S0-BDM-177

MUSEUMS OF THE WORLD
2011

MUSEUMS OF THE WORLD 2011

18th Edition

Volume 1

Afghanistan – United Arab Emirates

Volume 2

United Kingdom – Zimbabwe
Museum Associations
Indices

De Gruyter Saur

MUSEUMS OF THE WORLD 2011

18th Edition

Volume 2

United Kingdom – Zimbabwe
Museum Associations
Indices

De Gruyter Saur

Editors:
Ruth Lochar
Alexandra Meinhold
Hildegard Toma
Mies-van-der-Rohe-Str. 1
80807 München
Tel: +49 89 76902-316
Fax: +49 89 76902-333
Email: muse@gruyter.com

Advertising Representatives Worldwide:
DE GRUYTER
Mies-van-der-Rohe-Str. 1
80807 München
Tel: +49 89 76902-372
Fax: +49 89 76902-350
Email: anzeigen@degruyter.com
http://www.degruyter.com

ISBN 978-3-11-023485-5
e-ISBN 978-3-11-023486-2

Bibliographic information published by the Deutsche Nationalbibliothek
The Deutsche Nationalbibliothek lists this publication in the Deutsche Nationalbibliografie; detailed bibliographic data are available in the Internet at http://dnb.d-nb.de.

Typesetting: bsix information exchange GmbH, Braunschweig
Printing: Strauss GmbH, Mörlenbach
∞ Printed on acid-free paper

Printed in Germany

www.degruyter.com

Contents

Volume 1

Suggestions for Use . . . vii

Afghanistan . . . 3
Albania . . . 3
Algeria . . . 3
American Samoa . . . 4
Andorra . . . 4
Angola . . . 4
Argentina . . . 4
Armenia . . . 14
Australia . . . 15
Austria . . . 34
Azerbaijan . . . 61

Bahamas . . . 62
Bahrain . . . 62
Bangladesh . . . 62
Barbados . . . 62
Belarus . . . 63
Belgium . . . 63
Belize . . . 79
Benin . . . 79
Bermuda . . . 79
Bhutan . . . 80
Bolivia . . . 80
Bosnia and Herzegovina . . . 81
Botswana . . . 81
Brazil . . . 81
Brunei . . . 92
Bulgaria . . . 92
Burkina Faso . . . 98
Burundi . . . 98

Cambodia . . . 98
Cameroon . . . 98
Canada . . . 98
Cape Verde . . . 138
Central African Republic . . . 138
Chad . . . 138
Chile . . . 138
China . . . 139
Colombia . . . 154
Comoros . . . 159
Congo, Democratic Republic . . . 159
Congo, Republic . . . 159
Costa Rica . . . 159
Côte d'Ivoire . . . 159
Croatia . . . 160
Cuba . . . 163
Curaçao . . . 167
Cyprus . . . 167
Czech Republic . . . 168

Denmark . . . 179
Dominican Republic . . . 187

East Timor . . . 187
Ecuador . . . 187
Egypt . . . 189
El Salvador . . . 190
Equatorial Guinea . . . 190
Eritrea . . . 190
Estonia . . . 190
Ethiopia . . . 192

Faroe Islands . . . 192
Fiji . . . 192
Finland . . . 192
France . . . 206
French Guiana . . . 282
French Polynesia . . . 283

Gabon . . . 283
Gambia . . . 283
Georgia . . . 283
Germany . . . 284
Ghana . . . 404
Gibraltar . . . 404
Greece . . . 404
Greenland . . . 411
Grenada . . . 412
Guadeloupe . . . 412
Guam . . . 412
Guatemala . . . 412
Guinea . . . 412
Guinea-Bissau . . . 413
Guyana . . . 413

Haiti . . . 413
Honduras . . . 413
Hungary . . . 413

Iceland . . . 425
India . . . 426
Indonesia . . . 433
Iran . . . 435
Iraq . . . 437
Ireland . . . 437
Isle of Man . . . 441
Israel . . . 441
Italy . . . 445

Jamaica . . . 494
Japan . . . 494
Jordan . . . 517

Kazakhstan . . . 518
Kenya . . . 519
Korea, Democratic People's Republic . . . 520
Korea, Republic . . . 520
Kuwait . . . 523
Kyrgyzstan . . . 523

Laos . . . 523
Latvia . . . 523
Lebanon . . . 526
Lesotho . . . 527
Liberia . . . 527
Libya . . . 527
Liechtenstein . . . 527
Lithuania . . . 527
Luxembourg . . . 532
Macedonia . . . 533
Madagascar . . . 533
Malawi . . . 533
Malaysia . . . 534
Maldives . . . 535
Mali . . . 535
Malta . . . 535
Martinique . . . 535
Mauritania . . . 535
Mauritius . . . 536
Mexico . . . 536
Moldova . . . 548
Monaco . . . 548
Mongolia . . . 548
Montenegro . . . 549
Morocco . . . 549
Mozambique . . . 550
Myanmar . . . 550

Namibia . . . 550
Nepal . . . 551
Netherlands . . . 551
New Caledonia . . . 572
New Zealand . . . 572
Nicaragua . . . 577
Niger . . . 578
Nigeria . . . 578
Niue . . . 578
Norfolk Island . . . 578
Northern Cyprus . . . 578
Northern Mariana Islands . . . 579
Norway . . . 579

Oman . . . 590

Pakistan . . . 590
Palestine . . . 591
Panama . . . 591
Papua New Guinea . . . 592
Paraguay . . . 592
Peru . . . 592
Philippines . . . 595
Poland . . . 599
Portugal . . . 615
Puerto Rico . . . 624

Qatar . . . 625

Réunion . . . 625
Romania . . . 626
Russia . . . 636
Rwanda . . . 662
St. Lucia . . . 662
St. Pierre and Miquelon . . . 662

Samoa . . . 662
San Marino . . . 662
Sao Tome and Principe . . . 662
Saudi Arabia . . . 662
Senegal . . . 662
Serbia . . . 662
Seychelles . . . 665

Sierra Leone 665
Singapore 665
Slovakia 665
Slovenia 669
Solomon Islands 672
Somalia 672
South Africa 672
Spain 677
Sri Lanka 703
Sudan 704
Suriname 704
Swaziland 704
Sweden 704
Switzerland 715
Syria 736
Taiwan 736
Tajikistan 741
Tanzania 742
Thailand 742
Togo 743
Tonga 743
Trinidad and Tobago 743
Tunisia 743
Turkey 744
Turkmenistan 749
Uganda 749
Ukraine 749
United Arab Emirates 753

Volume 2

United Kingdom 755
United States of America 811
Uruguay 974
Uzbekistan 975
Vanuatu 976
Vatican City 976
Venezuela 976
Vietnam 977
Yemen 978
Zambia 978
Zimbabwe 978
Museum Associations 979
Index of Distinguished Persons 989
Alphabetical Index to Museums 1005
Index of Museum Staff 1209
Subject Index 1313

Museums of the World

Volume 2

United Kingdom

Abercrave

Dan-yr-Ogof Showcaves Museum, Glyntawe, Upper Swansea Valley, Abercrave • T: +44 1639 730284 • F: 730293 •
Dir.: *Ian Gwilim* •
Natural History Museum – 1964 ... 43231

Aberdare

Cynon Valley Museum and Gallery, Depot Rd, Gadys, Aberdare CF44 8DL • T: +44 1685 886729 • F: 886730 • cvm@rhondda-cynon-raff.gov.uk •
Cur.: *Chris Wilson* •
Local Museum / Public Gallery
Over 200 yrs history of Cynon valley ... 43232

Aberdeen

Aberdeen Art Gallery, Schoolhill, Aberdeen AB10 1FQ • T: +44 1224 523700 • F: 632133 • info@aagm.co.uk • www.aagm.co.uk •
Dir.: *Christine Rew* •
Fine Arts Museum – 1885
Fine and applied arts (18th-20th c), decorative art since 6th c – library ... 43233

Aberdeen Arts Centre, 33 King St, Aberdeen AB24 5AA • T: +44 1224 635208 • F: 626390 • enquiries@aberdeenartscentre.org.uk • www.aberdeenartscentre.org.uk •
Dir.: *Paula Gibson* •
Public Gallery ... 43234

Aberdeen Maritime Museum, Shiprow, Aberdeen AB11 5BY • T: +44 1224 337700 • F: 213066 • info@aagm.co.uk • www.aagm.co.uk •
Dir.: *John Edwards* •
Historical Museum – 1984/97
North Sea oil and gas industry, shipping and fishing exhibitions, multi-media and hands-on exhibits .. 43235

Aberdeen University Natural History Museum, c/o Department of Zoology, Tillydrone Av, Aberdeen AB9 2TZ • T: +44 1224 272850 • F: 272396 • m.gorman@abdn.ac.uk • www.abdn.ac.uk/zoology •
Hon. Cur.: *Dr. Martyn Gorman* •
University Museum / Natural History Museum
Ornithology, entomology, invertebrates/vertebrates coll ... 43236

Aberdeen University Natural Philosophy Museum, c/o Dept. of Physics, Fraser Noble Bldg, Aberdeen AB24 3UE • T: +44 1224 272507 • F: 272497 • j.s.reid@abdn.ac.uk • www.abdn.ac.uk/~nph126 •
Dir.: *Dr. John S. Reid* •
Science&Tech Museum / University Museum – 1971
Coll of apparatus used during the past 200 yrs for teaching and demonstrating the traditional branches of physics ... 43237

Aberdeen University Pathology and Forensic Medicine Collection, c/o Department of Pathology, Foresterhill, Aberdeen AB25 2ZD • T: +44 1224 553792 • F: 663002 • g.l.murray@abdn.ac.uk • www.abdn.ac.uk/museums •
Cur.: *Graeme Murray* •
University Museum / Natural History Museum ... 43238

Anatomy Museum, c/o Marischal College, University of Aberdeen, Broad St, Aberdeen AB10 1YS • T: +44 1224 274320 • F: 274329 • i.stewart@abdn.ac.uk • abdn.ac.uk •
Dir.: *Dr. Ian Stewart* •
University Museum / Natural History Museum
Human anatomical specimens and models ... 43239

The Blairs Museum, South Deeside Rd, Blairs, Aberdeen AB12 5YQ • T: +44 1224 863767 • F: 869424 • manager@blairsmuseum.com • www.blairsmuseum.com •
Manager/Curator: *Mandy Murray* •
Fine Arts Museum / Religious Arts Museum
Pictures of Mary Queen of Scots, Stuart portraits and memorabilia, 16th c Church robes, altarware, embroidergd vestments ... 43240

Geology Department Museum, c/o University of Aberdeen, Meston Bldg, Aberdeen AB24 3UE • T: +44 1224 273448 • F: 272785 • n.trewin@abdn.ac.uk •
Natural History Museum ... 43241

Gordon Highlanders Museum, Saint Luke's, Viewfield Rd, Aberdeen AB15 7XH • T: +44 1224 311200 • F: 319323 • museum@gordonhighlanders.com • www.gordonhighlanders.com •
Dir.: *Ian Lakin* • Cur.: *Sarah MacKay* •
Military Museum – 1937
History of Gordon Highlanders regiment, uniforms, medals, pictures, trophies ... 43242

Grays School of Art Gallery and Museum, c/o Faculty of Design, The Robert Gordon University, Garthdee Rd, Aberdeen AB9 2QD • T: +44 1224 263506/600 •
Public Gallery / University Museum ... 43243

Marischal Museum, Marischal College, University of Aberdeen, Broad St, Aberdeen AB10 1YS • T: +44 1224 274301 • F: 274302 • museum@abdn.ac.uk • www.abdn.ac.uk/marischal_museum •
Head: *Neil Curtis* • Cur.: *Dr. Jennifer Downes* (Science Colls) •
University Museum / Local Museum
Local archaeology, art, antiquities, non-western ethnography ... 43244

Peacock Visual Arts, 21 Castle St, Aberdeen AB11 5BQ • T: +44 1224 639539 • F: 627094 • info@peacockvisualarts.co.uk • www.peacockvisualarts.co.uk •
Dir.: *Lindsay Gordon* •
Public Gallery
Printmaking, photography, digital imaging, video 43245

Provost Skene's House, off Broad St, Guestrow, Aberdeen AB10 1AS • T: +44 1224 641086 • F: 632133 • info@aagm.co.uk • www.aagm.co.uk •
Dir.: *Christine Rew* •
Decorative Arts Museum – 1953
Scottish furniture, costume, archaeology, painted chapel ceiling (17th c), local history ... 43246

Robert Gordon University Collection, university Library, Garthdee Rd, Aberdeen AB10 7QE • T: +44 1224 263474 • library@rgu.ac.uk • www.rgu.ac.uk/library •
Head: *Ross Hayworth* (Collections) •
University Museum
Art cool, museum coll, items from the needlework development scheme ... 43247

Satrosphere, c/o The Tramsheds, 179 Constitution St, Aberdeen AB24 5TU • T: +44 1224 640340 • F: 622211 • satrosphere@satrosphere.net • www.satrosphere.net •
Dir.: *Dr. Alistair M. Flett* •
Science&Tech Museum
Scientific instruments, computers ... 43248

The Tolbooth Museum, Castle Street, Aberdeen AB10 1EX • T: +44 1224 621167 • F: 523666 • info@aagm.co.uk • www.aagm.co.uk •
Historical Museum / Historic Site – 1995
Housed in the former wardhouse, or prison for Aberdeen, a unique complex of 17th and 18th century gaol cells. Featured displays deal with local history and the evolution of crime and punishment over the centuries ... 43249

Aberfeldy

Castle Menzies, Weem, Aberfeldy PH15 2JD • T: +44 1887 820982 • menziesclan@tesco.net • www.menzies.org •
Pres.: *G.M. Menzies* •
Historical Museum
Small Clan-Museum in a 15th c building, Claymore sword, Broad swords, botanist Archibald Menzies ... 43250

Aberford

Lotherton Hall, Leeds Museums and Galleries, Aberford LS25 3EB • T: +44 113 2813259 • F: 2812100 • michael.thaw@kids.gov.uk • www.leeds.gov.uk •
Dir.: *John Roles* • Cur.: *Michael Thaw* •
Decorative Arts Museum / Fine Arts Museum – 1969
English and Continental paintings (17th-20th c), furniture, silver, British and oriental ceramics, costume, sculpture ... 43251

Abergavenny

Abergavenny Museum, Castle, Castle St, Abergavenny NP7 5EE • T: +44 1873 854282 • F: 736004 • abergavennymuseum@monmouthshire.gov.uk • www.abergavennymuseum.co.uk •
Dir.: *Rachael Rogers* •
Local Museum – 1959
History and industries of the town and district, kitchen, rural craft, tools, Roman relics from Gobannium site, grocers shop ... 43252

Aberlady

Myreton Motor Museum, In the Countryside, Aberlady EH32 0PZ • T: +44 1875 870288 • F: +44 1368 860199 •
Dir.: *Alex P. Dale* • Cur.: *Sara Schwab* •
Science&Tech Museum – 1966
Cars, motorcycles, bicycles, commercials, period advertising, posters, enamal signs ... 43253

Abernethy

Abernethy Museum, Cherrybank, Abernethy PH2 9LW • T: +44 131 3177300 •
Local Museum ... 43254

Abertillery

Abertillery Museum, Metropole Theater, Market St, Abertillery NP3 1TE • T: +44 1495 213806 •
Cur.: *Don Bearcroft* •
Local Museum ... 43255

Aberystwyth

Aberystwyth Arts Centre, c/o University College of Wales, Penglais, Aberystwyth SY23 3DE • T: +44 1970 622882, 622460 • F: 622883, 622461 • etr@aber.ac.uk • www.aber.ac.uk/~arcwww •
Dir.: *Alan Hewson* •
University Museum / Fine Arts Museum
ceramics gallery ... 43256

Ceramics Gallery, c/o Arts Centre, University of Wales, Penglais, Aberystwyth SY23 1NE • T: +44 1970 622460 • F: 622461 • neh@aber.ac.uk • www.aber.ac.uk •
Dir.: *Moira Vincentelli* •
Decorative Arts Museum – 1876
Studio Ceramics ... 43257

Ceredigion Museum, Coliseum, Terrace Rd, Aberystwyth SY23 2AQ • T: +44 1970 633088 • F: 633084 • museum@ceredigion.gov.uk • museum.ceredigion.gov.uk •
Dir.: *Michael Freeman* •
Local Museum – 1972
Alfred Worthington paintings, work of Hutchings, taxidermist ... 43258

School of Art Gallery and Museum, c/o University of Wales, Buarth Mawr, Aberystwyth SY23 1NG • T: +44 1970 622460 • F: 622461 • neh@aber.ac.uk • www.aber.ac.uk/museum •
Dir.: *Robert Meyrick* (Head of School) • Cur.: *Neil Holland* • *Moira Vincentelli* (Art history) •
Fine Arts Museum / Public Gallery / University Museum – 1876
Graphic art since 15th c, art in Wales since 1945, 20th c Italian and British photography, pottery (early 20th /contemporary British), slipware, Swansea and Nantgarw porcelain, Oriental ceramics – ceramic archive (documents, publications, sound and film) ... 43259

Abingdon

Abingdon Museum, Old County Hall, Market Pl, Abingdon OX14 3HG • T: +44 1235 523703 • F: 536814 • cherry.gray@oxfordshire.gov.uk •
Cur.: *Cherry Gray* •
Local Museum – 1925
Local hist, archaeology, contemporary crafts, 17th c building ... 43260

Abinger Common

Mesolithic Museum, Abinger Manor, Abinger Common RH5 6JD • T: +44 1306 730760 •
Dir.: *Robert Clarke* •
Historical Museum – 1953
Hut build, flint coll ... 43261

Accrington

Haworth Art Gallery, Haworth Park, Manchester Rd, Accrington BB5 2JS • T: +44 1254 233782 • F: 301954 • jennifer.renni@hyndburnbc.gov.uk • www.hyndburnbc.gov.uk •
Cur.: *Jennifer A. Rennie* •
Public Gallery – 1921
19th-20th c paintings, watercolours, largest Tiffany glass coll in Europe ... 43262

Acton Scott

Acton Scott Historic Working Farm, Wenlock Lodge, Acton Scott SY6 6QN • T: +44 1694 781306 • F: 781569 • acton.scott.museum@shropshire-cc.gov.uk • www.actonscottmuseum.co.uk •
Dir.: *Mike Barr* • Cur.: *Nigel Nixon* •
Open Air Museum / Agriculture Museum – 1975
Stock, crops and machinery (1870-1920), agriculture, rural crafts (wheelwright, farrier, blacksmith), dairy, period cottage – library, technical workshops, education ... 43263

Alcester

Coughton Court, Alcester B49 5JA • T: +44 1789 400777 • F: 765544 • carol@throckmortons.co.uk •
Cur.: *C. Throckmorton* •
Decorative Arts Museum / Local Museum
Family portraits, furniture ... 43264

Ragley Hall, Alcester B49 5NJ • T: +44 1789 762090 • F: 764791 • ragley@ragleyhall.com • www.ragleyhall.com •
Owner: *Marquess of Hertford* •
Fine Arts Museum / Decorative Arts Museum
Art 18th c, porcelain, furniture, baroque plasterwork ... 43265

Aldborough

Aldborough Roman Site, Main St, Aldborough YO51 3EP • T: +44 1423 322768 • customers@english-heritage.org.uk • www.english-heritage.org.uk/yorkshire •
Archaeology Museum / Historic Site ... 43266

Aldeburgh

Aldeburgh Museum, The Moot Hall, Aldeburgh IP15 5DS • T: +44 1728 454666 • aldeburgh@museum.freeserve.co.uk • www.aldeburghmuseum.org.uk •
Local Museum – 1911
1862 Anglo-Saxon Ship Burial finds, local hist ... 43267

Alderney

Alderney Railway - Braye Road Station, POB 75, Alderney, GY9 3DA • T: +44 1481 822978 • www.alderneyrailway.com •
Dir.: *Tony Le Blanc* •
Science&Tech Museum – 1978
Two industrial diesel locomotives, 4 petrol Wickham railcars, 2 1959 London transport driving trailer stock ... 43268

Alderney Society Museum, The Old School, High St, Alderney GY9 3TG • T: +44 1481 823222 • F: 824979 • alderneymuseum@alderney.net • www.alderneysociety.org •
Pres.: *Louis Jean* •
Local Museum – 1966
Local hist, arts and crafts, hist of the German occupation 1940-1945, Iron Age pottery finds, Elizabethan wreck ... 43269

Aldershot

Airborne Forces Museum, Browning Barracks, Aldershot GU11 2BU • T: +44 1252 349619 • F: 349203 • airborneforcesmuseum@army.mod.uk.net •
Cur.: *Tina Pittock* •
Military Museum – 1969
WW II operational briefing models, aircraft models, equip, vehicles, guns ... 43270

Aldershot Military Museum and Rushmoor Local History Gallery, Queens Av, Aldershot GU11 2LG • T: +44 1252 314598 • F: 342942 • musmim@hants.gov.uk • www.hants.gov.uk/museum/aldershot •
Cur.: *Sally Day* • Cons.: *Graham Smith* •
Military Museum / Local Museum – 1984
Borough of Rushmoor local history coll, vehicle coll ... 43271

Army Physical Training Corps Museum, Queens Av, Aldershot GU11 2LB • T: +44 1252 347168 • F: 340785 • regtsec@aptc.org.uk • www.aptc.org.uk •
Dir.: *R.J. Kelly* •
Military Museum
History of the Corps since 1860, achievements and personalities – library ... 43272

Royal Army Dental Corps Historical Museum, Evelyn Woods Rd, Aldershot GU11 2LS • T: +44 1252 347976 • F: 347726 •
Dir.: *C.D. Parkinson* •
Military Museum / Science&Tech Museum – 1969
Hist of the Corps, developments in dental techniques, connection between dentistry and the Army from the origins of the British Army, following the restoration of the monarchy in 1660 to the present ... 43273

Aldwarke Rotherham

Sheffield Bus Museum → South Yorkhire Transport Museum

South Yorkhire Transport Museum, Unit 9 Waddington Way South, Aldwarke Rotherham S65 3SH, South Yorkshire • T: +44 114 2553010 • info@sytm.co.uk • www.sytm.co.uk •
Dir.: *Douglas Miller* • *David Tummon* • *David Taylor* • *Stuart Lodge* •
Science&Tech Museum
Buses, tramcars, Commercial vehicles, cars plus transport artefacts ... 43274

Alexandria

Tobias Smollett Museum, Castle Cameron, Alexandria G83 8QZ • T: +44 1389 56226 •
Special Museum ... 43275

Alford, Aberdeenshire

Alford Heritage Centre, Mart Rd, Alford, Aberdeenshire AB33 8BZ • T: +44 19755 62906 •
Local Museum ... 43276

Grampian Transport Museum, Alford, Aberdeenshire AB33 8AE • T: +44 1975 562292 • F: 562180 • info@gtm.org.uk • www.gtm.org.uk •
Cur.: *Mike Ward* •
Science&Tech Museum
200 vehicle exhibt ... 43277

Alford, Lincolnshire

Alford Manor House Museum, West St, Alford, Lincolnshire LN13 9HT • T: +44 1507 462143, 463073 •
Local Museum
Hist of Alford Manor House, Victorian Drawing Room, Shops from the past (Pharmaca, Veterinary Surgery, Bootmaker) Prison Cell, Victorian Schoolroom, coll of Roman finds, displays from the Salt Works and Sweetmakers, costumes and photographs from past local life, Alford's connection with America ... 43278

Alfriston

Clergy House, The Tye, Alfriston BN26 5TL • T: +44 1323 870001 • F: 871318 • alfriston@nationaltrust.org.uk • www.nationaltrust.org.uk •
Dir.: *Tina Maher* •
Historic Site ... 43279

Allerford

West Somerset Rural Life Museum, Old School, Allerford TA24 8HN • T: +44 1643 862529 • www.allerfordwebsite.ic24.net •
Head: *Duncan Osborne* •
Local Museum
Victorian kitchen, laundry, dairy, school desks, books, toys ... 43280

Alloa

Aberdona Gallery, Aberdona Mains, Alloa FK10 3QP • T: +44 1259 752721 • F: 750276 •
Public Gallery ... 43281

Clackmannshire Council Museum, Speirs Centre, 29 Primrose St, Alloa FK10 1JJ • T: +44 1259 216913 • F: 721313 • smills@clacks.gov.uk • www.clacksweb.org.uk •
Head: *Susan Mills* •
Local Museum
Alloa pottery, textiles and woolen industry, social and indistrial history ... 43282

Alloway

Burns Cottage Museum, Alloway KA7 4PY • T: +44 1292 441215 • F: 441750 • info@burnsheritagepark.com • www.burnsheritagepark.com •
Dir.: *Nat Edwards* • Cur.: *John Manson* •
Special Museum – 1880
Cottage where poet Robert Burns (1759-1796) was born, memorabilia, manuscripts ... 43283

Tam O'Shanter Experience, Murdoch's Lone, Alloway • T: +44 1292 443700 • F: 441750 • info@burnsheritagepark.com • www.burnsheritagepark.com •
Dir.: *N. Edwards* •
Local Museum ... 43284

Alness

Clan Grant Museum, Grants of Dalvey, Alness IV17 0XT • T: +44 1349 884111 • F: 884100 •
Historical Museum ... 43285

Alnwick

Fusiliers Museum of Northumberland, Abbots Tower, Alnwick Castle, Alnwick NE66 1NG • T: +44 1665 602152 • F: 605257 • fusnorthld@aol.com • www.northumberlandfusiliers.org.uk •
Chm.: *Tony Adamson* •
Military Museum – 1929
Artefacts, uniforms, medals, weapons ... 43286

Museum of Antiquities, Alnwick Castle, Alnwick NE66 1NQ • T: +44 1665 510777 • F: 510876 • enquiries@alnwickcastle.com • www.alnwickcastle.com •
Cur.: *Clare Baxter* •
Historic Site / Museum of Classical Antiquities ... 43287

Museum of the Percy Tenantry Volunteers 1798-1814, Estate Office, Alnwick Castle, Alnwick NE66 1NQ • T: +44 1665 510777 • F: 510876 • enquiries@alnwickcastle.com • www.alnwickcastle.com •
Cur.: *Clare Baxter* •
Historic Site / Military Museum
Arms and accoutrements of Percy Tenantry Volunteers formed mainly from people on northern estates of the Duke of Northumberland ... 43288

Alresford

Mid Hants Railway 'Watercress Line', Railway Station, Alresford SO24 9JG • T: +44 1962 733810 • F: 735448 • www.watercressline.co.uk •
Science&Tech Museum
Historic steam and diesel trains ... 43289

Alston

South Tynedale Railway Preservation Society, Railway Station, Alston CA9 3JB • T: +44 1434 381696 • www.strps.org.uk •
Dir.: *M.J. LeMarie* •
Science&Tech Museum
Steam and diesel narrow-gauge locomotives ... 43290

Althorp

Althorp Museum, The Stables, Althorp NN7 4HQ • T: +44 1604 770107 • F: 770042 • mail@althorp.com • www.althorp.com •
Special Museum
Costume, memorabilia of Diana, Princess of Wales ... 43291

Alton

Allen Gallery, 10-12 Church St, Alton GU34 2BW • T: +44 1420 82802 • F: 84227 • musmtc@hants.gov.uk • www.hants.gov.uk/museums •
Dir.: *Tony Cross* •
Decorative Arts Museum
Works by W.H. Allen, ceramics from 1550 to the present day, silver – gallery, herb garden ... 43292

Curtis Museum, 3 High St, Alton GU34 1BA • T: +44 1420 82802 • F: 84227 • musmtc@hants.gov.uk • www.hants.gov.uk/museum/curtis •
Dir.: *Tony Cross* •
Local Museum – 1855
Local history, archaeology, geology, toys, hop growing, brewing ... 43293

Altrincham

Dunham Massey Hall, Altrincham WA14 4SJ • T: +44 161 9411025 • F: 9297508 • dunhammassey@nationaltrust.org.uk • www.nationaltrust.org.uk/main/w-dunhammassey •
Dir.: *Stephen Adams* •
Fine Arts Museum / Decorative Arts Museum
Fine coll of paintings, furniture and Huguenot silver – library ... 43294

Alva

Mill Trail Visitor Centre, West Stirling St, Alva FK12 5EN • T: +44 1259 769696 • F: 763100 •
Man.: *Anne Tindell* •
Science&Tech Museum
History of weaving industry, spinning wheels and looms ... 43295

Alyth

Alyth Museum, Perth Museum and Art Gallery, Commercial St, Alyth PH11 8AF • T: +44 1738 632488 • F: 443505 • museum@pkc.gov.uk • www.pkc.gov.uk/ah •
Local Museum
Local hist ... 43296

Amberley

Amberley Museum and Heritage Centre, Houghton Bridge, Amberley BN18 9LT, West Sussex • T: +44 1798 831370 • F: 831831 • office@amberleymuseum.co.uk • www.amberleymuseum.co.uk •
Dir.: *Dr. Claire Seymour* •
Science&Tech Museum / Open Air Museum – 1979
Milne electrical coll, Southdown omnibus coll, narrow gauge railways, printing, radio, TV and telephones, roads and roadmaking, industrial buildings – archives ... 43297

Ambleside

Hill Top, nr Sawrey, Hawkshead, Ambleside LA22 0LF • T: +44 15394 36269 • F: 36811 • hilltop@nationaltrust.org.uk • www.nationaltrust.org.uk •
Dir.: *John Moffat* •
Special Museum
Beatrix Potter wrote many "Peter Rabbit" books here, furniture, china and other possessions ... 43298

Museum of Ambleside, Rydal Rd, Ambleside LA22 9BL • T: +44 15394 31212 • F: 31313 • info@armitt.com • www.armitt.com •
Cur.: *Michelle Kelly* •
Local Museum – 1958
Social and economic history, archaeology, geology, natural history, photography, fine art – Library ... 43299

Rydal Mount, Ambleside LA22 9LU • T: +44 15394 33002 • F: 31738 • Rydalmount@aol.com •
Dir.: *Peter Elkington* • *Marian Elkington* •
Local Museum – 1970
Portraits, furniture, personal possessions ... 43300

Amersham

Amersham Museum, 49 High St, Amersham HP7 0DP • T: +44 1494 725754 • F: 725754 • amershammuseum@tiscali.co.uk • www.amersham.org.uk/museum •
Cur.: *Jane Bowen* •
Local Museum
Local hist, fossil and archeaological finds, wall paintings,lace making, strawplait work, brewing, postal services, agriculture ... 43301

Andover

Andover Museum, 6 Church Close, Andover SP10 1DP • T: +44 1264 366283 • F: 339152 • musmda@hants.gov.uk •
Cur.: *Jenny Stevens* •
Local Museum
Local hist, archaeology, natural science ... 43302

Museum of the Iron Age, 6 Church Close, Andover SP10 1DP • T: +44 1264 366283 • F: 339152 • musmda@hants.gov.uk • www.hants.gov.uk/museum/ironagem •
Cur.: *Jenny Stevens* •
Archaeology Museum
Finds from the excavations, hist of southern Britain in the Iron Age ... 43303

Annaghmore

Ardress Farmyard Museum, Ardress House, Annaghmore BT62 1SQ • T: +44 28 38851236 • F: 38851236 •
Decorative Arts Museum / Agriculture Museum – 1977
Fine furniture and pictures ... 43304

Annan

Historic Resources Centre, c/o Annan Museum, Bank St, Annan DG12 6AA • T: +44 1461 201384 • F: 205876 • info@dumfriesmuseum.demon.co.uk • www.dumfriesmuseum.demon.co.uk •
Local Museum
Ethnography, local history, relics, contemporary art, biographical essays ... 43305

Anstruther

Scottish Fisheries Museum, Saint Ayles, Habourhead, Anstruther KY10 3AB • T: +44 1333 310628 • F: 310628 • info@scotfishmuseum.org • www.scotfishmuseum.org •
Dir.: *Andrew Fox* •
Special Museum – 1969
History of Scotland's fishing industry and related trades, fishing gear, ship models and gear, fishermen's ethnography, whaling, aquarium, paintings and photographs with marine themes ... 43306

Antrim

Clotworthy Arts Centre, Castle Gardens, Randalstown Rd, Antrim BT41 4LH • T: +44 28 94481338 • F: 94481344 • clotworthy@antrim.gov.uk • www.antrim.gov.uk •
Fine Arts Museum / Decorative Arts Museum
Art exhibits, scale model of Antrim castle, 17th c Anglo Dutch water garden ... 43307

Appledore, Devon

North Devon Maritime Museum, Odun House, Odun Rd, Appledore, Devon EX39 1PT • T: +44 1237 422064 • appledore@devonmuseums.net • www.devonmuseums.net/appledore •
Dir.: *A.E. Grant* •
Science&Tech Museum
Scale models of ships using North Devon ports from the 17th c onwards, paintings of ships and maritime scenes, and old photogr of the district and its people ... 43308

Arbigland

John Paul Jones Birthplace Museum, John Paul Jones Cottage, Arbigland DG2 8BQ • T: +44 1387 880613 • F: 265081 • postmaster@dumfriesmuseum.demon.co.uk • www.jpj.demon.co.uk •
Chm.: *Alf Hannay* •
Special Museum
Navy history ... 43309

Arborfield

REME Museum of Technology, Royal Electrical and Mechanical Engineers, Isaac Newton Rd, Arborfield RG2 9NJ • T: +44 118 9763375 • F: 9762017 • enquiries@rememuseum.org.uk • www.rememuseum.org.uk •
Dir.: *I.W.J. Cleasby* •
Science&Tech Museum / Military Museum – 1958
History of the Corps of Royal Electrical and Mechanical Engineers, the British Army's equipment repair corps, documents, photos, equipment and memorabilia, weapons, small arms, electronics, aeronautics engineering, uniforms, medals – pictural archive 43310

Arbroath

Abbot's House, Arbroath Abbey, Arbroath DD11 1EG • T: +44 1241 2443101 •
Archaeology Museum ... 43311

Arbroath Art Gallery, Library, Hill Terrace, Arbroath DD11 1AH • T: +44 1241 872248 • F: 434396 • arbroath.library@angus.gov.uk • www.angus.gov.uk/history •
Libr.: *Teresa Roby* •
Fine Arts Museum – 1898
Local artists' works, 2 paintings by Breughel ... 43312

Arbroath Museum, Signal Tower, Ladyloan, Arbroath DD11 1PU • T: +44 1241 875598 • F: 439263 • signal.tower@angus.gov.uk • www.angus.gov.uk/history •
Local Museum – 1843
Bell Rock lighthouse artifacts, fishing ... 43313

Saint Vigeans Museum, Saint Vigeans, Arbroath • T: +44 1241 878756, 2443101 •
Religious Arts Museum – 1960
Picts, stone carvings of particular symbols ... 43314

Arlington

Arlington Court, Arlington EX31 4LP • T: +44 1271 850296 • F: 851108 • arlingtoncourt@nationaltrust.org.uk •
Dir.: *Ana Chylak* •
Special Museum
Coll of horse drawn carriages ... 43315

Armadale

Armadale Community Museum, West Main St, Armadale EH48 2JD • T: +44 1501 678400 • armadale.library@westlothian.org.uk • www.wlonline.org •
Man.: *Elizabeth Hunter* •
Local Museum ... 43316

Armagh

Armagh County Museum, The Mall East, Armagh BT61 9BE • T: +44 28 37523070 • F: 37522631 • acm.info@magni.org.uk • www.magni.org.uk •
Dir.: *Catherine McCullough* •
Local Museum – 1935
Irish art, hist and natural hist of Armagh County since prehistoric times, local folklore, applied art incl some works and manuscripts by George Russell – library ... 43317

Hayloft Gallery, Palace Stable Heritage Centre, Palace Demesne, Armagh BT60 4EL • T: +44 28 37529629 • F: 37529630 • stables@armagh.gov.uk • www.armagh.gov.uk •
Public Gallery ... 43318

Navan Centre, 81 Killylea Rd, Armagh BT60 4LD • T: +44 1861 525550, +44 28 37521801 • F: +44 1861 522323, +44 28 37520180 • info@armagh.gov.uk • www.armagh.gov.uk •
Local Museum
History, archaeology and mythology of Navan Fort (Emain Macha) ... 43319

Royal Irish Fusiliers Museum, Sovereign's House, The Mall East, Armagh BT61 9DL • T: +44 28 37522911 • F: 37522911 • rylirfusiliermus@aol.com •
Military Museum
Hist of the regiment, uniforms, photos ... 43320

Saint Patrick's Trian, 40 English St, Armagh BT61 7BA • T: +44 28 37521801 • F: 37510180 • info@saintpatrickstrian.com • www.armaghvisit.com •
Local Museum
Local hist from prehistory, Celts, Vikings, Brian Boru, Georgian times ... 43321

Arnol

Arnol Blackhouse Museum, Arnol PA86 9DB • T: +44 131 5568400 •
Archaeology Museum ... 43322

Arundel

Arundel Museum and Heritage Centre, 61 High St, Arundel BN18 9AJ • T: +44 1903 885708 •
Cur.: *Alan Chapman* •
Local Museum – 1963 ... 43323

Arundel Toy Museum Dolls House, 23 High St, Arundel BN18 9AD • T: +44 1903 883101 •
Dir.: *Diana Henderson* •
Special Museum – 1978
Britains model toy soldiers and animals/farms, small militaria, dolls houses and contents, games, tintoys, puppets, curiosities ... 43324

Ashburton

Ashburton Museum, 1 West St, Ashburton TQ13 7AB • T: +44 1364 653278 •
Cur.: *Jamie Broughton* •
Local Museum – 1954
Local history, American Indian antiques, costumes, arrows, tools ... 43325

Ashby-de-la-Zouch

Ashby-de-la-Zouch Museum, North St, Ashby-de-la-Zouch LE65 2HU • T: +44 1530 560090 • paul.monk@virgin.net • www.ashbydelazouchmuseum.org •
Dir.: *Kenneth Hillier* •
Local Museum – 1982 ... 43326

Ashford

Ashford Museum, Church Yard, Ashford TN23 1QG • T: +44 1233 631511 • F: 502599 • a.d.terry@excite.co.uk •
Historical Museum
Hist of Ashford and the surrounding area, with exhibits reflecting the social and domestic life of the past 150 yrs, and the occupations carried on in the district ... 43327

Swanton Mill, Lower Mersham, Ashford TN25 7HS • T: +44 1233 720223 •
Owner: *John Bickel* •
Science&Tech Museum ... 43328

Ashington

Woodhorn, Northumberland Museum, Archives and Country Park, Queen Elizabeth II Park, Ashington NE63 9YF • T: +44 1670 528080 • dtate@woodhorn.org.uk • www.experiencewoodhorn.com •
Dir.: *Neil Anderson* • Asst. Dir: *Margaret Barnfather* • Sc. Staff: *Jo Raw* • Head of Colls.: *Sue Wood* •
Local Museum
Coal-mining and social hist of SE Northumberland – study centre ... 43329

Ashton, Oundle

National Dragonfly Biomuseum, Ashton Wold, Ashton, Oundle PE8 5LZ • T: +44 1832 272427 • ndmashton@aol.com • www.natdragonflymuseum.org.uk •
Natural History Museum / Science&Tech Museum
Larva-feeding sessions, dragonflies in art, live larvae in tanks, Victorian hydropower station, vintage farm machinery coll, blacksmith's forge, local crafts, fish coll ... 43330

Ashton-under-Lyne

Central Art Gallery, Old St, Ashton-under-Lyne OL6 7SG • T: +44 161 3422650 • F: 3422650 • central.artgallery@tameside.gov.uk • www.tameside.gov.uk •
Dir.: *Marie Hedegaard Knudsen* •
Public Gallery ... 43331

Museum of the Manchester Regiment, Town Hall, Market Pl, Ashton-under-Lyne OL6 6DL • T: +44 161 3422254 • museum.manchesters@tameside.gov.uk •
Man.: *Emma Varnam* •
Historical Museum – 1987
Social and regimental hist ... 43332

Portland Basin Museum, 1 Portland Pl, Ashton-under-Lyne OL7 0QA • T: +44 161 3432878 • F: 3432869 • portland.basin@mail.gov.uk • www.tameside.gov.uk •
Dir.: *Dr. Alan Wilson* •
Local Museum ... 43333

Ashurst, Southampton

National Dairy Museum, Longdown Dairy Farm, Deerleap Ln, Ashurst, Southampton SO40 4UH • T: +44 23 80293326 • F: 80293376 • annette@longdown.uk.com •

Dir.: *Hannah Field* •
Agriculture Museum
Coll of horse-drawn and hand-pushed milk prams, history of dairy industry 43334

Ashwell

Ashwell Village Museum, Swan St, Ashwell SG7 5NY • T: +44 1462 742956 • peter.greener@care4free.net •
Cur.: *Peter Greener* •
Local Museum – 1930
Rural village life from the Stone Age to the present, a Tudor building, snuff and tinder boxes, straw-plaiting tools, lace-making tools, objects formerly used in farming and everyday life, leather eyeglasses 43335

Augrès

Sir Francis Cook Gallery, Rte de la Trinité, Augrès JE3 5JN • T: +44 1534 863333 • F: 864437 • museum@jerseyheritagetrust.org • www.jerseyheritagetrust.org •
Dir.: *Jon Carter* •
Fine Arts Museum
Work of Sir Francis Cook 43336

Auldgirth

Ellisland Farm, Hollywood Rd, Auldgirth DG2 0RP • T: +44 1387 740426, 256885 • friends@ellislandfarm.co.uk • scotlandvacations.com/Ellisland%20Farm.htm •
Agriculture Museum
Exhibits concerning the poet Robert Burns and his family, farming life 43337

Avebury

Alexander Keiller Museum, High St, Avebury SN8 1RF • T: +44 1672 539250 • F: 538038 • avebury@nationaltrust.org.uk •
Cur.: *Rosamund Cleal* •
Archaeology Museum – 1938
Prehistoric artifacts from the world heritage site of Avebury, pottery, flint implements, animal bone, bone implements 43338

Aviemore

Strathspey Railway, Aviemore Station, Dalfaber Rd, Aviemore PH22 1PY • T: +44 1479 810725 • F: 811022 • strathtrains@strathspeyrailway.co.uk • www.strathspeyrailway.co.uk •
Head: *Laurence Grant* •
Science&Tech Museum
British locomotives and rolling stocks, mainly 1950's-1960's 43339

Axbridge

King John's Hunting Lodge, The Square, Axbridge BS26 2AP • T: +44 1934 732012 • F: +44 1278 444076 • museums@sedgemoor.gov.uk •
Dir.: *Sarah Harbige* •
Local Museum
Archaeology 43340

Axminster

Axminster Museum, Old Court House, Church St, Axminster EX13 5LL • T: +44 1297 34137 • F: 32929 • axminster-museum@ukf.net • www.devonmuseums.net •
Cur.: *Sheila Hughes* •
Historical Museum 43341

Aylesbury

Buckinghamshire County Museum, Church St, Aylesbury HP20 2QP • T: +44 1296 331441 • F: 485590 • museum@buckscc.gov.uk • www.buckscc.gov.uk/museum •
Cur.: *Sarah Gray* • Sc. Staff: *Mike Palmer* (Natural History) • *Catherine Weston* (Social History) • *Alexandra MacCulloch* (Art, Clothing, Textiles) • *Brett Thorn* (Archaeology) • *David Erskine* (Education, Exhibitions) •
Local Museum – 1847
Buckinghamshire artefacts, prints, watercolours, drawings and paintings, studio pottery 43342

Ayot-Saint-Lawrence

Shaw's Corner, Ayot-Saint-Lawrence AL6 9BX • T: +44 1438 820307 • F: 820307 • shawscorner@nationaltrust.org.uk • www.nationaltrust.org.uk/shawscorner •
Special Museum
Former home of George Bernard Shaw (1856-1950) from 1906 until his death, memorabilia including Nobel Prize related literature of 1925 and objects concerning the Oscar-winning work 'Pygmalion' 43343

Ayr

Ayrshire Yeomanry Museum, Rozelle House, Monument Rd, Ayr KA7 4NQ • T: +44 1292 264091 •
Military Museum 43344

Dunaskin Open Air Museum, Waterside, Dalmellington Rd, Ayr KA6 7JF • T: +44 1292 531144 • F: 532314 • dunaskin@btconnect.com •
Open Air Museum / Science&Tech Museum
Ironworks, coal mining, brickworks, social and industrial history 43345

Kyle and Carrick District Library and Museum, 12 Main St, Ayr KA8 8ED • T: +44 1292 269141 ext 5227 • F: 611593 •
Dir.: *David T. Roy* •
Library with Exhibitions / Local Museum – 1934
Local history, rotating art coll 43346

Maclaurin Art Gallery, Rozelle Park, Monument Rd, Ayr KA7 4NQ • T: +44 1292 443708 • F: 442065 •
Man.: *Elizabeth I. Kwasnik* • Cur.: *Michael Bailey* •
Fine Arts Museum – 1976
Maclaurin coll of contemporary art, fine and applied art, sculpture 43347

Rozelle House Galleries, Rozelle Park, Monument Rd, Ayr KA7 4NQ • T: +44 1292 445447, 443708 • F: 442065 • rozelle.house@south-ayrshire.gov.uk • www.south-ayrshire.gov.uk/galleries •
Man.: *Elizabeth I. Kwasnik* •
Local Museum / Fine Arts Museum – 1982
Local hist, military exhibit, fine and applied art, Tom O'Shanter paintings 43348

Bacup

Natural History Society and Folk Museum, 24 Yorkshire St, Bacup OL13 9AE • T: +44 1706 873961 •
Pres.: *Arnold Barcroft* • Cur.: *Ken Simpson* •
Natural History Museum / Local Museum – 1878
Geology, Neolothic objects, local nature, household objects 43349

Baginton

Lunt Roman Fort, Coventry Rd, Baginton CV8 3AJ • T: +44 24 76303567 • F: 76832410 • luntromanfort@coventry.gov.uk • www.theherbert.org •
Dir.: *Roger Vaughan* •
Historical Museum
Roman cavalry fort 43350

Midland Air Museum, Coventry Airport, Baginton CV8 3AA • T: +44 24 76301033 • F: 76301033 •
Science&Tech Museum
Sir Frank Whittle Jet Heritage Centre, Wings over Coventry 43351

Bagshot

Archaeology Centre, 4-10 London Rd, Bagshot GU19 5HN • T: +44 1276 451181 • geoffreycole@shaat.netscapeonline.co.uk •
Dir.: *Geoffrey Cole* •
Archaeology Museum
Prehistoric, Roman, Saxon, medieval and post-medieval archaeological artefacts 43352

Baildon

Bracken Hall Countryside Centre, Glen Rd, Baildon BD17 5EA • T: +44 1274 584140 •
Dir.: *John Dallas* •
Natural History Museum – 1981
Wildlife, enviroment 43353

Bakewell

Chatsworth House, Bakewell DE45 1PP • T: +44 1246 565300 • F: 583536 • visit@chatsworth.org • www.chatsworth.org •
Decorative Arts Museum 43354

Old House Museum, Cunningham Pl, Bakewell DE45 1DD • T: +44 1629 813165, 815294 • bakerwellmuseum@tiscali.co.uk • www.oldhousemuseum.org.uk •
Local Museum – 1959
Costume, craftmen's tools, farming equip, lacework, toys, cameras 43355

Bala

Bala Lake Railway, Station Llanuwchllyn, Bala LL23 7DD • T: +44 1678 540666 • F: 540535 • www.bala-lake-railway.co.uk •
Science&Tech Museum 43356

Canolfan Y Plase, Plassey St, Bala LL23 7SW • T: +44 1678 520320 •
Chm.: *D.A. Jones* •
Local Museum
Artwork and local heritage 43357

Ballasalla

Rushen Abbey, Manx National Heritage, Ballasalla IM9 3DB • T: +44 1624 648000 • F: 648001 • enquiries@mnh.gov.im • www.storyofmann.com •
Dir.: *Stephen Harrison* •
Religious Arts Museum
Cistercian community life, archaeology 43358

Ballindalloch

Glenfarclas Distillery Museum, Ballindalloch AB37 9BD • T: +44 1807 500245 • F: 500234 • j&ggrant@glenfarclas.demon.co.uk •
Dir.: *J. Grant* •
Science&Tech Museum – 1973 43359

Ballycastle

Ballycastle Museum, 59 Castle St, Ballycastle BT54 6AS • T: +44 2820 762225 • F: 762515 • kmcgarry@moyle-council.org • www.moyle-council.org •
Historical Museum
Social hist, decorative, applied and fine art 43360

Larrybane and Carrick-A-Rede, Ballintoy, Ballycastle, Co. Antrim • T: +44 12657 62178, 31159 •
Local Museum 43361

Ballygrant

Finlaggan Centre, The Cottage, Ballygrant PA45 7QL • T: +44 1496 810629 • F: 810856 • LynMags@aol.com • www.islay.com •
Local Museum 43362

Ballymena

Arthur Cottage, 1b Church St, Ballymena BT43 6DF, mail addr: c/o Tourist Information, 76 Church St, Ballymena BT43 6DF • T: +44 28 25880781, 25638494 • F: 25638495 • tourist.information@ballymena.gov.uk • www.ballymena.gov.uk •
Dir.: *Sam Fleming* • Cur.: *May Kirkpatrick* •
Historical Museum 43363

Ballymena Museum, The Braid, 1-29 Bridge St, Ballymena BT43 5EJ • T: +44 28 25657161 • www.ballymena.gov.uk/tourism/ballymenamuseum.asp •
Dir.: *William Blair* •
Local Museum / Folklore Museum
Mixed local hist, folklife coll, farm machinery 43364

Royal Irish Regiment Museum, RIR Headquarter, Saint Patrick's Barracks, Ballymena BT43 7NX • T: +44 28 25661383 • F: 25661378 • hqririsħ@royalirishregiment.co.uk • www.royalirishregiment.co.uk •
Dir.: *M. Hegan* •
Military Museum
Military hist since 1689 43365

Ballymoney

Ballymoney Museum, c/o Ballymoney Town Hall, Townhead St, Ballymoney BT53 6BE • T: +44 28 27660245 • F: 27667659 • museum@ballymoney.gov.uk • www.ballymoneyancestry.com •
Man.: *Keith Beattie* •
Local Museum
Local hist since the earliest settlement in Ireland, motorcycle races, military hist, popular culture, art, crafts 43366

Leslie Hill Open Farm, Macfin Rd, Ballymoney BT53 6QL • T: +44 28 65663109, 65666803 • F: 65666803 • www.lesliehillopenfarm.co.uk •
Dir.: *James Leslie* •
Agriculture Museum
Horse-drawn farm machines 43367

Bamburgh

Bamburgh Castle, Bamburgh NE69 7DF • T: +44 1668 214515 • F: 214060 • bamburghcastle@aol.com • www.bamburghcastle.com •
Dir.: *Francis Watson-Armstrong* •
Historical Museum 43368

Grace Darling Museum, Radcliffe Rd, Bamburgh NE69 7AE • T: +44 1668 214465 • F: 214465 •
Cur.: *Christine Bell* •
Historical Museum – 1938 43369

Banbury

Banbury Museum, Spiceball Park Rd, Banbury OX16 2PQ • T: +44 1295 259855 • F: 672652 • banburymuseum@cherwell-dc.gov.uk • www.cherwell-dc.gov.uk/banburymuseum •
Dir.: *Simon Townsend* • Sc. Staff: *Natalie Chambers* (Education) • *Dale Johnson* (Exhibitions) •
Local Museum – 1958
Local hist since 1600, costume, canals and boat bldg – archives, library 43370

Upton House and Gardens, Banbury OX15 6HT • T: +44 1295 670266 • F: 671144 • uptonhouse@nationaltrust.org.uk • www.nationaltrust.org.uk •
Dir.: *Julie Smith* •
Fine Arts Museum / Decorative Arts Museum / Historic Site – 1948
Brussels tapestries, porcelain from Sèvres, Chelsea figurines, furniture, coll of internationally important paintings, 2nd Viscount Bearstead – gardens 43371

Banchory

Banchory Museum, Bridge St, Banchory AB31 5SX • T: +44 1771 622807 • F: 623558 • museums@aberdeenshire.gov.uk • www.aberdeenshire.gov.uk/heritage •
Local Museum – 1977
Hist of the Banchory area, social and domestic life, life and work of the local musician Scott Skinner, natural history, royal commemorative porcelain 43372

Crathes Castle, Crathes, Banchory AB31 5QJ • T: +44 1330 844525 • F: 844797 • crathes@nts.org.uk • www.nts.org.uk •
Man.: *Alexander Gordon* •
Decorative Arts Museum – 1951
Furnishings and decoration, painted ceiling from the end of the 16th c – garden 43373

Banff

Banff Museum, High St, Banff AB45 1AE • T: +44 1771 622807 • F: 623558 • museums@aberdeenshire.gov.uk • www.aberdeenshire.gov.uk/heritage •
Cur.: *William K. Mine* • Sc. Staff: *Andrew F. Hill* •
Local Museum – 1828
Banff silver, arms and armour 43374

Bangor, County Down

North Down Heritage Centre, Town Hall, Bangor Castle, Bangor, County Down BT20 4BT • T: +44 28 91270371, 91271200 • F: 91271370 • heritage@northdown.gov.uk • www.northdown.gov.uk/heritage •
Dir.: *I.A. Wilson* •
Local Museum – 1984
Local hist 43375

Bangor, Gwynedd

Gwynedd Museum and Art Gallery, Amgueddfa ac Oriel Gwynedd, Ffordd Gwynedd, Bangor, Gwynedd LL57 1DT • T: +44 1248 353368 • F: 370426 • gwynedd@gwynedd.gov.uk • www.gwynedd.gov.uk/museums •
Cur.: *Esther Roberts* •
Local Museum / Public Gallery – 1884
Archaeology, furniture, costume, textiles, prints 43376

Natural History Museum, c/o School of Biological Sciences, University of Wales, Deiniol Rd, Bangor, Gwynedd LL57 2UW • T: +44 1248 382315 • F: 371644 • c.bishop@bangor.ac.uk •
Cur.: *Dr. C. Bishop* •
Natural History Museum – 1900
Zoology 43377

Penrhyn Castle, Bangor, Gwynedd LL57 4HN • T: +44 1248 353084 • F: 371281 • penrhyncastle@nationaltrust.org.uk •
Dir.: *Joan Bayliss* •
Fine Arts Museum / Decorative Arts Museum – 1952
19th c furnished castle, neo-Norman architecture, industrial railway, dolls, Dutch and Italian masters 43378

Bardon Mill

Vindolanda Fort and Museum Chesterholm, Bardon Mill NE47 7JN • T: +44 1434 344277 • F: 344060 • info@vindolanda.com • www.vindolanda.com •
Dir.: *Robin Birley* •
Historical Museum / Archaeology Museum
Roman leather, textilles and wooden objects 43379

Barmouth

Barmouth Sailors' Institute, Ty Gwyn and Ty Crwn, The Quay, Barmouth LL42 1ET • T: +44 1341 241333 •
Historical Museum
Maritime heritage 43380

Barnard Castle

The Bowes Museum, Newgate, Barnard Castle DL12 8NP • T: +44 1833 690606 • F: 637163 • info@thebowesmuseum.org.uk • www.thebowesmuseum.org.uk •
Dir.: *Adrian Jenkins* • Sc. Staff: *Howard Coutts* (Ceramics) • *Joanna Hashagen* (Textiles) • *Emma House* (Art) • Head of Exhibitions: *Vivien Reid* • Development, Marketing: *Ruth Robson* •
Decorative Arts Museum / Fine Arts Museum / Local Museum – 1892
European painting, works by El Greco, Goya, Tiepolo, Boudin, Fragonard, pottery and porcelain, especially French, textiles, sculpture, furniture, other decorative, European Antiquities arts 43381

Barnet

Museum of Domestic Design and Architecture, Middlesex University, Cat Hill, Barnet EN4 8HT • T: +44 20 84115244 • F: 84115271 • moda@mdx.ac.uk • www.moda.mdx.ac.uk •
University Museum / Decorative Arts Museum
Silver studio coll, Sir J.M. Richards library, Peggy Angus archive; Charles Hassler coll, British and American domestic design archive, crown wallpaper archive 43382

Barnoldswick

Bancroft Mill Engine Trust, Gillians Ln, Barnoldswick BB8 5QR • T: +44 1282 865626 • www.pendletourism.com •
Dir.: *J. D. Gill* •
Science&Tech Museum
Corliss textile mill engine, weaving demo, hand fired boiler, textile related items 43383

Barnsley, South Yorkshire

Cannon Hall Museum, Bark House Ln, Barnsley, South Yorkshire S75 4AT • T: +44 1226 790270 • F: 792117 • cannonhall@barnsley.gov.uk • www.barnsley.gov.uk •
Man.: *Lynn Dunning* •
Fine Arts Museum 43384

U

Cooper Gallery, Church St, Barnsley, South Yorkshire S70 2AH • T: +44 1226 242905 • F: 297283 • coopergallery@barnsley.gov.uk • www.barnsley.gov.uk •
Public Gallery – 1980
17th, 18th and 19th c European paintings and English drawings and watercolours ... 43385

Worsbrough Mill Museum, Worsbrough Bridge, Barnsley, South Yorkshire S70 5LJ • T: +44 1226 774527 • F: 774527 • worsbroughmill@barnsley.co.uk •
Open Air Museum / Science&Tech Museum – 1976
Working 17th c water powered corn mill, working 19th c oil engine powered corn mill ... 43386

Barnstaple

Museum of Barnstaple and North Devon, The Square, Barnstaple EX32 8LN • T: +44 1271 346747 • F: 346407 • alison_mills@northdevon.gov.uk • www.devonmuseums.net/barnstaple •
Dir.: *Alison Miller* •
Local Museum – 1888
Natural and human hist of Northern Devon, industry, decorative arts, archaeology, geology, social hist, militaria – inc. the Royal Devon Yeomanry Museum ... 43387

Barrow-in-Furness

Dock Museum, North Rd, Barrow-in-Furness LA14 2PW • T: +44 1229 876400 • F: 811361 • dockmuseum@barrowbc.gov.uk • www.dockmuseum.org.uk •
Dir.: *Sabine Skae* •
Local Museum – 1906
Ship models, Vickers glass photographic negative archive, maritime, boats, social history, archaeology, geology, fine art, taxidermy ... 43388

Furness Abbey, Abbey Approach, Barrow-in-Furness LA13 0TJ • T: +44 1229 823420 • F: 823420 • furness.abbey@english-heritage.org.uk • www.english-heritage.org.uk •
Historic Site
Architectural stonework, sculpture ... 43389

Barrowford

Pendle Heritage Centre, Park Hill, Barrowford BB9 6JQ • T: +44 1282 661704 • F: 611718 •
Dir.: *E.M.J. Miller* •
Local Museum
Photos, doc and artefact reflecting the social hist 43390

Barton-upon-Humber

Baysgarth House Museum, Baysgarth Leisure Park, Caistor Rd, Barton-upon-Humber DN18 6AH • T: +44 1652 632318 • F: 636659 •
Local Museum
18th cent mansion house, porcelain, local hist, archeology and geology ... 43391

Basildon

The Motorboat Museum, Wat Tyler Country Park, Pitsea Hall Ln, Basildon SS16 4UH • T: +44 1268 550077 • F: 584207 • julie.graham@basildon.gov.uk • www.motorboatmuseum.org.uk •
Dir.: *Steven Prewer* •
Science&Tech Museum – 1986
Carstairs coll, Bert Savidge coll, motorboating (sports, leisure), in- and outboard motors, over 35 crafts – library ... 43392

Potland Museum, Langdon Visitor Centre, Third Av, Lower Dunton Rd, Basildon SS16 6EB • T: +44 1268 419103 • F: 546137 • langdon@essexwt.org.uk • www.essexwt.org.uk •
Local Museum – 1933 ... 43393

Basingstoke

Basing House, Redbridge Ln, Basingstoke RG24 7HB • T: +44 1256 467294 • musmat@hants.gov.uk • www.hants.gov.uk/museum/basingho •
Dir.: *Alan Turton* •
Archaeology Museum
Archaeological material from Basing House ... 43394

Milestones: Hampshire's Living History Museum, Basingstoke Leisure Park, Churchill Way W, Basingstoke RG21 6YR • T: +44 1256 477766 • F: 477784 • milestones.museum@hants.gov.uk • www.milestones-museum.com •
Cur.: *Gary Wragg* •
Historical Museum
Thornycroft and Tasker coll, Hampshire social and domestic hist coll, AA vehicle and memorabilia coll ... 43395

The Vyne, Sherborne Saint John, Basingstoke RG24 9HL • T: +44 1256 881337 • F: 881720 • thevyne@ntrust.org.uk •
Dir.: *J. Ingram* •
Decorative Arts Museum – 1958
Tudor house with renovations (1650-1760), Flemish glass, majolica tiles, panelling, furniture of Charles II, Queen Anne and Chippendale periods, rococo decoration ... 43396

Willis Museum, Old Town Hall, Market Pl, Basingstoke RG21 7QD • T: +44 1256 465902 • F: 471455 • sue.tapliss@hants.gov.uk • www.hants.gov.uk/museum/willis •
Dir.: *Sue Tapliss* •
Local Museum – 1930
Local hist, archaeology, natural sciences ... 43397

Bath

American Museum in Britain, Claverton Manor, Bath BA2 7BD • T: +44 1225 460503 • F: 469160 • info@americanmuseum.org • www.americanmuseum.org •
Dir.: *Sandra Barghini* •
Decorative Arts Museum – 1961
17th-19th c American decorative art and social hist, furnished period rooms, Indian section, 18th c tavern, folk art gallery, New Gallery housing the Dallas Pratt Coll of maps – library ... 43398

Bath Abbey Heritage Vaults, 13 Kingston Bldgs, Bath BA1 1LT • T: +44 1225 303314 • F: 429990 • heritagevaults@bathabbey.org •
Religious Arts Museum
Hist of the three churches ... 43399

Bath Postal Museum, 8 Broad St, Bath BA1 5LJ • T: +44 1225 460333 • F: 460333 • info@bathpostalmuseum.org • www.bathpostalmuseum.org •
Dir.: *Patrick Cassels* • Cur.: *Janine Marriott* •
Special Museum – 1979
4,000 years of communication, fron Sumarian tablets 2,000 BC to email ... 43400

Bath Royal Literary and Scientific Institution, 16-18 Queen Sq, Bath BA1 2HN • T: +44 1225 312084 • F: 442460 • enquiries@brlsi.org • www.brlsi.org •
Dir.: *Matt Williams* •
Association with Coll – 1824
Moore coll of fossils incl Type Specimens, Jenyns coll of natural hist, material-herbaria, Darwin letters and others from eminent naturalists, geology, ethnology, art, archaeology, mineralogy, early photography – library, archves ... 43401

Beckford's Tower and Museum, Lansdown Rd, Bath BA1 9BH • T: +44 1225 460705 • F: 481850 • beckford@bptrust.demon.co.uk • www.bath-preservation-trust.org.uk •
Local Museum – 1977/2000 ... 43402

Building of Bath Museum, The Countess of Huntingdon's Chapel, The Vineyards, Bath BA1 5NA • T: +44 1225 333895 • F: 445473 • enquiries@bathmuseum.co.uk • www.bathmuseum.co.uk •
Cur.: *Cathryn Spence* •
Special Museum ... 43403

The Herschel Museum of Astronomy, 19 New King St, Bath BA1 2BL • T: +44 1225 311342, 446865 • F: 446865 • admin@herschelpbt.fsnet.org.uk • www.bath-preservation-trust.org.uk •
Cur.: *Debbie James* •
Science&Tech Museum / Music Museum – 1978
Astronomical and 18th c musical instr, life in Georgian Bath – auditorium ... 43404

Holburne Museum of Art, Great Pulteney St, Bath BA2 4DB • T: +44 1225 466669 • F: 333121 • holburne@bath.ac.uk • www.bath.ac.uk/holburne •
Dir.: *Alexander Sturgis* • Cur.: *Amina Wright* •
Decorative Arts Museum / Fine Arts Museum – 1893
Decorative and fine art, furniture, silver, miniatures, porcelain, majolica, bronzes, netsukes, paintings by Stubbs, Gainsborough, Guardi and 18th c Old Masters together with 20th c work by leading British artist-craftspeople embracing textiles, ceramics, furniture, calligraphy, sculpture – library ... 43405

Museum of Bath at Work, Camden Works, Julian Rd, Bath BA1 2RH • T: +44 1225 318348 • F: 318348 • mobaw@hotmail.co.uk • www.bath-at-work.org.uk •
Dir.: *Stuart Burroughs* •
Science&Tech Museum – 1978
Coll of working machinery, hand-tools, brasswork, patterns, bottles and documents of all kinds, displayed as realistically as possible to convey an impression of the working life, local engineering, power generation, Horstmann car production hist – archive ... 43406

Museum of Costume, Bennett St, Bath BA1 2QH • T: +44 1225 477173 • F: 444743 • costume_enquiries@bathnes.gov.uk • www.museumofcostume.co.uk •
Dir.: *R. Harden* •
Special Museum – 1963
Some pre-18th c costumes, mainly 18th c to present day dress for men, women and children, incl haute couture pcs – study centre ... 43407

Museum of East Asian Art, 12 Bennett St, Bath BA1 2QJ • T: +44 1225 464640 • F: 461718 • info@meaa.org.uk • www.meaa.org.uk •
Cur.: *Michel Lee* •
Decorative Arts Museum – 1993
Ceramics, jades, bronzes, bamboo carvings and natural wood sculptures coming from China, Japan, Korea and Southeast Asia, Tibetan Buddhist bronzes, Chinese Imperial ceramics ... 43408

Number 1 Royal Crescent Museum, 1 Royal Crescent, Bath BA1 2LR • T: +44 1225 428126 • F: 481850 • no1museum@bptrust.org.uk • www.bath-preservation-trust.org.uk/museums/no1 •
Cur.: *Victoria Barwell* •
Historical Museum – 1970
Georgian Town House, example of Palladian architecture redecorated and furnished ... 43409

Roman Baths Museum, Pump Room, Stall St, Bath BA1 1LZ • T: +44 1225 477774 • F: 477243 • stephen_clews@bathnes.gov.uk • www.romanbaths.co.uk •
Cur.: *Stephen Clews* • *Susan Fox* (Collections) •
Archaeology Museum – 1897
Archaeological history of Bath in prehistoric, Roman, Saxon, medieval and later times ... 43410

The Royal Photographic Society, Fenton House, 122 Wells Rd, Bath BA2 3AH • T: +44 1225 325720 • lesley@rps.org • www.rps.org •
Dir.: *John Page* •
Special Museum – 1853/2005
Monthly exhibits of members of the Royal Photographic Society – library, archives ... 43411

Sally Lunn's Refreshment House and Kitchen Museum, 4 North Parade Passage, Bath BA1 1NX • T: +44 1225 811400, 461634 • F: 447090 • info@sallylunns.co.uk • www.sallylunns.co.uk •
Dir.: *Jonathan Overton* • *Julian Abraham* •
Historical Museum – 1680
Bakery and kitchen equip ... 43412

Victoria Art Gallery, Bridge St, Bath BA2 4AT • T: +44 1225 477233 • F: 477231 • jon_benington@bathnes.gov.uk • www.victoriagal.org.uk •
Man.: *Jon Benington* •
Fine Arts Museum / Decorative Arts Museum / Public Gallery – 1900
British paintings, sculptures, drawings and prints 18-20th c, ceramics and glass, European and British watches, British miniatures and silhouettes, craft, photography ... 43413

Bathgate

Bennie Museum, 9-11 Mansefield St, Bathgate EH48 4HU • T: +44 1506 634944 • bennie@benniemuseum.wanadoo.co.uk • www.benniemuseum.homestead.com •
Cur.: *William I. Millan* •
Local Museum
Local hist, early photos, glass from Bathgate glassworks and Victoriana ... 43414

Batley

Bagshaw Museum, Wilton Park, Batley WF17 0AS • T: +44 1924 326155 • F: 326164 • bagshaw.museum@kirklees.gov.uk • www.kirklees.gov.uk •
Dir.: *Kathryn White* • Cur.: *Brian Haigh* •
Local Museum – 1911
Antiquities of Britain and other areas, relics, ethnography, local hist, industries, natural hist, geology, textile industry, Oriental ceramics and finds, egyptology, tropical rainforest ... 43415

Batley Art Gallery, Market Pl, Batley WF17 5DA • T: +44 1924 326021 • F: 326308 • stephen.mcgrath@kirklees.gov.uk • www.kirklees.gov.uk •
Dir.: *Jenny Hall* •
Fine Arts Museum – 1948
Local arts and crafts, children's art, 19th-20th c art, works by Francis Bacon and Max Ernst – library 43416

Battle

Battle Abbey, Battle TN33 0AD • T: +44 14246 773792 • F: 775059 •
Historic Site ... 43417

Battle Museum of Local History, The Almonry, High St, Battle TN33 0EA • T: +44 1424 775955 • F: 772827 • enquiries@battlemuseum.org.uk • www.battlemuseum.org.uk •
Cur.: *Yvonne Cleland* •
Local Museum – 1956
Local hist, Romano-British ironwork, gunpowder industry, battle artifacts, diorama of the Battle of Hastings and reproductions of the Bayeux Tapestry ... 43418

Buckleys Yesterday's World, 89-90 High St, Battle TN33 0AQ • T: +44 1424 775378, 774269 • F: 893316 • info@yesterdaysworld.co.uk • www.yesterdaysworld.co.uk •
Dir.: *Annette Buckley* •
Special Museum
Advertising material and the contents of old shops, Victorian and Edwardian social hist ... 43419

Battlesbridge

Battlesbridge Motorcycle Museum, Maltings Rd, Battlesbridge SS11 7RF • T: +44 1268 575000 • F: 575001 • jim@battlesbridge.com •
Dir.: *Jim Gallie* •
Science&Tech Museum ... 43420

Beaminster

Beaminster Museum, Whitcombe Rd, Beaminster DT8 3NB • T: +44 1308 863623, 862880 • chemeng@btinternet.com •
Local Museum ... 43421

Beamish

Beamish, the North of England Open Air Museum, Regional Resource Centre, Beamish DH9 0RG • T: +44 191 3704000 • F: 3704001 • museum@beamish.org.uk • www.beamish.org.uk •
Dir.: *Miriam Harte* •
Open Air Museum / Local Museum – 1971
Period areas (1820s and 1913), early railways (1825), Town street, colliery village and farm (1913), manor house (1825) – photographic archive, reference library, trade catalogues ... 43422

Beaulieu

Buckler's Hard Village Maritime Museum, Buckler's Hard, Beaulieu SO42 7XB • T: +44 1590 616203 • F: 616283 •
Dir.: *Mike Lucas* •
Science&Tech Museum – 1963
Shipping, 18th c naval vessels, model of Hard in 1803, memorabilia on Nelson, Sir Francis Chichester, display cottages 18th c, recreation of vinage life, with models ... 43423

National Motor Museum, John Montagu Bldg, Brockenhurst, Beaulieu SO42 7ZN • T: +44 1590 614650 • F: 612655 • nmmt@beaulieu.co.uk • www.beaulieu.co.uk •
Dir.: *Marion Barnes* •
Science&Tech Museum – 1952
Cars, motorcycles, commercial vehicles, racing cars, photos, books, film – archives ... 43424

Beaumaris

Beaumaris Castle, Beaumaris LL58 8AB • T: +44 1248 810361 •
Archaeology Museum ... 43425

Beaumaris Gaol and Courthouse, Steeple Ln, Castle St, Beaumaris LL58 8ED • T: +44 1248 810921, 724444 • F: 750282 • agxlh@anglesey.gov.uk • www.anglesey.gov.uk •
Head: *Alun Gruffydd* •
Special Museum ... 43426

Museum of Childhood Memories, 1 Castle St, Beaumaris LL58 8AP • T: +44 1248 712498 • www.aboutbritain.com/museumofchildhoodmemories.htm •
Dir.: *Robert Brown* •
Special Museum – 1973
Audio and visual entertainment, pottery and glass, trains, cars, clockwork toys, ships and aeroplanes, art, educational toys, dolls, games, money boxes ... 43427

Beccles

Beccles and District Museum, Leman House, Beccles NR34 9ND • T: +44 1502 715722 • www.becclesmuseum.org.uk •
Cur.: *James Woodrow* •
Local Museum
Natural hist, geology, archaeology ... 43428

Beckenham

Bethlem Royal Hospital Archives and Museum, Monks Orchard Rd, Beckenham BR3 3BX • T: +44 20 32284307, 32284227 • F: 32284045 • museum@bethlemheritage.org.uk • www.bethlemheritage.org.uk •
Cur.: *J. Michael Phillips* •
Special Museum – 1967
Paintings and drawings by artists who have suffered from mental disorder, incl Richard Dadd, Louis Wain, Vaslav Nijinsky, William Kurelek, historical exhibits relating to Bethlem Hospital (the original 'Bedlam'), and to the history of psychiatry, mental health ... 43429

Bedale

Bedale Museum, Bedale Hall, Bedale DL8 1AA • T: +44 1677 423797 • F: 425393 •
Local Museum
Local hist, clothing, toys, craft tools, hand-drawn fire engine from 1748 ... 43430

Beddgelert

Sygun Museum of Wales, Sygun Copper Mine, Beddgelert LL55 4NE • T: +44 1766 890595 • F: 890595 • sygunruine@hotmail.com • www.syguncoppermine.co.uk •
Cur.: *S. Ward* •
Science&Tech Museum ... 43431

Beddington

Carew Manor and Dovecote, Church Rd, Beddington SM6 7NH • T: +44 20 87704781 • F: 87704777 • sutton.museum@ukonline.co.uk • www.sutton.gov.uk •
Man.: *Valary Murphy* •
Historic Site ... 43432

Bedford

Bedford Central Library Gallery, Harpur St, Bedford MK40 1PG • T: +44 1234 350931 •
Public Gallery ... 43433

Bedford Museum, Castle Ln, Bedford MK40 3XD • T: +44 1234 353323 • bedford.museum@bedford.gov.uk • www.bedfordmuseum.org •
Cur.: *Rosemary Brind* •
Local Museum – 1959
Local archaeological finds, hist, natural hist, medieval tile pavement, iron age bronze mirror, lace bobbins ... 43434

Cecil Higgins Art Gallery (temporarily closed), Castle Close, Castle Ln, Bedford MK40 3RP • T: +44 1234 211222 • F: 327149 • chag@bedford.gov.uk • www.cecilhigginsartgallery.org •
Cur.: *Victoria Partridge* •
Fine Arts Museum / Decorative Arts Museum – 1949
Fine and decorative arts, English watercolours,

international prints, English and continental porcelain, glass, furniture, silver, sculptures, lace, costume, Handley-Read coll of Victorian and Edwardian decorative arts, William Burges room 43435

Elstow Moot Hall, Church End, Elstow, Bedford MK42 9XT • T: +44 1234 266889 • countrysidesites@bedscc.gov.uk •
Historical Museum – 1951
17th c cultural history, local history, architecture 43436

John Bunyan Museum, Mill St, Bedford MK40 3EU • T: +44 1234 213722 • F: 213722 • bmeeting@dialstart.net •
Dir.: *Doreen Watson* •
Special Museum – 1946
"The Pilgrim's Progress" in over 170 languages and dialects, artefacts relating to John Bunyan's life and times – library 43437

Beetham

Heron Corn Mill and Museum of Papermaking, Waterhouse Mills, Beetham LA7 7AR • T: +44 15395 65027 • F: 65033 • info@heronmill.org •
Man.: *Audrey Steeley* •
Science&Tech Museum – 1975/1988
18th c working water driven corn mill, machinery, displays tell the story of 900 yrs of milling, displays of papermaking separate from Corn Mill in restored carter's barn – Corn mill, Museum of Papermaking 43438

Belfast

Crown Liquor Saloon, Great Victoria St, Belfast, Co. Antrim • T: +44 1232 249476 •
Decorative Arts Museum
Victorian ornamentation, woodwork, glass and tiles 43439

Fernhill House, The People's Museum, Glencairn Park, Belfast BT13 3PT • T: +44 28 90713810 • F: 90713810 • historydetective@hotmail.com • www.fernhillhouse.co.uk •
Historical Museum
Social, economic and military hist 43440

Lagan Lookout Centre, 1 Donegall Quay, Belfast BT1 3EA • T: +44 28 90315444 • F: 90311955 • lookout@laganside.com • www.laganside.com •
Science&Tech Museum
Hist of Belfast 43441

Natural History Museum, c/o School of Biology and Biochemistry, Queen's University, Medical and Biological Centre, 97 Lisburn Rd, Belfast BT9 7BL • T: +44 28 90975786 • F: 90975877 • sobb.office@qub.ac.uk • www.qub.ac.uk/bb/ •
Dir.: *Prof. W.I. Montgomery* •
Natural History Museum / University Museum
Educational zoology and geology collections 43442

Old Museum Arts Centre, 7 College Sq N, Belfast BT1 6AR • T: +44 28 90235053 • F: 90322912 • info@oldmuseumartscentre.freeserve.co.uk • www.oldmuseumartscentre.org •
Dir.: *Anne McReynolds* •
Public Gallery 43443

Ormeau Baths Gallery, 18a Ormeau Av, Belfast BT2 8HQ • T: +44 28 90321402 • F: 90312232 • admin@obgonline.net • www.obgonline.net •
Dir.: *Hugh Mulholland* •
Fine Arts Museum
Contemporary Irish and international art 43444

The People's Museum, Fernhill House, Glencairn Park, Belfast BT13 3PT • T: +44 28 90715599 • F: 90715582 •
Dir.: *Thomas G. Kirkham* •
Local Museum
Social and military coll 43445

The Police Museum, Brooklyn, Knock Rd, Belfast BT5 6LE • T: +44 28 90650222 ext 22499 • F: 90700124 • museum@psni.police.uk • www.psni.police.uk •
Man.: *Hugh Forrester* •
Military Museum
Includes uniforms and equipment, firearms, medals and other items relating to policing in Ireland from the early 19th c to date, database of RIC service records 1822-1922 43446

Royal Ulster Rifles Regimental Museum, The Royal Irish Rangers, 5 Waring St, Belfast BT1 2EW • T: +44 28 90232086 • F: 90232086 • rurmuseum@yahoo.co.uk • rurmuseum.tripod.com •
Cur.: *J. Knox* •
Military Museum – 1935
Uniforms, medals, history of the regiment 43447

Ulster Museum, Botanic Gardens, Belfast BT9 5AB • T: +44 28 90383000 • www.nmni.com/um/ •
Dir.: *John Wilson* • Sc. Staff: *S.B. Kennedy* (Fine and Applied Art) • *R. Warner* (Human History) • *P.S. Doughty* (The Sciences) •
Local Museum – 1892
International art, Irish painting, contemporary Irish art, silver ceramics, Williamite glass, Irish and European antiquities, treasure from Spanish Armada Galleass Girona, Irish botany, zoology, geology, wildlife art, industrial technology and history, numismatics, ethnography 43448

Whowhatwherewhenwhy W5, Odyssey, 2 Queen's Quay, Belfast BT3 9QQ • T: +44 28 90467700 • Montgomery@w5online.co.uk • www.w5online.co.uk •
Dir.: *Dr. Sally Montgomery* •
Science&Tech Museum
Sciences 43449

Benburb

Blackwater Valley Museum, 89 Milltown Rd, Tullymore Etra, Benburb BT71 7LZ • T: +44 28 90311156 • F: 90311264 • thelma@ulsterprovident.fsenet.co.uk •
Local Museum
Set of machinery, spinning, winding, warping, beaming, weaving, forge 43450

Benenden

Mervyn Quinlan Museum, Benenden Hospital, Goddards Green Rd, Benenden TN17 4AX • T: +44 1580 240333 • F: 241877 • jabbott@benenden.star.co.uk •
Cur.: *M. Quinlan* •
Historical Museum
Hist of the hospital 43451

Benson

Benson Veteran Cycle Museum, 61 Brook St, Benson OX10 6LH • T: +44 1491 838414 •
Science&Tech Museum
600 veteran and vintage cycles (1818-1930) 43452

Berkeley

Berkeley Castle, Berkeley GL13 9BQ • T: +44 1453 810332 • F: 512995 • info@berkeley-castle.com • www.berkeley-castle.com •
Man.: *E.J. Halls* •
Historical Museum
Paintings, furniture, tapestries, porcelain and silver 43453

Jenner Museum, The Chantry, Church Ln, Berkeley GL13 9BH • T: +44 1453 810631 • F: 811690 • manager@jennermuseum.com • www.jennermuseum.com •
Dir.: *D. Mullin* •
Special Museum – 1985
Illustration of the life and career of Edward Jenner, coll of memorabilia and personal possessions, computerized exhibition on immunology, medical science as founded by E. Jenner 43454

Berkhamsted

Dacorum Heritage, Museum Store, Clarence Rd, Berkhamsted HP4 3YL • T: +44 1442 879525 • F: 879525 • cpeet@dacht2.freeserve.co.uk • www.dacorumheritage.org.uk •
Cur.: *Catherine Peet* •
Local Museum
Hist of the borough of Dacorum 43455

Berkswell

Berkswell Village Museum, Lavender Hall Ln, Berkswell CV7 7BJ • T: +44 1676 533242 • www.baisley.co.uk •
Dir.: *D. Tracey* •
Local Museum
Local artefacts (19th-20th c), Saxon Parish of Berkswell 43456

Bernera

Bernera, Urras Eachdraibh Sgire Bhearnaraidh, Bernera PA86 9LZ • T: +44 1851 612331 •
Local Museum 43457

Berwick-upon-Tweed

Berwick Barracks, The Parade, Ravensdowne, Berwick-upon-Tweed TD15 1DE • T: +44 1289 304493 • F: 601999 • customers@english-heritage.org.uk • www.english-heritage.org.uk •
Cur.: *Andrew Morrison* •
Military Museum
Exhib on the hist of the British infantry, 1660-1880, barrack-room of the 1780s, an Army schoolroom of the 1860s with period figures, and other tableaux 43458

Berwick Borough Museum and Art Gallery, Clock Block, Berwick Barracks, Ravensdowne, Berwick-upon-Tweed TD15 1DQ • T: +44 1289 301869 • F: 330540 • museum@berwick-upon-tweed.gov.uk •
Dir.: *Chris Green* •
Local Museum / Fine Arts Museum – 1867
Local hist, decorative objects in ceramics, bronze and brass, paintings, natural history, archaeology – library 43459

Borough Museum and Art Gallery, Clock Block, Berwick Barracks, Parade, Berwick-upon-Tweed TD15 1DQ • T: +44 1782 301869 • F: 330540 • museum@berwick-upon-tweed.gov.uk • www.berwick-upon-tweed.gov.uk/leisure/museum.htm •
Cur.: *Chris Green* •
Local Museum / Decorative Arts Museum / Fine Arts Museum / Public Gallery – 1943
Staffordshire pottery, weapons, clocks, toys, local and civic hist, English paintings from 18th c to present 43460

King's Own Scottish Borderers Regimental Museum, The Barracks, Ravensdowne, Berwick-upon-Tweed TD15 1DG • T: +44 1289 307426 • F: 331928 • kosbmus@milnet.uk.net • www.kosb.co.uk/museum.htm •
Dir.: *W.N. Foster* •
Military Museum – 1951
Uniforms, medals, trophies, weapons, paintings and drawings, hist of the regiment – archives 43461

Lady Waterford Hall, Ford Village, Berwick-upon-Tweed TD15 2QG • T: +44 1890 820503 • F: 820384 • tourism@ford-and-etal.co.uk • www.fordetal.co.uk/ford.asp •
Cur.: *Dorien Irving* •
Fine Arts Museum – Mural paintings 43462

Paxton House Grannery, Paxton, Berwick-upon-Tweed TD15 1SZ • T: +44 1289 386291 • F: 386660 • info@paxtonhouse.com • www.paxtonhouse.com •
Dir.: *John Malden* •
Special Museum
country park 43463

Bessbrook

Derrymore House, Bessbrook, Co. Down • T: +44 1693 830353 •
Historical Museum
Local history 43464

Bettyhill

Strathnaver Museum, Clachan, Bettyhill KW14 7SS • T: +44 1641 521418 • projectmanager@strathnavermuseum.org.uk • www.strathnavermuseum.org.uk •
Local Museum / Archaeology Museum
Traditional collection of artefacts relating to the domestic and working lives of the people of Strathnaver Province, archaeological artefacts, a pictish stone in the garden. 43465

Betws-y-Coed

Betws-y-coed Motor Museum, Betws-y-Coed LL24 0AH • T: +44 1690 710760 •
Science&Tech Museum
Vintage and classic vehicles 43466

Conwy Valley Railway Museum, Old Goods Yard, Betws-y-Coed LL24 0AL • T: +44 1690 710568 • F: 710132 •
Cur.: *Colin M. Cartwright* •
Science&Tech Museum / Local Museum – 1973
Historic maps of Conwy Valley, memorabilia from narrow, gauge railways, hist of L.N.W.R. Railway, signalling working models all gauges 43467

Beverley

Beverley Art Gallery, Champney Rd, Beverley HU17 9BQ • T: +44 1482 392780 • F: 392779 • sally.hayes@eastriding.gov.uk •
Dir.: *Sally Hayes* • *Alison Brice* •
Public Gallery – 1906/1928
Regional art, contemporary art and craft, old paintings, prints, drawings and photographs, Fred Elwell coll 43468

Bewdley

Bewdley Museum, The Shambles, Load St, Bewdley DY12 2AE • T: +44 1299 403573 • F: 405306 • bewdley.museum@wyreforestdc.gov.uk • bewdleymuseum.tripod.com •
Dir.: *Monica Rees* •
Historical Museum / Local Museum – 1972
Crafts and industries of the Wyre Forest, pewter brass 43469

Severn Valley Railway, Railway Station, Bewdley DY12 1BG • T: +44 1299 403816 • F: 400839 • www.svr.co.uk •
Science&Tech Museum
Steam locomotives and pre-nationalisation coaches 43470

Bexhill-on-Sea

Bexhill Museum, Egerton Rd, Bexhill-on-Sea TN39 3HL • T: +44 1424 787950 • F: 787950 • museum@rother.gov.uk • www.bexhillmuseum.co.uk •
Cur.: *Julian Porter* •
Local Museum – 1914
Regional, natural and social history, geology, antiquities, ethnography 43471

Bexhill Museum of Costume and Social History, Manor Gardens, Upper Sea Rd, Bexhill-on-Sea TN40 1RL • T: +44 1424 210045 •
Special Museum 43472

Bexley

Hall Place, Bourne Rd, Bexley DA5 1PQ • T: +44 1322 526574 • F: 522921 • info@hallplace.org • www.hallplace.com •
Local Museum 43473

Bibury

Arlington Mill Museum, Arlington, Bibury GL7 5NL • T: +44 1285 740368, 740199 • F: 740368 •
Folklore Museum / Science&Tech Museum
Mill machinery, Victorian way of life 43474

Bickenhill

National Motorcycle Museum, Coventry Rd, Bickenhill B92 0EJ • T: +44 1675 443311 • F: 443310 • sales@nationalmotorcyclemuseum.co.uk • www.nationalmotorcyclemuseum.co.uk •
Dir.: *W.R. Richards* •
Science&Tech Museum – 1984
Ca. 800 machines covering the 60 golden years of British motorcycle manufacturing 43475

Bickleigh

Bickleigh Castle, Bickleigh EX16 8RP • T: +44 1884 855363 •
Owner: *Michael Boxall* •
Historical Museum
Agricultural objects and toys, model ships, civil war arms and armour 43476

Bideford

Burton Art Gallery and Museum, Kingsley Rd, Bideford EX39 2QQ • T: +44 1237 471455 • F: 473813 • burtonartgallery@torridge.gov.uk • www.burtonartgallery.co.uk •
Cur.: *John Butler* •
Public Gallery – 1951/1994
Watercolour paintings Hubert Coop and Ackland, Edwards, oils by Hunt, Opie, Fisher et al, ceramics, pewter, porcelain, Napoleonic ship models, local artefacts, touring exhibits 43477

Hartland Quay Museum, Hartland Quay, Bideford EX39 6DU • T: +44 1288 331353 •
Local Museum
History of the quay, coastal trades and industries, shipwreck, life-saving, Hartland Point Lighthouse, smuggling, geology and natural history 43478

Biggar

Biggar Gasworks Museum, c/o Biggar Museums Trust, Moat Park, Biggar ML12 6 DT • T: +44 1899 221050 • F: 221050 •
Science&Tech Museum
Original 1914 equiq of gasworks, early hist of the gas industry 43479

Brownsbank Cottage, Moat Park, Biggar ML12 6DT • T: +44 1899 221050 • F: 221050 •
Special Museum – 1992
Home of Hugh MacDiarmid, 1951-1978 43480

Gladstone Court Museum, North Back Rd, Biggar ML12 6DT • T: +44 1899 221050 • F: 221050 • margaret@bmtrust.freeserve.co.uk • www.biggar-net.co.uk •
Dir.: *Margaret Brown* •
Local Museum – 1968
Reconstructed street with various shops and trades 43481

Greenhill Covenanter's House, Burn Braes, Biggar ML12 6DT • T: +44 1899 221050 • F: 221050 • margaret@bmtrust.freeserve.co.uk • www.biggar-net.co.uk •
Dir.: *Margaret Brown* •
Historical Museum – 1981
17th c furniture and artefacts 43482

John Buchan Centre, Broughton, Biggar ML12 6HQ • T: +44 1899 830223, 221050 •
Special Museum
Carreer and achievements of John Buchan as a novelist, lawyer, politician, soldier, historian, biographer and Governor-General of Canada, photographs, books, personal possessions of John Buchan and his family 43483

Moat Park Heritage Centre, Kirkstyle, Biggar ML12 6DT • T: +44 1899 221050 • F: 221050 • margaret@bmtrust.freeserve.co.uk • www.biggar-net.co.uk •
Dir.: *Margaret Brown* •
Local Museum – 1988
Embroidery, archaeology – local archives 43484

Billericay

Barleylands Farm Museum, Barleylands Rd, Billericay CM11 2UD • T: +44 1268 290229 • F: 290222 • farmcentre@barleylands.co.uk • www.barleylands.co.uk •
Dir.: *Chris Philpot* •
Agriculture Museum – 1984
Tractors, steam engines, threshing machines, potato farming, diry equip 43485

Cater Museum, 74 High St, Billericay CM12 9BS • T: +44 1277 622023 •
Cur.: *Christine Brewster* •
Local Museum – 1960
Local and regional history, Victorian room settings, WWII exhibition, Zeppelin airship L32 which was destroyed near Billericay in 1916 43486

Billingham

Billingham Art Gallery, Queensway, Billingham TS23 2LN • T: +44 1642 397590, 527911 • F: 397594, 527912 • billinghamartgallery@stockton.gov.uk • www.stockton.gov.uk/museums •
Dir.: *Graham Bowen* • Sc. Staff: *Mark Rowland Jones* • *Kirstie Garbutt* •
Public Gallery 43487

U

Birchington

Pembroke Lodge Patterson Heritage Museum and Family History Centre, 4 Station Approach, Birchington CT7 9RD • T: +44 1843 841649 • pattersonheritage@tesco.net • www.pattersonheritage.co.uk •
Head: *J.J. Patterson-O'Regan* •
Historical Museum
Social Hist ... 43488

Quex Museum, House and Gardens, Quex Park, Birchington CT7 0BH • T: +44 1843 842168 • F: 846661 • enquiries@quexmuseum.org • www.quexmuseum.org •
Dir.: *Julia Walton* • Cur.: *Malcolm Harman* •
Ethnology Museum / Natural History Museum / Decorative Arts Museum – 1896
African and Asian natural hist and ethnography, primates, bovidae, dioramas of African and Asian animals, archaeology, firearms, oriental fine arts, European & English ceramics, Chinese imperial porcelain – library ... 43489

Birkenhead

Birkenhead Priory and Saint Mary's Tower, Priory St, Birkenhead CH41 5JH • T: +44 151 6661249 • F: 6661249 •
Religious Arts Museum
Hist of the bldg and its development ... 43490

Shore Road Pumping Station, Hamilton St, Birkenhead CH41 6DN • T: +44 151 6501182 • F: 6663965 •
Science&Tech Museum
Giant restored steam engine ... 43491

Williamson Art Gallery and Museum, Slatey Rd, Birkenhead CH43 4UE • T: +44 151 6524177 • F: 6700253 • willamsonartgallery@wirral.gov.uk •
Fine Arts Museum / Decorative Arts Museum – 1928
Paintings, sculpture, etchings, pottery, porcelain, glass, silver, furniture, local history, geology, numismatics, porcelain, shipping history, motor vehicle hist – archive ... 43492

Wirral Museum, Town Hall, Hamilton Sq, Birkenhead CH41 5BR • T: +44 151 6664010 • F: 6663965 •
Dir.: *Colin Simpson* •
Local Museum – 2001
Social, industrial and commercial hist ... 43493

Wirral Transport Museum and Bierkenhead Tramways, 1 Taylor St, Birkenhead CH41 1BG • T: +44 151 6472128 •
Man.: *Robert Tennant* •
Science&Tech Museum
Transport related working coll incl trams, buses, motorcycles and cars – library ... 43494

Birmingham

Aston Hall, Birmingham Museums, Trinity Rd, Aston, Birmingham B6 6JD • T: +44 121 3270062 • F: 3277162 • www.bmag.org.uk •
Head: *Nick Ralls* •
Decorative Arts Museum – 1864
Jacobean house dating from 1616-1635, period furniture and rooms ... 43495

Aston Manor Transport Museum, Old Tram Depot, 208-216 Witton Ln, Birmingham B6 6QE • T: +44 121 3223398 • F: 4494768 • info@amrtm.org • www.amrtm.org •
Science&Tech Museum
Over 100 vehicles, two unrestored tramcars, scale model tram, trolleybus layout ... 43496

Barber Institute of Fine Arts, c/o University of Birmingham, Edgbaston Park Rd, Birmingham B15 2TS • T: +44 121 4147333 • F: 4143370 • info@barber.org.uk • www.barber.org.uk •
Dir.: *Prof. Richard Verdi* • Sen. Cur.: *Dr. P. Spencer-Longhurst* •
Fine Arts Museum – 1932
European art up to the 20th c, works by Bellini, Veronese, Rubens, Rembrandt, Frans Hals, Gainsborough, Reynolds, Degas, Gauguin ... 43497

Biological Sciences Collection, c/o School of Biological Sciences, University of Birmingham, Birmingham B15 2TT • T: +44 121 4145465 • F: 4145925 •
Cur.: *James Hamilton* •
Natural History Museum
herbarium ... 43498

Birmingham Institute of Art and Design, c/o University of Central England, Gosta Green, Birmingham B4 7DX • T: +44 121 3315878 • F: 3317876 • sharon.shakespeare@uce.ac.uk • www.biad.uce.ac.uk •
Dean: *Prof. Mick Durman* •
Fine Arts Museum ... 43499

Birmingham Museum and Art Gallery, Chamberlain Sq, Birmingham B3 3DH • T: +44 121 3032834 • F: 3031394 • bmag-enquiries@birmingham.gov.uk • www.bmag.org.uk •
Dir.: *Rita McLean* • Cur.: *Martin Ehis* •
Fine Arts Museum / Decorative Arts Museum / Local Museum / Ethnology Museum / Archaeology Museum – 1885
British and European painting and sculpture, Pre.Raphaelite paintings and drawings, contemporary art, drawings and watercolours, ceramics, metalwork, costume and textiles, jewellery and stained glass, coll from India, China, Japan and the Far East, archaeology and ethnography, local history – picture library ... 43500

Blakesley Hall, Blakesley Rd, Yardley, Birmingham B25 8RN • T: +44 121 4648193 • F: 4640400 • bmag.enquiries@bcc.gov.uk • www.bmag.org.uk •
Man.: *Jean McArdle* •
Historical Museum – 1935
Historic and archaeological objects, period rooms in Tudor house ... 43501

Bournville Centre for Visual Arts, Linden Rd, Birmingham B30 1JX • T: +44 121 3315777 • F: 3315779 • tom.jones@uce.ac.uk • www.uce.ac.uk •
Dir.: *Prof. Tom Jones* •
University Museum / Public Gallery ... 43502

Chamberlain Museum of Pathology, c/o Medical School, University of Birmingham, Edgbaston, Birmingham B15 2TT • T: +44 121 4144014 • F: 4144019 • e.l.jones@bham.ac.uk •
Special Museum
Hist of medical education, human organs, anatomical coll ... 43503

Danford Collection of West African Art and Artefacts, c/o Centre of West African Studies, University of Birmingham, Edgbaston, Birmingham B15 2TT • T: +44 121 4145128 • F: 4143228 • cwas@bham.ac.uk • www.cwas.bham.ac.uk •
Cur.: *Dr. S. Brown* • *Dr. James Hamilton* •
Ethnology Museum / University Museum
Carvings, metalwork, textiles, paintings, domestic objects ... 43504

Ikon Gallery, 1 Oozells Sq, Brindleyplace, Birmingham B1 2HS • T: +44 121 2480708 • F: 2480709 • art@ikon-gallery.co.uk • www.ikon-gallery.co.uk •
Dir.: *Jonathan Watkins* • Dep. Dir.: *Graham Halstead* • Cur.: *Michael Stanley* • *Deborah Kermode* • *Nigel Prince* • *Saira Holmes* •
Public Gallery ... 43505

Lapworth Museum of Geology, c/o Earth Sciences, University of Birmingham, Edgbaston Park Rd, Birmingham B15 2TT • T: +44 121 4147294 • F: 4144942 • lapworth@contacts.bham.ac.uk • www.lapworth.bham.ac.uk •
Dir.: *Prof.Dr. M.P. Smith* • Cur.: *J.C. Clatworthy* •
University Museum / Natural History Museum – 1880
Palaeozoic fossils of the West Midlands and Welsh Borders, UK, fossil fish, minerals, stone implements – archive ... 43506

MAC, Midlands Art Centre, Cannon Hill Park, Egbaston, Birmingham B12 9QH • T: +44 121 4403838 • F: 4464372 • enquiries@macarts.co.uk • www.macarts.co.uk •
Dir.: *Dorothy Wilson* •
Public Gallery ... 43507

Museum of the Jewellery Quarter, 75-79 Vyse St, Hockley, Birmingham B18 6HA • T: +44 121 5543598 • F: 5549700 • www.bmag.org.uk •
Cur.: *Victoria Emmanuel* •
Special Museum
Old family run jewellery making factory, history of Birmingham jewellery industry ... 43508

Patrick Collection Motor Museum, 180 Lifford Ln, Birmingham B30 3NK • T: +44 121 4863399 • F: 4863388 •
Dir.: *J.A. Patrick* •
Science&Tech Museum – 1985
Motor vehicles since 1904 – reference library ... 43509

Sarehole Mill, Colebank Rd, Hall Green, Birmingham B13 0BD • T: +44 121 7776612 • F: 4640400 • www.bmag.co.uk •
Science&Tech Museum – 1969
Restored water mill (1760), hist of rural life, corn production, connections with Matthew Bovlton and J.R.R. Tolkien ... 43510

Selly Manor Museum, Cnr Maple and Sycamore Rds, Birmingham B30 2AE • T: +44 121 4720199 • F: 4156410 • sellymanor@bvt.org.uk • www.bvt.org.uk/sellymanor •
Man.: *Gillian Ellis* •
Decorative Arts Museum
Laurence Cadbury coll of Vernacular furniture (1500-1730) ... 43511

Soho House Museum, Soho Av, Hardsworth, Birmingham B18 5LB • T: +44 121 5549122 • F: 5545929 • www.bmag.org.uk •
Science&Tech Museum
Industrial pioneer Matthew Boulton (1766-1809), Lunar Society, scientists, engineers and thinkers, 18th c metalwork, ceramics & furniture ... 43512

Thinktank, The Museum of Science and Discovery, Millennium Point, Curzon St, Birmingham B4 7XG • T: +44 121 2022222 • F: 2022280 • findout@thinktank.ac • www.thinktank.ac •
Science&Tech Museum – 2001
Science, hist ... 43513

Tyseley Loco Works Visitor Centre, 670 Warwick Rd, Tyseley, Birmingham B11 2HL • T: +44 121 7084962 • F: 7084963 • office@vintagetrains.co.uk • www.vintagetrains.co.uk •
Dir.: *C.M. Whitehouse* •
Science&Tech Museum / Open Air Museum – 1969
Steam locomotives, passenger trains driving exp. ... 43514

University Archaeology Museum, c/o Institute of Archaeology and Antiquity, University of Birmingham, Edgbaston, Birmingham B15 2TT • T: +44 121 4145497 • F: 4143595 • g.b.shepherd@bham.ac.uk • www.iaa.bham.ac.uk/museum/museum.htm •
Cur.: *James Hamilton* •
Archaeology Museum
Study coll, about 1,700 Greek, Mycenean, Roman and Egyptian artefacts ... 43515

Weoley Castle, Selly Oak, Birmingham B29 5RX, mail addr: c/o Birmingham Museum and Art Gallery, Chamberlain Sq, Birmingham B3 3DH • T: +44 121 3031675, 3034698 • bmag_enquiries@birmingham.gov.uk •
Historical Museum ... 43516

West Midlands Police Museum, 641 Stratford Rd, Sparkhill, Birmingham B11 6EA • T: +44 121 6267181 • F: 6267066 • d.a.cross@west-midlands.police.uk •
Historical Museum
Police memorabilia ... 43517

Birsay

Orkney Farm and Folk Museum, Kirbister Hill, Birsay KW17 2LR • T: +44 1856 771268 • F: 771268 • www.orkneyislands.com/guide/museums/farm.html •
Cur.: *Katrina Mainland* •
Agriculture Museum / Folklore Museum ... 43518

Birstall

Oakwell Hall Country Park, Nutter Ln, Birstall WF17 9LG • T: +44 1924 326240 • F: 326249 • oakwell.hall@kirklees.gov.uk • www.oakwellhallcounrypark.co.uk •
Dir.: *Eric Brown* •
Historic Site – 1928
Elizabethan Manor House displayed as a family home of the 1690s, 17th c furniture, decorations – gardens ... 43519

Bishop Auckland

Binchester Roman Fort, Bishop Auckland DL14 8DJ, mail addr: c/o Durham County Council, Archaeology Dept., County Hall, Durham DH1 5UB • T: +44 191 3708712 • F: +44 191 3708897 • archaeology@durham.gov.uk • www.durham.gov.uk/binchester •
Keeper: *Niall Hammond* (Archaeology) •
Archaeology Museum
Roman fort of Vinovia ... 43520

Bishopbriggs

Thomas Muir Museum, Library, 170 Kirkintilloch Rd, Bishopbriggs G64 2LX • T: +44 141 7751185, 7724513 •
Historical Museum
Life of the 18th c Scottish radical and reformer Th. Muir ... 43521

Bishop's Castle

House on Crutches Museum, Tan House, Church St, Bishop's Castle SY9 5AA • T: +44 1588 630007 • hocmuseum@tan-house.demon.co.uk •
Local Museum
Agricultural tools, household objects, clothing, shop goods ... 43522

Bishop's Stortford

Bishop's Stortford Museum, South Rd, Bishop's Stortford CM23 3JG • T: +44 1279 641746 • F: 467171 • museum@rhodesbishopsstortford.org.uk • www.hertsmuseums.org.uk •
Local Museum
Local trades, archaeology, architecture, local and family hist ... 43523

Rhodes Memorial Museum and Commonwealth Centre, South Rd, Bishop's Stortford CM23 3JG • T: +44 1279 651746 • F: 467171 • rhodesmuseum@freeuk.com • www.hertsmuseums.org.uk •
Cur.: *Hannah Kay* •
Special Museum – 1938
Life and work of statesman Cecil John Rhodes, his birthplace, memorabilia, photos, documents, African ethnography and artwork ... 43524

Bishop's Waltham

Bishops Waltham Museum, Brook St, Bishop's Waltham SO32 1EX • T: +44 1489 895407 •
Cur.: *E. Hiscock* •
Local Museum
Local hist, Victorian pottery, brick making, brewing, local trades, shops ... 43525

Blackburn, Lancashire

Blackburn Museum and Art Gallery, Museum St, Blackburn, Lancashire BB1 7AJ • T: +44 1254 667130 • F: 685541 • paul.flintoff@blackburn.gov.uk • www.blackburn.gov.uk/museum •
Dir.: *S. Rigby* • Sc. Staff: *Paul Flintoff* • *Nick Harling* •
Fine Arts Museum / Historical Museum – 1874
Natural hist, geology, ethnography, applied and fine arts, Hart coll of coins, philatelics, medieval illuminated manuscripts, books, local hist, Lewis coll of Japanese prints, ornithology ... 43526

Lewis Textile Museum, 3 Exchange St, Blackburn, Lancashire BB1 7JN • T: +44 1254 667130 • F: 685541 • paul.flintoff@blackburn.gov.uk • www.blackburn.gov.uk/museum •
Dir.: *S. Rigby* • Sc. Staff: *Paul Flintoff* • *Nick Harling* •
Decorative Arts Museum / Science&Tech Museum – 1938
18th and 19th c cotton machinery, art gallery, hist of textile production ... 43527

Blackgang

Blackgang Sawmill and Saint Catherine's Quay, Blackgang PO38 2HN • T: +44 1983 730330 • F: 731267 • info@blackgangchine.com • www.blackgangchine.com •
Dir.: *Simon Dabell* •
Historical Museum
Local coastal history and country life ... 43528

Blackpool

Grundy Art Gallery, Queen St, Blackpool FY1 1PX • T: +44 1253 478170 • F: 478172 • grundyartgallery@blackpool.gov.uk • www.blackpool.gov.uk •
Cur.: *Stuart Tulloch* •
Fine Arts Museum – 1911
Contemporary paintings, watercolors, ivory, porcelain, Victorian and Edwardian oils, modern British painting ... 43529

Ripley's Believe It or Not Museum, 5-6 Ocean Blvd, Blackpool Pleasure Beach, Blackpool FY4 1EZ. • T: +44 870 4445566 • F: +44 1253 401098 • info@ripleysblackpool.com • www.ripleysblackpool.com •
Dir.: *Andy Ansell* •
Special Museum ... 43530

Blackridge

Blackridge Community Museum, Library, Craig Inn Centre, Blackridge EH48 3RJ • T: +44 1506 776347 • F: 776190 • museums@westlothian.org.uk • www.wlonline.org •
Local Museum ... 43531

Blaenau Ffestiniog

Gloddfa Ganol Slate Mine, Blaenau Ffestiniog LL41 3NB • T: +44 1766 830664 •
Science&Tech Museum ... 43532

Llechwedd Slate Caverns, Blaenau Ffestiniog LL41 3NB • T: +44 1766 830306 • F: 831260 • quarrytours@aol.com • www.llechwedd-slate-caverns.co.uk •
Science&Tech Museum – 1972
The earliest slate extraction, sawing and dressing equipment, coll of narrow gauge railway stock ... 43533

Blaenavon

Big Pit National Coal Museum of Wales, Blaenavon NP4 9XP • T: +44 1495 790311 • F: 792618 • bigpit@nmgw.ac.uk • www.nmgw.ac.uk/bigpit •
Man.: *Peter Walker* • Cur.: *Ceri Thomson* •
Science&Tech Museum – 1983
Preserved colliery shaft, winding engine and underground areas giving public guess & a former working colliery ... 43534

Pontypool and Blaenavon Railway, 13a Broad St, Blaenavon NP4 9ND • T: +44 1495 792263 • pbrsec@aol.com • www.pontypool-and-blaenavon.co.uk •
Science&Tech Museum
Locomotives, coaches ... 43535

Blair Atholl

Atholl Country Life Museum, Old School, Blair Atholl PH18 5SP • T: +44 1796 481232 • John.museum@virgin.net • www.athollcountrylifemuseum.org •
Cur.: *John Cameron* •
Local Museum – 1982
Caledonian Challenge Shield for rifle shooting, nature display (birds & animals of the area); postal, road & rail communications; Horse Harness ... 43536

Blair Castle, Blair Atholl PH18 5TL • T: +44 1796 481207 • F: 481487 • office@blair-castle.co.uk • www.blair-castle.co.uk •
Historical Museum
Furniture, paintings, arms and armour, china, lace and embroidery, masonic regalia and other treasures 43537

Blairgowrie

Errol Station Railway Heritage Centre, c/o Errol Station Trust, 48 Moyness Park Dr, Blairgowrie PH10 6LX • T: +44 15754 222 •
Science&Tech Museum
Station from 1847 on the main Aberdeen-Glasgow line ... 43538

Blandford Forum

Blandford Forum Museum, Bere's Yard, Market Pl, Blandford Forum DT11 7HQ • T: +44 1258 450388 • pjandrews@btinternet.com • www.blandfordmuseum.org •
Cur.: *Dr. Peter Andrews* • Dep. Cur.: *Michael le Bas* •
Historical Museum
Illustration of the life and occupations of people in Blandford and its neighbourhood from prehistoric times to the present day, good coll of domestic equipment, local militaria and Victorian and Edwardian costumes, Model Railway ... 43539

Cavalcade of Costume, Lime Tree House, The Plocks, Blandford Forum DT11 7AA • T: +44 1258 453006 • cavalcademuseum@tiscali.co.uk • www.cavalcadeofcostume.co.uk •
Special Museum
Costumes and accessories from 18th c to 1960s 43540

Royal Signals Museum, Blandford Camp, Blandford Forum DT11 8RH • T: +44 1258 482248 • F: 482084 • info@royalsignalsmuseum.com • www.royalsignalsmuseum.com •
Dir.: *Cliff Walters* • Cur.: *Stella McIntyre* •
Military Museum – 1935
Communication equipment, uniforms, prints and photos relating to the Royal Signals and Army communications 43541

Blantyre

David Livingstone Centre, 165 Station Rd, Blantyre G72 9BY • T: +44 1698 823140 • F: 821424 • kcarruthers@nts.org.uk •
Dir.: *Karen Carruthers* •
Special Museum – 1929
Memorabilia on explorer David Livingstone in the house of his birth 43542

Bletchley

Bletchley Park, Wilton Av, Bletchley MK3 6EB • T: +44 1908 640404 • F: 274381 • info@bletchleypark.org.uk • www.bletchleypark.org.uk •
Dir.: *Christine Large* •
Science&Tech Museum / Historical Museum
Cryptography, WW 2 codebreaking computer, radar and electronics, WW 2 memorabilia 43543

Blindley Heath

Times Past - Bygones of Yester Year, Bannisters Bakery, Eastbourne Rd, Blindley Heath RH7 6LQ • T: +44 1342 832086 • enquiries@bannistersbakery.co.uk • www.bannistersbakery.co.uk •
Owner: *R. Bannister* •
Special Museum
Household and rural bakery 43544

Bloxham

Bloxham Village Museum, Courthouse, Church St, Bloxham OX15 4ET • T: +44 1295 721256, 720801 • peter.barwell@btinternet.com •
Chm.: *John Phillips* •
Local Museum
Past life in this village 43545

Bodelwyddan

Bodelwyddan Castle Museum, Castle, Bodelwyddan LL18 5YA • T: +44 1745 584060 • F: 584563 • enquiries@bodelwyddan-castle.co.uk • www.bodelwyddan-castle.co.uk •
Dir.: *Dr. Kevin Mason* •
Fine Arts Museum – 1988
19th c colls of portraits, sculptures and furniture from the National Portrait Gallery, the Victoria & Albert Museum and the Royal Academy 43546

Bodmin

Bodmin and Wenford Railway, General Station, Bodmin PL31 1AQ • T: +44 1208 73666 • F: 77963 • manager@bodminandwenfordrailway.co.uk • www.bodminandwenfordrailway.co.uk •
Dir.: *Roger J. Webster* •
Science&Tech Museum
Various locomotives, carriages, wagons and steam crane 43547

Bodmin Town Museum, Mount Folly Sq, Bodmin PL31 2HQ • T: +44 1208 77067 • F: 264764 • bodmin.museum@ukonline.co.uk •
Historical Museum / Natural History Museum
Artefacts, docs, photos and text illustrating Bodmin's Local und social hist 43548

Lanhydrock, Bodmin PL30 5AD • T: +44 1208 265950 • F: 265959 • lanhydrock@nationaltrust.org.uk • www.nationaltrust.org.uk •
Dir.: *Rebecca Brookes-Sullivan* •
Historical Museum
18th c and Brussels and Mortlake tapestries, family portraits painted by Kneller and Romney, books, furniture, kitchens, nurseries, interiors 43549

Military Museum, The Keep, Bodmin PL31 1EG • T: +44 1208 72810 • F: 269516 • dclimus@talk21.com •
Cur.: *Trevor Stipling* •
Military Museum – 1923
Armaments, uniforms, military hist of the regiment, small arms, medals, badges, pictures 43550

Pencarrow House, Washaway, Bodmin PL30 3AG • T: +44 1208 841369 • F: 841722 • pencarrow@aol.com • www.pencarrow.co.uk •
Dir.: *Lady Iona Molesworth-Saint Aubyn* •
Fine Arts Museum / Decorative Arts Museum
Family portraits, including works by Reynolds, Northcote, Raceburn, Hudson and Devis, paintings by Samuel Scott and Richard Wilson, furnitures of the 18th and early 19th c 43551

Bognor Regis

Bognor Regis Museum, 69 High St, Bognor Regis PO21 1RY • T: +44 1243 865636 • d.jennings@freenet.co.uk •
Cur.: *David Jennings* •
Local Museum
Social and local hist (19th-20th c), wireless and constructional toy coll 43552

Bolton

Bolton Museum and Art Gallery, Le Mans Crescent, Bolton BL1 1SE • T: +44 1204 332211 • F: 332241 • museums@bolton.gov.uk • www.boltonmuseums.org.uk •
Dir.: *Steve Garland* • Cur.: *J. Shaw* (Art) • *Angela Thomas* (Human History) •
Natural History Museum / Fine Arts Museum / Local Museum / Ethnology Museum / Archaeology Museum / Decorative Arts Museum / Historical Museum – 1893
Natural hist, geology, fossils, archaeology, Egyptian coll, early English watercolors, British sculpture with works by Epstein, Moore, Hepworth, costumes, papers of inventor Samuel Crompton, local and social hist 43553

Bolton Steam Museum, Mornington Rd, Bolton BL1 4EU, mail addr: John Phillp, 84 Watkin Rd, Clayton-le-Woods PR6 7PX • T: +44 1257 265003 • john.phillp@btinternet.com • www.nmes.org •
Head: *John Phillp* •
Science&Tech Museum – 1992
Stationary steam engines from textile industry, cotton mill 43554

Hall i'th' Wood Museum, Green Way, off Crompton Way, Bolton BL1 8UA • T: +44 1204 332370 • F: 332241 • museum@bolton.gov.uk • www.boltonmuseums.org.uk •
Head: *Elizabeth Shaw* •
Historical Museum – 1902
Memorabilia of inventor Samuel Crompton, furniture, place where S. Compton inverted the spinning mule 43555

Smithills Hall Museum, Bolton Museum, Smithills Dean Rd, Bolton BL1 7NP • T: +44 1204 332377 • www.boltonmuseums.org.uk •
Man.: *Liz McNabb* •
Historical Museum – 1962
800 years old manor – garden/ parkland 43556

Bolventor

Daphne du Maurier's Smugglers's at Jamaica Inn, Old School, Jamaica Inn, Bolventor PL15 7TS • T: +44 1566 86250 • F: 86177 • enquiry@jamaicainn.co.uk • www.jamaicainn.co.uk •
Cur.: *Rose Mullins* •
Special Museum – 1861
Museum founded by Walter Potter, Victorian naturalist and taxidermist, Daphne Du Maurier's novel 'Jamaica Inn', smuggling history 43557

Bo'ness

Bo'ness Heritage Area, 17-19 North St, Bo'ness EH51 9AQ • T: +44 1506 825855 • F: 828766 •
Local Museum 43558

Kinneil Museum and Roman Fortlet, Duchess Anne Cottages, Kinneil Estate, Bo'ness EH51 0PR • T: +44 1506 778530 • callendar.house@falkirk.gov.uk • www.falkirkmuseum.org •
Dir.: *Maria Gillespie* •
Historical Museum / Archaeology Museum – 1976
Estate hist, medieval village and church, excavations 43559

Scottish Railway Museum, Bo'ness Station, off Union St, Bo'ness EH51 9AQ • T: +44 1506 822298 • F: 828766 • museum@srps.org.uk • www.srps.org.uk •
Dir.: *John Burnie* •
Science&Tech Museum
Scottish railways history, steam & diesel locomotives, coaches & wagons, equipment – archives, working railway 43560

Boosbeck

Margrove - South Cleveland Heritage Centre, Margrove Park, Boosbeck TS12 3BZ • T: +44 1287 610368 • F: 610368 •
Local Museum / Natural History Museum
Moors, valley and coast of South Cleveland, historic maps 43561

Boston

Boston Guildhall Museum, South St, Boston PE21 6HT • T: +44 1205 365954 • heritage@originalboston.freeserve.co.uk •
Cur.: *Andrew Crabtree* •
Local Museum – 1926
15th c building with medieval kitchen, cells used by Pilgrim Fathers pending trial in 1607, Council Chamber, Courtroom & Banqueting Hall Displays on archaeology, costume & textiles, ceramics, militaria, coins & tokens & Boston's maritime heritage 43562

Botley, Hampshire

Hampshire Farm Museum, Country Park, Brook Ln, Botley, Hampshire SO3 2ER • T: +44 1489 787055 •
Dir.: *Barbara Newbury* •
Agriculture Museum – 1984
Tools, implements machinery, domestic buildings 43563

Bourn

Wysing Arts Centre, Fox Rd, Bourn CB3 7TX • T: +44 1954 718881 • F: 718500 • info@wysingarts.org • www.wysingarts.org •
Dir.: *Donna Lynas* •
Fine Arts Museum
Exhibits of sculpture, ceramics, pottery, glass, textiles, carving, painting, photography, video, installation – studios 43564

Bournemouth

Bournemouth Natural Science Society Museum, 39 Christchurch Rd, Bournemouth BH1 3NS • T: +44 1202 553525 • administrator@bnss.org.uk • www.bnss.org.uk •
Chm.: *R. Broadey* • Cur.: *J. Cresswell* •
Natural History Museum / Archaeology Museum – 1903
Local natural hist, archaeology, zoology, herbarium (Dorset plants) entomology, geology, egyptology – botanical garden, library 43565

Bournemouth Transport Museum, Transport Depot, Mallard Rd, Bournemouth BH8 9PN •
Science&Tech Museum
Coll of busses 43566

Russell-Cotes Art Gallery and Museum, East Cliff, Bournemouth BH1 3AA • T: +44 1202 451800 • F: 451851 • diane.edge@bournemouth.gov.uk • www.russell-cotes.bournemouth.gov.uk •
Dir.: *Linda Constable* •
Fine Arts Museum / Ethnology Museum – 1922
Victorian painting, Japanese coll, ethnography from Oceania, New Zealand and Africa 43567

Bourton-on-the-Water

Cotswold Motoring Museum and Toy Collection, Old Mill, Bourton-on-the-Water GL5 2BY • T: +44 1451 821255 • motormuseum@csma-netlink.co.uk • www.cotswold-motor-museum.com •
Dir.: *Michael Tambini* •
Science&Tech Museum – 1978
Vintage enamel advertising signs, toys, cars 43568

Bovey Tracey

Devon Guild of Craftsmen Gallery, Riverside Mill, Bovey Tracey TQ13 9AF • T: +44 1626 832223 • F: 834220 • devonguild@crafts.org.uk • www.crafts.org.uk •
Dir.: *Alexander Murdin* •
Historical Museum
Craftwork, changing exhibits invited contributors work in the gallery, guilds 43569

Bovington

Tank Museum, Bovington BH20 6JG • T: +44 1929 405096 • F: 405360 • info@tankmuseum.co.uk • www.tankmuseum.co.uk •
Dir.: *John Woodward* •
Military Museum – 1923
Hist of tank units, small arms and ammunition 43570

Bowness-on-Windermere

Blackwell - The Arts and Crafts House, on B5360, Bowness-on-Windermere LA23 3JT • T: +44 15394 46139 • F: 88486 • info@blackwell.org.uk • www.blackwell.org.uk •
Dir.: *Edward King* •
Decorative Arts Museum – 2001
Rare surviving house from the arts and crafts movement, original interior, furniture, applied arts, crafts – library 43571

Windermere Steamboat Museum, Rayrigg Rd, Bowness-on-Windermere LA23 1BN • T: +44 15394 46139 • www.steamboat.co.uk •
Dir.: *Frances Snowden* •
Science&Tech Museum 43572

Bracknell

Look Out Discovery Centre, Nine Mile Ride, Bracknell RG12 7QW • T: +44 1344 354400 • F: 354422 • thelookout@bracknell-forest.gov.uk • www.bracknell-forest.gov.uk/lookout •
Science&Tech Museum 43573

South Hill Park Arts Centre, Bracknell RG12 7PA • T: +44 1344 484858 • F: 411427 • visualarts@southhillpark.org.uk • www.southhillpark.org.uk •
Public Gallery 43574

Bradford

Bolling Hall Museum, Bowling Hall Rd, Bradford BD4 7LP • T: +44 1274 431826 • F: 726220 • www.bradfordmuseums.org •
Dir.: *Mark Suggitt* •
Local Museum – 1915
House with historic furniture, 17th and 18th c plaster ceilings, 17th-18th c furniture, local history 43575

Bradford Industrial Museum and Horses at work, Moorside Rd, Eccleshill, Bradford BD2 3HP • T: +44 1274 435900 • F: 636362 • industrial.museum@bradford.gov.uk • www.bradfordmuseums.org •
Head: *Mark Suggitt* •
Science&Tech Museum / Historical Museum
Hist of local industry, industrial archaeology, wool, textiles, transport, horses, victorian mill 43576

Cartwright Hall Art Gallery, Lister Park, Bradford BD9 4NS • T: +44 1274 431212 • F: 481045 • cartwright.hall@bradford.gov.uk • www.bradfordmuseums.org •
Head: *Maggie Pedley* •
Fine Arts Museum – 1904
19th and 20th c British art, colls of intern contemporary prints and contemporary South Asian art and crafts 43577

Colour Museum, 1 Providence St, Bradford BD1 2PW • T: +44 1274 390955 • F: 392888 • museum@sdc.org.uk • www.sdc.org.uk •
Dir.: *Sarah Burge* •
Special Museum – 1978
Textile industry 43578

Gallery II, c/o University of Bradford, Chesham Bldg, off Great Horton Rd, Bradford BD7 1DP • T: +44 1274 383365 • F: 305340 •
Fine Arts Museum / University Museum 43579

Impressions Gallery of Photography, Centenary Sq, Bradford BD1 1SD • T: +44 845 0515882 • enquiries@impressions-gallery.com • www.impressions-gallery.com •
Dir.: *Anne E. McNeill* •
Public Gallery
Contemporary photography 43580

National Media Museum, Bradford BD1 1NQ • T: +44 1274 202030 • F: 723155 • talk@nationalmediamuseum.org.uk • www.nationalmediamuseum.org.uk •
Head: *Colin Philpott* •
Science&Tech Museum – 1983
Art, history and science of photography, film, television, digital imaging and new media; BBC Heritage Coll, William Henry Fox Talbot Coll, Kodak Coll, The Royal Photographic Society Coll – Pye Photographic Archive, Daily Herald Archive 43581

National Museum of Photography, Film and Television → National Media Museum

Peace Museum, 10 Piece Hall Yard, Bradford BD1 1PJ • T: +44 1274 434009 • peacemuseum@bradford.gov.uk • www.peacemuseum.org.uk •
Cur.: *Peter Nias* •
Historical Museum
Peace, non-violence and conflict resolution 43582

Bradford-on-Avon

Bradford-on-Avon Museum, Bridge St, Bradford-on-Avon BA15 1BY • T: +44 1225 863280 • www.bradfordmuseum.com •
Cur.: *Roger Clark* •
Local Museum
Local hist 43583

Bradford-on-Tone, Taunton

Sheppy's Farm and Cider Museum, Three Bridges Farm, Bradford-on-Tone, Taunton TA4 1ER • T: +44 1823 461233 • F: 461712 • louisa@sheppyscider.com • www.sheppyscider.com •
Dir.: *Louisa Sheppy* •
Agriculture Museum
Old cidermaking equip, agricultural implements, coopers' and other craftsmen's tools, video of cidermaker's year 43584

Brading

Aylmer Military Collection, Nunwell House, Church Ln, Brading PO36 0JQ • T: +44 1983 407240 •
Cur.: *John Alymer* •
Military Museum
Local family related militaria, uniforms, pictures 43585

Brading Roman Villa, Morton Old Rd, Brading PO36 0EN • T: +44 1983 406223 • F: 407924 • info@bradingromanvilla.org.uk • www.bradingromanvilla.org.uk •
Dir.: *Paul Knowlson* •
Archaeology Museum 43586

Isle of Wight Wax Works, 46 High St, Brading PO36 0DQ • T: +44 1983 407286 • F: 402112 • info@iwwaxworks.co.uk • www.iwwaxworks.co.uk •
Science&Tech Museum – 1965
Wax tableaux of Island since the Roman times, incl the oldest house on the Island (1066), instruments of torture from the Royal castle of Nuremberg, animal world, mechanical world (cars, steam engines etc) 43587

Lilliput Antique Doll and Toy Museum, High St, Brading PO36 0DJ • T: +44 1983 407231 • lilliput.museum@btconnect.com • www.lilliputmuseum.com •
Dir.: *Graham K. Munday* •
Special Museum – 1974
Over 2,000 antique dolls & toys from 2,000 BC to 1945 AD 43588

Braintree

Braintree District Museum, Manor St, Braintree CM7 3HW • T: +44 1376 325266 • F: 344345 • museum@braintree.gov.uk • www.braintree.gov.uk •
Head: *Jean Grice* •

U

Local Museum
Textile and engineering industries, archaeology, social hist, textiles, large photographic coll – Textile archive ... 43589

Bramhall

Bramall Hall, Bramhall Park, Bramhall SK7 3NX • T: +44 845 8330974 • F: +44 161 4866959 • bramall.hall@stockport.gov.uk • www.stockport.gov.uk/heritageattractions •
Man.: *Caroline Egan* •
Historic Site
14th to 16th c half-timbered Hall ... 43590

Brandon

Brandon Heritage Museum, George St, Brandon IP27 0BX • T: +44 1225 863280 • F: +44 1204 391352 •
Local Museum – 1991
Flint, fur and forestry industries, local history ... 43591

Braunton

Braunton and District Museum, Bakehouse Centre, Caen St, Braunton EX33 1AA • T: +44 1271 816688 • braunton@devonmuseums.net • www.devonmuseums.net/braunton •
Local Museum
Maritime, agriculture and village life ... 43592

Breamore

Breamore Countryside Museum, Upper St, Breamore SP6 2DF • T: +44 1725 512468 • F: 512858 • breamore@ukonline.co.uk • www.breamorehouse.com •
Dir.: *Sir Edward Hulse* • Cur.: *Philip Crouter* •
Agriculture Museum / Open Air Museum – 1972
Hand tools, tractors, steam engines, blacksmith's and wheel-wright's shops, brewery, dairy, farmworkerscottage, harnessmakers, boot makers, village general store, village school, laundry, coopers', baker's and clockmaker's shop, village garage ... 43593

Breamore House Museum, near Fordingbridge, Breamore SP6 2DF • T: +44 1725 512233 • F: 512858 • breamore@ukonline.co.uk • www.breamorehouse.com •
Dir.: *E. Hulse* •
Fine Arts Museum / Decorative Arts Museum / Folklore Museum – 1953
17th and 18th c furniture and paintings (mainly Dutch), tapestries, porcelain, 14 paintings of the intermingling of Indian races (c 1725), development of agricultural machinery, farm implements, aspects of a self-sufficient village ... 43594

Brechin

Brechin Museum, Town House, 28 High St, Brechin DD9 6EU • T: +44 1307 625536 • F: 462590 • brechin.museum@angus.gov.uk • www.angus.gov.uk •
Local Museum
Local history, antiquities ... 43595

Glenesk Folk Museum, The Retreat, Glenesk, Brechin DD9 7YT • T: +44 1356 648070 • visit@gleneskretreat.wanadoo.co.uk •
Cur.: *Anne Taylor* •
Folklore Museum – 1955
Local rural life, period rooms, Victoriana ... 43596

Brecon

Brecknock Museum and Art Gallery, Captain's Walk, Brecon LD3 7DS • T: +44 1874 624121 • F: 624121 • brecknock.museum@powys.gov.uk •
Cur.: *David Moore* •
Local Museum – 1928
Natural and local hist, archaeology, ceramics, costumes, early Christian monuments, art – art gallery ... 43597

Howell Harris Museum, Trefeca, Brecon LD3 0PP • T: +44 1874 711423 • F: 712212 • post@trefeca.org.uk • www.trefeca.org.uk •
Local Museum – 1957
Local and religious hist, Howell Harris (1714-1773) ... 43598

Regimental Museum of The Royal Welsh (Brecon), The Barracks, Brecon LD3 7EB • T: +44 1874 613310 • F: 613275 • swb@rrw.org.uk • www.rrw.org.uk •
Cur.: *Martin Everett* •
Military Museum – 1934
Regimental hist, medals, armaments, Anglo-Zulu War 1879 – archive ... 43599

Brentwood

Brentwood Museum, Cementery Lodge, Lorne Rd, Brentwood CM14 5HH • T: +44 1277 224012 • enquiries@brentwood.gov.uk • www.brentwood-council.gov.uk •
Cur.: *John Fryer* • *Audrey Puddy* • *Mary Humphreys* •
Local Museum
Social and domestic items, memorabilia from world wars, toys, games, furniture and costumes ... 43600

Brewood

Boscobel House, Bishop's Wood, Brewood ST19 9AR • T: +44 1902 850244 • www.english-heritage.org.uk •
Head: *Peter Trickett* •
Local Museum
Furniture, textiles, agricultural machinery, forge, dairy and salting room ... 43601

Bridgend

South Wales Police Museum, Cowbridge Rd, Bridgend CF31 3SU • T: +44 1656 655555 • julia.mason@south-wales.pnn.police.uk • www.south-wales.police.uk/museum •
Cur.: *Julia Mason* •
Special Museum – 1950
Policing in Glamorgan from the Celts to the present day, Edwardian police station, wartime tableau ... 43602

Bridgnorth

Bridgnorth Northgate Museum, Burgess Hall, High St, Bridgnorth WV16 4ER • T: +44 1746 762830 • www.northgatemuseum.org.uk •
Cur.: *Emma-Kate Burns* •
Local Museum
Local hist, firemarks, tokens, coins, agriculture ... 43603

Childhood and Costume Museum, Newmarket Bldg, Postern Gate, Bridgnorth WV16 4AA • T: +44 1746 764636 •
Cur.: *Jayne Griffiths* •
Folklore Museum – 1951
Clothes and accessories (1860-1960), toys, games, Victorian nursery, textiles, mineral coll ... 43604

Midland Motor Museum, Stanmore Hall, Stourbridge Rd, Bridgnorth WV15 6DT • T: +44 1746 762992 •
Science&Tech Museum ... 43605

Bridgwater

Blake Museum, Blake St, Bridgwater TA6 3NB • T: +44 1278 435599 • F: 436423 • museums@sedgemoor.gov.uk • www.sedgemoor.gov.uk/museums •
Officer: *Jessica Vale* • Asst.: *Carolyn Cudbill* •
Historical Museum – 1926
Local hist (18th c), archaeology, work of Admiral Blake (1598-1657) in the supposed place of his birth, relics on the Battle of Sedgemoor, photos ... 43606

Somerset Brick and Tile Museum, East Quay, Bridgwater TA6 3NB • T: +44 1278 426088 • F: 320229 • county-museum@somerset.gov.uk • www.somerset.gov.uk/museums •
Dir.: *David Dawson* •
Historical Museum
Brick and tile industry ... 43607

Westonzoyland Pumping Station Museum, Hoopers Ln, Westonzoyland, Bridgwater, mail addr: c/o W. Jewell, Gayton, Keens Ln, Othery TA7 0PU • T: +44 1823 698532 • webmaster@wzlet.org • www.wzlet.org •
Science&Tech Museum – 1977
Land drainage, steam pumps and engines ... 43608

Bridlington

Bayle Museum, Bayle Gate, Bridlington YO16 4PZ • T: +44 1262 424099 • F: 674308 • bayle.museum@btinternet.com • www.bayle.bridlington.net •
Cur.: *Ruth McDonald* •
Local Museum – 1920
Victorian kitchen, Green Howard's Regemental room, courtroom of 1388, the monastic gatehouse of the Augustinian Priory ... 43609

Bridlington Harbour Museum, Harbour Rd, Bridlington YO15 2NR • T: +44 1262 670148 • F: 602041 • secretary@bscps.com • www.bscps.com •
Historical Museum / Public Gallery ... 43610

Sewerby Hall Art Gallery and Museum, Museum of East Yorkshire, Church Ln, Sewerby, Bridlington YO15 1EA • T: +44 1262 677874 • F: 674265 • sewerby.hall@eastriding.gov.uk • www.bridlington.net/sewerby •
Local Museum / Fine Arts Museum – 1936
19th c marine paintings by Henry Redmore of Hull, early 19th c local watercolours, local archaeology, farm implements, vintage motorcycles, horse-drawn vehicles, military uniforms with local associations, fossils from the nearby chalk cliffs, the various seabirds and waders found in the area, memorabilia on air-woman Amy Johnson ... 43611

Bridport

Bridport Museum, South St, Bridport DT6 3NR • T: +44 1308 422116 • F: 420659 • curator@bridportmuseum.co.uk • www.bridportmuseum.co.uk •
Local Museum
Local hist, archaeology, natural hist, geology, trades, costumes, Victorian relics, net weaving ... 43612

Harbour Life Exhibition, Salt House, West Bay, Bridport DT6 4SA • T: +44 1308 422807 • F: 421509 •
Local Museum
Industry, history of Bridport Harbour ... 43613

Museum of Net Manufacture, Bridgeacre, Uploders, Bridport DT6 4PF • T: +44 1308 485349 • F: 485621 • aumjasq@aol.com •
Cur.: *Jacquie Summers* •
Historical Museum
Documents, ledgers, traditional tools, samples of natural fibres, historic trade ... 43614

Brierfield

Neville Blakey Museum of Locks, Keys and Safes, Burnley Rd, Brierfield BB9 5AD • T: +44 1282 613593 • F: 617550 • sales@blakeys.fsworld.co.uk •
Dir.: *W.H. Neville Blakey* •
Special Museum ... 43615

Brighouse

Smith Art Gallery, Halifax Rd, Brighouse HD6 2AF • T: +44 1484 719222 • F: 719222 • karen.belshaw@calderdale.gov.uk • www.calderdale.gov.uk •
Public Gallery – 1907
Paintings, English watercolors, 19th c woodcarvings ... 43616

Brightlingsea

Brightlingsea Museum, 1 Duke St, Brightlingsea CO7 0EA • T: +44 1206 303286 •
Cur.: *Margaret Stone* •
Local Museum
Maritime and oyster industry, local hist ... 43617

Brighton

Barlow Collection of Chinese Ceramics, Bronzes and Jades, c/o University of Sussex, Falmer, Brighton BN1 9QL • T: +44 1273 873506 • barlow@sussex.ac.uk • www.sussex.ac.uk/barlow •
Decorative Arts Museum
Chinese jades (since Han dynasty), bronzes (Shang & Zhou dynasties), porcelain and ceramics (Han until Ming dynasies) ... 43618

Booth Museum of Natural History, 194 Dyke Rd, Brighton BN1 5AA • T: +44 1273 292777 • F: 292778 • boothmuseum@brighton-hove.gov.uk •
Keeper: *Dr. G. Legg* (Natural Sciences) •
Natural History Museum – 1874
British birds, dioramas, butterflies, nymphalid butterflies, fossils, vertebrate skeletons, coll of Edward Thomas Booth – library ... 43619

Brighton Fishing Museum, 201 King's Rd Arches, Brighton BN1 1NB • T: +44 1273 723064 • F: 723064 • brightonfishingmuseum@btinternet.com •
Special Museum – 1994
Boats, artifacts, pictures – film and historical archive ... 43620

Brighton Museum and Art Gallery, Royal Pavilion Gardens, Brighton BN1 1EE • T: +44 1273 290900 • F: 292841 • museums@brighton-hove.gov.uk • www.brighton.virtualmuseum.info •
Dir.: *Pauline Scott-Garrett* (Heritage) •
Fine Arts Museum / Decorative Arts Museum / Local Museum – 1873
Fine coll of local and practical importance, Art Nouveau and Art Deco, fine art and non-Western art, costumes, Willett coll of pottery and porcelain – My Brighton – interactive local hist project ... 43621

Fabrica, 40 Duke St, Brighton BN1 1AG • T: +44 1273 778646 • F: 778646 • info@fabrica.org.uk • www.fabrica.org.uk •
Association with Coll
Artist-led space promoting the understanding of contemporary art in a regency church, site specific commissions with four exhibitions a year: craft, material based, lens based and digital ... 43622

Preston Manor, Preston Drove, Brighton BN1 6SD • T: +44 3000 290900 • F: +44 1273 292771 • visitor.services@brighton-hove.gov.uk • www.brighton-hove-museums.org.uk •
Decorative Arts Museum – 1933
Macquoid bequest English and Continental period furniture, silver, pictures, panels of stained glass, 16th-19th c original 17th leather wall-hangings – archives ... 43623

Rottingdean Grange Art Gallery and Museum, The Green, Rottingdean, Brighton BN2 7HA • T: +44 1273 301004 • F: 307197 •
Dir.: *B.J. Wholey* •
Public Gallery / Local Museum
Local history and arts ... 43624

Royal Pavilion, 4-5 Pavilion Bldgs, Brighton BN1 1EE • T: +44 1273 290900 • F: 292871 • visitor.services@brighton-hove.gov.uk • www.royalpavilion.org.uk •
Dir.: *David Beevers* •
Fine Arts Museum / Decorative Arts Museum – 1851
Seaside palace of George IV - designed by John Nash, exterior - Indian design - interiors in Chinese taste, items on loan from H.M. The Queen, 18th c gardens ... 43625

Stanmer Rural Museum, Stanmer Park, Brighton BN1 7HN • T: +44 1273 509563 •
Head: *N. Stopps* •
Folklore Museum – 1975
Rare horse trave, rural tools, local archaeological finds, agricultural implements ... 43626

University of Brighton Gallery, Grand Parade, Brighton BN2 0JY • T: +44 1273 643012 • F: 643038 • c.l.matthews@bton.ac.uk • www.bton.ac.uk/gallery-theatre/ •
Dir.: *Julian Crampton* •
Public Gallery ... 43627

Bristol

Arnolfini, 16 Narrow Quay, Bristol BS1 4QA • T: +44 117 9172300 • F: 9172303 • info@arnolfini.org.uk • www.arnolfini.org.uk •
Dir.: *Tom Trevor* • Dep. Dir.: *Polly Cole* •
Fine Arts Museum / Performing Arts Museum
Contemporary and 20th c art ... 43628

Ashton Court Visitor Centre, Stable Block, Ashton Court Estate, Long Ashton, Bristol BS41 9JN • T: +44 117 9639174 • F: 9532143 •
Dir.: *Cellan Rhys-Michael* •
Local Museum
History of this landscape, temporary exhibits ... 43629

At-Bristol, Anchor Rd, Harbourside, Bristol BS1 5DB • T: +44 845 3451235 • F: +44 117 9157200 • information@at-bristol.org.uk • www.at-bristol.org.uk •
Natural History Museum / Science&Tech Museum
Nature and science ... 43630

Blaise Castle House Museum, Henbury Rd, Henbury, Bristol BS10 7QS • T: +44 117 9039818 • F: 9039820 • general.museum@bristol.gov.uk • www.bristol-city.gov.uk/museums •
Dir.: *David J. Eveleigh* •
Local Museum – 1947
Regional social hist, agricultural hist, domestic life, costumes, toys, dairy ... 43631

Bristol's City Museum and Art Gallery, Queen's Rd, Bristol BS8 1RL • T: +44 117 9223571 • F: 9222047 • general.museum@bristol.gov.uk • www.bristol.gov.uk/museums •
Dir.: *Kate Bindley* • Sc. Staff: *Dr. Tim Ewin* (Geology) • *Samantha Trebilcok* (Biology) • *David J. Eveleigh* (Social History) • *Andy King* (Industrial and Maritime History) • *Sheena Stoddard* (Fine Art) • *Karin Walton* (Applied Art) • *Sue Giles* (Ethnography) • *Gail Boyle* • *Les Good* (Archaeology) • *Kate Newnham* (Eastern Art) •
Local Museum / Fine Arts Museum – 1905
Archaeology, geology, natural hist, transportation and technology of the area, ethnography, Classical and Egyptian finds, numismatics, local hist, fine art, ceramics, Eastern art, marine history ... 43632

Bristol's Industrial Museum (closed until 2011 for redevelopment), Princes Wharf, Wapping Rd, Bristol BS1 4RN • T: +44 117 9251470 • F: 9297318 • andy_king@bristol-city.gov.uk • www.bristol-city.gov.uk/museums •
Cur.: *Andy King* • Cons.: *Chris West* •
Science&Tech Museum – 1978
Models of early ships, flight deck of Concorde, buses, cars and motorcycles, ships, cranes and railway stock, slave trade ... 43633

British Empire and Commonwealth Museum, Clock Tower Yard, Temple Meads, Bristol BS1 6QH • T: +44 117 9254980 • F: 9254983 • staff@empiremuseum.co.uk • www.empiremuseum.co.uk •
Dir.: *Dr. G. Griffiths* •
Historical Museum
Social history, photographies, costume, film and moving images, weapons, art, personal affects – library, oral history archive, image archives ... 43634

Georgian House, 7 Great George St, Bristol BS1 5RR • T: +44 117 9211362 • F: 9222047 • general.museum@bristol.gov.uk • /www.bristol.gov.uk •
Cur.: *Karin Walton* •
Decorative Arts Museum – 1937
Period furnishings ... 43635

Guild Gallery, 68 Park St, Bristol BS1 5JY • T: +44 117 9265548 • F: 9255659 • info@bristolguild.co.uk •
Dir.: *Simone Andre* •
Public Gallery
Paintings, prints, crafts, sculptures a.o. visual arts ... 43636

Kings Weston Roman Villa, c/o City Museum and Art Gallery, Long Cross, Bristol BS11 0LP • T: +44 117 9223571 • F: 9222047 • general.museum@bristol.gov.uk • www.bristol.gov.uk •
Archaeology Museum
Farmhouse with bathhouse ... 43637

Red Lodge, Park Row, Bristol BS1 5LJ • T: +44 117 9211360 • F: 9222047 • general.museum@bristol.gov.uk • www.bristol.gov.uk •
Cur.: *Karin Walton* •
Decorative Arts Museum
Elizabethan house, modified 18th c, furnishings ... 43638

SS Great Britain Museum, Maritime Heritage Centre, Great Western Dockyard, Gasferry Rd, Bristol BS1 6TY • T: +44 117 9260680 • F: 9255788 • admin@ssgreatbritain.org • www.ssgreatbritain.org •
Dir.: *Matthew Tanner* •
Science&Tech Museum
Iron steam-ship 'Great Britain' and its designer I.K. Brunel, iron ship-building, paddle ships 'Great Western'-'Great Eastern', passenger emigration to Australia ... 43639

University of Bristol Theatre Collection, 21 Park Row, Bristol BS1 5LY, mail addr: Department of Drama, Cantocks Close, Bristol BS8 1UP • T: +44 117 3315086 • F: 3315082 • theatre-collection@bristol.ac.uk • www.bris.ac.uk/theatrecollection •

Dir.: *Jo Elsworth* •
Performing Arts Museum / University Museum
Written and graphic material, professional British theatre hist. – archives, library, live art archives 43640

Brixham

Brixham Heritage Museum, Old Police Stn, New Rd, Bolton Cross, Brixham TQ5 8LZ • T: +44 1803 856267 • mail@brixhamheritage.org.uk • www.brixhamheritage.org.uk •
Cur.: *Dr. Philip L. Armitage* • Cust.: *Brian Ogle* •
Local Museum – 1958
Fishing, shipping, maritime and local hist, archaeology 43641

Golden Hind Museum, The Quay, Habour, Brixham TQ5 8AW • T: +44 1803 856223 • postmaster@goldenhind.co.uk • www.goldenhind.co.uk •
Special Museum
Full sized replica of Sir Francis Drake's ship (16th c) 43642

Broadclyst

Paulise de Bush Costume Collection, Killerton House, Broadclyst EX5 3LE • T: +44 1392 881345 • F: 883112 • killerton@nationaltrust.org.uk • www.nationaltrust.org.uk/main/w-killerton •
Cur.: *Shelley Tobin* •
Decorative Arts Museum – 1978
Coll of costumes, which is displayed in a series of rooms furnished in different periods, from the 18th c to the present day 43643

Broadstairs

Bleak House, Fort Rd, Broadstairs CT10 1HD • T: +44 1843 862224 •
Dir.: *L.A. Longhi* •
Special Museum – 1958
Memorabilia on writer Charles Dickens (1811-1870) in the house where he wrote 'David Copperfield' and 'Bleak House' 43644

Crampton Tower Museum, Broadway, Broadstairs CT10 2AB • T: +44 1843 864446 • www.cramptontower.co.uk •
Cur.: *Robert Tolhurst* •
Science&Tech Museum – 1978
Thomas Russell Crampton memorabilia, designer of locomotives, submarine telegraphic cable, gas and water works, local transport 43645

Dickens House Museum Broadstairs, 2 Victoria Parade, Broadstairs CT10 1QS • T: +44 1843 863453, 861232 • F: 863453 • leeault@tiscali.co.uk • www.dickenshouse.co.uk •
Dir.: *Lee Ault* •
Special Museum – 1973
Dickens memorabila, costume and Victoriana 43646

Broadway

Broadway Magic Experience, 76 High St, Broadway WR12 7AJ • T: +44 1386 858323 • janicelonghi@hotmail.com •
Dir.: *Janice Longhi* •
Special Museum
Automated childhood scenes, Teddy bears 43647

Snowshill Manor, Snowshill, Broadway WR12 7JU • T: +44 1386 852410 • F: 842822 • snowshillmanor@nationaltrust.org.uk • www.nationaltrust.org.uk •
Sc. Staff: *Jennifer Rowley* •
Decorative Arts Museum – 1951
Chinese and Japanese objects, nautical instruments, Japanese Samurai armour, musical instruments, costumes, toys, spinning and weaving tools, bicycles, farm wagon models 43648

Brockenhurst

Beaulieu Abbey Exhibition of Monastic Life, Beaulieu, Brockenhurst SO42 7ZN • T: +44 1590 612345 • F: 612624 • susan.tomkins@beaulieu.co.uk • www.beaulieu.co.uk •
Chm.: *Lord Montagu of Beaulieu* •
Religious Arts Museum – 1952
Story of daily life for the Cistercian Monks, who founded the Abbey in 1204 43649

Brodick

Arran Heritage Museum, Rosaburn, Brodick KA27 8DP • T: +44 1770 302636 • F: 302185 • tom.macleod@arranmuseum.co.uk • www.arranmuseum.co.uk •
Local Museum
Agricultural machinery, geology, archaeology, smiddy, naval photos 43650

Brodick Castle, Brodick KA27 8HY • T: +44 1770 302202 • F: 302312 • brodickcastle@nts.org.uk • www.nts.org •
Dir.: *Ken Thorburn* •
Fine Arts Museum / Decorative Arts Museum
Beckford silver, porcelain, paintings, Duke of Hamilton and Earl of Rochford sport photos and trophies – garden 43651

Brokerswood

Phillips Countryside Museum, Woodland Park, Brokerswood BA13 4EH • T: +44 1373 823880 • F: 858474 •
Dir.: *Barry Capon* •
Natural History Museum – 1971
Natural history and forestry, botany, ornithology 43652

Bromham

Bromham Mill and Art Gallery, Bridge End, Bromham MK43 8LP • T: +44 1234 824330 • F: 228531 • wilemans@deed.bedfordshire.gov.uk • www.bedfordshire.gov.uk •
Dir.: *M. Kenworthy* •
Public Gallery
Art, industrial heritage, natural environment 43653

Bromsgrove

Avoncroft Museum of Historic Buildings, Redditch Rd, Stoke Heath, Bromsgrove B60 4JR • T: +44 1527 831886 • F: 876934 • avoncroft1@compuserve.com • www.avoncroft.org.uk •
Dir.: *Angela Gill* •
Open Air Museum – 1969
16th c inn, 19th c post mill, granary and barn, 15th c merchant's house, hand-made nail and chain workshops, medieval guesten roof, 17th c cockpit, 18th c ice-house, perry mill, National Telephone Kiosk coll, 20th c prefab 43654

Bromsgrove Museum, 26 Birmingham Rd, Bromsgrove B61 0DD • T: +44 1527 831809 • F: 831809 • k.spry@bromsgrove.gov.uk • www.bromsgrove.gov.uk •
Historical Museum – 1972
Social and industrial hist of the area, with exhibits covering glass, button, nail and salt-making, social hist of the 19th and early 20th c 43655

Bronwydd

Gwili Steam Railway, Bronwydd Arms Station, Bronwydd SA33 6HT • T: +44 1267 230666 • company@gwili-railway.co.uk • www.gwili-railway.co.uk •
Science&Tech Museum
Steam and diesel locomotives, carriages and wagons 43656

Brook

Agricultural Museum Brook, The Street, Brook TN25 5PF • T: +44 1304 824969 • F: 824969 • brianwimsett@hotmail.com • www.agricultural-museumbrook.org.uk •
Cur.: *Brian Wimsett* •
Agriculture Museum
Old farm equipment, medieval barn and oast house 43657

Broseley

Clay Tobacco Pipe Museum, The Flat, Duke St, Broseley TF8 7AW • T: +44 1952 884391 • F: 432204 • info@ironbridge.org.uk • www.ironbridge.org.uk •
Cur.: *Michael Vanns* •
Special Museum
Clay tobacco pipe factory 43658

Buckfastleigh

Buckfast Abbey, Buckfastleigh TQ11 0EE • T: +44 1364 642519 • F: 643891 •
Local Museum 43659

South Devon Railway, The Station, Buckfastleigh TQ11 0DZ • T: +44 1364 642338, 643536 • F: 642170 • info@southdevonrailway.org • www.southdevonrailway.org •
Man.: *Richard Elliott* •
Science&Tech Museum
17 steam locomotives, 3 heritage diesels, 27 carriages, 35 freight wagons 43660

The Valiant Soldier, 79 Fore St, Buckfastleigh TQ11 0BS • T: +44 1364 644522 • enquiries@valiantsoldier.org.uk •
Head: *Stuart Barker* •
Local Museum
Pub fittings, furniture, bar ephemera, promotional and domestic items, history coll 43661

Buckhaven

Buckhaven Museum, Fife Council Libraries & Museums, c/o Buckhaven Library, 3 College St, Buckhaven KY8 1LD • T: +44 1592 583211 • fife.museums@fife.gov.uk • www.fifedirect.org.uk/museums •
Co-ord.: *Dallas Mechan* •
Historical Museum
History and techniques of the local fishing industry, life of families who earned a living from it, recreations of interiors of home of past generations of Buckhaven people 43662

Buckie

Buckie Drifter Fishing Heritage Centre, Freuchny Rd, Buckie AB56 1TT • T: +44 1542 834646 • F: 835995 • buckie.drifter@moray.gov.uk •
Historical Museum
History of the town of Buckie, fishing industry, boats and techniques employed, navigation, achievements of the local lifeboat 43663

Peter Anson Gallery, Town House West, Cluny Pl, Buckie AB56 1HB • T: +44 1542 832121 • F: +44 1309 675863 • museums@moray.gov.uk • www.moray.org/museums •
Fine Arts Museum
Watercolours by marine artist Peter Anson 43664

Buckingham

Buckingham Movie Museum, Printers Mews, Market Hill, Buckingham MK18 1JX • T: +44 1280 816758 •
Special Museum
Coll of 200 projectors and 100 cine-cameras and accessories 43665

Old Gaol Museum, Market Hill, Buckingham MK18 1JX • T: +44 1280 823020 • F: 823020 • www.mkheritage.co.uk/ogb •
Man.: *John Roberts* •
Local Museum
Story of Buckingham, military hist 43666

Bude

Bude-Stratton Museum, The Castle Heritage Centre & Gallery, The Castle, Bude Cornwall EX23 8LG • T: +44 1288 357300 • F: 353576 • jboxter@bude-stratton.gov.uk • www.bude-stratton.gov.uk •
Historical Museum – 1976
Hist and working of the canal, by means of early photographs, models, drawings, plans and maps, development of Bude, shipwrecks and hist of the local lifeboat, railway, military and social hist – archive of documents relevant to North Cornwall 43667

Budleigh Salterton

Fairlynch Museum, 27 Fore St, Budleigh Salterton EX9 6NP • T: +44 1395 442666 • admin@fairlynchmuseum.org.uk • www.devonmuseums.net/fairlynch •
Pres.: *Priscilla Hull* •
Local Museum
Prehistory, art, local hist, ecology of Lower Otter Valley, geology 43668

Otterton Mill Centre and Working Museum, Fore Otterton St, Budleigh Salterton EX9 7HG • T: +44 1395 568521 • F: 568521 • escape@ottertonmill.com • www.ottertonmill.com •
Dir.: *Dr. Desna Greenhow* •
Science&Tech Museum – 1977
Use of water to drive mill machinery, production processes of flour, displays of crafts, fine arts and historical items – gallery, lace exhibition 43669

Bulwell

South Notts Hussars Museum, TA Centre, Hucknall Ln, Bulwell NG6 8AQ • T: +44 115 9272251 • F: 975420 •
Cur.: *G.E. Aldridge* •
Military Museum
Regimental uniforms, models of vehicles and guns from WW II, paintings and trophies 43670

Bunbury

Bunbury Watermill, Mill Ln, off Bowes Gate Rd, Bunbury CW6 9PP • T: +44 1829 261422 • F: 261422 • www.bunbury-mill.org •
Dir.: *Dennis Buchanan* •
Science&Tech Museum 43671

Bungay

Bungay Museum, Waveney District Council Office, Broad St, Bungay NR35 1EE • T: +44 1986 893155, 894463 • F: 892176 • www.bungay-suffolk.co.uk/activities/museum.asp •
Cur.: *Christopher Reeve* •
Local Museum
Local hist, fossils, flint implements, coins, J.B. Scott memorabilia 43672

Burford

Tolsey Museum, 126 High St, Burford OX18 4QU • T: +44 1993 823196 • cgwalker100@hotmail.com • www.tolseymuseumburford.com •
Dir.: *Christopher Walker* •
Local Museum – 1960
Local governmental hist, local crafts, social hist 43673

Burghead

Burghead Museum, 16-18 Grant St, Burghead IV36 1PH • T: +44 1309 673701 • alasdair.joyce@techleis.moray.gov.uk •
Local Museum 43674

Burnham-on-Crouch

Burnham on Crouch and District Museum, Coronation Rd, Burnham-on-Crouch CM0 8HW • T: +44 1621 783444 • pfhammond@btinternet.com •
Local Museum – 1981
archives 43675

Mangapps Railway Museum, Mangapps Farm, Burnham-on-Crouch CM0 8QQ • T: +44 1621 784898 • F: 783833 • www.mangapps.co.uk •
Science&Tech Museum
Good wagons, carriages, steam and diesel locomotives, railway artefacts 43676

Burnley

Museum of Local History, Towneley Holmes Rd, Burnley BB11 3RQ • T: +44 1282 424213 • F: 436138 • towneleyhall@burnley.gov.uk • www.towneleyhall.org.uk •
Historical Museum
Social and industrial hist, textiles, street scene, home life, a pub 43677

Queen Street Mill Textile Museum, Queen St, Harle Syke, Burnley BB10 2HX • T: +44 1282 412555 • F: 430220 • queenstreet.mill@mus.lancscc.gov.uk • www.lancsmuseums.gov.uk •
Dir.: *Edmund Southworth* •
Science&Tech Museum
Steam powered textile mill 43678

Towneley Hall Art Gallery and Museums, Towneley Holmes Rd, Burnley BB11 3RQ • T: +44 1282 424213 • F: 436138 • towneleyhall@burnley.gov.uk • www.towneleyhall.org.uk •
Cur.: *Susan Bourne* •
Fine Arts Museum / Local Museum – 1902
Paintings, water colours, furnishings, period house, archaeology, local hist, natural hist, decorative arts, military 43679

Weavers' Triangle Visitor Centre, 85 Manchester Rd, Burnley BB11 1JZ • T: +44 1282 452403 • www.weaverstriangle.co.uk •
Dir.: *Brian Hall* •
Historical Museum – 1980
Burnley's cotton industry and its workers, Victorian industrial area 43680

Burntisland

Burntisland Edwardian Fair Museum, 102 High St, Burntisland KY3 9AS, mail addr: Kirkcaldy Museum, Abbotshall Road, Kirkcaldy KY1 1YG • T: +44 1592 412860 • F: 412870 • kirkcaldy.museum@fife.gov.uk •
Co-ord.: *Dallas Mechan* •
Local Museum 43681

Museum of Communication, 131 High St, Burntisland KY3 9AA • T: +44 1592 874836 • mocenquiries@tiscali.co.uk • www.mocft.co.uk •
Pres.: *Prof. Heinz Wolff* •
Science&Tech Museum
18th-21st c communications, coll of telephones, radio, telegraphy, radar, computing, televion covering 20th c 43682

Bursledon

Bursledon Windmill, Windmill Ln, Bursledon SO31 8BG • T: +44 23 80404999 • musmgb@hants.gov.uk •
Cur.: *Dr. Gavin Bowie* •
Science&Tech Museum 43683

Manor Farm, Pylands Ln, Bursledon SO31 1BH • T: +44 1489 787055 • www.hants.gov.uk/countryside/manorfarm •
Man.: *Barbara Newbury* •
Agriculture Museum
Farm with machinery, woodworking, blacksmithing, farmhouse, living history displays 43684

Burston

Burston Strike School, Diss Rd, Burston IP22 5TP • T: +44 1379 741565 • www.burstonstrikeschool.org •
Historical Museum
Events in 1914 which lead to the longest strike in hist 43685

Burton Agnes

Burton Agnes Hall, Estate Office, Burton Agnes YO25 4NB • T: +44 1262 490324 • F: 490513 • burton.agnes@farmline.com • www.burton-agnes.com •
Dir.: *E.S. Cunliffe-Lister* •
Fine Arts Museum / Decorative Arts Museum – 1946
Elizabethan house with ceilings, overmantels, paintings by Reynolds, Gainsborough, Cotes, Marlow, Reinagle, also modern French paintings 43686

Burton Bradstock

Bredy Farm Old Farming Collection, Bredy Farm, Burton Bradstock DT6 4ND • T: +44 1308 897229 • jmallinson@bredyfarm.freeserve.co.uk •
Cur.: *Justin Mallinson* •
Agriculture Museum
A coll of yeaster-year farm 43687

U

Burton-upon-Trent

Museum of Brewing, Coors Visitor Centre, Horninglow St, Burton-upon-Trent DE14 1YQ • T: +44 1283 532880 • info@nationalbrewerycentre.co.uk • www.nationalbrewerycentre.co.uk •
Dir.: *Peter Orgill* • Cur.: *Roger Sellick* •
Science&Tech Museum – 1977
Hist of brewing in Burton, associated industries 43688

Burwash

Bateman's Kipling's House, Batemaris Ln, Burwash TN19 7DS • T: +44 1435 882302 • F: 882811 • batemans@nationaltrust.org.uk •
Man.: *Elaine Francis* •
Special Museum – 1939
Work and living rooms of writer Rudyard Kipling (1865-1936) in his former home, built 1634, manuscripts, Kiplings Rolls Royce, working water mill – Library 43689

Burwell

Burwell Museum, Mill Close, Burwell CB5 0HL • T: +44 1638 605544, 741713 • F: +44 871 6615378 • museum@burwellmuseum.org.uk • www.burwellmuseum.org.uk •
Local Museum
Local hist 43690

Bury, Lancashire

Bury Art Gallery and Museum, Moss St, Bury, Lancashire BL9 0DR • T: +44 161 2535878 • F: 2535915 • artgallery@bury.gov.uk • www.bury.gov.uk •
Cur.: *Richard Burns* (Art Gallery) • *Ronan Brindley* (Museum) • Asst.: *Katherine McClung-Oakes* (Arts, Colls) • Sc. Staff: *Alison Green* (Museum) •
Fine Arts Museum / Decorative Arts Museum / Local Museum
British paintings and drawings, local hist, natural hist, ornithology, archaeology, Wedgwood and local pottery, Bronze Age finds 43691

Bury Transport Museum, Castlecrof Rd, off Bolton St, Bury, Lancashire BL9 0LN • T: +44 1225 333895 • F: +44 1424 775174 •
Science&Tech Museum
Road and rail transport, buses, lorries, traction engines 43692

Fusiliers' Museum Lancashire, Wellington Barracks, Bolton Rd, Bury, Lancashire BL8 2PL • T: +44 161 7642208 • F: 7642208 • fusilierslancshq@btinternet.com • www.fusiliersmuseum-lancashire.org.uk •
Cur.: *Mike Glover* •
Military Museum – 1934
History of the regiment since 1688, uniforms, medals, militaria, Napoleonic relics 43693

Bury Saint Edmunds

Bury Saint Edmunds Art Gallery, Market Cross, Bury Saint Edmunds IP33 1BT • T: +44 1284 762081 • F: 750774 • enquiries@burystedmundsartgallery.org • www.burystedmundsartgallery.org •
Dir.: *Barbara Taylor* •
Public Gallery 43694

Moyse's Hall Museum, Cornhill, Bury Saint Edmunds IP33 1DX • T: +44 1284 757160 • F: 765373 • moyseshall@stedsbc.gov.uk • www.stedmundsbury.gov.uk •
Dir.: *Margaret Goodger* •
Local Museum – 1899
Prehistoric-medieval finds, regional natural hist, local hist, 12th c Norman building, Suffolk Regiment 43695

Suffolk Regiment Museum, The Keep, Gibraltar Barracks, Newmarket Rd, Bury Saint Edmunds IP33 3NR • T: +44 1284 752394 • F: 752026 • chief-clerk@anglian.army.mod.uk • www.suffolkregiment.org •
Cur.: *A.G.B. Cobbold* •
Military Museum – 1935
Regimental history 1685-1959 43696

West Stow Anglo-Saxon Village, The Visitor Centre, Icklingham Rd, West Stow, Bury Saint Edmunds IP28 6HG • T: +44 1284 728718 • F: 728277 • weststow@stedsbc.gov.uk • www.stedmundsbury.gov.uk/weststow.htm •
Man.: *Alan Baxter* •
Archaeology Museum
Early Anglo-Saxon archaeology from the West Stow settlement (c 450-650 AD) and cemetery, along with other local material from the same period, also mesolithic, Bronze Age, Iron Age and Roman material, reconstructed Anglo-Saxon village on original site 43697

Buscot

Buscot House, Buscot SN7 8BU • T: +44 1367 20786 •
Fine Arts Museum
Coll of paintings 43698

Bushey

Bushey Museum and Art Gallery, Rudolph Rd, Bushey WD23 3HW • T: +44 20 84204057, 89503233 • F: 84204923 • busmt@bushey.org.uk • www.busheymuseum.org •
Dir.: *David Whorlow* • Cur.: *Bryen Wood* •
Local Museum / Fine Arts Museum – 1993
Local history, Monro circle of artists, Herkomer and Kemp-Welch art schools – Local Study Centre 43699

Bushmead

John Dony Field Centre, Hancock Dr, Bushmead LU2 7SF • T: +44 1582 486983 • F: 422805 • trevor.tween@luton.gov.uk • www.luton.gov.uk •
Dir.: *Trevor Tween* •
Natural History Museum
Natural history, enviroment 43700

Bushmills

Causeway School Museum, 52 Causeway Rd, Bushmills BT57 8SU • T: +44 28 20731777 • F: 20732142 •
Head: *Vivienne Quinn* •
Historical Museum 43701

Giant's Causeway and Bushmills Railway, Runkerry Rd, Bushmills BT57 8SZ • T: +44 28 20732844 • F: 20732844 • infogcbr@btconnect.com • www.freewebs.com/giantscausewayrailway •
Dir.: *D.W. Laing* •
Science&Tech Museum 43702

Giant's Causeway Visitor Centre, 44a Causeway Rd, Bushmills, Co. Antrim • T: +44 12657 31582 •
Natural History Museum
World Heritage Centre and National Nature Reserve 43703

Bute Town

Drenewydd Museum, 26-27 Lower Row, Bute Town NP2 5QH • T: +44 1443 864224 • F: 864228 • morgacl@caerphilly.gov.uk • www.caerphilly.gov.uk •
Dir.: *Chris Morgan* •
Local Museum
Ironworker's cottages, domestic and household objects 43704

Buxton

Buxton Museum and Art Gallery, Peak Bldgs, Terrace Rd, Buxton SK17 6DA • T: +44 1298 24658 • F: 79394 • buxton.museum@derbyshire.gov.uk • www.derbyshire.gov.uk/azserv/libh024.htm •
Man.: *Ros. Westwood* •
Fine Arts Museum / Natural History Museum / Local Museum – 1891
Prehistoric cave finds, geology, mineralogy, archaeology, fine and decorative arts, craft, local hist 43705

Cadbury

Fursdon House, Cadbury EX5 5JS • T: +44 1392 860860 • F: 860126 • enquiries@fursdon.co.uk • www.fursdon.co.uk •
Dir.: *E.D. Fursdon* •
Special Museum
Coll of family possessions, furniture, portraits, everyday items, family costumes, coll of Iron Age and Roman material, old estate maps 43706

Caerleon

National Roman Legionary Museum, High St, Caerleon NP18 1AE • T: +44 1633 423134 • F: 422869 • post@nmgw.ac.uk • www.nmgw.ac.uk/rlm •
Cur.: *Bethan Lewis* •
Archaeology Museum – 1850
Finds from excavations at the fortress of the Second Augustan Legion 43707

Caernarfon

Caernarfon Maritime Museum, Victoria Dock, Caernarfon LL55 1SR • T: +44 1248 752083 •
Pres.: *F. Whitehead* •
Local Museum
History of Caernarfon as a harbour and port since Roman times, maritime education, trade and industry of the port, steam engine Seiont II, HMS Conway exhibits 43708

Gwynedd Education and Culture, Archives, Museums and Galleries, County Offices, Caernarfon LL55 1SH • T: +44 1286 679088, 678098 • F: 679637 • museums@gwynedd.gov.uk • www.gwynedd.gov.uk •
Ass. Dir.: *Gareth H. Williams* •
Historical Museum
Social hist 43709

Royal Welch Fusiliers Regimental Museum, Queen's Tower, The Castle, Caernarfon LL55 2AY • T: +44 1286 673362 • F: 677042 • rwfusiliers@callnetuk.com • www.rwfmuseum.org.uk •
Military Museum – 1960
Hist of the regiment since 1689, doc, portraits, medals, weapons, uniforms, badges and pictures 43710

Segontium Roman Museum, Beddgelert Rd, Caernarfon LL55 1AT • T: +44 1286 675625 • F: 678416 • info@segontium.org.uk • www.segontium.org.uk •
Cust.: *D. Jones* •
Archaeology Museum – 1937
Finds from excavations at the former Roman fort Segontium, inscriptions, medals, urns, tools, armaments, pottery, household objects 43711

Caerphilly

Caerphilly Castle, Caerphilly CF2 1UF • T: +44 29 20883143 •
Historical Museum
Full scale replica, working medieval siege engines 43712

Caldicot

Caldicot Castle, Church Rd, Caldicot NP26 4HU • T: +44 1291 420241 • F: 435094 • www.caldicotcastle.co.uk •
Man.: *Sarah Finch* •
Historic Site 43713

Callander

Rob Roy and Trossachs Victor Centre, Acaster Sq, Callander FK17 8ED • T: +44 1877 330342 • F: 330784 •
Man.: *Janet Michael* •
Historical Museum
Story of the highland folk hero Rob Roy MacGregor 43714

Calne

Atwell-Wilson Motor Museum, Stockley Ln, Calne SN11 0NF • T: +44 1249 813119 • F: 813119 • avmmcalne@aol.com • www.atwellwilson.org.uk •
Dir.: *Richard Atwell* •
Science&Tech Museum
Coll of post vintage and classic vehicles, motorcycles 43715

Bowood House and Gardens, Derry Hill, Calne SN11 0LZ • T: +44 1249 812102 • F: 821757 • houseandgardens@bowood.org • www.bowood.org •
Cur.: *Dr. Kate Fielden* •
Decorative Arts Museum / Fine Arts Museum / Historical Museum
Fine and decorative arts, paintings, watercolours, furniture, sculptures, architecture – library, archive 43716

Calverton

Calverton Folk Museum, Main St, Calverton NG14 6FQ • T: +44 115 9654843 • pressoc@ntlworld.com • www.welcome.to/calverton •
Folklore Museum
Hosiery industry, framework knitting machine – Library 43717

Camberley

National Army Museum, Sandhurst Outstation, Royal Military Academy Sandhurst, Camberley GU15 4PQ • T: +44 1276 63344 ext 2457 • F: 686316 • info@national-army-museum.ac.uk • www.national-army-museum.ac.uk •
Military Museum
Indian army memorial room, study coll of uniforms and equipment 43718

Royal Military Academy Sandhurst Collection, Royal Military Academy, Camberley GU15 4PQ • T: +44 1276 412489 • F: 412595 • drthwaites@rmas.mod.uk •
Cur.: *Dr. Peter Thwaites* •
Military Museum
History of the academy 43719

Surrey Heath Museum, Surrey Heath House, Knoll Rd, Camberley GU15 3HD • T: +44 1276 707284 • F: 671158 • museum@surreyheath.gov.uk • www.surreyheath.gov.uk/leisure •
Dir.: *Sharon Cross* •
Local Museum – 1935
Local 19th and 20th c social hist, costume, natural hist, local photographs 43720

Cambo

Wallington, The National Trust, Cambo NE61 4AR • T: +44 1670 773600/05 • F: 774420 • wallington@nationaltrust.org.uk • www.nationaltrust.org.uk •
Man.: *Lloyd Langley* •
Decorative Arts Museum / Fine Arts Museum – 1688
House rebuilt in 1688 on site of Peel Tower, 18th-19th c furnishings, rococo plaster decorations, 19th c hall decorated by William Bell Scott, Ruskin a.o., Trevelyan family, Chinese and European porcelain, paintings by Reynolds, Romney and Gainsborough 43721

Camborne

Trevithick Cottage, Penponds, Camborne TR14 0QG • T: +44 1209 612154 • trevithick1771@aol.com •
Cust.: *L. Humphrey* •
Historical Museum / Fine Arts Museum
Oil paintings of himself, wife and family 43722

Cambridge

Cambridge and County Folk Museum, 2-3 Castle St, Cambridge CB3 0AQ • T: +44 1223 355159 • info@folkmuseum.org.uk • www.folkmuseum.org.uk •
Cur.: *Cameron Hawke-Smith* • Ass. Cur.: *Rebecca Proctor* •
Local Museum – 1936
Hist of Cambridge and Cambridgeshire county since the 17th c, looking at the everyday life of people in the area, wide-ranging colls, strengths, domestic items, toys, tools 43723

Cambridge Museum of Technology, Old Pumping Station, Cheddars Ln, Cambridge CB5 8LD • T: +44 1223 368650 • cmt_master@hotmail.com • www.museumoftechnology.com •
Dir.: *David Gates* •
Science&Tech Museum
Local industry, steam engines, printing, scientific instr 43724

Cambridge University Collection of Air Photographs, Mond Bldg, Free School Ln, Cambridge CB2 3RF • T: +44 1223 334578 • F: 763300 • aerial-photography@lists.cam.ac.uk • www.aerial.cam.ac.uk •
Dir.: *Dr. B.J. Devereux* •
Special Museum
Air-photographs of the British Isles record both the natural environment and the effects of human activity from prehistoric times to the present day, substantial Irish coll and smaller collections of photographs taken over northern France, Denmark and the Netherlands 43725

Cavendish Laboratory, c/o University of Cambridge, Dept. of Physics, Madingley Rd, Cambridge CB3 0HE • T: +44 1223 337420, 337200 • F: 766360 • clh29@phy.cam.ac.uk • www.phy.cam.ac.uk •
Science&Tech Museum
Scientific instruments 43726

The Fitzwilliam Museum, c/o University of Cambridge, Trumpington St, Cambridge CB2 1RB • T: +44 1223 332900 • F: 332923 • fitzmuseum-enquiries@lists.cam.ac.uk • www.fitzmuseum.cam.ac.uk •
Dir.: *Dr. Timothy Potts* • Sc. Staff: *Dr. L. Burn* (Antiquities) • *J. Dawson* (Antiquities, Conservation) • *Dr. S. Ashton* (Antiquities) • *J.E. Poole* (Applied Arts) • *M.A.S. Blackburn* (Coins and Medals) • *Dr. A. Popescu* (Coins and Medals) • *Dr. S. Panayotova* (Manuscripts and Printed Books) • *D.E. Scrase* (Paintings and Drawings) • *J.A. Munro* (Paintings and Drawings) • *C. Hartley* (Prints) • Cons.: *Rebecca Proctor* • *B. Clarke* •
University Museum – 1816
Masterpieces of painting from 14th c to today, drawings and prints, sculpture, furniture, armour, pottery and glass, oriental art, illuminated manuscripts, coins and medals and antiquities from Egypt, the Ancient Near East, Greece, Rome and Cyprus, coll Richard VII Viscount Fitzwilliam of Merrion (the founder of the museum) – Hamilton Kerr Institute for the Conservation of Paintings 43727

Kettle's Yard, c/o University of Cambridge, Castle St, Cambridge CB3 0AQ • T: +44 1223 352124 • F: 324377 • mail@kettlesyard.co.uk • www.kettlesyard.co.uk •
Dir.: *Michael Harrison* •
Fine Arts Museum / University Museum – 1957
20th c art incl Ben Nicholson, Alfred Wallis, David Jones, Christopher Wood, Henri Gaudier-Brzeska 43728

Lewis Collection, Corpus Christi College, Cambridge CB2 1RH • T: +44 1223 338000 • F: 338061 • info@corpus.cam.ac.uk • www.corpus.cam.ac.uk •
Dir.: *Dr. Christopher Kelly* (Greek and Roman Art, Archaeology, Numismatics) •
Archaeology Museum / Decorative Arts Museum / University Museum
Classical and neo-classical gems, Greek and Roman coins, Greek and Roman figurines and pottery 43729

Museum of Classical Archaeology, c/o University of Cambridge, Sidgwick Av, Cambridge CB3 9DA • T: +44 1223 335153 • F: 335409 • jd125@cam.ac.uk • www.classics.cam.ac.uk/ark.html •
Cur.: *Henry Hurst* • Asst. Cur.: *John Donaldson* •
Museum of Classical Antiquities – 1884 43730

New Hall Art Collection, Huntingdon Rd, Cambridge CB3 0DF • T: +44 1223 762297 • F: 763110 • art@newhall.cam.ac.uk • www-art.newhall.cam.ac.uk •
Admin.: *Sarah Greaves* • Cur.: *Amy Jones* •
Fine Arts Museum
Works by contemporary women artists 43731

Scott Polar Research Institute Museum, University of Cambridge, Lensfield Rd, Cambridge CB2 1ER • T: +44 1223 336540, 336552 • F: 336549 • museum@spri.cam.ac.uk • www.spri.cam.ac.uk •
Dir.: *Prof. J.A. Dowdeswell* • Cur.: *H.E. Lane* •
Natural History Museum – 1920
Polar exploration and research, material from Antarctic expeditions of Captain Scott and from other historical British polar expeditions, exploration of the NW-Passage, relics, manuscripts, paintings, photographs, Eskimo art, ethnological coll – library, archives, picture library 43732

Sedgwick Museum of Earth Sciences, c/o University of Cambridge, Dept. of Earth Sciences, Downing St, Cambridge CB2 3EQ • T: +44 1223 333456 • F: 333450 • sedgwickmuseum@esc.cam.ac.uk • www.sedgwickmuseum.org •
Dir.: *Dr. David B. Norman* • Cur.: *Dr. R.B. Rickards* • *Dr. M.A. Carpenter* • *M. Dorling* •
Natural History Museum / University Museum – 1728/1904
Local, British, foreign rocks, fossils, minerals, memorabilia of scientist Adam Sedgwick, John Woodward coll (oldest intact geological coll), Darwin's Rocks – conservation laboratory, Sedgwick library 43733

University Museum of Archaeology and Anthropology, c/o University of Cambridge, Downing St, Cambridge CB2 3DZ • T: +44 1223 333516 • F: 333517 • cumaa@hermes.cam.ac.uk • museum.archanth.cam.ac.uk •

Dir.: *Prof. David W. Phillipson* •
Archaeology Museum / University Museum / Ethnology Museum – 1884
Prehistoric archaeology of all parts of the world, the archaeology of the Cambridge region from the paleolithic period to recent times, ethnographic material from all regions of the world with a particular emphasis on Oceania 43734

University Museum of Zoology, c/o University of Cambridge, Downing St, Cambridge CB2 3EJ • T: +44 1223 336650 • umzc@zoo.cam.ac.uk • www.zoo.cam.ac.uk/museum •
Dir.: *Prof. Michael Akam* •
Natural History Museum – 1814
Zoology: birds, fossil tetrapods, invertebrates, mammals, fish, reptiles and amphibians 43735

Whipple Museum of the History of Science, Free School Ln, Cambridge CB2 3RH • T: +44 1223 330906 • F: 334554 • hps-whipple-museum@lists.cam.ac.uk • www.hps.cam.ac.uk/whipple.html •
Dir.: *Liba Taub* (Early Scientific Instruments and Cosmology) •
Historical Museum / University Museum – 1944
Old scientific instruments, 16th-20th c books 43736

Camelford

British Cycling Museum, The Old Station, Camelford PL32 9TZ • T: +44 1840 212811 • F: 212811 • www.chycor.co.uk •
Owner: *J.W. Middleton* • Dir.: *S.I. Middleton* •
Science&Tech Museum
Cycling history, 400 cycles, over 1,000 cycling medals, enamel signs and posters, lighting – library 43737

North Cornwall Museum and Gallery, The Clease, Camelford PL32 9PL • T: +44 1840 212954 • F: 212954 • manager@camelfordtic.eclipse.co.uk •
Dir.: *Sally Holden* •
Local Museum / Folklore Museum 43738

Campbeltown

Campbeltown Museum, Public Library, Hall St, Campbeltown PA28 6BS • T: +44 1586 552366 • F: 552938 • manager@auchindrain.museum.org.uk •
Dir.: *Mary van Helmond* •
Local Museum – 1898 43739

Canterbury

Bargrave Collection, Canterbury Cathedral Archives and Library, Cathedral House, 11 The Precincts, Canterbury CT1 2EH • T: +44 1227 865330, 865287 • F: 865222 • library@canterbury-cathedral.org • www.canterbury-cathedral.org •
Head: *Keith M.C. O'Sullivan* •
Library with Exhibitions
Cabinets of curiosities – library, archives 43740

Buffs Museum, National Army Museum (currently closed), 18 High St, Canterbury CT1 2RA • T: +44 1227 452747 • F: 455047 • museums@canterbury.gov.uk • www.canterbury-museums.co.uk •
Military Museum / Fine Arts Museum
Royal East Kent Regiment story, uniforms, weapons, medals, sporting and leisure pursuits 43741

Canterbury Heritage Museum, Poor Priests' Hospital, Stour St, Canterbury CT1 2RA • T: +44 1227 452747 • F: 455047 • www.canterbury-museum.co.uk •
Dir.: *K.G.H. Reedie* •
Local Museum
History of Canterbury's 2000 years 43742

Roman Museum, Longmarket, Butchery Ln, Canterbury CT1 2RA • T: +44 1227 785575 • F: 455047 • museums@canterbury.gov.uk • www.canterbury-museums.co.uk •
Dir.: *K.G.H. Reedie* •
Archaeology Museum / Historical Museum / Historic Site
Round a Roman site, excavated finds, mosaic floors, hands-on area 43743

Royal Museum and Art Gallery, Canterbury City Museums (currently closed), 18 High St, Canterbury CT1 2RA • T: +44 1227 452747 • F: 455047 • museums@canterbury.gov.uk • www.canterbury-artgallery.co.uk •
Fine Arts Museum / Decorative Arts Museum / Public Gallery – 1899
Decorative arts, picture coll, cattle paintings, gallery of cattle painter Thomas Sidney Cooper (1803-1902) 43744

Saint Augustine's Abbey, Longport, Canterbury CT1 1TF • T: +44 1227 767345 • F: 767345 • dan.dennis@english-heritage.org.uk • www.english-heritage.org.uk •
Cur.: *Jan Summerfield* •
Religious Arts Museum
Religious art 43745

West Gate Towers, Saint Peters St, Canterbury • T: +44 1227 452747 • F: 455047 •
Historical Museum – 1906
British and foreign armaments, relics in 14th c medieval gate 43746

Canvey Island

Castle Point Transport Museum, 105 Point Rd, Canvey Island SS8 7TP • T: +44 1268 684272 •
Science&Tech Museum
Busses, coaches, commercial, military & emergency vehicles (1944-1972) 43747

Cardiff

Amgueddfa Cymru - National Museum of Wales, Cathays Park, Cardiff CF10 3NP • T: +44 29 20397951 • F: 20373219 • post@museumwales.ac.uk • www.museumwales.ac.uk •
Dir.: *Michael Tooby* • Sc. Staff: *R. Brewer* (Archaeology) • *Dr. M.G. Bassett* (Geology) • *Dr. Graham Oliver* (Biodiversity and systematic Biology) • *Oliver Fairclough* (Art) •
Archaeology Museum / Natural History Museum / Fine Arts Museum – 1907
Geology, mineralogy, fossils, botany, plant ecology, Welsh zoology, mollusks, entomology, Welsh archaeology up to the 16th c, numismatics, industrial history, European painting, sculpture, applied art, decorative objects, French art 43748

Cardiff Castle, Castle St, Cardiff CF10 3RB • T: +44 29 20878100 • F: 20231417 • cardiffcastle@cardiff.gov.uk • www.cardiffcastle.com •
Cur.: *Matthew Williams* •
Fine Arts Museum / Historical Museum
Architectural drawings, ceramics, painting, furniture, Burges coll, De Morgan coll 43749

Howard Gardens Gallery, University of Wales Institute, Howard Gardens, Cardiff CF24 0SP • T: +44 29 20416608 • F: 20416678 • rcox@uwic.ac.uk •
Dir.: *Richard Cox* •
Public Gallery 43750

IBRA Collection, International Bee Research Association, 18 North Rd, Cardiff CF10 3DT • T: +44 29 20372409 • F: 20665522 • mail@ibra.org.uk • www.ibra.org.uk •
Dir.: *Richard Jones* •
Agriculture Museum
Artifacts from around the world related to bees and beekeeping – library 43751

Regimental Museum of 1st The Queen's Dragoon Guards, Cardiff Castle, Cardiff CF10 2RB • T: +44 29 20781271 • F: 20781384 • curator@qdg.org.uk • www.qdg.org.uk •
Cur.: *Clive Morris* •
Military Museum 43752

Saint David's Hall, The Hayes, Cardiff CF1 2SH • T: +44 29 20878500 • F: 20878599 • st.davidshallreception@@cardiff.gov.uk • www.stdavidshallcardiff.co.uk •
Public Gallery 43753

Techniquest, Stuart St, Cardiff CF10 5BW • T: +44 29 20475475 • F: 20482517 • info@techniquest.org • www.techniquest.org •
C.E.O.: *Peter Trevitt* •
Science&Tech Museum
Science discovery centre – Planetarium 43754

The Welch Regiment Museum (41st/ 69th Foot) of the Royal Regiment of Wales, Black and Barbican Towers, Cardiff Castle, Cardiff CF10 2RB • T: +44 29 20229367 • welch@ukonline.co.uk • www.rrw.org.uk •
Cur.: *John Dart* •
Military Museum 43755

Cardigan

Cardigan Heritage Centre, Teifi Wharf, Castle St, Cardigan SA43 3AA • T: +44 1239 614404 • F: 614404 • heritagecentre@hanesaberteifi.freeserve.couk •
Local Museum
Town hist 43756

Carlisle

Border Regiment and King's Own Royal Border Regiment Museum, Queen Mary's Tower, Castle, Carlisle CA3 8UR • T: +44 1228 532774 • F: 545435 • korbrmuseum@aol.com • www.kingsownbordermuseum.btik.com •
Cur.: *Stuart Eastwood* •
Military Museum – 1932
Uniforms, badges, medals, pictures, silver, arms, dioramas, documents, trophies, militaria, photogr – archives 43757

Cathedral Treasury Museum, Carlisle Cathedral, 7 The Abbey, Carlisle CA3 8TZ • T: +44 1228 548151 • F: 547049 • office@carlislecathedral.org.uk • www.carlislecathedral.org.uk •
Dir.: *Canon Warden* •
Religious Arts Museum / Archaeology Museum
Articles connected with the site from Roman times to the present day, 15th to 20th c Church plate 43758

Guildhall Museum, Greenmarket, Carlisle CA2 4PW • T: +44 1228 618718 • F: 810249 • enquiries@tulliehouse.co.uk • www.tulliehouse.co.uk •
Dir.: *Hilary Wade* •
Local Museum
History of the medieval Guilds and on the building itself, medieval iron-bound muniment chest and town bell, two 1599 silver bells, medieval measures and pottery, civic regalia, the pillory and stocks, wearing, local history 43759

Tullie House - City Museum and Art Gallery, Castle St, Carlisle CA3 8TP • T: +44 1228 534781 • F: 810249 • enquiries@tulliehouse.co.uk • www.tulliehouse.co.uk •
Dir.: *Hilary Wade* •
Local Museum – 1892
Regional antiquities, natural history, Lake District specimens, pre-Raphaelite, 19th c painting, modern paintings, English porcelain, Roman finds, ornithology, historic railways, local history 43760

Carlton Colville

East Anglia Transport Museum, Chapel Rd, Carlton Colville NR33 8BL • T: +44 1502 518459 • F: 584658 • enquiries@eatm.org.uk • www.eatm.org.uk •
Science&Tech Museum
Trams, trolleybuses and a narrow gauge railway 43761

Carmarthen

Carmarthenshire County Museum, Abergwili, Carmarthen SA31 2JG • T: +44 1267 228696 • F: 223830 • museums@carmarthenshire.gov.uk • www.carmarthenshire.gov.uk •
Dir.: *Chris Delaney* • Sc. Staff: *A.G. Dorsett* (Arts) • *G.H. Evans* (Archaeology) •
Archaeology Museum / Agriculture Museum / Local Museum – 1905
Folklore, archaeology – library 43762

Carnforth

Leighton Hall, Carnforth LA5 9ST • T: +44 1524 734474 • F: 720357 • leightonhall@yahoo.co.uk • www.leightonhall.co.uk •
Owner: *R.G. Reynolds* •
Local Museum
Manor house dating from the 13th c with old and modern painting, antique furniture, Gillow family portraits 43763

Steamtown Railway Museum, Warton Rd, Carnforth LA5 9HX • T: +44 1524 732100 • F: 735518 •
Dir.: *W.D. Smith* •
Science&Tech Museum 43764

Carrbridge

Landmark Forest Heritage Park, Main St, Carrbridge PH23 3AJ • T: +44 1479 841613 • F: 841384 • landmarkcentre@compuserve.com • www.landmark-centre.co.uk •
Dir.: *David Hayes* •
Natural History Museum – 1970
Hist, natural environment and wildlife of the Highlands, nature trails, microscopy, steampowered sawmill, working horse and woodcraft centre 43765

Carrickfergus

Flame!, The Gasworks Museum of Ireland, 44 Irish Quarter W, Carrickfergus BT38 8AT • T: +44 28 93369575 • info@flamegasworks.co.uk • www.flamegasworks.co.uk •
Dir.: *Sam Gault* •
Science&Tech Museum
Old gas appliences and artefacts 43766

Castle Cary

Castle Cary and District Museum, Market House, High St, Castle Cary BA7 7AL • T: +44 1663 350680 •
Local Museum 43767

Castle Combe

Castle Combe Museum, Combe Cottage, Castle Combe SN14 7HU • T: +44 1249 782250 • F: 782250 • adrian@bishop60.fsnet.co.uk • www.castle-combe.com •
Cur.: *Adrian Bishop* •
Local Museum
Artefacts and historical displays 43768

Castle Donington

Donington Grand Prix Collection, Donington Park, Castle Donington DE74 2RP • T: +44 1332 811027 • F: 812829 • enquiries@doningtoncollection.co.uk • www.doningtoncollection.com •
Head: *Gary Rankin* •
Special Museum – 1973
Coll of Grand Prix Racing cars, McLarren, BRM, Lotus, Ferrari, Vanwalls, Williams, amazing coll of helmets, Formula 1 racing 43769

Castle Douglas

Castle Douglas Art Gallery, Marekt St, Castle Douglas DG7 1AE • T: +44 1556 331643 • F: 331643 • davidd@dumgal.gov.uk • www.dumgal.gov.uk/museums •
Public Gallery
Paintings by Ethel S.G. Bristowe, local artists 43770

Castleford

Castleford Museum Room, Library, Carlton St, Castleford WF10 2BB • T: +44 1924 722085 • F: 305353 • museumsandarts@wakefield.gov.uk • www.wakefield.gov.uk/cultureandleisure •
Dir.: *Colinn MacDonald* •
Historical Museum – 1935
Archaelogy in Castlefort 43771

Castlerock

Downhill Castle, Mussenden Rd, Castlerock, Co. Antrim • T: +44 12658 848728 •
Historical Museum 43772

Castleton

Castleton Village Museum, Methodist Schoolroom, Buxton Rd, Castleton S33 8WX • T: +44 1433 620950 •
Cur.: *Christine Thorpe* •
Local Museum
Local crafts, lead mining, rope-making, agriculture, Garland Ceremony and Blue John Caverns 43773

Castletown

Castle Rushen, Manx National Heritage, The Quay, Castletown IM9 1AB • T: +44 1624 648000 • F: 648001 • enquiries@mnh.gov.im • www.storyofmann.com •
Dir.: *Stephen Harrison* •
Historical Museum
Story of the Kings and Lords of Man, castle keep, draw bridge and dungeons 43774

Dunnet Pavilion, Dunnet Bay Caravan Site, Castletown KW14 8XD • T: +44 1847 821531 • F: 821531 • mary.legg@highland.gov.uk •
Natural History Museum 43775

Nautical Museum, Manx National Heritage, Castletown IM9 1AB • T: +44 1624 648000 • F: 648001 • enquiries@mnh.gov.im • www.storyofmann.com •
Dir.: *Stephen Harrison* •
Historical Museum
Maritime life and trade in the days of sail, armed yacht "Peggy" (1791), sailmakers loft, ship model, photos 43776

Old Grammar School, Manx National Heritage, Quay Ln, Castletown IM9 1LE • T: +44 1624 648000 • F: 648001 • enquiries@mnh.gov.im • www.storyofmann.com •
Dir.: *Stephen Harrison* •
Historical Museum
Hist of Manx education, Victorian school life 43777

Old House of Keys, Manx National Heritage, Parliament Sq, Castletown IM9 1LA • T: +44 1624 648000 • F: 648001 • enquiries@mnh.gov.im • www.storyofmann.com •
Dir.: *Stephen Harrison* •
Historical Museum
Assembly and debating chamber for Manx parliament, Manx politics 43778

Câtel Guernsey

Folk and Costume Museum, Saumarez Park, Route de Cobo, Câtel Guernsey GY5 7UJ • T: +44 1481 255384 • folkmuseumntgsy@cwgsy.net • www.nationaltrust-gsy.org.gg •
Dir.: *Paul Le Pelley* •
Folklore Museum – 1968
Agriculture tools of the island, horse drawn machinery, farmhouse interior, domestic and Guernsey fisherman's life (19th-20th c), trades, tools; traditional Guernsey costumes 1750 to the present day 43779

Caterham

East Surrey Museum, 1 Stafford Rd, Caterham CR3 6JG • T: +44 1883 340275 • F: 340275 • es@emuseums.freeserve.co.uk • www.tandridgedc.gov.uk •
Cur.: *Elizabeth Amias* •
Local Museum
Local Hist 43780

Cawdor

Cawdor Castle, Cawdor IV12 5RD • T: +44 1667 404401 • F: 404674 • info@cawdorcastle.com • www.cawdorcastle.com •
Decorative Arts Museum 43781

Cawthorne

Regimental Museum of the 13th/18th Royal Hussars and Kent Dragoons, Cannon Hall Museum, Cawthorne S75 4AT • T: +44 1226 790270 • F: 792117 • cannonhall@barnsley.gov.uk •
Dir.: *Dr. J. Whittaker* •
Military Museum
library 43782

Victoria Jubilee Museum, Taylor Hill, Cawthorne S75 4HQ • T: +44 1226 790545 • www.aboutbritain.com/cawthornemuseum.htm •
Pres.: *Barry Jackson* •
Local Museum – 1884
Local history, example of Cruck formation timbered building, Victoriana, domestic bygones 43783

Cenarth

National Coracle Centre, Cenarth Falls, Cenarth SA38 9JL • T: +44 1239 710980 • F: 710980 • martinfowler@btconnect.com •
Cur.: *Martin Fowler* •
Science&Tech Museum
Boat coll of coracles from Wales and other parts of the world 43784

Ceres

Fife Folk Museum, Weigh House, High St, Ceres KY15 5NF, mail addr: High Street, Ceres KY15 5YS • T: +44 1334 828180 • info@fifefolkmuseum.org • www.fifefolkmuseum.org •
Dir.: *Winifred Harley* •

Folklore Museum / Historical Museum / Agriculture Museum – 1968
Tools, weights & measures, agricultural items, domestic items, costume, ceramics, textiles 43785

Chalfont Saint Giles

Chiltern Open Air Museum, Newland Park, Gorelands Ln, Chalfont Saint Giles HP8 4AD • T: +44 1494 871117, 872163 • F: 872774 • coamuseum@netscape.net • www.coam.org.uk •
Dir.: *Sue Shave* •
Open Air Museum – 1976
Historic bldgs, iron age house, Victorian farm 43786

Milton's Cottage, Deanway, Chalfont Saint Giles HP8 4JH • T: +44 1494 872313 • F: 873936 • info@miltonscottage.org • www.miltonscottage.org •
Dir.: *E.A. Dawson* •
Special Museum – 1887
Life and times of writer John Milton (1608-1674), editions of Milton's work, portraits and engravings of Milton, memorabilia 43787

Chapel Brampton

Northampton and Lamport Railway, Pitsford and Brampton Station, Pitsford Rd, Chapel Brampton NN6 8BA • T: +44 1604 820327 • info@nlr.org.uk • www.nlr.org.uk •
Science&Tech Museum
Steam trains, diesel locomotives, signal boxes 43788

Chappel

East Anglian Railway Museum, Chappel and Wakes Colne Station, Chappel CO6 2DS • T: +44 1206 242524 • F: 242524 • info@earm.co.uk • www.earm.co.uk •
Science&Tech Museum 43789

Chard

Chard and District Museum, Godworthy House, 15 High St, Chard TA20 1QL • T: +44 1460 65091 • www.southsomerset.gov.uk •
Dir.: *P. Wood* •
Local Museum – 1970
John Stringfellow Pioneer of powered flight, James Gillingham maker of artificial limbs 43790

Forde Abbey, Chard TA20 4LU • T: +44 1460 220231 • F: 220296 • forde.abbay@virgin.net •
Dir.: *Mark Roper* •
Religious Arts Museum 43791

Hornsbury Mill, Hornsbury Hill, Chard TA20 3AQ • T: +44 1460 63317 • F: 63317 • hornsburymill@btclick.com • www.hornsburymill.co.uk •
Dir.: *S.J. Lewin* •
Science&Tech Museum – 1970
Milling machinery and equipment, agricultural implements, craftsmen's tools, domestic equipment and clothing 43792

Charlbury

Charlbury Museum, Corner House Community Centre, Market St, Charlbury OX7 3PN • T: +44 1608 810060 •
Cur.: *R. Prew* •
Local Museum
Traditional crafts and industry 43793

Chatham

Fort Amherst Heritage Museum, Dock Rd, Chatham ME4 4UB • T: +44 1634 847747 • F: 830612 • info@fortamherst.com • www.fortamherst.com •
Historical Museum
Napoleonic fortress, WW II, military artefacts 43794

Historic Dockyard Chatham, Chatham ME4 4TZ • T: +44 1634 823800/07 • F: 823801 • info@chdt.org.uk • www.thedockyard.co.uk •
Historical Museum – 1984
Georgian dockyard, warships incl HMS Cavalier, submarine Ocelot, HMS Gannet, ropery, lifeboats coll 43795

Kent Police Museum, Dock Rd, Chatham ME4 4TZ • T: +44 1634 403260 • F: 403260 • info@kent-police-museum.co.uk • www.kent-police-museum.co.uk •
Cur.: *John Endicott* •
Historical Museum
Hist of the Kent County Constabulary, uniforms, equipment, photos, crime exhibits, paintings, police vehicles 43796

Medway Heritage Centre, Dock Rd, Chatham ME4 4SH • T: +44 1634 408437 •
Cur.: *R. Ingle* •
Local Museum
Artefacts and photographs to River Medway 43797

Royal Engineers Museum, Brompton Barracks, Chatham ME4 4UG • T: +44 1634 822839 • F: 822006 • mail@re-museum.co.uk • www.royalengineers.org.uk •
Dir.: *Richard Dunn* • Cur.: *Rebecca Cheney* •
Science&Tech Museum / Military Museum – 1873
Hist of engineers' work, military engineering science, memorabilia on General 'Chinese' Gordon and Field Marshal Lord Kitchener of Khartoum 43798

Chawton

Jane Austen's House, Winchester Rd, Chawton GU34 1SD • T: +44 1420 83262 • F: 83262 • www.jane-austen-house-museum.org.uk •
Head: *T.F. Carpenter* • Man.: *Ann Channon* •
Special Museum – 1949
Memorabilia of writer Jane Austen (1775-1817) 43799

Cheam

Whitehall, 1 Malden Rd, Cheam SM3 8QD • T: +44 20 86431236 • F: 86431236 • whitehallcheam@btconnect.com • www.sutton.gov.uk •
Cur.: *Julia Howard* • Dir.: *Helen Henkens-Lewis* •
Historical Museum – 1978
Arts, crafts and local history centre, displays of material from the excavations at Nonsuch Palace, Cheam Pottery and Cheam School, works by or featuring William Gilpin (caricatured by Rowlandson as Dr. Syntax) inside a timber-framed house build c 15th 43800

Cheddar

Cheddar Man & the Cannibals, Cheddar Caves & Gorge, Cheddar BS27 3QF • T: +44 1934 742343 • F: 744637 • caves@cheddarcaves.co.uk • www.cheddarcaves.co.uk •
Archaeology Museum / Natural History Museum – 1934
Archaeology and zoology of Cheddar Gorge area, paleolithic pleistocene finds, artefacts from local caves 43801

Cheddleton

Cheddleton Flint Mill, Cheadle Rd, Cheddleton ST13 7HL • T: +44 1782 502907 • robert.copeland2@btopenworld.com • www.ex.ac.uk/~akoutram/chddleton-mill •
Head: *Ted Royle* •
Science&Tech Museum – 1967
Prime movers for the grinding of raw materials for the pottery industry, steam engines, transportation, grinding equipment 43802

Chelmsford

Chelmsford Museum, Oaklands Park, Moulsham St, Chelmsford CM2 9AQ • T: +44 1245 605700 • F: 268545 • oaklands@chelmsford.gov.uk • www.chelmsford.gov.uk/museums •
Man.: *N.P. Wickenden* • Sc. Staff: *A. Lutyens-Humfrey* (Art) • *T. Walentowicz* (Natural Sciences) • *G. Bowles* (Science and Industry) • *M.D. Bedenham* (Social History) •
Local Museum – 1835
Natural hist, geology, archaeology, local hist, coins,decorative arts, paintings, glass, costume, military hist, developing science and technology museum 43803

Engine House Project, Chelmsford Museums, Sandford Mill, Chelmsford CM2 6NY • T: +44 1245 475498 • F: 475498 • engine.house@chelmsfordbc.gov.uk • www.chelmsfordbc.gov.uk/museum/ •
Dir.: *G. Bowles* (Science and Industry) •
Science&Tech Museum
Industrial coll 43804

Essex Police Museum, Headquarters Springfield, Chelmsford CM2 6DA • T: +44 1245 457150 • F: 452456 • museum@essex.pnn.police.uk • www.essex.police.uk/museum •
Cur.: *Becky Latchford* •
Special Museum
Police history since 1836 43805

Essex Regiment Museum, Oaklands Park, Moulsham St, Chelmsford CM2 9AQ • T: +44 1245 605701 • F: 262428 • oaklands@chelmsford.gov.uk • www.chelmsford.gov.uk/museums •
Man.: *N.P. Wickenden* • Cur.: *I.D. Hook* •
Military Museum
Regimental hist 43806

Cheltenham

Axiom Centre for Arts, 57 Winchombe St, Cheltenham GL52 2NE • T: +44 1242 253183 • F: 253183 •
Public Gallery 43807

Bugatti Trust, Prescott Hill, Gotherington, Cheltenham GL52 9RD • T: +44 1242 677201 • F: 674191 • trust@bugatti.co.uk • www.bugatti.co.uk/trust •
Cur.: *Richard Day* •
Science&Tech Museum
Bugatti artefacts, technical drawings – photographic archive 43808

Chedworth Roman Villa Museum, Yanworth, Cheltenham GL54 3LJ • T: +44 1242 890256 • F: 890909 • chedworth@nationaltrust.org.uk •
Archaeology Museum – 1865
Ruins of Roman villa, finds 43809

Cheltenham Art Gallery and Museum, Clarence St, Cheltenham GL50 3JT • T: +44 1242 237431 • F: 262334 • artgallery@cheltenham.gov.uk • www.cheltenham.artgallery.museum •
Museum and Arts Manager: *Jane Lillystone* • Coll Manager: *Helen Brown* • Exhibitions Manager: *Sophia Wilson* • Education, Outreach: *Virginia Adsett* • *Faye Little* •
Fine Arts Museum / Decorative Arts Museum – 1907
Dutch, Flemish and English paintings, watercolours, local prints, English pottery and porcelain, geology, Cotswold area history, Chinese porcelain, memorabilia on Edward Wilson, internationally important coll of the Arts and Crafts Movement, furniture, metalwork, textiles 43810

Holst Birthplace Museum, 4 Clarence Rd, Pittville, Cheltenham GL52 2AY • T: +44 1242 524846 • F: 580182 • holstmuseum@btconnect.com • www.holstmuseum.org.uk •
Cur.: *Dr. Joanna Archibald* •
Music Museum
Musical instruments, Victorian hist, life and music of Gustav Holst, composer of The Planets 43811

Chepstow

Chepstow Museum, Gwy House, Bridge St, Chepstow NP16 5EZ • T: +44 1291 625981 • F: 635005 • chepstowmuseum@monmouthshire.gov.uk •
Dir.: *Anne Rainsbury* •
Local Museum – 1949
Local and social hist – education resource centre, paper and object conservation (social history, archaeology & ethnography) 43812

Tintern Abbey, Tintern, Chepstow NP6 6SE • T: +44 1291 689251 • cadw@wales.gsi.gov.uk • www.cadw.wales.gov.uk •
Religious Arts Museum
Ruins 43813

Chertsey

Chertsey Museum, The Cedars, 33 Windsor St, Chertsey KT16 8AT • T: +44 1932 565764 • F: 571118 • enquiries@chertseymuseum.org.uk • www.chertseymuseum.org.uk •
Cur.: *Emma Warren* •
Local Museum / Decorative Arts Museum – 1965
Olive Matthews coll of English dress and accessories, local hist, archaeology, ancient Greek pottery, horology 43814

Chesham

Chesham Museum-The Stables, Gamekeepers Lodge, Bellingdon Rd, Chesham HP5 2NN • T: +44 1494 793491 • info@cheshammuseum.org.uk • www.cheshammuseum.org.uk •
Chair: *Sue Gordon* •
Local Museum
Historical items 43815

Chester

Cheshire Military Museum, The Castle, Chester CH1 2DN • T: +44 1244 327617 • F: 401700 • r.corners@chester.ac.uk • www.chester.ac.uk/militarymuseum •
Cur.: *N.A. Hine* •
Military Museum – 1924
Memorabilia on the 22nd Cheshire Regiment, Cheshire Yeomanry, Battle of Meeanee 1843, memorabilia of Sir Charles Napier, Eaton Hall Officer Cadet School – archives 43816

Dewa Roman Experience, Pierpoint Ln, Off Bridge St, Chester CH1 1NL • T: +44 1244 343407 • F: 347737 • www.dewaromanexperience.co.uk •
Man.: *Lesley Wilson* •
Archaeology Museum
Streets of Roman Chester 43817

Grosvenor Museum, 27 Grosvenor St, Chester CH1 2DD • T: +44 1244 402033 • grosvenormuseum@cheshirewestandchester.gov.uk • www.grosvenormuseum.co.uk •
Dir.: *Susan Hughes* • Keeper: *Peter Boughton* (Art, Architecture) • *Hannat Crowdy* (Local and Social History) • *Dan Robinson* (Archaeology) • *Kate Riddington* (Natural History) •
Local Museum
Roman archaeology, local and natural hist, art, clocks, coins, silver, costume 43818

Chester-le-Street

Anker's House Museum, The Parish Centre, Church Chare, Chester-le-Street DH3 3QB • T: +44 191 3883295 •
Dir.: *Mike Rutter* •
Archaeology Museum
Roman and Anglo-Saxon artefacts 43819

Chesterfield

Chesterfield Museum and Art Gallery, Saint Mary's Gate, Chesterfield S41 7TD • T: +44 1246 345727 • F: 345720 • museum@chesterfieldbc.gov.uk • www.chesterfield.gov.uk •
Dir.: *A.M. Knowles* •
Historical Museum / Public Gallery – 1994
Locally produced salt glazed pottery (19th-20th c) 43820

Hardwick Hall, Doe Lea, Chesterfield S44 5QJ • T: +44 1246 850430 • F: 858424 • hardwickhall@nationaltrust.org.uk • www.nationaltrust.org.uk •
Man.: *Denise Edwards* •
Decorative Arts Museum – 1597
Furniture, tapestry, needlework, and portraits of the Cavendish family, late 16th c house, greatest of all Elizabethan houses and garden, corn mill and parkland 43821

Peacock Heritage Centre, Low Pavement, Chesterfield S40 1PB • T: +44 1246 345777/78 • F: 345770 • publicity@chesterfieldbc.gov.uk •
Local Museum 43822

Revolution House, Hight St, Old Whittington, Chesterfield • T: +44 1246 345727 • F: 345720 • museum@chesterfield.gov.uk • www.chesterfield.gov.uk •
Historical Museum – 1938
Restored 17th c bldg with 17th c furnishings, hist of the Revolution of 1688 43823

Chichester

Cathedral Treasury, Royal Chantry, Cathedral Cloisters, Chichester PO19 1PX • T: +44 1243 782595 • F: 812499 • visitors@chichestercathedral.org.uk • www.chichestercathedral.org.uk •
Head: *M.J. Moriarty* •
Religious Arts Museum
Historical ecclesiastical plate from diocese and cathedral 43824

Chichester District Museum, 29 Little London, Chichester PO19 1PB • T: +44 1243 784683 • F: 776766 • districtmuseum@chichester.gov.uk • www.chichester.gov.uk/museum •
Dir.: *Dr. I. Friel* • Sc. Staff: *Rosemary Gilmour* • *Simon Kitchin* • *Jane Seddon* •
Local Museum – 1964
Local archaeology and social hist 43825

Fishbourne Roman Palace and Museum, The Sussex Archaeological Society, Salthill Rd, Fishbourne, Chichester PO19 3QR • T: +44 1243 785859 • F: 539266 • adminfish@sussexpast.co.uk • www.sussexpast.co.uk •
Dir.: *David J. Rudkin* •
Archaeology Museum – 1968
Roman ruins and finds, Britains largest coll of 'in-situ' Roman mosaic floors and other architectural and artefactural remains, formal Roman garden planted to its original plan 43826

Goodwood House, Goodwood, Chichester PO18 0PX • T: +44 1243 755000 • F: 755005 • housevisiting@goodwood.co.uk • www.goodwood.co.uk •
Dir.: *Earl of March and Kinrara* • Cur.: *James Peill* •
Fine Arts Museum / Decorative Arts Museum – 1697
Family portraits, Sèvres porcelain, French & English furniture, Gobelins tapestries, paintings by Van Dyck, Canaletto (Giovanni Antonio Canal) & George Stubbs 43827

Guildhall Museum, Priory Park, Chichester PO19 1PB, mail addr: 29 Little London, Chichester PO19 1PB • T: +44 1243 784683 • F: 776766 • districtmuseum@chichester.gov.uk • www.chichester.gov.uk/museum •
Dir.: *Dr. I. Friel* • Sc. Staff: *Simon Kitchin* • *Rosemary Gilmour* • Asst. Cur.: *Jane Seddon* •
Historical Museum – 1930
Hist of the site/building, large stones from Roman Chichester 43828

Mechanical Music and Doll Collection, Church Rd, Portfield, Chichester PO19 4HN • T: +44 1243 372646 • F: 370299 •
Dir.: *Lester Jones* • *Clive Jones* •
Music Museum – 1983
Disc musical boxes, mechanical musical instruments, Lester Jones renaissance discs, replica discs for musical boxex 43829

Pallant House Gallery, 9 North Pallant, Chichester PO19 1TJ • T: +44 1243 774557 • F: 536038 • info@pallant.org.uk • www.pallant.org.uk •
Dir.: *Stefan Van Raay* • Cur.: *Frances Guy* •
Fine Arts Museum – 1982
20th c British art 43830

Royal Military Police Museum, Roussillon Barracks, Broyle Rd, Chichester PO19 4BL • T: +44 1243 534225 • F: 534288 • museum@rhqrmp.freeserve.co.uk • www.rhqrmp.freeserve.co.uk •
Dir.: *P.H.M. Squier* •
Military Museum – 1946
Military police, uniforms, medals, hist, books, pictures, arms – archives 43831

Weald and Downland Open Air Museum, Singleton, Chichester PO18 0EU • T: +44 1243 811348 • F: 811475 • office@wealddown.co.uk • www.wealddown.co.uk •
Dir.: *Richard Harris* •
Open Air Museum / Agriculture Museum – 1967
45 historic buildings, plumbing, carpentry, wheelwrighting, blacksmithing and shepherding equipment, tudor kitchen, working watermill, medieval farmstead – library 43832

Chicksands

Museum Military Intelligence, Intelligence Corps Museum and Medmenham Collection, Defence Intelligence and Security Centre, Chicksands SG17 5PR • T: +44 1462 752896 • F: 752341 • musarchdint-d@disc.mod.uk • www.army.mod.uk/intelligencecorps/chicksands.htm •
Cur.: *J.D. Woolmore* •
Military Museum
Hist of Intelligence Corps, hist of air photography, Y service 1940-45 43833

Chiddingstone

Chiddingstone Castle, Chiddingstone TN8 7AD • T: +44 1892 870347 •
Sc. Staff: *Ruth Eldridge* •
Historical Museum / Fine Arts Museum – 1956
Stuart royal letters and memorabilia, Jacobite coll,

U

portraits, manuscripts, medals, Japanese boxes, swords, lacquer, armaments and metal objects, Buddhistic art, Egyptian objects 4000 BC-Roman times 43834

Chippenham

Chippenham Museum and Heritage Centre, 10 Market Pl, Chippenham SN15 3HF • T: +44 1249 705020 • F: 705025 • heritage@chippenham.gov.uk • www.chippenham.gov.uk •
Cur.: *Mike Stone* •
Local Museum – 2000 43835

Chipping Norton

Chipping Norton Museum of Local History, Westgate Centre, High St, Chipping Norton OX7 5AD • T: +44 1608 643779 • museum@cn2001.fsnet.co.uk •
Local Museum
Agricultural tools and horse harnesses, local industries, tweed, brewing and agricultural engineering, baseball 43836

Chirk

Castell y Waun (Chirk Castle), Chirk Wrexham LL14 5AF • T: +44 1691 777701 • F: 774706 • chirkcastle@nationaltrust.org.uk •
Dir.: *M.V Wynne* •
Decorative Arts Museum / Fine Arts Museum / Historic Site
Paintings, decorative art, history, 16th, 17th, 18th and early 19th c decorations 43837

Chittlehampton

Cobbaton Combat Collection, Chittlehampton EX37 9RZ • T: +44 1769 540740 • info@cobbatoncombat.co.uk • www.cobbatoncombat.co.uk •
Dir.: *Preston Isaac* •
Military Museum – 1981
Over 70 WW2 mainly British tanks, trucks, artillery guns and armoured cars, restored to running order, thousands of smaller items 43838

Chollerford

Clayton Collection Museum, Chesters Roman Fort, Humshaugh, Chollerford NE46 4EP • T: +44 1434 681379 • www.english-heritage.org.uk •
Dir.: *Georgina Plowright* •
Archaeology Museum – 1903
Roman coll from sites along central sector of hadrian's wall, Roman inscriptions, sculpture, weapons, tools and ornaments from several forts 43839

Chorley

Astley Hall Museum and Art Gallery, Astley Park, Chorley PR7 1NP • T: +44 1257 515927 • F: 515923 • astley.hall@chorley.gov.uk • www.chorley.gov.uk/astleyhall •
Cur.: *Louise McCall* •
Fine Arts Museum / Decorative Arts Museum – 1924
Jacobean furniture, paintings and etchings, Flemish tapestries, Leeds pottery, 18th c English glass, magnificent plaster ceilings 43840

Christchurch

Museum of Electricity, Old Power Station, Bargates, Christchurch BH23 1QE • T: +44 1202 480467 • F: 480468 • www.scottish-southern.co.uk/museum •
Cur.: *Eric Jones* •
Science&Tech Museum
Working models of orginal experiments by Faraday, Richie and Barlow 43841

Red House Museum and Art Gallery, Quay Rd, Christchurch BH23 1BU • T: +44 1202 482860 • F: 481924 • musmjh@hants.gov.uk • www.hants.gov.uk/museums •
Cur.: *Jim Hunter* •
Local Museum – 1951
Paleolithic to medieval local archaeology, local hist, geology and natural hist, costumes, toys, dolls, domestic equipment, agricultural equipment, fashion plates, Victoriana, paintings, drawings, prints and photographs of local interest – library, herb garden 43842

Saint Michael's Loft Museum, The Christchurch Priory, Quay Rd, Christchurch BH23 1BU • T: +44 1202 485804 • uk.edlunds@ntlworld.com • www.christchurchpriory.org •
Cur.: *D. Edlund* •
Religious Arts Museum
Stonework and carving from Saxon to Tudor times, church memorabilia, map and models, tomb of Knight Tempian Stephen de Strapplebrugge, watercolours 43843

Chudleigh

Ugbrooke House, Chudleigh TQ13 0AD • T: +44 1626 852179 • F: 853322 •
Man.: *Lord Thomas Clifford* •
Military Museum / Decorative Arts Museum
Private military coll, portraits 43844

Cirencester

Cirencester Art & Local Studies Collection, Bingham Library Trust Art Collection, 3 Dyer St, Cirencester GL7 2PP • T: +44 1285 655646 • F: 643843 • clerk@cirencester.gov.uk • binghamlibrarytrust.org.uk/artcollection •
Cur.: *David Viner* •
Fine Arts Museum – 1975
Oil paintings, watercolours, prints and drawings 43845

Cirencester Lock-Up, Trinity Rd, Cirencester GL7 1BR • T: +44 1285 655611 • F: 643286 • museums@cotswold.gov.uk • www.cotswold.gov.uk •
Local Museum 43846

Corinium Museum, Park St, Cirencester GL7 2BX • T: +44 1285 655611 • john.paddock@cotswold.gov.uk • www.cotswold.gov.uk •
Dir.: *Dr. J.M Paddock* •
Archaeology Museum / Local Museum – 1856
Romano-British antiquities from the site of Corinium Dobunnorum, archaeology, local history, military 43847

Clare

Clare Ancient House Museum, High St, Clare CO10 8NY • T: +44 1787 277662 • F: 277662 • clareah@btinternet.com • www.clare-ancient-house-museum.co.uk •
Dir.: *David J. Ridley* •
Local Museum
Clare people and industries, exhibits from prehistoric to Victorian times 43848

Clatteringshaws

Galloway Deer Museum, c/o Forest Wildlife Centre, New Galloway, Castle Douglas, Clatteringshaws DG7 3SQ • T: +44 1671 402420 • F: 403708 •
Man.: *Ann Smith* •
Natural History Museum
Red deer, roe deer, wild goats, local wildlife, geology of the district 43849

Claydon

Bygones Museum, Butlin Farm, Claydon OX17 1EP • T: +44 1295 690258 • F: 690114 • bygonesmuseum@yahoo.com • www.bygonesmuseum.moonfruit.com •
Dir.: *Catherine Fox* •
Agriculture Museum
Domestic equipment, agricultural implements, tools once used by rural craftsmen 43850

Cleethorpes

Cleethorpes Discovery Centre, The Lakeside, Kings Rd, Cleethorpes DN35 0AG • T: +44 1472 323232 • F: 323233 • lynne.conlan@nelincs.gov.uk • www.cleethorpesdiscoverycentre.co.uk •
Man.: *Lynne Conlan* •
Historical Museum 43851

Clevedon

Clevedon Court, Tickenham Rd, Clevedon BS21 6QU • T: +44 1275 872257 •
Decorative Arts Museum
Eltonware ceramics, Nailsea glass, prints of bridges 43852

Clifton

Royal West of England Academy New Gallery, Queen's Rd, Clifton BS8 1PX • T: +44 117 9735129 • F: 9237874 • info@rwa.org.uk • www.rwa.org.uk •
Dir.: *Dragana Smart* •
Public Gallery / Fine Arts Museum
Fine art since the 2nd half of the 19th c, contemporary works 43853

Clitheroe

Castle Museum, Castle, Clitheroe BB7 1BA • T: +44 1200 424635 • F: 424568 • museum@ribblevalley.gov.uk • www.ribblevalley.gov.uk •
Cust.: *D. Mary Hornby* •
Local Museum
20 000 Geology specimens: International coll of rocks, minerals and fossils, local hist – earth science educational project 43854

North West Sound Archive, Old Steward's Office, Castle Crounds, Clitheroe BB7 1AZ • T: +44 1200 427897 • F: 427897 • nwsa@ed.lancscc.gov.uk •
Music Museum
100,000 recordings on all aspects of north west life, specialist coll incl folk music, radio astronomy, the survey of English dialects 43855

Clun

Clun Local History Museum, Town Hall, The Square, Clun SY7 8JA • T: +44 1588 640681 • F: 640681 • ktomey@fish.co.uk • www.clun.org.uk •
Dir.: *J.K. Tomey* •
Archaeology Museum / Local Museum – 1934
Over 1000 flint artefacts Palaeolithic to Bronze Age 43856

Clydebank

Clydebank Museum, Old Town Hall, Dumbarton Rd, Clydebank G81 1UE • T: +44 1389 738702 • F: 608044 • curator@clydebankmuseum.sol.co.uk •
Local Museum
Local industry and social hist, with special coll of ship models and sewing machines, shipbuilders hist 43857

Coalbrookdale

Darby Houses, Rosehill and Dale House, Darby Rd, Coalbrookdale TF8 7EL • T: +44 1952 432551 • F: 432551 • tic@ironbridge.org.uk • www.ironbridge.org.uk •
Cur.: *John Challen* •
Historical Museum
Quaker ironmasters, Abraham Darby III, builder of the Iron Bridge 43858

Long Warehouse, Coalbrookdale TF8 7EL • T: +44 1952 432141 • F: 432751 • coalbrokdale@ironbridge.org.uk • www.ironbridge.org.uk •
Man.: *Joanne Smith* •
Fine Arts Museum / Historical Museum
Elton coll of industrial art – library, archive 43859

Museum of Iron and Darby Furnace, Coalbrookdale TF8 7NR • T: +44 1952 433418 • F: 432237 • tic@ironbridge.org.uk • www.ironbridge.org.uk •
Cur.: *John Challen* •
Science&Tech Museum 43860

Coalport

Coalport China Museum, Ironbridge Gorge Museums, High St, Coalport TF8 7HT • T: +44 1952 580650 • F: 686540 • tic@ironbridge.org.uk • www.ironbridge.org.uk •
Cur.: *Sophie Heath* •
Decorative Arts Museum – 1976
History of China manufacturesince 1776, coll of Coalport and Caughley china 43861

Tar Tunnel, Coalport Rd, Coalport TF8 7HS • T: +44 1952 580627 • F: 432204 • tic@ironbridge.org.uk • www.ironbridge.org.uk •
Cur.: *Jennifer Thompson* •
Science&Tech Museum
18th c mining tunnel, natural bitumen 43862

Coalville

Snibston Discovery Park, Ashby Rd, Coalville LE67 3LN • T: +44 1530 278444 • F: 813301 • snibston@leics.gov.uk • www.leics.gov.uk/museum •
Cur.: *Iain Gordon* •
Science&Tech Museum
Transport, engineering, fashion, science, colliery, toys 43863

Coatbridge

Summerlee Heritage Park, Heritage Way, Coatbridge ML5 1QD • T: +44 1236 431261 • F: 440429 • www.summerlee.co.uk •
Local Museum / Historic Site – 1988
Social and industrial hist of Scotland's iron and steel and heavy engineering industries, coll of machinery and equipment, a steam winding engine, railway locomotives, two steam cranes, underground coalmine, recreated miner's cottages, Scotland's only electric tramway, art gallery 43864

Coate

Richard Jefferies Museum, Marlborough Rd, Coate SN3 6AA • T: +44 1793 783040 • r.jefferies_society@tiscali.co.uk • richardjefferiessociety.blogspot.com •
Dir.: *Jean Saunders* •
Special Museum – 1962
Memorabilia on Richard Jefferies, nature and countryside writer, manuscripts, attic and study restored as the writer described it 43865

Cobham

Cobham Bus Museum, The London Bus Preservation Trust, Redhill Rd, Cobham KT11 1EF • T: +44 1932 868665 • cobhambusmuseum@aol.com • www.lbpt.org •
Science&Tech Museum
London buses from 1925 to 1973 43866

Cockermouth

Cumberland Toy and Model Museum, Bank's Court, Market Pl, Cockermouth CA13 9NG • T: +44 1900 827606 • rod@toymuseum.co.uk • www.toymuseum.co.uk •
Dir.: *Rod Moore* •
Special Museum – 1989
British toys from 1900 to present 43867

Printing House Museum, 102 Main St, Cockermouth CA13 9LX • T: +44 1900 824984 • F: 823124 • printinghouse@uk2.net • www.printinghouse.co.uk •
Dir.: *David R. Winkworth* •
Science&Tech Museum – 1979
Hist of printing 15th c to 1970 (letterpress era), iron presses, platen presses, automatic presses, type founding machinery, stone and etching press – reference library 43868

Wordsworth House, Main St, Cockermouth CA13 9RX • T: +44 1900 824805 • F: 820883 • wordsworthhouse@nationaltrust.org.uk • www.wordsworthhouse.org.uk •
Cust.: *Louise Horsfield* •
Special Museum – 1937
Fine Georgian townhouse, birthplace and childhood home of William Wordsworth 43869

Cockley Cley

Iceni Village and Museums, Estate Office, Cockley Cley PE37 8AG • T: +44 1760 721339, 24588 • F: 721339 •
Dir.: *Sir Samuel Roberts* •
Local Museum / Agriculture Museum – 1971
Carriage and farm implements, engines 43870

Coggeshall

Coggeshall Heritage Centre and Museum, Saint Peter's Hall, Stoneham St, Coggeshall C06 1QG • T: +44 1376 563003 • spratcliffe@btinternet.com • www.btinternet.com/~coggeshall/ •
Cur.: *Shirley Ratcliffe* •
Local Museum
Local hist 43871

Colchester

Castle Museum, c/o Museum Resource Centre, 14 Ryegate Rd, Colchester CO1 1YG • T: +44 1206 282939 • F: 282925 • philip.wise@colchester.gov.uk • www.colchestermuseums.org.uk •
Dir.: *Peter Berridge* • Sc. Staff: *J. Bowdery* (Natural History) • *Tom Hodgson* (Social History) • *Philip Wise* (Archaeology) •
Archaeology Museum – 1860
Material from Iron Age and Roman Colchester, Roman military tombstones, Roman pottery, bronze figure of Mercury, civil war – library 43872

Firstsite at the Minories Art Gallery, 74 High St, Colchester CO1 1UE • T: +44 1206 577067 • F: 577161 • info@firstsite.uk.net • www.firstsite.uk.net •
Dir.: *Katherine Wood* •
Public Gallery – 1956/ 1992
Contemporary visual arts 43873

Hollytrees Museum, High St, Colchester CO1 1UG • T: +44 1206 282940 • F: 282925 • tom.hodgson@colchester.gov.uk • www.colchestermuseums.org.uk/hollytrees/holly_index.html •
Dir.: *Tom Hodgson* •
Historical Museum – 1928
Georgian town house built in 1718 housing domestic displays of toys, costumes and decorative arts from the 18th, 19th and early 20th c – library 43874

Natural History Museum, High St, All Saint's Church, Colchester CO1 1DN • T: +44 1206 282941 • F: 282925 • jerry.bowdrey@colchester.gov.uk • www.colchestermuseums.org.uk/NatHist_html/default_NatHist.html •
Dir.: *Jeremy Bowdrey* •
Natural History Museum – 1958
Natural hist of Essex 43875

Tymperleys Clock Museum, Trinity St, Colchester CO1 1JN • T: +44 1206 282943 • F: 282925 • tom.hodgson@colchester.gov.uk • www.colchestermuseums.org.uk •
Science&Tech Museum – 1987
216 clocks and 12 watches 43876

University of Essex Exhibition Gallery, Wivenhoe Park, Colchester CO4 3SQ • T: +44 1206 872074, 873333 • F: 873598 • kennj@essex.ac.uk • www.essex.ac.uk •
Dir.: *Jessica Kenny* •
Fine Arts Museum / University Museum
150 works with regional focus, sited sculpture, contemporary Latin American art 43877

Coldstream

Coldstream Museum, 12 Market Sq, Coldstream TD12 4BD • T: +44 1890 882630 • F: 882631 • museums@scotborders.gov.uk •
Cust.: *Margaret Dixon* •
Local Museum – 1971
Social and domestic hist of the Coldstream area, hist of the Coldstream guards regiment, raised in the town, 1659 43878

Hirsel Homestead Museum, The Hirsel, Coldstream TD12 4LP • T: +44 1573 224144 • F: 226313 • rogerdodd@btinternet.com • www.hirselcountrypark.co.uk •
Dir.: *Roger Dodd* •
Historical Museum
Archaeology, farming, forestry, the estate workshops, the laundry, the gardens, the dovecote, and natural history 43879

Coleraine

Hezlett House, 107 Sea Rd, Castlerock, Coleraine, Co. Derry • T: +44 28 70848728 •
Historical Museum / Agriculture Museum
Victorian style, farm implements 43880

Combe Martin

Combe Martin Museum, The Parade, Seaside, Combe Martin EX34 0AP • T: +44 1271 882441 •
Chm.: *Jean Griffiths* •
Local Museum
Local industries, silver/lead mining, lime quarrying and burning, maritime, social history, photographs, documents – archives 43881

Compton, Surrey

Watts Gallery, Down Ln, Compton, Surrey GU3 1DQ • T: +44 1483 810235 • www.wattsgallery.org.uk •
Cur.: *Perdita Hunt* •
Fine Arts Museum – 1903
Paintings and sculptures by G.F. Watts 43882

Congleton

Little Moreton Hall, Congleton CW12 4SD • T: +44 1260 272018 • F: 292802 • littlemoretonhall@nationaltrust.org.uk •
Dir.: *Dean V. Thomas* •
Decorative Arts Museum – 1937
Elizabethan house with woodwork, plaster decorations, oak furniture, pewter objects 43883

Coniston

Brantwood, Coniston LA21 8AD • T: +44 15394 41396 • F: 41263 • enquiries@brantwood.org.uk • www.brantwood.org.uk •
Dir.: *Howard Hull* •
Fine Arts Museum / Decorative Arts Museum
Former Ruskin's home, filled with Ruskin's watercolours and drawings, works by some of his friends, incl T.M. Rooke, Sir Edward Burne-Jones and William Holman Hunt, furniture, his personal possessions, books – library 43884

Ruskin Museum, Coniston Institute, Yewdale Rd, Coniston LA21 8DU • T: +44 15394 41164 • F: 41132 • vjm@ruskinmuseum.com • www.ruskinmuseum.com •
Dir.: *V.A.J. Slowe* • Cur.: *J. Dawson* •
Fine Arts Museum / Natural History Museum / Local Museum – 1900
Memorabilia of John Ruskin (1819-1900), incl watercolours, drawings, sketchbooks geology, slate and coppermines, Malcolm & Donald Campbell personalia and Bluebird memorabilia, water speed record 43885

Conwy

Aberconwy House, 2 Castle St, Conwy LL32 8AY • T: +44 1492 592246 • F: 860233 •
Dir.: *Linda Thorpe* •
Local Museum
Medieval house that dates from the 14th c, now houses the Conwy Exhibition, depicting the life of the borough from Roman times to the present day 43886

Plas Mawr (Great Hall), High St, Conwy LL32 8DE • T: +44 1492 580167 • www.conwy.com/plasmawr.html •
Historic Site
Plas Mawr is an Elizabethan mansion, built 1577-80, in virtually its original condition 43887

Royal Cambrian Academy of Art, Academi Frenhinol Gymreig, Crown Ln, Conwy LL32 8AN • T: +44 1492 593413 • F: 593413 • rca@rcaconwy.org • www.rcaconwy.org •
Cur.: *Gill Bird* •
Fine Arts Museum – 1881
Paintings, sculpture 43888

Cookham

Stanley Spencer Gallery, King's Hall, High St, Cookham SL6 9SJ • T: +44 1628 523484, 471885 • F: 526627 • www.stanleyspencer.org.uk •
Fine Arts Museum – 1962
Paintings and drawings by Stanley Spencer, reference works 43889

Cookstown

Wellbrook Beetling Mill, 20 Wellbrook Rd, Corkhill, Cookstown BT80 9RY • T: +44 28 86748210, 86751735 • uspest@smtp.ntrust.org.uk •
Science&Tech Museum
Last working mill 43890

Corbridge

Corbridge Roman Site and Museum, Roman Site, Corbridge NE45 5NT • T: +44 1434 632349 • F: 633168 • georgina.plowright@english-heritage.org.uk • www.english-heritage.org.uk •
Cur.: *Georgina Plowright* • *Sarah Lawrance* •
Archaeology Museum – 1906
Roman inscriptions and other finds from the Roman town of Corstopitum 43891

Corfe Castle

Corfe Castle Museum, West St, Corfe Castle BH20 5HB • T: +44 1929 480974 • k.woolaston@btinternet.com •
Historical Museum
Early agricultural and clay working implements, clay and stone industries 43892

Cornhill-on-Tweed

Heatherslaw Mill, Ford and Etal Estates, Cornhill-on-Tweed TD12 4TJ • T: +44 1890 820338 • F: 820384 • miller@hatherslaw.org.uk •
Head: *Julia Nolan* •
Science&Tech Museum
Traditional milling of wheat 43893

Cornwall

Museum of Witchcraft, Boscastle, The Harbour, Cornwall PL35 0HD • T: +44 1840 250111 • museumwitchcraft@aol.com • www.museumofwitchcraft.com •
Dir.: *Graham King* •
Special Museum
Displays on cursing, charms and spells, stone circles and ancient sacred sites, modern witchcraft, the persecution of witches – library 43894

Corsham

Corsham Court, Corsham SN13 0BZ • T: +44 1249 701610 • F: 701610 • jmethc@corc.fsbusiness.co.uk •
Dir.: *James Methuen-Campbell* •
Fine Arts Museum / Decorative Arts Museum
Paintings by various old masters incl Filippo Lippi, Rubens, van Dyck, Reynolds, Guido Reni, furniture by Chippendale, Adam, Cobb, Johnson 43895

Cottesmore

Rutland Railway Museum, Ashwell Rd, Cottesmore LE15 7BX • T: +44 1572 813203 •
Science&Tech Museum
Industrial railways, locomotives and wagons associated with local iron ore quarries 43896

Cotton

Mechanical Music Museum and Bygones, Blacksmith Rd, Cotton IP14 4QN • T: +44 1449 613876 • clarkeqff@aol.com • www.visitbritain.com •
Dir.: *Phyllis Keeble* •
Music Museum – 1982
Music boxes, polyphons, organettes, street pianos, barrel organs, pipe organs, fair organs, gigantic cafe organ, the Wurlitzer theatre pipe organ and many unusual items 43897

Cousland

Cousland Smiddy, 21 Hadfast Rd, Cousland EH22 2NZ • T: +44 131 6631058 • F: 6632730 • smiddy@cousland99.freeserve.co.uk •
Local Museum
Blacksmithing tools, farm machinery and tools 43898

Coventry

Coventry Toy Museum, Whitefriars Gate, Much Park St, Coventry CV1 2LT • T: +44 24 76227560 •
Dir.: *Ron Morgan* •
Special Museum
Coll of dolls, toys, games dating from 1750 to 1960, a display of amusement machines 43899

Coventry Transport Museum, Millennium Pl, Hales St, Coventry CV1 1PN • T: +44 24 76234270 • F: 76234284 • enquiries@transport-museum.com • www.transport-museum.com •
Dir.: *Gary Hall* •
Science&Tech Museum – 1980
230 bicycles, 75 motorcycles, more than 220 motor vehicles, a wide range of equipment and accessories – archives 43900

Coventry Watch Museum, POB 1714, Coventry CV3 6ZS • T: +44 24 76402331, 76502916 • covwatchmus@aol.com •
Science&Tech Museum
Coventry made clocks and watches, tools, watch industry, factories, and people at work 43901

Herbert Art Gallery and Museum, Jordan Well, Coventry CV1 5QP • T: +44 24 76832386 • F: 76834089 • info@theherbert.org • www.theherbert.org •
Dir.: *Roger Vaughan* (Arts, Heritage) • Sc. Staff: *Robin Johnson* (Lifelong Learning) • *Allison Taylor* (Outreach) • *Rosie Addenbrooke* (Exhibitions) • *Ron Clarke* (Visual Arts) • *Martin Roberts* (Social History) • *Huw Jones* (Industry) • *Paul Thompson* (Archaeology) • *Steve Lane* (Natural History) • Cons.: *Judi Browes* (Works on Paper) • *Gillian Irving* (Easel Paintings) • *Martin Grahn* (Objects) •
Fine Arts Museum / Local Museum / Archaeology Museum / Natural History Museum – 1960
Sutherland sketches used in making Coventry Cathedral tapestry, local art, topography, local natural hist, silk ribbons, archaeology, social and industrial hist – Lunt Roman Fort 43902

Jaguar Daimler Heritage Trust, Jaguar Cars, Browns Ln, Coventry CV5 9DR • T: +44 24 76402121 • F: 76202777 • smoore50@jaguar.com • www.jdht.com •
Dir.: *John Maries* • Cur.: *Tony O'Keeffe* •
Science&Tech Museum
Vehicles covering Jaguar, Daimler and Lanchester Cars 43903

Mead Gallery, c/o Warwick Arts Centre, University of Warwick, Gibbet Hill Rd, Coventry CV4 7AL • T: +44 24 76522589 • F: 76572664 • meadgallery@warwick.ac.uk • www.warwickartscentre.co.uk •
Cur.: *Sarah Shalgosky* •
Fine Arts Museum
British post-war paintings, photographs, prints, sculpture 43904

Coverack

Poldowrian Museum of Prehistory, Poldowrian, Gwenter Rd, Coverack TR12 6SL • T: +44 1326 280468 • vjhadley@tiscali.co.uk •
Dir.: *V.J. Hadley* •
Archaeology Museum
Flints, stone tools, pottery, bronce-age roundhouse 43905

Cowbridge

Cowbridge and District Museum, Town Hall Cells, Cowbridge CF71 7AD • T: +44 1446 775139 •
Cur.: *Pamela Robson* •
Local Museum
Local hist, archaeology of Cowbridge 43906

Cowes

Cowes Maritime Museum, Beckford Rd, Cowes PO31 7SG • T: +44 1983 823433 • F: 293341 • corina.westwood@iow.gov.uk • www.iwight.com •
Science&Tech Museum – 1975
Items and archives from J. Samuel White & Co Ltd. shipbuilders, 5 small boats designed by Uffa Fox, yachting photos by Kirk of Cowes – library 43907

Sir Max Aitken Museum, 83 High St, Cowes PO31 7AJ • T: +44 1983 295144, 293800 • F: 200253 •
Special Museum
Paintings of maritime subjects, half models and full models of ships, including a number of Sir Max Aitken's own yachts, and souvenirs of former royal yachts 43908

Coxwold

Byland Abbey, Coxwold YO61 4BD • T: +44 1347 868614 • www.english-heritage.org.uk •
Religious Arts Museum 43909

Shandy Hall, Coxwold YO61 4AD • T: +44 1347 868465 • F: 868465 • shandy.hall@dsl.pipex.com • www.asterisk.org.uk •
Chm.: *Dr. Nicolas Barker* • Cur.: *Patrick Wildgust* •
Special Museum
Place where Laurence Sterne wrote his 2 novels, relevant prints, paintings, books 43910

Cradley Heath

Haden Hall House, Off Barrs Rd, Cradley Heath B64 7JU • T: +44 1384 569444 • F: 412623 • alison_hyatt@sandwell.gov.uk • www.lea.sandwell.gov.uk/museums •
Historical Museum
Paintings, books etc 43911

Craigavon

Craigavon Museum, Pinebank House, 2 Tullygally Rd, Craigavon, Co. Armagh • T: +44 1762 341635 •
Agriculture Museum
Hist farming and a Georgian farmhouse 43912

Crail

Crail Museum, 62-64 Marketgate, Crail KY10 3TL • T: +44 1333 450869 • F: 450869 • www.crailmuseum.org.uk •
Cur.: *Susan Bradman* •
Local Museum – 1979
H.M.S. Jackdaw (world war II fleer air arm station) 43913

Cranbrook

Cranbrook Museum, Carriers Rd, Cranbrook TN17 3JX • T: +44 1580 712516 • www.localmuseum.freeserve.co.uk •
Cur.: *Andrew Saunders* •
Local Museum
Local trades, social, domestic and community exhibits, Boyd Alexander Bird Coll – archives 43914

Crawfordjohn

Crawfordjohn Heritage Venture, Crawfordjohn Church, Main St, Crawfordjohn ML12 6SS • T: +44 1864 504265 • F: 504206 • crawfordjohn-heritage@culturalprojects.co.uk •
Cur.: *Robert A. Clark* •
Local Museum
Traditional way of life, hill farms and community activity 43915

Crawley

Crawley Museum, Goffs Park House, Old Horsham Rd, Southgate, Crawley RH11 8PE • T: +44 1293 539088 • office@crawleymuseum.fsnet.co.uk • www.crawleymuseum.moonfruit.com •
Cur.: *Helen Poole* • *Jessica Hadfield* •
Historical Museum
Objects and photographs relating to local social history, e.g. the 'shoemakers workshop', 'Rex Forecar' and 'Victorian Kitchen' 43916

Ifield Watermill, Hydr Dr, Ifield, Crawley RH11 8PE • T: +44 1293 539088 • office@crawleymuseum.fsnet.co.uk • www.ifieldwatermill.moonfruit.com/ •
Dir.: *Nick Sexton* • Head of Watermill: *Edward Henbery* •
Cur.: *Helen Poole* •
Historic Site
Watermill restored to working order, coll of milling machinery, tools and other elements of local social history 43917

Creetown

Creetown Gem Rock Museum, Chain Rd, Creetown DG8 7HJ • T: +44 1671 820357 • F: 820554 • gem.rock@btinternet.com • www.gemrock.net •
Dir.: *Tim Stephenson* •
Natural History Museum – 1971
Scottish agate coll, British minerals, dinosaur egg fossil, cut gemstones, meteorite 43918

Creetown Heritage Museum, 91 Saint John's St, Creetown DG8 7JE • T: +44 1671 820471 • www.creetown-heritage-museum.com •
Chm.: *Andrew MacDonald* •
Local Museum 43919

Cregneash

Cregneash Folk Village, Cregneash IM9 5PX • T: +44 1624 648000 • F: 648001 • enquiries@mnh.gov.im • www.gov.im/mnh •
Man.: *Peddyr Cubberley* •
Folklore Museum 43920

Cremyll

Mount Edgcumbe House, Cremyll PL10 1HZ • T: +44 1752 822236 • F: 822199 • mt.edgcumbe@plymouth.gov.uk • www.mountedgcumbe.gov.uk •
Man.: *Ian Berry* •
Special Museum
Paintings, furniture, Plymouth porcelain 43921

Crewe

Museum of Primitive Methodism, Englesea Brook, Englesea Brook Methodist Chapel, Crewe CW2 5QW • T: +44 1270 820836 • engleseabrook-methodist-museum@supanet.com • www.engleseabrook-museum.org.uk •
Dir.: *Stephen Hatcher* •
Religious Arts Museum
Early pipe organ, printing press and chest of drawers pulpit, pottery, banners 43922

Railway Age, Vernon Wy, Crewe CW1 2DB • T: +44 1270 212130 •
Science&Tech Museum
railway archives 43923

Crewkerne

Crewkerne and District Museum and Heritage Centre, Market Sq, Crewkerne TA18 7JU • T: +44 1460 77079 •
Local Museum
Social and industrial history 43924

Cricciert

Lloyd George Museum, Llanystumdwy, Cricciert LL52 0SH • T: +44 1766 522071 • amgueddfalloydgeorge@gwynedd.gov.uk • www.gwynedd.gov.uk/museums •
Dir.: *Iwan Trefor Jones* •
Historical Museum – 1948
Caskets, original punch cartoons, original archives concerning Versailles Peace Treaty, Lloyd George memorabilia (prime minister during WWI) 43925

Criccieth

Criccieth Castle, off A497, Criccieth LL52 0AA • T: +44 1766 522227 •
Archaeology Museum 43926

Cromarty

Cromarty Courthouse Museum, Church St, Cromarty IV11 8XA • T: +44 1381 600418 • F: 600418 • info@cromarty-courthouse.org.uk • www.cromarty-courthouse.org.uk •
Local Museum
Cromarty's rule in the hist of Scottland 43927

Hugh Miller's Cottage, Church St, Cromarty IV11 8XA • T: +44 1381 600245 • F: 600796 • mgostwick@nts.org.uk • www.hughmiller.org •
Dir.: *Martin Gostwick* •
Special Museum – 1890
Birthplace of Scottish geologist Hugh Miller (1802-1852), memorabilia, equipment, letters, fossil coll 43928

Cromer

Cromer Museum, East Cottages, Tucker St, Cromer NR27 9HB • T: +44 1263 513543 • F: 511651 • cromer.museum@norfolk.gov.uk •
Cur.: *Alistair Murphy* •
Local Museum – 1978 43929

Henry Blogg Lifeboat Museum, The Gangway, Cromer NR27 9HE • T: +44 1263 511294 • F: 511294 • jumbo@csma.netlink.co.uk •
Cur.: *Frank H. Muirhead* •
Historical Museum
Houses the lifeboat, H F Bailey 777, hist of the lifeboat station from 1804, models, WW I Northsea convoys 43930

Cromford

The Arkwright Society, Sir Richard Arkwright's Cromfort Mill, Mill Ln, Cromford DE4 3RQ • T: +44 1629 823256 • F: 823256 • info@cromfordmill.co.uk • www.arkwrightsociety.org.uk •
Dir.: *Dr. Christopher Charlton* •
Historic Site ... 43931

Cromford Mill, Mill Ln, Cromford DE4 3RQ • T: +44 1629 824297 • F: 823256 • info@arkwrightsociety.org.uk • www.arkwrightsociety.org.uk •
Dir.: *Dr. Christopher Charlton* • Pres.: *Sir George Kenyon* •
Science&Tech Museum – 1979
Water powered cotton spinning mill ... 43932

Crosby-on-Eden

Solway Aviation Museum, Aviation House, Carlisle Airport, Crosby-on-Eden CA6 4NW • T: +44 1228 573823 • F: 573823 • info@solway-aviation-museum.org.uk • www.solway-aviation-museum.org.uk •
Dir.: *David Kirkpatrick* •
Science&Tech Museum
Local aviation heritage and the Blue Streak Project at Spadeadam ... 43933

Croydon

Croydon Museum, Croydon Clocktower, Katherine St, Croydon CR9 1ET • T: +44 20 82531022 • F: 82531003 • museum@croydon.gov.uk • www.croydon.gov.uk •
Local Museum / Decorative Arts Museum
Local people's lives, Riesco coll of Chinese porcelain ... 43934

Croydon Natural History and Scientific Society, 96a Brighton Rd, Croydon CR2 6AD • T: +44 20 86691501 •
Dir.: *John Greig* •
Association with Coll
Archaeology, geology, social history, ethnography ... 43935

Cuckfield

Cuckfield Museum, Queen's Hall, High St, Cuckfield RH17 5EL • T: +44 1444 454276 • frances_stenlake@yahoo.com • www.cuckfield-society.org.uk/museum_home •
Cur.: *Frances Stenlake* •
Local Museum
Local hist ... 43936

Cullyhanna

Cardinal Ó Fiaich Heritage Centre, Áras an Chairdiméil Ó Fiaich, Slatequarry Rd, Cullyhanna BT35 0JH • T: +44 1693 868757 • cregganhistory@yahoo.co.uk •
Man.: *Michael McShane* •
Religious Arts Museum ... 43937

Culross

Culross Palace, Sandhaven, Royal Burgh of Culross, Culross KY12 8JH • T: +44 1383 880359 • F: 882675 • cwilson@nts.org.uk • www.culross.org/palace.htm •
Cur.: *Christine Wilson* •
Decorative Arts Museum – 1932
17th c wood panelling, painted ceilings and furniture ... 43938

Town House, Sandhaven, Royal Burgh of Culross, Culross KY12 8HT • T: +44 1383 880359 • F: 882675 • cwilson@nts.org.uk • www.culross.org/townhouse.htm •
Head: *Robin Pellen* • Cur.: *Christine Wilson* •
Local Museum ... 43939

Cumbernauld

Cumbernauld Museum, 8 Allander Walk, Cumbernauld G67 1EE • T: +44 1236 725664 • F: 458350 •
Local Museum ... 43940

Cumnock

Baird Institute Museum, 3 Lugar St, Cumnock KA18 1AD • T: +44 1290 421701 • F: 421701 • baird@east-ayrshire.gov.uk • www.east-ayrshire.gov.uk •
Dir.: *Michael Harkiw* •
Decorative Arts Museum / Local Museum – 1891
Cumnock pottery, Mauchline boxware, Ayrshire embroidery and the "Lochnorris Collection" of artefacts mining equipment, photos, local and social history ... 43941

Cupar

Hill of Tarvit Mansion House, Cupar KY15 5PB • T: +44 1334 653127 • F: 653127 • hilloftarvit@nts.org.uk • www.nts.org.uk •
Dir.: *K. Caldwell* •
Decorative Arts Museum / Fine Arts Museum / Historic Site
French, Chippendale and vernacular furniture, Dutch paintings and pictures by Scottish artists Raeburn and Ramsay, Flemish tapestries, Chinese porcelain and bronzes ... 43942

Dagenham

Valence House Museum and Art Gallery, Becontree Av, Dagenham RM8 3HT • T: +44 20 82275293 • F: 82275293 • valencehousemuseum@hotmail.com • www.lbbd.gov.uk •
Dir.: *Sue Curtis* • Cur.: *Mark Watson* •
Local Museum / Fine Arts Museum – 1938
Local hist, art gallery, herb garden – archives ... 43943

Dalbeattie

Dalbeattie Museum, Southwick Rd, Dalbeattie DG5 4bs • T: +44 1556 610437, 611657 •
Local Museum
Local granite quarrying industy, shipping, the railway, mills history ... 43944

Dalkeith

Dalkeith Arts Centre, White Hart St, Dalkeith EH22 1OAE • T: +44 131 6636986, 2713970 • F: 4404635 • alan.reid@midlothian.gov.uk •
Man.: *Alan Reid* •
Public Gallery ... 43945

Dalmellington

Doon Valley Museum, Cathcartston, Dalmellington KA6 7QY • T: +44 1292 550633 • F: 550937 • elaine.jones@east-ayrshire.gov.uk •
Officer: *Stanley Sarsfield* •
Local Museum / Historical Museum
Hand-loom weaving, local coal, Iron and brockwork industries ... 43946

Scottish Industrial Railway Centre, Dunaskin Ironworks, Dalmellington KA6 7JF • T: +44 1292 269260 • F: 313579 • agcthoms@aol.com • www.arpg.org.uk •
Science&Tech Museum – 1980
Steam and diesel locomotives, rolling stock and cranes ... 43947

Darlington

Darlington Arts Centre and Myles Meehan Gallery, Vane Terrace, Darlington DL3 7AX • T: +44 1325 348843, 348845 • F: 365794 • wendy.scott@darlington.gov.uk • www.darlingtonarts.co.uk •
Dir.: *Wendy Scott* •
Fine Arts Museum
Chiefly 19th c topographical and 20th c works ... 43948

Darlington Railway Centre and Museum, North Road Station, Darlington DL3 6ST • T: +44 1325 460532 • F: 287746 • museum@darlington.gov.uk • www.drcm.org.uk •
Man.: *Robert Clark* •
Science&Tech Museum – 1975
Material relating to the Stockton and Darlington Railway, Ken Hoole coll, North Eastern Railway Association Library – Ken Hoole Study Centre ... 43949

Raby Castle, Staindrop, Darlington DL2 3AY • T: +44 1833 660202 • F: 660169 • admin@rabycastle.com • www.rabycastle.com •
Local Museum
Dutch,Italian and English paintings; coll. of five Meissen porcelain birds by Kaendler and Kirschner; Lord Barnard's Family Home ... 43950

Dartford

Dartford Borough Museum, Market St, Dartford DA1 1EU • T: +44 1322 343555 • F: 343209 • museum@dartford.gov.uk • www.dartford.gov.uk/community/museum •
Man.: *Chris Baker* (Archaeology) •
Local Museum – 1908
Natural hist, geology, prehistoric finds, Roman, Saxon and medieval finds, Saxon finds from Horton Kirby site ... 43951

Dartmouth

Dartmouth Castle, Dartmouth TQ6 0JN • T: +44 1803 833588 • F: 834445 • customer@english-heritage.org.uk • www.english-heritage.org.uk •
Cur.: *Tony Musty* •
Military Museum
Historic artillery and military items ... 43952

Dartmouth Museum, 6 The Butterwalk, Dartmouth TQ6 9PZ • T: +44 1803 832923 • darmouth@devonmuseums.net • www.devonmuseums.net •
Local Museum / Historical Museum
Local and maritime hist, ship models, 17th c panelling ... 43953

Newcomen Engine House, Mayors Av, Dartmouth TQ6 9YY • T: +44 1803 834224 • F: 835631 • holidays@discoverdartmouth.com • www.discoverdartmouth.com •
Science&Tech Museum
Pressure/steam engine of 1725 ... 43954

Darwen

Sunnyhurst Wood Visitor Centre, off Earnsdale Rd, Darwen BB3 • T: +44 1254 701545 • F: 701545 • countryside@blackburn.gov.uk •
Public Gallery ... 43955

Daventry

Daventry Museum, Moot Hall, Market Sq, Daventry NN11 4BH • T: +44 1327 302463, 71100 • F: 706035 • vgabbitas@daventrydc.gov.uk •
Cur.: *Victoria Gabbitas* •
Local Museum – 1986 ... 43956

Dawlish

Dawlish Museum, The Knowle, Barton Terrace, Dawlish EX7 9QH • T: +44 1626 888557 • dawlish@devonmuseums.net • www.devonmuseums.net •
Dir.: *Brenda French* •
Local Museum – 1969
Local hist, railway, industry, military, toys, clothes ... 43957

Deal

Deal Castle, Victoria Rd, Deal CT14 7BA • T: +44 1304 372762 • F: 372762 • sue.pearson@english-heritage.org.uk • www.english-heritage.org.uk •
Cur.: *Rowena Shepherd* •
Historical Museum
Herny VIII's fortifications and castle artillery ... 43958

Deal Maritime and Local History Museum, 22 Saint George's Rd, Deal CT14 6BA • T: +44 1304 381344 • F: 373684 • dealmuseum@lineone.net • home.freeuk.com/deal-museum •
Historical Museum / Local Museum
Local history, people, boats, life jackets, Deals maritime history, pilots, social history, fishing, trade, ship-to-shore communications ... 43959

Timeball Tower, Victoria Parade, Beach St, Deal CT14 7BP • T: +44 1304 360897, 201200 •
Science&Tech Museum
Telegraphy and semaphores ... 43960

Dedham

Sir Alfred Munnings Art Museum, Castle House, Dedham CO7 6AZ • T: +44 1206 322127 • F: 322127 • www.siralfredmunnings.co.uk •
Admin.: *Diane Roe* •
Fine Arts Museum – 1970
Coll of paintings, sketches, studies of racehorses and of racing, equestrian portraits and hunting scenes, house, studios and grounds where Sir Alfred Munnings lived and painted for 40 yrs, large coll of his works ... 43961

Deepcut

Royal Logistic Corps Museum, Princess Royal Barracks, Deepcut GU16 6RW • T: +44 1252 833371 • F: 833484 • rlcmuseum@btconnect.com • www.army-rlc.co.uk/museum •
Man: *Sue Lines* • Cur.: *Sebastian Puncher* •
Military Museum
Military artefacts ... 43962

Denbigh

Denbigh Castle, Bull Ln, Denbigh LL16 3SN • T: +44 1745 813385 • debbie.snow@denbighshire.gov.uk •
Historical Museum
History of Denbigh, methods of castle warfare, the campaigns and castles of Edward I in North Wales, Richard Dudley, Earl of Leicester ... 43963

Denbigh Museum and Gallery, Hall Sq, Denbigh LL16 3NU • T: +44 1745 816313 • F: 816427 • rose.mcmahon@denbighshire.gov.uk • www.denbighshire.gov.uk •
Local Museum / Fine Arts Museum
Social, industrial and economical growth, cultural heritage ... 43964

Derby

Derby Museum and Art Gallery, The Strand, Derby DE1 1BS • T: +44 1332 716659 • F: 716670 • diana.peake@derby.gov.uk • www.derby.gov.uk/museums •
Dir.: *Anneke Bambery* • Cur.: *Elizabeth Spencer* • *Janine Derbyshire* (Decorative Arts) • *Sarah Allard* (Fine Art) • *Bill Grange* • *Nick Moyes* (Natural History) • *Roger Shelly* (Social History) • *Louise Dunning* • *Maggie Cullen* (Exhibitions) • *Jonathan Wallis* (Colls) •
Fine Arts Museum / Local Museum – 1879
Paintings of Joseph Wright, history, natural history, porcelain, paintings, drawings, engravings, numismatics ... 43965

The Eastern Museum, Kedleston Hall, Derby DE22 5JH • T: +44 1332 842191 • F: 841972 • kedlestonhall@nationaltrust.org.uk •
Ethnology Museum – 1927
Eastern silver, textiles, ceramics, metalworks, ivory, furniture ... 43966

Pickford's House Museum, 41 Friar Gate, Derby DE1 1DA • T: +44 1332 255363 • F: 255277 • museums@derby.gov.uk • www.derby.gov.uk/museums •
Dir.: *Janine Derbyshire* • *Elizabeth Spencer* •
Decorative Arts Museum – 1988
Georgian town house, historic costumes, toy theatres, decorative art, domestic life ... 43967

Regimental Museum of the 9th/12th Royal Lancers, The Strand, Derby DE1 1BS • T: +44 1332 716656 • F: 716670 • museum@derby.gov.uk • www.derby.gov.uk/museums •
Cur.: *A. Tarnowski* •
Military Museum ... 43968

Royal Crown Derby Museum, 194 Osmaston Rd, Derby DE23 8JZ • T: +44 1332 712800 • F: 712899 • enquiries@royal-crown-derby.co.uk • www.royal-crown-derby.co.uk •
Dir.: *Hugh Gibson* • Cur.: *Jaqueline Smith* •
Decorative Arts Museum – 1969
Hist of ceramics production in Derby since 1748 ... 43969

Silk Mill - Derby's Museum of Industry and History, Silk Mill Ln, off Full St, Derby DE1 3AR • T: +44 1332 255308 • F: 716670 • roger.shelley@derby.gov.uk • www.derby.gov.uk/museums •
Dir.: *Roger Shelley* •
Science&Tech Museum – 1974
Rolls-Royce aero engine coll, items relating to the Midland Railway and especially to Railway Research carried out in Derby ... 43970

Dereham

Bishop Bonner's Cottage Museum, Saint Withburga Ln, Dereham NR19 1ED • T: +44 1362 695397 • katelynn@tiscali.co.uk •
Head: *Kitty Lynn* •
Local Museum
Local history, health and medicine, pictures, inns and taverns ... 43971

Hobbies Museum of Fretwork, 34-36 Swaffham Rd, Dereham NR19 2QZ • T: +44 1362 692985 • F: 699145 • enquire@hobbies-dereham.co.uk • www.hobbies-dereham.co.uk •
Special Museum
Machines and tools, fretwork articles ... 43972

Derry

Harbour Museum, Harbour Sq, Derry BT48 6AF • T: +44 28 71377331 • F: 71377633 • museums@derrycity.gov.uk • www.derrycity.gov.uk •
Dir.: *Roisin Doherty* •
Local Museum
History of the city, maritimes ... 43973

Tower Museum, Union Hall Pl, Derry BT48 6LU • T: +44 28 71372411 • F: 71366018 • tower.museum@derrycity.gov.uk •
Dir.: *Roisin Doherty* •
Local Museum – 1992
Local history ... 43974

Upperlands Eco-Museum, 55 Kilrea Rd, Upperlands, Derry, Co. Derry • T: +44 1648 43265 •
Historical Museum / Open Air Museum
History ... 43975

Workhouse Museum, Glendermott Rd, Waterside, Derry BT47 6BG • T: +44 28 71318328 • F: 71377633 •
Dir.: *Dermot Francis* •
Historical Museum
Protection of North Atlantic convoys during WW II, Workhouse Guardians, comparisons between Irish and African famines ... 43976

Dervaig

The Old Byre Heritage Centre, Dervaig PA75 6QR • T: +44 1688 400229 • theoldbyre@lineone.net • www.old-byre.co.uk •
Dir.: *J.M. Bradley* •
Natural History Museum / Local Museum
Natural hist and hist of Mull, geology, wildlife ... 43977

Devizes

Kennet and Avon Canal Trust Museum, Devizes Wharf, Couch Ln, Devizes SN10 1EB • T: +44 1380 721279, 729489 • F: 727870 • administrator@katrust.org • www.katrust.org •
Dir.: *Clive Hackford* • Cur.: *Warren Berry* •
Science&Tech Museum
History of the Kennet and Avon Canal from 1788 ... 43978

Wiltshire Heritage Museum and Gallery, 41 Long St, Devizes SN10 1NS • T: +44 1380 727369 • F: 722150 • wanhs@wiltshireheritage.org.uk • www.wiltshireheritage.org.uk •
Dir.: *Dr. Paul Robinson* • Sc. Staff: *Lisa Webb* (Natural Sciences) • *Dr. Lorna Haycock* (Library) •
Local Museum / Natural History Museum – 1853
Archaeology, Stourhead Bronze age coll, geological fossils, paintings by John Buckler, etchings by Robin Tanner – library ... 43979

Dewsbury

Dewsbury Museum, Crow Nest Park, Heckmondwike Rd, Dewsbury WF13 0AS • T: +44 1924 325100 • F: 325109 • dewsbury.museum@kirklees.gov.uk • www.dewsburymuseum.co.uk •
Dir.: *Christine Hebblethwaite* •
Local Museum
Local hist and childhood ... 43980

Didcot

Didcot Railway Centre, Didcot OX11 7NJ • T: +44 1235 817200 • F: 510621 • didrlyc@globalnet.co.uk • www.didcotrailwaycentre.org.uk •
Dir.: *M. Dean* •
Science&Tech Museum – 1967
Great Western Railway locomotives and carriages interactive exhibition ... 43981

U

Dingwall

Dingwall Museum, Town House, High St, Dingwall IV15 9RY • T: +44 1349 865366 •
Local Museum – 1975
Local hist, military exhibits 43982

Disley

Lyme Hall, Lyme Park, Disley SK12 2NX • T: +44 1663 762023 • F: 765035 • lymepark@nationaltrust.org.uk • www.nationaltrust.org.uk •
Dir.: *David Morgan* •
Decorative Arts Museum – 1946
16th c house, remodelled in 18th c in Palladian style interiors of many architectural fashions, Mortlake tapestries, limewood carvings, Heraldic glass, English clocks 43983

Diss

100th Bomb Group Memorial Museum, Common Rd, Diss IP21 4PH • T: +44 1379 740708 • 100bgmm@tiscali.co.uk • www.100bgmus.org.uk •
Dir.: *S.P. Hurry* •
Military Museum 43984

Bressingham Steam Museum, Low Rd, Bressingham, Diss IP22 2AB • T: +44 1379 686900 • F: 686907 • info@bressingham.co.uk • www.bressingham.co.uk •
Dir.: *Chris Leah* • *Dr. Howard Stephens* • *Alasdair Baker* •
Science&Tech Museum – 1961
Coll of steam engines, road and industrial engines, a steam-operated Victorian roundabout, steamhauled narrow gauge railways, locomotives, historic railway carriages, mechanical organ, Dad's Army exhibition 43985

Diss Museum, The Shambles, Market Pl, Diss IP22 3AB • T: +44 1379 650618 • F: 643848 • dissmuseum@lineone.net •
Cur.: *Alison Molnos* •
Local Museum
History and prehistory of the market town 43986

Ditchling

Ditchling Museum, Church Ln, Ditchling BN6 8TB • T: +44 1273 844744 • F: 844744 • info@ditchling-museum.com • www.ditchling-museum.com •
Dir.: *Hilary Williams* •
Local Museum / Fine Arts Museum – 1985
Calligraphy, engraving, typography 43987

Dobwalls

The John Southern Gallery, Dobwalls Family Adventure Park, Dobwalls PL14 6HB • T: +44 1579 320325, 321129 • F: 321345 • dobwallsad-venturepark@hotmail.com •
Cur.: *John Southern* •
Fine Arts Museum
Steven Townsend (born 1955), Britain's dog painter, his limited prints, Carl Brenders (born 1937) nature paintings and his limited prints 43988

Dollar

Dollar Museum, Castle Campbell Hall, 1 High St, Dollar FK14 7AY • T: +44 1259 742895, 743239 • F: 742895 • curator.dollarmuseum@btopenworld.com •
Cur.: *Janet Carolan* •
Local Museum
Local hist, Dollar academy, Devon Valley railway, castle hist 43989

Dolwyddelan

Ty Mawr Wybrnant, Dolwyddelan LL25 0HJ •
Special Museum
Birthplace of Bishop William Morgan who translated the Bible into Welsh which was the foundation of modern Welsh literature 43990

Doncaster

Brodsworth Hall & Gardens, Brodsworth, Doncaster DN5 7XJ • T: +44 1302 722598 • F: 337165 • michael.constantine@english-heritage.org.uk • www.english-heritage.org.uk •
Dir.: *Peter Gordon-Smith* • Cur.: *Caroline Carr-Whitworth* • *Dr. Crosby Stevens* •
Historical Museum
Paintings, Italian sculpture, furnishings, kitchen equipment – gardens 43991

Doncaster Museum and Art Gallery, Chequer Rd, Doncaster DN1 2AE • T: +44 1302 734293 • F: 735409 • museum@doncaster.gov.uk • www.doncaster.gov.uk/museums •
Dir.: *Geoff Preece* • Sc. Staff: *Carolyn Dalton* •
Local Museum / Fine Arts Museum – 1909
Regional hist, archaeology, natural hist, fine and decorative art, temporary exhibitions, Koyli Regimental Museum 43992

King's Own Yorkshire Light Infantry Regimental Museum, Chequer Rd, Doncaster DN1 2AE • T: +44 1302 734293 • F: 735409 • museum@doncaster.gov.uk • www.doncaster.gov.uk/museums •
Dir.: *Mjr. Alan Howarth* •
Military Museum – 1950
History of King's Own Yorkshire Light Infantry, medals, armaments, uniforms 43993

Museum of South Yorkshire Life, Cusworth Hall, Cusworth Ln, Doncaster DN5 7TU • T: +44 1302 782342 • F: 782342 • museum@doncaster.gov.uk • www.doncaster.gov.uk •
Dir.: *Frank Carpenter* •
Historical Museum – 1967
Social history of South Yorkshire, coll illustrate life over last 200 years (domestic life, costume, childhood, transport) 43994

Trolleybus Museum at Sandtoft, Belton Rd, Sandtoft, Doncaster DN8 5SX • T: +44 1724 711391 • F: 711846 • www.sandtoft.org.uk •
Head: *Brian Maguire* •
Science&Tech Museum – 1969
Trolleybuses, motorbus, coll of small exhibits relating to trolleybus, cycle shop, lawnmower coll – archive 43995

Donington-le-Heath

Manor House, Manor Rd, Donington-le-Heath LE67 2FW • T: +44 1530 831259 • F: 831259 • museums@leics.gov.uk • www.leics.gov.uk/lcc/tourism/attractions_museums.html •
Dir.: *Peter Liddle* •
Decorative Arts Museum 43996

Dorchester

Athelhampton House, Athelhampton, Dorchester • T: +44 1305 848363 • F: 848135 • enquiry@athelhampton.co.uk • www.athelhampton.co.uk •
Dir.: *Patrick Cooke* •
Historic Site – 1957
House with 15th-16th c furniture – architectural garden, library 43997

Dinosaur Museum, Icen Way, Dorchester DT1 1EW • T: +44 1305 269880, 269741 • F: 268885 • info@thedinosaurmuseum.com • www.thedinosaurmuseum.com •
Dir.: *Jackie D. Ridley* • Cur.: *Tim Batty* •
Natural History Museum – 1984 43998

Dorset County Museum, High West St, Dorchester DT1 1XA • T: +44 1305 262735 • F: 257180 • dorsetcountymuseum@dor-mus.demon.co.uk • www.dorsetcountymuseum.org •
Dir.: *Judy Lindsay* • Cur.: *Peter Woodward* •
Local Museum – 1846
County archaeology, geology, biology, local and regional hist, memorabilia, literature incl Thomas Hardy coll – library 43999

Dorset Teddy Bear Museum, Eastgate, High East St, Dorchester DT1 1JU • T: +44 1305 266040, 26974 • F: 268885 • info@teddybearmuseum.co.uk • www.teddybearmuseum.co.uk •
Dir.: *Jackie D. Ridley* •
Special Museum 44000

Military Museum of Devon and Dorset, The Keep, Bridport Rd, Dorchester DT1 1RN • T: +44 1305 264066 • F: 250373 • curator@keepmilitarymuseum.org • www.keepmilitarymuseum.org •
Cur.: *Charles E. Cooper* • Cust.: *John Murphy* •
Military Museum – 1927
Uniforms, weapons, militaria, medals, exhibits of Devon Regiment, Dorset Regiment, local militia and volunteers, Queens Own Dorset Yeomanry and Devonshire and Dorset Regiment – library, archives 44001

Old Crown Court and Cells, 58-60 High West St, Dorchester DT1 1UZ • T: +44 1305 252241 • F: 257039 • tourism@westdorset-dc.gov.uk • www.westdorset.com •
Historical Museum
Law and order (Rough Justice), Tolpuddle Martyrs, 200 yrs gruesome crime & punishment, waiting sits for prisoners for their appearance before the judge 44002

Terracotta Warriors Museum, High East St, Dorchester DT1 1JU • T: +44 1305 266040 • F: 268885 • info@terracottawarriors.co.uk • www.terracottawarriors.co.uk •
Dir.: *Jacqeline Ridley* • Cur.: *Tim Batty* •
Archaeology Museum – 2002 44003

Tutankhamun Exhibition, High West St, Dorchester DT1 1UW • T: +44 1305 269571, 269741 • F: 268885 • info@tutankhamun-exhibition.co.uk • www.tutankhamun-exhibition.co.uk •
Dir.: *Dr. Michael Ridley* •
Archaeology Museum – 1987 44004

Dorchester-on-Thames

Dorchester Abbey Museum, Abbey Guest House, Dorchester-on-Thames OX10 7HH • T: +44 1865 340054 • john.metcalfe@fritillary.demon.co.uk • www.dorchester-abbey.org.uk •
Chm.: *John Metcalfe* •
Local Museum
Local history since prehistoric time 44005

Dorking

Dorking and District Museum, 62a West St, Dorking RH4 1BS • T: +44 1306 876591 •
Local Museum
Toys, domestic and agricultural displays, fossil coll, minerals, photographes – archives 44006

Polesden Lacey House, Great Bookham, Dorking RH5 6BD • T: +44 1372 452048 • F: 452023 • polesdenlacey@nationaltrust.org.uk • www.nationaltrust.org.uk/polesdenlacey •
Man.: *Andrea Selley* •
Decorative Arts Museum
Fine paintings, pictures, furniture, tapestries, art objects, porcelain, silver 44007

Dornie

Eilean Donan Castle Museum, Dornie IV40 8DX • T: +44 1599 555202 • F: 555262 •
Local Museum 44008

Dornoch

Dornoch Heritage Museum, Struie Rd, Ardoch, Dornoch • T: +44 1862 810754 •
Local Museum 44009

Douglas, Lanarkshire

Douglas Heritage Museum, Saint Sophia's Chapel, Bells Wynd, Douglas, Lanarkshire ML11 0QH • T: +44 1555 851243 •
Local Museum
Village history, Douglas family 44010

Dover

Crabble Corn Mill, Lower Rd, River, Dover CT17 0UY • T: +44 1304 823292, 362569 • F: 826040 • miller@ccmt.org.uk • www.ccmt.org.uk •
Chm.: *Henry Reid* •
Science&Tech Museum 44011

Dover Castle, Stone Hut, Dover CT16 1HU • T: +44 1304 241892 • F: 241892 • rowena.willard-wright@english-heritage.org.uk • www.english-heritage.org.uk •
Cur.: *Rowena Willard-Wright* •
Historical Museum
Medieval fortress 44012

Dover Museum, Market Sq, Dover CT16 1PB • T: +44 1304 201066 • F: 241186 • museum@dover.gov.uk • www.dover.gov.uk/museum •
Cur.: *Jon Iveson* •
Local Museum / Archaeology Museum / Public Gallery – 1836
Local history, ceramics, geology, transportation, embroidery, cameras, natural history, bronze age archaeology 44013

English Heritage South East Region, Stone Hut, Dover CT16 1HU • T: +44 1304 241892 • F: 241892 • www.english-heritage.org.uk •
Cur.: *Rowena Willard-Wright* •
Association with Coll
Archaeological artefacts and furnishings 44014

Grand Shaft, South Military Rd, Dover CT17 9BZ • T: +44 1304 201200, 201066 • F: 201200 • www.dover.gov.uk/museum/g_shaft/home.htm •
Dir.: *Trevor Jones* •
Military Museum
Vast military fortification 44015

Princess of Wales's Royal Regiment and Queen's Regiment Museum, 5 Keep Yard, Dover Castle, Dover CT16 1HU • T: +44 1304 240121 • F: 240121 • pwrrqueensmuseum@tinyworld.co.uk • www.123pwrr.co.uk •
Cur.: *J.-C. Rogerson* •
Military Museum 44016

Roman Painted House, New St, Dover CT17 9AJ • T: +44 1304 203279 •
Dir.: *B.J. Philp* •
Archaeology Museum 44017

Downe

Home of Chales Darwin - Down House, Luxted Rd, Downe BR6 7JT • T: +44 1689 859119 • F: 862755 • www.english-heritage.org.uk •
Dir.: *Margaret Speller* •
Natural History Museum – 1929/1998
Memorabilia of biologist Charles Darwin, his study, portraits 44018

Downpatrick

Castle Ward, Strangford, Downpatrick BT30 7LS • T: +44 1396 44881204 • F: 44881729 • castleward@nationaltrust.org.uk •
Historical Museum
18th c house, fortified tower, cornmill 44019

Down County Museum, The Mall, Downpatrick BT30 6AH • T: +44 28 44615218 • F: 44615590 • museum@downdc.gov.uk • www.downcountymuseum.com •
Cur.: *Michael D. King* •
Local Museum – 1981 44020

Downpatrick Railway Museum, Railway Station, Market St, Downpatrick BT30 6LZ • T: +44 28 44615779 • F: 44615779 • downtrains@yahoo.co.uk • www.downpatricksteamrailway.co.uk •
Dir./ Cur.: *G. Cochrane* •
Science&Tech Museum
Steam and diesel locomotives, historic carriages and wagons, signal box, water tower, goods shed, travelling Post Office 44021

Draperstown

Plantation of Ulster Visitor Centre, 50 High St, Draperstown BT45 7AD • T: +44 28 79627800 • F: 79627732 • info@theflightoftheearlsexperience.com • www.theflightoftheearlsexperience.com •
Man.: *Frank Larey* •
Agriculture Museum
Story of plantation of Ulster, Irish history from 1595 44022

Driffield

Sledmere House, Sledmere, Driffield YO25 3XG • T: +44 1377 236637 • F: 236500 • info@sledmerehouse.com • www.sledmerehouse.com •
Dir.: *Baronet Tatton Sykes* •
Fine Arts Museum 44023

Droitwich

Droitwich Spa Heritage Centre, Saint Richard's House, Victoria Sq, Droitwich WR9 8DS • T: +44 1905 774312 • F: 794226 • heritage@droitwichspa.gov.uk •
Dir.: *Diane Malley* •
Local Museum
Local artefacts from Roman times to the 17th c 44024

Hanbury Hall, School Rd, Hanbury, Droitwich WR9 7EA • T: +44 1527 821214 • F: 821251 • hanburyhall@nationaltrust.org.uk • www.nationaltrust.org.uk •
Dir.: *Fiona Reynolds* •
Fine Arts Museum
gardens 44025

Drumnadrochit

Loch Ness 2000 Exhibition, Official Loch Ness Centre, Drumnadrochit IV63 6TU • T: +44 1456 450573 • F: 450770 • info@loch-ness-scotland.com • www.loch-ness-scotland.com •
Dir.: *R.W. Bremner* • Cur.: *Gary Richardson* •
Special Museum 44026

Urquhart Castle, Drumnadrochit IV3 6TX • T: +44 1456 450551 •
Cust.: *Cameron McKenzie* •
Historical Museum
Ruins 44027

Drumoak

Drum Castle, Drumoak AB31 5EY • T: +44 844 4932161 • F: 4932162 • drum@nts.org.uk • www.nts.org.uk/Property/24/ •
Dir.: *Alec Gordon* •
Historical Museum 44028

Dudley

Black Country Living Museum, Tipton Rd, Dudley DY1 4SQ • T: +44 121 5579643 • F: 5574242 • info@bclm.co.uk • www.bclm.co.uk •
Dir.: *Ian N. Walden* •
Historical Museum – 1975
Social and industrial history of Black Country 44029

Dudley Museum and Art Gallery, Saint James's Rd, Dudley DY1 1HU • T: +44 1384 815575 • F: 815576 • museum.pls@dudley.gov.uk • www.dudley.gov.uk •
Dir.: *Charles R. Hajdamach* • Keeper: *Roger C. Dodsworth* (Glass and Fine Art) • Sc. Staff: *Graham Worton* (Geology) •
Natural History Museum / Fine Arts Museum 44030

Royal Brierley Crystal Museum, Tipton Rd, Dudley DY1 4SH • T: +44 121 5305600 • F: 5305590 • sales@royalbrierley.com • www.royalbrierley.com •
Dir.: *T.G. Westbrook* •
Decorative Arts Museum 44031

Dufftown

Dufftown Whisky Museum, Fife St, Dufftown AB55 4AL • T: +44 1340 820507 • ramon@dufftown.co.uk • www.dufftown.co.uk/dufftown_whisky_museum.htm •
Cur.: *René Ramon* •
Science&Tech Museum – 2000
Many of the distillery ledgers, coopers' tools, and distillery equipment used in the past 44032

Dumbarton

Denny Ship Model Experiment Tank, Scottish Maritime Museum, Castle St, Dumbarton G82 1QS • T: +44 1389 763444 • F: 743093 • www.scottishmaritimemuseum.org •
Dir.: *David Thomson* •
Science&Tech Museum
First ship model testing tank by William Froude (1883), drawing office, clay model moulding beds and shaping machinery 44033

Dumfries

Crichton Museum (closed) 44034

Dumfries and Galloway Aviation Museum, Former Control Tower, Heathhall Industrial Estate, Dumfries DG2 3PH • T: +44 1387 251623, 259546 • F: 256680 • info@dumfriesaviationmuseum.com • www.dumfriesaviationmuseum.com •
Cur.: *David Reid* •
Science&Tech Museum
Engines and aeronautica 44035

Dumfries Museum & Camera Obscura, Observatory, Corberry Hill, Dumfries DG2 7SW • T: +44 1387 253374 • F: 265081 • elainek@dumgal.gov.uk • www.dumgal.gov.uk/museums •
Dir.: *David Lockwood* • Cur.: *Elaine Kennedy* •

U

Local Museum – 1835
Geology, paleontology, mineralogy, botany, zoology, ethnography, costumes, photos, archaeology, social history 44036

Gracefield Arts Centre, 28 Edinburgh Rd, Dumfries DG1 1NW • T: +44 1387 262084 • F: 255173 • dawnh@dumgal.gov.uk • www.artandcraft-southwestscotland.com •
Dir.: *D. Henderby* (Arts) •
Fine Arts Museum
Coll of Scottish paintings (19th-20th c), programme of contemporary art and craft exhibitions 44037

Old Bridge House Museum, Mill Rd, Dumfries DG2 7BE • T: +44 1387 256904 • F: 265081 • dumfriesmuseum@dumgal.gov.uk • www.dumgal.gov.uk/museums •
Dir.: *Siobhan Ratchford* • Cur.: *David Morrison* •
Decorative Arts Museum – 1959/1960
Period furnishings, interiors, various rooms, ceramics, costumes, toys, books 44038

Robert Burns Centre, Mill Rd, Dumfries DG2 7BE • T: +44 1387 264808 • F: 264808 • elainek@dumgal.gov.uk • www.dumgal.gov.uk/museums •
Dir.: *David Lockwood* • Cur.: *Jennifer Taylor* •
Special Museum – 1986
Exhibitions, including an audio-visual presentation on Robert Burns and his life in Dumfries, and on the growth of interest in his life and poetry 44039

Robert Burns House, Burns St, Dumfries DG1 2PS • T: +44 1387 255297 • F: 265081 • elainek@dumgal.gov.uk • www.dumgal.gov.uk/museums •
Man.: *David Lockwood* • Attendant: *Paul Cowley* •
Special Museum
House in which Robert Burns lived for three years prior to his death 44040

Dunbar

Dunbar Town House Museum, High St, Dunbar EH42 1ER • T: +44 1368 863734 • F: 828201 • elms@dunbarmuseum.org • www.dunbarmuseum.org •
Local Museum
16th c house, archeology, photographs of old Dunbar 44041

John Muir's Birthplace, 126 High St, Dunbar EH42 1JJ • T: +44 1386 865899 • info@jmbt.org.uk • www.jmbt.org.uk •
Special Museum
Father of the conservation movement John Muir 44042

Dunbeath

Laidhay Croft Museum, Laidhay, Dunbeath KW6 6EH • T: +44 1593 731244 • F: 721548 •
Dir.: *Elizabeth Cameron* •
Agriculture Museum – 1974
Examples of early agricultural machinery, household artefacts 44043

Dunblane

Dunblane Museum, 1 The Cross, Cathedral Sq, Dunblane FK15 0AQ • T: +44 1786 823440, 825691 • curator@dunblanemuseum.org.uk • www.dunblanemuseum.org.uk •
Cur.: *Marjorie Davies* •
Local Museum – 1943
Paintings, prints, medieval carvings, letters of Bishop Robert Leighton, Communion Tokens – library 44044

Dundee

Broughty Castle Museum, Broughty Ferry, Dundee DD5 2TF • T: +44 1382 436916 • F: 436951 • broughty@dundeecity.gov.uk • www.dundeecity.gov.uk/broughtycastle •
Historical Museum / Local Museum – 1969
Local history, fishing, Orchar Art Gallery, the wildlife of the Tay estuary, weapons and armour, military history 44045

H.M. Frigate Unicorn, Victoria Dock, Dundee DD1 3BP • T: +44 1382 200900 • F: 200923 • frigateunicorn@hotmail.com •
Chm.: *Earl of Dalhousie* • Man.: *Bob Hovell* •
Military Museum
Oldest British-built warship still afloat 'HMS Unicorn', Royal Navy coll, 46 gun frigate for the Royal Navy 44046

McManus Galleries, Albert Sq, Dundee DD1 1DA • T: +44 1382 432350 • F: 432369 • mcmanus.galleries@dundeecity.gov.uk • www.mcmanus.co.uk •
Heritage Manager: *J. Stewart-Young* • Heritage Section: *R. Brinklow* • Cur.: *C. Young* (Art) • *A. Robertson* (Art) • *R. Neave* (Collections) • *F. Sinclair* (Heritage) • Sc. Staff: *C. Millar* • *J. Sage* •
Fine Arts Museum / Local Museum – 1873
19th c Scottish fine art, Dundee & Edinburgh silver, contemporary Scottish art & photography & crafts, local hist, civil, church & history material, archaeology, whaling material, etc. natural hist of the region incl taywhale skeleton, botany, geology, entomology, zoology 44047

Mills Observatory Museum, Balgay Park, Glamis Rd, Dundee DD2 2UB • T: +44 1382 435846 • F: 435962 • mills.observatory@dundeecity.gov.uk • www.dundeecity.gov.uk/mills •
Dir.: *Dr. Bill Samson* •
Science&Tech Museum
Astronomy, space flight – planetarium 44048

North Carr Lightship, South Victoria Dock Rd, Dundee DD1 3BP • T: +44 1382 542516 • F: 542516 • info@northcarr.org.uk • www.northcarr.org.uk •
Dir.: *Robert Richmond* •
Science&Tech Museum
Last manned light vessel North Carr (1933-1974) with contemporary furniture, fittings and machinery 44049

Royal Research Ship Discovery, Discovery Point, Discovery Quay, Dundee DD1 4XA • T: +44 1382 201245 • F: 225891 • info@dundeeheritage.co.uk • www.rrsdiscovery.com •
Dir.: *Gill Poulter* •
Special Museum
Ship used by Robert Falcon Scott on the Antarctic Expedition of 1901-1904 44050

University of Dundee Exhibitions, Copper Gallery, Matthew Gallery, Lower Foyer and Lamb Gallery, c/o Duncan of Jordanstone College of Art and Design, 13 Perth Rd, Dundee DD1 4HT • T: +44 1382 345330 • F: 345192 • exhibitions@dundee.ac.uk • www.dundee.ac.uk/exhibitions •
Public Gallery 44051

University of Dundee Museum Collections, University of Dundee, Dundee DD1 4HN • T: +44 1382 384310 • F: 385523 • museum@dundee.ac.uk • www.dundee.ac.uk/museum/ •
Cur.: *Matthew Jarron* •
Fine Arts Museum / University Museum / Natural History Museum / Science&Tech Museum
Scientific and medical equipment, natural hist., historic and contemporary fine art and design – D'Arcy Thompson Zoology Museum; Tayside Medical History Museum; Tower Foyer & Lamb Galleries 44052

Verdant Works, 27 West Henderson's Wynd, Dundee DD1 5BT • T: +44 1382 225282 • F: 221612 • info@dundeeheritage.co.uk • www.verdantworks.com •
Dir.: *Gill Poulter* •
Science&Tech Museum – 1996
Museum of Dundee's textile industries, 19th c flax and jute mill, historic textile machinery 44053

Dunfermline

Andrew Carnegie Birthplace Museum, Moodie St, Dunfermline KY12 7PL • T: +44 1383 724302 • F: 721862 • info@carnegiebirthplace.com • www.carnegiebirthplace.com •
Dir.: *Lorna Owers* •
Special Museum – 1928
Birthplace of industrialist-philanthropist Andrew Carnegie with memorabilia, costumes, decorative arts 44054

Dunfermline Museum, Viewfield Terrace, Dunfermline KY12 7HY • T: +44 1383 313838 • F: 313837 • lesley.botten@fife.gov.uk • www.fifedirect.org.uk/museums •
Local Museum – 1961
Local industrial, social and natural history, significant local linen coll 44055

Pittencrieff House Museum, Pittencrieff Park, Dunfermline KY12 8QH • T: +44 1383 722935, 313838 • F: 313837 • dunfermline.museum@fife.gov.uk • www.fifedirect.org.uk •
Local Museum
Costume 19th-20th c, local hist, major arts, photographics 44056

Dungannon

The Argory, 144 Derrycaw Rd, Moy, Dungannon BT70 • T: +44 28 87784753 •
Historical Museum 44057

United States Grant Ancestral Homestead, Dergenagh Rd, Dungannon BT70 1TW • T: +44 28 87767259 • F: 87767911 • killymaddy@dstbc.org • www.inthedistrict.com/tourism-grant-ancestoral.shtml •
Man.: *Libby McLean* •
Historical Museum
19th c agricultural implements, Ulysses Simpson Grant, 18th President of the USA 44058

Dunkeld

Dunkeld Cathedral Chapter House Museum, Cathedral St, Dunkeld PH8 0AW • T: +44 1350 728732 • F: 728732 •
Dir.: *Ted Carr* •
Local Museum
Ecclesiastical artefacts, an cross slab from the 9th-c, the Pictish Apostles Stone from the 9th c, temporary displays reflecting different aspect of social history – Archives 44059

Little Houses, High St, Dunkeld PH8 0AN • T: +44 13502 727460 •
Dir.: *Gillian Kelly* •
Local Museum 44060

Dunoon

Castle House Museum, Castle Gardens, Dunoon PA23 7HH • T: +44 1369 701422 • info@castlehousemuseum.org.uk • www.castlehousemuseum.org.uk •
Local Museum 44061

Dunrossnes

Shetland Croft House Museum, Voe, Boddam, Dunrossnes ZE2 9JG • T: +44 1595 695057 • F: 696729 • museum@sic.shetland.gov.uk • www.shetland-museum.org.uk •
Dir.: *T. Watt* •
Agriculture Museum – 1972 44062

Duns

Biscuit Tin Museum, Manderston, Duns TD11 3PP • T: +44 1361 883450 • F: 882010 • palmer@manderston.co.uk • www.manderston.co.uk •
Dir.: *Lord Palmer* •
Decorative Arts Museum 44063

Duns Area Museum, 49 Newtown St, Duns TD11 3AU • T: +44 1361 884114 • F: 883961 • museums@scotborders.gov.uk •
Local Museum 44064

Jim Clark Room, 44 Newtown St, Duns TD11 3AU • T: +44 1361 883960 • museums@scotborders.gov.uk •
Historical Museum – 1969
Trophies and other memorabilia, world champion of motor racing 44065

Dunster

Dunster Castle, Dunster TA24 6SL • T: +44 1643 821314 • F: 823000 • dunstercastle@nationaltrust.org.uk •
Dir.: *William Wake* •
Historical Museum 44066

Dunvegan

Dunvegan Castle, Dunvegan IV55 8WF • T: +44 1470 521206 • F: 521205 • info@dunvegancastle.com • www.dunvegancastle.com •
Dir.: *John Lambert* •
Historical Museum 44067

Dunwich

The Museum, Saint James' St, Dunwich IP17 3EA • T: +44 1728 648796 •
Dir.: *M. Caines* •
Historical Museum – 1972
Local and natural hist of the seashore, marshes, heathland and woodland 44068

Durham

Durham Cathedral Treasures of Saint Cuthbert, Cloisters, off the Cathedral, Durham DH1 3EH • T: +44 191 3864266 • F: 3864267 • enquiries@durhamcathedral.co.uk • www.durhamcathedral.co.uk •
Dir.: *Dr. M. Kitchen* •
Religious Arts Museum – 1978
17th and 18th c altar plate, opus anglicanum embroidery, manuscripts, medieval bronze work, Saint Cuthbert relics 44069

Durham Heritage Centre and Museum, Saint Mary-le-Bow, North Bailey, Durham DH1 3ET • T: +44 191 3868719, 3845589 •
Cur.: *Josephine Jones* •
Historical Museum – 1975 44070

Durham Light Infantry Museum and Durham Art Gallery, Aykley Heads, Durham DH1 5TU • T: +44 191 3842214 • F: 3861770 • dli@durham.gov.uk • www.durham.gov.uk/dli •
Man.: *Steve Shannon* • Officer: *Dennis Hardingham* (Exhibitions, Collections) •
Military Museum / Fine Arts Museum – 1969/2000
Medals, uniforms, weapons, docs, arts 44071

Monks' Dormitory Durham Cathedral, The Chapter Library, The College, Durham DH1 3EH • T: +44 191 3864266 • F: 3864267 • enquiries@durham.cathedral.co.uk •
Religious Arts Museum 44072

Museum of Archaeology, c/o Durham University, Old Fulling Mill, The Banks, Durham DH1 3EB • T: +44 191 3341823 • F: 3341824 • fulling.mill@durham.ac.uk • www.dur.ac.uk/fulling.mill •
Cur.: *Craig Barclay* •
Archaeology Museum – 1833
Oswald Prideaux coll of Samian pottery, Roman stone inscriptions from Hadrian's Wall 44073

Oriental Museum Durham, c/o University of Durham, Elvet Hill, Durham DH1 3TH • T: +44 191 3345694 • F: 3345694 • oriental.museum@durham.ac.uk • www.dur.ac.uk/oriental.museum •
Cur.: *Craig Barcley* •
Fine Arts Museum / Decorative Arts Museum – 1960
Chinese ceramics, ancient Egyptian antiquities, Indian paintings and sculptures, Tibetan paintings and decorative art, hist of writing 44074

Duxford

Imperial War Museum Duxford, Duxford CB2 4QR • T: +44 1223 835000 • F: 837267 • duxford@iwm.org.uk • www.iwm.org.uk •
Dir.: *Richard Ashton* •
Military Museum – 1976
Largest coll of civil and military aircraft in EU also tanks, guns, naval exhibits, aircraft ranging from WWI biplanes, spitfires to concorde 44075

Royal Anglian Regiment Museum, Imperial War Museum, Land Warfare Hall, Duxford CB2 4QR • T: +44 1223 497298 • info@royalanglianmuseum.org.uk • www.royalanglianmuseum.org.uk •
Head: *George Boss* •
Military Museum
Military coll of the East Anglian Regiments and the Royal Anglian Regiment, customs, uniforms, operations, weapons 44076

Dymchurch

New Hall, New Hall Close, Dymchurch TN29 0LF • T: +44 1303 873897 • F: 874788 •
Local Museum
Local artefacts, pottery, muskets, cannon balls, rifles and general miscellany 44077

Earby

Museum of Yorkshire Dales Lead Mining, Old Grammar School, School Ln, Earby BB8 • T: +44 1282 841422 • F: 841422 • earby.leadmines@bun.com •
Dir.: *Peter R. Hart* •
Science&Tech Museum
Mining, miners' tools and equipment, tubs, mine plans, models, photos 44078

Eardisland

Burton Court, Eardisland HR6 9DN • T: +44 1544 388231 • helenjsimpson@hotmail.com •
Dir.: *Helen J. Simpson* •
Local Museum – 1967
Oriental and European costumes dating from the 16th c, natural history, ship and railway models, local history 44079

Earls Barton

Earls Barton Museum of Local Life, 26 The Square, Earls Barton NN6 0HT •
Chm.: *I.D. Flanagan* •
Local Museum
Local cottage industries, lace, boot and shoe making 44080

Easdale Island

Easdale Island Folk Museum, Easdale Island PA34 4TB • pwithall@aol.co.uk • www.slate.co.uk •
Cur.: *P. Jones* •
Folklore Museum – 1980
Industrial and domestic life of the Slate Islands during the 19th c 44081

East Budleigh

Countryside Museum, Bicton Park Botanical Gardens, East Budleigh EX9 7BJ • T: +44 1395 568465 • F: 568374 • museum@bictongardens.co.uk • www.bictongardens.co.uk •
Dir.: *Simon Listen* •
Local Museum – 1967
Agricultural and estate memorabilia 44082

East Carlton

East Carlton Steel Heritage Centre, Park, East Carlton LE16 8YF • T: +44 1536 770977 • F: 770661 •
Science&Tech Museum
Living room of 1930's steel worker, model of Corby Steel Works 44083

East Cowes

Osborne House, East Cowes PO32 6JX • T: +44 1983 200022 • F: 297281, 281380 • www.english-heritage.org.uk •
Cur.: *Michael Hunter* •
Fine Arts Museum / Decorative Arts Museum – 1954
Queen Victoria's seaside home, build in 1845, mementoes of royal travel abroad, Indian plaster decoration, furniture, 15.000 objects, pictures 44084

East Grinstead

Town Museum, East Court, College Ln, East Grinstead RH19 3LT • T: +44 1342 712087, 302233 • rhurstegm@yahoo.co.uk • www.hodgers.com/eastgrinsteadmuseum •
Cur.: *Rachel Hurst* •
Historical Museum
Coll of local handicrafts, 19th and 20th-c pottery 44085

East Hendred

Champs Chapel Museum, Chapel Sq, East Hendred OX12 8JT • T: +44 1235 833471 • F: 833312 • jstevehendred@aol.com • www.hendred.org.uk •
Local Museum
Material of local interest 44086

East Kilbride

Calderglen County Park, Strathaven Rd, East Kilbride G75 0QZ • T: +44 1355 236644 • F: 247618 • mike.taylor@southlanarkshire.gov.uk • www.southlanarkshire.gov.uk •
Historical Museum
Calderglen is made of two former estates of Torrance and Calderwood and opened in 1982 as a country park, Torrance house, which dates back to the 17th c and its courtyard (18th c) are now used as a combined private residence and visitor centre 44087

U

Hunter House, Maxwellton Rd, Calderwood, East Kilbride G74 3LU • T: +44 1355 261261 • lowparksmuseum@southlanarkshire.gov.uk •
Dir.: *Robert Smith* •
Special Museum
Exhibition portraying the lives of John and William Hunter, 18th c scientists 44088

Museum of Scottish Country Life, National Museums of Scotland, Philipshill Rd, Wester Kittochside, East Kilbride G76 9HR • T: +44 1355 224181, 2474377 • F: 571290 • kittochside@nms.ac.uk • www.nms.ac.uk/countrylife •
Man.: *Duncan Dornan* •
Agriculture Museum – 2001
History of farming and rural life throughout Scotland 44089

East Linton

Preston Mill and Phantassie Doocot, East Linton • T: +44 1620 860426 •
Science&Tech Museum 44090

East Looe

Old Guildhall Museum, Higher Market St, East Looe PL13 1BP • T: +44 1503 263709 • F: 265674 • eastlooetowntrust@yahoo.com •
Cur.: *Barbara Birchwood-Harper* •
Local Museum
Looe hist, fishing, smuggling, stocks, courthouse, ceramics, model boats 44091

East Molesey

Embroiderers' Guild Museum Collection, Apt 41, Hampton Court Palace, East Molesey KT8 9AU • T: +44 20 89431229 • F: 89779882 • administrator@embroiderersguild.com • www.embroiderersguild.com •
Dir.: *Michael Spender* • Cur.: *Lynn Szygenda* •
Special Museum 44092

Hampton Court Palace, Historic Royal Palaces, East Molesey KT8 9AU • T: +44 20 87819500 • F: 87819509 • www.hrp.org.uk •
Dir.: *Michael Day* • Cur.: *Lucy Worsley* •
Fine Arts Museum – 1514
Royal palace, original furnishings, Italian, Flemish, German, Dutch, Spanish paintings 44093

Mounted Branch Museum, Imber Court, East Molesey KT8 0BY • T: +44 20 82475480 •
Special Museum
Uniforms and equipment, saddles, harness, paintings and photographs 44094

East Tilbury

Thameside Aviation Museum, Coalhouse Fort, East Tilbury, Essex • T: +44 1277 655170 • F: +44 113 2470219 •
Science&Tech Museum
Aviation archaeology 44095

East Winterslow

New Art Centre, Roche Court, Sculpture Park and Gallery, East Winterslow SP5 1BG • T: +44 1980 862244 • F: 862447 • nac@sculpture.uk.com • www.sculpture.uk.com •
Dir.: *Madeleine Bessborough* •
Open Air Museum 44096

Eastbourne

Eastbourne Heritage Centre, 2 Carlisle Rd, Eastbourne BN21 4JJ • T: +44 1323 411189, 721825 •
Chm.: *Owen Boydell* •
Local Museum
Local hist about the last 100 yrs 44097

How we lived then - Museum of Shops, 20 Cornfield Tce, Eastbourne BN21 4NS • T: +44 1323 737143 • howwelivedthen@btconnect.com • www.how-we-lived-then.co.uk •
Owner: *Graham Upton* • *Jan Upton* •
Special Museum
Victorian-style shops 44098

Lifeboat Museum, King Edward's Parade, Eastbourne BN21 4BY • T: +44 1323 730717 • F: 730717 •
Cur.: *Alan Bennett* •
Special Museum – 1937
Hist of Eastbourne lifeboats from 1853 to present, lifeboat models, many photogr 44099

Queen's Royal Irish Hussars Museum, c/o Sussex Combined Services Museum, Redoubt Fortress, Royal Parade, Eastbourne BN22 7AQ • T: +44 1323 410300 • F: 732240 • eastbourne_museums@breathmail.net •
Cur.: *Richard Callaghan* •
Military Museum
Medals, uniforms, prints and memorabilia 44100

Redoubt Fortress - Military Museum of Sussex, Royal Parade, Eastbourne BN21 4BP • T: +44 1323 410300 • F: 438827 • redoubtmuseum@eastbourne.gov.uk • www.eastbournemuseums.co.uk •
Head: *Fran Stovold* •
Military Museum
Uniforms, medals, weapons, equipment and campaign medals illustrate the history of the former Royal Sussex Regiment and the Queen's Royal Irish Hussars, photos, books and documentary material 44101

Sussex Combined Services Museum, Redoubt Fortress, Royal Parade, Eastbourne BN22 7AQ • T: +44 1323 410300 • F: 732240 • eastbourne_museums@breathemail.net •
Cur.: *Richard Callaghan* •
Military Museum
Uniforms, medals and other military artefacts 44102

Towner Art Gallery and Local Museum, High St, Old Town, Eastbourne BN20 8BB • T: +44 1323 417961, 411688 • F: 648182 • townergallery@eastbourne.gov.uk • www.eastbourne.gov.uk •
Cur.: *Matthew Rowe* •
Fine Arts Museum / Historical Museum – 1923
History of Eastbourne from prehistory onwords to the original station clock and an original Victorian kitchen British art of mainly 19th and 20th cc, coll of contemporary art 44103

Eastleigh

Eastleigh Museum, The Citadel, 25 High St, Eastleigh SO50 5LF • T: +44 23 80643026 • F: 80653582 • musmst@hants.gov.uk • www.hants.gov.uk/museum/eastlmus/ •
Cur.: *Sue Tapliss* •
Local Museum
Local hist, railway, art gallery 44104

Royal Observer Corps Museum, 8 Roselands Close, Fair Oak, Eastleigh SO50 8GN • T: +44 2380 693823 • xrocmunc@hants.gov.uk • www.therocmuseum.org.uk •
Cur.: *Neville A. Cullingford* •
Military Museum
National coll of ROC equipment and memorabilia until the 1990s – ROC archive at couty record office 44105

Easton

Easton Farm Park, Easton IP13 0EQ • T: +44 1728 746475 • F: 747861 • easton@eastonfarmpark.co.uk • www.eastonfarmpark.co.uk •
Dir.: *John Kerr* •
Agriculture Museum
Farm buildings, machinery and animals 44106

Easton-on-the-Hill

Priest's House, 40 West St, Easton-on-the-Hill PE9 3LS •
Chm.: *Paul Way* •
Local Museum
Local artefacts 44107

Eastwood

D.H. Lawrence Birthplace Museum, 8a Victoria St, Eastwood NG16 3AW • T: +44 1773 717353 • F: 713509 • culture@broxtowe.gov.uk • www.broxtowe.gov.uk •
Dir.: *Sally Rose* • *Emma Herrity* •
Special Museum – 1975
9 original watercolours by D.H. Lawrence, original items of furniture owners by the Lawrence family, birthplace of D.H. Lawrence 44108

Durban House Heritage Centre, D.H. Lawrence Birthplace Museum, Mansfield Rd, Eastwood NG16 3DZ • T: +44 1773 717353 • F: 713509 • culture@broxtowe.gov.uk • www.broxtowe.gov.uk •
Dir.: *Sally Rose* •
Historical Museum
Local coal owners, D.H. Lawrence, town history 44109

Ecclefechan

Thomas Carlyle's Birthplace, The Arches House, High St, Ecclefechan DG11 3DG • T: +44 1576 300666 • www.nts.org.uk •
Head: *Fiona Auchterlonie* •
Special Museum
Birthplace of writer Thomas Carlyle, coll of his belongings and copy manuscript letters, local artisan dwelling, 44110

Edenbridge

Eden Valley Museum, Church House, 72 High St, Edenbridge TN8 5AR • T: +44 1732 868102 • curator@evmt.org.uk • www.evmt.org.uk •
Cur.: *Jane Higgs* •
Local Museum
Changing fashions, social structures, economic history of the Eden Valley, coll photographs, paintings and engravings 44111

Hever Castle and Gardens, Hever, Edenbridge TN8 7NG • T: +44 1732 865224 • F: 866796 • mail@hevercastle.co.uk • www.hevercastle.co.uk •
Dir.: *Robert Pullin* •
Decorative Arts Museum
Furniture, paintings, tapestries, 16th c portraits 44112

Kent and Sharpshooters Yeomanry Museum, Hever Castle, Edenbridge TN8 7DB • T: +44 1732 865224 • ksymuseum@aol.com • www.ksymuseum.org.uk •
Cur.: *Boris Mollo* •
Military Museum 44113

Edinburgh

Brass Rubbing Centre, Trinity Apse, Chalmers Close, High St, Royal Mile, Edinburgh EH1 1TG • T: +44 131 5564664 • F: 5583103 • www.cac.org.uk •
Special Museum 44114

Camera Obscura and World of Illusions, Castlehill, The Royal Mile, Edinburgh EH1 2LZ • T: +44 131 2263709 • F: 2254239 • info@camera-obscura.co.uk • www.camera-obscura.co.uk •
Dir.: *Andrew Johnson* •
Special Museum
Optical illusions, telescopes, hands-on science exhibits, optical toys, photographs of old Edinburgh 44115

City Art Centre, 2 Market St, Edinburgh EH1 1DE • T: +44 131 5293993 • F: 5293977 • www.cac.org.uk •
Sc. Staff: *Ian O'Riordan* •
Public Gallery – 1980
19th and 20th c artists, mostly Scottish 44116

Cockburn Museum, c/o Dept of Geology & Geophysics, University of Edinburgh, King's Bldg, West Mains Rd, Edinburgh EH9 3JF • T: +44 131 6508527 • F: 6683184 • heather.hooker@ed.ac.uk •
Natural History Museum / University Museum
Rocks, fossils and minerals 44117

The Danish Cultural Institute, 3 Doune Terrace, Edinburgh EH3 6DY • T: +44 131 2257189 • F: 2206162 • dci@dancult.demon.co.uk • www.dancult.demon.co.uk •
Dir.: *Kim Caspersen* •
Public Gallery
Danish paintings, prints, photography and craft 44118

Dean Gallery, 73 Belford Rd, Edinburgh EH4 3DS • T: +44 131 6246200 • F: 3433250 • enquiries@nationalgalleries.org • www.nationalgalleries.org •
Cur.: *Richard Calvocoressi* •
Fine Arts Museum 44119

Edinburgh Printmakers Workshop and Gallery, Wash-House, 23 Union St, Edinburgh EH1 3LR • T: +44 131 5572479 • F: 5588418 • printmakers@ednet.co.uk • www.edinburgh-printmakers.co.uk •
Dir.: *David Watt* •
Public Gallery
Lithography, screenprinting, relief, etching fine art prints 44120

Edinburgh Scout Museum, 7 Valleyfield St, Edinburgh EH3 9LP • T: +44 131 2293756 • F: 2219905 •
Special Museum
Hist of scout movement, memorabilia of movement in Edinburgh and worldwide 44121

Edinburgh University Anatomy Museum, c/o Department of Anatomy, University Medical School, Teviot Pl, Edinburgh EH8 9AG •
Natural History Museum 44122

Edinburgh University Natural History Collections, c/o ICAPB, Ashworth Laboratories, King' Bldgs, West Mains Rd, Edinburgh EH9 3JT • T: +44 131 6501000 • F: 6506564 • pat.preston@ed.ac.uk • www.nhc.ed.ac.uk •
Cur.: *Dr. B.E. Matthews* • *Dr. P.M. Preston* •
Natural History Museum 44123

Fruitmarket Gallery, 45 Market St, Edinburgh EH1 1DF • T: +44 131 2252383 • F: 2203130 • info@fruitmarket.co.uk • www.fruitmarket.co.uk •
Dir.: *Fiona Bradley* •
Public Gallery
Contemporary art 44124

The Georgian House, National Trust of Scotland, 7 Charlotte Sq, Edinburgh EH2 4DR • T: +44 131 2252160, 2263318 • F: 2263318 • thegeorgianhouse@nts.org.uk • www.nts.org.uk •
Man.: *Dr. Shéonagh Martin* •
Decorative Arts Museum / Historical Museum – 1975
Georgian house, furniture, life in Edinburgh (late 18th/early 19th c) 44125

The Grand Lodge of Scotland Museum, 96 George St, Edinburgh EH2 3DH • T: +44 131 2255304 • F: 2253953 • info@grandlodgescotland.com • www.grandlodgescotland.com •
Libr.: *Robert L.D. Cooper* •
Special Museum
Ceramics, glasware, artefacts relating to Freemasonry, manuscripts – library 44126

Heriot-Watt University Museum, Mary Burton Centre, Riccarton, Edinburgh EH14 4AS • T: +44 131 4513218 • F: 4513164 • archive@hw.ac.uk • www.hw.ac.uk/archive •
Archives: *Ann Jones* • Cur.: *Angela Edgar* •
University Museum / Historical Museum / Fine Arts Museum – 1982
History of the University, scientific and technical education in Scotland, art and artefacts, artists of the Edinburgh School, Scottish textile heritage – archive 44127

Historic Scotland, 3 Stenhouse Mill Ln, Edinburgh EH11 3LR • T: +44 131 4435635 • F: 4558260 • robert.wilmot@scotland.gsi.gov.uk • www.historic-scotland.gov.uk •
Man.: *Robert Wilmot* • Cons.: *Dr. Craig Kennedy* •
Historical Museum 44128

John Knox's House, 45 High St, Edinburgh EH1 1SR • T: +44 131 5569579 • F: 5575224 •
Dir.: *Dr. Donald Smith* •
Special Museum – 1849
Memorabilia on John Knox, memorabilia on James Mossman, goldsmith to Mary, Queen of Scots, memorabilia on Mary, information on 16th c Edinburgh 44129

Landings Gallery, Over-Seas House, 100 Princes St, Edinburgh EH2 3AB • T: +44 131 2251501 • F: 2263936 • culture@rosl.org.uk • www.roslarts.org.uk •
Public Gallery
Drawings, paintings, prints and photographs by hospitalfield in Arbroath artists 44130

Lauriston Castle, 2a Cramond Rd South, Edinburgh EH4 5QD, mail addr: 2 Market St, Edinburgh EH1 1DE • T: +44 131 3362060 • F: 3127165 • lauriston.castle@tiscali.co.uk • www.cac.org.uk •
Dir.: *David Scarratt* •
Decorative Arts Museum – 1926
"Blue John" Derbyshire Spar, wool mosaics, castle with period furnishings, tapestries, decorative arts 44131

Museum of Childhood, 42 High St, Royal Mile, Edinburgh EH1 1TG • T: +44 131 5294142 • F: 5583103 • cac.admin@edinburgh.gov.uk • www.cac.org.uk •
Dir.: *John Heyes* •
Special Museum – 1955
Social hist of childhood, hobbies, education, toys, dolls, games 44132

The Museum of Edinburgh, 142 Canongate, Edinburgh EH8 8DD • T: +44 131 5294143 • F: 5573346 • www.cac.org.uk •
Local Museum / Decorative Arts Museum – 1932
Local history, Scottish pottery, local glass and silver, paintings, prints, temporary exhibits 44133

Museum of Fire, c/o Lothian & Borders Fire and Rescue Service, Lauriston Pl, Edinburgh EH3 9DE • T: +44 131 2282401 • F: 2298359 • ian.mcmurtrie@lbfire.org.uk • www.lbfire.org.uk •
Cur.: *I. McMurtrie* •
Science&Tech Museum – 1966
Fire engines and fire-fighting equipment dating from the early 19th c to the present day, coll of badges, medals, photographs and records relating to fire-fighting – stables 44134

Museum of Lighting, 59 Saint Stephen St, Edinburgh EH3 5AG • T: +44 131 5564503 •
Cur.: *W.M. Purves* •
Science&Tech Museum – 1972
All types of lighting equipment, hand-operated machines from the 19th-c lampmakers, Bocock – archive 44135

Museum of Scotland, National Museums of Scotland, Chambers St, Edinburgh EH1 1JF • T: +44 131 2474422 • F: 2204819 • info@nms.ac.uk • www.nms.ac.uk •
Dir.: *Dr. Gordon Rintoul* •
Historical Museum 44136

Museum on the Mound, HBOS, The Mound, Edinburgh EH1 1YZ • T: +44 131 5291288 • F: 5291307 • info@museumonthemound.com • www.museumonthemound.com •
Man.: *Doug MacBeath* •
Special Museum 44137

National Gallery of Scotland, The Mound, Edinburgh EH2 2EL • T: +44 131 6246200 • F: 2200917 • enquiries@nationalgalleries.org • www.nationalgalleries.org •
Gen. Dir.: *John Leighton* • Sc. Staff: *Michael Clarke* • *Christopher Baker* • *Helen Smailes* • *Aidan Weston-Lewis* • *Emilie Gordenker* • *Valerie Hunter* • Libr.: *Penelope Carter* •
Fine Arts Museum – 1850
Paintings, drawings and prints by the greatest artists from the Renaissance to Post-Impressionism, incl Velazquez, El Greco, Rembrandt, Vermeer, Turner, Constable, Monet and Van Gogh, Botticelli, Titian, Rubens, Van Dyck, Goya, Poussin, Raphael 44138

National Museums of Scotland, Chambers St, Edinburgh EH1 1JF • T: +44 131 2474422 • F: 2204819 • info@nms.ac.uk • www.nms.ac.uk •
Dir.: *Dr. Gordon Rintoul* •
1781 44139

National War Museum of Scotland, National Museums of Scotland, The Castle, Edinburgh EH1 2NG • T: +44 131 2257534 • F: 2253848 • www.nms.ac.uk •
Cur.: *A.L. Carswell* •
Military Museum – 1930
Military history of Scotland and the experiance of war and military service, uniforms, weapons, paintings, sculpture, ptints 44140

Nelson Monument, 32 Calton Hill, Edinburgh EH7 5AA, mail addr: CAC, 2 Market St, Edinburgh EH1 1DE • T: +44 131 5562716 • F: 2205057 • www.cac.org.uk •
Historical Museum
Artefacts relating to Admiral Lord Nelson 44141

Newhaven Heritage Museum, 24 Pier Pl, Edinburgh EH6 4LP • T: +44 131 5514165 • F: 5573346 • www.cac.org.uk •
Historical Museum – 1994 44142

Palace of Holyroodhouse, Royal Collection, The Royal Mile, Edinburgh EH8 8DX, mail addr: The Official Residences of The Queen, London SW1A 1AA • T: +44 131 5565100 • F: +44 20 79309625 • bookinginfo@royalcollection.org.uk • www.royalcollection.org.uk •
Fine Arts Museum
Royal residence 44143

U

People's Story Museum, 163 Canongate, Royal Mile, Edinburgh EH8 8BN • T: +44 131 5294057 • F: 5563439 • www.cac.org.uk •
Historical Museum – 1989
Social history material representing the life and work of Edinburgh's people over the past 200 years, coll of Trade Union banners, oral tape recordings, photos ... 44144

The Queen's Gallery, Royal Collection, Palace of Holyroodhouse, The Royal Mile, Edinburgh EH8 8DX, mail addr: The Official Residences of The Queen, London SW1A 1AA • T: +44 131 5565100 • F: +44 20 79309625 • bookinginfo@royalcollection.org.uk • www.royalcollection.org.uk •
Fine Arts Museum
Royal coll ... 44145

Regimental Museum of the Royal Scots Dragoon Guards, The Castle, Edinburgh EH1 2YT • T: +44 131 2204387 • F: 3105101 •
Military Museum ... 44146

Reid Concert Hall Museum of Instruments, Edinburgh University Collection of Historic Musical Instruments, Bristo Sq, Edinburgh EH8 9AG • T: +44 131 6502422 • F: 6502425 • euchmi@ed.ac.uk • www.music.ed.ac.uk/euchmi/rch •
Dir./Cur.: *Dr. Arnold Myers* • Cur.: *Darryl Martin* •
Music Museum / University Museum – 1859
John Donaldson coll of over 1,000 musical instr, 400 years of hist of folk and domestic music, bands and orchestras – sound lab ... 44147

Royal Museum, National Museums of Scotland, Chambers St, Edinburgh EH1 1JF • T: +44 131 2474422 • F: 2204819 • info@nms.ac.uk • www.nms.ac.uk •
Dir.: *Dr. Gordon Rintoul* •
Archaeology Museum / Natural History Museum / Ethnology Museum / Science&Tech Museum – 1855
Archaeology, natural history, geology, technology, Oriental arts, ethnography of Plains Indians, Eastern Island and Maori cultures, Baule figures, minerals , fossils, mining and metallurgy, power industries, shipping, aeronautics, radiation, primitive arts, British birds – library ... 44148

Royal Pharmaceutical Society Late Victorian Pharmacy, 36 York Pl, Edinburgh EH1 3HU • T: +44 131 5564386 • F: 5588850 • scotinfo@rpsgb.org •
Dir.: *Lyndon Braddick* •
Historical Museum ... 44149

The Royal Scots Regimental Museum, The Castle, Edinburgh EH1 2YT • T: +44 131 3105014 • F: 3105019 • royalscotarchive@aol.com • www.theroyalscots.co.uk •
Dir.: *R.P. Mason* • Sc. Staff: *D. Murphy* •
Military Museum – 1951
History, uniforms, medals, weapons, trophies of the regiment, militaria ... 44150

Royal Scottish Academy, The Mound, Edinburgh EH2 2EL • T: +44 131 2256671 • F: 2206016 • info@royalscottishacademy.org • www.royalscottishacademy.org •
Dir.: *Dr. Ian McKenzie Smith* •
Fine Arts Museum / Public Gallery – 1826
Scottish painting & sculpture of the 19th & 20th c, drawings, photographs ... 44151

Royal Scottish Academy Collections, c/o Dean Gallery, 73 Belford Rd, Edinburgh EH4 3DS • T: +44 131 6246277 • F: 2206016 • info@royalscottishacademy.org • www.royalscottishacademy.org •
Cur.: *Joanna Soden* •
Fine Arts Museum – 1826
Visual arts, fine art – library and archives ... 44152

Saint Cecilia's Hall Museum of Instruments, Edinburgh University Collection of Historic Musical Instruments, Saint Cecilia's Hall, Niddry St, Cowgate, Edinburgh EH1 1LJ • T: +44 131 6502422 • F: 6502425 • euchmi@ed.ac.uk • www.music.ed.ac.uk/euchmi/sch •
Dir.: *Arnold Myers* •
Music Museum – 1968
Raymond Russell coll of early keyboard instr, harpsichords, virginals, spinets, clavichords, harps, lutes, guitars, chamber organs and early pianos 44153

Scotch Whisky Heritage Centre, 354 Castlehill, The Royal Mile, Edinburgh EH1 2NE • T: +44 131 2200441 • F: 2206288 • enquiries@whisky-heritage.co.uk • www.whisky-heritage.co.uk •
Special Museum ... 44154

Scott Monument, Princes St, Edinburgh EH2 2EJ • T: +44 131 5294068 • F: 2205057 • www.cac.org.uk •
Fine Arts Museum
National monument to Sir Walter Scott ... 44155

Scottish National Gallery of Modern Art Gallery, 75 Belford Rd, Edinburgh EH4 3DR • T: +44 131 6246200 • F: 3432802, 3433250 • enquiries@nationalgalleries.org • www.nationalgalleries.org •
Gen. Dir.: *John Leighton* • Dir.: *Richard Calvocoressi* • Cur.: *Philip Long* • *Fiona Pearson* (Paolozzi Collection) • *Alice Strang* • *Patrick Elliott* • *Keith Hartley* • *Daniel Hermann* • *Ann Simpson* • *Jane Furness* • *Andrea Ross* •
Public Gallery / Fine Arts Museum – 1960
20th c paintings, sculpture and graphic art incl works by Vuillard, Matisse, Kirchner, Picasso, Magritte, Miró, Dali and Ernst, coll of 20th c Scottish art – sculpture garden, conservation ... 44156

Scottish National Portrait Gallery, 1 Queen St, Edinburgh EH2 1JD • T: +44 131 6246200 • F: 5583691 • pginfo@nationalgalleries.org • www.nationalgalleries.org •
Dir.: *James Holloway* • Dep. Dir.: *Nicola Kalinsky* • Sc. Staff: *Sara Stevenson* • *Stephen Lloyd* • *Julie Lawson* • *Duncan Forbes* • *Imogen Gibbon* •
Fine Arts Museum / Historical Museum – 1882
Portraits in various media illustrating the hist of Scotland from the 16th c up to the present day, portraits of Scots by Scottish artists, and other artists including van Dyck, Gainsborough, Copley, Thorvaldsen, Rodin and Kokoschka ... 44157

Scottish Rugby Union Museum, Murrayfield Stadium, Edinburgh EH12 5PJ • T: +44 131 3465073 • F: 3465001 • library@sru.org.uk • www.sru.org.uk •
Dir.: *William S. Watson* •
Special Museum
History and development of the game and the Scottish Rugby Union, jerseys, caps, balls, trophies, photos ... 44158

Stills Gallery, 23 Cockburn St, Edinburgh • T: +44 131 6226200 • F: 6226201 • info@stills.org • www.stills.org •
Dir.: *Kate Tregaskis* •
Public Gallery ... 44159

Surgeons' Hall Museum, 18 Nicolson St, Edinburgh EH8 9DW • T: +44 131 5271649 • F: 5576406 • museum@rcsed.ac.uk • www.edinburgh.surgeonshall.museum •
Dir.: *Dawn Kemp* • Cur.: *Andrew Connell* •
Historical Museum / University Museum – 1805
Surgery, pathology and medicine, dentistry ... 44160

Talbot Rice Gallery, c/o University of Edinburgh, Old College, South Bridge, Edinburgh EH8 9YL • T: +44 131 6502211 • F: 6502213 • info.talbotrice@ed.ac.uk • www.trg.ed.ac.uk •
Dir.: *Pat Fisher* •
Fine Arts Museum / University Museum / Public Gallery
Coll of bronzes and paintings, esp Dutch and Italian contemporary Scottish art ... 44161

The Writers' Museum, Lady Stair's House, Lady Stair's Close, Royal Mile, Edinburgh EH1 2PA • T: +44 131 5294901 • F: 2205057 • enquiries@writersmuseum.demon.co.uk • www.cac.org.uk •
Dir.: *Herbert Coutts* •
Special Museum – 1907
Memorabilia of the writers Robert Burns, Sir Walter Scott, Robert Louis Stevenson ... 44162

Egham

Egham Museum, Literary Institute, High St, Egham TW20 9EW • T: +44 1344 843047, +44 1784 434483 •
Cur.: *John Mills* •
Historical Museum / Archaeology Museum
Archaeology and hist of Egham, Thorpe and Virginia Water, displays relating to Magna Carta, which was sealed nearby at Runnymede ... 44163

Royal Holloway College Picture Gallery, Egham Hill, Egham • T: +44 1784 434455 • F: 437520 • m.cowling@rhul.ac.uk • www.rhul.ac.uk •
Cur.: *Dr. Mary Cowling* •
Fine Arts Museum / University Museum – 1887
Paintings by Frith, Fildes, Maclise, Landseer, Millais and others ... 44164

Egremont

Florence Mine Heritage Centre, Egremont CA22 2NR • T: +44 1946 820683 • F: 820683 • info@florencemine.co.uk • www.florencemine.co.uk •
Man.: *Fiona Paine* •
Science&Tech Museum
Mining artefacts and geological displays ... 44165

Elgin

Elgin Museum, 1 High St, Moray, Elgin IV30 1EQ • T: +44 1343 543675 • F: 543675 • curator@elginmuseum.org.uk • www.elginmuseum.org.uk •
Cur.: *David Addison* • Cust.: *Mary McGarrie* •
Archaeology Museum – 1843
Local antiquities, Triassic and Permian reptile fossils, Devonian fish fossils, archaeology – scientific archive ... 44166

Ellesmere Port

Hooton Park Exhibition Centre, Hangars, South Rd, Hooton Park Airfield, Ellesmere Port CH65 1BQ • T: +44 151 3274701 • F: 3274701 •
Dir.: *D. Beckett* • *R. Copson* • *J. Graham* •
Science&Tech Museum
Classic vehicles (military and civil), aeronautical items, aircraft ... 44167

National Waterways Museum - Ellesmere Port, South Pier Rd, Ellesmere Port CH65 4FW • T: +44 151 3555017 • F: 3554079 • sophie.fowler@thewaterwaystrust.org • www.thewaterwaystrust.org •
Head: *Sophie Fowler* (Collections) •
Science&Tech Museum
60 canal and river craft, icebreakers, traditional boat-building techniques in progress, story of canal building, life of dock workers, working engines, canal hist – archive ... 44168

Ellon

Museum of Farming Life, Pitmedden House, Ellon AB41 7PD • T: +44 1651 842352 • F: 843188 • sburgess@nts.org.uk • nts.org.uk •
Dir.: *Susan M. Burgess* •
Agriculture Museum – 1982
Coll of farm tools and domestic equipment, a furnished farmhouse and bothy, restored outbuildings, hist of Scottish farming ... 44169

Elsham

Clocktower Museum, Elsham Hall, Elsham DN20 0QZ • T: +44 1652 688698 • F: 688240 •
Fine Arts Museum / Decorative Arts Museum
Resident artists and working craft ... 44170

Elvington

Yorkshire Air Museum, Halifax Way, Elvington YO41 4AU • T: +44 1904 608595 • F: 608246 • museum@yorkshireairmuseum.co.uk • www.yorkshireairmuseum.co.uk •
Dir.: *Ian Reed* •
Science&Tech Museum / Military Museum
Coll of WWII memorabilia, historical airframes and replica's (early days of aviation to modern military jets) – library ... 44171

Ely

Ely Museum, The Old Gaol, Market St, Ely CB7 4LS • T: +44 1353 666655 • F: 659259 • elymuseum@freeuk.com • www.elymuseum.org.uk •
Cur.: *Polly Hodgson* •
Local Museum
Social and natural hist of the Isle of Ely, local archeology ... 44172

Oliver Cromwell's House, 29 Saint Mary's St, Ely CB7 4HF • T: +44 1353 662062 • F: 668518 • tic@eastcambs.co.uk • www.elyeastcambs.co.uk •
Special Museum
Rooms incl kitchen, palour, Civil War exhibition, Cromwell's study and a haunted room, fen drainage ... 44173

Stained Glass Museum, South Triforium, Ely Cathedral, Ely CB7 4DN • T: +44 1353 660347 • F: 665025 • curator@stainedglassmuseum.com • www.stainedglassmuseum.com •
Dir.: *Juliet Short* • Cur.: *Susan Mathews* •
Decorative Arts Museum – 1972
100 windows from medieval to modern times, a photographic display of the styles and techniques of medieval glass, and models of a modern stained glass workshop, 19th-c glass, with examples of work from all the leading studios and designers – library ... 44174

Emsworth

Emsworth Museum, 10b North St, Emsworth PO10 7DD • T: +44 1243 378091 • www.emsworthmuseum.org •
Local Museum
Local hist, sea-faring families, oyster fishermen and coastal trade in grain and coal, P.G. Woodhouse, Sir Peter Blake memorabilia ... 44175

Enfield

Forty Hall Museum, Forty Hill, Enfield EN2 9HA • T: +44 20 83638196 • F: 83639098 • jan.metcalfe@enfield.gov.uk • www.enfield.gov.uk/museum •
Man.: *Val Munday* •
Decorative Arts Museum – 1955
Ceramics, hist childhood, Middlesex maps, furniture, paintings, teaching ... 44176

Enniskillen

Castle Coole, Enniskillen • T: +44 1365 322690 •
Historical Museum
Furniture and textiles ... 44177

Fermanagh County Museum at Enniskillen Castle, Castle Barracks, Enniskillen BT74 7HL • T: +44 28 66325000 • F: 66327342 • castle@fermanagh.gov.uk • www.enniskillencastle.co.uk •
Dir.: *Helen Lanigan Wood* •
Local Museum
Regional history, archaeology, fine and applied art ... 44178

Florence Court House, Florence Court, Enniskillen BT92 1DB • T: +44 28 66348249 • F: 66348873 • florencecourt@nationaltrust.org.uk • www.nationaltrust.org.uk •
Local Museum
18th cent. house and Irish estate with Rococo decoration and fine Irish furniture ... 44179

Royal Inniskilling Fusiliers Regimental Museum, The Castle, Enniskillen BT74 7HL • T: +44 28 66323142 • F: 66320359 • info@inniskillingsmuseum.com • www.inniskillingsmuseum.com •
Dir.: *J.M. Dunlop* •
Military Museum – 1931
Hist of the Royal Inniskilling Fusiliers (1689-1968) and associated Fermanagh, Tyrone and Donegal militia regiments, displays on 6th Inniskilling Dragoons (cavalry regiment) – library ... 44180

Sheelin Antiques and Irish Lace Museum, Bellanaleck, Enniskillen • T: +44 1365 348052 • info@irishlacemuseum.com • www.irishlacemuseum.com •
Dir.: *Rosemary Cathcart* •
Special Museum
Coll of Irish Lace ... 44181

Epping

North Weald Airfield Museum and Memorial, Ad Astra House, Hurricane Way, North Weald, Epping CM16 6AA • T: +44 1922 523010 • info@policeaviationnews.com • www.fly.to/northweald •
Chm.: *Stephen Wagstaffe* •
Science&Tech Museum – 1989
Historic and vintage flying aircraft in a 1940 period setting ... 44182

Epsom

College Museum, Epsom College, Epsom KT17 4QJ • T: +44 1372 728862 •
Sc. Staff: *M.D. Hobbs* •
University Museum / Natural History Museum ... 44183

Epworth

Epworth Old Rectory, 1 Rectory St, Epworth DN9 1HX • T: +44 1427 872268 • curator@epwortholdrectory.org.uk • www.epwortholdrectory.org.uk •
Cur.: *Alison Bodley* •
Historical Museum / Religious Arts Museum
Portraits, Prints, Furniture, Wesley family items – library ... 44184

Erith

Erith Museum (closed) ... 44185

Errol

Megginch Castle, Errol PH1 • T: +44 1821 2222 •
Historical Museum ... 44186

Etchingham

Bateman's, Kipling's House, Burwash, Etchingham TN19 7DS • T: +44 1435 882302 • F: 882811 • batemans@nationaltrust.org.uk •
Dir.: *Elaine Francis* •
Special Museum
Home of the Rudyard Kipling from 1902-36: Kipling's rooms and study are left, his Rolls-Royce ... 44187

Haremere Hall, Etchingham TN19 7QJ • T: +44 1580 81145 • members.tripod.com/~harmer1/reunion78c.html •
Decorative Arts Museum – 1978 ... 44188

Eton

Eton College Natural History Museum, Eton College, Keate's Lane, Eton SL4 6EW, mail addr: Eton College, SL4 6DW Windsor • T: +44 1753 671288 • F: 671159 • g.fussey@etoncollege.org.uk • www.etoncollege.com/NatHistMus.aspx •
Cur.: *Dr. George D. Fussey* • Sc. Staff: *Roddy Fisher* •
Natural History Museum / University Museum – 1875
Fossils, insects, molluscs, birds, mammals, fishes, plants – Herbarium ... 44189

Myers Museum of Egyptian and Classical Art, Brewhouse Gallery, Eton College, Eton SL4 6DW • T: +44 1753 801538 • F: 801538 • n.reeves@etoncollege.org.uk • www.etoncollege.org.uk •
Cur.: *Dr. Nicholas Reeves* (Egyptian and Classical Art) •
Archaeology Museum / Museum of Classical Antiquities
Coll of Egyptian archaeology by Major W.J. Myers (1883-1899), objects dating from Predynastic times to the Coptic period, faience, wooden models from tombs and a preserved panel portrait from a mummy of the 2nd c A.D. ... 44190

Evesham

Almonry Heritage Centre, Abbey Gate, Evesham WR11 4BG • T: +44 1386 446944 • F: 442348 • tic@almonry.ndo.co.uk • www.evesham.uk.com •
Local Museum / Archaeology Museum
Hist of the Vale of Evesham from prehistoric times to the present day, Romano-British, Anglo-Saxon and medieval items, together with material illustrating the history of the monastery, coll of agricultural implements ... 44191

Ewell

Bourne Hall Museum, Spring St, Ewell KT17 1UF • T: +44 20 83941734 • F: 87867265 • j.harte@bournehall.free-online.co.uk • www.epsom.townpage.co.uk •
Dir.: *Jeremy Harte* •
Decorative Arts Museum – 1969
Coll of some of the earliest wallpapers ever made (c 1690) and the Ann Hull Grundy coll of costume jewellery, prehistoric material found in the vicinity, Victorian and more modern costumes and accessories, and early cameras and radio sets, Roman archaeology, tools – library ... 44192

Exeter

Devon and Cornwall Constabulary Museum, Devon & Cornwall Constabulary Headquarters, Exeter EX2 7HQ • T: +44 1392 203025 • F: 426496 • force.museum-dcc@btconnect.com • www.devon-cornwall.police.uk •

Dir.: *Angela Sutton-Vane* •
Special Museum
The coll comprises of docs and artefacts relating to the hist of policing throughout Devon & Cornwall, incl photos, uniforms, equipment and memorabilia ... 44193

Guildhall, High St, Exeter EX4 3EB • T: +44 1392 277888, 665500 • F: 201329 • guildhall@exeter.gov.uk • www.exeter.gov.uk •
Local Museum ... 44194

Royal Albert Memorial Museum and Art Gallery (reopens in 2011), Queen St, Exeter EX4 3RX • T: +44 1392 665858 • F: 421252 • ramm@exeter.gov.uk • www.exeter.gov.uk •
Dir.: *Camilla Hampshire* • Sc. Staff: *John Madin* (Art) • *Tom Cadbury* (Antiquities) • *David Bolton* (Natural History) • *Tony Eccles* (Ethnography) •
Fine Arts Museum / Local Museum – 1865
Natural hist, butterflies, birds, local geology, fine art, local artists, applied arts, Exeter silver, Devon pottery, Devon lace, ethnography, Polynesian, N. American Indian material, Yoruba sculptures, 18th and early 19th c Eskimo and N.W. Coast material, Benin head and Staff mount, 18th c Tahitian morning dress, flute glass 'Exeter flute' c 1660, harpsichord by Vincentius Sodi ... 44195

Underground Passages, Roman Gate Passageway, off High St, Exeter EX4 3PZ • T: +44 1392 265206 • F: 265695 • upassages@exeter.gov.uk • www.exeter.gov.uk/visiting •
Science&Tech Museum
Introductory exhibition and video, scheduled ancient monument, vaulted medieval passages build in the 14th c unterground tour shows how water was brought into the city ... 44196

Exmouth

Exmouth Museum, Sheppards Row, Exeter Rd, Exmouth EX8 1PW • exmouth@devonmuseum.net • www.devonmuseums.net/exmouth •
Cur.: *Tom Haynes* •
Local Museum ... 44197

World of Country Life, Sandy Bay, Exmouth EX8 5BU • T: +44 1395 274533 • F: 273457 • worldofcountrylife@hotmail.com • www.worldofcountrylife.co.uk •
Folklore Museum / Science&Tech Museum / Agriculture Museum
Vintage cars, classic motorbikes and steam traction engines, Fire engines, moving farm machinery, gypsy caravan ... 44198

Eyam

Eyam Museum, Hawkhill Rd, Eyam S32 5QP • T: +44 1433 631371 • F: 631371 • www.eyam.org.uk •
Local Museum – 1994
Village heroism when in 1665 they had Bubonic Plague and quarantined themselves – C. Daniel Collection ... 44199

Eyemouth

Eyemouth Museum, Auld Kirk, Manse Rd, Eyemouth TD14 5JE • T: +44 1890 750678 •
Hon. Cur.: *Jean Bowle* •
Local Museum
History of the Eyemouth area and its people, with a special emphasis on the fishing industry, farming and the traditional trades and handicrafts ... 44200

Eynsford

Lullingstone Roman Villa, Eynsford DA4 0JA • T: +44 1322 863467 • F: 863467 • karen.hawkins@english-heritage.org.uk • www.english-heritage.org.uk •
Cur.: *Jo Gray* •
Archaeology Museum
Artefacts from the excavations of the 1960s ... 44201

Fair Isle

Auld Sköll, Utra, Fair Isle ZE2 9JU • T: +44 1595 760209 •
Historical Museum ... 44202

Fakenham

Fakenham Museum of Gas and Local History, Hempton Rd, Fakenham NR21 7LA • T: +44 1328 863507 • gaslhmus@hotmail.com • www.fakenham.org.uk •
Chm.: *Dr. E.M. Bridges* •
Science&Tech Museum / Local Museum
Local history, coal gas industry and trade, industrial archaeology ... 44203

Thursford Collection, Thursford Green, Fakenham NR21 0AS • T: +44 1328 878477 • F: 878415 • admin@thursfordcollection.co.uk •
Dir.: *J.R. Cusming* •
Science&Tech Museum / Music Museum – 1947
Steam and locomotives including showman's traction and ploughing engines, steam wagons, mechanical musical organs, Wurlitzer theater organ, concert organ, German organ Karl Frei ... 44204

Falkirk

Callendar House, Callendar Park, Falkirk FK1 1YR • T: +44 1324 503770 • F: 503771 • callendar.house@falkirk.gov.uk •
Dir.: *Sue Selwyn* •
Historical Museum – 1991
Mansion, working kitchen of 1825, social coll, archaeology – Falkirk Council Archives ... 44205

Park Gallery, Callendar Park, Falkirk FK1 1YR • T: +44 1324 503789 • artsandcrafts@falkirk.gov.uk •
Dir.: *Stephen Palmer* •
Fine Arts Museum
Coll of contemporary visual art and craft incl work by national, international and local artists ... 44206

Falkland

Falkland Palace and Garden, High St, Falkland KY15 7BV • T: +44 1337 857397 • F: 857980 • www.nts.org.uk •
Man.: *Karen Caldwell* •
Historical Museum / Historic Site – 1952
Chapel Royal, Royal Tennis Court (1539), Flemish tapestries ... 44207

Falmouth

Falmouth Art Gallery, Municipal Buildings, The Moor, Falmouth TR11 2RT • T: +44 1326 313863 • F: 318608 • info@falmouthartgallery.com • www.falmouthartgallery.com •
Dir.: *Brian Stewart* •
Fine Arts Museum / Public Gallery – 1978
Late 19th and early 20th c paintings (mainly British) donated by Alfred de Pass, photographs ... 44208

National Maritime Museum Cornwell, Discovery Quay, Falmouth TR11 3QY • T: +44 1326 313388 • F: 317878 • enquiries@nmmc.co.uk • www.nmmc.co.uk •
Dir.: *Jonathan Griffin* •
Historical Museum / Science&Tech Museum / Local Museum – 2002
Maritime history of Cornwall and the Falmouth Packet Service, models, photographs, navigational instruments, ship-building tools, coll of 120 historic boats – Database of Cornish built vessels 1786-1914, Bartlett library ... 44209

Pendennis Castle, Falmouth TR11 4LP • T: +44 1326 316594 • F: 319911 • customers@english-heritage.org.uk • www.english-heritage.org.uk •
Man.: *Jane Kessel* •
Historical Museum
Arms and uniforms, artillery, furnished interiors and discovery centre ... 44210

Fareham

Royal Armouries at Fort Nelson, Fort Nelson, Portsdown Hill, Down End Rd, Fareham PO17 6AN • T: +44 1329 233734 • F: 822092 • fnenquiries@armouries.org.uk • www.royalarmouries.org •
Cur.: *Philip Magrath* •
Historical Museum / Military Museum – 1988
19th c fortress home of the Royal armouries coll of artillery (worldwide 15th-20th c), regular exhibitions and firing days – Conservation education centre ... 44211

Westbury Manor Museum, 84 West St, Fareham PO16 0JJ • T: +44 1329 824895 • F: 825917 • musmop@hants.gov.uk • www.hants.gov.uk/museum/westbury •
Cur.: *Oonagh Palmer* •
Local Museum ... 44212

Faringdon

Buscot Park House, Faringdon SN7 8BU • T: +44 1367 240786 • F: 241794 • estbuscot@aol.com • www.buscot-park.com •
Dir.: *Lord Faringdon* •
Fine Arts Museum
Georgian house in park, containing Faringdon Coll of the English and Continental schools (Rembrandt, Murillo and Reynolds), 18th c furniture, porcelain, objets d'art ... 44213

Farleigh Hungerford

Farleigh Hungerford Castle, Farleigh Hungerford BA3 6RS • T: +44 1225 754026 • F: 754026 • customers@english-heritage.org.uk • www.english-heritage.org.uk •
Cur.: *Tony Musty* •
Archaeology Museum
Ruins of a fortified manor house containing a furnished chapel and a small collection of architectural fragments ... 44214

Farndon

Stretton Water Mill, Stretton, Farndon SY14 7RS • cheshiremuseums@cheshire.gov.uk •
Cur.: *Tom Jones* •
Science&Tech Museum ... 44215

Farnham

Farnhem Maltings Gallery, Farnham Maltings, Bridge Sq, Farnham GU9 7QR • T: +44 1252 726234 • F: 718177 • FarnMAlt@aol.com • www.farnhammaltings.com •
Dir.: *Tozzy Bridger* •
Public Gallery ... 44216

Foyer / James Hockey Gallery, University College for the Creative Arts, Campus, Falkner Rd, Farnham GU9 7DS • T: +44 1252 892668/46 • F: 892667 • galleries@ucreative.ac.uk • www.ucreative.ac.uk/galleries •
Man.: *Christine Kapteijn* •
Public Gallery / University Museum ... 44217

Museum of Farnham, Willmer House, 38 West St, Farnham GU9 7DX • T: +44 1252 715094 • F: 715094 • fmuseum@waverley.gov.uk • www.waverley.gov.uk/museumoffarnham •
Cur.: *Anne Jones* •
Decorative Arts Museum – 1960
House (1718) with English decorative arts, local history, archaeology, 18th c furniture, costumes, William Cobbett memorabilia – Reference library ... 44218

Rural Life Centre, Old Kiln Museum, Reeds Rd, Tilford, Farnham GU10 2DL • T: +44 1252 795571 • F: 795571 • rural.life@lineone.net • www.rural-life.org.uk •
Dir.: *Henry Jackson* •
Agriculture Museum – 1972
Waggons, farm implements, hand tools, dairy, kitchen, trades and crafts, rural life 1800-1950, arboretum, wheel whight ... 44219

Faversham

Belmont, Belmont Park, Throwley, Faversham ME13 0HH • T: +44 1795 890202 • F: 890042 • belmondadmin@btconnect.com • www.belmont-house.org •
Cur.: *Jonathan Betts* •
Science&Tech Museum
Clocks and watches coll ... 44220

Fleur de Lis Heritage Centre, 10-13 Preston St, Faversham ME13 8NS • T: +44 1795 534542 • F: 533261 • faversham@btinternet.com • www.faversham.org/society/tic.asp •
Dir.: *Arthur Percival* •
Local Museum – 1977
Telephones, explosives hist ... 44221

Maison Dieu, Water Lane, Ospringe, Faversham ME13 8TW • T: +44 1795 533751 • F: 533261 • www.english-heritage.co.uk •
Historical Museum – 1925
Medieval bldg, once part of complex which served as Royal lodge, pilgrims' hostel, hospital and almshouse for retired Royal retainers, now housing Roman finds and new displays illustrating Ospringe's eventful hist 44222

Ferniegair

Chatelherault, Ferniegair ML3 7UE • T: +44 1698 426213 • F: 421532 • tom.mccormack@southlanarkshire.gov.uk •
Historical Museum
Historic park, ruined Cadzow Castle, remains of coal mines ... 44223

Museums of South Lanarkshire, 116 Cadzow St, Ferniegair ML3 7UE • T: +44 1698 452168 • F: 452265 • customer.services@southlanarkshire.gov.uk • www.southlanarkshire.gov.uk •
Local Museum ... 44224

Fetlar

Fetlar Interpretive Centre, Beach of Houbie, Fetlar ZE2 9DJ • T: +44 1957 733206 • F: 733219 •
Local Museum ... 44225

Filching

Motor Museum, Filching Manor, Filching BN26 5QA • T: +44 1323 487838 • F: 486331 •
Science&Tech Museum
Wealden Hall house, car coll, 1893-1997 karting circuit ... 44226

Filey

Filey Museum, 8-10 Queen St, Filey YO14 9HB • T: +44 1723 515013 •
Cur.: *Margaret Wilkins* •
Local Museum
Fishing, lifeboats, the seashore, rural and domestic objects ... 44227

Filkins

Swinford Museum, Filkins GL7 3JQ • T: +44 1367 860209, 860504 •
Cur.: *Diane Blackett* •
Agriculture Museum / Folklore Museum – 1931
Household and farm tools, stone craft ... 44228

Firle

Charleston Trust, Charleston, Firle BN8 6LL • T: +44 1323 811626, 811265 • F: 811628 • info@charleston.org.uk • www.charleston.org.uk •
Dir.: *Colin McKenzie* • Cur.: *Dr. Wendy Hitchmough* •
Decorative Arts Museum – 1981
Domestic decorative art of Vanessa Bell and Duncan Grant ... 44229

Firle Place, Firle BN8 6LP • T: +44 1273 858335 • F: 858188 • gage@firleplace.co.uk •
Decorative Arts Museum
Home of Viscount Gage, old masters, Sevres and English porcelain, fine French and English furniture ... 44230

Fishguard

West Wales Arts Centre, 16 West St, Fishguard SA65 9AE • T: +44 1348 873867 • westwalesarts@btconnect.com • home.btconnect.com/WEST-WALES-ARTS •
Dir.: *Myles Pepper* •
Public Gallery
Paintings, sculpture, ceramics, jewllery and glass ... 44231

Fleetwood

Fleetwood Museum, Queen's Terrace, Fleetwood FY7 6BT • T: +44 1253 876621 • F: 878088 • fleedwood.museum@mus.lancscc.gov.uk • www.lancashire.com/lcc/museums •
Cur.: *Simon Hayhow* •
Natural History Museum – 1974
County Museum Service biology collections ... 44232

Flintham

Flintham Museum, Inholms Rd, Flintham NG23 5LF • T: +44 1636 525111 • flintham.museum@lineone.net • www.flintham-museum.org.uk •
Local Museum
Rural life, toys, postcards, pharmeceuticals, tobacco products, bike accessories, haberdashery and hardware – archive ... 44233

Flixton

Norfolk and Suffolk Aviation Museum, East Anglia's Aviation Heritage Centre, Buckeroo Way, The Street, Flixton NR35 1NZ • T: +44 1986 896644 • nsam.flixton@virgin.net • www.aviationmuseum.net •
Cur.: *Alan Hague* •
Science&Tech Museum – 1972
Coll of over 36 aircraft and items from Pre-WW I to present day, civil and military ... 44234

Fochabers

Fochabers Folk Museum, Pringle Church, High St, Fochabers IV32 7DU • T: +44 1343 821204 • F: 821291 •
Dir.: *Gordon A. Christie* •
Folklore Museum – 1984
Hist of the village of Fochabers since the late 18th c, coll of horse gigs and carts ... 44235

Folkestone

Elham Valley Railway Museum, Peene Yard, Peene, Newington, Folkestone CT18 8BA • T: +44 1303 273690 • F: 873939 • evlt.museum@btopenworld.com • www.elhamvalleylinetrust.org •
Chm.: *George Wright* •
Science&Tech Museum
Winston Churchill brought the Bochbuster railway gun during the war ... 44236

Folkestone Museum and Art Gallery, 2 Grace Hill, Folkestone CT20 1HD • T: +44 1303 256710 • F: 256710 • elizabeth.ritchie@kent.gov.uk • www.kent-museums.org.uk/folkston.html •
Cur.: *J. Adamson* •
Historical Museum – 1868
Natural hist, geology, local archaeology ... 44237

Metropole Gallery, The Leas, Folkestone CT20 2LS • T: +44 1303 244706 • F: 851353 • info@metropole.org.uk • www.metropole.org.uk •
Dir.: *Nick Ewbank* •
Public Gallery ... 44238

Fordingbridge

Rockbourne Roman Villa, Rockbourne, Fordingbridge SP6 3PG • T: +44 1725 518541 • musuijh@hants.gov.uk • www.hants.gov.uk/museum/rockbourne •
Cur.: *Jenny Stevens* •
Archaeology Museum
Mosaics and hypocaust exposed ... 44239

Fordyce

Fordyce Joiner's Visitor Centre, West Church St, Fordyce AB45 2SL • T: +44 1261 843322 • F: 623558 • heritage@aberdeenshire.gov.uk • www.aberdeenshire.gov.uk/heritage •
Special Museum
Rural joiner's workshop, woodworking tools ... 44240

Forest

German Occupation Museum, Les Houards, Forest GY8 0BG • T: +44 1481 238205 •
Dir.: *R.L. Heaume* •
Military Museum / Historical Museum – 1966
The museum tells the story of the German occupation with audio-visual tableaux including an occupation street, kitchen and bunker rooms etc, also restored German fortifications can be visited ... 44241

Forfar

Museum and Art Gallery, Meffan Institute, 20 West High St, Forfar DD8 1BB • T: +44 1307 464123, 467017 • F: 468451 • the.meffan@angus.gov.uk • www.angus.gov.uk/history •
Dir.: *Margaret H. King* •
Local Museum / Fine Arts Museum – 1898
Local hist, archaeology, art esp. pictish stones ... 44242

Forncett Saint Mary

Forncett Industrial Steam Museum, Low Rd, Forncett Saint Mary NR16 1JJ • T: +44 1508 488277 • www.phoenixbooks.org •
Dir.: *Dr. R. Francis* •
Science&Tech Museum
147 hp Cross Compound engine which was used to open Tower Bridge, 85t triple expansion water pump 600HP, 2 beam engines 44243

Forres

Brodie Castle, Brodie, Forres IV36 2TE • T: +44 1309 641371 • F: 641600 • brodiecastle@nts.org.uk •
Dir.: *Fiona Dingwall* •
Decorative Arts Museum / Fine Arts Museum – 1979
French furniture, English, Continental and Chinese porcelain and an important coll of 17th-c Dutch and 18th-c English paintings 44244

Falconer Museum, Tolbooth St, Forres IV36 1PH • T: +44 1309 673701 • F: 673701 • museums@moray.gov.uk • www.moray.org/museums •
Local Museum – 1871
Specialized herbaria, local history, natural history, Peter Anson, Hugh Falconer archive – library, archive 44245

Nelson Tower, Grant Park, Forres IV36 1PH • T: +44 1309 673701 • F: 675863 • museums@moray.gov.uk • www.moray.gov.uk/museums •
Fine Arts Museum 44246

Fort William

Glenfinnan Station Museum, Glenfinnan Railway Station, Fort William PH37 4LT • T: +44 1397 722295 • F: 722291 •
Science&Tech Museum 44247

Treasures of the Earth, Carpach by Fort William, Fort William PH33 7JL • T: +44 1397 772283 • F: 772133 •
Natural History Museum
Coll of crystals, gemstones and fossils, simulation of cave, cavern and mining scene 44248

West Highland Museum, Cameron Sq, Fort William PH33 6AJ • T: +44 1397 702169 • F: 701927 • info@westhighlandmuseum.org.uk • www.westhighlandmuseum.org.uk •
Dir.: *Fiona C. Marwick* •
Local Museum – 1922
Many aspects of the district and its hist, including geology, wildlife, tartans, maps and a coll of items connected with the 1745 Jacobite rising 44249

Foxton

Foxton Canal Museum, Middle Lock, Gumley Rd, Foxton LE16 7RA • T: +44 1858 466185 • info@fipt.org.uk • www.fipt.org.uk •
Cur.: *Mike Beech* •
Historical Museum
Canal items, 1793-1950, boat life and locks 44250

Framlingham

Lanman Museum, Castle, Framlingham IP13 9BP • T: +44 1728 724189 • F: 724324 •
Cur.: *A. Pickup* •
Local Museum
Objects of historical, educational or artistic interest connected with the town of Framlingham and the surrounding villages 44251

Parham Airfield Museum and Museum of the Resistance Organisation Museum, Parham Airfield, Framlingham IP13 9AF • T: +44 1473 621373 • 390bgmam@suffolkonline.net • www.parhamairfieldmuseum.co.uk •
Military Museum
Engines and other components of crashed U.S. and Allied aircraft, and many photographs and mementoes of the wartime life of both airmen and civilians, reconstructions of a radio hut and of a wartime office 44252

Fraserburgh

Fraserburgh Museum, 37 Gralton Pl, Fraserburgh AB51 7LD •
Local Museum 44253

Museum of Scottish Lighthouses, Kinnaird Head, Fraserburgh AB43 9DU • T: +44 1346 511022 • info@lighthousemuseum.org.uk • www.lighthousemuseum.org.uk •
Dir.: *Virginia Mayes-Wright* •
Special Museum / Historic Site / Science&Tech Museum
Coll. of artefacts, images and archives relating to lighthouses in Scotland and the work of the Northern Lighthouse Board; Robert Stevenson (engineer, lighthouse builder) – library, archive 44254

Freshwater Bay

Dimbola Lodge Museum, Terrace Ln, Freshwater Bay PO40 9QE • T: +44 1983 756814 • F: 755578 • administrator@dimbola.co.uk • www.dimbola.co.uk •
Cur.: *Dr. Brian Hinton* •
Decorative Arts Museum / Fine Arts Museum
Original Cameron images, Isle ogf Wight Pop festival, photography, applied and visual arts, Julia Margaret Cameron coll – library 44255

Frodsham

Castle Park Arts Centre, Fountain Ln, Frodsham WA6 6SE • T: +44 1928 735832 • www.castle-park-arts.co.uk •
Fine Arts Museum 44256

Frome

Frome Museum, 1 North Parade, Frome BA11 1AT • T: +44 1373 454611 •
Local Museum
History of local industries, including cloth-making, metal-casting, printing – library 44257

Fyvie

Fyvie Castle, Fyvie AB53 8JS • T: +44 1651 891266 • F: 891107 •
Local Museum
Coll of portraits including works by Batoni, Raeburn, Gainsborough, arms and armour, 16th c tapestries 44258

Gainsborough

Gainsborough Old Hall, Parnell St, Gainsborough DN21 2NB • T: +44 1427 612669 • F: 612779 • gainsborougholdhall@lincolnshire.gov.uk • www.lincolnshire.gov.uk •
Local Museum – 1974
In building dating from 1460, medieval room settings & kitchens, 17th c furniture 44259

Gairloch

Gairloch Heritage Museum, Achtercairn, Gairloch IV21 2BP • T: +44 1445 712287 • info@gairlochheritagemuseum.org.uk • www.gairlochheritagemuseum.org.uk •
Cur.: *Nicola Tayler* •
Local Museum – 1977
Archaeology, wildlife, agriculture, dairying, wood processing and the domestic arts, family hist section – archive, library 44260

Galashiels

Old Gala House, Scott Crescent, Galashiels TD1 3JS • T: +44 1750 720096 • F: 723282 • ssinclair@scotborckrs.gov.uk •
Cust.: *Keith Watt* •
Historical Museum
Home of Lairds of Gala, painted ceiling (1635) and wall (1988) 44261

Garlogie

Garlogie Mill Power House Museum, Garlogie AB32 6RX • T: +44 1771 622906 • F: 622884 • heritage@aberdeenshire.gov.uk • www.aberdeenshire.gov.uk/heritage •
Dir.: *Dr. David M. Bertie* •
Science&Tech Museum
19th c beam engine, 20th c turbine 44262

Gateshead

Bowes Railway Heritage Museum, Springwell Village, Gateshead NE9 7QJ • T: +44 191 4161847, 4193349 • alan.milburn2@virgin.net • www.bowesrailway.co.uk •
Man.: *Alan Milburn* •
Science&Tech Museum – 1976
Coll of historic colliery rolling stock, rope haulage equipment, coll of historic colliery buildings dating from 19th c 44263

The Gallery, Prince Consort Rd, Gateshead Central Library, Gateshead NE8 4LN • T: +44 191 4338400 • F: 4777454 • libraries@gateshead.gov.uk • www.gateshead.gov.uk/libraries •
Public Gallery – 1885 44264

Shipley Art Gallery, Prince Consort Rd, Gateshead NE8 4JB • T: +44 191 4771495 • F: 4787917 • shipley@tyne-wear-museums.org.uk • www.twmuseums.org.uk •
Fine Arts Museum / Decorative Arts Museum / Local Museum – 1917
17th-20th c British painting, also French, Flemish, German, Dutch and Italian painting, Schufelein Altar, contemporary craft, local glass, local industrial hist 44265

Gaydon

Heritage Motor Centre, Banbury Rd, Gaydon CV35 0BJ • T: +44 1926 641188 • F: 641555 • enquiries@heritage-motor-centre.co.uk • www.heritage-motor-centre.co.uk •
Dir.: *Julie Tew* •
Science&Tech Museum – 1993
Coll of over 200 historic cars, incl prototypes, concept vehicles, record breakers – archive, photographic library 44266

Gerston

Castletown Museum, Mingulay, Gerston KW12 6XQ •
Local Museum 44267

Gillingham, Dorset

Gillingham Museum, Chantry Fields, Gillingham, Dorset SP8 4UA • T: +44 1747 854018 • gillinghammuseum@waitrose.com • www.brwebsites.com/gillingham.museum/ •
Cur.: *Bill Moore* •
Local Museum – 1958
Articles manufactured, used or found in Gillingham and documents from the 13th century to the present 44268

Gillingham, Kent

Royal Engineers Museum, Prince Arthur Rd, Gillingham, Kent ME4 4UG • T: +44 1634 822839 • F: 822371 • mail@re-museum.co.uk • www.royalengineers.org.uk •
Dir.: *Richard Dunn* •
Ethnology Museum
Lives and works, worldwide, of Britain's soldier-engineers from 1066, scientific and engineering equipment, medals, weapons and uniforms, ethnography, decorative arts, esp. Chinese and Sudanese material 44269

Girvan

McKechnie Institute, 40a Dalrymple St, Girvan KA26 9AE • T: +44 1465 713643 • F: 713643 •
Local Museum / Fine Arts Museum
Local hist, community arts centre in a octagonal tower 44270

Penkill Castle, Girvan KA26 9TQ • T: +44 1465 871219 • F: 871215 • psbfd2@aol.com •
Dir.: *P.S.B.F. Dromgoole* •
Historic Site – 1978
Pre Raphaelite 44271

Glamis

Angus Folk Museum, Kirkwynd Cottages, Glamis DD8 1RT • T: +44 1307 840288 • F: 840233 •
Man.: *Mary Brownlow* •
Folklore Museum – 1957
Rural life in Angus last 200 yrs, agricultural implements, domestic artefacts 44272

Glamis Castle, Glamis DD8 1RJ • T: +44 1307 840393 • F: 840733 • enquiries@glamis-castle.co.uk • www.glamis-castle.co.uk •
Admin.: *David Adams* •
Historical Museum
Remodelled castle (17th c), historic pictures, furniture, porcelain and tapestries 44273

Glandford

Glandford Shell Museum, Glandford NR25 7JR • T: +44 1263 740081 • www.shellmuseum.org.uk •
Cur.: *Wendy Gill* •
Special Museum – 1915
Sir Alfred Jodrell shell collection, jewelry, pottery, tapestry 44274

Glasgow

Burrell Collection, Pollok Country Park, 2060 Pollokshaws Rd, Glasgow G43 1AT • T: +44 141 2872550 • F: 2872597 • museum@cls.glasgow.gov.uk • www.glasgowmuseums.com •
Man.: *Muriel King* •
Fine Arts Museum / Decorative Arts Museum / Museum of Classical Antiquities – 1983
Paintings, watercolors by Glasgow artists, ceramics, pottery, tapestries, antiquities, Islamic art, stained glass, sculpture, arms and armour, Oriental art, furniture, textiles 44275

CCA-Centre for Contemporary Arts, 350 Sauchiehall St, Glasgow G2 3JD • T: +44 141 3524900 • F: 3323226 • gen@cca-glasgow.com • www.cca-glasgow.com •
Dir.: *Francis McKee* • Sc. Staff: *Kerri Moogan* •
Public Gallery 44276

Clydebuilt, Scottish Maritime Museum, King's Inch Rd, Glasgow G51 4BN • T: +44 141 8861013 • F: 8861015 • clydebuilt@scotmaritime.org.uk • www.scottishmaritimemuseum.org •
Dir.: *David Thomson* •
Historical Museum
Story of the river Clyde, ships 44277

Collins Gallery, c/o University of Strathclyde, 22 Richmond St, Glasgow G1 1XQ • T: +44 141 5482558 • F: 5524053 • collinsgallery@strath.ac.uk • www.strath.ac.uk/collections •
Cur.: *Laura Hamilton* •
Science&Tech Museum / Public Gallery / University Museum – 1973
Anderson coll of scientific instruments, contemporary Scottish art, prints 44278

Eastwood House, Eastwood Park, Giffnoclk, Glasgow G46 6UG • T: +44 141 6381101 •
Public Gallery 44279

Fossil Grove, Victoria Park, Glasgow G14 1BN • T: +44 141 2872000 • museums@cls.glasgow.gov.uk • www.glasgowmuseums.com •
Man.: *Caroline Barr* •
Local Museum
330 million year old fossilised tree stumps 44280

Gallery of Modern Art, Royal Exchange Sq, Glasgow G1 3AH • T: +44 141 2291996 • F: 2045316 • museums@cls.glasgow.gov.uk • www.glasgowmuseums.com •
Man.: *Victoria Hollows* •
Fine Arts Museum – 1996
Paintings, sculptures and installations from around the world, artists represented incl T. Paterson, C. Borland and R. Sinclair 44281

Glasgow Print Studio, 22-25 King St, Glasgow G1 5QP • T: +44 141 5520704 • F: 5522919 • gallery@gpsart.co.uk • www.gpsart.co.uk •
Dir.: *John Mackechnie* •
Public Gallery
Work of Paolozzi, Bellany, Blackadder, Howson, Wiszniewski – printmaking workshop, Three Gallery 44282

Glasgow School of Art - Mackintosh Collection, 167 Renfrew St, Glasgow G3 6RQ • T: +44 141 3534500 • F: 3534746 • info@gsa.ac.uk • www.gsa.ac.uk •
Dir.: *Prof. Seona Reid* • Cur.: *Peter Trowles* •
University Museum – 1845
Coll of furniture, watercolours and architectural designs and drawings by Charles Rennie Mackintosh and other GSA-artists 44283

Hunterian Art Gallery, c/o University of Glasgow, 82 Hillhead St, Glasgow G12 8QQ • T: +44 141 3305431 • F: 3303618 • hunter@museum.gla.ac.uk • www.hunterian.gla.ac.uk •
Dep. Dir.: *Mungo Campbell* • Cur.: *Pamela B. Robertson* (Charles Rennie Mackintosh and Scottish Art) • *Peter Black* •
Fine Arts Museum / University Museum – 1980
Paintings, drawings, prints, furniture, sculpture, works by Charles Rennie Mackintosh, James McNeill Whistler 44284

Hunterian Museum, c/o University of Glasgow, University Av, Glasgow G12 8QQ • T: +44 141 3304221 • F: 3303617 • hunter@museum.gla.ac.uk • www.hunterian.gla.ac.uk •
Dir.: *Evelyn Silber* •
Ethnology Museum / Natural History Museum / Archaeology Museum / University Museum – 1807
Coll of the celebrated physician and anatomist Dr. William Hunter, geology, rocks, minerals and fossils, incl 'Bearsden shark', archaeology - early civilizations in Scotland, ethnography, Greek, Roman and Scottish coins, scientific instruments 44285

Hunterian Museum, Zoology Section, c/o University of Glasgow, Graham Kerr Bldg, Glasgow G12 8QQ • T: +44 141 3304772 • F: 3305971 • mreilly@museum.gla.ac.uk • www.hunterian.gla.ac.uk •
Cur.: *Margaret T. Reilly* •
Natural History Museum
Animal kingdom from microscopic creatures to elephants and whales 44286

Kelvingrove Art Gallery and Museum, Kelvin Park, Argyle St, Glasgow G3 8AG • T: +44 141 2769599 • F: 2769540 • museums@cls.glasgow.gov.uk • www.glasgowmuseums.com •
Dir.: *Mark O'Neill* •
Fine Arts Museum / Local Museum – 1902
British and European paintings including pictures by Rembrandt and Giorgione, 19th c French Impressionists, Scottish painting, sculpture, costume, silver, pottery and porcelain, arms and armour, wildlife and story of man in Scotland 44287

McLellan Galleries, 270 Sauchiehall St, Glasgow G2 3EH • T: +44 141 5654100 • F: 5654111 • museums@cls.glasgow.gov.uk • www.glasgowmuseums.com •
Dir.: *Bridget McConnell* •
Public Gallery 44288

Mitchell Library Exhibition, North St, Glasgow G3 7DN • T: +44 141 2872999 • F: 2872815 • lil@cls.glasgow.gov.uk •
Public Gallery
Paintings exhibits 44289

Museum of Anatomy, c/o IBLS, University of Glasgow, Laboratory of Human Anatomy, Glasgow G12 8QQ • T: +44 141 3305871 • F: 3304299 • a.payne@bio.gla.ac.uk • www.hunterian.gla.ac.uk •
Cur.: *Prof. Anthony Payne* •
Natural History Museum 44290

Museum of Piping, National Piping Centre, 30-34 McPhater St, Cowcaddens, Glasgow G4 0HW • T: +44 141 3530220 • F: 3531570 • reception@thepipingcentre.co.uk • www.thepipingcentre.co.uk •
Dir.: *Roddy McLeod* •
Music Museum
Bagpipes, piping artefacts 44291

Museum of Transport, 1 Bunhouse Rd, Glasgow G3 8DP • T: +44 141 2872720 • F: 2872692 • museums@cls.glasgow.gov.uk • www.glasgowmuseums.com •
Man.: *Lawrence Fitzgerald* •
Science&Tech Museum – 1964
Railroad locomotives, bicycles, motorcycles, streetcars, automobiles, mobile museums, fire engines, baby carriages, steam road vehicles, horse-drawn carriages 44292

Pearce Institute, 840 Govan Rd, Glasgow G51 3UT • T: +44 141 4451941 •
Public Gallery 44293

People's Palace Museum, Glasgow Green, Glasgow G40 1AT • T: +44 141 2712951 • F: 2712960 • museums@cls.glasgow.gov.uk • www.glasgowmuseums.com •
Cur.: *Fiona Hayes* •
Historical Museum – 1898
Glasgow's social, economic and political hist since 1750, exhibits on popular culture, public art 44294

Pollok House, 2060 Pollokshaws Rd, Glasgow G43 1AT • T: +44 141 6166410 • F: 6166521 • pollokhouse@nts.org.uk • www.nts.org.uk •
Dir.: *Robert S. Ferguson* •
Fine Arts Museum / Decorative Arts Museum – 1968
Stirling Maxwell coll incl Spanish school of painting El Greco, Goya, Murillo, British school, principally William Blake, decorative arts, furniture, pottery, porcelain, silver, bronzes 44295

Provan Hall House, Auchinlea Park, Auchinlea Rd, Glasgow G34 • T: +44 141 7711538 •
Decorative Arts Museum – 1983/2000 44296

Provand's Lordship, 3 Castle St, Glasgow G4 0RB • T: +44 141 5528819 • F: 5524744 • museums@cls.glasgow.gov.uk • www.glasgowmuseums.com •
Dir.: *Harry Dunlop* •
Historical Museum
Medieval building of 1471 44297

Royal Glasgow Institute Kelly Gallery, 118 Douglas St, Glasgow G2 4ET • T: +44 141 2486386 • F: 2210417 • rgi@robbferguson.co.uk • www.rgiscotland.co.uk •
Public Gallery 44298

Royal Highland Fusiliers Regimental Museum, 518 Sauchiehall St, Glasgow G2 3LW • T: +44 141 3320961 • F: 3531493 • assregseg@rhf.org.uk • www.rhf.org.uk •
Military Museum – 1969
History of the regiment since 1678, medals, uniforms, regimental silver, militaria, Role of Honour 1914-1918 and 1939-1945 – library 44299

Ruchill Centre for Contemporary Arts, 158 Brassey St, Glasgow G20 9HL • T: +44 78 03894508 • info@rcca.co.uk • www.rcca.co.uk •
Public Gallery – 1999 44300

Saint Mungo Museum of Religious Life and Art, 2 Castle St, Glasgow G4 0RH • T: +44 141 5532557 • F: 5524744 • museums@cls.glasgow.gov.uk • www.glasgowmuseums.com •
Man.: *Harry Dunlop* •
Religious Arts Museum – 1993
Objects associated with different religious faiths – Japanese Zen garden 44301

Scotland Street School Museum, 225 Scotland St, Glasgow G5 8QB • T: +44 141 2870500 • F: 2870515 • museums@cls.glasgow.gov.uk • www.glasgowmuseums.com •
Man.: *Caroline Barr* •
Historical Museum – 1990
Hist of education in Scotland from 1872, several historical classrooms, Edwardian cookery room, school hist, architectural hist (designed by Charles Rennie Mackintosh) 44302

Scottish Football Museum, c/o Museum of Transport, Kelvin Hall, 1 Bunhouse Rd, Glasgow G3 9DP • T: +44 141 2872697 • F: 2872692 • www.scottishfa.co.uk •
Special Museum
Football memorabilia coll from SFA since 1873 .. 44303

Scottish Jewish Museum, Garnethill Synagogue, 127 Hill St, Glasgow G3 6UB • T: +44 141 9562973 •
Historical Museum
archives 44304

Tenement House, 145 Buccleuch St, Glasgow G3 6QN • T: +44 141 3330183 • F: 3329368 • tenementhouse@nts.org.uk • www.nts.org.uk •
Dir.: *Lorna Hepburn* •
Local Museum
Typical lower middle class Glasgow tenement flat of the turn of the century with original furnishings and fittings. 44305

Glastonbury

Glastonbury Abbey Museum, Abbey Gatehouse, Magdalene St, Glastonbury BA6 9EL • T: +44 1458 832267 • F: 832267 • info@glastonburyabbey.com • www.glastonburyabbey.com •
Religious Arts Museum – 1908
Coll of tiles, carved stonework and miscellaneous artefacts from the site of the ruined Abbey, together with a model showing how the Abbey may have looked in 1539, before the Dissolution 44306

Glastonbury Lake Village Museum, The Tribunal, 9 High St, Glastonbury BA6 9DP • T: +44 1458 832954 • F: 832949 • glastonbury.tic@ukonline.co.uk • www.glastonburytic.co.uk •
Local Museum
Local history 44307

Somerset Rural Life Museum, Abbey Farm, Chilkwell St, Glastonbury BA6 8DB • T: +44 1458 831197 • F: 834684 • mcgryspeerdt@somerset.gov.uk • somerset.gov.uk/museums •
Dir.: *David A. Walker* •
Local Museum – 1976
Agricultural machinery, tools, waggons and carts, baskets and basket making tools, cider making equipment, chedder cheese making presses and equipment, kitchen and general domestic colls, Somerset furniture, contrymen's smocks 44308

Glenavy

Ballance House, 118a Lisburn Rd, Glenavy BT29 4NY • T: +44 28 92648492 • F: 92648098 • ballancenz@aol.com • www.ballance.utvinternet.com •
Historical Museum / Ethnology Museum
John Ballance (New Zealand premier 1891-93), ethnographical material, fine art, journals, books, furniture 44309

Glencoe

Glencoe and North Lorn Folk Museum, Glencoe PH49 4HS • T: +44 1855 811664 •
Pres.: *Rachel Grant* • Vice Pres.: *Arthur Smith* •
Local Museum
MacDonald and Jacobite relicts, domestic bygones, weapons, embroidery, costumes, agricultural items, Ballachulish slate industry, ethnology, military ... 44310

Glenesk

Glenesk Folk Museum, The Retreat, Glenesk DD9 7YT • T: +44 1356 648070 • F: 648084 • retreat@angusglens.co.uk • www.angusglens.co.uk •
Cur.: *Muriel McIntosh* •
Local Museum
Local hist, costume coll 44311

Glenluce

Glenluce Motor Museum, Glenluce DG8 0NY • T: +44 1581 300534 • F: 300258 •
Science&Tech Museum
Vintage and classic cars, motor cycles memorabilia and old type garage 44312

Glenrothes

Corridor Gallery, Viewfield Rd, Glenrothes KY6 2RB • T: +44 1592 415700 •
Public Gallery 44313

Glossop

Glossop Heritage Centre, Bank House, Henry St, Glossop SK13 8BW • T: +44 1457 869176 • www.glossopheritage.co.uk •
Man.: *Tony Murphy* •
Local Museum
Local newspapers, local hist 44314

Gloucester

City Museum and Art Gallery, Brunswick Rd, Gloucester GL1 1HP • T: +44 1452 396131 • F: 410898 • city.museum@gloucester.gov.uk • www.mylife.gloucester.gov.uk •
Dir.: *Christopher Morris* • Cur.: *Louise Allen* • Sc. Staff: *Sue Byrne* • *David Rice* •
Archaeology Museum / Fine Arts Museum / Decorative Arts Museum / Natural History Museum – 1860
Archaeology, geology, botany, natural hist, English pottery, glass and silver objects, numismatics, art 44315

Gloucester Folk Museum, 99-103 Westgate St, Gloucester GL1 2PG • T: +44 1452 396868/69 • F: 330495 • folk.museum@gloucester.gov.uk • www.livinggloucester.co.uk/histories/museums/folk_museum •
Head: *C.I. Morris* • Cur.: *Nigel R. Cox* (Social history) •
Historical Museum – 1935
Local hist since 1500, crafts, trades, farming tools, fishing, shops and workshops, toys and games, laundry and household equipment, steam engines, model aircraft, pewter, glass, pottery 44316

National Waterways Museum, Llanthony Warehouse, Gloucester Docks, Gloucester GL1 2EH • T: +44 1452 318200 • F: 318202 • bookingsnwm@thewaterwaystrust.org • www.nwm.org.uk •
Science&Tech Museum / Historical Museum – 1988
Inland waterways and its transport – archive 44317

Nature in Art, Wallsworth Hall, Tewkesbury Rd, Gloucester GL2 9PA • T: +44 1452 731422, 4500233 • F: 730937 • ninart@globalnet.co.uk • www.nature-in-art.org.uk •
Dir.: *Dr. Simon H. Trapnell* •
Fine Arts Museum
Fine, decorative and applied art inspired by nature, any period, and country in all media 44318

Soldiers of Gloucestershire Museum, Custom House, Gloucester Docks, Gloucester GL1 2HE • T: +44 1452 522682 • F: 311116 • regimental_secretary@ngbw.army.mod.uk • www.gloster.org •
Cur.: *George C. Streatfield* •
Military Museum – 1989
Uniforms, medals, badges, paintings, arts and crafts – archives 44319

Glyn Ceiriog

Ceiriog Memorial Institute, High St, Glyn Ceiriog LL20 7EH • T: +44 1691 718910, 718383 •
Special Museum 44320

Godalming

Charterhouse School Museum, Charterhouse, Godalming GU7 2DX • T: +44 1483 421006, 291515 • F: 291594 • dsh@charterhouse.org.uk •
Local Museum
Local hist, ethnography, archaeology, natural hist 44321

Godalming Museum, 109a High St, Godalming GU7 1AQ • T: +44 1483 426510 • F: 523495 • museum@godalming.ndo.co.uk • www.godalming-museum.org.uk •
Cur.: *Alison Pattison* •
Local Museum – 1920
Local hist and industries, Gertrude Jekyll and Edwin Lutyens coll – local studies library 44322

Golcar

Colne Valley Museum, Cliffe Ash, Golcar HD7 4PY • T: +44 1484 659762 •
Local Museum
Restored 19th c weavers' cottages. Demonstration of hand-loom weaving, working spinning jenny, spinning on saxony wheel and clog-making in gas-lit clogger's workshop 44323

Goldhanger

Maldon and District Agricultural and Domestic Museum, 47 Church St, Goldhanger CM9 8AR • T: +44 1621 788647 • F: 788647 •
Agriculture Museum / Folklore Museum
Coll of vintage farm machinery and tools, domestic bygones, printing machinery, photographs 44324

Golspie

Dunrobin Castle Museum, Dunrobin Castle, Golspie KW10 6SF • T: +44 1408 633177 • F: 634081 • info@dunrobincastle.net • www.highlandescape.com •
Dir.: *Keith Jones* •
Decorative Arts Museum
Victorian castle, formerly a summerhouse, collections of game trophies and Pictish stones, local history, ornithology, geology, ethnography 44325

Gomersal

Red House Museum, Oxford Rd, Gomersal BD19 4JP • T: +44 1274 335100 • F: 335105 • www.kirkleesmc.gov.uk/community/museums •
Dir.: *Helga Hughes* •
Historical Museum
Regency house with Brontë connections 44326

Gomshall

Gomshall Gallery, Station Rd, Gomshall GU5 9LB • T: +44 1483 203795 • F: 203282 •
Public Gallery 44327

Goodwood

Cass Sculpture Foundation, Sculpture Estate, Goodwood PO18 0QP • T: +44 1243 538449 • F: 531853 • info@sculpture.org.uk • www.sculpture.org.uk •
Founder: *Wilfred Cass* • Dir.: *Kate Simms* •
Public Gallery
Contemporary mid-scale sculptures, drawings and prints 44328

Goole

Goole Museum and Art Gallery, Carlisle St, Goole DN14 5BG, mail addr: Eryc Offices, Church St, Goole DN14 5BG • T: +44 1482 392777 • F: 392782 • janet.tierney@eastriding.gov.uk •
Janet Tierney •
Local Museum / Fine Arts Museum – 1968
Local hist, shipping, paintings 44329

Yorkshire Waterways Museum, Sobriety Project, Dutch River Side, Goole DN14 5TB • T: +44 1405 768730 • F: 769868 • waterwaysmuseum@btinternet.com • www.waterwaysmuseumandadventurecentre.co.uk •
Dir.: *Bob Watson* •
Historical Museum
Photos, docs, industrial and waterways heritage, Aire and Calder navigation, Goole port, Yorkshire waterways 44330

Gordon

Mellerstain House, Mellerstain, Gordon TD3 6LG • T: +44 1573 410225 • F: 410636 • mellerstain.house@virgin.net •
Dir.: *Earl of Haddington* •
Local Museum – 1952
Robert Adam architecture, old master paintings, antique dolls – Library 44331

Gosport

Explosion! The Museum of Naval Firepower, Priddy's Hard, Gosport PO12 4LE • T: +44 23 92505600 • F: 92505605 • info@explosion.org.uk • www.explosion.org.uk •
Cur.: *Jo Lawler* •
Science&Tech Museum
Small arms, cannon and large guns, shells and munitions, mines, torpedoes and modern missiles 44332

Fort Brockhurst, Gunner's Wy, Elson, Gosport PO12 4DS • T: +44 1424 775705 • www.english-heritage.org.uk •
Cur.: *Pam Braddock* •
Military Museum
Artillery coll 44333

Gosport Museum, Walpole Rd, Gosport PO12 1NS • T: +44 23 92588035 • F: 92501951 • oonagh.palmer@hants.gov.uk • www.discoverycentres.co.uk/gosport •
Cur.: *Oonagh Palmer* •
Local Museum – 1974
Local hist, geology 44334

Hovercraft Museum, Argus Gate, Chark Ln, Gosport PO13 9NY • T: +44 23 92552090 • enquiries@hovercraft-museum.org • www.hovercraft-museum.org •
Dir.: *Warwick Jacobs* •
Historical Museum / Science&Tech Museum – 1986
66 hovercraft and history – photo archives 44335

Royal Navy Submarine Museum, Haslar Jetty Rd, Gosport PO12 2AS • T: +44 2392 510354 • F: 511349 • rnsubs@rnsubmus.co.uk • www.rnsubmus.co.uk •
Dir.: *J.J. Tall* • Cur.: *Bob Mealings* •
Military Museum – 1963
Models of all British submarines from 1879 to the present day, H.M. Submarine Alliance, Holland I – library, photographic archive, diving museum 44336

SEARCH-Centre, 50 Clarence Rd, Gosport PO12 1BU • T: +44 23 92501957 • F: 92501921 • janet.wildman@hants.gov.uk • www.hants.gov.uk/museum/search •
Man.: *Janet Wildman* •
Historical Museum / Natural History Museum
History and natural hist 44337

Gott

Tingwall Agricultural Museum, 2 Veensgarth, Gott ZE2 9SB • T: +44 1595 840344 •
Agriculture Museum
18th c farming 44338

Goudhurst

Finchcocks, Living Museum of Music, Goudhurst TN17 1HH • T: +44 1580 211702 • F: 211007 • katrina@finchcocks.co.uk • www.finchcocks.co.uk •
Dir.: *Richard Burnett* •
Music Museum – 1971
Coll of historical keyboard instruments, chamber organs, and a wide range of early pianos, musical furniture 44339

Gower

Gower Heritage Centre, Y Felin Ddwr, Parkmill, Gower SA3 2EH • T: +44 1792 371206 • F: 371471 • info@gowerheritagecentre.co.uk • www.gowerheritagecentre.co.uk •
Local Museum / Agriculture Museum
Milling and wheelwrights equipment, agriculture and farming 44340

Grange-over-Sands

Lakeland Motor Museum, Holker Hall, Cark-in-Cartmel, Grange-over-Sands LA11 7PL • T: +44 15395 58509 • F: 58509 • info@lakelandmotormuseum.co.uk • www.lakelandmotormuseum.co.uk •
Dir.: *D.J. Sidebottom* •
Science&Tech Museum – 1978
Classic and vintage cars, motor-cycles, cycles, tractors, engines and automobilia, 1920's garage re-creation, Campbell Bluebird exhibition, ... 44341

Grangemouth

Grangemouth Museum, Bo'ness Rd, Grangemouth FK3 8AG • T: +44 1324 504699 • callendar.house@falkirk.gov.uk • www.falkirkmuseums.org •
Staff: *Janet Paxton* •
Local Museum
Scotlands earliest planned industrial town, petroleum and chemical industries 44342

Museum Workshop and Stores, 7-11 Abbotsinch Rd, Abbotsinch Industrial Estate, Grangemouth FK3 9UX • T: +44 1324 504689 • niamh.conlon@falkirk.gov.uk •
Local Museum
Extensive industrial collection, machines, hand tools, transport items 44343

Grantham

Belton House, Grantham NG32 2LS • T: +44 1476 566116 • F: 579071 • belton@nationaltrust.org.uk •
Man.: *Duncan Bowen* •
Fine Arts Museum / Decorative Arts Museum
Plasterwork ceilings by Edward Goudge, wood carvings of the Grinling Gibbons School, portraits, furniture, tapestries, oriental porcelain 44344

Grantham Museum, Saint Peter's Hill, Grantham NG31 6PY • T: +44 1476 568783 • F: 592457 • grantham.museum@lincolnshire.gov.uk • www.lincolnshire.gov.uk/granthammuseum •
Keeper: *Anna Brearley* •
Local Museum – 1926
Local history from prehistoric to modern times, two special exhibitions are devoted to the town's famous personalities - Isaac Newton and Margaret Thatcher 44345

The Queen's Royal Lancers Regimental Museum, Belvoir Castle, Grantham NG32 1PD • T: +44 115 9573295, 870909 • F: 9573195 • hhqandmuseumqrl@ukonline.co.uk • www.qrl.uk.com •

U

Cur.: *J.Mick Holtby* • Asst. Cur.: *Terence Brighton* •
Military Museum – 1963
Arms, uniforms, paintings, silver, medals – archives 44346

Grantown-on-Spey

Grantown Museum, Burnfield House, Burnfield Av, Grantown-on-Spey PH26 3HH • T: +44 1479 872478 • F: 872478 • gosmuseum@btconnect.com • www.grantownmuseum.com •
Dir.: *Angus Miller* •
Local Museum 44347

Grasmere

Dove Cottage and the Wordsworth Museum, Centre for British Romanticism, Town End, Grasmere LA22 9SH • T: +44 15394 35544 • F: 35748 • enquiries@wordsworth.org.uk • www.wordsworth.org.uk •
Dir.: *Dr. Robert Woof* •
Special Museum – 1890/1935
Former home of William Wordsworth (1799-1808) with memorabilia of the poet, books and manuscripts of major Romantic figures, lake district fine art and topography 44348

Grassington

Upper Wharfedale Folk Museum, 6 The Square, Grassington BD23 5AQ • T: +44 1756 752801 •
Folklore Museum
History of the Dales, with a special emphasis on farming, lead mining and domestic life 44349

Gravesend, Kent

Gravesham Museum, High St, Gravesend, Kent DA12 0BQ • T: +44 1474 365600 •
Local Museum
Local archaeology and hist, paintings 44350

Grays

Thurrock Museum, Thameside Complex, Orsett Rd, Grays RM17 5DX • T: +44 1375 382555 • F: 392666 • jcatton@thurrock.gov.uk • www.thurrock.gov.uk/museum •
Dir.: *Jonathan Catton* • Cur.: *Terry J. Carney* •
Local Museum – 1956
Local hist and archaeology, maritime – Borough archives 44351

Great Ayton

Captain Cook Schoolroom Museum, 101 High St, Great Ayton TS9 6NB • T: +44 1642 724296 • www.captaincookschoolroommuseum.co.uk •
Historical Museum – 1928 44352

Great Bardfield

Great Bardfield Cage, Bridge St, Great Bardfield CM7 4UA • T: +44 1371 810516 • stella_herbert@hotmail.com •
Dir.: *Peter Cott* •
Local Museum
19th c village 44353

Great Bardfield Cottage Museum, Dunmow Rd, Great Bardfield CM7 4UA • T: +44 1371 810516 • stella_herbert@hotmail.com •
Chm.: *Charles Holden* •
Local Museum
19th and 20th c domestic and agricultural artefacts, rural crafts 44354

Great Grimsby

National Fishing Heritage Centre, Alexandra Dock, Great Grimsby DN31 1UZ • T: +44 1472 323345 • F: 323555 • nfhc@nelincs.gov.uk • www.nelincs.gov.uk/museums •
Man.: *Tracy Welsh* •
Special Museum
Fishing artefacts, ship models, marine paintings, folk life 44355

Great Malvern

Malvern Museum, Abbey Gateway, Abbey Rd, Great Malvern WR14 3ES • T: +44 1684 567811 • cora@malvernspa.com •
Cur.: *M.J. Hebden* • *F. Renger* •
Local Museum – 1979
Geology, water-cure, radar 44356

Great Yarmouth

Elizabethan House, 4 South Quay, Great Yarmouth NR30 2QH • T: +44 1493 855746 • yarmouth.museums@norfolk.gov.uk • www.museums.norfolk.gov.uk •
Dir.: *Dr. Sheila Watson* •
Historical Museum – 1975
Tudor merchant's house of 1596 with displays of social hist and domestic life, Victorian kitchen and scullery 44357

Great Yarmouth Museums, Tolhouse St, Great Yarmouth NR30 2SH • T: +44 1493 745526 • F: 745459 • yarmouth.museums@norfolk.gov.uk • www.museums.norfolk.gov.uk •
Public Gallery – 1961 44358

Old Merchant's House and Row 111 Houses, South Quay, Great Yarmouth NR30 2RQ • T: +44 1493 857900 • www.english-heritage.org.uk •
Cur.: *Sara Lunt* • Cust.: *Sara Lunt* •
Local Museum
Two 17th c town houses, original fixtures and displays of local architectural fittings salvaged from bombing in 1942/43 44359

Time & Tide - Museum of Great Yarmouth Life, 16 Blackfriars Rd, Great Yarmouth NR30 3BX • T: +44 1493 743930 • F: 743940 • yarmouth.museums@norfolk.gov.uk • www.museums.norfolk.gov.uk •
Dir.: *James Stewart* •
Historical Museum / Local Museum – 1967
Maritime hist of the district, fishing, historic racing lateener (1829), life saving, reconstructed Row houses, boats, prehistory, fine art, temporary exhibitions – library, archive 44360

The Tolhouse, Tolhouse St, Great Yarmouth NR30 2SH • T: +44 1493 858900 • yarmouth.museums@norfolk.gov.uk • www.museums.norfolk.gov.uk •
Cust.: *Robert Warnes* •
Historical Museum – 1976
Victorian prison cells 44361

Greenfield

Greenfield Valley Museum, Basingwerk House, Heritage Park, Greenfield CH8 7QB • T: +44 1352 714172 • F: 714791 • info@greenfieldvalley.com • www.greenfieldvalley.com •
Science&Tech Museum / Agriculture Museum
Farm machinery and industrial heritage 44362

Greenford

London Motorcycle Museum, Ravenor Farm, 29 Oldfield Ln S, Greenford UB6 9LB • T: +44 20 85756644 • thelmm@hotmail.com • www.london-motorcycle-museum.org •
Dir.: *Philippa Crosby* •
Science&Tech Museum – 1999
80 motorcycles (incl. many prototypes and Crosby Coll of Triumphs) 44363

Greenhead

Roman Army Museum, Carvoran, nr B6318, Greenhead CA6 7JB • T: +44 16977 47485 • F: 47487 • vindolandatrust@btinternet.com • www.vindolanda.com •
Military Museum 44364

Greenock

McLean Museum and Art Gallery, 15 Kelly St, Greenock PA16 8JX • T: +44 1475 715624 • F: 715626 • museum@inverclyde.gov.uk • www.inverclyde.gov.uk •
Cur.: *Valerie N.S. Boa* •
Local Museum – 1876
Caird and McKellar coll of pictures, shipping, objects of J. Watt 44365

Gressenhall

Roots of Norfolk at Gressenhall, Gressenhall NR20 4DR • T: +44 1362 860563 • F: 860385 • gressenhall.museum@norfolk.gov.uk • www.museums.norfolk.gov.uk •
Man.: *Stuart Gillis* •
Open Air Museum / Agriculture Museum – 1976
Rural and Norfolk life, working traditional farm 44366

Grimsby

Welholme Galleries (closed) 44367

Grouville

La Hougue Bie Museum, Grouville JE2 3NF • T: +44 1534 853823 • F: 856472 • museum@jerseyheritagetrust.org • www.jerseyheritagetrust.org •
Historical Museum / Archaeology Museum
Main coll of geological and archaeological artefacts 44368

Guildford

Guildford Cathedral Treasury, Stag Hill, Guildford GU2 7UP • T: +44 1483 547860 • F: 303350 • www.guildford-cathedral.org •
Cur.: *Ann Wickham* •
Religious Arts Museum
Cathedral plate and other artefacts 44369

Guildford House Gallery, 155 High St, Guildford GU1 3AJ • T: +44 1483 444742 • F: 444742 • guildfordhouse@guildford.gov.uk • www.guildfordhouse.co.uk •
Public Gallery – 1959
Topographical paintings, drawings and prints from the 18th to 20th centuries including pastel and oil portraits by local artist John Russell, 20th century craft collection, textiles, glass, ceramics – library 44370

Guildford Museum, Castle Arch, Guildford GU1 3SX • T: +44 1483 444751 • F: 532391 • museum@guildford.gov.uk • www.guildfordmuseum.co.uk •
Dir.: *Sue Roggero* •
Local Museum – 1898
County archaeology, history, needlework, ethnography, Saxon cemetery finds from Guildtown 44371

Lewis Elton Gallery, c/o University of Surrey, Guildford GU2 7XH • T: +44 1483 689167 • F: 300803 • gallery@surrey.ac.uk • www.surrey.ac.uk •
Cur.: *Patricia Grayburn* •
Public Gallery / University Museum 44372

Loseley House, Estate Office, Loseley Park, Guildford GU3 1HS • T: +44 1483 304440 • F: 302036 • enquiries@loseley-park.com • www.loseley-park.com •
Owner: *Michael More-Molyneux* •
Decorative Arts Museum
House (1562) with Elizabethan architecture, works of art, furniture, paintings, ceilings, tapestries 44373

Queen's Royal Surrey Regiment Museum, Clandon Park, Guildford GU4 7RQ • T: +44 1483 223419 • F: 223419 • www.queensroyalsurreys.org.uk •
Military Museum
Militaria, medals, photos and reference material relating to the Infantry Regiments of Surrey 44374

Haddington

Jane Welsh Carlyle's House, Lodge St, Haddington EH41 3DX • T: +44 1620 823738 • F: 825147 •
Dir.: *Garth Morrison* •
Local Museum
House from the middle of the 18th century 44375

Lennoxlove House, Lennoxlove Estate, Haddington EH41 4NZ • T: +44 1620 823720 • F: 825112 • events@lennoxlove.com • www.lennoxlove.com •
Cur.: *Duke of Hamilton* •
Fine Arts Museum 44376

Hagley

Hagley Hall, Hagley Worcestershire DY9 9LG • T: +44 1562 882408 • F: 882632 • info@hagleyhall.org • www.hagleyhall.org •
Historic Site / Fine Arts Museum – 1756
Pictures – library 44377

Hailes

Cistercian Museum, Hailes Abbey, Hailes GL54 5PB • T: +44 1242 602398 • F: 602398 • customers@english-heritage.org.uk • www.englishheritage.org.uk •
Cur.: *Tony Musty* •
Religious Arts Museum
Medieval sculpture, coll of 13th c roof bosses, medieval tiles, manuscripts, pottery and iron work 44378

Hailsham

Hailsham Heritage Centre, Blackman's Yard, Market St, Hailsham BN27 2AE • T: +44 1323 840947 •
Dir.: *M. Alder* •
Local Museum – 1962 44379

Michelham Priory, Upper Dicker, Hailsham BN27 3QS • T: +44 1323 844224 • F: 844030 • adminmich@sussexpast.co.uk • www.sussexpast.co.uk •
Dir.: *Henry Warner* • Officer: *Helen Poole* •
Decorative Arts Museum / Agriculture Museum / Historical Museum – 1960
Medieval and Tudor-Jacobean furniture, 17th c tapestries, 16th-17th c ironwork, glass, 18th c paintings, agricultural and wheelwright tools and equipment, buildings date from 1229 and include 16th c Great Barn and working watermill 44380

Halesworth

Halesworth and District Museum, Railway Station, Station Rd, Halesworth IP19 8BZ • T: +44 1986 873030 •
Cur.: *Michael Fordham* •
Local Museum
Local geology and archaeology 44381

Halifax

Bankfield Museum, Akroyd Park, Boothtown Rd, Halifax HX3 6HG • T: +44 1422 354823, 352334 • F: 349020 • bankfield-museum@calderdale.gov.uk • www.calderdale.gov.uk •
Ass. Dir.: *Rosemary Crook* • Sc. Staff: *June Hill* •
Local Museum / Military Museum / Decorative Arts Museum – 1887
Textiles, crafts, militaria, toys, ethnography 44382

The Dean Clough Galleries, Dean Clough, Halifax HX3 5AX • T: +44 1422 250250 • F: 341148 • linda@designdimension.org • www.deanclough.com •
Dir.: *Roger Standen* • Cur.: *Doug Binder* •
Public Gallery 44383

Eureka! The Museum for Children, Discovery Rd, Halifax HX1 2NE • T: +44 1422 330069 • F: 330275 • info@eureka.org.uk • www.eureka.org.uk •
Dir.: *Leigh-Anne Stradeski* •
Science&Tech Museum – 1992 44384

Museum of The Duke of Wellington's Regiment, Boothtown Rd, Halifax HX3 6HG • T: +44 1422 352334 • F: 349020 • bankfield-museum@calderdale.gov.uk • www.calderdale.gov.uk •
Dir.: *Philippa Mackenzie* •
Military Museum – 1959
History of the regiment, memorabilia on the first Duke of Wellington (1769-1852) 44385

Piece Hall Art Gallery, Piece Hall, Halifax HX1 1RE • T: +44 1422 358300 • F: 300878 • piece.hall@calderdale.gov.uk • www.calderdale.gov.uk •
Dir.: *Rosie Crook* •
Public Gallery
Temporary exhibition program, regular artists in residence, complementary workshops, educational programs 44386

Quilters' Guild Resource, c/o Room 190, Dean Clough, Halifax HX3 5AX • T: +44 1422 247669 • F: 345017 • resource@qghalifax.org.uk • www.quiltersguild.org.uk •
Association with Coll 44387

Shibden Hall Museum and West Yorkshire Folk Museum, Shibden Hall, Lister's Rd, Halifax HX3 6XG • T: +44 1422 352246 • F: 348440 • shibden.hol@calderdale.gov.uk • www.calderdale.gov.uk •
Folklore Museum / Historical Museum – 1937
15th c house with 16th-19th c furnishings, 17th c aisled barn, farm tools, vehicles, musical instruments, local pottery – workshops 44388

Hallaton

Hallaton Museum, Hog Ln, Hallaton LE16 8UE • T: +44 1858 555388 •
Chm.: *C. Kirby* •
Local Museum
Local hist, village life, agricultural artefacts 44389

Halstead

Brewery Chapel Museum, Adams Court, off Trinity St, Halstead CO9 1LF • T: +44 1787 478463 •
Local Museum 44390

Halton

County Museum Technical Centre, Tring Rd, Halton HP22 5PJ • T: +44 1296 696012 • F: 694519 • museum@buckscc.gov.uk • www.buckscc.gov.uk/museum •
Local Museum 44391

Hamilton

Hamilton Low Parks Museum, 129 Muir St, Hamilton ML3 6BJ • T: +44 1698 328232 • F: 328412 • lowparksmuseum@southlanarkshire.gov.uk • www.southlanarkshire.gov.uk •
Dir.: *Kirsty Lingstadt* •
Military Museum / Local Museum – 1996
History of the Cameronians regiment, costumes, weaving, industrial and local history, musician gallery 44392

Handsworth

Handsworth Saint Mary's Museum, Handsworth Parish Centre, Rectory Grounds, Handsworth S13 9BZ • T: +44 114 2692537 •
Local Museum
Coll of photographs, artefacts and documents of the local community 44393

Harewood

Harewood House, Moor House, Harewood LS17 9LQ • T: +44 113 2181010 • F: 2181002 • info@harewood.org • www.harewood.org •
Dir.: *Terence Suthers* •
Decorative Arts Museum 44394

Harlech

Harlech Castle, Castle Sq, Harlech LL46 2YH • T: +44 1766 780552 • F: 780552 •
Historic Site 44395

Harlow

Museum of Harlow, Muskham Rd, Off First Av, Harlow CM20 2LF • T: +44 1279 454959 • TMOH@harlow.gov.uk • www.harlow.gov.uk •
Man.: *Chris Lydamore* • Ass. Cur.: *George Luckey* •
Local Museum – 1973
Roman, post-medieval pottery – library 44396

Harray

Orkney Farm and Folk Museum, Corrigall, Harray KW15 1DH • T: +44 1856 771411 • F: 771411 • www.orkneyislands.com/guide/ •
Cust.: *Harry Flett* •
Local Museum
Series of restored Orkney farm buildings, visitors can see the traditional use of local resources - stone, straw, heather and horn, poultry, native sheep 44397

Harrington

Carpetbagger Aviation Museum, Sunnyvale Farm, off Lamport Rd, Harrington NN6 9PF • T: +44 1604 686608 • cbaggermuseum@aol.com • www.harringtonmuseum.org.uk •
Chm.: *Ron Clarke* •
Science&Tech Museum / Military Museum
Memorabilia of 801st-492nd, BG. USAAF Royal Observer Corps, Thor Rocket site, Harrington airfield history 44398

Harrogate

Gascoigne Gallery, Royal Parade, Harrogate HG2 0QA • T: +44 1423 525000 • www.thegascoignegallery.com •
Public Gallery
Contemporary paintings 44399

Harlow Carr Museum of Gardening, Royal Horticultural Society, Crag Ln, Harrogate HG3 1QB • T: +44 1423 565418 • F: 530663 • education-harlowcarr@rhs.org.uk • www.rhs.org.uk •
Agriculture Museum – 1989
Old gardening equipment, gardening ephemera, large coll available to students 44400

Mercer Art Gallery, Swan Rd, Harrogate HG1 2SA • T: +44 1423 566188 • F: 556130 • lg12@harrogate.gov.uk • www.harrogate.gov.uk/museums •
Dir.: *Ceryl Evans* •
Fine Arts Museum – 1991
Paintings, watercolors, lithographs, drawings, prints, reproductions 44401

Royal Pump Room Museum, Crown Pl, Harrogate HG1 2RY • T: +44 1423 566130, 556188 • F: 556130 • LG12@harrogate.gov.uk • www.harrogate.gov.uk/museums •
Dir.: *Ceryl Evans* •
Local Museum – 1953
History of sulphur and mineral wells in Harrogate, costumes, pottery, archaeology, local social life 44402

War Room and Motor House Collection, 30 Park Parade, Harrogate HG1 5AG • T: +44 1423 500704 •
Keeper: *Brian Jewell* •
Historical Museum / Science&Tech Museum
Exhibits from two WW's transport and architectural models 44403

Harrow

Harrow Museum, Headstone Manor, Pinner View, Harrow HA2 6PX • T: +44 20 88612626 • F: 88636407 • museum@harrowarts.com • www.harrowarts.com •
Cur.: *Mary Hesling* •
Local Museum
History of Harrow 44404

Hartland

Hartland Quay Museum, Hartland Quay, Hartland EX39 6DU • T: +44 1288 331353 •
Dir.: *Mark Myers* • *M. Nix* •
Local Museum – 1980
Geological and natural hist, a marine aquarium, maritime hist, shipwrecks, local hist, coastal industry – aquarium 44405

Hartlepool

Hartlepool Art Gallery, Church Sq, Hartlepool TS24 7EQ • T: +44 1429 869706 • customer.service@hartlepool.gov.uk • www.hartlepool.gov.uk •
Fine Arts Museum
19th and 20th c paintings, oriental antiquities, displays of local hist and the hist of Christ Church – Japanese gallery 44406

HMS Trincomalee, Jackson Dock, Hartlepool TS24 0SQ • T: +44 1429 223193 • F: 864385 • office@hms-trincomalee.co.uk • www.hms-trincomalee.co.uk •
Science&Tech Museum – 1817
Historic warship 44407

Museum of Hartlepool, Jackson Dock, Hartlepool TS24 0SQ • T: +44 1429 860077 • F: 867332 • arts-museum@hartlepool.gov.uk • www.destinationhartlepool.co.uk •
Local Museum – 1995
Local history stone age to present, natural history, fishing, shipbuilding, archaeology, local sailing craft 44408

Tees Archaeology, Sir William Gry House, Clarence Rd, Hartlepool TS24 8BT • T: +44 1429 523455 • F: 523477 • tees-archaeology@hartlepool.gov.uk •
Dir.: *R. Daniels* •
Archaeology Museum 44409

Harwich

Harwich Maritime Museum and Harwich Lifeboat Museum, Harwich Green, Harwich CO12 3NL • T: +44 1255 503429 • F: 503429 • info@harwich-society.com • www.harwich-society.com •
Dir.: *T. Beirne* • *C. Farnell* •
Science&Tech Museum 44410

Harwich Redoubt Fort, Main Rd, Harwich CO12 3LT • T: +44 1255 503429 • F: 503429 • info@harwich-society.com • www.harwich-society.com •
Dir.: *A. Rutter* •
Historical Museum
180ft diameter circular fort built 1808 against Napoleonic invasion, 11 guns/battlements 44411

National Vintage Wireless and Television Museum, The High Lighthouse, Harwich CO12 3DT • T: +44 7796 280980 •
Cur.: *Anthony O'Neill* •
Special Museum
Large coll of vintage wirelesses, televisions and broadcasting items, tracing the history of broadcasting from Baird and Marconi to the present day 44412

Haslemere

Haslemere Educational Museum, 78 High St, Haslemere GU27 2LA • T: +44 1428 642112 • F: 645234 • haslemeremuseum@aol.com • www.haslemeremuseum.co.uk •
Cur.: *Julia Tanner* •
Local Museum – 1888
Natural history, geology, ornithology, botany, prehistory, local history, peasant art 44413

Hastings

Fishermen's Museum, Rock-a-Nore Rd, Hastings TN34 3DW • T: +44 1424 461446 • F: 461446 •
Keeper: *Philip Ornsby* •
Historical Museum
Hastings fishing industry, maritime hist 44414

Hastings Museum and Art Gallery, Bohemia Rd, Hastings TN34 1ET • T: +44 1424 781155 • F: 781165 • museum@hastings.gov.uk • www.hmag.org.uk •
Cur.: *Victoria Williams* •
Fine Arts Museum / Ethnology Museum – 1890
Paintings, Oriental art, Pacific and American Indian ethnography, geology, zoology, archaeology, history, folklore, topography, Durbar Hall, Hawaiian feather cloak, majolica dish of 1594, Sussex pottery – library, archives, two North American Indian galleries 44415

Museum of Local History, Old Town Hall, High St, Hastings TN34 1EW • T: +44 1424 781166 • F: 781165 • oldtownmuseum@hastings.gov.uk • www.hmag.org.uk •
Dir.: *Victoria Williams* •
Historical Museum – 1949
Local archaeology and history, topography 44416

Shipwreck Heritage Centre, Rock-a-Nore Rd, Hastings TN34 3DW • T: +44 1424 437452 • F: 437452 •
Dir.: *Peter Marsden* •
Archaeology Museum – 1986
Objects from historic and ancient shipwrecks around southern England, prehistoric, Roman, medieval and later 44417

Hatfield

Art and Design Gallery, c/o Faculty of Art & Design, University of Hertfordshire, College Ln, Hatfield AL10 9AB • T: +44 1707 285376 • F: 285310 • m.b.shaul@herts.ac.uk • www.herts.ac.uk/artdes •
Dir.: *Chris McIntyre* •
Public Gallery
Fine and applied arts, design and craft 44418

Mill Green Museum and Mill, Mill Green, Hatfield AL9 5PD • T: +44 1707 271362 • F: 272511 • museum@welhar.gov.uk • www.welhat.gov.uk/museum •
Dir.: *Dr. John Beckerson* •
Local Museum / Science&Tech Museum – 1973
18th c brick-built, three-storey watermill, with early 19th c wood and iron machinery, 18th-c miller's house, the social hist of the Welwyn-Hatfield area, archaeology, domestic life, the railways, industry, farming and wartime, Belgic and Roman pottery from Welwyn, pottery from Hatfield – library 44419

Havant

Havant Museum, 56 East St, Havant PO9 1BS • T: +44 23 92451155 • F: 92498707 • musmcp@hants.gov.uk • www.hants.gov.uk/museum/havant •
Cur.: *Dr. Oonagh Palmer* •
Local Museum – 1977
Vokes Coll of sporting firearms, Havant local industries and local hist 44420

Havenstreet

Isle of Wight Steam Railway and Isle of Wight Railway Heritage Museum, Railway Station, Havenstreet PO33 4DS • T: +44 1983 882204 • F: 884515 • hugh@iwsteamrailway.co.uk • www.iwsteamrailway.co.uk •
Dir.: *Hugh Boynton* •
Science&Tech Museum – 1971
Locomotives and rolling stock from early Isle of Wight-railways, oldest vehicle built in 1864, oldest locomotive built in 1876 – Documentary and photographic archive 44421

Haverfordwest

Scolton Manor Museum, Bethlehem, Haverfordwest SA62 5QL • T: +44 1437 731328 • F: 731457 • mark.thomas@pembrokeshire.gov.uk •
Local Museum – 1967
Social hist, costume, photographs, military hist, fine art, decorative art, natural hist, coins & numismatics, geology, archaeology 44422

Haverhill

Haverhill and District Local History Centre, Town Hall Arts Centre, High St, Haverhill CB9 8AR • T: +44 1440 714962 •
Local Museum
Photographs of local area, newspaper on microfilm 44423

Hawes

Dales Countryside Museum, Station Yard, Hawes DL8 3NT • T: +44 1969 666210 • F: 666239 • dcm@yorkshiredales.co.uk •
Cur.: *Fiona Rosher* •
Local Museum – 1979
History, agriculture, industry, local crafts transport, communications , telling the story of the people and landscape of the Dales 44424

Hawick

Drumlanrig's Tower, 1 Tower Knowe, Hawick TD9 9EN • T: +44 1450 377615 • F: 378506 •
Head: *Andrea Hardie* •
Local Museum / Fine Arts Museum
Borders hist, local traditions, Tom Scott Gallery of late Victorian watercolours 44425

Hawick Museum, Wilton Lodge Park, Hawick TD9 7JL • T: +44 1450 373457 • F: 378506 • museums@scotborders.gov.uk •
Local Museum / Fine Arts Museum – 1856/1975
Scottish and natural hist, militaria, coins, knitwear industry, 19th-20th c paintings 44426

The Scott Gallery, Hawick Museum, Wilton Lodge Park, Hawick TD9 7JL • T: +44 1450 373457 • F: 378506 • fionacolton@hotmail.com •
Cur.: *Fiona Colton* •
Public Gallery 44427

Hawkinge

Kent Battle of Britain Museum, Aerodrome Rd, Hawkinge CT18 7AG • T: +44 1303 893140 • kentbattleofbritainmuseum@btinternet.com • www.kentbattleofbritainmuseum.org.uk •
Cur.: *Mike Llewellyn* •
Military Museum
Battle of Britain artefacts, aircraft, vehicles, weapons, flying equipment, prints and relics from over 600 crashed aircraft 44428

Hawkshead

Beatrix Potter Gallery, The National Trust, Main St, Hawkshead LA22 0NS • T: +44 15394 36355 • F: 36187 • beatrixpottergallery@nationaltrust.org.uk • www.nationaltrust.org.uk •
Dir.: *John Moffat* •
Fine Arts Museum
Beatrix Potter's original drawings and illustrations of her children's storybooks 44429

Haworth

Brontë Parsonage Museum, Church St, Haworth BD22 8DR • T: +44 1535 642323 • F: 674131 • bronte@bronte.org.uk • www.bronte.info •
Dir.: *Alan Bentley* •
Special Museum – 1895
Books, manuscripts, letters, paintings, drawings, furniture and personal treasures of the Brontë family 44430

Keighley and Worth Valley Railway Museum, Station, Haworth BD22 8NJ • T: +44 1535 645214, 647777 • F: 647317 • kwvr@hotmail.com • www.kwvr.co.uk •
Admin.: *Jim Shipley* •
Science&Tech Museum
Engines and rolling stock, Pullman train, vintage carriages and steam locomotives, station signs 44431

Headcorn

Lashenden Air Warfare Museum, Aerodrome, Headcorn TN27 9HX • T: +44 1622 890226, 206783 • F: 206783 • lashairwar@blueyonder.co.uk •
Man.: *D. Campbell* •
Science&Tech Museum
Aviation relics, uniforms, civilians at war, prisoners of war 44432

Heathfield

Sussex Farm Museum, Horam Manor, Heathfield TN21 0JB • T: +44 1435 813352 • F: 812597 • vfmroberts@tiscali.co.uk •
Dir.: *F. Goulden* • Cur.: *V. Roberts* •
Agriculture Museum / Local Museum – 1985
Farming and farm life since 1900. Large items in farmyard, rest in ancient barns in room settings, also toys, tools and hobbies 44433

Hebden Bridge

Automobilia Transport Museum, Billy Ln, Old Town, Wadsworth, Hebden Bridge HX7 8RY • T: +44 1422 844775 • F: 842884 •
Science&Tech Museum
Coll of cars, motorcycles, bicycles hr 44434

Heckington

Heckington Windmill, Hale Rd, Heckington NG34 9JJ • T: +44 1529 461919 • www.visitlincolnshire.com •
Science&Tech Museum
Ancillary machinery 44435

Hednesford

Museum of Cannock Chase, Valley Rd, Hednesford WS12 1TD • T: +44 1543 877666 • F: 428272 • museum@cannockchase.gov.uk • www.cannockchasedc.gov.uk/museum •
Dir.: *Adrienne Whitehouse* •
Local Museum
Social history, industry, coal mining 44436

Hedon

Hedon Museum, Town Hall Complex, Saint Augustine's Gate, Hedon HU12 8EX • T: +44 1482 890908 • hedon_museum@yahoo.co.uk • www.hedon-museum.d.tec.org •
Local Museum
Docs, photos and artefacts to the hist of the town 44437

Helensburgh

Hill House, Upper Colquhoun St, Helensburgh G84 9AJ • T: +44 844 4932208 • F: 4932209 • lhepburn@nts.org.uk • www.nts.org.uk •
Cur.: *Lorna Hepburn* •
Decorative Arts Museum
Domestic architecture of Charles Rennie Mackintosh, furniture, interior design, new domestic design (annual exhibition) 44438

Portico Gallery, 78 W Clyde St, Helensburgh G84 0AB • T: +44 1436 671821 • F: 677553 •
Fine Arts Museum 44439

Helmsdale

Timespan Heritage Centre and Art Gallery, Dunrobin St, Helmsdale KW8 6JX • T: +44 1431 821327 • F: 821058 • enquiries@timespan.org.uk • www.timespan.org.uk •
Man.: *Lorna Jappy* •
Local Museum / Public Gallery
Highland's people, natural hist, rare and medicinical plants – garden 44440

Helmshore

Helmshore Mills Textile Museum, Holcombe Rd, Helmshore BB4 4NP • T: +44 1706 226459 • F: 218554 • helmshore.museum@mus.lancscc.gov.uk • www.lancsmuseums.gov.uk •
Science&Tech Museum – 1967
Platt coll of early textile machinery, Lancs textile industry 44441

Helston

Flambards Experience, c/o Culdrose Manor, Clodgey Ln, Helston TR13 0QA • T: +44 1326 573404 • F: 573344 • info@flambards.co.uk • www.flambards.co.uk •
Dir.: *James K. Hale* •
Science&Tech Museum – 1976
Flying machines, several aircraft, motorcycles motorcars and vehicles, historic streets, shops, houses, coaches, dresses, Victorian village, Britain in the Blitz historical aviation pioneers 44442

Flambards Victorian Village and Gardens, Culdrose Manor, Helston TR13 0QA • T: +44 1326 573404 • F: 573344 • info@flambards.co.uk • www.flambards.co.uk •
Dir.: *James Kingsford Hale* •
Science&Tech Museum
Flambards Victorian Village and Britain in The Blitz, life-sized recreations, aeropark coll, aviation in peace and war 44443

Helston Folk Museum, The Old Butter Market, Market Pl, Helston TR13 8TH • T: +44 1326 564027 • F: 569714 • enquiries@helstonmuseum.org.uk • www.helstonmuseum.org.uk •
Man.: *J. Spargo* •
Local Museum – 1949
Local history, trades & industries of the Lizard Peninsula, railway, costume, toys, domesatic – art gallery 44444

National Museum of Gardening, Trevarno Manor, Trevarno, Sithney, Helston TR13 0RU • T: +44 1326 574274 • F: 574282 • enquiry@trevarno.co.uk • www.trevarno.co.uk •
Dir.: *Mike Sagin* •
Special Museum
Garden tools, implements, requisites, memorabilia and ephemera, cultivation, watering, cutting, trimming 44445

Poldark Mine and Heritage Complex, Wendron, Helston TR13 0ER • T: +44 1326 573173, 563166 • F: 563166 • info@poldark-mine.co.uk • www.poldark-mine.co.uk •
Dir.: *Richard Williams* •
Science&Tech Museum – 1971
18th c tin mine heritage, Cornish beam engine, machinery 44446

Hemel Hempstead

Old Town Hall Arts Centre, High St, Hemel Hempstead HP1 3AE • T: +44 1442 228091 • F: 228091 • www.oldtownhall.co.uk •
Fine Arts Museum 44447

Henfield

Henfield Museum, Henfield Hall, Coopers Wy, Henfield BN5 9DB • T: +44 1273 492507 • F: 494898 • office@henfield.gov.uk • www.henfield.gov.uk •
Cur.: *Marjorie W. Carreck* •
Local Museum – 1948
Local hist, costume, agricultural and domestic bygones, local paintings and photographs, uniforms 44448

Henley-on-Thames

River and Rowing Museum, Mill Meadows, Henley-on-Thames RG9 1BF • T: +44 1491 415600 • F: 415601 • museum@rrm.co.uk • www.rrm.co.uk • Dir.: *Paul E. Mainds* •
Historical Museum
History, ecology and archaeology of the River Thames, development of the international sport of rowing and the town of Henley-on-Thames, "The Wind of the Willows" famous children's book 44449

Stonor Park, Henley-on-Thames RG9 6HF • T: +44 1491 638587 • F: 638587 • administrator@stonor.com • www.stonor.com •
Dir.: *Lord Camoys* •
Fine Arts Museum 44450

Henllan

Teifi Valley Railway, Station Yard, Henllan SA44 5TD • T: +44 1559 371077 • F: 371077 • teifivr@f9.co.uk •
Science&Tech Museum
Steam locomotive, Motorail diesel units 44451

Heptonstall

Heptonstall Museum, Church Yard Bottom, Heptonstall HX7 7PL • T: +44 1422 843738 • shibden.hall@calderdale.gov.uk • www.calderdale.gov.uk •
Local Museum 44452

Hereford

Cider Museum and King Offa Distillery, Pomona Pl, Whitecross Rd, Hereford HR4 0LW, mail addr: 21 Ryelands St, Hereford HR4 0LW • T: +44 1432 354207 • F: 371641 • enquiries@cidermuseum.co.uk • www.cidermuseum.co.uk •
Dir.: *Margaret Thompson* •
Agriculture Museum – 1981
Farm cider making, the evolution of the modern cider factory 44453

Hereford Museum and Art Gallery, Broad St, Hereford HR4 9AU • T: +44 1432 260692 • F: 342492 •
Local Museum / Fine Arts Museum – 1874
Roman remains, agricultural implements, natural hist, social histpaintings and prints, glass and china, militaria, coins 44454

Herefordshire Regiment Museum, TA Centre, Harold St, Hereford HR1 2QX • T: +44 1432 359917 •
Military Museum
History of the Regiment, uniforms, weapons, medals, Colours and documents 44455

Mappa Mundi and Chained Library, 5 College Cloisters, Cathedral Close, Hereford HR1 2NG • T: +44 1432 374209 • F: 374220 • exhibition@herefordcathedral.org • www.herefordcathedral.org •
Man.: *Ken Tomkins* • Libr.: *James Anthony* •
Library with Exhibitions – 1996
Medieval world map, chained medieval books, Anglo-Saxon gospels, Hereford breviary and the Wycliffe Cider Bible – library 44456

Old House, High Town, Hereford HR1 2AA • T: +44 1432 260694 • F: 342492 • herefordmuseums@herefordshire.gov.uk • www.herefordshire.gov.uk •
Ass.: *Alexia Clark* •
Historical Museum – 1928
17th c furniture, model of the city (1640), Jacobean domestic architecture 44457

Piano Collection, c/o Royal National College for the Blind, College Rd, Hereford HR1 1EB • T: +44 1432 265725 • F: 376628 • info@rncb.ac.uk • www.rncb.ac.uk •
Dir.: *Roisin Burge* •
Music Museum
Early keyboard instruments 44458

Saint John and Coningsby Medieval Museum, Coningsby Hospital, Widemarsh St, Hereford HR4 9HN • T: +44 1432 358134 •
Historical Museum
History of the Order of St John and the wars during the 300 years of the Crusades, armour and emblazons, also models, in period dress and bandages, of the Coningsby pensioners who used the hospital 44459

Waterworks Museum - Hereford, Broomy Hill, Hereford HR4 0JS, mail addr: Llancraugh Cottage, Marstow, Ross-on-Wye HR9 6EH • T: +44 1432 344062 • F: +44 1600 890009 • info@waterworksmuseum.org.uk • www.waterworksmuseum.org.uk •
Dir.: *Dr. Noel Meeke* •
Science&Tech Museum
Hist of drinking water supplies, wide range of working pumping engines (steam, diesel, gas) incl a triple-expansion condensing steam engine by Worth, McKenzie & Co (1895) 44460

Herne Bay

Herne Bay Museum and Gallery, 12 William St, Herne Bay CT6 5EJ • T: +44 1227 367368 • F: 742560 • museums@canterbury.gov.uk • www.hernebay-museum.co.uk •
Local Museum / Fine Arts Museum
Displays about the town, about Reculver Roman Fort, archaeology and fossils 44461

Hertford

Hertford Museum, 18 Bull Plain, Hertford SG14 1DT • T: +44 1992 582686 • F: 552100 • helen@hertfordmuseum.org • www.hertford.net/museum •
Cur.: *Helen Gurney* •
Local Museum – 1902
Geology, natural history, archaeology, history of East Hertfordshire, militaria, Saxon coins, ethnography 44462

Hexham

Hexham Old Gaol, The Old Gaol, Hallgate, Hexham NE46 3NH • T: +44 1434 652351 • F: 652364 • museum@tynedale.gov.uk • wwwtynedale.gov.uk •
Dir.: *Janet Goodridge* •
Historical Museum – 1980
Weapons and armour 44463

Heywood

The Corgi Heritage Centre, 53 York St, Heywood OL10 4NR • T: +44 1706 365812 • F: 627811 • corgi@zen.co.uk • www.zen.co.uk/home/page/corgi •
Special Museum
Model vehicle, history of Corgi toys 44464

High Wycombe

Hughenden Manor, High Wycombe HP14 4LA • T: +44 1494 755565, 755573 • F: 474284 • hughenden@nationaltrust.org.uk • http://www.nationaltrust.org.uk/main/w-hughendenmanor •
Special Museum – 1948
Home of Prime Minister Benjamin Disraeli (1848-1881), containing much of his furniture, pictures, books 44465

Wycombe Museum, Castle Hill House, Priory Av, High Wycombe HP13 6PX • T: +44 1494 421895 • F: 421897 • museum@wycombe.gov.uk • www.wycombe.gov.uk/museum •
Dir.: *Sally Ackroyd* •
Local Museum / Historical Museum – 1932
Windsor chair making industry, lace, local hist, domestic furniture – library 44466

Higher Bockhampton

Hardy's Cottage, off A35, Higher Bockhampton DT2 8QJ • T: +44 1305 262366 •
Special Museum
Memorabilia of writer Thomas Hardy (1840-1928) 44467

Hillsborough

The Art Gallery, White Image, 34 Lisburn St, Hillsborough BT26 6AB • T: +44 28 92689896 • info@whiteimage.com • www.whiteimage.com •
Fine Arts Museum
Irish art, contemporary pieces, watercolours, oils, etchings, drawings 44468

Hillside

Sunnyside Museum, Sunnyside Royal Hospital, Hillside DD10 9JP • T: +44 1674 830361 • F: 830251 •
Cur.: *Dr. P. Thompson* •
Special Museum
Administrative records from 1797, clinical records from 1815, photographs taken in Victorian and Edwardian times, medical instruments, a straitjacket, firefighting uniforms, replicas of nursing uniforms, examples of patients' craftwork 44469

Himley

Himley Hall, Himley Park, Himley DY3 4DF • T: +44 1902 326665 • F: 894163 • himley.pls@mbc-dudley.gov.uk • www.dudley.gov.uk •
Dir.: *Charles Hajdamach* •
Decorative Arts Museum – 1995
Temporary exhibitions programme 44470

Hinckley

Hinckley and District Museum, Framework Knitters' Cottage, Lower Bond St, Hinckley LE10 1QX • T: +44 1455 251218 • www.hinckleydistrictmuseum.org.uk •
Chm.: *H.A. Beavin* •
Local Museum
Hosiery and boot and shoe industries of Hinckley and district, machines, social history, WW I and II, pickering coll, local archaeological finds 44471

Hindley

Hindley Museum, Market St, Hindley WN2 3AU • T: +44 1942 55287 •
Local Museum 44472

Hitchin

British Schools Museums, 41 Queen St, Hitchin SG4 9TS • T: +44 1462 420144 • F: 420144 • brsch@britishschools.freeserve.co.uk • www.britishschools.org.uk •
Special Museum
Teaching coll 44473

Hitchin Museum and Art Gallery, Paynes Park, Hitchin SG5 1EQ • T: +44 1462 434476 • F: 431316 • hitchin.museum@north-herts.gov.uk • www.north-herts.gov.uk •
Cur.: *David Nodges* • Asst. Cur.: *Jennifer Hammond* •
Local Museum – 1939
Local history, militaria, costumes, paintings by Samuel Lucas, Victorian chemist shop and physic garden 44474

North Hertfordshire Museums Resources Centre, Burymead Rd, Hitchin SG5 1RT • T: +44 1462 422946, 434896 • F: 434883 • k.matthews@north-herts.gov.uk • www.north-herts.gov.uk •
Historical Museum 44475

Hoddesdon

Lowewood Museum, High St, Hoddesdon EN11 8BH • T: +44 1992 445596 • museum.leisure@broxbourne.gov.uk • homepages.tesco.net/~hdp •
Cur.: *Neil Robbins* •
Local Museum – 1947
Archaeology, geology and social history 44476

Holm

Norwood Museum, Graemshall, Holm KW17 2RX • T: +44 1856 78217 •
Local Museum
Coll of Norris Wood 44477

Holsworthy

Holsworthy Museum, Manor Office, Holsworthy EX22 6DJ • T: +44 1409 259337 • holsworthy@devonmuseum.net • www.devonmuseums.net/holsworthy •
Local Museum
Local history, social, domestic and agricultural items 44478

Holt

Muckleburgh Collection, Weybourne Military Camp, Holt NR25 7EG • T: +44 1263 588210 • F: 588425 • info@muckleburgh.co.uk • www.muckleburgh.co.uk •
Dir.: *Michael Saray* •
Military Museum
Suffolk and Norfolk Yeomanry, RAF reconnaissance, air sea rescue and marine craft, Norfolk Dunkirk veterans, tanks and military vehicles 44479

Holy Island

Lindisfarne Priory, Holy Island TD15 2RX • T: +44 1289 389200 • customers@english-heritage.org.uk • www.english-heritage.org.uk •
Cur.: *Susan Harrison* •
Archaeology Museum
Anglo-Saxon sculpture and archaeological finds from the Priory 44480

Holyhead

Holyhead Maritime Museum, 8 Llainfain Estate, Llaingoch, Holyhead LL65 1NF • T: +44 1407 764374 • F: 769745 • jonncave4@aol.com •
Dir.: *William B. Carroll* •
Historical Museum
Photographs, plans, ship-models, marine tools 44481

Holywood

Ulster Folk and Transport Museum, Cultra, Holywood BT18 0EU • T: +44 28 90428428 • F: 90428728 • uftm.info@magni.org.uk • www.uftm.org.uk •
Dir.: *Marshall McKee* •
Folklore Museum / Science&Tech Museum – 1958
History and social life of Northern Ireland, aviation, maritime, road and rail transport, reconstructed buildings, homes and shops, paintings, rural life, agriculture, fishing, crafts, music, textiles – archive, library 44482

Honiton

Allhallows Museum of Lace and Antiquities, High St, Honiton EX14 1PG • T: +44 1404 44966 • F: 46591 • info@honitonmuseum.co.uk • www.honitonmuseum.co.uk •
Local Museum – 1945
Honiton Pottery, antiquities, fossils, local history, industry, Doll's House, Honiton lace 44483

Hornsea

Hornsea Museum of Village Life, 11-13 Newbegin, Hornsea HU18 1AB • T: +44 1964 533443 • www.hornseamuseum.com •
Local Museum
Representations of the dairy, kitchen, parlour and bedroom of a c ago, tools of local craftsmen, photographs and postcards of local scenes, local industry, Hornsea Brick and Tile Works (c 1868-96), Hull and Hornsea Railway (1864-1964) 44484

Horringer

Ickworth House, The Rotunda, Horringer IP29 5QE • T: +44 1284 735270 • F: 735175 •
Man.: *Kate Carver* •
Decorative Arts Museum – 1957
House dating from 1795-1830 with 18th c French furniture, silver, paintings 44485

Horsforth

Horsforth Village Museum, 3-5 The Green, Horsforth LS18 5JB • T: +44 113 2819877 • horsforthmuseum@hotmail.com • www.yourhorsforth.co.uk/history.htm •
Local Museum
Local hist 44486

Second World War Experience Centre, 5 Feast Field, Horsforth LS18 4TJ • T: +44 113 2584993 • F: 2582557 • enquiries@war-experience.org • www.war-experience.org •
Dir.: *Dr. Peter Liddle* •
Historical Museum
Soldiers, sailors, airmen and civilians daily life experience of men, women and children during wartime – international archive 44487

Horsham

Horsham Arts Centre, North St, Horsham RH12 1RL • T: +44 1403 259708, 268689 • F: 211502 •
Public Gallery 44488

Horsham Museum, Causeway House, 9 Causeway, Horsham RH12 1HE • T: +44 1403 254959 • F: 282594 • museum@horsham.gov.uk •
Cur.: *Jeremy Knight* •
Local Museum – 1893
Costumes, toys, old bicycles, local crafts and industries, reconstructed shops, in 16th c house, Shelley, geology, saddlery, farming – archives 44489

Horsham Saint Faith

City of Norwich Aviation Museum, Old Norwich Rd, Horsham Saint Faith NR10 3JF • T: +44 1603 893080 • dnam.marketin@btinternet.com • www.cnam.co.uk •
Man.: *Barry C. Holdstock* • Cust.: *Rodney Scott* •
Science&Tech Museum / Military Museum
Aviation history of Norfolk, Vulcan bomber and some military and civil aircraft 44490

Housesteads

Housesteads Roman Fort and Museum, Haydon Bridge, Bardon Mill, Housesteads NE47 6NN • T: +44 1434 344363 • customers@english-heritage.org.uk • www.english-heritage.org.uk •
Dir.: *Georgina Plowright* •
Archaeology Museum – 1936
Inscriptions, sculptures, small finds, arm from Housesteads Roman Fort Arms 44491

Hove

British Engineerium, Nevill Rd, Hove BN3 7QA • T: +44 1273 559583 • F: 566403 • info@britishengineerium.com • www.britishengineerium.com •
Dir.: *Dr. Jonathan E. Minns* •
Science&Tech Museum
Restored Victorian pumping station 44492

Hove Museum and Art Gallery, 19 New Church Rd, Hove BN3 4AB • T: +44 1273 290200 • F: 292827 • museums@brighton-hove.gov.uk • www.hove.virtualmuseum.info •
Dir.: *Pauline Scott-Garrett* •
Fine Arts Museum / Local Museum – 1927
Early film material, toys, contemporary and decorative arts (all media), 20th c English art 44493

West Blatchington Windmill, Holmes Av, Hove BN3 7LF • T: +44 1273 776017 • visitor.services@brighton-hove.gov.uk • www.brighton-hove-museums.org.uk •
Dir.: *Peter Hill* •
Science&Tech Museum – 1979
Artefacts of milling and agricultural hist 44494

Hoy

Scapa Flow Museum, Hoy KW16 3NT • T: +44 1856 791300 • F: 871560 • museum@orkney.gov.uk • www.orkneyheritage.com •
Cust.: *Lewis Munro* •
Military Museum
Large military vehicles, boats, cranes, field artillery and railway rolling stock 44495

Huddersfield

Huddersfield Art Gallery, Princess Alexandra Walk, Huddersfield HD1 2SU • T: +44 1484 221964 • F: 221952 • info.galleries@kirklees.gov.uk • www.kirklees.gov.uk/art •
Head: *Robert Hall* •
Fine Arts Museum – 1898
British paintings, drawings, prints and sculpture since 1850 (Bacon, Hockney), changing exhibitions of contemporary art by regional and national artists in all media 44496

Tolson Memorial Museum, Ravensknowle Park, Wakefield Rd, Huddersfield HD5 8DJ • T: +44 1484 223830 • F: 223843 • tolson.museum@kirklees.gov.uk • www.kirklees.gov.uk •
Dir.: *Jenny Salton* • *J.H. Rumsby* • Cur.: *S. Cooper* •
Local Museum / Folklore Museum – 1920
Local and natural hist, mineralogy, geology, botany, zoology, local life in Stone Age, Bronze Age, Iron Age, Roman and medieval periods, dolls, English glass, woolens, old vehicles, money scales, Ronnie The Raven's puzzlepath – Kirklees image archive 44497

Victoria Tower, Castle Hill, Lumb Ln, Almondbury, Huddersfield HD5 8DJ • T: +44 1484 223830 • F: 223843 • tolson.museum@kirklees.gov.uk • www.kirklees.gov.uk/museums •
Historical Museum
Prehistoric hill-fort and medieval earthworks 44498

Hull

Hull Maritime Museum, Queen Victoria Sq, Hull HU1 3DX • T: +44 1482 300300 • F: 613710 • museums@hullcc.gov.uk • www.hullcc.gov.uk/museums •
Dir.: *Arthur G. Credland* •
Historical Museum – 1975
Whaling, shipbuilding, weapons, gear, maritime history, marine paintings, shipmodels, scrimshaw work – library 44499

Huntingdon

Cromwell Museum, Grammar School Walk, Huntingdon PE29 3LF • T: +44 1480 375830 • F: 459563 • john.goldsmith@cambridgeshire.gov.uk • www.cambridgeshire.gov.uk/cromwell •
Dir.: *John Goldsmith* •
Historical Museum – 1962
Documents on Cromwell's time, portraits, memorabilia on Oliver Cromwell (1599-1658) 44500

Hinchingbrooke House, Brampton Rd, Huntingdon PE18 6BN • T: +44 1480 451121, 345700 •
Cur.: *T.R. Wheeley* •
Historic Site
Cromwell family and the Earls of Sandwich, portraits 44501

Ramsey Rural Museum, Wood Ln, Ramsey, Huntingdon PE26 2XD • T: +44 1487 815715 • F: 814030 • mikenixon@btopenworld.com • www.ramseyruralmuseum.co.uk •
Historical Museum
Victorian kitchen, sitting room, bedroom, schoolroom, farm machinery 44502

Huntly

Brander Museum, The Square, Huntly AB54 8AE • T: +44 1771 622906, 622807 • F: 622884, 623558 • heritage@aberdeenshire.gov.uk • www.aberdeenshire.gov.uk •
Dir.: *Ian Bonner* •
Local Museum
Communion tokens, author George MacDonald memorabilia 44503

Leith Hall, Kennethmont, Huntly AB54 4NQ • T: +44 844 4932164 • F: 4932176 • leithhall@nts.org.uk • www.nts.org.uk/Property/42/Contact/ •
Dir.: *Robin Pellew* •
Historic Site / Military Museum
Exhibit 'For Crown and Country', Leith Hay family's military memorabilia 44504

Hutton-le-Hole

Ryedale Folk Museum, Hutton-le-Hole YO62 6UA • T: +44 1751 417367 • F: 417367 • info@ryedalefolkmuseum.co.uk • www.ryedalefolkmuseum.co.uk •
Man.: *Mike Benson* •
Open Air Museum / Folklore Museum – 1966
Tools, spinning, weaving, reconstructions of a 15th c Cruck house, 16th c manor house, 18th c cottage, 16th c glass furnace, agricultural coll of folk life 44505

Hynish, Isle of Tiree

Skerryvore Museum, Story of Skerryvore Lighthouse, Lower Sq, Morton Boyd Hall, Hynish, Isle of Tiree PA77 6UG • T: +44 1879 220726, 1865 311468 • F: +44 1865 311468 • info@hebrideantrust.org • www.hebrideantrust.org •
Local Museum
Lighthouse story, pictures 44506

Hythe, Kent

Hythe Local History Room, 1 Stade St, Oaklands, Hythe, Kent CT21 6BG • T: +44 1303 266152/53 • F: 262912 • admin@hythe-kent.com • www.hythe-kent.com •
Local Museum – 1933
Local and military hist, archaeology, paintings 44507

Ilchester

Ilchester Museum, Town Hall and Community Centre, High St, Ilchester BA22 8NQ • T: +44 1935 841247 •
Local Museum
Prehistory, Roman, Saxon, Medieval, Post Medieval history 44508

Ilford

Redbridge Museum, Central Library, Clements Rd, Ilford IG1 1EA • T: +44 20 87082317 • F: 87082431 • redbridge.museum@redbridge.gov.uk • www.redbridge.gov.uk/leisure/museum.cfm •
Man.: *Gerard Greene* •
Local Museum
Local and social hist of the London Borough of Redbridge 44509

Ilfracombe

Ilfracombe Museum, Wilder Rd, Ilfracombe EX34 8AF • T: +44 1271 863541 • ilfracombe@devonmuseums.net • www.devonmuseums.net •
Man.: *Sue Pullen* •
Local Museum – 1932
Local history, archaeology, ethnography, natural history, fauna, Victorian relics, shipping, armaments, militaria 44510

Ilkeston

Erewash Museum, High St, Ilkeston DE7 5JA • T: +44 115 9071141 • F: 9329264 • museum@erewash.gov.uk • www.erewash.gov.uk •
Man.: *Victoria J. Geddes* • Cur.: *Julie Biddlecombe* •
Local Museum – 1981/82
Listed Georgian and Victorian town house with kitchen and scullery display and other rooms containing civic, local and social hist collections, Long Eaton Town Hall 44511

Ilkley

Manor House Art Gallery and Museum, Castle Yard, Ilkley LS29 9DT • T: +44 1943 600066 • F: 817079 • gavinedwards@btconnect.com • www.bradfordmuseums.org •
Local Museum / Public Gallery – 1961
Local hist, Roman finds from Olicana Roman Fort, temporary art exhibitions in Elizabethan Manor House 44512

Ilminster

Perry's Cider Mills, Dowlish Wake, Ilminster TA19 0NY • T: +44 1460 52681 • F: 55195 • info@perryscider.co.uk • www.perryscider.co.uk •
Folklore Museum
Old farm tools, wagons, cider-making equipment, stone jars and related country living items 44513

Immingham

Immingham Museum, Margaret St, Immingham DN40 1LE • T: +44 1469 577066 • immingham@bmummery.freeserve.co.uk • www.ukwebwizard.co.uk •
Dir.: *Brian Mummery* •
Historical Museum
Coll of the Great Central Railway Society, creation of a port – archive 44514

Ingatestone

Ingatestone Hall, Hall Ln, Ingatestone CM4 9NR • T: +44 1277 353010 • F: 248979 •
Decorative Arts Museum – 1989
Essex history, furniture, china 44515

Ingrow

Museum of Rail Travel, Railway Centre, Ingrow BD22 8NJ • T: +44 1535 680425 • F: 610796 • admin@vintagecarriagestrust.org • www.vintagecarriagestrust.org •
Cur.: *Michael W. Cope* •
Science&Tech Museum
Coll of historic railway coaches 44516

Innerleithen

Traquair House, Innerleithen EH44 6PW • T: +44 1896 830323 • F: 830639 • enquiries@traquair.co.uk • www.traquair.co.uk •
Dir.: *Catherine Maxwell Stuart* •
Decorative Arts Museum / Historical Museum – 1963
16th-17th c embroidery, Jacobite glass, silver, manuscripts, books, household objects, furniture, musical instruments, memorabilia of Mary, Queen of Scotland, 18th c Brew House 44517

Inveraray

Auchindrain Museum, Auchindrain, Inveraray PA32 8XN • T: +44 1499 500235 • F: 500235 • joanne@auchindrain-museum.org.uk • www.auchindrain-museum.org.uk •
Cur.: *Joanne Howdle* •
Open Air Museum
West Highland farming township buildings, agricultural, domestic and social history 44518

Inveraray Castle, Cherry Park, Inveraray PA32 8XE • T: +44 1499 302203 • F: 302421 • enquiries@inveraray-castle.com • www.inveraray-castle.com •
Dir.: *A. Montgomery* •
Decorative Arts Museum
Home of the Duke and Duchess of Argyll, armoury coll, French tapestries, fine exemples of Scottish and European furniture – Clan room 44519

Inveraray Jail, Church Sq, Inveraray PA32 8TX • T: +44 1499 302381 • F: 302195 • inverarayjail@btclick.com • www.inverarayjail.co.uk •
Dir.: *Jim Linley* •
Special Museum
19th c Scottish prison 44520

Inveraray Maritime Museum, The Pier, Inveraray PA32 8UY • T: +44 1499 302213 •
Historical Museum
Clyde maritime memorabilia, three masted schooner 44521

Inverkeithing

Inverkeithing Museum, The Friary, Queen St, Inverkeithing KY11 1LS • T: +44 1383 313594 • lesley.botton@fife.gov.uk •
Local Museum
Social and local history 44522

Inverness

Culloden Battlefield and Visitor Centre, Culloden Moor, Inverness IV2 5EU • T: +44 844 4932159 • F: 4932160 • culloden@nts.org.uk • www.nts.org.uk •
Man.: *Deirdre Smyth* •
Historical Museum – 1970
Historical display of the Battle of Culloden (1745/46), Jacobite exhibition – Old Leanach Cottage 44523

Eden Court Art Gallery, Eden Court Theatre, Bishops Rd, Bishop's Road Inverness IV3 5SA • T: +44 1463 234234 • ecmail@cali.co.uk •
Public Gallery 44524

Inverness Museum and Art Gallery, Castle Wynd, Inverness IV2 3EB • T: +44 1463 237114 • contact@invernessmuseum.com • http://inverness.highland.museum •
Cur.: *Catharine Niven* •
Local Museum – 1825
Local hist, archaeology, paleontology, geology, bagpipes, local prints and paintings, old kitchen, silver, uniforms, costumes, weapons 44525

Regimental Museum The Highlanders, The Queen's Own Highlanders Collection, Fort George, Inverness IV2 7TD • T: +44 131 3108701 • info@thehighlandersmuseum.com • www.thehighlandersmuseum.com/contact-us.html •
Military Museum
Chronological display of uniforms, pictures, equipment, medals, pipe banners 44526

Inverurie

Carnegie Museum (closed) 44527

Ipswich

Christchurch Mansion, Soane St, Ipswich IP4 2BE • T: +44 1473 433554 • F: 433564 • christchurch.mansion@ipswich.gov.uk • www.ipswich.gov.uk •
Fine Arts Museum / Decorative Arts Museum / Public Gallery – 1896
English pottery, porcelain, glass, paintings (Gainsborough, Constable, Wilson Steer), prints, sculptures, furniture, in 16th c house 44528

HMS Ganges Museum, Old Sail Loft, Shotley Marina, Shotley Gate, Ipswich IP9 1QJ, mail addr: George Barnham, 48 Colchester Road, Bures, Suffolk CO8 5AE • T: +44 1787 228417 • secretary@hmsgangesmuseum.org.uk • www.hmsgangesmuseum.org •
Sec.: *George Barnham* •
Fine Arts Museum / Military Museum – 1982
Photographs, artefacts, memorabilia of The Royal Navy and HMS Ganges - the premier boy entrants training ship 1905-1976 44529

Ipswich Museum, High St, Ipswich IP1 3QH • T: +44 1473 433550/53 • F: 433558 • museums.service@ipswich.gov.uk • www.ipswich.gov.uk •
Dir.: *Tim Heyburn* • Sc. Staff: *David Jones* (Human History) • *D. Lampard* (Botany) • *S. Dummer* (Collections, Registration) • *R. Entwistle* (Cons.) • *T. Butler* (Public Services) • *M.R. Hayward* (Design) • *E. Dodds* (Design) • *R. Weaver* (Exhibitions) • *Catherine Richardson* (Art Education) • *M. Swift* (Christchurch Mansion) • *E. Richardson* (Ipswich Museum) •
Local Museum / Natural History Museum – 1881
Archaeology, geology, natural history, ethnology, wildlife 44530

Ipswich Transport Museum, Old Trolleybus Depot, Cobham Rd, Ipswich IP3 9JD • T: +44 1473 715666 • F: 832260 • enquiries@ipswichtransportmuseum.co.uk • www.ipswichtransportmuseum.co.uk •
Science&Tech Museum
Road transport, built or used in the Ipswich area, on air, water and rail transport – archive, library 44531

Ironbridge

Ironbridge Tollhouse, Ironbridge TF8 7AW • T: +44 1952 884391 • F: 884391 • tic@ironbridge.org.uk • www.ironbridge.org.uk •
Cur.: *David de Haan* •
Science&Tech Museum
First iron bridge, erected in 1779 44532

Teddy Bear Shop and Museum, Dale End, Ironbridge TF8 7AQ • T: +44 1952 433029 • info@ironbridge.org.uk • www.merrythought.co.uk •
Man.: *Barbara Stewart* •
Special Museum
Famous teddy bear factory, modern Merrythought products, historical coll of soft toys 44533

Irvine

Irvine Burns Club Museum, Fairburn, 5 Burns St, Irvine KA12 8RW • T: +44 1294 274511 •
Special Museum
Flax dresser works, letters from Honorary Members Dickens, Garibaldi, Tennyson 44534

Scottish Maritime Museum, Laird Forge Bldgs, Gottries Rd, Irvine KA12 8QE • T: +44 1294 278283 • F: 313211 • www.scottishmaritimemuseum.org • www.scottishmaritimemuseum.org •
Dir.: *David Thomson* •
Science&Tech Museum
Traditional maritime skills, shipbuilding, equipment, tools, boats, a Puffer 'Spartan', a tug 'Garnock', steam yacht 'Carola' – archives 44535

Vennel Gallery, 4-10 Glasgow Vennel, Irvine KA12 0BD • T: +44 1294 275059 • F: 275059 • vennel@north-ayrshire.gov.uk • www.north-ayrshire.gov.uk/museums •
Fine Arts Museum
Robert Burns' heckling shop and lodging house where he worked and lived, international art 44536

Isle-of-Barra

Dualchas-Museum Bharraigh Agus Bhatarsaidh, Castlebay, Isle-of-Barra HS9 5XD • T: +44 1871 810413 • F: 810413 •
Chm.: *Malcolm MacNeil* •
Local Museum 44537

Isle-of-Iona

Iona Abbey and Nunnery, Iona Abbey, Isle-of-Iona PA76 6SQ • T: +44 1681 700512 • F: 640217 • hs.ionaabbey@scotland.gsi.gov.uk • www.historic-scotland.gov.uk •
Religious Arts Museum – 1938
Celtic gravestones and crosses, Celtic artefacts 44538

Iona Heritage Centre, Isle-of-Iona PA76 6SJ • T: +44 1681 700576 • F: 700580 • heritage@ionagallery.com • www.isle-of-iona.com •
Cur.: *Mary Hay* •
Historical Museum
Social history 44539

Isle-of-Lewis

Dell Mill, North Dell, Ness, Isle-of-Lewis PA86 0SN •
Science&Tech Museum
Equipment, machinery, local grain production 44540

Isle-of-Skeye

Museum of the Isles, Clan Donald Visitor Centre, Armadale Castle, Isle-of-Skeye IV45 8RS • T: +44 1599 534454 • F: +44 1471 844735 • library@cland.demon.co.uk • www.clandonald.com •
Cur.: *Ann. M MacKinnon* •
Historical Museum
Story of 1300 yrs of Clan Donald's hist and of the Lordship of the Isles 44541

Isleworth

Osterley Park House, Jersey Rd, Isleworth TW7 4RB • T: +44 20 82325050 • F: 82325080 • osterley@nationaltrust.org.uk • www.nationaltrust.org.uk/osterley •
Man.: *Sian Harrington* •
Decorative Arts Museum – 1949
18th c villa, neo-classical interior decoration and furnishings designed by Robert Adam, landscape park and gardens 44542

Ivinghoe

Ford End Watermill, Station Rd, Ford End Farm, Ivinghoe LU6 3QB • T: +44 1582 600391 • www.fordendwatermill.co.uk •
Man.: *David Lindsey* •
Science&Tech Museum
Milling machines and artefacts 44543

Jackfield

Jackfield Tile Museum, Jackfield TF8 7JX • T: +44 1952 882030 • F: 884124 • tic@ironbridge.org.uk • www.ironbridge.org.uk •
Cur.: *Michael Vanns* •
Decorative Arts Museum
Decorative wall and floor tiles 44544

Jarrow

Bede's World, Church Bank, Jarrow NE32 3DY • T: +44 191 4892106 • F: 4282361 • visitor.info@bedesworld.co.uk • www.bedesworld.co.uk •
Dir.: *Keith Merrin* • Cur.: *Laura M. Sole* •
Archaeology Museum – 1974
Finds from the Anglo-Saxon and medieval monastic site of St. Paul's, Jarrow, Anglo-Saxon demonstration farm – library 44545

Jedburgh

Harestanes Countryside Visitor Centre, Harestanes, Ancrum, Jedburgh TD8 6UQ • T: +44 1835 830306 • F: 830734 • mascott@scotborders.gov.uk •
Natural History Museum
Environmental education, natural science, craft 44546

Jedburgh Castle Jail Museum, Castlegate, Jedburgh TD8 6QD • T: +44 1835 864750 • F: 864750 • museums@scotborders.gov.uk •
Historical Museum – 1965
19th c prison life, local history coll 44547

Mary Queen of Scots' House, Queen St, Jedburgh TD8 6EN • T: +44 1835 863331 • F: 863331 • museums@scotborders.gov.uk •
Cur.: *Fiona Colton* •
Special Museum
Facility for the 400th anniversary of the death of Mary Stuart ... 44548

Kegworth

Kegworth Museum, 52 High St, Kegworth DE74 2DA • T: +44 1509 670137 •
Local Museum
Village life and trades ... 44549

Keighley

Cliffe Castle, Spring Gardens Ln, Keighley BD20 6LH • T: +44 1535 618231 • F: 610536 • alison.armstrong@bradford.gov.uk • www.bradfordmuseums.org •
Dir.: *Jane Glaister* • Keeper: *Alison C. Armstrong* (Geology and Education) •
Local Museum / Natural History Museum – 1899
Paintings, sculpture, applied arts, household and farm tools, natural history, geology, archaeology, reconstructed craft shops, militaria, sports, toys, costumes ... 44550

East Riddlesden Hall, Bradford Rd, Keighley BD20 5EL • T: +44 1535 607075 • F: 691462 • eastriddlesdenhall@ntrust.org.uk • www.nationaltrust.org.uk •
Dir.: *Lloyd Taylor* •
Decorative Arts Museum
Coll of furniture, textiles, embroidery, pweter ... 44551

Ingrow Loco Museum, South St, Keighley BD21 5AX, mail addr: 296 Didsbury Rd, Stockport SK4 3JH • T: +44 1535 690739 • www.bahamas45596.co.uk •
Science&Tech Museum – 1968
Locomotive Museum, Coaltank Locomotiv No 1054 ... 44552

Kelevedon

Feering and Kelvedon Local History Museum, Aylett's School, Maldon Rd, Kelevedon CO5 9BA • T: +44 1376 571206 • F: 573163 • gh.wh@virgin.net • www.kelvedon.org.uk •
Dir.: *G.H. Wheldon* •
Local Museum
Local and manorial history, Roman remains, agricultural and domestic, schools, transport, post ... 44553

Kelmscott

Kelmscott Manor, Kelmscott GL7 3HJ • T: +44 1367 252486 • F: 253754 • admin@kelmscottmanor.co.uk • www.kelmscottmanor.co.uk •
Manager: *Jane Milne* • *Tristan Molloy* •
Decorative Arts Museum
Home of William Morris, coll of textiles, furniture and drawings by Morris and his associates ... 44554

Kelso

Floors Castle, Roxburghe Estate's Office, Kelso TD5 7SF • T: +44 1573 223333 • F: 226056 • marketing@floorscastle.com • www.floorscastle.com •
Dir.: *Duke of Roxburghe* •
Fine Arts Museum ... 44555

Kelvedon Hatch

Kelvedon Hatch Secret Nuclear Bunker, Kelvedon Hall Ln, Kelvedon Hatch CM14 5TL • T: +44 1277 364883 • F: 365260 • bunker@japar.demon.co.uk • www.japar.demon.co.uk •
Cur.: *M. Parrish* •
Historical Museum
Cold war artefacts ... 44556

Kendal

Abbot Hall Art Gallery, Kendal LA9 5AL • T: +44 1539 722464 • F: 722494 • info@abbothall.org.uk • www.abbothall.org.uk •
Dir.: *Edward King* •
Fine Arts Museum – 1962
18th c furnished rooms, paintings, sculpture, pottery, portraits by Kendal painters, Romney, works by Schwitters, Ruskin, Riley, Turner and Nicholson, watercolours of the English Lake District ... 44557

Kendal Museum, Station Rd, Kendal LA9 6BT • T: +44 1539 815597 • F: 737976 • info@kendalmuseum.org.uk • www.kendalmuseum.org.uk •
Cur.: *Morag Clement* •
Local Museum – 1796
Local history, archaeology, geology, natural history, British birds and mounted mammals ... 44558

Museum of Lakeland Life, Abbot Hall, Kendal LA9 5AL • T: +44 1539 722464 • F: 722494 • info@lakelandmuseum.org.uk • www.lakelandmuseum.org.uk •
Dir.: *Edward King* •
Local Museum – 1970
Local cultural history of Lake District area, industry, period rooms, costumes, trades, farming, printing presses, weaving equipment, photogr coll, Simpson furniture, Arthur Ransome Society ... 44559

Kenton

Powderham Castle, Kenton EX6 8JQ • T: +44 1626 890243 • F: 890729 • castle@powderham.co.uk • www.powderham.co.uk •
Man.: *Clare Crawshaw* •
Fine Arts Museum
Earl of Devon, channon book cases, music room, state bedroom & dinning room – libraries ... 44560

Keswick

Cars of the Stars Motor Museum, Standish St, Keswick CA12 5LS • T: +44 17687 73757 • F: 72090 • cotsmm@aol.com • www.carsofthestars.com •
Cur.: *John Nelson* • *Philip Nelson* •
Science&Tech Museum
Celebrity TV and film vehicles ... 44561

Cumberland Pencil Museum, Southey Works, Greta Bridge, Keswick CA12 5NG • T: +44 17687 73626 • F: 74679 • museum@acco-uk.co.uk • www.pencils.co.uk •
Dir.: *D.J. Sharrock* • Staff: *Y. Gray* •
Special Museum – 1981
Process of pencil manufacture, history of pencils using ... 44562

Keswick Mining Museum, Otley House, Otley Rd, Keswick CA12 5LE • T: +44 17687 80055 • coppermaid@aol.com • www.keswickminingmuseum.co.uk •
Dir.: *Ian Tyler* •
Science&Tech Museum – 1988
Minerals, geological and industrial hist, explosives, mining and quarring since the 16th c, minaeral coll – private archive ... 44563

Keswick Museum and Art Gallery, Fitz Park, Station Rd, Keswick CA12 4NF • T: +44 17687 73263 • F: 80390 • keswick.museum@allerdale.gov.uk • www.allerdale.gov.uk/keswick-museum •
Cur.: *Philip Crouch* •
Local Museum / Fine Arts Museum – 1873
Keswick's social and natural hist, archaeology, lit coll, geology ... 44564

Mirehouse, Underskiddaw, Keswick CA12 4QE • T: +44 17687 72287 • info@mirehouse.com • www.mirehouse.com •
Dir.: *Janaki Spedding* •
Fine Arts Museum
Pictures, furniture, works by Constable, Turner, De Wint, Hearne, Girtin, Romney, Morland, coll of papers and books of Francis Bacon ... 44565

Kettering

Alfred East Art Gallery, Sheep St, Kettering NN16 0AN • T: +44 1536 534274 • F: 534370 • museum@kettering.gov.uk • www.kettering.gov.uk •
Dir.: *Su Davies* •
Public Gallery – 1913
English paintings, watercolors, drawings and prints since 1800, Sir Alfred East, Thomas Cooper-Gotch ... 44566

Boughton House, Kettering NN14 1BJ • T: +44 1536 515731 • F: 417255 • llt@boughtonhouse.org.uk • www.boughtonhouse.org.uk •
Dir.: *Gareth Fitzpatrick* • Cust.: *Micheal Crick* •
Fine Arts Museum
Paintings of Van Dyck, English and Flemish tapestries, armoury ... 44567

Manor House Museum, Sheep St, Kettering NN16 0AN • T: +44 1536 534219 • F: 534370 • museum@kettering.gov.uk • www.kettering.gov.uk •
Dir.: *Su Davies* •
Local Museum
Hist of Kettering Borough, geology, archaeology, social and industrial hist ... 44568

Kew

Kew Bridge Steam Museum, Green Dragon Ln, Brentford, Kew TW8 0EN • T: +44 20 85684757 • F: 85699978 • info@kbsm.org • www.kbsm.org •
Dir.: *Lesley Bossine* •
Science&Tech Museum – 1975
Cornish beam engines dating from 1820 plus other stationary steam engines, all housed in a Victorian waterworks, 'Water for Life' exhibition on London's water supply ... 44569

Kew Palace Museum & Queen Charlotte's Cottage, Historic Royal Palaces, Royal Botanic Gardens, Kew TW9 3AB • T: +44 20 87819500 • F: 89481197 • www.hrp.org.uk •
Dir.: *Prof. Peter Crane* •
Decorative Arts Museum / Natural History Museum – 1759 ... 44570

Museum - Treasures from the National Archives, Ruskin Av, Kew TW9 4DU • T: +44 20 83925279 • F: 83925345 • events@nationalarchives.gov.uk • www.nationalarchives.gov.uk •
Man.: *Sue Laurence* •
Special Museum – 1902
Millenium treasure ... 44571

Kidderminster

Kidderminster Railway Museum, Station Approach, Comberton Hill, Kidderminster DY10 1QX • T: +44 1562 825316 • F: 861039 • krm@krm.org.uk • www.krm.org.uk •
Science&Tech Museum
Railways of the British Isles, signalling equipment ... 44572

Worcestershire County Museum, Hartlebury Castle, Kidderminster DY11 7XZ • T: +44 1299 250416 • F: 251890 • Museum@worcestershire.gov.uk • www.worcestershire.gov.uk/museum •
Dir.: *Robin Hill* •
Local Museum – 1966
Horse-drawn vehicles including gypsy caravans, glass, crafts, toys, costumes, folk life, archaeology, coll from the Victorian period, including costume, transport, domestic life, agriculture and crafts and trades – library ... 44573

Kidwelly

Kidwelly Industrial Museum, Broadford, Kidwelly SA17 4LW • T: +44 1554 891078 •
Science&Tech Museum
Local history, tinplate industry, local coal industry, gear and steam winding engine from Morlais colliery, geology, ... 44574

Kilbarchan

Weaver's Cottage, The Cross, Kilbarchan PA10 2JG • T: +44 1505 705588 • cmacleod@nts.org.uk •
Man.: *Chris Macleod* •
Historical Museum – 1954
18th c cottage, handloom weaver, looms, weaving equipment, domestic utensils, local history ... 44575

Kildonan, Isle of South Uist

Kildonan Museum, Museum Chill Donnan, South Uist, Kildonan, Isle of South Uist HS8 5RZ • T: +44 1878 710343 •
Local Museum / Ethnology Museum
Artefacts, photos, genealogical information, local archaeology ... 44576

Killin

Breadalbane Folklore Centre, Falls of Dochart, Killin FK21 8XE • T: +44 1567 820254 • F: 820764 • killin@visitscotlandcom.com • www.breadalbanefolk-lorecentre.com •
Man.: *Anne Tindall* •
Folklore Museum
Folklore, legends and myths of Breadalbane, contains the healing stones of St Fillan, exhibition on clan heritage, area wildlife and audio visual presentations ... 44577

Kilmarnock

Dean Castle, Dean Rd, off Glasgow Rd, Kilmarnock KA3 1XB • T: +44 1563 554702 • F: 554720 • jason.sutcliffe@east-ayrshire.gov.uk •
Historical Museum – 1976
Medieval arms and armour, musical instruments, tapestries ... 44578

Dick Institute Museum and Art Gallery, 14 Elmbank Av, Kilmarnock KA1 3BU • T: +44 1563 554343 • F: 554344 • jason.sutcliffe@east-ayrshire.gov.uk • www.east-ayrshire.gov.uk •
Dir.: *Adam Geary* •
Natural History Museum / Fine Arts Museum – 1901
Geology, ethnography, archaeology, conchology, armaments, numismatics, church and trade relics, paintings, etchings, sculptures, old bibles, incunabula, fine art, local and social history ... 44579

Kilmartin

Museum of Ancient Culture, Kilmartin House, Kilmartin PA31 8RQ • T: +44 1546 510278 • F: 510330 • museum@kilmartin.org • www.kilmartin.org •
Dir.: *Colin Schafer* •
Historical Museum / Archaeology Museum
Archaeology, landscape interpretation, prehistory, early Scottish hist, neolithic, bronze age, Dalriada, rock art, ancient monuments ... 44580

Kilsyth

Colzium Museum, Colzium House, Colzium-Lennox Estate, Kilsyth G65 0PY • T: +44 1236 735077 • F: 781407 •
Local Museum
Local hist, battle of Kilsyth – garden, woodland walk, chilren's zoo ... 44581

Kilsyth's Heritage, Library, Burngreen, Kilsyth G65 0HT • T: +44 1236 735077 • F: 781407 •
Local Museum
History of Kilsyth from the 18th c ... 44582

Kilwinning

Dalgarven Mill, Museum of Ayrshire Country Life and Costume, Dalry Rd, Kilwinning KA13 6PL • T: +44 1294 552448 • F: 552448 • admin@dalgarvenmill.org.uk • www.dalgarvenmill.org.uk •
Dir.: *Moira Ferguson* •
Science&Tech Museum / Folklore Museum
Water-driven flour mill, agricultural and domestic implements, costume ... 44583

Kilwinning Abbey Tower, Main St, Kilwinning KA13 6AN, mail addr: c/o North Ayrshire Museum, Manse St, Saltcoats KA21 5AA • T: +44 1294 464174 • F: 464174 • namuseum@north-ayrshire.gov.uk • www.north-ayrshire.gov.uk/museums •
Local Museum
Local and social history ... 44584

King's Lynn

Houghton Hall Soldier Museum, Houghton Hall, King's Lynn PE31 6UE • T: +44 1485 528569 • F: 528167 • administrator@houghtonhall.com • www.houghtonhall.com •
Military Museum – 1978 ... 44585

King's Lynn Arts Centre, 29 King St, King's Lynn PE30 1HA • T: +44 1553 779095 • F: 766834 • lfalconbridge@west-norfolk.gov.uk • www.kingslynnarts.org.uk •
Man.: *Liz Falconbridge* •
Public Gallery ... 44586

King's Lynn Museum, Market St, King's Lynn PE30 1NL • T: +44 1553 775001 • F: 775001 • lynn.museum@norfolk.gov.uk • www.museums.norfolk.gov.uk •
Dir.: *Robin Hanley* • Cur.: *Timothy Thorpe* •
Local Museum – 1904
Natural hist, archaeology, local hist, geology, temporary exhibition gallery ... 44587

Tales of the Old Gaol House, Saturday Market Pl, King's Lynn PE30 5DQ • T: +44 1553 774297 • F: 772361 • gaolhouse@west-norfolk.gov.uk • www.west-norfolk.gov.uk •
Historical Museum
History of crime and punishment in Lynn, King John Cup (1340), King John Sword ... 44588

Town House Museum of Lynn Life, 46 Queen St, King's Lynn PE30 5DQ • T: +44 1553 773450 • F: 775001 • townhouse.museum@norfolk.gov.uk • www.museums.norfolk.gov.uk •
Dir.: *Robin Hanley* •
Folklore Museum – 1992
Ceramic and glass, furniture, Victoriana ... 44589

True's Yard Fishing Museum, North St, King's Lynn PE30 1QW • T: +44 1553 770479 • F: 765100 • info@truesyard.co.uk • www.truesyard.co.uk •
Man.: *Joanna Barrett* •
Historical Museum ... 44590

Kingsbridge

Cookworthy Museum of Rural Life in South Devon, c/o Old Grammar School, 108 Fore St, Kingsbridge TQ7 1AW • T: +44 1548 853235 • wcookworthy@talk21.com • www.devonmuseums.net •
Chair: *Margaret Lorenz* •
Local Museum – 1971
Cookworthy's porcelain and work as a chemist, costumes, craft tools, photos, agricultural machinery, Edwardian pharmacy, Victorian kitchen – library, archives ... 44591

Kingston-upon-Hull

Arctic Corsair, 25 High St, Kingston-upon-Hull HU1 3DX, mail addr: c/o Ferens Art Gallery, Queen Victoria Sq, Kingston-upon-Hull HU1 3RA • T: +44 1482 613902 • F: 613710 • museums@hullcc.gov.uk •
Dir.: *Arthur G. Credland* •
Science&Tech Museum
Deep sea trawler (1960) ... 44592

Burton Constable Hall, Skirlaugh, Kingston-upon-Hull HU11 4LN • T: +44 1964 562400 • F: 563229 • helendewson@btconnect.com • www.burtonconstable.com •
Dir.: *Dr. D.P. Connell* •
Decorative Arts Museum
Paintings, furniture ... 44593

Ferens Art Gallery, Hull City Museum & Art Gallery, Queen Victoria Sq, Kingston-upon-Hull HU1 3RA • T: +44 1482 613902 • F: 613710 • museums@hullcc.gov.uk • www.hullcc.gov.uk/museums •
Keeper: *Kirsten Simister* (Art) • Asst. Keeper: *Laura Turner* (Collections) • *Christine Brady* (Exhibitions) •
Fine Arts Museum – 1927
Coll of old masters (British, Dutch, Flemish, French, Italian), portraits (16th c to present), marine painting, 19th & 20th c British art, contemporary British painting, sculpture and photography ... 44594

Guildhall Collection, Queen Victoria Square, Kingston-upon-Hull HU1 3RA, mail addr: c/o Hull Museums Monument Buildings, Alfred Gelder Street • T: +44 1482 300300 • F: 613710 • museums@hullcc.gov.uk •
Cur.: *Robin Diaper* •
Fine Arts Museum / Decorative Arts Museum / Historic Site – 2004
Paintings, sculptures, antiques, silver, Hull Tapestry ... 44595

Hands on History, South Church Side, Kingston-upon-Hull HU1 1RR, mail addr: Queen Victoria Sq, Kingston-upon-Hull HU1 3RA • T: +44 1482 613902 • F: 613710 • museums@hullcc.gov.uk • www.hullcc.gov.uk/museums •

U

Dir.: *Lydia Saul* •
Historical Museum – 1988
Hull and its people, Victorian Britain, ancient Egypt 44596

Hull and East Riding Museum, 36 High St, Kingston-upon-Hull HU1 1NQ, mail addr: Queen Victoria Sq, Kingston-upon-Hull HU1 3RA • T: +44 1482 613902 • F: 613710 • museums@hullcc.demon.co.uk • www.hullcc.gov.uk/museums •
Keeper: *Martin Foreman* (Archaeology) •
Archaeology Museum / Local Museum – 1925
Prehistoric and early hist of Humberside, mosaics found at the Roman villa at Horkstow, archaeology, geology and natural history 44597

Hull University Art Collection, Middleton Hall, University Cottingham Rd, Kingston-upon-Hull HU6 7RX • T: +44 1482 465035 • F: 465192 • j.g.bernasconi@hull.ac.uk • www.hull.ac.uk/artcoll •
Dir.: *John G. Bernasconi* •
Public Gallery / Decorative Arts Museum – 1963
Art in Britain 1890-1940, the Thompson Collections of Chinese porcelain 44598

Spurn Lightship, Castle St, Hull Marina, Kingston-upon-Hull HU1 3TJ, mail addr: Queen Victoria Sq, Kingston-upon-Hull HU1 3RA • T: +44 1482 613902 • F: 613710 • museum@hullcc.gov.uk • www.hullcc.gov.uk/museums •
Dir.: *Arthur G. Credland* •
Science&Tech Museum 44599

Streetlife Museum, Hull Museum of Transport, 26 High St, Kingston-upon-Hull HU1 1PS, mail addr: Queen Victoria Sq, Kingston-upon-Hull HU1 3RA • T: +44 1482 613902 • F: 613710 • museums@hullcc.gov.uk • www.hullcc.gov.uk •
Keeper: *Michael Hughes* •
Science&Tech Museum
Railway displays, street scene, carriage gallery 44600

Wilberforce House, 25 High St, Kingston-upon-Hull HU1 1NQ, mail addr: Queen Victoria Sq, Kingston-upon-Hull HU1 3RA • T: +44 1482 613902 • F: 613710 • museums@hullcc.gov.uk • www.hullcc.gov.uk •
Dir.: *Vanessa Salter* (Social History) •
Historical Museum – 1906
Silver, costume, dolls, slavery, slave trade and abolition, items relating to William Wilberforce, born here in 1759 – library 44601

Kingston-upon-Thames

Kingston Museum, Wheatfield Way, Kingston-upon-Thames KT1 2PS • T: +44 20 85476460 • F: 85476747 • king.mus@rbk.kingston.gov.uk • www.kingston.gov.uk/museum •
Head: *Anne McCormack* •
Local Museum – 1904
Archaeology, oral hist, social hist, borougn archives, Eadweard Muybridge colls, Martin Brothers (art pottery) 44602

The Stanley Picker Gallery, Middle Mill Island, Knights Park, Kingston-upon-Thames KT1 2QJ • T: +44 20 85478074 • F: 85478068 • picker@kingston.ac.uk • www.kingston.ac.uk/picker •
Dir.: *Prof. Bruce Russell* •
Public Gallery 44603

Kingswinford

Broadfield House Glass Museum, Compton Dr, Kingswinford DY6 9NS • T: +44 1384 812745 • glass.museum@dudley.gov.uk • www.glassmuseum.org.uk •
Decorative Arts Museum – 1980
Coll of British glass from the 17th c to the present 44604

Kingussie

Kingussie Museum, Highland Folk Museum, Duke St, Kingussie PH21 1JG • T: +44 1540 661307 • F: 661631 • highland.folk@highland.gov.uk • www.highlandfolk.com •
Cur.: *Robert Powell* •
Open Air Museum – 1934
Agriculture, costume, furniture, tartans, folklore, social history, vehicles of Scotland 44605

Kinnesswood

Michael Bruce Cottage Museum, The Cobbles, Kinnesswood KY13 9HL • T: +44 1592 840203 •
Dir.: *Dr. David M. Munro* •
Special Museum – 1906
Manuscripts by poet Michale Bruce, local articles 44606

Kinross

Kinross Museum, Loch Leven Community Campus, Kinross KY13 7FQ • T: +44 1738 632488 • F: 443505 • info@kinrossmuseum.co.uk • www.kinrossmuseum.co.uk •
Local Museum 44607

Kirkcaldy

John McDouall Stuart Museum, Rectory Ln, Dysart, Kirkcaldy KY1 2TP • T: +44 1592 412860 • F: 412870 • kirkcaldy.museum@fife.gov.uk • www.fifedirect.org.uk/museums •
Cur.: *Dallas Machan* •
Historical Museum
Australian exploration by John McDouall, hist of Dysart 44608

Kirkcaldy Museum and Art Gallery, War Memorial Gardens, Abbotshall Road, Kirkcaldy KY1 1YG • T: +44 1592 412860 • F: 412870 • kirkcaldy.museum@fife.gov.uk •
Co-ord.: *Dallas Mechan* •
Local Museum – 1926
Local ceramics (ware), paintings by W.M. McTaggart and S.J. Peploe, Adam Smith items, archaeology, history, natural sciences 44609

Kirkcudbright

Broughton House and Garden, 12 High St, Kirkcudbright DG6 4JX • T: +44 1557 330437 • F: 330437 • seastgate@nts.org.uk • www.nts.org.uk •
Man.: *S. Eastgate* •
Special Museum – 1950
18th c town house, photographs, paintings by Edward Atkinson Hornel (1864-1933), Dumfries and Galloway history and literature – library, Robert Burns archives 44610

The Stewartry Museum, 6 Saint Mary St, Kirkcudbright DG6 4AQ • T: +44 1557 331643 • F: 331643 • davidd@dumgal.gov.uk • www.dumgal.gov.uk/museums •
Dir.: *David Devereux* •
Local Museum – 1881
Items concerning the Stewartry of Kirkcudbright, natural hist, geology, archaeology, fine art 44611

Tolbooth Art Centre, High St, Kirkcudbright DG6 4JL • T: +44 1557 331556 • F: 331643 • davidd@dumgal.gov.uk • www.dumgal.gov.uk/museums •
Dir.: *David Devereux* •
Public Gallery
Paintings from the art colony, contemporary art 44612

Kirkintilloch

Auld Kirk Museum, East Dunbartonshire Museums, Cowgate, The Cross, Kirkintilloch G66 1AB • T: +44 141 5780144 • F: 5780140 • peter.mccormack@eastdunbarton.gov.uk •
Cur.: *Peter McCormack* •
Local Museum
Local hist, art and craft living conditions of working-class families (early 20th c), weaving, mining, boatbuilding and ironfounding industries, displays of tools, equipment and photographs 44613

Kirkoswald

Souter Johnnie's Cottage, Main Rd, Kirkoswald KA19 8HY • T: +44 1655 760603 • F: 760671 •
Special Museum
Home of John Daidson, village cobbler and original Souter (shoemaker) 44614

Kirkwall

Orkney Museum, Tankerness House, Broad St, Kirkwall KW15 1DH • T: +44 1856 873191 • F: 871560 • museum@orkney.gov.uk • www.orkneyheritage.com •
Dir.: *Janette A. Park* •
Archaeology Museum / Historical Museum – 1968
Extensive archaeological coll, social hist 44615

Orkney Wireless Museum, Kiln Corner, Junction Rd, Kirkwall KW15 1LB • T: +44 1856 871400, 874272 • www.owm.org.uk •
Dir.: *Peter MacDonald* •
Science&Tech Museum – 1983
Coll of wartime radio and defences used at Scapa Flow, also hist of domestic wireless in Orkney 44616

Kirriemuir

Barrie's Birthplace & Camera Obscura, 9 Brechin Rd, Kirriemuir DD8 4BX • T: +44 1575 572646 • F: 575461 •
Man.: *Mary Brownlow* •
Special Museum – 1963
Life and work of writer Sir James M. Barrie (1860-1937), memorabilia, manuscripts 44617

Kirriemuir Gateway to the Glens Museum, The Townhouse, 32 High St, Kirriemuir DD8 4BB • T: +44 1575 575479 • F: 462590 • kirriegateway@angus.gov.uk • www.angus.gov.uk/history •
Cur.: *Fiona Guest* •
Local Museum
Archaeology and social hist town and western angus glens, wildlife and multimedia programme 44618

Knaresborough

Knaresborough Castle and Old Courthouse Museum, Castle Yard, Knaresborough HG5 8AS • T: +44 1423 556188 • F: 556130 • lg12@harrogate.gov.uk • www.harrogate.gov.uk/museums •
Local Museum / Historic Site
Medieval castle with King's Tower, underground sallyport, tudor courthouse, civil war gallery, local history, site archaeology, art 44619

Saint Robert's Cave, Abbay Rd, Knaresborough HG5 8AS • T: +44 1423 556188 • F: 556130 • lg12@harrogate.gov.uk • www.harrogate.gov.uk/museums •
Dir.: *Ceryl Evans* •
Religious Arts Museum
Medival heritage with living area and chapel remains 44620

Knebworth

Knebworth House, Park, Knebworth SG3 6PY • T: +44 1438 812661 • F: 811908 • info@knebworthhouse.com • www.knebworthhouse.com •
Dir.: *Martha Lytton Cobbold* •
Decorative Arts Museum
Tudor to Victorian, Gothic and 20th c interior, wallpapers, rustic wall paintings, literary coll 44621

Knutsford

First Penny Farthing Museum, 92 King St, Knutsford WA16 6ED • T: +44 1565 653974 •
Dir.: *Glynn Stockdale* •
Science&Tech Museum
Largest coll of penny farthing bicycle in the UK 44622

Knutsford Heritage Centre, 90a King St, Knutsford WA16 6ED • T: +44 1565 650506 • www.knutsfordheritage.com •
Man.: *Valerie Dawson* •
Local Museum
Millennium tapestry 44623

Tabley House, Tabley Ln, Knutsford WA16 0HB • T: +44 1565 750151 • F: 653230 • tableyhouse@btconnect.com • www.tableyhouse.co.uk •
Cur.: *Peter Cannon-Brookes* •
Fine Arts Museum / Historic Site
Coll of British paintings (1790's to 1826) formed by Sir John-Fleming Leicester Bart 44624

Tatton Park, Knutsford Cheshire WA16 6QN • T: +44 1625 374400 • F: 374403 • tatton@cheshire.gov.uk • www.tattonpark.org.uk •
Man.: *Brendan Flanagan* •
Decorative Arts Museum
Coll Gillow furniture, pictures by Canaletto, Carracci, De Heem, Poussin, Nazzari, Van Dyck, silver by Paul Storr, porcelain, European, English and Chinese, carpet 44625

Kyle

Raasay Heritage Museum, 6 Osgaig Park, Raasay, Kyle W40 8PB • T: +44 1478 660207 • F: +44 870 1227170 • osgaig@lineone.net •
Chm.: *F. Maclennan* •
Local Museum 44626

Lacock

Fox Talbot Museum, Lacock SN15 2LG • T: +44 1249 730459 • F: 730501 • roger.watson@nationaltrust.org.uk • www.nationaltrust.org.uk/lacock •
Cur.: *Roger Watson* •
Science&Tech Museum – 1975
Coll of manuscripts, documents, correspondence and photographs relating to the invention and discovery of the positive/negative photographic process by William Henry Fox Talbot (1800-1877), photographic technology – research and documentation facilities 44627

Lackham Museum of Agriculture and Rural Life, Wiltshire College, Lackham, Lacock SN15 2NY • T: +44 1249 466800 • sheena.awdry@wiltshire.ac.uk • www.lackhamcountrypark.co.uk •
Cur.: *Andrew Davies* •
Agriculture Museum – 1946
Agricultural tools, machinery, rural life displays, historic farm buildings 44628

Lanark

Lanark Museum, 8 Westport, Lanark ML11 9HD • T: +44 1555 666681 •
Local Museum
Lanark history, 19th c costume, photographs, maps and plans 44629

Lancaster

Cottage Museum, 15 Castle Hill, Lancaster LA1 1YS • T: +44 1524 64637 • F: 841692 • lancaster.citymuseum@mus.lancscc.gov.uk • www.lancashire.gov.uk/education/museums/ •
Dir.: *Edmund Southworth* •
Decorative Arts Museum – 1978
Artisan cottage, furnished in the style of the 1820s 44630

Judges' Lodgings, Church St, Lancaster LA1 1YS • T: +44 1524 32808, 846315 • F: 846315 • judgeslodgings.lcc@btinternet.com • www.lancsmuseums.gov.uk •
Dir.: *Edmund Southworth* • Keeper: *Anthea Dennett* •
Decorative Arts Museum
Local hist, toys, dolls, games, costumes, coll of Gillow furniture (18th-20th c) 44631

King's Own Royal Regiment Museum, Market Sq, Lancaster LA1 1HT • T: +44 1524 555619 • F: 841692 • kingsownmuseum@iname.com • www.kingsownmuseum.plus.com •
Dir.: *Edmund Southworth* • Cur.: *Peter Donnelly* •
Military Museum – 1929
Archives, medals, photographs, uniforms, memorabilia 44632

Lancaster City Museum, Old Town Hall, Market Sq, Lancaster LA1 1HT • T: +44 1524 64637 • F: 841692 • lancaster.citymuseum@mus.lancscc.gov.uk •
Dir.: *Edmund Southworth* • Sc. Staff: *P. Donelly* (Military History) • Asst. Keeper: *H. Dowler* •
Local Museum / Decorative Arts Museum / Military Museum / Archaeology Museum – 1923
Local hist, decorative arts, Roman finds, weights and measures, Gillow furniture, drawings, local paintings 44633

Lancaster Maritime Museum, Custom House, Saint George's Quay, Lancaster LA1 1RB • T: +44 1524 382264 • F: 841692 • lancaster.citymuseum@mus.lancscc.gov.uk •
Dir.: *Edmund Southworth* •
Science&Tech Museum – 1985
Ship models, ship fittings, navigational aids, customs equipment, charts 44634

Peter Scott Gallery, c/o University of Lancaster, Lancaster LA1 4YW • T: +44 1524 593057 • F: 592603 • galleryenquiries@lancaster.ac.uk • www.peterscottgallery.com •
Dir.: *Mary Gavagan* •
Fine Arts Museum / Decorative Arts Museum / University Museum
Art, pottery, ceramics, sculptures 44635

Roman Bath House, Vicarage Field, Lancaster LA1 1HN • T: +44 1524 64637 • F: 841692 • lancaster.citymuseum@mus.lancscc.gov.uk •
Dir.: *Edmund Southworth* •
Archaeology Museum 44636

Ruskin Library, Lancaster University, Bailrigg, Lancaster LA1 4YH • T: +44 1524 593587 • F: 593580 • ruskin.library@lancaster.ac.uk • www.lancs.ac.uk/users/ruskinlib •
Dir.: *S.G. Wildman* •
Library with Exhibitions / Fine Arts Museum / University Museum – 1997
Drawings and waretcolours bythe writer John Ruskin, paintings, drawings by artists associated with him, photographs, letters a.o. manuscripts – library 44637

Langbank

The Dolly Mixture, Finlaystone Country Estate, Langbank PA14 6TJ • T: +44 1475 540505 • F: 540285 • info@finlaystone.co.uk • www.finlaystone.co.uk •
Dir.: *Arthur MacMillan* •
Special Museum – 1996
Finlaystone Doll Collection - The Dolly Mixture - All sorts of dolls from around the world 44638

Langholm

Clan Armstrong Museum, Lodge Walk, Langholm DG13 0NY • T: +44 13873 81610 • info@armstrongclan.org • www.armstrongclan.org •
Cur.: *Miriam Bibby* •
Historical Museum
Armstrong clan artefacts, maps, documents, portraits, weapons, armour, ancient Reiver clan of the Scottish border – library, archives 44639

Langton Matravers

Langton Matravers Museum, Saint George's Close, Langton Matravers BH19 3HZ • T: +44 1929 423168 • F: 427534 • localhistory@langtonia.org.uk • www.langtonia.gov.uk •
Cur.: *Reg Saville* •
Science&Tech Museum – 1974
History of the local stone industry from Roman times to the present day, quarrying equipment, stone-masons' tools, model of a section of underground working, test pieces, photographs, social history exhibits 44640

Lapworth

Packwood House, Solihull, Lapworth B94 6AT • T: +44 1564 783294 • F: 782912 •
Decorative Arts Museum – 1941
17th c country house with tapestries, needlework, English antique furniture, topiary 44641

Largs

Largs Museum, Main St, Kirkgate House, Manse Court, Largs KA30 8AW • T: +44 1475 687081 • kathryn.valentine@virgin.net • freespace.virgin.net/mike.mackenzie2/LDHSoc.htm •
Local Museum – 1973
Local hist 44642

Larne

Larne Museum, 2 Victoria Rd, Carnegie Bldg, Larne BT40 1RN • T: +44 28 28279482 • F: 28260660 • admin@larne.gov.uk • www.larne.gov.uk •
Head: *Joan Morris* •
Local Museum
Local history colls incl country kitchen, smithy, agricultural implements, old photographs, old newspapers, maps 44643

Latheron

Clan Gunn Museum, Old Parish Church, Latheron KW1 4DD • T: +44 1593 721325 • F: 721325 •
Local Museum
Hist of Orkney and the North 44644

Lauder

Thirlestane Castle, Lauder TD2 6RU • T: +44 1578 722430 • F: 722761 • admin@thirlestanecastle.co.uk • www.thirlestanecastle.co.uk •
Head: *I.I.D. Garner* •
Historical Museum 44645

U

Laugharne

Dylan Thomas Boathouse, Dylan's Walk, Laugharne SA33 4SD • T: +44 1994 427420 • F: 427420 • boathouse@carmarthenshire.gov.uk •
Special Museum
Dylan Thomas' writing hut, memorabilia 44646

Launceston

Launceston Steam Railway, Saint Thomas Rd, Launceston PL15 8DA • T: +44 1566 775665 •
Dir.: *Bowman* •
Science&Tech Museum
Locomotives, vintage cars and motorcycles, stationary steam engines, machine tools 44647

Lawrence House Museum, 9 Castle St, Launceston PL15 8BA • T: +44 1566 773277 •
Cur.: *Jean Brown* •
Local Museum
Local hist, WW II and home front, coins, feudal dues, rural and domestic Victorian life, costume 44648

Laxey

Museum Cheann A'Loch (Kinloch Historical museum), Laxay Hall, Laxay HS2 9PJ • T: +44 1851 830778 •
Local Museum 44649

Laxey

Great Laxey Wheel and Mines Trail, Manx National Heritage, Laxey IM4 7AH • T: +44 1624 648000 • F: 648001 • enquiries@mnh.gov.im • www.storyofmann.com •
Dir.: *Stephen Harrison* •
Science&Tech Museum
Largest working water wheel in the world, pump water from the lead and zinc mines, Victorian engineering 44650

Murray's Motorcycle Museum, The Bungalow, Snaefell, Laxey IM4 • T: +44 1624 861719 •
Science&Tech Museum
Over 120 motorcycles (1902-1961), motoring memorabilia 44651

Laxfield

Laxfield and District Museum, The Guildhall, Laxfield IP13 8DU • T: +44 1986 798026 •
Local Museum
Rural domestic and working life, local archaeology, costumes and natural hist 44652

Leadhills

Leadhills Miners' Library Museum, 15 Main St, Leadhills ML12 6XP • T: +44 1659 74242 • F: +44 1864 54206 • leadhills@miners-library.fsnet.co.uk •
Chm.: *Harry Shaw* •
Library with Exhibitions
Minerals, orig. book stock, library holds objects, photos 44653

Leamington Spa

Leamington Spa Art Gallery and Museum, Royal Pump Rooms, The Parade, Leamington Spa CV32 4AA • T: +44 1926 742700 • F: 742705 • prooms@warwickdc.gov.uk • www.warwickdc.gov.uk/wdc/royalpumprooms •
Dir.: *Jeffrey Watkin* •
Fine Arts Museum / Decorative Arts Museum – 1928
British, Dutch, Flemish paintings (16th-20th c), ceramics from Delft, Wedgwood-Whieldon, Worchester, 18th c drinking glasses, costumes, photography, ethnography, social hist 44654

Leatherhead

Leatherhead Museum of Local History, c/o Hampton Cottage, 64 Church St, Leatherhead KT22 8DP • T: +44 1372 386348 • leatherheadmuseum@localhistory.free-online.co.uk • www.leatherheadlocalhistory.org.uk •
Cur.: *Graham Evans* •
Local Museum
Local hist, coll of Art Deco Ashtead pottery 44655

Ledaig

Lorn Museum, An Sailean, Ledaig PA37 1QS • T: +44 1631 720282 •
Dir.: *Catherine MacDonald* •
Local Museum 44656

Ledbury

Ledbury Heritage Centre, Church Ln, Ledbury HR8 1DN • T: +44 1531 635680 • F: +44 1432 342492 • herefordmuseums@herefordshire.gov.uk • www.herefordshire.gov.uk •
Local Museum / Science&Tech Museum 44657

Leeds

Abbey House Museum, Leeds Museums and Galleries, Kirkstall Rd, Leeds LS5 3EH • T: +44 113 2305492 • F: 2305499 • abbeyhouse.museum@virgin.net • www.leeds.gov.uk •
Cur.: *Samantha Flavin* •
Decorative Arts Museum / Local Museum – 1926
Reconstructions of 3 street scenes (19th c) cottages, Victorian city, childhood, Kirkstall abbey 44658

Art Library, Municipal Bldgs, Leeds LS1 3AB • T: +44 113 2478247 • F: 2478247 • artlibrary@leedslearning.net • www.leeds.gov.uk •
Dir.: *Catherine Blanshard* •
Library with Exhibitions 44659

City Art Gallery, Leeds Museums and Galleries, The Headrow, Leeds LS1 3AA • T: +44 113 2478248 • F: 2449689 • city.art.gallery@leeds.gov.uk • www.leeds.gov.uk/artgallery •
Dir.: *John Roles* • Cur.: *Nigel Walsh* (Exhibitions) • *Matthew Whitey* (Sculpture) • *Amanda Phillips* (Education) •
Fine Arts Museum – 1888
Victorian paintings, early English watercolours, British 20th c paintings and sculpture 44660

Henry Moore Institute, 74 The Headrow, Leeds LS1 3AH • T: +44 113 2343158, 2467467 • F: 2461481 • hmi@henry-moore.ac.uk • www.henry-moore-fdn.co.uk •
Cur.: *Penelope Curtis* •
Fine Arts Museum
Sculpture production and reception, drawings, letters – research facilities 44661

Leeds Industrial Museum, Leeds Museums and Galleries, Armley Mills, Canal Rd, Leeds LS12 2QF • T: +44 113 2637861, 2244362 • F: 2244365 • armley.mills@leeds.gov.uk • www.leeds.gov.uk/armleymills •
Dir.: *John Roles* • Cur.: *Neil Dowlan* (Engineering) •
Science&Tech Museum
Large woollen mill, textile, printing, engineering, locomotives 44662

Leeds Metropolitan University Gallery, Woodhouse Ln, Leeds LS1 3HE • T: +44 113 2833130 • F: 2835999 • m.innes@lmu.ac.uk • www.lmu.ac.uk/arts •
Cur.: *Moira Innes* •
Public Gallery 44663

Leeds Museum, Leeds Institute, Millenium Sq, Leeds LS10 1AA • T: +44 113 2478275 • F: 2342300 • evelyn.silber@leeds.gov.uk •
Dir.: *John Roles* •
Local Museum – 1820
Natural science, archaeology, anthropology, local history 44664

Royal Armouries Museum, Armouries Dr, Leeds LS10 1LT • T: +44 113 2201940, 2201999 • F: 2201934 • enquiries@armouries.org.uk • www.armouries.org.uk •
Chief Ex.: *Paul Evans* (Master of the Armouries) • Acad. Dir.: *Greame Rimer* • Dep. Dir.: *Steve Burt* (Collections, Education) • Keeper: *Thom Richardson* (European Armour and Oriental Ars and Armour) • Head: *Philip Abbott* (Information, Research) • Sc. Staff: *Suzanne Kitto* (Collection Care) •
Military Museum – 1996
Arms and armour coll, five galleries describe the evolution of arms and armour in war, tournament, oriental, self-defence and hunting, computer programmes and films on this topic – Conservation, library, education centre 44665

Temple Newsam House, Leeds Museums and Galleries, Leeds LS15 0AE • T: +44 113 2647321 • F: 2602285 • john.roles@leeds.gov.uk • www.leeds.gov.uk •
Dir.: *John Rules* • Sc. Staff: *Anthony Wells-Cole* (Historic Interiors, Crafts) • *James Lomax* (Silver) • *Adam White* (Ceramics) • *Polly Putnam* (Textiles) •
Fine Arts Museum / Decorative Arts Museum – 1922
English and Continental paintings (16th-20th c), furniture, silver, ceramics, country house from 1490 44666

Tetley's Brewery Wharf, The Waterfront, Leeds LS1 1QG • T: +44 113 2420666 • F: 2594125 •
Man.: *Ian Glenholme* •
Science&Tech Museum
History of Joshua Tetley & Son Brewery, brewing and allied trades 44667

Thackray Museum, Beckett St, Leeds LS9 7LN • T: +44 113 2444343 • F: 2470219 • info@thackraymuseum.org • www.thackraymuseum.org •
Dir.: *Almut Grüner* • Cur.: *Joanne Stewardson* • Libr.: *Alan Humphries* •
Historical Museum – 1997
History of surgery, medicine, health care and medical supply trades – library 44668

Thwaite Mills Watermill, Thwaite Ln, Stourton, Leeds LS10 1RP • T: +44 113 2762887 • F: 2776737 • www.leeds.gov.uk/thwaitemills •
Science&Tech Museum
Working water-powered mill 44669

University Gallery Leeds, Parkinson Bldg, Woodhouse Ln, Leeds LS2 9JT • T: +44 113 3432777 • F: 3435561 • gallery@leeds.ac.uk • www.leeds.ac.uk/gallery •
Head: *Dr. Hilary Diaper* •
Public Gallery / University Museum
British paintings, drawings and prints (16th-20th c), Bloomsbury group 44670

Leek

Out of the Blue Gallery, 30 Stockwell St, Leek ST13 6DW • T: +44 1538 483732 • F: 483733 •
Public Gallery – 1884 44671

Leicester

Abbey Pumping Station Museum, Corporation Rd, off Abbey Ln, Leicester LE4 5PX • T: +44 116 2995111 • F: 2995125 • www.leicestermuseums.ac.uk •
Cur.: *Nick Ladlow* •
Science&Tech Museum
Horse drawn vehicles, bicycles, motorcycles, motor vehicles, 1891 Beam pumping engines 44672

Belgrave Hall Museums and Gardens, Church Rd, off Thurcaston Rd, Leicester LE4 5PE • T: +44 116 2666590 • museums@leicester.gov.uk • www.leicestermuseums.ac.uk •
Dir.: *Sarah Leritt* • Cur.: *Nick Ladlow* •
Decorative Arts Museum
Queen Anne House with 18th-19th c furniture, botanical/ historical gardens 44673

The City Gallery, 90 Granby St, Leicester LE1 1DJ • T: +44 116 2232060 • F: 2232069 • city.gallery@leicester.gov.uk • www.leicester.gov.uk/citygallery •
Public Gallery 44674

Guildhall, Guildhall Ln, Leicester LE1 5FQ • T: +44 116 2532569 • F: 2539626 • www.leicestermuseums.ac.uk •
Dir.: *Nick Ladlow* •
Historical Museum
Medieval timber construction 44675

Jewry Wall Museum, Leicester Museums, Saint Nicholas Circle, Leicester LE1 4LB • T: +44 116 2254971 • museums@leicester.gov.uk • www.leicestermuseums.ac.uk •
Dir.: *S. Levitt* •
Archaeology Museum – 1966
Roman wall dating from the 2nd c, local archaeology, Roman milestone, mosaics and painted wallplaster 44676

Leicestershire CCC Museum, County Ground, Grace Rd, Leicester LE2 8AD • T: +44 116 2832128 • F: 2440363 • enquiries@leicestershireccc.co.uk • www.leicestershireccc.co.uk •
Chief Exec.: *Paul Mayland-Mason* •
Special Museum
Coll of cricket memorabilia to the Leicestershire County Cricket Club 44677

Museum of the Royal Leicestershire Regiment, Magazine Gateway, Oxford St, Leicester LE2 7BY • T: +44 116 2473222 • F: 2470403 •
Dir.: *S. Levitt* • Cur.: *J.A. Legget* •
Military Museum – 1969
Hist of the Royal Leicestershire Regiment, in building from 1400, medieval gateway 44678

The National Gas Museum, 195 Aylestone Rd, Leicester LE2 7QH • T: +44 116 2503190 • F: 2503190 • information@gasmuseum.co.uk • www.gasmuseum.co.uk •
Cur.: *M. Martin* •
Science&Tech Museum – 1977
Gas production (tools and equipment), gas lighting, meters, gas appliances (heating, cooking etc.) ... 44679

National Space Centre, Exploration Dr, Leicester LE4 5NS • T: +44 116 2582111 • F: 2582100 • info@spacecentre.co.uk • www.spacecentre.co.uk •
Dir.: *George Barnett* •
Science&Tech Museum 44680

New Walk Museum and Art Gallery, 53 New Walk, Leicester LE1 7EA • T: +44 116 2254900 • F: 2254927 • museums@leicester.gov.uk • www.leicester.gov.uk/museums •
Dir.: *Sarah Levitt* • Cur.: *Simon Lake* • Sc. Staff: *Jane May* • *Mark Evans* •
Fine Arts Museum / Decorative Arts Museum / Natural History Museum – 1849
Geology, natural hist, mammalogy, ornithology, ichthyology, biology, English painting, German expressionists, French painting (19th-20th c), old masters, English ceramics, silver, egyptology – library 44681

Newarke Houses Museum, The Newarke, Leicester LE2 7BY • T: +44 116 2473222 • F: 2254982 • www.leicestermuseums.ac.uk •
Dir.: *Nick Ladlow* • Cur.: *Jane May* (Decorative Art) • Cons.: *Libby Finney* • *Fiona Graham* •
Music Museum / Science&Tech Museum – 1940
Clock making, mechanical musical instruments, reconstructed 19th c street, reconstructed workshop of 18th c clockmaker Samuel Deacon, sewing accessories, tokens, toys, Gimson furniture 44682

Record Office for Leicestershire, Leicester and Rutland, Long St, Leicester LE18 2AH • T: +44 116 2571080 • F: 2571120 • recordoffice@leics.gov.uk • www.leics.gov.uk •
Fine Arts Museum
Official, public and private archives and local studies library, from medieval times to date, serves Leicestershire, the City of Leicester and Rutland, jointly supported by Leicestershire County Council, Leicester City Council and Rutland County Council 44683

William Carey Museum, Central Baptist Church, Charles St, Leicester LE1 1LA • T: +44 116 2873864, 2550413 • F: 2766862 • keithw.harrison@virgin.net • www.central-baptist.org.uk •
Cur.: *Keith Harrison* •
Religious Arts Museum – 1960
Taleaux of scenes from life of William Carey (1762-1834), modern Overseas missionary, 4 vols Voyages of James Cook (1784) 44684

Leigh, Lancashire

Turnpike Gallery, Civic Sq, Leigh, Lancashire WN7 1EB • T: +44 1942 404469 • F: 404447 • turnpikegallery@wlct.org • www.wlct.org •
Officer: *Martyn Lucas* •
Public Gallery
20th c prints, contemporary art 44685

Leigh-on-Sea

Leigh Heritage Centre, 13a High St, Old Town, Leigh-on-Sea SS9 2EN • T: +44 1702 470834 • carole.pavitt@btopenworld.com •
Fine Arts Museum
Leigh photographs and artefacts 44686

Leighton Buzzard

Ascott House, Ascott Estate, Office Wing, Leighton Buzzard LU7 0PS • T: +44 1296 688242 • F: 681904 • info@ascottestate.co.uk • www.ascottestate.co.uk •
Decorative Arts Museum
French and Chippendale furniture, original needlework, paintings by Rubens, Hogarth, Gainsborough, Hobbema, Oriental porcelain, former possession of Anthony de Rothschild 44687

Leighton Buzzard Railway, Page's Park Station, Billington Rd, Leighton Buzzard LU7 4TN • T: +44 1525 373888 • F: 377814 • info@buzzrail.co.uk • www.buzzrail.co.uk •
Chm.: *Mervyn Leah* •
Science&Tech Museum
Over 50 unique and historic locomotives 44688

Leiston

Long Shop Museum, Main St, Leiston IP16 4ES • T: +44 1728 832189 • F: 832189 • longshop@care4free.net • www.longshop.care4free.net •
Man.: *Stephen Mael* •
Science&Tech Museum – 1980
History of Richard Garrett eng works for over 200 years 1778-1980, display of Leiston air base occupied by 357th fighter group USAAF World War II 44689

Leominster

Croft Castle, Leominster HR6 9PW • T: +44 1568 780246 •
Fine Arts Museum 44690

Leominster Folk Museum, Etnam St, Leominster HR6 8AL • T: +44 1568 615186 •
Cur.: *Lynne Moult* •
Folklore Museum / Local Museum
Local trades and industry, cider house, dairy, costume, social hist 44691

Lerwick

Böd of Gremista Museum, Gremista, Lerwick ZE1 0PX • T: +44 1595 694386 • F: 696729 • museum@sic.shetland.gov.uk • www.shetland-museum.org.uk •
Dir.: *T. Watt* •
Historical Museum
Birthplace of Arthur Anderson 44692

Shetland Museum, Hay's Dock, Lerwick ZE1 0EL • T: +44 1595 695057 • F: 696729 • museum@sic.shetland.gov.uk • www.shetland-museum.org.uk •
Dir.: *T. Watt* •
Local Museum – 1970 44693

Letchworth

First Garden City Heritage Museum, 296 Norton Way South, Letchworth SG6 1SU • T: +44 1462 482710 • F: 486056 • fgchm@letchworth.com • www.letchworth.com •
Cur.: *Josh Tidy* •
Historical Museum – 1977 44694

Letchworth Museum and Art Gallery, Broadway, Letchworth SG6 3PF • T: +44 1462 685647 • F: 481879 • letchworth.museum@north-herts.gov.uk • www.north-herts.gov.uk •
Dir.: *Rosamond Allwood* •
Local Museum / Fine Arts Museum – 1914
Archaeology, natural hist, local hist, fine and decorative art, costume 44695

North Hertfordshire Museums, North Herts District Council, Gernon Rd, Letchworth SG6 3JF • T: +44 1462 474274 • F: 474500 •
Association with Coll 44696

Lewes

Anne of Cleves House Museum, 52 Southover High St, Lewes BN7 1JA • T: +44 1273 474610 • F: 486990 • anne@sussexpast.co.uk • www.sussexpast.co.uk •
Cur.: *Steve Watts* •
Decorative Arts Museum / Folklore Museum
Furniture, domestic equipment and a wide range of material illustrating the everyday life of local people during the 19th and early 20th c 44697

Lewes Castle and Barbican House Museum, 169 High St, Lewes BN7 1YE • T: +44 1273 486290 • F: 486990 • castle@sussexpast.co.uk • www.sussexpast.co.uk •

Man.: *Sally White* •
Historical Museum
Built on the orders of William the Conqueror soon after the conquest of 1066, history of the area ... 44698

Military Heritage Museum, 7-9 West St, Lewes BN7 2NJ • T: +44 1273 480208 • F: 476562 • www.wallisandwallis.co.uk •
Cur.: *S.R. Butler* •
Military Museum – 1977
Roy Butler Coll – Hist of the British army 1650-1914 ... 44699

Museum of Sussex Archaeology, Barbican House, 169 High St, Lewes BN7 1YE • T: +44 1273 486290 • F: 486990 • castle@sussexpast.co.uk • www.sussexpast.co.uk •
Dir.: *Emma O'Connor* •
Archaeology Museum
Prehistoric, Roman and Saxon archaeology, watercolors ... 44700

Leyland

British Commercial Vehicle Museum, King St, Leyland PR5 1LE • T: +44 1772 451011 • F: 623404 • enquiries@commercialvehiclemuseum.co.uk • www.commercialvehiclemuseum.co.uk •
Dir.: *Andrew Buchan* •
Science&Tech Museum – 1983
Historic commercial vehicles and buses, truck and bus building ... 44701

South Ribble Museum and Exhibition Centre, Old Grammar School, Church Rd, Leyland PR25 3FJ • T: +44 1772 422041 • F: 625363 •
Cur.: *Dr. D.A. Hunt* •
Local Museum
Archaeological material from excavations at the Roman site at Walton-le-Dale, and a wide range of smaller items relating to the history of the town and the area ... 44702

Lichfield

Friary Art Gallery, The Friary, Lichfield WS13 6QG • T: +44 1543 510700 • F: 510716 • lichfield.library@staffordshire.gov.uk •
Fine Arts Museum / Decorative Arts Museum – 1859
Crafts, textiles ... 44703

Hanch Hall, Lichfield WS13 8HH • T: +44 1543 490308 •
Dir.: *Colin Lee* • *Linda Lee* •
Local Museum ... 44704

Lichfield Heritage Centre, Market Sq, Lichfield WS13 6LG • T: +44 1543 256611 • F: 414749 • info@lichfieldheritage.org.uk • www.lichfieldheritage.org.uk •
Local Museum ... 44705

Samuel Johnson Birthplace Museum, Breadmarket St, Lichfield WS13 6LG • T: +44 1543 264972 • F: 258441 • sjmuseum@lichfield.gov.uk • www.samueljohnsonbirthplace.org.uk •
Special Museum – 1901
Memorabilia of Dr. Samuel Johnson (1709-1784), paintings, association books, documents, letters of Johnson, Boswell, Anna Seward – library ... 44706

Staffordshire Regiment Museum, Whittington Barracks, Lichfield WS14 9PY • T: +44 121 434390 • F: 434391 • museum@rhqstaffords.fsnet.co.uk • www.armymuseums.org.uk •
Cur.: *Erik Blakeley* •
Military Museum – 1963
Hist of the regiment, medals, armaments, uniforms, exterior WW I trench, Anderson shelters ... 44707

Wall Roman Site and Museum - Letocetum, off A5, Watling St, Wall, Lichfield WS14 0AW, mail addr: POB 569, Swindon SN2 2YP • T: +44 1543 480768 • customers@english-heritage.org.uk • www.english-heritage.org.uk •
Cur.: *Sara Lunt* • Cust.: *Kevin Lynch* •
Archaeology Museum – 1912
Roman finds, pottery, coins, tools, , Roman military road to North Wales (Watling St) ... 44708

Lifton

Dingles Steam Village - Fairground Museum, Lifton PL16 0AT • T: +44 1566 783425 • F: 783584 • info@fairground-heritage.org.uk • www.fairground-heritage.org.uk •
Science&Tech Museum
Steam and other early road rollers, lorries, fairground engines and industrial engines, old road signs, fairground art, belt driven machinery ... 44709

U

Lincoln

City and County Museum, 12 Friars Ln, Lincoln LN2 5AL • T: +44 1522 530401 • F: 530724 • librarian@lincolncathedral.com • www.lincolnshire.gov.uk/thecollection •
Dir.: *Dr. Nicholas Bennett* •
Historical Museum
Perhistoric, Roman and medievial Lincolnshire, armoury ... 44710

Doddington Hall, Doddington, Lincoln LN6 4RU • T: +44 1522 694308 • F: 682584 • info@doddingtonhall.com • www.doddingtonhall.com •
Dir.: *Antony Jarvis* •
Association with Coll
China, glass, furniture, textiles, pictures ... 44711

Lincoln Cathedral Treasury, Cathedral, Lincoln • T: +44 1522 552222 ext 2805, 811308 •
Dir.: *O.T. Griffin* •
Religious Arts Museum ... 44712

Lincolnshire Road Transport Museum, Whisby Rd, off Doddington Rd, Lincoln LN6 3QT • T: +44 1522 689497 •
Science&Tech Museum – 1959/93
Vehicles dating from 1920s to 1960s, mostly restored and with local connections ... 44713

Museum of Lincolnshire Life, Burton Rd, Lincoln LN1 3LY • T: +44 1522 528448 • F: 521264 • lincolnshire-life_museum@lincolnshire.gov.uk • www.lincolnshire.gov.uk/visiting/museums/museum-of-lincolnshire-life •
Dir.: *Jon Finch* •
Local Museum – 1969
Agricultural and industrial machinery, horse-drawn vehicles, displays devoted to domestic, community and commercial life within the county, crafts and trades, Royal Lincolnshire Regiment – windmill archive . 44714

Royal Lincolnshire Regiment Museum, Burton Rd, Lincoln LN1 3LY • T: +44 1522 528448 • F: 521264 • lincolnshirelife.museum@lincolnshire.gov.uk •
Dir.: *Jon Finch* •
Military Museum – 1986
Hist of the regiment, uniforms, medals, war trophies, armaments, social history coll relating to Lincolnshire ... 44715

Sir Joseph Banks Conservatory, Lincoln's Lawn, Union Rd, Lincoln LN1 3BL • T: +44 1522 560330 • daisy.arnold@lincoln.gov.uk •
Science&Tech Museum
aquarium ... 44716

Usher Gallery, Lindum Rd, Lincoln LN2 1NN • T: +44 1522 527980 • F: 560165 • usher.gallery@lincolnshire.gov.uk • www.thecollection.lincoln.museum/ •
Fine Arts Museum / Decorative Arts Museum – 1927
Paintings, watercolours, watches, miniatures, silver, porcelain, manuscripts and memorabilia of Alfred Lord Tennyson (1809-1892) ... 44717

Linford

Walton Hall Museum, Walton Hall Rd, Linford SS17 0RH • T: +44 1375 671874 • F: 641268 • www.waltonhallmuseum.com •
Science&Tech Museum
Historic farm machinery, gypsy caravan, motor rollers ... 44718

Linlithgow

House of the Binns, Linlithgow EH49 7NA • T: +44 1506 834255 •
Dir.: *Kathleen Dalyell* •
Historical Museum / Decorative Arts Museum – 1944
Historic home of the Dalyells, 17th c moulded plaster ceilings, paintings, porcelain, furniture ... 44719

Linlithgow Palace, Edinburgh Rd, Linlithgow EH49 6QS • T: +44 1506 842896 •
Decorative Arts Museum ... 44720

Linlithgow Story, Annet House, 143 High St, Linlithgow EH49 7JH • T: +44 1506 670677 •
Local Museum ... 44721

Linlithgow Union Canal Museum, Canal Basin, Manse Rd, Linlithgow EH49 6AJ • T: +44 1506 671215 • info@lucs.org.uk • www.lucs.org.uk •
Science&Tech Museum / Historic Site ... 44722

Lionacleit

Museum Nan Eilean, Sgoil Lionacleit, Lionacleit HS7 5PJ • T: +44 1870 602864 • F: 602864 • dmacphee1a@fnes.net • www.cne-siar.gov.uk •
Cur.: *Dana MacPhee* •
Local Museum
Hist and culture of the Islands ... 44723

Liphook

Hollycombe Steam Collection, Iron Hill, Liphook GU30 7UP • T: +44 1428 724900 • F: 723682 • info@hollycombe.co.uk • www.hollycombe.co.uk •
Dir.: *Brian Gooding* •
Science&Tech Museum ... 44724

Lisburn

Irish Linen Centre and Lisburn Museum, Market Sq, Lisburn BT28 1AG • T: +44 28 92663377 • F: 92672624 • irishlinencentre@lisburn.gov.uk • www.lisburn.gov.uk •
Cur.: *Brian Mackey* •
Special Museum
Linen, archaeology, social hist, art and industry of the Lagan valley – specialist library ... 44725

Lisnaskea

Folklife Display, Library, Lisnaskea BT92 0AD • T: +44 28 67721222 •
Local Museum ... 44726

Litcham

Litcham Village Museum, Fourways, Mileham Rd, Litcham PE32 2NZ • T: +44 1328 701383 •
Chm.: *R.W. Shaw* •
Local Museum
Local artefacts ... 44727

Little Walsingham

Shirehall Museum, Common Pl, Little Walsingham NR22 6BP • T: +44 1328 820510 • F: 820098 • jackie@walsingham-estate.co.uk • www.walsingham.uk.com •
Historical Museum
Local and village history, pilgrimages to Walsingham, court ... 44728

Littlehampton

Littlehampton Museum, Manor House, Church St, Littlehampton BN17 5EW • T: +44 1903 738100 • F: 731690 • littlehamptonmuseum@littlehampton-tc.gov.uk • www.littlehampton-tc.gov.uk •
Local Museum – 1928
Nautical and maritime exhibits, photography ... 44729

Liverpool

Anfield Museum, c/o LFC, Anfield Rd, Liverpool L4 0TH • T: +44 151 2606677, 2632361 • F: 2608813 •
Special Museum
Football history in Liverpool ... 44730

The Beatles Story, Britannia Vaults, Albert Dock, Liverpool L3 4AD • T: +44 151 7091963 • F: 7080039 • info@beatlesstory.com • www.beatlesstory.com •
Music Museum – 1990 ... 44731

Conservation Centre Presentation, Whitechapel, Liverpool L1 6HZ • T: +44 151 4784980 • F: 4784990 • info@liverpoolmuseums.org.uk • www.conservationcentre.org.uk •
Special Museum
Works of museum conservators ... 44732

Croxteth Hall, Croxteth Hall Ln, Liverpool L12 0HB • T: +44 151 2336910 • F: 2282817 • croxtethcountrypark@liverpool.gov.uk • www.croxteth.co.uk •
Dir.: *Irene Vickers* • Cur.: *Julia Carder* •
Decorative Arts Museum
Edwardian furnishings ... 44733

The Garstang Museum of Archaeology, University of Liverpool, School of Archaeology, Classics and Egyptology, 14 Abercromby Sq, Liverpool L69 7WZ • T: +44 151 7942467 • F: 7942226 • winkerpa@liverpool.ac.uk • www.liv.ac.uk/sace •
Head: *Dr. Steven R. Snape* •
Archaeology Museum
Objects from the excavations in Egypt and the Near East, classical Aegean and prehistoric antiquities, coins ... 44734

Her Majesty Customs and Excise National Museum, c/o Merseyside Maritime Museum, Albert Dock, Liverpool L3 4AQ • T: +44 151 4784499 • F: 4784590 • custumsandexcise@liverpoolmuseums.org.uk • www.customsandexcisemuseum.org.uk •
Cur.: *Karen Bradbury* •
Special Museum
Battle between smugglers and duty men since the 1700's ... 44735

Lark Lane Motor Museum, 1 Hesketh St, Liverpool L17 8XJ •
Science&Tech Museum
Cars, motorcycles and a wide range of mechanical devices ... 44736

Lennon House, 251 Menlove Av, Liverpool L25 6ET • T: +44 151 7088574 •
Music Museum ... 44737

Liverpool Scottish Regimental Museum, Forbes House, Botanic Rd 139-141, Liverpool L7 5PZ, mail addr: c/o Scot Riley, 51a Common Ln, Culcheth WA3 4EY • T: +44 151 6455717 • ilriley@liverpoolscottish.org.uk • www.liverpoolscottish.org.uk •
Cur.: *Ian Riley* •
Military Museum
Uniforms, weapons, equipment, photographs and documents, history of the Regiment from 1900 to the present day ... 44738

McCartney House, 20 Forthlin Rd, Liverpool L18 9TN • T: +44 151 7088574 •
Music Museum ... 44739

Merseyside Maritime Museum, Albert Dock, Liverpool L3 4AQ • T: +44 151 4784499 • F: 4784590 • maritime@liverpoolmuseums.org.uk • www.merseysidemaritimemuseum.org.uk •
Dir.: *Michael Stammers* •
Science&Tech Museum – 1980
Coll of full-size craft, incl the pilot boat, Edmund Gardner (1953), coll of models, paintings, marine equipment, the history of cargo-handling in the Port of Liverpool, the develop and operation of the encl dock system ... 44740

Museum of Dentistry, c/o School of Dental Surgery, University of Liverpool, Pembroke Pl, Liverpool L69 3BX • T: +44 151 7065279, 7062000 • F: 7065809 •
Cur.: *John Cooper* •
Special Museum – 1880
Hist of dentistry, dentistal education, equipment 44741

Museum of Liverpool Life, New Museum of Liverpool (reopening 2011), Pier Head, Liverpool L3 1PZ • T: +44 151 4784060 • F: 4784090 • liverpoollife@liverpoolmuseums.org.uk • www.museumofliverpoollife.org.uk •
Dir.: *Graham Boxer* • Keeper: *Mike Stammers* •
Historical Museum – 1993
Displays on Making a Living, Demanding a Voice and Mersey Culture, forthcoming developments: Phase 2 - Homes and Communities, Phase 3 - The King's Regiment ... 44742

Open Eye Gallery, 28-32 Wood St, Liverpool L1 4AQ • T: +44 151 7099460 • F: 7093059 • info@openeye.org.uk • www.openeye.co.uk •
Dir.: *Patrick Henry* •
Public Gallery ... 44743

Speke Hall, The Walk, Liverpool L24 1XD • T: +44 151 4277231 • F: 4279860 •
Dir.: *S. Osborne* •
Decorative Arts Museum – 1970
Decorative art, social history, interior design (16th-19th c), furniture (17th-19th c), Victorian furnishings (19th c) ... 44744

Sudley House, Mossley Hill Rd, Aigburth, Liverpool L18 8BX • T: +44 151 7243245 • F: 7290531 • sudley@liverpoolmuseums.org.uk • www.sudleyhouse.org.uk •
Dir.: *Julian Treuherz* •
Fine Arts Museum – 1986
British paintings ... 44745

Tate Liverpool, Albert Dock, Liverpool L3 4BB • T: +44 151 7027400 • F: 7027401 • liverpoolinfo@tate.org.uk • www.tate.org.uk/liverpool •
Dir.: *Christoph Grunenberg* • Head: *Simon Groom* (Exhibitions) •
Public Gallery – 1988
National coll of 20th c art ... 44746

University of Liverpool Art Gallery, Art and Heritage Collections, Victoria Gallery and Museum, Ashton Str, Liverpool L69 3BX • T: +44 151 7942348 • F: 7944787 • artgall@liv.ac.uk • www.liv.ac/vgm •
Fine Arts Museum / University Museum – 1977
Watercolours by Turner, Girtin, Cozens, Cotman, early English porcelain, oil paintings by Turner, Audubon, Joseph Wright of Derby, Augustus John, sculpture by Epstein ... 44747

The Walker Art Gallery, William Brown St, Liverpool L3 8EL • T: +44 151 4784199 • F: 4784190 • walker@nmgm.org • www.thewalker.org.uk •
Keeper: *Julian Treuherz* •
Fine Arts Museum – 1877
European painting, sculpture, drawings, watercolors and prints since 1300, Italian, Dutch and Flemish paintings (13th-17th c), British paintings (17th-19th c), 19th c Liverpool art, contemporary painting ... 44748

Western Approaches, 1 Rumford St, Liverpool L2 8SZ • T: +44 151 2272008 • F: 2366913 • www.liverpoolwarmuseum.co.uk •
Historical Museum
Vist the labyrinth in the 50,000 sq ft headquarters beneath the streets of Liverpool, incl main operations room, Anderson Shelter, teleprinters, educational centre, admiral's office etc ... 44749

World Museum Liverpool, William Brown St, Liverpool L3 8EN • T: +44 151 4784393 • F: 4784390 • themuseum@liverpoolmuseums.org.uk • www.worldmuseumliverpool.org.uk •
Dir.: *Dr. David Fleming* • Keeper: *John Millard* • Sc. Staff: *Edmund Southworth* (Antiquities) • *R. Cowell* (Archaeology) • *L. Stumpe* (Ethnology) • *I. Wallace* (Zoology) • *J. Edmondson* (Botany) • *Margaret Warhurst* (Antiquities) •
Local Museum – 1851
Archaeology, astronomy, botany, ethnology, geology, ceramics, decorative arts, zoology, musical instruments, vivarium and aquarium, natural history ... 44750

Livingston

Almond Valley Heritage Centre, Millfield, Kirkton North, Livingston EH54 7AR • T: +44 1506 414957 • F: 497771 • rac@almondvalley.co.uk • www.almondvalley.co.uk •
Dir.: *Dr. Robin Chesters* •
Local Museum – 1990
Agriculture, Scottish shale oil industry, local & social hist ... 44751

Llanberis

Electric Mountain - Museum of North Wales, Amgueddfa'r Gogledd, Llanberis LL55 4UR • T: +44 1286 870636 • info@electricmountain.co.uk • www.fhc.co.uk/electric_mountain.htm •
Man.: *Dorothy Lamb* •
Local Museum
Art, temporary exhibitions, underground trips to Dinorwig pumped storage power station ... 44752

Llanberis Lake Railway, Gilfach Ddu, Llanberis LL55 4TY • T: +44 1286 870549 • F: 870549 • info@lake-railway.co.uk • www.lake-railway.co.uk •
Science&Tech Museum ... 44753

National Slate Museum, Pardarn Country Park, Llanberis LL55 4TY • T: +44 1286 870630 • F: 871906 • slate@nmgw.ac.uk • www.nmgw.ac.uk/wsm •
Dir.: *Dr. Dafydd Roberts* • Cur.: *Cadi Iolen* •
Natural History Museum – 1971
Geology, industry, slate, engineering, victorian technology, social history, mining, quarrying ... 44754

Llandrindod Wells

Museum of Local History and Industry, Temple St, Llandrindod Wells LD1 5DL • T: +44 1686 412605 • Cur.: *Chris Wilson* •
Local Museum / Archaeology Museum
Castell Collen Roman Fort, archaeology, local history and industry, dolls, life in Llanidloes 19th-20th c 44755

National Cycle Collection, Automobile Palace, Temple St, Llandrindod Wells LD1 5DL • T: +44 1597 825531 • F: 825531 • cycle.museum@care4free.net • www.cyclemuseum.org.uk •
Cur.: *David Higman* •
Science&Tech Museum
Bicycle since 1818, photographs, early lighting art, racing stars 44756

Radnorshire Museum, Temple St, Llandrindod Wells LD1 5DL • T: +44 1597 824513 • F: 825781 • radmus@powys.gov.uk •
Cur.: *Will Adams* •
Local Museum 44757

Llandudno

Great Orme Tramway, Victoria Station, Church Walks, Llandudno LL30 2NB • T: +44 1492 879306 • tramwayenquiries@conwy.gov.uk • www.greatormetramway.com •
Science&Tech Museum
Cable-hauled street tramway (only one in the world outside San Francisco), 4 tramcars working in 2 pairs, trams built by Hurst Nelson in 1902/3 44758

Llandudno Museum and Art Gallery, Chardon House, 17-19 Gloddaeth St, Llandudno LL30 2DD • T: +44 1492 876517 • F: 876517 • llandudno.museum@lineone.net • www.llandudno-tourism.co.uk/museum •
Cust.: *Richard Ll. Hughes* •
Local Museum / Fine Arts Museum
Archaeology, hist and early industry of the area, with a special section on early 20th-c tourism and entertainment 44759

Oriel Mostyn Gallery, 12 Vaughan St, Llandudno LL30 1AB • T: +44 1492 879201, 870875 • F: 878869 • post@mostyn.org • www.mostyn.org •
Dir.: *Martin Barlow* •
Public Gallery 44760

Llandysul

National Wool Museum, National Museum Wales, Dre-fach Felindre, Llandysul SA44 5UP • T: +44 1559 370929 • F: 371592 • wool@museumwales.ac.uk • www.nmgw.ac.uk/mwwi •
Man.: *Ann Whittall* •
Science&Tech Museum
Hist of the Welsh woollen ind, former Cambrian mill, produced shirts and shawls, blankets and bedcovers, woollen stockings for the local marked, flannell makers, textiles machinery, gallery of the variety of Welsh textiles (blankets to unusual 19th c knitted socks) 44761

Llanelli

Parc Howard Museum and Art Gallery, Parc Howard, Felinfoel Rd, Llanelli SA15 3LJ • T: +44 1554 772029 • F: +44 1267 223830 • chrisdelaney@carmarthenshire.gov.uk • www.carmarthenshire.gov.uk •
Cur.: *Ann Dorsett* •
Decorative Arts Museum / Fine Arts Museum / Local Museum – 1912
Pottery, art 44762

Public Library Gallery, Vaughan St, Llanelli SA15 3AS • T: +44 1554 773538 • F: 750125 •
Dir.: *D.F. Griffiths* •
Public Gallery 44763

Llanfair Caereinion

Welshpool and Llanfair Light Railway, The Station, Llanfair Caereinion SY21 0SF • T: +44 1938 810441 • F: 810861 • info@wllr.org.uk • www.wllr.org.uk •
Dir.: *Terry Turner* •
Science&Tech Museum
Narrow gauge steam railway 44764

Llanfairpwll

Plas Newydd, Llanfairpwll LL61 6DQ • T: +44 1248 714795 • F: 713673 • plasnewydd@nationaltrust.org.uk • www.nationaltrust.org.uk •
Dir.: *Paul Carr-Griffin* •
Historical Museum
Wall paintings, relics of the 1st Marquess of Anglesey and the Battle of Waterloo, and the Ryan coll of military uniforms and headdresses 44765

Llangefni

Oriel Ynys Môn (Anglesey Heritage Gallery), Rhosmeirch, Llangefni LL77 7TQ • T: +44 1248 724444 • F: 750282 • agxlh@anglesey.gov.uk • www.anglesey.gov.uk •
Head: *Alun Gruffydd* •
Local Museum
Tunnicliffe wildlife art coll, Welsh art, archaeology, geology, social hist 44766

Llangernyw

Sir Henry Jones Museum, Y Cwm, Llangernyw LL22 8PR • T: +44 1492 575371 • F: 513664 • info@sirhenryjones-museums.org • www.sirhenryjones-museums.org •
Chm.: *John Hughes* •
Local Museum
Furniture and domestic items, craft tools, books, photographs and costume relating to Sir Henry Jones and village life 44767

Llangollen

Canal Museum, The Wharf, Llangollen LL20 8TA • T: +44 1978 860702 • F: 860799 • sue@horsedrawnboats.co.uk • www.horsedrawnboats.co.uk •
Dir.: *D.R. Knapp* • *S. Knapp* •
Historical Museum – 1974
Canals and canals boats 44768

Y Capel, Castle St, Llangollen LL20 8NU • T: +44 1978 860828 • kim.dewsbury@denbighshire.gov.uk •
Public Gallery
Contemporary paintings 44769

Llangollen Motor Museum, Pentrefelin, Llangollen LL20 8EE • T: +44 1978 860324 • llangollen-motormuseum@hotmail.com • www.llangollen-motormuseum.co.uk •
Owner: *Ann Owen* • Dir.: *Gwilym Owen* •
Science&Tech Museum
Cars, motorcycles, memorabilia from 1910-1970's, history of Britain's canals 44770

Llangollen Railway, The Station, Abbey Rd, Llangollen LL20 8SN • T: +44 1978 860979 • F: 869247 • www.llangollen-railway.co.uk •
Dir.: *Gordon Heddon* •
Science&Tech Museum
Locomotive, passenger stock and goods vehicles in its possession and care 44771

Plas Newydd, Hill St, Llangollen LL20 8AW • T: +44 1978 861314 • F: 708258 • rose.mcmahon@denbighshire.gov.uk •
Dir.: *Rose MacMahon* •
Special Museum
Home of the Ladies of Llangollen 44772

Llanidloes

Llanidloes Museum, Town Hall, Great Oak St, Llanidloes SY18 6BU, mail addr: c/o Powysland Museum, Canal Wharf, Welshpool SY21 7AQ • T: +44 1938 554656 • F: 554656 • powysland@powys.gov.uk • powysmuseums.powys.gov.uk •
Dir.: *Eva B. Bredsdorff* •
Local Museum – 1933 44773

Llantrisant

Model House Gallery, Bull Ring, Llantrisant CF72 8EB • T: +44 1443 237758 • F: 224718 •
Public Gallery 44774

Royal Mint Museum, The Royal Mint, Llantrisant CF72 8YT • T: +44 1443 623061 • information.office@royalmint.gov.uk • www.royalmint.gov.uk/RoyalMint/web/site/Corporate/Corp_museum •
Special Museum 44775

Llwynypia

Rhondda Museum, Glyncornel Environmental Study Centre, Llwynypia CF40 2HT • T: +44 1443 431727 •
Local Museum 44776

Lockerbie

Rammerscales, Lockerbie DG11 1LD • T: +44 1387 811988 •
Fine Arts Museum
20th c paintings, tapestries, sculpture 44777

London

198 Gallery, 194-198 Railton Rd, Herne Hill, London SE24 0LU • T: +44 20 79788309 • F: 77375315 • gallery@198gallery.co.uk • www.198gallery.co.uk •
Dir.: *Lucy Davies* •
Public Gallery
Contemporary art by artists from diverse cultural backgrounds – education centre 44778

AJEX Military Museum → Jewish Military Museum

Alexander Fleming Laboratory Museum, Saint Mary's Hospital, Praed St, London W2 1NY • T: +44 20 78866528 • F: 78866739 • kevin.brown@st-marys.nhs.uk • www.st-marys.nhs.uk/about/fleming_museum.htm •
Dir.: *Kevin Brown* •
Science&Tech Museum – 1993
Reconstruction of the laboratory in which Fleming discovered Penicillin in 1928 44779

Alfred Dunhill Museum, 48 Jermyn St, Saint James, London SW1Y 6LX • T: +44 20 78388233 • F: 78388304 • peter.tilley@dunhill.com • www.dunhill.com •
Cur.: *Peter Tilley* •
Decorative Arts Museum
Dunhill antique products, incl. motoring accessories, pens, compendiums, lighters, watches – archive 44780

All Hallows Undercroft Museum, Byward St, London EC3R 5BJ • T: +44 20 74812928 • F: 74883333 • office@hallows-tower.swinternet.co.uk • www.allhallowsbythetower.org.uk •
Cur.: *T. Barber* •
Religious Arts Museum
Roman pavement in situ, classical tombstones, a model of the Roman city, ancient registers, a Crusader Altar 44781

The Anaesthesia Heritage Centre, The Charles King Collection, 21 Portland Pl, London W1B 1PY • T: +44 20 76318806/811 • F: 76314352 • heritage@aagbi.org • www.aagbi.org •
Dir.: *Patricia Willis* •
Historical Museum
Historic anaesthetic equipment 44782

Anaesthetic Museum, c/o Dept. of Anaesthesia, Saint Bartholomew's Hospital, London EC1A 7BE • T: +44 20 76017518 •
Owner: *Dr. D.J. Wilkinson* •
Historical Museum
Encompasses apparatus general, local anaesthetics since 1800's 44783

Apsley House, 149 Piccadilly, Hyde Park Cnr, London W1J 7NT • T: +44 20 74995676 • F: 74936576 • www.apsleyhouse.org.uk •
Cur.: *Susan Jenkins* •
Decorative Arts Museum / Fine Arts Museum – 1952
Art coll, decorative objects, paintings, silver, sculpture, porcelain 44784

Architecture Foundation, 2a Kingsway Pl, Sans Walk, London EC1R 0LS • T: +44 20 72533334 • F: 72533335 • mail@architecturefoundation.org.uk • www.architecturefoundation.org.uk •
Dir.: *Rowan Moore* •
Public Gallery 44785

Arsenal Museum, Arsenal Football Club, Emirates Stadium, London N5 1BU • T: +44 20 77044000 • F: 77044001 • info@arsenal.co.uk •
Special Museum – 1993/ 2006
Football based coll, dating from 1886 to the present day 44786

Baden-Powell House, 65-67 Queen's Gate, London SW7 5JS • T: +44 20 75847031 • F: 75905103 • www.towntocountry.co.uk/bpHouse/ •
Special Museum – 1961
Life and work of Lord Robert Baden-Powell (1857-1941), founder of the Boy Scouts 44787

Bank of England Museum, Threadneedle St, London EC2R 8AH • T: +44 20 76015545 • F: 76015808 • museum@bankofengland.co.uk • www.bankofengland.co.uk/museum •
Cur.: *John Keyworth* •
Special Museum – 1988
Banknotes, coins and medals, photographs, pictures, prints and drawings, statuary, artefacts, Roman and medieval 44788

Bankside Gallery, 48 Hopton St, London SE1 9JH • T: +44 20 79287521 • F: 79282820 • info@banksidegallery.com • www.banksidegallery.com •
Dir.: *Angela Parker* •
Public Gallery 44789

Banqueting House, Historic Royal Palaces, Whitehall Palace, London SW1A 2ER • T: +44 203 1666184 • F: 1666159 • www.hrp.org.uk •
Cust.: *Irma Hay* • Dep. Cust.: *SimonWhite Lumley* •
Fine Arts Museum
Historic building, completed in 1622 to a design by Inigo Jones, ceiling paintings by Rubens installed in 1635 44790

Barbican Art Gallery, Barbican Centre, Silk St, London EC2Y 8DS • T: +44 20 76384141, 73827105 • F: 73822308 • artinfo@barbican.org.uk • www.barbican.org.uk •
Dir.: *Kate Bush* (Head of Art Galleries) •
Public Gallery 44791

Barnet Museum, 31 Wood St, Barnet, London EN5 4BE • T: +44 20 84408066 • www.barnetmuseum.co.uk •
Dir.: *Dr. Gilian Gear* • Cur.: *Sandra Lea* •
Local Museum – 1935
Barnet district history, costume photogr – library 44792

Ben Uri Gallery, London Jewish Museum of Art, 108a Boundary Rd, Saint Johns Wood, London NW8 0RH • T: +44 20 76043991 • F: 76043992 • info@benuri.org.uk • www.benuri.org.uk •
Public Gallery / Fine Arts Museum
Contemporary art by Jewish artists, paintings, drawings, sculptures 44793

Benjamin Franklin House, 36 Craven St, London WC2N 5NF • T: +44 20 79309121 • F: 79309124 • benjaminfranklinhouse@msn.com • www.benjaminfranklinhouse.org •
Historical Museum 44794

Bexley Museum, Hall Place, Bourne Rd, Bexley, London DA5 1PQ • T: +44 1322 526574 • F: 522921 • fau@hallplace.co.uk • www.hallplace.com •
Historical Museum – 1934
Local geology, archaeology, fauna and flora, minerals, Roman finds, local history and industries 44795

BFI Southbank Gallery, South Bank, Waterloo, London SE1 8XT • T: +44 20 74012636, 79283535 • F: 79287938 • Caroline.Ellis@bfi.org.uk • www.bfi.org.uk/museum •
Dir.: *Elisabetta Fabrizi* •
Historical Museum – 1988
Early optical toys, costumes from modern science fiction films, Charlie Chaplin's hat and cane, Fred Astaire's tail coat, the IBA coll of period television sets, the world's first fourscreen unit 44796

Black Cultural Archives, 1 Othello Close, London SE11 4RE • T: +44 20 77384591 • F: 77387168 • info@aambh.org.uk •
Dir.: *Sam Walker* •
Local Museum
Biographies, music manuscripts, slave documents, photographs, British parliamentary records, volumes of "Crisis" and "Messenger" 44797

Bramah Tea and Coffee Museum, 40 Southwark St, London SE1 1UN • T: +44 20 74035650 • F: 74035650 • bramah@btconnect.com • www.bramahmuseum.co.uk •
Dir.: *Edward Bramah* •
Special Museum – 1992
Large coll of teapots and coffee making machinery 44798

Brent Museum, 95 High Rd, Willesden Green, London NW10 2SF • T: +44 20 82046870 • F: 82047681 • customer.services@brent.gov.uk • www.brent.gov.uk •
Cur.: *Victoria Barlow* •
Historical Museum – 1977
Archive of the British Empire exhibition held at Wembley 1924, historical displays relating to Brent people 44799

Britain at War Experience, 64-66 Tooley St, London SE1 2TF • T: +44 20 74033171 • F: 74035104 • info@britainatwar.co.uk • www.britainatwar.co.uk •
Historical Museum / Military Museum
2nd World War 44800

British Dental Association Museum, 64 Wimpole St, London W1G 8YS • T: +44 20 75634549 • F: 79356492 • museum@bda.org • www.bda.org/museum •
Head: *Rachel Bairsto* •
Historical Museum
Hist, art and science of dental surgery 44801

British Library, 96 Euston Rd, London NW1 2DB • T: +44 20 74127332 • F: 74127340 • visitor-services@bl.uk • www.bl.uk •
Chief Ex.: *Lynne Brindley* •
Library with Exhibitions
Permanent display: Magna Carta, first folio of Shakespeare's works, Gutenberg Bible 44802

The British Museum, Great Russell St, London WC1B 3DG • T: +44 20 73238000 • F: 73238616 • information@thebritishmuseum.ac.uk • www.thebritishmuseum.ac.uk •
Dir.: *Neil MacGregor* • Chm. Trustees: *Sir John Boyd* • Sc. Staff: *Dr. W.V. Davies* (Ancient Egypt and Sudan) • *J. Cribb* (Coins and Medals) • *Dr. D. Williams* (Greek and Roman) • *A. Griffiths* (Prints and Drawings) • *R. Knox* (Asia) • *Dr. J. Curtis* (Ancient Near East) • *Dr. David Saunders* (Conservation, Documentation, Science) • *Jonathan King* (Africa, Oceania and The Americas) • *L. Webster* (Prehistory, Europe) •
Museum of Classical Antiquities – 1753
Coins and medals, ancient Egypt, Sudan, Africa, Oceania and Americas ethnography, Greek and Roman antiquities, Japanese antiquities, prehistory and Europe, prints & drawings, Ancient Near East 44803

British Optical Association Museum, College of Optometrists, 42 Craven St, London WC2N 5NG • T: +44 20 77664353 • F: 78396800 • museum@college-optometrists.org • www.college-optometrists.org/museum •
Cur.: *Neil Handley* •
Historical Museum / Science&Tech Museum – 1901
Optical and ophthalmic items, incl. spectacles, pince-nez, lorgnettes, spyglasses, lenses and artificial eyes, diagnostic instruments, fine art, ceramics, oil paintings – library 44804

The British Postal Museum & Archive, Freeling House, Phoenix Pl, London WC1X 0DL • T: +44 20 72392570 • F: 72392576 • info@postalheritage.org.uk • www.postalheritage.org.uk •
CEO: *Tony Conder* •
Special Museum – 1965
Philatelics, British stamps since 1840, world coll since 1878, special display on the creation of the One Penny Black 1840, Postal services and artefacts 44805

British Red Cross Museum and Archives, 44 Moorfields, Moorfields House, London EC2Y 9AL • T: +44 20 78777058 • F: 75622056 • enquiry@redcross.org.uk • www.redcross.org.uk/museum&archives •
Dir.: *Sir Nicholas Young* •
Special Museum
Uniforms, medals, badges, textiles, medical equipment and humanitarian aid – archives 44806

Bruce Castle Museum, Lordship Ln, Tottenham, London N17 8NU • T: +44 20 88088772 • F: 88084118 • museum.services@haringey.gov.uk • www.haringey.gov.uk •
Dir.: *Deborah Hedgecock* •
Local Museum – 1906
Haringey Libraries, archives 44807

Brunei Gallery, School of Oriental and African Studies, University of London, Thornhaugh St, Russell Sq, London WC1H 0XG • T: +44 20 78984915, 78984046 • F: 78984259 • gallery@soas.ac.uk • www.soas.ac.uk/gallery •
Dir.: *John Hollingworth* • Asst. Dir.: *Joy Onyejiako* •

Public Gallery / University Museum – 1996
Temporary exhibition programme and a permanent coll of historical and contemporary works of and from Africa, Asia and the Middle East 44808

The Brunel Museum, Railway Av, Rotherhithe, London SE16 4LF • T: +44 20 72313840 • robert.hulse@brunel-museum.org.uk • www.brunel-museum.org.uk •
Dir.: *Robert Hulse* •
Historic Site – 1974
Brunel's Thames tunnel, models, watercolours, peepshows, memorabilia, Great Eastern steamship, electro lumine scence 44809

Buckingham Palace, Royal Collections, London SW1A 1AA, mail addr: Stable Yard House, Saint James's Palace, London SW1A 1JR • T: +44 20 73212233 • F: 79309625 • buckinghampalace@royalcollection.org.uk • www.royalcollection.org.uk •
Fine Arts Museum / Decorative Arts Museum
State appartments, royal coll 44810

Cafe Gallery Projects, Southwalk Park, Bermondsey, London SE16 2UA • T: +44 20 72371230 • www.cafegalleryprojects.org •
Public Gallery 44811

Camden Arts Centre, Arkwright Rd, London NW3 6DG • T: +44 20 74725500 • F: 74725501 • info@camdenartscentre.org • www.camdenartscentre.org •
Dir.: *Jenni Lomax* •
Public Gallery
Contemporary art, painting and pottery 44812

Canada House Gallery, Trafalgar Sq, London SW1Y 5BJ • T: +44 20 72586493 • F: 72586434 • www.international.gc.ca/london •
Dir.: *J. Rennie* •
Public Gallery 44813

Carlyle's House, 24 Cheyne Row, London SW3 5HL • T: +44 20 73527087 • F: 73525108 • carlyleshouse@nationaltrust.org.uk • www.nationaltrust.org.uk/carlyleshouse •
Cust.: *Lin Skippings* •
Special Museum – 1895
Home of Thomas and Jane Carlyle with furniture, books, letters, personal effects and portraits – library 44814

A Celebration of Immigration, 19 Princelet St, London E1 6QH • T: +44 20 72475352 • F: 73751490 • information@19princeletstreet.org.uk • www.19princeletstreet.org.uk •
Historical Museum
Various waves of immigration through the East End of London, religious textiles of the Jewish life – archive 44815

Centre for the Magic Arts, 12 Stephenson Wy, Euston, London NW1 2ND • T: +44 20 73872222 • F: 73875114 • henry.lewis@classictm.net • www.themagiccircle.co.uk •
Fine Arts Museum
Original posters ephemera, apparatus of conjurours and illusionists 44816

The Charles Dickens Museum, 48 Doughty St, London WC1N 2LX • T: +44 20 74052127 • F: 78315175 • info@dickensmuseum.com • www.dickensmuseum.com •
Dir.: *Andrew Xavier* •
Special Museum / Historic Site – 1925
Former home of the writer Charles Dickens (1812-1870), memorabilia, manuscripts, letters, first editions, furniture – library 44817

Chartered Insurance Institute's Museum, 20 Aldermanbury, London EC2V 7HY • T: +44 20 74174412 • F: 77260131 • customer.service@cii.co.uk • www.cii.co.uk •
Special Museum – 1934
History of insurance, insurance company fire brigades, fire marks, early firefighting equipment – library 44818

Chisenhale Gallery, 64 Chisenhale Rd, London E3 5QZ • T: +44 20 89814518 • F: 89807169 • mail@chisenhale.org.uk • www.chisenhale.org.uk •
Dir.: *Simon Wallis* •
Public Gallery 44819

Chiswick House, Burlington Ln, Chiswick, London W4 2RP • T: +44 20 89950508, 87421978 • F: 87423104 • www.english-heritage.org.uk •
Cur.: *Cathy Power* •
Decorative Arts Museum
18th c furniture and interiors, paintings coll from 17th-19th c 44820

Church Farmhouse Museum, Greyhound Hill, Hendon, London NW4 4JR • T: +44 20 83593942 • gerrard.roots@barnet.gov.uk • www.churchfarmhousemuseum.co.uk •
Dir.: *Gerrard Roots* • Archivist: *Hugh Petrie* •
Local Museum – 1955
Period furniture, local hist 44821

Churchill Museum and Cabinet War Rooms, Clive Steps, King Charles St, London SW1A 2AQ • T: +44 20 79306961 • F: 78395897 • cwr@imw.org.uk • cwr.iwm.org.uk •
Dir.: *Philip Reed* •
Historical Museum – 1984
Churchill's underground Headquarter, 30 historic rooms incl the cabinet room, Churchill's sirensuit and his Noble Prize, original materials 44822

Cinema Museum, 2 Dugard Way, Kennington, London SE11 4TH • T: +44 20 78402200 • F: 78402299 • martin@cinemamuseum.org.uk •
Dir.: *Martin Humphries* • Cur.: *Ronald Grant* •
Performing Arts Museum
Hist of cinemas since 1896 44823

City of London Police Museum, 37 Wood St, London EC2P 2NQ • T: +44 20 76012747 •
Cur.: *Roger Appleby* •
Historical Museum
Police and the policing of the city, insignia truncheons, tipstaves, uniforms and photographs 44824

Clink Prison Museum, 1 Clink St, London SE1 9DG • T: +44 20 73781558, 74030900 • F: 74035813 • museum@clink.co.uk • www.clink.co.uk •
Owner: *Ray Rankin* •
Local Museum
Coll of torture equipment 44825

Clipper Ship Cutty Sark, King William Walk, London SE10 9HT • T: +44 20 88583445 • F: 88533589 • info@cuttysark.org.uk • www.cuttysark.org.uk •
Historical Museum – 1957
Restored ship, launched 1869, with figureheads, ship hist 44826

Clockmakers' Museum, Guildhall Library, Aldermanbury, London EC2P 2EJ • T: +44 20 73321865 • keeper@clockmakers.org • www.clockmakers.org •
Keeper: *Sir George White* •
Science&Tech Museum – 1813
Old clocks, marine chronometers, watch keys 44827

College Art Collections → UCL Art Collections

Commonwealth Institute, Kensington High St, London W8 6NQ • T: +44 20 76034535 • F: 76034525 • information@commonwealth.org.uk • www.commonwealth.org.uk •
Dir.: *David French* • Sc. Staff: *Karen Peters* (Information and Resources) •
Public Gallery – 1962
Exhibitions of commonwealth countries, commonwealth artists, culture, literature, books, artefacts from commonwealth countries – library, educational programmes 44828

County Hall Gallery, Riverside Bldg, Westminster Bridge Rd, London SE1 7PB • T: +44 20 74507612, 76202720 • F: 76203120 • gallery@countyhallgallery.com • www.countyhallgallery.com •
Public Gallery
Dali universe of sculptures, graphics, illustration, crystal and furniture 44829

Courtauld Institute Gallery, Courtauld Institute of Art, Somerset House, Strand, London WC2R 0RN • T: +44 20 78482526 • F: 78482589 • galleryinfo@courtauld.ac.uk • www.courtauld.ac.uk •
Dir.: *Dr. Ernst Vegelin Van Claerbergen* • Cur.: *Dr. Caroline Campbell* (Paintings) • *Dr. Ioanna Selborne* (Drawings, Prints) • *Dr. Alexandra Gerstein* (Sculpture, Decorative Arts) • *Dr. Barnaby Wright* (Exhibitions) •
Fine Arts Museum – 1932
French Impressionists, Post-Impressionists and European Old Masters (14th-18th c), Old Masters & 19th c drawings, prints, sculpture and decorative arts 44830

Crafts Council Gallery, 44a Pentonville Rd, Islington, London N1 9BY • T: +44 20 72787700 • F: 78736891 • reference@craftscouncil.org.uk • www.craftscouncil.org.uk •
Decorative Arts Museum – 1971
Crafts – gallery, education workshop, picture and reference libraries 44831

The Crossness Engines, Old Works Thames Water, Belvedere Rd, Abbey Wood, London SE2 9AQ • T: +44 20 83113711 • F: 83036723 • mail@crossness.org.uk • www.crossness.org.uk •
Science&Tech Museum
Four 1865 James Watt rotative beam engines, sanitation ware 44832

Crystal Palace Museum, Anerley Hill, London SE19 2BA • T: +44 20 86760700 • F: 86760700 • museum@thecrystalpalace.fsnet.co.uk •
Dir.: *K. Kiss* • *B. McKay* •
Historical Museum
Oil paintings, ceramics, medals, papers, 3-dimensional objects, history of the Crystal Palace 44833

Cuming Museum, 155-157 Walworth Rd, London SE17 1RS • T: +44 20 75252163 • F: 75252345 • cuming.museum@southwark.gov.uk • www.southwark.gov.uk •
Dir.: *Catherine Hamilton* •
Archaeology Museum / Historical Museum / Decorative Arts Museum – 1902
Roman and medieval remains from archaeological excavations, coll of London superstitions, worldwide coll of the Cuming family 44834

De Morgan Centre, 38 West Hill, Wandsworth, London SW18 1RZ • T: +44 20 88711144 • info@demorgancentre.org.uk • www.demorgan.org.uk •
Pres.: *Jon Catleugh* •
Fine Arts Museum – De Morgan, Willam; De Morgan, Evelyn
Paintings and drawings by Evelyn De Morgan, ceramics by William De Morgan 44835

Delfina, 50 Bermondsey St, London SE1 3UD • T: +44 20 73576600 • F: 73570250 •
Public Gallery
Modern and contemporary art 44836

Design Museum, Shad Thames, London SE1 2YD • T: +44 870 8339955 • F: 9091909 • info@designmuseum.org • www.designmuseum.org •
Dir.: *Alice Rawsthorn* •
Fine Arts Museum – 1989 44837

Diorama Arts, 3-7 Euston Centre, London NW1 3JG • T: +44 20 79165467 • admin@diorama-arts.org.uk • www.diorama-arts.org.uk •
Public Gallery 44838

Dr. Johnson's House, 17 Gough Sq, London EC4A 3DE • T: +44 20 73533745 • F: 73533745 • curator@drjohnsonshouse.org • www.drjohnsonshouse.org •
Chm.: *Lord Harmsworth* • Cur.: *Stephanie Pickford* •
Special Museum – 1912
Memorabilia on Samuel Johnson (1709-1784) 44839

Dulwich Picture Gallery, Gallery Rd, Dulwich, London SE21 7AD • T: +44 20 86935254 • F: 82998700 • info@dulwichpicturegallery.org.uk • www.dulwichpicturegallery.org.uk •
Dir.: *Ian Dejardin* •
Fine Arts Museum – 1811
Paintings by Poussin, Gainsborough, Canaletto Rembrandt, Rubens, building by Sir John Soane 44840

Ealing College Gallery, The Green, London W5 5EW • T: +44 20 82316303 • F: 82316013 • www.etc.ac.uk •
University Museum 44841

East Ham Nature Reserve, Norman Rd, London E6 4HN • T: +44 20 84704525 •
Man.: *Barry Graham* •
Local Museum
Local and natural hist 44842

Eltham Palace, Court Yard, Eltham, London SE9 5QE • T: +44 20 82942548 • F: 82942621 • www.english-heritage.org.uk •
Decorative Arts Museum / Historical Museum – 1999
Rare 1930's Art Deco house, build for Edvard IV and medieval Great Hall build for Edvard IV and Henry VIII – gardens 44843

Estorick Collection of Modern Italian Art, 39a Canonbury Sq, Islington, London N1 2AN • T: +44 20 77049522 • F: 77049531 • curator@estorickcollection.com • www.estorickcollection.com •
Dir.: *Roberta Cremoncini* •
Fine Arts Museum
Italian modern art 44844

Fan Museum, 12 Crooms Hill, Greenwich, London SE10 8ER • T: +44 20 88587879, 83051441 • F: 82931889 • admin@fan-museum.org • www.fan-museum.org •
Dir.: *Helene Alexander* •
Decorative Arts Museum – 1985
European 17th and 18th c fans, fan leaves, important oriental export fans 44845

Fashion and Textile Museum, 83 Bermondsey St, London SE1 3XF • T: +44 20 74078664 • info@ftmlondon.org • www.ftmlondon.org •
Dir.: *Gity Monsef* •
Decorative Arts Museum 44846

Fenton House, Hampstead Grove, Hampstead, London NW3 6SP • T: +44 20 74353471 • fentonhouse@nationaltrust.org.uk • www.nationaltrust.org.uk/main/w-fentonhouse •
Cust.: *Jane Ellis* •
Music Museum / Decorative Arts Museum – 1953
Benton-Fletcher coll of early musical instruments, Binning coll of porcelain, Georgian furniture, needlework 44847

Firepower - Royal Artillery Museum, Number One St, Royal Arsenal, Woolwich, London SE18 6ST • T: +44 20 88557755 • F: 88557100 • info@firepower.org.uk • www.firepower.org.uk •
Cur./Dir.: *Mark Smith* •
Military Museum – 2001
Fuses, ammunition, artillery instruments, rockets, edge weapons, display of artillery from the 14th c to present day 44848

Florence Nightingale Museum, Gassiot House, 2 Lambeth Palace Rd, London SE1 7EW • T: +44 20 76200374 • F: 79281760 • info@florence-nightingale.co.uk • www.florence-nightingale.co.uk •
Dir.: *Alex Attewell* •
Historical Museum – 1989
Nightingale memorabilia 44849

Foundation for Women's Art, 11 Northburgh St, London EC1V 0AH • T: +44 20 72514881 • F: 72514882 • admin@fwa-uk.org • www.fwa-uk.org •
Dir.: *Monica Petzal* •
Fine Arts Museum
archives 44850

Foundling Museum, 40 Brunswick Sq, London WC1N 1AZ • T: +44 20 78413600 • F: 78413601 • enquiries@foundlingmuseum.org.uk • www.foundlingmuseum.org.uk •
Dir.: *Rhian Harris* •
Fine Arts Museum / Historical Museum – 2004
London's first art gallery, works by Hogarth, Gainsborough and Reynolds, social history, Gerald Coke Handel coll 44851

Freud Museum, 20 Maresfield Gardens, Hampstead, London NW3 5SX • T: +44 20 74352002 • F: 74315452 • info@freud.org.uk • www.freud.org.uk •
Dir.: *Michael Molnar* •
Special Museum – 1986
Sigmund Freud's complete working environment, including the famous couch, his library and coll of 2000 Greek, Roman and Egyptian antiquities 44852

Fusiliers London Volunteers' Museum, 213 Balham High Rd, London SW17 7BQ • T: +44 20 86721168 •
Cur.: *A.H. Mayle* •
Military Museum 44853

Geffrye Museum, Kingsland Rd, Shoreditch, London E2 8EA • T: +44 20 77399893 • F: 77295647 • info@geffrye-museum.org.uk • www.geffrye-museum.org.uk •
Dir.: *David Dewing* •
Decorative Arts Museum – 1914
English domestic interiors from 1600-2000, walled herb garden and period garden rooms fom 17th to 20th c – library 44854

Geology Collections, University College London, c/o Dept. of Earth Sciences, Gower St, London WC1E 6BT • T: +44 20 76797900 • F: 73874166 • geolcoll@ucl.ac.uk • www.geology.museum.ucl.ac.uk •
Cur.: *Wendy Kirk* • Man.: *Jayne Dunn* •
Natural History Museum
Geological specimens, rocks, minerals and fossils, planetary images 44855

Gilbert Collection (closed) 44856

Golden Hinde Educational Trust, Saint Mary Overie Dock, Cathedral St, London SE1 9DE • T: +44 20 74030123 • F: 74075908 • info@goldenhinde.co.uk • www.goldenhinde.co.uk •
Chm.: *Christopher Allen* •
Historical Museum
16th c warship in which Sir Francis Drake circumnavigated the world between 1577 and 1580 44857

Goldsmiths' Hall, Foster Lane, London EC2V 6BN • T: +44 20 76067010 • F: 76061511 • the.clerk@thegoldsmiths.co.uk • www.thegoldsmiths.co.uk •
Dir.: *R.G. Melly* • Cur.: *Rosemary Ransome Wallis* •
Decorative Arts Museum
English silver, modern jewelry since 1960 – library 44858

Gordon Museum, Saint Thomas St, London SE1 9RT • T: +44 20 79554358 • william.edwards@kcl.ac.uk •
Keeper: *William Edwards* •
Special Museum
Specimens of human disease, wax models of anatomical dissection and skin diseases 44859

Government Art Collection, Queen's Yard, 179A Tottenham Court Road, London W1T 7PA • T: +44 20 75809120 • F: 75809130 • gac@culture.gsi.gov.uk • www.gac.culture.gov.uk •
Dir.: *Penny Johnson* •
Fine Arts Museum – 1898
British art from the 16th c to the present day 44860

Grange Museum of Community History → Brent Museum

Grant Museum of Zoology and Comparative Anatomy, c/o Biology Department, University College London, Darwin Bldg, Gower St, London WC1E 6BT • T: +44 20 76792647 • F: 76797096 • zoology.museum@ucl.ac.uk • www.grant.museum.ucl.ac.uk •
Dir.: *Dr. H.J. Chatterjee* •
Natural History Museum / University Museum – 1828
Robert Edmund Grant, T.H. Huxley, E. Ray Lankester, DMS Watson, J. Cloudsly Thompson colls, Quagga skeleton and many rare, endangered and extinct animals, extensive skeletal and fossil coll 44861

Great Ormond Street Hospital for Children Museum, c/o Great Ormond Street Hospital for Children NHS Trust, Great Ormond St, London WC1N 3JH • T: +44 20 74059200 ext 5920 • BaldwiN@gosh.nhs.uk • www.ich.ucl.ac.uk •
Cur.: *Nicholas Baldwin* •
Historical Museum – 1989
Hospital hist, books, photographs, artworks; medical, surgical and other equipment – archives 44862

Greenwich Heritage Centre, The Royal Arsenal, Woolwich, London SE18 4DX • T: +44 20 88542452 • F: 88542490 • heritage.centre@greenwich.gov.uk • www.greenwich.gov.uk •
Cur.: *Beverley Burford* •
Local Museum – 1919
Local history, natural history 44863

Greenwich Local History Centre, 90 Mycenae Rd, Blackheath, London SE3 7SE • T: +44 20 88584631 • F: 82934721 • local.history@greenwich.gov.uk •
Man.: *Julian Watson* •
Local Museum – 1972
A.R. Martin coll of documents relating to Greenwich, Blackheath and Charlton 44864

The Guards Museum, Wellington Barracks, Birdcage Walk, London SW1E 6HQ • T: +44 20 74143271 • F: 74143429 •
Cur.: *Andrew Wallis* •
Military Museum
Uniforms, weapons, memorabilia, Royal items, 350 yrs of the fire regiments of Foot Guards 44865

Guildhall Art Gallery, Guildhall Yard, off Gresham St, London EC2P 2EJ • T: +44 20 73323700 • F: 73323342 • guildhall.artgallery@cityoflondon.gov.uk • www.guildhall-art-gallery.org.uk •
Dir.: *David Bradbury* • Cur.: *V.M. Knight* •
Fine Arts Museum – 1886
London paintings, 19th c works of art, portraits, sculpture, Sir Matthew Smith Studio coll 44866

U

Gunnersbury Park Museum, Gunnersbury Park, Popes Ln, London W3 8LQ • T: +44 20 89921612 • F: 87520686 • gp-museum@cip.org.uk • www.cip.org.uk •
Cur.: *Jill Draper* •
Local Museum / Archaeology Museum / Historic Site – 1929
Local archaeology, Sadler coll of flints, Middlesex maps and topography, costumes, toys and dolls, transportation, local crafts and industries – library 44867

Hackney Museum, 1 Reading Ln, Hackney, London E8 1GQ • T: +44 20 83563500 • F: 89857600, 83562563 • hmuseum@hackney.gov.uk • www.hackney.gov.uk/hackneymuseum •
Cur.: *Sue McAlpine* • Man.: *Jane Sarre* • *Sophie Perkins* •
Local Museum
History of Hackney, local people and their 1000 years of immigration, social history, industry, housing, local cultures, art and archaeology 44868

Hamilton's Gallery, 13 Carlos Pl, London W1Y 5AG • T: +44 20 74999493/94 • F: 76299919 •
Fine Arts Museum 44869

Hampstead Museum, Burgh House, New End Sq, London NW3 1LT • T: +44 20 74310144 • F: 74358817 • hmcurator@btinternet.com • www.burghhouse.org.uk •
Cur.: *Marilyn Greene* •
Local Museum – 1979
Local history, Helen Allingham coll, Isokon furniture 44870

Handel House Museum, 25 Brook St, London W1K 4HB • T: +44 20 74951685 • F: 74951759 • mail@handelhouse.org • www.handelhouse.org •
Dir.: *Sarah Bardwell* •
Music Museum
Portraits, prints, 18th c furniture, manuscripts, printed scores, musical instruments 44871

Hayward Gallery, South Bank Centre, Belvedere Rd, London SE1 8XZ • T: +44 20 79210813 • F: 74012664 • customer@southbankcentre.co.uk • www.hayward.org.uk •
Dir.: *Susan Ferleger Brades* •
Fine Arts Museum – 1968
Arts Council coll, National Touring exhibitions 44872

Her Majesty Tower of London, Historic Royal Palaces, Tower Hill, London EC3N 4AB • T: +44 870 7566060 • F: +44 20 74805543 • www.tower-of-london.org.uk •
Chief.: *Michael Day* •
Historical Museum / Decorative Arts Museum / Historic Site
Royal palace and fortress, White Tower, towers and remparts, royal lodgings, Chapel Royal of Saint Peter at Vincula, crown jewels, Royal amouries 44873

Hermitage Rooms, South Bldg, Somerset House, Strand, London WC2R 1LA • T: +44 20 78454630, 78454631 • F: 78454637 • www.hermitagerooms.com •
Dir.: *Dr. Deborah Swallow* •
Public Gallery
Rotating coll from the State Hermitage Museum in St Petersburg 44874

HMS Belfast, Morgans Ln, off Tooley St, London SE1 2JH • T: +44 20 79406300 • F: 74030719 • www.iwm.org.uk/hmsbelfast •
Dir.: *Brad King* •
Military Museum – 1971
Armoured warship of WWII, Royal Navy 44875

Hogarth's House, Hogarth Ln, Great West Rd, Chiswick, London W4 2QN • T: +44 20 89946757 • info@cip.org.uk • www.hounslow.info/hogarthshouse •
Fine Arts Museum – 1904
Prints, engravings 44876

Honeywood Heritage Centre, Honeywood Walk, Carshalton, London SM5 3NX • T: +44 20 87704297 • F: 87704297 • ibshoneywood@ukonline.co.uk • www.sutton.gov.uk/leisure/heritage •
Cur.: *Jane Howard* •
Local Museum
Hist of London borough Sutton 44877

Horniman Museum, 100 London Rd, Forest Hill, London SE23 3PQ • T: +44 20 86991872 • F: 82915506 • enquiry@horniman.ac.uk • www.horniman.ac.uk •
Dir.: *Janet Vitmayer* •
Local Museum / Ethnology Museum / Music Museum – 1890
Anthropology, natural history, musical instruments – library, gardens, Living Waters aquarium, conservation 44878

House Mill, Three Mill Ln, Bromley-by-Bow, London E3 3DU • T: +44 20 89804626 • F: 89800725 • rltmt@boos.demon.co.uk •
Dir.: *John Haggerty* •
Science&Tech Museum
Original machinery incl millstones and waterwheels, mill owners' info 44879

Hunterian Museum, 35-43 Lincoln's Inn Fields, London WC2A 3PE • T: +44 20 78696000 • F: 78696005 • museums@rcseng.ac.uk • www.rcseng.ac.uk/services/museums •
Special Museum – 1813
Anatomy, physiology and pathology from the coll of John Hunter (1728-1793) historical surgical instruments 44880

Imperial War Museum, Lambeth Rd, London SE1 6HZ • T: +44 20 74165000 • mail@iwm.org.uk • www.iwm.org.uk •
Gen. Dir.: *D. Lees* • Sc. Staff: *A. Richards* (Dept. Documents) • *R. McAlister* (Dept. Exhibits) • *S. Bardgette* (Dept. Research) •
Historical Museum – 1917
Wars since 1914, armaments, documents, books involving Britain and the Commonwealth, films, records, photographs, 20th cent. art collection, drawings, paintings, sculpture and special exhibitions, Holocaust exhibition – Explore History Centre, Library, Archives 44881

Inns of Court and City Yeomanry Museum, 10 Stone Bldgs, Lincoln's Inn, London WC2A 3TG • T: +44 20 74058112 • F: 74143496 •
Cur.: *D. Durkin* •
Military Museum
Hist of 2 regiments, the Inns of Court Regiment and City of London Yeomanry from 1780 to date 44882

Institute of Contemporary Arts, The Mall, London SW1Y 5AH • T: +44 20 79303647 • F: 78730051 • info@ica.org.uk • www.ica.org.uk •
Dir.: *Elton Eshun* •
Public Gallery 44883

Islington Education Artefacts Library, Block D, Barnsbury Complex, Barnsbury Park, London N1 1QG • T: +44 20 75275524 • F: 75275564 • els@islington.gov.uk • www.objectlessons.org •
Historical Museum
History, science, natural history, religion and multiculture, costume coll, the National Curriculum – library, delivers a loan box handling coll 44884

Islington Museum, Foyer Gallery, Islington Town Hall, Upper St, London N1 2UD • T: +44 20 73549442 • F: 75273049 • valerie.munday@islington.gov.uk • www.islington.gov.uk/localinfo/leisure/museum •
Dir.: *Alison Lister* •
Local Museum
Local and social hist of Islington borough, popular culture 44885

J.M.W. Turner House, 153 Cromwell Rd, London SW5 0TQ • T: +44 20 73735560 • F: 73735560 • selbywhittingham@hotmail.com • www.jmwturner.org •
Fine Arts Museum 44886

Jewel Tower, Abingdon Rd, Westminster, London SW1P 3JY • T: +44 20 72222219 • F: 72222219 • www.english-heritage.org.uk •
Local Museum 44887

Jewish Military Museum, Shield House, Harmony Way, off Victoria Rd, Hendon, London NW4 2BZ • T: +44 20 82015656 • F: 82029900 • visit@thejmm.org.uk • www.ajex.org.uk •
Cur.: *Henry Morris* •
Military Museum – 1996
Documents, medals, memorabilia, uniforms, photographs relating to the contribution made by British Jews to the British Armed Forces – Archive 44888

Jewish Museum, Raymond Burton House, 129-131 Albert St, London NW1 7NB • T: +44 20 72841997 • F: 72679008 • admin@jewishmuseum.org.uk • www.jewishmuseum.org.uk •
Dir.: *Rickie Burman* •
Religious Arts Museum / Historical Museum – 1932
Jewish ceremonial art and hist 44889

Jewish Museum Finchley, 80 East End Rd, London N3 2SY • T: +44 20 83491143 • F: 83432162 • enquiries@jewishmuseum.org.uk • www.jewishmuseum.org.uk •
Cur.: *Sarah Jillings* •
Religious Arts Museum / Historical Museum
Hist of Jewish immigration and settlement in London 44890

The John Wesley House and The Museum of Methodism, 47 City Rd, London EC1Y 1AU • T: +44 20 72532262 • F: 76083825 • museum@wesleyschapel.org.uk • www.wesleyschapel.org.uk •
Dir.: *Christian Dettlaff* •
Special Museum – 1898
John Wesley's library, personal memorabilia, largest coll of Wesleyana in the world 44891

Keats House, Keats Grove, Wentworth Pl, London NW3 2RR • T: +44 20 74352062 • F: 74319293 • keatshouse@corpoflondon.org.uk • www.keatshouse.org.uk •
Head: *Dr. Deborah Jenkins* •
Special Museum – 1925
Life and work of the poet John Keats (1795-1821), manuscripts, relics, furniture – library 44892

Kensington Palace State Apartments, Historic Royal Palaces, Kensington Palace, London W8 4PX • T: +44 20 79379561 • F: 73760198 • www.hrp.org.uk •
Decorative Arts Museum 44893

Kenwood House, The Iveagh Bequest, Hampstead Ln, London NW3 7JR • T: +44 20 83481286 • F: 87933891 • www.english-heritage.org.uk •
VSM: *Susan Coventry* • Cur.: *Laura Houliston* •
Fine Arts Museum – 1927
Paintings by Rembrandt, Vermeer, Gainsborough, Reynolds and others, 18th c furniture and sculpture 44894

Kingsgate Gallery, 110-116 Kingsgate Rd, London NW6 2JG • T: +44 20 73287878 • F: 73287878 • mail@kingsgateworkshops.org.uk • www.kingsgateworkshops.org.uk •
Dir.: *Bridget Eadie* •
Public Gallery
Contemporary art 44895

Kirkaldy Testing Museum, 99 Southwark St, Southwark, London SE1 0JF •
Science&Tech Museum
David Kirkaldy's original all-purpose testing machine 44896

Kufa Gallery, 26 Westbourne Grove, London W2 5RH • T: +44 20 72291928 • F: 72438513 • info@kufagallery.co.uk • www.kufagallery.co.uk •
Man.: *Walid Atiyeh* •
Public Gallery
Islamic and Middle Eastern Art and architecture 44897

Lauderdale House, Waterlow Park, Highgate Hill, London N6 5HG • T: +44 20 83488716 • F: 84429099 • admin@lauderdale.org.uk • www.lauderdalehouse.co.uk •
Dir.: *Carolyn Naish* •
Local Museum
Fine art, photography, local hist 44898

Leighton House Museum, 12 Holland Park Rd, Kensington, London W14 8LZ • T: +44 20 76023316 • F: 73712467 • leightonhousemuseum@rbkc.gov.uk • www.rbkc.gov.uk/leightonhousemuseum •
Cur.: *Daniel Robbins* •
Special Museum – 1898
Leighton drawings coll, Leighton letters and archive 44899

Lewisham Local Studies Centre, c/o Lewisham Library, 199-201 Lewisham High St, London SE13 6LG • T: +44 20 82970682 • F: 82971169 • local.studies@lewisham.gov.uk • www.lewisham.gov.uk/localStudies •
Library with Exhibitions
Coll of archives, printed materials (incl newspapers), maps, pictorial materials and objects 44900

Linley Sambourne House, 18 Stafford Terrace, Kensington, London W8 7BH • T: +44 20 76023316 • F: 73712467 • museum@rbkc.gov.uk • www.rbkc.gov.uk/linleysambournehouse •
Decorative Arts Museum
Edward Linley Sambourne, political cartoonist at punch magazine, funishings 44901

Little Holland House, 40 Beeches Av, Carshalton, London SM5 3LW • T: +44 20 87704781 • F: 87704777 • valary.murphy@sutton.gov.uk • www.sutton.gov.uk •
Dir.: *Valary Murphy* •
Decorative Arts Museum – 1974
Furniture, paintings and craft by Frank R. Dickinson 44902

Livesey Museum for Children, 682 Old Kent Rd, London SE15 1JF • T: +44 20 76395604 • F: 72775384 • livesey.museum@southwark.gov.uk • www.liveseymuseum.org.uk •
Dir.: *Theresa Dhaliwal* •
Special Museum – 1974
Interactive childrens museum 44903

London Brass Rubbing Centre, The Crypt, Saint Martin-in-the-Fields Church, Trafalgar Sq, London WC2N 4JJ • T: +44 20 79309306 • F: 79309306 • www.visitlondon.co.uk •
Dir.: *A. Dodwell* • *P. Dodwell* •
Decorative Arts Museum
Coll of facsimile monumental church brasses from the medieval and Tudor periods portraying kings, knights, merchants and families, designs from Celtic sources 44904

London Canal Museum, 12-13 New Wharf Rd, King's Cross, London N1 9RT • T: +44 20 77130836 • F: 76896679 • info@canalmuseum.org.uk • www.canalmuseum.org.uk •
Chm.: *Martin Sach* •
Historical Museum – 1992
Inland waterways of the London region 44905

London Fire Brigade Museum, 94a Southwark Bridge Rd, Southwark, London SE1 0EG • T: +44 20 75872894 • F: 75872878 • museum@london-fire.gov.uk • www.london-fire.gov.uk •
Cur.: *Esther Mann* •
Historical Museum
Fire appliances, equipment, clothing, medals and miscellanea (17th-20th c), fire-related art 44906

London Scottish Regimental Museum, 95 Horseferry Rd, London SW1P 2DX • T: +44 20 76301639 • F: 72337909 • smallpipes@aol.com •
Military Museum
History of the regiment 44907

London Sewing Machine Museum, 292-312 Balham High Rd, Tooting Bec, London SW17 7AA • T: +44 20 86827916 • F: 87674726 • wimbledonsewingmachinecoltd@btinternet.com • www.sewantique.com •
Cur.: *Ray Rushton* •
Science&Tech Museum
Industrial and domestic sewing machines from 1850-1950 44908

London's Transport Museum, Covent Garden Piazza, 39 Wellington St, London WC2E 7BB • T: +44 20 73796344 • F: 75657254 • contact@ltmuseum.co.uk • www.ltmuseum.co.uk •
Dir.: *Sam Mullins* •
Science&Tech Museum – 1980
London transport history, coll of historic vehicles and artefacts – photographic library, reference library 44909

Madame Tussaud's, Marylebone Rd, London NW1 5LR • T: +44 870 4003000 • F: +44 20 74650862 • firstname.lastname@madame-tussauds.com • www.madame-tussauds.com •
Man.: *Peter Philipson* • PR: *Diane Moon* •
Special Museum – 1770
Wax figures, planetarium – Planetarium 44910

Mall Galleries, The Mall, 17 Carlton House Terrace, London SW1Y 5BD • T: +44 20 79306844 • F: 78397830 • info@mallgalleries.com • www.mallgalleries.org.uk •
Dir.: *Lewis McNaught* •
Public Gallery – 1971 44911

Mander and Mitchenson Theatre Collection, c/o Jerwood Library, Trinity College of Music, Old Royal Naval College, King Charles Court, London SE10 9JF • T: +44 20 83054426 • F: 83059426 • rmangan@tcm.ac.uk • www.tcm.ac.uk •
Dir.: *Richard Mangan* •
Performing Arts Museum
Theatre, music hall, pantomime, circus, opera, ballet, incl. posters, playbills, programmes, engravings, photographs, paintings, manuscripts, china and unique ephemera 44912

Markfield Beam Engine and Museum, Markfield Rd, South Tottenham, London N15 4RB • T: +44 20 88020680 • F: 88020680 •
Dir.: *A.J. Spackman* •
Science&Tech Museum
Beam pumping engine, public health engineering 44913

Matt's Gallery, 42-44 Copperfield Rd, London E3 4RR • T: +44 20 89831771 • F: 89831435 • info@mattsgallery.org • www.mattsgallery.org •
Dir.: *Robin Klassnik* •
Public Gallery – 1979 44914

MCC Museum, Lord's Cricket Ground, Saint John's Wood, London NW8 8QN • T: +44 20 76168595/96 • F: 74321062, 76168659 • museum@mcc.org.uk • www.lords.org •
Cur.: *Adam Chadwick* •
Historical Museum – 1953
History of cricket, pictures, relics, trophies, art objects – library 44915

Metropolitan Police Historical Collection, Empress State Building, Lillie Rd, West Brompton, London SW6 1TR • T: +44 20 71611234 • historicstore@met.police.uk • www.met.police.uk •
Head: *Neil Paterson* •
Historical Museum
Beat duty equipment, uniforms, books, documents, photos, medals, paintings, police station furnishings, officer records from 1829 44916

Michael Faraday's Laboratory and Museum, The Royal Institution of Great Britain, 21 Albemarle St, London W1S 4BS • T: +44 20 74092992 • F: 76293569 • ri@ri.ac.uk • www.rigb.org •
Head: *Frank James* •
Science&Tech Museum – 1972
Unique coll of Faraday's original apparatus, diaries 44917

Millwall FC Museum, Millwall Football Club, Zampa Rd, London SE16 3LN • T: +44 20 86980793 • F: 86980793 •
Cur.: *Chris Bethell* •
Special Museum
Football and social history from 1885 to date 44918

Morley Gallery, Art Centre, Morley College, 61 Westminster Bridge Rd, London SE1 7HT • T: +44 20 74509226 • F: 79284074 • janet.browne@morleycollege.ac.uk •
Sc. Staff: *Jane Hartwell* (Exhibitions Organisation) •
Public Gallery / University Museum 44919

Museum and Study Collection, Central Saint Martins College of Art and Design, Southampton Row, London WC1B 4AP • T: +44 20 75147146 • F: 75147024 • c.pound@csm.arts.ac.uk • www.csm.arts.ac.uk/museum •
Fine Arts Museum / Decorative Arts Museum 44920

Museum in Docklands, West India Quay, Hertsmere Rd, London E14 4AL • T: +44 870 4443856 • F: 4443858 • info@museumindocklands.org.uk • www.museumindocklands.org.uk •
Dir.: *Prof. Jack Lohman* • *David Spence* •
Historical Museum
Cargo handling, port trades, navigational aids such as ship's compasses , sextants and wheels – archives, library 44921

Museum of Contemporary Art London, 113 Bellenden Rd, Project Space, London SE15 4QY • T: +44 20 77719778 • mocalondon@yahoo.com •
Fine Arts Museum 44922

Museum of Freemasonry, Freemasons Hall, Great Queen St, London WC2B 5AZ • T: +44 20 73959257 • F: 74047418 • libmus@ugle.org.uk • www.freemasonry.london.museum •
Dir.: *Diane Clements* •
Special Museum – 1837
Coll of class, porcelain, silver, furniture, paintings – Library 44923

Museum of Fulham Palace, Bishop's Av, London SW6 6EA • T: +44 20 77363233 • F: 77363233 • fulhampalace@waitrose.com •
Cur.: *Miranda Poliakoff* •
Historical Museum
Archaeology, paintings, stained glass, history of the Palace 44924

Museum of Garden History, Saint Mary at Lambeth, Lambeth Palace Rd, London SE1 7LB • T: +44 20 74018865 • F: 74018869 • info@museumgardenhistory.org • www.museumgardenhistory.org •
Exec. Dir.: *Victoria Farrow* •
Historical Museum – 1977
Garden tools, 17th c garden reproduction,John Tradescants, UK garden history, plant hunting, horticulture, 44925

Museum of Installation, 171 Deptford High St, London SE8 3NU • T: +44 20 84694140 • moi@dircon.co.uk •
Dir.: *Nicolas de Oliveira* • *Nicola Oxley* • *Michael Petry* • *Jeremy Wood* •
Fine Arts Museum
Installations – archive 44926

Museum of Instruments, Royal College of Music, c/o Centre for Performance History, Prince Consort Rd, South Kensington, London SW7 2BS • T: +44 20 75914346 • F: 75897740 • museum@rcm.ac.uk • www.rcm.ac.uk •
Cur.: *Jenny Nex* •
Music Museum – 1883
Old stringed, wind and keyboard instruments 44927

Museum of London, London Wall, London EC2Y 5HN • T: +44 870 4443851 • F: 4443853 • info@museumoflondon.org.uk • www.museumoflondon.org.uk •
Dir.: *Jack Lohman* • Sc. Staff: *Roy Stephenson* (Early London Dept.) • *Dr. Cathy Ross* (Later London Dept.) • *Kate Starling* (Cur. Division) •
Historical Museum – 1912
Prehistoric, Roman and medieval antiquities, costumes, decorative art objects, topography, history and social life of London, royal relics, Parliament and legal history, Cromwellian period relics, objects relating to the Great Fire of London, toys, fire engines, glass 44928

Museum of Methodism, 49 City Rd, London EC1Y 1AU • T: +44 20 75322262 • F: 76083825 • museum@wesleyschapel.org.uk • www.wesleyschapel.org.uk •
Cur.: *Heather Carson* •
Religious Arts Museum – 1984
Wesleyana, paintings, manuscript letters, social history 44929

Museum of the Order of Saint John, Saint John's Gate, Saint John's Ln, Clerkenwell, London EC1M 4DA • T: +44 20 73244005, 73244070 • F: 73360587 • museum@nhq.sja.org.uk • www.museumstjohn.org.uk •
Cur.: *Pamela Willis* •
Religious Arts Museum / Historical Museum / Fine Arts Museum / Decorative Arts Museum – 1915
Silver, coins, armour, medals, furniture, documents on the order of Saint John, manuscripts, uniforms, glass, ceramics, paintings, prints, cartography, textiles, St John Ambulance, hist of religious military order, Rhodes, Malta, European estates – Library, archive 44930

Museum of the Royal College of Physicians, 11 Saint Andrews Pl, Regent's Park, London NW1 4LE • T: +44 20 79351174 ext 510 • F: 74863729 • info@rcplondon.ac.uk • www.rcplondon.ac.uk •
Dir.: *Prof. C.M. Black* • Cur.: *Emma Shepley* •
Fine Arts Museum
History of the college and the physician's profession, portraits of fellows a.o., Symon's coll of medical instr, William Harvey's demonstration rod, anatomical tables 44931

Museum of the Royal College of Surgeons of England, 35-43 Lincoln's Inn Fields, London WC2A 3PE • T: +44 20 78696560 • F: 78696564 • museums@rcseng.ac.uk • www.rcseng.ac.uk •
Historical Museum 44932

Museum of the Royal Pharmaceutical Society of Great Britain, 1 Lambeth High St, London SE1 7JN • T: +44 20 75722210 • F: 75722499 • museum@rpsgb.org • www.rpsgb.org •
Keeper: *Briony Hudson* (Collections) •
Special Museum – 1841
History of pharmacy, 17th-18th c drugs, English Delft drug jars, mortars, retail and dispensing equipment 44933

Musical Museum, 399 High St, Brentford, London TW8 0DU • T: +44 20 85608108 • www.musicalmuseum.co.uk •
Cur.: *Richard Cole* •
Music Museum – 1963
Automatic musical instruments in working order, pianos, music rolls, Welte Philharmonic Reproducing Pipe Organ Model, Wurlitzer Theater Organ, Double Mills Violano Virtuoso, Hupfeld Phonoliszt Violina, Hupfeld Animatic Clavitist Sinfonie-Jazz and orchestra, Edison Phonograph, Hupfeld Piano, Street Barrel Pianos and Organs, Broadwood Grand Piano 44934

Narwhal Inuit Art Gallery, 55 Linden Gardens, Chiswick, London W4 2EH • T: +44 20 87421268 • F: 87421268 • narwhalman@blueyonder.co.uk • www.niaef.com •
Dir.: *Ken Mantel* •
Ethnology Museum
Canadian, Russian and Greenlandic Inuit art 44935

National Army Museum, Royal Hospital Rd, London SW3 4HT • T: +44 20 77300717 • F: 78236573 • info@national-army-museum.ac.uk • www.national-army-museum.ac.uk •
Dir.: *Dr. A.J. Guy* • Asst. Dir.: *Dr. P. Boyden* • Sc. Staff: *D. Smurthwaite* •
Military Museum – 1960
Hist of the British Army 1415-2000, Indian Army until 1947, colonial forces, uniforms, weapons, medals, British military painting, early photographs – library 44936

National Gallery, Trafalgar Sq, London WC2N 5DN • T: +44 20 77472885 • F: 77472423 • information@ng-london.org.uk • www.nationalgallery.org.uk •
Dir.: *Nicholas Penny* • Cur.: *Dr. Susan Foister* (Dir. of Collections) • *Dr. Dawson Carr* • *Dr. Dillian Gordon* • *David Jaffe* • *Christopher Riopelle* • *Dr. Humphrey Wine* • *Lorne Campbell* • *Sarah Herring* • *Betsy Wieseman* • *Carol Plazotta* • *Xavier Bray* • *Minna Moore Ede* • *Anne Robbins* • *Luke Syson* • *Simona Di Nepi* • Sc. Staff: *Dr. Ashok Roy* (Dir. of Scientific Research) • *Jo Kirby* • *Marika Spring* • *Joe Padfield* • *Helen Howard* • *Ali Alsam* • *David Peggie* • Conservation: *Martin Wyld* (Dir.) • *Jill Dunkerton* • *Paul Ackroyd* • *Larry Keith* •
Fine Arts Museum – 1824
Western European paintings from 1250-1900 incl works by Leonardo, Rembrandt, Velázquez and others, British painters from Hogarth to Turner, 19th c French painting such as Cézanne, Van Gogh, Manet, Monet 44937

National Hearing Aid Museum, Royal Throat, Nose and Ear Hospital, Grays Inn Rd, London WC1X 1DA • T: +44 20 79151390 • F: 79151646 • p.turner@ucl.ac.uk •
Cur.: *Peggy Chalmers* •
Science&Tech Museum
Hearing aids and test equipment, the first cochlea implant device to be fitted in the United Kingdom 44938

National Maritime Museum, Park Row, Greenwich, London SE10 9NF • T: +44 20 88584422 • F: 83126632 • www.nmm.ac.uk •
Dir.: *Roy Clare* •
Historical Museum – 1934
Maritime hist of Britain, Navy, Merchant Service, yachting, ship models and plans 1700-1990, oil paintings, Nelson memorabilia, astronomical instruments (17th-20th c), globes since 1530, rare books from 1474, manuscripts, charts, atlases – library 44939

National Portrait Gallery, Saint Martin's Pl, London WC2H 0HE • T: +44 20 73060055 • archiveenquiry@npg.org.uk • www.npg.org.uk •
Dir.: *Sandy Nairne* •
Fine Arts Museum – 1856
National coll of portraits (15th-21st c), incl paintings, sculptures, miniatures, engravings and photographs – archive, library; Montacute House, Beningbrough Hall, Bodelwyddan Castle 44940

Natural History Museum, Cromwell Rd, London SW7 5BD • T: +44 20 79425000 • information@nhm.ac.uk • www.nhm.ac.uk •
Dir.: *Dr. Michael Dixon* • Sc. Staff: *Prof. P. Rainbow* (Zoology) • *A. Wheeler* (Entomology) • *N. MacLeod* (Palaeontology) • *Prof. A. Fleet* (Mineralogy) • *Prof. Richard Bateman* (Botany) •
Natural History Museum – 1881
Botany, entomology, mineralogy, palaeontology, zoology, ornithology 44941

Nelson Collection at Lloyd's, 1 Lime St, London EC3M 7HA • T: +44 20 76237100 • F: 73276400 •
Decorative Arts Museum
Silver, documents, letters, objects of art associated with Admiral Lord Nelson and his contemporaries 44942

New Realms, 111 Lexington Bldg, Fairfield Rd, London E3 2UE • T: +44 20 89818225 • contact@newrealms.org.uk • www.newrealms.org.uk •
Public Gallery
Contemporary paintings, prints and sculpture by Eastern European, Baltic, Russian and Central Asian artists 44943

New Realms at Adams Street, 9 Adam St, London WC2N 6AA • T: +44 20 89818225 • contact@newrealms.org.uk • www.newrealms.org.uk •
Public Gallery
Contemporary art by Eastern European, Baltic and Central Asian artists 44944

Newham Heritage Centre, 292 Barking Rd, London E6 3BA • T: +44 20 84306393 • F: 84306392 • sean-sherman@newham.gov.uk • www.newham.gov.uk •
Man.: *Sean Sherman* •
Local Museum – 2002
Archaeology, local hist, Bow porcelain, work of Madge Gill (local outsider artist) – museum education 44945

North Woolwich Old Station Museum, Pier Rd, North Woolwich, London E16 2JJ • T: +44 20 74747244 • F: 74736065 • kathy.taylor@newham.gov.uk • www.newham.gov.uk •
Man.: *Kathy Taylor* •
Science&Tech Museum 44946

Odontological Museum, c/o Royal College of Surgeons of England, 35-43 Lincoln's Inn Fields, London WC2A 3PE • T: +44 20 778696562 • F: 78696564 • museums@rcseng.ac.uk •
Special Museum
Coll of 6,000 animal skulls, of which 2,000 are pathological, 3,500 human casts, skulls and jaws, 200 dentures, 300 dental instruments 44947

Old Operating Theatre, Museum and Herb Garret, London's Museums of Health and Medicine, 9a Saint Thomas St, London SE1 9RY • T: +44 20 79554791 • F: 73788383 • curator@thegarret.org.uk • www.thegarret.org.uk •
Dir.: *Kevin Flude* • Cur.: *Karen Howell* •
Special Museum – 1957
Medicine, surgery, hospitals, Florence Nightingale, herbal medicine 44948

Old Speech Room Gallery, Harrow School, High St, Harrow, London HA1 3HP • T: +44 20 88728021, 88728205 • F: 84233112 • wdriver@harrowschool.org.uk • www.harrowschool.org.uk/osrg •
Cur.: *Carolyn R. Leder* •
Fine Arts Museum / Museum of Classical Antiquities
School's coll (Egyptian and Greek antiquities, watercolours, printed books) 44949

Palace of Westminster, Houses of Parliament, London SW1A 0AA • T: +44 20 72196218, 72195503 • F: 72194250 • curator@parliament.uk • www.parliament.uk •
Cur.: *Malcolm Hay* •
Decorative Arts Museum
Coll of Victorian wall paintings depicting scenes from British history, coll of political portraits, and works of art, coll of Gothic Revival furniture 44950

Parasol Unit, Foundation for Contemporary Art, 14 Wharf Rd, London N1 7RW • T: +44 20 74907373 • F: 74907775 • info@parasol-unit.org • www.parasol-unit.org •
Public Gallery 44951

Pathological Museum, Saint Bartholomew's, Royal London School of medicine Smithfield, London EC1A 7BE • T: +44 20 76018537 • F: 76018530 • d.g.lowe@mds.qmw.ac.uk • www.mds.qmw.ac.uk/ •
Cur.: *Dr. D. Lowe* •
Special Museum 44952

Petrie Museum of Egyptian Archaeology, University College London, Malet Pl, London WC1E 6BT • T: +44 20 76792884 • F: 76792886 • petrie.museum@ucl.ac.uk • www.petrie.ucl.ac.uk •
Dir.: *S. MacDonald* • Cur.: *Stephen Quirke* (Papyri, Ostraca, Hieroglyphic Texts) •
Archaeology Museum / University Museum – 1913
Archaeology, hieroglyphic texts, Sir William M. Flinders Petrie (1853-1942) 44953

The Photographers' Gallery, 58 Great Newport St, London WC2H 7HY • T: +44 20 78311772 • F: 78369704 • info@photonet.org.uk • www.photonet.org.uk •
Dir.: *Brett Rogers* •
Public Gallery 44954

Pitzhanger Manor-House and Gallery, Walpole Park, Mattlock Ln, Ealing, London W5 5EQ • T: +44 20 85671227 • F: 85670595 • pmgallery&house@ealing.gov.uk •
Dir.: *Helen Walker* • *Carol Swords* •
Decorative Arts Museum / Local Museum
Contemporary art exhibitions, Martinware pottery 44955

Polish Cultural Institute, 34 Portland Pl, London W1B 1HQ • T: +44 20 76366032 • F: 76372190 •
Dir.: *Pawel Potorczyn* •
Public Gallery
Polish paintings, graphics, photography, posters, tapestry, folk art 44956

Pollock's Toy Museum, 1 Scala St, London W1T HTL • T: +44 20 76363452 • info@pollocksmuseum.co.uk • www.pollocksmuseum.co.uk •
Head: *Eddy Fawdry* •
Historical Museum – 1956
Old toys and games, toytheatres 44957

POSK Gallery, 238-246 King St, London W6 0RF • T: +44 20 87411940 • F: 87463798 • admin@posk.org • www.posk.org •
Dir.: *Joanna Ciechanowska* •
Association with Coll – 1980
POSK coll. charting the history of Poles in GB -paintings, photos, archives, engravings, maps, tin plate, armour, sculptures – archives 44958

Pump House Gallery, Battersea Park, London SW11 4NJ • T: +44 20 88717572 • F: 72289062 • pumphouse@wandsworth.gov.uk • www.wandsworth.gov.uk/gallery •
Dir.: *Susie Gray* •
Public Gallery
Contemporary art 44959

Pumphouse Educational Museum, Lavender Pond and Nature Park, Lavender Rd, Rotherhithe, London SE16 5DZ • T: +44 20 72312976 • F: 72312976 • c.marais@btclick.com • www.pumphouse.org.uk •
Dir.: *Caroline Marais* •
Local Museum
Dockers tools, life in Rotherhithe, replica 6ft Queen Elizabeth II wedding cake 44960

Queen Elizabeth's Hunting Lodge, Rangers Rd, Chingford, London E4 7QH • T: +44 20 85296681 • F: 85298209 • sophie.lillington@corpoflondon.gov.uk • www.cityoflondon.gov.uk •
Head: *Sophie Lillington* •
Historical Museum – 1895
Local history, archaeology, forestry, zoology, botany 44961

The Queen's Gallery, Royal Collection, Buckingham Palace, London SW1A 1AA, mail addr: Stable Yard House, Saint James's Palace, London SW1A 1JR • T: +44 20 77667301 • F: 79309625 • bookinginfo@royalcollection.org.uk • www.royalcollection.org.uk •
Fine Arts Museum – 1962
Paintings, drawings, artworks from Royal coll, furniture, clocks, porcelain, silver, scientific instruments, books, miniatures, gems 44962

Ragged School Museum, 46-50 Copperfield Rd, London E3 4RR • T: +44 20 89806405 • F: 89833481 • enquiries@raggedschoolmuseum.org.uk • www.raggedschoolmuseum.org.uk •
Dir.: *Eleanor Clark* •
Special Museum – 1990
Victorian school room, social hist 44963

RIBA Drawings and Archives Collections, Royal Institute of British Architects, c/o Victoria and Albert Museum, Cromwell Rd, London SW7 2RL • T: +44 20 73073708 • F: 75893175 • drawings&archives@inst.riba.org • www.architecture.com •
Cur.: *Charles Hind* •
Fine Arts Museum
Over 500,000 architectural drawings c1480 to the present together with engravings, models, portraits, mauscripts 44964

Royal Academy of Arts, Burlington House, Piccadilly, London W1J 0BD • T: +44 20 73008000 • F: 73008001 • www.royalacademy.org.uk •
Fine Arts Museum – 1768
Works of art are mainly by British artists from the 18th century to the present day, including paintings, sculptures, drawings, prints, historic books, archives, historic photographs and plaster casts. 44965

Royal Academy of Arts, Burlington House, Piccadilly, London W1J 0BD • T: +44 20 73008000 • F: 73008001 • events.lectures@royalacademy.org.uk • www.royalacademy.org.uk •
Pres.: *Sir Nicholas Grimshaw* •
Public Gallery – 1768
British art 18th-21st c, creation, enjoyment and appreciation of the visual arts, temporary exhibits of painting, sculpture, architecture and engraving 44966

Royal Air Force Museum, Grahame Park Way, London NW9 5LL • T: +44 20 82052266 • F: 82001751, 83584981 • london@rafmuseum.com • www.rafmuseum.com •
Dir. Gen.: *Dr. Michael A. Fopp* •
Military Museum – 1972
Sister Museum at Cosford, reserve coll at Cosford, one of the best WW II fighter coll in the world, 70 aircraft on permanent display 44967

Royal Armouries, Tower of London, London EC3N 4AB • T: +44 20 74806358 • F: 74812922 • enquiries@armouries.org.uk • www.armouries.org.uk •
Dir.: *Paul Evans* (Master of the Armouries) • Keeper: *Dr. Geoffrey Parnell* (Tower History) • Collections South: *Bridget Clifford* •
Historical Museum / Military Museum / Decorative Arts Museum
Eight galleries tell the story of the evolution of the Tower and the Armouries itself over the centuries, incl weapons of the historic Tower arsenal, Royal Tudor and Stuart armours, coll 16th-19th c – Education centre, archives 44968

Royal British Society of Sculptors, 108 Old Brompton Rd, London SW7 3RA • T: +44 20 73738615 • F: 73703721 • info@rbs.org.uk • www.rbs.org.uk •
Pres.: *Prof. Brian Falconbridge* •
Public Gallery 44969

Royal Ceremonial Dress Collection, Kensington Palace, London W8 4PX • T: +44 20 79379561 • F: 73760198 • www.hrp.org.uk •
Dir.: *Nigel Arch* •
Special Museum
Royal and ceremonial dress 44970

Royal College of Art, Kensington Gore, London SW7 2EU • T: +44 20 75904444 • F: 75904124 • info@rca.ac.uk • www.rca.ac.uk •
Dir.: *Prof. Christopher Frayling* •
Public Gallery 44971

Royal College of Obstetricians and Gynaecologists Collection, 27 Sussex Pl, Regent's Park, London NW1 4RG • T: +44 20 77726309 • F: 72628331 • museum@rcog.org.uk • www.rcog.org.uk •
Cur.: *Ian L.C. Fergusson* •
Special Museum – 1929
Obstetric, midwifery, surgical and gynaecological instruments and artefacts – library, archive 44972

Royal Fusiliers Museum, Tower of London, London EC3N 4AB • T: +44 20 74885610 • F: 74811093 • royalfusiliers@fsmail.net •
Dir.: *C.P. Bowes-Crick* •
Military Museum – 1949
History of regiment 1685-1968, silver, documents, armaments, uniforms 44973

Royal Hospital Chelsea Museum, Royal Hospital Rd, Chelsea, London SW3 4SR • T: +44 20 78815203 • F: 78815463 • info@chelsea.pensioners.org.uk • www.chelsea.pensioners.org.uk •
Historical Museum
Royal hospital artefacts, medals and uniforms, veteran soldiers 44974

Royal London Hospital Archives and Museum, Newark St, Whitechapel, London E1 2AA • T: +44 20 73777608 • F: 73777413 • jonathan.evans@bartsandthelondon.nhs.uk • www.bartsandthelondon.nhs.uk •
Head: *Jonathan Evans* •
Historical Museum – 1989
Medical science hist 44975

The Royal Mews, Royal Collection, Buckingham Palace, London SW1A 1AA, mail addr: Stable Yard House, Saint James's Palace, London SW1A 1JR • T: +44 20 77667302 • F: 79309625 • buckinghampalace@royalcollection.org.uk • www.royalcollection.org.uk •
Dir.: *Hugh Roberts* •
Historical Museum – 1825
The Monarch's magnificent gilded State carriages and coaches, incl the unique Gold State Coach are housed here together with their horses and State liveries 44976

Royal Observatory Greenwich, National Maritime Museum, Greenwich, London SE10 9NF • T: +44 20 88584422 • F: 83126632 • www.nmm.ac.uk •
Dir.: *Roy Clare* •
Science&Tech Museum 44977

Saatchi Gallery, c/o Duke of York's HQ, Sloane Sq, Chelsea, London SW3 4RY • T: +44 20 78232332 • F: 78232334 • www.saatchi-gallery.co.uk •
Public Gallery
Contemporary art 44978

Saint Bartholomew's Hospital Museum, West Smithfield, London EC1A 7BE • T: +44 20 76018152 • barts.archives@bartsandthelondon.nhs.uk • www.bartsandthelondon.nhs.uk/museums •
Historical Museum – 1997
Archives 44979

Saint Bride's Crypt Museum, Saint Bride's Church, Fleet St, London EC4Y 8AU • T: +44 20 74270133 • F: 75834867 • info@stbrides.com • www.stbrides.com •
Dir.: *James Irving* •
Religious Arts Museum – 1957
History of Fleet Street, Roman ruins, former church ruins 44980

Saint Thomas' Hospital Mini Museum, Lambeth Palace Rd, London SE1 7EH • T: +44 20 71887188 • www.guysandthomas.nhs.uk •
Dir.: *Karen Sarkissian* •
Historical Museum
800 years hospital hist 44981

Salvation Army International Heritage Centre (closed, reopening September 2011), House 14, William Booth College, Denmark Hill, London SE5 8BQ • T: +44 20 77373327 • F: 77374127 • heritage@salvationarmy.org • www.salvationarmy.org •
Dir.: *Stephen Grinsted* •
Historical Museum
library, archives 44982

Science Museum, The National Museum of Science & Industry, Exhibition Rd, South Kensington, London SW7 2DD • T: +44 870 8704868 • F: +44 20 79424447 • sciencem@sciencemuseum.org.uk • www.sciencemuseum.org.uk •
Dir.: *Prof. Martin J. Earwicker Freng* •
Science&Tech Museum / Historical Museum – 1857
History of science and industry, agriculture, astronomy, air, sea and land transport, civil, electrical, marine and mechanical engineering, jet engines, geophysics, telcommunications, domestic appliances, spacetechnology, chemistry, medicine, interactive galleries 44983

Serpentine Gallery, Kensington Gardens, London W2 3XA • T: +44 20 72981515, 74026075 • F: 74024103 • information@serpentinegallery.org • www.serpentinegallery.org •
Dir.: *Julia Peyton-Jones* •
Public Gallery – 1970 44984

Shakespeare's Globe Exhibition and Tour, 21 New Globe Walk, Bankside, London SE1 9DT • T: +44 20 79021500 • F: 79021515 • exhibit@shakespearesglobe.com • www.shakespeares-globe.org •
Gen. Dir.: *Peter Kyle* • Dir.: *Dominic Dromgoole* •
Fine Arts Museum
The largest exhibition in the world dedicated to Shakespeare in his workplace, incl a tour of today's working theatre 44985

Sherlock Holmes Museum, 221b Baker St, London NW1 6XE • T: +44 20 79358866 • F: 77381269 • info@sherlock-holmes.co.uk • www.sherlock-holmes.co.uk •
Dir.: *Grace Riley* •
Special Museum – 1990
Victorian memorabilia, artefacts to Arthur Conan Doyle 44986

Sir John Soane's Museum, 13 Lincoln's Inn Fields, London WC2A 3BP • T: +44 20 74052107 • F: 78313957 • jbrock@soane.og.uk • www.soane.org •
Dir.: *Tim Knox* •
Fine Arts Museum / Decorative Arts Museum – 1833
Sir John Soane's private art and antiquities coll, paintings (Hogarth, Turner), architectural and other drawings – library 44987

South London Gallery, 65 Peckham Rd, London SE5 8UH • T: +44 20 77036120 • F: 72524730 • mail@southlondongallery.org • www.southlondongallery.org •
Dir.: *Margot Heller* • Cur.: *Kit Hammonds* •
Public Gallery – 1891
Work of mid-carrer British artists and emerging and establ. international artists in an annual programme of contemporary and live art 44988

Southside House, Wimbledon Common, London SW19 4RJ • T: +44 20 89467643 • F: 89467643 • info@southsidehouse.com •
Dir.: *Adam Munthe* •
Decorative Arts Museum 44989

Spencer House, 27 Saint James's Pl, London SW1A 1NR • T: +44 20 74998620, 75141964 • F: 74092952 • www.spencerhouse.co.uk •
Dir.: *Jane Rick* •
Fine Arts Museum / Decorative Arts Museum
Built in 1756-66, London's finest surviving private palace, 18th c furniture and paintings 44990

Stephen Lawrence Gallery, University of Greenwich, Queen Mary Court, Old Royal Naval College, Park Row, London SE10 9LS • T: +44 79 41125901 • k.oreilly@gre.ac.uk •
Public Gallery / University Museum 44991

Stephens Collection, Av House, East End Rd, Finchley, London N3 3QE • T: +44 20 83463072 • stephens.collection@avenuehouse.org.uk • www.london-northwest.com •
Historical Museum
History of the Stephens Ink Company 44992

Strang Print Room, University College London, South Cloisters, Gower St, London WC1E 6BT • T: +44 20 76792540 • www.art.museum.ucl.ac.uk •
Fine Arts Museum / University Museum
Coll of fine german drawings,old master prints, plaster models, drawings by Flaxman, early english watercolours and english prints, japanese ukiyo-e prints, history of the slade school 44993

Syon House, Syon Park, Brentford, London TW8 8JF • T: +44 20 85600882 • F: 85680936 • info@syonpark.co.uk • www.syonpark.co.uk •
Historic Site – 1415
Robert Adam furniture and decoration, famous picture coll, home of the Dukes of Northumberland 44994

Tate Britain, Millbank, London SW1P 4RG • T: +44 20 78878000 • F: 78878007 • information@tate.org.uk • www.tate.org.uk •
Dir.: *Sir Nicholas Serota* • *Stephen Deuchar* • Head: *Celia Clear* (Publications) • Sc. Staff: *Sheena Wagstaff* (Exhibitions and Displays) • *Richard Humphreys* (Education and Interpretation) • *Peter Wilson* (Building and Gallery Services) •
Fine Arts Museum – 1897
National art coll of British Art from 15th c to the present, Turner bequest, Pre-Raphaelites – library and archive of 20th c art 44995

Tate Modern, Bankside, London SE1 9TG • T: +44 20 78878000 • F: 74015129 • information@tate.org.uk • www.tate.org.uk/modern •
Dir.: *Dercon Chris* • Head: *Celia Clear* (Publications) • Sc. Staff: *Sheena Wagstaff* (Exhibitions and Displays) • *Toby Jackson* (Education and Interpretation) • *Peter Wilson* (Building and Gallery Services) •
Fine Arts Museum – 2000
International 20th c art incl major works by some of the most influential artists of this century such as Picasso, Matisse, Dalí, Duchamp, Moore, Bacon, Gabo, Giacometti and Warhol 44996

Thames Police Museum, Wapping Police Station, 98 Wapping High St, London E1W 2NE • T: +44 20 72754421 • F: 72754490 • thames.metpol@gtnet.gov.uk • thamespolicemuseum.org.uk •
Dir.: *Bob Jeffries* •
Special Museum
Uniforms and items 44997

Tower Bridge Exhibition, Tower Bridge, London SE1 2UP • T: +44 20 74033761 • F: 73577935 • enquiries@towerbridge.org.uk • www.towerbridge.org.uk •
Dir.: *David White* •
Science&Tech Museum – 1982
Original Victorian steam machinery, glass walkways obove the Thames 44998

UCL Art Collections, The Strang Print Room, South Cloisters, University College London, Gower St, London WC1 6BT • T: +44 20 76792540 • college.art@ucl.ac.uk • www.ucl.ac.uk •
Association with Coll
18th and 19th cent British works of art (Wright, Turner, Constable et al.); Flaxman coll (sculptor); Old Masters Prints (Dürer, Mantegna, Rembrandt et al., Old Masters Drawings (Hans Burgkmair, Sigmund Holbein et al.); student prize works of Slade school of Fine Art 44999

UCL Museums and Collections, University College London, c/o Institute of Archaeology, 31-34 Gordon Sq, London WC1H 0PT • T: +44 20 76797495 • F: 73832572 • curator@ucl.ac.uk • www.museum.ucl.ac.uk •
Local Museum / University Museum
Art, ethnography, archaeology, natural hist, scientific instr 45000

VA Museum of Childhood, Bethnal Green, Cambridge Heath Rd, London E2 9PA • T: +44 20 89835200 • F: 89835225 • moc@vam.ac.uk • www.vam.ac.uk/moc/your_visit/index.html •
Dir.: *Diane Lees* •
Special Museum – 1872
Toys, dolls, games, puppets, children's books, doll's houses, teddy bears, toy soldiers, trains, nursery coll incl children's costume and furniture, baby equipment 45001

Vestry House Museum, Vestry Rd, Walthamstow, London E17 9NH • T: +44 20 84964391 • vhm.enquiries@walthamforest.gov.uk • www.walthamforest.gov.uk/vestry-house •
Man.: *Vicky Carroll* •
Local Museum – 1931
Bremer car, built between 1892-94 – local studies library and archives 45002

Veterinary Museum, c/o Library, Royal Veterinary College, Royal College St, London NW1 0TU • T: +44 20 74685162 • F: 74685162 • www.rvc.ac.uk •
Historical Museum – 1992
Development of veterinary education and science, veterinary instruments, print and manuscript items, pictures, history of veterinary medicine in the UK 45003

Victoria and Albert Museum, Cromwell Rd, South Kensington, London SW7 2RL • T: +44 20 79422000 • F: 79422496 • vanda@vam.ac.uk • www.vam.ac.uk •
Dir.: *Mark Jones* • Dep. Dir.: *Beth McKillop* • Sc. Staff: *Jane Lawson* • *Moira Gemmill* • *Damien Whitmore* • *Jo Prosser* • *Sean Williams* •
Decorative Arts Museum – 1852
Books, ceramics, drawings, fashion, furniture, glass, jewellery, manuscripts, metalwork, miniatures, paintings, photographs, prints, sculpture, silver, tapestries, textiles, watercolours, woodwork, theatre and performance coll's – National Art Library 45004

Vintage Wireless Museum, 23 Rosendale Rd, West Dulwich, London SE21 8DS • T: +44 20 86703667 •
Cur.: *Gerald Wells* •
Science&Tech Museum / Historical Museum
Broadcast receiving history, TV and radio 45005

The Wallace Collection, Hertford House, Manchester Sq, London W1U 3BN • T: +44 20 75639500 • F: 72242155 • visiting@wallacecollection.org • www.wallacecollection.org •
Dir.: *Rosalind J. Savill* •
Fine Arts Museum – 1900
Paintings of all European schools, 17th c Dutch painting, incl works by Titian, Rubens, Van Dyck, Rembrandt, 18th c French art, furniture and ceramics, sculpture, goldsmith art, porcelain, maiolica, European and Oriental armaments 45006

Wandsworth Museum, 38 West Hill, Wandsworth, London SW18 1RZ • T: +44 208 8706060 • contact@wandsworthmuseum.co.uk • www.wandsworthmuseum.co.uk •
Cur.: *Patricia Astley Cooper* •
Local Museum – 1986 45007

Wellcome Library, 210 Euston Rd, London NW1 2BE • T: +44 20 76118722 • F: 76118369 • library@wellcome.ac.uk • library.wellcome.ac.uk •
Libr.: *Dr. Simon Chaplin* •
Library with Exhibitions
Asian coll, manuscripts, iconographics, rare books – archives 45008

Wellcome Museum of Anatomy and Pathology, c/o Royal College of Surgeons of England, 35-43 Lincoln's Inn Fields, London WC2A 3PE • T: +44 20 778696562 • F: 78696564 • museums@rcseng.ac.uk •
Special Museum / University Museum
Human anatomy and systemic pathology, histological and histopathological microscope slides 45009

Wellington Arch, Hyde Park Cnr, London W1J 7JZ • T: +44 20 79302726 • F: 79251019 • www.english-heritage.org.uk •
Historical Museum
Arch and Buckingham palace history 45010

Wernher Collection at Ranger's House, Chesterfield Walk, Blackheath, London SE10 8QX • T: +44 20 88530035 • F: 88530090 • www.english-heritage.org.uk •
Cur.: *Tori Reeve* •
Decorative Arts Museum – 1688
Sevres porcelain, renaissance jewellery, limoges, enamels 45011

Westminster Abbey Museum, Chapter House, East Cloister, 20 Dean's Yard, London SW1P 3PE • T: +44 20 72225897 • F: 72220960 • info@westminster-abbey.org • www.westminster-abbey.org •
Fine Arts Museum – 1987
Medieval sculture, wall paintings 45012

Westminster Dragoons Collection, 87 Fulham High St, London SW6 3JS • T: +44 20 73844201 • ry-wsqnpsao@tanet.mod.uk • www.westminsterdragoons.co.uk •
Dir.: *Christopher Sayer* •
Military Museum
Uniforms and artefacts of the Westminster Dragoons 45013

Whitechapel Art Gallery, 80-82 Whitechapel High St, London E1 7QX • T: +44 20 75227878 • F: 73771685 • info@whitechapel.org • www.whitechapel.org •
Dir.: *Iwona Blazwick* •
Public Gallery – 1901 45014

William Morris Gallery and Brangwyn Gift, Lloyd Park, Forest Rd, London E17 4PP • T: +44 20 84964390 • wmg.enquiries@walthamforest.gov.uk • www.walthamforest.gov.uk/william-morris •
Dir.: *Vicky Carroll* •
Decorative Arts Museum – 1950
19th/early 20th c English arts & crafts, decorative arts, life & work of socialist poet William Morris 1834-1896, incl work by Morris & Co. and the Pre-Raphaelites – library, archive 45015

The William Morris Society, Kelmscott House, 26 Upper Mall, London W6 9TA • T: +44 20 87413735 • F: 87485207 • william.morris@care4free.net • www.morrissociety.org •
Cur.: *Helen Elletson* •
Association with Coll 45016

Wimbledon Lawn Tennis Museum, All England Lawn Tennis Club, Church Rd, Wimbledon, London SW19 5AE • T: +44 20 89466131 • F: 89446497 • museum@aeltc.com • www.wimbledon.org/museum •
Cur.: *Honor Godfrey* •
Special Museum – 1977
Prints, photos, pictures on Wimbledon, objects, pictures etc. relating to the history of tennis – Kenneth Ritchie Wimbledon Library 45017

Wimbledon Society Museum of Local History, 22 Ridgway, Wimbledon, London SW19 4QN • T: +44 20 82969914 • wimbledonmuseum@yahoo.co.uk • www.wimbledonmuseum.org.uk •
Dir.: *Charles Toase* •
Historical Museum – 1916
Flints, artefacts, prints, watercolours, maps, photographs, manuscripts, natural hist, ephemera 45018

Wimbledon Windmill Museum, Windmill Rd, Wimbledon Common, London SW19 5NR • T: +44 20 89472825 • F: 89472825 • www.wimbledonwindmillmuseum.org.uk •
Science&Tech Museum – 1976 45019

Women's Art Library - Make, c/o Goldsmith College, Rutherford Bldg, New Cross, London SE14 6NW • T: +44 20 77172295 • F: 79197165 • a.greenan@gold.ac.uk • www.goldsmiths.ac.uk/make •
Head: *Althea Greenan* •
Association with Coll
Photographs and slides, postcards, catalogues, posters, press cuttings and ephemera, coll. Pauline Boty, coll. Monica Sjoo 45020

Wyndham Lewis Society, Olympia National Hall, Hammersmith Rd, London W14 8UX • T: +44 870 1219994 • matthew@mhall.info • www.wyndhamlewis.com •
Public Gallery – 1991 45021

Young's Brewery Museum, The Ram Brewery, Wandsworth High St, London SW18 4JD • T: +44 20 88757000 • F: 88757100 • brewerytap@youngs.co.uk • www.youngs.co.uk •
Dir.: *John Young* •
Science&Tech Museum
Brewing equip 19th-20th c, drinking vessels 16th-20th c 45022

Londonderry

Orchard Gallery, 21 Orchard St, Londonderry BT48 6EG, mail addr: 98 Strand Rd, Londonderry BT48 7NN • T: +44 28 71269675 • F: 71267273 • www.derrycity.gov.uk •
Dir.: *Brendan McMenamin* •
Fine Arts Museum 45023

Long Eaton

Long Eaton Town Hall, Derby Rd, Long Eaton NG10 1HU • T: +44 115 9071141 • F: 9329264 • museum@erewash.org.uk • www.erewash.gov.uk •
Local Museum / Fine Arts Museum
Paintings, mostly 18th and 19th c oils, local history 45024

Long Hanborough

Combe Mill Beam Engine and Working Museum, Blenheim Sawmill, Long Hanborough OX8 8ET • T: +44 1865 455029 •
Dir.: *Dr. W.A. Parker* •
Science&Tech Museum
Working forge, 19th c steam beam engine, powering line shafting driving various machinery, small steam engines, 19th c artifacts, tools, crafts 45025

Oxford Bus Museum Trust, Old Station Yard, Main Rd, Long Hanborough OX29 8LA • T: +44 1993 883617 • F: 881662 • colin@print-rite.u-net.com • www.oxfordbusmuseum.org.uk •
Science&Tech Museum
Public transport from Oxford and Oxfordshire, local buses, horse trams, vehicles 45026

Long Melford, Sudbury, Suffolk

Melford Hall, The National Trust, Long Melford, Sudbury, Suffolk CO10 9AA • T: +44 1787 379228 • F: 379228 • melford@nationaltrust.org.uk •
Decorative Arts Museum
Coll of furniture and porcelain, Beatrix Potter memorabilia 45027

Long Wittenham

Pendon Museum, Long Wittenham OX14 4QD • T: +44 1865 407365, 408143 • www.pendonmuseum.com •
Chm.: *Chris Webber* •
Historical Museum – 1954
The Vale Scene, based on the Vale of White Horse, indoor model village & railway illustrating a cross-section of local building styles and materials over the years ... 45028

Lossiemouth

Lossiemouth Fisheries and Community Museum, Pitgaveny St, Lossiemouth IV31 6AA • T: +44 1343 543221 •
Local Museum ... 45029

Lostwithiel

Lostwithiel Museum, 16 Fore St, Lostwithiel PL22 0BW • T: +44 1208 872079 •
Chm.: *Jeanne Jones* •
Local Museum
Former town jail, houses local domestic and agricultural exhibits ... 45030

Loughborough

Charnwood Museum, Granby St, Loughborough LE11 3DU • T: +44 1509 233754, 233737 • F: 268140 • museum@leics.gov.uk • www.charnwoodbc.gov.uk •
Cur.: *Susan Cooke* •
Local Museum
Natural history of Charnwood borough, social and cultural hist ... 45031

Great Central Railway Museum, Great Central Station, Loughborough LE11 1RW • T: +44 1509 230726 • F: 239791 • museum@mlst.org • www.mlst.org •
Dir.: *R. Stephens* •
Special Museum
Station signs, loco's name and number plates, cast iron notices, silverware, signalling equipment, maps and diagrams, models, photos ... 45032

John Taylor Bellfoundry Museum, Freehold St, Loughborough LE11 1AR • T: +44 1509 233414, 212241 • F: 263305 • museum@taylorbells.co.uk • www.taylorbells.co.uk •
Cur.: *Robert Bracegirdle* •
Special Museum
Bells, bellfounding, Talor family history ... 45033

Old Rectory Museum, Rectory Pl, Loughborough LE11 1UW • T: +44 1509 233754 • F: 268140 •
Archaeology Museum
Local archaeological finds ... 45034

War Memorial Carillon Tower and Military Museum, Queen's Park, Loughborough LE11 2TT • T: +44 1509 263370, 634704 • F: 262370 • carillonmuseum@tiscali.co.uk •
Military Museum ... 45035

Loughgall

Dan Winters House - Ancestral Home, 9 Derryloughan Rd, The Diamond, Loughgall BT61 8PH • T: +44 28 38851344 • winter@orangenet.org • www.orangenet.org/winter •
Owner: *Hilda Winter* •
Historical Museum / Historic Site
17th c family chair, Rushlamp, thatching needle, 1903 pram, guns and lead shot, sword from The Battle of the Diamond 1795, churn, milk separator ... 45036

Louth

Louth Museum, 4 Broadbank, Louth LN11 0EQ • T: +44 1507 601211 • www.louth.org.uk •
Pres.: *D.N. Robinson* • Cur.: *Jean Howard* •
Local Museum – 1910
Impressions of seals collected by G.W. Gordon, architect James Fowler (1828-92), woodcarver Thomas Wilkinson Wallis (1821-1903) ... 45037

Lower Broadheath

The Elgar Birthplace Museum, Crown East Ln, Lower Broadheath WR2 6RH • T: +44 1905 333224 • F: 333426 • birthplace@elgarmuseum.org • www.elgarmuseum.org •
Dir.: *Catherine Sloan* •
Music Museum – 1938
Memorabilia of composer Edward Elgar (1857-1934), manuscripts, clippings, photos ... 45038

Lower Methil

Lower Methil Heritage Centre, 272 High St, Lower Methil KY8 3EQ • T: +44 1333 422100 • F: 422101 • www.virtual-pc.com/museum/methil •
Local Museum ... 45039

Lower Stondon

Stondon Museum, Station Rd, Lower Stondon SG16 6JN • T: +44 1462 850339 • F: 850824 • transportmuseum@supanet.com • www.transportmuseum.co.uk •
Dir.: *Maureen Hird* •
Science&Tech Museum
Coll of transport, replica of Captain Cooks ship HM Bark Endeavour, over 100 years of motoring from 1896-1996 ... 45040

Lowestoft

Lowestoft and East Suffolk Maritime Museum, Sparrows Nest Park, Whapload Rd, Lowestoft NR32 1XG • T: +44 1502 561963 •
Dir.: *P. Parker* •
Historical Museum – 1968
Hist of local fishing (1900-1995), evolution of R.N. lifeboats, marine art gallery, ship and boat models ... 45041

Lowestoft Museum, Broad House, Nicholas Everitt Park, Oulton Broad, Lowestoft NR33 9JR • T: +44 1502 511457, 5133795 •
Local Museum – 1972
Lowestoft porcelain, archaeology, hist ... 45042

Royal Naval Patrol Service Association Museum, Europa Room, Sparrows Nest, Lowestoft NR32 1XG • T: +44 1502 586250 •
Cur.: *A. Muffet* •
Historical Museum / Military Museum
Naval uniforms, memorabilia, records, log books, photos, working models ... 45043

Ludham

Toad Hole Cottage Museum, How Hill, Ludham NR29 5PG • T: +44 1692 678763 • F: 678763 • todeholeinfo@broads-authority.gov.uk • www.broads-authority.gov.uk •
Local Museum
Home and working life on the Broads marshes ... 45044

Ludlow

Ludlow Museum, Shropshire County Museum, Castle St, Ludlow SY8 1AS • T: +44 1584 875384, 873857 • F: 872019 • ludlow.museum@shropshire-cc.gov.uk • www.shropshireonline.gov.uk/museums.nsf •
Dir.: *Daniel Lockett* •
Local Museum ... 45045

Luton

Bedfordshire and Hertfordshire Regiment Association Museum Collection, Wardown Park Museum, Old Bedford Rd, Luton LU2 7HA • T: +44 1582 546722 • F: 546763 • museum.gallery@luton.gov.uk • www.lutonline.gov.uk/museums •
Cur.: *Elizabeth Adey* •
Military Museum
Medals, memorabilia, uniforms, equipment – library ... 45046

Stockwood Park Museum, Farley Hill, Luton LU1 4BH • T: +44 1582 738714 • F: 746763 • museum.gallery@luton.gov.uk • www.luton.gov.uk •
Local Museum
Horse-drawn vehicles, rural life, crafts and trades ... 45047

Wardown Park Museum, Old Bedford Rd, Luton LU2 7HA • T: +44 1582 746722 • F: 746763 • museum.gallery@luton.gov.uk • www.lutonline.gov.uk/museums •
Dir.: *Maggie Appleton* • Cur.: *Dr. Elizabeth Adey* (Local History) • *Dr. Paul Hyman* (Natural Sciences, Archaeology) • *Karen Perkins* (Stockwood) • Sc. Staff: *Siobhan Kirrane* • *Helen Close* • *Eleanor Markland* (Education) • *Karen Grunzweil* (Community Outreach) • *Fiona Morton* (Events) • *Trevor Tween* (Nature Conservation) • *Marian Nichols* (Collections, Interpretation) • *Chris Grabham* (Photography) • *Neil Evans* (Design, Exhibitions) •
Local Museum / Fine Arts Museum – 1928
Regional social hist, archaeology, industrial hist, local art, lace, textiles, costumes, toys, rural trades, crafts, porcelain, glass, furniture, household objects ... 45048

Lutterworth

Percy Pilcher Museum, Stanford Hall, Lutterworth LE17 6DH • T: +44 1788 860250 • F: 860870 • enquiries@stanfordhall.co.uk • www.stanfordhall.co.uk •
Dir.: *Robert G. Thomas* •
Science&Tech Museum
First man in England to fly using unpowered flight, replica 1898 fliying machine, rare photos ... 45049

Sherrier Resources Centre, Church St, Lutterworth LE17 4AG • T: +44 1455 552834 • F: 552845 • museums@leics.gov.uk •
Keeper: *Eleanor Thomas* •
Local Museum
Natural, life, working, archaeology, domestic life, artworks ... 45050

Lydd

Lydd Town Museum, Old Fire Station, Queens Rd, Lydd TN29 9DF, mail addr: 4 Varne Mews, Littlestone TN28 8N5 • T: +44 1797 366566 • boxallas@onetel.net • www.lyddtownmuseum.co.uk •
Local Museum
Old fire engine, horse bus and unique beach cart, the army, agriculture, fishing and the fire service ... 45051

Lydney

Dean Forest Railway Museum, c/o Norchard Railway Centre, Forest Rd, New Mill, Lydney GL15 4ET • T: +44 1594 843423, 845840 • F: 845840 • museum@deanforestrailway.co.uk • www.deanforestrailway.co.uk •
Cur.: *John Metherall* •
Science&Tech Museum
Last remaining part of original Severn and Wye railway, railway artefacts since 1809 ... 45052

Lyme Regis

Dinosaurland, Coombe St, Lyme Regis DT7 3PY • T: +44 1297 443541 •
Dir.: *Steve Davies* • *Jenny Davies* •
Natural History Museum – 1995
Fossils, models and live animals, coll of Jurassic fossils ... 45053

Lyme Regis Philpot Museum, Bridge St, Lyme Regis DT7 3QA • T: +44 1297 443370 • info@lymeregismuseum.co.uk • www.lymeregismuseum.co.uk •
Hon. Cur.: *Max Habditch* • Cur.: *Jo Draper* •
Local Museum – 1922
Geology, local history, literary ... 45054

Lymington

Saint Barbe Museum and Art Gallery, New St, Lymington SO41 9BH • T: +44 1590 676969 • F: 679997 • office@stbarbe-museum.org.uk • www.stbarbe-museum.org.uk •
Cur.: *Steven Marshall* •
Local Museum / Fine Arts Museum
Boat building, fishing, engineering, Local and social history, archaeology ... 45055

Lyndhurst

New Forest Museum and Visitor Centre, High St, Lyndhurst SO43 7NY • T: +44 23 80283444 • F: 80284236 • office@newforestmuseum.org • www.newforestmuseum.org •
Dir.: *Joanne Benson* •
Natural History Museum
History, traditions, character and wildlife of New Forest ... 45056

Lynton

Lyn and Exmoor Museum, Saint Vincent's Cottage, Market St, Lynton EX35 6AF • T: +44 1598 752219 • www.devonmuseums.net/lynton •
Chm.: *John Pedder* •
Local Museum
Arts, crafts, implements depicting hist and life, railway, lifeboat, flood features, Victorian doll's house ... 45057

Lyth

Lyth Arts Centre, Wick, Caithness, Lyth KW1 4UD • T: +44 1955 641270 • www.lytharts.org.uk •
Dir.: *William Wilson* •
Fine Arts Museum
Coll of visual art ... 45058

Lytham Saint Anne's

Lytham Hall, Ballam Rd, Lytham Saint Anne's FY8 4LE • T: +44 1253 736652 • F: 737656 •
Dir.: *E.M.J. Miller* •
Decorative Arts Museum
Portraits of the Clifton family, landscape pictures, Gillow furniture ... 45059

Lytham Heritage Centre, 2 Henry St, Lytham Saint Anne's FY8 5LE • T: +44 1253 730787 • F: 730767 • thecentre@lythamheritage.fsnet.co.uk • www.lythamheritage.fsnet.co.uk •
Chm.: *Alan Ashton* •
Local Museum
Lytham hist, local artists and art societies ... 45060

Lytham Lifeboat Museum, Old Lifeboat House, Lytham Saint Anne's FY8 5EQ • T: +44 1253 730155 • www.legendol.freeserve.co.uk/museum.html •
Cur.: *Frank Kilroy* •
Historical Museum – 1886 ... 45061

Lytham Windmill Museum, 2 Henry St, Lytham Saint Anne's FY8 5LE • T: +44 1253 794879 • thecentre@lythamheritage.fsnet.co.uk • www.lythamheritage.fsnet.co.uk •
Chm.: *Alan Ashton* •
Science&Tech Museum
History of mills and milling ... 45062

Macclesfield

Gawsworth Hall, Church Ln, Macclesfield SK11 9RN • T: +44 1260 223456 • F: 223469 • gawsworth@lineone.net • www.gawsworth.com •
Decorative Arts Museum ... 45063

Jodrell Bank Visitor Centre, Lower Withington, Macclesfield SK11 9DL • T: +44 1477 571339 • F: 571695 • visitorcentre@jb.man.ac.uk • www.jb.man.ac.uk/scicen •
Man.: *Linda Bennett* •
Science&Tech Museum – 1880
Lovell radio telescope, displays on astronomy, space, energy and satellites, planetarium, arboretum ... 45064

Macclesfield Silk Museum, Heritage Centre, Roe St, Macclesfield SK11 6XD • T: +44 1625 613210 • F: 617880 • info@macclesfield.silk.museum • www.macclesfield.silk.museum •
Dir.: *Richard de Peyer* •
Historical Museum / Local Museum – 1987
Hist of the building and of the Macclesfield Sunday School, religious, educational and social life of the town, silk industry, social and industrial change ... 45065

Paradise Mill Silk Industry Museum, Paradise Mill, Old Park Ln, Macclesfield SK11 6TJ • T: +44 1625 612045 • F: 612048 • info@macclesfield.silk.museum • www.macclesfield.silk.museum •
Dir.: *Ricard de Peyer* •
Science&Tech Museum – 1984/2002
Silk industry, machinery, photos, textiles, silk mill, jacquard handlooms ... 45066

West Park Museum, Prestbury Rd, Macclesfield SK10 3BJ • T: +44 1625 613210 • F: 617880 • info@macclesfield.silk.museum • www.macclesfield.silk.museum •
Dir.: *Louanne Collins* •
Local Museum / Decorative Arts Museum / Fine Arts Museum – 1898
Art, Egyptian antiquities, local coll ... 45067

Machynlleth

Corris Railway Museum, Station Yard, Corris, Machynlleth SY20 9SH • T: +44 1654 761624 • alfo@corris.co.uk • www.corris.co.uk •
Science&Tech Museum – 1970
Locomotives, brake van, coaches, a number of wagons, hist and photogr of the railway ... 45068

MOMA Wales, Y Tabernacl, Heol Penrallt, Machynlleth SY20 8AJ • T: +44 1654 703355 • F: 702160 • info@momawales.org.uk • www.momawales.org.uk •
Head: *Ruth Lambert* •
Fine Arts Museum
British/Welsh paintings 20th c ... 45069

Magherafelt

Bellaghy Bawn, Castle St, Bellaghy, Magherafelt BT45 8LA • T: +44 28 79386812 • F: 79386556 • bellaghy.bawn@doeni.gov.uk •
Local Museum
Plantation coll, local artefacts, literature ... 45070

Springhill House, 20 Springhill Rd, Moneymore, Magherafelt, Co. Derry • T: +44 16487 48210 •
Decorative Arts Museum
Furniture, paintings and an important costume coll ... 45071

Maidstone

Dog Collar Museum, Leeds Castle, Maidstone ME17 1PL • T: +44 1622 765400 • F: 767838 • enquiries@leeds-castle.co.uk • www.leeds-castle.com •
Special Museum
Coll of dog collars, of iron, brass, silver and leather from 16th c ... 45072

Maidstone Museum and Bentlif Art Gallery, Saint Faith's St, Maidstone ME14 1LH • T: +44 1622 602838/39 • F: 685022 • museuminfo@maidstone.gov.uk • www.museum.maidstone.gov.uk •
Dir.: *Simon Lace* • Sc. Staff: *F. Woolley* (Fine and Applied Art) • *Prof. Dr. E. Jarzembowski* (Natural History) • *Giles Guthrie* (Human History) •
Local Museum / Fine Arts Museum / Decorative Arts Museum – 1858
Elizabethan furnishings, archaeology, fine arts, ceramics, oriental art and ethnography, natural history, local history, costume ... 45073

Queen's Own Royal West Kent Regimental Museum, Saint Faith's St, Maidstone ME14 1LH • T: +44 1622 602842 • F: 685022 • qorwkmuseum@maidstone.gov.uk •
Dir.: *H.B.H. Waring* •
Military Museum – 1961
Regiment's artifacts ... 45074

Tyrwhitt-Drake Museum of Carriages, Archbishop's Stables, Mill St, Maidstone ME15 6YE • T: +44 1622 602835 • F: 685022 • museuminfo@maidstone.qov.uk • www.museum.maidstone.gov.uk •
Dir.: *Simon Lace* •
Science&Tech Museum – 1946
Horse-drawn coaches and carriages (17th to 19th c) ... 45075

Maldon

Maldon District Museum, 47 Mill Rd, Maldon CM9 5HX • T: +44 1621 842688 • bygones@maldonmuseum.fsnet.co.uk •
Dir.: *M. Derek Maldon Fitch* •
Local Museum
Social history ... 45076

Mallaig

Mallaig Heritage Centre, Station Rd, Mallaig PH41 4PY • T: +44 1687 462085 • F: 462085 • info@mallaigheritage.org.uk • www.mallaigheritage.org.uk •
Man.: *Malcolm Poole* •
Local Museum
Local and social history of West Lochaber, railway, steamers, fishing, agriculture ... 45077

U

Malmesbury

Athelstan Museum, Town Hall, Cross Hayes, Malmesbury SN16 9BZ • T: +44 1666 829258 • F: 829258 • athelstanmuseum@northwilts.gov.uk • www.northwilts.gov.uk •
Cur.: *J.R. Prince* •
Local Museum – 1979
Malmesbury lacemaking industry, Malmesbury branch railway, costume, local engineering company, bicycles, coins and the new educational/hands-on activity facility for children and their families, prints, drawings 45078

Malton

Eden Camp, Modern History Theme Museum, Junction of A64 and A169, Malton YO17 6RT • T: +44 1653 697777 • F: 698243 • admin@edencamp.co.uk • www.edencamp.co.uk •
Dir.: *N.B. Hill* •
Historical Museum – 1987
History of the British wartime, series of reconstructed scenes, story of civilian life during World War II 45079

Malton Museum, Old Town Hall, Market Pl, Malton YO17 7LP • T: +44 1653 695136, 692610 •
Archaeology Museum
Roman finds, coins, Samian ceramics, iron, bronze and stone objects 45080

Manchester

Chethams's Hospital and Library, Long Millgate, Manchester M3 1SB • T: +44 161 8347961 • F: 8395797 • librarian@chethams.org.uk • www.chethams.org.uk •
Head: *Michael Powell* •
Library with Exhibitions
Printed books, manuscripts, ephemera, newspapers, photographs, prints, maps, drawings, paintings, furniture 45081

Gallery of Costume, Manchester City Galleries, Platt Hall, Rusholme, Manchester M14 5LL • T: +44 161 2245217 • F: 2563278 • www.manchestergalleries.org.uk •
Dir.: *Dr. Miles Lambert* •
Special Museum – 1947
History of costume since 1700, housed in a 1760s building – library 45082

Greater Manchester Police Museum, Newton St, Manchester M1 1ES • T: +44 161 8563287 • F: 8563286 • police.museum@gmp.police.uk • www.gmp.police.uk •
Cur.: *Duncan Broady* • Asst. Cur.: *David Tetlow* •
Historical Museum – 1981 45083

Heaton Hall, Manchester City Galleries, Heaton Park, Prestwich, Manchester M25 5SW • T: +44 161 7731231 • F: 2367369 • r.shrigley@manchester.gov.uk • www.manchestergalleries.org •
Dir.: *Virginia Tandy* • Cur.: *Ruth Shrigley* •
Fine Arts Museum / Decorative Arts Museum – 1906
Neo-classical interiors, furniture, ceramics, paintings, plasterwork, musical instruments 45084

Holden Gallery, c/o Faculty of Art and Design, Manchester Metropolitan University, Cavendish St, Manchester M15 6BR • T: +44 161 2476225 • F: 2476870 •
Fine Arts Museum 45085

Imperial War Museum North, The Quays, Trafford Wharf Rd, Manchester M17 1TZ • T: +44 161 8364000 • F: 8364090 • iwmnorth@iwm.org.uk • www.iwm.org.uk/north •
Gen. Dir.: *Diane Lees* • Dir.: *Jim Forrester* •
Historical Museum – 2002
Material showing people's experience of war and the impact of war on society from 1914 to the present day 45086

John Rylands University Library-Special Collections, 150 Deansgate, Manchester M3 3EH • T: +44 161 8345343, 2753764 • F: 8345574 • special.collections@manchester.ac.uk • www.rylibweb.man.ac.uk/spcoll •
Dir.: *William G. Simpson* •
Library with Exhibitions
Coll of rare books and manuscripts – Library 45087

Manchester Art Gallery, Mosley St, Manchester M2 3JL • T: +44 161 2358888 • F: 2358899 • k.gowland@notes.manchester.gov.uk • www.manchestergalleries.org •
Dir.: *Virginia Tandy* • Asst. Dir.: *Moira Stevenson* • Cur.: *Anthea Jarvis* (Costume) • *Sandra Martin* (Fine Art) • *Ruth Shrigley* (Decorative Arts) • *Tim Wilcox* (Exhibitions) • Cons.: *Amanda Wallace* •
Fine Arts Museum / Decorative Arts Museum – 1882
British art, painting, drawing, sculptures, decorative arts 45088

Manchester Jewish Museum, 190 Cheetham Hill Rd, Manchester M8 8LW • T: +44 161 8349879 • F: 8349801 • info@manchesterjewishmuseum.com • www.manchesterjewishmuseum.com •
Admin.: *Don Rainger* •
Historical Museum / Religious Arts Museum
Hist of Manchester's Jewish community over the past 250 years 45089

The Manchester Museum, The University of Manchester, Oxford Rd, Manchester M13 9PL • T: +44 161 2752634 • F: 2752676 • museum@manchester.ac.uk • www.manchester.ac.uk/museum •
Dir.: *Dr. Nicholas Merriman* • Sc. Staff: *Dr. A.J.N.W. Prag* • *B. Sitch* (Archaeology) • *L. Wolstenholme* (Botany) • *H. McGhie* (Zoology) • *Dr. D. Logunov* (Entomology) • *D. Gelsthorpe* • *Dr. P. Manning* (Paleontology) • *L. Harris* (Anthropology) • *Dr. D. Green* (Mineralogy) • *K. Sugden* (Numismatics) • *Wendy Hodkinson* (Archery) • *A. Gray* (Herpetology) •
Archaeology Museum / Natural History Museum / University Museum / Ethnology Museum
Egyptology, zoology, entomology, herpetology, botany, anthropology, ethnology, mineralogy, numismatics, archaeology, archery 45090

Manchester United Museum and Tour Centre, Sir Matt Busby Way, Old Trafford, Manchester M16 0RA • T: +44 810 4421994 • F: 8688861 • tours@manutd.co.uk •
Dir.: *Mike Maxfield* •
Special Museum – 1986
British Football – education dept 45091

Museum of Science and Industry in Manchester, Liverpool Rd, Castlefield, Manchester M3 4FP • T: +44 161 8322244 • F: 8331471 • marketing@msim.org.uk • www.msim.org.uk •
Dir.: *Ian Griffin* •
Science&Tech Museum – 1983
World's oldest passenger railway station, steam engines, electricity, gas, air and space, textiles – library, restoration workshop, coll centre 45092

Museum of Transport - Greater Manchester, Boyle St, Cheetham, Manchester M8 8UW • T: +44 161 2052122 • F: 2052122 • gmtsenquire@btinternet.com • www.gmts.co.uk •
Dir.: *Dennis Talbot* •
Science&Tech Museum – 1979
Transports, buses, trams 45093

North West Film Archive, c/o Manchester Metropolitan University, Minshull House, 47-49 Chorlton St, Manchester M1 3EU • T: +44 161 2473097 • F: 2473098 • n.w.filmarchiv@mmu.ac.uk • www.nwfa.mmu.ac.uk •
Dir.: *Marion Hewitt* •
Performing Arts Museum
Film coll 45094

Pankhurst Centre, 60-62 Nelson St, Chorlton-on-Medlock, Manchester M13 9WP • T: +44 161 2735673 • F: 2743525, 2744979 • pankhurst@zetnet.co.uk • www.thepankhurstcentre.org.uk •
Dir.: *Yvonne Edge* •
Historical Museum
Women's suffrage movement 45095

People's History Museum, Left Bank, Spinningfields, Manchester M3 3ER • T: +44 161 8389190 • F: 8389190 • info@phm.org.uk • www.phm.org.uk •
Dir.: *Nicholas Mansfield* •
Historical Museum
Working class life and institutions, labour and women's movements, banners, photographs, badges, tools, regalia, paintings 45096

Portico Library and Gallery, 57 Mosley St, Manchester M2 3HY • T: +44 161 2366785 • F: 2366803 • librarian@theportico.org.uk • www.theportico.org.uk •
Public Gallery 45097

University of Manchester Medical School Museum, Medical Admin., Stopford Bldg, Oxford Rd, Manchester M13 9PT, mail addr: The Curator, room 3.24, Stopford Bldg, Oxford Rd, Manchester M13 9PT • T: +44 161 2755027 • peter.mohr@manchester.ac.uk • www.medicine.manchester.ac.uk/museum •
Cur.: *Dr. Peter Mohr* • Ass.: *Julie Mohr* •
Historical Museum / University Museum – 1973
Medical, surgical and pharmaceutical equipment, instruments, trade catalogues, photographs 45098

The Whitworth Art Gallery, c/o The University of Manchester, Oxford Rd, Manchester M15 6ER • T: +44 161 2757450 • F: 2757451 • whitworth@manchester.ac.uk • www.manchester.ac.uk/whitworth •
Dir.: *Dr. Maria Balshaw* • Deputy Dir.: *Dr. Jennifer Harris* (Textiles) • Sc. Staff: *David Morris* (Head of collections) • *Jo Beggs* (Head of Development) • *Esmé Ward* (Head of Learning and Interpretation) • *Nicola Walker* (Head of Collections Care and Access) • *Andrea Hawkins* (Head of Public Engagement) • Cur.: *Mary Griffiths* (Modern Art) • Cur.: *Heather Birchall* (Historic Art) • Cur.: *Frances Pritchard* (Textiles) • Cur.: *Christine Woods* (Wallpapers) • Conservator: *Dionysia Christoforou* (Works on paper) • *Ann French* (Textiles) •
Fine Arts Museum / University Museum – 1889
Watercolors and drawings, prints, textiles, contemporary art, wallpapers 45099

Wythenshawe Hall, Manchester City Art Gallery, Wythenshawe Park, Northenden, Manchester M23 0AB • T: +44 161 9982331 • F: 9454403 • r.shrigley@notes.manchester.gov.uk • www.manchestergalleries.org •
Dir.: *Virginia Tandy* •
Decorative Arts Museum / Fine Arts Museum – 1930
17th-19th c furniture and pictures housed in country manor 45100

Mansfield

Mansfield Museum and Art Gallery, Leeming St, Mansfield NG18 1NG • T: +44 1623 463088 • F: 412922 • mansfield@museum.gov.uk • www.mansfield-dc.gov.uk •
Cur.: *Elizabeth Weston* •
Local Museum / Fine Arts Museum – 1904
Natural hist, local hist, art works, ceramics, archaeology, Buxton watercolors, Wedgwood and Pinxton China 45101

Marazion

Giant's Castle, Saint Michael's Mount, Marazion TR17 0HT • T: +44 1736 710507 • F: 719930 • godolphin@manor-office.co.uk •
Special Museum 45102

March

March and District Museum, High St, March PE15 9JJ • T: +44 1354 655300 • F: 653714 • march.museum@freeuk.com •
Local Museum
Coll of domestic and agricultural artefacts 45103

Margam

Abbey and Stones Museum, Port Talbot, Margam • T: +44 1639 891548 • F: 891548 •
Religious Arts Museum 45104

Margate

Lifeboat House, The Rendezvous, Margate Harbour, Margate CT9 1AA • T: +44 1843 221613 •
Special Museum 45105

Margate Old Town Hall Local History Museum, Market Pl, Margate CT9 1ER • T: +44 1843 231213 • F: 582359 • margatemuseum@bonkers16.freeserve.co.uk •
Cur.: *Bob Bradley* •
Local Museum 45106

Tudor House (closed) 45107

Market Bosworth

Bosworth Battlefield Heritage Centre and Country Park, Sutton Cheney, Market Bosworth CV13 0AD • T: +44 1455 290429 • F: 292841 • bosworth@leics.gov.uk • www.bosworthbattlefield.com •
Dir.: *Alan Morrison* •
Historical Museum / Historic Site – 1973
Archaeological artefacts from the battle, including a large collection of medieval cannonballs and The Bosworth Boar 45108

Market Harborough

Harborough Museum, Council Offices, Adam and Eve St, Market Harborough LE16 7AG • T: +44 1858 821085 • F: 821086 • harboroughmuseum@leics.gov.uk • www.leics.gov.uk •
Local Museum 45109

Rockingham Castle, Market Harborough LE16 8TH • T: +44 1858 770240 • F: +44 1536 771692 • estateoffice@rockinghamcastle.com • www.rockinghamcastle.com •
Fine Arts Museum
Fine pictures, furniture 45110

Market Lavington

Market Lavington Village Museum, Church St, Market Lavington SN10 4DP • T: +44 1380 818736 • F: 816222 •
Cur.: *Peggy Gye* •
Local Museum
Life in the village 45111

Maryport

Maryport Maritime Museum, 1 Senhouse St, Maryport CA15 5AB • T: +44 1900 813738 • F: 326320 • www.allerdale.gov.uk •
Historical Museum / Science&Tech Museum – 1976
Maryport local hist, maritime hist 45112

Senhouse Roman Museum, Sea Brows, Maryport CA15 6JD • T: +44 1900 816168 • F: 816168 • romans@senhouse.freeserve.co.uk • www.senhousemuseum.co.uk •
Man.: *Jane Laskey* •
Archaeology Museum 45113

Matlock

Caudwell's Mill & Craft Centre, Rowsley, Matlock DE4 2EB • T: +44 1629 734374 • F: 734374 • raymarjoran@compuserve.com • caudwellsmill.museum.com •
Man.: *Lesley Wyld* •
Science&Tech Museum – 1982
Flour mill machinery 45114

National Tramway Museum, Crich Tramway Village, Matlock DE4 5DP • T: +44 1773 854321 • F: 854320 • enquiry@tramway.co.uk • www.tramway.co.uk •
Chm.: *Colin Heaton* •
Science&Tech Museum – 1959
Coll of tramcars, built between 1873 and 1969, from Britain and abroad – library, restoration workshop 45115

Peak District Mining Museum, The Pavilion, South Parade, Matlock DE4 3NR • T: +44 1629 583834 • mail@peakmines.co.uk • www.peakmines.co.uk •
Dir.: *Robin Hall* •
Science&Tech Museum – 1978
Artifacts relating to the history of lead mining from the Roman period to present day, mining tools, water pressure engine built in 1819 45116

Peak Rail Museum, Station, Matlock DE4 3NA • T: +44 1629 580381 • F: 760645 • peakrail@peakrail.co.uk • www.peakrail.co.uk •
Dir.: *J.T. Starham* •
Science&Tech Museum
Standard gauge, railway 45117

Mauchline

Burns House Museum, Castle St, Mauchline KA5 5BZ • T: +44 1290 550045 • F: 550045 • jason.sutcliffe@east-ayrshire.gov.uk • www.east-ayrshire.gov.uk •
Special Museum – 1969
Memorabilia on Poet Robert Burns, Mauchline Boxware, curling stones 45118

Maud

Maud Railway Museum, Station, Maud AB42 5LY • T: +44 1771 622807 • F: 623558 • museums@aberdeenshire.gov.uk • www.aberdeenshire.gov.uk/heritage/ •
Science&Tech Museum 45119

Maybole

Culzean Castle and Country Park, Culzean Castle, Maybole KA19 8LE • T: +44 1655 884455 • F: 884503 • culzean@nts.org.uk • www.nts.org.uk •
Dir.: *Michael L. Tebbutt* •
Decorative Arts Museum – 1945
Interiors, fittings designed by Robert Adam, staircase, plaster ceilings, 18th c castle built by Robert Adam 45120

Measham

Measham Museum, High St, Measham DE12 7HZ • T: +44 1530 273956 • F: 273986 •
Chm.: *Keith Elliott* •
Local Museum
Local hist, Measham ware pottery, mining artefacts, Hart coll, pictures 45121

Meigle

Meigle Museum of Sculptured Stones, Dundee Rd, Meigle PH12 8SB • T: +44 1828 640612 • F: 640612 • hs.explorer@scotland.gsi.gov.uk • www.historic-scotland.gov.uk •
Archaeology Museum – ca 1890
Pictish and early Christian stone objects 45122

Melbourne

Melbourne Hall, Church Sq, Melbourne DE73 8EN • T: +44 1332 862502 • F: 862263 • melbhall@globalnet.co.uk • www.melbournehall.com •
Cur.: *Gill Weston* •
Fine Arts Museum
Fine art, furniture, William Bamb (2nd Viscount Melbourne, gave his name to Melbourne, Australia) 45123

Melrose

Abbotsford House, Melrose TD6 9BQ • T: +44 1896 752043 • F: 752916 • enquiries@sirwalterscottsabbotsford.co.uk • www.sirwalterscottsabbotsford.co.uk •
Historical Museum
Relics of Sir Walter Scott – library 45124

Melrose Abbey Museum, off A7 or A68, Melrose TD6 9LG • T: +44 1896 822562 • hs.website@scotland.gsi.gov.uk •
Religious Arts Museum
Abbey pottery, architecture, stone sculpture 45125

Melton Mowbray

Melton Carnegie Museum, Thorpe End, Melton Mowbray LE13 1RB • T: +44 1664 569946 • F: 564060 • museums@leics.gov.uk • www.leics.gov.uk/museum •
Dir.: *Jenny Dancey* •
Local Museum – 1977
Local hist, hunting pictures 45126

Meopham

Meopham Windmill, Wrotham Rd, Meopham DA13 0QA • T: +44 1474 813779 • F: 813779 • dustymillerll@waitrose.com • www.meopham.org •
Science&Tech Museum 45127

Mere

Mere Museum, Barton Ln, Mere BA12 6JA • T: +44 1747 861444 • meremuseum@onetel.com •
Cur.: *Dr. D. Longbourne* •
Local Museum 45128

Merthyr Tydfil

Brecon Mountain Railway, Pant Station, Dowlais, Merthyr Tydfil CF4 2UP • T: +44 1685 722988 • F: 384854 • www.breconmountainrailway.co.uk •
Science&Tech Museum
Locomotives from three continents built between 1894 and 1930 45129

Cyfarthfa Castle Museum and Art Gallery, Brecon Rd, Merthyr Tydfil CF47 8RE • T: +44 1685 723112 • F: 723112 • museum@merthyr.gov.uk • www.museums.merthyr.gov.uk •
Cur.: *Scott Reid* •
Decorative Arts Museum / Fine Arts Museum – 1910
Paintings, ceramics, silver, geology, ethnology, local, natural and industrial history ... 45130

Joseph Parry's Ironworkers Cottage, 4 Chapel Row, Merthyr Tydfil CF48 1BN • T: +44 1685 721858, 723112 • F: 723112 • museum@merthyr.gov.uk • www.museums.merthyr.gov.uk •
Cur.: *Scott Reid* •
Music Museum – 1979
Story of the Welsh composer Joseph Parry ... 45131

Merton

Barometer World Museum, Quicksilver Barn, Merton EX20 3DS • T: +44 1805 603443 • F: 603344 • curator@barometerworld.co.uk • www.barometerworld.co.uk •
Cur.: *Philip Collins* •
Special Museum
Barometers ... 45132

Methil

Methil Heritage Centre, 272 High St, Methil KY8 3EQ • T: +44 1333 422100 • F: 422101 • www.fifedirect.org.uk/museums •
Local Museum
Social, industrial and natural hist of Levenmourth ... 45133

Methlick

Haddo House, Methlick AB41 7EQ • T: +44 1651 851440 • F: 851888 • haddo@nts.gov.uk • www.nts.org.uk •
Dir.: *Lorraine Hesketh-Campbell* •
Decorative Arts Museum
Earls of Aberdeen mansion ... 45134

Mevagissey

Folk Museum, Frazier House, East Quay, Mevagissey PL26 6PP • T: +44 1726 843568 • widgerywinks@yahoo.co.uk • www.geocities.com/mevmus •
Cur.: *Gordon Kane* •
Folklore Museum ... 45135

Mickley

Thomas Bewick's Birthplace, Station Bank, nr Stocksfield, Mickley NE43 7DD • T: +44 1661 843276 • cherryburn@nationaltrust.org.uk •
Cur.: *Hugh Dixon* •
Fine Arts Museum – 1990
Engravings and prints from Bewick's wood blocks ... 45136

Middle Claydon

Claydon House, Middle Claydon MK18 2EY • T: +44 1296 730349 • F: 738511 • todgen@smtp.ntrust.org.uk • www.nationaltrust.org.uk •
Decorative Arts Museum
Rococo State rooms with carving and decoration by Luke Lightfoot and plasterwork by Joseph Rose the younger ... 45137

Middle Wallop

Museum of Army Flying, Middle Wallop SO20 8DY • T: +44 1980 674421 • F: +44 1264 781694 • alison@flying-museum.org.uk • www.flying-museum.org.uk •
Cur.: *David Patterson* •
Military Museum
British Army's flying history, balloons in Bechuanaland, helicopters in the Falklands campaign of 1982, Army helicopters and fixed-wing aircraft, a captured Huey helicopter, British WWII assault gliders ... 45138

Middlesbrough

Captain Cook Birthplace Museum, Stewart Park, Marton, Middlesbrough TS7 8AT • T: +44 1642 311211 • F: 515659 • captcookmuseum@middlesbrough.gov.uk • www.captcook-ne.co.uk •
Cur.: *Phil Philo* •
Special Museum – 1978
The life of Cook, a reconstruction of the below-deck accomodation in his famous ship, the Endeavour, galleries illustrating countries he visited until his death in Hawaii in 1779, ethnographical material ... 45139

Dorman Memorial Museum, Linthorpe Rd, Middlesbrough TS5 6LA • T: +44 1642 813781 • F: 358100 • dormanmuseum@middlesborough.gov.uk • www.dormanmuseum.co.uk •
Dir.: *Godfrey Worsdale* • Sc. Staff: *Ken Sedman* •
Local Museum – 1904
Natural, industrial and social hist, pottery, conchology, archaeology, world cultures ... 45140

Middlesbrough Institute of Modern Art, Centre Sq, Middlesbrough TS1 2AZ • T: +44 1642 726720 • F: 726722 • mima@middlesbrough.gov.uk • www.visitmima.co.uk •
Dir.: *Kate Brindley* • Cur.: *James Beighton* •
Public Gallery – 2007
Mima coll of fine and applied art from 1900 to the present day ... 45141

Newham Grange Leisure Farm Museum, Coulby Newham, Middlesbrough TS1 • T: +44 1642 300261 • F: 300276 •
Agriculture Museum
History of farming in Cleveland, reconstructions of a farmhouse kitchen, a veterinary surgeon's room, a saddler's shop, coll of farm tools, implements and equipment ... 45142

Middleton-by-Wirksworth

Middleton Top Engine House, Middleton-by-Wirksworth DE4 4LS • T: +44 1629 823204 • F: 825336 • middletontop@derbyshire.gov.uk • www.derbyshire.gov.uk/countryside •
Science&Tech Museum – 1974
Restored Beam Winding Engine of 1830 ... 45143

Mildenhall, Suffolk

Mildenhall and District Museum, 6 King St, Mildenhall, Suffolk IP28 7EX • T: +44 1638 716970 • webmaster@mildenhallmuseum.co.uk • www.mildenhallmuseum.co.uk •
Cur.: *Chris Mycock* •
Local Museum
Local history, history of RAF Mildenhall, archaeology ... 45144

Milford Haven

Milford Haven Museum, Old Customs House, Sybil Way, The Docks, Milford Haven SA73 3AF • T: +44 1646 694496 • F: 699454 •
Local Museum
History of the town and waterway, fishing port ... 45145

Milford, Staffordshire

Shugborough Estate Museum, Milford, Staffordshire ST17 0XB • T: +44 1889 881388 • F: 881323 • shugborough.promotions@stafordshire.gov.uk • www.shugborough.org.uk •
Dir.: *Richard Kemp* •
Local Museum – 1966
The social and agricultural history of rural Staffordshire, stables, laundry, ironing room, brewhouse, coachhouse, coll of horse-drawn vehicles ranges from family coaches and carriages to farm carts ... 45146

Millom

Folk Museum, The Station Building, Station Rd, Millom LA18 4DD • T: +44 1229 774819 • F: 772555 • www.visitcumbria.com/wc/milmmus.htm •
Cur.: *B. Bennison* •
Folklore Museum ... 45147

Millport

Museum of the Cumbraes, Garrison House, Garrison Grounds, Millport KA28 0DG • T: +44 1294 464174 • namuseum@north-ayrshire.gov.uk • www.futuremuseum.co.uk/PartnerMuseums.aspx/Show/museum_of_the_cumbraes/ •
Dir.: *John Travers* • Manager: *Morag McNicol* • *Debra L. Keasal* • Asst. Man.: *Margaret Weir* •
Local Museum
Local hist in a reconstructed wash house ... 45148

Robertson Museum, c/o University Marine Biological Station, Millport KA28 0EG • T: +44 1475 530581 • F: 530601 • june.thomsom@millport.gla.ac.uk • www.gla.ac.uk/centres/marinestation/facilities/museum.html •
Dir.: *Dr. Rupert Ormond* •
Natural History Museum / University Museum – 1900
Local marine life – aquarium ... 45149

Milngavie

Lillie Art Gallery, East Dunbartonshire Museums, Station Rd, Milngavie G62 8BZ • T: +44 141 5788847 • F: 5700244 • museums@eastdunbarton.gov.uk •
Dir./ Cur.: *Hildegarde Berwick* •
Fine Arts Museum – 1962
Paintings 20th c, paintings water colors and etchings by Robert Lillie, drawings by Joan Eardley ... 45150

Milnrow

Ellenroad Engine House, Elizabethan Way, Milnrow OL16 4LG • T: +44 1706 881952 • F: +44 161 6880634 • ellenroad@aol.com • www.ellenroad.org.uk •
Science&Tech Museum ... 45151

Milton Abbas

Park Farm Museum, Park Farm, Blanford, Milton Abbas DT11 0AX • T: +44 1258 880216 •
Agriculture Museum ... 45152

Milton Keynes

Milton Keynes Gallery, 900 Midsummer Blvd, Milton Keynes MK9 3QA • T: +44 1908 676900 • F: 558308 • info@mk-g.org • www.mk-g.org •
Dir.: *Michael Stanley* •
Public Gallery ... 45153

Milton Keynes Museum, McConnell Dr, Wolverton, Milton Keynes MK12 5EL • T: +44 1908 316222 • F: 319148 • enquiries@mkmuseum.org.uk • www.mkmuseum.org.uk •
Dir.: *Bill Griffiths* •
Local Museum
Industry, agriculture, social hist, retailing, railways ... 45154

Warner Archive, Bradbourne Dr, Tilbrook, Milton Keynes MK7 8BE • T: +44 1908 658021 • F: 658020 • sue_kerry@walkergreenbank.com •
Historical Museum / Decorative Arts Museum
Textiles, designs, Warner Fabrics, Wilson, Keith, Norris, Walters, Scott Richmond and Helios ... 45155

Minehead

West Somerset Railway, Station, Minehead TA24 5BG • T: +44 1643 704996, 707650 • F: 706349 • info@west-somerset-railway.co.uk • www.west-somerset-railway.co.uk •
Dir.: *Mark Smith* •
Science&Tech Museum
Colls of Great Western Railway artefacts, Somerset and Dorset Railway artefacts ... 45156

Minera

Minera Lead Mines, Wern Rd, Minera LL11 3DU • T: +44 1978 261529 • F: 361703 • museum@wrexham.gov.uk • www.wrexham.gov.uk/heritage •
Man.: *David Bowers* •
Science&Tech Museum
19th c lead mine, beam engine house, geology ... 45157

Minster-in-Sheppey

Minster Abbey Gate House Museum, Union Rd, Minster-in-Sheppey MC12 2HW • T: +44 1795 872303 • www.swale.gov.uk •
Historical Museum ... 45158

Mintlaw

Aberdeenshire Farming Museum, Station Rd, Mintlaw AB42 5FQ • T: +44 1771 622807 • F: 623558 • museums@aberdeenshire.gov.uk • www.aberdeenshire.gov.uk/heritage/ •
Sc. Staff: *Dr. David M. Bertie* •
Open Air Museum / Agriculture Museum
Agricultural colls, hist of farming in NE Scotland ... 45159

Mitcham

Wandle Industrial Museum, Vestry Hall Annexe, London Rd, Mitcham CR4 3UD • T: +44 20 86480127 • F: 86850249 • curator@wandle.org • www.wandle.org •
Science&Tech Museum
Life and industries of the River Wandle ... 45160

Moffat

Moffat Museum, The Neuk, Church Gate, Moffat DG10 9EG • T: +44 1683 220868 •
Local Museum ... 45161

Moira

Moira Furnace, Furnace Ln, Moira DE12 6AT • T: +44 1283 224667 • F: 224667 • wendyjr@tiscali.co.uk •
Man.: *Wendy Robinson* •
Science&Tech Museum
Cast iron goods produced at the furnace foundary, foundryman's tools, social hist – boat trips ... 45162

Mold

Mold Museum, Earl Rd, Mold CH7 1AP • T: +44 1352 754791 • F: 707207 • mold.library@flintshire.gov.uk • www.flintshire.gov.uk •
Man.: *Nia Jones* (Library) •
Library with Exhibitions / Public Gallery
Archaeology, social and industrial hist, life of Welsh novelist Daniel Owen – library, gallery ... 45163

Oriel Gallery, Clwyd Theatr Cymru, Mold CH7 1YA • T: +44 1352 756331 • F: 701558 • oriel@clwyd-theatr-cymru.co.uk • www.clwyd-theatr-cymru.co.uk •
Cur.: *Jonathan Le Vay* •
Public Gallery ... 45164

Moneymore

Springhill Costume Museum, Magherafelt, 20 Springhill Rd, Moneymore BT45 7NQ • T: +44 28 86747927 • F: 86747927 • helen.mcaneney@nationaltrust.org.uk •
Cur.: *Helen McAneney* •
Special Museum ... 45165

Moniaive

James Paterson Museum, Meadowcroft, North St, Moniaive DG3 4HR • T: +44 1848 200583 •
Cur.: *Anne Paterson* •
Special Museum ... 45166

Monmouth

Castle and Regimental Museum, The Castle, Monmouth NP25 3BS • T: +44 1600 772175 • F: 711428 • curator@monmouthcastlemuseum.org.uk • www.monmouthcastlemuseum.org.uk •
Dir.: *P. Lynesmith* •
Military Museum ... 45167

Nelson Museum and Local History Centre, Priory St, Monmouth NP25 3XA • T: +44 1600 710630 • F: 775001 • monmouthmuseum@monmouthshite.gov.uk • www.nelsonmuseum.org.uk •
Cust.: *Valerie Eveleigh* •
Historical Museum – 1924
Documents and material concerning Nelson, local coll including the Hon. Charles Stuart Rolls (of Rolls Royce Fame) – commercial conservation service ... 45168

Montacute

National Portrait Gallery, Montacute House, Montacute TA15 6XP • T: +44 1935 823289 • F: 826921 • wmogen@smtp.ntrust.org.uk •
Decorative Arts Museum
Furniture, tapestries, wood panelling, amorial glass, portraits of Tudor and Jacobean courts ... 45169

Montgomery

Old Bell Museum, Arthur St, Montgomery SY15 6RA • T: +44 1686 668313 • curator@oldbellmuseum.org.uk • www.oldbellmuseum.org.uk •
Local Museum
Local history, railway, workhouse ... 45170

Montrose

House of Dun, Montrose DD10 9LQ • T: +44 1674 810264 • F: 810722 • www.nts.org.uk/Property/32 •
Historic Site
Georgian house built in 1730, exhibition on the architecture of house and garden ... 45171

Montrose Air Station Museum, Waldron Rd, Broomfield, Montrose DD10 9BB • T: +44 1674 678222 • rafmontrose@aol.com • www.rafmontrose.org.uk •
Chm.: *David Butler* •
Military Museum
Aircrafts, wartime artefacts ... 45172

Montrose Museum and Art Gallery, Panmure Pl, Montrose DD10 8HF • T: +44 1674 673232 • F: +44 1307 462590 • montrose.museum@angus.gov.uk • www.angus.gov.uk •
Cur.: *Rachel Benvie* •
Local Museum / Fine Arts Museum – 1836
Social history, ethnography, natural science, fossils, watercolours, sculpture and etchings, geology, molluscs ... 45173

William Lamb Memorial Studio, Montrose Museum, 24 Market St, Montrose DD10 8NB, mail addr: Panmure Pl, Montrose, DD10 8HF • T: +44 1674 673232 • F: +44 1307 462590 • montrose.museum@angus.gov.uk • www.angus.gov.uk •
Cur.: *Rachel Benvie* •
Fine Arts Museum
Sculptures, wood carvings, etchings, drawings, watercolours ... 45174

Moreton-in-Marsh

Wellington Aviation Museum, British School House, Broadway Rd, Moreton-in-Marsh GL56 0BG • T: +44 1608 650323 • F: 650323 •
Dir.: *Gerry V. Tyack* •
Local Museum / Military Museum
Royal Air Force treasures, local history ... 45175

Moreton Morrell

Warwickshire Museum of Rural Life, c/o College of Agriculture, Horticulture & Equine Studies, Moreton Hall, Moreton Morrell CV35 9BL • T: +44 1926 651367 • F: 419840 •
Agriculture Museum
Implements, equipments, farming like ploughing, sowing, harvesting and dairying, locally made ploughs ... 45176

Morpeth

Morpeth Chantry Bagpipe Museum, Bridge St, Morpeth NE61 1PD • T: +44 1670 500717 • F: 500710 • anne.moore@castlemorphet.gov.uk •
Cur.: *Anne Moore* •
Music Museum – 1987
W.A. Cocks Coll of bagpipes and manuscripts ... 45177

Morwellham

Morwellham Quay Historic Port & Copper Mine, Morwellham PL19 8JL • T: +44 1822 832766 • F: 833808 • enquiries@morwellham-quay.co.uk • www.morwellham-quay.co.uk •
Dir.: *Peter Kenwright* •
Open Air Museum – 1970
George and Charlotte Copper Mine, industry, archaeology, local hist, River Port ... 45178

Motherwell

Motherwell Heritage Centre, High St, Motherwell ML1 3HU • T: +44 1698 251000 • F: 268867 • heritage@mhc158.freeserve.co.uk • www.northlan.gov.uk •
Local Museum – 1996
Heritage from the Roman period, the rise and fall of heavy industry, present days – study room, exhibition gallery ... 45179

Mottistone

Bembridge Windmill Museum, Strawberry Ln, Mottistone PO30 4EA • T: +44 1983 873945 •
Science&Tech Museum 45180

Mouldsworth

Mouldsworth Motor Museum, Smithy Ln, Mouldsworth CH3 8AR • T: +44 1928 731781 • www.mouldsworthmotormuseum.com •
Owner: *James Peacop* •
Science&Tech Museum – 1971
Automobilia, motoring, art, enamel signs, motorcars & motorbicycles, sportscars, early bicycles, coll of early teapots 45181

Much Hadham

Forge Museum, The Forge, High St, Much Hadham SG10 6BS • T: +44 1279 843301 • F: 843301 • christinaharrison@btopenworld.com • www.hertsmuseums.org.uk •
Cur.: *Christina Harrison* •
Science&Tech Museum
Tools, blacksmith's shop and works, beekeeping equipment 45182

Henry Moore Foundation, Dane Tree Houses, Perry Green, Much Hadham SG10 6EE • T: +44 1279 843333 • F: 843647 • www.henry-moore-fdn.co.uk •
Head: *David Mitchinson* • Cur.: *Anita Feldman Bennet* •
Fine Arts Museum
Sculpture production and reception, drawings, letters 45183

Much Marcle

Hellens Manor, off A449, Much Marcle HR8 2LY • T: +44 1531 660504 • F: 660416 • info@hellensmanor.com • www.hellensmanor.com •
Dir.: *Nicolas Stepens* •
Decorative Arts Museum 45184

Much Wenlock

Much Wenlock Museum, The Square, High St, Much Wenlock TF13 6HR • T: +44 1952 727773 • much.wenlock.museum@shropshire-cc.gov.uk • www.shropshiremuseums.co.uk •
Local Museum
Modern Olympics 45185

Mullach Ban

Mullach Ban Folk Museum, Tullymacrieve, Mullach Ban BT35 9XA • T: +44 28 30888278 • F: 30888100 • micealmccoy@ireland.com •
Folklore Museum / Agriculture Museum
Past agricultural way of life 45186

Mytchett

Army Medical Services Museum, Keogh Barracks, Ash Vale, Mytchett GU12 5RQ • T: +44 1252 868612 • F: 868832 • museum@keogh72.freeserve.co.uk •
Cur.: *Peter Starling* •
Military Museum – 1953
History of Army Medical Service since pre-Tudor times, Crimean relics, memorabilia on Napoleon I (1769-1821) and Wellington (1769-1852), Falklands and Gulfwar military uniforms 45187

Basingstoke Canal Exhibition, Mytchett Place Rd, Mytchett GU16 6DD • T: +44 1252 370073 • F: 371758 • info@basingstoke-canal.co.uk • www.basingstoke-canal.org.uk/cancent.htm •
Historical Museum
200 year history of the Basingstoke Canal 45188

Nairn

Nairn Fishertown Museum, Viewfield House, Viewfield Dr, Nairn IV12 4EE • T: +44 1667 456798, 456791 • F: 455399 • manager@nairnmuseum.freeserve.co.uk • www.nairnmuseum.co.uk •
Local Museum
Domestic life of the fishertown, fishing industry around Moray Firth 45189

Nantgarw

Nantgarw China Works Museum, Tyla Gwyn, Nantgarw CF15 7TB • T: +44 1443 841703 • F: 841826 •
Dir.: *Gerry Towell* •
Decorative Arts Museum
William Billingsley, 19th c porcelain maker and decorator 45190

Nantwich

Nantwich Museum, Pillory St, Nantwich CW5 5BQ • T: +44 1270 627104 • nantwich.museum@origin.net • www.nantwich.org.uk •
Cur.: *Susan Pritchard* •
Local Museum 45191

Narberth

Blackpool Mill, Canaston Bridge, Narberth SA67 8BL • T: +44 1437 541233 •
Science&Tech Museum 45192

Narberth Museum, 13 Market Sq, Narberth SA67 7AX • T: +44 1834 861719 • F: 861719 • narbethmuseum@yahoo.co.uk •
Cur.: *Pauline Griffiths* •
Local Museum
Local hist, brewery items, local shops and busi- nesses, costume, toys, postcards and domestic items 45193

Naseby

Battle and Farm Museum, Purlieu Farm, Naseby NN6 7DD • T: +44 1604 740241 • F: 740800 • www.hillyer.demon.co.uk/museum.htm •
Military Museum / Agriculture Museum – 1975
Battle of 1645, agriculture 45194

Neath

Cefn Coed Colliery Museum, Neath Rd, Crynant, Neath SA10 8SN • T: +44 1639 750556 • F: 750556 •
Historical Museum – 1980
Mining of coal, story of men and machines 45195

Neath Museum, Gwyn Hall, Orchard St, Neath SA11 1DT • T: +44 1639 645741, 645726 • F: 645726 •
Local Museum
Natural and local history, archaeology, art gallery 45196

Nefyn

Lleyn Historical and Maritime Museum, Old Saint Mary's Church, Church St, Nefyn LL53 6LB • T: +44 1758 720270 •
Special Museum
Model ships, lifeboats, maritime paintings, local hist and items on the fishing industry 45197

Nelson

British in India Museum, Hendon Mill, Hallam Rd, Nelson BB9 8AD • T: +44 1282 870215, 613129 • F: 870215 •
Dir.: *Henry Nelson* •
Historical Museum / Military Museum – 1972
Model soldiers, photographs, documents, paintings, postage stamps, letters, military uniforms – library 45198

Llancaiach Fawr Living History Museum, Llancaiach Fawr Manor, Nelson CF46 6ER • T: +44 1443 412248 • F: 412688 • walked1@caerphilly.gov.uk • www.caerphilly.gov.uk/visiting •
Dir.: *Diane Walker* •
Local Museum / Historic Site
Social history, decorative arts, archaeology 45199

Nenthead

Nenthead Mines Heritage Centre, Nenthead CA9 3PD • T: +44 1434 382037 • F: 382294 • info@npht.com • www.npht.com •
Science&Tech Museum
History and geology of Nenthead lead mining area 45200

Nether Stowey

Coleridge Cottage, 35 Lime St, Nether Stowey TA5 1NQ • T: +44 1278 732662 • stc1798@aol.com •
Cust.: *Derrick Woolf* •
Special Museum – 1909
Memorabilia of poet Samuel Taylor Coleridge (1772-1834) 45201

New Abbey

Shambellie House Museum of Costume, National Museums of Scotland, Shambellie House, New Abbey DG2 8HQ • T: +44 1387 850375 • F: 850461 • info@nms.ac.uk • www.nms.ac.uk/costume •
Dir.: *Dr. Gordon Rintoul* • Cur.: *Fiona Anderson* •
Special Museum – 1977
Costumes from 1850 to 1950, European fashionable dress 45202

New Lanark

New Lanark Visitor Centre, New Lanark Mills, New Lanark ML11 9DB • T: +44 1555 661345 • F: 665738 • visit@newlanark.org • www.newlanark.org •
Dir.: *J. Arnold* •
Historical Museum – 1785
Millennium experience, 19th c textile machinery, period housing and shop 45203

New Mills

New Mills Heritage and Information Centre, Rock Mill Ln, New Mills SK22 3BN • T: +44 1663 746904 • F: 746904 • www.newmillsheritage.com •
Local Museum
Local material describing the Torrs, the pre-industrial history incl Domesday and the royal forest of Peak, story of the New Mill, coal mining, textile industry, engraver's workshop, model of the town in 1884 45204

New Milton

Sammy Miller Motorcycle Museum, Bashley Cross Road, New Milton BH25 5SZ • T: +44 1425 620777 • F: 619696 • info@sammymiller.co.uk • www.sammymiller.co.uk •
Dir.: *Sammy Miller* •
Science&Tech Museum
Motorcycles of the world, rare and exotic phototypes, over 300 world's rarest bikes 45205

New Pitsligo

Northfield Farm Museum, New Pitsligo AB43 6PX • T: +44 1771 653504 •
Agriculture Museum
Coll of historic farm machinery and household impl 45206

New Romney

Romney Toy and Model Museum, Romney Hythe and Dymchurch Railway, New Romney Station, New Romney TN28 8PL • T: +44 1797 362353 • F: 363591 • danny.martin2@btinternet.com • www.rhdr.demon.co.uk •
Dir.: *Danny Martin* •
Special Museum
World's smallest public railway, traditional toys 45207

New Tredegar

Elliot Colliery Winding House, White Rose Way, New Tredegar NP24 6DF • T: +44 1443 864224, 822666 • F: 838609 • museum@caerphilly.gov.uk • www.caerphilly.gov.uk •
Dir.: *Chris Morgan* •
Science&Tech Museum
Thornewill and Warham steam engine, hist of mining in Caerphilly county borough 45208

Newark-on-Trent

Castle Story, Gilstrap Heritage Centre, Castlegate, Newark-on-Trent NG24 1BG • T: +44 1636 611908 • F: 612274 • gilstrap@newark-sherwooddc.gov.uk • www.newark-sherwooddc.gov.uk •
Local Museum 45209

Gilstrap Heritage Centre, Castlegate, Newark-on-Trent NG24 1BG • T: +44 1636 655765 • F: 655767 • museums@nsdc.info • www.newark-sherwooddc.gov.uk •
Historical Museum 45210

Newark Air Museum, Airfield Winthorpe, Newark-on-Trent NG24 2NY • T: +44 1636 707170 • F: 707170 • newarkair@lineone.net • www.newarkairmuseum.co.uk •
Dir.: *H.F. Heeley* •
Science&Tech Museum
Aircraft, instruments, avionics, memorabilia, uniforms 45211

Newark Museum, Appleton Gate, Newark-on-Trent NG24 1JY • T: +44 1636 655740 • F: 655745 • museums@nsdc.info • www.newark-sherwooddc.gov.uk •
Local Museum – 1912
Archaeology, social hist, art, natural science, photogr, coins, documents 45212

Newark Town Treasures and Art Gallery, Market Pl, Newark-on-Trent NG24 1DU • T: +44 1636 680333 • F: 680350 • post@newark.gov.uk • www.newarktowntreasures.co.uk •
Cur.: *Patty Temple* •
Historical Museum / Fine Arts Museum
Civic plate, insignia, regalia, gifts and fine art 45213

Newart's Millgate Museum, 48 Millgate, Newark-on-Trent NG24 4TS • T: +44 1636 655730 • F: 655735 • museums@nsdc.info • www.newark-sherwooddc.gov.uk •
Dir.: *Melissa Hall* •
Folklore Museum – 1978
Items mid-19th c-1950's relating to everyday life 45214

Saint Mary Magdalene Treasury, Parish Church, Newark-on-Trent • T: +44 1636 706473 •
Religious Arts Museum 45215

The Time Museum, Upton Hall, Upton, Newark-on-Trent NG23 5TE • T: +44 1636 813795 • F: 812258 • clocks@bhi.co.uk • www.bhi.co.uk •
Man.: *Martin Taylor* •
Special Museum – 1858 45216

Vina Cooke Museum of Dolls and Bygone Childhood, Old Rectory, Cromwell, Newark-on-Trent NG23 6JE • T: +44 1636 821364 •
Dir.: *Vina Cooke* •
Special Museum – 1984
Dolls, costume, general items associated with childhood 45217

Newburgh

Laing Museum, High St, Newburgh KY14 6DX • T: +44 1334 412690 • F: 412691 •
Cur.: *Lin Collins* •
Local Museum
Local antiquities, geology, shells, ethnography 45218

Newburn

Newburn Hall Motor Museum, 35 Townfield Gardens, Newburn NE15 8PY • T: +44 191 2642977 • F: 2642977 •
Dir.: *D.M. Porrelli* •
Science&Tech Museum
Veteran, vintage and classic cars since the 1920s, model cars, 1947 Invita Black Prince 45219

Newbury

British Balloon Museum, c/o West Berkshire Museum, The Wharf, Newbury RG14 5AS • T: +44 1635 30511 • tjthafb@aol.com • www.britishballoonmuseum.org.uk •
Cur.: *John Baker* •
Science&Tech Museum
Cloud-hopper balloon, two baskets, model of hot air burner built 1904, balloon envelopes, baskets and related items – library 45220

West Berkshire Museum, The Wharf, Newbury RG14 5AS • T: +44 1635 30511 • F: 38535 • heritage@westberks.gov.uk • www.westberkshiremuseum.org.uk •
Cur.: *Jane Burrell* • *Viv Wilson* •
Local Museum – 1904
Archaeology, medieval hist, natural hist, Bronze Age objects, local pottery, costume, pewter, local and social hist 45221

Newby Bridge

Stott Park Bobbin Mill, Low Stott Park, nr Ulverston, Newby Bridge LA12 8AX • T: +44 15395 31087 • F: 530849 • www.english-heritage.org.uk/northwest •
Man.: *Michael Nield* •
Science&Tech Museum – 1983
Bobbin turning with Victorian machinery, working Victorian static steam engine 45222

Newcastle-upon-Tyne

A Soldier's Life, Blandford Sq, Newcastle-upon-Tyne NE1 4JA • T: +44 191 2326789, 2772262 • F: 2302614 • ralph.thompson@twmuseums.org.uk • www.lightdragoons.org.uk •
Cur.: *A. Wilson* •
Military Museum
Regimental Museum of the 15th/19th The King's Royal Hussars and Northumberland Hussars and Light Dragoons 45223

Bessie Surtees House, 41-44 Sandhill, Newcastle-upon-Tyne NE1 3JF • T: +44 191 2691200 • F: 2611130 • www.english-heritage.org.uk •
Archaeology Museum
Archaeological artefacts and furnishings 45224

Castle Keep Museum, Castle Garth, Newcastle-upon-Tyne NE1 1RQ • T: +44 191 2327938 • www.castlekeep-newcastle.org.uk •
Local Museum 45225

Discovery Museum, Blandford House, Blandford Sq, Newcastle-upon-Tyne NE1 4JA • T: +44 191 2326789 • F: 2302614 • discovery@twmuseums.org.uk • www.twmuseums.org.uk •
Cur.: *Graham Bradshaw* •
Local Museum
Local and maritime hist, science and technology, fashion and textiles, 45226

Hancock Museum, Barras Bridge, Newcastle-upon-Tyne NE2 4PT • T: +44 191 2226765 • F: 2226753, 2617537 • hancock.museum@ncl.ac.uk • www.twmuseums.org.uk/hancock •
Dir.: *Alec Coles* • Cur.: *Steve McLean* (Geology) • Keeper: *L. Jessop* (Biology) • Asst. Keeper: *E. Morton* (Biology) • *Sylvia Humphrey* (Geology) • Sc. Staff: *G. Mason* (Education) •
Natural History Museum / University Museum – 1829
Natural hist, mounted birds, Hutton fossilised plants, Hancock and Atthey fossilised vertebrates, N.J. Winch herbarium 45227

Hatton Gallery, c/o University of Newcastle, The Quadrangle, Newcastle-upon-Tyne NE1 7RU • T: +44 191 2226059 • F: 2223454 • hatton-gallery@ncl.ac.uk • www.ncl.ac.uk/hatton •
Cur.: *Liz Ritson* •
Fine Arts Museum / University Museum – 1926
14th-18th c European paintings, 16th-18th c Italian art, contemporary English drawings, Kurt Schwitters' Elterwater 'Merzbau' 45228

Laing Art Gallery, New Bridge St, Newcastle-upon-Tyne NE1 8AG • T: +44 191 2327734 • F: 2220952 • julie.milne@twmuseums.org.uk • www.twmuseums.org.uk •
Dir.: *Alec Coles* • Cur.: *Julie Milne* • Sc. Staff: *Sarah Richardson* • Asst.: *Natalie Frost* (Exhibition) • *Ruth Trotter* (Fine and Decorative Art) • *Julie Watson* (Education) •
Fine Arts Museum / Decorative Arts Museum – 1904
British paintings and water colours, pottery, porcelain, silver, glass and pewter 45229

Museum of Antiquities, c/o University of Newcastle-upon-Tyne, The Quadrangle, Newcastle-upon-Tyne NE1 7RU • T: +44 191 2227846/9 • F: 2228561 • m.o.antiquities@ncl.ac.uk • www.ncl.ac.uk/antiquities •
Dir.: *Lindsay Allason-Jones* •
Museum of Classical Antiquities / Archaeology Museum / University Museum – 1813
Roman inscriptions, sculptures, small finds, prehistoric and Anglo Saxon, antiquities from north Britain (from Paleolithic period to Tudors and Stuarts) – library 45230

Shefton Museum of Greek Art and Archaeology, c/o University of Newcastle, Armstrong Bldg, Newcastle-upon-Tyne NE1 7RU • T: +44 191 2228996 • F: 2228561 • m.o.antiquities@ncl.ac.uk • www.ncl.ac.uk/shefton-museum •
Dir.: *Lindsay Allason-Jones* •
Fine Arts Museum / Archaeology Museum / University Museum – 1956

U

Classical art and archaeology, Corinthian bronze helmet, Apulian helmet, 6th c Laconian handles, Greek bronze hand mirrors, 7th c B.C. Etruscan bucchero, Pelike by Pan Painter, Italic late archaic carved amber, Tarantine and Sicilian terra cotta ... 45231

Trinity House, 29 Broad Chare, Quayside, Newcastle-upon-Tyne NE1 3DQ • T: +44 191 2328226 • F: 2328448 • ncl_trinityhouse@hotmail.com • www.newcastletrinityhouse.org.uk •
Science&Tech Museum ... 45232

The University Gallery, c/o Northumbria University, Sandyford Rd, Newcastle-upon-Tyne NE1 8ST • T: +44 191 2274424 • F: 2274718 • mara-helen.wood@unn.ac.uk • www.northumbria.ac.uk/universitygallery •
Dir.: *Mara-Helen Wood* •
Public Gallery
1977 ... 45233

Newent

Shambles Museum, 16-24 Church St, Newent GL18 1PP • T: +44 1531 822144 • F: 821120 •
Dir.: *Jim Chapman* •
Historical Museum
Complete Victorian town layout with shops, houses, gardens ... 45234

Newhaven

Local and Maritime Museum, Paradise Leisure Park, Avis Rd, Newhaven BN9 9EE • T: +44 1273 612530 • secretary@newhavenmuseum.co.uk • www.newhavenmuseum.co.uk •
Pres.: *Lord Greenway* • Cur.: *Peter Bailey* •
Local Museum / Science&Tech Museum ... 45235

Newhaven Fort, Fort Rd, Newhaven BN9 9DL • T: +44 1273 517622 • F: 512059 • info@newhavenfort.org.uk • www.newhavenfort.org.uk •
Military Museum
WW I and WW II exhibition ... 45236

Planet Earth Museum, Paradise Park, Avis Rd, Newhaven BN9 0DH • T: +44 1273 512123, 616011 • F: 616005 • enquiries@paradisepark.co.uk • www.paradisepark.co.uk •
Natural History Museum
Fossils, minerals and crystals from around the world ... 45237

Newmarket

BSAT Gallery, British Sporting Art Trust, 99 High St, Newmarket CB8 8LU • T: +44 1264 710344 • F: 710114 • bsatrust@btconnect.com • www.bsat.co.uk •
Fine Arts Museum ... 45238

National Horseracing Museum, 99 High St, Newmarket CB8 8JL • T: +44 1638 667333 • F: 665600 • www.nhrm.co.uk •
Dir.: *Graham Snelling* •
Special Museum – 1983
Archive, Hands on gallery ... 45239

Newport, Gwent

Museum and Art Gallery, John Frost Sq, Newport, Gwent NP20 1PA • T: +44 1633 840064 • F: 222615 • museum@newport.gov.uk •
Bruce Campbell (Collections and Natural Science) •
Local Museum / Decorative Arts Museum / Fine Arts Museum – 1888
Local archaeology, Roman caerwent, natural sciences, geology, zoology, local history, social & industrial, art, 18th & 19th British watercolours, 20th c British oil paintings, ceramics including teapots – Newport medieval ship project ... 45240

Tredegar House and Park, Coedkernew, Newport, Gwent NP10 8YW • T: +44 1633 815880 • F: 815895 • tredegar.house@newport.gov.uk •
Archaeology Museum ... 45241

Newport, Isle of Wight

Calbourne Water Mill and Rural Museum, Calbourne, Newport, Isle of Wight PO30 4JN • T: +44 1983 531227 • F: 531227 • www.calbournewatermill.co.uk •
Dir.: *S. Chaucer* •
Local Museum / Science&Tech Museum
War and country life ... 45242

Carisbrooke Castle Museum, Carisbrooke Castle, Newport, Isle of Wight PO30 1XY • T: +44 1983 523112 • carismus@lineone.net • www.carisbrooke-castlemuseum.org.uk •
Cur.: *Rosemary Cooper* •
Historical Museum – 1898
Isle of Wight hist, material associated with Charles I and the Civil War, objects owned by Alfred Lord Tennyson – library ... 45243

Museum of Island History, The Guildhall, High St, Newport, Isle of Wight PO30 1TY • T: +44 1983 823366 • F: 823841 • museums@iow.gov.uk • www.iwight.com •
Dir.: *Dr. Michael Bishop* • Cur.: *Corinna Westwood* (Human History) •
Local Museum
Archaeology, geology, palaeontology, social history and fine art ... 45244

Newport Roman Villa, Cypress Rd, Newport, Isle of Wight PO30 1HE • T: +44 1983 823828 • museums@iow.gov.uk • www.iwight.com •
Dir.: *Dr. Michael Bishop* •
Archaeology Museum / Historic Site – 1963
Roman finds, villa ruins, mosaics ... 45245

Newport Pagnell

Beatty Museum, Chicheley Hall, Newport Pagnell MK16 9JJ • T: +44 1234 391252 • F: 391388 • enquiries@chicheleyhall.co.uk • www.chicheleyhall.co.uk •
Military Museum
Memorabilia and photographs of Admiral Beatty 45246

Newport Pagnell Historical Society Museum, Chandos Hall, Silver St, Newport Pagnell MK16 0EW • T: +44 1908 614490 • www.mkheritage.co.uk/nphs •
Local Museum
Archaeological, domestic and trade artefacts ... 45247

Newquay

Lappa Valley Steam Railway, Saint Newlyn East, Newquay TR8 5HZ • T: +44 1872 510317 • steam@lappa.freeserve.co.uk • www.lappa-railway.co.uk •
Owner: *Amanda Booth* •
Science&Tech Museum ... 45248

Trerice, Kestle Mill, Newquay TR8 4PG • T: +44 1637 875404 •
Local Museum ... 45249

Newry

Newry and Mourne Arts Centre and Museum, 1a Bank Parade, Newry BT35 6HP • T: +44 28 30266232 • F: 30266839 • noreen.cunningham@newryandmourne.gov.uk • www.newryandmourne.gov.uk •
Dir.: *Mark Hughes* • Cur.: *Noreen Cunningham* •
Public Gallery / Local Museum
Artefacts, photos and docs which record the archaeology, industrial heritage, social and political hist of Newry and Mourne ... 45250

Newton Abbot

Newton Abbot Town and Great Western Railway Museum, 2a Saint Paul's Rd, Newton Abbot TQ12 2HP • T: +44 1626 201121 • F: 201119 • museum@newtonabbot-tc.gov.uk • www.devonmuseums.net •
Cur.: *Felicity Cole* •
Local Museum / Science&Tech Museum
Local social history, Aller Vale Art pottery coll, Mapleton Butterfly Coll, signal box, railway memorabilia ... 45251

Newton Stewart

The Museum, York Rd, Newton Stewart DG8 6HH • T: +44 1671 402472 • jmclaymus@btinternet.com •
Local Museum – 1978 ... 45252

Newtongrange

Scottish Mining Museum, Lady Victoria Colliery, Newtongrange EH22 4QN • T: +44 131 6637519 • F: 6541618 • enquiries@scottishminingmuseum.com • www.scottishminingmuseum.com •
Dir.: *Fergus Waters* •
Science&Tech Museum
Machinery and artefacts relating to 900 yrs of Scottish coal mining, historic 'A-listed' Victorian colliery – Library ... 45253

Newtonmore

Clan Macpherson House and Museum, Main St, Newtonmore PH20 1DE • T: +44 1540 673332 • F: 673332 • museum@clan-macpherson.org • www.clan-macpherson.org •
Cur.: *C. Morag Hunter Carsch* •
Special Museum
Family hist incl Cluny of the '45, relics associated with Prince Charles Edward Stuart (Bonnie Prince Charlie) and 'Ossian' Macpherson – Library ... 45254

Newtonmore Museum, Highland Folk Museum, Aultlarie Croft, Newtonmore PH20 1AY, mail addr: Kingussie Road, Newtonmore PH20 1AY • T: +44 1540 67351 • F: 673693 • highland.folk@highland.gov.uk • www.highlandfolk.com •
Cur.: *Robert Powell* • Asst. Cur.: *Rachel Chisholm* •
Open Air Museum – 1934
Working farm, historic buildings ... 45255

Newtown, Powys

Newtown Textile Museum, 5-7 Commercial St, Newtown, Powys SY16 2BL, mail addr: c/o Powysland Museum, Canal Wharf, Welshpool SY21 7AQ • T: +44 1938 554656 • F: 554656 • powysland@powys.gov.uk • powysmuseums.powys.gov.uk •
Dir.: *Eva B. Bredsdorff* •
Special Museum – 1964
Crafts, Newtown woollen industry 1790-1910 ... 45256

Oriel Davies Gallery, The Park, Newtown, Powys SY16 2NZ • T: +44 1686 625041 • F: 623633 • samantha@orieldavies.org • www.orieldavies.org •
Dir.: *Amanda Farr* •
Public Gallery – 1985
Contemporary visual art and craft from Wales ... 45257

Robert Owen Memorial Museum, Broad St, Newtown, Powys SY16 2BB • T: +44 1686 626345 • johnd@robert-owen.midwales.com • www.robert-owen.midwales.com •
Dir.: *J. Hatton Davidson* •
Special Museum – 1929
Life of socialist Robert Owen (1771-1858) ... 45258

W.H. Smith Museum, c/o W.H. Smith & Son Ltd., 24 High St, Newtown, Powys SY16 2NP • T: +44 1686 626280 •
Cur.: *Hilary Hyde* •
Historical Museum
Original oak shop front, tiling and mirrors, plaster relief decoration, models, photographs and a variety of historical momentoes, history of W.H. Smith from its beginning in 1792 until the present day ... 45259

Newtownards

Ards Art Centre, Town Hall, Conway Sq, Newtownards BT23 4DB • T: +44 28 91810803 • F: 91823131 • arts@ards-council.gov.uk • www.ards-council.gov.uk •
Dir.: *Eilis O'Baoill* •
Public Gallery ... 45260

Mount Stewart House, Portaferry Rd, Greyabbey, Newtownards BT22 2AD • T: +44 28 42787800 • F: 42788569 • mountstewart@nationaltrust.org.uk •
Man.: *Debbie McCamphill* •
Historical Museum
Home of Lord Castlereagh, statues, plants, furniture, paintings – Temple of the Winds, garden ... 45261

Somme Heritage Centre, 233 Bangor Rd, Newtownards BT23 7PH • T: +44 28 91823202 • F: 91823214 • sommeassociation@btconnect.com • www.irishsoldier.org •
Dir.: *Carol Walker* •
Military Museum
WW I through 10th and 16th (Irish) Divisions and 36th (Ulster) Division, Battle of the Somme, uniforms and artefacts ... 45262

Newtownbutler

Crom Estate, Newtownbutler, Co. Fermanagh • T: +44 139657 38118 •
Historical Museum / Archaeology Museum
History, archaeology ... 45263

North Berwick

National Museum of Flight, National Museums of Scotland, East Fortune Airfield, North Berwick EH39 5LF • T: +44 1620 897240 • F: 880355 • info@nms.ac.uk • www.nms.ac.uk/flight •
Dir.: *Dr. Gordon Rintoul* •
Science&Tech Museum – 1971
Protected WWIWWII airfield, flying machines, Concorde Alpha-Alpha ... 45264

North Berwick Museum, School Rd, North Berwick EH39 4JU • T: +44 1620 828203 • F: 828201 • elms@northberwickmuseum.org • www.northberwickmuseum.org •
Local Museum – 1957
Local and domestic hist, natural hist, archaeology, hist of the Royal Burgh of North Berwick ... 45265

North Leverton

North Leverton Windmill, Mill Ln, North Leverton DN22 0AB • T: +44 1427 880662, 880573 •
Man.: *Keith Barlow* •
Science&Tech Museum
Coll of stationary barn engines ... 45266

North Shields

Stephenson Railway Museum, Middle Engine Lane, West Chirton, North Shields NE29 8DX • T: +44 191 2007146 • F: 2007144 • john.pratt@twmuseums.org.uk • www.twmuseums.org.uk •
Man.: *John Pratt* •
Science&Tech Museum
Railways in Tyne and Wear from 18th c to the present day ... 45267

Northampton

Abington Park Museum, Abington Park, Northampton NN1 5LW • T: +44 1604 838110 • F: 238720 • museums@northampton.gov.uk • www.northampton.gov.uk/museums •
Dir.: *Peter Field* •
Local Museum
Social hist, costums, military ... 45268

Museum of the Northamptonshire Regiment, Abington Park, Northampton NN1 5LW • T: +44 1604 838110 • F: 838720 • museums@northamton.gov.uk • www.northampton.gov.uk/museums •
Dir.: *William Brown* •
Military Museum
Military coll ... 45269

Northampton Museum and Art Gallery, Guildhall Rd, Northampton NN1 1DP • T: +44 1604 838111 • F: 838720 • museums@northampton.gov.uk • www.northampton.gov.uk/museums •
Dir.: *Peter Field* • Sc. Dir.: *W. Brown* (Information, Resources) • Sc. Staff: *S. Constable* (Shoe Heritage) • *A. Cowling* (Art, Exhibitions) • *J. Minchinton* (Records, Resoures) • *S. Pevereit* (Senior Community Heritage, Education) • *K. Grünzweil* (Community History) •
Local Museum / Fine Arts Museum – 1865
Northampton archaeology, geology, natural history, ethnography, Italian and modern English paintings, decorative arts, boots and shoes, leathercraft ... 45270

Northleach

Keith Harding's World of Mechanical Music, Oak House, High St, Northleach GL54 3ET • T: +44 1451 860181 • F: 861133 • keith@mechanicalmusic.co.uk • www.mechanicalmusic.co.uk •
Dir.: *Keith Harding* •
Music Museum – 1977
Self-playing musical instruments, clocks – restoration workshops ... 45271

Northwich

Lion Salt Works, Ollershaw Ln, Marston, Northwich CW9 6ES • T: +44 1606 41823 • F: 41823 • afielding@lionsalt.demon.co.uk • www.lionsaltworkstrust.co.uk •
Dir.: *Andrew Fielding* •
Science&Tech Museum ... 45272

Salt Museum, 162 London Rd, Northwich CW9 8AB • T: +44 1606 41331 • F: 350420 • cheshiremuseums@cheshire.gov.uk • www.cheshire.gov.uk/saltmuseum •
Cur.: *Matt Wheeler* •
Special Museum
Salt making in Cheshire ... 45273

Norwich

Blickling Hall, Buckling, Norwich NR11 6NF • T: +44 1263 738030 • F: 731660 • abgusr@smtp.ntrust.org.uk • www.nationaltrust.org.uk •
Man.: *Peter Griffiths* •
Local Museum
Jacobian architecture, plaster ceilings, tapestries, furniture, pictures ... 45274

Bridewell Museum, Bridewell Alley, Norwich NR2 1AQ • T: +44 1603 629127 • museums@norfolk.gov.uk • www.museums.norfolk.gov.uk •
Cur.: *John Renton* •
Historical Museum – 1925
Textiles, shoes, timepieces, printing, brewing and iron founding, pharmacy, manufacturing industries ... 45275

Castle Museum and Art Gallery, Castle, Norwich NR1 3JQ • T: +44 1603 493625 • F: 495909 • museums@norfolk.gov.uk • www.norfolk.gov.uk •
Decorative Arts Museum / Fine Arts Museum – 1894
Fine and applied arts, archaeology, geology, natural hist, social hist, Norwich paintings, porcelain, silver, coins ... 45276

East Anglian Film Archive, University of East Anglia, Archive Centre, Martineau Ln, Norwich NR1 2DQ • T: +44 1603 592664 • F: 761885 • eafa@uea.ac.uk • www.uea.ac.uk/eafa •
Dir.: *Richard Taylor* •
Performing Arts Museum
Film and videotape of East Anglian life (official regional film archive for the counties of Bedfordshire, Cambridgeshire, Essex, Hertfordshire, Norfolk and Suffolk) plus museum of film-making and projection equipment ... 45277

Felbrigg Hall, Roughton, Norwich NR11 8PR • T: +44 1263 837444 • F: 837032 • felbrigg@nationaltrust.org.uk •
Decorative Arts Museum
Historic country house with complete furnishings and fittings ... 45278

Inspire - Hands-on Science Centre, Saint Michael's Church, Coslany St, Norwich NR3 3DT • T: +44 1603 612612 • F: 616721 • inspire@science-project.org • www.science-project.org •
Dir.: *Ian Simmons* •
Science&Tech Museum ... 45279

John Jarrold Printing Museum, Whitefriars, Norwich NR3 1SH • T: +44 1603 660211 • F: 630162 • www.johnjarroldprintingmuseum.org.uk •
Science&Tech Museum – 1982 ... 45280

Mustard Shop Museum, 15 Royal Arcade, Norwich NR2 1NQ • T: +44 1603 627889 • F: 762142 • www.colmansmustardshop.com •
Dir.: *Avril Houseago* •
Special Museum – 1973
Colman's mustard manufaturing history ... 45281

Norwich Gallery, Saint George St, Norwich NR3 1BB, mail addr: c/o Norwich School of Art and Design, Francis House, 3-7 Redwell St, Norwich NR2 4SN • T: +44 1603 610561, 756248 • F: 615728 • info@norwichgallery.co.uk • www.norwichgallery.co.uk •
Cur.: *Lynda Morris* •
University Museum / Public Gallery ... 45282

Royal Air Force Air Defence Radar Museum, RAF Neatishead, Norwich NR12 8YB • T: +44 1692 633309 • F: 633214 • curator@radarmuseum.co.uk • www.radarmuseum.co.uk •
Man.: *Doug Robb* •
Science&Tech Museum / Military Museum
History of radar and air defence since 1935, radar engineering, military communications systems, Cold War operations room, Royal Observer Corps rooms, ballistic missile and space defence, Bloodhound missile systems, radar vehicles ... 45283

Royal Norfolk Regimental Museum, Shirehall, Market Av, Norwich NR1 3JQ • T: +44 1603 493649 • F: 493623 • museums@norfolk.gov.uk • www.rnrm.org.uk •
Cur.: *Kate Thaxton* •
Military Museum – 1933
Hist of regiment, medals, uniforms, hist of British Army 45284

Sainsbury Centre for Visual Arts, c/o University of East Anglia, Norwich NR4 7TJ • T: +44 1603 593199 • F: 259401 • scva@uea.ac.uk • www.uea.ac.uk/scva •
Dir.: *Nichola Johnson* •
Fine Arts Museum / Association with Coll / University Museum – 1978
Robert and Lisa Sainsbury coll, Anderson coll of Art Nouveau, coll of abstract/constructivist art and Design 45285

Saint Peter Hungate Church Museum, Princes St, Norwich NR3 1AE • T: +44 1603 667231 • F: 493623 •
Religious Arts Museum – 1936
In 15th c church, hist of parish life, illuminated manuscripts, church musical instruments 45286

Strangers Hall Museum of Domestic Life, Charing Cross, Norwich NR2 4AL • T: +44 1603 667229 • F: 765651 • cathy.terry@norfolk.gov.uk • www.museums.norfolk.gov.uk •
Cur.: *Cathy Terry* • Ass. Cur.: *Helen Renton* •
Decorative Arts Museum – 1922
House with parts dating from 14th c, period rooms, toys, transportation, tapestries 45287

Nottingham

Angel Row Gallery, Central Library Bldg, 3 Angel Row, Nottingham NG1 6HP • T: +44 115 9152869 • F: 9152860 • angelrow.info@nottighamcity.gov.uk • www.nottinghamcity.gov.uk/sitemap/angelrow_gallery •
Man.: *Deborah Dean* •
Public Gallery
Temporary exhibitions 45288

Brewhouse Yard Museum, Castle Blvd, Nottingham NG7 1FB • T: +44 115 9153600, 9153640 • F: 9153601 • kaveandali@ncmg.demon.co.uk • www.nottinghamcity.gov.uk •
Dir.: *Michael Williams* • Asst. Dir.: *Peter Milton* • Cur.: *Suella Postles* •
Local Museum – 1977
Daily life material in Nottingham, local working, domestic, community, personal life from 1750 to present 45289

Djanogly Art Gallery, University of Nottingham, c/o Lakeside Arts Centre, University Park, Nottingham NG7 2RD • T: +44 115 9513192 • F: 9513194 • lakeside-marketing@nottingham.ac.uk • www.lakesidearts.org.uk •
Dir.: *Shona Powell* • Sc. Staff: *Neil Walker* (Exhibitions) •
Fine Arts Museum / University Museum – 1956
Duke of Newcastle coll: a bequest of family portraits from the 17th to 19th c, Glen Bott Bequest: a coll of watercolours and drawings of the British School (19th-20th c) 45290

Galleries of Justice, Shire Hall, High Pavement, Lace Market, Nottingham NG1 1HN • T: +44 115 9520555 • F: 9939828 • info@galleriesofjustice.org.uk • www.galleriesofjustice.org.uk •
Cur.: *Paul Baker* •
Historical Museum
Gaol, court rooms and 1905 police station, crime and punishment, prisons 45291

Green's Windmill and Science Centre, Windmill Ln, Sneinton, Nottingham NG2 4QB • T: +44 115 9156878 • F: 9156875 • enquiries@greenmill.org.uk • www.greensmill.org.uk •
Man.: *Jo Kemp* •
Science&Tech Museum – 1985
Mathematics, physics, mills and milling, science 45292

Industrial Museum, Courtyard Bldgs, Wollaton Park, Nottingham NG8 2AE • T: +44 115 9153900 • F: 9153932 • wollaton@ncmg.demon.co.uk • www.wollatonhall.org.uk •
Man.: *R. Dewsbury* •
Science&Tech Museum – 1971
Lace machinery, bicycles, telecommunications, steam engines 45293

Museum of Costume and Textiles, 51 Castle Gate, Nottingham NG1 6AF • T: +44 115 9153500 • F: 9153599 •
Special Museum
Dresses, laces, textiles 45294

Natural History Museum, Wollaton Hall, Nottingham NG8 2AE • T: +44 115 9153900 • F: 9153932 • wollaton@ncmg.demon.co.uk • www.wollatonhall.org.uk •
Sc. Staff: *G.P. Walley* (Team Leader Natural Sciences) • Keeper: *Dr. Sheila Wright* (Biology) • *C.D. Paul* (Biology) • *Neil Turner* (Geology) •
Natural History Museum – 1867
Mounted British and foreign birds, birds eggs, herbaria, European Diptera, lepidoptera, British aculeates, British beetles, mounted African big game heads, paleontology, fossils, Nottinghamshire plant and animal coll – library, Nottinghamshire Enironmental Archive, City Wildlife Database 45295

NCCL Galleries of Justice, High Pavement, The Lace Market, Nottingham NG1 1HN • T: +44 115 9520555 • F: 9939828 • info@galleriesofjustice.org.uk • www.galleriesofjustice.org.uk •
Cur.: *Tim Desmond* •
Special Museum 45296

Newstead Abbey Historic House and Gardens, Newstead Abbey Park, Nottingham NG15 8NA • T: +44 1623 455900 • F: 455904 • sallyl@newsteadabbey.org.uk • www.newsteadabbey.org.uk •
Cur.: *Haidee Jackson* •
Special Museum – 1170
Roe-Byron Coll: Poet's possessions and furniture, manuscripts, letters and first editions, 18th and 19th c furniture – library 45297

Nottingham Castle, off Friar Ln, Lenton Rd, Nottingham NG1 6EL • T: +44 115 9153700 • F: 9153653 • castle@ncmg.demon.co.uk • www.nottinghamcity.gov.uk/museums •
Dir.: *Michael Williams* • Asst. Dir.: *Brian Ashley* •
Decorative Arts Museum / Fine Arts Museum – 1878
Fine and applied arts, archaeology, militaria, medieval alabaster, English glass and silver, ceramics 45298

Nottingham University Museum of Archaeology, University Park, Nottingham NG7 2RD • T: +44 115 9514820, 9514813 • F: 9514812 • lloyd.laing@nottingham.ac.uk •
Dir.: *Lloyd Laing* •
University Museum / Archaeology Museum
Prehistoric, Roman and medieval artefacts from Eastern England, Bronze Age and Greek Mediterranean ceramics 45299

Surface Gallery, 7 Mansfield Rd, Nottingham NG1 3FB • T: +44 115 9348435 • admin@surfacegallery.org • surfacegallery.org •
Public Gallery – 1999/2002 45300

William Booth Birthplace Museum, 14 Notintone Pl, Sneiton, Nottingham NG2 4QG • T: +44 115 9503927 • F: 9598604 • djepson@salvationarmy.org.uk • www.salvationarmy.org.uk •
Dir.: *David Jepson* •
Special Museum
Birthplace William Booth (1829-1912) founder of the Salvation Army 45301

The Worcestershire and Sherwood Foresters Regimental Museum, Castle, Nottingham NG1 6EL, mail addr: Foresters House, Chetwynd Barracks, Chilwell NG9 5HA • T: +44 115 9465415 • F: 9469853 • Rhqmercian.notts@btconnect.com • www.wfrmuseum.org.uk •
Cur.: *M. J. Green* •
Military Museum – 1964
History of Worcestershire and Sherwood Foresters Regiment since 1741, armaments, medals, uniforms 45302

Yard Gallery, Wollaton Park, Nottingham NG8 2AE • T: +44 115 9153920 • F: 9153932 •
Public Gallery 45303

Nuneaton

George Eliot Hospital Museum, College St, Nuneaton CV10 7DJ • T: +44 870 8552540 • www.geh.nhs.uk •
Cur.: *Ann Cahill* •
Special Museum
Coll of medical equipment, photographs and archival documents 45304

Museum and Art Gallery, Coton Rd, Riversley Park, Nuneaton CV11 5TU • T: +44 24 76350720 • F: 76343559 • museum@nuneatonandbedworthbc.gov.uk • www.nuneatonandbedworth.gov.uk •
Local Museum – 1917
Roman to late-medieval archaeology, ethnography, George Eliot Coll, Baffin Land Inuit material, miniatures, oil painting and watercolours 45305

Nunnington

Carlisle Collection of Miniature Rooms, Nunnington Hall, Attic Rm, Nunnington YO62 5UY • T: +44 1439 748283 • F: 748284 • nunningtonhall@ntrust.org.uk •
Special Museum – 1981 45306

Oakham

Normanton Church Museum, Rutland Water, Oakham LE15 8PX • T: +44 1572 653026/27 • F: 653027 • tic@anglianwater.co.uk • www.anglianwaterleisure.co.uk •
Local Museum – 1985
Fossils, Anglo Saxon artefacts 45307

Oakham Castle, Rutland County Museum, Market Pl, Oakham LE15 6DT • T: +44 1572 758440 • F: 758445 • museum@rutland.gov.uk • www.rutnet.co.uk/rcc/rutlandmuseums •
Dir.: *Simon Davies* •
Historical Museum
12th c Great Hall of Norman Castle, coll of horseshoes 45308

Rutland County Museum, Catmose St, Oakham LE15 6HW • T: +44 1572 758440 • F: 758445 • museum@rutland.gov.uk • www.rutnet.co.uk/rcc/rutlandmuseums •
Dir.: *Simon Davies* •
Local Museum – 1967
Local hist and archaeology, Anglo-Saxon and other finds, farm tools, rural trades 45309

Oban

Oban War and Peace Museum, Old Oban Times Bldg, Corran Esplanade, Oban PA34 5PX • T: +44 1631 570007 • obanwarandpeace@aol.com • www.obanmuseum.org.uk •
Historical Museum / Local Museum – 1995
Local hist, local activities during the war years 45310

Okehampton

Finch Foundry Working Museum, Sticklepath, Okehampton EX20 2NW • T: +44 1837 840046 • F: 840046 • www.nationaltrust.org.uk •
Cust.: *Roger Boney* •
Science&Tech Museum – 1966
Working water powered tilt hammers 45311

Museum of Dartmoor Life, 3 West St, Okehampton EX20 1HQ • T: +44 1837 52295 • F: 659300 • dartmoormuseum@eclipse.co.uk • www.museumofdartmoorlife.eclipse.co.uk •
Cur.: *Maurie Webber* •
Folklore Museum 45312

Old Warden

Shuttleworth Collection, Aerodrome, Old Warden SG18 9EP • T: +44 1767 627927 • F: 627949 • enquiries@shuttleworth.org • www.shuttleworth.org •
Science&Tech Museum
Classic grass aerodrome, historic aeroplanes, aviation from a 1909 Bleriot XI to a 1942 Spitfire, veteran and vintage cars, motorcycles and horse-drawn carriages 45313

Oldham

Gallery Oldham, Greaves St, Oldham OL1 1AL • T: +44 161 9114653 • F: 9114669 • ecs.galleryoldham@oldham.gov.uk • www.galleryoldham.org.uk •
Head: *Sheena Macfarlane* (Heritage, Libraries, Arts) • Man.: *Stephen Whittle* (Development) • Coord.: *Dinah Winch* (Exhibits, Colls) • Sc. Staff: *Sean Baggaley* (Social History) • *Alan Price* (Natural History) •
Public Gallery / Fine Arts Museum / Local Museum – 1883/2002
Social, industrial and natural history of the region, 19th-20th c British paintings and sculpture, English watercolors, glass, Oriental art objects 45314

Oldland

Artemis Archery Collection, 29 Batley Court, Oldland BS30 8YZ • T: +44 117 9323276 • F: 9323276 • bogaman@btinternet.com •
Dir.: *Hugh David Hewitt-Soar* •
Special Museum
Traditional British longbows and arrows (since 18th c), related artifacts and ephemera – research library 45315

Olney

Cowper and Newton Museum, Orchardside, Market Pl, Olney MK46 4AJ • T: +44 1234 711516 • F: +44 870 1640662 • museum@onley.co.uk • www.cowperandnewtonmuseum.org.uk •
Cust.: *Joan McKillop* •
Special Museum – 1900
Memorabilia of writer William Cowper (1731-1800), memorabilia of the hymn writer, antislave campaigner and ev. preacher Reverend John Newton (1725-1807), lace industry 45316

Omagh

Ulster-American Folk Park, 2 Mellon Rd, Castle Town, Omagh BT78 5QY • T: +44 28 82243292 • F: 82242241 • info@uafp.co.uk • www.folkpark.com •
Cur.: *Dr. P. Mowat* • *P. O'Donnell* • *L. Corry* • Sc. Staff: *J. Bradley* • *E. Cardwell* (Education) •
Open Air Museum / Folklore Museum / Historical Museum – 1976
Buildings representing life in Ulster and North America 18th-19th c, emigration and the Atlantic crossing, ship and dockside gallery, reconstruction of an early 19th c sailing brig – library, migration studies centre 45317

Ulster History Park, Cullion, Omagh BT79 7SU • T: +44 28 81648188 • F: 81648011 • uhp@omagh.gov.uk • www.ballynagarrick.net/historicireland/HI901.htm •
Man.: *Anthony Candon* •
Historical Museum
Models of homes and monuments (17th c) 45318

Onchan

Groudle Glen Railway, Groudle Glen, Onchan, mail addr: 29 Hawarden Av, Douglas IM1 4BP • T: +44 1624 670453 • tbeard@manx.net •
Science&Tech Museum
Narrow gauge steam railway, steam locomotive Sea Lion, built in 1896, battery electric locomotive Polar Bear 45319

Orpington

Crofton Roman Villa, Crofton Rd, Orpington BR6 8AF • T: +44 20 84624737 • F: 84624737 • croftonvilla@aol.com • www.bromley.gov.uk •
Man.: *Brian Philp* •
Archaeology Museum 45320

London Borough of Bromley Museum, The Priory, Church Hill, Orpington BR6 0HH • T: +44 1689 873826 • bromley.museum@bromley.gov.uk •
Cur.: *Dr. Alan Tyler* •
Local Museum – 1965
Avebury coll 45321

Ospringe

Maison Dieu, Water Ln, Ospringe ME13 9DW • T: +44 1795 533751 • F: 533261 • faversham@btinternet.com • www.english-heritage.org.uk •
Staff: *Margaret Slythe* •
Historical Museum
Roman pottery and other finds from the local cemetry, finds from the medieval hospital 45322

Oswestry

Oswestry Transport Museum, Oswald Rd, Oswestry SY11 1RE • T: +44 1691 671749 • hignetts@enterprise.net • www.cambrianline.co.uk •
Science&Tech Museum
Railway, coll of 105 bicycles 45323

Otley

Otley Museum, Civic Centre, Cross Green, Otley LS21 1HP • T: +44 1943 461052 •
Local Museum – 1962
Archives relating to the printing machine industry in Otley, large coll of flints from the region 45324

Oundle

Oundle Museum, The Courthouse, Mill Rd, Oundle PE8 4BW • T: +44 1832 272741, 273871 • F: 273871 • oundle@globalnet.co.uk •
Local Museum 45325

Owermoigne

Mill House Cider Museum and Dorset Collection of Clocks, Owermoigne DT2 8HZ • T: +44 1305 852220 • F: 854760 • contact@millhousecider.com • www.millhousecider.com •
Dir.: *D.J. Whatmoor* •
Special Museum
Horology, cider mills and presses, long case and Turret clocks 45326

Oxford

Ashmolean Museum of Art and Archaeology, Oxford University, Beaumont St, Oxford OX1 2PH • T: +44 1865 278000 • F: 278018 • www.ashmol.ox.ac.uk •
Dir.: *Dr. Christopher Brown* • Chief Cons.: *Mark Normann* •
Archaeology Museum / Fine Arts Museum / University Museum / Decorative Arts Museum / Historical Museum / Religious Arts Museum – 1683
Archaeology of Britain, Europe, the Mediterranean, Egypt, the Near East, Italian, French, Dutch, Flemish and English paintings, Old Master drawings, modern drawings, watercolours, prints, miniatures, European ceramics, sculpture, bronze and silver, engraved portraits, numismatics, Oriental art, Indian and Islamic arts and crafts 45327

Bate Collection of Musical Instruments, Faculty of Music, Saint Aldatés, Oxford OX1 1DB • T: +44 1865 276139 • F: 76128 • bate.collection@music.ox.ac.uk •
Cur.: *Dr. Hélèna LaRue* •
Music Museum / University Museum
European woodwind, brass and percussion instruments, many early keyboards 45328

Christ Church Cathedral Treasury, Christ Church, Oxford OX1 1DP • T: +44 1865 201971 • F: 276277 • edward.evans@christ-church.ox.ac.uk •
Cur.: *Edward Evans* •
Religious Arts Museum
Church plate 45329

Christ Church Picture Gallery, Christ Church, Canterbury Quadrangle, Oriel Sq, Oxford OX1 1DP • T: +44 1865 202429 • F: 202429 • picturegallery@chch.ox.ac.uk • www.chch.ox.ac.uk •
Dir.: *Dr. Richard Rutherford* • Cur.: *Jacqueline Thalmann* •
Public Gallery – 1968
Old Master paintings and drawings 1300-1750 45330

The Crypt - Town Hall Plate Room, Town Hall, Saint Aldate's, Oxford OX1 1BX • T: +44 1865 252195 • www.oxford.gov.uk •
Decorative Arts Museum – 1897
Historic civic plate, large silver gilt mace, two Sear-geant's maces, gold porringer, cups and covers 45331

Modern Art Oxford, 30 Pembroke St, Oxford OX1 1BP • T: +44 1865 722733, 813830 • F: 722573 • info@modernartoxford.org.uk • www.modernartoxford.org.uk •
Dir.: *Andrew Nairne* • Sen. Cur.: *Suzanne Cotter* • Officer: *Sara Dewsbery* (Press, Marketing) • *Kirsty Kelso* (Marketing) •
Fine Arts Museum – 1965
Temporary exhibitions 45332

Museum of Oxford, Saint Aldates, Oxford OX1 1DZ • T: +44 1865 252761, 815539 • F: 252555 • museum@oxford.gov.uk • www.oxford.gov.uk/museum •
Dir.: *Vanessa Lea* • Cur.: *Jacqueline Hunt* •
Local Museum – 1975
Objects relating to the hist of the city and University of Oxford 45333

U

Museum of the History of Science, c/o University of Oxford, Broad St, Oxford OX1 3AZ • T: +44 1865 277280 • F: 277288 • museum@mhs.ox.ac.uk • www.mhs.ox.ac.uk •
Dir.: *James A. Bennett* • Asst. Keeper: *Stephen Johnston* •
University Museum – 1924
Historic scientific instruments (astrolabes, armillary spheres, sundials, clocks and watches, microscopes and telescopes, various apparatus) – library 45334

The Oxford Story, Broad St, Oxford OX1 3AJ • T: +44 1865 728822 • F: 791716 • info@oxfordstory.co.uk • www.oxfordstory.co.uk •
Historical Museum
900 years of University history 45335

Oxford University Museum of Natural History, Park Rd, Oxford OX1 3PW • T: +44 1865 272950, 272966 • F: 272970 • info@oum.ac.uk • www.oum.ox.ac.uk •
Dir.: *Prof. W.J. Kennedy* • Cur.: *Prof. E. Seiffert* (Geology) • *Dr. D. Waters* (Mineralogy) • *Dr. T.S. Kemp* (Zoology) • *Dr. G. McGavin* •
University Museum / Natural History Museum – 1860
Entomology, geology, mineralogy, zoology 45336

Oxford University Press Museum, Great Clarendon St, Oxford OX2 6DP • T: +44 1865 353527 • F: 267908 • martin.maw@oup.com •
Cur.: *Dr. Martin Maw* •
Special Museum
OUP's history from the middle ages to the age of technology, historical printing and typographical artefacts – archives 45337

Pitt Rivers Museum, c/o University of Oxford, South Parks Rd, Oxford OX1 3PP • T: +44 1865 270927 • F: 270943 • prm@pitt-rivers-museum.oxford.ac.uk • www.prm.ox.ac.uk •
Dir.: *Dr. Michael O'Hanlon* • Cur.: *Dr. Hélène LaRue* (Ethnomusicology) • *Dr. Laura Peers* (Social Anthropology, Central and South America) • *Dr. Chris Gosden* (European and Pacific Archaeology and Archaeological Method/Theory, Central Asia and ironage sites in Oxfordshire) • *Dr. Peter Mitchell* (African Archaeology) • *Dr. C. Harris* (Art, Anthropology, Asia esp. Tibet) •
Ethnology Museum / University Museum – 1884
Prehistoric archaeology, musical instruments, Captain Cook's material collected 1773-1774, African art, North American, Arctic, Pacific material, masks, textiles, arms and armour, amulets, charms, costumes, pottery, lamps, jewellery and bodyadornment – archives, Balfour Library 45338

Regimental Museum of the Oxfordshire and Buckinghamshire Light Infantry, T.A. Centre, Slade Park, Headington, Oxford OX3 7JJ • T: +44 1865 780128 • oxfoff@royalgreenjackets.co.uk •
Military Museum 45339

Padiham

Gawthorpe Hall, Rachel Kay Shuttleworth Textile Collections, nr Burnley, Padiham BB12 8UA • T: +44 1282 771004 • F: 770178 • gawthorpehall@mus.lancscc.gov.uk • www.lancsmuseums.gov.uk •
Keeper: *Rachel Bentley* •
Decorative Arts Museum
Embroideries from 17th c including goldwork, whitework, canvas work, quilts, lacework, costumes from 1750 – library 45340

Padstow

Padstow Museum, The Institute, Market Pl, Padstow PL28 8AD • T: +44 1841 532470 • www.padstowmuseum.com •
Cur.: *Peta Venner* •
Local Museum 45341

Paignton

Compton Castle, Marldon, Paignton TQ3 1TA • T: +44 1803 875740 • F: 875740 • www.nationaltrust.org.uk •
Historic Site
Historic house and garden 45342

Kirkham House, Kirkham St, Paignton TQ3 3AX • T: +44 117 9750700 • F: 9750701 • www.english-heritage.org.uk •
Cur.: *Tony Musty* •
Decorative Arts Museum
Coll of reproduction furniture 45343

Paignton and Dartmouth Steam Railway, Queen's Park Station, Torbay Rd, Paignton TQ4 6AF • T: +44 1803 555872 • F: 664313 • pdsr@talk21.com • www.paignton-steamrailway.co.uk •
Science&Tech Museum 45344

Paisley

Paisley Museum and Art Galleries, High St, Paisley PA1 2BA • T: +44 141 8893151 • F: 8899240 • museum.els@renfrewshire.gov.uk • www.renfrewshire.gov.uk •
Dir.: *Andrea J. Kerr* • Sc. Staff: *Jane Kidd* (Art) • *Shona Allan* (Natural History) • *Valerie A. Reilly* (Textiles) • *Chris Lee* (Social History) •
Local Museum / Fine Arts Museum – 1870
Art, local and natural history, local weaving art, Paisley Shawl Coll – observatory 45345

Pateley Bridge

Nidderdale Museum, Council Offices, King St, Pateley Bridge HG3 5LE • T: +44 1423 711225 • museum@nidderdale.fsnet.co.uk • www.nidderdalemuseum.com •
Local Museum – 1975
Domestic, farming and industrial material, costumes, a cobbler's shop, solicitor's office, general store, a Victorian parlour 45346

Paulerspury

Sir Henry Royce Memorial Foundation, The Hunt House, Paulerspury NN12 7NA • T: +44 1327 811048 • F: 811797 • shrmf@rrec.co.uk • www.henry-royce.org •
Cur.: *Philip Hall* •
Science&Tech Museum
Life and work of Henry Royce, Rolls-Royce & Bentley cars, photographs, historic docs, artifacts 45347

Peebles

Cornice Museum of Ornamental Plasterwork, Innerleithen Rd, Peebles EH45 8BA • T: +44 1721 720212 • F: 720212 •
Special Museum 45348

Tweeddale Museum, c/o Chambers Institute, High St, Peebles EH45 8AJ • T: +44 1721 724820 • F: 724424 • rhannay@scotborders.gov.uk •
Dir.: *Rosemary Hannay* •
Natural History Museum / Archaeology Museum / Local Museum – 1859
Geology coll and local prehistoric material 45349

Peel

House of Manannan, Manx National Heritage, Peel IM5 1TA • T: +44 1624 648000 • F: 648001 • enquiries@mnh.gov.im • www.storyofmann.com •
Dir.: *Stephen Harrison* •
Historical Museum
Celtic, Viking and maritime traditions, art' technology 45350

Peel Castle, Manx National Heritage, Peel IM5 1TB • T: +44 1624 648000 • F: 648001 • enquiries@mnh.gov.im • www.storyofmann.com •
Dir.: *Stephen Harrison* •
Religious Arts Museum / Historic Site
History, mystery and suspence of Peel castle, Viking heritage 45351

Penarth

Turner House Gallery, Plymouth Rd, Penarth CF64 3DM • T: +44 29 20708870 • turnerhouse@ffotogallery.org • www.ffotogallery.org •
Dir.: *C. Coppock* •
Public Gallery
Photography 45352

Pendeen

Geevor Tin Mining Museum, Geevor Tin Mine, Pendeen TR19 7EW • T: +44 1736 788662 • F: 786059 • pch@geevor.com • www.geevor.com •
Dir.: *W.G. Lakin* •
Science&Tech Museum
History of tin mining in Cornwall, geology, Cornish minerals, mining artefacts 45353

Pendeen Lighthouse, Penzance, Pendeen TR19 7ED • T: +44 1736 788418 • F: 786059 • info@trevithicktrust.com • www.trevithicktrust.com •
Science&Tech Museum
Engine room with sounder 45354

Pendine

Museum of Speed, Pendine SA33 4NY • T: +44 1994 453488 • F: +44 1267 223830 • museums@carmarthenshire.gov.uk • www.carmarthenshire.gov.uk •
Science&Tech Museum
Hist of Pendine sands 45355

Penrith

Dalemain Historic House, Penrith CA11 0HB • T: +44 17684 86450 • admin@dalemain.com • www.dalemain.com •
Historical Museum 45356

Penrith Museum, Robinson's School, Middlegate, Penrith CA11 7PT • T: +44 1768 212228 • F: 891754 • museum@eden.gov.uk • www.eden.gov.uk •
Cur.: *Judith Clarke* • Sc. Staff: *Dr. Sydney T. Chapman* •
Local Museum – 1990 45357

Penshurst

Penshurst Place and Toy Museum, Penshurst TN11 8DG • T: +44 1892 870307 • F: 870866 • enquiries@penshurstplace.com • www.penshurstplace.com •
Dir.: *Viscount de L'Isle* •
Special Museum 45358

Penzance

Cornwall Geological Museum, Saint John's Hall, Alverton St, Penzance TR18 2QR • T: +44 1736 332400 • F: 332400 • honsec@geological.org.uk • www.geological.nildram.co.uk •
Pres.: *David Freegman* •
Natural History Museum – 1914
Mineralogy, petrography, paleontology, Cornwall geology 45359

Newlyn Art Gallery, 4 Causewayhead, Penzance TR18 2SN • T: +44 1736 363715 • F: 331578 • mail@newlynartgallery.co.uk • www.newlynartgallery.co.uk •
Dir.: *Elizabeth Knowles* •
Public Gallery – 1895 45360

Penlee House Gallery and Museum, Morrab Rd, Penzance TR18 4HE • T: +44 1736 363625 • F: 361312 • info@penleehouse.org.uk • www.penleehouse.org.uk •
Dir.: *Alison Bevan* •
Fine Arts Museum / Local Museum / Archaeology Museum
Archaeology, local and natural history, Bronze Age pottery, Newlyn school paintings 45361

Trinity House National Lighthouse Centre, Wharf Rd, Penzance TR18 4BN • T: +44 1736 360077 • F: 364292 • www.lighthousecentre.org.uk •
Man.: *Alan Renton* •
Science&Tech Museum – 1990
Coll of lighthouse equipment 45362

Perranporth

Perranzabuloe Folk Museum, Oddfellows Hall, Ponsmere Rd, Perranporth TR6 0BW •
Local Museum
History of mining, fishing, farming 45363

Perth

The Black Watch Castle & Museum, Balhousie Castle, Hay St, Perth PH1 5HR • T: +44 1738 638152 • F: +44 1738 643245 • museum@theblackwatch.co.uk • www.theblackwatch.co.uk •
Man.: *Emma Halford-Forbes* •
Military Museum – 1924
History of Black Watch, related regiments, paintings, royal relics, uniforms, trophies from 1740 to the present 45364

Fergusson Gallery, Marshall Pl, Perth PH2 8NS • T: +44 1738 441944 • F: 621152 • museum@pkc.gov.uk • www.pkc.gov.uk •
Head: *Michael Taylor* • Sc. Staff: *J. Kinnear* (Arts Office) •
Fine Arts Museum
Oil paintings, watercolours, drawings, sculpture, sketchbooks by Scottish colourist John Duncan Fergusson – archive 45365

Perth Museum and Art Gallery, 78 George St, Perth PH1 5LB • T: +44 1738 632488 • F: 443505 • museum@pkc.gov.uk • www.pkc.gov.uk/ah •
Dir.: *M.A. Taylor* • Sc. Staff: *Robin Rodger* (Heritage) • *Fiona Slattery* (Applied Art) • *Susan Payne* (Human History) • *Robert Jarvie* (Design) • *Barbara Hamilton* (Education) • *Mark Hall* (Archaeology) • *Mark Simmons* (Natural Sciences) •
Local Museum / Fine Arts Museum / Natural History Museum / Public Gallery – 1935
Scottish and other paintings, applied art (Perth silver and glass), regional natural history, ethnography, archaeology, geology, antiquities 45366

Regimental Museum of the Black Watch → The Black Watch Castle & Museum

Scone Palace, Perth PH2 6BD • T: +44 1738 552300 • F: 552588 • visits@scone-palace.co.uk • www.scone-palace.co.uk •
Head: *Elspeth Bruce* •
Decorative Arts Museum
Finest French furniture, china, clocks, ivories, needlework, vases, several historic rooms, halls and galleries, rare art 45367

Peterborough

Peterborough Museum and Art Gallery, Priestgate, Peterborough PE1 1LF • T: +44 1733 343329 • F: 341928 • museum@peterborough.gov.uk • www.peterboroughheritage.org.uk •
Dir.: *Glenys Wass* • Cur.: *Julia Habeshaw* •
Local Museum – 1880
Archaeology, history, geology, marine reptils, natural history, ceramics, glass, portraits, paintings, finds from Castor, Napoleonic P.O.W. coll from Norman cross, bone carving, straw marquetry 45368

Railworld, Oundle Rd, Peterborough PE2 9NR • T: +44 1733 344240 • F: 319362 • railworld@aol.com • www.railworld.net •
Dir.: *Richard Paten* •
Science&Tech Museum – 1993
Exhibits on worldwide modern trains, transport environmental factors, Maglev trains, the age of steam, 21st c rail 45369

Southwick Hall, Main St, Southwick, Peterborough PE8 5BL • T: +44 1832 274064 • enquiries@southwickhall.co.uk • www.southwickhall.co.uk •
Man.: *Greg Bucknill* •
Local Museum
Development of the English manor house comprising medieval, Elizabethan, Georgian and 19th-c architectural styles, incl coll of Victorian and Edwardian artefacts and exhibitions of local hist 45370

Peterhead

Arbuthnot Museum, Saint Peter St, Peterhead AB42 1QD • T: +44 1771 622807 • F: 622884, 623558 • heritage@aberdeenshire.gov.uk • www.aberdeenshire.gov.uk/heritage/ •
Dir.: *Dr. David M. Bertie* •
Local Museum – 1850
Local hist, fishing, shipping and whaling, coin and medal coll, 19th c Inuit ethnography 45371

Petersfield

Bear Museum, 38 Dragon St, Petersfield GU31 4JJ • T: +44 1730 265108 • F: 266119 • www.bearmuseum.co.uk •
Dir.: *Judy Sparrow* • *John Sparrow* •
Special Museum – 1984
The world's first museum of teddy bears 45372

Flora Twort Gallery, Church Path, 21 The Square, Petersfield GU32 1HS • T: +44 1730 260756 • musmtc@hants.gov.uk • www.hants.gov.uk/museum/floratwo •
Cur.: *Tony Cross* •
Public Gallery
Works by Flora Twort (1893-1985) 45373

Petersfield Museum, The Old Courthouse, Saint Peters Rd, Petersfield GU32 3HX • T: +44 1730 262601 • petersfieldmuseum@hantsweb.org.uk • www.hants.org.uk/petersfieldmuseum •
Local Museum 45374

Uppark Exhibition, South Harting, Petersfield GU31 5QR • T: +44 1730 825415 • F: 825873 • uppark@nationaltrust.org.uk • www.nationaltrust.org.uk/uppark •
Man.: *Philippa Rawlinson* •
Special Museum
The fire of 1989 and subsequent restoration work 45375

Petworth

Coultershaw Beam Pump, Station Rd, Coultershaw Mill, Petworth GU28 0JE • T: +44 1798 865774 • F: 865774 • rlwconsult@compuserve.com • www.coultershaw.co.uk •
Cur.: *Robin Wilson* •
Science&Tech Museum / Historic Site
Waterwheel-driven pump, other pumps incl hydraulic ram 45376

The Petworth Cottage Museum, 346 High St, Petworth GU28 0AU • T: +44 1798 342100 • F: 343467 • stevensonguk@yahoo.co.uk • www.petworthcottagemuseum.co.uk •
Cur.: *Peter Jerrome* •
Historical Museum
Restored as an estate cottage occupied by a seamstress employed by the Leconfield estate (1910) 45377

Petworth House and Park, Church St, Petworth GU28 0AE • T: +44 1798 342207 • F: 342963 • petworth@nationaltrust.org.uk • www.nationaltrust.org.uk/petworth •
Fine Arts Museum
Old Masters (Turner, van Dyck, Laguerre), ancient and neo-classical sculpture 45378

Pevensey

Court House and Museum, High St, Pevensey BN24 5LF • T: +44 1323 762309 •
Cur.: *G.J.H. Dent* •
Local Museum 45379

Pewsey

Pewsey Heritage Centre, Whatleys Old Foundry, Avonside Works, High St, Pewsey SN9 5AF • T: +44 1672 56240, 562617 • F: 562617 • mikeasbury@tiscali.co.uk • www.pewsey-heritage-centre.org.uk •
Man.: *Michael J. Asbury* •
Local Museum
Farming, commerce, the home, engineering 45380

Pickering

Beck Isle Museum of Rural Life, Bridge St, Pickering YO18 8DU • T: +44 1751 473653 • F: 475996 • info@beckislemuseum.co.uk • www.beckislemuseum.co.uk •
Dir.: *Gordon Clitheroe* •
Local Museum – 1967
Reconstructed rooms and shops, photographic work of the late "Sidney Smith", works of Rex Whistler 45381

North Yorkshire Moors Railway, Station, Pickering YO18 7AJ • T: +44 1751 472508 • F: 476970 • customerservices@nymr.fsnet.co.uk • www.northyorkshiremoorsrailway.com •
Science&Tech Museum 45382

Pinxton

John King Workshop Museum, Victoria Rd, Pinxton NG16 6LR • T: +44 1773 811667 •
Science&Tech Museum
Mining, engineering, railway artifacts 45383

Pitlochry

Clan Donnachaidh Museum, Bruar, Pitlochry PH18 5TW • T: +44 1796 483264 • F: 483338 •
Dir.: *Ann McBay* •
Historical Museum
Items related to the Jacobite rising of 1745 45384

Pass of Killiecrankie Visitor Centre, Killiecrankie, Pitlochry PH16 5LG • T: +44 1796 473233 • F: 473233 • Man.: *Ben Notley* • Historical Museum / Natural History Museum Historic battle of 1689, natural hist 45385

Pitlochry Festival Theatre Art Gallery, Port-na-Craig, Pitlochry PH16 5DR • T: +44 1796 484600 • F: 484616 • boxoffice@pitlochry.org.uk • Dir.: *Roy Wilson* • Performing Arts Museum – 1951 45386

Pitstone

Pitstone Green Farm Museum, Vicarage Rd, Pitstone LU7 9EY • T: +44 1582 605464 • norman.groom@lineone.net • website.lineone.net/~pitstonemus • Local Museum WW2 collection of radio & radar with a replica section of a Avro Lancaster Bomber 45387

Pittenweem

Kellie Castle, Pittenweem KY10 2RF • T: +44 1333 720271 • F: 330132 • kelliecastle@nts.org.uk • www.nts.org.uk • Head: *M. Pirnie* • Decorative Arts Museum 16th c castle 45388

Plumridge

Sperrin Heritage Centre, 274 Glenelly Rd, Plumridge BT79 8LS • T: +44 28 81648142 • F: 71381348 • shc@strabanedc.com • www.strabanedc.com • Supervisor: *Anne Bradley* • Local Museum Ghosts, gold and poteen, local environment 45389

Plymouth

City Museum and Art Gallery, Drake Circus, Plymouth PL4 8AJ • T: +44 1752 304774 • F: 304775 • museum@plymouth.gov.uk • www.plymouthmuseum.gov.uk • Cur.: *Nicola Moyle* • Man.: *Jonathan Wilson* (Collections) • *David Paget-Woods* (Learning) • Local Museum / Fine Arts Museum – 1897 Porcelain, Cottonian coll, paintings, drawings, prints, early printed books, 17th c portraits, ceramics, personalia of Sir Francis Drake, archaeology, ethnography, model ships, Oceanic coll, tokens, photography, minerals – herbarium 45390

Merchant's House Museum, 33 Saint Andrew's St, Plymouth PL1 2AA • T: +44 1752 304774 • F: 304775 • museum@plymouth.gov.uk • www.plymouthmuseum.gov.uk • Dir.: *N. Moyle* • Historical Museum – 1977 Social hist of Plymouth 45391

Plymouth Arts Centre, 38 Looe St, Plymouth PL4 8AJ • T: +44 1752 221450 • Public Gallery 45392

Saltram House, Plympton, Plymouth PL7 1UH • T: +44 1752 333500 • F: 336474 • saltram@nationaltrust.org.uk • www.nationaltrust.org.uk • Man.: *Susan M. Baumbach* • Fine Arts Museum / Decorative Arts Museum / Historic Site 18th c living 45393

Pocklington

Stewart Collection, Burnby Hall Gardens, Pocklington YO4 2QF • T: +44 1759 302068 • burnbyhallgardens@hotmail.com • www.burnbyhallgardens.co.uk • Admin.: *Brian Petrie* • Local Museum 45394

Point Clear Bay

Museum of the 40's, Martello Tower, Point Clear Bay CO16 8NG • Military Museum / Science&Tech Museum Navy in WW II 45395

Polegate

Windmill and Milling Museum, Park Croft, Polegate BN26 5LB • T: +44 1323 734496 • Dir.: *Lawrence Stevens* • Science&Tech Museum 45396

Pontefract

Pontefract Castle Museum, Castle Chain, Pontefract WF8 1QH • T: +44 1977 723440 • museumsandarts@wakefield.gov.uk • www.wakefield.gov.uk/cultureandleisure • Dir.: *Colin MacDonald* • Historical Museum 45397

Pontefract Museum, 5 Salter Row, Pontefract WF8 1BA • T: +44 1977 722740 • F: 722742 • museumsandarts@wakefield.gov.uk • www.wakefield.gov.uk/cultureandleisure • Local Museum Archaeology, local hist, Knottingley glass and topography 45398

Ponterwyd

Llywernog Silver-Lead Mine Museum, Ponterwyd SY23 3AB • T: +44 1970 890620 • F: +44 1545 570823 • silverrivermine@aol.com • www.silverminetours.co.uk • Dir.: *Peter Lloyd Harvey* • Science&Tech Museum Coll of mining artefacts 45399

Pontypool

Amgueddfa Pontypool Museums, Park Bldgs, Pontypool NP4 6JH • T: +44 1495 752036 • F: 752043 • pontypoolmuseum@hotmail.com • www.pontypoolmuseum.org.uk • Cur.: *Deborah Wildgust* • Local Museum – 1978 45400

Junction Cottage, Lower Mill, off Fontain Rd, Pontymoile, Pontypool NP4 0RF • T: +44 800 5422663 • F: +44 1495 755877 • junctioncottage@messages.co.uk • www.junctioncottage.co.uk • Special Museum 45401

Torfaen Museum, Park Bldgs, Pontypool NP4 6JH • T: +44 1495 752036 • F: 752043 • Dir.: *Martin Buckridge* • Local Museum Social and industrial history of Torfaen 45402

Pontypridd

Pontypridd Museum, Bridge St, Pontypridd CF37 4PE • T: +44 1443 409748 • F: 490746 • bdavies@pontypriddmuseum.org.uk • www.pontypriddmuseum.org.uk • Cur.: *Brian Davies* • Local Museum Social hist, industry, agriculture, culture and recreation 45403

Pool

Camborne School of Mines Geological Museum and Art Gallery, c/o University of Exeter, Pool TR15 3SE • T: +44 1209 714866 • F: 716977 • l.atkinson@csm.ex.ac.uk • Cur.: *Lesley Atkinson* • Natural History Museum / Public Gallery Rocks and minerals, incl. fluorescent, radioactive, gem and ore minerals, art exhibitions 45404

Poole

Lighthouse Gallery, Poole Centre for the Arts, Kingland Rd, Poole BH15 1UG • T: +44 870 0668701 • F: 0668076 • oliviah@lighthousepoole.co.uk • www.lighthousepoole.co.uk • Dir.: *Olivia Hamilton* • Public Gallery Digital art 45405

Old Lifeboat Museum, East Quay, Poole • www.rnli.org.uk • Historical Museum 1938 lifeboat Thomas Kirk Wright 45406

Royal National Lifeboat Institution Headquarters Museum, West Quay Rd, Poole BH15 1HZ • T: +44 1202 663000 • F: 662243 • www.rnli.org.uk • Dir.: *Andrew Freemantle* • Historical Museum Lifeboat, models, paintings, fundraising artifacts 45407

Scaplen's Court Museum, High St, Poole BH15 1DA • T: +44 1202 262600 • F: 262622 • museums@poole.gov.uk • www.boroughofpoole.com • Historical Museum – 1932 45408

The Study Gallery, Bournemouth and Poole College, North Rd, Poole BH14 0LS • T: +44 1202 205200 • F: 205240 • info@thestudygallery.org • www.thestudygallery.org • Public Gallery / Fine Arts Museum – 2000 45409

Waterfront Museum, 4 High St, Poole BH15 1BW • T: +44 1202 262600 • F: 262622 • museums@poole.gov.uk • www.boroughofpoole.com • Dir.: *Clive Fisher* • Historical Museum – 1989 Boats and shipping, local hist, archaeology 45410

Port Charlotte

Museum of Islay Life, Port Charlotte PA48 7UN • T: +44 1496 850358 • F: 850358 • imt@islaymuseum.freeserve.co.uk • www.islaymuseum.freeserve.co.uk • Cur.: *Louise Airlie* • Local Museum – 1977 Victorian domestic items, industry, carved stone coll 6th to 16th c – lapidarium, library 45411

Port-of-Ness

Ness Historical Society Museum, Factory Habost, Port-of-Ness HS2 0TG • T: +44 1851 810377 • Local Museum History of the Ness area ofthe Isle of Lewis, photos, docs, fishing equipment, domestic utensils, implements used in crofting 45412

Port Saint Mary

Cregneash Village Folk Museum, Manx National Heritage, Cregneash, Port Saint Mary IM9 5PT • T: +44 1624 648000 • F: 648001 • enquiries@mnh.gov.im • www.storyofmann.com • Dir.: *Stephen Harrison* • Folklore Museum Manx crofting and fishing 19th-20th c, traditional cooking and spinning, farming 45413

Sound Visitor Centre, Manx National Heritage, The Sound, Port Saint Mary IM9 5PT • T: +44 1624 648000, 838123 • F: 648001 • enquiries@mnh.gov.im • www.storyofmann.com • Dir.: *Stephen Harrison* • Natural History Museum Natural wonders of the Sound and the Calf of Man 45414

Port Sunlight

Lady Lever Art Gallery, Lower Rd, Port Sunlight CH62 5EQ • T: +44 151 4784136 • F: 4784140 • ladylever@liverpoolmuseums.org.uk • www.liverpoolmuseums.org.uk/ladylever • Head: *Sandra Penketh* • Fine Arts Museum – 1922 British and other paintings, watercolors, sculpture, porcelain, furniture 45415

Port Sunlight Heritage Centre, 95 Greendale Rd, Port Sunlight CH62 4XE • T: +44 151 6446466 • F: 6448973 • info@portsunlightvillage.com • www.portsunlightvillage.com • Head: *Lionel Bolland* • Historical Museum Story of William Hesketh Lever, his soap factory, period soap packaging, village hist 45416

Port Talbot

Margam County Park, Margam, Port Talbot SA13 2TJ • T: +44 1639 881635 • F: 895897 • Ass. Dir.: *Raymond Butt* • Natural History Museum Over 800 acres of parkland, orangery, castle, sculpture coll, nursery rhyme village for children under 8, large hedge maze 45417

South Wales Miner's Museum, Afan Argoed Country Park, Cynonville, Port Talbot SA13 3HG • T: +44 1639 850564 • F: 850446 • www.southwalesminersmuseum.co.uk • Dir.: *Mair Boast* • Science&Tech Museum 45418

Portchester

Portchester Castle, Castle St, Portchester PO16 9QW • T: +44 23 92378291 • www.english-heritage.org.uk • Cur.: *Pam Braddock* • Local Museum 45419

Porthcawl

Porthcawl Museum, Old Police Station, John St, Porthcawl CF36 3BD • T: +44 1656 782111 • Local Museum Local and maritime hist, military hist of 49th Recce Regiment, costume, photographs – archives 45420

Porthcurno

Porthcurno Telegraph Museum, Eastern House, Porthcurno TR19 6JX • T: +44 1736 810966, 810478 • F: 810966 • info@porthcurno.org.uk • www.porthcurno.org.uk • Dir.: *Mary Godwin* • Science&Tech Museum Working submarine telegraphy equipment, underground WW II tunnels, local hist – archive 45421

Porthmadog

Porthmadog Maritime Museum, The Harbour, Porthmadog LL49 9LU • T: +44 1766 513736 • Historical Museum Models, paintings, pictures, docs, materials for shipbldg, navigation aids and seafaring 45422

Welsh Highland Railway Museum, Tremadog Rd, Porthmadog LL49 9DY • T: +44 1766 513402 • F: 514024, 513402 • info@whr.co.uk • www.whr.co.uk • Chm.: *James Hewett* • Science&Tech Museum – 1980 Hist of Welsh Highland and allied railway systems, steam locomotives and quarry wagons, tramway horse, hearse van 45423

Portland

Portland Castle, Castletown, Portland DT5 1AZ • T: +44 1305 820539 • F: 860853 • customers@english-heritage.org.uk • www.english-heritage.org.uk • Cur.: *Tony Musty* • Local Museum 45424

Portland Museum, 217 Wakeham, Portland DT5 1HS • T: +44 1305 821804 • F: 789922 • tourism@weymouth.gov.uk • www.weymouth.gov.uk • Natural History Museum – 1930 Cunnington, fossil coll (c 1925) of national importance being the best coll of faunal fossils from the Portlandian rocks 45425

Portslade-by-Sea

Foredown Tower, Foredown Rd, Portslade-by-Sea BN41 2EW • T: +44 3000 290900 • F: 1273 292092 • visitor.services@brighton-hove.gov.uk • www.brighton-hove-museums.org.uk • Natural History Museum – 1909 Camera obscura, astronomy, gallery of surrounding countryside 45426

Portsmouth

Charles Dickens Birthplace Museum, 393 Old Commercial Rd, Portsmouth PO1 4QL • T: +44 23 92827261 • F: 875276 • david.evans@portsmouthcc.gov.uk • www.portsmouthmuseums.co.uk • Dir.: *David Evans* • Special Museum – 1904 Birthplace of writer Charles Dickens (1812-1870), furniture, memorabilia, prints 45427

City Museum and Records Office, Museum Rd, Portsmouth PO1 2LJ • T: +44 23 92827261 • F: 875276 • david.evans@portsmouthcc.gov.uk • www.portsmouthmuseums.co.uk • Dir.: *David Evans* • Local Museum / Decorative Arts Museum / Fine Arts Museum / Historical Museum – 1972 Local history, ceramics, sculpture, paintings, wood engravings, prints, furniture, glass 45428

D-Day Museum and Overlord Embroidery, Clarence Esplanade, Southsea, Portsmouth PO5 3NT • T: +44 23 92827261 • F: 875276 • david.evans@portsmouthcc.gov.uk • www.ddaymuseum.co.uk • Dir.: *David Evans* • Military Museum – 1984 "The Overlord embroidery" D-day military hist 45429

Eastney Industrial Museum, Henderson Rd, Eastney, Portsmouth PO4 9JF • T: +44 23 92827261 • F: 92875276 • david.evans@portsmouthcc.gov.uk • www.portsmouthmuseums.co.uk • Dir.: *Sarah Quail* • Science&Tech Museum 45430

Fort Widley, A333, Portsdown Hill, Portsmouth PO6 3LS • T: +44 23 92324553 • Military Museum Bunker, tunnels, the Great Ditch 45431

HMS Victory, HM Naval Base, Portsmouth PO1 3NH • T: +44 23 92723111 • F: 92723171 • victorypa@a.dii.mod.uk • www.flagship.org.uk • Head: *J. Scivier* • Military Museum Flagship of Lord Nelson at the Battle of Trafalgar 45432

HMS Warrior 1860, Victory Gate, HM Naval Base, Portsmouth PO1 3QX • T: +44 23 92778600 • F: 92778601 • info@hmswarrior.org • www.hmswarrior.org • Dir.: *David Newbery* • Military Museum – 1987 Historic warship 45433

Mary Rose Museum, HM Naval Base, College Rd, Portsmouth PO1 3LX • T: +44 23 92750521 • F: 92870588 • mail@maryrose.org • www.maryrose.org • Dir.: *John Lippiett* • Military Museum Objects recovered from the wreck of Henry VIII's warship, the 'Mary Rose', surviving portion of the ship's hull 45434

Natural History Museum, Cumberland House, Eastern Parade, Southsea, Portsmouth PO4 9RF • T: +44 23 92827261 • F: 92875276 • info@portsmouthnaturalhistory.co.uk • www.portsmouthmuseums.co.uk • Dir.: *Dr. Jane Mee* • Natural History Museum 45435

Royal Marines Museum, Eastney, Southsea, Portsmouth PO4 9PX • T: +44 23 92819385 • F: 92838420 • info@royalmarinesmuseum.co.uk • www.royalmarinesmuseum.co.uk • Dir.: *Chris Newbery* • Military Museum – 1958/1975 Documents, paintings, weapons, uniforms, campaign relics since 1664, band and music paraphernalia 45436

Royal Naval Museum Portsmouth, HM Naval Base, PP66, Portsmouth PO1 3NH • T: +44 23 92727562 • F: 92727575 • information@royalnavalmuseum.org • www.royalnavalmuseum.org • Dir.: *H. Campbell McMurray* • Dep. Dir.: *C.S. White* • Historical Museum – 1911 Figureheads, ship models, ship furniture, memorabilia of Lord Nelson, panorama of the Battle of Trafalgar, history of the Royal Navy from earliest times to present – library, full research facilities 45437

Southsea Castle, Clarence Esplanade, Southsea, Portsmouth PO5 3PA • T: +44 23 92827261 • F: 875276 • david.evans@portsmouthcc.gov.uk • www.portsmouthmuseums.co.uk • Dir.: *David Evans* • Military Museum – 1967 History of Portsmouth fortress, tudor, civilwar, Victorian military hist 45438

Treadgolds Museum, 1 Bishop St, Portsmouth PO1 3DA • T: +44 23 92824745 • F: 92837310 • www.hants.gov.uk/museums •
Man.: *Peter Lawton* •
Special Museum
Machine and hand tools, materials, business archives, Treadgolds, ironmongers and engineers ... 45439

Potterne

Wiltshire Fire Defence and Brigades Museum, Worton Rd, Potterne SN10 5PP • T: +44 1380 723601 • F: 727000 • www.wiltshirefirebrigade.com •
Science&Tech Museum
Coll of 18th- 20th c fire-fighting equip ... 45440

Potters Bar

Potters Bar Museum, Wyllotts Pl, Darkes Ln, Potters Bar EN6 4HN • T: +44 1707 645005 ext 20 •
Cur.: *Arnold Davey* •
Local Museum
History and archaeology of Potters Bar, telephones, Potters Bar Zeppelin ... 45441

Prescot

Prescot Museum, 34 Church St, Prescot L34 3LA • T: +44 151 4307787 • F: 4307219 • prescot.museum.dlcs@knowsley.gov.uk • www.knowsley.gov.uk •
Cur.: *R. John Griffiths* •
Science&Tech Museum – 1982
Horology, earthenware ... 45442

Presteigne

Judge's Lodging, Broad St, Presteigne LD8 2AD • T: +44 1544 260650/51 • F: 260652 • info@judgeslodging.org.uk • www.judgeslodging.org.uk •
Cur.: *Gabrielle Rivers* •
Local Museum – 1980
Victorian domestic life, law and order, local/regional hist ... 45443

Preston

Harris Museum and Art Gallery, Market Sq, Preston PR1 2PP • T: +44 1772 258248 • F: 886764 • harris.museum@preston.gov.uk •
Dir.: *Alexandra Walker* • Sc. Staff: *James Green* (Exhibition) • *Francis Marshall* (Fine Art) • *Emma Heslewood* (Social History) • *Amanda Draper* (Decorative Art) •
Fine Arts Museum / Decorative Arts Museum / Local Museum – 1893
Fine art, Devis Coll of 18th c paintings, Newsham Bequest of 19th c British paintings, Haslam Bequest of 19th c watercolors, decorative art, costumes, ceramics, porcelain, social hist, archaeology, skeleton of Mesolithic elk, photogr ... 45444

Museum of Lancashire, Stanley St, Preston PR1 4YP • T: +44 1772 264075 • F: 264079 • museum.enquiries@mus.lancscc.co.uk • www.lancsmuseums.gov.uk •
Dir.: *Edmund Southworth* •
Local Museum / Natural History Museum
Geology, archaeology, social, military and natural hist ... 45445

National Football Museum, Sir Tom Finney Wy, Deepdale, Preston PR1 6RU • T: +44 1772 908442 • F: 908433 • enquiries@nationalfootballmuseum.com • www.nationalfootballmuseum.com •
Dir.: *Kevin Moore* • Cur.: *Malcom MacCallum* •
Special Museum
Hist of football ... 45446

Queen's Lancashire Regiment Museum, Fulwood Barracks, Watling St Rd, Preston PR2 8AA • T: +44 1772 260362 • F: 260583 • rhq.qlr@btconnect.com • www.army.mod.uk/qlr/museum_archives/index.htm •
Dir.: *John Downham* • Cur.: *Jane Davies* •
Military Museum – 1926
Regimental history, uniforms, silver, trophies, weapons – library, archive ... 45447

Prestongrange

Prestongrange Museum, Morison's Haven, Prestongrange EH32 9RX • T: +44 131 6532904 • F: +44 1620 828201 • elms@prestongrangemuseum.org • www.prestongrangemuseum.org •
Head: *Peter Gray* •
Science&Tech Museum
History of Prestongrange colliery and brickworks, colliery locomotives, 800 years mining history ... 45448

Prickwillow

Prickwillow Drainage Engine Museum, Main St, Prickwillow CB7 4UN • T: +44 1353 688360 • F: 723456 • www.prickwillow-engine-museum.co.uk •
Chm.: *Les Walton* •
Science&Tech Museum – 1982
Memorabilia, drainage artefacts, diesel engines e.g. Mirlees Bickerton & Day, Vickers Petter 2stroke, Ruston GMXR, Allen ... 45449

Pudsey

Moravian Museum, 55-57 Fulneck, Pudsey LS28 8NT • T: +44 113 2564147 •
Special Museum
Moravian memoabilia, lace and emboidery, ethnography, oldest working fire engine in England, Victorian parlour ... 45450

Pulborough

Roman Villa Museum at Bignor, Bignor Ln, Pulborough RH20 1PH • T: +44 1798 869259 • F: 869259 • bignorromanvilla@care4free.net •
Dir.: *J. Tupper* •
Archaeology Museum – 1960
Samian pottery, tiles, finds from 1811 excavations, mosaics ... 45451

Purton

Purton Museum, Library, High St, Purton SN5 4AA • T: +44 1793 770648 • curator@purtonmuseum.com • www.purtonmuseum.com •
Cur.: *Nigel Manfield* •
Local Museum
Local hist, archaeological material, agricultural hand tools ... 45452

Quainton

Buckinghamshire Railway Centre, Quainton Road Station, Quainton HP22 4BY • T: +44 1296 655720 • F: 655720 • bucksrailcentre@btopenworld.com • www.bucksrailcentre.org.uk •
Cur.: *R. Miller* •
Science&Tech Museum
Steam/Diesel locomotives, carriages, wagons, railway artefacts ... 45453

Quatt

Dudmaston House, Quatt WV15 6QN • T: +44 1746 780866 • F: 780744 • mduefe@smtp.ntrust.org.uk •
Decorative Arts Museum
Dutch flower paintings, fine furniture ... 45454

Radstock

Radstock Museum, Market Hall, Wateroo Rd, Radstock BA3 3EP • T: +44 1761 437722 • F: 420470 • info@radstockmuseum.co.uk • www.radstockmuseum.co.uk •
Chm.: *Richard Maggs* •
Local Museum – 1986
Local history, industrial archaeology, mining ... 45455

Rainham

Rainham Hall, The Broadway, Rainham RM13 9YN • T: +44 1494 528051 • F: 463310 •
Decorative Arts Museum
Plasterworks, carved porch and interor ... 45456

Ramsgate

Ramsgate Maritime Museum, Pier Yard, Royal Harbour, Ramsgate CT11 8LS • T: +44 1843 570622 • F: 582359 • curator@greatstorm.fsnet.co.uk • www.ekmt.fsnet.co.uk •
Dir.: *Charles Payton* •
Science&Tech Museum – 1984
Artifacts from H.M.S. Stirling Castle wrecked on the Goodwin sands in the great storm of 1703, historic ship coll incl Dunkirk little ship motor yacht Sundowner ... 45457

Ramsgate Museum, Guildford Lawn, Ramsgate CT11 9AY • T: +44 1843 593532 • F: 852692 • www.kent.gov.uk •
Dir.: *Beth Thomson* • Sc. Staff: *Sheena Watson* •
Local Museum – 1912
library ... 45458

Spitfire and Hurricane Memorial Museum, The Airfield, Manston Rd, Ramsgate CT12 5DF • T: +44 1843 821940 • F: 821940 • spitfire752@btconnect.com • www.spitfire.memorial.museum •
Man.: *Peter Turner* •
Science&Tech Museum / Military Museum
Spitfire and hurricane fighters, aviation-related artefacts relating to WW II period (battle of Britain) ... 45459

Ravenglass

Muncaster Castle, Ravenglass CA18 1RQ • T: +44 1229 717614 • F: 717010 • info@muncaster.co.uk • www.muncaster.co.uk •
Dir.: *Peter Frost-Pennington* •
Fine Arts Museum ... 45460

Muncaster Watermill, Ravenglass CA18 1ST • T: +44 1229 717232 •
Science&Tech Museum ... 45461

Railway Museum, c/o Ravenglass and Eskdale Railway Co. Ltd., Ravenglass CA18 1SW • T: +44 1229 717171 • F: 717011 • pvanzeller@onetel.net.uk • www.ravenglass-railway.co.uk •
Cur.: *Peter van Zeller* •
Science&Tech Museum – 1978
History of the Ravenglass and Eskdale Railway since 1875 and its effect on the area ... 45462

Ravenshead

Gordon Brown Collection, Longdale Craft Centre, Longdale Ln, Ravenshead NG15 9AH • T: +44 1623 794858 • longdale@longdale.co.uk • www.longdale.co.uk •
Dir.: *Gordon Brown* •
Decorative Arts Museum ... 45463

Longdale Craft Centre and Museum, Longdale Ln, Ravenshead NG15 9AH • T: +44 1623 794858 • F: 794858 • www.longdalecraftcentre.co.uk •
Dir.: *Gordon Brown* •
Folklore Museum – 1972
Recreated Victorian Craft Village ... 45464

Papplewick Pumping Station, off Longdale Ln, Ravenshead NG15 9AJ • T: +44 115 9632938 • F: 9638153 • director@papplewickpumpingstation.co.uk • www.papplewickpumpingstation.co.uk •
Dir.: *Ian Smith* •
Science&Tech Museum
Boilers, beam engines, working forge, colliery winding engine, 7.25" gauge passenger carrying steam railway ... 45465

Rawtenstall

Rossendale Footware Heritage Museum, Greenbridge Works, Fallbarn Rd, Rawtenstall BB4 7NY • T: +44 1706 235160 • F: 235129 •
Special Museum ... 45466

Reading

Cole Museum of Zoology, c/o School of Animal and Microbial Science, University of Reading, Whiteknights, Reading RG6 6AJ • T: +44 118 3786355 • a.callaghan@rdg.ac.uk • www.ams.rdg.ac.uk/zoology/colemuseum •
Dir.: *Dr. Amanda Callaghan* •
Natural History Museum / University Museum – 1906
Form and function in the animal kingdom ... 45467

Museum of English Rural Life, c/o University of Reading, Redlands Rd, Reading R61 5EX • T: +44 118 3788660 • F: 3785632 • merl@reading.ac.uk • www.merl.org.uk •
Agriculture Museum – 1951
Farming development, history of local rural life – library ... 45468

Museum of Reading, Town Hall, Blagrave St, Reading RG1 1QH • T: +44 118 9399800 • F: 9399881 • mail@readingmuseum.org • www.readingmuseum.org •
Dir.: *Karen Hull* •
Decorative Arts Museum / Archaeology Museum / Local Museum – 1993
Britain's only full-size replica of the Bayeux Tapestry, the Silchester coll (Roman artefacts from Callera Atrebatum, now Silchester) ... 45469

Reading Abbay Gateway, The Forbury, Reading RG1 1QH • T: +44 118 9399809/05 • F: 9566719 •
Religious Arts Museum ... 45470

Riverside Museum at Blake's Lock, Gas Works Rd, off Kenavon Dr, Reading RG1 3DH • T: +44 118 9399800 • F: 9399881 • blakes@readingmuseum.org.uk • www.readingmuseum.org.uk •
Dir.: *Karen Hull* •
Historical Museum – 1985
Waterways (rivers Thames and Kennet), traders and industries of Reading ... 45471

Ure Museum of Greek Archaeology, c/o Dept. of Classics, School of Humanities, University of Reading, HumSS, Whiteknights, Reading RG6 6AA • T: +44 118 3786990 • F: 3786661 • ure@reading.ac.uk • www.rdg.ac.uk/ure •
Dir.: *Dr. Amy C. Smith* •
Archaeology Museum / Museum of Classical Antiquities / University Museum – 1922
Greek & Egyptian antiquities; Attic, Corinthian, Boeotian and South Italian vases, Coptic textiles – excarvation archives of Rhitsona ... 45472

Redcar

Kirkleatham Museum, Kirkleatham, Redcar TS10 5NW • T: +44 1642 479500 • F: 474199 • museum_services@redcar-cleveland.gov.uk • www.redcar-cleveland.gov.uk/museums •
Dir.: *P. Philo* •
Local Museum – 1981
Local 19th c artists (the staithes group incl Dame Laura Knight), commondale pottery, poster coll (some international from Poland and Germany), photographic coll, maritime hist incl lifeboats of the north east coast, iron stone mining, iron making, fishing ... 45473

Zetland Lifeboat Museum, 5 King St, Redcar TS10 3DT • T: +44 1642 494311 • zetland.museum@yahoo.co.uk •
Historical Museum – 1969
Shipping and fishing, industrial development, sea rescue, oldest lifeboat, ship models, marine paintings, scientific instruments ... 45474

Redditch

Forge Mill Needle Museum and Bordesley Abbey Visitor Centre, Needle Mill Ln, Riverside, Redditch B98 8HY • T: +44 1527 62509 • museum@redditchbc.gov.uk • www.redditchbc.gov.uk •
Man.: *Liz Shiftins* •
Science&Tech Museum / Archaeology Museum – 1983
Needles and fishing tackle from the Redditch needle-making district – Bordesley Abbey archaeological archive ... 45475

Redruth

Cornish Mines, Engines and Cornwall Industrial Discovery Centre, Pool, Redruth TR15 3NP • T: +44 1209 315027 • F: 315027 • info@trevithicktrust.com • www.trevithicktrust.com •
Science&Tech Museum
Cornish beam-engines used for mine winding and pumping ... 45476

Geological Museum and Art Gallery, University of Exeter, Camborne School of Mines, Redruth TR15 3SE • T: +44 1209 714866 • F: 716977 • scamm@csm.ex.ac.uk •
Dir.: *Simon Camm* •
Natural History Museum – 1888
Robert Hunt Coll of minerals, flurorescent minerals, ore specimens, Cornish minerals, fossils, worldwide coll of rocks and minerals, shows by local artists – library ... 45477

Tolgus Tin Mill and Streamworks, c/o Trevithick Trust, Portreath Rd, Portreath, Redruth TR15 3NP • T: +44 1209 210900 • F: 210900 • info@cornishgoldsmiths.com • www.trevithicktrust.com •
Man.: *Dervla Jarratt* •
Science&Tech Museum – 1921
Tin-streaming machinery in operation incl Cornish stamps (water powered), tin-mining – library ... 45478

Reeth

Swaledale Museum, The Green, Reeth DL11 6QT • T: +44 1748 884118 • museum@swaledale.org • www.swaledalemuseum.org •
Cur.: *R.A. Bainbridge* • Sc. Staff: *H.M. Bainbridge* •
Folklore Museum – 1975
Folk exhibits, social hist and traditions of the Dale, sheep farming, lead mining ... 45479

Reigate

Fire Brigades Museum, Saint David's, 70 Wray Park Rd, Reigate RH2 0EJ • T: +44 1737 242444 • F: 224095 •
Man.: *Derek Chinery* •
Science&Tech Museum
A coll of vintage fire engines, photographs and general information related to fire fighting in Surrey from 1754 to modern times ... 45480

Museum of the Holmesdale Natural History Club, 14 Croydon Rd, Reigate RH2 0PG • www.hnhc.co.uk •
Dir.: *H. Collier* •
Natural History Museum / Local Museum – 1857
Fossils, local archaeological finds, coll of British birds, photographs and postcards, local history, geology – herbarium ... 45481

Reigate Priory Museum, Reigate Priory, Bell St, Reigate RH2 7RL • T: +44 1737 222550 • www.reigateprioryrmuseum.org.uk •
Cur.: *Eileen Wood* •
Historical Museum
The coll incl domestic memorabilia, local hist and costume ... 45482

Renishaw

Renishaw Hall Museum and Art Gallery, Renishaw Park, Renishaw S21 3WB • T: +44 1246 432310 • F: 430760 • info2@renishaw-hall.co.uk • www.sitwell.co.uk •
Owner: *Sir Reresby Sitwell* •
Performing Arts Museum
Art, John Piper music and theatre exhibits, Sitwell memorabilia ... 45483

Retford

Bassetlaw Museum and Percy Laws Memorial Gallery, Amcott House, 40 Grove St, Retford DN22 6JD • T: +44 1777 713749 • enquiries@bassetlawmuseum.org.uk • www.bassetlawmuseum.org.uk •
Cur.: *Samantha Glasswell* •
Local Museum – 1986
Local archaeology, civic, social and agricultural history, fine and applied art ... 45484

Rhuddlan

Rhuddlan Castle, Rhuddlan LL18 5AE • T: +44 1745 590777 •
Archaeology Museum ... 45485

Rhyl

Rhyl Library, Museum and Arts Centre, Church St, Rhyl LL18 3AA • T: +44 1745 353814 • F: 331438 • rose.mcmahon@denbighshire.gov.uk •
Cur.: *Rose Mahon* •
Local Museum
Town's role, firstly as a fishing village and later as a seaside resort, maritime and social history ... 45486

Ribchester

Ribchester Roman Museum, Riverside, Bremetennacum, Ribchester PR3 3XS • T: +44 1254 878261 • F: 820803 • ribchestermuseum@btconnect.com • www.ribchestermuseum.org •
Cur.: *Patrick Tostevin* •
Archaeology Museum – 1914
Prehistoric and Roman finds from the Ribble Valley and the fort and civilian settlerment at Ribchester ... 45487

Richmond, North Yorkshire

Georgian Theatre Royal Museum, Victoria Rd, Richmond, North Yorkshire DL10 4DW • T: +44 1748 823710, 825252 • F: 823303 • admin@georgiantheatreroyal.co.uk • www.georgiantheatreroyal.co.uk •
Dir.: *Vaughn Curtis* •
Performing Arts Museum
Coll of original playbills from 1792 to the 1840s, the oldest and largest complete set of painted scenery in Britain, dating from 1836, displays of model theatres, photographs 45488

Green Howards Regimental Museum, Trinity Sq, Richmond, North Yorkshire DL10 4QN • T: +44 1748 826561 • F: 826561 • museum@greenhowards.org.uk • www.greenhowards.org.uk •
Dir.: *J.R. Chapman* • Cur.: *David Tetlow* •
Military Museum – 1973
History of the regiment since 1688, uniforms, medals, equipment, paintings, photographs & memorabilia 45489

Richmondshire Museum, Ryder's Wynd, Richmond, North Yorkshire DL10 4JA • T: +44 1748 825611 •
Cur.: *Julie Dodd* •
Local Museum
Development of the Richmond area since 1071, earlier archaeological material, Anglo-Saxon carved stones, farming and craftsmen's tools and equipment, leadmining relics, costumes, needlework, reconstructions of a carpenter's and a blacksmith's shop, model of Richmond railway station, old photographs and prints 45490

Richmond, Surrey

Economic Botany Collection/Museum No. 1, Royal Botanic Gardens, Kew, Richmond, Surrey TW9 3AE • T: +44 20 83325386 • F: 83325771 • ecbot@kew.org • www.rbgkew.org.uk/collections/ecbot •
Head: *Mark Nesbitt* •
Natural History Museum / Folklore Museum – 1847
Economic botany coll, useful plants and their products worldwide, wood coll 45491

Ham House, Ham St, Petersham, Richmond, Surrey TW10 7RS • T: +44 20 89401950 • F: 83326903 • hamhouse@nationaltrust.org.uk • www.nationaltrust.org.uk/hamhouse •
Local Museum
19th-c kitchen, textiles, paintings and tapestries 45492

Museum of Richmond, Old Town Hall, Whittaker Av, Richmond, Surrey TW9 1TP • T: +44 20 83321141 • F: 89487570 • museumofrichmond@btconnect.co.uk • www.museumofrichmond.com •
Cur.: *Stephanie Kirkpatrick* •
Local Museum
Local history 45493

Rickmansworth

Batchworth Lock Canal Centre, 99 Church St, Rickmansworth WD3 1JD • T: +44 1923 778382 • F: 710903 •
Science&Tech Museum
Visitor centre, maps, information, books, photographs of old Batchworth, working boats on canal, canal trips on summer sunday afternoons 45494

Three Rivers Museum of Local History, Basing House, Rickmansworth WD3 1HP • T: +44 1923 727333 • info@trmt.org.uk •
Local Museum
Coll. of photographs and artefacts relating to the Three Rivers area 45495

Rievaulx

Rievaulx Abbay, Rievaulx YO6 5LB • T: +44 1439 798340 • F: 798480 • customers@english-heritage.org.uk • www.english-heritage.org.uk •
Cur.: *Susan Harrison* •
Decorative Arts Museum / Religious Arts Museum
English landscape design in the 18th c, medieval tiles, everyday items 45496

Ripley, Derbyshire

Midland Railway Butterley, Butterley Station, Ripley, Derbyshire DE5 3QZ • T: +44 1773 747674 • F: 570721 • mrc@rapidial.co.uk • www.midlandrailwaycentre.co.uk •
Cur.: *Dudley Fowkes* •
Science&Tech Museum 45497

Ripley, North Yorkshire

Ripley Castle, Ripley, North Yorkshire HG3 3AY • T: +44 1423 770152 • F: 771745 • enquiries@ripleycastle.co.uk • www.ripleycastle.co.uk •
Dir.: *Sir Thomas C.W. Ingilby* •
Historical Museum – 1307
Civil War armour, Elizabethan panelling 45498

Ripon

Newby Hall, Ripon HG4 5AE • T: +44 1423 322583 • F: 324452 • info@newbyhall.com • www.newbyhall.com •
Dir.: *Richard Compton* •
Fine Arts Museum / Decorative Arts Museum
Classical statuary, Chippendale furniture, chamber pots, Gobelin tapestries 45499

Norton Conyers, off A1, Ripon HG4 5EQ • T: +44 1765 640333 • F: 640333 • norton.conyers@bronco.co.uk •
Dir.: *Sir James Graham* • *Lady Graham* •
Historical Museum
Coll of family portraits from 1580-1903; medieval to 18th c furniture, costumes 45500

Ripon Prison and Police Museum, Saint Marygate, Ripon HG4 1LX • T: +44 1765 690799, 603006 • info@riponmuseums.co.uk • www.riponmuseums.co.uk •
Cur.: *Ralph Lindley* • Man.: *Christine Orsler* •
Historical Museum
Police mementoes and equipment from the 17th c to the present day, illustration of the 17th to 19th-c methods of confinement and punishment 45501

Ripon Workhouse Museum, Sharow View, Allhallowgate, Ripon HG4 1LX •
Cur.: *Ralph Lindley* •
Historical Museum
Victorian poor law artefacts 45502

Robertsbridge

Bodiam Castle, Robertsbridge TN32 5UA • T: +44 1580 830436 • F: 830398 • bodiamcastle@nationaltrust.org.uk • www.nationaltrust.org.uk/places/bodiamcastle •
Historical Museum – 1926
Relics connected with Bodiam castle 45503

Robin Hood's Bay

Robin Hood's Bay and Fylingdale Museum, Fisherhead, Robin Hood's Bay YO22 4ST • T: +44 1947 880097 •
Local Museum
Local social history, geology 45504

Rochdale

Rochdale Art and Heritage Centre, c/o Touchstones Rochdale, The Esplanade, Rochdale OL16 1AQ • T: +44 1706 641085 • F: 712723 •
Dir.: *Andrew Pearce* •
Fine Arts Museum / Local Museum – 1903
British paintings and watercolours, contemporary paintings, local social history – archives, museum and art gallery colls 45505

Rochdale Pioneers Museum, 31 Toad Ln, Rochdale OL12 0NU • T: +44 1706 524920 • F: +44 161 2462946 • museum@co-op.ac.uk • museum.co-op.ac.uk •
Cur.: *Gillian Lonergan* •
Historical Museum – 1931
Formation of the Rochdale Equitable Pioneers Society, which marked the beginning of the worldwide co-operative movement, Co-operative Wholesale Society, celebrations (1944 and 1994) celebrations, international social reformer and co-operator Robert Owen 45506

Rochester

Guildhall Museum, High St, Rochester ME1 1PY • T: +44 1634 848717 • F: 832919 • guildhall.museum@medway.gov.uk •
Dir.: *Peter Boreham* • Sc. Staff: *Steve Nye* (Archaeology, Collections) • *Joni Brobbel* (Museum Support) • *A. Freeman* (Display and Technic) • *Jeremy Clarke* (Education) •
Historical Museum
18th-19th c prison hulks, prisoner-of-war ship models, Charles dickens relates books & objects 45507

Medway Towns Gallery, Civic Centre, Strood, Rochester ME20 4AW • T: +44 1634 727777 •
Public Gallery 45508

Strood Library Gallery, 32 Bryant Rd, Strood, Rochester ME2 3EP • T: +44 1634 718161 • F: 718161 •
Public Gallery 45509

Rolvenden

C.M. Booth Collection of Historic Vehicles, c/o Falstaff Antiques, 63-67 High St, Rolvenden TN17 4LP • T: +44 1580 241234 •
Dir.: *C.M. Booth* •
Science&Tech Museum – 1972
Morgan 3-wheel cars 45510

Romney Marsh

Brenzett Aeronautical Museum Trust, Ivychurch Rd, Brenzett, Romney Marsh TN29 0EE • T: +44 1797 344747 • www.brenzettaero.co.uk •
Chm.: *Frank J. Beckley* •
Military Museum / Historical Museum – 1972
Former Womens Land Army Hostel, coll incl engines and other relicts from British ,American and German wartime aircraft, the 4 1/2 ton dam-buster bomb, designed by Sir Barnes Wallis in 1943, Vampire MKII, Dakota Nose, Canberra 45511

Romsey

Mountbatten Exhibition, Broadlands, Romsey SO51 9ZD • T: +44 1794 505010 • F: 518605 • admin@broadlands.net • www.broadlands.net •
Historical Museum
Lives and careers of Lord and Lady Mountbatten 45512

Rosemarkie

Groam House Museum, High St, Rosemarkie IV10 8UF • T: +44 1381 620961 • F: 621730 • groamhouse@ecosse.net •
Cur.: *Susan Seright* •
Archaeology Museum – 1989
13 Pictish sculptured stones & Rosemarkie Pictish cross-slab, artistic impressions of Ross and Cromarty Pictish stones, photographs of all the Pictish carved stones in Scotland, celtic art by George Bain 45513

Roslin

Rosslyn Chapel, Roslin EH25 9PU • T: +44 131 4402159 • F: 4401979 • rosslych@aol.com • www.rosslynchapel.org.uk •
Dir.: *Colin Glynne-Percy* •
Historic Site 45514

Rossendale

Rossendale Museum, Whitaker Park, Haslingden Rd, Rossendale BB4 6RE • T: +44 1706 217777, 244682 • F: 250037 • rossendalemuseum@btconnect.com •
Cur.: *Sandra Cruise* •
Local Museum – 1902
Local and natural hist, fine and decorative arts 45515

Rothbury

Cragside House, Rothbury NE65 7PX • T: +44 1669 620150 • F: 620066 • ncrvmx@smtp.ntrust.org.uk •
Decorative Arts Museum
Original furniture and fittings, stained glass, earliest wallpaper, Lord Armstrong's first hydro-electric lighting 45516

Rotherham

Clifton Park Museum, Clifton Ln, Rotherham S65 2AA • T: +44 1709 336633 • F: 336628 • cliftonparkmuseum@rotherham.gov.uk • www.rotherham.gov.uk •
Dir.: *Di Billups* •
Local Museum – 1893
Rockingham porcelain, Roman archaeology 45517

Rotherham Art Gallery, Walker Pl, Rotherham S65 1JH • T: +44 1709 336633 • F: 336628 • wendyfoster@rotherham.gov.uk • www.roterham.gov.uk •
Dir.: *David Gilbert* •
Public Gallery – 1893
Fine and decorative art 45518

York and Lancaster Regimental Museum, Arts Centre, Walker Pl, Rotherham S65 1JH • T: +44 1709 336633 • F: 336628 • steve.blackbourn@rotherham.gov.uk • www.rotherham.gov.uk •
Dir.: *Di Billups* •
Military Museum – 1930
Uniform, orders, medals and effects of field marshal Viscount H.C.O. Plumer of Messines 45519

Rothesay

Bute Museum, Stuart St, Rothesay PA20 0EP • T: +44 1700 505067 • ivor@butemuseum.fsnet.co.uk •
Cur.: *Ivor Gibbs* •
Local Museum – 1905
Local archaeology, geology, mineralogy, natural and social hist 45520

Royston

Royston and District Museum, Lower King St, Royston SG8 5AL • T: +44 1763 242587 • curator@roystonmuseum.org.uk • www.roystonmuseum.org.uk •
Cur.: *Carole Kaszak* •
Local Museum
Local archaeology, social history, local art 45521

Wimpole Hall and Home Farm, Arrington, Royston SG8 0BW • T: +44 1223 206000 • wimpolehall@nationaltrust.org.uk • www.wimpole.org •
Local Museum / Agriculture Museum 45522

Ruddington

Nottingham Transport Heritage Centre, Mere Way, Ruddington NG11 6NX • T: +44 115 9405705 • F: 9405905 • www.nthc.org.uk •
Dir.: *Alan Kemp* •
Science&Tech Museum
Steam railway and bus, signal boxes and sundry railwayana 45523

Ruddington Framework Knitters' Museum, Chapel St, Ruddington NG11 6HE • T: +44 115 9846914 • F: 9841174 • jack@smirfitt.demon.co.uk • www.rfkm.org •
Pres.: *Prof. Stanley Chapman* •
Special Museum – 1971
Textile machinery 45524

Ruddington Village Museum, Saint Peters Rooms, Church St, Ruddington NG11 6HA • T: +44 115 9146645 • enquiries@ruddingtonlms.org.uk • www.ruddingtonlms.org.uk •
Cur.: *Gavin Walker* •
Local Museum – 1968
Village history, fish and chips shop, pharmacy 45525

Rufford

Rufford Old Hall, nr Ormskirk, Rufford L40 1SG • T: +44 1704 821254 • F: 821254 • rrufoh@smtp.ntrust.org.uk • www.nationaltrust.org.uk •
Man.: *Nick Brooks* •
Historic Site – 1936
Antique furniture, tapestries, armaments 45526

Rugby

Rugby Art Gallery and Museum, Little Gborow St, Rugby CV21 3BZ • T: +44 1788 533201 • F: 533204 • rugbyartgallery&museum@rugby.gov.uk • www.rugbygalleryandmuseum.org.uk •
Dir.: *Wendy Parry* •
Local Museum – 2000
Rugby coll, British art (20th c), Tripontium 45527

Rugby School Museum, 10 Little Church St, Rugby CV21 3AW • T: +44 1788 556109 • F: 556228 • museum@rugbyschool.net •
Man.: *Rusty MacLean* •
Historical Museum
School hist and artefacts, early Rugby Football memorabilia, art colls 45528

Rugeley

Brindley Bank Pumping Station and Museum, Wolseley Rd, Rugeley WS15 2EU • T: +44 1922 38282 •
Science&Tech Museum
Flowmeters, pumps, maps, documents and other items illustrating the history of the Waterworks and the public water supply in the area 45529

Runcorn

Norton Priory Museum, Manor Park, Tudor Rd, Runcorn WA7 1SX • T: +44 1928 569895 • F: 589743 • info@nortonpriory.org • www.nortonpriory.org •
Dir.: *Steven Miller* •
Religious Arts Museum – 1975
Medieval decorated floor tiles, medieval carved stonework, remains of medieval priory, medieval mosaic tile floor (70 sq.m.), statue of St. Christopher, 12th c undercroft, contemporary sculpture – priory remains, gardens 45530

Rustington

Rustington Heritage Exhibition Centre, 34 Woodlands Av, Rustington BN16 3HB • T: +44 1903 784792 •
Local Museum
Social history 45531

Ruthin

Ruthin Goal, 46 Clwyd St, Yr Hen Garchar, Ruthin LL15 1HP • T: +44 1824 708281 • F: 708258 • debbie.snow@denbighshire.gov.uk •
Special Museum
Prison from 1654 to 1946 45532

Ruthwell

Savings Banks Museum, Ruthwell DG1 4NN • T: +44 1387 870640 • info@savingsbanksmuseum.co.uk • www.savingsbanksmuseum.co.uk •
Cur.: *Linda Williams* •
Special Museum – 1974
History of savings banks movement, coll of money boxes, bank memorabilia, familiy history, Runic stone – archive 45533

Ryde, Isle of Wight

Cothey Bottom Heritage Centre, Westridge, Brading Rd, Ryde, Isle of Wight PO33 1QS • T: +44 1983 814875 • F: 823841 • paul.simpson@iow.gov.uk •
Cons.: *Paul Simpson* •
Local Museum 45534

Rye

Rye Art Gallery, Ockman Ln and East St, Rye TN31 7JY • T: +44 1797 223218, 222433 • F: 225376 • www.ryeartgallery.co.uk •
Man.: *Sarah Money* •
Fine Arts Museum / Decorative Arts Museum / Public Gallery
Contemporary fine art and craft 45535

Rye Castle Museum, 3 East St, Rye TN31 7JY • T: +44 1797 226728 •
Cur.: *Allan Downend* •
Local Museum – 1928
Medieval Rye pottery, modern Rye pottery, prints and drawings local scenes, costumes, cinque ports regalia, 18th c fire engine, toys, maritime history of Rye 45536

Rye Heritage Centre Town Model, Strand Quay, Rye TN31 7AY • T: +44 1797 226696 • F: 223460 • ryetic@rother.gov.uk • www.visitrye.co.uk •
Dir.: *J. Arkley* •
Local Museum 45537

Saffron Walden

Audley End House, Saffron Walden CB11 4JF • T: +44 1799 522842 • F: 521276 • www.english-heritage.org.uk •
Cur.: *Gareth Hughes* •

Decorative Arts Museum
House history, Howard, Neville and Cornwallis coll, early neo-classical furniture, British Portraits, natural history, silver coll 45538

Fry Art Gallery, Bridge End Gardens, Castle St, Saffron Walden CB10 1BD • T: +44 1799 513779 • F: 520212 • gcummings@totalise.co.uk • www.fryartgallery.org •
Dir.: *Nigel Weaver* •
Fine Arts Museum – 1985
Works by 20th c artists from the area of North West Essex, incl Ravilious, Bawden, Rothenstein, Aldridge, Rowntree, Cheese, Hoyle a.o. 45539

Saffron Walden Museum, Museum St, Saffron Walden CB10 1JL • T: +44 1799 510333 • F: 510334 • museum@uttlesford.gov.uk • www.saffronwaldenmuseum.org.uk •
Cur.: *Carolyn Wingfield* • Sc. Staff: *Sarah Kenyon* (Natural History) • Cons.: *Lynn Morrison* •
Local Museum – 1832
Local archaeology, hilt of a 6th c ring sword, unique viking pendant necklace from saxon walden cemetery, social and natural hist, geology, ethnography, ceramics, glass, costumes, furniture, dolls 45540

Saint Albans

Clock Tower, Market Pl, Saint Albans AL3 5DR • T: +44 1727 751810, 751820 • F: 859919 • museum@stalbans.gov.uk • www.stalbansmuseums.org.uk •
Historical Museum
A 15th-c town belfry 45541

De Havilland Aircraft Heritage Centre, Mosquito Aircraft Museum, Salisbury Hall, London Colney, Saint Albans AL2 1EX • T: +44 1727 822051, 826400 • F: 826400 • wa050@dhamt.freeserve.co.uk • www.dehavillandmuseum.co.uk •
Dir.: *Ralph Steiner* •
Science&Tech Museum – 1959
Displays of de Havilland aircraft, Mosquito prototype – Library, photo archive 45542

Hypocaust, Verulamium Park, Saint Albans AL3 4SW • T: +44 1727 751810 • F: 859919 • museum@stalbans.gov.uk • www.stalbansmuseums.org.uk •
Dir.: *Chris Green* •
Historic Site
Roman hypocaust system and mosaic floor 45543

Kingsbury Watermill Museum, Saint Michaels Village, Saint Albans AL3 4SJ • T: +44 1727 853502 • F: 730459 •
Science&Tech Museum / Agriculture Museum
Working watermill, coll of old fram implements 45544

Kyngston House Museum, Inkerman Rd, Saint Albans AL1 3BB • T: +44 1727 819338 • F: 836282 • museum@stalbans.gov.uk • www.stalbansmuseums.org.uk •
Local Museum / Archaeology Museum
Social hist coll, archaeology 45545

Margaret Harvey Gallery, c/o Faculty of Art & Design, University of Herfordshire, 7 Hatfield Rd, Saint Albans AL1 3RS • T: +44 1707 285376 • F: 285310 • s.moore@herts.ac.uk • perseus.herts.ac.uk/schools/art/uhgalleries •
Dir.: *Chris McIntyre* •
Public Gallery 45546

Museum of Saint Albans, Hatfield Rd, Saint Albans AL1 3RR • T: +44 1727 819340 • F: 837472 • museum@stalbans.gov.uk • www.stalbansmuseums.org.uk •
Dir.: *Chris Green* •
Local Museum – 1898
Salaman coll of trade and craft tools, social and local history, natural sciences, temporary exhibitions 45547

Roman Theatre of Verulamium, Gorhambury Dr, off Bluehouse Hill, Saint Albans AL3 6AH • T: +44 1727 835035 • F: 843675 • stalbans@struttandparker.co.uk • www.romantheatre.co.uk •
Cust.: *Dinah Ryle Bagnall* •
Performing Arts Museum
Theatre discovered on the Gorhambury estate in 1847, fully excavated by Dr Kathleen Kenyon in 1935, first constructed in AD 160, the theatre was altered and modified during 2 centuries and was used for religious processions, ceremonies and plays 45548

Saint Albans Organ Museum, 320 Camp Rd, Saint Albans AL1 5PE • T: +44 1727 869693, 768652 • F: 768651 • keithpinner@fsbdial.co.uk • www.stalbansorgantheatre.org.uk •
Chm.: *Keith Pinner* •
Music Museum – 1959
Two theatre pipe organs, four Belgian dance organs, music boxes, mills violano virtuoso, player pianos, player reed organs 45549

Verulamium Museum, Saint Michael's, Saint Albans AL3 4SW • T: +44 1727 751810 • F: 859919 • museum@stalbans.gov.uk • www.stalbansmuseums.org.uk •
Dir.: *Chris Green* •
Archaeology Museum – 1939
Iron Age & Roman, excavated at Verulamium, c. 100 BC-AD 450, Roman hypocaust 45550

Saint Andrews

Bell Pettigrew Museum, c/o University of Saint Andrews, Bute Medical Bldg, Queens Terrace, Saint Andrews KY16 9TS • T: +44 1334 463608 • F: 462401 • mjm5@st-and.ac.uk • biology.st-and.ac.uk/sites/bellpet •
Cur.: *Dr. M.J. Milner* •
Natural History Museum / University Museum 45551

British Golf Museum, Bruce Embankment, Saint Andrews KY16 9AB • T: +44 1334 460046 • F: 460064 • judychance@randa.org • www.britishgolfmuseum.co.uk •
Cur.: *Angela Howe* •
Special Museum
Golf hist in Britain since the middle ages, clubs, balls, trophies, costume, paintings 45552

Crawford Arts Centre (closed) 45553

Saint Andrews Cathedral Museum, The Pends, South St, Saint Andrews KY16 9QU • T: +44 1334 472563 •
Religious Arts Museum – 1950
Early Christian crosses, medieval cathedral relics, pre-Reformation tomb stones, 9th-10th c sarcophagus 45554

Saint Andrews Museum, Doubledykes Rd, Saint Andrews KY16 9DP • T: +44 1334 412690 • F: 413214 •
Cur.: *Gavin Grant* •
Local Museum
Social hist, archaeology 45555

Saint Andrews Preservation Museum, 12-16 North St, Saint Andrews KY16 9PW • T: +44 1334 477629 • trust@standrewspreservationtrust.org.uk • www.standrewspreservationtrust.co.uk •
Cur.: *Rachel Clerke* •
Local Museum
Social and industrial hist 45556

Saint Andrews University Museum Collections, University of Saint Andrews, Saint Andrews KY16 9TR • T: +44 1334 462417 • F: 462401 • hcr1@st-andrews.ac.uk • www.st-and.ac.uk/services/muscoll •
Keeper: *Prof. Ian A. Carradice* • Cur.: *Helen Rawson* •
Natural History Museum / University Museum
Fine and applied art, silver, furniture, anatomy, pathology, chemistry, ethnography, geology, psychology, zoology, natural history, Amerindians, scientific apparatus and instruments 45557

Scotland's Secret Bunker, Troywood, Saint Andrews KY16 8QH • T: +44 1333 310301 • F: 312040 • mod@secretbunker.co.uk • www.secretbunker.co.uk •
Man.: *Brian Light* •
Historical Museum
Underground secret nuclear command bunker 45558

Saint Asaph

Saint Asaph Cathedral Treasury, The Cathedral, Saint Asaph LL19 0RD • T: +44 1745 583429 • F: 583566 • karenwilliams@churchinwales.org.uk • www.stasaphcathedral.org.uk •
Religious Arts Museum
Relics relating to the Clwyd area 45559

Saint Austell

China Clay Country Park Mining and Heritage Centre, Wheal Martyn, Carthew, Saint Austell PL26 8XG • T: +44 1726 850362 • F: 850362 • info@chinaclaycountry.co.uk • www.chinaclaycountry.co.uk •
Dir.: *David Owens* • Cur.: *Elisabeth Chald* •
Historical Museum
China clay industry, communuties arround the quarries, machinery, Victorian clay works, water wheels – trails 45560

Saint Brelade

Noirmont Command Bunker, Noirmont Point, Saint Brelade JE3 8JA, mail addr: M.P. Costard, Rte de Vinchelez, Saint Ouen JE3 2DJ • T: +44 1534 746795 • F: 838001 • m.costard@spoor.co.uk • wwww.ciosjersey.org.uk •
Dir.: *Matthew P. Costard* •
Historical Museum / Military Museum – 1977
German artillery command bunker 45561

Saint Dominick, Cornwall

Cotehele House, Saint Dominick, Cornwall PL12 6TA • T: +44 1579 351346 • F: 351222 • cotehele@nationaltrust.org.uk • www.nationaltrust.org.uk •
Special Museum
Armour, tapestries and furniture 45562

Shamrock and Cotehele Quay Museum, National Maritime Museum Outstation, Cotehele Quay, Saint Dominick, Cornwall PL12 6TA • T: +44 1579 350830 • www.nmm.ac.uk •
Keeper: *Peter Allington* •
Science&Tech Museum – 1979
Historic sailing barge Shamrock 45563

Saint Dominick, Jersey

Jersey Heritage Trust, Clarence Rd, Saint Dominick, Jersey JE3 5HQ • T: +44 1534 833300 • F: 833301 • archives@jerseyheritagetrust.org • www.jerseyheritagetrust.org •
Association with Coll – 1993
Royal Court, H E Lieutenant Governor, parishes, churches, societies and individuals relating to the island 45564

Saint Fagans

Museum of Welsh Life, Saint Fagans CF5 6XB • T: +44 29 20573500 • F: 20573490 • mwl@nmgw.ac.uk • www.nmgw.ac.uk/mwl •
Dir.: *John Williams-Davies* • Keeper: *Dr. Beth Thomas* •
Open Air Museum / Folklore Museum / Agriculture Museum / Ethnology Museum – 1948
Farmhouses, cottages, Victorian shop complex, tollhouse, tannery, smithy, corn mill, woollen mill, bakehouse, pottery 45565

Saint Helens

The Citadel, Waterloo St, Saint Helens WA10 1PX • T: +44 1744 735436 • F: 20836 • marketing@citadel.org.uk • www.citadel.org.uk •
Dir.: *Morag Murchison* •
Public Gallery 45566

North West Museum of Road Transport, 51 Hall St, Saint Helens WA10 1DU • T: +44 1744 451681 • general@hallstreetdepot.info • www.hallstreetdepot.info •
Head: *G. Sandford* •
Science&Tech Museum
Historic buses, cars, lorries, vans a.o. historic road vehicles, former steam tram shed 45567

World of Glass, Chalon Way E, Saint Helens WA10 1BX • T: +44 1744 22766 • F: 616966 • info@worldofglass.com • www.worldofglass.com •
Cur.: *Joanna Hayward* •
Decorative Arts Museum / Local Museum
History of glass manufacture and design, local glass industry, local hist, oils and waterclours 45568

Saint Helier

Barreau Le Maistre Art Gallery, Jersey Museum, The Weighbridge, Saint Helier JE2 3NF • T: +44 1534 633300 • F: 633301 • museum@jerseyheritagetrust.org • www.jerseyheritagetrust.org •
Dir.: *Jon Carter* •
Fine Arts Museum 45569

Elizabeth Castle, Saint Aubins Bay, Saint Helier JE2 6QN • T: +44 1534 723971 • F: 610338 • museum@jerseyheritagetrust.org • www.jerseyheritagetrust.org •
Dir.: *Jonathan Carter* •
Fine Arts Museum / Local Museum / Military Museum / Historic Site – 1923
Jersey militia, numismatic coll, silver, German occupation, regional hist, hist. tableaux, paintings by Jersey Artists 45570

Jersey Museum, The Weighbridge, Saint Helier JE2 3NF • T: +44 1534 633300 • F: 633301 • museum@jerseyheritagetrust.org.uk • www.jerseyheritagetrust.org •
Dir.: *Jonathan Carter* •
Local Museum – 1992
Archaeology, German occupation, prison hist, rural life, finance, art 45571

Jersey Photographic Museum, Hôtel de France, Saint Saviour Rd, Saint Helier JE1 7PX • T: +44 1543 634894 • rolleiclubjersy@localdial.com •
Science&Tech Museum
Cameras, images, processing, history of photography since 1840 45572

Occupation Tapestry Gallery and Maritime Museum, New North Quay, Saint Helier JE2 3ND • T: +44 1534 811043 • F: 874099 • museum@jerseyheritagetrust.org • www.jerseyheritagetrust.org •
Dir.: *Jon Carter* •
Decorative Arts Museum – 1996
Twelve-panel tapestry depicting the occupation of Jersey during WWII by Germany 45573

Saint Ives, Cambridgeshire

Norris Museum, 41 The Broadway, Saint Ives, Cambridgeshire PE17 5BX • T: +44 1480 497314 • bob@norrismuseum.fsnet.co.uk •
Cur.: *R.I. Burn-Murdoch* •
Local Museum – 1931
Archaeology of all periods, French prisoner-of-war material from Norman Cross, ice-skates, 18th c fire engine, local prints, paintings, newspapers, paleolithic finds, lace production – library 45574

Saint Ives, Cornwall

Barbara Hepworth Museum and Sculpture Garden, Barnoon Hill, Saint Ives, Cornwall TR26 1TG • T: +44 1736 796226 • F: 792189 • visiting.stives@tate.org.uk • www.tate.org.uk/stives •
Dir.: *Martin Clark* • Press: *Arwen Fitch* • Learning: *Susan Lamb* •
Fine Arts Museum – 1976
A selection of sculptures devoted to the work of Barbara Hepworth (1903-75) 45575

Penwith Galleries, Back Rd W, Saint Ives, Cornwall TR26 1NL • T: +44 1736 795579 •
Cur.: *Kathleen Watkins* •
Fine Arts Museum – 1949
Paintings, sculpture, pottery 45576

Saint Ives Museum, Wheal Dream, Saint Ives, Cornwall TR26 1PR • T: +44 1736 796005 •
Cur.: *B. Stevens* •
Local Museum – 1951
Local industries, arts, crafts, fishing, mining, paintings, folklore 45577

Saint Ives Society of Artists Members Gallery (Norway Gallery) and Mariners Gallery, Old Mariners Church, Norway Sq, Saint Ives, Cornwall TR26 1NA • T: +44 1736 795582 • gallery@stisa.co.uk • www.stisa.co.uk •
Dir.: *Nicola Tilley* • Man.: *April Brooks* •
Public Gallery – 1926
Paintings, sculpture, cards, prints 45578

Tate Saint Ives, Porthmeor Beach, Saint Ives, Cornwall TR26 1TG • T: +44 1736 796226 • F: 792189 • visiting.stives@tate.org.uk • www.tate.org.uk/stives •
Dir.: *Martin Clark* • Press: *Arwen Fitch* • Learning: *Susan Lamb* •
Fine Arts Museum – 1993
Intern Modern and contemporary art including works from the Tate Coll 45579

Saint Lawrence

Hamptonne Country Life Museum, Rue de la Patente, Saint Lawrence JE3 1HG • T: +44 1534 863955 • F: 863935 • museum@jerseyheritagetrust.org • www.jerseyheritagetrust.org •
Dir.: *Jon Carter* •
Agriculture Museum – 1993
Restored farm complex tracing its hist through centuries of development 45580

Jersey War Tunnels, Les Charrières Malorey, Meadowbank, Saint Lawrence JE3 1FU • T: +44 1534 860808 • F: 860886 • info@jerseywartunnels.com • www.jerseywartunnels.com •
Dir.: *Paul Simmonds* •
Military Museum
Wartime occupation, arms 45581

Saint Margaret's Bay

Saint Margaret's Museum, Beach Rd, Saint Margaret's Bay CT15 6DZ • T: +44 1304 852764 • F: 853626 • www.baytrust.org.uk •
Cur.: *Colette Fronty* •
Local Museum
Marine and local hist, ships' badges 45582

Saint Martin

Mont Orgueil Castle, Le Mont de Gouray, Saint Martin JE3 6ET • T: +44 1534 853292 • F: 854303 • museum@jerseyheritagetrust.org • www.jerseyheritagetrust.org •
Dir.: *Jon Carter* •
Historic Site / Archaeology Museum – 1907
Medieval castle, local archaeological finds 45583

Saint Mary's

Isles of Scilly Museum, Church St, Saint Mary's TR21 0JT • T: +44 1720 422337 • F: 422337 • info@iosmuseum.org • www.iosmuseum.org •
Cur.: *Amanda Martin* •
Local Museum
Scilly archaeology, hist, shipwrecks, birds and lichens 45584

Saint Neots

Saint Neots Museum, 8 New St, Old Court, Saint Neots PE19 1AE • T: +44 1480 388921 • F: 388791 • stneotsmuseum@tiscali.co.uk •
Cur.: *Anna Mercer* •
Local Museum
Local hist, local archaeology, crafts and trades, home and community life 45585

Saint Osyth

East Essex Aviation Society and Museum of the 40's, Martello Tower, Point Clear, Saint Osyth CO15 4JD, mail addr: c/o R. Barrell, 37 Brookland Rd, Brantham CO11 1RP •
Historical Museum
Military hist 45586

Saint Ouen

Channel Islands Military Museum, The Five Mile Rd, Saint Ouen JE2 4SL • T: +44 1534 723136 • damienhorn@jerseymail.co.uk •
Owner: *Damian Horn* • *Ian Cabot* •
Historical Museum / Military Museum
Civilian and military items (June 1940-May 1945), WW2 German optical equipment, weapons, enigma decoding machine, Axis military motorcycles 45587

Jersey Battle of Flowers Museum, La Robeline, Mont des Corvees, Saint Ouen JE3 2ES • T: +44 1534 482408 •
Cur.: *Florence Bechelet* •
Special Museum
Animals and birds, made with dried wildflowers 45588

Saint Peter Port

Castle Cornet Military and Maritime Museums, Castle Emplacement, Saint Peter Port GY1 1AU • T: +44 1481 721657 • F: 714021 • admin@museum.guernsey.net • www.museums.gov.gg •

Dir.: *Peter M. Sarl* •
Military Museum / Historical Museum / Historic Site – 1950
Royal militia coll 45589

Coach-House Gallery, Les Islets, Saint Peter Port GY7 9ES • T: +44 1481 265339 •
Public Gallery
Works by contemporary British artists 45590

Guernsey Museum and Art Gallery, Candie Gardens, Saint Peter Port GY1 1UG • T: +44 1481 726518 • F: 715177 • admin@museum.guernsey.net • www.museum.guernsey.net •
Dir.: *Peter M. Sarl* • Sc. Staff: *Matt Harvey* (Social History) • *Heather Sebire* (Archaeology) • *Alan C. Howell* (Natural History) • *Paul Le Tissier* (Design Development) • *Helen Conlon* (Fine Arts) • *Lynne Ashton* (Education) •
Archaeology Museum / Local Museum / Fine Arts Museum
Natural hist, local watercolours, Frederick Corbin Lukis, Joshua Gosselin 45591

Hauteville House, Musées de la Ville de Paris, 38 Hauteville, Saint Peter Port GY1 1DG • T: +44 1481 721911 • F: 715913 • hugohouse@cwgsy.net •
Dir.: *Danielle Molinari* • Admin.: *Odile Blanchette* •
Special Museum
Exile House of Victor Hugo (1802-1885), in which he lived from 1856 to 1870, decorated in his very own unique style, using tapestries, Delft tiles carvings, chinoiseries, etc – garden 45592

Royal Air Force Museum 201 Squadron, Castle Cornet, Saint Peter Port GY1 1AU • T: +44 1481 721657 • F: 714021 • admin@museum.guernsey.net • www.museums.gov.gg •
Dir.: *Peter M. Sarl* •
Military Museum – 2001
Hist of Guernsey's own Squadron 45593

Story of Castle Cornet, Castle Cornet, Saint Peter Port GY1 1AU • T: +44 1481 721657 • F: 714021 • admin@museum.guernsey.net • www.museums.gov.gg •
Dir.: *Peter M. Sarl* • Cur.: *Helen Colon* •
Fine Arts Museum – 1997
Objects and paintings depicting the hist of the Castle 45594

Saint Pierre du Bois

Fort Grey and Shipwreck Museum, Rocquaine Bay, Rte de la Lague, Saint Pierre du Bois GY7 9BY • T: +44 1481 265036 • F: 263279 • admin@museum.guernsey.net • www.museum.guernsey.net •
Dir.: *P.M. Sarl* •
Military Museum / Historical Museum – 1976
Shipwrecks, Martello tower 45595

Salcombe

Overbeck's Museum, Sharpiter, Salcombe TQ8 8LW • T: +44 1548 842893 • F: 845020 • overbecks@nationaltrust.org.uk • www.nationaltrust.org.uk •
Man.: *Victoria Pepper* •
Historical Museum – 1938
Secret childrens room, doll coll, Britain's lead soldier coll, local shipbuilding tools and other nautical artefacts 45596

Salcombe Maritime Museum, Town Hall, Market St, Salcombe TQ8 8DE • T: +44 1548 843080 •
Resp.: *Yvonne Paul* •
Historical Museum – 1974
Shipwreck, shipbuilding, smuggling 45597

Salford

Chapman Gallery, University of Salford, Crescent, Salford M5 4WT • T: +44 161 7455000 ext 3219 •
Public Gallery 45598

The Lowry Gallery, Quays, Pier 8, Salford M50 3AZ • T: +44 161 8762020 • F: 8762021 • info@thelowry.com • www.thelowry.com •
Dir.: *Lindsay Brooks* • Cur.: *Ruth Salisbury* •
Fine Arts Museum
Coll of paintings and drawings by L.S. Lowry 45599

Ordsall Hall Museum, Ordsall Ln, Salford M5 3AN • T: +44 161 8720251 • F: 8724951 • ordsall.hall@salford.gov.uk • www.salford.gov.uk/museums •
Dir.: *Samantha Smith* •
Folklore Museum – 1972
Folk life, leather figures, sword made in Solingen 45600

Salford Museum and Art Gallery, The Crescent, Peel Park, Salford M5 4WU • T: +44 161 7780800 • F: 7459490 • salford.museum@salford.gov.uk • www.salford.gov.uk/museums •
Dir.: *John Sculley* • Cur.: *Nicola Power* (Heritage Service) • Sc. Staff: *Julie Allsop* (Heritage Development) • *Peter Ogilive* (Collections) • *Meg Ashworth* (Exhibitions) • *K. Craven* (Research) • *Jo Clarke* (Learning) • *A. Monaghan* (Outreach) •
Local Museum / Fine Arts Museum – 1850
Fine and applied arts, social and local hist 45601

Salisbury

Edwin Young Collection, c/o Salisbury Library and Galleries, Market Pl, Salisbury SP1 1BL • T: +44 1722 324145, 343277 • F: 413214 • peterriley@wiltshire.gov.uk •
Cur.: *Peter Riley* •
Fine Arts Museum
Victorian and Edwardian oils and watercolours, contemporary paintings 45602

John Creasey Museum, c/o Salisbury Library and Galleries, Market Pl, Salisbury SP1 1BL • T: +44 1722 324145 • F: 413214 • peterriley@wiltshire.gov.uk •
Cur.: *Peter Riley* •
Fine Arts Museum
John Creasey's work, a local author - all editions, all languages, all pseudonyms (23 in all), contemporary art 45603

Mompesson House, The Close, Salisbury SP1 2EL • T: +44 1722 420980, 335659 • F: 321559 • mompessonhouse@nationaltrust.org.uk • www.nationaltrust.org.uk •
Man.: *Karen Rudd* •
Fine Arts Museum / Decorative Arts Museum
18th c drinking glasses, furniture, ceramics 45604

Rifles (Berkshire and Wiltshire) Museum, The Wardrobe, 58 The Close, Salisbury SP1 2EX • T: +44 1722 419419 • F: 421626 • curator@thewardrobe.org.uk • www.thewardrobe.org.uk •
Cur.: *M.J. Cornwell* • *Jackie Dryden* •
Military Museum – 1982
Hist of the County Infantry Regiments of Berkshire and Wiltshire covering more than 200 yrs, including uniforms, weapons, equip, campaign relics, medals and Regimental silver 45605

Salisbury and South Wiltshire Museum, The Kings House, 65 The Close, Salisbury SP1 2EN • T: +44 1722 332151 • museum@salisburymuseum.org.uk • www.salisburymuseum.org.uk •
Dir.: *Adrian Green* • Cur: *Jane Ellis-Schön* •
Local Museum / Archaeology Museum / Decorative Arts Museum – 1860
Archaeology, ceramics, English china, glass, Pitt Rivers coll, Brixie Jarvis coll of Wedgwood, Lace, Salisbury Giant – library 45606

Saltash

Saltash Heritage Centre, 17 Lower Fore St, Saltash PL12 6JQ • T: +44 1752 848466 • www.saltash-heritage.org.uk •
Cur.: *David Kent* •
Local Museum
Local hist 45607

Saltcoats

North Ayrshire Museum, Manse St, Kirkgate, Saltcoats KA21 5AA • T: +44 1294 464174 • F: 464234 • namuseum@north-ayrshire.gov.uk • www.north-ayrshire.gov.uk/museums •
Cur.: *Dr. Martin Bellamy* •
Local Museum – 1957
Local hist, stone carvings, old kitchen 45608

Sandal

Sandal Castle, Manygates Ln, Sandal WF1 5PD • T: +44 1924 249779 • museumsandarts@wakefield.gov.uk • www.wakefield.gov.uk/cultureandleisure •
Historical Museum 45609

Sandford Orcas

Manor House, Sandford Orcas DT9 4SB • T: +44 1963 220206 •
Dir.: *Sir Mervyn Medlycott* •
Fine Arts Museum / Decorative Arts Museum
Queen Anne and Chippendale furniture, 17th c Dutch and 18th c English pictures, coll medieval stained glass 45610

Sandhurst, Berkshire

National Army Museum Sandhurst Departments, Royal Military Academy, Sandhurst, Berkshire GU15 4PQ • T: +44 1276 63344 ext 2457 •
Dir.: *I.G. Robertson* • Cur.: *S.K. Hopkins* •
Military Museum 45611

Sandling

Museum of Kent Life, Lock Ln, Sandling ME14 3AU • T: +44 1622 763936 • F: 662024 • enquiries@museum-kentlife.co.uk • www.museum-kentlife.co.uk •
Dir.: *Nigel Chew* •
Open Air Museum – 1983
Agricultural rural coll mainly from Kent: hopping industry darling buds of May exhibit, life in Kent exhibit, hop, herb and kitchen gardens, oast, granary, barn, farmhouse, farmyard, village hall 45612

Sandown

Dinosaur Isle, Culver Parade, Sandown PO36 8QA • T: +44 1983 404344 • F: 407502 • dinosaur@iow.gov.uk • www.dinosaurisle.com •
Dir.: *Martin Munt* •
Natural History Museum – 1913
Fossils, dinosaurs, local rocks 45613

Sandringham

Sandringham House Museum, Sandringham PE35 6EN • T: +44 1553 772675 • F: +44 1485 541571 • www.sandringhamestate.co.uk •
Dir.: *G. Pattinson* •
Local Museum / Science&Tech Museum
Photogr, commemorative china, Royal Daimler cars, fire engine 45614

Sandwich

Guildhall Museum, Cattle Market, Sandwich CT13 9AN • T: +44 1304 617197 • F: 620170 • sandwichtowncouncil@btopenworld.com • www.sandwichtowncouncil.co.uk •
Cur.: *Ray Harlow* •
Historical Museum – 1930
Victorian photogr by a Sandwich photographer 1869 to 1897, 50 Nazi propaganda photogr of the German Army in the field, as distributed to their Embassies worldwide in the 1939-45 War 45615

Richborough Castle - Roman Fort, Richborough Rd, Sandwich CT13 9JW • T: +44 1304 612013 • www.english-heritage.org.uk •
Dir.: *Tracey Wahdan* • Cur.: *Jo Gray* •
Archaeology Museum – 1930
Roman finds, pottery, tools, coins, ornaments 45616

White Mill Rural Heritage Centre, Ash Rd, Sandwich CT13 9JB • T: +44 1304 612076 • www.sandwich-kent.uk.net •
Head: *R.G. Barber* •
Science&Tech Museum
Agricultural implements, artefacts and craft workers' tools, domestic artefacts, blacksmith forge, models 45617

Sanquhar

Sanquhar Tolbooth Museum, High St, Sanquhar DG4 5BN • T: +44 1659 50186 • info@dumfriesmuseum.demon.co.uk • www.dumfriesmuseum.demon.co.uk •
Dir.: *Robert Martin* •
Local Museum
1735 town house, local hist, geology knitting 45618

Sauchen

Castle Fraser, Sauchen AB51 7LD • T: +44 1330 833463 • F: 833819 • castlefraser@nts.org.uk •
Man.: *Nicola Batrd* •
Historic Site 45619

Saxtead

Saxtead Green Post Mill, Saxtead Green, Saxtead IP13 9QQ • T: +44 1728 685789 • www.english-heritage.org.uk •
Cust.: *Anita Baldry* •
Science&Tech Museum
Milling equipment, tools 45620

Scalloway

Scalloway Museum, Main St, Scalloway ZE1 0TR •
Local Museum
Local objects from Neolithic times to the present, Scalloway's role in the WWII – photographic archive 45621

Scarborough

Crescent Arts, The Crescent, Scarborough YO11 2PW • T: +44 1723 351461 • info@crescentarts.co.uk • www.crescentarts.co.uk •
Head: *Rachel Massey* •
Public Gallery 45622

Rotunda Museum of Archaeology and Local History, Museum Terrace, Vernon Rd, Scarborough YO11 2NN • T: +44 1723 374839 • F: 376941 • scab.meg@pop3.poptel.org.uk • www.scarboroughmuseums.org.uk •
Dir.: *Karen Snowden* •
Archaeology Museum / Local Museum – 1829
Regional archaeological coll, esp mesolithic, Bronze Age and medieval, fine surviving example of early purpose-built museums 45623

Scarborough Art Gallery, The Crescent, Scarborough YO11 2PW • T: +44 1723 374753 • F: 376941 • www.scarboroughmuseums.org.uk •
Dir.: *Louise Marr* •
Fine Arts Museum – 1947
Paintings (17th-20th c), local paintings (19th c), contemporary prints 45624

Wood End Museum, Londesborough Lodge, The Crescent, Scarborough YO11 2PW • T: +44 1723 367326 • F: 376941 • www.scarboroughmuseums.org.uk •
Dir.: *Karen Snowden* •
Natural History Museum – 1951
Regional fauna, flora and geology, entomology, conchology, vertebrates 45625

Scunthorpe

Normanby Hall, Normanby Hall Country Park, Scunthorpe DN15 9HU • T: +44 1724 720588 • normanby.hall@northlincs.gov.uk • www.northlincs.gov.uk/museums •
Dir.: *Susan Hopkinson* •
Decorative Arts Museum – 1964
Period furniture, decorations, costume galleries – Normanby park farming museum 45626

Normanby Park Farming Museum, Normanby Hall Country Park, Scunthorpe DN15 9HU • T: +44 1724 720588 • normanby.hall@northlincs.gov.uk • www.northlincs.gov.uk/museums •
Dir.: *Susan Hopkinson* •
Agriculture Museum – 1989
Rural crafts and industry, transport, agriculture – Normanby Hall 45627

North Lincolnshire Museum, Oswald Rd, Scunthorpe DN15 7BD • T: +44 1724 843533 • F: 270474 • steve.thompson@northlincs.gov.uk • www.northlincs.gov.uk/museums •
Cur.: *K.A. Leahy* • Keeper: *Steve Thompson* • Sc. Staff: *Adam Smith* •
Local Museum – 1909
Geology, hist and natural hist, archaeology 45628

Seaford

Seaford Museum of Local History, Martello Tower, Esplanade, Seaford BN25 1JH • T: +44 1323 898222 • museumseaford@tinyonline.co.uk • www.seafordmuseum.org •
Local Museum – 1970
Housing register, huge photographic coll, Connie Brewer coll, television and radio 45629

Seaton

Seaton Delaval Hall, Estate Office, Seaton NE26 4QR • T: +44 191 2371493 •
Owner: *Lord Hastings* •
Decorative Arts Museum
Pictures and documents 45630

Seething

448th Bomb Group Memorial Museum, Airfield, Seething NR2 4HB • T: +44 1603 614041 •
Military Museum
Photos, memorabilia, artefacts, a restored WW II control tower 45631

Selborne

Gilbert White's House and the Oates Museum, The Wakes, Selborne GU34 3JH • T: +44 1420 511275 • F: 511040 • info@gilbertwhiteshouse.org.uk • www.gilbertwhiteshouse.org.uk •
Dir.: *Maria Newbery* •
Special Museum – 1955
Restored 18th c home of Gilbert White, travels of Frank Oates in Africa, Cpt. L. Oates Antarctic fame – library 45632

Romany Folklore Museum, Selborne GU34 3JW • T: +44 1420 511486 •
Folklore Museum
Coll of vehicles, the early history, language, music, dress and crafts of the gypsies 45633

Selkirk

Bowhill Collection, Selkirk TD7 5ET • T: +44 1750 722204 • F: 722204 • bht@buccleuch.com • www.heritageontheweb.co.uk •
Cur.: *Duke of Buccleuch* •
Decorative Arts Museum / Fine Arts Museum
Art coll, French furniture, silver, porcelain, relicts of Duke of Monmouth, Queen Victoria ans Sir Walter Scott 45634

Halliwell's House Museum, Halliwell's Close, Market Pl, Selkirk TD7 4BC • T: +44 1750 20096, 20054 • F: 23282 • museums@scotborders.gov.uk •
Local Museum – 1984
Late 19th/early 20thc domestic ironmongery 45635

James Hogg Exhibition, Bowhill, Selkirk TD7 5ET • T: +44 1750 722204 • F: 722204 • bht@buccleuch.com •
Cur.: *Duke of Buccleuch* •
Special Museum
Life and work of the Scottish writer James Hogg 45636

Robson Gallery, Halliwells House Museum, Market Pl, Selkirk TD7 4BC • T: +44 1750 20096 • F: 23282 • museums@scotborders.gov.uk •
Dir.: *Ian Brown* •
Public Gallery 45637

Sir Walter Scott's Courtroom, Town Hall, Market Pl, Selkirk TD7 4JX • T: +44 1750 720096 • F: 723282 • museums@scotborders.gov.uk •
Dir.: *Fiona Colton* •
Local Museum – 1994
Civic hist, Sir Walter Scott, Sheriff of the County 45638

Selsey

Selsey Lifeboat Museum, Kingsway, Selsey PO20 0DL • T: +44 1243 602387 • F: 607790 • lauramariner@btinternet.com • www.selseylifeboats.co.uk •
Historical Museum
Lifeboat memorabilia 45639

U

Settle

Museum of North Craven Life, The Folly, Settle BD24 9EY • T: +44 15242 51388 • earead@clara.co.uk • www.ncbpt.org.uk/folly •
Local Museum 45640

Sevenoaks

Knole, Sevenoaks TN15 0RP • T: +44 1732 462100, 450608 • F: 465528 • knole@nationaltrust.org.uk • www.nationaltrust.org.uk •
Decorative Arts Museum
State rooms, paintings, furniture, tapestries, silver, built in 1456 45641

Sevenoaks Museum and Gallery, Library, Buckhurst Ln, Sevenoaks TN13 1LQ • T: +44 1732 605210 • F: 605221 • elizabeth.ritchie@kent.gov.uk •
Cur.: *Susan Gosling* •
Local Museum
Local geology, history, archaeology, local fossils, paintings 45642

Shackerstone

Shackerstone Railway Museum, Shackerstone Station, Shackerstone CV13 6NW • T: +44 1827 880754 • F: 881050 • www.battlefield-line-railway.co.uk •
Pres.: *John C. Jacques Mbe* •
Science&Tech Museum
Coll of railway items, signalbox equipment, station signs, timetables from 1857 and dining-car crockery, eighteen paraffin lamps, a coll of WWII-railway posters 45643

Shaftesbury

Shaftesbury Abbey Museum and Garden, Park Walk, Shaftesbury SP7 8JR • T: +44 1747 852910 • F: 852910 • user@shaftesburyabbey.fsnet.co.uk • www.shaftesburyabbey.co.uk •
Dir.: *Anna McDowell* •
Archaeology Museum – 1951
Tiles, stone finds, Saxon relics, reconstruction of old church 45644

Shaftesbury Town Museum, Gold Hill, Shaftesbury SP7 8JW • T: +44 1747 852157 • sdhs@waitrose.com •
Local Museum 45645

Shallowford

Izaak Walton's Cottage, Worston Ln, Shallowford ST15 0PA • T: +44 1785 760278, 619131 • info@staffordbc.gov.uk • www.staffordbc.gov.uk/izaak-waltons-cottage •
Man.: *Gillian Bould* •
Local Museum
Local history, fishing, material on Izaak Walton (1593-1683) author of 'The Compleat Angler' 45646

Sheerness

Minster Abbey Gatehouse Museum, Minster, Union Rd, Sheerness ME12 2HW • T: +44 1795 872303, 661119 •
Chm.: *Jonathan Fryer* •
Local Museum
Local memorabilia, fossils, tools, costume, telephones, radios, photographs, toys 45647

Sheffield

Abbeydale Industrial Hamlet, Abbeydale Rd S, Sheffield S7 2QW • T: +44 114 2367731 • F: 2353196 • postmaster@simt.co.uk • www.simt.co.uk •
Dir.: *John Hamshere* •
Science&Tech Museum – 1970
A fully restored 18th c water-powered skythe and steel works, four working water wheels, craftsmen on site, science, production of steel tools, development of steel manufacturing 45648

Bishops' House Museum, Norton Lees Ln, Sheffield S8 9BE • T: +44 114 2782600 • F: 2782604 • info@sheffieldgalleries.org.uk • www.sheffieldgalleries.org.uk •
Dir.: *Dr. Gordon Rintoul* •
Special Museum – 1976
16th and 17th c oak furniture 45649

Fire Police Museum, 101-109 West Bar, Sheffield S3 8PT • T: +44 114 2491999 • info@firepolicemuseum.org.uk •
Special Museum
Fire fighting, police coll, Charlie Peace exhibition 45650

Graves Art Gallery, Surrey St, Sheffield S1 1XZ • T: +44 114 2782600 • F: 2782705 • info@sheffieldgalleries.org.uk • www.sheffieldgalleries.org.uk •
Dir.: *Dr. Gordon Rintoul* • Sc. Staff: *Anne Goodchild* (Visual Art) •
Fine Arts Museum 45651

Kelham Island Museum, Alma St, off Corporation St, Sheffield S3 8RY • T: +44 114 2722106 • F: 2757847 • postmaster@simt.co.uk • www.simt.co.uk •
Dir.: *John Hamshere* •
Historical Museum / Science&Tech Museum – 1982
Story of Sheffield, industry and life, working steam engine, workshops, working cutlers and craftspeople – library; lecture room 45652

Millennium Galleries, Arundel Gate, Sheffield S1 2PP • T: +44 114 2782600 • F: 2782604 • info@sheffieldgalleries.org.uk • www.sheffieldgalleries.org.uk •
Dir.: *Dr. Gordon Rintoul* • Sc. Staff: *Caroline Krzesinska* (Exhibitions, Collections) • *Dorian Church* (Decorative Art) •
Fine Arts Museum / Decorative Arts Museum – 2001
Metalworks, Ruskin coll, craft, design, visal art 45653

National Fairground Archive, University of Sheffield, Main Library, Western Bank, Sheffield S10 2TN • T: +44 114 2227231 • F: 2227290 • fairground@sheffield.ac.uk • www.shef.ac.uk/nfa •
Library with Exhibitions – 1994
Photographs, works drawings and social history, posters 45654

Traditional Heritage Museum, 605 Ecclesall Rd, Sheffield S11 8PR • T: +44 114 2226296 • j.widdowson@sheffield.ac.uk • www.shef.ac.uk/english/natcect •
Cur.: *Prof. J.D.A. Widdowson* •
University Museum
Life and work in Sheffield area, Muir Smith puppet coll, handcrafted trades – archives 45655

Weston Park Museum, Weston Park, Sheffield S1 4DU • T: +44 114 2782600 • F: 2782604 • info@sheffieldgalleries.org.uk • www.sheffieldgalleries.org.uk/westonpark •
Dir.: *Nick Dodd* • Sc. Staff: *Paul Richards* (Natural History) • *Gill Woolrich* (Archaeology and Ethnography) • *Kim Streets* (Social History) •
Local Museum / Fine Arts Museum – 1875
Archaeology, antiquities, armaments, applied arts, natural hist, relics, ethnography, geology, glass, entomology, numismatics, porcelain, paintings (16th-19th c) 45656

Shepton Mallet

East Somerset Railway, Cranmore Station, Shepton Mallet BA4 4QP • T: +44 1749 880417 • F: 880764 • info@eastsomersetrailway.org • www.eastsomersetrailway.org •
Science&Tech Museum
Steam and diesel locomotives 45657

Shepton Mallet Museum, Great Ostry, Shepton Mallet BA4 •
Local Museum / Natural History Museum / Archaeology Museum – 1900
Geology, fossils, Roman kiln and jewelry 45658

Sherborne

Sherborne Castle, New Rd, Sherborne DT9 5NR • T: +44 1935 813182 • F: 816727 • enquiries@sherbornecastle.com • www.sherbornecastle.com •
Cur.: *Ann Smith* • Man.: *Ian Pollard* (Events) •
Fine Arts Museum / Decorative Arts Museum / Historic Site – 1594
Oriental and European porcelain, painting 'Procession of Elizabeth I' attributed to Peake the Elder, furniture, pictures, miniatures – library 45659

Sherborne House, Regional Centre for Visual Arts, Newland, Sherborne DT9 3JG • T: +44 1935 816426 • admin@sherbornehouse.org.uk • www.sherbornehouse.org.uk •
Public Gallery 45660

Sherborne Museum, Abbeygate House, Church Ln, Sherborne DT9 3BP • T: +44 1935 812252 • admin@shermus.fsnet.co.uk •
Cur.: *N. Darling-Finan* •
Local Museum – 1966
Local history, prehistoric finds, geology, Roman and medieval relics, silk production, 20th c fiberglass, local architecture and 14th c wall painting 45661

Sheringham

North Norfolk Railway Museum, Station, Sheringham NR26 8RA • T: +44 1263 820800 • F: 820801 • enquiry@mandgn.co.uk • www.mandgn.co.uk •
Science&Tech Museum
Steam locomotives, rolling stock, station buildings 45662

Sheringham Museum, Station Rd, Sheringham NR26 8RE • T: +44 1263 821871 • F: 825741 • info@sheringhammuseum.co.uk • www.sheringhammuseum.co.uk •
Dir.: *Mary Blyth* •
Local Museum
Local hist, customs and development, local families, boat building, lifeboats, fishing industry, geology, art gallery 45663

Shifnal

Royal Air Force Museum, Cosford, Shifnal TF11 8UP • T: +44 1902 376200 • F: 376211 • cosford@rafmuseum.com • www.rafmuseum.co.uk •
General Manager: *Alex Medhurst* • Cur.: *Al McLean* •
Military Museum – 1979
Military and civil aircraft, aviation coll with 70 historic aircraft on display in three wartime hangars and within the National Cold War Exhibition, including the Vulcan, Victor and Valiant bombers, the Hastings, the York, the Bristol 188, the last airworthy Britannia and other airliners used by British Airways, Missiles and aero engines also on display 45664

Shildon

Locomotion - The National Railway Museum at Shildon, Hackworth Close, Shildon DL4 1PQ • T: +44 1388 777999 • F: 771448 • s.joyce@locomotion.uk.com • www.locomotion.uk.com •
Man.: *George Muirhead* •
Science&Tech Museum – 1975
Sans Pareil locomotive, railway vehicles 45665

Shipley

Museum of Victorian Reed Organs and Harmoniums, Victoria Hall, Victoria Rd, Saltaire, Shipley BD18 3JS • T: +44 1274 585601 • phil@harmoniumservice.demon.co.uk •
Dir.: *Phil Fluke* • *Pam Fluke* •
Music Museum – 1986
Coll of around 100 instruments, Harmoniums of all types and styles 45666

Shoreham

Shoreham Aircraft Museum, High St, Shoreham TN14 7TB • T: +44 1959 524416 • F: 524416 • mail@shoreham-aircraft-museum.co.uk • www.shoreham-aircraft-museum.co.uk •
Cur.: *Geoff Nutkins* •
Historical Museum / Military Museum
Aviation relics from crashed British and German aircrafts (shot down over Southern England 1940), flying equipment, home front memorabilia 45667

Shoreham-by-Sea

Marlipins Museum, 36 High St, Shoreham-by-Sea BN43 5DA • T: +44 1273 462994 • F: +44 1323 844030 • smomich@sussexpast.co.uk • www.sussexpast.co.uk •
Cur.: *Helen E. Poole* •
Local Museum – 1922/1928
Local history, paintings of ships, ship models, in a 12th-14th c building 45668

Shoreham Airport Visitor Centre, Airport, Shoreham-by-Sea BN15 9EW • T: +44 1273 441061 • F: 440146 • sama.archive@virgin.net • www.thearchiveshoreham.co.uk •
Dir.: *David Dunstall* •
Science&Tech Museum
Hist of Shoreham Airport, aircraft restoration projects – library 45669

Shotts

Shotts Heritage Centre, Benhar Rd, Shotts ML27 5EN • T: +44 1501 821556 •
Local Museum
Iron and coal industries and social history, local studies 45670

Shrewsbury

Attingham Park, Atcham, Shrewsbury SY4 4TP • T: +44 1743 708162, 708123 • F: 708175 • attingham@nationaltrust.org.uk • www.nationaltrust.org.uk •
Man.: *Jane Alexander* •
Decorative Arts Museum
Italian furniture (c 1810), silver and pictures, printed pottery 45671

Clive House Museum, College Hill, Shrewsbury SY1 1LT • T: +44 1743 354811 • F: 358411 • museums@shrewsbury-atcham.gov.uk • www.shrewsburymuseums.com •
Local Museum
Archaeology, geology of Shropshire, paintings and prints, prehistory, local and regional social history, natural history 45672

Coleham Pumping Station, Longden Coleham, Shrewsbury SY3 7DB • T: +44 1743 361196 • F: 358411 • museums@shrewsbury-atcham.gov.uk • www.shrewsburymuseums.com •
Science&Tech Museum 45673

Radbrook Culinary Museum, Centre for Catering and Management Studies, Radbrook Rd, Shrewsbury SY3 9BL • T: +44 1743 232686 • F: 271563 • mail@s-cat.ac.uk •
Head: *Sue Warton* • Library: *F. Bates* •
Special Museum – 1985
Food, nutrition, housecraft, education 45674

Shrewsbury Castle, Castle St, Shrewsbury SY1 2AT • T: +44 1743 358516 • F: 270023 • museums@shrewsbury.gov.uk • www.shrewsburymuseums.com •
Cust.: *L. Cliffe* •
Decorative Arts Museum 45675

Shrewsbury Museum and Art Gallery, Rowley's House, Barker St, Shrewsbury SY1 1QH • T: +44 1743 361196 • F: 358411 • museums@shrewsbury.gov.uk • www.shrewsburymuseums.com •
Man.: *Mary White* • Sc. Staff: *Mike Stokes* (Archaeology) • *Peter Boyd* (Collections) •
Local Museum / Archaeology Museum / Decorative Arts Museum – 1935
Main repository of excavated material from Viroconium, Shropshire geology and prehist, natural hist incl Shropshire herbarium, costume, decorative arts incl Shropshire manufacturers 45676

Shropshire Regimental Museum, The Castle, Castle St, Shrewsbury SY1 2AT • T: +44 1743 262292, 358516 • F: 270023 • shropsrm@zoom.co.uk •
Cur.: *Peter Duckers* •
Military Museum
Coll of the 53rd & 8th Regiments, the King's Shropshire Light Infantry & Yeomanny , Shropeshire Royal Horse Artillery 45677

Sibsey

Trader Windmill, off A16, W of Sibsey, Sibsey PE22 0UT, mail addr: c/o Ian Ansell, Pear Tree Cottage, Pitcher Row Ln, Algarkirk PE20 2LJ • T: +44 1205 750036 • traderwindmill@sibsey.fsnet.co.uk •
Science&Tech Museum
Flour milling related machinery and bygones 45678

Sidmouth

Sid Vale Heritage Centre → Sidmouth Museum

Sidmouth Museum, Hope Cottage, Church St, Sidmouth EX10 8LY • T: +44 1395 516139 • sidmouth@devonmuseums.net • www.sidvaleassociation.org.uk •
Cur.: *Dr. Robert F. Symes* •
Local Museum
Local hist, Jurassic coast, geological development 45679

Silsoe

Wrest Park, Silsoe MK45 4HS • T: +44 1525 860152 • www.english-heritage.org.uk •
Cur.: *Gareth Hughes* •
Local Museum
18th c sculpture in a landscaped garden 45680

Sittingbourne

Court Hall Museum, High St, Milton Regis, Sittingbourne ME10 1RW • T: +44 1795 478446 •
Local Museum
Documents, objects and photographs 45681

Dolphin Sailing Barge Museum, Crown Quay Ln, Sittingbourne ME10 3SN • T: +44 1795 424132, 421549 •
Dir.: *Peter J. Morgan* •
Science&Tech Museum – 1969
Sailing barge artefacts, photographs, maps, documents, brickmaking tools, sample bricks, moored in our basin is the Thames Sailingbarge Cambria 45682

Sittingbourne and Kemsley Light Railway, Mill Way, Sittingbourne ME7 4QL, mail addr: POB 300, Sittingbourne ME10 2DZ • T: +44 871 135033 • info@sklr.net • www.sklr.net •
Science&Tech Museum
Narrow gauge locomotives and wagons, photographs, industrial archaeology, paper-making 45683

Sittingbourne Heritage Museum, 67 East St, Sittingbourne ME10 4BQ • T: +44 1795 423215 •
Local Museum
Archaeological finds, local hist 45684

Skegness

Church Farm Museum, Church Rd S, Skegness PE25 2ET • T: +44 1754 766658 • F: 898243 • www.lincolnshire.gov.uk/visiting/museums/church-farm-museum/ •
Dir.: *Heather Cummins* •
Open Air Museum / Agriculture Museum – 1976 45685

Skidby

Skidby Windmill and Museum of East Riding Rural Life, Beverley Rd, Skidby HU16 5TF • T: +44 1482 848405 • F: 842889 • janet.tierney@eastriding.gov.uk •
Dir.: *Janet Tierney* • *Jane Bielby* •
Historical Museum / Agriculture Museum – 1974
Alex West (agricultural exhibits) 45686

Skinningrove

Cleveland Ironstone Mining Museum, Deepdale, Skinningrove TS13 4AP • T: +44 1287 642877 • F: 642970 • visits@ironstonemuseum.co.uk • www.ironstonemuseum.co.uk •
Science&Tech Museum
Social history, tools, minestone face 45687

Skipton

Craven Museum and Gallery, Town Hall, High St, Skipton BD23 1AH • T: +44 1756 706407 • F: 706412 • museum@cravendc.gov.uk • www.cravendc.gov.uk •
Dir.: *Andrew Mackay* •
Local Museum / Public Gallery – 1928 45688

Embsay Bolton Abbey Steam Railway, Bolton Abbey Station, Skipton BD23 6AF • T: +44 1756 710614 • F: 710720 • embsay.steam@btinternet.com • www.embsayboltonabbeyrailway.org.uk •
Dir.: *Stephen Walker* •
Science&Tech Museum – 1968
Coll of industrial steam locomotives in Britain, items of passenger and freight rolling stock, photogr and documents illustrates the hist of the line and of the Trust which operates it 45689

Slough

Slough Museum, 278-286 High St, Slough SL1 1NB • T: +44 1753 526422 • F: 526422 • info@sloughmuseum.co.uk • www.sloughmuseum.co.uk •
Cur.: *Eleanor Pulfer* •
Local Museum ... 45690

Smethwick

Avery Historical Museum, c/o Avery Weigh-Tronix, Foundry Ln, Smethwick B66 2LP • T: +44 870 9034343 • info@averyweigh-tronix.com • www.averyweigh-tronix.com •
Cur.: *John Doran* •
Historical Museum – 1928
Coll of scales, weights and records relating to the hist of weighing ... 45691

Soudley

Dean Heritage Museum, Camp Mill, Soudley GL14 2UB • T: +44 1594 822170 • F: 823711 • deanmuse@btinternet.com • www.deanheritagemuseum.com •
Dir.: *Deborah Cook* • Cur.: *Dawn Heywood* •
Local Museum – 1982
Forest of dean, foresty, royal hunting, clocks, mining, geology, archaeology, social/industrial history – reference library ... 45692

South Molton

South Molton and District Museum, Guildhall, South Molton EX36 3AB • T: +44 1769 572951 • F: 574008 • curatorsouthmolton@lineone.net • www.devonmuseums.net/southmolton •
Local Museum – 1951
Local hist, farming and mining, pewter, fire engines, cider presses, local painting, pottery, original royal charters ... 45693

South Queensferry

Dalmeny House, South Queensferry EH30 9TQ • T: +44 131 3311888 • F: 3311788 • events@dalmeny.co.uk • www.dalmeny.co.uk •
Decorative Arts Museum
French furniture, tapestries, porcelain from Mentmore, Napleonic coll, British portraits ... 45694

Hopetoun House, South Queensferry EH30 9SL • T: +44 131 3312451 • F: 3191885 •
Decorative Arts Museum / Fine Arts Museum – 1700
Paintings, mid-18th c furniture, sculpture, manuscripts, costumes, tapestries, architecture ... 45695

Queensferry Museum, 53 High St, South Queensferry EH30 9HP • T: +44 131 3315545 • F: 5573346 • museums&galleries@edinburgh.gov.uk • www.cac.org.uk •
Historical Museum – 1951
History of South Queensferry and Dalmeny ... 45696

South Shields

Arbeia Roman Fort and Museum, Baring St, South Shields NE33 2BB • T: +44 191 4561369 • F: 4276862 • liz.elliott@twmuseums.org.uk • www.twmuseums.org.uk •
Dir.: *Alec Coles* • Cur.: *Paul Bidwell* (Archaeology) •
Archaeology Museum – 1953
Roman finds from the fort ... 45697

South Shields Museum and Art Gallery, Ocean Rd, South Shields NE33 2JA • T: +44 191 4568740 • F: 4567850 • alisdair.wilson@tyne-wear-museums.org.uk • www.twmuseums.org.uk •
Local Museum / Fine Arts Museum – 1876
Glass, natural hist, local hist, watercolours – Catherine Cookson Gallery ... 45698

South Witham

Geeson Brothers Motor Cycle Museum, 4-6 Water Ln, South Witham NG33 5PH • T: +44 1572 767280, 768195 •
Science&Tech Museum
85 British bikes, automobilia ... 45699

Southall

Collection of Martinware Pottery, Southall Library, Osterley Park Rd, Southall UB2 4BL • T: +44 20 85743412 • F: 85717629 • libuser@ealing.gov.uk • www.ealing.gov.uk/libraries •
Head: *Neena Sohal* •
Decorative Arts Museum
library ... 45700

Southampton

Ancient Order of Foresters, College Pl, Southampton SO15 2FE • T: +44 23 80229655 • F: 80229657 • mail@foresters.ws •
Religious Arts Museum / Historical Museum
Regalia, certificates, photographs, badges, banners, memorabilia and court room furniture, murual self-help ... 45701

Bitterne Local History Centre, 225 Peartree Av, Bitterne, Southampton SO19 7RD • T: +44 23 80490948 • jim@jimbrown.me.uk • www.bitterne.net •
Local Museum
Lives of the people, gentry and traders of Bitterne ... 45702

Hawthorns Urban Wildlife Centre, Hawthorns Centre, Southampton Common, Southampton SO15 7NN • T: +44 23 80671921 • F: 80676859 • l.hand@southampton.gov.uk • www.southampton.gov.uk/leisure •
Natural History Museum – 1980
Biological records ... 45703

John Hansard Gallery, University of Southampton, Highfield, Southampton SO17 1BJ • T: +44 23 80592158 • F: 80594192 • info@hansardgallery.org.uk • www.hansardgallery.org.uk •
Dir.: *Stephen Foster* •
Public Gallery ... 45704

Solent Sky, Albert Rd S, Southampton SO1 3FR • T: +44 23 80635830 • F: 80223383 • aviation@spitfireonline.co.uk • www.spitfireonline.co.uk •
Dir.: *Alan Jones* •
Science&Tech Museum – 1982
History of Solent aviation, story of 26 aircraft companies ... 45705

Southampton City Art Gallery, Civic Centre, Southampton SO14 7LP • T: +44 23 80834563 • F: 80832153 • art.gallery@southampton.gov.uk • www.southampton.gov.uk/leisure/arts •
Dir.: *Tim Craven* (Collections) • Sc. Staff: *Esta Mion-Jones* (Exhibitions) • *Charlotte Baber* (Cons.) • *Rosie Shirley* (Education) •
Public Gallery – 1939
18th-20th c English paintings, Continental Old Masters 14th-18th c, modern French paintings, incl the Impressionist, also sculpture and ceramics, paintings and drawings, contemporary British art ... 45706

Southampton Maritime Museum, Wool House, Town Quay, Bugle St, Southampton SO14 2AR • T: +44 23 80635904, 80223941 • F: 80339601 • museums@southampton.gov.uk • www.southampton.gov.uk •
Cur.: *Alastair Arnott* •
Science&Tech Museum – 1962
Models of the great liners and shipping ephemera, Titanic to Queen Mary, docks model showing the port c 1938 and associated displays, in 14th c warehouse ... 45707

Tudor House Museum, Bugle St, Southampton SO14 2AD • T: +44 23 80635904, 80332513 • F: 80339601 • museums@southampton.gov.uk • www.southampton.gov.uk •
Dir.: *Sian Jones* • Cur.: *Alastair Arnott* (Local Collections) •
Local Museum – 1912
Furniture, domestic artefacts, paintings and drawings of Southampton, costumes and decorative arts ... 45708

Southborough

Salomons Museum, Memento Rooms, David Salomons House, Broomhill Rd, Southborough TN3 0TG • T: +44 1892 515152 • F: 539102 • enquiries@salomons.org.uk • www.salomonscentre.org.uk •
Head: *Prof. Tony Lavender* •
Science&Tech Museum / Local Museum
David Salomons, first Jewish Lord Mayor of London; inventions of David Lionel Salomons, scientist and engineer ... 45709

Southend-on-Sea

Beecroft Art Gallery, Station Rd, Westcliff-on-Sea, Southend-on-Sea SS0 7RA • T: +44 1702 347418 • F: 347681 • www.beecroft-art-gallery.co.uk •
Dir.: *Clare Hunt* • Keeper: *C.F. Leming* (Art) •
Fine Arts Museum – 1953
Permanent coll by artists as Constable, Molenaer, Bright, Epstein, Weight, Lear and Seago, Thorpe-Smith coll of local works, Todman coll, drawings of Nubia by Alan Sorrell ... 45710

Focal Point Gallery, Southend Central Library, Victoria Av, Southend-on-Sea SS2 6EX • T: +44 1702 534108 ext 207 • F: 469241 • focalpointgallery@southend.gov.uk • www.focalpoint.org.uk •
Dir.: *Lesley Farrell* •
Public Gallery ... 45711

Prittlewell Priory Museum, Priory Park, Victoria Av, Southend-on-Sea SS1 2TF • T: +44 1702 342878 • F: 349806 • southendmuseum@hotmail.com • www.southendmuseums.co.uk •
Dir.: *John Skinner* •
Local Museum – 1922
Natural history of South East Essex, medieval life, printing, radios and television ... 45712

Southchurch Hall Museum, Southchurch Hall Close, Southend-on-Sea • T: +44 1702 467671 • F: 349806 • soutendmuseum@hotmail.com • www.southendmuseums.co.uk •
Dir.: *John Skinner* •
Historical Museum – 1974
Medieval manor house with 17th c furnishings, built in 1340 ... 45713

Southend Central Museum, Victoria Av, Southend-on-Sea SS2 6EW • T: +44 1702 215131, 434449 • F: 349806 • southendmuseum@hotmail.com • www.southendmuseums.co.uk •
Dir.: *J.F. Skinner* • Keeper: *K.L. Crowe* (Human History) • *C. Hunt* (Art) • Asst. Keeper: *R.G. Payne* (Natural History) • *D.I. Mitchell* (Human History) • Cons.: *C. Reed* •
Local Museum
Natural and human history of south-east Essex, Planetarium ... 45714

Southend Pier Museum, Southend Pier, Western Esplanade, Southend-on-Sea SS1 2EQ • T: +44 1702 611214, 614553 • peggy@spmfh.fsnet.co.uk • www.southendpiermuseum.co.uk •
Dir.: *Peggy Dowie* •
Historical Museum – 1989
History of the pier, buildings, staff, illuminations, pleasure boats, war years, lifeboats, disasters, pier transport and railway ... 45715

Southport

Atkinson Art Gallery, Lord St, Southport PR8 1DH • T: +44 151 9342110 • F: 9342109 • atkinson.gallery@leisure.sefton.gov.uk • www.atkinson-gallery.co.uk •
Man.: *Joanna Jones* •
Public Gallery – 1878
Paintings, drawings, watercolours, prints, sculptures ... 45716

Botanic Gardens Museum, Churchtown, Southport PR8 7NB • T: +44 1704 227547 • F: 224112 • www.seftonarts.co.uk •
Natural History Museum – 1938
Toys, Liverpool porcelain, natural hist, Victorian artefacts ... 45717

British Lawnmower Museum, 106-114 Shakespeare St, Southport PR8 5AJ • T: +44 1704 501336 • F: 500564 • info@lawnmowerworld.com • www.lawnmowerworld.com •
Cur.: *Brian Radam* •
Special Museum
Garden machinery, 200 lawnmowers ... 45718

Southwick

Manor Cottage Heritage Centre, Southwick St, Southwick BN42 4TE • T: +44 1273 465164 • nigeldivers@ntlworld.com •
Historical Museum / Decorative Arts Museum
Coll of modern, specially worked needlework panels depicting the history of Southwick ... 45719

Southwold

Southwold Museum, 9-11 Victoria St, Southwold IP18 6HZ • T: +44 1502 723374 • F: 723379 • southwold.museum@virgin.net • www.southwoldmuseum.org •
Cur.: *David de Kretser* •
Local Museum – 1933
Local hist, Southwold railway relics, natural hist, battle of Sole Bay, tokens, coins, watercolours ... 45720

Spalding

Ayscoughfee Hall Museum, Churchgate, Spalding PE11 2RA • T: +44 1775 725468 • F: 762715 • museum@sholland.gov.uk • www.sholland.gov.uk •
Dir.: *Richard Davies* •
Local Museum
Medieval manor house, local and social history ... 45721

Pinchbeck Marsh Engine and Land Drainage Museum, Pinchbeck Marsh of West Marsh Rd, Spalding PE11 3UW • T: +44 1775 725861, 725468 • F: 711054 • museum@shetland.gov.uk • www.sholland.gov.uk •
Science&Tech Museum
Drain pipes, pumps, gas and oil engines, a dragline excavator, hand tools, digging tools, turf cutters, weed cutters, fishing spears, plank road equipment, lamps, steam engine, punt ... 45722

Spalding Gentlemen's Society Museum, Broad St, Spalding PE11 1TB • T: +44 1775 724658 •
Pres.: *J. W. Belsham* •
Local Museum – 1710
Local history, ceramics, glass, coins, medals, trade tokens ... 45723

Sparkford

Haynes International Motor Museum, Castle Cary Rd, Sparkford BA22 7LH • T: +44 1963 440804 • F: 441004 • mike@haynesmotormuseum.co.uk • www.haynesmotormuseum.co.uk •
Dir.: *John H. Haynes* • Cur.: *Michael Penn* •
Science&Tech Museum – 1985
1897 Daimler Station Wagonette,1905 Daimler detachable top limousine, the 1965 AC Cobra, 400 motoring and motorcycling items of historical and cultural interest, American coll, Red coll of sports cars, William Morris Garages, 1929 model J. Deusenberg, Derham bodied tourster, V16 Cadillac, speedway motorcycles, British Motorcycles – library ... 45724

Spean Bridge

Clan Cameron Museum, Achnacarry, Spean Bridge PH34 4EJ • T: +44 1397 712090 • museum@achnacarry.fsnet.co.uk • www.clan-cameron.org •
Dir.: *Sir Donald Cameron of Lochiel* •
Historical Museum
Hist of Cameron Clan and their involvement in the Bonnie Prince Charlie rising ... 45725

Spey Bay

Moray Firth Wildlife Centre, Tugnet Ice House, Spey Bay IV32 7PS • T: +44 1343 820339 • F: 829065 • wildlifecentre@wdcs.org • www.wdcs.org/wildlifecentre •
Natural History Museum
Local wildlife incl whales, dolphins and porpoises ... 45726

Staffin

Staffin Museum, 6 Ellishadder, Staffin IV51 9JE • T: +44 1470 562321 •
Dir.: *Dugald Alexander Ross* •
Archaeology Museum
Fossils from Skye (Jurassic age), incl the only dinosaur bones from Scotland (these represent five species incl the world's oldest recorded Stegosaur), 18th c implements and furniture, old glass and eartheware bottles, geological and mineralogical specimens, horse-drawn implements of the 19th c, archaeological exhibits incl neolithic flint arrowheads and pottery ... 45727

Stafford

Ancient High House, Greengate St, Stafford ST16 2JA • T: +44 1785 619131 • F: 619132 • ahh@staffordbc.gov.uk • www.staffordbc.gov.uk/ancienthighhouse •
Dir.: *Jill Fox* •
Historic Site
England's largest timber-framed town house built in 1595, period room settings ... 45728

Royal Air Force Museum Reserve Collection, RAF Stafford, Beaconside, Stafford ST18 0AQ • T: +44 1785 258200 • F: 220080 • ken.hunter@rafmuseum.com • www.rafmuseum.com •
Keeper: *Ken Hunter* •
Military Museum / Science&Tech Museum ... 45729

Shire Hall Gallery, Market Sq, Stafford ST16 2LD • T: +44 1785 278345 • F: 278327 • shirehallgallery@staffordshire.gov.uk • www.staffordshire.gov.uk/sams •
Dir.: *Kim Tudor* •
Fine Arts Museum / Decorative Arts Museum – 1927
Works by Staffordshire artists and of subjects in the county, contemporary jewellery coll by British makers ... 45730

Stafford Castle and Visitor Centre, Newport Rd, Stafford ST16 1DJ • T: +44 1785 257698 • F: 257698 • castlebc@btconnect.com •
Dir.: *Nicholas Thomas* •
Historical Museum / Archaeology Museum
Archaeological display ... 45731

Staines

Spelthorne Museum, 1 Elmsleigh Rd, Staines TW18 4PM • T: +44 1784 461804 • staff@spelthorne.free-online.co.uk • www.spelthornemuseum.org.uk •
Cur.: *Ralph Parsons* •
Local Museum
Archaeology, local and social history, Roman town model, Victorian kitchen, 1738 fire engine ... 45732

Stalybridge

Astley-Cheetham Art Gallery, Trinity St, Stalybridge SK15 2BN • T: +44 161 3386767 • F: 3431732 • astley.cheetham@tameside.gov.uk • www.tameside.gov.uk •
Public Gallery
Art, craft, photography ... 45733

The People's Gallery, 30 Melbourne St, Stalybridge SK15 2JJ • T: +44 161 3388333 • www.thepeoplesgallery.net •
Public Gallery ... 45734

Stamford

Burghley House, Stamford PE9 3JY • T: +44 1780 752451 • F: 756057 • burghley@burghley.co.uk • www.burghley.co.uk •
Dir.: *P. Gompertz* •
Fine Arts Museum / Decorative Arts Museum / Historic Site
Italian paintings 17th c, fine oriental ceramics, European furniture, works of art ... 45735

Stamford Brewery Museum, All Saints St, Stamford PE9 2PA • T: +44 1780 752186 •
Science&Tech Museum
Victorian steam brewery ... 45736

Stamford Museum, Broad St, Stamford PE9 1PJ • T: +44 1780 766317 • F: 480363 • stamford_museum@lincolnshire.gov.uk • www.lincolnshire.gov.uk/stamfordmuseum •
Keeper: *Tracey Crawley* •
Local Museum – 1961
Archaeology, medieval pottery, social history relating to town of Stamford ... 45737

Stansted Mountfitchet

House on the Hill Museums Adventure, Stansted Mountfitchet CM24 8SP • T: +44 1279 813567 • F: 816391 • moutfitchitcastle1066@btinternet.com • www.moutfitchitcastle.com •
Dir.: *Jeremy Goldsmith* •
Special Museum – 1991
Action man coll, star wars coll, toys, books, games since late Victorian times ... 45738

Stansted Mountfitchet Windmill, Millside, Stansted Mountfitchet CM24 8AT • T: +44 1279 813964 • F: 816391 •
Dir.: *Peggy Honour* •
Science&Tech Museum
Tower mill, last type of mill to evolve in England ... 45739

U

Staplehurst

Brattle Farm Museum, Five Oak Ln, Staplehurst TN12 0HE • T: +44 1580 891222 • F: 891222 •
Owner: *Brian Thompson* •
Agriculture Museum / Historical Museum
Vintage cars, motorcycles, bicycles, tractors, wagons, horse-drawn machinery, hand tools and crafts, war items, laundry and dairy ... 45740

Stevenage

Boxfield Gallery, Stevenage Arts and Leisure Centre, Stevenage SG1 1LZ • T: +44 1438 766644 • F: 766675 •
Public Gallery ... 45741

Stevenage Museum, Saint George's Way, Stevenage SG1 1XX • T: +44 1438 218881 • F: 218882 • museum@stevenage.gov.uk • www.stevenage.gov.uk/museum •
Cur.: *Claire Hill* •
Local Museum – 1954
Local hist, archaeology, natural hist, geology ... 45742

Stevington

Stevington Windmill, Stevington MK43 7QB • T: +44 1234 824330 • F: 228531 • www.bedfordshire.gov.uk/bromhammill •
Resp.: *Ed Burnett* •
Science&Tech Museum
Post mill with several features worthy of note ... 45743

Stewarton

Stewarton and District Museum, Council Chambers, Avenue Sq, Stewarton KA3 5AB, mail addr: 17 Grange Terrace, Kilmarnock KA1 2JR • T: +44 1563 524748 • ianhmac@aol.com • www.stewarton.org •
Dir.: *I.H. Macdonald* •
Local Museum – 1980
Local industry, bonnet-making ... 45744

Steyning

Steyning Museum, Church St, Steyning BN44 3YB • T: +44 1903 813333 • steyningmuseum.org.uk •
Cur.: *Chris Tod* •
Historical Museum
Hist of Steyning from Saxon times to the present day, photographs, documents and objects ... 45745

Stibbington

Nene Valley Railway, Wansford Station, Stibbington PE8 6LR • T: +44 1780 784444 • F: 784440 • nvrorg@aol.com • www.nvr.org.uk •
Science&Tech Museum
Steam and diesel locomotives and rolling stock, railwayana ... 45746

Stirling

Argyll and Sutherland Highlanders Regimental Museum, The Castle, Stirling FK8 1EH • T: +44 1786 475165 • F: 446038 • museum@argylls.co.uk • www.argylls.co.uk •
Cur.: *C.A. Campbell* •
Military Museum – 1961
History of regiment since 1794, medals, pictures, silver armours, uniforms – archives ... 45747

Art Collection, c/o University of Stirling, Stirling FK9 4LA • T: +44 1786 466050 • F: 466866 • art.collection@stir.ac.uk • www.stir.ac.uk/artcol •
Cur.: *Jane Cameron* •
Fine Arts Museum
Paintings, prints, sculptures, tapestries, J.D. Fergusson ... 45748

Bannockburn Heritage Centre, Glasgow Rd, Stirling FK7 0LJ • T: +44 1786 812664 • F: 810892 • bannockburn@nts.org.uk • www.nts.org.uk •
Historical Museum – 1987
Scottish hist ... 45749

National Wallace Monument, Abbey Craig, Hillfoots Rd, Stirling FK9 5LF • T: +44 1786 472140 • F: 461332 •
Man.: *Eleanor Muir* •
Historical Museum
Scotlands National hero Sir William Wallace, famous Scots, countryside, history ... 45750

Stirling Old Town Jail, Saint John St, Stirling FK8 1EA • T: +44 1786 450050 • F: 471301 • otjua@visitscotland.com • www.oldtownjail.com •
Man.: *Neil Craig* •
Historical Museum – restored 19th c jail with original cells ... 45751

Stirling Smith Art Gallery and Museum, 40 Albert Pl, Dumbarton Rd, Stirling FK8 2RQ • T: +44 1786 471917 • F: 449523 • museum@smithartgallery.demon.co.uk • www.smithartgallery.demon.co.uk •
Dir.: *Elspeth King* •
Local Museum / Fine Arts Museum – 1874
Local hist, archaeology, ethnology, geology, paintings, watercolors ... 45752

Stockport

Hat Works, Museum of Hatting, Wellington Mill, Wellington Rd S, Stockport SK3 0EU • T: +44 161 3557770 • F: 4808735 • linn.holmstrom@stockport.gov.uk • www.hatworks.org.uk •
Dir.: *John Baker* •
Historical Museum – 2000
Hatting coll, machinery, historical and contemporary hats ... 45753

Stockport Art Gallery, Wellington Rd S, Stockport SK3 8AB • T: +44 161 4744453/54 • F: 4804960 • stockport.artgallery@stockport.gov.uk • www.stockport.gov.uk/tourism/artgallery •
Man.: *Jackie Mellor* •
Fine Arts Museum – 1924
Paintings, sculptures, bronze of Yehudi Menuhin by Epstein ... 45754

Stockport Museum, Vernon Park, Turncroft Ln, Stockport SK1 4AR • T: +44 161 4744460 • F: 4744449 • www.stockport.gov.uk/heritageattractions •
Local Museum – 1860
Local history, natural history, applied arts, example of a window made from 'Blue John' fluorspar ... 45755

Stockton-on-Tees

Green Dragon Museum, Theatre Yard, off Silver St, Stockton-on-Tees TS18 1AT • T: +44 1642 527982 • F: 527037 • greendragon@stockton.gov.uk • www.stockton.gov.uk/museums •
Dir.: *Alan Cracket* •
Local Museum – 1973
Social, industrial and administrative development, ship models, coll of local pottery ... 45756

Preston Hall Museum, Preston Park, Yarm Rd, Stockton-on-Tees TS18 3RH • T: +44 1642 781184 • F: 788729 • prestonhall@stockton.gov.uk • www.stockton.gov.uk/museums •
Dir.: *Paul Lake* •
Local Museum – 1953
Arms, working crafts, Victorian life, toys ... 45757

Stoke Bruerne

Canal Museum, Stoke Bruerne NN12 7SE • T: +44 1604 862229 • F: 864199 • stokebruerne@thewaterwaystrust.org.uk • www.stokebruerne-canalmuseum.org.uk •
Gen. Man.: *David Henderson* • Man.: *Louise Stockwin* •
Historical Museum – 1963
Canal history, 200 years of inland waterways ... 45758

Stoke-on-Trent

Etruria Industrial Museum, Lower Bedford St, Etruria, Stoke-on-Trent ST4 7AF • T: +44 1782 233144 • F: 233145 • etruria@swift.stoke.gov.uk • www.stoke.gov.uk/museums •
Science&Tech Museum – 1991
Ceramic processing, steam beam engine (1820), grinding machinery (1856), blacksmithing coll ... 45759

Ford Green Hall, Ford Green Rd, Smallthorne, Stoke-on-Trent ST6 1NG • T: +44 1782 233195 • F: 233194 • www.stoke.gov.uk/fordgreenhall •
Dir.: *Angela Graham* •
Decorative Arts Museum – 1952
In 17th c framed house, period furniture, household objects ... 45760

Gladstone Pottery Museum, Uttoxeter Rd, Longton, Stoke-on-Trent ST3 1PQ • T: +44 1782 237777 • gladstone@stoke.gov.uk • www.stoke.gov.uk/gladstone •
Man.: *Sally Coleman* •
Special Museum
Story of the British pottery industry, the history of the industry in Staffordshire, social history, tiles and tilemaking, sanitary ware, colour and decoration 45761

The Potteries Museum and Art Gallery, Bethesda St, Stoke-on-Trent ST1 3DW • T: +44 1782 232323 • museums@stoke.gov.uk •
Dir.: *Deb Clarke* •
Archaeology Museum / Fine Arts Museum / Decorative Arts Museum / Natural History Museum – 1956
Staffordshire pottery and porcelain, Continental and Oriental pottery, sculpture, natural history, local history, 18th c watercolors, English paintings, contemporary art ... 45762

Spode Museum, Church St, Stoke-on-Trent ST4 1BX • T: +44 1782 744011 • F: 572526 • spodemuseum@spode.co.uk • www.spode.co.uk •
Cur.: *Pam Woolliscroft* •
Decorative Arts Museum – 1938
Spode & Copeland ceramics since 1780, bone china and transfer printed wares ... 45763

Wedgwood Museum, Barlaston, Stoke-on-Trent ST12 9ES • T: +44 1782 282818 • F: 223315 • info@wedgwoodmuseum.org.uk • www.wedgwoodmuseum.com •
Dir.: *Gaye Blake Roberts* •
Decorative Arts Museum – 1906
Wedgwood family hist, ceramics, paintings and manuscripts, mason's ironstone, historical pottery, comprehensive coll of the works of Wedgwood – library, archives ... 45764

Stonehaven

Tolbooth Museum, The Harbour, Stonehaven AB39 2JU • T: +44 1771 622906, 622807 • F: 622884, 623558 • museums@aberdeenshire.gov.uk • www.aberdeenshire.ov.uk/heritage/ •
Local Museum – 1963 ... 45765

Stornoway

An Lanntair Art Centre, Kenneth St, Stornoway HS1 2DS • T: +44 1851 703307 • F: 703307 • info@lanntair.com • www.lanntair.com •
Dir.: *Roddy Murray* •
Public Gallery ... 45766

Museum Nan Eilean, Steornabhagh, Francis St, Stornoway HS1 2NF • T: +44 1851 703773 ext 266 • F: 706318 • rlanghorne@cne-siar.gov.uk • www.cne-siar.gov.uk •
Cur.: *Richard Langhorne* •
Archaeology Museum / Historical Museum
Archeology of Lewis and Harris, history of the islands, maritime, fishing, crofting coll ... 45767

Storrington

Parham House & Gardens, Parham Park, Storrington RH20 4HS • T: +44 1903 744888 • F: 746557 • enquiries@parhaminsussex.co.uk • www.parhaminsussex.co.uk •
Fine Arts Museum – 1577
Historical portraits, needlework, Equestrian portrait of Henry Frederick, Prince of Wales, Kangaroo and Dingo by Stubbs ... 45768

Stourbridge

Red House Glass Cone Museum, High St, Wordsley, Stourbridge DY8 4AZ • T: +44 1384 812750 • F: 812751 • www.dudley.gov.uk/redhousecone •
Special Museum
A glassmaking site dating from the late 18th c and in use until 1936, a 90' high glassmaking cone, built in 1790 and one of only four left in the UK, is surrounded by buildings related to the Victorian glass industry ... 45769

Stourton

Stourhead House, Stourton BA12 6QH • T: +44 1747 842020 • F: 842005 • emily.blanshard@nationaltrust.org.uk •
Man.: *Emily Blanshard* •
Fine Arts Museum / Decorative Arts Museum – 1720
Furniture, paintings, porcelain, furniture designed by Thomas Chippendale, the Younger ... 45770

Stow-on-the-Wold

The Toy Museum, Park St, Stow-on-the-Wold GL54 1AQ • T: +44 1451 830159 • info@thetoymuseum.co.uk • www.thetoymuseum.co.uk •
Special Museum
Dolls ... 45771

Unicorn Trust Equestrian Centre, Netherswell, Stow-on-the-Wold GL54 1JZ • T: +44 1451 832317 • F: 832126 • sydney.smith@dialin.net •
Science&Tech Museum
Equestrian sports, carriage driving ... 45772

Stowmarket

Museum of East Anglian Life, Abbot's Hall, Stowmarket IP14 1DL • T: +44 1449 612229 • F: 672307 • enquiries@eastanglianlife.org.uk • www.eastanglianlife.org.uk •
Dir.: *Tony Butler* •
Historical Museum – 1965
Rural life in East Anglia, household equipment, farm and craft tools, re-erected watermill, wind pump, smithy ... 45773

Strabane

Gray's Printing Press, 49 Main St, Strabane, Co. Tyrone • T: +44 1504 884094 •
Science&Tech Museum
18th c printing press, 19th c printing machines ... 45774

Strachur

Strachur Smiddy Museum, The Clachan, Strachur PA27 8DG • T: +44 1369 860565 •
Dir.: *Cathie Montgomery* •
Science&Tech Museum ... 45775

Stranraer

Castle of Saint John, Castle St, Stranraer DG9 7RT • T: +44 1776 705088 • F: 705835 • JohnPic@dumgal.gov.uk • www.dumgal.gov.uk/museums •
Dir.: *John Pickin* •
Historical Museum
Castle hist, town jail in the 18th and 19th c ... 45776

Stranraer Museum, 55 George St, Stranraer DG9 7JP • T: +44 1776 705088 • F: 705835 • JohnPic@dumgal.gov.uk • www.dumfriesmuseum.demon.co.uk •
Dir.: *John Pickin* •
Local Museum – 1939
Social and local history, archaeology, art, numismatics, costume, photos – archive ... 45777

Stratfield Saye

Stratfield Saye House and Wellington Exhibition, Stratfield Saye RG27 2BZ • T: +44 1256 882882 • F: 881466 • stratfield-saye@btconnect.com • www.stratfield-saye.co.uk •
Historical Museum
Home of Dukes of Wellington, paintings, captured from Joseph Bonaparte, and the Great Duke's personal belongings – Library ... 45778

Stratford-upon-Avon

Anne Hathaway's Cottage, Shakespeare Birthplace Trust, Shottery, Stratford-upon-Avon CV37 9HH • T: +44 1789 292100, 297240 • F: 296083 • info@shakespeare.org.uk • www.shakespeare.org.uk •
Dir.: *Roger Pringle* •
Special Museum
Thatched farmhouse, home of Anne Hathaway before her marriage to Shakespeare in 1582, period furniture since 16th c with traditional English cottage garden and orchard ... 45779

Hall's Croft, Shakespeare Birthplace Trust, Old Town, Stratford-upon-Avon CV37 6BG • T: +44 1789 292107 • F: 296083 • info@shakespeare.org.uk • www.shakespeare.org.uk •
Dir.: *Roger Pringle* •
Special Museum – 1847
Home of Shakespeare's daughter Susanna and her husband Dr. John Hall, Tudor and Jacobean furniture, period medical Exhibition ... 45780

Harvard House, Shakespeare Birthplace Trust, 26 High St, Stratford-upon-Avon CV37 6AU • T: +44 1789 204507 • F: 296083 • info@shakespeare.org.uk • www.shakespeare.org.uk •
Dir.: *Roger Pringle* •
Special Museum
Neish Pewter coll, architecture ... 45781

Mary Arden's House and the Countryside Museum, Shakespeare Birthplace Trust, Wilmcote, Stratford-upon-Avon CV37 9UN • T: +44 1789 293455 • F: 296083 • info@shakespeare.org.uk • www.shakespeare.org.uk •
Dir.: *Roger Pringle* •
Special Museum
Tudor farmhouse, home of Mary Arden (Shakespeare's mother) before her marriage to John Shakespeare, exhibitions and displays of life and work on the land over the centuries ... 45782

Nash's House and New Place, Shakespeare Birthplace Trust, Chapel St, Stratford-upon-Avon CV37 6EP • T: +44 1789 292325 • F: 296083 • info@shakespeare.org.uk • www.shakespeare.org.uk •
Dir.: *Roger Pringle* •
Special Museum
Nash's House, former home of Thomas Nash, husband of Elizabeth Hall (Shakespeare's grand-daughter), adjacent to stunning Elizabethan-style Knott garden set in foundations of New Place, Shakespeare's home from 1597 until he died in 1616 ... 45783

Royal Shakespeare Company Collection, Royal Shakespeare Theatre, Waterside, Stratford-upon-Avon CV37 6BB • T: +44 1789 296655 • F: 262870 • david.howells@RSC.org.uk •
Cur.: *David Howells* •
Performing Arts Museum – 1881
Relics of famous actors and actresses, portraits of actors, paintings of scenes from Shakespeare's plays, sculptures, scenery and costume designs, costume coll plus props, production photographs, programmes, posters, stage equipment – archive ... 45784

Shakespeare Birthplace Trust, Shakespeare Centre, Henley St, Stratford-upon-Avon CV37 6QW • T: +44 1789 204016 • F: 296083 • info@shakespeare.org.uk • www.shakespeare.org.uk •
Dir.: *Roger Pringle* •
Special Museum – 1847
16th c timbered house, Shakespeare's birthplace, memorabilia, ceramics, books, documents, pictures – specialist Shakespeare library, records office, education department ... 45785

The Teddy Bear Museum, 19 Greenhill St, Stratford-upon-Avon CV37 6LF • T: +44 1789 293160 • F: 87413454 • info@theteddybearmuseum.com • www.theteddybearmuseum.com •
Dir.: *Sylvia Coote* •
Special Museum – 1988 ... 45786

Strathaven

John Hastie Museum, 8 Threestanes Rd, Strathaven ML10 6DX • T: +44 1357 521257 • lowparksmuseum@southlanarkshire.gov.uk • www.southlanarkshire.gov.uk •
Resp.: *Robert Smith* •
Historical Museum / Decorative Arts Museum
History of Strathaven, with special displays on the weaving industry, Covenanting and the Radical Uprising, coll of porcelain ... 45787

Strathpeffer

Highland Museum of Childhood, The Old Station, Strathpeffer IV14 9DH • T: +44 1997 421031 • F: 421031 • info@highlandmuseumofchildhood.org.uk • www.highlandmuseumofchildhood.org.uk •
Cur.: *Jennifer Maxwell* •
Special Museum – 1992
Dolls, toys, children's costume ... 45788

Strathpeffer Spa Pumping Room Exhibition, Park House Studio, The Square, Strathpeffer IV14 9DL • T: +44 1997 420124 •
Science&Tech Museum ... 45789

Street

Shoe Museum, c/o C. & J. Clark Ltd., 40 High St, Street BA16 0YA • T: +44 1458 842169 • F: 842226 • linda.stevens@clarks.com •
Dir.: *John Keery* •
Special Museum – 1974 ... 45790

Stretham

Stretham Old Engine, Green End Ln, Stretham CB6 3LE • T: +44 1223 860895 •
Science&Tech Museum
Pumping station (1831) 45791

Stromness

Pier Arts Centre, 28-30 Victoria St, Stromness KW16 3AA • T: +44 1856 850209 • F: 851462 • info@pierartscentre.com • www.pierartscentre.com •
Dir.: *Neil Firth* •
Public Gallery / Fine Arts Museum – 1979
20th c art, paintings, Barbara Hepworth, Naum Gabo, Ben Nicholson, Alfred Wallis – temporary exhibition of contemporary art; reference library 45792

Stromness Museum, 52 Alfred St, Stromness KW16 3DF • T: +44 1856 850025 •
Cur.: *B. Wilson* •
Local Museum – 1837
Orkney maritime and local history, natural history, birds and fossils, Stone Age finds, ship models, Invit carvings, fishing 45793

Stroud

The Museum in the Park, Stratford Park, Stratford Rd, Stroud GL5 4AF • T: +44 1453 763394 • F: 752400 • museum@stroud.gov.uk • www.stroud.gov.uk •
Cur.: *Susan P. Hayward* •
Local Museum – 1899
Geology, archaeology, fine and decorative arts, social hist 45794

Styal

Quarry Bank Mill & Styal Estate, Quarry Bank Rd, Styal SK9 4LA • T: +44 1625 527468 • F: 539267 • quarrybankmill@nationaltrust.org.uk • www.quarrybankmill.org.uk •
Head: *Andrew Backhouse* •
Science&Tech Museum / Historic Site – 1976
Working textile museum, hist of cotton, skilled machine demonstration, steam engines, most powerful waterwheel, water turbine 45795

Sudbury, Derbyshire

Museum of Childhood, Sudbury Hall, Sudbury, Derbyshire DE6 5HT • T: +44 1283 585305 •
F: 585139 • sudburyhall@nationaltrust.org.uk •
Dir.: *Carolyn Aldridge* •
Special Museum – 1974
Betty Cadbury coll of playthings past – education dept. 45796

Sudbury, Suffolk

Gainsborough's House, 46 Gainsborough St, Sudbury, Suffolk CO10 2EU • T: +44 1787 372958 •
F: 376991 • mail@gainsborough.org • www.gainsborough.org •
Dir.: *Diane Perkins* •
Fine Arts Museum – 1958
Paintings, drawings and prints by Thomas Gainsborough 45797

Sunderland

Military Vehicle Museum, c/o North East Land Sea and Air Museum's, Old Washington Rd, Sunderland SR5 3HZ • T: +44 191 5190662 • info@neam.org.uk • www.military-museum.org.uk •
Military Museum – 1983
Military vehicles and artefacts since 1900, soldiers life. North East at War Display – Library and archives (by prior appointment) 45798

Monkwearmouth Station Museum, North Bridge St, Sunderland SR5 1AP • T: +44 191 5677075 •
F: 5109415 • juliet.horsley@tyne-wear-museums.org.uk • www.twmuseums.org.uk •
Dir.: *Juliet Horsley* •
Science&Tech Museum – 1973
Land transport in North East England, rail exhibits and other land transport items 45799

National Glass Centre, Liberty Way, Sunderland SR6 0GL • T: +44 191 5155555 • F: 5155556 • info@nationalglasscentre.com • www.nationalglasscentre.com •
Exec. Dir.: *Katherine Pearson* •
Decorative Arts Museum / Public Gallery 45800

North East Aircraft Museum, Old Washington Rd, Sunderland SR5 3HZ • T: +44 191 5190662 • www.neam.co.uk •
Dir.: *William Fulton* •
Science&Tech Museum
30 aircraft, incl. Vulcan bomber 45801

Northern Gallery for Contemporary Art, City Library and Arts Centre, Fawcett St, Sunderland SR1 1RE •
T: +44 191 5141235 • F: 5148444 • ngca@sunderland.gov.uk • www.ngca.co.uk •
Dir.: *Alistair Robinson* •
Public Gallery 45802

Ryhope Engines Museum, Pumping Station, Ryhope, Sunderland SR2 0ND • T: +44 191 5210235 • ryhope.engines@virgin.net • www.ryhopeengines.org.uk •
Science&Tech Museum – 1973
Mechanical engineering design and construction, beam engines and pumps, history of water supply, pair of beam engines (1868) complete with boilers, chimneys etc. 45803

Saint Peter's Church, Vicarage, Saint Peter's Way, Sunderland SR6 0DY • T: +44 191 5160135 • monkwearmouth.parish@durham.anglican.org • www.parishofmonkwearmouth.co.uk •
Religious Arts Museum / Historic Site 45804

Sunderland Museum, Burdon Rd, Sunderland SR1 1PP • T: +44 191 5532323 • F: 5537828 • sheryl.muxworthy@tyne-wear-museums.org.uk • www.twmuseums.org.uk •
Dir.: *Alec Coles* • Cur.: *Juliet Horsley* (Fine and Applied Art) • *M.A. Routledge* (Social History) • Sc. Staff: *Jo Cunningham* • *H.M. Sinclair* (Education) •
Decorative Arts Museum / Fine Arts Museum / Natural History Museum – 1846
Pottery and glass, Wearside paintings, prints, etchings and photographs, Wearside maritime history, British silver, zoology, botany, geology, industries, archaeology 45805

Sutton, Norwich

Sutton Windmill and Broads Museum, Mill Rd, Sutton, Norwich NR12 9RZ • T: +44 1692 581195 •
F: 583214 •
Cur.: *Chris Nunn* •
Local Museum
Social history, tricycles, engines, pharmacy, tobacco 45806

Sutton-on-the-Forest

Sutton Park, Main St, Sutton-on-the-Forest YO61 1DP • T: +44 1347 810249 • F: 811251 • suttonpark@fsbdial.co.uk • www.statelyhome.co.uk •
Dir.: *Sir Reginald Sheffield* •
Fine Arts Museum / Decorative Arts Museum
Furniture, paintings, porcelain, plasterwork by Cortese 45807

Swaffham, Norfolk

Swaffham Museum, 4 London St, Swaffham, Norfolk PE37 7DQ • T: +44 1760 721230 • enquiries@swaffhammuseum.co.uk • www.swaffhammuseum.co.uk •
Sc. Staff: *David Wickerson* •
Local Museum – 1985
Local and social history, hand crafted figures from English literature 45808

Swalcliffe

Swalcliffe Barn, Shipston Rd, Swalcliffe OX15 5ET •
T: +44 1295 788278 •
Local Museum
Agricultural and trade vehicles, local history 45809

Swanage

Swanage Museum, Tithe Barn Museum and Art Centre, Church Hill, Swanage BH19 1HU • T: +44 1929 427174 • mail@swanagemuseum.org.uk • www.swanagemuseum.org.uk •
Cur.: *David Haysom* •
Local Museum / Fine Arts Museum / Archaeology Museum – 1976
Fossils of the area, examples of the various beds of Purbeck stone, exhibitions of archaeological, architectural and social history material relating to the district, paintings and sculpture 45810

Swanage Railway, Station House, Swanage BH19 1HB • T: +44 1929 425800 • F: 426680 • general@swanrail.freeserve.co.uk • www.swanagerailway.co.uk •
Man.: *Michael Scott* •
Science&Tech Museum
Coll of goods vehicles, locomotives 45811

Swansea

Egypt Centre, c/o University of Wales Swansea, Singleton Park, Swansea SA2 8PP • T: +44 1792 295960 • F: 295739 • c.a.graves-Brown@swansea.ac.uk • www.swansea.ac.uk/egypt •
Cur.: *Carolyn Graves-Brown* • Dep. Cur.: *Wendy Goodridge* •
University Museum / Historical Museum / Archaeology Museum – 1998 45812

Glynn Vivian Art Gallery, Alexandra Rd, Swansea SA1 5DZ • T: +44 1792 516900 • F: 516903 • glynn.vivian.gallery@swansea.gov.uk • www.glynnviviangallery.org •
Dir.: *Jenni Spencer-Davies* •
Fine Arts Museum – 1911
20th c Welsh art, Swansea and Nantgarw pottery and porcelain, glass 45813

Mission Gallery, Gloucester Pl, Maritime Quarter, Swansea SA1 1TY • T: +44 1792 652016 •
F: 652016 • missiongallery@btconnect.com • www.missiongallery.co.uk •
Dir.: *Jane Phillips* •
Public Gallery
Contemporary arts and crafts 45814

National Waterfront Museum, Maritime and Industrial Museum, Maritime Quarter, Oystermouth Rd, Swansea SA1 1SN • T: +44 1792 638950 • F: 638956 • waterfrontmuseum@nmgw.ac.uk • www.nmgw.ac.uk/nwms/ •
Science&Tech Museum 45815

Swansea Museum, Victoria Rd, Maritime Quarter, Swansea SA1 1SN • T: +44 1792 653763 •
F: 652585 • swansea.museum@swansea.gov.uk • www.swansea.gov.uk •
Cur.: *Garethe El-Tawab* •
Local Museum – 1835
Social history, archaeology, ceramics, natural history, Welsh ethnography, numismatic, maps, topography, photography, manuscripts – library 45816

Swindon

Lydiard House, Lydiard Tregoze, Swindon SN5 3PA •
T: +44 1793 770401 • F: 877909, 770968 • lydiardpark@swindon.gov.uk • www.swindon.gov.uk/lydiardhouse •
Dir.: *Sarah Finch-Crisp* •
Fine Arts Museum / Decorative Arts Museum – 1955
Palladian mansion, ancestral home of the Viscount of Bolingbroke, furniture, family portrait coll – 17th c parish church 45817

Royal Wiltshire Yeomanry Museum, Yeomanry House, Church Pl, Swindon SN1 5EH • T: +44 1793 523865 • F: 529350 •
Military Museum
Military uniforms, militaria 45818

Steam Museum of the Great Western Railway, Kemble Dr, Swindon SN2 2TA • T: +44 1793 466646 •
F: 466615 • steampostbox@swindon.gov.uk • www.steam-museum.org.uk •
Cur.: *Felicity Ball* •
Science&Tech Museum – 1962/2000
Hist of railway, hist locomotives, models, illustrations 45819

Swindon Museum and Art Gallery, Bath Rd, Swindon SN1 4BA • T: +44 1793 466556 • F: 484141 • libraries@swindon.gov.uk • www.swindon.gov.uk/heritage •
Dir.: *Robert Dickinson* • Cur.: *Isobel Thompson* • Art Gallery: *Rosalyn Thomas* •
Local Museum / Fine Arts Museum – 1920
Numismatics, natural hist, ethnography, 20th-21th c British paintings, local archaeology and social history 45820

Symbister

Hanseatic Booth or The Pier House, Symbister ZE2 9AA • T: +44 1806 566240 •
Dir.: *J.L. Simpson* •
Local Museum 45821

Tain

Tain and District Museum, Tower St, Tain IV19 1DY •
T: +44 1862 894089 • info@tainmuseum.org.uk • www.tainmuseum.org.uk •
Head: *Margaret Urquhart* •
Local Museum – 1966
Clan Ross, Tain silver, local artefacts, photos – archive 45822

Tamworth

Tamworth Castle Museum, The Holloway, Tamworth B79 7NA • T: +44 1827 709626 • F: 709630 • heritage@tamworth.gov.uk • www.tamworthcastle.co.uk •
Local Museum – 1899
Local hist, archaeology, Saxon and Norman coins, costumes, heraldry, arms – archives 45823

Tangmere

Military Aviation Museum, Tangmere Airfield, Tangmere PO20 2ES • T: +44 1243 775223 •
F: 789490 • admin@tangmere-museum.org.uk • www.tangmere-museum.org.uk •
Cur.: *Alan Bower* •
Military Museum
British world speed record aircraft-Meteorand Hunter – Library 45824

Tarbert

Mercantile Arts Centre Kintyre, An Tairbeart, Campbeltown Rd, Tarbert PA29 6SX • T: +44 1880 821212 • info@mackandmags.co.uk • www.mackandmags.co.uk •
Dir.: *Kate MacDonald* •
Public Gallery
Scottish artists and craftspeople 45825

Tarbolton

Bachelor's Club, Sandgate St, Tarbolton KA5 5RB •
T: +44 1292 541940 •
Man.: *David Rodger* •
Special Museum – 1938
Memorabilia of Robert Burns and his friends, period furnishings, free masonery 45826

Tarporley

Beeston Castle, Tarporley CW6 9TX • T: +44 1829 260464 • F: 260464 • customers@english-heritage.org.uk • www.english-heritage.org.uk •
Cur.: *Andrew Morrison* •
Archaeology Museum
Archeological finds 45827

Tattershall

Guardhouse Museum, Tattershall Castle, Tattershall LN4 4LR • T: +44 1526 342543 • F: 348826 • tattershallcastle@nationaltrust.org.uk • www.nationaltrust.org.uk •
Archaeology Museum
Model of the Castle (17th c), Lord Curzon artefacts, fossils, axe-heads, archaeological material, pottery, glass and metal objects 45828

Taunton

Somerset County Museum, The Castle, Castle Green, Taunton TA1 4AA • T: +44 1823 320200 •
F: 320229 • county-museums@somerset.gov.uk • www.somerset.gov.uk/museums •
Local Museum
Archaeology, geology, biology, pottery, costume and textiles and other applied crafts 45829

Somerset Cricket Museum, 7 Priory Av, Taunton TA1 1XX • T: +44 1823 275893 • somersetcricket.museum@btinternet.com •
Special Museum
Cricket memorabilia – library 45830

Somerset Military Museum, Taunton Castle, Taunton TA1 4AA • T: +44 1823 320201 • F: 320229 • info@sommilmuseum.org.uk • www.sommilmuseum.org.uk •
Dir.: *D. Eliot* •
Military Museum – 1921
Militaria 45831

Teignmouth

Teignmouth and Shaldon Museum, 29 French St, Teignmouth TQ14 8ST • T: +44 16267 777041 • enquiries@teignmuseum.org.uk •
Dir.: *Carol Chambers* •
Local Museum / Historical Museum – 1978
Late 16th c Venetian trading vessel, various artefacts; local maritime history (Admiral Pellew, marine artist Thomas Luny); Haldon aerodrome, Brunel's athmospheric railway 45832

Telford

Blists Hill Victorian Town, Legges Way, Madeley, Telford TF8 7AW • T: +44 1952 601010 • F: 588016 • tic@ironbridge.org.uk • www.ironbridge.org.uk •
Dir.: *Traci Dix-Williams* •
Open Air Museum
Working Victorian town: foundry, candle factory, saw mill, printing shop 45833

Templepatrick

Patterson's Spade Mill, 751 Antrim Rd, Templepatrick BT39 • T: +44 13967 51467 • colin.dawson@nationaltrust.org.uk •
Dir.: *Ruth Laird* •
Science&Tech Museum
Last surviving water driven spade mill in Ireland 45834

Tenbury Wells

Tenbury and District Museum, Goff's School, Cross St, Tenbury Wells WR15 8EF • T: +44 1299 832143 •
Cur.: *John Greenhill* •
Local Museum
Local history, bath and fountain from spa, farming, Victorian kitchen 'Tedbury Advertisers' coll 45835

Tenby

Tenby Museum and Art Gallery, Castle Hill, Tenby SA70 7BP • T: +44 1834 842809 • F: 842809 • tenbymuseum@hotmail.com • www.tenbymuseum.free-online.co.uk •
Cur.: *Jon Beynon* •
Fine Arts Museum / Local Museum / Natural History Museum – 1878
Geology, archaeology, natural an social hist of Pembrokeshire, local hist of Tenby 45836

Tenterden

Colonel Stephens Railway Museum, Kent and East Sussex Railway, Tenterdon Town Stn, Station Rd, Tenterden TN30 6HE • T: +44 1580 765350 •
F: 763468 • colonel@hfstephens-museum.org.uk • www.hfstephens-museum.org.uk •
Cur.: *John Miller* •
Special Museum
Story of LtCol Holman F. Stephens, his family backround and carrier as soldier and engineer 45837

Ellen Terry Memorial Museum, Smallhythe Pl, Tenterden TN30 7NG • T: +44 1580 762334 •
F: 761960 • smallhytheplace@ntrust.org.uk • www.nationaltrust.org.uk •
Performing Arts Museum – 1929
Memorabilia on actress Dame Ellen Terry, costumes, Edith Craig archive 45838

Kent and East Sussex Railway, Tenterden Town Stn, Station Rd, Tenterden TN30 6HE • T: +44 1580 765155 • F: 765654 • kesroffice@aol.com • www.kesr.org.uk •
Man.: *Graham Baldwin* •
Science&Tech Museum – 1974
1900 railway with 5 stations, steam and diesel locomotives, carriages, various goods vehicles 45839

U

Tenterden and District Museum, Station Rd, Tenterden TN30 6HN • T: +44 1580 764310 • F: 766648 • www.tenterdentown.co.uk • Cur.: *Debbie Greaves* • Local Museum – 1976
Hist of Tenterden, weights and measures of the former Borough, scale model of the town as it was in the mid 19th c, domestic and business life of Tenterden in the 18th and 19th c ... 45840

Tetbury

Tetbury Police Museum, 63 Long St, Tetbury GL8 8AA • T: +44 1666 504670 • F: 504670 • tetbury.council@virgin.net • www.tetbury.org • Cur.: *Gwen Bruist* • Historical Museum
Cells, artefacts, uniforms, history of the Gloucestershire force ... 45841

Tewkesbury

John Moore Countryside Museum, 41 Church St, Tewkesbury GL20 5SN • T: +44 1684 297174 • simonlawton@btconnect.com • www.gloster.demon.co.uk/jmcm/index.html • Cur.: *Simon R. Lawton* • Special Museum
Row of medieval cottages, furnishings, domestic environment, British woodland and wetland wildlife, writer and naturalist John Moore ... 45842

Little Museum, 45 Church St, Tewkesbury GL20 5SN • T: +44 1684 297174 • Dir.: *Simon R. Lawton* • Local Museum ... 45843

Tewkesbury Museum, 64 Barton St, Tewkesbury GL20 5PX • T: +44 1684 292901 • F: 292277 • info@tewkesburymuseum.org • www.tewkesburymuseum.org • Cur.: *Maggie Thurnton* • Local Museum / Archaeology Museum – 1962
Hist, trades and industries of Tewkesbury, a diorama of the Battle of Tewkesbury in 1471, models of fairground machines, coll of woodworking tools ... 45844

Thaxted

Guildhall, Town St, Thaxted CM6 24D • T: +44 1371 831339 • F: 830418 • www.thaxted.co.uk • Chm.: *P.G. Leeder* • Local Museum
Thaxted archives coll and old Thaxted photographs ... 45845

Thetford

Ancient House Museum, White Hart St, Thetford IP24 1AA • T: +44 1842 752599 • F: 752599 • ancient.house.museum@norfolk.gov.uk • www.museums.norfolk.gov.uk • Cur.: *Oliver Bone* • Local Museum – 1924
Early tudor house, local history and industries ... 45846

Euston Hall, Thetford IP24 2QP • T: +44 1842 766366 • www.eustonhall.co.uk • Dir.: *Duke of Grafton* • Fine Arts Museum ... 45847

Thetford Collection, c/o Library, Raymond St, Thetford IP24 2EA • T: +44 1842 752048 • F: 750125 • thetford.lib@norfolk.gov.uk • Fine Arts Museum / Local Museum
Duleep Singh Picture coll, Thomas Paine coll local life and history – archives ... 45848

Thirsk

Thirsk Museum, 14-16 Kirkgate, Thirsk YO7 1PQ • T: +44 1845 527707, 524510 • thirskmuseum@supanet.com • thirskmuseum.org • Local Museum – 1975
Local life, hist and industry, exhibits of cobblers' and blacksmith's tools, veterinary and medical equipment, Victorian clothing, esp underwear, children's games, archaeological finds, medieval pottery, cameras and cricket mementoes, agricultural implements, pharmaceutical products and equipment, photographs ... 45849

Thornbury, South Gloucestershire

Thornbury and District Museum, 4 Chapel St BS35 2BJ, mail addr: c/o The Town Hall, 35 High Street, BS35 2AR Thornbury, South Gloucestershire • T: +44 1454 857774 • enquiries@thornburymuseum.org.uk • www.thornburymuseum.org.uk • Local Museum – 1986
Local social and landscape history, local crafts and industry, farming, local archaeology – archive ... 45850

Thorney

Thorney Heritage Museum, The Tankyard, Station Rd, Thorney PE6 0QE • T: +44 1733 270908 • dot.thorney@tesco.net • www.thorney-museum.org.uk • Local Museum
History of the village ... 45851

Thornhaugh

Sacrewell Farm and Country Centre, Sacrewell, Thornhaugh PE8 6HJ • T: +44 1780 782254 • info@sacrewell.fsnet.co.uk • www.sacrewell.org.uk • Dir.: *Peter Thompson* • Agriculture Museum – 1987 ... 45852

Thornhill, Central

Farmlife Centre, Dunaverig, Ruskie, Thornhill, Central FK8 3QW • T: +44 1786 850277 • F: 850404 • Dir.: *Sarah Stewart* • Agriculture Museum
A small mixed farm, the story of the parish of Ruskie from ancient times to the present day, models, photographs, charts, implements ... 45853

Thornhill, Dumfriesshire

Alex Brown Cycle History Museum, Drumlanrig Castle, Thornhill, Dumfriesshire DG3 4AQ • T: +44 1848 31555 • alex@calderbrown.com • Dir.: *Alex Brown* • Science&Tech Museum ... 45854

Drumlanrig Castle, Thornhill, Dumfriesshire DG3 4AQ • T: +44 1848 330248 • F: 331682 • bre@drumlanrigcastle.org.uk • www.drumlanrigcastle.org.uk • Dir.: *Claire Fisher* • Decorative Arts Museum ... 45855

Thurgoland

Wortley Top Forge, Forge Ln, Thurgoland S35 7DN • T: +44 114 2887576 • www.topforge.co.uk • Resp.: *Gordon Parkinson* • Science&Tech Museum
Iron works, water wheels, hammers, cranes, stationary steam engine coll ... 45856

Thurso

Thurso Heritage Museum, Town Hall, High St, Thurso KW14 8AG • T: +44 1847 62459 • www.caithness.org/community/museums/thursoheritage • Dir.: *E. Angus* • Local Museum – 1970
Pictish stones ... 45857

Tilbury

Tilbury Fort, 2 Office Block, The Fort, Tilbury RM18 7NR • T: +44 1375 858489 • www.english-heritage.org.uk • Cur.: *Jan Summerfield* • Cust.: *Bernard Truss* • Military Museum ... 45858

Tiptree

Tiptree Museum, Tiptree CO5 0RF • T: +44 1621 814524 • F: 814555 • tiptree@tiptree.com • www.tiptree.com • Local Museum
Jam making equipment, story of village life ... 45859

Tiverton

Tiverton Museum of Mid Devon Life, Beck's Sq, Tiverton EX16 6PJ • T: +44 1884 256295 • curator04@tivertonmuseum.org.uk • www.tivertonmuseum.org.uk • Dir.: *Patrick Brooke* • Cur.: *Judith Elsdon* • Science&Tech Museum / Local Museum / Agriculture Museum – 1960
Wagons, agricultural implements, railway, clocks, industries, crafts – library ... 45860

Tobermory

Mull Museum, Columbia Bldg, Main St, Tobermory PA75 6NY • T: +44 1688 302493 • F: 302454 • olivebrown@msn.com • Cur.: *Dr. W.H. Clegg* • Local Museum
Isle of Mull history – library, archives ... 45861

Toddington

Gloucestershire Warwickshire Railway, Railway Station, Toddington GL54 5DT • T: +44 1242 621405 • enquiries@gwsr.com • www.gwsr.com • Science&Tech Museum
Stand and gauge steam and diesel locomotives, carriages, wagons ... 45862

Tolpuddle

Tolpuddle Martyrs Museum, Main Rd, Tolpuddle DT2 7EH • T: +44 1305 848237 • F: 848237 • jpickering@tuc.org.uk • www.tolpuddlemartyrs.org.uk • Head: *Nigel Costely* • Historical Museum
Story of the six agricultural workers transported to Australia in 1834 after forming a trade union ... 45863

Tomintoul

Tomintoul Museum, The Square, Tomintoul AB3 9ET • T: +44 1807 673701 • F: 673701 • museums@moray.gov.uk • www.moray.org/museums • Local Museum
Local and natural hist, geology ... 45864

Topsham

Topsham Museum, Holman House, 25 The Strand, Topsham EX3 0AX • T: +44 1392 873244 • museum@topsham.org • www.devonmuseums.net/topsham • Local Museum – 1967
Local history, crafts, shipbuilding, lacemaking, fishing ... 45865

Torquay

Torquay Museum, 529 Babbacombe Rd, Torquay TQ1 1HG • T: +44 1803 293975 • F: 294186 • ros.palmer@ukonline.co.uk • www.torquaymuseum.org • Cur.: *Ros Palmer* • Local Museum – 1844
Geology, entomology, botany, folk life, social hist – local studies, library ... 45866

Torre Abbey Historic House and Gallery, The King's Dr, Torquay TQ2 5JE • T: +44 1803 293593 • F: 215948 • torre-abbey@torbay.gov.uk • www.torre-abbey.org.uk • Dir.: *Dr. Michael Rhodes* • Cur.: *Leslie Retallick* • Fine Arts Museum / Historic Site – 1930
Paintings, furniture, glass, archaeology, historic building, the most complete medieval monastery in Devon and Cornwall ... 45867

Torrington

Dartington Crystal Glass Museum, Torrington EX38 7AN • T: +44 1805 626242 • F: 626263 • tours@dartington.co.uk • www.dartington.co.uk • Dir.: *Jon Courtiour* • Decorative Arts Museum – 1987
Dartington Crystal production from 1967 to present, non Dartington Crystal, historical glass from 1650 to today ... 45868

Torrington Museum, Town Hall, High St, Torrington EX38 8HN • T: +44 1805 624324 • Chm.: *Valerie Morris* • Local Museum
Local and social history, trade – archive ... 45869

Totnes

British Photographic Museum, Bowden House, Ashprington Rd, Totnes TQ9 7PW • T: +44 1803 863664 • www.bowdenhouse.co.uk • Cur.: *Christopher Petersen* • Science&Tech Museum – 1974
Moving-picture photographic equipment of the period 1875 to 1960 ... 45870

Totnes Costume Museum, Bogan House, 43 High St, Totnes TQ9 5NP • T: +44 1803 862857 • Cur.: *Julie Fox* • *Val Oatley* • Special Museum
Coll of period costume since mid 18th c ... 45871

Totnes Elizabethan House Museum, 70 Fore St, Totnes TQ9 5RU • T: +44 1803 863821 • Info@totnesmuseum.co.uk • www.devonmuseums.net/totnes • Dir.: *Kristin Saunders* • Local Museum – 1961
Period furniture, costumes, farm and household objects, archaeology, toys, computers – library, archive ... 45872

Trehafod

Rhondda Heritage Park, Lewis Merthyr Colliery, Coed Cae Rd, Trehafod CF37 7NP • T: +44 1443 682036 • F: 687420 • info@rhonddaheritagepark.com • www.rhonddaheritagepark.com • Dir.: *John Harrison* • Local Museum
Social, cultural and industrial history of the Rhondda and South Wales Valleys ... 45873

Tresco

Gallery Tresco, New Grimsby, Tresco TR24 0QF • T: +44 1720 424925 • F: 423105 • gallery@tresco.co.uk • www.gallerytresco.co.uk • Public Gallery
Artworks by local, Cornish and visiting artists ... 45874

Valhalla Museum, National Maritime Museum, Tresco Abbey Garden, Tresco TR24 0PU • T: +44 1720 424105 • F: 422868 • mikenelhams@tresco.co.uk • www.tresco.co.uk • Cur.: *Mike Nelhams* • Special Museum
Figureheads and ships' carvings from wrecks ... 45875

Tring

Natural History Museum, Akeman St, Tring HP23 6AP • T: +44 20 79426158 • F: 79426150 • r.prys-jones@nhm.ac.uk • www.nhm.ac.uk • Man.: *Dr. R. Prys-Jones* • Natural History Museum
Coll of birds ... 45876

Walter Rothschild Zoological Museum, Akeman St, Tring HP23 6AP • T: +44 20 79426171 • F: 79426150 • terw@nhm.ac.uk • www.nhm.ac.uk • Dir.: *Teresa Wild* • Natural History Museum – 1892
Ornithological coll of the Natural History Museum, London SW (research coll, not open to the public), zoological specimens, most of which were collected by Walter Rothschild (open to the public) ... 45877

Trowbridge

Trowbridge Museum, The Shires, Court St, Trowbridge BA14 8AT • T: +44 1225 751339 • F: 754608 • staff@trowbridgemuseum.co.uk • www.trowbridgemuseum.co.uk • Cur.: *Clare Lyall* • Science&Tech Museum – 1976
Textile machinery, West of England woollen industry ... 45878

Truro

Royal Cornwall Museum, River St, Truro TR1 2SJ • T: +44 1872 272205 • F: 240514 • enquiries@royalcornwallmuseum.org.uk • www.royalcornwallmuseum.org.uk • Dir.: *Hilary Bracegirdle* • Local Museum – 1818
Philip Rashleigh coll of minerals, John Opie, artist, Newlyn School paintings, old master drawings – Courtney library and archives ... 45879

Saint Mawes Castle, Saint Mawes, Truro TR2 3AA • T: +44 1326 270526 • F: 270526 • customers@english-heritage.org.uk • www.english-heritage.org.uk • Head Cust: *Keith Robson* • Local Museum
A well preserved, clover leaf design Tudor coastal fort situated in a strategic position overlooking the entrance to the Carrick Rd anchorage ... 45880

Tunbridge Wells

Tunbridge Wells Museum and Art Gallery, Civic Centre, Mount Pleasant, Tunbridge Wells TN1 1JN • T: +44 1892 554171 • F: 554131 • museum@tunbridgewells.gov.uk • www.tunbridgewells.gov.uk/museum • Man.: *Caroline Ellis* • Cur.: *Dr. Ian Beavis* • Local Museum / Public Gallery – 1885
Victorian paintings, local hist, natural hist, prints, drawings, dolls, toys, Tunbridge ware, archaeology, historic costume ... 45881

Turriff

Delgatie Castle, Turriff AB53 5TD • T: +44 1888 563479 • F: 563479 • Dir.: *Joan Johnson* • Historical Museum / Fine Arts Museum / Decorative Arts Museum
Arts, arms and armour, paintings, furniture, history ... 45882

Old Post Office Museum, 24 High St, Turriff AB53 4DS • T: +44 1888 562631 • Dir.: *Deirdre S. McAinsh* • Local Museum – 1999
Local goverment and history ... 45883

Session Cottage Museum, Session Close, Castle St, Turriff AB53 4AS • T: +44 1888 562631 • Dir.: *Deirdre S. McAinsh* • Historical Museum – 1979
Lighting, domestic irons ... 45884

Turton Bottoms

Turton Tower, Chapeltown Rd, Turton Bottoms BL7 0HG • T: +44 1204 852203 • F: 853759 • turton.tower@mus.lancscc.gov.uk • www.lancsmuseums.goc.uk • Decorative Arts Museum – 1952
National coll, furniture, 16th, 17th and 19th c, arms and armour, metalwork, paintings ... 45885

Twickenham

Marble Hill House, Richmond Rd, Twickenham TW1 2NL • T: +44 20 88925115 • F: 86079976 • www.english-heritage.org.uk • Cur.: *Cathy Power* • Fine Arts Museum – 1966
Overdoor and overmantel paintings by G.P. Painini ... 45886

Museum of Rugby → World Rugby Museum

Museum of the Royal Military School of Music, Kneller Hall, Twickenham TW2 7DU • T: +44 20 88985533 • F: 88987906 • armymusichq@aol.com • Cur.: *R.G. Swift* • Music Museum – 1935
Old music instruments ... 45887

Orleans House Gallery, Riverside, Twickenham TW1 3DJ • T: +44 20 88316000 • F: 87440501 • artsinfo@richmond.gov.uk • www.richmond.gov.uk/arts • Cur.: *Mark de Novellis* • Local Museum / Fine Arts Museum – 1972
Nearly 2.400 topographical paintings, prints and photographs of Richmond, Twickenham Hampton court area, Sir Richard Burton coll (paintings, portraits, photos, letters, personal objects, 1821-1890) ... 45888

Twickenham Museum, 25 The Embankment, Twickenham TW1 3DU, mail addr: Clarence Cottage, Hampton Court Rd, East Molesey KT8 9BY • T: +44 20 84080070, 89775198 • Local Museum ... 45889

World Rugby Museum, Twickenham Stadium, Rugby Rd, Twickenham TW1 1DZ • T: +44 20 88928877 • F: 88922817 • museum@rfu.com • www.rfu.com/museum • Cur.: *M. Rowe* •

U

Special Museum – 1996
World's largest coll of rugby football and pre-code football artefacts from 19th, 20th, and 21st c from all rugby-playing nations 45890

Tywyn

Narrow Gauge Railway Museum, Wharf Station, Tywyn LL36 9EY • T: +44 1654 710472 • F: 711755 • enquiries@ngrm.org.uk • www.talyllyn.co.uk/ngrm • Dir.: *Christopher White* • Cur.: *Brian Owen* •
Science&Tech Museum – 1956
Locomotives and wagons of Narrow Gauge Railways of Britain and Ireland 45891

Uckfield

Bluebell Railway, Sheffield Park Station, Uckfield TN22 3QL • T: +44 1825 720800 • F: 720804 • info@bluebell-railway.co.uk • www.bluebell-railway.co.uk •
Cur.: *John Potter* •
Science&Tech Museum – 1960
Train tickets, badges, timetables 45892

Heaven Farm Museum, Heaven Farm, Danehill, Uckfield TN22 3RG • T: +44 1825 790226 • F: 790881 • www.heavenfarm.co.uk •
Cur.: *John Butler* •
Agriculture Museum
A large coll of small farm tools and implements displayed in several Victorian farm buildings incl a barn, cowshed and oast house 45893

Uffculme

Coldharbour Mill, Working Wool Museum, Uffculme EX15 3EE • T: +44 1884 840960 • F: 840858 • info@coldharbourmill.org.uk • www.coldharbourmill.org.uk •
Cur.: *Jill E. Taylor* •
Science&Tech Museum – 1982
Turn-of-the-c worsted spinning machinery, spinning mule and looms, steam engine, new world tapestry 45894

Uffington

Tom Brown's School Museum, Broad St, Uffington SN7 7RA • T: +44 1367 820259 • museum@uffington.net • www.uffington.net/museum •
Cur.: *Sharon Smith* •
Historical Museum – 1984
150 editions of Tom Brown's schoolday by Thomas Hughes 45895

Ullapool

Ullapool Museum, 7-8 West Argyle St, Ullapool IV26 2TY • T: +44 1854 612987 • F: 612987 • admin@ullapoolmuseum.co.uk • www.ullapoolmuseum.co.uk •
Head: *Alex Eaton* •
Local Museum – 1988/96
Village tapestry and quilt, natural hist, education, genealogy, archaeology, social hist, fishing, religion, emigration – archive 45896

Ulverston

Laurel and Hardy Museum, 4c Upperbrook St, Ulverston LA12 7BH • T: +44 1229 582292 • F: 582292 •
Cur.: *Marion Grave* •
Special Museum
Memorabilia of Stan Laurel and Lucille Hardy, waxwork models 45897

Upminster

Upminster Mill, Saint Mary's Ln, Upminster RM14 1AU • T: +44 1708 226040 • bobsharp@ukonline.co.uk • www.upminsterwindmill.co.uk •
Head: *R.W.D. Sharp* •
Science&Tech Museum 45898

Upminster Tithe Barn Museum of Nostalgia, Hall Ln, Upminster RM14 1AU, mail addr: 123 Victor Close, Hornchurch RM12 4XU • T: +44 78 55633917 •
Cur.: *M.L. Cullen* •
Local Museum – 1978
Agriculture, horticulture, social history 45899

Uppermill

Saddleworth Museum and Art Gallery, High St, Uppermill OL3 6HS • T: +44 1457 874093 • F: 870336 • curator@saddleworthmuseum.co.uk • www.saddleworthmuseum.co.uk •
Cur.: *Peter Fox* •
Science&Tech Museum / Fine Arts Museum – 1962
Textile machinery, transport, art – archive 45900

Usk

Usk Rural Life Museum, Malt Barn, New Market St, Usk NP5 1AU • T: +44 1291 673777 • uskrurallife.museum@virgin.net • uskmuseum.members.easyspace.com •
Agriculture Museum – 1966
Agricultural and rural artefacts, vintage machinery 45901

Uttoxeter

Uttoxeter Heritage Centre, 34-36 Carter St, Uttoxeter ST14 8EU • T: +44 1889 567176 • F: 568426 • info@uttoxetercouncil.co.uk •
Local Museum
Local history 45902

Uxbridge

Beldam Gallery, Brunel University, Wilfred Brown Bldg, Cleveland Rd, Uxbridge UB8 3PH • T: +44 1895 273482 • F: 269735 • artcentre@brunel.ac.uk • www.brunel.ac.uk/artcentre •
Public Gallery / University Museum 45903

Ventnor

Ventnor Heritage Museum, 11 Spring Hill, Ventnor PO38 1PE • T: +44 1983 855407 • ventnorheritage@aol.com • www.ventnorheritage.org.uk •
Cur.: *Graham Bennett* •
Local Museum
Local history, old photographs, prints 45904

Waddesdon

Waddesdon Manor, The Rothschild Collection, Waddesdon HP18 0JH • T: +44 1296 653226 • F: 653212 • waddesdonmanor@nationaltrust.org.uk • www.waddesdon.org.uk •
Exec. Dir.: *Fabia Bromovsky* • Head: *Pippa Shirley* (Collections) •
Decorative Arts Museum / Fine Arts Museum – 1959
French decorative arts, English portraits, Dutch and Flemish painting, majolica, design drawings, textiles, glass, Renaissance jewellery, European arms, French, Italian and Flemish illuminated manuscripts, French books and bindings 45905

Wainfleet All Saints

Magdalen College Museum, Saint John St, Wainfleet All Saints PE24 4DL • T: +44 1754 880343 • doreenreg@aol.com •
Historical Museum / Local Museum 45906

Wakefield

Clarke Hall, Aberford Rd, Wakefield WF1 4AL • T: +44 1924 302700 • F: 302701 • clarkehall1@wakefield.gov.uk • www.wakefieldmuseum.org •
Head: *David Stockdale* •
Historic Site / Historical Museum
17th c gentleman farmer's house 45907

The Hepworth Wakefield (opening spring 2011), 1 Wentworth Terrace, Wakefield WF1 3QW • T: +44 1924 305900 • hepworthwakefield@wakefield.gov.uk • www.hepworthwakefield.com •
Fine Arts Museum – 2011
Works of Barbara Heoworth and Henry Moore, coll of british art 1900 til today, Wakefield and Yorkshire in pictures 45908

National Arts Education Archive, University od Leeds, c/o Lawrence Batley Centre & Gallery, Bretton Hall, West Bretton, Wakefield WF4 4LG • T: +44 113 3439134 • F: 3433913 • skielty@bretton.ac.uk • naea.leeds.ac.uk •
Dir.: *Prof. Ron George* •
Fine Arts Museum
Paintings, drawings, books, sculpture, ceramics, magazines, papers and letters, games, puzzles, videos, audiotapes, films, photographs and slides 45909

National Coal Mining Museum for England, Caphouse Colliery, New Rd, Overton, Wakefield WF4 4RH • T: +44 1924 848806 • F: 844567 • info@ncm.org.uk • www.ncm.org.uk •
Dir.: *Dr. Margaret Faull* •
Science&Tech Museum – 1988
Colliery site buildings incl original twin-cylinder steam winding engine, mining in the English coalfields, mining literature, photos, mine safety lamps, mining related coll – research library, photographic coll 45910

Nostell Priory, Doncaster Rd, Wakefield WF4 1QE • T: +44 1924 863892 • F: 865282 • www.nationaltrust.org.uk/nostellpriory •
Decorative Arts Museum
Chippendale furniture, 18th c silver, old master and British pictures, James Paine and Robert Adam's interiors 45911

Stephen G. Beaumont Museum, Fieldhead Hospital, Ouchthorpe Ln, Wakefield WF1 3SP • T: +44 1924 328654 • F: 327340 •
Cur.: *Michael J. McCarthy* •
Historical Museum
Originally West Riding Pauper Lunatic Asylum, plans, records, documents, scale model of 1818, surgical and medical equipment, tradesmen's tools 45912

Wakefield Art Gallery, Wentworth Terrace, Wakefield WF1 3QW • T: +44 1924 305796, 305900 • F: 305770 • museumsandarts@wakefield.gov.uk • www.wakefield.gov.uk/CultureAndLeisure/Galleries/WakefieldArtGallery •
Dir.: *Antonino Vella* •
Public Gallery – 1934
20th c British and other paintings, sculpture 45913

Wakefield Museum, Wood St, Wakefield WF1 2EW • T: +44 1924 305351 • F: 305353 • museumsandarts@wakefield.gov.uk • www.wakefield.gov.uk/cultureandleisure/museums/wakefield •
Dir.: *Colin MacDonald* •
Local Museum – 1920
Local history, natural history, costumes, photogr – Resource Centre 45914

Wallingford

Wallingford Museum, Flint House, High St, Wallingford OX10 0DB • T: +44 1491 835065 •
Dir.: *Stuart Dewey* • Cur.: *Judy Dewey* •
Local Museum
History of the town, items date from the 19th and 20th c, Civil War relics, a model of Wallingford Castle 45915

Wallsend

Buddle Arts Centre, 258b Station Rd, Wallsend NE28 8RH • T: +44 191 2007132 • F: 2007142 • buddle@ntynearts.demon.co.uk •
Public Gallery
Prints 45916

Segedunum Roman Fort, Baths and Museum, Buddle St, Wallsend NE28 6HR • T: +44 191 2369347 • F: 2955858 • segedunum@twmuseums.org.uk • www.twmuseums.org.uk •
Cur.: *Geoff Woodward* •
Archaeology Museum 45917

Walmer

Walmer Castle, Kingsdown Rd, Walmer CT14 7LJ • T: +44 1304 364288 • F: 364826 • sally.mewton-hynds@english-heritage.org.uk • www.english-heritage.org.uk •
Cur.: *Jo Gray* • Cust.: *Sally Mewton-Hynds* • Sc. Staff: *Chris Pascall* •
Decorative Arts Museum
Wellington memorabilia 45918

Walsall

Jerome K. Jerome Birthplace Museum, Belsize House, Bradford St, Walsall WS1 1PN • T: +44 1922 627686, 629000 • F: 721065, 632824 • museum@walsall.gov.uk • www.jeromekjerome.com •
Pres.: *Jeremy Nicholas* •
Special Museum
Story of the life and work of the famous author 45919

The New Art Gallery Walsall, Gallery Sq, Walsall WS2 8LG • T: +44 1922 654400 • F: 654401 • info@artatwalsall.org.uk • www.artatwalsall.org.uk •
Dir.: *Stephen Snoddy* • Cur.: *Jo Digger* (Fine Art) • *Deborah Robinson* (Exhibitions) •
Public Gallery – 1887/2000
Garman Ryan coll: works by Matisse, Modigliani, Pissarro, Manet, Constable, Van Gogh, prints and sculpture; non-Western artefacts, social and local hist of Walsall 45920

Walsall Leather Museum, Littleton St W, Walsall WS2 8EQ • T: +44 1922 721153 • F: 725827 • leathermuseum@walsall.gov.uk • www.walsall.gov.uk/leathermuseum •
Dir.: *Michael Glasson* •
Special Museum – 1988
Designer leatherwork, lethergoods, saddlery, harness trades – library, photo archive 45921

Walsall Local History Centre, Essex St, Walsall WS2 7AS • T: +44 1922 721305 • F: 634954 • localhistorycentre@walsall.gov.uk •
Local Museum
Archives 45922

Walsall Museum, Lichfield St, Walsall WS1 1TR • T: +44 1922 653116 • F: 632824 • museum@walsall.gov.uk • www.walsall.gov.uk/museums •
Cur.: *Stuart Warburton* •
Local Museum
Locally made goods, social history, costume, contemporary colls, local industry 45923

Waltham Abbey

Epping Forest District Museum, 39-41 Sun St, Waltham Abbey EN9 1EL • T: +44 1992 716882 • F: 700427 • museum@eppingforestdc.gov.uk • www.eppingforestdistrictmuseum.org.uk •
Dir.: *Tony O'Connor* •
Local Museum
Objects, documents, photographs and pictures relating to the social history of the area, an bi-annual 'Art in Essex' exhibition of work by county artists 45924

Royal Gunpowder Mills, Beaulieu Dr, Waltham Abbey EN9 1JY • T: +44 1992 707370 • F: 707372 • info@royalgunpowdermills.com • www.royalgunpowdermills.com •
Dir.: *Trevor Knapp* •
Science&Tech Museum
Gunpowder a.o. explosives production since the 1660s, largest heronry in Essex, producers, transport equipment for the internal canal and railway network 45925

Walton-on-the-Naze

Walton Maritime Museum, Old Lifeboat House, East Tce, Walton-on-the-Naze CO14 8AA • T: +44 1255 678259 •
Resp.: *Margaret Sandell* •
Historical Museum / Local Museum
Geology, fossils, flints, military history, local lifeboats, piers, old mills 45926

Wanlockhead

Museum of Lead Mining, Wanlockhead ML12 6UT • T: +44 1659 74387 • F: 74481 • miningmuseum@hotmail.com • www.leadminingmuseum.co.uk •
Dir.: *Brian Montgomery* •
Science&Tech Museum / Natural History Museum – 1974
Rare mineral coll, unique mining engines – library 45927

Wantage

Charney Bassett Mill, 7 Clement Close, Wantage OX12 7ED • T: +44 1235 868677, 210612 • bruce.hedge@virgin.net •
Science&Tech Museum
Early 19th c watermill 45928

Vale and Downland Museum, Church St, Wantage OX12 8BL • T: +44 1235 771447 • F: 760991 • museum@wantage.com • www.wantage.com/museum •
Cur.: *Tony Hadland* •
Local Museum – 1983
17th c merchants house 45929

Ware

Ware Museum, The Priory Lodge, Ware SG12 9AL • T: +44 1920 487848 • www.waremuseum.org.uk •
Local Museum
Local and historical objects, archaeology, malting industry, railcar manufacturers 45930

Wareham

Wareham Town Museum, Town Hall, East St, Wareham BH20 4NS • T: +44 1929 553448 • F: 553521 • info@wtm.org.uk • www.wtm.org.uk •
Cur.: *Michael O'Hara* •
Local Museum 45931

Warminster

Dewey Museum, Library, Three Horseshoes Mall, Warminster BA12 9BT • T: +44 1985 216022 • F: 846332 • deweywar@ukonline.co.uk •
Cur.: *Alwyn Hardy* •
Local Museum – 1972
History of Warminster and the surrounding area, the geology of the area 45932

Infantry and Small Arms School Corps Weapons Collection, Land Warfare Centre, Warminster BA12 0DJ • T: +44 1985 222487 • F: 222972 • regsecsasc@aol.com • www.sasc-comradeo.org •
Military Museum – 1853
Coll of infantry firearms, dating from the 16th c to the present day, and ranging from small pocket pistols to heavy anti-tank weapons – Travers library 45933

Longleat House, Warminster BA12 7NW • T: +44 1985 844400 • F: 844885 • enquiries@longleat.co.uk • www.longleat.co.uk •
Dir.: *Marquess of Bath* •
Historical Museum / Fine Arts Museum – 1541
6th Marquess of Bath's coll of Churchilliana and Hitleriana and children's books; incunabula, medieval manuscripts, porcelain, Italian/Dutch paintings, English portraits, French/Italian/English furniture – library, archives 45934

Warrington

Walton Hall Heritage Centre, Walton Hall Gardens, Walton Lea Rd, Warrington WA4 6SN • T: +44 1925 601617 • www.warrington.gov.uk/waltongardens •
Dir.: *Keith Webb* •
Local Museum – 1995 45935

Warrington Museum and Art Gallery, Museum Street, Cultural Quarter, Warrington WA1 1JB • T: +44 1925 442733 • F: 443257 • museum@warrington.gov.uk • www.warrington.gov.uk/museum •
Man.: *Janice Hayel* •
Local Museum / Public Gallery – 1848
Local hist, Romano-British finds from Wilderspool, prehistory, ethnology, natural hist, local industries, early English watercolours, ceramics, local longcase clocks, temporary exhibits 45936

Warwick

The Queen's Own Hussars Museum, Lord Leycester Hospital, High St, Warwick CV34 4BH • T: +44 1926 492035 • F: 492035 • trooper@qohm-fsnet.co.uk • www.army.mod.uk •
Dir.: *P.J. Timmons* •
Military Museum – 1966
Regimental ans social history of the 3rd King's own Hussars and the 7th Queen's own Hussars since 1685, uniforms, medals – library 45937

Royal Regiment of Fusiliers Museum, Royal Warwickshire, Saint John's House, Warwick CV34 4NF • T: +44 1926 491653 • F: +44 1869 257633 • areasecretary@rrfmuseumwarwick.demon.co.uk • www.warwickfusiliers.com •
Head: *R.G. Mills* •
Military Museum
County Infantry Regiment related items from 1674 onwards 45938

U

St John's House, St John's, Warwick CV34 4NF • T: +44 1926 412132 • museum@warwickshire.gov.uk • www.warwickshire.gov.uk •
Manager: *Michelle Alexander* •
Local Museum – 1961
Social hist, costume, toys and games ... 45939

Warwick Castle, Castle Hill, Warwick CV34 4QU • T: +44 870 4422000 • F: +44 1926 401692 • custumer.information@warwick-castle.com • www.warwick-castle.co.uk •
Gen. Man.: *S.E. Montgomery* •
Historic Site
Over 1000 items of armour & weaponry, "A Royal Weekend Party 1898", award-winning wax exhibit by Madame Tussauds depicting a weekend party, "Kingmaker-Preparation for Battle 1471" from Richard Neville ... 45940

Warwick Doll Museum (closed) ... 45941

Warwickshire Museum, Market Pl, Warwick CV34 4SA • T: +44 1926 412500 • F: 419840 • museum@warwickshire.gov.uk • www.warwickshire.gov.uk/museum •
Dir.: *Dr. Helen Maclagan* •
Local Museum – 1836
Archaeology, geology, natural history, musical instruments, tapestry map ... 45942

Warwickshire Yeomanry Museum, Court House, Jury St, Warwick CV34 4EW • T: +44 1926 492212 • F: 411694 • wtc.admin@btclick.com •
Military Museum
History of the Yeomanry from 1794 to 1968, uniforms, swords, firearms, pictures and paintings relating to the Regiment, campaign relics ... 45943

Warwickshire

Compton Verney, Compton Verney, Warwickshire CV35 9HZ • T: +44 1926 645500 • F: 645501 • info@comptonverney.org.uk • www.comptonverney.org.uk •
Dir.: *Steven Pariessien* •
Fine Arts Museum / Folklore Museum – 2004
Old master paintings, Chinese coll., British folk art ... 45944

Washford

Somerset and Dorset Railway Trust Museum, Railway Station, Washford TA23 0PP • T: +44 1984 640869 • info@sdrt.org • www.sdrt.org •
Cur.: *John Smith* •
Science&Tech Museum
Locomotives, rolling stock, railway buildings, signs, small artefacts, plans, railway signal box ... 45945

Washington

Arts Centre, Biddick Ln, Fatfield, Washington NE38 8AB • T: +44 191 2193455 • F: 2193466 •
Public Gallery ... 45946

Washington Old Hall, The Avenue, District 4, Washington NE38 7LE • T: +44 191 4166879 • F: 4192065 • washington.oldhall@nationaltrust.org.uk • www.washington.co.uk •
Special Museum
Ancestral home of the first President of the United States ... 45947

Watchet

Watchet Market House Museum, Market St, Watchet TA23 0AN • T: +44 1984 631209 •
Cur.: *Roger Wedlake* •
Local Museum – 1979
Maritime history of the port of Watchet, archaeology, Saxon Mint, industry, railways, mining ... 45948

Waterbeach

Farmland Museum and Denny Abbey, Ely Road, Waterbeach CB5 9PQ • T: +44 1223 860988 • F: 860988 • f.m.denny@tesco.net • www.dennyfarmlandmuseum.org.uk •
Cur.: *Kate Brown* • Curatorial Officer: *Corinna Bower* • Education: *Jeremy Ressiter* •
Agriculture Museum / Historic Site – 1997
Interactive displays on farming, esp village life in the 1940/50s, Denny Abbey ... 45949

Watford

Watford Museum, 194 High St, Watford WD17 2DT • T: +44 1923 232297 • F: 244772 • info@watfordmuseum.org.uk • www.watfordmuseum.org.uk •
Cur.: *Marion Duffin* •
Fine Arts Museum / Local Museum – 1981
Fine art coll (17th-18th c), North European genre painting, 19th c English landscape, sculpture by Sir Jacob Epstein, African art coll, local hist, archaeology ... 45950

Weardale

Killhope, the North of England Lead Mining Museum, Cowshill, Weardale DL13 1AR • T: +44 1388 537505 • F: 537617 • info@killhope.org.uk • www.killhope.org.uk •
Man.: *Mike Boase* •
Open Air Museum
Lead mining, waterwheels, minerals of the North Pennines, spar box coll. ... 45951

Weardale Museum, Ireshopeburn, Weardale DL14 • T: +44 1388 537417 • dtheaherington@argonet.co.uk • www.weardalemuseum.co.uk •
Chm.: *Ken Ward* •
Local Museum
Typical Weardale cottage room, geology, farming, railways, water resources and early settlement ... 45952

Wednesbury

Wednesbury Museum and Art Gallery, Holyhead Rd, Wednesbury WS10 7DF • T: +44 121 5560683 • F: 5051625 • rachel_smith@sandwell.gov.uk • lea.sandwell.gov.uk/museums •
Dir.: *Raj Pal* • Cur.: *Emma Cook* •
Local Museum / Fine Arts Museum – 1891
William Howson-Taylor-Ruskin pottery, Helen Caddick ethnography coll, George Robbins geological coll, English 19th c oil paintings and watercolours, ethnography, local hist, geology, decorative arts ... 45953

Welbeck

Creswell Crags Museum and Education Centre, off Crags Rd, Welbeck S80 3LH • T: +44 1909 720378 • F: 724726 • info@creswell-crags.org.uk • www.creswell-crags.org.uk •
Cur.: *Ian Wall* •
Archaeology Museum
Ecofacts and artefacts of Pleistocene research ... 45954

Wellesbourne

Wellesbourne Wartime Museum, Airfield, Control Tower Entrance, Wellesbourne CV35 9EU • T: +44 121 7773518 •
Man.: *Derek Powell* •
Historical Museum
Underground bunker, roll of honour 316 airmen killed while serving at Wellesbourne 1941-45, Vampire TII, Piston Pruvost airframes, cockpits of Seavixen and Vulcan ... 45955

Wellingborough

Irchester Narrow Gauge Railway Museum, Irchester Country Park, Wellingborough NN29 7DL, mail addr: Gypsy Lane, Irchester, Wellingborough NN29 7DL • T: +44 1604 675368 • kadams7000@btinternet.com • www.ingrt.freeuk.com •
Science&Tech Museum – 1984
9 steam and diesel locomotives ... 45956

Wellingborough Heritage Centre, Dulley's Baths, 12 Castle Way, Wellingborough NN8 1XB • T: +44 1933 276838 • F: 222740 • wellingboroughmuseum@msn.com • www.wellingboroughmuseum.co.uk •
Local Museum – 1985 ... 45957

Wells

Wells Museum, 8 Cathedral Green, Wells BA5 2UE • T: +44 1749 673477 • F: 675337 • wellsmuseum@ukonline.co.uk • www.wellsmuseum.org.uk •
Cur.: *Estelle Gilbert* •
Local Museum / Archaeology Museum – 1893
Archaeology, natural history, geology, speleology, social history, needlework samplers ... 45958

Wells-next-the-Sea

Bygones Museum at Holkham, Holkham Park, Wells-next-the-Sea NR23 1AB • T: +44 1328 710227 • F: 711707 • enquiries@holkham.co.uk • www.holkham.co.uk •
Historical Museum
Agricultural Bygones, craft, domestic items, steam and motor vehicles, farming ... 45959

Wells Walsingham Light Railway, Stiffkey Rd, Wells-next-the-Sea NR23 1QB • T: +44 1328 711630 •
Dir.: *R.W. Francis* •
Science&Tech Museum
Garratt steam locomotive, 2 diesel locomotives, carriages ... 45960

Welshpool

Andrew Logan Museum of Sculpture, Berriew, Welshpool SY21 8PJ • T: +44 1686 640689 • F: 640764 • info@andrewlogan.com • www.andrewlogan.com •
Dir.: *Anne Collins* •
Decorative Arts Museum
Sculpture, Alternative Miss World memorabilia ... 45961

Powis Castle, Welshpool SY21 8RF • T: +44 1938 551929 • F: 554336 • powiscastle@nationaltrust.org.uk • www.nationaltrust.org.uk •
Dir.: *Anna Orton* •
Decorative Arts Museum / Historic Site – 1987
Plasterwork and panelling, paintings, tapestries, sculpture, early Georgian furniture, Clive Museum of Indian artifacts; historic garden with surviving 17th/18th-century terraces; 'Clive of India' coll. – historic garden ... 45962

Powysland Museum and Montgomery Canal Centre, Canal Wharf, Welshpool SY21 7AQ • T: +44 1938 554656 • F: 554656 • powysland@powys.gov.uk • powysmuseums.powys.gov.uk •
Dir.: *Eva B. Bredsdorff* •
Local Museum – 1874
Archiological finds, social history ... 45963

Welwyn

Welwyn Roman Baths Museum, Welwyn AL6 0BL, mail addr: c/o Mill Green Museum, Mill Green, Hatfield AL9 5PD • T: +44 1707 271362 • F: 272511 • museum@welhat.gov.uk • www.welhat.gov.uk/museum •
Dir.: *Dr. John Beckerson* • Cur.: *Caroline Rawle* •
Archaeology Museum – 1975
Remains of a 3rd c AD bath house ... 45964

Wensleydale

Yorkshire Museum of Carriages and Horse Drawn Vehicles, Yore Bridge, Aysgarth Falls, Wensleydale DL8 3SR • T: +44 1969 663399 • F: 663399 •
Dir.: *David Kiely* • *Ann Kiely* • Cur.: *Elizabeth Shaw* •
Science&Tech Museum
Coll of both everyday and more lavish horse-drawn vehicles in Britain today ... 45965

West Bretton

Yorkshire Sculpture Park, West Bretton WF4 4LG • T: +44 1924 830302, 832631 • F: 832600 • info@ysp.co.uk • www.ysp.co.uk •
Dir.: *Peter Murray* • *Clare Lilley* •
Public Gallery / Fine Arts Museum / Open Air Museum
Works by Moore, Hepworth, Gormley, Caro, Le Witt and Nash, temporary exhibits by international artists – auditorium, workshop ... 45966

West Bromwich

Oak House Museum, Oak Rd, West Bromwich B70 8HJ • T: +44 121 5530759 • F: 5255167 • oakhouse@sandloell.gov.uk • www.museums.sandwell.gov.uk •
Dir.: *Frank Caldwell* • Cur.: *Emma Cook* •
Historical Museum – 1898
Fine coll of oak furniture of 16th c and 17th c, 17th c interiors, Tudor Yeoman farmers house with period furniture ... 45967

West Clandon

Queen's Royal Surrey Regiment Museum, Clandon Park, West Clandon GU4 7RQ • T: +44 1483 223419 • F: 223419 • queenssurrey@care4free.net • www.queensroyalsurreys.org.uk •
Military Museum ... 45968

West Hoathly

Priest House, North Ln, West Hoathly RH19 4PP • T: +44 1342 810479 • F: +44 1273 486990 • priest@sussexpast.co.uk • www.sussexpast.co.uk •
Cur.: *Antony Smith* •
Local Museum – 1908
Vernacular furniture, domestic implements, embroidery ... 45969

West Kilbride

West Kilbride Museum, Public Hall, 1 Arthur St, West Kilbride KA23 9EN • T: +44 1294 822102 •
Local Museum ... 45970

West Malling

Kent and Medway Museum, KCC, Gibson Dr, Kings Hill, West Malling ME19 4AL • T: +44 1622 605213 • F: 605221 • joanna.follett@kent.gov.uk • www.kent.gov.uk/museums •
Historical Museum
Models and replicas which illustrate the history, natural history and development of Kent and of the wider world ... 45971

West Mersea

Mersea Island Museum, High St, West Mersea CO5 8QD • T: +44 1206 385191 • www.mersea-museum.org.uk •
Dir.: *David W. Gallifant* •
Local Museum ... 45972

West Walton

Fenland and West Norfolk Aviation Museum, Old Lynn Rd, West Walton PE14 7DA • T: +44 1945 463996, 461771 • bill@welbourne.freeserve.co.uk • www.fawnaps.co.uk •
Chm.: *Richard Mason* •
Science&Tech Museum / Military Museum
Aircraft crash sites, Boeing 747 flight deck, general aviation memorabilia, aircraft, lighting, Vampire Provost ... 45973

West Wycombe

West Wycombe Motor Museum, Cockshoot Farm, Chorley Rd, West Wycombe HP14 •
Science&Tech Museum ... 45974

West Wycombe Park House, West Wycombe HP14 3AJ • T: +44 1494 52441 •
Fine Arts Museum
Contemporary frescos, painted ceilings ... 45975

Westbury

Woodland Heritage Museum, Woodland Park, Brokerswood, Westbury BA13 4EH • T: +44 1373 823880, 822238 • F: 858474 • woodland.park@virgin.net •
Dir.: *S.H. Capon* •
Natural History Museum
Wildlife diorama, a bird wall and exhibits on forestry, bird eggs, natural history ... 45976

Westerham

Chartwell, Mapleton Rd, Westerham TN16 1PS • T: +44 1732 866368, 868381 • F: 868193 • chartwell@nationaltrust.org.uk • www.nationaltrust.org.uk •
Man.: *Carole Kenwright* •
Historic Site
Sir Winston Churchill's family house ... 45977

Quebec House, National Trust Chartwell, Quebec Sq, Westerham TN16 1TD • T: +44 1959 562206 • www.nationaltrust.org.uk •
Hon. Cur.: *David Boston* •
Historical Museum
General Wolfe's early years, battle of Quebec ... 45978

Squerryes Court, Goodley Stock Rd, Westerham TN16 1SJ • T: +44 1959 562345 • F: 565949 • enquiries@squerryes.co.uk • www.squerryes.co.uk •
Dir.: *John St. Andrew Warde* • Admin.: *Pam White* •
Fine Arts Museum
Paintings (Dutch 17th c, English 18th) tapestries, furniture, porcelain, General James Wolfe ... 45979

Weston-super-Mare

The Helicopter Museum, The Heliport, Locking Moor Rd, Weston-super-Mare BS24 8PP • T: +44 1934 635227 • F: 645230 • office@helimuseum.fsnet.co.uk • www.helicoptermuseum.co.uk •
Chm.: *Elfan ap Rees* • Man.: *Wendy Cowlin* •
Science&Tech Museum – 1979
App. 90 helicopters incl Bristol Belvedere, Cierva C-30A, Fairey Rotodyne, Mil Mi-1, PZL Swidnik SM-2, Westland Lynx 3, Sud Super Frelon, MBB Bo-102, Mil M.24D, Queens Flight Wessex XV 733 – library, archive ... 45980

North Somerset Museum, Burlington St, Weston-super-Mare BS23 1PR • T: +44 1934 621028 • F: 612526 • museum.service@n-somerset.gov.uk • www.n-somerset.gov.uk/museum •
Dir.: *Nick Goff* •
Local Museum – Archaeology, hist, wildlife and geology, cameras, temporary exhib. ... 45981

Time Machine, Burlington St, Weston-super-Mare BS23 1PR • T: +44 1934 621028 • F: 612526 • museum.service@n-somerset.gov.uk • www.n-somerset.gov.uk/museum •
Archaeology Museum / Local Museum
Archaeology, hist (esp seaside holidays), wildlife and geology of north Somerset ... 45982

Weston-under-Lizard

Weston Park, Weston-under-Lizard TF11 8LE • T: +44 1952 850207 • F: 850430 • enquiries@weston-park.com • www.weston-park.com •
Dir.: *Colin Sweeney* •
Decorative Arts Museum / Fine Arts Museum – 1671
House built 1671 with English, Flemish and Italian paintings, furniture, books, tapestries, art objects ... 45983

Wetheringsett

Museum of the Mid-Suffolk Light Railway, Brockford Station, Wetheringsett IP14 5PW • T: +44 1449 766899 • secretary@mslr.org.uk • www.mslr.org.uk •
Science&Tech Museum
Artefacts, railway station, photographs of mild-Suffolk railway ... 45984

Weybridge

Brooklands Museum, Brooklands Rd, Weybridge KT13 0QN • T: +44 1932 857381 • F: 855465 • info@brooklandsmuseum.com • www.brooklandsmuseum.com •
Dir.: *Allan Winn* •
Science&Tech Museum
British motorsport and aviation, Brooklands cars, aircraft, motorcycles, cycles, John Cobb's 24-litre Napier-Railton, Wellington bomber of Loch Ness, Concorde G-BBDG – library, archive ... 45985

Elmbridge Museum, Church St, Weybridge KT13 8DE • T: +44 1932 843573 • F: 846552 • info@elm-mus.datanet.co.uk • www.elmbridge.history.museum •
Local Museum – 1909/1996
Local history, archaeology, excavated material from Oatlands palace, costume, natural history ... 45986

Weymouth

Deep Sea Adventure and Diving Museum, Custom House Quay, Weymouth • T: +44 1305 760690 • F: 760690 •
Special Museum ... 45987

Nothe Fort Museum of Coastal Defence, Barrack Rd, Weymouth DT4 8UF • T: +44 1305 766626 • F: 766465 • fortressweymouth@btconnect.com • www.fortressweymouth.co.uk •
Cur.: *Alisdair Murray* •
Military Museum – 1979
Coastal fortress, coll of guns, weapons and equipment, military badges, torpedos, uniforms ... 45988

U

Tudor House, Weymouth Civic Society, 3 Trinity St, Weymouth DT4 8TW • T: +44 1305 779711 • info@weymouthcivicsociety.org • www.weymouthcivicsociety.org •
Chm.: *Derek Cope* •
Historical Museum – 1977
17th c life ... 45989

Water Supply Museum, Sutton Poyntz, Pumping Station, Weymouth DT3 6LT • T: +44 1305 832634 • F: 834287 • museum@wessexwater.co.uk • www.wessexwater.co.uk •
Cur.: *John Willows* •
Science&Tech Museum
Water turbine pump from 1857, pumping plant and artefacts ... 45990

Weymouth Museum, Timewalk and Brewery Museum, Brewer's Quay, Hope Sq, Weymouth DT4 8TR • T: +44 1305 777622, 752323 • F: 761680 •
Historical Museum
Local and social history, costume, prints, paintings, Bussell coll – local archives ... 45991

Whaplode

Museum of Entertainment, Rutland Cottage, Millgate, Whaplode PE12 6SF • T: +44 1406 540379 •
Cur.: *Iris Tunnicliff* •
Performing Arts Museum
History of entertainment, mechanical music, the fairground, radio, TV, cinema, theatre ... 45992

Whitburn

Whitburn Community Museum, Library, Union Rd, Whitburn EH47 0AR • T: +44 1506 776347 • F: 776190 • museums@westlothian.org.uk • www.wlonline.org •
Local Museum
Whitburn and its mining past ... 45993

Whitby

Captain Cook Memorial Museum, Grape Ln, Whitby YO22 4BA • T: +44 1947 601900 • F: 601900 • cookmuseum@tiscali.co.uk • www.cookmuseumwhitby.co.uk •
Dir.: *Dr. Sophie Forgan* •
Historical Museum
House where Captain Cook lodged as an apprentice 1746-49, paintings,drawings, artifacts from the voyages, documents ... 45994

Museum of Victorian Whitby, 4 Sandgate, Whitby YO22 4DB • T: +44 1947 601221 •
Man.: *T.J. Ruff* •
Local Museum
Life in Victorian Whitby, whaling, jet mining and jewellery manufacturing ... 45995

Sutcliffe Gallery, 1 Flowergate, Whitby YO21 3BA • T: +44 1947 602239 • F: 820287 • photographs@sutcliffe-gallery.fsnet.co.uk • www.sutcliffe-gallery.co.uk •
Fine Arts Museum
Prints, photography, negative coll of works by Frank Meadow (19th c) ... 45996

Whitby Abbey Museum, Whitby YO22 4JT • T: +44 1947 603568 • F: 825561 • customers@english-heritage.org.uk • www.english-heritage.org.uk/yorkshire •
Cur.: *Andrew Morrison* •
Archaeology Museum – 2001
Architectural material from the abbey ... 45997

Whitby Museum, Pannett Park, Whitby YO21 1RE • T: +44 1947 602908 • F: 897638 • keeper@whitbymuseum.org.uk • www.whitbymuseum.org.uk •
Local Museum – 1823
Local history, archaeology, geology, natural history, shipping, fossils, paintings, ceramics, costumes, Scoresby, Captain Cook, toys and dolls, militaria – library, archives ... 45998

Whitchurch

Whitchurch Silk Mill, 28 Winchester St, Whitchurch RG28 7AL • T: +44 1256 892065 • F: 893882 • silkmill@btinternet.com • www.whitchurchsilkmill.org.uk •
Man.: *Stephen Bryer* •
Science&Tech Museum
Textile watermill, 19th-c machinery, weavers of silk a.o. ... 45999

Whitehaven

The Beacon, Whitehaven, West Strand, Whitehaven CA28 7LY • T: +44 1946 592302 • F: 598150 • thebeacon@copelandbc.gov.uk • www.thebeacon-whitehaven.co.uk •
Dir.: *Sue Palmer* •
Local Museum – 1996
Maritime hist, coal mining, pottery, social hist – library ... 46000

Haig Colliery Mining Museum, Solway Rd, Kells, Whitehaven CA28 9BG • T: +44 1946 599949 • F: 599949 • root@haigpit.com • www.haigpit.com •
Man.: *Pamela Telford* •
Historical Museum / Historic Site
Restored Bever Dorling steam winding engine, coal mining hertiage of West Cumbria ... 46001

Whitehead

Railway Preservation Society of Ireland, Whitehead Excursion Station, Castleview Rd, Whitehead BT39 9NA • T: +44 28 28260803 • F: 28260803 • rpsitrains@hotmail.com • www.rpsi-online.org •
Association with Coll
9 steam locomotives, 2 diesel locomotives, over 25 coaches, wagons, steam crane ... 46002

Whitfield

Dover Transport Museum, Willingdon Rd, Port Zone, Whitfield CT16 2HJ • T: +44 1304 822409 • F: 822409 • jhnlines@aol.com • www.dovertransportmuseum.co.uk •
Chm.: *David Atkins* •
Science&Tech Museum
Local transport hist by land, sea and air ... 46003

Whithorn

Whithorn - Cradle of Christianity, 45-47 George St, Whithorn DG8 8NS • T: +44 1988 500508 • F: 500508 •
Religious Arts Museum ... 46004

Whithorn Priory and Museum, Bruce St, Whithorn DG8 8P4 • T: +44 131 24431010 •
Dir.: *Michael Lyons* •
Archaeology Museum – 1908
Early Christian medieval stonework, Latins Stone 450 AD to mid 5th c, St Peters Cross, 7th c, 10th c Monrieith Cross, fine coll of other crosses ... 46005

Whitstable

Ethnic Doll and Toy Museum, Castlegate House, 1 Tankerton Rd, Whitstable CT5 2AB • T: +44 1983 407231 •
Special Museum ... 46006

Whitstable Museum and Gallery, Oxford St, Whitstable CT5 1DB • T: +44 1227 276998 • F: 772379 • museums@canterbury.gov.uk • www.whitstable-museum.co.uk •
Cur.: *Kenneth Reedie* •
Historical Museum / Fine Arts Museum
Seafaring tradition (oyster industry, divers), fine coll of maritime art ... 46007

Whitstable Oyser and Fishery Exhibition, Harbour, East Quay, Whitstable CT5 1AB • T: +44 1227 280753 • F: 264829 • bayes@seasalter.evnet.co.uk •
Man.: *Norman Goodman* •
Special Museum
Traditional oyster and fishing industry, oyster farming, science and technology ... 46008

Whittlesey

Whittlesey Museum, Town Hall, Market St, Whittlesey PE7 1BD • T: +44 1733 840968 •
Cur.: *Maureen Watson* •
Local Museum
Local trades and industry ... 46009

Whitworth

Whitworth Historical Society Museum, North St, Whitworth OL12 8RE • T: +44 1706 343370 •
Local Museum
Coll of quarry tools, photographs, books, local council records ... 46010

Wick

Wick Heritage Centre, 20 Bank Row, Wick KW1 5EY • T: +44 1955 605393 • F: 605393 • museum@wickheritage.org • www.wickheritage.org •
Local Museum / Fine Arts Museum
Coll of photographs of 115 years history, art gallery ... 46011

Widnes

Catalyst, Science Discovery Centre, Gossage Bldg, Mersey Rd, Widnes WA8 0DF • T: +44 151 4201121 • F: 4952030 • info@catalyst.org.uk • www.catalyst.org.uk •
Dir.: *C. Allison* •
Science&Tech Museum – 1989 ... 46012

Wigan

History Shop, Library St, Wigan WN1 1NU • T: +44 1942 828128 • F: 827645 • heritage@wiganmbc.gov.uk • www.wlct.org/culture/heritage •
Dir.: *A. Gillies* •
Local Museum – 1992
Local and family hist, archaeology, numismatics, local art, industries ... 46013

Wigan Pier, Wallgate, Wigan WN3 4EF • T: +44 1942 323666 • F: 701927 • wiganpier@wiganmbc.gov.uk •
Dir.: *Carole Tydesley* •
Local Museum
Social and industrial life in Wigan in 1900 ... 46014

Willenhall

Locksmith's House, Willenhall Lock Museum, 54-56 New Rd, Willenhall WV13 2DA • T: +44 121 5208054 • www.bclm.co.uk/lhhome.htm •
Cur.: *Kathleen Howe* •
Science&Tech Museum – 1987
17th-20th c locks and keys, early 20th c womens childrens's clothing ... 46015

Willenhall Museum, Willenhal Historic Map Room and Museum, Above Willenhall Library, Walsall St, Willenhall WV13 2EX • T: +44 1922 653116 • F: 632824 • museum@walsall.gov.uk • www.walsall.gov.uk/museums •
Man.: *Louise Troman* •
Local Museum
History of this Midland town ... 46016

Williton

Bakelite Museum, Orchard Mill, Williton TA4 4NS • T: +44 1984 632133 • F: 632322 • info@bakelitemuseum.com • www.bakelitemuseum.co.uk •
Cur.: *Patrick Cook* •
Decorative Arts Museum
Design ... 46017

Wilton, Marlborough

Wilton Windmill, Wilton, Marlborough SN8 3SP • T: +44 1672 870266 • F: 871105 • peter.lemon@talk21.com • www.wiltonwindmill.com •
Chm.: *Dick Marchant Smith* •
Science&Tech Museum
Working windmill ... 46018

Wilton, Salisbury

Wilton House, Wilton, Salisbury SP2 0BJ • T: +44 1722 746720 • F: 744447 • tourism@wiltonhouse.com • www.wiltonhouse.com •
Historic Site
Paintings, interior, art coll, marble bustos ... 46019

Wimborne Minster

Priest's House Museum, 23-27 High St, Wimborne Minster BH21 1HR • T: +44 1202 882533 • F: 882533 • priestshouse@eastdorset.gov.uk •
Cur.: *Emma Ayling* • Ass. Cur.: *Caitlin Griffiths* •
Librarian: *Julia Wenham* •
Local Museum – 1962
Local hist, archaeology, Bronze Age and Romano-British finds, toys ... 46020

Wincanton

Wincanton Museum, 32 High St, Wincanton BA9 9JF • T: +44 1963 34063 •
Cur.: *H.C. Rodd* •
Local Museum ... 46021

Winchcombe

Sudeley Castle, Winchcombe GL54 5JD • T: +44 1242 602308 • F: 602959 • marketing@sudeley.org.uk • www.sudeleycastle.co.uk •
Dir.: *Kevin Jones* •
Historical Museum
Relics and art, paintings by Turner, Van Dijk, Rubens ... 46022

Winchcombe Folk and Police Museum, Old Town Hall, High St, Winchcombe GL54 5LJ • T: +44 1242 609151 • www.winchcombemuseum.org.uk •
Cur.: *M.H. Simms* •
Local Museum / Historical Museum – 1983
Local hist, police uniforms and equipment ... 46023

Winchcombe Railway Museum, 23 Gloucester St, Winchcombe GL54 • T: +44 1242 620641 •
Dir.: *Tim Petchey* •
Science&Tech Museum – 1968
Tickets, labels, horsedrawn railway road vehicles, mechanical signalling, lineside notices ... 46024

Winchelsea

Winchelsea Museum, Court Hall, Winchelsea TN36 4EN • T: +44 1797 226382 •
Cur.: *G. Alexander* •
Local Museum – 1945
Cinque Ports, town history, maps, seals, coins, pictures, pottery, clay pipes, model of the town in 1292 ... 46025

Winchester

Adjutant General's Corps Museum, Romsey Rd, Winchester SO23 8TS • T: +44 1962 877826 • F: 887690 • agc.museum@milnet.uk.net •
Dir.: *G. McMeekin* •
Military Museum
Uniforms, badges and medals, old office machines, documents related to Royal Army Pay Corps, Educational Corps, Military Provost Corps, Army Legal Corps, Women' Royal Army Corps ... 46026

Balfour Museum of Hampshire Red Cross History, Red Cross House, Stockbridge Rd, Winchester SO22 5JD • T: +44 1962 865174 • F: 869721 • balfourmuseum@redcross.org.uk • www.redcross.org.uk •
Historical Museum
History of the British Red Cross, Hampshire branch – archives ... 46027

Guildhall Gallery, Broadway, Winchester SO23 9LJ • T: +44 1962 848289 • F: 848299 • museums@winchester.gov.uk • www.winchester.gov.uk •
Dir.: *Geoffrey Denford* • Cons.: *Christopher Bradbury* •
Public Gallery – 1969 ... 46028

Gurkha Museum, Peninsula Barracks, Romsey Rd, Winchester SO23 8TS • T: +44 1962 842832, 843659 • F: 877597 • curator@thegurkhamuseum.co.uk • www.thegurkhamuseum.co.uk •
Cur.: *Gerald Davies* •
Military Museum / Ethnology Museum
Story of Gurkha since 1815, ethnographic objects from Nepal, India, Tibet and Afghanistan – library, archives ... 46029

The King's Royal Hussars Museum, Peninsula Barracks, Romsey Rd, Winchester SO23 8TS • T: +44 1962 828539 • F: 828538 • beresford@krhmuseum.freeserve.co.uk • www.hants.gov.uk/leisure/museum/royalhus •
Dir.: *P. Beresford* •
Military Museum – 1980
Story of the Regiment from the raising of the 10th and 11th Hussars (originally Light Dragoons) in 1715, armoured vehicles, uniforms, equipment, medals, photographs and campaign relics ... 46030

Light Infantry Museum, Peninsula Barracks, Romsey Rd, Winchester SO23 8TS • T: +44 1962 828550, 828530 • F: 828534 • www.army.doc.uk/army/press/museums/details/m158ligh.html •
Man.: *Patrick Kirkby* •
Military Museum
Modern military museum, elite regiment, Berlin wall, Gulf war, Bosnia exhibition ... 46031

Royal Green Jackets Museum, Peninsula Barracks, Romsey Rd, Winchester SO23 8TS • T: +44 1962 828549 • F: 828500 • museum@royalgreenjackets.co.uk • www.royalgreenjackets.co.uk •
Cur.: *K. Gray* •
Military Museum – 1926
Regimental hist, compaigns over 5 continents, silver, medals, uniforms, weapons, Battle of Waterloo-diorama ... 46032

Royal Hampshire Regiment Museum, Serle's House, Southgate St, Winchester SO23 9EG • T: +44 1962 863658 • F: 888302 • serleshouse@aol.com • www.royalhampshireregimentmuseum.co.uk •
Cur.: *R.M.O. Webster* •
Military Museum – 1933
History of the regiment – archive ... 46033

Textile Conservation Centre, University of Southampton, Winchester Campus, Park Av, Winchester SO23 8DL • T: +44 23 80597100 • F: 80597101 • tccuk@soton.ac.uk • www.soton.ac.uk/~wsart/tcc.htm •
Dir.: *Nell Hoare* •
Special Museum
Karen Finch Library ... 46034

Westgate Museum, High St, Winchester • T: +44 1962 848269 • F: 848299 • museums@winchester.gov.uk • www.winchester.gov.uk •
Dir.: *Dr. G.T. Denford* •
Local Museum – 1898
Medieval building: coll of weights and standards 46035

Winchester Cathedral Triforium Gallery, Cathedral, South Transept, Winchester SO23 9LS • T: +44 1962 857223 • F: 841519, 857201 • john.hardacre@winchester-cathedral.org.uk • www.winchester-cathedral.org.uk •
Dir.: *John Hardacre* •
Religious Arts Museum
Artistic and archeological material from the Cathedral, stone and wooden sculpture ... 46036

Winchester City Museum, The Square, Winchester SO23 9EX • T: +44 1962 848269 • F: 848299 • museum@winchester.gov.uk • www.winchester.gov.uk •
Dir.: *Kenneth Qualmann* •
Archaeology Museum / Local Museum – 1847
Archeology and history of Winchester ... 46037

Winchester College Treasury, College St, Winchester SO23 9NA • T: +44 1962 866079 • F: 843005 •
Dir.: *Victoria Hebron* •
Fine Arts Museum
Historic documents, antique silver, ceramics, Egyptian and Central American objects, antique scientific instruments, modern crafts ... 46038

Windermere

Windermere Steamboat Museum, Rayrigg Rd, Windermere LA23 1BN • T: +44 15394 45565 • F: 48769 • info@steamboat.co.uk • www.steamboat.co.uk •
Dir.: *Tony Jackson* •
Science&Tech Museum – 1977
Coll of 35 steam, sail and motorboats undercover in a purpose built wet dock made on Lake Windermere – art gallery, library ... 46039

Windsor

Dorney Court, Dorney, Windsor SL4 6QP • T: +44 1628 604638 • F: 665772 • palmer@dorneycourt.co.uk • www.dorneycourt.co.uk •
Dir.: *Peregrine Palmer* •
Historical Museum – 1440 ... 46040

Frogmore House, Royal Collections, Home Park, Windsor SL4 1NJ, mail addr: Stable Yard House, Saint James's Palace, London SW1A 1JR • T: +44 20 77667305 • F: 79309625 • windsorcastle@royalcollection.org.uk • www.royalcollection.org.uk •
Fine Arts Museum / Decorative Arts Museum
Royal coll ... 46041

Household Cavalry Museum, Combermere Barracks, Windsor SL4 3DN • T: +44 1753 755112 • F: 755161 • homehq@householdcavalry.co.uk •
Sc. Staff: *S.F. Sibley* • *K.C. Hughes* •
Military Museum – 1952
Regimental history, uniforms, armaments, pictures ... 46042

Museum of Eton Life, c/o Eton College, Windsor SL4 6DB • T: +44 1753 671177 • F: 671265, 671029 • r.hunkin@etoncollege.com • www.etoncollege.com •
Cur.: *P. Hatfield* •
Historical Museum – 1985
Daily life and duties, food and living conditions, punishments, uniforms, and school books and equipment, history of rowing at the College and the Officer Training Corps ... 46043

Royal Berkshire Yeomanry Cavalry Museum, Territorial Army Centre, Bolton Rd, Windsor SL4 3JG • T: +44 1753 860600 • F: 854946 •
Cur.: *A.P. Verey* •
Military Museum ... 46044

Royal Borough Museum Collection, Windsor Library, Bachelors Acre, Windsor SL4 1ER, mail addr: Tinkers Depot, Tinkers Ln, Dedworth, Windsor SL4 4LR • T: +44 1628 796829 • F: 796859 • museum.collections@rbwm.gov.uk • www.rbwm.gov.uk •
Head: *Caroline McCutcheon* •
Local Museum
Royal Borough of Windsor and Maidenhead history ... 46045

Town and Crown Exhibition, Royal Windsor Information Centre, 24 High St, Windsor SL4 1LH • T: +44 1628 796829 • F: 796859 • museum.collections@rbwm.gov.uk • www.rbwm.gov.uk/museum •
Cur.: *Judith Hunter* •
Local Museum
History of the town of Windsor ... 46046

Windsor Castle, Royal Collections, Windsor SL4 1NJ, mail addr: Stable Yard House, Saint James's Palace, London SW1A 1JR • T: +44 20 77667300 • F: 79309625 • windsorcastle@royalcollection.org.uk • www.rocyalcollection.org.uk •
Fine Arts Museum / Decorative Arts Museum
Royal coll, state apartments – Royal library ... 46047

Wisbech

Peckover House, North Brink, Wisbech PE13 1JR • T: +44 1945 583463 • F: 583463 • aprigx@smtp.ntrust.org.uk •
Decorative Arts Museum ... 46048

Wisbech and Fenland Museum, 5 Museum Sq, Wisbech PE13 1ES • T: +44 1945 583817 • F: 589050 • info@wisbechmuseum.org.uk • www.wisbechmuseum.org.uk •
Cur.: *Dr. Jane Hubbard* •
Local Museum / Decorative Arts Museum – 1835
Pottery and porcelain, figures, local photographs, manuscripts by Dickens and Lewis, calotypes by Samuel Smith, early Madagascar photos, parish records, african ethnography, geology, Fenland hist and folklife, coins, applied art, fossils – library, archives ... 46049

Withernsea

Withernsea Lighthouse Museum, Hull Rd, Withernsea HU19 2DY • T: +44 1964 614834 •
Admin.: *Janet Standley* •
Science&Tech Museum ... 46050

Witney

Bishop of Winchester's Palace Site, Mount House, Church Green, Witney OX28 6AZ • T: +44 1865 300557 • museums.resource.centre@oxfordshire.gov.uk • www.oxfordshire.gov.uk •
Historic Site
12th c palace ... 46051

Cogges Manor Farm Museum, Church Ln, Cogges, Witney OX28 3LA • T: +44 1993 772602 • F: 703056 • info@cogges.org • www.cogges.org •
Dir.: *Clare Pope* •
Agriculture Museum – 1976
Historic Manor House with Victorian period room displays, 17th c study with original painted panelling, agricultural history, tools, wagon coll ... 46052

Witney and District Museum and Art Gallery, Gloucester Court Mews, High St, Witney OX28 6LX • T: +44 1993 775915 • witney.museum@virgin.net • www.witneymuseum.com •
Chm.: *Ian Petty* •
Local Museum
Social hist, fine art, military hist, photographic coll, Woollen/ Blanket hist ... 46053

Woburn

Woburn Abbey, Woburn MK17 9WA • T: +44 1525 290666 • F: 290271 • enquiries@woburnabbay.co.uk •
Cur.: *Lavinia Wellicome* •
Historical Museum
Coll of English and French furniture (18th c), paintings by Canaletto, Reynolds, Van Dyck ... 46054

Woking

The Lightbox, 1 Crown Sq, Woking GU21 6HR • T: +44 1483 725517 • F: 725501 • info@thelightbox.org.uk • www.thelightbox.org.uk •
Dir.: *Marilyn Scott* •
Local Museum
Life in Woking past and present, arts and crafts ... 46055

Wollaston

Wollaston Museum, 102-104 High St, Wollaston NN29 7RJ • T: +44 1933 664468 •
Cur.: *Joan Lewry* •
Local Museum
Farming implements, archaeological finds from the area, village life, paintings by local artists, footware manufacture, lace making ... 46056

Wolverhampton

Bantock House, Bantock Park, Finchfield Rd, Wolverhampton WV3 9LQ • T: +44 1902 552195 • F: 552196 • www.wolverhamptonart.org.uk •
Cur.: *Helen Steatham* •
Decorative Arts Museum / Local Museum
Enamels, steel jewellery, japanned ware ... 46057

Bilston Craft Gallery, Mount Pleasant, Bilston, Wolverhampton WV14 7LU • T: +44 1902 552507 • BilstonCraftGallery@wolverhampton.gov.uk • www.wolverhamptonart.org.uk •
Cur.: *Emma Daker* • Sc. Staff: *Val Plummer* • *Anne Hardy* • *Natalie Cole* •
Public Gallery – 1937
Local hist, enamels, contemporary crafts, glass, jewellry, sculptures ... 46058

Wightwick Manor, Wightwick Bank, Wolverhampton WV6 8EE • T: +44 1902 761400 • F: 764663 • wightwickmanor@nationaltrust.org.uk • www.nationaltrust.org.uk •
Cur.: *T.A. Clement* •
Local Museum ... 46059

Wolverhampton Art Gallery, Lichfield St, Wolverhampton WV1 1DU • T: +44 1902 552055 • F: 552053 • info@wolverhamptonart.org.uk • www.wolverhamptonart.org.uk •
Cur.: *Julia Bigham* •
Fine Arts Museum – 1884
British 18th-20th c paintings and sculpture, British and American Pop Art and contemporary paintings, geology, interactive display "ways of seeing" ... 46060

Woodbridge

Suffolk Punch Heavy Horse Museum, Market Hill, Woodbridge IP12 4LU • T: +44 1394 380643 • F: 610005 • sec@suffolkhousesociety.org.uk • www.suffolkhousesociety.org.uk •
Cur.: *P. Ryder-Davis* •
Special Museum ... 46061

Woodbridge Museum, 5a Market Hill, Woodbridge IP12 4LP • T: +44 1394 380502 •
Cur.: *John Hampton* •
Local Museum – 1981 ... 46062

Woodchurch

Woodchurch Village Life Museum, Susans Hill, Woodchurch TN26 3RE • T: +44 1233 860240 • F: 860653 • curator@woodchurchmuseum.com • www.woodchurchmuseum.fsnet.com •
Cur.: *G. Loynes* •
Local Museum
Social history of the village, vehicles, agricultural equipment, ceramics, coins, household items, books, papers and photographs ... 46063

Woodhall Spa

Woodhall Spa Cottage Museum, Bungalow, Iddesleigh Rd, Woodhall Spa LN10 6SH • T: +44 1526 353775 • ddfradford@hotmail.com •
Local Museum
History of Woodhall Spa, geology ... 46064

Woodley

Museum of Berkshire Aviation, Mohawk Way, Woodley RG5 4UE • T: +44 118 9448089 • museumberksav@gmail.com • home.comcast.net/~aero51/html/index.htm •
Cur.: *Ken Fostekew* •
Science&Tech Museum – 1992
Aircraft, models, archives, photographs ... 46065

Woodstock

Blenheim Palace, Woodstock OX20 1PX • T: +44 1993 811091 • F: 813527 • admin@blenheimpalace.com • www.blenheimpalace.com •
Dir.: *Julian Blades* •
Fine Arts Museum / Decorative Arts Museum
Tapestries, paintings, sculpture, porcelain, fine furniture ... 46066

The Oxfordshire Museum, Fletcher's House, Park St, Woodstock OX20 1SN • T: +44 1993 811456 • F: 813239 • oxon.museum@oxfordshire.gov.uk • www.tomocc.org.uk •
Cur.: *Cherry Gray* •
Local Museum – 1965
Archaeology, history and natural environment of Oxfordshire, county's innovative industry ... 46067

Wookey Hole

Wookey Hole Cave Diving and Archaeological Museum, Wookey Hole BA5 1BB • T: +44 1749 672243 • F: 677749 • witch@wookey.co.uk • www.wookey.co.uk •
Dir.: *Gerry W. Cottle* •
Natural History Museum / Archaeology Museum – 1980
Archaeological finds, cave exploration, geology, cave diving, myth and legends ... 46068

Woolpit

Woolpit and District Museum, The Institute, Woolpit IP30 9QH • T: +44 1359 240822 •
Dir.: *John Wiley* •
Local Museum – 1983
Coll of items from the village and its immediate surroundings ... 46069

Worcester

Commandery Museum, Sidbury, Worcester WR1 2HU • T: +44 1905 361821 • F: 361822 • thecommandery@cityofworcester.gov.uk • www.cityofworcester.gov.uk •
Local Museum – 1977
Local hist, english Civil War, 15th c wall paintings ... 46070

Museum of the Worcestershire Regiment, Foregate St, Worcester WR1 1DT • T: +44 1905 25371 • F: 616979 • rhq_wfr@army.mod.uk •
Cur.: *R. Prophet* •
Military Museum
History of the regiment, medals, uniforms, equipment ... 46071

Museum of Worcester Porcelain, Severn St, Worcester WR1 2NE • T: +44 1905 746000 • F: 617807 • www.worcesterporcelainmuseum.org •
Dir.: *Amanda Savidge* • Chm.: *T.G. Westbrook* • Cur.: *Wendy Cook* •
Decorative Arts Museum – 1946
Worcester porcelain since 1751 ... 46072

Worcester City Museum and Art Gallery, Foregate St, Worcester WR1 1DT • T: +44 1905 25371 • F: 616979 • artgalleryandmuseum@cityofworcester.gov.uk • www.worcestercitymuseums.org.uk •
Local Museum / Fine Arts Museum – 1833
Area hist, geology, natural hist, archaeology, glass, ceramics, local prints, watercolors ... 46073

Worcestershire Yeomanry Cavalry Museum, Foregate St, Worcester WR1 1DT • T: +44 1905 25371 • F: 616979 • tbridges@cityofworcester.gov.uk • www.worcestercitymuseums.org.uk •
Head: *Tim Bridges* •
Military Museum – 1929
Hist of regiment, medals, uniforms, decorations, documents and prints ... 46074

Workington

Helena Thompson Museum, Park End Rd, Workington CA14 4DE • T: +44 1900 326255 • F: 326320 • whghtm@hotmail.co.uk • www.helenathompsonmuseum.co.uk/Home.html •
Dir.: *Philip Crouch* • Cur.: *Janet Baker* •
Local Museum – 1940
Workington's social and industrial hist, costume, ceramics, other decorative art ... 46075

Worksop

Harley Gallery, Welbeck, Worksop S80 3LW • T: +44 1909 501700 • F: 488747 • info@harley-welbeck.co.uk • www.harleygallery.co.uk •
Dir.: *Lisa Gee* •
Fine Arts Museum / Decorative Arts Museum
Contemporary visual arts and crafts, Portland coll of fine and decorative arts ... 46076

Mr. Straw's House, 7 Blyth Grove, Worksop S81 0JG • T: +44 1909 482380 • mrstrawshouse@nationaltrust.org.uk • www.nationaltrust.org.uk •
Manager: *Philippa Caurie* •
Historical Museum
Family ephemera, costume, furniture and household objects from 19th and 20th c ... 46077

Worksop Museum, Memorial Av, Worksop S80 2BP • T: +44 1777 713749 • F: 713749 •
Cur.: *Malcolm J. Dolby* •
Local Museum
Local history, archaeology, natural history, Stone Age finds ... 46078

Worthing

Worthing Museum and Art Gallery, Chapel Rd, Worthing BN11 1HP • T: +44 1903 239999 ext 1150 • F: 236277 • museum@worthing.gov.uk • www.worthing.gov.uk •
Man.: *Diana Smart* (Antiquities) • Cur.: *Kate Rose* (Historic Colls) • *Emma Walder* (Arts, Exhibitions) •
Local Museum / Decorative Arts Museum / Fine Arts Museum – 1908
Archaeology, geology, local history, costumes, paintings, decorative arts, toys ... 46079

Wotton-under-Edge

Wotton Heritage Centre, The Chipping, Wotton-under-Edge GL12 7AD • T: +44 1453 521541 • wootonhs@freeuk.com • www.wottonheritage.com •
Cur.: *Beryl A. Kingan* •
Local Museum
Local and family history – library ... 46080

Wrexham

Bersham Heritage Centre and Ironworks, Bersham, Wrexham LL14 4HT • T: +44 1978 261529 • F: 361703 • museum@wrexham.gov.uk • www.wrexham.gov.uk •
Science&Tech Museum / Local Museum – 1983/1992 ... 46081

Erddig, c/o J. Cragg, Wrexham LL13 0YT • T: +44 1978 355314 • F: 313333 • erddig@ntrust.org.uk • www.nationaltrust.org.uk •
Dir.: *Jamie Watson* •
Decorative Arts Museum – 1977
Early 18th c furniture, servants portraits, state rooms and servants quarters, Chinese wallpaper, horse-drawn carriage – garden ... 46082

King's Mill Visitor Centre, King's Mill Rd, Wrexham LL13 0NT • T: +44 1978 358916 •
Science&Tech Museum
Restored mill featuring life and work in an 18th c mill, waterwheel, by appointment ... 46083

Wrexham Arts Centre, Rhosddu Rd, Wrexham LL11 1AU • T: +44 1978 292093 • F: 292611 • arts.centre@wrexham.gov.uk • www.wrexham.gov.uk •
Dir.: *Tracy Simpson* •
Public Gallery ... 46084

Wrexham County Borough Museum, County Bldgs, Regent St, Wrexham LL11 1RB • T: +44 1978 317970 • F: 317982 • museum@wrexham.gov.uk • www.wrexham.gov.uk/heritage •
Dir.: *Steve Grenter* • Cur.: *Joanne Neri* •
Local Museum
Local hist, iron and coal industry (18th-19th c), bronze age skeleton of Brymbo man, archeology ... 46085

Wroughton

Science Museum Wroughton, The National Museum of Science & Industry, Hachten Ln, Wroughton SN4 9NS • T: +44 1793 846221 • F: 815413 • janet.marshall@nmsi.ac.uk • www.sciencemuseum.org.uk/wroughton •
Dir.: *Lindsay Sharp* •
Science&Tech Museum ... 46086

Wroxeter

Viroconium Museum, Wroxeter Roman City, Roman Site, Wroxeter SY5 6PH • T: +44 1743 761330 • www.english-heritage.org.uk •
Cur.: *Sara Lunt* •
Archaeology Museum
Archeological finds, pottery, coins, metalwork ... 46087

Wylam

Railway Museum, Falcon Centre, Falcon Terrace, Wylam NE41 8EE • T: +44 1661 852174 •
Cur.: *Philip R.B. Brooks* •
Science&Tech Museum – 1981
Railway hist. ... 46088

Wymondham

Wymondham Heritage Museum, 10 The Bridewell, Norwich Rd, Wymondham NR18 0NS • T: +44 1953 600205 • www.wymondham-norfolk.co.uk •
Man.: *Adrian D. Hoare* •
Local Museum / Historic Site
Local history ... 46089

Y Bala

Canolfan Y Plase, Plassey St, Y Bala LL23 7YD • T: +44 1678 520320 •
Chm.: *D. A. Jones* •
Local Museum
Exhibitions of artwork and local heritage topics ... 46090

Yanworth

Chedworth Roman Villa, Yanworth GL54 3LJ • T: +44 1242 890256 • F: 890544 • chedworth@nationaltrust.org.uk • www.nationaltrust.org.uk •
Dir.: *Philip Bethell* •
Archaeology Museum – 1865
Roman walls, mosaic floors, bath suites, wooden buildings, heating in Roman houses, Roman gardens, religion in Roman times, archaeology, classical history ... 46091

Yarmouth

Sunken History Exhibition, Fort Victoria Museum, Fort Victoria Country Park, off Westhill Ln, Yarmouth PO41 0RR • T: +44 1983 761214 • F: +44 2380 593052 • info@hwtma.org.uk • www.hwtma.org.uk •
Cur.: *Paul Blake* •
Historical Museum ... 46092

Yell

Old Haa, Burravoe, Yell ZE2 9AY • T: +44 19577 22339 •
Dir.: *A. Nisbet* •
Local Museum
Arts and crafts, local history, tapes of Shetland music and folklore, Bobby Tulloch collection, genealogy 46093

Yelverton

Buckland Abbey, Yelverton PL20 6EY • T: +44 1822 853607 • F: 855448 • bucklandabbey@nationaltrust.org.uk • www.nationaltrust.org.uk •
Man.: *Michael Coxson* • Cur.: *Nicola Moyle* •
Historic Site – 1952
Sir Francis Drakes' drum, standards, portraits and family pictures, furniture and ship navigation equipment ... 46094

Yelverton Paperweight Centre, 4 Buckland Tce, Leg O'Mutton Cnr, Yelverton PL20 6AD • T: +44 1822 854250 • F: 854250 • paperweightcentre@btinternet.com • www.paperweightcentre.co.uk •
Fine Arts Museum
Work of glass artists and others ... 46095

Yeovil

Museum of South Somerset, Hendford, Yeovil BA20 1UN • T: +44 1935 424774, 462855 • F: 477107 • heritageservices@southsomerset.gov.uk • www.southsomersetmuseums.org.uk •
Local Museum – 1928
Archaeology, industries, household and farm tools, topography, armaments, glass, costumes ... 46096

Yeovilton

Fleet Air Arm Museum - Concorde, The National Museum of Science & Industry, D6, Royal Naval Air Station, Yeovilton BA22 8HT • T: +44 1935 840565 • F: 842630 • enquires@fleetairarm.com • www.fleetairarm.com •
Dir.: *Graham Mottram* •
Military Museum – 1964
Helicopter transported, replica of the aircraft carrier HMS Ark Royal, flight deck experience, nuclear bomb, 1st British build Concord, naval aircraft coll ... 46097

York

Beningbrough Hall, Beningbrough, York YO30 1DD • T: +44 1904 470666 • F: 470002 • ybblmb@smtp.ntrust.org.uk •
Man.: *Ray Barker* •
Fine Arts Museum – 1979
Restored early Georgian country house, National portrait gallery ... 46098

Castle Museum, Eye of York, nr Clifford's Tower, Coppergate, York YO1 9RY • T: +44 1904 687687 • F: 671078 • castle.museum@ymt.org.uk • www.york.trust.museum •
Dir.: *Keith Matthews* •
Ethnology Museum / Historical Museum – 1938
Yorkshire ethnology, period rooms, household and farm tools, militaria, costumes, crafts ... 46099

DIG, Saint Saviour Church, Saint Saviourgate, York YO1 8NN • T: +44 1904 543403 • F: 627097 • info@digyork.co.uk • www.digyork.co.uk •
Archaeology Museum
City's archaeological heritage ... 46100

Divisional Kohima Museum, Imphal Barracks, Fulford Rd, York YO1 4AU • T: +44 1904 662381, 665806 • F: 662655 • thekohimamuseum@hotmail.com • www.armymuseums.org.uk •
Cur.: *Nigle Magrane* •
Military Museum ... 46101

Fairfax House Museum, Castlegate, York YO1 9RN • T: +44 1904 655543 • F: 652262 • peterbrown@fairfaxhouse.co.uk • www.fairfaxhouse.co.uk •
Dir.: *Peter B. Brown* •
Decorative Arts Museum / Fine Arts Museum – 1984
18th c English furniture, clocks, dining room silver, early York silver, paintings and glass, elaborate stucco ceilings ... 46102

Jorvik Museum, Coppergate, York YO1 9WT • T: +44 1904 615505 • F: 627097 • jorvik@yorkat.co.uk • www.viking-jorvik-centre.co.uk •
Dir.: *John Walker* •
Archaeology Museum / Historical Museum
Viking life and trade ... 46103

Merchant Adventurers' Hall, Fossgate, York YO1 9XD • T: +44 1904 654818 • F: 616150 • enquiries@theyorkcompany.co.uk • www.theyorkcompany.co.uk •
Dir.: *James Finlay* •
Decorative Arts Museum
Early furniture, silver and portraits, pottery and other objects ... 46104

National Railway Museum, The National Museum of Science and Industry, Leeman Rd, York YO26 4XJ • T: +44 1904 621261 • F: 611112 • nrm@nmsi.ac.uk • www.nrm.org.uk •
Head: *Andrew Scott* • Dep. Head: *Janice Murray* • Sc. Staff: *Su Matthewman* (Marketing) • *Camilla Harrison* (Press, PR) • *Jim Rees* (Engineering) • *Ed Bartholomew* (Photographic Archive, Arts Colls) •
Science&Tech Museum – 1975
Mallard - the world's fastest steam locomotive, Queen Victoria's Royal Carriage, Working Replica of Stephenson's Rocket, Japanese Bullet train, Chinese 4-8-4 locomotive, the Flying Scotsman – library, archives and picture archives ... 46105

Regimental Museum of the Royal Dragoon Guards and the Prince of Wales's Own Regiment of Yorkshire, 3a Tower St, York YO1 9SB • T: +44 1904 642036 • F: 642036 • hhq@rdgmuseum.org.uk • www.rdgmuseum.org.uk •
Cur.: *W.A. Henshall* •
Military Museum
Uniforms, paintings, pictures, medals, horse furniture, weapons, soldiers' personal memorabilia, regiments history ... 46106

Treasurer's House, Minster Yard, York YO1 7JL • T: +44 1904 624247 • F: 647372 • treasurershouse@nationaltrust.org.uk •
Dir.: *Jane Whitehead* •
Decorative Arts Museum – 1930
Furniture 17th-18th c, pictures, glass, ceramics 46107

Viking City of Jorvik → Jorvik Museum

York Art Gallery, Exhibition Sq, York YO1 7EW • T: +44 1904 551861 • F: 551866 • art.gallery@ymt.org.uk • www.york.trust.museums •
Dir.: *Richard Green* • Sc. Staff: *Lara Goodband* (Exhibitions and Publicity) • *Victoria Osborne* (Art) •
Fine Arts Museum – 1879
European painting 1350-1930, topographical prints and drawings, Italian paintings, English stoneware pottery – library ... 46108

York Minster Undercroft Treasury and Crypt, York Minster, Deangate, York YO1 7JF • T: +44 1904 557216 • F: 557218 • visitors@yorkminster.org • www.yorkminster.org •
Man.: *Louise Hampson* •
Religious Arts Museum
Coll of diocesan church plate (13th-20th c), archaeological artefacts ... 46109

York Racing Museum, Racecourse, The Knavesmire, York YO23 1EX • T: +44 1904 620911 • F: 611071 • info@yorkracecourse.co.uk • www.yorkracecourse.co.uk •
Dir.: *Scott Brown* • Cur.: *Dede Marks* •
Special Museum – 1965
Horce racing ... 46110

York Story, Saint Mary Castlegate, York YO1 9RN • T: +44 1904 628632 •
Local Museum
Social and architectural history of York ... 46111

Yorkshire Museum, Museum Gardens, York YO1 7FR • T: +44 1904 687687 • F: 687662 • yorkshire.museum@ymt.org.uk • www.york.trust.museum •
Cur.: *Paul Howard* •
Local Museum – 1822
Archaeology, decorative art, pottery, biology, mammals, entomology, herbarium, geology, fossils, minerals, numismatics, figures – library ... 46112

Yorkshire Museum of Farming, Murton Park, York YO19 5UF • T: +44 1904 489966 • F: 489159 • amanda@murtonpark.co.uk • www.murtonpark.co.uk •
Agriculture Museum – 1982
Viking village, Roman fort, Celtic houses, 200 years farm machinery and equipment ... 46113

Zennor

Wayside Folk Museum, Old Mill House, Zennor TR26 3DA • T: +44 1736 796945 •
Cur.: *Th. Priddle* •
Folklore Museum – 1935
Mining, household objects, milling, fishing, archaeology, wheelwrights and carpenters shop, blacksmiths, agriculture, millers cottage kitchen, childhood memories, the sea, village dairy, cobblers shop working water wheels, photographs ... 46114

United States of America

Abercrombie ND

Fort Abercrombie Historic Site, 816 Broadway St, Abercrombie, ND 58001 • T: +1 701 5538513 • histsoc@state.nd.us • www.discovernd.com/hist •
Head: *June Singer* •
Historical Museum – 1905
Relics from pioneer days, early fort hist ... 46115

Aberdeen SD

Dacotah Prairie Museum and Lamont Art Gallery, 21 S Main St, Aberdeen, SD 57402-0395 • T: +1 605 6267117 • F: +1 605 6264026 • dpm@brown.sd.us • www.brown.sd.us/museum •
Dir.: *Sue Gates* • Cur.: *Lora Schaunaman* (Exhibits) • Education: *Sherri Rawstern* •
Local Museum – 1964
Local hist, American Indian art – Ruth Bunker Memorial library ... 46116

Northern Galleries, Northern State University, 1200 S Jay St, Aberdeen, SD 57401 • T: +1 605 6262596 • F: +1 605 6262263 • mulvaner@wolf.northern.edu •
Dir.: *Rebecca Mulvaney* •
Fine Arts Museum – 1902
Drawings, paintings, prints, photography, sculpture ... 46117

Aberdeen Proving Ground MD

United States Army Ordnance Museum, c/o U.S. Army Ordnance Center and School, 2601 Aberdeen Blvd, Aberdeen Proving Ground, MD 21005 • T: +1 410 2783602, 2782396 • F: +1 410 2787473 • museum@ocs.apg.army.mil • www.ordmusfound.org •
Dir./Cur.: *William F. Atwater* •
Military Museum – 1919
Military history ... 46118

Abilene KS

Dickinson County Heritage Center, 412 S Campbell St, Abilene, KS 67410 • T: +1 785 2632681 • F: +1 785 2630380 • heritagecenterdk@sbcglobal.net • www.heritagecenterdk.com •
Dir.: *Jeff Sheets* •
Local Museum – 1928
Local history – telephone hist ... 46119

Dwight D. Eisenhower Library-Museum, 200 SE 4th St, Abilene, KS 67410 • T: +1 913 2636700 • F: +1 913 2636718 • eisenhower.library@nara.gov • www.eisenhower.archives.gov •
Dir.: *Daniel D. Holt* • Cur.: *Dennis H.J. Medina* •
Library with Exhibitions – 1962
Presidential library ... 46120

Greyhound Hall of Fame, 407 S Buckeye, Abilene, KS 67410 • T: +1 785 2633000 • F: +1 785 2632604 • info@greyhoundhalloffame.com • www.greyhoundhalloffame.com •
Pres.: *Vey O. Weaver* • Dir.: *Edward Scheele* • Exec. Asst.: *Kathryn Lounsbury* •
Special Museum – 1963
Greyhound sports and animal ... 46121

Museum of Independent Telephony, 412 S Campbell, Abilene, KS 67410 • T: +1 785 2632681 • F: +1 785 2630380 • mitmuseum@sbcglobal.net • www.mitmuseum.com •
Dir.: *Robin Black* • Cur.: *Janet Groninga* •
Science&Tech Museum – 1973
Telephonic history ... 46122

Western Museum, 201 SE 6th, Abilene, KS 67410 • T: +1 785 2634612 •
Head: *B.G. French* •
Local Museum
Local history, housed in Rock Island Railroad Stn 46123

Abilene TX

The Grace Museum, 102 Cypress St, Abilene, TX 79601 • T: +1 325 6734587 • F: +1 325 6755993 • info@thegracemuseum.org • www.thegracemuseum.org •
Exec. Dir.: *Judith Godfrey* •
Local Museum / Fine Arts Museum – 1937
20th c American paintings and prints, local history, T&P railway coll ... 46124

Ryan Fine Arts Center, c/o McMurry University, 514th and Sayles Blvd, Abilene, TX 79697 • T: +1 915 7933823 • F: +1 915 7934662 • www.mcm.edu •
University Museum / Fine Arts Museum – 1970 . 46125

Abingdon VA

William King Regional Arts Center, 415 Academy Dr, Abingdon, VA 24212 • T: +1 276 6285005 • F: +1 276 6283922 • info@wkrac.org • www.wkrac.org •
Dir.: *Betsy K. White* • Dep. Dir.: *Suzanne G. Brewster* •
Cur.: *Heather McBride* •
Fine Arts Museum – 1979 ... 46126

Abington MA

Dyer Memorial Library, 28 Centre Av, Abington, MA 02351, mail addr: POB 2245, Abington, MA 02351 • T: +1 781 8788480 • info@dyerlibrary.org • www.dyerlibrary.org •
Cur.: *Stephanie McDonough* •
Library with Exhibitions – 1932
Civil war monographs, local hist ... 46127

Abiquiu NM

Florence Hawley Ellis Museum of Anthropology, Ruth Hall Museum of Paleontology, Ghost Ranch Conference Center, Abiquiu, NM 87510-9601, mail addr: HC 77 Box 11, Abiquiu, New Mexico 87510 • T: +1 505 6854333 ext 118 • F: +1 505 6854519 • cherylm@ghostranch.org • www.ghostranch.org •
Dir./Cur.: *Cheryl L. Muceus* (Anthropology) • Cur.: *Alex Downs* (Paleontology) •
Ethnology Museum / Archaeology Museum / Natural History Museum – 1980
Anthropology, Paleontology ... 46128

Piedra Lumbre Visitors Center, Ghost Ranch Living Museum, US Hwy 84, Abiquiu, NM 87510 • T: +1 505 6854312 •
Dir.: *Ray Martinez* •
Fine Arts Museum – 1959
Paintings and prints related to natural resources 46129

Ruth Hall Museum of Paleontology, Ghost Ranch Conference Center, Abiquiu, NM 87510-9601 • T: +1 505 6854333 ext 118 • F: +1 505 6854519 • www.newmexico-ghostranch.org •
Dir./Cur.: *Cheryl Muceus* (Anthropology) • Cur.: *Alex Downs* (Paleontology) •
Natural History Museum – 1980
Paleontology, geology ... 46130

Accokeek MD

The Accokeek Foundation, 3400 Bryan Point Rd, Accokeek, MD 20607 • T: +1 301 2832113 • F: +1 301 2832049 • accofound@accokeek.org • www.accokeek.org •
Pres.: *Wilton C. Corkern* •
Agriculture Museum – 1958
Agriculture ... 46131

Acme MI

The Music House Museum, 7377 US-31 N, Acme, MI 49610 • T: +1 231 9389300 • F: +1 231 9383650 • musichouse@coslink.net • www.musichouse.org •
Dir.: *David L. Stiffler* •
Music Museum – 1983
Music, musical instrument, automatic instruments, music boxes, early phonographs, radios and TVs – library ... 46132

Acton MA

The Discovery Museums, 177 Main St, Acton, MA 01720 • T: +1 978 2644200 • F: +1 978 2640210 • staff@discoverymuseums.org • www.discoverymuseums.org •
Dir.: *Michael Judd* (Exec.) • *Geoff Nelson* (Exhibits) • *Denise LeBlanc* (Science Education) • *Lauren Kotkin* (Children's Education) • Assoc. Dir.: *Maria Conley* (Science Education) • *Amy Leonard* (Children's Education) •
Science&Tech Museum – 1981
Children's hands-on museums, Sea of Clouds, sciences ... 46133

Ada MN

Memorial Museum & Pioneer Museum, East and First Sts, Thorpe Av, Ada, MN 56510-1519 • T: +1 218 7847311 • lnelson25@cox.net •
Pres.: *Rude DeFloren* •
Local Museum – 1957
Prairie village museum ... 46134

Ada OK

University Gallery, East Central University, 1100 E 14th St, Ada, OK 74820 • T: +1 580 3105353 • F: +1 580 4363329 • bjessop@mailclerk.ecok.edu • www.ecok.edu •
Public Gallery ... 46135

Addison VT

Chimney Point, 7305 VT Rte 125, Addison, VT 05491 • T: +1 802 7592412 • F: +1 802 7592547 • chimneypoint@historicvermont.org • www.historicvermont.org •
Head: *Elsa Gilbertson* •
Local Museum – 1968
Historic building ... 46136

John Strong Mansion, 6656 Rte 17 W, Addison, VT 05491-8893 • T: +1 802 7592309 • www.northshirecomputer.com/VTDAR •
Cur.: *Jane V. Roberts* •
Local Museum – 1934
China, cut glass, samplers, furniture ... 46137

Adrian MI

Klemm Gallery, Siena Heights University, 1247 Siena Heights Dr, Adrian, MI 49221 • T: +1 517 2647860/63 • F: +1 517 2647739 • creising@sienahts.edu • www.sienahts.edu/~art •
Dir.: *Dr. Peter J. Barr* •
Public Gallery / University Museum – 1919
Work by national artists, cultural expression such as Japanese kimonos, Peruvian arpilleras, Australian fibers, Classical Indian artwork, and contemporary Native American art – studios ... 46138

Stubnitz Gallery, Adrian College, Downs Hall, 110 S Madison St, Adrian, MI 49221 • T: +1 517 2643902 • F: +1 517 2643331 • croyer@adrian.edu • www.adrian.edu •
Dir.: *Brian Steele* •
Public Gallery / University Museum ... 46139

Aiken SC

Aiken County Historical Museum, 433 Newberry St SW, Aiken, SC 29801 • T: +1 803 6422017 • F: +1 803 6422016 • acmuseum@duesouth.net • www.duesouth.net •
Dir.: *Carolyn W. Miles* • C.E.O.: *Owen Clary* •
Local Museum – 1970
Regional hist ... 46140

Aiken Thoroughbred Racing Hall of Fame and Museum, Whiskey Rd and Dupree Pl, Aiken, SC 29801 • T: +1 803 6427758, 6427650 • F: +1 803 6427639 • halloffame@aiken.net • www.aiken.net •
Special Museum – 1979
Horse races ... 46141

Ainsworth NE

Coleman House Museum, 456 Old Highway #7, Ainsworth, NE 69210 •
Local Museum – 1973 ... 46142

Aitkin MN

Aitkin County Museum, 20 Pacific St SW, Aitkin, MN 56431, mail addr: POB 215, Aitkin, MN 56431 • T: +1 218 9273348 • achs@mlec.mn.net • www.cpinternet.com/~achs •
Local Museum ... 46143

Ajo AZ

Organ Pipe Cactus Museum, 10 Organ Pipe Dr, Ajo, AZ 85321-9626 • T: +1 520 3876894 • F: +1 520 3877144 • orpi_information@nps.gov • www.nps.gov/orpi •
C.E.O.: *Kathy Billings* •
Natural History Museum – 1937
Natural and history specimens ... 46144

Akron CO

Washington County Museum, 34445 Hwy 63, Akron, CO 80720 • T: +1 970 3456446 •
Cur.: *Mildred Starlin* •
Local Museum – 1958
Local history ... 46145

Akron OH

Akron Art Museum, 70 E Market St, Akron, OH 44308-2084 • T: +1 330 3769185 • F: +1 330 3761180 • mail@akronartmuseum.org • www.akronartmuseum.org •
Dir.: *Mitchell Kahan* • Chief Cur.: *Barbara Tannenbaum* • Ass. Cur.: *Kathryn Wat* •
Fine Arts Museum – 1922
Modern art, paintings and sculpture, photography ... 46146

National Inventors Hall of Fame, 221 S Broadway, Akron, OH 44308 • T: +1 330 7626565 • F: +1 330 7626313 • museum@invent.org • www.invent.org •
C.E.O.: *David G. Fink* •
Science&Tech Museum – 1973
Science ... 46147

Stan Hywet Hall and Gardens, 714 N Portage Path, Akron, OH 44303-1399 • T: +1 330 8365533 • F: +1 330 8362680 • wbohan@stanhywet.org • www.stanhywet.org •
C.E.O.: *Harry P. Lynch* •
Decorative Arts Museum / Historical Museum – 1957
16th-18th c European tapestries, 18th-20th c American and British fine art, 19th-20th c glass, ceramics, textiles ... 46148

Summit County Historical Society Museum, 550 Copley Rd, Akron, OH 44320 • T: +1 330 5351120 • F: +1 330 5350250 • schs@summithistory.org • www.summithistory.org •
Pres.: *Donald Fair* • Exec.Dir.: *Paula G. Moran* •
Historical Museum – 1924
Transportation, pottery, costumes, glass, manuscripts ... 46149

University Art Galleries, 150 E Exchange St, Akron, OH 44325 • T: +1 330 9725950 • F: +1 330 9725960 • bengston@uakron.edu • www.uakron.edu •
Dir.: *Rod Bengston* •
University Museum / Fine Arts Museum – 1974
Contemporary photography, Southeast Asian ceramics and artifacts ... 46150

Alamogordo NM

New Mexico Museum of Space History, Top of New Mexico, Hwy 2001, Alamogordo, NM 88310 • T: +1 505 4372840 • F: +1 505 4342245 • spacepr@zianet.com • www.spacefame.org •
C.E.O.: *Mark Santiago* • Cur.: *George M. House* • Ass. Cur.: *Kimberly Merrill* •
Science&Tech Museum – 1973
Objects documenting the history of astronautics, astronomy ... 46151

Tularosa Basin Historical Society Museum, 1301 White Sands Blvd, Alamogordo, NM 88310 • T: +1 505 4344438 • tbhs@zianet.com • www.alamogordomuseum.org •
Pres.: *Dr. Rick Miller* •
Historical Museum – 1971
Indian artifacts, fossils, temporary exhibitions ... 46152

Alamosa CO

Adams State College Luther Bean Museum, Richardson Hall, Alamosa, CO 81102 • T: +1 719 5877011 • F: +1 719 5877522 • lsrelyea@adams.edu •
Historical Museum – 1968
Anthropology, ethnology, Indian, history ... 46153

Albany GA

Albany Civil Rights Movement Museum, 326 Whitney Av, Albany, GA 31701 • T: +1 229 4321698 • F: +1 229 4322150 • mtzion@surfsouth.com •
Dir.: *Mirma A. Johnson* •
Historical Museum – 1994
Photographs, artifacts – Archives ... 46154

Albany Museum of Art, 311 Meadowlark Dr, Albany, GA 31707 • T: +1 912 4398400 • F: +1 912 4391332 • curator@albanymuseum.com • www.albanymuseum.com •
Dir.: *Aaron Berger* • Cur.: *Kathryn Delee* •
Fine Arts Museum – 1964
Archaeology, ethnography, costumes and textiles, decorative arts, paintings, photographs, prints, drawings, graphic arts, sculpture ... 46155

Thronateeska Heritage Foundation, 100 W Roosevelt Av, Albany, GA 31701 • T: +1 912 4326955 • F: +1 912 4351572 • info@heritagecenter.org • www.heritagecenter.org •
Pres.: *Leslie Hudson* • Dir.: *Tommy Gregors* •
Local Museum – 1974
Historical Indian and South West Georgia artifacts, railroad cars, locomotives, trains – planetarium ... 46156

Albany NY

Albany Institute of History and Art, 125 Washington Av, Albany, NY 12210-2296 • T: +1 518 4634478 • F: +1 518 4621522 • information@albanyinstitute.org • www.albanyinstitute.org •
Dir.: *Christine M. Miles* •
Fine Arts Museum / Historical Museum – 1791
Fine and decorative arts, photographs, 18th - 20zh c portraits, paintings and documents ... 46157

Historic Cherry Hill, 523 1/2 S Pearl St, Albany, NY 12202 • T: +1 518 4344791 • F: +1 518 4344806 • info@historiccherryhill.org • www.historiccherryhill.org •
C.E.O. & Dir.: *Liselle LaFrance* • Cur.: *Erin Crissman* •
Decorative Arts Museum / Historical Museum – 1964
Furnishings, textiles, ceramics, paintings, silver, personal pers and belongings of the Van Rensselaer-Rankin family ... 46158

Irish American Museum, 991 Broadway, Albany, NY 12204 • T: +1 518 4326598 • F: +1 518 4492540 • irishamermuseum@cs.com • www.irishamericanheritagemusem.org •
Historical Museum – 1986
History ... 46159

Museum of Prints and Printmaking, POB 6578, Albany, NY 12206 • T: +1 518 4494756 • bolonda@nycap.rr.com • www.pcaprint.org •
Cur.: *Charles Semowich* •
Fine Arts Museum – 1990
About 13,000 prints since the 16th c – library, archives ... 46160

New York State Museum, 3099 Cultural Education Center, Empire State Plaza, Albany, NY 12230 • T: +1 518 4745877 • F: +1 518 4863696 • cryan@mail.nysed.gov • www.nysm.nysed.gov •
Dir.: *Clifford A. Siegfried* • Sc. Staff: *William Kelly* (Geology) • *Jeanine Grinage* • *John Hart* (Collections) •
Local Museum – 1858
New York historical artfacts, New Yoirk Indian ethnology, African American hist, geology, archaeology, dec art, fire engines, transportation, agricultural, industrial and domestic technology ... 46161

Schuyler Mansion, 32 Catherine St, Albany, NY 12202 • T: +1 518 4340834 • F: +1 518 4343821 • marcy.shaffer@oprhp.state.ny.us • www.nysparks.com/hist •
Dir.: *Marcy Shaffer* •
Historical Museum – 1911
18th c furnishings and decorative arts, Schuyler family memorabilia, books, glass, silver ... 46162

Shaker Heritage Society Museum, 875 Watervliet-Shaker Rd, Albany, NY 12211-1051 • T: +1 518 4567890 • F: +1 518 4527348 • shakerwv@crisny.org • crisny.org/not-for-profit/shakerwv •
Dir.: *Starlyn D'Angelo* •
Historical Museum / Religious Arts Museum – 1977
1848 Shaker Meeting House, Shaker artifacts and hist ... 46163

Ten Broeck Mansion, 9 Ten Broeck Pl, Albany, NY 12210 • T: +1 518 4369826 • F: +1 518 4361489 • history@tenbroeck.org •
C.E.O. & Pres.: *Gordon Lattey* •
Decorative Arts Museum / Historical Museum – 1947
Local hist, decorative arts ... 46164

University Art Museum, State University of New York at Albany, 1400 Washington Av, Albany, NY 12222 • T: +1 518 4424035 • F: +1 518 4425075 • www.albany.edu/museum •
Interim Dir.: *J. Jason Stewart* •
Fine Arts Museum / University Museum – 1967
Late modern and contemporary art ... 46165

Albany OR

Albany Regional Museum, 136 Lyon St SW, Albany, OR 97321 • T: +1 541 9677122 • armuseum@peak.org • www.armuseum.org •
Local Museum – 1980
Culture, memorabilia ... 46166

Albany TX

The Old Jail Art Center, 201 S Second St, Albany, TX 76430 • T: +1 325 7622269 • F: +1 325 7622260 • ojac@camalott.com •
Exec. Dir.: *Margaret Blagg* •
Fine Arts Museum – 1977
European and American Art from 19th to 21st c ... 46167

Albemarle NC

Morrow Mountain State Park Museum, 49104 Morrow Mountain Rd, Albemarle, NC 28001 • T: +1 704 9824402 • F: +1 704 9825323 • morrow.mountain@ncmail.net • www.ncsparks.net •
Head: *Timothy McCree* •
Natural History Museum – 1962
Mountain area geology, plants and animals, exhibit of Indian civilization ... 46168

Stanly County Historic Museum, 245 E Main St, Albemarle, NC 28001 • T: +1 704 9863777 • F: +1 704 9863778 • evarner@co.stanly.nc.us • www.co.stanly.nc.us •
Dir.: *Christine M. Dwyer* •
Historical Museum – 1973
Period furniture, clothing, textiles ... 46169

Albert Lea MN

Freeborn County Museum, 1031 Bridge Av, Albert Lea, MN 56007-2205 • T: +1 507 3738003 • fchm@smig.net • www.smig.net/fchm •
Exec. Dir.: *Bev Jackson* •
Local Museum / Open Air Museum – 1948
Regional history, pioneer village, Eddie Cochran, rorerunner in the Rock'n Roll era, Marion Ross – libraryCochran archives ... 46170

Albion IN

Old Jail Museum, 215 W Main St, Albion, IN 46701 • T: +1 219 6362428 • www.noblehistoricalsociety.org •
Historical Museum – 1968 ... 46171

Albion MI

Bobbitt Visual Arts Center, Albion College Print Collection and Art Exhibitions, 805 Cass St, Albion, MI 49224 • T: +1 517 6290246 • F: +1 517 6290752 • bwickre@albion.edu • www.albion.edu/art/PrintCollection.asp •
Fine Arts Museum / University Museum – 1835
Contempotary artists, prints ... 46172

Brueckner Museum of Starr Commonwealth, 13725 Starr Commonwealth Rd, Albion, MI 49224-9580 • T: +1 517 6295591 • F: +1 517 6294650 • info@starr.org •
Pres.: *Dr. Arlin E. Ness* •
Fine Arts Museum – 1956
Paintings, sculpture, drawings, prints, ethnic coll, minerals ... 46173

Gardner House Museum, 509 S Superior St, Albion, MI 49224 • T: +1 517 6295100 • history@forks.org • www.forks.org/history •
C.E.O.: *Marorie Ulbrich* •
Local Museum – 1968
Local history, domestic and commercial items – library, archives ... 46174

Albuquerque NM

The Albuquerque Museum of Art and History, 2000 Mountain Rd NW, Albuquerque, NM 87104 • T: +1 505 2437255 • F: +1 505 7646546 • jmoore@cabq.gov • www.cabq.gov/museum •
Dir.: *James C. Moore* • Sc. Staff: *Douglas Fairfield* (Art) • *Deb Slaney* (History) • *Thomas C. Lark* (Collections) • *Chris Steiner* (Education) •
Fine Arts Museum / Historical Museum – 1967
Decorative arts, regional fine arts and crafts, costumes, photography, Southwestern hist, Indian jewelry ... 46175

Anderson/Abruzzo Albuquerque International Balloon Museum, 400 Marquette Av NW, Albuquerque, NM 87102 • T: +1 505 7683555 • F: +1 505 7682846 • bjsmith@cabq.gov • www.cabq.quu.balloon •
Dir.: *Dr. Beej Nierengarten-Smith* •
Science&Tech Museum – 2000
Hist, science and art of ballooning, gondolas, balloons, personal artifacts ... 46176

Explora Science Center and Children's Museum, 2100 Louisiana Blv, Albuquerque, NM 87110-5410 • T: +1 505 8421537 • F: +1 505 8425915 • explorainfo@esccma.org • www.explora.mus.nm.us •
C.E.O.: *Paul Tatter* •
Science&Tech Museum – 1996
Science ... 46177

Indian Pueblo Cultural Center, 2401 12th St NW, Albuquerque, NM 87104 • T: +1 505 8437270 • F: +1 505 8426959 • preck@indianpueblo.org • www.indianpueblo.org •
Dir.: *Ron Solimon* • Cur.: *Pat Reck* •
Ethnology Museum / Historical Museum – 1976
Indian hist ... 46178

Institute of Meteoritics Meteorite Museum, c/o University of New Mexico, Albuquerque, NM 87131 • T: +1 505 2772747 • F: +1 505 2773577 • epswww.unm.edu/iom •
Dir.: *Dr. James J. Papike* • Chief Cur.: *Rhian Jones* •
University Museum / Natural History Museum – 1944
Mineralogy and petrology of meteorites ... 46179

Jonson Gallery, University of New Mexico Art Museum, 1909 Las Lomas NE, Albuquerque, NM 87131 • T: +1 505 2774967 • F: +1 505 2773188 • jonsong@unm.edu • www.unm.edu/~jonsong •
Cur.: *Robert Ware* •
Fine Arts Museum / University Museum – 1950
Works by 20th c and contemporary artists ... 46180

Maxwell Museum of Anthropology, c/o University of New Mexico, Albuquerque, NM 87131-1201 • T: +1 505 2774405 • F: +1 505 2771547 • gbawden@unm.edu • www.unm.edu/~maxwell •
Dir.: *Dr. Garth L. Bawden* • Chief Cur.: *Mari Lyn Salvador* •
University Museum / Ethnology Museum / Archaeology Museum / Folklore Museum – 1932
Archaeology, ethnology, osteology, anthropology, Navajo and other Southwestern weaving, silver, musical instruments, Pakistani, South American textiles and jewelry ... 46181

Museum of Southwestern Biology, c/o Biology Department, University of New Mexico, Albuquerque, NM 87131 • T: +1 505 2772604 • F: +1 505 2776079 • tlowrey@unm.edu • www.unm.edu/~museum •
Dir.: *Thomas F. Turner* (Fishes) • Cur.: *Dr. Christopher C. Witt* (Birds) • *Dr. Howard L. Snell* (Reptiles and Amphibians) • *Dr. Kelly B. Miller* (Arthropods) • *Dr. Joseph A. Cook* (Div. of Genomic Resources; Mammals) • *Dr. Timothy K. Lowrey* (Herbarium) • *Dr. Eric S. Loker* (Parasitology) •
University Museum / Natural History Museum – 1930
Birds, mammals, plants, arthropods, fishes, reptiles, amphibians – frozen tissues coll ... 46182

National Atomic Museum, 1905 Mountain Rd NW, Albuquerque, NM 87104 • T: +1 505 2843243 • F: +1 505 2843244 • info@atomicmuseum.com • www.atomicmuseum.com •
Dir.: *Jim Walther* • Ass. Dir.: *Merri Lewis* • Sc. Staff: *Sam Bono* (History) •
Science&Tech Museum – 1969
Energy, science, weapons, history, aircraft, robotics, medicine ... 46183

National Hispanic Art Museum, 1701 4th St W, Albuquerque, NM 87102 • T: +1 505 2462261 • F: +1 505 2462613 • webmaster@nhccnm.org • www.nhccnm.org •
Dir.: *Thomas E. Chavez* •
Fine Arts Museum
Folk and decorative art from US, Latin American and Spain ... 46184

New Mexico Museum of Natural History and Science, 1801 Mountain Rd NW, Albuquerque, NM 87104-1375 • T: +1 505 8412823 • F: +1 505 8412866 • mtanner@nmmnh.state.nm.us • www.nmnaturalhistory.org •
Pres.: *Ken Hueter* •
Natural History Museum – 1986
Natural hist materials of New Mexico and southwest, botany, geology, paleontology ... 46185

Rattlesnake Museum, 202 San Felipe, Albuquerque, NM 87104-1426 • T: +1 505 2426569 • F: +1 505 2426569 • zoomuseum@aol.com • www.rattlesnakes.com •
Dir.: *Bob Myers* •
Natural History Museum – 1990
Herpetology, American artifacts ... 46186

Telephone Pioneer Museum of New Mexico, 110 Fourth St NW, Albuquerque, NM 87103 • T: +1 505 8422937 • www.nmculture.org •
Dir.: *Neal Roch* •
Science&Tech Museum – 1961
Telephones ... 46187

Turquoise Museum, 2107 Central NW, Albuquerque, NM 87104 • T: +1 505 2478650 • F: +1 505 2478765 •
Historical Museum / Archaeology Museum
Turquoise coll, hist, mining, geology, cutting, marketing, archaeology ... 46188

University Art Museum, University of New Mexico, UNM Center for the Arts, Albuquerque, NM 87131-1416 • T: +1 505 2774001 • F: +1 505 2777315 • lbahm@unm.edu • www.unmartmuseum.unm.edu •
Dir.: *Linda Bahm* • Cur.: *Lee Savary* (Exhibits) •
Fine Arts Museum / University Museum – 1963
19th to 20th c art, sculpture, Beaumont Newhall and Harold Edgerton coll, – Johnson Gallery ... 46189

Very Special Arts Gallery, 4904 Fourth St NW, Albuquerque, NM 87107 • T: +1 505 3452872 • F: +1 505 3452896 • info@vfranm.org • www.collectorsguide.com/vsgallery •
Dir.: *Beth Rudolph* •
Fine Arts Museum – 1994
Drawings, paintings, ceramics, crafts, folk art ... 46190

Alden KS

Atchinson, Topeka and Santa Fe Depot, POB 158, Alden, KS 67512 • T: +1 316 5342425 • prflrcraft@aol.com •
Dir.: *Sara Fair Sleeper* •
Science&Tech Museum – 1970
Transportation ... 46191

Alden NY

Alden Historical Society Museum, 13213 Broadway, Alden, NY 14004 • T: +1 716 9377606 • Dir.: *Ronald Savage* • Cur.: *Ruth Davis* • Asst.Cur.: *Laura Airey* •
Historical Museum – 1965
Farm tools, kitchen utensils, town hist 46192

Aledo IL

Essley-Noble Museum, Mercer County Historical Society, 1406 SE 2nd Av, Aledo, IL 61231 • T: +1 309 5844820 •
Pres.: *Alicia Ives* • Cur.: *Shirley Crawford* •
Local Museum – 1959
Genealogy, local hist 46193

Aledo TX

Living Word National Bible Museum, 3909 Snow Creek Dr, Aledo, TX 76008 • T: +1 817 2444504 • F: +1 817 2444504 • johnhellstern@charter.net •
C.E.O.: *Dr. John R. Hellstern* •
Religious Arts Museum – 1991
Ancient bibles & scolls, full-scale Gutenberg press 46194

Alexandria LA

Alexandria Museum of Art, 933 Main St, Alexandria, LA 71301 • T: +1 318 4433458 • F: +1 318 4430545 • frances@themuseum.org • www.themuseum.org •
Dir.: *Frances Morrow* • Cur.: *Cheryy Davis* (Art) •
Fine Arts Museum – 1977
Art, housed in a old Bank Bldg 46195

University Gallery, c/o Louisiana State University at Alexandria, Student Ctr, Hwy 71 S, Alexandria, LA 71302 • T: +1 318 4736449 •
Cur.: *Roy V. de Ville* •
Fine Arts Museum / University Museum – 1960
Louisiana contemporary art 46196

Alexandria MN

Runestone Museum, 206 N Broadway, Alexandria, MN 56308 • T: +1 320 7633160 • F: +1 320 7639705 • bigole@rea-alp.com • www.runestonemuseum.org •
Dir.: *LuAnn W. Patton* •
Local Museum – 1958
History, story of the Kensington Runestone, norse hist, pioneer life, Native Americans, Fort Alexandria replica 46197

Alexandria VA

Alexandria Archaeology Museum, 105 N Union St, 327, Alexandria, VA 22314 • T: +1 703 8384399 • F: +1 703 8386491 • archaeology@ci.alexandria.va.us • www.alexandriaarchaeology.org •
Dir.: *Dr. Pamela J. Cressey* •
Archaeology Museum – 1977
18th & 19th c artifacts from archaeological digs in Alexandria 46198

Alexandria Black History Resource Center, 638 N Alfred St, Alexandria, VA 22314 • T: +1 703 8384356 • F: +1 703 7063999 • black.history@ci.alexandria.va.us • www.alexblackhistory.org •
Dir.: *Louis Hicks* • Ass. Dir.: *Audrey P. Davis* • Sc. Staff: *Lillian Patterson* •
Local Museum – 1983
African American hist 46199

The Athenaeum, 201 Prince St, Alexandria, VA 22314 • T: +1 703 5480035 • F: +1 703 7687471, 5480456 • nvfaa@nvfaa.org • www.alexandria-athenaeum.org •
Exec. Dir.: *Mary Gaissert Jackson* •
Fine Arts Museum – 1964
American artists 46200

Carlyle House, 121 N Fairfax St, Alexandria, VA 22314 • T: +1 703 5492997 • F: +1 703 5495738 • mrcoleman@juno.com • www.carlylehouse.org •
Dir.: *Mary Ruth Coleman* • Cur.: *Heidi Miller* •
Local Museum – 1976
18th c decorative arts, archaeological coll 46201

Collingwood Library and Museum on Americanism, 8301 E Boulevard Dr, Alexandria, VA 22308 • T: +1 703 7651652 • F: +1 703 7658390 • clma1@erols.com • collingwoodlibrary.com •
Cur.: *Warren L. Baker* • Ass. Cur.: *Barbara J. Baker* •
Historical Museum – 1977
Indian artifacts, coins, china, Revolutionary War items 46202

Fort Ward Museum and Historic Site, 4301 W Braddock Rd, Alexandria, VA 22304 • T: +1 703 8384848 • F: +1 703 6717350 • fort.ward@ci.alexandria.va.us • www.fortward.org •
Cur.: *Susan G. Cumbey* •
Military Museum – 1964
Military hist, Civil War coll and Union Fort 46203

Frank Lloyd Wright's Pope-Leighey House, 9000 Richmond Hwy, Alexandria, VA 22309 • T: +1 703 7804000 • F: +1 703 7808509 • woodlawn@nthp.org • www.popeleighey1940.org •
Dir.: *Ross Randall* •
Local Museum / Fine Arts Museum – 1964 46204

Friendship Firehouse, 107 S Alfred St, Alexandria, VA 22314 • T: +1 703 8383891 • F: +1 703 8384997 • jean.federico@ci.alexandria.va.us • ci.alexandria.va.us/oha/ •
Dir.: *Jean Federico* •
Science&Tech Museum – 1993
Hist of Friendship Fire Company & local firefighting hist, uniforms, letters 46205

Gadsby's Tavern Museum, 134 N Royal St, Alexandria, VA 22314 • T: +1 703 8384242 • F: +1 703 8384270 • gadsbys.tavern@ci.alexandria.va.us • www.gadsbystavern.org •
Dir.: *Gretchen M. Bulova* •
Local Museum – 1976
Historic buildings 46206

Gallery West, 205 S Union St, Alexandria, VA 22314 • T: +1 703 5497359 •
Dir.: *Craig Snyder* •
Fine Arts Museum – 1979 46207

George Washington Masonic National Memorial, 101 Callahan Dr, Alexandria, VA 22301 • T: +1 703 6832007 • F: +1 703 5199270 • gseghers@gwmemorial.org • www.gwmemorial.org •
Dir.: *George D. Seghers* •
Historical Museum – 1910
George Washington & his family, Masonic affiliations of Washington and his associates 46208

The John Q. Adams Center for the History of Otolaryngology - Head and Neck Surgery, 1 Prince St, Alexandria, VA 22314 • T: +1 703 8364444 • F: +1 703 6835100 • museum@entnet.org • www.entnet.org/museum •
C.E.O.: *David R. Nielsen* •
Special Museum – 1990
Medical hist, doc of hte hist of otolarygology 46209

Lee-Fendall House Museum, 614 Oronoco St, Alexandria, VA 22314 • T: +1 703 5481789 • F: +1 703 5480931 • staff@leefendallhouse.org • www.leefendallhouse.org •
Dir.: *Kristin Miller* •
Local Museum – 1974
Lee family heirlooms, decorative arts 46210

The Lyceum - Alexandria's History Museum, 201 S Washington St, Alexandria, VA 22314 • T: +1 703 8384994 • F: +1 703 8384997 • lyceum@alexandriava.us • www.alexandriahistory.org •
Dir.: *Jim Mackay* • Ass. Dir.: *Kristin B. Lloyd* •
Historical Museum – 1974
18th to 20th c artifacts relating to the local area 46211

Ramsay House, 221 King St, Alexandria, VA 22314 • T: +1 703 8385005 • F: +1 703 8384683 • jmitchell@funside.com • www.funside.com •
Dir.: *Joanne Mitchell* • Dep. Dir.: *Laura Overstreet* •
Local Museum – 1962
Historic house 46212

Stabler-Leadbeater Apothecary Museum, 105-107 S Fairfax St, Alexandria, VA 22314 • T: +1 703 8363713 • F: +1 703 8363713 • apothecary@starpower.net • www.apothecary.org •
Dir.: *Sara Becker* •
Special Museum – 1792
Pharmaceutical equipment and artifacts, apothecary bottles 46213

Torpedo Factory Art Center, 105 N Union St, Alexandria, VA 22314 • T: +1 703 8384199, 8384565 • F: +1 703 8383603 • director@torpedofactory.org • www.torpedofactory.org •
C.E.O.: *Trudi VanDyke* •
Fine Arts Museum / Folklore Museum – 1974
Hist of the torpedo factory, U.S. Naviy loaned torpedoes, art coll 46214

Woodlawn Plantation, 9000 Richmond Hwy, Alexandria, VA 22309 • T: +1 703 7804000 • F: +1 703 7808509 • woodlawn@nthp.org • www.woodlawn1805.org •
Dir.: *Ross Randall* •
Local Museum – 1951
19th c decorative arts, furniture, paintings, china 46215

Alfred NY

The Schein-Joseph International Museum of Ceramic Art, New York State College of Ceramics at Alfred University, Alfred, NY 14802 • T: +1 607 8712421 • F: +1 607 8712615 • ceramicsmuseum@alfred.edu • www.ceramicsmuseum.alfred.edu •
Dir./Chief Cur.: *Dr. Margaret Carney* •
Decorative Arts Museum – 1900
Ceramics, pottery, porcelain 46216

Algodones NM

Legends of New Mexico Museum, 601 Frontage Rd, Algodones, NM 87001 • T: +1 505 8678600 • F: +1 505 8361700 • www.legendmuseum.com •
Dir.: *Furber Marthalie* •
Historical Museum – 2003
Personal artifacts, photographs, decorative arts, costumes 46217

Alice TX

South Texas Museum, 66 S Wright St, Alice, TX 78332 • T: +1 512 6688891 • F: +1 512 6643327 • stmuseum@sbcglobal.com •
Chm.: *Mary Dru Burns* •
Local Museum – 1975
Firearms, Civil War items, farm & ranch tools, saddles, antique furniture 46218

Aliceville AL

Aliceville Museum, 104 Broad St, Aliceville, AL 35442 • T: +1 205 3732363 • museum@nctv.com • www.cityofaliceville.com •
Dir.: *Paluzzi Mary Bess K.* •
Military Museum – 1993
Military hist, Coca-Cola exhibit 46219

Aline OK

Sod House Museum, Rte 1, Aline, OK 73716 • T: +1 580 4632441 • www.ok-history.mus.ok.us •
C.E.O.: *DeAnn McGahen* •
Historical Museum – 1968
Household and farm equipment 46220

Allaire NJ

Historic Allaire Village, Allaire State Park, Rte 524, Allaire, NJ 07727-3715 • T: +1 732 9193500, 9382253 • F: +1 732 9383302 • allairevillage@bytheshore.com • www.allairevillage.org •
Dir.: *John Curtis* •
Open Air Museum / Local Museum – 1957
Farm tools, textiles, decorative arts, furniture, trade artifacts 46221

Allegan MI

Allegan County Historical and Old Jail Museum, Old Jail Museum, 113 Walnut St, Allegan, MI 49010 • T: +1 616 6738292 • F: +1 616 6734853 • kharrier@datawise.net • www.historicallegan.org •
Dir.: *Marguerite Miller* •
Local Museum – 1952
Regional history, former county jail and sheriff's home – library 46222

Alleghany CA

Underground Gold Miners Museum, 356 Main St, Alleghany, CA 95910 • T: +1 530 2873330 • info@undergroundgold.com • www.undergroundgold.com •
C.E.O.: *Rae Bell* •
Science&Tech Museum / Historical Museum – 1995
Mining technology, history of Allgehany 46223

Allentown NJ

Artful Deposit Galleries, 1 Church St, Allentown, NJ 08505 • T: +1 609 2593234 • www.theartfuldeposit.com •
Owner: *C.J. Mugavero* •
Fine Arts Museum – 1986
Works by Patrick Antonelle, Joseph Dawley, Hanneke de Neve and Ken McIndoe 46224

Allentown PA

Allentown Art Museum, Fifth and Court Sts, Allentown, PA 18105 • T: +1 610 4324333 ext 10 • F: +1 610 4347409 • membership@allentownartmuseum.org • www.allentownartmuseum.org •
Pres.: *Gary L. Millenbruch* • Dep. Dir.: *David R. Brigham* • Dir.: *Lisa Miller* (Admin) • Cur.: *Christine L. Oaklander* (Coll. & Exhebitions) • *Ruta Saliklis* (Textiles) • Sc. Staff: *Lisa Dubé* (Education) • *Carl Schafer* (Registrar) •
Fine Arts Museum – 1934
European and American paintings, sculpture, prints, drawings, rextiles, decoratives art 46225

Allied Air Force Museum, Queen City Airport, 1730 Vultee St, Allentown, PA 18103 • T: +1 610 7915122 • F: +1 610 7915453 •
Dir./Cur.: *Joseph B. Fillman* •
Military Museum / Science&Tech Museum – 1984
Aeronautics 46226

Lehigh County Historical Society Museum, Hamilton at Fifth St, Allentown, PA 18101 • T: +1 610 4351074 • F: +1 610 4359812 • lchs@voicenet.com • www.lehighcountyhistoricalsociety.org •
Dir.: *Bernard P. Fishman* • Cur.: *Andree Miller* (Collections) • Education: *Sarah E. Nelson* •
Historical Museum – 1904
Reninger house, Elverson house, Gruber house, Trout hall (1770), Haines mill (1760), One-room-schoolhouse 46227

Martin Art Gallery, c/o Muhlenberg College, 2400 Chew St, Allentown, PA 18104-5586 • T: +1 484 6643467 • F: +1 484 6643633 • gallery@muhlenberg.edu • www.muhlberg.edu •
Dir.: *Dr. Lori Verderame* •
Fine Arts Museum – 1976
Master prints, Rembrandt, Durer, Whistler, Goya, 20th c American contemporary art 46228

Museum of Indian Culture, 2825 Fish Hatchery Rd, Allentown, PA 18103-9801 • T: +1 610 7972121 • F: +1 610 7972801 • lenape@epixi.net • www.lenape.org •
C.E.O.: *James Albany* • Cur.: *Alexander Nome* •
Ethnology Museum – 1981
Lenni Lenape Indians, early Pennsylvania German artifacts 46229

Alliance NE

Knight Museum, 908 Yellowstone, Alliance, NE 69301 • T: +1 308 7622384, 7625400 ext 261 • F: +1 308 7627848 • museum@panhandle.net • www.cityofalliance.com •
C.E.O.: *Richard Cayer* • Dir.: *Becci Thomas* •
Local Museum – 1965
Indian artifacts, gun collection 46230

Alliance OH

Mabel Hartzell Historical Home, Mount Union College, 840 N Park Av, Alliance, OH 44601 • T: +1 330 8234115 •
Pres.: *Gordon Harrison* •
Special Museum – 1939
Original furnishings, Renaissance revival furniture 46231

Allison IA

Butler County Historical Museum, Little Yellow Schoolhouse, Fair Grounds, 714 Elm St, Allison, IA 50602 • T: +1 319 2672255 •
Pres.: *Deb Bochmann* •
Local Museum – 1956
Local history 46232

Allison Park PA

Depreciation Lands Museum, 4743 S Pioneer Rd, Allison Park, PA 15101 • T: +1 412 4860563 •
Historical Museum – 1974
Hist of the Indian Wars of 1784-1795, financial Problems of the War 46233

Alma CO

Alma Firehouse and Mining Museum, 1 W Buckskin Rd, Alma, CO 80420-0336 • T: +1 719 8363413, 8363117 •
Dir.: *Don Gostisha* •
Historical Museum – 1976
Fire fighting 46234

Alma KS

Wabaunsee County Historical Museum, Missouri and Third Sts, Alma, KS 66401, mail addr: POB 387, Alma, KS 66401 • T: +1 785 7652200 •
C.E.O./Pres.: *Gertrude Gehrt* • Cur.: *Rayonna Mock* •
Local Museum – 1968
Indian artifacts 46235

Almond NY

Hagadorn House Museum, Almond Historical Society, 7 N Main St, Almond, NY 14804 • T: +1 607 2766781 • dodomo@infoblvd.net •
C.E.O. & Pres.: *Charlotte Baker* •
Historical Museum – 1965
Genaological files, , local artifacts 46236

Alpena MI

Jesse Besser Museum, 491 Johnson St, Alpena, MI 49707-1496 • T: +1 989 3562202 • F: +1 989 3563133 • jbmuseum@northland.lib.mi.us • www.bessermuseum.org •
Exec. Dir.: *Dr. Janice V. McLean* • Cur.: *Richard Clute* •
Local Museum – 1962
Fine and decorative arts, history and science of Northeast Michigan, anthropology, natural history – library, planetarium 46237

Alpine TX

Museum of the Big Bend, Sul Ross State University, Alpine, TX 79832 • T: +1 915 8378143 • F: +1 915 8378381 • francell@sulross.edu • www.sulross.edu/~museum/ •
Dir.: *Larry Francell* • Cur.: *Mary Bridges* •
Local Museum – 1926
Late 19th - early 20th c hist materials, archaeological artifacts from paleo-man to modern ethnology 46238

Altenburg MO

Perry County Lutheran Historical Society, Church St, Altenburg, MO 63732 • T: +1 573 8245542 •
Pres.: *Leonard A. Kuehnert* •
Association with Coll / Religious Arts Museum – 1910
Religious books, bibles, sermon books, pictures, records 46239

Alton IL

Alton Museum of History and Art, 2809 College Av, Alton, IL 62002 • T: +1 618 4622763 • F: +1 618 4626390 • altonmuseum@618connect.com • www.altonweb.com/museum/index.html •
Pres.: *David A. Culp* •
Fine Arts Museum / Historical Museum – 1971
Local paintings, sculpture, artifacts from local railroads, photgraphs 46240

National Great Rivers Museum, POB 337, Alton, IL 62002 • T: +1 618 4626979 • F: +1 618 4627650 • www.mvs.usace.army.mil •
Science&Tech Museum – 2000
Mississippi river hist and workings of the locks and dams system 46241

Altoona PA

Railroader's Memorial Museum, 1300 Ninth Av, Altoona, PA 16602 • T: +1 814 9460834 • F: +1 814 9469457 • admin@railroadcity.com • www.railroadcity.com •
Dir.: *R. Cummins McNitt* • Cur.: *Janice Hartman* •
Science&Tech Museum – 1980
Railroading artifacts, paintings, photographs, RPO car, dining car, K-4 & 0-4-0 steam locomotives ... 46242

Southern Alleghenies Museum of Art, 1210 11th Av, Altoona, PA 16603 • T: +1 814 9464464 • F: +1 814 9463131 • sama@penn.com • www.sama-art.org •
Dir.: *Barbara J. Hollander* •
Fine Arts Museum
American painting, sculpture, drawing and prints 46243

Alturas CA

Modoc County Historical Museum, 600 S Main St, Alturas, CA 96101 • T: +1 530 2339944 • contactus@alturaschamber.org • www.alturaschamber.org •
Dir.: *Paula Murphy* •
Local Museum – 1967
Regional history, period firearms ... 46244

Altus OK

Museum of the Western Prairie, 1100 Memorial Dr, Altus, OK 73521 • T: +1 580 4821044 • F: +1 580 4820128 • muswestpr@ok-history.mus.ok.us • www.museumwesternprairie.org •
C.E.O.: *Dr. Bob Blackburn* • Dir.: *Bart McClenny* •
Historical Museum – 1970
Agricultural and ranching tools, Native American collo, local hist, fossils ... 46245

Alva OK

Cherokee Strip Museum, 901 14 St, Alva, OK 73717 • T: +1 580 3272030 • ffkirk@email.com •
C.E.O. & Pres.: *Kirk Trekell* •
Local Museum – 1961
Indian sculptures with bow and headdress, farm machinary and equip, telephone coll, dolls, dishes, war uniforms, costumes ... 46246

Northwestern Oklahoma State University Museum, Jesse Dunn Hall, 709 Oklahoma Blvd, Alva, OK 73717 • T: +1 580 3271700 ext 8513 • F: +1 580 3271881 • vnpowders@nwosu.edu • www.nwosu.edu •
Dir.: *Dr. Vernon Powders* •
University Museum – 1902
Indian artifacts; natural science, mounted birds and mammals and study skin, pleistocene fossils, geology, mineralogy, paleontology ... 46247

Amagansett NY

Mabel and Victor D'Amico Studio and Archive, POB 1266, Amagansett, NY 11930 • T: +1 631 2673123 • F: +1 631 2672901 • damico@hamptons.com • www.theartbarge.com •
Dir.: *Christopher Kohan* •
Fine Arts Museum / Open Air Museum – 1960
Early modernist furnishings, found art and objects – archive, sculpture garden ... 46248

Amana IA

Museum of Amana History, 4310 220th Trail, Amana, IA 52203 • T: +1 319 6223567 • F: +1 319 6226481 • amherit@juno.com • www.amanaheritage.org •
Dir.: *Lanny Haldy* •
Local Museum / Folklore Museum – 1969
Local hist ... 46249

Amarillo TX

Amarillo Museum of Art, 2200 S Van Buren, Amarillo, TX 79109 • T: +1 806 3715050 • F: +1 806 3739235 • amoa@actx.edu • www.amarilloart.org •
Exec. Dir.: *Tom R. Toperzer* •
Fine Arts Museum – 1972
20th c American paintings, prints, drawings, photographs, sculpture, Asian art ... 46250

American Quarter Horse Heritage Center and Museum, 2601 I-40, E, Amarillo, TX 79104 • T: +1 806 3765181 • F: +1 806 3761005 • museum@aqha.org • www.aqha.org •
C.E.O.: *Billy Brewer* • Dir.: *Ron Marshall* •
Special Museum – 1991
Equine, Hall of Fame ... 46251

Don Harrington Discovery Center, 1200 Streit Dr, Amarillo, TX 79106 • T: +1 806 3559547 ext 20 • F: +1 806 3555703 • g.ganpat@dhdc.org • www.dhdc.org •
Exec. Dir.: *Ganesh Ganpat* • Dir.: *David Sargus* (Exhibits) •
Science&Tech Museum – 1968
Science and technology ... 46252

Texas Pharmacy Museum, c/o Texas Tech School of Pharmacy, 1300 S Coulter St, Amarillo, TX 79106 • T: +1 806 3564000 ext 268 • F: +1 806 3564017 • paul.katz@ttuhsc.edu • www.texas.pharmacy.museum •
Cur.: *Paul Katz* •
Historical Museum – 1968
Pharmacy in Texas, the US and Western Europe, tools for prescription compounding and delivery systems, pharmaceutical products, pharmacy art, Texas practitioners, laboratory glassware, antique bottles ... 46253

Ambridge PA

Old Economy Village Museum, 14-16 and Church Sts, Ambridge, PA 15003 • T: +1 724 2664500 • F: +1 724 2667506 • mlandis@state.pa.us • www.oldeconomyvillage.org •
Dir.: *Mary Ann Landis* •
Open Air Museum / Historical Museum – 1919
18 buildings of the original town of Economy (now Ambridge), PA, built between 1824 and 1831 ... 46254

American Falls ID

Massacre Rocks State Park Museum, 3592 Park Lane, American Falls, ID 83211-5555 • T: +1 208 5482672 • F: +1 208 5482671 • mas@idpr.state.id.us • www.idahoparks.org •
Natural History Museum – 1967
Idian artifacts, natural hist, pioneer and Oregon trail hist ... 46255

Ames IA

Brunnier Art Museum, Iowa State University, 290 Scheman Bldg, Iowa State University, Ames, IA 50011 • T: +1 515 2943342 • F: +1 515 2947070 • museums@adp.iastate.edu • www.museums.iastate.edu •
Dir.: *Lynette Pohlman* •
Decorative Arts Museum / Fine Arts Museum / University Museum – 1975
Decorative and fine arts ... 46256

Farm House Museum, Iowa State University, Knoll Rd, Iowa State University Campus, Ames, IA 50011 • T: +1 515 2943342, 2947426 • F: +1 515 2947070 • eostend@iastate.edu • www.museums.iastate.edu •
Dir.: *Lynette Pohlman* • Cur.: *Eleanor Ostendorf* •
University Museum / Agriculture Museum – 1976
C 1860 - 1910 decorative arts, c 1860 - 1864 The Farm House ... 46257

Gallery 181, Iowa State University, 134 College of Design Bldg, Iowa State University, Ames, IA 50011 • T: +1 515 2947428 • F: +1 515 2949755 • isutoded@iastate.edu • www.design.iastate.edu •
Dir.: *Dean Mark. Engelbrecht* •
Fine Arts Museum / University Museum – 1978
Art ... 46258

The Octagon Center for the Arts, 427 Douglas Av, Ames, IA 50010-6281 • T: +1 515 2325331 • F: +1 515 2325088 • directorart@isunet.net • www.octagonarts.org •
Fine Arts Museum – 1966
Art, Feinberg Mask collection ... 46259

Amesbury MA

Bartlett Museum, 270 Main St, Amesbury, MA 01913 • T: +1 978 3884528 •
Local Museum – 1968
Local hist, natural hist, carriage trade, Indian artifacts, tools and implements, paintings ... 46260

John Greenleaf Whittier Home, 86 Friend St, Amesbury, MA 01913 • T: +1 978 3881337 •
Pres.: *Sally Ann Lavery* • Cur.: *Stephen McDonough* •
Special Museum – 1898
Home of poet and abolitionist John Greenleaf Whittier – library ... 46261

Newburyport Maritime Society Museum, 25 Water St, Amesbury, MA 01950 • T: +1 978 3380162 • F: +1 978 8340762 • info@themaritimesociety.org • www.themaritimesociety.org •
Historical Museum – 1793
Fleet of dories, patterns, handtools, buiseness papers ... 46262

Amherst MA

Amherst College Museum of Natural History, College St, Amherst, MA 01002 • T: +1 413 5422165 • F: +1 413 5422713 • llthomas@amherst.edu • www.amherst.edu/~pratt •
Dir.: *Peter D. Crowley* • Cur.: *Linda L. Thomas* •
University Museum / Natural History Museum – 1848
Natural history, paleontology, mineralogy, petrology, osteology, anthropology, taxidermy – library ... 46263

Amherst History Museum, Strong House, 67 Amity St, Amherst, MA 01002 • T: +1 413 2560678 • F: +1 413 2560678 • info@amhersthistory.org • www.amhersthistory.org •
Dir.: *Fiona Russell* •
Local Museum – 1899
Local hist, decorative and folk arts, tools, weapons ... 46264

Emily Dickinson Museum, 280 Main St, Amherst, MA 01002 • T: +1 413 5428161 • F: +1 413 5422152 • info@emilydickinsonmuseum.org • www.emilydickinsonmuseum.org •
Dir.: *Cindy Dickinson* • *Jane Wald* •
Special Museum – 1965
Literature, decorative arts ... 46265

Eric Carle Museum of Picture Book Art, 125 W Bay Rd, Amherst, MA 01002 • T: +1 4136581113 • F: +1 413 6581139 • www.picturebookart.org •
Dir.: *H. Nicholas B. Clark* •
Library with Exhibitions – 2001
Picture book art – library ... 46266

Herter Art Gallery, University of Massachusetts, 125A Herter Hall, 151 Presidents Dr, Amherst, MA 01003-9330 • T: +1 413 5450976 • F: +1 413 5451189 • www.umass.edu/art/exhibitions/herter-schedule.html •
University Museum / Fine Arts Museum
Paintings, sculpture and works on paper ... 46267

Mead Art Museum, Amherst College, Amherst, MA 01002-5000 • T: +1 413 5422335 • F: +1 413 5422117 • mead@amherst.edu • www.amherst.edu/~mead •
Dir.: *Jill Meredith* • Cur.: *Trinkett Clarck* (American Art) • *Jill Meredith* (European Art) • *Carol Solomon Kiefer* (Prints) •
Fine Arts Museum / University Museum – 1821
American fine and decorative arts, Asian, pre-Columbian, Mexican and African art – Fairchild gallery ... 46268

Museum of Natural History, University of Massachusetts, Dept. of Biology, Amherst, MA 01003 • T: +1 413 5452902 • F: +1 413 5453243 • wbemis@bio.umass.edu • bcrc.bio.umass.edu/ummnh •
Dir.: *William Bemis* • Cur.: *Alan Richmond* (Herpetology) • *Douglas G. Smith* (Invertebrates) • *Elizabeth R. Dumont* (Mammals) • *Margery C. Coombs* (Vertebrate Paleontology) • *Laurie Godfrey* (Anthropology) • *Donald E. Kroodsma* (Birds) • *Karen B. Searcy* (Herbarium) • *Ben Normark* (Insects) •
University Museum / Natural History Museum – 1863
Zoology, fishes, amphibians and reptiles, invertebrates, mammals, birds – herbarium ... 46269

National Yiddish Book Center, Weinberg Bldg, 1021 West St, Amherst, MA 01002-3375 • T: +1 413 2564900 ext 124 • F: +1 413 2564700 • yiddish@bikher.org • www.yiddishbookcenter.org •
Pres.: *Aaron Lansky* •
Historical Museum / Religious Arts Museum – 1980
Yiddish culture and history – library ... 46270

University Gallery, University of Massachusetts at Amherst, Fine Arts Center, 151 Presidents Dr, Amherst, MA 01003-9331 • T: +1 413 5453670 • F: +1 413 5452018 • ugallery@acad.umass.edu • www.umass.edu/fac/universitygallery •
Dir.: *Loretta Yarlow* • Cur.: *Regina Coppola* •
Fine Arts Museum / Public Gallery / University Museum – 1975
20th c American photography, prints and drawings ... 46271

Amherst NY

Amherst Museum, 3755 Tonawanda Creek Rd, Amherst, NY 14228 • T: +1 716 6891440 • F: +1 716 6891409 • amhmuseum@adelphia.net • www.amherstmuseum.org •
Dir.: *Lynn S. Beman* • Cur.: *Jessica A. Norton* • *Kristine Rhoback* (Textiles) • *Jean W. Neff* (Education) •
Local Museum – 1970
Local history and culture, historic buildings ... 46272

Amherst VA

Amherst County Museum, 154 S Main St, Amherst, VA 24521 • T: +1 804 9469068 • F: +1 804 9469068 • achmuseum@aol.com • members.aol.com/achmuseum/achmhis.htm •
Dir.: *Meghan Wallace* •
Local Museum / Folklore Museum – 1973
Aspects of former Amherst life-styles ... 46273

Amityville NY

Lauder Museum, Amitville Historical Society, 170 Broadway, Amityville, NY 11701-0764 •
C.E.O.: *William T. Lauder* •
Local Museum – 1973
Hist of the town ... 46274

Amory MS

Amory Regional Museum, 715 Third St, Amory, MS 38821 • T: +1 662 2562761 • F: +1 662 2562761 •
Dir.: *Lynn R. Millender* •
Local Museum – 1976
Archaeological aartifacts, Chickasaw Indian artifacts ... 46275

Amsterdam NY

Walter Elwood Museum, 300 Guy Park Av, Amsterdam, NY 12010-2228 • T: +1 518 8435151 • F: +1 518 8436098 •
Dir.: *Mary Margaret Gage* •
Local Museum – 1940
Early American and Indian material, hist, natural hist ... 46276

Anaconda MT

Copper Village Museum and Arts Center, 401 E Commercial, Anaconda, MT 59711 • T: +1 406 5632422 •
Dir.: *Carol Jette* •
Fine Arts Museum / Historical Museum – 1971
Local hist artifacts, miners, industry, barbed wire coll, smelting process, workers – library ... 46277

Anacortes WA

Anacortes Museum, 1305 8th, Anacortes, WA 98221 • T: +1 360 2931915 • F: +1 360 2931929 • coa.museum@cityofanacortes.org • www.anacorteshistorymuseum.org •
Dir.: *Garry Cline* • Cur.: *Judy Hakins* (Collections) • *James Stover* (Maritime History) •
Local Museum – 1957
Local hist ... 46278

Anadarko OK

Anadarko Philomathic Museum, 311 E Main St, Anadarko, OK 73005 • T: +1 405 2473240 •
C.E.O. & Pres.: *Linda Taylor* • Cur.: *Lee Reeves* •
Local Museum – 1936
Pioneer furnishings and clothing; military exhibits, railroad memorabilia and ticket office ... 46279

Indian City U.S.A., Hwy 8, 2 1/2 miles SE of Anadarko, Anadarko, OK 73005 • T: +1 405 2475661 • F: +1 405 2472467 • indiancity@aol.com • www.indiancityusa.com •
Pres.: *Linda Poolaw* •
Historical Museum / Natural History Museum / Ethnology Museum – 1955
Indian artifacts, dance costumes, pottery, cradles, bags; Indian dolls and paintings ... 46280

National Hall of Fame for Famous American Indians, Hwy 62 E, Anadarko, OK 73005 • T: +1 405 2475555 • F: +1 405 2475571 • dailynews@tanet.net •
C.E.O.: *Joe McBride* •
Historical Museum – 1952 ... 46281

Southern Plains Indian Museum, 715 E Central Blvd, Anadarko, OK 73005, mail addr: POB 749, Anadarko, OK 73005 • T: +1 405 2476221 • F: +1 405 2477593 •
Ethnology Museum / Folklore Museum – 1947
Native American art in all mediums, exhibitions of contemporary art ... 46282

Anaheim CA

Anaheim Museum, 241 S Anaheim Blvd, Anaheim, CA 92805 • T: +1 714 7783301 • F: +1 714 7786740 • info@anaheimmuseum.com • www.anaheimmuseum.com •
Dir./Cur.: *Joyce Franklin* •
Local Museum / Historical Museum – 1982
Local history, culture, art works ... 46283

Anaktuvuk Pass AK

Simon Paneak Memorial Museum, 341 Mekiana Rd, Anaktuvuk Pass, AK 99721 • T: +1 907 6613413 • F: +1 907 6613414 • grant.spearman@north-slope.org • www.north-slope.org/nsb/55.htm •
Dir.: *George N. Ahmaogak* • Cur.: *Grant Spearman* •
Local Museum – 1986
Local history, ethnography – library ... 46284

Anchorage AK

Alaska Aviation Museum, 4721 Aircraft Dr, Anchorage, AK 99502 • T: +1 907 2485325 • F: +1 907 2486391 • info@alaskaairmuseum.org • www.alaskaairmuseum.org •
Dir.: *Norm Lagasse* •
Science&Tech Museum – 1988
Ca 30 aircrafts, aviation hist – theater ... 46285

Alaska Heritage Museum at Wells Fargo, 301 W Northern Lights Blvd, Anchorage, AK 99503 • T: +1 907 2652834 • historicalservices@wellsfargo.com • www.wellsfargohistory.com/museums •
Dir.: *Beverly Smith* • Cur.: *Artemis Bona Dea* •
Fine Arts Museum / Ethnology Museum – 1968
Alaskian native ethnology and contemporary crafts, Regional fine arts – library ... 46286

Alaska Museum of Natural History, 201 N Bragaw Rd, Anchorage, AK 99508 • T: +1 907 2742400 • webcontact@alaskamuseum.org • www.alaskamuseum.org •
Dir.: *Tom Bennett* •
Natural History Museum – 1992
Exhibits of rocks, minerals and exciting rare fossils ... 46287

Anchorage Museum at Rasmuson Center, 625 C Street, Anchorage, AK 99501 • T: +1 907 3436172, 3434326 • F: +1 907 3436149 • museum@ci.anchorage.ak.us • www.anchoragemuseum.org •
Dir.: *James Pepper Henry* • Dep. Dir.: *Suzi Jones* • Cur.: *Walter Van Horn* (Collections) • *David Nicholls* (Exhibits) • *Jocelyn Young* (Public Art) • *Marylin Knapp* (History) • Libr.: *Kathleen Hertel* •
Fine Arts Museum / Historical Museum / Ethnology Museum / Science&Tech Museum – 1968
Alaskan art, history, archaeology, ethnology and science – Atwood Resource Center (library and archives) 46288

Imaginarium Discovery Center, Anchorage Museum, 625 C St, Anchorage, AK 99501 • T: +1 907 3436172, 3434326 • F: +1 907 3436149 • museum@ci.anchorage.ak.us • www.anchoragemuseum.org •
Exec. Dir.: *Christopher B. Cable* •
Science&Tech Museum – 1987
Science, technology ... 46289

Oscar Anderson House Museum, 420 M St, Anchorage, AK 99501 • T: +1 907 2742336 • F: +1 907 2743600 •
Man.: *Mary A. Flaherty* •
Historical Museum – 1982
Founding of the city of Anchorage ... 46290

U

Anderson IN

Alford House-Anderson Fine Arts Center, 32 W 10th St, Anderson, IN 46016-1406 • T: +1 765 6491248 • F: +1 765 6490199 • info.taca@sbcglobal.net • www.andersonart.org/index.html •
Pres.: *Rick Moore* • C.E.O.: *Deborah McBratney-Stapleton* •
Fine Arts Museum – 1966
American art 46291

Gustav Jeeninga Museum of Bible and Near Eastern Studies, Theology Bldg, 1123 Anderson University Blvd, Anderson, IN 46012-3495 • T: +1 765 6499071 • F: +1 765 6413005 • dneidert@anderson.edu • www.anderson.edu •
Dir.: *David Neidert* •
University Museum / Archaeology Museum / Religious Arts Museum – 1963
Archaeology, religion 46292

Anderson SC

Anderson County Arts Center, 405 N Main St, Anderson, SC 29621 • T: +1 864 2248811 • F: +1 864 2248864 • info@andersonartscenter.org • www.andersonartscenter.org •
Exec. Dir.: *Kimberly Spears* •
Public Gallery – 1972 46293

Anderson County Museum, 202 E Greenville St, Anderson, SC 29624 • T: +1 864 2604737 • F: +1 864 2604044 • acm@andersoncountysc.org • www.andersoncountysc.org/museum.htm •
Dir.: *Paula Reel* • Cur.: *Alison Hinman* •
Military Museum – 1983
Local hist 46294

Andersonville GA

Andersonville Prison, 496 Cemetery Rd, Andersonville, GA 31711 • T: +1 912 9240343 • F: +1 912 9289640 • www.nps.gov/ande •
Historical Museum – 1971
P.O.W. camps, Civil War and later conflicts, monuments and sculptures 46295

Andover MA

Addison Gallery of American Art, Phillips Academy, 180 Main St, Andover, MA 01810 • T: +1 978 7494015 • F: +1 978 7494025 • addison@andover.edu • www.addisongallery.org •
Dir.: *Brian T. Allen* • Dir./Cur.: *Susan Faxon* • Cur.: *Allison Kemmerer* •
Fine Arts Museum – 1931
American paintings, sculpture, drawings and prints, photography since 17th c 46296

Andover Historical Society Museum, 97 Main St, Andover, MA 01810 • T: +1 978 4752236 • F: +1 978 4702741 • info@andhist.org • www.andhist.org •
C.E.O.: *Elaine Clements* • Cur.: *Juliet Mofford* (Education) •
Local Museum – 1911
Local history, furnishings, costumes, agriculture, trade – library 46297

Robert S. Peabody Museum of Archaeology, Phillips Academy, 175 Main St, Andover, MA 01810 • T: +1 978 7494490 • F: +1 978 7494495 • rpeabody@andover.edu • www.andover.edu •
Dir./Cur.: *Malinda S. Blustain* •
Archaeology Museum / University Museum – 1901
Archaeology of North America, emphasis, Highland Maya ethnology, anthropology – library 46298

Angleton TX

Brazoria County Historical Museum, 100 E Cedar, Angleton, TX 77515 • T: +1 979 8641208 • F: +1 979 8641217 • bchm@bchm.org • www.bchm.org •
Dir.: *Jackie Haynes* • Pres.: *Marcus Stephenson* •
Historical Museum – 1983
Hist from prhistoric to modern times – archives 46299

Angola IN

General Lewis B. Hershey Museum, Trine University, Ford Hall, 318 S. Darling Street, Angola, IN 46703, mail addr: 1 University Avenue, Angola, IN 46703 • T: +1 260 6654161 • F: +1 260 6654283 • brewerk@tristate.edu • www.trine.edu •
University Museum / Military Museum – 1970
Military 46300

Angwin CA

Rasmussen Art Gallery, c/o Pacific Union College, 1 Angwin Av, Angwin, CA 94508 • T: +1 707 9656311 • www.puc.edu/puc-life/arts/rag-gallery •
Public Gallery 46301

Ann Arbor MI

Ann Arbor Art Center, 117 W Liberty, Ann Arbor, MI 48104 • T: +1 734 9948004 ext 110 • F: +1 734 9943610 • kjohnson@annarborartcenter.org • www.annarborartcenter.org •
Dir.: *Deborah Campbell* •
Fine Arts Museum – 1909
Multicultural exhibits 46302

Ann Arbor Hands-on Museum, 220 E Ann St, Ann Arbor, MI 48104 • T: +1 734 9955439 • F: +1 734 9951188 • museum@aahom.org • www.aahom.org •
Exec. Dir.: *Mel Drumm* • Dep. Dir.: *Carol Knauss* (Operations) • Sc. Staff: *John Bowditch* (Exhibits) •
Science&Tech Museum – 1982
Science, technology, central firehouse 46303

Artrain USA, 1100 N Main St, Ste 106, Ann Arbor, MI 48104 • T: +1 734 7478300 • F: +1 734 7428530 • laura.drew@artrainusa.org • www.artrainusa.org •
CEO: *Debra Polich* •
Fine Arts Museum – 1971
Traveling art, housed in five railroad cars, paintings, drawings, sculptures, fiber art 46304

Exhibit Museum of Natural History, University of Michigan, 1109 Geddes Av, Ann Arbor, MI 48109-1079 • T: +1 734 7634190, 7640478 • F: +1 734 6472767 • dmadaj@umich.edu • www.exhibits.lsa.umich.edu •
Dir.: *Amy S. Harris* •
University Museum / Natural History Museum – 1881
Evolution, paleontology, paleobotany, ethnology, anthropology, human anatomym geology, astromony, Michigan wildlife and ecology – planetarium 46305

Jean Paul Slusser and Warren Robbins Galleries, University of Michigan, School of Art and Design, 2000 Bonisteel Blvd, Ann Arbor, MI 48109-2069 • T: +1 734 9362082 • F: +1 734 6156761 • slussergallery@umich.edu • www.art-design.umich.edu •
Dir.: *Todd M. Cashbaugh* •
University Museum / Fine Arts Museum – 1975
Contemporary art & design, architecture – library 46306

Kelsey Museum of Archaeology, 434 S State St, Ann Arbor, MI 48109 • T: +1 734 7649304 • F: +1 734 7638976 • www.lsa.umich.edu/kelsey •
Dir./Cur.: *Sharon C. Herbert* (Research) • Cur.: *Suzanne Davis* (Conservation) • *Margaret Root* (Greece and Near East) • *Lauren Talalay* • *Thelma Thomas* (Late Antique and Textiles Post Classical) • *Janet Richards* (Dynastic Egypt) • *Terry Wilfong* (Graeco-Roman Egypt) • *Robin Meador-Woodruff* (Slides, Photographs) •
University Museum / Archaeology Museum / Museum of Classical Antiquities – 1928
Archaeology, Hellenic, Roman and Egypt antiques, photos – library 46307

Museum of Anthropology, University of Michigan, 4009 Ruthven Museums Bldg, 1109 Geddes Av, Ann Arbor, MI 48109 • T: +1 734 7640485 • F: +1 734 7637783 • anthro-museum@umich.edu • www.umma.lsa.umich.edu •
Dir./Cur.: *Richard I. Ford* (Ethnology, Ethnobotany) • Cur.: *Robert Whallon* (Mediterranean Archaeology) • *Carla Sinopoli* (South and Southeast Archaeology) • *John D. Speth* (North American Archaeology) • *Jeffrey R. Parsons* (Latin American Archaeology) • *C. Loring Brace* (Physical Anthropology) • *Kent V. Flannery* (Environmental Archaeology) • *John O'Shea* (Great Lakes Archaeology) • *Henry T. Wright* (Old World Civilization) • *Joyce Marcus* (Mayan and Central American Archaeology) • *Dr. William R. Farrand* (Analytical Colls) •
University Museum / Ethnology Museum / Archaeology Museum – 1922
Anthropology, archaeology, old world civilization, ethnology, ethnobotany, human osteology – labs 46308

Museum of Art, University of Michigan, 525 S State St, Ann Arbor, MI 48109-1354 • T: +1 734 7638662, 7640395 • F: +1 734 7643731 • umma.info@umich.edu • www.umich.edu/~umma/ •
Dir.: *James Steward* • Sc. Staff: *Carole McNamara* (Western Art) • *Jacob Proctor* (Modern and Contemporary Art) • *Ruth Slavin* (Education) •
Fine Arts Museum / University Museum – 1946
Western, Asian, African Art – Asian Art Conservation Laboratory 46309

Museum of Zoology, University of Michigan, 1109 Geddes Av, Ann Arbor, MI 48109-1079 • T: +1 734 7640476 • F: +1 734 7634080 • wfink@umich.edu • www.ummz.lsa.umich.edu •
Dir.: *Prof. David P. Mindell* • Assoc. Dir.: *Prof. R.A. Nussbaum* • Sc. Staff: *Prof. Reeve M. Bailey* • *Prof. Richard D. Alexander* • *Prof. Thomas E. Moore* • *Prof. Robert B. Payne* • *Prof. Robert W. Storer* • *Prof. Arnold G. Kluge* • *Prof. William L. Fink* • *Prof. Philip Myers* • *Prof. Barry M. O'Connor* • *Prof. Priscilla K. Tucker* • *Prof. David P. Mindell* • *Prof. Diarmaid O'Foighil* •
University Museum / Natural History Museum – 1838
Zoology, insects, mammals, birds, fishes, reptiles and amphibians, molluscs 46310

Stearns Collection of Musical Instruments, University of Michigan, School of Music, Ann Arbor, MI 48109-2085 • T: +1 734 9362891 • F: +1 734 6471897 • jclam@umich.edu • www.music.umich.edu/research/stearns_collection •
Dir.: *Joseph Lam* •
Music Museum – 1899
Over 2,500 historical and contemporary musical instruments from all over the world 46311

Work Gallery, University of Michigan, 306 S State St, Ann Arbor, MI 48104 • T: +1 734 9986178 • www.art-design.umich.edu •
Dir.: *Gregory Steel* •
University Museum / Fine Arts Museum
Students present work (business side of art) 46312

Annandale MN

Minnesota Pioneer Park Museum, 725 Pioneer Park Trail, Annandale, MN 55302 • T: +1 320 2748489 • F: +1 320 2749612 • pioneerp@pioneerpark.org • www.pioneerpark.org •
Pres.: *Carol Anderson* •
Historical Museum – 1972
Memorabilia from 1850-1910, life, customs and cultures 46313

Annandale-on-Hudson NY

Center for Curatorial Studies, Bard College, 33 Garden Rd, Annandale-on-Hudson, NY 12504-5000 • T: +1 845 7587598 • F: +1 845 7582442 • ccs@bard.edu • www.bard.edu/ccs •
Dir.: *Norton Batkin* •
Fine Arts Museum – 1990
Contemporary art 46314

Annapolis MD

Cardinal Gallery, 801 Chase St, Annapolis, MD 21401 • T: +1 410 2635544 • F: +1 410 2635114 • mdhall@annap.infi.net • www.mdhallarts.org •
Exec. Dir.: *Linnell R. Bowen* •
Fine Arts Museum – 1979
Drawings, photography, sculpture, watercolors, woodcarvings, woodcuts, Afro-American art 46315

Charles Carroll House of Annapolis, 107 Duke of Gloucester St, Annapolis, MD 21401 • T: +1 410 2691737 • F: +1 410 2691746 • ccarroll@toad.net •
Dir.: *Sandria B. Ross* •
Historic Site – 1987 46316

Chesapeake Children's Museum, 25 Silopanna Road, Annapolis, MD 21403 • T: +1 410 9901993 • info@theccm.org • www.theccm.org •
Special Museum – 1994 46317

Elizabeth Myers Mitchell Art Gallery, Saint John's College, 60 College Av, Annapolis, MD 21401 • T: +1 410 2632371, +1 410 6262556 • F: +1 410 6262886 • hyde.schaller@stjohnscollege.edu • www.stjohnscollege.edu •
Dir.: *Hydee Schaller* •
Fine Arts Museum / University Museum – 1989
16th c Italien art, 14th c Chinese art, 17th c Dutch art, 20th c Japanese art 46318

Hammond-Harwood House Museum, 19 Maryland Av, Annapolis, MD 21401 • T: +1 410 2634683 • F: +1 410 2676891 • clively@hammondharwoodhouse.org • www.hammondharwoodhouse.org •
Pres.: *David Morrow* • Exec. Dir.: *Carter C. Lively* •
Historical Museum / Folklore Museum – 1938
18th c furniture, decorative arts and paintings 46319

Historic Annapolis Foundation, 18 Pinkney St, Annapolis, MD 21401 • T: +1 410 2677619 • F: +1 410 2676189 • www.annapolis.org •
Pres./C.E.O.: *Greg Stgiverson* •
Local Museum – 1952
Local history, 1715 Shiplap House 46320

United States Naval Academy Museum, 118 Maryland Av, Annapolis, MD 21402-5034 • T: +1 410 2932108 • F: +1 410 2935220 • jsharmon@usna.edu • www.usna.edu/museum •
Dir.: *Dr. J. Scott Harmon* • Cur.: *Robert F. Sumrall* (Ship Models) • *Sigrid Trumpy* (Beverly R. Robinson Collection) •
Fine Arts Museum / Historical Museum – 1845
Naval hist 46321

Anniston AL

Anniston Museum of Natural History, 800 Museum Dr, Anniston, AL 36207 • T: +1 256 2376766 • info@annistonmuseum.org • www.annistonmuseum.org •
Dir.: *Cheryl H. Bragg* • Cur.: *Daniel Spaulding* (Collections) •
Natural History Museum – 1930
North American and African animals, Egyptian mummies, anthropology, ethnology, entomology, paleontology, geology 46322

Anoka MN

Anoka County Museum, 2135 Third Av N, Anoka, MN 55303 • T: +1 763 4210600 • F: +1 761 3230218 • bonnie@ac-hs.org • www.ac-hs.org •
Exec. Dir.: *Bonnie McDonald* •
Local Museum – 1934
Regional history, period furniture & clothing, dolls, rugs – library 46323

Ansted WV

Hawks Nest State Park, Rte 60, Ansted, WV 25812 • T: +1 304 6585212, 6585196 • F: +1 304 6584549 • hawksnest@citynet.net • www.hawknestsp.com •
Head: *Thomas L. Shriver* •
Local Museum – 1935
Local hist 46324

Antigo WI

Langlade County Historical Society Museum, 404 Superior St, Antigo, WI 54409 • T: +1 715 6274464 • lchs@dwave.net • langladehistory.com •
Local Museum – 1929
Local hist, medical and musical equipment 46325

Antwerp OH

Ehrhart Museum, 118 N Main St, Antwerp, OH 45813 • T: +1 419 2582665 • F: +1 419 2581875 •
Pres.: *Randy Shaffer* •
Natural History Museum – 1963
Archaeology, ethnology, geology, mountain birds and animals, insects 46326

Apalachicola FL

John Gorrie Museum, 46 6th St and Avenue D, Apalachicola, FL 32329, mail addr: POB 267, Apalachicola, FL 32329-0267 • T: +1 850 6539347 • www.baynavigator.com •
Dir.: *Fran P. Mainella* •
Science&Tech Museum – 1955
Inventor of machine tomate ice, fore-runner of refrigerator and air conditioning 46327

Apple Valley CA

Victor Valley Museum and Art Gallery, 11873 Apple Valley Rd, Apple Valley, CA 92307 • T: +1 760 2402111 • F: +1 760 2405290 • info@vvmuseum.com • www.vvmuseum.com •
Pres.: *Calvin Camara* • C.E.O.: *Carl Mason* •
Fine Arts Museum / Local Museum – 1976
Natural hist, art, archaeology, geography 46328

Appleton WI

Appleton Art Center, 111 W College Av, Appleton, WI 54911 • T: +1 920 7334089 • F: +1 920 7334149 • info@appletonartcenter.org • www.appletonartcenter.org •
Pres.: *Craig Uhlenbrauck* • Dir.: *Tracey Jenks* •
Fine Arts Museum – 1960
Art 46329

Hearthstone Historic House Museum, 625 W Prospect Av, Appleton, WI 54911 • T: +1 920 7308204 • F: +1 920 7308266 • hearthstonemuseum@athenet.org • www.hearthstonemuseum.org •
Dir.: *Christine Cross* •
Historical Museum / Science&Tech Museum – 1986
First residence in the world to be lighted from a central hydroelectric powerplant using the Edison system in 1882, original Edison light fixtures and switches, early hist of electric lighting, 19thc furnishings, dec art 46330

Outagamie Museum, 330 E College Av, Appleton, WI 54911 • T: +1 920 7338445 • F: +1 920 7338636 • ochs@foxvalleyhistory.org • www.foxvalleyhistory.org •
Dir.: *Terry Bergen* • Cur.: *Matthew Carpenter* •
Local Museum – 1872
Hist of technology, regional hist, tools, life and career of A.K.A. Houdini 46331

Wriston Art Center Galleries, Lawrence University, 613 E College Av, Appleton, WI 54912-0599 • T: +1 920 8326890 • F: +1 920 8327362 • nadine.wassman@lawrence.edu •
Cur.: *Frank C. Lewis* •
University Museum / Fine Arts Museum – 1989
19th and 20th c paintings and prints, coll of German Expressionist art, Japanese prints and drawings 46332

Appomattox VA

Appomattox Court House, POB 218, Appomattox, VA 24522 • T: +1 804 3528987 • F: +1 804 3528330 • jow_williams@nps.gov • www.nps.gov/apco •
Local Museum – 1940
Civil War, historic buildings 46333

Arapahoe NE

Furnas-Gosper Historical Society and Museum, 401 Nebraska Av, Arapahoe, NE 68922 •
C.E.O. & Pres.: *Robert Trosper* •
Local Museum – 1966
Clothing, books glassware, piano, sewing machines, crockery 46334

Arcade NY

Arcade Historical Museum, 331 W Main St, Arcade, NY 14009 • T: +1 716 4924466 •
Pres.: *Sandra Dutton* •
Local Museum – 1956
Crafts, local hist, household items, toys 46335

Arcadia CA

Ruth and Charles Gilb Arcadia Historical Museum, 380 W Huntington Dr, Arcadia, CA 91006 • T: +1 626 5745440 • F: +1 626 8219057 • www.ci.arcadia.ca.us •
Dir.: *Janet Sporleder* •
Local Museum – 2001
Arcadia hist from 1903 46336

Arcadia MI

Arcadia Historical Museum, Lake St, Arcadia, MI 49613 • T: +1 616 8894360 • LMatt613@msn.com • www.arcadiami.com •
Dir.: *Virgil Schneider* •
Local Museum – 2000
Lumbertown, Sawmill furniture factory, mirror works 46337

U

Arcata CA

Natural History Museum, c/o Humboldt State University, 1315 G St, Arcata, CA 95521 • T: +1 707 8264479 • F: +1 707 8264477 • www.humboldt.edu/~natmus •
Dir.: *Melissa L. Zielinski* •
University Museum / Natural History Museum – 1989
NW Pacific natural hist, fossils 46338

Reese Bullen Gallery, c/o Humboldt State University, 1 Harpst St, Arcata, CA 95521 • T: +1 707 8265802 • F: +1 707 8263628 • mm1@axe.humboldt.edu • www.humboldt.edu/~rbg •
Dir.: *Martin Morgan* • Asst. Dir.: *Nancy Clark* •
University Museum / Fine Arts Museum – 1970
20th c American and African art 46339

Archbold OH

Historic Sauder Village, 22611 State Rte 2, Archbold, OH 43502 • T: +1 419 4462541 • F: +1 419 4455251 • info@saudervillage.org • www.saudervillage.com •
Exec. Dir.: *D. Sauder David* •
Historical Museum – 1971
Farm equip, woodworking tools, household, smith, potter, glassblower, spinning and weaving, broommaking, basket maker, grist mill 46340

Archer City TX

Archer County Museum, Old County Jail, Archer City, TX 76351 • T: +1 940 4236426 •
Cur.: *Jack Loftin* •
Local Museum – 1974
Lizard fossils, art and western paintings, old clothing, farm machinery, vehicles, Indian artifacts, early pictures 46341

Arco ID

Craters of the Moon, 18 mi SW of Arco on Hwy 26, Arco, ID 83213, mail addr: POB 29, Arco, ID 83213 • T: +1 208 5273257 • F: +1 208 5273073 • crmo_resource_management@nps.gov • www.nps.gov/crmo •
Natural History Museum – 1924
Local geology, hist 46342

Ardmore OK

Charles B. Goddard Center for Visual and Performing Arts, 401 First Av and D St SW, Ardmore, OK 73401 • T: +1 580 2260909 • F: +1 580 2268891 • godart@sbcglobal.net • www.godart.org •
Dir.: *Mort Hamilton* •
Fine Arts Museum / Performing Arts Museum – 1969
20th c art, western painting and sculpture, native American painting 46343

Eliza Cruce Hall Doll Museum, 320 East St NW, Ardmore, OK 73401 • T: +1 580 2238290 •
Dir.: *Carolyn J. Franks* •
Special Museum – 1971
Dolls dating back to 1700s, original dolls of the French Court, English pedlars, French fashion dolls, Lenci, Kruse, Kestner a.o. 46344

Tucker Tower Nature Center Park Museum, 3310 S Lake Murray Dr, Ardmore, OK 73401 • T: +1 405 2232109 • F: +1 405 2232109 • markttnc@aol.com • www.oklahomaparks.com •
Natural History Museum – 1952
Fossils, minerals, meteorite, nature exhibits 46345

Argonia KS

Salter Museum, 220 W Garfield, Argonia, KS 67004, mail addr: POB 116, Argonia, KS 67004 • T: +1 316 4356376 •
Pres.: *Laura High* • Chief Cur.: *Helen Nafziger* •
Historical Museum – 1961
Home of America's first woman mayor Susanna M. Salter 46346

Arkansas City

Cherokee Strip Land Rush Museum, 31639 Hwy 77 S, Arkansas City, mail addr: Box 778, Arkansas City, KS 67005 • T: +1 316 4426750 • F: +1 316 4414332 • museum@arkcity.org • www.arkcity.org/csm.html •
Historical Museum – 1966
Cherokee Strip Run, Indian and military coll, Hardy Jail 46347

Arlington MA

Arlington Historical Society Collection, 7 Jason St, Arlington, MA 02476 • T: +1 781 6484300 • arlhistsoc@aol.com • www.arlingtonhistorical.org •
Pres.: *Howard Winkler* •
Historical Museum – 1897
Local history, weapons, furnishings, textiles, dolls – archives 46348

Old Schwamb Mill, 17 Mill Ln, Arlington, MA 02476-4189 • T: +1 781 6430554 • F: +1 781 6488809 • schwambmill@aol.com • www.oldschwambmill.org •
Science&Tech Museum – 1969
Industrial history, waterpowered mill 46349

Arlington TX

Arlington Museum of Art, 201 W Main St, Arlington, TX 76010 • T: +1 817 2754600 • F: +1 817 8604800 • ama@arlingtonmuseum.org • www.arlingtonmuseum.org •
C.E.O.: *Reecanne Joeckel* • Dir./Cur.: *Anne Allen* •
Fine Arts Museum – 1987
Contemporary Texas art 46350

The Gallery at UTA, c/o University of Texas at Arlington, 502 S Cooper St, Fine Arts Bldg, Arlington, TX 76019 • T: +1 817 2725658, 2723143 • F: +1 817 2722805 • bhuerta@uta.edu • www.uta.edu/gallery •
Dir.: *Benito Huerta* •
Fine Arts Museum / University Museum – 1976
Contemporary art 46351

Legends of the Game Baseball Museum, 1000 Ballpark Way, Ste 400, Arlington, TX 76011 • T: +1 817 2735059 • F: +1 817 2735093 • museum@texasrangers.com • museum.texasrangers.com •
Dir.: *Amy E. Polley* •
Special Museum – 1994
National baseball hist, Texas League, Texas Rangers, Negro Leagues 46352

River Legacy Living Science Center, 703 NW Green Oaks Blvd, Arlington, TX 76006 • T: +1 817 8606752 • F: +1 817 8601595 • rlegacy@earthlink.net • www.riverlegacy.org •
C.E.O.: *Phyllis Snider* •
Natural History Museum – 1996
Animals, botany, archaeology, geology 46353

Tarrant County College Southeast The Art Corridor, 2100 Southeast Pkwy, Arlington, TX 76018 • T: +1 817 5153522 • F: +1 817 5153189 • dcerrick.white@tccd.net •
Dir.: *Derrick White* •
Fine Arts Museum / University Museum
Contemporary art 46354

Arlington VA

Arlington Arts Center, 3550 Wilson Blvd, Arlington, VA 22201 • T: +1 703 7974574 • F: +1 703 5275445 • artcenter@erols.com • www.arlingtonartscenter.org •
Dir.: *Carole Sullivan* • Cur.: *Gerry Rogers* •
Fine Arts Museum – 1976
Visual arts 46355

Arlington Historical Museum, 1805 S Arlington Ridge Rd, Arlington, VA 22202 • T: +1 703 8924204, 8129479 • arlhistory@aol.com • www.arlingtonhistoricalsociety.org •
Local Museum – 1956
Local hist artifacts from English exploration period to present 46356

Arlington House - The Robert E. Lee Memorial, Arlington National Cemetery, Arlington, VA 22211 • T: +1 703 5351530 • F: +1 703 5351546 • gwmp-arlingtonhouse@nps.gov • www.nps.gov/arho •
Dir.: *Terry Carustrom* •
Local Museum – 1925
Historic house, residence of Gen. Robert E. Lee 46357

Newseum, c/o Freedom Forum Newseum, 1101 Wilson Blvd, Arlington, VA 22209 • T: +1 703 2843700, 8886397386 • F: +1 703 2843777 • newseum@freedomforum.org • www.newseum.org •
Dir.: *Joe Urschel* •
Historical Museum – 1997
Hist and present time 46358

US Patent and Trademark Office - Museum, 2121 Crystal Dr, Ste 0100, Arlington, VA 22202 • T: +1 703 3058341 • F: +1 703 3085258 • mscott@invent.org • www.uspto.gov •
C.E.O.: *Richard Maulsby* • Cur.: *Ruth Ann Nyblod* •
Science&Tech Museum – 1995
Hist of the patent and trademark systems, models 46359

Arlington Heights IL

Arlington Heights Historical Museum, 110 W Fremont St, Arlington Heights, IL 60004 • T: +1 847 2551225 • F: +1 847 2551570 • susan@ahpd.org • www.ahmuseum.org •
Cur.: *Mickey Horndasch* •
Local Museum – 1957
Local history 46360

Armonk NY

North Castle Historical Society Museum, 440 Bedford Rd, Armonk, NY 10504 • T: +1 914 2734510 •
C.E.O.: *Robbie Morris* •
Local Museum – 1971
Farm implements, tools, home styles , fashions 46361

Armour SD

Douglas County Museum Complex, Courthouse Grounds, Armour, SD 57313 • T: +1 605 7242129 •
Dir.: *Laverne Vanderwerff* • Pres./ Cur.: *Sharon A. Wiese* • Ass. Dir.: *Dot Hoveng* •
Local Museum – 1958
Local hist 46362

Arnolds Park IA

Iowa Great Lakes Maritime Museum, 243 W Broadway, Arnolds Park, Arnolds Park, IA 51331 • T: +1 712 3322183 • F: +1 712 3322186 • www.okobojimuseum.org •
Dir.: *Steven R. Anderson* • Cur.: *Mary Kennedy* •
Historical Museum – 1987
Maritime artifacts, photos 46363

Arrow Rock MO

Arrow Rock State Historic Site, 4th and Van Buren, Arrow Rock, MO 65320 • T: +1 660 8373330 • F: +1 660 8373300 • dsparro@mail.dnr.state.mo.us • www.arrowrock.org •
C.E.O.: *Michael Dickey* •
Historical Museum / Open Air Museum – 1923
Paintings, manuscripts textiles, artifacts 46364

Artesia NM

Artesia Historical Museum and Art Center, 505 Richardson Av, Artesia, NM 88210 • T: +1 505 7482390 • F: +1 505 7463886 • ahmac@pvtnetworks.net •
Dir.: *Nancy Dunn* • Registrar: *Merle Rich* •
Historical Museum – 1970
Local and area hist 46365

Arthurdale WV

Arthurdale Heritage Museum, Q and A Rds, Arthurdale, WV 26520 • T: +1 304 8643959 • F: +1 304 8644602 • ahi1934@aol.com • www.arthurdaleheritage.org •
Pres.: *Natasha Showalter* • CEO: *Amanda Griffith* •
Local Museum – 1987
Local hist, tools and equipment, artisan crafts 46366

Arvada CO

Arvada Center for the Arts and Humanities, 6901 Wadsworth Blvd, Arvada, CO 80003 • T: +1 303 4313080 • F: +1 303 4313083 • kathy-a@arvadacenter.org • www.arvadacenter.org •
Exec. Dir.: *Deborah L. Ellerman* • Cur.: *Debra Sanders* •
Fine Arts Museum – 1976
19th c photography, contemporary art and crafts, outdoor sculpture 46367

Asheville NC

Asheville Art Museum, 2 S Pack Sq, Asheville, NC 28801, mail addr: PO Box 1717, Asheville, NC 28802 • T: +1 828 2533227 • F: +1 828 2574503 • mailbox@ashevilleart.org • www.ashevilleart.org •
Dir.: *Pamela L. Myers* • Cur.: *Frank E. Thomson* •
Fine Arts Museum – 1948
20th c American art of all media, works of significatnce to Western North Carolina's cultural heritage including Studio Craft, Black Mountain College and Cherokee artists. – Art Library, Education Dept., Curatorial 46368

Biltmore Estate, 1 N Pack Sq, Asheville, NC 28801 • T: +1 828 2551776 • F: +1 828 2551111 • rking@biltmore.com • www.biltmore.com •
C.E.O.: *William A.V. Cecil* •
Local Museum – 1930
1895 house, paintings, tapestries, sculpture, furniture 46369

Black Mountain College Museum and Arts Center, POB 18912, Asheville, NC 28814 • T: +1 828 2999306 • F: +1 828 2299306 • bmcmac@bellsouth.net • www.blackmountaincollege.org •
C.E.O.: *Connie Bostic* •
Historical Museum / Fine Arts Museum – 1993
Artwork, photographs, books 46370

Colburn Gem and Mineral Museum, 2 S Pack Sq, Asheville, NC 28801 • T: +1 828 2547162 • F: +1 828 2574505 • ddmowrey@yahoo.com • www.main.nc.us/colburn •
Pres.: *Kempton Roll* • Exec. Dir.: *Debbie Mowrey* • Cur.: *Christian Richart* •
Natural History Museum – 1960
Minerals, gems, rocks, mining artifacts 46371

Estes-Winn Antique Automobile Museum, 111 Grovewood Rd, Asheville, NC 28804 • T: +1 828 2537651 • F: +1 828 2542489 • automuseum@grovewood.com • www.grovewood.com •
Pres.: *S.M. Patton* •
Science&Tech Museum – 1970 46372

Grovewood Gallery, 111 Grovewood Rd, Asheville, NC 28804 • T: +1 828 2537651 • F: +1 828 2542489 • homespun@grovewood.com • grovewood.com •
Pres.: *S.M. Patton* •
Special Museum – 1901
Textiles, 1920 homespun tools, looms, machinery 46373

Health Adventure, 2 S Pack Sq, Asheville, NC 28802-0180 • T: +1 828 2546373 • F: +1 828 2574521 • info@thehealthadventure.org • www.thehealthadventure.org •
C.E.O.: *Todd Boyette* •
Special Museum – 1968 46374

Smith-McDowell House Museum, 283 Victoria Rd, Asheville, NC 28801 • T: +1 828 2539231 • F: +1 828 2535518 • smh@wnchistory.org • www.wnchistory.org •
Pres.: *Stephen Jones* • Exec. Dir.: *Rebecca B. Lamb* •
Dir.: *Sharon Withrow* •
Historic Site – 1981
19th c decorative arts and artifacts 46375

Southern Highland Craft Guild at the Folk Art Center, Milepost 382, Blue Ridge Pkwy, Asheville, NC 28815 • T: +1 828 2987928 • F: +1 828 2987962 • shcg@buncombe.main.nc.us • www.southernhighlandguild.org •
Man.Dir.: *Tom Bailey* •
Folklore Museum – 1930
Traditional and contemporary crafts and trades – Robert W. Gray library and archives 46376

Thomas Wolfe Memorial, 52 N Market St, Asheville, NC 28801 • T: +1 828 2538304 • F: +1 828 2528171 • contactus@wolfememorial.com • www.wolfememorial.com •
Dir.: *Steve Hill* •
Special Museum – 1949
Life and work of the author, memorabilia, 20th c boarding house furnishings 46377

Ashford WA

Mount Rainier National Park Museum, Tahoma Woods, Star Rte, Ashford, WA 98304 • T: +1 360 5692211 • F: +1 360 5692187 • www.nps.gov/mora •
Natural History Museum – 1940
Botany, zoology, geology, hist, Indian artifacts 46378

Ashland KS

Pioneer-Krier Museum, 430 W 4th St, Ashland, KS 67831 • T: +1 620 6352227 • F: +1 620 6352227 • pioneer@ucom.net •
C.E.O: *Mike Harden* •
Local Museum – 1968
Kitchen equip, dolls and toys, bits and spurs, guns, musical instruments, furniture, military, pioneer pictures, farm machinery, fossils, aerobatic airplanes 46379

Ashland KY

Highlands Museum and Discovery Center, 1620 Winchester Av, Ashland, KY 41101 • T: +1 606 3298888 • F: +1 606 3243218 • highlandsmuseum@yahoo.com • www.highlandsmuseum.com •
C.E.O.: *Dr. Kenneth Hauswald* •
Local Museum – 1984
Clothing, transportation, communication, industry, country music, medical and military since WW 2 46380

Ashland MA

Ashland Historical Society Museum, 2 Myrtle St, Ashland, MA 01721 • T: +1 508 8818183 •
Cur.: *Catherine Powers* •
Local Museum – 1905
Local history, furniture, implements, portraits, china, glassware, docs 46381

Ashland ME

Ashland Logging Museum, POB 866, Ashland, ME 04732 • T: +1 207 4356039 • F: +1 207 4356579 •
Cur.: *Ed Chase* •
Ethnology Museum – 1964
Reproduction of an early logging camp, Lombard steam andgas log haules, log hauler sleds, bateau, tote wagon, pine log 46382

Ashland NE

Strategic Air & Space Museum, 28210 W Park Hwy, Ashland, NE 68003 • T: +1 402 9443100 • F: +1 402 9443160 • staff@strategicairandspace.com • www.strategicairandspace.com •
Dir.: *Scott Hazelrigg* • Dep. Dir.: *Steve Prall* • Cur.: *Brian York* •
Military Museum – 1959
Aircraft, missiles, aviation artifacts 46383

Ashland NH

Ashland Railroad Station Museum, 69 Depot St, Ashland, NH 03217 •
Science&Tech Museum – 1999 46384

Pauline E. Glidden Toy Museum, Pleasant St, Ashland, NH 03217-9401 • T: +1 603 9687289 •
Dir.: *Shirley Splaine* •
Special Museum – 1990
Toys 46385

Whipple House Museum, 14 Pleasant St, Ashland, NH 03217-0175 • T: +1 603 9687716 • F: +1 603 9687716 • www.ashlandnh.org •
Pres.: *David Ruell* • Cur.: *Sandra Ray* •
Local Museum / Historical Museum – 1970
Local history, home of George Hoyt Whipple, Nobel Prize for medicine 46386

Ashland OH

The Coburn Gallery, Ashland College Arts & Humanities Gallery, 401 College Ave, Ashland, OH 44805 • T: +1 419 2894142 •
Dir.: *Larry Schiemann* •
Fine Arts Museum / Public Gallery
Contemporary works 46387

Ashland OR

Schneider Museum of Art, Southern Oregon University, 1250 Siskiyou Blvd, Ashland, OR 97520 • T: +1 541 5526245 • F: +1 541 5528241 • schneider_museum@sou.edu • www.sou.edu/sma •
Dir.: *Mary N. Gardiner* •
University Museum / Fine Arts Museum – 1986
Contemporary art, Native American baskets, preColumbia artifacts 46388

Southern Oregon University Museum of Vertebrate Natural History, 1250 Siskiyou Blvd, Ashland, OR 97520 • T: +1 541 5526341 • F: +1 541 5526415 • stonek@sou.edu •
Cur.: *Karen Stone* •
University Museum / Natural History Museum – 1969
Bird skins, mammal skins, reptiles, amphibians, fish 46389

Ashland PA

Museum of Anthracite Mining, Pine and 17 Sts, Ashland, PA 17921 • T: +1 570 8754708 • F: +1 570 8753732 • ddubick@state.pa.us • www.phmc.state.pa.us •
Dir.: *David Dubick* • Cur.: *Chester Kulesa* •
Science&Tech Museum / Historical Museum – 1971
Geology, mining technology 46390

Ashland WI

Ashland Historical Society Museum, 509 W Main, Ashland, WI 54806 • T: +1 715 6824743 • ashlandhistory@centurytel.net • ashlandhistory.com •
Local Museum – 1954
Costumes, hist, military 46391

Ashley ND

McIntosh County Historical Society Museum, 117 3rd Av, Ashley, ND 58413 • T: +1 701 2883388 •
Pres.: *Ronald J. Meidinger* •
Local Museum – 1977
Local hist, sod house, outdoor baking oven, domestic and farm equipment, clothing, blacksmith equipment 46392

Ashley Falls MA

Colonel Ashley House, Cooper Hill Rd, Ashley Falls, MA 01222, mail addr: POB 792, Stockbridge, MA 01262 • T: +1 413 2983239 • F: +1 413 2985239 • wgarrison@ttor.org • www.thetrustees.org •
Exec. Dir.: *Andrew Kendall* •
Historical Museum – 1972
Colonial furnishings, pottery, tools 46393

Ashtabula OH

Ashtabula Arts Center, 2928 W 13 St, Ashtabula, OH 44004 • T: +1 440 9643396 • F: +1 440 9643396 • aac@suite224.net • www.ashartscenter.org •
Pres.: *Judy Robson* • Exec. Dir.: *Elizabeth Koski* •
Fine Arts Museum – 1953
Contemporary art 46394

Ashville OH

Slate Run Living Historical Farm, Metro Parks, 9130 Marcy Rd, Ashville, OH 43103 • T: +1 614 8331880 • F: +1 614 8341220 •
C.E.O.: *John O'Meara* •
Local Museum / Agriculture Museum – 1976
Local hist, agricultural implements and machinery, household goods 46395

Askov MN

Pine County Historical Society Museum, 3851 Glacier Rd, Askov, MN 55704 • T: +1 320 8383792 • lizesp@juno.com •
Pres.: *Elizabeth Espointour* •
Local Museum – 1948
Agriculture, manuscripts, folklore – archives, library 46396

Aspen CO

Aspen Art Museum, 590 N Mill St, Aspen, CO 81611 • T: +1 970 9258050 • F: +1 970 9258054 • info@aspenartmuseum.org • www.aspenartmuseum.org •
Dir./Chief Cur.: *Heidi Zuckerman Jacobson* •
Fine Arts Museum – 1979 46397

Aspen Historical Society Museum, 620 W Bleeker St, Aspen, CO 81611 • T: +1 970 9253721 • F: +1 970 9255347 • info@heritageaspen.org • www.heritageaspen.org •
Dir.: *Georgia Hanson* • Cur.: *Sarah Oates* •
Local Museum – 1969
Local history, Wheeler-Stallard house 46398

Astoria NY

American Museum of the Moving Image, 35th Av at 36th St, Astoria, NY 11106 • T: +1 718 7844520 • F: +1 718 7844681 • www.ammi.org •
Dir.: *Rochelle Slovin* • Chief Cur.: *David Schwartz* (Film and Video) • Cur.: *Carl Goodman* (Digital Media) • *Dana Sergent Nemeth* (Popular Culture) •
Special Museum – 1977
Art, production, technology, craft, business and history of motion pictures 46399

Astoria OR

Captain George Flavel House Museum, 441 Eighth St, Astoria, OR 97103 • T: +1 503 3252203 • F: +1 503 3257727 • cchs@seasurf.net • www.clatsophistoricalsociety.org •
Exec.Dir.: *McAndrew Burns* • Cur.: *Liisa Penner* (Collections) •
Historical Museum – 1951
Furniture, art, Victorian era furnishings, household objects 46400

Columbia River Maritime Museum, 1792 Marine Dr, Astoria, OR 97103 • T: +1 503 3252323 • F: +1 503 3252331 • information@crmm.org • www.crmm.org •
Dir.: *Jerry Ostermiller* • Cur.: *Dave Pearson* •
Historical Museum / Science&Tech Museum – 1962 46401

Fort Clatsop National Memorial, 92343 Fort Clatsop Rd, Astoria, OR 97103 • T: +1 503 8612471 • F: +1 503 8612585 • focl_administration@nps.gov • www.nps.gov/focl •
Open Air Museum / Military Museum – 1958
Material conc Lewis and Clark expedition 46402

The Heritage Museum, Clatsop Historical Society, 1618 Exchange St, Astoria, OR 97103 • T: +1 503 3384849 • F: +1 503 3386265 • cchs@seasurf.net • www.clatsophistoricalsociety.org •
Exec.Dir.: *McAndrew Burns* • Cur.: *Liisa Penna* (Collections) •
Local Museum – 1985
Local history, natural hist, geology, Native American artifacts, early immigration and settlement 46403

Uppertown Firefighters Museum, 30th and Marine Dr, Astoria, OR 97103 • T: +1 503 3252203 • F: +1 503 3257727 • cchs@seasurf.net • www.clatsophistoricalsociety.org •
Dir.: *McAndrew Burns* • Cur.: *Liisa Penner* (Collections) •
Historical Museum – 1990
Fire-fighting equipment 46404

Atascadero CA

Atascadero Historical Society Museum, 6500 Palma Av, Atascadero, CA 93422 • T: +1 805 4668341, +1 805 7817939 • F: +1 805 4610606 • acornejo@thetribunenews.com • www.atascaderohistoricalsociety.org •
Dir./Cur.: *Mike Lindsay* • Asst. Cur.: *Marjorie R. Mackey* • *Bill Smith* •
Local Museum – 1965
Local history, E.G. Lewis the founder of the town, Ralph Holmes paintings, Hooper coll incl WWII items, elephants, household articles, furniture 46405

Atchison KS

Amelia Earhart Birthplace Museum, 223 North Tce, Atchison, KS 66002 • T: +1 913 3674217 • aemuseum@lvnworth.com • www.ameliaearhartmuseum.org •
Special Museum – 1984
Women pilots 46406

Atchison County Historical Society, 200 S 10th St, Santa Fe Depot, Atchison, KS 66002 • T: +1 913 3676238 • gowest@atchisonhistory.org • www.atchisonhistory.org •
Association with Coll – 1966
Local history, westward expansion, Lewis O'Clark, Amelia Barhart 46407

The Muchnic Gallery, 704 N 4th, Atchison, KS 66002 • T: +1 913 3674278 • F: +1 913 3672939 • atchart@ponyexpress.net • www.atchison-art.org •
Dir./Cur.: *Gloria Conkle Davis* •
Public Gallery – 1970 46408

Athens GA

Center Art Gallery, University of Georgia, 153 Tate Student Center, Athens, GA 30602 • T: +1 706 5426396 • F: +1 706 5425584 •
Dir.: *Kevin McKee* •
University Museum / Fine Arts Museum 46409

Church-Waddel-Brumby House Museum, 280 E Dougherty St, Athens, GA 30601 • T: +1 706 3531820 • F: +1 706 3531770 • athenswc@negia.net • www.visitathensga.com •
Dir.: *Mary Beth Justus* •
Historical Museum – 1968
Photographs, period furnishings 46410

Georgia Museum of Art, University of Georgia, 90 Carlton St, Athens, GA 30602-6719 • T: +1 706 5424662 • F: +1 706 5421051 • buramsey@arches.uga.edu • www.uga.edu/gamuseum •
Dir.: *William U. Eiland* • Cur.: *Romita Ray* (Prints and Drawings) • *Ashley Brown* (Decorative Arts) •
Fine Arts Museum / University Museum – 1945
Prints, drawings, paintings, photographs, graphics 46411

Georgia Museum of Natural History, University of Georgia, Natural History Bldg, Athens, GA 30602-1882 • T: +1 706 5421663 • F: +1 706 5423920 • musinfo@arches.uga.edu • museum.nhm.uga.edu/ •
Dir./Cur.: *B.J. Freeman* (Zoology) •
University Museum / Natural History Museum – 1977
Zooarchaeology, botany, paleontology, geology, mycology, palynology, entomology 46412

Lyndon House Art Center, 293 Hoyt St, Athens, GA 30601 • T: +1 706 6133623 • F: +1 706 6133627 •
Dir.: *Claire Benson* • Cur.: *Nancy Lukasiewicz* •
Fine Arts Museum / Decorative Arts Museum – 1973 46413

Sed Gallery, University of Georgia, c/o School of Environmental Design, G14 Caldwell Hall, Athens, GA 30602 • T: +1 706 5428292 • F: +1 706 5424485 • rds@uga.edu • www.sed.uga.edu/ •
C.E.O.: *Rene D. Shoemaker* •
Fine Arts Museum / University Museum – 1993
Architecture 46414

Taylor-Grady House, 634 Prince Av, Athens, GA 30601 • T: +1 706 5498688 • F: +1 706 6130860 • jlathens@aol.com •
Pres.: *Laura K. Brown* •
Historical Museum – 1968
1844 Henry W. Grady home 46415

United States Navy Supply Corps Museum, 1425 Prince Av, Athens, GA 30606-2205 • T: +1 706 3547349 • F: +1 706 3547239 • dan.roth@nscs.com •
Cur.: *Dan Roth* •
Military Museum – 1974
Naval Museum housed in 1910 Carnegie Library 46416

Athens OH

Kennedy Museum of Art, Ohio University, 117 Lin Hall, Athens, OH 45701-2979 • T: +1 740 5931304 • F: +1 740 5931305 • kennedymuseum@ohio.edu • www.ohiou.edu/museum •
Dir.: *James Wyman* • Cur.: *Sally Delgado* (Education) • *Jennifer McLerran* •
Fine Arts Museum / University Museum – 1993
Southwest native American art, 20th c prints, African art 46417

Seigfred Gallery, Ohio University, School of Art, Seigfred Hall 528, Athens, OH 45701 • T: +1 740 5934290 • F: +1 740 5930457 •
Dir.: *Jenita Landrum-Bittle* •
Fine Arts Museum – 1993 46418

Athens PA

Tioga Point Museum, 724 S Main St, Athens, PA 18810 • T: +1 717 8887225 • tpointmuseum@stny.rr.com • home.stny.rr.com/tpointmuseum •
Pres.: *Dr. Donald W. Hunt* • Dir./Cur.: *Jaime A. Anderson* •
Local Museum – 1895
Native american art, archaeology 46419

Athens TN

McMinn County Living Heritage Museum, 522 W Madison Av, Athens, TN 37303 • T: +1 423 7450329 • F: +1 423 7450329 • livher@usit.net • www.livingheritagemuseum.com •
Dir.: *Ann Davis* •
Historical Museum – 1982
Local hist 46420

Atlanta GA

Apex Museum, 135 Auburn Av NE, Atlanta, GA 30303 • T: +1 404 5232739 • F: +1 404 5233248 • apexmuseum@aol.com • www.apexmuseum.org •
Pres.: *Dan Moore* • Dir.: *Dan Moore jr.* •
Fine Arts Museum / Historical Museum – 1978
African-American art, artifacts 46421

Art Gallery, c/o Georgia State University, 10 Peachtree Center Av, Rm 117, Atlanta, GA 30303 • T: +1 404 6512257 • F: +1 404 6511779 •
Dir.: *Teri Williams* •
Fine Arts Museum / University Museum 46422

The Atlanta College of Art, 1280 Peachtree St, NE, Atlanta, GA 30309 • T: +1 404 7335001 • F: +1 404 7335201 • acainfo@woodruffcenter.org • www.aca.edu •
Pres.: *Ellen L. Meyer* •
Fine Arts Museum / Library with Exhibitions – 1928
Artist's books 46423

Atlanta Contemporary Art Center, 535 Means St NW, Atlanta, GA 30318 • T: +1 404 6881970 • F: +1 404 5775856 • gallery@thecontemporary.org • www.thecontemporary.org •
Exec. Dir.: *Rob Smulian* • Dir.: *Helena Reckitt* •
Fine Arts Museum – 1973
Contemporary Art 46424

The Atlanta Cyclorama, 800-C Cherokee Av SE, Atlanta, GA 30315 • T: +1 404 6587625 • F: +1 404 6587045 • atlcyclorama@mindspring.com • www.webguide.com/cyclorama.html •
Dir.: *Pauline M. Smith* • Ass. Dir.: *Keith G. Lauer* •
Fine Arts Museum / Historical Museum – 1898
Art, Civil War, c.1886 Cyclorama that depicts the July 22, 1864 Battle of Atlanta 46425

Atlanta History Museum, 130 W Paces Ferry Rd NW, Atlanta, GA 30305-1366 • T: +1 404 8144000 • F: +1 404 8144186 • jbruns@atlantahistorycenter.com • www.atlantahistorycenter.com •
Pres.: *James Bruns* •
Historical Museum / Military Museum / Open Air Museum – 1926
Local history, historic houses, gardens – library 46426

Atlanta International Museum of Art and Design, 285 Peachtree Center Av, Atlanta, GA 30303 • T: +1 404 6882467 ext 307 • F: +1 404 5219311 • info@atlantainternationalmuseum.org • www.atlantainternationalmuseum.org •
Dir.: *John V. Johnes* •
Fine Arts Museum – 1989 46427

Atlanta Museum, 537-39 Peachtree St, NE, Atlanta, GA 30308 • T: +1 404 8728233 •
Dir.: *J.H. Elliott* • Cur.: *Mary Gene Elliott* •
Local Museum – 1938
General Museum 46428

Callanwolde Fine Arts Center, 980 Briarcliff Rd NE, Atlanta, GA 30306 • T: +1 404 8725338 • F: +1 404 8725175 • info@callanwolde.org • www.callanwolde.org •
Dir.: *Laurie Allan* (Gallery) •
Fine Arts Museum – 1973
Personal artifacts, paintings, prints, sculpture, pottery and jewelry 46429

Center for Puppetry Arts, 1404 Spring St, Atlanta, GA 30309 • T: +1 404 8733089 ext 110 • F: +1 404 8739907 • info@puppet.org • www.puppet.org •
Exec. Dir.: *Vincent Anthony* •
Special Museum – 1978
Puppetry, poster and toys coll 46430

City Gallery at Chastain, 135 W Wieuca Rd NW, Atlanta, GA 30342 • T: +1 404 2571804 • F: +1 404 8511270 •
Dir.: *Lynn Ho* •
Local Museum
Contemporary art by local, regional and national artists and designers 46431

City Gallery East, 675 Ponce de Leon Av NE, Atlanta, GA 30308 • T: +1 404 8176981 • F: +1 404 8176827 • citygalleryeast@mindspring.com •
Dir.: *Karen Comer* •
Fine Arts Museum
Contemporary fine art 46432

Clark Atlanta University Art Galleries, Trevor Arnett Hall, Atlanta, GA 30314, mail addr: 223 James P Brawley Dr SW, Atlanta, GA 30314 • T: +1 404 8806102 • F: +1 404 8806968 • tdunkley@cau.edu • www.cau.edu/art_gallery/art_gal_dir_right.html •
Pres.: *Susan E. Neugent* •
Fine Arts Museum – 1942
African-American and world art 46433

Fernbank Museum of Natural History, 767 Clifton Rd NE, Atlanta, GA 30307-1221 • T: +1 404 9296300, 9296400 • F: +1 404 3788140, 3708087 • fernbankmail@fc.dekalb.k12.ga.us • www.fernbank.edu/museum •
Pres.: *Susan E. Neugent* •
Natural History Museum / Archaeology Museum – 1992
Jewelry, costumes, textiles, Georgia minerals, gemstones, plants, shells, animals, dinosaurs, argentinosaurus, gigantosaurus 46434

Fernbank Science Center, 156 Heaton Park Dr NE, Atlanta, GA 30307-1398 • T: +1 404 3784311 • F: +1 404 3701336 • fernbankmail@fc.dekalb.k12.ga.us • fsc.fernbank.edu •
Dir.: *William M. Sudduth* •
Natural History Museum – 1967
Skins, hides, entomology, ornitology, geology, minerals, skeletal, insects – Library, planetarium 46435

Folk Art and Photography Galleries, High Museum of Art, 133 Peachtree St NE, Atlanta, GA 30303 • T: +1 404 5776940, 7334400 • F: +1 404 6530916 • dean.walcott@woodruffcenter.org • www.high.org •
Dir.: *Dr. Michael E. Shapiro* •
Fine Arts Museum / Folklore Museum – 1986
Photographic art, folk art 46436

Georgia Capitol Museum, Georgia State Capitol Rm 431, Atlanta, GA 30334 • T: +1 404 6516996 • F: +1 404 6573801 • dolson@sos.state.ga.us • www.sos.state.ga.us •
Dir.: *Dorothy Olson* •
Historical Museum – 1895
Portraits, sculpture, historic flags 46437

Georgia State University Art Gallery, University Plaza, 10 Peachtree Center Av, Atlanta, GA 30303 • T: +1 404 6512257, 6513424 • F: +1 404 6511779 • artgallery@gsu.edu • www.gsu.edu/~wwwart/ •
Dir.: *Teri Williams* • Cur.: *Ann England* •
Public Gallery / Fine Arts Museum / University Museum – 1970
American contemporary paintings, prints, photographs and crafts 46438

Gilbert House, 2238 Perkerson Rd SW, Atlanta, GA 30315 • T: +1 404 7669049 • F: +1 404 7652806 •
Dir.: *Erin Bailey* •
Decorative Arts Museum / Historical Museum – 1984 46439

Global Health Odyssey, 1600 Clifton Rd, NE-MS A14, Atlanta, GA 30333 • T: +1 404 6390830 • F: +1 404 6390834 • global@cdc.gov • www.cdc.gov/global •
Dir.: *Judy M. Gantt* •
Historical Museum – 1996
Hist of Public Health Service and Disease Control, protection of health 46440

High Museum of Art, 1280 Peachtree St NE, Atlanta, GA 30309 • T: +1 404 7334000 • F: +1 404 7334529 • annualfund@woodruffcenter.org • www.high.org •
Dir.: *Michael Shapiro* •

Fine Arts Museum – 1926
Audiovisual and film, decorative arts, paintings, photographs, prints, drawings, graphic arts, sculpture 46441

Margaret Mitchell House and Museum, 990 Peachtree St, Atlanta, GA 30309-3964 • T: +1 404 2497012 • F: +1 404 2499388 • ashley_burdell@gwtw.org • www.gwtw.org •
C.E.O.: *Mary Rose Taylor* •
Historical Museum – 1990
History of Atlanta 46442

Martin Luther King jr. Center and Preservation District, Birth Home Museum, Ebenezer Baptist Church Museum, 449-450 Auburn Av NE, Atlanta, GA 30312 • T: +1 404 3315190, 5268900 • F: +1 404 7303112 • information@thekingcenter.org • www.nps.gov/malu •
Dir.: *Frank Catroppa* •
Historical Museum / Historic Site – 1980
Neighborhood in which Dr. Martin Luther King, Jr. grew up, incl birthplace, boyhood home, church and gravesite 46443

Michael C. Carlos Museum, Emory University, 571 S Kilgo St, Atlanta, GA 30322 • T: +1 404 7274282 • F: +1 404 7274292 • carlos@emory.edu • www.carlos.emory.edu •
Dir.: *Bonnie Speed* • Cur.: *Dr. Peter Lacovara* (Ancient Art) •
Fine Arts Museum / University Museum / Archaeology Museum – 1920
Arts, archaeology 46444

The Museum of Contemporary Art, 1447 Peachtree St, Atlanta, GA 30309 • T: +1 404 8811109 • F: +1 404 8811138 • info@mocaga.org • www.mocaga.org •
Dir.: *Annette Cone Skelton* •
Fine Arts Museum
Paintings, prints, sculpture, viedeo 46445

Museum of the Jimmy Carter Library, 441 Freedom Pkwy, Atlanta, GA 30307 • T: +1 404 8657101 • F: +1 404 8657102 • carter.library@nara.gov • www.jimmycarterlibrary.org •
Dir.: *Jay E. Hakes* • Cur.: *Sylvia Mansour Naguib* •
Library with Exhibitions – 1986
Presidential library 46446

Oglethorpe University Museum of Art, 4484 Peachtree Rd NE, Atlanta, GA 30319 • T: +1 404 3648555 • F: +1 404 3648556 • museum@oglethorpe.edu • museum.oglethorpe.edu •
Dir.: *Lloyd Nick* •
Fine Arts Museum / University Museum – 1993 46447

Photographic Investments Gallery, 3977 Briarcliff Rd, Atlanta, GA 30345 • T: +1 404 3201012 • F: +1 404 3203465 • ecsymmes@aol.com •
Dir.: *Edwin C. Symmes* •
Fine Arts Museum – 1979
Photography Art 46448

Robert C. Williams American Museum of Papermaking, 500 10th St NW, Atlanta, GA 30318 • T: +1 404 8947840 • F: +1 404 8944778 • cindy.bowden@ipst.edu • www.ipst.edu/amp •
Dir.: *Cindy Bowden* • Cur.: *Teri Williams* •
Science&Tech Museum – 1936
Paper technology 46449

Salvation Army Southern Historical Museum, 1032 Metropolitan Pkwy SW, Atlanta, GA 30310-3488 • T: +1 404 7527578 • F: +1 404 7531932 • historical_center@uss.salvationarmy.org • www.salvationarmysouth.org/museum.htm •
Dir.: *Michael Nagy* •
Religious Arts Museum – 1986
Religious hist 46450

Science and Technology Museum of Atlanta, 395 Piedmont Av NE, Atlanta, GA 30308 • T: +1 404 5225500 ext 202 • F: +1 404 5256906 • tsmith@scitrek.org • www.scitrek.org •
Pres./ C.E.O.: *Lewis Massey* •
Natural History Museum / Science&Tech Museum – 1988
Exhibits of science and technology 46451

Southeast Arts Center, 215 Lakewood Way SE, Atlanta, GA 30315 • T: +1 404 6586036 • F: +1 404 6240746 •
Dir.: *Alberta Ward* •
Fine Arts Museum
Photography, ceramics, handbuilt pottery and Raku Firing, jewelry-making, sewing, art for youth 46452

Southern Arts Federation, 1800 Peachtree St NW, Atlanta, GA 30309 • T: +1 404 8747244 • F: +1 404 8732148 • jtalbott@southarts.org • www.southarts.com •
Association with Coll – 1975 46453

Spelman College Museum of Fine Art, 350 Spelman Ln SW, Atlanta, GA 30314 • T: +1 404 6813643 • F: +1 404 2237665 • museum@spelman.edu • www.museum.spelman.edu •
C.E.O.: *Andrea D. Barnwell* •
Fine Arts Museum – 1996
African American and European art, African sculpture, textiles, crafts 46454

Spruill Gallery, 4681 Ashford Dunwoody Rd, Atlanta, GA 30338 • T: +1 770 3944019 • F: +1 770 3943987 • sprgalry@bellsouth.net • www.spruillarts.org •
Exec. Dir.: *Sandra Bennett* •
Public Gallery – 1975 46455

Swan Coach House Gallery, 3130 Slaton Dr NW, Atlanta, GA 30305 • T: +1 404 2662636 • F: +1 404 2332012 • gallery@swancoachhouse.com • www.swancoachhouse.com •
Cur.: *Marianne Lambert* •
Public Gallery 46456

William Breman Jewish Heritage Museum, 1440 Spring St NW, Atlanta, GA 30309 • T: +1 404 8731661 • F: +1 404 8814009 • reinstein@thebreman.org; • www.thebreman.org •
Dir.: *Jane Leavey* •
Historical Museum – 1996
Georgia Jewish hist since 1845, Holocaust 46457

Wren's Nest House Museum, 1050 R.D. Abernathy Blvd SW, Atlanta, GA 30310-1812 • T: +1 404 7537735 ext 1 • F: +1 404 7538535 • wrensnest@aol.com •
Exec. Dir.: *Sharon Crutchfield* •
Special Museum – 1909
Home of Joel Chandler Harris, creator of Uncle Remus and chronicler of stories 46458

Atlantic Beach NC

Fort Macon, E Fort Macon Rd, Atlantic Beach, NC 28512 • T: +1 252 7263775 • F: +1 252 7262497 • fort.macon@ncmail.net • www.clis.com/friends •
Head: *Jody A. Merritt* •
Historical Museum – 1924
1834-1944 military artifacts, soldier and garrison life relics, hist buildings 46459

Atlantic City NJ

Atlantic City Historical Museum, New Jersey Av and The Boardwalk, Garden Pier, Atlantic City, NJ 08401 • T: +1 609 3475839, 3441943 • F: +1 609 3475284 • princetn@earthlink.net • www.acmuseum.org •
Dir.: *Vicki Gold-Levi* •
Local Museum – 1982
Furnishings, personal artifacts, drawings, praphic arts, textiles 46460

Historic Gardner's Basin, 800 N New Hampshire Av, Atlantic City, NJ 08401 • T: +1 609 3482880 • F: +1 609 3454238 • fea_more@yahoo.com • www.oceanlifecenter.com •
Exec. Dir.: *Jack Keith* •
Local Museum – 1976
Water craft, maritime artifacts, paintings, sculpture 46461

Attleboro MA

Attleboro Area Industrial Museum, 42 Union St, Attleboro, MA 02703-2911 • T: +1 508 2223918 • info@industrialmuseum.com • www.industrialmuseum.com •
Exec. Dir.: *George Shelton* •
Historical Museum – 1975
Local hist & ind, jewelry ind – library 46462

Attleboro Museum-Cerner for the Arts, 86 Park St, Attleboro, MA 02703 • T: +1 508 2222644 ext 5 • F: +1 508 2264401 • museum@naisp.net • www.attleboromuseum.org •
Exec. Dir.: *Dore Van Dyke* •
Fine Arts Museum / Public Gallery – 1929
Regional American prints, paintings, civil war coll 46463

Atwater CA

Castle Air Museum, 5050 Santa Fe Dr, Atwater, CA 95301 • T: +1 209 7232178 • F: +1 209 7230323 • www.castleairmuseum.org •
Dir.: *Jean Astorino* •
Military Museum / Science&Tech Museum – 1981
Aircraft history, 40 aircrafts esp. SR-71, B-17, strategic bombardment 46464

Auburn AL

Jule Collins Smith Museum of Fine Arts, c/o Auburn University, 901 S College St, Auburn, AL 36849 • T: +1 334 8441484 • jcsm@auburn.edu • jcsm.auburn.edu/the_museum •
Dir.: *Miah Michaelsen* •
University Museum / Fine Arts Museum – 2003
American art, Tibetan Bronzes coll, colour prints by John J. Audubon 46465

Auburn IN

Auburn-Cord-Duesenberg Museum, 1600 S Wayne St, Auburn, IN 46706 • T: +1 219 9251444 • F: +1 219 9256266 • www.acdmuseum.org •
Pres.: *Robert Pass* •
Science&Tech Museum – 1973
100 antique and classic cars, photographs 46466

National Automotive and Truck Museum of the United States, 1000 Gordon M. Buehrig Pl, Auburn, IN 46706-0686 • T: +1 219 9259100 • F: +1 219 9258695 • info@natmus.com • www.natmus.com •
Science&Tech Museum – 1989
Automobiles, trucks, models, toys, gas and oil pumps 46467

Auburn ME

Androscoggin Historical Society Museum, County Bldg, 2 Turner St, Auburn, ME 04210-5978 • T: +1 207 7840586 • androhs@verizon.net • www.rootsweb.com/~meandrhs •
Pres.: *David C. Young* •
Local Museum / Historical Museum – 1923
County history 46468

Auburn NY

Cayuga Museum, 203 Genesee St, Auburn, NY 13021 • T: +1 315 2538051 • F: +1 315 2539829 • cayugamuseum@cayuganet.org • www.cayuganet.org/cayugamuseum •
Dir.: *Eileen McHugh* • Cur.: *Pat Elkovitch* •
Local Museum – 1936
Indian artefacts, history 46469

Schweinfurth Memorial Art Center, 205 Genesee St, Auburn, NY 13021 • T: +1 315 2551553 • F: +1 315 2550871 • smac@baldcom.net • www.cayuganet.org/smac •
Dir.: *Donna Lamb* • Cur.: *Stephanie Bielejec* •
Fine Arts Museum – 1975 46470

Seward House, 33 South St, Auburn, NY 13021 • T: +1 315 2533351 • F: +1 315 2533351 • info@sewardhouse.org • www.sewardhouse.org •
Exec. Dir.: *Peter A. Wisbey* • Cur.: *Jeniffer Haines* •
Coll.Mgr: *Paul McDonald* •
Local Museum – 1951
Originla furnishings, Life of Seward, Civil War material, Alaskan memorabilia and artifacts 46471

Ward O'Hara Agricultural Museum of Cayuga County, E Lake Rd, Auburn, NY 13021 • T: +1 315 2527644 • F: +1 315 2527644 • mrr423@dreamscape.com • www.cayuganet.org •
C.E.O.: *Norman Riley* •
Agriculture Museum – 1975
Agricultural machinery and implements 46472

Auburn WA

White River Valley Museum, 918 H St SE, Auburn, WA 98002 • T: +1 253 2887433 • F: +1 253 9313098 • www.wrvmuseum.org •
C.E.O.: *Patricia Cosgrove* •
Local Museum – 1959
Life of Native Americans and settlers, railroading, farming 46473

Auburn Hills MI

Walter P. Chrysler Museum, 1 Chrysler Dr, CIMS 488.00.00, Auburn Hills, MI 48327-2766 • T: +1 248 9440432 • F: +1 248 9440460 • bd28@daimlerchrysler.com • www.chryslerheritage.com •
Man.: *Barry Dressel* •
Science&Tech Museum / Historical Museum – 1998
Founder of the automotive industry, historical vehicles 46474

Augusta GA

Augusta Museum of History, 560 Reynolds St, Augusta, GA 30901 • T: +1 706 7228454 • F: +1 706 7245192 • amh@csra.net • www.augustamuseum.org •
Dir.: *Scott W. Loehr* •
Historical Museum – 1937
The past of Augusta, Georgia and its environs 46475

Ezekiel Harris House, 1822 Broad St, Augusta, GA 30904 • T: +1 706 7372820 •
Historical Museum – 1965
Furnishings 46476

Fort Discovery, 1 Seventh St, Augusta, GA 30901 • F: +1 706 8210648 • info@nscdiscovery.org • www.nationalsciencecenter.org •
Science&Tech Museum – 1997 46477

Gertrude Herbert Institute of Art, 506 Telfair St, Augusta, GA 30901 • T: +1 706 7225495 • F: +1 706 7223670 • ghia@ghia.org • www.ghia.org •
Fine Arts Museum – 1937
Paintings, sculptures, graphics 46478

Lucy Craft Laney Museum of Black History, 1116 Philips St, Augusta, GA 30901 • T: +1 706 7243576 • F: +1 706 7243576 • info@lucycraftlaneymuseum.com • www.lucycraftlaneymuseum.com •
Exec. Dir.: *Christine Miller-Betts* •
Ethnology Museum / Folklore Museum – 1991
African Americans from the Augusta and Central Savannah River area education, religion, science sports, entertainment and medicine 46479

Meadow Garden Museum, 1320 Independence Dr, Augusta, GA 30901-1038 • T: +1 706 7244174 •
Historical Museum
1791-1804 residence of George Walton, signer of the Declaration of Independence 46480

Morris Museum of Art, 1 Tenth St, Augusta, GA 30901-0100 • T: +1 706 7247501 • F: +1 706 7247612 • mormuse@themorris.org • www.themorris.org •
Cur.: *Jay Williams* •
Fine Arts Museum – 1985
Paintings 46481

Augusta KS

Augusta Historical Museum, 303 State, Augusta, KS 67010 • T: +1 316 7755655 • augustahm@earthlink.net •
Pres.: *Virginia Belt* •
Local Museum – 1938
Pioneer relics, items from 18th c 46482

Augusta ME

Blaine House, 192 State St, Augusta, ME 04330 • T: +1 207 2872121 •
Head: *Sue J. Plummer* •
Local Museum
Blaine House, governer's mansion of Maine 46483

Jewett Hall Gallery, c/o University of Maine at Augusta, 46 University Dr, Augusta, ME 04330 • T: +1 207 613243 • F: +1 207 6213293 • kareng@maine.edu •
Dir.: *Karen Gilg* •
Fine Arts Museum – 1970
Drawings, painting, outdoor sculpture 46484

Maine State Museum, 83 State House Station, State House Complex, Augusta, ME 04333-0083 • T: +1 207 2872301 • F: +1 207 2876633 • museum@state.me.us • www.mainestatemuseum.org •
Dir.: *Joseph R. Phillips* • Cur.: *Deanna Bonner* •
Local Museum – 1837
History, natural history, archaeology, fine arts, graphics 46485

Old Fort Western, 16 Cony St, Augusta, ME 04330 • T: +1 207 6262385 • F: +1 207 6262304 • oldfort@oldfortwestern.org • www.oldfortwestern.org •
Dir./Cur.: *Jay Adams* •
Historical Museum – 1922
Historic house 46486

Auriesville NY

Kateri Galleries, National Shrine of the North American Martyrs, off Rte 5S, Auriesville, NY 12016 • T: +1 518 8533033 • F: +1 518 8533051 • office@martyshrine.org •
Dir.: *John G. Marzolf* • Ass. Dir. & Cur.: *John J. Paret* •
Cur.: *Owen Smith* •
Religious Arts Museum – 1950
American Indian contemporary handicrafts and local artifacts, paintings and charts, Geology, anthropology 46487

Aurora CO

Aurora History Museum, 15001 E Alameda Dr, Aurora, CO 80012-1547 • T: +1 303 7396660 • F: +1 303 7396657 • museum@auroragov.org • www.aurora-museum.org •
Dir.: *Gordon Davis* • Cur.: *Mary Ellen Schoonover* • *Matthew Chasansky* • *Michelle Bahe* •
Local Museum – 1979
History 46488

Aurora IL

Aurora Historical Museum, Cedar and Oak Sts, Aurora, IL 60506 • T: +1 630 8979029 • F: +1 630 9060657 • ahs@aurorahistoricalsociety.org • www.aurorahistoricalsociety.org •
Dir.: *John R. Jaros* • Cur.: *Dennis Buck* •
Local Museum – 1906
Local hist and genealogy 46489

Aurora Public Art Commission, 20 E Downer Pl, Aurora, IL 60507 • T: +1 630 8443623, 9060654 • F: +1 630 8920741 • rchurch@ci.aurora.il.us •
Dir./Cur.: *Rena J. Church* •
Fine Arts Museum – 1996
Watercolours, photography, sculpture, intaglio 46490

Blackberry Farm-Pioneer Village, 100 S Barnes Rd, Aurora, IL 60506 • T: +1 630 2647408 • F: +1 630 8921661 • ibeasley@fvdp.net • www.foxvalleyparkdistrict.org •
C.E.O.: *Robert Vaughan* • Cur.: *Carol Nordbrock* •
Historical Museum / Agriculture Museum – 1969
Village, agriculture 46491

Grand Army of the Republic Memorial and Veteran's Military Museum, 23 E Downer Pl, Aurora, IL 60505 • T: +1 630 8977221 •
Pres.: *Charles Gates* •
Military Museum – 1875
Civil, Spanish American, Korean, Vietnam wars, WW 1 and 2 colls 46492

Schingoethe Center for Native American Cultures, Aurora University, Dunham Hall, 347 S Gladstone, Aurora, IL 60506-4892 • T: +1 630 8445402 • F: +1 630 8448884 • museum@aurora.edu • www.aurora.edu/museum •
Dir.: *Michael P. Sawdey* •
University Museum / Ethnology Museum / Folklore Museum – 1989
Native American art 46493

Scitech, 18 W Benton, Aurora, IL 60506 • T: +1 630 8593434 ext 210 • F: +1 630 8598692 • ronen@scitech.mus.il.us • scitech.mus.il.us •
Exec. Dir.: *Dr. Ronen Mir* •
Science&Tech Museum / Natural History Museum – 1988
Weather, light and color, heat, mathematics, sound and music, magnets and electricity, physics, chemistry, biology, astronomy and nuclear physics 46494

Aurora IN

Hillforest House Museum, 213 Fifth St, Aurora, IN 47001 • T: +1 812 9260087 • F: +1 812 9261075 • hillforest@seidata.com • www.hillforest.org •
Pres.: *Tamara Zoller* •
Historical Museum – 1956
Victorian Ohio River Valley mansion 46495

Aurora NE

Edgerton Explorit Center, 208 16th St, Aurora, NE 68818 • T: +1 402 6944032 • F: +1 402 6944035 • edgerton@hamilton.net • www.edgerton.org •
Science&Tech Museum – 1995
Hands-on science exhib 46496

Plainsman Museum, 210 16th St, Aurora, NE 68818 • T: +1 402 6946531 • www.plainsmanmuseum.org •
Pres.: *Bruce Ramsour* • Dir./Cur.: *Megan Sharp* •
Local Museum – 1935 46497

Aurora OH

Aurora Historical Society Museum, 115 E Pioneer Trail, Aurora, OH 44202 • T: +1 330 5626502 • lsc5128446@aol.com •
Pres.: *Lynn Hubach* •
Local Museum – 1968
Home furnishings, farm implements, cheese-making equipment 46498

Austin MN

SPAM Museum, 1101 N Main St, Austin, MN 55912 • T: +1 507 4346721 • F: +1 507 4346721 • spammuseum@hormel.com • www.spam.com •
Dir./Cur.: *Tom Rypka* •
Special Museum – 2001
Photographs, audios, videos and displays depicting the hist of the SPAM brand and the Hormel Foods Corp. 46499

Austin TX

Arthouse at the Jones Center, 700 Congress Av, Austin, TX 78701 • T: +1 512 4535312 • F: +1 512 4594830 • info@arthousetexas.org • www.arthousetexas.org •
Dir.: *Sue Graze* •
University Museum / Fine Arts Museum – 1911
Contemporary art by Texas artists 46500

Austin Children's Museum, 201 Colorado St, Austin, TX 78701 • T: +1 512 4722499 • F: +1 512 4722495 • info@austinkids.org • www.austinkids.org •
Dir.: *Mike Nellis* •
Special Museum – 1983 46501

Austin History Center, 810 Guadalupe St, Austin, TX 78701, mail addr: PO BOX 2287, Austin, TX 78768-2287 • T: +1 512 9747480 • F: +1 512 9747483 • ahc_reference@ci.austin.tx.us • www.cityofaustin.org/library/ahc •
Head: *Mike Miller* •
Local Museum – 1955
Archives & manuscripts, local hist books, photographs, maps, ephemera – library, archives 46502

Austin Museum of Art, 823 Congress Av, Ste 100, Austin, TX 78701 • T: +1 512 4959224 • F: +1 512 4959029 • info@amoa.org • www.amoa.org •
Dir.: *Dana Friis-Hansen* • Cur.: *Judith Sims* (Film and Video) • Education: *Eva Buttacavoli* •
Fine Arts Museum – 1961
Modern & contemporary art, paintings, sculpture 46503

Austin Museum of Art Laguna Gloria, 3809 W 35th St, Austin, TX 78703 • T: +1 512 4588191 • F: +1 512 4581571 • info@amoa.org • www.amoa.org •
Dir.: *Judith Sims* •
Fine Arts Museum – 1961
Art 46504

Austin Nature and Science Center, 301 Nature Center Dr, Austin, TX 78746 • T: +1 512 3278180 ext 25 • F: +1 512 3278745 •
Natural History Museum – 1960
Native Texas mammals, fish, birds, reptiles and insects, microscope tables, Eco-Detective 46505

Center for American History, c/o University of Texas, Sid Richardson Hall, Unit 2, Austin, TX 78712 • T: +1 512 4954515 • F: +1 512 4954542 • cahref@uts.cc.utexas.edu • www.cah.utexas.edu •
Dir.: *Dr. Don Carleton* • Cur.: *Lynn Bell* •
Historical Museum
Special coll & programs relating to hist of Texas, the South, Rocky Mountain West and select national areas 46506

Elisabet Ney Museum, 304 E 44th St, Austin, TX 78751 • T: +1 512 4582255 • F: +1 512 4722174 • enm@ci.austin.tx.us • www.ci.austin.tx.us/elisabetney •
Dir./Cur.: *Mary Collins Blackmon* •
Decorative Arts Museum – 1911
508 piece collection of Ney portrait sculptures and personal memorabilia on permanent loan from the Harry Ransom Humanities Research Center of The University of Texas at Austin 46507

Fine Arts Exhibitions, c/o Saint Edward's University, 3001 S Congress Av, Austin, TX 78704 • T: +1 512 4888400 ext 1338 • F: +1 512 4488492 •
Dir.: *Stan Irvin* •
University Museum / Public Gallery – 1961 46508

The French Legation Museum, 802 San Marcos St, Austin, TX 78702 • T: +1 512 4728180 • F: +1 512 4729457 • dubois@french-legation.mus.tx.us •
Head: *Pete Wehmeyer* •
Historical Museum – 1956
Antique French & early American furnishings, artifacts reflecting life in early days of Republicof Texas 46509

Harry Ransom Humanities Research Center, c/o University of Texas at Austin, 21st & Guadelupe, Austin, TX 78705 • T: +1 512 4718944 • F: +1 512 4719646 • info@hrc.utexas.edu • www.hrc.utexas.edu •
Dir.: *Dr. Thomas F. Staley* • Assoc. Dir.: *Sally Leach* • *Mary Beth Bigger* • *Sue Murphy* • Cur.: *Roy Flukinger* (Photography) • *Peter Mears* (Art) • Librarian: *Dr. Richard W. Oram* •
Library with Exhibitions / Public Gallery
Manuscripts of works by Graham Greene, Lillian Hellman, D.H. Lawrence, Carson McCullers, George Bernard Shaw; Pforzheimer Library of English Literature (1450-1700); libraries of James Joyce, Evelyn Waugh and E.E. Cummings; hist photographs; temporary exhibitions 46510

Jack S. Blanton Museum of Art, c/o University of Texas at Austin, 23rd and San Jacinto, Austin, TX 78712-1205 • T: +1 512 4717324 • F: +1 512 4717023 • blantonmuseum@utexas.edu • www.blantonmuseum.org •
Dir.: *Jessie Otto Hite* • Cur.: *Jonathan Bober* (Prints and Drawings, European Painting) • *Annette DiMeo Carlozzi* (American Art) • Education: *Anne Manning* •
Fine Arts Museum / University Museum – 1963
Antiquites, Renaissance & Baroque art, 19th to 20th c American Art 46511

Jourdan-Bachman Pioneer Farm, 11418 Sprinkle Cut Off Rd, Austin, TX 78754 • T: +1 512 8371215 • F: +1 512 8374503 • john.hirsch@ci.austin.tx.us •
Dir.: *John Hirsch* •
Agriculture Museum – 1975
Historical agriculture 46512

Julia C. Butridge Gallery, 1110 Barton Springs Rd, Dougherty Arts Center, Austin, TX 78704 • T: +1 512 3971455 • F: +1 512 3971460 • megan.weiler@ci.austin.tx.us • www.ci.austin.tx.us/dougherty/gallery.htm •
Public Gallery 46513

Lyndon Baines Johnson Museum, 2313 Red River St, Austin, TX 78705-5702 • T: +1 512 7210200 • F: +1 512 7210171 • johnson.library@nara.gov • www.lbjlib.utexas.edu •
Dir.: *Betty Sue Flowers* • Cur.: *Sandor Cohen* •
Library with Exhibitions – 1971
Presidential library 46514

Mexic-Arte Museum, 419 Congress St, Austin, TX 78701 • T: +1 512 4809373 • F: +1 512 4808626 • director@mexic-artemuseum.org • www.mexic-artemuseum.org •
Exec. Dir.: *Sylvia Orozco* •
Fine Arts Museum – 1984
Art, focus on Mexican and Latin American culture 46515

Neill-Cochran Museum House, 2310 San Gabriel St, Austin, TX 78705 • T: +1 512 4782335 • F: +1 512 4781865 • ncmuseum@flash.net •
Head: *Kathleen Peterson-Moussaid* •
Historical Museum – 1956
Historic house, furnishings of the period 1855-1878 46516

O. Henry Museum, 409 E 5th St, Austin, TX 78701 • T: +1 512 4721903 • F: +1 512 4727102 • valerie.bennett@ci.austin.tx.us • www.ci.austin.tx.us/parks/ohenry.htm •
Dir.: *Valerie Bennett* •
Local Museum – 1934
Historic house, containing artifacts and archives of the famous short story writer William Sidney Porter, known as O. Henry. 46517

Texas Governor's Mansion, 1010 Colorado, Austin, TX 78701 • T: +1 512 4635518, 4635516 • F: +1 512 4631850 • www.txfgm.org •
Historical Museum / Historic Site
Furnishings, American Federal & Empire periods, decorative arts 46518

Texas Memorial Museum, 2400 Trinity, Austin, TX 78705 • T: +1 512 4711604 • F: +1 512 4714794 • tmmweb@uts.cc.utexas.edu • www.texasmemorialmuseum.org •
Dir.: *Edward C. Theriot* • Cur.: *Dean A. Hendrickson* (Ichthyology) • *Chris J. Durden* (Entomology) • *James R. Reddell* (Arthropods) • *David C. Cannatella* (Herpetology) •
Historical Museum / Ethnology Museum / Natural History Museum – 1936
Geology, mineralogy, meteorites, tektites, paleobotany, entomology, arachnology 46519

Texas Military Forces Museum, 2200 W 35th St, Austin, TX 78703 • T: +1 512 7825659 • F: +1 512 7826750 • museum@agd.state.tx.us • www.kwanah.com/txmilmus •
Cur.: *John C.L. Scribner* •
Military Museum – 1992
Paintings, military hist of Texas, vehicles, equip, uniforms 46520

Texas Music Museum, 1109 E 11th St, Austin, TX 78761 • T: +1 512 4728891 • F: +1 512 4719600 • cshorkey@mail.utexas.edu • www.texasmusicmuseum.org •
Music Museum – 1984
Music instrument-makers, recording and reproduction devices, music art 46521

Umlauf Sculpture Garden and Museum, 605 Robert E. Lee Rd, Austin, TX 78704 • T: +1 512 4455582 • F: +1 512 4455583 • info@umlaufsculpture.org • www.umlaufsculpture.org •
Dir.: *Nelie Plourde* •
Fine Arts Museum – 1991
Sculpture 46522

Women and Their Work, 1710 Lavaca St, Austin, TX 78701 • T: +1 512 4771064 • F: +1 512 4771090 • wtw@eden.com • www.womenandtheirwork.org •
Dir.: *Chris Cowden* • *Kathryn Davidson* •
Fine Arts Museum – 1978 46523

Avalon CA

Catalina Island Museum, Casino Bldg, Avalon, CA 90704, mail addr: POB 366, Avalon, CA 90704 • T: +1 310 5102414 • F: +1 310 5102780 • catalinaislmuseum@catalinaisp.com • www.catalinamuseum.org •
Dir.: *Stacey A. Otte* •
Historical Museum / Archaeology Museum – 1953
Regional history, ship models, Indian artifacts, natural hist 46524

Avella PA

Meadowcroft Museum of Rural Life, Historical Society of Western Pennsylvania, 401 Meadowroft Rd, Avella, PA 15312 • T: +1 724 5873412 • mcroft@cobweb.net • www.meadowcroftmuseum.org •
Local Museum
History of Life on the land in Western Pennsylvania, archeological site, 19th c village 46525

Avon CT

Avon Historical Society Museum, 8 E Main St, Avon, CT 06001 • T: +1 860 6787621 • oakeshoward@aol.com • www.avonct.com •
C.E.O.: *Nora Howard* •
Local Museum
Local hist – archives 46526

Aztec NM

Aztec Museum and Pioneer Village, 125 N Main Av, Aztec, NM 87410-1923 • T: +1 505 3349829 • F: +1 505 3347648 • aztecmuseum@cyberport.com • www.aztecnm.com/museum/museum_index.htm •
Historical Museum / Natural History Museum – 1963
Lobato cool of rocks, minerals, fossils, artifacts, prehistoric coll 46527

Aztec Ruins, Ruins Rd, Aztec, NM 87410 • T: +1 505 3346174 • F: +1 505 3346372 • azru_curatorial@nps.gov • www.nps.gov/azru •
Archaeology Museum – 1923
Archaeological materials of Chaco and Mesa Verde Anasazi 46528

Bahama NC

Historic Stagville, 5825 Old Oxford Hwy, Bahama, NC 27503 • T: +1 919 6200120 • F: +1 919 6200422 • stagvill@sprynet.com • www.historicstagvillefoundation.org •
Agriculture Museum – 1977
Period furnishings, farm implements, plantation house, slave cabins, barn 46529

Bailey NC

The Country Doctor Museum, 6642 Peele Rd, Bailey, NC 27807 • T: +1 919 2354165 • F: +1 252 2352372 • anderson@mail.ecu.edu • www.countrydoctormuseum •
Historical Museum – 1967
Practice of medicine 18th-19th c 46530

Bainbridge OH

Dr. John Harris Dental Museum, 208 W Main St, Bainbridge, OH 45612 • T: +1 740 6342228 • F: +1 937 9813218 • djhdm1940@yahoo.com •
Special Museum – 1939 46531

Baird TX

Callahan County Pioneer Museum, 100 W 4th, B-1, Baird, TX 79504 • T: +1 325 8541718 • F: +1 325 8541227 • callahancl@bitstreet.com •
C.E.O.: *Joyce McAuley* • Cur.: *Sonia Walker* •
Local Museum – 1940
Local pioneer articles 46532

Baker MT

O'Fallon Historical Museum, 723 S Main St, Baker, MT 59313, mail addr: P.O. Box 285, Baker, Mt. 59313-0285 • T: +1 406 7783265 • F: +1 406 778 • ofmuseum@midrivers.com •
Pres.: *Ken Griffith* • Dir.: *Lora Heyen* •
Historical Museum – 1968
Local homestead articles from 1908, Indian artifacts 46533

Baker City OR

National Historic Oregon Trail Interpretive Center, 22267 Flagstaff Hill, Oregon Hwy 86, Baker City, OR 97814 • T: +1 541 5231845 • F: +1 541 5231834 • nhotic@or.blm.gov •
Dir.: *Sarah LeCompte* •
Local Museum – 1992 46534

Oregon Trail Regional Museum, 2180 Grove St, Baker City, OR 97814 • T: +1 541 5239308 • F: +1 541 5230244 • cmotrm@oregontrail.net • www.bakercounty.org •
Dir.: *Cherry Mires* •
Historical Museum – 1982
Original settlers artifacts 46535

Bakersfield CA

Bakersfield Museum of Art, 1930 R St, Bakersfield, CA 93301 • T: +1 661 3237219 • F: +1 661 3237266 • ggilfoy@bmoa.org • www.bmoa.org •
Dir.: *Bernard J. Herman* • Ass. Dir.: *David Gordon* • Cur.: *Emily Falke* •
Fine Arts Museum – 1987
Paintings, sculpture, graphics, complete Marion Osborn Cunningham coll, Phil Paradise serigraphs – library 46536

Kern County Museum, 3801 Chester Av, Bakersfield, CA 93301 • T: +1 661 8525000 • F: +1 661 3226415 • kcmuseum@kern.org • www.kcmuseum.org •
Dir.: *Carola Rupert Enriquez* • Cur.: *Jeff Nickell* •
Local Museum / Open Air Museum – 1945
Local history, culture, tools & equipment, firearms, photo coll – Lori Brock children's discovery center 46537

Todd Madigan Gallery, c/o California State University, 9001 Stockdale Hwy, Bakersfield, CA 93311-1022 • T: +1 661 6542238 • kplunkett@csub.edu • www.csub.edu/art/gallery •
Public Gallery 46538

Baldwin NY

Baldwin Historical Society Museum, 1980 Grand Av, Baldwin, NY 11510 • T: +1 516 2236900 •
C.E.O. & Pres.: *Constance Grando* • C.E.O.: *Jack Bryck* • Cur.: *Gary R. Hammond* •
Local Museum – 1971 46539

Baldwin City KS

Old Castle Museum, Baker University, 515 Fifth St, Baldwin City, KS 66006, mail addr: POB 65, Baldwin City, KS 66006-0065 • T: +1 785 5946809 • F: +1 785 5942522 • day@harvey.bakeru.edu • www.bakeru.edu •
Dir.: *Brenda Day* • Asst. Dir.: *Jane Richards* • Pres.: *Dan Lambert* •
Historical Museum – 1953
Original Santa Fe Trail Post Office, replica of Kibbee cabin 46540

William A. Quayle Bible Collection, Baker University, Collins Library, Spencer Quayle Wing, 8th and Grove Sts, Baldwin City, KS 66006 • F: +1 913 5946721 •
Dir.: *Dr. John M. Forbes* •
University Museum / Religious Arts Museum / Library with Exhibitions – 1925
Rare Bibles 46541

Ballston Spa NY

Brookside Museum, 6 Charlton St, Ballston Spa, NY 12020 • T: +1 518 8854000 • F: +1 518 8857085 • info@brooksidemuseum.org • www.brooksidemuseum.org •
Exec.Dir.: *Joy Houle* • Cur.: *Linda Gorham* •
Local Museum – 1962 46542

Balsam Lake WI

Polk County Museum, 120 Main St, Balsam Lake, WI 54810 • T: +1 715 4833979 • darose@centurytel.net • www.co.polk.wi.us/museum •
C.E.O.: *Rosalie Kittleson* • Cur.: *Willis D. Erickson* •
Local Museum – 1960
Local hist, arts and crafts 46543

Baltimore MD

The Albin O. Kuhn Gallery, University of Maryland-Baltimore County, 1000 Hilltop Circle, Baltimore, MD 21250 • T: +1 410 4552270 • F: +1 410 4551153 • www.umbc.edu/library •
Chief Cur.: *Tom Beck* • Cur.: *Cynthia Wayne* •
University Museum / Fine Arts Museum – 1975
19th and 20th c photographs, biological sciences, popular culture 46544

American Visionary Art Museum, 800 Key Hwy, Baltimore, MD 21230 • T: +1 410 2441900 • F: +1 410 2445858 • www.avam.org •
Pres.: *Rebecca Hoffberger* • Dir.: *Mark Ward* •
Fine Arts Museum – 1995
Sculpture, works on paper, textiles, folk art 46545

Archaeological Collection, c/o Johns Hopkins University, 3400 North Charles St, Baltimore, MD 21218 • T: +1 410 5167561 •
Dir.: *Dr. Betsy Bryan* (Near Eastern and Egyptian Art) •
Cur.: *Dr. Eunice Maguire* •
Archaeology Museum / University Museum – 1876
Egyptian, Roman antiquities 46546

The B & O Railroad Museum, Smithsonian Institution, 901 W Pratt St, Baltimore, MD 21223 • T: +1 410 7522490 • F: +1 410 7522499 • info@borail.org • www.borail.org •
Dir.: *Courtney B. Wilson* • Chief Cur.: *Edward Williams* •
Historical Museum / Science&Tech Museum – 1953
Transport, site of c.1830 B & O Railroad's Mt Clare shops, site of first common carrier rail service in the US, which received SFB Morse's first long distance telegraph message in 1844 ... 46547

Babe Ruth Birthplace & Museum, 216 Emory St, Baltimore, MD 21230 • T: +1 410 7271539 • F: +1 410 7271652 • lauriew@baberuthmuseum.com • www.baberuthmuseum.com •
Dir.: *Michael L. Gibbons* • Cur.: *Gregory Schwalenberg* •
Fine Arts Museum – 1974
Sports, photographs, fine arts ... 46548

Baltimore Civil War Museum, 601 President St, Baltimore, MD 21202 • T: +1 410 3855188 • F: +1 410 3852105 • Museum@mdhs.org • www.mdhs.org •
C.E.O: *Dennis Fiori* •
Historical Museum / Military Museum – 1995
Civil War artifacts ... 46549

Baltimore Maritime Museum, Pier 3, Pratt St, Baltimore, MD 21202 • T: +1 410 3963453 • F: +1 410 3963393 • admin@baltomaritimemuseum.org • www.baltomaritimemuseum.org •
Dir.: *John Kellett* •
Historical Museum – 1982
Coast Guard cutter, WW II submarine, lightship, lighthouse ... 46550

The Baltimore Museum of Art, 10 Art Museum Dr, Baltimore, MD 21218-3898 • T: +1 410 3967100 • F: +1 410 3967153 • amannix@artbma.org • www.artbma.org •
Dir.: *Doreen Bolger* • Sen. Cur.: *Jay Fisher* (Cultural Affairs, Cur. Prints, Drawings, Photographs) • *Sona Johnston* (Painting, Sculpture before 1900) •
Fine Arts Museum – 1914
Arts of America, Africa, Asia, Oceania, decorative arts, drawings, prints, photography – sculpture gardens ... 46551

Baltimore Museum of Industry, 1415 Key Hwy, Inner Harbour S, Baltimore, MD 21230 • T: +1 410 7274808 • F: +1 410 7274869 • chboyd@thebmi.org • www.thebmi.org •
Exec. Dir.: *Paul Cypher* •
Historical Museum / Science&Tech Museum – 1981
History, industry, housed in a water front oyster cannery ... 46552

Baltimore Public Works Museum, 751 Eastern Av, Baltimore, MD 21202 • T: +1 410 3961509, 3965565 • F: +1 410 5456781 • bpwm@erols.com • www.baltimorepublicworksmuseum.org •
Exec. Dir.: *Mari B. Ross* • Dir.: *Vince Pompa* •
Historical Museum / Science&Tech Museum – 1982
Urban environmental history, sewage pumping stn ... 46553

Baltimore Streetcar Museum, 1901 Falls Rd, Baltimore, MD 21211 • T: +1 410 5470264 • F: +1 410 5470264 • www.baltimoremd.com/streetcar •
Pres.: *John J. O'Neill* •
Science&Tech Museum – 1966
Transport, RR Terminal ... 46554

Baltimore's Black American Museum, 1767 Carswell St, Baltimore, MD 21218 • T: +1 410 2439600 •
Ethnology Museum / Historical Museum ... 46555

The Benjamin Banneker Museum, 300 Oella Av, Baltimore, MD 21228 • T: +1 410 8871081 • F: +1 410 2032747 • slee@co.ba.md.us • www.thefriendsofbanneker.org •
Dir.: *Steven Lee* •
Fine Arts Museum / Natural History Museum – 1998
Traditional and contemporary art from Africa and the Americas, plants ... 46556

Community College Art Gallery, 2901 Liberty Heights Av, Baltimore, MD 21215 • T: +1 410 4628000 • F: +1 410 4627614 •
Fine Arts Museum – 1965 ... 46557

Contemporary Museum, 100 W Centre St, Baltimore, MD 21201 • T: +1 410 7835720 • F: +1 410 7835722 • info@contemporary.org • www.contemporary.org •
Dir.: *Leslie Shaffer* •
Fine Arts Museum – 1989
Audiovisual, paintings, photographs, prints, sculpture ... 46558

Cylburn Nature Museum, 4915 Greenspring Av, Baltimore, MD 21209 • T: +1 410 3672217 • F: +1 410 3677112 • caa4915@bcpl.net • www.cylburnorganization.com •
Pres.: *Jane Baldwin* • Dir.: *Patsy Perlman* •
Natural History Museum – 1954
Natural history, horticulture ... 46559

Decker, Meyerhoff and Pinkard Galleries, Maryland Institute, College of Art, 1300 W Mount Royal Ave, Baltimore, MD 21217 • T: +1 410 6699200 • whipps@mica.edu • www.mica.edu •
Fine Arts Museum ... 46560

Dr. Samuel D. Harris National Museum of Dentistry, 31 S Greene St, Baltimore, MD 21201 • T: +1 410 7068314 • F: +1 410 7068313 • rfetter@dentalmuseum.umaryland.edu • www.dentalmuseum.org •
Exec. Dir.: *Rosemary Fetter* • Cur.: *Dr. John M. Hyson* •
Special Museum – 1840
Dentistry ... 46561

Edgar Allan Poe House and Museum, 203 N Amity St, Baltimore, MD 21223 • T: +1 410 3967932 • F: +1 410 3965662 • eapo@baltimorecity.gov • www.eapoe.org •
Cur.: *Jeff Jerome* •
Special Museum – 1923
Home of Edgar Allan Poe ... 46562

Eubie Blake National Jazz Museum and Cultural Center, 847 N Howard St, Baltimore, MD 21201 • T: +1 410 2253110 • F: +1 410 2253139 • eubieblake@erols.com • www.eubieblake.org •
Pres.: *James Crokett* • Exec. Dir.: *Kenny Murphy* •
Music Museum – 1983 ... 46563

Evergreen House, Johns Hopkins University, 4545 N Charles St, Baltimore, MD 21210 • T: +1 410 5160341 • F: +1 410 5160864 • rssarnio@jhu.edu • www.jhu.edu/historichouses •
Dir.: *Robert Saarnio* •
Historical Museum – 1952
Historic House, purchased for the Garrett family in 1878 ... 46564

Fort McHenry, End of E Fort Av, Baltimore, MD 21230-5393 • T: +1 410 9624290 • F: +1 410 9622500 • fomc_superintendent@nps.gov • www.nps.gov/fomc •
Sc. Staff: *Scott Sheads* (History) •
Historic Site / Historical Museum / Military Museum – 1933
Site of bombardment which inspired Francis Scott Key to write "The Star-Spangled Banner", 1861-65 site of Union Prison Camp, 1917-25 U.S. Army General Hospital 2 ... 46565

Great Blacks in Wax Museum, 1601-03 E North Av, Baltimore, MD 21213-1409 • T: +1 410 5633404 • F: +1 410 6755040 • jmartin@greatblacksinwax.org • www.greatblacksinwax.org •
Dir.: *Dr. Joanne M. Martin* • Dep. Dir.: *Trevis Henson* •
Historical Museum / Decorative Arts Museum – 1983
History, wax figures ... 46566

Heritage Museum, Hamlet Court, 4509 Prospect Circle, Baltimore, MD 21216 • T: +1 410 6646711 • F: +1 410 6646711 • heritagemuseum@startpower.net •
Dir.: *Steven Lee* •
Historical Museum – 1991
Art from Africa and America, records, stamps, artifacts ... 46567

Homewood House Museum, Johns Hopkins University, 3400 N Charles St, Baltimore, MD 21218 • T: +1 410 5165589 • F: +1 410 5167859 • rsaarnio@jhu.edu • www.jhu.edu/historichouses •
Dir.: *Robert Saarnio* •
University Museum / Historical Museum – 1987
Furnishings, porcelain, ceramics, textiles, furniture, prints, paintings, kitchen equipments ... 46568

Intercultural Museum and Art Gallery, 128 W North Av, Baltimore, MD 21201-5813 • T: +1 410 3473288 • F: +1 410 3473298 • imag128@aol.com • www.imag128.org •
Dir.: *Jimmie Jokulo Cooper* •
Public Gallery – 1996 ... 46569

James E. Lewis Museum of Art, Morgan State University, 245 Carl Murphy Fine Arts Center, 1700 E Coldspring Ln, Baltimore, MD 21251 • T: +1 410 3193030 • F: +1 410 3194024 • gtenabe@moac.morgan.edu • www.morgan.edu •
Dir.: *Gabriel S. Tenabe* •
Fine Arts Museum / University Museum – 1955
19th and 20th c American and European sculpture, graphics, paintings, decorative arts, archaeology, African and New Guinean sculpture ... 46570

Jewish Museum of Maryland, 15 Lloyd St, Baltimore, MD 21202 • T: +1 410 7326400 • F: +1 410 7326451 • info@jewishmuseummd.org • www.jewishmuseummd.org •
Dir.: *Avi Y. Decter* • Cur.: *Melissa Martens* •
Historical Museum – 1960
Local and American Jewish hist, Jewsih art – research archives, library ... 46571

The Johns Hopkins University Archaeological Collection, 129 Gilman Hall, 34th and Charles, Baltimore, MD 21218 • T: +1 410 5168402 • F: +1 410 5165218 • emaguire@jhu.edu •
Cur.: *Dr. Betsy Bryan* (Classical, Near Eastern and Egyptian Antiquities) • *Dr. Eunice Daughterman-Maguire* •
Museum of Classical Antiquities / University Museum / Archaeology Museum – 1884
Classical art, archaeology ... 46572

The Lacrosse Museum and National Hall of Fame, 113 W University Pkwy, Baltimore, MD 21210 • T: +1 410 2356882 ext 122 • F: +1 410 3666735 • info@lacrosse.org • www.lacrosse.org •
Dir.: *Susan Koches* •
Folklore Museum – 1959
National Hall of Fame, Photographs, art, vintage equipments and uniforms, sculpture, trophies, artifacts, medals, prints ... 46573

Lovely Lane Museum, 2200 Saint Paul St, Baltimore, MD 21218 • T: +1 410 8894458 • F: +1 410 8891501 • lovinmus@bcpl.net •
Pres.: *Dennis Whitmore* • C.E.O.: *Suni Johnson* •
Religious Arts Museum – 1855
Religion, methodist history, Strawbridge shrine ... 46574

Maryland Art Place, 8 Market Pl, Ste 100, Baltimore, MD 21202 • T: +1 410 9628565 • F: +1 410 2448017 • map@mdartrplace.org • www.mdartplace.org •
Dir.: *Julie Ann Cavnor* •
Public Gallery – 1982
Temporary exhibitions ... 46575

Maryland Historical Society Museum, 201 W Monument St, Baltimore, MD 21201 • T: +1 410 6853750 • F: +1 410 3852105 • webcomments@mdhs.org • www.mdhs.org •
Local Museum – 1844
Maryland portraits and landscapes, furniture, silver, China, glass, costumes, clocks, uniforms, textiles, Chesapeake Bay maritime coll, Civil War, prints, architectural drawings ... 46576

Maryland Institute Museum, College of Art Exhibitions, 1300 Mount Royal Av, Baltimore, MD 21217 • T: +1 410 2252280 • F: +1 410 2252396 • vhecht@mica.edu • www.mica.edu •
Pres.: *Fred Lazarus* • Dir.: *Will Hipps* •
Fine Arts Museum / University Museum – 1826
George A. Lucas coll of 19th c paintings ... 46577

Maryland Science Center, 601 Light St, Baltimore, MD 21230 • T: +1 410 6855225 • F: +1 410 5455974 • whannon@mdsci.org • www.mdsci.org •
Exec. Dir.: *Gregory Paul Andorfer* •
Science&Tech Museum – 1797
Science, technology – planetarium ... 46578

Meredith Gallery, 805 N Charles St, Baltimore, MD 21201 • T: +1 410 8373575 • F: +1 410 8373577 • www.meredithgallery.com •
Dir.: *Judith Lippman* •
Decorative Arts Museum – 1977 ... 46579

Mount Clare Museum House, 1500 Washington Blvd, Baltimore, MD 21230 • T: +1 410 8373262 • F: +1 410 8370251 • Mountclaremuseum@aol.com • www.mountclare.org •
Dir.: *Jane D. Woltereck* •
Historical Museum / Decorative Arts Museum – 1917
18th c and earley 19th c silver, furniture, china, glassware ... 46580

Mount Vernon Museum of Incandescent Lighting, 717 Washington Pl, Baltimore, MD 21201 • T: +1 410 7528586 •
Dir.: *Hugh Francis Hicks* •
Science&Tech Museum – 1963
Science, technology, development of the electronic light bulb (1878-now) ... 46581

Museum of Maryland History, Maryland Historical Society, 201 W Monument St, Baltimore, MD 21201 • T: +1 410 6853750 • F: +1 410 3852105 • mray@mdhs.org • www.mdhs.org •
Dir.: *Dennis A. Fiori* • Cur.: *Nancy Davis* •
Local Museum – 1844
Regional history – library ... 46582

National Museum of Ceramic Art and Glass, 2406 Shelleydale Dr, Baltimore, MD 21209 • T: +1 410 7641042 • F: +1 410 7641042 •
Pres.: *Richard Taylor* •
Decorative Arts Museum – 1989
Ceramic art, glass ... 46583

Port Discovery Children's Museum, 35 Market Place, Baltimore, MD 21202 • T: +1 410 7278120 • F: 7273042 • info@portdiscovery.org • www.portdiscovery.org •
Special Museum – 1990 ... 46584

Port Discovery - Children's Museum in Baltimore, 35 Market Pl, Ste 905, Baltimore, MD 21202 • T: +1 410 8642700, 7278120 • F: +1 410 7273042 • info@portdiscovery.org • www.portdiscovery.org •
Pres./C.E.O.: *Bryn Parchman* •
Ethnology Museum – 1977 ... 46585

Reginald F. Lewis Museum of Maryland African American History and Culture, 830 E Pratt St, Baltimore, MD 21202 • T: +1 443 2631800 • F: +1 410 3331138 • emailus@maamc.org • www.africanamericanculture.org •
Dir.: *David Taft Terry* •
Historical Museum – 1998
Resource center, oral history, folk art ... 46586

Rosenberg Gallery, Goucher College, 1021 Dulaney Valley Rd, Baltimore, MD 21204 • T: +1 410 3376073 • F: +1 410 3376405 • lburns@goucher.edu • www.goucher.edu/rosenberg •
Dir.: *Laura Burns* •
University Museum / Fine Arts Museum – 1885
Contemporary and modern art, photography, prints, Asian art, Mexican ceramics ... 46587

Star-Spangled Banner Flag House and 1812 Museum, 844 E Pratt St, Baltimore, MD 21202 • T: +1 410 8371793 • F: +1 410 8371812 • info@flaghouse.org • www.flaghouse.org •
Dir.: *Sally Johnston* •
Historical Museum – 1927
Home of Mary Pickersgill, maker of the 30'x42' banner which flew over Fort McHenry during the War of 1812 ... 46588

Steamship Collection, University of Baltimore, Library, 1420 Maryland Av, Baltimore, MD 21201-5779 • T: +1 410 8374334 • F: +1 410 8374330 • ghaitsuka@ubmail.ubalt.edu •
Pres.: *Timothy J. Dacey* •
University Museum – 1940
Steamship and navigation-related items ... 46589

USS Constellation Museum, Pier 1, 301 E Pratt St, Baltimore, MD 21202-3134 • T: +1 410 5391797 • F: +1 410 5396238 • www.constellation.org •
Exec. Dir.: *Christopher Rowsom* • Cur.: *John Pentangelo* •
Military Museum – 1999
Civil War, US naval hist ... 46590

Walters Art Museum, 600 N Charles St, Baltimore, MD 21201-5185 • T: +1 410 5479000 • F: +1 410 7837969, 7524797 • info@thewalters.org • www.thewalters.org •
Dir.: *Dr. Gary Vikan* • Cur.: *Joaneath Spicer* (Renaissance/ Baroque Art) •
Fine Arts Museum / Museum of Classical Antiquities – 1931
Decorative arts, paintings, sculpture, arms and armor, jewelry, manuscripts, rare books ... 46591

Bancroft NE

John G. Neihardt Center, Nebraska State Historical Society, 306 W Elm and Washington, Bancroft, NE 68004 • T: +1 402 6483388 • F: +1 402 6483388 • neihardt@gpcom.net • www.neihardt.com •
Dir.: *Nancy S. Gillis* •
Historical Museum – 1976
Literature – library ... 46592

Bandera TX

Frontier Times Museum, 510 31th St, Bandera, TX 78003 • T: +1 830 7963864 • www.frontiertimesmuseum.org •
Dir.: *Bob Perry* •
Historical Museum – 1933
Texas' early pioneer days, western paintings, antiques, photographs ... 46593

Bandon OR

Coquville River Museum, Bandon Historical Society, 270 Fillmore & Hwy 101, Bandon, OR 97411 • T: +1 541 3472164 • F: +1 541 3472164 • bandonhistoricalmuseum@yahoo.com • bandonhistoricalmuseum.org •
Dir.: *Judy Knox* • Cur.: *Reg Pullen* •
Historical Museum – 1977
Native American artifacts, pioneer family artifacts, natural history, Bandon's 1914 and 1936 fires, farming, industry, fishing, timber ... 46594

Bangor ME

Bangor Historical Society Museum, 159 Union St, Bangor, ME 04401 • T: +1 207 9425766 • F: +1 207 9410266 • info@bangormuseum.org • www.bangormuseum.org •
C.E.O: *Linda Jaffe* •
Local Museum – 1864
Local history ... 46595

Cole Land Transportation Museum, 405 Perry Rd, Bangor, ME 04401 • T: +1 207 9903600 • F: +1 207 9902653 • mail@colemuseum.com • www.colemuseum.org •
Pres.: *Garrett Cole* •
Science&Tech Museum – 1990
Vehicle hist, World War II veternas memorial ... 46596

Isaac Farrar Mansion, 17 Second St, Bangor, ME 04401 • T: +1 207 9412808 • F: +1 207 9412812 •
Dir.: *Lynda Clyve* • *Peggy Wentworth* •
Local Museum – 1972
History of the 19th c Bangor, paintings, period furnishings, ... 46597

Orono Museum of Art, University of Maine, 40 Harlow St, Bangor, ME 04401-5102 • T: +1 207 5613350 • F: +1 207 5613351 • umma@umit.maine.edu • www.umma.umaine.edu •
Dir.: *Wally Mason* •
Fine Arts Museum
Coll of American and European prints and paintings ... 46598

Banning CA

Malki Museum, 11795 Fields Rd, Morongo Indian Reservation, Banning, CA 92220 • T: +1 951 8497289 • F: +1 951 8493549 • MalkiMuseumMail@gmail.com • www.malkimuseum.org •
Pres.: *Katherine Saubel* •
Historical Museum / Ethnology Museum / Folklore Museum – 1964
Indian tribe artifacts e.g. Cahuilla, anthropology, archaeology and ethnology – ethno-botanical garden ... 46599

Bar Harbor ME

Abbe Museum at Downtown, 26 Mount Desert St, Bar Harbor, ME 04609 • T: +1 207 2883519 • F: +1 207 2888979 • info@abbemuseum.org • www.abbemuseum.org •
Dir.: *Diane Kopec* • Cur.: *Rebecca Cole-Will* •
Ethnology Museum / Archaeology Museum – 2001
Archaeology, anthropology, ethnology ... 46600

U

Abbe Museum at Sieur de Monts Spring, Sieur de Monts Spring, Acadia National Park, Bar Harbor, ME 04609 • T: +1 207 2883519 • F: +1 207 2888979 • abbe@midmaine.com • www.abbemuseum.org •
Ethnology Museum / Archaeology Museum – 1928
Archaeology, anthropology, ethnology ... 46601

Bar Harbor Historical Society Museum, 33 Ledgelawn Av, Bar Harbor, ME 04609 • T: +1 207 2883807, 2880000 • bhhistorical@acadia.net • www.barharborhistorical.org •
Cur.: *Deborah Dyer* •
Local Museum – 1946
Local history ... 46602

George B. Dorr Museum of Natural History, College of the Atlantic, 105 Eden St, Bar Harbor, ME 04609 • T: +1 207 2885015 • F: +1 207 2882917 • museum@ecology.coa.edu •
Dir.: *Dr. Stephen Katona* • Cur.: *Dianne Clendaniel* (Program) •
Natural History Museum – 1982
Natural history ... 46603

Baraboo WI

Circus World Museum, 550 Water St, Baraboo, WI 53913 • T: +1 608 3568341 • F: +1 608 3561800 • ringmaster@circusworldmuseum.com • www.circusworldmuseum.com •
C.E.O.: *Lawrence A. Fisher* •
Performing Arts Museum – 1959
Circus artifacts, wagons, railroad cars, archives, site of 1884-1918 original winter quarters of Ringling Bros. ... 46604

International Crane Foundation Museum, E 11376 Shady Lane Rd, Baraboo, WI 53913-0447 • T: +1 608 3569462 • F: +1 608 3569465 • cranes@savingcranes.org • www.savingcranes.org •
Pres.: *James T. Harris* • Sc. Staff: *Jeb Barzen* (Ecology) • Cur.: *Michael Putnam* •
Natural History Museum – 1973
Ornithology ... 46605

Sauk County Historical Museum, 531 4th Av, Baraboo, WI 53913 • T: +1 608 3561001 • history@saukcounty.com • www.saukcounty.com/schs •
Dir.: *Peter Shrake* • Cur.: *Mary Farrell Stieve* •
Local Museum – 1906
Local hist, Indian artifacts, circus mementoes ... 46606

Bardstown KY

My Old Kentucky Home, 501 E Stephen Foster Av, Bardstown, KY 40004 • T: +1 502 3483502, 3237803 • F: +1 502 3490054 • www.kystateparks.com •
Head: *Alice Heaton* •
Special Museum – 1922
Home of Judge John Rowan, where Stephen Foster wrote My Old Kentucky Home ... 46607

Oscar Getz Museum of Whiskey History, Bardstown Historical Museum, 114 N 5th St, Bardstown, KY 40004-1402 • T: +1 502 3482999 • F: +1 502 3482999 • bourbonmuseum@cube3.net •
Cur.: *Mary Ellin Hamilton* •
Local Museum – 1984
Memorabilia & artifacts pertaining to the hist of the American whiskey industry since its beginnings in the mid-18th c, through the Prohibition Era ... 46608

Barkers Island WI

S.S. Meteor Maritime Museum, 300 Marina Dr, Barkers Island, WI 54880 • T: +1 715 3925742, 3945712 • F: +1 715 3943810 • www.superiorpublicmuseums.org •
C.E.O.: *Susan K. Anderson* •
Science&Tech Museum – 1973
Whaleback, boat models, ship equip and building hist ... 46609

Barnesville OH

Belmont County Victorian Mansion, 532 N Chestnut St, Barnesville, OH 43713 • T: +1 740 4252926 •
C.E.O. & Pres.: *Lester R. Milton* • Dir.: *Jane Barker* •
Historical Museum – 1966
Glass, china, furniture, clothing, Native American artifacts ... 46610

Barnet VT

Barnet Historical Society, Goodwille House, Barnet, VT 05821 • T: +1 802 6332611, 2563 •
Pres.: *Hazel McLaren* •
Association with Coll – 1967
Early pictures of the area, household & farm equipment, store ledgers ... 46611

Barnstable MA

Donald G. Trayser Memorial Museum, 3353 Main St, Barnstable, MA 02630, mail addr: 200 Main St, Hyannis, MA 02601 • T: +1 617 3622092 •
Local Museum – 1960
Local history, old Customs House, marine, ship models, Indian artifacts ... 46612

Olde Colonial Courthouse, Rendevouz Ln and Rte 6a, Barnstable, MA 02630 • T: +1 508 3629056 • F: +1 508 3629056 •
Pres.: *Carol Clarke Di Vico* •
Historical Museum – 1949
Sachem Iyanough's gravesite dedicated to early Indians who befriended Pilgrims ... 46613

Barnwell SC

Barnwell County Museum, Hagood and Marlboro Aves, Barnwell, SC 29812 • T: +1 803 2591916, 2593277 • F: +1 803 2591916 • barnwellmuseum@barnwellsc.com •
Cur.: *Pauline Zidlick* •
Local Museum – 1978
Local hist ... 46614

Barre MA

Barre Historical Museum, 18 Common St, Barre, MA 01005 • T: +1 978 3554067 •
Cur.: *Bertyne Smith* •
Local Museum – 1955
Local history, first parish records, United Methodist church ... 46615

Barrington IL

Barrington Area Historical Museum, 212 W Main St, Barrington, IL 60010 • T: +1 708 3811730 • F: +1 708 3811766 •
Exec. Dir.: *Michael J. Harkins* •
Local Museum – 1969
Local history, housed in two Folk Victorian Houses ... 46616

Bartlesville OK

Bartlesville Area History Museum, 401 S Johnstone, Bartlesville, OK 74005 • T: +1 918 3375336 • F: +1 918 3375338 • kkswoods@bartlesville.org • www.cityofbartlesville.org •
Dir. & Cur.: *Karen Smith Woods* •
Local Museum – 1964
Early settlers and Native Indian tribes ... 46617

Frank Phillips Home, Oklahoma Historical Society, 1107 S Cherokee, Bartlesville, OK 74003 • T: +1 918 3362491 • F: +1 918 3363529 • fphillipshome@ok-history.mus.ok.us •
C.E.O.: *Dr. Bob Blackburn* • Cur.: *Susan J. Lacey* •
Local Museum – 1973
1930s furniture ... 46618

Price Tower Arts Center, 510 S Dewey, Bartlesville, OK 74003 • T: +1 918 3364949 • F: +1 918 3367117 • onfo@pricetower.org • www.pricetower.org •
Exec. Dir.: *Richard P. Townsend* •
Fine Arts Museum / Decorative Arts Museum – 1985
Frank Lloyd Wright and Bruce Goff, architectural and design objects 20th and 21st c ... 46619

Woolaroc Museum, State Hwy 123, Bartlesville, OK 74003 • T: +1 918 3360307 ext 10 • F: +1 918 3360084 • woolaroc1@aol.com • www.woolaroc.org •
Interim Dir. & Cur. Coll.: *Kenneth Meek* • Cur.: *Linda Stone* (Art) • *Kenneth Meek* (Collections) •
Fine Arts Museum / Historical Museum – 1929
Indian artifacts, paintings, sculpture, ethnology, archaeology, anthropology, gun coll ... 46620

Barton VT

Crystal Lake Falls Historical Association, 97 Water St, Barton, VT 05822 • T: +1 802 5253583, 5256251 • F: +1 802 5253583 • jbrown@kingcon.com •
Pres.: *Avis Harper* •
Association with Coll – 1984
Artifacts from education in Barton ... 46621

Bartow FL

Polk County Historical Museum, 100 E Main St, Bartow, FL 33830 • T: +1 863 5344386 • F: +1 863 5344387 • museum@polk-county.net • www.polk-county.net/museum •
Dir./Cur.: *Brenda McLain* •
Local Museum – 1998
Political hist of Polk County, prehistory, agriculture, industrial hist ... 46622

Basking Ridge NJ

Environmental Education Center, 190 Lord Stirling Rd, Basking Ridge, NJ 07920 • T: +1 908 7662489 • F: +1 908 7662687 • cschrein@parks.co.somerset.nj.us • www.park.co.somerset.nj.us •
Head: *Catherine Schrein* •
Natural History Museum – 1970
Natural hist ... 46623

Bastrop LA

Snyder Museum and Creative Arts Center, 1620 E Madison Av, Bastrop, LA 71220 • T: +1 318 2818760 •
Pres.: *Madeline Herring* • Dir.: *Helen T. Landress* •
Fine Arts Museum / Local Museum – 1974
Local history, art ... 46624

Batavia IL

Batavia Depot Museum, 155 Houston, Batavia, IL 60510 • T: +1 630 4065253 • F: +1 630 8799537 • carlah@batpkdist.org • www.bataviahistoricalsociety.org •
Dir.: *Carla Hill* •
Local Museum – 1960
Local history, housed in 1854 Batavia Depot, one of the oldest railroad station on the Burlington Line ... 46625

Batavia NY

Holland Land Office Museum, 131 W Main St, Batavia, NY 14020 • T: +1 716 3434727 • F: +1 716 3450023 • info@hollandoffice.com • www.hollandoffice.com •
Pres.: *Don Burkel* • Cur.: *Patrick Weissend* •
Local Museum – 1894
Medical and military coll, glass and china, early furniture, musical instruments ... 46626

Batesville AR

Old Independence Regional Museum, 380 S Ninth St, Batesville, AR 72501 • T: +1 870 7932121 • F: +1 870 7932101 • oirm@oirm.org • www.oirm.org •
Cur.: *Twyla Wright* •
Local Museum – 1994
Regional history of North Central Arkansas ... 46627

Bath ME

Maine Maritime Museum, 243 Washington St, Bath, ME 04530 • T: +1 207 4431316 • F: +1 207 4431665 • drumm@bathmaine.com • www.bathmaine.com/ •
Dir.: *Thomas R. Wilcox* • Cur.: *Nathan Lipfert* •
Historical Museum – 1964
Maritime, historic shipyard ... 46628

Bath NC

Historic Bath, 207 Carteret St, Bath, NC 27808 • T: +1 252 9233971 • F: +1 252 9230174 • bath@ncmail.net • www.bath.nchistoricsites.org •
C.E.O.: *Patricia Samford* •
Historic Site – 1963 ... 46629

Bath OH

Hale Farm and Village, 2686 Oak Hill Rd, Bath, OH 44210 • T: +1 330 6663711 • F: +1 330 6669497 • mtramontine@wrhs.org • www.wrhs.org •
C.E.O.: *Margaret Tromantine* •
Open Air Museum / Agriculture Museum – 1957
Agriculture, crafts ... 46630

Baton Rouge LA

Baton Rouge Gallery, 1442 City Park Av, Baton Rouge, LA 70808-1037 • T: +1 225 3831470 • F: +1 225 3360943 •
Dir.: *Kathleen Pheney* •
Fine Arts Museum – 1966
Contemporary art ... 46631

Enchanted Mansion Doll Museum, 190 Lee Dr, Baton Rouge, LA 70808 • T: +1 225 7690005 • F: +1 225 7666822 • termansion@bellsouth.net • www.enchantedmansion.org •
CEO: *Rosemary Sedberry* •
Decorative Arts Museum – 1995
Doll coll, Victorian doll house ... 46632

Louisiana Arts and Science Center, 100 S River Rd, Baton Rouge, LA 70802 • T: +1 225 3445272, 3449478 • F: +1 225 3449477 • lasm@lasm.org • www.lasm.org •
Dir.: *Carol Sommerfeldt Gikas* • Ass. Dir.: *Sam Losavio* • Cur.: *Elizabeth Chubbuck Weinstein* •
Fine Arts Museum / Science&Tech Museum – 1960
Art, science, former Illinois Central Railroad Station ... 46633

Louisiana Naval War Memorial U.S.S. Kidd, 305 S River Rd, Baton Rouge, LA 70802 • T: +1 225 3421942 • F: +1 225 3422039 • kidd661@aol.com • www.usskidd.com •
Dir./Cur.: *H. Maury Drummond* •
Military Museum – 1981
Maritime, historic ship ... 46634

Louisiana Old State Capitol, Center for Political and Governmental History, 100 North Blvd, Baton Rouge, LA 70801 • T: +1 225 3420500 • F: +1 225 3420316 • osc@sec.state.la.us • www.sec.state.la.us •
Dir.: *Mary Louise Prudhomme* • Ass. Dir.: *Sailor Jackson* (Film, Video) • Cur.: *Dr. Florent Hardy* •
Historical Museum – 1990
Louisiana State history ... 46635

Louisiana State University Museum of Art, Memorial Tower, Baton Rouge, LA 70803 • T: +1 225 5784003 • F: +1 225 5789288 • indiana@lsu.edu • www.lsumoa.com •
Exec. Dir.: *Dr. Laura F. Lindsay* •
Fine Arts Museum / University Museum – 1959
Decorative arts, sculpture, drawings, paintings, ceramics, silver coll, Charels E .Craig jun. coll, ... 46636

Magnolia Mound Plantation, 2161 Nicholson Dr, Baton Rouge, LA 70802 • T: +1 504 3434955 • F: +1 504 3436739 • marketmmp@aol.com • magnoliamound.org •
Pres.: *Suzette Tannehill* • Dir.: *Dorinda Hilbun* •
Open Air Museum / Local Museum – 1968
Plantation house and outbuildings ... 46637

Museum of Natural Science, 119 Foster Hall, LSU, Baton Rouge, LA 70803 • T: +1 225 5782855 • F: +1 225 5783075 • namark@lsu.edu • www.museum.lsu.edu •
Dir.: *Dr. Mark S. Hafner* (Mammals) • Cur.: *Dr. J. Michael Fitzsimons* (Fishes) • *Dr. J.V. Remsen* (Birds) • *Dr. Jim A. McGuire* (Reptiles) •
University Museum / Natural History Museum – 1936
Paleontology, anthropology, mammals, birds, fishes, reptiles ... 46638

Robert A. Bogan Fire Museum, 427 Laurel St, Baton Rouge, LA 70801 • T: +1 225 3448558 • F: +1 225 3447777 • gnadler@acgbr.com •
Pres.: *Hank Saurage* • Dir.: *Genny Nadler Thomas* •
Historical Museum – 1924
Fire fighting ... 46639

Rural Life Museum and Windrush Gardens, 4600 Essen Ln, Baton Rouge, LA 70809 • T: +1 225 7652437 • F: +1 225 7652639 • rulife1@lsu.edu • rurallife.lsu.edu/ •
Dir.: *David Floyd* •
Local Museum – 1970
Local hist ... 46640

School of Art Gallery, c/o Louisiana State University, 123 Art Bldg, Baton Rouge, LA 70803-0001 • T: +1 225 5785411 • F: +1 225 5785424 • mdaughi@isu.edu •
Dir.: *Michael Crespo* •
Fine Arts Museum / University Museum – 1934
Contemporary graphic works, prints, drawings ... 46641

Union Art Gallery, c/o Louisiana State University, Nicholson Dr, Baton Rouge, LA 70803 • T: +1 225 3885162, 3885117 • www.lib.lsu.edu •
Dir.: *Judith R. Stahl* •
Fine Arts Museum / University Museum – 1964 ... 46642

Battle Creek MI

Art Center of Battle Creek, 265 E Emmett St, Battle Creek, MI 49017 • T: +1 616 9629511 • F: +1 616 9693838 • artcenterofbc@yahoo.com •
Exec. Dir.: *Linda Holderbaum* •
Public Gallery – 1948
Art gallery of Michigan artists ... 46643

Kimball House Museum, 196 Capital Av NE, Battle Creek, MI 49017 • T: +1 616 9652613 • F: +1 616 9662495 • bchist@net-link.net • www.kimballhouse.org •
CEO: *Michael Evans* • Sc. Staff: *Mary Butler* (Research) •
Local Museum – 1966
83-year tradition of medical service (3 generations Kimball), Sojourner Truth items, cereal pioneers C.W. Post and W.K. Kellogg ... 46644

Kingman Museum of Natural History, W Michigan Av at 20th St, Battle Creek, MI 49017 • T: +1 616 9655117 • F: +1 616 9625610 •
Cur.: *D. Thomas Johnson* (Exhibits) • *Kathy Ward* (Collections) •
Natural History Museum – 1869
Natural history ... 46645

Battle Ground IN

Museum at Prophetstown, 3549 Prophetstown Tr, Battle Ground, IN 47920, mail addr: POB 331, Battle Ground, IN 47920 • T: +1 765 5674700 • F: +1 765 5674736 • farmdirector@prophetstown.org • www.prophetstown.org •
Pres.: *Nola Gentry* •
Local Museum / Agriculture Museum – 1996
Agriculture, personal artifacts, art, botany ... 46646

Tippecanoe Battlefield, 200 Battle Ground Av, Battle Ground, IN 47920 • T: +1 765 5672147 • F: +1 765 5672149 • info@tcha.mus.in.us • www.tcha.mus.in.us •
Pres.: *George Hanna* •
Historic Site – 1972
Kethtppecanuk, a lonf forgotten settlement, Tippecanoe and Tyler, Two ... 46647

Baudette MN

Lake of the Woods County Museum, 119 Eighth Av SE, Baudette, MN 56623 • T: +1 218 6341200 • lowhsociety@wiklel.com •
Cur.: *Marlys Hirst* •
Local Museum – 1978
County history – library, archives ... 46648

Baxter Springs KS

Baxter Springs Heritage Center and Museum, 740 East Av, Baxter Springs, KS 66713 • T: +1 620 8562385 • heritagectr@4state.com • home.4state.com/~heritagectr/ •
Cur.: *Phyliss Abbott* •
Local Museum / Military Museum – 1962
Civil War coll, Oct 1863 Baxter Massacre of General Blunt's men, quilt coll, mining ... 46649

Bay City MI

Historical Museum of Bay County, 321 Washington Av, Bay City, MI 48708 • T: +1 989 8935733 • F: +1 989 8935741 • bchsco@mindspring.com • www.bchsmuseum.org •
Dir.: *Gay McInerney* •
Historical Museum – 1919
County, Michigan & Great Lakes history, sugar beet industry – library ... 46650

Bay City TX

Matagorda County Museum, 2100 Av F, Bay City, TX 77414-0851 • T: +1 979 2457502 • F: +1 979 2451233 • mcma@matagordacountmyuseum.org • www.matagordacountymuseum.org •
C.E.O.: *Sarah Higgins* •
Local Museum – 1965
Time of the Karankawa Indians, Civil War, cattle ranching into the 20th c ... 46651

U

Bay View MI

Bay View Historical Museum, Bay View Association Encampment 1715, Bay View, MI 49770 • T: +1 231 3476225, 3478182 • F: +1 231 3474337 • bvarod@freeway.net • www.bayviewassoc.com •
Cur.: *William D. Green* •
Local Museum – 1970
Local hist ... 46652

Bay Village OH

Baycrafters, 28795 Lake Rd, Bay Village, OH 44140 • T: +1 440 8716543 • F: +1 440 8710452 • info@baycrafters.com • www.baycrafters.com •
Dir.: *Sally Irwin Price* •
Fine Arts Museum / Science&Tech Museum – 1948
Coll of paintings, Nickel Plate Railroad Station, Norfolk and Western Caboose ... 46653

Lake Erie Nature and Science Center, 28728 Wolf Rd, Bay Village, OH 44140 • T: +1 440 8712900 • F: +1 440 8712901 • larry@lensc.org • www.lensc.org •
Dir.: *Larry D. Richardson* •
Natural History Museum – 1945
Astronomy, botany, geology, physical science, zoology, ecology ... 46654

Rose Hill Museum, 27715 Lake Rd, Bay Village, OH 44140 • T: +1 440 8717338 • mail@bayhistorical.com • www.bayhistorical.com •
Pres.: *Eric Eakin* •
Historical Museum – 1960
Clothing, children's toys, furniture ... 46655

Bayfield WI

Apostles Island National Lakeshore Maritime Museum, 415 W Washington Av, Bayfield, WI 54814 • T: +1 715 7797007 • F: +1 715 7793049 • susan_mackreth@nps.gov • www.nps.gov/apis •
Special Museum – 1970
Hist buildings, fishery ... 46656

Bayside NY

QCC Art Gallery, 222-05 56th Av, Bayside, NY 11364-1497 • T: +1 718 6316396 • F: +1 718 6316620 • qccgallery@aol.com • www.qcc.cuny.edu/ArtGallery •
Dir.: *Faustino Quintanilla* •
University Museum / Fine Arts Museum – 1966
Role of art in cultural hist of people, American artists after 1950 ... 46657

Beachwood OH

Temple Museum of Religious Art, 26000 Shaker Blvd, Beachwood, OH 44122 • T: +1 216 8313233 ext 108 • F: +1 216 8314216 • makeck@hotmail.com • www.ttti.org/museum.asp •
Dir.: *Susan Koletsky* •
Religious Arts Museum – 1950
Antique Torah hangings, Holyland pottery, Israel stamps, paintings, sculpture, stained glass, Torah ornaments ... 46658

Beacon NY

Beacon Historical Society Museum, 477 Main St, Beacon, NY 12508, mail addr: POB 89, Beacon, NY 12508 • T: +1 914 8311877 •
Cur.: *Joan Van Voorhis* •
Local Museum – 1976 ... 46659

Dia:Beacon, Riggio Galleries, Dia Art Foundation, 3 Beekman St, Beacon, NY 12508 • T: +1 845 4400100 • F: +1 845 4400092 • info@diaart.org • www.diaart.org •
Dir.: *Michael Govan* • Cur.: *Lynne Cooke* • Sc. Staff: *Patrick Heilman* (Digital Media) • *Laura Fields* (Graphics) • *Steven Evans* (Gallery) •
Fine Arts Museum – 1974
Contemporary American/European art ... 46660

The Madam Brett Homestead, 50 Van Nydeck Av, Beacon, NY 12508 • T: +1 845 8316533 •
C.E.O.: *Pamela Barrack* • Cur.: *Linda Shedd* •
Local Museum – 1954
Original furnishings of Brett family, doll coll, local hist ... 46661

Bear Mountain NY

Bear Mountain Trailside Museums Wildlife Center, Bear Mountain State Park, Bear Mountain, NY 10911 • T: +1 845 7862701 ext 263 • F: +1 845 7867157 •
Dir.: *John Focht* •
Natural History Museum – 1927
Local coll of animals, plants, minerals and rocks, Indian artifacts, ... 46662

Beatrice NE

Gage County Historical Museum, 101 N Second, Beatrice, NE 68310 • T: +1 402 2281679 • gagecountymuseum@beatricene.com • www.beatricene.com/gagecountymuseum •
Dir.: *Lesa Arterburn* • Ass. Dir./Cur.: *Rita Clawson* •
Historical Museum – 1971
Railroad artifacts, agriculture implements, medical equipments ... 46663

Homestead National Monument of America, 8523 W State Hwy, Beatrice, NE 68310 • T: +1 402 2233514 • F: +1 402 2284231 • home_interpretation@nps.gov • www.nps.gov •
Agriculture Museum – 1936
Agriculture hist, paintings ... 46664

Beaufort NC

Beaufort Historic Site, 138 Turner St, Beaufort, NC 28516 • T: +1 252 7285225 • F: +1 252 7284966 • bha@clis.com • www.historicbeaufort.com •
Exec. Dir.: *Patricia Suggs* •
Local Museum – 1960
Hist buildings, period furnishings ... 46665

North Carolina Maritime Museum, 315 Front St, Beaufort, NC 28516 • T: +1 252 7287317 • F: +1 252 7282108 • maritime@ncmail.net • www.ah.dcr.state.nc.us/sections/maritime •
Dir.: *Dr. George Ward Shannon* • Cur.: *Jeannie W. Kraus* • Sc. Staff: *Patricia Hay* (Natural Science) • *David Moore* (Nautic Archaeology) • *Paul Fontenoy* (Watercraft and Maritime Research) • *Jerry Heiser* (Exhibits) • *JoAnne Powell* (Education) •
Natural History Museum / Science&Tech Museum – 1975
Ship models, sea shells, marine artifacts ... 46666

Beaufort SC

Beaufort Museum, 713 Craven St, Beaufort, SC 29902 • T: +1 843 5257077, 5257005 • F: +1 843 3793331 • wedlerbf@hargray.com •
Cur.: *Jacque Wedler* •
Fine Arts Museum / Local Museum – 1939
Local hist ... 46667

University of South Carolina Beaufort Art Gallery, 801 Carteret St, Beaufort, SC 29902 • T: +1 843 5214145 • info@beaufortarts.com • www.beaufortarts.com •
Exec. Dir.: *Eric V. Holowacz* •
University Museum / Fine Arts Museum – 1990 . 46668

The Verdier House, 801 Bay St, Beaufort, SC 29902 • T: +1 843 3793331 • F: +1 843 3793371 • histbft@hargray.com • www.historic-beaufort.org •
Exec. Dir.: *Jefferson Mansell* •
Local Museum – 1977
Historic house ... 46669

Beaumont TX

Art Museum of Southeast Texas, 500 Main St, Beaumont, TX 77701 • T: +1 409 8323432 • F: +1 409 8328508 • info@amset.org • www.amset.org •
Exec. Dir.: *Lynn P. Castle* • Cur.: *Sandra Laurette* (Education) •
Fine Arts Museum – 1950
19th-21st c American paintings, sculpture, photography & folk art ... 46670

The Art Studio, 720 Franklin St, Beaumont, TX 77701-4424 • T: +1 409 8385393 • F: +1 409 8384695 • artstudio@artstudio.org • www.artstudio.org •
Dir.: *Greg Busceme* • Ass. Dir.: *Tracy Danna* •
Public Gallery – 1983
Bob Willis Memorial Ceramic coll, Memorial Art Library ... 46671

Babe Didrikson Zaharias Museum, I-10 W, MLK Blvd, Exit 854, Beaumont, TX 77701 • T: +1 409 8803749, 8334622 • F: +1 409 8803750 • klewis@ci.beaumont.tx.us • www.beaumontcvb.com •
Cur.: *Karen Lewis* •
Special Museum – 1976
Sports trophies, golf clubs, Olympic medals ... 46672

Beaumont Art League, 2675 Gulf St, Fairgrounds, Beaumont, TX 77703 • T: +1 409 8334179 • F: +1 409 8321563 • abarr91511@aol.com •
Pres.: *Frank Gerrietts* •
Fine Arts Museum – 1943
Art ... 46673

Dishman Art Gallery, 1030 E Lavaca, Beaumont, TX 77705 • T: +1 409 8808959 • F: +1 409 8801799 • alicia.hargreaves@lamar.edu •
Dir.: *Dr. Lynne Lokensgard* •
Fine Arts Museum / Public Gallery – 1983
Art ... 46674

Edison Museum, 350 Pine St, Beaumont, TX 77701 • T: +1 409 9813089 • F: +1 409 8382361 • info@edisonmuseum.org • www.edisonmuseum.org •
Chm.: *Michael Barnhill* •
Local Museum – 1980
Focus is on inventions & innovations of Thomas A. Edison ... 46675

Fire Museum of Texas, 400 Walnut at Mulberry, Beaumont, TX 77701 • T: +1 409 8803927 • F: +1 409 8803914 • firemuseum@ci.beaumont.tx.us • www.firemuseumoftexas.org •
Pres.: *Allison Golias* •
Science&Tech Museum – 1986
Firefighting services in Texas, international firefighting memorabilia, fire pumper ... 46676

John Jay French House, 3025 French Rd, Beaumont, TX 77706 • T: +1 409 8980348 • F: +1 409 8988487 • jjfrench@sbcglobal.net • www.jjfrench.com •
Dir.: *Nell Truman* • Ass. Dir.: *Carol White* •
Local Museum – 1968
Decorative arts, Pre-Civil War era Furnishing & accessories ... 46677

McFaddin-Ward House, 1906 McFaddin Av, Beaumont, TX 77701 • T: +1 409 8321906 • F: +1 409 8323483 • info@mcfaddin-ward.org • www.mcfaddin-ward.org •
Dir.: *Metthew White* • Cur.: *Nathan Campbell* •
Local Museum – 1982
American factory-made furniture, arts and crafts, furniture and decorative arts, oriental rugs, paintings ... 46678

Spindletop and Gladys City Boomtown Museum, Hwy 69 at University Dr, Beaumont, TX 77705 • T: +1 409 8350823 • F: +1 409 8389107 • marinocc@hal.lamar.edu • www.spindletop.org •
Cur.: *Christy Marino* •
Local Museum – 1975
Oilfield machinery and tools, furniture, textiles, glass items ... 46679

Texas Energy Museum, 600 Main St, Beaumont, TX 77701 • T: +1 409 8335100 • F: +1 409 8334282 • smithtem@msn.com • www.texasenergymuseum.org •
Dir.: *D. Ryan Smith* •
Science&Tech Museum – 1987
Petroleum geology and technology, hist of the petroleum industry, especially relating to southeast Texas .. 46680

Beaver Dam WI

Dodge County Historical Society Museum, 105 Park Av, Beaver Dam, WI 53916 • T: +1 920 8871266 • dchs@powercom.net •
Pres.: *Roger Noll* • Cur.: *Thom Neuman* •
Local Museum – 1938
Local hist, arrowheads ... 46681

Beaver Island MI

Beaver Island Historical Museum, Old Mormon Print Shop Museum, Marine Museum and Protar Home, 26275 Main St, Beaver Island, MI 49782 • T: +1 616 4482254 • F: +1 616 4482106 • news@beaverisland.net • www.beaverisland.net/history •
Dir.: *William Cashman* •
Local Museum – 1957
Local history, Irish immigration, fishing, James J. Strang – archives ... 46682

Beaverdam VA

Scotchtown, 16120 Chiswell Ln, Beaverdam, VA 23015 • T: +1 804 2273500 • F: +1 804 2273559 • scotchtown@apva.org •
C.E.O.: *Susan Nepomuceno* •
Decorative Arts Museum – 1958
18th c furnishings, Henry Map table ... 46683

Becker MN

Sherburne County Historical Museum, 13122 First St, Becker, MN 55308 • T: +1 763 2614433 • F: +1 763 2614437 • schs@sherbtel.net • www.rootsweb.com/~mnschs •
Exec. Dir.: *Kurt K. Kragness* •
Local Museum – 1972
Regional history, farming, costumes, decorative arts – library ... 46684

Beckley WV

Youth Museum of Southern West Virginia & Beckley Exhibition Coal Mine, New River Park, Beckley, WV 25802 • T: +1 304 2523730 • F: +1 304 2523764 • ymswv@citynet.net • beckleymine.com •
Exec. Dir.: *Sandi Parker* •
Science&Tech Museum – 1977 ... 46685

Bedford IN

Antique Auto and Race Car Museum, 3348 16th St, Bedford, IN 47421 • T: +1 812 2750556 • F: +1 812 2791916 • retail@autoracemuseum.com • www.autoracemuseum.com •
Cur.: *Eddie L. Evans* •
Science&Tech Museum – 1994
Antique, race and special interest cars ... 46686

Lawrence County Historical Museum, 931 15th St, Bedford, IN 47421 • T: +1 812 2788575 • F: +1 812 2788583 • lchgs@hpcisp.com •
Cur.: *Helen Burchard* •
Local Museum – 1928
Local history ... 46687

Bedford NY

Museum of the Bedford Historical Society, 38 Village Green, Bedford, NY 10586 • T: +1 914 2349751 • F: +1 914 2345461 • bedhist@bestweb.net • www.bedfordhistoricalsociety.org •
Exec.Dir.: *Lynne Ryan* •
Local Museum – 1916
Local hist, Nativa American artifacts, manuscript coll ... 46688

Bedford OH

Bedford Historical Society Museum, 30 Squire Pl, Bedford Commons, Bedford, OH 44146 • T: +1 440 2320796 • www.bedfordohiohistory.org •
C.E.O.: *Janet Caldwell* •
Local Museum – 1955
Americana, archaeology, period furniture, glass, local and Ohio hist, Indian artifacts, railway – library .. 46689

Bedford PA

Fort Bedford Museum, Fort Bedford Dr, Bedford, PA 15522 • T: +1 814 6238891 • F: +1 814 6232011 • fbm@nb.net •
Local Museum – 1958
Indian artifacts, anique vehicles, antiques ... 46690

Bedford VA

Bedford City/County Museum, 201 E Main St, Bedford, VA 24523 • T: +1 540 5864520 • bccm-info@bedfordvamuseum.org • www.bedfordvamuseum.org •
Dir.: *Ellen A. Wandrei* • Cur.: *Bernice Sizemore* •
Local Museum – 1932
Indian relics, tolls & implements ... 46691

Beeville TX

Beeville Art Museum, 401 E Fannin, Beeville, TX 78104 • T: +1 361 3588615 • F: +1 361 3580743 • BeevilleArtMuseumr@Beeville.Net • beeville.net/BeevilleArtmuseum/ •
Dir.: *Brenda Fleming-Pawelek* •
Fine Arts Museum
Art ... 46692

Belchertown MA

Stone House Museum, 20 Maple St, Belchertown, MA 01007 • T: +1 413 2892010 • stonehousemuseum@yahoo.com • www.stonehousemuseum.org •
Dir./Cur.: *Caren Anne Harrington* •
Historical Museum – 1903
Historic house, china, furniture, jewelry, costumes, toys, musical instr, farm equipment – small library ... 46693

Belcourt ND

Turtle Mountain Chippewa Heritage Center, Hwy 5, Belcourt, ND 58316 • T: +1 701 4772639 • F: +1 701 4770065 • tmchc@utma.com •
Exec.Dir.: *Sheldon Williams* •
Fine Arts Museum / Historical Museum – 1985
Chippewa Indian history, art, contemp art ... 46694

Belhaven NC

Belhaven Memorial Museum, 211 E Main St, Belhaven, NC 27810 • T: +1 252 9436817 • F: +1 252 9432357 • www.beaufort-county.com/Belhaven/museum •
Pres.: *Ed Harris* •
Historical Museum – 1965
19th and 20th c regionalhist, early phonographs, Indian artifacts, marine ... 46695

Bella Vista AR

Bella Vista Historical Museum, 1885 Bella Vista Way, Bella Vista, AR 72714 • T: +1 501 8552335 •
Pres.: *Carole Westby* •
Local Museum – 1976
History and genealogy of Bella Vista ... 46696

Bellefontaine OH

Logan County Historical Society Museum, 521 E Columbus Av, Bellefontaine, OH 43311 • T: +1 937 5937557 • logancomuseum@earthlink.net • www.logancountymuseum.org •
Dir.: *Todd McCormick* •
Local Museum – 1945
Household appliances, tools, clothing, cameras, music, firearms, uniforms, railroad, military and pioneer costumes, geology, Warren Cushman art ... 46697

Bellefonte PA

Centre County Library Historical Museum, 203 N Allegheny St, Bellefonte, PA 16823 • T: +1 814 3551516 • F: +1 814 3552700 • paroom@centrecountylibrary.org • www.ccfpl.org •
Dir.: *Charlene K. Brungard* •
Local Museum – 1939
Furniture, china, county artifacts ... 46698

Belleville IL

Saint Clair County Historical Museum, 701 E Washington St, Belleville, IL 62220 • T: +1 618 2340600 • F: +1 618 2343060 •
Pres.: *Robert Fietsam* •
Local Museum – 1905
Period furniture, costumes, trools, memorabilia, manuscripts coll ... 46699

Belleville KS

Republic County Historical Society Museum, 2726 Hwy 36, Belleville, KS 66935 • T: +1 785 5275971 • repcomuse@nckcn.com • www.nckcn.com/homepage/republic_co/repmus.htm •
Cur.: *Patricia Walter* •
Local Museum – 1985
Pioneer historic tools, railroad caboose ... 46700

Belleville MI

Belleville Area Museum, 405 Main St, Belleville, MI 48111 • T: +1 734 6971944 • F: +1 734 6971944 • dwilson@provide.net • www.vanburen-mi.org/hot_topics/belleville_area_museum.html •

U

Dir.: *Diane Wilson* •
Local Museum – 1989
Local history, small-scale replicas of historical buildings 46701

Yankee Air Museum, Willow Run Airport, H-2041 A St, Belleville, MI 48111, mail addr: POB 590, Belleville, MI 48112-0590 • T: +1 734 4834030 • F: +1 734 4835076 • mark.ernst@yankeeairmuseum.org • www.yankeeairmuseum.org •
Pres.: *Richard Stewart* • Dep. Dir./Cur.: *Gayle Drews* •
Military Museum / Science&Tech Museum – 1981
WWII Aviation, WWII Home Front & Arsenal of Democracy, Ford Willow Run bomber plant airplane hangar, 1940-45 aircrafts – library, archives, annual air show, education 46702

Bellevue NE

Sarpy County Historical Museum, 2402 Clay St, Bellevue, NE 68005-3932 • T: +1 402 2921880 • F: +1 402 2926045 • sarpymuseum@prolinx.net •
Dir.: *Gary Iske* •
Local Museum – 1970 46703

Bellevue OH

Historic Lyme Village, 5001 State Rte 4, Bellevue, OH 44811 • T: +1 419 4834949, 4836052 • www.lymevillage.com •
Pres.: *Ray Parker* • Cur.: *Alvina Schaeffer* •
Historical Museum – 1972 46704

Bellevue PA

John A. Hermann jr. Memorial Art Museum, 318 Lincoln Av, Bellevue, PA 15202 • T: +1 412 7618008 • www.borough.bellevue.pa.us/hermann/hermann.html •
Dir.: *Donald F. Gust* •
Fine Arts Museum – 1940 46705

Bellevue WA

Bellevue Art Museum, 510 Bellevue Way NE, Bellevue, WA 98004, mail addr: POB 1705, Bellevue, WA 98004-1705 • T: +1 425 5190770 • F: +1 425 6371799 • bam@bellevueart.org • www.bellevueart.org •
Dir.: *Kathleen Harleman* • Cur.: *Greg Ginger* •
Fine Arts Museum – 1975
Art 46706

Eastside Heritage Center, Key Center, 601 108th Av NE, Bellevue, WA 98015 • T: +1 425 4501049 • F: +1 425 4501050 • director@eastsideheritagecenter.org • www.eastsideheritagecenter.org •
Historical Museum – 1965
Material and artifacts from the region from 1850s onwards 46707

Kidsquest Children's Museum, 4055 Factoria Mall SE, Bellevue, WA 98006 • T: +1 425 6378100 • info@kidsquestmuseum.org • www.kidsquestmuseum.org •
Exec. Dir.: *Putter Bert* •
Science&Tech Museum / Natural History Museum – 1997 46708

Rosalie Whyel Museum of Doll Art, 1116 108th Av, NE, Bellevue, WA 98004 • T: +1 425 4551116 • F: +1 425 4554793 • dollart@dollart.com • www.dollart.com •
Dir.: *Rosalie Whyel* • Cur.: *Jill Gorman* •
Special Museum – 1989
Dolls, games, books, photographs, costumes 46709

Bellflower IL

Bellflower Genealogical and Historical Society Museum, Latcha St, Bellflower, IL 61724 • T: +1 309 7223757, 7223467 •
Pres.: *Dorothy Woliung* • *Phyllis Kumler* •
Local Museum – 1976
Early 1900 Illinois Central Railroad Depot 46710

Bellingham WA

American Museum of Radio and Eletricity, 1312 Bay St, Bellingham, WA 98225 • T: +1 360 7383886 • F: +1 360 7383472 • jwinter@americanradiomuseum.org • www.americanradiomuseum.org •
Science&Tech Museum – 1982
Hist of radio and electricity, radios, televisions, recording devices and players 46711

Viking Union Gallery, Western Washington University, 516 High St, Bellingham, WA 98225 • T: +1 360 6503450 • F: +1 360 6506507 • aspart@cc.wwu.edu • www.as.www.edu/asp/gallery •
Dir.: *Caroline Knebelsberger* •
Fine Arts Museum / University Museum – 1899/1950
Art 46712

Western Gallery, Western Washington University, Bellingham, WA 98225-9068 • T: +1 360 6503900 • F: +1 360 6506878 • sarah.clarklangager@wwu.edu • www.westerngallery.wwu.edu •
Dir.: *Sarah Clark-Langager* • Sc. Staff: *Paul Brower* (Preparator) •
University Museum / Fine Arts Museum – 1950
Fine and performing art 46713

Whatcom Museum of History and Art, 121 Prospect St, Bellingham, WA 98225 • T: +1 360 7788930 • F: +1 360 7788931 • museuminfo@cob.org • www.whatcommuseum.org •
Dir.: *Patricia Leach* • Cur.: *Janis Olson* (Collections) • *Richard Vanderway* (History) •
Local Museum – 1940
Art, history 46714

Bellows Falls VT

Adams Old Stone Grist Mill, Mill St, Bellows Falls, VT 05101 • T: +1 802 4633734 • facades@sover.net •
Folklore Museum – 1965
19th c milling equipment, locally manufactured machinery 46715

Rockingham Free Public Library and Museum, 65 Westminster St, Bellows Falls, VT 05101 • T: +1 802 4634270 • rockingham@dol.state.vt.us • www.rockingham.lib.vt.us •
Historical Museum – 1909
Photographs, period toys, Civil War artifacts, local hist., steropticon view 46716

Bellport NY

Bellport-Brookhaven Historical Society Museum, 31 Bellport Ln, Bellport, NY 11713 • T: +1 516 2860888 • robinastarbuck@aol.com •
Pres.: *Carol Blesser* •
Historical Museum – 1963
Local hist 46717

Belmont CA

The Wiegand Gallery, c/o Notre Dame de Namur University, 1500 Ralston Av, Belmont, CA 94002 • T: +1 650 5083595 • pr@ndnu.edu • www.ndnu.edu •
Dir.: *Charles Strong* • Cur.: *Robert Poplack* •
Public Gallery / University Museum – 1970 46718

Belmont NY

Allegany County Museum, 11 Wells St, Belmont, NY 14813 • T: +1 716 2689293 • F: +1 716 2689446 • historian@alleganyco.com •
Dir.: *Craig R. Braack* •
Local Museum – 1970
19th c tools and furenishings, maps 46719

Americana Manse, Whitney-Halsey Home, 39 South St, Belmont, NY 14813 • T: +1 716 2685130 •
C.E.O.: *Ruth Czankus* •
Local Museum – 1964
Victorian art and furniture, household items 46720

Belmont VT

Mount Holly Community Historical Museum, Tarbellville Rd, Belmont, VT 05730 • T: +1 802 2593722 • www.vermonthistory.org •
Cur.: *Janice Bamforth* •
Local Museum – 1968
Quilts, clothing, old farmtools, early records, apothecary 46721

Beloit KS

Little Red Schoolhouse-Living Library, Roadside Park, N Walnut and Hwy 24, Beloit, KS 67420 • T: +1 785 7385301 •
Chm.: *Mildred Peterson* •
Historical Museum – 1976
Education, one-room school 46722

Beloit WI

Beloit Historical Society Museum, 845 Hackett St, Beloit, WI 53511 • T: +1 608 3657835 • F: +1 608 3655999 • beloiths@ticon.net • www.ticon.net/~beloiths •
Dir.: *Paul K. Kerr* •
Local Museum – 1910
Period furnishings, 1857 rural school house 46723

Logan Museum of Anthropology, 700 College St, Beloit, WI 53511 • T: +1 608 3632677 • F: +1 608 3632248 • www.beloit.edu/~museum/logan •
Dir.: *Dr. William Green* (North American Archaeology) •
Sc. Staff: *Karla Wheeler* • *Nicolette Meister* •
University Museum / Ethnology Museum / Archaeology Museum – 1892
Anthropology 46724

Wright Museum of Art, Beloit College, 700 College St, Beloit, WI 53511 • T: +1 608 3632095 • F: +1 608 3632248 • newlandj@beloit.edu • www.beloit.edu/~museum/wright/ •
Dir.: *Thomas H. Wilson* • Cur.: *Judy Newland* •
Fine Arts Museum / University Museum – 1892
European and American pointings, hist and contemp photography, culpture, Asian art 46725

Belton SC

South Carolina Tennis Hall of Fame, Main St, Belton, SC 29627 • T: +1 864 3387751 • F: +1 864 3384034 • rexmaynard@statecom.net • www.beltonsc.com •
Cur.: *Rex Maynard* •
Special Museum – 1984
Tennis 46726

Belton TX

Bell County Museum, 201 N Main St, Belton, TX 76513 • T: +1 254 9335243 • F: +1 254 9335756 • museum@co.bell.tx.us • www.bellcountytx.com •
Dir.: *Stephanie Turnham* •
Local Museum – 1975
Local hist items 46727

Belvidere IL

Boone County Historical Society Museum, 311 Whitney Blvd, Belvidere, IL 61008 • T: +1 815 5448391 • F: +1 815 5471691 • info@boonecountyhistoricalmuseum.org • www.boonecountyhistoricalmuseum.org •
Pres.: *Patrick Murphy* •
Local Museum – 1968
Civil War, natural hist, sewing and washing machines, – library 46728

Bement IL

Bryant Cottage, 146 E Wilson Av, Bement, IL 61813 • T: +1 217 6788184 •
Head: *Marilyn L. Ayers* •
Historical Museum – 1925
1856 Bryant Cottage where the Lincoln-Douglas Debates were verbally agreed to be part of the 1858 Senate campaigns 46729

Bemidji MN

Beltrami County Historical Center, 130 Minnesota Av SW, Bemidji, MN 56601 • T: +1 218 4443376 • F: +1 218 4443377 • depot@paulbunyan.net • www.paulbunyan.net/users/depot •
Exec. Dir.: *Wanda Hoyum* •
Local Museum – 1952
Regional hist, Indian artifacts, log school furnishings – archives 46730

Harlow Ceramics Collection & Kleven Print Collection, Bemidji State University, College of Arts and Letters, 1500 Birchmond Dr, Bemidji, MN 56601-2699 • T: +1 218 7553708 • F: +1 218 7553735 • sskaul@bemidjistate.edu • www.bemidjistate.edu •
Dir.: *Cecilia Lieder* •
University Museum / Fine Arts Museum
20 th c Japanese and American prints, contemporary ceramic vessels 46731

Bend OR

Deschutes County Historical Society Museum, 63090 Sherman Rd, Bend, OR 97708 • T: +1 541 3891813 • F: +1 541 3179345 • info@deschutes.historical.museum • deschutes.historical.museum •
C.E.O. & Pres.: *John Sholes* •
Local Museum – 1975
County hist exhibits 46732

High Desert Museum, 59800 S Hwy 97, Bend, OR 97702-7963 • T: +1 541 3824754 • F: +1 541 3825256 • info@highdesertmuseum.org • www.highdesertmuseum.org •
Pres.: *Forrest B. Rodgers* • Cur.: *Robert Boyd* (Western History) •
Ethnology Museum / Natural History Museum / Historical Museum – 1974
Great basin prehistory and cultural hist, Native American ethnographic objects, Columbia River Plateau clothing and bags, natural hist 46733

Benicia CA

Benicia Historical Museum, 2060 Camel Rd, Benicia, CA 94510 • T: +1 707 7455435 • F: +1 707 7455869 • info@beniciahistoricalmuseum.org • www.beniciahistoricalmuseum.org •
Dir.: *Julia Bussinger* • Cur.: *Harry Wassmann* •
Historical Museum – 1985
Local history 46734

Benjamin TX

Knox County Museum, Courthouse, Benjamin, TX 79505 • T: +1 940 4592229 • F: +1 940 4592229 • kchc@nts-online.net • www.knoxcountytexas.com •
C.E.O.: *Mary Jane Young* •
Local Museum – 1966
Photographs and books relating to life in Knox County from prehist period to early 20th c 46735

Benkelman NE

Dundy County Historical Society Museum, 522 Araphahoe, Benkelman, NE 69021 • T: +1 308 4232750 •
Pres.: *Betty Deyle* • Dir.: *Dee Fries* •
Local Museum – 1970
Indian artifacts, railroad depot, doll coll, bank and post office artifacts 46736

Bennettsville SC

Jennings-Brown House Female Academy, 119 S Marlboro St, Bennettsville, SC 29512 • T: +1 843 4795624 • marlborough@mecsc.net •
Exec. Dir.: *Lucille Carabo* •
Local Museum – 1967
Historic house 46737

Marlboro County Historical Museum, 123 S Marlboro St, Bennettsville, SC 29512 • T: +1 843 4795624 • marlborough@mecse.net •
Pres.: *Marty Rankin* • Dir.: *Lucille Carabo* •
Local Museum – 1970
Local hist 46738

Bennington VT

The Bennington Museum, 75 Main St, Bennington, VT 05201 • T: +1 802 4471571 • F: +1 802 4428305 • dborges@benningtonmuseum.org • www.benningtonmuseum.com •
Dir.: *Richard Borges* •
Fine Arts Museum / Decorative Arts Museum / Local Museum – 1875
Art, decorative arts, local hist 46739

Southern Vermont College Art Gallery, 982 Mansion Dr, Bennington, VT 05201 • T: +1 802 4425427 • F: +1 802 4474695 •
Dir.: *Greg Winterhalter* •
Public Gallery / University Museum – 1979
Photo coll of 1910 Everett Mansion construction 46740

Benson MN

Swift County Historical Museum, 2135 Minnesota Av, Bldg 2, Benson, MN 56215 • T: +1 320 8434467 • historical.society@co.swift.mn.us •
Exec. Dir.: *Marlys M. Gallagher* •
Local Museum – 1929
Regional hist – newspapers library 46741

Benton WI

Swindler's Ridge Museum, 25 W Main St, Benton, WI 53803 • T: +1 608 7595182 •
Dir.: *Peg Roberts* •
Historical Museum – 1993 46742

Benton Harbor MI

Morton House Museum, 501 Territorial Rd, Benton Harbor, MI 49022 • T: +1 616 9257011 • smith.arnold@gmail.com • www.parrett.net/~morton •
Pres.: *Chuck Jager* •
Local Museum – 1966
Local hist, costumes, foreign small boxes 46743

Bentonville AR

Peel Mansion Museum, 400 S Walton Blvd, Bentonville, AR 72712 • T: +1 479 2739664 • F: +1 479 2739688 • flower@arkmola.net • www.peelmansion.org •
Dir.: *Leah Whitehead* •
Local Museum – 1992
Genealogy, county hist 46744

Benzonia MI

Benzie Area Historical Museum, 6941 Traverse Av, Benzonia, MI 49616-0185 • T: +1 231 8825539 • F: +1 231 8825539 • museum@t-one.net •
Dir.: *Debbra Eckhout* •
Local Museum – 1969
County hist, Bruce Catton, Gwen Frostic, AA car ferries, railroads, agriculture, civil war – library 46745

Berea KY

Berea College Burroughs Geological Museum, Main St, Berea, KY 40404 • T: +1 859 9853893 • F: +1 859 9853303 • larry_lipchinsky@berea.edu • www.berea.edu/ •
Cur.: *Zelek L. Lipchinsky* •
University Museum / Natural History Museum – 1920
Geology 46746

Berea College Doris Ulmann Galleries, Corner of Chestnut and Elipse St, Berea, KY 40404 • T: +1 859 9869341 ext 5530 • F: +1 859 9853541 • robert_boyce@berea.edu • www.berea.edu/ •
Public Gallery – 1975
Contemporary arts 46747

Berea OH

Fawick Art Gallery, Baldwin-Wallace College, 275 Eastland Rd, Berea, OH 44017 • T: +1 216 8262152 • F: +1 216 8263380 • www.baldwinwallacecollege.com •
Dir.: *Prof. Paul Jacklitch* •
Fine Arts Museum
Drawings and prints from 16th - 20th c, paintings, sculptures by American Midwest artists 46748

Bergen NY

Bergen Museum of Local History, 7547 Lake Rd, Bergen, NY 14416 • T: +1 585 4941121 • F: +1 585 4941488 •
Pres.: *Raymond MacConnell* • Sc. Staff: *Tracy Miller* •
Local Museum – 1964
Photographs and documents, dec art, folklore, pottery; farm, trde and home articles 46749

Berkeley CA

Badè Museum of Biblical Archaeology, Pacific School of Religion, 1798 Scenic Av, Berkeley, CA 94709 • T: +1 510 8498286 • F: +1 510 8458948 • bade@psr.edu • bade.psr.edu •
Dir.: *Aaron Brody* •

Religious Arts Museum / Archaeology Museum – 1926
Tell en-Nasbeh (archaeological site), Judah, Israel/Palestine, religion, rare bibles – Howell Bible Collection 46750

Berkeley Art Center, 1275 Walnut St, Berkeley, CA 94709 • T: +1 510 6446893 • F: +1 510 5400343 • info@berkeleyartcenter.org • www.berkeleyartcenter.org •
Dir.: *Robbin Henderson* • Cur.: *Patrice Wagner* •
Fine Arts Museum – 1967
Contemporary work by emerging regional artists 46751

Berkeley Art Museum and Pacific Film Archive, c/o University of California, 2625 Durant Avenue, Ste 2250, Berkeley, CA 94720-2250 • T: +1 510 6420808 • bampfa@berkeley.edu • www.bampfa.berkeley.edu •
Dir.: *Kevin E. Consey* • Dep. Dir.: *Stephen Gong* • Ass. Dir.: *Lucinda Barnes* (Programs) • Cur.: *Constance Lewallen* (Exhibitions) • *Terri Cohn* (Matrix) • *Dara Solomon* (Collections) •
Fine Arts Museum / University Museum – 1965
Art, film, video 46752

Essig Museum of Entomology, c/o University of California, 1170 Valley Life Sciences Bldg, Berkeley, CA 94720 • T: +1 510 6429895 • cbarr@nature.berkeley.edu • essig.berkeley.edu •
Dir.: *Rosemarie G. Gillespie* • Cur.: *Cheryl B. Barr* • *Robert Zuparko* •
University Museum / Natural History Museum – 1939
Entomology, esp arachnids, insects and other terrestrial arthropods 46753

Hall of Health, Children's Hospital and Research Center (closed) 46754

Judah L. Magnes Museum, 2911 Russell St, Berkeley, CA 94705 • T: +1 510 5496950 • info@magnes.org • www.magnes.org •
Dir.: *Terry Pink Alexander* • Cur.: *Alla Efimova* • *Elayne Grossbard* (Judaica) • *Francesco Spagnolo* (Music) • *Aaron T. Kornblum* (Western Jewish History Center) • Education: *Carin Jacobs* •
Fine Arts Museum / Religious Arts Museum – 1962
Judaica, prints, drawings, paintings, sculptures, Western US Jewish history – Blumenthal library, archives 46755

Kala Gallery, c/o Kala Art Institute, 2990 San Pablo Av, Berkeley, CA 94702 • T: +1 510 8417000 • kala@kala.org • www.kala.org •
Dir.: *Archana Horsting* • Artistic Dir.: *Yuzo Nakano* •
Public Gallery – 1974
Artists who specialize in a printmaking, photographic, or digital process 46756

Lawrence Hall of Science, c/o University of California, 1 Centennial Dr, Berkeley, CA 94720-5200 • T: +1 510 6425132 • lhsweb@berkeley.edu • www.lawrencehallofscience.org •
Dir.: *Dr. Elisabeth K. Stage* • Dep. Dir.: *Susan Gregory* • *Brooke Smith* (Exhibits) •
Natural History Museum / Science&Tech Museum
Science, technology 46757

Museum of Paleontology, c/o University of California, Berkeley, 1101 Valley Life Sciences Bldg, Berkeley, CA 94720-4780 • T: +1 510 6421821 • F: +1 510 6421822 • ucmpwebmaster@berkeley.edu • www.ucmp.berkeley.edu •
Dir./Cur.: *Roy Caldwell* • Sc. Staff: *Jere Lipps* • *Antony Barnosky* • *Carole Hickman* • *Kevin Padian* • *Leslea Hlusko* • *Sheila Patek* • *William Clemens* • *David R. Lindberg* • *James Valentine* • *Joseph Gregory* •
Natural History Museum – 1921
Vertebrates, plants, invertebrates, microfossils, paleobotany – library, labs 46758

Museum of Vertebrate Zoology, c/o University of California, Berkeley, 3101 Valley Life Sciences Bldg, Berkeley, CA 94720-3160 • T: +1 510 6423567 • F: +1 510 6438238 • judithf@berkeley.EDU • mvz.berkeley.edu •
Dir.: *Craig Moritz* • Sc. Staff: *Christopher J. Conroy* (Herpetology, Mammals) • *Carla Cicero* (Birds) • *Eileen A. Lacey* • *James L. Patton* (Mammalogy) • *Jimmy A. McGuire* (Herpetology) •
University Museum / Natural History Museum – 1908
Zoology, herpetology, mammology – library 46759

Phoebe Apperson Hearst Museum of Anthropology, c/o University of California, Berkeley, 103 Kroeber Hall, Berkeley, CA 94720-3712 • T: +1 510 6423682 • F: +1 510 6426271 • pahma@berkeley.edu • hearstmuseum.berkeley.edu •
Dir.: *Douglas Sharon* • Dep. Dir.: *Sandra Harris* • Cons.: *Madeleine Fang* • Cur.: *Leslie Freund* •
University Museum / Ethnology Museum / Archaeology Museum – 1901
Anthropology, specimens worldwide 46760

Berkeley Springs WV

Homeopathy Works, 33 Fairfax Str, Berkeley Springs, WV 25411 • T: +1 304 2582541 • F: +1 304 2586335 • info@homeopathyworks.com •
Special Museum
Homeopathy artifacts from 1800's topresent 46761

Museum of the Berkeley Springs, POB 99, Berkeley Springs, WV 25411 • T: +1 304 2583743 •
Head: *Hettie G. Hawvermale* •
Local Museum – 1984
Hist of town, region and mineral springs 46762

Berlin MA

Berlin Art and Historical Collections, 4 Woodward Av, Berlin, MA 01503, mail addr: Box 35, Berlin, MA 01503 • T: +1 978 8382502 • historical@townofberlin.com • www.townofberlin.com/historical •
Dir.: *Barry W. Eager* •
Historical Museum / Local Museum / Decorative Arts Museum – 1950
Local hist, military, costumes, folklore, genealogy – archives 46763

Berlin WI

Berlin Historical Society Museum of Local History, 111 S Adams Av, Berlin, WI 54923 • T: +1 920 3611274 • F: +1 920 3615439 • berlinc@vbe.com • www.1berlin.com •
Pres.: *Jack Wahlers* • Vice Pres.: *Jim Petersen* • Cur.: *Dan Freimark* •
Local Museum – 1962
Local hist 46764

Bermuda LA

Beau Fort Plantation Home, 4078 Hwy 494 and Hwy 119, Bermuda, LA 71456 • T: +1 318 3529580, 3528352 • F: +1 318 3527280 • beaufort@worldnetla.net •
Folklore Museum – 1790
1790 Creole one and one-half cottage type building 46765

Berrien Springs MI

Courthouse Museum, 313 N Cass St, Berrien Springs, MI 49103 • T: +1 616 4711202 • F: +1 616 4717412 • bcha@berrienhistory.org • www.berrienhistory.org •
Dir.: *Leo J. Goodsell* •
Local Museum – 1967
County hist, court records, clark equipment company – library 46766

Siegfried H. Horn Archaeological Museum, Andrews University (by appointment only), Apple Valley Plaza, Old US-31 and Garland Ave, Berrien Springs, MI 49104-0990 • T: +1 616 4713273 • F: +1 616 4713619 • hornmusm@andrews.edu • www.andrews.edu/archaeology •
Dir.: *Dr. Randall W. Younker* • Cur.: *Constance Gane* •
Archaeology Museum / University Museum – 1970
Middle East archaeology, pottery, tools, coins, papyrii, textiles, cuneiform tablets – library 46767

Berryville AR

Heritage Center Museum, Old Court House on Square, Berryville, AR 72616 • T: +1 870 4236312 • www.rootsweb.com/~arcchs •
Local Museum – 1960
County hist, genealogy 46768

Saunders Memorial Museum, 113-115 E Madison Av, Berryville, AR 72616 • T: +1 870 4232563 •
Man.: *Rose M. Garrett* •
Museum of Classical Antiquities – 1955
Gun coll, period furnishings, arts and crafts 46769

Berryville VA

Clarke County Historical Museum, 32 E Main St, Berryville, VA 22611 • T: +1 540 9552600 • F: +1 540 9550285 • ccha@visuallink.com • www.clarkehistory.org •
Pres./Cur.: *Don Cady* •
Local Museum – 1939
Portraits, photos, early fairfax land grants 46770

Bessemer AL

Bessemer Hall of History, 1905 Alabama Av, Bessemer, AL 35020 • T: +1 205 4261633 • F: +1 205 4261633 • www.bhamrails.info/Bess_Hall_Hist/bess_hall_hist_02 •
Dir.: *Dominga N. Toner* •
Local Museum – 1970
local hist in old railroad terminal 46771

Bethany WV

Historic Bethany, Bethany College, Bethany, WV 26032 • T: +1 304 8294285 • F: +1 304 8294287 • historic@mail.bethanywv.edu • www.bethanywv.edu •
C.E.O.: *Dr. G.T. Smith* •
University Museum / Historical Museum – 1840
Campbell family house items, books and papers 46772

Bethel AK

Yupiit Piciryarait Museum, 420 Chief Eddie Hoffman Hwy, Bethel, AK 99559, mail addr: POB 219, Bethel, AK 99559 • T: +1 907 5431819 • F: +1 907 5431885 • museum@yupiitpiciryaraitmuseum.org • ypmuseum.org •
Local Museum – 1967
Native culture, lifestyle of the Central Yupik Eskimos of the Yukon-Kuskokwin delta 46773

Bethel ME

Bethel Historical Society's Regional History Center, 10-14 Broad St, Bethel, ME 04217-0012 • T: +1 207 8242908 • F: +1 207 8240882 • info@bethelhistorical.org • www.bethelhistorical.org •
Exec. Dir.: *Dr. Stanley Russell Howe* •
Historical Museum – 1974
Regional history of Northern New England, federal style house – Research library, period house museum, 46774

Bethesda MD

DeWitt Stetten Jr., Museum of Medical Research, National Institutes of Health, Bldg 45, Rm. 3AN38 MSC 6330, Bethesda, MD 20892-2092 • T: +1 301 4966610 • F: +1 301 4021434 • museum@nih.gov • http://history.nih.gov •
Dir.: *Robert Martensen* • Cur.: *Michele Lyons* •
Historical Museum – 1987
20th and 21st century biomedical research instruments; non-scientific NIH-related objects; heart valves – Library and NIH-focused archives 46775

Bethlehem NH

Crossroads of America, 6 Trudeau Rd, Bethlehem, NH 03574 • T: +1 603 8693919 • Roger.Hinds@forthBBS •
C.E.O.: *Roger Hinds* •
Special Museum – 1981
Model railroad, toy 46776

Bethlehem PA

Burnside Plantation, 1461 Schoenersville Rd, Bethlehem, PA 18018 • T: +1 610 8685044 • F: +1 610 8825044 • historicbethlehem@hotmail.com • www.historicbethlehem.org •
Exec. Dir.: *Charlene Donchez Mowers* •
Decorative Arts Museum – 1986
Farming, domestic crafts and decorative arts 46777

Colonial Industrial Quarter, 459 Old York Rd, Bethlehem, PA 18018 • T: +1 610 8820450 • F: +1 610 8820460 • histbeth2@hotmail.com • www.historicbethlehem.org •
Exec. Dir.: *Charlene Donchez Mowers* •
Historical Museum – 1957
Historic artifacts from 1700-1885, industrial hist 46778

Kemerer Museum of Decorative Arts, 427 N New St, Bethlehem, PA 18018 • T: +1 610 8686868 • F: +1 610 3322459 • historicbethlehem@hotmail.com • www.historicbethlehem.org •
Exec. Dir.: *Charlene Donchez Mowers* • Sc. staff: *Barbara L. Schafer* (Coll.) •
Decorative Arts Museum – 1954
18th - 19th c American decorative arts, photographs, prints and stereographyc, regional 19th c oil paintings 46779

Lehigh University Art Galleries/Museum, Zoellner Arts Center, 420 E Packer Av, Bethlehem, PA 18015 • T: +1 610 7583615 • F: +1 610 7584580 • rv02@lehigh.edu • www.luag.org •
Dir./Cur.: *Ricardo Viera* • Asst. Dir.: *Denise Stangl* • Cur./Prep.: *Robert Lopata* •
Fine Arts Museum / University Museum – 1864
Paintings, photography, prints, artist books, artifacts, American fok art 46780

Moravian Museum of Bethlehem, 66 W Church St, Bethlehem, PA 18018 • T: +1 610 8670173 • F: +1 610 6940960 • historicbethlehem@hotmail.com • www.historicbethlehem.org •
Exec. Dir.: *Charlene Donchez Mowers* • Sc. Staff: *Barbara L. Schafer* •
Local Museum / Religious Arts Museum – 1938
Religious items, tools, textiles 46781

National Museum of Industrial History, 530 E Third St, Bethlehem, PA 18015 • T: +1 610 6946644 • F: +1 610 6946641 • nmih@fast.net • www.nmih.org •
Pres./C.E.O.: *Stephen G. Donches* •
Historical Museum – 1997
Textile equipment, papermaking machine, steel making models, industrial development of the 19th - 20th c, steel machinery and equipment, slate industry machinery – Archives 46782

Payne Gallery, Moravian College, Main and Church Sts, Bethlehem, PA 18018 • T: +1 610 8611675, 8611680 • F: +1 610 8611682 • medjr01@moravian.edu • www.moravian.edu •
Dir.: *Dr. Diane Radycki* •
Fine Arts Museum – 1982
Drawings, paintings, photography, prints, watercolors, woodcuts 46783

Bettendorf IA

Family Museum of Arts and Science, 2900 Learning Campus Dr, Bettendorf, IA 52722 • T: +1 563 3444106 • F: +1 563 3444164 • familymuseum@bettendorf.org • www.familymuseum.org •
Dir.: *Tracey K. Kuehl* •
Fine Arts Museum / Natural History Museum – 1974
Children's museum 46784

Beverly MA

Beverly Historical Museum, 117 Cabot St, Beverly, MA 01915 • T: +1 508 9221186 • F: +1 508 9227387 • info@beverlyhistory.org • www.beverlyhistory.org •
Exec. Dir.: *Mary-Ellen Smiley* •
Local Museum – 1891
Local history, John Cabot mansion, John Balch house, Hale farm – library, archive 46785

Endicott College Museum, 376 Hale St, Beverly, MA 01915 • T: +1 978 2322257 • F: +1 978 2322231 • www.endicott.edu •
University Museum / Local Museum 46786

Beverly OH

Oliver Tucker Historic Museum, State Rte 60, Beverly, OH 45715 • T: +1 740 9842489 •
Pres.: *Greg Holdren* • Cur.: *Mary A. Irvin* •
Local Museum – 1971
Furniture, clothes, tools 46787

Beverly WA

Wanapum Dam Heritage Center, Hwy 243, Beverly, WA 98823, mail addr: POB 878, Ephrata, Wa98823 • T: +1 509 7543541 • F: +1 509 7542522 • sjohns1@gcpud.org •
Dir.: *Angela Buck* •
Local Museum / Folklore Museum – 1966
Local hist 46788

Beverly WV

Randolph County Museum, Main St, Beverly, WV 26253 • T: +1 304 6360841 • dlrice@citynet.net •
Dir.: *Randy Allan* • Cur.: *Donald Rice* •
Local Museum – 1924
Local hist 46789

Beverly Hills CA

The Academy Gallery, c/o Academy of Motion Picture Arts and Sciences, 8949 Wilshire Blvd, Beverly Hills, CA 90211-1972 • T: +1 310 2473000 • F: +1 310 8599619 • gallery@oscars.org • www.oscars.org •
Dir.: *Robert Smolkin* •
Public Gallery – 1970 46790

California Museum of Ancient Art, POB 10515, Beverly Hills, CA 90213 • T: +1 818 7625500 • cmaa@earthlink.net •
Pres.: *John D. Hofbauer* • Dir./Cur.: *Jerôme Berman* •
Archaeology Museum – 1983
Near Eastern art and archaeology (Egypt, Mesopotamia, the Holy Land) 46791

Museum of Television and Radio → The Paley Center for Media

The Paley Center for Media, 465 N Beverly Dr, Beverly Hills, CA 90210 • T: +1 310 7861000 • tebright@paleycenter.org • www.paleycenter.org •
Dir.: *Barbara Dixon* •
Science&Tech Museum – 1975
Communication, TV and radio programming, ads, shows 46792

Bexley OH

Bexley Historical Society Museum, 2242 E Main St, Bexley, OH 43209-2319 • T: +1 614 2358694 • F: +1 614 2353420 •
Pres.: *Barbara Hysell* • Cur.: *Debbie McCue* •
Local Museum – 1974 46793

Big Cypress FL

Ah-Tah-Thi-Ki Museum, County Rd 833 and W Boundary Rd, Big Cypress, FL 33440 • T: +1 863 9021115 • F: +1 863 9021117 • museum@semtribe.com • www.seminoletribe.com/enterprises/bigcypress/museum.shtml •
Dir.: *Billy L. Cypress* •
Historical Museum – 1989
Seminole Indian reservation 46794

Big Horn WY

Bradford Brinton Memorial, 239 Brinton Rd, Big Horn, WY 82833 • T: +1 307 6723173 • F: +1 307 6723258 • kls_bbm@vcn.com • www.bradfordbrintonmemorial.com •
Dir./Cur.: *Kenneth L. Schuster* • Cur.: *Dwight Layton* •
Fine Arts Museum – 1961
Western paintings, prints, drawings and sculpture 46795

Big Pool MD

Fort Frederick, 11100 Fort Frederick Rd, Big Pool, MD 21711 • T: +1 301 8422155 • F: +1 301 8420028 • fortfrederick@erols.com • www.dnr.state.md.us •
Fine Arts Museum / Military Museum – 1922
Indian relics, artifacts, hist of Fort Frederick, Civil War rifle 46796

Big Rapids MI

Jim Crow Museum, Ferris State University, 820 Campus Dr, Big Rapids, MI 49307 • T: +1 231 5915873 • F: +1 231 5912541 • thorj@ferris.edu • www.ferris.edu/jimcrow •
Dir.: *John P. Thorp* •
Historical Museum / University Museum – 1996
Early to mid 20th c artifacts, Ku Klux Klan artifacts, civil rights artifacts, civil rights movement 46797

Mecosta County Historical Museum, 129 S Stewart St, Big Rapids, MI 49307-1839 • T: +1 616 5925091 •
Dir.: *Maxine Safoulis* • *Anna Young* •
Local Museum – 1957
Agriculture, costumes, folklore, Indian artifacts, numismatic, logging and lumbering, toys – children's museum 46798

Big Spring TX

Heritage Museum and Potton House, 510 Scurry, Big Spring, TX 79720 • T: +1 915 2678255 • F: +1 915 2678255 • heritage@crcom.net • www.bigspringmuseum.com •
Cur.: *Tammy Schrecengost* •
Local Museum – 1971
Paintings by Harvey W. Caylor, early artifacts of the area, pioneer settlement, cattle industry railroad 46799

Big Stone Gap VA

Southwest Virginia Museum, 10 W First St N, Big Stone Gap, VA 24219 • T: +1 276 5231322 • F: +1 276 5236616 • swvamuseum@dcr.state.va.us • www.dcr.state.va.us/parks/swvamus.htm •
Historical Museum – 1948
Early hist and pioneer period of Southwest Virginia, industrial develop, coal bboom 46800

Big Timber MT

Crazy Mountain Museum, Exit 367 Cemetery Rd, Hwy I-90, Big Timber, MT 59011 • T: +1 406 9325126 •
Dir.: *Fran Elgen* •
Local Museum – 1990
Hist of Sweet Grass County 46801

Bigfork MT

Butter Pat Museum, 265 Eagle Bend Dr, Bigfork, MT 59911-6235 •
Dir.: *Mary Dessoie Stephen Weppner* •
Decorative Arts Museum
2000 butter pats (miniature plates used to hold an individual portion of butter), 19th c China 46802

Biglerville PA

National Apple Museum, 154 W Hanover St, Biglerville, PA 17307-0656 • T: +1 717 6774556 • info@nationalapplemuseum.com • www.nationalapplemuseum.com •
Pres.: *Dick Mountfort* •
Agriculture Museum – 1990
Artifacts, photo coll, agriculture hist 46803

Billings MT

Moss Mansion Museum, 914 Division St, Billings, MT 59101 • T: +1 406 2565100 • F: +1 406 2520091 • mossmansion@mossmansion.com • www.mossmansion.com •
C.E.O.: *Kristin Gravatt* •
Decorative Arts Museum – 1986
Clothing & personal effects, city directories 1949-1998 46804

Northcutt Steele Gallery, Montana State University Billings, 1500 N 30th St, Billings, MT 59101-0298 • T: +1 406 6572324 • F: +1 406 6572187 • njussila@msu-b.edu •
Dir.: *Neil Jussila* •
Fine Arts Museum
Visual art 46805

Peter Yegen Jr. Yellowstone County Museum, 1950 Terminal Circle, Logan Field, Billings, MT 59105 • T: +1 406 2566811 • F: +1 406 2566031 • ycm@pyjrycm.org • www.pyjrycm.org •
Dir.: *Robin Urban* •
Local Museum – 1953
Indian artifacts, Western memorabilia, military items, dinosaur bones, Leory Greene paintings 46806

Western Heritage Center, 2822 Montana Av, Billings, MT 59101 • T: +1 406 2566809 ext 21 • F: +1 406 2566850 • heritage@ywhc.org • www.ywhc.org •
Dir.: *Kevin Kooistra-Manning* • Cur.: *Dorla Brunner* •
Historical Museum – 1971
Material culture of the Yellowstone vally 46807

Yellowstone Art Museum, 401 N 27 St, Billings, MT 59101 • T: +1 406 2566804 • F: +1 406 2566817 • artinfo@artmuseum.org • yellowstone.artmuseum.org •
Dir.: *Robert Knight* •
Fine Arts Museum – 1964
Contemporary art and prints, abstract expresionists art 46808

Biloxi MS

Beauvoir, Jefferson Davis Home and Presidential Library, 2244 Beach Blvd, Biloxi, MS 39531 • T: +1 228 3884400 • F: +1 228 3887800 • rtrowbridge@beauvoir.org • www.beauvoir.org •
Dir.: *Richard V. Forte* •
Historical Museum / Military Museum / Historic Site – 1941
19th c period furnishings & decorative, Civil War militaria 46809

Mardi Gras Museum, 119 Rue Magnolia, Biloxi, MS 39530 • T: +1 228 4356245 • F: +1 228 4356246 • museums@biloxi.ms.us • www.biloxi.ms.us •
Dir.: *Lolly Barnes* •
Historical Museum
History Museum, housed in the restored antebellum 1847 Magnolia Hotel 46810

Maritime and Seafood Industry Museum (temporarily closed), 115 First St, Biloxi, MS 39530, mail addr: POB 1907, Biloxi, MS 39533 • T: +1 228 4356320 • F: +1 228 4356309 • info@maritimemuseum.org • www.maritimemuseum.org •
Dir.: *Robin Krohn-David* •
Historical Museum / Science&Tech Museum – 1986
Maritime crafts, fishing &boatbuilding implements, early photographs of industry & watercraft 46811

Museum of Biloxi → Maritime and Seafood Industry Museum

The Ohr-O'Keefe Museum of Art, 1596 Glenn Swetman Street, Biloxi, MS 39530 • T: +1 228 3745547 • F: +1 228 4363641 • info@georgeohr.org • www.georgeohr.org •
Dir.: *Denny Mecham* •
Fine Arts Museum – 1994
George Ohr pottery, ceramics & regional and national pottery 46812

Tullis-Toledano Manor, 360 Beach Blvd, Biloxi, MS 39530 • T: +1 228 4356293, 4356308 • F: +1 228 4356246 • museums@biloxi.ms.us • www.biloxi.ms.us •
Dir.: *Ed Miles* •
Decorative Arts Museum 46813

Binghamton NY

Broome County Historical Society Museum, 185 Court St, Binghamton, NY 13901 • T: +1 607 7783572 • F: +1 607 7718905 • broomehistory@bclibrary.info • www.bclibrary.inf/history.htm •
Pres.: *David J. Dixon* •
Local Museum – 1919 46814

Discovery Center of the Southern Tier, 60 Morgan Rd, Binghamton, NY 13903 • T: +1 607 7738750 • F: +1 607 7738019 • info@thediscoverycenter.org • www.thediscoverycenter.org •
C.E.O.: *Margaret S. Crocker* • Dir.: *Elaine Kelly* (Exhibits) •
Special Museum – 1983
Fossils, minerals, shells, African artifacts, textiles, 46815

Roberson Museum and Science Center, 30 Front St, Binghamton, NY 13905 • T: +1 607 7720660 • F: +1 607 7718905 • www.roberson.org •
Exec.Dir.: *Susan I. Sherwood* • Cur.: *Catherine Schwoeffermann* (Folklife) •
Fine Arts Museum / Historical Museum / Science&Tech Museum – 1954
Decorative arts, American art, ethnol and archaeol coll, mounted birds and mammals 46816

University Art Museum, State University of New York at Binghamton, Binghamton, NY 13902-6000 • T: +1 607 7772634 • F: +1 607 7772613 • hogan@binghamton.edu •
Dir.: *Lynn Gamwell* • Ass. Dir.: *Jacqueline Hogan* • Ass. Cur.: *Silvia Ivanova* •
Fine Arts Museum / University Museum – 1967
Fine and dec art, Wedgwood ceramics, Asian and African art,ancient pottery 46817

Birdsboro PA

Daniel Boone Homestead, 400 Daniel Boone Rd, Birdsboro, PA 19508 • T: +1 610 5824900 • F: +1 610 5821744 • jlewars@state.pa.us • www.danielboonehomestead.org •
Pres.: *Dan Berg* •
Local Museum / Folklore Museum / Open Air Museum – 1937
Arts, furniture, folklore, agricultural and blacksmithing tools 46818

Birmingham AL

Alabama Museum of the Health Science, Lister Hall Library, 1700 University Blvd, Birmingham, AL 35294-0013, mail addr: c/o University of Alabama, 1530 3rd Av S - 300 LHL, Birmingham, AL 35294-0013 • T: +1 205 9344475 • F: +1 205 9758476 • www.uab.edu/historical/museum.htm •
Dir.: *Flannery Michael A.* •
Historical Museum – 1981
Medicine hist 46819

Arlington Antebellum Museum, 331 Cotton Av, Birmingham, AL 35211 • T: +1 205 7805656 • www.informationbirmingham.com/arlington/index •
Dir.: *Daniel F. Brooks* •
Folklore Museum – 1953
19th c decorative arts 46820

Birmingham Civil Rights Institute Gallery, 520 16th St N, Birmingham, AL 35203 • T: +1 205 3289696 • bcri@bcri.org • www.bcri.org •
Exec. Dir.: *Dr. Lawrence J. Pijeaux* •
Public Gallery – 1992
Civil rights movement, African American life 46821

Birmingham Museum of Art, 2000 8th Av N, Birmingham, AL 35203 • T: +1 205 2542565 • F: +1 205 2542714 • museum@artsbma.org • www.artsbma.org •
Dir.: *Gail Andrews* • Cur.: *Dr. Donald A. Wood* (Asian Art) • *Dr. Anne Forschler* (Decorative Arts) • *Jeannine O'Grody* (European Art) • *Emily Hanna* (Arts of Africa and The Americas) • *Ron Platt* (Contemporary and Modern Art) • *Graham C. Boettcher* (American Art) • Education: *Samantha Kelley* • Libr.: *Tantum Preston* •
Fine Arts Museum – 1951
American, Asian, European, African and African American, Modern, Contemporary, Decorative Arts (Wedgwood and Lamprecht collections), Pre-Columbian, Native American, Folk Art – Hanson library 46822

McWane Science Center, 200 19th St N, Birmingham, AL 35203 • T: +1 205 7148300 • www.mcwane.org •
Dir.: *Timothy Ritchie* •
Science&Tech Museum – 1997
Science, technology 46823

Sloss Furnaces, 20 32nd St N, Birmingham, AL 35222 • T: +1 205 3241911 • F: +1 205 3246758 • info@slossfurnaces.com • www.slossfurnaces.com •
Dir.: *Robert R. Rathburn* • Cur.: *Karen Utz* • *Paige Wainwright* •
Science&Tech Museum – 1983
Industry, ironmaking plant incl blast furnaces, blowing engines, power house, boilers 46824

Southern Museum of Flight, 4343 N 73rd St, Birmingham, AL 35206-3642 • T: +1 205 8338226 • F: +1 205 8362439 • southernmuseumofflight@yahoo.com • www.southernmuseumofflight.org •
Dir.: *Dr. J. Dudley Pewitt* • Ass. Dir.: *Dr. Donald B. Dodd* •
Science&Tech Museum – 1965
Aviation, civilian, experimental and military aircrafts 46825

Space One Eleven, 2409 Second Av N, Birmingham, AL 35203-3809 • T: +1 205 3280553 • F: +1 205 2546176 • www.spaceoneeleven.org •
Dir.: *Anne Arrasmith* • *Peter Prinz* •
Public Gallery – 1988
Arts education 46826

Visual Arts Gallery, c/o University of Alabama, 113 Humanities Bldg, 900 13th St S, Birmingham, AL 35294-1260 • T: +1 205 9340815 • F: +1 205 9752836 • www.uab.edu •
Dir.: *Brett Levine* •
Public Gallery – 1972
Contemporary art 46827

Birmingham MI

Birmingham Bloomfield Art Center, 1516 S Cranbrook Rd, Birmingham, MI 48009 • T: +1 248 6440866 • F: +1 248 6447904 • jtorno@globalbiz.net • www.bbartcenter.org •
Dir.: *Janet E. Torno* • *Deborah Calahan* (Education) •
Public Gallery – 1956 46828

Birmingham Historical Museum & Park, 556 W Maple Rd, Birmingham, MI 48009 • T: +1 248 5301928 • F: +1 248 5301681 • museum@ci.birmingham.mi.us • www.ci.birmingham.mi.us •
Dir.: *William K. McElhone* •
Historical Museum – 2000
Local hist, furnishings, hist buildings – John West Hunter House, Allen House, and historic park 46829

Bisbee AZ

Bisbee Mining and Historical Museum, 5 Copper Queen Plaza, Bisbee, AZ 85603, mail addr: PO Box 14, Bisbee, AZ 85603 • T: +1 520 4327071 • F: +1 520 4327800 • carrie@bisbeemuseum.org • www.bisbeemuseum.org •
Dir.: *Carrie Gustavson* • Collections Manager: *Annie Larkin* • Office Manager: *Anna Garcia* •
Historical Museum / Science&Tech Museum – 1971
Local hist, mining, AZ Territorial history – Research library (non-circulating); extensive photographic collection 46830

Bishop CA

Laws Railroad Museum, Silver Canyon Rd, Bishop, CA 93515 • T: +1 760 8735950 • lawsmuseum@aol.com • www.lawsmuseum.org •
CEO: *Denton Sonke* •
Science&Tech Museum / Local Museum – 1966
Railroad depot and 22 other bldgs, engines, cars and wagons, paintings 46831

Bishop Hill IL

Bishop Hill Colony, POB 104, Bishop Hill, IL 61419 • T: +1 309 9273345 • F: +1 309 9273345 • www.bishophill.com •
Man.: *Martha J. Downey* •
Local Museum – 1946
Colony artifacts-agricultural items, furnitures, household items, textiles and tools, Olof Krans coll of folk art paintings 46832

Bishop Hill Heritage Museum, 103 N Bishop Hill St, Bishop Hill, IL 61419 • T: +1 309 9273899 • F: +1 309 9273010 • bhha@winco.net • www.bishophill.com •
Pres.: *Warren Schulz* • Dir.: *Michael G. Wendel* •
Local Museum – 1962
Local history, Swedish communal settlement founded in 1846 by religious dissenters 46833

Bishopville SC

South Carolina Cotton Museum, 121 W Cedar Ln, Bishopville, SC 29010 • T: +1 803 4844497 • F: +1 803 4845203 • scottonmus@ftc-i.net • www.sccotton.org •
Exec. Dir.: *Janson L. Cox* •
Special Museum – 1993
Economic, social and political impact of cotton 46834

Bismarck ND

De Mores State Historic Site, c/o North Dakota Heritage Center, 612 E Boulevard Av, Bismarck, ND 58505-0830 • T: +1 701 3282666 • F: +1 701 3283710 • www.state.nd.us/hist/ •
Dir.: *Merl Paaverud* •
Historical Museum – 1936 46835

Elsa Forde Galleries, Bismarck State College, 1500 Edwards Ave, Bismarck, ND 58501 • T: +1 701 2245520 • F: +1 701 2245550 • michelle.lindblom@bsc.nodak.edu • www.bismarckstate.com •
Dir.: *Michelle Lindblom* •
Public Gallery 46836

Former Governors' Mansion, 320 Av B East, Bismarck, ND 58501 • T: +1 701 2553819, 3282666 • F: +1 701 3283710 • www.discovernd.com/hist •
Dir.: *Merlan E. Paaverud* •
Historic Site – 1895
Historic house, original furnishing, gubernatorial memorabilia 46837

Gateway to Science, Frances Leach High Prairie Arts and Science Complex, 1810 Schafer St, Bismarck, ND 58503 • T: +1 701 2581975 • F: +1 701 2581975 • gscience@gscience.org • www.gscience.org •
Exec. Dir.: *Elisabeth Demke* •
Science&Tech Museum – 1994 46838

State Historical Society of North Dakota Museum, c/o North Dakota Heritage Center, 612 E Blvd Av, Bismarck, ND 58505 • T: +1 701 3282666 • F: +1 701 3283710 • histsoc@state.nd.us • www.discovernd.com/hist •
Dir.: *Merlan E. Paaverud* •
Historical Museum – 1895 46839

Black Hawk CO

Lace House Museum, 161 Main St, Black Hawk, CO 80422 • T: +1 303 5825221 • F: +1 303 5820429 •
Historical Museum – 1976 46840

Black River Falls WI

Jackson County Historical Society Museum, 13 S First and 321 Main St, Black River Falls, WI 54615-0037 • T: +1 715 2845314 • www.blackriverfalls.com •
Pres.: *Eugene Gutknecht* •
Local Museum – 1916
Costumes, tools, school artifacts, glass plates, furniture, dairy 46841

Blacksburg SC

Kings Mountain National Military Park, 2625 Park Rd, Blacksburg, SC 29702 • T: +1 864 9367921 • F: +1 864 9369897 •
Military Museum – 1931
Military hist 46842

Blacksburg VA

Armory Art Gallery, c/o Virgina Polytechnic Institute and State University, 201 Draper Rd, Blacksburg, VA 24061 • T: +1 540 2314859, 2315547 • F: +1 540 2317826 •
Fine Arts Museum / University Museum – 1969 46843

Historic Smithfield, 1000 Smithfield Plantation Rd, Blacksburg, VA 24060 • T: +1 540 2313947 • F: +1 540 2313006 • smithfield.plantation@vt.edu • www.civic.bev.net/smithfield •
Local Museum – 1964
Furnishings from the Colonial & Federal periods 46844

Museum of the Geological Sciences, Virginia Polytechnic Institute and State University, Derring Hall, Blacksburg, VA 24061 • T: +1 540 2316029 • F: +1 540 2313386 • seriksn@vt.edu •
Dir./Cur.: *Susan C. Eriksson* •
University Museum / Natural History Museum – 1969
Geology, mineralogy 46845

Perspective Gallery, c/o Virgina Polytechnic Institute and State University, Squires Student Center, Blacksburg, VA 24061 • T: +1 540 2315431, 2316040 • F: +1 540 2315430 •
Dir.: *Tom Butterfield* •
Fine Arts Museum / University Museum – 1969 46846

Blackwell OK

Top of Oklahoma Historical Museum, 303 S Main St, Blackwell, OK 74631 • T: +1 580 3630209 • F: +1 580 3630209 •
Pres.: *George Glaze* •
Local Museum – 1972
Pioneers, artifacts, equip and household items; local school hist, church; military, hist of local ind and agriculture 46847

Blairsville GA

Union County Historical Society Museum, 1 Town Sq, Blairsville, GA 30512 • T: +1 706 7455493 • F: +1 706 7811899 • history1@alltel.net • www.unioncountyhistory.org •
Dir.: *Steve Oakley* •
Local Museum – 1988
Local history 46848

U

Blakely GA

Kolomoki Mounds State Park Museum, Indian Mounds Rd, off US-Hwy 27 and Kolomoki Rds, Blakely, GA 31723 • T: +1 229 7242150/51 • F: +1 229 7242152 • kolomoki@alltel.net • www.gastateparks.org •
Archaeology Museum – 1951
13th c Indian burial mound and village ... 46849

Blanding UT

Edge of the Cedars State Park Museum, 660 West, 400 N, Blanding, UT 84511 • T: +1 435 6782238 • F: +1 435 6783348 • edgeofthecedars@utah.gov •
Cur.: *Deborah Westfall* •
Ethnology Museum – 1978
Remains of 700 AD to 1220 AD structures, ancient dwellings of the Anasazi Indian culture ... 46850

Bloomfield NJ

Historical Society of Bloomfield Museum, 90 Broad St, Bloomfield, NJ 07003 • T: +1 973 5666220 • F: +1 973 5666220 • bloomfhist@aol.com •
Pres.: *Ina Campbell* •
Local Museum – 1966
Furniture, tools, household articles, paintings, toys, newspapers, postcards ... 46851

Bloomfield NM

San Juan County Museum, 6131 US Hwy 64, Bloomfield, NM 87413 • T: +1 505 6322013 • F: +1 505 6321707 • sreducation@sisna.com • www.salmonruins.com •
Exec. Dir.: *Larry L. Baker* •
Local Museum – 1973
History, archaeology, anthropology ... 46852

Bloomfield NY

A.W.A. Electronic-Communication Museum, 2 South Av, Bloomfield, NY 14469 • T: +1 585 3923088 • F: +1 716 3923088 • egable@rochester.rr.com • www.antiquewireless.org •
Dir.: *Thomas Peterson* • Cur.: *Edward Gable* •
Science&Tech Museum – 1953
Radio and communications, elctricity, electronics 46853

Bloomfield Academy Museum, 8 South Av, Bloomfield, NY 14443 • T: +1 585 6577244 • F: +1 585 6577244 • ebhs1838@hotmail.com •
Pres.: *Charles Thomas* • C.E.O.: *Carl J. Elsbree* •
Historical Museum – 1967
Local artifacts, clothing, furniture, household furnishings, agricultural tools and artifacts ... 46854

Bloomfield Hills MI

Cranbrook Art Museum, 39221 Woodward Av, Bloomfield Hills, MI 48303-0801 • T: +1 248 6453361 • F: +1 248 6453324 • artmuseum@cranbrook.edu • www.cranbrookart.edu/museum •
Dir.: *Gregory M. Wittkopp* • Cur.: *Joe Houston* (Exhibits) • Sc. Staff: *Elena Ivanova* (Education) •
Fine Arts Museum – 1927
20th c art, architecture and design – library, auditorium ... 46855

Cranbrook House, 380 Lone Pine Rd, Bloomfield Hills, MI 48303-0801 • T: +1 248 6453147 • F: +1 248 6453151 • mkrygier@cranbrook.edu • www.cranbrook.edu/housegardens •
Decorative Arts Museum – 1971
Home of George Gough and Ellen Scripps Booth, hand-carved woodworking, English arts and crafts-style antiques – gardens ... 46856

Cranbrook Institute of Science - Natural History Collections, 39221 Woodward, Bloomfield Hills, MI 48303-0801 • T: +1 248 6453200 • F: +1 248 6453050 • dbrose@cranbrook.edu • science.cranbrook.edu •
Dir.: *Dr. David S. Brose* • Cur.: *Mark Uhen* (Science) •
Natural History Museum / Science&Tech Museum – 1932
Sciences, mineralogy, geology, zoology, botany, anthropology – library, planetarium, auditorium, herbarium, nature center ... 46857

U

Bloomington IL

Children's Discovery Museum of Central Illinois, 716 E Empire, Bloomington, IL 61701 • T: +1 309 8296222 • F: +1 309 8292292 • staff@cdmci.org • www.cdmci.org •
CEO: *Shari Spaniol Buckellew* •
Natural History Museum / Science&Tech Museum – 1994 ... 46858

The David Davis Mansion, 1000 E Monroe Dr, Bloomington, IL 61701 • T: +1 309 8281084 • F: +1 309 8283493 • davismansion@yahoo.com • www.davismansion.org •
Dir.: *Dr. Marcia Young* • Cur.: *Jeff Saulsbery* •
Historical Museum – 1960
David Davis' Second Empire Italianate brick mansion (Abraham Lincoln's campaign manager) ... 46859

McLean County Arts Center, 601 N East St, Bloomington, IL 61701 • T: +1 309 8290011 • F: +1 309 8294928 • djohnson@mcac.org • www.mcac.org •
Exec. Dir.: *Douglas Johnson* • Cur.: *Alison Hatcher* •
Fine Arts Museum / Public Gallery – 1922 ... 46860

McLean County Museum of History, 200 N Main, Bloomington, IL 61701 • T: +1 309 8270428 • F: +1 309 8270100 • mcmh@mchistory.org • www.mchistory.org •
Exec. Dir.: *Greg Koos* • Cur.: *Susan Hartzold* •
Local Museum – 1892
Prehistoric, costumes, textiles, Civil War artifacts, – Library and Archives ... 46861

Merwin and Wakeley Galleries, Illinois Weslayn University, Bloomington, IL 61702-2900 • T: +1 309 5563077 • F: +1 309 5563976 • jlapham@titan.iwu.edu •
Dir.: *Jennifer Lapham* •
Fine Arts Museum / University Museum
Paintings, sculptures, printmaking, drawings, photraphy ... 46862

Bloomington IN

Elizabeth Sage Historic Costume Collection, Memorial Hall E 232, Indiana University, Bloomington, IN 47405 • T: +1 812 8554627, 8555223 • F: +1 812 8550362 • rowold@indiana.edu • www.indiana.edu/~amid •
Dir.: *Nelda M. Christ* • Cur.: *Kathleen L. Rowold* •
Special Museum – 1935
Textiles, costume ... 46863

Indiana University Art Museum, 1133 E Seventh St, Bloomington, IN 47405 • T: +1 812 8555445 • F: +1 812 8551023 • iuartmus@indiana.edu • www.indiana.edu/~iuam •
Dir.: *Adelheid M. Gealt* •
Fine Arts Museum – 1941
Egyptian, Greek, Roman sculpture; vases, jewelery, coins, glass, 14th - 21th c European and American art ... 46864

Monroe County Historical Society, 202 E Sixth St, Bloomington, IN 47408 • T: +1 812 3322517 • F: +1 812 3555593 • mchm@kiva.net • www.kiva.net/~mchm/museum.htm •
C.E.O: *Kari Price* • Cur.: *Carrie Herz* •
Local Museum – 1980
County history ... 46865

School of Fine Arts Gallery, Indiana University Bloomington, 1201 E Seventh St, 123 Fine Arts Bldg, Bloomington, IN 47405 • T: +1 812 8558490 • F: +1 812 8557498 • sofa@indiana.edu • www.fa.indiana.edu •
Dir.: *Betsy Stirratt* •
Public Gallery – 1987
Paintings, graphics, drawings, sculptures, decorative arts ... 46866

William Hammond Mathers Museum, 416 N Indiana Av, Bloomington, IN 47405 • T: +1 812 8556873 • F: +1 812 8550205 • mathers@indiana.edu • www.indiana.edu/~mathers/ •
Dir.: *Geoffrey Conrad* • Cur.: *Ellen Sieber* •
Folklore Museum / Historical Museum / Ethnology Museum / University Museum – 1963
Anthropology, history, folklore ... 46867

Wylie House Museum, 307 E 2nd St, Bloomington, IN 47401, mail addr: 317 E 2nd St, Bloomington, IN 47401 • T: +1 812 8556224 • libwylie@indiana.edu • www.indiana.edu/~libwylie •
Dir.: *Jo Burgess* •
Historical Museum
Indiana University's first president ... 46868

Bloomington MN

Bloomington Art Center, Greenberg Gallery & Atrium Gallery, 1800 W Old Shakopee Rd, Bloomington, MN 55431 • T: +1 952 5638587 • F: +1 952 5638576 • info@bloomingtonartcenter.com • www.bloomingtonartcenter.com •
Dir.: *Susan M. Anderson* • *Rachel Daly Flentje* (Exhibitions) • *Judith Yerhot* (Education) • Cur.: *Elizabeth R. Greenbaum* •
Public Gallery – 1976
Multi-disciplinary art, paintings ... 46869

Bloomington Historical Museum, 10200 Penn Av, S, Bloomington, MN 55431 • T: +1 952 9484327 •
Pres.: *Vonda Kelly* •
Local Museum – 1964
Local hist, manuscripts ... 46870

Bloomsburg PA

Columbia County Museum, 225 Market St, Bloomsburg, PA 17815-0360 • T: +1 570 7841600 • research@colcohist-gensoc.org • www.colcohist-gensoc.org •
Exec. Dir.: *Bonnie Farver* • Dir.: *Julia Driskell* •
Local Museum – 1914
Local hist, Indian artefacts, agriculture ... 46871

Haas Gallery of Art, c/o Bloomsburg University of Pennsylvania, Arts Dept., Old Science Hall, Bloomsburg, PA 17815 • T: +1 570 3894646 • F: +1 570 3894459 • rhuber@bloomu.edu • www.bloomu.edu •
Dir.: *Vincent Hron* •
University Museum / Fine Arts Museum – 1966 . 46872

Blue Earth MN

Wakefield House Museum, 405 E Sixth St, Blue Earth, MN 56013 • T: +1 507 5265421 • becity@becity.org •
Cur.: *Gerald Schaal* •
Local Museum – 1948
Period furnishings, rural school & life, farm machinery ... 46873

Blue Hill ME

Parson Fisher House, Jonathan Fisher Memorial, 44 Mines Rd, Blue Hill, ME 04614 • T: +1 207 3742159 • parsonfisher@hypernet.com • www.jonathanfisherhouse.org •
Pres.: *Eric Linnel* •
Folklore Museum / Local Museum – 1954
Paintings, art work and furniture by Jonathan Fisher ... 46874

Blue Mounds WI

Little Norway, 3576 Hwy JG N, Blue Mounds, WI 53517 • T: +1 608 4378211 • F: +1 608 4377827 • www.littlenorway.com •
Pres.: *Scott Winner* •
Historical Museum – 1926
Farmstead buildings, Norwegian and pioneer artifacts ... 46875

Blue Mountain Lake NY

Adirondack Museum, Rts 28N and 30, Blue Mountain Lake, NY 12812-0099 • T: +1 518 3527311 ext 101 • F: +1 518 3527653 • dswanson@adkmuseum.org • www.adkmuseum.org •
Exec.Dir.: *David L. Pamperin* • Chief Cur.: *Caroline Welsh* • Cur.: *Hallie Bond* •
Local Museum – 1952
Regional wood and mining industry, boats and boating, transportation, rustic furniture, ... 46876

Bluefield WV

Science Center of West Virginia, 500 Bland St, Bluefield, WV 24701 • T: +1 304 3258855 • F: +1 304 3240513 • info@sciencecenterwv.com • www.sciencecenterwv.com •
Pres.: *Patty Wilkinson* • Dir.: *Thomas R. Willmitch* •
Natural History Museum / Science&Tech Museum – 1994
Interactive science exhibitions ... 46877

Bluffton IN

Wells County Historical Museum, 420 W Market St, Bluffton, IN 46714 • T: +1 219 8249956 • jcsturgeon@adamswells.com • www.parlorcity.com •
Pres.: *James Sturgeon* •
Local Museum – 1935
Culture, industry, genealogy, family hist ... 46878

Blunt SD

Mentor Graham Museum, 103 N Commercial Av, Blunt, SD 57522 •
Local Museum – 1950
Local hist ... 46879

Boalsburg PA

Boal Mansion Museum, Business Rte 322, Boalsburg, PA 16827 • T: +1 814 4666210 • F: +1 814 4669266 • office@boalmuseum.com • www.boalmuseum.com •
Pres.: *Christopher Lee* • *Mathilde Boal Lee* •
Fine Arts Museum / Historical Museum – 1952
Hist for the World War I, decorative arts ... 46880

Pennsylvania Military Museum and 28th Division Shrine, Rtes 322 and 45, Boalsburg, PA 16827 • T: +1 814 4666263 • F: +1 814 4666618 • wleech@state.pa.us • www.psu.edu/dept/aerospace/museum •
Pres.: *Dr. Charles Yesalis* • Dir./Cur.: *William J. Leech* •
Military Museum – 1969
Military hist ... 46881

Boca Raton FL

Boca Raton Historical Society Museum, 71 N Federal Hwy, Boca Raton, FL 33432 • T: +1 561 3956766 • F: +1 561 3954049 • info@bocahistory.org • www.bocahistory.org •
Dir.: *Mary Csar* •
Historical Museum – 1972
History of Balm Beach county cities ... 46882

Boca Raton Museum of Art, 501 Plaza Real, Mizner Park, Boca Raton, FL 33432 • T: +1 561 3922500 • F: +1 561 3916410 • info@bocamuseum.org • www.bocamuseum.org •
Dir.: *George S. Bolge* • Cur.: *Courtney P. Curtiss* (Exhibitions) • *Richard J. Frank* (Education) •
Fine Arts Museum – 1950 ... 46883

Children's Museum of Boca Raton, 498 Crawford Blvd, Boca Raton, FL 33432 • T: +1 561 3686875 • F: +1 561 3957764 •
Exec. Dir.: *Poppi Mercier* •
Special Museum – 1979
World of play toys, Sophia S. Kuzmick dolls, Korvetz Shell coll – library ... 46884

International Museum of Cartoon Art, 201 Plaza Real, Boca Raton, FL 33432 • T: +1 561 3912200 • F: +1 561 3912721 • correspondance@cartoon.org • www.cartoon.org •
Dir.: *Abigail Roeloffs* • Cur.: *Charla Stephen* •
Fine Arts Museum – 1974
Art, cartoons, comic strips and books, magazine illustration – Library ... 46885

Jewish Institute for the Arts, 9557 Islamorade Tce, Boca Raton, FL 33496 • T: +1 561 8835023 • F: +1 561 8835019 • jiarts@aol.com •
Cur.: *Shalom Goldberg* •
Fine Arts Museum – 1966
Paintings, sculpture and works on paper ... 46886

University Gallery, c/o Florida Atlantic University, 777 Glades Rd, Boca Raton, FL 33431-0991 • T: +1 561 2972966, 2973406 • F: +1 561 2972166 • wfaulds@fau.edu • www.fau.edu/galleries •
Dir.: *W. Rod Faulds* •
Fine Arts Museum – 1970
Artinian coll, Albert Binny Backus paintings ... 46887

Boerne TX

Agricultural Heritage Museum, 102 City Park Rd, Boerne, TX 78006, mail addr: POB 1076, Boerne, TX 78006 • T: +1 830 2496007 • info@agmuseum.org • www.agmuseum.org •
Pres./ Dir.: *Kristy Watson* • Dir.: *Darvan Alexander* •
Agriculture Museum – 1986
Ranching and farming life in Texas, machinery and tools, blacksmith shop ... 46888

Kuhlmann King Historical House and Museum, 402 E Blanco, Boerne, TX 78006 • T: +1 830 2492030 • edmond@qvtc.com •
Local Museum – 1970
Furniture, clothing, toys, kitchen utensils ... 46889

Boise ID

Boise Art Museum, 670 S Julia Davis Dr, Boise, ID 83702 • T: +1 208 3458330 • F: +1 208 3452247 • comments@boiseartmuseum.org • www.boiseartmuseum.org •
Exec. Dir.: *Timothy Close* • Cur.: *Sandy Harthorn* (Exhibitions) •
Fine Arts Museum – 1931
20th c American art, Nortwest artists, ceramics, prints, photographs, European art, James Castle coll ... 46890

Discovery Center of Idaho, 131 Myrtle St, Boise, ID 83702 • T: +1 208 3439895 • F: +1 208 3430105 • info@scidaho.org • www.scidaho.org •
Exec. Dir.: *Janine Boire* •
Science&Tech Museum – 1986 ... 46891

Idaho Museum of Mining and Geology, 2455 Old Penitentiary Rd, Boise, ID 83712 • T: +1 208 3433315 • idahomuseum@hotmail.com •
Pres.: *Edward D. Fields* •
Natural History Museum / Science&Tech Museum – 1989
Mining, mineralogy ... 46892

Idaho State Historical Museum, 610 N Julia Davis Dr, Boise, ID 83702 • T: +1 208 3342120 • F: +1 208 3344059 • jochoa@shs.state.id.us • www.state.id.us/ishs/ •
Dir.: *Steve Guerber* • Cur.: *Joe Toluse* •
Historical Museum – 1881
Historical objects, Indian artifacts, maps, photographs ... 46893

Bolinas CA

Bolinas Museum, 48 Wharf Rd, Bolinas, CA 94924 • T: +1 415 8680330 • F: +1 415 8680607 • info@bolinasmuseum.org • www.bolinasmuseum.org •
Dir.: *Dolores Richards* •
Fine Arts Museum / Local Museum – 1982
Marin county history and art ... 46894

Bolivar OH

Fort Laurens State Memorial, 11067 Fort Laurens Rd NW, Bolivar, OH 44612 • T: +1 330 8742059 • F: +1 330 8742936 • zoar@cannet.com • ohiohistory.org/places/ftlaurens •
Dir.: *Kathleen M. Fernández* •
Military Museum – 1972
Archaeological artifacts from the fort, weapons, uniforms ... 46895

Bolton MA

Bolton Historical Museum, Sawyer House, 676 Main St, Bolton, MA 01740 • T: +1 978 7796392 •
Pres.: *Tim Fiehler* •
Local Museum – 1962
Local history, farm/barn blacksmith shop ... 46896

Bolton Landing NY

Marcella Sembrich Opera Museum, 4800 Lake Shore Dr, Bolton Landing, NY 12814-0417 • T: +1 518 6442492 • F: +1 518 6442191 • sembrich@webtv.net • www.operamuseum.org •
Pres.: *Hugh Allen Wilson* • Cur.: *Richard Wargo* •
Music Museum – 1937
Golden age of Opera, opera hist, life and career of Marcella Sembrich ... 46897

Bonham TX

Fort Inglish, Hwy 56 and Chinner St, Bonham, TX 75418 • T: +1 903 5833943, 6402228 •
Pres.: *Wayne Moore* •
Military Museum – 1976
Concentration on northeast Texas hist ... 46898

Sam Rayburn House Museum, West Hwy 82/56, Bonham, TX 75418, mail addr: POB 308, Bonham, TX 75418 • T: +1 903 5835558 • F: +1 903 6400800 • srhmdir@texoma.net • www.thc.stete.tx.us/samrayhouse/srhoefault.html •
Dir.: *Carole Stanton* • Cur.: *Anne Carlson* •
Local Museum – 1975
Sam Rayburn's personal belongings, household furnishings and vehicles ... 46899

The Sam Rayburn Museum, Center for American History of the University of Texas at Austin, 800 W Sam Rayburn Dr, Bonham, TX 75418 • T: +1 903 5832455 • F: +1 903 5837394 • www.cah.utexas.edu •
Asst.Dir.: *Dr. Patrick Cox* •
Local Museum – 1957
Paintings, Sam Rayburn's personal papers and his mementoes on microfilm ... 46900

Bonner Springs KS

The National Agricultural Center and Hall of Fame, 630 Hall of Fame Dr, Bonner Springs, KS 66012 • T: +1 913 7211075 • F: +1 913 7211202 • info@aghalloffame.com • www.aghalloffame.com •
C.E.O.: *Cathi Hahner* • Chm.: *Robert Carlson* •
Agriculture Museum / Science&Tech Museum / Historical Museum – 1958
Agriculture, harvesting & planting equipment, steam traction engines ... 46901

Wyandotte County Museum, 631 N 126th St, Bonner Springs, KS 66012 • T: +1 913 7211078 • F: +1 913 7211394 • pschurkamp@wycokck.org •
Dir.: *Trish Schurkamp* •
Local Museum – 1889
Emigrant Tribal material, textiles and crafts ... 46902

Boone IA

Boone County Historical Center, 602 Story St, Boone, IA 50036 • T: +1 515 4321907 • bchs@opencominc.com •
Pres.: *James Caffrey* • Dir.: *Charles W. Irwin* •
Folklore Museum / Historical Museum / Natural History Museum – 1990
Miltary domestic, agriculture and industrial hist ... 46903

Iowa Railroad Historical Museum, 225 10th St, Boone, IA 50036 • T: +1 515 4324249 • F: +1 515 4324253 • www.scenic-valley.com •
Pres.: *Ken Barkwill* • C.E.O.: *Fenners W. Stevenson* •
Science&Tech Museum – 1983
Railroad cars and engines, photographs, railroad hist ... 46904

Kate Shelley Railroad Museum, 1198 232nd St, Boone, IA 50036 • T: +1 515 4321907 • bchs@opencominc.com •
Pres.: *James Caffrey* • Dir.: *Charles W. Irwinnson* •
Science&Tech Museum / Historic Site – 1976
Railroad hist, working telegraphs ... 46905

Mamie Doud Eisenhower Birthplace, 709 Carroll St, Boone, IA 50036 • T: +1 515 4321896 • F: +1 515 4323097 • mamiedoud@opencominc.com • www.booneiowa.com •
C.E.O./Cur.: *Larry Adams* •
Historic Site – 1970 ... 46906

Boonesboro MO

Boone's Lick Site, State Rd 187, Arrow Rock, Boonesboro, MO 65233 • T: +1 660 8373330 • F: +1 660 8373300 • dsparro@mail.dnr.state.mo.us • www.mostateparks.com/booneslick.htm •
Dir.: *Michael Dickey* •
Archaeology Museum – 1960
Salt kettles, tools, wooden buckests, water wheel drive shaft excavated on site ... 46907

Boonsboro MD

Boonsborough Museum of History, 113 N Main St, Boonsboro, MD 21713-1007 • T: +1 301 4326969 • F: +1 301 4162222 •
Dir.: *Douglas G. Bast* •
Historical Museum – 1975
Local and state hist, Russian icons, ceramics, glassware, Civil War artifacts ... 46908

Boonville IN

Warrick County Museum, 217 S First St, Boonville, IN 47601 • T: +1 812 8973100 • F: +1 812 8976104 • wcmuseum@aol.com •
Pres.: *Ted Kroeger* • Dir.: *Virginia S. Allen* •
Local Museum – 1976
Life in Warrick county from earley 1800 ... 46909

Boothbay ME

Boothbay Railway Village, 586 Wiscasset Rd, Boothbay, ME 04537 • T: +1 207 6334727 • F: +1 207 6334733 • staff@railwayvillage.org • www.railwayvillage.org •
Dir.: *Robert Ryan* •
Science&Tech Museum / Open Air Museum / Historical Museum – 1962
Railroad, locomotives, vehicles ... 46910

Boothbay Harbor ME

Boothbay Region Art Foundation, 7 Townsend Av, Boothbay Harbor, ME 04538 • T: +1 207 6332703 • localart@gwi.net • www.boothbayartists.org •
Pres.: *Linda Yarmosh* •
Fine Arts Museum – 1964 ... 46911

Boothwyn PA

Real World Computer Museum, c/o US Ikon Naaman Creek Center, 7 Creek Pkwy, Boothwyn, PA 19061 • T: +1 610 4949000 • F: +1 610 4942090 • museum@phila.usconnect.com •
Head: *Craig Collins* •
Science&Tech Museum – 1990
Computer ... 46912

Bordentown NJ

Bordentown Historical Society Museum, 211 Crosswicks St, Bordentown, NJ 08505 • T: +1 609 2981740 •
Cur.: *L.A. LeJambre* •
Local Museum – 1930 ... 46913

Borger TX

Hutchinson County Museum, 618 N Main, Borger, TX 79007 • T: +1 806 2730130 • F: +1 806 2730128 • museum@nts-online.net • www.hutchinson-countymuseum.org •
Dir.: *Edward Benz* •
Local Museum – 1977
Local hist of the county from perhistoric times to the present, emphasizing the Oil Boom of 1926 ... 46914

Borrego Springs CA

Anza-Borrego Desert Park Museum, 200 Palm Canyon Dr, Borrego Springs, CA 92004 • T: +1 760 7675311 • info@parks.ca.gov • www.anzaborrego.statepark.org •
Head: *David Van Cleve* •
Natural History Museum / Archaeology Museum – 1979
Archaeology, paleontology, plio-pleistocene mammals, birds and reptiles ... 46915

Boston MA

Ancient and Honorable Artillery Company Museum, Armory, Faneuil Hall, Boston, MA 02109 • T: +1 617 2271638 • F: +1 617 2277221 • vze28mhb@verizon.net • www.ahacsite.org •
Cur.: *John F. McCauley* •
Military Museum
Military weapons, war relicts, paintings, portraits, – archives ... 46916

The Art Institute of Boston Main Gallery, 700 Beacon St, Boston, MA 02215 • T: +1 617 2621223 • F: +1 617 4371226 • robinson@aiboston.edu • www.aiboston.edu •
C.E.O.: *Bonnell Robinson* •
Public Gallery / University Museum – 1909
Visual fine arts ... 46917

Barbara and Steven Grossman Gallery, School of the Museum of Fine Arts, 230 The Fenway, Boston, MA 02115-5596 • T: +1 617 3693718 • F: +1 617 3693856 • www.smfa.edu/News_Exhibitions/Exhibitions •
Fine Arts Museum / University Museum
Faculty and student work of the school ... 46918

Boston Athenæum, Fine Arts Collection and Library, 10 1/2 Beacon St, Boston, MA 02108 • T: +1 617 2270270 • F: +1 617 2275266 • reference@bostonathenaeum.org • www.bostonathenaeum.org •
Dir.: *Richard Wendorf* • Ass. Dir.: *John Lannon* • Cur.: *David Dearinger* (Paintings, Sculpture) • *Sally Pierce* (Prints, Photographs) • *Stanley E. Cushing* (Rare Books) • Cons.: *James Reid-Cunningham* •
Fine Arts Museum / Library with Exhibitions – 1807
Library with art coll of paintings,prints, drawings, sculpture, photographs, daguerrotypes – archives ... 46919

Boston Children's Museum, Museum Wharf, 300 Congress St, Boston, MA 02210-1034 • T: +1 617 4266500 • F: +1 617 4261944 • info@bostoncildrensmuseum.org • www.bostoncildrensmuseum.org •
Pres./CEO: *Louis B. Casagrande* • Dir.: *Leslie Swartz* •
Ethnology Museum – 1913
Indian America, NE Native Americans, multi-cultural period, ethnic dolls, science, industrial art, crafts – theater ... 46920

Boston Fire Museum, 344 Congress St, Boston, MA 02210 • T: +1 617 4821344 • info@bostonfiremuseum.org • www.bostonfiremuseum.org •
Chm.: *Theodore Gerber* • Cur.: *John Vahey* •
Historical Museum – 1977
Fire fighting history, Congress Street Fire Station 46921

Boston National Historical Park, Charlestown Navy Yard, Boston, MA 02129 • T: +1 617 2425642 • F: +1 617 2418650 • marty_blatt@nps.gov • www.nps.gov/bost/ •
Dir.: *Martin Blatt* • Sc. Staff: *Phillip Hunt* •
Historic Site / Archaeology Museum – 1974
library, archives ... 46922

Boston Public Library, 700 Boylston St, Boston, MA 02116, mail addr: POB 286, Boston, MA 02117 • T: +1 617 5365400 • F: +1 617 2364306 • fineartsref@bpl.org; musicref@bpl.org; printref@bpl.org; rarebooksref@bpl.org • www.bpl.org •
Dir.: *Amy Ryan* • Cur.: *Janice Chadbourne* (Visual arts, architecture, and decorative arts reference and books/special collections) • *Susan Glover* (Prints, Rare Books, Manuscripts and Archives) • *Kimberly Reynolds* (Manuscripts) • *Diane Ota* (Music and dance reference and books/special collections) • Ass. Cur.: *Karen Shafts* (Prints) • Cons.: *Steward Walker* (Books) •
Fine Arts Museum / Library with Exhibitions – 1852
Public and research library with art collections (mural decorations, paintings, prints, lithographs, drawings, photographs, sculpture, dioramas) and reference departments with non-circulating research collections on the arts and architecture (books, periodicals, microforms, architectural drawings, manuscripts, rare books, musical scores, special collections and archives.) – Fine Arts Department; Music Department; Prints and Photographs Department; Rare Books and Manuscripts Department; Special Collections ... 46923

Boston Sculptors Gallery, 486 Harrison Av, Boston, MA 02118 • T: +1 617 4827781 • F: +1 617 4827781 • bostonsculptors@yahoo.com • www.bostonsculptors.com •
Dir.: *Anna Shapiro* •
Public Gallery – 1992 ... 46924

Boston University Art Gallery, 855 Commonwealth Av, Boston, MA 02215 • T: +1 617 3534672 • F: +1 617 3534509 • gallery@bu.edu • www.bu.edu/art •
Dir.: *Stacey McCarroll* •
Public Gallery / University Museum – 1960
Paintings ... 46925

Bromfield Art Gallery, 450 Harrison Av, Boston, MA 02118-2436 • T: +1 617 4513605 • bromfield@iname.com • www.bromfieldartgallery.com •
Public Gallery – 1974 ... 46926

Commonwealth Museum, 220 William T. Morrissey Blvd, Boston, MA 02125 • T: +1 617 7272816 • F: +1 617 2288429 • stephen.kenney@sec.state.ma.us • www.sec.state.ma.us/mus/museum •
Dir.: *Maxine Trost* • *Stephen Kenney* •
Historical Museum – 1986
State history ... 46927

Dillaway-Thomas House, Roxbury Heritage State Park, 183 Roxbury St, John Eliot Sq, Boston, MA 02119 • T: +1 617 4453399 • F: +1 617 4455883 • www.mass.gov/dcr/parks/metroboston/rxhp.htm •
Dir.: *Antonio Menefee* •
Local Museum – 1901
Local history ... 46928

Federal Reserve Bank of Boston Collection, POB 2076, Boston, MA 02106-2076 • T: +1 617 9733454, 9733368 • F: +1 617 9734272 • www.bos.frb.org •
Fine Arts Museum – 1978
Coll on US art since the mid-1950s ... 46929

Gallery Nature and Temptation, Kaji Aso Studio, 40 Saint Stephen St, Boston, MA 02115 • T: +1 617 2471719 • F: +1 617 2477564 • administrator@kajiasostudio.com • www.kajiasostudio.com •
Dir.: *Gary Tucker* •
Fine Arts Museum – 1973
Watercolor, oil and Sumi Painting ... 46930

Gibson House Museum, 137 Beacon St, Boston, MA 02116 • T: +1 617 2676338 • F: +1 617 2675121 • gibsonmuseum@aol.com • www.thegibsonhouse.org •
Exec. Dir.: *Barbara Thibault* •
Historical Museum – 1957
Historic house, decorative art, paintings, sculptures, Victorian furniture, clothing ... 46931

Historic New England, Harrison Gray Otis House, 141 Cambridge St, Boston, MA 02114 • T: +1 617 2273956 • F: +1 617 2279204 • pzea@spnea.org • www.historicnewengland.com •
Pres.: *Carl Nold* •
Decorative Arts Museum / Historic Site – 1910
New England antiquities, decorative arts, textiles, photographs – library, archives ... 46932

Institute of Contemporary Art, 955 Boylston St, Boston, MA 02115-3194 • T: +1 617 2665152 • F: +1 617 2664021 • info@icaboston.org • www.icaboston.org •
Dir.: *Jill Medvedow* • Cur.: *Nicholas Baume* •
Public Gallery – 1936
Contemporary artworks in various media ... 46933

Isabella Stewart Gardner Museum, 280 The Fenway, Boston, MA 02115 • T: +1 617 5661401 • F: +1 617 2785167 • Information@isgm.org • www.gardnermuseum.org •
Dir.: *Anne Hawley* • Cur.: *Alan Chong* • *Pieranna Cavalchini* (Contemporary Art) • *Margaret Burchenal* (Education) • Cons.: *Valentine Talland* • *Kathy Francis* •
Fine Arts Museum / Decorative Arts Museum – 1903
15th c Venetian style bldg, paintings, prints, sculpture, tapestries, stained glass, furniture, decorative arts – library ... 46934

John F. Kennedy Presidential Library-Museum, Columbia Point, Boston, MA 02125 • T: +1 617 5141600 • F: +1 617 5141652 • library@kennedy.nara.gov • www.jfklibrary.org •
Dir.: *Deborah Leff* • Cur.: *Frank Rigg* •
Library with Exhibitions – 1979
History, presidential library, documents, film, photos ... 46935

Mary Baker Eddy Library for the Betterment of Humanity, 200 Massachusetts Av, Boston, MA 02115-3017 • T: +1 617 4507000 • F: +1 617 4507048 • www.marybakereddy.org •
Cur.: *Alan K. Lester* •
Library with Exhibitions – 2000
Books, manuscripts, letters and correspondence, personal artifacts, photographs, films, paintings 46936

Massachusetts Historical Society, 1154 Boylston St, Boston, MA 02215 • T: +1 617 6460500 • F: +1 617 8590074 • library@masshist.org • www.masshist.org •
Pres.: *Amalie Kass* • Dir.: *Dennis Fiori* •
Historical Museum – 1791
American and New England history, paintings, sculpture, manuscript colls – library, archives ... 46937

Mobius Gallery, 374 Congress St, Boston, MA 02210 • T: +1 617 5427416 • F: +1 617 4512910 • mobius@mobius.org • www.mobius.org •
Dir.: *Nancy Adams* •
Public Gallery – 1977 ... 46938

Museum of Afro-American History, 46 Joy St, Boston, MA 02114 • T: +1 617 7250022 • F: +1 617 7205225 • history@afroammuseum.org • www.afroammuseum.org •
Exec. Dir.: *Beverly Morgan-Welch* •
Ethnology Museum / Folklore Museum – 1966
Afro-Americans in New England, papers on civil rights and civic organizations, Black family papers, sculptures, Civil War, artworks – library ... 46939

Museum of Fine Arts, Boston, 465 Huntington Av, Boston, MA 02115-5519 • T: +1 617 2679300 • F: +1 617 2670280 • webmaster@mfa.org • www.mfa.org •
Dir.: *Malcolm Rogers* • Dep. Dir./Cur.: *Katherine Getchell* • Cur.: *Rita E. Freed* (Ancient Egyptian, Nubian and Near Eastern Art) • Sc. Staff: *Elliot Davis* (Art Americas) • *Clifford S. Ackley* (Prints, Drawings, Photographs) • *Jane Portal* (Asiatic Art) • *George T.M. Shackelford* (European Arts) • *Edward Saywell* (Contemporary Art) • *Pamela A. Parmal* (Textiles, Fashion arts) • *Darcy Kuronen* (Musical Instruments) • Cons.: *Matthew Siegal* •
Fine Arts Museum – 1870
Ancient Egyptian, Nubian and Near Eastern art, decorative art, Asiatic art, drawings, photography, European and American paintings, contemporary art – library ... 46940

Museum of Science, Science Park, Boston, MA 02114-1099 • T: +1 617 7232500 • F: +1 617 7422246 • information@mos.org • www.mos.org •
Pres./Dir.: *Ioannis Miaoulis* •
Natural History Museum / Science&Tech Museum – 1830
Science technology, natural hist – library, planetarium, theatre of electricity, IMAX ... 46941

Museum of the National Center of Afro-American Artists, 300 Walnut Av, Roxbury, Boston, MA 02119 • T: +1 617 4428614 • F: +1 617 4455525 • bgaither@mfa.org • www.ncaaa.org/museum.html •
Dir./Cur.: *Barry Gaither* •
Fine Arts Museum – 1969
Afro-American, Caribbean, Latin American, contemporary and traditional African art, coll and criticism of black visual arts heritage worldwide 46942

Nichols House Museum, 55 Mount Vernon St, Boston, MA 02108 • T: +1 617 2276993 • F: +1 617 7238026 • nhm@earthlink.net • www.nicholshousemuseum.org •
Dir.: *Flavia Cigliano* •
Decorative Arts Museum / Historical Museum – 1961
Decorative arts, former Beacon Hill home of Rose Standish Nichols, paintings, sculpture ... 46943

Old South Meeting House, 310 Washington St, Boston, MA 02108 • T: +1 617 4826439 • F: +1 617 4829621 • rdeblosi@osmh.org • www.oldsouthmeetinghouse.org •
Pres.: *Catherine E. C. Henn* • Exec. Dir.: *Emily Curran* •
Historical Museum / Historic Site – 1877
Early American political and religious history, Boston Tea Party and the American Revolution ... 46944

Old State House, 206 Washington St, Boston, MA 02109 • T: +1 617 7201713 • F: +1 617 7203289 • bostoniansociety@bostonhistory.org • www.bostonhistory.org •
Exec. Dir.: *Brian Lemay* •
Historical Museum – 1881
Revolutionary and colonial artifacts – library ... 46945

Paul Revere House, 19 North Sq, Boston, MA 02113 • T: +1 617 5232338 • F: +1 617 5231775 • staff@paulreverehouse.org • www.paulreverehouse.org •
Dir.: *Nina Zannieri* • Cur.: *Edith Steblecki* •
Historical Museum – 1907
History, c.1680 Paul Revere house (Boston's oldest) ... 46946

Photographic Resource Center, 832 Commonwealth Av, Boston, MA 02215 • T: +1 617 9750600 • F: +1 617 9750606 • prc@bu.edu • www.prcboston.org •
Exec. Dir.: *Terrence Morash* • Cur.: *Leslie Brown* •
Public Gallery – 1976
library ... 46947

Scollay Square Gallery, 1 City Hall Plaza, Boston, MA 02201 • T: +1 617 6353245 • F: +1 617 6353031 • sarah.hutt@cityofboston.gov • www.cityofboston.gov/arts/galleries.asp •
Dir.: *Sarah Hutt* •
Fine Arts Museum
Temporary exhibitions ... 46948

Shirley-Eustis House, 33 Shirley St, Boston, MA 02119 • T: +1 617 4422275 • F: +1 617 4422270 • ataaffe@netway.com • www.shirleyeustishouse.org • Pres.: *Brian Pfeiffer* • Exec. Dir.: *Tamsen E. George* • Decorative Arts Museum – 1913
Decorative arts, Gov. William Eustis inventory, carriage house ... 46949

Society of Arts and Crafts-Exhibition Gallery, 175 Newbury St, Boston, MA 02116 • T: +1 617 2661810 • F: +1 617 2665654 • exhibitiongallery@societyofcrafts.org • www.societyofcrafts.org • Man.: *Diane Abbott* • Public Gallery – 1897 ... 46950

The Sports Museum of New England, TD Banknorth Garden, 1 Fleet Pl, Boston, MA 02114 • T: +1 617 6241105, 6241234 • F: +1 617 6241326 • mgormley@dncboston.com • www.sportsmuseum.org • Exec. Dir.: *Michelle Gormley* • Cur.: *Brian Codagnone* • Special Museum – 1977 ... 46951

Tremont Gallery, c/o Chinese Culture Institute, 276 Tremont St, Boston, MA 02116 • T: +1 617 5424599 • F: +1 617 3384274 • CCIBoston@yahoo.com • Dir.: *Doris Chu* • Fine Arts Museum – 1980 ... 46952

UrbanArts Institute, c/o Massachusetts College of Art, 621 Huntington Av, Boston, MA 02115-5801 • T: +1 617 8797973 • F: +1 617 8797969 • ricardo.barreto@massart.edu • www.urbanartsinstitute.org • Dir.: *Ricardo D. Barreto* • Fine Arts Museum
Architecture ... 46953

USS Constitution Museum, Charlestown Navy Yard, Bldg 22, Boston, MA 02129, mail addr: POB 1812, Boston, MA 02129 • T: +1 617 4261812 • F: +1 617 2420496 • info@ussconstitutionmuseum.org • www.ussconstitutionmuseum.org • Dir.: *Burt Logan* • Cur.: *Sarah Watkins* • Military Museum / Historical Museum – 1972
Historic ship, old ironsides, launched 1797, world's oldest commissioned warship afloat ... 46954

Boulder CO

Boulder History Museum, 1206 Euclid Av, Boulder, CO 80302 • T: +1 303 4493464 • F: +1 303 9388322 • info@boulderhistory.org • www.boulderhistorymuseum.org • Exec. Dir.: *Nancy Geyer* • Local Museum – 1944
Local history ... 46955

Boulder Museum of Contemporary Art, 1750 13th St, Boulder, CO 80302 • T: +1 303 4432122 • F: +1 303 4471633 • info@bmoca.com • www.bmoca.com • Dir.: *Brandi Mathis* • Fine Arts Museum
Art ... 46956

Colorado University Art Galleries, 318 UCB, Sibell Wolle Fine Arts Bldg, Boulder, CO 80309-0318 • T: +1 303 4928300 • F: +1 303 7354197 • gilmore@colorado.edu • www.colorado.edu/cuartgalleries • Dir.: *Jerry Gilmore* • Fine Arts Museum – 1939
Colorado coll ... 46957

Colorado University Heritage Center, Campus, Old Main St, Boulder, CO 80309 • T: +1 303 4926329 • F: +1 303 4926799 • oltmans_k@cufund.colorado.edu • Dir.: *Kay Oltmans* • Ass. Dir.: *Nancy Lee Miller* • University Museum / Local Museum – 1985
Local hist ... 46958

Cu Heritage Center, University of Colorado, Boulder Campus, Old Main St, Boulder, CO 80309 • T: +1 303 4926329 • F: +1 303 4926799 • oltmans_k@cufund.colorado.edu • Dir.: *Kay Oltmanns* • University Museum – 1985
University hist, space exploration and Colorado astronauts, Glenn Miller's gold records ... 46959

Leanin' Tree Museum of Western Art, 6055 Longbow Dr, Boulder, CO 80301 • T: +1 303 5301442 ext 299 • F: +1 303 5812152 • artmuseum@leanintree.com • www.leanintree.com • Pres.: *Thomas E. Trumble* • Dir.: *Edward P. Trumble* • Fine Arts Museum – 1974
Art of the American West ... 46960

University of Colorado Museum, Broadway, between 15th and 16th St, Boulder, CO 80309 • T: +1 303 4926892 • F: +1 303 4924195 • cumuseum@colorado.edu • www.colorado.edu/cumuseum • Dir.: *Prof. Linda S. Cordell* • Cur.: *John R. Rohner* (Museography) • *Deborah Confer* (Anthropology) • *Stephen Lakson* • *Tim Hogen* (Botany) • *Nan Lederer* • University Museum / Natural History Museum / Local Museum – 1902
Natural history, anthropology, botany, zoology, geology, entomology, osteology, textiles, archaeology ... 46961

Boulder UT

Anasazi State Park Museum, 460 North Hwy 12, Boulder, UT 84716 • T: +1 435 3357308 • F: +1 435 3357352 • nrdpr.ansp@tate.ut.us • www.stateparks.utah.gov • Dir.: *Cortland Nelson* • Cur.: *Bill Latady* • Archaeology Museum – 1970
1050-1200 AD, excavated Anasazi Indian Village ... 46962

Boulder City NV

Hoover Dam Museum, 1305 Arizona St, Boulder City, NV 89005 • T: +1 702 2941988 • F: +1 702 2944380 • bcmha@yahoo.com • www.bcmha.org • Pres.: *Cheryl Ferrence* • Science&Tech Museum – 1981 ... 46963

Bountiful UT

Davis Art Center, 745 S Main St, Bountiful, UT 84010 • T: +1 801 2920367 • F: +1 801 2927298 • bdac@networld.com • www.bdac.org • Dir.: *Arley Curtz* • Public Gallery
Art ... 46964

Bourbonnais IL

Exploration Station, A Children's Museum, 1095 W Perry St, Bourbonnais, IL 60914 • T: +1 815 9339905 • F: +1 815 9335468 • rose@btpd.org • www.btpd.org • Science&Tech Museum – 1987
Aircraft, NASA replicas ... 46965

Bourne MA

Aptucxet Trading Post Museum, 24 Aptucxet Rd, Bourne, MA 02532-0795 • T: +1 508 7598167 • info@bournehistoricalsoc.org • www.bournehistoricalsoc.org • C.E.O.: *Judith McAlister* • Cur.: *Eleanor A. Hammond* • Historical Museum – 1921
Trading post, reconstructed on original 1627 site, railroadsalt work ... 46966

Bowie MD

Belair Mansion, 12207 Tulip Grove Dr, Bowie, MD 20715 • T: +1 301 8093089 • F: +1 301 8092308 • museums@cityofbowie.org • www.cityofbowie.org • Cur.: *Stephen E. Patrick* • Decorative Arts Museum – 1968
Furniture and decorative artcs c 1730-1957 ... 46967

Belair Stable Museum, 2835 Belair Dr, Bowie, MD 20715 • T: +1 301 8093089 • F: +1 301 8092308 • museums@cityofbowie.org • www.cityofbowie.org • Dir.: *Stephen E. Patrick* • Science&Tech Museum / Agriculture Museum – 1968
Thoroughbred racing, farming, carriage coll ... 46968

Bowie Railroad Station and Huntington Museum, 8614, Chestnut Av, Bowie, MD 20715 • T: +1 301 8093089 • F: +1 301 8092308 • museums@cityofbowie.org • www.cityofbowie.org • Dir.: *Stephen E. Patrick* • Science&Tech Museum / Local Museum – 1994
Railroad artifacts ... 46969

Bowie's Heritage and Children's Museum, Welcome Center, Bowie, MD 20715 • T: +1 301 5752488 • F: +1 301 8092308 • museums@cityofbowie.org • www.cityofbowie.org • Dir.: *Stephen E. Patrick* • Ass. Dir.: *Pamela Williams* • Local Museum – 1968
Local hist ... 46970

Radio-Television Museum, 2608 Mitchellville Rd, Bowie, MD 20716 • T: +1 301 8093088 • F: +1 301 8092308 • kmellgren3@comcast.net • www.radiohistory.org • Dir.: *Stephen E. Patrick* • Science&Tech Museum – 1999
Broadcast hist 1900-present ... 46971

Bowling Green FL

Paynes Creek Historic State Park, 888 Lake Branch Rd, Bowling Green, FL 33834 • T: +1 863 3754717 • F: +1 863 3754510 • paynes.creek@dep.state.fl.us • www.floridastateparks.org • Historical Museum / Folklore Museum – 1981
Fort Chokonikla, Seminole Indian War Fort ... 46972

Bowling Green KY

The Kentucky Museum, Western Kentucky University, 1 Big Red Way, Bowling Green, KY 42101 • T: +1 502 7456258 • F: +1 502 7454878 • nancy.baird@wku.edu • www.wku.edu/library/museum/ • Cur.: *Donna Parker* (Exhibits) • *Sandra Staebell* (Collections) • *Laura Harper Lee* (Education) • University Museum / Local Museum – 1931
Period furniture, Shaker decorative arts, toys musical instruments ... 46973

National Corvette Museum, 350 Corvette Dr, Bowling Green, KY 42101 • T: +1 270 7817973 • F: +1 270 7815286 • bobbiejo@corvettemuseum.com • www.corvettemuseum.com • Dir.: *Wendell Strode* • Science&Tech Museum – 1989
75 Corvette automobiles ... 46974

Riverview at Hobson Grove, 1100 W Main Av, Bowling Green, KY 42102-4859 • T: +1 270 8435565 • F: +1 270 8435557 • rivervw@bowlinggreen.net • www.bgky.org/riverview.htm • Decorative Arts Museum / Historical Museum / Historic Site – 1972 ... 46975

Western Kentucky University Gallery, Rm 441, Ivan Wilson Center for Fine Arts, Bowling Green, KY 42101 • T: +1 270 7453944 • F: +1 270 7455932 • artdept@wku.edu • www.wku.edu/ • Head: *Kim Chalmers* • Fine Arts Museum / Public Gallery / University Museum – 1973
Contemporary works ... 46976

Bowling Green OH

Fine Arts Center Galleries, Bowling Green State University, Fine Arts Center, Bowling Green, OH 43403-0211 • T: +1 419 3728525 • F: +1 419 3722544 • jnathan@bgsu.edu • www.gallery.bgsu.edu • Dir.: *Jaqueline S Nathan* • Fine Arts Museum / University Museum – 1960
Contemporary Prints and Regional Artists, Changing exhibitions ... 46977

Wood County Historical Center, 13660 County Home Rd, Bowling Green, OH 43402 • T: +1 419 3520967 • F: +1 419 3526220 • director@woodcountyhistory.org • www.woodcountyhistory.org • Dir.: *Christie Raber* • Local Museum – 1955
Regional hist, poor farm hist, oil boom hist, farm machinery, drilling eqipment, dec art ... 46978

Boyertown PA

Boyertown Museum of Historic Vehicles, 85 S Walnut St, Boyertown, PA 19512-1415 • T: +1 610 3672090 • F: +1 610 3679712 • museum@enter.net • www.boyertownmuseum.org • Pres.: *Bernard Hofmann* • Exec. Dir.: *Kenneth D. Wells* • Science&Tech Museum – 1968
Vehicle building in Southeastern Pennsylvania ... 46979

Boys Town NE

Boys Town Hall of History & Father Flanagan House, 14057 Flanagan Blvd, Boys Town, NE 68010 • T: +1 402 4981185 • F: +1 402 4981159 • lyncht@boystown.org • www.boystown.org • C.E.O.: *Thomas J. Lynch* • Historic Site – 1986
History ... 46980

Bozeman MT

Helen E. Copeland Gallery, Montana State University, c/o MSU School of Art, 242 Haynes Hall, Bozeman, MT 59717 • T: +1 406 9944501 • F: +1 406 9943680 • dungan@montana.edu • www.montana.edu/wwwart • Dir.: *Erica Howe Dungan* • Fine Arts Museum / University Museum – 1974
Japanese patterns, native American ceramics, prints ... 46981

Museum of the Rockies, 600 W Kagy, Bozeman, MT 59717 • T: +1 406 9945283 • F: +1 406 9942682 • sfischer@montana.edu • www.museumoftherockies.org • Dir.: *Sheldon McKamey* • Cur.: *Dr. Leslie B. Davis* (Archaeology and Ethnology) • *John R. Horner* (Paleontology) • *Steven B. Jackson* (Art and Photography) • *Margaret M. Woods* (Textiles and History) • Local Museum – 1956
American Western and Indian art, drawings, anthropology, archaeology, ethnology, decorative arts ... 46982

School of Art Gallery, Montana State University, 213 Haynes Hall, Bozeman, MT 59717 • T: +1 406 9944501 • F: +1 406 9943680 • art@montana.edu • www.montana.edu/art • Dir.: *Richard Helzer* • Public Gallery ... 46983

Bradenton FL

Art League of Manatee County, 209 Ninth St W, Bradenton, FL 34205 • T: +1 941 7462862 • F: +1 941 7462319 • artleague@almc.org • www.almc.org • Dir.: *Diane Shelly* • Fine Arts Museum – 1937
Temporary exhibitions ... 46984

Desoto National Memorial, 75th St NW, Bradenton, FL 34209, mail addr: POB 15390, Bradenton, FL 34280-5390 • T: +1 941 7920458 • F: +1 941 7925094 • deso_ranger_activities@nps.gov • www.nps.gov/deso • Historical Museum – 1948
6th-c European military artifacts, pre-historic Native American artifacts ... 46985

Manatee Village Historical Park Museum, 604 15th St E, Bradenton, FL 34208 • T: +1 941 7497165 • F: +1 941 7085924 • cathy.slusser@manateeclerk.com • www.manateeclerk.com • Historic Site – 1974
County hist from 1841 - 1914 ... 46986

South Florida Museum, 201 10th St W, Bradenton, FL 34205 • T: +1 941 7464131 ext 14 • F: +1 941 7462556 • info@southfloridamuseum.org • www.southfloridamuseum.org • Dir.: *Alan P. Haines* • Local Museum / Natural History Museum – 1946
Archaeology, astronomy, geology and natural hist, Indian artifacts ... 46987

Bradford OH

Bradford Ohio Railroad Museum, 501 E Main St, Bradford, OH 45308 • C.E.O.: *Marilyn K. Kosica* • Special Museum
Historic railyards, railroad tower ... 46988

Bradford PA

Zippo and Case Visitors Center Company Museum, 1932 Zippo Dr, Bradford, PA 16701, mail addr: 33 Barbour St, Bradford, PA 16701 • T: +1 814 3681932, 3682711 • F: +1 814 3682874 • lmeabon@zippo.com • www.zippo.com • Dir.: *Patrick Grandy* • Cur.: *Linda Meabon* • Historical Museum – 1994
Zippo lighters and other Zippo products, company hist, case knives ... 46989

Bradford VT

Bradford Historical Society, Town Hall, Main St, Bradford, VT 05033 • T: +1 802 2224727, 2229026 • Cur.: *Phyllis Lavelle* • Association with Coll – 1959
Civil War artifacts, scrapbooks, store signs, china, 19th c costumes, jewelry, cemetery survey ... 46990

Brainerd MN

Crow Wing County Historical Museum, 320 W Laurel St, Brainerd, MN 56401 • T: +1 218 8293268 • F: +1 218 8284434 • history@brainerd.net • Exec. Dir.: *Mary Lou Moudry* • Local Museum – 1927
Local hist, archaeology, industry, Indian artifacts – library, archives ... 46991

Braintree MA

Braintree Historical Society Museum, 31 Tenney Rd, Braintree, MA 02184-6512 • T: +1 781 8481640 • F: +1 781 3800731 • genthayer@aol.com • www.braintreehistorical.org • Dir.: *Brian A. Kolner* • Cur.: *Dr. Robert H. Downey* • *Jennifer Potts* • Local Museum – 1930
History, decorative arts, antique furniture, American and Japanese fans – library ... 46992

Branchville SC

Branchville Railroad Shrine and Museum, 7505 Freedom Rd, Branchville, SC 29432 • T: +1 803 2748821 • F: +1 803 2748760 • Pres.: *Luther Folk* • C.E.O.: *Johnny Norris* • Science&Tech Museum – 1969
Transportation ... 46993

Brandon OR

Coquille River Museum, 270 Fillmore and Hwy 101, Brandon, OR 97411 • T: +1 541 3472164 • F: +1 541 3472164 • jnrknox@harborside.com • Dir.: *Judy Knox* • Local Museum – 1977
Native Americans, fires, period clothing, businesses and schools ... 46994

Branford CT

Harrison House, 124 Main St, Branford, CT 06405 • T: +1 203 4884828 • branfordhistory.org • Pres.: *Joseph Chadwick* • Local Museum – 1960
Local hist, farm tools ... 46995

Branson MO

Hollywood Wax Museum, 3030 W Hwy-76, Branson, MO 65616 • T: +1 417 3378277 • branson@hollywoodwax.com • www.hollywoodwax.com • Special Museum ... 46996

Roy Rogers-Dale Evans Museum, 3950 Green Mountain Dr, Branson, MO 65616 • T: +1 417 3391900 • administrator@royrogers.com • www.royrogers.com • C.E.O.: *Roy Dusty Rogers jr* • Performing Arts Museum / Ethnology Museum – 1967
Costumes, cowboy and western memorabilia, awards, gun coll – Theater ... 46997

Brattleboro VT

Brattleboro Museum and Art Center, 10 Vernon St, Brattleboro, VT 05301 • T: +1 802 2570124 • F: +1 802 2589182 • info@brattleboromuseum.org • www.brattleboromuseum.org • Dir.: *Konstantin von Krusenstiern* • Fine Arts Museum / Local Museum – 1972
Changing visual arts exhib, art of our times ... 46998

Brazosport TX

Brazosport Museum of Natural Science, 400 College Dr., Brazosport, TX 77566 • T: +1 979 2657831 • Pres.: *J.H. McIver* • Natural History Museum / Science&Tech Museum – 1962
Natural science ... 46999

Brea CA

City of Brea Gallery, 1 Civic Center Circle, Brea, CA 92821-5732 • T: +1 714 9907730 • breagallery@cityofbrea.net • www.cityofbrea.net • Dir.: *Dianna Rivera* • Public Gallery – 1980
Large sculptures ... 47000

Breckenridge CO

Fine Arts Gallery, c/o Colorado Mountain College, 103 S Harris St, Breckenridge, CO 80424 • T: +1 970 4536757 • F: +1 970 4512209 •
Public Gallery – 1980 47001

Breckenridge MN

Wilkin County Historical Museum, 704 Nebraska Av, Breckenridge, MN 56520 • T: +1 218 6431303 •
Pres./CEO: *Gordon Martinson* •
Local Museum – 1965
Local hist, maps/ manuscripts 47002

Breckenridge TX

Swenson Memorial Museum of Stephens County, 116 W Walker, Breckenridge, TX 76424 • T: +1 254 5598471 •
Dir.: *Freda Mitchell* •
Local Museum – 1970
Farming, business, ranching, oil equipment 47003

Bremerton WA

Bremerton Naval Museum, 402 Pacific Av, Bremerton, WA 98337 • T: +1 360 4797447 • F: +1 360 3774186 • bremnavmuseum@aol.com •
C.E.O.: *Charleen Zettl* •
Military Museum – 1954
Naval hist 47004

Kitsap County Historical Museum, 280 4th St, Bremerton, WA 98337, mail addr: POB 903, Bremerton, WA 98337 • T: +1 360 4796226 • F: +1 360 4159294 • kchsm@telebyte.net • www.waynes.net/kchsm/ •
Cur.: *Gail Campbell-Ferguson* •
Historical Museum – 1949
Local hist 47005

Brenham TX

Texas Baptist Historical Center Museum, 10405 FM 50, Brenham, TX 77833 • T: +1 979 8365117 • F: +1 979 8362929 •
Dir./Cur.: *D.H. Strickland* •
Religious Arts Museum – 1965
Church artifacts, manuscript coll 47006

Brevard NC

Spiers Gallery, Brevard College, c/o Sims Art Center, 400 N Broad St, Brevard, NC 28712 • T: +1 828 8838292 ext 2325 • F: +1 828 8843790 • bbyers@brevard.edu • www.brevard.edu •
Dir.: *Bill Byers* •
Fine Arts Museum / University Museum
Changing exhibitions of temporary American art 47007

Brewster MA

Cape Cod Museum of Natural History, 869 Rte 6a, Brewster, MA 02631-7710 • T: +1 508 8963867 • F: +1 508 8968844 • info@ccmnh.org • www.ccmnh.org •
Dir.: *Connie Boyce* •
Natural History Museum – 1954
Natural history, marine aquaria, local archaeology – herbarium 47008

New England Fire and History Museum, 1439 Main St, Brewster, MA 02631 • T: +1 508 8965711, 4322450 •
Pres./Dir.: *Joan Frederici* •
Local Museum – 1972
Fire fighting, local fire-relatad history, paintings, lithographs 47009

Brewster NY

Southeast Museum, 67 Main St, Brewster, NY 10509 • T: +1 845 2797500 • F: +1 845 2791992 • sem@bestweb.net • www.southeastmuseum.org •
Exec. Dir.: *Amy Campanero* • Cur.: *Joan Crawford* •
Local Museum / Science&Tech Museum – 1963
McLane Railroad, hist of the Harlem Line, Borden Condensed Milk Factory coll, local economic and social hist, early circus items 47010

Brewton AL

Thomas E. McMillan Museum, c/o Jefferson Davis College, 220 Alco Dr, Brewton, AL 36426 • T: +1 251 8091528 • F: +1 251 8091528 • museum@jdcc.edu • museum.jdcc.edu •
Decorative Arts Museum / Archaeology Museum – 1978
Archaeology, Indian artifacts, trading wares, Civil war 47011

Bridgehampton NY

Bridgehampton Historical Society Museum, 2368 Main St, Bridgehampton, NY 11932-0977 • T: +1 631 5371088 • F: +1 631 5374225 • bhhs@hamptons.com • www.hamptons.com/bhhs •
Pres.: *Leonard Davenport* •
Historical Museum – 1956
Agriculture, tolls, antique engines and early farm machines; period furniture, costumes 47012

Dan Flavin Art Institute, Corwith Av, Bridgehampton, NY 11932 • T: +1 631 5371476 • www.diaart.org •
Fine Arts Museum
Works by Dan Flavin 47013

Bridgeport AL

Russell Cave Museum, 3729 County-Rd 98, Bridgeport, AL 35740 • T: +1 256 4952672 • F: +1 256 4959220 • ruca_ranger_activities@nps.gov • nps.gov/ruca •
Head: *Bill Springer* •
Archaeology Museum – 1961
Archaeology 47014

Bridgeport CA

Bodie Park Museum, Hwy 395, Bridgeport, CA 93517, mail addr: POB 515, Bridgeport, CA 93517 • T: +1 760 6476445 • info@parks.ca.gov • www.parks.ca.gov •
Cur.: *Shirley J. Kendall* • *Judith K. Polanich* •
Historic Site / Open Air Museum – 1962
Gold rush mining boom town 47015

Bridgeport CT

Barnum Museum, 820 Main St, Bridgeport, CT 06604 • T: +1 203 3311104 • F: +1 203 3394341 • www.barnum-museum.org •
Cur.: *Kathleen Maher* •
Performing Arts Museum – 1893
Circus history 47016

The Discovery Museum, 4450 Park Av, Bridgeport, CT 06604 • T: +1 203 3723521 • F: +1 203 3741929 • hamilton@discoverymuseum.org • www.discoverymuseum.org •
Pres.: *Paul Audley* • Dir.: *Leslie E. Birkmaier* •
Fine Arts Museum / Science&Tech Museum – 1958
Art, science – planetarium 47017

Housatonic Museum of Art, 900 Lafayette Blvd, Bridgeport, CT 06604-4704 • T: +1 203 3325000 • F: +1 203 3325123 • rzella@hcc.commnet.edu • www.hcc.commnet.edu •
Dir./Cur.: *Robbin Zella* •
Fine Arts Museum / University Museum – 1967
European and American art 19th-20th c, contemporary Latinamerican and Connecticut art 47018

University Art Gallery, Bernhard Arts and Humanities Center, 84 Iranistan Av, Bridgeport, CT 06601-2449 • T: +1 203 5764402, 5764419 • F: +1 203 5764051 • tjuliusb@cr2.nai.net •
Dir.: *Thomas Juliusburger* •
Fine Arts Museum / University Museum – 1972
Art gallery 47019

Bridgeton NJ

New Sweden Farmstead Museum, City Park, Bridgeton, NJ 08302 • T: +1 609 6531271 • www.biderman.net/nj/new.htm •
C.E.O.: *Michael Tartaglio* •
Historical Museum / Agriculture Museum – 1983
Reproduction of 17th c farmstead 47020

Woodruff Museum of Indian Artifacts, 150 E Commerce St, Bridgeton, NJ 08302 • T: +1 856 4512620 •
Ethnology Museum – 1976 47021

Bridgewater VA

Reuel B. Pritchett Museum, Bridgewater College, E College St, Bridgewater, VA 22812 • T: +1 540 8285462, 8285414 • F: +1 540 8285482 • tbarkley@bridgewater.edu • www.bridgewater.edu •
Cur.: *Terry Barkley* •
Decorative Arts Museum / University Museum / Ethnology Museum – 1954
Rare books & bibles, manuscripts coll, coin & currency, guns, glass 47022

Bridgton ME

Bridgton Historical Museum, Gibbs Av, Bridgton, ME 04009 • T: +1 207 6473699 • bhs@megalink.net • www.megalink.net/~bhs •
Pres.: *Brandon Woolley* •
Local Museum / Folklore Museum – 1953
Artifacts, decorative arts, mechanical implements, photographs 47023

Brigham City UT

Brigham City Museum-Gallery, 24 N 300 W, Brigham City, UT 84302 • T: +1 435 7236769 • F: +1 435 7236769 • www.brigham-city.org •
Dir.: *Larry Douglass* •
Local Museum – 1970
Art, historical coll focussing the Mormon Communitarian Society in Brigham City 1865-1881 47024

Golden Spike National Historic Site, POB 897, Brigham City, UT 84302 • T: +1 435 4712209 • F: +1 435 4712341 • gosp_interpretation@nps.gov • www.nps.gov/gosp •
Head: *Mary Risser* •
Local Museum – 1965
Material relating to the railrod, constrution, workers, manuscript coll 47025

Brighton CO

Adams County Museum, 9601 Henderson Rd, Brighton, CO 80601-8100 • T: +1 303 6597103 • F: +1 303 6597103 • dnvrdixie@aol.com •
Pres.: *Virginia Eppinger* •
Local Museum / Agriculture Museum – 1987
Prehistoric life and geology, native Americans, early Western settlement 47026

Bristol CT

American Clock and Watch Museum, 100 Maple St, Bristol, CT 06010 • T: +1 860 5836070 • F: +1 860 5831862 • info@clockmuseum.org • www.clockmuseum.org •
Dir.: *Donald Muller* • Sc. Staff: *Chris H. Bailey* (Horology) •
Special Museum – 1952
Horology 47027

New England Carousel Museum, 95 Riverside Av, Bristol, CT 06010 • T: +1 860 5855411 • F: +1 860 3140483 • info@thecarouselmuseum.com • www.thecarouselmuseum.com •
Exec. Dir.: *Louise L. DeMars* •
Special Museum – 1989 47028

Bristol IN

Elkhart County Historical Museum, 304 W Vistula St, Bristol, IN 46507 • T: +1 574 5356458, 8484322 • F: +1 584 8485703 • museum@elkhartcountyparks.org • www.elkhartcountyparks.org •
Dir.: *Tina Mellott* •
Local Museum – 1896
Agriculture, railroad hist, furnishings, apparel, military, Native Americans 47029

Bristol RI

Asri Environmental Education Center, 1401 Hope St, Bristol, RI 02809 • T: +1 401 2457500 • F: +1 401 2459339 • eec@asri.org • www.asri.org •
C.E.O.: *Lee C. Schisler jr* •
Special Museum – 1897
Flora, fauna and natural hist of Rhode Island 47030

Bristol Historical and Preservation Society Museum, 48 Court St, Bristol, RI 02809 • T: +1 401 2537223 • F: +1 401 2537223 •
Cur.: *Ray Battcher* •
Local Museum – 1936
Local hist 47031

Coggeshall Farm Museum, Rte 114, Poppasquash Rd, Bristol, RI 02809 • T: +1 401 2539062 • coggmuseum@aol.com •
Exec. Dir.: *Robert L. Shermann* •
Historical Museum – 1968
Rural life on a RI coastal farm 47032

Haffenreffer Museum of Anthropology, Brown University, Mount Hope Grant, 300 Tower St, Bristol, RI 02809-4050 • T: +1 401 2538388, 2531287 • F: +1 401 2531198 • kathleen_luke@Brown.edu • www.haffenreffermuseum.org •
Dir.: *Shepard Krech* • Dep. Dir. & Chief Cur.: *Kevin P. Smith* • Cur./Res.: *Douglas Anderson* • Ass. Cur.: *Thierry Gentis* • *David Gregg* •
University Museum / Ethnology Museum – 1956
Anthropology 47033

Herreshoff Marine Museum/America's Cup Hall of Fame, 1 Burnside St, Bristol, RI 02809 • T: +1 401 2535000 • F: +1 401 2536222 • b.knowles@herreshoff.org • www.herreshoff.org •
C.E.O. & Pres.: *Halsey C. Herreshoff* • Cur.: *John Palmeiri* •
Historical Museum – 1971
Maritime hist 47034

Brockport NY

Tower Fine Arts Gallery, SUNY Brockport, Tower Fine Arts Bldg, Brockport, NY 14420 • T: +1 585 3952209 • F: +1 585 3952588 • llonnen@po.brockport.edu • www.brockport.edu/art •
Chm.: *Debra Fisher* • Dir.: *Timothy Massey* •
University Museum / Public Gallery – 1964
Changing exhibitions, works of E.E. Cummings – Rainbow Gallery with students' works 47035

Brockton MA

Fuller Craft Museum, 455 Oak St, Brockton, MA 02301 • T: +1 508 5886000 • F: +1 508 5876191 • director@fullercraft.org • www.fullermuseum.org •
Dir.: *Gretchen Keyworth* •
Fine Arts Museum – 1969
Contemporary crafts, art – library 47036

Broken Bow NE

Custer County Historical Society Museum, 455 S 9th St, Broken Bow, NE 68822 • T: +1 308 8722203 • custer.county.history@navix.net • www.rootsweb.com/~necuster •
Local Museum – 1962
Local hist 47037

Broken Bow OK

Oklahoma Forest Heritage Center Forestry Museum, US-259A, Beavers Bend Resort Park, Broken Bow, OK 74728, mail addr: POB 157, Broken Bow, OK 74728 • T: +1 580 4946497 • F: +1 580 4946689 • fhc@beaversbend.com • www.beaversbend.com •
Dir.: *Michelle Finch* •
Natural History Museum – 1976
Forestry tools, Caddo Indian papermaking, lumbering 47038

Bronx NY

Bartow-Pell Mansion Museum, 895 Shore Rd, Pelham Bay Park, Bronx, NY 10464 • T: +1 718 8851461 • F: +1 718 8859164 • bartowpell@aol.com • www.bartowpellmansionmuseum.org •
Dir.: *Robert Engel* •
Local Museum – 1914
1842 Greek revival house with carriage house and garden, Empire furniture 47039

Bronx Museum of the Arts, 1040 Grand Concourse, Bronx, NY 10456 • T: +1 718 6816000 • F: +1 718 6816181 • bxarts@bronxview.com • www.bronxview.com/museum •
Sen. Cur.: *Lydia Yee* • Asst. Cur.: *Amy Rosenblum Martin* •
Fine Arts Museum – 1971
Modern and contemporary African, Latin American and Asian art 47040

Bronx River Art Center, 1087 E Tremont Av, Bronx, NY 10460 • T: +1 718 5895819 • F: +1 718 8608303 • www.bronxriverart.org •
Dir.: *Gail Nathan* •
Public Gallery
Changing exhibitions of modern and contemporary art of all media 47041

Brooklyn Arts Council Gallery, 195 Cadman Plaza W, Bronx, NY 11201 • T: +1 718 6250080 •
Dir.: *Pamela Billig* •
Association with Coll
Junior Children's Museum of Art 47042

City Island Nautical Museum, 190 Fordham St, Bronx, NY 10464 • T: +1 718 8850008 • F: +1 718 8850507 • cihs@cityislandmuseum.org • www.cityislandmuseum.org •
Pres. & Cur.: *Tom Nye* •
Historical Museum – 1964
Hist artifacts, paintings and photographs, from Indian times to present, City Island and yachting industry 47043

Edgar Allan Poe Cottage, Bronx County Historical Society, Poe Park, Grand Concourse, E Kingsbridge Rd, Bronx, NY 10458 • T: +1 718 8818900 • F: +1 718 8814827 • kmcauley@bronxhistoricalsociety.org • www.bronxhistoricalsociety.org •
Dir.: *Dr. Gary D. Hermalyn* • Cur.: *Kathleen A. McAuley* •
Special Museum – 1955
Historic house, final residence of E.A. Poe, paintings, photographs, original furniture 47044

Glyndor Gallery and Wave Hill House Gallery, 675 W 252nd St, Bronx, NY 10471 • T: +1 718 5493200 • F: +1 718 8848952 • arts@wavehill.org • www.wavehill.org/arts •
Public Gallery
Art concerning the interaction between human beings and the natural environment 47045

Hall of Fame for Great Americans, Bronx Community College, University Av and W 181 St, Bronx, NY 10453 • T: +1 718 2895161 • F: +1 718 2896496 • www.bcc.cuny.edu/halloffame •
Dir.: *Ralph Rourke* •
Historical Museum – 1900
98 original bronzes of notable Americans in art, sciences, humanities, government, business and labor 47046

The Judaica Museum, The Hebrew Home, 59601 Palisade Av, Bronx, NY 10471 • T: +1 718 5811787 • F: +1 718 5811009 • judaicamuseum@aol.com • www.hebrewhome.org/museum •
C.E.O.: *Daniel Reingold* • Dir.: *Karen S. Franklin* • Cur.: *Bella Vilsker* •
Fine Arts Museum / Religious Arts Museum
Objects, paintings and textiles from Jewish religion, arts and culture, objects from the Bezalel School of Art and Crafts in Jerusalem (1906-1929), Ralph and Leuba Baum Collection, over 100 pre-war European, contemporary and Israeli crafted mezuzot from the Rosenbaum Foundation 47047

Lehman College Art Gallery, 250 Bedford Park Blvd W, Bronx, NY 10468-1589 • T: +1 718 9608731 • F: +1 718 9606991 • susan@lehman.cuny.edu • ca80.lehman.cuny.edu/gallery •
Dir.: *Susan Hoeltzel* • Ass. Dir.: *Mary Ann Siano* •
Fine Arts Museum / University Museum – 1985
Contemporary art, interactive media in the visual arts 47048

Maritime Industry Museum at Fort Schuyler, SUNY Maritime College, 6 Pennyfield Av, Bronx, NY 10465 • T: +1 718 4097218 • F: +1 718 4096130 • pperez@sunymaritime.edu • maritimeindustrymuseum.org/about.htm •
Dir.: *Eric J. Johansson* • Cur.: *William Sokol Jr.* •
Natural History Museum – 1986
Marine hist, ship models, Tufnell watercolor coll, Frank Cronican models, Maritime College hist, navigation instruments 47049

Museum of Bronx History (Bronx County Historical Society), 3266 Bainbridge Av, Bronx, NY 10467 • T: +1 718 8818900 • F: +1 718 8814827 • kmcauley@bronxhistoricalsociety • www.bronxhistoricalsociety.org •
Dir.: *Dr. Gary D. Hermalyn* • Cur.: *Kathleen A. McAuley* • Sc. Staff: *Prof. Lloyd Ultan* (History) • *Dr. Stephen Stertz* (Research) •
Local Museum – 1897
Bronx, NY City and Westchester hist 1639 to present, 18th-19th c furnishings, domestic artifacts 47050

U

Museum of Migrating People, Harry S. Truman High School, 750 Baychester Av, Bronx, NY 10475 • T: +1 718 9045400 •
Cur.: *Peter Lerner* •
Local Museum – 1974
Material on the immigration experiences of Americans 47051

North Wind Undersea Institute, 610 City Island Av, Bronx, NY 10464 • T: +1 718 8850701 • F: +1 718 8851008 • www.northwind.org •
Dir.: *Michael Sandlofer* • *Irene Sullivan* • Cur.: *Lee Clodfelter* •
Natural History Museum – 1976
Marine hist 47052

Van Cortlandt House Museum, Van Cortlandt Park, Bronx, NY 10471 • T: +1 718 5433344 • F: +1 718 5433315 • vancortlandthouse@juno.com •
Pres.: *Ana Duff* • Dir.: *Laura Carpenter Correa* •
Historical Museum – 1896
House built by Frederick Van Cortlandt, 17th-18th c decorative and useful arts 47053

Bronxville NY

Eastchester Historical Society Museum, 388 California Rd, Bronxville, NY 10708 • T: +1 914 7931900 • eastchesterhistoricalsociety@juno.com •
Pres.: *Saul Radin* •
Local Museum – 1959
1835 Marble Schoolhouse, toys, costumes, photos, furniture, school artifacts 47054

Brookfield CT

Brookfield Craft Center, 286 Whisconier Rd, Brookfield, CT 06804, mail addr: POB 122, Brookfield, CT 06804 • T: +1 203 7754526 • F: +1 203 7407815 • craftcenter@charter.net • www.brookfield-craftcenter.org •
Exec. Dir.: *John I. Russell* •
Decorative Arts Museum – 1954
Craft, old grist mill 47055

Brookfield VT

Historical Society of Brookfield, 113 Ridge Rd, Brookfield, VT 05036 • T: +1 802 2763277 •
Cur.: *Jacalin W. Wilder* •
Association with Coll – 1935
19th c furnishings & artifacts 47056

Marvin Newton House, 1133 Ridge Rd, Brookfield, VT 05036 • T: +1 802 2763277 •
Cur.: *Jacalin W. Wilder* •
Local Museum – 1935
Historic house 47057

Brookings SD

Brookings Arts Council, 524 Fourth St, Brookings, SD 57006 • T: +1 605 6924177 • F: +1 605 6928298 • arts@brookings.net • www.mjts.com/bac •
Dir.: *Vicki Schuster* •
Fine Arts Museum / Folklore Museum – 1977
Cultural arts 47058

South Dakota Art Museum, Harvey Dunn St and Medary Av, Brookings, SD 57007-0999 • T: +1 605 6885423 • F: +1 605 6884445 • sdsu_sdam@sdstate.edu • sdartmuseum.sdstate.edu •
Dir.: *M. Lynn Verschoor* • Cur.: *Lisa Scholten* (Collections) • *John Rychtarik* (Exhibitions) • Sc. Staff: *David Merhib* (Education) •
Fine Arts Museum – 1969
Harvey Duinn Native American tribal art, American artists, Marghab Linen collection 47059

State Agricultural Heritage Museum, South Dakota State University, 925 11th St, Brookings, SD 57007 • T: +1 605 6886226 • F: +1 605 6886303 • sdsu_agmuseum@sdstate.edu • www.agmuseum.com •
Dir.: *John C. Awald* • Sc. Staff: *William D. Lee* • *Carrie Lavarnway* • *Michelle Glanzer* • *Dawn Stephens* •
Agriculture Museum / University Museum – 1967
Agricultural hist 47060

Brookline MA

Brookline Historical Society Museum, 347 Harvard St, Brookline, MA 02446 • T: +1 617 5665747 • w_eayrs@yahoo.com •
Cur.: *Walter F. Eayrs* •
Local Museum – 1901
Local history, period furnishings 47061

Frederick Law Olmsted Collection, 99 Warren St, Brookline, MA 02445 • T: +1 617 5661689 • F: +1 617 2323964 • olmsted_archives@nps.gov • www.nps.gov/frla •
Head: *Myra Harrison* •
Historic Site / Decorative Arts Museum – 1979
Home and office of Frederick Law Olmsted, landscape plans and drawings – archives 47062

John Fitzgerald Kennedy House, 83 Beals St, Brookline, MA 02446 • T: +1 617 5667937 • F: +1 617 2323964 • frla_kennedy_nhs@nps.gov • www.nps.gov/jofi •
Head: *Myra Harrison* •
Historical Museum – 1969
Birthplace of John F. Kennedy 47063

Larz Anderson Auto Museum, 15 Newton St, Brookline, MA 02445 • T: +1 617 5226547 • F: +1 617 5240170 • curator@mot.org • www.mot.org •
Dir.: *John Sweeney* • Cur.: *Evan Ide* •
Science&Tech Museum – 1952
Gas, electric and steam cars before 1942, horse-drawn carriages and sleights – library 47064

Brooklyn CT

Museum Necca and New England Center for Contemporary Art, 248 Pomfret Rd, Brooklyn, CT 06234 • T: +1 860 7748899 • F: +1 860 7799291 • museum@museum-necca.org • www.museum-necca.org •
Dir.: *Henry Riseman* • Cur.: *John Roth* • *Xue Jian Xin* (Chinese Art) •
Fine Arts Museum – 1975
Contemporary art 47065

Brooklyn NY

Art Gallery, Kingsborough Community College, CUNY, 2001 Oriental Blvd, Art & Science Bldg, Brooklyn, NY 11235 • T: +1 718 3685449 • F: +1 718 3684872 • kccgallery@gmail.com • www.kingsborough.edu/academicDepartments/art/gallery/index.html •
Dir.: *Peter Malone* •
Fine Arts Museum
Changing exhibitions of modern and contemporary art 47066

Brooklyn Children's Museum, 145 Brooklyn Av, Brooklyn, NY 11213 • T: +1 718 7354400 • F: +1 718 6047442 • acanty@bchildmus.org • www.brooklynkids.org •
Pres.: *Carol Enseki* • Dir.: *Beth Alberty* (Collections) •
Special Museum – 1899
Ethnology and natural history, folk crafts 47067

Brooklyn Historical Society Museum, 128 Pierrepont St, Brooklyn, NY 11201-2711 • T: +1 718 2224111 • F: +1 718 2223794 • www.brooklynhistory.org •
Pres.: *Deborah Schwartz* • Vice Pres.: *Marilyn Pettit* (Collections) • Sc. Staff: *Kate Fernoile* (Exhibitions and Education) •
Historical Museum – 1863
Brooklyn hist 17th c to present, photographs, costumes, books, paintings, graphics – library 47068

Brooklyn Museum, 200 Eastern Pkwy, Brooklyn, NY 11238-6052 • T: +1 718 6385000 • F: +1 718 5016136 • information@brooklynmuseum.org • www.brooklynmuseum.org •
Chm. Board of Trustees: *Norman M. Feinberg* • Dir.: *Arnold L. Lehman* • Chief Cons. & Dir.: *Kenneth Moser* (Collections) • Deputy Dir. for Art: *Charles Desmarais* • Chief Cur.: *Kevin Stayton* • Cur.: *Edward Bleiberg* (Egyptian, African, Asian, and Islamic Art) • *Edna Russmann* (Egyptian, African, Asian, and Islamic Art) • *Barry Harwood* (Decorative Arts) • *Teresa Carbone* (American Art) • *Joan Cummins* (Asian Art) • *Richard Aste* (European Art) • *Nancy Rosoff* (Arts of the Americas) • *Eugenie Tsai* (Contemporary Art) • *Catherine J. Morris* (Elizabeth A. Sackler Center for Feminist Art) • Associate Cur.: *Ladan Akbarnia* (Islamic Art) • *Patrick Amsellem* (Photography) • Asst. Cur.: *Kevin D. Dumouchelle* (Arts of Africa and Pacific Islands) • Man.Cur.: *Judy Kim* (Exhibitions) •
Fine Arts Museum – 1823
Fine and decorative arts worldwide and from ancient to contemporary art – art reference library (Wilbour Library of Egyptology), Educ. Division 47069

The Chassidic Art Institute, 375 Kingston Av, Brooklyn, NY 11213 • T: +1 718 7749149 • F: +1 718 7747540 •
Dir.: *Zev Markowitz* •
Fine Arts Museum
Jewish art 47070

The Doll and Toy Museum of NYC, 157 Montague Str, Brooklyn, NY 11201 • T: +1 718 2430820 • www.dtmnyc.org •
Exec.Dir.: *Marlene Alperin-Hochman* •
Special Museum – 1999
Early 1900s dollhouses and accessoires, tin toys, construction toys, hist and modern dolls, paper dolls 47071

Dumbo Arts Center, 30 Washington St, Brooklyn, NY 11201 • T: +1 718 6243772 • F: +1 718 6940867 •
Dir.: *Joy Glidden* •
Public Gallery
Changing exhibitions of modern and contemporary art – archive 47072

Franklin Furnace Archive, James A. Davis Arts Bldg, Ste 301, 80 Hanson Pl, Brooklyn, NY 11217-1506 • T: +1 718 3987255 • F: +1 718 3987256 • mail@franklinfurnace.org • www.franklinfurnace.org •
Dir.: *Martha Wilson* •
Association with Coll – 1976
Works by modern and contemporary artists 47073

Galapagos Art Space, 70 N 6th St, Brooklyn, NY 11211 • T: +1 718 7825188 • www.galapagosartspace.com •
Dir.: *Robert Elmes* •
Public Gallery
Contemporary art of all media 47074

Harbor Defense Museum of New York City, Bldg 230, Fort Hamilton, Brooklyn, NY 11252-5701 • T: +1 718 6304349 • F: +1 718 6304888 • Info@harbordefensemuseum.com • www.harbordefensemuseum.com/ •
Dir.: *Paul Morando* • Cur.: *Richard Cox* (Technic) •
Military Museum – 1966
17th c to present US military coll, coast defense 1800-1950, arms and uniforms 47075

Holland Tunnel Art, 61 S first St, Brooklyn, NY 11211 • T: +1 718 3845738 • hollandtunnel@yahoo.com • www.hollandtunnelart.com •
Dir.: *Paulien Lethen* •
Fine Arts Museum
Art 47076

The Kurdish Museum, 144 Underhill Av, Brooklyn, NY 11238 • T: +1 718 7837930 • F: +1 718 3984365 • kurdishlib@aol.com •
Dir.: *Dr. Vera Beaudin Saeedpour* •
Ethnology Museum / Folklore Museum – 1988
Ethnology, costumes and accessoires, jewelry, rugs 47077

Lefferts Historic House, 95 Prospect Park W, Brooklyn, NY 11215 • T: +1 718 7892822 • F: +1 718 7894724 • www.prospectpark.org •
Dir.: *Maria Cobo* • Ass. Dir.: *Erica Catlin* •
Local Museum – 1918
Period furniture, hist doc, household items 47078

Lesbian Herstory Educational Foundation, 14th St, Brooklyn, NY 11215, mail addr: POB 1258, New York, NY 10116 • T: +1 718 7683953 • F: +1 718 7684663 • www.lesbianherstoryarchives.org/ •
Special Museum – 1974
Lesbian hist and culture 47079

New York Transit Museum, Boerum Pl and Schermerhorn St, Brooklyn, NY 11201, mail addr: 130 Livingston St, Brooklyn, NY 11201 • T: +1 718 6941600 • F: +1 718 6941791 • gashubert@nyct.com • www.mta.info/museum •
Dir.: *Gabrielle Shubert* • Dep. Dir.: *Carlos Gutierrez-Solana* • Sen. Cur.: *Charles L. Sachs* •
Science&Tech Museum – 1976
Urban transportation 47080

Rotunda Gallery, 33 Clinton St, Brooklyn, NY 11201 • T: +1 718 8754047 • F: +1 718 4880609 • rotunda@brooklynx.org • www.brooklynx.org/rotunda •
Dir.: *Janet Riker* •
Public Gallery – 1981
Contemporary works of all media by Brooklyn artists 47081

Rubelle and Norman Schafler Gallery, Pratt Institute, 200 Willoughby Av, Brooklyn, NY 11205 • T: +1 718 6363517 • F: +1 718 6363785 • exhibits@pratt.edu • www.pratt.edu/exhibitions •
Dir.: *Loretta Yarlow* •
Fine Arts Museum / University Museum – 1967
19th - 21st c European and American paintings, sculptures, prints, graphics and decorative art 47082

Sonya Art, 305 Vandebilt Av, Brooklyn, NY 11205 • T: +1 718 7892545 • F: +1 718 7892545 • www.sonyany.com •
Public Gallery
Contemporary art 47083

Waterfront Museum and Showboat Barge, 699 Columbia St, Brooklyn, NY 11231 • T: +1 718 6244719 • F: +1 718 6244719 • dsharps@waterfrontmuseum.org • www.waterfrontmuseum.org •
Pres. & C.E.O.: *David Sharps* •
Historical Museum – 1986
Maritime artifacts 47084

The Wyckoff Farmhouse Museum, 5816 Clarendon Rd, Brooklyn, NY 11203 • T: +1 718 6295400 • F: +1 718 6293125 • info@wyckoffassociation.org • wyckoffassociation.org •
C.E.O.: *Sean Sawyer* •
Decorative Arts Museum – 1982
Colonial and early American furniture 47085

Brooklyn OH

Brooklyn Historical Society Museum, 4442 Ridge Rd, Brooklyn, OH 44144-3353 • T: +1 216 9410160 • groundhogsgarden@wowway.com •
Pres.: *Barbara Stepic* •
Local Museum – 1970 47086

Brookneal VA

Red Hill-Patrick Henry National Memorial, 1250 Red Hill Rd, Brookneal, VA 24528 • T: +1 804 3762044 • F: +1 804 3762647 • redhill@redhill.org • www.redhill.org •
Dir.: *Dr. Jon Kukla* •
Historical Museum – 1944
Local hist 47087

Brooks OR

Pacific Northwest Truck Museum, 3995 Brooklake Rd NE, Brooks, OR 97305 • T: +1 503 4638701, +1 503 3120039 • delh@integrity.com • www.pacificnwtruckmuseum.org •
Pres.: *Terry Dovre* •
Science&Tech Museum – 1989
Truck, manufacturing and transportation industries, photographs, drawings, patents 47088

Brooks Air Force Base TX

Edward H. White II Memorial Museum, 311 ABG/MU, Brooks Air Force Base, TX 78235-5329 • T: +1 210 5362203 • F: +1 210 2403224 • museum@platinum.brooks.af.mil •
Dir./Cur.: *Ulysses S. Rhodes* •
Science&Tech Museum / Historical Museum – 1966
Aeronautics 47089

Brookville IN

Franklin County Museum, 5th and Mill St, Brookville, IN 47012 • T: +1 765 6475182 • mick@sur-seal.com • www.franklinchs.com •
Pres.: *Mick Wilz* •
Local Museum – 1969
Local history, housed in a original Franklin County Seminary bldg 47090

Brookville NY

Hillwood Art Museum, Long Island University, C.W. Post Campus, 720 Northern Blvd, Brookville, NY 11548 • T: +1 516 2994073 • F: +1 516 2992787 • museum@cwpost.liu.edu • www.liu.edu/museum •
Pres.: *Dr. David Steinberg* • C.E.O.: *Barry Stern* •
Fine Arts Museum / University Museum – 1973
Precolumbian and African art, contemporary photography, prints 47091

Browning MT

Museum of the Plains Indian, Junction of Hwys 2 and 89 W, Browning, MT 59417 • T: +1 406 3382230 • F: +1 406 3387404 • mpi@3rivers.net • www.iacb.doi.gov •
Cur.: *Loretta Pepion* •
Folklore Museum – 1941
American Indian artefacts and crafts 47092

Brownington VT

The Old Stone House Museum, Orleans County Historical Society, 28 Old Stone House Rd, Brownington, VT 05860 • T: +1 802 7542022 • information@oldstonehousemuseum.org • oldstonehousemuseum.org •
Dir.: *Jack Sumberg* •
Local Museum – 1916
Period furniture, paintings, early farm 47093

Browns Valley MN

Sam Brown Log House, W Broadway, Browns Valley, MN 56219-0013 • T: +1 320 6952110 •
Historical Museum – 1932
Log cabin, 1-room country school, Indian artifacts, clothing, guns, arts and crafts 47094

Brownsville OR

Linn County Historical Museum and Moyer House, 101 Park Av, Brownsville, OR 97327 • T: +1 541 4663390, 4463070 • F: +1 541 4665312 • jmoyer@peak.org •
C.E.O.: *Brian Carrol* •
Local Museum / Folklore Museum – 1962
Local hist, industry, commerce, domestic arts, furnishings, doc 47095

Brownsville TX

Brownsville Museum of Fine Art, 230 Neale Dr, Brownsville, TX 78520 • T: +1 956 5420941 • F: +1 956 5427094 • balart@prodigy.net • brownsvilleartleague.netfirms.com •
Pres.: *Tencha Sloss* •
Fine Arts Museum – 1935
Fine art incl oil paintings, watercolors, collages, prints, sculptures 47096

Brownville NE

Brownville Historical Society Museum, Main St, Brownville, NE 68321 • T: +1 402 8256001 • F: +1 402 8256581 • brownvillemills@alltel.net •
Local Museum – 1956 47097

Meriwether Lewis Dredge Museum, RR 1, Brownville State Recreation Area, Brownville, NE 68321 • T: +1 402 8253341 •
Cur.: *Clay W. Kennedy* •
Local Museum – 1977 47098

Brownville NY

General Jacob Brown Historical Society Museum, 216 Brown Blvd, Brownville, NY 13615 • T: +1 315 7824508 • F: +1 315 7861178 • bville@gisco.net •
Pres.: *Constance G. Hoard* •
Military Museum – 1978
Tools, weapons, furnishings 47099

Brunswick GA

Hofwyl-Broadfield Plantation, 5556 US-Hwy 17 N, Brunswick, GA 31525 • T: +1 912 2647333 • F: +1 912 2623346 • hofwyl@darientel.net •
Agriculture Museum – 1974
Historic plantation 47100

Brunswick MD

Brunswick Railroad Museum, 40 W Potomac St, Brunswick, MD 21716 • T: +1 301 8347100 • moppy6@juno.com • www.brrm.org •
Pres.: *Brian H. Durbin* • Cur.: *Vicki Dearing* • *Suzette Richwagen* •
Science&Tech Museum / Historical Museum – 1974
Railroad social hist and town life ca 1880-1920 47101

Brunswick ME

Bowdoin College Museum of Art, Walker Art Bldg, 9400 College Station, Brunswick, ME 04011-8494 • T: +1 207 7253275 • F: +1 207 7253762 • artmuseum@bowdoin.edu • www.bowdoin.edu/artmuseum •
Dir.: *Katy Kline* • Cur.: *Alison Ferris* •
Fine Arts Museum / University Museum – 1811
Assyrian sculpture, Greek and Roman antiques, European and American paintings, prints, drawings, sculpture and decorative arts 47102

The Peary-MacMillan Arctic Museum, Bowdoin College, 9500 College Stn, Brunswick, ME 04011-8495 • T: +1 207 7253416, 7253062 • F: +1 207 7253499 •
Dir.: *Dr. Susan A. Kaplan* • Cur.: *Dr. Geneviève LeMoine* •
Ethnology Museum / University Museum – 1967
Ethnology, archaeology, natural history 47103

Pejepscot Historical Society Museum, 159 Park Row, Brunswick, ME 04011 • T: +1 207 7296606 • F: +1 207 7296012 • pejepscot@curtislibrary.com • www.curtislibrary.com/pejepscot.htm •
Dir.: *Deborah A. Smith* • Cur.: *Kate Higgins* •
Local Museum – 1888
Regional history 47104

Bryan TX

Brazos Valley Museum of Natural History, 3232 Briarcrest Dr, Bryan, TX 77802 • T: +1 979 7762195 • F: +1 979 7740252 • bvmnh@myriad.net • bvmuseum.myriad.net •
Dir.: *Thomas F. Lynch* • Ass. Cur.: *Amy Witte* (Collections, Exhibitions) •
Natural History Museum – 1961
Paleontology, anthropology, archaeology, gology,and mineralogy, mammalogy, ornithology, entomology 47105

Bryant Pond ME

Woodstock Historical Society Museum, 70 S Main St, Bryant Pond, ME 04219 • T: +1 207 6652450 •
Cur.: *Billings Larry* •
Local Museum – 1979
Paintings and local artifacts 47106

Bryn Athyn PA

Glencairn Museum, 1001 Cathedral Rd, Bryn Athyn, PA 19009-0757 • T: +1 215 9382600 • F: +1 215 9142986 • glencairnmuseum@newchurch.edu • www.glencairnmuseum.org •
Dir.: *Stephen H. Morley* • Cur.: *C. Edward Gyllenhaal* •
Religious Arts Museum – 1878
Furniture, sculpture, jewelery, arms and amor, religious art 47107

Bryson City NC

Smoky Mountain Trains, 100 Greenlee St, Bryson City, NC 28713 • T: +1 828 4885200 • F: +1 828 4883162 • info@smokymountaintrains.com • www.smokymountaintrains.com •
C.E.O. & Cur.: *Timothy O. Cooper* •
Science&Tech Museum – 2002 47108

Buckhead GA

Steffen Thomas Museum of Art, 4200 Bethany Rd, Buckhead, GA 30625 • T: +1 706 3427557 • F: +1 706 3424348 • info@steffenthomas.org • www.steffenthomas.org •
Pres.: *Linda Thomas* • Dir.: *Mary Nell McLauchlin* •
Fine Arts Museum – 1998
Sculpture, paintings, mosaics, furniture, works on paper by expressionist S.Thomas – library, archives 47109

Bucksport ME

Bucksport Historical Museum, Main St, Bucksport, ME 04416 • T: +1 207 4692464 •
Pres.: *Frances D. Beamis* •
Local Museum – 1964
Local history, Old Maine Central Railroad Station 47110

Bucyrus OH

Bucyrus Historical Society Museum, 202 S Walnut St, Bucyrus, OH 44820 • T: +1 419 5626386 •
Dir.: *Dr. John K. Kurtz* • Cur.: *Linda Blicke* •
Local Museum – 1969
Furniture and furnishings 47111

Buena Park CA

Buena Park Historical Society Museum, 6631 Beach Blvd, Buena Park, CA 90621 • T: +1 714 5623570 • info@historicalsociety.org • www.historicalsociety.org •
Cur.: *Jane Mueller* •
Local Museum – 1967
Period furnishings 47112

Buffalo MN

Wright County Historical Museum, 2001 Hwy-25 N, Buffalo, MN 55313 • T: +1 763 6827323 • F: +1 763 6828945 •
Cur.: *Maureen Galvin* •
Local Museum – 1942
Regional hist, culture, ethic, social hist, crafts, agriculture, T Ford model by Humphrey – library, archives 47113

Buffalo NY

Albright-Knox Art Gallery, 1285 Elmwood Av, Buffalo, NY 14222-1096 • T: +1 716 8828700 • F: +1 716 8821958 • corlick@albrightknox.org • www.albrightknox.org •
Dir.: *Louis Grachos* • Sr.Cur.: *Douglas Dreishpoon* • Cur.: *Kenneth Wayne* (Modern Art) • Assoc.Cur.: *Claire Schneider* (Contemporary Art) • Sc. Staff: *Mariann Smith* (Education) • *Susana Tejada* (Library) •
Fine Arts Museum – 1862
Painting, sculpture, photography, works on paper 47114

Benjamin and Dr. Edgar R. Cofeld Judaic Museum of Temple Beth Zion, 805 Delaware Av, Buffalo, NY 14209 • T: +1 716 8366565 • F: +1 716 8311126 • tbz@webt.com • www.tbz.org •
Dir.: *Mortimer Spiller* • Dep. Dir.: *Harriet Spiller* •
Religious Arts Museum – 1981
Judaic artifacts from 10th c to present, coins, medallions, folk art and textiles; holocaust remembrances, hist memorabilia 47115

Buffalo and Erie County Historical Society Museum, 25 Nottingham Ct, Buffalo, NY 14216 • T: +1 716 8739644 • F: +1 716 8738754 • bechs@buffnet.net • www.bechs.org •
Pres.: *Richard A. Wiesen* • Exec. Dir.: *William Siener* • Dir.: *Walter Mayer* (Collections) •
Local Museum – 1862
Archaeology, ethnology; agriculture, household items, transportation, crafts 47116

Buffalo and Erie County Naval and Military Park, One Naval Park Cove, Buffalo, NY 14202 • T: +1 716 8471773 • F: +1 716 8476405 • navalpark@ch.ci.buffalo.ny.us • www.buffalonavalpark.org •
C.E.O.: *Patrick J. Cunningham* •
Military Museum – 1979
Military hist, artifacts of all armed forces, tanks, aircrafts 47117

Buffalo Museum of Science, 1020 Humboldt Pkwy, Buffalo, NY 14211 • T: +1 716 8965200 • F: +1 716 8976723 • dchesebrough@sciencebuff.org • www.buffalomuseumofscience.org •
Pres. & C.E.O.: *David Chesebrough* • Dir.: *Dr. John Grahan* (Science and collections) • Cur.: *Dr. Richard S. Laub* (Geology) • *Dr. Elizabeth Pena* (Anthropology) •
Natural History Museum – 1861
Anthropology, botany, zoology, geology 47118

Burchfield-Penney Art Center, Buffalo State College, 1300 Elmwood Av, Buffalo, NY 14222 • T: +1 716 8786012 • F: +1 716 8786003 • burchfld@buffalostate.edu • www.burchfield-penney.org •
Dir.: *Ted Pietrzak* • *Nancy Weekly* (Curatorial Sevices) •
Fine Arts Museum – 1966
Western New York art and craft, watercolors by Charles E. Burchfield 47119

Center for Exploratory and Perceptual Art, 617 Main St, Buffalo, NY 14203 • T: +1 716 8562717 • F: +1 716 2700184 • cepa@cepagallery.com • cepagallery.com •
Dir.: *Lawrence F. Brose* •
Public Gallery – 1974
Photography 47120

Hallwalls Arts Center, 2495 Main St, Buffalo, NY 14214 • T: +1 716 8357362 • F: +1 716 8357364 • www.hallwalls.org •
Dir.: *Edmund Cardoni* • *John Massier* •
Public Gallery
Art 47121

Karpeles Manuscript Museum, 453 Porter Av, Buffalo, NY 14201 • T: +1 716 8854139 • www.karpeles.com •
Dir.: *Chris Kelly* •
Fine Arts Museum
Art 47122

El Museo Francisco Oller y Diego Rivera, 91 Allen St, Buffalo, NY 14202 • T: +1 716 8849693 • F: +1 716 8849362 • elmuseubuffalo@aol.com • www.buffalo.com/elmuseobuffalo •
Fine Arts Museum 47123

Theodore Roosevelt Inaugural National Historic Site, 641 Delaware Av, Buffalo, NY 14202 • T: +1 716 8840095 • F: +1 716 8840330 • thri_administration@nps.gov • www.nps.gov/thri/ •
Dir.: *Molly Quackenbush* • Cur.: *Leonora Henson* •
Historical Museum – 1971
Buffalo Barracks, the site of the inauguration of Pres. Theodore Roosevelt in 1901 47124

University at Buffalo Anderson Art Gallery, State University of New York, Martha Jackson Pl, Buffalo, NY 14214 • T: +1 716 8293754 • F: +1 716 8293757 • artgallery.buffalo.edu •
Dir.: *Sandra H. Olsen* •
Fine Arts Museum / University Museum
Paintings, sculpture – archives 47125

University at Buffalo Art Galleries, 201A Center for the Arts, Buffalo, NY 14260-6000 • T: +1 716 6456912 ext 1423 • F: +1 716 6456753 • kriemer@buffalo.edu • www.artgallery.buffalo.edu •
Dir.: *Sandra H. Olsen* • Cur.: *Sandra Firmin* •
Fine Arts Museum / University Museum – 1994
Modern and contemporary art, early 19th c works 47126

Buffalo WY

Jim Gatchell Memorial Museum, 100 Fort St, Buffalo, WY 82834 • T: +1 308 6849331 • F: +1 307 6840354 • jmuseum@trib.com • www.jimgatchell.com •
C.E.O.: *John Gavin* • Cur.: *Robert Wilson* •
Local Museum – 1957
Military, pioneer, Indian hist, mineralogy, paintings, old vehicles 47127

Buford GA

Lanier Museum of Natural History, 2601 Buford Dam Rd, Buford, GA 30518 • T: +1 770 9324460 • F: +1 770 9323055 • mark.patterson@gwinnettcounty.com • www.gwinnettcounty.com •
Dir.: *Dr. Mark A. Patterson* •
Natural History Museum – 1978
Botany, zoology, geology, paleontology 47128

Burfordville MO

Bollinger Mill, 113 Bollinger Mill Rd, Burfordville, MO 63739 • T: +1 573 2434591 • F: +1 573 2435385 • www.mostateparks.com •
Open Air Museum / Local Museum – 1967
Mill machinery 47129

Burley ID

Cassia County Museum, East Main and Highland Av, Burley, ID 83318 • T: +1 208 6787172 •
Local Museum – 1972
Train car, farm machinery, county covered wagon 47130

Burlington IA

The Apple Trees Museum, 1616 Dill St, Burlington, IA 52601 • T: +1 319 7532449 • dmcohist@interl.net •
Pres.: *Judy Johnson* •
Historical Museum / Agriculture Museum – 1972
Charles E. Perkins mansion 47131

Art Guild of Burlington Gallery, Seventh and Washington, Burlington, IA 52601 • T: +1 319 7548069 • F: +1 319 7544731 • arts4living@aol.com • www.artguildofburlington.org •
Pres.: *Burton Prugh* • Dir.: *Lois Rigdon* •
Public Gallery – 1966
Paintings and sculptures by regional artists 47132

Hawkeye Log Cabin, 2915 S Main St, Burlington, IA 52601 • T: +1 319 7532449 • dmcohist@interl.net •
Pres.: *Judy Johnson* •
Special Museum – 1971
1909 log cabin, furnishings, pioneer tools 47133

Phelps House, 521 Columbia, Burlington, IA 52601 • T: +1 319 7535880 • dmcohist@interl.net •
Pres.: *Judi Johnson* •
Decorative Arts Museum – 1974
1851 Victorian mansion 47134

Burlington KY

Dinsmore Homestead, 5656 Burlington Pike, Burlington, KY 41005 • T: +1 859 5866117 • F: +1 859 3343690 • info@dinsmorefarm.org • www.dinsmorefarm.org •
Exec. Dir.: *Marty McDonald* •
Historical Museum – 1986
Furniture, books, paintings, textiles. African-American slavery 47135

Burlington MA

Burlington Historical Museum, Town hall, 29 Center st, Burlington, MA 01803 • T: +1 781 2701049, 2720606 • F: +1 781 2701608 • archives@burlmass.org •
C.E.O.: *Norman Biggart* •
Local Museum – 1970
Historical and todays artifacts, primitive agriculture, costumes, dolls, furniture, WW 1, toys, paintings, town hist – library 47136

Burlington ME

Stewart M. Lord Memorial Museum, Burlington, ME 04448 • T: +1 207 7324121 •
Cur.: *Fern P. Cummings* •
Historical Museum / Folklore Museum – 1968
History , restored tavern, housing wagons, surry boat 47137

Burlington NC

Alamance Battleground, 5803 S NC Rte 62, Burlington, NC 27215 • T: +1 336 2274785 • F: +1 336 2274787 • alamance@ncmail.net • www.alamancebattleground.nchistoricsites.org •
Head: *Bryan Dalton* •
Open Air Museum / Military Museum – 1955
Hist house, military hist 47138

Burlington NJ

Burlington County Historical Society Museum, 451 High St, Burlington, NJ 08016 • T: +1 609 3864773 • F: +1 609 3864828 • bchsnj@earthlink.net •
Dir.: *Douglas E. Winterich* • Cur.: *Lisa Fox* •
Historical Museum – 1915
Decorative arts, clocks, quilts, Delaware River Decoys 47139

Colonial Burlington Foundation, 213 Wood St, Burlington, NJ 08016 • T: +1 609 2392266 • F: +1 609 3863415 • jacques@colorite-resins.com •
Historical Museum – 1939
Hist house, furnishings, gardens 47140

Hoskins House, 202 High St, Burlington, NJ 08016 • T: +1 609 3860200 • F: +1 609 3860214 • www.tourburlington.org •
Local Museum / Folklore Museum – 1975
Local history 47141

Burlington VT

The Children's Discovery Museum of Vermont, 1 College St, Burlington, VT 05401 • T: +1 802 8641848 • F: +1 802 8646832 • lcbsc@together.net • www.uvm.edu/~lcbsc/ •
Dir.: *Betsy Rosenbluth* •
Special Museum – 1995
Children's museum 47142

Francis Colburn Gallery, c/o University of Vermont, Art Department, Williams Hall, Burlington, VT 05405 • T: +1 802 6562014 • F: +1 802 6562064 •
Public Gallery / University Museum – 1975
Various forms of art media exhib by students, faculty and visiting artists 47143

Robert Hull Fleming Museum, c/o University of Vermont, 61 Colchester Av, Burlington, VT 05405-0064 • T: +1 802 6560750 • F: +1 802 6568059 • fleming@uvm.edu • www.flemingmuseum.org •
Dir.: *Janie Cohen* • Cur.: *Evelyn Hankins* •
Fine Arts Museum / University Museum / Ethnology Museum – 1931
American & European paintings, sculpure, prints, & drawings 47144

Burnet TX

Fort Croghan Museum, 703 Buchanan Dr, Burnet, TX 78611 • T: +1 512 7568281 • F: +1 512 7562548 • piejoy@tstar.net • www.fortcroghan.org •
Dir.: *Carole Goble* •
Local Museum / Historic Site – 1957
Stone and log cabins from the 1850s and 1860s 47145

Burns OR

Harney County Historical Museum, 18 W D St, Burns, OR 97720 • T: +1 503 5731461 • F: +1 541 5737225 • dpurdy@centurytel.net • www.burnsmuseum.com •
C.E.O. & Pres.: *Dorothea Purdy* • Cur.: *Kathy Elsbury* •
Local Museum – 1960
Archaeology, natural hist, geology and mineralogy, Indian artifacts, Indian headdress, butterfly coll, gun coll 47146

Burton OH

Century Village Museum, 14653 E Park St, Burton, OH 44021 • T: +1 440 8344012 • F: +1 440 8344012 • gchs@jmzcomputer.com • www.geaugahistorical.org •
Pres.: *Kurt Updegraff* •
Open Air Museum / Historical Museum – 1938
Recreated 19th c village, hist houses, railroad station, furniture and artifacts 47147

Burwell NE

Fort Hartsuff, RR 1, Burwell, NE 68823 • T: +1 308 3464715 • F: +1 308 3464715 • fortharsuff@nctc.net • www.ngpc.state.ne.us •
Military Museum – 1874
Period furniture, arch. coll, Great Plains military hist – Military Museum 47148

Bushnell FL

Dade Battlefield, 7200 County Rd 603, Bushnell, FL 33513 • T: +1 352 7934781 • F: +1 352 7934230 • dbshs@sum.net • www.floridastateparks.org •
Dir.: *Barbara Webster* •
Historic Site / Historical Museum – 1921
Site of a battle between Seminole and US soldiers on Dec. 28, 1835 47149

U

Butte MT

Butte Silver Bow Arts Chateau, 321 W Broadway, Butte, MT 59701 • T: +1 406 7237600 • F: +1 406 7235083 • glenv@in-tch.com • www.artschateau.org • Dir.: *Glenn Bodish* •
Fine Arts Museum – 1977
Contemporary regional art 47150

Copper King Mansion, 219 W Granite, Butte, MT 59701 • T: +1 406 7827580 •
Head: *John Thompson* •
Fine Arts Museum / Decorative Arts Museum – 1966
Victorian furniture, crystal, cut glass, silver, porcelains, art 47151

Mineral Museum, Montana Tech., 1300 W Park St, Butte, MT 59701-8997 • T: +1 406 4964414 • F: +1 406 4964451 • dberg@mtech.edu • www.mbmg.mtech.edu/museum.htm •
Cur.: *Dr. Richard B. Berg* •
University Museum / Natural History Museum / Science&Tech Museum – 1900
1,300 mineral specimens 47152

World Museum of Mining, 155 Museum Way, Butte, MT 59703, mail addr: PO Box 33, Butte, MT 59703 • T: +1 406 7237211 • F: +1 406 7232398 • director@miningmuseum.org • www.miningmuseum.org •
Dir.: *Tina Green* •
Science&Tech Museum / Natural History Museum / Historic Site / Local Museum – 1964
Mining & Cultural history of Butte Montana and the surrounding region during the turn of the century – Photo Archives 47153

Byron NY

Byron Historical Museum, 6407 Town Line Rd, Byron, NY 14422 • T: +1 585 5489008 • byron-historian@juno.com •
Pres.: *Alvin Shaw* • Sc. Staff: *Beth Wilson* • *Bob Wilson* (History) •
Local Museum – 1967
General coll 47154

Cable WI

Cable Natural History Museum, 13470 County Hwy M and Randysek Rd, Cable, WI 54821 • T: +1 715 7983890 • F: +1 715 7983828 • mail@cablemuseum.org • www.cablemuseum.org •
Exec. Dir.: *Michelle L. Gostomski* •
Natural History Museum – 1968
Natural hist 47155

Cabot VT

Cabot Historical Society, 193 McKinistry St, Cabot, VT 05647 • T: +1 802 5632547 •
Pres.: *Leonard H. Spencer* • Cur.: *Eric Ginette* •
Association with Coll – 1966
Artifacts, pictures, documents, manuscripts coll 47156

Cadillac MI

Wexford County Historical Museum, 127 Beech St, Cadillac, MI 49601 • T: +1 231 7751717, 3892615 •
Local Museum – 1978
Local hist exhibits 47157

Cahokia IL

Cahokia Courthouse, 107 Elm St, Cahokia, IL 62206 • T: +1 618 3321782 • F: +1 618 3321737 • cahokiacourt@peaknet.net •
Head: *Molly McKenzie* •
Historical Museum – 1940
Local hist, French colonial architecture, artifacts, furniture 47158

Cairo IL

Magnolia Manor Museum, 2700 Washington Av, Cairo, IL 62914 • T: +1 618 7340201 • F: +1 618 7340201 •
Cur.: *Tim Slapinski* •
Local Museum – 1952
19th c furnishings, Cairo hist, Civil War items 47159

Caldwell ID

Kiwanis Van Slyke Museum Foundation, Caldwell Municipal Park, Grant and Kimball Sts, Caldwell, ID 83605 • T: +1 208 4591597 • F: +1 208 4591608 •
Chm.: *Mac McCann* •
Local Museum / Agriculture Museum – 1958
Agriculture, hist, farm implements 47160

Rosenthal Gallery of Art, Albertson College of Idaho, 2112 Cleveland Blvd, Caldwell, ID 83605 • T: +1 208 4595321 • F: +1 208 4542077 • gclaassen@collegeofidaho.edu •
Dir.: *Garth Claassen* •
Fine Arts Museum / University Museum – 1891
Prints, paintings, sculptures 47161

Caldwell NJ

Grover Cleveland Birthplace, 207 Bloomfield Av, Caldwell, NJ 07006 • T: +1 973 2260001 • F: +1 973 2261810 •
Local Museum – 1913 47162

Visceglia Art Gallery, Caldwell College, 9 Ryerson Av, Caldwell, NJ 07006 • T: +1 973 6183457 •
Dir.: *Kendall Baker* •
Fine Arts Museum – 1970 47163

Caldwell TX

Burleson County Historical Museum, Burleson County Courthouse, Caldwell, TX 77836 • T: +1 979 5677196 • F: +1 979 5670366 •
Chm.: *Bill Giesenschlag* •
Local Museum – 1968
Fort Tenoxtitlan artifacts, antique toys and children's furniture 47164

Caledonia NY

Big Springs Museum, 3089 Main St, Caledonia, NY 14423 • T: +1 585 5389880 •
Pres.: *Jeanne Guthrie* • Cur.: *Patty Garret* • Cur.Asst.: *Betty DeNoon* •
Local Museum – 1936
History, early farm inplements, Native American coll, Civil War, WW I and II, weaving equip, household items 47165

Calera AL

Heart of Dixie Railroad Museum, 1919 Ninth St, Calera, AL 35040 • T: +1 800 9434490 • F: +1 205 6689900 • www.hodrrm.org •
Pres.: *Honeycutt Ralph* •
Science&Tech Museum – 1963
History of railroads in Alabama 47166

Calhoun GA

New Echota, 1211 Chatsworth Hwy NE, Calhoun, GA 30701 • T: +1 706 6241321 • F: +1 706 6241323 • n_echoda@innerx.net • www.gastateparks.org •
Historic Site / Open Air Museum
1825 capital town of Cherokee Nation 47167

Calistoga CA

Sharpsteen Museum, 1311 Washington St, Calistoga, CA 94515 • T: +1 707 9425911 • F: +1 707 9426325 • sharpsteenmuseum@att.net • www.sharpsteen-museum.org •
Pres.: *Jo Noble* •
Local Museum – 1978
City and upper Napa Valley hist 47168

Calumet MI

Coppertown U.S.A. Mining Museum, 25815 Red Jacket Rd, Calumet, MI 49913 • T: +1 906 3374354 • www.uppermichigan.com/coppertown •
Pres.: *Richard Dana* •
Science&Tech Museum – 1973
Restored former Calumet and Hecla Mining Co. 47169

Calverton NY

Grumman Memorial Park, Rte 25 and 25A, Calverton, NY 11933, mail addr: POB 147, Calverton, NY 11933 • T: +1 631 3699488 • F: +1 631 3699489 • grummanpk@aol.com • www.grummanpark.org •
Chm.: *Joe Vande Wetering* •
Military Museum
Military aircraft 47170

Camarillo CA

Studio Channel Islands Art Center, 2221 Ventura Blvd, Camarillo, CA 93010 • T: +1 805 8381368 • F: +1 805 4829168 • sciartcenter@aol.com • www.studiochannelislands.org •
Dir.: *Magee Kildee* •
Fine Arts Museum / University Museum
Art by regional and national artists 47171

Cambridge MA

Arthur M. Sackler Museum, 485 Broadway, Cambridge, MA 02138 • T: +1 617 4952397 • www.fas.harvard.edu/~hoart/ •
Fine Arts Museum 47172

Botanical Museum of Harvard University, 26 Oxford St, Cambridge, MA 02138 • T: +1 617 4952326 • F: +1 617 4955667 • hmnh@oeb.harvard.edu • www.hmnh.harvard.edu •
Cur.: *Andrew H. Knoll* (Paleobotany) •
University Museum / Natural History Museum – 1858
Botany, narcotic and medical plants, rubber-yielding plants, fossil woods, orchids – library 47173

Cambridge Historical Museum, 159 Brattle St, Cambridge, MA 02138 • T: +1 617 5474252 • F: +1 617 6611623 • camhistory@aol.com • www.cambridgehistory.org •
Cur.: *Lindsay Leard Coolidge* •
Local Museum – 1905
Local history 47174

Gallery 57, City Hall Annex at 344 Broadway, Cambridge, MA 02139 • T: +1 617 3494380 • F: +1 617 3494669 • cac@ci.cambridge.ma.us • www.cambridgeartscouncil.org/exhibitions_gallery.html •
Dir.: *Jason Weeks* •
Public Gallery – 1973 47175

Harvard Museum of Natural History, 26 Oxford St, Cambridge, MA 02138 • T: +1 617 4953045 • F: +1 617 4968206 • hmnh@oeb.harvard.edu • www.hmnh.harvard.edu •
University Museum / Natural History Museum – 1995
University's 3 natural history institutions: Botanical Museum, Museum of Comparative Zoology, Mineralogical Museum 47176

Harvard University Art Museums, Fogg Art Museum, Arthur M. Sackler Museum, Busch-Reisinger Museum, 32 Quincy St, Cambridge, MA 02138 • T: +1 617 4959400 • F: +1 617 4969762 • huam@fas.harvard.edu • www.artmuseums.harvard.edu •
Dir.: *Thomas W. Lentz* • Cur.: *Mary McWilliams* (Islamic and Later Indian Art) • *Theodore E. Stebbins* (American Art) • *Robert Mowry* (Chinese Art) • *Harry Cooper* (Modern Art) • *Linda Norden* (Contemporary Art) • *David G. Mitten* (Ancient Art) • *Deborah M. Kao* (Photography) • *Marjorie B. Cohn* (Prints) • *Ivan Gaskell* (Paintings, Sculpture, Decorative Arts) • *William W. Robinson* (Drawings) • *Peter Nisbet* (Busch-Reisinger) •
Fine Arts Museum / University Museum – 1895
Sackler Museum: Asian, Islamic and Indian art, Busch-Reisinger Museum: European art since late medivial, Fogg Museum: European and American fine and decorative arts – Lyonel Feininger archive 47177

Harvard University Semitic Museum, 6 Divinity Av, Cambridge, MA 02138 • T: +1 617 4954631 • F: +1 617 4968904 • stager@fas.harvard.edu • www.fas.harvard.edu/~semitic/ •
Dir.: *Prof. Lawrence Stager* • Cur.: *Prof. Piotr Steinkeller* (Cuneiform Coll) •
University Museum / Archaeology Museum – 1889
Ancient and medieval Near East archaeology 47178

List Visual Arts Center, Massachusetts Institute of Technology, 20 Ames St, Wiesner Bldg E15-109, Cambridge, MA 02139 • T: +1 617 2534680, 2534400 • F: +1 617 2587265 • hiroco@mit.edu • web.mit.edu/lvac •
Dir.: *Jane Farver* • Cur.: *Bill Arning* •
Fine Arts Museum / University Museum – 1963
Sculpture, paintings, photographs, drawings 47179

Longfellow House, 105 Brattle St, Cambridge, MA 02138 • T: +1 617 8764491 • F: +1 617 8766014 • frla_longfellow@nps.gov • www.nps.gov/long •
Cur.: *Janice Hodson* •
Historical Museum – 1972
Home of Henry Wadsworth Longfellow, headqaurters of George Washington 1775-1776 – library 47180

Margaret Hutchinson Compton Gallery, 77 Massachusetts Av, Bldg 10, Cambridge, MA 02139-4307 • T: +1 617 4522111 • F: +1 617 2538994 • web.mit.edu/museum •
Dir.: *Beryl Rosenthal* •
Fine Arts Museum / Science&Tech Museum – 1978
Technology 47181

Mineralogical Museum of Harvard University, 24 Oxford St, Cambridge, MA 02138 • T: +1 617 4954758 • F: +1 617 4958839 • francis@eps.harvard.edu • www.fas.harvard.edu/~geomus •
Cur.: *Dr. Carl Francis* • Ass. Cur.: *William C. Metropolis* •
Natural History Museum – 1784
Scientific referance coll, rocks 47182

MIT Museum, 265 Massachusetts Av, Cambridge, MA 02139-4307 • T: +1 617 2535927 • F: +1 617 2538994 • museum@mit.edu • web.mit.edu/museum •
Dir.: *John Durant* • Cur.: *Kurt Hasselbalch* (Hart Nautical Coll) • *Gary Van Zante* (Architecture, Design) • *Deborah Douglas* (Science, Technology) •
University Museum / Science&Tech Museum – 1971
Science & technology, architecture & design, holography, naval architecture & marine engineering – holography lab 47183

Museum of Comparative Zoology, Harvard University, 26 Oxford St, Cambridge, MA 02138 • T: +1 617 4955891 • F: +1 617 4958308 • mdevanney@oeb.harvard.edu • www.mcz.harvard.edu •
Dir.: *James Hanken* •
Natural History Museum / University Museum – 1859
Biological oceanography, entomology, herpetology, ichthyology, malacology, mammology, vertebrates and invertebrates, ornithology 47184

Peabody Museum of Archaeology and Ethnology, Harvard University, 11 Divinity Av, Cambridge, MA 02138 • T: +1 617 4961027 • F: +1 617 4957535 • peabody@fas.harvard.edu • www.peabody.harvard.edu •
Dir.: *Dr. William L. Fash* • Cur.: *Dr. Diane Loren* • *Dr. Michele Morgan* • *Dr. Patricia Capone* • Cons.: *Rose Holdcraft* •
University Museum / Ethnology Museum / Archaeology Museum – 1866
Anthropology, archaeology, ethnology mainly The Americas and the Old World – library, labs 47185

Cambridge MD

Dorchester County Historical Society Museum, 902 LaGrange Av, Cambridge, MD 21613 • T: +1 410 2287953 • F: +1 410 2282947 • dchs@fastol.com • www.bluecrab.org/dchs/index.html •
Pres.: *Gregory J. Papin* •
Local Museum – 1953
History and genealogy, agriculture, furnishings, dolls, potraits 47186

Cambridge MN

Isanti County Museums, 33525 Flanders St NE, Cambridge, MN 55008, mail addr: POB 525, Cambridge, MN 55008 • T: +1 763 6894229 • F: +1 763 5520740 • ichs@nsatel.net • www.ichs.ws •
Exec. Dir.: *Kathleen J. McCully* •
Local Museum – 1964
Pioneer artifacts, school, church, household and farm equip 47187

Cambridge NE

Cambridge Museum, 612 Penn St, Cambridge, NE 69022 • T: +1 308 6974385 •
Pres.: *Marilyn Kester* • Cur.: *Marjorie Ridpath* •
Fine Arts Museum / Local Museum – 1938
Local history 47188

Cambridge OH

Cambridge Glass Museum, 812 Jefferson Av, Cambridge, OH 43725 • T: +1 614 4323045 •
Dir.: *Harold D. Bennett* • Dep. Dir.: *Dorthy E. Bennett* •
Decorative Arts Museum – 1973
Cambridge glass and art pottery of the first of the 20th c 47189

Degenhart Paperweight and Glass Museum, 65323 Highland Hills Rd, Cambridge, OH 43725 • T: +1 614 4322626 • degmus@clover.net •
Pres.: *James Caldwell* • Cur.: *Jennifer Hatcher* •
Decorative Arts Museum – 1978
Ohio Valley midwestern pattern glass, 20th-c paperweights, artglass, decorative items 47190

Guernsey County Museum, 218 N Eighth St, Cambridge, OH 43725 • T: +1 740 4395884 •
Cur.: *Calvin Rice* •
Local Museum – 1964
Regional hist, county hall of fame, military, glass, china, pottery 47191

Cambridge City IN

Huddleston Farmhouse Inn Museum, 838 National Rd, Mt Auburn, Cambridge City, IN 47327 • T: +1 765 4783172 • F: +1 765 4783410 • huddleston@historiclandmarks.org • www.historiclandmarks.org •
Pres.: *J. Reid Williamson* • Dir.: *Brady Kress* •
Historical Museum / Agriculture Museum – 1977
Federal style brick, 3-story farmhouse 47192

Camden AR

McCollum-Chidester House, 926 Washington St NW, Camden, AR 71701 • T: +1 870 8369243, 8363610 • www.ouachitacountyhistoricalsociety.org •
Pres.: *Clara Freeland* •
Historical Museum – 1847
Stage coach house, Union general quartered here during battle at Poison Springs, furniture and personal effects of family (1863-1963) – Leake-Ingham library 47193

Camden ME

Old Conway Homestead Museum and Mary Meeker Cramer Museum, US Rte 1, Camden-Rockport Ln, Camden, ME 04843, mail addr: POB 747, Rockport, ME 04856 • T: +1 207 2362257 • crmuseum@midcoast.com • www.crmuseum.org •
C.E.O.: *Marlene Hall* •
Local Museum – 1962
Paintings, books, photographs, costumes, folklore, local hist, marine, musical instruments, agriculture 47194

Camden NJ

Camden County Historical Society Museum, 1900 Park Blvd and Euclid Av, Camden, NJ 08103 • T: +1 856 9643333 • F: +1 856 9640378 • cchsnj@voicenet.com • www.cchsnj.org •
Dir.: *John R. Seitter* • Cur.: *Judith Synder* • Libr.: *Joanne Diogo* •
Local Museum – 1899
Social, cultural and economic hist of New Jersey 47195

Stedman Art Gallery, Rutgers State University of New Jersey, Camden, NJ 08102 • T: +1 856 2256245 • F: +1 609 2256330 • arts@camdenrutgers.edu • rcca.camden.rutgers.edu •
Dir.: *Virginia Oberlin Steel* • Cur.: *Nancy Maguire* (Exhibits) • Sc. Staff: *Noreen Scott Garrity* (Education) •
Fine Arts Museum – 1975
Modern and Contemporary art 47196

Walt Whitman House, 328 Mickle Blvd, Camden, NJ 08103 • T: +1 609 9645383 • F: +1 856 9641088 • whitmanshs@snip.net •
Cur.: *Leo D. Blake* •
Historical Museum – 1946
Furniture, photographs, manuscripts 47197

Camden NY

Carriage House Museum, 2 N Park St, Camden, NY 13316 • T: +1 315 2454652 • historycamden@a-znet.com •
Pres. & Cur.: *Elaine Norton* •
Historical Museum – 1975
Hist of Camden, tools, photographs, furniture, clothing 47198

Camden SC

Camden Archives and Museum, 1314 Broad St, Camden, SC 29020-3535 • T: +1 803 4256050 • F: +1 803 4244053 • www.mindspring.com/~camdenarchives/index.html •
Dir.: *Agnes Corbett* •
Local Museum – 1973
Local hist 47199

U

Fine Arts Center of Kershaw County, 810 Lyttleton St, Camden, SC 29020 • T: +1 803 4257676 • F: +1 803 4257679 • DEdwards@fineartscenter.org • www.fineartscenter.org •
Dir.: *Dianne Pantoja* •
Fine Arts Museum – 1976
Art ... 47200

Historic Camden Revolutionary War Site, 222 Broad St, Camden, SC 29020 • T: +1 803 4329841 • F: +1 803 4323815 • hiscamden@camden.net • www.historic-camden.net •
Dir.: *Joanna Craig* •
Local Museum – 1967
Local hist ... 47201

Kershaw County Historical Society Museum, 811 Fair St, Camden, SC 29020 • T: +1 803 4251123 • kchistory@mindspring.com • www.mindspring.com/~kchistory •
Dir.: *Kathleen P. Stahl* •
Local Museum – 1954
Local hist ... 47202

Cameron TX

Milam County Museum and Annex, Main and Fannin Sts, Cameron, TX 76520 • T: +1 254 6974770, 6978963 • F: +1 254 6977028 • milamco@tlab.net • www.cameron-tx.com •
Dir.: *Charles King* •
Local Museum – 1977
Original Sam Houston and Bejamin Bryant documents ... 47203

Cameron WI

Barron County Historical Society's Pioneer Village Museum, 1 1/2 mile west of Cameron on Museum Rd, Cameron, WI 54822 • T: +1 715 4582080 •
Pres.: *Ken Mosenstine* •
Local Museum / Open Air Museum – 1960
State and local hist, agricultural, logging and household items, Ameriucan Indian artifacts, dental, medical, veterinary tools – farmstead, depot, dentists office, post office a.o. hist. buildings, shops and offices ... 47204

Camp Douglas WI

Wisconsin National Guard Memorial Library and Museum, 101 Independence Dr, Camp Douglas, WI 54618 • T: +1 608 4271280 • F: +1 608 4271399 • eric.lent@dva.state.wi.us •
C.E.O.: *Raymond Boland* • Cur.: *Eric Lent* •
Military Museum – 1984
Military hist ... 47205

Camp Verde AZ

Fort Verde, Fort Verde State Historic Park, Camp Verde, AZ 86322, mail addr: POB 397, Camp Verde, AZ 86322 • T: +1 928 5673275 • azstateparks.com/Parks/FOVE/map.html •
Military Museum – 1956
Local military hist ... 47206

Campbell CA

Ainsley House, 300 Grant St, Campbell, CA 95008 • T: +1 408 8662119 • F: +1 408 8662795 • kerryd@cityofcampbell.com • www.ci.campbell.ca.us •
Dir.: *Gloria Chun Hoo* • Cur.: *Kerry Dickie* •
Historical Museum – 1964 ... 47207

Campbell Historical Museum, 51 N Central Av, Campbell, CA 95008 • T: +1 408 8662119 • F: +1 408 8662795 • karenB@ci.campbell.ca.us • www.ci.campbell.ca.us •
Dir.: *Gloria Chun Hoo* • Cur.: *Kerry Dickie* •
Historical Museum – 1964
Local history ... 47208

Campbellsport WI

Henry S. Reuss Ice Age Visitor Center, DNR, Kettle Moraine State Forest N 2875 Hwy 67, Campbellsport, WI 53010, mail addr: N 1765, Hwy G, Campbellsport, WI 53010 • T: +1 920 5338322, +1 262 6262116 • F: +1 262 6262117 • jackie.scharfenberg@dnr.state.wi.us • www.dnr.state.wi.us •
Natural History Museum – 1980
Glacial geology ... 47209

Canaan NY

Canaan Historical Society Museum, Warner's Crossing Rd, Canaan, NY 12029 • T: +1 518 7814228 •
Cur.: *Anna Mary Dunton* •
Local Museum – 1963
Artifacts from the Shakers, Civil War artifacts, Canaan militia during Civil War, furniture, glassware ... 47210

Canajoharie NY

Canajoharie Library and Art Gallery, 2 Erie Blvd, Canajoharie, NY 13317 • T: +1 518 6732314 • F: +1 518 6735243 • etrahan@sals.edu • www.clag.org •
Dir.: *Eric Trahan* • Cur.: *James Crawford* •
Fine Arts Museum / Library with Exhibitions – 1929
American art, American impressionists, Ash Can artists, American watercolors ... 47211

Canal Fulton OH

Canal Fulton Heritage Society Museum, 103 Tuscarawas St, Canal Fulton, OH 44614 • F: +1 330 8542902 • canaler@ohio-eriecanal.com •
Pres.: *Rochelle Rossi* •
Local Museum – 1968
Canal artifacts, furnishings, maps ... 47212

Canandaigua NY

Granger Homestead Society Museum, 295 N Main St, Canandaigua, NY 14424 • T: +1 716 3941472 • F: +1 716 3946958 • ghomestead@aol.com • www.grangerhomestead.org •
Dir.: *Saralinda Hooker* •
Local Museum – 1946
Decorative arts, furniture, silver, glass, china; 1820-1930 horse drawn vehicles ... 47213

Ontario County Historical Society Museum, 55 N Main St, Canandaigua, NY 14424 • T: +1 716 3944975 • F: +1 716 3949351 • director@ochs.org • www.ochs.org •
Dir.: *Edward Varno* • Cur.: *Wilma T. Townsend* •
Historical Museum – 1902
Hist and genealogy of the region ... 47214

Sonnenberg Mansion, 151 Charlotte Str, Canandaigua, NY 14424 • T: +1 585 3944922 • F: +1 585 3942192 • info@sonnenberg.org • www.sonnenberg.org •
Historical Museum – 1970
Queen Anne style mansion with furnishings and decorative art ... 47215

Canby OR

Canby Depot Museum, 888 NE Fourth Av, Canby, OR 97013, mail addr: POB 160, Canby, OR 97013 • T: +1 503 2666712 • F: +1 503 2669775 • depotmuseum@canby.com •
Pres.: *Nora Clark* •
Local Museum – 1984
Hist of Canby, caboose, speeder car, railroading, agriculture and daily living ... 47216

Caney KS

Caney Valley Historical Society, 310 W 4th St, Caney, KS 67333, mail addr: POB 354, Caney, KS 67333 • T: +1 620 8795198 • F: +1 620 8795198 •
Local Museum / Association with Coll – 1984
Education, medical, post office, military, municipal, religious, industry and domestic items ... 47217

Cannon Falls MN

Cannon Falls Area Historical Museum, 208 W Mill St, Cannon Falls, MN 55009 • T: +1 507 2634080 • cfmuseum@citlink.net •
Dir.: *Heidi Holmes Helgren* •
Local Museum – 1979
Regional history ... 47218

Canon City CO

Canon City Municipal Museum, 612 Royal Gorge Blvd, Canon City, CO 81212 • T: +1 719 2765279 • F: +1 719 2699017 • museum@canoncity.org •
Dir.: *Ann Swim* •
Local Museum – 1928
Local history ... 47219

Canterbury CT

Prudence Crandall Museum, Rts 14 and 169, Canterbury Green, Canterbury, CT 06331-0058 • T: +1 860 5469916 • F: +1 860 5467803 • crndll@snet.net •
Cur.: *Kazimiera Kozlowski* •
Historical Museum – 1984
New England life in 19th c, abolitionism, education ... 47220

Canterbury NH

Canterbury Shaker Village, 288 Shaker Rd, Canterbury, NH 03224 • T: +1 603 7839511 • F: +1 603 7839152 • info@shakers.org • www.shakers.org •
Pres.: *Scott T. Swank* • Chief Cur.: *Shery N. Hack* •
Open Air Museum / Historical Museum / Historic Site – 1969
Artifacts, manuscripts and photographs ... 47221

Canton MA

Canton Historical Museum, 1400 Washington St, Canton, MA 02021 • T: +1 781 8288537 • F: +1 781 8215780 • www.canton.org •
Pres.: *James Roche* •
Local Museum – 1893
Local history ... 47222

Canton NY

Pierrepont Museum, 864 State Hwy 68, Canton, NY 13617 • T: +1 315 3868311 • F: +1 315 3790415 •
Sc. Staff: *Charlotte Ragan* (History) •
Local Museum – 1977
Farm tools of pre-machine era, kitchen artifacts, rural school setting, clothing 1800-1875 ... 47223

Richard F. Brush Art Gallery and Permanent Collection, Saint Lawrence University, 23 Romoda Dr, Canton, NY 13617 • T: +1 315 2295174 • F: +1 315 2297445 • ctedford@stlawu.edu • www.stlawu.edu/gallery •
Dir.: *Catherine L. Tedford* •
Fine Arts Museum / University Museum – 1967
19th and 20th c American and European works on paper, photography ... 47224

Silas Wright House, Saint Lawrence County Historical Association, 3 E Main St, Canton, NY 13617 • T: +1 315 3868133 • F: +1 315 3868134 • slcha@northnet.org • www.slcha.org •
C.E.O.: *Trent Trulock* •
Local Museum – 1947
Local hist artifacts, early 19th c furniture, household utensils, costumes – archive ... 47225

Canton OH

Canton Classic Car Museum, 555 Market Av, Canton, OH 44702 • T: +1 330 4553603 • F: +1 330 4550363 • cccm@cantonclassiccar.org • www.cantonclassiccar.org •
Dir.: *Robert C. Lichty* •
Science&Tech Museum – 1978
45 rare and unusual classic and special interest automobiles ... 47226

The Canton Museum of Art, 1001 Market Av N, Canton, OH 44702 • T: +1 330 4537666 • F: +1 330 4524477 • staff@cantonart.org • www.cantonart.org •
Pres.: *David Baker* • C.E.O.: *M.J. Albacete* •
Fine Arts Museum – 1935
19th-20th c American art, contemporary ceramics, European art ... 47227

National Football Museum, 2121 George Halas Dr NW, Canton, OH 44708 • T: +1 330 4568207 • F: +1 330 4568175 • jaikens@profootballhof.com • www.profootballhof.com •
Pres. & Exec. Dir.: *John Bankert* •
Special Museum – 1963 ... 47228

William McKinley Presidential Library and Museum, 800 McKinley Monument Dr NW, Canton, OH 44708 • T: +1 330 4557043 • F: +1 330 4551137 • mmuseum@neo.rr.com • www.mckinleymuseum.org •
Dir.: *Joyce Yut* • Cur.: *Kimberley Kenney* •
Historical Museum / Science&Tech Museum / Natural History Museum – 1946
Astronomy, natural and physical sciences, architecture, ceramics, glass, decorative arts – planetarium, archives ... 47229

Canyon TX

Panhandle-Plains Historical Museum, 2401 Fourth Av, Canyon, TX 79015 • T: +1 806 6512244 • F: +1 806 6512250 • museum@wtamu.edu • www.panhandleplains.org •
Dir.: *Walter R. Davis* • Ass. Dir.: *Lisa Shippee Lambert* • Cur.: *Michael R. Grauer* (Art) • *Dr. William E. Green* (History) • *Dr. Jeff Indeck* (Archaeology) •
Local Museum – 1921
Geology, plains Indian ethnology, firearms ... 47230

Canyon City OR

Grant County Historical Museum, 101 S Canyon City Blvd, Canyon City, OR 97820 • T: +1 541 5750362, 5751993 • museum@ortelco.net •
Pres.: *Ellen D. Stull* •
Local Museum – 1953
Pioneer antiques and relics, Indian artifacts ... 47231

Cape Coral FL

The Children's Science Center, 2915 NE Pine Island Rd, Cape Coral, FL 33909 • T: +1 239 9970012 • F: +1 239 9977215 • scicenter@hotmail.com •
C.E.O.: *Jeff Rodgers* •
Science&Tech Museum – 1989
Artifacts, technology ... 47232

Cape Girardeau MO

Glenn House, 325 S Spanish, Cape Girardeau, MO 63701 •
Decorative Arts Museum – 1972
Structures, furnishings ... 47233

Southeast Missouri State University Museum, 1 University Pl, Cape Girardeau, MO 63701 • T: +1 573 6512260 • F: +1 573 6515909 • museum@semovm.semo.edu • www5.semo.edu/museum •
Dir.: *Stanley I. Grand* • Cur.: *James Phillips* •
University Museum – 1976
Fine art, American military hist, archaeology ... 47234

Cape May NJ

Mid-Atlantic Center for the Arts, 1048 Washington St, Cape May, NJ 08204-0340 • T: +1 609 8845404 • F: +1 609 8842006 • mac4arts@capemaymac.org • www.capemaymac.org •
Dir.: *B. Michael Zuckerman* • Cur.: *Elizabeth Bailey* •
Fine Arts Museum – 1970 ... 47235

Cape May Court House NJ

Cape May County Historical Museum, 504 Rte 9 N, Cape May Court House, NJ 08210 • T: +1 609 4653535 • F: +1 609 4654274 • museum@co.cape-may.nj.us • www.cmcmuseum.org •
Pres.: *James Waltz* • Cur.: *Rachel A. Rodgers* •
Local Museum – 1927
Native American and Victorian artifacts, China, glass and furniture, military and maritime exhibits ... 47236

Leaming's Run Garden and Colonial Farm, 1845 Rte 9 N, Cape May Court House, NJ 08210 • T: +1 609 4655871 • www.leamingsrungardens.com •
Dir.: *Gregg Aprill* •
Agriculture Museum – 1978
Agriculture ... 47237

Cape Vincent NY

Cape Vincent Historical Museum, James St, Cape Vincent, NY 13618 • T: +1 315 6544400 •
Pres.: *J. Thompson* • Dir.: *Mary Hamilton* • *Peter Margrey* (Town History) •
Local Museum – 1968 ... 47238

Capitola CA

Capitola Historical Museum, 410 Capitola Av, Capitola, CA 95010 • T: +1 831 4640322 • cswift@ci.capitola.ca.us • www.capitolamuseum.org •
Local Museum – 1967
Historic photo coll ... 47239

Carbondale IL

Children's Science Center, 1237 E Main, Carbondale, IL 62901 • T: +1 618 5295431 • F: +1 618 5295431 • fun@yoursciencecenter.org • www.yoursciencecenter.com •
C.E.O.: *Pamela Madden* •
Science&Tech Museum – 1994 ... 47240

University Museum, Southern Illinois University, 2469 Faner Hall N, Carbondale, IL 62901-4508 • T: +1 618 4535388 • F: +1 618 4537409 • museum@siu.edu • www.museum.siu.edu/ •
Dir.: *Dona R. Bachman* • Cur.: *Lorilee C. Huffman* •
University Museum / Local Museum – 1869
University history ... 47241

Caribou ME

Nylander Museum, 657 Main St, Caribou, ME 04736 • T: +1 207 4934209 • nylander@mfx.net • www.nylandermuseum.org •
Dir.: *Jeanie L. McGowan* •
Natural History Museum – 1938
Fossils, shells, mineralogy, wetland exhibit ... 47242

Carlisle PA

Hamilton Library and Two Mile House, 21 N Pitt St, Carlisle, PA 17013 • T: +1 717 2497610 • F: +1 717 2589332 • info@historicalsociety.com • www.historicalsociety.com •
Dir.: *Linda F. Witmer* • Libr.: *Cara Holtry* •
Historical Museum / Library with Exhibitions – 1874
Papers, books, genealogy – Photo archives ... 47243

Trout Gallery, Dickinson College, W High St, Carlisle, PA 17013, mail addr: POB 1773, Carlisle, PA 17013-2896 • T: +1 717 2451344 • F: +1 717 2458929 • trout@dickinson.edu • www.dickinson.edu/trout •
C.E.O.: *Dr. Phillip J. Earenfight* •
Fine Arts Museum – 1983
Prints, Oriental and decorative arts, African art ... 47244

United States Army Heritage and Education Center, 22 Ashburn Dr, Carlisle, PA 17013-5008 • T: +1 717 2453611 • F: +1 717 2454370 • usamhi@carlisle.army.mil • carlisle-www.army.mil/usamhi •
Dir.: *Robert Dallesandro* •
Military Museum – 1967
Military hist ... 47245

Carlsbad CA

Children's Discovery Museum of North San Diego (closed) ... 47246

Museum of Making Music, 5790 Armanda Dr, Carlsbad, CA 92008 • T: +1 760 4385996 • F: +1 760 4388964 • museum@museumofmakingmusic.org • www.museumofmakingmusic.org •
Exec. Dir.: *Carolyn Grant* •
Music Museum – 1998
Musical instruments, hist of music making ... 47247

Carlsbad NM

Carlsbad Caverns National Park, 3225 National Parks Hwy, Carlsbad, NM 88220 • T: +1 505 7853116 • F: +1 505 7852317 • david_kayser@nps.gov • www.nps.gov •
Archaeology Museum / Natural History Museum – 1923
Hist of the caverns, botany, entomology, geology, paleontology, archaeology, herpetology ... 47248

Carlsbad Museum and Art Center, 418 W Fox St, Carlsbad, NM 88220 • T: +1 505 8870276 • F: +1 505 8858809 • virginia_dodier@carlsbadmuseum.org • www.carlsbadmuseum.org •
Dir.: *Virginia Dodier* •
Fine Arts Museum / Archaeology Museum / Historical Museum / Science&Tech Museum – 1931
Archaeology, regional hist, New Mexico art, Pueblo pottery, Peruvian antiques, temporary exhibits ... 47249

Carmel CA

Center for Photographic Art, San Carlos & Nineth Sts, Carmel, CA 93921, mail addr: POB 1100, Carmel, CA 93921 • T: +1 831 6255181 • F: +1 831 6255199 • info@photography.org • www.photography.org •
Dir.: *Dennis High* •
Fine Arts Museum – 1992 ... 47250

U

Mission San Carlos Borromeo del Rio Carmelo, 3080 Rio Rd, Carmel, CA 93923 • T: +1 831 6241271 • www.carmelmission.org •
Cur.: *James Griffin* •
Historical Museum / Religious Arts Museum – 1770 ... 47251

Carmel IN

National Association of Miniature Enthusiasts, 130 N Rangeline Rd, Carmel, IN 46032 • T: +1 317 5718094 • F: +1 317 5718105 • name@miniatures.org • www.miniatures.org •
Exec. Dir.: *John K. Purcell* •
Association with Coll – 1972 ... 47252

Carmi IL

White County Historical Museums, Ratcliff Inn, L.Haas Store, Matsel Cabin, Robinson-Stewart House, 203 N Church St, Carmi, IL 62821 • T: +1 618 3828425 • cbconly@midwest.net • www.rootsweb.com/~ilwcohs •
Pres.: *Jim Pumphrey* •
Local Museum – 1957
Local history ... 47253

Carpinteria CA

Carpinteria Valley Museum of History, 956 Maple Av, Carpinteria, CA 93013 • T: +1 805 6843112 • info@carpinteriahistoricalmuseum.org • www.carpinteriahistoricalmuseum.org •
Dir./Cur.: *David W. Griggs* •
Historical Museum – 1959
Chumash Indians, early pioneers items – library, archives ... 47254

Carrollton MS

Merrill Museum, Rte 1, Carrollton, MS 38917 • T: +1 662 2379254 •
Pres.: *Kay Slocum* •
Local Museum – 1961
Local history ... 47255

Carrollton OH

Civil War Museum, McCook House, Public Sq, Carrollton, OH 44615 • T: +1 330 6273345 • F: +1 330 6275366 • ohswww@winslo.ohio.gov • www.ohiohistory.org/places/mcookhse/ •
Pres.: *Thomas Konst* •
Historical Museum – 1963
Local hist, local pottery, military items, medical instruments and textiles ... 47256

Carson CA

International Printing Museum, 315 Torrance Blvd, Carson, CA 90745 • T: +1 310 5157166 • mail@printmuseum.org • www.printmuseum.org •
Pres.: *John Hedlund* • Dir.: *Mark Barbour* •
Science&Tech Museum – 1988
Typography, antique printing machinery – library 47257

University Art Gallery, c/o California State University, Dominguez Hills, 1000 E Victoria St, Carson, CA 90747 • T: +1 310 2433334 • F: +1 310 2176967 • kzimmerer@csudh.edu • cah.csudh.edu/art_gallery •
Dir.: *Kathy Zimmerer* •
University Museum / Fine Arts Museum – 1978 . 47258

Carson City NV

Bowers Mansion, 4005 US 395 N, Carson City, NV 89704 • T: +1 702 8490201 • F: +1 775 8499568 •
Cur.: *Betty Hood* •
Historical Museum – 1946 ... 47259

Nevada State Museum, 600 N Carson St, Carson City, NV 89701 • T: +1 775 6874810 • F: +1 775 6874168 • www.nevadaculture.org •
Dir.: *Jim Barmore* • Cur.: *George Baumgardner* (Natural History) •
Historical Museum / Natural History Museum – 1939
History, natural and prehistory ... 47260

Nevada State Railroad Museum, 2180 S Carson St, Carson City, NV 89701 • T: +1 702 6876953 • F: +1 702 6878294 • pdbarton@clan.lib.nv.us • www.nsrm-friends.org •
Cur.: *Daniel P. Thielen* •
Science&Tech Museum – 1980
Virginia and Truckee railway coll, cars and locomotives, railroad equip ... 47261

Stewart Indian Cultural Center, 1329 Stanford Dr, Carson City, NV 89701 • T: +1 702 8821808 •
Folklore Museum / Ethnology Museum ... 47262

Cartersville GA

Bartow History Center, 13 N Wall St, Cartersville, GA 30120 • T: +1 770 3823818 • F: +1 770 3839314 • micheler@bartowhistorycenter.org • www.bartowhistorycenter.org •
Dir.: *Michele Rodgers* •
Historical Museum – 1987
Personal artifacts, farming tools, photographs, textiles ... 47263

Etowah Indian Mounds, 813 Indian Mounds Rd SE, Cartersville, GA 30120 • T: +1 770 3873747 • F: +1 770 3873972 • etowahl@innerx.net •
Head: *Elizabeth Bell* •
Ethnology Museum / Archaeology Museum – 1953
Indian, archaeology ... 47264

Rose Lawn Museum, 224 W Cherokee Av, Cartersville, GA 30120 • T: +1 770 3875162 • F: +1 770 3861527 • roselawn@mindspring.com • www.roselawnmuseum.com •
Dir.: *Jane Drew* •
Religious Arts Museum – 1973
Victorian mansion, former home of evangelist Samuel Porter Jones ... 47265

Carterville IL

John A. Logan College Museum, 700 Logan College Rd, Carterville, IL 62918 • T: +1 618 9853741 • F: +1 6189852248 • museum@jalc.edu • www.jal.cc.il.us/museum •
Fine Arts Museum / Decorative Arts Museum – 1967
Contemporary regional art and craft, international ethnic crafts, relating to the life of General John A Logan memorabilia ... 47266

Carthage IL

Hancock County Historical Museum, 306 Walnut, Carthage, IL 62321 •
Pres.: *Karen Taylor* •
Local Museum – 1968
1906-8 Stone Courthouse, on historical site in Carthage ... 47267

Carthage MO

Powers Museum, 1617 Oak St, Carthage, MO 64836 • T: +1 417 3582667 • F: +1 417 3599627 • info@powersmuseum.com • www.powersmuseum.com •
Dir.: *Michele Hansford* •
Local Museum – 1982
Late 19th-20th c decorative arts, textiles & costumes, photograph coll ... 47268

Casa Grande AZ

Casa Grande Valley History Museum, 110 W Florence Blvd, Casa Grande, AZ 85222-4033 • T: +1 520 8362223 • F: +1 520 8365065 • info@cgvhs.org • www.cgvhs.org •
Man.: *Sybil Rossiter* •
Local Museum – 1964
Local history – library ... 47269

Cashmere WA

Cashmere Museum & Pioneer Village, 600 Cotlets Way, Cashmere, WA 98815 • T: +1 509 7823230 • F: +1 509 7828905 • cchspvm@aol.com • www.cashmeremuseum.com •
Dir.: *Angelina LeRoy* •
Local Museum – 1956
Local hist ... 47270

Casper WY

Fort Caspar Museum, 4001 Fort Caspar Rd, Casper, WY 82604 • T: +1 307 2358462 • F: +1 307 2358464 • ftcaspar@cityofcasperwy.com • www.fortcasparwyoming.com •
C.E.O.: *Richard L. Young* • Cur.: *Dana Schaar* •
Military Museum – 1936
Civil War, Indian and pioneer artifacts, fort buildings ... 47271

Nicolaysen Art Museum, 400 East Collins Dr, Casper, WY 82601 • T: +1 307 2355247 • F: +1 307 2350923 • info@thenic.org • www.thenic.org •
Dir.: *Joe Ellis* • Cur.: *Joe Ramirez* •
Fine Arts Museum – 1967
Contemporary artists ... 47272

Tate Geological Museum, Casper College, 125 College Dr, Casper, WY 82601 • T: +1 307 2682447 • F: +1 307 2683026 • dbrown@caspercollege.edu • www.caspercollege.edu/tate/webpage.htm •
Dir.: *David Brown* •
Natural History Museum – 1979
Paleontological and mineralogical items from Wyoming ... 47273

Caspian MI

Iron County Museum, Brady at Museum Rd, Caspian, MI 49915 • T: +1 906 2652617 • icmuseum@up.net • www.ironcountymuseum.com •
Dir.: *Harold O. Bernhardt* • *Shirley Carlson* (Art) • Cur.: *Marcia A. Bernhardt* •
Local Museum – 1962
Lumbering, mining and farm tools, miniature lumbering, glass mining diorama, mining tools – Le Blanc Memorial Art Gallery, Bernhardt Art Gallery ... 47274

Cassville WI

Stonefield Historic Site, 12195 County Rd V V, Cassville, WI 53806 • T: +1 608 7255210 • F: +1 608 7255919 • stonefld@pcii.net •
Dir.: *Allen Schroeder* • Cur.: *Judith Meyerdierks* •
Agriculture Museum / Open Air Museum – 1952
Agriculture ... 47275

Castalian Springs TN

Historic Cragfont, 200 Cragfront Rd, Castalian Springs, TN 37031 • T: +1 615 4527070 • historiccragfont@hotmail.com • www.srlab.net/cragfont/ •
Historic Site – 1958
Historic house, home of General James Winchester ... 47276

Castile NY

William Pryor Letchworth Museum, 1 Letchworth State Park, Castile, NY 14427 • T: +1 585 4933600 • F: +1 716 4935272 • brian.scriven@oprhp.state.ny.us • www.nysparks.state.ny.us •
Dir.: *Brian Scriven* •
Local Museum – 1913
Indian artifacts, pioneer artifacts, Civilian Conservation Corps hist ... 47277

Castine ME

Castine Scientific Society Museum, 107 Perkins St, Castine, ME 04421 • T: +1 207 3269247 • F: +1 207 3268545 • admin@wilsonmuseum.org • www.wilsonmuseum.org •
Dir.: *Norman Doudiet* • Cur.: *Patricia Hutchins* •
Natural History Museum / Historical Museum – 1921
Science ... 47278

Castleton VT

Castleton Historical Society, The Higley Homestead, Main St, Castleton, VT 05735-0219 • T: +1 802 4685523 •
Association with Coll – 1973
Local hist ... 47279

Catasauqua PA

George Taylor House, Lehigh and Poplar Sts, Catasauqua, PA 18032 • T: +1 610 4354664 • F: +1 610 4359812 •
Pres.: *Robert M. McGovern* • Exec. Dir.: *Andree Miller* •
Local Museum – 1904 ... 47280

Catawissa PA

Catawissa Railroad Museum, 119 Pine St, Catawissa, PA 17820 • T: +1 570 3562675 • F: +1 570 3567876 • waltgosh@ptd.net • www.caboosenut.com •
Science&Tech Museum
Railroad cabooses, track, railroad memorabilia .. 47281

Cathlamet WA

Wahkiakum County Historical Museum, 65 River St, Cathlamet, WA 98612 • T: +1 360 7953954 •
Pres.: *Ralph Keyser* •
Local Museum – 1956
Local hist ... 47282

Catonsville MD

Alumni Museum, Spring Grove Hospital Center, 55 Wade Av, Garrett Bldg, Catonsville, MD 21228 • T: +1 410 4027786, 4027856 • F: +1 410 4027050 • helseld@dhmh.state.md.us • www.springgrove.com/history.html •
Historical Museum
Tools, furniture, photos ... 47283

Cattaraugus NY

Cattaraugus Area Historical Center, 23 Main St, Cattaraugus, NY 14719 • T: +1 716 2579500 •
Dir.: *Dawn D. Waite* •
Local Museum – 1955
Agriculture, costumes,, Indian artifacts, hist of the local school ... 47284

Cavalier ND

Pioneer Heritage Center, 13571 Hwy 5, Cavalier, ND 58220 • T: +1 701 2654561 • F: +1 701 2654443 • isp@state.nd.us •
Pres.: *Alice Olson* •
Local Museum – 1989
Local hist, settlement story (1870-1920) ... 47285

Cazenovia NY

Chapman Art Center Gallery, Cazenovia College, Cazenovia, NY 13035 • T: +1 315 6557162 • F: +1 315 6552190 • jaistars@cazcollege.edu • www.cazcollege.edu •
Dir.: *John Aistars* •
Public Gallery / University Museum – 1978 ... 47286

Lorenzo State Historic Site, 17 Rippleton Rd, Cazenovia, NY 13035 • T: +1 315 6553200 • F: +1 315 6554304 • lincklaen@juno.com •
Head: *Pam Ellis* • Cur.: *Jackie Vivirito* •
Historic Site – 1968
History, furniture, 19th c decorative art, carriages, fine art ... 47287

Rothschild Petersen Patent Model Museum, 4796 W Lake Rd, Cazenovia, NY 13035 • museum@patentmodel.org • www.patentmodel.org •
Special Museum – 1998
4,000 patent models and related documents ... 47288

Cedar City UT

Braithwaite Fine Arts Gallery, c/o Southern Utah University, 351 W Center St, Cedar City, UT 84720 • T: +1 435 5865432 • F: +1 435 8658012 • museums@suu.edu • www.suu.edu/pva/artgallery •
Dir.: *Lydia Johnson* •
Public Gallery / University Museum – 1976
18th-20th c American art ... 47289

Iron Mission Museum, 635 N Main St, Cedar City, UT 84720 • T: +1 435 5869290 • F: +1 435 86656830 • ironmission@utah.gov • www.stateparks.utah.gov •
Dir.: *Todd Prince* • Cur.: *Ryan Paul* •
Historical Museum – 1973
Pioneer hist ... 47290

Cedar Falls IA

Cedar Falls Historical Museum, 308 Franklin St, Cedar Falls, IA 50613 • T: +1 319 2665149, 2778817 • F: +1 319 2681812 • cfhs_schmitz@cfu.net • www.cedarfallshistorical.org •
Pres.: *Dr. Thomas Connors* •
Local Museum – 1962
Local history ... 47291

Gallery of Art, University of Northern Iowa, Kamerick Art Bldg, 1601 W 27th St, Cedar Falls, IA 50614-0362 • T: +1 319 2732077 • F: +1 319 2737333 • galleryofart@uni.edu • www.uni.edu/artdept •
Dir.: *Darrell Taylor* •
Fine Arts Museum / University Museum – 1978
20th c American and European art ... 47292

James and Meryl Hearst Center for the Arts, 304 W Seerley Blvd, Cedar Falls, IA 50613 • T: +1 319 2738641 • F: +1 319 2738659 • mary.huber@cedarfalls.com • www.hearstartscenter.com •
Dir.: *Mary Huber* • Cur.: *Emily Drennan* •
Fine Arts Museum / Open Air Museum / Public Gallery – 1989
Arts, Children's Book Illustration coll, Public Art coll, Local Artist coll – sculpture garden ... 47293

University of Northern Iowa Museum Museums & Collections, 3219 Hudson Rd, Cedar Falls, IA 50614-0199 • T: +1 319 2732188 • F: +1 319 2736924 • doris.mitchell@uni.edu • www.uni.edu/museum •
Dir.: *Sue Grosboll* •
University Museum – 1892
Ethnological and zoological coll ... 47294

Cedar Grove NJ

Cedar Grove Historical Society Museum, 903 Pompton Av, Cedar Grove, NJ 07009 • T: +1 973 2395414 •
Local Museum – 1968
Farming, sports equip, textiles ... 47295

Cedar Key FL

Cedar Key Historical Society Museum, 2nd St at State Rd 24, Cedar Key, FL 32625 • T: +1 352 5435549 • ckhistory@inetw.net •
Cur.: *Peggy Rix* •
Local Museum – 1979
Fossils, shells, pencil and lumber industry, railroad hist, maps, Indian hist, commercial fishing industry ... 47296

Cedar Key State Park Museum, 12231 SW 166 Court, Cedar Key, FL 32625 • T: +1 352 5435350, 5435567 • F: +1 352 5436315 • charles.neese@dep.state.fl.us • www.dep.state.fl.us/parks •
Exec. Dir.: *David B. Struhs* •
Local Museum / Natural History Museum – 1962
History, shell coll, natural hist ... 47297

Cedar Lake IN

Lake of the Red Cedars Museum, 7900 Constitution Av, Cedar Lake, IN 46303 • T: +1 219 3746157 • F: +1 219 3746157 •
Pres.: *Tim Brown* • Dir.: *Anne Zimmerman* •
Local Museum – 1977
Local history; prehistoric-modern, industrial and farming artifacts; photographs ... 47298

Cedar Rapids IA

African American Historical Museum, 55 12th Av SE, Cedar Rapids, IA 52401 • T: +1 319 8622101 • F: +1 319 8622105 • blackiowa.org •
Exec. Dir./Cur.: *Thomas Moore* • Cur.: *Susan Kuecker* •
Historical Museum – 1994
library ... 47299

Brucemore, 2160 Linden Dr SE, Cedar Rapids, IA 52403 • T: +1 319 3627375 • F: +1 319 3629481 • mail@brucemore.org • www.brucemore.org •
Dir.: *Peggy Whitworth* •
Historical Museum / Historic Site – 1981
Historic Site, culture, housed in 21-room Queen Anne-style mansion ... 47300

Cedar Rapids Museum of Art, 410 Third Av, SE, Cedar Rapids, IA 52401 • T: +1 319 3667503 • F: +1 319 3664111 • crma@earthlink.net • www.crma.org •
Dir.: *Terence Pitts* •
Fine Arts Museum – 1905
20th c modern and regionalist paintings, works by Grant Wood, Marvin Cone, James Swann, Malvina Hoffman, Bertha Jaques, Mauricio Lasansky, 19th and 20th c sculpture, paintings, prints, photographs ... 47301

Coe College Collection of Art, Stewart Memorial Library, 1220 first Av NE, Cedar Rapids, IA 52402 • T: +1 319 3998023 • F: +1 319 3998019 • rdoley@coe.edu • www.coe.edu •
Fine Arts Museum
Art – Library 47302

Eaton-Buchan Gallery and Marvin Cone Gallery, c/o Coe College, 1220 First Av NE, Cedar Rapids, IA 52402 • T: +1 319 3998217 • F: +1 319 3998557 • www.coe.edu •
Dir.: *Delores Chance* •
Fine Arts Museum – 1942
Art works, paintings, contemporary art 47303

Iowa Masonic Library and Museum, 813 1st Av, SE, Cedar Rapids, IA 52402 • T: +1 319 3651438 • F: +1 319 3651439 • gliowa@qwest.net • www.gl-iowa.org •
Local Museum – 1845
Local history – library 47304

Linn County Historical Society Museum, 615 First Av, Cedar Rapids, IA 52401-2022 • T: +1 319 3621501 • F: +1 319 3626790 • martha@historycener.org • www.historycenter.org •
Dir.: *Martha Aldridge* • Cur.: *Bryan Preston* •
Local Museum – 1969
Iowa history from prehistoric-presents, history of Cedar Rapid 47305

National Czech and Slovak Museum, 30 16th Av SW, Cedar Rapids, IA 52404 • T: +1 319 3628500 • F: +1 319 3632209 • www.ncsml.org •
Cur.: *Edith Blanchard* •
Special Museum – 1974
Ethnic museum, 1880-1900 restored Czech immigrant home 47306

Science Station Museum, 427 First St, SE, Cedar Rapids, IA 52401 • T: +1 319 3660968 • F: +1 319 3664590 • nolte@mcleodusa.net • www.sciencestation.org •
Pres.: *Terry Bergen* • Exec. Dir.: *Joseph Nolte* •
Science&Tech Museum – 1986
Science, housed in a fire station bldg 47307

Cedarburg WI

Ozaukee Art Center, W 62 N 718 Riveredge Dr, Cedarburg, WI 53012 • T: +1 414 3778230 • ozaukeeartcenter@pauljyank.com •
Pres.: *Lon Horton* • Dir.: *Paul Yank* •
Fine Arts Museum – 1971
Paintings, sculpture, ptints, jewelry, glass, ceramics 47308

Cedartown GA

Polk County Historical Museum, 205 N College St, Cedartown, GA 30125 • T: +1 404 7480073 •
Local Museum – 1974
Local artifacts and history 47309

Celina OH

Mercer County Historical Museum, The Riley Home, 130 E Market, Celina, OH 45822-0512 • T: +1 419 5866065 • histalig@bright.net •
Exec. Dir.: *Joyce L. Alig* •
Historical Museum – 1959
Agriculture, carpnetrs' and blacksmiths' tools, furniture, costumes, glassware – archives 47310

Centennial CO

B'S Ballpark Museum, 8611 E Otero Pl, Centennial, CO 80112 • T: +1 720 3510665 • F: +1 303 6946761 •
Dir.: *Bruce S. Hellerstein* •
Special Museum – 1999
Baseball hist 47311

Centennial WY

Nici Self Museum, 2734 Hwy 130, 28 miles west of Laramie, Centennial, WY 82055 • T: +1 307 7427763 • F: +1 307 6344955 • rgoldie@wyomail.com •
Pres.: *Jim Chase* •
Local Museum – 1974
Local hist, railroad equipment, gold and platinum mining 47312

Center ND

Fort Clark Trading Post, 1074 27th Av SW, Center, ND 58530 • T: +1 701 3282666, 7948832 • F: +1 701 3283710 • histsoc@state.nd.us • www.state.nd.us/hist/ftClark •
Special Museum – 1965
Mandan Indian villages, fur trading 47313

Center Sandwich NH

Sandwich Historical Society Museum, 4 Maple St, Center Sandwich, NH 03227 • T: +1 603 2846269 • F: +1 603 2846269 • shistory@worldpath.net • www.sandwichhistorical.org •
Pres.: *Peter Booty* • Dir.: *Marcene Molinaro* •
Local Museum / Fine Arts Museum – 1917
Paintings, furniture 47314

Centerport NY

Suffolk County Vanderbilt Museum, 180 Little Neck Rd, Centerport, NY 11721 • T: +1 516 8545579 • F: +1 516 8545527 • vanderbilt@webscope.com • www.vanderbiltmuseum.org •
Exec. Dir.: *J. Lance Mallamo* • Pres.: *Steve Gittelman* •
Historic Site / Natural History Museum – 1950
Marine biology and wildlife, butterfly, bird and shell coll; Spanish Revival mansion, furniture, medieval till 19th c decorative arts, fine art; antique and classic automobiles 47315

Centerville MA

Centerville Historical Museum, 513 Main St, Centerville, MA 02632 • T: +1 508 7750331 • F: +1 508 8629211 • info@centervillehistoricalmuseum.org • www.centervillehistoricalmuseum.org •
Cur.: *Britt Beedenbender* •
Local Museum – 1952
Costumes, quilts, military, marine, tools, toys, Dodge MacKnight watercolors, decorative arts – library 47316

Centerville OH

Walton House Museum, 89 W Franklin St, Centerville, OH 45459 • T: +1 937 4330123 • F: +1 937 4244629 • cwhistorical@sbcglobal.net • www.mvcc.net/centerville/histsoc •
Pres.: *Sue Hufnagle* •
Local Museum – 1967
Hist of the area 47317

Central AK

Central Museum, 128 Mile Steese Hwy, Central, AK 99730 • T: +1 907 5201893 • F: +1 907 5201893 •
Cur.: *Julie Cooper* •
Local Museum – 1977
Gold mining hist, living conditions 47318

Central City CO

Central City Opera House Museum, 124 Eureka St, Central City, CO 80427 • T: +1 303 2926500 • F: +1 303 2924958 • marketing@centralcityopera.org • www.centralcityopera.org •
Gen. Dir.: *Pelham G. Pearce* •
Performing Arts Museum – 1932
Theater museum, housed in Teller House Hotel Museum 47319

Gilpin History Museum, 228 E High St, Central City, CO 80427-0247 • T: +1 303 5825283 • F: +1 303 5825283 • gchs@wispertel.net • www.gilpinhistory.org •
Pres.: *Linda Jones* • Dir.: *James J. Prochaska* •
Local Museum – 1971
Regional history 47320

Chadds Ford PA

Brandywine Battlefield, US Rte 1, Chadds Ford, PA 19317 • T: +1 610 4593342 • F: +1 610 4599586 • erump@state.pa.us • www.ushistory.org/brandywine •
Head: *Elisabeth Rump* •
Military Museum / Historic Site / Open Air Museum – 1947
Military hist 47321

Brandywine River Museum, US Rte 1 at PA Rte 100, Chadds Ford, PA 19317 • T: +1 610 3882700 • F: +1 610 3881197 • inquiries@brandywine.org • www.brandywinemuseum.org. •
Dir.: *James H. Duff* • Cur.: *Gene E. Harris* • Ass. Cur.: *Virginia Herrick O'Hara* • *Christine B. Podmaniczky* (N.C. Wyeth Collections) •
Fine Arts Museum – 1971
American art 47322

Christian C. Sanderson Museum, Rte 100 N, Chadds Ford, PA 19317 • T: +1 610 3886545 • F: +1 302 7982567 • racford@cu-access.net • www.sandersonmuseum.org •
Pres. & C.E.O.: *Richard A. McLellan* •
Local Museum – 1967
Personal and family items, hist artifacts 47323

Chadron NE

Chadron State College Main Gallery, 1000 Main St, Chadron, NE 69337 • T: +1 308 4326326 • F: +1 308 4323561 • rbird@csc.edu •
Dir.: *Richard Bird* •
Fine Arts Museum – 1967 47324

Dawes County Historical Society Museum, 341 Country Club Rd, Chadron, NE 69337 • T: +1 308 4324999, 4322309 • www.chadron.com/ •
C.E.O. & Pres.: *Rolin Curd* •
Local Museum – 1964
Local history 47325

Eleanor Barbour Cook Museum of Geology, Chadron State College, Math & Science Bldg, 1000 Main St, Chadron, NE 69337 • T: +1 308 4326293 • F: +1 308 4326434 • mleite@csc1.csc.edu • www.csc.edu •
Dir.: *Prof. Michael Leite* •
University Museum / Natural History Museum – 1939
Paleontology, archaeology, geology 47326

Museum of the Fur Trade, 6321 E Hwy 20, Chadron, NE 69337 • T: +1 308 4323843 • F: +1 308 4325943 • museum@furtrade.org • www.furtrade.org •
C.E.O. & Dir.: *Gail DeBuse Potter* •
Historical Museum – 1955
American, Indian and Alaskan objects 47327

Chagrin Falls OH

Chagrin Falls Historical Society Museum, 21 Walnut St, Chagrin Falls, OH 44022 • T: +1 440 2474695 •
Cur.: *Pat E. Zalba* • Libr.: *Dorothy Pauly* •
Local Museum – 1949
Doll coll, household articles, costumes, glassware, regional art and artifacts, Chagrin Falls' industry hist 47328

Chamberlain SD

Akta Lakota Museum, Saint Joseph's Indian School, Chamberlain, SD 57325 • T: +1 605 7343452 • F: +1 605 7343388 • aktalakota@stjo.org • www.stjo.org •
Dir.: *Dixie Thompson* •
Historical Museum / Folklore Museum – 1991 47329

South Dakota Hall of Fame, 1480 S Main, Chamberlain, SD 57325 • T: +1 605 7344216 • F: +1 605 7344216 • sdhof@midstatesd.net • www.sdhalloffame.com •
C.E.O: *Ron Frame* •
Historical Museum – 1974
Biographical info, artwork, pioneer and Indian days 47330

Chambersburg PA

Fort Loudoun, 1720 Brooklyn Rd, Chambersburg, PA 17021 • T: +1 717 3693473 • F: +1 717 7831073 • www.fortloudoun-pa.com •
Historical Museum – 1968
Historical of the fort occupation 47331

Old Brown's Mill School, 1051 S Coldbrook Av, Chambersburg, PA 17201 • T: +1 717 7050559 • F: +1 717 7831073 •
Local Museum – 1956 47332

Champaign IL

Champaign County Historical Museum, 102 E University Av, Champaign, IL 61820 • T: +1 217 3561010 • director@champaignmuseum.org • www.champaignmuseum.org •
Dir.: *Paul Idleman* •
Historical Museum – 1972
Furnishings, furniture, County hist 47333

I Space, University of Illinois at Urbana-Champaign, 230 W Superior St, Champaign, IL 60610 • T: +1 312 5879976 • F: +1 312 5879978 • mantonak@uiuc.edu • www.ispace.uiuc.edu •
Dir.: *Mary Antonakos* •
University Museum / Fine Arts Museum
Contemporary art 47334

Krannert Art Museum, University of Illinois, 500 E Peabody Dr, Champaign, IL 61820 • T: +1 217 3331861 • F: +1 217 3330883 • schumach@uiuc.edu • www.art.uiuc.edu/kam/ •
Dir.: *Karen Hewitt* • Cur.: *Michael Conner* •
Fine Arts Museum – 1961
Cultures of Africa, Asia, Europe and Americas 47335

Parkland College Art Gallery, 2400 W Bradley, Champaign, IL 61821 • T: +1 217 3512485, 2200 • F: +1 217 3733899 • lcostello@parkland.edu • www.parkland.edu/gallery •
Dir.: *Lisa Costello* • Ass. Dir.: *Cindy Smith* •
Fine Arts Museum – 1980
Contemporary art, drawings, paintings, photography, crafts, sculpture, ethnic arts 47336

Champion NE

Champion Mill, POB 117, Champion, NE 69023 • T: +1 308 8825860, 3945118 • anderssra@ngpc.state.ne.us •
Science&Tech Museum – 1969
Milling machinery, tools 47337

Chandler AZ

Chandler Museum, 178 E Commonwealth Av, Chandler, AZ 85225 • T: +1 480 7822717 • F: +1 480 7822765 • chandlermuseum@aol.com • www.chandlermuseum.org •
Pres.: *Michael E. Larson* •
Local Museum – 1969
Local history, Indian artifacts 47338

Chandler OK

Lincoln County Historical Society Museum of Pioneer History, 717 Manvel Av, Chandler, OK 74834 • T: +1 405 2582425 • F: +1 405 2582809 • lchs@brightok.net •
C.E.O. & Pres.: *Ronald Brasel* •
Local Museum – 1954
Artifacts, memorabilia, archival mat 47339

Chanute KS

Imperato Collection of West African Artifacts, Safari Museum, 111 N Lincoln Av, Chanute, KS 66720 • T: +1 620 4312730 • F: +1 620 4313848 • osajohns@safarimuseum.com • www.safarimuseum.com •
Dir.: *Conrad G. Froehlich* •
Ethnology Museum – 1974
West African sculpture, masks, ancestor figures and ritual objects, household items, musical instruments 47340

Johnson Collection of Photographs, Movies and Memorabilia, Safari Museum, 111 N Lincoln Av, Chanute, KS 66720 • T: +1 620 4313848 • F: +1 620 4313848 • osajohns@safarimuseum.com • www.safarimuseum.com •
Dir.: *Conrad G. Froehlich* •
Special Museum 47341

Martin and Osa Johnson Safari Museum, 111 N Lincoln Av, Chanute, KS 66720 • T: +1 620 4312730 • F: +1 620 4313848 • osajohns@safarimuseum.com • www.safarimuseum.com •
Dir.: *Conrad G. Froehlich* • Cur.: *Barbara E. Henshall* • *Jaquelyn L. Borgeson* •
Special Museum – 1961
Biography, ethnography, film/photo, Santa Fe Train Depot 47342

Selsor Gallery of Art, Safari Museum, 111 N Lincoln Av, Chanute, KS 66720 • T: +1 620 4312730 • F: +1 620 4313848 • osajohns@safarimuseum.com • www.safarimuseum.com •
Dir.: *Conrad G. Froehlich* •
Fine Arts Museum – 1981 47343

Chapel Hill NC

Ackland Art Museum, University of North Carolina at Chapel Hill, 9302 Columbia and Franklin Sts, Chapel Hill, NC 27599 • T: +1 919 9665736 • F: +1 919 9661400 • ackland@email.unc.edu • www.ackland.org •
C.E.O.: *Gerald D. Bolas* • Asst.Dir.: *Timothy A. Riggs* • Cur.: *Barbara Matilsky* (Exhibitions) • Cons.: *Lyn Koehnline* • Education: *Carolyn Wood* •
Fine Arts Museum – 1958
European and American art, 15th-20th c prints, photographs, NC folk art, Indian miniatures and sculptures 47344

Chapel Hill Museum, 523 E Franklin St, Chapel Hill, NC 27514 • T: +1 919 9671400 • F: +1 919 9676230 • info@chapelhillmuseum.org • www.chapelhillmuseum.org •
Fine Arts Museum 47345

Chappaqua NY

Horace Greeley House, New Castle Historical Society, 100 King St, Chappaqua, NY 10514 • T: +1 914 2384666 • F: +1 914 2381296 • newcastlehs@aol.com • www.newcastlehistoricalsociety.org •
Pres.: *Albert Hutin* • Exec. Dir.: *Betsy Towl* •
Local Museum – 1966
Hist of New Castle, Quakers coll, Horace Greeley memorabilia 47346

Chappell NE

Chappell Art Gallery, 289 Babcock St, Chappell, NE 69129 • T: +1 308 8742626 •
Fine Arts Museum – 1935
Aaron Pyle coll, art works from many countries – Memorial library 47347

Chappell Hill TX

Chappell Hill Historical Society Museum, 9220 Poplar St, Chappell Hill, TX 77426 • T: +1 979 8366033 • F: +1 979 8367438 • chmuseum@alpha1.net • www.chappellhillmuseum.org •
Dir.: *Ladonna Vest* •
Local Museum / Historical Museum – 1964
Confederate, early Texashost of Chappell Hill, Johnnie Swearingen folk art – archives 47348

Charles City IA

Floyd County Historical Museum, 500 Gilbert St, Charles City, IA 50616-2738 • T: +1 641 2281099 • F: +1 641 2281157 • fchs@fiai.net • www.floydcountymuseum.org •
Pres.: *Jody Flint* • Dir.: *Mary Ann Townsend* •
Local Museum / Historical Museum – 1961
Archaeology, medical, pharmagology, military, 1850 - 1900 period rooms 47349

Charles City VA

Berkeley Plantation, 12602 Harrison Landing Rd, Charles City, VA 23030 • T: +1 804 8296018 • F: +1 804 8296757 • www.berkeleyplantation.com •
Head: *Malcom E. Jamieson* •
Local Museum – 1619
Period artifacts, Indian & Civil War relics 47350

Sherwood Forest Plantation, 14501 John Tyler Memorial Hwy, Charles City, VA 23030 • T: +1 804 8295377 • F: +1 804 8292947 • ktylerl@sherwoodforest.org • www.sherwoodforest.org •
Pres.: *Frances P.B. Tyler* • Dir.: *Kay Montgomery Tyler* •
Historic Site – 1975
Historic house, home of President John Tyler 47351

Shirley Plantation, 501 Shirley Plantation Rd, Charles City, VA 23030 • T: +1 804 8295121 • F: +1 804 8296322 • info@shirleyplantation.com • www.shirleyplantation.com •
Head: *Charles Hill Carter* •
Local Museum – 1613/1952
Hist of the Hill & Carter families 47352

Westover House, 7000 Westover Rd, Charles City, VA 23030 • T: +1 804 8292882 • F: +1 804 8295528 •
Man.: *F.S. Fisher* •
Historical Museum 47353

U

Charleston IL

Tarble Arts Center, Eastern Illinois University, S 9th St at Cleveland Av, Charleston, IL 61920-3099 • T: +1 217 5812787 • F: +1 217 5817138 • cfmw@eiu.edu • www.eiu.edu/~tarble •
Dir.: *Michael Watts* •
Fine Arts Museum / University Museum – 1982
20th c Illinois folk arts, prints, paintings 47354

Charleston MO

Mississippi County Historical Society, 403 N Main, Charleston, MO 63834 • T: +1 314 6834348 •
Pres.: *Tom Graham* •
Association with Coll – 1966
Archaeology, costumes, military, Indian artifacts 47355

Charleston SC

American Military Museum, 360 Concord St, Charleston, SC 29403, mail addr: POB 31351, Charleston, Sc 29417 • T: +1 803 7239620 • ammilmus@aol.com •
Dir./Cur.: *George E. Meagher* •
Military Museum – 1987
Military hist 47356

Avery Research Center for African American History and Culture, College of Charleston, 125 Bull St, Charleston, SC 29424 • T: +1 843 9537609 • F: +1 843 9537607 •
Dir.: *W. Marvin Dulaney* • *Curtis J. Franks* (Education and Exhibits) •
Historical Museum – 1985
History 47357

Charles Towne Landing 1670, 1500 Old Town Rd, Charleston, SC 29407 • T: +1 843 8524200 • F: +1 803 8524205 • charles_towne_landing_sp@scprt.com • www.southcarolinaparks.com •
Head: *Ron Fischer* •
Historic Site / Historical Museum – 1970
1670 site of the first permanent English settlement in South Carolina 47358

The Charleston Museum, 360 Meeting St, Charleston, SC 29403-6297 • T: +1 803 7222996 • F: +1 843 7221784 • jbrumgardt@charlestonmuseum.org • www.charlestonmuseum.org •
Dir.: *Dr. John R. Brumgardt* • Ass. Dir.: *Carl P. Borick* • Cur.: *Dr. Albert E. Sanders* (Natural History) • *Dr. William Post* (Ornithology) • *Martha Zierden* (Historical Archaeology) • *J. Grahame Long* (History) • Libr.: *Sharon Bennett* •
Local Museum – 1773
Local hist – library 47359

The Citadel Archives and Museum, The Military College of South Carolina, 171 Moultrie St, Charleston, SC 29409 • T: +1 843 9536846 • F: +1 843 9536956 • yatesj@citadel.edu • www.citadel.edu/archivesandmuseum •
Dir.: *Jane M. Yates* •
Historical Museum – 1842
History 47360

City Hall Council Chamber Gallery, 80 Broad St, Charleston, SC 29401 • T: +1 843 7243799, 5776970 • F: +1 843 7203827 •
C.E.O: *Joseph P. Riley jr.* •
Fine Arts Museum / Historical Museum – 1818
Art, history 47361

Drayton Hall, 3380 Ashley River Rd, Hwy 61, Charleston, SC 29414 • T: +1 843 7692600 • F: +1 803 7660878 • dhmail@draytonhall.org • www.draytonhall.org •
Dir.: *George W. McDaniel* • Ass. Dir.: *Wade Lawrence* •
Historic Site – 1974
Historic house 47362

Elizabeth O'Neill Verner Studio Museum, 38 Tradd St, Charleston, SC 29401 • T: +1 803 7224246 • F: +1 803 7221763 • info@vernergallery.com •
C.E.O. & Pres.: *David Hamilton* • Cur.: *Elizabeth Verner Hamilton* •
Fine Arts Museum – 1970
Works by E. O'Neill Verner 47363

Gibbes Museum of Art, 135 Meeting St, Charleston, SC 29401 • T: +1 843 7222706 • F: +1 843 7201682 • mloftus@gibbesmuseum.org • www.gibbesmuseum.org •
Dir./Cur.: *Angela Mack* •
Fine Arts Museum – 1858
Art of the American South from the colonial era to present day, painting, miniature portraiture, sculpture, photographs – library 47364

Halsey Gallery, College of Charleston, School of the Arts, 66 George St, Charleston, SC 29424-0001 • T: +1 843 9535680 • F: +1 843 9537890 • sloanm@cofc.edu • www.cofc.edu •
Dir.: *Mark Sloan* •
Fine Arts Museum / University Museum – 1978
Contemporary art in all mediums 47365

Heyward-Washington House, 87 Church St, Charleston, SC 29401 • T: +1 843 7220354 • F: +1 843 7221784 • www.charlestonmuseum.com •
Historical Museum – 1929 47366

Historic Charleston Foundation, 40 E Bay, Charleston, SC 29401 • T: +1 843 7231623 • F: +1 843 5772067 • krobinson@historiccharleston.org • historiccharleston.org •
C.E.O.: *Katharine S. Robinson* • Dir.: *Jonathan Poston* • *Stephen Hanson* •
Local Museum – 1947
Local hist 47367

John Rivers Communications Museum, College of Charleston, 58 George St, Charleston, SC 29424 • T: +1 843 9535810 • F: +1 843 9535815 • evansc@cofc.edu • www.cofc.edu/~jrmuseum •
Dir.: *Cathy Evans* • Cur.: *Rick Zender* •
Science&Tech Museum / University Museum – 1989
Communications 47368

Joseph Manigault House, 350 Meeting St, Charleston, SC 29403 • T: +1 843 7232926 • F: +1 843 7221784 • www.charlestonmuseum.com •
Decorative Arts Museum – 1773
Charleston silver, furniture, textiles, fossils, porcelain 47369

Karpeles Manuscript Museum, 68 Spring St, Charleston, SC 29403 • T: +1 843 8534651 • www.karpeles.com •
Fine Arts Museum
Changing exhibitions of illustrated books, manuscripts, documents, contemporary art 47370

Macaulay Museum of Dental History, Medical University of South Carolina, 175 Ashley Av, Charleston, SC 29425 • T: +1 843 7922288 • F: +1 843 7928619 • waring.library.musc.edu •
Cur.: *Jane Brown* •
Special Museum – 1975
Dental hist 47371

Magnolia Plantation, 3550 Ashley River Rd, Charleston, SC 29414 • T: +1 843 5711266 • F: +1 843 5715346 • tours@magnoliaplantation.com • www.magnoliaplantation.com •
Dir.: *Taylor Nelson* •
Fine Arts Museum / Local Museum / Natural History Museum / Historic Site – 1676
Historic house 47372

Middleton Place Foundation, 4300 Ashley River Rd, Charleston, SC 29414 • T: +1 843 5566020 • F: +1 843 7664460 • ttodd@middletonplace.org • www.middletonplace.org •
Pres.: *Charles H.P. Duell* •
Decorative Arts Museum / Historical Museum – 1974
American hist 47373

Old Exchange and Provost Dungeon, 122 East Bay St, Charleston, SC 29401 • T: +1 843 7272165 • F: +1 843 7272163 • oldexchange@infoave.net • www.oldexchange.com •
Exec. Dir.: *Frances McCarthy* •
Historic Site – 1976
Local hist 47374

Powder Magazine, 79 Cumberland St, Charleston, SC 29401 • T: +1 843 7231623 • F: +1 843 5772067 • www.historiccharleston.org •
Pres.: *Harold Pratt Thomas* • Exec. Dir.: *Katharine S. Robinson* •
Local Museum – 1713
Local hist 47375

South Carolina Historical Society Museum, 100 Meeting St, Charleston, SC 29401 • T: +1 843 7233225 • F: +1 843 7238584 • info@schistory.org • www.schistory.org •
Dir.: *David O. Percy* • Ass. Dir.: *Daisy R. Bigda* • Libr.: *Ashley Yandle* •
Historical Museum – 1855
Regional history, Robert Mills Bldg 47376

University Medical Museum, Medical University of South Carolina, 175 Ashley Av, Charleston, SC 29425 • T: +1 843 7922288 • F: +1 843 7928619 • waring.library.musc.edu •
Dir.: *W. Curtis Washington jr* •
Historical Museum / University Museum – 1966
Doctors' saddle bags, medicine chests, amputation kits, electro-therapeutic machines, bleeding instruments, obstatrical specula and forceps, pharmaceutical items 47377

Charleston WV

Avampato Discovery Museum (Sunrise Museum), 300 Leon Sullivan Way, Charleston, WV 25301 • T: +1 304 3448035, 5613575 • F: +1 304 3472313 • info@avampatodiscoverymuseum.org • www.avampatodiscoverymuseum.org •
Dir.: *Judith L. Wellington* •
Fine Arts Museum – 1961
19th - 20th c American paintings, drawings, prints and works on paper, sculptures 47378

Frankenberger Art Gallery, University of Charleston, 2300 MacCorkle Av SE, Charleston, WV 25304 • T: +1 304 3574870 • swatts@ucwv.edu • www.ucwv.edu •
Public Gallery 47379

West Virginia State Museum, 1900 Kanawha Blvd, Charleston, WV 25305 • T: +1 304 5580220 • F: +1 304 5582779 • charles.morris@wvculture.org • www.wvculture.org •
Dir.: *Stan Baumgardner* • *Fredrick Armstrong* (Archives and History) • C.E.O.: *Nancy Herholdt* • Cur.: *James Mitchell* • Sc. Staff: *Richard Ressmeyer* (Art) • *Stephanie Lilly* (Exhibitions) • *Charles Morris* (Collections) •
Local Museum – 1905
Local hist and culture of West Virginia, hist costumes and textiles, firearms, edged weapons, early industrial tools and products 47380

Charlestown MA

Bunker Hill Museum, 43 Monument Sq, Charlestown, MA 02129-4543 • T: +1 617 2425642 • F: +1 617 2426006 • www.nps.gov/bost •
Pres.: *Arthur L. Hurley* •
Historical Museum – 1975
Bunker Hill Monument grounds 47381

Charlestown NH

Old Fort Number 4 Associates, Springfield Rd, Rte 11, Charlestown, NH 03603 • T: +1 603 8265700 • F: +1 603 8263368 • fortat4@cyberportal.net •
C.E.O.: *Roberto M. Rodriguez* •
Local Museum – 1948
American artifacts, colonial furnishings, tools, weapons, French-Indian War 47382

Charlotte MI

Courthouse Square Museum, 100 W Lawrence Av, Charlotte, MI 48813 • T: +1 517 5436999 • F: +1 517 5436999 • preserve@ia4u.net • www.visitcourthousesquare.org •
Dir.: *Christa Christensen* •
Local Museum – 1993
Military, textiles, government records, political coll, agriculture – library 47383

Charlotte NC

Afro-American Cultural Center, 401 N Myers St, Charlotte, NC 28202 • T: +1 704 3741565 • F: +1 704 3749273 • www.aacc.charlotte.org •
Exec. Dir.: *John Moore* •
Folklore Museum – 1974
African and African American art and culture 47384

Carolinas Aviation Museum, 4108 Airport Dr, Charlotte, NC 28208 • T: +1 704 3598442 • F: +1 704 3590429 • chacboss@aol.com • www.chacweb.com •
Pres. & C.E.O.: *Floyd S. Wilson* •
Science&Tech Museum – 1991
Aviation history and artifacts, Wright Brothers first flight, WW II military aircraft, civil aircraft 47385

Charlotte Museum of History and Hezekiah Alexander Homesite, 3500 Shamrock Dr, Charlotte, NC 28215 • T: +1 704 5681774 • F: +1 704 5661817 • info@charlottemuseum.org • www.charlottemuseum.org •
Pres. & C.E.O.: *Pamela Meister* •
Local Museum – 1976
Furnishings from 1774-1820, regional artifacts and objects 47386

Charlotte Nature Museum, 1658 Sterling Rd, Charlotte, NC 28209 • T: +1 704 3726261 ext. 605 • F: +1 704 3338948 • www.discoveryplace.org •
C.E.O. & Pres.: *John L. Mackay* •
Natural History Museum – 1947
Teaching coll of biological material 47387

Discovery Place, 301 N Tryon St, Charlotte, NC 28202 • T: +1 704 3726261 • F: +1 704 3372670 • www.discoveryplace.org •
Pres. & C.E.O.: *John L. Mackay* •
Natural History Museum / Science&Tech Museum – 1981
Ethnology, geology, archaeology, herpetology, mineralogy, ornithology 47388

Historic Rosedale, 3427 N Tryon St, Charlotte, NC 28206 • T: +1 704 3350325 • F: +1 704 3350384 • roseplan@bellsouth.net •
C.E.O.: *Elaine Wood* •
Decorative Arts Museum – 1989
Decorative arts of the Catawba River Valley, family hist 47389

Levine Museum of the New South, 200 E Seventh St, Charlotte, NC 28202 • T: +1 704 3331887 • F: +1 704 3331896 • www.museumofthenewsouth.org •
Exec.Dir.: *Emily F. Zimmern* • Dir. Exhibits: *John Hilarides* • Cur.: *Jean Johnson* •
Historical Museum – 1991
Hist of the American South from Civil War to present 47390

The Light Factory, 345 N College St, Charlotte, NC 28202 • T: +1 704 3339755 • F: +1 704 3335910 • info@lightfactory.org • www.lightfactory.org •
C.E.O. & Exec.Dir.: *Marcie Kelso* •
Performing Arts Museum – 1972
Light-generated media incl photography, video and film 47391

Mint Museum of Art, 2730 Randolph Rd, Charlotte, NC 28207 • T: +1 704 3372000 • F: +1 704 3372101 • nrider@mintmuseum.org • www.mintmuseum.org •
C.E.O.: *Phil Kline* • Dep.Dir.: *Mark Leach* • Dir.: *Charles Mo* (Collections and Exhibitions) • *Cheryl Palmer* (Education) • Cur.: *Michael Whittington* (Pre-Columbian and African Art) • *Dr. Barbara Perry* (Decorative Arts) • Libr.: *Sara Wolf* •
Fine Arts Museum – 1933
American and European paintings, furniture and dec art, African, pre-Columbian and Spanish Colonial art, porcellain, pottery, glass, regional crafts, hist costumes 47392

Mint Museum of Craft and Design, 220 N Tryon St, Charlotte, NC 28202 • T: +1 704 3372000 • F: +1 704 3372101 • nrider@mintmuseum.org • www.mintmuseum.org •
C.E.O.: *Phil Kline* • Dir.: *Mark Leach* • *Mary Beth Ausman* (Education) • Cur.: *Melissa G. Post* •
Decorative Arts Museum – 1999 47393

Charlotte Harbor FL

Charlotte County Historical Center, 22959 Bayshore Rd, Charlotte Harbor, FL 33980 • T: +1 941 6297278 • F: +1 941 7433917 • museum@sunline.net •
Local Museum – 1969
Fossils, shells, local and state hist 47394

Charlottesville VA

Ash Lawn-Highland, College of William and Mary, 1000 James Monroe Pkwy, Charlottesville, VA 22902 • T: +1 434 2939539 • F: +1 434 2938000 • info@ashlawnhighland.org • www.ashlawnhighland.org •
Dir.: *Carolyn C. Holmes* •
Decorative Arts Museum / Historic Site – 1930
Monroe American & French furnishings & other objects of the period 1790-1825 47395

Children's Health Museum, Primary Care Center, Lee St and Park Pl, Charlottesville, VA 22908 • T: +1 434 9241593 • F: +1 434 9825872 • epv4p@virginia.edu •
Dir.: *Ellen Vaughan* •
Historical Museum – 1982
Children's health 47396

Michie Tavern Ca. 1784, 683 Thomas Jefferson Pkwy, Rte 53, Charlottesville, VA 22902 • T: +1 434 9771234 • F: +1 434 2967203 • info@michietavern.com • www.michietavern.com •
Cur.: *Cynthia Conte* •
Local Museum – 1928
18th - 19th c Southern furniture & artifacts 47397

Monticello, Home of Thomas Jefferson, Rte 53 Thomas Jefferson Pkwy, Charlottesville, VA 22902 • T: +1 434 9849808 • F: +1 434 9777751 • publicaffairs@monticello.org • www.monticello.org •
Cur.: *Susan R. Stein* •
Historical Museum – 1923
Art objects, manuscripts, personal items, slavery artifacts, American independence declaration 47398

The Rotunda, University of Virginia, Charlottesville, VA 22903 • T: +1 434 9247969, 9241019 • F: +1 434 9243817 • rotunda@virginia.edu •
University Museum / Local Museum – 1819
Historic buildings, site of Thomas Jefferson's original academical village 47399

University of Virginia Art Museum, 155 Rugby Rd, Charlottesville, VA 22904-4119 • T: +1 434 9243592 • F: +1 434 9246321 • jhartz@virginia.edu • www.virginia.edu/artmuseum •
Dir.: *Jill Hartz* • Cur.: *Stephen N. Margulies* (Prints) • *Suzanne Foley* •
Fine Arts Museum / University Museum – 1935
15th - 20th c American & European paintings, sculpture & works on paper 47400

The Virginia Discovery Museum, 524 E Main St, Charlottesville, VA 22902 • T: +1 434 9771025 • F: +1 434 9779681 • vdm@ntelos.net • www.vadm.org •
C.E.O.: *Peppy G. Linden* •
Special Museum – 1981
400 brass animals, puppets, nigerian coll, exhib. 47401

Charlottesville VT

Second Street Gallery, 115 Second St, Charlottesville, VT 22902 • T: +1 434 9777284 • F: +1 434 9799793 • ssg@secindstreetgallery.org • www.secindstreetgallery.org •
Dir.: *Leah Stoddard* •
Public Gallery – 1973
Contemporary art space featuring a revolfing series of exhib of paintings 47402

Chase City VA

MacCallum More Museum, 603 Hudgins St, Chase City, VA 23924 • T: +1 804 3720502, 3723120 • F: +1 804 3723483 • mmmg@meckcom.net • www.mmmg.org •
Pres.: *Gay Butler* • C.E.O.: *Brenda Arriaga* •
Ethnology Museum – 1991
Indian artifacts exhib, 1st c Roman bust, Spanish cloister, eight fountains 47403

Chatham MA

Atwood House Museum, 347 Stage Harbor Rd, Chatham, MA 02633 • T: +1 508 9452493 • F: +1 508 9451205 • chathamhistoricalsociety@verizon.net • www.chathamhistoricalsociety.org •
Cur.: *Dr. Ernest A. Rohdenburg* •
Decorative Arts Museum / Historical Museum – 1923
Harold Brett, Harold Dunbar and Frederick Wright paintings, sandwich glass, 17th-18th c furnishings, antique tools, China 47404

Chatsworth GA

Vann House, 82 Hwy 225 N, Chatsworth, GA 30705 • T: +1 706 6952598 • F: +1 706 5174255 • vannhouse@alltel.net •
Pres.: *Laura Greeson* •
Historical Museum – 1952
Furniture, genealogy, Cherokee Indian hist 47405

Chattanooga TN

Chattanooga African Museum, Bessie Smith Hall, 200 E Martin Luther King Blvd, Chattanooga, TN 37403 • T: +1 423 2668658 • F: +1 423 2671076 • caami@bellsouth.net • www.caamhistory.com •
Dir.: *Vilma S. Fields* •
Historical Museum – 1983 47406

Chattanooga Regional History Museum, 400 Chestnut St, Chattanooga, TN 37402 • T: +1 423 2653247 • F: +1 423 2669280 • office@chattanoogahistory.com • www.chattanoogahistory.com •
Dir.: *R. Brit. Brantley* • Cur.: *Patrice Glass* •
Local Museum – 1978
10.000 y of human hist in SE Tennessee 47407

Creative Discovery Museum, 321 Chestnut St, Chattanooga, TN 37402 • T: +1 423 7562738 • F: +1 423 2679344 • hhs@cdmfun.org • www.cdmfun.org •
C.E.O.: *Henry H. Schulson* •
Special Museum – 1995 47408

Cress Gallery of Art, 355 Fine Arts Center, Chattanooga, TN 37403, mail addr: c/o University of Tennessee at Chattanooga, Dept. of Art, 615 McCallie Av, Chattanooga, TN 37403 • T: +1 423 4254178 • F: +1 423 4252101 • ruth-grover@utc.edu • www.utc.edu/~artdept •
Cur.: *Ruth Grover* •
Fine Arts Museum / University Museum – 1952
Graphics, paintings, student's works 47409

Houston Museum of Decorative Arts, 201 High St, Chattanooga, TN 37403 • T: +1 423 2677176 • houston@chattanooga.net • www.chattanooga.net/houston •
C.E.O.: *Amy H. Frierson* •
Decorative Arts Museum – 1949
American decorative arts 1750-1930 47410

Hunter Museum of American Art, 10 Bluff View, Chattanooga, TN 37403-1197 • T: +1 423 2670968 • F: +1 423 2679844 • robkret@huntermuseum.org • www.huntermuseum.org •
Dir.: *Robert A. Kret* • Cur.: *Ellen Simak* (Collections) •
Fine Arts Museum – 1952
American paintings, photography, prints, sculpture, glass 47411

National Medal of Honor Museum of Military History, Northgate Mall, Hwy 153, Chattanooga, TN 37403 • T: +1 423 4230710 • www.mohm.org •
Dir.: *Patty Parks* •
Military Museum – 1987
Uniforms, pre-Civil War to present, weapons, military prints and paintings, medals, badges, Nuremburg coll, Confederate coll, currency coll, helmets 47412

Siskin Museum of Religious and Ceremonial Art, 101 Carter St, Chattanooga, TN 37403 • T: +1 423 6481700 • F: +1 423 6481717 •
Religious Arts Museum – 1950
Religious art 47413

Tennessee Valley Railroad Museum, 4119 Cromwell Rd, Chattanooga, TN 37421 • T: +1 423 8948028 • F: +1 423 8948029 • info@tvrail.com • www.tvrail.com •
Pres. & C.E.O.: *Robert M. Soule* •
Science&Tech Museum – 1961
Railroad 47414

Chattsworth GA

Fort Mountain, 181 Fort Mountain Park Rd, Chattsworth, GA 30705 • T: +1 706 6952621 • F: +1 706 5173520 • www.gastateparks.org •
Archaeology Museum – 1938
Pre-historic stone wall 47415

Chazy NY

Alice T. Miner Colonial Collection, 9618 Main St, Chazy, NY 12921 • T: +1 518 8467336 • F: +1 518 8468771 • minermuseum@westelcom.com • www.minermuseum.org •
C.E.O., Pres. & Chm.: *Joan T. Burke* • Dir./Cur.: *Frederick G. Smith* •
Local Museum – 1924
Paintings, dec art, furnishings, glass, Indian artifacts 47416

Chehalis WA

Lewis County Historical Museum, 599 NW Front Way, Chehalis, WA 98532 • T: +1 360 7480831 • F: +1 360 7405646 • lchm@lewiscountymuseum.org • www.lewiscountymuseum.org •
Dir.: *Rianne Perry* •
Local Museum – 1965
Local hist 47417

Chelmsford MA

Chelmsford Historical Society Museum, 40 Byam Rd, Chelmsford, MA 01824 • T: +1 978 2562311 • martis@netway.com • www.chelmhist.org •
Dir.: *Donald Patterschal* • Cur.: *Judy Fichtenbaum* •
Local Museum – 1930
Coll of Chelmsford glass, clothing, model of country store – library 47418

Cheraw SC

Cheraw Lyceum Museum, 200 Market St, Cheraw, SC 29520 • T: +1 843 5378425 • F: +1 843 5378407 • townofcherawtour@mindspring.com • www.cheraw.com •
Dir.: *J. William Taylor* • Cur.: *Sarah C. Spruill* •
Local Museum – 1962
Local hist 47419

Cherokee IA

Joseph A. Tallman Museum, Mental Health Institute, 1205 W Cedar Loop, Cherokee, IA 51012 • T: +1 712 2256922 • F: +1 712 2256969 • larmstrl@dhs.state.IA.us •
Head: *Dr. Tom Deiker* •
Historical Museum – 1961
Psychiatry 47420

Sanford Museum, 117 E Willow St, Cherokee, IA 51012 • T: +1 712 2253922 • F: +1 712 2250446 • sanford@cherokee.k12.ia.us •
Dir.: *Linda Burkhart* • Ass. Dir.: *Michele Deiber-Kumm* •
Local Museum – 1941
Archaeology, history, geology, paleontology, zoology, ethnology – planetarium 47421

Cherokee NC

Mountain Farm Museum, 1194 Newfound Gap Rd, Cherokee, NC 28719 • T: +1 828 4971900 • F: +1 828 4971910 • grsm_smokies_information@nps.gov • www.nps.gov/grsm •
Agriculture Museum – 1953 47422

Museum of the Cherokee Indian, 589 Tsali Blvd, Cherokee, NC 28719 • T: +1 828 4973481 • F: +1 828 4974985 • littlejohn@cherokeemuseum.org • www.cherokeemuseum.org •
Dir.: *Ken Blankenship* •
Historical Museum / Ethnology Museum – 1948
Artifacts, relics, 16th c Cherokee homestead – archive 47423

Cherokee County TX

Caddoan Mounds, State Hwy 21 W, Cherokee County, TX 75925 • T: +1 936 8583218 • F: +1 936 8583227 • www.tpwd.state.tx.us/park/caddoan/ •
Local Museum – 1983 47424

Cherry Valley CA

Edward-Dean Museum, 9401 Oak Glen Rd, Cherry Valley, CA 92223 • T: +1 909 8452626 • F: +1 909 8452628 • btoor@rivcoeda.org • www.edward-deanmuseum.org •
Dir.: *Teresa Gallavam* •
Decorative Arts Museum / Fine Arts Museum – 1958
Decorative arts, paintings, prints, sculpture, ceramics, glass, textiles, furniture and other miscelanious objects from 16th through 19th century – reference Library 47425

Cherry Valley NY

Cherry Valley Museum, 49 Main St, Cherry Valley, NY 13320 • T: +1 607 2643303 • F: +1 607 2649320 • museum@celticart.com •
Pres.: *James Johnson* •
Local Museum – 1958
Farm and village life in Cherry Valley since late 18th c 47426

Cherryville NC

C. Grier Beam Truck Museum, 111 N Mountain St, Cherryville, NC 28021 • T: +1 704 4353072 • F: +1 704 4459010 • jdismukes1@juno.com • www.beamtruckmuseum.com •
Pres.: *Sandra B. Dismukes* • C.E.O.: *Michael N. Beam* •
Science&Tech Museum – 1982
14 trucks, hist of Carolina trucking company 47427

Chesapeake Beach MD

Chesapeake Beach Railway Museum, 4155 Mears Av, Chesapeake Beach, MD 20732 • T: +1 410 2573892 • www.cbrm.org •
Science&Tech Museum / Local Museum – 1979
Early railroad era of 1900-1935, local hist 47428

Chesnee SC

Cowpens National Battlefield, POB 308, Chesnee, SC 29323 • T: +1 864 4612828 • F: +1 864 4617077 • cowp-information@nps.gov • www.nps.gov/cowp/ •
Historic Site / Military Museum – 1933
Military hist, site of the 1781 Battle of Cowpens 47429

Chester MT

Liberty County Museum, Second St E and Madison, Chester, MT 59522 • T: +1 406 7595256 •
C.E.O.: *Betty Marshall* •
Local Museum – 1969
Agriculture, archaeology, costumes, Indian artifacts 47430

Liberty Village Arts Center and Gallery, 400 Main St, Chester, MT 59522 • T: +1 406 7595652 • F: +1 406 7595652 • lvac@mtintouch.net •
Folklore Museum – 1976
Paintings, quilts by local artists 47431

Chester PA

Widener University Art Collection and Gallery, 14th and Chestnut Sts, Chester, PA 19013 • T: +1 610 4991189 • F: +1 610 4994425 • rebecca.m.warda@widener.edu • www.widener.edu/campuslife.edu •
Head: *Rebecca M. Warda* •
Fine Arts Museum / University Museum – 1970
19th c and 20th c American paintings and sculpture 47432

Chester SC

Chester County Museum, 107 McAliley St, Chester, SC 29706 • T: +1 803 3852330, 5814354 •
Dir.: *Gary Roberts* •
Historical Museum – 1959
Local hist 47433

Chester VT

Chester Art Guild, The Green, Main St, Chester, VT 05143 • T: +1 802 8753767 • doadiem@vermontel.net •
Pres.: *Doris Ingram* •
Fine Arts Museum – 1960
Paintings by local artists 47434

Chesterfield MA

Bisbee Mill Museum, 66 East St, Chesterfield, MA 01012 • T: +1 413 2964750 • cabjr18@webtv.net • www.bisbeemillmuseum.org •
Agriculture Museum / Science&Tech Museum – 1995
Agriculture and industrial history 47435

Edwards Memorial Museum, 3 North Rd, Chesterfield, MA 01012 • T: +1 413 2964750 • F: +1 413 2964394 •
Cur.: *Christopher Smith* •
Folklore Museum – 1950
Woodworking and agricultural tools 47436

Chesterfield MO

Thornhill Historic Site and 19th Century Village, 15185 Olive Blvd, Chesterfield, MO 63017-1805 • T: +1 636 5321030 • F: +1 638 5320604 • jfoley@stlouisco.com • www.st-louiscountyparks.com •
Dir.: *Jim Foley* • Cur.: *Jesse Francis* •
Local Museum – 1968 47437

Chesterfield VA

Chesterfield Historical Society of Virginia, 10201 Iron Bridge Rd, Chesterfield, VA 23832 • T: +1 804 7779663 • F: +1 804 7779643 • dpfarmerchs@aol.com • www.chesterfieldhistory.com •
Dir.: *Dennis P. Farmer* •
Association with Coll – 1961
Museum housed in replica, genealogy 47438

Chestertown MD

Historical Society of Kent County Museum, 101 Church Alley, Chestertown, MD 21620 • T: +1 410 7783499 • F: +1 410 7883747 • hskcmd@friend.ly.net • www.hskcmd.com •
Historical Museum
Indian artifacts, furniture, paintings, maps 47439

Chestnut Hill MA

Longyear Museum, 1125 Boylston St, Chestnut Hill, MA 02467 • T: +1 617 2789000 • F: +1 617 2789003 • letters@longyear.org • www.longyear.org •
Exec. Dir.: *John Baehrend* • Cur.: *Stephen R. Howard* •
Religious Arts Museum – 1923
Life of Mary Baker Eddy, Christians science – library 47440

McMullen Museum of Art, Boston College, Devlin Hall, 140 Commonwealth Av, Chestnut Hill, MA 02467 • T: +1 617 5528587 • F: +1 617 5528577 • netzer@bc.edu • www.bc.edu/artmuseum •
Dir.: *Dr. Nancy Netzer* • Cur.: *Alston Conley* •
Fine Arts Museum / Public Gallery / University Museum – 1986
European art, Japanese prints, American paintings 47441

Chetopa KS

Chetopa Historical Museum, 419 Maple St, Chetopa, KS 67336 • T: +1 620 2367121 •
Cur.: *Fannie Bassett* •
Local Museum – 1881
Local history 47442

Cheyenne OK

Black Kettle Museum, Oklahoma Historical Society, 101 S LL, Intersection of US 283 and State Hwy 47, Cheyenne, OK 73628, mail addr: POB 252, Cheyenne, OK 73628 • T: +1 580 4973929 • F: +1 580 4973537 • bkmus@ok-history.mus.ok.us •
Historical Museum / Military Museum – 1958
Hist of the Battle of Washita, attack of G. Custer on Black Kettle' Village; Native American arts and crafts 1860-1880 47443

Cheyenne WY

Cheyenne Frontier Days Old West Museum, 4610 N Carey Av, Cheyenne, WY 82001 • T: +1 307 7787290 • F: +1 307 7787288 • www.oldwestmuseum.org •
Dir.: *Wayne Hansen* • Cur.: *Bob Gant* •
Local Museum – 1978
Western art coll, horse-drawn vehicles, rodeo coll 47444

Historic Governors' Mansion, 300 E 21st St, Cheyenne, WY 82001 • T: +1 307 7777878 • F: +1 307 6357077 • sphs@state.wy.us • www.wyoparks.state.wy.us •
Dir.: *Bill Gentle* • Cur.: *Timothy J. White* •
Local Museum – 1904
Historic building, hist fruniture, photographs, art 47445

Warren ICBM and Heritage Museum, 7405 Marne Loop, 90th SW/MU, Frances E. Warren Air Force Base, Cheyenne, WY 82005 • T: +1 307 7732980 • F: +1 307 7732791 • paula.taylor@warren.af.mil • www.pawnee.com/fewmuseum •
Dir./Cur.: *Paula Bauman Taylor* • Cur.: *Cary Wiesner* •
Military Museum – 1967
Military uniforms and equipment 1840-present, memorabilia of the 90th Bomb Group-The Jolly Rogersof WWII 47446

Wyoming Arts Council Gallery, 2320 Capitol Av, Cheyenne, WY 82002 • T: +1 307 7777742 • F: +1 307 7775499 • lfranc@state.wy.us • wyoarts.state.wy.us •
Dir.: *John G. Coe* •
Fine Arts Museum – 1976
Changing exhibitions of Wyoming artists 47447

Wyoming State Museum, Barrett Bldg, 2301 Central Av, Cheyenne, WY 82002 • T: +1 307 7777022 • F: +1 307 7775375 • wsm@state.wy.us • wyomuseum.state.wy.us •
Dir.: *Marie Wilson-McKee* •
Historical Museum – 1895
State hist of Wyoming, textiles, firearms, household artifacts, Native American artifacts 47448

Chicago IL

A.R.C. Gallery, 734 N Milwaukee Av, Chicago, IL 60622 • T: +1 312 7332787 • F: +1 312 7332787 • pam@arcartistindex.com • www.arcgallery.com •
Pres.: *Carolyn King* •
Public Gallery – 1973
American paintings, sculpture 47449

American Police Center and Museum, 1926 S Canalport Av, Chicago, IL 60616-1071 • T: +1 312 4558770 •
CEO: *Joseph P. Pecoraro* •
Historical Museum – 1974
Uniforms, weapons, photos, documents 47450

Anchor Graphics, 119 W Hubbard St, 5W, Chicago, IL 60610 • T: +1 312 5959598 • F: +1 312 5959895 • print@anchorgraphics.org • www.anchorgraphics.org •
Dir.: *David Jones* •
Fine Arts Museum
Graphics, art 47451

The Art Institute of Chicago, 111 S Michigan Av, Chicago, IL 60603-6110 • T: +1 312 4433600 • F: +1 312 4430849 • www.artic.edu •
Pres./Dir.: *James N. Wood* • Asst. Dir.: *Dorothy Schroeder* (Exhibition) • Cur.: *James Rondeau* (Art) •
Fine Arts Museum – 1879
Architecture, paintings, drawings, prints, photos, decorative art, art from America, Asia, Europe – Ryerson and Burnham libraries, art school 47452

U

Artemisia Gallery, 700 N Carpenter, Chicago, IL 60622 • T: +1 312 2267323 • F: +1 312 2267756 • artemisi@enteract.com •
Pres.: *Marji Vecchio* •
Public Gallery – 1973 47453

The Arts Club of Chicago, 201 E Ontario St, Chicago, IL 60611 • T: +1 312 7873997 • F: +1 312 7878664 • artsclub@mindspring.com •
Association with Coll 47454

Australian Exhibition Center, 114 W Kinzie St, Chicago, IL 60610 • T: +1 312 6451948 • F: +1 312 2224119 • exhibit@aec-chicago.com.au •
Public Gallery 47455

Balzekas Museum of Lithuanian Culture, 6500 S Pulaski Rd, Chicago, IL 60629 • T: +1 773 5826500 • F: +1 773 5825133 • president@lithuanian-museum.org • www.lithaz.org/museums/balzekas •
Dir.: *Stanley Balzekas* • *Robert Balzekas* (Genealogy) • *Edward Mankus* (Audio Visual Department) • *Val Martis* (Periodicals Collection) • *Karile Vaitkute* (Education) •
Cur.: *Frank Zapolis* (Folk Art) •

Historical Museum / Folklore Museum / Fine Arts Museum – 1966
Ethnology, folklore, numismatics, folk art – library, archive 47456

Beverly Art Center, 2407 W 111th St, Chicago, IL 60655 • T: +1 773 4453838 • www.beverlyartcenter.org •
Public Gallery 47457

Boulevard Arts Center, 1525 W 60th St, Chicago, IL 60636 • T: +1 773 4764900 • F: +1 773 4765951 •
Dir.: *Pat Devine-Reed* • *Marti Price* •
Fine Arts Museum – 1986
Visual arts, wood carving, stone carving 47458

Carlson Tower Gallery, c/o North Park University, 3225 W Foster, Chicago, IL 60625 • T: +1 773 2446200 • F: +1 773 2445230 • www.northpark.edu •
Dir.: *Tim Lowly* •
Fine Arts Museum / University Museum
Contemporary Christian, Illinois and Scandinavian art 47459

Center for Intuitive and Outsider Art, 756 N Milwaukee Av, Chicago, IL 60622 • T: +1 312 2439088 • F: +1 312 2439089 • intuit@art.org • www.outsider.art.org •
Exec. Dir.: *Jeff Cory* •
Ethnology Museum / Fine Arts Museum – 1991 47460

Charnley-Persky House Museum, 1365 N Astor St, Chicago, IL 60610-2144 • T: +1 312 5731365 • F: +1 312 5731141 • psaliga@sah.org • www.sah.org •
C.E.O: *Lynn Osmond* •
Local Museum – 1998
Local hist 47461

Chicago Architecture Foundation, 224 S Michigan Av, Chicago, IL 60604-2507 • T: +1 312 9223432 • F: +1 312 9222607 • losmond@architecture.org • www.architecture.org •
Pres.: *Lynn Osmond* •
Fine Arts Museum – 1966
Architecture tours, exhibits, education programs 47462

The Chicago Athenaeum, Museum of Architecture and Design, 307 N Michigan Av, Chicago, IL 60601 • T: +1 815 7774444 • F: +1 815 7772471 • info@chi-athenaeum.org • www.chi-athenaeum.org •
Dir.: *Christian K. Warkiewicz Laine* •
Fine Arts Museum – 1988
Industrial and product design, graphics, architectural drawings, models, decorative art 47463

Chicago Children's Museum, 700 E Grand Av, Ste 127, Navy Pier, Chicago, IL 60611 • T: +1 312 5271000 • F: +1 312 5279082 • www.chichildrensmuseum.org •
C.E.O: *Peter England* •
Special Museum – 1982 47464

Chicago Cultural Center, 78 E Washington St, Chicago, IL 60602 • T: +1 312 7446630 • F: +1 312 7447482 • culture@cityofchicago.org • www.chicagoculturalcenter.org •
Dir.: *Gregory Knight* • Cur.: *Lanny Silverman* • Asst. Cur.: *Sofia Zutautas* •
Public Gallery – 1897
Paintings, sculptures photographs, ceramics, graphics, design, mixed media, installations 47465

Chicago Historical Society, Clark St at North Av, Chicago, IL 60614-6099 • T: +1 312 6424600 • F: +1 312 2662077 • bunch@chicagohistory.org • www.chicagohistory.org •
Pres.: *Lonnie G. Bunch* •
Historical Museum – 1856
Chicago, American and Illinois hist, Revolutionary and Civil War, architectural drawings, models, fragments, paintings, sculpture, manuscripts 47466

Clarke House Museum, 1821 S Indiana Av, Chicago, IL 60616 • T: +1 312 7450040 • F: +1 312 7450077 • clarkehouse@interaccess.com • www.cityofchicago.org/culturalaffairs/ •
Cur.: *Eduard M. Maldonado* •
Local Museum – 1984
Greek Revival building, oldest home in Chicago 47467

Columbia College Art Gallery, 72 E Eleventh St, Chicago, IL 60605-2312 • T: +1 312 3346156 • F: +1 312 3448067 •
Dir.: *Denise Miller* • *Dr. John Mulvany* • Ass. Dir.: *Nancy Fewkes* •
Fine Arts Museum / University Museum – 1984 47468

Contemporary Art Workshop, 542 W Grant Pl, Chicago, IL 60614 • T: +1 773 4724004 • F: +1 773 4724505 • info@contemporaryartworkshop.org • www.contemporaryartworkshop.org •
Pres.: *John Kearney* • Dir.: *Lynn Kearney* •
Public Gallery 47469

DePaul University Art Gallery, 2350 N Kenmore, Chicago, IL 60614 • T: +1 773 3257506 • F: +1 773 3254506 • lfatemi@depaul.edu • www.museums.depaul.edu/artwebsite •
Dir./Cur.: *Louise Lincoln* •
Fine Arts Museum / University Museum – 1987 47470

Dix Art Mix, 2068 N Leavitt, Chicago, IL 60647 • T: +1 773 3845142 • F: +1 773 3845142 • thompeter@earthlink.net •
Public Gallery
Art 47471

Dusable Museum of African-American History, 740 E 56th Pl, Chicago, IL 60637 • T: +1 773 9470600 • F: +1 773 9470677 • awright@dusablemuseum.org • www.dusablemuseum.org •
Chief Cur.: *Selean Holmes* •
Historical Museum / Ethnology Museum – 1961
African, Afro-American art hist, Chicago and Illinois hist 47472

Field Museum of Natural History, 1400 S Lake Shore Dr, Chicago, IL 60605 • T: +1 312 9229410 • F: +1 312 9220741 • jmcarter@fmnh.org • www.fieldmuseum.org •
Pres.: *John W. McCarter* •
Natural History Museum – 1893
Natural history, zoology, geology, botany, anthropology 47473

First National Bank of Chicago Art Collection, 1 Bank One Plaza, Chicago, IL 60670 • T: +1 312 7325935 •
Dir./ Cur.: *Lisa K. Erf* •
Fine Arts Museum / Archaeology Museum – 1968
Art from Africa, America, Asia, Australia, the Caribbean Basin, Europe, Latin America, Near East and the South Seas since 16th c 47474

Frederick C. Robie House, 5757 S Woodlawn Av, Chicago, IL 60637 • T: +1 773 8341847 • F: +1 773 8341538 • mercuri@wrightplus.org • www.wrightplus.org •
Pres.: *Joan B. Mercuri* •
Historical Museum – 1974
Residence of Frederick C. Robie 47475

Gallery 2, c/o School of the Art Institute of Chicago, 847 W Jackson Blvd, Chicago, IL 60607 • T: +1 312 5635162 • F: +1 312 5630510 • www.artic.edu/saic/art/galleries/gallery2.html •
Dir.: *Anthony Wight* •
Fine Arts Museum / University Museum – 1988 47476

Gallery 400 at the University of Illinois at Chicago, c/o Art & Design Hall, 400 South Peoria Street, Chicago, IL 60607-7034 • T: +1 312 9966114 • F: +1 312 3553444 • uicgallery400@gmail.com • gallery400.aa.uic.edu •
Dir.: *Lorelei Stewart* • Asst. Dir.: *Anthony Elms* •
Fine Arts Museum / University Museum – 1983 47477

Glessner House Museum, 1800 S Prairie Av, Chicago, IL 60616-1333 • T: +1 312 3261480 • F: +1 312 3261397 • glessnerhouse@earthlink.net • www.glessnerhouse.org •
Dir./Cur.: *Corina Carusi* • Pres.: *Penelope Boardman* •
Historical Museum – 1994
Home designed by H.H. Richardson, English arts and crafts, furniture, decorative arts 47478

Hellenic Museum and Cultural Center, 168 N Michigan Av, Chicago, IL 60610 • T: +1 312 7261234 • F: +1 312 6551221 • info@hellenicmuseum.org • www.hellenicmuseum.org •
Pres.: *John Marks* • Exec. Dir.: *Elaine Kollintzas Drikakis* •
Folklore Museum / Historical Museum / Museum of Classical Antiquities / Archaeology Museum – 1987 47479

Hyde Park Art Center, 5020 S Cornell Av, Chicago, IL 60615 • T: +1 773 3245520 • F: +1 773 3246641 • info@hydeparkart.org • www.hydeparkart.org •
Exec. Dir.: *Chuck Thurow* •
Fine Arts Museum – 1939
Oldest alternative exhibition space in the city 47480

Illinois Art Gallery, 100 W Randolph, Ste 2-100, Chicago, IL 60601 • T: +1 312 8145322 • F: +1 312 8143471 • ksmith@museum-state.il.us • www.museum.state.il.us/ismsites/chicago/ •
Dir.: *Kent Smith* • Ass. Dir.: *Jane Stevens* •
Fine Arts Museum / Public Gallery – 1985
Paintings, drawings, printmaking, ceramics, photography, textiles 47481

International Museum of Surgical Science, 1524 N Lake Shore Dr, Chicago, IL 60610 • T: +1 312 6426502 • F: +1 312 6429516 • info@imss.org • www.imss.org •
Chm.: *Raymond A. Dieter Jr.* •
Special Museum – 1953
Medical imaging/radiology, orthopedics, pain management/anesthesia, ophthalmology, and nursing, fine arts coll – Library, manuscript archives 47482

Jane Addams' Hull-House Museum, University of Illinois Chicago, 800 S Halsted St, Chicago, IL 60607-7017 • T: +1 312 4135353 • F: +1 312 4132092 • jahh@uic.edu • www.uic.edu/jaddams/hull/ •
Dir.: *Margaret Strobel* •
Historical Museum – 1967
Hull Mansion occupied by Jane Addams in 1889, serving as the first settlement building of Hull House complex 47483

John H. Vanderpoel Art Gallery, 9625 S Longwood Dr, Ridge Park, Chicago, IL 60643 • T: +1 312 2398320 •
Fine Arts Museum
About 500 paintings, prints and works of sculpture 47484

Leather Museum, 6418 N Greenview Av, Chicago, IL 60626 • T: +1 773 7619200 • F: +1 773 3814657 • archives@leatherarchives.org • www.leatherarchives.org •
Pres.: *Chuck Renslow* • Dir.: *Rick Storer* •
Special Museum – 1996
Art and prints, club colors, pins and ephemera, leathers and boots, manuscripts, oral histories, periodicals, photography, posters and banners, shirts and fibers, titleholders and sashes, toys and devices, uniforms and clothing, International Mr. Leather – archives 47485

The Lithuanian Museum, 5620 S Claremont Av, Chicago, IL 60636-1039 • T: +1 773 4344545 • F: +1 773 4349363 • lithuanianresearch@ameritech.net •
Dir.: *Skirmante Miglinas* • *Milda Budrys* (Medicine) •
Cur.: *Peter Paul Zansitis* •
Fine Arts Museum / Historical Museum – 1989
Lithuanian art, culture and history, Lithuanian immigration to the U.S. 47486

The Martin D'Arcy Museum of Art, Loyola University Museum of Medieval, Renaissance and Baroque Art, 6525 N Sheridan Rd, Chicago, IL 60626 • T: +1 773 5082679 • F: +1 773 5082993 • www.luc.edu/luma •
Dir.: *Dr. Sally Metzler* • Ass. Dir.: *Rachel Baker* •
Fine Arts Museum / University Museum – 1969
European art 47487

Marwen Foundation, 833 N Orleans St, Chicago, IL 60610 • T: +1 312 9442418 • F: +1 312 9446696 • acontro@marwen.org • www.marwen.org •
Dir.: *Antonia Contro* • Asst. Dir.: *Jesse McClelland* •
Public Gallery – 1987 47488

Mexican Fine Arts Center Museum, 1852 W 19th St, Chicago, IL 60608 • T: +1 312 7381503 • F: +1 312 7389740 • carlost@mfacmchicago.org • www.mfacmchicago.org •
Exec. Dir.: *Carlos Tortolero* •
Fine Arts Museum / Folklore Museum – 1982
Mexican art, folk art, photography, graphics art 47489

Museum of Ancient Artifacts, 2528 N Luna St, Chicago, IL 60639 • T: +1 773 3859840 •
Cur.: *Michael J. Kurban* •
Natural History Museum – 1990
Rare photographs and gems, tapestries, artwork – botanical garden 47490

The Museum of Broadcast Communications, 78 E Washington St, Chicago, IL 60602-9837 • T: +1 312 6296000 • F: +1 312 6296009 • archives@museum.tv • www.museum.tv •
Pres.: *Bruce DuMont* •
Special Museum – 1987
Historic and contemporary radio and TV memorabilia – archives 47491

Museum of Contemporary Art, 220 E Chicago Av, Chicago, IL 60611-2604 • T: +1 312 2802660 • F: +1 312 3974095 • webmaster@machicago.org • www.mcachicago.org •
Dir.: *Robert Fitzpatrick* • Chief Cur.: *Elizabeth Smith* •
Fine Arts Museum – 1967
Post World War II paintings, sculpture, graphics, photography, conceptual works 47492

The Museum of Contemporary Photography, Columbia College Chicago, 600 S Michigan Av, Chicago, IL 60605-1901 • T: +1 312 6635554 • F: +1 312 3448067 • mocp@aol.com • www.mocp.org •
C.E.O: *Bert Gall* • Pres.: *Warrick L. Carter* •
University Museum – 1976
Photography 47493

Museum of Cosmetology, c/o National Cosmetology Assn, 401 N Michigan Av, Chicago, IL 60611 • T: +1 312 5276765 •
Exec. Dir.: *Gordon Miller* •
Special Museum – 1990
History of American hair and cosmetics fashion 47494

Museum of Decorative Art, 4611 N Lincoln Av, Chicago, IL 60625 • T: +1 773 9894310 •
Decorative Arts Museum 47495

Museum of Holography, 1134 W Washington Blvd, Chicago, IL 60607 • T: +1 312 8292292 • F: +1 312 8299636 • hologram@maritech.net • holographiccenter.com •
Dir.: *Loren Billings* • *John Hoffmann* (Research) •
Special Museum – 2001
Holography 47496

Museum of Science and Industry, 57th S Lake Shore Dr, Chicago, IL 60637 • T: +1 773 6849844 • F: +1 773 6847141 • www.msichicago.org •
Pres.: *David R. Mosena* • Dir.: *Dr. Barry Aprison* •
Natural History Museum / Science&Tech Museum – 1926
Science, technology, housed in classic Greek structure constructed as the Palace of Fine Arts for the World's Fair Columbian Exposition of 1893 in Chicago 47497

NAB Gallery, 1117 W Lake, Chicago, IL 60607 • T: +1 312 7381620 •
Dir.: *Craig Anderson* • *Robert Horn* •
Fine Arts Museum – 1974 47498

The National Time Museum, 57th St and Lake Shore Dr, Chicago, IL 60637-2093 • T: +1 312 7421412 •
Dir.: *Michael Lash* •
Science&Tech Museum – 2001
Horology 47499

Northeastern Illinois University Art Gallery, 5500 N Saint Louis Av, Chicago, IL 60625 • T: +1 773 5834050 • F: +1 773 4424920 • www.neiu.edu •
Fine Arts Museum / University Museum – 1973 47500

Northern Illinois University Art Gallery in Chicago, 215 W Superior St, Chicago, IL 60610 • T: +1 312 6426010 • F: +1 312 6429635 • hweber@niu.edu • www.vpa.niu.edu/museum •
Dir.: *Julie Charmelo* •
Fine Arts Museum / University Museum – 1984 47501

Oriental Institute Museum, University of Chicago, 1155 E 58th St, Chicago, IL 60637 • T: +1 773 7029514 • F: +1 773 7029853 • oi-museum@uchicago.edu • oi.uchicago.edu •
Dir.: *Geoff Emberling* • Cur.: *Raymond D. Tindel* •
Fine Arts Museum / Archaeology Museum / University Museum – 1894
Archaeology, ancient history, art 47502

The Palette and Chisel Academy, 1012 N Dearborn St, Chicago, IL 60610 • T: +1 312 6424149 • F: +1 312 6424317 • fineart1012@sbcglobal.net • www.paletteandchisel.org •
Exec. Dir.: *William Ewers* •
Fine Arts Museum / Public Gallery – 1895
Paintings, sculptures 47503

Palette & Chisel Academy of Fine Arts Gallery Exhibition, 1012 N Dearborn Av, Chicago, IL 60610 • T: +1 312 6424400 • F: +1 312 6424317 • fineart1012@sbcglobal.net • www.paletteandchisel.org •
Public Gallery
Drawings, paintings and sculpture exhibits 47504

The Peace Museum, 100 N Central Park Av, Chicago, IL 60610-1912 • T: +1 773 6386450 • F: +1 312 4401267 • virginia@peacemuseum.org • www.peacemuseum.org •
Dir.: *Bruce Doblin* •
Fine Arts Museum – 1981
Exhibits on Martin Luther King and bombing of Hiroshima, 20th c anti-war-poster, Veitnem War, folk art and art 47505

Peggy Notebaert Nature Museum, Chicago Academy of Sciences, 2430 N Cannon Dr, Chicago, IL 60614 • T: +1 773 7555100 • F: +1 773 7555199 • webmaster@naturemuseum.org • www.naturemuseum.org •
C.E.O.: *Joseph Shacter* • Cur.: *Dr. Douglas Taron* (Biology) •
Natural History Museum – 1857
Natural hist 47506

Polish Museum of America, 984 N Milwaukee Av, Chicago, IL 60622-4104 • T: +1 773 3843352 and 3731 • F: +1 773 3843799 • pma@prcua.org • www.prcua.org/pma •
Dir.: *Jan Lorys* • Pres.: *Joan Kosinski* • Ass. Cur.: *Bohdan Gorczynski* •
Folklore Museum / Ethnology Museum – 1937
Polonica, Paderewski memorabilia, world fair 1939, art 47507

Polvo Art Studio, 1458 W 18th St, Chicago, IL 60608 • T: +1 773 3441944 • polvoarte@yahoo.com • www.polvo.org •
Head: *Miguel Cortez* •
Public Gallery – 2003 47508

Pullman Museum, Historic Pullman Foundation, 11141 S Cottage Grove Av, Chicago, IL 60628-4614 • T: +1 773 7858181 • F: +1 773 7858182 • foundation@pullmanil.org • www.pullmanil.org •
Pres.: *Cynthia McMahon* •
Historical Museum – 1973
Historic District, first planned model industrial community in 1880 47509

The Renaissance Society at the University of Chicago, 5811 S Ellis Av, Chicago, IL 60637 • T: +1 773 7028670 • F: +1 773 7029669 • info@renaissancesociety.org • www.renaissancesociety.org •
Pres.: *Timothy Flood* • Dir.: *Susanne Ghez* •
Fine Arts Museum – 1915
Temporary exhibitions 47510

Roy Boyd Gallery, 739 N Wells St, Chicago, IL 60610 • T: +1 312 6421606 • F: +1 312 6422143 •
Dir.: *Roy Bod* •
Fine Arts Museum
Contemporary American paintings, Russian and Baltic photography, sculpture and works on paper 47511

School of the Art Institute of Chicago Gallery, 847 W. Jackson Blvd, Chicago, IL 60603 • T: +1 312 8995219 • F: +1 312 8991840 • www.artic.edu/saic •
Dir.: *Rhoda Rosen* •
Public Gallery 47512

Smart Museum of Art, University of Chicago, 5550 S Greenwood Av, Chicago, IL 60637 • T: +1 773 7020200 • F: +1 773 7023121 • smart-museum@uchicago.edu • smartmuseum.uchicago.edu •
Dir.: *Anthony G. Hirschel* • Cur.: *Stephanie Smith* •
Fine Arts Museum / University Museum – 1974 47513

Spertus Museum, 618 S Michigan Av, Chicago, IL 60605 • T: +1 312 3221747 • F: +1 312 9223934 • musm@spertus.edu • www.spertus.edu •
Dir.: *Rhoda Rosen* • Cur.: *Susan Marcus* •
Archaeology Museum / Religious Arts Museum – 1968
Culture, art, archaeology 47514

Stephen A. Douglas Tomb, 636 E 35th St, Chicago, IL 60616 • T: +1 312 2252620 • F: +1 312 2257855 • stephenadouglas@aol.com •
Head: *Michael Carson* •
Historic Site – 1865
1866-1881 Stephen A. Douglas Tomb 47515

Swedish American Museum Association of Chicago Museum, 5211 N Clark St, Chicago, IL 60640 • T: +1 312 7288111 • F: +1 312 7288870 • museum@samac.org • www.samac.org •
Pres.: *Philip J. Anderson* •
Folklore Museum – 1976
Swedish history 47516

TBA Exhibition Space, c/o Thomas Blackman Assoc., 230 W Huron St, Ste 3E, Chicago, IL 60610 • T: +1 312 5873300 • F: +1 312 5873304 •
Dir.: *Heather Hubbs* • Cur.: *Pedro Velez* •
Public Gallery 47517

Terra Museum of American Art, 666 N Michigan Av, Chicago, IL 60611 • T: +1 312 6643939 ext 1233 • F: +1 312 6642052 • terra@terramuseum.org • www.terramuseum.org •
Dir.: *Elizabeth Glassman* • Cur.: *Elizabeth Kennedy* •
Fine Arts Museum – 1979 47518

Ukrainian Institute of Modern Art, 2320 W Chicago Av, Chicago, IL 60622 • T: +1 773 2275522 • www.brama.com/uima •
Dir.: *Oleh Kowerko* •
Fine Arts Museum
Contemporary art 47519

Ukrainian National Museum, 721 N Oakley Blvd, Chicago, IL 60612 • T: +1 312 4218020 • F: +1 773 7722883 • hankewych@msn.com • www.ukrntlmuseum.org •
C.E.O.: *Jaroslaw J. Hankewych* •
Folklore Museum – 1952
Ukrainian folk art, textiles, artists, paintings sculpture, photos, music – Archiv 47520

Union Street Gallery, 1655 Union St, Chicago, IL 60411 • T: +1 708 7542601 • F: +1 708 7548779 • info@unionstreetgallery.org • www.unionstreetgallery.org •
Fine Arts Museum
Contemporary art 47521

Virginia Groot A. Foundation, 215 W Superior St, Chicago, IL 60610 • T: +1 312 3374658 •
Public Gallery
Sculptures 47522

Woman Made Gallery, 685 N Milwaukee Av, Chicago, IL 60642 • T: +1 312 7380400 • F: +1 312 7380404 • gallery@womanmade.org • www.womanmade.org •
Public Gallery – 1992 47523

Chico CA

1078 Gallery, 820 Broadway, Chico, CA 95928 • T: +1 530 3431973 • www.1078gallery.org •
Dir.: *Lynette Krehe* • *John Ferrell* •
Public Gallery 47524

Bidwell Mansion, 525 Esplanade, Chico, CA 95926 • T: +1 530 8956144 • F: +10 530 8956699 •
Historical Museum – 1964
Victorian furnishings 47525

Chico Museum, 141 Salem St, Chico, CA 95928 • T: +1 530 8914336 • F: +1 530 8914336 • www.chicomuseum.org •
Pres.: *Beulah Robinson* • Cur.: *Anne Seiler* •
Local Museum – 1980
Local history since 1840's, Chinese Taoist temple – library 47526

Janet Turner Print Museum, California State University, Chico, 400 W First St, Chico, CA 95929-0820 • T: +1 530 8984476 • F: +1 530 8985581 • csullivan@exchange.csuchico.edu • www.csuchico.edu/art/galleries/turnergallery.html •
Cur.: *Catherine Sullivan Sturgeon* •
University Museum / Fine Arts Museum – 1981
Fine art prints 47527

Museum of Anthropology, California State University, Chico, 400 W First St, Chico, CA 95929-0400 • T: +1 530 8985397 • F: +1 530 8986143 • anthromuseum@csuchico.edu • www.csuchico.edu/anth/museum •
Dir.: *Dr. Stacy Schaefer* • *Dr. Georgia Fox* • Cur.: *Adrienne Scott* •
University Museum / Ethnology Museum – 1969
Caöofornia prehistory and history, worldwide archaeological and ethnographical materials 47528

University Art Gallery, c/o California State University, 111 Alva Taylor Hall, First and Normal Sts, Chico, CA 95929-0820 • T: +1 530 8985864 • jtannen@csuchico.edu • www.csuchico.edu/art/galleries/univgallery.html •
Cur.: *Jason Tannen* •
Fine Arts Museum / University Museum 47529

Childress TX

Childress County Heritage Museum, 210 Third St, Childress, TX 79201 • T: +1 940 9372261 • F: +1 940 9373188 • cchm@srcaccess.net •
C.E.O.: *Johnann Schaefer* • *Shawna Murphy* •
Local Museum – 1976
Hist of the area from Paleolithic era, Republic of Texas 47530

Chillicothe MO

George W. Somerville Historical Library, Livingston County Library, 450 Locust, Chillicothe, MO 64601 • T: +1 660 6460547 • F: +1 660 6465504 • ugy001@mail.connect.more.net • www.livcolibrary.org •
Pres.: *Patricia Henry* • Dir.: *Karen Hicklin* •
Library with Exhibitions – 1966 47531

Grand River Historical Society & Museum, 1401 Forest Dr, Chillicothe, MO 64601 • T: +1 660 6463430 • drfstark@cmuonline.net •
Pres.: *Dr. Frank Stark* • Cur.: *Dr. John R. Neal* •
Historical Museum – 1959 47532

Chillicothe OH

Adena State Memorial, Home of Thomas Worthington, Adena Rd, Chillicothe, OH 45601 • T: +1 740 7721500 • F: +1 740 7752746 • jrupp@ohiohistory.org • www.ohiohistory.org/places/adena/ •
C.E.O.: *Jane E. Rupp* •
Local Museum – 1946
American furniture and furnishings 1760-1830 47533

Hopewell Culture National Historic Park, 16062 State Rte 104, Chillicothe, OH 45601-8694 • T: +1 740 7741125 • F: +1 740 7741140 • hocu_superintendent@nps.gov • www.nps.gov/hocu •
Sc. Staff: *Jennifer Pederson* (Archaeology) •
Ethnology Museum – 1923
200 B.C.-500 A.D. archaeological artifacts, Hopewell Indian mound enclosure 47534

Lucy Hayes Heritage Center, 90 W Sixth St, Chillicothe, OH 45601 • T: +1 740 7755829 • F: +1 740 7755829 • www.ohiotourism.com •
Pres.: *Shirley Thacker* •
Historical Museum – 1996
Restored birthplace of First Lady Lucy Hayes, pictures, postcards, letters, White House invitations 47535

Pump House Center for the Arts, POB 1613, Chillicothe, OH 45601-5613 • T: +1 740 7725783 • F: +1 740 7725783 • pumpart@bright.net • www.bright.net/~pumpart •
Exec. Dir.: *Beverly J. Mullen* •
Decorative Arts Museum / Fine Arts Museum – 1986
Baskets, fine art, designer crafts, glass, prints, quilts, woodcarving 47536

Ross County Historical Society Museum, 45 W Fifth St, Chillicothe, OH 45601 • T: +1 740 7721936 • info@rosscountyhistorical.org • www.rosscountyhistorical.org •
Pres.: *Eric Picciano* • C.E.O.: *Thomas G. Kuhn* •
Local Museum – 1896
Prehistoric artifacts, pioneer tools, household furnishings 47537

Chiloquin OR

Collier Memorial State Park and Logging Museum, 46000 Hwy 97 N, 30 miles north of Klamath Falls, Chiloquin, OR 97624 • T: +1 541 7832471 • F: +1 541 7832707 •
Cur.: *Lowell N. Jones* •
Historical Museum – 1945
Pioneer surveying instruments, locomotives, lumber and logging trucks and equip, Indian artifacts 47538

Chincoteague VA

The Oyster and Maritime Museum of Chincoteague, 7125 Maddox Blvd, Chincoteague, VA 23336 • T: +1 757 3366117 •
C.E.O.: *Gerald C. West* •
Natural History Museum / Science&Tech Museum / Historical Museum – 1966
Natural science, maritime hist, Fresnel lens 47539

Chino CA

Yorba and Slaughter Families Adobe Museum, 17127 Pomona Rincon Rd, Chino, CA 91710 • T: +1 909 5978332 • F: +1 909 3070748 • creyes@sbcm.sbcounty.gov • www.co.san-bernardino.ca.us •
Dir.: *Robert L. McKernan* • Cur.: *Dr. Ann Deegan* •
Historical Museum – 1976
Decorative arts, farm equipment, photos, local hist 47540

Chinook MT

Blaine County Museum, 501 Indiana, Chinook, MT 59523 • T: +1 406 3572590 • F: +1 406 3572199 • blmuseum@mtintouch.net • www.chinookmontana.com •
Dir.: *Stuart C. MacKenzie* • Cur.: *Jude Sheppard* •
Local Museum – 1977
Paleontology, old west artifacts, Indian culture 47541

Chinook WA

Fort Columbia House Museum, Fort Columbia State Park, Hwy 101, Chinook, WA 98614 • T: +1 360 7778221 • F: +1 360 6424216 • lcic@parks.wa.gov •
Military Museum – 1954
Commander's house, artillery barracks, American Indian and pioneer artifacts, period furniture, appliances 47542

Chiriaco Summit CA

General Patton Memorial Museum, 62-510 Chiriaco Rd, Chiriaco Summit, CA 92201 • T: +1 760 2273483 • F: +1 760 2273483 • Contact@GeneralPattonMuseum.com • www.generalpattonmuseum.com •
Pres.: *Jan Roberts* •
Military Museum – 1988
Military, headquarters of WW II desert training areas 47543

Chisholm MN

Ironworld Discovery Center, 801 SW Hwy-169, Chisholm, MN 55719 • T: +1 218 2547959 • F: +1 218 2547971 • marketing@ironworld.com • www.ironworld.com •
Dir.: *Marianne Bouska* •
Local Museum / Science&Tech Museum – 1977
Mining, regional hist, social & civic organizations – library 47544

Chittenango NY

Chittenango Landing Canal Boat Museum, 7010 Lakeport Rd, Chittenango, NY 13037 • T: +1 315 6873801 • F: +1 315 6873801 • www.chittenagolandingcanalboatmuseum.com •
Pres.: *Joan DiChristina Hager* • Cur.: *Barbara Richardson* •
Science&Tech Museum – 1986 47545

Choteau MT

Old Trail Museum, 823 Main St, Choteau, MT 59422 • T: +1 406 4665332 • F: +1 406 4665874 • otm@3rivers.net • www.oldtrailmuseum.org •
Dir.: *Christine Hoyt* • Cur.: *Todd Crowell* •
Local Museum – 1985
Antiques, Indian artifacts, fossils, grizzly cabin, paleontology antechamber 47546

Christiansburg VA

Montgomery Museum and Lewis Miller Regional Art Center, 300 S Pepper St, Christiansburg, VA 24073 • T: +1 540 3825644 • F: +1 540 3829127 • www.montgomerymuseum.org •
Dir.: *Shearon Campbell* •
Local Museum / Fine Arts Museum – 1983
Photographs, genealogical, New River Valley artifacts 47547

Christmas FL

Fort Christmas, 1300 Fort Christmas Rd, Christmas, FL 32709 • T: +1 407 5684149 • F: +1 407 5686629 • trudy.trask@ocfl.net • www.parks.onetgov.net •
Cur.: *Vickie Prewett* •
Military Museum – 1977
Military artifacts 47548

Church Rock NM

Red Rock Museum, Red Rock State Park, Church Rock, NM 87311 • T: +1 505 8631337 • F: +1 505 8631297 • rrsp@ci.gallup.nm.us • www.ci.gallup.nm.us/rrsp/00182_redrock.html •
Head: *Joe Athens* • Sc. Staff: *Maxine Armstrong Touchine* •
Ethnology Museum / Natural History Museum / Public Gallery – 1951
Human and natural hist, fine arts, crafts and artifacts of the prehistoric 47549

Cimarron NM

Old Mill Museum, 220 W 17th St, Rte 1, Cimarron, NM 87714 • T: +1 505 3762913 • F: +1 505 3762417 •
Cur.: *F.E. Morse* •
Local Museum – 1967
New Mexico hist, espallicy northeastern New Mexico, manuscripts 47550

The Philmont Museum, Seton Memorial Library and Kit Carson Museum, Philmont Scout Ranch, 17 Deer Run Rd, Cimarron, NM 87714 • T: +1 505 3762281 ext 46 • F: +1 505 3762602 • jschubert@philmontscoutranch.org •
Dir.: *Jason Schubert* • Libr.: *Seth McFarland* •
Fine Arts Museum / Local Museum – 1967
Art, archaeology, ethnology, history od the Southwest 47551

Cincinnati OH

American Classical Music Hall of Fame and Museum, 4 W Fourth St, Cincinnati, OH 45202 • T: +1 513 6213263 • F: +1 513 6211563 • contact@americanclassicalmusic.org • www.americanclassicalmusic.org •
Exec. Dir.: *Stefan A. Skirtz* •
Music Museum – 1995
Classical music in US 47552

Betts House Research Center, 416 Clark St, Betts-Longworth Historic District, Cincinnati, OH 45203-1420 • T: +1 513 6510734 • F: +1 513 6512143 • bettshouseerc@fuse.net • www.bettshouse.org •
Pres.: *Robert W. Dorsey* • Exec.Dir.: *J. Duncan Muir* •
Local Museum – 1995
Local hist 47553

Cary Cottage, 7000 Hamilton Av, Cincinnati, OH 45231 • T: +1 513 5223860 • F: +1 513 7283946 • clovernook@clovernook.org • www.clovernook.org •
C.E.O.: *Jeff Brasie* •
Special Museum – 1903
1832 home of poets Alice and Phoebe Cary, furnishings, literatura 47554

Cincinnati Art Museum, 953 Eden Park Dr, Cincinnati, OH 45202-1596 • T: +1 513 6392995 • F: +1 513 6392888 • information@cincyart.org • www.cincinnatiartmuseum.org •
Dir.: *Timothy Rub* • Cur.: *Anita Ellis* (Decorative Arts) • *Kristin Spangenberg* (Prints, Drawings) • *Betsy Wieseman* (European Art) • *Glenn Markoe* (Classical, Near Eastern) • *Julie Aronson* (American Art) • Ass. Cur.: *Dennis Kiel* (Photography) • Libr./Archive: *Mona Chapin* •
Fine Arts Museum / Decorative Arts Museum – 1881
Ancient art, near and far Eastern art, medieval art, decorative arts, musical instruments, African and Native American art, contemporary art 47555

Cincinnati Fire Museum, 315 W Court St, Cincinnati, OH 45202 • T: +1 513 6215553 • F: +1 513 6211456 • www.cincyfiremuseum.com •
Pres.: *Rick Jahnigen* • Exec. Dir.: *Barbara M. Hammond* •
Special Museum – 1979
Fire-fighting artifacts mid 19th c to present 47556

Cincinnati Museum Center, 1301 Western Av, Cincinnati, OH 45203 • T: +1 513 2877000 • F: +1 513 2877029 • www.cincymuseum.org •
Pres.: *Douglass W. McDonald* •
Ethnology Museum / Local Museum / Natural History Museum / Science&Tech Museum – 1835
Nature and natural sciences, hist of nature, minerals, geology, biology 47557

Contemporary Arts Center, 44 E Sixth St, Cincinnati, OH 45202-9834 • T: +1 513 3458400 • F: +1 513 7217418 • admin@spiral.org • www.spiral.org •
Dir.: *Linda Shearer* • Cur.: *Matt Distel* • *Lisa Buck* •
Fine Arts Museum – 1939 47558

DAAP Galleries, University of Cincinnati, 2624 Clifton Av, Cincinnati, OH 45221-0016 • T: +1 513 5564933 • F: +1 513 5563288 • daapgalleries@uc.edu • www.daap.uc.edu/gallery/gallery.html •
Dir.: *Anne Timpano* •
Fine Arts Museum / University Museum – 1993
Art of ancient Greece, Europa, Asia, American art of the late 19th and early 20th c 47559

Fine Arts Department Gallery, Xavier University, 3800 Victory Pkwy, Cincinnati, OH 45207-7311 • T: +1 513 7453811 • F: +1 513 7451098 • www.xu.edu •
Dir.: *Kitty Uetz* •
Public Gallery 47560

Hillel Jewish Student Center Gallery, 2615 Clifton Av, Cincinnati, OH 45220-2885 • T: +1 513 2216728 • F: +1 513 2217134 • www.hillelcincinnati.org •
Exec. Dir.: *Abie Ingber* •
Fine Arts Museum – 1982
Jewish art, antique Judaica 47561

Indian Hill Historical Society Museum, 8100 Given Rd, Cincinnati, OH 45243 • T: +1 513 8911873 • F: +1 513 8911873 • ihhist@one.net • www.indianhill.org •
Pres.: *Margaret Gillespie* •
Local Museum – 1973 47562

Skirball Museum, Hebrew Union College-Jewish Institute of Religion, 3101 Clifton Av, Cincinnati, OH 45220 • T: +1 513 2211875 • F: +1 513 2210316 • jlucas@cnhuc.edu • www.huc.edu •
Cur.: *Judith S. Lucas* •
Religious Arts Museum – 1913
Ceremonial objects, biblical archaeology, , textiles, decorative arts, prints 47563

Studio San Giuseppe Art Gallery, College of Mount Saint Joseph, 5701 Delhi Rd, Dorothy Meyer Ziv Art Bldg, Cincinnati, OH 45233-1670 • T: +1 513 2444314 • F: +1 513 2444942 • Jerry_Bellas@mail.msj.edu • www.msj.edu •
Dir.: *Gerald M. Bellas* •
Fine Arts Museum / University Museum / Public Gallery – 1962 47564

The Taft Museum of Art, 316 Pike St, Cincinnati, OH 45202-4293 • T: +1 513 2410343 • F: +1 513 2417762 • taftmuseum@taftmuseum.org • www.taftmuseum.org •
Dir. & C.E.O.: *Phillip C. Long* • Chief Cur.: *Lynne Ambrosini* •
Fine Arts Museum – 1932
Furnishings, coll of Duncan Phyfe furniture, paintings, old masters, Chinese porcelains, Limoges enamels 47565

Trailside Nature Center and Museum, Brookline Dr, Burnet Woods Park, Cincinnati, OH 45220 • T: +1 513 3216070 • F: +1 513 7513679 • vivian.wagner@cinparks.rcc.org • www.cinci-parks.org •
Dir.: *Willie F. Carden* • *Vivian Wagner* •
Natural History Museum – 1930
Local birds, insects a.o., rocks and fossils, tree and plant specimens 47566

William Howard Taft National Historic Site, 2038 Auburn Av, Cincinnati, OH 45219 • T: +1 513 6843262 • F: +1 513 6843627 • wiho_superintendent@nps.gov •
Man.: *E. Ray Henderson* •
Historical Museum – 1969
Furnishings, memorabilia of President Taft and his family 47567

Circle MT

McCone County Museum, 1507 Av B, Circle, MT 59215 • T: +1 406 4852414 •
Cur.: *Donald Zahn* •
Local Museum – 1953
County newspapers, animal husbandry, geology, Indians 47568

City of Industry CA

Workman and Temple Family Homestead Museum, 15415 E Don Julian Rd, City of Industry, CA 91745-1029 • T: +1 626 9688492 • F: +1 626 9682048 • info@homesteadmuseum.org • www.

homesteadmuseum.org •
Dir.: *Karen Graham Wade* •
Historical Museum – 1981
Decorative arts and furnishings 47569

Clackamas OR

Oregon Military Museum, Camp Withycombe, 10101 SE Clackamas Rd, Clackamas, OR 97015 • T: +1 503 5575359 • F: +1 503 5576713 • tracy.thoennes@us.army.mil •
Cur.: *Tracy Buckley* •
Military Museum – 1974
Archives, artwork, clothing, documents, emblems, equipage, flags, furnishings, insignia, library volumes ordnance, subsistence, vehicles, and trophies - Oregon National Guard, US Army 47570

Claremont CA

Clark Humanities Museum, Scripps College, Humanities Bldg, 1030 Columbia Av, Claremont, CA 91711 • T: +1 909 6073606 • F: +1 909 6077143 • www.scrippscollege.edu •
Head: *Dr. Eric T. Haskell* •
Fine Arts Museum / University Museum – 1970
Paintings, prints,cloisonne, African sculpture, drawings – Denison library 47571

Kenneth G. Fiske Museum, c/o Claremont Colleges, 450 N College Way, Claremont, CA 91711-4491 • T: +1 909 6218307 • arrice@rocketmail.com • www.cuc.claremont.edu •
Cur.: *Dr. Albert R. Rice* •
Music Museum – 1986
Musical instruments 47572

Petterson Museum of Intercultural Art, 730 Plymouth Rd, Claremont, CA 91711 • T: +1 909 3995544, 6219581 • F: +1 909 3995508 • cgil@pilgrimplace.org • www.pilgrimplace.org •
Head: *Carol Bowdion Gil* •
Fine Arts Museum / Decorative Arts Museum – 1968
International folk and fine art – library 47573

Pomona College Museum of Art, Montgomery Art Center, 330 N College Way, Claremont, CA 91711-6344 • T: +1 909 6218283 • F: +1 909 6218989 • museuminfo@pomona.edu • www.pomona.edu/museum •
Dir.: *Kathleen Howe* • Ass. Dir.: *Steve Comba* • Cur.: *Rebecca McGrew* •
Fine Arts Museum / University Museum – 1958
Kress Renaissance paintings, prints and drawings, photographs, basketry, ceramics 47574

Raymond M. Alf Museum of Paleontology, 1175 W Baseline Rd, Claremont, CA 91711 • T: +1 909 6242798 • F: +1 909 6242798 • dlofgren@webb.org • www.alfmuseum.org •
Dir.: *Donald Lofgren* •
Natural History Museum – 1937
Rocks, minerals, paleontology, osteology, archaeology 47575

Ruth Chandler Williamson Gallery, Scripps College, 1030 N Columbia Av, Claremont, CA 91711-6344 • T: +1 909 6073517 • mmacnaug@scrippscollege.edu • www.scrippscollege.edu/williamson-gallery •
Dir.: *Mary Davis MacNaughton* •
Fine Arts Museum / University Museum
American paintings, Japanese prints, sculptures 47576

Claremont NH

Claremont Museum, 26 Mulberry St, Claremont, NH 03743 • T: +1 603 5431400 •
Pres.: *Colin J. Sanborn* •
Local Museum – 1966
Paintings, furnishings, tools, artifacts 47577

Claremore OK

J.M. Davis Arms and Historical Museum, 333 N Lynn Riggs Blvd, Claremore, OK 74017 • T: +1 918 3415707 • F: +1 918 3415771 • ktthompson@sbcglobal.net • www.thegunmuseum.com •
Dir.: *Duane Kyler* • Cur.: *Dennis Duval* •
Historical Museum / Military Museum – 1965
20.000 firearms, western memorabilia, Native American artifacts 47578

Will Rogers Memorial Museums, 1720 W Will Rogers Blvd, Claremore, OK 74018 • T: +1 918 3410719 • F: +1 918 3438119 • wrinfo@willrogers.com • www.willrogers.com •
Dir.: *Michelle Lefebvre-Carter* • Cur.: *Gregory Malak* •
Historical Museum – 1938
Life of Will Rogers, cowboy, newspaperman, radio moderator, memorabilia, manuscripts; saddle coll 47579

Clarence NY

Town of Clarence Museum, Historical Society of the Town of Clarence, 10465 Main St, Clarence, NY 14031 • T: +1 716 7598575 •
Pres.: *Lorna Helm* • Cur.: *Alicia L. Braaten* •
Local Museum – 1954
Geealogy, Indian artifacts, garage and workshop of the invention of the implantable pacemaker 47580

Clarion IA

4-H Schoolhouse Museum, 302 S Main, Clarion, IA 50525 • T: +1 515 5322256 • F: +1 515 5322511 • clchamb@goldfieldaccess.net • www.clarion-iowa.com •
Exec. Dir.: *Cheryl Ketelsen* •
Historical Museum – 1955 47581

Clarion PA

Sutton-Ditz House Museum, 17 S Fifth Av, Clarion, PA 16214 • T: +1 814 2264450 • F: +1 814 2267106 • cchs@csonline.net • www.csonline.net/cchs/ •
Dir./Cur.: *Lindsley A. Dunn* •
Local Museum – 1955
Local hist manuscripts and photographs 47582

University Galleries, Clarion University of Pennsylvania, c/o Carlson Library A-4, Clarion, PA 16214 • T: +1 814 3932523, 3932412 • F: +1 814 3932168 • gallery@clarion.edu •
Interim Dir.: *Joe Thomas* •
Public Gallery / University Museum – 1982
Contemporary art 47583

Clark CO

Hahns Peak Area Historical Museum, Hahns Peak Village, Clark, CO 80428 • T: +1 970 8796781 •
Cur.: *Mallory Pollock* •
Local Museum – 1972
Regional history, Old Horse Barn, one room school house 47584

Clark NJ

Dr. William Robinson Plantation, 593 Madison Hill Rd, Clark, NJ 07066 • T: +1 732 3883600 •
Pres.: *Eleanor Warren* •
Local Museum – 1974 47585

Clark SD

Beauvais Heritage Center, Hwy 212, Clark, SD 57225 • T: +1 605 5323722, 5325216 •
C.E.O. & Pres.: *Ailene Luckhurst* •
Local Museum – 1978
Cultural hist, pioneer home, depot, store, military, religion, machine, tool 47586

Clarksdale MS

Delta Blues Museum, 114 Delta Av, Clarksdale, MS 38614, mail addr: POB 459, Clarksdale, MS 38614 • T: +1 662 6276820 • F: +1 662 6277263 • info@deltabluesmuseum.org • www.deltabluesmuseum.org •
C.E.O.: *Shelley Ritter* •
Music Museum – 1979
American music & Black hist 47587

Rock 'n Roll & Blues Heritage Museum, 113 E Second St, Clarksdale, MS 38614 • T: +1 901 6058662 • frog@rockmuseum.biz • www.rockmuseum.biz •
Dir.: *Theo P. Dasbach* •
Music Museum – 2006
Music, dance, Elvis movie memorabilia 47588

Clarkson NE

Clarkson Historical Museum, 221 Pine St, Clarkson, NE 68629 • T: +1 402 8923854 • F: +1 402 8923059 •
Pres.: *Ruth D. Waters* •
Local Museum – 1967
Artifacts, musical instruments, photographs 47589

Clarkston MI

Yesteryear House-Central Mine, 7995 Perry Lake Rd, Clarkston, MI 48348 • T: +1 248 6254296 •
Owner: *Nicholas Bell* • *Theresa Rekawek* •
Science&Tech Museum – 1860
Local copper mining hist 47590

Clarkston WA

Valley Art Center, 842 6th St, Clarkston, WA 99403 • T: +1 509 7588331 • valleyarts@clarkston.com •
Exec. Dir.: *Pat Rosenberger* •
Fine Arts Museum – 1968
Art 47591

Clarksville TN

Clarksville-Montgomery County Museum-Customs House Museum, 200 S Second St, Clarksville, TN 37040 • T: +1 931 6485780 • F: +1 931 5535179 • info@customshousemuseum.org • www.customshousemuseum.org •
Dir.: *Ned Crouch* • Ass. Dir.: *Linda Maki* • Cur.: *Steven Stewart* •
Local Museum – 1984
Local hist 47592

Mabel Larson Fine Arts Gallery, Austin Peay State University, POB 4677, Clarksville, TN 37044 • T: +1 615 6487333 •
Fine Arts Museum – 1994
Drawings 47593

Margaret Fort Trahern Gallery, College and Eighth Sts, Clarksville, TN 37044 • T: +1 931 2217333 • F: +1 931 2217432 • holteb@apsu02.apsu.edu •
C.E.O.: *Bettye Holte* •
Public Gallery – 1974
Art 47594

Clarksville VA

Prestwould Foundation, 429 Prestwould Dr, Clarksville, VA 23927 • T: +1 804 3748672 • F: +1 434 3743060 •
Dir.: *Dr. Julian D. Hudson* •
Local Museum – 1963
Costumes, original furniture, original French scenic wallpaper 47595

Clawson MI

Clawson Historical Museum, 41 Fisher Ct, Clawson, MI 48017 • T: +1 248 5889169 •
Cur.: *Deloris M. Kumler* •
Local Museum – 1973
Local history, household articles 47596

Clay City KY

Red River Historical Society Museum, 4541 Main St, Clay City, KY 40312 • T: +1 606 6632555 •
Dir.: *Larry G. Meadows* • Cur.: *John Faulkner* (Archaeology) • *Steve Abner* (Photography) • *Jim Spencer* (History) •
Local Museum – 1966
Local history, archaeology, photography, genealogy 47597

Clayton ID

Custer Museum, Yankee Fork Ranger District Museum, Clayton, ID 83227 • T: +1 208 8382201 • F: +1 208 8383329 •
Pres.: *Craig Wolford* •
Historical Museum – 1961
Located on site of 1870 Gold Rush 47598

Clayton MO

Hanley House, 7600 Westmoreland Rd, Clayton, MO 63105 • T: +1 314 2908516 • F: +1 314 2908517 •
Decorative Arts Museum
Period furnishings 47599

Clayton NY

Antique Boat Museum, 750 Mary St, Clayton, NY 13624 • T: +1 315 6864104 • F: +1 315 6862775 • info@abm.org • www.abm.org •
Exec.Dir.: *John MacLean* • Dir.: *John Summers* • Cur.: *Rebecca Hopfinger* •
Historical Museum – 1964
Boats, boat building molds and tools, nautical artifacts 47600

Handweaving Museum and Arts Center, 314 John St, Clayton, NY 13624 • T: +1 315 6864123 • F: +1 315 6863459 • info@hm-ac.org • www.hm-ac.org •
Dir.: *Beth Conlon* • Cur.: *Sonja Wahl* •
Decorative Arts Museum – 1966 47601

Thousand Islands Museum of Clayton, 312 James St, Clayton, NY 13624 • T: +1 315 6865794 • F: +1 315 6864867 • info@timuseum.org • www.timuseum.org •
Dir.: *Linda Schleher* •
Historical Museum – 1964
Decoy carving, muskie fishing, Indian artifacts, hist of Clayton and the islands 47602

Cle Elum WA

Carpenter Home Museum, 221 E 1st St, Cle Elum, WA 98922 • T: +1 509 6745702 •
Pres.: *Sarah Engdahl* •
Historical Museum – 1990
Home, with ballroom and a special maid's room 47603

Cle Elum Telephone Museum, 221 E 1st St, Cle Elum, WA 98922 • T: +1 509 6745702 •
Pres.: *Sarah Engdahl* •
Science&Tech Museum – 1967
History of the phone 47604

Clear Lake IA

Kinney Pioneer Museum, 9184 G 265th St, Clear Lake, IA 50428 • T: +1 641 4231258 •
Dir.: *Bonnie Frenz* •
Local Museum – 1964
Local hist, Indian artifacts, fossil coll, dolls and toys 47605

Clear Lake WI

Clear Lake Area Historical Museum, 450 Fifth Av, Clear Lake, WI 54005 • T: +1 715 2633050, 2632042 •
Pres./Cur.: *Charles T. Clark* •
Local Museum – 1977
Local hist 47606

Clearfield PA

Clearfield County Historical Society Museum, 104 E Pine St, Clearfield, PA 16830 • T: +1 814 7656125 •
Cur.: *Harry Reickart* •
Local Museum – 1955 47607

Clearmont WY

Big Red Barn Gallery, 30 Big Red Ln, Clearmont, WY 82835 • T: +1 307 7372291 • F: +1 307 7372322 • ucross@wyoming.com •
Exec. Dir.: *Sharon Dynak* •
Decorative Arts Museum – 1981
Quilts by Linda Behar 47608

Ucross Foundation Art Gallery, 30 Big Red Ln, Clearmont, WY 82835 • T: +1 307 7372291 • F: +1 307 7372322 • ucross@wyoming.com •
Dir.: *Elizabeth Guheen* •
Public Gallery
Art 47609

Clearwater FL

Napoleonic Society of America, 1115 Ponce de Leon Blvd, Clearwater, FL 33756 • T: +1 727 5861779 • F: +1 727 5812578 • napoleonic1@juno.com • www.napoleonic-society.com •
Pres.: *Robert Timlia* •
Historical Museum / Military Museum – 1983
Books and prints of the life of Napoleon Bonaparte, artifacts – library 47610

Ruth Eckerd Hall, 1111 McMullen-Booth d, Clearwater, FL 33759 • T: +1 727 7917060 • F: +1 727 7916020 • www.rutheckerdhall.net •
Exec. Dir.: *Robert A. Freedman* •
Public Gallery – 1978
Traveling exhibitions 47611

Cleburne TX

Layland Museum, 201 N Caddo, Cleburne, TX 76031 • T: +1 817 6450940 • F: +1 817 6414161 • museum@cleburne.net • www.cleburne.net •
Cur.: *Julie P. Backer* •
Historical Museum – 1964
History of domestic life, photographs, archives .. 47612

Clemson SC

Bob Campbell Geology Museum, Clemson University, 103 Garden Trail, Clemson, SC 29634-0130 • T: +1 864 6564600 • F: +1 864 6566230 • dcheech@clemson.edu • www.clemson.edu/geomuseum •
Cur.: *David Cicimurri* •
Natural History Museum – 1989
Minerals, gems, fossils, meteorites, mining artifacts 47613

Fort Hill - The John C. Calhoun House, Clemson University, Fort Hill St, Clemson, SC 29634 • T: +1 864 6562475 • F: +1 864 6561026 • hiottw@clemson.edu • www.clemson.edu/welcome/history/forthill.htm •
Dir.: *William D. Hiott* •
Local Museum – 1889
Historic house, home of John C. Calhoun 47614

Hanover House, Clemson University, c/o South Carolina Botanical Garden,, Perimeter Rd, Clemson, SC 29634 • T: +1 864 6562241, 6562475 • F: +1 803 6561026 • hiottw@clemson.edu • www.clemson.edu/welcome/history/hanover.htm •
Dir.: *William D. Hiott* •
Local Museum – 1941
Historic house 47615

Rudolph E. Lee Gallery, College of Architecture, Arts and Humanities Lee Hall, Clemson University, Clemson, SC 29634 • T: +1 864 6563883 • F: +1 864 6567523 •
Dir.: *David Houston* •
Fine Arts Museum / University Museum – 1956
Architecture, graphics, paintings 47616

Clermont IA

Montauk, POB 372, Clermont, IA 52135 • T: +1 319 4237173 • F: +1 319 4237378 • montauk@acegroup.cc •
Historical Museum – 1968
Home of William Larrabee, Iowa's 12th Governor 47617

Clermont NY

Clermont State Historic Site, 1 Clermont Av, Clermont, NY 12526 • T: +1 518 5374240 • F: +1 518 5376240 • fof@valstar.net • www.friendsofclermont.org •
Dir. & C.E.O.: *Bruce E. Naramore* • Cur.: *Travis Bowman* (Collections) • Education: *Heidi L. Hill* •
Historic Site / Local Museum – 1962
18th - 20th c furnishings and decorative arts, photographic and manuscript coll – Library 47618

Cleveland GA

White County Historical Society Museum, Cleveland Sq, Cleveland, GA 30528 • T: +1 706 8653225 • www.georgiamagazine.com •
Pres.: *Janet Cox* •
Local Museum – 1965
Civil War, local pictures, historic papers 47619

Cleveland MS

Fielding L. Wright Art Center, Delta State University, Cleveland, MS 38733 • T: +1 662 8464720 • F: +1 662 8464016 • artinfo@deltastate.edu • www.deltastate.edu •
Dir.: *Patricia Brown* •
Fine Arts Museum / University Museum – 1924
Southern art & folk art 47620

Cleveland OH

African American Museum, 1765 Crawford Rd, Cleveland, OH 44106 • T: +1 216 7911700 • F: +1 216 7911774 • ourstory@aamcleveland.org • www.aamcleveland.org •
C.E.O.: *Nancy Nolan Jones* •
Folklore Museum – 1953
African American artifacts and culture 47621

The Children's Museum of Cleveland, 10730 Euclid Av, Cleveland, OH 44106-2200 • T: +1 216 7917114 • F: +1 216 7918838 • info@clevelandchildrensmuseum.org • clevelandchildrensmuseum.org •
Pres.: *Richard J. Maicki* •
Special Museum – 1981 47622

Cleveland Institute of Art Gallery, 11141 East Blvd, Cleveland, OH 44106 • T: +1 216 4217000 • F: +1 216 4217438 • admiss@gate.cia.edu • www.cia.edu •
Public Gallery 47623

The Cleveland Museum of Art, 11150 East Blvd, Cleveland, OH 44106 • T: +1 216 4217350, +1 888 2420033 • F: +1 216 4210411 • info@clevelandart.org • www.clevelandart.org •
Dir.: *Katharine Lee Reid* • Dep. Dir.: *Charles L. Venable* (Collections) • *Janet Ashe* (Admin., Treasurer) • Chief Cur.: *Karel Paukert* (Musical Arts) • Cur.: *Sylvain Bellenger* (Paintings) • *Jane Glaubinger* (Prints) • *Michael Bennett* (Greek and Roman Art) • *Holger A. Klein* (Medieval Art) • *Stanislaw Czuma* (Southeast Asian Art and Indian Art) • *Ju-Hsi Chou* (Chinese Art) • *Louise W. Mackie* (Textiles and Islamic Art) • *William H. Robinson* (Modern European Art) • *Tom E. Hinson* (Photography) • Assoc.Cur.: *Susan Bergh* (Art of the Ancient Americas) • *Stephen N. Fliegel* (Medieval Art) • *Jeffrey Grove* (Contemporary Art) • Asst.Cur.: *Constantine Petridis* (African Art) • *Heather Lemonedes* (Prints & Drawings) •
Fine Arts Museum – 1913
Egyptian, early Christian and Byzantine art, Renaissance armor, Asian coll 47624

The Cleveland Museum of Natural History, 1 Wade Oval Dr, Cleveland, OH 44106-1767 • T: +1 216 2314600 ext 3235 • F: +1 216 2315919 • info@cmnh.org • www.cmnh.org •
Dir.: *Dr. Bruce Latimer* • Dep. Dir. & Cur.: *Dr. Brian Redmond* (Science, Archaeology) • *James Bissell* (Conservation, Botany) • *Bonnie Cummings* (Education) • *Harvey Webster* (Wildlife Resources) • Cur.: *Dr. Joseph Hannibal* (Invertebrate Paleontology) • *Dr. Timothy Matson* (Vertebrate Zoology) • *Dr. Shya Chitaley* (Paleobotany) • *Dr. Michael Ryan* (Vertebrate Paleontology) • *Dr. Joe Keiper* (Invertebrate Zoology) • *Dr. David Saja* (Mineralogy) •
Natural History Museum – 1920
Archaeology, ethnography, physical anthropology, botany, paleontology, paleobotany – archives 47625

Cleveland Police Museum, 1300 Ontario St, Cleveland, OH 44113-1600 • T: +1 216 6235055 • F: +1 216 6235145 • museum@clevelandpolicemuseum.org • www.clevelandpolicemuseum.org •
Dir.: *David C. Holcombe* • Cur.: *Anne T. Kmieck* •
Special Museum – 1983
Police hist 1866 to present, criminology, selected criminal cases, memorabilia, uniforms, firearms, police motorcycles 47626

Cleveland State University Art Gallery, 2307 Chester Av, Cleveland, OH 44114-3607 • T: +1 216 6872103 • F: +1 216 6879340 • artgallery@csuohio.edu • www.csuohio.edu/art/gallery •
Dir. & Cur.: *Robert Thurmer* • Pres.: *Michael Schwartz* •
Fine Arts Museum / University Museum – 1973
African artifacts, mostly from 19th c, African-Carribean and African American artifacts 47627

Crawford Auto-Aviation Museum, 10825 East Blvd, Cleveland, OH 44106 • T: +1 216 7215722 • F: +1 216 7210645 • www.wrhs.org •
Cur.: *Chris Dawson* •
Science&Tech Museum – 1963
Autos and aircraft transportation documents 47628

Dittrick Museum of Medical History, Case Western Reserve University, 11000 Euclid Av, Cleveland, OH 44106-1714 • T: +1 216 3686391 • F: +1 216 3680165 • jme3@po.cwru.edu • www.cwru.edu/artsci/dittrick •
Dir./Cur.: *Dr. Jim Edmonson* •
Historical Museum – 1926
Medical past of the area, microscopy and surgical instr, endoscopy – library, archives 47629

Dunham Tavern Museum, 6709 Euclid Av, Cleveland, OH 44103 • T: +1 216 4311060 • dunhamtavern@sbcglobal.net • www.dunhamtavern.org •
Pres.: *Garrit Wamelink* • C.E.O.: *Raymond L. Cushing* •
Local Museum – 1939
Folklore, glass, textiles, pewter, early American artifacts, Queen Ann and Chippendale furniture 47630

Great Lakes Science Center, 601 Erieside Av, Cleveland, OH 44114 • T: +1 216 6942000 • F: +1 216 6962140 • pamblancom@glsc.org • www.greatscience.com •
Exec.Dir. & Pres: *Linda Abraham-Silver* •
Natural History Museum / Science&Tech Museum – 1988 47631

Healthspace Cleveland, 8911 Euclid Av, Cleveland, OH 44106 • T: +1 216 2315010 • F: +1 216 2315129 • horvath@healthspacecleveland.org • healthspacecleveland.org •
Chm.: *Theodore J. Castele* •
Special Museum – 1936
Anatomy, physiology, Robert L. Dickinson Collection on human reproduction 47632

International Women's Air and Space Museum, 1501 N Marginal Rd, Cleveland, OH 44114 • T: +1 216 6231111 • F: +1 216 6231113 • cluhta@iwasm.org • www.iwasm.org •
Pres.: *Caroline Luhta* •
Science&Tech Museum – 1976
History of women in air and space, photos, uniforms, clothing and costumes, personal papers 47633

Museum of Contemporary Art, 8501 Carnegie Av, Cleveland, OH 44106 • T: +1 216 4218671 • F: +1 216 4210737 • info@mocacleveland.org • www.contemporaryart.org •
Dir.: *Jill Snyder* • Cur.: *Margo Crutchfield* •
Fine Arts Museum – 1968
Visual arts, changing exhibitions 47634

NASA Glenn Research Center's Visitor Center, 21000 Brookpark Rd, MS 8-1, Cleveland, OH 44135 • T: +1 216 4332846 • F: +1 216 4339834 • david.lowenfeld@lere.nasa.gov •
Man.: *David Lowenfeld* •
Science&Tech Museum – 1976 47635

Oberlin College Gallery, 1305 Euclid Av, Cleveland, OH 44115 • T: +1 216 8610748 • www.oberlin.edu/herehere •
Public Gallery
Contemporary art, sculpture, installations 47636

Reinberger Galleries, Cleveland Institute of Art, 11141 East Blvd, Cleveland, OH 44106 • T: +1 216 4217407 • F: +1 216 4217438 • www.cia.edu •
Dir.: *Bruce Checefsky* •
Public Gallery
Exhibitions of contemporary art 47637

Rock 'n' Roll Hall of Fame and Museum, 1 Key Plaza, Cleveland, OH 44114-1022 • T: +1 216 7817625 • F: +1 216 7811832 • director@rockhall.org • www.rockhall.com •
Chief Cur.: *James Henke* •
Music Museum – 1985
Rock 'n' Roll music 47638

Romanian Ethnic Art Museum, 3256 Warren Rd, Cleveland, OH 44111 • T: +1 216 9415550 • F: +1 216 9413068 • dobreaart@aol.com • www.smroc.org •
Dir. & Pres.: *George Dobrea* •
Ethnology Museum – 1960
Folk art, Romanian art, costumes, ceramics, painters, rugs, silver & woodwork, 47639

Spaces, 2220 Superior Viaduct, Cleveland, OH 44113 • T: +1 216 6212314 • F: +1 216 6212314 • spaces@apk.net • www.spacesgallery.org •
Dir.: *Susan R Channing* •
Public Gallery
Special exhibitions of contemporary artists 47640

Steamship William G. Mather Museum, 1001 E Ninth St Pier, Cleveland, OH 44114 • T: +1 216 5749053 • F: +1 216 5742536 • wgmather@aol.com • www.wgmather.org •
Exec.Dir.: *Holly Holcombe* • Cur.: *Loucinda Holt* •
Science&Tech Museum – 1991
Freighter and flagship, Great Lakes shipping 47641

The Temple Museum of Religious Art, 1855 Ansel Rd, Cleveland, OH 44106 • T: +1 216 8313233 • F: +1 216 8314216 • www.ttti.org •
C.E.O.: *Jeremy Handler* • Dir.: *Sue Koletsky* •
Religious Arts Museum – 1950
Religious objects, ritual slver, paintings, sculpture, graphics, decorative art 47642

Ukrainian Museum, 1202 Kenilworth Av, Cleveland, OH 44113 • T: +1 216 7814329 • F: +1 216 7815844 • staff@umacleveland.org • www.umacleveland.org •
C.E.O.: *Andrew Fedynsky* •
Fine Arts Museum / Folklore Museum / Historical Museum – 1952
Ukrainian culture and history, artifacts, documents, books, art 47643

Western Reserve Historical Society Museum, 10825 East Blvd, Cleveland, OH 44106-1788 • T: +1 216 7215722 • F: +1 216 7218934 • www.wrhs.org •
Exec.Dir.: *Patrick H. Reymann* •
Historical Museum – 1867
Costumes, textiles, glassware and porcelain, early aircraft, 1895-1976 automobiles, 1770-1920 furnished rooms, hist buildings 47644

Cleveland TN

Red Clay State Historical Park, 1140 Red Clay Park Rd, SW, Cleveland, TN 37311 • T: +1 423 4780339 •
Head: *Lois I. Osborne* •
Historic Site – 1979
Historic site, seat of the former Cherokee government and site of eleven general councils national affairs 47645

Clewiston FL

Clewiston Museum, 112 S Commercio St, Clewiston, FL 33440-3706 • T: +1 941 9832870 • www.clewiston.org/museum.htm •
Dir./Cur.: *Gretchen C. DeBree* •
Local Museum – 1984 47646

Clifton KS

Clifton Community Historical Society, 108 Clifton St, Clifton, KS 66937-0223 •
Association with Coll – 1976
Military and scout clothing, photos, schools records, China, tools 47647

Clifton NJ

Hamilton van Wogener Museum, 971 Valley Rd, Clifton, NJ 07013 • T: +1 973 7445707 • F: +1 973 7445732 • hamilton971@att.net •
Cur.: *Genevieve Generalli* •
Local Museum – 1974
Local hist, period furnishings of the 19th c 47648

Clifton Forge VA

Alleghany Highlands Arts and Crafts Center, 439 E Ridgeway St, Clifton Forge, VA 24422 • T: +1 540 8624447 •
Pres.: *Annette Anderson* • Exec. Dir.: *Nancy Newhard-Farrar* •
Decorative Arts Museum – 1984
Works produced by Highlands and other regional artists 47649

Clinton IA

Clinton County Museum, 601 S First St, Clinton, IA 52732 • T: +1 563 2421201 •
Pres.: *Donald Dethmann* •
Local Museum – 1965 47650

River Arts Center, 229 Fifth Av S, Clinton, IA 52733-0132 • T: +1 319 2423300, 2428055 • gwenwes@webtv.net •
Dir.: *Gwen Chrest* •
Fine Arts Museum – 1968
Painting, beaded loin cloth, photographs, lithography, engraving, sculptures, etching, prints, pottery, fabric, pencil, wood 47651

Clinton IL

C.H. Moore Homestead, 219 E Woodlawn St, Clinton, IL 61727 • T: +1 217 9356066 • F: +1 217 9350553 • chmoure@davesword.net • www.chmoorehomestead.org •
C.E.O.: *Shirley Strange* •
Historic Site – 1967
General museum housed in C.H. Moore home, dolls, agriculture, railroad, period furnishing 47652

Clinton MD

Surratt House Museum, 9118 Brandywine Rd, Clinton, MD 20735 • T: +1 301 8681121 • F: +1 301 8688177 • laurie.verge@pgks.com • www.surratt.org •
Dir.: *Laurie Verge* •
Historical Museum – 1976
Historic house, Civil Warand Lincoln assassination 47653

Clinton MI

Wings of Love, 12803 E Michigan Av, Clinton, MI 49236 • T: +1 517 4568800 • F: +1 517 4568800 •
Dir.: *Dr. Khristina Smith-Speelman* •
Fine Arts Museum
Healing arts, photographs 47654

Clinton MO

Henry County Museum and Cultural Arts Center, 203 W Franklin St, Clinton, MO 64735 • T: +1 660 8858414 • F: +1 660 8902228 • hcmus@mid-america.net •
Dir.: *Alta Dulaban* •
Decorative Arts Museum / Local Museum – 1974
Architecture, drawings, archaeology, ethnology, costumes, decorative art, Eskimo art, dolls, Oriental, Persian and Greek antiquities 47655

Clinton MS

Mississippi Baptist Historical Commission, Mississippi College, Library-College St, Clinton, MS 39058 • T: +1 601 9253434 • F: +1 601 9253435 • mbhc@mc.edu • www.mc.edu •
C.E.O.: *Edward McMillan* •
Religious Arts Museum / University Museum – 1887
Baptist hist, old church minute books 47656

Clinton NJ

Hunterdon Museum of Art, 7 Lower Center St, Clinton, NJ 08809 • T: +1 908 7358415 • F: +1 908 7358416 • info@hunterdonmuseumofart.org • www.hunterdonmuseumofart.org •
Exec. Dir.: *Marjorie Frankel Nathanson* • Dir.: *Donna Gustafson* (Exhibitions) •
Fine Arts Museum – 1952
Print coll 47657

Red Mill Museum Village, 56 Main St, Clinton, NJ 08809-0005 • T: +1 908 7354101 • F: +1 908 7350914 • hhmredmill@yahoo.com •
Pres. & C.E.O.: *Jo-an' Van Doren* • Dir.: *Dr. Charles Speierl* •
Local Museum – 1960
18th - 19th and 20th c social, industrial and agriculture hist 47658

Clinton NY

Emerson Gallery, Hamilton College, 198 College Hill Rd, Clinton, NY 13323 • T: +1 315 8594396 • F: +1 315 8594687 • emerson@hamilton.edu • www.hamilton.edu/college/emerson_gallery •
Dir.: *David Nathans* • Cur.: *Bill Salzillo* (Hamilton Collections) • *Susanna White* •
Fine Arts Museum / University Museum – 1982
15th - 20th c works on paper, 19th and 20th c paintings, Naive American objects, Asian painting, sculpture and ceramics, Beinecke Coll of 16th - 19th c prints, drawings and maps of West Indies 47659

Kirkland Art Center, East Park Row, Clinton, NY 13323, mail addr: POB 213, Clinton, NY 13323 • T: +1 315 8538871 • F: +1 315 8532076 • kacinc@adelphia.net •
Public Gallery
Art 47660

Clinton OK

Oklahoma Route 66 Museum, Oklahoma Historical Society, 2229 Gary Blvd, Clinton, OK 73601 • T: +1 580 3237866 • F: +1 580 3232870 • rt66mus@ok-history.mus.ok.us • www.route66.org •
Dir.: *Pat A. Smith* • Cur.: *Andy Watson* •
Local Museum – 1967 47661

Clintonville WI

Four Wheel Drive Foundation, E 11th St, Clintonville, WI 54929 • T: +1 715 8232141 • F: +1 715 8235768 • rich@seagrave.com •
Science&Tech Museum – 1948
Racing and passenger cars, trucks, fire engines 47662

Cloquet MN

Carlton County Historical Museum and Heritage Center, 406 Cloquet Av, Cloquet, MN 55720-1750 • T: +1 218 8791938 • F: +1 218 8791938 • cchs@cpinternet.com • www.carltoncountyhs.org •
Dir.: *Marlene Wisuri* •
Historical Museum – 1949
Regional hist, Indians, pioneers, agriculture, dairy, railroad, lumbering, fires of 1918 47663

Cloudcroft NM

Sacramento Mountains Historical Museum, 1000 Hwy 82, Cloudcroft, NM 88317 • T: +1 505 6822932 •
C.E.O. & Pres.: *Patrick Rand* •
Local Museum – 1977
Newspaper, photographs, history of the Sacramento Mountains 47664

Cloutierville LA

The Kate Chopin House and Bayou Folk Museum, 243 Hwy 495, Cloutierville, LA 71416 • T: +1 318 3792233 • F: +1 318 3790055 •
Pres.: *Julia Hildebran* • Dir.: *Amanda Chenault* •
Special Museum – 1965
Home of Creole writer Kate Chopin from 1880-1884 47665

Clute TX

Brazosport Museum of Natural Science, 400 College Dr, Clute, TX 77531 • T: +1 979 2657831 • F: +1 979 2656022 • bmns@bcfas.org • bcfas.org/museum •
Natural History Museum – 1962
Archaeology, botany, geology, malacology, marine, mineralogy, paleontology, zoology 47666

Cochran GA

Peacock Gallery, Middle Georgia College, 103 Sarah Street, Russell Hall, Cochran, GA 31014, mail addr: 1100 Second St SE, Cochran, GA 31014 • T: +1 478 9343043 • F: +1 478 9343517 • cagnew@mgc.edu • www.mgc.edu/academic/humanities/art/gallery.cfm •
Public Gallery / University Museum – 1970
Students artworks 47667

Cockeysville MD

Baltimore County Historical Museum, 9811 Van Buren Ln, Cockeysville, MD 21030 • T: +1 410 6661876 and 1878 • F: +1 410 6665276 • bchistory@msn.com • www.hsobc.org •
Exec. Dir.: *Kristi Alexander* •
Local Museum – 1959
Regional history 47668

Cocoa FL

Brevard Museum, 2201 Michigan Av, Cocoa, FL 32926-5618 • T: +1 321 6321830 • F: +1 321 6317551 • bmhs@brevardmuseum.com • www.brevardmuseum.com •
Dir.: *JaNeen Kniprath Smith* • Cur.: *Kay Davis* (History) • *Winnifred Mendenhall* (Science) •
Local Museum / Natural History Museum – 1969
History, natural history 47669

Coconut Grove FL

Barnacle Historic State Park, 3485 Main Hwy, Coconut Grove, FL 33133 • T: +1 305 4489445 • F: +1 305 4487484 • www.floridastateparks.org/thebarnacle •

Historical Museum – 1973
Memorabilia of pioneer Commodore Ralph Middleton Munroe, marine, photographs, cameras, furniture 47670

Cody WY

Buffalo Bill Historical Center, 720 Sheridan Av, Cody, WY 82414 • T: +1 307 5874771 ext 0 • F: +1 307 5784066 • thomh@bbhc.org • www.bbhc.org •
Dir.: *Dr. Robert E. Shimp* • Assoc. Dir.: *Eugene Reber* • *Dr. Judi A. Winchester* (Buffalo Bill Museum) • *Simeon Stoddart* (Cody Firearms Museum) • *Dr. Sarah E. Boehme* (Whitney Gallery of Western Art) • *Emma Hansen* (Plains Indian Museum) • *Dr. Charles R. Preston* (Draper Museum of Natural History) •
Historical Museum / Fine Arts Museum – 1917
American history coll, Western memorabilia, posessions of Buffalo Bill Cody and his Wild West Show, Western art, – Plains Indian Museum, Draper Museum of Natural History 47671

Coeur d'Alene ID

Museum of North Idaho, 115 NW Blvd, Coeur d'Alene, ID 83814 • T: +1 208 6643448 • dd@museumni.org • www.museumni.org •
Dir.: *Dorothy Dahlgren* •
Local Museum – 1968
Region's hist, logging, mining and steamboat, textiles, firearms, Native American artifacts, hist of Fort Sherman 47672

Coffeyville KS

Brown Mansion, 2019 Walnut, Coffeyville, KS 67337 • T: +1 620 2510431 • F: +1 620 2515448 • chamber@coffeyville.com • www.coffeyville.com •
Pres.: *Lue Barndollar* •
Special Museum – 1904
Period furniture, carpets, china, silver; Tiffany 47673

Dalton Defenders Museum, 113 E 8th, Coffeyville, KS 67337 • T: +1 620 2515944 • F: +1 620 2515448 • chamber@coffeyville.com • www.coffeyville.com •
Pres.: *Lue Barndollar* •
Historical Museum – 1954
Mementos from the Dalton Raid and from Dalton Gang member 47674

Cohasset MA

Captain John Wilson House, 4 Elm St, Cohasset, MA 02025 • T: +1 781 3831434 • cohassethistory@yahoo.com • www.cohassethistoricalsociety.org •
Cur.: *David H. Wadsworth* •
Historical Museum – 1936
Decorative arts, costumes, maritimes – archive 47675

Maritime Museum, 4 Elm St, Cohasset, MA 02025 • T: +1 781 3831434 • cohassethistory@yahoo.com • www.cohassethistoricalsociety.org •
Cur.: *David H. Wadsworth* •
Special Museum – 1928
Maritime hist, shipwreck, lifesaving, ship building and models, fishing 47676

Pratt Building, 106 S Main St, Cohasset, MA 02025 • T: +1 781 3831434 • cohassethistory@yahoo.com • www.cohassethistoricalsociety.org •
Historical Museum – 1928 47677

South Shore Art Center, 119 Ripley St, Cohasset, MA 02025 • T: +1 781 3832787 • F: +1 781 3832964 • info@ssac.org • www.ssac.org •
Exec. Dir.: *Sarah Hannan* •
Public Gallery
Exhibits of contemporary paintings, prints and sculptures 47678

Cokato MN

Cokato Museum and Akerlund Photography Studio, 175 W Fourth St, Cokato, MN 55321-0686 • T: +1 320 2862427 • F: +1 320 2865876 • cokatomuseum@cmgate.com • www.cokato.mn.us/cmhs •
Dir.: *Mike Worcester* •
Local Museum / Fine Arts Museum – 1976
Businesses, Finnish culture, agriculture, immigration artifacts, period furnishings and women's clothing – library, photo archive 47679

U

Colby KS

The Prairie Museum of Art and History, 1905 S Franklin, Colby, KS 67701 • T: +1 913 4624590 • F: +1 785 4624592 • prairiem@colby.ixks.com • www.prairiemuseum.org •
Dir.: *Sue Ellen Taylor* •
Local Museum – 1959
Dolls, signed porcelain, silver, books, textiles – archives 47680

Cold Spring NY

Putnam County Historical Society and Foundry School Museum, 63 Chestnut St, Cold Spring, NY 10516 • T: +1 845 2654010 • F: +1 845 2652884 • office@pchs-fsm.org • www.pchs-fsm.org •
Exec.Dir.: *Shannon M. Risk* • Cur.: *Trudie Alexis Grace* •
Local Museum – 1906
Hist and articles of the West Point Foundry, 19th c paintings, genealogy records 47681

Cold Spring Harbor NY

Cold Spring Harbor Whaling Museum, Main St, Cold Spring Harbor, NY 11724 • T: +1 516 3673418 • F: +1 516 6927037 • cshwm@optonline.net • www.cshwhalingmuseum.org •
Exec.Dir.: *Paul B. DeOrsay* • Cur.: *Daniel S. Trachtenberg* •
Special Museum – 1936
Scrimshaw coll, ship models, maritime paintings, 19thc whaleboat, whaling implements, Long Island whaling industry 47682

SPLIA Gallery, 161 Main St, Cold Spring Harbor, NY 11724 • T: +1 631 6924664 • F: +1 631 6925265 • splia@aol.com • splia.org •
Dir.: *Robert B. MacKay* •
Local Museum – 1948
Decorative arts, textiles, ceramics, glass, photos, paintings, prints, manuscripts 47683

Colfax IA

Trainland U.S.A., 3135 Hwy 117 N, Colfax, IA 50054 • T: +1 515 6743813 • F: +1 515 6743813 • judy@trainland-usa.com • www.trainlandusa.com •
Pres.: *Leland Atwood* •
Decorative Arts Museum – 1981
Toys 47684

College Park MD

The Art Gallery, University of Maryland, 1202 Art-Sociology Bldg, College Park, MD 20742 • T: +1 301 4052763 • F: +1 301 3147774 • theartgallery@umd.edu • www.artgallery.umd.edu •
Dir.: *Scott Habes* •
Fine Arts Museum / University Museum – 1966
20th c American paintings and prints, African art, Japanese prints, photography 47685

College Park Aviation Museum, 1985 Cpl. Frank Scott Dr, College Park, MD 20740-3836 • T: +1 301 8646029 • F: +1 301 9276472 • aviationmuseum@pgparks.com • www.collegeparkaviationmuseum.com •
Dir.: *Catherine Allen* • Cur.: *Anne Smallman* •
Science&Tech Museum / Historical Museum – 1982
Aviation, oldest continually operating airport in the world 47686

University College Gallery, University of Maryland, University Blvd, College Park, MD 20742-1600 • T: +1 301 9857822 • F: +1 301 9857678 •
Dir.: *Diana Crosson* •
University Museum
Temporary exhibitions 47687

College Station TX

George Bush Presidential Library and Museum, 1000 George Bush Dr W, College Station, TX 77845-3906 • T: +1 979 2609552 • F: +1 979 2609557 • library@bush.nara.gov • bushlibrary.tamu.edu •
C.E.O.: *Dr. David E. Alsobrook* •
Historical Museum – 1992
Personal and family memorabilia, 80.000 artifacts 47688

J. Wayne Stark University Center Galleries, Texas A&M University, Mem. Student Center, Rm 193, College Station, TX 77844 • T: +1 979 8456081 • F: +1 979 8623381 • uart@stark.tamu.edu • stark.tamu.edu •
Cur.: *Catherine A. Hastedt* •
Public Gallery / University Museum – 1973
Works by Texas artists of the 20th c, incl E.M "Buck" Schiwetz, Micheal Frary, Russell Waterhouse 47689

MSCC Forsyth Center Galleries, Texas A & M University, Room 227, Memorial Student Center, Joe Routt Blvd, College Station, TX 77843-1237 • T: +1 979 8459251 • F: +1 979 8455117 • ncurtis@mscc.tamu.edu • www.forsyth.tamu.edu •
Dir.: *Nan Curtis* •
Fine Arts Museum / Decorative Arts Museum / University Museum – 1989
19th c English and French Cameo Glass; Mt. Washington/Pairpoint Glass; Paperweights; American Cut Glass; Tiffany Glass; American Impressionist paintings; American Southwest paintings; The Eight; American Quilts; Guatemalan Maya Textiles 47690

Collegeville PA

Philip and Muriel Berman Museum of Art, Ursinus College, 601 East Main St, Collegeville, PA 19426-1000 • T: +1 610 4093500 • F: +1 610 4093664 • lhanover@ursinus.edu • www.ursinus.edu •
Dir./Cur.: *Lisa Tremper Hanover* • Head: *Andrea Cooper* (Collections) •
Fine Arts Museum / University Museum – 1987
Art 47691

Collierville TN

Biblical Resource Center & Museum, 140 Mulberry E, Collierville, TN 38017 • T: +1 901 8459578 • F: +1 901 8549883 • info@biblical-museum.org • www.biblical-museum.org •
C.E.O.: *Donald E. Bassett* • Dir.: *Nancy W. Bassett* •
Religious Arts Museum / Archaeology Museum
Biblical, Egyptian & ancient Near Eastern archaeological artifacts – library, theater, bible lab 47692

Collingswood NJ

Film Forum - Film Archives, 579A Haddon Av, Collingswood, NJ 08108-1445 • T: +1 856 8543221 •
Dir.: *David J. Grossman* •
Performing Arts Museum – 1989
Color design, social issues on film 47693

Collinsville IL

Cahokia Mounds, 30 Ramey St, Collinsville, IL 62234 • T: +1 618 3465160 • F: +1 618 3465162 • cahokiamounds@ezl.com • www.cahokiamounds.com •
C.E.O.: *Cheryl Jett* •
Ethnology Museum / Archaeology Museum – 1930
Archaeology, 800-1500 A.D., site of largest prehistoric Indian city in North America 47694

Collinsville OK

Collinsville Depot Museum, 115 S 10th St, Collinsville, OK 74021 • T: +1 918 3713540 •
Pres.: *William T. Thomas* •
Historical Museum – 1975
Early 1800 living room and kitchen; local and railroad artifacts; WW I & II uniforms 47695

Coloma CA

Gold Discovery Museum, c/o Marshall Gold Discovery State Historic Park, 310 Back St, Coloma, CA 95613 • T: +1 530 6223470 • www.parks.ca.gov •
Pres.: *Conrad Tracy* •
Historical Museum – 1890 47696

Colorado City TX

Heart of West Texas Museum, 340 E 3rd St, Colorado City, TX 79512 • T: +1 915 7288285 • F: +1 915 7288944 • ccmuseum@wtxs.net • www.coloradocity.org/MuseumHeartOfWestTexas •
Pres.: *Jay McCollum* • Dir.: *Beth Purcell* •
Local Museum – 1960
Pioneer memorabiblia, old coaches, period artifacts, china 47697

Colorado Springs CO

Carriage House Museum, 10 Lake Circle, Colorado Springs, CO 80906 • T: +1 719 6347711 ext 5353 •
C.E.O: *Robert J. Hiller* • Cur.: *Franeis Pittz* •
Science&Tech Museum – 1941
Transportation 47698

Colorado Springs Fine Arts Center, Taylor Museum, 30 W Dale St, Colorado Springs, CO 80903 • T: +1 719 6345581 • F: +1 719 6340570 • info@csfineartscenter.org • www.csfineartscenter.org •
Pres./C. E. O: *Michael DeMarsche* •
Fine Arts Museum – 1936
Historic and contemporary art, Southwestern U.S. culture 47699

Colorado Springs Museum, 215 S Tejon, Colorado Springs, CO 80903 • T: +1 719 3855990 • F: +1 719 3855645 • cosmuseum@springsgov.com • www.cspm.org •
Dir.: *Matthew Mayberry* • Cur.: *Katie Davis Gardner* •
Local Museum – 1937
Local history 47700

Gallery of Contemporary Art, University of Colorado, 1420 Austin Bluffs Pkwy, Colorado Springs, CO 80918 • T: +1 719 2623567 • F: +1 719 2623183 • griggs@uccs.edu • www.galleryuccs.org •
Dir.: *Gerry Riggs* •
Fine Arts Museum / University Museum – 1981
Contemporary art 47701

McAllister House Museum, 423 N Cascade Av, Colorado Springs, CO 80903 • T: +1 719 6357925 • F: +1 719 5285869 • harrisonbl@aol.com • oldcolo.com/~mcallister •
Cur.: *Patric Fox* • *Barbara Gately* •
Historical Museum – 1961
Major Henry McAllister home, Victorian furnishings 47702

May Natural History Museum and Museum of Space Exploration, 710 Rock Creek Canyon Rd, Colorado Springs, CO 80926-9799 • T: +1 719 5760450 • F: +1 719 5763644 • maymusem2001@yahoo.com • www.maymuseum-camp-rvpark.com •
Pres./ Dir.: *John M. May* •
Science&Tech Museum / Natural History Museum – 1941
Natural history, entomology, space exploration 47703

Museum of the American Cowboy, 101 Pro Rodeo Dr, Colorado Springs, CO 80907 • T: +1 719 5484874 • phildebrand@prorodeo.com • www.prorodeo.com •
C.E.O.: *Steven J. Hatchell* •
Special Museum
Rodeo hist 47704

Museum of the American Numismatic Association, 818 N Cascade Av, Colorado Springs, CO 80903 • T: +1 719 6322646 • F: +1 719 6344085 • ana@money.org • www.money.org •
Dir.: *Christopher Cipoletti* • Cur.: *Douglas Mudd* •
Special Museum – 1967
Numismatics 47705

Pro Rodeo Hall of Fame and Museum of the American Cowboy, 101 Pro Rodeo Dr, Colorado Springs, CO 80919 • T: +1 719 5284761 • F: +1 719 5484874 • phildebrand@prorodeo.com • www.prorodeo.com •
C.E.O.: *Steven J. Hatchell* •
Special Museum – 1979
Rodeo and Cowboy history 47706

Western Museum of Mining and Industry, 125 N Gate Rd, Colorado Springs, CO 80921 • T: +1 719 4880880 • F: +1 719 4889261 • westernmuseum@aol.com • www.wmmi.org •
Dir.: *Linda D. LeMieux* • Cur.: *Terry Girouard* •
Science&Tech Museum – 1970
Mining, technology history 47707

World Figure Skating Museum and Hall of Fame, 20 First St, Colorado Springs, CO 80906 • T: +1 719 6355200 • F: +1 719 6359548 • museum@usfsa.org • www.worldskatingmuseum.org •
Dir.: *John LeFevre* • Cur.: *Beth Davis* •
Special Museum – 1964
Sports 47708

Colter Bay Village WY

Colter Bay Indian Arts Museum, Visitor Center, Grand Teton National Park, Colter Bay Village, WY 83012 • T: +1 307 7393399 • F: +1 307 7393504 • danna_kinsey@nps.gov • www.nps.gov/grte •
Cur.: *Danna Kinsey* •
Fine Arts Museum – 1972 47709

Colton CA

Colton Area Museum, 380 N La Cadena Dr, Colton, CA 92324, mail addr: POB 1648, Colton, CA 92324 • T: +1 909 8248814 • info@coltonmuseum.net • www.coltonmuseum.net •
Pres.: *Patricia S. Barlow* •
Local Museum – 1984
Indian artifacts, the Earps, local hist, citrus ind, Southern Pacific Railroad 47710

Colton Point MD

Saint Clements Island-Potomac River Museum, 38370 Pointbreeze Rd, Colton Point, MD 20626 • T: +1 301 7692222 • F: +1 301 7692225 • mdbeginshere@excite.com • www.co.saint-marys.md.us/recreate/museums/stclementsisland.asp •
Dir.: *Debra L. Pence* • Ass. Dir.: *Lydia Wood* •
Historic Site / Historical Museum / Archaeology Museum – 1975
Archaeology, history, 1634 landing site of Maryland colonists, first Roman Catholic mass in English colonies 47711

Columbia MD

African Art Museum of Maryland, 5430 Vantage Point Rd, Columbia, MD 21044-0105 • T: +1 410 7307105 • F: +1 410 7153047 • africanartmuseum@aol.com • www.africanartmuseum.org •
Dir.: *Doris H. Ligon* •
Fine Arts Museum / Ethnology Museum – 1980
African art 47712

Howard County Center of African American Culture, 5434 Vantage Point Rd, Columbia, MD 21044 • T: +1 410 7151921 • F: +1 410 7158755 • hccaacmd@Juno.com •
Dir.: *Wylene Sims-Burch* •
Historical Museum – 1987
Artifacts, murals 47713

Columbia MO

Bingham Gallery, University of Missouri, A126 Fine Arts Center, Columbia, MO 65211-6090 • T: +1 573 8823555 • F: +1 314 8846807 • binghamgallery@missouri.edu • binghamgallery.missouri.edu •
Public Gallery / University Museum 47714

Davis Art Gallery, Stephens College, Columbia, MO 65215 • T: +1 573 8764267 • F: +1 573 8767248 • irene@stephens.edu • www.stephens.edu •
C.E.O: *Robert Friedman* • Cur.: *Irene O. Alexander* •
Public Gallery – 1962
Modern paintings & graphics, Melanesian sculpture 47715

Lewis, James and Nellie Stratton Gallery, Stephens College, 1200 E Broadway, Columbia, MO 65215 • T: +1 573 8767267 • F: +1 573 8767248 • www.stephens.edu •
Dir.: *Robert Friedman* • Cur.: *Irene O. Alexander* •
Fine Arts Museum
Modern graphics and paintings, primitive sculpture 47716

Museum of Anthropology, University of Missouri, 104 Swallow Hall, Columbia, MO 65211 • T: +1 573 8823573 • F: +1 573 8841435 • frenchme@missouri.edu • coas.missouri.edu/anthromuseum •
Dir.: *Michael J. O'Brien* • Cur.: *Mary French* •
University Museum / Ethnology Museum – 1949
Prehistoric Missouri archaeological artifacts 47717

Museum of Art and Archaeology, University of Missouri-Columbia, 1 Pickard Hall, University Av and S 9th St, Columbia, MO 65211 • T: +1 573 8823591 • F: +1 573 8844039 •
Dir.: *Marlene Perchinske* • Assoc. Cur.: *Joan Stack* (European and American Art) •
Fine Arts Museum / University Museum / Archaeology Museum – 1957
Ancient art & archaeology incl Egypt, Palestine, Iran, Cyprus, Greece, Etruria, rome early Christian & Byzantine art 47718

State Historical Society of Missouri, 1020 Lowry St, Columbia, MO 65201-7298 • T: +1 573 8827083 • F: +1 573 8844950 • shsofmo@umsystem.edu • www.system.missouri.edu/shs •
Dir.: *Dr. James W. Goodrich* • Cur.: *Sidney Larson* •
Fine Arts Museum – 1898
Paintings, sculpture, drawings, prints, photographs, maps, cartoon ... 47719

Columbia PA

National Watch and Clock Museum, 514 Poplar St, Columbia, PA 17512-2130 • T: +1 717 6848261 • F: +1 717 6840878 • charris@nawcc.org • www.nawcc.org •
Dir.: *Carolin M. Stuckert* • Cur.: *Carter Harris* • Libr.: *Sharon Gordon* •
Science&Tech Museum – 1971
Clocks, watches, horological tools ... 47720

Wright's Ferry Mansion, 38 S Second St, Columbia, PA 17512 • T: +1 717 6844325 •
Dir.: *Thomas Cook* • Cur.: *Elizabeth Meg Schaefer* •
Local Museum – 1974
Philadelphia furniture, English ceramics, glass and textiles ... 47721

Columbia SC

Columbia Fire Department Museum, 1800 Laurel St, Columbia, SC 29201 • T: +1 803 7338350 • F: +1 803 7338311 • www.columbia.net/city/fire.htm •
C.E.O.: *John D. Jansen jr* •
Historical Museum – 1996
Metropolitan horse drawn steamer, hist of fire service ... 47722

Columbia Museum of Art, Main and Hampton Sts, Columbia, SC 29201, mail addr: POB 2068, Columbia, SC 29202 • T: +1 803 7992810 • F: +1 803 3432150 • rshirley@colmusart.org • www.columbiamuseum.org •
Pres.: *Carroll Heyward* •
Fine Arts Museum – 1950
Fine and decorative art, photography – Kress Collection of Renaissance and Baroque old masters ... 47723

Edventure, 211 Gervais St, Columbia, SC 29201 • T: +1 803 7793100 ext 10 • F: +1 803 7793144 • info@edventure.org • www.edventure.org •
Natural History Museum / Science&Tech Museum – 1994 ... 47724

Governor's Mansion, 800 Richland St, Columbia, SC 29201 • T: +1 803 7371710 • F: +1 803 7373860 • nbunch@gov.state.us • www.scgovernorsmansion.org •
Dir.: *Ryan Zimmerman* • Cur.: *Nancy B. Bunch* •
Historic Site – 1855
Historic house ... 47725

Hampton-Preston Mansion, 1615 Blanding St, Columbia, SC 29201 • T: +1 803 2521770 • F: +1 803 9297695 •
Pres.: *Dr. John G. Sproat* • Exec. Dir.: *Arrington Cox* •
Historical Museum – 1961
Historic house ... 47726

Historic Columbia Foundation, 1601 Richland St, Columbia, SC 29201 • T: +1 803 2527742 • F: +1 803 9297695 • rwaites@historiccolumbia.org • www.historiccolumbia.org •
Pres.: *Belinda Gergel* • Dir.: *Robin Waites* •
Local Museum – 1961
Local hist ... 47727

McKissick Museum, University of South Carolina, Pendleton and Bull Sts, Columbia, SC 29208 • T: +1 803 7777251 • F: +1 803 7772829 • robertsonl@gwm.sc.edu • www.cla.sc.edu/mcks/index.html •
Dir.: *Lynn Robertson* • Chief Cur.: *Jay Williams* (Exhibitions) • *Saddler Taylor* (Folklife and Research) • Cur.: *Jason Shaiman* (Traveling Exhibitions) • *Karen Swager* (Collections) • *Alice Booknight* (Education) •
Folklore Museum / University Museum – 1976
Folk art, natural sciences, material culture ... 47728

Mann-Simons Cottage, 1403 Richland St, Columbia, SC 29201 • T: +1 803 2521770 • F: +1 803 2525001 •
Pres.: *Dr. John G. Sproat* • C.E.O.: *Arrington Cox* •
Historical Museum – 1961
House dedicated to free African American Celia Mann and her family who lived here until 1970 ... 47729

Robert Mills Historic House and Park, Historic Columbia Foundation, 1601 Richland St, Columbia, SC 29201 • T: +1 803 2521770 • F: +1 803 9297695 •
Pres.: *Roger D. Poston* • Exec. Dir.: *Cynthia Moses Nesmith* •
Local Museum – 1961
Local hist ... 47730

South Carolina Confederate Relic Room and Museum, 301 Gervais St, Columbia, SC 29201 • T: +1 803 7378095 • F: +1 803 7378099 • sschoon@crr.state.sc.us • www.crr.sc.gov •
Dir.: *W. Allen Roberson* •
Historical Museum – 1896
History ... 47731

South Carolina Law Enforcement Officers Hall of Fame, 5400 Broad River Rd, Columbia, SC 29212 • T: +1 803 8968199 • F: +1 803 8968067 • www.scdps.org/cja/halloffame •
Admin.: *Marsha T. Ardila* •
Historical Museum – 1979
Law enforcement artifacts ... 47732

South Carolina Military Museum, 1 National Guard Rd, Columbia, SC 29201-4766 • T: +1 803 8064440 • F: +1 803 8062103 •
Dir.: *Ewell G. Sturgis jr* •
Military Museum – 1982
SC military and National Guard, artillery ... 47733

South Carolina State Museum, 301 Gervais St, Columbia, SC 29201 • T: +1 803 8984921 • F: +1 803 8984969 • publicrelations@scmuseum.org • www.southcarolinastatemuseum.org •
Dir.: *William Calloway* • Cur.: *Fritz Hamer* (History) • *Jim Knight* (Natural History) • *Tom Falvey* (Science/Technology) • *Paul Matheny* (Art) • *Elaine Nichols* (Cultural History) •
Local Museum – 1973
Local hist, culture, art, natural history, science, technology ... 47734

Columbia TN

James K. Polk Ancestral Home, 301 W 7th St, Columbia, TN 38401, mail addr: POB 741, Columbia, TN 38402 • T: +1 931 3882354 • F: +1 931 3885971 • jameskpolk@bellsouth.net • www.jameskpolk.com •
Dir.: *John C. Holtzapple* • Cur.: *Thomas E. Price* •
Historic Site – 1924
Historic house ... 47735

Columbia City IN

Whitley County Historical Museum, 108 W Jefferson, Columbia City, IN 46725 • T: +1 219 2446372 • F: +1 219 2446384 • wcmuseum@whitleynet.org • historical.whitleynet.org/ •
Dir.: *Ruth Kirk* • Cur.: *Susan Richey* •
Local Museum – 1963
County history, genealogy, archaeology, pioneer ... 47736

Columbia Falls ME

Ruggles House, 1/4m off U.S. Rte 1, Columbia Falls, ME 04623 • T: +1 207 4834637, 5467429 • etenan@rugleshouse.com •
Pres.: *Linda Long* •
Historical Museum – 1949
Gnealogy, records, photographs, newfiles, maps, glass, childrens toys and furniture ... 47737

Columbiana OH

Log House Museum, 10 E Park Av, Columbiana, OH 44408 • T: +1 330 4822983 •
Cur.: *Nora Salmen* • Sc. Staff: *Ada Wilhelm* •
Local Museum – 1953
Civil War and WW I artifacts, local hist, period furniture, clothing, wooden tools, pottery ... 47738

Columbus GA

The Columbus Museum, 1251 Wynnton Rd, Columbus, GA 31906 • T: +1 706 6490713 • F: +1 706 6491070 • information@columbusmuseum.com • www.columbusmuseum.com •
Dir.: *Charles Thomas Butler* •
Fine Arts Museum / Historical Museum – 1952
Archaeology, costumes and textiles, decorative arts, paintings, photos, prints, drawings, graphics, sculpture, furnishings, personal artifacts ... 47739

Port Columbus National Civil War Naval Center, 1002 Victory Dr, Columbus, GA 31902 • T: +1 706 3247334 • F: +1 706 3247225 • cwnavy@portcolumbus.org • www.portcolumbus.org •
Exec. Dir.: *Bruce Smith* •
Military Museum – 1962
Civil War hist ... 47740

Columbus IN

Bartholomew County Historical Museum, 524 Third St, Columbus, IN 47201 • T: +1 812 3723541 • F: +1 812 3723113 • bchs@hsonline.net • www.barthist.com •
Pres.: *Chuck Baker* • Cur.: *Sandy Greenlee* •
Local Museum – 1921
Local History, genealogy ... 47741

Indianapolis Museum of Art - Columbus Gallery, 390 The Commons, Columbus, IN 47201-6764 • T: +1 812 3762597 • F: +1 812 3752724 • jhandley@imacg.org • www.artcom.com/Museums/newones/47201-41.htm •
Pres.: *Em Rodway* • Dir.: *Jerry Handley* •
Fine Arts Museum – 1974
Temporary exhibitions ... 47742

Columbus KY

Columbus-Belmont Civil War Museum, 350 Park Rd, Columbus, KY 42032 • T: +1 270 6772327 • F: +1 270 6774013 • cindy.lynch@mail.state.ky.us •
Head: *Cindy Lynch* •
Historical Museum / Military Museum – 1934
History, civil war, military ... 47743

Columbus MS

Blewitt-Harrison-Lee Museum, 316 Seventh St N, Columbus, MS 39701 • T: +1 662 3293533 •
Cur.: *Caroline Neault* •
Local Museum – 1960
Local artifacts, furniture, jewelry, pictures, portraits ... 47744

Fine Arts Gallery, c/o Mississippi University for Women, Fine Arts Bldg, 1100 College St, Columbus, MS 39701 • T: +1 662 3297341, 2416976 • F: +1 662 2417815 •
Dir.: *Lawrence Feeney* •
Fine Arts Museum / University Museum – 1948
American art, paintings, sculpture, photographs, drawings, ceramics, prints ... 47745

Florence McLeod Hazard Museum, 316 7th St N, Columbus, MS 39701 • T: +1 662 3278888 • www.historic-columbus.org •
Pres.: *Libba Johnson* • Dir.: *Heather Roland* •
Historical Museum – 1959
Clothing, portraits, silver, furniutre, china, glass, guns ... 47746

Mississippi University for Women Museum, Fant Library, Columbus, MS 39701 • T: +1 662 3297332 • F: +1 662 3297348 • pmattes@muw.edu • www.muw.edu •
University Museum – 1978 ... 47747

Columbus NJ

Georgetown Museum, 4 Fitzgerald Ln, Columbus, NJ 08022-2383 • T: +1 609 2984174 •
Local Museum – 1976
Township hist ... 47748

Columbus OH

ACME Art Gallery, 1129 N High St, Columbus, OH 43201 • T: +1 614 2994003 • www.acmeart.com •
Dir.: *Melesa Klosek* •
Fine Arts Museum ... 47749

Canzani Center Galleries, Columbus College of Art and Design, Cleveland Av, Gay St, Columbus, OH 43215 • T: +1 614 2224022 • communications@ccad.edu • www.ccad.edu •
Dir.: *Natalie Marsh* •
Public Gallery ... 47750

Columbus Cultural Arts Center, 139 W Main St, Columbus, OH 43215 • T: +1 614 6457047 • F: +1 614 6455862 • jljohnson@cmhmetro.net •
Pres.: *Richard Wissler* • Dir.: *Jennifer L. Johnson* •
Fine Arts Museum / Folklore Museum / Historical Museum – 1978
Paintings ... 47751

Columbus Museum of Art, 480 E Broad St, Columbus, OH 43215 • T: +1 614 2216801 • F: +1 614 2210226 • info@cmaohio.org • www.columbusmuseum.org •
Pres.: *Jeff Edwards* • Exec. Dir.: *Nannette V. Maciejunes* • Chief Cur.: *Catherine Evans* (Photography) • Cur.: *Annegreth T. Nill* (20th-C European and Contemporary Art) • *Mark Cole* (American Art) •
Fine Arts Museum – 1878
Coll of European and American modernism, photography, 19th c American textiles, contemp art, 20th c folk art, ... 47752

COSI Columbus, 333 W Broad St, Columbus, OH 43215-2738 • T: +1 614 2282674 • F: +1 614 2286363 • cosi@cosi.org • www.cosi.org •
Pres.: *Kathryn D. Sullivan* •
Science&Tech Museum – 1964
Science, technology, industry, hist ... 47753

Historic Costume & Textiles Collection, Ohio State University, College of Human Ecology, 1787 Neil Av, Columbus, OH 43210-1295 • T: +1 614 2923090 • F: +1 614 6888133 • costume@osu.edu • www.hec.ohio-state.edu/cts/collect/buckel •
Cur.: *Gayle Strege* •
Decorative Arts Museum ... 47754

Kelton House Museum, 586 E Town St, Columbus, OH 43215-4888 • T: +1 614 4642022 • F: +1 614 4643346 • Keltonhouse@cs.com • www.keltonhouse.com •
Cur.: *Joseph Lowe* •
Folklore Museum – 1976
Historic house, period furniture, china and ceramics, decorative arts, music boxes, Victorian toys and clothing ... 47755

Northwood Art Space, 2231 N High St, Columbus, OH 43201 • T: +1 614 2945113 • F: +1 614 2945114 •
Dir: *Dianne Efsic* •
Public Gallery
Paintings, drawings, prints, ceramics and installations of local artists ... 47756

Ohio Historical Society Museum, 1982 Velma Av, Columbus, OH 43211-2497 • T: +1 614 2972300 • F: +1 614 2972318 • ohsref@ohiohistory.org • www.ohiohistory.org •
C.E.O.: *William K. Laidlaw* • Cur.: *Martha Otto* (Archaeology) • *William Gates* (History) • *Robert Glotzhober* (Natural History) •
Historical Museum – 1885
Archaeology, history, natural history ... 47757

Orton Geological Museum, Ohio State University, 155 S Oval Mall, Columbus, OH 43210 • T: +1 614 2926896 • F: +1 614 2921496 • gnidovec.1@osu.edu •
Dir.: *Prof. Bill Ausich* • Cur.: *Dale Gnidovec* •
University Museum / Natural History Museum – 1892
Fossils, minerals, rocks, meteorites ... 47758

Riffe Gallery, Ohio Arts Council, 77 S High St, Columbus, OH 43215 • T: +1 614 6449624 • F: +1 614 7283939 • mary.gray@oac.state.oh.us •
Dir.: *Mary H. Gray* •
Public Gallery
Changing exhibitions ... 47759

Schumacher Gallery, Capital University, 2199 E Main St, Columbus, OH 43209 • T: +1 614 2366319 • F: +1 614 2366490 • www.capital.edu/cc/gall/index.shtml •
Dir.: *Dr. Cassandra Tellier* •
Fine Arts Museum / University Museum – 1964
Asian art, inuit art, ethnic arts, cont painting and sculpture, graphics ... 47760

Wexner Center for the Arts, Ohio State University, 1871 N High St, Columbus, OH 43210-1393 • T: +1 614 2920330 • F: +1 614 2923369 • wexner@cgrg.ohio-stete.edu • www.wexarts.org •
Dir.: *Sherri Geldin* • Cur.: *Jeffrey Kipnis* (Exhibitions) • *Charles Helm* (Performing Arts) • *Patricia Trumps* (Education) • *William Horrigan* (Media Arts) •
Fine Arts Museum / University Museum / Performing Arts Museum – 1989 ... 47761

Comfrey MN

Jeffers Petroglyphs Historic Site, 27160 County Rd 2, Comfrey, MN 56019 • T: +1 507 6285591 • F: +1 507 6285593 • www.jefferspetrolyphs.com •
Historic Site / Historical Museum – 1966
American Indian hist and spirituality ... 47762

Commack NY

Long Island Culture History Museum, Hoyt Farm Park, New Hwy, Commack, NY 11725, mail addr: POB 1542, Stony Brook, NY 11790 • T: +1 631 9298725 • F: +1 631 9296967 • gaystone@optonline.net •
Dir.: *Gaynell Stone* •
Historical Museum / Archaeology Museum / Open Air Museum – 1986
History, archaeology ... 47763

Commerce TX

University Gallery, Texas A & M University, POB 3011, Commerce, TX 75429 • T: +1 903 8865208 • F: +1 214 8865415 • BarbaraFrey@tamu-commerce.edu •
Dir.: *Barbara Frey* •
Fine Arts Museum – 1979 ... 47764

Comstock TX

Seminole Canyon State Historical Park, U.S. Hwy 90 West, Comstock, TX 78837 • T: +1 432 2924464 • F: +1 432 2924596 • www.tpwd.state.tx.us •
Archaeology Museum – 1980
Archaeology site, containing 4,000-year old pictographs ... 47765

Concord MA

Concord Art Association, 37 Lexington Rd, Concord, MA 01742 • T: +1 508 3692578 • F: +1 508 3712496 • gallery@concordart.org • www.concordart.org •
Exec. Dir.: *Deborah Disston* •
Fine Arts Museum – 1922
Early American portraits, paintings, sculpture, graphics, decorative arts – sculpture garden ... 47766

Concord Museum, 200 Lexington Rd, Concord, MA 01742 • T: +1 978 3699763 • F: +1 978 3699660 • cm1@concordmuseum.org • www.concordmuseum.org •
Exec. Dir.: *Desiree Caldwell* • Cur.: *David Wood* •
Local Museum / Decorative Arts Museum – 1886
Local history, American decorative arts, paintings, graphics – library ... 47767

Minute Man, National Historical Park, 174 Liberty St, Concord, MA 01742 • T: +1 978 3696993 • F: +1 978 3187800 • mima_info@nps.gov • www.nps.gov/mima •
Cur.: *Teresa Wallace* •
Historical Museum / Military Museum / Open Air Museum / Historic Site – 1959
Battles of Lexington and Concord, Alcott and Lothrop family home, New England literature 19th c ... 47768

The Old Manse, 269 Monument St, Concord, MA 01742 • T: +1 978 3693909 • F: +1 978 2876154 • oldmanse@ttor.org • www.thetrustees.org/pages/346_old_manse.cfm •
Man.: *Deborah Kreiser-Francis* •
Historical Museum / Military Museum – 1939
Site of the 1st major skirmish of the Revolutionary War 1775, home to Ralph Waldo Emerson and Nathaniel Hawthorne ... 47769

Orchard House, 399 Lexington Rd, Concord, MA 01742 • T: +1 978 3694118 • F: +1 978 3691367 • info@louisamayalcott.org • www.louisamayalcott.org •
Dir.: *Jan Turnquist* •
Special Museum – 1911
House where Louisa May Alcott wrote Little Women, Bronson Alcott's School of Philosophy, American history, art ... 47770

Ralph Waldo Emerson House, 28 Cambridge Turnpike, Concord, MA 01742 • T: +1 978 3692236 • www.fiddlersgreen.net/buildings/new-england/emerson-hse/info/info.htm •
Dir.: *Barbara Mongan* •
Special Museum – 1930
Home of the essayist, philosopher and poet ... 47771

U

Concord MI

Historic Mann House, 205 Hanover St, Concord, MI 49237 • T: +1 517 5248943, 3731979 • F: +1 517 3737597 • www.michiganhistory.org •
Historian: *Patrick Murphy* •
Historical Museum – 1970
Decorative arts, furnishings 47772

Concord NH

Art Center In Hargate, Saint Paul's School, 325 Pleasant St, Concord, NH 03301 • T: +1 603 2294644 • F: +1 603 2295696 • ccallaha@sps.edu •
Dir.: *Colin J. Callahan* •
Fine Arts Museum – 1967
Drawings, graphics, paintings, sculpture, school gifts – Ohrstrom library 47773

McLane Center at Silk Farm, Audubon Society of New Hampshire, 3 Silk Farm Rd, Concord, NH 03301 • T: +1 603 2249909 • F: +1 603 2260902 • asnh@nhaudubon.org • www.nhaudubon.org •
C.E.O. & Pres.: *Richard A. Minard jr.* •
Association with Coll / Natural History Museum – 1914
Stuffed birds – Library 47774

Museum of New Hampshire History, 6 Eagle Sq, Concord, NH 03301 • T: +1 603 2286688 • F: +1 603 2286308 • jdesmarais@nhhistory.org • www.nhhistory.org •
Pres.: *Sherilyn B. Young* • Dir.: *Joan Desmarais* •
Decorative Arts Museum / Historical Museum – 1823
Decorative and fine arts, New Hampshire artifacts, furniture, silver, paintings, ceramics, costumes and textiles 47775

Pierce Manse, 14 Penacook St, Concord, NH 03301 • T: +1 603 2245959 •
C.E.O.: *Florence Holden* •
Historical Museum – 1966
1842-1848 home of President Franklin Pierce 47776

Concordia KS

Cloud County Historical Society Museum, 635 Broadway, Concordia, KS 66901 • T: +1 785 2432866 • cchsm@dustdevil.com •
Pres.: *Don Kaufmann* • Cur.: *Linda Palmquist* •
Local Museum – 1959
Prisoner of war camp, miltary casket flags, clothing, dolls, medical 47777

Condon OR

Depot Museum Complex, Gilliam County Historical Society, Hwy 19 at Burns Park, Condon, OR 97823 • T: +1 541 3844233 •
Pres.: *June Mikkalo* •
Local Museum – 1985
County hist, farm machinery 47778

Conneaut OH

Conneaut Railroad Museum, Depot St, Conneaut, OH 44030 • T: +1 440 5997878 •
Science&Tech Museum – 1962
Engines, hopper car, wood caboose, Red River Line, Ashtabula disaster 1876, models 47779

Connersville IN

Henry H. Blommel Historic Automotive Data Collection, 427 E County Rd, 215 S, Connersville, IN 47331 • T: +1 317 8259259 •
Cur.: *Henry H. Blommel* •
Science&Tech Museum – 1928
Automotive industry 47780

Constantine MI

John S. Barry Historical Society Museum, 300 N Washington, Constantine, MI 49042 • T: +1 616 4355825, 4355385 •
Local Museum – 1945
Local history, 1835-1847 Governor Barry house, needlework, lamps, dolls, musical instruments 47781

Conway AR

Baum Gallery, University of Central Arkansas, McCastlain Hall, L.A. Niven Dr, Conway, AR 72035, mail addr: 201 Donaghey Av, Conway, AR 72035 • T: +1 501 4505793 • barbaras@uca.edu • www.uca.edu/cfac/baum •
Dir./Cur.: *Barbara Satterfield* •
Public Gallery – 1994
Contemporary American and European art 47782

Faulkner County Museum, 805 Locust St, Conway, AR 72032 • T: +1 501 3295918 • fcm@conwaycorp.net • www.faulknercountymuseum.org •
Dir.: *Lyntia Langley-Ware* •
Local Museum – 1992
Local history, Indien artifacts, cadron settlement, agriculture, medicine 47783

Conway SC

Horry County Museum, 428 Main St, Conway, SC 29526 • T: +1 803 2481282 • F: +1 843 2481854 • hcmuseum@sccoast.net • www.horrycountymuseum.org •
C.E.O.: *Stewart J. Pabst* • Dep. Dir.: *Terri Hooks* •
Historical Museum / Ethnology Museum / Archaeology Museum – 1979
Anthropology, history, archaeology 47784

Cookeville TN

Cookeville Depot Museum, 116 W Broad St, Cookeville, TN 38501 • T: +1 931 5288570 • F: +1 931 5261167 • depotmuseum@frontiernet.net •
C.E.O.: *Judy Duke* •
Science&Tech Museum – 1984
Transportation, railroads 47785

Coolidge AZ

Casa Grande Ruins, 1100 W Ruins Dr, Coolidge, AZ 85228 • T: +1 520 7233172 • F: +1 520 7237209 • www.nps.gov/cagr •
Historic Site / Archaeology Museum – 1892
Archaeology, located on Hohokam village approx. 500-1450 AD 47786

Coolspring PA

Coolspring Power Museum, Main St, Coolspring, PA 15730 • T: +1 814 8496883 • F: +1 814 8495495 • gaild@penn.com • www.coolspringpowermuseum.org •
Dir.: *John Wilcox* • *Weldon Forman* • *Preston Foster* • *Edward Kuntz* • *Vance Packard* • Dir./Cur.: *Paul Harvey* •
Science&Tech Museum – 1985
Internal combustion engine technology, electric generators, oil field technology 47787

Cooperstown NY

The Farmers' Museum, Lake Rd, Cooperstown, NY 13326 • T: +1 607 5471450 • F: +1 607 5471499 • info@nysha.org • www.farmersmuseum.org •
Acting C.E.O. & Pres.: *Esther Nelson* • Chief Cur.: *Paul D'Ambrosio* • Cons.: *C.R. Jones* •
Open Air Museum / Agriculture Museum – 1943
Early 19th c Village, agricultural tools and implements 47788

Fenimore Art Museum, New York State Historical Associaton, Lake Rd, Cooperstown, NY 13326 • T: +1 607 5471400 • F: +1 607 5471404 • info@nysha.org • www.nysha.org •
Acting Pres. & C.E.O.: *Esther Nelson* • Chief Cur.: *Paul D'Ambrosio* • Cons.: *C.R. Jones* •
Fine Arts Museum – 1899
Fine and decorative arts, folk art, American Indian art 47789

Hyde Hall, 1 Mill Rd, Cooperstown, NY 13326 • T: +1 607 5475098 • F: +1 607 5478462 • hydehall@wpe.com • www.hydehall.org •
Pres.: *Christopher Holbrook* • Exec. Dir.: *Lorraine Joseph* •
Historical Museum – 1964
Family artifacts dating from 1820 47790

National Baseball Hall of Fame and Museum, 25 Main St, Cooperstown, NY 13326 • T: +1 607 5477200 • F: +1 607 5472044 • info@baseballhalloffame.org • www.baseballhalloffame.org •
Pres.: *Dale A. Petroskey* • Chief Cur.: *Ted Spencer* •
Cur.: *Peter P. Clark* (Collections) • Libr.: *Jim Gates* •
Special Museum – 1936 47791

The Smithy-Pioneer Gallery, 55 Pioneer St, Cooperstown, NY 13326 • T: +1 607 5478671 •
Public Gallery
Sculpture, art 47792

Coopersville MI

Coopersville Area Historical Society Museum, 363 Main St, Coopersville, MI 49404 • T: +1 616 8376978 • budphoto@i2k.com • www.coopersville.com/museum •
Cur.: *Jim Budzynski* •
Local Museum
Railroad, gold records, Del Shannon an early Rock-n-Roll star 47793

Coos Bay OR

Coos Art Museum, 235 Anderson Av, Coos Bay, OR 97420 • T: +1 541 2673901 • F: +1 541 2674877 • info@coosart.org • www.coosart.org •
Pres.: *Lynda Shapiro* •
Fine Arts Museum – 1950 47794

Copper Center AK

George I. Ashby Memorial Museum, c/o Copper Center Lodge, Mile 101 Old Richardson Hwy Loop Rd, Copper Center, AK 99573 • T: +1 907 8223245 • info@coppercenterlodge.com • www.coppercenterlodge.com •
Pres.: *Fred Williams* •
Local Museum 47795

Copper Harbor MI

Fort Wilkins Historic Complex, Fort Wilkins, Copper Harbor, MI 49918 • T: +1 906 2894215 • F: +1 906 2894939 • www.michiganhistory.org •
Local Museum – 1923
Military hist, coppermining, Pittsburgh and Boston Co. Mine, decorative arts, lighthouse – auditorium 47796

Coral Gables FL

Center for Visual Communication, 4021 Laguna St, Coral Gables, FL 33146 • T: +1 305 4466811 • F: +1 305 4463456 •
Public Gallery
Contemporary art photography 47797

College of Arts and Sciences Gallery, University of Miami, 1210 Stanford Dr, Coral Gables, FL 33146 • T: +1 305 2842792 • F: +1 305 2842115 • tracey@miami.edu • www.as.miami.edu/art •
Public Gallery
Art 47798

Coral Gables Merrick House, 907 Coral Way, Coral Gables, FL 33134 • T: +1 305 4605361, 4605095 • F: +1 305 4605097 • www.citybeautiful.net •
Decorative Arts Museum – 1976
Plantation house belonging to family of founder of Coral Gables 47799

Creatabilitoys! - Museum of Advertising Icons, 1550 Madruga Av, Coral Gables, FL 33146 • T: +1 305 6637374 • F: +1 305 6690092 • info@creatability.com • www.toymuseum.com •
Dir.: *Ritchie Lucas* • Cur.: *Carmen Rodriguez* •
Decorative Arts Museum – 1995
Toys and dolls 47800

The Florida Museum of Hispanic and Latin American Art, 4006 Aurora St, Coral Gables, FL 33126-1414 • T: +1 305 4447060 • F: +1 305 2616996 • Hispmuseum@aol.com •
Pres.: *Raul M. Oyuela* •
Fine Arts Museum – 1991
Art 47801

Lowe Art Museum, University of Miami, 1301 Stanford Dr, Coral Gables, FL 33124-6310 • T: +1 305 2843603 • F: +1 305 2842024 • bdursum@miami.edu • www.miami.edu/lowe •
Dir./Chief Cur.: *Brian A. Dursum* • Cur.: *Margaret Jackson* (Pre-Columbian Art) • *Marcilene Wittmer* (African Art) • *Perri L. Roberts* (Renaissance) •
Fine Arts Museum / University Museum – 1950 47802

Coral Springs FL

Coral Springs Museum of Art, 2855 Coral Springs Dr, Coral Springs, FL 33065 • T: +1 954 3405000 • F: +1 954 3464424 • ctbok@coral-springs.com • www.csmart.org •
Dir.: *Barbara K. O'Keefe* •
Fine Arts Museum – 1997
Visual arts of Florida residents 47803

Coralville IA

Johnson County Historical Society Museum, 310 Fifth St, Coralville, IA 52241 • T: +1 319 3515738 • F: +1 319 3515310 • joctyhistsociety@yahoo.com •
Pres.: *Kristin Summerwill* •
Local Museum – 1973
Local history 47804

Cordele GA

Georgia Veterans Memorial Museum, 2459-A Hwy 280 W, Cordele, GA 31015 • T: +1 912 2762371 • F: +1 912 2762372 • gavets@sowega.net •
Head: *Charles Luther* •
Military Museum – 1962
Historic aircraft, fighting vehicles, uniforms, weapons 47805

Cordova AK

Cordova Historical Museum, 622 First St, Cordova, AK 99574, mail addr: POB 391, Cordova, AK 99574 • T: +1 907 4246665 • infoservices@cityofcordova.net • www.cordovamuseum.org •
Dir.: *Cathy R. Sherman* •
Local Museum – 1966
Fishing artifacts, railway, lighthouse, Native Americans 47806

Corinth MS

Jacinto Courthouse, Corinth, MS 38835-1174 • T: +1 601 2868662 • F: +1 601 2866777 •
Historical Museum – 1966 47807

Northeast Mississippi Museum, 204 E 4th St, Corinth, MS 38834 • T: +1 662 2873120 • F: +1 662 6650241 • nemma@corinth.ms •
Cur.: *Kristy White* •
Local Museum – 1980
Fossils, American Indian, Civil War, railroad, local arts and crafts 47808

Corning NY

Benjamin Patterson Inn Museum Complex, 59 W Pulteney St, Corning, NY 14830-2212 • T: +1 607 9375281 • F: +1 607 9375281 • cpphs@aol.com •
Pres.: *Sheri Golder* • C.E.O.: *Roger Grigsby* •
Historical Museum – 1976
Hist. buildings, furnishings and artifacts of the hist inn, costume coll, local hist 47809

Corning Museum of Glass, 1 Museum Way, Corning Glass Center, Corning, NY 14830-2253 • T: +1 607 9375371 • F: +1 607 9748470 • cmg@cmog.org • www.cmog.org •
Exec.Dir.: *David B. Whitehouse* • Cur.: *Jane S. Spillman* • *Dedo von Kerssenbrock-Krosigk* • *Tina Oldknow* •
Decorative Arts Museum – 1951
Glass vessels, archeological remains, glass hist 47810

Rockwell Museum of Western Art, 111 Cedar St, Corning, NY 14830-2694 • T: +1 607 9375386 • F: +1 607 9744895 • info@rockwellmuseum.org • www.rockwellmuseum.org •
Dir.: *Kristin Swain* • Sc. Staff: *Joyce Penn* •
Fine Arts Museum / Decorative Arts Museum – 1976
19th and 20th c art of the North American West, Native American art, firearms 47811

Cornish NH

Saint Gaudens National Historic Site, 139 Saint Gaudens Rd, Cornish, NH 03745 • T: +1 603 6752175 • F: +1 603 6752701 • saga@nps.gov • www.nps.gov/saga/index.html •
Dir.: *John Dryfhout* • Cur.: *Henry J. Duffy* •
Fine Arts Museum – 1926
Sculpture, American art 47812

Cornwall CT

Cornwall Historical Museum, 7 Pine St, Cornwall, CT 06753 •
Pres.: *Michael R. Gannett* •
Local Museum – 1964
Local history 47813

Cornwall PA

Cornwall Iron Furnace, Rexmont Rd at Boyd St, Cornwall, PA 17016 • T: +1 717 2729711 • F: +1 717 2720450 •
Head: *Stephen G. Somers* •
Science&Tech Museum – 1931
Industrial hist 47814

Cornwall-on-Hudson NY

Museum of the Hudson Highlands, The Boulevard, Cornwall-on-Hudson, NY 12520 • T: +1 845 5345506 • F: +1 845 5344581 • sbaxter@museumhudsonhighlands.org • www.museumhudsonhighlands.org •
Exec.Dir.: *Jaqueline Grant* • Chm.: *Ed Hoyd* •
Fine Arts Museum / Natural History Museum – 1962
Preserved fishes, reptiles, amphibians, ichthyology coll from Hudson River tributaries 47815

Corona CA

Fender Museum of Music and the Arts, 365 N Main St, Corona, CA 92880 • T: +1 909 7352440 • bacosta@fendermuseum.com • www.fendermuseum.com •
Dir.: *Debbie Shuck* •
Music Museum – 1997
Guitars, amps, factory equipment, music memorabilia 47816

Corpus Christi TX

Art Center of Corpus Christi, 100 N Shoreline, Corpus Christi, TX 78401 • T: +1 361 8846406 • F: +1 361 8848836 • artctrcc@sbcglobal.net •
Dir.: *Debbie Luce* •
Public Gallery
Local art 47817

Art Museum of South Texas, South Texas Institute for the Arts, 1902 N Shoreline, Corpus Christi, TX 78401 • T: +1 361 8253500 • F: +1 361 8253520 • stiaweb@mail.tamucc.edu • www.stia.org •
Dir.: *Dr. William G. Otton* • Ass. Dir.: *Marilyn Ramey* •
Ass. Cur.: *Michelle Locke* •
Fine Arts Museum – 1960
Paintings, drawings, sculpture, graphics, photographs 47818

Asian Cultures Museum, 1809 N Champarral St, Corpus Christi, TX 78401 • T: +1 361 8822641 • F: +1 361 8825718 • asiancm@yahoo.com • www.geocities.com/asiancm •
C.E.O.: *Dongwol Kim Robertson* •
Decorative Arts Museum – 1973
Decorative arts of Japan, China, India & Korea, incl Hakata dolls, porcelains, metal ware 47819

Corpus Christi Museum of Science and History, 1900 N Chaparral, Corpus Christi, TX 78401 • T: +1 361 8264667 • F: +1 361 8847392 • www.ccmuseum.com •
Dir.: *Richard R. Stryker* • Cur.: *Donald P. Zuris* •
Local Museum – 1957
Hist, marine and earth science, Texas mollusks, Columbus replica ships 47820

Joseph A. Cain Memorial Art Gallery, Del Mar College, 101 Baldwin, Corpus Christi, TX 78404-3897 • T: +1 361 6981216 • F: +1 361 6981511 • elambert@delmar.edu • www.delmar.edu •
Dir.: *Randy Flowers* •
Fine Arts Museum – 1932
Drawings and small sculpture 47821

USS Lexington Museum on the Bay, 2914 N Shoreline Blvd, Corpus Christi, TX 78402 • F: +1 361 8838361 • www.usslexington.com •
Exec. Dir.: *Frank Montesano* • Dir.: *M. Charles Reustle* •
Military Museum – 1991
Naval military hist 47822

Weil Art Gallery, Texas A&M University, 6300 Ocean Dr, Corpus Christi, TX 78412 • T: +1 361 8255752 • F: +1 361 9946097 • anderson@falcon.tamucc.edu •
Dir.: *Dorothy Lumb* •
Fine Arts Museum / University Museum – 1979
Contemporary Texas and Mexican art 47823

Corry PA

Corry Area Historical Society Museum, 937 Mead Av, Corry, PA 16407 • T: +1 814 6644749 • www.tbscc.com •
Pres.: *James R. Nelson* •
Local Museum – 1965
Transpotation, costumes, agriculture, music 47824

Corsicana TX

Navarro County Historical Society Museum, 912 W Park Av, Corsicana, TX 75110 • T: +1 903 6544846 • F: +1 903 6544983 • byoung@ci.corsicana.tx.us •
Cur.: *Bobbie Young* •
Local Museum – 1958
8 log buildings contructed in Navarro County during 1838-1865 47825

Pearce Collections Museum, Navarro College, 3100 Collin St W, Corsicana, TX 75110 • T: +1 903 8757438 • F: +1 903 8757593 • www.navarrocollege.edu/library/archives •
C.E.O. & Cur.: *Julie Holcomb* •
Local Museum – 1996
Artifacts, photographs, memorabilia, graphic, American Western art, hist of Navarro College 47826

Cortez CO

Crow Canyon Archaeological Center, 23390 County Rd K, Cortez, CO 81321 • T: +1 970 5658975 • F: +1 970 5654859 • cleoncini@crowcanyon.org • www.crowcanyon.org •
C.E.O.: *Ricky Lightfoot* •
Archaeology Museum – 1983
Archaeology 47827

Cortland NY

1890 House-Museum and Center for Arts, 37 Tompkins St, Cortland, NY 13045-2555 • T: +1 607 7567551 • F: +1 607 7567551 • the1890house@usadatanet.net •
Pres.: *Joseph Puzo* • C.E.O.: *Teresa Roberts* •
Decorative Arts Museum / Fine Arts Museum – 1975
Victorian chateauesque style mansion 47828

Bowers Science Museum, State University of New York College at Cortland, Bowers Hall, SUNY, Cortland, NY 13045 • T: +1 607 7532900 • F: +1 607 7532927 • www.cortland.edu •
Dir./Cur.: *Dr. Peter K. Ducey* •
University Museum / Natural History Museum / Science&Tech Museum – 1964
Anatomy, local plants, birds and mammals, teaching skins, mounted fish, human biology, geology, fungi 47829

Cortland County Historical Society Museum, 25 Homer Av, Cortland, NY 13045 • T: +1 607 7566071 • F: +1 607 7535728 • graffa@cortland.edu •
Dir.: *Mary Ann Kane* •
Local Museum – 1925
Hist artifacts, local and county hist, local industry – library 47830

Dowd Fine Arts Gallery, State University of New York, Suny Cortland, Prospect Terr & Graham Av, Cortland, NY 13045 • T: +1 607 7534216 • F: +1 607 7535728 • graffa@cortland.edu • www.cortland.edu/art/html/gallery.html •
Dir.: *Allison Graff* •
Fine Arts Museum / University Museum – 1967
19th-20th c European and American paintings, drawings, prints and sculpture 47831

Corvallis OR

Corvallis Art Center, Oregon State University, 700 SW Madison, Corvallis, OR 97331-4501 • T: +1 541 7541551 •
Public Gallery 47832

Fairbanks Gallery, c/o Oregon State University, 106 Fairbank Hall, Corvallis, OR 97331-4501 • T: +1 541 7375009 • F: +1 541 7375372 • drussell@orst.edu •
Dir.: *Douglas Russell* •
Fine Arts Museum / University Museum 47833

Giustina Gallery, c/o Oregon State University, La Sells Stewart Center, 875 SW 26th St, Corvallis, OR 97331-4501 • T: +1 541 7372402 • F: +1 541 7373187 •
Dir.: *Melanie Fahrenbruch* •
Fine Arts Museum / University Museum 47834

Greensward Gallery, Oregon State University, 968 NW Circle Blvd, Corvallis, OR 97331-4501 • T: +1 541 7540204 •
Public Gallery 47835

Memorial Union Art Gallery, c/o Oregon State University, Memorial Union, Corvallis, OR 97331-4501 • T: +1 541 7378511 • F: +1 541 7371565 • kent.sumner@mu.orst.edu •
Dir.: *Kent Sumner* •
Fine Arts Museum / University Museum
Paintings, sculptur, prints 47836

Corydon IA

Prairie Trails Museum of Wayne County, Hwy 2 E, Corydon, IA 50060 • T: +1 641 8722211 • F: +1 641 8722664 • ptmuseum@grm.net • www.prairetrailmuseum.org •
Pres.: *Hal Greenlee* •
Local Museum – 1942
General museum 47837

Coshocton OH

Johnson-Humrickhouse Museum, Roscoe Village, 300 N Whitewoman St, Coshocton, OH 43812 • T: +1 740 6228710 • F: +1 740 6228710 • JHMuseum@sbcglobal.net • www.jhmuseum.org •
Dir.: *Patti Malenke* •
Local Museum – 1931
American Indian, Japanese, Chinese, American and European decorative arts, weapon, Ohio history 47838

Roscoe Village Foundation, 600 North Whitewoman St, Coshocton, OH 43812 • T: +1 740 6229310 • F: +1 614 6236555 • www.roscoevillage.com •
C.E.O.: *Joel L. Hampton* • Dir.: *Thomas Mackie* (History and Education) • Sc. Staff: *Wilma Hunt* (History) •
Historic Site / Open Air Museum – 1968
Historic buildings, furniture and related artifacts, Ohio's canal-era period 47839

Costa Mesa CA

Orange County Museum of Art - Orange Lounge, 3333 Bristol St, Costa Mesa, CA 92626 • T: +1 949 7591122 ext 272 • orangelounge@ocma.net • www.ocma.net •
Fine Arts Museum 47840

Cottage Grove OR

Cottage Grove Museum, 147 High St and Birch Av, Cottage Grove, OR 97424 • T: +1 541 9423963 •
Head: *Isabelle S. Woolcott* •
Local Museum – 1961
Local hist, pioneer artifacts, Indian artifacts, mining tools, Civil War items, Italian stained glass windows 47841

Cottonwood AZ

Clemenceau Heritage Museum, 1 N Willard, Cottonwood, AZ 86326, mail addr: POB 511, Cottonwood, AZ 86326 • T: +1 520 6342868 • clemenceauheritagem@questoffice.net • www.clemenceaumuseum.org •
C.E.O.: *Jim McMeekin* •
Historical Museum – 1989
History of the Verde Valley 47842

Cottonwood ID

The Historical Museum at Saint Gertrude, Keuterville Rd, Cottonwood, ID 83522-9408 • T: +1 208 9627123 • F: +1 208 9628647 • museum@velocitus.net • www.historicalmuseumatstgertrude.com •
Dir.: *Lyle Wirtanen* • Cur.: *Mary Cay Henry* •
Local Museum – 1931
Local hist 47843

Cottonwood Falls KS

Chase County Historical Society, 301 Broadway, Cottonwood Falls, KS 66845 • T: +1 620 2738500 • skyways.lib.ks.us/genweb/society/cottonwd •
Cur.: *Patty J. Donelson* •
Association with Coll – 1934
Old tools, household items 47844

Roniger Memorial Museum, 315 Union, Cottonwood Falls, KS 66845 • T: +1 620 2736310, 2736412 • F: +1 620 2736671 • dcroy@kansas.net •
Pres./Cur.: *David E. Croy* •
Historical Museum – 1959
Indian artifacts, stuffed native animals 47845

Cotuit MA

Cahoon Museum of American Art, 4676 Falmouth Rd, Cotuit, MA 02635 • T: +1 508 4287581 • F: +1 508 4203709 • cmaa@cahoonmuseum.org • www.cahoonmuseum.org •
C.E.O.: *Cindy Nickerson* •
Fine Arts Museum – 1984
American paintings, period furnishings – library 47846

Cotulla TX

Brush Country Museum, 201 S Stewart, Cotulla, TX 78014 • T: +1 830 8792429 • www.historicdistrict.com/museum •
Cur.: *Elizabeth Seidel* • *Nita Gierisch* •
Local Museum – 1982
Photographs, artifacts, hist of La Salle County and the Brush Country of South Texas 47847

Coudersport PA

Potter County Historical Society Museum, 308 N Main St, Coudersport, PA 16915 • T: +1 814 2744410 • pottercohist@adelphia.net • www.pottercountypa.net/history •
Pres.: *Leon B. Reed* •
Local Museum – 1919
Pioneer artifacts, prints, paintings, books, local industry coll 47848

Coulee City WA

Dry Falls Interpretive Center, Sun Lakes State Park, 34875 Park Lake Rd NE, Coulee City, WA 99115 • T: +1 509 6325583 • F: +1 509 6325971 •
Dir.: *Mike Sternback* •
Local Museum – 1965
Local geological events, archaeology, anthropology, ethnology, Indian artifacts 47849

Council Bluffs IA

Historic General Dodge House, 605 3rd St, Council Bluffs, IA 51503 • T: +1 712 3222406 • F: +1 712 3223504 • generaldodgehouse@alltel.net • www.councilbluffsiowa.com •
Exec. Dir.: *Kori L. Nielsen* •
Local Museum / Decorative Arts Museum – 1961
Historic house 47850

Union Pacific Railroad Museum, 200 Pearl St, Council Bluffs, IA 51503 • T: +1 712 32198307 • F: +1 402 2716460 • upmuseum@up.com • www.up.com •
Dir.: *Beth Lindquist* • Cur.: *Keely Rennie* •
Historical Museum / Science&Tech Museum – 1921
Western railroad hist up to modern technology, American hist 47851

Council Grove KS

Kaw Mission, 500 N Mission, Council Grove, KS 66846 • T: +1 620 7675410 • F: +1 620 7675816 • kawmission@cgtelco.net • www.kawmission.org •
Cur.: *Ron Parks* •
Ethnology Museum – 1951
19th c America, plains Indians 47852

Coupeville WA

Fort Casey Interpretive Center, 1280 Engle Rd, Coupeville, WA 98239 • T: +1 360 6797391, 6784519 • F: +1 360 6797327 •
Historical Museum / Military Museum
Coast artillery, natural history, sea life 47853

Gallery at the Wharf, Coupeville Arts Center, Coupeville, WA 98239, mail addr: POB 171, Coupeville, WA 98239 • T: +1 360 6783396 • F: +1 360 6787420 • cac@whidbey.net •
Exec. Dir.: *Judy G. Lynn* •
Fine Arts Museum – 1989
Paintings, photography 47854

Island County Historical Society Museum, 908 NW Alexander St, Coupeville, WA 98239 • T: +1 360 6783310 • F: +1 360 6781702 • ichscpvl@whidbey.net • www.islandhistory.org •
Local Museum – 1949
Indian artifacts, military hist in Island County 47855

Courtland VA

Rawls Museum Arts, 22376 Linden St, Courtland, VA 23837 • T: +1 757 6530754 • F: +1 757 6530341 • rma@beldar.com • www.rawlsart.cjb.net •
Dir.: *Barbara Easton-Moore* •
Fine Arts Museum – 1958
Contemporary regional painting, glass, Indian artifacts, seashells 47856

Coventry RI

Western Rhode Island Civic Museum, 7 Station St, Coventry, RI 02816 • T: +1 401 8214095 •
Pres.: *Lynda Hawkins* •
Local Museum – 1945
Local hist 47857

Covington KY

Behringer-Crawford Museum, 1600 Montague Rd, Devou Park, Covington, KY 41012 • T: +1 859 4914003 • F: +1 859 4914006 •
Dir.: *Laurie Risch* •
Local Museum – 1950
Paleontology, prehistoric, mineralogy, ceramics 47858

The Railway Exposition, 315 W Southern Av, Covington, KY 41015 • T: +1 513 7613500 •
Science&Tech Museum – 1975
Railroad passenger cars & locomotives 47859

Covington LA

Saint Tammany Art Gallery, 129 N New Hampshire St, Covington, LA 70433 • T: +1 985 8928650 • F: +1 985 8980976 • staa1@juno.com •
Public Gallery 47860

Covington TN

Tipton County Museum, 751 Bert Johnston Av, Covington, TN 38019 • T: +1 901 4760242 • F: +1 901 4760261 •
Dir.: *Alice Fisher* •
Local Museum
Military hist, environmental education 47861

Coweta OK

Mission Bell Museum, 201 South Av B, Coweta, OK 74429 • T: +1 918 4862513 • F: +1 918 4862513 • CowetaCham@aol.com •
Local Museum – 1977 47862

Coxsackie NY

Bronck Museum, Rte 9 W, Coxsackie, NY 12051, mail addr: 90 County Rte 42, Coxsackie, NY 12051 • T: +1 518 7314960 • F: +1 518 7317672 • gchsbm@mhtable.com • www.gchistory.org •
C.E.O. & Pres.: *Robert Hallock* •
Local Museum – 1929
Local hist and genealogy, art, furniture, silver, pottery, carriages, early tools – Vedder library 47863

Cozad NE

The 100th Meridian Museum, 206 E Eighth St, Cozad, NE 69130 • T: +1 308 7841100, 7842704 •
Pres.: *Howard McGinnis* •
Local Museum – 1994
Farm equipment, china silver, historical photographs, piano, local hist 47864

Craig CO

Museum of Northwest Colorado, 590 Yampa Av, Craig, CO 81625 • T: +1 970 8246360 • F: +1 970 8247175 • musnwco@quik.com • www.museumnwco.org •
Dir.: *Dan Davidson* • Ass. Dir.: *Jan Gerber* •
Local Museum – 1964
Regional history 47865

Cranbury NJ

Cranbury Historical and Preservation Society Museum, 4 Park Pl E, Cranbury, NJ 08512 • T: +1 609 3950702 • historycenter@comcast.net • www.cranbury.org/history/museum.htm •
Pres.: *Karl Kusserow* • Cur.: *Don Jo Swanagan* •
Historical Museum / Local Museum – 1967
History – library 47866

Cranford NJ

Tomasulo Gallery, Union County College, 1033 Springfield Av, Cranford, NJ 07016-1599 • T: +1 908 7097155 • F: +1 908 7091599 • www.ucc.edu/tomasulo_art_gallery/htm •
Dir.: *Valeri Lanko* •
Fine Arts Museum
Contemporary fine arts 47867

Crawford NE

Fort Robinson Museum, 3200 US Hwy 20, Crawford, NE 69339 • T: +1 308 6652919 • F: +1 308 6652917 • fortrob@bbc.net •
Cur.: *Thomas R. Buecker* •
Military Museum / Ethnology Museum – 1956
Indian artifacts, military items, ethnology, archaeology, anthropology 47868

Crawfordsville IN

General Lew Wallace Museum, 200 Wallace Av, Crawfordsville, IN 47933, mail addr: POB 662, Crawfordsville, IN 47933 • T: +1 765 3625769 • F: +1 765 3625769 • study@ben-hur.com • www.ben-hur.com •
Dir.: *Chinnamon Catlin-Legutko* •
Historical Museum – 1905
General Lew Wallace Study, author of 'Ben-Hur', movies and broadway play 47869

The Lane Place, 212 S Water St, Crawfordsville, IN 47933 • T: +1 765 3623416 • mchs@wico.net • www.lane-mchs.org •
Pres.: *Karen Zech* • C.E.O.: *Tamara Hemmerlein* •
Historical Museum / Decorative Arts Museum – 1911
Home of Henry S. Lane, governor and senator of Indiana 47870

The Old Jail Museum, c/o Montgomery County Cultural Foundation, 225 N Washington, Crawfordsville, IN 47933 • T: +1 765 3625222 • F: +1 765 3625222 • oldjail@tctc.com • crawfordsville.org/jail.html •
Pres.: *Sam Smith* •
Historical Museum – 1975
1882 Old Jail only remaining rotating circular jail still in working condition 47871

Crawfordville GA

Confederate Museum, 456 Alexander St, Crawfordville, GA 30631 • T: +1 706 4562221, 4562602 • F: +1 706 4562396 • ah__stephens_park@mail.dnr.state.ga.us • www.gastateparks.org •
Historical Museum – 1952
Political hist of Civil War, life of A. H. Stephens, furnishings 47872

Crazy Horse SD

Indian Museum of North America, Av of the Chiefs, Crazy Horse, SD 57730-9506 • T: +1 605 6734681 • F: +1 605 6732185 • memorial@crazyhorse.org • www.crazyhorse.org •
Dir.: *Anne Ziolkowski* •
Ethnology Museum – 1972
Indian culture 47873

Crescent City CA

Del Norte County Historical Society Museum → Main Museum

Main Museum, 577 H St, Crescent City, CA 95531 • T: +1 707 4643922 • dnhissoc@northcoast.com • www.delnortehistory.org •
Pres.: *Loren Bommerlyn* •
Local Museum – 1951
Local archaeology, history, lumber ind, agriculture, mining and shipping 47874

Redwood Museum, 1111 Second St, Crescent City, CA 95531 • T: +1 707 4646101 • F: +1 707 4641812 • redw_superintendent@nps.gov • www.nps.gov/redw •

U

Cur.: *James B. O'Barr* •
Natural History Museum / Ethnology Museum – 1968
Natural history, cultural hist of northcoast California 47875

Cresson TX

Pate Museum of Transportation, Hwy 377, Cresson, TX 76035, mail addr: 1227 W Magnolia Av, Ste 420, Fort Worth, TX 76104 • T: +1 817 9229504, 3964305 • F: +1 817 9229536 •
Cur.: *Sue Baker* •
Science&Tech Museum – 1969
Antique and classic autos, aircraft, railroad artifacts, space exhibits 47876

Crestline OH

Crestline Shunk Museum, 211 N Thoman St, Crestline, OH 44827 • T: +1 419 6833410 •
Pres.: *Nancy Everly* •
Local Museum – 1947
Railroad artifacts, glassware, silver, china, toys, dolls 47877

Creston IA

Union County Historical Complex, McKinley Park, Creston, IA 50801 • T: +1 515 7824247 •
Pres.: *Paul Roeder* •
Local Museum / Open Air Museum – 1966
General and village museum, 10 pioneer bldgs 47878

Crestview KY

TM Gallery, c/o Thomas More College, 333 Thomas More Pkwy, Crestview, KY 41017 • T: +1 606 3443419/20 • F: +1 606 3443345 •
Dir.: *Barb Rauf* •
Fine Arts Museum
Drawings, graphics, photography, sculpture, ceramics 47879

Crestwood MO

Thomas Sappington House Museum, 1015 S Sappington Rd, Crestwood, MO 63126 • T: +1 314 8228171 • F: +1 314 1294794 • lblumer@ci.crestwood.mo.us • www.ci.crestwood.mo.us •
Pres.: *Betty Oberhaus* •
Historical Museum – 1967
Decorative arts, Federal period furnishings 47880

Creswell NC

Somerset Place, 2572 Lake Shore Rd, Creswell, NC 27928 • T: +1 919 7974560 • F: +1 919 7974171 • somerset@coastalnet.com •
Head: *Dorothy Spruill Redford* •
Local Museum – 1969
Agriculture, several buildings 47881

Cripple Creek CO

Cripple Creek District Museum, 500 E Bennett Av, Cripple Creek, CO 80813 • T: +1 719 6899540 • F: +1 719 6890512 • ccdistrictmuseum@aol.com • www.cripple-creek.co.us •
Dir.: *Erik Swanson* •
Fine Arts Museum / Local Museum – 1953
Local history, historic camp where gold was discovered in 1890 47882

Critz VA

Reynolds Homestead, 463 Homestead Ln, Critz, VA 24082 • T: +1 276 6947181 • F: +1 276 6947183 • registrar.reynolds.homestead@vt.edu • www.reynoldshomestead.vt.edu •
Dir.: *Carolyn Beale* •
Fine Arts Museum / Local Museum – 1970
Photograph album, paintings, silver 47883

Crockett TX

Discover Houston County Visitors Center-Museum, 303 S First St, Crockett, TX 75835 • T: +1 936 5449520 • F: +1 936 5448053 •
Cur.: *Eliza H. Bishop* •
Local Museum – 1983
State & local hist exhi; photograph coll, WW I & II listings 47884

Croghan NY

American Maple Museum, POB 81, Croghan, NY 13327 • T: +1 315 3461107 •
Pres.: *Vernon Lyndaker* • Cur.: *Laurie Walseman* •
Agriculture Museum – 1977
Hist, products and tools of maple industry 47885

Crookston MN

Polk County Museum, 719 E Robert st, Crookston, MN 56716 • T: +1 218 2811038 • polkcomuseum@rru.net • www.visitcrookston.com/attractions.htm •
Pres./Cur.: *Allen Brolsma* •
Local Museum – 1930
Regional hist, agriculture, anthropology, Indian artifacts, industry, medicine, folklore, costumes, fire equip – archives 47886

Crosby MN

Cuyuna Range Museum, 101 First St NE, Crosby, MN 56441 • T: +1 218 5465435 •
Pres.: *Ray Nelson* •
Local Museum – 1970
Local hist, mining, logging, tools, clothing, pictures 47887

Crosby ND

Divide County Historical Society Museum, Pioneer Village, Crosby, ND 58730 • T: +1 701 9656297 •
Pres.: *Perry E. Rosenquist* •
Local Museum – 1969
Regional history, pioneer artifacts, farm implements, vehicles, stationary gas engines 47888

Crosbyton TX

Crosby County Pioneer Memorial Museum, 101 W Main intersection U.S. 82 and F.M. 651, Crosbyton, TX 79322 • T: +1 806 6752331, 6752906 • F: +1 806 6757012 • ccpmm@door.net • www.crosbycountymuseum.com •
Dir.: *Verna Anne Wheeler* •
Local Museum – 1958
Agriculture, home & family artifats, Texas plains area artifacts, cowboy memorabilia – archives 47889

Cross Creek FL

Marjorie Kinnan Rawlings Historic State Park, 187000 S County Rd 325, Cross Creek, FL 32640 • T: +1 352 4669273 • F: +1 352 4664743 • mkrshs@bellsouth.net •
Historical Museum / Agriculture Museum – 1970
1930's citrus farm and home of Marjorie Kinnan Rawlings, a rambling Cracker farmhouse 47890

Cross River NY

Trailside Nature Museum, Ward Pound Ridge Reservation, 11 Reservation Rd, Cross River, NY 10518, mail addr: POB 236, Cross River, NY 10518 • T: +1 914 8647322 • F: +1 914 8647323 •
Cur.: *Brenda Bates* •
Natural History Museum – 1937
Geology, local hist, mounted birds, mammals, insects 47891

Crow Agency MT

Little Bighorn Battlefield Museum, I-90 and Hwy 212, Crow Agency, MT 59022, mail addr: POB 39, Crow Agency, MT 59022-0039 • T: +1 406 6382621 ext 110 • F: +1 406 6382623 • darrell_cook@nps.gov • www.nps.gov/libi •
Head: *Darrell J. Cook* •
Historical Museum / Military Museum – 1940
Indian and military items, Sioux War 1876-90 47892

Crowley LA

Crowley Art Association and Gallery, 220 N Parkerson, Crowley, LA 70527-2003 • T: +1 337 7833747 • F: +1 337 7833747 •
Fine Arts Museum / Decorative Arts Museum – 1980
Art gallery, crafts 47893

The Rice Museum, 6428 Airport Rd, Crowley, LA 70527-1176 • T: +1 337 7836417 • F: +1 318 7880123 • crpwright@cs.com • www.crystalrice.com •
Pres.: *Diane Hoffpauer* •
Special Museum – 1970
Furniture and implements, pictures, dolls, hist of the rice industry 47894

Crown Point NY

Crown Point State Historic Site, 739 Bridge Rd, Crown Point, NY 12928 • T: +1 518 5973666 • F: +1 518 5974668 • williamfarrar@oprhp.state.ny.us •
Head: *Bill Farar* •
Military Museum – 1910
1734 remains of French fort, 1759 British Fort Crown Point and Outwork fortifications controlling Lake Champlain 47895

Crystal River FL

Coastal Heritage Museum, 532 Citrus Av, Crystal River, FL 34429 • T: +1 352 7951755, 3416428 • F: +1 352 3416445 • webmaster@cccourthouse.org • www.cccourthouse.org •
Pres.: *Sharon Padgett* •
Local Museum – 1986
Citrus County hist, Aboriginal and Seminole Indian artifacts, archaeology 47896

Crystal River State Archaeological Site, 3400 N Museum Pt, Crystal River, FL 34428 • T: +1 352 7953817 • F: +1 352 7956061 • crsas@flanet.com • www.floridastateparks.org •
Archaeology Museum – 1965
Archaeology, Florida Indian artifacts 47897

Cuddebackville NY

Neversink Valley Area Museum, D and H Canal Park, Hoag Rd, Cuddebackville, NY 12729 • T: +1 845 7548870 • F: +1 845 7548870 • nvame@frontiernet.net • www.neversinkmuseum.org •
C.E.O. & Pres.: *Stephen. Skye* •
Local Museum / Natural History Museum – 1964
Local hist of Delaware and Hudson Canal, 19th c life in the Valley 47898

Cuero TX

DeWitt County Historical Museum, 312 E Broadway, Cuero, TX 77954 • T: +1 361 2756322 • www.cuero.org •
Local Museum – 1973
Furnishings, traveling exhib 47899

Culbertson MT

Culbertson Museum, Hwy 2 E, Culbertson, MT 59218, mail addr: POB 95, Culbertson, MT 59218-0095 • T: +1 406 7876320, 7875337 •
Local Museum – 1990
Hist in northeastern Montana 47900

Northeastern Montana Threshers and Antique Association, POB 12, Culbertson, MT 59218 • T: +1 406 7875265 • elk1@nemontel.net •
Pres.: *David Krogedal* •
Association with Coll – 1964
Agriculture, pioneer House, steam & gasoline tractors, small engines 47901

Cullman AL

Cullman County Museum, 211 Second Av NE, Cullman, AL 35055 • T: +1 256 7391258 • F: +1 256 7378782 • Curator@cullmancountymuseum.com • www.cullmancountymuseum.com •
Cur.: *Elaine L. Fuller* •
Local Museum – 1975
Local history, housed in replica of 1873 home of Col. John G. Cullman, founder of Cullman 47902

Cullowhee NC

Mountain Heritage Center, Western Carolina University, 150 H. F. Robinson Building, Cullowhee, NC 28723 • T: +1 828 2277129 • www.wcu.edu/mhc •
Dir.: *H. Tyler Blethen* • Cur.: *Suzanne McDowell* •
Historical Museum / University Museum – 1975
Cherokee and pioneer artifacts, tools, folklore recordings 47903

Culver City CA

Museum of Jurassic Technology, 9341 Venice Blvd, Culver City, CA 90232 • T: +1 310 8366131 • F: +1 310 2872267 • info@mjt.org • www.mjt.org •
Dir.: *David Wilson* •
Natural History Museum
Ecclectic coll of relicts and artifacts from Lower Jurassic 47904

Cumberland MD

Allegany County Historical Museum, 218 Washington St, Cumberland, MD 21502 • T: +1 301 7778678 • F: +1 301 7778678 ext 51 • hhouse@allconet.org • www.historyhouse.allconet.org •
Dir.: *Sharon Nealis* • Cur.: *Wilma Thompson* •
Local Museum – 1937
Historic house, regional history 47905

Cumberland Theatre Arts Gallery, 101-103 N Johnson St, Cumberland, MD 21502 • T: +1 301 7594990 • F: +1 301 7777092 • ctdon@mindspring.com • www.cumberlandtheatre.com •
Performing Arts Museum – 1987 47906

George Washington's Headquarters, Greene St, Cumberland, MD 21502 • T: +1 301 7596636 • F: +1 301 7593223 • djohnson@allconet.org • www.ci.cumberland.md.us •
Dir.: *Diane Johnson* •
Historical Museum – 1925
1755 log cabin built during the French and Indian war 47907

Cummington MA

Kingman Tavern Historical Museum, 41 Main St, Cummington, MA 01026 • T: +1 413 6345527 • www.virtual-valley.com/kingman •
Head: *Stephen Howes* •
Local Museum – 1967
Local history, frame bldg used as a post office, Masonic Lodge meeting hall and tavern 47908

William Cullen Bryant Homestead, 207 Bryant Rd, Cummington, MA 01026 • T: +1 413 6342244 • F: +1 413 6340376 • bryanthomestead@ttor.org • www.thetrustees.org/pages/285_bryant_homestead.cfm •
C.E.O.: *Andy Kendall* •
Special Museum – 1928
Boyhood home and stuff summer residence of literary figure William Cullen Bryant 47909

Cupertino CA

Cupertino Historical Museum, 10185 N Stelling Rd, Cupertino, CA 95014 • T: +1 408 9731495 • info@cupertinohistoricalsociety.org • www.cupertinohistoricalsociety.org •
Pres.: *Darryl Stowe* •
Local Museum – 1990
Local history, farming, blacksmith tools, furniture, clothing – archives 47910

Euphrat Museum of Art, De Anza College, 21250 Stevens Creek Blvd, Cupertino, CA 95014 • T: +1 408 8648836 • rindfleischjanet@fhda.edu • www.deanza.edu/euphrat •
Dir.: *Jan Rindfleisch* •
University Museum / Fine Arts Museum – 1971
Contemporary paintings, prints and photographs, sculpture 47911

Currie MN

End-O-Line Railroad Museum, 440 N Mill St, Currie, MN 56123-1004 • T: +1 507 7633708, 7633113 • F: +1 507 7633708 • EndOLine@co.murray.mn.us • www.endoline.com •
Dir.: *Louise Gervais* •
Science&Tech Museum – 1972
Chicago & nortwestern railroad, watertower, wooden Caboose, diesel switcher, steam engine, HO model train 47912

Currie NC

Moores Creek National Battlefield, 40 Patriots Hall Dr, Currie, NC 28435 • T: +1 910 2835591 • F: +1 910 2835351 • anw_childress@nps.gov • www.nps.gov/mocr •
Dir.: *Ann Childress* •
Open Air Museum / Military Museum – 1926
Weapons, diorama 47913

Cushing OK

Cimarron Valley Railroad Museum, South Kings Hwy, Cushing, OK 74023 • T: +1 918 2253936, 2251657 •
Cur.: *Robert F. Read* •
Science&Tech Museum – 1970
Railroad artifacts 47914

Custer SD

Custer County 1881 Courthouse Museum, 411 Mount Rushmore Rd, Custer, SD 57730 • T: +1 605 6732443 • F: +1 605 6732443 •
C.E.O.: *Dan McPherson* •
Local Museum – 1974
Local hist 47915

National Museum of Woodcarving, Hwy 16 W, Custer, SD 57730 • T: +1 605 6734404 • F: +1 605 6733843 • woodcarv@gwtc.net • www.blackhills.com/woodcarving •
C.E.O.: *Dale Schaffer* •
Fine Arts Museum – 1966
Woodcarvings 47916

Cuyahoga Falls OH

Richard Gallery and Almond Tea Gallery, 2250 Front St, Cuyahoga Falls, OH 44221 • T: +1 330 9291575 • jrichard@gallery.com •
Dir.: *Jack Richard* •
Fine Arts Museum – 1960
Local, regional and national works of art 47917

Cypress CA

Cypress College Fine Arts Gallery, 9200 Valley View St, Cypress, CA 90630 • T: +1 714 4847133, 8265593 • F: +1 714 5278238 •
Dir.: *Betty Disney* •
Fine Arts Museum / University Museum – 1969 47918

Dade City FL

Pioneer Florida Museum, 15602 Pioneer Museum Rd, Dade City, FL 33523 • T: +1 352 5670262 • F: +1 352 5671262 • curator@pioneerfloridamuseum.org • www.pioneerfloridamuseum.org •
Pres.: *John Falls* •
Historical Museum / Agriculture Museum – 1961
Agriculture and town hist 47919

Dahlonega GA

Dahlonega Courthouse Gold Museum, 1 Public Sq, Dahlonega, GA 30533 • T: +1 706 8642257 • F: +1 706 8648370 • dgmgold@alltel.net • www.ngeorgia.com/parks/dahlonega.html •
Head: *Sharon Johnson* •
Historical Museum / Science&Tech Museum – 1966
Tools and equipment, personal artefacts, furnishings 47920

Dakota City IA

Humboldt County Historical Association Museum, 905 First Av N, Dakota City, IA 50529 • T: +1 515 3325280, 3323449 •
Dir.: *Bette Newton* •
Local Museum – 1962
General Indian coll, local county hist 47921

Dalhart TX

Dallam-Hartley XIT Museum, 108 E 5th St, Dalhart, TX 79022 • T: +1 806 2445390 • F: +1 806 2443031 • xitmusm@xit.net • www.xitmuseum.com •
Dir./Cur.: *Nicky L. Olson* •
Local Museum / Public Gallery – 1975
XIT Ranch hist, art gallery with changing exhi 47922

Dallas NC

Gaston County Museum of Art and History, 131 W Main St, Dallas, NC 28034 • T: +1 704 9227681 • F: +1 704 9227683 • museum@co.gaston.nc.us • www.gastoncountymuseum.org •
C.E.O. & Dir.: *Barbara H. Brose* •
Fine Arts Museum / Local Museum – 1975
Carriage and sleigh coll, textiles, 19th c parlors, regional hist 47923

Dallas TX

African American Museum, 3536 Grand Av, Dallas, TX 75210, mail addr: POB 150153, Dallas, TX 75315 • T: +1 214 5659026 • F: +1 214 4218204 •
Cur.: *Phillip Collins* •
Folklore Museum / Historical Museum – 1974
African American fine art and folk art, Texas Black hist 47924

Age of Steam Railroad Museum, 1105 Washington St, Fair Park, Dallas, TX 75210 • T: +1 214 4280101 • F: +1 214 4261937 • info@dallasrailwaymuseum.com • www.dallasrailwaymuseum.com •
Exec. Dir.: *Robert LaPrelle* •
Science&Tech Museum – 1963
Small rail vehicles, equipment typical of the 20-50 era 47925

American Museum of the Miniature Arts, 2201 N Lamar St, Ste 100, Dallas, TX 75202 • T: +1 214 9695502, 9695997 • F: +1 214 9695997 • minimuseum@aol.com •
Exec. Dir.: *Janet Wilhite* •
Decorative Arts Museum – 1988
Dolls, house designs, miniature room boxes, vignettes, antique toys, militaria scenes 47926

Biblical Arts Center, 7500 Park Ln, Dallas, TX 75225 • T: +1 214 6914662 • F: +1 214 6914752 • rjmach329@aol.com • www.biblicalarts.org •
Dir.: *Greg Vaughn* • Cur.: *R.J. Machacek* • *Scott Peck* •
Religious Arts Museum – 1966
16th c to present art in mediums incl. performing arts pertaining to the Bible 47927

The Dallas Center for Contemporary Art, 2801 Swiss Av, Dallas, TX 75204-5925 • T: +1 214 8212522 • F: +1 214 8219103 • info@thecontemporary.net • www.thecontemporary.net •
Exec. Dir.: *Joan Davidow* •
Fine Arts Museum / Public Gallery – 1981
Contemporary art 47928

Dallas Historical Society Museum, Hall of State, Fair Park, 3939 Grand Av, Dallas, TX 75315 • T: +1 214 4214500 • F: +1 214 4217500 • staff@dallashistory.org • www.dallashistory.org •
Local Museum – 1922
Manuscripts coll, costume coll, archive material relating to Southwestern & U.S.. hist 47929

Dallas Holocaust Museum, 211 N Record St, Ste 100, Dallas, TX 75230 • T: +1 214 7417500 • F: +1 214 7417510 • info@dallasholocaustmuseum.org • www.dallasholocaustmuseum.org •
Pres.: *Michael Schiff* • Exec. Dir.: *Elliott Dlin* •
Historical Museum – 1984
Holocaust memorial – library 47930

Dallas Museum of Art, 1717 N Harwood St, Dallas, TX 75201 • T: +1 214 9221200, 9540234 • F: +1 214 9221350 • jconner@dallasmuseumofart.org • www.dallasmuseumofart.org •
Dir.: *Eugene McDermott* • Cur.: *Charles Wylie* (Contemporary Art) • *Dr. Dorothy Kosinski* (European Art) • *Dr. Anne Bromberg* (Ancient Art) • *Carol Robbins* (New World and Pacific Cultures) • *Suzanne Weaver* (Contemporary Art) • Deputy Director: *Bonni Pitman* •
Fine Arts Museum – 1903
European & American paintings, decorative arts & sculpture 47931

Dallas Museum of Natural History, Fair Park, 3535 Grand Av, Dallas, TX 75210 • T: +1 214 4213466 • F: +1 214 4284356 • info@dmnhnet.org • www.dallasdino.org •
C.E.O.: *Nicole G. Small* • Cur.: *Avelino Segura* (Exhibits) • *Alex Barker* (Archaeology) • *Anthony Fiorillo* (Earth Sciences) •
Natural History Museum – 1935
Paleontology, mammalogy, ornithology, herpetology, geology 47932

Frontiers of Flight Museum, 8008 Cedar Springs Blvd, Dallas, TX 75235 • T: +1 214 3503600, 3501651 • F: +1 214 3510101 • info@flightmuseum.com • www.flightmuseum.com •
Dir.: *Dan Hamilton* • Cur.: *Knox Bishop* •
Science&Tech Museum – 1988
Aviation hist from pre-Wright brothers through modern space age 47933

International Museum of Cultures, 7500 W Camp Wisdom Rd, Dallas, TX 75236 • T: +1 972 7087537 • F: +1 972 7087341 • kamm@internationalmuseumofcultures.org • www.internationalmuseumofcultures.org •
Dir.: *Mary Fae Kamm* •
Ethnology Museum – 1974
Ethnographic coll from S America, Papua New Guinea, Africa SE Asia 47934

Meadows Museum, Southern Methodist University, 5900 Bishop Blvd, Dallas, TX 75205 • T: +1 214 7682516 • F: +1 214 7681688 • jcarrell@smu.edu • www.meadowsmuseumdallas.org •
Dir.: *Dr. Edmund P. Pillsbury* • Cur.: *Dr. Mark Roglán-Kelly* •
Fine Arts Museum / University Museum – 1965
Spanish art 47935

Museum of Geometric and Madi Art, 3109 Carlisle, Dallas, TX 75204 • T: +1 214 8557802 • F: +1 214 8555479 • www.madimuseumdallas.org •
Dir./Cur.: *Dorothy Masterson* •
Fine Arts Museum – 2003
Works by founder Carmelo Arden Quin, Rothfuss, Blaszko, Verdanega, Herrera etc 47936

Old City Park - The Historical Village of Dallas, 1717 Gano St, Dallas, TX 75215 • T: +1 214 4215162 • F: +1 214 4286351 • dagns@airmail.net • www.oldcitypark.org •
Dir.: *Gary N. Smith* • *Jennifer Bransom* • Cur.: *Hal Simon* •
Local Museum – 1966
Historic village 47937

Pollock Gallery, Southern Methodist University, Meadow School of the Arts, mail addr: POB 750356, Dallas, TX 75205-0356 • T: +1 214 7684439 • F: +1 214 7684257 • pvankeur@mail.smu.edu •
Public Gallery / University Museum
Amrican art 47938

Sixth Floor Museum at Dealey Plaza, 411 Elm St, Ste 120, Dallas, TX 75202 • T: +1 214 7476660 • F: +1 214 7476662 • jfk@jfk.org • www.jfk.org •
Cur.: *Gary Mack* •
Historical Museum – 1989
John F. Kennedy and the memory of a nation 47939

The Southland Art Collection, 2711 N Haskell, Dallas, TX 75221-0711 • T: +1 214 8287011 •
Cur.: *Richard Allen* •
Fine Arts Museum
Paintings, sculpture, photography, works on paper 47940

Southwest Museum of Science and Technology, 1318 Second Av in Fair Park, Dallas, TX 75210 • T: +1 214 4285555 • F: +1 214 4282033 • dhueter@scienceplace.org • www.scienceplace.org •
C.E.O.: *Diana Hueter* •
Science&Tech Museum – 1946
Hands-on physics, robot dinosaurs, medical & health, Kidsplace, dental, astronomy 47941

Trammell and Margaret Crow Collection of Asian Art, 2010 Flora St, Dallas, TX 75201-2335 • T: +1 214 9796430/32 • F: +1 214 9796439 • education@crowcollection.com • crowcollection.org •
Dir.: *Amy E. Lewis* •
Fine Arts Museum – 1998
Arts from Japan, China, Cambodia, India, Thailand, Myanmar and Tibet 47942

The Women's Museum, 3800 Parry Av, Dallas, TX 75226 • T: +1 214 9150887 • F: +1 214 9150870 • bbooker@thewomensmuseum.org • www.thewomensmuseum.org •
Historical Museum – 1998
Paper, books, buttons, poster, garments, personal artifacts relating to women's hist 47943

The Dalles OR

The Dalles Art Association Gallery, 220 E Fourth St, The Dalles, OR 97058 • T: +1 541 2964759 •
Dir.: *Carolyn Wright* •
Public Gallery – 1967 47944

Fort Dalles Museum, 500 W 15th St, The Dalles, OR 97058 • T: +1 541 2964547 •
Head: *Sam Woolsey* •
Military Museum – 1951
Military and local hist 47945

Wasco County Historical Museum, 5000 Discovery Dr, The Dalles, OR 97058 • T: +1 541 2968600 • F: +1 541 2988660 • perry@gorgediscovery.org • www.gorgediscovery.org •
C.E.O. & Exec.Dir.: *Ken Karsmizki* •
Local Museum – 1997
Geological hist, wildlife and natural hist of the Columbia River Gorge, Native American artifacts, local industry and pioneer artifacts 47946

Dalton GA

Creative Arts Guild, 520 W Waugh St, Dalton, GA 30722-1485 • T: +1 706 2780168 • F: +1 706 2786996 • cagarts@creativeartsguild.org • www.creativeartsguild.org •
Pres.: *Sam Turner* •
Fine Arts Museum – 1963
Paintings, prints, drawings, photos, sculptures 47947

Crown Gardens Museum, 715 Chattanooga Av, Dalton, GA 30720 • T: +1 706 2780217 •
Pres.: *Marvin Sowder* • Dir.: *Marcelle White* •
Local Museum – 1977
1848 Blunt House, 1908 Wright Hotel-Chatsworth, 1840 John Hamilton House-Chatsworth Depot – archives 47948

Dalton MA

Crane Museum of Papermaking, Rtes 8 and 9, Dalton, MA 01226 • T: +1 413 6846481 • F: +1 413 6840817 • customerservice@crane.com • www.crane.com •
Cur.: *Charles Wellspeak* • *Carol Shea* •
Historical Museum / Science&Tech Museum – 1930
Hist of papermaking in a 1844 stone mill 47949

Damariscotta ME

Round Top Center for the Arts - Arts Gallery, POB 1316, Damariscotta, ME 04543 • T: +1 207 5631507 • rtca@lincoln.midcoast.com •
Fine Arts Museum – 1988 47950

Dana IN

Ernie Pyle House, 120 W Briarwood Av, Dana, IN 47847-0338 • T: +1 765 6653633 • F: +1 765 6659312 • erniepyle@abcs.com • www.in.gov/ism/historicsites/erniepyle/historic.asp •
Cur.: *Rick Bray* •
Historical Museum – 1976
Birthplace of Ernie Pyle, noted WW II journalist 47951

Danbury CT

Danbury Museum, 43 Main St, Danbury, CT 06810 • T: +1 203 7435200 • F: +1 203 7431131 • dmhs@danburyhistorical.org • www.danburyhistorical.org •
Exec. Dir.: *Brigid Durkin* •
Local Museum – 1942
Local history, Charles Ives birthplace 47952

Military Museum of Southern New England, 125 Park Av, Danbury, CT 06810 • T: +1 203 7909277 • F: +1 203 7900420 • military@sbcglobal.net • www.usmilitarymuseum.org •
Exec. Dir.: *Kevin Mara* •
Military Museum – 1985
Military hist 47953

Danbury WI

Forts Folle Avoine Historic Park, 8500 County Rd U, Danbury, WI 54830 • T: +1 715 8668890 • F: +1 715 8668081 • fahp@centuryitel.net •
Dir.: *Kevin Klucas* •
Local Museum – 1945
Fur trade 47954

Dania Beach FL

South Florida Museum of Natural History, 481 S Federal Hwy, Dania Beach, FL 33004 • T: +1 954 9257770 • F: +1 954 9257064 • digit@sfmuseumnh.org •
Pres.: *Robert P. Kelley* •
Archaeology Museum / Natural History Museum – 1980
Archaeology, paleontology, natural history 47955

Danvers MA

Danvers Historical Society Exhibition, 9 Page St, Danvers, MA 01923 • T: +1 978 7771666 • F: +1 978 7775028 • dhs@danvershistory.org • www.danvershistory.org •
Historical Museum – 1889
Local history – archives 47956

Rebecca Nurse Homestead, 149 Pine St, Danvers, MA 01923 • T: +1 978 7748799 • president@rebeccanurse.org • www.rebeccanurse.org •
Cur.: *Richard B. Trask* •
Historical Museum – 1974
1678 home of Rebecca Nurse, hanged as a witch in 1692 47957

Danville CA

Blackhawk Museum, 3700 Blackhawk Plaza Circle, Danville, CA 94506 • T: +1 925 7362277 • F: +1 925 7364818 • museum@blackhawkmuseum.org • www.blackhawkmuseum.org •
Dir.: *Don Dunn* •
Science&Tech Museum – 1988
Automotive art 47958

Eugene O'Neill Home, POB 280, Danville, CA 94526 • T: +1 925 8380249 • F: +1 925 8389471 • euon_interpretation@nps.gov • www.nps.gov/euon •
Special Museum – 1976
Playwright E. O'Neill, clothing, jewelry, paintings, letters, photos, furnishings 47959

Danville IL

Vermilion County Museum, 116 N Gilbert St, Danville, IL 61832-8506 • T: +1 217 4422922, 4422001 • F: +1 217 4422001 • webmaster@vermilioncountymuseum.org • www.vermilioncountymuseum.org •
Pres.: *Donald Richter* •
Local Museum – 1964
History, housed in 1855 Dr. Fithian Home, doctor's residence, often visited by Abraham Lincoln 47960

Danville KY

Constitution Square, 134 S Second St, Danville, KY 40422 • T: +1 859 2397089 • F: +1 859 2397894 • brenda.willoughby@mail.state.ky.us •
Head: *Brenda Willoughby* •
Historic Site – 1937
Historic site 47961

McDowell House Museum, 125 S Second St, Danville, KY 40422 • T: +1 859 2362804 • F: +1 859 2362804 • mcdhse@kih.net • www.mcdowellhouse.org •
Dir.: *Carol J. Senn* •
Historical Museum / Historic Site – 1939
McDowell House 1800, 1795 Apothecary shop 47962

Danville VA

Danville Museum of Fine Arts and History, 975 Main St, Danville, VA 24541 • T: +1 804 7935644 • F: +1 804 7996145 • dmfah@gamewood.net • www.danvillemuseum.org •
Exec. Dir.: *Lynne Bjarnesen* •
Fine Arts Museum / Historical Museum – 1974
Paintings, prints, textiles, American costumes, hist documents and artifacts 47963

Danville Science Center, 677 Craghead St, Danville, VA 24541 • T: +1 804 7915160 • F: +1 804 7915168 • www.dsc.smv.org •
Science&Tech Museum – 1974
Astronomy, chemistry, physics, rocks and minerals, zoology 47964

Daphne AL

American Sport Art Museum, 1 Academy Dr, Daphne, AL 36526-7055 • T: +1 251 6263303 • F: +1 251 6251035 • asama@ussa.edu • www.asama.org •
Dir.: *Dr. Thomas P. Rosandich* • Cur.: *Kay Daughdrill* •
Fine Arts Museum / University Museum – 1985
Bronze sculptures, paintings, lithographs – archives 47965

Darien CT

Bates-Scofield Homestead, 45 Old King's Hwy, N, Darien, CT 06820 • T: +1 203 6559233 • F: +1 203 6563892 • darien.historical@snet.net •
Dir.: *Judith Geoppa* • Cur.: *Babs White* (Cosumes) • *Lois Hofmann* (Quilts) •
Historical Museum – 1954 47966

Darien GA

Fort King George, Fort King George Dr, Darien, GA 31305 • T: +1 912 4374770 • F: +1 912 4375479 • ftkgeo@darientel.net • www.darientel.net/~ftkgeo •
Head: *Ken Akins* •
Historic Site / Historical Museum / Military Museum – 1961
Archaeology, photos, ethnography 47967

Darlington SC

Darlington Raceway Stock Car Museum, 1301 Harry Byrd Hwy, Darlington, SC 29532, mail addr: POB 500, Darlington, SC 29540 -0500 • T: +1 803 3958821 • F: +1 803 3933911 • www.darlingtonraceway.com •
Science&Tech Museum – 1965
Stock cars 47968

Joe Weatherly Museum → Darlington Raceway Stock Car Museum

Davenport IA

Davenport Museum of Art, 1737 W 12th St, Davenport, IA 52804 • T: +1 563 3267804 • F: +1 563 3267876 • dma@ci.davenport.ia.us • www.art-dma.org •
Dir.: *Linda Downs* • Cur.: *Michelle Robinson* •
Fine Arts Museum – 1925
Art 47969

Putnam Museum of History and Natural Science, 1717 W 12th St, Davenport, IA 52804 • T: +1 563 3241933 • F: +1 563 3246638 • museum@putnam.org • www.putnam.org •
Dir.: *Christopher J. Reich* • Chief Cur.: *Eunice Schlicting* •
Historical Museum / Ethnology Museum / Natural History Museum – 1867
History, natural science, anthropology 47970

Davenport WA

Fort Spokane Visitor Center and Museum, 44150 District Office Ln N, Davenport, WA 99122 • T: +1 509 7252715, 6333836 • F: +1 509 6333834 •
Military Museum – 1965
Military hist 47971

Lincoln County Historical Museum, 7th and Park, Davenport, WA 99122 • T: +1 509 7256711 •
Local Museum – 1972
Early agricultural tools, combine harvester, steam engine, cameras, printing tools, costumes, guns, Indian artifacts, railroad 47972

Davidson NC

William H. Van Every jr. and Edward M. Smith Galleries, Davidson College, c/o Belk Visual Arts Center, 315 N Main St, Davidson, NC 28036-1720 • T: +1 704 8942519 • F: +1 704 8942691 • brthomas@davidson.edu • www.davidson.edu •
Dir.: *Brad Thomas* •
Fine Arts Museum – 1962
15th - 20th c graphics, photography, painting, sculpture 47973

Davie FL

Young at Art Children's Museum, 11584 W State Rd 84, Davie, FL 33325 • T: +1 954 4240085 ext 21 • F: +1 954 3705057 • visitorservices@youngatartmuseum.org • www.youngatartmuseum.org •
Dir.: *Mindy Shrago* • Ass. Dir.: *Esther Shrago* •
Special Museum – 1985
Hands-On Children's museum 47974

U

Davis CA

Davis Art Center, 1919 F St, Davis, CA 95617 • T: +1 530 7564100 • F: +1 530 7563041 • davisart@dcn.org • www.davisartcenter.org •
Dir.: *Jackie Steven* •
Fine Arts Museum – 1959
Contemporary art ... 47975

The Nelson, Richard L. Nelson Gallery and Fine Arts Collection, c/o University of California, One Shields Av, Art Bldg, Room 124, Davis, CA 95616 • T: +1 530 7528500 • F: +1 530 7549112 • nelsongallery@ucdavis.edu • www.nelsongallery.info •
Dir.: *Renny Pritikin* •
Fine Arts Museum / University Museum – 1976
16th c present paper, prints, drawings and paintings, contemporary art, SE Asian, European and American art, ceramics – library ... 47976

Pence Gallery, 212 D St, Davis, CA 95616 • T: +1 530 7583370 • F: +1 530 7584670 • penceassistant@sbcglobal.net • www.pencegallery.org •
Dir.: *Natalie Nelson* • Asst. Dir.: *Eileen Hendren* •
Public Gallery – 1975
Californian art and history, contemporary art ... 47977

R.M. Bohart Museum of Entomology, University of California, 1124 Academic Surge, Davis, CA 95616-8584 • T: +1 530 7520493 • F: +1 530 7529464 • bmuseum@ucdavis.edu • bohart.ucdavis.edu •
Dir.: *Dr. Lynn S. Kimsey* • Cur.: *Philip S. Ward* • *Peter S. Cranston* • *Penny J. Gullan* •
University Museum / Natural History Museum – 1946
7 million arthropod specimens – library ... 47978

Daviston AL

Horseshoe Bend National Military Park, 11288 Horseshoe Bend Rd, Daviston, AL 36256 • T: +1 256 2347111 • www.nps.gov/hobe •
Historic Site – 1959
Horseshoe Bend Battlefield, location of final battle of the Creek Indian War of 1813-1814 ... 47979

Davisville MO

Dillard Mill State Historic Site, 142 Dillard Mill Rd, Davisville, MO 65456 • T: +1 573 2443120 • F: +1 573 2445672 • dillmill@misn.com • www.dnr.state.mo.us/dsp •
Head: *Jerry Wilson* •
Historical Museum – 1975
Mill machinery ... 47980

Dawson Springs KY

Dawson Springs Museum and Art Center, 127 S Main St, Dawson Springs, KY 42408 • T: +1 270 7973503, 7973891 •
Pres./Cur.: *Claude A. Holeman* •
Fine Arts Museum / Decorative Arts Museum / Historical Museum – 1986 ... 47981

Dayton OH

Boonshoft Museum of Discovery, 2600 DeWeese Pkwy, Dayton, OH 45414 • T: +1 937 2757431 • F: +1 937 2755811 • info@boonshoftmuseum.org • www.boonshoftmuseum.org •
Dir. & CEO: *Mark Meister* • Dir.: *Cheryl Adams* (Astronomy) • Cur.: *Tammery Olsen* (Live Animals) • *Lynn Simonelli* (Anthropology) • Asst. Cur.: *William Kennedy* •
Archaeology Museum / Natural History Museum / Science&Tech Museum – 1893
Archaeology, natural history, ethnology ... 47982

Carillon Historical Park, 1000 Carillon Blvd, Dayton, OH 45409 • T: +1 937 2932841 ext 100 • F: +1 937 2935798 • bkress@carillonpark.org • www.carillonpark.org •
Exec. Dir.: *Brady Kress* •
Historical Museum – 1950
Wright brothers' 1905 airplane, old automobiles and bicycles, train car, trolley bus ... 47983

Dayton Art Institute, 456 Belmonte Park N, Dayton, OH 45405-4700 • T: +1 937 2235277 • F: +1 937 2233140 • info@daytonartinstitue.org • www.daytonartinstitute.org •
Dir.: *Alexander Lee Nyerges* • Chief Cur.: *Michael Komanecky* • Cur.: *Li Jian* (Asian Art) •
Fine Arts Museum – 1919
Asian, African, pre-Columbian, Oceanic and European paintings, sculptures, graphic and decorative art 47984

Dayton Visual Arts Center, 118 N Jefferson St, Dayton, OH 45402 • T: +1 937 2243822 • dvac@daytonvisualarts.org • www.daytonvisualarts.org •
Dir.: *Jane A. Black* •
Public Gallery – 1991
Regional contemp art ... 47985

Montgomery County Historical Society Center, 224 North St.Clair St, Dayton, OH 45402 • T: +1 937 2286271 • F: +1 937 3317160 • mchs@daytonhistory.org • www.daytonhistory.org •
Exec.Dir.: *Brian Hackett* • Ass. Dir.: *Claudia Watson* •
Historical Museum – 1896
Miami Valley hist, decorative arts, textiles, hand tools – archive ... 47986

Patterson Homestead, 1815 Brown St, Dayton, OH 45409 • T: +1 937 2229724 • F: +1 937 2220345 • www.pattersonhome.com •
Dir.: *Denise L. Darling* • Cur.: *Ray Shook* •
Decorative Arts Museum / Historical Museum – 1953
Period furniture, memorabilia of the Patterson family ... 47987

Paul Laurence Dunbar State Memorial, 219 N Paul Laurence Dunbar St, Dayton, OH 45407 • T: +1 937 2247061 • F: +1 937 2244256 • paul33@sbcglobal.net • www.ohiohistory.org •
Head: *LaVerne Sci* •
Historical Museum / Historic Site – 1936
Paintings, photographs and memorabilia of the Dunbar family ... 47988

Sunwatch Indian Village - Archaeological Park, 2301 W River Rd, Dayton, OH 45418 • T: +1 937 2688199 • F: +1 937 2681760 • asawyer@sunwatch.org • www.sunwatch.org •
Exec. Dir.: *Mark Meister* •
Historical Museum / Archaeology Museum – 1988
Prehistory, Fort Ancient Indian culture ... 47989

Wright State University Art Galleries, 3640 Colonel Glenn Hwy, Dayton, OH 45435 • T: +1 937 7752896 • F: +1 937 7754082 • cmartin@desire.wright.edu • www.wright.edu/artgalleries/ •
Fine Arts Museum / University Museum – 1974
Contemporary art ... 47990

Dayton VA

Heritage Center, Harrisonburg-Rockingham Historical Society, 382 High St, Dayton, VA 22821 • T: +1 540 8792616 • F: +1 540 8792616 • heritag1@shentel.net • www.heritagecenter.com •
Pres.: *Shelvie Carr* •
Local Museum – 1895
Local folklore, art of the Society, electric map of Sonewall Jacksons' Valley campaign ... 47991

Dayton WA

Dayton Historical Depot Society Museum, 222 E Commercial St, Dayton, WA 99328 • T: +1 509 3822026 • F: +1 509 3822640 •
Local Museum – 1974 ... 47992

Daytona Beach FL

Halifax Historical Museum, 252 S Beach St, Daytona Beach, FL 32114 • T: +1 904 2556976 • F: +1 904 2557605 • mail@halifaxhistorical.org • www.halifaxhistorical.org •
Dir.: *Suzanne Heddy* •
Local Museum – 1949
American artifacts, military and local hist ... 47993

The Museum of Arts and Sciences, 1040 Museum Blvd, Daytona Beach, FL 32114 • T: +1 904 2550285 • F: +1 904 2555040 • info@moas.org • www.moas.org •
Exec. Dir.: *Gary R. Libby* • Dep. Dir.: *Patricia Thalheimer* • Cur.: *Richard Lussky* •
Fine Arts Museum / Local Museum / Natural History Museum – 1971
American, European, African and Cuban fine, folk and decorative art ... 47994

Southeast Museum of Photography, Daytona Beach Community College, 1200 W International Speedway Blvd, Daytona Beach, FL 32114 • T: +1 386 5063080 • F: +1 386 5064487 • millerm@daytonastate.edu • www.smponline.org •
Interim Dir.: *Kevin R. Miller* • Cur.: *Trish Thompson* •
Fine Arts Museum / Science&Tech Museum – 1979
Photography ... 47995

De Land FL

De Land Museum of Art, 600 N Woodland Blvd, De Land, FL 32720-3447 • T: +1 386 7344371 • F: +1 386 7347697 • info@delandmuseum.com • www.delandmuseum.com •
Dir.: *Jennifer Coolidge* •
Fine Arts Museum – 1951
20th c American fine arts, crafts, photography, Florida and wildlife art ... 47996

Duncan Gallery of Art, Stetson University, 421 N Woodland Blvd, De Land, FL 32723 • T: +1 386 8227266 • F: +1 386 8227268 • cnelson@stetson.edu • www.stetson.edu/departments/art •
Dir.: *Dan Gunderson* •
Public Gallery / University Museum – 1965
Drawings, prints, watercolors, oils, ceramics, sulptures, photographs ... 47997

Gillespie Museum of Minerals, Stetson University, 234 E Michigan Av, De Land, FL 32720 • T: +1 386 8227330 • F: +1 386 8227328 • hvanter@stetson.edu • www.gillespiemuseum.stetson.edu •
Dir.: *Dr. Robert S. Chauvin* • Cur.: *Dr. Bruce C. Bradford* •
University Museum / Natural History Museum – 1958
Mineralogy ... 47998

De Pere WI

Oneida Nation Museum, W892 County Trunk EE, De Pere, WI 54115 • T: +1 920 8692768 • F: +1 920 8692959 • rlara@oneidanation.org • www.oneidanation.org •
Dir.: *Rita Lara* •
Ethnology Museum – 1979
Oneida Nation and Iroquois Confederacy, Indian artifacts, Civil War ... 47999

White Pillars Museum, 403 N Broadway, De Pere, WI 54115 • T: +1 920 3363877 •
Dir.: *Pete Safford* •
Local Museum – 1970
Local hist ... 48000

De Smet SD

De Smet Depot Museum, 104 Calumet Ave NE, De Smet, SD 57231 • T: +1 605 8543991 •
C.E.O.: *Gary Wolkow* •
Local Museum – 1965
Local hist ... 48001

Deadwood SD

Adams Museum, 54 Sherman St, Deadwood, SD 57732 • T: +1 605 5781714, 5781928 • F: +1 605 5781194 • director@adamsmuseumandhouse.org • www.adamsmuseumandhouse.org •
Dir.: *Mary A. Kopco* • Cur.: *Darrel Nelson* (Exhibits) • *Arlette Hansen* •
Local Museum – 1930
Local hist ... 48002

Adams Museum and Historic Adams House, 54 Sherman St and 22 Van Buren Av, Deadwood, SD 57732, mail addr: POB 252, Deadwood, SD 57732 • T: +1 605 5781928 • F: +1 605 5781194 • amhdirector@rushmore.com • www.adamsmuseumandhouse.org •
Dir.: *Mary A. Kopco* •
Historical Museum
American hist ... 48003

House of Roses, Senator Wilson Home, 15 Forest Av, Deadwood, SD 57732 • T: +1 605 7221879 •
Pres./Dir.: *Michael Bockwoldt* •
Historical Museum – 1976
Historic house, antique Victorian furniture, paintings, old prints ... 48004

Dearborn MI

Automotive Hall of Fame, 21400 Oakwood Blvd, Dearborn, MI 48124 • T: +1 313 2404000 • F: +1 313 2408641 • www.automotivehalloffame.org •
Dir.: *Rod Alberts* •
Science&Tech Museum – 1939
Histories of companies, corporations and peoples – library ... 48005

Commandant's Quarters, Dearborn Historical Museum, 21950 Michigan Av, Dearborn, MI 48124-2322 • T: +1 313 5650844 • F: +1 313 5654848 • mmacdonald@ci.dearborn.mi.us • www.cityofdearborn.org/departments/historicalmuseum •
Cur.: *Mary V. MacDonald* •
Historical Museum ... 48006

Dearborn Historical Museum, McFadden Ross House, 915 Brady St, Dearborn, MI 48124-2322 • T: +1 313 5653000 • F: +1 313 5654848 • mmacdonald@ci.dearborn.mi.us • www.cityofdearborn.org/departments/historicalmuseum •
Cur.: *Mary V. MacDonald* •
Local Museum – 1950
Local history, ex powder magazine, agriculture, ,military, costumes, decorative arts – archives, library ... 48007

Henry Ford Estate, Universty of Michigan-Dearborn, 4901 Evergreen Rd, Dearborn, MI 48128 • T: +1 313 5935590 • F: +1 313 5935243 • grodgers@umd.umich.edu • www.henryfordestate.org •
Decorative Arts Museum / Historical Museum – 1957
Former home of Henry Ford, decorative arts, mechanical equip – library ... 48008

Henry Ford Museum, 20900 Oakwood Blvd, Dearborn, MI 48124 • T: +1 313 9826001 • F: +1 313 9826250 • barbh@thehenryford.org • www.hfmgv.org •
Dir.: *Christian Overland* •
Historical Museum / Science&Tech Museum – 1929
Agricultural equip, power machinery, precursor to his model T-wagons and 'horseless carriages' – library, IMAX ... 48009

Death Valley CA

Death Valley National Park Museum, on Hwy 190, Death Valley, CA 92328, mail addr: POB 579, Death Valley, CA 92328 • T: +1 760 7863200 • F: +1 760 7863283 • blair-davenport@nps.gov • home.nps.gov/applications/museum •
Head: *James T. Reynolds* •
Ethnology Museum / Archaeology Museum / Historical Museum – 1933
Natural and local hist, anthropology and Archaeology – herbarium ... 48010

Decatur AL

Department of Fine Arts Gallery, c/o Calhoun Community College, Hwy 31 N, Fine Arts Bldg, Decatur, AL 35609-2216 • T: +1 256 3062695, 2891479 • alanshero@gmail.com • www.calhoun.cc.al.us •
Dir.: *Kristine Beadle* •
Fine Arts Museum / University Museum – 1965
American and European graphics, student art and photographs ... 48011

Decatur GA

Dalton Gallery, Agnes Scott College, E College Av, Decatur, GA 30030 • T: +1 404 4716000 • F: +1 404 4715369 • lalembik@agnesscott.edu • www.agnesscott.edu •
Chm.: *Dr. Donna Sadler* •
Fine Arts Museum / University Museum – 1957
Decorative art, paintings, sculpturs ... 48012

DeKalb Historical Society Museum, 101 E Court Sq, Decatur, GA 30030 • T: +1 404 3731088 • F: +1 404 3738287 • dhs@dekalbhistory.org • www.dekalbhistory.org •
Exec. Dir.: *Sue Ellen Owens* •
Local Museum – 1947
Local history, Old Courthouse, personal artefacts, costumes, photos ... 48013

Decatur IL

Birks Museum, c/o Millikin University, 1184 W Main, Decatur, IL 62522 • T: +1 217 4246337 • F: +1 217 4243993 • ewalker@mail.millikin.edu • www.millikin.edu •
Dir.: *Edwin G. Walker* •
Decorative Arts Museum / University Museum – 1981
China, glass, paper weights ... 48014

Kirkland Fine Arts Center-Perkinson Gallery, c/o Millikin University, 1184 W Main St, Decatur, IL 62522 • T: +1 217 4246227 • F: +1 217 4243993 • jschietinger@mail.millikin.edu •
Dir.: *James Schietinger* • Ass. Dir.: *Lyle Salmi* •
Fine Arts Museum / University Museum – 1969
19th - 20th c art, graphics, decorative art, paintings, sculpture ... 48015

Macon County Museum Complex, 5580 N Fork Rd, Decatur, IL 62521 • T: +1 217 4224919 • F: +1 217 4224773 • info@mchsdecatur.org • www.mchsdecatur.org •
Pres.: *Roger Rosenkranz* •
Local Museum – 1973
Historical artifacts from Macon County, photoghraphs, tools ... 48016

Perkinson Gallery, Millikin University, c/o Millikin University, 1184 W Main St, Decatur, IL 62522 • T: +1 217 4246227 • F: +1 217 4243993 • jschietinger@mail.millikin.edu • www.millikin.edu •
Dir.: *Jim Schitinger* •
Fine Arts Museum / University Museum – 1970
Drawings, painting, prints, sculpture, watercolors ... 48017

Decatur IN

Adams County Historical Museum, 420 W Monroe St, Decatur, IN 46733 • T: +1 219 7242341 • goble@adamswells.net •
Pres.: *Rebecca Goble* •
Local Museum – 1965
Local history, in Charles Dugan Home ... 48018

Decatur MI

Historic Newton Home, 20689 Marcellus Hwy, Decatur, MI 49045 • T: +1 616 4459016 •
Historical Museum – 1974 ... 48019

Decatur TX

Wise County Heritage Museum, 1602 S Trinity, Decatur, TX 76234 • T: +1 940 6275586 • wisemuseum@ntws.net •
C.E.O.: *Rosalie Gregg* •
Local Museum – 1967
Hist artifacts, manuscript coll ... 48020

Decorah IA

Luther College Fine Arts Collection, 700 College Dr, Luther College Library, Decorah, IA 52101-1042 • T: +1 563 3871195 • F: +1 563 3871657 • kempjane@luther.edu • finearts.luther.edu •
Head: *Jane Kemp* • Gallery Coord.: *David Kamm* •
Fine Arts Museum / University Museum
Paintings, prints, sculpture, Marcks Gerhard drawings and prints, Wildenhain Marguerite pottery and drawings, Skandinavian immigrant paintings ... 48021

Vesterheim Norwegian-American Museum, 523 W Water St, Decorah, IA 52101-0379 • T: +1 563 3829681 • F: +1 319 3828828 • vesterheim@vesterheim.org • www.vesterheim.org •
Dir.: *Janet Blohm Pultz* • Chief Cur.: *Darrell D. Henning* • Cur.: *Laurann Gilbertson* (Textiles) • *Tova Brandt* • *Steven Johnson* (Jacobson Farmstead) • Libr.: *Carol Hasvold* •
Ethnology Museum / Folklore Museum – 1877
Ethnology, genealogy ... 48022

Dedham MA

Dedham Historical Museum, 612 High St, Dedham, MA 02027-0215, mail addr: Box 215, Dedham, MA 02027-0215 • T: +1 781 3261385 • F: +1 781 3265762 • society@dedhamhistorical.org • www.dedhamhistorical.org •
Exec. Dir.: *Ronald F. Frazier* •
Fine Arts Museum / Decorative Arts Museum / Local Museum – 1859
Local history, pottery, needlework and costume, Katharine Pratt silver, fine arts, furniture, industry – library ... 48023

Fairbanks House, 511 East St, Dedham, MA 02026 • T: +1 781 3261170 • F: +1 781 3262147 • fairbankshouse@aol.com • www.fairbankshouse.org • Pres.: *Lynn Fairbank* • Cur.: *Julie Letendre* • Historical Museum – 1903 Oldest surviving timber frame house in North America, built for Fairbanks family, furnishings 48024

Museum of Bad Art, 580 High St, Dedham, MA 02026, mail addr: 73 Parker Rd, Needham, MA 02494 • T: +1 781 4446757 • F: +1 781 4339991 • moba@museumofbadart.org • www.museumofbadart.org • C.E.O.: *Louise R. Sacco* • Cur.: *Scott Wilson* • Fine Arts Museum – 1993 Works of exuberant, although crude, execution by artists barely in control of the brush 48025

Deer Isle ME

Deer Isle-Stonington Historical Society Museum, Rte 15A, Deer Isle, ME 04627 • T: +1 207 3482897 • becke01@prexar.com • Pres.: *Becke Thompson* • Historical Museum – 1959 Local history 48026

Haystack Mountain School of Crafts Gallery, 89 Haystack School Dr, Deer Isle, ME 04627-0518 • T: +1 207 3482306 • F: +1 207 3482307 • www.haystack-mtn.org • Dir.: *Stuart J. Kestenbaum* • Historical Museum – 1950 48027

Deer Lodge MT

Grant-Kohrs Ranch, 266 Warren Ln, Deer Lodge, MT 59722 • T: +1 406 8462070 • F: +1 406 8463962 • laura_rotegard@nps.gov • www.nps.gov/grko • Dir.: *Darlene Koontz* • Cur.: *Chris Ford* • Historic Site – 1972 Horse drawn vehicles, tack, bunkhouse furnishings, textiles 48028

Montana Auto Museum, 1106 Main St, Deer Lodge, MT 59722 • T: +1 406 8463111 • F: +1 406 8463156 • oldprisonmuseums@in-tch.com • Dir.: *Andrew C. Towe* • Cur.: *James R. Haas* • Historical Museum – 1980 48029

Old Prison Museum, 1106 Main St, Deer Lodge, MT 59722 • T: +1 406 8463111 • F: +1 406 8463156 • oldprisonmuseums@in-tch.com • www.pcmaf.org • Dir.: *John O'Donnell* • Cur.: *James R. Haas* • Historical Museum – 1980 Contraband exhib & display areas 48030

Powell County Museum, 1106 Main St, Deer Lodge, MT 59722 • T: +1 406 8461694 • F: +1 406 8463156 • oldprisonmuseums@in-tch.com • www.pcmaf.org • Dir.: *Andrew C. Towe* • Cur.: *James R. Haas* • Local Museum – 1964 Mastodon fossils, trapping & mining tools, railroad exhib 48031

Deerfield MA

Historic Deerfield, The Street, Deerfield, MA 01342, mail addr: POB 321, Deerfield, MA 01342 • T: +1 413 7745581 • F: +1 413 7757220 • tours@historic-deerfield.org • www.historic-deerfield.org • Dir.: *Donald R. Friary* • Cur.: *Amanda E. Lange* • *Edward F. Maeder* (Textiles) • Sc. Staff: *Joshua W. Lane* • Decorative Arts Museum / Local Museum – 1952 Local history and life,decorative arts, silver, metalwork – library 48032

Memorial Hall Museum, 8 Memorial St, Deerfield, MA 01342 • T: +1 413 7743768 • F: +1 413 7747070 • info@old-deerfield.org • www.old-deerfield.org • Dir.: *Timothy C. Neumann* • Cur.: *Suzanne Flynt* • Decorative Arts Museum / Local Museum – 1870 Local hist, decorative arts, Deerfield Academy, musical instr, American paintings – library 48033

Deerfield Beach FL

Deerfield Beach Historical Society Museum, 380 E Hillsboro Blvd, Deerfield Beach, FL 33443 • T: +1 954 4290378 • F: +1 954 4290378 • fdallen49@earthlink.net • Exec. Dir.: *Dale Allen* • Local Museum – 1973 Photographs, oral histories, costumes, period furnishings – archives, library 48034

Defiance MO

Historic Daniel Boone Home and Boonesfield Village, Lindenwood University, 1868 Hwy F, Defiance, MO 63341 • T: +1 636 7982005 • F: +1 636 7982914 • www.lindenwood.edu • Dir.: *Pam Jensen* • *Greta Maxheimer* • Open Air Museum / Local Museum – 1803 Authenic documents, furniture and artifacts of the Boone family 48035

Defiance OH

Au Glaize Village, 12296 Krouse Rd, Defiance, OH 43512, mail addr: 15806 Campbell Rd, Defiance, OH 43512 • T: +1 419 7840107 • www.defiance-online.com/auglaize • C.E.O.: *Lynn Lantz* • Local Museum – 1966 Local hist, farming, household, clothing, military uniforms, natural hist, archaeology, model trains 48036

DeKalb IL

Anthropology Museum, Northern Illinois University, DeKalb, IL 60115 • T: +1 815 7530230 • F: +1 815 7537027 • awparsons@niu.edu • www.niu.edu/anthro_museum • Exec. Dir.: *Winifred Creamer* • Dir.: *Ann Wright-Parsons* • University Museum / Ethnology Museum / Archaeology Museum – 1964 Anthropology, archaeology 48037

Ellwood House Museum, 509 N First St, DeKalb, IL 60115 • T: +1 815 7564609 • F: +1 815 7564645 • ellwoodhouse@hcnet.com • www.ellwoodhouse.org • Dir.: *Gerald J. Brauer* • Cur.: *April Arrecis* • Historical Museum – 1965 Ellwood family and DeKalb county hist, decorative arts 48038

NIU Art Museum, Northern Illinois University, Altgeld Hall, DeKalb, IL 60115 • T: +1 815 7531936 • F: +1 815 7537897 • pdoherty@niu.edu • www.vpa.niu.edu/museum • Dir.: *Peggy M. Doherty* • Ass. Dir.: *Jo Burke* • Fine Arts Museum / University Museum – 1970 Contemporary and modern paintings, prints, sculptures and photographs, Burmese art, native American art 48039

Del Rio TX

Whitehead Memorial Museum, 1308 S Main St, Del Rio, TX 78840 • T: +1 830 7747568 • F: +1 830 7680223 • director@whitehead-museum.com • www.whitehead.museum.com • C.E.O.: *Lee Lincoln* • Local Museum – 1962 Jersey Lily Saloon replica, pioneer log cabin 48040

Delafield WI

Hawks Inn, 426 Wells St, Delafield, WI 53018 • T: +1 414 6464794 • www.hawksinn.org • Pres.: *Mary Daniel* • Local Museum – 1960 Historic building 48041

Saint John's Northwestern Military Academy Museum, 1101 N Genesee St, Delafield, WI 53018 • T: +1 414 6467118 • F: +1 414 6467155 • alumni@sjnma.org • www.sjnma.org • C.E.O.: *Margaret H. Koller* • Military Museum – 1984 Military hist 48042

Delano CA

Delano Heritage Park Museum, 330 Lexington and Garces Hwy, Delano, CA 93215 • T: +1 661 7256730 • F: +1 559 7572344 • Pres.: *James Sevier* • Local Museum / Natural History Museum – 1961 Local history, argriculture – arboretum 48043

Delaware City DE

Fort Delaware Society Museum, 122 Washington St, Delaware City, DE 19706, mail addr: POB 553, Delaware City, DE 19706 • T: +1 302 8341630 • ftdsociety@del.net • www.del.net/org/fort • Pres.: *William G. Robelen* • Military Museum – 1950 Military hist 48044

Delray Beach FL

Cornell Museum, 51 N Swinton Av, Delray Beach, FL 33444 • T: +1 561 2437922 • F: +1 561 2437022 • museum@oldschool.org • www.oldschool.org • Dir.: *Gloria Rejune Adams* • Fine Arts Museum – 1990 Fine art, sculpture 48045

The Morikami Museum and Japanese Gardens, 4000 Morikami Park Rd, Delray Beach, FL 33446 • T: +1 561 4950233 • F: +1 561 4992557 • morikami@co.palm-beach.fl.us • www.morikami.org • Dir.: *Larry Rosensweig* • Cur.: *Thomas Gregersen* • Ethnology Museum / Folklore Museum – 1977 Ethnology, Japanese culture 48046

Palm Beach Photographic Center, 55 NE Second Av, Delray Beach, FL 33444 • T: +1 561 2769797 • F: +1 561 2761932 • info@workshop.org • www.workshop.org • Man.: *Art NeJame* • Public Gallery 48047

Delta CO

Delta County Museum, 251 Meeker St, Delta, CO 81416 • T: +1 970 8748721 • deltamuseum@aol.com • Dir.: *James K. Wetzel* • *Bernice Musser* • Local Museum – 1964 Regional history 48048

Fort Uncompahgre History Museum, 205 Gunnison River Dr, Delta, CO 81416 • T: +1 970 8748349 • F: +1 970 8741353 • fortunc@doci.net • www.deltafort.org • C.E.O.: *Wilma Erven* • Cur.: *Paul D. Suppes* • Historical Museum – 1990 Fur trade period items ranging from firearms and trape to tools and livestock 48049

Delta UT

Great Basin Historical Society & Museum, 328 W 100 N, Delta, UT 84624 • T: +1 435 8645013 • F: +1 435 8642446 • greatbasin@hubwest.com • Dir.: *Charlotte K. Morrison* • Cur.: *Sindy McMichael* • Local Museum – 1988 Geological specimens, personal artifacts, tools & implements for geology 48050

Delton MI

Bernard Historical Museum, 7135 W Delton Rd, Delton, MI 49046 • T: +1 616 6235451 • Pres./CEO: *Richard Martin* • Historical Museum – 1962 Local history, farming, pioneer family – small library 48051

Demarest NJ

Center Gallery, Old Church Cultural Center School, 561 Piermont Rd, Demarest, NJ 07627 • T: +1 201 7677160 • F: +1 201 7670497 • gallery@occcartschool.org • www.occcartschool.org • Dir: *Laura Bundesen* • *Paula Madawick* • Public Gallery Temporary exhibitions 48052

Deming NM

Luna Mimbres Museum, 301 S Silver St, Deming, NM 88030 • T: +1 505 5462382 • F: +1 505 5440121 • dlm-museum@zianet.com • www.cityofdeming.org/museum.html • Dir.: *Sharon Lein* • Local Museum / Decorative Arts Museum – 1955 Local hist, vintage clothing, Mimbres Indian artifacts, American Indian art, anthropology, folk art, decorative art 48053

Demopolis AL

Bluff Hall Museum, 405 N Commissioners Av, Demopolis, AL 36732 • T: +1 334 2899644 • Local Museum Costumes, history, furnishings 48054

Gaineswood Mansion, 805 S Cedar Av, Demopolis, AL 36732 • T: +1 334 2894846 • gaineswd@bellsouth.net • www.preserveala.org • Dir.: *Matthew D. Hartzell* • Historical Museum / Historic Site – 1975 Antebellum mansion 48055

Denison TX

Eisenhower Birthplace, 208 E Day, Denison, TX 75020 • T: +1 903 4658908 • F: +1 903 4658988 • eisenhower@texoma.net • www.eisenhowerbirthplace.org • Pres.: *Wayne Jamison* • Local Museum – 1946 Historic house where Dwight D. Eisenhower was born 48056

Dennis MA

Cape Museum of Fine Arts, 60 Hope Ln, Dennis, MA 02638 • T: +1 508 3854477 • F: +1 508 3857933 • cmfa@capecod.net • www.cmfa.org • Exec. Dir.: *Elizabeth Ives Hunter* • Cur.: *Michael Giaquinto* • Fine Arts Museum / Public Gallery – 1981 20th c art – auditorium, archives 48057

Dennison OH

Dennison Railroad Depot Museum, 400 Center St, Dennison, OH 44621 • T: +1 740 9226776 • F: +1 740 9224929 • depot@tusco.net • www.dennisondepot.org • C.E.O.: *Wendy R. Zucal* • Science&Tech Museum / Local Museum – 1984 Railroad, WW 2 Canteen site, local hist, model train, Keystone exhibit, roliing stock 48058

Denton MD

Museum of Rural Life, 16 N Second St, Denton, MD 21629, mail addr: POB 514, Denton, MD 21629 • T: +1 410 4792055 • F: +1 410 4794513 • Historical Museum Personal artifacts, Maryland hist, photographs 48059

Denton TX

Dar Museum First Ladies of Texas Historic Costumes Collection, Texas Woman's University, Denton, TX 76204 • T: +1 940 8983350 • F: +1 940 8983306 • www.twu.edu • Pres.: *Ann Stuart* • Special Museum – 1940 Inaugural Ball gowns of Texas Ist ladies, incl. wives of the presidents of the Republik of Texas and the governors of the State 48060

Denton County Historical Museum, 5800 North I-35, Denton, TX 76201 • T: +1 940 3800877 • F: +1 940 3801699 • dchminc@earthlink.net • www.dentoncountyhistoricalmuseum.com • Dir.: *Judy Selph* • Local Museum – 1977 Local hist, incl community, school, church, early settlers & cementeries 48061

Texas Woman's University Art Galleries, 1200 Frame St, Denton, TX 76204 • T: +1 940 8982530 • F: +1 940 8982496 • visualarts@twu.edu • www.twu.edu • Head: *John Weinkein* • Fine Arts Museum / University Museum – 1901 Art 48062

University of North Texas Art Gallery and Cora Stafford Gallery, School of Visual Arts, 1201 W Mulberry at Welch, Denton, TX 76203, mail addr: POB 305100, Denton, TX 76203-5100 • T: +1 940 5654005 • F: +1 940 5654717 • block@art.unt.edu • www.art.unt.edu • Dir.: *Diana Block* • University Museum / Fine Arts Museum – 1972 20th c contemporary art most of which is by former students & faculty 48063

Denver CO

Black American West Museum and Heritage Center, 3091 California St, Denver, CO 80205 • T: +1 303 2922566 • F: +1 303 3821981 • bawmhc@aol.com • www.coax.net/people/lwf/bawmus.htm • Dir.: *Ottawa Harris* • Historical Museum – 1971 African American hist 48064

Byers-Evans House Museum, 1310 Bannock St, Denver, CO 80204 • T: +1 303 6204933 • F: +1 303 6204795 • byer@rmi.net • Dir.: *Kevin Gramer* • Historical Museum – 1990 48065

Center for the Visual Arts, c/o Metropolitan State College of Denver, 1734 Wazee St, Denver, CO 80202 • T: +1 303 2945207 • F: +1 303 2945210 • andrewka@mscd.edu • www.mscd.edu/news/cva • C.E.O.: *Sally L. Perisho* • Dir.: *Amy Banker* • Public Gallery / University Museum – 1991 Temporary exhibitions of contemporary art 48066

Center for the Visual Arts, Metropolitan State College of Denver, 1734 Wazee St, Denver, CO 80202 • T: +1 303 2945207 • F: +1 303 2945210 • andrewka@mscd.edu • www.mscd.edu/news/cva • Public Gallery 48067

Children's Museum of Denver, 2121 Children's Museum Dr, Denver, CO 80211 • T: +1 303 4337444 • F: +1 303 4339520 • lindaf@cmdenver.org • www.cmdenver.org • Dir.: *Dr. Linda E. Farley* • Special Museum – 1973 48068

Colorado Historical Society Museum, 1300 Broadway, Denver, CO 80203 • T: +1 303 8663682 • F: +1 303 8665739 • information@chs.state.co.us • www.coloradohistory.org • Dir.: *Georgianna Contiguglia* • Cur.: *Moya Hansen* (Decorative and Fine Arts) • *Eric Paddock* (Photography) • Historical Museum – 1879 State history 48069

Colorado Photographic Arts Center, 1513 Boulder St, Denver, CO 80211 • T: +1 303 4558999 • F: +1 303 2783693 • Pres.: *R. Skip* • Public Gallery 48070

Core, 2045 Larimer St, Denver, CO 80205 • T: +1 303 2978428 • rgarriott@electricstores.com • Dir.: *Dave Griffin* • Public Gallery 48071

Denver Art Museum, 100 W 14th Av Pkwy, Denver, CO 80204 • T: +1 720 8655000 • F: +1 720 9130001 • info@denverartmuseum.org • www.denverartmuseum.org • Dir.: *Lewis I. Sharp* • Dep. Dir./Cur.: *Christoph Heinrich* (Modern/Contemp. Art) • Chief Cur.: *Margaret Young-Sanchez* (New World, Pre-Columbian) • Cur./AIGA Ass. Cur.: *Darrin Alfred* (Architecture, Design, Graphics) • Cur.: *Ronald Otsuka* (Asian Art) • *Gwen Chanzit* (Modern/Contemp. Art, Bayer Coll and Archive) • *Nancy Blomberg* (Native Arts) • *Donna Pierce* (New World, Spanish Colonial Art) • *Timothy Standring* (Paintings, Sculpture) • *Eric Paddock* (Photography/New Media) • Ass. Cur.: *Polly Nordstrand* (Native Arts, American Indian art) • *Angelica Daneo* (Paintings, Sculpture) • *Alice Zrebiec* (Textile Art) • Cur./Dir.: *Peter H. Hassrick* (Petrie Institute of Western American Art) • Sr. Scholar: *Joan Carpenter Troccoli* (Petrie Institute of Western American Art) • Fine Arts Museum – 1893 Native art, modern and contemporary art, pre-Columbian and Spanish Colonial, architecture, paintings, graphics, sculptures – Bayer Collection and Archive, AIGA Archive 48072

Denver Firefighters Museum, 1326 Tremont Pl, Denver, CO 80204 • T: +1 303 8921436 • F: +1 303 8921436 • info@denverfirefightersmuseum.org • www.denverfirefightersmuseum.org • Exec. Dir.: *Carey Southwell* • Special Museum – 1978 Firefighting artifacts, hand-drawn to motorized fire fighting vehicles, Photographs, books 48073

Denver Museum of Miniatures, Dolls and Toys, 1880 Gaylord St, Denver, CO 80206 • T: +1 303 3221053 • F: +1 303 3223704 • bazyl57@yahoo.com • www.dmmdt.com • Dir./Cur.: *Abigail L. Bentz* • Special Museum – 1981 Dolls and puppets 48074

U

Denver Museum of Nature and Science, 2001 Colorado Blvd, Denver, CO 80205 • T: +1 303 3706357 • F: +1 303 3316492 • ngire@dmns.org • www.dmns.org •
Pres.: *Tom Swanson* • Cur.: *Dr. Kirk R. Johnson* (Earth Sciences, Paleontology) • *Dr. Ella Maria Ray* (Anthropology) • *Joyce L. Herold* (Ethnology) • *Dr. Cheri A. Jones* (Mammalogy) • *Dr. Jack A. Murphy* (Geology) • *Dr. Stephen Holen* (Archaeology) • *Dr. Paula E. Cushing* (Entomology, Arachnology) •
Natural History Museum – 1900
Natural history, archaeology, paleontology, mammals, entomology, earth and space sciences ... 48075

Emmanuel Gallery, Auraria Campus, 10th and Lawrence Sts, Denver, CO 80217 • T: +1 303 5568337 • F: +1 303 5562335 •
Dir.: *Jennifer Garner* •
University Museum / Fine Arts Museum ... 48076

Forney Transportation Museum, 4303 Brighton Blvd, Denver, CO 80216 • T: +1 303 2971113 • F: +1 970 4989505 • museum@forneymuseum.com • www.forneymuseum.com •
Pres.: *Jack D. Forney* •
Science&Tech Museum – 1961
Transport ... 48077

Ginny Williams Family Foundation, 299 Fillmore St, Denver, CO 80206 • T: +1 303 3214077 •
Fine Arts Museum ... 48078

Grant-Humphreys Mansion, 770 Pennsylvania St, Denver, CO 80203 • T: +1 303 8942505 • F: +1 303 8942508 • gran@rmi.net •
Dir.: *Kevin Gramer* •
Historical Museum – 1976
Historic Beaux-Arts style home ... 48079

Kirkland Museum of Fine and Decorative Art, 1311 Pearl St, Denver, CO 80203 • T: +1 303 8328576 • F: +1 303 8328404 • info@kirklandmuseum.org • www.kirklandmuseum.org •
Dir.: *Hugh Grant* • Cons.: *Dean Sartori* • Man.: *Gerald Horner* (Admin.) • Sc. Staff: *Christopher Herron* (Collections, Registration) •
Fine Arts Museum – 1932
Decorative Art, Vance Kirkland works, artworks of Art Nouveau, Wiener Werkstätten, de Stijl, Bauhaus, Art Deco and Modern, 600 works of Colorado painters, sculptures, ceramists and furniture design e.g. Frank Lloyd Wright, Russel and Mary Wright, Eva Zeisel – library ... 48080

Mizel Museum, 400 S Kearney St, Denver, CO 80224 • T: +1 303 3949993 • F: +1 303 3941119 • ellen@mizelmuseum.org • www.mizelmuseum.org •
Dir.: *Ellen Premack* • Dep. Dir.: *Janelle Mock* (Education) •
Religious Arts Museum – 1982
Religious art and culture – Babi yar Park in Denver ... 48081

Molly Brown House Museum, 1340 Pennsylvania St, Denver, CO 80203 • T: +1 303 8324092 ext 16 • F: +1 303 8322340 • admin@mollybrown.org • www.mollybrown.org •
Dir.: *Kerri Atter* •
Historical Museum – 1970 ... 48082

Museo de Las Americas, 861 Santa Fe Dr, Denver, CO 80204 • T: +1 303 5714401 • F: +1 303 6079761 • gloria@museo.org • www.museo.org •
Dir.: *Pita Martinez* •
Fine Arts Museum / Historical Museum – 1991
Art and history ... 48083

Museum of Anthropology, c/o University of Denver, 2000 Asbury Av, Ste 146, Denver, CO 80208 • T: +1 303 8712406 • F: +1 303 8712437 • ckreps@du.edu • www.du.edu/ •
Dir.: *Dr. Christina Kreps* •
University Museum / Ethnology Museum – 1932
Anthropology, ethnology, textiles ... 48084

Museum of Contemporary Art Denver, 1275 19th St, Denver, CO 80202 • T: +1 303 298754 • F: +1 303 2987553 • scotta@mocadenver.com •
Dir.: *Sidney Payton* •
Fine Arts Museum – 1997 ... 48085

Pirate Contemporary Art Oasis, 1370 Verbana, Denver, CO 80220 • T: +1 303 4586058 •
Dir.: *Phil Bender* •
Public Gallery ... 48086

Rocky Mountain College of Art and Design Galleries, 6875 E Evans Av, Denver, CO 80224 • T: +1 303 7536046 • F: +1 303 7594970 • lspival@rmcad.edu • www.rmcad.edu •
Dir.: *Steven Steele* • Cur.: *Lisa Spivak* •
Fine Arts Museum / University Museum – 1963 ... 48087

Rocky Mountain Conservation Center, University of Denver, 2420 S University Blvd, Denver, CO 80208 • T: +1 303 7332508 • F: +1 303 7332508 • lmellon@du.edu •
Dir.: *Lori A. Mellon* •
Special Museum
Photography, textiles, archaeologie, ethnology, maps ... 48088

School of Art and Art History Galleries, Victoria H. Myhren Gallery, Gallery 023, c/o University of Denver, 2121 E Asbury Av, Denver, CO 80210 • T: +1 303 8712846 • F: +1 303 8714112 • www.du.edu/art •
Dir.: *Shannen Hill* •
Fine Arts Museum / University Museum – 1940
European and regional masters art ... 48089

Spark Gallery, 1535 Platte St, Denver, CO 80202 • T: +1 303 4554435 •
Public Gallery
Drawings, photography, prints, sculptures ... 48090

Trianon Museum and Art Gallery, 335 14th St, Denver, CO 80202 • T: +1 303 6230739 •
Fine Arts Museum / Public Gallery ... 48091

Denville NJ

Blackwell Street Center for the Arts, POB 808, Denville, NJ 07834 • T: +1 201 3372143 • F: +1 201 3372143 • wblakeart@nac.net • www.blackwell-st-artists.org •
Dir.: *Annette Adrian Hanna* •
Public Gallery – 1983
Professional gallery for NJ/NY artists ... 48092

Derry NH

The Children's Metamorphosis, Children's Museum, 6 W Broadway, #24, Derry, NH 03038 • T: +1 603 4252560 • info@childrensmet.org • childrensmet.org •
Exec. Dir.: *Betsy Anderson* •
Special Museum – 1991 ... 48093

Robert Frost Farm, N.H. Rte 28, Derry, NH 03038 • T: +1 603 4323091 •
Pres.: *Laura Burnham* •
Special Museum – 1968
1900-1909 home of the poet and author Robert Frost (1875-1963), 1900's furniture, personal belongings ... 48094

Des Arc AR

Lower White River Museum, 2009 W Main St, Des Arc, AR 72040 • T: +1 870 2563711 • lowerwhiterivermuseum@arkansas.com • www.arkansasstateparks.com •
Dir.: *Neva Boatright* •
Local Museum – 1971
County history, farming ... 48095

Des Moines IA

Des Moines Art Center, 4700 Grand Av, Des Moines, IA 50312 • T: +1 515 2774405 • F: +1 515 2710357 • marketing@desmoinesartcenter.org • www.desmoinesartcenter.org •
Dir.: *Susan Lubowsky Talbott* • Sen. Cur.: *Jeff Fleming* •
Fine Arts Museum – 1933
19th-20th c European and American art ... 48096

Hoyt Sherman Place, 1501 Woodland Av, Des Moines, IA 50309 • T: +1 515 2440507 • F: +1 515 2373582 • barcus@hoytsherman.org • www.hoytsherman.org •
Dir.: *Leisha Barcus* •
Fine Arts Museum – 1907
17th - 19th c paintings, furniture, decorative arts, artifacts ... 48097

Polk County Heritage Gallery, 111 Court Av, Des Moines, IA 50309 • T: +1 515 2863215 • F: +1 515 2863082 • www.co.polk.ia.us •
Pres.: *Mel Shivvers* • C.E.O.: *Angela Connolly* •
Public Gallery – 1980
Art ... 48098

Polk County Historical Society, 317 SW 42nd St, Des Moines, IA 50312 • T: +1 515 2556657 •
Pres.: *Robert Simon* •
Local Museum – 1938
Local history, Fort Des Moines II ... 48099

Salisbury House, 4025 Tonawanda Dr, Des Moines, IA 50312 • T: +1 515 2741777 • F: +1 515 2740184 • contactus@salisburyhouse.org • www.salisburyhouse.org •
Dir.: *Scott Brunscheen* •
Historical Museum – 1993
14th - 20th c furnishings and tapetries, manuscript coll, paintings ... 48100

Science Center of Iowa, 4500 Grand Av, Greenwood-Ashworth Park, Des Moines, IA 50312-2499 • T: +1 515 2744138 • F: +1 515 2743404 • info@sciowa.org • www.sciowa.org •
Exec. Dir.: *Mary B. Sellers* •
Science&Tech Museum – 1965
Science, technology ... 48101

State Historical Society of Iowa Museum, 600 E Locust St, Des Moines, IA 50319 • T: +1 515 2816412 • F: +1 515 2820502 • deirdre.giesler@iowa.gov • www.iowahistory.org •
C.E.O.: *Anita Walker* • Chief Cur.: *Michael O. Smith* •
Local Museum – 1892
Regional history, natural history ... 48102

Terrace Hill Historic Site and Governor's Mansion, 2300 Grand Av, Des Moines, IA 50312 • T: +1 515 2817205 • F: +1 515 2817267 • david.cordes@igov.state.ia.us • www.terracehill.org •
Pres.: *Jackie Devine* • C.E.O.: *David Cordes* •
Local Museum – 1971 ... 48103

Des Plaines IL

Des Plaines Historical Museum, 789 Pearson St, Des Plaines, IL 60016-4506 • T: +1 847 3915399, 4741 • F: +1 847 2971710 • dphslibrary@juno.com •
Dir.: *Joy A. Matthiessen* • Cur.: *Lynne Mickle-Smaczny* •
Local Museum – 1967
Furnishings, items of local significance, costumes, photographs, pioneer tools ... 48104

Desert Hot Springs CA

Cabot's Pueblo Museum, 67-616 E Desert View Av, Desert Hot Springs, CA 92240 • T: +1 760 3297610 • cabotsmuseum@roadrunner.com • www.cabotsmuseum.org •
Man.: *Edna Wells* •
Fine Arts Museum – 1968 ... 48105

Detroit MI

Black Legends of Professional Basketball Museum, 8900 E Jefferson, Detroit, MI 48202, mail addr: POB 02384, Detroit, MI 48214 • T: +1 313 8228208 • F: +1 313 8228227 • blpbf@cs.com •
Pres.: *Dr. John Kline* •
Special Museum – 1997
Photos, posters, uniforms 1904-1959 ... 48106

Casa de Unidad (Unity House), Boggs Center, 1920 Scotten, Detroit, MI 48209 • T: +1 313 8439598 • F: +1 313 8437307 • www.boggscenter.org/network/casa.shtml •
Dir.: *Marta E. Lagos* • *David W. Conklin* •
Public Gallery – 1981
Works of Latino artists ... 48107

Center Galleries, College for Creative Studies, 301 Frederick Douglas Av, Detroit, MI 48202-4034 • T: +1 313 6647800 • F: +1 313 6647880 • admissions@ccscad.edu • www.ccscad.edu •
Dir.: *Michelle M. Perron* •
Public Gallery / University Museum – 1989 ... 48108

Charles H. Wright Museum of African American History, 315 E Warren Av, Detroit, MI 48202 • T: +1 313 4945800 • F: +1 313 4945855 • dhamm@maah-detroit.org • www.maah-detroit.org •
Pres.: *Christy Matthews* • Cur.: *Patrina Chatman* •
Folklore Museum / Historical Museum – 1965
History, ethnography, folk art ... 48109

Children's Museum, 6134 Second Av, Detroit, MI 48202 • T: +1 313 8738100 • F: +1 313 8733384 • www.detroitchildrensmuseum.org •
Dir.: *Dwight R. Levens* •
Ethnology Museum – 1917
Ethnology, natural hist, science, folk arts, costumes & textiles, toys, dolls, musical instr – library ... 48110

Community Arts Gallery, Wayne State University, Dept. of Arts, 450 Reuther Mall, 150 Community Arts Bldg, Detroit, MI 48202 • T: +1 313 5772423, 5772980 • F: +1 313 5778935 • s.dupret@wayne.edu • www.art.wayne.edu •
Cur.: *Sandra Dupret* •
Public Gallery / University Museum – 1958
American and European graphics, paintings, sculpture ... 48111

Detroit Artists Market, 4719 Woodward Av, Detroit, MI 48201 • T: +1 313 8328540 • F: +1 313 8328543 • info@detroitartistsmarket.org • www.detroitartistsmarket.org •
Exec. Dir.: *Aaron Timlin* • Cur.: *Christine Stamas* •
Fine Arts Museum – 1932
Paintings, photographs, prints, drawings, sculpture, ceramics, glass ... 48112

Detroit Historical Museum, 5401 Woodward Av, Detroit, MI 48202-4009 • T: +1 313 8331805 • F: +1 313 8335342 • dhswebmaster@hist.ci.detroit.mi.us • www.detroithistorical.org •
Dir.: *Dr. Dennis Zembala* • Cur.: *Jill Grannan* (Social History) • *John Polacsek* (Marine, Military and Industrial History) • Sc. Staff: *Henry Amick* (Education) • *James Conway* (Programs) • *Catherine Klingman* (Special Exhibits) •
Historical Museum / Local Museum – 1928
Local hist, military, social and maritime hist, industry, automotive hist, Woodlands Indian, decorative and fine arts, costumes – library ... 48113

The Detroit Institute of Arts, 5200 Woodward Av, Detroit, MI 48202 • T: +1 313 8337900 • F: +1 313 8332357 • ebruss@dia.org • www.dia.org •
Dir.: *Graham W.J. Beal* • Cur.: *Laurie Barns* (Middle Eastern Islamic and Asian Art) • *David Penney* (Native American Art) • *MaryAnn Wilkinson* (Modern Art) • *Jim Tottis* (American Art) • *Alan P. Darr* (European Sculpture, Decorative Arts) • *George S. Keyes* (European Paintings) • *Nancy Sojka* (Graphic Arts) • *Valerie Mercer* (African-American Art) • *Elliot Wilhelm* (Film Theatre, Video) • *Tara Robinson* (Exhibits) • Cons.: *Barbara Heller* •
Fine Arts Museum – 1885
American, European, African, Middle Eastern, Islamic, Oceanic and Asien art, modern and contemporary art and design, New World culture, decorative art – library, auditorium, conservation lab ... 48114

Detroit Repertory Theatre Gallery, 13103 Woodrow Wilson, Detroit, MI 48238-3686 • T: +1 313 8681347 • F: +1 313 2598242 • DetRepTh@aol.com • www.detroitreptheatre.com/gallery.htm •
Cur.: *Gilda Snowden* •
Fine Arts Museum / Decorative Arts Museum – 1957 ... 48115

Dossin Great Lakes Museum, Detroit Historical Museum, 100 Strand Dr, Belle Isle, Detroit, MI 48207 • T: +1 313 2978366, 8524051 • F: +1 313 8335342 • polacsekj@hist.ci.detroit.mi.us • www.glmi.org •
Cur.: *John F. Polacsek* •
Historical Museum – 1948
Great Lakes maritime hist, model ships, nauticalia, photographs, paintings – small library ... 48116

Elaine L. Jacob Gallery, Wayne State University, Dept. of Arts, 480 W Hancock, Detroit, MI 48202 • T: +1 313 5772423 • F: +1 313 5778935 • ac1370@wayne.edu • www.art.wayne.edu •
Dir.: *Sandra Dupret* •
Public Gallery / University Museum
Contemporary art ... 48117

Ellington White Project-Art Exhibitions, 18100 Meyers Rd, Detroit, MI 48235 • T: +1 313 3421000 • F: +1 313 3426733 • ewp@provide.net • www.ellington-white.com •
Exec. Dir.: *Dwight Smith* •
Public Gallery – 1997
Creative visual arts experiences ... 48118

The Gallery & Beyond Words Gallery, Marygrove College, 8425 W McNichols, Liberal Arts Bldg, Detroit, MI 48221-2599 • T: +1 313 9271336 • bhsmith@marygrove.edu • www.marygrove.edu/facilities/art_gallery.asp •
Dir.: *Nelson Smith* •
Public Gallery / University Museum
Temporary exhibits of works from the students and guests ... 48119

Heritage Museum of Fine Arts for Youth, 110 E Ferry Av, Detroit, MI 48202 • T: +1 313 8711667 •
Dir.: *Josephine Harreld Love* •
Decorative Arts Museum / Fine Arts Museum – 1969
Fine arts for Youth, toys, games, puppets, children's music – library ... 48120

International Gospel Music Hall of Fame and Museum, 18301 W McNichols St, Detroit, MI 48219 • T: +1 313 5920017 • F: +1 313 5928762 • gmhfm@cs.com • www.igmhf.org •
Pres.: *David Gough* • C.E.O.: *Phyllis Siders* • Chief Cur.: *Sherry Dupree* •
Music Museum – 1995
Photos, tapes, sound recordings, songbooks, artist biographies, musical instr, clothing – library, auditorium ... 48121

International Institute of Metropolitan Detroit, 111 E Kirby St, Detroit, MI 48202 • T: +1 313 8718600 • F: +1 313 8711651 • info@iimd.org • www.iimd.org •
Pres.: *David Newman* •
Fine Arts Museum / Folklore Museum – 1919
International and ethnic folk art – auditorium ... 48122

Michigan Chapter Gallery, National Conference of Artists, NW Activities Center, 18100 Meyers, Detroit, MI 48235 • T: +1 313 3421000 • info@ncamich.org • www.ncamich.org •
Dir.: *Esther Vivian Brewer* • Cur.: *Celia Woodson* • Sc. Staff: *Geneva Brown* •
Fine Arts Museum
African Americam artists ... 48123

Michigan Sports Hall of Fame, Cobo Center, 1 N Washington Blvd, Detroit, MI 48226, mail addr: POB 1073, Farmington Hill, MI 48332 • T: +1 248 4730656 • F: +1 248 4730674 • mshof@twmi.rr.com • www.michigansportshof.org •
Pres.: *William F. McLaughlin* •
Special Museum – 1955
State sports ... 48124

Motown Historical Museum, 2648 W Grand Blvd, Detroit, MI 48208-1237 • T: +1 313 8752264 • F: +1 313 8752267 • rterry@motownmuseum.org • www.motownmuseum.org •
C.E.O.: *Esther Gordy Edwards* •
Music Museum – 1988
Hist of Motown Records, recording artists, founder Berry Gordy – archives ... 48125

Museum of Anthropology, Wayne State University, 4841 Cass Av, Detroit, MI 48202 • T: +1 313 5772598 • F: +1 313 5779759 • ad4844@wayne.edu • www.anthro.wayne.edu/museum/homepage.html •
Dir.: *Tamara Bray* •
University Museum / Ethnology Museum / Archaeology Museum – 1958
Anthropology, ethnography, archaeology – library ... 48126

Museum of Natural History, Wayne State University, Biological Sciences Bldg, 5047 Gullen Mall, Detroit, MI 48202 • T: +1 313 5772872 • F: +1 313 5776891 • wmoore@biology.biosci.wayne.edu • www.biosci.wayne.edu/outreach/natural_history.html •
Dir.: *William S. Moore* • Cur.: *Jesheskel Shoshani* (Vertebrates) • *Stanley K. Gangwere* (Insects) • *D. Carl Freeman* (Herbarium) •
University Museum / Natural History Museum – 1972
Natural hist, plants, birds, mammals, molluscs, insects, vertebrates, invertebrates, fossils – herbarium ... 48127

New Detroit Science Center, 5020 John R. St, Detroit, MI 48202 • T: +1 313 5778400 ext 440 • F: +1 313 8321623 • info@sciencedetroit.org • www.sciencedetroit.org •
Pres./CEO: *Shawn M. Kahle* •
Science&Tech Museum – 1970
IMAX, planetarium ... 48128

Pewabic Pottery Museum, 10125 E Jefferson St, Detroit, MI 48214 • T: +1 313 8220954 ext 110 • F: +1 313 8226266 • pewabic@pewabic.com • www.pewabic.com •
Exec. Dir.: *Terese Ireland* •
Decorative Arts Museum – 1903
Pottery factory, blueprints, art pottery, ceramics and tiles by Mary Chase Stratton, drawings, photographs, arts and crafts furniture ... 48129

Swords Into Plowshares Peace Center and Gallery, 33 E Adams St, Detroit, MI 48226 • T: +1 313 9637575 • F: +1 313 9632569 • createvisionsofpeace@yahoo.com •
Special Museum – 1985
Peace art coll ... 48130

Detroit Lakes MN

Becker County Historical Museum, 714 Summit Av, Detroit Lakes, MN 56501 • T: +1 218 8472938 • F: +1 218 8475048 • mail@beckercountyhistory.org • www.beckercountyhistory.org •
Dir.: *Erin McMillan* • *Joann Splonskowski* •
Local Museum – 1924
Local hist, implements, tools, pictures – library ... 48131

Devils Tower WY

Devils Tower Visitor Center, State Hwy 110, Bldg 170, Devils Tower, WY 82714 • T: +1 307 4675283 • F: +1 307 4675350 • christine_czazasty@nps.gov • www.nps.gov/deto •
Cur.: *Jim Cheatham* •
Local Museum – 1906
Geology, hist, botany, Indian artifacts ... 48132

Dewey OK

Dewey Hotel, 801 N Delaware, Dewey, OK 74029 • T: +1 918 5324416 •
Local Museum – 1967
Furniture and furnishings 1890-1910, Delaware, Cherokee and Osage Indian artifacts, photograph coll 1880-1930 ... 48133

Tom Mix Museum, Oklahoma Historical Society, 721 N Delaware, Dewey, OK 74029 • T: +1 918 5341555 •
C.E.O.: *Dr. Bob Blackburn* •
Local Museum – 1973
Life of Tom Mix ... 48134

Dexter ME

Dexter Historical Society Museum, 3 Water St, Dexter, ME 04930 • T: +1 207 9243043, 9245721 • dexhist@panax.com • www.dexterhistoricalsociety.com •
Dir./Cur.: *Richard Whitney* •
Local Museum – 1966
Local history, Grist mill, Millers house, Carr school ... 48135

Dexter MI

Dexter Area Museum, 3443 Inverness, Dexter, MI 48130 • T: +1 734 4262519 • dexmuseum@aol.com • www.hvcn.org/info/dextermuseum •
Dir.: *Nina Doletsky-Rackham* •
Local Museum – 1976
Local hist, Civil war, costumes, local business ... 48136

Diamond MO

George Washington Carver House, 5646 Carver Rd, Diamond, MO 64840 • T: +1 417 3254151 • F: +1 417 3254231 • gwca_superintendent@nps.gov •
Science&Tech Museum / Agriculture Museum – 1943
Scientist George Washington Carver ... 48137

Dickinson ND

Dakota Dinosaur Museum, 200 E Museum Dr, Dickinson, ND 58601 • T: +1 701 2253466 • F: +1 701 2270534 • info@dakotadino.com • www.dakotadino.com •
C.E.O.: *Alice League* •
Science&Tech Museum – 1991
Paleontology, geology, minerals, triceratops and dinosaur skeletons, seashells ... 48138

Gallery of Art, Dickinson State University, Dickinson, ND 58601-4896 • T: +1 701 4832312 • F: +1 701 4832006 • sharon_linnehan@dsu.nodak.edu •
Dir.: *Sharon Linnehan* •
Fine Arts Museum
Contemporary graphics, Zoe Bailer paintings, American art ... 48139

Dighton KS

Lane County Historical Museum, 333 N Main St, Dighton, KS 67839 • T: +1 620 3975652 •
Cur.: *Virginia Johnston* •
Local Museum – 1976
Early Lane County photos, manusript coll, sod house ... 48140

Dillingham AK

Samuel K. Fox Museum, Seward and D Sts, Dillingham, AK 99576, mail addr: POB 273, Dillingham, AK 99576 • T: +1 907 8424831 • samfoxmuseum@nushtel.net • www.ci.dillingham.ak.us •
Dir.: *Tim Troll* •
Ethnology Museum – 1974
Alaskan Native and Indian museum, Central Yup'ik objects – library ... 48141

Dillon MT

Beaverhead County Museum, 15 S Montana, Dillon, MT 59725-2433 • T: +1 406 6835027 • bvhdmuseum@bmt.net •
Pres.: *Lynn Westad* •
Historical Museum – 1947
Indian artifacts, relics of mining, natural hist ... 48142

Western Art Gallery and Museum, The University of Montana, 710 S Atlantic St, Dillon, MT 59725-3598 • T: +1 406 6837232 • F: +1 406 6837493 • R_Horst@umwestern.edu • www.umwestern.edu •
Dir.: *Randy Horst* •
Fine Arts Museum / University Museum – 1986
Seidensticker wildlife trophy coll – Lucy Carson Library ... 48143

Dillon SC

James W. Dillon House Museum, W Main St, Dillon, SC 29536 • T: +1 843 7749051 • F: +1 843 7745521 • jru@multi-forms.com •
Dir.: *Don Barclay* •
Local Museum
Local hist ... 48144

Dixon IL

John Deere House, 8393 S Main, Grand Detour, Dixon, IL 61021-9406 • T: +1 815 6524551 • F: +1 815 6523835 • timmermanlynna@johndeere.com •
Pres./Dir.: *James H. Collins* •
Archaeology Museum / Historic Site – 1953
Archaeology ... 48145

Ronald Reagan Boyhood Home, 816 S Hennepin Av, Dixon, IL 61021 • T: +1 815 2883830 • F: +1 815 2886757 •
Chm.: *Norm Wymbs* •
Historic Site – 1980
Historic house ... 48146

Dodge City KS

Boot Hill Museum, Front St, Dodge City, KS 67801 • T: +1 316 2278188 • F: +1 316 2277673 • frontst@pld.com • www.boothill.org •
C.E.O.: *Tammy Moody* • Chm.: *Gwen Nelson* • Chief Cur.: *David N. Kloppenborg* • Cur.: *Susan Dame* (Exhibitions) •
Historical Museum – 1947
Western history, located on Boot Hill ... 48147

Home of Stone - a Ford County Museum, Av A and Vine St, Dodge City, KS 67801 • T: +1 620 2276791 • georgedchs@yahoo.com • www.ukans.edu/kansas/ford/ •
Local Museum – 1965
Kerosene lamps, clocks, picture coll ... 48148

The Kansas Teachers' Hall of Fame, 603 5th Av, Dodge City, KS 67801 • T: +1 316 2257311 •
Dir.: *Nancy Tramer* •
Historical Museum – 1977
Pictures of outstanding techaers inducted into the Hall ... 48149

Dolores CO

Anasazi Heritage Center, 27501 Hwy 184, Dolores, CO 81323 • T: +1 970 8824811 • F: +1 970 8827035 • marilynn_eastin@co.blm.gov • www.co.blm.gov/ahc/index.htm •
Dir.: *LouAnn Jacobson* •
Ethnology Museum / Archaeology Museum – 1988
Archaeology ... 48150

Donner Pass CA

Western Ski Sport Museum, I-80, Boreal Ridge Ski Area, Donner Pass, CA 95728, mail addr: POB 829, Soda Springs, CA 95728 • T: +1 530 4263313 • F: +1 530 4263501 • info@assoc.auburnskiclub.org • www.auburnskiclub.org •
Dir.: *Bill Clark* •
Special Museum – 1969
Skiing hist ... 48151

Dora AL

Alabama Mining Museum, 120 East St, Dora, AL 35062 • T: +1 205 6482442 •
Dir.: *Bonnie Sue Groves* • Cur.: *Florence Wiley* •
Historical Museum / Science&Tech Museum – 1982
Mining, 1905 steam locomotive, ore car and caboose; c.1900 Oakman, Alabama railroad depot bldg, c.1930 U.S. post office ... 48152

Dorchester NE

Saline County Historical Society Museum, Hwy 33, S of Main St, Dorchester, NE 68343 • T: +1 402 9472911 • F: +1 402 9472911 •
Pres.: *Norma Knoche* •
Local Museum – 1956
Agriculture, machinery, paintings by local artists, glass ... 48153

Dothan AL

Alabama Agricultural Museum, Landmark Park, Hwy 431 N, Dothan, AL 36302, mail addr: POB 6362, Dothan, AL 36302 • T: +1 334 6780171 • F: +1 334 6777229 • parkinfo@landmarkpark.com • www.landmarkpark.com, www.archives.state.al.us •
Dir./Cur.: *Karen Hill* • CEO: *William M. Holman* •
Agriculture Museum – 1976
State agriculture from prehistory to 1960s, historic farm ... 48154

Wiregrass Museum of Art, 126 Museum Av, Dothan, AL 36303-1624 • T: +1 334 7943871 • www.wiregrassmuseum.org •
Exec. Dir.: *Esther Hockett* • Cur.: *Ron Parrish* • Sr. Scholar: *Erin Schovel* •
Public Gallery – 1988
Regional art 19th-21st c, decorative art ... 48155

Douglas AZ

Slaughter Ranch Museum, 6153 Geronimo Trail, Douglas, AZ 85608, mail addr: POB 438, Douglas, AZ 85608 • T: +1 520 5582474 • sranch@vtc.net • www.vtc.net/~sranch •
Dir.: *Harvey Finks* • Cur.: *John Lavanchy* • Sc. Staff: *Dr. Reba Grandrud* (History) •
Local Museum – 1982
Local hist, costumes, furniture ... 48156

Douglas MI

Keewatin Maritime Museum, Steamship Keewatin, Harbour Village, Union St & Blue Star Hwy, Douglas, MI 49406, mail addr: POB 638, Douglas, MI 49406 • T: +1 616 8572464, 8577235 • drydock@modempool.com • www.keewatinmaritimemuseum.com •
Cur.: *Bob Zimmerman* • *Cindy Zimmerman* •
Historical Museum / Science&Tech Museum – 1965
History of the Keewatin, former Great Lakes passenger steamship of the Canadian Pacific Railroad ... 48157

Douglas WY

Fort Fetterman, 752 Hwy 93, Douglas, WY 82633 • T: +1 307 3582864, 6847629 • F: +1 307 3582864 • rwilso@missc.state.wy.us •
C.E.O.: *Pat Green* • Head: *Bill Gentle* •
Military Museum – 1963
Military hist, life of the native Americans (Dakota, Cheyenne, Shoshone) ... 48158

Wyoming Pioneer Memorial Museum, Wyoming State Fairgrounds, 400 W Center St, Douglas, WY 82633 • T: +1 307 3589288 • F: +1 307 3589293 • aeklanstate.wy.us •
Dir./Cur.: *Arlene E. Earnst* •
Local Museum – 1956
Local hist, pioneer relics, Indian artifacts, textiles, agriculture, hist buildings ... 48159

Douglass KS

Douglass Historical Museum, 318 S Forest, Douglass, KS 67039 • T: +1 316 7462319, 7462166 •
Dir./Cur.: *Francis Renfro* •
Historical Museum – 1949
Pioneer museum ... 48160

Dover DE

Biggs Museum of American Art, 406 Federal St, Dover, DE 19901, mail addr: POB 711, Dover, DE 19903 • T: +1 302 6742111 • F: +1 302 6745133 • biggs@delaware.net • www.biggsmuseum.org •
Dir.: *Karol A. Schmiegel* • Cur.: *Ryan Grover* •
Fine Arts Museum – 1989
Art Museum housed in 1858 County Office Building ... 48161

Delaware Agricultural Museum, 866 N Dupont Hwy, Dover, DE 19901 • T: +1 302 7341618 • F: +1 302 7340457 • damv@dol.net • www.agriculturalmuseum.org •
Dir.: *Linda Chatfield* • Cur.: *Jennifer Griffin* •
Local Museum / Agriculture Museum – 1974
Rural life, agricultural hist and technology ... 48162

Delaware State Museums, Rose Cottage, 102 S State St, Dover, DE 19901 • T: +1 302 7395316 • F: +1 302 7396712 • jim.stewart@state.de.us • www.destatemuseums.org •
Cur.: *Claudia F. Leister* (Collections Management) • *Madeline Dunn* (Education) •
Local Museum – 1931
Local hist ... 48163

Dover MA

Caryl House and Fisher Barn, 107 Dedham St, Dover, MA 02030 • T: +1 508 7851832 • F: +1 508 7850789 • DHS1895@aol.com • www.doverhistoricalsociety.org/caryl.html •
C.E.O.: *Paul H. Tedesco* • Cur.: *Priscilla Pitt Jones* •
Historical Museum / Agriculture Museum – 1920
Historic house of minister, doctor and families, agriculture ... 48164

Sawin Museum, 80 Dedham St, Dover, MA 02030 • T: +1 508 7851832 • DHS1895@aol.com • www.doverhistoricalsociety.org/sawin.html •
C.E.O.: *Paul H. Tedesco* • Cur.: *Shirley McGill* • *Glenda Mattes* •
Local Museum – 1895
Regional history – small library ... 48165

Dover NH

Annie E. Woodman Institute, 182 Central Av, Dover, NH 03821-0146 • T: +1 603 7421038 •
Ass. Cur.: *Llyr E. Couture* •
Local Museum – 1916
Indian artifacts, mineralogy, geology ... 48166

Dover OH

J.E. Reeves Home and Museum, 325 E Iron Av, Dover, OH 44622 • T: +1 330 3437040 • F: +1 330 3436290 • reeves@tusco.net • web.tusco.net/reeves •
Pres.: *James D. Nixon* • Dir.: *Chris Nixon* •
Local Museum – 1958
Agriculture, archaeology, geology, Civil War artifacts, furnishings, vehicles, carriages – Library ... 48167

Warther Museum, 331 Karl Av, Dover, OH 44622 • T: +1 330 3437513 • info@warthers.com • www.warthers.com •
Pres./Dir.: *David R. Warther* •
Decorative Arts Museum / Agriculture Museum – 1936
Works by Ernest warther, ivory, ebony, pearls, ... 48168

Dover TN

Fort Donelson National Battlefield Museum, 120 Fort Donelson Rd, Dover, TN 37058 • T: +1 931 2325706 • F: +1 931 2326331 • fodo_ranger_-activities@nps.gov •
Military Museum – 1928
Civil War ... 48169

Dover, Air Force Base DE

Air Mobility Command Museum, 1301 Heritage Rd, Dover, Air Force Base, DE 19902-5301 • T: +1 302 6775938 • F: +1 302 6775940 • museum@dover.af.mill • www.amcmuseum.org •
C.E.O.: *Michael D. Leister* • Cur.: *James R. Leech* •
Military Museum / Science&Tech Museum – 1986
Military airlift and air refuelling aircraft e.g. C-7A, C-54M, C-119G, C-130E, C-141AB, F-101, F-106, B-17G,TG-4A, BT-13, T-33, UH-1 – library, theater 48170

Dover-Foxcroft ME

Blacksmith Shop Museum, 107 Dawes Rd, Dover-Foxcroft, ME 04426 • T: +1 207 5648618 • dblockwood@midmaine.com •
Cur.: *Dave Lockwood* •
Historical Museum – 1963 ... 48171

Dowagiac MI

Heddon Museum, 414 West St, Dowagiac, MI 49047, mail addr: 204 W. Telegraph St., Dowagiac, MI 49047 • T: +1 269 7825698 • heddonmuseum@lyonsindustries.com •
Dir.: *Joan Lyons* • *Don Lyons* •
Historical Museum / Historic Site – 1995
Heddon factory, Over 1500 Heddon lures, 160 reels and 215 rods are on display plus, violin bows, ski poles, box kites, military antennas, aviation. At one time the worlds largest manufacturer of fishing tackle – Research library ... 48172

Museum at Southwestern Michigan College, 58900 Cherry Grove Rd, Dowagiac, MI 49047 • T: +1 269 7821374 • F: +1 269 7821460 • museum@swmich.edu • www.swmich.edu/museum •
Dir.: *Ann Thompson* • Cur.: *Steve Arseneau* (History) • *Tom Caskey* (Science Exhibits) • *Judy Van der Linden* (Education) •
University Museum / Historical Museum / Science&Tech Museum – 1982
Local history and industry, Heddon fishing lures, Round Oak Stove Co. a.o. ... 48173

Downers Grove IL

The Downers Grove Park District Museum, 831 Maple Av, Downers Grove, IL 60515 • T: +1 630 9631309 • F: +1 630 9630496 • mharmon@xnet.com • www.dgparks.org •
Dir.: *Mark S. Harmon* •
Local Museum – 1968
Furniture, decorative arts, textiles, toys, tools, photographs, personal documents ... 48174

Downey CA

Downey Museum of Art, 10419 S Rives Av, Downey, CA 90241 • T: +1 562 8610419 • F: +1 562 8610419 • www.downeyca.org •
Dir.: *Kate Davies* •
Fine Arts Museum – 1957
Paintings, sculpture, graphics, photographs ... 48175

Downieville CA

Downieville Museum, 330 Main St, Downieville, CA 95936 • T: +1 530 2893423 •
Dir.: *Kevel Jane Gutman* •
Local Museum – 1932
Gold rush, Cinese and Indian artifacts – library ... 48176

Doylestown PA

Fonthill Museum of the Bucks County Historical Society, E Court St, Doylestown, PA 18901 • T: +1 215 3489461 • F: +1 215 3489462 • info@fonthillmuseum.org • www.fonthillmuseum.org •
C.E.O.: *Douglas C. Dolan* •
Decorative Arts Museum / Historical Museum – 1930
Home of Henry C. Mercer (1856-1930), noted archeologist, arts and crafts ... 48177

Heritage Conservancy, 85 Old Dublin Pike, Doylestown, PA 18901 • T: +1 215 3457020 • F: +1 215 3454328 • cdavid@heritageconservancy.org • www.heritageconservancy.org •

Pres.: *Clifford C. David* •
Fine Arts Museum / Decorative Arts Museum – 1958
Artwork of local artists, decorative arts, photographs 48178

James A. Michener Art Museum, 138 S Pine St, Doylestown, PA 18901 • T: +1 215 3409800 • F: +1 215 3409807 • jamam1@mitchenerartmuseum.org • www.michenerartmuseum.org •
Dir.: *Bruce Katsiff* • Sen. Cur.: *Brian Peterson* • Cur.: *Adrienne Neszmelyi* (Education) • *Zoriana Siokalo* (Public Programs) •
Fine Arts Museum – 1987
Pennsylvania impressionism 20th c 48179

Mercer Museum of the Bucks County Historical Society, 84 S Pine St, Doylestown, PA 18901-4999 • T: +1 215 3450210 • F: +1 215 2300823 • info@mercermuseum.org • www.mercermuseum.org •
Dir.: *Douglas C. Dolan* • Ass. Dir.: *Molly W. Lowell* • Cur.: *Cory Amsler* • Libr.: *Betsy Smith* •
Folklore Museum / Historical Museum / Science&Tech Museum – 1916
Folk art, local hist, artifacts, technology 48180

Dragoon AZ

Amerind Foundation, 2100 N Amerind Rd, Dragoon, AZ 85609, mail addr: POB 400, Dragoon, AZ 85609 • T: +1 520 5863666 • F: +1 520 5864679 • amerind@amerind.org • www.amerind.org •
Dir.: *Dr. John A. Ware* •
Ethnology Museum / Archaeology Museum – 1937
Archaeology, ethnology, oils, sculpture, santos, ivory and scrimshaw – library 48181

Dresden ME

Pownalborough Court House, Courthouse Rd, Rte 128, Dresden, ME 04342 • T: +1 207 8826817 • lcha@wiscasset.net • www.lincolncountyhistory.org •
Dir.: *Margaret M. Shiels* •
Historical Museum – 1954
Archaeology, furniture, military hist 48182

Drummond Island MI

Drummond Island Historical Museum, Water St, Drummond Island, MI 49726 • T: +1 906 4935746 •
Dir.: *Harry Ropp* • *John Lowe jr.* • *Michael McDonald* • *Rosalie Sasso* • *Catherine Ashley* • Cur.: *Audrey Moser* •
Local Museum – 1961
Local history, Indian artifacts, lumbering and farming, early snowmobile and marine engines, sleds, fossils, geology – archives, library 48183

Duarte CA

Justice Brothers Racing Museum, 2734 E Huntington Dr, Duarte, CA 91010 • T: +1 626 3599174 • gencontact@justicebrothers.com • www.justicebrothers.com •
Pres.: *Ed jr. Justice* •
Science&Tech Museum – 1985
Racing vehicles, motorcycles, passenger cars 48184

Dublin GA

Dublin-Laurens Museum, 311 Academy Av, Dublin, GA 31021 • T: +1 478 2729242 • history@nlamerica.com •
Dir.: *Scott Thompson* •
Special Museum – 1979
Archaeology, geology, paintings, photos, artefacts, housed in 1904 restored Carnegie Library 48185

Dubois WY

Dubois Museum, 909 West Ramshorn, Dubois, WY 82513 • T: +1 307 4552284 • F: +1 307 4552912 • dmuseum@dteworld.com • duboismuseum.org •
Dir.: *Norma Williamson* •
Local Museum – 1976
Local hist, natural hist and archaeology, Indian artifacts – Lucius Burch Center for Western Traditions 48186

Dubuque IA

Dubuque Museum of Art, 701 Locust St, Dubuque, IA 52001 • T: +1 563 5571851 • F: +1 319 5577826 • dbqartmuseum@mcleodusa.net • www.dbqartmuseum.com •
Exec. Dir.: *Geri Shafer* •
Fine Arts Museum / Folklore Museum – 1874
Art 48187

Mathias Ham House, 2241 Lincoln Av, Dubuque, IA 52001 • T: +1 563 5579545 • F: +1 563 5831241 • rivermuse@mwci.net • www.mississippirivermuseum.com •
Dir.: *Jerome A. Enzler* •
Historical Museum – 1964
Local history 48188

Mississippi River Museum, 350 E 3rd St, Dubuque, IA 52001 • T: +1 563 5579545 • F: +1 563 5831241 • rivermuseum@mwci.net • www.mississippirivermuseum.com •
Dir.: *Jerome A. Enzler* • Cur.: *Tacie N. Campbell* •
Historical Museum – 1950
Maritime and naval history, Mississippi Riverboats – archives 48189

Dugger IN

Dugger Coal Museum, 8178 E Main St, Dugger, IN 47848 •
Cur.: *Martha Marlow* •
Science&Tech Museum – 1980
Deep coal mines and strip mines, tools, clothing, protective gear, old mining reports 48190

Duluth GA

Jacqueline Casey Hudgens Center for the Arts, 6400 Sugarloaf Pkwy, Bldg 300, Duluth, GA 30097 • T: +1 770 6236002 • F: +1 770 6233555 • elliott@hudgenscenter.org • www.hudgenscenter.org •
Dir.: *Nancy Gullickson* • Cur.: *Lucy Elliott* •
Fine Arts Museum – 1981
Works by Picasso, Kandinsky, Miro and Rauschenberg 48191

Southeastern Railway Museum, 3595 Peachtree Rd, Duluth, GA 30096 • T: +1 770 4762013 • F: +1 770 6227269 • admin@srmduluth.org • www.srmduluth.org •
Pres.: *Jamie Reid* •
Science&Tech Museum / Historical Museum – 1968
Rail cars and locomotives 1910 - present, rail-related artefacts 48192

Duluth MN

Duluth Art Institute, 506 W Michigan St, Duluth, MN 55802 • T: +1 218 7337560 • F: +1 218 7337506 • getart@duluthartinstitute.org • www.duluthartinstitute.org •
Exec. Dir.: *Samantha Gibb Roff* •
Public Gallery – 1896
Regional art 48193

Duluth Children's Museum, 506 W Michigan St, Duluth, MN 55802 • T: +1 218 7337544 • F: +1 218 7337547 • explore@duluthchildrensmuseum.org • www.duluthchildrensmuseum.org •
Dir./Cur.: *Bonnie A. Cusick* •
Local Museum – 1930
Childhood, cultural anthropology, Saint Louis county heritage, arts 48194

Glensheen-Historic Congdon Estate, University of Minnesota Duluth, 3300 London Rd, Duluth, MN 55804 • T: +1 218 7268910 • F: +1 218 7268911 • glen@d.umn.edu • www.d.umn.edu/glen •
CEO: *Jack Bowman* •
Historical Museum – 1979
Historic neo-Jacobean mansion, sleighs, carriages, original furnishings, glass art 48195

Karpeles Manuscript Museum, 902 E 1st St, Duluth, MN 55805 • T: +1 218 7280630 • www.rain.org/~karpeles/dulfrm.html •
Dir.: *Lee R. Fadden* • Cur.: *Karen Fadden* •
Library with Exhibitions
Contemporary art, manuscripts, illustrations and documents 48196

Lake Superior Maritime Visitors Center, 600 Lake Av S, Duluth, MN 55802 • T: +1 218 7272497 • F: +1 218 7205270 • info@lsmma.com • www.lsmma.com •
Dir.: *Thomas Holden* •
Historical Museum / Science&Tech Museum – 1973
Marine hist, navigation hist, ships, cargoes, wrecks, harbor development 48197

Lake Superior Railroad Museum, 506 W Michigan St, Duluth, MN 55802 • T: +1 218 7337590 • F: +1 218 7337596 • museum@lsrm.org • www.lsrm.org •
Dir.: *Ken Buehler* • Cur.: *Thomas Gannon* •
Science&Tech Museum – 1974
Railroad, Duluth Union depot bldg, locomotives, cars, steam wrecking crane, china, silver, timetables – library 48198

Saint Louis County Historical Museum, 506 W Michigan St, Duluth, MN 55802 • T: +1 218 7337580 • F: +1 218 7337585 • history@chartemi.net • www.vets-hall.org •
Exec. Dir.: *JoAnne Coombe* •
Local Museum – 1922
Regional history, E. Johnson coll of drawings and paintings, Ojibwe and Sioux beadwork, quillwork, furniture 48199

Tweed Museum of Art, University of Minnesota Duluth, 1201 Ordean Ct, Duluth, MN 55812 • T: +1 218 7268222 • F: +1 218 7268503 • tma@d.umn.edu • www.d.umn.edu/tma •
Dir./Cur.: *Peter Spooner* •
Fine Arts Museum / University Museum – 1950
American prints, paintings, sculpture, modern & contemporary art, ceramics – library, archives 48200

Dumas AR

Desha County Museum, Hwy 165 E, Dumas, AR 71639, mail addr: POB 141, Dumas, AR 71639 • T: +1 870 3822222 •
Dir.: *Ann Brigham* • Cur.: *Charlotte Shexnayder* •
Local Museum – 1979
Local history 48201

Dumas TX

Moore County Historical Museum, 1820 S Dumas Av, Dumas, TX 79029-4329 • T: +1 806 9353113 • F: +1 806 9343621 •
C.E.O.: *Kurt Stallwitz* • Dir.: *Terri George* •
Local Museum – 1976
Local wild life, hist items, indian artifacts, oral tapes – archives 48202

Dumfries VA

Weems-Botts Museum, 3944 Cameron St, Dumfries, VA 22026 • T: +1 703 2212218 • F: +1 703 2212218 • weemsbotts@msn.com • www.geocities.com/HDVINC •
Pres.: *Barry K. Ward* • Cur.: *Kimberley D. Ward* •
Local Museum – 1974
Colonial hist, tabacco farming, Civil War items 48203

Duncan OK

Chisholm Trail Heritage Center Museum, 1000 N 29th St, Duncan, OK 73534 • T: +1 580 2526692, 2526563 • F: +1 580 2526567 • statue@texhoma.net •
Exec. Dir.: *Dr. Chris Jefferies* •
Historical Museum
Cowboy, cavalry, Native American clothing, cattle trails, sculptures 48204

Stephens County Historical Society Museum, Hwy 81 and Beech, Fuqua Park, Duncan, OK 73533 • T: +1 580 2520717 • F: +1 580 2513195 •
Dir.: *PeeWee Cary* • Ass. Dir.: *Vickie Zimmerman* • *Pat Hale* • *Marge Rigdon* • *Sharleen Johns* • *Teresa Fritts* •
Local Museum – 1971
Anthropology, archaeology, fine and dec art, ethnology, American Indian artifacts, hist of oil ind 48205

Dundee IL

Dundee Township Historical Society Museum, 426 Highland Av, Dundee, IL 60118 • T: +1 847 4286996 •
Local Museum – 1964
Local history 48206

Dunedin FL

David L. Mason Children's Art Museum, 1143 Michigan Blvd, Dunedin, FL 34698 • T: +1 727 2983322 • F: +1 727 2983326 • info@dfac.org • www.dfac.org/childmuseum.html •
Dir.: *Todd Still* •
Fine Arts Museum 48207

Dunedin Fine Arts Center, 1143 Michigan Blvd, Dunedin, FL 34698 • T: +1 727 2983322 • F: +1 727 2983326 • info@dfac.org • www.dfac.org •
Exec. Dir.: *George Ann Bissett* •
Public Gallery 48208

Dunedin Historical Society Museum, 349 Main St, Dunedin, FL 34697-2393 • T: +1 727 7361176 • F: +1 727 7381871 • dunhist@compuserve.com • www.ci.dunedin.fl.us •
Dir.: *Vincent Luisi* •
Local Museum – 1969
Local history, 1920 Atlantic Coastline Passenger Station, Old Freight Warehouse 48209

Dunkirk IN

The Glass Museum, 309 S Franklin, Dunkirk, IN 47336 • T: +1 765 7686872 • F: +1 765 7686872 • marynewsome@netscape.net • www.dunkirkpubliclibrary.com •
Dir.: *Ailesia Franklin* • Cur.: *Mary Newsome* •
Decorative Arts Museum – 1979
Glass 48210

DuPont WA

DuPont Historical Museum, 207 Barksdale Av, DuPont, WA 98327 • T: +1 253 9642399, 9643492 • F: +1 253 9643554 • info@dupontmuseum.com • www.dupontmuseum.com •
Pres.: *Lorraine Overmyer* •
Local Museum – 1976
Local hist 48211

Durango CO

Animas Museum, 3065 W 2nd Av, Durango, CO 81301 • T: +1 970 2592402 • F: +1 970 2594749 • animasmuseum@frontier.net • www.frontier.net/~animasmuseum •
Dir.: *Robert McDaniel* • Cur.: *Charles A. DiFerdinando* •
Local Museum – 1978
Local hist 48212

Strater Hotel, 699 Main Av, Durango, CO 81301 • T: +1 970 2474431, 3757121 • F: +1 970 2592208 • info@strater.com • www.strater.com •
C.E.O.: *Rod Barker* •
Historical Museum – 1887
Walnut antiques 48213

Durant OK

Fort Washita, Oklahoma Historical Society, 15 miles east of Madill on SH 199, Durant, OK 74701 • T: +1 580 9246502 • F: +1 580 9246502 • ftwashita@ok-history.mus.ok.us •
C.E.O.: *Dr. Bob Blackburn* •
Historic Site / Open Air Museum / Military Museum – 1967
Ruins and restored buildings 48214

Durham NC

Bennett Place, 4409 Bennett Memorial Rd, Durham, NC 27705 • T: +1 919 3834345 • F: +1 919 3834349 • bennettplace@ncmail •
Man.: *Davis Waters* •
Local Museum – 1962
Military hist, log kitchen 48215

Duke Homestead, 2828 Duke Homestead Rd, Durham, NC 27705 • T: +1 919 4775498 • F: +1 919 4797092 • dukehomestead@verizon.net • www.dukehomestead.nchistoricsites.org •
Pres.: *Walker S. Stone* •
Historical Museum – 1974
Tobacco hist, tobacco barn, 1870 tobacco factory, 1852 homestead with furnishing 48216

Duke University Union Museum and Brown Art Gallery, Bryan Center, Science Dr, Durham, NC 27708 • T: +1 919 6842911 • F: +1 919 6848395 • www.duke.edu/web/duu/ •
Dir.: *Regan Bedouin* •
University Museum / Fine Arts Museum 48217

Durham Art Guild, 120 Morris St, Durham, NC 27701 • T: +1 919 5602713 • F: +1 919 5602704 • artguild1@yahoo.com • www.durhamartguild.org •
Dir.: *Lisa Morton* •
Public Gallery 48218

Historic Stagville, 5828 Old Oxford Hwy, Durham, NC 27712 • T: +1 919 6200120 • F: +1 919 6200422 • stagvill@sprynet.com •
Pres.: *Judith Fortson* •
Local Museum – 1977
Hist buildings, furniture and furnishings, farm implements, tools 48219

History of Medicine Collections, Duke University, Medical Center Library Rm 102, Erwin Rd & Fulton St, Durham, NC 27710 • T: +1 919 6601143 • F: +1 919 6817599 • porte004@mc.duke.edu • www.mclibrary.duke.edu •
Cur.: *Suzanne Porter* •
Historical Museum – 1956
Medical instruments and artifacts, books, journals, manuscripts and photographs 48220

Nasher Museum of Art at Duke University, Buchanan Blvd at Trinity, Durham, NC 27708 • T: +1 919 6845135 • F: +1 919 6818624 • wendy.hower@duke.edu • www.duke.edu/duma/ •
Dir.: *Dr. Kimerly Rorschach* • Cur.: *Dr. Sarah W. Schroth* • Ass. Cur.: *Dr. Anne Schroder* •
Fine Arts Museum / University Museum – 1968
Medieval and renaissance sculpture and decorative art, Greek and Roman antiquities, American and European paintings, drawings and sculpture, contemporary art 48221

North Carolina Central University Art Museum, 1801 Fayetteville St, Durham, NC 27707 • T: +1 919 5606211 • F: +1 919 5605649 • Krodgers@wpo.nccu.edu • www.nccu.edu/artmuseum/ •
Dir.: *Kenneth G. Rodgers* •
Fine Arts Museum / University Museum – 1971
Traditional African sculpture and artifacts, contemporary art esp. of African American artists 48222

North Carolina Museum of Life and Science, 433 Murray Av, Durham, NC 27704 • T: +1 919 2205429 • F: +1 919 2205575 • contactus@ncmls.org • www.ncmls.org •
C.E.O.: *Dr. Thomas H. Krakauer* •
Science&Tech Museum / Natural History Museum – 1946
Wildlife and flora of NC, exotic butterflies, biology, geology, physics 48223

Durham NH

Durham Historic Association Museum, Newmarket Rd and Main St, Durham, NH 03824 • T: +1 603 8685436 •
Pres.: *Marion E. James* • C.E.O.: *Alexander Amell* •
Local Museum – 1851
American Indian artifacts, history of Durham 48224

Museum of Art, University of New Hampshire, Paul Creative Arts Center, 30 Academic Way, Durham, NH 03824-2617 • T: +1 603 8623712 • F: +1 603 8622191 • Museum.of.Art@unh.edu • www.unh.edu/moa •
Interim Dir.: *W. Weston LaFountain* • Ass. Dir.: *Astrida Schaeffer* •
Fine Arts Museum / University Museum – 1960
20th c works on paper, 19th c American paintings 48225

Duxbury MA

Alden House, 105 Alden St, Duxbury, MA 02332 • T: +1 781 9349092 • F: +1 781 9349149 • curator@alden.org • www.alden.org •
Dir.: *Alden Ringquist* • Cur.: *James W. Baker* •
Historical Museum – 1906
Pilgrims John Alden and Priscilla Mullins, founders of Duxbury, antique furnihings 48226

Art Complex Museum, 189 Alden St, Duxbury, MA 02331 • T: +1 781 9346634 • F: +1 781 9345117 • info@artcomplex.org • www.artcomplex.org •
Dir.: *Charles A. Weyerhaeuser* • Cur.: *Catherine Mayes* •

Sc. Staff: *Alice Hyland* • *Craig Bloodgood* •
Fine Arts Museum – 1967
European and American paintings, prints, Shaker furniture, Asian art, Native American art – library 48227

Duxbury Ma

Duxbury Rural and Historical Society, 479 Washington St, Duxbury, Ma 02331 • T: +1 781 9346106 • F: +1 781 9345730 • pbrowne@duxburyhistory.org • www.duxburyhistory.org •
Exec. Dir.: *Patrick Browne* •
Local Museum – 1883
Decorative arts, textiles, relics of 1869 French-American cable, Weston shipyards, farms, wharves, mill and ropewalk – Nathaniel Winsor jr. house, King Caesar house, Gershom Bradford house, library, archives 48228

Dyersville IA

National Farm Toy Museum, 1110 16th Av Ct SE, Dyersville, IA 52040 • T: +1 563 8752727 • F: +1 563 8758467 • farmtoys@dyersville.com • www.nftmonline.com •
Cur.: *Anne Reitzler* •
Special Museum – 1986
Trucks, life size toy tractor display 48229

Eagle WI

Old World Wisconsin, S103 W37890 Hwy 67, Eagle, WI 53119 • T: +1 262 5946300 • F: +1 262 5946342 • oww@whs.wisc.edu • www.oldworldwisconsin.org •
Dir.: *Peter S. Arnold* • Ass. Dir.: *John W. Reilly* • Cur.: *Martin Perkins* (Collections) • *Carol Smith* •
Agriculture Museum / Historical Museum / Historic Site – 1976
Ethnology, 10 ethnic farmsteads and 1870's rural village with over 65 original buildings from 19th and early 20th c 48230

Eagle City AK

Eagle Historical Society and Museums, Third and Chamberlain Rd, Eagle City, AK 99738, mail addr: POB 23, Eagle City, AK 99738 • T: +1 907 5472325 • F: +1 907 5472232 •
Pres.: *Elva Scott* •
Local Museum – 1961
Local military and Indians history 48231

Eagle Harbor MI

Eagle Harbor Lighthouse and Museums, Keweenaw County Historical Society, 670 Lighthouse St, M-26, Eagle Harbor, MI 49950 • T: +1 906 2894990, 2962561 • www.keweenawhistory.org •
Pres.: *David H. Thomas* •
Historical Museum / Science&Tech Museum – 1982
Marine hist, copper mining, fire fighting 48232

East Aurora NY

Aurora Historical Museum, Aurora Historical Society, Old Town Hall, East Aurora, NY 14052, mail addr: POB 472, East Aurora, NY 14052 • T: +1 716 6524735 •
Local Museum – 1951
Local hist, paintings, Indian artifacts, coll of Indian arrowheads 48233

Elbert Hubbard-Roycroft Museum, Aurora Historical Society, 363 Oakwood Av, East Aurora, NY 14052, mail addr: POB 472, East Aurora, NY 14052-0472 • T: +1 716 6524735 • ebert@earthlink.net • www.roycrofter.com/museum.htm •
Sc. Staff: *M. Fisher* • *Geneviève Steffen* • *Bruce F. Bland* • *Maryann Myers* •
Local Museum / Decorative Arts Museum – 1962
Books, leathercraft, metal work, furniture, artglass, arts and crafts movement, local hist 48234

Explore & More ... a Children's Museum, 300 Gleed Av, East Aurora, NY 14052 • T: +1 716 6555131 • F: +1 716 6555466 • mail@exploreandmore.org • www.exploreandmore.org •
C.E.O. & Dir.: *Barbara Park Leggett* •
Special Museum – 1994
Hands-on exhibits on building and architecture, food and nutrition, world culture, brain teasers 48235

Millard Fillmore House, Aurora Historical Society, 24 Shearer Av, East Aurora, NY 14052 • T: +1 716 6528875 •
Pres.: *Diane Meade* • Sc. Staff: *Marie Schnurr* • *Lyn Chemera* • *Rachelle Francis* • *Vivian Pikett* • *Joce Teach* •
Local Museum – 1951
Period furnishing; 13th U.S. Pres. Millard Fillmore 48236

East Brunswick NJ

East Brunswick Museum, 16 Maple St, East Brunswick, NJ 08816 • T: +1 732 2572313, 2571508 •
Pres.: *Mark Nonesteid* •
Local Museum – 1978
Local and regional hist 48237

East Durham NY

Durham Center Museum, Star Rte 145, East Durham, NY 12423, mail addr: POB 192, East Durham, NY 12423 • T: +1 518 2398461 • F: +1 518 2394081 • dougsancie@aol.com •
C.E.O.: *Dan Clifton* • Exec.Dir & Asst.Cur.: *Sancie Thomsen* • Cur.: *Douglas Thomsen* •
Local Museum – 1960
Folk art, local and Green County artifacts, Civil War, genealogy 48238

Irish American Heritage Museum, 2267 Rte 145, East Durham, NY 12423 • T: +1 518 6347497 • F: +1 518 6347497 • irishamermuseum@cs.com • www.irishamericanheritagemuseum.org •
C.E.O.: *Joseph J. Dolan* •
Folklore Museum – 1986
Ethnic museum 48239

East Greenwich RI

James Mitchell Varnum House and Museum, 57 Peirce St, East Greenwich, RI 02818 • T: +1 401 8841776 • k8bcm@home.com • www.varnumcontinentals.org •
Dir.: *Bruce C. MacGunnigle* •
Local Museum – 1938
Historic house 48240

New England Wireless and Steam Museum, 1300 Frenchtown Rd, East Greenwich, RI 02818 • T: +1 401 8850545 • F: +1 401 8840683 • newsm@ids.net • users.ids.net/~newsm •
Pres.: *Robert W. Merriam* •
Science&Tech Museum – 1964
Engineering 48241

Varnum Memorial Armory and Military Museum, 6 Main St, East Greenwich, RI 02818 • T: +1 401 8844110 • k8bcm@cox.net • www.varnumcontinentals.org •
Dir.: *Bruce C. MacGunnigle* • Cur.: *Donald Marcum* •
Military Museum – 1913
Military hist 48242

East Hampton NY

Clinton Academy Museum, East Hampton Historical Society, 151 Main St, East Hampton, NY 11937 • T: +1 631 3246850 • F: +1 631 3249885 • ehhs@optonline.net • www.easthamptonhistory.org •
Exec.Dir.: *Maralyn Rittenour* •
Historic Site – 1784
Decorative and fine arts, textiles, tools, equipment, hist of education 48243

East Hampton Historical Society Museum, 101 Main St, East Hampton, NY 11937 • T: +1 516 3246850 • F: +1 516 3249885 • ehhs@optonline.net • www.easthamptonhistory.org •
Exec.Dir.: *Maralyn Rittenour* •
Local Museum – 1921
Hist structures and landscapes, dec and fine arts, textiles, tools and equipment 48244

Guild Hall Museum, 158 Main St, East Hampton, NY 11937 • T: +1 631 3240806 • F: +1 631 3292722 • museum@guildhall.org • guildhall.org •
Exec.Dir.: *Ruth Appelhof* • Cur.: *Christina Moissaides* • Ass. Cur.: *Maura Doyle* •
Fine Arts Museum – 1931
Works by 19th and 20th c regional artists 48245

Home Sweet Home Museum, 14 James Ln, East Hampton, NY 11937 • T: +1 613 3240713 • F: +1 613 3240713 • www.easthampton.com/homesweethome •
Dir.: *Hugh R. King* •
Decorative Arts Museum / Local Museum / Historic Site – 1928
Lustreware, Staffordshire and English ceramics of late 18th/early 19th c, 17th - 19th American furniture 48246

Mulford House and Farm, East Hampton Historical Society, 10 James Ln, East Hampton, NY 11937 • T: +1 631 3246850 • F: +1 631 3249885 • ehhs@optonline.net • www.easthamptonhistory.org/pages/mulford.html •
Exec.Dir.: *Maralyn Rittenour* •
Agriculture Museum / Open Air Museum – 1948
Farm house with several buildings and 18th period furnishings, farm implements, crafts, domestic equip, 48247

Osborn-Jackson House, East Hampton Historical Society, 101 Main St, East Hampton, NY 11937 • T: +1 631 3246850 • F: +1 631 3249885 • ehhs@optonline.net • www.easthamptonhistory.org •
Exec.Dir.: *Maralyn Rittenour* •
Decorative Arts Museum / Local Museum – 1979
18th c decorative and fine arts, textiles, costumes and fashion 48248

Town House, East Hampton Historical Society, Clinton Academy Museum, 149 Main St, East Hampton, NY 11937 • T: +1 631 3246850 • F: +1 631 3249885 • ehhs@optonline.net • www.easthamptonhistory.org •
Historic Site – 1958
1731 one-room schoolhouse, 18th c schooling 48249

East Haven CT

Shore Line Trolley Museum, 17 River St, East Haven, CT 06512-2519 • T: +1 203 4676927 • F: +1 203 4677635 • sltmbera@sbcglobal.net • www.bera.org •
Dir.: *George Boucher* • *Theodore Eickmann* (Vehicle Collection) • *Frederick Sherwood* (Exhibits) •
Science&Tech Museum – 1945
Transport technology, operating railway 48250

East Islip NY

Islip Art Museum, 50 Irish Ln, East Islip, NY 11730 • T: +1 631 2245402 • F: +1 631 2245417 • info@islipartmuseum.org • www.islipartmuseum.org •
Dir.: *Mary Lou Cohalan* • *Susan Simmons* • Cur.: *Karen Shaw* • *Janet Goleas* •
Fine Arts Museum – 1973
Contemporary and avant-garde art 48251

East Jordan MI

East Jordan Portside Art and Historical Museum, 1787 S M-66 Hwy, East Jordan, MI 49727 • T: +1 231 5363416, 5363381 • portsideartsfair@hotmail.com •
Cur.: *Laura R. Hansen* •
Fine Arts Museum / Local Museum – 1976
Local hist, lumbering, agriculture, contemporary coll of Purchase Prize winners (arts fair) – archives 48252

East Lansing MI

Kresge Art Museum, Michigan State University, Auditorium and Physics Rds, East Lansing, MI 48824-1119 • T: +1 517 3539834 • F: +1 517 3556577 • kamuseum@msu.edu • artmuseum.msu.edu •
Dir.: *Dr. Susan J. Bandes* • Cur.: *April Kingsley* • Sc. Staff: *Elizabeth K. Whiting* (Education) •
Fine Arts Museum / University Museum – 1959
Drawings, graphics, bronzes, African art, ethnology, ceramics, etchings and engravings, Afro-American art, Eskimo art, decorative arts, baroque art, antiquities, German woodcuts 48253

Michigan State University Museum, W Circle Dr, East Lansing, MI 48824-1045 • T: +1 517 3552370 • F: +1 517 4322846 • dewhurs1@msu.edu • museum.msu.edu •
Dir./Cur.: *Dr. C. Kurt Dewhurst* (Folk Arts) • Sc. Staff: *Dr. Barbara Lundrigan* • *Dr. Pamela C. Rasmussen* (Mammalogy, Ornithology) • *Dr. William A. Lovis* (Anthropology) • *Dr. Michael D. Gottfried* (Vertebrate Paleontology) • *Dr. Jodie A. O'Gorman* (Great Lakes Archaeology) • *Val Roy Berryman* (History) • *Dr. Marsha MacDowell* (Folk Arts) • *Dr. Yvonne R. Lockwood* (Folklife) • *Juan S. Alvarez* (Exhibits) • *Dr. Kristine Morrissey* (Interpretation) •
University Museum / Local Museum – 1857
Science, cultural history, folk art, history, archaeology, ethnology, anthropology, paleontology, mammalogy, ornitology 48254

East Liverpool OH

Lou Holtz and Upper Ohio Valley Hall of Fame, 120 E Fifth St, East Liverpool, OH 43920 • T: +1 330 3865443 • F: +1 330 3820244 • gulutz@raex.com • www.louholtzhalloffame.com •
Pres.: *Frank C. Dawson* •
Local Museum – 1998 48255

Museum of Ceramics, 400 E Fifth St, East Liverpool, OH 43920 • T: +1 330 3866001 • F: +1 330 3860488 • ceramics@clover.net • www.ohiohistory.org •
Cur.: *Mark Twyford* •
Decorative Arts Museum – 1980
Ohio ceramics, local relics 48256

East Meredith NY

Hanford Mills Museum, County Routes 10 and 12, East Meredith, NY 13757 • T: +1 607 2785744 • F: +1 607 2786299 • hanford1@hanfordmills.org • www.hanfordmills.org •
Dir.: *Elizabeth Callahan* • Asst.Dir. & Cur.: *Caroline de Marais* •
Science&Tech Museum – 1973
Sawmilling, woodworking, gristmilling equip 48257

East Poultney VT

East Poultney Museum, The Green, East Poultney, VT 05741 • T: +1 802 2875268, 2330 •
Local Museum – 1954
Early Poultney hist, costumes, farming 48258

East Springfield NY

Boswell Museum of Music, 5748 US Hwy 20, East Springfield, NY 13333 • T: +1 607 2643321 • F: +1 607 2643321 • boswell@telenet.net • www.boswellmuseum.org •
C.E.O.& Pres.: *Vet Boswell Minnerly* • Cur.: *Gretchen Sorin* •
Music Museum – 1990
Recorded sound, clothing, furnishings 48259

Hyde Hall, Mill Rd, Near Glimmerglass State Park, East Springfield, NY 13333, mail addr: POB 721, Cooperstown, NY 13326 • T: +1 607 5475098 • F: +1 607 5478462 • hydehall@wpe.com • www.hydehall.org •
Historical Museum – 1964 48260

East Tawas MI

Iosco County Historical Museum, 405 W Bay St, East Tawas, MI 48730 • T: +1 989 3628911 • klenowr@alpena.cc.mi.us • www.ioscomuseum.com •
Pres./CEO: *Rosemary E. Klenow* •
Local Museum – 1976
Local history, Detroit and Mackinaw Railway, tools, clothing, fire fighting, Indian birch canoe, military hist 48261

East Troy WI

East Troy Electric Railroad Museum, POB 943, East Troy, WI 53120-0943 • T: +1 262 6423263 • F: +1 262 6423197 • JFTrolley@aol.com • www.easttroyrr.org •
Science&Tech Museum – 1975
Trolley cars, electric railway cars 48262

East Windsor CT

Connecticut Fire Museum, 58 North Rd, East Windsor, CT 06088 • T: +1 860 6234732 • F: +1 860 6276510 • ctfiremuseum@hotmail.com • www.ceraweb.org/firemuseum •
Pres.: *Bert Johnson* •
Historical Museum – 1968
Antique fire apparatus and equipment 48263

Connecticut Trolley Museum, 58 North Rd, East Windsor, CT 06088-0360 • T: +1 860 6276540 • F: +1 860 6276510 • office@ceraweb.org • www.ceraonline.org •
Pres.: *Louis A. Grimaldi* •
Science&Tech Museum – 1940
Trolley transport system 48264

Scantic Academy Museum, 115 Scantic Rd, East Windsor, CT 06088 • T: +1 860 6235327 • eastwindsorhistory@att.net • www.eastwindsorhistory.home.att.net •
Pres.: *Michael Hunt* •
Local Museum – 1965
Industry, agriculture, transpotation, paintings 48265

Eastlake OH

Croatian Heritage Museum, 34900 Lakeshore Blvd, Eastlake, OH 44095 • T: +1 440 9462044 • F: +1 216 9912310 • banjelacic@worldnet.att.net •
C.E.O. & Pres.: *Robert M. Jerin* • Cur.: *Branka Malinar* •
Ethnology Museum / Folklore Museum – 1983
Croatian textiles, folk costumes, wood carvings, sculpture, metal work, leather work, paintings – library 48266

Easton MD

Academy Art Museum, 106 South St, Easton, MD 21601 • T: +1 410 8222787 • F: +1 410 8225997 • info@art-academy.org • www.art-academy.org •
Dir.: *Christopher J. Brownawell* •
Fine Arts Museum – 1958
Multi-disciplinary arts 48267

Historical Museum of Talbot County, 25 S Washington St, Easton, MD 21601 • T: +1 410 8220773 • F: +1 410 8227911 • director@hstc.org • www.hstc.org •
Pres.: *Albert Smith* • Dir.: *Glenn Uminowicz* •
Local Museum – 1954
Regional history 48268

Easton PA

Lafayette College Art Gallery, Hamilton and High St, Easton, PA 18042-1768 • T: +1 610 3305361 • F: +1 610 3305642 • artgallery@lafayette.edu • www.lafayette.edu •
Dir./Cur.: *Michiko Okaya* •
Fine Arts Museum / Public Gallery / University Museum – 1983
Art 48269

National Canal Museum, 30 Centre Sq, Easton, PA 18042-7743 • T: +1 610 5596613 • F: +1 610 5596686 • ncm@canals.org • canals.org •
Exec. Dir.: *Robert Rudd* • Sc. Staff: *Lance Metz* (History) • Cur.: *Sonya Dollins-Colton* •
Science&Tech Museum – 1970
Technology 48270

Northampton County Historical and Genealogical Society Museum, 107 S Fourth St, Easton, PA 18042 • T: +1 610 2531222 • F: +1 610 2531224 • director@northamptonctymuseum.org • www.northamptonctymuseum.org •
Pres.: *Dorothy Cieslicki* • Exec. Dir.: *Colleen Cunningham Lavdar* •
Local Museum – 1906
Native American artifacts, dolls, military equipment and uniforms, costumes, folk art – Library 48271

U

Eastsound WA

Orcas Island Historical Museum, 181 North Beach Rd, Eastsound, WA 98245 • T: +1 360 3764849 • F: +1 360 3762994 • orcasmuseum@rockisland.com • www.orcasisland.org •
Pres.: *Linda Tretheway* • Dir.: *Jennifer Vollmer* •
Local Museum – 1950
Local hist 48272

Eatonton GA

Uncle Remus Museum, 360 Oak St, Hwy 441 S, Eatonton, GA 31024 • T: +1 706 4856856 •
Pres./Cur.: *Norma Watterson* •
Historical Museum – 1963
House c.1820 slave cabin, furnishings 48273

Eatonville FL

The Zora Neale Hurston Museum, 227 E Kennedy Blvd, Eatonville, FL 32751 • T: +1 407 6473307 • F: +1 407 6473559 •
Dir.: *N. Y. Nathiri* •
Fine Arts Museum
African American art ... 48274

Eatonville WA

Pioneer Farm Museum and Ohop Indian Village, 7716 Ohop Valley Rd E., Eatonville, WA 98328 • T: +1 360 8326300 • F: +1 360 8324533 • pioneer@mashell.com • www.pioneerfarmmuseum.org •
Head: *Dean Carpenter* •
Local Museum / Agriculture Museum – 1975
Agriculture, local hist ... 48275

Eau Claire WI

Chippewa Valley Museum, 1204 Carson Park Dr, Eau Claire, WI 54703 • T: +1 715 8347871 • F: +1 715 8346624 • info@cvmuseum.com • www.cvmuseum.com •
Dir.: *Susan M. McLeod* •
Local Museum – 1966
Hist of the Chippewa Valley, hist buildings ... 48276

Foster Gallery, University of Wisconsin-Eau Claire, 121 Water St, Eau Claire, WI 54702-4004 • T: +1 715 8362328 • F: +1 715 8364882 • wagenetk@uwec.edu • www.uwec.edu •
Dir.: *Thomas K. Wagner* •
Fine Arts Museum / University Museum – 1970 . 48277

Paul Bunyan Logging Camp, 1110 Carson Park Dr, Eau Claire, WI 54703 • T: +1 715 8356200 • F: +1 715 8356293 • info@paulbunyancamp.org • www.paulbunyancamp.org •
Pres.: *Gordon Wall* •
Science&Tech Museum – 1934
Logging and lumbering ... 48278

Ebensburg PA

Cambria County Historical Society Museum, 615 N Center, Ebensburg, PA 15931 • T: +1 814 4726674 • info@cambriacountyhistorical.com • www.cambriacountyhistorical.com •
C.E.O. & Pres.: *Fremont McKenrick* • Cur.: *Kathy Jones* •
Local Museum – 1925
History, agriculture, Indian artifacts ... 48279

Echo OR

Chinese House Museum, 230 W Bridge and 33208 Marble, Echo, OR 97826 • T: +1 541 3768411 • F: +1 541 3768218 • ecpl@centurytel.net • www.echo-oregon.com •
C.E.O.: *Diane Berry* •
Religious Arts Museum – 1987 ... 48280

Echo Historical Museum, 76502 Echo Meadows Rd, Echo, OR 97826 • T: +1 541 3768150 •
Cur.: *Elaine Attum* •
Local Museum – 1981
Umatilla County hist, guns, musical instruments, Indian artifacts, dolls ... 48281

Fort Henrietta, 10 W Main, Echo, OR 97826 • T: +1 541 3768411 • F: +1 541 3768218 • ecpl@centurytel.net • www.echo-oregon.com •
C.E.O.: *Diane Berry* •
Historical Museum – 1985
Fire equipment, covered wagons, chemical wagon, hook and ladder, 1st Umatilla County jail ... 48282

Eckley PA

Eckley Miners' Village, Main St, Eckley, PA 18255 • T: +1 570 6362070 • F: +1 570 6362938 • www.phmc.state.pa.us •
Dir.: *David Dubick* • Cur.: *Chester Kulesa* •
Historical Museum – 1969
54 houses built in the 1850s as coal patch town including Roman Catholic Church, Episcopal Church, doctor's office, coal breaker, visitor's center and mule barn ... 48283

Edenton NC

Historic Edenton, 108 N Broad St, Edenton, NC 27932 • T: +1 252 4822637 • F: +1 252 4823499 • edentonshs@inteliport.com •
Man.: *Linda Jordan Eure* •
Historical Museum – 1951
Buildings, period furnishings, kitchen utensils ... 48284

Edgartown MA

Martha's Vineyard Exhibition, Cooke and School Sts, Edgartown, MA 02539, mail addr: POB 1310, Edgartown, MA 02539 • T: +1 508 6274441 • F: +1 508 6274436 • proth@mvmuseum.org • www.marthasvineyardhistory.org •
Dir.: *Matthew Stackpole* • Cur.: *Jill Bouck* •
Open Air Museum / Local Museum – 1922
Thomas Cooke House, 2nd Captain Francis Pease House, archarology, history and industry, decorative arts – library, archives, herbarium ... 48285

Edgerton KS

Lanesfield School, 18745 S Dillie Rd, Edgerton, KS 66021 • T: +1 913 8936645 • F: +1 913 6316359 •
Dir.: *Mindi Love* •
Historical Museum – 1967
One-room schoolhouse restored to 1904 appearance ... 48286

Edgerton WI

Albion Academy Historical Museum, 605 Campus Ln, Edgerton, WI 53534 • T: +1 608 8843940 •
Pres.: *Clifford Townsend* •
Local Museum – 1959
Local hist ... 48287

Edgewater MD

Historic London Town, 839 Londontown Rd, Edgewater, MD 21037 • T: +1 410 2221919 • F: +1 410 2221918 • londntown@historiclondontown.com • www.historiclondontown.com •
Exec. Dir.: *Donna M. Ware* •
Historical Museum / Archaeology Museum – 1971
Significant archaeological site ... 48288

Edina MO

Knox County Historical Society Museum, Court House, 107 N 4th St, Edina, MO 63537 • T: +1 660 3972349 • F: +1 660 3973331 •
Pres.: *Brent Karhoff* •
Historical Museum – 1967 ... 48289

Edinboro PA

Bruce Gallery of Art, Edinboro University of Pennsylvania, Doucette Hall, Edinboro, PA 16444 • T: +1 814 7322513 • F: +1 814 7322629 • wmathie@edinboro.edu • www.edinboro.edu •
Dir.: *William Mathie* •
Public Gallery
Contemporary art in all mediums ... 48290

Edinburg TX

Museum of South Texas History, 121 E McIntyre St, Edinburg, TX 78541 • T: +1 956 3836911 • F: +1 956 3816911 • mosthistory@mosthistory.org • www.mosthistory.org •
Dir.: *Shan Rankin* • Asst. Dir.: *Thomas A. Fort* •
Local Museum – 1967
Hist & culture of the Rio Grande Valley ... 48291

University Galleries, Art Dept., University of Texas Pan American, 1201 W University Dr, Edinburg, TX 78539 • T: +1 210 3812655 • F: +1 210 3845072 • galleries@panam.edu •
Dir.: *Dindy Reich* •
Fine Arts Museum ... 48292

Edisto Island SC

Edisto Island Historic Preservation Society Museum, 8123 Chisolm Plantation Rd, Edisto Island, SC 29438 • T: +1 843 8691954 • F: +1 843 8691954 •
Dir.: *Karen Nickles* •
Archaeology Museum – 1986
Prehistoric low country ... 48293

Edmond OK

Edmond Historical Society Museum, 431 S Blvd, Edmond, OK 73034 • T: +1 405 3400078 • F: +1 405 3402771 • edmondhistory@coxinet.net • www.edmondhistory.org •
Pres.: *Shelley Codomi* • Dir.: *Brenda Granger* • *Iris Muno Jordan* •
Local Museum – 1983
Local hist ... 48294

Edmonds WA

Edmonds Arts Festival Museum, 700 Main St, Edmonds, WA 98020, mail addr: POB 699, Edmonds, WA 98020 • T: +1 425 7716412 • F: +1 425 6737700 • hardarmc@cmc.net • www.edmondsartsfestival.com •
Cur.: *Darlene McLellan* •
Fine Arts Museum – 1979
Fine arts ... 48295

Edmonds South Snohomish County Historical Society Museum, 118 Fifth Av N, Edmonds, WA 98020 • T: +1 425 7740900 • F: +1 425 7746507 • jonisein@yahoo.com • www.historicedmonds.org •
Dir.: *Joni L. Sein* • Cur.: *Mike Wilcox* •
Local Museum – 1973
Local hist ... 48296

Edna KS

Edna Historical Museum, 100 S Delaware, Edna, KS 67342 •
Cur.: *Ronald Neidigh* •
Local Museum – 1978 ... 48297

Edna TX

Texana Museum, 403 N Wells, Edna, TX 77957 • T: +1 361 7825431 •
Dir.: *Mary Sayles* •
Local Museum – 1967
Hist artifacts, doc ... 48298

Edwards CA

Air Force Flight Test Center Museum, Air Force Base, 95 ABW/MU, 405 S Rosamond Blvd, Edwards, CA 93524-1850, mail addr: POB 57, Edwards, CA 93523 • T: +1 661 2778050 • F: +1 661 2778051 • afftc.edwards.af.mil •
Dir./Cur.: *Doug Nelson* •
Military Museum / Science&Tech Museum – 1986
Aeronautics, hist of flight testing – library ... 48299

Edwardsville IL

Madison County Historical Museum, 715 N Main St, Edwardsville, IL 62025-1111 • T: +1 618 6567562 • www.madisoncountyhistory.org/ •
Dir.: *Evelyn Bowles* •
Local Museum – 1924
Local history, housed in John H. Weir home – library ... 48300

The University Museum, Southern Illinois University Edwardsville, Edwardsville, IL 62026 • T: +1 618 6502996 • F: +1 618 6502995 • ebarnet@siue.edu •
Dir.: *Eric B. Barnett* • Cur.: *Michael E. Mason* •
Local Museum / University Museum – 1959 ... 48301

Egg Harbor WI

Chief Oshkosh Native American Arts, 7631 Egg Harbor Rd, Egg Harbor, WI 54209 • T: +1 920 8683240 •
Dir.: *Coleen Bins* •
Fine Arts Museum – 1975
Native American art ... 48302

Cupola House, 7836 Egg Harbor Rd, Egg Harbor, WI 54209 • F: +1 920 8681710 • cupolahouse@itol.com • www.cupolahouse.com •
C.E.O.: *Gloria Hanson* •
Decorative Arts Museum – 1871
Period furnishings, pottery, Amish quilts ... 48303

Eglin Air Force Base FL

Air Force Armament Museum, 100 Museum Dr, Eglin Air Force Base, FL 32542-1497 • T: +1 850 8824062, 8824189 • F: +1 850 8823990 • sneddon@eglin.af.mil • www.wg53.eglin.af.mil •
C.E.O.: *Russell C. Sneddon* •
Military Museum – 1985
Military ... 48304

Egypt PA

Troxell-Steckel House and Farm Museum, 4229 Reliance St, Egypt, PA 18052 • T: +1 610 4351074 • F: +1 610 4359812 •
Cur.: *Andree Miller* •
Local Museum / Agriculture Museum – 1904 ... 48305

Ekalaka MT

Carter County Museum, 306 N Main St, Ekalaka, MT 59324 • T: +1 406 7756886 • ccmuseum@midrivers.com •
Dir.: *Warren O. White* •
Local Museum / Ethnology Museum / Natural History Museum / Folklore Museum – 1936
Mounted dinosaur skeletons, anatosaurus, triceratops ... 48306

El Cajon CA

Heritage of the Americas Museum, 12110 Cuyamaca College Dr W, El Cajon, CA 92019 • T: +1 619 6705194 • F: +1 619 6705198 • hofam@sbcglobal.net • www.cuyamaca.net/museum •
Dir.: *Cheryl Minshew* •
Natural History Museum / Historical Museum – 1993
Natural and Human history of the Americas ... 48307

Hyde Art Gallery, Grossmont College, 8800 Grossmont College Dr, El Cajon, CA 92020-1799 • T: +1 619 6447299 • F: +1 619 6447922 • www.grossmont.edu/artgallery •
Cur.: *Ben Aubert* • Asst. Cur.: *Teresa Markey* •
University Museum / Fine Arts Museum – 1961
Prints, ceramics, photography ... 48308

El Campo TX

El Campo Museum of Natural History, 2350 N Mechanic St, El Campo, TX 77437 • T: +1 979 5436885 • F: +1 979 5435788 • museum@wcnet.net • www.elcampomuseum.com •
Pres.: *Phyllis Laughlin* • Cur.: *Denise Prochazka* •
Natural History Museum – 1978
More than 100 big game trophy specimes obtained world-wide, sea shell coll, varved, wooden coll ... 48309

El Dorado AR

South Arkansas Arts Center, 110 E Fifth St, El Dorado, AR 71730 • T: +1 870 8625474 • info@saac-arts.org • www.saac-arts.org •
Exec. Dir.: *Beth James* •
Fine Arts Museum – 1962
Art gallery, paintings, wood block prints, children's Oriental coll – library, theater ... 48310

El Dorado KS

Butler County Society Museum and Kansas Oil Museum, 383 E Central Av, El Dorado, KS 67042 • T: +1 316 3219333 • F: +1 316 3213619 • bchs@powwwer.net • www.skyways.org/museums/kom •
Dir.: *Rebecca Matticks* • Cur.: *Debbie Amend* (Education) • *Brad Amend* (Collections) • *Dale Wilson* (Project Dir.) •
Local Museum / Science&Tech Museum – 1956
Kansas oil history, local history ... 48311

Coutts Memorial Museum of Art, 110 N Main St, El Dorado, KS 67042-0001 • T: +1 316 3211212 • F: +1 316 3211215 • coutts@southwind.com • skyways.lib.ks.us/museums/coutts/ •
Pres.: *Rhoda Hodges* • Dir.: *Terri Scott* •
Fine Arts Museum – 1970
Fine art, Frederic Remington sculpture coll ... 48312

El Monte CA

El Monte Historical Society Museum, 3150 Tyler Av, El Monte, CA 91731 • T: +1 626 5802232 • www.californiahistoricalsociety.org •
Cur.: *Donna Crippen* •
Local Museum – 1958
Furnishings, clothing, guns, musical instr, Indian and Mexican artifacts, natural hist – library, archive . 48313

El Paso CO

United States Air Force Academy Museum, 2346 Academy Dr, USAF Academy, El Paso, CO 80840-9400 • T: +1 719 3332569 • F: +1 719 3334402 •
University Museum / Military Museum – 1961
Military ... 48314

El Paso TX

Americana Museum, 5 Civic Center Plaza, El Paso, TX 79901-1153 • T: +1 915 5420394 • F: +1 915 5424511 •
Cur.: *William Kwiecinski* • Ass. Cur.: *Victor Gomez* •
Folklore Museum / Historical Museum ... 48315

Bridge Center for Contemporary Art, 1 Union Fasion Center, El Paso, TX 79902 • T: +1 915 5326707 •
Dir.: *Richard Baron* •
Public Gallery – 1986 ... 48316

The Centennial Museum at the University of Texas at El Paso, University Av, Wiggins Rd, El Paso, TX 79968-0533 • T: +1 915 7475565 • F: +1 915 7475411 • museum@utep.edu • www.utep.edu/museum/ •
Dir.: *Florence Schwein* • Cur.: *Scott Cutler* •
University Museum / Natural History Museum – 1936
Natural & human hist, arhaeology, botany, geology ... 48317

Chamizal National Memorial, 800 S San Marcial, El Paso, TX 79905 • T: +1 915 5327273 • F: +1 915 5327240 • www.nps.gov/cham •
Local Museum – 1967
History, art, ethnology ... 48318

El Paso Holocaust Museum, 310 N Mesa, Ste 906, El Paso, TX 79901 • T: +1 915 3510048 • F: +1 915 3510048 • info@elpasoholocaustmuseum.org • www.elpasoholocaustmuseum.org •
Exec. Dir.: *Leslie Novick* •
Historical Museum – 1992
Artifacts, dramatic displays, pictures, authentic hist of Europe during the Nazia era ... 48319

El Paso Museum of Archaeology at Wilderness Park, 4301 Transmountain Rd, El Paso, TX 79924 • T: +1 915 7554332 • F: +1 915 7596824 •
Dir.: *Marc Thompson* •
Archaeology Museum – 1977
Prehist artifacts of the American southwest & northern Mexico, ceramics, textiles, stone tools ... 48320

El Paso Museum of Art, 1 Arts Festival Plaza, El Paso, TX 79901 • T: +1 915 5321707 • F: +1 915 5321010 • Arts@elpasotexas.gov • www.elpasoartmuseum.org •
Dir.: *Becky Duval Reese* • Cur.: *Christian Gerstheimer* •
Fine Arts Museum – 1930
19th and 20th c American art, Mexican colonial art – Kress collection of 13th-18th c European art ... 48321

El Paso Museum of History, 12901 Gateway West, El Paso, TX 79928 • T: +1 915 8581928 • F: +1 915 8584591 • cityhistorymuseum@elpasotexas.gov •
Dir.: *René Harris* • Cur.: *Barbara J. Angus* •
Historical Museum – 1974
Guns, horse gear, hist clothing, hist of El Paso ... 48322

Glass Gallery and Main Gallery, University of Texas at El Paso, Department of Art, 500 W University Av, El Paso, TX 79968 • T: +1 915 7475181 • F: +1 915 7976749 • bonansin@utep.edu • www.utep.edu/arts •
Dir.: *Kate Bonansinga* •
Public Gallery – 1940 ... 48323

Insights - El Paso Science Center, 505 N Santa Fe, El Paso, TX 79901 • T: +1 915 5340000 • F: +1 915 5327416 • contact@insightsmuseum.org • www.insightsmuseum.org •
Science&Tech Museum – 1979
Participatory science exhib on perception & energy, human anatomy ... 48324

International Museum of Art, 1211 Montana St, El Paso, TX 79902 • T: +1 915 5436747 • F: +1 915 5439222 •
Dir.: *Michael Kirkwood* •
Fine Arts Museum
Western, African and Heritage gallery, William Kolliker Gallery ... 48325

Magoffin Home, 1120 Magoffin Av, El Paso, TX 79901 • T: +1 915 5335147 • F: +1 915 5444398 • Pres.: *Prestene Dehrkoop* • Local Museum – 1976 Historic house ... 48326

National Border Patrol Museum and Memorial Library, 4315 Transmountain Rd, El Paso, TX 79924 • T: +1 915 7596060 • F: +1 915 7590992 • info@borderpatrolmuseum.com • www.borderpatrolmuseum.com • Cur.: *Brenda Tisdale* • Historical Museum – 1985 ... 48327

People's Gallery, 2 Civic Center Plaza, El Paso, TX 79901 • T: +1 915 5414481 • F: +1 915 5414902 • drewaj@ci.el-paso.tx.us • www.artsresources.org • Dir.: *Alejandrina Drew* • Fine Arts Museum – 1979 ... 48328

The Stanley and Gerald Rubin Center for the Visual Arts, University of Texas at El Paso, 500 W University Av, El Paso, TX 79968 • T: +1 915 7475181 • F: +1 915 7476749 • bonansin@utep.edu • www.utep.edu/arts/ • Dir.: *Kate Bonansinga* • University Museum / Public Gallery – 1936 Changing exhibitions of cont art ... 48329

El Reno OK

Canadian County Historical Museum, 300 S Grand, El Reno, OK 73036 • T: +1 405 2625121 • Pres.: *Vicki Proctor* • Cur.: *Pat Reuter* • Local Museum – 1969 Indian artifacts; railway exhibits, Fort Reno, pioneer exhibits, tools, vehicles, farm machinery ... 48330

Elberta AL

Baldwin County Heritage Museum, 25521 Hwy 98 E, Elberta, AL 36530, mail addr: POB 356, Elberta, AL 36530 • T: +1 251 9868375 • F: +1 251 9868375 • bchm@gulftel.com • www.baldwincountyheritagemuseum.com • Head: *June Taylor* • Local Museum / Agriculture Museum – 1981 Rural hist, farming, forestry and associated crafts ... 48331

Elberton GA

Elberton Granite Museum, 1 Granite Plaza, Elberton, GA 30635 • T: +1 706 2832551 • F: +1 706 2836380 • granite@egaonline.com • www.egaonline.com • Dir.: *Thomas A. Robinson* • Local Museum – 1981 Local hist, granite monuments ... 48332

Elbow Lake MN

Grant County Historical Museum, Hwy 79 E, Elbow Lake, MN 56531 • T: +1 218 6854864 • gcmnhist@runestone.net • Dir./Cur.: *Patricia Benson* • Historical Museum – 1944 Regional hist, Indian artifacts, pioneers life, agriculture, fossil remains – library ... 48333

Elgin IL

Elgin Area Museum, 360 Park St, Elgin, IL 60120 • T: +1 847 7424248 • F: +1 847 9316199 • elginhistory@foxvalley.net • www.elginhistory.org • Dir.: *Elizabeth Marston* • Local Museum – 1961 Local city hist, architecture ... 48334

Elgin Public Museum, 225 Grand Blvd, Elgin, IL 60120 • T: +1 847 7416655 • F: +1 847 9316787 • epm@mc.net • www.elginpublicmuseum.org • Pres.: *Marty Kellams* • Ethnology Museum / Natural History Museum – 1904 Natural hist ... 48335

Elgin ND

Grant County Museum, 119 Main St, Elgin, ND 58533 • T: +1 701 5842900 • gcn@westriv.com • Head: *Duane Schatz* • Local Museum – 1970 Railroad equipment, pioneer items, printing equipment ... 48336

Elizabeth WV

Beauchamp Newman Museum, Court St, Elizabeth, WV 26143 • T: +1 304 2756534 • Cur.: *Jean Daley* • Decorative Arts Museum – 1950 Furniture, costumes, domestic artifacts ... 48337

Elizabeth City NC

Museum of the Albemarle, 501 S Water St, Elizabeth City, NC 27909 • T: +1 252 3351453 • F: +1 252 3350637 • moa@ncdcr.gov • www.museumofthealbemarle.com • Admin.: *Ed Merrell* • Cur.: *Tom Butchko* • Historical Museum – 1963 Indian artifacts, agricultural exhibits, lumbering items, farming, fishing ... 48338

Elizabethtown KY

Coca-Cola Memorabilia Museum of Elizabethtown, 321 Peterson Dr, Elizabethtown, KY 42701-9375 • T: +1 270 7693320 ext 328 • F: +1 270 7693323 • Cur.: *Channing C. Hardy* • Science&Tech Museum – 1977 Company museum ... 48339

Elizabethtown NY

Adirondack History Center, 7590 Court St, Elizabethtown, NY 12932 • T: +1 518 8736466 • echs@northnet.org • adkhistorycenter.org • Pres.: *James Rogers* • Dir.: *Margaret Gibbs* • Local Museum – 1954 Farm implements, horse-drawn vehicles, mining equip, surveying equip, tools, household items, exhibits of camping, hiking and hunting ... 48340

Elk City OK

National Route 66 Museum and Old Town Museum, 2717 W Hwy 66, Elk City, OK 73644 • T: +1 405 2256266 • F: +1 580 2253234 • Cur.: *Lucy Stansberry* • Local Museum – 1966 Pioneer artifacts, Western Rodeo, Indian artifacts, hist buildings ... 48341

Elk Grove Village IL

Elk Grove Farmhouse Museum, 399 Biesterfield Rd, Elk Grove Village, IL 60007 • T: +1 847 4379494 • F: +1 847 2283508 • cfearing@parks.elkgrove.org • www.elkgrove.org/egpd • Dir.: *Chris Fearing* • Cur.: *Peggy Rogers* • *Karen David* • Local Museum – 1976 ... 48342

Elk Horn IA

Danish Immigrant Museum, 2212 Washington St, Elk Horn, IA 51531 • T: +1 712 7647001 • F: +1 712 7647002 • info@danishmuseum.org • www.danishmuseum.org • C.E.O: *John Mark* • Ethnology Museum – 1983 Culture and diversity of Danish-American immigrants ... 48343

Elk River MN

Oliver H. Kelley Farm, 15788 Kelley Farm Rd, Elk River, MN 55330 • T: +1 612 4416896 • F: +1 612 4416302 • kelleyfarm@mnhs.org • www.mnhs.org/kelleyfarm • Head: *Bob M. Quist* • Agriculture Museum – 1961 Agricultural hist of Minnesota ... 48344

Elkhart IN

CTS Turner Museum, 905 N West Blvd, Elkhart, IN 46514 • T: +1 574 2937511 ext 296 • F: +1 574 2936146 • turnermuseum@ctscorp.com • www.ctscorp.com • Dir.: *Joseph Carlson* • Science&Tech Museum – 1979 Company history, technology ... 48345

Midwest Museum of American Art, 429 S Main St, Elkhart, IN 46515, mail addr: POB 1812, Elkhart, IN 46516 • T: +1 219 2936660 • F: +1 219 2936660 • info@midwestmuseum.us • midwestmuseum.us • Dir.: *Jane Burns* • Ass. Cur.: *Brian Byrn* • Fine Arts Museum – 1978 American impressionists, Chicago and pop art, prints, drawings, sculpture and photography, works by Robert Reid, Charles Hawthorne, Earnest Lawson, JohnSinger Sargent , Grandma Moses, Grant Wood and Thomas Hart Benton ... 48346

National New York Central Railroad Museum, 721 S Main St, Elkhart, IN 46516 • T: +1 219 2943001 • F: +1 219 2959434 • artscul@michiana.org • www.nycrrmuseum.org • C.E.O.: *David M. Bird* • Science&Tech Museum – 1987 Railroads ... 48347

Ruthmere House Museum, 302 E Beardsley Av, Elkhart, IN 46514 • T: +1 219 2640330 • F: +1 219 2660474 • dwiegand@ruthmere.org • www.ruthmere.org • Dir.: *Robert Beardsley* • Special Museum – 1969 Louis XV revival furniture, Dresden and Sevres porcelain; Wedgwood pottery; art glass by Steuben and Tiffany; sculpture by A. Rodin, C. Claudel, W.O. Patridge; paintings by Samuel F. B. Morse, Alma-Tadema and other 19th c American and European artists and craftmens ... 48348

RV/MH Heritage Foundation, 801 Benham Av, Elkhart, IN 46516-3369 • T: +1 219 2932344 • F: +1 219 2933466 • rvmhhall@aol.com • rv-mh-hall-of-fame.org • Pres.: *Carl A. Ehry* • Science&Tech Museum – 1972 Industrial hist ... 48349

S. Ray Miller Auto Museum, 2130 Middlebury St, Elkhart, IN 46516 • T: +1 219 5220539 • F: +1 219 5220358 • millermuseum@earthlink.net • CEO: *Linda L. Miller* • Science&Tech Museum / Folklore Museum – 1987 Automobile, vintage clothing, antiques ... 48350

Elkhorn WI

Webster House Museum, 9 E Rockwell, Elkhorn, WI 53121 • T: +1 262 7234248 • walcohistsoc@elknet.net • Pres.: *Doris M. Reinke* • Decorative Arts Museum / Local Museum – 1955 Local hist ... 48351

Elkins PA

Tyler Galleries, Temple University, Tyler School of Art, Beech and Penrose Av, Elkins, PA 19127 • T: +1 215 7822776 • F: +1 215 7822879 • nany.lewis@temple.edu • Fine Arts Museum Contemporary and visual culture ... 48352

Elkins Park PA

Temple Judea Museum, American Jewish Museums, 8339 Old York Rd, Elkins Park, PA 19027-1597 • T: +1 215 8872027 • F: +1 215 8871070 • tjmuseum@aol.com • www.kenesethisrael.org/mus.htm • Dir./Cur.: *Rita Rosen Poley* • Religious Arts Museum – 1984 800 Jewish artifacts from around the world, fabrics crafted for religious ceremonial, folk and cultural use, prints, paintings – archives, library ... 48353

Elko NV

Northeastern Nevada Museum, 1515 Idaho St, Elko, NV 89801 • T: +1 775 7383418 • F: +1 775 7789318 • info@museumelko.org • www.museumelko.org • C.E.O.: *Kim Steninger* • Dir.: *Claudia Wines* • Fine Arts Museum / Historical Museum / Natural History Museum – 1968 History, natural history, wildlife, art – library, archives ... 48354

Elkton MD

Historical Museum of Cecil County, 135 E Main St, Elkton, MD 21921 • T: +1 410 3981790 • history@cchistory.org • www.cchistory.org • Pres.: *Paula H. Newton* • Local Museum – 1931 Regional history ... 48355

Ellensburg WA

Central Washington Gallery, 313 N Pearl St, Ellensburg, WA 98926 • T: +1 509 9331919 • F: +1 509 9332547 • reuben@ambyedinger.com • www.centralwashingtongallery.org • Dir.: *Reuben Edinger* • Public Gallery Paintings, sculpture, photgraphy ... 48356

Clymer Museum of Art, 416 N Pearl St, Ellensburg, WA 98926 • T: +1 509 9626416 • F: +1 509 9626424 • clymermuseum@charter.net • www.clymermuseum.com • Dir.: *Diana Tasker* • Cur.: *Randall Sharp* • Fine Arts Museum – 1988 ... 48357

CWU Anthropology Department Collection, c/o Central Washington University, Farrell Hall, Ellensburg, WA 98926-7544 • T: +1 509 9633201 • F: +1 509 9633215 • anthro@cwu.edu • www.cwu.edu/~anthro • Ethnology Museum / University Museum – 1972 Ethnography from New Guinea, Mexico, Western Plateau, San Blas Islands, Panama, Africa, SW and NW Coast US ... 48358

Gallery One, 408 1/2 N Pearl St, Ellensburg, WA 98926 • T: +1 509 9252670 • F: +1 509 9252674 • Dir.: *Mary Frances* • Fine Arts Museum – 1968 Oil paintings, enamel paintings, watercolors, wood sculpture ... 48359

Olmstead Place State Park, 921 N Ferguson Rd, Ellensburg, WA 98926 • T: +1 509 9251943 • F: +1 509 9251955 • Local Museum / Agriculture Museum – 1968 Local hist, agricultural equipment ... 48360

Sarah Spurgeon Gallery, c/o Central Washington University, 400 E Eighth Av, Ellensburg, WA 98926-7564 • T: +1 509 9632665 • F: +1 509 9631918 • chinnm@cwu.edu • www.cwu.edu/~art • Dir.: *James Sahlstrand* • University Museum / Public Gallery – 1970 ... 48361

Ellenton FL

The Judah P. Benjamin Confederate Memorial at Gamble Plantation, 3708 Patten Av, Ellenton, FL 34222 • T: +1 941 7234536 • F: +1 941 7234538 • william.wolbert@dep.state.fl.us • www.fcn.state.fl.us • Local Museum / Agriculture Museum – 1926 Home of Robert Gamble, main house of the Gamble sugar plantation ... 48362

Ellerbe NC

Rankin Museum, 131 Church St, Ellerbe, NC 28338 • T: +1 910 6526378 • www.rankinmuseum.com • Cur.: *P.R. Rankin jr* • Fine Arts Museum / Historical Museum Southeastern Indian artifacts, Arts of Amazonia, Melanesian art ... 48363

Ellicott City MD

B & O Railroad Station Museum, 2711 Maryland Av, Ellicott City, MD 21043 • T: +1 410 4611945 • ecbostation@aol.com • www.borail.org • Pres.: *Lisa Mason-Chaney* • Science&Tech Museum – 1976 Transport, railroad stn, first terminus in the U.S ... 48364

Bagpipe Music Museum, 840 Oella Av, Ellicott City, MD 21043 • T: +1 410 3139311 • C.E.O.: *James Coldren* • Music Museum – 1997 ... 48365

Firehouse Museum, 3829 Church Rd, Ellicott City, MD 21043 • T: +1 410 4650232 • Dir.: *Carolyn Klein* • Science&Tech Museum – 1991 Original fire equipment ... 48366

Howard County Historical Society Museum, 8328 Court Av, Ellicott City, MD 21041, mail addr: POB 109, Ellicott, MD 21041 • T: +1 410 7500370 • F: +1 410 7500370 • hchs@clark.net • Dir.: *Phyllis Knill* • Local Museum – 1957 County's hist, furniture, weapons ... 48367

Ellis KS

Walter P. Chrysler Boyhood Home and Museum, 102 W 10th, Ellis, KS 67637 • T: +1 785 7263636 • F: +1 785 7263636 • chrysler@eaglecom.net • Dir.: *Karen Day* • Special Museum – 1954 Personal artifacts, jewelry, books ... 48368

Ellis Grove IL

Pierre Menard Home, 4230 Kaskaskta Rd, Ellis Grove, IL 62241-9702, mail addr: 4372 Park Rd, Ellis Grove, IL 62241-9704 • T: +1 618 8593031 • F: +1 618 8593741 • menrdhom@midwest.net • www.state.il.us/hpa/hs/menard.htm • C.E.O.: *Robert Coomer* • Historical Museum – 1927 Home of Pierre Menard, the first Lt. Governor of Illinois and U.S. Agent of Indian Affairs ... 48369

Ellsworth KS

Hodgen House Museum Complex, 104 W South Main, Ellsworth, KS 67439 • T: +1 785 4723059 • echs@informatics.net • Pres.: *Patricia L. Bender* • Local Museum – 1961 Folklore, Indian, agriculture, archaeology, costumes ... 48370

Rogers House Museum and Gallery, 102 E Main S, Ellsworth, KS 67439 • T: +1 785 4725674 • Dir.: *Robert Rogers* • Fine Arts Museum / Public Gallery – 1968 Paintings, prints, art ... 48371

Ellsworth ME

Telephone Museum, 166 Winkumpaugh Rd, Ellsworth, ME 04605 • T: +1 207 6679491 • www.thetelephonemuseum.org • Pres./Dir.: *Galley Sandra* • Science&Tech Museum – 1983 History of communication ... 48372

Woodlawn Museum, 19 Black House Dr, Ellsworth, ME 04605 • T: +1 207 6678671 • F: +1 207 6677950 • info@woodlawnmuseum.com • www.woodlawnmuseum.com • Exec. Dir.: *Joshua C. Torrance* • Local Museum – 1928 Decorative and fine art, personal arifacts, social hist of Down East, local hist ... 48373

Elmhurst IL

Elmhurst Art Museum, 150 Cottage Hill Av, Elmhurst, IL 60126 • T: +1 630 8340202 • F: +1 630 8340234 • info@elmhurstartmusesum.org • www.elmhurstartmuseum.org • Dir.: *D. Neil Bremer* • Ass. Dir.: *Heather Pastore* • Cur.: *Melissa Ganje* • Fine Arts Museum – 1981 Artwork by'regional American and European artists ... 48374

Elmhurst Historical Museum, 120 E Park Av, Elmhurst, IL 60126 • T: +1 630 8331457 • F: +1 630 8331326 • ehm@elmhurst.org • www.elmhurst.org • Dir.: *Brian F. Bergheger* • Cur.: *Marcia Lautanen-Raleigh* • Local Museum – 1952 Local history, artifacts, manuscripts, literature on local hist, photographs ... 48375

Lizzadro Museum of Lapidary Art, 220 Cottage Hill Av, Elmhurst, IL 60126 • T: +1 630 8331616 • F: +1 630 8331225 • info@lizzadromuseum.org • www.lizzadromuseum.org • Exec. Dir.: *John S. Lizzadro* • Dir.: *Dorothy J. Asher* • Special Museum – 1962 Chinese jades, hardstone carvings, gemstones, minerals, earth science ... 48376

Milano Model and Toy Museum, 116 Park Av, Elmhurst, IL 60126 • T: +1 630 2794422 • www.toys-n-cars.com • CEO: *Dean Milano* • Special Museum – 2002 Model cars, 1930s-1960s vintage toys 48377

Elmira NY

Arnot Art Museum, 235 Lake St, Elmira, NY 14901 • T: +1 607 7343697 • F: +1 607 7345687 • jdo@arnotartmuseum.org • www.arnotartmuseum.org • Pres.: *Laurie Liberatore* • Dir./Cur.: *John D. O'Hern* • Cur.: *Karen Kucharski* • Sc.Staff: *Michael Sampson* (Registrar) • Fine Arts Museum – 1913 48378

Chemung Valley History Museum, 415 E Water St, Elmira, NY 14901 • T: +1 607 7344167 • F: +1 607 7341565 • cchs@chemungvalleymuseum.org • www.chemungvalleymuseum.org • Pres.: *Douglas C. Tifft* • Dir.: *Amy H. Wilson* • Local Museum – 1923 Civil war and Indian artifacts, Mark twain coll, local hist, industry and agriculture 48379

National Soaring Museum, Harris Hill, 51 Soaring Hill Dr, Elmira, NY 14903-9204 • T: +1 607 7343128 • F: +1 607 7326745 • nsm@soaringmuseum.org • www.soaringmuseum.org • Pres.: *Robert Gaines* • Exec. Dir.: *Peter W. Smith* • Science&Tech Museum – 1969 Gliders and sailplanes – archive, library 48380

Elsah IL

School of Nations Museum, Principia College, Elsah, IL 62028 • T: +1 618 3142100, 3745236 • F: +1 618 2753519 • bkh@prin.edu • Dir.: *Dan Hanna* • University Museum / Folklore Museum – 1930 American Indian and pre-Columbian artifacts, Asian art, decorative arts, textiles, costumes, dolls, pottery 48381

Village of Elsah Museum, 26 LaSalle, Elsah, IL 62028 • T: +1 618 3745238 • F: +1 618 3742625 • voemayor@empowering.com • www.elsah.org • C.E.O.: *Marjorie A. Doerr* • Local Museum – 1978 Local history housed in 1857 school, archaeological, geological, photography, social and industrial components 48382

Elverson PA

Hopewell Furnace Hopewell Furnace National Historic Site, 2 Mark Bird Ln, Elverson, PA 19520 • T: +1 610 5828773 • F: +1 610 5822768 • hofu_superintendent@nps.gov • www.nps.gov/hofu/index.html • Head: *Edie Shean-Hammond* (Superintendent) • Open Air Museum – 1938 Daybooks of the iron furnace operation during the 19th century. Archives of the Civilian Conservation Corps work in the 1930's 48383

Ely NV

East Ely Railroad Depot Museum, 1100 Avenue A, Ely, NV 89301 • T: +1 775 2891663 • F: +1 775 2891664 • esm@the-onramp.net • Science&Tech Museum / Historical Museum – 1992 Mining and transportation, industrial development, right of way maps 48384

White Pine Public Museum, 2000 Aultman St, Ely, NV 89301 • T: +1 775 2894710 • F: +1 775 2894710 • wpmuseum@idsely.com • www.idsely.com/~wpmuseum • Head: *Tom Puckett* • Local Museum / Natural History Museum – 1957 Indian artifacts, dolls, antiques, historic costumes, natural hist, trains, railroads 48385

Elyria OH

Hickories Museum, 509 Washington Av, Elyria, OH 44035 • T: +1 216 3223341 • F: +1 440 3222817 • thehickories@alltel.net • lchs.org • Pres.: *George Strom* • Exec.Dir.: *William Bird* • Historical Museum – 1889 Costumes and textiles 1860-1950, household furnishings, glassware, toys 48386

Spirit of '76 Museum, 509 Washington Av, Elyria, OH 44035 • T: +1 216 6474367 • Cur.: *Diane Stanley* • Historical Museum – 1970 Archibald M. Willard memorabilia 48387

Elysian MN

Le Sueur County Historical Museum, NE Frank and Second Sts, Elysian, MN 56028 • T: +1 507 2674620, 3628350 • F: +1 507 2674750 • museum@lchs.mus.mn.us • Pres.: *Pat Nusbaum* • CEO: *Shirley Zimprich* • Local Museum – 1966 County hist, artworks by local artists, life in 1900, agriculture, American Indians, Red Wing pottery, WWII & Civil war items 48388

Emmaus PA

Shelter House, 601 S Fourth St, Emmaus, PA 18049 • T: +1 610 9659258 • Pres.: *R. Mitchell Freed* • Local Museum – 1951 History 48389

Empire MI

Empire Area Heritage Group Museum, 11544 S La Core, Empire, MI 49630 • T: +1 231 3265568 • Pres.: *David Taghon* • Local Museum – 1972 Local hist, logging, lumbering, sailing, railroads, horse-drawn vehicles, woodworking, fire fighting 48390

Emporia KS

Emporia State University Geology Museum, 1200 Commercial St, Emporia, KS 66801 • T: +1 316 3415330 • F: +1 316 3416055 • Dir.: *Dr. Michael Morales* • Natural History Museum – 1983 Geology 48391

Lyon County Historical Society and Museum, 118 E 6th Av, Emporia, KS 66801 • T: +1 620 3420933 • lycomu@osprey.net • Pres.: *Lisa Goldstein* • Dir.: *J. Greg Jordon* • Chm.: *Annette Rice* • Sc. Staff: *Laura Dodge* (Education) • *Jake Dalton* (Registrar) • *Carolyn Eckstrom* (Asst. Registrar) • Local Museum – 1937 County history, Carnegie library and Richard Howe home 48392

Norman R. Eppink Art Gallery, Emporia State University, 1200 Commercial St, Emporia, KS 66801 • T: +1 316 3415246 • F: +1 316 3416246 • schwarml@emporia.edu • www.emporia.edu/m/www/art/eppink.htm • C.E.O.: *Kay Schallenkamp* • Dir.: *Larry W. Schwarm* • Public Gallery – 1939 Contemporary drawings & prints, paintings, sculpture 48393

Richard H. Schmidt Museum of Natural History, Emporia State University, 1200 Commercial St, Emporia, KS 66801 • T: +1 316 3415311 • F: +1 316 3415607 • mooredwi@emporia.edu • www.emporia.edu/smnh/ • Dir.: *Dr. Gregory Smith* • University Museum / Natural History Museum – 1959 Skins & mounted specimens of birds, mammals, ornithology 48394

Encampment WY

Grand Encampment Museum, 817 Barnett Av, Encampment, WY 82325 • T: +1 307 3275558, 3275308 • F: +1 307 3275427 • gemuseum@aol.com • Pres.: *Mary Martin* • Cur.: *Rick Martin* • Science&Tech Museum – 1965 Mining, Indian artifacts 48395

Encinitas CA

San Dieguito Heritage Museum, 450 Quail Gardens Dr, Encinitas, CA 92024 • T: +1 760 6329711 • F: +1 760 6325695 • sdheritage@sbcglobal.net • www.sdheritage.org • CEO: *Mary Fran Riggs* • Local Museum – 1988 District history and development 48396

Enfield NH

Enfield Historical Society Museum, Rte 4A, Enfield, NH 03749 • T: +1 603 6327740 • Pres.: *John Goodwin* • Local Museum – 1991 Local hist 48397

Enfield Shaker Museum, 24 Caleb Dyer Ln, Enfield, NH 03748 • T: +1 603 6324346 • F: +1 603 6324346 • chosen.vale@valley.net • www.shakermuseum.org • Dir.: *Michael J. O'Connor* • Local Museum / Historical Museum – 1986 Historic village, furniture, photographs 48398

Lockehaven Schoolhouse Museum, Lockehaven Rd, Enfield, NH 03748 • T: +1 603 6327740 • C.E.O.: *John Goodwin* • Local Museum – 1947 48399

Englewood CO

The Museum of Outdoor Arts, 1000 Englewood Pkwy, Ste 2-230, Englewood, CO 80110 • T: +1 303 8060444 • F: +1 303 8060504 • cmleitner@moaonline.org • www.moaonline.org • Exec. Dir.: *Cynthia Leitner* • Fine Arts Museum – 1982 Art 48400

Enid OK

Museum of the Cherokee Strip, Oklahoma Historical Society, 507 S Fourth St, Enid, OK 73701 • T: +1 580 2371907 • F: +1 580 2422874 • mcs1@onenet.net • www.ok-history.mus.ok.us • C.E.O.: *Dr. Bob Blackburn* • Cur.: *Steve Dortch* • Historical Museum – 1951 Native American coll, farm implements, regional hist 48401

Ephraim WI

Ephraim Foundation Museums, Thomas Goodletson House, Pioneer School House, Anderson Store Museum, Anderson Barn Museum, 3060 Anderson Ln, Ephraim, WI 54211 • T: +1 920 8549688 • F: +1 920 8547232 • efoundation@itol.com • www.ephraim.org • Dir.: *Sally Jacobson* • Local Museum – 1949 Hist buildings with original furniture 48402

Ephrata PA

Cocalico Valley Museum, 249 W Main St, Ephrata, PA 17522 • T: +1 717 7331616 • Pres.: *Clarence Spohn* • Historical Museum – 1957 History of Cocalico Valley 48403

Ephrata Cloister, 632 W Main St, Ephrata, PA 17522-1717 • T: +1 717 7336600 • F: +1 717 7334364 • tocollins@state.pa.us • www.state.pa.us • Pres.: *Dr. William Ives* • Dir.: *Toni I. Collins* • Open Air Museum / Religious Arts Museum / Historic Site – 1732 Religious Village Museum comprising of 12 mid-18th c bldgs of Germanic architectural style, located on original site of a celibate religious community 48404

Ephrata WA

Grant County Historical Museum, 742 Basin St NW, Ephrata, WA 98823 • T: +1 509 7543334 • F: +1 509 7542148 • grantcomuseum@vib.tv • Local Museum – 1951 48405

Epping ND

Buffalo Trails Museum, Main St, Epping, ND 58843 • T: +1 701 8594361 • F: +1 701 8594361 • buffalotrails@epping.govoffice.com • Pres.: *Duane Syverson* • Local Museum – 1966 Indian artifacts, geology, mineralogy, archaelogy 48406

Epworth IA

Father Weyland Gallery, c/o Divine Word College, 102 Jacoby Dr SW, Epworth, IA 52045-0380 • T: +1 319 8763353 • F: +1 319 8763353 • Fine Arts Museum – 1985 Art of Africa and Papua New Guinea 48407

Erie KS

Mem-Erie Historical Museum, 225 S Main St, Erie, KS 66733 • T: +1 620 2443218 • F: +1 620 2443332 • ruthmc@cox.net • Pres.: *Ruth McKinney-Tandy* • Local Museum – 1994 Furnishings, photographs, personal artifacts 48408

Erie PA

Erie Art Museum, 411 State St, Erie, PA 16501 • T: +1 814 4595477 • F: +1 814 4521744 • contact@erieartmuseum.org • www.erieartmuseum.org • Pres.: *Dr. Kirk W. Steehler* • Exec. Dir.: *John L. Vanco* • Fine Arts Museum – 1898 Indian bronze and stone sculpture, Chinese porcelain, American ceramics 48409

Erie History Center, 417-419 State St, Erie, PA 16501 • T: +1 814 4541813 • F: +1 814 4521744 • echs@eriecountyhistory.org • www.eriecountyhistory.org • Dir.: *Dr. C. Raymond* • Local Museum – 1903 Nortwestern Pensylvania and Great Lakes hist 48410

Experience Children's Museum, 420 French St, Erie, PA 16507 • T: +1 814 4533743 • F: +1 814 4599735 • junep@erie.net • www.eriechildrensmuseum.org • Exec. Dir.: *June Pintea* • Ethnology Museum Children,Youth 48411

United States Brig Niagara, Homeport Erie Maritime Museum, 150 E Front St, Ste 100, Erie, PA 16507 • T: +1 814 4522744 • F: +1 814 4556760 • sail@brigniagara.org • www.brigniagara.org • Dir.: *Walter P. Rybka* • *John Beebe* • Military Museum – 1943 Military hist 48412

Watson-Curtze Mansion, Erie Historical Museum, 356 W Sixth St, Erie, PA 16507 • T: +1 814 8715790 • F: +1 814 8790988 • echs2@velocity.net • www.eriecountyhistory.org • Dir.: *Dr. David Frew* • Local Museum – 1899 48413

Escanaba MI

Delta County Historical Society Museum, 16 Waterplant Rd, Escanaba, MI 49829 • T: +1 906 7863428, 7896790 • F: +1 906 7863428 • Pres.: *Peter Strom* • Dir.: *Mrs. Robert Ham* • Local Museum – 1947 Area hist, agriculture, marine, military hist, logging, Indian artifacts, anthropology – archives 48414

Sandpoint Lighthouse, Sandpoint, Ludington Park, Escanaba, MI 49829-4052 • T: +1 906 7863763 • Pres.: *Peter Strom* • Dir.: *Mrs. Robert Mosenfelder* • Historical Museum – 1990 Latern room, maps, pictures 48415

William Bonifas Fine Arts Center, 700 First Av S, Escanaba, MI 49829 • T: +1 906 7863833 • F: +1 906 7863840 • pasqua@bonifasarts.org • www.bonifasarts.org • Dir.: *Pasqua Warstler* (Visual Arts, Education) • Cust.: *Jesse Farkas* • Fine Arts Museum / Public Gallery – 1974 Regional arts from the Great Lakes – theater, studio 48416

Escondido CA

Auto Museum, Deer Park Winery, 29013 Champagne Blvd, Escondido, CA 92026 • T: +1 760 2981666 • www.deerparkwinery.com • C.E.O.: *Clark Knapp* • Science&Tech Museum – 1979 Automoblile hist 48417

California Center for the Arts Museum, 340 N Escondido Blvd, Escondido, CA 92025 • T: +1 760 8394120 • marketing@artcenter.org • www.artcenter.org • Dir.: *Mary-Catherine Ferguson* • Fine Arts Museum – 1992 California and Latin American contemporary art – library, art classes 48418

Heritage Walk Museum, c/o Escondido History Center, 321 N Broadway, Escondido, CA 92025 • T: +1 760 7438207 • ehc@escondidohistory.org • www.escondidohistory.com • Exec. Dir.: *Wendy Barker* • Historical Museum – 1956 Local history, railroad car with model train layout, furnishing, blacksmith shop – library 48419

Mingei International Museum, North County Satellite, 155 W Grand Av, Escondido, CA 92025 • T: +1 760 7355533 • F: +1 760 7353306 • mingei@mingei.org • www.mingei.org • Dir.: *Martha W. Longenecker* • Fine Arts Museum Art, folk art, craft and design 48420

Essex CT

Connecticut River Museum, 67 Main St, Essex, CT 06426 • T: +1 860 7678269 • F: +1 860 7677028 • crm@ctrivermuseum.org • www.ctrivermuseum.org • Dir.: *Stuart Parnes* • Cur.: *Alison Guinness* • Historical Museum / Folklore Museum – 1974 Maritime and River Museum: housed in 1878 wooden warehouse 48421

Essex Historical Museum, 22 Prospect St, Essex, CT 06426 • T: +1 860 7670681 • www.essexhistory.org • Pres.: *Frank Flores* • Local Museum – 1955 Local history 48422

Essex MA

Essex Shipbuilding Museum, 66 Main St, Essex, MA 01929 • T: +1 978 7687541 • F: +1 978 7682541 • info@essexshipbuildingmuseum.org • www.essexshipbuildingmuseum.org • Cur.: *Jim Witham* • Science&Tech Museum – 1976 Hearse house, maritime history, old Story Yard site where launched 300 fishing schooners – library, archives, drydock 48423

Essington PA

Governor Printz Park, Second and Taylor Aves, Essington, PA 19029 • T: +1 610 5837221 • F: +1 610 4599586 • C.E.O.: *Brent D. Glass* • Local Museum – 1937 48424

Estero FL

Koreshan Historic Site, US 41 at Corkscrew Rd, Estero, FL 33928 • T: +1 941 9920311 • F: +1 941 9921607 • catherine.ohnemus@dep.state.fl.us • www.floridastateparks.org • Special Museum – 1961 Historic buildings, 1894-1977 utopian settlement 48425

Estes Park CO

Estes Park Area Historical Museum, 200 Fourth St, Estes Park, CO 80517 • T: +1 970 5866256 • F: +1 970 5866909 • bkilsdonk@estes.org • www.estesnet.com/museum • Dir.: *Betty Kilsdonk* • Cur.: *Lisel Goetze* • Local Museum – 1962 General museum, regional history 48426

Rocky Mountain National Park Museum, 1000 Hwy 36, Estes Park, CO 80517 • T: +1 970 5861340 • F: +1 970 5861310 • christine_baker@nps.gov • www.nps.gov • Natural History Museum – 1915 Natural history, archaeology, botany, zoology, history 48427

Estherville IA

H.G. Albee Memorial Museum, 1720 Third Av, S, Estherville, IA 51334 • T: +1 712 3622750 •
Pres.: *David Kaltved* •
Special Museum – 1964
General coll pertaining to Emmet county and country schools ... 48428

Eufaula AL

Shorter Mansion Museum, 340 N Eufaula Av, Eufaula, AL 36027 • T: +1 334 6873793 • F: +1 334 6871836 • info@eufaulapilgrimage.com • www.eufaulapilgrimage.com •
Dir.: *Calvin Wingo* •
Local Museum – 1965
Local hist, Admiral Thomas H. Moorer room ... 48429

Eugene OR

Adell McMillan Gallery, University of Oregon, EMU, Eugene, OR 97403-1228 • T: +1 541 3464373 • F: +1 541 3464400 •
University Museum / Fine Arts Museum – 1981 . 48430

Jordan Schnitzer Museum of Art at the University of Oregon, 1430 Johnson Ln, Eugene, OR 97403 • T: +1 541 3463027 • F: +1 541 3460976 • uoma@darkwing.uoregon.edu • uoma.uoregon.edu •
Dir.: *David Turner* • Ass. Dir.: *Lawrence Fong* •
Fine Arts Museum / University Museum – 1932
American and regional art,European, Korean, Chinese and Japanese art, Russian icons ... 48431

Lane Community College Art Gallery, 4000 E 30th Av, Eugene, OR 97405 • T: +1 541 7474501 • F: +1 541 7444185 •
Dir.: *Nancy LaVelle* •
University Museum / Fine Arts Museum – 1970
Contemporary art ... 48432

Lane County Historical Museum, 740 W 13th Av, Eugene, OR 97402 • T: +1 541 6824242 • F: +1 541 6827361 • lchm@efn.org • www.lchmuseum.org •
Dir.: *Robert L. Hart* •
Historical Museum – 1951
1850-1950 textiles, tools, vehicles, household furnishings; early settlement period; trades 1920s-1930s – Gallery ... 48433

Maude I. Kerns Art Center, 1910 E 15th Av, Eugene, OR 97403 • T: +1 541 3451571 • F: +1 541 3456248 • staff@mkartcenter.org • www.mkartcenter.org •
Dir.: *Karen Marie Pavelec* •
Fine Arts Museum / Public Gallery – 1962
Changing exhibitions ... 48434

Museum of Natural and Cultural History, University of Oregon, 1680 E 15th Av, Eugene, OR 97403-1224 • T: +1 541 3463024 • F: +1 541 3465334 • mnh@uoregon.edu • natural-history.uoregon.edu •
Dir.: *C. Melvin Aikens* • *Patricia Krier* (Program) • *Thomas Connolly* (Research for Archaeology) • *Pamela Endzweig* (Collections) •
Natural History Museum / Ethnology Museum / Archaeology Museum – 1935
Oregon natural and cultural history and archaeology, worldwide ethnology ... 48435

New Zone Virtual Gallery, 2740 Onyx St, Eugene, OR 97403 • T: +1 541 3435651 • ross@newzone.org • www.newzone.org •
Dir.: *Jerry Ross* •
Fine Arts Museum – 1978
Etchings, abstract figurative, oils, landscape, portraits ... 48436

Oregon Air and Space Museum, 90377 Boeing Dr, Eugene, OR 97402 • T: +1 541 4611101 • F: +1 541 6893593 • Thomasewinn@cs.com •
Pres.: *Michael Holcomb* •
Science&Tech Museum – 1987
Aviation, space ... 48437

The Science Factory Children's Museum, 2300 Leo Harris Pkwy, Eugene, OR 97401 • T: +1 541 6827888 • F: +1 541 4849027 • info@sciencefactory.org • www.sciencefactory.org •
Exec.Dir. & C.E.O.: *Edward Gerdes* •
Science&Tech Museum – 1960 ... 48438

Eureka CA

Clarke Historical Museum, 240 E St, Eureka, CA 95501 • T: +1 707 4431947 • F: +1 707 4430290 • clarkehistorical@att.net • www.clarkemuseum.org •
Dir./Cur.: *Pam Service* •
Local Museum – 1960
Native American beskerty and ceremonial regalia, natural hist, decorative arts, textiles – library, photo archive ... 48439

Morris Graves Museum of Art, 636 F St, Eureka, CA 95501 • T: +1 707 4420278 • F: +1 707 4422040 • jemima@humboldtarts.org • www.humboldtarts.org •
Pres./Dir.: *Sally Arnot* • Cur.: *Jemima Harr* •
Fine Arts Museum ... 48440

Eureka KS

Greenwood County Historical Society, 120 W 4th St, Eureka, KS 67045-1445 • T: +1 620 5836682 • gwhistory@correct-connect.com •
Pres.: *James H. Francis* •
Association with Coll – 1973
Dolls, tolls, pictures, 1868-2000, newspaper, dishes, bottles, genealogy ... 48441

Eureka SD

Eureka Pioneer Museum of McPherson County, 1210 J Av, Eureka, SD 57437 • T: +1 605 2842987 • www.glpta.org/eurekamuseum.htm •
Dir./Cur.: *Edmund Opp* •
Local Museum – 1978
Local hist ... 48442

Eureka UT

Tintic Mining Museum, POB 218, Eureka, UT 84628 • T: +1 435 4336842 •
Dir.: *J.L. McNulty* • Ass. Dir.: *E.C. McNulty* •
Historical Museum – 1974
Mining artifacts, complete mineral display ... 48443

Eureka Springs AR

Gay Nineties Button and Doll Museum, 338 Onyx Cave Ln, Eureka Springs, AR 72632 • T: +1 479 2539321 •
Pres.: *Muriel H. Schmidt* •
Special Museum – 1971
Dolls, household accessoires, glassware ... 48444

Eustis FL

Eustis Historical Museum, 536 N Bay St, Eustis, FL 32726-3439 • T: +1 352 4830046 • EustisHist@aol.com • www.eustis.org •
Cur.: *Emily Hayes* •
Historical Museum – 1983
Florida hist and culture ... 48445

Evanston IL

Charles Gates Dawes House, Evanston Historical Society Museum, 225 Greenwood St, Evanston, IL 60201 • T: +1 847 4753410 • F: +1 847 4753599 • evanstonhs@nortwestern.edu • www.evanstonhistorical.org •
Pres.: *Marilyn Lindemann* •
Local Museum – 1898
Restored 1894 national historic landmark, home of former Vice President and Nobel laureate Dawes, original 1920s furnishings ... 48446

Dittmar Memorial Gallery, Northwestern University, 1999 S Campus Dr, Evanston, IL 60208 • T: +1 847 4912348 • F: +1 847 4914333 •
Fine Arts Museum – contemporary art ... 48447

Eavanston Ecology Center, 2024 McCormick Blvd, Evanston, IL 60201 • T: +1 847 8645181 • F: +1 847 4488805 • ecologycenter@cityofevanston.org • www.laddarboretum.org •
Cur.: *Alexis Carlson-Reynolds* •
Natural History Museum – 1975
Natural hist of region, environment ... 48448

Evanston Art Center, 2603 Sheridan Rd, Evanston, IL 60201 • T: +1 847 4755300 • F: +1 847 4755330 • www.evanstonartcenter.org •
Dir.: *Alan Leder* •
Fine Arts Museum – 1929
Contemporary art ... 48449

Frances Willard House Museum, WCTU Museum, 1730 Chicago Av, Evanston, IL 60201-4585 • T: +1 847 3287500 • F: +1 847 3287500 • willardhouse@earthlink.net • www.franceswillardhouse.org •
Pres.: *Lori Osborne* •
Special Museum – 1946
Home of Frances E. Willard (1865-1898) ... 48450

The Kendall College Mitchell Museum of the American Indian, 2600 Central Park, Evanston, IL 60201 • T: +1 847 4951030 • F: +1 847 4950911 • mitchellmuseum@mindsprung.com • www.mitchellmuseum.org •
Dir.: *Janice B. Klein* •
University Museum / Historical Museum / Ethnology Museum – 1977 ... 48451

Levere Memorial Temple, 1856 Sheridan Rd, Evanston, IL 60201-3837 • T: +1 847 4751856 • F: +1 847 4752250 • lburrows@sae.net • www.sae.net •
Pres.: *Thomas Bower* • C.E.O.: *Thomas Goodale* •
Historical Museum – 1929 ... 48452

Mary and Leigh Block Museum of Art, Northwestern University, 40 Arts Circle Dr, Evanston, IL 60208-2410 • T: +1 847 4914000 • F: +1 847 4912261 • block-museum@nortwestern.edu • www.blockmuseum.northwestern.edu •
Dir.: *David Alan Roiberts* •
Fine Arts Museum / University Museum – 1851
Prints, drawings and photographs, architectural renderings by Walter Burley, and Marion Mahony Griffin, textiles by Theo Leffman – Sculpture garden ... 48453

Mitchell Museum of the American Indian, 2600 Central Park, Evanston, IL 60201 • T: +1 847 4751030 • F: +1 847 4750911 • mitchellmuseum@mindspring.com • www.mitchellmuseum.org •
Dir.: *Janice B. Klein* •
Ethnology Museum / Folklore Museum – 1977
Cultures of the Woodlands, Plains, Arctic, Southwest and Northwest coast ... 48454

Evansville IN

Angel Mounds, 8215 Pollack Av, Evansville, IN 47715 • T: +1 812 8533956 • F: +1 812 8587686 • curator@angelmounds.org • www.angelmounds.org •
Cur.: *Mike Linderman* • Ass. Cur.: *William Salvador Spellazza* •
Archaeology Museum / Folklore Museum – 1972
Archaeology, pre-historic middle Mississippian Indian site ... 48455

Evansville Museum of Arts and Science, 411 SE Riverside Dr, Evansville, IN 47713 • T: +1 812 4252406 • F: +1 812 4217506 • mary@emuseum.org • www.emuseum.org •
Dir.: *John W. Streetman* • Cur.: *Thomas R. Lonnberg* (History) •
Local Museum / Fine Arts Museum – 1926
Paintings, sculpture, prints and drawings, transpotation, arms and armor – planetarium ... 48456

Reitz Home Museum, 224 SE First St, Evansville, IN 47713 • T: +1 812 4261871 • F: +1 812 4262179 • reitz@evansville.net • www.reitzhome.evansville.net •
Dir.: *Tess C. Grimm* • Cur.: *Mary Bower* •
Local Museum – 1974 ... 48457

Evansville WY

Reshaw Exhibit, 235 Curtis, Evansville, WY 82636 • T: +1 307 2346530 • F: +1 307 2665109 •
Chm.: *Joyce Hill* •
Science&Tech Museum – 1963
Indian and pioneer artifacts ... 48458

Eveleth MN

United States Hockey Hall of Fame, 801 Hat Trick Av, Eveleth, MN 55734 • T: +1 218 7445167 • F: +1 218 7442590 • sersha@ushockeyhall.com • www.ushockeyhall.com •
Exec. Dir.: *Tom Sersha* •
Special Museum – 1969
Cleve Bennewitz skate coll, Olympic & women's hockey – archive ... 48459

Evergreen CO

Hiwan Homestead Museum, 4208 S Timbervale Dr, Evergreen, CO 80439 • T: +1 303 6746262 • F: +1 303 6707746 • jsteinle@co.jefferson.co.us • www.co.jefferson.co.us/ext/dpt/comm_res/openspac/hiwan.htm •
Cur.: *Angela Rayne* •
Ethnology Museum / Historical Museum / Historic Site – 1974
Local history, Camp Neosho, later renamed Hiwan Ranch Inetrnational antique Doll collection, Southwestern native american pottery – Archive of local hist ... 48460

Ewing NJ

The College Art Gallery, College of New Jersey, 2000 Pennington Rd, Ewing, NJ 08628 • T: +1 609 7712633 • F: +1 609 6375193 • masterjp@tcnj.edu • www.tcnj.edu/~artmain •
Dir.: *Judy Masterson* •
Public Gallery
Photography, printmaking, paintings, drawings, sculpture ... 48461

Excelsior MN

Excelsior-Lake Minnetonka Historical Museum, 305 Water St, Excelsior, MN 55331 • T: +1 952 2214766 • elmhs@stribmail.com •
Pres.: *Tad Shaw* •
Local Museum – 1972
Local history, maps, pictures ... 48462

Exeter NH

American Independence Museum, 1 Governors Ln, Exeter, NH 03833-2420 • T: +1 603 7722622 • F: +1 603 7720861 • info@independencemuseum.org • www.independencemuseum.org •
Exec. Dir.: *Funi Burdick* • Cur.: *Laura Gowing* •
Historical Museum – 1991
17th to 19th c decorative arts ... 48463

Gilman Garrison House, 12 Water St, Exeter, NH 03833 • T: +1 603 4363205 • F: +1 617 2279204 • GilmanGarrisonHouse@HistoricNewEngland.org • www.historicnewengland.org •
Pres.: *Carl Nold* •
Local Museum – 1965 ... 48464

The Lamont Gallery, Phillips Exeter Academy, Frederic R. Mayer Bldg, 20 Main St, Exeter, NH 03833 • T: +1 603 7773461 • F: +1 603 7774371 • gallery@exeter.edu • www.exeter.edu •
Dir.: *Karen Burges-Smith* •
Fine Arts Museum / University Museum – 1953 . 48465

Exira IA

Courthouse Museum, Washington St, Exira, IA 50076 • T: +1 712 5633984 • aced@netins.net •
Local Museum – 1960
Indian hist, flood memorabilia ... 48466

Exton PA

Thomas Newcomen Museum, 412 Newcomen Rd, Exton, PA 19341 • T: +1 610 3636600 • F: +1 610 3630612 • mewcomen@libertynet.org •
Pres.: *C. Daniel Hayes* • C.E.O.: *Edward Kottcamp* •
Science&Tech Museum – 1923 ... 48467

Fabius NY

Pioneer Museum, Highland Forest Rte 80, Fabius, NY 13063 • T: +1 315 4536767 • F: +1 315 4536762 • stemarie@nysnet.net • www.ongov.net/parks •
Head: *Valerie Bell* •
Local Museum – 1959 ... 48468

Fairbanks AK

Alaska Centennial Center for the Arts, 2300 Airport Way, Fairbanks, AK 99701, mail addr: c/o Fairbanks Arts Association, POB 72786, Fairbanks, AK 99707 • T: +1 907 4566485 • june@fairbanksarts.org • www.fairbanksarts.org •
Public Gallery
Contemporary arts ... 48469

Dog Mushing Museum, 250 Cushman St, Fairbanks, AK 99701, mail addr: POB 80136, Fairbanks, AK 99708 • T: +1 907 4566874 • F: +1 907 4566874 •
Dir.: *Elizabeth Smith* •
Historical Museum – 1987
Dog mushing, with photo displays, sleds, clothing, harnesses, trophies and racing info ... 48470

Fairbanks Community Museum, 410 Cushman St, Fairbanks, AK 99701 • T: +1 907 4573669, 4528671 • glake@alaska.net • fairbanks-alaska.com/fairbanks-museum.htm •
Local Museum ... 48471

Museum of the North, 907 Yukon Dr, Fairbanks, AK 99775-6960, mail addr: c/o University of Alaska, POB 756960, Fairbanks, AK 99775 • T: +1 907 4747505 • F: +1 907 4745469 • museum@uaf.edu • www.uaf.edu/museum •
Dir.: *Aldona Jonaitis* • Cur.: *Roland Gangloff* (Earth Sciences) • *Gordon Haas* (Ichthyology) • *Link Olson* (Mammalogy) • *Molly Lee* (Ethnology, History) • *Daniel Odess* (Archaeology) • *Kevin Winker* (Ornithology) •
Coord.: *Barry McWayne* (Fine Arts) • *Wanda Chin* (Exhibits) • *Leonard Kamerling* (Alaska Center for Docu Film) •
Fine Arts Museum / University Museum / Natural History Museum / Ethnology Museum – 1929
Natural history, fine art, photography, ethnograpgy, archaeology, paleontology, geology, ornitology, mammals – herbarium ... 48472

Pioneer Air Museum, Airport Way and Peger Rd, Fairbanks, AK 99701, mail addr: POB 70437, Fairbanks, AK 99707-0437 • T: +1 907 4510037 • www.akpub.com/akttt/aviat.html •
Pres.: *George C. Clayton* •
Science&Tech Museum – 1983
Aviation artifacts pertaining Arctic Alaska, aircraft, pictures, engines, radios, memorabilia of bush pilots and airlines ... 48473

Pioneer Museum, Pioneer Park, Airport Way, Fairbanks, AK 99707-6970, mail addr: POB 70176, Fairbanks, AK 99707 • T: +1 907 4568579 • F: +1 907 4796970 • fairbanks-alaska.com/pioneer-museum-big-stampede.htm •
Historical Museum
Historical coll of photos and artifacts of Fairbanks ... 48474

Wickersham House Museum, 535 Second Av, Fairbanks, AK 99708 • T: +1 907 4576165 • F: +1 907 4576165 • tyhs@ptialaska.net • www.ptialaska.net •
Pres.: *Renee Blahuta* •
Historical Museum – 1959
Alaskan hist ... 48475

Fairbury NE

Rock Creek Station State Historic Park, 57425 710 Rd, Fairbury, NE 68352 • T: +1 402 7295777 • rockcreek@ngpc.state.ne.us • www.ngpc.state.ne.us •
Head: *Wayne Brandt* •
Local Museum / Open Air Museum – 1980 ... 48476

Fairfax VA

Fairfax Museum and Visitor Center, 10209 Main St, Fairfax, VA 22030 • T: +1 703 3858414 • F: +1 703 3858692 • sgray@ci.fairfax.va.us • www.ci.fairfax.va.us •
C.E.O.: *Susan Inskeep Gray* •
Historical Museum / Natural History Museum – 1992
Personal & archaeological artifacts, photographs 48477

National Firearms Museum, 11250 Waples Mill Rd, Fairfax, VA 22030 • T: +1 703 2671620 • F: +1 703 2673913 • nfmstaff@nrahq.org • www.nrahq.org •
Cur.: *Doug Wicklund* •
Science&Tech Museum – 1871
3,500 Firearms both period & modern ... 48478

Fairfield CO

The Gallery of Contemporary Art, Sacred Heart University, 5151 Park Av, Fairfield, CO 06825-1000 • T: +1 203 3657650 • F: +1 203 3968361 • gevass@sacredheart.edu • artgallery.sacredheart.edu •
Dir.: *Sophia Gevas* •
Public Gallery ... 48479

Fairfield CT

Connecticut Audubon Birdcraft Museum, 314 Unquowa Rd, Fairfield, CT 06430 • T: +1 203 2590416 • F: +1 203 2591344 • birdcraft@ctaudubon.org • www.ctaudubon.org •
Dir.: *Milan Bull* •
Natural History Museum – 1914
Bird life, eggs, insects ... 48480

Ogden House, 636 Old Post Rd, Fairfield, CT 06430 • T: +1 203 2591598 • F: +1 203 2552716 • asaintpierre@fairfieldhs.org • www.fairfieldhistoricalsociety.org •
Dir.: *Steve Young* • Cur.: *Adrienne Saint-Pierre* •
Local Museum – 1902
Local history – library ... 48481

Thomas J. Walsh Art Gallery and Regina A. Quick Center for the Arts, Fairfield University, 1073 N Benson Rd, Fairfield, CT 06430 • T: +1 203 2544242 • F: +1 203 2544113 • dmille@fairfield.edu • www.fairfield.edu/quick •
Dir.: *Dr. Diana Dimodica Mille* • *Thomas V. Zingarelli* •
Fine Arts Museum / University Museum – 1990
Art ... 48482

Fairfield IA

Carnegie Historical Museum, 114 S Court, Fairfield, IA 52556 • T: +1 641 4722485 •
Local Museum – 1870
Indian, Iowa pioneer artifacts, mounted birds, mammals, rocks ... 48483

Fairfield Art Museum, POB 904, Fairfield, IA 52556 • T: +1 515 4729424 • fpac@lisco.com • www.fairfieldpublicaccess.org/art_association.htm •
Pres.: *Diane Boltesilverman* •
Fine Arts Museum – 1964
Paintings and sculpture by Iowa and Midwest artists ... 48484

Fairfield TX

Freestone County Historical Museum, 302 E Main St, Fairfield, TX 75840 • T: +1 903 3893738 • fcmuseum@airmail.net • www.fairfieldtx.com •
Cur.: *Molly Fryer* •
Historical Museum – 1967
Local hist, telephones ... 48485

Fairfield UT

Camp Floyd, 18035 W 1540 N, Fairfield, UT 84013 • T: +1 801 7688932 • F: +1 801 7681731 • marktrotter@utah.gov • parks.state.ut.us •
Head: *Mark A. Trotter* •
Local Museum – 1855
Historic site, former Army of Utah camp ... 48486

Fairfield VT

President Chester A. Arthur Historic Site, Chester Arthur St, Fairfield, VT 05455 • T: +1 802 8283051 • F: +1 802 8283206 • john.dumville@state.vt.us • www.historicvermont.org •
Head: *John P. Dumville* •
Local Museum – 1953
Memorabilia ... 48487

Fairhope AL

Eastern Shore Art Center, 401 Oak St, Fairhope, AL 36532 • T: +1 334 9282228 • F: +1 334 9285188 • esac@esartcenter.com • www.easternshoreartcenter.com •
Dir.: *Robin Fitzhugh* •
Fine Arts Museum – 1958
Contemporary American paintings ... 48488

Fairmont MN

Martin County Pioneer Museum, 303 E Blue Earth Av, Fairmont, MN 56031 • T: +1 507 2355178 • F: +1 507 2355179 • mch@frontiernet.net • www.co.martin.mn.us/mchs •
Exec.Dir.: *Lenny Tvedten* • Cur.: *Muriel Malliet* •
Local Museum – 1929
Regional history ... 48489

U

Fairmont WV

Pricketts Fort, Rte 3, Fairmont, WV 26554 • T: +1 304 3633030 • F: +1 304 3633857 • pfort@host.dmsc.net • www.prickettsfort.org •
C.E.O.: *Melissa May Dobbins* •
Local Museum – 1976 ... 48490

Fairplay CO

South Park City Museum, 100 Fourth, Fairplay, CO 80440 • T: +1 719 8362387 • www.southparkcity.org •
Pres.: *Raymond Dellacroce* • Dir.: *Carol A. Davis* •
Open Air Museum / Local Museum – 1957
Historic village, Colorado mining town ... 48491

Fairport NY

Fairport Historical Museum, Perinton Historical Society, 18 Perrin St, Fairport, NY 14450-2122 • T: +1 716 2233989 • F: +1 716 4251962 • www.angelfire.com/ny5/fairporthistmuseum •
Dir.: *Matson Ewell* •
Local Museum – 1935
Local lore, Native Americans, 19th c industry, home, farm and village exhibits ... 48492

Fairport Harbor OH

Fairport Marine Museum, 129 Second St, Fairport Harbor, OH 44077 • T: +1 440 3544825 • fhhs@ncweb.com • www.fairportlighthouse.com •
Pres.: *Valerie Laczko* •
Special Museum – 1945
Navigation instruments, marine charts, ship carpenters' tools, ship models ... 48493

Fairview UT

Fairview Museum of History and Art, 85 N 100 E, Fairview, UT 84629 • T: +1 435 4279216 • www.sanpete.com •
Dir.: *Ron Staker* •
Historical Museum / Fine Arts Museum – 1966
Pioneer relics, miniature carvings, arts & crafts, Indian bead & leather, engines, Columbian Mammoth, paintings & prints ... 48494

Fairway KS

Grinter Place State Historical Site, c/o Shawnee Indian Mission, 3403 W 53rd St, Fairway, KS 66205 • T: +1 913 2990373 • F: +1 913 7888046 • grinter@kshs.org • www.kshs.org •
Historical Museum – 1971
1860's-1890's household furnishings ... 48495

Falcon Heights MN

Gibbs Farm Museum, Museum of Pioneer and Dakota Life, 2097 W Larpenteur Av, Falcon Heights, MN 55113 • T: +1 651 6468629 • F: +1 651 2238539 • admin@rchs.com • www.rchs.com •
C.E.O.: *Priscilla Farnham* •
Agriculture Museum / Folklore Museum – 1949
Gibbs family artifacts, Dakotah Native Americans, agricultural and woodworking items ... 48496

Falfurrias TX

The Heritage Museum at Falfurrias, 415 N Saint Mary's St, Falfurrias, TX 78355 • T: +1 512 3252907 • www.heritagemuseum-falfurrias.com •
Cur.: *E. Cabrera* •
Local Museum – 1965
Texas Ranger Room, Dryden photo negative coll 48497

Fall River MA

Battleship Cove - Maritime Heritage Museums, Battleship Cove, Fall River, MA 02721 • T: +1 508 6781100 • F: +1 508 6745597 • battleship@battleshipcove.org • www.battleshipcove.org •
Exec. Dir.: *Ernst E. Cummings* • Cur.: *Christopher J. Nardi* •
Historical Museum / Military Museum / Historic Site – 1965
Historic ships, model aircraft coll, WWII ... 48498

Fall River Historical Society Museum, 451 Rock St, Fall River, MA 02720 • T: +1 508 6791071 • F: +1 508 6755754 • www.lizzieborden.org •
Cur.: *Michael Martins* •
Local Museum – 1921
Local history, Lizzie Borden murder trail, decorative arts, guns, marine – library ... 48499

Marine Museum at Fall River, 70 Water St, Fall River, MA 02721 • T: +1 508 6743533 • F: +1 508 6743534 • staff@marinemuseum.org • www.marinemuseum.org •
CEO: *Donna J. Futoransky* •
Historical Museum – 1968
Marine, restored machine shop, steam ship models, photos, prints, glass slides – library ... 48500

Fall River Mills CA

Fort Crook Museum, Fort Crook Av and Hwy 299, Fall River Mills, CA 96028, mail addr: POB 397, Fall River Mills, CA 96028 • T: +1 530 3365110 • fortcrook@frontiernet.net • www.ftcrook.com •
Cur.: *Dorothy Mason* •
Historical Museum – 1934
Local hist, agriculture, industry, Indians ... 48501

Fallon NV

Churchill County Museum and Archives, 1050 S Maine St, Fallon, NV 89406 • T: +1 775 4233677 • F: +1 775 4233662 • ccmuseum@phonewave.net • ccmuseum.org •
Dir./Cur.: *Jane Pieplow* •
Local Museum – 1967
Anthropology, archaeology, costumes, geology, glass, Indian artifacts ... 48502

Falls Village CT

Canaan Historical Society, 44 Railroad St, Falls Village, CT 06031 • T: +1 860 8247235 • F: +1 860 8244506 • www.betweenthelakes.com/canaan/hist_society.htm •
Cur.: *Dorris Longaven* •
Local Museum – 1952
Costumes, local history items ... 48503

Fallsington PA

Historic Fallsington, 4 Yardley Av, Fallsington, PA 19054 • T: +1 215 2956567 • F: +1 215 2956567 • info@historicfallsington.org • www.historicfallsington.org •
Pres.: *Robert L.B. Harman* • Exec. Dir.: *Erica Armour* •
Open Air Museum / Local Museum – 1953
Local hist coll, ... 48504

Falmouth MA

Falmouth Historical Museum, 55-65 Palmer Av, Falmouth, MA 02540 • T: +1 508 5484857 • F: +1 508 5400968 • fhs@cape.com • www.falmouthhistoricalsociety.org •
Exec. Dir.: *Carolyn T. Powers* • Cur.: *Judith McAlister* •
Local Museum – 1900
Local history, French furnishings, china, paintings, whaling era, Katherine Lee Bates, farming – small library ... 48505

Far Hills NJ

Golf Museum, 77 Liberty Corner Rd, Far Hills, NJ 07931 • T: +1 908 2342300 • F: +1 908 2340319 • museum@usga.org • www.usga.org •
Dir.: *David Fay* • Cur.: *Rand Jerris* •
Special Museum – 1935
Golf ... 48506

Fargo GA

Stephen C. Foster State Park, Rte 1, Fargo, GA 31631, mail addr: POB 131, Fargo, GA 31631 • T: +1 912 6375274 • F: +1 912 6375587 • scfost@planttel.net •
Natural History Museum – 1954
Folk culture, personal artefacts, lumbering, turpentining ... 48507

Fargo ND

Children's Museum at Yunker Farm, 1201 28th Av N, Fargo, ND 58102 • T: +1 701 2326102 • F: +1 701 2324605 • info@childrensmuseum-yunker.org • www.childrensmuseum-yunker.org •
C.E.O.: *Yvette Nasset* •
Special Museum – 1989 ... 48508

Fargo Air Museum, 1609 19th Av N, Fargo, ND 58102 • T: +1 701 2938043 • F: +1 701 2938103 • www.fargoairmuseum.org •
Science&Tech Museum – 2001
Civilian and miltary aircrafts, aviation artifacts ... 48509

Memorial Union Gallery, North Dakota State University, Memorial Union, Fargo, ND 58105-5476 • T: +1 701 2318239 • F: +1 701 2318043 • peg-furshong@ndsu.nodak.edu • www.ndsu.nodak.edu/memorial_union/gallery •
Dir.: *Peg Furshong* •
University Museum / Fine Arts Museum – 1975
Contemporary American artworks ... 48510

Plains Art Museum, 704 First Av N, Fargo, ND 58102 • T: +1 701 2323821 ext. 101 • F: +1 701 2931082 • museum@plainsart.org • www.plainsart.org •
Chm.: *John Q. Paulsen* • Pres. & C.E.O.: *Edward E. Pauley* •
Fine Arts Museum – 1973
Contemporary American, American Indian and West African art, fine arts, Woodland and Plain Indian art ... 48511

Faribault MN

Rice County Museum of History, 1814 NW Second Av, Faribault, MN 55021 • T: +1 507 3322121 • F: +1 507 3322121 • rchs@rchistory.org • www.rchistory.org •
Exec. Dir.: *Susan Garwood* •
Local Museum – 1926
Local history, agriculture, pioneer & Indians artifacts, picture coll – library ... 48512

Farmington CT

Hill-Stead Museum, 35 Mountain Rd, Farmington, CT 06032 • T: +1 860 6774787 • F: +1 860 6770174 • hillstead@hillstead.org • www.hillstead.org •
Dir.: *Linda Stagleder* •
Fine Arts Museum – 1946
French impressionist painting, prints, sculpture, porcelain, furniture ... 48513

Museum of the University of Connecticut Health Center, c/o School of Dental Medicine, 263 Farmington Av, Farmington, CT 06030 • T: +1 860 6793211 •
Head: *Jay Christian* •
University Museum / Special Museum – 1979
Dental antiques ... 48514

Stanley-Whitman House, 37 High St, Farmington, CT 06032 • T: +1 860 6779222 • F: +1 860 6777758 • lisa@stanleywhitman.org • www.stanleywhitman.org •
C.E.O: *Lisa Johnson* •
Local Museum – 1936
Interpretation of 18th-c Farmington ... 48515

Farmington ME

Nordica Homestead Museum, 116 Nordica Ln, Farmington, ME 04938 • T: +1 207 7782042 •
Pres.: *Tom Sawyer* •
Local Museum – 1927
Regional history ... 48516

Farmington NM

Farmington Museum, 3041 E Main St, Farmington, NM 87402 • T: +1 505 5991174 • F: +1 505 3267572 • rwelch@fmtn.org • www.farmingtonmuseum.org •
Dir.: *Richard Welch* • Cur.: *Bart Wilsey* •
Ethnology Museum / Local Museum – 1980
Costumes, oil and gas industry, living hist farm and orchards ... 48517

Farmington PA

Fort Necessity National Battlefield, 1 Washington Pkwy, Farmington, PA 15437 • T: +1 724 3295512 • F: +1 724 3298682 • lawren_dunn@nps.gov • www.nps.gov/fone/ •
Dir.: *Joanne Hanley* • Cur.: *Lawren Dunn* •
Open Air Museum / Military Museum – 1931
Military histotry ... 48518

Touchstone Center for Crafts, RR 1, Box 60, Farmington, PA 15437-9707 • T: +1 724 3291370 • F: +1 724 3291371 • tcc@hhs.net • www.touchstonecrafts.com •
Decorative Arts Museum – 1972
Ceramics, jewelry ... 48519

Farmington UT

Pioneer Village, 375 N Lagoon Ln, Farmington, UT 84025 • T: +1 801 4518050 • F: +1 801 4518015 •
Dir.: *Peter Freed* •
Local Museum / Open Air Museum – 1954
Carriages, guns, toys, coins & silver ... 48520

Farmington Hills MI

Zekelman Family Holocaust Memorial Center, 28123 Orchard Lake Rd, Farmington Hills, MI 48334-3788 • T: +1 248 5532400 • F: +1 248 5532433 • info@holocaustcenter.org • www.holocaustcenter.org •
CEO/ Founder: *Charles Rosenzveig* •
Historical Museum
Films, artifacts, portraits, biographies, culture of European Jews, emigration, postwar pogroms in Poland and Yedwabne case, migration to Israel and USA – library ... 48521

Farmville NC

May Museum and Park, 213 S Main St, Farmville, NC 27828 • T: +1 252 7535814 • F: +1 252 7533190 • maymuseum@farmville.ng.com •
Dir.: *Clifford P. Kendall* • Cur.: *John S. Rossi* •
Local Museum – 1991
Local history, decorative art, textiles ... 48522

Farmville VA

Longwood Center for the Visual Arts, 129 N Main St, Farmville, VA 23901 • T: +1 434 3952206 • F: +1 434 3926441 • kjbowles@longwood.edu • web.longwood.edu/lcva •
Dir.: *Kay Johnson Bowles* •
Fine Arts Museum / University Museum – 1971
19th c American art, African & Chinese art ... 48523

Fayette AL

Fayette Art Museum, 530 Temple Av N, Fayette, AL 35555 • T: +1 205 9328727 • www.fayette.net/chamber/locart.htm •
Dir./Cur.: *Jack Black* •
Fine Arts Museum – 1969
20th c art, folk art ... 48524

Fayette MO

Ashby-Hodge Gallery of American Art, 411 CMC Sq, Fayette, MO 65248 • T: +1 660 2486324 • F: +1 660 2482622 • jgeist@coin.org • www.centralmethodist.edu •
Cur.: *Dr. Joseph E. Geist* •
Fine Arts Museum / University Museum – 1994
Midwestern regionalist artists ... 48525

Stephens Museum, Central Methodist College, Fayette, MO 65248 • T: +1 660 2486370 • F: +1 660 2482622 • delliot@cmc.edu •
Dir./Cur.: *Dana R. Elliott* •
Natural History Museum / Religious Arts Museum / University Museum – 1879 ... 48526

Fayetteville AR

Arkansas Air Museum, 4290 S School Av, Fayetteville, AR 72701 • T: +1 479 5214947 • F: +1 479 5214947 • arkairmus@aol.com • www.arkairmuseum.org •
C.E.O.: *Rick McKinney* •
Historical Museum / Science&Tech Museum – 1986
Classic aircraft, WW II naval aviation ... 48527

University of Arkansas Collections, Fulbright College of Arts and Sciences, Old Main 525, Biomass Bldg, Fayetteville, AR 72701 • T: +1 501 5753684 • msuter@uark.edu • www.uark.edu/~arsc/collections •
Dir.: *Dr. John Hehr* • Cur.: *Mary Suter* (Collections) •

Ass. Cur.: *Dr. Nancy G. McCartney* (Zoology) • University Museum / Ethnology Museum / Natural History Museum / Archaeology Museum – 1873
Ethnology, anthropology, zoology – herbarium ... 48528

Fayetteville NC

Airborne and Special Operations Museum, 100 Bragg Blvd, Fayetteville, NC 28301 • T: +1 910 4833003 ext 221 • F: +1 910 4838232 • duvalljs@bragg.army.mil • www.asomf.org •
Dir.: *John S. Duvall* •
Military Museum – 2000
Military artifacts, uniforms, weapons, WW 2 airplane, helicopter ... 48529

Arts Council of Fayetteville/Cumberland County, 301 Hay St, Fayetteville, NC 28302 • T: +1 910 3231776 • F: +1 910 3231727 • admin@theartscouncil.com • www.theartscouncil.com •
Exec. Dir.: *Deborah Martin Mintz* •
Association with Coll / Public Gallery – 1974 ... 48530

Fayetteville Museum of Art, 839 Stamper Rd, Fayetteville, NC 28303 • T: +1 910 4855121 • F: +1 910 4855233 • mima@fay-moa.org • www.fayettevillemuseumart.org •
Pres.: *Virginia Oliver* • Dir.: *Tom Grubb* •
Fine Arts Museum – 1971
Contemporary NC art, contemp Russian art, African artifacts ... 48531

Museum of the Cape Fear Historical Complex, Corner of Bradford and Arsenal Aves, Fayetteville, NC 28305 • T: +1 910 4861330 • F: +1 910 4861585 • mcfhc@ncmail.net • www.ncmuseumofhistory.org •
Pres.: *V. Weyher Dawson* •
Historical Museum – 1985
History and culture of southern NC ... 48532

Fenton MI

Pioneer Memorial Association of Fenton and Mounty Townships, 2436 N Long Lake Rd, Fenton, MI 48430 • T: +1 810 6297748 •
Pres.: *Joyce Simmons* •
Local Museum – 1967
Pioneer life, genealogy, furniture, furnishings ... 48533

Fergus Falls MN

Otter Tail County Museum, 1110 Lincoln Av W, Fergus Falls, MN 56537 • T: +1 218 7366038 • F: +1 218 7393075 • otchs@prtel.com •
Dir.: *Chris Schuelke* • Cur.: *Kathy M. Lindberg* •
Local Museum – 1927
Regional hist, American Indians, Scandinavians, agriculture, clippings – library, archives ... 48534

Fernandina Beach FL

Amelia Island Museum of History, 233 S Third St, Fernandina Beach, FL 32034 • T: +1 904 2617378 • F: +1 904 2619701 • aimuseum@bellsouth.net • www.ameliaislandmuseumofhistory.org •
Pres.: *Susan Little* • Dir.: *Carmen Godwin* •
Historical Museum – 1986
Photographs, railroads, Civil War, local hist, spanish colonial hist ... 48535

Fort Clinch, 2601 Atlantic Av, Fernandina Beach, FL 32034 • T: +1 904 2777274 • F: +1 904 2777249 • www.floridastateparks.org •
C.E.O.: *Virginia Wetherell* •
Historical Museum – 1935
Living hist ... 48536

Ferrisburgh VT

Rokeby Museum, Rowland Evans Robinson Memorial Association, 4334 Rte 7, Ferrisburgh, VT 05456-9711 • T: +1 802 8773406 • F: +1 802 8773406 • rokeby@adelphia.net • www.rokeby.org •
Dir.: *Jane Williamson* •
Historic Site – 1962
Historic house ... 48537

Ferrum VA

Blue Ridge Institute and Museum, c/o Ferrum College, 20 Museum Dr, Ferrum, VA 24088 • T: +1 540 3654416 • F: +1 540 3654419 • bri@ferrum.edu • www.blueridgeinstitute.org •
Dir.: *J. Roderick Moore* • Ass. Dir.: *Vaughan Webb* •
Folklore Museum – 1971
Items from the Blue Ridge Mountains from the late 18th c to te present ... 48538

Fessenden ND

Wells County Museum, 305 First St S, Fessenden, ND 58450 • T: +1 701 5473100, 5473415 •
Cur.: *Lorraine Rau* •
Local Museum – 1972
Antique agricultural implements, period automobiles, horse drawn vehicles, ethnic artifacts ... 48539

Fifield WI

Old Town Hall Museum, W 7213 Pine St, Fifield, WI 54524 • T: +1 715 3392254 • matrojak@pctcnet.net •
Pres.: *Therese Trojak* • Cur.: *Patricia Schroeder* •
Local Museum / Historical Museum – 1969
Local hist, logging and farming tools, clothing, books, household ... 48540

Fillmore CA

Fillmore Historical Museum, 350 Main St, Fillmore, CA 93016 • T: +1 805 5240948 • fillmore.museum@sbcglobal.net • www.fillmorehistoricalmuseum.com •
Local Museum – 1971 ... 48541

Fillmore UT

Territorial Statehouse State Museum, 50 W Capitol Av, Fillmore, UT 84631 • T: +1 435 7435316 • F: +1 435 7434723 •
Head: *Gordon Chatland* •
Historical Museum – 1930
Pioneer furniture, tools, pictures, handicrafts, music, cotton ... 48542

Findlay OH

Dudley and Mary Marks Lea Gallery, c/o University of Findlay, 1000 N Main St, Findlay, OH 45840 • T: +1 419 4244577 • F: +1 419 4244757 •
Dir.: *Doug Salveson* •
Fine Arts Museum / University Museum – 1962 ... 48543

Hancock Historical Museum, 422 W Sandusky St, Findlay, OH 45840 • T: +1 419 4234433 • F: +1 419 4232154 • www.hancockhistoricalmuseum.org •
Pres.: *Ed Ingold* • C.E.O.: *Sue Tucker* •
Local Museum – 1970
1875-1900 art glass, 1880s pattern glass, 1916 Grant car, dolls, hist buildings ... 48544

Fish Creek WI

Eagle Bluff Lighthouse, Peninsula State Park, Fish Creek, WI 54212, mail addr: 7698 W Kangaroo Lake Dr, Bailey's Harbor, WI 54202 • T: +1 920 8392377 •
Local Museum – 1964
Maritime coll, period furnishings ... 48545

Fishers IN

Conner Prairie Living History Museum, 13400 Allisonville Rd, Fishers, IN 46038 • T: +1 317 7766000 • F: +1 317 7766014 • info@connerprairie.org • www.connerprairie.org •
Pres.: *Ellen Rosenthal* •
Historical Museum – 1964
William Conner home; living history, natural history area ... 48546

Fishkill NY

Van Wyck Homestead Museum, 504 Rte 9, Fishkill, NY 12524 • T: +1 845 8969560 • vanwyckhomestead@hotmail.com •
Dir.: *Roy Jorgensen* • Chair: *Susan Dexter* •
Historical Museum / Local Museum – 1962
Artefacts of Rev. war, early Dutch settler artifacts, Hudson valley painting ... 48547

Fitchburg MA

Fitchburg Art Museum, 185 Elm St at Merriam Pkwy, Fitchburg, MA 01420 • T: +1 978 3454207 • F: +1 978 3452319 • info@fitchburgartmuseum.org • www.fitchburgartmuseum.org •
Dir.: *Peter Timms* • Sc. Staff: *Roger Dell* (Education) • *Pamela Russell* • *Kristina Durocher* •
Fine Arts Museum – 1925
Pre-Columbian, Asian, and African art, European and American paintings (18th-20th c), Egyptian and Classical antiquities, photographs, drawings, prints, decorative arts, illustrated books – sculpture garden ... 48548

Fitchburg Historical Society Museum, 50 Grove St, Fitchburg, MA 01420 • T: +1 978 3451157 • F: +1 978 3452229 • fitchburghistory@aol.com • fitchburghistory.fsc.edu •
Pres.: *Helen Obermeyer Simmons* • Exec. Dir.: *Betsy Hannula* •
Historical Museum – 1892
Local hist, US wars colls, local portraits, glass and china, dolls, musical instr – library, auditorium ... 48549

Flagstaff AZ

Arizona Historical Society Pioneer Museum, 2340 N Fort Valley Rd, Flagstaff, AZ 86001 • T: +1 520 7746272 • F: +1 520 7741596 • AHSFlagstaff@azhs.gov • www.ahs.state.az.us •
Dir.: *Joseph M. Meehan* • Cur.: *Susan Wilcox* •
Historical Museum – 1953
Industrial, social, istitutuional and transportion history of the area – archives ... 48550

Museum of Northern Arizona, 3101 N Fort Valley Rd, Flagstaff, AZ 86001 • T: +1 928 7745213 • F: +1 928 7791527 • info@mna.mus.az.us • www.musnaz.org •
Dir.: *Robert G. Breunig* • Cur.: *David D. Gillette* (Geology) • *Kelley Hays-Gilpin* (Anthropology) • *Larry Stevens* (Ecology & Conservation) •
Local Museum – 1928
Local history, anthropology and prehistory, geology and paleontology, biological specimen, fine art, Indian artifacts – herbarium, library ... 48551

NAU Art Museum, Northern Arizona University, Knoles and McMullen Circle, Bldg 10, North Campus, Flagstaff, AZ 86011-6021, mail addr: POB 6021, Flagstaff, AZ 86011-6021 • T: +1 928 5233471 • F: +1 928 5231424 • art.museum@nau.edu • www.nau.edu/art_museum •
Dir.: *John Burton* • Cur.: *Kathleen Battali* •
University Museum / Fine Arts Museum / Public Gallery – 1961
Paintings, graphics, ceramics, sculpture, furniture ... 48552

Riordan Mansion, 409 West Riordan Rd, Flagstaff, AZ 86001 • T: +1 928 7794395 • www.pr.state.az.us •
Decorative Arts Museum – 1983
Local hist, coll of original craftsman furnishings ... 48553

Flandreau SD

Moody County Museum, 706 East Pipestone Av, Flandreau, SD 57028 • T: +1 605 9973191 • duncana@iw.net • www.rootsweb.com/~sdmoody •
Dir.: *Roberta Williamson* •
Local Museum – 1965
Local hist ... 48554

Flat Rock NC

Carl Sandburg Home, 81 Carl Sandburg Ln, Flat Rock, NC 28731 • T: +1 704 6934178 • F: +1 704 6934179 • carl_administration@nps.gov • www.nps.gov/carl •
C.E.O./Cur.: *Connie Backlund* •
Local Museum – 1969
The author's working library, letters, photographs ... 48555

Fleming CO

Fleming Historical Museum, Heritage Museum Park, Fleming, CO 80728 • T: +1 970 2653721, 2652591 •
C.E.O.: *Wanda West* •
Local Museum – 1965
Local history ... 48556

Flemington NJ

Doric House, 114 Main St, Flemington, NJ 08822 • T: +1 908 7821091 •
Pres.: *Richard H. Stothoff* •
Local Museum – 1885
Agriculture, Indian artifacts, genealogy, local hist 48557

Flint MI

Buckham Gallery, 134 1/2 W Second St, Flint, MI 48502 • T: +1 810 2396334 • gfac@prodigy.net • www.gfn.org/buckham •
Fine Arts Museum – 1984
Contemporary arts since 1984 ... 48558

Flint Institute of Arts, 1120 E Kearsley St, Flint, MI 48503 • T: +1 810 2341695 • F: +1 810 2341692 • info@flintarts.org • www.flintarts.org •
Dir.: *John B. Henry* • Cur.: *Monique Desormeau* (Education) • *Lisa Baylis Ashby* (Exhibits, Colls) •
Fine Arts Museum – 1928
African, American, Asian and European art, decorative arts – library, auditorium ... 48559

Sloan Museum & Longway Planetarium, 1221 E Kearsley St, Flint, MI 48503 • T: +1 810 2373450 • F: +1 810 2373451 • tshickles@flintcultural.org • www.sloanlongway.org •
Dir.: *Tim Shickles* • Cur.: *Jeff Taylor* (Colls) • *Laurie Bone* (Programs) • Reg.: *Jane McIntosh* •
Historical Museum / Science&Tech Museum – 1966
Local origin, natural hist, carriages, vintage & prototypes of Buick automobiles – library, auditorium ... 48560

Flora MS

Mississippi Petrified Forest, 124 Forest Park Rd, Flora, MS 39071 • T: +1 601 8798189 • F: +1 601 8798165 • mspforest@aol.com • www.mspetrifiedforest.com •
Dir.: *C. J. McNamara* •
Natural History Museum – 1963 ... 48561

Floral Park NY

Queens County Farm Museum, 73-50 Little Neck Pkwy, Floral Park, NY 11004 • T: +1 718 3473276 ext 10 • F: +1 718 3473243 • afischetti@queensfarm.org • www.queensfarm.org •
Dir.: *Amy Fischetti* •
Agriculture Museum – 1975
Hist orchard, 19th and 20th c farm tools and household artifacts ... 48562

Florence AL

Kennedy-Douglass Center for the Arts, 217 E Tuscaloosa St, Florence, AL 35630 • T: +1 256 7606379 • bbroach@florenceal.org • www.kennedydouglasscenter.org •
Dir.: *Barbara K. Broach* •
Fine Arts Museum – 1976
library ... 48563

Pope's Tavern Museum, 203 Hermitage Dr, Florence, AL 35630 • T: +1 256 7606439 • bbroach@florenceal.org • www.florenceal.org •
Cur.: *Jo Parkhurst* •
Local Museum – 1968
One-time stagecoach stop, tavern and inn, hospital by Confederate and Union forces ... 48564

University Art Gallery, University of North Alabama, 1 Harrison Plaza, Visual Arts Bldg, Florence, AL 35632-0001, mail addr: POB 5006, Florence, AL 35632-0001 • T: +1 256 7654384 • F: +1 256 7654511 • kmauldin@una.edu • www.una.edu •
Public Gallery / University Museum
Art by local artist, professional artist, and students ... 48565

W.C. Handy Home Museum and Library, 620 W College St, Florence, AL 35630, mail addr: 217 E Tucsaloosa St, Florence, AL 35630 • T: +1 256 7606434 • F: +1 256 7606382 • bbroach@florenceal.org • www.florenceal.org •
Dir.: *Barbara K. Broach* • Cur.: *Sandra Ford* •
Historical Museum / Music Museum – 1968
Large collection of personal memorabilia of W.C. Handy including his piano, trumpet, sheet music, photos, awards, log cabin home – library ... 48566

Florence AZ

Pinal County Historical Museum, 715 S Main St, Florence, AZ 85232, mail addr: POB 815, Florence, AZ 85232-0815 • T: +1 520 8684382 •
Local Museum – 1958
Agriculture, ranching, mining, Indian, Spanish, Mexican and Anglo contributions from residents of Pinal County, artifacts – archives ... 48567

Florence CO

Florence Price Pioneer Museum, Pikes Peak Av and Front St, Florence, CO 81226 • T: +1 719 7843157 •
C.E.O.: *Frank W. Tedesko* • Cur.: *Leona Blackman* •
Local Museum – 1964
Industrial, folklore, Indian artifacts, mineralogy ... 48568

Florence KS

Harvey House Museum, 221 Marion, Florence, KS 66851 • T: +1 316 8784296 • members.tripod.com/harvey_house •
Pres.: *Kristal Timm* •
Special Museum – 1971
Antique furniture, pictures, dishes, tools, baggage wagon ... 48569

Florence SC

Florence Museum of Art, Science and History, 558 Spruce St, Florence, SC 29501 • T: +1 843 6623351 • flomus@bellsouth.net • www.florenceweb.com/museum.htm •
Dir.: *Betsy Olsen* •
Fine Arts Museum / Natural History Museum / Historical Museum – 1936
Art, science, history ... 48570

Florissant MO

Florissant Valley Historical Society, 1896 S Florissant Rd, Florissant, MO 63031 • T: +1 314 5241100 • fredmary2@aol.com • www.florissantoldtown.com •
Association with Coll – 1958 ... 48571

Old Saint Ferdinand's Shrine, 1 Rue Saint François, Florissant, MO 63031 • T: +1 314 8314237 • F: +1 314 8393829 • Bill052946@sbcglobal.net •
Pres.: *Bill Bray* •
Religious Arts Museum – 1958 ... 48572

Flovilla GA

Indian Springs State Park Museum, Hwy 42, 5 Miles S of Jackson, Flovilla, GA 30216, mail addr: 678 Lake Clark Rd, Flovilla, GA 30216 • T: +1 770 5042277 • F: +1 770 5042178 • www.georgiastateparks.org •
Head: *Don Coleman* •
Natural History Museum – 1825
History, photos ... 48573

Flushing NY

Art Center, Queens College, 65-30 Kissena Blvd, Flushing, NY 11367 • T: +1 718 9975770 • F: +1 718 9973753 • ssimor@qc1.qc.edu •
Public Gallery ... 48574

Bowne House, 37-01 Bowne St, Flushing, NY 11354 • T: +1 718 3590528 • F: +1 718 3590873 •
Pres.: *Douglas F. Bauer* • Exec. Dir.: *Evangeline T. Egglezos* •
Local Museum – 1945
17th to 19th c furnishings, paintings and documents ... 48575

Godwin-Ternbach Museum, Queens College, 65-30 Kissena Blvd, Flushing, NY 11367 • T: +1 718 9974747 • F: +1 718 9974734 • amy_winter@qc.edu • www.qc.edu/art/gtmus.html •
C.E.O.: *Dr. Amy Winter* •
Fine Arts Museum / University Museum – 1957
Ancient Near-Eastern art, Asian, Western and contemporary art ... 48576

Kingsland Homestead, Queens Historical Society, Weeping Beech Park, 143-35 37th Av, Flushing, NY 11354 • T: +1 718 9390647 ext 17 • F: +1 718 5399885 • info@queenshistoricalsociety.org • www.queenshistoricalsociety.org •
Pres.: *Stanley Cogan* • Exec. Dir.: *Mitchell Grubler* •
Historical Museum / Historic Site – 1968
Photographs, postcards, maps, personal papers, architectural renderings, ... 48577

New York Hall of Science, 47-01 111th St, Flushing, NY 11368 • T: +1 718 6990005 • F: +1 718 6991341 • wbrez@nyscience.org • www.nyhallsci.org •
Dir.: *Dr. Alan J. Friedman* • Sc. Staff: *Harold Chapnick* (Internal Affairs) • *Marilyn Hoyt* (External Affairs) • *Eric*

Siegel (Planning, Programs) • *Martin Weiss* (Biology) • *Preeti Gupta* (Education) • PR: *Wendy Brez* •
Science&Tech Museum – 1964
library 48578

Queens Historical Society, 143-35 37th Av, Flushing, NY 11354 • T: +1 718 9390647 • F: +1 718 5399885 • info@queenshistoricalsociety.org • www.queenshistoricalsociety.org •
Pres.: *Stanley Cogan* • Exec. Dir.: *Joyce A. Cook* •
Association with Coll – 1968
Local history, Kingsland homestead 48579

Queens Museum of Art, New York City Bldg, Flushing Meadows Corona Park, Flushing, NY 11368-3398 • T: +1 718 5929700 • F: +1 718 5925778 • www.queensmuseum.org •
Dir.: *Laurene Buckley* • Cur.: *Sharon Vatksy* •
Fine Arts Museum – 1972
Paintings, sculpture, prints, photography 48580

Folsom CA

Folsom History Museum, 823 Sutter St, Folsom, CA 95630 • T: +1 916 9852707 • info@FolsomHistoryMuseum.org • www.folsomhistorymuseum.org •
Dir.: *Karen Mehring* •
Local Museum – 1960
Hist of the area, Pony Express, railroad, gold mining, electricity – library, archives 48581

Folsom NM

Folsom Museum, Main St, Folsom, NM 88419 • T: +1 505 2783616 • bkthompson@bacavalley.com • folsommuseum.netfirms.com •
Pres.: *Vinita Brown* •
Archaeology Museum – 1967
Local artifacts, archaeology 48582

Fond du Lac WI

Galloway House and Village-Blakely Museum, 336 Old Pioneer Rd, Fond du Lac, WI 54935 • T: +1 920 9220991 • F: +1 920 9220991 • fdlhistorical@dotnet.com •
Chief Cur.: *Jim Glassel* • Cur.: *William J. Weinshrott* •
Local Museum – 1955
Historic village 48583

Fonda NY

Native American Exhibit, National Kateri Shrine, Rte 5, 1/2 mile West of Fonda, Fonda, NY 12068 • T: +1 518 8533646 • F: +1 518 8533371 • kkenny@nycap.rr.com • www.katerishrine.com •
Dir.: *Kevin Kenny* • Ass. Dir.: *Florence Guiffre* •
Historical Museum / Archaeology Museum / Folklore Museum / Religious Arts Museum – 1949
1666-1693 staked out Mohawk Indian castle and 1666-1676 residence of Kateri Tekakwitha 48584

Forest Grove OR

Pacific University Museum, Pacific University, 2043 College Way, Forest Grove, OR 97116 • T: +1 503 3592211 • F: +1 503 3592252 • ur@pacificu.edu • www.pacificu.edu •
Pres.: *Dr. Philip D. Creighton* •
University Museum / Historical Museum / Folklore Museum – 1949
University hist, Asian cultures, missionary activities 48585

Forked River NJ

Lacey Historical Society Museum, Rte 9, Old Schoolhouse, Forked River, NJ 08731 • T: +1 609 9710467 •
Local Museum – 1962
Local artifacts 48586

Forrest City AR

Saint Francis County Museum, 603 Front St, Forrest City, AR 72335 • T: +1 870 2611744 • info@sfcmuseum.org • www.sfcmuseum.org •
Dir.: *Laura Mazzanti* •
Local Museum – 1995
Military hist, native Americans, Afro-Americans, medical hist 48587

Forsyth MT

Rosebud County Pioneer Museum, 1335 Main St, Forsyth, MT 59327 • T: +1 406 3567547 •
Pres.: *Cal MacConnel* •
Local Museum – 1966
Pioneer memorabilia 48588

Fort Apache AZ

Nohwike' Bagowa, White Mountain Apache Cultural Center and Museum, Fort Apache, AZ 85926, mail addr: POB 507, Fort Apache, AZ 85926 • T: +1 928 3384625 • www.wmat.us •
Dir.: *Dr. Karl Hoerig* •
Local Museum – 1969
Apache culture, Fort Apache artifacts 48589

Fort Atkinson WI

Hoard Historical Museum, 407 Merchant Av, Fort Atkinson, WI 53538 • T: +1 920 5637769 • F: +1 920 5683203 • hartwick@hoardmuseum.org • www.hoardmuseum.org •
Dir.: *Sue Hartwick* • Cur.: *Karen O'Connor* •
Local Museum – 1933
Local hist 48590

Fort Benning GA

National Infantry Museum, c/o US Army Infantry School, Bldg 396, Baltzell Av, Fort Benning, GA 31905-5593 • T: +1 706 5452958 • F: +1 706 5455158 • hannerz@benning.army.mil • www.benningmwr.com •
Dir.: *Frank Hanner* •
Military Museum – 1959
Military history 48591

Fort Benton MT

Fort Benton Museum of the Upper Missouri, 1810 Front St, Fort Benton, MT 59442 • T: +1 406 6225316 • F: +1 406 6223725 • fbmusuems@hotmail.com •
Dir.: *John G. Lepley* •
Local Museum – 1957
Artifacts, maps, models, pictures 48592

Museum of the Northern Great Plains, 20th and Washington, Fort Benton, MT 59442 • T: +1 406 6225133 • F: +1 406 6223725 • fbmuseums@hotmail.com •
Dir.: *John G. Lepley* •
Agriculture Museum – 1989
Hist of agriculture in the region 48593

Fort Bliss TX

Fort Bliss Museum, Bldig 1735, Marshall Rd, Fort Bliss, TX 79916 • T: +1 915 5686940 • F: +1 915 5686941 • geery@bliss.army.mil • www.bliss.army.mil •
Military Museum – 1954
Military hist, local installation hist 48594

United States Army Air Defense Artillery Museum, Bldg 1735, Marshall Rd, Fort Bliss, TX 79916 • T: +1 915 5685412 • F: +1 915 5687166 • rossd@emhio.bliss.army.mil •
Cur.: *David A. Ross* •
Military Museum – 1975
Military hist 48595

Fort Bragg CA

Guest House Museum, 343 Main St, Fort Bragg, CA 95437, mail addr: POB 71, Fort Bragg, CA 95437 • T: +1 707 9644251 • info@fortbragghistory.com • www.fortbragghistory.org •
Pres./C.E.O.: *Mark Ruedrich* •
Science&Tech Museum / Historical Museum – 1980
Mill artifacts, logging tools 48596

Triangle Tattoo and Museum, 356B N Main St, Fort Bragg, CA 95437 • T: +1 707 9648814 • info@triangletattoo.com • www.triangletattoo.com •
Special Museum – 1992
Tattoo designs and implements worldwide, photographs, tattoo history, different tattooing exhibits 48597

Fort Bragg NC

82nd Airborne Division War Memorial Museum, Gela and Ardennes Sts, Bldg C-6841, Fort Bragg, NC 28310 • T: +1 910 4325307 • F: +1 910 4361642 • aarsenj@bragg.army.mil •
Chief Cur.: *John W. Aarsen* •
Military Museum – 1945
Hist of WW I and II, hist of the division, weapons, uniforms, vehicles, aircraft 48598

JFK Special Warfare Museum, Ardennes and Marion Sts, Bldg D-2502, Fort Bragg, NC 28307 • T: +1 910 4321533 • F: +1 910 4324062 • merrittr@soc.mil •
Cur.: *Roxanne M. Merritt* •
Military Museum – 1963
Unconventionel warfare, military and ethnogr coll 48599

Fort Bridger WY

Fort Bridger State Museum, 37,000 Business Loop I-80, Fort Bridger, WY 82933 • T: +1 307 7823842 • F: +1 307 7827181 • lnewma@missc.state.wy • www.wyobest.org •
Dir.: *Bill Gentle* • C.E.O.: *Phil Noble* • Cur.: *Cecil Sanderson* •
Historical Museum / Historic Site – 1843
Local hist, military life, westwards expansion exhibits on Indian mountain men, mormon pioneers 48600

Fort Calhoun NE

Fort Atkinson State Historical Park, 1 mile E of Hwy 75, Fort Calhoun, NE 68023 • T: +1 402 4685611 • F: +1 402 4685066 • ftatkin@ngpc.state.ne.us • www.ngpc.state.ne.us •
Head: *John Slader* •
Military Museum – 1963
Military hist, artifacts from Fort period 48601

Washington County Historical Association Museum, 14th and Monroe St, Fort Calhoun, NE 68023 • T: +1 402 4685740 • F: +1 402 4685741 • info@newashcohist.org • www.newashcohist.org •
Local Museum – 1938
Fort Atkinson relicts, Indian artifacts, furniture, toys, hand crafts, musical instruments 48602

Fort Campbell KY

Don F. Pratt Memorial Museum, 5702 Tennessee Av, Fort Campbell, KY 42223-5335 • T: +1 270 7984986 • F: +1 270 7982605 • boggsr@emh2.army.mil • www.campbell.army.mil/pratt/ •
Dir./Cur.: *Rex Boggs* • Sc. Staff: *John O'Brian* (History) •
Military Museum – 1956
Airborne military, history of 101st ABN Division 48603

Fort Carson CO

Third Cavalry Museum, Bldg 2160, 6797 Barkeley Rd, Fort Carson, CO 80913-5000 • T: +1 719 5261404, 5262028 • F: +1 719 5266573 • paul.martin@carson.army.mil • www.3acr.com •
Dir.: *Paul Martin* •
Military Museum – 1959
Military hist 48604

Fort Collins CO

Curfman Gallery and Duhesa Lounge, Colorado State University, Fort Collins, CO 80523 • T: +1 970 4916444, 4915838 • F: +1 970 4913746 • curfman@lamar.colostate.edu • www.curfman.colostate.edu •
Dir.: *Miriam B. Harris* •
Fine Arts Museum / Public Gallery – 1968 48605

Discovery Center Science Museum, 703 E Prospect Rd, Fort Collins, CO 80525 • T: +1 970 4723990 • F: +1 970 4723997 • dcsm@psd.k12.co.us • www.dcsm.org •
Exec. Dir.: *Anette Geiselman* •
Science&Tech Museum – 1989
Science 48606

Fort Collins Museum, 200 Matthews, Fort Collins, CO 80524 • T: +1 970 2216738 • F: +1 970 4162236 • bhiggins@fcgov.com • www.fcgov.com/museum •
Local Museum – 1940
Local hist 48607

Gustafson Gallery, Colorado State University, 314 Gifford St, Fort Collins, CO 80523 • T: +1 970 4911983 • F: +1 970 4914376 • carlson@cahs.colostate.edu • dm.cahs.colostate.edu/Pages/Gustafson/index.aspx •
Cur.: *Linda Carlson* •
Special Museum / University Museum – 1986
Historical costume and textile 48608

Hatton Gallery, Colorado State University, Visual Arts Bldg, Fort Collins, CO 80523 • T: +1 970 4916774 • F: +1 970 4910505 • linda.frickman@colostate.edu • www.colostate.edu/Depts/Art/hg/ •
Dir.: *Linda Frickman* •
Fine Arts Museum / University Museum – 1970
Art gallery, Japanese prints, contemporary posters, African art 48609

Museum of Contemporary Art, 201 S College Av, Fort Collins, CO 80524 • T: +1 970 4822787 • F: +1 970 4820804 • fcmoca@frii.com • www.fcmoca.com •
Dir.: *Jeanne Shoaff* •
Fine Arts Museum – 1990 48610

Fort Davis TX

Fort Davis Museum, Hwy 17-118, Fort Davis, TX 79734 • T: +1 915 4263224 ext 27 • F: +1 915 4263122 • foda_superintendent@nps.gov • www.nps.gov/foda •
Military Museum / Historical Museum – 1963
Military hist, Indians wars Western frontier history 48611

Neill Museum, 7th and Court St, Fort Davis, TX 79734 • T: +1 915 4263838, 4263969 •
Dir.: *Teda W. Neill* •
Local Museum – 1960
Dolls, bottles, period furnishings, hist House 48612

Fort DeRussy HI

United States Army Museum of Hawaii, Battery Randolph, Kalia Rd, Fort DeRussy, HI 96815 • T: +1 808 4382821 • F: +1 808 4382819 • mclaughlinje@shafter.army.mil •
Dir./Cur.: *Thomas M. Fairfull* •
Historical Museum / Military Museum – 1976
Military history, Battery Randolph, a Pre-WWI coast artillery defense bastion 48613

Fort Dix NJ

Fort Dix Museum, AFRC-FA-KPS, Fort Dix, NJ 08640 • T: +1 609 5626983 • F: +1 609 5622164 • Dixweb2@conus.army.mil • www.dix.army.mil •
Dir.: *Daniel W. Zimmerman* •
Military Museum – 1984
Firearms, uniforms, military and personal equipment, photograph coll 48614

Fort Dodge IA

Blanden Art Museum, 920 Third Av S, Fort Dodge, IA 50501 • T: +1 515 5732316 • F: +1 515 5732317 • www.blanden.org •
Pres.: *Dr. Kenneth Adams* • Dir.: *Margaret A. Skove* •
Fine Arts Museum – 1930
American and european paintings and sculpture, 16th - 20th c prints, drawings, 17th - 20th c Asian art 48615

Fort Dodge Historical Museum, Museum Rd, Fort Dodge, IA 50501 • T: +1 515 5734231 • F: +1 515 5734231 • thefort@frontiernet.net • www.fortmuseum.com •
Pres.: *Dr. E. Ryan* • Exec. Dir.: *David Parker* •
Military Museum / Local Museum – 1962
Local and military history 48616

Fort Douglas UT

Fort Douglas Military Museum, 32 Potter St, Fort Douglas, UT 84113 • T: +1 801 5811710 • F: +1 801 5819846 • admin@fortdouglas.org • www.fortdouglas.org •
Dir.: *Robert S. Voyles* • Cur.: *Ephriam D. Dickson III* •
Military Museum – 1974
Military hist, weapons, equipment, uniforms 48617

Fort Duchesne UT

Cultural Rights and Protection/Ute Indian Tribe, Hwy 40, Fort Duchesne, UT 84026 • T: +1 435 7224992 • F: +1 435 7222083 • betsyc@ubta.com •
Dir.: *Betsy Chapoose* •
Ethnology Museum – 1976
Indian hist 48618

Fort Edward NY

Old Fort House Museum, 29 Lower Broadway, Fort Edward, NY 12828 • T: +1 518 7479600 • F: +1 518 7477790 • oldfort@localnet.com • www.ftedward.com •
Pres.: *Lawrence Konopka* • C.E.O.: *R. Paul McCarty* •
Local Museum – 1925
Arcahaeology, hist, Indian and colonial war artifacts, Colonial-Victorian furniture, glass, local pottery 48619

Fort Eustis VA

United States Army Transportation Museum, 300 Washington Blvd, Besson Hall, Fort Eustis, VA 23604 • T: +1 757 8781115 • F: +1 757 8785656 • barbara.bower@eustis.army.mil • www.eustis.army.mil/DPTMSEC/museum.htm •
Dir.: *Barbara A. Bower* • Cur.: *David S. Hanselman* •
Military Museum / Science&Tech Museum – 1959
Military transportation 48620

Fort Garland CO

Old Fort Garland, S of US-160, on Hwy 159, Fort Garland, CO 81133 • T: +1 719 3793512 • F: +1 719 3793479 • rick.manzanares@state.co.us • coloradohistory.org •
Dir.: *Rick Manzanares* •
Historical Museum / Military Museum – 1858
Military, pioneer and Indian artifacts 48621

Fort Gibson OK

Fort Gibson Historic Site, Oklahoma Historical Society, 907 N Garrison Av, Fort Gibson, OK 74434 • T: +1 918 4784088 • F: +1 918 4784089 • fortgibson@ok-history.mus.ok.us •
C.E.O.: *Dr. Bob Blackburn* • Dir.: *Chris Morgan* •
Military Museum – 1936
Military, acoutrements, weapons 48622

Fort Gordon GA

United States Army Signal Corps and Fort Gordon Museum, Bldg 29807, Fort Gordon, GA 30905-5735 • T: +1 706 7912818, 7913856 • F: +1 706 7916069 • wiset@gordon.army.mil • www.gordon.army.mil/ •
Dir.: *Robert Anzuoni* • Cur.: *Michael W. Rodgers* •
Military Museum / Science&Tech Museum – 1965
Military, technology, photos 48623

Fort Hood TX

4th Infantry Division Museum, Bldg 418, 761st Tank Bn Av and 27th St, Fort Hood, TX 76544 • T: +1 254 2878811 • F: +1 254 2873833 • strattonc@hood.emh3.army.mil •
Dir.: *Ceilia M. Stratton* •
Military Museum / Science&Tech Museum – 1949
Military vehicles, tank destroyers, III corps 48624

First Cavalry Division Museum, 56th & 761 Tank Bn Av, Fort Hood, TX 76545 • T: +1 254 2873626 • F: +1 254 2876423 • steven.draper@hood.army.mil •
Dir./Cur.: *Steven C. Draper* •
Military Museum – 1971
Military hist 48625

Fort Huachuca AZ

Fort Huachuca Museum, Boyd and Grierson Sts, Fort Huachuca, AZ 85613-6000 • T: +1 520 5333898 • F: +1 520 5335736 • www.nps.gov •
Dir.: *Tim Phillips* • Cur.: *Barbara Tuttle* •
Historical Museum / Military Museum – 1960
Military history of the Indian Wars 48626

Fort Hunter NY

Schoharie Crossing, 129 Schoharie St, Fort Hunter, NY 12069-0140 • T: +1 518 8297516 • F: +1 518 8297491 •
Historical Museum – 1966
Hist of transportation in the Mohawk Valley 48627

Fort Jackson SC

Fort Jackson Museum, 4442 Jackson Blvd, Fort Jackson, SC 29207-5325 • T: +1 803 7517419 • F: +1 803 7514434 • williamsb1@jackson.army.mill •
Cur.: *Judith M. Matteson* •
Military Museum – 1974
Military hist 48628

United States Army Chaplain Museum, 10100 Lee Rd, Fort Jackson, SC 29207-7090 • T: +1 803 7518827 • F: +1 803 7518740 • mcmanusm@usaches.army.mil • www.usachcs.army.mil •
Dir./Cur.: *Marcia McManus* •
Military Museum – 1957
Military hist 48629

United States Army Finance Corps Museum, ATSG-FSM, Bldg 4392, Fort Jackson, SC 29207 • T: +1 803 7513771 • F: +1 803 7511749 • carnesw@jackson.army.mil •
Cur.: *Henry D. Howe* •
Military Museum – 1954
Military 48630

Fort Johnson NY

Old Fort Johnson, 14 Tessiero Dr, Fort Johnson, NY 12070, mail addr: POB 196, Fort Johnson, NY 12070 • T: +1 518 8430300 • museum@oldfortjohnson.org • www.oldfortjohnson.org •
Pres.: *Dennis Drenzek* •
Local Museum – 1905
Home of Sir William Johnson during French and Indian war, original furnishings, Mohawk valley Indian artifacts, ... 48631

Fort Jones CA

Fort Jones Museum, 11913 Main St, Fort Jones, CA 96032-0428 • T: +1 530 4685568 •
Dir.: *Cecelia Reuter* •
Local Museum – 1947
Indian beads, baskets, guns, tools and clothing 48632

Fort Knox KY

Patton Museum of Cavalry and Armor, 4554 Fayette Av, Fort Knox, KY 40121-0208 • T: +1 502 6243812 • F: +1 502 6242364 • museum@knox.army.mil • www.generalpatton.org •
Dir.: *John Purdy* • Cur.: *C.R. Lemons* • Libr.: *Candy Fuller* •
Military Museum – 1948
Tanks, armored vehicles, firearms 48633

Fort Laramie WY

Fort Laramie, 965 Gray Rocks Rd, Fort Laramie, WY 82212 • T: +1 307 8372221 • F: +1 307 8372120 •
Cur.: *Tammy Benson* •
Historic Site – 1938
19th c military hist, plains Indian artifacts and culture, hist buildings 48634

Fort Lauderdale FL

Bonnet House, 900 N Birch Rd, Fort Lauderdale, FL 33304-3326 • T: +1 954 5635393 • F: +1 954 5614174 • director@bonnethouse.com • www.bonnethouse.com •
Dir.: *William Marks* • Cur.: *Tom Edmonds* •
Local Museum – 1987
Collection of Frederic C. and Evelyn F. Barlett, hist artwork, rare China coll 48635

Broward County Historical Museum, 151 SW Second St, Fort Lauderdale, FL 33301 • T: +1 954 7654670 • F: +1 954 7654437 • ceck@broward.org • www.broward.org/history.htm •
Cur.: *Denyse M. Cunningham* •
Local Museum – 1972
Hist books, maps, microfilm, prints, artifacts of South Florida 48636

International Swimming Hall of Fame, One Hall of Fame Dr, Fort Lauderdale, FL 33316 • T: +1 954 4626536 • F: +1 954 5254031 • ishof@ishof.org • www.ishof.org •
Cur.: *Bob Duenkel* •
Special Museum – 1965
Hist of swimming sports 48637

Museum of Art, 1 E Las Olas Blvd, Fort Lauderdale, FL 33301 • T: +1 954 5255500 • F: +1 954 5246011 • museumofart@hotmail.com • www.museumofart.org •
Dir.: *Devell Lynn* • Cur.: *Jorge Santis* •
Fine Arts Museum – 1958
20th c European and American art 48638

Museum of Discovery and Science, 401 SW Second St, Fort Lauderdale, FL 33312-1707 • T: +1 954 4676637 • F: +1 954 4670046 • kcavendish@mods.net • www.mods.org •
Pres.: *Kim L. Cavendish* • Dir.: *Joe Cytacki* •
Science&Tech Museum – 1976
Science, technology, minerals, corals, aviation – childhood education 48639

Old Fort Lauderdale Museum of History, 219 SW Second Av, Fort Lauderdale, FL 33301 • T: +1 954 4634431 • F: +1 954 5236228 • villageinfo@oldfortlauderdale.org • www.oldfortlauderdale.org •
Dir.: *Joan Mikus* •
Local Museum – 1962
Local history 48640

Stranahan House, 335 SE Sixth Av, Fort Lauderdale, FL 33301 • T: +1 954 5244736 • F: +1 954 5252838 • stranahanl@aol.com • www.stranahanhouse.com •
Dir.: *Barbara W. Keith* •
Local Museum – 1981
Historic house 48641

Fort Leavenworth KS

Frontier Army Museum, 100 Reynolds Av, Fort Leavenworth, KS 66027-2334 • T: +1 913 6843767, 6843553 • F: +1 913 6843192 • web-dptm-museum@leavenworth.army.mil •
Dir.: *Stephen J. Allie* • Sc. Staff: *George Moore* (Exhibit) • *Richard W. Frank* (Collection) •
Military Museum – 1938
Evolution of military technology 48642

Post Museum, Fort Leavenworth Historical Society, Reynolds and Gibbons Sts, Fort Leavenworth, KS 66027 • T: +1 913 6517440 •
Science&Tech Museum – 1950
Local history, postal hist 48643

Fort Lee NJ

Fort Lee Historic Park and Museum, Hudson Terrace, Fort Lee, NJ 07024 • T: +1 201 4611776 • F: +1 201 4617275 • flhp@intac.com • www.njpalisades.org •
Pres.: *Jim Hall* • Dir.: *John Muller* •
Military Museum – 1976
Revolutionary War military hist 48644

Fort Lee VA

The United States Army Quartermaster Museum, Bldg 5218, A Av, Fort Lee, VA 23801-1601 • T: +1 804 7344203 • F: +1 804 7344359 • www.qmmuseum.lee.army.mil •
Dir.: *Tim O'Gorman* • Cur.: *Luther D. Hanson* •
Military Museum – 1957
Uniforms, flags, insignia, equestrian equipment, weapons hist military dioramas 48645

United States Army Women's Museum, 2100 Adams Av, Fort Lee, VA 23801 • T: +1 804 7344327 • F: +1 804 7344337 • burgessj@lee.army.mil • www.awm.lee.army.mil •
Pres.: *Vickie Longenecker* • C.E.O.: *Jerry G. Burgess* •
Military Museum – 1955
Women military history 48646

Fort Leonard Wood MO

United States Army Engineer Museum, Bldg 1607, Fort Leonard Wood, MO 65473 • T: +1 573 5960131 • F: +1 573 5960169 • combsk@wood.army.mil • www.wood.army.mil/museum •
Dir.: *Robert K. Combs* • Cur.: *Wilfried W. H. Rust* • *Stephen D. Wells* •
Military Museum – 1972 48647

United States Army Military Police Corps Museum, 495 Dakota Av, Fort Leonard Wood, MO 65473 • T: +1 573 5960604 • F: +1 573 5960603 • wilsonj@wood.army.mil • www.wood.army.mil/museum •
Dir.: *Scott L. Norton* •
Military Museum – 1960
Uniforms, weapons, flags, insignia, badges, military artifacts from allied countries 48648

Fort Lewis WA

Fort Lewis Military Museum, Bldg 4320, Fort Lewis, WA 98433-5000 • T: +1 256 9677206 • F: +1 253 9663029 • archamba@lewis.army.mil •
Dir./Cur.: *Alan H. Archambault* •
Military Museum – 1970
Military hist 48649

Fort McKavett TX

Fort McKavett State Historic Park, POB 68, Fort McKavett, TX 76841 • T: +1 325 3962358 • F: +1 325 3962818 • mckavett@airmail.net • www.tpwd.state.tx.us •
Head: *Michael A. Garza* •
Military Museum – 1968
Military hist 48650

Fort McNair DC

United States Army Center of Military History, Museum Division, 103 Third Av, Fort McNair, DC 20319-5058 • T: +1 202 6852453 • F: +1 202 6852113 • terry.dougherty@hqda.army.mil • www.army.mil/cmh-pg •
Dir.: *J. Terry Dougherty* •
Military Museum – 1946
Military hist 48651

Fort Madison IA

North Lee County Historic Center and Santa Fe Depot Museum Complex, 10th and Av H, Fort Madison, IA 52627 • T: +1 319 3727661, 3727363 • F: +1 319 3727363 •
Pres.: *Tom Barr* • Dir.: *Sheila Sallen* •
Local Museum / Fine Arts Museum – 1962
Historic Center, located in Santa Fe Railroad depot, caboose, country school, Brush College, 1993 Flood museum, Louis Koch Gallery of historic paintings, furniture 48652

Fort Meade MD

Fort George G. Meade Museum, 4674 Griffin Av, Fort Meade, MD 20755-5094 • T: +1 301 6776966, 6777054 • F: +1 301 6772953 • johnsonr@emh1.ftmeade.army.mil • www.ftmeade.army.mil/museum/index.htm •
Cur.: *Robert S. Johnson* •
Military Museum – 1963
Military hist 48653

National Cryptologic Museum, Colony 7 Rd, Fort Meade, MD 20755-6000 • T: +1 301 6885849 • F: +1 301 6885847 • museum@nsa.gov • www.nsa.gov/museum •
Cur.: *Jack E. Ingram* •
Military Museum – 1993
Cripher machines, cryptologic equipment, historical computers 48654

Fort Meade SD

Old Fort Meade Museum, 55 Fort Meade, Fort Meade, SD 57741, mail addr: POB 164, Fort Meade SD 57741 • T: +1 605 3479822 • ftmeade@rapidnet.com • www.fortmeademuseum.org •
Pres.: *Peg Aplan* • Dir.: *Charles Rambow* •
Military Museum – 1964
Military hist 48655

Fort Mitchell KY

Vent Haven Museum, 33 W Maple, Fort Mitchell, KY 41011 • T: +1 859 3410461, 3319500 • F: +1 859 3410461 • venthaven@home.com • www.venthaven.org •
Cur.: *Lisa Sweasy* •
Performing Arts Museum – 1973
Ventriloguial figures and memorabilia 48656

Fort Monmouth NJ

United States Army Communications-Electronics Museum, Kaplan Hall, Bldg 275, Fort Monmouth, NJ 07703 • T: +1 732 5321682 • F: +1 732 5322637 •
Dir.: *Mindy Rosewitz* •
Military Museum / Science&Tech Museum – 1976
International communication and electronic equipment, military 48657

Fort Monroe VA

Casemate Museum, 20 Bernard Rd, Fort Monroe, VA 23651-0341 • T: +1 757 7883391 • F: +1 757 7883886 • mroczkod@monroe.army.mil •
Dir.: *Dennis P. Mroczkowski* •
Military Museum – 1951
Restored cell where Jefferson Davis was imprisoned in 1865, Civil War 48658

Fort Morgan CO

Fort Morgan Museum, 414 Main City Park, Fort Morgan, CO 80701 • T: +1 970 8676331 • F: +1 970 5423008 • ftmormus@ftmorganmus.org • www.ftmorganmus.org/ •
Dir.: *Marne Jurgemeyer* •
Local Museum / Music Museum – 1969
General museum, music Glen Miller 48659

Fort Myer VA

The Old Guard Museum, 204 Lee Av, 3rd US Infantry, Fort Myer, VA 22211-1199 • T: +1 703 6966670, 6964168 • F: +1 703 6964256 • bogana@fmmc.army.mil • www.mdw.army.mil/oldguard/ •
Dir.: *Alan T. Bogan* • Sc. Staff: *John Manes* • *Kirk Heflin* •
Military Museum – 1962
Military univorms & weapons, accoutrements of the 3rd INF Regiment 1784 to the present 48660

Fort Myers FL

Edison and Ford Winter Estates, 2350 McGregor Blvd, Fort Myers, FL 33901 • T: +1 239 3347419 • F: +1 239 3326684 • estateinfo@cityftmyers.com • www.edison-ford-estate.com •
Dir.: *Chris Pendleton* •
Historic Site – 1947
1886 wooden vernacular, pre-cut in Maine, brought to Florida by schooner, early light bulbs, early phonographs laboratory with equipment, Edison papers/rubber research 48661

Edison Community College Gallery of Fine Art, 8099 College Pkwy SW, Fort Myers, FL 33919, mail addr: POB 60210, Fort Myers, FL 33906 • T: +1 941 4899313 • F: +1 941 4899482 • rbishop@edison.edu • www.edison.edu/ •
Cur.: *Ron Bishop* •
Fine Arts Museum / Public Gallery / University Museum – 1979
Art centre 48662

Imaginarium Hands-On Museum, 2000 Cranford Av, Fort Myers, FL 33916 • T: +1 941 3373332 • F: +1 941 3372109 • gbuck@cityftmyers.com • www.cityftmyers.com •
Science&Tech Museum – 1989
Science and technology 48663

Southwest Florida Museum of History, 2300 Peck St, Fort Myers, FL 33901 • T: +1 941 3325955 • F: +1 941 3326637 • mjohnson@cityftmyers.com • www.cityftmyers.com •
Local Museum / Historical Museum – 1982
Local history, in a 1924 ACL Railroad Depot 48664

Fort Oglethorpe GA

Chickamauga-Chattanooga National Military Park, 3370 LaFayette Rd, US 27 S, Fort Oglethorpe, GA 30742 • T: +1 706 8669241 • F: +1 423 7525215 • www.nps.gov/chch •
Military Museum – 1890
Civil War artifacts, American military 48665

Fort Peck MT

Fort Peck Museum, Fort Peck Power Plant, Fort Peck, MT 59223 • T: +1 406 5263431 • F: +1 406 5263477 • John.E.Daggett@usace.army.mil •
Dir.: *John E. Daggett* •
Natural History Museum 48666

Fort Pierce FL

Harbor Branch Oceanographic Institution, 5600 US 1 N, Fort Pierce, FL 34946 • T: +1 561 4653400 ext 306 • F: +1 561 4680757 • hanisak@hboi.edu • www.hboi.edu •
Dir.: *Richard Herman* • Cur.: *Dr. Dennis Hanisak* •
Natural History Museum / Science&Tech Museum – 1975
Natural hist 48667

Saint Lucie County Historical Museum, 414 Seaway Dr, Fort Pierce, FL 34949 • T: +1 561 4621795 • F: +1 561 4621877 • maddoxi@co.st-lucie.fl.us • www.st-lucie.lib.fl.us/museum.htm •
C.E.O.: *Iva Jean Maddox* •
Local Museum – 1965
Local hist sites – Library 48668

UDT-SEAL Museum, 3300 N State Rd A1A, Fort Pierce, FL 34949-8520 • T: +1 561 5955845 • F: +1 561 5955847 • www.udt-sealmuseum.org •
Pres.: *James H. Barnes* • Exec. Dir.: *H.T. Aldhizer* •
Military Museum – 1985
Military, site where the Navy first trained Frogmen (underwater demolition teams), artifacts, weapons 48669

Fort Plain NY

Fort Plain Museum, 389 Canal St, Fort Plain, NY 13339 • T: +1 518 9932527 •
Dir.: *Glenadore Wetterau* •
Local Museum – 1963
Indian hist and culture, Hist, anthropology 48670

Fort Polk LA

Fort Polk Military Museum, Bldg 917, S Carolina Av, Fort Polk, LA 71459-0916 • T: +1 337 5317905, 5314840 • F: +1 337 5314202 • binghamd@polk.army.mil •
Dir./Cur.: *David S. Bingham* •
Military Museum – 1972
Military 48671

Fort Ransom ND

Bjarne Ness Gallery, Bear Creek Hall, Fort Ransom, ND 58033 • T: +1 701 9734461 •
Fine Arts Museum
American painting, woodcarvings, paintings of Bjarne Ness 48672

Ransom County Historical Society Museum, 101 Mill Rd SE, Fort Ransom, ND 58033-9740 • T: +1 701 6782045 • F: +1 701 6782045 • rbower@northpro.net • www.members.tripod.com/rchsmuseum •
Pres.: *Richard Birklid* •
Local Museum – 1972
Regional hist 48673

Fort Recovery OH

Fort Recovery Museum, 1 Fort Site St, Fort Recovery, OH 45846 • T: +1 419 3754649 • F: +1 419 3754629 • bmeiring@columbus.rr.com •
Dir.: *Barbara Meiring* •
Military Museum / Ethnology Museum – 1982
History and relics of the battle of St. Clair's defeat (1791) and of Wayne's victory (1794), Indian artifact, fort buildings 48674

Fort Riley KS

First Territorial Capitol of Kansas, Bldg 693, Huebner Rd, K-18, Fort Riley, KS 66442 • T: +1 785 7845535 • www.kshs.org •
Cur.: *Gary R. Dierking* •
Historical Museum – 1928
Civil War calvery accoutrements, weapons, Indian artifacts 48675

United States Cavalry Museum, Bldg. 205, Fort Riley, KS 66442 • T: +1 785 2392737 • F: +1 785 2396243 • mckalew@riley.army.mil • www.kshs.org •
Dir.: *William McKale* • Sc. Staff: *Don Rush* (Exhibitions) •

U

Military Museum – 1957
Military, housed in a old hospital 1855-1890 (post headquarters 1890-1948), 1st Infantry and 1st Armored division ... 48676

Fort Rucker AL

United States Army Aviation Museum, Bldg 6000, Fort Rucker, AL 36362-0610, mail addr: POB 6206 10-0610, Fort Rucker, AL 36330 • T: +1 334 2553036 • F: +1 334 2553054 • giftshop@armyavnmuseum.org • www.armyavnmuseum.org •
Dir.: *R.S. Maxham* • Cur.: *Harford Edwards jr.* •
Military Museum / Science&Tech Museum – 1962
Aircraft, fixed wing and rotary wing, army aviation – library ... 48677

Fort Sam Houston TX

Fort Sam Houston Museum, 1210 Stanley Rd, Fort Sam Houston, TX 78234-5002 • T: +1 210 2211886 • F: +1 210 2211311 • john.manguso@amedd.army.mil • www.cs.amedd.army.mil/rlbc •
Dir.: *John M. Manguso* • Sc. Staff: *Jacqueline B. Davis* • *Martin L. Callahan* •
Military Museum / Historical Museum / Historic Site – 1967
Military hist ... 48678

United States Army Medical Department Museum, Bldg 1046, 2310 Stanley Rd, Fort Sam Houston, TX 78234 • T: +1 210 2216358 • F: +1 210 2216781 • thomas.mcmasters@amedd.army.nil • www.ameddmuseumfoundation.com •
Dir.: *Thomas O. McMasters* •
Military Museum / Historical Museum – 1955
Military medicine ... 48679

Fort Scott KS

Fort Scott National Historic Site, Old Fort Blvd, Fort Scott, KS 66701 • T: +1 620 2230310 • F: +1 620 2230188 • fosc_superintendent@nps.gov • www.nps.gov/fosc •
Historic Site – 1978
Restored and reconstructed Fort Scott ... 48680

Historic Preservation Association of Bourbon County, 117 S Main, Fort Scott, KS 66701 • T: +1 316 2231557 • ldrdevon@ckt.net •
Pres.: *Arnold Schofield* •
Local Museum / Military Museum – 1973
Religious items, hist artifacts, furnishings, documents ... 48681

Fort Sill OK

Fort Sill Museum, 437 Quanah Rd, Fort Sill, OK 73503-5100 • T: +1 580 4425123 • F: +1 580 4428120 • spiveyt@sill.army.mil •
Dir./Cur.: *Towana Spivey* •
Military Museum / Historic Site – 1934
U.S. Army field artillery, miltary exhibits, uniforms, horse furnishings, vehicles – archives ... 48682

Fort Smith AR

The Fort Smith, National Historic Site, 301 Parker Av, Fort Smith, AR 72901, mail addr: POB 1406, Fort Smith, AR 72902 • T: +1 501 7833961 • F: +1 501 7835307 • www.nps.gov./fosm •
Head: *William N. Black* • Cur.: *Emily Lovick* •
Historic Site – 1961
1817-24 Fort Smith, became the Federal courthouse and jail 1871-1896 ... 48683

Fort Smith Art Center, 423 N Sixth St, Fort Smith, AR 72901 • T: +1 501 7842787 • F: +1 501 7849071 • info@fortsmithartcenter.org • www.fortsmithartcenter.com •
Dir.: *Kathy Williams* • *Michael Richardson* •
Fine Arts Museum – 1948
Contemporary Americanart, decorative arts ... 48684

Fort Smith Museum of History, 320 Rogers Av, Fort Smith, AR 72901 • T: +1 501 7837841 • www.fortsmithmuseum.com •
Cur.: *Leslie Przybylek* •
Local Museum – 1910
Local and military hist ... 48685

Patent Model Museum, 400 N 8th St, Fort Smith, AR 72901-2204 • T: +1 501 7829014 •
Special Museum – 1976
85 working patent models 1836 to the 1870s ... 48686

Fort Stewart GA

Fort Stewart Museum, 2022 Frank Cochran Dr, Fort Stewart, GA 31314 • T: +1 912 7677885 • F: +1 912 7672121 • www.stewart.army.mil •
Military Museum – 1977
History of World War I and II, Korea and Desert Storm ... 48687

Fort Stockton TX

Annie Riggs Memorial Museum, 301 S Main St, Fort Stockton, TX 79735 • T: +1 432 3362167 • F: +1 432 3362402 • annieriggs@ftstockton.net •
Cur.: *Martha King* •
Local Museum – 1955
Hist textiles, Native American artifacts, cowboy & ranch implements, kitchen utensils ... 48688

Historic Fort Stockton, 300 E Third, Fort Stockton, TX 79735 • T: +1 915 3362400 • F: +1 915 3362402 • hfs@ci.fort-stockton.tx.us •
Dir./Cur.: *LeAnna S. Biles* •
Historic Site / Open Air Museum / Military Museum – 1990
Military & supporting contractors directly relaiting to Fort Stockton ... 48689

Fort Sumner NM

Billy The Kid Museum, 1435 E Sumner Av, Fort Sumner, NM 88119 • T: +1 575 3552380 • F: +1 575 3551380 • btkmuseum@plateautel.net •
Special Museum – 1953
Old West local and Billy the Kid memorabilia ... 48690

Fort Sumner, Billy the Kid Rd, Fort Sumner, NM 88119-0356 • T: +1 505 3552573 • F: +1 505 3552573 • hweeldi@plateautel.net •
Historical Museum – 1968 ... 48691

Fort Totten ND

Fort Totten State Historic Museum, Pioneer Daughters Museum, Bldg 14, Fort Totten, ND 58335 • T: +1 701 7664441 • F: +1 701 7661382 • histsoc@state.nd.us •
Local Museum / Open Air Museum – 1960
Hist buildings, local hist ... 48692

Fort Towson OK

Fort Towson Military Park, Oklahoma Historical Society, HC 63, Box 1580, Fort Towson, OK 74735-9273 • T: +1 580 8732634 • F: +1 580 8739385 • johndavis@ok-history.mus.ok.us • www.ok-history.mus.ok.us •
Dir.: *William Vandever* • Cur.: *Keith Tolman* •
Military Museum / Archaeology Museum – 1972 48693

Fort Valley GA

A.L. Fetterman Educational Museum, 100 Massee Ln, Fort Valley, GA 31030 • T: +1 912 9672358, 9672722 • F: +1 912 9672083 • acs@alltel.net • www.camellias-acs.com •
Pres.: *Arthur Gonos* • Dir.: *Ann Walton* •
Decorative Arts Museum – 1989
Porcelain, paintings, prints, books, sculpture ... 48694

Gallery of Art, Fort Valley State University, 1005 State University Dr, Fort Valley, GA 31030-4313 • T: +1 478 8256387 • F: +1 478 8256132 • www.fvsu.edu •
Dir.: *Bobby Dickey* •
Fine Arts Museum / University Museum
Paintings, graphics, sculpture, photography ... 48695

Fort Walton Beach FL

Indian Temple Mound Museum, 139 Miracle Strip Pkwy SE, Fort Walton Beach, FL 32548 • T: +1 850 8339595 • F: +1 850 8339675 • www.fwb.org •
Dir.: *Anna M. Peele* •
Historical Museum / Ethnology Museum – 1962
Anthropology, ethnology of Indian, Indian Temple Mound of Fort Walton Culture, prehistoric Native American cultures of local area ... 48696

Fort Washakie WY

Shoshone Tribal Cultural Center, 31 First St, Fort Washakie, WY 82514 • T: +1 307 3329106 • F: +1 307 3323055 • rteran@mail.washakie.net •
C.E.O.: *Reba Jo Teran* •
Folklore Museum – 1988
Indian culture, history and art ... 48697

Fort Washington MD

Fort Washington, 13551 Fort Washington Rd, Fort Washington, MD 20744 • T: +1 301 7634600 • F: +1 301 7631389 • nace_fort_washington_parks@nps.gov • www.nps.gov/fowa •
Military Museum – 1940
Military, Commandant's house ... 48698

Fort Washington PA

Childventure Museum, 430 Virginia Dr, Fort Washington, PA 19034 • T: +1 215 6439906 •
Dir.: *Nina Kardon* • *Beverly Levine* •
Special Museum – 1989 ... 48699

Fort Washington Museum, 473 Bethlehem Pike, Fort Washington, PA 19034 • T: +1 215 6466065 •
Pres.: *Lewis Keen* •
Historical Museum / Historic Site – 1935
1801 Clifton House ... 48700

Highlands, 7001 Sheaff Ln, Fort Washington, PA 19034 • T: +1 215 6412687 • F: +1 215 6412556 • highlandshistorical@earthlink.net •
Dir.: *Margaret Bleecker Blades* • Cur.: *Jeniffer P. April* •
Local Museum – 1975 ... 48701

Hope Lodge and Mather Mill, 553 S Bethlehem Pike, Fort Washington, PA 19034 • T: +1 215 6461595 • F: +1 215 6289471 • hkrueger@state.pa.us • www.ushistory.org/hope •
Pres.: *William Lutz* • C.E.O.: *Hilary Folwell Krueger* • Cur.: *Jennifer Glass* •
Local Museum / Historic Site / Open Air Museum – 1957 ... 48702

Fort Wayne IN

Allen County-Fort Wayne Historical Society Museum, 302 E Berry St, Fort Wayne, IN 46802 • T: +1 219 4262882 • F: +1 219 4244419 • histsociety@comcast.net • www.fwhistorycenter.com •
Cur.: *Walter Font* •
Local Museum – 1921
Military artifacts, manuscripts, local and Indian hist ... 48703

Fort Wayne Firefighters Museum, 226 W Washington Blvd, Fort Wayne, IN 46802 • T: +1 219 4260051 •
Science&Tech Museum
Hist of the Fort Wayne Fire Department, apparatus, equip, patch coll ... 48704

Fort Wayne Museum of Art, 311 E Main St, Fort Wayne, IN 46802 • T: +1 219 4226467 • F: +1 219 4221374 • mail@fwmoa.org • www.fwmoa.org •
Dir.: *Charles A. Shepard* • Cur.: *Sachi Yanari* •
Fine Arts Museum – 1922
19th and 20th c American and European prints and paintings, sculpture, works on paper, contemporary art ... 48705

John E. Weatherhead Gallery, University of Saint Francis, 2701 Spring St, Fort Wayne, IN 46808 • T: +1 219 4343235 • F: +1 219 4343194 • rcartwright@sf.edu • www.sf.edu/art •
Public Gallery ... 48706

The Lincoln Museum, 200 E Berry, Fort Wayne, IN 46801-7838 • T: +1 219 4553864 • F: +1 219 4556922 • thelincolnmuseum@lnc.com • www.thelincolnmuseum.org •
C.E.O.: *Joan L. Flinspach* •
Historical Museum – 1928
History, civil war ... 48707

Fort Worth TX

American Airlines C.R. Smith Museum, 4601 Hwy 360 at FAA Rd, Fort Worth, TX 76155 • T: +1 817 9675905, 9675904 • F: +1 817 9675737 • info.crsmitmuseum@aa.com • www.crsmithmuseum.org •
Cur.: *Ben Kristy* •
Science&Tech Museum / Historical Museum – 1993
American Airlines hist ... 48708

Amon Carter Museum, 3501 Camp Bowie Blvd, Fort Worth, TX 76107-2695 • T: +1 817 7381933 • F: +1 817 9895099 • carol.noel@cartermuseum.org • www.cartermuseum.org •
Dir.: *Rick Stewart* • Dep. Dir.: *Robert Workman* • Chief Cur.: *Jane Myers* • Cur.: *Barbara McCandless* • *Patricia Junker* • Ass. Cur.: *John Rohrbach* • *Rebecca Lawton* •
Fine Arts Museum / Folklore Museum – 1961
American paintings, sculpture, works on paper, Carter Coll of Remington & Russel works ... 48709

Atrium Gallery, c/o University of North Texas Health Science Center, 3500 Camp Bowie Blvd, Fort Worth, TX 76109-2699 • T: +1 817 7352000 • jsager@hsc.unt.edu • www.hsc.unt.edu/artcompetition •
Head: *Judy Sager* •
Fine Arts Museum – 1986
Watercolors ... 48710

Bank One Fort Worth Collection, 500 Throckmorton, Fort Worth, TX 76102 • T: +1 817 8844000 • F: +1 817 8702454 •
Fine Arts Museum – 1974
Sculpture, drawings, graphics, paintings, prints and tapestries ... 48711

Cattle Raisers Museum, 1301 W 7th St, Fort Worth, TX 76102-2665 • T: +1 817 3328551 • F: +1 817 3328749 • museum@texascattleraisers.org • www.cattleraisersmuseum.org •
Dir.: *Rachel L. Donahue* •
Local Museum – 1981
Hist of the cattle industry in Texas ... 48712

Fort Worth Museum of Science and History, 1501 Montgomery St, Fort Worth, TX 76107 • T: +1 817 2559300 • F: +1 817 7327635 • webmaster@fwmshz.org • www.fortworthmuseum.org •
Dir.: *James P. Diffily* • *Dennis Gabbard* • Chief Cur.: *William J. Voss* (Science) • Ass. Cur.: *Renee Tucker* (History) •
Historical Museum / Science&Tech Museum – 1941
Botany, entomology, malacology, ethnology, herpetology, loc hist ... 48713

Fort Worth Public Library Arts and Humanities, Fine Arts Section, 500 W Third Rd, Fort Worth, TX 76102 • T: +1 817 8717737 • F: +1 817 8717734 • tstone@fortworthlibrary.org • www.fortworthlibrary.org •
Man.: *Thelma Stone* •
Library with Exhibitions – 1902 ... 48714

Kimbell Art Museum, 3333 Camp Bowie Blvd, Fort Worth, TX 76107-2792 • T: +1 817 3328451 ext 224 • F: +1 817 8771264 • mriesenberg@kimbellmuseum.org • www.kimbellart.org •
Dir.: *Timothy Potts* • Cur.: *Malcolm Warner* • *Jennifer R. Casler* (Asian and Non-Western Art) • *Nancy E. Edwards* (European Art) • *Patricia C. Loud* (Architecture) • Libr.: *Chia-Chun Shih* •
Fine Arts Museum – 1972
Western Eurpean paintings, sculpture from antiquity to 20th c, pre-Columbian objects, Asian sculpure, sceens ... 48715

Log Cabin Village, 2100 Log Cabin Village Ln, Fort Worth, TX 76109 • T: +1 817 3925881 • F: +1 817 3927610 • director@logcabinvillage.org • www.logcabinvillage.org •
Dir.: *Kelli L. Pickard* • Cur.: *Dulce Ivette Ray* •
Local Museum / Open Air Museum – 1966
Historic village ... 48716

Modern Art Museum of Fort Worth, 3200 Darnell St at University Dr, Fort Worth, TX 76107 • T: +1 817 7389215 • F: +1 817 7351161 • info@themodern.org • www.themodern.org •
Dir.: *Dr. Marla J. Price* • Chief Cur.: *Michael Auping* • Ass. Cur.: *Andrea Karnes* •
Fine Arts Museum – 1892
Post World War II international art in all media ... 48717

National Cowgirl Museum and Hall of Fame, 1720 Gendy St, Fort Worth, TX 76107 • T: +1 817 3364475 • F: +1 817 3362470 • susan@cowgirl.net • www.cowgirl.net •
Dir.: *Pat Riley* • Cur.: *Tricia Taylor Dixon* •
Historical Museum – 1975/2002
Hist of Western American women ... 48718

Sid Richardson Collection of Western Art, 309 Main St, Fort Worth, TX 76102 • T: +1 888 3326554 • F: +1 817 3328671 • info@sidrmuseum.org • www.sidrmuseum.org •
Dir.: *Jan Brenneman* • Ass. Dir.: *Monica Herman* •
Fine Arts Museum – 1982
Paintings by Frederic Remington and Charles W. Russel ... 48719

University Art Gallery, Texas Christian University, University Dr & Cantey, Fort Worth, TX761209, mail addr: c/o Dept. of Art and Art History, POB 298000, Fort Worth, TX 76129 • T: +1 817 2577643 • F: +1 817 2577399 • r.watson@tcu.edu •
Dir.: *Ronald Watson* •
University Museum / Fine Arts Museum – 1874
Works on paper by traditional as well as contemporary artists ... 48720

Forty Fort PA

Nathan Denison House, 35 Denison St, Forty Fort, PA 18704-4390 • T: +1 717 2885531 •
Pres.: *Louise Robinson* •
Local Museum – 1970 ... 48721

Fossil OR

Fossil Museum, First and Main Sts, Fossil, OR 97830 • T: +1 541 7632113 • F: +1 541 7632026 •
Local Museum – 1966 ... 48722

Fostoria OH

Fostoria Area Historical Museum, 123 W North St, Fostoria, OH 44830 •
C.E.O.: *Leonard Skonecki* •
Local Museum – 1972
Furniture, fire equipment ... 48723

Fountain MN

Fillmore County Historical Center, 202 County Rd No. 8, Fountain, MN 55935 • T: +1 507 2684449 • F: +1 507 2684492 • fillmorehistory@earthlink.net • www.rootsweb.com/~mnfillmo/society.htm •
Dir.: *Jerry D. Henke* •
Local Museum – 1934
Rural life, agriculture, school, Civil War – archives, library ... 48724

Fountain City IN

Levi Coffin House, 113 U.S. 27 North, Fountain City, IN 47341 • T: +1 765 8472432 • F: +1 765 8472498 • coffinhs@infocom.com • www.waynet.org •
Pres.: *Janice McGuire* •
Historical Museum – 1967
Furnishings and furniture of the period ... 48725

Four Oaks NC

Bentonville Battleground, 5466 Harper House Rd, Four Oaks, NC 27524 • T: +1 910 5940789 • F: +1 910 5940027 • bentonville@ncmail.net •
Man.: *Donald B. Taylor* •
Military Museum / Open Air Museum – 1961 ... 48726

Fox Island WA

Fox Island Historical Society Museum, 1017 Ninth Av, Fox Island, WA 98333 • T: +1 253 5492835 • F: +1 253 5492461 • museum@foxisland.net • www.foxisland.net •
Pres.: *David McHughs* •
Local Museum – 1895
Local hist ... 48727

Fox Lake WI

Fox Lake Historical Museum, 211 Cordelia St and S College Av, Fox Lake, WI 53933 • T: +1 920 2960254 • F: +1 920 9283754 •
Local Museum / Science&Tech Museum – 1970
Railroad, local hist ... 48728

U

Framingham MA

Danforth Museum of Art, 123 Union Av, Framingham, MA 01702 • T: +1 508 6200050 ext 13 • F: +1 508 8725542 • dmadev@conversent.net • www.danforthmuseum.org •
Dir.: *Hope Coolidge* • Cur.: *Laura McCarty* •
Fine Arts Museum – 1975
Visual fine arts of America – library 48729

Framingham Historical Museum, 16 Vernon St, Framingham, MA 01701, mail addr: POB 2032, Framingham, MA 01703-2032 • T: +1 508 8723780 • F: +1 508 8723780 • office@framinghamhistory.org • www.framinghamhistory.org •
Dir.: *Tom Harris* • Cur.: *Dana Dautermann Ricciardi* •
Local Museum – 1888
Local history, culture, agriculture, domestic and industrial items, fine and decorative arts – library, archives 48730

Franconia NH

New England Ski Museum, Franconia Notch State Park, Pkwy Exit 2, Franconia, NH 03580 • T: +1 603 8237177 • F: +1 603 8239505 • staff@skimuseum.org • www.skimuseum.org •
Pres.: *Glenn Parkinson* • Exec. Dir.: *Jeffrey R. Leich* •
Special Museum – 1977
Ski artifacts, posters, photographs 48731

Frankenmuth MI

Frankenmuth Historical Museum, 613 S Main St, Frankenmuth, MI 48734 • T: +1 517 6529701 • F: +1 517 6529390 • frankenmuthmuseum@yahoo.com • frankenmuth.michigan.museum •
C.E.O.: *Sally D. Van Ness* • Cur.: *Michaela McInerney* (Education) •
Local Museum / Folklore Museum – 1963
Regional history, ethnic & early pioneer artifacts 48732

Frankfort IN

Clinton County Museum, 301 E Clinton St, Frankfort, IN 46041 • T: +1 765 6592030, 6594079 • F: +1 765 6547773 • cchsm@geetel.net • www.cchsm.org •
Dir.: *Nancy Hart* •
Local Museum – 1980
Local history, housed in Old Stoney, former Frankfort High School building 48733

Frankfort KY

The Executive Mansion, 704 Capitol Av, Frankfort, KY 40601 • T: +1 502 5648004 • F: +1 502 5645022 • rlyons@mail.state.ky.us •
Dir.: *Rex Lyons* •
Local Museum – 1914
Historic house, residence of 22 of Kentucky's governors 48734

The Governor's Mansion, 704 Capitol Av, Frankfort, KY 40601 • T: +1 502 5648004 • F: +1 502 5645022 • kenny.bishop@ky.gov •
Dir.: *Kenny Bishop* •
Historical Museum – 1914
1914 Beau-Arts style residence of 23 of Kentucky governors 48735

Jackson Hall Gallery, c/o Kentucky State University, Frankfort, KY 40601 • T: +1 502 5975994/95 • jalexandra@qwmail.kysu.edu •
Head: *John Bater* •
Fine Arts Museum / University Museum – 1886
African art 48736

Kentucky Historical Society, 100 W Broadway, Frankfort, KY 40601 • T: +1 502 5643016 • F: +1 502 5644701 •
Dir.: *Kent Whitworth* •
Association with Coll – 1836
History, housed in the Kentucky History Center and 1869 Old State Capitol Annex 48737

Kentucky Military History Museum, 125 E Main St, Frankfort, KY 40601 • T: +1 502 5641792, ext. 4498 • F: +1 502 5644054 • bill.bright@mail.state.ky.us • www.kyhistory.org •
Cur.: *Bill Bright* •
Military Museum – 1974
Ordnance, flags, uniforms, personal items 48738

Kentucky State Capitol, 700 Capitol Av, Frankfort, KY 40601 • T: +1 502 5643000 • F: +1 502 5646505 • paul.gannoe@ky.gov • finance.ky.gov/properties/capitol.htm •
Cur.: *Lou Karibo* •
Decorative Arts Museum
First Ladies Doll coll, oil paintings of chief justices, period furnishings 48739

Liberty Hall, 218 Wilkinson St, Frankfort, KY 40601 • T: +1 502 2272560 • F: +1 502 2273348 • libhall@dcr.net • www.libertyhall.org •
Dir.: *Sara Farley-Harger* •
Decorative Arts Museum – 1937
Formerly the home of the Kentucky Senator John Brown 48740

The Old Governor's Mansion, 420 High St, Frankfort, KY 40601 • T: +1 502 5645500 • F: +1 502 5644099 • Frankie.McNulty@mail.state.ky.us •
Cur.: *F.M. McNulty* •
Local Museum – 1798
Emphasizing historic furniture & artifacts, than belonged to former Governors 48741

Orlando Brown House, Liberty Hall Museum, 220 Wilkinson St, Frankfort, KY 40601 • T: +1 502 8754952 • F: +1 502 2273348 • www.libertyhall.org •
Dir.: *Sara Harger* •
Fine Arts Museum / Decorative Arts Museum – 1956
Paul Sawyier paintings, original furnishings 48742

Vest-Lindsey House, 401 Wapping St, Frankfort, KY 40601 • T: +1 502 5643000, 5646980 • F: +1 502 5646505 • helen.evans@mail.state.ky.us •
Dir.: *Helen H. Evans* • Cur.: *Lou Karibo* •
Local Museum – 1978
Early 188th c furnishings 48743

Frankfort SD

Fisher Grove Country School, 17290 Fishers Ln, Frankfort, SD 57440 • T: +1 605 4721212 •
Local Museum – 1884
Historic building 48744

Franklin AL

Alabama River Museum, Claiborne Lock and Dam, Franklin, AL 36444, mail addr: POB 1637, Monroeville, AL 36461 • T: +1 251 2824206 • F: +1 251 5752513 • mchm@frontiernet.net • www.tokillamockingbird.com •
C.E.O.: *Jane E. Clark* •
Historical Museum – 2000
Tools, points and ceremonial from native Americans of Alabama, Alabama river steamboats artifacts, fossils 48745

Franklin IN

Johnson County History Museum, 135 N Main St, Franklin, IN 46131 • T: +1 317 7364655 • F: +1 317 7365451 • map@netdirect.net • www.johnsoncountymuseum.org •
Dir.: *Mary Ann Plummer* • Cur.: *Jill Hasprunar* •
Local Museum – 1931
Textiles, furniture, tools, artifacts 48746

Franklin ME

Franklin Historical Society Museum, Rte 200, Franklin, ME 04634 • T: +1 207 5652223 •
Cur.: *Helen Cantor* •
Local Museum – 1960
Uniforms, pictures, farming tools, granite work tools, local hist 48747

Franklin NJ

Franklin Mineral Museum, 32 Evans St, Franklin, NJ 07416 • T: +1 973 8273481 • F: +1 973 8270149 • rockman@tellurian.com • www.franklinmineralmuseum.com •
Pres.: *Steven Phillips* • Cur.: *John Cianciulli* •
Natural History Museum / Science&Tech Museum – 1965
Geology, mining 48748

Franklin OH

Harding Museum, 302 Park Av, Franklin, OH 45005 • T: +1 513 7435832 •
Pres.: *Patrick LeVangie* •
Local Museum – 1965
Franklin area hist, General Harding's personal artifacts 48749

Franklin TN

Carnton Plantation, 1345 Carnton Ln, Franklin, TN 37064 • T: +1 615 7940903 • F: +1 615 7946563 • carnton@mindspring.com • www.carnton.org •
Exec. Dir.: *Angela Calhoun* •
Historic Site – 1977
Historic house 48750

The Carter House, 1140 Columbia Av, Franklin, TN 37065 • T: +1 615 7911861 • F: +1 615 7941327 • carterhouse1864@aol.com • www.carter-house.org •
Exec. Dir./Cur.: *Thomas Cartwright* •
Local Museum – 1951
Artifacts from the Battle of Franklin 48751

Franklin Center PA

Franklin Mint Museum, Franklin Mint, Rte 1, Franklin Center, PA 19091 • T: +1 610 4596881 ext 6348 • F: +1 610 4596463 • csiano@franklinmint.com • www.franklinmint.com •
Pres.: *Bruce Newman* •
Science&Tech Museum – 1973 48752

Franklinville NY

Ischua Valley Historical Society, 9 Pine St, Franklinville, NY 14737 • T: +1 716 6765651 •
Pres.: *Gertrude H. Schnell* •
Local Museum – 1966
Hist houses, local artifacts and doc 48753

Frederick MD

Barbara Fritchie House Museum, 154 W Patrick St, Frederick, MD 21701 • T: +1 301 6980630 • F: +1 301 6988994 •
Pres.: *Margaret Kline* •
Decorative Arts Museum – 1927
Furnishings, personal, distribution and transportation artifacts, paintings 48754

The Children's Museum of Rose Hill Manor Park, 1611 N Market St, Frederick, MD 21701-4304 • T: +1 301 6941648 • F: +1 301 6942595 • parksandrecreation@fredco-md.net • www.rosehillmuseum.com •
Historical Museum – 1972
19th c house and farm 48755

Historical Museum of Frederick County, 24 E Church St, Frederick, MD 21701 • T: +1 301 6631188 • F: +1 301 6630526 • mhudson@hsfcinfo.org • www.hsfcinfo.org •
Dir.: *Mark S. Hudson* • Cur.: *Heidi Campbell-Shoaf* •
Local Museum – 1888
Decorative and fine arts, historical artifacts, photographs, textiles, genealogy 48756

Monocacy National Battlefield, 4801 Urbana Pike, Frederick, MD 21704 • T: +1 301 6623515, 4327677 • F: +1 301 6623420 • cathy_beeler@nps.gov • www.nps.gov/mono/mo_visit.htm •
Historic Site – 1907
Dedicated to soldiers who fought in the Battle of Monocacy, July 9, 1864 48757

National Museum of Civil War Medicine, 48 E Patrick St, Frederick, MD 21701 • T: +1 301 6951864 • F: +1 301 6956823 • museum@civilwarmed.org • www.CivilWarMed.org •
Pres.: *Robert E. Gearinger* •
Special Museum – 1990
Medical hist, Civil War artifacts 48758

Rose Hill Manor Park & Children's Museum, 118 North Market St, Frederick, MD 21701 • T: +1 301 6001646 • F: 6002595 • parksandrecreation@fredco-md.net • www.rosehillmuseum.com •
Historical Museum / Special Museum
Rose Hill Manor, the retirement home of Thomas Johnson, Maryland's first elected Governor, serves as a hands-on history museum for children of all ages 48759

Schifferstadt Architectural Museum, 1110 Rosemont Av, Frederick, MD 21701 • T: +1 301 6633885 • F: +1 301 6634807 • fredcolandmark@aol.com • www.fredericklandmarks.org •
Pres.: *Bea Smith* •
Fine Arts Museum – 1974
Architecture 48760

Fredericksburg TX

Admiral Nimitz National Museum of the Pacific War, 340 E Main St, Fredericksburg, TX 78624 • T: +1 830 9974379 ext 225 • F: +1 830 9978220 • jeff.hunt@tpwd.state.tx.us • www.nimitz-museum.org •
Dir.: *Joe Cavanaugh* •
Military Museum – 1967
Pacific WW II related artifacts incl uniforms, personal equipment, weapons, flags & souvenirs 48761

Pioneer Museum and Vereins Kirche, Gillespie County Historical Society, 309 W San Antonio St, Fredericksburg, TX 78624 • T: +1 830 9972835 • F: +1 830 9973891 • sbuford@pioneermuseum.com • www.pioneermuseum.com •
Dir.: *Paul Camfield* •
Local Museum – 1936
Historic Houses, early settlement of Fredericksburg and Gillespie county 48762

Fredericksburg VA

Belmont, Gari Melchers Estate and Memorial Gallery, 224 Washington St, Fredericksburg, VA 22405 • T: +1 540 6541015 • F: +1 540 6541785 • belmont@mwc.edu • www.mwc.edu/belm •
Dir.: *David S. Berreth* • Cur.: *Joanna D. Catron* •
Public Gallery – 1975
1871-1932 paintings & sketches by Gari Melchers, European antiquites, china 48763

Fredericksburg and Spotsylvania National Military Park, 120 Chatham Ln, Fredericksburg, VA 22405 • T: +1 540 3710802 • F: +1 540 3711907 •
Sc. Staff: *Robert K. Krick* • *Donald C. Pfanz* (History) •
Military Museum – 1927
Military artifacts, manuscripts, pre-Civil War Innis House 48764

Fredericksburg Area Museum and Cultural Center, 907-911 Princess Anne St, Fredericksburg, VA 22401 • T: +1 540 3715668, 3713037 • F: +1 540 3736569 • famoffice@earthlink.net • www.famcc.org •
Dir.: *Edwin W. Watson* • Cur.: *Mary H. Dellinger* •
Local Museum – 1985
Native American artifacts, Civil War items, decorative arts, glass, ceramics 48765

Historic Kenmore, 1201 Washington Av, Fredericksburg, VA 22401 • T: +1 540 3733381 • F: +1 540 3716066 • raudenbush@gwffoundation.org • www.kenmore.org •
Pres.: *W. Vernon Edenfield* •
Local Museum – 1922
Local hist 48766

Hugh Mercer Apothecary Shop, 1020 Caroline St, Fredericksburg, VA 22401 • T: +1 540 3733362 • F: +1 540 3731569 • www.apva.org •
Dir.: *Gail G. Braxton* •
Special Museum
Hist apothecary shop from 1761, pharmaceutical and medical implements 48767

James Monroe Museum and Memorial Library, 908 Charles St, Fredericksburg, VA 22401 • T: +1 540 6541043 • F: +1 540 6541106 • jmmuseum@mwc.edu •
Dir.: *John N. Pearce* • Ass. Dir./Cur.: *David B. Voelkel* •
Historical Museum / Library with Exhibitions / Historic Site – 1928
Presidential library, local hist, James Monroe (1785 - 1831, 5th president of the USA) 48768

Mary Washington College Galleries, College Av at Seacobeck St, Fredericksburg, VA 22401-5358 • T: +1 540 6541013 • F: +1 540 6541171 • gallery@mwc.edu •
Dir.: *Dr. Thomas P. Somma* •
Public Gallery – 1956
European, American and Asian art 48769

Mary Washington House, 1200 Charles St, Fredericksburg, VA 22401 • T: +1 540 3731569 • F: +1 540 3731569 • www.apva.org •
Dir.: *Gail G. Braxton* •
Local Museum – 1772
18th c furnishings 48770

Ridderhof Martin Gallery, Mary Washington College, College Av at Seacobeck St, Fredericksburg, VA 22401-5358 • T: +1 540 6541013 • F: +1 540 6541171 • gallery@mwc.edu • www.mwc.edu •
Dir.: *Dr. Thomas P. Somma* •
Fine Arts Museum / University Museum – 1956 48771

Rising Sun Tavern, 1304 Caroline St, Fredericksburg, VA 22401 • T: +1 540 3711494 • F: +1 540 3711494 • www.apva.org •
Dir.: *Gail G. Braxton* •
Decorative Arts Museum – 1760
Tavern owned by Charles Washington 48772

Saint James' House, 1300 Charles St, Fredericksburg, VA 22401 • T: +1 703 3731569 • www.apva.org •
Dir.: *Gail G. Braxton* •
Local Museum – 1760
Private coll of the donores William H. Tollerton and Daniel J. Breslin 48773

Fredonia KS

Stone House Gallery, 320 N 7th St, Fredonia, KS 66736 • T: +1 620 3782052 • stonehouse@twinmounds.com •
Pres.: *Joyce Fulghum* •
Public Gallery – 1967
Visual art coll 48774

Wilson County Historical Society Museum, 420 N 7th, Fredonia, KS 66736 • T: +1 620 3783965 •
Local Museum – 1961
Local history, located in old county jail 48775

Fredonia NY

D.R. Barker Historical Museum, 20 E Main St, Fredonia, NY 14063 • T: +1 716 6722114 • F: +1 716 6793547 • BarkerMu@netsync.net • www.netsync.net/users/barkermu •
Pres.: *Peter Clark* • Cur.: *Nancy Brown* •
Library with Exhibitions / Historical Museum – 1884
Books, newspapers, documents and letters, records of the Fredonia Academy 1821-1867, portraits, photographs, furniture, costumes, uniforms 48776

Gallery of the State University College at Fredonia, State University College at Fredonia, Central Av, Fredonia, NY 14063 • T: +1 716 6733537 • F: +1 716 6734990 • visualarts.newmedia@fredonia.edu • www.fredonia.edu •
Public Gallery
Drawings, paintings, photographs, Oriental and Far Eastern art 48777

Michael C. Rockefeller Arts Center Gallery, State University College, Fredonia, NY 14063 • T: +1 716 6734897 • F: +1 716 6733397 • gaaschc@fredonia.edu • www.fredonia.edu •
Dir./Cur.: *Cynnie Gaasch* •
Fine Arts Museum / University Museum – 1826
Contemporary sculpture, prints and drawings 48778

Freehold NJ

Monmouth County Historical Museum, 70 Court St, Freehold, NJ 07728 • T: +1 908 4621466 • F: +1 908 4628346 • mcha@monmouth.com • www.monmouthhistory.org •
Dir.: *Lee Ellen Griffith* •
Local Museum – 1898
History of the county, decorative arts, folk art, paintings, silver, furniture, glass, ceramics, toys 48779

Freeman SD

Heritage Hall Museum and Archives, 748 S Main St, Freeman, SD 57029 • T: +1 605 9254237 • F: +1 605 9254271 • archives@freemanmuseum.org • www.freemanmuseum.org •
Cur.: *Cleon Graber* •
Local Museum / Ethnology Museum – 1976
Local history, ethnology – archives 48780

Freeport IL

Freeport Arts Center, 121 N Harlem Av, Freeport, IL 61032 • T: +1 815 2359755 • F: +1 815 2356015 • artscenter@aeroinc.net •
Pres.: *Jim Lee* •

U

Fine Arts Museum – 1975
19th c paintings, sculptures, prints and Florentine mosaics from the W.T. Rawleight coll, jewellry, glass, Kacina dolls and pre-Columbian artifacts 48781

Silvercreek Museum, 2954 Walnut Rd, Freeport, IL 61032 • T: +1 815 2322350 • janicekay.klever@gte.net • www.thefreeportshow.com •
Pres.: *Larry Buttel* • Dir.: *Rose Reeter* •
Local Museum / Agriculture Museum – 1988
Regional history, in 1906 county poor farm 48782

Stephenson County Historical Society Museum, 1440 S Carroll Av, Freeport, IL 61032 • T: +1 815 2328419 • F: +1 815 2970313 • director@stephcohs.org • www.stephcohs.org •
Historical Museum – 1944
Local and regional history, Jane Addams, Abraham Licoln, underground railroad, school, agriculture 48783

Freeport ME

Freeport Historical Society Museum, 45 Main St, Freeport, ME 04032 • T: +1 207 8653170 • F: +1 207 8659055 • info@freeporthistoricalsociety.org • www.freeporthistoricalsociety.org •
C.E.O.: *Randall Wade Thomas* •
Library with Exhibitions – 1969
Manuscripts, photographs, artifacts, furnishings 48784

Fremont CA

Olive Hyde Art Gallery, 123 Washington Blvd, Fremont, CA 94539 • T: +1 510 7914357, 4944228 • F: +1 510 4944753 • programs@olivehydeartguild.org • www.olivehydeartguild.org •
Cur.: *Sandra Hemsworth* •
Public Gallery – 1964 48785

Fremont NC

Charles B. Aycock Birthplace, 264 Governor Aycock Rd, Fremont, NC 27830 • T: +1 919 2425581 • F: +1 919 2426668 • aycock@ncmail.net •
Head: *Leigh V. Strickland* •
Local Museum – 1959
19th c farm life and education 48786

Fremont NE

Louis E. May Museum, 1643 N Nye, Fremont, NE 68025 • T: +1 402 7214515 • F: +1 402 7218354 • maymuseum@juno.com • www.connectfremont.org •
Cur.: *Patty Manhart* •
Historical Museum – 1969
Furniture, photographs, farm implements, dolls, general store 48787

Fremont OH

Rutherford B. Hayes Presidential Center, Spiegel Grove, Fremont, OH 43420-2796 • T: +1 419 3322081 • F: +1 419 3324952 • admin@rbhayes.org • www.rbhayes.org •
Exec. Dir.: *Thomas J. Culbertoon* • Cur.: *Nan Card* (Manuscript) • *Gilbert Gonzalez* (Photography) • Libr.: *Becky Hill* •
Historical Museum / Library with Exhibitions – 1916
1859 Hayes residence, books, manuscripts, photographic prints, family memorabilia, President's carriage, White House china – research library 48788

Fresno CA

Discovery Center, 1937 N Winery Av, Fresno, CA 93703 • T: +1 209 2515533 • F: +1 209 2515531 • office@thediscoverycenter.net • www.thediscoverycenter.net •
Pres.: *Gary Pigg* • Exec. Dir.: *Janet Berry* •
Historical Museum / Natural History Museum / Science&Tech Museum – 1956
Science exhibits, local life – cactus garden 48789

Fresno Art Museum, 2233 N First St, Fresno, CA 93703-9955 • T: +1 559 4414221 • F: +1 559 4414227 • info@fresnoartmuseum.org • www.fresnoartmuseum.org •
Dir.: *Eva Torres* • Cur.: *Jacquelin Pilar* • Asst. Cur: *Rebecca Shepard* •
Fine Arts Museum – 1949
Pre-Columbian, Mexican, Peruvian and Californian works of art – library, sculpture garden 48790

Fresno Metropolitan Museum, 1555 Van Ness Av, Fresno, CA 93721 • T: +1 559 4411444 • F: +1 559 4418607 • DanaT@fresnomet.org • www.fresnomet.org •
Dir.: *Dana Thorpe* • Cur.: *Kristina Hornback* •
Fine Arts Museum / Decorative Arts Museum – 1984 48791

Kearney Mansion Museum, 7160 W Kearney Blvd, Fresno, CA 93706 • T: +1 559 4410862 • F: +1 559 4411372 • frhistsoc@aol.com • www.valleyhistory.org •
Exec. Dir.: *Jill Moffat* •
Local Museum – 1919
Native American items, furnishings, costume,s agriculture 48792

Meux Home Museum, 1007 R St, Fresno, CA 93710, mail addr: POB 70, Fresno, CA 93707 • T: +1 559 2338007 • F: +1 559 2332331 • judi@meux.mus.ca.us • www.meux.mus.ca.us •
Pres.: *Sharon Dunn* •
Historical Museum – 1979
Victorian lifestyle, costumes, textiles, furniture, family artifacts, Dr. Meux'Civil War uniform 48793

Friday Harbor WA

San Juan Historical Society Museum, 405 Price St, Friday Harbor, WA 98250 • T: +1 360 3783949 • F: +1 360 3783949 • curator@sjmuseum.org • www.sjmuseum.org •
Dir.: *Kris Day Vincent* •
Local Museum – 1961
Island hist, town jail, log cabin, farmhouse, milk house 48794

San Juan Island National Historical Park, 125 Spring St, Friday Harbor, WA 98250 • T: +1 360 3782240 • F: +1 360 3782615 • sajh@nps.gov •
Military Museum – 1966
Pig War 1859-72, US Army and British Royal Marines 48795

Whale Museum, 62 First St N, Friday Harbor, WA 98250 • T: +1 360 3784710 • F: +1 360 3785790 • oap@whalemuseum.org • www.whalemuseum.com •
Dir. & Science Cur.: *Richard Osborne* •
Natural History Museum – 1978
Skeletoms, biological specimens 48796

Frisco CO

Frisco Historical Society Museum, 120 Main St, Frisco, CO 80443 • T: +1 970 6683428 • F: +1 970 6683428 • friscohistorical@aol.com •
Dir.: *Rita Bartram* •
Local Museum – 1983
Local hist 48797

Frisco NC

Frisco Native American Museum and Natural History Center, 53536 Hwy 12, Frisco, NC 27936 • T: +1 252 9954440 • F: +1 252 9954030 • bfriend1@mindspring.com • www.nativeamericanmuseum.org •
Exec. Dir.: *Carl Bornfriend* •
Ethnology Museum – 1986
Native American life, hist and culture, esp of Algonquian tribe 48798

Fritch TX

Lake Meredith Aquatic and Wildlife Museum, 101 N Robey St, Fritch, TX 79036 • T: +1 806 8572458 • F: +1 806 8573229 • shelda@amaonline.com •
Dir.: *Shelda Pingree* •
Natural History Museum – 1976
Six dioramas displaying vegetation and wildlife be LaNelle Poling 48799

Fromberg MT

Little Cowboy Bar and Museum, 105 W River, Fromberg, MT 59029 • T: +1 406 6689502 • diehard@imt.net • www.littlecowboy.com •
Dir.: *Shirley Smith* •
Local Museum – 1990
Local hist and culture 48800

Front Royal VA

Warren Rifles Confederate Museum, 95 Chester St, Front Royal, VA 22630 • T: +1 540 6366982, 6352219 • F: +1 540 6352219 • silwood@rmaonline.net •
Dir.: *Suzanne Silek* •
Military Museum – 1959
Relics of the War between the States incl guns, uniforms 48801

Frostburg MD

Stephanie Ann Roger Gallery, c/o Frostburg State University, 1011 Braddock Rd, Frostburg, MD 21532 • T: +1 301 6874797 • F: +1 301 6873099 • ddavis@frostburg.edu •
Chm.: *Dustin P. Davis* •
University Museum / Fine Arts Museum / Decorative Arts Museum – 1972
Folk, art, prints 48802

Thrasher Carriage Museum, 19 Depot St, Frostburg, MD 21532 • T: +1 301 6893380 • F: +1 301 6893882 • info@thrashercarriagemuseum.com • www.thrashercarriagemuseum.com •
Science&Tech Museum – 1992
Horse-drawn conveyances 48803

Fryeburg ME

The Fryeburg Fair Farm Museum, Rte 5 N, Fryeburg, ME 04037 • T: +1 207 9353268 • F: +1 207 9353662 • info@fryeburgfair.com •
Cur.: *Edward Jones* •
Agriculture Museum – 1970
Agriculture 48804

Fullerton CA

Fullerton Museum Center, 301 N Pomona Av, Fullerton, CA 92832 • T: +1 714 7386545 • F: +1 714 7383124 • DannielleM@ci.fullerton.ca.us • www.ci.fullerton.ca.us/depts/museum •
Dir.: *Dannielle Mauk* •
Local Museum – 1971
Leo Fender coll of musical instr, history, science, and arts exhibits 48805

Main Art Gallery, California State University, 800 N State College Blvd, Fullerton, CA 92834-6850 • T: +1 657 2787750 • mmcgee@fullerton.edu • www.fullerton.edu/arts/art/gallery_info.asp •
Dir.: *Prof. Mike McGee* •
Public Gallery / University Museum – 1967
Contemporary prints, sculpture – library 48806

Muckenthaler Gallery and Mansion Museum, 1201 W Malvern Av, Fullerton, CA 92833 • T: +1 714 7386366 • F: +1 714 7386366 • info@TheMuck.org • www.themuck.org •
Dir.: *Patricia House* •
Public Gallery – 1966 48807

Museum of Anthropology, California State University, Dept. of Anthropology, McCarthy Hall, Fullerton, CA 92834-6846 • T: +1 714 2782844, 2783564 • F: +1 714 2787046 • anthropology@exchange.fullerton.edu • anthro.fullerton.edu/museum.htm •
Dir.: *Susan Parman* •
Ethnology Museum – 1970
Southern California anthropology, prehistory, ethnography from South Pacific, Near East, Mexico ans South America – library, lab 48808

Fulton MO

William Woods University Art Gallery, 1 University Av, Fulton, MO 65251 • T: +1 573 5924245 • F: +1 573 5921623 • tmartin@williamswoods.edu • www.williamswoods.edu •
Dir.: *Terry Martin* •
Fine Arts Museum 48809

Winston Churchill Memorial and Library in the United States, Westminster College, 501 Westminster Av, Fulton, MO 65251 • T: +1 573 5925369 • F: +1 573 5925222 • henslej@jaynet.wcmo.edu • www.churchillmemorial.org •
Dir.: *Dr. Jerry Morelock* •
University Museum / Historical Museum – 1962
Churchill memorabilia, Wren architecture, rare map coll, paintings, graphics 48810

Gadsden AL

Gadsden Museum of Fine Arts, 515 Broad St, Gadsden, AL 35901 • T: +1 256 5467365 • F: +1 256 5467365 • stemple@cityofgadsden.com • www.gadsdenmuseum.com •
Dir./Cur.: *Steve Temple* • Cur.: *Jim Loftin* •
Fine Arts Museum – 1965
Oils, acrylics and watercolours, lithographs, antiques, decorative arts, porcelain – library 48811

Gaffney SC

Winnie Davis Museum of History, Limestone College, 1115 College Dr, Gaffney, SC 29340 • T: +1 864 4888399 • F: +1 864 4877151 • chayward@saint.limestone.edu • www.limestone.edu •
Historical Museum / University Museum – 1976
History 48812

Gainesville FL

Florida Museum of Natural History, SW 34th St and Hull Rd, Gainesville, FL 32611-7800, mail addr: POB 117800, Gainesville, FL 32611-7800 • T: +1 352 3921721, 8462000 • F: +1 352 3928783, 8460253 • cosmith@flmnh.ufl.edu • www.flmnh.ufl.edu •
Dir.: *Dr. Douglas S. Jones* •
University Museum / Natural History Museum – 1917
Paleontology, archaeology, ornitology, botany, mammalogy, herpetlogy, art 48813

Matheson Museum, 513 E University Av, Gainesville, FL 32601 • T: +1 352 3782280 • F: +1 352 3781246 • info@mathesonmuseum.org • www.mathesonmuseum.org •
Dir.: *Lisa B. Auel* •
Local Museum – 1994
Local culture and hist of Gainesville, Alachua County, and North Central Florida region 48814

Morningside Nature Center, 3540 E University Av, Gainesville, FL 32641 • T: +1 352 3342170 • F: +1 352 3342248 • www.natureoperations.org •
C.E.O: *Gary A. Paul* •
Agriculture Museum / Open Air Museum / Natural History Museum – 1972
Natural hist – Library 48815

Samuel P. Harn Museum of Art, University of Florida, 34 Th St and Hull Rds, Gainesville, FL 32611-2700 • T: +1 352 3929826 • F: +1 352 3923892 • coral@harn.ufl.edu • www.harn.ufl.edu •
Dir.: *Rebecca Nagy* •
Fine Arts Museum / University Museum – 1981
American, pre-Columbian, African and Asien art 48816

Santa Fe Gallery, Santa Fe Community College, 3000 NW 83rd St, Gainesville, FL 32606 • T: +1 352 3955000 • F: +1 352 3954432 • leslie.lambert@sfcc.edu • www.sfcc.edu/~vpa •
Fine Arts Museum – 1978 48817

Thomas Center Galleries, 302 NE 6th Av, Gainesville, FL 32601 • T: +1 352 3345064 • F: +1 352 3342314 • cultural@ci.gainesville.fl.us •
Dir.: *Erin H. Friedberg* •
Public Gallery 48818

University Gallery, University of Florida, Gainesville, FL 32611, mail addr: POB 115803, Gainesville, FL 32611-5803 • T: +1 352 3920201 • F: +1 352 8460266 • amyv@ufl.edu • www.arts.ufl.edu/galleries •
Dir.: *Amy Vigilante* •
Public Gallery / University Museum – 1965
Master of fine arts student coll, contemporary art 48819

Gainesville GA

Brenau University Galleries, 1 Centennial Circle, Gainesville, GA 30501 • T: +1 770 5346299 • F: +1 770 5384599 • gallery@lib.brenau.edu • www.brenau.edu •
Dir./Cur.: *Jean Westmacott* •
Fine Arts Museum / University Museum – 1983
Paintings, drawings, prints, sculpture, photos, craft works 48820

Georgia Mountains History Museum, c/o Brenau University, 311 Green St SE, Gainesville, GA 30501 • T: +1 770 5360889 • F: +1 770 5349488 • gamthist@bellsouth.net • www.gamtshistorymuseum.org •
Historical Museum / University Museum – 1981
Textile, poultry, medical, African-American, Indian, firefighting, arts and crafts, Ed Dodd, Mark Trail 48821

Simmons Visual Center, Brenau University, 1 CCentennial Circle, Gainesville, GA 30501 • T: +1 770 3546263 • F: +1 770 5384599 •
Dir.: *Jean Westmacott* •
University Museum / Fine Arts Museum
Contemporary exhibitions 48822

Gainesville TX

Morton Museum of Cooke County, 210 S Dixon St, Gainesville, TX 76240 • T: +1 940 6688900 • F: +1 940 6680533 • mortonmuseum@nortexinfo.net • www.mortonmuseum.org •
Cur.: *Shana Powell* •
Local Museum – 1968
Preservation project, Cooke County, Texas artifacts and manusripts – library 48823

Gaithersburg MD

The London Brass Rubbing Centre in Washington D.C., 11808 Silent Valley Ln, Gaithersburg, MD 20878 • T: +1 301 2797046 • me2you@starpower.net •
C.E.O.: *Karen L. Hammond* •
Decorative Arts Museum – 1977
Arts, crafts 48824

Galax VA

Jeff Matthews Memorial Museum, 606 W Stuart Dr, Galax, VA 24333 • T: +1 540 2367874 •
Cur.: *Rita Edwards* •
Local Museum – 1974
10,000 Indian artifacts, knife coll 48825

Galena IL

The Chicago Athenaeum - Galena, 601 S Prospect St, Galena, IL 61036 • T: +1 815 7774444 • F: +1 815 7772471 • info@chi-athenaeum.org • www.chi-athenaeum.org •
Fine Arts Museum – 1988
Architecture and design 48826

Galena-Jo Daviess County Historical Museum, 211 S Bench St, Galena, IL 61036 • T: +1 815 7779129 • F: +1 815 7779131 • ghmuseum@galenalink.net • www.galenahistorymuseum.org •
Exec. Dir.: *Daryl Watson* •
Local Museum – 1938
Local history, civil war, lead mining, clothing, geology, dolls and toys 48827

Old Market House, Market Sq, Galena, IL 61036 • T: +1 815 7773310 • F: +1 815 7773310 • granthome@granthome.com • www.granthome.com •
Historical Museum – 1947
Greek Revival-style Market House 48828

Sculpture Galena, The Chicago Athenaeum, c/o Ulysses S. Grant Home, 500 Bouthillier St, Galena, IL 61036 • T: +1 815 7774444 • F: +1 815 7772471 • info@chi-athenaeum.org • www.chi-athenaeum.org •
Fine Arts Museum
Contemporary sculpture 48829

United States Grant's Home, 500 Bouthillier St, Galena, IL 61036 • T: +1 815 7773310 • F: +1 815 7773310 • granthome@granthome.com • www.granthome.com •
Man.: *Terry J. Miller* •
Historical Museum – 1932
Italianate bracketed style house presented to Gen. Grant in 1865 48830

Galena KS

Galena Mining and Historical Museum, 319 W 7th St, Galena, KS 66739 • T: +1 316 7832192 •
Science&Tech Museum / Historical Museum – 1984
Lead and zinc mine artifacts, tools, lamps, pictures, paintings, mineral, locomotive, caboose, helicopter, tank 48831

U

Galesburg IL

Carl Sandburg State Historic Site, 331 E 3rd St, Galesburg, IL 61401 • T: +1 309 3422361 • F: +1 309 3422141 • carl@sandburg.org • www.sandburg.org •
Man.: *Steve Holden* •
Historical Museum – 1945
Immigrant railroad worker's cottage ... 48832

Galesburg Civic Art Center, 114 E Main St, Galesburg, IL 61401 • T: +1 309 3427415 • F: +1 309 3432650 • info@gallesburgarts.org • www.galesburgarts.org •
C.E.O.: *Heather L. Norman* •
Public Gallery – 1923
Piantings ... 48833

Illinois Citizen Soldier Museum, 1001 Michigan Av, Galesburg, IL 61401 • T: +1 309 3421181 •
Cur.: *Kenneth R. Johnson* •
Military Museum – 1988
Artifacts and military memorabile of Illinois ... 48834

Galesville MD

Carrie Weedon Natural Science Museum, 911 Galesville Rd, Galesville, MD 20765 • T: +1 410 2221625 • F: +1 410 8670588 •
Dir.: *Dotty Cheney* •
Natural History Museum / Science&Tech Museum – 1988
Natural history and science ... 48835

Galeton PA

Pennsylvania Lumber Museum, 5600 U.S Rte 6, Galeton, PA 16922 • T: +1 814 4352652 • F: +1 814 4356361 • www.lumbermuseum.org •
Pres.: *Robert Currin* •
Historical Museum / Natural History Museum – 1970 ... 48836

Galion OH

Brownella Cottage, 132 S Union St, Galion, OH 44833 • T: +1 419 4689338 •
Pres.: *Jerry A. Lantz* •
Local Museum – 1981
Local hist items, furniture, personal belongings ... 48837

Galion Historical Museum, 132 S Union St, Galion, OH 44833 • T: +1 419 4689338 •
Dir.: *Dr. Bernard M. Mansfield* •
Local Museum – 1956
Area hist, Indian artifacts, pioneer items, local industrial items ... 48838

Gallipolis OH

French Art Colony, 530 First Av, Gallipolis, OH 45631 • T: +1 740 4463834 • F: +1 740 4463834 • facart@zoomnet.net •
Dir.: *Mary Bea McCalla* • Cur.: *Janice M. Thaler* •
Fine Arts Museum – 1971
Fine art and antiques ... 48839

Our House State Memorial, 432 First Av, Gallipolis, OH 45631 • T: +1 740 4460586 •
Cur.: *Janice Layne* •
Historical Museum – 1933
Period furnishings ... 48840

Gallitzin PA

Allegheny Portage Railroad, 110 Federal Park Rd, Gallitzin, PA 16641 • T: +1 814 8866116 • F: +1 814 8840206 • nancy_smith@nps.gov • www.nps.gov/alpo •
Cur.: *Lawren Dunn* •
Science&Tech Museum – 1964
Pennsylvania Mainline Canal, transportation, social and economic hist ... 48841

Galveston TX

Ashton Villa, 2328 Broadway, Galveston, TX 77550-2014 • T: +1 409 7623933 • F: +1 409 7621904 • foundation@galvestonhistory.org • www.galvestonhistory.org •
C.E.O.: *Marsh Davis* •
Local Museum – 1974
Historic house, home of the James Moreau Braun ... 48842

The Bishop's Palace, 1402 Broadway, Galveston, TX 77550 • T: +1 409 7622475 • F: +1 409 7621810 •
Cur.: *Joseph Fiorenza* •
Local Museum – 1886
Historic building, Victorian furniture ... 48843

Galveston Arts Center, 2127 Strand, Galveston, TX 77550 • T: +1 409 7632403 • F: +1 409 7630531 • galartsctr@aol.com •
Exec. Dir.: *Clint Willour* •
Fine Arts Museum ... 48844

Galveston County Historical Museum, 2219 Market St, Galveston, TX 77550 • T: +1 409 7662340 • F: +1 409 7952157 • christy.carl@galvestonhistory.org • www.galvestonhistory.org •
Dir.: *Christine S. Carl* • Cur.: *Robin Munson* •
Historical Museum – 1972
Costumes, photos, clothings, artifacts from Galveston county hist, lighthouse lens ... 48845

Harris Art Gallery, James M. Lykes Maritime Gallery and Hutchings Gallery, c/o Rosenberg Library, 2310 Seally Av, Galveston, TX 77550 • T: +1 409 7638854 • F: +1 409 7630275 • child@rosenberg-library.org • www.rosenberg-library.org •
Exec. Dir.: *Nancy Milnor* • Cur.: *Elisabeth Darst* •
Fine Arts Museum / Local Museum
19th-20th c painting, sculpture and graphic, Russian icons ... 48846

John Sydnor's 1847 Powhatan House, 3427 Avenue O, Galveston, TX 77550 • T: +1 409 7630077 • F: +1 409 7441456 • evangelinewhorton@yahoo.com •
C.E.O.: *Patrick McGuffey* •
Local Museum – 1938
Furnishings of antebellum & Victorian perod ... 48847

Lone Star Flight Museum/Texas Aviation Hall of Fame, 2002 Terminal Dr, Galveston, TX 77554 • T: +1 409 7407722 • F: +1 409 7407612 • flight@lsfm.org • www.lsfm.org •
Dir.: *Ralph Royce* • Cur.: *Darla Harmon* •
Military Museum / Science&Tech Museum – 1986
Aeronautics ... 48848

The Moody Mansion Museum, 2618 Broadway, Galveston, TX 77550 • T: +1 409 7627668, 7659770 • F: +1 409 7627055 • b.massey@northenendowment.org • www.moodymansion.org •
Pres.: *Edward L. Protz* •
Historical Museum / Decorative Arts Museum – 1991
History, decorative arts, costume, textiles ... 48849

Ocean Star Offshore Drilling Rig & Museum, 20th St at Harborside Dr, 2002 Wharf C, Galveston, TX 77553-2040, mail addr: 200 Dairy Ashford Rd N, Houston, TX 77079 • T: +1 281 6798040 • F: +1 281 5442441 • osmuseum@aol.com • www.oceanstaroec.com •
Exec. Dir.: *Sandra Morton* •
Science&Tech Museum – 1997
Offshore energy, drilling equipment, geology, seismic ... 48850

Texas Seaport Museum, Pier 21, 8, Galveston, TX 77550 • T: +1 409 7631877 • F: +1 409 7633037 • elissa@galvestonhistory.org • www.tsm-elissa.org •
C.E.O.: *Marsh Davis* • Dir.: *Kurt Voss* •
Historical Museum – 1982
Maritime hist ... 48851

Gambier OH

Olin Art Gallery, c/o Kenyon College, Olin Library, Gambier, OH 43022 • T: +1 740 4275000 • F: +1 740 4275272 •
Fine Arts Museum ... 48852

Ganado AZ

Hubbell Trading Post, Hwy 264, Ganado, AZ 86505, mail addr: POB 150, Ganado, AZ 86505 • T: +1 928 7553475 • F: +1 928 7553405 • www.nps.gov/hutr/ •
Cur.: *Edward M. Chamberlin* •
Historical Museum / Historic Site – 1967
Native American arts and crafts, ethnology, agriculture ... 48853

Garden City KS

Finney County Kansas Historical Museum, Finnup Park, 403 S 4th, Garden City, KS 67846-0796 • T: +1 316 2723664 • F: +1 316 2723664 • fico.historical@gcnet.com •
Dir.: *Mary Regan* • Cur.: *Laurie Oshel* •
Historical Museum – 1949
Pioneer hist, sugar beet & cattle industry of county – library ... 48854

Garden City NY

Cradle of Aviation Museum, 1 Davis Av, Garden City, NY 11530 • T: +1 516 5724111 • F: +1 516 5724079 • www.cradleofaviation.org •
Cur.: *Joshua Stoff* •
Science&Tech Museum – 1979
Aircrafts and spacecraft, aerospace heritage of Long Island ... 48855

Firehouse Art Gallery, Nassau Community College, 1 Education Dr, Garden City, NY 11530 • T: +1 516 5727165 • F: +1 516 5727302 •
Dir.: *Lynn R. Casey* • Cur.: *Meg Oliveri* •
University Museum / Fine Arts Museum – 1965
16th c to present fine art ... 48856

Long Island Children's Museum, 11 Dvis Av, Garden City, NY 11530 • T: +1 516 2245800 • F: +1 516 3028188 • licm1@aol.com • www.licm.org •
Pres.: *Robert Lemle* • Dir.: *Bonnie Dixon* •
Special Museum – 1990 ... 48857

Long Island Museum of Science and Technology, Museum Row, Mitchel Field, Garden City, NY 11530, mail addr: 1 Davis Av, Garden City, NY 11530 • T: +1 516 6810345 • bparris@limsat.org • www.limsat.org •
C.E.O.: *Benjamin J. Parris* •
Science&Tech Museum – 1982
Physics, computer information science, telecommunications, biochemistry ... 48858

Gardiner NY

Locust Lawn and Terwilliger House, 400 Rte 32 S, Gardiner, NY 12525 • T: +1 845 2551660 • F: +1 845 2550376 • hhsoffice@hhs-newpaltz.org • www.hhs-newpaltz.org •
Dir.: *John Braunlein* • Cur.: *Leslie LeFevre Stratton* •
Decorative Arts Museum / Historical Museum – 1894
Queen Anne-early Victorian furnishings, China, fabrics, laces, textiles, toys, paintings, coaches, tools ... 48859

Mohonk Preserve, 3197 Rte 44/55, Gardiner, NY 12525-0715, mail addr: POB 715, New Paltz, NY 12561 • T: +1 845 2550919 • F: +1 845 2555646 • info@mohonkpreserve.org • mohonkpreserve.org •
Exec.Dir.: *Glenn D. Hoagland* •
Historical Museum / Natural History Museum – 1963
Natural hist, Northern Shawanqunk Ridge hist ... 48860

Gardner MA

East Wing Art Gallery, Mount Wachusett Community College, 444 Green St, Gardner, MA 01440 • T: +1 978 6309168 • www.mwcc.mass.edu/tour/gallery.html •
Fine Arts Museum – 1971
Color art posters and reproductions, prints, ceramic, student coll ... 48861

Gardner Museum, 28 Pearl St, Gardner, MA 01440 • T: +1 978 6323277 •
Pres.: *Muriel Caouette* •
Fine Arts Museum / Local Museum – 1978
Local history and development, silver tools and pieces ... 48862

Gardnerville NV

Carson Valley Museum, Douglas County Historical Society, 1477 Hwy 395 N, Gardnerville, NV 89410-5214 • T: +1 775 7822555 • F: +1 775 7838802 •
C.E.O.: *Mary Ellen Conaway* •
Historical Museum – 1961
Washoe Indian hist, Carson Valley hist 1850-present, early settlement, farming ... 48863

Garnavillo IA

Garnavillo Historical Museum, 205 N Washington, Garnavillo, IA 52049 • T: +1 563 9642607 • gmaomie@alpine.com.net •
Pres.: *Kurt Kuenzel* •
Local Museum – 1965
Local history, housed in a church ... 48864

Garnett KS

Anderson County Historical Museum, W 6th St, Garnett, KS 66032-0217 • T: +1 785 8672966, 4485740 • ancohiso@ecksor.net •
Pres.: *Dorothy L. Lickteig* •
Local Museum / Folklore Museum – 1968
Local hist, home and carriage house of Dr. Harris, Longfellow school building ... 48865

Walker Art Collection, c/o Garnett Public Library, 125 W 4th Av, Garnett, KS 66032 • T: +1 913 4483388 • F: +1 913 4483936 • garnettlibrary@yahoo.com • www.garnettks.net •
Chm.: *Terry J. Solander* •
Fine Arts Museum – 1965
19th - 20th c oil-paintings & works on paper, regional artists ... 48866

Garrett IN

Garrett Historical Museum, Heritage Park, 300 N Randolph St, Garrett, IN 46738 • T: +1 219 3575575 • jmohre@locl.net • www.garretthistoricalsociety.org •
Cur.: *Robert Parker* •
Local Museum / Science&Tech Museum – 1971
Local history, 3 railroad cars, diesel locomotive ... 48867

Garrison NY

Boscobel Restoration, 1601 Rte 9D, Garrison, NY 10524 • T: +1 845 2653638 • F: +1 845 2654405 • info@boscobel.org • www.boscobel.org •
Dir.: *Charles T. Lyle* •
Decorative Arts Museum / Historical Museum – 1955
NY neo-classical furniture, federal period decorative art and paintings, china, glass and silver purchased by Dyckman in London ... 48868

Garryowen MT

Custer Battlefield Museum, Town Hall, Garryowen, MT 59031-0200 • T: +1 406 6381876 • F: +1 406 6382019 • chris@custermuseum.org • www.custermuseum.org •
C.E.O.: *Christopher Kortlander* • Cur.: *James Thompson* •
Military Museum – 1994
Paintings, rare books, Battle of the Little Bighorn ... 48869

Gastonia NC

Schiele Museum of Natural History, 1500 E Garrison Blvd, Gastonia, NC 28054-5199 • T: +1 704 8666908 • F: +1 704 8666041 • dbrose@schielemuseum.org • www.schielemuseum.org •
Dir.: *Dr. V. Ann Tippitt* •
Natural History Museum / Ethnology Museum – 1960
Natural hist, ethnography, colonial hist, mammals, birds, bird eggs, reptiles and amphibians, fossils, minerals and gems ... 48870

Gate OK

Gateway to the Panhandle, Main St, Gate, OK 73844 • T: +1 405 9343133 • emaphet@@direcway.com •
Dir.: *L. Ernestine Maphet* • Dep. Dir.: *Karen Bond* •
Charlene Husted • *Louise Hein* •
Local Museum / Historical Museum – 1975
Farm implements and equip, household items, dishes, Civil War items ... 48871

Gates Mills OH

Gates Mills Historical Society Museum, 7580 Old Mill Rd, Gates Mills, OH 44040 • T: +1 440 4234808 • F: +1 440 4234808 •
C.E.O. & Pres.: *Harriet Leedy* •
Local Museum – 1946
Interior furnishings, kitchen utensils, Native Indian ... 48872

Gatlinburg TN

Arrowmont School of Arts and Crafts Collection, 556 Pkwy, Gatlinburg, TN 37738 • T: +1 865 4365860 • F: +1 865 4304101 • info@arrowment.org • www.arrowmont.org •
Dir.: *David Willard* • Cur.: *Billi R.S. Rothove* •
Fine Arts Museum / Decorative Arts Museum – 1945
Arts and crafts from 1950 to present ... 48873

Gaylord MI

Call of the Wild - Wildlife Museum, 850 S Wisconsin Av, Gaylord, MI 49735 • T: +1 989 7324336 • F: +1 989 7323749 • call@thenewsiewriter.com • www.gocallofthewild.com •
Dir.: *William C. Johnson* • *Judy Fleet* •
Natural History Museum – 1957
Various animals and habitats ... 48874

Geddes SD

Geddes Historic District Village, POB 97, Geddes, SD 57342 • T: +1 605 3372501 • F: +1 605 3373535 • dufsdfek@midstate.net • www.geddes.org •
Pres.: *Mike R. Dufek* •
Local Museum – 1969
Local hist ... 48875

Geneseo NY

Bertha V.B. Lederer Arts Gallery, SUNY at Geneseo, Fine Arts Bldg, 1 College Circle, Geneseo, NY 14454 • T: +1 716 2455814 • F: +1 716 2455815 •
C.E.O.: *Carl Shanahan* •
Fine Arts Museum / University Museum – 1967 ... 48876

Livingston County Historical Society Museum, 30 Center St, Geneseo, NY 14454 • T: +1 716 2432281 • www.livingstoncountyhistoricalsociety.org •
Pres.: *David W. Parish* •
Historical Museum – 1876 ... 48877

Geneva IL

Fabyan Villa Museum and Dutch Windmill, Rtes 31 and 25 at Fabyan Pkwy, Geneva, IL 60134, mail addr: 1925 S Batavia Av, Geneva, IL 60134 • T: +1 630 2324811 • F: +1 630 3776424 • fabyanvilla@aol.com •
Cur.: *Diana Malakar* •
Decorative Arts Museum / Science&Tech Museum – 1941
Oriental porcelains, Fabyan hist, Frank Lloyd Wright furniture ... 48878

Geneva IN

Limberlost State Historic Site, 200 East 6th St, Geneva, IN 46740 • T: +1 219 3687428 • F: +1 219 3687007 • limberlost368@embarqmail.com • www.genestrattonporter.net •
Pres.: *Fran Austin* • Cur.: *Randy Lehmann* •
Special Museum – 1947
Home of Gene Stratton-Porter, author and naturalist ... 48879

Geneva NY

Prouty-Chew Museum, Geneva Historical Society, 543 S Main St, Geneva, NY 14456 • T: +1 315 7895151 • F: +1 315 7890314 • info@genevahistoricalsociety.com • www.genevahistoricalsociety.com •
C.E.O.: *Kenneth Shefsiek* • Cur.: *John C. Marks* (Collections) • Sc. Staff: *Anne F. Dealy* (Education) •
Historical Museum – 1883
Federal, Empire and Vistorian furnishings and accessoires, photograph coll, hist exhibits – archives ... 48880

Rose Hill Mansion, Geneva Historical Society, Rte 96A, Geneva, NY 14456 • T: +1 315 7895151 • F: +1 315 7890314 • info@genevehistoricalsociety.com • www.genevahistoricalsociety.com •
C.E.O.: *Charles C. W. Bauder* •
Historical Museum – 1968
Historic house with furniture and interior ... 48881

Geneva OH

Platt R. Spencer Memorial Archives and Special Collections Area, 860 Sherman St, Geneva, OH 44041-9101 • T: +1 440 4664521 ext 107 • F: +1 440 4660162 • acgs@ashtabulagen.org • www.ashtabula.lib.oh.us/geneva.htm •
Dir.: *William Tokarczyk* •
Special Museum – 1988
Local hist files, genealogy, family memorabilia, Edith M. Thomas poetry coll – Archive ... 48882

Shandy Hall, 6333 S Ridge W, Geneva, OH 44041 • T: +1 440 4663680 • www.wrhs.org •
Cur.: *Byron Robertson* •
Historical Museum – 1937
Harper family's original furnishings, clothes, books, toys 48883

Geneva-on-the-Lake OH

Ashtabula County Historical Society Museum, 5685 Lake Rd, Geneva-on-the-Lake, OH 44041 • T: +1 440 4667337 • www.ashtcohs.com •
Pres.: *David Lepard* •
Local Museum – 1838
Period furnishings 1850-1900, office furnishings, manuscript collection 48884

Genoa NV

Genoa Courthouse Museum, 2304 Main St, Genoa, NV 89411, mail addr: Douglas County Historical Society, 1477 Hwy 395 N, Gardnerville, NV 89410-5214 • T: +1 775 7824355 •
C.E.O.: *Mary Ellen Conaway* • Cur.: *Cindy Southerland* •
Historical Museum – 1971
Original courthouse and jail, artifacts 1850-1900, geneaology 48885

Mormon Station Museum, 2295 Main St, Genoa, NV 89411 • T: +1 775 7822590, 6874379 • F: +1 775 6878972 • mormonstation@parks.nv.gov • parks.nv.gov •
Historical Museum – 1947
Site of Nevada's first permanent nonnative settlement, replica of 1851 Mormon station 48886

Gentryville IN

Colonel William Jones House, 3/4 m W of U.S. Hwy 231, on Boone St, Gentryville, IN 47537 • T: +1 812 9372802 • F: +1 812 9377038 • coljones@psci.net •
Cur.: *Peggy Brooks* •
Historical Museum – 1976 48887

Georgetown CO

Historic Georgetown, 305 Argentine, Georgetown, CO 80444 • T: +1 303 5692840 • F: +1 303 5692111 • preservation@historicgeorgetown.org • www.historicgeorgetown.org •
Dir.: *Dana Abrahamson* • Cur.: *Deirdre Baldwin* •
Open Air Museum / Local Museum – 1970
William A. Hamill house and office bldg, stable and carriage house, Bowman-White house; Tucker-Rutherford house; Miner's cottage, log cabin 48888

Hotel de Paris Museum, 409 Sixth Av, Georgetown, CO 80444 • T: +1 303 5692311 • F: +1 303 7568768 • mrc6118@aol.com • www.hoteldeparismuseum.org •
Historical Museum – 1875
Furniture, dishes and kitchen utensils – library 48889

Georgetown DE

Treasures of the Sea Exhibit, Delaware Technical and Community College, Rte 18/404, Seashore Hwy, Georgetown, DE 19947-0610, mail addr: POB 610, Georgetown, DE 19947-0610 • T: +1 302 8565700 • F: +1 302 8585462 • treasures@dtcc.edu • www.treasuresofthesea.org •
Dir.: *Robert Hearn* •
Historical Museum – 1988
Artifacts from the 1622 shipwreck of the Spanish galleon Nuestra Senora de Atocha 48890

Georgetown KY

Georgetown College Gallery, Mullberry and College Sts, Georgetown, KY 40324 • T: +1 502 8638106, 8638399 • F: +1 502 8688888 • Christine_Huskisson@georgetowncollege.edu •
Dir.: *Christine Huskisson* •
Fine Arts Museum – 1959
Contemporary graphics, painting and sculpture, crafts 48891

Ward Hall, 1782 Frankfort Pike, Georgetown, KY 40324 • T: +1 859 2330525 •
Cur.: *Frances Susong Jenkins* •
Decorative Arts Museum – 1979
Period furnishings 48892

Georgetown SC

Hopsewee Plantation, 494 Hopsewee Rd, Georgetown, SC 29440 • T: +1 843 5467891 • mail@hopsewee.com • www.hopsewee.com •
Dir.: *Helen B. Maynard* •
Local Museum – 1970
Historic house, birthplace of Thomas Lynch Jr., signer of the Declaration of Independence 48893

Kaminski House Museum, 1003 Front St, Georgetown, SC 29440 • T: +1 843 5467706 • F: +1 843 5454062 • klawrimore@cityofgeorgetownsc.com • www.cityofgeorgetownsc.com •
C.E.O.: *Katrina P. Lawrimore* • Pres.: *Alice Williams* •
Local Museum – 1973
Historic house 48894

The Rice Museum, 633 Front St, Georgetown, SC 29440 • T: +1 843 5467423 • F: +1 843 5459093 • thericemuseum@sc.rr.com • www.ricemuseum.org •
Dir.: *James A. Fitch* • Man.: *Lu Hook* •
Historical Museum / Fine Arts Museum / Historic Site – 1968
History, contemporary art 48895

Gering NE

North Platte Valley Museum, 11 and J Sts, near Hwys 92 and 71, Gering, NE 69341 • T: +1 308 4365411 • F: +1 308 4362592 • npvm@actcom.net • www.npum.org •
Local Museum – 1969
Furnishings, photographs, vehicles, military, Indian artifacts 48896

Oregon Trail Museum, Scotts Bluff National Monument, Hwy 92 W, Gering, NE 69341 • T: +1 308 4364340, 4362975 • F: +1 308 4367611 • www.nps.gov/scbl •
Historical Museum – 1919
Geology, prehistory, archaeology, ethnological history, history of Western migration 48897

Scotts Bluff National Monument, 190276 Hwy 92, Gering, NE 69341-0027 • T: +1 308 4364340 • F: +1 308 4367611 • www.nps.gov/scbl •
Archaeology Museum – 1919
Oregon trail hist, prehistoric and historic objects, watercolors 48898

Germantown TN

P.T. Boat Museum, 1384 Cordova Rd, Germantown, TN 38138, mail addr: POB 38070, Germantown, TN 38138 • T: +1 901 7558440 • F: +1 901 7510522 • ptboats@pop.net • www.ptboats.org •
Dir.: *Alyce N. Guthrie* •
Historical Museum / Military Museum – 1946
Maritime, naval history, at Battleship Cove, WW II – archives 48899

Gettysburg PA

Adams County Historical Society Museum, 111 Seminary Ridge, Gettysburg, PA 17325 • T: +1 717 3344723 • F: +1 717 3344723 • info@achs-pa.org • www.achs-pa.org •
Dir.: *Arthur Weaner* •
Local Museum – 1934
Art, Indian artifacts, genealogy 48900

Eisenhower National Historic Site, 97 Taneytown Rd, Gettysburg, PA 17325-2804 • T: +1 717 3389114 • F: +1 717 3380821 • eise_site_manager@nps.gov • www.nps.gov/eise •
Dir.: *Dr. John A. Latschar* • Cur.: *Michael R. Florer* •
Historian: *Carol A. Hegeman* •
Historical Museum – 1967
Presidential and retirement home of Dwight D. Eisenhower 48901

Gettysburg National Military Park, 97 Taneytown Rd, Gettysburg, PA 17325 • T: +1 717 3341124 • F: +1 717 3341891 • gett_curator@nps.gov • www.nps.gov/gett •
Dir.: *Dr. John A. Latschar* • Cur.: *Greg Goodell* •
Historic Site / Military Museum / Open Air Museum – 1895
Hist of the Civil War 48902

Lincoln Train Museum, 425 Steinwehr Av, Gettysburg, PA 17325 • T: +1 717 3345678 • F: +1 717 3349100 • www.gettysburgbattlefieldtours.com •
Special Museum
Model train of Abe Lincoln, railroad dioramas 48903

Soldiers National Museum, 777 Baltimore St, Gettysburg, PA 17325 • T: +1 717 3344890 • www.gettysburgbattlefieldtours.com •
Military Museum – 1950
Military, confederate encampment 48904

Ghent NY

Art Omi, 59 Letter South Rd, Ghent, NY 12075 • T: +1 518 3927656 • F: +1 518 3922181 • artomi55@aol.com • www.artomi.org •
Dir.: *Katleen Heike Triem* •
Fine Arts Museum
Art 48905

Parker-O'Malley Air Museum, 1571 Rte 66, Ghent, NY 12075 • T: +1 518 3927200 • F: +1 518 3922227 • jmcmahon@parkeromalley.org • www.parkeromalley.org •
C.E.O.: *James E. McMahon* •
Science&Tech Museum – 1991
Aeronautics, aircraft from 1920s-1940s 48906

Gig Harbor WA

Gig Harbor Peninsula Historical Society Museum, 4218 Harborview Dr, Gig Harbor, WA 98332 • T: +1 253 8586722 • F: +1 253 8534211 • info@gigharbormuseum.org • www.gigharbormuseum.org •
Historical Museum – 1962
Local hist 48907

Gillett AR

Arkansas Post Museum, 5530 Hwy 165 S, Gillett, AR 72055 • T: +1 870 5482634 • arkansaspostmuseum@arkansas.com • www.arkansasstateparks.com/arkansaspostmuseum •
Dir.: *Thomas E. Jordan* •
Local Museum – 1960
Local history, archaeology and agriculture, Quapaw Indians, military hist – Arkansas Post National Memorial 48908

Gillette WY

Campbell County Rockpile Museum, 900 W Second St, Gillette, WY 82716 • T: +1 307 6825723 • F: +1 307 6868528 • rockpile@vcn.com •
Dir.: *Robert J. Kothe* •
Local Museum – 1974
Local hist, agricultural equipment, Native American weapons and tools 48909

Gills Rock WI

Door County Maritime Museum, 12724 Wisconsin Bay Rd, Gills Rock, WI 54210 • T: +1 920 8541844, 7435958 • F: +1 920 7439483 • info@dcmm.com • www.dcmm.org •
Pres.: *Dan Austad* • Exec. Dir.: *Brian Kelsey* •
Historical Museum – 1969
Sailing, Coast Guard, commercial fishing, shipbuilding industry 48910

Gilmanton NH

Carpenter Museum of Antique Outboard Motors, POB 459, Gilmanton, NH 03237-0459 • T: +1 603 5247611 • F: +1 603 5247611 • amc@cyberportal.net •
Science&Tech Museum – 1976
Outboard motors, marine 48911

Gilroy CA

City of Gilroy Museum, 195 Fifth St, Gilroy, CA 95020 • T: +1 408 8460446 • gilroy.museum@ci.gilroy.ca.us • www.ci.gilroy.ca.us •
Head: *Cathy Mirelez* •
Local Museum – 1958
Local and Santa Clara valley history, Ohlone Indians 48912

Gavilan College Art Gallery, 5055 Santa Teresa Blvd, Gilroy, CA 95020 • T: +1 408 8464974 • art4change@mac.com • www.gavilan.edu •
Public Gallery – 1967 48913

Girard PA

Battles Museums of Rural Life, 436 Walnut St, Girard, PA 16417 • T: +1 814 7744788 • F: +1 814 4521744 • echs@eriecountyhistory.org • www.eriecountyhistory.org •
Historical Museum – 1989
Battles family 48914

Glasgow MT

Valley County Pioneer Museum, Hwy 2 W, Glasgow, MT 59230 • T: +1 406 2288692 • vcmuseum@nemontel.net • www.valleycountymuseum.com •
Dir.: *Lenore Hinerman* • Cur.: *Carol Cotton* •
Historical Museum – 1964
Pioneer memorabilia, wildlife exhib, farm machinery 48915

Glastonbury CT

Glastonbury Museum, 1944 Main St, Glastonbury, CT 06033 • T: +1 860 6336890 • F: +1 860 6336890 • hsglastonbury@netzero.net •
Dir.: *James Bennett* •
Historical Museum – 1936 48916

Glen Allen VA

Meadow Farm Museum, 3400 Mountain Rd, Glen Allen, VA 23060, mail addr: POB 27032, Richmond, VA 23273 • T: +1 804 5015520 • F: +1 804 5015284 • www.co.henrico.va.us/rec •
Decorative Arts Museum / Folklore Museum – 1981
19th-20th c folk art 48917

Glen Echo MD

Clara Barton National Historic Site, 5801 Oxford Rd, Glen Echo, MD 20812 • T: +1 301 3201410 • F: +1 301 3201415 • gwmp_clara_barton_nhs@nps.gov • www.nps.gov/clba •
Historic Site – 1975
Home with personal items of Clara Barton, founder of the American Red Cross 48918

Glen Ellen CA

House of Happy Walls, Jack London State Historic Park, 2400 London Ranch Rd, Glen Ellen, CA 95442 • T: +1 707 9385216 • www.jacklondons.net/museum •
Head: *Sheryl Lawton* •
Special Museum – 1959
History, paintings, sculpture, London's typewriter and dictaphone, manuscripts – Wolf house 48919

Glen Ellyn IL

Gahlberg Gallery, College of DuPage, 425 Fawell Blvd, Glen Ellyn, IL 60137-6599 • T: +1 630 9422321 • F: +1 630 9806 •
Dir.: *Barbara Wiesen* •
Fine Arts Museum / University Museum – contemporary art 48920

Stacy's Tavern Museum, 557 Geneva Rd, Glen Ellyn, IL 60137, mail addr: POB 283, Glen Ellyn, IL 60138 • T: +1 630 8588696 • F: +1 630 8588696 • www.glen-ellyn.com/historical/ •
Pres.: *John Costersian* • Dir.: *Jan Langford* •
Local Museum – 1976
1846 Moses Stacy House and Inn, local Glen Ellyn history – archive 48921

Glen Rose TX

Barnard's Mill Art Museum, 307 SW Barnard St, Glen Rose, TX 76043 • T: +1 817 8972611 •
Dir.: *Richard H. Moore* • Cur.: *Hollis Taylor* •
Fine Arts Museum – 1989
200 paintings, bronzes & etchings, works of artist Amy Miears Jackson, Lester Hughes 48922

Dinosaur Valley State Park, FM 205, Glen Rose, TX 76043 • T: +1 817 8974588 • F: +1 817 8973409 • www.tpwd.state.tx.us •
Head: *Billy P. Baker* •
Natural History Museum – 1969
Paleontology 48923

Somervell County Museum, Elm & Vernon Sts, Glen Rose, TX 76043 • T: +1 254 8974529 •
Dir.: *Jeanne P. Mack* •
Local Museum – 1966
Agriculture, archaeology, geology, Indian artifacts – archives 48924

Glencoe MO

Wabash Frisco and Pacific Association, 199 Grand Av, Glencoe, MO 63038, mail addr: 1569 Ville Angela Ln, Hazelwood, MO 63042-1630 • T: +1 636 5873538, +1 314 3519385 • it1569djn@earthlink.net • www.wfprr.com •
Science&Tech Museum / Association with Coll – 1939
Miniature steam locomotives, trank, hopper and gondola cars, wooden benched flat cars and gondolas, passenger cars, milling machines 48925

Glendale AZ

The Bead Museum, 5754 W Glenn Dr, Glendale, AZ 85301 • T: +1 623 9312737 • F: +1 623 6926013 • info@beadmuseumaz.org • www.beadmuseumaz.org •
Dir.: *Kelly Norton* •
Fine Arts Museum – 1985
Ancient and modern beads, beaded art, contemporary artists – research library, tours 48926

Deer Valley Rock Art Center, 3711 W Deer Valley Rd, Glendale, AZ 85308 • T: +1 623 5828007 • F: +1 623 5828831 • dvrac@asu.edu • www.asu.edu/clas/anthropology/dvrac •
C.E.O.: *Dr. Peter Welsh* •
Archaeology Museum – 1994
Rock art recording and interpretation – library 48927

Glendale CA

Brand Library Art Galleries, 1601 W Mountain St, Glendale, CA 91201-1200 • T: +1 818 5482051 • F: +1 818 5485079 • info@brandlibrary.org • www.ci.glendale.ca.us/library/brand_galleries.asp •
Dir.: *Alyssa Resnick* •
Library with Exhibitions / Fine Arts Museum – 1956
Prints, art and music books, records, slides, video & DVD 48928

Casa Adobe de San Rafael, 1330 Dorothy Dr, Glendale, CA 91202 • T: +1 818 5485400 • F: +1 818 5483789 • www.ci.glendale.ca.us/parks/casa_adobe.asp •
Dir.: *Merry Franzen* •
Historical Museum – 1867
Early California furniture 48929

Forest Lawn Museum, 1712 S Glendale Av, Glendale, CA 91205 • T: +1 323 2543131 • slawrence@forestlawn.com • www.forestlawn.com •
Dir.: *Alison Bruesehoff* •
Special Museum – 1951
American hist since 1770, paintings, sculptures, gems – theater 48930

Glendive MT

Frontier Gateway Museum, I-94, Exit 215 State St, Glendive, MT 59330, mail addr: PO Box 1181, Glendive, MT 59330 • T: +1 406 3778168 •
Cur.: *Louise Cross* •
Local Museum – 1963
Indian artifacts; natural hist, fossils, mammoth; fashion, photographs – library 48931

Glenford OH

Flint Ridge State Memorial Museum, 7091 Brownsville Rd SE, Glenford, OH 43739-9609 • moundbuilders@cgate.net •
Man.: *James Kingery* •
Natural History Museum – 1933
Geological exhibits, flint deposits 48932

Glenn Dale MD

Marietta House Museum, 5626 Bell Station Rd, Glenn Dale, MD 20769 • T: +1 301 4645291 • F: +1 301 4565654 • susan.wolfe@pgparks.com • www.pgparks.com •
Pres.: *Susan Wolfe* •
Local Museum – 1978
Federal style brick home, 19th c history 48933

Glens Falls NY

Chapman Historical Museum, 348 Glen St, Glens Falls, NY 12801 • T: +1 518 7932826 • F: +1 519 7932831 • chapman@capital.net • www.chapmanmuseum.org •
Dir.: *Timothy Weidner* • Cur.: *Rebecca Gereau* •
Historical Museum – 1967
Social and industrial hist of Glenn Falls, Queensbury and the Southern Adirondacks, Stoddard Coll of paintings, photographs, published works and memorabilia 48934

The Hyde Collection, 161 Warren St, Glens Falls, NY 12801 • T: +1 518 7921761 • F: +1 518 7929197 • info@hydecollection.org • www.hydecollection.org •
Dir.: *Randall Suffolk* • Cur.: *Erin M. Coe* •
Fine Arts Museum – 1952
European and American art by Raphael, Da Vinci, Rubens, Rembrandt, Cézanne, Renoir, Van Gogh, Picasso, Whistler, Eakins, Homer 48935

Glenside PA

Arcadia University Art Gallery, Church and Easton Rds, Glenside, PA 19038 • T: +1 215 5722131 • F: +1 215 8818774 • torchia@beaver.edu • www.beaver.edu •
Dir.: *Richard Torchia* •
University Museum / Fine Arts Museum – 1853
Paintings, sculpture, prints 48936

Glenview IL

Glenview Area Historical Museum, 1121 Waukegan Rd, Glenview, IL 60025 • T: +1 847 7242235 • F: +1 847 7242235 •
Pres.: *Beverly Dawson* •
Local Museum – 1965
Local history, housed in original 1864 farm house – library 48937

Hartung's Auto and License Plate Museum, 3623 W Lake St, Glenview, IL 60025 • T: +1 847 7244354 •
C.E.O.: *Lee Hartung* •
Science&Tech Museum – 1971
Automotive, over 150 antique cars, trucks, tractors and motorcycles 48938

Kohl Children's Museum of Greater Chicago, 2100 Patriot Blvd, Glenview, IL 60026 • T: +1 847 8326600 • F: +1 847 7246469 • info@kohlchildrensmuseum.org • www.kohlchildrensmuseum.org •
C.E.O.: *Sheridan Turner* • Vice Pres. of Programs: *Mary Pinon* • Dir. of Exhibits: *Richard Lyon* • Dir. of Education: *Erika Miller* •
Ethnology Museum – 1985
Dominick's Supermarket, Water Works, Pet Vet, Car Care 48939

Glenville NY

Empire State Aerosciences Museum, 250 Rudy Chase Dr, Glenville, NY 12302 • T: +1 518 3772191 • F: +1 518 3771959 • esam@ensam.org • www.esam.org •
Pres.: *Dan Silva* •
Science&Tech Museum – 1984
Aerosciences and airplane 48940

Glenwood IA

Mills County Museum, Glenwood Lake Park, Glenwood, IA 51534, mail addr: POB 255, Glenwood, IA 51534 • T: +1 712 5279221, 7129533 • carriemerritt@hotmail.com •
Pres.: *Carrie Merrit* •
Local Museum – 1957
Machinery hall, dolls, guns, household furniture, glass and china, Indian artifacts 48941

Glenwood MN

Pope County Historical Museum, 809 S Lakeshore Dr, Glenwood, MN 56334-1115 • T: +1 320 6343293 • F: +1 320 6343293 • pcmuseum@runestone.net •
Dir.: *Merlin Peterson* •
Local Museum / Open Air Museum – 1931
Regional history, native American crafts, Indian artifacts, church records, country schools – library, biographical archive 48942

Globe AZ

Gila County Historical Museum, 1330 N Broad St, Globe, AZ 85502, mail addr: POB 2891, Globe, AZ 85502 • T: +1 928 4257385 •
Man.: *Dr. Bill Haak* •
Local Museum / Archaeology Museum – 1972
Indian pottery, furniture, kitchen ware, glass coll 48943

Gloucester MA

Beauport-Sleeper-McCann House, 75 Eastern Point Blvd, Gloucester, MA 01930, mail addr: 141 Cambridge St, Boston, MA 02114 • T: +1 978 2830800 • F: +1 978 2834484 • beauporthouse@spnea.org • www.historicnewengland.org •
Man.: *Carl Nold* •
Decorative Arts Museum – 1942
American, European and Oriental decorative arts 48944

Cape Ann Historical Museum, 27 Pleasant St, Gloucester, MA 01930 • T: +1 978 2830455 • F: +1 978 2834141 • judithmcculloch@verizon.net • www.cape-ann.com/historical-museum •
Dir.: *Judith McCulloch* •
Historical Museum / Fine Arts Museum – 1873
Maritime hist, fishing ind, paintings, sculpture, drawings, silver, china, granite quarrying exhibit – library 48945

Hammond Castle Museum, 80 Hesperus Av, Gloucester, MA 01930 • T: +1 508 2837673, 2832080 • F: +1 508 2831643 • www.hammondcastle.org •
Dir./Cur.: *John W. Pettibone* •
Historical Museum – 1930
Medieval-style built in 1926-29, Roman, medieval and Renaissance artifacts 48946

The North Shore Arts Association, 197 E Main St at Pirate's Ln, Gloucester, MA 01930 • T: +1 978 2831857 • F: +1 978 2829819 • arts@gis.net • www.northshoreartsassoc.org •
Dir.: *Trudy J. Allen* •
Association with Coll – 1922
Contemporary regional and US art 48947

Sargent House Museum, 49 Middle St, Gloucester, MA 01930 • T: +1 978 2812432 • F: +1 978 2812432 • sargenthouse@cove.com • sargenthouse.org •
Dir.: *Martha Oaks* •
Fine Arts Museum / Decorative Arts Museum – 1919
Furniture, decorative and fine arts, coll of works by John Singer Sargent 48948

Gloucester Point VA

Virginia Institute of Marine Science, Fish Collection, Rte 1208, Great Rd, Gloucester Point, VA 23062 • T: +1 804 6847000 • F: +1 804 6847327 • jmusick@vims.edu • www.vims.edu •
Cur.: *Melanie M. Harbin* •
Agriculture Museum – 1969
Preserved fishes, scientific ichthyology coll 48949

Glover VT

Bread & Puppet Museum, 753 Heights Rd, Rte 122, Glover, VT 05839 • T: +1 802 5253031, 5256972 • F: +1 802 5253618 •
Dir.: *Peter Schumann* •
Decorative Arts Museum – 1975
Puppets 48950

Gnadenhutten OH

Gnadenhutten Historical Park and Museum, 352 S Cherry St, Gnadenhutten, OH 44629 • T: +1 740 2544143 • F: +1 740 2544992 • gnadmuse@tusco.net • www.gnaden.tusco.net •
C.E.O.: *Jack J. McKeown* •
Local Museum – 1963
18th-19th c Indian artifacts, regional hist 48951

Goehner NE

Seward County Historical Society Museum, I-80 Exit 373, Goehner, NE 68364 • T: +1 402 6434935, 5234055 • drouss//@connect.ccsn.edu •
Pres.: *Larry Hansen* •
Local Museum / Agriculture Museum – 1978
Local history, agriculture 48952

Goessel KS

Mennonite Heritage Museum, 200 N Poplar St, Goessel, KS 67053 • T: +1 620 3678200 • mhmuseum@futureks.net •
Pres.: *Dwight Schmidt* • Dir./Cur.: *Darlene Schroeder* •
Historical Museum – 1974
Artifacts, manuscripts coll, clothing, household goods, books, farm machinery 48953

Gold Beach OR

Curry Historical Society Museum, 29419 Ellensburg Av, Gold Beach, OR 97444 • T: +1 541 2476113 • hissoc@gb.wave.net • curryhistory.com •
Pres.: *Darlene McCarthy* •
Local Museum – 1974
Local hist, animal life, local beach ecosystem, Rogue River Indian artifacts 48954

Golden CO

Astor House Hotel Museum, 822 12th St, Golden, CO 80401 • T: +1 303 2783557 • F: +1 303 2788916 • info@astorhousemuseum.org • www.astorhousemuseum.org •
C.E.O.: *Kim Lennoy* •
Historical Museum – 1972
Decorative art 48955

Buffalo Bill Memorial Museum, 987.5 Lookout Mountain Rd, Golden, CO 80401 • T: +1 303 5260747 • F: +1 303 5260197 • buffalobill.museum@ci.denver.co.us • www.buffalobill.org •
Dir.: *Steve Friesen* •
Historical Museum – 1921
Founded by Johnny Baker, a close friend of Buffalo Bill and an important member of Buffalo Bill's Wild West Show 48956

Colorado Railroad Museum, 17155 W 44th Av, Golden, CO 80403 • T: +1 303 2794591 • F: +1 303 2794229 • info@crrm.org • www.crrm.org •
Dir.: *William Gould* •
Science&Tech Museum – 1958
Rolling stock of Colorado railroads, interior 48957

Colorado School of Mines Geology Museum, 1301 Maple St, Golden, CO 80401-1887 • T: +1 303 2733823, 2733815 • F: +1 303 2733859 • pbartos@mines.edu •
Cur.: *Paul I. Bartos* •
University Museum / Natural History Museum – 1874
Geology, mineralogy 48958

Foothills Art Center, 809 15th St, Golden, CO 80401 • T: +1 303 2793922 • F: +1 303 2799470 • fac@foothillsartcenter.org • www.foothillsartcenter.org •
Exec. Dir.: *Jennifer Cook* • Cur.: *Michael Chavez* •
Fine Arts Museum – 1968
American art, ethnic art, sculptures 48959

Golden Pioneer Museum, 923 10th St, Golden, CO 80401 • T: +1 303 2787151 • F: +1 303 2782755 • goldenpm@comcast.net • www.goldenpioneermuseum.com •
Dir.: *Elinor Packard* • Cur.: *Michelle Zupan* •
Historical Museum – 1938
Golden and Colorado history, prehistory – library 48960

Rocky Mountain Quilt Museum, 1111 Washington Av, Golden, CO 80401 • T: +1 303 2770377 • F: +1 303 2151636 • rmqm@att.net • www.rmqm.org •
Dir.: *Janet Finley* •
Folklore Museum – 1982
Quilts, fabric and quilt-related items 48961

Goldendale WA

Klickitat County Historical Society Museum, 127 W Broadway, Goldendale, WA 98620 • T: +1 509 7734303 •
C.E.O. & Pres.: *Bonny Beeks* •
Local Museum – 1958
Local hist 48962

Maryhill Museum of Art, 35 Maryhill Museum Dr, Goldendale, WA 98620 • T: +1 509 7733733 • F: +1 509 7736138 • maryhill@maryhillmuseum.org • www.maryhillmuseum.org •
Dir.: *Colleen Schafroth* •
Fine Arts Museum – 1923
Watercolors and sculptures by Rodin, North American Native art and culture, European and American art 48963

Goldsboro NC

Arts Council of Wayne County - Museum, 2406 E Ash St, Goldsboro, NC 27534 • T: +1 919 7363300 • F: +1 919 7363335 • arts@esn.net •
Dir.: *Alice Strickland* •
Fine Arts Museum – 1963
Contemp art 48964

Goleta CA

South Coast Railroad Museum, 300 N Los Carneros Rd, Goleta, CA 93117 • T: +1 805 9643540 • F: +1 805 9643549 • www.goletadepot.org •
Dir.: *Gary B. Coombs* • Asst. Dir.: *Phyllis J. Olsen* •
Science&Tech Museum – 1983
Southern Pacific railroad depot 48965

Stow House, Museum of Goleta Valley History, 304 N Los Carneros Rd, Goleta, CA 93117 • T: +1 805 9644407 • staff@goletahistory.org • www.stowhouse.com •
Historical Museum / Local Museum – 1967
18th-19th c local hist, farm machinery, wedding gown coll – library 48966

Goliad TX

Goliad State Historical Park, 108 Park Rd, Goliad, TX 77963 • T: +1 361 6453405 • F: +1 361 6458538 • elizabeth.livingston@tpwd.state.tx.us • www.tpwd.state.tx.us •
Local Museum / Folklore Museum – 1931
Archaeological artifacts, Spanish Colonial & early Texas Parks 48967

Presidio La Bahia, Refugio Hwy, 1 mile south of Goliad on Hwy 183, Goliad, TX 77963 • T: +1 512 6453752 • F: +1 512 6451706 • presidiolabahia@goliad.net • www.presidiolabahia.org •
Dir.: *Newton M. Warzecha* •
Military Museum / Historic Site / Archaeology Museum – 1966
Artifacts showing 9 levels of civilization 48968

Gonzales TX

Gonzales Memorial Museum, E Saint Lawrence St, Gonzales, TX 78629 • T: +1 830 6726350 •
Cur.: *Mary Bea Arnold* •
Local Museum – 1936
Original documents & relics dating to 1660 dealing with Texas, Mexico & Great Southwest 48969

Goochland VA

Goochland County Museum, 2875 River Rd West, Goochland, VA 23063, mail addr: POB 602, Goochland, VA 23063 • T: +1 804 5563966 • F: +1 804 5564617 • gchs1admin@earthlink.net • www.goochlandhistory.org •
Local Museum – 1968
Land grant parents (1700's) 48970

Goodland KS

High Plains Museum, 1717 Cherry, Goodland, KS 67735 • T: +1 785 8994595 •
Dir.: *Linda Holton* •
Historical Museum – 1959
Plains Indians artifacts, pioneer life, farming tools 48971

Goodwell OK

No Man's Land Historical Museum, Oklahoma Historical Society, 207 W Sewell St, Goodwell, OK 73939 • T: +1 580 3496697 • F: +1 580 3492670 • nmlhs@ptsi.net •
Dir.: *Dr. Kenneth R. Turner* • Pres.: *Gerald Dixon* •
Historical Museum – 1932
Hal Clark and William B. Baker Indian artifacts, Duckett Alabaster carvings, regional hist 48972

Gordonsville VA

Exchange Hotel Civil War Museum, 400 S Main St, Gordonsville, VA 22942 • T: +1 540 8322944 • tcouch@ns.gemlink.com • www.hgiexchange.org •
Historical Museum / Military Museum – 1971
Civil war, railroad hotel used as a Confederate receiving hospital 48973

Gore OK

Cherokee Courthouse Museum, Hwy 64, 2 1/2 miles SE of Gore, Gore, OK 74435 • T: +1 918 4895663 • F: +1 918 4892217 • fgilliam@twinterritories.com • www.twinterritories.com •
Historical Museum – 1973
Hist of early-day Cherokees, replicas of the original capital of the Cherokee Nation 48974

Gorham ME

Baxter House Museum, 69 South St, Gorham, ME 04038 • T: +1 207 8395031 • F: +1 207 8397749 •
Pres.: *Linda M. Frinsko* •
Local Museum – 1908
Furniture, historical items and records of Gorham 48975

USM Art Gallery, Campus, 37 College Av, Gorham, ME 04038 • T: +1 207 7805008 • F: +1 207 7805759 • cexler@usm.maine.edu • www.usm.maine.edu/~gallery •
C.E.O.: *Carolyn Eyler* •
Fine Arts Museum / University Museum – 1965
Paintings, sculpture, graphics, photographs 48976

Goshen CT

Goshen Historical Society Museum, 21 Old Middle Rd, Goshen, CT 06756 • T: +1 860 4919610 • jvnkuq@juno.com • www.goshenhistoricalsociety.org •
Pres.: *Margaret K. Wood* • Cur.: *Henrietta Horray* •
Historical Museum – 1955 48977

Goshen NY

Harness Racing Museum and Hall of Fame, 240 Main St, Goshen, NY 10924 • T: +1 845 2946330 • F: +1 845 2943463 • hrm@frontiernet.net • www.harnessmuseum.com •
Dir.: *Gail C. Cunard* •
Special Museum – 1951 48978

Gothenburg NE

Sod House Museum, I-80 and Hwy 47, Gothenburg, NE 69138 • T: +1 308 5372680 • sodlady2001@hotmail.com •
Dir.: *Merle Block* •
Local Museum – 1988
Indian artifacts, photographs, sculpture, pony express hist 48979

Gouverneur NY

Gouverneur Historical Association Museum, 30 Church St, Gouverneur, NY 13642 • T: +1 315 2870570 •
Pres./Cur.: *Joseph Laurenza* • Cur.: *Jean Tyler* (Acquisitions) •
Historical Museum – 1974
History, Victorian furnishings, rock and mineral coll, farm tools,Western Indians coll 48980

Grafton MA

Willard House and Clock Museum, 11 Willard St, Grafton, MA 01536 • T: +1 508 8393500 • F: +1 508 8393599 • seth@willardhouse.org • www.willardhouse.org •
Dir.: *John R. Stephens* •
Science&Tech Museum – 1971
Clocks by The Willard's, family portraits, patents, tools 48981

Grafton VT

Grafton Historical Society, 147 Main St, Grafton, VT 05146 • T: +1 802 8432489, 8431010 • grafhist@sover.net • www.graftonhistory.org •
Cur.: *Elisha Prouty* •
Association with Coll – 1962
Writing accessories, soapstone, photographs, hist 48982

Nature Museum at Grafton, 186 Townshend Rd, Grafton, VT 05146 • T: +1 802 8432111 • F: +1 802 8431164 • staff@nature-museum.org • www.nature-museum.org •
Cur.: *Betsy Bennett* •
Natural History Museum – 1987
Emphasis on the mammals, birds & geology of New England 48983

Granby CT

Salmon Brook Historical Society Museum, 208 Salmon Brook St, Granby, CT 06035 • T: +1 860 6539713 • www.salmonbrookhistorical.org •
Pres.: *Carol Bressor* •
Local Museum – 1959 48984

Grand Canyon AZ

Grand Canyon Museum Collection, Albright Maintenance Area, South Rim, Grand Canyon, AZ 86023, mail addr: POB 129, Grand Canyon, AZ 86023 • T: +1 928 6387769 • F: +1 928 6387769 • grca_Museum_Collection@nps.gov • www.nps.gov/grca/photos/museum.htm •
Dir.: *Joseph Alston* •
Natural History Museum – 1919
Deep gorge of the Colorado River, 277 mi long, 1-18 mi wide, 1 mi deep, natural history, geology, prehistory – herbarium, archives 48985

Grand Forks ND

Dakota Science Center, POB 5023, Grand Forks, ND 58206 • T: +1 701 7958500 • F: +1 701 7758484 • mckinnon@dakota-science.org • dakota-science.org •
Pres.: *Ann Porter* •
Science&Tech Museum – 1993 48986

Grand Forks County Historical Society Museum, 2405 Belmont Rd, Grand Forks, ND 58201 • T: +1 701 7752216 • gfhistory@wiktel.com • grandforkshistory.com •
Pres.: *Suellen Bateman* • Dir.: *Leah Byzewski* •
Historical Museum – 1970 48987

Hughes Fine Arts Center, c/o Univerisity of North Dakota, Dept of Visual Arts, Rm 127, Grand Forks, ND 58202-7099 • T: +1 701 7772257 •
University Museum / Fine Arts Museum – 1979
Prints, drawings 48988

North Dakota Museum of Art, Centennial Dr, Grand Forks, ND 58202 • T: +1 701 7774195 • F: +1 701 7774425 • ndmoa@ndmoa.com • www.ndmoa.com •
Pres.: *Ann Brown* • Dir. & Chief Cur.: *Laurel J. Reuter* •
Asst. Dir.: *Brian Lofthus* •
Fine Arts Museum – 1970
Contemporary art 48989

University of North Dakota Zoology Museum, Dept of Biology, University of North Dakota, Grand Forks, ND 58202 • T: +1 701 7772621 • F: +1 701 7772623 • www.und.edu/dept/biology/undergrad/bio_undergrad.html •
Cur.: *Dr. Richard Crawford* (Vertebrates) • *Dr. Rick Sweitzer* (Vertebrates) • *Dr. W.J. Wrenn* (Invertebrates) •
University Museum / Natural History Museum – 1883
Birds, fishes, mammals, insects, reptiles and amphibians 48990

Grand Haven MI

Tri-Cities Museum, 200 Washington Av, Grand Haven, MI 49417 • T: +1 616 8420700 • F: +1 616 8423698 • tcmuseum@grandhaven.com • www.tri-citiesmuseum.org •
Dir.: *Dennis W. Swartout* •
Local Museum – 1962
Regional history, agriculture, medicine, Indian artifacts, lumbering, shipping, costumes, coast guard, railroads – Depot Museum 48991

Grand Island NE

Stuhr Museum of the Prairie Pioneer, 3133 W Hwy 34, Grand Island, NE 68801 • T: +1 308 3855316 • F: +1 308 3855028 • info@stuhrmuseum.org • www.stuhrmuseum.org •
C.E.O.: *Joe Black* •
Local Museum – 1961
Frunishings, pioneer crafts, Indian and Old West artifacts, farm machinery 48992

Grand Junction CO

Doozoo Children's Museum, 421 Colorado Av, Grand Junction, CO 81501 • T: +1 970 2415225 • F: +1 970 2418510 • www.doozoo.com •
Dir.: *Valerie Johns* •
Special Museum – 1984 48993

Museum of Western Colorado, 462 Ute Av, Grand Junction, CO 81502-5020 • T: +1 970 2420971 • F: +1 970 2423960 • info@westcomuseum.org • www.wcmuseum.org •
Exec. Dir.: *Michael L. Perry* • Cur.: *David Bailey* •
Local Museum – 1966
Cultural and natural history, paleontology, farming 48994

Western Colorado Center for the Arts, 1803 N Seventh, Grand Junction, CO 81501 • T: +1 970 2437337 • F: +1 970 2432482 • info@gjartcenter.org • www.gjartcenter.org •
C.E.O: *Claudia Crowell* •
Fine Arts Museum / Decorative Arts Museum – 1953
Arts, crafts 48995

Grand Lake CO

Grand Lake Area Historical Museum, Kauffman House, Lake Av at Pitkin St, Grand Lake, CO 80447-0656 • T: +1 970 6279644 • F: +1 970 6273693 • glhistory@rkymtnhi.com •
Local Museum – 1973
Regional history 48996

Grand Marais MN

Cook County Historical Museum, 5 S Broadway, Grand Marais, MN 55604 • T: +1 218 3872883 • cchistsoc@boreal.org •
Dir.: *Patricia Zankman* •
Local Museum – 1966
Regional hist, trapping, logging, schools, textiles, folklore, industry, military, agriculture, fishing, arts – Lightkeepers house 48997

Johnson Heritage Post Art Gallery, 115 W Wisconsin St, Grand Marais, MN 55604 • T: +1 218 3872314 • cchristsoc@boreal.org •
Dir.: *Joanne Krause* •
Fine Arts Museum – 1966
Coll of Anna C. Johnson, contemporary art exhibits 48998

Grand Portage MN

Grand Portage National Monument, 170 Mile Creek Rd, Grand Portage, MN 55605-0426, mail addr: PO Box 426, Grand Portage, MN 55605-0426 • T: +1 218 4750123 • F: +1 218 4750123 • grpo_resource_-management@nps.gov • www.nps.gov/grpo •
Head: *Dr- Timothy Cochrane* •
Historical Museum – 1958
18th c reconstructed NW Co. fur trading depot, Chippewa (Ojibwe)Indian history and culture – small library 48999

Grand Rapids MI

106 South Division Gallery, Calvin College, 106 S Division Av, Grand Rapids, MI 49503, mail addr: 3201 Burton St SE, Grand Rapids, MI 49546-4388 • T: +1 616 5266271 • F: +1 616 5268551 • jhz2@calvin.edu • www.calvin.edu/centerartgallery/106southdivision •
Dir.: *Joel Zwart* •
Fine Arts Museum / University Museum
Students and classes exhibits 49000

Center Art Gallery, Calvin College, 3201 Burton St SE, Spoelhof Center, Grand Rapids, MI 49546-4388 • T: +1 616 5266271 • F: +1 616 5268551 • jhz2@calvin.edu • www.calvin.edu/centerartgallery •
Dir.: *Joel Zwart* •
Fine Arts Museum / University Museum – 1974
Dutch 17th-19th c paintings & drawings, 20th prints, paintings, drawings & sculpture – library 49001

Gerald R. Ford Museum, 303 Pearl St NW, Grand Rapids, MI 49504 • T: +1 616 2540400 • F: +1 616 2540386 • ford.museum@nara.gov • www.ford.utexas.edu •
Cur.: *Donald Holloway* •
Library with Exhibitions / Historical Museum – 1980
Hist of the post-WW II era, G.R. Ford presidency (1974-77) – Presidential library 49002

Grand Rapids Art Museum, 155 Division N, Grand Rapids, MI 49503-3154 • T: +1 616 8311000 • F: +1 616 5590422 • pr@gr-artmuseum.org • www.gramonline.org •
Dir.: *Celeste M. Adams* • Cur.: *Richard H. Axsom* (Prints, Photographs) • Asst. Cur.: *Sarah Holian* • Cons.: *Dan Van de Steeg* •
Fine Arts Museum – 1910
European Renaissance to 20th c paintings, American 19th-20th c art, 20th c design & decorative arts – library 49003

Grand Rapids Children's Museum, 22 Sheldon Av NE, Grand Rapids, MI 49503 • T: +1 616 2354726 ext 100 • F: +1 616 2354728 • tthome@grcm.org • www.grcm.org •
Exec. Dir.: *Teresa L. Thome* •
Special Museum – 1992
Hands-on discovery 49004

Kendall Gallery, Kendall College of Art and Design, Ferris State University, 17 Fountain St NW, Grand Rapids, MI 49503 • F: +1 616 4519867 • josephs@ferris.edu • www.kcad.edu •
Dir.: *Sarah Joseph* •
Public Gallery / University Museum 49005

Urban Institute for Contemporary Arts, 41 Sheldon Blvd SE, Grand Rapids, MI 49503 • T: +1 616 4547000 • F: +1 616 4599395 • info@uica.org • www.uica.org •
Exec. Dir.: *Jeffrey Meeuwsen* • Cur.: *Israel Davis* (Ceramics) •
Public Gallery – 1977
theater 49006

Van Andel Museum Center, Public Museum of Grand Rapids, 272 Pearl St NW, Grand Rapids, MI 49504-5371 • T: +1 616 4563977 • F: +1 616 4563873 • staff@grmuseum.org • www.grmuseum.org •
Dir.: *Timothy J. Chester* • Sc. Staff: *Paula Gangopadhyay* (Education) • *David L. DeBruyn* (Astronomy) • *Christian G. Carron* (Collections) • *Veronica Kandl* (History) • *Tom Bantle* (Exhibits) •
Local Museum – 1854
Local history, ethnology, natural history, astronomy – library, auditorium, planetarium, nature center 49007

Grand Rapids MN

Forest History Center, 2609 County Rd 76, Grand Rapids, MN 55744 • T: +1 218 3274482 • F: +1 218 3274483 • foresthistory@mnhs.org • www.mnhs.org/foresthistory •
Man.: *Robert M. Drake* •
Natural History Museum – 1978
Forest artifacts, logging camp, log drives wanigan & batteau – small library, nature trails 49008

Itasca Heritage Center Museum, 10 Fifth St NW, Grand Rapids, MN 55744-0664 • T: +1 218 3266431 • F: +1 218 3267083 • ichs@paulbunyan.net • www.itascahistorical.com •
CEO: *Lilah J. Crowe* •
Local Museum – 1948
County hist, papermaking, life & career of Judy Garland – small library 49009

Judy Garland Museum, 2727 Hwy 169 S, Grand Rapids, MN 55744, mail addr: POB 724 • T: +1 218 3279276 • www.judygarlandmuseum.com •
Special Museum
Judy Garland Birthplace Historic House, shoes of the Ruby Slippers from Wizard of Oz 49010

Grand Ronde OR

Tribal Museum, Confederated Tribes of Grand Ronde, 9615 Grand Ronde Rd, Grand Ronde, OR 97347 • T: +1 503 8792248 • F: +1 503 8792126 • lindy.trolan@grandronde.org • www.grandronde.org/culture/ikanum •
Ethnology Museum – 1995
Hist and culture of Grand Ronde tribal community, traditional Indian arts and crafts 49011

Grandview MO

Harry's Truman Farm Home, 12301 Blue Ridge Blvd, Grandview, MO 64030 • T: +1 816 2542720 • F: +1 816 2544491 • james_a_sanders@nps.gov • www.nps.gov/hstr •
Agriculture Museum / Historical Museum – 1994
Life of Harry S. Truman, local agriculture and rural life 49012

Grandview WA

Ray E. Powell Museum, 313 S Division, Grandview, WA 98930 • T: +1 509 8822070 •
Chm.: *Jack Norting* •
Local Museum – 1969
Local hist 49013

Granite Falls MN

Yellow Medicine County Historical Museum, Hwy 67 and 23, Granite Falls, MN 56241, mail addr: POB 145, Granite Falls, MN 56241 • T: +1 320 5644479 • F: +1 320 5644146 • ymchs@kilowatt.net • www.kilowatt.net/ymchs •
Dir./Cur.: *Melanie Gatchell* •
Local Museum – 1952
Geology, archaeology, pioneer artifacts, bison bones – historic chapel 49014

Grant NE

Perkins County Historical Society Museum, Central Av, Grant, NE 69140 • T: +1 308 3524019 • F: +1 308 3522346 •
Pres.: *Brenda Styskal* •
Local Museum – 1964
Agriculture, costumes, religouis artifacts, local hist 49015

Grants Pass OR

Grants Pass Museum of Art, 229 SW G St, Grants Pass, OR 97526 • T: +1 541 4793290 • F: +1 541 4790508 • museum2@grantspass.net •
Dir.: *Don Brown* •
Fine Arts Museum / Folklore Museum / Public Gallery – 1979
Contemporary regional art 49016

Schmidt House Museum, 508 SW 5th St, Grants Pass, OR 97526 • T: +1 541 4797827 • F: +1 541 4728928 • jchs@terragon.com • www.webtrail.com/jchs •
C.E.O.: *Rose Scott* •
Decorative Arts Museum – 1960
Furniture, toys, clothes, photos 49017

Wiseman and Firehouse Galleries, Rogue Community College, 3345 Redwood Hwy, Grants Pass, OR 97527 • T: +1 541 9567339 • F: +1 541 4713588 • tdrake@roguecc.edu • www.roguecc.edu/galleries •
Dir.: *Tommi Drake* •
Fine Arts Museum / University Museum – 1988
Contemporary art 49018

Granville NY

Pember Museum of Natural History, 33 W Main St, Granville, NY 12832 • T: +1 518 6421515 • F: +1 518 6423097 • pember@adelphia.net • www.pembermuseum.com •
Dir.: *Patricia E. Bailey* • Pres.: *Dan Wilson* •
Natural History Museum – 1909
Birds, mammals, egg sets and nests, insects, herbarium mounts,minerals, fossils 49019

Slate Valley Museum, 17 Water St, Granville, NY 12832 • T: +1 518 6421417 • F: +1 518 6421417 • mail@slatevalleymuseum.org • www.slatevalleymuseum.org •
Dir.: *Mary Lou Willits* •
Science&Tech Museum / Local Museum – 1994
Slate quarrying and slate industry, geological specimens, prints, paintings 49020

Granville OH

Denison University Art Gallery, Burke Hall, Granville, OH 43023 • T: +1 740 5876255 • F: +1 740 5875701 • vanderheijde@cc.denison.edu • www.denison.edu/art/artgallery •
Dir.: *Merijn Van Der Heijden* •
Fine Arts Museum / University Museum – 1946
Asian art, esp. Burmese art, Cuna Indian artifacts, European and American prints, paintings and drawings 49021

Granville Historical Museum, 115 E Broadway, Granville, OH 43023 • T: +1 740 5873951 • office@granvillehistory.org • www.granvillehistory.org •
Pres.: *Lance Clarke* • Cur.: *Cynthia Cort* •
Historical Museum – 1885
Local artifacts, family and business papers, carpenter tools, costumes, glass, dec arts 49022

Robbins Hunter Museum, Avery-Downer House, 221 E Broadway, Granville, OH 43023 • T: +1 740 5870430 • F: +1 740 5870430 •
Pres.: *David Neel* •
Historical Museum – 1981
1842 hist house, period furnishings, 19th c life in Granville, Ohio 1840-1870 49023

Grass Lake MI

Waterloo Area Farm Museum, 9998 Waterloo-Munith Rd, Grass Lake, MI 49240 • T: +1 517 5962254 • info@waterloofarmmuseum.org • www.waterloofarmmuseum.org •
Agriculture Museum – 1962 49024

Grass Valley CA

North Star Mining Museum, Mill St at Allison Ranch Rd, Grass Valley, CA 95959 • T: +1 916 2734255 • info@nevadacountyhistory.org • www.nevadacountyhistory.org •
Dir.: *Glenn Jones* •
Science&Tech Museum
World largest Pelton wheel, models, machinery, tools 49025

Video History Museum, Central off Colfax Av, Memorial Park, Grass Valley, CA 95959 • T: +1 916 2741126 •
Dir.: *Ron Sturgell* •
Science&Tech Museum
theater 49026

Great Bend KS

Barton County Historical Society Village and Museum, 85 S Hwy 281, Great Bend, KS 67530 • T: +1 316 7935125 • F: +1 316 7935125 • director@bartoncountymuseum.org • www.bartoncountymuseum.org •
Chm.: *Beverly Komarek* •
Local Museum / Open Air Museum – 1963
Historic Village. Dolls, wedding dresses, period artifacts, general store 49027

Great Falls MT

C.M. Russell Museum, 400 13 St N, Great Falls, MT 59401 • T: +1 406 7278787 • F: +1 406 7272402 • www.cmrussell.org •
Dir.: *Inez S. Wolins* •
Fine Arts Museum – 1953
Latin American, Mexican and American Indian art, decorative arts, archaeology, ethnology, folk art 49028

Fine Arts Gallery, University of Great Falls, 1301 20th St S, Great Falls, MT 59405 • T: +1 406 7915375 • F: +1 406 7915395 •
Dir.: *Julia Becker* •
Fine Arts Museum / University Museum
Contemporary art 49029

High Plains Heritage Center, Cascade County Historical Museum & Society, 422 2nd St S, Great Falls, MT 59405 • T: +1 406 4523462 • F: +1 406 7613805 • hphc@highplainsheritage.org • www.highplainsheritage.org •
Dir.: *Cindy Kittredge* • Cur.: *Heather Dorociak* •
Local Museum – 1976
archive 49030

Paris Gibson Square Museum of Art, 1400 First Av N, Great Falls, MT 59401 • T: +1 406 7278255 • F: +1 406 7278256 • info@the-square.org • www.the-square.org •
Pres.: *Bill Preston* • Exec. Dir.: *Dr. Lynne Spriggs* • Cur.: *Jessica Hunter Larsen* (Art) • Sc. Staff: *Susan Thomas* (Education) •
Fine Arts Museum – 1976
American art, folk art, decorative art 49031

Greeley CO

Centennial Village Museum, 1475 A St, Greeley, CO 80631 • T: +1 970 3509224 • F: +1 970 3509570 • museums@ci.greeley.co.us • www.greeleymuseums.com •

Pres.: *Gene Bellamy* •
Open Air Museum / Historical Museum – 1976
28 historic structures illustrating the hist of Colorado 49032

John Mariani Art Gallery, c/o University of Northern Colorado, Eighth Av and 18 St, Greeley, CO 80639 • T: +1 970 3512184 • F: +1 970 3512299 •
C.E.O: *Charlotte Nichols* • Ass. Dir.: *Katie Cooper* •
Fine Arts Museum / University Museum – 1972
Art 49033

Meeker Home Museum, 1324 9th Av, Greeley, CO 80631 • T: +1 970 3509220 • F: +1 970 3509570 • museums@ci.greeley.co.us • www.greeleymuseums.com •
Dir.: *Chris Dill* • Cur.: *Peggy A. Ford* (Education) •
Historical Museum – 1929
Home of Nathan Cook Meeker, founder of Greeley 49034

Plumb Farm Museum, 955 39th Av, Greeley, CO 80634 • T: +1 970 3509220 • F: +1 970 3509570 • museum@ci.greeley.co.us • www.greeleymuseums.com •
Dir.: *Chris Dill* •
Agriculture Museum – 1997
Farming history 49035

Green Bay WI

The Children's Museum of Green Bay, Port Plaza Mall, 320 N Adams St, Green Bay, WI 54301 • T: +1 920 4324397 • F: +1 920 4324566 • gbcmuseum@yahoo.com •
C.E.O.: *Bobbie Schuette* •
Special Museum – 1989 49036

Green Bay Packers Hall of Fame, 1265 Lombardi Av, Green Bay, WI 54304 • T: +1 920 4994281 • F: +1 920 4055564 •
Pres.: *Mike Gage* • Dir.: *Kelly Schiltz* •
Special Museum – 1969
Sports 49037

Heritage Hill State Park, 2640 S Webster Av, Green Bay, WI 54301 • T: +1 920 4485150 ext 10 • F: +1 920 4485127 • www.heritagehillgb.org •
Dir.: *Doug Brown* • Cur.: *Randy Klemm* (Collections) • *Jack Moga* •
Local Museum – 1976
Local hist, hist buildings and furnishings 49038

Lawton Gallery, University of Wisconsin-Green Bay, 2420 Nicolet Dr, Green Bay, WI 54311-7001 • T: +1 920 4652916 • F: +1 920 4652890 • perkinss@uwgb.edu • www.uwgb.edu •
Cur.: *Stephen Perkins* •
Public Gallery 49039

National Railroad Museum, 2285 S Broadway, Green Bay, WI 54304-4832 • T: +1 920 4377623 • F: +1 920 4371291 • staff@nationalrrmuseum.org • www.nationalrrmuseum.org •
Exec. Dir.: *Peter Chapman* •
Science&Tech Museum – 1957
Railroad, steam and diesel-electric locomotives, Eisenhower equipment, 19th c maps 49040

Neville Public Museum of Brown County, 210 Museum Pl, Green Bay, WI 54303 • T: +1 920 4484460 ext 0 • F: +1 920 4484458 • bc_museum@co.brown.wi.us • www.nevillepublicmuseum.org •
Dir.: *Eugene Umberger* • Cur.: *Nan Curtis* (Art) • *Trevor Jones* (History) • *Matt Welter* (Education) • *John Jacobs* (Science) • *Louise Pfotenhauer* (Collections) •
Local Museum – 1915
Local hist 49041

Green Lane PA

Goschenhoppen Folklife Museum, Rte 29, Red Men's Hall, Green Lane, PA 18054-0476 • T: +1 610 3678286 • redmens_hall@goschenhoppen.org • www.goschenhoppen.org •
C.E.O.: *George Spotts* •
Ethnology Museum / Folklore Museum / Historical Museum – 1965
Folk art and love, local history 49042

Green River WY

Sweetwater County Historical Museum, 3 E Flaming Gorge Way, Green River, WY 82935 • T: +1 307 8726435 • F: +1 307 8723234 • swchm@sweetwater.net • www.sweetwatermuseum.org •
Dir.: *Ruth Lauritzen* •
Local Museum – 1967
Indian and pioneer artifacts, local hist 49043

Greenbelt MD

Greenbelt Museum, 15 Crescent Rd, Greenbelt, MD 20770 • T: +1 301 4741936, 5076582 • F: +1 301 4418248 • greenbeltmuseum@ci.greenbelt.md.us • www.greenbeltmuseum.org •
Cur.: *Katie Scott-Childress* •
Historical Museum – 1987
Local history, culture of the 1930s, planned communities 49044

Greenbush WI

Wade House Stagecoach Inn and Wesley Jung Carriage Museum, W 7747 Plank Rd, Greenbush, WI 53026 • T: +1 920 5263271 • F: +1 920 5263626 • wadehous@dotnet.com •
Dir.: *Jeffrey Schultz* • Cur.: *Jeffrey Murray* •
Science&Tech Museum – 1953
Transportation, horse and hand-drawn vehicles, restored hist buildings 49045

Greencastle IN

DePauw University Anthropology Museum, Harrison Hall, Rm 206, Greencastle, IN 46135 • T: +1 765 6584336 • F: +1 765 6586552 • kajohnson@depauw.edu • www.depauw.edu/acad/art/galleries •
Dir./Cur.: *Kaytie Johnson* •
University Museum / Ethnology Museum – 1984
Sculpture from Africa, ceramics 49046

Richard E. Peeler Art Center, DePauw University, 10 W Hauna St, Greencastle, IN 46135 • T: +1 765 6584336 • F: +1 765 6586552 • kajohnson@depauw.edu • www.depauw.edu/galleries •
Cur.: *Kaytie Johnson* •
Fine Arts Museum – 1984
Changing exhibitions 49047

William Weston Clarke Emison Museum of Art, DePauw University, 10 W Hanna St, Greencastle, IN 46135 • T: +1 765 6584336 • F: +1 765 6586552 • kajohnson@depauw.edu • www.depauw.edu/museum •
Dir./ Cur.: *Kaytie Johnson* •
Fine Arts Museum – 2004
African, American and Asian arts, ethnographic art 49048

Greencastle PA

Allison-Antrim Museum, 365 S Ridge Av, Greencastle, PA 17225-1157 • T: +1 717 5979325 • aamuseum@greencastlemuseum.org • www.greencastlemuseum.org •
Historical Museum – 1995
20th-c paintings, historic artifacts, Henry P Fletcher US diplomat 49049

Greeneville TN

Andrew Johnson National Historic Site, 101 N College St, Greeneville, TN 37743 • T: +1 423 6383551, 6393711 • F: +1 423 6389194, 7980754 • jim_small@nps.gov • www.nps.gov/anjo/ •
Dir.: *Jim Small* • Cur.: *Elaine R. Clark* • Superint.: *Mark Corey* •
Historical Museum – 1942
Memorabilia of Andrew Johnson (US-President), history, civil war, furnishings – Tailor Shop Museum 49050

Doak House Museum, Tusculum College, Greeneville, TN 37743 • T: +1 423 6368554, 6367348 • F: +1 423 6387166 • clucas@tusculum.edu •
Assoc. Dir. & Cur.: *Cynthia L. Lucas* •
University Museum / Historical Museum
Located in home of Samuel W. Doak 49051

President Andrew Johnson Museum, Tusculum College, Greeneville, TN 37743 • T: +1 423 6367348 • F: +1 423 6387166 • gcollins@tusculum.edu •
Dir.: *George Collins* • Assoc. Dir. & Cur.: *Cynthia L. Lucas* •
University Museum / Historical Museum – 1993
Political memoabilia of Johnson family 49052

Greenfield IA

Iowa Aviation Museum, 2251 Airport Rd, Greenfield, IA 50849 • T: +1 641 3437184 • aviation@mddc.com • www.flyingmuseum.com •
Dir.: *Lee Ann Nelson* • Pres.: *Ron Havens* •
Science&Tech Museum – 1990
Aviation 49053

Greenfield IN

James Whitcomb Riley Birthplace and Museum, Riley Home, 250 W Main St, Greenfield, IN 46140 • T: +1 317 4628539 • F: +1 317 4628556 • parks_rec@greenfieldin.org • www.hccn.org/parks/rileyhouse.htm •
Historical Museum – 1937
Furniture, relics, manuscripts, paintings, china, crafts, newspaper coll 49054

Old Log Jail and Chapel Museums, Rte 40 and Apple St, Greenfield, IN 46140, mail addr: POB 375, Greenfield, IN 46140 • T: +1 317 4627780, 7875474 •
Dir.: *Greg Roland* •
Ethnology Museum – 1966
Indian arrowhead coll, coverlets, pictures, agriculture 49055

Greenfield MA

Greenfield Museum, Corner of Church & Union Sts, Greenfield, MA 01302-0415 • T: +1 413 7743663 •
Cur.: *Tim Blagg* •
Local Museum – 1907
Local history, antiques, photos – small library 49056

Greenfield WI

Greenfield Historical Society Museum, 5601 W Layton Av, Greenfield, WI 53220 • T: +1 414 5451117 • greenfieldhissoc@hotmail.com • www.greenfieldlibrary.org/greenfield_history.htm •
Cur.: *Carol Pingel* •
Local Museum – 1965
Local hist, furniture, tools, housewares 49057

Greenport NY

East and Seaport Museum, 3rd St at Ferry Dock, Greenport, NY 11944 • T: +1 631 4772100 • F: +1 631 4773422 • eseaport@aol.com • www.eastendseaport.org •
Chm. & Pres.: *Leueen Miller* •
Science&Tech Museum – 1990
Shipbuilding in Greenport during WW 2, tools, models, lighthouse 49058

Greensboro AL

Magnolia Grove House Museum, 1002 Hobson St, Greensboro, AL 36744 • T: +1 334 6248618 • maggrov2@bellsouth.net • www.preserveal.org •
Dir.: *Eleanor W. Cunningham* •
Historic Site – 1943
Birthplace of Rear Admiral Richmond Pearson Hobson 49059

Greensboro NC

Green Hill Center for North Carolina Art, 200 N Davie St, Greensboro, NC 27401 • T: +1 336 3337460 • F: +1 336 3332612 • info@greenhillcenter.org • www.greenhillcenter.org •
Dir.: *Jennifer Moore* •
Fine Arts Museum – 1974
Visual arts of NC 49060

Greensboro Artists' League Gallery, 200 N Davie St, Greensboro, NC 27401 • T: +1 336 3337485 • F: +1 336 3732617 • info@greensboroart.org • www.greensboroart.org •
Dir.: *Tamara Brown* • Cur.: *Corrie Lisk-Hurst* •
Fine Arts Museum – 1956
Changing exhibitions 49061

Greensboro Historical Museum, 130 Summit Av, Greensboro, NC 27401-3016 • T: +1 336 3732043 • F: +1 336 3732204 • fred.goss@ci.greensboro.nc.us • www.greensborohistory.org •
Dir.: *Fred Goss* • Cur.: *Jennifer McRae* (Exhibits) • Sc. Staff: *Jon Zachman* (Collections) • *Bbetty K. Phipps* (Education) • *Susan Webster* (Registrar) •
Historical Museum – 1924
Hist of transportation, military, pharmacy and medical artifacts, decorative arts, American historical glass 49062

Guilford College Art Gallery, 5800 W Friendly Av, Greensboro, NC 27410 • T: +1 336 3162438 • F: +1 336 3162950 • thammond@guilford.edu • www.guilford.edu/artgallery •
Dir.: *Theresa N. Hammond* •
University Museum / Fine Arts Museum – 1990
Changing exhibitions, permanent fine art coll 49063

Guilford Courthouse National Military Park, 2332 New Garden Rd, Greensboro, NC 27410-2355 • T: +1 336 2881776 • F: +1 336 2822296 • guco_administration@nps.gov • www.nps.gov/guco •
Military Museum – 1917
Revolutionary War weapons, RW equipments and artifacts 49064

Irene Cullis Gallery, Greensboro College, 815 W Market St, Greensboro, NC 27401-1875 • T: +1 336 2727102 ext 301 • F: +1 336 2716634 • bkowski@gborocollege.edu •
Dir.: *Robert Kowski* •
Fine Arts Museum / University Museum – 1838
Monthly exhibitions of contemp artists 49065

Mattye Reed African Heritage Center, N.C.A. & T. State University, Greensboro, NC 27411 • T: +1 336 3343209 • F: +1 336 3344378 • ndegec@ncat.edu •
Dir.: *Conchita F. Ndege* • Cur.: *Vandorn Hinnant* •
Decorative Arts Museum / University Museum / Historical Museum / Folklore Museum – 1968
African, New Guinea and Haiti arts and crafts 49066

The Natural Science Center of Greensboro, 4301 Lawndale Dr, Greensboro, NC 27455 • T: +1 336 2883769 • F: +1 336 2880545 • info@natsci.org • www.natsci.org •
Dir.: *Glenn Dobrogosz* • Cur.: *Richard G. Bolling* • *Patricia Day* • *Peggy V. Ferebee* • *Kenneth A. Schneidmiller* • *Roger D. Joyner* • *Richard A. Betton* •
Natural History Museum – 1957
Geology, ornithology, entomology, mammals, reptiles 49067

Weatherspoon Art Museum, University of North Carolina at Greensboro, Spring Garden and Tate St, Greensboro, NC 27402-6170 • T: +1 336 3345770 • F: +1 336 3345907 • weatherspoon@uncg.edu • weatherspoon.uncg.edu •
Dir.: *Nancy Doll* • Cur.: *Ronald Platt* • *Will South* •
Fine Arts Museum / University Museum – 1942
American modern and contemp art, works on paper, Claribel and Etta Cone Collection including Matisse lithographs and bronzes 49068

Greensburg KS

Big Well, 315 S Sycamore, Greensburg, KS 67054 •
Natural History Museum – 1937
World's largest Pallasite meteorite 49069

Greensburg PA

Westmoreland Museum of American Art, 221 N Main St, Greensburg, PA 15601-1898 • T: +1 724 8371500 • F: +1 724 8372921 • info@wmuseumaa.org • www.wmuseumaa.org •
Dir.: *Judith H. O'Toole* • Cur.: *Barbara L. Jones* •
Fine Arts Museum – 1949
Paintings, sculpture, drawings, prints, furniture 49070

Greentown IN

Greentown Glass Museum, 112 N Meridian, Greentown, IN 46936 • T: +1 317 6286206 • www.eastern.k12.in.us/gpl/greentownglassmuseum.htm •
Head Cur.: *Norma Jean David* • Cur.: *Annette LaRowe* •
Decorative Arts Museum – 1969
Glass 49071

Greenville IL

Hoiles-Davis Museum, 318 W Winter St, Greenville, IL 62246 • T: +1 618 6641590 • duke36prod@hotmail.com •
Local Museum – 1955
Artifacts, antiques, newspaper articles 49072

Greenville MS

Old Firehouse Museum, 340 Main St, Greenville, MS 38701 • T: +1 662 3781538 • F: +1 662 3781564 • ctyofgvl@tecinfo.com •
C.E.O.: *Paul C. Artman jr* •
Historical Museum – 1994
Firefighting equip 49073

Greenville NC

Greenville Museum of Art, 802 S Evans St, Greenville, NC 27834 • T: +1 252 7581946 • F: +1 252 7581946 • info@gmoa.org • www.gmoa.org •
C.E.O. & Dir.: *C. Barbour Strickland* • Gallery Asst.: *Christopher Daniels* •
Fine Arts Museum – 1956
Paintings, sculpture, ceramics, 20th c American art 49074

Wellington B. Gray Gallery, East Carolina University, Jenkins Fine Arts Center, E 5th St and Jarvis St, Greenville, NC 27858-4353 • T: +1 252 3286336 • F: +1 252 3286441 • leebrickg@mail.ecu.edu • www.ecu.edu/graygallery •
Dir.: *Gilbert Leebrick* •
Fine Arts Museum / University Museum – 1978
African art, Baltic ceramics 49075

Greenville OH

Garst Museum, 205 N Broadway, Greenville, OH 45331-2222 • T: +1 937 5485250 • F: +1 937 5487645 • www.garstmuseum.org •
Dir.: *Judy Logan* •
Historical Museum – 1903
Hist artifacts from Anthony Wayne army, Lohmann Telescope collection, James Young stereoscope collection 49076

Greenville PA

Weyers-Sampson Art Gallery, c/o Thiel College, 75 College Av, Greenville, PA 16125 • T: +1 724 5892095 • F: +1 724 5892021 •
Fine Arts Museum – 1971
Paintings, prints 49077

Greenville SC

Bob Jones University Museum & Gallery, Inc., 1700 Wade Hampton Blvd, Greenville, SC 29614 • T: +1 864 7701331 • F: +1 864 7701306 • contact@bjumg.org • www.bjumg.org •
Dir.: *Erin Jones* • Cur.: *John Nolan* •
Fine Arts Museum / Religious Arts Museum – 1951
Coll of European Old Masters, icons, antiquities, furniture, tapestries 49078

Greenville County Museum of Art, 420 College St, Greenville, SC 29601 • T: +1 864 2717570 • F: +1 864 2717579 • info@greenvillemuseum.org • www.greenvillemuseum.org •
Dir.: *Thomas W. Styron* • Cur.: *Martha R. Severens* •
Fine Arts Museum – 1963
Career surveys of Andrew Wyeth and Jasper Johns; Southern Collection, which traces the history of American art using Southern-related examples from Colonial times to the present – Center for Museum Education 49079

Greenville TX

Audie Murphy/American Cotton Museum, 600 I-30 E, Greenville, TX 75401 • T: +1 903 4504502, 4541990 • F: +1 903 4541990 • staff@cottonmuseum.com • www.cottonmuseum.com •
Dir.: *Adrien Witkofsky* •
Historical Museum – 1987
Cotton tools and machinary, cotton production hist 49080

Greenwich CT

Bruce Museum of Arts and Science, One Museum Dr, Greenwich, CT 06830 • T: +1 203 8690376 • F: +1 203 8690963 • webmaster@brucemuseum.com • www.brucemuseum.com •
Dir.: *Peter C. Sutton* •
Fine Arts Museum / Decorative Arts Museum – 1912
American paintings, sculpture and decorative art 49081

Historical Society Museum of the Town of Greenwich, 39 Strickland Rd, Greenwich, CT 06807 • T: +1 203 8696899 • F: +1 203 8619720 • dmecky@hstg.org • www.hstg.org •
C.E.O.: *Debra Walker Mecky* •
Local Museum – 1931
Local history – archives 49082

Putnam Cottage, 243 E Putnam Av, Greenwich, CT 06830 • T: +1 203 8699697 •
Chm.: *Eugene Fox* •
Historical Museum – 1900
Local his 49083

Greenwich NJ

Cumberland County Historical Society Museum, 960 YeGreate St, Greenwich, NJ 08323 • T: +1 856 4554055 • F: +1 856 4558580 • lummis2@juno.com • www.co.cumberland.nj.us •
Pres.: *Sara C. Watson* •
Local Museum – 1908
Agriculture, costumes, Civil War artifacts, furnishings, fossil and Indian artifacts 49084

Greenwood MS

Cottonlandia Museum, 1608 Hwy 82 W, Greenwood, MS 38930 • T: +1 662 4530925 • F: +1 662 4557556 • rperson@tecinfo.info • www.cottonlandia.org •
Exec. Dir.: *Robin Seage Person* •
Local Museum – 1969
Mississippi archaeology 49085

Florewood State Park, 1999 County Rd 145, Greenwood, MS 38930 • T: +1 662 4553821 • F: +1 662 4532459 • florew@tecinfo.com • www.mdwfp.com •
Dir.: *Myra Castle* •
Local Museum – 1976
Furniture, household accessories, food preparations utensils, blacksmith and carpentry tools, farm implements, steam engines, mill, wagons 49086

Greenwood SC

The Museum, 106 Main St, Greenwood, SC 29646 • T: +1 864 2297093 • F: +1 864 2299317 • themuseum@greenwood.net • www.themuseum-greenwood.org •
Pres.: *Mack Baltzegar* • Dir.: *M. Lyda Carroll* •
Local Museum – 1968
Local hist 49087

Gresham OR

Gresham History Museum, 410 N Main Av, Gresham, OR 97030 • T: +1 503 6610347 •
Pres.: *Debra Insley* •
Local Museum – 1976
Local history 49088

Greybull WY

Greybull Museum, 325 Greybull Av, Greybull, WY 82426 • T: +1 307 7652444 •
Dir.: *Wanda L. Bond* •
Local Museum – 1967
Local hist, Indian artifacts, geology, fossils and minerals 49089

Grinnell IA

Faulconer Gallery at Grinnell College, 1108 Park St, Grinnell, IA 50112 • T: +1 641 2694660 • F: +1 641 2694626 • strongdj@grinnell.edu • www.grinnell.edu/faulconergallery •
Dir.: *Lesley Wright* • Assoc. Dir.: *Dan Strong* • Cur.: *Kay Wilson Jenkins* •
Fine Arts Museum / University Museum – 1998
Art on paper, African ceramics 49090

Grinnell College Art Gallery, Grinnell, IA 50112 • T: +1 641 2693371 • F: +1 641 2694283 • jenkins@grinnell.edu •
Cur.: *Kay Wilson* •
Fine Arts Museum – 1983
Works of art on paper 49091

Grinnell Historical Museum, 1125 Broad St, Grinnell, IA 50112 • T: +1 641 2363252 •
Chm.: *Jeff Garland* •
Historical Museum – 1954
Historic artifacts 49092

Grosse Ile MI

Grosse Ile Historical Museum, Railroad Depot, E River Rd and Grosse Ile Pkwy, Grosse Ile, MI 48138, mail addr: POB 138, Grosse Ile, MI 48138 • T: +1 734 6751250 •
Pres.: *Sarah Lawrence* •
Local Museum – 1967
Local history, clothing, furniture 49093

Grosse Pointe MI

Edsel and Eleanor Ford House, 1100 Lake Shore Rd, Grosse Pointe, MI 48236 • T: +1 313 8844222 • F: +1 313 8845977 • info@fordhouse.org • www.fordhouse.org •
Pres.: *John F. Miller* • Cur.: *Josephine Shea* •
Decorative Arts Museum / Historic Site – 1978
Historic house, decorative art, design, paintings, graphics, automotive, architecture, landscape design 49094

Groton CT

Submarine Force Museum and Historic Ship Nautilus, 1 Crystal Lake Rd, Groton, CT 06349-5571 • T: +1 860 4493558 • F: +1 860 4494150 • www.ussnautilus.org •
Cur.: *Stephen Finnigan* •
Military Museum / Science&Tech Museum – 1954
Historic submarine 49095

Groton MA

Groton Historical Museum, 172 Main St, Groton, MA 01450 • T: +1 978 4489476 • F: +1 781 2463335 • encarter@aol.com •
Cur.: *Earl N. Carter* •
Historical Museum – 1894
Local hist, Revolutionary and Civil war items 49096

Grove OK

Har-Ber Village, 4404 W 20th St, Grove, OK 74344 • T: +1 918 7866446, 7863488 • F: +1 918 7876213 • harbervil@aol.com • www.har-bervillage.com •
Pres.: *Gary Smith* • Dir.: *Jan Norman* •
Local Museum – 1968
Pottery, China, toys, natural hist, furniture, lamps, dolls, farm machinery, stagecoaches, prairie schooner, steam engines, wagons, guns, musical instruments, clothing 49097

Groveland CA

Groveland Yosemite Gateway Museum, 18990 Main St, Groveland, CA 95321 • T: +1 209 9620300 • information@grovelandmuseum.org • www.grovelandmuseum.org •
Cur.: *Robert Worthington* •
Historical Museum – 2000
History of southern Tuolumne county 49098

Groveport OH

Motts Military Museum, 5075 S Hamilton Rd, Groveport, OH 43125 • T: +1 614 8361500 • F: +1 614 8365110 • info@mottsmilitarymuseum.org • www.mottsmilitarymuseum.org •
Dir.: *Warren E. Motts* •
Military Museum – 1988
Military items from the Civil War, WW 1 and 2, Korea, Vietnam and Desert Storm 49099

Groversville NY

Fulton County Museum, 237 N Kingsboro Av, Groversville, NY 12078 • T: +1 518 7252203 •
C.E.O. & Pres.: *William C. Clizbe* •
Historical Museum – 1891
Local history, glove and leather industry, weavings, machinery, leather tanning, weapons 49100

Guernsey WY

Lake Guernsey Museum, U.S. 26 to State Hwy 317, Guernsey, WY 82214 • T: +1 307 8362334, 8362900 • F: +1 307 8363088 •
Interim Dir.: *John Keck* •
Local Museum – 1936
Agriculture, ethnology, local hist, industry, military 49101

Guilford CT

Henry Whitfield State Historical Museum, 248 Old Whitfield St, Guilford, CT 06437 • T: +1 203 4532457 • F: +1 203 4537544 • whitfieldmuseum@snet.net • www.chc.state.ct.us •
Dir.: *Michael A. McBride* •
Historical Museum – 1899
First state-owned museum in Connecticut (1899) 49102

Hyland House, 84 Boston St, Guilford, CT 06437 • T: +1 203 4539477 • cfmacken@aol.com • www.hylandhouse.com •
Cur.: *Pamela Besse* •
Decorative Arts Museum – 1918
17th to early 18th c furniture, decorative arts 49103

Thomas Griswold House and Museum, 171 Boston St, Guilford, CT 06437 • T: +1 203 4533176 • info@guilfordkeepingsociety.com • www.guilfordkeepingsociety.com •
C.E.O.: *Sandra L. Rux* •
Historical Museum – 1947
19th-20th c glass and photographs, docs, deeds, genealogical data from 1693, furnishings 49104

Gulf Shores AL

Gulf Shores Museum, 244 W 19 th Av, Gulf Shores, AL 36561, mail addr: POB 299, Gulf Shores, AL 36547 • T: +1 251 9681473 • F: +1 251 99681175 • museum@gulfshoresal.gov • www.cityofgulfshores.org/pages_2006/museum •
Natural History Museum – 2000
Local hist 49105

Gunnison CO

Gunnison County Pioneer and Historical Museum, S Adams St and Hwy 50, Gunnison, CO 81230 • T: +1 970 6414530 •
Dir.: *C.J. Miller* •
Local Museum – 1930
Regional history, geology, minerals, costume, railroad, agriculture 49106

Gurdon AR

Hoo-Hoo International Forestry Museum, 207 Main St, Gurdon, AR 71743 • T: +1 870 3534997 • F: +1 870 3534151 • info@hoo-hoo.org • www.hoo-hoo.org •
Ethnology Museum / Fine Arts Museum – 1981
Logging, lumber, tools, woodcarvings 49107

Guthrie OK

Oklahoma Territorial Museum, Oklahoma Historical Society, 402 E Oklahoma Av, Guthrie, OK 73044 • T: +1 405 2821889 •
C.E.O.: *Dr. Bob Blackburn* • Dir.: *Michael Bruce* • Cur.: *Helen Stiefmiller* • Ass. Cur.: *Nathan Turner* •
Historical Museum – 1970
Oklahoma 1889 Land run 49108

State Capital Publishing Museum, Oklahoma Historical Society, 301 W Harrison Av, Guthrie, OK 73044 • T: +1 405 2824123 • guthriecomplex@ok-history.mus.ok.us •
C.E.O.: *Dr. Bob Blackburn* • Dir.: *Michael Bruce* • Cur.: *Helen Stiefmiller* • Ass. Cur.: *Nathan Turner* •
Science&Tech Museum – 1976
Printing and publishing artifacts, vintage printing press, manuscript coll 49109

Hackensack NJ

Edward Williams Gallery, Fairleigh Dickinson University, 150 Kotte Pl, Hackensack, NJ 07601 • T: +1 201 6922449 •
Dir.: *Rachel Friedberg* •
University Museum / Fine Arts Museum 49110

New Jersey Naval Museum, 78 River St, Hackensack, NJ 07601 • T: +1 201 3423268 • F: +1 201 3423268 • njnavalmus@aol.com • www.njnm.com •
C.E.O.: *Chris Buermeyer* •
Military Museum – 1974
Naval military hist 49111

Hackettstown NJ

Hackettstown Historical Society Museum, 106 Church St, Hackettstown, NJ 07840 • T: +1 908 8528797 •
Cur.: *Helen G. Montfort* •
Local Museum – 1975
Local history, genealogy – library 49112

Haddon Township NJ

Hopkins House Gallery, 250 S Park Dr, Haddon Township, NJ 08108 • T: +1 856 8580040 • F: +1 856 8693548 • cccandhc@co.camden.nj.us • arts.camden.lib.nj.us •
Dir.: *Ruth Bogutz* •
Public Gallery – New Jersay and Camden art 49113

Haddonfield NJ

Haddonfield Museum, 343 Kings Hwy E, Haddonfield, NJ 08033 • T: +1 856 4297375 • hadhistsoc@netcarrier.com • www.08033.com •
Dir.: *Robert A. Marshall* • Cur.: *Dianna H. Snodgrass* •
Historical Museum – 1914
History of New Jersey and Haddonfield – library 49114

Indian King Tavern Museum, 233 King's Hwy E, Haddonfield, NJ 08033 • T: +1 856 4296792 • www.indiankingtavern.com •
Cur.: *William J. Mason* •
Historic Site – 1903
Historical items 49115

Markeim Art Center, Lincoln Av & Walnut Sts, Haddonfield, NJ 08033 • T: +1 856 4298585 • F: +1 856 4298585 • markeimartcenter@msn.com •
Dir.: *Lisa Hamil* •
Fine Arts Museum
Art 49116

Hadley MA

Hadley Farm Museum, 147 Russell St, Hadley, MA 01035 • T: +1 413 5861812 •
Cur.: *Norman Barstow* •
Agriculture Museum – 1930
Farm equipment, plows, winnowing machines, corn sheller, stage coach, sleighs, crafts 49117

Porter-Phelps-Huntington House Museum, 130 River Dr, Hadley, MA 01035 • T: +1 413 5844699 • www.pphmuseum.org •
Dir.: *Susan J. Lisk* •
Historical Museum – 1955
Womens history studies 49118

Hagerstown IN

Wilbur Wright Birthplace and Interpretive Center, 1525 N County Rd 750 E, Hagerstown, IN 47346 • T: +1 765 3322495 • F: +1 765 3322805 • wilbur@nltc.net •
Historical Museum – 1929
Furniture, local hist 49119

Hagerstown MD

Jonathan Hager House and Museum, 110 Key St, Hagerstown, MD 21740 • T: +1 301 7398393 • hagerhouse@hagerstownmd.org • www.hagerhouse.org •
Cur.: *John Nelson* •
Fine Arts Museum – 1962
18th c furnishings, pottery, glass, china, 18th c coins 49120

Miller House Museum, 135 W Washington St, Hagerstown, MD 21740 • T: +1 301 7978782 • F: +1 301 7979509 • histsoc@earthlink.net • www.rootsweb.com/~mdwchs •
Exec. Dir.: *Melinda Marsden* • Cur.: *Jeniffer Dintaman* •
Local Museum – 1911
Antiques, local history 49121

Washington County Museum of Fine Arts, City Park, 91 Key St, Hagerstown, MD 21741, mail addr: PO Box 423, Hagerstown, MD 21741 • T: +1 301 7395764 • F: +1 301 7453741 • rmlane@wcmfa.org • www.washcomuseum.org •
Dir.: *Jean Woods* • Cur.: *Amy Metzger* •
Fine Arts Museum – 1929
19th and 20th c american art 49122

Hailey ID

Blaine County Historical Museum, 218 N Main St, Hailey, ID 83333 • T: +1 208 7264226 • teddie1@mindsprin.com • www.bchistoricalmuseum.org •
C.E.O: *Teddie Daley* •
Local Museum – 1964
Regional history 49123

Haines AK

Alaska Indian Arts, Historic Bldg 13, Fort Seward, Haines, AK 99827, mail addr: POB 271, Haines, AK 99827 • T: +1 907 7662160 • www.alaskaindianarts.com •
Dir.: *Lee D. Heinmiller* • Asst. Dir.: *Charles Jimmie* (Culture) • *John Hagen* (Art) •
Fine Arts Museum / Ethnology Museum – 1957
Indian living village, Indian arts 49124

Sheldon Museum, 11 Main St, Haines, AK 99827, mail addr: POB 269, Haines, AK 99827 • T: +1 907 7662366 • F: +1 907 7662368 • museumdirector@aptalaska.net • www.sheldonmuseum.org •
Dir.: *Cynthia L. Jones* •
Historical Museum / Ethnology Museum – 1924
History, ethnography, Tlingit Indian culture, Eldred Rock Lighthouse Lense – library, archives 49125

Haines OR

Eastern Oregon Museum on the Old Oregon Trail, Third and Wilcox, Haines, OR 97833 • T: +1 503 8563233 • mrifer@eoni.com • www.hainesoregon.com/eomuseum.html •
Pres.: *Coral Rose* •
Local Museum / Historical Museum – 1958
Farm implements, blacksmith shop, parlor, buggies, cutters 49126

Haledon NJ

American Labor Museum, 83 Norwood St, Haledon, NJ 07508 • T: +1 973 5957953 • F: +1 973 5957291 • labormuseum@aol.com • www.geocities.com/labormuseum •
Dir.: *Angelica M. Santomauro* •
Historical Museum – 1982
Textile artifacts, photographs. local hist 49127

Halifax NC

Historic Halifax, Saint David and Dobb Sts, Halifax, NC 27839 • T: +1 252 5837191 • F: +1 252 5839241 • histhalifax@ncolnet.com • www.ah.dcr.state.nc.us/sections/hs/halifax/halifax.htm •
Local Museum – 1955
Archaeology, history, hist houses 49128

Hallowell ME

Harlow Gallery, Kennebec Valley Art Association, 160 Water St, Hallowell, ME 04347 • T: +1 207 6223813 • kvaa@harlowgallery.org • www.harlowgallery.org •
Pres.: *Penny Markley* • C.E.O.: *Linda Murray* •
Fine Arts Museum / Public Gallery – 1963
Art 49129

Halstead KS

Kansas Health Museum, Kansas Learning Center for Health, 505 Main St, Halstead, KS 67056 • T: +1 316 8352662 • F: +1 316 8352755 • sweesner@learningcenter.org • www.learningcenter.org •
Dir.: *Megan E. Evans* •
Historical Museum – 1965
Health, education 49130

Hamden CT

Eli Whitney Museum, 915 Whitney Av, Hamden, CT 06517 • T: +1 203 7771833 • F: +1 203 7771229 • wb@eliwhitney.org • www.eliwhitney.org •
Pres.: *Beverly Hodgson* • Dir.: *William Brown* •
Science&Tech Museum – 1976 49131

Hamilton MT

Ravalli County Museum, 205 Bedford, Hamilton, MT 59840 • T: +1 406 3633338 • F: +1 406 3633338 • rcmuseum@cybernet1.com • www.cybernet1.com/rcmuseum •
Dir.: *Helen Ann Bibler* •
Local Museum – 1955
Early life of Bitterroot Valley, Indian, railroads 49132

Hamilton NJ

Grounds for Sculpture, 18 Fairgrounds Rd, Hamilton, NJ 08619 • T: +1 609 5860616 • F: +1 609 5860968 • info@groundsforsculpture.org • www.groundsforsculpture.org •
Dir. & Cur.: *Brooke Barrie* •
Fine Arts Museum – 1992
Sculptures 49133

International Sculpture Center, 14 Fairground Rd, Ste B, Hamilton, NJ 08619-3447 • T: +1 609 6891051 • F: +1 609 6891061 • Kristine@sculpture.org • www.sculpture.org •
Public Gallery / Association with Coll – 1960
Temporary coll of sculpture 49134

John Abbott II House, 2200 Kuser Rd, Hamilton, NJ 08690 • T: +1 609 5851686 • Thomglo@aol.com •
Pres.: *Thomas Glover* •
Historical Museum – 1976
Colonial and Victorian furnishings, textiels. local hist 49135

Kuser Farm Mansion, 390 Newkirk Av, Hamilton, NJ 08610 • T: +1 609 8903630 • F: +1 609 8903632 • comments@hamiltonnj.com • www.hamiltonnj.com •
Cur.: *Denise Dale Zemlansky* •
Historical Museum – 1979
Local hist 49136

Hamilton NY

Picker Art Gallery, Colgate University, 13 Oak Dr, Hamilton, NY 13346 • T: +1 315 2287634 • F: +1 315 2287932 • pickerart@mail.colgate.edu • picker.colgate.edu •
Fine Arts Museum – 1966
Modern and contemporary art, pre-Columbian art 49137

Hamilton OH

Butler County Museum, 327 N Second St, Hamilton, OH 45011 • T: +1 513 8969930 • F: +1 513 8969936 • bcomuseum@fuse.net • home.fuse.net/butlercountymuseum/ •
Pres.: *Tom Stander* • Dir.: *Carol A. Gabriel* •
Historical Museum – 1934
Glassware, china, musical instruments, toys, dolls, Indian artifacts 49138

Pyramid Hill Sculpture Park and Museum, 1763 Rte 128, Hamilton, OH 45013 • T: +1 513 8688336 • F: +1 513 8683585 • pyramid@pyramidhill.org • www.pyramidhill.org •
Dir.: *Harry T. Wilks* •
Fine Arts Museum – 1997
Sculptures, art 49139

Hamlet NC

National Railroad Museum and Hall of Fame, 2 Main St, Hamlet, NC 28345 • T: +1 910 5822382 •
C.E.O. & Pres.: *Bill Williams* •
Science&Tech Museum – 1976
Maps, model railroad, rolling stock, SAL locomotive 49140

Hammond IN

Hammond Historical Museum, 564 State St, Hammond, IN 46320 • T: +1 219 9315100, 8522255 • F: +1 219 9313474 • longs@hammond.lib.in.us • www.hammondindiana.com •
Dir.: *Margaret Evans* •
Local Museum – 1960
Local hist – library 49141

Purdue University Calumet Library Gallery, POB 2590, Hammond, IN 46323-2094 • T: +1 219 9892249 • F: +1 219 9892070 •
Decorative Arts Museum – 1976
Chinese scroll coll 49142

Hammondsport NY

Glenn H. Curtiss Museum, 8419 Rte 54, Hammondsport, NY 14840 • T: +1 607 5692160 • F: +1 607 5692040 • curtiss@linkny.com • www.linkny.com/curtissmuseum •
C.E.O. & Dir.: *Trafford Doherty* • Cur.: *Kirk House* •
Historical Museum / Science&Tech Museum – 1961
Aviation, local hist, hist of Glenn Curtiss aircraft and engines, Curtiss motorcycles 49143

Wine Museum of Greyton H. Taylor, 8843 Greyton H. Taylor Memorial Dr, Hammondsport, NY 14840 • T: +1 607 8684814 • F: +1 607 8683205 • bullyhil@ptd.net • www.bullyhill.com •
Pres.: *Lillian Taylor* •
Agriculture Museum – 1967
Hist of wine, winemaking equip, wine bottles and labels 49144

Hammonton NJ

Batsto Village, 4110 Nesco Rd, Hammonton, NJ 08037 • T: +1 609 5610024 • F: +1 609 5678116 • whartonsf@netzero.com • www.batstovillage.org •
Historic Site / Decorative Arts Museum – 1954
33 historic buildings built in the 1800s 49145

Hampden-Sydney VA

The Esther Thomas Atkinson Museum, College Rd, Hampden-Sydney, VA 23943 • T: +1 804 2236134, 2236000 • F: +1 804 2236344 • lmastemaker@hsc.edu • www.hsc.edu/museum/ •
Cur.: *Lorie Mastemaker* •
University Museum / Historical Museum – 1968
College hist from 18th c to the present 49146

Hampton NH

Tuck Museum, 40 Park Av, Hampton, NH 03842 • T: +1 603 9262543 • info@hamptonhistoricalsociety.org • www.hamptonhistoricalsociety.org •
Pres.: *Bennett Moore* •
Historical Museum – 1925
Period artifacts and documents pertaining to colonial Hampton 49147

Hampton SC

Hampton County Historical Museum, 702 1st West, 601 South, Hampton, SC 29924 • T: +1 803 9435484 • wmowen1@earthlink.net •
C.E.O. & Pres.: *Kevin Brown* •
Local Museum / Folklore Museum – 1979
Local hist 49148

Hampton VA

Air Power Park and Museum, 413 W Mercury Blvd, Hampton, VA 23666 • T: +1 757 7271163 •
Dir.: *Jim Wilson* •
Military Museum / Science&Tech Museum – 1964
Military aircraft 49149

Charles H. Taylor Arts Center, 4205 Victoria Blvd, Hampton, VA 23669 • T: +1 757 7271490 • F: +1 757 7271167 • artscom@hampton.gov • www.theamericantheatre.com/taylor •
Dir.: *Michael P. Curry* •
Fine Arts Museum – 1989
Regional and national artists 49150

Hampton History Museum, 120 Old Hampton Llane, Hampton, VA 23669 • T: +1 757 7271610 • F: +1 757 7276712 • www.hampton.gov/history_museum •
Dir.: *James W. Hollomon* • Cur.: *Michael Cobb* •
Historical Museum – 2003
Area hist from 1607 to present 49151

Hampton University Museum, Hampton University, Hampton, VA 23668 • T: +1 757 7275308 • F: +1 757 7275170 • museum@hamptonu.edu • www.hamptonu.edu •
Cur.: *Nancy Challender* •
University Museum – 1868
Traditional Africa, Asian, Oceanic & American Indian art 49152

Saint John's Church and Parish Museum, 100 W Queensway, Hampton, VA 23669 • T: +1 757 7222567 • F: +1 757 7220641 • office@stjohns-hampton.org • www.stjohns-hampton.org •
Cur.: *Beverly F. Gundry* •
Local Museum / Historic Site – 1976
16th - 18th c paryer books & bible, ceramics, memorabilia 49153

Virginia Air and Space Center, 600 Settlers Landing Rd, Hampton, VA 23669 • T: +1 757 7270900 • F: +1 757 7270898 • tbridgfo@vasc.org • www.vasc.org •
Dir.: *Todd C. Bridgford* •
Science&Tech Museum – 1991
Aerospace, military & civilian aircraft, NASA spacecraft 49154

Hamtramck MI

Ukrainian-American Museum, 11756 Charest St, Hamtramck, MI 48212-3027 • T: +1 313 3669764 •
Dir./Cur.: *Danylo Dmytrykiw* •
Folklore Museum / Local Museum / Historical Museum – 1958
archives, library 49155

Hana, Maui HI

Hana Cultural Center, 4974 Uakea Rd, Hana, Maui, HI 96713 • T: +1 808 2488622 • F: +1 808 2488620 • hccm@aloha.net •
Exec. Dir.: *Coila Eade* •
Local Museum – 1971
Hawaii artifacts, natural hist, paintings, geological, archaeology, ethnology, photographs 49156

Hancock NH

Hancock Historical Society Museum, 7 Main St, Hancock, NH 03449-6008 • T: +1 603 5259379 • hancockhistsoc@monaj.net • www.mv.com/ipusers/hancocknh/hhs/home.htm •
Cur.: *Eleanor Amidon* •
Local Museum / Historical Museum – 1903
18th - 20th c archives, photographs 49157

Hancock's Bridge NJ

Hancock House, 3 Front St, Hancock's Bridge, NJ 08038 • T: +1 856 9354373 • F: +1 856 9257818 • fortmott@snip.net •
Historical Museum
18th - 19th c regional furnishings 49158

Hannibal MO

Mark Twain Home and Museum, 208 Hill St, Hannibal, MO 63401 • T: +1 573 2219010 • F: +1 573 2217975 • museumoffice@marktwainmuseum.org • www.marktwainmuseum.org •
Dir.: *Henry Sweets* • Pres.: *Herbert S. Parham* •
Historical Museum – 1936
Mark Twain memorabilia, historic restaurations 49159

Hanover KS

Hollenberg Pony Express Station, 2889 23rd Rd, Kansas Hwy 243, Hanover, KS 66945-9634 • T: +1 913 3372635 • hollenberg@ksk.org •
Cur.: *Duane R. Durst* •
Historic Site
1857 Pony Express stn, comprised of general store, tavern and stage stn 49160

Hanover MI

Conklin Reed Organ and History Museum, 105 Fairview, Hanover, MI 49241 • T: +1 517 5638927 • F: +1 517 5638927 • info@conklinreedorganmuseum.org • www.conklinreedorganmuseum.org •
Pres.: *Jim Mann* •
Local Museum / Music Museum – 1977
Lee Conklin's reed organs, parlor and chapel organs, teaching hist, local hist, printing equip – auditorium, nature trail 49161

Hanover NH

Hood Museum of Art, Dartmouth College, Hanover, NH 03755 • T: +1 603 6462808 • F: +1 603 6461400 • hood.museum@dartmouth.edu • www.hoodmuseum.dartmouth.edu •
Dir.: *Derrick R. Cartwright* • Sc. Staff: *T. Barton Thurber* (European Art) • *Lesley Wellman* (Education) • *Barbara Thompson* (African, Oceanic and Native American Art) • *Evelyn Marcus* (Exhibitions) • *Katherine Hart* (Academic Programming) • *Barbara J. MacAdam* (American Art) • *Margaret E. Spicer* (Costumes) •
Fine Arts Museum / University Museum – 1772
Fine arts, anthropology 49162

Hanover VA

Hanover Historical Society Museum, Hwy 301, Court Green, Hanover, VA 23069 • T: +1 804 5375815 •
Cur.: *W.C. Wickham* •
Local Museum – 1967
Local hist material, portraits in courthouse 49163

Harbor Springs MI

Andrew J. Blackbird Museum, 368 E Main St, Harbor Springs, MI 49740 • T: +1 231 5260612 • F: +1 231 5262705 •
Man.: *Joyce Shagonaby* •
Historical Museum / Ethnology Museum – 1952
Odawa Indian clothing, tools, art 49164

Hardin MT

Big Horn County Historical Museum, RR 1 Box 1206A, Hardin, MT 59034 • T: +1 406 6651671 • F: +1 406 6653068 • bhcmuseum@bhwi.net • www.museumonthebighorn.org •
Dir.: *Diana Scheidt* • Cur.: *Bonnie Stark* •
Historical Museum – 1979
Farm equipment, horse drawn equip, farmhouse, barn; Crow Indians and Northern Cheyenne Indians hist and culture; settlement hist 49165

Hardwick MA

Hardwick Historical Museum, On Hardwick Common, Hardwick, MA 01037 • T: +1 413 4776635, 4778734 •
Cur.: *Leon Thresher* •
Local Museum – 1959
Local history, Indian artifacts, costumes, china, tools, portraits 49166

Hardy VA

Booker T. Washington National Monument, 12130 Booker T. Washington Hwy, Hardy, VA 24101 • T: +1 540 7212094 • F: +1 540 7218311, 7215128 • www.nps.gov/bowa •
C.E.O.: *Rebecca Harriett* •
Historic Site – 1956
Birthplace & early home of Booker T. Washington 49167

Harleysville PA

Mennonite Heritage Center, 565 Yoder Rd, Harleysville, PA 19438 • T: +1 215 2563020 • F: +1 215 2563023 • info@mhep.org • www.mhep.org/ •
Dir.: *Sarah Heffner* • Cur.: *Joel D. Alderfer* •
Local Museum / Religious Arts Museum – 1974
Local hist – library 49168

Harlingen TX

Lon C. Hill Home, Rio Grande Valley Museum, Boxwood at Raintree, Harlingen, TX 78550 • T: +1 956 4308500 • F: +1 956 4308502 • rgvmuse@hiline.net •
Dir.: *Linn Keller* •
Local Museum – 1905
Historic house 49169

Rio Grande Valley Museum, Boxwood at Raintree, Harlingen, TX 78550 • T: +1 956 4308500 • F: +1 956 4308502 • rgvmuse@hiline.net • hiline.net/rgvmuse •
Dir.: *Linn Keller* •
Local Museum – 1967
Valley hist incl Republic of Texas, natural hist, period clothing 49170

Harlowtown MT

Upper Musselshell Museum, 11 S Central, Harlowtown, MT 59036 • T: +1 406 6325519 • museum@mtintouch.net • harlowtonmuseum.com •
Local Museum – 1985
Local area items dating back to the dinosaur age 49171

Harper KS

Harper City Historical Museum, 804 E 12th St, Harper, KS 67058 • T: +1 620 8962824 •
Pres.: *Gail Bellar* •
Local Museum – 1959
Local history, housed in a German Apostolic Church 49172

Harpers Ferry IA

Effigy Mounds Museum, 151 Hwy 76, Harpers Ferry, IA 52146 • T: +1 563 8733491 • F: +1 563 8733743 • efmo_superintendent@nps.gov • www.nps.gov/efmo •
Head: *Phyllis Ewing* •
Archaeology Museum – 1949
Indian artifacts, ethnology, archaeology 49173

Harpers Ferry WV

Harpers Ferry National Historical Park, Fillmore St, Harpers Ferry, WV 25425 • T: +1 304 5356224 • F: +1 304 5356244 • www.nps.gov/hafe/home.htm •
Dir.: *Terry Carlstrom* • Cur.: *Marsha B. Starkey* •
Local Museum – 1944
56 restored buildings and Civil War fortifications 49174

John Brown Wax Museum, 168 High St, Harpers Ferry, WV 25425 • T: +1 304 5356342 • www.johnbrownwaxmuseum.com •
Pres.: *D.D. Kilham* •
Decorative Arts Museum – 1964 49175

Harpursville NY

Colesville and Windsor Museum at Saint Luke's Church, Maple Str, Harpursville, NY 13787 • T: +1 607 6553174 •
Pres.: *Fran Bromley* • Cur.: *Marjory Hinman* •
Historical Museum – 1971
American Indian artifacts, period furnishings, early agriculture and industry 49176

Harrisburg NE

Banner County Historical Museum, 200 N Pennsylvania Av, Harrisburg, NE 69156 • T: +1 308 7831660 •
Pres.: *Harold L. Brown* • C.E.O.: *George Penick* • Cur.: *Norris Leafdale* •
Local Museum / Agriculture Museum – 1969
Regional history 49177

Harrisburg PA

Art Association of Harrisburg, 21 N Front St, Harrisburg, PA 17101 • T: +1 717 2361432 • F: +1 717 2366631 • carrie@artassocofhbg.com • www.artassocofhbg.com •
Pres.: *Carrie Wissler-Thomas* •
Public Gallery – 1926
Paintings, graphics 49178

Fort Hunter Mansion, 5300 N Front St, Harrisburg, PA 17110 • T: +1 717 5995751 • F: +1 717 5995838 • www.forthunter.org •
Dir.: *Carl A. Dickson* • Cur.: *Julia Hair* •
Decorative Arts Museum / Local Museum – 1933
Historic house 49179

John Harris/Simon Cameron Mansion, 219 S Front St, Harrisburg, PA 17104 • T: +1 717 2333462 • F: +1 717 2336059 • www.dauphincountyhistoricalsociety.org • Dir.: *Robert D. Hill* (Collection) • Local Museum – 1869 Decorative arts, household items, tools, toys, textiles, Indian artifacts, ceramics, glass, John Harris and Simon Cameron furniture – library ... 49180

State Museum of Pennsylvania, 300 North St, Harrisburg, PA 17120-0024 • T: +1 717 7874980 • F: +1 717 7834558 • museum@statemuseumpa.org • www.statemuseumpa.org • Dir.: *Anita D. Blackaby* • Cur.: *Dr. Robert M. Sullivan* (Paleontology, Geology) • *N. Lee Stevens* (Art colls) • *Albert G. Mehring* (Natural Science) • *John Zwierzyna* (Industry and Technology, Military and Political History) • *Stephen G. Warfel* (Archaeology) • *Curt Miner* (Popular Culture) • *Dr. Walter Mehaska* (Zoology, Botany) • Local Museum / Historical Museum – 1905 Archaeology, fine and decorative arts, natural science, paleontology, geology, industry, technology, military history, mammals, native American village, civil war – planetarium, dino lab ... 49181

Susquehanna Art Museum, 301 Market St, Harrisburg, PA 17101 • T: +1 717 2338668 • F: +1 717 2338155 • www.squart.org • Pres.: *William Lehr* • Dir.: *Rusty Baker* • Cur.: *Jonathon Van Dyke* • Fine Arts Museum – 1989 ... 49182

Whitaker Center for Science and the Arts, 222 Market St, Kunkel Bldg, Harrisburg, PA 17101 • T: +1 717 2218201 • F: +1 717 2218208 • info@whitakercenter.org • www.whitakercenter.org • C.E.O.: *Byron Quann* • Natural History Museum / Fine Arts Museum – 1993 Physical, natural and life science ... 49183

Harrison OH

American Watchmakers-Clockmakers Institute, 701 Enterprise Dr, Harrison, OH 45030 • T: +1 513 3679800 • F: +1 513 3671414 • jlubic@awi-net.org • www.awi-net.org • C.E.O.: *James E. Lubic* • Special Museum – 1960 Watches, clocks, tools ... 49184

Village Historical Society Museum, 10659 New Biddinger Rd, Harrison, OH 45030 • T: +1 513 3679379 • C.E.O.: *Betty Cookendorfer* • Local Museum – 1962 1804 hist house ... 49185

Harrisonburg VA

D. Ralph Hostetter Museum of Natural History, c/o Eastern Mennonite University, 1200 Park Rd, Harrisonburg, VA 22802-2462 • T: +1 540 4324400, 4324000 • F: +1 540 4324488 • mellinac@emu.edu • www.emu.edu/ • Dir.: *Christine C. Hill* • Cur.: *A. Clair Mellinger* • University Museum / Natural History Museum – 1968 Geology, mineralogy, zoology, anthropology, paleontology ... 49186

Sawhill Gallery, Duke Hall, James Madison University, Main and Grace Sts, Harrisonburg, VA 22807 • T: +1 540 5686407 • F: +1 540 5686598 • Dir.: *Stuart Downs* • Public Gallery – 1967 Ernest Staples coll, Indonesian works, Sawhill coll, pre-classical & classical items, modern works of art . 49187

Harrodsburg KY

Morgan Row Museum, 220 S Chiles St, Harrodsburg, KY 40330 • T: +1 859 7345985 • F: +1 859 7345985 • Pres.: *George Neal* • Local Museum – 1907 Genealogy, early Kentucky hist ... 49188

Old Fort Harrod Museum, US-Hwy 68, Harrodsburg, KY 40330, mail addr: POB 156, Harrodsburg, KY 40330 • T: +1 859 7343314 • F: +1 859 7340794 • joan.huffman@mail.state.ky.us • www.oldfortharrod.com • Dir.: *Joan Huffman* • Local Museum – 1925 Greek revival mansion, replica of original fort, costumes, decorative arts, glass, hist, Indian artifacts, music ... 49189

Shaker Village of Pleasant Hill, 3501 Lexington Rd, Harrodsburg, KY 40330 • T: +1 859 7345411 • F: +1 859 7347278 • lcurry@shakervillageky.org • www.shakervillageky.org • Dir.: *Larrie Spier Curry* • Sc. Staff: *Ralph E. Ward* (Historic Farming) • Local Museum / Open Air Museum – 1961 34 hist early 19th c bldgs ... 49190

Harrogate TN

Abraham Lincoln Museum, Lincoln Memorial University, Hwy 25 E, Cumberland Gap Pkwy, Harrogate, TN 37752 • T: +1 423 8696235 • F: +1 423 8696350 • museum@lmunet.edu • www.lmunet.edu/museum • Dir.: *Dr. Charles M. Hubbard* • Cur.: *Steven Wilson* • Historical Museum – 1897 Abraham Lincoln & Civil War materials, 20.000 artifacts – Library ... 49191

Hartford CT

Antiquarian and Landmarks Society Museum, 66 Forest St, Hartford, CT 06105-3204 • T: +1 860 2478996 • F: +1 860 2494907 • www.hartnet.org/als • Dir.: *William Hosley* • Historical Museum – 1936 ... 49192

Connecticut Historical Society Museum, 1 Elizabeth St, Hartford, CT 06105 • T: +1 860 2365621 • F: +1 860 2362664 • ask_us@chs.org • www.chs.org • Dir.: *David M. Kahn* • Fine Arts Museum / Decorative Arts Museum / Local Museum / Historical Museum – 1825 ... 49193

Harriet Beecher Stowe Center, 77 Forest St, Hartford, CT 06105 • T: +1 860 5229258 • F: +1 860 5229259 • stowelib@hartnet.org • www.harrietbeecherstowe.org • Dir.: *Katherine Kane* • Cur.: *Dawn C. Adiletta* • Historical Museum – 1941 ... 49194

Mark Twain House, 351 Farmington Av, Hartford, CT 06105-4498 • T: +1 860 2470998 • F: +1 860 2788148 • info@marktwainhouse.org • www.marktwainhouse.org • Dir.: *John Vincent Boyer* • Dep. Dir.: *Debra Petke* • Cur.: *Diane Forsberg* • Historical Museum – 1929 ... 49195

Menczer Museum of Medicine and Dentistry, 230 Scarborough St, Hartford, CT 06105 • T: +1 860 2365613 • F: +1 860 2368401 • library.uchc.edu/hms • Cur.: *Bernard Kosta* • Special Museum – 1974 ... 49196

Museum of Connecticut History, Connecticut State Library, 231 Capitol Av, Hartford, CT 06106 • T: +1 860 7576535 • F: +1 860 7576533 • www.cslib.org • Cur.: *Howard L. Miller* • *David J. Corrigan* • Historical Museum – 1910 ... 49197

Old State House, 800 Main St, Hartford, CT 06103 • T: +1 860 5226766 • F: +1 860 5222812 • info@sharect.org • www.ctosh.org • Pres.: *James Williams* • Fine Arts Museum / Historical Museum – 1975 . 49198

Wadsworth Atheneum, 600 Main St, Hartford, CT 06103-2990 • T: +1 860 2782670 • F: +1 860 5270803 • info@wadsworthatheneum.org • www.wadsworthatheneum.org • Dir.: *Willard Holmes* • Cur.: *Elizabeth Mankin Kornhauser* (American Painting and Sculpture) • *Eric Zafran* (European Painting and Sculpture) • *Linda H. Roth* (European Decorative Arts) • *Carol Dean Krute* (Costume, Textiles) • *Eugene R. Gaddis* (Austin House) • *Joanna Marsh* (Contemporary Art) • *Rena Hoisington* (European Art) • *John Teahan* (Special Book coll) • Chief. Kons.: *Ulrich Brinkmaier* (Paintings) • Fine Arts Museum – 1842 ... 49199

Widener Gallery, c/o Trinity College, Austin Arts Center, 300 Summit St, Hartford, CT 06106 • T: +1 860 2975232 • F: +1 860 2975349 • felice.caivano@mail.trincoll.edu • www.trincoll.edu • Dir.: *Joseph Byrne* • Cur.: *Felice Caivano* • Fine Arts Museum / University Museum – 1964 . 49200

Hartford WI

Wisconsin Automotive Museum, 147 N Rural St, Hartford, WI 53027 • T: +1 262 6737999 • info@wisconsinautomuseum.com • www.wisconsinautomuseum.com • Dir.: *Dale W. Anderson* • Science&Tech Museum – 1982 Vintage cars, trucks, fire engines, 1906-1931 Kissel automobiles ... 49201

Hartsville SC

Cecelia Coker Bell Gallery, Coker College, Art Dept., 300 E College Av, Hartsville, SC 29550 • T: +1 843 3838156 • F: +1 843 3838048 • lmerriman@coker.edu • www.coker.edu/art/gallery.html • Dir.: *Larry Merriman* • University Museum / Public Gallery – 1983 Graphiocs, sculpture, photography, installation, painting ... 49202

Hartsville Museum, 222 N Fifth St, Hartsville, SC 29550 • T: +1 843 3833005 • F: +1 843 3832477 • info@hartsvillemuseum.org • www.hartsvillemuseum.org • Dir.: *Patricia J. Wilmot* • Local Museum / Fine Arts Museum – 1980 Local hist, art ... 49203

Jacob Kelley House Museum, 2585 Kellytown Rd, Hartsville, SC 29551 • T: +1 843 3326401, 3399093 • F: +1 843 3328017 • DCHC1968@aol.com • Dir.: *Doris Gandy* • Decorative Arts Museum Handcrafted furniture, decorative arts ... 49204

Robinson Visitors Center, 3581 West Entrance Rd, Hartsville, SC 29550 • T: +1 843 8575291 • F: +1 843 8571319 • mike.mccracken@cpk.com • Science&Tech Museum – 1968 Nuclear and coal generation of electricity ... 49205

Harvard IL

Greater Harvard Area Historical Society, 308 N Hart Blvd, Harvard, IL 60033 • T: +1 815 9436770, 9436141 • C.E.O.: *Brian Schultz* • Cur.: *Elaine Fideucci* • Local Museum – 1977 Miltary uniforms and artifacts, local hist ... 49206

Harvard MA

Fruitlands Museums, 102 Prospect Hill Rd, Harvard, MA 01451 • T: +1 978 4563924 ext 237 • F: +1 978 4568078 • atomicdesign@charter.net • www.fruitlands.org • Pres.: *Martin Fay* • Cur.: *Michael Volmar* (Archaeology) • Fine Arts Museum / Local Museum – 1914 Regional history, decorative and fine arts, folklore, memorabilia of Thoreau, Emerson, Lane and Margaret Fuller – library, nature trails ... 49207

Harvard Historical Society Collection, 215 Still River Rd, Harvard, MA 01451 • T: +1 508 4568285 • curator@harvardhistory.org • www.harvardhistory.org • Cur.: *Joyce Verrando* • Historical Museum – 1897 ... 49208

Harwich MA

Brooks Academy Museum, 80 Parallel St, Harwich, MA 02645 • T: +1 508 4328089 • info@harwichhistoricalsociety.org • www.harwichhistoricalsociety.org • Dir./Cur.: *James A. Brown* • Local Museum – 1954 Local and Cape Cod hist, marine hist, cranberry ind, lace coll, Indian arrow heads ... 49209

Hastings MI

Historic Charlton Park Museum, 2545 S Charlton Park Rd, Hastings, MI 49058-8102 • T: +1 269 9453775 • F: +1 269 9450390 • gshannon@iserv.net • www.charltonpark.org • Dir.: *George W. Shannon* • Cur.: *Hollie Bonnema* • Local Museum / Open Air Museum – 1936 Regional history, rural life, agriculture, industry, decorative arts – library, historic village ... 49210

Hastings NE

Hastings Museum, 1330 N Burlington, Hastings, NE 68901 • T: +1 402 4612399 • F: +1 402 4612379 • hastingsmuseum@alltel.com • Dir.: *Terry Hunter* • Cur.: *Teresa Kreutzer* • Local Museum / Natural History Museum – 1926 North American mammals and birds, minerals, glassware, pioneer items, regional hist ... 49211

Hattiesburg MS

HAHS Museum, Hattiesburg Area Historical Society, 723 Main St, Hattiesburg, MS 39401 • T: +1 601 5825460 • Local Museum – 1970 Area ist, education, patriotsm ... 49212

Turner House Museum, 500 Bay, Hattiesburg, MS 39401 • T: +1 601 5821771 • Dir./Cur.: *David Sheley* • Fine Arts Museum / Decorative Arts Museum – 1970 18th c furniture, silver, Persian rugs, paintings (old masters), tapestries ... 49213

Hatton ND

Hatton-Eielson Museum, 405 Eielson St, Hatton, ND 58240 • T: +1 701 5433726 • F: +1 701 5434013 • C.E.O. & Pres.: *Arden Johnson* • Cur.: *Eileen Mork* • Historical Museum – 1973 ... 49214

Haverford PA

Main Line Art Center, Old Buck Rd and Lancaster Av, Haverford, PA 19041 • T: +1 610 5250272 • F: +1 610 5255036 • jherman@mainlineart.org • www.mainlineart.org • Exec. Dir.: *Judy Herman* • Public Gallery – 1937 ... 49215

Haverhill MA

The Buttonwoods Museum, 240 Water St, Haverhill, MA 01830 • T: +1 978 3744626 • F: +1 978 5219176 • blangenau@haverhillhistory.org • www.haverhillhistory.org • Dir.: *Joanne Sullivan* • Local Museum – 1897 Local history, lower Merrimack Valley archaeology ... 49216

Havertown PA

Haverford Township Historical Society Museum, Karakung Dr, Powder Mill Valley Park, Havertown, PA 19083 • T: +1 610 7895169 • nitrehall@aol.com • www.haverfordhistoricalsociety.org • Cur.: *Carolyn Joseph* • Local Museum – 1939 Costumes, glass photographic plates of railroads, engines ... 49217

Havre MT

H. Earl Clack Museum, 306 Third Av, Havre, MT 59501 • T: +1 406 2654000 • F: +1 406 2655487 • www.theheritagecenter.com • Archaeology Museum – 1964 Dinosaur fossils, Indian artifacts & hist of Chippewa-Crree ... 49218

Havre de Grace MD

Havre de Grace Decoy Museum, 215 Giles St, Havre de Grace, MD 21078 • T: +1 410 9393739 • F: +1 410 9393775 • decoymuseum@aol.com • www.decoymuseum.com • Pres.: *Dr. Patrick Vincenti* • Dir.: *Debra Pence* • Cur.: *Diane Rees* • Folklore Museum – 1983 ... 49219

Steppingstone Museum, 461 Quaker Bottom Rd, Havre de Grace, MD 21078 • T: +1 410 9392299, +1 888 4191762 • F: +1 410 9392321 • steppingstonemuseum@msn.com • www.steppingstonemuseum.org • Dir.: *Linda M. Noll* • Decorative Arts Museum / Agriculture Museum – 1968 Arts and crafts, agriculture hist ... 49220

Susquehanna Museum of Havre de Grace, 817 Conesteo St, Havre de Grace, MD 21078 • T: +1 410 9395780 • F: +1 410 9395780 • susqmuseum@erols.com • users.erols.com/susqmuseum • Local Museum – 1970 Local hist ... 49221

Hays KS

Ellis County Historical Society Museum, 100 W 7th St, Hays, KS 67601 • T: +1 785 6282624 • F: +1 785 6280386 • office@elliscountyhistoricalmuseum.org • www.elliscountyhistoricalmuseum.org • Dir.: *Sharon Behrman* • *Janet Johannes* • Cur.: *Elisha Beck* • Historical Museum – 1971 Area hist from 1867 to present, hist of Wild West, Volga-German immigration – archives ... 49222

Fort Hays State Historic Site, 1472 Hwy 183, Alt., Hays, KS 67601-9212 • T: +1 785 6256812 • F: +1 785 6256812 • thefort@kshs.org • www.kshs.org/places/forthays/index • Dir.: *Robert Wilhelm* • Military Museum / Historic Site – 1965 Uniforms, utensils, Indian artifacts, weapons ... 49223

Moss-Thorns Gallery of Arts, c/o Fort Hays State University, 600 Park St, Hays, KS 67601 • T: +1 785 6284247 • F: +1 785 6284087 • ctaylor@fhsu.edu • www.fhsu.edu • Chm.: *Leland Powers* • Fine Arts Museum / University Museum – 1953 Contemporary prints, small paintings, prints and drawings, Oriental scroll coll ... 49224

Sternberg Museum of Natural History, Fort Hays State University, 3000 Sternberg Dr, Hays, KS 67601-2006 • T: +1 785 6285516 • F: +1 785 6284518 • jchoate@fhsu.edu • www.fhsu.edu/sternberg/ • Dir.: *J.R. Choate* (Mammals) • Chief Cur.: *R.J. Zakrzewski* • Cur.: *J.R. Thomasson* (Plants) • *J.R. Choate* (Mammals) • Asst. Cur.: *G. Farley* (Birds) • *R. Packauskas* (Insects) • *William Stark* (Fishes) • Coll. Mgr.: *M. Eberle* (Mammals) • Assoc. Cur.: *C. J. Schmidt* (Herptiles) • *Travis Taggart* (Herptiles) • University Museum / Natural History Museum – 1926 Mammals, birds, insects, plants, fossils, amphibians, reptiles, fishes – library ... 49225

University Gallery, Fort Hays State University, 600 Park St, Hays, KS 67601 • T: +1 785 6284247 • F: +1 785 6284087 • ctaylor@fhsu.edu • www.fhsu.edu/art • Public Gallery ... 49226

Hayward CA

C.E. Smith Museum of Anthropology, California State University, Meiklejohn Hall, 25800 Carlos B Blvd, Hayward, CA 94542-3039 • T: +1 510 8857414, 8853104 • F: +1 510 8853353 • george.miller@csueastbay.edu • class.csueastbay.edu/anthropologymuseum • Dir./Cur.: *Prof. George Miller* • Ethnology Museum / University Museum – 1975 Jack Lee kachina coll, baskets, Indians artifacts, African art – library ... 49227

Hayward Area Museum, 22701 Main St, Hayward, CA 94541 • T: +1 510 5810223 • F: +1 510 5810217 • info@haywardareahistory.org • www.haywardareahistory.org • C.E.O.: *Myron Freedman* • Local Museum – 1956 Local hist ... 49228

McConaghy House, 18701 Hesparian Blvd, Hayward, CA 94580 • T: +1 510 5810223 • F: +1 510 5810217 • info@haywardareahistory.org • www.haywardareahistory.org • C.E.O.: *Jim DeMersman* • Historical Museum – 1976 Victorian furnishings and decorative arts ... 49229

Sun Gallery, Hayward Area Forum of the Arts, 1015 E St, Hayward, CA 94541 • T: +1 510 5814050 • F: +1 510 5813384 • sungallery@comcast.net • www.sungallery.org •

Dir./Cur.: *Veronica Dondero* •
Fine Arts Museum – 1975
Creative contemporary art from Northern California 49230

Hayward WI

National Fresh Water Fishing Hall of Fame, 10630 Hall of Fame Dr, Hayward, WI 54843 • T: +1 715 6344440 • F: +1 715 6344440 • fishhall@win.bright.net •
Pres.: *Harold Tiffany* •
Special Museum – 1960
Fishing, outboard motors, mounted fish, reels, lures, rods 49231

Hazelwood MO

Museum of the Western Jesuit Missions, 700 Howdershell Rd, Hazelwood, MO 63031 • T: +1 314 8373525 •
Dir.: *William B. Faherty* • Cust.: *Mark Fischer* •
Religious Arts Museum – 1971 49232

Hazleton PA

Greater Hazleton Historical Society Museum, 55 N Wyoming St, Hazleton, PA 18201 • T: +1 570 4558576 • F: +1 570 4558576 • hazletonmuseum@yahoo.com • www.hazletonhistory.8m.com •
C.E.O. & Pres.: *Jean Coll Gormley* •
Local Museum – 1983
Local hist 49233

Healdsburg CA

Tile Heritage Foundation, POB 1850, Healdsburg, CA 95448 • T: +1 707 4318453 • F: +1 707 4318455 • foundation@tileheritage.org • www.tileheritage.org •
Pres.: *Joseph A. Taylor* • Dir.: *Sheila A. Menzies* • Coll. Man.: *Brechelle Ware* •
Decorative Arts Museum – 1987
Coll of historic and contemporary ceramic tiles, mosaic – library, archives 49234

Healdton OK

Healdton Oil Museum, Oklahoma Historical Society, 315 E Main St, Healdton, OK 73438 • T: +1 580 2290900 • F: +1 580 2290900 •
C.E.O.: *Dr. Bob Blackburn* • Cur.: *Claude N. Woods* •
Science&Tech Museum – 1973
Oil field rigs, equipmet and tools, pohtographs and documents 49235

Heavener OK

Peter Conser House, Oklahoma Historical Society, 47114 Conser Creek Rd, Hodgens off Hwy 59, Heavener, OK 74937 • T: +1 918 6532493 • hawhope@clnk.com •
C.E.O.: *Dr. Bob Blackburn* • Dir.: *Blake Wade* •
Historical Museum – 1970
Furnishings and artifacts 1894-1910 49236

Helena AR

Delta Cultural Center, 141 Cherry St, Helena, AR 72342 • T: +1 870 3384350 • info@deltaculturalcenter.com • www.deltaculturalcenter.com •
Dir.: *Katie Harrington* •
Folklore Museum – 1990
Folk culture, Arkansas Delta hist 49237

Phillips County Museum, 623 Pecan St, Helena, AR 72342 • T: +1 870 3387790 •
Cur.: *Danielle Burch* •
Local Museum / Military Museum – 1929
County hist, military hist, typical furnishings 49238

Helena MT

Archie Bray Foundation, 2915 Country Club Av, Helena, MT 59602 • T: +1 406 4433502 • F: +1 406 4430934 • archiebray@archiebray.org • www.archiebray.org •
Dir.: *Josh De Weese* •
Fine Arts Museum
Temporary exhibitions 49239

Helena Gallery, 1215 Beaverhead Dr, Helena, MT 59602-7602 •
Pres.: *Paul Finkster* •
Public Gallery
Contemporary regional art 49240

Holter Museum of Art, 12 E Lawrence St, Helena, MT 59601 • T: +1 406 4426400 • F: +1 406 4422404 • msquires@holtermuseum.org • www.holtermuseum.org •
Cur.: *Katie Knight* •
Fine Arts Museum – 1987
Contemporary regional art 49241

Montana Historical Society, 225 N Roberts, Helena, MT 59620, mail addr: POB 201201, Helena, MT 59620-1201 • T: +1 406 4442694 • F: +1 406 4442696 • gashmore@state.mt.us • www.montanahistoricalsociety.org •
Dir.: *Dr. Arnold Olsen* • Cur.: *Kirby Lambert* •
Association with Coll – 1865
Human hist, archaeology, native American material culture – library 49242

Hellertown PA

Gilman Museum, At the Cave, POB M, Hellertown, PA 18055 • T: +1 610 8388767 • F: +1 610 8382961 • info@lostcave.com • www.lostcave.com •
Dir.: *Robert G. Gilman* • Chief Cur.: *Beverly L. Rozewicz* •
Local Museum – 1955
Natural hist, antique weapons 49243

Helvetia WV

Helvetia Museum, POB 42, Helvetia, WV 26224 • T: +1 304 9246435 •
Mgr.: *Eleanor F. Mailloux* •
Local Museum – 1969
Swiss immigration, furniture of Swiss design, agricultural tools of the region 49244

Hemet CA

Hemet Museum, Santa Fe State Depot, Florida Av & State St, Hemet, CA 92543, mail addr: POB 2521, Hemet, CA 92546 • T: +1 909 9294409 • www.cityofhemet.org •
Cur.: *Bill Jennings* • *Anne B. Jennings* •
Local Museum – 1973
Local hist 49245

Hempstead NY

Hofstra Museum, 112 Hofstra University, Hempstead, NY 11549-1120 • T: +1 516 4635672 • F: +1 516 4634743 • elgkta@hofstra.edu • www.hofstra.edu/museum •
Dir.: *David C. Christman* • Cur.: *Eleanor Rait* • *Karen T. Albert* •
Fine Arts Museum / University Museum – 1963
19th/20th c art, pre-Columbian art, outdoor sculpture 49246

Henderson KY

John James Audubon Museum, Audubon State Park, U.S. Hwy 41 N, Henderson, KY 42419-0576 • T: +1 270 8271893 • F: +1 270 8262286 • curator@henderson.net • www.state.ky.us/agencies/parks/parkhome.htm •
Cur.: *Don Boarman* •
Special Museum – 1938
Nr migratory bird route, coll on life and work of John James Audubon 49247

Henderson MN

Sibley County Historical Museum, 700 Main St W, Henderson, MN 56044 • T: +1 507 2483434, 2483818 • schs@prairie.lakes.com • www.history.sibley.mn.us •
Cur.: *Sharon Haggenmiller* •
Historical Museum – 1940
Regional hist 18th-19th c, household, agriculture, music, tools, industry – library 49248

Henderson NV

Clark County Museum, 1830 S Boulder Hwy, Henderson, NV 89015 • T: +1 702 4557955 • F: +1 702 4557948 • ryz@co.clark.nv.us • www.co.clark.nv.us •
Admin.: *Mark Ryzdynski* • Cur.: *Dawna Jolliff* (Exhibitions) •
Historical Museum / Science&Tech Museum – 1968
Prehistoric, southwest Indians, miners, cultural hist 49249

Howard W. Cannon Aviation Museum, 1830 S Boulder Hwy, Henderson, NV 89015 • T: +1 702 4557968 • F: +1 702 4557948 • mhp@co.clark.nv.us • www.accessclarkcounty.com •
C.E.O. & Dir.: *Mark P. Hall-Patton* •
Science&Tech Museum – 1993
History of aviation of southern Nevada 49250

Henderson TX

The Depot Museum Complex, 514 N High St, Henderson, TX 75652 • T: +1 903 6574303 • F: +1 903 6572679 • sweaver@depotmuseum.com • www.depotmuseum.com •
Dir.: *Susan Weaver* •
Historical Museum – 1979
Costumes, tools, photograph coll, rural southern, communication artifacts 49251

Howard-Dickinson House Museum, 501 S Main St, Henderson, TX 75654 • T: +1 903 6576925 • F: +1 903 6579283 • lslov@worldnet.att.net •
C.E.O.: *Louise Slover* •
Agriculture Museum / Historical Museum – 1964
Furnishings, hist books,papers, paintings, manuscript coll, clothing 49252

Hendersonville NC

Mineral and Lapidary Museum of Henderson County, 400 N Main St, Hendersonville, NC 28792 • T: +1 828 6981977 • F: +1 828 6981977 • minlap@henderson.main.nc.us • www.minmuseum.org •
Chm.: *Larry Hauser* •
Natural History Museum – 1996
Minerals, fossils 49253

Hendricks MN

Lincoln County Pioneer Museum, 610 W Elm, Hendricks, MN 56136 • T: +1 507 2753537 •
Dir.: *Pearl Johnson* •
Local Museum – 1969
Regional hist, train depot, farming, military uniforms, doll colls, opera items, medical instr 49254

Henniker NH

New England College Gallery, 7 Main St, Henniker, NH 03242 • T: +1 603 4282329 • F: +1 603 4282266 • dfurtkamp@nec.edu • www.nec.edu •
Dir.: *Jan Hodges Baer* •
Fine Arts Museum / University Museum – 1988
Paintings, photographs, works on paper 49255

Hereford AZ

Coronado National Memorial Museum, 4101 E Montezuma Canyon Rd, Hereford, AZ 85615 • T: +1 520 3665515 ext 23 • www.nps.gov/coro •
Head: *Dale Thompson* •
Historical Museum – 1952
Tools and equipment for science and technology, costumes, paintings 49256

Hereford TX

Deaf Smith County Museum, 400 Sampson, Hereford, TX 79045 • T: +1 806 3637070 • deafsmithmuseum@wtrt.net • www.deafsmithcountymuseum.org •
Dir.: *Paula Edwards* •
Local Museum – 1966
1905 household items, ranching gear, Santa Fe caboose 49257

Herington KS

Tri-County Historical Society and Museum, 800 S Broadway, Herington, KS 67449-3060 • T: +1 785 2582842 • trimusda@ikansas.com •
Dir.: *Jolene Bradford* •
Local Museum / Science&Tech Museum – 1975
Local hist, railroad hist 49258

Herkimer NY

Herkimer County Historical Society Museum, 400 N Main St, Herkimer, NY 13350 • T: +1 315 8666413 • herkimerhistory@yahoo.com • www.rootsweb.com/~nyhchs •
Pres.: *Jeffrey Steele* •
Historical Museum – 1896 49259

Hermann MO

Deutschheim State Historic Site, 109 W Second St, Hermann, MO 65041 • T: +1 573 4862200 • F: +1 573 4862249 • moparks@dnr.mo.gov • www.mostateparks.com/deutschheim.htm •
Historical Museum – 1979
German Heritage Museum 49260

Historic Hermann Museum, 312 Schiller St, Hermann, MO 65041 • T: +1 573 4862017 • F: +1 573 4862017 • www.hermanmo.com •
C.E.O.: *Mollye C. Mundwiller* •
Local Museum – 1956
Wine making & hist of wineries, textiles, marine, Indian artifacts 49261

Hermitage TN

The Hermitage - Home of President Andrew Jackson, 4580 Rachel's Ln, Hermitage, TN 37076 • T: +1 615 8892941 • F: +1 615 8899289 • info@thehermitage.com • www.thehermitage.com •
Exec. Dir.: *Patricia Leach* • Dir. & Chief Cur.: *Marsha A. Mullin* • Cur.: *B. Anthony Guzzi* •
Local Museum – 1889
Andrew Jackson and his family 49262

Hershey PA

Hershey Museum, 170 W Hersheypark Dr, Hershey, PA 17033 • T: +1 717 5205722 • F: +1 717 5348940 • info@hersheymuseum.org • www.hersheymuseum.org •
Exec. Dir.: *James McMahon* • Cur.: *Amy Bischof* • *Jennifer Hammond* •
Decorative Arts Museum / Historical Museum / Local Museum – 1933
History 49263

Museum of Bus Transportation, 161 Museum Dr, Hershey, PA 17033 • T: +1 717 5667100 • F: +1 717 5667300 •
Pres./C.E.O.: *J. Thomas Collins* •
Science&Tech Museum
14 buses from 1912-1977, photos, models, signs 49264

Hibbing MN

Hibbing Historical Museum, 400 23rd St, Hibbing, MN 55746 • T: +1 218 2638522 • hibbhist@uslink.net •
Man.: *Heather Jo McLaughlin* •
Local Museum – 1958
Local hist, mining, logging, children's items, clothing, Greyhound Bus hist – library 49265

Hickory NC

Catawba Science Center, 243 Third Av NE, Hickory, NC 28601 • T: +1 828 3228169 ext 300 • F: +1 828 3221585 • msinclair@catawbascience.org • www.catawbascience.org •
Dir.: *Mark E. Sinclair* • Asst. Dir.: *Tricia Little* •
Natural History Museum / Science&Tech Museum – 1975 49266

Hickory Museum of Art, 243 Third Av NE, Hickory, NC 28601 • T: +1 828 3278576 • F: +1 828 3277281 • hma@w3link.com • www.hickorymuseumofart.org •
Acting Exec.Dir.: *Andrea Maricich* • Pres.: *Mary Elizabeth Geitner* •
Fine Arts Museum – 1944
1850 to present American art, pottery, outsider art, NC glass 49267

Propst House and Marple Grove, 542 5th Av NE, Hickory, NC 28601 • T: +1 828 3224731 • F: +1 828 3279096 • info@hickorylandmarks.org • hickorylandmarks.org •
Decorative Arts Museum – 1968
3 Victorian House museums, Victorian period furniture 49268

Hickory Corners MI

Gilmore Car Museum, 6865 W Hickory Rd, Hickory Corners, MI 49060 • T: +1 269 6715089 • F: +1 269 6715843 • info@gilmorecarmuseum.org • www.gilmorecarmuseum.org •
CEO: *Michael J. Spezia* •
Science&Tech Museum – 1964
Period and classic cars, pre- and post-war Cadillacs, Packards, Corvettes and Rolls-Royces 49269

Hicksville NY

Hicksville Gregory Museum, Heitz Pl, Hicksville, NY 11801 • T: +1 516 8227505 • F: +1 516 8223227 • gregorymuseum@earthlink.net • gregorymuseum.org •
Dir.: *William P. Bennett* • Cur.: *Donald Curran* •
Historian: *Richard Evers* •
Historical Museum / Natural History Museum – 1963
Indian artifacts, minerals, fossils, sea shells, butterflies, local hist 49270

Hidalgo TX

Hidalgo Pumphouse Heritage, 902 S Second St, Hidalgo, TX 78557 • T: +1 956 8438686 • F: +1 956 8436519 • chuck_snyder@hotmail.com • www.hidalgotexas.com •
Science&Tech Museum – 1999
Pumps driven by two double action, steam engines, diesel engine and pump 49271

Hiddenite NC

Hiddenite Center, 316 Church St, Hiddenite, NC 28636 • T: +1 828 6326966 • F: +1 828 6325756 • hidnight@aol.com • www.hiddenitecenter.com •
Exec. Dir.: *Dwaine C. Coley* •
Fine Arts Museum / Local Museum – 1981
Visual arts, dolls, gems and minerals, pottery, basketry 49272

Higginsville MO

Confederate Memorial State Historic Site, 211 W First St, Higginsville, MO 64037 • T: +1 660 5842853 • F: +1 660 5845134 • www.mostateparks.com •
Local Museum – 1925
Documents, photographs 49273

High Falls NY

Delaware and Hudson Canal Museum, 23 Mohonk Rd, High Falls, NY 12440 • T: +1 845 6879311 • F: +1 845 6872340 • info@canalmuseum.org • www.canalmuseum.org •
C.E.O.: *Jeanne M. Bollendorf* • Pres.: *Gretchen Reed* •
Historical Museum / Science&Tech Museum – 1966 49274

High Point NC

High Point Museum, 1859 E Lexington Av, High Point, NC 27262 • T: +1 336 8851859 • F: +1 336 8833284 • www.highpointmuseum.org •
C.E.O.: *Barbara E. Taylor* • Cur.: *Eve Weipert* • *Edith Brady* •
Historical Museum – 1966
Hist buildings, 18th and 19th c regeional hist 49275

Springfield Museum of Old Domestic Life, 555 E Springfield Rd, High Point, NC 27263 • T: +1 910 8894911 •
Dir.: *Brenda Haworth* •
Historical Museum – 1935
Early settlers' woodworking and shoemaking tools, cooking utensils, loom, Indian artifacts 49276

Highland KS

Native American Heritage Museum at Highland Mission, 1737 Elgin Rd, Highland, KS 66035 • T: +1 785 4423304, 3374 • nahm@kshs.org • www.kshs.org •
Pres.: *Robert Novrie* • Cur.: *Suzette McCord-Rogers* •
Historical Museum / Folklore Museum – 1943
House serving as a mission, dormitory and school for the Iowa, Sac and Fox Indians 49277

Highland NY

Highland Cultural Arts Gallery, Highland Cultural Center, 257 South Riverside Rd, Highland, NY 12528 • T: +1 845 6916008 • F: +1 845 6916148 • info@hcc-arts.org • www.hcc-arts.org •
Dir.: *Elisa Pritzker* •
Public Gallery
Art ... 49278

Highland Heights KY

Museum of Anthropology, Northern Kentucky University, University Dr, Highland Heights, KY 41099-6210 • T: +1 859 5725259 • F: +1 859 5725566 • www.nku.edu/~anthro/ •
Dir.: *Dr. James F. Hopgood* •
Ethnology Museum / Archaeology Museum – 1976
Anthropology ... 49279

Northern Kentucky University Art Galleries, Nunn Dr, Highland Heights, KY 41099 • T: +1 859 5725421 • F: +1 859 5726501 • knight@nku.edu • www.nku.edu/~art •
Dir.: *David J. Knight* •
Public Gallery – 1968
Regional & national artists and varied media ... 49280

Highland Park IL

Highland Park Historical Museum, 326 Central Av, Highland Park, IL 60035 • T: +1 847 4327090 • F: +1 847 4327307 • hphistoricalsociety@worldnet.att.net • www.highlandpark.org/ •
C.E.O.: *Ellsworth Mills* • Pres.: *Frederic I. Orkin* •
Local Museum – 1966
Local history, slides, artifacts, tool coll ... 49281

Hill Air Force Base UT

Hill Aerospace Museum, 7961 Wardleigh Rd, Hill Air Force Base, UT 84506-5842 • T: +1 801 7776868 • F: +1 801 7776386 • www.hill.af.mil/museum •
Dir.: *Leslie Peterson* •
Military Museum / Science&Tech Museum – 1985
Aircraft from late 1930 to modern air force type . 49282

Hill City SD

Black Hills Museum of Natural History, 217 Main St, Hill City, SD 57745 • T: +1 605 5744505 • F: +1 605 5742518 • ammoniteguy@bhigr.com •
Dir./Cur.: *Neal L. Larson* • Pres.: *Joe Harris* • Cur.: *Robert A. Farrar* •
Natural History Museum – 1990
Natural hist ... 49283

Hillsboro KS

Hillsboro Museums, 501 S Ash St, Hillsboro, KS 67063 • T: +1 620 9473775 • hillsboro_museums@yahoo.com •
C.E.O.: *Stan R. Harden* •
Local Museum / Agriculture Museum – 1958
Local history, pioneer adobe house, typical Mennonite dwelling ... 49284

Hillsboro ND

Traill County Historical Society Museum, 306 Caledonia W Av, Hillsboro, ND 58045 • T: +1 701 4365571 •
Pres.: *John Wright* • Cur.: *Aagot Nysveen* •
Historical Museum – 1965 ... 49285

Hillsboro OH

Fort Hill Museum, 13614 Fort Hill Rd, Hillsboro, OH 45133 • T: +1 937 5883221 • www.ohiohistory.org •
Head: *Keith Bengtson* •
Historical Museum
Hopewell culture, natural science ... 49286

Highland House Museum, 151 E Main St, Hillsboro, OH 45133 • T: +1 937 3933392 •
C.E.O. & Pres.: *Jean Wallis* • Dir.: *M. Van Frank* •
Historical Museum – 1965
1844 Highland house, furnishings, china, glassware, tools, Indian relics, memorabilia of the 1874 Crusade Against Intoxicating Liquid ... 49287

Hillsboro TX

Texas Heritage Museum, c/o Hill College, 112 Lamar Dr, Hillsboro, TX 76645 • T: +1 254 5822555 ext 258 • F: +1 254 5825740 • museum@hillcollege.edu • www.hillcollege.edu •
Dir.: *Dr. T. Lindsay Baker* •
Military Museum – 1963
Hood's Texas Brigade, Civil War, WW I & WW II miltary guns ... 49288

Hillsboro WI

Hillsboro Area Historical Society Museum, Maple St, Hillsboro, WI 54634 • T: +1 608 4893192 •
Pres. & C.E.O.: *Betty Havlik* •
Local Museum – 1958
Local hist ... 49289

Hillsboro WV

Pearl S. Buck Birthplace, Rte 219 N, Hillsboro, WV 24946, mail addr: POB 126, Hillsboro, WV 24946 • T: +1 304 6534430 •
Special Museum
Memorabilia of the wrighter and his family ... 49290

Hillsborough NC

Orange County Historical Museum, 201 N Churton St, Hillsborough, NC 27278 • T: +1 919 7322201 • F: +1 919 7326322 • ochm@mail.com • www.orangecountymuseum.org •
Cur.: *Evans McKinney* • *Elvan Cobb* •
Local Museum – 1957
Early settlers' crafts, Indian artifacts ... 49291

Hillsborough NH

Franklin Pierce Homestead, Second N.H. Turnpike, Hillsborough, NH 03244 • T: +1 603 4781081, 4783165 • F: +1 603 4645401 • c_chadwick@conknet.com • www.conknet.com/~hillsboro/pierce •
Historical Museum – 1804
Childhood home of Franklin Pierce, 14th U.S. President ... 49292

Hilo HI

Lyman House Memorial Museum, 276 Haili St, Hilo, HI 96720 • T: +1 808 9355021 • F: +1 808 9697685 • info@lymanmuseum.org • www.lymanmuseum.org •
Dir.: *Dr. Marie D. Strazar* •
Local Museum – 1931
Hawaiin and missionary artifacts, local culture, world class minerals, Chinese art, natural hist ... 49293

Hilton Head Island SC

Coastal Discovery Museum, 100 William Hilton Pkwy, Hilton Head Island, SC 29926-3497 • T: +1 843 6896767 • F: +1 843 6896769 • info@coastaldiscovery.org • www.coastaldiscovery.org •
Pres. & C.E.O.: *Michael J. Marks* •
Historical Museum / Science&Tech Museum – 1985
History and science ... 49294

Hinckley ME

L.C. Bates Museum, c/o Good Will-Hinckley Home For Boys and Girls, Rte 201, Hinckley, ME 04944 • T: +1 207 2384250 • F: +1 207 2384007 • lcbates@gwh.org • www.gwh.org •
Dir./ Cur.: *Deborah Staber* •
Local Museum – 1911
Native American and natural history coll, archaeological stone, art, mineralogy, military and local pioneer artifacts, ... 49295

Hinckley MN

Hinckley Fire Museum, 106 Old Hwy 61, Hinckley, MN 55037 • T: +1 320 3847338 • F: +1 320 3847338 • hfire@ecenet.com • www.hinckleyfire.org •
Dir.: *Jeanne Coffey* •
Historical Museum – 1976
Fire fighting, Saint Paul & Duluth railroad depot, clothing 19th c ... 49296

Hingham MA

Bare Cove Fire Museum, Bare Cove Park, 19 Fort Hill St, Hingham, MA 02043, mail addr: POB 262, Hingham, MA 02043 • T: +1 781 7490028 • www.barecovefiremuseum.org •
Cur.: *Andrew Addoms* •
Historical Museum – 1974
Hand and horse motored antique fire engines, equip ... 49297

Old Ordinary House Museum, 21 Lincoln St, Hingham, MA 02043 • T: +1 781 7490013, 7497721 • info@hinghamhistorical.org • www.hinghamhistorical.org •
Historical Museum – 1922
18th/19th c furnishings – small library ... 49298

Hinsdale IL

Hinsdale Historical Society Museum, 15 S Clay St, Hinsdale, IL 60521 • T: +1 630 7892600 • www.hinsdalehistory.org •
Cur.: *Kim Morrison* •
Local Museum / Historical Museum – 1981 ... 49299

Robert Crown Center for Health Education, 21 Salt Creek Ln, Hinsdale, IL 60521 • T: +1 630 3251900 • F: +1 630 3253970 • rcche@robertcrown.org • www.robertcrown.org •
Pres.: *Kathleen M. Burke* •
Local Museum – 1958
Human biology, physiology ... 49300

Hinton WV

Bluestone Museum, Rte 87, Hinton, WV 25951 • T: +1 304 4661454 •
Natural History Museum – 1972 ... 49301

Ho-Ho-Kus NJ

The Hermitage, 335 N Franklin Turnpike, Ho-Ho-Kus, NJ 07423 • T: +1 201 4458311 • F: +1 201 4450437 • info@thehermitage.org • www.thehermitage.org •
Dir.: *Joeen Ciannella* •
Historical Museum – 1972
Clothing ant textiles ... 49302

Hobart IN

Hobart Historical Society Museum, 706 E Fourth St, Hobart, IN 46342 • T: +1 219 9420970 •
Dir.: *Dorothy Ballantyne* •
Local Museum – 1968
Life in the Hobart are, wheelwright and woodworking tools, agriculture ... 49303

Hobbs NM

Lea County Cowboy Hall of Fame and Western Heritage Center, 5317 Lovington Hwy, Hobbs, NM 88240 • T: +1 505 3921275, 6576260 • F: +1 505 3925871 • lburnett@nmjc.cc.nm.us •
Dir.: *La Jean Burnett* •
Historical Museum – 1978
History, ranching and rodeo artifacts ... 49304

New Mexico Wing-Commemorative Air Force, U.S. Hwy 62-180, Hobbs, NM 88240 • T: +1 505 3936696 • F: +1 505 3921441 • hafnau@aol.com •
C.E.O.: *Harold Brown* • Exec. Officer: *Phill Ross* •
Military Museum – 1968
World War II aircraft ... 49305

Hobson MT

Utica Museum, HC 81, Hobson, MT 59452 • T: +1 406 4235531 •
Pres.: *Katherine Hodge* •
Historical Museum – 1965 ... 49306

Hodgenville KY

Abraham Lincoln Birthplace, 2995 Lincoln Farm Rd, Hodgenville, KY 42748 • T: +1 270 3583137 • F: +1 270 3583874 • abli_superintendent@nps.gov • www.nps.gov/abli •
Head: *Kenneth Apschnikat* •
Special Museum – 1916
Lincoln family homestead ... 49307

Hogansburg NY

Akwesasne Museum, 321 State Rte 37, Hogansburg, NY 13655-3114 • T: +1 518 3582240, 3582461 • F: +1 518 3582649 • info@akwesasneculturalcenter.org • www.akwesasneculturalcenter.org •
C.E.O.: *Glory Cole* •
Special Museum – 1972
Mohawk traditional artifacts, Akwesasne Mohawk culture, Haudenosaunee (Iroquois) ethnology ... 49308

Hohenwald TN

Meriwether Lewis National Monument, 189 Meriwether Lewis Park, Hohenwald, TN 38462 • T: +1 931 7962675 • F: +1 931 7965417 •
Historical Museum – 1936
Death and burial site of Meriwether Lewis (1774-1809) ... 49309

Holbrook NY

Sachem Historical Society Museum, 59 Crescent Circle, Holbrook, NY 11741 • T: +1 516 5883967 •
Historical Museum – 1963 ... 49310

Holderness NH

Squam Lakes Natural Science Center, Rte 113, Holderness, NH 03245 • T: +1 603 9687194 • F: +1 603 9682229 • info@nhnature.org • www.nhnature.org •
Dir.: *Will Abbott* •
Natural History Museum – 1966
700 birds and mammals ... 49311

Holdrege NE

Nebraska Prairie Museum, Phelps County Historical Society, N Burlington Hwy 183, Holdrege, NE 68949-0164 • T: +1 308 9955015 • F: +1 308 9953955 • pchs@nebi.com • www.nebraskaprairie.org •
Local Museum – 1966
Local history – library ... 49312

Holland MI

Cappon House Museum, 228 W 9th St, Holland, MI 49423 • T: +1 616 3926740 • F: +1 616 3944756 • hollandmuseum@hollandmuseum.org • www.hollandmuseum.org •
Historical Museum – 1986
Home of Holland's first mayor ... 49313

DePree Art Center and Gallery, Hope College, 275 Columbia Av, Holland, MI 49422-9000 • T: +1 616 3957500 • F: +1 616 3957499 • admissions@hope.edu • www.hope.edu/tour/academic/depree.html •
Dir.: *Dr. John Hanson* •
University Museum / Public Gallery – 1982
Painting, drawing, print-making, silkscree/lithography, photography, ceramics and sculpture ... 49314

First Michigan Museum of Military History, U.S. 31 and New Holland St, Holland, MI 49424 • T: +1 616 3991955 •
Pres.: *Craig DeSeyter* •
Military Museum – 1953
Military history ... 49315

Holland Area Arts Council, 150 W Eighth St, Holland, MI 49423 • T: +1 616 3963278 • F: +1 616 3966298 •
Exec. Dir.: *Jason Kalajainen* •
Public Gallery – 1967
Local arts and art competition exhibits ... 49316

Holland Museum, 31 W 10th St, Holland, MI 49423 • T: +1 616 3941362 • F: +1 616 3944756 • hollandmuseum@hollandmuseum.org • www.hollandmuseum.org •
Dir.: *Paula Dunlap* • Cur.: *Joel Lefever* •
Local Museum – 1937
Dutch heritage and decorative arts, local history – archive, library ... 49317

Windmill Island Municipal Park Museum, 1 Lincoln Av, Holland, MI 49423 • T: +1 616 3551030 • F: +1 616 3551035 • ad@windmillisland.org • www.windmillisland.org •
Dir.: *Ad Van den Akker* •
Special Museum – 1965
Restored Dutch windmills, Little Netherlands exhibit ... 49318

Holland VT

Holland Historical Society Museum, Gore Rd, Holland, VT 05830 • T: +1 802 8952917 • F: +1 802 8954440 •
Dir.: *Russell Sykes* • *Cyril Worth* •
Local Museum – 1972
Pews, lecterns, local hist artifacts, 19th c paintings & portraits ... 49319

Hollister CA

San Benito County Museum, 498 Fifth St, Hollister, CA 95024-0357 • T: +1 831 6350335 • info@sbchistoricalsociety.org • www.sbchistoricalsociety.org •
Pres.: *Bill Heidrich* •
Local Museum – 1956
Local history ... 49320

Holly Springs MS

Marshall County Historical Museum, 111 Van Doren Av, Holly Springs, MS 38635 • T: +1 662 2523669 •
C.E.O. & Cur.: *Lois Swanee* •
Historical Museum – 1970
Farm tools, relics from Civil War, War of 1812, Spanish-American War, WWI & II ... 49321

Hollywood FL

Art and Culture Center of Hollywood, 1650 Harrison St, Hollywood, FL 33020 • T: +1 954 9213274 • F: +1 954 9213273 • info@artandculturecenter.org • www.artandculturecenter.org •
Dir.: *Cynthia Berman-Miller* •
Fine Arts Museum / Folklore Museum – 1976
Multi-disciplinary Arts Center ... 49322

Hollywood MD

Sotterley Plantation Museum, 44300 Sotterley Ln, Rte 245, Hollywood, MD 20636 • T: +1 301 3732280 • F: +1 301 3738474 • officemanager@sotterley.org • www.sotterley.com •
C.E.O.: *Catherine Kelly Elder* • Pres.: *Dr. R. Fulton* •
Historic Site – 1961
1710 Manor House, outbuildings, illustrating Tidewater Plantation culture ... 49323

Holmdel NJ

Longstreet Farm, Holmdel Park, Longstreet Rd, Holmdel, NJ 07733 • T: +1 732 9463758, 8424000 • F: +1 732 9460750 • info@monmouthcountyparks.com • www.monmouthcountyparks.com •
Agriculture Museum – 1967
Living hist, 19th c agriculture ... 49324

Vietnam Era Educational Center, 1 Memorial Ln, Holmdel, NJ 07733 • F: +1 732 3351107 • klwatts@njvmf.org • www.njvmf.org •
Exec. Dir.: *Kelly Watts* •
Military Museum / Historical Museum – 1998
Photos, letters, personal items, hist Vietnam War 49325

Holyoke CO

Phillips County Museum, 109 S Campbell Av, Holyoke, CO 80734 • T: +1 970 8542129 • F: +1 970 8543811 • statz@henge.com • www.rootsweb.com/~cophilli/resource.htm •
Pres.: *Diane Rahe* •
Local Museum – 1967
Regional history, agriculture, doll coll, weapons . 49326

Holyoke MA

Children's Museum, 444 Dwight St, Holyoke, MA 01040 • T: +1 413 5367048 • F: +1 413 5332999 • kmorneau@childrensmuseumholyoke.org • www.childrensmuseumholyoke.org •
Exec. Dir.: *Kathy McCreary* • Dir.: *Elizabeth Barton* •
Special Museum – 1981
Cultural and natural hist, health education ... 49327

Wistariahurst Museum, 238 Cabot St, Holyoke, MA 01040 • T: +1 413 3225660 • F: +1 413 5342344 • wistariahurst@ci.holyoke.ma.us • www.holyoke.org/wistariahurst.htm •
CEO: *Carol Constant* •
Local Museum / Historical Museum – 1959
Mansion and carriage house, silk manufacture, costumes 49328

Homer AK

Bunnell Street Gallery, 106 W Bunnel, Ste A, Homer, AK 99603 • T: +1 907 2352662 • F: +1 907 2359427 • asia@bunnellstreetgallery.org • www.bunnellstreetgallery.org •
Public Gallery – 1994
Contemporary arts 49329

Pratt Museum, 3779 Bartlett St, Homer, AK 99603 • T: +1 907 2358635 • F: +1 907 2352764 • director@prattmuseum.org • www.prattmuseum.org •
Pres.: *Mel Strydom* • Dir.: *Diane Converse* • Cur.: *Pete Lundskow* (Colls) • *Holly Cusack-McVeigh* (Exhibits) •
Local Museum / Fine Arts Museum – 1968
Natural history, Alaskan history, ethnography, archaeology, local art – library 49330

Homer NY

Homeville Antique Fire Department, 32 Center St, Homer, NY 13077 • T: +1 607 7494466 • F: +1 607 7494466 • MIJfire@aol.com • members.aol.com/MIJfire/home.htm •
Pres.: *Mahlon Irish Jr.* •
Special Museum – 1979
American LaFrance fire apparatuses from 1900s to 1970s 49331

Homeville Museum, 49 Clinton St, Homer, NY 13077-1024 • T: +1 607 7493105 •
Head: *Kenneth M. Eaton* •
Historical Museum / Military Museum – 1976
Military accoutrements, local railroad memorabilia 49332

Hominy OK

Drummond Home, Oklahoma Historical Society, 305 N Price, Hominy, OK 74035 • T: +1 918 8852374 •
Dir.: *Blake Wade* •
Historical Museum – 1986
Original furnishing of Drummond family, clothing documents 49333

Honesdale PA

Wayne County Museum, 810 Main St, Honesdale, PA 18431 • T: +1 570 2533240 • F: +1 717 2535204 • wchs@ptd.net-director • www.waynehistorypa.org •
Dir.: *Sally Talaga* •
Historical Museum / Historic Site – 1917
History, hist of the Delaware and Hudson Canal – Research library, Farm Museum, room of a county school 49334

Honolulu HI

Bernice Pauahi Bishop Museum, 1525 Bernice St, Honolulu, HI 96817-2704 • T: +1 808 8473511 • F: +1 808 8418968 • museum@bishopmuseum.org • www.bishopmuseum.org •
Chm.: *Marcus Polivka* •
Local Museum / Ethnology Museum / Natural History Museum – 1889
Cultury, natural history 49335

The Contemporary Museum, 2411 Makiki Heights Dr, Honolulu, HI 96822 • T: +1 808 5261322 • F: +1 808 5365973 • info@tcmhi.org • www.tcmhi.org •
Dir.: *Georgiana Lagoria* • Chief Cur.: *James Jensen* •
Fine Arts Museum – 1961
Contemporary fine art, historic house, gardens 49336

The Contemporary Museum at First Hawaiian Center, 999 Bishop St, Honolulu, HI 96813, mail addr: 2411 Makiki Heights Dr, Honolulu, HI 96822 • T: +1 808 5261322 • F: +1 808 5365973 • info@tcmhi.org • www.tcmhi.org •
Dir.: *Georgiana Lagoria* • Chief Cur.: *James Jensen* •
Fine Arts Museum – 1989
Contemporary art 49337

East-West Center, 1601 East-West Rd, Honolulu, HI 96848 • T: +1 808 9447111 • F: +1 808 9447070 • feltzb@ewc.hawaii.edu • www.ewc.hawaii.edu •
Cur.: *Jeanette Bennington* •
Fine Arts Museum / Historical Museum / Natural History Museum / Folklore Museum – 1961 49338

Hawaii Children's Discovery Center, 111 Ohe St, Honolulu, HI 96813 • T: +1 808 5925437 • F: +1 808 5925400 • info@discoverycenterhawaii.org • www.discoverycenterhawaii.org •
Pres.: *Loretta Yajima* •
Special Museum – 1985 49339

Hawaii Maritime Center, Pier 7, Honolulu, HI 96813 • T: +1 808 5236151 • F: +1 808 5361519 • bmoore@bishopmuseum.org •
C.E.O.: *Dr. William Brown* •
Historical Museum – 1988
Maritime history 49340

Hawaii Pacific University Gallery, 1164 Bishop St, Honolulu, HI 96813 • T: +1 808 5440200 • F: +1 808 5441136 • nellis@hpu.edu • www.hpu.edu •
Dir.: *Sanit Khewhok* •
University Museum / Fine Arts Museum – 1983 49341

Hawaii State Foundation on Culture and the Arts, 250 S Hotel St, Honolulu, HI 96813 • T: +1 808 5860300 • F: +1 808 5860308 • sfca@sfca.state.hi.us • www.state.hi.us/sfca •
Dir.: *Lisa Yoshihara* •
Public Gallery – 1965
Contemporary & traditional Hawaiian arts, music 49342

Honolulu Academy of Arts Museum, 900 S Beretania St, Honolulu, HI 96814 • T: +1 808 5328700 • F: +1 808 5328787 • info@honoluluacademy.org • www.honoluluacademy.org •
Dir.: *Stephen Little* • Cur.: *Julia White* (Asian Art) •
Fine Arts Museum – 1922
Paintings, sculpture, ceramics, bronze, furniture, James A. Michener Japanese prints, traditional arts of Oceania, Islamic art – library 49343

Iolani Palace, King and Richards Sts, Honolulu, HI 96813 • T: +1 808 5220822 • F: +1 808 5321051 • kanaina@iolanipalace.org • www.iolanipalace.org •
Dir.: *Deborah F. Dunn* •
Historical Museum / Historic Site – 1966
Iolani Palace c. 1982 erected as state residence for Hawaii's last king Kalakaua, Iolani Barracks of the Royal Guard, paintings 49344

Japanese Cultural Center of Hawaii, 2454 S Beretania St, Honolulu, HI 96826 • T: +1 808 9457633 • F: +1 808 9441123 • jcch@lava.net • www.jcch.com •
Pres.: *Susan Kodani* •
Historical Museum – 1987
Japanes-American hist from 1885, literature on the Japanes American hist and Japanes culture 49345

King Kamehameha V - Judiciary History Center, 417 S King St, Honolulu, HI 96813 • T: +1 808 5394999 • F: +1 808 5394996 • jhchawaii@yahoo.com • jhchawaii.org •
Exec. Dir.: *Paris K. Chai* •
Historical Museum – 1989
History, Ali'iolani Hale, Hawaii Supreme Court Bldg 49346

KOA Art Gallery, University of Hawaii, Kapiolani Community College, 4303 Diamond Head Rd, Honolulu, HI 96816 • T: +1 808 7349374 • F: +1 808 7349151 •
Public Gallery 49347

Lassen Museum Waikiki, 2250 Kalakaua Av, Honolulu, HI 96815 • T: +1 808 9233435 •
Fine Arts Museum 49348

Mission Houses Museum, 553 S King St, Honolulu, HI 96813 • T: +1 808 5310481 • F: +1 808 5452280 • mhm@lava.net • www.missinhouses.org •
Cur.: *Margo Vittereli* •
Historical Museum – 1920
Missionary and Polynesia artifacts, furnishings 49349

Moanalua Gardens Foundation, 1352 Pineapple Pl, Honolulu, HI 96819 • T: +1 808 8395334 • F: +1 808 8393658 • mgf@pixi.com • mgf-hawaii.com •
Dir.: *Marilyn Schoenke* •
Archaeology Museum – 1970
Archaeology 49350

Queen Emma Gallery, 1301 Punchbowl St, Honolulu, HI 96802-0861 • T: +1 808 5474397 • F: +1 808 5474646 • efubuda@queens.org • www.queens.org •
Dir.: *Masa Morioka Tira* •
Decorative Arts Museum – 1977
Nicholas Bleecker sculptures, Hawaiian moon calender, Queen Emma's medallion, linocut prints, batik hangings, glass spheres 49351

Queen Emma Summer Palace, 2913 Pali Hwy, Honolulu, HI 96817 • T: +1 808 5953167 • F: +1 808 5954395 • doh1903@hawaii.rr.com • www.daughtersofhawaii.org •
Historical Museum / Decorative Arts Museum – 1915
Former home of Queen Emma and King Kamehameha IV 49352

Ramsay Museum, 1128 Smith St, Honolulu, HI 96817-5194 • T: +1 808 5372787 • F: +1 808 5316873 • ramsey@lava.net • www.ramsaymuseum.org •
Dir.: *Russ Sowers* •
Fine Arts Museum
Drawings, painting-American, photography, prints, sculpture 49353

Tennent Art Foundation Gallery, 203 Prospect St, Honolulu, HI 96813 • T: +1 808 5311987 •
Dir.: *Elaine Tennent* •
Fine Arts Museum – 1954
Oil, watercolors and drawings 49354

University of Hawaii Art Gallery, 2535 The Mall, Honolulu, HI 96822 • T: +1 808 9566888 • F: +1 808 9569659 • gallery@hawaii.edu • www.hawaii.edu/artgallery •
Dir.: *Tom Klobe* •
Fine Arts Museum / Public Gallery / University Museum – 1976
Japanese and Polish posters 49355

USS Arizona Memorial, 1 Arizona Memorial Pl, Honolulu, HI 96818 • T: +1 808 4222771 • F: +1 808 4838608 • www.nps.gov/usar •
Cur.: *Marshall Owens* • Sc. Staff: *Daniel Martinez* (History) •
Military Museum – 1980
Military, WW II Memorial 49356

USS Bowfin Submarine Museum, 11 Arizona Memorial Dr, Honolulu, HI 96818 • T: +1 808 4231341 • F: +1 808 4225201 • info@bowfin.org • www.bowfin.org •
Dir.: *Gerald Hofwolt* • Cur.: *Nancy Richards* •
Historical Museum / Military Museum – 1978
Military ship 49357

Hood VA

Roaring Twenties Antique Car Museum, Rte 230 W, Hood, VA 22623 • T: +1 540 9486290 • F: +1 540 9486290 • info@roaring-twenties.com • www.roaring-twenties.com •
C.E.O.: *Clarissa Dudley* •
Science&Tech Museum – 1967
Classic cars of 1920s & 1930s with emphasis on rare body styles & models 49358

Hood River OR

Hood River County Historical Museum, 300 E Port Marina Dr, Hood River, OR 97031 • T: +1 541 3866772 • F: +1 541 3866722 • hrchm@gorge.net • www.co.hood-river.or.us/museum •
Chm.: *Caroll Faull* •
Local Museum – 1907
Hood River hist 49359

Hope ND

Steele County Museum, 301 Steele Av, Hope, ND 58046 • T: +1 701 9452394 • F: +1 701 9452394 • scmuseum@invisimax.com • www.steelecomuseum.com •
Pres.: *Dorothy Moore* • Dir.: *Jaff Matson* •
Local Museum – 1966
Clothing, furniture, farm machinery, Indian artifacts 49360

Hope NJ

Hope Historical Society Museum, High St, Rte 519, Hope, NJ 07840 • T: +1 908 6374120 • F: +1 908 6374120 • cleopatra@nac.net •
Historical Museum – 1950 49361

Hopewell NJ

Hopewell Museum, 28 E Broad St, Hopewell, NJ 08525 • T: +1 609 4660103 •
Pres.: *David M. Mackey* • Cur.: *Beverly Weidl* •
Local Museum – 1924
Colonial furniture, Indian handicrafts, antique china, glass, silver and pewter, parlors, early needlework 49362

Hopewell VA

Flowerdew Hundred Foundation, 1617 Flowerdew Hundred Rd, Hopewell, VA 23860 • T: +1 804 5418897 • F: +1 804 4587738 • flowerdew@firstsaga.com • www.flowerdew.org •
Cur.: *Karen K. Shriver* •
Archaeology Museum – 1978
17th - 19th c artifacts and data from 65 archaeology sites, incl prehistoric 49363

Hopkinsville KY

Pennyroyal Area Museum, 217 E 9th St, Hopkinsville, KY 42240 • T: +1 270 8874270 • F: +1 270 8874271 • museum@hopkinsville.net • www.hopkinsville.net/~museum •
Exec. Dir.: *Donna K. Stone* •
Local Museum – 1975
Pioneer bedroom, household & farm tools, Black heritage objects 49364

Hopkinton NH

New Hampshire Antiquarian Society Museum, 300 Main St, Hopkinton, NH 03229 • T: +1 603 7463825 • nhas@tds.net • www.nhantiquarian.org •
Dir.: *Elaine P. Loft* •
Local Museum – 1859 49365

Hoquiam WA

Polson Park and Museum, 1611 Riverside Av, Hoquiam, WA 98550 • T: +1 360 5335862 • jbl@polsonmuseum.org • www.polsonmuseum.org •
Dir.: *John Larson* •
Local Museum – 1976
Local hist 49366

Horicon WI

Satterlee Clark House, 322 Winter St, Horicon, WI 53032 • T: +1 414 4852011, 4854028 •
Pres.: *Alferna Dobbratz* •
Local Museum – 1972
Historic house, furnishings, Indian artifacts, red wing pottery 49367

Hornby NY

Hornby Museum, County Rte 41, Hornby, NY 14813, mail addr: 185 South Av, Corning, NY 14830 • T: +1 607 9624471 •
Pres.: *Susan J. Moore* •
Local Museum – 1958
19th c household and farm items 49368

Horseheads NY

National Warplane Museum, 17 Aviation Dr, Horseheads, NY 14845 • T: +1 607 7398200 ext 221 • F: +1 607 7398374 • nwm@warplane.org • www.warplane.org •
Exec. Dir. & C.E.O.: *Michael S. Hall* • Dir.: *Ed Kittner* (Restoration) • Cur.: *Edward Flesch* •
Science&Tech Museum / Military Museum – 1983
WWII-era aircraft, nonflying jet and propeller driven aircraft, engins, instruments, uniforms 49369

Regional Science and Discovery Center, 3300 Chambers Rd, Horseheads, NY 14844 • T: +1 607 7395297 • F: +1 607 7395297 • scenter@stny.rr.com • www.sdcsciencecenter.org •
C.E.O. & Chm.: *Dale Wexell* •
Natural History Museum – 1996
Physical and biological sciences 49370

Horsham PA

Graeme Park/Keith Mansion, 859 County Line Rd, Horsham, PA 19044 • T: +1 215 3430965 • F: +1 215 3432223 • pmousley@state.pa.us • www.ushistory.org/graeme •
Dir.: *Patricia K. Mousley* •
Local Museum – 1958
Local hist 49371

Hot Springs AR

Fine Arts Center of Hot Springs, 610-A Central Av, Hot Springs, AR 71902 • T: +1 501 6240489 • hsfac610@sbcglobal.net • www.hsfac.com •
Dir.: *Nan Wayne* •
Public Gallery – 1947
Local contemporary artists 49372

Mid-America Science Museum, 500 Mid-America Blvd, Hot Springs, AR 71913 • T: +1 501 7673461 • F: +1 501 7671170 • www.midamericamuseum.org •
Dir.: *Glenda Eshenroder* •
Science&Tech Museum – 1971
Sciences, energy, life, matter and human perception – aquarium, laser theater 49373

Hot Springs SD

Fall River County Historical Museum, Rte 1, Hot Springs, SD 57747, mail addr: POB 361, Hot Springs, SD 57747 • T: +1 605 7455147 •
Dir./Cur.: *Paul Hickock* •
Local Museum / Decorative Arts Museum – 1961
Local hist 49374

Mammoth Site of Hot Springs, 1800 Highway 18, Bypass, Hot Springs, SD 57747 • T: +1 605 7456017 • F: +1 605 7453038 • mammoth@mammothsite.com • www.mammothsite.com •
C.E.O.: *Joe Muller* •
Natural History Museum – 1975
Paleontology 49375

Hot Sulphur Springs CO

Grand County Museum, 110 E Byers, Hot Sulphur Springs, CO 80451 • T: +1 970 7253939 • F: +1 970 7250129 • gcha@rkymtnhi.com • www.grandcountymuseum.com •
Dir.: *Don Woster* •
Local Museum – 1974
Regional history, settlement 49376

Houghton MI

A.E. Seaman Mineral Museum, Michigan Technological University, 1400 Townsend Dr, EERC, Houghton, MI 49931 • T: +1 906 4872572 • F: +1 906 4873027 • sjdyl@mtu.edu • www.geo.mtu.edu/museum/ •
Dir.: *Dr. Theodore J. Bornhorst* • Cur.: *Dr. George W. Robinson* •
University Museum / Natural History Museum – 1902
Minerals, crystals, native copper, silver and iron ores – small library 49377

Houma LA

Terrebonne Museum, 1208 Museum Dr, Houma, LA 70360 • T: +1 985 8510154 • F: +1 985 8681476 • southdown@mobiltel.com • www.southdownmuseum.org •
Dir.: *Karen Hart* •
Folklore Museum – 1975
Hist and culture of Terrebonne Parish 49378

Housatonic MA

Museum of Teenage Art, POB 556, Housatonic, MA 01236 • T: +1 413 2749913 •
Fine Arts Museum
Temporary mulicultural exhibitions of art 49379

Houston TX

Art League of Houston, 1953 Montrose Blvd, Houston, TX 77006-1243 • T: +1 713 5239530 • F: +1 713 5234053 • artleagh@neosoft.com • www.artleaguehouston.org •
Exec. Dir.: *Claudia Solis* • Pres.: *Margaret Poissant* •
Fine Arts Museum – 1948
Various forms of art media 49380

Bayou Bend Collection, 1 Westcott, Houston, TX 77007 • T: +1 713 6397750 • F: +1 713 6397770 • pmarzio@mfah.org • www.mfah.org •
Dir.: *Peter C. Marzio* •
Fine Arts Museum / Decorative Arts Museum – 1956
American decorative arts, paintings, English ceramics 49381

The Children's Museum of Houston, 1500 Binz, Houston, TX 77004-7112 • T: +1 713 5221138 ext 215/200 • F: +1 713 5225747 • lraschke@cmhouston.org • www.cmhouston.org •
Dir.: *Tammie Kahn* • Dept. Dir.: *Cheryl McCallum* • Cur.: *Niobe Ngozi* (Arts Education) •
Special Museum – 1981
Participatory exhib & programs in art, science, cultures, technology and enviroment 49382

Contemporary Arts Museum, 5216 Montrose Blvd, Houston, TX 77006-6598 • T: +1 713 2848250 • F: +1 713 2848275 • www.camh.org •
Dir.: *Marti Mayo* • Asst. Dir.: *Michael Reed* • Cur.: *Lynn M. Herbert* • *Valerie Cassell* • *Paola Morsiani* •
Fine Arts Museum – 1948 49383

Cy Twombly Gallery, The Menil Collection, 1501 Branard St, Houston, TX 77006 • T: +1 713 5259450 • F: +1 713 5259444 • public_affairs@menil.org • www.menil.org •
Dir.: *Paul Winkler* •
Fine Arts Museum – 1995
Cy Twombly's paintings, sculpture and works on paper 49384

Diverse Works, 1117 E Freewy, Houston, TX 77002 • T: +1 713 2238346 • F: +1 713 2234608 • info@diverseworks.org • www.diverseworks.org •
Dir.: *Diane Barber* •
Performing Arts Museum – 1983 49385

The Heritage Society Museum, 1100 Bagby St, Houston, TX 77002 • T: +1 713 6551912 • F: +1 713 6557527 • info@heritagesociety.org • www.heritagesociety.org •
C.E.O.: *Alice Collette* • Cur.: *Wallace Saage* •
Local Museum – 1954 49386

Holocaust Museum Houston, 5401 Caroline St, Houston, TX 77004 • T: +1 713 9428000 ext 100 • F: +1 713 9427953 • mrosen@hmh.org • www.hmh.org •
C.E.O.: *Susan Llanes-Myers* •
Historical Museum – 1996
Diaries, photographs, maps, prison uniforms, bricks from Auschwitz and the Warsaw ghetto 49387

Houston Center for Photography, 1441 W Alabama, Houston, TX 77006-4103 • T: +1 713 5294755 • F: +1 713 5299248 • hcphoto@insync.net • www.hcponline.org •
Dir.: *Jean Caslin* • *Diane Griffin Gregory* •
Association with Coll 49388

Houston Fire Museum, 2403 Milam St, Houston, TX 77006 • T: +1 713 5242526 • F: +1 713 5207566 • hfmi@houstonfiremuseum.org • www.houstonfiremuseum.org •
Dir.: *Emily Ponte* •
Historical Museum – 1982
Fire fighting tools & equipment, fire trucks, badges 49389

Houston Museum of Natural Science, 1 Hermann Circle Dr, Houston, TX 77030-1799 • T: +1 713 6394601, 6394614, 6394629 • F: +1 713 5234125 • www.hmns.org •
C.E.O.: *Joel A. Bartsch* •
Natural History Museum / Science&Tech Museum – 1909
Natural science, anthropology, astronomy, gems & minerals, entomology, malacology 49390

Houston Police Museum, 17000 Aldine Westfield Rd, Houston, TX 77073 • T: +1 281 2302360 • F: +1 281 2302334 •
Dir.: *Denny G. Hair* •
Historical Museum – 1981
1841 to the present day items pertaining to the hist of the police dept. 49391

John P. McGovern Museum of Health and Medical Science, 1515 Hermann Dr, Houston, TX 77004-7126 • T: +1 713 9427054 • F: +1 713 5261434 • info@mhms.org • www.mhms.org •
Dir.: *Randy W. Ray* •
Science&Tech Museum – 1969
Amazing Body, a walk-through interactive exhib . 49392

Lawndale Art Center, 4912 Main St, Houston, TX 77002 • T: +1 713 5285858 • F: +1 713 5284140 • askus@lawndaleartcenter.org • www.lawndaleartcenter.org •
Public Gallery 49393

The Menil Collection, 1515 Sul Ross, Houston, TX 77006 • T: +1 713 5259400 • F: +1 713 5259444 • info@menil.org • www.menil.org •
Cur.: *Matthew Drutt* • Cons.: *Elizabeth Lunning* • Registrar: *Anne Adams* •
Fine Arts Museum – 1987
Art, antiquities, art books 49394

Midtown Art Center, 3414 La Branch St, Houston, TX 77004-3841 • T: +1 713 5218803 • F: +1 713 5218803 •
Dir.: *Ida Thompson* •
Public Gallery – 1982 49395

Museum of American Architecture and Decorative Arts, c/o Houston Baptist University, 7502 Fondren Rd, Houston, TX 77074 • T: +1 281 6493311, 3000 • F: +1 281 6493489 •
Dir.: *Lois Miller* •
Decorative Arts Museum – 1964
Schissler miniature furniture, household goods and decorative arts of the ethnic groups who etablished the Republic of Texas 49396

The Museum of Fine Arts, 1001 Bissonnet, Houston, TX 77265 • T: +1 713 6397300 • F: +1 713 6397395 • pr@mfah.org • www.mfah.org •
Dir.: *Peter C. Marzio* • Assoc. Dir.: *David B. Warren* (Bayou Bend) • *Joseph Havel* (Glassell School of Art) • *Gwendolyn H. Goffe* (Finance, Admin.) • Cur.: *Barry Walker* (Prints, Drawings) • *Tina Llorente* (Textiles, Costumes) • *James Clifton* (Renaissance, Baroque Painting and Sculpture, Blaffer coll) • *Edgar Peters Bowron* (European Painting and Sculpture) • *Alison de Lima Greene* • *Alivia Wardlaw* (Modern and Contemporary Art) • *Anne W. Tucker* (Photography, Gus and Lyndall Wortham) • *Katherine S. Howe* (Decorative Arts) • *Michael K. Brown* (Bayou Bend Coll) • *Marian Lutz* (Film and Video) • *Emily Neff* (American Paintings and Sculpture) • *Frances Marzio* (Glassell coll) • *Mari Carmen Ramirez* (Latin American Art) • Cons.: *Wynne Phelan* •
Fine Arts Museum – 1900
Multicultural art, European ancient art, American and European decorative art, impressionist and post-impressionist art, African gold, prints, pre-columbian drawings, photos 49397

Museum of Printing History, 1324 W Clay, Houston, TX 77019 • T: +1 713 5224652 • F: +1 713 5228342 • donpiercy@hotmail.com • www.printingmuseum.org •
Exec. Dir.: *Elizabeth P. Griffin* • Dir.: *Sarah McNett* •
Science&Tech Museum – 1983
Hist of communication, tools and machines 49398

Musical Box Society International Museum, 1102 Heights Blvd, Houston, TX 77008-6916 • T: +1 713 8693332 • F: +1 713 8809771 • biesboech@mindspring.com • www.mbsi.org •
Music Museum – 1949
Mechanical musical instruments, primarily 1840-1925 49399

National Museum of Funeral History, 415 Barren Springs Dr, Houston, TX 77090 • T: +1 281 8763063 • F: +1 281 8764403 • info@nmfh.org • www.nmfh.org •
Dir.: *Georg Favell* •
Historical Museum – 1992
Concentration on funeral-related artifacts & memorabilia, from mid-19th c through mid-20th c America 49400

Rice University Art Gallery, 6100 Main St, Sewall Hall, Houston, TX 77005 • T: +1 713 3486069 • F: +1 713 3485980 • ruag@rice.edu • www.rice.edu/ruag •
Dir.: *Kimberly Davenport* •
University Museum / Public Gallery – 1971
Contemporary art, installation 49401

Robert A. Vines Environmental Science Center, 8856 Westview Dr, Houston, TX 77055 • T: +1 713 3654175 • F: +1 713 3654178 • harrisp@springbranchisd.com • www.springbranchisd.com •
Dir.: *Randell A. Beavers* •
Natural History Museum
Natural hist of Houston, Texas region 49402

Rothko Chapel, 1409 Sul Ross, Houston, TX 77006 • T: +1 713 5249839 • F: +1 713 5247461 • info@rothkochapel.org • www.rothkochapel.org •
Dir.: *K.C. Eynatten* •
Religious Arts Museum – 1971
Religious art, paintings by Mark Rothko 49403

Sarah Campbell Blaffer Foundation, 5601 Main St, Houston, TX 77006 • T: +1 713 6397741 • F: +1 713 6397742 • jclifton@mfah.org • www.rice.edu •
Dir.: *Dr. James Clifton* • Pres.: *Charles W. Hall* •
Fine Arts Museum – 1964
Traveling art coll, Old Master paintings 49404

Sarah Campbell Blaffer Gallery, Art Museum of the University of Houston, 120 Fine Arts Bldg, Houston, TX 77204-4018 • T: +1 713 7439521 • F: +1 713 7439525 • tsultan@uh.edu • www.blaffergallery.org •
Dir.: *Terrie Sultan* •
Fine Arts Museum / University Museum – 1973
Blaffer Gallery is a non-collecting institution that preseents a series of changing exhib 49405

Space Center Houston, 1601 NASA Road One, Houston, TX 77058-3696 • T: +1 281 2442105 • F: +1 281 2837724 • www.spacecenter.org •
Pres.: *Richard Allen* •
Science&Tech Museum – 1992
Space technology, artifacts & exhib pertaining to America's manned space program 49406

Howell NJ

Howell Historical Society & Committee Museum, 427 Lakewood-Farmingdale Rd, Howell, NJ 07731 • T: +1 732 9382212 • howellhist@aol.com • www.howellnj.com/historic/ •
Pres.: *Steve Meyer* •
Historical Museum – 1971
Local hist 49407

Howes Cave NY

Iroquois Indian Museum, Caverns Rd, Howes Cave, NY 12092 • T: +1 518 2968949 • F: +1 518 2968955 • info@iroquoismuseum.org • www.iroquoismuseum.org •
C.E.O. & Dir.: *Erynne Ansell* • Cur.: *Neal B. Keating* • *Mike Tarbell* •
Ethnology Museum / Folklore Museum – 1980
Contemporary Iroquois art and craft, ethnology, archaeology 49408

Hubbardton VT

Hubbardton Battlefield Museum, Monument Hill Rd, Hubbardton, VT 05749 • T: +1 802 7592412, 8283051 • F: +1 802 8283206 • john.dumville@state.vt.us • www.historicvermont.org •
Head: *John P. Dumville* •
Special Museum – 1948
Military, Revolutionary War 49409

Hudson NY

American Museum of Fire Fighting, 125 Harry Howard Av, Hudson, NY 12534 • T: +1 518 8281875 • F: +1 518 8288520 • fasnyfiremuseum@mhcable.com • www.fasnyfiremuseum.org •
Cur.: *R. Dennis Randall* •
Science&Tech Museum – 1925
Fire-fighting apparatus, tools and equip, photographs, prints 49410

Olana State Historic Site, 5720 Rte 9-G, Hudson, NY 12534 • T: +1 518 8280135 • F: +1 518 8286742 • linda.mclean@oprhp.state.ny.us • www.olana.org •
Dir.: *Linda E. McLean* • Cur.: *Evelyn Trebilcock* • Assoc. Cur.: *Valerie Balint* •
Fine Arts Museum / Historical Museum – 1966
1870 Persian style villa, 19th c dec arts, 19th c American art and photograph coll, Old Masters Coll 49411

Hudson OH

Historical Society Museum, 22 Aurora St, Hudson, OH 44236-2947 • T: +1 330 6536658 • F: +1 330 6504693 • archives3@hudson.lib.oh.us • www.hudsonlibrary.org •
Dir. & Cur.: *E. Leslie Polott* •
Historical Museum – 1910
Hist of Hudson township, John Brown family papers 49412

Hudson WI

The Octagon House, 1004 Third St, Hudson, WI 54016 • T: +1 715 3862654 • octagonhousemuseum@juno.com •
Dir.: *Faye Kangas* •
Local Museum – 1948
Local hist, furniture, books, doll coll 49413

Hugo CO

Headlund Museum, 617 Third Av, Hugo, CO 80821 • T: +1 719 7432485 • F: +1 719 7432447 • twbndee@yahoo.com •
Head: *Garald Ensign* •
Local Museum – 1972
Local hist 49414

Hugo OK

Choctaw Museum, 309 N B St, Hugo, OK 74743 • T: +1 580 3266630 • F: +1 580 3266686 • npence@sailfish.net •
Pres.: *Noel Pence* •
Science&Tech Museum – 1978
Railroad hist and items, local hist 49415

Hugoton KS

Stevens County Gas and Historical Museum, 905 S Adams, Hugoton, KS 67951 • T: +1 620 5448751 • svcomus@pld.com •
Pres.: *Richard Barnes* • Cur.: *Gladys Renfro* •
Historical Museum / Science&Tech Museum – 1961
Original A.T.S.F. depot country store, incl barber shop, grocery store, agricultural and automotive displays 49416

Hull MA

Hull Lifesaving Museum, The Museum of Boston Harbor Heritage, 1117 Nantasket Av, Hull, MA 02045 • T: +1 781 9255433 • F: +1 781 9250992 • lifesavingmuseum@comcast.net • www.lifesavingmuseum.org •
Dir.: *Corinne Leung* • Cur.: *Paula Morse* •
Special Museum – 1978
Prints, clothing, lifesaving vessels, logs, fleet of traditional small crafts – library 49417

Humboldt IA

Humboldt County Historical Association Museum, POB 162, Humboldt, IA 50548 • T: +1 515 3325280, 3323449 •
Pres.: *Tim Smith* • Dir.: *Bette Newton* •
Local Museum – 1962
County history 49418

Humboldt KS

Humboldt Historical Museum, Second and Charles Sts, Humboldt, KS 66748 • T: +1 620 4732250, 3658267 • rrthompson504@yahoo.com • www.usd258.net/~humbmuseum •
Local Museum – 1966
Farm equip, Civil War cannon, original jail cell, horse-drawn adult and infant hearse 49419

Huntersville NC

Historic Latta Plantation, 5225 Sample Rd, Huntersville, NC 28078 • T: +1 704 8752312 • F: +1 704 8751724 • www.lattaplantation.org •
Dir.: *Duane Smith* •
Historical Museum – 1972
Agricultural artifacts, farm buildings, African American hist and culture, Federal period furniture 49420

Huntingdon PA

Huntingdon County Museum, 106 4th St, Huntingdon, PA 16652 • T: +1 814 6435449 • F: +1 814 6432711 • mail@huntingdonhistory.org • www.huntingdonhistory.org •
Dir.: *Sandra L. Carowick* • Pres.: *Michael B. Husband* •
Local Museum – 1937
Local hist 49421

Juniata College Museum of Art, Moore and 17th Sts, Huntingdon, PA 16652 • T: +1 814 6413505 • F: +1 814 6413607 • siegel@juniata.edu • www.juniata.edu/museum •
Dir.: *Nancy Siegel* •
Fine Arts Museum – 1998
American portrait miniatures, contemporary works on paper, paintings, prints 49422

Swigart Museum, Museum Park, Rte 22 E, Huntingdon, PA 16652 • T: +1 814 6430885 • F: +1 814 6432857 • tours@swigartmuseum.com • www.swigartmuseum.com •
Science&Tech Museum – 1927
Automobiles 49423

Huntington IL

Robert E. Wilson Art Gallery, Merillat Centre for the Arts, Huntington, IL 46750 • T: +1 219 3594261 •
Dir.: *Margaret Roush* •
Fine Arts Museum – paintings, sculpture 49424

Huntington IN

Dan Quayle Center and Museum, 815 Warren St, Huntington, IN 46750 • T: +1 219 3566356 • F: +1 219 3561455 • info@quaylemuseum.org • www.quaylemuseum.org •
CEO: *Daniel Johns* •
Historical Museum – 1993
Artifacts relating to all U.S. Vice Presidents and Vice Presidents from Indiana 49425

Huntington County Historical Society Museum, 315 Court St, Huntington, IN 46750 • T: +1 219 3567264 • F: +1 219 3567264 •
Dir.: *James C. Taylor* • Cur.: *Helen Shock* •
Historical Museum – 1932
Local history 49426

Wings of Freedom Museum, 1365 Warren Rd, Huntington, IN 46750 • T: +1 219 3561945 • F: +1 219 3561315 •
Pres.: *John Shuttleworth* •
Military Museum / Science&Tech Museum – 1996
Aeronautics, military hist 49427

Huntington NY

David Conklin Farmhouse, Huntington Historical Society, 2 High St, Huntington, NY 11743 • T: +1 631 4277045 • F: +1 631 4277056 • barbra@huntington-historicalsociety.org • www.huntingtonhistoricalsociety.org •
C.E.O.: *Barbra Wells Fitzgerald* • Dir./Cur.: *Judith Estes* •
Decorative Arts Museum – 1903
Furnished in Colonial, Empire and Victorian styles, dols, toys, pottery 49428

Dr. Daniel W. Kissam House, Huntington Historical Society, 434 Park Av, Huntington, NY 11743 • T: +1 631 4277045 • F: +1 631 4277056 • barbra@huntingtonhistoricalsociety.org • www.huntingtonhistoricalsociety.org •
C.E.O.: *Barbra Wells Fitzgerald* • Dir./Cur.: *Judith Estes* •
Historical Museum – 1903
Early physicians home 49429

Heckscher Museum of Art, 2 Prime Av, Huntington, NY 11743-7702 • T: +1 631 3513250 • F: +1 631 4232145 • info@heckscher.org • www.heckscher.org •
C.E.O.: *Ph.D. Erik H. Neil* • Cur.: *Ph.D. Kenneth Wayne* •
Fine Arts Museum – 1920
16th to 20th c European and American painting, sculpture and works on paper 49430

Huntington Trade School, Huntington Historical Society, 209 Main St, Huntington, NY 11743 • T: +1 516 4277045 • F: +1 516 4277056 • barbra@huntingtonhistoricalsocietymuseum.org • www.huntingtonhistoricalsociety.org •
C.E.O.: *Barbra Wells Fitzgerald* • Dir./Cur.: *Judith Estes* •
Local Museum – 1903
History 49431

Soldiers and Sailors Memorial Building, Huntington Historical Society, 228 Main St, Huntington, NY 11743 • T: +1 516 4277045 • F: +1 516 4277056 • barbra@huntingtonhistoricalsocietymuseum.org • www.huntingtonhistoricalsociety.org •
C.E.O.: *Barbra Wells Fitzgerald* • Dir./Cur.: *Judith Estes* •
Local Museum – 1903
Local history, costumes and textiles, dec arts, paintings, prints ... 49432

Huntington VT

Birds of Vermont Museum, 900 Sherman Hollow Rd, Huntington, VT 05462 • T: +1 802 4342167 • F: +1 802 4342167 • museum@birdsofvermont.org • www.birdsofvermont.org •
Pres.: *Craig Reynolds* • C.E.O.: *Robert N. Spear* •
Natural History Museum – 1986 ... 49433

Huntington WV

Geology Museum, Marshall University, 3rd Av & Hal Greer Blvd, Huntington, WV 25755 • T: +1 304 6966720 • F: +1 304 6962600 •
Dir.: *Dr. Dewey D. Sanderson* • Sc. Staff: *Dr. Richard B. Bonnett* • *Dr. Ronald L. Martino* •
University Museum / Natural History Museum / Science&Tech Museum – 1837
Geology, minerals, rocks, fossils ... 49434

Huntington Museum of Art, 2033 McCoy Rd, Huntington, WV 25701 • T: +1 304 5292701 • F: +1 304 5297447 • info@hmoa.org • www.hmoa.org •
Exec. Dir.: *Margaret Mary Lane* • Dep. Dir.: *Katherine Cox* (Education) • *Mike Beck* (Horticulture) • Chief Cur.: *Jenine Culligan* •
Fine Arts Museum – 1947
Appalachian folk art, American decorative art and furniture, Ohio River Valley glass, American and European studio glass, American and European 19th and 20th cc paintings, drawings and prints, English Georgian silver, firearms coll, Near Eastern dec art and furniture, pre-Columbian artifacts ... 49435

Huntington Railroad Museum, 14th St W, River Park, Huntington, WV 25705 • T: +1 304 4531641 • F: +1 304 4536120 • newrivertrain@aol.com • www.newrivertrain.com •
Science&Tech Museum – 1959
Steam locomotive caboose, diesel locomotive cab, working pump handcar on track ... 49436

Museum of Radio and Technology, 1640 Florence Av, Huntington, WV 25701 • T: +1 304 5258890 • www.wrtwv.org •
Science&Tech Museum – 1992 ... 49437

Huntington Beach CA

Huntington Beach Art Center, 538 Main St, Huntington Beach, CA 92648 • T: +1 714 3741650 • F: +1 714 3745304 • deangel@surfcity-hb.org • www.surfcity-hb.org/Visitors/art_center.cfm •
Dir.: *Kate Hoffman* • Cur.: *Darlene DeAngelo* •
Public Gallery – 1995
Contemporary art ... 49438

Huntington Beach International Surfing Museum, 411 Olive Av, Huntington Beach, CA 92648 • T: +1 714 9603483 • F: +1 714 9601434 • www.surfingmuseum.org •
Dir.: *Natalie Kotsch* •
Special Museum – 1987
Surfing boards, clothing, records, sculpture, art ... 49439

Huntington Station NY

Walt Whitman Birthplace, 246 Old Walt Whitman Rd, Huntington Station, NY 11746-4148 • T: +1 631 4275240 • F: +1 631 4275247 • www.waltwhitman.org •
Dir.: *Barbara M. Bart* •
Historical Museum – 1949
Birthplace and early home of Walt Whitman (1819-1892), life and work of the author, memorabilia, 130 portraits of Whitman, original letters and manuscripts, first editions ... 49440

Huntsville AL

Art Galleries at UAH, c/o University of Alabama in Huntsville, Roberts Hall 313, Dept of Art and Art History, Huntsville, AL 35899 • T: +1 256 8246114 • art@uah.edu • www.uah.edu/colleges/liberal/art •
Head: *Kristy From-Brown* •
University Museum / Public Gallery – 1975
Print coll ... 49441

Burritt on the Mountain, 3101 Burritt Dr, Huntsville, AL 35801 • T: +1 256 5362882 • F: +1 256 5321784 • pat.robertson@hsvcity.com • www.burrittonthemountain.com •
Dir.: *James L. Powers* • Asst. Dir.: *Pat Robertson* •
Local Museum / Historical Museum / Natural History Museum – 1955
Rural life in North Alabama and South Tennesse in 19th c – library ... 49442

Earlyworks Museums, 404 Madison St, Huntsville, AL 35801 • T: +1 256 5648100 • F: +1 256 5648151 • info@earlyworks.com • www.earlyworks.com •
Dir.: *Bart Williams* •
Historical Museum – 1982
Hands-on exhibits, Alabama constitution – library ... 49443

Huntsville Museum of Art, 300 Church St S, Huntsville, AL 35801 • T: +1 256 5354350 • info@hsvmuseum.org • www.hsvmuseum.org •
Exec. Dir.: *Deborah Taylor* • Cur.: *Peter J. Baldaia* •
Fine Arts Museum – 1970
19th-20th c American art, contemporary crafts and glass – library ... 49444

United States Space and Rocket Center, 1 Tranquility Base, Huntsville, AL 35805, mail addr: POB 070015, Huntsville, AL 35807-7015 • T: +1 256 7217199 • www.spacecamp.com •
C.E.O.: *Larry Capps* •
Science&Tech Museum – 1968
Aviation, early space history to space shuttle, missiles, Apollo artefacts – archives, auditorium, IMAX ... 49445

Huntsville TX

Sam Houston Memorial Museum, 1836 Sam Houston Av, Huntsville, TX 77341, mail addr: POB 2057 SHSU, Huntsville, TX 77341 • T: +1 936 2941832 • F: +1 409 2943670 • SMM_PBN@shsu.edu • www.samhouston.memorial.museum •
Dir.: *Patrick B. Nolan* • Cur.: *Mac Woodward* • Sc. Staff: *Michael Sproat* • *Helen Belcher* •
Historical Museum – 1927
Memorabilia of Sam Houston (1793 - 1863), histof Texas 1836-1865, American Western Frontier 1793-1865 – library, archives ... 49446

Hurley NY

Hurley Patentee Manor, 464 Old Rte 209, Hurley, NY 12443 • T: +1 845 3315414 • F: +1 845 3315414 •
Dir.: *Carolyn M. Waligurski* • *Stephen S. Waligurski* •
Historical Museum – 1968 ... 49447

Hurley WI

Old Iron County Courthouse Museum, Wisconsin State Historical Society, 303 Iron St, Hurley, WI 54534 • T: +1 715 5612244 •
Pres. & C.E.O.: *Gene Cisewski* •
Local Museum / Historic Site – 1976
Local hist, iron mining, loging, early family living, historical records ... 49448

Hurleyville NY

Sullivan County Historical Society Museum, 265 Main St, Hurleyville, NY 12747-0247 • T: +1 845 4348044 • F: +1 845 4368620 • schs@warwick.net • www.sullivancountyhistory.org •
Pres.: *Allan Dampman* •
Local Museum – 1898
O and W Railroad, Anti-Rent War artifacts, Sullivan County maps and postcards, ledgers, Stephen Crane collection ... 49449

Huron SD

Dakotaland Museum, State Fair Grounds, Huron, SD 57350 • T: +1 605 3524626 •
Pres.: *K.O. Kauth* • Dir.: *Ruby Johannsen* •
Local Museum – 1960
Bird and animals (stuffed), Local hist ... 49450

Hurricane UT

Hurricane Valley Heritage Park Museum, 35 W State, Hurricane, UT 84737 • T: +1 435 6353245 • F: +1 435 6354696 • hurricanemuseum@hotmail.com •
C.E.O.: *Gregory Lawton* •
Local Museum – 1989
Local pioneer and Indian culture, dools, tools, family histories, town military men ... 49451

Hurricane WV

Museum in the Community, 3 Valley Park Dr, Hurricane, WV 25526 • T: +1 304 5620484 • F: +1 304 5624733 • info@museuminthecommunity.org • www.museuminthecommunity.org •
Dir.: *Kelly S. Burns* • Pres.: *John E. Arthur* •
Fine Arts Museum / Folklore Museum / Ethnology Museum – 1983
Art ... 49452

Hutchinson KS

Hutchinson Art Center, Hutchinson Art Association, 405 N Washington, Hutchinson, KS 67501 • T: +1 620 6631081 • F: +1 620 6636367 • director@hutchinsonartcenter.com • www.hutchinsonartcenter.com •
C.E.O.: *Brett Beatty* •
Association with Coll
19th -20th c American & European art ... 49453

Kansas Cosmosphere and Space Center, 1100 N Plum, Hutchinson, KS 67501-1499 • T: +1 620 6622305 • F: +1 620 6623693 • info@cosmo.org • www.cosmo.org •
Pres.: *Jeff Ollenburger* •
Science&Tech Museum – 1962
Coll of US & Soviet space artifacts incl Mercury, Gemini & Apollo spacecrafts ... 49454

Reno County Museum, 100 S Walnut, Hutchinson, KS 67501-0664 • T: +1 620 6621184 • F: +1 620 6620236 • rcmado@renocomuseum.org • renocomuseum.org •
Dir.: *Jay Smith* • Cur.: *Barbara Ulrich-Hicks* •
Local Museum – 1961
Artifacts relating to the local hist ... 49455

Hyannis NE

Grant County Museum, Grant County Courthouse, 105 E Harrison, Hyannis, NE 69350-9706 • T: +1 308 4582371 • F: +1 308 4582485 •
C.E.O.: *Leroy Evans* •
Local Museum – 1963
Genealogy, local hist ... 49456

Hyde Park NY

Eleanor Roosevelt National Historic Site, Rte 9 G, Hyde Park, NY 12538, mail addr: 4097 Albany Post Rd, Hyde Park, NY 12538 • T: +1 914 2299115 • F: +1 914 2290739 • rova-superintendent@nps.gov • www.nps.gov/elro •
Local Museum – 1977
Home of Eleanor Roosevelt (1884-1962) from 1945-1962, original furnishings ... 49457

Franklin D. Roosevelt Presidential Library-Museum, Rte 9, Hyde Park, NY 12538, mail addr: 4079 Albany Post Rd, Hyde Park, NY 12538 • T: +1 845 4867770 • F: +1 845 4861147 • library@roosevelt.nara.gov • www.fdrlibrary.marist.edu •
Dir.: *Cynthia M. Koch* • Cur.: *Herman Eberhardt* •
Historical Museum / Library with Exhibitions – 1940
Life and work of U.S. President F.D. Roosevelt (1882-1945) and of his wife Eleanor (1884-1962), personal and family memorabilia, Pres. Roosevelt's coll on naval hist, culture coll, manuscripts, books, film, photographs, White House records ... 49458

Home of Franklin D. Roosevelt, Rte 9, Hyde Park, NY 12538, mail addr: Hyde Park, NY 12538 • T: +1 845 2299115 • F: +1 845 2290739 • rova_webmaster@nps.gov • www.nps.gov/fofr •
Local Museum – 1946
Original furnishings and memorabilia of U.S. President Roosevelt (1882-1945) ... 49459

Vanderbilt Mansion National Museum, Rte 9, Hyde Park, NY 12538, mail addr: 4079 Albany Post Rd, Hyde Park, NY 12538 • T: +1 845 2299115 • F: +1 845 2290739 • rova_webmaster@nps.gov • www.nps.gov/vama •
Historical Museum – 1940
1896/99 Beaux-Arts style mansion, Beaux arts interiors ... 49460

Idabel OK

Museum of the Red River, 812 E Lincoln Rd, Idabel, OK 74745 • T: +1 508 2863616 • F: +1 508 2863616 • motrr@hotmail.com • www.museumoftheredriver.org •
Dir.: *Henry Moy* • Sen.Cur.: *Mary Herron* • Cur.: *Daniel Vick* • *Mario Rivera* •
Ethnology Museum / Archaeology Museum / Fine Arts Museum / Natural History Museum – 1974
American Indian coll, local Indian hist, art and archaeology, regional natural hist ... 49461

Idaho Falls ID

Museum of Idaho, 200 N Eastern Av, Idaho Falls, ID 83402 • T: +1 208 5221400 • F: +1 208 5245060 • david.pennock@museumofidaho.org • www.museumofidaho.org •
Exec. Dir.: *David Pennock* •
Local Museum – 1985
Regional natural and cultural hist, personal artifacts, furnishing, archaeology ... 49462

Idaho Springs CO

Argo Gold Mine, Mill and Museum, 2350 Riverside Dr, Idaho Springs, CO 80452 • T: +1 303 5672421 • F: +1 303 5679304 • argo2350@aol.com • www.historicargotours.com •
Dir.: *Robert N. Maxwell* •
Science&Tech Museum – 1977
Mining history ... 49463

Heritage Museum, 2060 Colorado Blvd, Idaho Springs, CO 80452 • T: +1 303 5674382 • F: +1 303 5674605 • jabowland@earthlink.net • www.historicidahosprings.com •
Cur.: *Chee Chee Bell* •
Local Museum – 1964
Local hist ... 49464

Underhill Museum, 2060 Miner St, Idaho Springs, CO 80452 • T: +1 303 5674709 • F: +1 303 5674605 • jabowland@earthlink.net • www.historicidahosprings.com •
Cur.: *Chee Chee Bell* •
Science&Tech Museum
Mining, U.S. Mineral Surveyor James Underhill ... 49465

Ignacio CO

Southern Ute Museum and Cultural Center, 14826 Hwy 172 N, Ignacio, CO 81137, mail addr: POB 737CO 81137, Ignacio • T: +1 970 5639583 • F: +1 970 5634641 • www.southernutemuseum.org •
Dir.: *Cheryl A. Frost* • Cur.: *Liz Kent* •
Historical Museum / Ethnology Museum / Folklore Museum – 1972
Native American history, ethnology, Indian reservation ... 49466

Ilion NY

Remington Firearms Museum, 14 Hoefler Av, Ilion, NY 13357 • T: +1 315 8953200 • www.remington.com •
Cur.: *Fred Supry* •
Science&Tech Museum – 1959
Remington firearms and artifacts 1820 to present, wildlife and related art ... 49467

Ilwaco WA

Ilwaco Heritage Museum, 115 SE Lake St, Ilwaco, WA 98624-0153 • T: +1 360 6423446 • F: +1 360 6424615 • ihm@willapabay.org • www.ilwacoheritagemuseum.org •
Dir.: *Hobe Kytr* • Man.: *Barbara Minard* (Collections) •
Fine Arts Museum / Historical Museum – 1983
History, art ... 49468

Independence CA

Eastern California Museum, 155 Grant St, Independence, CA 93526 • T: +1 760 8780364 • F: +1 760 8780412 • ecmuseum@inyocounty.us • ecmuseum.inyocounty.us •
Historical Museum / Ethnology Museum / Natural History Museum – 1928
Paiute and Shoshone Indians cultural artifacts, pioneer mining and farming – library, herbarium ... 49469

Independence KS

Independence Historical Museum, 8th and Myrtle, Independence, KS 67301 • T: +1 316 3313515 • museum@comgen.com • www.comgen.com/museum •
Dir.: *Julie Gosell* •
Local Museum – 1882
Fred Hudiburg coll, coll of Christmas plates, glass, hist, Indian artifacts, paintings, black pottery ... 49470

Independence MO

1859 Jail-Marshal's Home and Museum, 217 N Main St, Independence, MO 64050 • T: +1 816 2521892 • F: +1 816 4617792 • jail@jchs.org • www.jchs.org •
Dir.: *Susan Church* •
Local Museum – 1958
Jackson county history, furnishings of mid-19th c ... 49471

Community of Christ, 1001 W Walnut, The Temple, Independence, MO 64051 • T: +1 816 8331000 • F: +1 816 5213089 • mscherer@cofchrist.org • www.cofchrist.org •
Dir.: *Mark A. Scherer* •
Religious Arts Museum – 1956
Artifacts, manuscript & pictorial coll ... 49472

Harry S. Truman Home, 223 N Main St, Independence, MO 64050 • T: +1 816 2542720 • F: +1 816 2544491 • james_a_sanders@nps.gov • www.nps.gov/hstr/ •
Dir.: *James A. Sanders* • Cur.: *Carol J. Dage* •
Historical Museum – 1982
Former farm home young Harry S. Truman ... 49473

Harry S. Truman Office and Courtroom, Main and Lexington St, Independence, MO 64050 • T: +1 816 7958200 ext 1260 • F: +1 816 7957938 • juligor@gw.co.jackson.mo.us • www.co.jackson.mo.us •
Historical Museum – 1973
Judge & Harry S. Truman office & courtroom, ... 49474

Harry S. Truman Presidential Library and Museum, 500 W Hwy 24, Independence, MO 64050 • T: +1 816 2688200 • F: +1 816 2688295 • truman.library@nara.gov • www.trumanlibrary.org •
Cur.: *Clay R. Bauske* •
Historical Museum – 1957
Personal papers, goverment records, still photographs, film – Archives ... 49475

Independence Visitors Center, 937 W Walnut St, Independence, MO 64050 • T: +1 816 8363466 • F: +1 816 252 6256 • vcindepend@ldschurch.org • www.ldschurch.org •
Dir.: *George G. Romney* •
Folklore Museum – 1971
Art, Mormon life in Missouri from 1831-1839 ... 49476

John Wornall House Museum, 6115 Wornall Rd, Independence, MO 64113 • T: +1 816 4441858 • F: +1 816 3618165 • candice@wornallhouse.org • www.wornallhouse.org •
Historical Museum / Decorative Arts Museum – 1972
Decorative arts, furniture, architecture ... 49477

Mormon Visitors Center, 937 W Walnut St, Independence, MO 64050 • T: +1 816 8363466 • F: +1 816 2526256 • vcindepend@ldschurch.org • www.ldschurch.org •
Dir.: *Barrie McKay* •
Religious Arts Museum – 1970
30 ft mural of Christ, painting, reproductions ... 49478

National Frontier Trails Center, 318 W Pacific, Independence, MO 64050 • T: +1 816 3257575 • F: +1 816 3257579 • www.frontiertrailscenter.com •
Dir.: *John Mark Lambertson* • Cur.: *David Aamodt* •
Historical Museum – 1990
Trails hist 1800-1870, trades & craft tools, decorative arts, costumes ... 49479

Vaile Mansion - Dewitt Museum, 1500 N Liberty St, Independence, MO 64050 • T: +1 816 3257111 • F: +1 816 3257400 • www.ci.independence.mo.us • C.E.O. & Pres.: *Sandra Dougherty* • Cur.: *Cathy Offutt* • Local Museum / Historic Site – 1983
1880s furnishings 49480

Indian Rocks Beach FL

Gulf Beach Art Center, 1515 Bay Palm Blvd, Indian Rocks Beach, FL 37785 • T: +1 727 5964331 • Dir.: *Betsy Choetf* • Public Gallery – 1978
Temporary art exhibitions for children – Library 49481

Indiana PA

Historical and Genealogical Society of Indiana County, 200 S Sixth St, Indiana, PA 15701-2999 • T: +1 724 4639600 • F: +1 724 4639899 • ichs@ptd.net • www.rootsweb.com/~paicgs/ • Pres.: *Scott E. Decker* • Exec. Dir.: *Coleen B. Chambers* • Local Museum – 1938
Local hist 49482

The Jimmy Stewart Museum, 835 Philadelphia St, Indiana, PA 15701 • T: +1 724 3496112 • F: +1 724 3496140 • curator@jimmy.org • www.jimmy.org • Dir.: *Elizabeth Salome* • Performing Arts Museum – 1994
Cinematography 49483

Kipp Gallery, Indiana University of Pennsylvania, 470 Sprowls Hall, 11th St, Indiana, PA 15705 • T: +1 412 3577930 • F: +1 412 3577778 • www.arts.iup.edu/museum • Dir.: *Dr. Richard Field* • Fine Arts Museum / University Museum – 1971 49484

The University Museum, Indiana University of Pennsylvania, Sutton Hall, 1011 South Dr, Indiana, PA 15705-1087 • T: +1 724 3577930 • F: +1 724 3577778 • pacarone@iup.edu • www.arts.iup.edu/museum • Pres.: *Margaret Purdy* • Dir.: *Michael Hood* • Fine Arts Museum / University Museum – 1981
Art 49485

Indianapolis IL

Bona Thompson Memorial Center, 5350 University Av, Indianapolis, IL 46219-7009 • T: +1 317 3532662 • Dir.: *Steve Barnett* • Public Gallery
Art – archives 49486

Indianapolis IN

Center for Agricultural Science and Heritage, 1201 E 38th St, Indianapolis, IN 46205 • T: +1 317 9252410 • F: +1 317 9257277 • thebarn@iquest.net • www.centerforag.com • Exec. Dir.: *Betsy Kranz* • Agriculture Museum – 1994
Agriculture 49487

The Children's Museum of Indianapolis, 3000 N Meridian St, Indianapolis, IN 46208 • T: +1 317 3344000 • F: +1 317 9214122 • tcmi@childrensmuseum.org • www.childrensmuseum.org • Pres.: *Dr. Jeffrey Patchen* • Ethnology Museum / Local Museum / Natural History Museum – 1925
Archaeology, ethnology, geology, natural hist, railroads 49488

Clowes Fund Collection, 1200 W 38th St, Indianapolis, IN 46208 • T: +1 317 9231331 • F: +1 317 9311978 • ima@ima-art.org • Dir.: *Bret Waller* • Fine Arts Museum
Italian painting 49489

Eiteljorg Museum of American Indians and Western Art, 500 W Washington St, Indianapolis, IN 46204 • T: +1 317 6369378 • F: +1 317 2641724 • museum@eiteljorg.org • www.eiteljorg.org • Pres.: *John Vanausdall* • Cur.: *James H. Nottage* • Fine Arts Museum – 1989
Native American art and cultural artifacts, bronzes, paintings and drawings 49490

Emil A. Blackmore Museum of the American Legion, 700 N Pennsylvania St, Indianapolis, IN 46204 • T: +1 317 6301356 • F: +1 317 6301241 • library@legion.org • www.legion.org/national/divisions/libraryandmuseum • Cur.: *Howard C. Trace* • Military Museum
Military hist 49491

Eugene and Marilyn Glick Indiana History Center, 450 W Ohio St, Indianapolis, IN 46202-3269 • T: +1 317 2321882 • F: +1 317 2333109 • webmaster@indianahistory.org • www.indianahistory.org • C.E.O: *John Herbst* • Historical Museum – 1830
Hist of Indiana and the Old North West territory, photographs, manuscripts, artifacts – library 49492

Herron Gallery, Indiana University - Purdue University Indianapolis, 1701 N Pennsylvania St, Indianapolis, IN 46202 • T: +1 317 9202416 • F: +1 317 9202401 • Herrart@iupui.edu • www.herron.iupui.edu • Public Gallery 49493

Historic Landmarks Foundation of Indiana, 340 W Michigan St, Indianapolis, IN 46202-3204 • T: +1 317 6394534 • F: +1 317 6396734 • www.historiclandmarks.org • Pres.: *J. Reid Williamson* • Historical Museum – 1960
Historic Landmarks Headquarters, Morris-Butler House, Huddleston Farmhouse Inn Museum, Cambridge City, Kemper House Wedding Cake House 49494

Hoosier Salon Gallery, 714 E 65th St, Indianapolis, IN 46220-1610 • T: +1 317 2535340 • F: +1 317 2535468 • info@hoosiersalon.org • www.hoosiersalon.org • Exec. Dir.: *Amy Kindred* • Public Gallery – 1926
Paintings, sculpture, prints 49495

Indiana Medical History Museum, 3045 W Vermont St, Indianapolis, IN 46222 • T: +1 317 6357329 • F: +1 317 6357349 • edenharter@aol.com • www.imhm.org • Dir.: *Virginia L. Terpening* • Historical Museum – 1969
Medicine, housed in c.1896 pathology laboratory 49496

Indiana State Museum, 650 West Washington Street, Indianapolis, IN 46204 • T: +1 317 2321637 • F: +1 317 2327090 • museumcommunication@dnr.in.gov • www.indianamuseum.org • C.E.O/Pres.: *Barry Dressel* • Chief Cur.: *Dale Ogden* (Cultural history) • *Rex Garniewicz* (Collections) • *Ron Richards* (Natural History) • Local Museum / Historical Museum / Natural History Museum – 1869
General museum, science and culture, anthropology, natural history 49497

Indiana War Memorials Museum, 431 N Meridian St, Indianapolis, IN 46204 • T: +1 317 2327615 • F: +1 317 2334285 • iwm@iwm.state.in.us • www.state.in.us/iwm • Pres.: *Brian Regan* • Historic Site / Military Museum – 1927
Military, war history, housed in 1927 Indiana War Memorial 49498

Indianapolis Art Center, Churchman-Fehsenfeld-Gallery, 820 E 67th St, Indianapolis, IN 46220 • T: +1 317 2552464 • F: +1 317 2540486 • info@indplsartcenter.org • www.indplsartcenter.org/ • Pres. & Dir.: *Joyce A. Sommers* • Fine Arts Museum – 1934
Contemporary visual arts – Art Teaching Center 49499

Indianapolis Motor Speedway Hall of Fame Museum, 4790 W 16th St, Indianapolis, IN 46222 • T: +1 317 4818500 • F: +1 317 4846449 • ebireley@brickyard.com • www.brickyard.com • Dir.: *Ellen K. Bireley* • Science&Tech Museum – 1956
Cars, art, trophy coll. racing and automotive memorabilia 49500

Indianapolis Museum of Art, 4000 Michigan Rd, Indianapolis, IN 46208-3326 • T: +1 317 9231331 • F: +1 317 9311978 • ima@imamuseum.org • www.ima-art.org • Dir.: *Diane DeGrazia* • Chief Cons.: *Martin J. Radecki* • Fine Arts Museum / Historic Site / Folklore Museum – 1883
Contemporary art, decorative art, paintings, sculptures, drawings, prints, photos, textiles and costumes – library 49501

Iupui Cultural Arts Gallery, 815 W Michigan St, Indianapolis, IN 46202-5164 • T: +1 317 2743931 • F: +1 317 2747099 • life.iupui.edu/campcomm • Fine Arts Museum
Paintings, sculpture 49502

James Whitcomb Riley Museum, 528 Lockerbie St, Indianapolis, IN 46202 • T: +1 317 6315885 • F: +1 317 6344478 • rileyhome@rileymem.org • Dir.: *Sandra Crain* • Historical Museum – 1922
Art woorks, furnishings 49503

Morris-Butler House Museum, 1204 N Park Av, Indianapolis, IN 46202 • T: +1 317 6365409 • F: +1 317 6362630 • mbhouse@historiclandmarks.org • www.historiclandmarks.org • Pres.: *J. Reid Williamson* • Decorative Arts Museum – 1969
1850 - 1886 Victorian decorative arts 49504

National Art Museum of Sport, University Pl, 850 W Michigan St, Indianapolis, IN 46202-5198 • T: +1 317 2743627 • F: +1 317 2743878 • arein@iupui.edu • www.namos.iupui.edu • Pres.: *John D. Short* • Fine Arts Museum – 1959
Sports art, paintings,sculpture, graphics, photographs of sporting subjects 49505

President Benjamin Harrison Home, 1230 N Delaware St, Indianapolis, IN 46202-2598 • T: +1 317 6311888 • F: +1 317 6325488 • harrison@presidentbenjaminharrison.org • www.pbhh.org • Dir.: *Phyllis Geeslin* • Cur.: *Jennifer Capps* • Historical Museum – 1966
Furniture, furnishings, gifts to the president, personal papers 49506

Indianola IA

Farnham Galleries, Simpson College, 701 N C St, Indianola, IA 50125 • T: +1 515 9616251, 9611561 • F: +1 515 9611498 • jnostra@simpson.edu • www.simpson.edu • Pres.: *Dr. Kevin Lagree* • Dir.: *Justin Nostrala* • Fine Arts Museum / University Museum – 1982 49507

Indio CA

Coachella Valley Museum, 82-616 Miles Av, Indio, CA 92201 • T: +1 760 3426651 • F: +1 760 3426651 • www.coachellavalleymuseum.org • Dir.: *Carolyn Cooke* • Cur.: *Connie Cowan* • Local Museum – 1984
Farm tools, machinery, Cahuilla Indian artifacts 49508

S.C.R.A.P. Gallery, 46-350 Arabia St, Indio, CA 92201 • T: +1 760 8637777 • F: +1 760 8638973 • scrapgallery@earthlink.net • www.nonprofitpages.com/scrapgallery • C.E.O.: *Karen Riley* • Public Gallery – 1996
Art objects 49509

Ingalls KS

Santa Fe Trail Museums of Gray County, 204 S Main St, Ingalls, KS 67853 • T: +1 620 3355220 • dlmkwend@ucom.net • Dir.: *Kyleen Lacy* • Cur.: *Candice Wendel* • Local Museum – 1973
Exhib pertaining to the area incl: china, glass, silver, antiques 49510

Ingram TX

Duncan-McAshan Visual Arts Center, Hwy 39, Ingram, TX 78025, mail addr: POB 1169, Ingram, TX 78025 • T: +1 830 3675121, 3675120 • F: +1 830 3675725 • artdept@hcaf.com • Exec. Dir.: *George Eychner* • Fine Arts Museum / Public Gallery – 1958
Contemporary paintings, prints, photographs and sculpture 49511

Interlaken NY

Interlaken Historical Society Museum, Main St, Interlaken, NY 14847 • T: +1 607 5324213 • Pres.: *Allan Buddle* • Cur.: *Diane Bassette Nelson* • Historical Museum – 1960
Tools, pictures, geneaology 49512

International Falls MN

Koochiching County Historical Museum & Bronko Nagurski Museum, 214 Sixth Av, International Falls, MN 56649 • T: +1 218 2834316 • F: +1 218 2838243 • Exec. Dir.: *Edgar S. Oerichbauer* • Local Museum – 1958
Regional history, paintings, football memorabilia 49513

Inverness FL

Old Courthouse Heritage Museum, 532 Citrus Av, Crystel River, Inverness, FL 34450 • T: +1 352 3416428/29 • F: +1 352 3416445 • Dir.: *Kathy Turner* • Local Museum – 1985
Regional history 49514

Iola KS

Allen County Historical Society, 20 S Washington Av, Iola, KS 66749 • T: +1 620 3653051 • www.frederickfunston.org • Cur.: *Michael Anderson* • Association with Coll – 1956
Local memorabilia 49515

The Major General Frederick Funston Boyhood Home and Museum, 20 S Washington Av, Iola, KS 66749 • T: +1 316 3653051, 3656728 • www.frederickfunston.org • Dir.: *Michael Anderson* • Local Museum – 1995 49516

Iowa City IA

Hospitals and Clinics Medical Museum, University of Iowa, 200 Hawkins Dr, 8014 RCP, Iowa City, IA 52242 • T: +1 319 3561616, 3567106 • adrienne-drapkin@uiowa.edu • www.uihealthcare.com/depts/medmuseum.html • Dir.: *Adrienne Drapkin* • Historical Museum – 1989
Medicine 49517

Old Capitol, University of Iowa, 24 Old Capitol, Iowa City, IA 52242 • T: +1 319 3350548 • F: +1 319 3350558 • Ann-Smothers@uiowa.edu • www.uiowa.edu/~oldcap • Dir.: *Ann E. Smothers* • Historical Museum – 1976
Period furniture and furnishings – library 49518

Plum Grove Historic Home, 1030 Carroll St, Iowa City, IA 52240 • T: +1 319 3376846 • F: +1 319 3515310 • johctyhistscty@yahoo.com • www.uiwoa.edu/~plumgrov • Exec. Dir.: *Margaret Wieting* • Historical Museum – 1944
Home of Robert Lucas, first governor of Territory of Iowa 49519

Project Art-University of Iowa and Clinics, 200 Hawkins Dr, Iowa City, IA 52242-1009 • T: +1 319 3561616 • F: +1 319 3563862 • uihc-projectart@uiowa.edu • www.uihealthcare.com/projectart • Dir.: *Adrienn Drapkin* • Fine Arts Museum – 1967
American art, historic and modern Native American art 49520

University of Iowa Museum of Art, 150 N Riverside Dr, Iowa City, IA 52242 • T: +1 319 3351727 • F: +1 319 3353677 • uima@uiowa.edu • www.uiowa.edu/uima • Dir.: *Howard Collinson* • Cur.: *Pam Trimpe* (Painting, Sculpture) • Fine Arts Museum – 1967
International art 49521

University of Iowa Museum of Natural History, 10 Macbride Hall, Iowa City, IA 52242 • T: +1 319 3350481 • F: +1 319 3350653 • mus-nat-hist@uiowa.edu • www.uiowa.edu/~nathist • Dir.: *Willard L. Boyd* • University Museum / Natural History Museum – 1858
Natural history 49522

Iowa Falls IA

Calkins Field Museum, 18335 135th St, Iowa Falls, IA 50126-8511 • T: +1 641 6489878 • F: +1 641 6489878 • hccb@iowaconnect.com • www.hardincounty.net • Dir.: *Duane Rieken* • University Museum / Natural History Museum – 1890
Natural science, history 49523

Ipswich MA

Ipswich Historical Society, Heard House Museum & Whipple House Museum, 54 S Main St, Ipswich, MA 01938-2322 • T: +1 978 3562811 • F: +1 978 3562817 • info@ipswichmuseum.net • www.ipswichmuseum.net • Dir.: *Bonnie H. Smith* • Cur.: *Amanda Nelson* • Local Museum – 1890
Local history – library 49524

Ipswich SD

J.W. Parmely Historical Home Museum, Hwy 12 and Hwy 45, Ipswich, SD 57451 • T: +1 605 4266949 • C.E.O. & Pres.: *Phyllis M. Herrick* • Local Museum – 1931
Local hist 49525

Iraan TX

Iraan Museum, Alley Oop Fantasy Land Park, West City Limits Off Hwy 190, Iraan, TX 79744 • T: +1 432 6392522 • Cur.: *Morine Collett* • Asst. Cur.: *Laurisa Cash* • Historical Museum / Archaeology Museum – 1965
Stone tools & weapons, historical objects of Spanish period 49526

Ironton OH

Lawrence County Gray House Museum, 506 S Sixth St, Ironton, OH 45638-0073 • T: +1 614 5321222 • lchsmus@cloh.net • Pres.: *Naomi Deer* • Local Museum / Historic Site – 1925/1988
Charcoal iron furnaces, stone cutting tools, Indian cooking utensils, vintage clothing, furniture, local and personal items and coll 49527

Ironwood MI

Ironwood Area Historical Museum, 150 N Lowell St, Ironwood, MI 49938 • T: +1 906 9320287 • Dir.: *Gary Harrington* • Cur.: *Ray Maurin* • Local Museum – 1970
Local hist, iron mining, railroads, early sports 49528

Irvine CA

Beall Center for Art and Technology, Claire Trevor School of the Arts, University of California, 712 Arts Plaza, Irvine, CA 92967-2775 • T: +1 949 8244543 • estewart@uci.edu • beallcenter.uci.edu • Dir.: *Nohema Fernandez* • Asst. Dir.: *Indi McCarthy* • Fine Arts Museum / Science&Tech Museum / University Museum – 1965
New relationships between the arts, sciences, and engineering 49529

Irvine Fine Arts Center, 14321 Yale Av, Irvine, CA 92604-1901 • T: +1 949 7246880 • rmcgraw@cityofirvine.org • www.cityofirvine.org/depts/cs/finearts • Dir.: *Toni McDonald Pang* • Public Gallery – 1980
workshops 49530

Irvine Museum, 18881 Von Karman Av, Irvine, CA 92612 • T: +1 949 4760294 • www.irvinemuseum.org • Dir.: *Jean Stern* • Fine Arts Museum – 1992
California paintings 1870-1930 49531

Irving TX

Beatrice M. Haggerty Gallery, University of Dallas, 1845 E Nothgate Dr, Irving, TX 75062 • T: +1 972 7215087 • F: +1 972 7215319 •
Dir.: *Christine Bisetto* •
Fine Arts Museum / University Museum
Temporary exhibitions ... 49532

Irving Arts Center, 3333 N MacArthur Blvd, Irving, TX 75062 • T: +1 972 2527558 • F: +1 972 5704962 • minman@ci.irving.tx.us • www.ci.irving.tx.us/arts •
Exec. Dir.: *Richard E. Huff* •
Fine Arts Museum – 1990
Sculpture garden, works by Jesus Bautista Moroles and Michael Manjarris, paintings, pottery, woodcuts ... 49533

The National Museum of Communications, 6305 N O'Connor, Ste 123, Irving, TX 75062 • T: +1 972 6903636 • F: +1 972 8892329 • bill@yesterdayusa.com • www.yesterdayusa.com •
Special Museum – 1979
Vintage radio equipment, master control console from Voice of America in Woshington, DC ... 49534

National Scouting Museum, 1329 W Walnut Hill Ln, Irving, TX 75015-2079 • T: +1 972 5802100 • F: +1 972 5802020 • nsmuseum@netbsa.org • www.bsamuseum.org •
Dir.: *Connie Adams* •
Special Museum – 1959
History of the Boy Scouts of America, uniforms, Norman Rockwell oils, patches ... 49535

Irvington NY

Judaica Museum of Central Synagogue, 43 Mallard Rise, Irvington, NY 10533 • T: +1 212 8385122 • F: +1 212 6442168 •
Dir.: *Livia Thompson* •
Folklore Museum / Religious Arts Museum – 1962
Coins, medals, textiles, ritual objects from home and synagoge, sculpture, ephemera, memorabilia, artefacts ... 49536

Irvington VA

Foundation for Historic Christ Church, State Rtes 646 & 709, Irvington, VA 22480, mail addr: POB 24, Irvington, VA 22480 • T: +1 804 4386855 • F: +1 804 4385186 • fhcc@crosslink.net • www.christchurch1735.org •
C.E.O. & Exec.Dir.: *Robert E. Cornelius* •
Association with Coll – 1958
Church and social hist, books, documents – Library ... 49537

Ishpeming MI

United States National Ski Hall of Fame and Museum, 610 Palms, Ishpeming, MI 49849 • T: +1 906 4856323 • F: +1 906 4864570 • skihall@uplogon.com • www.skihall.com •
Pres.: *Richard Goetzman* • Cur.: *Raymond Leverton* •
Special Museum – 1954
library ... 49538

Island Falls ME

John E. and Walter D. Webb Museum of Vintage Fashion, Sherman St, Rte 2, Island Falls, ME 04747 • T: +1 207 8623797, 4632404 •
C.E.O.: *Frances R. Stratton* •
Special Museum – 1983 ... 49539

Isle La Motte VT

Isle La Motte Historical Society, 283 School St, Isle La Motte, VT 05463 • T: +1 802 9283173 • F: +1 802 9283342 • camilm@surfglobal.net •
Cur.: *Gloria McEwen* •
Association with Coll – 1925
18th-19th c artifacts, Indian culture ... 49540

Isleford ME

Isleford Historical Museum, Little Cranberry Island, Isleford, ME 04646 • T: +1 207 2885463 • F: +1 207 2885507 • brooke_childrey@nps.gv • www.nps.gov/acad •
Cur.: *Brooke Childrey* •
Local Museum
Maritime hist and settlement ... 49541

Islesboro ME

Islesboro Historical Museum, 388 Main Rd, Islesboro, ME 04848 • T: +1 207 7346733 •
Local Museum – 1964
Local hist ... 49542

Sailor's Memorial Museum, Grindle Point, Islesboro, ME 04848-0076 • T: +1 207 7342253 • F: +1 207 7348394 •
Historical Museum – 1936
Maritime, 1850 Lighthouse ... 49543

Issaquah WA

Gilman Town Hall Museum, Issaquah History Museums, 165 SE Andrews St, Issaquah, WA 98027 • T: +1 425 3923500 • F: +1 425 3923500 • info@issaquahhistory.org • www.issaquahhistory.org •
Dir.: *Erica S. Maniez* •
Historical Museum – 1972 ... 49544

Issaquah Depot Museum, Issaquah History Museums, 50 Rainier Blvd, Issaquah, WA 98027, mail addr: POB 695, Issaquah, WA 98027 • T: +1 425 3923500 • F: +1 425 3924236 • info@issaquahhistory.org • www.issaquahhistory.org/depot •
Historical Museum – 1972 ... 49545

Ithaca NY

DeWitt Historical Society Museum, 401 E State St, Ithaca, NY 14850 • T: +1 607 2738284 • F: +1 607 2736107 • mbraun@tompkinscountyhistory.org • www.tompkinscountyhistory.org •
Dir.: *Matthew Braun* •
Historical Museum – 1935
19th-20th c county hist, photograph coll, esp 1897-1945, ... 49546

Handwerker Gallery of Art, Ithaca College, 1170 Gannett Ctr, Ithaca, NY 14850-7276 • T: +1 607 2743018 • F: +1 607 2741774 • ckramer@ithaca.edu • www.ithaca.edu/handwerker •
Dir.: *Prof. Cheryl Kramer* •
Fine Arts Museum / University Museum – 1977
Contemporary art of faculty and local artists ... 49547

Herbert F. Johnson Museum of Art, Cornell University, Ithaca, NY 14853 • T: +1 607 2556464 • F: +1 607 2559940 • museum@cornell.edu • www.museum.cornell.edu •
Dir.: *Franklin W. Robinson* • Sen.Cur.: *Nancy Green* (Prints, Drawings and Photographs) • Chief Cur.: *Ellen Avril* (Asian Art) • Cur.: *Cathy Klimaszewski* (Education) • *Andrea Inselmann* (Modern and Contemporary Art) •
Fine Arts Museum / University Museum – 1973
Asian art coll, European and American painting, sculpture, drawings, prints, photographs ... 49548

Hinckley Foundation Museum, 410 E Seneca St, Ithaca, NY 14850 • T: +1 607 2737053 •
Dir.: *Nicole M. Carrier* • Pres.: *Tim Bumgardner* •
Historical Museum / Folklore Museum – 1972 ... 49549

Sciencenter, 601 First St, Ithaca, NY 14850 • T: +1 607 2720600 • F: +1 607 2777469 • info@sciencenter.org • www.sciencenter.org •
Dir.: *Charles H. Trautmann* • Dept. Dir.: *Kerry Flannery* (Visitor Services & Operations) •
Science&Tech Museum – 1983
More than 250 interactive, hands-on science and technology exhibits – Outdoor science park, science-themed miniature golf course ... 49550

State of the Art Gallery, 120 W Sate St, Ithaca, NY 14850 • T: +1 607 2771626 • www.soag.org/index.html •
Dir.: *Diana Ozolins* •
Public Gallery
Art, photography ... 49551

Ivoryton CT

The Hollycroft Foundation, POB 278, Ivoryton, CT 06442-0278 • T: +1 860 7672624 • hollyc.javanet@ron.com •
Dir.: *William C. Bendig* •
Fine Arts Museum
Sculpture, art ... 49552

Museum of Fife and Drum, POB 277, Ivoryton, CT 06442 • T: +1 860 7672237 • F: +1 860 7679765 • www.fifedrum.com/thecompany •
Cur.: *Ed Olsen* •
Music Museum – 1965
Musical instruments, uniforms, documents – archives ... 49553

Jacksboro TX

Fort Richardson, Hwy 281 S, Jacksboro, TX 76458 • T: +1 940 5673506 • F: +1 940 5675488 •
Local Museum – 1968
Items pertaining to Fort Richardson, Jack County & Lost Battalion ... 49554

Jackson CA

Amador County Museum, 225 Church St, Jackson, CA 95642, mail addr: 810 Court St, Jackson, CA 95642-2132 • T: +1 209 2236368 • F: +1 209 2230749 • museum@volcano.net • www.co.amador.ca.us/depts/museum •
Cur.: *Georgia Fox* •
Local Museum – 1949
Western and gold rush hist, gold mining and mine models, hardrock mining, Native-American and Chinese artifacts, 1859 house ... 49555

Jackson MI

Ella Sharp Museum, 3225 Fourth St, Jackson, MI 49203 • T: +1 517 7872320 • F: +1 517 7872933 • info@ellasharp.org • www.ellasharp.org •
CEO: *Lynnea Loftis* • Dep. Dir.: *Jim Zuleski* (Colls, Exhibits) • Sc. Staff: *Amy Wellington* (Art Education) • *LeeAnn Kendall* (Development) •
Local Museum – 1964
County hist, Merriman-Sharp family ... 49556

Jackson MS

Chimneyville Crafts Gallery, c/o Mississippi Agriculture and Forestry Museum, 1150 Lakeland Dr, Jackson, MS 39216 • T: +1 601 9812499 • F: +1 601 9810488 • mscrafts@mscraftsmensguild.org • www.mscraftsmensguild.org •
Exec. Dir.: *V.A. Patterson* •
Historical Museum – 1989
Choctaw Indian culture ... 49557

Manship House Museum, 420 E Fortification St, Jackson, MS 39202-2340 • T: +1 601 9614724 • F: +1 601 3546043 • manship@mdah.state.ms.us • www.mdah.state.ms.us •
C.E.O.: *Elbert R. Hillard* • Dir.: *Marilyn Jones* •
Historical Museum – 1980
19th c Mississippi furnishings & decorative arts ... 49558

Mississippi Governor's Mansion, 300 E Capitol, Jackson, MS 39201 • T: +1 601 3593175 • F: +1 601 3596473 • www.mdah.state.ms.us •
Dir.: *Elbert R. Hilliard* • Cur.: *Mary Lohrenz* •
Historic Site – 1842 ... 49559

Mississippi Museum of Art, 201 E Pascagoula St, Jackson, MS 39201 • T: +1 601 9601515 • F: +1 601 9601505 • mmart@netdoor.com • www.msmuseumart.org •
Dir.: *Betsy Bradley* • Cur.: *Rene Paul Barilleaux* •
Fine Arts Museum – 1911
19th & 20th c American & Mississippi art ... 49560

Mississippi Museum of Natural Science, 2148 Riverside Dr, Jackson, MS 39202 • T: +1 601 3547303 • F: +1 601 3547227 • libby.hartfield@mmns.state.ms.us • www.mdwfp.state.ms.us/museum •
Dir.: *Libby Hartfield* • Sc. Staff: *Georgia Spencer* (Project WILD) • *Dr. Robert Jones* (Zoology) •
Natural History Museum – 1934
Herpetology, ichthyology, mammalogy, ornithology – library ... 49561

Mississippi Agriculture and Forestry/National Agricultural Aviation Museum, 1150 Lakeland Dr, Jackson, MS 39216 • T: +1 601 7133365 • F: +1 601 9824292 • robert@mdac.state.us.ms.org • www.mdac.state.us.ms.org •
C.E.O.: *Charlie Dixon* • Cur.: *Robert Gembroll* •
Agriculture Museum – 1973
Hist of agriculture, forestry & agriculture aviation in Mississippi ... 49562

Museum of the Southern Jewish Experience, 4915 I-55 N, Ste 204B, Jackson, MS 39206 • T: +1 601 3626357 • F: +1 601 3666293 • information@msje.org • www.msje.org •
C.E.O.: *Macy B. Hart* •
Religious Arts Museum – 1989
Silver ritual Judaica & ceremonial items ... 49563

Oaks House Museum, 823 N Jefferson St, Jackson, MS 39202 • T: +1 601 3539339 •
Dir.: *Dr. Kathryn Breese* •
Local Museum – 1960
Family life in Jackson ... 49564

Old Capitol Museum of Mississippi History, 100 S State, Jackson, MS 39201 • T: +1 601 3596920 • F: +1 601 3596981 • ocmuseum@mdah.state.ms.us • www.mdah.state.ms.us •
Dir.: *Lucy Allen* • Dept. Dir.: *Clay Williams* (Exhibits) • Sc. Staff: *John Gardner* • *Michael Wright* (Collections) • *Carol Rietvelt* •
Historical Museum – 1961
State history, household objects, textiles (quilts, clothing), folk art, craft, civil war flags, swords ... 49565

Smith Robertson Museum, 528 Bloom St, Jackson, MS 39202-4005 • T: +1 601 9601457 • F: +1 601 9602070 •
Cur.: *Turry Flucker* •
Historical Museum – 1977
Public funds for Blacks in Jackson ... 49566

Jackson NJ

Post-Archaeological Institute, 981 Woodlane Rd, Jackson, NJ 08527 • T: +1 732 3641964 •
Dir.: *Amber Ramhorn* •
Special Museum
Civilizations and other pseudo archeological curiosities ... 49567

Jackson TN

Casey Jones Home and Railroad Museum, Casey Jones Village, 30 Casey Ln, Jackson, TN 38305 • T: +1 731 6681222 • F: +1 731 6647782 • casey@caseyjones.com • www.caseyjonesvillage.com •
C.E.O.: *J. Lawrence Taylor* •
Science&Tech Museum – 1956
Railroads, home of Casey Jones ... 49568

University Art Gallery, Union University, 1050 Union University Dr, Jackson, TN 38305 • T: +1 901 6615289 • F: +1 901 6615175 • mmallard@uu.edu •
Dir.: *Michael Mallard* •
Public Gallery ... 49569

Jackson WY

Art West Gallery, Community Art Association, 260 W Pearl St, Jackson, WY 83001 • T: +1 307 7336379 •
Public Gallery
Temporary exhibitions ... 49570

Jackson Hole Historical Museum, 105 Mercill Av, Jackson, WY 83001 • T: +1 307 7332414 • F: +1 307 7399019 • jhhsm@wyom.net • www.jacksonholehistory.org •
C.E.O. & Dir.: *Lokey Lytjen* • Cur.: *Fay Bisbee* • *Mary Gores* •
Local Museum – 1958
Archaeology, regional prehistory, Plains Indians, fur trade, game heads, regional hist ... 49571

Muse Gallery, 745 W Broadwy, Jackson, WY 83001 • T: +1 307 7330555 • F: +1 307 7330555 • www.jhmusegallery.com •
Dir.: *Tayloe Pigott* •
Public Gallery
Paintings, sculpture, drawings and prints ... 49572

National Museum of Wildlife Art, 2820 Rungius Rd, Jackson, WY 83001 • T: +1 307 7335771, 7335328 • F: +1 307 7335787 • info@wildlifeart.org • www.wildlifeart.org •
Dir.: *Dr. Francine Carraro* • Cur.: *Jane Lavino* (Education) • *Adam Duncan Harris* (Art) •
Fine Arts Museum – 1987
Art ... 49573

Jacksonport AR

Courthouse Museum, 205 Avenue St, Jacksonport, AR 72075 • T: +1 870 5232143 • F: +1 870 5234620 • jacksonport@arkansas.com •
Dir.: *Mark Ballard* • Cur.: *Angela Jackson* •
Historical Museum / Local Museum – 1965 ... 49574

Jacksonville AL

Dr. Francis Medical and Apothecary Museum, 310 Church Av SE, Jacksonville, AL 36265, mail addr: 204 Ladiga St SE, Jacksonville, AL 36265 • T: +1 256 4357611 • F: +1 256 4354103 •
Dir.: *Denise Rucker* •
Historical Museum – 1968
Medical instr, cases, books and journals, apothecaty items, medical and apothecary memorabilia ... 49575

Jacksonville FL

Alexander Brest Museum, 2800 University Blvd N, Jacksonville, FL 32211 • T: +1 904 7457371 • F: +1 904 7457375 • dlauder@ju.edu •
Dir.: *David Lauderdale* •
Fine Arts Museum / University Museum – 1977
Decorative arts, pre-Columbian ivory coll, Steuben and Tiffany glass, Chinese and Boehm porcelains, Persian carpets ... 49576

Cummer Museum of Art, 829 Riverside Av, Jacksonville, FL 32204 • T: +1 904 3566857 • F: +1 904 3534101 • info@cummer.org • www.cummer.org •
Dir.: *Dr. Maarten Van de Guchte* • Chief Cur.: *Jeanette Toohey* •
Fine Arts Museum – 1958
European and American paintings and sculpture, Meissen porcelain, Asian and ancient art, works on paper and decorative art – Library ... 49577

Fort Caroline Memorial Museum, 13165 Mount Pleasant Rd, Jacksonville, FL 32225 • T: +1 904 6417155 • F: +1 904 6413798 • timu_operations@nps.gov • www.nps.gov/timu •
Dir.: *Barbara Goodman* •
Historical Museum / Natural History Museum – 1988
History, natural history, American culture, plantation period and slavery ... 49578

Jacksonville Fire Museum, 146 Gator Bowel Blvd, Jacksonville, FL 32202 • T: +1 904 6300618 • F: +1 904 6304202 • lindat@coj.net • www.jacksonville-firemuseum.com •
Cur.: *John M. Peavy* •
Historical Museum – 1974
Antique fire apparatus, fire fighting hist ... 49579

Jacksonville Museum of Modern Art, 333 N Laura St, Jacksonville, FL 32202 • T: +1 904 3666911 • F: +1 904 3666901 • jmomapres@mindspring.com • www.jmoma.org •
Dir.: *Jane Craven* •
Fine Arts Museum – 1948
American painting, sculpture, graphics, contemporary and modern art since 1920 ... 49580

Karpeles Manuscript Museum, 101 W First St, Jacksonville, FL 32206 • T: +1 904 3562992 • www.karpeles.com •
Dir.: *Cheryl Allerman* •
Fine Arts Museum
Contemporary art, exhibitions of manuscripts, illustrations and documents ... 49581

Kent Campus Museum and Gallery, c/o Florida Community College, 3939 Roosevelt Blvd, Jacksonville, FL 32205 • T: +1 904 3813400, 3813674 • F: +1 904 3813690 • kwarren@fccj.org • www.fccj.org •
Public Gallery / University Museum – 1971
Temporary exhibitions of art ... 49582

Kingsley Plantation, 11676 Palmetto Av, Jacksonville, FL 32226 • T: +1 904 2513537 • F: +1 904 2513577 • timu_interpretation@nps.gov • www.nps.gov/timu •
Historical Museum / Historic Site – 1955
Historic site ... 49583

Museum of Science and History of Jacksonville, 1025 Museum Circle, Jacksonville, FL 32207-9053 • T: +1 904 3967062, 3966674 • F: +1 904 3965799 • director@themosh.org • www.themosh.org •
Dir.: *Margo Dundon* •
Historical Museum / Science&Tech Museum – 1941
Archaeology, ethnology, geology, natural hist, natural health and physical science, decorative arts, costumes, toys, antique dolls – Planetarium ... 49584

Museum of Southern History, 4304 Herschel St, Jacksonville, FL 32210 • T: +1 904 3883574 • www.museumsouthernhistory.com •
C.E.O./Cur.: *Van Seagraues* •
Historical Museum – 1975
Black soldiers in Confederate service 49585

University Art Gallery, 4567 Saint Johns Bluff Rd S, Jacksonville, FL 32224 • T: +1 904 6202534 •
Public Gallery 49586

Jacksonville IL

David Strawn Art Gallery, Art Association of Jacksonville, 331 W College Av, Jacksonville, IL 62651-1213 • T: +1 217 2439390 • F: +1 217 2457445 • strawn@myhtn.net • www.japl.lib.il.us/community/strawn •
Pres.: *Boyd Ferguson* •
Fine Arts Museum / Public Gallery – 1873
Mississippi Indian pottery, Miriam Cougar Allen coll of antique and collectible dolls, Charles Prentice Thompson classical coll 49587

Jaffrey NH

Jaffrey Civic Center, 40 Main St, Jaffrey, NH 03452 • T: +1 603 5326527 •
Dir.: *Scott Cunningham* • *Kim Cunningham* •
Folklore Museum / Local Museum – 1964 49588

Jamaica NY

Chung-Cheng Art Gallery, Saint John's University, 8000 Utopia Pkwy, Sun Yat Sen Hall, Jamaica, NY 11439 • T: +1 718 9906250, 9907476 • F: +1 718 9902075 • artdept@stjohns.edu • www.stjohns.edu •
Public Gallery 49589

Jamaica Center for Arts, 161-04 Jamaica Av, Jamaica, NY 11432 • T: +1 718 6587400 • F: +1 718 6587922 • visual@jcal.org • www.jcal.org •
Dir.: *S. Zurie McKie* •
Fine Arts Museum / Decorative Arts Museum – 1972 49590

King Manor, 150-03 Jamaica Av, King Park, Jamaica, NY 11432 • T: +1 718 2060545 • F: +1 718 2060541 • kingmanor1@earthlink.net •
C.E.O.: *Mary Anne Mrozinski* •
Historical Museum – 1900
Rufus King, signer of the US constitution, senator from NY state and ambasador to Great Britain under four presidents 49591

Queens Library Gallery, 89-11 Merrick Blvd, Jamaica, NY 11432 • T: +1 718 9908665 • F: +1 718 2918936 • mkrazmien@queenslibrary.org • www.queenslibrary.org/gallery •
Sc. Staff: *Mindy Krazmien* •
Public Gallery – 1995
Art and hist exhibits 49592

Jamestown ND

Frontier Village Museum, 17th St SE, Jamestown, ND 58401 • T: +1 701 2527492, 2528089 • F: +1 701 2528089 •
Pres.: *Charles Tanata* •
Local Museum – 1959
Local hist, early Pioneer hist 49593

Stutsman County Memorial Museum, 321 Third Av SE, Jamestown, ND 58401-1002 • T: +1 701 2526741 •
Head: *Leah Mitchell* •
Local Museum / Historical Museum – 1964 49594

Jamestown NY

Art Gallery, James Prendergast Library, 509 Cherry St, Jamestown, NY 14701 • T: +1 716 4847135 • F: +1 716 4871148 • reference@cclslib.org • www.cclslib.org/prendergast/ •
Dir.: *Catherine A. Way* •
Fine Arts Museum – 1880
19th c American and European paintings, reproductions 49595

CCC Weeks Gallery, 525 Falconer St, Jamestown, NY 14701-0020 • T: +1 716 6659188 • F: +1 716 6657023 • jimcolby@mail.sunyjcc.edu • www.sunyjcc.edu/gallery •
Dir.: *James Colby* •
Public Gallery – 1969 49596

Fenton History Center-Museum and Library, 67 Washington St, Jamestown, NY 14701 • T: +1 716 6646256 • F: +1 716 4837524 • information@fentonhistorycenter.org • www.fentonhistorycenter.org •
Dir.: *Joni Blackman* • *Sara W. Reale* (Education) • Cur.: *Wendy Chadwick-Case* (Exhibits) • *Norman P. Carlson* (Collections) • *Lina Cowan* (Exhibtion design) •
Registrar: *Sylvia Lobb* • Libr.: *Karen E. Livsey* •
Historical Museum – 1964 49597

Roger Tory Peterson Institute of Natural History Museum, 311 Curtis St, Jamestown, NY 14701 • T: +1 716 6652473 • F: +1 716 6653794 • webmaster@rtpi.org • www.rtpi.org •
Pres.: *James M. Berry* •
Natural History Museum – 1984
Environment, wildlife art and photography, naturalists' artifacts, work of R.T. Peterson 49598

Jamestown RI

Jamestown Museum, 92 Narragansett Av, Jamestown, RI 02835 • T: +1 401 4230784 •
Pres.: *William Burgin* •
Local Museum – 1972
Local hist 49599

Watson Farm, 455 North Rd, Jamestown, RI 02835 • T: +1 401 4230005, +1 617 2273956 • F: +1 617 2279204 • pzea@spnea.org • www.historicnewengland.org •
Vice Pres.: *Philip Zea* • Head: *Don Minto* • *Heather Minto* •
Folklore Museum / Agriculture Museum
Historic farm 49600

Jamestown VA

Jamestown Visitor Center & Museum, Jamestown Island, Jamestown, VA 23081 • T: +1 757 2291733 • F: +1 757 2294273 • colo_interpretation@nps.gov • www.nps.gov/colo •
Dir.: *Alec Gould* • Historian: *Diane G. Stallings* •
Local Museum – 1930
Local hist, first permanent English settlement at Jamestown 49601

Janesville WI

The Lincoln-Tallman Restorations, 440 N Jackson St, Janesville, WI 53545 • T: +1 608 7524519, 7564509 • F: +1 608 7419596 • rchs@rchs.us • www.rchs.us •
C.E.O.: *Anamari Golf* • Pres.: *Stuart Charland* •
Local Museum – 1951
Historic house, 19th c decorative arts and furnishings, Tallman ethnological coll, Tallman family artifacts 49602

Rock County Historical Society Museum, 426 N Jackson St, Janesville, WI 53545 • T: +1 608 7564509 • F: +1 608 7419596 • rchs@ticon.net • www.rchs.us •
C.E.O.: *Anamari Golf* • Pres.: *Stuart Charland* •
Local Museum – 1948
Local hist, 19th c decorative arts and furniture, vehicles, manuscript and iconographic coll, hist buildings 49603

Jeannette PA

Bushy Run Battlefield Museum, Rte 993, Jeannette, PA 15644 • T: +1 724 5275584 • F: +1 724 5275610 • davmiller@state.pa.us • www.bushyrunbattlefield.com •
Military Museum – 1933
18th-c military 49604

Jefferson GA

Crawford W. Long Museum, 28 College St, Jefferson, GA 30549 • T: +1 706 3675307 • F: +1 706 3675307 • dbutler@crawfordlong.org • www.crawfordlong.org •
Dir.: *Tina B. Harris* •
Historical Museum – 1957
19th c medical artifacts, documents, photographs 49605

Jefferson IA

Jefferson Telephone Museum, 105 W Harrison, Jefferson, IA 50129 • T: +1 515 3862626, 3864141 • F: +1 515 3862600 •
Dir.: *James Daubendiek* •
Science&Tech Museum – 1960
History and evolution of thelephony 49606

Jefferson OH

Dezign House, 4090 Lenox-New Lyme Rd, Jefferson, OH 44047 • T: +1 440 2942778 •
Dir. & C.E.O.: *Ramon Jan Elias* • Cur.: *Margery M. Elias* •
Decorative Arts Museum – 1962
Western decorative art and furniture 49607

Jefferson TX

Excelsior House, 211 W Austin St, Jefferson, TX 75657 • T: +1 903 6652513 • F: +1 903 6659389 • excelsior@jeffersontx.com • texasmonthly.com/mag/1991/jun/excelsior.html •
Decorative Arts Museum – 1850 49608

Jefferson Historical Museum, 223 W Austin, Jefferson, TX 75657 • T: +1 903 6652775 •
Chm.: *Bill McCay* •
Local Museum – 1948
Civil War artifacts, dolls, glass, medical items, paintings, farm implements 49609

Jefferson WI

Aztalan Museum, N 6264 Hwy Q, Jefferson, WI 53549 • T: +1 920 6484362 •
Pres.: *Dr. Cheryl D. Peterson* • Cur.: *Steve Steigerwald* • *Ethel Mae Wagner* •
Local Museum – 1941
Local hist, Indian Artifacts, Hansen Granary Tool coll 49610

Jefferson City MO

Cole County Historical Museum, 109 Madison, Jefferson City, MO 65101 • T: +1 573 6351850 •
Pres: *Natalie Tackett* •
Historical Museum – 1941
Inaugural ball gowns of governors' wives 49611

Elizabeth Rozier Gallery, State Capitol, Rm B-2, Jefferson City, MO 65101 • T: +1 573 7514210 • F: +1 573 5262927 •
Dir.: *Molly Strode* •
Public Gallery – 1981 49612

Jefferson Landing State Historic Site, 100 Jeffersen St, Jefferson City, MO 65101 • T: +1 573 7512854 • F: +1 573 5262927 • dsp.state.museum@dnr.mo.gov • www.mostateparks.com/jefferson.htm •
Cur.: *Juli Kemper* • *Katherine Keil* •
Historical Museum – 1976
Artifacts and documents about Missouri State Capitol and the settlement of Jefferson City 49613

Missouri State Museum, c/o Missouri State Capitol, 201 Capitol Av, Jefferson City, MO 65101 • T: +1 573 7512854 • F: +1 573 5262927 • www.mostateparks.com •
Dir.: *Molly O'Donnell* • Cur.: *L.T. Shelton* • *Gina Danna* •
Historical Museum – 1919
Paintings, artifacts, documents 49614

Jefferson City TN

Omega Gallery, Carson-Newman College, Warren Bldg, Jefferson City, TN 37760, mail addr: Box 71995, Jefferson City, TN 37760 • T: +1 865 4713572 • F: +1 573 5262927 •
Dir.: *David Underwood* •
University Museum / Fine Arts Museum
Art in all mediums 49615

Jeffersonville IN

Howard Steamboat Museum, 1101 E Market St, Jeffersonville, IN 47130, mail addr: POB 606, Jeffersonville, IN 47130-0606 • T: +1 812 2833728 • F: +1 812 2836049 • hsmsteam@aol.com • www.steamboatmuseum.org •
Pres.: *David Reinhardt* •
Special Museum – 1958
Home of Edmonds J. Howard, son of James E. Howard, founder of the Howard Shipyards 1834, steamboat construction tools, models 49616

Jekyll Island GA

Jekyll Island Museum, Stable Rd, Jekyll Island, GA 31520, mail addr: 381 Riverview Dr, Jekyll Island, GA 31527 • T: +1 912 6352119, 6352236 • F: +1 912 6354420 • jhunter@jekyllisland.com • www.jekyllisland.com •
Dir.: *F. Warren Murphey* •
Local Museum – 1954
Furnishings, photographs, documents 49617

Jemez Springs NM

Jemez State Monument, 1 mile N of Jemez Springs, State Hwy 4, Jemez Springs, NM 87025 • T: +1 505 8293530 • F: +1 505 8293530 • giusewa@sulphurcanyon.com • www.nmmonuments.org •
Dir.: *Jose Cisneros* •
Local Museum – 1935
500 years old Indian village, Catholic mission, ruins of Giusewa 49618

Jenkintown PA

Abington Art Center, 515 Meetinghouse Rd, Jenkintown, PA 19046 • T: +1 215 8874882 • F: +1 215 8871402 • info@abingtonartcenter.org • www.abingtonartcenter.org •
Dir.: *Laura Burnham* •
Fine Arts Museum – 1939 49619

Old York Road Historical Society, 460 Old York Rd, Jenkintown, PA 19046 • T: +1 215 8868590 • info@oyrhs.org • www.oyrhs.org •
Pres./C.E.O.: *David B. Rowland* •
Local Museum – 1936
Local hist items 49620

Jennings LA

The Zigler Museum, 411 Clara St, Jennings, LA 70546 • T: +1 318 8240114 • F: +1 318 8240120 • zigler-museum@charter.net • www.jeffdavis.org •
Cur.: *Dolores Spears* • Pres.: *Gregory Marcantel* •
Fine Arts Museum – 1963
Art American and Europesn, wildlife and natural history art 49621

Jericho Corners VT

Jericho Historical Society, Rte 15, Jericho Corners, VT 05465 • T: +1 802 8993225 • www.jerichohistoricalsociety.com •
Pres.: *Ann Squires* •
Association with Coll – 1971 49622

Jerome ID

Jerome County Historical Society, 220 N Lincoln, Jerome, ID 83338 • T: +1 208 3245641 • F: +1 208 3247694 • info@historicaljeromecounty.com • www.historicaljeromecounty.com •
C.E.O.: *Francis Egbert jr* •
Local Museum / Agriculture Museum – 1981
Antique farm machinery, hist of Jerome County 49623

Jersey City NJ

Afro-American Historical Society Museum, 1841 Kennedy Blvd, Jersey City, NJ 07305 • T: +1 201 5475262 • F: +1 201 5475392 •
Dir.: *Neal E. Brunson* • Cons.: *Theodore Brunson* •
Historical Museum – 1977
New Jersey African American hist, musical instruments, black dolls police and firemen, artifacts 49624

The C.A.S.E Museum, 80 Grand St, Jersey City, NJ 07302 • T: +1 201 3325200 •
Fine Arts Museum – Russian art 49625

Courtney and Lemmerman Galleries, New Jersey City University, 2039 Kennedy Blvd, Jersey City, NJ 07305 • T: +1 201 2003214 • F: +1 201 2003224 • hbastidas@njcu.edu •
Dir.: *Hugo Xavier Bastidas* •
Fine Arts Museum
Prints, drawings 49626

Jersey City Museum, 350 Montgommery St, Jersey City, NJ 07302 • T: +1 201 4130303 • F: +1 201 4139922 • info@jerseycitymuseum.org • www.jerseycitymuseum.org •
Exec. Dir.: *Marion Grzesiak* • Cur.: *Rocio Aranda-Alvarado* •
Fine Arts Museum / Local Museum – 1901
Local artifacts, regional art (19th-20th c) 49627

Saint Peter's College Art Gallery, 2641 Kennedy Blvd, Jersey City, NJ 07306 • T: +1 201 9159000 • F: +1 201 4131669 •
Dir.: *Oscar Magnan* •
Fine Arts Museum – 1971
Different art trends 49628

Jewell KS

Palmer Museum, POB 314, Jewell, KS 66949 •
Local Museum / Historical Museum – 1991
Printing machines, city hist 49629

Jim Thorpe PA

Asa Packer Mansion, Packer Hill, Jim Thorpe, PA 18229-0108 • T: +1 717 3253229 • F: +1 717 3258154 • www.asapackermansion.com •
Dir.: *Ron Sheehan* •
Local Museum – 1913
Local hist, 1860 Victorian mansion 49630

Johnson City TN

Carroll Reece Museum, East Tennessee State University, Gilbreath Dr, Johnson City, TN 37614 • T: +1 423 4394392 • F: +1 423 4394283 • whiteb@etsu.edu • cass.etsu.edu/museum •
Dir./Cur.: *Blair H. White* •
Fine Arts Museum / University Museum / Historical Museum – 1965
Contemporary regional art, B. Carool Reece memorial coll 49631

Elizabeth Slocumb Galleries, Carrol Reece Museum, c/o Dept. of Art, East Tennessee State University, POB 70708, Johnson City, TN 37614-0708 • T: +1 423 4395315 • F: +1 423 4394393 • ropp@access.etsu-tn.edu •
Dir.: *Ann Ropp* •
Fine Arts Museum / University Museum – 1965
Art 49632

Hands On! Regional Museum, 315 E Main St, Johnson City, TN 37601 • T: +1 423 9286508, 9286509 • F: +1 423 9286915 • handson@handsonmuseum.org • www.handsonmuseum.org •
Pres.: *Duffie Jones* •
Science&Tech Museum – 1986
Children's museum 49633

Museum at Mountain Home, c/o East Tennessee State University, POB 70693, Johnson City, TN 37614-0693 • T: +1 423 4398069 • F: +1 423 4317025 • whaleym@etsu.edu •
Cur.: *Martha Whaley* •
Historical Museum / Military Museum / University Museum – 1994
Medical and military artifacts 49634

Slocumb Galleries, East Tennessee State University, Gilbreath Dr, Johnson City, TN 37614-0708 • T: +1 423 4397078 • F: +1 423 4394393 • ropp@access.etsu-tn.edu • www.etsu-tenn-st.edu/design/ •
Dir.: *Ann Ropp* •
Public Gallery / University Museum – 1965
Contemporary art in all mediums 49635

Tipton-Haynes Museum, 2620 S Roan St, Johnson City, TN 37604 • T: +1 423 9263631 • information@tipton-haynes.org • www.tipton-haynes.org •
C.E.O.: *Penny McLaughlin* •
Agriculture Museum / Local Museum – 1965
Agriculture, hist, folklore 49636

Johnson City TX

Lyndon B. Johnson National Historical Park, 100 Ladybird Ln, Johnson City, TX 78636 • T: +1 830 8687128 • F: +1 830 8680810 • lyjo_superintendent@nps.gov • www.nps.gov/lyjo •
Dir.: *Leslie Starr Hart* • Cur.: *Virginia Kilby* •
Local Museum – 1969
Furnishings, farm & ranch equipment 49637

Johnstown NY

Johnson Hall, Hall Av, Johnstown, NY 12095 • T: +1 518 7628712 • F: +1 518 7622330 • wanda.burch@oprhp.state.ny.us • www.nysparks.com •
Head: *Wanda Burch* •
Historical Museum – 1763
Period furnishings, family possessions 49638

Johnstown Historical Society Museum, 17 N William St, Johnstown, NY 12095 • T: +1 518 7627076 •
Pres.: *Catherine Levee* • Cur.: *James F. Morrison* •
Historical Museum – 1892
Local hist, personalities of Johnstown 49639

Johnstown PA

Flood Museum, 304 Washington St, Johnstown, PA 15907, mail addr: POB 1889, Johnstown, PA 15907 • T: +1 814 5391889 • F: +1 814 5351931 • info@jaha.org • www.jaha.org •
Dir.: *Richard A. Burkert* • Cur.: *Daniel Ingram* •
Local Museum – 1971
Local hist 49640

Johnstown Area Heritage Association, 304 Washington St, Johnstown, PA 15901, mail addr: POB 1889, Johnstown, PA 15907 • T: +1 814 5391889 • F: +1 814 5351931 • info@jaha.org • www.jaha.org •
Dir.: *Richard A. Burkert* • Cur.: *Daniel Ingram* •
Association with Coll – 2001
Local hist 49641

Southern Alleghenies Museum of Art at Johnstown, c/o University of Pittsburgh at Johnstown, Pasquerilla Performing Arts Center, Johnstown, PA 15904 • T: +1 814 2697234 • F: +1 814 2697240 • ncward@pitt.edu • www.sama-art.org •
Dir.: *Michael Tomor* • Cur.: *Nancy Ward* •
Fine Arts Museum – 1982 49642

Joliet IL

Joliet Area Historical Museum, 204 N Ottawa St, Joliet, IL 60432 • T: +1 815 7235201 • F: +1 815 7239039 • susan.english@jolietmuseum.org • www.jolietmuseum.org •
Pres.: *Gregory Repetti* •
Local Museum – 1981/1999
Local hist, Joliet industries, photos 49643

Laure A. Sprague Art Gallery, c/o Joliet Junior College, J Bldg, 1215 Houbolt Rd, Joliet, IL 60436-8938 • T: +1 815 7299020 ext 2423 • F: +1 815 7445507 • www.jjc.cc.il.us •
Dir.: *Joe B. Milosevich* •
Fine Arts Museum – 1978 49644

Jolon CA

San Antonio Missions Museum, Mission Creek Rd, Jolon, CA 93928 • T: +1 831 3854478 • F: +1 831 3869332 • home.nps.gov/applications/museum •
Religious Arts Museum – 1771
Indian relicts, wine press of Spanish colonial period, manuscripts – archive 49645

Jonesboro AR

Arkansas State University Art Gallery, Caraway Rd, Jonesboro, AR 72467, mail addr: POB 1920, ASU, Jonesboro, AR 72467 • T: +1 870 9723055 • F: +1 870 9723932 • jlockhar@astate.edu • union.astate.edu/artgallery •
Chm.: *Curtis Steele* •
Fine Arts Museum / University Museum – 1967
Prints, drawings, paintings, sculpture by students 49646

Arkansas State University Museum, Dean B. Ellis Library building, Jonesboro, AR 72401, mail addr: POB 490, ASU, Jonesboro, AR 72467 • T: +1 870 9722074 • F: +1 870 9722793 • museum@astate.edu • www2.astate.edu/a/museum •
Dir.: *Dr. Marti Allen* •
University Museum / Historical Museum – 1933
University hist, state hist, archaeology, ethnology, military and natural hist – library 49647

Jonesborough TN

Jonesborough-Washington County History Museum, 117 Boone St, Jonesborough, TN 37659 • T: +1 423 7539580 • F: +1 423 7531020 • dmontanti@jwcheritagealliance.org • www.jwcheritagealliance.org •
Dir.: *Deborah Montanti* •
Local Museum – 1982
Local artifacts from 1770 - present 49648

Joplin MO

Dorothea B. Hoover Historical Museum, 504 Schifferdecker Park, Joplin, MO 64802 • T: +1 417 6231180 • F: +1 417 6236393 • jopmusm@ipa.net • www.joplinmuseum.org •
Dir.: *Brad Belk* • Cur.: *Christopher Wiseman* •
Local Museum – 1973
Textiles, dolls, miniature circus 49649

George A. Spiva Center for the Arts, 222 W Third St, Joplin, MO 64801 • T: +1 417 6230183 • F: +1 417 6233805 • spiva@spivaarts.com • www.spivaarts.com •
Dir.: *Jo Mueller* •
Fine Arts Museum – 1957 49650

Tri-State Mineral Museum, 504 Schifferdecker Av, Joplin, MO 64801 • T: +1 417 6231180 • F: +1 417 6236393 • jopmusm@ipa.net • www.joplinmuseum.org •
Dir.: *Brad Belk* • Cur.: *Christopher Wiseman* •
Natural History Museum – 1931
Lead & zinc ore dating back to Joplin's mining days 49651

Julesburg CO

Depot Museum and Fort Sedgwick Museum, 201 W First St, Julesburg, CO 80737 • T: +1 970 4742061 • F: +1 970 4742061 • history@kcl.net • www.kci.net/-history •
Local Museum – 1940
Union Pacific Railroad depot, agriculture, archaeology, folklore, military, local and natural history, Indian artifacts 49652

Juliette GA

Jarrell Plantation Georgia, 711 Jarrell Plantation Rd, Juliette, GA 31046-2515 • T: +1 912 9865172 • F: +1 912 9865919 • jarrell_planatation_park@mail.dnr.state.ga.us • www.dnr.state.ga.us/dnr/parks •
Agriculture Museum – 1974
Georgia agriculture and farm life 49653

Junction TX

Kimble County Historical Museum, 101 N 4th St, Junction, TX 76849 • T: +1 915 4464219 • F: +1 915 4462871 • burtwyatt@ctesc.net •
Cur.: *Frederica Wyatt* •
Local Museum – 1966
Farm & ranch tools, household items, guns, clothing, photographs 49654

Junction City OR

Junction City Historical Society Museum, 655 Holly St, Junction City, OR 97448 • T: +1 541 9982924 • F: +1 541 9982924 • cgoodin@aol.com • www.junctioncity.com/history •
Cur.: *Kitty Goodin* •
Historical Museum – 1971
Furnishings, medical and dental tools, Indian artifacts 49655

Juneau AK

Alaska State Museum, 395 Whittier St, Juneau, AK 99801-1718 • T: +1 907 4652901 • F: +1 907 4652976 • Bruce.kato@alaska.gov • www.museums.state.ak.us •
Cur.: *Bruce Kato* • Sc. Staff: *Scott Carrlee* (Services) • *Bob Banghart* (Exhibits) • *Steve Henrikson* (Colls) • *Lisa Golisek* (Services) •
Local Museum / Folklore Museum – 1900
General State museum – library 49656

House of Wickersham, 213 Seventh St, Juneau, AK 99801, mail addr: Alaska State Parks, 400 Willoughby Av, Suite 500, Juneau, AK 99811 • T: +1 907 4654563, +1 907 5869001 • dnr.alaska.gov/parks/units/wickrshm.htm •
Cur.: *Ellen Carrlee* (Collection, Exhibits) • *Jane Lindsey* (Education, Program) •
Historical Museum / Historic Site – 1984
Alaskan memorabilia, native crafts 49657

Juneau-Douglas City Museum, 114 W Fourth St, Juneau, AK 99801, mail addr: 155 S Seward St, Juneau, AK 99801 • T: +1 907 5863572 • Jane_Lindsey@ci.juneau.ak.us • www.juneau.lib.ak.us/parksrec •
Dir.: *Jane Lindsey* •
Local Museum – 1976
Local history, culture, gold mining tools and equipment, fine and decorative arts 49658

Jupiter FL

Florida History Museum, 805 N US Hwy 1, Jupiter, FL 33477 • T: +1 561 7476639 • F: +1 561 5753292 • visit@lrhs.org • www.lrhs.org •
Historical Museum – 1971
General History, lighthouse, historic homes 49659

Hibel Museum of Art, 5353 Parkside Dr, Jupiter, FL 33458 • T: +1 561 6225560 • F: +1 561 6224881 • info@hibelmuseum.org • www.hibelmuseum.org •
Dir.: *Nancy Walls* • Chief Cur.: *Edna Hibel* •
Fine Arts Museum – 1977
Paintings, sculptures, drawings, graphics, porcelain art, crystal art and dolls by Edna Hibel, furniture, paperweights and snuff bottle colls 49660

Kadoka SD

Jackson-Washabaugh Historical Museum, S Main St, Kadoka, SD 57543 • T: +1 605 8372229 • F: +1 605 8371262 • kadokacity@wcenet.com •
Local Museum – 1980
Local hist 49661

Kailua-Kona HI

Hulihee Palace, 75-5718 Alii Dr, Kailua-Kona, HI 96740 • T: +1 808 3291877 • F: +1 808 3291321 • hulihee@ilhawaii.net • www.huliheepalace.org •
Decorative Arts Museum / Natural History Museum – 1928
Residence for Hawaiian royalty 49662

Kalamazoo MI

Alamo Township Museum - John E. Gray Memorial Museum, 8119 N Sixth St, Kalamazoo, MI 49009 • T: +1 269 3442107 •
Pres.: *Mary Gobel* •
Ethnology Museum / Local Museum – 1969
Church, school and township records, farm implements 49663

Gallery II, Western Michigan University, College of Fine Arts, 1903 W Michigan Av, Kalamazoo, MI 49008-5421 • T: +1 269 3872455, 3872436 • F: +1 269 3872477 • art-exhibitions@wmich.edu • www.wmich.edu/cfa •
Dir.: *Jacquelyn Ruttinger* • Cur.: *Holly Stephenson* •
Public Gallery / University Museum – 1975
Print coll, sculptures, contemporary art, American and European 19th-20th c fine and decorative arts 49664

Kalamazoo Aviation History Museum, 3101 E Milham Rd, Kalamazoo, MI 49002 • T: +1 269 3826555 • F: +1 269 3821813 • airzoo@airzoo.org • www.airzoo.org •
Dir.: *Robert E. Ellis* • Cur.: *Bill Painter* • Rest.: *Dick Schaus* •
Military Museum / Historical Museum – 1977
Aviation hall of fame, Guadalcana memorial, ca. 70 aircrafts – library, simulation station 49665

Kalamazoo Institute of Arts, 314 S Park St, Kalamazoo, MI 49007-5102 • T: +1 269 3497775 ext 3001 • F: +1 269 3499313 • museum@kiarts.org • www.kiarts.org •
Dir.: *James A. Bridenstine* • Cur.: *Don Desmett* • *Susan Eckhardt* •
Fine Arts Museum – 1924
American and European art, graphics, ceramics, small sculture, works on paper, photography – school, library, studios 49666

Kalamazoo Valley Museum, Kalmazoo Valley Community College, 230 N Rose St, Kalamazoo, MI 49003, mail addr: POB 4070, Kalamazoo, MI 49003-4070 • T: +1 269 3737988 • F: +1 269 3737997 • Rnorris@kvcc.edu • www.kalamazoomuseum.org •
Dir.: *Dr. Patrick Norris* • Asst.Dir.: *Paula Metzner* (Collection) • *Elspeth Inglis* (Education) •
Historical Museum / Science&Tech Museum – 1927
Local hist, science, technology, decorative arts, ethnology – Digistar 4 Planetarium, Challenger Learning Center, Digital Video Theater 49667

Kalaupapa HI

Kalaupapa Historical Park, 7 Puahi St, Kalaupapa, HI 96742, mail addr: POB 2222, Kalaupapa, HI 96742 • T: +1 808 5676802 • F: +1 808 5676729 • kala-superintendent@nps.gob • www.nps.gov/kala/gov •
Head: *Tom Workmann* •
Historic Site – 1980
Site of the 1886-present Molokai Island leprosy settlement 49668

Kalida OH

Putnam County Historical Society Museum, 201 E Main St, Kalida, OH 45853 • T: +1 419 5323008 • pchs@bright.net • www.bright.net/~pchs •
Pres.: *Joe Balbaugh* • Cur.: *Carol Wise* •
Local Museum – 1873
Local artifacts, genealogical material 49669

Kalispell MT

Conrad Mansion Museum, Between Third and Fourth Sts on Woodland Av, Kalispell, MT 59901 • T: +1 406 7552166 • F: +1 406 7552176 • paivaa@altavista.net • www.conradmansion.com •
Chm.: *Rita Fitzsimmons* •
Historical Museum – 1975
Clothing, toys, furniture 49670

Hockaday Museum of Arts, 302 Second Av E at Third St, Kalispell, MT 59901 • T: +1 406 7555268 • F: +1 406 7552023 • information@hockadayartmuseum.org • www.hockadayartmuseum.org •
Pres.: *Mark Norley* • Exec. Dir.: *Linda Engh-Grady* •
Fine Arts Museum – 1968
Contemporary & hist art and culture of Montana 49671

Kalkaska MI

Kalkaska County Historical Museum, 335 S Cedar St, Kalkaska, MI 49646, mail addr: POB 1178, Kalkaska, MI 49646 • T: +1 231 2589719, 2587840 •
Dir.: *Donald F. Bellinger* •
Local Museum – 1967
Regional history, Grand Rapids and Indiana Railroad depot, 1898 Elmer car, a 1929 Durant car – library 49672

Kampsville IL

Center for American Archeology, Hwy 100, Kampsville, IL 62053 • T: +1 618 6534316 • F: +1 618 6534232 • csutton@caa-archeology.org • www.caa-archeology.org •
Dir.: *Cynthia Sutton* •
Archaeology Museum – 1954/1970
Archaeology 49673

Kane PA

Thomas L. Kane Memorial Chapel, 30 Chestnut St, Kane, PA 16735 • T: +1 814 8379729 • F: +1 814 8372213 •
Religious Arts Museum – 1970
Religious hist, historic chapel 49674

Kaneohe HI

Gallery Lolani, Windward Community College, 54-720 Keaahala Rd, Kaneohe, HI 96744 • T: +1 808 2639155 • F: +1 808 2475362 •
Dir.: *Antoinette Martin* •
University Museum / Fine Arts Museum
Art 49675

Kankakee IL

Kankakee County Historical Society Museum, 801 South Eighth Av, Kankakee, IL 60901 • T: +1 815 9325279 • F: +1 815 9325204 • museum@daily-journal.com • www.kankakeecountymuseum.com •
C.E.O.: *Norman S. Stevens* •
Local Museum – 1906
Native American artifacts, firearms, textiles, decorative arts, agriculture items 49676

Kannapolis NC

Cannon Village Visitor Center, 200 West Av, Kannapolis, NC 28081 • T: +1 704 9383200 • F: +1 704 9324188 • www.cannonvillage.com •
Local Museum – 1974
Textile art, historic artifacts, antique handloom 49677

Kanopolis KS

Fort Harker Museum Complex, 309 W Ohio St, Kanopolis, KS 67454 • T: +1 785 4723059 • echs@informatics.net •
Military Museum – 1961
Documents, files, 1870-present, gunds & material, uniforms, Indian artifacts 49678

Kansas City KS

The Children's Museum of Kansas City, 4601 State Av, Kansas City, KS 66102 • T: +1 913 2878888 • F: +1 913 2878332 • childrensmuseumofkc@hotmail.com • www.kidmuzm.org •
Dir.: *Marty Porter* • Sc. Staff: *Elizabeth Schwartz* (Exhibitions) •
Special Museum – 1984
Over 40 hands-on exhib for 2-8 yr olds 49679

Clendening History of Medicine Library and Museum, University of Kansas Medical Center, 3901 Rainbow Blvd, Kansas City, KS 66160-7311 • T: +1 913 5887244 • F: +1 913 5887060 • nhulston@kumc.edu • www.kumc.edu •
Chm.: *Robert L. Martensen* • Cur.: *Nancy Hulston* •
University Museum / Historical Museum – 1945
Egyptian amultets, surgical instruments, diagnostic doll, 19th c American medical artifacts 49680

Grinter Place, 1420 S 78th St, Kansas City, KS 66111 • T: +1 913 2990373 • F: +1 913 7888046 • grinter@kshs.org • www.kshs.org •
Cur.: *Eric Page* •
Historical Museum – 1968
Historic house, built by Moses Grinter, culture, domestic life 49681

Kansas City MO

American Jazz Museum, 1616 E 18th St, Kansas City, MO 64108-1610 • T: +1 816 4748463 ext 204 • F: +1 816 4740074 • jmoore@kcjazz.org • www.americanjazzmuseum.com •
Exec. Dir.: *Juanita Moore* •
Music Museum – 1994
Jazz masters Louis Armstrong, Duke Ellington, Ella Fitzgerald and Charlie Parker 49682

American Royal Museum, 1701 American Royal Court, Kansas City, MO 64102 • T: +1 816 2219800 • F: +1 816 2218189 • nancyp@americanroyal.com • www.americanroyal.com •
Dir.: *Nancy Perry* •
Agriculture Museum – 1992
Coll from American Royal Livestock, Rodeo & Horse Shows 49683

American Truck Historical Society, 10380 N Ambassador Dr, Kansas City, MO 64153 • T: +1 816 8919900 • F: +1 816 8919903 • aths@mindspring.com • www.aths.org •
Pres.: *George Crouse* •
Association with Coll – 1971
Artifacts and literature about trucks 49684

U

Belger Arts Center, University of Missouri-Kansas City, 2100 Walnut St, Kansas City, MO 64108 • T: +1 816 4743250 • F: +1 816 2211621 • info@belgerartscenter.org • www.belgerartscenter.org •
University Museum
Temporary exhibitions 49685

Commerce Bancshares Fine Art Collection, 1000 Walnut St, Kansas City, MO 64199 • T: +1 816 7607885 • F: +1 816 2342356 • robin.trafton@commercebank.com •
Dir.: *Laura Kemper Fields* •
Fine Arts Museum – 1964
Contemporary American realist art, paintings and drawings 49686

Creative Arts Center, 4525 Oak St, Kansas City, MO 64111 • T: +1 816 7511236 • F: +1 816 7617154 •
Dir.: *Marc F. Wilson* •
Fine Arts Museum 49687

Kaleidoscope, 2501 McGee, Kansas City, MO 64108 • T: +1 816 2748301, 2748934 • F: +1 816 2743148 • lavery1@hallmark.com • www.hallmarkkaleidoscope.com •
Admin.: *Regi Ahrens* •
Special Museum – 1969
Children's participatory art exhibit 49688

Kansas City Art Institute Gallery, 16 E 43 St, Kansas City, MO 64111 • T: +1 816 5615563 • F: +1 816 5615993 • www.kcai.edu •
Dir.: *Raechell Smith* •
University Museum
Temporary art 49689

Kansas City Museum, 3218 Gladstone Blvd, Kansas City, MO 64123 • T: +1 816 4838300 •
Local Museum
Hist 49690

Kemper Museum of Contemporary Art, 4420 Warwick Blvd, Kansas City, MO 64111-1821 • T: +1 816 7535784 • F: +1 816 7535806 • info@kemperart.org • www.kemperart.org •
Dir.: *Rachael Blackburn* • Cur.: *Dana Self* •
Fine Arts Museum – 1994
Contemporary art 49691

Leedy-Voulke's Art Center, 2012 Baltimore Av, Kansas City, MO 64108 • T: +1 816 4741919 • F: +1 816 4741919 •
Dir.: *Sherry Leedy* •
Public Gallery
Contemporary arts and crafts 49692

Liberty Memorial Museum of World War One, 100 W 26th St, Kansas City, MO 64108 • T: +1 816 7841918 • F: +1 816 7841929 • staff@libertymemorialmuseum.org • www.libertymemorialmuseum.org •
Dir.: *Eli Paul* •
Historical Museum – 1919
World War I memorabilia – archives 49693

The Money Museum, Federal Reserve Bank of Kansas City, 925 Grand Blvd, Kansas City, MO 64198 • T: +1 816 8812683 • F: +1 816 8812569 • www.kc.frb.org •
C.E.O.: *Thomas M. Hoenig* •
Special Museum
Banking hist to present day, role of money 49694

The Nelson-Atkins Museum of Art, 4525 Oak St, Kansas City, MO 64111-1873 • T: +1 816 7511278 • F: +1 816 5617154 • www.nelson-atkins.org •
Dir.: *Marc F. Wilson* (Oriental Art) • Chief Cur.: *Deborah Emont Scott* • Cur.: *Ian Kennedy* (European Art) • *Christina Futter* (European and American Decorative Arts) • *Dr. Robert Cohon* (Ancient Art) • *Margaret Conrads* (American Art) • *Gaylord Torrence* (American Indian Art) • *Xiaoneng Yang* (Early Chinese Art) • *Jan Schall* (20th c Art) • *George L. McKenna* (Prints and Photographs) • Cons.: *Elisabeth Batchelor* • *Scott Heffley* (Paintings) • *Kathleen Garland* (Objects) •
Fine Arts Museum – 1933
English pottery, Oriental ceramics, paintings, sculpture, contemporary artworks, Oriental art and furniture 49695

Print Consortium, 6121 NW 77th St, Kansas City, MO 64151 • T: +1 816 5871986 • eickhorst@griffon.nwic.edu •
Exec. Dir.: *Dr. William S. Eickhorst* •
Fine Arts Museum – 1983
Prints by established professional artists 49696

Roger Guffey Gallery, Federal Reserve Bank of Kansas City, 925 Grand Blvd, Kansas City, MO 64198 • T: +1 816 8812554 • F: +1 816 8812007 • www.kc.frb.org •
Pres.: *Thomas M. Hoenig* •
Fine Arts Museum / Public Gallery – 1985 49697

Shoal Creek Living History Museum, 7000 NE Barry Rd, Kansas City, MO 64156 • T: +1 816 7922655 • F: +1 816 7923469 • shoalcreeklhm@aol.com • www.kcmo.org •
Open Air Museum – 1975
Architecture, material culture 19th c 49698

Society for Contemporary Photography, 2016 Baltimore Av, Kansas City, MO 64108, mail addr: POB 32284, Kansas City, MO 64171 • T: +1 816 4712115 • F: +1 816 4712462 • spc@sky.net • www.scponline.org •
Dir.: *Kathy Aron* •
Fine Arts Museum – 1984 49699

Steamboat Arabia Museum, 400 Grand Blvd, Kansas City, MO 64106 • T: +1 816 4711856 • F: +1 816 4711616 • info@1856.com • www.1856.com •
Cur.: *Greg Hawley* •
Fine Arts Museum – 1991
Tools, textiles, leather, glassware, China, tin, international goods, equip from the steamboat 49700

Thomas Hart Benton Home and Studio, 3616 Belleview, Kansas City, MO 64111 • T: +1 816 9315722 • F: +1 816 9315722 • dspbhome@mail.dnr.state.mo.us • www.mostateparks.com/benton.htm •
Historical Museum – 1977 49701

Thornhill Gallery, Avila College, 11901 Wornall Rd, Kansas City, MO 64145 • T: +1 816 9428400 ext 2443 • F: +1 816 8464726 • chrismanga@mail.avila.edu • www.avila.edu •
Dir.: *Dr. Carol Coburn* • Cur.: *Lisa Ann Sugimoto* •
Fine Arts Museum / University Museum – 1978
Japanese woodblock prints 49702

Toy and Miniature Museum of Kansas City, 5235 Oak St, Kansas City, MO 64112 • T: +1 816 3339328 • F: +1 816 3332055 • toynmin@swbell.net • www.umkc.edu/tmm •
Dir.: *Roger Berg* • Cur.: *Mary Wheeler* •
Decorative Arts Museum – 1981
Toys 49703

UMKC Gallery of Art, University of Missouri-Kansas City, 203 Fine Arts, 51th and Rockhill Rds, Kansas City, MO 64110-2499 • T: +1 816 2351502 • F: +1 816 2356528 • sublerc@umkc.edu •
Dir.: *Craig A. Subler* •
Fine Arts Museum / University Museum – 1975
Contemporary art (20th-21st c) 49704

Union Station Kansas City, 30 W Pershing Rd, Kansas City, MO 64123-1199 • T: +1 816 4602000 • F: +1 816 4602260 • info@sciencecity.com • www.kcmuseum.com •
Ethnology Museum / Historical Museum / Natural History Museum / Science&Tech Museum – 1939
Archaeology, costumes, general hist 49705

Kapaa HI

Kauai Children's Discovery Museum, 4-831 Kuhio Hwy, Kapaa, HI 96746 • T: +1 808 8238222 • F: +1 808 8212558 • robin@kcdm.org • www.kcdm.org •
Dir.: *Robin Mazor* •
Special Museum – 1994
Science, cultures, technology and nature 49706

Katonah NY

Caramoor Center for Music and the Arts, Girdle Ridge Rd, Katonah, NY 10536 • T: +1 914 2325035 • F: +1 914 2325521 • museum@caramoor.com • www.caramoor.com •
Exec.Dir & C.E.O.: *Michael Barrett* •
Fine Arts Museum / Decorative Arts Museum / Performing Arts Museum – 1946
Eastern and Western art and furnishings, jade coll 49707

John Jay Homestead, 400 Rte 22, & Jay St, Katonah, NY 10536 • T: +1 914 2325651 • F: +1 914 2328085 • julia.warger@oprhp.state.ny.us • www.nysparks.org •
Pres.: *Wendy Ross* •
Historical Museum – 1958
Home and farm of chief justice John Jay, portraits, farming 49708

Katonah Museum of Art, Rte 22 at Jay St, Katonah, NY 10536 • T: +1 914 2329555 • F: +1 914 2323128 • info@katonahmuseum.org • www.katonahmuseum.org •
Acting Dir.: *Nancy Wallach* (Exhibitions) •
Fine Arts Museum – 1953
Changing exhibitons 49709

Kauai HI

Koke'e Natural History Museum, Kokee State Park, Kauai, HI 96752, mail addr: POB 100, Kekaha, HI 96752 • T: +1 808 3359975 • F: +1 808 3356131 • kokee@aloha.net • www.aloha.net/~kokee •
Exec. Dir.: *Marsha Erickson* •
Natural History Museum – 1952
Kaua'i's ecology, geology, climatology 49710

Kaukauna WI

Charles A. Grignon Mansion, 1313 Augustine St, Kaukauna, WI 54130 • T: +1 920 7359370, ext 103 • F: +1 920 7338636 • ochs@foxvalleyhistory.org • www.foxvalleyhistory.org •
C.E.O.: *Terry Bergen* •
Local Museum / Folklore Museum – 1837
Historic house, original interior 49711

Kawaihae HI

Puukohola Heiau Ruins, 62-3601 Kawaihae Rd, Kawaihae, HI 96473, mail addr: POB 44340, Kawaihae, HI 96473 • T: +1 808 8827218 • F: +1 808 8821215 • ernest-young@nps.gov • www.nps.gov/puhe •
Cur.: *Ernest Young* •
Archaeology Museum – 1972
Ruins of Puukohola Heiau, Temple on the hill of the Puukohla Whale, war temple built 1790-1791 49712

Kearney MO

Jesse James Farm and Museum, 21216 Jesse James Farm Rd, Kearney, MO 64060 • T: +1 816 6286065 • F: +1 816 6286676 • jessejames@claycogov.com •
Head: *Elizabeth Gilliam-Beckett* •
Local Museum – 1978
Birthplace of Jesse James 49713

Kearney NE

Fort Kearney Museum, 131 S Central Av, Kearney, NE 68847 • T: +1 308 2345200 •
Dir.: *Marlo L. Johnson* •
Military Museum – 1950
Indian and pioneer relics, tools and arms, European and Oriental material, Egyptian and African objects 49714

Fort Kearny, 1020 V Rd, Kearney, NE 68847 • T: +1 308 8655305 • F: +1 308 8655306 • ftkrny@ngpsun.ngpc.state.ne.us •
Dir.: *Eugene A. Hunt* •
Military Museum / Historic Site – 1929
1848 -1871military records, microfilm 49715

Kearney Area Children's Museum, 2013 Av A, Kearney, NE 68847 • T: +1 308 6892228 • F: +1 308 6892229 • kearneychildrensmuseum@alltel.net •
C.E.O.: *Peggy Abels* •
Special Museum – 1989
Technology, railroads, hot air ballon 49716

Museum of Nebraska Art, 2401 Central Av, Kearney, NE 68847 • T: +1 308 8658559 • F: +1 308 8658104 • monet@unk.edu • monet.unk.edu/mona •
Dir.: *Audrey S. Kauders* • Cur.: *Josephine Martins* •
Fine Arts Museum – 1986
Art and artists from earley 19th c to present 49717

Keene NH

Cheshire County Museum, 246 Main St, Keene, NH 03431 • T: +1 603 3521895 • F: +1 603 3529226 • hscc@cheshire.net • www.hsccnh.org •
Dir.: *Alan F. Rumrill* •
Local Museum – 1927
Local hist, furniture, glass, pottery, maps, photos, silver, toys, paintings 49718

Horatio Colony House Museum, 199 Main St, Keene, NH 03431 • T: +1 603 3520460 • colonymuseum@webryders.com • www.horatiocolonymuseum.org •
Decorative Arts Museum
19th - 20th c furniture, furnishings, decorative arts, portraits, photos, silver 49719

Thorne-Sagendorph Art Gallery, Keene State College, Wyman Way, Keene, NH 03435-3501 • T: +1 603 3582720 • F: +1 603 3582238 • thorne_education@yahoo.com • www.keene.edu/tsag •
Dir.: *Maureen Ahern* •
Fine Arts Museum – 1965
Paintings, prints, drawings, sculpture, crafts, regional and national artists 49720

Kellogg IA

Kellogg Historical Society Museum, 218 High St, Kellogg, IA 50135 •
Local Museum – 1980
Pitcher coll, farming 49721

Kelly WY

Murie Museum, 1 Ditch Creek Rd, Kelly, WY 83011 • T: +1 307 7334765 • F: +1 307 7399388 • info@tetonscience.org • www.tetonscience.org •
Dir.: *John C. Shea* •
Natural History Museum – 1973
Mammals and birds 49722

Kelso WA

Cowlitz County Historical Museum, 405 Allen St, Kelso, WA 98626 • T: +1 360 5773119 • F: +1 360 4239987 • freeced@co.cowlitz.wa.us • www.co.cowlitz.wa.us/museum •
Pres.: *Travis Cavens* • Dir.: *David W. Freece* •
Local Museum – 1953
Local hist 49723

Kemmerer WY

Fossil Country Museum, 400 Pine, Kemmerer, WY 83101 • T: +1 307 8776551 • F: +1 307 8776552 • museum@hamsfork.net • www.hamsfork.net/~museum •
Dir.: *Hilary Barton Billman* •
Local Museum – 1989
Local hist, coal mine, genuine whiskey stills 49724

Kenilworth IL

Kenilworth Historical Society Museum, 415 Kenilworth Av, Kenilworth, IL 60043-1134 • T: +1 847 2512565 • F: +1 847 2512565 • kenilworthhistory@earthlink.net • www.kenilworthhistory.org •
Pres.: *Ray Drexler* •
Local Museum – 1922
Costumes, local documents, artifacts, architectural 49725

Kenly NC

Tobacco Farm Life Museum, Hwy 301 N, 709 Church St, Kenly, NC 27542 • T: +1 919 2843431 • F: +1 919 2849788 • director@tobaccofarmlifemuseum.org • www.tobaccofarmlifemuseum.org •
C.E.O.: *Lynn Wagner* • Cur.: *Harold Jacobson* •
Local Museum – 1983
Agriculture, heritage, Eastern NC farm family 49726

Kennebunk ME

Brick Store Museum, 117 Main St, Kennebunk, ME 04043 • T: +1 207 9854802 • F: +1 207 9856887 • info@brickstoremuseum.org • www.brickstoremuseum.org •
Dir.: *Tracy L. Baetz* •
Historical Museum – 1936
Fine and decorative arts, furniture, archival materials, textiles, maritime-related objects – archives 49727

Kennebunkport ME

Kennebunkport Historical Museum, 125 North St, Kennebunkport, ME 04046 • T: +1 207 9672751 • F: +1 207 9671205 • kporths@gwi.net • www.kporthistory.org •
C.E.O.: *Beverly Keough* •
Local Museum – 1952
Local crafts, art, geneaology, photographs 49728

Seashore Trolley Museum, 195 Log Cabin Rd, Kennebunkport, ME 04046-1690 • T: +1 207 9672712 • F: +1 207 9672800 • carshop@gwi.net • www.trolleymuseum.org •
Science&Tech Museum – 1939
Electric railway history and technology, steetcars, mass transit 49729

Kennesaw GA

Kennesaw Civil War Museum, 2829 Cherokee St, Kennesaw, GA 30144 • T: +1 770 4272117 • F: +1 770 4294538 • groups@southernmuseum.org • www.southernmuseum.org •
Dir.: *Dr. Jeffrey A. Drobney* •
Historic Site / Historical Museum / Science&Tech Museum – 1972
Civil war, trains, site where the Great Locomotive chase began, the locomotive General 49730

Kennesaw Mountain National Battlefield Park, 900 Kennesaw Mountain Dr, Kennesaw, GA 30152 • T: +1 770 4274686 • F: +1 770 5288399 • www.nps.gov/kemo •
Cur.: *Retha Stephens* •
Historical Museum / Military Museum – 1917
Civil war history 49731

Kennett MO

Dunklin County Museum, 122 College, Kennett, MO 63857 • T: +1 573 8886620 • cbrown@sheltonbbs.com •
Dir.: *Charles B. Brown* •
Historical Museum – 1976 49732

Kenosha WI

Kenosha County Museum, 2202 51st Pl, Kenosha, WI 53140 • T: +1 262 6545770 • F: +1 262 6541730 • kchs@acronet.net • www.kenoshahistorycenter.org •
Pres.: *Elise Shelley* • Exec. Dir.: *Tom Schleif* • Cur.: *Cindy J. Nelson* •
Local Museum – 1878
Local hist, folk art and portraits, American decorative arts, Kenosha automotive manufacturing 49733

Kenosha Public Museum, 5500 First Av, Kenosha, WI 53140 • T: +1 262 6534140 • F: +1 262 6534437 • mpaulat@kenosha.org • www.kenosha.org •
Dir.: *Paula Touhey* • Cur.: *Dan Joyce* • *Nancy Mathews* •
Fine Arts Museum / Decorative Arts Museum / Natural History Museum – 1933
Natural hist, Asian art, ivory carvings, Native American, Oceanian and African ethnographic coll 49734

Kensington CT

New Britain Youth Museum at Hungerford Park, 191 Farmington Av, Kensington, CT 06037 • T: +1 860 8279064 • F: +1 860 8271266 • newb@snet.net • www.newbritainyouthmuseum.org •
Dir.: *Ann F. Peabody* •
Special Museum – 1984
Farming, geological and agricultural tools 49735

Kent CT

Kent Art Association, 21 S Main St, Kent, CT 06757 • T: +1 860 9273989 • kent.art.assoc@snet.net • www.kentart.org •
Pres.: *Constance Horton* •
Public Gallery – 1923 49736

Sloane-Stanley Museum and Kent Furnace, Rte 7, Kent, CT 06757, mail addr: POB 917, Kent, CT 06757 • T: +1 860 9273849 • F: +1 860 9272152 • sloanestanley.museum@snet.net • www.chc.state.ct.us •
Dir.: *Karin Peterson* •
Science&Tech Museum – 1969
Iron industry, paintings 49737

Kent OH

The Kelso House Museum, 1106 Old Forge Rd, Kent, OH 44240 • T: +1 330 6731058 • kelsohouse1@aol.com •
Dir.: *Edgar L. McCormick* • Cur.: *Michelle Wardle* •
Historical Museum – 1963
Hist house, furniture, household items, tools, local hist 49738

Kent State University Art Galleries, School of Art, Kent State University, Kent, OH 44242-0001 • T: +1 330 6722444, 6727853 • F: +1 330 6724729 • fsmith@kent.edu • dept.kent.edu/art/galleries.html •
Dir.: *Fred T. Smith* •
Fine Arts Museum / University Museum – 1950
20th c prints, paintings, sculptures, phoztographs, African art coll – James A. Mitchener Collection, Downtown Gallery, Eells Gallery 49739

Kent State University Museum, Rockwell Hall, Corner of E Main & S Lincoln Sts, Kent, OH 44242-0001, mail addr: P.O. Box 5190, 515 Hilltop Drive, Kent, Ohio 44242 • T: +1 330 6723450 • F: +1 330 6723218 • museum@kent.edu • www.kent.edu/museum/ •
Dir.: *Jean Druesedow* • Cur.: *Anne Bissonnette* •
Decorative Arts Museum – 1981
Costumes and decorative arts, 19th and 20th c American glass, ceramics, furniture 49740

Kentfield CA

College of Marin Art Gallery, 835 College Av, Kentfield, CA 94904 • T: +1 415 4859494 • www.marin.edu •
Dir.: *Francis Arnold* •
Fine Arts Museum / University Museum – 1970 49741

Kenton OH

Hardin County Historical Museum, 223 N Main St, Kenton, OH 43326 • T: +1 419 6737147 • F: +1 419 6753547 • hchm@dbscorp.net • www.hardinmuseums.org •
Historical Museum – 1991
Local hist, Native American artifacts, military artifacts, farming implements, pioneer artifacts, china and glass 49742

Kenyon MN

Gunderson House, 107 Gunderson Blvd, Kenyon, MN 55946 • T: +1 507 7895936 •
Pres.: *Lois Easton* •
Historical Museum – 1895
Victorian arcitecture & furnishings 49743

Keokuk IA

Keokuk Art Center, 300 Main St, Keokuk, IA 52632 • T: +1 319 5248354 •
Dir.: *Thomas Seabold* •
Fine Arts Museum – 1954
Paintings, sculptures 49744

Keokuk River Museum, Foot of Johnson St, Victory Park, Keokuk, IA 52632 • T: +1 319 5244765 • F: +1 319 5242642 • www.geomverity.org •
Dir.: *Charles Pietscher* •
Historical Museum – 1962
Maritime history, housed in Geo. M. Verity Mississippi River Steamboat 49745

Keosauqua IA

Van Buren County Historical Society Museum, First St, Keosauqua, IA 52565 • T: +1 319 2933211 •
Pres.: *Vel Luse* •
Local Museum – 1960
Local history 49746

Keota OK

Overstreet-Kerr Historical Farm, 29186 Kerr-Overstreet Rd, Keota, OK 74941 • T: +1 948 9663396 • F: +1 948 9663396 • okhfarm@crosstel.net • www.kerrcenter.com •
Dir.: *Alan Ware* • Cur.: *Jim Combs* •
Agriculture Museum – 1990
Lifestyle of the Native American and Pioneer farmer from 1871 to 1952, 1890-1940 farm equipment 49747

Kerby OR

Kerbyville Museum, 24195 Redwood Hwy, Kerby, OR 97531 • T: +1 541 5925252 • F: +1 541 5925252 • kerbyvillemuseum@mail.com • www.kerbyvillemuseum.com •
C.E.O. & Pres.: *Sandra J.. Hare* •
Local Museum – 1959
Military items of WW I & II; Indian artifacts, pioneer home furnishings; mining, farming 49748

Kerrville TX

Hill Country Museum, 226 Earl Garrett St, Kerrville, TX 78028-5305 • T: +1 830 8968633 •
Dir.: *Griffiths Carnes* •
Local Museum – 1983
Local hist of Hill Country from 1850s to 1930s 49749

L.D. Brinkman Art Foundation, 444 Sidney Baker St S, Kerrville, TX 78028 • T: +1 830 2572000 • F: +1 830 2572030 •
Dir.: *L.D. Brinkman* • *Pam Stone* • *Charles C. Thomas* •
Fine Arts Museum – 1985
Western art 49750

The Museum of Western Art, 1550 Bandera Hwy, Kerrville, TX 78028 • T: +1 830 8962553 • F: +1 830 8962556 • richard.assunto@americanwesternart.org • www.americanwesternart.org •
Exec. Dir.: *Richard Assunto* •
Fine Arts Museum – 1983
Paintings, drawings, sculpture 49751

Ketchikan AK

Tongass Historical Museum, 629 Dock St, Ketchikan, AK 99901 • T: +1 907 2255600 • F: +1 907 2255602 • michaeln@city.ketchikan.ak.us • www.city.ketchikan.ak.us/departments/museums •
Dir.: *Michael Naab* • Cur.: *Chris Hanson* • *Richard H. Van Cleave* •
Local Museum – 1961
Regional history 49752

Totem Heritage Center, 601 Deermount St, Ketchikan, AK 99901, mail addr: 629 Dock St, Ketchikan, AK 99901 • T: +1 907 2255900 • F: +1 907 2255901 • michaeln@city.ketchikan.ak.us • www.city.ketchikan.ak.us/departments/museums •
Dir.: *Michael Naab* •
Local Museum – 1976
Alaska totems, contemporary NW-coast Indian art 49753

Ketchum ID

Sun Valley Center for the Arts, 191 5th St E, Ketchum, ID 83340, mail addr: POB 656, Sun Valley, ID 83353 • T: +1 208 7269491 • F: +1 208 7262344 • information@sunvalleycenter.org • www.sunvalleycenter.org •
Cur.: *Courtney Gilbert* •
Public Gallery / Fine Arts Museum 49754

Kewaunee WI

Kewaunee County Historical Museum, Court House Sq, 613 Dodge St, Kewaunee, WI 54216 • T: +1 920 3887176 • gandolfoo@hotmail.com • www.rootsweb.com/~wikewaun •
Dir.: *Gerald Abitz* • Cur.: *Tom Schuller* •
Local Museum – 1970
Local hist, 1876 Sheriff's residence with interior, Indian and early settlers artifacts 49755

Key West FL

Audubon House, 205 Whitehead St, Key West, FL 33040 • T: +1 305 2942116 • F: +1 305 2944513 • johnaudubon@earthlink.net • www.audubonhouse.com •
Cur.: *Tom Greenwood* •
Historical Museum – 1960
Early 19th c home of Capt. John H. Geiger which commemorates John James Audubon visit to Key West in 1832 – gardens 49756

Donkey Milk House, 613 Eaton St, Key West, FL 33040 • T: +1 305 2961866 • F: +1 305 2960922 •
C.E.O.: *Denison Tempal* •
Decorative Arts Museum – 1992
Period frunishings and furniture from 1830 - 1930 c 49757

East Martello Museum, 3501 S Roosevelt Blvd, Key West, FL 33040 • T: +1 305 2963913 • F: +1 305 2966206 • staffmembername@kwahs.org • www.kwahs.com •
Dir.: *Claudia L. Pennington* •
Historical Museum – 1951
Mario Sanchez and Stanley Papio folk art coll, costumes, WPA and local paintings, period furnishing 49758

Ernest Hemingway House Museum, 907 Whitehead, Key West, FL 33040 • T: +1 305 2941136, 2941575 • F: +1 305 2942755 • hemingwayhome@prodigy.net • www.hemingwayhome.com •
Dir.: *Mike Morawski* •
Special Museum / Historic Site – 1964
1931-1961 Ernest Hemingway Home in Spanish Colonial Style 49759

Fort Zachary Taylor, End of Southard St, Truman Annex, Key West, FL 33040, mail addr: POB 6560, Key West, FL 33041 • T: +1 305 2926713 • F: +1 305 2926881, 2926850 • foft1985@yahoo.com •
Military Museum – 1985
Civil War, Spanish-American war, WW 12 49760

Harry S. Truman Little White House Museum, 111 Front St, Key West, FL 33040 • T: +1 305 2949911 • F: +1 305 2949988 • bwolz@historictours.com • www.trumanlittlewhitehouse.com •
Exec. Dir.: *Robert J. Wolz* •
Historical Museum – 1991
Temporary exhibitions 49761

Key West Lighthouse Museum, 938 Whitehead, Key West, FL 33040 • T: +1 305 2963913 • F: +1 305 2966206 • communications@kwahs.org • www.kwahs.com •
Dir.: *Claudia L. Pennington* •
Science&Tech Museum – 1966
Housed in 1887 lighthouse keepers home, 1846 lighthouse 49762

Key West Museum of Art and History, 281 Front St, Key West, FL 33040 • T: +1 305 2956616 • F: +1 305 2956649 • www.kwahs.com •
Dir.: *Claudia L. Pennington* •
Fine Arts Museum / Historical Museum – 1951
Fine art, folk art, historic artifacts, documents, photographs from 1860 49763

Mel Fisher Maritime Heritage Museum, 200 Greene St, Key West, FL 33040 • T: +1 305 2942633 • F: +1 305 2945671 • info@melfisher.org • www.melfisher.org •
Dir.: *Madeleine H. Burnside* •
Historical Museum – 1982
17th and 18th c coins, decorative arts, ceramics, Slave Trade and maritime artifacts 49764

The Oldest House Museum, 322 Duval St, Key West, FL 33040 • T: +1 305 2949502 • F: +1 305 2944509 • oirf@bellsouth.net • www.oirf.org •
Dir.: *Frances A. Marchbank* •
Historical Museum / Folklore Museum – 1975
Maritime, Conch house, home of local sea captain and wrecker 49765

Ripley's Believe it or not!, 527 Duval St, Key West, FL 33040 • T: +1 305 2959686 • F: +1 305 2939709 • www.ripleys.com •
Cur.: *James R. McGarry* •
Museum of Classical Antiquities / Decorative Arts Museum – 1927
Asian and primitive tribal artifacts, human oddities, arcane art 49766

Keyport NJ

Steamboat Dock Museum, POB 312, Keyport, NJ 07735 • T: +1 732 2647822 •
Cur.: *Angela Jeandron* •
Local Museum – 1976
Business ledgers, photographs of commercial, industrial and domestic life 49767

Keyport WA

Naval Undersea Museum, Garnett Way, Keyport, WA 98345-7610 • T: +1 360 3964148 • F: +1 360 3967944 • underseainfo@kpt.nuwc.navy.mil •
Dir.: *Bill Galvani* • Cur.: *Barbara Moe* •
Science&Tech Museum / Military Museum – 1979 49768

Keystone SD

Big Thunder Gold Mine, 604 Blair, Keystone, SD 57751 • T: +1 605 6664847 • F: +1 605 6664566 • www.bigthundergoldmine.com •
C.E.O.: *Charles McLain* •
Science&Tech Museum – 1958
1880s gold mine's equipment 49769

Keystone Area Museum, 410 3rd St, Keystone, SD 57751 • T: +1 605 6664494 • F: +1 605 6664824 • keygadfly@aol.com • www.keystonehistory.com •
Pres.: *Betty Jo Sagdalen* •
Local Museum – 1983
Local hist 49770

Mount Rushmore National Memorial, Hwy 244, Keystone, SD 57751 • T: +1 605 5742523 • F: +1 605 5742307 • jim_popovich@nps.gov • www.nps.gov/moru/ •
Cur.: *Bruce Weisman* •
Historical Museum – 1925
Massive granite sculpture, carved into a mountainside, memorializing the likenesses of the Presidents Washington, Jefferson, Theodore Roosevelt & Lincoln 49771

National Presidential Wax Museum, Highway 609 16-A, Keystone, SD 57751 • T: +1 605 6664455 • F: +1 605 6664455 •
Dir.: *Mary Ann Riordan* •
Decorative Arts Museum – 1970
Wax figures 49772

Keytesville MO

General Sterling Price Museum, 412 Bridge St, Keytesville, MO 65261 • T: +1 660 2883204 •
Pres.: *Janet Weaver* •
Local Museum – 1964
China, glass, silver, geology, furniture 49773

Kilgore TX

East Texas Oil Museum at Kilgore College, Hwy 259 at Ross St, Kilgore, TX 75662 • T: +1 903 9838531, 9838295 • F: +1 903 9838600 • www.easttexasoilmuseum.com •
Dir.: *Joe L. White* •
Science&Tech Museum / University Museum – 1980
Largest oil field in the world at the time 49774

Kill Devil Hills NC

Wright Brothers National Memorial, Virginia Dare Trail-US 158, Kill Devil Hills, NC 27948 • T: +1 919 4417430 • F: +1 919 4417730 • www.nps.gov/wrbr •
C.E.O.: *Larry Belli* •
Science&Tech Museum – 1927
Reconstructed Wright camp buildings and 1903 flyer, 1902 glider and wind tunnel, tools 49775

Kimberly OR

John Day Fossil Beds National Monument, 32651 Hwy 19, Kimberly, OR 97848 • T: +1 541 9872333 • F: +1 541 9872336 • joda_interpretation@nps.gov • www.nps.gov/joda/ •
Cur.: *T. Fremd* (Paleontology) •
Natural History Museum / Archaeology Museum – 1975
Fossils, geology, cenozoic 49776

Kinderhook NY

Columbia County Historical Society Museum, 5 Albany Av, Kinderhook, NY 12106-0311 • T: +1 518 7589265 • F: +1 518 7582499 • cchs@cchsny.org • www.cchsny.org •
Dir.: *Sharon S. Palmer* • Cur.: *Helen M. McLallen* •
Historical Museum / Local Museum – 1916
NY Federal and Dutch furnishings, dec and fine arts, county hist documents and artifacts 49777

Martin Van Buren Historic Site, Rte 9H, Kinderhook, NY 12106 • T: +1 518 7589689 • F: +1 518 7586986 • superintendent@nps.gov/mava • www.nps.gov/mava •
C.E.O.: *Dan Dattilio* • Cur.: *Dr. Patricia West* •
Historical Museum – 1974
Home of Pres. Martin van Buren (1782-1862), personal effects, furnishings 49778

King WI

Wisconsin Veterans Museum King, Wisconsin Veterans Home, Hwy QQ, King, WI 54946 • T: +1 608 2661009 • F: +1 608 2647615 •
Chm.: *Don Heiliger* • Cur.: *Mary Bade* •
Military Museum – 1935
Military hist, WW I and WW II artifacts, uniforms, arms, Korean and Vietnam Wars' exhibits 49779

Kingfield ME

Stanley Museum, 40 School St, Kingfield, ME 04947 • T: +1 207 2652729 • F: +1 207 2654700 • maine@stanleymuseum.org • www.stanleymuseum.org •
Pres.: *Susan S. Davis* •
Historical Museum / Science&Tech Museum – 1981
History, transportation, steam car 49780

Kingfisher OK

Chisholm Trail Museum, Oklahoma Historical Society, 605 Zellers Av, Kingfisher, OK 73750 • T: +1 405 3755176 • F: +1 405 3755176 +1 call first • reneem@ok-history.mus.ok.us • www.ok-history.mus.ok.us/ •
Cur.: *Renee Mitchell* •
Local Museum – 1970
Agriculture, archaeology, cowboy and old farm equip, Indian artifacts 49781

Governor Seay Mansion, 605 Zellers Av, Kingfisher, OK 73750 • T: +1 405 3755176 • F: +1 405 3755176 • reneem@ok-history.mus.ok.us • www.ok-history.mus.ok.us/ •
C.E.O.: *Dr. Bob Blackburn* • Cur.: *Renee Mitchell* •
Historical Museum – 1967
Period furniture 49782

Kingman AZ

Mohave Museum of History and Arts, 400 W Beale St, Kingman, AZ 86401 • T: +1 520 7533195 • www.mohavemuseum.org •
Dir.: *Shannon Rossiter* •
Local Museum / Fine Arts Museum – 1961
Regional hist, archaeology, art, Indian artifacts, military hist, mining – library, photo archives 49783

Kingman KS

Kingman County Historical Museum, 400 N Main, Kingman, KS 67068 • T: +1 316 5322627 •
C.E.O.: *Ted Giesert* • Cur.: *June Walker* •
Local Museum – 1969
County history, housed in 1888 City Hall 49784

Kings Point NY

American Merchant Marine Museum, United States Merchant Marine Academy, Kings Point, NY 11024-1699 • T: +1 516 7735515 • F: +1 516 4825340 • museum@usmma.edu • www.usmma.edu •
Exec.Dir.: *Dennis Fanucci* •
Historical Museum / Science&Tech Museum – 1978
American Merchant Marine hist, ship models, paintings and artifacts; nautical instrument coll 49785

Kingsport TN

Netherland Inn House Museum, 2144 Netherland Inn Rd, Kingsport, TN 37660 • T: +1 615 2474043 • jgibson@naxs.net •
Local Museum – 1966
Costumes, guns, musical instruments, toys, farm equip, tools, flatboat replica, railroading, forts, antique dolls, antique furniture, wagons 49786

Kingston MA

Major John Bradford House, Maple St and Landing Rd, Kingston, MA 02364 • T: +1 781 5856300 •
Pres.: *Norman P. Tucker* •
Historical Museum – 1921
Halfhulls a.o. maritime artifacts, looms, doll coll 49787

Kingston NJ

Rockingham, POB 496, Kingston, NJ 08528 • T: +1 609 9218835 • peggi@superlink.net • www.rockingham.net •
Cur.: *Margaret Carlsen* •
Local Museum – 1896
Hist house, period furnishings, paintings 49788

Kingston NY

Fred J. Johnston House Museum, Friends of Historic Kingston, 63 Main St, Kingston, NY 12401, mail addr: POB 3763, Kingston, NY 12402 • T: +1 845 3390720 • twothings@hvc.rr.com • www.fohk.org •
Historical Museum – 1965
Local hist, period rooms 49789

Hudson River Maritime Museum, 50 Rondout Landing, Kingston, NY 12401 • T: +1 845 3380071 • F: +1 845 3380583 • hrmm@ulster.net • www.hrmm.org •
C.E.O.: *Gregory Bell* • Cur.: *Allynne Lange* •
Historical Museum – 1980 49790

Senate House, 296 Fair St, Kingston, NY 12401 • T: +1 914 3382786 • F: +1 914 3348173 • www.nysparks.com •
Dir.: *Jane Kellar* •
Fine Arts Museum / Historic Site / Historical Museum – 1887
Hist houses, period furnishings, paintings 49791

Trolley Museum of New York, 89 E Strand, Kingston, NY 12401 • T: +1 845 3313399 • info@tmny.org • tmny.org •
Pres.: *Michael Hanna* •
Science&Tech Museum – 1955
Trolley, subway cars, photos 49792

Volunteer Firemen's Mall and Museum of Kingston, 265 Fair St, Kingston, NY 12401 • T: +1 845 3310866, 3381247 •
Pres.: *Ronald Matthews* •
Historical Museum – 1980
Personal and firemanic artifacts, fire engines, hose carriages 49793

Kingston RI

Pettaquamscutt Museum, 2636 Kingstown Rd, Kingston, RI 02881 • T: +1 401 7831328 •
Dir.: *Christpher P. Bickford* •
Local Museum – 1958
Local hist 49794

University of Rhode Island Fine Arts Center Galleries, University of Rhode Island, 105 Upper College Rd, Ste 1, Kingston, RI 02881-0820 • T: +1 401 8742775 • F: +1 401 8742007 • jtolnick@uri.edu • www.uri.edu/artgalleries •
Dir.: *Judith E. Tolnick* •
University Museum / Fine Arts Museum – 1968
20 th c American and European art 49795

Kingsville TX

John E. Conner Museum, c/o Texas A&M University, 905 W Santa Gertrudis St, Kingsville, TX 78363 • T: +1 361 5932810 • F: +1 361 5932112 •
Dir.: *Hal Hau* • Cur.: *Kate Hogue* •
Local Museum – 1925
South Texas & University hist 49796

Kinsley KS

Edwards County Historical Museum, POB 64, Kinsley, KS 67547 • T: +1 316 6592420 •
Cur.: *Maxine Blank* •
Local Museum – 1967
Agriculture, costumes, Inidan artifacts, farm implements 49797

Kinzers PA

Rough and Tumble Engineers Museum, Rte 30 1/2 E of Kinzers, Kinzers, PA 17535 • T: +1 717 4424249 • www.roughandtumble.org •
C.E.O. & Pres.: *Lewis Frantz* •
Agriculture Museum / Science&Tech Museum – 1948
Agricultural and mechanical technology and hist 49798

Kirksville MO

Adair County Historical Society Museum, 211 Elson St, Kirksville, MO 63501 • T: +1 660 665.6502 • peeve@cableone.net • www.webmoondance.com/ACHS/ACHS.htm •
C.E.O. & Pres.: *Pat Ellebrecht* • Cur.: *Walter Davison* •
Historical Museum – 1976
Adair county hist, Indian and early settlers artifacts 49799

E.M. Violette Museum (closed) 49800

Still National Osteopathic Museum, 800 W Jefferson, Kirksville, MO 63501 • T: +1 660 6262359 • F: +1 660 6262984 • museum@kcom.edu • www.kcom.edu/museum •
Dir.: *Jason Haxton* •
Special Museum – 1978 49801

Kirtland OH

Kirtland Temple Historic Center, 9020 Chillicothe Rd, Kirtland, OH 44094 • T: +1 440 2563318 • F: +1 440 2561929 • temple2@ncweb.com • www.kirtlandtemple.org •
C.E.O.: *Barbara B. Walden* •
Historic Site
1830s tools and furnishings 49802

Kirtland Hills OH

Lake County Historical Society Museum, 8610 Mentor Rd, Kirtland Hills, OH 44060 • T: +1 440 2558979 • F: +1 440 2558980 • www.lakehistory.org •
Exec.Dir.: *Kathie Purmal* •
Local Museum – 1936
19th and 20th c decorative arts, costumes, photographs, tools 49803

Kit Carson CO

Kit Carson Historical Society Museum, 202 W Hwy, Kit Carson, CO 80825 • T: +1 719 9623306 • www.kcdrl.org •
Dir.: *Polly Johnson* •
Local Museum – 1968
Local history 49804

Kittery ME

Kittery Historical and Naval Museum, Rogers Rd, Kittery, ME 03904, mail addr: POB 453, Kittery, ME 03904 • T: +1 207 4393080 • F: +1 207 4393080 •
Dir.: *Wayne Manson* •
Local Museum / Historical Museum – 1976
Local hist, ship models, decorative arts, crafts, regional archaeology, shipbuilding 49805

Klamath Falls OR

Favell Museum of Western Art and Indian Artifacts, 125 W Main St, Klamath Falls, OR 97601 • T: +1 541 8829996 • favellmuseum@favellmuseum.org • www.favellmuseum.org •
C.E.O.: *Patsy H. McMillan* •
Fine Arts Museum / Ethnology Museum / Archaeology Museum – 1972
Indian artifacts: arrowheads, stonework, bonework, pottery; contemporary Western art; miniature firearms 49806

Klamath County Museum, 1451 Main St, Klamath Falls, OR 97601 • T: +1 541 8834208 • F: +1 541 8835710 • tourklco@cdsnet.net •
Dir.: *Kim Bellavia* •
Historical Museum – 1953
Anthropology, archaeology, botany, hist 49807

Senator George Baldwin Hotel, Klamath County Museum, 31 Main St, Klamath Falls, OR 97601 • T: +1 541 8834208 • F: +1 541 8835710 • tourklco@cdsnet.net •
Historical Museum – 1954
Furnishings, personal artifacts 49808

Knoxville IA

National Sprint Car Hall of Fame and Museum, 1 Sprint Capital Pl, Knoxville, IA 50138 • T: +1 641 8426176 • F: +1 641 8426177 • sprintcarhof@sprintcarhof.com • www.sprintcarhof.com •
Dir.: *Bob Baker* • Cur.: *Thomas J. Schmeh* •
Science&Tech Museum – 1986
Sprint car racing memorabilia, sprint cars, big cars, midget cars, championship cars, supermodified cars – library, research center, theatre 49809

Knoxville IL

Knox County Museum, c/o City Hall, Public Sq, Knoxville, IL 61448 • T: +1 309 2892814 • F: +1 309 2898825 • tbould@wmco.net •
C.E.O.: *Peg Bivens* • Cur.: *Sally Hutchcrot* •
Local Museum – 1954
Abingdon pottery 49810

Knoxville TN

Beck Cultural Exchange Center, 1927 Dandridge Av, Knoxville, TN 37915-1997 • T: +1 865 5248461 • F: +1 865 5248462 • beckcec@korrnet.org • www.korrnet.org/beckcec •
Exec. Dir.: *Avon William Rollins* • Pres.: *William V. Powell* •
Folklore Museum / Library with Exhibitions / Historical Museum – 1975
Local Black history 49811

Blount Mansion, 200 W Hill Av, Knoxville, TN 37902 • T: +1 865 5252375 • F: +1 865 5465315 • info@blountmansion.org • www.blountmansion.org •
C.E.O.: *C.L. Creech* •
Local Museum / Historic Site – 1926
1792 home and office of William Blount 49812

C. Kermit Ewing Gallery of Art and Architecture, University of Tennessee, Art and Architecture Bldg, 1715 Volunteer Blvd, Knoxville, TN 37996 • T: +1 865 9743200 • F: +1 865 9743198 • spangler@utk.edu • sunsite.utk.edu/ewing_gallery •
Dir.: *Sam Yates* •
University Museum / Fine Arts Museum – 1981
Contemporary arts and architecture, Japanese prints and woodblocks 49813

Confederate Memorial Hall-Bleak House, 3148 Kingston Pike SW, Knoxville, TN 37919 • T: +1 865 5222371 •
Pres.: *Florence E. Hillis* •
Local Museum – 1959
Confederate hist, relics of Civil War – archives .. 49814

Crescent Bend/Armstrong-Lockett House and William P. Toms Memorial Gardens, 2728 Kingston Pike, Knoxville, TN 37919 • T: +1 865 6373163 • F: +1 865 6371709 • tennlaw@awgd.com •
Cur.: *William Beall* •
Decorative Arts Museum – 1975
1720 - 1820 American and English furniture, paintings, mirrors 49815

East Tennessee Discovery Center, 516 N Beaman, Chilhowee Park, Knoxville, TN 37914-0204 • T: +1 423 5941494 • F: +1 423 5941469 • etdc@bellsouth.net • www.etdiscovery.org •
Pres.: *Margaret Maddox* •
Science&Tech Museum / Natural History Museum – 1960
Science Museum with Planetarium/Kama Health Discovery 49816

East Tennessee Historical Society Museum, 600 Market St, Knoxville, TN 37902 • T: +1 865 2158824 • F: +1 865 2158819 • eths@east-tennessee-history.org • www.east-tennessee-history.org •
Dir.: *Cherel Henderson* • Pres.: *Ginny Rogers* • Cur.: *Lisa Oakley* •
Local Museum – 1834
Concentration on East Tennessee hist 49817

Frank H. McClung Museum, University of Tennessee, 1327 Circle Park Dr, Knoxville, TN 37996-3200 • T: +1 865 9742144 • F: +1 865 9743827 • museum@utk.edu • mcclungmuseum.utk.edu •
Dir.: *Dr. Jefferson Chapman* • Cur.: *Elaine A. Evans* • *Dr. Gary Crites* (Paleoethnobotany) • *Dr. Lynne Sullivan* (Archaeology) • *Dr. Paul Parmalee* (Malacology) •
University Museum / Local Museum / Natural History Museum / Archaeology Museum – 1961
Local hist, natural hist, SE U.S. archaeology 49818

Knoxville Museum of Art, 1050 World's Fair Park Dr, Knoxville, TN 37916-1653 • T: +1 865 5256101 • F: +1 865 5463635 • kma@esper.com • www.knoxart.org •
Chm.: *Shirley Brown* • Cur.: *Dana Self* •
Fine Arts Museum – 1961
Modern and contemporary art in all media, Mexican folk masks, Thorne miniature, paintings, sculpture, photography, works on paper 49819

Mabry-Hazen House, 1711 Dandridge Av, Knoxville, TN 37915 • T: +1 865 5228661 • F: +1 865 5228471 • mabryhazen@yahoo.com • www.korrnet.org/mabry •
Pres.: *Pat Austin* •
Local Museum – 1992
Historic house 49820

Marble Springs State Historic Farmstead Governor John Sevier Memorial, 1220 West Gov. John Sevier Hwy, Knoxville, TN 37920 • T: +1 865 5735508 • F: +1 865 5739768 • marble@vic.net •
Exec. Dir.: *Karma S. King* • Dir. Maint.: *Harry King* •
Local Museum – 1941
Historic house, home of Tennessee's first governor John Sevier 49821

Ramsey House Museum Plantation, 2614 Thorngrove Pike, Knoxville, TN 37914 • T: +1 865 5460745 • F: +1 865 5461851 • ramhse@aol.com • www.ramseyhouse.org •
Dir.: *Fiona McAnally* •
Local Museum – 1952
18th c textiles, furnishings of 1797-1820 49822

Kodiak AK

Alutiiq Museum and Archaeological Repository, 215 Mission Rd; Suite 101, Kodiak, AK 99615 • T: +1 907 4867004 • F: +1 907 4867048 • www.alutiiqmuseum.com •
Dir.: *Sven Haakanson* • Dep. Dir.: *Amy Steffian* • Cur.: *Patrick Saltonstall* •
Ethnology Museum / Archaeology Museum – 1995
Coastal Alaska archaeology 49823

Baranov Museum, Erskine House, 101 Marine Way, Kodiak, AK 99615 • T: +1 907 4865920 • F: +1 907 4863166 • baranov@ak.net • www.baranovmuseum.org •
Dir.: *Katie Oliver* • Cur.: *Ellen Lester* •
Local Museum – 1954
Local history, Alexander Baranov (chief manager of Russian American Company), Handwoven grass baskets, Russian American occupation, Kodiak and Aleutian Island hist – library 49824

Kokomo IN

Elwood Haynes Museum, 1915 S Webster St, Kokomo, IN 46902-2040 • T: +1 765 4567500 •
Dir.: *Kay J. Frazer* • Cur.: *Nancy Kennedy* •
Special Museum – 1967
Home of the inventor Elwood Haynes, Haynes memorabilia, 4 Haynes cars and inventions, exhib. about industrial hist 49825

Howard County Historical Museum, 1200 W Sycamore St, Kokomo, IN 46901 • T: +1 317 4524314 • F: +1 765 4524581 • director@howardcountymuseum.org • www.howardcountymuseum.org •
C.E.O.: *Kelly Thompson* •
Historic Site / Local Museum – 1916
Local history 49826

Kotzebue AK

NANA Museum of the Arctic → Northwest Arctic Heritage Center

Northwest Arctic Heritage Center, 100 Shore Av, Kotzebue, AK 99508, mail addr: POB 1029, Kotzebue, AK 99752 • T: +1 907 4423890, +1 907 4423760 • F: +1 907 4428316 • www.nps.gov/kova/contacts.htm •
C.E.O.: *Michael Scott* •
Local Museum – 1967
Local history 49827

Kure Beach NC

Fort Fisher, 1610 Fort Fisher Blvd, Kure Beach, NC 28449 • T: +1 910 4585538 • F: +1 910 4580477 • fisher@ncmail.net • www.fortfisher.nchistoricsites.org •
Man.: *Barbara G. Hoppe* •
Historical Museum – 1961
Underwater archaeology, military accoutrements, memorabilia, Civil War artifacts and fort ruins 49828

Kutztown PA

Pennsylvania German Cultural Heritage Center, c/o Kutztown University, Luckenbill Rd, Kutztown, PA 19530 • T: +1 610 6831589 • culture@fast.net • www.kutztown.edu/community/PGCHC •
Dir.: *Dr. David Valuska* •
Folklore Museum / Ethnology Museum – 1992
Folk items and art, agriculture, costumes, colonial period 49829

Sharadin Art Gallery, Kutztown University of Pennsylvania, W Main St & College Blvd, Kutztown, PA 19530 • T: +1 610 6834546 • F: +1 610 6834547 • talley@kutztown.edu • www.kutztown.edu/acad/artgallery •
Dir.: *Dan R. Talley* •
Public Gallery
Contemporary art in all mediums 49830

La Conner WA

Museum of Northwest Art, 121 S First St, La Conner, WA 98257 • T: +1 360 4664446 • F: +1 360 4667431 • marketing@museumofnwart.org • www.museumofnwart.org •
C.E.O.: *Kris Molesworth* •
Fine Arts Museum – 1981
Art 49831

Skagit County Historical Museum, 501 S 4th St, La Conner, WA 98257-0818 • T: +1 360 4663365 • F: +1 360 4661611 • museum@co.skagit.wa.us • www.skagitcounty.net/museum •
Dir.: *Karen Marshall* • Library: *Mari C. Anderson-Densmore* • Cur.: *Patricia A. Doran* •
Local Museum / Folklore Museum – 1959
Local hist 49832

La Crosse KS

Barbed Wire Museum, W 1st St, La Crosse, KS 67548 • T: +1 785 2229900 • www.rushcounty.org/barbedwiremuseum •
Chm.: *Bradley Penka* •
Historical Museum – 1971
Tools and items relatedto barbed wire 49833

Post Rock Museum, Rush Co Historical Society, 202 W First St, La Crosse, KS 67548-0473 • T: +1 785 2222719, 2223508 •
Pres.: *James Jecha* •
Natural History Museum – 1962
Geology, post rock is limestone bedrock used in building, churches, homes 49834

La Crosse WI

Hixon House, 429 N 7th St, La Crosse, WI 54601 • T: +1 608 7821980 • F: +1 608 7931359 • bjordan_lchs@yahoo.com •
Dir.: *Rick Brown* • Cur.: *Brenda Jordan* •
Historic Site – 1898
Historic house with originla furniture and interior, upper class 19th c life style 49835

Pump House Regional Arts Center, 119 King St, La Crosse, WI 54601 • T: +1 608 7851434 • F: +1 608 7851432 • info@thepumphouse.org • www.thepumphouse.org •
Exec. Dir.: *Toni Asher* •
Fine Arts Museum – 1980
Regional art, hist photographs 49836

Riverside Museum, 410 E Veterans Memorial Dr, La Crosse, WI 54601 • T: +1 608 7821980 • F: +1 608 7931359 • bjordan_lchs@yahoo.com •
C.E.O. & Dir.: *Carl R. Miller* • Cur.: *Brenda Jordan* • *Lynda Flanety* •
Local Museum – 1990
Local hist 49837

Swarthout Memorial Museum, 112 S 9th St, La Crosse, WI 54601 • T: +1 608 7821980 • F: +1 608 7931359 • bjordan_lchs@yahoo.com •
C.E.O. & Dir.: *Rick Brown* • Cur.: *Brenda Jordan* •
Local Museum – 1898
Local hist, lumbering, brewing, period costumes, Hixon coll of 19th c lifestyle 49838

U

La Fargeville NY

Agricultural Museum at Stone Mills, Rte 180 at Stone Mills, La Fargeville, NY 13656 • T: +1 315 6582353 • agstonemills@usadatanet.net • home.usadatanet.net/~agstonemills •
Dir.: *Maguerite Raineri* •
Agriculture Museum – 1968
Early farm and home equip, early milk-handling, hist houses 49839

La Grange GA

Chattahoochee Valley Art Museum, 112 Lafayette Pkwy, La Grange, GA 30240 • T: +1 706 8823267 • F: +1 706 8822878 • info@cvam-online.org • www.cvam-online.org •
Dir.: *Keith Rasmussen* •
Fine Arts Museum / Decorative Arts Museum – 1963
Decorative arts, paintings, prints, drawings, graphics, sculpture 49840

Lamar Dodd Art Center, c/o La Grange College, 302 Forrest Av, La Grange, GA 30240 • T: +1 706 8808000 • F: +1 706 8127329 • doddartc@mentor.lgc.peachnet.edu •
Dir.: *John Lawrence* •
Fine Arts Museum / University Museum – 1982
Southwestern, American Indian, Lamar Dodd paintings and drawings; 20th c photography 49841

La Grange KY

Oldham County History Center, 106 N Second Av, La Grange, KY 40031 • T: +1 502 2220826 • F: +1 502 2227115 • ochstryctr@aol.com • www.oldhamcounty-historicalsociety.org •
Exec. Dir.: *Nancy Theiss* •
Local Museum 49842

La Grange TX

Fayette Heritage Museum, 855 S Jefferson, La Grange, TX 78945 • T: +1 979 9686418 • F: +1 979 9685357 • library@fais.net • lagrange.fais.net/library/museum.html •
Dir.: *Kathy Carter* •
Historical Museum – 1978
Local hist items, photos, business memorabilia 49843

Nathaniel W. Faison Home and Museum, 232 Clear Lake Dr, La Grange, TX 78945-5330 • T: +1 979 2423566 •
Pres.: *Sue Jones* •
Local Museum – 1960
Paintings, hist of Texas, original furniture, documents 49844

La Habra CA

Children's Museum, 301 S Euclid St, La Habra, CA 90631 • T: +1 562 9059693 • april_baxter@lahabracity.com • www.lahabracity.com •
Dir.: *Kimberly Albarian* • Cur.: *April Baxter* •
Special Museum – 1977
History, taxidermied animals, railroad cars, gems and minerals 49845

La Jolla CA

Museum of Contemporary Art San Diego - La Jolla, 700 Prospect St, La Jolla, CA 92037 • T: +1 858 4543541 • F: +1 858 4546985 • info@mcasd.org • www.mcasd.org •
Dir.: *Hugh M. Davies* • Dep. Dir.: *Charles E. Castle* • *Anne Farrell* (External Affairs) • Sen. Cur.: *Stephanie Hanor* • Cur.: *Robin Clark* •
Fine Arts Museum – 1941 49846

Stuart Collection, University of California, San Diego, 9500 Gilman Dr, La Jolla, CA 92093-0010 • T: +1 858 5342117 • F: +1 858 5349713 • mbeebe@ucsd.edu • stuartcollection.ucsd.edu •
Dir.: *Mary L. Beebe* •
Fine Arts Museum / University Museum – 1981
Contemporary sculptures 49847

University Art Gallery, University of California, San Diego, 9500 Gilman Dr, La Jolla, CA 92093-0327 • T: +1 858 5340419 • uag@ucsd.edu • universityartgallery.ucsd.edu •
Cur.: *Isabelle Lutterodt* •
Public Gallery / University Museum – 1967 49848

La Junta CO

Bent's Old Fort, 35110 Hwy 194 E, La Junta, CO 81050 • T: +1 719 3835010 • F: +1 719 3832129 • www.nps.gov/beol •
Cur.: *Melissa Bechhoefer* •
Historical Museum – 1963
Living history 49849

Koshare Indian Museum, 115 W 18th, La Junta, CO 81050 • T: +1 719 3844411 • F: +1 719 3848836 • koshare@ria.net • www.koshare.org •
Exec. Dir.: *Linda Powers* •
Ethnology Museum – 1949
Indian arts and crafts 49850

La Mirada CA

Biola University Art Gallery, Marshburn Hall, 13800 Biola Av, La Mirada, CA 90639 • T: +1 562 9034807 ext 5655 • F: +1 562 9034748 • art@biola.edu • www.biola.edu •
Public Gallery / University Museum 49851

La Pointe WI

Madeline Island Historical Museum, Woods Av and Main St, La Pointe, WI 54850 • T: +1 715 7472415 • F: +1 715 7476985 • madeline@whs.wisc.edu •
Dir.: *Alicia L. Goehring* • Cur.: *Victoria A. Lock* •
Historical Museum – 1958
Local hist, native american hist, English and French exploration and fur trade, maritime, 19th c. trades 49852

La Porte IN

La Porte County Historical Society Museum, 2405 Indiana Av, La Porte, IN 46350 • T: +1 219 3266808 ext 276 • F: +1 219 3249029 • lpcohist@csinet.net • www.lapcohistsoc.org/ •
Cur.: *James Rodgers* • Asst. Cur.: *Susie Richter* •
Historical Museum – 1906
Local history, Jones coll of firearms 49853

La Porte TX

Battleship Texas, 3527 Battleground Rd, La Porte, TX 77571 • T: +1 281 4792431 • F: +1 281 4794197 • barry.ward@tpwd.state.tx.us • www.tpwd.state.tx.us •
Dir.: *Barry J. Ward* •
Historic Site / Military Museum – 1948
Hist of the Battleship, only surviving WW I era Dreadnought 49854

San Jacinto Museum of History, 1 Monument Circle, La Porte, TX 77571 • T: +1 281 4792421 • F: +1 281 4796619 • sjm@sanjacinto-museum.org • sanjacinto-museum.org •
Pres.: *George J. Donnelly* •
Historical Museum – 1938
Hist of region & Texas, artifacts, manuscripts, visual arts 49855

La Veta CO

Francisco Fort Museum, 306 Main St, La Veta, CO 81055 • T: +1 719 7425501 • F: +1 719 7425501 • lvcc@ruralwideweb.com •
Dir.: *Katharine Emsden* •
Local Museum – 1956
Local history 49856

Lac du Flambeau WI

George W. Brown jr. Ojibwe Museum, 603 Peace Pipe, Lac du Flambeau, WI 54538 • T: +1 715 5883333, 5882139 • F: +1 715 5882355 • deanna@ojibwe.com •
Folklore Museum – 1988
Cultural history of The Lake Superior Ojibwe 49857

Lackawaxen PA

Zane Grey Museum, Scenic Dr, Lackawaxen, PA 18435 • T: +1 570 6854871 • F: +1 570 6854874 • upde_interpretation@nps.gov • www.nps.gov/upde •
Special Museum – 1978 49858

Lackland Air Force Base TX

History and Traditions Museum, 2051 George Av, 37 TRW/MU, Bldg 5206, Lackland Air Force Base, TX 78236-5218 • T: +1 210 6713055 • F: +1 210 6710347 • henry.valdez@lackland.af.mil •
Dir.: *E.A. Valdez* •
Military Museum – 1956
Rare aeronautical equipment, 40 static aircraft to represent note pilots & squadrons, engines 49859

Laclede MO

General John J. Pershing Boyhood Home, POB 141, Laclede, MO 64651 • T: +1 660 9632525 • F: +1 660 9632520 •
Dir.: *Dr. Douglas Eiken* •
Historical Museum – 1960
WW I General Pershing 49860

Laconia NH

The Belknap Mill Museum, The Mill Plaza, Laconia, NH 03246 • T: +1 603 5248813 • F: +1 603 5281228 • belknap@metrocast.net •
Dir.: *Mary Rose Boswell* • Cur.: *Roger Gibbs* •
Fine Arts Museum / Local Museum – 1970
Art, history 49861

Ladysmith WI

Rusk County Historical Society Museum, W 7891 Old 8 Rd, Ladysmith, WI 54848 • T: +1 715 5325615 •
Cur.: *Janet Platteter* •
Local Museum – 1961
Local hist, hist tools 49862

Lafayette CO

Lafayette Miners Museum, 108 E Simpson St, Lafayette, CO 80026-2322 • T: +1 303 6657030 • minersmuseum@cityoflafayette.com •
Dir.: *Glenda L. Chermak* •
Science&Tech Museum – 1976
Coal mining 49863

Lafayette IN

Art Museum of Greater Lafayette, 102 S 10th St, Lafayette, IN 47905 • T: +1 765 7421128 • F: +1 765 7421120 • glma@glmart.org • www.dcwi.com/~glma •
Pres.: *Dr. Rob Lindsey* • Cur.: *David Kwashigroh* •
Fine Arts Museum – 1909
19th and 20th c paintings, prints and drawings, historical and contemporary Indiana art, Latin American prints, American art pottery 49864

Tippecanoe County Historical Museum, 909 South St, Lafayette, IN 47901 • T: +1 765 4768411 • F: +1 765 4768414 • mail@tcha.mus.in.us • www.tcha.mus.in.us •
Dir.: *Kevin O'Brien* •
Local Museum / Historical Museum – 1925
General museum, housed in Moses Fowler Home, nr the site of 1717-91 Fort Ouiatenon, the first fortified European settlement in Indiana 49865

Lafayette LA

Lafayette Museum - Alexandre Mouton House, 1122 Lafayette St, Lafayette, LA 70501 • T: +1 337 2342208 • F: +1 337 2342208 • lafayettemuseum@aol.com •
Pres.: *Anne Marie Hightower* •
Local Museum – 1954
Furniture, hisrorical documents, portraits, Indian artifacts, Mardi Gras costumes and photographs 49866

Lafayette Natural History Museum, 637 Girard Park Dr, Lafayette, LA 70503 • T: +1 337 2685544 • F: +1 337 2685464 • members@lnhm.org • lnhm.org •
Cur.: *David Hostetter* •
Natural History Museum – 1969
Natural history – planetarium, library 49867

University Art Museum, University of Louisiana at Lafayette, 710 E Saint Mary Blvd, Lafayette, LA 70503 • T: +1 337 4825326 • F: +1 337 4825907 • artmuseum@louisiana.edu • www.louisiana.edu/uam •
Dir.: *Herman Mhire* •
Fine Arts Museum – 1968
19th to 20th c paintings, drawings, phtographs and sculpture, 19th to 20th c Japanes prints, naive art 49868

Lafox IL

Garfield Farm Museum, 3N016 Garfield Rd, Box 403, Lafox, IL 60147 • T: +1 630 5848485 • F: +1 630 5848522 • info@garfieldfarm.org • www.garfieldfarm.org •
Dir.: *Jerome Martin Johnson* •
Agriculture Museum – 1977
Historic farm and Inn 49869

Lagrange IN

Machan Museum, 405 S Poplar St, Lagrange, IN 46761 • T: +1 219 4633232 •
Local Museum – 1966
Local artifacts, history 49870

Laguna Beach CA

Laguna Art Museum, 307 Cliff Dr, Laguna Beach, CA 92651 • T: +1 949 4948971 ext 200 • info@lagunaartmuseum.org • www.lagunaartmuseum.org •
Dir.: *Bolton Colburn* •
Fine Arts Museum – 1918
American art, with focus on California painting and sculptures 49871

Lahaina HI

Lahaina Restoration Foundation, 120 Dickenson, Lahaina, HI 96761 • T: +1 808 6613262 • F: +1 808 6619309 • lrf@hawaii.pp.com • www.lahainarestoration.org •
Dir.: *George W. Freeland* •
Folklore Museum / Open Air Museum – 1962
Historical Houses, Baldwin Home, Masters Reading Rm, Hale Pai Printing House, Wohing Chinese Temple, Hale Paahao Prison, Hale Aloha Church, depicting the life of the Lahaina during the Hawaiian Monarchy (1820-1893) 49872

Laie HI

Polynesian Cultural Center, 55-370 Kamehameha Hwy, Laie, HI 96762 • T: +1 808 2933005 • F: +1 808 2933022 • clawsone@polynesia.com •
Decorative Arts Museum / Ethnology Museum – 1963
Decorative arts, ethnic material, graphics, paintings, sculpture 49873

Lake Bronson MN

Kittson County History Center Museum, 332 E Main St, Lake Bronson, MN 56734 • T: +1 218 7544100 • history@wiktel.com •
Dir.: *Cindy Adams* •
Local Museum – 1973
Regional hist since the settlement – library 49874

Lake Buena Vista FL

Epcot, Walt Disney World Resort, Lake Buena Vista, FL 32830, mail addr: POB 10000, Lake Buena Vista, FL 32830 • T: +1 407 8244321 • F: +1 407 5604987 • www.disneyworld.com •
Pres.: *Al Weiss* •
Fine Arts Museum – 1971 49875

Lake Charles LA

Children's Museum of Lake Charles, 925 Enterprise Blvd, Lake Charles, LA 70601 • T: +1 318 4339420 • F: +1 318 4330144 • dan@child-museum.org • www.child-museum.org •
Dir.: *Dan Ellender* •
Special Museum – 1988 49876

Gibson Barham Gallery, 204 W Sallier St, Lake Charles, LA 70601 • T: +1 337 4393797 • F: +1 337 4396040 •
Fine Arts Museum – 1963 49877

Imperial Calcasieu Museum, 204 W Sallier St, Lake Charles, LA 70601 • T: +1 337 4393797 • F: +1 337 4396040 •
Exec. Dir.: *Laura Winford* •
Fine Arts Museum / Local Museum – 1963
Furnishings, pharmacy, Civil War and World War I coll, dolls and toys, musical instrumenst, Indian artifacts 49878

Lake City FL

Columbia County Historical Museum, 105 S Hernando St, Lake City, FL 32025 • T: +1 904 7559096 • F: +1 904 7556605 • histmusm@isgroup.net •
Dir.: *John Shoemaker* •
Local Museum – 1984
Civil war, hist of area 49879

Lake City SC

Browntown Museum, Hwy 341, Lake City, SC 29560, mail addr: 414 Main St, Hemingway, SC 29554 • T: +1 843 5582355 •
Cur.: *Nell Morris* •
Agriculture Museum
Cotton gin, corn crib, smokehouse, farm equip 49880

Lake City SD

Fort Sisseton, 11907 434th Av, Lake City, SD 57247-6142 • T: +1 605 4485701 • F: +1 605 4485572 • fortsisseton@state.sa.us •
Local Museum – 1972
Local hist 49881

Lake Forest IL

Lake Forest-Lake Bluff Historical Society Museum, 361 E Westminster, Lake Forest, IL 60045 • T: +1 847 2345253 • F: +1 847 2345236 • info@lflbhistory.org • www.lflbhistory.org •
Exec. Dir.: *Janice Hack* •
Local Museum – 1972
Temporary exhibitions 49882

Lake George NY

Fort William Henry Museum, Canada St, Lake George, NY 12845 • T: +1 518 6685471 • F: +1 518 6684926 • info@fwh.com • www.fwhmuseum.com •
Pres.: *Robert Flacke* • Cur.: *Gerald Bradfield* •
Military Museum – 1952
Artifacts, Colonial documents, 1750-1790 weapons, manuscripts 49883

Lake George Historical Association Museum, Old Country Courthouse, Canada St, Lake George, NY 12845 • T: +1 518 6685044 • www.lakegeorge-historical.org •
Pres.: *Stephen Miller* •
Fine Arts Museum / Local Museum / Historical Museum – 1964
Paintings, local hist, natural hist, Indian artifacts, tramp art, boats 49884

Lake Jackson TX

Lake Jackson Historical Museum, 249 Circle Way, Lake Jackson, TX 77566 • T: +1 979 2971570 • F: +1 979 2850043 • ljmuseum@compatron.net • www.lakejacksonmuseum.org •
Historical Museum – 1981
Hist of Lake Jackson, art, clothing, textiles, documents, newspaper 49885

Lake Junaluska NC

World Methodist Museum, 575 N Lakeshore Dr, Lake Junaluska, NC 28745 • T: +1 828 4569432 • F: +1 828 4569433 • wmc6@juno.com • www.worldmethodistcouncil.org •
Dir.: *Dr. George Freeman* • Cur.: *Jeanette W. Roberson* •
Religious Arts Museum – 1881
Hist of early Methodism, letters by Wesley, Asbury and Coke 49886

Lake Linden MI

Houghton County Historical Museum, 5500 Hwy M-26, Lake Linden, MI 49945 • T: +1 906 2964121, 2969911 • dmkorman@houghtonhistory.org • www.houghtonhistory.org •
Dir.: *Leo W. Chaput* •
Local Museum – 1961
Regional history, copper lithographs, steam engine, wagons, mining, firefighting – library 49887

Lake Norden SD

South Dakota Amateur Baseball Hall of Fame, 519 Main Av, Lake Norden, SD 57248 • T: +1 605 7853553, 7853884 • F: +1 605 7853315 •
Cur.: *Rusty Antonen* •
Special Museum – 1976
Sports 49888

Lake Oswego OR

Oregon Electric Railway Historical Society Museum, 311 N State St, Lake Oswego, OR 97034 • T: +1 503 2222226 • www.trainweb.org/oerhs •
Dir.: *Greg Bonn* •
Science&Tech Museum – 1957 49889

Lake Placid NY

John Brown Farm, John Brown Rd, Lake Placid, NY 12946 • T: +1 518 5233900 • F: +1 518 5234952 • brendan.mills@oprhp.state.ny.us • www.nysparks.com •
Sc. Staff: *Brendon Mills* •
Agriculture Museum – 1895
Home and grave of abolitionist John Brown (1800-1859), period furnishings, personal possessions of Brown and his family 49890

Lake Placid-North Elba Historical Society Museum, Averyville Rd, Lake Placid, NY 12946 • T: +1 518 5231608 •
Pres.: *John B. Huttlinger* • C.E.O. & Dir.: *Gary Francois* •
Historical Museum – 1967
Regional hist and memorabilia, general store replica, railroad memorabilia, Dewey decimal system coll 49891

Lake Tomahawk WI

Northland Historical Society Museum, 7247 Kelly Dr, Lake Tomahawk, WI 54539 • T: +1 715 2773123, 2772261 • F: +1 715 2773123 • lkhbowen@newnorth.net •
Pres.: *Linda Kay Houghton-Bowen* •
Local Museum – 1957
Local hist 49892

Lake Village AR

Museum of Chicot County Arkansas, 614 Cokley St, Lake Village, AR 71653 • T: +1 870 2652868 • chicotcountymuseum.com •
C.E.O.: *Judge Fred Zieman* •
Local Museum – 1994
Medical instr, nursery equip 49893

Lake Waccamaw NC

Lake Waccamaw Depot Museum, 201 Flemington Av, Lake Waccamaw, NC 28450 • T: +1 910 6461992 • lwdm@whitevillenc.net • www.lakewaccamaw.com •
Chm.: *Nancy Sigmon* • Cur.: *Michele Gore* •
Historical Museum / Natural History Museum / Science&Tech Museum – 1977
Deeds, artifacts, fossils, tools, railroad and logging industry 49894

Lake Wales FL

Lake Wales Depot Museum, 325 S Scenic Hwy, Lake Wales, FL 33853 • T: +1 863 6784209 • F: +1 863 6784299 • thedepot@cityoflakewales.com • www.cityoflakewales.com •
Dir.: *Mimi Reid Hardman* •
Local Museum – 1976
Local history artifacts, railroad, cars, photograhs 49895

Lake Worth FL

Museum of the City of Lake Worth, 414 Lake Av, Lake Worth, FL 33460 • T: +1 561 5861700 • F: +1 561 5861651 •
C.E.O./Cur.: *Beverly Mustaine* •
Local Museum – 1982
Photograph coll, tools, office equipment, cameras, kitchen equipments 49896

National Museum of Polo and Hall of Fame, 9011 Lake Worth Rd, Lake Worth, FL 33467 • T: +1 561 9693210 • F: +1 561 9648299 • polomuseum@att.net • www.polomuseum.com •
Dir.: *George DuPont* •
Special Museum – 1984
Polo sports hist 49897

Palm Beach Institute of Contemporary Art, 601 Lake Av, Lake Worth, FL 33460 • T: +1 561 5820006 • F: +1 561 5820504 • info@palmbeachica.org • www.palmbeachica.org •
Dir.: *Dr. Michael Rush* •
Fine Arts Museum – 1989
Art, housed in 1939 art deco movie theatre 49898

Lakefield MN

Jackson County Historical Museum, 307 N Hwy 86, Lakefield, MN 56150-0238 • T: +1 507 6625505 • jchs@rconnect.com •
Man.: *Madonna M. Kellen* •
Local Museum – 1931
Local history 49899

Lakeland FL

International Sport Aviation Museum, 4175 Medulla Rd, Lakeland, FL 33811 • T: +1 863 6440741 • F: +1 863 6489264 • museum@airmuseum.org • www.airmuseum.org •
Pres.: *John Burton* •
Science&Tech Museum – 1989
Aviation 49900

Melvin Art Gallery, Florida Southern College, Dept. of Art, 111 Lake Hollingsworth Dr, Lakeland, FL 33801 • T: +1 863 6804220 • F: +1 863 6804147 • www.flsouthern.edu •
Public Gallery 49901

Polk Museum of Art, 800 E Palmetto, Lakeland, FL 33801-5529 • T: +1 941 6887743 • F: +1 941 6882611 • info@polkmuseumofart.org • www.polkmuseumofart.com •
Dir.: *Daniel E. Stetson* • Pres.: *Richard Powers* • Dep. Dir.: *Judith M. Barger* • Cur.: *Todd Behrens* •
Fine Arts Museum – 1966
15th to 18th c European arts, pre-Columbian and American art, photography 49902

Lakeport CA

Historic Courthouse Museum, 255 N Main St, Lakeport, CA 95453 • T: +1 707 2634555 • F: +1 707 2637918 • museum@co.lake.ca.us • www.co.lake.ca.us •
Cur.: *Linda Lake* •
Historical Museum – 1936
Native American artifacts, Pomo baskets, stone tools – library 49903

Lakeview OR

Schminck Memorial Museum, 128 S E St, Lakeview, OR 97630 • T: +1 541 9473134 • F: +1 541 9473134 • jenglenn@centurytel.com •
Cur.: *Sherrain Glenn* •
Local Museum – 1936
Glass, graphics, costumes, tools, furniture, china, saddles, guns, Indian artifacts 49904

Lakewood CO

Lakewood's Heritage Culture and the Arts Galleries, 797 S Wadsworth Blvd, Lakewood, CO 80226 • T: +1 303 9877850 • F: +1 303 9877851 • beckos@lakewood.org •
Dir.: *Kris Anderson* • Cur.: *Winifred Filipunas* • *Elizabeth Nosek* •
Local Museum / Public Gallery – 1976
20th c Lakewood history 49905

Lakewood NJ

M. Christina Geis Gallery, Georgian Court College, 900 Lakewood Av, Lakewood, NJ 08701-2697 • T: +1 732 3642200 ext 348 • F: +1 732 9058571 • velasquez@georgian.edu •
Dir.: *Dr. Geraldine Velasquez* •
Fine Arts Museum – 1964 49906

Lakewood OH

Cleveland Artists Foundation at Beck Center for the Arts, 17801 Detroit Av, Lakewood, OH 44107 • T: +1 216 2279507 • F: +1 216 2286050 • laurenhansgen@clevelandartists.org • www.clevelandartists.org •
Fine Arts Museum – 1984
Contemporary art incl acrylics, collages, etchings, oils, sculpture, watercolors 49907

Oldest Stone House Museum, 14710 Lake Av, Lakewood, OH 44107 • T: +1 216 2217343 • F: +1 216 2210320 • lkwdhist@bge.net • www.lkwdpl.org/histsoc/ •
Dir.: *Mazie M. Adams* •
Historical Museum / Local Museum – 1952
Period furnishings of the 1830s and 40s, slide coll of early photographs 49908

Lamar CO

Big Timbers Museum, 7515 US-Hwy 50, Lamar, CO 81052 • T: +1 719 3362472 • F: +1 719 3362472 •
Cur.: *Jeanne Clark* •
Historical Museum – 1966
General hist, WW I posters coll, Worth Charles Frederick gown exhibit, Indian arrowhead coll, furniture – library 49909

Lamar MO

Harry S. Truman Birthplace, 1009 Truman Av & 11th St, Lamar, MO 64759 • T: +1 417 6822279 • F: +1 417 6826304 • www.mostateparks.com •
Dir.: *Pam Myers* •
Historical Museum – 1959 49910

Osage Historic Village, 1009 Truman, Lamar, MO 64759 • T: +1 417 6822279 • F: +1 417 6826304 • www.mostateparks.com •
C.E.O.: *Pam Myers* •
Archaeology Museum – 1984 49911

Lamesa TX

Dal-Paso Museum, 310 S First St, Lamesa, TX 79331 • T: +1 806 8725007 • F: +1 806 8725700 •
C.E.O.: *Wayne C. Smith* •
Local Museum – 1988
General & historical interest to local viewers 49912

Lamoni IA

Liberty Hall Historic Center, 1300 W Main St, Lamoni, IA 50140 • T: +1 515 7846133, 7846408 • libhall@grm.net •
Pres.: *Pamela Clinefelter* • Dir.: *Martha McKain* •
Local Museum – 1976
Local hist 49913

Lancaster CA

Antelope Valley Indian Museum, Av M between E 150th & 170th Sts, Lancaster, CA 93534 • T: +1 661 9420662 • www.avim.parks.ca.gov •
Cur.: *Edra Moore* •
Historical Museum – 1928
American Indian culture 49914

Lancaster Museum / Art Gallery, 44801 W Sierra Hwy, Lancaster, CA 93535 • T: +1 661 9481322 • F: +1 661 9481322 •
Dir.: *Lyle Norton* • Cur.: *Norma Gurba* •
Fine Arts Museum / Local Museum – 1984
Lancaster hist, prehistory, decorative arts, household items 49915

Western Hotel Museum, 557 W Lancaster Blvd, Lancaster, CA 93534 • T: +1 661 7236250 •
Cur.: *Norma Gurba* •
Historical Museum
Antelope Valley Native American and pioneer history 49916

Lancaster MA

Fifth Meeting House, Town Common, Lancaster, MA 01523 • T: +1 978 3652427 • F: +1 978 3680194 •
Religious Arts Museum
1st church of Christ items 49917

Lancaster NH

Lancaster Historical Society Museum, 226 Main St, Lancaster, NH 03584 • T: +1 603 7883004, 7884629 •
Pres.: *Myra Emerson* •
Historical Museum – 1964
Architecture, local artists, local hist 49918

Lancaster OH

The Decorative Arts Center of Ohio, 145 E Main St, Lancaster, OH 43130 • T: +1 740 6811423 • F: +1 740 6812713 • brookover@decartsohio.org • www.decartsohio.org •
C.E.O.: *John Thomas La Porte* •
Decorative Arts Museum – 1997 49919

The Georgian, 105 E Wheeling St, Lancaster, OH 43130 • T: +1 740 6549923 • F: +1 740 6549890 • fairfieldheritage@greenapple.com • www.fairfieldheritage.org •
Dir.: *Peggy Smith* •
Historical Museum / Local Museum – 1976
Furniture, silver, costumes, tools, glass, Indcian artifacts 49920

The Sherman House, 137 East Main St, Lancaster, OH 43130 • T: +1 740 6875891 • F: +1 740 6549890 • fairfieldheritage@greenapple.com • www.shermanhouse.com •
C.E.O. & Pres.: *Diane Eversole* • Dir.: *Connie Leitnaker* •
Local Museum
Furniture, silver, costumes, tools, glass, Indian artifacts, hist Civil War, Sherman family 49921

Lancaster PA

Hands-on House, Children's Museum of Lancaster, 721 Landis Valley Rd, Lancaster, PA 17601-4888 • T: +1 717 5695437 • F: +1 717 5819283 • info@handsonhouse.org • www.handsonhouse.org •
Exec. Dir.: *Lynne Morrison* •
Natural History Museum / Science&Tech Museum – 1987 49922

Heritage Center of Lancaster County, 13 W King St, Lancaster, PA 17603-3813 • T: +1 717 2996440 • F: +1 717 2996916 • heritage@paonline.com • www.lancasterheritage.com •
Dir.: *Peter S. Seibert* •
Fine Arts Museum / Decorative Arts Museum / Local Museum – 1976
History, decorative art 49923

Historic Rock Ford, 881 Rock Ford Rd, Lancaster, PA 17602 • T: +1 717 3927223 • F: +1 717 3927283 • hhbecker@rockfordplantation.org • www.rockfordplantation.org •
Pres.: *Merrill Spahn* •
Local Museum – 1958
Local hist 49924

James Buchanan Foundation for the Preservation of Wheatland, 1120 Marietta Av, Lancaster, PA 17603 • T: +1 717 3928721 • F: +1 717 2958825 • jbwwheatland@aol.com • www.wheatland.org •
Pres.: *Michael J Minney* •
Historical Museum – 1936
Restored 1828 Federal mansion, residence of President James Buchanan 49925

Lancaster County Museum, 230 N President Av, Lancaster, PA 17603-3125 • T: +1 717 3924633 • F: +1 717 2932739 • lchs@ptd.net • www.lancasterhistory.org •
Exec. Dir.: *Thomas R. Ryan* •
Local Museum – 1886
Local hist 49926

Lancaster Museum of Art, 135 N Lime St, Lancaster, PA 17602 • T: +1 717 3943497 • F: +1 717 3940101 • lmart@mindspring.com • www.lancastermuseumart.com •
Dir.: *Cindi Morrison* • Asst. Dir.: *Leigh Mackow* •
Fine Arts Museum – 1965
Contemporary art 49927

Landis Valley Museum, 2451 Kissel Hill Rd, Lancaster, PA 17601 • T: +1 717 5690401 ext 200 • F: +1 717 5602147 • rswody@state.pa.us • www.landisvalleymuseum.org •
Dir.: *Russell Swody* • Cur.: *Bruce Bomberger* (Agricultural History) •
Local Museum / Agriculture Museum – 1925
Rural life and culture 49928

The North Museum of Natural History and Science, 400 College Av, Lancaster, PA 17604 • T: +1 717 2913941 • F: +1 717 3584504 • www.northmuseum.org •
Cur.: *Dr. Steven Kirsch* (Geology, Mineralogy) • *Allison Eichelberger* (Entomology) • *John Coolidge* (Paleontology) •
Natural History Museum – 1901
Natural history 49929

Pennsylvania College of Art and Design Galleries, 204 N Prince St, Lancaster, PA 17603 • T: +1 717 3967833 • F: +1 717 3961339 • admissions@pcad.edu • www.pcad.edu •
Public Gallery
Works of nationally and regionally known artists and designers 49930

The Phillips Museum of Art, 700 College Av, Lancaster, PA 17604 • T: +1 717 2913879 • F: +1 717 3584437 • www.fandm.edu/departments/phillipsmuseum •
Dir./Cur.: *Carol E. Faill* •
Fine Arts Museum – 2000
18th - 20th c American paintings, 20th c photographs, American decoratvie arts, Pennsylvania folk art, African art 49931

Lancaster VA

Mary Ball Washington Museum, 8346 Mary Ball Rd, Lancaster, VA 22503 • T: +1 804 4627280 • F: +1 804 4626107 • history@rivnet.net •
C.E.O.: *Valencia Keeve* •
Historical Museum – 1958
Textiles, furniture, paintings, family and regional memorabilia, folklife, Civil War 49932

Lander WY

Fremont County Pioneer Museum, 630 Lincoln St, Lander, WY 82520 • T: +1 307 3324137 • F: +1 307 3326498 •
Dir.: *Todd Guenther* •
Local Museum – 1908
Local hist, pioneer and Indian artifacts, agricultural and transportation implements 49933

Landing NJ

Lake Hopatcong Historical Museum, Hopatcong State Park, Landing, NJ 07850, mail addr: POB 668, Landing, NJ 07850 • T: +1 973 3982616 • F: +1 973 3618987 • lhhistory@worldnet.att.net • www.hopatcong.org/museum •
C.E.O.: *Rosemarie Winfield* •
Local Museum – 1955
Native American, Morris Canal, hist, Lake Hopatcong 49934

Landisville PA

Amos Herr House, 1756 Nissley Rd, Landisville, PA 17538 • T: +1 717 8988822 • info@herrhomestead.org • www.herrhomestead.org •
Cur.: *Catherine Glass* •
Decorative Arts Museum – 1990
Furnishings, farm equip 49935

Langhorne PA

Historic Langhorne Museum, 160 W Maple Av, Langhorne, PA 19047 • T: +1 215 7571888, 7576158 • F: +1 215 7415767 • hla.buxcom.net •
Cur.: *Ruth E. Irwin* (Artifacts) •
Historical Museum – 1965
History 49936

Langston OK

Melvin B. Tolson Black Heritage Center, c/o Langston University, POB 907, Langston, OK 73050 • T: +1 405 4662231 • F: +1 405 4662979 •
Cur.: *Bettye Black* •
Historical Museum / University Museum – 1959
African American art and artifacts, paintings, photographs 49937

Langtry TX

Judge Roy Bean Visitor Center, Hwy 90, W, Loop 25, Langtry, TX 78871 • T: +1 432 2913340 • F: +1 432 2913366 • lytic@dot.state.tx.us • www.dot.state.tx.us • Head: *Vernon N. Billings* •
Local Museum – 1939
Relics pertaining to Judge Roy Bean, Law West of the Pecos ... 49938

Lanham-Seabrook MD

Howard B. Owens Science Center, 9601 Greenbelt Rd, Lanham-Seabrook, MD 20706 • T: +1 301 9188750 • F: +1 301 9188753 • owens2@erols.com • www.pgcps.org •
Natural History Museum – 1978
Science – planetarium ... 49939

Lansing IL

Lansing Veterans Memorial Museum, Lansing Municipal Airport, Lansing, IL 60438 • T: +1 708 8951321 • F: +1 708 4740798 •
Dir.: *David Schwolger* •
Military Museum – 1991
Coll of photos, maps, models an other items related to Farword Air controllers in Vietnam ... 49940

Lansing MI

Impression 5 Science Center, 200 Museum Dr, Lansing, MI 48933 • T: +1 517 4858116 • F: +1 517 4858125 • contompasis@impression5.org • www.impression5.org •
C.E.O.: *Ellen Sprouls* •
Science&Tech Museum – 1972 ... 49941

Lansing Art Gallery, 113 S Washington Sq, Lansing, MI 48933 • T: +1 517 3746400 • lansingartgallery@yahoo.com • www.lansingartgallery.org •
Dir.: *Catherine Babcock* •
Public Gallery – 1965
Visual arts by Michigan artists ... 49942

Michigan Historical Museum, 702 W Kalamazoo, Lansing, MI 48909-8240 • T: +1 517 3730515 • F: +1 517 2414738 • kwiatkowskip@michigan.gov • www.michiganhistory.org •
Dir.: *Phillip C. Kwiatkowski* •
Historical Museum / Local Museum – 1879
State and NW-Territory history, settlements, industry, technology, ethnic items, agriculture, folklore, decorative arts ... 49943

Museum of Surveying, 220 Museum Dr, Lansing, MI 48933 • T: +1 517 4846605 • F: +1 517 4843711 • museumofsurvey@acd.net • www.surveyhistory.org •
Dir.: *Lisa D. Jacobs* •
Special Museum – 1989
Land surveying hist and instr ... 49944

R.E. Olds Transportation Museum, 240 Museum Dr, Lansing, MI 48933 • T: +1 517 3720529 • F: +1 517 3722901 • autos@reoldsmuseum.org • www.reoldsmuseum.org •
Dir.: *Deborah Horstik* •
Science&Tech Museum – 1981
REO speedwagons, wheels, aviation, oldsmobile 49945

Laona WI

Camp Five Museum, 5480 Connor Farm Rd, Laona, WI 54541 • T: +1 715 6743414 • F: +1 715 6747400 • www.camp5museum.org •
Science&Tech Museum – 1969
Logging, early farm tools, railroad artifacts ... 49946

Laramie WY

The Geological Museum, University of Wyoming, Laramie, WY 82071-3006 • T: +1 307 7664227 • F: +1 307 7666679 • uwgeoms@uwyo.edu •
Dir.: *Brent H. Breithaupt* •
University Museum / Natural History Museum – 1887
Geology, vertebrate and invertebrate paleontology, rocks, minerals ... 49947

Laramie Plains Museum, 603 Ivinson Av, Laramie, WY 82070-3299 • T: +1 307 7424448 • laramiemuse@vcn.com • www.laramiemuseum.org •
C.E.O. & Dir.: *Daniel A. Nelson* • Cur.: *Jonel Wilmot* •
Local Museum – 1966
Local hist, pioneer life, Indian artifacts ... 49948

University of Wyoming Anthropology Museum, Anthropology Bldg, 14th and Ivinson St, Laramie, WY 82071 • T: +1 307 7665136 • F: +1 307 7662473 • arrow@uwyo.edu •
Dir.: *Charles A. Reher* •
University Museum / Ethnology Museum / Archaeology Museum – 1966
Anthropology, ethnology ... 49949

University of Wyoming Art Museum, 2111 Willett Dr, Laramie, WY 82071-3807 • T: +1 307 7666620 • F: +1 307 7663520 • uwartmus@uwyo.edu • www.uwyo.edu/artmuseum •
Dir./ Cur.: *Susan Moldenhauer* • Cur.: *Scott Boberg* •
Fine Arts Museum / University Museum – 1968
Contemporary art, photography ... 49950

Wyoming Territorial Prison, 975 Snowy Range Rd, Laramie, WY 82070 • T: +1 307 7456161 • F: +1 307 7458620 • prison@lariat.org • www.wyoprisonpark.org •
C.E.O. & Dir.: *Tom Lindmier* • Cur.: *Teresa Beyer* •
Local Museum – 1986
Local hist, hist buildings ... 49951

Laredo TX

Laredo Children's Museum, c/o Laredo Community College, P-56 Fort Macintosh, Laredo, TX 78040 • T: +1 956 7252299 • F: +1 956 7257776 • lcmuseum1@yahoo.com •
Pres.: *Melissa Cigarroa* • C.E.O.: *Evelyn Smietana* •
Historical Museum – 1988
Large playground equipment for children 5 & up, toddler area for infants to 3 yrs ... 49952

Largo FL

Heritage Village - Pinellas County Historical Museum, 11909-125 St N, Largo, FL 33774 • T: +1 727 5822123 • F: +1 727 5822211 • plighthi@co.pinellas.fl.us • www.pinellascounty.org/heritage •
Dir.: *Jan Luth* • Cur.: *Alison Giesen* •
Historical Museum / Open Air Museum – 1961
Living history ... 49953

Larned KS

Fort Larned, Rte 3, Larned, KS 67550 • T: +1 620 2856911 • F: +1 620 2853571 • george_elmore@nps.gov • www.nps.gov/fols •
Dir.: *Steve Linderer* • Cur.: *George Elmore* •
Historical Museum / Military Museum – 1964
Historic fort on Santa Fe Trail ... 49954

Santa Fe Trail Museum, Rte 3, Larned, KS 67550 • T: +1 620 2852054 • F: +1 620 2857491 • trailctr@larned.net • www.larned.net/trailctr/ •
Dir.: *Ruth Olson Peters* • Cur.: *Betsy Crawford-Gore* • *Beverly Howell* •
Historical Museum – 1974
Prehistoric & hist artifacts – library ... 49955

Las Animas CO

Kit Carson Museum, 425 Carson St, Las Animas, CO 81054 • T: +1 719 4562005 •
Cur.: *Ann Smith* •
Local Museum – 1959
Indian coll, cattle industry items, railroad, agriculture, Fort Lyon, Santa Fee room, Llewellyn Thompson memorial room, history ... 49956

Las Cruces NM

Branigan Cultural Center, 500 N Water, Las Cruces, NM 88001 • T: +1 505 5412155 • F: +1 505 5253645 • bcc1@zianet.com • www.lascruces-culture.org •
Dir.: *Sharon Bode-Hempton* •
Fine Arts Museum / Natural History Museum – 1981
Natural hist, art ... 49957

Las Cruces Museum of Natural History, 700 S Telshor Blvd, Las Cruces, NM 88011 • T: +1 505 5223120 • F: +1 505 5223744 • mnh@las-cruces.org • www.nmsu.edu/~museum •
Natural History Museum – 1988
Chihuhuan desert animals ... 49958

Museum of Fine Art, 500 N Water St, Las Cruces, NM 88001 • T: +1 505 5412221 • F: +1 505 5253645 • bcc1@zianet.com • www.lascruces-culture.org •
Pres.: *Bobbie Provencio* • Dir.: *Sharon Bode-Hempton* •
Fine Arts Museum
20th c art ... 49959

New Mexico Farm and Ranch Heritage Museum, 4100 Dripping Springs Rd, Las Cruces, NM 88011 • T: +1 505 5224100 • F: +1 505 5223085 • cmassey@frh.state.nm.us • www.frhm.org •
Dir.: *Mac R. Harris* •
Agriculture Museum / Historical Museum – 1991
Agriculture hist in New Mexico, prehistoric and modern items ... 49960

New Mexico State University Art Gallery, University Av E. of Solano, Las Cruces, NM 88003-8001 • T: +1 505 6461705 • F: +1 505 6468036 • artglry@nmsu.edu • www.nmsu.edu/~artgal •
Dir.: *Mary Anne Redding* •
Fine Arts Museum / University Museum – 1973
19th c Mexican retablos, contemporary prints, photographs, painting and work on paper ... 49961

New Mexico State University Museum, Kent Hall, University Av, Las Cruces, NM 88003 • T: +1 505 6463739 • F: +1 505 6461419 • estaski@nmsu.edu • www.nmsu.edu/~museum •
Dir.: *Edward Staski* • Cur.: *Dr. Terry Reynolds* •
University Museum / Ethnology Museum – 1959
Soutwestern U.S. and Nortwestern Mexico archaeology, ethnology, history, science ... 49962

Las Vegas NM

City of Las Vegas and Rough Riders Memorial Museum, 727 Grand Av, Las Vegas, NM 87701 • T: +1 505 4541401 ext 283 • F: +1 505 4257335 •
Dir.: *Melanie LaBorwit* •
Local Museum – 1960
Local history, Spanish-American war ... 49963

The Ray Drew Gallery, New Mexico Highlands University, Donelly Library, National Av, Las Vegas, NM 87701 • T: +1 505 4543338 • F: +1 505 4540026 • bread_77@hotmail.com •
Dir.: *Bob Read* •
University Museum / Fine Arts Museum – 1956
Art of the American Southwest-Hispanic and American Indian ... 49964

Las Vegas NV

Charleston Heights Art Center Gallery, City of Las Vegas, 800 South Brush St, Las Vegas, NV 89107 • T: +1 702 2294674 • www.artlasvegas.org •
Dir.: *Nancy Sloan* •
Public Gallery
Contemporary art exhibitions ... 49965

Donna Beam Fine Art Gallery, University of Nevada, Las Vegas, 4505 S Maryland Pkwy, Las Vegas, NV 89154-5002 • T: +1 702 8953893 • F: +1 702 8953751 • schefcij@nevada.edu • www.unlv.edu/main/museums.html •
Dir.: *Jerry A. Schefcik* •
Fine Arts Museum / University Museum – 1962
US works of all media of art (2nd half of 20th c) ... 49966

Flamingo Gallery, c/o Clark County Library, 1401 E Flamingo Rd, Las Vegas, NV 89119 • T: +1 702 7337810 • F: +1 702 7327271 • www.lvccld.org •
Dir.: *Denise Shapiro* •
Public Gallery – 1970 ... 49967

George L. Sturman Museum, University of Nevada at Las Vegas, 4505 Maryland Pkwy, Las Vegas, NV 89154 • T: +1 702 8953893 • F: +1 702 8953751 • www.unlv.edu/main/indexpage.html#dlist •
Dir.: *Jerry A. Schefcik* •
Fine Arts Museum / University Museum
Temporary art coll ... 49968

Guggenheim Hermitage Museum (closed) ... 49969

Guinness World of Records Museum, 2780 Las Vegas Blvd S, Las Vegas, NV 89109 • T: +1 702 7923766 • F: +1 702 7920530 • guinness@wizard.com •
Cur.: *Oli Lewis* •
Special Museum – 1990 ... 49970

Las Vegas Art Museum, 9600 W Sahara Av, Las Vegas, NV 89117 • T: +1 702 3608000 • F: +1 702 3608080 • lvam@earthlink.net •
Exec. Dir.: *Marianne Lovenz* • Cur.: *Dr. James Mann* •
Fine Arts Museum – 1950
Contemporary international art ... 49971

Las Vegas International Scouting Museum, 2915 W Charleston, Las Vegas, NV 89102 • T: +1 702 8787268 • F: +1 702 8222020 • luismdirector@aol.com •
Dir.: *J.C. Ellis* •
Special Museum ... 49972

Las Vegas Natural History Museum, 900 Las Vegas Blvd N, Las Vegas, NV 89101 • T: +1 702 3843466 • F: +1 702 3845343 • dino@lvnhm.org • www.lvnhm.org •
C.E.O.: *Marilyn Gillespie* •
Natural History Museum – 1989
Natural history, fossils, animals, prints & culture 49973

Liberace Museum, 1775 E Tropicana, Las Vegas, NV 89119 • T: +1 702 7985595 ext 20 • F: +1 702 7987386 • info@liberace.org • www.liberace.org •
Pres.: *Joel Strote* •
Music Museum – 1979
Pianos, jewelry, cars, photographs, performing and creative arts, costumes ... 49974

Lied Discovery Children's Museum, 833 Las Vegas Blvd N, Las Vegas, NV 89101 • T: +1 702 3823445 • F: +1 702 3820592 • mail@ldcm.org • www.ldcm.org •
Dir.: *Suzanne LeBlanc* •
Special Museum – 1984 ... 49975

Museum of Ancient Artifacts, 3019 McLeod Dr, Las Vegas, NV 89121 • T: +1 702 4571377 •
C.E.O. & Cur.: *Michael J. Kurban* •
Local Museum – 1990
Photographs and gems, tapestries, artwork, books and artifacts from 550 to present ... 49976

Nevada State Museum, 700 Twin Lakes Dr, Las Vegas, NV 89107 • T: +1 702 4865205 • F: +1 702 4865172 • www.nevadaculture.net •
Exec. Dir.: *Greta Brunschwyler* • Cur.: *David Millman* (Collections) •
Local Museum – 1982
Southern Nevada hist, natural and local hist ... 49977

Reed Whipple Cultural Center, City of Las Vegas, 821 Las Vegas Blvd, Las Vegas, NV 89101 • T: +1 702 2294674 • www.artlasvegas.org •
Dir.: *Nancy Sloan* •
Public Gallery
Temporary exhibitions of art ... 49978

UNLV Barrick Museum, 4505 Maryland Pkwy, Las Vegas, NV 89154-4012 • T: +1 702 8953011 • F: +1 702 8953094 • barrickmuseum@ccmail.nevada.edu • hrc.nevada.edu/museum •
Dir.: *Dr. Klaus Stetzenbach* • Cur.: *Aurore Giguet* •
University Museum / Natural History Museum – 1967 ... 49979

The Walker African American Museum, 705 W van Buren Av, Las Vegas, NV 89106 • T: +1 702 6492238, 3995400 •
Pres., Dir. & Cur.: *Gwendolyn Walker* •
Historical Museum – 1993
Hist of African Americans in nevade from 1800's to present, books, documents, artifacts ... 49980

Wynn Collection, 3145 Las Vegas Blvd S, Las Vegas, NV 89109 • T: +1 702 7334100 •
Fine Arts Museum
13 paintings by Steve Wynn ... 49981

Laurel MD

The Laurel Museum, 817 Main St, Laurel, MD 20707 • T: +1 301 7257975 • F: +1 301 7257975 • laurelmuseum@netscape.net • www.laurelhistory.org •
Pres.: *Shari E. Pollard* •
Local Museum – 1996
Local hist ... 49982

Montpelier Cultur Arts Center, 12826 Laural-Bowie Rd, Laurel, MD 20708 • T: +1 301 9531993 • F: +1 301 2069682 • montpeliermansion@smart.net •
Dir.: *Richard Zandler* •
Decorative Arts Museum – 1979 ... 49983

Montpellier Mansion, Rte 197 and Muirkirk Rd, Laurel, MD 20708 • T: +1 301 9531376 • F: +1 301 6992544 • ann.wagner@pgparks.com • www.pgparks.com •
Decorative Arts Museum – 1976
18th - 19th c decorative arts and furnishings ... 49984

Laurel MS

Lauren Rogers Museum of Art, 565 N 5th Av, Laurel, MS 39440 • T: +1 601 6496374 • F: +1 601 6496379 • lrma@c-gate.net • www.lrma.org •
Dir.: *George Bassi* • Registrar: *Tommie Rodgers* •
Fine Arts Museum – 1923
19th c European art, 18th c Englisg Georgian silver, 18th-19th c Japanese Ukiyo-e Woodblock prints, 19th-20th c American paintings – library ... 49985

Laurens IA

Pocahontas County Iowa Historical Society Museum, 272 N Third, Laurens, IA 50554 • T: +1 712 8452577 •
Cur.: *Joseph Sobotka* •
Local Museum – 1977
Regional history, photos, medicine ... 49986

Laurens SC

James Dunklin House, 544 W Main St, Laurens, SC 29360 • T: +1 864 9844735 • hackster@backroads.net •
Decorative Arts Museum – 1972 ... 49987

Laurinburg NC

Indian Museum of the Carolinas, 607 Turnpike Rd, Laurinburg, NC 28352 • T: +1 910 2765880 •
Dir.: *Dr. Margaret Houston* •
Archaeology Museum – 1969
Indian artifacts ... 49988

Lava Hot Springs ID

South Bannock County Historical Center, 110 E Main St, Lava Hot Springs, ID 83246 • T: +1 208 7765254 • F: +1 208 7765254 •
Dir.: *Ruth Ann Olson* • Cur.: *Dee Forsythe* •
Local Museum – 1980
Artifactsm photographs, family hist, native American ... 49989

Lawrence KS

Museum of Anthropology, University of Kansas, Lawrence, KS 66045 • T: +1 913 8644245 • F: +1 913 8645243 •
Dir.: *Alfred E. Johnson* • Cur.: *Anta Montet-White* • *Brad Logan* • *Mary J. Adair* • *John Hoopes* • *Jack Hofman* • *Robert J. Smith* •
University Museum / Ethnology Museum – 1975
Anthropology ... 49990

Natural History Museum, University of Kansas, 1345 Jayhawk Blvd, Lawrence, KS 66045-7561 • T: +1 785 8644540 • F: +1 785 8645335 • kunhm@ku.edu • www.nhm.ku.edu •
Dir.: *Dr. Leonard Krishtalka* • Asst. Dir.: *W. Bradley M. Kemp* • *Jordan Yochim* • *James Beach* • Cur.: *Robert Timm* • *Norman Slade* (Mammalogy) • *Richard O. Prum* • *Townsend Peterson* (Ornithology) • *James S. Ashe* • *Michael Engel* (Entomology) • *Roger L. Kaesler* (Invertebrate Paleontology) • *Daphne Fautin* (Invertebrate Zoology) • *Mark Mort* • *Craig C. Freeman* (Botany) • *Linda Trueb* (Herpetology) • *E.O. Wiley* • *Walter Dimmick* (Ichthyology) • *Larry Martin* (Vertebrate Paleontology) • *Tom Taylor* • *Edith Taylor* (Paleobotany) •
Natural History Museum – 1866
Natural history, zoology, ornitology, icthyology, mammalogy, entomology, paleontology, botany, herpetology – biodiversity research center ... 49991

Spencer Museum of Art, University of Kansas, 1301 Mississippi St, Lawrence, KS 66045-7500 • T: +1 785 8644710 • F: +1 785 8643112 • spencerart@ku.edu • www.ku.edu/~sma •
Ass. Dir.: *Carolyn Chinn Lewis* • Cur.: *Susan Earle* (European and American Arts) • *Stephen Goddard*

U

(Prints, Drawings) • *John Pultz* (Photography) •
Fine Arts Museum / University Museum – 1928
American, European and Asian art, prints, photos, drawings 49992

University of Kansas Ryther Printing Museum, 2425 W 15th St, Lawrence, KS 66049-3903 • T: +1 785 8644341 • F: +1 785 8647356 • jgs@ku.edu •
Dir.: *John G. Sayler* •
Science&Tech Museum – 1955
Antique printing equipment 49993

Watkins Community Museum of History, 1047 Massachusetts St, Lawrence, KS 66044 • T: +1 785 8414109 • F: +1 785 8419547 • wcmhist@sunflower.com • www.dchsks.org •
Dir.: *Rebecca J. Phipps* • Cur.: *Alison Miller* (Collections) •
Local Museum – 1972
Scrapbooks, maps & pamphlets, textiles, tools & agricultural equipment 49994

Lawrence MA

Immigrant City Museum, 6 Essex St, Lawrence, MA 01840 • T: +1 978 6869230 • F: +1 978 9752154 • lawrencehistorycenter@conversent.net • www.lawrencehistorycenter.org •
Exec. Dir.: *A. Patricia Jaysane* •
Local Museum – 1978
City hist, immigrant labor – archives 49995

Lawrenceville GA

Gwinnett History Museum, 455 S Perry St, Lawrenceville, GA 30045 • T: +1 770 8825178 • F: +1 770 2375612 • gwinnetthistorymuseum@gwinnettcounty.com •
Dir.: *Karen Tyler* •
Local Museum – 1974
County's history 49996

Lawrenceville NJ

Rider University Art Gallery, 2083 Lawrenceville Rd, Lawrenceville, NJ 08648-3099 • T: +1 609 8965168 • F: +1 609 8965232 • lnaar@rider.edu • www.rider.edu •
Dir.: *Harry I. Naar* •
Fine Arts Museum – 1970 49997

Lawson MO

Watkins Woolen Mill, 26600 Park Rd N, Lawson, MO 64062 • T: +1 816 2963357 • F: +1 816 5803784 • nrwatki@mail.dnr.state.mo.us •
Head: *Michael D. Beckett* •
Agriculture Museum / Science&Tech Museum – 1964 49998

Lawtell LA

Matt's Museum, McClelland St and Hwy 190, Lawtell, LA 70550 • T: +1 337 5437223 •
Dir.: *Edvin Matt* •
Local Museum – 1968
Firearms, crafts, coins, glassware, Civil War artifacts 49999

Lawton OK

Museum of the Great Plains, 601 NW Ferris Av, Lawton, OK 73507 • T: +1 580 5813460 • F: +1 580 5813458 • mgp@museumgreatplains.org • www.museumgreatplains.org •
Dir.: *John Hernandez* • Sc. Staff: *Deborah Baroff* • *Jim Whiteley* • *Timothy Poteete* • *Maurianna Johnson* • *Brian Smith* • *Joe Anderson* (Archaeology) •
Local Museum / Historical Museum – 1961
Documents and catalogs conc hist of Great Plains, agricultural machinery 50000

Le Claire IA

Buffalo Bill Museum of Le Claire, Iowa, 200 N River Dr, Le Claire, IA 52753 • T: +1 319 2895580 • F: +1 319 2894989 • museum@buffalobillmuseumleclaire.com • www.buffalobillmuseumleclaire.com •
Cur.: *Gloria Byrd* •
Local Museum / Historical Museum – 1957
Indian preservation 50001

Le Roy NY

Le Roy House and Jell-o Gallery, 23 E Main St, Le Roy, NY 14482 • T: +1 585 7687433 • F: +1 585 7687579 • www.jellomuseum.com •
Dir.: *Lynne J. Belluscio* •
Local Museum – 1940
Paintings and dec arts, agriculture, textiles, costumes, glass 50002

Le Sueur MN

City of Le Sueur Museum, 970 Ferry St, Le Sueur, MN 56058 • T: +1 507 6652050, 6656588 •
Pres.: *Helen Meyer* • Cur.: *Ann Little* •
Local Museum – 1968
Local hist, etchings of George T. Plowman, hist of canning, veterinary medicine, early pharmacy – library 50003

Lead SD

Black Hills Mining Museum, 323 W Main, Lead, SD 57754 • T: +1 605 5841605 • bhminmus@mato.com • www.mining-museum.blackhills.com •
Dir.: *John Meade* • Cur.: *Donald D. Toms* •
Science&Tech Museum – 1986 50004

Leadville CO

Healy House and Dexter Cabin, 912 Harrison Av, Leadville, CO 80461 • T: +1 719 4860487 • F: +1 719 4862557 • www.coloradohistory.org •
Dir.: *Maureen Scanlon* •
Historical Museum – 1947 50005

National Mining Hall of Fame and Museum, 120 W 9th St, Leadville, CO 80461 • T: +1 719 4861229 • F: +1 719 4863927 • www.mininghalloffame.org •
Dir.: *William Skala* •
Science&Tech Museum – 1953
Matchless mine historical items 50006

Tabor Opera House Museum, 308 Harrison Av, Leadville, CO 80461 • T: +1 719 4861206 • F: +1 719 4860474 • sharon@taboroperahouse.net • taboroperahouse.net •
C.E.O.: *Sharon Bland* •
Performing Arts Museum – 1955
Paintings, theater history, original scenery 50007

Leavenworth KS

Carroll Mansion Home - Leavenworth County Historical Society Museum, 1128 5th Av, Leavenworth, KS 66048-3212 • T: +1 913 6827759 • F: +1 913 6822089 • lvcohistsoc@lvnworth.com • leavenworth-net.com/lchs •
Cur.: *Nicole Dupe* •
Local Museum – 1964
Furniture, silver, china, costumes 50008

Goppert Gallery, University of Saint Mary, 4100 S 4th St, Trafficway, Leavenworth, KS 66048-5082 • T: +1 913 7586151 • F: +1 913 7586140 • nelsons@stmary.edu • www.stmary.edu/acad_art/gallery2.asp •
Dir.: *Susan Nelson* •
Public Gallery 50009

Lebanon NH

Ava Gallery and Art Center, 11 Bank St, Lebanon, NH 03766 • T: +1 603 4483117 • F: +1 603 4484827 • ava@valley.net • www.avagallery.org •
Dir.: *Bente Torjusen* •
Fine Arts Museum
Contemporary exhibitions of art 50010

Lebanon OH

Glendower State Memorial, 105 Cincinnati Av, Lebanon, OH 45036 • T: +1 513 9321817 • F: +1 513 9328560 • wchs@go-concepts.com •
Historic Site – 1944
19th c furniture and furnishings 50011

Warren County Historical Society Museum, 105 S Broadway, Lebanon, OH 45036-1707 • T: +1 513 9321817 • F: +1 513 9328560 • wchs@go-concepts.com • www.wchsmuseum.com •
Dir.: *Mary Payne* • Cur.: *Mary Klei* •
Local Museum – 1940
Early artifacts from SW Ohio, Shaker furniture, household articles, Native American artifacts – Genealogy library 50012

Lebanon PA

The Stoy Museum of the Lebanon County Historical Society, 924 Cumberland St, Lebanon, PA 17042 • T: +1 717 2721473 • F: +1 717 2727474 • staff@lebanonhistory.org • www.lebanonhistory.org •
Pres.: *Philip H. Feather* •
Local Museum – 1898
Local hist 50013

Lebec CA

Fort Tejon, 4201 Fort Tejon Rd, Lebec, CA 93243 • T: +1 661 2486692 • tejon@forttejon.org • www.forttejon.org •
Military Museum – 1939 50014

Lecompton KS

Territorial Capital-Lane Museum, 393 N 1900 Rd, Lecompton, KS 66050 • T: +1 785 8876184 • F: +1 785 8876148 • lanemuseum@aol.com • www.lecomptonkansas.com •
Dir.: *Paul M. Bahnmaier* • Cur.: *Arlene Simmons* •
Historical Museum – 1969
State history 50015

Lee MA

Warehouse Gallery, c/o Council for Creative Projects, 17 Main St, Lee, MA 01238-1648 • T: +1 413 2438030 • F: +1 413 2438031 •
Dir.: *Dr. Gail Gelburd* •
Fine Arts Museum – 1989
Art 50016

Lees Summit MO

Missouri Town 1855, 8010 E Park Rd, Lees Summit, MO 64015 • T: +1 816 7958200 ext 1260 • F: +1 816 7957938 • juligor@gw.co.jackson.mo.us • www.co.jackson.mo.us •
Open Air Museum – 1963
Agrigulutre, costumes, decorative arts, folklore, general, hist 50017

Leesburg VA

Loudoun Museum, 14-16 Loudoun St, Leesburg, VA 20175 • T: +1 703 7778331, 7777427 • F: +1 703 7378873 • info@loudounmuseum.org • www.loudounmuseum.org •
Exec. Dir.: *Karen MacLeod* • Cur.: *Eric Larson* •
Local Museum – 1967
Decorative arts, local history 50018

Morven Park Museum, Old Waterford Rd, Leesburg, VA 20175, mail addr: POB 6228, Leesburg, VA 220178-921 • T: +1 703 7776034 • F: +1 703 7719211 • myork@morvenpark.org • www.morvenpark.com •
Local Museum – 1955
Carriages, fox hunting, agriculture, folklore 50019

Oatlands, 20850 Oatlands Plantation Ln, Leesburg, VA 20175 • T: +1 703 7773174 • F: +1 703 7774427 • oatlands@erols.com • www.oatlands.org •
Dir.: *David Y. Boyce* •
Local Museum – 1965
18th - 20th c decorative arts 50020

Lehi UT

John Hutchings Museum of Natural History, 55 N Center St, Lehi, UT 84043 • T: +1 801 7687180 •
C.E.O.: *Lamar Hutchings* •
Natural History Museum – 1955
Pionner, Johnstons Army from the Civil & Revolutionary Wars, undersea life from Puerto Rico & Pacific, minerals 50021

Leland MI

Leelanau Historical Museum, 203 E Cedar St, Leland, MI 49654 • T: +1 231 2567475 • F: +1 231 2567650 • info@leelanauhistory.org • www.leelanauhistory.org •
Dir.: *John Mitchell* • Cur.: *Laura J. Quackenbush* •
Historical Museum – 1957
Local hist, tradional handicrafts, local Odawa arts 50022

Lenhartsville PA

Pennsylvania Dutch Folk Culture Society Museum, Lenhartsville, PA 19534 • T: +1 610 8676705 •
Folklore Museum – 1965
Folk culture 50023

Lenox MA

Berkshire Scenic Railway Museum, Willow Creek Rd, Lenox, MA 01240 • T: +1 413 6372210, +1 518 3922620 • F: +1 518 3922225 • woodworks@taconic.net •
Cur.: *John Trowill* •
Science&Tech Museum – 1984
1920s-1960s railroad equip, vintage coaches 50024

Lerna IL

Lincoln Log Cabin, 400 S Lincoln Hwy Rd, Lerna, IL 62440 • T: +1 217 3456489, 3451845 • F: +1 217 3456472 • lincoln_loc@ihpa.state.il.us •
C.E.O.: *Maynard Crossland* •
Historical Museum – 1929
Reconstructed cabin and farm of Thomas and Sara Lincoln on original site 50025

Moore Home, 400 S Lincoln Hwy Rd, Lerna, IL 62440 • T: +1 217 3456489, 3451845 • F: +1 217 3456472 • lincol_log@ihpa.state.il.us •
Historical Museum – 1929
Home of Reuben and Matilda Moore, daughter of Sara Lincoln and stepsister of Abraham Lincoln 50026

Leverett MA

Leverett Historical Museum, Moore's Corner Schoolhouse, N Leverett Rd, Leverett, MA 01054 • T: +1 413 5489082, 3672800 •
Local Museum – 1963
Local history 50027

Lewes DE

Lewes Historical Society Museum, 110 Shipcarpenter St, Lewes, DE 19958 • T: +1 302 6457670 • F: +1 302 6452375 • info@historiclewes.org • www.historiclewes.org •
Dir.: *Michael DiPaolo* •
Local Museum – 1961
Local hist 50028

Lewisburg PA

Edward and Marthann Samek Art Gallery, Bucknell University, Elaine Langone Center, Lewisburg, PA 17837 • T: +1 570 5773792 • F: +1 570 5773215 • inthebag@bucknell.edu • www.bucknell.edu/SamekArtGallery •
Dir.: *Dan Mills* • Op.Mgr.: *Cynthia Peltier* •
Fine Arts Museum / University Museum – 1983
Contemporary art, 19th and early 20th c European and American art 50029

Packwood House Museum, 8 Market St, Lewisburg, PA 17837 • T: +1 570 5240323 • F: +1 570 5240548 • packwood@jdweb.com • www.Packwoodhousemuseum.com •
Dir.: *Sarah Phinney Kelley* •
Local Museum – 1976
Local hist 50030

Slifer House, Riverwoods, 1 River Rd, Lewisburg, PA 17837 • T: +1 570 5242245 • F: +1 570 5242245 • sliferhs@postoffice.ptd.net • www.albrightcare.org/slifer •
Dir.: *Gary W. Parks* •
Historic Site – 1976
Victorian furnishings and decoratives art 50031

Union County Museum, Courthouse, Second and Saint Louis Sts, Lewisburg, PA 17837 • T: +1 570 5248666 • F: +1 570 5248743 • hstoricl@ptd.net • www.rootsweb.com •
Pres.: *Jeannette Lasansky* • Dir.: *Gary Slear* •
Local Museum – 1906
History, genealogy, oral traditions 50032

Lewisburg WV

Fort Savannah Museum, 100 E Randolph St, Lewisburg, WV 24901 • T: +1 304 6454010 •
C.E.O.: *Rosalie S. Detch* •
Military Museum – 1962
Military hist 50033

North House Museum, 301 W Washington St, Lewisburg, WV 24901 • T: +1 304 6453398 • F: +1 304 6455201 • tammy@greenbrierhistorical.org • www.greenbrierhistorical.org •
Dir.: *Joyce Mott* • Pres.: *John B. Arbuckle* •
Local Museum / Historical Museum – 1963
Civil War, Indian artifacts, early tools and vehicles 50034

Lewiston ID

Nez Perce County Museum, 306 Third St, Lewiston, ID 83501 • T: +1 208 7432535 • F: +1 208 7432535 • registrar@npchistsoc.org •
Pres.: *Richard Riggs* •
Local Museum – 1960
Artifacts, photographs 50035

Lewiston ME

Atrium Gallery, University of Southern Main, c/o Lewiston-Auburn College, 51 Westminster St, Lewiston, ME 04240 • T: +1 207 7536500 • F: +1 207 7536555 • www.usm.maine.edu/lac/art •
Dir.: *Robin Holman* •
Fine Arts Museum / University Museum
Contemporary exhibitions 50036

Bates College Museum of Art, 75 Russell St, Lewiston, ME 04240 • T: +1 207 7866158 • F: +1 207 7868335 • museum@bates.edu • www.bates.edu/acad/museum •
Dir.: *Mark H. C. Bessir* • Asst. Cur.: *William Low* • Sc. Staff: *Liz Sheehan* •
Fine Arts Museum / University Museum – 1986
European and American paintings, prints, sculpture 50037

Dorothea C. Witham Gallery, 59 Canal St, Lewiston, ME 04240 • T: +1 207 7821369 • F: +1 207 7825931 • cpacm@cpacm.com • www.cpacm.pair.com •
Exec. Dir.: *J. Michel Patry* •
Fine Arts Museum – 1993
Photography 50038

Lewiston NY

Niagara Power Project Visitors' Center, 5777 Lewiston Rd, Lewiston, NY 14092 • T: +1 716 2866661 • F: +1 716 2866091 • www.nypa.gov/vc/niagara.htm •
Dir.: *Joanne Willmont* •
Science&Tech Museum 50039

Lewistown IL

Dickson Mounds Museum, 10956 N Dickson Mounds Rd, Lewistown, IL 61542 • T: +1 309 5473721 • F: +1 309 5473189 • wiant@museum.state.il.us • www.museum.state.il.us/ismsites/dickson/ •
Dir.: *Dr. Michael D. Wiant* • Asst. Cur.: *Alan D. Harn* •
Archaeology Museum / Historic Site – 1927
Archaeology, American Indian cultures of Illinois 50040

Lewistown MT

Central Montana Historical Association, 408 NE Main St, Lewistown, MT 59457 • T: +1 406 5385436 •
Pres.: *George Simonson* •
Historical Museum – 1955
Geology, Native American artifacts 50041

Lewistown Art Center, 801 W Broadway, Lewistown, MT 59457 • T: +1 406 5388278 • F: +1 406 5388278 • www.lewistown2000.com •
Dir.: *Nancy Hedrick* •
Fine Arts Museum – 1971
Art work from artists from Central Montana 50042

Lewistown PA

McCoy House, 17 N Main St, Lewistown, PA 17044 • T: +1 717 2421022, 2484711 • F: +1 717 2421022 • mchistory@acsworld.net • www.mccoyhouse.com • Pres.: *Ray Metthews* • Local Museum – 1921 Historic building ... 50043

Lexington KY

American Saddlebred Museum, 4093 Iron Works Pkwy, Lexington, KY 40511 • T: +1 859 2592746 • F: +1 859 2552909 • ashm@mis.net • www.american-saddlebred.com • Exec. Dir.: *Tolley Graves* • Special Museum – 1962 ... 50044

Ashland - Henry Clay Estate, 120 Sycamore Rd, Lexington, KY 40502 • T: +1 859 2668581 • F: +1 859 2687266 • ahmichel@henryclay.org • www.henryclay.org • Exec. Dir.: *Kelly B. Willis* • Cur.: *Eric Brooks* • Historical Museum – 1926 1811 estate of Henry Clay, 1857 house of James Clay ... 50045

Aviation Museum of Kentucky, 4316 Hangar Dr, Blue Grass Airport, Lexington, KY 40510 • T: +1 859 2311219 • F: +1 859 3818739 • sparker@aviationky.org • www.aviationky.org • Science&Tech Museum – 1995 Kentucky aviation hist, space and exploration ... 50046

Explorium of Lexington, 440 W Short St, Lexington, KY 40507 • T: +1 859 2583253 • F: +1 859 2583255 • lcmo@prodigy.net • www.lexingtonchildrensmuseum.com • Dir.: *Sara Nees Holcomb* • Special Museum – 1990 ... 50047

Headley-Whitney Museum, 4435 Old Frankfort Pike, Lexington, KY 40510 • T: +1 859 2556653 • F: +1 859 2558375 • hwmuseum@headley-whitney.org • www.headley-whitney.org • Dir.: *Diane C. Wachs* • Cur.: *Travis Robinson* • Decorative Arts Museum – 1973 Chinese porcelains, oriental textiles, ceramics & furniture, Asian, European & American textiles ... 50048

Hunt-Morgan House, 201 N Mill St, Lexington, KY 40507 • T: +1 859 2323290 • F: +1 859 2599210 • info@bluegrasstrust.org • www.bluegrasstrust.org • Local Museum – 1955 Period furniture, decorative objects, portraits ... 50049

International Museum of the Horse, 4089 Iron Works Pkwy, Lexington, KY 40511 • T: +1 859 2594231, 2334303 • F: +1 859 2254613 • bcooke@kyhorsepark.com • www.imh.org • Dir.: *Bill Cooke* • Cur.: *Jenifer Stermer* • Fine Arts Museum – 1978 Equine hist coll covering all breeds worldwide ... 50050

Lexington History Museum, 215 W Main St, Lexington, KY 40507 • T: +1 859 2540530 • F: +1 859 2548372 • info@lexingtonhistorymuseum.org • www.lexingtonhistorymuseum.org • C.E.O.: *E.T. Houlihan* • Chm.: *Dr. Thomas D. Clark* • Historical Museum – 1999 ... 50051

The Living Arts and Science Center, 362 N Martin Luther King Blvd, Lexington, KY 40508 • T: +1 859 2525222, 2552284 • F: +1 859 2557448 • lasc6898@aol.com • www.livingartsandscience.org • Dir.: *Franklin Thompson* • Cur.: *Jim Brancaccio* • Fine Arts Museum / Ethnology Museum – 1968 Children's art, science ... 50052

Mary Todd Lincoln House, 578 W Main, Lexington, KY 40507 • T: +1 859 2339999 • F: +1 859 2729601 • gwen@mtlhouse.org • www.mtlhouse.org • Dir.: *Gwen Thompson* • Decorative Arts Museum – 1968 Lincoln book coll, rare old leather bound books ... 50053

Morlan Gallery, c/o Transylvania University, Mitchell Fine Arts Ctr, 300 N Broadway, Lexington, KY 40508 • T: +1 859 2338210 • F: +1 859 2338797 • nwolsk@mail.transy.edu • Dir.: *Nancy Wolsk* • Decorative Arts Museum / Natural History Museum – 1978 19th c natural history and portraits, decorative arts ... 50054

Photographic Archives, University of Kentucky, M.I. King Library, Lexington, KY 40506-0039 • T: +1 859 2572654 • F: +1 859 2576311 • jasonf@uky.edu • www.uky.edu/libraries/specials/av • Special Museum / University Museum – 1978 Hist photographs ... 50055

Transylvania Museum, 300 N Broadway, Lexington, KY 40508 • T: +1 859 2338229 • F: +1 859 2358171 • jday@transy.edu • www.transy.edu • Dir.: *Dr. James Day* • Historical Museum – 1882 Early 19th c scientific appparatus, equipment used in teachering ... 50056

University of Kentucky Art Museum, Rose and Euclid Av, Lexington, KY 40506-0241 • T: +1 859 2575716 • F: +1 859 3231994 • jrbosw00@uky.edu • www.uky.edu/artmuseum • Dir.: *Kathleen Walsh-Piper* • Cur.: *Deborah Borrowdale-Cox* • Fine Arts Museum / University Museum – 1976 14th to 20th c paintings, sculpture, works on art on paper, decorative arts ... 50057

Waveland Museum, 225 Waveland Museum Ln, Lexington, KY 40515-1601 • T: +1 859 2723611 • F: +1 859 2454269 • ron.langdon@ky.us • www.kystateparks.com • Decorative Arts Museum – 1957 Furniture, portraits, China ... 50058

William S. Webb Museum of Anthropology, c/o University of Kentucky, 211 Lafferty Hall, Lexington, KY 40506-0024 • T: +1 859 2578208 • F: +1 859 3231968 • www.uky.edu/as/anthropology/museum/museum.htm • Dir.: *George M. Crothers* • Archaeology Museum / Ethnology Museum / University Museum – 1931 Archaeological and ethnographic materials ... 50059

Lexington MA

Lexington Historical Society, Buckman Tavern, Munroe Tavern, Hancock-Clarke House, Antique Fire Equipment Museum, 1 Bedford St, Hancock St, Lexington, MA 02420, mail addr: POB 514, Lexington, MA 02420 • T: +1 781 8621703 • F: +1 781 8624920 • info@lexingtonhistory.org • www.lexingtonhistory.org • Exec. Dir.: *George S. Comtois* • Historical Museum – 1886 Local history, American revolution, fire fighting ... 50060

National Heritage Museum, Van Gorden-Williams Library and Archives, 33 Marrett Rd, Lexington, MA 02421 • T: +1 781 8616559 ext 100 • F: +1 781 8619846 • info@monh.org • www.monh.org • Dir.: *John H. Ott* • *Hilary Anderson* (Collections, Exhibitions) • *Cynthia Robinson* (Education, Programs) • Cur.: *Mark Tabbert* (Masonic and Fraternal Colls) • Historical Museum – 1971 American prints, maps, paintings, decorative arts, costumes – library, auditorium ... 50061

Lexington MO

Battle of Lexington State Historic Site, John Shea Dr, Ext Hwy 13, Lexington, MO 64067 • T: +1 660 2594654 • F: +1 660 2592378 • moparks@dnr.mo.gov • www.mostateparks.com/lexington • C.E.O.: *Janae Fuller* • Military Museum / Historical Museum – 1959 ... 50062

Lexington NC

Davidson County Historical Museum, Old Courthouse, 2 S Main St, Lexington, NC 27292 • T: +1 336 2422035 • F: +1 336 2482871 • choffmann@co.davidson.nc.us • www.co.davidson.nc.us • C.E.O./Cur.: *Catherine Matthews Hoffmann* • Local Museum – 1976 Hist building, local hist ... 50063

Davidson County Museum of Art, 220 S Main St, Lexington, NC 27292 • T: +1 336 2492742 • F: +1 336 2496302 • www.co.davidson.nc.us/arts • Exec. Dir.: *Carla Copeland-Burns* • Fine Arts Museum – 1968 ... 50064

Lexington NE

Dawson County Historical Museum, 805 N Taft St, Lexington, NE 68850-0369 • T: +1 308 3245340 • F: +1 308 3245340 • dawcomus@alltel.net • Pres.: *Gail Hall* • Dir.: *Barbara Vondras* • Local Museum – 1958 Local hist items, pioneer household artifacts ... 50065

Lexington OH

Richland County Museum, 51 Church St, Lexington, OH 44904 • T: +1 419 8842230 • Pres.: *Jeffrey Mandeville* • Folklore Museum / Local Museum – 1966 Hist, agriculture, art, ... 50066

Lexington SC

Lexington County Museum, 231 Fox St, Lexington, SC 29072 • T: +1 803 3598369 • F: +1 803 8082160 • museum@lex-co.com • Dir.: *Horace E. Harmon* • Decorative Arts Museum / Local Museum – 1970 Local hist ... 50067

Lexington VA

American Work Horse Museum, 487 Maury River Rd, Lexington, VA 24450 • T: +1 540 4642950 • F: +1 540 4642999 • info@horsecenter.org • www.horsecenter.org • Dir.: *John Scott* • Special Museum – 1971 Work horse hist from 1493 to the present ... 50068

Dupont Gallery, Washington and Lee University, Lexington, VA 24450-0303 • T: +1 540 4638861 • F: +1 540 4638112 • olsonk@wlu.edu • www.wlu.edu • Dir.: *Kathleen Olson* • Public Gallery – 1954 Temporary exhib by regional as well as nationally known artists ... 50069

George C. Marshall Museum, VMI Parade Ground, Lexington, VA 24450 • T: +1 540 4637103 ext 225 • F: +1 540 4645229 • cushmanpn@marshallfoundation.org • www.marshallfoundation.org • Dir.: *Paula N. Cushman* • Historical Museum / Military Museum – 1953 Military and political history (20th c) ... 50070

Lee Chapel and Museum, Washington and Lee University, Lexington, VA 24450 • T: +1 540 4638768 • F: +1 540 4638741 • phobbs@wlu.edu • leechapel.wlu.edu • C.E.O.: *Patricia A. Hobbs* • Historical Museum – 1867 Traces the hist of the uni emphasizing the contributions ot he namesakes ... 50071

The Reeves Center, Washington and Lee University, Lexington, VA 24450 • T: +1 540 4638744 • F: +1 540 4638741 • hbailey@wlu.edu • Dir.: *Thomas V. Litzenburg* • Fine Arts Museum / University Museum – 1982 Chinese ceramics, European ceramics, paintings 50072

Rockbridge Historical Society, 101 E Washington St, Lexington, VA 24450 • T: +1 540 4641058 • rochist@hotmail.com • www.rockhist.org • C.E.O.: *George W. Warren* • Association with Coll – 1939 Books, documents, journals, photographs, works of art ... 50073

Stonewall Jackson House, 8 Washington St E, Lexington, VA 24450 • T: +1 540 4632552 • F: +1 540 4634088 • sjh1@rockbridge.net • www.stonewalljackson.org • Exec.Dir.: *Michael Anne Lynn* • Cur.: *Megan M. Newman* • Historical Museum – 1954 Furniture, furnishings and memorabilia of US General Thomas J. Jackson and his family ... 50074

Virginia Military Institute Museum, Virginia Military Institute, Jackson Memorial Hall, Lexington, VA 24450 • T: +1 540 4647232 • F: +1 540 4647112 • gibsonke@vmi.edu • www.vmi.edu/museum • Dir.: *Keith E. Gibson* • Military Museum – 1856 Artifacts, drawings, paintings and photographs relating to the hist of VMI ... 50075

Liberal KS

Mid-America Air Museum, 2000 W 2nd St, Liberal, KS 67901 • T: +1 316 6245263 • F: +1 316 6245454 • liberalcityam@swko.net • www.liberalairmuseum.com • Exec. Dir.: *Frank Yount* • Science&Tech Museum – 1987 Aviation ... 50076

Liberty MO

Clay County Historical Museum, 14 N Main, Liberty, MO 64068 • T: +1 816 7921849 • info@claycountymuseum.org • www.claycountymuseum.org • Cur.: *Kevin M. Fisher* • Historical Museum – 1965 Medical & druggist equipment, furniture, dating from 1800-1930s ... 50077

Historic Liberty Jail Visitors Center, 216 N Main, Liberty, MO 64068 • T: +1 816 7813188 • F: +1 816 7817311 • C.E.O.: *J. Derle Thorpe* • Religious Arts Museum – 1936 Religious hist ... 50078

Jesse James Bank Museum, 103 N Water, Liberty, MO 64068 • T: +1 816 7814458 • F: +1 816 6286676 • Head: *Elizabeth Gilliam-Beckett* • Historical Museum – 1966 Civil War banking,picutres & Documentspertaining to the James Gang ... 50079

Libertyville IL

Libertyville-Mundelein Historical Society Museum, 413 N Milwaukee Av, Libertyville, IL 60048-2280 • T: +1 847 3622330 • F: +1 847 3620006 • Pres.: *Jerrold L. Schulkin* • Local Museum – 1955 Pioneer artifacts, Civil War items, historical books, maps, pictures and paper ... 50080

Ligonier IN

Stone's Tavern Museum, 4946 N State Rd 5 and U.S. 33, Ligonier, IN 46767 • T: +1 219 8562871 • www.stonestrace.com • Pres.: *Dick Hursey* • Historical Museum – 1964 Tavern of 1839 ... 50081

Ligonier PA

Fort Ligonier Museum, 216 S Market St, Ligonier, PA 15658-1206 • T: +1 724 2389701 • F: +1 724 2389732 • Dir.: *J. Martin West* • Cur.: *Penelope A. West* • *Shirley McQuillis Iscrupe* • Historical Museum – 1946 History ... 50082

Lihu'e HI

Kaua'i Museum, 4428 Rice St, Lihu'e, HI 96766 • T: +1 808 2456931 • www.kauaimuseum.org • Dir.: *Carol Lovell* • Cur.: *Margaret Lovett* • Fine Arts Museum / Historical Museum – 1960 History, art ... 50083

Lima OH

Allen County Museum, 620 W Market St, Lima, OH 45801 • T: +1 419 2229426 • F: +1 419 2220649 • acmuseum@bright.net • www.allencountymuseum.org • Dir.: *Patricia F. Smith* • Cur.: *Anna B. Selfridge* (Manuscripts and archives) • *John Carnes* (Collections) • *Sarah Rish* (Education) • Asst. Cur.: *Charles Bates* • Local Museum – 1908 Indian relics, minerals. fossils, pioneer rooms and tools, furniture, fire-fight equipment ... 50084

Artspace-Lima, 65-67 Town Sq, Lima, OH 45802 • T: +1 419 2221721 • F: +1 419 2226587 • artspace@worcnet.gen.oh.us • Exec. Dir.: *Patricia Good* • Public Gallery – 1940 ... 50085

Limon CO

Limon Heritage Museum, 899 First St, Limon, CO 80828 • T: +1 719 7758605 • F: +1 719 7758808 • limonmuseum@hotmail.com • Dir.: *Vivian J. Lowe* • Local Museum – 1990 Local hist of the Colorado high plans, Union Pacific and Rock Island railroads, ranching and farming, sheepherder's wagons, period farm machinery ... 50086

Lincoln IL

Postville Courthouse Museum, 914 Fifth St, Lincoln, IL 62656 • T: +1 217 7328930 • Head: *Richard Schachtsiek* • Historic Site – 1953 ... 50087

Lincoln KS

Lincoln County Historical Society, 214 W Lincoln Av, Lincoln, KS 67455, mail addr: POB 85, Lincoln, KS 67455 • Cur.: *Loa Page* • Association with Coll – 1978 Farm equip, dolls, Indian artifacts, military, F.A. Cooper's paintings, drawings and metal etchings ... 50088

Lincoln MA

Codman Estate - The Grange, 36 Codman Rd, Lincoln, MA 01773 • T: +1 781 2598843 • F: +1 781 2598843 • CodmanEstate@HistoricNewEngland.org • www.historicnewengland.org/visit/homes/codman.htm • Decorative Arts Museum – 1969 1740 Georgian mansion, home of Codman family ... 50089

DeCordova Museum and Sculpture Park, 51 Sandy Pond Rd, Lincoln, MA 01773 • T: +1 781 2598355 • F: +1 781 2593650 • info@decordova.org • www.decordova.org • Dir.: *Paul Master-Karnik* • Cur.: *Rachel Rosenfield Lafo* • Fine Arts Museum – 1948 American arts, 20th c sculptures, photography, paintings, graphics – library, studios ... 50090

Drumlin Farm, 208 S Great Rd, Lincoln, MA 01773 • T: +1 781 2592200 • F: +1 781 2597917 • drumlinfarm@massaudubon.org • www.massaudubon.org • Dir.: *Christy Foote-Smith* • Agriculture Museum – 1955 Agriculture, aviary, environment – library ... 50091

Gropius House, 68 Baker Bridge Rd, Lincoln, MA 01773 • T: +1 781 2598098 • F: +1 781 2599722 • gropiushouse@spnea.org • www.historicnewengland.org/visit/homes/gropius.htm • Historical Museum – 1985 Modern architecture, designed by combining Bauhaus principles with New England building materials ... 50092

The Thoreau Institute at Walden Woods, 44 Baker Farm, Lincoln, MA 01773 • T: +1 781 2594730 • F: +1 781 2594730 • jeff.cramer@walden.org • www.walden.org/institute • Dir.: *Kathi Anderson* • Cur.: *Jeffrey S. Cramer* • Library with Exhibitions – 1991 Regional history, American literature – library ... 50093

Lincoln NE

Elder Art Gallery, Rogers Center for Fine Arts, c/o Nebraska Wesleyan University, 50th St and Huntington Av, Lincoln, NE 68504 • T: +1 402 4662371, 4652230 • F: +1 402 4652179 • dp@nebrwesleyan.edu • Dir.: *Dr. Donald Paoletta* • University Museum / Historical Museum / Fine Arts Museum – 1966 Campus coll, prints, paintings and sculpture ... 50094

Gallery of the Department of Art and Art History, University of Nebraska, 207 Nelle Cochrane Woods Hall, Lincoln, NE 68588-0114 • T: +1 402 4722631, 4725541 • F: +1 402 4729746 • www.unl.edu • Dir.: *Joseph M. Ruffo* • Fine Arts Museum – 1985 Student works ... 50095

U

Great Plains Art Collection in the Christlieb Gallery, c/o University of Nebraska-Lincoln, 1155 Q St, Hewit Pl, Lincoln, NE 68588-0250 • T: +1 402 4726220 • F: +1 402 4722960 • asummers3@unl.edu • www.unl.edu/plains/ •
Dir.: *James Stubbendieck* • Cur.: *Reece Summers* •
Fine Arts Museum / University Museum – 1980
Western art, sculpture, paintings, graphics, photographs ... 50096

Haydon Gallery, Nebraska Art Association, 335 N 8th St, Lincoln, NE 68508 • T: +1 402 4755421 • F: +1 402 4755422 • haydongallery@alltel.net •
Dir.: *Teliza V. Rodriguez* •
Public Gallery – Contemporary art ... 50097

The Lentz Center for Asian Culture, c/o University of Nebraska-Lincoln, 1155 Q St, Lincoln, NE 68588-0252 • T: +1 402 4725841 • F: +1 402 4728899 • bbanks@unl.edu • www.unl.edu/lentz •
Dir./Cur.: *Barbara Chapman Banks* •
Fine Arts Museum – 1986
Asian art collection, Tibetan ritual and secular art ... 50098

Lincoln Children's Museum, 1420 P St, Lincoln, NE 68508 • T: +1 402 4770128 • F: +1 402 4772004 • lbullcm@navix.net • www.lincolnchildrensmuseum.org •
Dir.: *Marilyn R. Gorham* •
Special Museum – 1989 ... 50099

Museum of American Historical Society of Germans from Russia, 631 D St, Lincoln, NE 68502-1199 • T: +1 402 4743363 • F: +1 402 4747229 • ahsgr@aol.com • www.ahsgr.org •
Historical Museum – 1968 ... 50100

Museum of Nebraska History, 131 Centennial Mall N, Lincoln, NE 68508 • T: +1 402 4714754 • F: +1 402 4713144 • museum@nebraskahistory.org • www.nebraskahistory.org •
Dir.: *Lawrence J. Sommer* •
Historical Museum – 1878
Art, antropology, archaeology, ethnology, Western hist ... 50101

National Museum of Roller Skating, 4730 South St, Lincoln, NE 68506 • T: +1 402 4837551 ext 16 • F: +1 402 4831465 • directorcurator@rollerskatingmuseum.com • www.rollerskatingmuseum.com •
Dir.: *James Vannurden* •
Special Museum – 1980
Roller skates, patents, trophies and medals, costumes ... 50102

Nebraska Conference United Methodist Historical Center, 5000 Saint Paul Av, Lincoln, NE 68504-2796 • T: +1 402 4645994 • F: +1 402 4646203 • nebrumchc@yahoo.comom • www.umcneb.org •
Dir./Cur.: *Maureen Vetter* •
Religious Arts Museum – 1968 ... 50103

Noyes Art Gallery, 119 S Ninth St, Lincoln, NE 68508 • T: +1 402 4751061 • rnoyes1348@aol.com • www.noyesart.com •
Dir.: *Julia Noyes* •
Decorative Arts Museum / Fine Arts Museum – 1994 ... 50104

Sheldon Memorial Art Gallery and Sculpture Garden, c/o University of Nebraska-Lincoln, 12th and R Sts, Lincoln, NE 68588-0300 • T: +1 402 4722461 • F: +1 402 4724258 • sheldon@unl.edu • www.sheldon.unl.edu/ •
Dir.: *Janice Driesbach* • Cur.: *Daniel Siedell* • *Karen Janovy* •
Fine Arts Museum / University Museum – 1963
19th to 21th c American art, sculpture, American modernism ... 50105

Thomas P. Kennard House, 1627 H St, Lincoln, NE 68501 • T: +1 402 4714764 • F: +1 402 4714764 • museum02@nebraskahistory.org • www.nebraskahistory.org •
C.E.O./Dir.: *Larry Sommers* •
Historical Museum – 1968
Period furnishing, local hist ... 50106

University of Nebraska State Museum, 307 Morrill Hall, 14 and U Sts, Lincoln, NE 68588-0338 • T: +1 402 4723779 • F: +1 402 4728899 • pgrewl@unl.edu • www.museum.unl.edu •
Dir.: *Dr. Priscilla C. Grew* • Cur.: *Dr. Judy Diamond* •
University Museum / Natural History Museum – 1871
Natural science exhibitions – planetarium, library 50107

University Place Art Center, 2601 N 48th St, Lincoln, NE 68504 • T: +1 402 4668692 • F: +1 402 4663786 • info@universityplaceart.com • www.universityplaceart.com •
C.E.O.: *Julie Diegel* •
Fine Arts Museum – 1978
Antique and collectible dolls, American prints, paperweights, quilts, buttons, paintings ... 50108

Lincoln NM

Historic Lincoln, Hwy 380, Lincoln, NM 88346 • T: +1 505 6534025 • F: +1 505 6534627 • moth@zianet.com • www.zianet.com/museum •
Dir.: *Bruce B. Eldredge* • Cur.: *Barbara Bertucio* •
Historical Museum – 1937
Local Hispanic and native American cultural artifacts, Lincoln county War and Billy the Kid coll, photographs, furniture, medical instruments ... 50109

Old Lincoln County Courthouse Museum, Lincoln State Monument, Hwy 380, Lincoln, NM 88338 • T: +1 505 6534372 • F: +1 505 6534372 •
Historical Museum – 1937
Indian artifacts, papers, furniture, clothing, tools, art objects ... 50110

Lincoln RI

Blackstone Valley Historical Society Museum, 1873 Old Louisquisett Pike, Lincoln, RI 02865 • T: +1 401 7252847 •
Cur.: *Ken Smith* •
Local Museum – 1958
Local hist ... 50111

Flanagan Valley Campus Art Gallery, Louisquisset Pike, Lincoln, RI 02865-4585 • T: +1 401 3337154 • F: +1 401 8252265 •
Dir.: *Tom Morrissey* •
Fine Arts Museum – 1974 ... 50112

Lincoln Park MI

Lincoln Park Historical Museum, 1335 Southfield Rd, Lincoln Park, MI 48146 • T: +1 313 3863137 • citylp@aol.com •
Cur.: *Muriel Lobb* •
Local Museum – 1972
Local and regional hist, people stories – library 50113

Lincolnton GA

Elijah Clark Memorial Museum, 2959 McCormick Hwy, Lincolnton, GA 30817 • T: +1 706 3593458 • F: +1 706 3595856 • eclark@nu-z.net • www.gastateparks.com •
Special Museum – 1961
Replica of the house of Elijah Clark ... 50114

Lincroft NJ

Monmouth Museum, College Campus, Newman Springs Rd, Lincroft, NJ 07738 • T: +1 732 7472266 • F: +1 732 7478592 • monmuseum@netlabs.net • www.monmouthmuseum.org •
Dir.: *Dorothy V. Morehouse* •
Local Museum – 1963
Culture, art, scientific phenomena ... 50115

Lind WA

Adams County Historical Society Museum, First St, Lind, WA 99341 • T: +1 509 6773393 • galeirma@ritzcom.net •
Dir.: *Irma E. Gfeller* •
Local Museum – 1963 ... 50116

Linden IN

Linden Railroad Museum, 215 N Main St, Linden, IN 47955-0061 • T: +1 765 3394896 • F: +1 765 3394896 • eutsler@tctc.com •
Dir.: *David White* • Cur.: *Therese Eutsler* •
Science&Tech Museum – 1986
Railroad ... 50117

Lindenhurst NY

Old Village Hall Museum, 215 S Wellwood Av, Lindenhurst, NY 11757-0296 • T: +1 516 9574385 •
Dir.: *Evelyn M. Ellis* • Asst. Dir.: *Mary Kollar* •
Historical Museum – 1958 ... 50118

Lindsay OK

Murray-Lindsay Mansion, POB 282, Lindsay, OK 73052 • T: +1 405 7562121 •
Pres.: *Shawn Bridwell* •
Historical Museum – 1971
Period furniture 1879/1880 ... 50119

Lindsborg KS

Birger Sandzén Memorial Gallery, 401 N First St, Lindsborg, KS 67456-0348 • T: +1 785 2272220 • F: +1 785 2274170 • fineart@sandzen.org • www.sandzen.org •
Dir.: *Larry L. Griffis* •
Fine Arts Museum – 1957
Oils, watercolors, prints, archives of letters & papers by Birger Sandzén ... 50120

McPherson County Old Mill Museum, 120 Mill St, Lindsborg, KS 67456 • T: +1 785 2273595 • F: +1 785 2272810 • oldmillmuseum@hotmail.com • www.lindsborg.org/oldmill •
Dir.: *Lorna Batterson* •
Historical Museum / Science&Tech Museum – 1959
Local history, mill, genealogy ... 50121

Mingenback Art Center Gallery, c/o Bethany College, 421 N First St, Lindsborg, KS 67456 • T: +1 785 2273311 ext 8145 • F: +1 785 2272004 • tubbsj@bethany.bethanylb.edu •
Dir.: *Caroline Kahler* •
Public Gallery
Oil paintings, watercolors, prints, etchings, lithographs, wood engravings, ceramics and sculpture ... 50122

Linthicum MD

Historical Electronics Museum, 1745 W Nursery Rd, Linthicum, MD 21090 • T: +1 410 7653803 • F: +1 410 7650240 • radarmus@erols.com • www.hem-usa.org •
Dir.: *Katherine Duer Marks* •
Science&Tech Museum – 1980
Electronics technology, radar systems, transmitters, receivers, antennas, satellites ... 50123

Lioncoln AR

Arkansas County Doctor Museum, 109 N Starr Av, Lioncoln, AR 72744 • T: +1 479 8244307 • F: +1 479 8244307 • acdm@pgtc.com • www.drmuseum.net •
Pres.: *Dr. Joe B. Hall* •
Special Museum – 1994
Medical instr, equip and furnishings ... 50124

Lisbon OH

Lisbon Historical Society Museum, 117-119 E Washington St, Lisbon, OH 44432 • T: +1 330 4249000 • F: +1 330 4241861 • lisbonhs@sky-access.com • www.lisbonhistory.org •
Pres.: *David Privette* • Cur.: *Gene Krotky* •
Local Museum – 1938 ... 50125

Lisle IL

Jurica Nature Museum, Benedictine University, 5700 College Rd, Lisle, IL 60532 • T: +1 630 8296545 • F: +1 630 8296546 • tsuchy@ben.edu • alt.ben.edu/resources/J-Museum/index.htm •
Cur.: *Theodore D. Suchy* •
University Museum / Natural History Museum – 1970
Natural history ... 50126

Lisle Station Park, 921 School St, Lisle, IL 60532 • T: +1 630 9680499 • F: +1 630 9680499 • jlozano@lisleparkdistrict.org • www.lisleparkdistrict.org •
Cur.: *Jacob Lozano* •
Historical Museum / Historic Site / Open Air Museum – 1978
Temporary exhibitions ... 50127

Litchfield CT

Litchfield Historical Society Museum, 7 South St, Litchfield, CT 06759 • T: +1 860 5674501 • F: +1 860 5673565 • cfields@litchfieldhistoricalsociety.org • www.litchfieldhistory.org •
Dir.: *Catherine Keene Fields* •
Historical Museum – 1856 ... 50128

White Memorial Conservation Center, 80 Whitehall Rd, Litchfield, CT 06759-0368 • T: +1 860 5670857 • info@whitememorialcc.org • www.whitememorialcc.org •
Dir.: *Keith R. Cudworth* •
Natural History Museum – 1964 ... 50129

Litchfield MN

Meeker County Museum and G.A.R. Hall, 308 N Marshall Av, Litchfield, MN 55355 • T: +1 320 6938911 • www.garminnesota.org •
Dir.: *Cheryl Bulau* •
Local Museum / Military Museum – 1961
Regional hist, Grand Army of the Republic memorabilia ... 50130

Lithopolis OH

Wagnalls Memorial, 150 E Columbus St, Lithopolis, OH 43136 • T: +1 614 8334767 ext 104 • F: +1 614 8370781 • jneff@wagnalls.org • www.wagnalls.org •
Dir.: *Jerry W. Neff* •
Special Museum – 1924
Paintings, poems, Mabel Wagnalls Jones publishing ... 50131

Little Compton RI

Little Compton Historical Society Museum, 548 W Main Rd, Little Compton, RI 02837 • T: +1 401 6354035 • F: +1 401 6354035 • lchistory@yahoo.com • www.littlecompton.org •
Dir.: *Carlton C. Brownell* •
Local Museum – 1937
Antique farming, "Peggotty" artists studio, Portuguese exhibit ... 50132

Little Falls MN

Charles A. Lindbergh House, 1620 S Lindbergh Dr, Little Falls, MN 56345 • T: +1 320 6165421 • F: +1 320 6165423 • lindbergh@mnhs.org • www.mnhs.org/places/sites/lh •
Man.: *Charles D. Paulter* •
Historical Museum
Flight that inspired an aviation revolution ... 50133

Charles A. Weyerhaeuser Memorial Museum, 2151 S Lindbergh Dr, Little Falls, MN 56345-0239 • T: +1 320 6324007 • F: +1 320 6328409 • mchs@littlefalls.net • www.morrisoncountyhistory.org •
Dir.: *Jan Warner* •
Local Museum – 1936
Local hist – library, archives ... 50134

Little Falls NJ

Yogi Berra Museum, c/o Montclair State University, 8 Quarry Rd, Little Falls, NJ 07424 • T: +1 973 6552378 • F: +1 973 6556894 • www.yogiberramuseum.org •
Dir.: *Dave Kaplan* •
Special Museum / University Museum – 1998
Yankees and baseball hist ... 50135

Little Falls NY

Herkimer Home, 200 State Rte 169, Little Falls, NY 13365 • T: +1 315 8230398 • F: +1 315 8230587 • thomas.kernan@oprhp.state.ny.us • www.nysparks.com •
Head: *Thomas J. Kernan* •
Historical Museum – 1913
Period furnishings, memorabilia of Herkimer family ... 50136

Little Falls Historical Museum, 319 S Ann St, Little Falls, NY 13365 • T: +1 315 8230643 • lfhistor@ntcnet.com • www.lfhistoricalsociety.org •
Pres. & C.E.O.: *Katharine Lyon* •
Historical Museum – 1962
Local memorabilia, genealogy – library ... 50137

Little Rock AR

Arkansas Arts Center, MacArthur Park, 9th and Commerce Sts, Little Rock, AR 72203-2137 • T: +1 501 3724000 • info@arkarts.com • www.arkarts.com •
Dir.: *Dr. Ellen Plummer* • Cur.: *Anne Gochenour* (Contemporary Craft) • *Brian Young* (Arts) •
Fine Arts Museum – 1937
American and European drawings, paintings, sculpture, prints and photographs. American, Asian and European decorative art – children's theater ... 50138

Historic Museum of Arkansas, 200 E Third St, Little Rock, AR 72201-1608 • T: +1 501 3249351 • F: +1 501 3249345 • o@historicarkansas.org • www.arkansashistory.com •
Dir.: *William B. Worthen jr.* • Dep. Dir.: *Starr Mitchell* (Education) • Cur.: *Swannee Bennett* • Cons.: *Andrew Zawacki* •
Historical Museum – 1941
State hist, 19th c decorative, mechanical and fine arts ... 50139

Museum of Discovery, Arkansas Museum of Science and History, 500 President Clinton Av, Little Rock, AR 72201 • T: +1 501 3967050 • F: +1 501 3967054 • bnelsen@amod.org • www.amod.org •
Dir.: *Nan Selz* • Dep. Dir.: *Carol Couser* (Technology) • Sc. Staff: *Susan Robinson* (Programs, Health Education) • *Marci Bynum Robertson* (Collections, Historical Research) • *Lance Nolley* (Exhibits, Facilities) • *Anne Carter* (Development) • Cust.: *Lynne Larsen* •
Historical Museum / Natural History Museum – 1927
Natural hist, science, South American and African anthropology, Indian and pioneer artifacts ... 50140

Old State House Museum, 300 W Markham St, Little Rock, AR 72201 • T: +1 501 3249685 • F: +1 501 3249688 • info@oldstatehouse.org • www.oldstatehouse.com •
Dir.: *Bill Gatewood* •
Historical Museum – 1952
Political state hist, first state capitol (1836-1911) ... 50141

University of Arkansas at Little Rock Art Galleries, c/o Dept. of Art, 2801 S University Av, Little Rock, AR 72204-1099 • T: +1 501 5698977 • F: +1 501 5698775 • www.ualr.edu/artdept •
Dir.: *Brad Cushman* • Cur.: *Nathan Larson* •
Fine Arts Museum / University Museum – 1972
20th c American art ... 50142

Littleton CO

Colorado Gallery of the Arts, Araphoe Community College, 5900 S Santa Fe Dr, Littleton, CO 80160 • T: +1 303 7975649 • F: +1 303 7975935 • www.arapahoe.com •
Fine Arts Museum / University Museum – 1979
Art gallery ... 50143

Littleton Historical Museum, 6028 S Gallup, Littleton, CO 80120 • T: +1 303 7953950 • F: +1 303 7309818 • curator@mail.littleton.org • www.littletongov.org •
Dir.: *Tim Nimz* • Cur.: *Bill Hastings* • *Lorena Donohue* •
Local Museum – 1969
Local history, farming ... 50144

Livermore IA

Humboldt County Old Settler's Museum, Old Settlers Park, Hwy 222, Livermore, IA 50558 • T: +1 515 3791642 • F: +1 515 3791472 •
Pres.: *Floyd Raney* •
Historical Museum / Museum of Classical Antiquities – 1885
Antiques, settlement ... 50145

Liverpool NY

Sainte Marie among the Iroquois, Onondaga Lake Pkwy, Liverpool, NY 13088 • T: +1 315 4536767 • F: +1 315 4536762 • stemarie@nysnet.net • www.ongov.net/parks •
Cur.: *Valerie Bell* •
Open Air Museum – 1933
17th c culture of the Haudenosaunee (Iroquois), daily life in the 1650's ... 50146

Salt Museum, Onondaga Lake Park, Liverpool, NY 13088 • T: +1 315 4536715 • F: +1 315 4536762 • stemarie@nysnet.net • www.ongov.net/parks •
Dir.: *Elaine Wisowaty* •
Special Museum – 1934
History of salt production till 1920 in Syracuse, kettles, wooden barrels and other equipment ... 50147

U

Livingston MT

Livingston Depot Center, 200 W Park, Livingston, MT 59047 • T: +1 406 2222300 • F: +1 406 2222401 • Pres.: *John Sullivan* • Dir.: *Diana L. Seider* •
Local Museum – 1985
Northern Pacific railroad stn 50148

Yellowstone Gateway Museum of Park County, 118 W Chinook, Livingston, MT 59047 • T: +1 406 2224184 • F: +1 406 2224184 • museum@yci.net • www.livingstonmuseums.org •
Dir.: *Brian Sparks* •
Local Museum – 1976
Local and regional history and archaeology, railroads 50149

Livingston TX

Alabama-Coushatta Indian Museum, U.S. Hwy 190, between Livingston and Woodville, Livingston, TX 77351 • F: +1 409 5634397 •
Head: *Roland A. Poncho* •
Ethnology Museum – 1965
Native American hist 50150

Polk County Memorial Museum, 514 W Mill St, Livingston, TX 77351 • T: +1 409 3278192 • F: +1 936 3278192 • museum@livingston.net • www.livingston.net/museum •
C.E.O.: *Ray Hill* • Cur.: *Wanda L. Bobinger* •
Local Museum – 1963
Agriculture, archaeology, folklore, glass, Indian artifacts, military 50151

Livingston Manor NY

Catskill Fly Fishing Center and Museum, 1031 Old Rte 17, Livingston Manor, NY 12758 • T: +1 845 4394810 • F: +1 845 4393387 • flyfish@catskill.net • www.cffcm.org •
Interim Dir.: *James Krul* •
Special Museum – 1981
Memorabilia and hist of flyfishing 50152

Loachapoka AL

Lee County Historical Society Museum, 6500 Stage Rd, Loachapoka, AL 36865, mail addr: POB 206, Loachapoka, AL 36865 • T: +1 334 8873007 • webmaster@leecountyhistoricalsociety.org • www.leecountyhistoricalsociety.org •
Pres.: *Carl Summers* •
Local Museum – 1968 50153

Lock Haven PA

Heisey Museum, 362 E Water St, Lock Haven, PA 17745 • T: +1 570 7487254 • F: +1 570 7481590 • heisey@clintoncountyhistory.com • www.clintoncountyhistory.com •
Exec. Dir.: *Anne M. McCloskey* •
Local Museum / Historical Museum – 1921
Hist furniture, artifacts 50154

Piper Aviation Museum, 1 Piper Way, Lock Haven, PA 17745 • T: +1 570 7488283 • F: +1 570 8938357 • piper@cub.kcnet.org • www.pipermuseum.com •
Cur.: *Dr. Ira Masemore* •
Science&Tech Museum – 1986
Piper aircraft hist 50155

Lockport IL

Illinois and Michigan Canal Museum, 803 S State St, Lockport, IL 60441 • T: +1 815 8385080 • salletgreen@aol.com •
Dir.: *Rose Bucciferro* •
Local Museum – 1964
Local history, pioneer settlement 50156

Lockport Gallery, Illinois State Museum, 201 W 10th St, Lockport, IL 60441 • T: +1 815 8387400 • F: +1 815 8387448 • jzimmer@museum.state.il.us • www.museum.state.il.us/ismsites/lockport/ •
Dir.: *Jim L. Zimmer* • Asst. Cur.: *Geoffrey Bates* •
Fine Arts Museum / Public Gallery – 1987
Fine, decorative, and ethnogrphic art of Illinois 50157

Lockport NY

Niagara County Historical Center, 215 Niagara St, Lockport, NY 14094-2605 • T: +1 716 4347433 • F: +1 716 4343309 • history1@localnet.com • www.niagaracounty.org •
C.E.O. & Cur.: *Melissa L. Dunlap* •
Local Museum / Historical Museum – 1947
Local history, military, pioneer, Tuscarora trades, Erie Cunnl 50158

Lodi CA

San Joaquin County Historical Museum, 11793 N Micke Grove Rd, Lodi, CA 95240 • T: +1 209 3312055 • info@sanjoaquinhistory.org • www.sanjoaquinhistory.org •
Dir.: *Michael W. Bennett* •
Historical Museum / Agriculture Museum – 1966
Anthropology, agricultural a.o. tools and implements, local hist, mining, Joseph Tope, C. Gregory Crampton, Charls M. Weber coll – library, botanical garden 50159

Logan KS

Dane G. Hansen Memorial Museum, 110 W Main St, Logan, KS 67646 • T: +1 785 6894846 • F: +1 785 6894892 • hansenmuseum@ruraltel.net • www.hansenmuseum.org •
Dir.: *Lee M. Favre* •
Fine Arts Museum – 1973
Oriental art, gun, coin & spoon coll, work by Kansas artists 50160

Logan UT

Nora Eccles Harrison Museum of Art, c/o Utah State University, 650 N 1100 E, Logan, UT 84322 • T: +1 435 7971414 • F: +1 435 7973423 • linda.pierson@usu.edu • www.artmuseum.usu.edu •
Dir.: *Victoria Rowe* •
Fine Arts Museum / University Museum – 1982
20th & 21st c West coast American paintings, sclupture & drawings, ceramic vessels, Native American art 50161

Logansport IN

Cass County Historical Society Museum, 1004 E Market St, Logansport, IN 46947 • T: +1 219 7533866 • F: +1 219 7229267 • cchistoricalsoc@myvine.com • www.casscountyin.tripod.com •
Pres.: *Peggy Wihebrink* •
Local Museum – 1907
Local history 50162

Loma MT

House of a Thousand Dolls, 106 First St, Loma, MT 59460 • T: +1 406 7394338 •
C.E.O.: *Marion Britton* •
Decorative Arts Museum – 1979
Dolls & toys from 1830 to present 50163

Lombard IL

Lombard Historical Museum, 23 W Maple, Lombard, IL 60148 • T: +1 630 6291885 • F: +1 630 6299927 • lombard_historical_society@netzero.com • www.villageoflombard.org/lhsweb/index.html •
Dir.: *Joel Van Haaften* •
Local Museum – 1970
Victorian crafts, artifacts and photogrpahs 50164

Sheldon Peck Museum, 355 E Parkside, Lombard, IL 60148 • T: +1 630 6291885 • F: +1 630 6299927 • lombard_historical_society@netzero.com • www.villageoflombard.org/lhsweb/index.html •
Dir.: *Joel Van Haaften* •
Fine Arts Museum – 1999
Reproduction artwork of Sheldon Peck 50165

Lompoc CA

Lompoc Museum, 200 S H St, Lompoc, CA 93436 • T: +1 805 7363888 • F: +1 805 7362840 • lompocmuseum@verizon.net • www.lompocmuseum.org •
Dir./Cur.: *Dr. Lisa Renken* (Anthropology) •
Ethnology Museum / Local Museum – 1969
Chumash Indians, development of the area, natural hist 50166

La Purisima Mission, 2295 Purisima Rd, Lompoc, CA 93436 • T: +1 805 7333713 • F: +1 805 7332497 • lpminfo@parks.ca.gov • www.lapurisimamission.org •
Head: *Michael Curry* •
Religious Arts Museum – 1935
Mission hist, docs, architectural studies 50167

London KY

Mountain Life Museum, 998 Levi Jackson Mill Rd, London, KY 40744 • T: +1 606 8788000 • F: +1 606 8643825 • www.kystateparks.com •
Dir.: *William Meadors* •
Historical Museum / Ethnology Museum – 1929
Indian artifacts, textiles, transportation 50168

Londonderry NH

The Children's Metamorphosis, 217 Rockingham Rd, Londonderry, NH 03053 • T: +1 603 4252560 • www.discoverthemet.com •
Dir.: *Betsy Anderson* •
Special Museum – 1991
Children's museum 50169

Lone Jack MO

Civil War Museum of Lone Jack, Jackson County, 301 S Bynum Rd, Lone Jack, MO 64070 • T: +1 816 6978833 •
Pres.: *Charlotte Remington* • Chm.: *Loeda Helmig* •
Military Museum – 1964
Military, local hist, Civil War, artifacts, electric map 50170

Nance Museum, POB 292, Lone Jack, MO 64070 • T: +1 816 6972526 • pjnmuseum@worldnet.att.net •
Dir.: *Paul J. Nance* •
Ethnology Museum – 1981
Traditional Saudi Arabia and Islam outfitted Bedouin tent, jewelry, costume, ethnographic, folk culture – library 50171

Long Beach CA

American Museum of Straw Art, 2324 Snowden Av, Long Beach, CA 90815 • T: +1 562 4313540 • F: +1 562 5980457 • curator@strawartmuseum.org • www.strawartmuseum.org •
Cur.: *Morgyn Owens-Celli* •
Decorative Arts Museum / Folklore Museum – 1989
Decorative and folk art – library 50172

Gatov Gallery, Alpert Jewish Community Center, 3801 E Willow St, Long Beach, CA 90815 • T: +1 562 4267601 • F: +1 562 4243915 • jantonoff@alpertjcc.org • www.alpertjcc.org •
Dir.: *Michael Witenstein* •
Public Gallery – 1960 50173

Long Beach Museum of Art, 2300 E Ocean Blvd, Long Beach, CA 90803 • T: +1 562 4392119 • F: +1 562 4393587 • megane@lbma.org • www.lbma.org •
Dir.: *Ron Nelson* • Cur.: *Sue Ann Robinson* (Collections) •
Fine Arts Museum – 1950
Southern Californian and European art since 1850, decorative arts, sculpture, ceramics – library, workshops for children 50174

Museum of Latin American Art, 628 Alamitos Av, Long Beach, CA 90802 • T: +1 562 4371689 • info@molaa.org • www.molaa.org •
Dir.: *Gregorio Luke* •
Fine Arts Museum – 1996
Contemporary art from Mexico, Central & South America, Spanish-speaking Caribbean – library 50175

Queen Mary Museum, 1126 Queen's Hwy, Long Beach, CA 90802-6390 • T: +1 562 4353511 • www.queenmary.com •
Pres.: *Joseph F. Prevratil* • Dir.: *Lovetta Kramer* •
Special Museum – 1971
Retired British luxury liner as museum ship 50176

Rancho Los Alamitos, 6400 Bixby Hill Rd, Long Beach, CA 90815 • T: +1 562 4313541 • info@rancholosalamitos.com • www.rancholosalamitos.com •
Dir.: *Pamela Seager* •
Historical Museum – 1970
Decorative arts, furniture, tools, farm implements, Indian artifacts 50177

Rancho Los Cerritos, 4600 Virginia Rd, Long Beach, CA 90807 • T: +1 562 5701755 • F: +1 562 5701893 • Ellen.Calomiris@longbeach.gov • www.rancholoscerritos.org •
Dir.: *Ellen Calomiris* • Cur.: *Stephen C. Iverson* •
Historical Museum – 1955
Furniture, tools, costumes, quilts, coverlets, archaeology – library 50178

University Art Museum, California State University, Long Beach, 1250 Bellflower Blvd, Long Beach, CA 90840-1901 • T: +1 562 9855761 • F: +1 562 9857602 • uam@csulb.edu • www.csulb.edu/uam •
Dir.: *Christopher Scoates* • Assoc. Dir.: *Ilee Kaplan* •
Cur.: *Mary-Kay Lombino* • *Marina Freeman* • *Liz Harvey* •
Fine Arts Museum / University Museum – 1949
Modern and contemporary art on paper, sculpture – Gordon F. Hampton coll, library, theater 50179

Long Beach WA

World Kite Museum and Hall of Fame, 3rd St NW, Long Beach, WA 98631 • T: +1 360 6424020 • F: +1 360 6424020 • info@worldkitemuseum.com • www.worldkitemuseum.com •
Dir.: *Kay Buesing* •
Special Museum – 1988
Japanese kites, other kites used in science, sport and religion 50180

Long Branch NJ

Long Branch Historical Museum, 1260 Ocean Av, Long Branch, NJ 07740 • T: +1 732 2290600 •
C.E.O. & Pres.: *Florence Dinkelspiel* •
Local Museum – 1953
1879 Saint James Episcopal Chapel, Church of the Presidents, civil war 50181

Long Island City NY

Isamu Noguchi Garden Museum, 32-37 Vernon Blvds, Long Island City, NY 11106 • T: +1 718 2047088 • F: +1 718 2782348 • museum@noguchi.org • www.noguchi.org •
Dir.: *Jenny Dixon* •
Fine Arts Museum – 1985
Sculpture, models, drawings by Noguchi, video documentation 50182

LaGuardia and Wagner Archives, LaGuardia Community College, 31-10 Thomson Av, Long Island City, NY 11101 • T: +1 718 4825065 • F: +1 718 4825069 • richardli@lagcc.cuny.edu • www.laguardiawagnerarchive.lagcc.cuny.edu •
Dir.: *Richard K. Lieberman* •
Local Museum – 1981
20th c New York City, papers of the Mayors, NY City housing authority records, Queens local hist, NY City council, hist of Steinway and sons 50183

Museum for African Art, 36-01 43rd Av, Long Island City, NY 11101 • T: +1 718 7847700 • F: +1 718 7847718 • museum@africanart.org • www.africanart.org •
Pres.: *Elsie Crum McCabe* • Cur.: *Laurie Farrell* •
Ethnology Museum / Fine Arts Museum – 1982 50184

P.S. 1 Contemporary Art Center, Museum of Modern Art, 22-25 Jackson Av, Long Island City, NY 11101 • T: +1 718 7842084 • F: +1 718 4829454 • mail@ps1.org • www.ps1.org •
Exec.Dir. & Pres.: *Alanna Heiss* •
Fine Arts Museum – 1971
Contemporary art 50185

Scalamandre, 37-24 24th St, Long Island City, NY 11101 • T: +1 718 3618500 • F: +1 718 3618311 • info@scalamandre.com • www.scalamandre.com •
Chm.: *Edwin Bitter* • Dir.: *Adriana Bitter* • Pres.: *Robert F. Scalamandre Bitter* • Sc. Staff: *Marc J. Scalamandre Bitter* •
Science&Tech Museum – 1947
Textile coll, wallcoverings and passementerie, carpeting, Scalamandre family and firm 50186

Sculpture Center, 44-19 Purves St, Long Island City, NY 11101 • T: +1 718 3611750 • F: +1 718 7869336 • info@sculpture-center.org • www.sculpture-center.org •
Dir.: *Mary Ceruti* •
Fine Arts Museum 50187

Socrates Sculpture Park, Broadway and Vernon Blvds, Long Island City, NY 11106 • T: +1 718 9561819 • F: +1 718 6261533 • info@socratessculpturepark.org • www.socratessculpturepark.org •
Dir.: *Alyson Baker* •
Fine Arts Museum – 1985
Sculpture garden, contemporary art 50188

Long Lake MN

3M Collection of Contemporary Western and Americana Art, c/o Old Utica Co., 1935 W Wayzata Blvd, Long Lake, MN 55356 • T: +1 952 4499611 • F: +1 952 2497813 • custserv@oldutica.com • www.oldutica.com/threemabout.html •
Fine Arts Museum – 1974 50189

Western Hennepin County Pioneers Museum, 1953 W Wayzata Blvd, Long Lake, MN 55356 • T: +1 952 4736557 • F: +1 952 4736557 • pioneer_museum@hotmail.com • www.amatalon.com/pioneermuseum •
Cur.: *Russ Ferrin* •
Local Museum – 1907
Native American relicts, farming, transportation, family histories, tools, clothing – archives 50190

Longboat Key FL

Longboat Key Center for the Arts, 6860 Longboat Dr S, Longboat Key, FL 34228 • T: +1 941 3832345 • F: +1 941 3837915 • director@longboatkeyartscenter.org •
Dir.: *Jeniffer Glassmoyer* •
Fine Arts Museum / Decorative Arts Museum – 1952 50191

Longmeadow MA

Richard Salter Storrs House, 697 Longmeadow St, Longmeadow, MA 01106 • T: +1 413 5673600, 5652920 • LCAsearch@aol.com • www.longmeadow.org •
C.E.O.: *Phillip Clark* • Cur.: *Linda C. Abrams* •
Local Museum – 1899
Local history, genealogy, decorative arts, early American furnishings 50192

Longmont CO

Longmont Museum, 400 Quail Rd, Longmont, CO 80501 • T: +1 303 6518374 • F: +1 303 6518590 • martha.clevenger@ci.longmont.co.us • www.ci.longmont.co.us/museum.htm •
Dir.: *Martha R. Clevenger* • Cur.: *Svein Edland* • *Jill Overlie* • *Cathey Dunn* • *Erik Mason* •
Local Museum – 1940/2002
Regional history 50193

Longview TX

Gregg County Historical Museum, 214 N Fredonia St, Longview, TX 75601 • T: +1 903 7535840 • F: +1 903 7535854 • ginia@gregghistorical.org • www.gregghistorical.org •
Exec. Dir.: *Ginia Northcutt* •
Local Museum – 1983
Artifacts, memorabilia & photographs illustrating the development of Gregg County 50194

Longview Museum of Fine Art, 215 E Tyler St, Longview, TX 75602 • T: +1 903 7538103 • F: +1 903 7538217 • director@lmfa.org • www.lmfa.org •
Dir.: *Renee Hawkins* •
Fine Arts Museum – 1970
Paintings, sculpture, graphics 50195

Longview WA

Lower Columbia College, The Arts Gallery, 1600 Maple St, Longview, WA 98632 • T: +1 360 4422510 • dbartlett@lowercolumbia.edu • www.lowercolumbia.edu/artgallery •
Dir.: *Diane Bartlett* •
Fine Arts Museum / University Museum – 1978
Art 50196

Lookout Mountain TN

Cravens House, Point Park Visitor Center, Lookout Mountain, TN 37350 • T: +1 423 8217786 • F: +1 423 8215129 •
C.E.O.: *Sam Weddle* •
Local Museum
Furniture datin from 1830 - 1890s ... 50197

Lopez Island WA

Lopez Island Historical Museum, 28 Washburn Pl, Lopez Village, Lopez Island, WA 98261 • T: +1 360 4683447, 4682049 • lopezmuseum@rockisland.com • www.rockisland.com/~lopezmuseum •
Dir.: *Mark Thompson-Klein* • Cur.: *Nancy McCoy* •
Historical Museum – 1966
Local hist ... 50198

Lorain OH

Black River Historical Society of Lorain Museum, 309 W 5th St, Lorain, OH 44052-1611 • T: +1 440 2452563 • F: +1 440 2453591 • brhsmoore@centurytel.net • www.loraincityhistory.org •
C.E.O. & Pres.: *Frank Sipkovsky* • Cur.: *Dee Trifiletti* •
Local Museum – 1981
Businesses and industries of Lorain ... 50199

Lordsburg NM

Shekspear Ghost Town Museum, 2 1/2 miles S of Mine St, Lordsburg, NM 88045, mail addr: POB 253, Lordsburg • T: +1 505 5429034 •
Dir.: *Hough Emanuel D.* • Cur.: *Hough Janaloo H.* •
Local Museum – 1970
Soutwest hist 1856-1935, furniture, period guns, custom made holsters, saddle and tack ... 50200

Loretto PA

Southern Alleghenies Museum of Art, Saint Francis University Mall, Loretto, PA 15940 • T: +1 814 4723920 • F: +1 814 4724131 • sama@sfcpa.edu • www.sama-art.org •
Dir.: *Dr. Michael Tomor* •
Fine Arts Museum – 1975
Visual art ... 50201

Los Alamos NM

Art Center at Fuller Lodge, 2132 Central Av, Los Alamos, NM 87544 • T: +1 505 6629331 • artful@losalamos.com • www.artfulnm.org •
Dir.: *Gloria Gilmore-House* • Asst. Dir.: *Craig Carmer* •
Fine Arts Museum – 1977
Art media ... 50202

Bandelier National Monument, State Rte 4, Los Alamos, NM 87544, mail addr: HCR 1, Box 1, Ste 15, Los Alamos, NM 87544-9701 • T: +1 505 6723861 ext 517 • F: +1 505 6729607 • band-visitor-center@nps.gov • www.nps.gov/band •
Archaeology Museum – 1916
Archaeology, ethnology of Pueblo Indians of the Pajarito Plateau ... 50203

Bradbury Science Museum, 15 and Central, Los Alamos, NM 87544 • T: +1 505 6653339 • F: +1 505 6656932 • museum@lanl.gov • bsm.lanl.gov •
Dir.: *John S. Rhoades* •
Natural History Museum – 1963 ... 50204

Los Alamos County Historical Museum, Fuller Lodge Cultural Center, 1921 Juniper St, Los Alamos, NM 87544 • T: +1 505 6626272 • F: +1 505 6626312 • historicalsociety@losalamos.com • www.losalamos.com/historicalsociety •
Dir.: *Hedy Dunn* • Pres.: *Nancy Bartlit* • Cur.: *Marianne Mortenson* •
Historical Museum – 1968
Manhattan Project history ... 50205

Upstairs Art Gallery, 2400 Central Av, Los Alamos, NM 87544 • T: +1 505 6628240 • F: +1 505 6628245 •
Dir.: *Charlie Kalogeros-Chattan* •
Fine Arts Museum
Local, regional and national art; architectural models, botanical illustrations, digital art, fiber arts, furniture, glasswork, jewelry, paintings, photographs, prints and sculpture ... 50206

U

Los Altos CA

Los Altos History Museum, 51 S San Antonio Rd, Los Altos, CA 94022 • T: +1 650 9489427 • lbajuk@losaltoshistory.org • www.losaltoshistory.org •
Dir.: *Madelyn Crawford* •
Historical Museum – 1977 ... 50207

Los Angeles CA

American Film Institute, 2021 N Western Av, Los Angeles, CA 90027 • T: +1 323 8567600 • F: +1 323 4674578 • www.afi.com •
Dir.: *Jean Firstenberg* • *James Hindman* •
Fine Arts Museum – 1967
Cinematography, film/ TV script coll – library ... 50208

Annette Green Perfume Museum, Fashion Institute of Design and Merchandising, 919 S Grand Av, Los Angeles, CA 90015-1421 • T: +1 213 6241200 • fidm.edu •
Special Museum
Historical and cultural role of fragrances ... 50209

Automobile Driving Museum, 610 Lairport St, Los Angeles, CA 90245 • T: +1 310 9090950 • jeffw@automobiledrivingmuseum.org • www.automobiledrivingmuseum.org •
Dir.: *Jeff Walker* • Cur.: *Earl Rubenstein* •
Science&Tech Museum – 2002
Vehicles from 1915-1970 ... 50210

Barnsdall Art Center and Junior Arts Center, 4800 Hollywood Blvd, Los Angeles, CA 90027 • T: +1 323 6446295 • www.culturela.org/artcenters •
Dir.: *Istiharoh Glasgow* •
Fine Arts Museum – 1966
library, theater ... 50211

Ben Maltz Gallery and Bolsky Gallery, Otis College of Art and Design, 9045 Lincoln Blvd, Los Angeles, CA 90045 • T: +1 310 6656905 • galleryinfo@otis.edu • www.otis.edu •
Dir.: *Meg Linton* • Coord.: *Jinger Heffner* (Exhibits) • Education: *Kathy MacPherson* •
Public Gallery – 1940
Work of art students ... 50212

CAFAM, Los Angeles Craft and Folk Art Museum, 5814 Wilshire Blvd, Los Angeles, CA 90036-4501 • T: +1 323 9374230 • info@cafam.org • www.cafam.org •
Dir.: *Maryna Hrushetska* •
Folklore Museum / Historical Museum – 1973
Contemporary American crafts and design, international folk art, masks of the worlds ... 50213

California African American Museum, 600 State Dr, Exposition Park, Los Angeles, CA 90037 • T: +1 213 7447432 • www.caamuseum.org •
Dir.: *Jai Henderson* • Cur.: *Redell Hearn* (History) • *Mar Hollingworth* (Education) •
Fine Arts Museum / Historical Museum / Ethnology Museum – 1981
Culture of African-Americans, fine art, sculpture, prints, costume, decorative arts – library ... 50214

California Science Center, 700 Exposition Park Dr, Los Angeles, CA 90037 • T: +1 323 7243623 • F: +1 213 7442034 • 4info@cscmail.org • www.californiasciencecenter.org •
Pres.: *Jeffrey N. Rudolph* • Sr. VP: *Dr. Diane Perlov* (Exhibits) • Deputy Dir.: *Dr. Ron Rohovit* (Education) • *Dr. Krisztina Eleki* (Life Science) • *Dr. Chuck Kopczak* (World of Ecology) • *Dr. Kenneth E. Phillips* (Aerospace) • *Dr. David Bibas* (Technology) •
Science&Tech Museum – 1998
Human innovation, life science, air and space – IMAX ... 50215

Center for the Study of Political Graphics, 8124 W Third St, Los Angeles, CA 90048 • T: +1 323 6534662 • F: +1 323 6536991 • cspgweb@politicalgraphics.org • www.politicalgraphics.org •
Dir./Cur.: *Carol A. Wells* •
Historical Museum / Fine Arts Museum – 1988
Political and protest posters from early 20th c to present ... 50216

Children's Museum of Los Angeles (closed for renovation), Hansen Dam Recreation Park, Los Angeles, CA 90012, mail addr: 14410 Sylvan St, Van Nuys, CA 91401 • T: +1 213 6878800, +1 818 7862656 • F: +1 213 6870319 • www.childrensmuseumla.org •
Dir.: *Candace Barrett* •
Special Museum – 1979 ... 50217

Corita Art Center Gallery, Immaculate Heart High School, 5515 Franklin Av, Los Angeles, CA 90028 • T: +1 323 4662157 • sasha@corita.org • www.corita.org •
Public Gallery – 1995
Serigraphs and watercolors of Corita Kent ... 50218

County of Los Angeles Fire Museum, 1320 N Eastern Av, Los Angeles, CA 90063-3294 • T: +1 323 8812411 • www.clafma.org/collection •
Dir.: *Paul Schneider* •
Historical Museum – 1974
Fire-fighting, historic equipment, helmets, badges ... 50219

FIDM Museum and Galleries, Fashion Institute of Design and Merchandising, 919 S Grand Av, Los Angeles, CA 90015-1421 • T: +1 213 6241200 • fidm.edu •
Dir.: *Robert Nelson* • Cur.: *Kevin L. Johns* • Sc.Staff: *Christina Johnson* •
Decorative Arts Museum – 1977
Costumes from 18th c to present, 20th c costume and ready-to-wear ... 50220

Fine Arts Gallery, California State University Los Angeles, 5151 State University Dr, Los Angeles, CA 90032 • T: +1 323 3434040 • Art-Gallery@calstatela.edu • www.calstatela.edu/academic/art •
Dir.: *Lamont Westmorland* •
Fine Arts Museum / University Museum – 1954 . 50221

Fisher Gallery, University of Southern California, 823 Exposition Blvd, Los Angeles, CA 90089-0292 • T: +1 213 7404561 • fmoa@usc.edu • www.usc.edu/fishergallery •
Dir.: *Dr. Selma Holo* • Assoc. Dir.: *Kay Allen* • Cur.: *Jennifer Jaskowiak* • Sc. Staff: *Max Schulz* •
Fine Arts Museum / University Museum – 1939
15th-20th c paintings, prints graphics and sculpture – library ... 50222

Fowler Museum of Cultural History, University of California, Los Angeles, POB 951549, Los Angeles, CA 90095-1549 • T: +1 310 8254361 • F: +1 310 2067007 • fowlerws@arts.ucla.edu • www.fowler.ucla.edu •
Dir.: *Marla C. Berns* • Asst. Dir.: *David Blair* • *Patricia Anawalt* (Center for Study Regional Dress) • *Betsy D. Quick* (Education) • Cur.: *Mary Nooter Roberts* • *Roy Hamilton* (Southeast Asia and Oceania) • *Wendy Teeter* (Archaeology) •
Fine Arts Museum / Decorative Arts Museum / Folklore Museum / Ethnology Museum / Archaeology Museum – 1963
Worldwide art, cultural and archaeological material – library ... 50223

Frederick R. Weisman Art Foundation, 275 N Carolwood Dr, Los Angeles, CA 90077 • T: +1 310 2775321 • F: +1 310 2775075 • billie@weismanfoundation.org • www.weismanfoundation.org •
Dir.: *Billie Milam Weisman* •
Fine Arts Museum – 1982
Works by European Modernists incl Cézanne, Picasso and Kandinsky; Surrealist works by Ernst, Miro and Magritte. postwar art by Giacometti, Noguchi, Calder, Rauschenberg, and Johns. abstract expressionist paintings by de Kooning, Francis, Still, and Rothko. Pop Art by Warhol, Lichtenstein, Oldenburg, and Rosenquist. Contemporary California works ... 50224

George J. Doizaki Gallery, Japanese American Cultural & Community Center, 244 S San Pedro St, Los Angeles, CA 90012-3895 • T: +1 213 6282725 • F: +1 213 6178576 • www.jaccc.org •
Dir.: *Chris Aihara* •
Public Gallery – 1980
Japanese treasures, Bugaku costumes ... 50225

Grier Musser Museum, 403 S Bonnie Brae St, Los Angeles, CA 90057 • T: +1 213 4131814 • griermusser@hotmail.com • griermussermuseum.com •
Dir.: *Susan Tejada* •
Public Gallery – 1984
Postcards of old Los Angeles ... 50226

Grunwald Center for the Graphic Arts, University of California, Hammer Museum, 10899 Wilshire Blvd, Los Angeles, CA 90024-4201 • T: +1 310 4437078 • www.hammer.ucla.edu •
Dir./Cur.: *Cynthia Burlingham* • Assoc. Cur.: *Claudine Dixon* • Asst. Cur.: *Carolyn Peter* •
Fine Arts Museum / University Museum – 1956
Graphic art, prints, drawings, photos ... 50227

Hammer Museum, University of California Los Angeles, 10899 Wilshire Blvd, Los Angeles, CA 90024 • T: +1 310 4437000 • www.hammer.ucla.edu •
Dir.: *Ann Philbin* • Dep. Dir.: *Lisa Whitney* •
Fine Arts Museum / University Museum – 1994
Sculpture garden, paintings, lithographs – library ... 50228

Heritage Square Museum, 3800 Homer St, Los Angeles, CA 90031 • T: +1 323 2252700 • jessica@heritagesquare.org • www.heritagesquare.org •
Dir.: *Berry H. Herlihy* • Cur.: *Dina Morgan* (Furniture, Furnishings) • *Steven Ormenyi* (Gardens) • *Justin Gershuny* (Architecture) •
Historical Museum – 1969
Victorian structure from Los Angeles and Pasadena, furniture and personal artifacts, textiles, medical, communication, transportation and buiseness equipments and artifacts ... 50229

Hollywood Bowl Museum, 2301 N Highland Av, Los Angeles, CA 90078 • T: +1 323 8502058 • museum@laphil.org • www.hollywoodbowl.org •
Dir./Cur.: *Dr. Carol Merrill-Mirsky* •
Music Museum / Performing Arts Museum – 1984
Music and performing arts, musical artists 20th c ... 50230

Hollywood Entertainment Museum, 1140 N Citrus Av, Los Angeles, CA 90038 • T: +1 323 4657900 • F: +1 323 4699576 • r.ayson@hollywoodmuseum.com • www.hollywoodmuseum.com •
Pres./C.E.O.: *Phyllis Caskey* •
Performing Arts Museum – 1996
Film, television, sound recording, new media ... 50231

Hollywood Guinness World of Records Museum, 6764 Hollywood Blvd, Los Angeles, CA 90028 • T: +1 323 4636433 • www.guinnessattractions.com •
Dir.: *Chanchil Sundher* •
Special Museum – 1991 ... 50232

Hollywood Heritage Museum, 2100 N Highland Av, Los Angeles, CA 90068 • T: +1 323 8742276 • membership@hollywoodheritage.org • www.hollywoodheritage.org •
Dir.: *Robert Nudelman* • Cur.: *Mark Wanamaker* •
Performing Arts Museum – 1982
Film industry hist – library ... 50233

Hollywood Wax Museum, 6767 Hollywood Blvd, Los Angeles, CA 90028 • T: +1 323 4625991 • lon@hollywoodwax.com • www.hollywoodwax.com •
Cur.: *Ken Horn* •
Special Museum – 1965 ... 50234

Holyland Exhibition, 2215 Lake View Av, Los Angeles, CA 90039 • T: +1 323 6643162 •
Dir./Cur.: *Betty J. Shepard* •
Religious Arts Museum – 1924
Porrty from Egypt, Persia, Palestine, Syria, Rome and Greece, mummy case, coins, oil lamps, furniture, tapestries, paintings ... 50235

I-5 Gallery, c/o Brewery Art Association, 2100 N Main St, Los Angeles, CA 90031 • T: +1 323 3420717 • gallery@breweryart.org • www.the-brewery.net •
Dir.: *Jill D' Agnencia* •
Public Gallery
Art ... 50236

J. Paul Getty Museum, 1200 Getty Center Dr 1200, Los Angeles, CA 90049-1687 • T: +1 310 4407300 • F: +1 310 4407751 • info@getty.edu • www.getty.edu •
Dir.: *Deborah Gribbon* • Cur.: *Marion True* (Antiquities) • *Gillian Wilson* (Research) • *Thomas Kren* (Manuscripts) • *Weston Naef* (Photographs) • *Scott Schaefer* (Paintings, Sculpture, Works of Art) • *Lee Hendrix* (Drawings) • Cons.: *Brian Considine* (Decorative Arts, Sculpture) • *Mark Leonard* (Paintings) • *Jerry Podany* (Antiquities) • *Marc Harnly* (Paper) •
Fine Arts Museum – 1953
Greek and Roman antiquities, Western European paintings, drawings; decorative arts, sculpture – conservation labs ... 50237

Japanese American National Museum, 369 E First St, Los Angeles, CA 90012 • T: +1 213 6250414 • F: +1 213 6251770 • ckomai@janm.org • www.janm.org •
Dir.: *Irene Hirano* •
Historical Museum / Fine Arts Museum – 1985
History, former 1925 Nishi Hongwanji Buddhist Temple – library ... 50238

José Drudis-Biada Art Gallery, c/o Mount Saint Mary's College, 12001 Chalon Rd, Los Angeles, CA 90049 • T: +1 310 9544360 • jbaral@msmc.edu • www.msmc.la.edu •
Dir.: *Jody Baral* •
Public Gallery
Gene Mako coll, recent paintings, works on paper ... 50239

Korean American Museum, 3727 W Sixth St, Los Angeles, CA 90020 • T: +1 213 3884229 • F: +1 213 3811288 • info@kamuseum.org • www.kamuseum.org •
Dir.: *Yujin Lim* •
Ethnology Museum / Fine Arts Museum – 1991
History, culture and achievements of the Korean American community ... 50240

Korean Cultural Center Art Gallery, 5505 Wilshire Blvd, Los Angeles, CA 90036 • T: +1 323 9367141 • F: +1 323 9365712 • www.kccla.org •
Dir.: *Jun Young Jae* •
Public Gallery
library ... 50241

Laband Art Gallery, Loyola Marymount University, 1 LMU Dr, Los Angeles, CA 90045-2659 • T: +1 310 3382880 • www.lmu.edu •
Dir.: *Gordon L. Fuglie* •
Fine Arts Museum / University Museum – 1985
Coll of figurative Flemish and Italian Baroque art, drawings and prints by Max Thalmann ... 50242

LACE - Los Angeles Contemporary Exhibitions, 6522 Hollywood Blvd, Los Angeles, CA 90028-6210 • T: +1 323 9571777 • F: +1 323 9579025 • info@welcometolace.org • www.welcometolace.org •
Dir.: *Irene Tsatsos* •
Public Gallery – 1977 ... 50243

Los Angeles County Museum of Art, 5905 Wilshire Blvd, Los Angeles, CA 90036 • T: +1 323 8576000 • publicinfo@lacma.org • www.lacma.org •
Dir.: *Andrea L. Rich* • Cons.: *Victoria Blyth-Hill* • Cur.: *Stephanie Barron* • *Howard Fox* (Modern and Contemporary Art) • *Stephen A. Markel* (Indian and Southeast Asian Art) • *J. Keith Wilson* (Far Eastern Art) • *Kevin Salatino* (Prints, Drawings) • *Sharon Takeda* (Costumes, Textiles) • *J. Patrice Marandel* (European Painting and Sculpture) • *Bruce Robertson* (American Art) • *Nancy K. Thomas* (Collections) • *Wendy Kaplan* (Decorative Arts) • *Robert Sobieszek* (Photography) • *Robert Singer* (Japanese Art) • Head: *Dorrance Stalvey* (Music) • *Ian Birnie* (Film) •
Fine Arts Museum – 1910
Asian paintings, sculpture, ceramics and works on paper, American, European and Islamic arts, decorative arts, textiles – library, conservation center, theater ... 50244

Los Angeles Municipal Art Gallery, Barnsdall Art Park, 4800 Hollywood Blvd, Los Angeles, CA 90027 • T: +1 323 6446269 • F: +1 323 6446271 • cadmag@earthlink.net • www.culturela.org/lamag/Home •
Cur.: *Mark Steven Greenfield* • *Scott Canty* • *Sara Cannon* •
Fine Arts Museum – 1954
Contemporary fine arts and design ... 50245

Los Angeles Museum of the Holocaust, 6435 Wilshire Blvd, Los Angeles, CA 90048 • T: +1 323 6513704 • F: +1 323 6513706 • www.lamuseumoftheholocaust.org •
Dir.: *Mark Rothman* •
Historical Museum – 1961/2005
Jewish art, children's art, personal memorabilia, pre war photos and decorative art – library ... 50246

Los Angeles Police Museum, 6045 York Blvd, Los Angeles, CA 90042 • T: +1 323 3449445 • F: +1 323 3449516 • laphs@earthlink.net • www.laphs.com •
Exec. Dir.: *Glynn B. Martin* •
Special Museum – 1989
Hist of LA Police Dept. ... 50247

Luckman Gallery, California State University Los Angeles, 5151 State University Dr, Los Angeles, CA 90032-8116 • T: +1 323 3436604 • info@luckmanarts.org • www.luckmanarts.org •
Fine Arts Museum / University Museum ... 50248

Lummis Home 'El Alisal', c/o Historical Society of Southern California, 200 E Av 43, Los Angeles, CA 90031 • T: +1 323 4605632 • hssc@socalhistory.org • www.socalhistory.org •
Exec. Dir.: *Thomas F. Andrews* •
Historical Museum
Archaeology, ethnology, history, Indian artifacts . 50249

MAK Center for Art and Architecture, 835 N Kings Rd, Los Angeles, CA 90069 • T: +1 323 6511510 • F: +1 323 6512340 • office@makcenter.org • www.makcenter.org •
Dir.: *Kimberli Meyer* •
Fine Arts Museum – 1994 50250

Museum of African American Art, 4005 Crenshaw Blvd, Macy's 3rd fl, Los Angeles, CA 90008 • T: +1 323 2947071 • F: +1 323 2947084 • info@maaala.org • www.maaala.org •
Pres.: *Belinda Fontenote-Jamerson* •
Fine Arts Museum – 1976
Arts of African and African-descendant peoples, traditional sculptures, paintings and prints – library 50251

Museum of Contemporary Art Los Angeles, 250 S Grand Av, Los Angeles, CA 90012 • T: +1 213 6266222 • F: +1 213 6208674 • www.moca.org •
Dir.: *Jeremy Strick* • *Olga Gerrard* • Cur.: *Paul Schimmel* • *Brian Gray* (Exhibits) • *Robert Hollister* (Collections) • Education: *Suzanne Isken* •
Fine Arts Museum – 1979
Art after 1940 50252

Museum of Death, 6031 Hollywood Blvd, Los Angeles, CA 90028 • T: +1 323 4668011 • F: +1 323 4668011 • www.museumofdeath.net •
Special Museum – 1995
Artwork and letters from serial murderers, caskets and coffins, execution devices, mortician tools 50253

Museum of Neon Art, 501 W Olympic Blvd, Los Angeles, CA 90015 • T: +1 213 4899918 • info@neonmona.org • www.neonmona.org •
Exec. Dir.: *Kim Koga* •
Fine Arts Museum – 1981
Contemporary fine electric and kinetic art, historic neon signs – small library 50254

Museum of the American West, Autry National Center, Griffith Park campus, 4700 Western Heritage Way, Los Angeles, CA 90027-1462 • T: +1 323 6672000 • F: +1 323 6605721 • vservices@autrynationalcenter.org • www.autrynationalcenter.org •
Pres.: *John L. Gray* • Cur.: *Michael Duchemin* (History) • *Amy Scott* (Visual Arts) • *Estella Chung* (Popular Culture) • *Bryson Strauss* (Music, Oral History) • Cons.: *Linda Strauss* •
Historical Museum – 1984
Multicultural American West, popular culture and fine arts – library, archives 50255

Museum of Tolerance, A Simon Wiesenthal Center Museum, 9786 West Pico Blvd, Los Angeles, CA 90035, mail addr: 1399 S Roxbury Dr, Los Angeles, CA 90035 • T: +1 310 5539036 • F: +1 310 5534521 • information@wiesenthal.net • www.museumoftolerance.com •
Special Museum – 1993
Holocaust hist, contemporary racism, anti-semitism – library 50256

Natural History Museum of Los Angeles County, 900 Exposition Blvd, Los Angeles, CA 90007 • T: +1 213 7633466 • info@nhm.org • www.nhm.org •
Dir.: *Dr. Jane G. Pisano* • Dep. Dir.: *Jural J. Garrett* • Sc. Staff: *Dr. John Heyning* (Research, Collections) •
Natural History Museum – 1910
Ethnology, archaeology, anthropology, botany, ichthyology, entomology, herpetology, paleontology, ornithology, mammalogy, invertebrate zoology, mineralogy – library, lab 50257

Petersen Automotive Museum, 6060 Wilshire Blvd, Los Angeles, CA 90036 • T: +1 323 9646356 • info@petersen.org • www.petersen.org •
Dir.: *Richard Messer* •
Special Museum – 1994
Works on paper, history of the automobile 50258

Pharmaka Gallery, 101 W Fifth St, Los Angeles, CA 90013 • T: +1 323 6897799 • info@pharmaka-art.org • www.pharmaka-art.org •
Public Gallery
Works of artists that live and work in Los Angeles 50259

Platt and Borstein Galleries, American Jewish University, 15600 Mulholland Dr, Los Angeles, CA 90077 • T: +1 310 4769777 • F: +1 310 4760347 • arts@ajula.edu • www.ajula.edu •
University Museum – 1985
Council artists, paintings, sculpture 50260

Plaza de la Raza - Boathouse Gallery, 3450 N Mission Rd, Los Angeles, CA 90031-3135 • T: +1 323 2232475 • F: +1 323 2231804 • information@plazadelaraza.org • www.plazadelaraza.org •
C.E.O.: *Rose Marie Cano* • Cur.: *Melissa Richardson Banks* •
Fine Arts Museum / Folklore Museum – 1969
Mexican-American folk art, Latino artworks, photos – Chicano coll 50261

El Pueblo de Los Angeles, 125 Paseo de la Plaza, Ste 400, Los Angeles, CA 90012 • T: +1 213 6281274, 6802525 • F: +1 213 4858238 • www.cityofla.org •
Man.: *Edward Navarro* •
Historical Museum – 1953
Manuscripts, architectural drawings, slides, archaeology – library 50262

Self Help Graphics and Art Gallery, 3802 Cesar E. Chavez Av, Los Angeles, CA 90063 • T: +1 323 8816444 • info@selfhelpgraphics.com • www.selfhelpgraphics.com •
Dir.: *Evonne Gallardo* •
Public Gallery – 1972 50263

Skirball Cultural Center, Hebrew Union College, 2701 N Sepulveda Av, Los Angeles, CA 90049 • T: +1 310 4404500 • info@skirball.org • www.skirball.org •
Dir.: *Lori Starr* • Cur.: *Grace Cohen Grossman* • *Barbara Gilbert* • *Erin Clancey* (Archaeology) • *Tal Gozani* •
Fine Arts Museum / Historical Museum – 1996
Jewish ceremonial art, archaeology, numismatics, Jewish fine and decorative arts 50264

Southwest Museum of the American Indian, Autry National Center, Mt Washington campus, 234 Museum Dr, Los Angeles, CA 90065 • T: +1 323 6672000 • F: +1 323 6605721 • vservices@autrynationalcenter.org • www.autrynationalcenter.org •
Dir.: *Dr. Duane H. King* • Dep. Dir.: *Barbara Arvi* (Education) • Asst. Cur.: *Byrn Potter* • Sc. Staff: *George Kritzman* •
Ethnology Museum – 1907
Art and culture of prehistoric, historic and contemporary Native Americans, decorative arts, folklore and folk art – library 50265

Travel Town Transportation Museum, 5200 Zoo Dr, Los Angeles, CA 90027 • T: +1 323 6625874 • travel.town@lacity.org • www.laparks.com/grifmet/tt/ •
Dir.: *Linda J. Barth* •
Science&Tech Museum – 1952
Railroad equipment, steam locomotives, wagons, model trains – library 50266

Watts Towers Arts Center, 1727 E 107th St, Los Angeles, CA 90002 • T: +1 213 8474646 • cadwattsctr@earthlink.net • www.trywatts.com/art_center •
Dir.: *Rosie Lee Hooks* •
Public Gallery – 1975
African art, contemporary art 50267

Weingart Gallery, Occidental College, Weingart 109, 1600 Campus Rd, Los Angeles, CA 90041 • T: +1 323 2592749 • F: +1 323 2592930 • bdillon@oxy.edu • www.oxy.edu •
Pres.: *Linda Lyke* •
Public Gallery – 1938 50268

Wells Fargo History Museum, 333 S Grand Av, Los Angeles, CA 90071 • T: +1 213 2537166 • WFMUSEUM.LA@wellsfargo.com • www.wellsfargohistory.com •
Cur.: *Juan Colato* •
Historical Museum – 1982
Banking and express history, gold scales, mining 50269

Zimmer Children's Museum, 6505 Wilshire Blvd, Los Angeles, CA 90048 • T: +1 323 7618989 • F: +1 323 7618990 • info@zimmermuseum.org • www.zimmermuseum.org •
Dir.: *Esther Netter* •
Historical Museum – 1991
Jewish culture, traditions and history 50270

Los Banos CA

Ralph Milliken Museum, Merced County Park, US-Hwy 152, Los Banos, CA 93635, mail addr: POB 2294, Los Banos, CA 93635 • T: +1 209 8265505 •
Folklore Museum / Local Museum – 1954
Agriculture, archaeology, military uniforms, Indian artifacts 50271

Los Gatos CA

Art Museum, 4 Tait Av, Los Gatos, CA 95030 • T: +1 408 3542646 • ED@MuseumsofLosGatos.org • www.museumsoflosgatos.org •
Pres.: *Sandy Decker* • Dir.: *Elke Groves* • Cur.: *Catherine Politopoulos* •
Fine Arts Museum – 1965 50272

History Museum, 75 Church St, Los Gatos, CA 95030 • T: +1 408 3957375 • ED@MuseumsofLosGatos.org • www.MuseumsofLosGatos.org •
Exec. Dir.: *Laura Bajuk* • Cur.: *Alice McCammon* (Natural History) • *Jade Bradbury* (History) •
Historical Museum / Natural History Museum – 1965 50273

Los Olivos CA

Wildling Art Museum, 2329 Jonata St, Los Olivos, CA 93441 • T: +1 805 6881082 • info@wildlingmuseum.org • www.wildlingmuseum.org •
Exec. Dir.: *Elizabeth P. Knowles* •
Fine Arts Museum – 1997
Paintings, prints, drawings, photographs, sculpture depicting flora and fauna 50274

Lost Creek WV

Watters Smith Memorial, Duck Creek Rd, Lost Creek, WV 26385 • T: +1 304 7453081 • F: +1 304 7453631 • watterssmith@aol.com •
Head: *Larry A. Jones* •
Local Museum / Agriculture Museum – 1964
Local hist, hist hose, farm and household machinery, tools and utensils 50275

Loudonville OH

Cleo Redd Fisher Museum, 203 E Main St, Loudonville, OH 44842 • T: +1 419 9944050 •
Cur.: *James Sharp* •
Local Museum – 1973
Local history, Indian relics, pioneer and Victorian rooms 50276

Louisa VA

Louisa County Historical Society, 200 Church Av, Courthouse Sq, Louisa, VA 23093 • T: +1 703 9672794 • lchs@firstva.com • www.trevilians.com •
Association with Coll – 1966
Regional history, in 1868 jail 50277

Louisburg NC

Louisburg College Art Gallery, 501 N Main St, Louisburg, NC 27549 • T: +1 919 4962521 • F: +1 919 4961788 • hintonwj@yahoo.com • www.louisburg.edu •
Dir. & Cur.: *William Hinton* •
Public Gallery / University Museum – 1957
American Impressionist art, Primitive art, contemp art 50278

Louisiana MO

Saint Luis University, 17405 Hwy UU, Louisiana, MO 63353 • T: +1 573 7546641 • F: +1 573 7543589 • hlay@nemonet.com •
Dir.: *Janet Couch* • *Nanette E. Boileau* •
University Museum / Fine Arts Museum
Sculptures 50279

Louisville KY

Allen R. Hite Art Institute, University of Louisville, Belknap Campus, Louisville, KY 40292 • T: +1 502 8526794 • F: +1 502 8526791 • john.begley@louisville.edu • www.art.louisville.edu •
Dir.: *Prof. James Grubola* •
Fine Arts Museum / University Museum – 1946
15th-20th c Eurpoean & American prints, drawings & paintings 50280

American Printing House for the Blind, Callahan Museum, 1839 Frankfort Av, Louisville, KY 40206 • T: +1 502 8952405 ext 365 • F: +1 502 8992363 • museum@aph.org • www.aph.org •
Dir.: *Carol B. Tobe* •
Historical Museum – 1994
Artifacts & interpretive material related to the hist of the educationof blind & visually impaired people 50281

Artswatch, 2337 Frankfort Av, Louisville, KY 40206-2467 • T: +1 502 8939661 • F: +1 502 8939661 • admin@frankave.win.net • www.frankfortave.com •
Exec. Dir.: *Andy Perry* •
Public Gallery – 1985 50282

Creative Arts Studio, Louisville Visual Art Association, 4121 Shelbyville Rd, Louisville, KY 40207 • T: +1 502 8953700 • F: +1 502 8963779 • troy@louisvillevisualart.org • www.artopia.org •
Dir.: *Elizabeth George* •
Public Gallery
Contemporary art 50283

Farmington Historic Home, 3033 Bardstown Rd, Louisville, KY 40205 • T: +1 502 4529920 • F: +1 502 4561976 • knic@bellsouth.net • www.farmington-historichome.org •
Dir.: *Carolyn Brooks* •
Historical Museum – 1958
1830 period furnishings 50284

The Filson Historical Society, 1310 S Third St, Louisville, KY 40208 • T: +1 502 6355083 • F: +1 502 6355086 • filson@filsonhistorical.org • www.filsonhistorical.org •
Dir.: *Mark V. Wetherington* •
Association with Coll – 1884
Local history 50285

The Joseph A. Callaway Archaeological Museum, Southern Baptist Theological Seminary, 2825 Lexington Rd, Louisville, KY 40280 • T: +1 502 8974011 • F: +1 502 8974056 • campinfo@sbts.edu • www.sbts.edu •
Cur.: *Joel F. Drinkard* •
Archaeology Museum / Religious Arts Museum – 1961
Biblical archaeology 50286

Kentucky Derby Museum, 704 Central Av, 1 Churchill Downs, Louisville, KY 40201 • T: +1 502 6371111 • F: +1 502 6365855 • info@derbymuseum.org • www.derbymuseum.org •
Dir.: *Lynn Ashton* • Cur.: *Chris Goodlett* • *Sandy Flaksman* •
Special Museum – 1985
Thoroughbred racing, equine art, history, science 50287

Kentucky Museum of Arts and Design, 715 W Main St, Louisville, KY 40202 • T: +1 502 5890102 • F: +1 502 5890154 • admin@kentuckyarts.org • www.kentuckyarts.org •
Dir.: *Mary Miller* • Cur.: *Brion Clinkingbeard* •
Decorative Arts Museum / Fine Arts Museum – 1981
Contemporary crafts & folk arts 50288

Locust Grove, 561 Blankenbaker Ln, Louisville, KY 40207 • T: +1 502 8979845 • F: +1 502 8970103 • lghh@locustgrove.org • www.locustgrove.org •
Dir.: *Julie Parke* •
Historic Site – 1964
1790 Georgian mansion, home of General George Rogers Clark 50289

Louisville Science Center, 727 W Main St, Louisville, KY 40202 • T: +1 502 5616100 • F: +1 502 5616145 • admin@lsclouienet.com • www.louisvillescience.org •
Dir.: *Gail R. Becker* •
Natural History Museum / Science&Tech Museum – 1872
Physical, life and natural science, industrial technology, ethnographic, costumes 50290

Louisville Slugger Museum, 800 W Main St, Louisville, KY 40202 • T: +1 502 5887228 • F: +1 502 5851179 • museum@slugger.com • www.sluggermuseum.org •
Exec. Dir.: *Anne Jewell* •
Special Museum – 1996
Baseball bats, baseball related artifacts 50291

Louisville Visual Art Museum, 3005 River Rd, Louisville, KY 40207 • T: +1 502 8962146 • F: +1 502 8962148 • feedback@louisvillevisualart.org • www.louisvillevisualart.org •
Exec Dir.: *Elizabeth George* • Pres.: *Sally Labaugh* •
Fine Arts Museum – 1909
Art gallery 50292

Morris Belknap Gallery, Dario Covi Gallery and SAL Gallery, c/o University of Louisville, Hite Art Institute, Louisville, KY 40292 • T: +1 502 8526794 • F: +1 502 8526791 • www.louisville.edu •
Chm.: *James Grubola* •
University Museum / Fine Arts Museum – 1935
Teaching coll, paintings, drawings, prints 50293

Muhammad Ali Center, 144 N Sixth St, Louisville, KY 40202 • T: +1 502 5849254 • F: +1 502 5894905 • info@alicenter.org • www.alicenter.org •
Pres./ CEO: *Greg Roberts* •
Special Museum
Muhammad Ali's life story, boxing career, humanitarian efforts, legacy and ideals of Ali to promote respect, hope and understanding – library, archives 50294

Portland Museum, 2308 Portland Av, Louisville, KY 40212 • T: +1 502 7767678 • F: +1 502 7769874 • pmuseum@iglou.com • www.goportland.org •
Dir.: *Nathalie Taft Andrews* •
Local Museum – 1978
Regional history, Beach Grove residence of William Skene 50295

Sons of the American Revolution Museum, 1000 S 4th St, Louisville, KY 40203 • T: +1 502 5891776 • F: +1 502 5891671 • www.sar.org •
Dir.: *William H. Roddis* • Cur.: *Fletcher L. Elmore* •
Historical Museum – 1876
National history 50296

Speed Art Museum, 2035 S Third St, Louisville, KY 40208 • T: +1 502 6342700 • F: +1 502 6362899 • info@speedmuseum.org • www.speedmuseum.org •
Dir.: *Peter Morrin* • Cur.: *Ruth Cloudman* •
Fine Arts Museum – 1925
Paintings, sculpture, graphics 50297

Thomas Edison House, 729-31 E Washington St, Louisville, KY 40202 • T: +1 502 5855247 • F: +1 502 5855231 • edisonhouse@edisonhouse.org • www.edisonhouse.org •
Exec. Dir.: *Heather Bain* •
Science&Tech Museum – 1978
Phonographs, kinetoscope and bulbs 50298

Tryart Gallery, Louisville Visual Art Association, 400 W Market St, Louisville, KY 40202 • T: +1 502 5855550 • lou@louisvillevisualart.org • www.louisvillevisualart.org •
Public Gallery 50299

Louisville MS

American Heritage "Big Red" Fire Museum, 332 N Church Av, Louisville, MS 39339 • T: +1 662 7733421 • F: +1 662 7737183 • www.taylorbigred.com •
Historical Museum – 1989
Early fire equip, hand pumpers, hose reels, horse drawn ladder trucks, Indian artifacts, art 50300

Loveland CO

Loveland Museum and Gallery, 503 N Lincoln, Loveland, CO 80537 • T: +1 970 9622410, 9622490 • F: +1 970 9622910 • isons@ci.loveland.co.us • www.ci.loveland.co.us •
Dir.: *Susan P. Ison* • Cur.: *Tom Katsimpalis* • *Jennifer Slichter* •
Local Museum / Public Gallery – 1937
Local history, art 50301

Loveland OH

Greater Loveland Historical Society Museum, 201 Riverside Dr, Loveland, OH 45140 • T: +1 513 6835692 • F: +1 513 6837409 • glhsm@fuse.net • www.lovelandmuseum.org •
Dir.: *Janet Beller* • Cur.: *Jenny Shives* •
Local Museum – 1975
Local hist, period furniture, costumes, dolls 50302

Lowell MA

American Textile History Museum, 491 Dutton St, Lowell, MA 01854-4221 • T: +1 978 4410400 • F: +1 978 4411412 • athm@netwayl.mdc.net • www.athm.org •
Dir.: *Michael J. Smith* • Cur.: *Karen Herbaugh* (Machinery, Textiles) • Cons.: *Deirdre Windsor* (Textiles) •
Historical Museum / Science&Tech Museum – 1960
Textile history, machinery and industry, tools, costumes, clothing – library, archive 50303

Brush Art Gallery, 256 Market St, Lowell, MA 01852 • T: +1 978 4597819 • info@thebrush.org • www.thebrush.org •
Fine Arts Museum – 1982
Nationally acclaimed, culturally significant and cutting edge exhibits – artists and crafts studios 50304

Middlesex Canal Collection, Center for Lowell History, 40 French St, Lowell, MA 01852 • T: +1 978 9344997 • F: +1 978 9344995 • martha_mayo@uml.edu • library.uml.edu/clh/MidCan.html •
Dir.: *Martha Mayo* •
Local Museum – 1971
Local history, paintings, maps, docs – archives .. 50305

New England Quilt Museum, 18 Shattuck St, Lowell, MA 01852 • T: +1 978 4524207 • F: +1 978 4525405 • director@nequiltmuseum.org • www.nequiltmuseum.org •
Dir.: *Jennifer Gilbert* •
Decorative Arts Museum – 1987
Traditional and contemporary quilts – library 50306

Revolving Museum, 22 Shattuck St, Lowell, MA 01852 • T: +1 978 9372787 • F: +1 978 9372788 • gallery@revolvingmuseum.org • www.revolvingmuseum.org •
Dir.: *Jerry Beck* • Cur.: *Janice Pokorski* •
Fine Arts Museum 50307

Whistler House Museum of Art, Lowell Art Association, 243 Worthen St, Lowell, MA 01852 • T: +1 978 4527641 • F: +1 978 4542421 • mlally@whistlerhouse.org • www.whistlerhouse.org •
Exec. Dir.: *Michael H. Lally* • Cur.: *Michele Auger* •
Fine Arts Museum – 1878
Birthplace of James McNeill Whistler, American art – Parker gallery 50308

Lowville NY

Lewis County Historical Society Museum, 7552 S State St, Lowville, NY 13367 • T: +1 315 3768957 • histsoc@northnet.org •
Dir.: *Lisa Becker* •
Local Museum – 1930
Archaeology, costumes, family heirlooms, geology, military, technology, local hist 50309

Lubbock TX

Buddy Holly Center, 1801 Av G, Lubbock, TX 79401 • T: +1 806 7672686 • F: +1 806 7670732 • cgibbons@mail.ci.lubbock.tx.us •
Music Museum / Fine Arts Museum – 1984
Buddy Holly memorabilia, contemporary art 50310

The Galleries of Texas Tech, Texas Tech University, 18th St and Flint Av, Lubbock, TX 79409-1971 • T: +1 806 7421947 • F: +1 806 7421971 • landmark@ttu.edu •
Dir.: *Joe R. Arredondo* •
Fine Arts Museum
Art 50311

Museum of Texas Tech University, 3301 4th St, Lubbock, TX 79409-3191 • T: +1 806 7422442 • F: +1 806 7421136 • museum.texastech@ttu.edu • www.museum.ttu.edu •
Exec. Dir.: *Gary Edson* • Cur.: *Dr. Eileen Johnson* (Anthropology) • *Mei Campbell* (Ethnology, Clothing, Textiles) • *Dr. Robert Baker* (Natural Science) • *Dr. Sankar Chatterjee* (Paleontology) • *Henry B. Crawford* (History) • *Dr. Robert Bradley* (Mammalogy) • *Dr. Nancy McIntyre* (Ornithology) • *Dr. Peter Briggs* (Art) • Sc. Staff: *James Cokendolpher* (Invertebrates) •
University Museum / Local Museum – 1929
History of Southwestern US, ethnology, natural science, paleontology, fine and contemporary art, Pre-Columbian art coll, SW Indian art coll, archaeology, mammalogy, biodiversity 50312

Science Spectrum, 2579 S Loop 289, Lubbock, TX 79423 • T: +1 806 7481040 • F: +1 806 7451115 • spectrum@door.net • www.sciencespectrum.com •
Dir.: *Cassandra L. Henry* •
Science&Tech Museum – 1986
Hands-on displays of science, focusing on physical science 50313

Lubec ME

Roosevelt Campobello, POB 129, Lubec, ME 04652 • T: +1 506 7522922 • F: +1 506 7526000 • quirk@fdr.net • www.fdr.net •
Head: *Christopher Roosevelt* •
Historical Museum – 1964
Summer home of Pres. Franklin D. Roosevelt 50314

Lucas OH

Malabar Farm State Park, 4050 Bromfield Rd, Lucas, OH 44843 • T: +1 419 8922784 • F: +1 419 8923988 • malabarfarm@earthlink.net • www.malabarfarm.org •
C.E.O.: *Louis M. Andres* •
Historical Museum / Agriculture Museum – 1939 50315

Ludington MI

Historic White Pine Village, 1687 S Lakeshore Dr, Ludington, MI 49431 • T: +1 231 8434808 • F: +1 231 8437089 • info@historicwhitepinevillage.org • www.historicwhitepinevillage.org •
Dir.: *Ronald M. Wood* •
Open Air Museum / Local Museum – 1937
Local hist, reconstr historical village, 21 bldgs incl Rose Hawley Museum, Maritime Museum, Abe Nelson Lumbering Museum, Scottville Clown Band's Museum of Music – library, archives 50316

Ludlow VT

Black River Academy Museum, High St, Ludlow, VT 05149 • T: +1 802 2285050 • F: +1 802 2287444 • llt44@ludl.tds.net •
Dir.: *Georgia Brehm* •
Local Museum – 1972
Paintings, portraits, photographs, hist of the area 50317

Lufkin TX

The Museum of East Texas, 503 N Second St, Lufkin, TX 75901 • T: +1 409 6394434 • F: +1 409 6394435 •
Dir.: *J.P. McDonald* • Cur.: *Michael Collins* • *Nancy Wilson* (Photographic Coll) •
Fine Arts Museum / Historical Museum – 1975
Regional hist artifacts through the present, fine arts coll of American, European & regional artists 50318

Texas Forestry Museum, 1905 Atkinson Dr, Lufkin, TX 75901 • T: +1 936 6329535, 6336248 • F: +1 936 6329543 • info@treetexas.com • www.treetexas.com •
C.E.O.: *Carol Riggs* •
Natural History Museum – 1972
Logging equipment, tools, logging train, chain saws, fire look out tower 50319

Lumberton NJ

Air Victory Museum, 68 Stacey Haines Rd, Lumberton, NJ 08048 • T: +1 609 2674488 • F: +1 609 7021852 • info@airvictorymuseum.org • www.airvictorymuseum.org •
Dir.: *Fred Koch* •
Science&Tech Museum / Historical Museum – 1989
Airdraft from Korea and following wars, artifacts, uniforms 50320

Lumpkin GA

Bedingfield Inn Museum, Cotton St on the Square, Lumpkin, GA 31815 • T: +1 229 8386419 • F: +1 229 8386134 • schc@sowega.net • www.bedingfieldinn.org •
Dir.: *Vickie Harville* •
Historical Museum – 1965
1836 Stagecoach Inn , costumes, paintings, decorative art, furnishings 50321

Westville Historic Handicrafts Museum, 1850 Martin Luther King Dr, Lumpkin, GA 31815 • T: +1 229 8386310 • F: +1 229 8384000 • director@westville.org • www.westville.org •
Dir.: *Matthew M. Moye* • Cur.: *L.M. Moye* •
Historical Museum / Open Air Museum – 1966
Historic Village 34 buildings and houses c.1850 50322

Lutherville MD

Fire Museum of Maryland, 1301 York Rd, Lutherville, MD 21093 • T: +1 410 3217500 • F: +1 410 7698433 • info@firemuseummd.org • www.firemuseummd.org •
Cur.: *Stephen G. Heaver* •
Historical Museum – 1971
Fire fighting apparatus and equipment, hist of fire service, artifacts, photographs 50323

Lynchburg VA

The Anne Spencer Memorial Foundation, 1313 Pierce St, Lynchburg, VA 24501 • T: +1 804 8451313 •
Chm.: *Hugh R. Jones* •
Local Museum – 1977
Photographs, writings of Anne Spence, book, stained glass window 50324

Daura Gallery, Lynchburg College, 1501 Lakeside Dr, Lynchburg, VA 24501-3199 • T: +1 804 5448100, 5448343 • F: +1 804 5448277 • rothermel@lynchburg.edu • www.lynchburg.edu/daura •
Dir.: *Barbara Rothermel* •
Public Gallery – 1974
Art of Catalan-American Modernis Pierre Daura . 50325

Lynchburg Fine Arts Center, 1815 Thomson Dr, Lynchburg, VA 24501 • T: +1 804 8468451 • F: +1 804 8463806 • finearts@lynchburg.net •
Exec. Dir.: *Mary Brumbauch* •
Fine Arts Museum – 1958
Area photography and area art 50326

Lynchburg Museum System, 901 Court St, Lynchburg, VA 24504 • T: +1 804 8471459 • F: +1 804 5280162 • museum@lynchburgva.gov • www.lynchburgmuseum.org •
Dir.: *Thomas G. Ledford* •
Historical Museum – 1976
Furniture, costumes, art exhib, machinery 50327

Maier Museum of Art, Randolph-Macon Woman's College, ONe Quinlan St, Lynchburg, VA 24503 • T: +1 434 9478136 • F: +1 434 9478726 • klawson@rmwc.edu • maiermuseum.rmwc.edu •
Dir.: *Karol Lawson* • Assoc. Dir.: *Ellen Schall Agnew* •
Fine Arts Museum / University Museum – 1920
American paintings, works on paper, photographs 50328

Point of Honor, 112 Cabell St, Lynchburg, VA 24504 • T: +1 434 8471459 • F: +1 434 5280162 • museum@lynchburgva.us • www.pointofhonor.org •
Pres.: *Karen Buchanan* •
Local Museum / Folklore Museum – 1976
Historic house, home from Patrick Henry and George Cabell 50329

South River Meeting House, 5810 Fort Av, Lynchburg, VA 24502 • T: +1 434 2392548 • F: +1 434 2396071 •
Chm.: *Jim Garrett* •
Local Museum – 1757
Photographs, artifacts from 1864 Civil War battle 50330

Lynden WA

Lynden Pioneer Museum, 217 W Front St, Lynden, WA 98264 • T: +1 360 3543675 • troy@lyndenpioneermuseum.com • www.lyndenpioneermuseum.com •
Cur.: *Troy Luginbill* •
Local Museum – 1976
Prehistory and settlement, historic travel and transportation 50331

Lyndon KS

Osage County Historical Society, 631 Topeka Av, Lyndon, KS 66451 • T: +1 785 8283477 • research@kanza.net • www.osagechs.org •
Association with Coll – 1963
Local history, family history, library 50332

Lynn MA

Lynn Museum, 125 Green St, Lynn, MA 01902 • T: +1 781 5922465 • F: +1 781 5920012 • lynnmuse@shore.net • www.lynnmuseum.org •
Exec. Dir.: *Connie Colom* • Cur.: *Jennifer Gaudio* •
Local Museum – 1897
Local history, Hyde-Mills house, shoes ,early shoe making tools, costumes – library 50333

Lyons CO

The Lyons Redstone Museum, 340 High St, Lyons, CO 80540, mail addr: POB 9, Lyons, CO 80540 • T: +1 303 8236692, 8235271 • F: +1 303 8238257 • lavern921@aol.com •
Head: *LaVern M. Johnson* •
Local Museum / Historic Site – 1973
Local history 50334

Lyons KS

Coronado-Quivira Museum, 105 W Lyon, Lyons, KS 67554 • T: +1 316 2573941 • cqmuseum@hotmail.com •
Dir.: *Janel Cook* •
Local Museum – 1927
Indian & pionneer artifacts, anthropology, archaeology 50335

Lyons NY

Wayne County Historical Society Museum, 21 Butternut St, Lyons, NY 14489 • T: +1 315 9464943 • F: +1 315 9460069 • info@waynehistory.org • www.waynehistory.org •
Dir.: *Joseph O'Toole* • Cur.: *Lisa Williams* •
Historical Museum – 1946
Local hist, agriculture, early criminology, glass and ceramics 50336

Lyons Falls NY

Lewis County Historical Society Museum, High St, Lyons Falls, NY 13368 • T: +1 315 3488089 • histsoc@northnet.org • web.northnet.org/lewiscounty-historicalsociety/home.html •
Dir.: *Lisa Becker* •
Historical Museum – 1930 50337

Mabel MN

Steam Engine Museum, Steam Engine Park, Mabel, MN 55954 • T: +1 507 4935350 •
Pres.: *Tom Mengis* •
Science&Tech Museum – 1990
Farm machinery of the steam-power era, gas engines, tractors 50338

McAllen TX

McAllen International Museum, 1900 Nolana, McAllen, TX 78504 • T: +1 956 6821564 • F: +1 956 6861813 • mim@hiline.net • www.mcallenmuseum.org •
Dir.: *John Mueller* • Cur.: *Vernon Weckbacher* •
Fine Arts Museum / Science&Tech Museum – 1969/1967
Science, geology, paleonology, fine & contemporary art, Mexican folk art 50339

McCalla AL

Iron and Steel Museum of Alabama, 12632 Confederate Pkwy, McCalla, AL 35111 • T: +1 205 4775711 • F: +1 205 4779400 • ironmuseum@bellsouth.net • www.tannehill.org •
Dir.: *Vicki Kes* •
Historical Museum / Science&Tech Museum – 1970
Tannehill iron works, iron working implements, iron industry 1700-1850s – library 50340

McClellanville SC

Hampton Plantation, 1950 Rutledge Rd, McClellanville, SC 29458 • T: +1 803 5469361 • F: +1 803 5274995 • hampton_plantataion_sp@scprt.com • www.southcarolinaparks.com •
Local Museum – 1971
Historic house 50341

McConnells SC

Historic Brattonsville, 1444 Brattonsville Rd, McConnells, SC 29726 • T: +1 803 6842327 • F: +1 803 6840149 • www.yorkcounty.org •
Dir.: *Van Shields* •
Open Air Museum / Historic Site – 1976
Local hist 50342

McCook NE

High Plains Museum, 421 Norris Av, McCook, NE 69001 • T: +1 308 3453661 •
Dir.: *Marilyn Hawkins* •
Historical Museum – 1969
German bibles, pharmacy, farm tools, railroad, musical instruments 50343

Senator George Norris State Historic Site, 706 Norris Av, McCook, NE 69001-3142 • T: +1 308 3458484 • F: +1 308 3458484 • norris@mccooknet.com • www.nebraskahistory.org •
Dir.: *Lynne Ireland* •
Historical Museum – 1969
Furnishings, local historical artifacts 50344

McCutchenville OH

McCutchen Overland Inn, Rte 53 N, McCutchenville, OH 44844 • T: +1 419 9812052 •
Pres.: *Robin Schuster* • Cur.: *Anna Bea Heilman* •
Local Museum – 1967
Historic house, Victorian furniture, kitchen items, quilts, china 50345

McDade TX

McDade Museum, Main St, McDade, TX 78650 • T: +1 512 2730044 • www.rootsweb.com/~bxmdhms/ •
Pres.: *Don Grissom* •
Local Museum – 1962 50346

Macedon NY

Macedon Historical Society Museum, 1185 Macedon Center Rd, Macedon, NY 14502 • T: +1 315 9864845 •
Pres.: *David Taber* • Cur.: *Blanch Kemp* •
Historical Museum – 1962
Quaker school artifacts 50347

McFarland WI

McFarland Historical Society Museum, 5814 Main St, McFarland, WI 53558 • T: +1 608 8383992 • bluebee@madtown.net •
Chm.: *Dale Marsden* • Cur.: *Harold Muenkel* •
Local Museum – 1964
Local hist 50348

Machias ME

Burnham Tavern, Hannah Weston Chapter, National Society Daughters of the American Revolution, Main St, Machias, ME 04654 • T: +1 207 2554432 • info@burnhamtavern.com • www.burnhamtavern.com •
Dir.: *Valdine C. Atwood* •
Historical Museum – 1910
Tavern from 1770, local hist 50349

U

Machiasport ME

Gates House, 344 Port Rd, Machiasport, ME 04655 • T: +1 207 2558461 • franklf@nemalne.com • C.E.O.: *Suzanne Sousa* • Historical Museum – 1964
Marine artifacts, photos, furniture, geneaology ... 50350

Mackinac Island MI

Fort Mackinac, Huron Rd, Mackinac Island, MI 49757 • T: +1 906 8473328, +1 231 4364100 • F: +1 906 8473815 • mackinacparks@michigan.gov • www.mackinacparks.com • Dir.: *Phil Porter* • Cur.: *Steven C. Brisson* • *Katherine Cederholm* • *Lynn Evans* (Archaeology) • Open Air Museum / Military Museum / Historic Site – 1895
Regional history, archaeology, military items, battlefield of 1814, Indian dormitory ... 50351

Stuart House City Museum, Market St, Mackinac Island, MI 49757 • T: +1 906 8478181 • Dir.: *Lewis D. Crusoe* • Cur.: *Daniel Seeley* • Historical Museum / Local Museum – 1930
American Fur Co., fur post, natural hist, Indian artifacts, fur and pelts trade, war of 1812 ... 50352

Mackinaw City MI

Great Lakes Lighthouse Museum, 100 Dock Dr, Mackinaw City, MI 49701-0712 • T: +1 231 4363333, 4365580 • F: +1 231 4367870 • C.E.O.: *Dr. Sandra Planisek* • Science&Tech Museum / Fine Arts Museum – 1998
Fine arts, archeology, lighthouse and maritime artifacts ... 50353

Historic Mill Creek, 9001 US-23 S, Mackinaw City, MI 49701 • T: +1 231 4364100 • F: +1 231 4364211 • mackinacparks@michigan.gov • www.mackinacparks.com • Dir.: *Phil Porter* • Natural History Museum / Science&Tech Museum – 1975
18th-c water-powered sawmill, natural science ... 50354

Old Mackinac Point Lighthouse & Colonial Michilimackinac, 102 W Straits Av, Mackinaw City, MI 49701 • T: +1 231 4364100 • F: +1 231 4364210 • mackinacparks@michigan.gov • www.mackinacparks.com • Science&Tech Museum / Historical Museum – 1909/2004 ... 50355

Teysen's Woodland Indian Museum, 300 E Central St, Mackinaw City, MI 49701 • T: +1 231 4367011, 4367519 • F: +1 231 4365932 • C.E.O.: *Kenneth Teysen* • Ethnology Museum / Folklore Museum – 1950
Local hist ... 50356

McKinney TX

Bolin Wildlife Exhibit and Antique Collection, 1028 N McDonald, McKinney, TX 75069 • T: +1 972 5622639 • Dir.: *Jerry Bolin* • Cur.: *Tammie Ragsdale* • Natural History Museum / Museum of Classical Antiquities – 1980
North American & African wildlife animals ... 50357

Collin County Farm Museum, 7117 County Rd 166, McKinney, TX 75070 • T: +1 972 5484793 • F: +1 972 5422265 • jlawrence@co.collin.tx.us • www.co.collin.tx.us • Man.: *John Lawrence* • Agriculture Museum – 1976
Farming ... 50358

Heard Natural Science Museum, 1 Nature Pl, McKinney, TX 75069-8840 • T: +1 972 5625566 • F: +1 972 5489119 • info@heardmuseum.org • www.heardmuseum.org • Exec. Dir.: *Elizabeth W. Bleiberg* • Cur.: *Kenneth Steigman* • Natural History Museum – 1963
Malachology, insects, herpetology, ornithology ... 50359

McLean TX

McLean-Alanreed Area Museum, 116 Main St, McLean, TX 79057 • T: +1 806 7792731 • Cur.: *Dorothy McKee* • Local Museum – 1969
Pioneer costumes, western exhib. incl saddles, ranch tools, plows & Indian artifacts ... 50360

McLean VA

CIA Museum, McLean, VA 22101, mail addr: 2430 E St NW, Washington, DC 20505 • T: +1 703 4820623 • F: +1 703 4821739 • www.cia.gov/cia/information/tour/cia_museum.html • Historical Museum – 1988 ... 50361

Emerson Gallery, 1234 Ingleside Av, McLean, VA 22101 • T: +1 703 7901953 • F: +1 703 7901012 • Dir.: *Nancy Perry* • *Deborah McLeod* • Public Gallery
Paintings, sculpture, photography, graphics ... 50362

McMinnville OR

Evergreen Aviation Museum, 500 NE Captain Michael King Smith Way, McMinnville, OR 97128 • T: +1 503 4344180 • F: +1 503 4344058 • darrell.kuelpman@sprucegoose.org • www.sprucegoose.org • Exec.Dir.: *Darrell Kuelpman* • Science&Tech Museum – 1992
Military and general aviation, warbirds, vintage aircraft, HK-1 Hughes Flying Boat ... 50363

Macomb IL

Western Illinois University Art Gallery, 1 University circle, Macomb, IL 61455 • T: +1 309 2981587 • F: +1 309 2982400 • JR-Graham1@wiu.edu • www.wiu.edu/artgallery • Cur.: *John R. Graham* • Fine Arts Museum / University Museum – 1899
Graphics, paintings, sculpture, glass, jewelry, ceramics, Native American pottery, WPA graphics and paintings ... 50364

Macon GA

Georgia Music Hall of Fame, 200 Martin Luther King Jr Blvd, Macon, GA 31201 • T: +1 478 7508555 • F: +1 478 7500350 • info@gamusichall.com • www.gamusichall.com • C.E.O.: *Elizabeth F. Garcia* • Music Museum – 1996
Georgia music from pre-colonial to the present ... 50365

Hay House Museum, 934 Georgia Av, Macon, GA 31201-6708 • T: +1 478 7428155 • F: +1 478 7454277 • hayhouse@georgiatrust.org • www.georgiatrust.org • Dir.: *Suzanne Jones Harper* • Historical Museum – 1973
Historic House: 1855-59 Italian Renaissance Revival Mansion ... 50366

Museum of Arts and Sciences, 4182 Forsyth Rd, Macon, GA 31210 • T: +1 478 4773232 • F: +1 478 4773251 • info@masmacon.com • www.masmacon.com • Dir.: *Sheila Stewart-Leach* • Fine Arts Museum / Science&Tech Museum – 1956
Archaeology, artifacts, exotic moths and butterflies, toys, paintings, drwaings, prints, sculpture, ceramics – plantearium ... 50367

Ocmulgee National Monument, 1207 Emery Hwy, Macon, GA 31201 • T: +1 478 7528257 ext 10 • F: +1 478 7528259 • www.nps.gov/ocmu • Archaeology Museum – 1936
Archaeology, Indian artifacts ... 50368

Sidney Lanier Cottage, Middle Georgia Historical Society, 935 High St, Macon, GA 31208-3358 • T: +1 912 7433851 • F: +1 912 7453132 • sidneylanier@bellsouth.net • www.sidneylaniercottage.org • Dir.: *Katherine C. Oliver* • *Lucia C. Carr* • Historical Museum / Local Museum – 1964
Regional History, birthplace of poet Sidney Lanier – library ... 50369

Tubman African-American Museum, 340 Walnut St, Macon, GA 31208 • T: +1 912 7438544 • F: +1 912 7439063 • cpickard@tubmanmuseum.com • www.tubmanmuseum.com • Pres.: *Virgil Adams* • Historical Museum / Folklore Museum – 1982
African-American art and folk ... 50370

McPherson KS

McPherson College Gallery, Friendship Hall, 1600 E Euclid, McPherson, KS 67460 • T: +1 316 2410731 • F: +1 316 2428443 • www.mcpherson.edu • Dir.: *Wayne Conyers* • Fine Arts Museum – 1960
Oils, original prints, watercolors ... 50371

McPherson Museum, 1130 E Euclid, McPherson, KS 67460 • T: +1 620 2418464 • F: +1 620 24184644 • Dir.: *David Flask* • Local Museum – 1890
Local history, prehistory, paleontology ... 50372

Madawaska ME

Tante Blanche Museum, U.S. 1, Madawaska, ME 04756 • T: +1 207 7284518 • Pres.: *Verna Fortin* • Local Museum – 1968
Local hist ... 50373

Madera CA

Madera County Museum, 210 W Yosemite Av, Madera, CA 93637 • T: +1 209 6730291 • Cur.: *Dorothy Foust* • Local Museum – 1955
Local Indians, hist, lumber, agriculture ... 50374

Madison CT

Madison Historical Society Museum, 853 Boston Post Rd, Madison, CT 06443 • T: +1 203 2454567, 4213050 • dtallenrd:@aol.com • www.madisonct.com • Pres.: *Donald T. Allen* • Local Museum – 1917 ... 50375

Madison FL

North Florida Community College Art Gallery, Turner Davis Dr, Madison, FL 32340 • T: +1 850 9732288 • F: +1 850 9732288 • Dir.: *William F. Gardner jr.* • Fine Arts Museum / University Museum – 1975
College Art Gallery ... 50376

Madison GA

Madison-Morgan Cultural Center Collection, 434 S Main St, Madison, GA 30650 • T: +1 706 3424743 • info@mmcc-arts.org • www.mmcc-arts.org • Exec. Dir.: *Tina Lilly* • Local Museum – 1976
History of the region, decorative arts, costumes, architectural fragments, tools, household utensils, Civil War ... 50377

Madison IN

Historic Madison House, 500 West St, Madison, IN 47250 • T: +1 812 2652967 • F: +1 812 2733941 • hmihmfi@seidata.com • www.historicmadisoninc.com • Pres.: *John E. Galvin* • Historical Museum / Science&Tech Museum – 1960 ... 50378

Jefferson County Historical Museum, 615 W First St, Madison, IN 47250 • T: +1 812 2652335 • F: +1 812 2735023 • jchs@seidata.com • www.jcohs.org • Pres.: *Robert Wolfschlag* • Local Museum – 1850
County history, Madison railroad stn ... 50379

Lanier Mansion, 511 W 1st St, Madison, IN 47250 • T: +1 812 2653526 • F: +1 812 2653501 • lanier1@seidata.com • Cur.: *Link Ludington* • Historic Site – 1925
Home of J.F.D. Lanier ... 50380

Shrewsbury Windle House, 301 W 1st St, Madison, IN 47250 • T: +1 812 2654481 • Dir.: *Ann S. Windle* • Historical Museum – 1948
Period furnishings, glass, china, silver ... 50381

Madison MN

Lac Qui Parle County Historical Museum, 250 8th Av S, Madison, MN 56256 • T: +1 612 5987678 • madville@frontiernet.net • Cur.: *Lorraine Connor* • Local Museum – 1948
Regional history, pioneer & Indian artifacts, military, farming, doll coll, poet Robert Bly – library ... 50382

Madison NJ

Elizabeth P. Korn Gallery, Drew University, Rte 24, Madison, NJ 07940 • T: +1 973 4083000 • Dir.: *Livio Saganic* • Fine Arts Museum / Archaeology Museum – 1968
Ancient Near-East archeaology, different arts ... 50383

History Center of the United Methodist Church, 36 Madison Av, Madison, NJ 07940 • T: +1 973 4083189 • F: +1 973 4083909 • cyrigoye@drew.edu • Religious Arts Museum – 1885
archives ... 50384

Museum of Early Trades and Crafts, Main St at Green Village Rd, Madison, NJ 07940 • T: +1 973 3772982 • F: +1 973 3777358 • metc@msn.com • www.rosenet.org/metc • Dir.: *Deborah Starker* • Cur.: *Peter Rothenberg* • Historical Museum – 1969
Regional hist ... 50385

Madison SD

Prairie Village, W Hwy 34, Madison, SD 57042 • T: +1 605 2563644 • F: +1 605 2569616 • Pres.: *Mike Tammen* • Local Museum / Open Air Museum – 1966
40 restored buildings ... 50386

Smith-Zimmermann Museum, 221 NE 8th St, Dakota State University Campus, Madison, SD 57042 • T: +1 605 2565308 • F: +1 605 2565643 • smith.zimmermann@dsu.edu • www.smith-zimmermann.dsu.edu • Pres.: *Lory Norby* • Historical Museum – 1952
History of lake County, Ethnic settlers artifacts, Chautaugua, musical instruments, transportation, farm implements ... 50387

Madison WI

Chazen Museum of Art, 800 University Av, Madison, WI 53706-1479 • T: +1 608 2632246 • F: +1 608 2638188 • ppowell@lvm.wisc.edu • chazen.wisc.edu • Dir.: *Dr. Russell Panczenko* • Cur.: *Maria Saffiotti Dale* (Paintings, Sculpture, Decorative Arts) • *Anne Lambert* (Education) • *Andrew Stevens* (Prints, Drawings) • Fine Arts Museum / University Museum – 1962
Paintings, sculpture, drawings and prints, decorative arts, Russian and Soviet paintings, Davies Collection of Russian icons, Watson Coll of Indian miniatures, Van Vleck Coll of Japanese prints, Hall Coll of European prints ... 50388

DeRicci Gallery, c/o Edgewood College, 855 Woodrow, Madison, WI 53711 • T: +1 608 2574861 • F: +1 608 6633291 • Fine Arts Museum – 1965
Local and regional arts ... 50389

Helen Louise Allen Textile Collection, 1300 Linden Dr, University of Wisconsin, Madison, WI 53706 • T: +1 608 2621162 • F: +1 608 2625009 • sohe.wisc.edu/hlatc • Cur.: *Mary Ann Fitzgerald* • Special Museum – 1968
Textiles and costume from 15th c onwoards, ethnographic textiles and costumes ... 50390

The James Watrous Gallery, Wisconsin Academy, 211 State St, Madison, WI 53703 • T: +1 608 2631692 • F: +1 608 2653039 • www.wisconsinacademy.org • Dir.: *Randall Berndt* • Fine Arts Museum
Art ... 50391

Madison Children's Museum, 100 State St, Madison, WI 53703 • T: +1 608 2566445 • F: +1 608 2563226 • mcm@kidskiosk.org • www.kidskiosk.org • C.E.O.: *Dan Stewart* • Special Museum – 1980 ... 50392

Madison Museum of Contemporary Art, 222 W Washington Av Ste 350, Madison, WI 53703 • T: +1 608 2570158 • F: +1 608 2575722 • info@mmoca.org • www.mmoca.org • Dir.: *Stephen Fleischman* • Cur.: *Sara Krajewski* • *Sheri Castelnuovo* • Fine Arts Museum – 1901
American paintings, sculpture and prints, photography, Mexican, Japanese and European artworks ... 50393

Steep and Brew Gallery, 544 State St, Madison, WI 53703 • T: +1 608 2562902 • mduerr@madison.k12.wi.us • Dir.: *Mark Duerr* • Fine Arts Museum – 1985 ... 50394

University of Wisconsin Zoological Museum, 250 North Mills St, Lowell E. Noland Bldg, Madison, WI 53706 • T: +1 608 2621051 • F: +1 608 2625395 • Dir.: *Dr. John A.W. Kirsch* • Cur.: *E. Elizabeth Pillaert* (Osteology) • *Paula M. Holahan* (Mammology, Ornithology) • *Dr. John D. Lyons* (Fish) • *Dr. Gregory C. Mayer* (Herpetology) • Cur.em.: *Dr. John E. Dallman* (Paleontology) • Registrar: *Allison A. Smith* • University Museum / Natural History Museum – 1887
Vertebrate zoology, osteology ... 50395

Wisconsin Historical Museum, 30 N Carroll St, Madison, WI 53703-2707 • T: +1 608 2646555 • F: +1 608 2646575 • museum@whs.wisc.edu • www.wisconsinhistory.org/museum • Dir.: *Ann Koski* (Archaeology) • Dep. Dir.: *Jennifer Kolb* • Dir. Museum: *Kelly Hamilton* (Archaeology) • Chief Cur.: *Paul Bourcier* • Cur.: *Joseph Kapler* (Domestic Life) • *Leslie Bellais* (Costumes, Textiles) • *Joan Garland* • *David Driscoll* (Business, Technology) • *Beth Kowalski* • *Denise Wiggins* • *Monica Harrison* (Public Programms) • Registrar: *Scott Roller* • Local Museum – 1846
Wisconsin hist, firearms, crafts, costumes, dolls, ethnology, anthropology, dec arts, glass, ceramics ... 50396

Wisconsin Union Gallery, University of Wisconsin-Madison, 800 Langdon St, Madison, WI 53706 • T: +1 608 2625969 • F: +1 608 2628862 • schmoldt@wisc.edu • Dir.: *Mark Guthier* • Dep.Dir.: *Ken Gibson* • Cur.: *Robin Schmoldt* • Fine Arts Museum / University Museum
Temporary exhibitions ... 50397

Wisconsin Veterans Museum Madison, 30 W Mifflin St, Madison, WI 53703 • T: +1 608 2661009 • F: +1 608 2647615 • C.E.O.: *John Scocos* • Dir.: *Dr. Richard Zeitlin* • Cur.: *William F. Brewster* • *Mary Bade* • *Jeffrey Kollath* • Military Museum – 1901
Aircraft, vehicles, arms, artifacts – archive ... 50398

Mahomet IL

Early American Museum, State Rte 47, Mahomet, IL 61853 • T: +1 217 5862612 • F: +1 217 5863491 • early@earlyamericanmuseum.org • www.earlyamericanmuseum.org • Dir.: *Cheryl Kennedy* • Cur.: *Barbara Oelschlaeger-Garvey* • Historical Museum / Agriculture Museum / Decorative Arts Museum – 1967
Tools, implements, furniture ... 50399

Mahwah NJ

Art Galleries of Ramapo College, 505 Ramapo Valley Rd, Mahwah, NJ 07430 • T: +1 201 6847587 • Dir.: *Shalom Gorewitz* • Fine Arts Museum – 1979
Haitian art, popular art, prints ... 50400

Maitland FL

Holocaust Memorial Center of Central Florida, 851 N Maitland Av, Maitland, FL 32751 • T: +1 407 6280555 ext 284 • F: +1 407 6281079 • eva@holocaustedu.org • www.holocaustedu.org • C.E.O.: *Tess Wise* • Cur.: *Anita Lam* • Historical Museum – 1983
Holocaust artifacts – Library ... 50401

Maitland Art Center, 231 W Packwood Av, Maitland, FL 32751-5596 • T: +1 407 5392181 • F: +1 407 5391198 • gshepp@maitlandartcenter.org • www.maitlandartcenter.org • Dir.: *James G. Shepp* • Cur.: *Richard D. Colvin* • Fine Arts Museum – 1972
Works of Andre Smith and Jerry Raidiger ... 50402

Maitland Historical Museums, Maitland Historical Museum, Telephone Museum, Historic Waterhouse Residence Museum, 820 Lake Lily Dr, Maitland, FL 32751 • T: +1 47 6442451 • F: +1 407 6440057 • maiths@mpinet.net •
C.E.O.: *Allison Chapman* •
Local Museum / Science&Tech Museum – 1970
Seminole wars, telephone technology, Maitland and Waterhouse family hist, furnishings, tools and textiles 50403

Makawao HI

Hui No'eau Visual Arts Center, 2841 Baldwin Av, Makawao, HI 96768 • T: +1 808 5726560 • F: +1 808 5722750 • info@huinoeau.com • www.huinoeau.com •
Exec. Dir.: *Martin Betz* •
Fine Arts Museum – 1934
Artifacts of Harry and Ethel Baldwin and Francis Cameron 50404

Malden MA

Malden Public Library Art Collection, 36 Salem St, Malden, MA 02148 • T: +1 781 3240218, 3880800 • F: +1 617 3244467 • dmalgeri@maldenpubliclibrary.org • www.maldenpubliclibrary.org •
Head: *Dina G. Malgeri* •
Fine Arts Museum – 1879
Library with fine art coll, oils, sculpture 50405

Malibu CA

Adamson House and Malibu Lagoon Museum, 23200 Pacific Coast Hwy, Malibu, CA 90265 • T: +1 310 4568432 • www.adamsonhouse.org •
Pres.: *John Ghini* •
Historical Museum – 1982
Local history since Chumash Indians – library, botanical garden 50406

Frederick R. Weisman Museum of Art, Pepperdine University Center for the Arts, 24255 Pacific Coast Hwy, Malibu, CA 90263 • T: +1 310 5064522 • todd.eskin@pepperdine.edu • www.pepperdine.edu/cfa/weismanmuseum.htm •
Michael Zakian •
Fine Arts Museum / University Museum – 1992
20th c American art 50407

Malone NY

House of History, Franklin County Historical and Museum Society, 51 Milwaukee St, Malone, NY 12953 • T: +1 518 4832750 • fcohms@northnet.org • www.franklinhistory.org •
Pres.: *Peggy Roulston* •
Historical Museum – 1903
Hist, agriculture, period rooms, headquarter papers of 16th CW Regiment of NY State Volunteers 50408

Malta NY

Bass Manor Museum, 30 Rte 9P, Malta, NY 12020 • T: +1 518 8045 • F: +1 212 4914 • information@bassmanor.com • www.bassmanor.com •
Dir.: *Cassandra Dooley* •
Fine Arts Museum
Paintings, sculpture, photography 50409

Malvern AR

The Boyle House, Hot Spring County Museum, 302 E Third St, Malvern, AR 72104 • T: +1 501 3374775 •
Dir.: *Janis West* • Cur.: *Dorothy Keith* •
Local Museum – 1981
County history 50410

Malvern PA

Wharton Esherick Museum, 1520 Horseshoe Trail, Malvern, PA 19355, mail addr: POB 595, Paoli, PA 19301-0595 • T: +1 610 6445822 • F: +1 610 6442244 • whartonesherickmuseum@netzero.net •
Exec. Dir.: *Robert Leonard* • Cur.: *Mansfield Bascom* •
Historical Museum – 1971
Furniture, paintings, prints, sculpture in wood, stone and ceramic, utensils and woodcuts 50411

Mamaroneck NY

Larchmont Historical Society Museum, 740 W Boston Post Rd, Mamaroneck, NY 10543 • T: +1 914 3812239, 8345136 • lhs@savvy.net • members.savvy.net/lhs •
Local Museum – 1980
Hist of village and postal district of Larchmont 50412

Mammoth Spring AR

Depot Museum, Mammoth Spring State Park, Mammoth Spring, AR 72554, mail addr: Box 36, Mammoth Spring, AR 72554 • T: +1 870 6257364 • mammothspring@arkansas.com • www.arkansasstateparks.com •
Historical Museum – 1971
Local history, Frisco railroad depot, train memorabilia, Frisco Caboose 50413

Manassas VA

Manassas Museum System, 9101 Prince William St, Manassas, VA 20110 • T: +1 703 3681873 • F: +1 703 2578406 • info@manassasmuseum.org • www.manassasmuseum.org •
C.E.O.: *Melinda Herzog* • Cur.: *Roxana Adams* •
Local Museum – 1995
Civil War, local hist, African-American hist 50414

Manassas National Battlefield Park, 6511 Sudley Rd, Rte 234, Manassas, VA 20109 • T: +1 703 3611339 • F: +1 703 3617106 • mana_superintendent@nps.gov • www.nps.gov/mana •
Sc. Staff: *Edmund Raus* •
Military Museum / Historic Site – 1940
Hist artifacts & documentary materials relating to 1st & 2nd Battles of Manassas 50415

Manchester CT

Cheney Homestead, 106 Hartford Rd, Manchester, CT 06040 • T: +1 860 6435588 • manchesterhistory@juno.com • www.ci.manchester.ct.us/cheney/historic.htm •
C.E.O.: *David Smith* •
Local Museum – 1969 50416

Lutz Children's Museum, 247 S Main St, Manchester, CT 06040 • T: +1 860 6492838 • reckert@lutzmuseum.org • www.lutzmuseum.org •
Dir.: *Bob Eckert* • Cur.: *Marcie Charest* • *Michelle LaGrave* •
Special Museum – 1953 50417

Old Manchester Museum, 126 Cedar St, Manchester, CT 06040 • T: +1 860 6479983 • manchesterhistory@juno.com • www.manchesterhistory.org •
Pres.: *David K. Smith* •
Historical Museum – 1965
History, genealogy 50418

Manchester MA

Manchester Historical Society Museum, 10 Union St, Manchester, MA 01944 • T: +1 978 5267230 • F: +1 978 5260060 • manchesterhistorical@verizon.net • www.manchesterhistorical.org •
Dir.: *John Huss* • Cur.: *William Walker* •
Decorative Arts Museum / Local Museum – 1886
Decorative arts, local history 50419

Manchester NH

Alva DeMars Megan Chapel Art Center, Saint Anselm College, 100 Saint Anselm Dr, Manchester, NH 03102 • T: +1 603 6417000 • F: +1 603 6417116 • imaclell@anselm.edu • www.anselm.edu •
Dir.: *Iain MacLellan* •
Fine Arts Museum / University Museum – 1967
Sculpture, paintings, graphics 50420

Currier Gallery of Art, 201 Myrtle Way, Manchester, NH 03104 • T: +1 603 6696144 • F: +1 603 6697194 • hjordan@currier.org • www.currier.org •
Dir.: *Susan E. Strickler* • Deputy Dir.: *Susan Leidy* • Cur.: *P. Andrew Spahr* •
Fine Arts Museum – 1929
Architecture, American paintings, drawings, American Indian art, folk art, textiles, woodcarvings, Oriental art, Asian art, porcelain, silver, tapestries, enamels, Baroque art – library 50421

Lawrence L. Lee Scouting Museum, 40 Blondin Rd, Manchester, NH 03109 • T: +1 603 6271492, 6698919 • F: +1 603 6416436 • administrator@scoutingmuseum.org • www.scoutingmuseum.org •
Dir.: *Al Lambert* • Cur.: *Edward Rowan* •
Folklore Museum – 1969
Scouting 50422

Manchester Historic Association Museum, 255 Commercial St, Manchester, NH 03101 • T: +1 603 6227531 • F: +1 603 6220822 • history@manchesterhistoric.org • www.manchesterhistoric.org •
Dir.: *Gail Nessell Colglazier* • Cur.: *Marylou Ashooh Lazos* • *Eileen O'Brien* •
Historical Museum – 1896
History of the Manchester area, hist textile industry 50423

New Hampshire Institute of Art, 148 Concord St, Manchester, NH 03104 • T: +1 603 6230313 • F: +1 603 6411832 • admissions@nhia.edu • www.nhia.edu •
C.E.O. & Pres.: *Andrew Jay Svedlow* •
Fine Arts Museum – 1898
Contemporary regional fine arts and traditional crafts 50424

See Science Museum, 200 Bedford St, Manchester, NH 03101 • T: +1 603 6690400 • F: +1 603 6690400 • info@see-sciencecenter.org •
Dir.: *Douglas Heuser* •
Science&Tech Museum
Technology 50425

Manchester VT

The American Museum of Fly Fishing, 1070 Main St, Manchester, VT 05254 • T: +1 802 3623300 • F: +1 802 3623308 • amff@sover.net • www.amff.com •
Dir.: *John Price* • Cur.: *Yoshi Akiyama* •
Special Museum – 1968
Presidential fishing tackle, fly fishing art 50426

Hildene, 1005 Hildene Rd, Manchester, VT 05254 • T: +1 802 3621788 • F: +1 802 3621564 • info@hildene.org • www.hildene.org •
Cur.: *Brian L. Knight* •
Local Museum – 1978
Lincoln family hist, fruniture, President Lincoln 50427

Southern Vermont Art Center, West Rd, Manchester, VT 05254 • T: +1 802 3621405 • F: +1 802 3623274 • info@svac.org • www.svac.org •
Dir.: *Christopher Madkour* •
Fine Arts Museum / Folklore Museum – 1929
Paintings, sculpture, graphics, photographs 50428

Mangum OK

Old Greer County Museum and Hall of Fame, 222 W Jefferson St, Mangum, OK 73554 • T: +1 580 7822851 • museum@itl.net.net • greercountyok.com •
Cur.: *Patsie Smith* •
Local Museum – 1972 50429

Manhasset NY

Science Museum of Long Island, Leeds Pond Preserve, 1526 N Plandome Rd, Manhasset, NY 11030 • T: +1 516 6279400 ext 10 • F: +1 516 3658927 • smli@compuserve.com • ourworld.compuserve.com/homepages/smli •
Dir.: *John Loret* •
Science&Tech Museum – 1963
Natural hist, physics, technology 50430

Manhattan KS

Goodnow Museum, 2301 Clafin Rd, Manhattan, KS 66502 • T: +1 785 5656490 • F: +1 785 5656491 • www.kshs.org •
Dir.: *D. Cheryl Collins* •
Historical Museum – 1969
Home of pioneer Kansas educator, Isaac Tichenor Goodnow 50431

Harold M. Freund American Museum of Baking, 1213 Bakers Way, Manhattan, KS 66502 • T: +1 785 5374750 • F: +1 785 5371493 • info@aibonline.org • www.aibonline.org •
Asst. Cur.: *Tammy L. Popejoy* •
Special Museum – 1982
Hist of baking and milling 50432

Hartford House Museum, 2309 Claflin Rd, Manhattan, KS 66502 • T: +1 785 5656490 • www.co.riley.ks.us •
Dir.: *D. Cheryl Collins* •
Historical Museum – 1974
Pre-fabricated house shipped on the Hartford Steamboat to Manhattan, KS 50433

Manhattan Arts Center, 1520 Poyntz Av, Manhattan, KS 66502 • T: +1 785 5374420 • F: +1 785 5393356 • director@manhattanarts.org • www.manhattanarts.org •
Dir.: *Penny Senften* •
Fine Arts Museum – 1996
Photograph coll, Grandma Layton drawing, Remington sculpture, Bronco Buster 50434

Marianna Kistler Beach Museum of Art, Kansas State University, 14th St & Anderson Av, Manhattan, KS 66506, mail addr: 701 Beach Lane, Manhattan, KS 66506 • T: +1 785 5327718 • F: +1 785 5327498 • Beachart@ksu.edu • www.ksu.edu/bma •
C.E.O.: *Lorne E. Render* •
Fine Arts Museum / University Museum – 1996
Art of Kansas and the Mountain Plains region with emphasis on printmaking, 20th- 21st c American printmaking 50435

Pioneer Log Cabin, City Park, 11th and Poyntz, Manhattan, KS 66502 • T: +1 785 5656490 • F: +1 785 5656491 • www.co.riley.ks.us •
Dir.: *D. Cheryl Collins* •
Historical Museum – 1914
Farm and shop tools 50436

Riley County Historical Museum, 2309 Claflin Rd, Manhattan, KS 66502 • T: +1 785 5656490 • F: +1 785 5656491 • www.co.riley.ks.us •
Dir.: *D. Cheryl Collins* •
Local Museum – 1914
County history 50437

Wolf House Museum, 630 Fremont, Manhattan, KS 66502 • T: +1 785 5656490 • F: +1 785 5656491 • www.co.riley.ks.us •
Dir.: *D. Cheryl Collins* • Cur.: *Edna Williams* •
Historical Museum – 1983
1868 boarding house 50438

Manistee MI

Manistee County Historical Museum, 425 River St, Manistee, MI 49660 • T: +1 231 7235531 • www.rootsweb.com/~mimanist/Page63.html •
Dir.: *Steve Harold* •
Local Museum – 1953
Regional history, A.H. Lyman Drug Co. – archive 50439

Manistique MI

Imogene Herbert Historical Museum, Pioneer Park, Deer St, Manistique, MI 49854, mail addr: POB 284, Manistique, MI 49854 • T: +1 906 3418131, 3415045 •
Dir.: *Carol Dixon* • Cur.: *Janet Hickey* •
Local Museum – 1963
Local hist, geology,Upper Peninsula of Michigan fiction, plat, Indian artifacts, period fire engine – archives 50440

Manitou Springs CO

Miramont Castle Museum, 9 Capitol Hill Av, Manitou Springs, CO 80829 • T: +1 719 6851011 •
Dir.: *Bob Yager* •
Historical Museum – 1976 50441

Manitowoc WI

Manitowoc County History Museum, 1701 Michigan Av, Manitowoc, WI 54220 • T: +1 920 6844445, 6845110 • F: +1 920 6840573 • mchistsoc@lakefield.net • www.mchistsoc.org •
Pres.: *Don Janda* •
Local Museum / Open Air Museum – 1970
Local hist, hist village – library 50442

Rahr West Art Museum, 610 N 8th St, Manitowoc, WI 54220 • T: +1 920 6834501 • F: +1 920 6835047 • rahrwest@manitowoc.org • www.rahrwestartmuseum.org •
Dir.: *Jan Smith* • Asst. Dir.: *Daniel Juchniewich* •
Fine Arts Museum – 1950
19th till 21st cc American paintings, 19th c American decorative art, Chinese ivory carvings 50443

Wisconsin Maritime Museum, 75 Maritime Dr, Manitowoc, WI 54220-6823 • T: +1 920 6840218 • F: +1 920 6840219 • museum@wisconsinmaritime.org • www.wisconsinmaritime.org •
Exec.Dir.: *Robert M. O'Donnell* • Asst.Dir. & Cur.: *Bill Thiesen* •
Historical Museum – 1968
Maritime hist, submarine artifacts, ship tools 50444

Mankato KS

Jewell County Historical Museum, 118 N Commercial St, Mankato, KS 66956 • T: +1 785 5456426 • lkboden@dustdevil.com •
Cur.: *Karen Boden* •
Local Museum / Agriculture Museum – 1961
Farm implements, county hist, geology, oral hist 50445

Mankato MN

Blue Earth County Museums, Heritage Center Museum & R. D. Hubbard House, 415 Cherry St, Mankato, MN 56001 • T: +1 507 3455566 • bechs@juno.com • www.rootsweb.com/~mnbechs •
Exec. Dir.: *James Lundgreen* •
Local Museum – 1901
Local hist, American Indians, settlers, fine & decorative arts, clothing, transportation hist – library 50446

Carnegie Art Center, Rotunda and Fireplace Galleries, 120 S Broad St, Mankato, MN 56001 • T: +1 507 6252730 •
Fine Arts Museum – 1981
Regional visual arts 50447

Manlius NY

Town of Manlius Museum, 101 Scoville Av, Manlius, NY 13104 • T: +1 315 6826660 • F: +1 315 6823252 • mhsdirector@aol.com • www.manliushistory.org •
Dir.: *Julian Tasick* •
Local Museum – 1976
Farm and carpentry tools, wheelwright impl and tools, colonial ironware, local hist artifacts 50448

Mansfield LA

Mansfield State Historc Site, 15149 Hwy 175, 3 mi SE of Mansfield, Mansfield, LA 71052 • T: +1 318 8721474, +1 888 6776267 • F: +1 318 8714345 • mansfield@crt.state.la.us • www.crt.state.la.us/ •
Historical Museum / Military Museum – 1957
Civil War artifacts and books 50449

Mansfield MO

Laura Ingalls Wilder-Rose Wilder Lane Historic Home and Museum, 3068 Hwy A, Mansfield, MO 65704 • T: +1 417 9243626 • F: +1 417 9248580 • liwhome@getgoin.net • www.lauraingallswilderhome.com •
Dir.: *Jean C. Coday* •
Historical Museum – 1957
Translations of Little House books 50450

Mansfield OH

Mansfield Art Center, Ohio State University, 700 Marion Av, Mansfield, OH 44906 • T: +1 419 7561700 • F: +1 419 7560860 • allen56@osu.edu •
Dir.: *H. Daniel Butts* •
Public Gallery / University Museum – 1946
Changing exhibitions 50451

Manteo NC

Fort Raleigh, US 64-264, 1401 National Park Dr, Manteo, NC 27954 • T: +1 252 4732111 • F: +1 252 4732595 • www.nps.gov/fora •
Natural History Museum – 1941
Relics from the period of the first colony, Indian artifacts 50452

Roanoke Island Festival Park, Homeport of the Elizabeth II, 1 Festival Park, Manteo, NC 27954 • T: +1 252 4751500 • F: +1 252 4751507 • rifp.information@ncmail.net • www.roanokeisland.com •
Dir.: *Deloris Harrell* •
Historical Museum – 1983
Local hist 50453

Mantorville MN

Dodge County Historical Museum, 615 N Main, Mantorville, MN 55955-0433 • T: +1 507 6355508 • dchgc@kmtel.com • www.dodgecohistorical.addr.com •
Dir.: *Idella M. Conwell* •
Local Museum – 1876
Regional history, civil war recruiting, military, agriculture, costumes, paintings, Indian artifacts – archive 50454

Restoration House, 540 Main St, Mantorville, MN 55955 • T: +1 507 6355140 •
Decorative Arts Museum – 1963
19th-c furniture, dishes, utensils, furnishings – library 50455

Maple Valley WA

Maple Valley Historical Museum, 23015 SE 216th Way, Maple Valley, WA 98038 • T: +1 425 4323470 • F: +1 425 4323470 • pilgrim.dsk-c@worldnet.att.net •
Pres.: *Mona Pickering* •
Local Museum – 1972
Local hist, fire engines 50456

Maquoketa IA

Jackson County Historical Museum, 1212 E Quarry, Fairgrounds, Maquoketa, IA 52060-1245 • T: +1 319 6525020 • F: +1 319 6525020 • museum@willinet.net • www.clintonengine.com •
C.E.O: *Asher Schroeder* •
Local Museum / Agriculture Museum – 1964
Local history, agriculture – library 50457

Marathon FL

Crane Point Hammock Museum, 5550 Overseas Hwy, Marathon, FL 33050 • T: +1 305 7437124, 7433900 • F: +1 305 7430429 • tropcranept@aol.com • www.cranepoint.org •
Exec. Dir.: *Deanna S. Lloyd* •
Natural History Museum – 1990
Photos, shipwrecks artifacts, railroad, shells, hardwoods, natural hist of Florida Keys 50458

Marble CO

Marble Historical Museum, 412 W Main St, Marble, CO 81623 • T: +1 970 9631710 • F: +1 970 9638435 •
C.E.O.: *Thomas Williams* •
Local Museum – 1977
Local history 50459

Marblehead MA

Jeremiah Lee Mansion, 161 Washington St, Marblehead, MA 01945 • T: +1 781 6311768 • F: +1 781 6310917 •
Dir.: *Pam Peterson* • Cur.: *Judy Anderson* •
Decorative Arts Museum – 1909
Decorative arts, paintings, maritime and folk art, musical instruments – archives 50460

King Hooper Mansion, 8 Hooper St, Marblehead, MA 01945 • T: +1 781 6312608 • F: +1 781 6397890 • maa@marbleheadarts.org • www.marbleheadarts.org •
Decorative Arts Museum – 1922
Period furnishings, pottery, prints 50461

Marblehead Museum and J.O.J. Frost Folk Art Gallery, 170 Washington St, Marblehead, MA 01945 • T: +1 781 6311768 • F: +1 781 6310917 • petermhd@hotmail.com • www.marbleheadmuseum.org •
Dir.: *Pam Peterson* • Cur.: *Karen MacInnis* • *Judy Anderson* •
Local Museum / Decorative Arts Museum – 1898
Local history, home of Col. Jeremiah Lee, decorative and folk arts, maritime paintings – archives 50462

Marbletown NY

Ulster County Historical Society Museum, 2682 Rte 209, Marbletown, NY, mail addr: POB 279, Stone Ridge, NY 12484-0279 • T: +1 845 3385614 •
Dir.: *Amanda C. Jones* •
Local Museum / Historical Museum – 1939
Furniture, ceramics, dec arts, paintings, Civil War artifacts, tools 50463

Marbury MD

Smallwood's Retreat, Smallwood State Park, 2750 Sweden Point Rd, Marbury, MD 20658 • T: +1 301 7437613 • F: +1 301 7439405 • park-smallwood@dnr.state.md.us • www.dnr.state.md.us •
Military Museum – 1954
State park museum, 18th c house and kitchen 50464

Marcellus NY

Marcellus Historical Society Museum, 6 Slocombe Av, Marcellus, NY 13108 • T: +1 315 6733112 •
Pres.: *Peg Nolan* •
Historical Museum – 1960
Household and farm utensils, treadmill, guns, china, agricultural implements 50465

Marfa TX

Chinati Foundation, 1 Cavalry Row, Marfa, TX 79843 • T: +1 432 7294362 • F: +1 432 7294597 • information@chinati.org • www.chinati.org •
Dir.: *Marianne Stockebrand* • Pres.: *Fredericka Hunter* •
Fine Arts Museum – 1986
Installations by contemporary artists 50466

Marietta GA

Cobb County Youth Museum, 649 Cheatham Hill Dr, Marietta, GA 30064 • T: +1 770 4272563 • F: +1 770 4271060 • youthmuseum@aol.com • www.cobbcountyyouthmuseum.org •
Dir.: *Anita S. Barton* •
Ethnology Museum – 1964
Youth, jet trainer, canoe, caboose 50467

Cobb Museum of Art, 30 Atlanta St NE, Marietta, GA 30060 • T: +1 770 5281444 • F: +1 770 5281440 • jfoxmcma@bellsouth.net • www.mariettasquare.com •
Fine Arts Museum – 1986
American art (19th-20th c) 50468

Marietta Museum of History, 1 Depot St, Marietta, GA 30060 • T: +1 770 5280430, 5280431 • F: +1 770 5280450 • dcox@city.marietta.ga.us • www.mariettahistory.org •
Exec. Dir.: *Dan Cox* •
Local Museum
History of Georgia Piedmont 50469

Museo Abarth, 1111 Via Bayless, Marietta, GA 30066 • T: +1 770 9281342 •
Science&Tech Museum – 1989
Abarth engine, model cars, new car brochures, emblems, art work 50470

Root House Museum, Marietta Pkwy and Polk St, Marietta, GA 30060 • T: +1 770 4264982 • F: +1 770 4999540 • clhs2@bellsouth.net • cobblandmarks.com •
Dir.: *Daryl Barksdale* •
Local Museum – 1972 50471

Marietta OH

Campus Martius Museum, 601 Second St, Marietta, OH 45750 • T: +1 740 3733750 • F: +1 740 3733680 • cmmoriv@ohiohistory.org • www.ohiohistory.org/places/campus •
Dir.: *Gary C. Ness* •
Historical Museum – 1919
19th and 20th c farming, transportation and industry, farmers migration 1880-1920, Appalachian migration 1917-1970 50472

The Castle, 418 Fourth St, Marietta, OH 45750 • T: +1 740 3734180 • F: +1 740 3734233 • www.mariettacastle.org •
Dir.: *Lynne Shuman* •
Local Museum – 1992
Hist house, Victorian furniture 50473

Grover M. Hermann Fine Arts Center, c/o Marietta College, 215 Fifth St, Marietta, OH 45750 • T: +1 740 3764696 • F: +1 740 3764529 • www.marietta.edu •
Fine Arts Museum – 1965
Contemporary American paintings, sculpture and crafts, African and pre-Columbian art 50474

Harmar Station, 220 Gilman St, Marietta, OH 45750 • T: +1 740 3743424 • F: +1 740 3737808 • jmoberg@charter.net • www.harmarstation.com •
C.E.O.: *Jack Moberg* •
Science&Tech Museum – 1996
Engines, cars and accesories, "0" gauge, larger scale toys 50475

Ohio River Museum, 601 Front St, Marietta, OH 45750 • T: +1 614 3733750 • F: +1 614 3733680 • cmmoriv@ohiohistory.org • www.ohiohistory.org/places/ohriver •
Dir.: *Gary C. Ness* •
Historical Museum – 1941
Ecological, recreational and commercial hist of the Ohio River 50476

Marilla NY

Marilla Historical Society Museum, 1810 Two Rod Rd, Marilla, NY 14102 • T: +1 716 6525396 •
Pres.: *Mary Beth Serafin* •
Local Museum – 1960 50477

Marineland FL

Marineland Ocean Museum, 9507 Ocean Shore Blvd, Marineland, FL 32086-9602 • T: +1 904 4711111 • F: +1 904 4610156 • mor@aug.com • www.marineland.com •
Dir.: *David Internoscia* •
Historical Museum – 1938 50478

Marinette WI

Marinette County Historical Museum, Stephenson Island, U.S. Hwy 41, Marinette, WI 54143 • T: +1 715 7320831 •
Pres.: *Frank Lauermann* •
Local Museum – 1962
Local hist, logging and woodland Indian artifacts, copper culture, fishing, agriculture 50479

Marion IA

Granger House, 970 Tenth St, Marion, IA 52302-0753 • T: +1 319 3776672 • grangerhouse@juno.com • community.marion.ia.us/granger/ •
Decorative Arts Museum – 1973
Victorian furnishings, wagons 50480

Marion Heritage Center, 590 Tenth St, Marion, IA 52302 • T: +1 319 4776377 • Marionheritage@juno.com •
Local Museum
Local hist 50481

Marion IL

Williamson County Historical Society Museum, 105 S Van Buren, Marion, IL 62959 • T: +1 618 9975863 • charla@thewchs.com • www.thewchs.com •
Local Museum – 1976
County history, housed former county jail and sheriff's home 50482

Marion IN

Marion Public Library Museum, Carnegie Bldg, 600 S Washington St, Marion, IN 46953 • T: +1 765 6682900 • F: +1 765 6682911 • jfelton@marion.lib.in.us • www.marion.lib.in.us •
Dir.: *Sue Israel* •
Local Museum – 1884
Local hist 50483

Quilters Hall of Fame, 926 S Washington St, Marion, IN 46953 • T: +1 765 6749602 • F: +1 765 6649333 • quilters@comteck.com • quiltershalloffame.org •
Pres.: *Hazel Carter* •
Special Museum – 1979 50484

Marion MA

Marion Art Center, 80 Pleasant St, Marion, MA 02738 • T: +1 508 7481266 • F: +1 508 7482759 • wendymac@gis.net • www.marionartcenter.org •
Dir.: *Wendy Bidstrup* •
Fine Arts Museum – 1957
Cecil Clark Davis (1877-1955), portrait paintings 50485

Marion NC

Historic Carson House, 1805 US Hwy 70 W, Marion, NC 28752 • T: +1 704 7244648 •
Pres.: *Nina Greenlee* •
Historical Museum – 1964
Pioneer artifacts, textiles, musical instruments 50486

Marion OH

Harding Home and Museum, 380 Mount Vernon Av, Marion, OH 43302 • T: +1 740 3879630 • F: +1 740 3879630 • hardinghome@marion.net • www.ohiohistory.org •
C.E.O.: *Melinda Gilpin* •
Historical Museum – 1925
President Warren G. Hardings life, memorabilia, original interior 50487

Marion County Historical Society Museum, 169 E Church St, Marion, OH 43302 • T: +1 740 3874255 • F: +1 740 3870117 • mchs@historymarion.org • www.historymarion.org •
Dir.: *Jane E. Rupp* •
Local Museum – 1989
County hist, Warren G. Harding presidential artifacts 50488

Stengel-True Museum, 504 S State St, Marion, OH 43302 • T: +1 740 3877150 •
Dir.: *J.C. Ballinger* •
Local Museum – 1973
Interior, paintings, glass, chine, Indian artifacts 50489

Marion VA

Smyth County Museum, 105 E Strother St, Marion, VA 24354 • T: +1 540 7837067, 7837286 •
Pres.: *Brenda Gwyn* • Dir.: *Joan T. Armstrong* •
Local Museum – 1961
Weaving, pictoial hist of county, medical items 50490

Mariposa CA

California State Mining and Mineral Museum, 5005 Fairgrounds Dr, Mariposa, CA 95338 • T: +1 209 7427625 • F: +1 209 9663597 • mineralmuseum@sti.net • www.parks.ca.gov •
Dir.: *Peggy Ronning* •
Natural History Museum / Science&Tech Museum – 1988 50491

Marksville LA

Tunica-Biloxi Native American Museum, 150 Melancon Rd, Marksville, LA 71351 • T: +1 318 2538174 • F: +1 318 2537711 • www.tunica.org •
Cur.: *Earl J. Barbry jr* •
Historical Museum – 1989
French and Indian artifacts 50492

Marlboro MA

Peter Rice Homestead, 377 Elm St, Marlboro, MA 01752 • T: +1 508 4854763, 4603499 •
Cur.: *Richard Wilcox* •
Historical Museum – 1962
Local history and industry 50493

Marlboro NY

Gomez Foundation for Mill House, Mill House Rd, Marlboro, NY 12542 • T: +1 914 2363126 • F: +1 914 2363365 • gomezmillhouse@juno.com • www.gomez.org •
Dir.: *Ruth K. Abrahams* •
Decorative Arts Museum / Local Museum – 1984
Historic house, dec arts, furniture 50494

Marlboro VT

Marlboro Historical Society, N Main St, Marlboro, VT 05344 • T: +1 802 4640329 • F: +1 802 4641275 • dhdaod@sover.net • www.marlboro.vt.us •
Pres.: *Alan O. Dann* •
Association with Coll – 1958
Household items, farm implements, furniture 50495

Southern Vermont Natural History Museum, Rte 9, Hogback Mt. Overlook, Marlboro, VT 05344 • T: +1 802 4640048 • F: +1 802 4640017 • metcalfe@sover.net • www.vermontmuseum.org •
Dir.: *Edward C. Metcalfe* •
Natural History Museum – 1962
Natural hist of the northeast U.S. 50496

Marlinton WV

Pocahontas County Museum, Seneca Trail, Marlinton, WV 24954 • T: +1 304 7994973 • F: +1 304 7996466 •
Pres.: *Mary Lou Dilley* • Cur.: *H.L. Sheets* •
Local Museum – 1962
Local hist 50497

Marquette MI

DeVos Art Museum, Northern Michigan University, Lee Hall, 1401 Preque Isle Av, Marquette, MI 49855 • T: +1 906 2271481 • F: +1 906 2272276 • wfrancis@nmu.edu • art.nmu.edu/department/Museum •
Dir.: *Wayne Francis* •
Fine Arts Museum
Contemporary prints ans sculpture, Japanese and American illustrations 50498

Marquette County History Museum, 213 N Front St, Marquette, MI 49855 • T: +1 906 2263571 • fporter@up.net • www.marquettecohistory.org •
Exec. Dir.: *Frances J. Porter* • Cur.: *Kaye Hiebel* •
Local Museum – 1918
Local hist, archaeology, geology, mining, ethnology, shipping and logging, technologies, decorative arts, folklore – library, archives 50499

Oasis Gallery, 227 W Washington, Marquette, MI 49855 • T: +1 906 2251377 • bsampson@up.net •
Dir.: *Bill Sampson* •
Public Gallery
Contemporary art 50500

Marshall IL

Clark County Museum, 4th and Maple Sts, Marshall, IL 62441 • T: +1 217 8266098 • dalannemiller@hotmail.com •
Pres.: *Ken Jensen* •
Local Museum – 1969
China, furniture, photos, documents, primitive kitchen equipment, quilts, hand tools, copies of local paper 1870-1986 50501

Marshall MI

American Museum of Magic, 107 E Michigan St, Marshall, MI 49068 • T: +1 269 7817674, 7817666 •
Owner: *Elaine H. Lund* •
Performing Arts Museum – 1978
Posters, coins & tokens, magican apparatus, toys, games, figures, costumes, memorabilia – library 50502

Honolulu House Museum, 107 N Kalamazoo, Marshall, MI 49068 • T: +1 269 7818544 • F: +1 269 7890371 • dircherie@cablespeed.com • www.marshallhistoricalsociety.org •
C.E.O.: *Mary Jo Byrne* •
Local Museum – 1962
19th c Midwest hist, folklore, decorative arts, military hist, early settlement – library 50503

Marshall MN

Anthropology Museum, Southwest State University, 1501 State St, Marshall, MN 56258 • fortier@southwest.msus.edu • www.southwestmsu.edu/program/list6.cfm •
Dir.: *Jana Fortier* •
University Museum / Ethnology Museum 50504

Lyon County Historical Society Museum, 114 N Third St, Marshall, MN 56258 • T: +1 507 5376580 • F: +1 507 5377699 •
Pres.: *Paul Henricksen* •
Local Museum – 1934
Regional hist, pioneer & Indian artifacts, natural hist, geology – library 50505

Museum of Natural History, Southwest State University, 1501 State St, Marshall, MN 56258 • T: +1 507 5376178 • F: +1 507 5377154 • desy@ssu.southwest.msus.edu • www.southwestmsu.edu/program/list6.cfm •
Dir.: *Dr. Elizabeth A. Desy* •
University Museum / Natural History Museum – 1972
SW Minnesota flora & fauna – planetarium 50506

William Whipple Art Gallery, Soutwest Minnesota State University, 1501 State St, Marshall, MN 56258 • T: +1 507 5377191 • F: +1 507 5376577 • edevans@southwest.msus.edu • www.southwestmsu.edu/wwag •
Dir.: *Edward Evans* •
Fine Arts Museum / University Museum – 1972
Contemporary artworks by alumni, regional and international artists 50507

Marshall TX

Franks Antique Doll Museum, 211 W Grand Av, Marshall, TX 75670 • T: +1 903 9353065, 9353070 •
Head: *Clara Franks* •
Decorative Arts Museum – 1960
1700s to the present dolls 50508

Harrison County Historical Museum, 707 N Washington, Marshall, TX 75670 • T: +1 903 9382680 • F: +1 903 9272534 • museum@marshalltx.com • www.cets.sfasu.edu/Harrison/ •
C.E.O.: *Carrol Fletcher* •
Historical Museum – 1965
Portaits, paintings, porcelains, jewelry, silverware, manuscraipts 50509

Michelson Museum of Art, 216 N Bolivar, Marshall, TX 75671 • T: +1 903 9359480 • F: +1 903 9351974 • leomich@shreve.net • www.michelsonmuseum.org •
Dir.: *Susan Spears* •
Fine Arts Museum – 1985
1887-1978 Russian-American works of Leo Michelson, Kronenberg Collection of Early 20th Century American Artists, Ward Collection of African Masks and Artifacts. 50510

Marshalltown IA

Central Iowa Art Museum, Fisher Center, 709 S Center St, Marshalltown, IA 50158 • T: +1 641 7539013 • F: +1 641 7539013 • ciaa@marshallnet.com •
Dir.: *Valerie Busse* •
Fine Arts Museum / Decorative Arts Museum – 1946
Art, ceramics 50511

Marshall County Museum, 202 E Church St, Marshalltown, IA 50158 • T: +1 515 7526664 • marshallhistory@adiis.net • www.marshallhistory.org/museum.html •
Pres.: *Jeffery Quam* • Dir.: *Gary L. Cameron* •
Local Museum – 1908
Local artifacts, costumes, natural hist, anthropology and geological artifacts 50512

Marshfield WI

New Visions Gallery, 1000 N Oak Av, Marshfield, WI 54449 • T: +1 715 3875562 • newvisions.gallery@verizon.net • www.newvisionsgallery.org •
Dir.: *Ann Waisbrot* •
Fine Arts Museum / Decorative Arts Museum – 1975
Japanese prints, West African masks and sculpture, Haitian paintings, Gothic prints 50513

Upham Mansion, 212 W Third, Marshfield, WI 54449 • T: +1 715 3873322 • F: +1 715 3873322 • upham@tznet.com • www.uphamansion.com •
Pres.: *Kelly Franklin* • Cur.: *Connie Buhrmann* •
Local Museum – 1952
Pioneer artifacts and furnishings, clothing, household goods, 50514

Marthaville LA

Louisiana Country Music Museum, 1260 State Hwy 1221, Marthaville, LA 71450 • T: +1 318 6773600 • F: +1 318 4729315 • rebel@crt.state.la.us • www.lastateparks.com •
Music Museum – 1981
Musical instrument-shaped bldg 50515

Martin TN

University Museum, Paul Meek Library, The University of Tennessee at Martin, Martin, TN 38238 • T: +1 731 8817094 • F: +1 731 8817074 • museum@utm.edu •
Dir.: *Richard Saunders* •
University Museum / Local Museum – 1981
Rotating and traveling exhi 50516

Martinez CA

Martinez Museum, 1005 Escobar St, Martinez, CA 94553 • T: +1 510 2288160 • webmaster@martinezhistory.org • www.martinezhistory.org •
Dir.: *Justine Sellick* •
Local Museum – 1974
Local hist 50517

Martinsburg WV

Boarman Arts Center, 208 S Queen St, Martinsburg, WV 25401 • T: +1 304 2630224 •
Fine Arts Museum – 1987
Arts and crafts incl photography, sculpture, oils, acrylics, watercolors by local artisans 50518

General Adam Stephen House, 309 East John St, Martinsburg, WV 25402 • T: +1 304 2674434 •
Pres.: *Martin Keesecker* • Cur.: *Keith E. Hammersla* •
Local Museum / Folklore Museum / Decorative Arts Museum – 1959
Historic house, period furniture 50519

Martinsville VA

Piedmont Arts Museum, Piedmont Arts Association, 215 Starling Av, Martinsville, VA 24112 • T: +1 540 6323221 • F: +1 540 6383963 • PAA@PiedmontArts.org • www.piedmontarts.org •
Dir.: *Toy L. Cobbe* •
Fine Arts Museum – 1961
Rotation exhib of national & regional artists and craftsmen 50520

Virginia Museum of Natural History, 1001 Douglas Av, Martinsville, VA 24112 • T: +1 276 6668600 • F: +1 276 6326487 • tgette@vmnh.org •
Exec Dir.: *Timothy J. Gette* • Cur.: *Dr. Richard S. Hoffman* (Recent Invertebrates) • *Dr. Lauck W. Ward* (Invertebrate Paleontology) • Asst. Cur.: *Dr. James S. Beard* (Earth Sciences) • *Dr. Nicholas C. Fraser* (Vertebrate Paleontology) • *Dr. Nancy D. Moncrief* (Mammalogy) •
Natural History Museum – 1984
Mesozoic & Cenozoic vertebrate & plant fossils, Virginia archaeology, insects, bryozoans 50521

Marylhurst OR

The Art Gym, Marylhurst University, 17600 Pacific Hwy 43, Marylhurst, OR 97036-0261 • T: +1 503 6996243 • F: +1 503 6369526 • artgym@marylhurst.edu • www.marylhurst.edu/ •
Dir./Cur.: *Terri M. Hopkins* •
Fine Arts Museum / University Museum – 1980
Contemporary northwest art 50522

Marysville OH

Union County Historical Society Museum, 246-254 W Sixth St, Marysville, OH 43040 • T: +1 937 6440568 •
Pres.: *Robert W. Parrott* •
Local Museum – 1949
Farm tools, furniture, china, glass, silver, clothing, Indian arrowheads, chisels and hammers 50523

Maryville MO

Olive DeLuce Art Gallery, Northwest Missouri State University, 800 University Dr, Maryville, MO 64468, mail addr: 219 Administration Building, 800 University Dr, Maryville, MO 64468 • T: +1 660 5621326 • F: +1 660 5621346 • abrown@nwmissouri.edu • www.nwmissouri.edu •
Cur.: *Philip Laber* •
Fine Arts Museum
Contemporary 50524

Maryville TN

Fine Arts Center Gallery, Maryville College, 502 E Lamar Alexander Pkwy, Maryville, TN 37804 • T: +1 865 9818000, 9818150 • F: +1 865 2738873 • karen.eldridge@maryvillecollege.edu • www.maryvillecollege.edu •
Chm.: *Mark Hall* •
Public Gallery 50525

Sam Houston Historical Schoolhouse, 3650 Old Sam Houston School Rd, Maryville, TN 37804 • T: +1 865 9831550 • samhoustonschool@aol.com • www.geocities.com/samhoustonschoolhouse •
Pres.: *Enixx Simerly* •
Local Museum – 1965
Early schol an dpioneer artifacts 50526

Mashantucket CT

Mashantucket Pequot Museum, 110 Pequot Trail, Mashantucket, CT 06338-3180 • F: +1 860 3967013 • dholahan@mptn.org • www.mashantucket.com •
Exec. Dir.: *Theresa H. Bell* •
Historical Museum – 1998
Personal artifacts, tribial hist, area and land hist, archaeology, ethnobotany 50527

Mason TX

Mason County Museum, 321 Moody St, Mason, TX 76856 • T: +1 915 3476137 •
Local Museum – 1965
Housed in 18'76 school bldg 50528

Mason City IA

Charles H. MacNider Museum, 303 2nd St, SE, Mason City, IA 50401-3666 • T: +1 641 4213666 • F: +1 641 4229612 • macnider@macniderart.org • www.macniderart.org •
Dir.: *Sheila Perry* •
Fine Arts Museum – 1964
American art incl. Bil Baird puppets 50529

Mason Neck VA

Gunston Hall Plantation, 10709 Gunston Rd, Mason Neck, VA 22079 • T: +1 703 5509220 • F: +1 703 5509480 • Historic@GunstonHall.org • www.gunstonhall.org •
Dir.: *Thomas A. Lainhoff* • Dep. Dir.: *Susan A. Borchardt* •
Local Museum – 1932
Furniture & fine arts of the 18th c, rare books, decorative arts, – archives 50530

Massillon OH

Massillon Museum, 121 Lincoln Way E, Massillon, OH 44646 • T: +1 330 8334061 • F: +1 330 8332925 • www.massillonmuseum.org •
Dir. & C.E.O.: *Christine Fowler Shearer* • Cur.: *Alexandra Nicholis* •
Fine Arts Museum / Historical Museum – 1933
American folk art, china, glass, pottery, metal, American and European fine and decorative arts, Indian artifacts 50531

Matawan NJ

Burrowes Mansion Museum, 94 Main St, Matawan, NJ 07747 • T: +1 732 5663817, 5665605 •
Local Museum – 1976
18th - 19th c furniture, postcards, costumes, photographs 50532

Thomas Warne Historical Museum, Madison Twp. Historical Society, 4216 Rte 516, Matawan, NJ 08816 • T: +1 732 5660348, 5662108 • F: +1 732 5666943 •
C.E.O. & Pres.: *Ann Miller* • Cur.: *Alvia D. Martin* •
Historical Museum – 1964 50533

Mathias WV

Lost River State Park Museum, 321 Park Dr, Mathias, WV 26812 • T: +1 304 8975372 • F: +1 304 8975325 • lrsp@hardynet.com • www.lostriversp.com •
Local Museum – 1968
Old documents, clothing, tools 50534

Mattapoisett MA

Mattapoisett Museum and Carriage House, 5 Church St, Mattapoisett, MA 02739 • T: +1 508 7582844 • mattapoisett.museum@verizon.net • www.mattapoisetthistoricalsociety.org •
Cur.: *Bette Roberts* •
Local Museum – 1959
Local history, Christian meeting house, decorative arts, farming, maritime and whaling items 50535

Maumee OH

Wolcott House Museum Complex, 1031 River Rd, Maumee, OH 43537 • T: +1 419 8939602 • F: +1 419 8933108 • mvhs@accesstoledo.com •
Pres.: *Suzanne Moesser* • Cur.: *Charles Jacobs* •
Historical Museum – 1961
Maumee Valley hist, house furnishings, clothing 50536

Mauston WI

The Boorman House, 211 N Union St, Mauston, WI 53948 • T: +1 608 4625931 • rclarkjco@hotmail.com •
Pres.: *Nancy McCullick* •
Local Museum – 1963
Local hist, genealogy 50537

Maxwell IA

Community Historical Museum, Main St, Maxwell, IA 50161 • T: +1 515 3852376 • jlengeli@prodigy.net •
Pres.: *Robert Swanson* • Cur.: *Jesse Parr* • Historian: *Mildred McIntosh* •
Local Museum – 1964
Local history 50538

Maysville KY

Museum Center, 215 Sutton St, Maysville, KY 41056 • T: +1 606 5645865 • F: +1 606 5644372 • museum@masoncountymuseum.org • www.masoncountymuseum.org •
Dir.: *Dawn C. Browning* • Cur.: *Sue Ellen Grannis* (Books, Art and Artifacts) •
Local Museum – 1876
Art, genealogy, local history 50539

Mayville MI

Mayville Area Museum of History and Genealogy, 2124 Ohmer Rd, Mayville, MI 48744 • T: +1 989 8437185, 8436389 •
Pres./Dir./Cur.: *Frank E. Franzel* •
Local Museum – 1972
Local hist, familiy records 50540

Mayville ND

Northern Lights Art Gallery, c/o Mayville State University, 330 NE Third St, Mayville, ND 58257 • T: +1 701 7862301 ext 4742 • F: +1 701 7864748 •
Dir.: *Cynthia Kaldor* •
University Museum / Fine Arts Museum – 1999 50541

Mayville WI

Mayville Historical Society Museum, 11 N German St, Mayville, WI 53050 • T: +1 920 3872420 • F: +1 920 3875944 • www.mayvillecity.com •
Pres.: *Ann Guse* •
Local Museum – 1968
Local hist, wagons, agricultural items, glassware, clothing 50542

Mazomanie WI

Mazomanie Historical Society Museum, 118 Brodhead St, Mazomanie, WI 53560 • T: +1 608 7952992, 7954533 • F: +1 608 7954576 • robem@charter.net •
Pres.: *Joan S. Moc* • Cur.: *Rita Frakes* •
Historical Museum – 1965
Local hist, tool coll 50543

Meade KS

Meade County Historical Society Museum, 200 E Carthage, Meade, KS 67864 • T: +1 620 8732359 • F: +1 620 8732359 • meademuseums@swko.net •
C.E.O.: *Norman Dye* •
Local Museum – 1969
American Indian artifacts 50544

Meadville PA

Baldwin-Reynolds House Museum, 639 Terrace St, Meadville, PA 16335 • T: +1 814 7246080 • F: +1 814 3338173 • cchs@ccfls.org • www.visitcrawford.org •
Exec. Dir.: *Laura W. Polo* •
Historical Museum – 1883
Historic house 50545

Bowman, Megahan and Penelec Galleries, Allegheny College, N Main St, Meadville, PA 16335 • T: +1 814 3324365 • F: +1 814 7246834 • robert.raczka@allegheny.edu • www.allegheny.edu •
Dir.: *Robert Raczka* •
Fine Arts Museum / University Museum – 1970
Contemporary American art 50546

Medford MA

Royall House, 15 George St, Medford, MA 02155 • T: +1 781 3969032 • F: +1 781 3957766 •
Exec. Dir.: *Fred J. Schlicher* •
Historical Museum – 1908
Colonial history, Georgian architecture, American slavery – library 50547

Tufts University Art Gallery, Aidekman Arts Center, 40 Talbot Av, Medford, MA 02155 • T: +1 617 6273518 • F: +1 617 6273121 • galleryinfo@tufts.edu • www.tufts.edu/as/gallery •
Dir.: *Amy Ingrid Schlegel* •
Fine Arts Museum / University Museum – 1955
Modern and contemporary art, Greek and Roman antiquities, pre-Columbian art 50548

Medford NJ

Air Victory Museum, 68 Stacy Haines Rd, Medford, NJ 08055 • T: +1 609 2674488 • F: +1 609 7021852 • www.airvictorymuseum.org •
Pres.: *Barbara C. Snyder* • Dir.: *Ray Pysher* •
Science&Tech Museum – 1989
Aeronautics 50549

Medford OK

Grant County Museum, Main and Cherokee Sts, Medford, OK 73759 • T: +1 580 3952822 • F: +1 580 3952343 •
Pres.: *Mariann Smrcka* •
Local Museum / Historical Museum – 1965 50550

Medford OR

Rogue Gallery and Art Center, 40 S Bartlett, Medford, OR 97501 • T: +1 541 7728118 • F: +1 541 7720294 • juday@roguegallery.org • www.roguegallery.org •
Dir.: *Judy Barnes* •
Public Gallery / Fine Arts Museum – 1959
Regional prints, paintings and artworks; Oregon sculpture 50551

Southern Oregon Historical Society Museum, 106 N Central Av, Medford, OR 97501-5926 • T: +1 541 7736536 • F: +1 541 7767994 • director@sohs.org • www.sohs.org •
Dir.: *John Enders* • Cur.: *Steve Wyatt* (Collections) • *Matt Watson* (Exhibits) •
Historical Museum / Open Air Museum – 1946
19th and 20th c photographic equipment, regional business and industries, furnishings, textiles – library, archive 50552

Media PA

Colonial Pennsylvania Plantation, Ridley Creek State Park, Media, PA 19063 • T: +1 610 5661725 • F: +1 610 5664252 • www.colonialplantation.org •
Pres.: *David Stitely* •
Local Museum – 1973
Local hist 50553

Medicine Lodge KS

Carry A. Nation Home Memorial, 209 W Fowler, Medicine Lodge, KS 67104 • T: +1 316 8863553 • F: +1 316 8865978 •
C.E.O.: *Dorothy Reed* •
Historical Museum – 1950
1880-1903 home of Carry A. Nation 50554

Medina NY

Medina Railroad Museum, 530 West Av, Medina, NY 14103 • T: +1 585 7986106 • F: +1 585 7981086 • office@railroadmuseum.net • www.railroadmuseum.net •
Dir.: *Martin C. Phelps* •
Science&Tech Museum – 1997
Railroad and firefighting artifacts, memorabilia, photographs, HO-scale model train layout 50555

Medina OH

America's Ice Cream and Dairy Museum, 1050 W Lafayette Rd, Medina, OH 44256 • T: +1 330 7649269, 7223839 • F: +1 330 7221326 • elmfarm1934@aol.com • www.elmfarm.com •
Pres.: *Sherry S. Abell* • Cur.: *Carl T. Abell* •
Special Museum – 1999
Ice cream, signs, advertising, photos, prints, antique toys, prototype scooper, trucks and bottles, cream separators, butter churns, milk wagon 50556

John Smart House, Medina County Historical Society, 206 N Elmwood St, Medina, OH 44256 • T: +1 330 7221341 • www.medinahistory.org •
Cur.: *Thomas D. Hilberg* •
Local Museum / Historic Site – 1922
Indian and Civil War artifacts, Victorian furnishings 50557

Portholes Into the Past, 4450 Poe Rd, Medina, OH 44256 • T: +1 330 7250402 • F: +1 330 7222439 • mmishne@nobleknights.com •
Pres./Dir.: *Merle H. Mishne* •
Fine Arts Museum / Public Gallery / Historical Museum / Science&Tech Museum – 1984
Vintage cars (Alfa Romeo, Bugatti, Lotus & Maserati a.o.), car art, blue prints and cutaway drawings, paintings, prints, phtotgraphs, advertising art 50558

Medora ND

Chateau de Mores State Historic Site, Medora, ND 58645, mail addr: POB 106, Medora, ND 58645 • T: +1 701 6234355 • F: +1 701 6234921 • drogness@state.nd.us • www.state.nd.us/hist/chateau/chateau.htm •
Dir.: *Merl Paaverud* •
Historical Museum – 1936
Chateau and park of Antoine de Vallombrosa, Marquis de Mores, French nobleman and entrepreneur founding Medora, memorabilia, original furnishings 50559

Meeker CO

White River Museum, 565 Park St, Meeker, CO 81641 • T: +1 970 8789982 •
Cur.: *Ardith Douglass* •
Folklore Museum / Local Museum – 1956
Regional history 50560

Meeteetse WY

Bank Museum, Meeteetse Museums, 1033 Park Av, Meeteetse, WY 82433 • T: +1 307 8682423 • F: +1 307 8682565 • www.meeteetsewy.com •
Local Museum – 1974
Photographs, artifacts – archives 50561

Charles J. Beldon Museum of Western Photography, Meeteetse Museums, 1947 State, Meeteetse, WY 82433 • T: +1 307 8682423 • F: +1 307 8682565 • www.meeteetsewy.com •
Special Museum – 1974
Photographs, Pitchford Ranch collections 50562

Hall Museum, Meeteetse Museums, 942 Mondell Ave, Meeteetse, WY 82433 • T: +1 307 8682423 • F: +1 307 8682565 • www.meeteetsewy.com •
Local Museum – 1974
Local hist, military from WW 1 to Vietnam War 50563

Melbourne FL

Brevard Museum of Art and Science, 1463-1520 Highland Av, Melbourne, FL 32935, mail addr: POB 360835, Melbourne, FL 32936-0835 • T: +1 321 2420737 • F: +1 321 2420798 • artmuseum@mindspring.com • www.artandscience.org •
Dir.: *Dane F. Pollei* • Cur.: *Jackie Borsanyi* •
Fine Arts Museum / Science&Tech Museum – 1978
Works on paper, Asian art, contemporary artists 50564

Melrose LA

Melrose Plantation Home Complex, Melrose General Delivery, Melrose, LA 71452 • T: +1 318 3790055 • F: +1 318 3790055 • melrose@worldnet.la.net •
Cur.: *Gail Vogel* •
Historical Museum / Folklore Museum – 1971
Early Louisiana type plantation, home of Marie Therese Coin-Coin, freed slave who became owner of plantation 50565

Memphis TN

Art Museum of the University of Memphis, 3750 Norriswood Av, CFA Bldg, Memphis, TN 38152-3200 • T: +1 901 6782224 • F: +1 901 6785118 • artmuseum@memphis.edu • www.amum.org •
Dir.: *Dr. Leslie Luebbers* • Asst. Dir.: *Lisa Francisco* •
Fine Arts Museum – 1981
Egyptian coll, antiquities 3.500 BC - 700 AD 50566

C.H. Nash Museum-Chucalissa Archaeological Museum, 1987 Indian Village Dr, Memphis, TN 38109 • T: +1 901 7853160 • F: +1 901 7850519 • dswan@memphis.edu • www.chucalissa.org •
Dir.: *Dr. Daniel Swan* •
University Museum / Ethnology Museum / Archaeology Museum – 1958
Archaeology, Native American Choctaw & Pre-historic Native Mississippian culture 50567

The Children's Museum of Memphis, 2525 Central Av, Memphis, TN 38104 • T: +1 901 4582678 • F: +1 901 4584033 • children@cmom.com • www.cmom.com •
Dir.: *Judy Caldwell* •
Special Museum – 1987 50568

Clough-Hanson Gallery, Rhodes College, Clough Hall, 2000 N Pkwy, Memphis, TN 38112 • T: +1 901 8433442 • F: +1 901 8433727 • dobbinsh@rhodes.edu •
Dir./Cur.: *Marina Pacini* •
Public Gallery – 1970
Local, regional and national art, textiles, woodcarvings 50569

Delta Axis Contemporaray Arts Center, Power House, 45 GE Patterson, Memphis, TN 38103 • T: +1 901 5785545 •
Public Gallery
New and neglected art in all mediums, performance and outsider art 50570

The Dixon Gallery and Gardens, 4339 Park Av, Memphis, TN 38117 • T: +1 901 7615250 • F: +1 901 6820943 • jkamm@dixon.org • www.dixon.org •
Dir.: *James J. Kamm* • Cur.: *Robert W. Torchia* • Asst. Cur.: *Vivian Kung-Haga* •
Fine Arts Museum – 1976
Imressionist art – library 50571

Fire Museum of Memphis, 118 Adams Av, Memphis, TN 38103 • T: +1 901 3205650 • F: +1 901 4528422 • brier@firemuseum.com • www.firemuseum.com •
Exec.Dir: *Brier Smith Turner* •
Special Museum – 1993
Fire fighting 50572

Graceland, 3764 Elvis Presley Blvd, Memphis, TN 38116 • T: +1 901 3323322 • F: +1 901 3443116 • graceland@memphisonline.com • www.elvis.com •
Pres.: *Jack Soden* •
Music Museum – 1982
Mansion occupied by Elvis Presley from 1957 until his death in 1977 50573

Magevney House, 198 Adams, Memphis, TN 38103 • T: +1 901 5264464 • F: +1 901 5268666 • www.memphismuseums.org •
Cur.: *Kate Dixon* •
Historic Site – 1941
Historic house 50574

Mallory-Neely House, 652 Adams Av, Memphis, TN 38105 • T: +1 901 5231484 • F: +1 901 5268666 • www.memphismuseums.org •
Cur.: *Kate Dixon* •
Historic Site – 1973
Historic House 50575

Memphis Belle B17 Flying Fortress, 5118 Park Av, Ste 110, Memphis, TN 38117 • T: +1 901 4128071 • F: +1 901 7674612 • doctorhar@aol.com • www.memphisbelle.com •
Pres.: *John Coy* •
Military Museum – 1964
B-17 aircraft 50576

Memphis Brooks Museum of Art, 1934 Poplar Av, Overton Park, Memphis, TN 38104 • T: +1 901 5446200 • F: +1 901 7254071 • www.brooksmuseum.org •
Dir.: *Kaywin Feldman* • Cur.: *Marina Pacini* •
Fine Arts Museum – 1913
Fine arts, paintings, sculpture, glass, textiles, porcelain – library 50577

Memphis College of Art Gallery, 1930 Poplar Av, Memphis, TN 38104 • T: +1 901 7264085, 2725100 • F: +1 901 2725104 • info@mca.edu • www.mca.edu •
Pres.: *Jeffrey D. Nesin* •
Public Gallery – 1936
Art – library 50578

Memphis Pink Palace Museum, 3050 Central Av, Memphis, TN 38111 • T: +1 901 3206398, 6320 • F: +1 901 3206391 • more_info@memphismuseums.org • www.memphismuseums.org •
Dir.: *Stephen J. Pike* • Pres.: *Fleet Abston* •
Historical Museum / Natural History Museum / Science&Tech Museum – 1928
Natural hist, science, cultural hist 50579

Mississippi River Museum at Mud Island River Park, 125 N Front St, Memphis, TN 38103 • T: +1 901 5767241 • F: +1 901 5767235 • giuntini@memphis.magibox.net • www.mudisland.com •
Cur.: *Trey Giuntini* •
Local Museum / Natural History Museum – 1978
Natural and cultural hist of the Lower Mississippi River 50580

National Civil Rights Museum, 450 Mulberry St, Memphis, TN 38103 • T: +1 901 5219699 • F: +1 901 5219740 • www.civilrightsmuseum.org •
Pres.: *J.R. Hyde* • Dir.: *Beverly Robertson* •
Historical Museum – 1991
Document of civil rights 50581

National Ornamental Metal Museum, 374 Metal Museum Dr, Memphis, TN 38106 • T: +1 901 7746380 • F: +1 901 7746382 • cf@mentalmuseum.org • www.metalmuseum.org •
Dir.: *James A. Wallace* • Asst. Dir.: *Charles Ferryman* •
Decorative Arts Museum – 1976
Historic and contemporary decorative & fine art metalwork, sculptures, jewellery 50582

Peabody Place Museum, 119 S Main St, Memphis, TN 38103 • T: +1 901 5232787 •
Fine Arts Museum – 1998
Asian art 50583

Rhodes College, 2000 N Pkwy, Memphis, TN 38112 • T: +1 901 8433833 • F: +1 901 8433727 • coonin@rhodes.edu • kelser.biology.rhodes.edu •
Pres.: *William Trout* •
Natural History Museum 50584

Mendham NJ

John Ralston Museum, 313 Rte 24, Mendham Rd, Mendham, NJ 07945 • T: +1 973 5436878 • F: +1 973 5431149 • pfr14@aol.com • www.ralstonmuseum.org •
Pres.: *Tracey Kinsell* •
Historical Museum – 1964
Local artifacts, local papers and documents, manuscript coll 50585

Mendocino CA

Kelley House Museum, 45007 Albion St, Mendocino, CA 95460 • T: +1 707 9375791 • F: +1 707 9372156 • staff@mendocinohistory.org • www.mendocinohistory.org •
Dir.: *Nancy Freeze* •
Local Museum – 1973
Local hist 50586

Mendocino Art Center Gallery, 45200 Little Lake St, Mendocino, CA 95460 • T: +1 707 9375818 • register@mendocinoartcenter.org • www.mendocinoartcenter.org •
Dir.: *Peggy Templer* • Cur.: *Judith Goodrich* •
Public Gallery – 1959
Graphics, paintings, sculpture 50587

Mendota MN

Sibley House, 1357 Sibley Memorial Hwy, Mendota, MN 55150 • T: +1 651 4521596 • F: +1 651 4521238 • sibleyhouse@mnhs.org •
Pres.: *Marveen Minish* •
Historical Museum – 1910
Henry Hastings Sibley, dwelling & trading, Minnesota's first state governor's house, Euro-American & Dakota Indian artifacts 50588

Menominee MI

Menominee County Heritage Museum, 904 11th Av, Menominee, MI 49858 • T: +1 906 8639000 • F: +1 906 8634611 •
Dir./Cur.: *William C. King* •
Local Museum – 1967
Regional history, Indian artifacts, geological and military colls, fishing – library 50589

Menomonee Falls WI

Old Falls Village, N 96 W 15791 County Line Rd, Menomonee Falls, WI 53051 • T: +1 262 2505096 • F: +1 262 2505097 •
Dir.: *Debra Zindler* •
Local Museum – 1966
Historic village, primitive and country furniture of 19th c, 18th/19th c dec art 50590

Menomonie WI

The Furlong Gallery, University of Wisconsin-Stout, Micheels Hall, 415 13th Av E, Menomonie, WI 54751 • T: +1 715 2322261 • F: +1 715 2321669 • hunts@uwstout.edu • www.furlonggallery.uwstout.edu •
Chm. & Cur.: *Susan Hunt* • Dir.: *Geoffrey Wheeler* •
Fine Arts Museum / University Museum – 1965
20th c paintings, grawings, sculpture and prints 50591

Mentor OH

James A. Garfield National Historic Site, 8095 Mentor Av, Mentor, OH 44060 • T: +1 440 2558722 • F: +1 440 2558545 • garfield@wrhs.org • www.wrhs.org •
Dir.: *Edith Serkownek* •
Historic Site – 1936
Garfield memorabilia, furniture, photographs, presidential library 50592

Mequon WI

Concordia University-Wisconsin Art Gallery, 12800 N Lake Shore Dr, Mequon, WI 53097 • T: +1 262 2434552 • F: +1 262 2434351 • gstone@bach.cuw.edu • www.cuw.edu •
Dir.: *Jeff Shawhan* •
Fine Arts Museum / University Museum
Russian bronzes, religious art, Chinese porcelain 50593

Crafts Museum, 11458 N Laguna Dr, 21 W, Mequon, WI 53092 • T: +1 262 2421571 •
C.E.O. & Dir.: *Bob Siegel* •
Special Museum – 1972
Arts and crafts, wood-carving - woodenshoes, handcarving, ice-harvesting (storage, iceboxes etc.) 50594

Merced CA

Merced County Courthouse Museum, 21st and N Sts, Merced, CA 95340 • T: +1 209 7232401 • F: +1 209 7238029 • info@mercedmuseum.org • www.mercedmuseum.org •
Dir.: *Sarah Lim* •
Local Museum – 1975
Local history, photos, clothes 50595

Mercer PA

Mercer County Museum, 119 S Pitt St, Mercer, PA 16137 • T: +1 724 6623490 • info@mchspa.org • www.mchspa.org •
Exec. Dir.: *William C. Philson* • Pres. & C.E.O.: *David M. Miller* •
Historical Museum – 1946
Local hist 50596

Mercerville NJ

Johnson Atelier, c/o Technical Institute of Sculpture, 60 Ward Av Extension, Mercerville, NJ 08619 • T: +1 609 8907777 • F: +1 609 8901816 • dwarner@atelier.org • www.atelier.org •
Dir.: *Gyuri Hollosy* •
Fine Arts Museum – 1977 50597

Meriden CT

Meriden Historical Society Museum, 1090 Hanover St, Meriden, CT 06451 • T: +1 203 2375079 •
Cur.: *Allen Weathers* •
Historical Museum / Local Museum / Folklore Museum – 1893 50598

Meridian MS

Jimmie Rodgers Museum, 1725 Highland Park Dr, Meridian, MS 39304 • T: +1 601 4851808 •
C.E.O.: *Betty Lou Jones* •
Music Museum – 1976
Father of country music 50599

Meridian Museum of Art, 25th Av and 7th St, Meridian, MS 39301 • T: +1 601 6931501 • F: +1 601 4853175 • meridianmuseum@aol.com •
Dir.: *Terence Heder* • Pres.: *Bob Bresnahan* •
Fine Arts Museum – 1969
Drawings, American paintings, prints, sculpture, pottery, photography, decorative art 50600

Merion PA

Barnes Foundation, 300 N Latch's Ln, Merion, PA 19066 • T: +1 610 6670290 • F: +1 610 6644026 • www.barnesfoundation.org •
C.E.O.: *Kimberly Camp* •
Fine Arts Museum / Decorative Arts Museum
American decorative arts, African sculpture, horticulture, Greek, Roman and Egyptian antiquities, paintings, metalwork 50601

Merrill WI

Merrill Historical Museum, 804 E Third St, Merrill, WI 54452 • T: +1 715 5365652 • merrillhs@aol.com •
C.E.O.: *William N. Heideman* • Cur.: *Beverly King* •
Historical Museum – 1978
Wisconsin hist since 1600s 50602

Merrillan WI

Thunderbird Museum, Hatfield N 9517 Thunderbird Ln, Merrillan, WI 54754-8033 • T: +1 715 3335841 • F: +1 715 3337214 •
Head: *Robert Flood* • *Ellen Flood* •
Local Museum – 1959
Local hist, Indian Artifacts 50603

Mesa AZ

Arizona Museum for Youth, 35 N Robson, Mesa, AZ 85201 • T: +1 480 6442468 • F: +1 480 6442466 • azmus4youth@cityofmesa.org • www.arizonamuseumforyouth.com •
Dir.: *Barbara Meyerson* • Cur.: *Rebecca Akins* •
Special Museum – 1980
Children fine arts 50604

Arizona Museum of Natural History, 53 N Macdonald St, Mesa, AZ 85201 • T: +1 480 6442230 • azmnh.info@mesaaz.gov • www.mesasouthwestmuseum.com •
Dir.: *Dr. Thomas Wilson* • Cur.: *Dr. Robert McCord* (Natural Sciences, Paleontology) • *Dr. Heide-Marie Johnson* (Paleontology) • *Brian Curtis* (Paleontology (Vertebrates)) • *Douglas Wolfe* (Paleontology)

U

(Invertebrates)) • *Leo Langland* (Geology) • *Dr. Larry Marshall* (Mammology, Paleontology) • *Keith Foster* (History, Social Sciences) • *Jerry Howard* (Anthropology) • *Dr. Susan Shaffer Nahmias* (Ethnology) • *Paula Liken* (Collections) • *Kate Litteral* (Paleo Education) •
Natural History Museum – 1977
Paleontology, archaeology, ethnology, natural and cultural hist – library 50605

Arizona Wing Commemorative Air Force Museum, 2017 N Greenfield Rd, Mesa, AZ 85215 • T: +1 480 9241940 • F: +1 480 9811954 • info@azcaf.org • www.azcaf.org •
Military Museum
Military aircrafts 50606

Mesa Contemporary Arts, Mesa Arts Center, 1 East Main St, Mesa, AZ 85211-1466 • T: +1 480 6442056, 6446501 • F: +1 480 6442901 • mesacontemporaryarts@mesaartscenter.com • www.mesaarts.com •
Cur.: *Patty Haberman* •
Public Gallery – 1981
Contemporary arts and crafts, prints 50607

Mesa Historical Museum, 2345 N Horne St, Mesa, AZ 85203, mail addr: POB 582, Mesa, AZ 85211 • T: +1 480 8357358 • info@mesahistoricalmuseum.org • www.mesaaz.org •
Dir.: *Shirley Huffaker* •
Local Museum – 1987
library 50608

Mesa Verde CO

Mesa Verde National Park Museum, Mesa Verde, CO 81330 • T: +1 970 5294465 • F: +1 970 5294637 • www.nps.gov/meve •
Dir.: *Larry Wiese* • Cur.: *Carolyn Landes* •
Archaeology Museum – 1917
Archaeology from AD 550-1300, ethnography 50609

Mesilla NM

Gadsden Museum, 1875 Boutz Rd, Mesilla, NM 88046 • T: +1 505 5266293 •
Cur.: *Mary Veitch Alexander* •
Historical Museum – 1931
Civil war coll, Indian artifacts, clothing, guns 50610

Mesquite NV

Mesquite Heritage Museum, 15 W Mesquite Blvd, Mesquite, NV 89024 • T: +1 702 3468417 •
Dir.: *Randy McArthur* •
Fine Arts Museum
Contemporary art 50611

Mesquite TX

Florence Ranch Homestead, 1424 Barnes Bridge Rd, Mesquite, TX 75150 • T: +1 972 2166468 • F: +1 972 2168109 • corr@ci.mesquite.tx.us • www.cityofmesquite.com •
Exec. Dir.: *Charlene Orr* •
Agriculture Museum – 1987
Artifacts from 1880-1920, ranch house, homestead 50612

Metamora IL

Metamora Courthouse, 113 E Partridge, Metamora, IL 61548 • T: +1 309 3674470 • jm20936@mtco.com •
Historical Museum – 1845
Courthouse located in the 8th Judicial Circuit that Abraham Lincoln traveled as a circuit lawyer 50613

Metamora IN

Whitewater Canal Historic Site, 19083 Clayborn St, Metamora, IN 47030 • T: +1 765 6476512 • F: +1 765 6472734 • wwcshe@cnz.com •
Cur.: *Jay Dishman* •
Science&Tech Museum – 1845
Milling machinery, transportation 50614

Metlakatla AK

Duncan Cottage Museum, 501 Tait St, Metlakatla, AK 99926 • T: +1 907 8868687, +1 907 8864441 • F: +1 907 8864436 •
Local Museum – 1975
Tsimpshians history and furnishings, ceremonial drum – library 50615

Metropolis IL

Fort Massac, 1308 E Fifth St, Metropolis, IL 62960 • T: +1 618 5249321, 5244712 • F: +1 618 5249321 • SMCREE@dnrmail.state.il.us •
Dir.: *Terry Johnson* •
Historical Museum – 1908
Located on the site of 1756-1814, Military Post 50616

Mexico MO

American Saddlebred Horse Museum, 501 S Muldrow, Mexico, MO 65265 • T: +1 573 5817155 • achs@swbell.net • www.audrain.org •
Dir.: *Kristine Smiley* •
Historical Museum – 1970
Ribobons, bits, bridles, saddles, paintings, George Ford Morris drawings 50617

Audrain Historical Museum, Graceland, 501 S Muldrow, Mexico, MO 65265 • T: +1 573 5813910 • F: +1 573 5817155 • achs@audrain.org • www.audrain.org •
Dir.: *Dana Keller* •
Local Museum – 1952
Furniture, paintings, piano, dolls, toys, antique glass, agriculture, costumes, hist, Indian artifacts 50618

Miami FL

Ancient Spanish Monastery of Saint Bernard de Clairvaux Cloisters, 16711 W Dixie Hwy, Miami, FL 33160 • T: +1 305 9451462 • F: +1 305 9456986 • monastery@earthlink.net • www.spanishmonastery.com •
Dir.: *Dr. Ronald Fox* •
Fine Arts Museum – 1952
Religious art, reconstruction of monastery, built in 1141 in Segovia Spain, with original stones brought to United States by William Randolph Hearst 50619

Art Gallery Kendall, c/o Dade Community College, 11380 NW 27th Av, Miami, FL 33167 • T: +1 305 2372322 •
Public Gallery 50620

The Art Museum at Florida International University, University Park, PC 110, Miami, FL 33199 • T: +1 305 3482890 • F: +1 305 3482762 • artinfo@fiu.edu • www.artmuseumatfiu.org •
Dir.: *Dahlia Morgan* •
Fine Arts Museum / University Museum – 1977
European, North and South American paintings, drawings, prints and sculpture; African, Oriental and pre-Columbian artifacts 50621

Black Heritage Museum, POB 570327, Miami, FL 33257-0327 • T: +1 305 2523535 • F: +1 305 2523535 • blkhermu@bellsouth.net •
Pres.: *Pricilla G. Stephens Kruize* • Cur.: *L.S. Houston* •
Historical Museum / Folklore Museum – 1987
Books, art, artifacts reated to the Black Heritage around the world 50622

Cuban Museum of Arts and Culture, 1300 SW 12 Av, Miami, FL 33129 • T: +1 305 8588006 • F: +1 305 8589639 •
Dir.: *Louis Jeffery Collette* •
Fine Arts Museum / Folklore Museum – 1985 50623

Gold Coast Railroad Museum, 12450 SW 152nd St, Miami, FL 33177 • T: +1 305 2530063 • F: +1 305 2334641 • gcrm@askchuck.com • www.goldcoast-railroad.org •
Dir.: *Michael D. Hall* •
Science&Tech Museum – 1957
Railroads hist 50624

Historical Museum of Southern Florida, 101 W Flagler St, Miami, FL 33130 • T: +1 305 3751492 • F: +1 305 3751609 • hmsf@hmsf.org • www.hmsf.org •
Chief Cur.: *Dr. Steve Stuempfle* • Cur.: *Rebecca A. Smith* •
Historical Museum – 1940
Historical artifacts of South Florida 50625

Holocaust Documentation and Education Center, 13899 Biscayne Blvd, Miami, FL 33181 • T: +1 305 9195690 • F: +1 305 9195691 • info@hdec.org • www.firn.edu/doe/holocaust •
Pres.: *Harry A. Levy* •
Historical Museum – 1979
Holocaust oral histories, artifacts, memorabilia 50626

International Art Center, 70 Miracle Mile, Coral Gables, Miami, FL 33134 • T: +1 305 5671750, 4717383 •
Public Gallery 50627

Kendall Campus Art Gallery, Miami-Dade Community College, 11011 SW 104th St, Miami, FL 33176-3393 • T: +1 305 2372322 • F: +1 305 2372901 • lfontana@kendall.mdcc.edu •
Dir.: *Lilia Fontana* •
Fine Arts Museum / University Museum – 1970
Paintings, prints, sculpture, photography, 16th to 19th c engravings 50628

Latin American Art Museum, 2206 SW 8th St, Miami, FL 33135-4914 • T: +1 305 6441127 • F: +1 305 2616996 • HispMuseum@aol.com • www.latinartmuseum.org •
C.E.O.: *Raul M. Oyuela* •
Fine Arts Museum – 1991
Latin American cultures 50629

Miami Art Central, 5960 Red Rd, Miami, FL 33143 • T: +1 305 4553333 • F: +1 305 4553334 • www.miamiartcentral.com •
Public Gallery
Contemporary exhibitions 50630

Miami Art Museum, 101 W Flagler St, Miami, FL 33130 • T: +1 305 3753000 • F: +1 305 3751725 • griera@miamidade.gov • www.miamiartmuseum.org •
Dir.: *Suzanne Delehanty* • Asst. Dir.: *José Garcia* • Cur.: *Peter Boswell* •
Fine Arts Museum – 1978
20th and 21th c international art 50631

Miami Children's Museum, 980 MacArthur Causeway, Miami, FL 33132 • T: +1 305 3735437 • F: +1 305 3735431 • info@miamichildrensmuseum.org • www.miamichildrensmuseum.org •
Dir.: *Deborah Spiegelman* •
Special Museum – 1984 50632

Miami Museum of Science, 3280 S Miami Av, Miami, FL 33129 • T: +1 305 6464200 • F: +1 305 6464300 • dsierra@miamisci.org • www.miamisci.org •
Pres.: *Gillian Thomas* •
Science&Tech Museum – 1949
Seashells, Entomology collection, Pre-Columbian collection, unique fossile heritage – planetarium 50633

Museum of Contemporary Art, 770 NE 125th St, Miami, FL 33161 • T: +1 305 8936211 • F: +1 305 8911472 • info@mocanomi.org • www.mocanomi.org •
Dir.: *Bonnie Clearwater* •
Fine Arts Museum – 1981
Visual contemporary art 50634

Vizcaya Museum, 3251 S Miami Av, Miami, FL 33129 • T: +1 305 2509133 ext 2221 • F: +1 305 2852004 • tmg@vizcayamuseum.com • www.vizcayamuseum.com •
Dir.: *Joel M. Hoffmann* •
Decorative Arts Museum / Historical Museum – 1952
Italian Renaissance-styled Villa, formal gardens – gardens 50635

Weeks Air Museum, 14710 SW 128 St, Miami, FL 33196 • T: +1 305 2335197 • F: +1 305 2324134 • www.weeksairmuseum.com •
Dir.: *Vincent Tirado* •
Science&Tech Museum – 1987
Aircraft, engines, instruments, propellers 50636

Wirtz Gallery, First National Bank of South Miami, 5750 Sunset Dr, Miami, FL 33143 • T: +1 305 6625414 • F: +1 305 6625413 • gallery@fnbsm.com • www.fnbsm.com/wirtz.htm •
Public Gallery – 1983 50637

Miami MO

Van Meter State Park, Rte 1, Hwy 122, Miami, MO 65344 • T: +1 660 8867537 • F: +1 660 8867512 • van.meter.state.park@dnr.mo.us • www.dnr.state.mo.us/dsp/homedsp.htm •
Archaeology Museum – 1932
Missouri Indian village archaeological site 50638

Miami TX

Roberts County Museum, 120 E Commercial, Miami, TX 79059 • T: +1 806 8683291 • F: +1 806 8683381 • robertscomuseum@amaonline.net • www.robertscountymuseum.org •
Exec. Dir.: *Cecil Gill* • Dir.: *Katie Underwood* •
Local Museum – 1979
East Room, medical toys, school memorabilia, American War artifacts 50639

Miami Beach FL

Art Center of South Florida, 924 Lincoln Rd, Miami Beach, FL 33139 • T: +1 305 6748278 • F: +1 305 6748772 • info@artcentersf.org • www.artcentersf.org •
Dir.: *Dena Bianchino* • *Isabel Block* •
Public Gallery – 1984 50640

Bass Museum of Art, 2121 Park Av, Miami Beach, FL 33139-1729 • T: +1 305 6737530 • F: +1 305 6737062 • info@bassmuseum.org • www.bassmuseum.org •
Dir.: *Diane W. Camber* •
Fine Arts Museum – 1964
Paintings, sculpture, textiles, artifacts, 20th c American graphics, prints, drawings, Latin and Haitian art 50641

Jewish Museum of Florida, 301 Washington Av, Miami Beach, FL 33139-6965 • T: +1 305 6725044 ext 10/18 • F: +1 305 6725933 • mzerivitz@jewishmuseum.com • www.jewishmuseum.com •
Historical Museum – 1995
Jewish hist of Florida 50642

Rubell Family Collection, 95 NW 29th St, Miami Beach, FL 33127 • T: +1 305 5736090 • F: +1 305 5736023 • rubellcollection@mindspring.com • www.rubellfamilycollection.com •
Dir.: *Marc Coetzee* •
Public Gallery
Contemporary art 50643

The Wolfsonian, Florida International University, 1001 Washington Av, Miami Beach, FL 33139 • T: +1 305 5311001 • F: +1 305 5312133 • leffc@fiu.edu • www.wolfsonian.org •
Dir.: *Cathy Leff* • Asst. Dir.: *Marianne Lamonaca* •
Decorative Arts Museum / Historical Museum – 1986
Design, decorative arts, architecture, historic Art Deco District – library 50644

Miami Lakes FL

Jay I. Kislak Foundation, 7900 Miami Lakes Dr W, Miami Lakes, FL 33016 • T: +1 305 3644208 • F: +1 305 8943209 • adunkelman@kislak.com • www.jayikislakfoundation.org •
Dir.: *Arthur Dunkelman* •
Fine Arts Museum – 1988
Historic maps, manuscripts, pre-Columbian art 50645

Michigan City IN

Lubeznik Center for the Arts, 720 Franklin Sq, Michigan City, IN 46360 • T: +1 219 8744900 • F: +1 219 8726829 • jgbartcenter@adsnet.com • www.lubeznikcenter.org •
Dir.: *Lelde Alida Kalmite* •
Fine Arts Museum – 1977
Art, housed in the Old Michigan City Library, an Beaux Arts structure of Indiana Limestone 50646

Old Lighthouse Museum, Heisman Harbor Rd, Washington Park, Michigan City, IN 46360 • T: +1 219 8726133 •
Pres.: *Fred Devries* •
Science&Tech Museum – 1973
Maritime history 50647

Middleborough MA

Middleborough Historical Museum, Jackson St, Middleborough, MA 02346 • T: +1 508 9471969 •
Pres.: *Cynthia McNair* •
Local Museum – 1960
Local history, folklore, G.A.R. and Thuumb colls 50648

Robbins Museum of Archaeology, 17 Jackson St, Middleborough, MA 02346-0700 • T: +1 508 9479005 • F: +1 508 9479005 • mas@bridgew.edu • webhost.bridgew.edu/mas •
Pres.: *Ron Dalton* •
Archaeology Museum – 1939
Native American archaeology, Paleo-Indian to present – library 50649

Middleburg PA

The Snyder County Historical Society Museum, 30 E Market St, Middleburg, PA 17842 • T: +1 510 8376191 • F: +1 510 8374282 • ikneep@snydercounty.org •
C.E.O. & Pres.: *Teresa J. Berger* •
Historical Museum – 1898
Local hist 50650

Middlebury VT

Frog Hollow Vermont State Craft Center, 1 Mill St, Middlebury, VT 05753 • T: +1 802 3883177 • F: +1 802 3885020 • info@froghollow.org • www.froghollow.org •
Exec. Dir.: *William F. Brooks jr.* •
Historical Museum – 1971 50651

The Henry Sheldon Museum of Vermont Histoy, 1 Park St, Middlebury, VT 05753 • T: +1 802 3882117 • F: +1 802 3882112 • info@henrysheldonmuseum.org • www.henrysheldonmuseum.org •
Man.: *Jon Mathewson* •
Fine Arts Museum / Local Museum – 1882
Art, history, furniture, textiles, documents 50652

The Middlebury College Museum of Art, Center for the Arts, Middlebury, VT 05753-6177 • T: +1 802 4435007 • F: +1 802 4432069 • www.middlebury.edu/~museum •
Dir.: *Richard H. Saunders* • Assoc. Dir.: *Emmie Donadio* • Cur.: *Sandra Olivo* (Education) •
Fine Arts Museum / University Museum – 1968
Prints, drawings, paintings, sculpture, photographs 50653

Vermont Folklife Center, Masonic Hall, 3 Court St, Middlebury, VT 05753 • T: +1 802 3884964 • F: +1 802 3881844 • www.vermontfolklifecenter.org •
Dir.: *Jane C. Beck* • Assoc. Dir.: *Meg Ostrum* •
Folklore Museum – 1983
Contemporary folk art 50654

Middlesboro KY

Cumberland Gap National Historical Park, U.S. 25 E, Middlesboro, KY 40965 • T: +1 606 2482817 • F: +1 606 2487276 • www.nps.gov/cuga •
Head: *Mark H. Woods* •
Historical Museum / Archaeology Museum – 1959
History, archaeology 50655

Middleton MA

Lura Watkins Museum, Pleasant St, Middleton, MA 01949 • T: +1 978 7748132 • F: +1 978 7773270 • flint3@comcast.net • www.flintlibrary.org/history.htm •
Dir.: *Henry Tragent* •
Local Museum – 1976
Local history, shoe making, farming, ladies' fashion – library 50656

Middletown CT

Davison Art Center, Wesleyan University, 301 High St, Middletown, CT 06459-0487 • T: +1 860 6852500 • F: +1 860 6852501 • swiles@wesleyan.edu • www.wesleyan.edu/dac •
Fine Arts Museum / University Museum – 1952 50657

Ezra and Cecile Zilkha Gallery, c/o Wesleyan University, Center for the Arts, Middletown, CT 06459-0442 • T: +1 860 6852695 • F: +1 860 6852061 • parnold@wesleyan.edu • www.wesleyan.edu/cfa/zilkha/home.html •
Cur.: *Nina Felshin* •
Fine Arts Museum / University Museum – 1973
Art 50658

Middlesex County Historical Society Museum, 151 Main St, Middletown, CT 06457 • T: +1 860 3460746 • F: +1 860 3460746 • middlesexhistory@wesleyan.edu • www.middlesexhistory.org •
Dir.: *Dione Longley* •
Local Museum – 1901 50659

U

Middletown MD

Middletown Valley Historical Society Museum, 305 W Main St, Middletown, MD 21769 • T: +1 301 3717582 • F: +1 301 3717582 • j.dwighthut@juno.com •
Pres.: *J. Dwight Hutchinson* •
Local Museum – 1976
Local artifacts 50660

Middletown NY

Historical Society of Middletown and the Wallkill Precinct, 25 East Av, Middletown, NY 10940 • T: +1 914 3420941 • enjine@aol.com •
Pres.: *Marvin H. Cohen* •
Historical Museum – 1923
Photographs, Indian artifacts, clothing, china, local genealogy 50661

Middletown OH

Middletown Fine Arts Center, AIM Bldg, 130 N Verity Pkwy, Middletown, OH 45042 • T: +1 513 4242417 • F: +1 513 4241682 • mfac@siscom.net • www.middletownfinearts.com •
Pres.: *Rick Davies* •
Public Gallery – 1963
library 50662

Middletown VA

Belle Grove Plantation, 336 Belle Grove Rd, Middletown, VA 22645 • T: +1 540 8692028 • F: +1 540 8699638 • info@bellegrove.org • www.bellegrove.org •
Dir.: *Elizabeth McClung* •
Local Museum – 1964
Decorative arts, architecture 50663

Middleville MI

Historic Bowens Mills and Pioneer Park, 200 Old Mill Rd, Middleville, MI 49333 • T: +1 616 7957530 • carleen@bowensmills.com • www.bowensmills.com •
Dir.: *Carleen Sabin* • *Owen Sabin* •
Science&Tech Museum / Local Museum – 1978
Water-powered grist and cider mill, farming, Indian artifacts, early settlement, tools, spinning & weaving 50664

Midland MI

Alden B. Dow Museum of Science and Art, Midland Center for the Arts, 1801 W Saint Andrews Dr, Midland, MI 48640 • T: +1 989 6315931 • F: +1 989 6317890 • winslow@mcfta.org • www.mcfta.org •
Fine Arts Museum / Science&Tech Museum 50665

Arts Midland Galleries, 1801 W Saint Andrews Rd, Midland, MI 48640 • T: +1 989 6315931 ext 1401 • F: +1 989 6317890 • winslow@mcfta.org • www.mcfta.org •
Dir.: *Bruce B. Winslow* •
Public Gallery – 1956
Great Lake regional art, local history photographs – art school 50666

Midland County Historical Museum, 3417 W Main St, Midland, MI 48640 • T: +1 989 6315930 ext 1302 • F: +1 989 8359120 • skory@mcfta.org • www.mcfta.org/historical_society •
Dir.: *Gary F. Skory* •
Local Museum – 1952
Regional history, lumbering, decorative arts, chemistry (Dow Chemical), carriages & sleights 50667

Midland NC

Reed Gold Mine, 9621 Reed Mine Rd, Midland, NC 28107 • T: +1 704 7214653 • F: +1 704 7214657 • reedmine1799@msn.com • www.reedmine.com •
Science&Tech Museum – 1971
Mining machinery, coins, steam engines, stamp mill 50668

Midland TX

American Airpower Heritage Museum, 9600 Wright Dr, Midland, TX 79711 • T: +1 432 5631000, 5673009 • F: +1 432 5673047 • director@aahm.org • www.airpowermuseum.org •
Dir.: *Tami O'Bannion* • Cons.: *Jeff Wood* •
Historical Museum / Military Museum / Science&Tech Museum – 1957
World War II, military aviation 50669

J. Evetts Haley History Center, 1805 W Indiana Av, Midland, TX 79701 • T: +1 915 6825785 • F: +1 915 6853512 • haley-mail@att.net • www.haleylibrary.com •
Local Museum – 1976
Local hist 50670

McCormick Gallery, Midland College, 3600 N Garfield, Midland, TX 79705 • T: +1 915 6854770 • F: +1 915 6854721 • mccormickgallery@midland.edu • www.midland.edu/mccormick •
Dir.: *J. Don Wallace* •
Public Gallery 50671

Midland County Historical Museum, 301 W Missouri, Midland, TX 79701 • T: +1 915 6822931, 6888947 •
Pres.: *Nancy R. McKinley* •
Local Museum – 1930
Archaeology, manuscripts, glass, original pictures of Midland – archives 50672

Museum of the Southwest, 1705 W Missouri Av, Midland, TX 79701-6516 • T: +1 915 6832882, 5707770 • F: +1 915 5707077 • tjones@museumsw.org • www.museumsw.org •
Dir.: *Thomas Jones* • Cur.: *Karin Winkler* • *Daniel Holeva* •
Fine Arts Museum / Local Museum – 1965
Southwestern art, archaeological coll of West Texas, contemporary and historic Soutwestern art with emphasis on Texas and New Mexico – Planetarium 50673

The Petroleum Museum, 1500 Interstate 20 W, Midland, TX 79701 • T: +1 432 6834403 • F: +1 432 6834509 • info@petroleummuseum.org • www.petroleummuseum.org •
Exec. Dir.: *Jane Phares* •
Historical Museum / Science&Tech Museum – 1967
Original paintings showing histsorical and oil industry subjects, period oil industry machingery – archive 50674

Taylor Brown and Sarah Dorsey House, 213 N Weatherford, Midland, TX 79701, mail addr: 2102 Community Ln, Midland, TX 79701 • T: +1 915 6822931 •
Decorative Arts Museum – 1899
Furniture & furnishings, historical artifacts 50675

Midway GA

Fort Morris, 2559 Fort Morris Rd, Midway, GA 31320 • T: +1 912 8845999 • F: +1 912 8845285 • ftmorris@clds.net • www.fortmorris.org •
Dir.: *Arthur C. Edgar jr.* •
Military Museum – 1978
Military history 50676

Midway Museum, US-Hwy 17, Midway, GA 31320, mail addr: POB 195, Midway, GA 31320 • T: +1 912 8845837 •
Cur.: *Joann Clark* •
Local Museum – 1957
Local history, furnishing 50677

Seabrook Village, 660 Trade Hill Rd, Midway, GA 31320 • T: +1 912 8847008 • F: +1 912 8847005 • wallstalk@cld.net • www.seabrookvillage.org •
Cur.: *Meredith Devendorf* •
Folklore Museum
Living hist of an African-American village 50678

Mifflinburg PA

Mifflinburg Buggy Museum, 598 Green St, Mifflinburg, PA 17844 • T: +1 570 9661355 • F: +1 570 9669231 • buggymuseum@jlink.net • www.buggymuseum.org •
Pres.: *Ronald A. Cohen* • Dir.: *Bronwen Sanders* •
Science&Tech Museum – 1978
Tools and equipments of the William A. Heiss Coachworks, Heiss family furnishings, docs, photos, buggy industry, transportation 50679

Milan OH

Milan Historical Museum, 10 Edison Dr, Milan, OH 44846 • T: +1 419 4992968 • F: +1 419 4999004 • museum@milanhist.org • www.milanhistory.org •
Dir.: *Debra S. Fisher* •
Local Museum / Decorative Arts Museum – 1930
Hist, folklor, costumes, doll coll, dec arts 50680

Thomas Edison Birthplace Museum, 9 Edison Dr, Milan, OH 44846 • T: +1 419 4992135 • F: +1 419 4993241 • edisonbp@accnorwalk.com • www.tomedison.org •
C.E.O. & Pres.: *Robert K.L. Wheeler* • Cur.: *Laurence J. Russell* •
Historical Museum – 1947
Edisons inventions, memorabilia, family furnishings from 1780-1870 50681

Miles City MT

Custer County Art Center, Water Plant Rd, Miles City, MT 59301 • T: +1 406 2320635 • F: +1 406 2320637 • ccartc@midrivers.com • www.ccac.milescity.org •
Exec. Dir.: *Mark Browning* •
Fine Arts Museum – 1975
Photographic coll, Montana arts 50682

Milford CT

Eells-Stow House, 34 High St, Milford, CT 06460 • T: +1 203 8742664 • F: +1 203 8745789 • mhsoc@hotmail.com •
Local Museum – 1930
Local hist 50683

Milford DE

Milford Historical Society Museum, 501 NW Front St, Milford, DE 19963 • T: +1 302 4224222 • genltorbert@aol.com •
Pres.: *Marvin Shelhouse* •
Local Museum – 1961
Local hist 50684

Milford Museum, 121 S Walnut St, Milford, DE 19963 • T: +1 302 4241080 •
C.E.O: *John Huntzinger* •
Local Museum – 1983
Local hist 50685

Milford MI

Milford Historical Museum, 124 E Commerce St, Milford, MI 48381 • T: +1 248 6857308 • milfordhistory@yahoo.com • www.milfordhistory.org •
Dir.: *Mary Lou Gharrity* •
Local Museum – 1976
Local hist – microfilm coll 50686

Milford NJ

Volendam Windmill Museum, 231 Adamic Hill Rd, Holland Township, Milford, NJ 08848 • T: +1 908 9954365 • ctbrown2@ron.com •
C.E.O.: *Charles T. Brown* •
Historical Museum – 1965
Antique mill machinery 50687

Milford OH

Promont House Museum, 906 Main St, Milford, OH 45150 • T: +1 513 2480324 • promonthouse@juno.com • promonthouse.org •
C.E.O. & Pres.: *Dan Shaw* •
Historical Museum / Local Museum – 1967
Regional hist, Indian items 50688

Milford PA

Pike County Museum, 608 Broad St, Milford, PA 18337 • T: +1 570 2968126 • pchs1@ptd.net •
Pres.: *Edwin Reynolds* •
Local Museum – 1930
Local history 50689

Mill Run PA

Fallingwater - Western Pennsylvania Conservancy, Rte 381 S, Mill Run, PA 15464-0167, mail addr: POB R, Mill Run, PA 15464 • T: +1 724 3298501 • F: +1 724 3290881 • fallingwater@paconserve.org • www.fallingwater.org •
Dir.: *Lynda Waggoner* •
Fine Arts Museum / Decorative Arts Museum – 1963
Historic house designed by Frank Lloyd Wright in 1935, coll of ceramics, decorative arts, furniture, glass, paintings and graphic works by Picasso, Diego Rivera, Japanese prints, sculptures, textiles 50690

Milledgeville GA

Museum and Archives of Georgia Education, Georgia College and State University, 131 S Clarke St, Milledgeville, GA 31061 • T: +1 478 4454391 • F: +1 478 4456847 • lamonica.sanford@gscu.edu • www.library.gcsu.edu •
Dir.: *Lamonica J. Sanford* •
Historical Museum – 1975
History, education 50691

Millersburg OH

Holmes County Historical Society Museum, 484 Wooster Rd, Millersburg, OH 44654 • T: +1 330 6740022 • F: +1 208 4398675 • hchs@valkyrie.net • www.victorianhouse.org •
C.E.O.: *Mark Boley* • Cur.: *Helen Smith* •
Local Museum – 1965
Furniture, early medical equip, war mementoes, early law office, pioneer tools 50692

Millersburg PA

Upper Paxton Township Museum, 330 Center St, Millersburg, PA 17061 • T: +1 717 6924084 • mbghist@epix.net •
Dir.: *Nancy L. Wert* •
Local Museum – 1980
Local history 50693

Millersville TN

Beverage Containers Museum, 1055 Ridgecrest Dr, Millersville, TN 37072 • T: +1 615 8595236 • F: +1 615 8595238 • www.gono.com •
Cur.: *Tom Bates* •
Special Museum – 1987
Beverage containers, openers, advertising signs, posters 50694

Milltown NJ

Eureka Fire Museum, 39 Washington Av, Milltown, NJ 08850 • T: +1 732 8287207 •
Cur.: *Edward S. Harto* •
Historical Museum – 1981
Fighting equip, fire patches, badges and insignia from around the world, uniforms, helmets, hand-drawn fire apparatus 50695

Millville NJ

Museum of American Glass at WheatonArts, 1501 Glasstown Rd, Millville, NJ 08332 • T: +1 856 8256800 ext 142 • F: +1 856 8252410 • gtaylor@wheatonvillage.org • www.wheatonarts.org/museumamericanglass •
Cur.: *Gay LeCleire Taylor* •
Decorative Arts Museum – 1968 50696

Millwood VA

Burwell-Morgan Mill, 15 Tannery Ln, Millwood, VA 22646 • T: +1 540 9552600 • F: +1 540 9550285 • ccha@visuallink.com • www.clarkehistory.org •
Pres.: *Don Cady* •
Science&Tech Museum – 1964
Wooden water wheel and gear train, grinding stones 50697

Milton MA

Blue Hills Trailside Museum, 1904 Canton Av, Milton, MA 02186 • T: +1 617 3330690 ext 0 • F: +1 617 3330814 • bluehills@massaudubon.org • www.massaudubon.org •
Dir.: *Norman Smith* •
Natural History Museum – 1959
Natural history, environment, archaeology – small library 50698

Captain Forbes House Museum, 215 Adams St, Milton, MA 02186 • T: +1 617 6961815 • F: +1 617 6961815 • forbeshousemuseum@verizon.net • www.forbeshousemuseum.org •
Dir.: *Christine M. Sullivan* •
Decorative Arts Museum / Historical Museum – 1964
Family coll, Asian porcelain, furniture and paintings, European fine and decorative arts, civil war 50699

Suffolk Resolves House, 1370 Canton Av, Milton, MA 02186 • T: +1 617 3339700 • mhs1904@aol.com •
Local Museum – 1904
Local history – archives 50700

Milton NH

New Hampshire Farm Museum, 1305 White Mountain Hwy, Milton, NH 03851 • T: +1 603 6527840 • F: +1 603 6527840 • info@farmmuseum.org • www.farmmuseum.org •
Pres.: *Curtis Johnston* •
Agriculture Museum – 1970
Farm tools, machinery, implements, photographs 50701

Milton VT

Milton Historical Museum, 13 School St, Milton, VT 05468 • T: +1 802 8932340 •
Dir.: *Jane Fitzgerald* •
Local Museum – 1979
Wedding gowns of early Milton brides, pictures, tools, mantle clock 50702

Milton WI

Milton House Museum, 18 S Janesville St, Milton, WI 53563 • T: +1 608 8687772 • F: +1 608 8681698 • miltonhouse@miltonhouse.org • www.miltonhouse.org •
Exec.Dir.: *Dr. David McKay* • Asst. Dir.: *Sue Schlueter* •
Historical Museum – 1948
Local hist, pioneer items, Civil War artifacts 50703

Milwaukee WI

America's Black Holocaust Museum, 2233 N Fourth St, Milwaukee, WI 53212 • T: +1 414 2642500 • F: +1 414 2640112 • abhmwi@aol.com • www.blackholocaustmuseum.com •
Dir.: *Jessie Leonard* •
Historical Museum – 1988
Slavery, civil rights 50704

Betty Brinn Children's Museum, 929 E Wisconsin Av, Milwaukee, WI 53202 • T: +1 414 3905437 ext 200 • F: +1 414 2910906 • www.bbcmkids.org •
C.E.O.: *Fern Shupeck* •
Special Museum – 1995
Walt Disney cell art, Warner animation cells 50705

Charles Allis Art Museum, 1630 E Royal Pl, Milwaukee, WI 53202 • T: +1 414 2788295 • F: +1 414 2780335 • eberry@cavtmuseums.org • www.cavtmuseums.org •
Dir.: *James D. Temmer* • Cur.: *Sarah Haberstroh* •
Fine Arts Museum – 1947
Chinese porcelain, painting, 16th-19th cbronzes 50706

Discovery World - The James Lovell Museum of Science, Economics and Technology, 815 N James Lovell St, Milwaukee, WI 53233 • T: +1 414 7659966 • F: +1 414 7650311 • hdq@discoveryworld.org • www.discoveryworld.org •
Dir.: *Paul J. Krajniak* • Pres.: *Michael J. Cudahy* •
Science&Tech Museum – 1984
Inetractive science and technology exhibits 50707

Gallery 218, Walker's Point Artists Association, 207 E Buffalo St, Ste 218, Milwaukee, WI 53204 • T: +1 414 6431732 • 218@gallery218.com • www.gallery218.com •
Public Gallery
Art 50708

Greene Memorial Museum, University of Wisconsin-Milwaukee, 3209 N Maryland Av, Milwaukee, WI 53211 • T: +1 414 2295067, 2294561 • F: +1 414 2295452 • www.geology.uwm.edu •
Cur.: *Rod Watkins* •
Natural History Museum / University Museum – 1913
Geology, paleontology, mineralogy, Devonian and Silurian fossils 50709

U

Institute of Visual Arts, University of Wisconsin-Milwaukee, 3253 N Downer Av, Milwaukee, WI 53211 • T: +1 414 2295070 • F: +1 414 2296785 • inova@uwm.edu • www.uwm.edu/soa/inova •
Cur.: *Marilu Knode* •
Public Gallery / University Museum – 1982
Contemporary art ... 50710

Milwaukee Art Museum, 700 N Art Museum Dr, Milwaukee, WI 53202 • T: +1 414 2243200 • F: +1 414 2717588 • mam@mam.org • www.mam.org •
Dir.: *David Gordon* • Sc. Staff: *Laurie Winters* (Earlier European Art) •
Fine Arts Museum – 1888
American and European art of 19th and 20th c, contemp. art, American decorative art, folk art ... 50711

Milwaukee County Historical Society Museum, 910 N Old World Third St, Milwaukee, WI 53203 • T: +1 414 2738288 • F: +1 414 2733268 • mchs@prodigy.net • www.milwaukeecountyhistsoc.org •
Dir.: *Robert T. Teske* • Cur.: *Steven L. Daily* • *William R. Frick* •
Local Museum – 1935
County hist, brewery materials, early settlers' material ... 50712

Milwaukee Public Museum, 800 W Wells St, Milwaukee, WI 53233 • T: +1 414 2782700 • F: +1 414 2786100 • smedley@mpm.edu • www.mpm.edu •
Pres. & C.E.O.: *Michael D. Stafford* •
Natural History Museum – 1882
Natural and human hist ... 50713

Milwaukee Very Special Art Center, 3211 S Lake Dr, Milwaukee, WI 53235-3702 • T: +1 414 4890539 • hplaney@yahoo.com •
Public Gallery ... 50714

Mitchell Gallery of Flight, c/o Mitchell International Airport, 5300 S Howell Av, Milwaukee, WI 53207-6189 • T: +1 414 7474503 • F: +1 414 7474525 • flymitchell@mitchellgallery.org • www.mitchellgallery.org •
Pres. & Dir.: *Charles Boie* •
Science&Tech Museum – 1984
1911 Curtiss Pusher aircraft, models, aviation engines, aviation art ... 50715

Mount Mary College Costume Museum, 2900 N Menomonee River Pkwy, Milwaukee, WI 53222 • T: +1 414 2560164 • F: +1 414 2560172 • gastone@mtmary.edu • www.mtmary.edu •
Cur.: *Elizabeth Gaston* •
University Museum / Special Museum – 1928
Costumes from 1750s onwoards, works of American and European designers ... 50716

Northwestern Mutual Art Gallery, Cardinal Stritch University, 6801N Yates Rd, Milwaukee, WI 53217 • T: +1 414 4104105 • F: +1 414 4104111 • bjknackert@stritch.edu •
Dir.: *Bruce J. Knackert* •
Fine Arts Museum / University Museum
Collection of African art ... 50717

The Patrick and Beatrice Haggerty Museum of Art, Marquette University, 13th and Clybourn, Milwaukee, WI 53233, mail addr: POB 1881, Milwaukee, WI 53201-1881 • T: +1 414 2887290 • F: +1 414 2885415 • haggertym@vms.csd.mu.edu • www.marquette.edu/haggerty •
Dir.: *Dr. Curtis L. Carter* • Asst. Dir.: *Lee Coppernoll* • Cur.: *Annemarie Sawkins* •
Fine Arts Museum / University Museum – 1984
European and American painting 15th - 20th cc., 20th c international art ... 50718

Union-Art Gallery, University of Wisconsin-Milwaukee, 2200 E Kenwood Blvd, Milwaukee, WI 53201 • T: +1 414 2296310 • F: +1 414 2296709 • art_gallery@aux.uwm.edu • www.aux.uwm.edu/union/artgal.htm •
Head: *Steven D. Jaeger* •
Public Gallery / University Museum – 1972
Changing exhibitions of contemp. art ... 50719

Villa Terrace Decorative Arts Museum, 2220 N Terrace Av, Milwaukee, WI 53202 • T: +1 414 2713656 • F: +1 414 2713986 • eberry@cavtmuseums.org • www.cavtmuseums.org •
Exec. Dir.: *James D. Temmer* • Cur.: *Sarah Haberstroh* •
Decorative Arts Museum – 1967
European decorative art 15th - 19th cc ... 50720

Walker's Point Center for the Arts, 911 W National Av, Milwaukee, WI 53204 • T: +1 414 6722787 • F: +1 414 6725399 • staff@wpca-milwaukee.org • www.wpca-milwaukee.org •
Public Gallery – 1987 ... 50721

Minburn IA

Voas Museum, 1930 Lexington Rd, Minburn, IA 50167 • T: +1 515 4653577 • F: +1 515 4653579 • info@dallas25.org •
CEO: *Mike Wallace* •
Natural History Museum – 1991
Rocks, fossils, minerals, rare native elements, quartz specimens ... 50722

Minden NE

Harold Warp Pioneer Village Foundation, 138 E Hwy 6, Minden, NE 68959-0068 • T: +1 308 8321181 • F: +1 308 8321181 • manager@pioneervillage.org • www.pioneervillage.org •
Pres.: *Harold G. Warp* •
Historical Museum – 1953
Transpotation, cars, farm machinery, musical instruments, toys, local hist art ... 50723

Kearney County Historical Museum, 530 N Nebraska Av, Minden, NE 68959 • T: +1 308 8321765 •
Dir.: *Wayne Bergsten* •
Local Museum – 1925
Local hist ... 50724

Mineral VA

North Anna Nuclear Information Center, Rte 700, 1022 Haley Dr, Mineral, VA 23117 • T: +1 804 7713200 • F: +1 540 8940379 • mike.duffey@dom.com • www.dom.com •
Science&Tech Museum – 1973
Nuclear energy ... 50725

Mineral Point WI

Pendarvis Historic Site, Wisconsin Historical Society, 114 Shake Rag St, PO box 270, Mineral Point, WI 53565 • T: +1 608 9872122 • F: +1 608 9873738 • pendarvis@wisconsinhistory.org • pendarvis.wisconsinhistory.org •
Dir.: *Allen Schroeder* • Cur.: *Tamara Funk* •
Local Museum / Historical Museum / Decorative Arts Museum / Open Air Museum / Historic Site – 1971
Historic houses, 1840-1940s household furnishings, 19th c mining tools ... 50726

Minneapolis MN

American Swedish Institute, 2600 Park Av, Minneapolis, MN 55407 • T: +1 612 8714907 • F: +1 612 8718682 • info@americanswedishinst.org • www.americanswedishinst.org •
Exec. Dir.: *Bruce N. Karstadt* • Cur.: *Curt Pederson* •
Ethnology Museum / Folklore Museum / Fine Arts Museum – 1929
Ethnic museum, Swedish-American & Swedish culture, clothing, antiques, fine arts, folklore – library ... 50727

Art Gallery, North Hennepin Community College, 7411 85th Av N, Brooklyn Park, Minneapolis, MN 55445 • T: +1 612 4240779 • F: +1 612 4240929 • wagar@nhcc.mnscu.edu • www.nhcc.edu/studentlife •
Dir.: *Will Agar* •
Public Gallery / University Museum
Student works and local artists ... 50728

The Bakken, Library and Museum of Electricity in Life, 3537 Zenith Av S, Minneapolis, MN 55416-4623 • T: +1 612 9263878 • F: +1 612 9277265 • info@thebakken.org • www.thebakken.org •
Dir.: *David J. Rhees* • Sc. Staff: *Beth Murphy* (Education) • *Ellen Kuhfeld* (Instruments) • *Paul Manuelas* (Exhibits) •
Science&Tech Museum – 1975
History of electricity in medicine, magnetism, culture – book coll ... 50729

Exhibitions @ Soap Factory, 110 Fifth Av SE at Second St, Minneapolis, MN 55458-1696 • T: +1 612 6239176 • info@soapfactory.org • www.soapfactory.org •
Public Gallery ... 50730

Frederick R. Weisman Art Museum, University of Minnesota, 333 E River Rd, Minneapolis, MN 55455 • T: +1 612 6259494 • F: +1 612 6259630 • benru001@umn.edu • www.weisman.umn.edu •
Dir./Cur.: *Lyndel King* • Cur.: *Diane Mullin* •
Fine Arts Museum / University Museum – 1934
18th-20th c paintings, drawings, prints by American artists, ceramics, Korean furniture ... 50731

Hennepin History Museum, 2303 Third Av S, Minneapolis, MN 55404 • T: +1 612 8701329 • F: +1 612 8701320 • hhmuseum@mtn.org • www.hhmuseum.org •
Exec. Dir.: *Jack A. Kabrud* •
Local Museum / Historical Museum – 1938
Minneapolis & Hennepin county hist ... 50732

Humphrey Forum, 301 19th Av S, Minneapolis, MN 55455 • T: +1 612 6245893 • F: +1 612 6253513 • ssandell@hhh.umn.edu • www.hhh.umn.edu/humphrey-forum •
Dir.: *Stephen Sandell* •
Special Museum – 1989
Hubert Humphrey's career up to US Vice President – library ... 50733

Intermedia Arts Minnesota-Galleries, 2822 Lyndale Av S, Minneapolis, MN 55408 • T: +1 612 8714444 • F: +1 612 8712769 • info@intermediaarts.org • www.intermediaarts.org •
Exec. Dir.: *Daniel Gumnit* • Cur.: *Sandy Agustin* •
Public Gallery – 1973
Multidisciplinary arts ... 50734

James Ford Bell Museum of Natural History, University of Minnesota, 10 Church St SE, Minneapolis, MN 55455 • T: +1 612 6247083, 6244112 • F: +1 612 6267704 • info@bellmuseum.org • www.bellmuseum.org •
Dir.: *Scott M. Lanyon* • Sc. Staff: *Anita Cholewa* (Vascular Plants) • *Bob Zink* (Ornithology) • *David McLaughlin* (Funghi, Lichens) • *Andrew Simons* (Ichthyology) • *Susan J. Weller* (Invertebrate Biology) • *Kendall W. Corbin* (Frozen Tissues) • *Gordon R. Murdock* • *Kevin Williams* (Education) • *Donald T. Luce* (Exhibits) •
Natural History Museum / University Museum – 1872
Regional coll of seeds, fossils, mollusks, birds, fishes, mammals, reptiles, amphibians, frozen tissues, flowering plants, lichens, mosses, nature art – herbarium, labs, auditorium ... 50735

Katherine Nash Gallery, University of Minnesota, Regis Center, 405 21st Av S, Minneapolis, MN 55455 • T: +1 612 6246518 • F: +1 612 6250152 • nash@tc.umn.edu • nash.umn.edu •
Dir.: *Nicholas B. Shank* •
Public Gallery / University Museum – 1973
Ceramics, paintings, prints, sculpture, metalwork, photography ... 50736

Lutheran Brotherhood Gallery, 625 Fourth Av S, Minneapolis, MN 55415 • T: +1 612 3407000 • F: +1 612 3408447 • cic@luthbro.com • www.thrivent.com/heritage/art •
Cur.: *Richard L. Hillstrom* •
Fine Arts Museum / Religious Arts Museum – 1982
Prints & drawings by European & American masters 15th-20th c ... 50737

MCAD Gallery, Minneapolis College of Art and Design, 2501 Stevens Av S, Minneapolis, MN 55404 • T: +1 612 8743700, 8743667 • F: +1 612 8743704 • brianszott@mn.mcad.edu • www.mcad.edu •
Dir.: *Larz Masen* • Cur.: *Kristin Makholm* (Programs) •
Fine Arts Museum / University Museum – 1886
New and innovative work by local, regional and national artists and designers ... 50738

Minneapolis Institute of Arts, 2400 Third Av S, Minneapolis, MN 55404 • T: +1 612 8703131 • F: +1 612 8703004 • miagen@artsmia.org • www.artsmia.org •
Dir.: *Patricia J. Grazzini* • Cur.: *Richard Campbell* (Prints, Drawings) • *Carroll T. Hartwell* (Photography) • *Christopher Monkhouse* (Decorative Arts, Sculpture) • *Robert Jacobsen* (Asian Art) • *Lotus Stack* (Textiles) • *Patrick Noon* (Paintings) •
Fine Arts Museum – 1883
European, American, Native American, Oriental, African, Oceanic, Islamic and ancient arts – library, auditorium, lab ... 50739

Museum of Russian Art, 5500 Stevens Av S, Minneapolis, MN 55419 • T: +1 612 8219045 • bshinkle@tmora.org • www.tmora.org •
Pres. & Dir.: *Bradford Shinkle* • Cur.: *Reed Fellner* •
Fine Arts Museum – 2002
Late 19th c to 1991 Russian art, 20th c Russian Socialist Realist art from the Soviet era, 19th c realist/itinerant school, past-1991 Russian paintings, laquer boxes ... 50740

Northern Clay Center Galleries, 2424 Franklyn Av, Minneapolis, MN 55406 • T: +1 612 3398007 • F: +1 612 3390592 • nccinfo@northernclaycenter.org • www.northernclaycenter.org •
Dir.: *Emily Galusha* •
Fine Arts Museum
Pottery and sculpture by regional and nation ceramic artists ... 50741

Soo Visual Arts Center, 2642 Lyndale Av S, Minneapolis, MN 55408 • T: +1 612 8712263 • suzy@soovac.org • www.soovac.org •
Dir.: *Suzy Greenberg* •
Public Gallery
Relationship with art to personal and cummunity health and vision ... 50742

Walker Art Center, 1750 Hennepin Av, Minneapolis, MN 55403 • T: +1 612 3757622 • F: +1 612 3757618 • info@walkerart.org • www.walkerart.org •
Dir.: *Kathy Halbreich* • Dep. Dir./Cur.: *Richard Flood* • Cur.: *Joan Rothfuss* (Permanent Coll) • *Philippe Vergne* • *Douglas Fogle* • *Siri Engberg* • Asst. Cur.: *Elizabeth Carpenter* •
Fine Arts Museum – 1879/1971
Contemporary paintings, sculpture, drawings, prints & films – sculpture garden ... 50743

Wells Fargo History Museum, Sixth and Marquette Av, Minneapolis, MN 55401, mail addr: 420 Mongomery St, San Francisco, CA 94163 • T: +1 612 6674210 • F: +1 612 3164361 • www.wellsfargohistory.com •
Historical Museum
Banking & express hist, Concord stagecoach, gold rush, gold specimens, coins ... 50744

Minot ND

Lillian and Coleman Taube Museum of Art, 2 Main St N, Minot, ND 58703 • T: +1 701 8384445 • F: 8386471 • taube@srt.com • www.taubemuseum.org •
Dir.: *Nancy F. Walter-Brown* •
Fine Arts Museum – 1970
Drawings, folk art, paintings, sculptures by local and national artists ... 50745

Northwest Art Center, Minot State University, 500 W University Av, Minot, ND 58707 • T: +1 701 8583264 • F: +1 701 8583894 • nac@minotstateu.edu •
Dir.: *Catherine Walker* •
Fine Arts Museum ... 50746

Ward County Historical Society Museum, North Dakota Fairgrounds, Minot, ND 58702 • T: +1 701 8397330 • wchs@minto.com • www.wardcountymuseum.org •
C.E.O.: *Mark Timbrook* •
Local Museum – 1951
Local hist, automobiles, farm implements, hist village houses ... 50747

Mishawaka IN

Hannah Lindahl Children's Museum, 1402 S Main St, Mishawaka, IN 46544 • T: +1 219 2544540 • F: +1 219 2544585 • pmarker@hlcm.org • www.hlcm.org •
Pres.: *Eva Jojo* • Dir.: *Peggy Marker* •
Ethnology Museum / Local Museum – 1946
Indian artifacts, archaeology, military items, tools, natural hist ... 50748

Mission Hills CA

Andres Pico Adobe, San Fernando Valley Historical Society Museum, 10940 Sepulveda Blvd, Mission Hills, CA 91345 • T: +1 818 3657810 • www.sfvhs.com •
Cur.: *Dr. Richard Doyle* •
Local Museum – 1943
Paintings, decorative arts, costumes, history, Indian artifacts ... 50749

Mississippi State MS

Cobb Institute of Archaeology Museum, POB AR, Mississippi State, MS 39762 • T: +1 662 3253826 • F: +1 662 3258690 • jds1@ra.msstate.edu • www.cobb.msstate.edu •
Dir.: *Joe Seger* • Cur.: *John O'Hear* (Artifacts) •
Archaeology Museum – 1972 ... 50750

Dunn-Seiler Museum, Mississippi State, MS 39762, mail addr: c/o MSU, Dept of Geosciences, POB 5448, Mississippi State, MS 39762 • T: +1 601 3253915 • F: +1 601 3252907 • visit@saffairs.msstate.edu • www.msstate.edu •
Cur.: *Dr. Chris Dawey* •
Natural History Museum / University Museum – 1947
Mesozoic and Cenozoic palaeontology, Upper Cretaceous lepadomorph barnacles, mineralogy, geology ... 50751

Missoula MT

Gallery of Visual Arts, University of Montana, Missoula, MT 59812 • T: +1 406 2432813 • F: +1 406 2434968 • www.umt.edu/art/gva.htm •
Fine Arts Museum ... 50752

Historical Museum at Fort Missoula, 322 Fort Missoula, Missoula, MT 59804 • T: +1 406 7283476 • F: +1 406 5436277 • ftmslamuseum@montana.com • www.montana.com/ftmslamuseum •
Dir.: *Robert M. Brown* • Cur.: *L. Jane Richards* • *Tamara King* •
Local Museum – 1975
Furnishings, textiles, clothing, tools, army uniforms, weapons ... 50753

Missoula Art Museum, 335 N Pattee, Missoula, MT 59802 • T: +1 406 7280447 • F: +1 406 5438691 • museum@missoulaartmuseum.org • www.missoulaartmuseum.org •
Dir.: *Laura J. Millin* • Sc. Staff: *Renee Taaffe* (Education) • *Stephen Glueckert* (Exhibitions) •
Fine Arts Museum – 1975
Architecture, ethnology, American Indian, Afo-American, Western art, African art, decorative arts ... 50754

Montana Museum of Art and Culture, University of Montana, Main Hall 006, Missoula, MT 59812 • T: +1 406 2432019 • F: +1 406 2432797 • museum@umontana.edu • www.umt.edu/montanamuseum •
Dir.: *Margaret Mudd* •
Fine Arts Museum
Contemporary ceramic sculpture and paintings ... 50755

Museum of Fine Arts, School of Fine Arts, University of Montana, Missoula, MT 59812 • T: +1 406 2434970 • F: +1 406 2435726 • m.mudd@selway.umt.edu • www.dnr.state.mi.us •
Dir.: *Margaret Mudd* •
Fine Arts Museum / University Museum – 1956 ... 50756

Paxson Gallery → Montana Museum of Art and Culture

Philip L. Wright Zoological Museum, University of Montana, Div. of Biological Sciences, 32 Campus Dr #4824, Missoula, MT 59812 • T: +1 406 2435222 • F: +1 406 2434184 • dave.dyer@mso.umt.edu •
Cur.: *Dr. Richard Hutto* (Ornithology) • *Dr. Kerry Foresman* (Mammalogy) • *David Dyer* (Collections) •
University Museum / Natural History Museum – 1909
Research coll – herbarium ... 50757

Missouri Valley IA

Steamboat Bertrand Museum, DeSoto National Wildlife Refuge, Missouri Valley, IA 51555 • T: +1 712 6424121 • F: +1 712 6425427 • r3bertrand@fws.gov •
Cur.: *Larry Klimek* •
Historical Museum – 1969 ... 50758

Mitchell GA

Hamburg State Park Museum, 6071 Hamburg State Park Rd, Mitchell, GA 30820 • T: +1 912 5522393 • F: +1 912 5531457 • hamburg@accucomm.net • www.gastateparks.org/info/hamburg •
Historical Museum / Science&Tech Museum
Industry, 1920 water turbine powered gin and milling complex ... 50759

U

Mitchell IN

Spring Mill State Park Pioneer Village and Grissom Memorial, Hwy 60 E, Mitchell, IN 47446 • T: +1 812 8494129 • F: +1 812 8494004 • spring@tima.com • Historical Museum / Open Air Museum – 1927
Village life, housed in a Grist Mill, flourishing pioneer village in the 1800s ... 50760

Virgil I. Grissom State Memorial, POB 376, Mitchell, IN 47446 • T: +1 812 8494129 • F: +1 812 8494004 • www.dnr.com •
Dir.: *Gerald J. Pagac* •
Special Museum – 1971
America's exploration of outer space, second American astronaut ... 50761

Mitchell SD

Enchanted World Doll Museum, 615 N Main, Mitchell, SD 57301 • T: +1 605 9969896 • F: +1 605 9960210 • vala@santel.net •
Pres.: *Tom Wudel* •
Decorative Arts Museum – 1977
Dolls ... 50762

Middle Border Museum of American Indian and Pioneer Life, 1300 E University St, Mitchell, SD 57301 • T: +1 605 9962122 • F: +1 605 9960323 • john.day@usd.edu • www.oscarhowe.com •
Dir.: *Lori Holmberg* •
Local Museum / Open Air Museum / Ethnology Museum – 1939
Historic village ... 50763

Oscar Howe Art Center, 119 W Third Av, Mitchell, SD 57301 • T: +1 605 9964111 • F: +1 605 9960323 • john.day@usd.edu • www.oscarhowe.org •
Dir.: *Joanita Kant Monteith* •
Fine Arts Museum / Folklore Museum – 1971
Local and regional artists, paintings by Sioux artists ... 50764

Moab UT

Dan O'Laurie Canyon Country Museum, 118 E Center, Moab, UT 84532 • T: +1 801 2597985 •
Pres.: *Tom Stengel* •
Local Museum – 1958
Local hist, geology, archaeology ... 50765

Mobile AL

Bragg-Mitchell Mansion, 1906 Springhill Av, Mobile, AL 36608 • T: +1 334 4716364 • F: +1 334 4783800 • info@braggmitchellmansion.com • www.braggmitchellmansion.com •
Dir.: *Virginia McKean* •
Museum of Classical Antiquities / Local Museum – 1987
Antiques, historic house ... 50766

Gulf Coast Exploreum, 65 Government St, Mobile, AL 36602 • T: +1 251 2086873 • F: +1 251 2086889 • hsheth@exploreum.com • www.exploreum.net •
Dir.: *Michael Sullivan* •
Science&Tech Museum – 1979
Interactive exhibits – IMAX ... 50767

Mobile Medical Museum, 1664 Springhill Av, Mobile, AL 36616 • T: +1 251 4151109 • F: +1 251 4151110 • med_museum@yahoo.com • www.mobilemedicalmuseum.com •
Pres.: *Byron E. Green jr.* •
Historical Museum – 1962
Early 17th c medical artifacts, 300 yrs practice of medicine – library ... 50768

Mobile Museum of Art, 4850 Museum Dr, Langan Park, Mobile, AL 36608-1917 • T: +1 251 2085200 • mivy@mobilemuseumofart.com • www.mobilemuseumofart.com •
Dir.: *Tommy A. McPherson* • Cur.: *Paul W. Richelson* • *Donan Klooz* (Exhibits) • *Kurtis Thomas* (Colls) •
Fine Arts Museum – 1964
Regional furniture and art, contemporary crafts and glass, ethnic art, European and American paintings and prints – library ... 50769

Museum of Mobile, 111 S Royal St, Mobile, AL 36652-2068, mail addr: POB 2068, Mobile, AL 36602 • T: +1 251 2087569 • F: +1 251 2087686 • www.museumofmobile.com •
Dir.: *George H. Ewert* • Asst. Dir.: *Shelia M. Flanagan* • Cur.: *Todd A. Kreamer* (History) • *Dave Morgan* (Collections) • *Terri Price* (Exhibits) •
Local Museum – 1962
Local history, French, British and Spanish colonial history, Indians, African Americans – library ... 50770

Oakleigh House Museum, 350 Oakleigh Pl, Mobile, AL 36604 • T: +1 251 4326161, 4321281 • hmps@bellsouth.net • www.historicmobile.org •
Dir.: *Marilyn Culpepper* •
Historical Museum – 1935
Local history, civil war, period furnishings ... 50771

USS Alabama Battleship Memorial Park, 2703 Battleship Pkwy, Mobile, AL 36601, mail addr: POB 65, Mobile, AL 36601-0065 • T: +1 251 4332703 • F: +1 251 4332777 • btunnell@ussalabama.com • www.ussalabama.com •
C.E.O.: *Bill Tunnell* •
Military Museum / Open Air Museum – 1963
Military ships, weapons and equipments, military aircrafts, uniforms, graphics ... 50772

Mobridge SD

Klein Museum, 1820 W Grand Crossing, Mobridge, SD 57601 • T: +1 605 8457243 • kleinmuseum@westriv.com •
Dir./Cur.: *Diane Kindt* •
Local Museum – 1975
Local hist, native American beadwork, Sitting Bul pictures ... 50773

Modesto CA

Great Valley Museum of Natural History, 1100 Stoddard Av, Modesto, CA 95350 • T: +1 209 5756196 • F: +1 209 5197050 • louise.crawford@ccc-infonex.edu •
Head: *Louise J. Crawford* •
Natural History Museum – 1973
Mammals, birds, invertebrates, fishes, shells, geology ... 50774

McHenry Museum, 1402 I St, Modesto, CA 95354 • T: +1 209 5775235 • museum@mchenrymusuem.org • www.mchenrymuseum.org •
Cur.: *Laura Mesa* •
Local Museum – 1965
Local history, gold mining, fire display, costumes – library ... 50775

Modjeska Canyon CA

Helena Modjeska Historic House, 29042 Modjeska Canyon Rd, Modjeska Canyon, CA 92676, mail addr: 25151 Serrano Rd, Lake Forest, CA 92630-2534 • T: +1 949 9232230 • ardenmodjeska@ocparks.com • www.ocparks.com/modjeskahouse •
Historical Museum – 1990
Furniture, personal artifacts ... 50776

Mohall ND

Renville County Historical Society Museum, 504 First St NE, Mohall, ND 58761 • T: +1 701 7566195 •
Pres.: *Dorothy Aalund* •
Local Museum / Historical Museum – 1978
Local items ... 50777

Moline IL

Deere Museum, One John Deere Pl, Moline, IL 61265 • T: +1 309 7654881 • F: +1 309 7654088 • www.deere.com •
Dir.: *James H. Collins* •
Science&Tech Museum – 1837
Agriculture, industrial, archaeology ... 50778

Rock Island County Historical Museum, 822-11 Av, Moline, IL 61266-0632 • T: +1 309 7648590 • F: +1 309 7644748 • richs@netexpress.net • www.netexpress.net/~richs/ •
Local Museum – 1905
Folklore, military, agriculture, costumes, Indian artifacts, carriage house ... 50779

Monhegan ME

The Monhegan Museum, 1 Lighthouse Hill, Monhegan, ME 04852 • T: +1 207 5967003 • monheganmuseum@earthlink.net •
Cur.: *Tralice Peck Bracy* •
Fine Arts Museum / Local Museum / Natural History Museum – 1968
Local history ... 50780

Monkton MD

Ladew Manor House, 3535 Jarrettsville Pike, Monkton, MD 21111 • T: +1 410 5579570, 5579466 • F: +1 410 5577763 • kbabcock@ladewgardens.com • www.ladewgardens.com •
C.E.O.: *Emily W. Emerick* •
Historic Site – 1977
Manor House, English furniture, fox hunting memorabilia and equestrian-inspired paintings ... 50781

Monmouth IL

Buchanan Center for the Arts, 64 Public Sq, Monmouth, IL 61462 • T: +1 309 7343033 • F: +1 309 7343554 • bca@misslink.net •
Dir.: *Mike A. Difuccia* •
Fine Arts Museum – 1990
Furnishing, personal artifacts, drawings, paintings, decorative arts, costumes, textiles ... 50782

Wyatt Earp Birthplace, 406 S 3rd St and Wyatt Earp Way, Monmouth, IL 61462-1435 • T: +1 309 7346419 • F: +1 309 7346419 • wyattearpbirthp@webtv.net • www.misslink.net/misslink/earp.htm •
Local Museum – 1986
Local hist ... 50783

Monmouth OR

Campbell Hall Gallery, Western Oregon University, 345 Monmouth Av, Monmouth, OR 97361 • T: +1 503 8388000 • curt@wou.edu • www.wou.edu •
Head: *Don Hoskisson* •
University Museum / Fine Arts Museum
Contemporary Northwestern visual art ... 50784

Jensen Arctic Museum, Western Oregon University, 590 W Church St, Monmouth, OR 97361 • T: +1 503 8388468, 8388281 • F: +1 503 8388289 • www.wou.edu/president/universityadvancement/jensen •
Cur.: *Keni Sturgeon* •
University Museum / Folklore Museum – 1985
Arctic culture ... 50785

Monroe CT

Monroe Historical Society Museum, 433 Barnhill Rd, Monroe, CT 06468-0212 • T: +1 203 2611383 • www.monroehistoricsociety.org •
Cur.: *Michelle Olfra* •
Local Museum / Historical Museum – 1959 ... 50786

Monroe LA

Biedenharn Museum and Gardens, Emy-Lou Biedenharn Foundation, 2006 Riverside Dr, Monroe, LA 71201 • T: +1 318 3875281 • F: +1 318 3878253 • bmuseum@bayou.com • www.bmuseum.org •
Pres.: *Murray Biedenharn* • Dir.: *Ralph Calhoun* •
Religious Arts Museum – 1971
Bible museum, home of Joseph A. Biedenharn, first bottler of Coca-Cola ... 50787

Bry Gallery, c/o University of Louisiana at Monroe, 700 University Av, Bry Hall, Monroe, LA 71209-0310 • T: +1 318 3421375 • F: +1 318 3421369 • aralexander@uln.edu •
Head: *Robert Ward* • *Cliff Tresner* •
University Museum / Fine Arts Museum – 1931 ... 50788

Masur Museum of Art, 1400 S Grand, Monroe, LA 71202 • T: +1 318 3292237 • F: +1 318 3292847 • masur@ci.monroe.la.us • www.masurmuseum.org •
Dir.: *Suzanne M. Prudhomme* •
Fine Arts Museum – 1963
Paintings, prints, sculpture, photographs ... 50789

Northeast Louisiana Delta African American Heritage Museum, 503 Plum St, Monroe, LA 71202 • T: +1 318 3231167 • F: +1 318 3231167 •
Dir.: *Nancy T. Johnson* •
Historical Museum / Folklore Museum – 1994
African art, visual art, furniture ... 50790

Monroe MI

Monroe County Historical Museum, 126 S Monroe St, Monroe, MI 48161 • T: +1 734 2407780 • F: +1 734 2407788 • ralph_naveaux@monroemi.org • www.monroemi.org/museum •
Dir.: *Ralph Naveaux* • Cur.: *James Ryland* •
Local Museum – 1939
Regional hist, Gen. G.A. Custer home, regional Indians, musical instr, costumes, tools, medical/ dental equip – library ... 50791

Monroe NC

Union County Public Library Union Room, 316 E Windsor St, Monroe, NC 28112 • T: +1 704 2838184 • F: +1 704 2820657 • www.union.lib.nc.us •
Dir.: *Dave Eden* •
Public Gallery ... 50792

Monroe NY

Museum Village, 1010 Rte 17 M, Monroe, NY 10950 • T: +1 845 7828248 • F: +1 845 7826432 • musvil@frontiernet.net • www.museumvillage.org •
Exec.Dir.: *Mary Ellen Count* •
Open Air Museum – 1950
25 buildings of 19th c village, life and work in Hudson Valley in 19th c ... 50793

Monroe Township NJ

Stone Museum, 608 Spotswood-Englishtown Rd, Monroe Township, NJ 08831 • T: +1 732 5212232 • F: +1 732 5213388 • displayworld@erols.com • www.thestonemuseum.com •
Dir.: *Gerald Kleiner* •
Natural History Museum
Fluorescent rock, rare seashells, minerals, fossils, rare plants ... 50794

Monroeville AL

Monroe County Heritage Museum, Old Courthouse, Downtown Sq, Monroeville, AL 36461, mail addr: POB 1637, Monroeville, AL 36461 • T: +1 251 5757433 • mchm@frontiernet.net • www.tokillamockingbird.com •
Dir.: *Kathy McCoy* •
Local Museum – 1990
Local hist, 1900-1950s clothing ... 50795

Montague MI

Montague Museum, Church and Meade Sts, Montague, MI 49437 • T: +1 231 8933055 • F: +1 231 8949955 • cityofmontaque@aol.com •
Pres.: *Henry E. Roesler jr.* •
Local Museum – 1964
Local hist, lumbering, farming, Indian artifacts, military, religious items, manikins – library ... 50796

Montauk Ny

Montauk Point Lighthouse Museum, 2000 Montauk Hwy, Montauk, Ny 11954 • T: +1 631 6682544 ext 22 • F: +1 631 6682546 • director@montauklighthouse.com •
Dir.: *Ann Shengold* •
Science&Tech Museum – 1987
Fresnel lens, paintings, photographs, maritime, lanterns ... 50797

Montclair NJ

Montclair Art Museum, 3 S Mountain Av, Montclair, NJ 07042-1747 • T: +1 973 7465555 • F: +1 973 7469118 • mail@montclair-art.com • www.montclair-art.com •
Pres.: *William H. Turner* • Dir.: *Patterson Sims* • Cur.: *Gail Stavitsky* •
Fine Arts Museum – 1914
American Indian art, costumes, silver, paintings, drwaings ... 50798

Montclair Historical Society Museum, 108 Orange Rd, Montclair, NJ 07042 • T: +1 973 7441796 • F: +1 973 7839419 • mail@montclairehistorical.org • www.montclairhistorical.org •
Pres.: *Kathleen Zaracki* • C.E.O.: *Alice Schatteman* •
Local Museum – 1965 ... 50799

Monterey CA

Colton Hall Museum, City Hall, Pacific St, Monterey, CA 93940 • T: +1 831 6465648 • F: +1 831 6463917 • catlin@ci.monterey.ca.us • monterey.org/museum •
Historical Museum – 1949
Constitutional Convention items ... 50800

Monterey History and Art Association, 5 Custom House Plaza, Monterey, CA 93940 • T: +1 831 3722608 • F: +1 831 6553054 • Pam@MontereyHistory.org • montereyhistory.org •
Association with Coll – 1931 ... 50801

Monterey Maritime and History Museum, 5 Custom House Plaza, Monterey, CA 93940 • T: +1 831 3722608 ext 17 • www.montereyhistory.org/museum.htm •
Cur.: *Tim Thomas* •
Historical Museum – 1970
Maritime artifacts, ship models, paintings, photos ... 50802

Monterey Museum of Art, 559 Pacific St, Monterey, CA 93940 • T: +1 831 3725477 • F: +1 831 3725680 • info@montereyart.org • www.montereyart.org •
Dir.: *Richard W. Gadd* • Cur.: *Mary Murray* •
Fine Arts Museum – 1959
California paintings, sculpture, works on paper and photography, Asian art, international folk art ... 50803

Monterey Museum of Art at La Mirada, 720 Via Mirada, Monterey, CA 93940 • T: +1 831 3723689 • info@montereyart.org • www.montereyart.org •
Fine Arts Museum – 1993
Asian coll, contemporary art ... 50804

Old Monterey Jail, Colton Hall Museum, City Hall, Pacific St, Monterey, CA 93940 • T: +1 831 6465640 • catlin@ci.monterey.ca.us • www.monterey.org/museum •
Historical Museum – 1949 ... 50805

Perry-Downer House and Costume Gallery, 201 Van Buren St, Monterey, CA 93940 • T: +1 831 3759182 • www.historicmonterey.org/?p=perry_downer_house •
Dir.: *Shauna Hershfield* •
Local Museum / Decorative Arts Museum – 1997
Local history, paintings, costume, artworks – Carriage house ... 50806

Robert Louis Stevenson House, 530 Houston St, Monterey, CA 93940 • T: +1 831 6497118 • F: +1 831 6476236 • www.mchsmuseum.com/stevensonhouse.html •
Cur.: *Kris N. Quist* •
Historical Museum / Open Air Museum – 1938
Native American coll, anthropology, costumes, fine arts – historic theater ... 50807

Monterey MA

Bidwell House Museum, 100 Art School Rd, Monterey, MA 01245-0537 • T: +1 413 5286888 • F: +1 413 6449997 • bidwellhouse@verizon.net • www.bidwellhousemuseum.org •
C.E.O.: *Brian O'Grady* •
Decorative Arts Museum / Local Museum – 1990
Decorative arts, redware, textiles, furniture, delph, domestic and agricultural tools – library ... 50808

Monterey VA

Highland Maple Museum, U.S. 220 S, Monterey, VA 24465 • T: +1 540 4682420, 4682550 • F: +1 540 4682551 • highcc@cfw.com • www.highlandcounty.org •
C.E.O.: *Carolyn Pohowsky* •
Agriculture Museum – 1983
Old & new tools, exhib demonstrating techniques of maple syrup & sugar production from Indian times to present ... 50809

Monterey Park CA

Vincent Price Art Museum, c/o East Los Angeles College, 1301 Av Cesar Chevez, Monterey Park, CA 91754 • T: +1 323 2658841 • F: +1 323 2658763 • vincentpriceartmuseum@elac.edu • www.elac.edu/collegeservices/vincentprice/gallery •

U

Dir.: *Thomas Silliman* •
Fine Arts Museum – 1958
Africa, Peruvian, Mexican artifacts, North Amercian Indian art 50810

Montevideo MN

Chippewa County Museum, 151 Pioneer Dr, Montevideo, MN 56265 • T: +1 320 2697636 • cchs.june@juno.com • www.montevideomn.com •
Pres.: *Larry Larson* •
Open Air Museum / Local Museum – 1936
Sioux uprising, dugout canoe, farmstead, civil war, Henry Gippe & Frank Stay colls 50811

Montezuma KS

Stauth Memorial Museum, 111 N Azetec St, Montezuma, KS 67867 • T: +1 620 8462527 • F: +1 620 8462810 • stauthm@ucom.net • stauthmemorialmuseum.org •
Dir.: *Kim Legleiter* •
Local Museum / Decorative Arts Museum – 1996
Decorative art, woodcarvings, sculpture, ivory carvings, musical instruments, weapons, hides, jewelry 50812

Montgomery AL

Alabama Department of Archives and History Museum, 624 Washington Av, Montgomery, AL 36130-0100 • T: +1 334 2424435 • www.archives.state.al.us/mv.html •
Dir.: *Edwin C. Bridges* •
Historical Museum – 1901
American Indian art, archaeology, textiles, ceramics, pottery, decorative arts, manuscripts, military art, local history 50813

First White House of the Confederacy, 644 Washington Av, Montgomery, AL 36130 • T: +1 334 2421861 •
Head: *John H. Napier* •
Historic Site – 1900
Historic building, confederate hist 50814

Montgomery Museum of Fine Arts, 1 Museum Dr, Montgomery, AL 36117 • T: +1 334 2404333 • museuminfo@mmfa.org • www.mmfa.org •
Dir.: *Mark M. Johnson* • Dep. Dir: *Andrea Carman* • Cur.: *Margaret Lynne Ausfeld* • *Irja Bonafede* • *Marisa Pascucci* (Collections) • *Tara Sartorius* • Education: *Samantha Kelly* •
Fine Arts Museum – 1930
American paintings and sculptures, Old Master prints, Southern regional art and decorative arts – library 50815

Montgomery NY

Brick House, Rte 17K, Montgomery, NY 12549, mail addr: 211 Rte 416, Montgomery, NY 12549 • T: +1 845 4574921 • F: +1 845 4574906 •
Cur.: *Susan A. Tucker* •
Historical Museum – 1979
Furnishings late 17th c - present, family papers 50816

Hill-Hold Museum, 211 Rte 416, Montgomery, NY 12549 • T: +1 845 4574905 • F: +1 914 4574906 • stucker@co.orange.ny.us •
Cur.: *Susan A. Tucker* •
Historic Site – 1976
Hudson Valley furnishings 17th c to 1830's 50817

Monticello AR

Drew County Historical Museum, 404 S Main St, Monticello, AR 71655 • T: +1 870 3677446 • www.monticelloedc.org •
Pres.: *Connie Mullis* •
Local Museum – 1970
Regional history, Indian artifacts, textiles, furniture 50818

Monticello IL

Monticello Railway Museum, Interstate 72 at Exit 166, Market St, Monticello, IL 61856 • T: +1 217 7629011 • F: +1 217 4223257 • mrm@prairienet.org • www.prairienet.org/mrm/ •
Pres.: *Kent McClure* • Dir.: *Donna McClure* •
Science&Tech Museum – 1966
Railways 50819

Piatt County Museum, 315 W Main, Monticello, IL 61856 • T: +1 217 7624731 • www.piattcountymuseum.com •
C.E.O.: *Blanche Stoller* •
Local Museum – 1965
Fine arts, Civil War relics, World War I an II relics 50820

Monticello IN

White County Historical Society Museum, 101 S Bluff St, Monticello, IN 47960 • T: +1 219 5833998 •
Local Museum – 1911
Local history 50821

Montour Falls NY

Schuyler County Historical Society Museum, 108 N Catharine, Montour Falls, NY 14865 • T: +1 607 5359741 • schuylermuseum@aol.com •
Dir.: *John Potter* •
Local Museum – 1960
Local hist, agriculture, costumes, art, artifacts of the 1900's 50822

Montpelier OH

Williams County Historical Museum, 611 E Main St, Montpelier, OH 43543 • T: +1 419 4858200 •
Cur.: *Jane McCaskey* •
Local Museum – 1956
American prehist and hist Indian artifacts, agricultural items, anthropology, Hopewell Indian Mounds, railroad artifacts 50823

Montpelier VT

T.W. Wood Gallery and Arts Center, c/o Vermont College, 36 College St, Montpelier, VT 05602 • T: +1 802 8288743 • F: +1 802 8288645 • woodartgallery@tui.edu • www.twwoodgallery.org •
Dir.: *Joyce Mandeville* •
Fine Arts Museum / Public Gallery – 1891
438 oils, drawings & watercolors by Thomas Waterman Wood & his contemporaries 50824

Vermont Historical Society Museum, 109 State St, Pavilion Bldg, Montpelier, VT 05609-0901 • T: +1 802 8282291 • F: +1 802 8283638, 4798510 • vhs@vhs.state.vt.us • www.vermonthistory.org •
C.E.O.: *J. Kevin Graffagnino* • Cur.: *Jacqueline Calder* •
Local Museum – 1838
State hist from 1600 to present, American Indian artifacts 50825

The Vermont State House, 115 State St, Montpelier, VT 05633 • T: +1 802 8282228 • F: +1 802 8282424 • www.leg.state.vt.us •
Pres.: *Mary Leahy* • Cur.: *David Schutz* •
Decorative Arts Museum / Local Museum – 1808
Historic building, costumes, textiles 50826

Montpelier Station VA

James Madison's Montpelier, 11407 Constitution Hwy, Montpelier Station, VA 22957 • T: +1 540 6722728 • F: +1 540 6720411 • education@mmtpelier.org • www.montpelier.org •
Local Museum – 1984
Historic house, home of President James Madison 50827

Montreat NC

Presbyterian Historical Society Museum, 318 Georgia Terrace, Montreat, NC 28757 • T: +1 828 6697061 • F: +1 828 6695369 • pcusadoh@montreat.edu • www.history.pcusa.org •
Religious Arts Museum – 1927
Hist of the Presbyterian Church, communion items, US and its missions 50828

Montrose CO

Montrose County Historical Museum, Main and Rio Grande, Montrose, CO 81402, mail addr: POB 1882, Montrose, CO 81402 • T: +1 970 2496135 • stepbackintime@montrose.net •
Cur.: *Marilyn Cox* •
Local Museum – 1974
Regional history 50829

Ute Indian Museum, 17253 Chipeta Dr, Montrose, CO 81401 • T: +1 970 2493098 • F: +1 970 2528741 • cj.brafford@state.co.us •
C.E.O.: *C.J. Brafford* •
Ethnology Museum / Historic Site – 1956
Indian history 50830

Montrose PA

Susquehanna County Historical Society, Two Monument Sq, Montrose, PA 18801 • T: +1 570 2781881 • F: +1 570 2789336 • suspulib@epix.net • www.susqcohistsoc.org •
Cur.: *Elizabeth A. Smith* • Asst. Cur.: *Debra Adleman* •
Local Museum – 1890
Local hist 50831

Montville NJ

Montville Township Historical Museum, Taylor Town Rd, Montville, NJ 07045, mail addr: P.O. Box 519, Montville, NJ 07045 • T: +1 973 3343665 •
Pres.: *Kathleen Fisher* •
Historical Museum – 1963
Agriculture, costumes, decorative arts, history, Indian artifacts 50832

Moorestown NJ

Perkins Center for the Arts, 395 Kings Hwy, Moorestown, NJ 08057 • T: +1 856 2356488 • F: +1 856 2356624 • create@perkinscenter.org • www.perkinscenter.org •
Dir.: *Alan Willoughby* • Asst. Dir.: *Lise Ragbir* •
Fine Arts Museum – 1977
Art 50833

Moorhead MN

Clay County Museum, 202 First Av N, Moorhead, MN 56561-0501 • T: +1 218 2995520 • F: +1 218 2995525 • mark.peihl@ci.moorhead.mn.us • www.info.co.clay.mn.us/history •
Dir.: *Lisa Vedaa* •
Local Museum / Historical Museum – 1932
County hist, glass plate negatives, clothing, guns, pioneer artifacts 50834

Comstock House, 506 Eighth St, Moorhead, MN 56560 • T: +1 218 2914211 • comstock@mnhs.org • www.mnhs.org/places/sites/ch •
Chm.: *Robert J. Loeffler* •
Historical Museum – 1975
Historic business, First National Bank, Red River Valley railroad, Moorhead State University 50835

Heritage Hjemkomst Interpretive Center, 202 First Av N, Moorhead, MN 56560 • T: +1 218 2995511 • F: +1 218 2995510 • charlotte.cox@ci.moorhead.mn.us • www.hjemkomst-center.com •
Exec. Dir.: *Dean Sather* •
Ethnology Museum – 1986
Viking ship, Red River Valley hist, wooden church replica 50836

Mora MN

Kanabec History Center, 805 W Forest Av, Mora, MN 55051 • T: +1 320 6791665 • F: +1 320 6791673 • center@kanabechistory.org • www.kanabechistory.org •
Dir.: *Sharon L. Vogt* •
Historical Museum – 1978
County hist 50837

Moraga CA

Hearst Art Gallery, Saint Mary's College, 1928 Saint Mary's Rd, Moraga, CA 94575-5110 • T: +1 925 6314379 • F: +1 925 3765128 • cbrewste@stmarys-ca.edu • www.stmarys-ca.edu/arts/art_gallery •
Head: *Carrie Brewster* •
Fine Arts Museum / University Museum – 1977
William Keith paintings, European icons, ceramics, sculpture, works by Morris Graves 50838

Moravia NY

Cayuga-Owasco Lakes Historical Society Museum, 14 W Cayuga, Moravia, NY 13118 • T: +1 315 4973096 •
Pres.: *Arlene Murphy* •
Local Museum – 1966
Photos, clothing, household and farm artifacts, family files 50839

Morehead KY

Claypool-Young Art Gallery, Morehead State University, 108 Clypool-Young Art Building, Morehead, KY 40351 • T: +1 606 7835446 • F: +1 606 7835048 • r.franzi@morehead-st.edu • www.morehead-st.edu •
Head: *Jennifer Reis* •
University Museum / Fine Arts Museum – 1922
Prints, lithographs 50840

Kentucky Folk Art Center, 102 W First St, Morehead, KY 40351 • T: +1 606 7832204 • F: +1 606 7835034 • g.barker@morehead-st.edu • www.kyfolkart.org •
Dir.: *Garry Barker* • Cur.: *Adrian Swain* •
Folklore Museum – 1985
Folk art 50841

Morehead City NC

The History Place, 1008 Arendell St, Morehead City, NC 28557 • T: +1 252 2477533 • F: +1 252 2472756 • historyplace@starfishnet.com • www.thehistoryplace.org •
Pres.: *Janet Eshleman* •
Local Museum – 1971
Local genealogy, Carteret County, pictures 50842

Morganton NC

Jailhouse Galleries, Burke Arts Council, 115 E Meeting St, Morganton, NC 28655 • T: +1 828 4337282 • F: +1 828 4337282 • burkearts@hci.net •
Fine Arts Museum / Decorative Arts Museum 50843

Morgantown WV

Comer Museum, MRB, Morgantown, WV 26506-6070, mail addr: Box 6070, Morgantown, WV 26506-6070 • T: +1 304 2934211 • F: +1 304 2936751 • jdean@cemewvu.edu •
Interim Dir.: *Jim Dean* •
Science&Tech Museum – 1990
History, social and technological aspects of coal, oil and gas industries of West Virginia 50844

Laura and Paul Mesaros Galleries, POB 6201, Morgantown, WV 26506 • T: +1 304 2934841 ext 3210 • F: +1 304 2935731 • kolson@wvu.edu • www.wvu.edu •
Cur.: *Kristina Olson* •
Performing Arts Museum – 1867
Costumes, music, paintings, theatre 50845

West Virginia University Mesaros Galleries, Creative Arts Center, Evansdale Campus, West Virginia University, Morgantown, WV 26506-6111 • T: +1 304 2932140 ext 3210 • F: +1 304 2935731 • bob.bridges@wvu.edu • www.wvu.edu/~ccarts •
Dir.: *J. Bernard Schultz* • Cur.: *Kristina Olson* • *Robert Cobridges* (Collection) •
University Museum / Fine Arts Museum – 1968 50846

Moriarty NM

Moriarty Historical Museum, 201 Broadway St, Moriarty, NM 87035 • T: +1 505 8320839 • F: +1 505 8325270 •
Cur.: *Joseph McComb* •
Agriculture Museum – 1981
19th to mid 20th c homestead/ranching area hist 50847

Morrilton AR

Museum of Automobiles, Petit Jean Mountain, 8 Jones Ln, Morrilton, AR 72110 • T: +1 501 7275427 • F: +1 501 7276482 • info@museumofautos.com • www.museumofautos.com •
Dir.: *Buddy Hoelzeman* •
Science&Tech Museum – 1964
Classic cars 50848

Morris MN

Stevens County Historical Society Museum, 116 W Sixth St, Morris, MN 56267 • T: +1 320 5891719 • carnegiemus@hometonsolutions.net •
Dir.: *Randee Hokanson* •
Local Museum – 1920
Local hist, clothing, war items, farm tools, furnishings, manuscript coll 50849

Morristown NJ

Acorn Hall House Museum, 68 Morris Av, Morristown, NJ 07960-4212 • T: +1 973 2673465 • F: +1 973 2678773 • acornhall@juno.com • www.acornhall.org •
C.E.O.: *Bonnie-Lynn Nadzeika* • Cur.: *Debra Westmoreland* •
Historical Museum – 1945
19th c and mid-Victorian period 50850

Fosterfields Living Historical Farm, 73 Kahdena Rd, Morristown, NJ 07960 • T: +1 973 3267644 • F: +1 973 6315023 • mtexel@morrisparks.net • www.morrisparks.net •
Dir.: *Mark Texel* • Cur.: *Rebecca Hoskins* •
Agriculture Museum – 1978
Machinery and tools, furniture and decorative arts, farm records 50851

Historic Speedwell, 333 Speedwell Av, Morristown, NJ 07960 • T: +1 973 5400211 • F: +1 973 5400476 • info@speedwell.org • www.speedwell.org •
Historical Museum – 1966
Genealogy, architecture, decoartive arts, industry, iron industry 50852

Macculloch Hall Historical Museum, 45 Macculloch Av, Morristown, NJ 07960 • T: +1 973 5382404 • F: +1 973 5389428 • maccullochhall@aol.com • www.maccullochhall.org •
Dir.: *David Breslauer* • Cur.: *Ryan Hyman* • *Jane Bedula* •
Historical Museum / Fine Arts Museum – 1950
18th and 19th English and American decorative arts, oriental rugs and china, Thomas Nast coll 50853

The Morris Museum, 6 Normandy Heights Rd, Morristown, NJ 07960 • T: +1 973 9713700 • F: +1 973 5380154 • morrismuseum@worldnet.att.net • www.morrismuseum.org •
Exec. Dir.: *Steve Miller* • Cur.: *Erin Dougherty* • *Jennifer Martin* •
Local Museum – 1913
American Indian, African and Asian art, paintings, sculpture, anthropology, archaeology, costumes, ceramics, crafts, pottery, primitive and decorative art 50854

Morristown National Historical Park, 30 Washington Pl, Morristown, NJ 07960 • T: +1 973 5392016 • F: +1 973 5398361 • Joni_Rowe@nps.gov • nps.gov/morr •
Historical Museum – 1933
18th c military weapons and equipment, decorative arts, paintings, furnishings 50855

Schuyler-Hamilton House, 5 Olyphant Pl, Morristown, NJ 07960 • T: +1 973 2674039 • F: +1 908 8521361 • abren85271@aol.com •
C.E.O.: *Anita Brennan* • Cur.: *Phyllis Sanftner* • *JoAnn Bowman* •
Decorative Arts Museum – 1923
1720 - 1850 period furnishings 50856

Morristown NY

Red Barn Museum, 518 River Rd E, Morristown, NY 13669 • T: +1 315 3756390 •
Pres./Dir.: *Lorraine B. Bogardus* •
Open Air Museum – 1971
Early settlers' memorabilia, farm tools, clothing, military memorabilia 50857

Morristown TN

Crockett Tavern Museum, 2002 E Morningside Dr, Morristown, TN 37814 • T: +1 423 5879900 • crockett@korrnet.org • www.korrnet.org/crockett •
Head: *Sally B. Bennett* •
Local Museum – 1958
Furniture, pioneer utensils 50858

Rose Center Museum, 442 W 2nd N St, Morristown, TN 37816 • T: +1 423 5814330 • F: +1 423 5814307 • rosecent@usit.net • www.rosecenter.org •
C.E.O.: *Zack Allen* •
Folklore Museum / Local Museum – 1976
Civic art and culture 50859

U

Morrisville PA

Pennsbury Manor, 400 Pennsbury Memorial Rd, Morrisville, PA 19067 • T: +1 215 9460400 • F: +1 215 2952936 • willpenn17@aol.com • www.pennsburymanor.org •
Dir.: *Douglas Miller* • Asst. Dir.: *Mary Ellyn Kunz* • Cur.: *Kimberly McCarty* •
Historic Site – 1939
1683 Pennsbury Manor, residence of William Penn, reconstructed in 1939 50860

Morro Bay CA

Morro Bay State Park Museum of Natural History, State Park Rd, Morro Bay, CA 93442 • T: +1 805 7722694 • info@morrobaymuseum.org • www.slostateparks.com •
Cur.: *Nancy Dreher* •
Natural History Museum – 1962
Chumash Indians, birds, mammals, reptiles, insects, plants, rocks and minerals – library 50861

Morton Grove IL

Morton Grove Historical Museum, 6240 Dempster St, Morton Grove, IL 60053 • T: +1 847 9650203 • F: +1 847 9657484 • mwalsh@mortongroveparks.com •
Cur.: *Mary Walsh* •
Local Museum – 1984
Local memorabilia and artifacts 50862

Moscow ID

The Appaloosa Museum and Heritage Center, 2720 W Pullman Rd, Moscow, ID 83843 • T: +1 208 8825578 ext 279 • F: +1 208 8828150 • museum@appaloosa.com • www.appaloosamuseum.org •
Dir.Cur.: *Sherry Caisley* • Pres.: *King Rockhill* •
Historical Museum / Ethnology Museum – 1973
Local history 50863

McConnell Mansion, 110 S Adams, Moscow, ID 83843 • T: +1 208 8821004 • F: +1 208 8820759 • lchsoffice@moscow.com • www.moscow.com/lchs •
Dir.: *Mary Reed* • Cur.: *Marilyn Sandmeyer* •
Historical Museum – 1968
Local history, Governor McConnell – library 50864

Prichard Art Gallery, University of Idaho, 414-416 S Main St, Moscow, ID 83843 • T: +1 208 8853586 • F: +1 208 8853622 •
Dir.: *Gail Siegel* •
Fine Arts Museum / University Museum
Contemporary exhibitions 50865

Moses Lake WA

Adam East Museum Art Center, 122 W Third, Moses Lake, WA 98837 • T: +1 509 7669395 • F: +1 509 7669392 •
Interim Dir.: *Terry Mulkey* •
Fine Arts Museum – 1957
Art 50866

Moses Lake Museum and Art Center, 228 W Third, Moses Lake, WA 98837 • T: +1 509 7669395 • F: +1 509 7669243 • museum@moses-lake.com • www.moses-lake.com •
Dir.: *Terry Mulkey* •
Local Museum / Fine Arts Museum – 1957
Native American cultures 50867

Moundsville WV

Grave Creek Mound Archaeology Complex, 801 Jefferson Av, Moundsville, WV 26041 • T: +1 304 8434128 • F: +1 304 8434131 • grave.creek@wvculture.org • www.wvculture.org •
Archaeology Museum – 1978
Archaeology of West Virginia 50868

Mount Airy NC

Gertrude Smith House, 708 N Main St, Mount Airy, NC 27030 • T: +1 336 7866856 • F: +1 336 7869193 • visitandy@tcia.net • www.visitmayberry.com •
Exec.Dir.: *Ann L. Vaughn* •
Decorative Arts Museum – 1984
Furniture, artwork 50869

Mount Airy Museum of Regional History, 301 N Main St, Mount Airy, NC 27030 • T: +1 336 7864478 • F: +1 336 7861666 • www.northcarolinamuseum.org •
Dir.: *Linda Blue Stanfield* • Cur.: *Amy Snyder* •
Local Museum 50870

Mount Clemens MI

The Art Center, 125 Macomb Pl, Mount Clemens, MI 48043 • T: +1 810 4698666 • www.theartcenter.org •
Exec. Dir.: *Jo-Anne Wilkie* •
Public Gallery – 1969
Michigan artists 50871

Crocker House, 15 Union St, Mount Clemens, MI 48043 • T: +1 586 4652488 •
Man.: *Sharon Retka* •
Local Museum – 1964
County hist, Victorian Italianate style 50872

Michigan Transit Museum, 200 Grand Av, Mount Clemens, MI 48046-5412 • T: +1 586 4631863 • F: +1 586 4639813 • info@michigantransitmuseum.org • www.michigantransitmuseum.org •
Pres./Dir.: *Timothy D. Backhurst* •
Science&Tech Museum – 1973
Transport equipment, Grand Trunk depot railroad 50873

Mount Desert ME

Somesville Museum, 373 Sound Dr, Mount Desert, ME 04660 • T: +1 207 2769323 • F: +1 207 2764024 • mdihistory@acadia.net • www.mdihistory.org •
Dir.: *Jaylene Roths* •
Historical Museum
Period artifacts, furniture, historic books, local hist 50874

Sound School House Museum, 373 Sound Dr, Mount Desert, ME 04660 • T: +1 207 2769323 • F: +1 207 2764024 • mdihistory@acadia.net •
Dir.: *Jaylene Roths* •
Historical Museum 50875

Mount Dora FL

Mount Dora Center for the Arts, 138 E Fifth Av, Mount Dora, FL 32757 • T: +1 352 3830880 • F: +1 352 3837753 • info@mountdoracenterforthearts.org • www.mountdoracenterforthearts.org •
Dir.: *Pat Huizing* •
Folklore Museum – 1985
Folk art 50876

Mount Gilead NC

Town Creek Indian Mound Historic Site, 509 Town Creek Mound Rd, Mount Gilead, NC 27306 • T: +1 910 4396802 • F: +1 910 4396441 • towncreek@carolina.net • www.towncreek.nchistoricsites.org •
Open Air Museum / Historical Museum – 1936
Anthropology, Indian artifacts, Indian mound 50877

Mount Holly NJ

Historic Burlington County Prison Museum, 128 High St, Mount Holly, NJ 08060 • T: +1 609 2655476 • F: +1 609 2655797 • parks@co.burlington.nj.us •
Pres.: *Janet L. Sozio* •
Historical Museum – 1966
Penology, prison items 50878

John Woolman Memorial, 99 Branch St, Mount Holly, NJ 08060 • T: +1 609 2673226 • www.woolmancentral.com •
Dir.: *Jack Walz* • *Carol Walz* •
Special Museum – 1915 50879

Mount Ida AR

Heritage House Museum of Montgomery County, 819 Luzerne St, Mount Ida, AR 71957, mail addr: POB 1362, Mount Ida, AR 71957-1362 • T: +1 870 8674422 • museum@hhmmc.org • hhmmc.org •
Pres.: *Dean Shaw* • Dir.: *Emilie Kinney* •
Local Museum / Science&Tech Museum – 1999
Genealogy of Montgomery County, family hist, mining industry, minerals, crystals 50880

Mount Olivet KY

Blue Licks Battlefield Museum, Blue Licks Battlefield Park, Hwy 68, Mount Olivet, KY 41064 • T: +1 859 2895507 • F: +1 859 2895409 • doug.price@mail.state.ky.us • www.state.ky.us/agencies/parks/bluelick.htm •
Local Museum – 1928
Artifacts from Fort Ancient, mastodon bones, pioneer items 50881

Mount Pleasant IA

Harlan-Lincoln House, 101 W Broad St, Mount Pleasant, IA 52641, mail addr: Iowa Wesleyan College, 601 N Main St, Mount Pleasant, IA 52641 • T: +1 319 3856320 • F: +1 319 3856324 • iwcarch@iwc.edu • www.iwc.edu •
Dir.: *Lynn Ellsworth* •
Historical Museum – 1959
Home of U.S. Senator James Harlan (1876-99) and the secondary residence of the Robert Todd Lincoln Family – Iowa Wesleyan College Archives 50882

Midwest Old Settlers and Threshers Association Museum, 405 E Threshers Rd, Mount Pleasant, IA 52641 • T: +1 319 3858937 • F: +1 319 3850563 • admin@oldthreshers.org • www.oldthreshers.org •
Pres.: *Earl Reynolds* • Cur.: *Dr. Mike Kramme* (Theatre) •
Agriculture Museum / Science&Tech Museum – 1950
Agricultural history, technic 50883

Mount Pleasant MI

Museum of Cultural and Natural History, Central Michigan University, Bellows St, 103 Rowe Hall, Mount Pleasant, MI 48859 • T: +1 989 7743829 • F: +1 989 7742612 • fauverLN@cmich.edu • www.museum.cmich.edu •
Dir.: *Lynn N. Fauver* • Cur.: *Joyce Salisbury* (History) •
University Museum / Local Museum – 1970
Archaeology, anthropology, Indian artifacts, rocks, minerals, paleontology, zoology, Native American art 50884

University Art Gallery, Central Michigan University, Dept. of Art, Cnr Preston and Franklin Sts, Wightman 132, Mount Pleasant, MI 48859 • T: +1 989 7743800 • F: +1 989 7742278 • guhee1em@cmich.edu • www.uag.cmich.edu •
Dir.: *Elizabeth Guheen* •
Public Gallery / University Museum 50885

Mount Pleasant OH

Mount Pleasant Historical Society Museum, 342 Union St, Mount Pleasant, OH 43939 • T: +1 740 7692893 • F: +1 740 7692804 • kaspenwa@1st.net • users.1st.net/gudzent •
Pres.: *Sherry Sawchuk* • Cur.: *Jim Aspenwall* •
Local Museum / Historical Museum – 1948
Books, tools, documents, period clothing 50886

Quaker Meeting House, 298 Market St, Mount Pleasant, OH 43939 • T: +1 740 7692893 • gudzent@1st.net • users.1st.net/gudzent •
Man.: *Mark Twyford* •
Religious Arts Museum – 1814
Religious items 50887

Mount Pleasant SC

Patriots Point Naval and Maritime Museum, 40 Patriots Point Rd, Mount Pleasant, SC 29464 • T: +1 803 8842727 • F: +1 803 8814232 • patriotspt@infoave.net • www.state.sc.us/patpt •
Exec. Dir.: *David Burnette* • Dir.: *Steve Ewing* (Exhibits, Historian) •
Military Museum – 1976
Museum housed in the aircraft carrier USS Yorktown, destroyer USS Laffey, submarine USS Clamagore, Coast Guard cutter Ingham 50888

Mount Prospect IL

Mount Prospect Historical Society Museums, 101 S Maple, Mount Prospect, IL 60056 • T: +1 847 3929006 • F: +1 847 3928995 • mphist@wowway.com • www.mtphistory.org •
Dir.: *Gavin Kleespies* •
Historical Museum – 1976
Decorative arts, costumes, photographs 50889

Mount Pulaski IL

Mount Pulaski Courthouse, 113 S Washington, Mount Pulaski, IL 62548 • T: +1 217 7923919 •
Historical Museum – 1936
Greek Revival County Court House, part of Illinois 8th Judicial Circuit 50890

Mount Vernon IA

Peter Paul Luce Gallery, Cornell College, 600 1st St W, Mount Vernon, IA 52314-1098 • T: +1 319 8954491 • F: +1 319 8955926 • scolemann@cornellcollege.edu • www.cornellcollege.edu •
Pres.: *Les Garner* •
Fine Arts Museum / University Museum – 1853
Sonnenschein coll of baroque drawings, drawings and prints of Thomas Nast, whiting coll of Phoenician glass, paintings by Karel Appel, lithography prints by Roy Lichtenstein, lithographs by Grant Wood – library 50891

Mount Vernon IL

Mitchell Museum at Cedarhurst, Richview Rd, Mount Vernon, IL 62864 • T: +1 618 2421236 • F: +1 618 2429530 • mitchellmuseum@cedahurst.org • www.cedarhurst.org •
Cur.: *Michael J. Beam* • *Heather Owens* (Craft Fair) •
Fine Arts Museum – 1973
Art, sculpture park 50892

Mount Vernon VA

George Washington's Mount Vernon, S End of George Washington Pkwy, Mount Vernon, VA 22121 • T: +1 703 7802000 • F: +1 703 7998654 • mvinfo@mountvernon.org • www.mountvernon.org •
Exec. Dir.: *James C. Rees* •
Local Museum – 1853
Decorative arts, graphic arts, textiles, archaeology, fine arts, Washington memorabilia 50893

Mountain Lake MN

Heritage Village, County Rd 1, Mountain Lake, MN 56159 • T: +1 507 4272709 • conductorron@swwnet.com •
Dir.: *Alvin Dick* •
Historical Museum – 1971
Mennonite people from Russia (Russian-German Lutherans) 50894

Mountain View AR

Ozark Folk Center, 1032 Park Av, Mountain View, AR 72560 • T: +1 870 2693851 • ozarkfolkcenter@arkansas.com • www.ozarkfolkcenter.com •
Dir.: *Bill Young* •
Folklore Museum – 1973
Folk arts and crafts of the Ozark mountain region – library 50895

Mountainair NM

Salinas Pueblo Missions National Monument, 120 Ripley St, Mountainair, NM 87036 • T: +1 505 8472585 • F: +1 505 8472441 • leslie_newkirk@nps.gov • www.nps.gov/sapu •
Head: *Glenn Fulfer* •
Archaeology Museum – 1909
Prehistoric pithouses c. 800 A.D.; prehistoric Indian ruins c. 1100-1670 A.D.; four Spanish mission ruins c. 1622-1672 50896

Mountainside NJ

Trailside Nature and Science Center, 452 New Providence Rd, Mountainside, NJ 07092 • T: +1 908 7893670 • F: +1 908 7893270 • pbertsch@ucnj.org •
Dir.: *Patricia Bertsch* •
Natural History Museum – 1941
Minerals, fossils, Indian artifacts 50897

Mountainville NY

Storm King Art Center, Old Pleasant Hill Rd, Mountainville, NY 10953-0280 • T: +1 845 5343115 • F: +1 845 5344457 • www.stormking.org •
Dir./Cur.: *David R. Collens* •
Fine Arts Museum – 1960
20th c American and international sculpture 50898

Mulberry FL

Mulberry Phospate Museum, 101 SE 1st St, Mulberry, FL 33860 • T: +1 863 4252823 • F: +1 863 4250188 • phaag@gte.net •
Archaeology Museum – 1985
Cenozoic fossils 50899

Mullens WV

Twin Falls Museum, Hwy Rte 97, Twin Falls State Park, Mullens, WV 25882 • T: +1 304 2944000 • F: +1 304 2945000 • www.twinfallsresort.com •
Head: *A. Scott Durham* •
Local Museum – 1976
Local hist, furniture, household items, tools 50900

Mumford NY

Genesee Country Village and Museum, 1410 Flint Hill Rd, Mumford, NY 14511 • T: +1 585 5386822 • F: +1 716 5382887 • onfo@gcv.org • www.gcv.org •
C.E.O. & Pres.: *Elizabeth Harrison* • Cur.: *Dan Barber* (Collections) • Sc. Staff: *Jill Roberts* (Gallery of Wildlife and Sporting Art) • Sen. Dir.: *Connie Bodner* (Programs) •
Fine Arts Museum / Historical Museum / Open Air Museum – 1966
67 villag and farm buildings of the 19th c, furnishings and tools, horse drawn vehicles 50901

Muncie IN

Ball State University Museum of Art, Riversite Av at Warwick Rd, Muncie, IN 47306 • T: +1 765 2855242 • F: +1 765 2854003 • nhuth@bsu.edu • www.bsu.edu/artmuseum •
Dir.: *Peter F. Blume* • Cur.: *Nancy Huth* •
Fine Arts Museum / University Museum – 1936
Italienrenaissance art, 17th - 19th c European paintings, 19th - 20th c American paintings, prints, drawings 50902

Biology Department Teaching Museum, 2000 University, Muncie, IN 47306 • T: +1 765 2891241, 2892641 • F: +1 765 2858804 • jetaylor@bsu.edu •
Chm.: *Dr. Carl E. Warnes* •
University Museum / Natural History Museum – 1918 50903

Muncie Children's Museum, 515 S High St, Muncie, IN 47305 • T: +1 765 2861660 • F: +1 765 2861662 • MUSEUM@munciechildrensmuseum.com • www.munciechildrensmuseum.com •
Dir.: *Lenette Freeman* •
Ethnology Museum – 1977 50904

National Model Aviation Museum, Academie of Model Aeronautics, 5151 E Memorial Dr, Muncie, IN 47302 • T: +1 765 2871256 • F: +1 765 2894248 • michaels@modelaircraft.org • www.modelaircraft.org •
Dir.: *Joyce Hager* • Cur.: *Michael Smith* •
Special Museum – 1936
Aeronautics – Library 50905

Munising MI

Alger County Heritage Center, 1496 Washington St, Munising, MI 49862 • T: +1 906 3874308 • F: +1 906 4643665 • algerchs@up.net •
Pres.: *Mary Jo Cook* •
Local Museum – 1966
Pioneer hist, portraits – library 50906

Grand Island Trader's Cabin, Washington St, Munising, MI 49862 • T: +1 906 3874308 •
Pres.: *Mark Louma* •
Historical Museum – 1972
Abraham Williams, 1st white settler in Alger County 50907

Munnsville NY

Fryer Memorial Museum, Williams St, Munnsville, NY 13409 • T: +1 315 4955395, 4956148 •
Dir.: *Olive S. Boylan* •
Historical Museum – 1977
Genealogical coll 50908

Munster IN

Northern Indiana Arts Association, 1040 Ridge Rd, Munster, IN 46321 • T: +1 219 8361839 • F: +1 219 8361863 • www.niaaonline.org •
Pres.: *Connie Skozen* •
Fine Arts Museum – 1969
Regional art 50909

Murdo SD

Pioneer Auto Museum, 503 E Fifth St, Murdo, SD 57559, mail addr: POB 76, Murdo, SD 57559 • T: +1 605 6692691 • F: +1 605 6693217 • pas@pioneerautoshow.com • www.pioneerautoshow.com •
Dir.: *David A. Geisler* • Cur.: *John Geisler* •
Science&Tech Museum – 1953
Transportation 50910

Murfreesboro AR

Ka-Do-Ha Indian Village Museum, 1010 Caddo Dr, Murfreesboro, AR 71958 • T: +1 870 2853736 • caddotc@windstream.net • www.caddotc.com •
Cur.: *Sherri Wright* •
Ethnology Museum / Archaeology Museum – 1964
Archaeology, Moundbuilders village 50911

Murfreesboro NC

William Rea Museum, William and Fourth Sts, Murfreesboro, NC 27855, mail addr: POB 3, Murfreesboro, NC 27855 • T: +1 252 3985922 • F: +1 252 3985871 • mha@murfreesboronc.org • /www.murfreesboronc.org •
C.E.O.: *Dale Neighbors* •
Historical Museum – 1967
Agriculture, general hist, Indian artifacts 50912

Murfreesboro TN

Baldwin Photo Gallery, Middle Tennessee State University, Learning Resources Center, Murfreesboro, TN 37132 • T: +1 615 8982085 • F: +1 615 8985682 • tjimison@mtsu.edu •
Cur.: *Tom Jimison* •
Fine Arts Museum / University Museum – 1961 . 50913

Oaklands Historic House Museum, 900 N Maney Av, Murfreesboro, TN 37130 • T: +1 615 8930022 • F: +1 615 8930513 • info@oaklandsmuseum.org • www.oaklandsmuseum.org •
Pres.: *William C. Ledbetter* •
Local Museum – 1959
Historic house, period furnishings, millitary, textiles, medical 50914

Stones River National Battlefield, 3501 Old Nashville Hwy, Murfreesboro, TN 37129-3094 • T: +1 615 8939501 • F: +1 615 8939508 • stri_information@nps.gov • www.nps.gov/stri •
Dir.: *Stuart Johnson* •
Military Museum / Historic Site – 1927
Civil War intems pertaining to the Battle of Stones River 50915

Murphy ID

Owyhee County Historical Museum, POB 67, Murphy, ID 83650 • T: +1 208 4952319 •
Pres.: *James Howard* • Dir.: *Byron Johnson* •
Local Museum – 1960
Agriculture, Indian artifacts, geology, mining 50916

Murphy NC

Cherokee County Historical Museum, 87 Peachtree St, Murphy, NC 28906 • T: +1 828 8376792 • F: +1 828 8376792 • cchm@webworkz.com • www.tib.com/cchm •
Dir.: *Wanda Stalcup* •
Historical Museum – 1977
Indian artifacts, early housewares, minerals, firearms 50917

Murray KY

University Art Galleries, Murray State University, 604 Fine Arts Center, Murray, KY 42071-3342 • T: +1 270 7623052 • F: +1 270 7623920 • sarah.henrich@murraystate.edu •
Dir.: *Sarah Henrich* •
Fine Arts Museum / University Museum – 1971
Permanent coll 50918

Wrather West Kentucky Museum, Murray State University, 100 Wrather Museum, Murray, KY 42071-0009 • T: +1 502 7624771 • F: +1 502 7624485 • www.murraystate.edu •
C.E.O.: *Kate A. Reeves* •
Local Museum – 1982
Guns, period furniture & tools, memorabilia 50919

Murrells Inlet SC

Brookgreen Gardens, 1931 Brookgreen Gardens Dr, Murrells Inlet, SC 29576 • T: +1 843 2356000 • F: +1 843 2371014 • info@brookgreen.org • www.brookgreen.org •
Pres.: *Robert Jewell* •
Fine Arts Museum – 1931
American sculpture 50920

Muscatine IA

Muscatine Art Center, 1314 Mulberry Av, Muscatine, IA 52761 • T: +1 319 2638282 • F: +1 319 2634702 • info@ci.muscatine.ia.us • www.muscatineartcenter.org •
Dir.: *Barbara Christensen* • Registrar: *Virginia Cooper* • Education Coordinator: *Maria Norton* •
Fine Arts Museum – 1965
Art, housed in Edwardian style Musser mansion, contemporaray gallery 50921

Pearl Button Museum, 117 W Second St, Muscatine, IA 52761 • T: +1 563 2631052 • www.pearlbuttoncapitol.com •
Dir.: *Jeffrey J. Kurtz* •
Special Museum – 1992
Clamming hist, button manufacturing 50922

Muskegon MI

Great Lakes Naval Memorial and Museum, 1346 Bluff St, Muskegon, MI 49441 • T: +1 231 7551230 • F: +1 231 7555883 • SS236@aol.com • www.silversides.org •
Cur.: *Robert G. Morin* •
Historical Museum / Military Museum – 1987
WW II navy sub USS Silversides, USCGC McLane W-146 50923

Hackley & Hume Houses, W Webster Av and Sixth St, Muskegon, MI 49440 • T: +1 231 7227578 • F: +1 231 7284119 • dawn@muskegonmuseum.org • www.muskegonmuseum.org/hh_site.asp •
Exec. Dir.: *John H. McGarry* • Cur.: *Dawn Willi* •
Historical Museum – 1971
Homes of lumber barons Charles H. Hackley and his business partner Thomas Hume – library 50924

Muskegon County Museum, 430 W Clay, Muskegon, MI 49440 • T: +1 231 7220278 • F: +1 231 7284119 • info@muskegonmuseum.org • www.muskegonmuseum.org •
Dir.: *John H. McGarry* • Cur.: *Melinda Johnson* (Exhibits) • Sc. Staff: *Melissa Horton* (Education) •
Local Museum – 1937
Regional hist, Indian, pioneer & lumbering artifacts, natural & maritime hist – library, archives 50925

Muskegon Museum of Art, 296 W Webster Av, Muskegon, MI 49440 • T: +1 231 7202570 • F: +1 231 7202585 • www.muskegonartmuseum.org •
Dir.: *Susan Talbot-Stanaway* • Cur.: *J. Houghton* •
Fine Arts Museum / Public Gallery – 1912
American and European paintings and graphics, Tiffany and modern studio glass, photography, sculpture – library, auditorium 50926

Muskogee OK

Ataloa Lodge Museum, 2299 Old Bacone Rd, Muskogee, OK 74403-1597 • T: +1 918 6834581 ext 283 • F: +1 918 6875913 • jtimothy@bacone.edu • www.bacone.edu •
Ethnology Museum / Folklore Museum – 1932
Native American material, basketry, pottery 50927

Five Civilized Tribes Museum, Agency Hill, Honor Heights Dr, Muskogee, OK 74401 • T: +1 918 6831701 • F: +1 918 6833070 • the5tribes@azalea.net • www.fivetribes.com •
Pres.: *Richard Bradley* • Dir.: *Clara Reekie* •
Historical Museum / Ethnology Museum / Folklore Museum – 1966
American Indian Museum and Art Gallery, Cherokee, Chickasaw, Choctaw, Creek and Seminole art and hist 50928

Muskogee War Memorial Park and Military Museum, 3500 Batfish Rd, Muskogee, OK 74401 • T: +1 918 6826294 • F: +1 918 6826294 • uss_batfish@okconnect.net • www.ussbatfish.com •
Cur.: *Charles Fletcher* •
Military Museum – 1972
Army, navy, marine and air force, WW 2, Civil War 50929

Thomas-Foreman Home, Oklahoma Historical Society, 1419 W Okmulgee, Muskogee, OK 74401 • T: +1 918 6826938 •
Pres.: *Jerry Marshall* •
Historical Museum – 1970
Private coll and furnishings 50930

Three Rivers Museum, 220 Elgin, Muskogee, OK 74401 • T: +1 918 6866624 • F: +1 918 6823477 • staff@3riversmuseum.com • www.3riversmuseum.com •
Dir.: *Linda Moore* •
Local Museum – 1985
Furnishings, clothing, military, railroad 50931

Myrtle Beach SC

Children's Museum of South Carolina, 2501 N Kings Hwy, Myrtle Beach, SC 29577 • T: +1 843 9469469 • F: +1 843 9467011 • cmsc@sccoast.net • www.cmsckids.org •
C.E.O.: *Pam Ross* •
Science&Tech Museum / Natural History Museum – 1993 50932

Franklin G. Burroughs-Simeon B. Chapin Art Museum, 3100 S Ocean Blvd, Myrtle Beach, SC 29577 • T: +1 843 2382510 • F: +1 843 2382910 • artmuse@sccoast.com • myrtlebeachartmuseum.org •
Dir.: *Patricia Goodwinl* •
Fine Arts Museum – 1989
Art 50933

Mystic CT

Mystic Art Center, 9 Water St, Mystic, CT 06355 • T: +1 860 5367601 • F: +1 860 5360610 • info@mysticarts.org • www.mysticarts.org •
Exec. Dir.: *Claire Matthews* • Dir.: *Tamara S. Rich* (Education) • *Karen Barthelson* (Exhibitions, Collections) •
Public Gallery – 1914
Creative arts 50934

Mystic Seaport, 75 Greenmanville Av, Mystic, CT 06355-0990 • T: +1 860 5720711 • F: +1 860 5725327 • info@mysticseaport.org • www.mysticseaport.org •
Dir.: *Douglas H. Teeson* • Cur.: *William Cogar* • *William Peterson* • *Mary Anne Stets* •
Special Museum – 1929 50935

Nacogdoches TX

SFA Galleries, Stephen F. Austin State University, 208 Griffith Fine Arts Bldg, Nacogdoches, TX 75962, mail addr: c/o Stephen F. Austin State University, POB 13041, Nacogdoches, TX 75962 • T: +1 936 4681131 • F: +1 936 4682938 • eadams@sfasu.edu • www.art.sfasu.edu •
Dir.: *Eloise Adams* •
Fine Arts Museum / University Museum 50936

Sterne-Hoya Museum, 211 S Lanana St, Nacogdoches, TX 75961 • T: +1 936 5605426 • F: +1 936 5699813 • brayb@ci.nacogdoches.tx.us • ci.nacogdoches.tx.us •
Local Museum – 1959
Texas artifacts, household fixtures – library 50937

Stone Fort Museum, Stephen F. Austin State University, Alumni & Griffith Blvds, Nacogdoches, TX 75962, mail addr: POB 6075 (SFASU), Nacogdoches, TX 75962 • T: +1 936 4682408 • F: +1 936 4687084 • cspears@sfasu.edu •
Dir.: *Dr. James E. Corbin* • Cur.: *Carolyn Spears* •
Local Museum / University Museum – 1936
Local hist, Spanish colonial and East Texas prior till 1900 50938

Nageezi NM

Chaco Culture National Historical Park, 1808 Rd 7950, Nageezi, NM 87037 • T: +1 505 7867014 ext 221 • F: +1 505 7867061 • chcu@nps.gov • www.nps.gov/chcu •
Cur.: *Wendy Bustard* •
Archaeology Museum – 1907
Archaeology, ethnology, prehistoric and historic site 50939

Nambe NM

Liquid Paper Museum, Nambe, NM 87501 • T: +1 505 4553848 • www.gihon.com •
Exec. Dir.: *Marcia J. Summers* •
Special Museum – 1980
Manufacturing tools and equipments 50940

Nampa ID

Land of the Yankee Fork Historical Museum, 2825 Myrtlewood Way, Nampa, ID 83686, mail addr: POB 1086, Challis, ID 83226 • T: +1 208 8382201 • F: +1 208 8383329 • louben3@aol.com •
Pres.: *Garvin McDaniels* •
Natural History Museum / Local Museum – 1961
Gold rush mining equip, household artifacts 50941

Nantucket MA

Artists Association of Nantucket Gallery, 19 Washington St, Nantucket, MA 02554 • T: +1 508 2280722, 2280294 • F: +1 508 3255251 • aan@nantucket.net • www.nantucketarts.org •
Dir.: *Robert Foster* •
Association with Coll – 1945
Early 20th c and contemporary art 50942

Coffin School Museum, Egan Institute of Maritime Studies, 4 Winter St, Nantucket, MA 02554 • T: +1 508 2282505 • F: +1 508 2287069 • egan@eganinstitute.com • www.eganinstitute.com •
Dir.: *Nathaniel Philbrick* • Cur.: *Margaret Moore Booker* •
Historical Museum – 1996
Coffin School memorabilia, Local history, portraits and marine paintings, ship models 50943

Hinchman House Natural Science Museum, Maria Mitchell Association, 7 Milk St, Nantucket, MA 02554 • T: +1 508 2289198, 2280898 • F: +1 508 2281031 • ahunt@mmo.org • www.mmo.org •
Dir.: *Dr. William Maple* • *Kathryn K. Pochman* • Sc. Staff: *Dr. Robert Kennedy* (National Science) • Cur.: *Mara Alper* •
Natural History Museum – 1902
Natural science, astronomy 50944

Mitchell House, Maria Mitchell Association, 1 Vestal St, Nantucket, MA 02554 • T: +1 508 2282896 • F: +1 508 2281031 • kpochman@mmo.org • www.mmo.org •
Dir.: *Dr. William Maple* • *Vladimir Strelnitski* (Astronomy) • Cur.: *Jascin Leonardo-Finger* •
Historical Museum / Natural History Museum – 1902
Women's history, Quakerism, Maria Mitchell's birthplace – observatories, aquarium, science library 50945

Nantucket Historical Association Museum, 15 Broad St, Nantucket, MA 02554-1016 • T: +1 508 2281894 • F: +1 508 2285618 • nhainfo@nha.org • www.nha.org •
CEO: *Dr. Frank D. Milligan* • Cur.: *Niles Parker* •
Local Museum – 1894
Local maritime history, whaling, crafts, tools, decorative arts – library, archives 50946

Nantucket Life-Saving Museum, 158 Polpis Rd, Nantucket, MA 02554-2320 • T: +1 508 2281885 • mo72506@nantucket.net • www.nantucketlifesavingmuseum.com •
Exec. Dir.: *Maurice E. Gibbs* •
Historical Museum 50947

Nantucket Lightship Basket Museum, 49 Union St, Nantucket, MA 02554 • T: +1 508 2281177 • F: +1 508 2287092 • adminoffice@nantucketlightshipbasketmuseum.org • www.nantucketlightshipbasketmuseum.org •
Dir.: *Jane Lee* •
Science&Tech Museum – 1997
Lightship baskets since 1850s, photos of early basketmakers 50948

Nanuet NY

Hudson Valley Children's Museum, Nanuet mall, 75 W Rte 159, Nanuet, NY 10954 • T: +1 845 7359720 • F: +1 845 7359770 • HVCM95@worldnet.att.net • www.HVCM.org •
Exec. Dir.: *Elizabeth Edgert* •
Science&Tech Museum / Natural History Museum – 1995 50949

Naperville IL

Naper Settlement Museum, 523 S Webster St, Naperville, IL 60540 • T: +1 630 4206010, 3055250 • F: +1 630 3055255 • towncrier@naperville.il.us • www.napersettlement.org •
Dir.: *Peggy Frank* • Cur.: *Louise Howard* • Asst. Cur.: *Debbie Grinnell* •
Open Air Museum – 1969
Furnishings 1830-1900, costumes, dolls, glass, china, folk, historical artifacts 50950

Naples FL

Collier County Museum, 3301 Tamiami Trail E, Naples, FL 34112 • T: +1 941 7748476 • F: +1 941 7748580 • curator@att.net • www.colliermuseum.com •
C.E.O.: *Ron D. Jamro* •
Local Museum – 1978
Local hist and archaeology, steam logging locomotive 50951

Naples Museum of Art, 5833 Pelican Bay Blvd, Naples, FL 34108-2740 • T: +1 239 5971900 • F: +1 239 2542753 • info@thephil.org • www.thephil.org •
Dir., Chief. Cur.: *Ph.D. Michael Culver* •
Public Gallery – 1989
Photography, sculpture, paintings, American and Mexican modernism, miniatures, glass 50952

Teddy Bear Museum of Naples, 2511 Pine Ridge Rd, Naples, FL 34109 • T: +1 941 5982711 • F: +1 941 5989239 • info@teddymuseum.com • www.teddymuseum.com •
Cur.: *George B. Black* •
Special Museum – 1990
Bear sculptures, manufactured bears 50953

The von Liiebig Art Center, 585 Park St, Naples, FL 34102 • T: +1 941 2626517 • F: +1 941 2625404 • info@naplesartcenter.org • www.naplesartcenter.org •
Pres.: *Pat Scoville* •
Fine Arts Museum – 1954
Paintings, sculpture, photography, works on paper 50954

Narragansett RI

South County Museum, POB 709, Narragansett, RI 02882 • T: +1 401 7835400 • F: +1 401 7830506 • www.southcountymuseum.org •
Exec.Dir.: *L.J. McElroy* •
Archaeology Museum / Historical Museum – 1933
Agriculture, technology, Yankee crafts, household items, rural life 1800-1930, farming and sea artifacts . 50955

Narrowsburg NY

Fort Delaware, Rte 97 N of Port Jervis, Narrowsburg, NY 12764, mail addr: POB 5012, Monticello, NY 12701-5192 • T: +1 845 2526660 • F: +1 845 7943459 • co.sullivan.ny.us •
Dir.: *Joanna Szakmary* •
Historical Museum – 1957 50956

Nashua IA

Chickasaw County Historical Society Museum, Bradford Village, 2729 Cheyenne Av, Nashua, IA 50658 • T: +1 641 4352567 • F: +1 641 4352567 •
C.E.O.: *Duane Tracy* •
Local Museum – 1953
Local history 50957

Nashua NH

Nashua Center for Arts, 14 Court St, Nashua, NH 03060 • T: +1 603 8831506 • F: +1 603 8827705 • Public Gallery 50958

Nashville IN

Brown County Art Gallery and Museum, 1 Artist Dr, Nashville, IN 47448 • T: +1 812 9884609 • brncagal@aol.com • www.geocities.com/brncagal • Pres.: *Dr. Emanuel Klein* • Fine Arts Museum – 1926 Oil paintings, pastels by Glen Cooper Henshaw, paintings, atchings, prints 50959

Brown County Historical Society Pioneer Museum, Museum Ln, Nashville, IN 47448 • T: +1 812 9886089 • bcarchive@hotmail.com • Pres.: *Ann Callahan* • Local Museum / Open Air Museum – 1957 Local history, museum village 50960

T.C. Steele State Historic Site, 4220 T.C. Steele Rd, Nashville, IN 47448 • T: +1 812 9882785 • F: +1 812 9888457 • tcsteele@bloomington.in.us • www.in.gov/ism/StateHistoricSites/index.aspx • Cur.: *Andrea Smith de Tarnowsky* • Fine Arts Museum / Historic Site – 1945 C.1907 T.C. Steele home, c.1916 artists studio 50961

Nashville TN

Belle Meade Plantation, 5025 Harding Rd, Nashville, TN 37205 • T: +1 615 3560501 • F: +1 615 3562336 • noburns@bellemeadeplantation.com • www.bellemeadeplantation.com • Exec. Dir.: *Norman O. Burns* • Pres.: *John Rochford* • Local Museum – 1953 Costumes, period furniture, civil war memorabilia 50962

Cheekwood Museum of Art, 1200 Forrest Park Dr, Nashville, TN 37205-4242 • T: +1 615 3568000 • F: +1 615 3530919 • info@cheekwood.org • www.cheekwood.org • Dir.: *Dr. Jack Becker* • Cur.: *Celia Walker* • Asst. Cur.: *Rusty Freeman* • *Terri Smith* • Fine Arts Museum – 1957 American art , contemporary art and sculpture, american and english decorative arts – library 50963

Country Music Hall of Fame and Museum, 222 5th Av of the Arts, Nashville, TN 37203 • T: +1 615 4162001 • F: +1 615 2552245 • membership@countrymusichalloffame.com • www.countrymusichalloffame.com • C.E.O.: *Kyle Young* • Pres.: *Vince Gill* • Cur.: *Mark Medley* • Music Museum – 1964 Country music and historic recording studio 50964

Cultural Museum at Scarritt-Bennett Center's, 1104 19th Av, Nashville, TN 37212 • T: +1 615 3407481 • F: +1 615 3407463 • museum@scarrittbennett.org • www.scarrittbennett.org • Dir.: *Herschell Parker* • Decorative Arts Museum / Ethnology Museum – 1992 Anthropology 50965

Cumberland Science Museum, 800 Fort Negley Blvd, Nashville, TN 37203 • T: +1 615 8625160 • F: +1 615 8625178 • www.csmisfun.com • C.E.O.: *Ralph J. Schulz* • Dir.: *Bert Matthews* (Physical Science) • Natural History Museum / Science&Tech Museum – 1944 Anthropology, archaeology, geology, transportation, general sciences – planetarium 50966

Disciples of Christ Historical Society Museum, 1101 19th Av S, Nashville, TN 37212-2196 • T: +1 615 3271444 • F: +1 615 3271445 • mail@dishistsoc.org • www.dishistsoc.org • Pres.: *Peter M. Morgan* • Religious Arts Museum – 1941 Religious hist 50967

Fisk University Galleries, Aaron Douglas Gallery and Carl Van Vechten Gallery, 1000 17th Av N, Nashville, TN 37208-3051 • T: +1 615 3298720 • F: +1 615 3298544 • galleries@fisk.edu • www.fisk.edu • C.E.O.: *Dr. Carolyn Reid-Wallace* • Fine Arts Museum – 1949 African and American art, paintings, drawings, folk art – library 50968

Fort Nashborough, 170 First Av N, Nashville, TN 37201 • T: +1 615 8628400 • F: +1 615 8625493 • jackie.jones@nashville.gov • www.nashville.gov/parks • C.E.O.: *Curt Garrigan* • Historical Museum – 1780 First settlement in Nashville 50969

Frist Center for the Visual Arts, 919 Broadway, Nashville, TN 37203 • T: +1 615 2443340 • F: +1 615 2443339 • mail@fristcenter.org • www.fristcenter.org • C.E.O.: *Susan Edwards* • Fine Arts Museum – 2001 Photographs, sculpture, paintings 50970

Fugitive Art Center, 440 Houston St, Nashville, TN 37203 • T: +1 615 2567067 • contact@fugitiveart.com • www.fugitiveart.com • Head: *Jack Dingo Ryan* • Public Gallery – 1999 50971

The Parthenon, 2600 W End Av, Nashville, TN 37203 • T: +1 615 8628431 • F: +1 615 8802265 • info@parthenon.org • www.parthenon.org • Dir.: *Wesley M. Paine* • Fine Arts Museum – 1897 Cowan coll, paintings, artworks 50972

Ruby Green Contemporary Arts Foundations, 514 Fifth Av S, Nashville, TN 37203 • T: +1 615 2447179 • www.rubygreen.org • Dir.: *Chris Campbell* • Public Gallery 50973

Tennessee Central Railway Museum, 220 Willow St, Nashville, TN 37210-2159 • T: +1 615 2449001 • F: +1 615 2442120 • hultman@nashville.com • www.tcry.org • C.E.O. & Dir.: *Terry L. Bebout* • Science&Tech Museum – 1990 Rolling stock, Nashville, Chattanooga and St Louis Ry 50974

Tennessee Historical Society Museum, War Memorial Bldg, Nashville, TN 37243-0084 • T: +1 615 7418934 • F: +1 615 7418937 • tnhissoc@tennesseehistory.org • www.tennesseehistory.org • C.E.O.: *Ann Toplovich* • Historical Museum – 1849 Paintings, manuscripts, artifacts – library, archives 50975

Tennessee State Museum, 505 Deaderick St, Nashville, TN 37243-1120 • T: +1 615 7412692 • F: +1 615 7417231 • info@tnmuseum.org • www.tnmuseum.org • Dir.: *Lois S. Riggins-Ezzell* • Asst. Dir.: *Dan E. Pomeroy* (Collections) • Education: *Patricia Rasbury* • Historical Museum – 1937 State history and art, military history – library 50976

Travellers Rest Plantation and Museum, 636 Farrell Pkwy, Nashville, TN 37220 • T: +1 615 8328197 • F: +1 615 8328169 • travellersrest@earthlink.net • www.travellersrestplantation.org • Exec. Dir.: *David Currey* • Cur.: *Rob DeHart* • Local Museum – 1955 Furniture, decorative arts and books, federal and early empire furniture 50977

Upper Room Chapel Museum, 1908 Grand Av, Nashville, TN 37212 • T: +1 615 3407206 • F: +1 615 3407293 • kkimball@upperroom.org • www.upperroom.org • Dir.: *Kathryn A. Kimball* • Religious Arts Museum / Fine Arts Museum – 1953 Bibles from 1577, woodcarving of Da Vinci's last supper, furniture, antiquites, religious art, African, Asian, Arfo-American and African Indian art 50978

Vanderbilt Kennedy Center Arts Gallery, 405 MRL Bldg, 21st and Magnolia Circle, Nashville, TN 37203, mail addr: Peabody Box 40230 Appleton Pl, Nashville, TN 37203 • T: +1 615 3228240, 3228147 • kc.vanderbilt.edu/kennedy/community/art.html • Fine Arts Museum / University Museum – 1994 Art by and about people with disabilities; changing exhibitions 50979

Vanderbilt University Fine Arts Gallery, 23rd & W End Avs, Nashville, TN 37203 • T: +1 615 3220605, 3431704 • F: +1 615 3433786 • www.vanderbilt.edu/gallery • Dir.: *Joseph S. Mella* • Asst. Cur.: *Krista Nance* • Fine Arts Museum / University Museum – 1961 Asian art, old master print, Italian renaissance painting, Oriental art, rare books; changing exhibitions of contemporary art – Arts library 50980

Vanderbilt University Sarratt Gallery, 207 Sarratt Student Center, Nashville, TN 37240-0026 • T: +1 615 3222471 • F: +1 615 3438081 • bridgette.kohnhorst@vanderbilt.edu • www.vanderbilt.edu/sarratt • Dir.: *Bridgette Kohnhorst* • Fine Arts Museum / Public Gallery Contemporary art of national and local artists, changing exhibitions 50981

Watkins Institute Art Gallery, 100 Powell Pl, Nashville, TN 37204 • T: +1 615 3834848 • F: +1 615 3834849 • Pres.: *Dr. Jim Brooks* • Public Gallery – 1885 Paintings, graphics and sculpture, etchings 50982

Nassau Bay TX

Arts Alliance Center at Clear Lake, 2000 NASA Rd, Nassau Bay, TX 77058 • T: +1 281 3357777 • F: +1 281 3338571 • www.taaccl.org • Dir.: *Kay Taylor Burnett* • Public Gallery – 1997 50983

Natchez MS

Grand Village of the Natchez Indians, 400 Jefferson Davis Blvd, Natchez, MS 39120 • T: +1 601 4466502 • F: +1 601 4466503 • gvni@bkbank.com • mdah.state.ms.us • Dir.: *James F. Barnett* • Ethnology Museum / Archaeology Museum – 1976 1700-1730 Natchez Indian ceremonial mound center 50984

Rosalie House Museum, 100 Orleans St, Natchez, MS 39120 • T: +1 601 4454555 • F: +1 601 3041376 • rosaliemansion@yahoo.com • www.rosaliemansion.com • Man.: *Rebecca J. Martin* • Historical Museum – 1898 Decorative arts, architecture, family life – historic garden 50985

Natchitoches LA

Bishop Martin Museum, Second and Church Sts, Natchitoches, LA 71457 • T: +1 318 3523422 • Dir.: *Kenneth J. Roy* • Cur.: *Kay Fortenberry* • Religious Arts Museum – 1839 Immaculate Conception Church and Rectory 50986

Fort Saint Jean Baptiste, 130 Moreau, Natchitoches, LA 71457 • T: +1 318 3573101 • F: +1 318 3577055 • ftstjean@crt.state.la.us • www.crt.sate.la.us • Cur.: *James Prudhomme* • Historic Site – 1982 Reconstruction of 1732 fort and several bldgs 50987

Lemee House, 310 Jeffersn St, Natchitoches, LA 71457 • T: +1 318 3577907 • F: +1 318 3522415 • Historical Museum – 1834 Paintings, lamps, 1792 map by French engineer 50988

Natick MA

Historical and Natural History Museum of Natick, 58 Eliot St, Natick, MA 01760 • T: +1 508 6474841 • F: +1 508 6517013 • info@natickhistory.com • www.natickhistory.com • Dir./Cur.: *Anne K. Schaller* • Local Museum – 1870 Local history, American Indian artifacts, natural history – library 50989

Naturita CO

Rimrocker Historical Museum of West Montrose County, 411 W Second Av, Naturita, CO 81422, mail addr: POB 913, Nucla, CO 81424 • T: +1 970 8652877, 8647438 • F: +1 970 8647564 • cookib@aol.com • www.Rimrocker.org • Pres.: *Mary Helen de Koevend* • Local Museum – 1980 Regional history 50990

Nauvoo IL

Joseph Smith Historic Center, 149 Water St, Nauvoo, IL 62354 • T: +1 217 4532246 • F: +1 217 4536416 • jshisctr@nauvoo.net • www.joseph-smith.com • C.E.O.: *Joyce A. Shireman* • Historical Museum – 1918 Joseph Smith Homestead, Joseph Smith Red Brick Store, Joseph Smith Mansion House 50991

Nauvoo Historical Society Museum, 980 S Bluff St, Nauvoo, IL 62354 • T: +1 217 4536355 • F: +1 217 4532512 • Pres.: *Mary Reed* • Local Museum – 1953 Geology, history, American Indian artifacts, 1850 wine cellar 50992

Nazareth PA

Moravian Historical Society Museum, 214 E Center St, Nazareth, PA 18064 • T: +1 610 7595070 • F: +1 610 7592461 • director@moravianhistoricalsociety.org • www.moravianhistoricalsociety.org • Exec.Dir.: *Susan M. Dreydoppel* • Religious Arts Museum / Historical Museum – 1857 Moravian Church hist – library 50993

Whitefield House Museum, Moravian Historical Society, 214 E Center St, Nazareth, PA 18064 • T: +1 610 7595070 • F: +1 610 7592461 • www.moravianhistoricalsociety.org • Dir.: *Susan M. Dreydoppel* • Cur.: *Mark A. Turdo* • Local Museum / Religious Arts Museum – 1857 Local hist 50994

Nebraska City NE

Arbor Lodge, 2600 Arbor Av, Nebraska City, NE 68410 • T: +1 402 8737222 • www.ngpc.state.ne.us • Dir.: *Rex Amack* • Historical Museum – 1923 Furniture, Tiffany glassware, silverware, paintings and other art items 50995

Wildwood Center, 420 S Steinhart Park Rd, Nebraska City, NE 68410 • T: +1 402 8736340 • Dir.: *Gail Wurtele* • Historical Museum / Fine Arts Museum – 1967 Victorian frame home, Victorian furnishings 50996

Neenah WI

Bergstrom-Mahler Museum, 165 N Park Av, Neenah, WI 54956 • T: +1 920 7514658, 7514670 • F: +1 920 7514755 • info@paperweightmuseum.com • www.paperweightmuseum.com • Dir.: *Alex D. Vance* • Cur.: *Jamie H. Severstad* • Fine Arts Museum / Decorative Arts Museum – 1954 Glass paperweights, Germanic and contemporary glass, Victorian glass baskets 50997

Hiram Smith Octagon House, 343 Smith St, Neenah, WI 54956 • T: +1 920 7254160 • neenahhistoricalsociety@powernetonline.com • Historical Museum – 1948 Local hist 50998

Neeses SC

Neeses Farm Museum, 6449 Savannah Hwy, Neeses, SC 29107 • T: +1 803 2475811 • F: +1 803 2475811 • Dir.: *Sonja Gleaton* • Agriculture Museum – 1976 Native American artifacts, pottery, clothing 50999

Neligh NE

Antelope County Historical Museum, 509 L St, Neligh, NE 68756 • T: +1 402 8874999 • Pres.: *Alta J. DeCamp* • Local Museum – 1965 Indian artifacts, genealogy, regional hist 51000

Neligh Mills, North St and Wylie Dr, Neligh, NE 68756, mail addr: POB 271, Neligh, NE 68756 • T: +1 402 8874303 • F: +1 402 8874303 • mill@bloomnet.com • www.nebraskahistory.org • Science&Tech Museum – 1971 500-barrel a day flour mill, elevators, plan sifters, purifiers, reel sifters, sackers, water powered turbine 51001

Nelsonville OH

Hocking Valley Museum of Theatrical History, 34 Public Sq, Nelsonville, OH 45764 • T: +1 740 7531924 • F: +1 740 7531982 • shirleystuart-soperahouse@yahoo.com • Pres.: *Frederick L. Oremus* • Performing Arts Museum – 1978 51002

Neodesha KS

Norman No. 1 Oil Well Museum, 106 S First St, Neodesha, KS 66757 • T: +1 316 3255316 • F: +1 316 3255316 • Dir.: *Jackie Clark* • Science&Tech Museum – 1967 Full sized derrick, oil well equip, farming, clown clothing, dolls, standard oil refinery Osage 51003

Neosho MO

Crowder College-Longwell Museum and Camp Crowder Collection, 601 La Clede, Neosho, MO 64850 • T: +1 417 4513223 ext 201 • F: +1 417 4514280 • www.crowder.edu • Dir.: *Lori L. Marble* • Fine Arts Museum / University Museum – 1970 Oil paintings, prints, Japanese wood block prints, Chinese paper cuts 51004

Neptune NJ

Neptune Township Historical Museum, 25 Neptune Blvd, Neptune, NJ 07754-1125 • T: +1 732 9885200 • F: +1 732 7741132 • nept_mus@hotmail.com • Cur.: *Evelyn Stryker Lewis* • Historical Museum – 1971 Cherokee based Sand Hill Indians, Antarctic explorer Schlossbach, regional history – library 51005

Nevada MO

Bushwhacker Museum, 231 N Main St, Nevada, MO 64772 • T: +1 417 6679602 • F: +1 417 6674571 • bushwhackerjail@sbcglobal.net • www.bushwhacker.org • C.E.O.: *Joe C. Kraft* • Local Museum / Historical Museum – 1964 Archaeology, Indian artifacts, medicine, local hist, primitive tools, military hist 51006

Nevada City CA

Firehouse No. 1 Museum, 215 Main St, Nevada City, CA 95959 • T: +1 916 2655468 • info@nevadacountyhistory • www.nevadacountyhistory.org • Historical Museum – 1947 Relicts of the Donner Party, mining, clothing, furnishings, Chinese exhibit, Maidu Native Americans – Searls library 51007

New Albany IN

Carnegie Center for Art & History, 201 E Spring St, New Albany, IN 47150 • T: +1 812 9447336 • F: +1 812 9813544 • info@carnegiecenter.org • www.carnegiecenter.org • Dir.: *Sally Newkirk* • Fine Arts Museum / Local Museum – 1971 Local history, art gallery 51008

Conway Fire Museum Aka Vintage Fire Engines, 402 Mount Tabor Rd, New Albany, IN 47150 • T: +1 812 9419901 • F: +1 812 9419901 • CEO: *Allen C. Conway* • Science&Tech Museum – 1990 Hand and horsedrawn, motorized firefighting equip 51009

Culbertson Mansion, 914 E Main St, New Albany, IN 47150 • T: +1 812 9449600 • F: +1 812 9496134 • guestservices@dnr.in.gov • www.indianamuseum.com • Cur.: *Joellen Bye* • Historical Museum – 1976 W.S. Culbertson Home, a 25-room Victorian mansion 51010

New Albany MS

Union County Heritage Museum, 114 Cleveland St, New Albany, MS 38652 • T: +1 662 5380014 • F: +1 662 5380019 • uchs@dixie-net.com • www.ucheritagemuseum.com •
Dir.: *Jill Smith* •
Local Museum – 1991
Memorabilia of the writer William Faulkner (1897-1962), scale model of his birthplace 51011

New Bedford MA

Free Public Library Special Collections, 613 Plesant St, New Bedford, MA 02740 • T: +1 508 9916275 • F: +1 508 9791481 • www.ci.new-bedford.ma.us •
Fine Arts Museum – 1852
Paintings, watercolors 51012

New Bedford Art Museum, 608 Plesant St, New Bedford, MA 02740 • T: +1 508 9613072 • info@newbedfordartmuseum.org • www.newbedfordartmuseum.org •
Exec. Dir.: *Karie C. Vincent* • Cur.: *Joan Backes* •
Fine Arts Museum – 1996 51013

New Bedford Whaling Museum, 18 Johnny Cake Hill, New Bedford, MA 02740 • T: +1 508 9970046 ext 101 • F: +1 508 9970018 • abrengle@whalingmuseum.org • www.whalingmuseum.org •
Exec. Dir.: *Anne B. Brengle* • Cons.: *Robert Hauser* •
Natural History Museum / Local Museum – 1903
International whaling history, Local social, political, economic and cultural history – library, auditorium 51014

Rotch-Jones-Duff House and Garden Museum, 396 County St, New Bedford, MA 02740 • T: +1 508 9971401 • F: +1 508 9976846 • info@rjdmuseum.org • www.rjdmuseum.org •
Dir.: *Kate Corkum* • Cur.: *Everett H. Hoag* •
Historical Museum – 1984
Period furnishings, horticulture 51015

University Art Gallery & Cabinet of Natural History, University of Massachusetts-Dartmouth, 715 Purchase St, New Bedford, MA 02740 • T: +1 508 9998555 • F: +1 508 9998912 • lantonsen@umassd.edu • www.umassd.edu/cvpa/universityartgallery •
Dir.: *Lasse B. Antonsen* •
University Museum / Fine Arts Museum / Natural History Museum 51016

New Berlin WI

The New Berlin Historical Society Museum, 19765 W National Av, New Berlin, WI 53146 • T: +1 262 5424773 • meidenbauer@juno.com •
Dir.: *Roy Meidenbauer* •
Historical Museum – 1965
Local hist, 1890's local home, school house room of 1863, 1850's Pioneer log home, carriage barn 51017

New Bern NC

Attmore-Oliver House, 510 Pollock St, New Bern, NC 28560 • T: +1 252 6388558 • F: +1 252 6385773 • nbhistoricalsoc@cconnect.net • www.newbernhistorical.org •
Dir.: *Joanne G. Ashton* •
Historical Museum – 1923
18th/19th c American furnishings, Civil War items 51018

Friends of Firemen's Museum, 408 Hancock St, New Bern, NC 28560 • T: +1 252 6364087 • F: +1 252 6364087 • firechief-nb@admin.ci.new-bern.nc.us •
C.E.O.: *Robert Aster* •
Special Museum – 1955
Fire engines, hors-drawn equipment, hist 51019

Tryon Palace, 610 Pollock St, New Bern, NC 28562 • T: +1 252 5144900 • F: +1 252 5144876 • kwilliams@tryonpalace.org • www.tryonpalace.org •
Dir.: *Kay P. Williams* • Cur.: *Nancy E. Richards* •
Historical Museum – 1945
Paintings of American artists, e.g. Thomas Gainsborough, Graphics, special exhibitions 51020

New Braunfels TX

Das Anwesen and Pharmacy Museum, 360 Millies Ln, New Braunfels, TX 78132 • T: +1 830 6255992 •
Historical Museum 51021

McKenna Children's Museum, 386 W San Antonio St, New Braunfels, TX 78130 • T: +1 830 6200939 • F: +1 830 6060440 • mkrause@mckenna.org • www.nbchildren.org •
Dir.: *Melissa Krause* •
Special Museum – 1986 51022

New Braunfels Conservation Society Museum, 1300 Church Hill, New Braunfels, TX 78131-0933, mail addr: POB 310933, New Braunfels, TX 78131-0933 • T: +1 210 6292943 • nbcs@axs4u.net • www.nbconservation.org •
Dir.: *Martha Rehler* •
Decorative Arts Museum / Local Museum – 1964
Handmade furniture, household items 51023

New Braunfels Museum of Art & Music, 1259 Gruene Rd, New Braunfels, TX 78130 • T: +1 830 6255636 • F: +1 830 6255636 • nbmuseum.org •
Dir.: *Charles R. Gallagher* •
Fine Arts Museum / Music Museum
Photographs, oral histories, recordings, film documentaries, folk art, music and crafts in Texas 51024

Sophienburg Museum, 401 W Coll St, New Braunfels, TX 78130 • T: +1 830 6291572, 6291900 • F: +1 830 6293906 • sophienburg@nbtx • www.nbtx.com/sophienburg •
Dir.: *Michelle Oatman* •
Local Museum – 1926
1840-1960 German immigration & Texas settlement, german and local artifacts 51025

New Bremen OH

Bicycle Museum of America, 7 W Monroe St, New Bremen, OH 45869 • T: +1 419 6292311 • F: +1 419 6293256 • annette.thompson@crown.com • www.bicyclemuseum.com •
C.E.O.: *James F. Dicke* •
Science&Tech Museum – 1997
200 bicycles from 1816 to present 51026

New Brighton PA

The Merrick Art Gallery, 1100 5th Av, New Brighton, PA 15066 • T: +1 724 8461130 • F: +1 724 8460413 • merrick@ccia.com • www.themerrick.org •
Dir.: *Cynthia A. Kundar* •
Fine Arts Museum – 1880
Art 51027

New Britain CT

Museum of Central Connecticut State University, Art Dept, Samuel S.T. Chen Art Center, New Britain, CT 06050 • T: +1 860 8322633 • F: +1 860 8322634 • www.art.ccsu.edu/gallery.html •
Dir.: *Dr. Cora Marshall* • *Mark Strathy* •
University Museum / Fine Arts Museum / Local Museum – 1965
Anthropology, archaeology, folklore, decorative arts, industry, paintings, graphics, sculpture 51028

New Britain Museum of American Art, 56 Lexington St, New Britain, CT 06052-1412 • T: +1 860 2290257 • F: +1 860 2293445 • nbmaa@nbmaa.org • www.nbmaa.org •
Dir.: *Laurene Buckley* • *Douglas Hyland* •
Fine Arts Museum – 1903
American art 51029

New Britain Youth Museum, 30 High St, New Britain, CT 06051 • T: +1 860 2253020 • F: +1 860 2294982 • marketing@newbritainyouthmuseum.org • www.newbritainyouthmuseum.org •
Dir.: *Deborah Pfeiffenberger* •
Special Museum – 1956
Circus items, clothing, toys, dolls, farming tools 51030

New Brunswick NJ

Buccleuch Mansion, Buccleuch Park, Easton Av, New Brunswick, NJ 08901 • T: +1 732 7455094 • skenen@telcordia.com •
Cur.: *Susan Kenen* •
Historical Museum / Decorative Arts Museum – 1915
Temporary exhibitions 51031

Henry Guest House, 58 Livingston Av, New Brunswick, NJ 08901 • T: +1 732 7455108 • F: +1 732 8460226 •
Dir.: *Robert Belvin* •
Historical Museum – 1760 51032

Jane Voorhees Zimmerli Art Museum, Rutgers State University of New Jersey, 71 Hamilton St, New Brunswick, NJ 08901-1248 • T: +1 732 9327237 • F: +1 732 9328201 • jujusant@rci.rutgers.edu • www.zimmerlimuseum.rutgers.edu •
Dir.: *Gregory P. Perry* • Cur.: *Julie Mellby* (Prints and Drawings) •
Fine Arts Museum / University Museum – 1966
Paintings, graphics, sculpture, American and European art – art library 51033

Mason Gross Art Galleries, RutgersUniversity, 33 Livingston Av, New Brunswick, NJ 07042 • T: +1 732 9322222 • F: +1 732 9322217 • rinac@rci.rutgers.edu •
Fine Arts Museum
Cotemporary art 51034

Museum of the American Hungarian Foundation, 300 Somerset St, New Brunswick, NJ 08903 • T: +1 732 8465777 • F: +1 732 2497033 • info@ahfoundation.org • www.ahfoundation.org •
Dir.: *August J. Molnar* • Cur.: *Patricia L. Fazekas* •
Fine Arts Museum / Folklore Museum – 1954
Hsitory of Hungarian immigration, art and life 51035

New Canaan CT

Glastonbury Gild Arts Center, 1037 Silvermine Rd, New Canaan, CT 06840 • T: +1 203 9666668 •
Public Gallery 51036

New Canaan Historical Society Museum, 13 Oenoke Ridge, New Canaan, CT 06840 • T: +1 203 9661776 • F: +1 203 9725917 • newcanaan.historical@snet.net • www.nchistory.org •
Dir.: *Janet Lindstrom* •
Local Museum / Historical Museum – 1889 51037

Silvermine Guild Arts Center, 1037 Silvermine Rd, New Canaan, CT 06840 • T: +1 203 9669700 • F: +1 203 9727236 • sgac@silvermineart.org • www.silvermineart.org •
Dir.: *Cynthia Clair* •
Fine Arts Museum – 1922 51038

New Castle DE

New Castle Historical Society Museum, 2 E Fourth St, New Castle, DE 19720 • T: +1 302 3222794 • F: +1 302 3228923 • nchistorical@aol.com • www.newcastlehistory.org •
Dir.: *Bruce Dalleo* •
Local Museum – 1934
Local hist 51039

New Castle IN

Henry County Historical Museum, 606 S 14th St, New Castle, IN 47362 • T: +1 765 5294028 • hchisoc@kiva.net • www.kiva.net/~hchisoc/museum.htm •
Cur.: *Marianne Hughes* •
Local Museum / Historical Museum – 1887
Home of Civil War General William Grose 51040

Indiana Basketball Hall of Fame, 408 Trojan Ln, New Castle, IN 47362 • T: +1 765 5291891 • F: +1 765 5290273 • mail@hoopshall.com • www.hoopshall.com •
Dir.: *Roger Dickinson* •
Special Museum – 1965
Sports hist, history of Indiana high schools 51041

New Castle PA

Hoyt Institute of Fine Arts, 124 E Leasure Av, New Castle, PA 16101 • T: +1 724 6522882 • F: +1 724 6578786 • hoyt@hoytartcenter.org • www.hoytartcenter.org •
Dir.: *Kimberly B. Koller-Jones* • Asst. Dir.: *Mary Ann Fazzone* • Cur.: *David Collins* •
Fine Arts Museum – 1965
Art 51042

New City NY

Rockland County Museum, 20 Zukor Rd, New City, NY 10956 • T: +1 914 6349629 • F: +1 914 6348690 • info@rocklandhistory.org • www.rocklandhistory.org •
C.E.O.: *Erin L. Martin* •
Local Museum – 1965
Furniture, dec arts, paintings, prints, photographs, genealogical documents 51043

New Glarus WI

Chalet of the Golden Fleece, 618 2nd St, New Glarus, WI 53574 • T: +1 608 5272614 • F: +1 608 5272062 •
C.E.O.: *Mark Eisenmann* •
Local Museum – 1955
Historic house, period Swiss and early American furniture, glass, china, jewelry, weapons 51044

Swiss Historical Village, 612 7th Av, New Glarus, WI 53574 • T: +1 608 5272317 • F: +1 608 5272302 • www.swisshistoricalvillage.com •
Pres.: *John F. Marty* •
Local Museum / Open Air Museum – 1938
Historical buildings, tools 51045

New Gloucester ME

Shaker Museum, 707 Shaker Rd, New Gloucester, ME 04260 • T: +1 207 9264597 • usshakers@aol.com • www.shaker.lib.me.us •
Dir.: *Leonard L. Brooks* • Cur.: *Michael S. Graham* •
Historic Site – 1931
Religion, Shaker religious community – library 51046

New Harbor ME

Colonial Pemaquid Historical Site, New Harbor, ME 04554 • T: +1 207 6772423 •
Archaeology Museum – 1970
Archaeological dig site 51047

New Harmony IN

Historic New Harmony, 506 1/2 Main St, New Harmony, IN 47631 • T: +1 812 6824488 • F: +1 812 6824313 • harmony@usi.edu • www.newharmony.org •
Dir.: *Connie Weinzapfel* • Cur.: *Jean Lee* •
Historic Site – 1973
American history, early natural science, early abolition, early women's movement 51048

New Harmony Gallery of Contemporary Art, University of Southern Indiana, 506 Main St, New Harmony, IN 47631 • T: +1 812 6823156 • F: +1 812 6823870 • avasherdea@usi.edu • www.nhgallery.com •
Dir.: *April Vasher-Dean* •
Fine Arts Museum – 1975
Contemporary art 51049

New Harmony State Historic Site, 510 Main St, New Harmony, IN 47631 • T: +1 812 6824488 • F: +1 812 6824313 • newharmonyshs@dynasty.net • www.ai.org/ism •
Historical Museum – 1937
Site of two early utopian experiments, the communal societies of New Harmony 51050

New Haven CT

John Slade Ely House, 51 Trumbull St, New Haven, CT 06510 • T: +1 203 6248055 • F: +1 203 6242306 • info@clyhouse.org •
Cur.: *Paul Clabby* •
Fine Arts Museum – 1960
Paintings, sculpture 51051

Knights of Columbus Museum, 1 State St, New Haven, CT 06511-6702 • T: +1 203 8650400 • F: +1 203 8650351 • www.kofc.org •
Dir.: *Larry Sowinski* • Cur.: *Mary Lou Cummings* •
Archivist: *Susan Brosnan* •
Historical Museum / Association with Coll – 1982
Fine and decorative art, paintings, prints, sculpture, costumes, Coll on Christopher Columbus, memorabilia and manuscripts relating to the history of the Knights – library 51052

New Haven Colony Historical Society Museum, 114 Whitney Av, New Haven, CT 06510 • T: +1 203 5624183 • F: +1 203 5622002 • dcarter@nhchs.org • www.nhchs.org •
Dir.: *Peter Thomas Lemothe* • Cur.: *Amy L. Trout* •
Local Museum – 1862
Decorative and inventive arts, photographs, economic, technological, political and ethnic-racial hist of NH 51053

Peabody Museum of Natural History, Yale University, 170 Whitney Av, New Haven, CT 06520-8118 • T: +1 203 4325050 • F: +1 203 4329816 • peabody.webmaster@yale.edu • www.peabody.yale.edu •
Dir. & Cur.: *Michael Donoghue* (Botany) • Sc. Staff: *Marcello A. Canuto* (Anthropology) • *Frank Hole* (Anthropology) • *Andrew Hill* (Anthropology) • *Richard L. Burger* (Anthropology) • *Leo J. Hickey* (Paleobotany) • *Elisabeth Vrba* (Vertebrate Zoology) • *Jacques A. Gauthier* (Vertebrate Zoology, Ornitology, Vertebrate Paleontology) • *Adolf Seilacher* (Invertebrate Paleontology) • *Leo W. Buss* (Invertebrates) • *Jay Ague* (Mineraology) • *Karl K. Turekian* (Meteorites, Planetary Science) • *David F. Musto* (Historical Scientific Instruments) • *Leonard Munstermann* (Entomology) • *Sean Rice* (Invertebrates) • *David Skelly* (Vertebrate Zoology) •
University Museum / Natural History Museum – 1866
Anthropology, paleontology, mineralogy, paleobotany, vertebrates, invertebrates, otnitology, entomology, zoology, botany, meteorits, historic scientific instr. 51054

Yale Center for British Art, 1080 Chapel St, New Haven, CT 06520-8280 • T: +1 203 4322850 • F: +1 203 4324538 • bacinfo@yale.edu • www.yale.edu/ycba •
Dir.: *Amy Meyers* • Dep. Dir.: *Constance Clement* • Cur.: *Angus Trumble* • *Cassandra Albinson* (Paintings and Drawings) • *Scott Wilcox* • *Gillian Forrester* (Prints and Drawings) • *Elisabeth Fairman* (Rare Books and Archives) • Chief Cons.: *Theresa Fairbanks* •
Fine Arts Museum / University Museum – 1977
British art from 16th c to present including paintings, drawings, prints, books, sculptures and manuscripts 51055

Yale University Art Gallery, 1111 Chapel St, New Haven, CT 06510 • T: +1 203 4320600 • artgalleryinfo@yale.edu • artgallery.yale.edu •
Dir.: *Jock Reynolds* • Cur.: *Patricia Kane* (American Decorative Arts) • *Helen Cooper* (American Painting) • *Susan Matheson* (Ancient Art) • *Jennifer Gross* (European and Contemporary Art) •
Fine Arts Museum / University Museum – 1832
Paintings, sculpture, drawings, photographs, Garvan Mabel Brady coll of the 17th-19th c American paintings and decorative arts, Jarves and Griggs coll of Italian renessaince paintings, Hobart and Edward Small Moore coll fo Near and Far Eastern art, Ordway Katharine coll of 19th and 20th c art, coll of the Societe Anonyme, Olson coll of Precolumbian art, Linton coll of African sculpture 51056

Yale University Collection of Musical Instruments, 15 Hillhouse Av, New Haven, CT 06520 • T: +1 203 4320822 • F: +1 203 4328342 • musinst@pantheon.yale.edu • www.yale.edu/musicalinstruments •
Dir.: *Richard Rephann* • Assoc. Cur.: *Nicholas Renouf* • Cur.: *Susan Thompson* •
University Museum / Music Museum – 1900
16th-18th c keyboard instruments, 17th-20th c stringed instruments, 15th-20th c American and European instruments, bells 51057

New Haven KY

Kentucky Railway Museum, 136 S Main St, New Haven, KY 40051 • T: +1 502 5495470 • F: +1 502 5495472 • kyrail@bardstown.com • kyrail.org •
C.E.O.: *Greg Mathews* •
Science&Tech Museum – 1954
Railroad engines, steam and diesel, passenger and freight car, blueprints 51058

New Holstein WI

Calumet County Historical Society Museum, 1704 Eisenhower St, New Holstein, WI 53061 • T: +1 920 8981333 •
Pres.: *Ronald Zarling* •
Local Museum / Agriculture Museum – 1963
Farming 51059

Pioneer Corner Museum, 2103 Main St, New Holstein, WI 53061 • T: +1 414 8985258 • F: +1 414 8989022 • wcramer@esls.lib.wi.us •
Cur.: *Wendy Cramer* •
Local Museum – 1961
Local hist ... 51060

New Hope PA

Lockhouse-Friends of the Delaware Canal, 145 S Main St, New Hope, PA 18938 • T: +1 215 8622021 • F: +1 215 8622021 • fodc@erols.com • www.fodc.org •
Dir.: *Susan H. Taylor* •
Local Museum – 1991
Artifacts, photographs ... 51061

Parry Mansion Museum, 45 S Main St, New Hope, PA 18938 • T: +1 215 8625652, 7945260 • F: +1 215 8628227 • newhopehs@verizon.net • www.newhopehs.org •
Pres.: *Lester Isbrandt* • Sc. Staff: *Terry McNealy* •
Decorative Arts Museum / Historical Museum – 1966
Decorative arts ... 51062

New Hyde Park NY

Goudreau Museum of Mathematics in Art and Science, Herricks Community Center, 999 Herricks Rd, New Hyde Park, NY 11040-1353 • T: +1 516 7470777 • F: +1 516 7470777 • info@mathmuseum.org • www.mathmuseum.org •
Dir.: *Beth J. Deaner* •
Special Museum – 1980
Math models and puzzles ... 51063

New Iberia LA

The Shadows-on-the-Teche, 317 E Main St, New Iberia, LA 70560 • T: +1 337 3696446 • F: +1 337 3655213 • shadows@shadowsontheteche.org • www.shadowsontheteche.org •
Dir.: *Patricia Kahle* •
Historic Site – 1961
Plantation home ... 51064

New Ipswich NH

Barrett House, Forest Hall, Main St, New Ipswich, NH 03071 • T: +1 603 8782517 • F: +1 617 2279204 • pgittlemann@spnea.org • www.historicnewengland.org •
Dir.: *Carl Nold* •
Historical Museum – 1948
18th - 19th c furniture, musical instruments ... 51065

New London CT

Lyman Allyn Museum of Art, 625 Williams St, New London, CT 06320 • T: +1 860 4432545 • F: +1 860 4421280 • crusan@lymanllyn.org • www.lymanallyn.org •
Dir.: *Ronald L. Crusan* •
Fine Arts Museum – 1930
American fine arts, decorative arts ... 51066

New London County Historical Society Museum, 11 Blinman St, New London, CT 06320 • T: +1 860 4431209 • F: +1 860 4431209 • nlchsinc@aol.com • www.newlondonhistory.org •
Pres.: *Patricia M. Schaefer* •
Local Museum / Historical Museum – 1870 ... 51067

United States Coast Guard Museum, c/o US Coast Guard Academy, 15 Mohegan Av, New London, CT 06320-4195 • T: +1 860 4448511 • F: +1 860 4448289 • cherrick@cga.uscg.mil • www.uscg.mil •
Historical Museum – 1967 ... 51068

New London NH

New London Historical Society Museum, Little Sunapee Rd, New London, NH 03257-0965 • T: +1 603 5266564 • info@newlondonhistoricalsociety.org • www.nlhs.net •
Pres.: *Henry S. Otto* •
Local Museum – 1954
Local and regional hist ... 51069

New London WI

New London Public Museum, 406 S Pearl St, New London, WI 54961 • T: +1 920 9828520 • F: +1 920 9828617 • museum@newlondonwi.org • www.newlondonwi.org •
Dir.: *Laurel Spencer Forsythe* •
Local Museum – 1917
Local hist, mounted birds, Native American artifacts, African folk art ... 51070

New Madrid MO

Hunter-Dawson State Historic Site, Dawson Rd, Hwy U, New Madrid, MO 63869 • T: +1 573 7485340 • F: +1 573 7487228 • hunter-dawson.state.historic.site@dar.mo.gov • www.mostateparks.com •
Head: *Michael Comer* •
Historic Site – 1967 ... 51071

New Market VA

New Market Battlefield State Historical Park, 8895 Collins Dr, New Market, VA 22844 • T: +1 540 7403101 • F: +1 540 7403033 • nmbjw@shentel.net • www.vmi.edu/museum/nm •
Dir.: *Scott H. Harris* • Exec. Dir.: *Keith E. Gibson* •
Historic Site / Military Museum – 1967
Civil War artifacts, personal items owned by Stonewall Jackson ... 51072

New Milford CT

New Milford Historical Society Museum, 6 Aspetuck Av, New Milford, CT 06776-0359 • T: +1 860 3543069 • F: +1 860 2100263 • nmhistorical@mail.com • www.nmhistorical.org •
Cur.: *Pamela Edwards* •
Historical Museum – 1915 ... 51073

New Orleans LA

Academy Gallery, New Orleans Academy of Fine Arts, 5256 Magazine St, New Orleans, LA 70115 • T: +1 504 8998111 • F: +1 504 8976811 • noafa@bellsouth.com • www.noafa.com •
Dir.: *Patsy Adams* •
Fine Arts Museum – 1978 ... 51074

Beauregard-Keyes House, 1113 Chartres St, New Orleans, LA 70116 • T: +1 504 5237257 • F: +1 504 5237257 •
Dir.: *Marion S. Chambon* •
Historical Museum – 1970
Louisiana raised cottage, home of Gen. P.G.T. Beauregard and Frances Parkinson Keyes ... 51075

Black Arts National Diaspora, 1530 N Claiborne Av, New Orleans, LA 70116 • T: +1 504 9490807 • F: +1 504 9496052 • blkarts@bellsouth.net • www.bandinc.org/ •
Exec. Dir.: *Dr. G. Jeannette Hodge* •
Fine Arts Museum
Contemporary African art from around the world 51076

Confederate Museum, 929 Camp St, New Orleans, LA 70130 • T: +1 504 5234522 • F: +1 504 5238595 • memhall@aol.com • www.confederatemuseum.com •
Cur.: *Pat Ricci* •
Military Museum – 1891
Military, housed in 1890 Memorial Hall ... 51077

Contemporary Arts Center, 900 Camp St, New Orleans, LA 70130 • T: +1 504 5283800/05 • F: +1 504 5283828 • mweber@cacno.org • www.cacno.org •
Dir.: *Jay Weigel* •
Fine Arts Museum / Public Gallery – 1976
Art ... 51078

Fine Arts Gallery, c/o University of New Orleans, 2000 Lake Shore Dr, New Orleans, LA 70148 • T: +1 504 2806493 • F: +1 504 2807346 • finearts@uno.edu • www.uno.edu •
Dir.: *Doyle Gertjejansen* •
University Museum / Fine Arts Museum – 1974 . 51079

Fort Pike, Rte 6, Box 194, New Orleans, LA 70129 • T: +1 504 6625703 • F: +1 504 6620147 • fortpike_mgr@crt.state.la.us • www.crt.state.la.us •
Military Museum – 1934
Armaments, dress and battle orders, Louisiana hist ... 51080

Gallier House, 1118-32 Royal St, New Orleans, LA 70116 • T: +1 504 5255661 • F: +1 504 5689735 • hgrimagallier@aol.com • www.gnofn.org/~hggh •
Dir.: *Stephen A. Moses* • Cur.: *Jan Bradford* •
Local Museum – 1969
Decorative arts ... 51081

Hermann-Grima House, 820 Saint Louis St, New Orleans, LA 70112 • T: +1 504 5255661 • F: +1 504 5689735 • hgrimagallier@aol.com • www.gnofn.org/~hggh •
Dir.: *Stephen A. Moses* • Cur.: *Jan Bradford* •
Local Museum – 1971
Early example of American influence on New Orleans architecture ... 51082

The Historic New Orleans Collection, 533 Royal St, New Orleans, LA 70130 • T: +1 504 5234662 • F: +1 504 5987108 • hnocinfo@hnoc.org • www.hnoc.org •
Dir.: *Priscilla O'Reilly-Lawrence* • *John H. Lawrence* •
Cur.: *Dr. Alfred Lemmon* •
Historical Museum – 1966
Hist and culture of Louisiana and Gulf South ... 51083

Jean Lafitte National Historical Park and Preserve, 419 Decatur St, New Orleans, LA 70130-1142 • T: +1 504 5893882 • F: +1 504 5893851 • kathy_lang@nps.gov • www.nps.gov/jela •
Local Museum / Historical Museum / Military Museum / Natural History Museum / Historic Site – 1978
Natural history specimens and associated field records; archeological objects systematically recovered from within the park's boundaries and associated field records; historic objects, tools and equipment, furnishings, and household items related to the Battle of New Orleans, the Barataria Preserve, and the Acadian Culture ... 51084

Longue Vue House, 7 Bamboo Rd, New Orleans, LA 70124 • T: +1 504 4885488 • F: +1 504 4867015 • lschmalz@longuevue.com • www.longuevue.com •
Exec. Dir.: *Bonnie Goldblum* • Cur.: *Lydia H. Schmalz* •
Fine Arts Museum / Decorative Arts Museum – 1968
Furniture, textile, needlework, wallapers, ceramics, contemporary and modern art – gardens ... 51085

Louisiana Children's Museum, 420 Julia St, New Orleans, LA 70130 • T: +1 504 5231357 • F: +1 504 5293666 • www.lcm.org •
C.E.O.: *Julia W. Bland* •
Special Museum – 1981 ... 51086

Louisiana State Museum, 751 Chartres St, New Orleans, LA 70116 • F: +1 504 5684995 • sschulkens@crt.state.la.us • lsm.crt.state.la.us •
Dir.: *James F. Sefcik* • Cur.: *Sam Rykels* •
Historical Museum – 1906
History, decorative arts, costume, textiles, maps, manuscripts, aviation ... 51087

National D-Day Museum, 945 Magazine St, New Orleans, LA 70130 • T: +1 504 5276012 • F: +1 504 5276088 • info@ddaymuseum.org • www.ddaymuseum.org •
C.E.O: *Dr. Gordon H. Mueller* •
Military Museum – 1991
American experience WW 2 ... 51088

New Orleans Artworks Gallery, 727 Magazine St, New Orleans, LA 70130 • T: +1 504 5297279 • F: +1 504 5395417 •
Pres.: *Gerlool Baronne* •
Decorative Arts Museum / Fine Arts Museum – 1990
Prints, sculpture, etchings and engravings, decorative arts, furniture, glass, metalwork, enamels ... 51089

New Orleans Fire Department Museum, 1135 Washington Av, New Orleans, LA 70130 • T: +1 504 8964756, 5657818 • F: +1 504 8964756 •
C.E.O: *Warren E. McDaniels* •
Historical Museum – 1992
History of fire fighting in New Orleans ... 51090

New Orleans Museum of Art, 1 Collins Diboll Circle, New Orleans, LA 70124 • T: +1 504 4882631 • F: +1 504 4846662 • jbullard@noma.org • www.noma.org •
Dir.: *E. John Bullard* •
Fine Arts Museum – 1910
Art, decorative art, paintings, photos ... 51091

New Orleans Pharmacy Museum, 514 Rue Chartres, New Orleans, LA 70130-2110 • T: +1 504 5658027 • F: +1 504 5658028 • nopharmsm@aol.com • www.pharmacymuseum.org •
Dir.: *Liz Good* • Cur.: *Jen Lushear* •
Historical Museum – 1950
Pharmacy, medicine, Louis Joseph Dufilho, Jr., first licensed pharmacist in the U.S ... 51092

Newcomb Art Gallery, c/o Newcomb College, Tulane University, Woldenberg Art Center, New Orleans, LA 70118-5698 • T: +1 504 8655328 • F: +1 504 8655329 • smain@tulane.edu • www.newcombartgallery.com •
Dir.: *Erik Neil* • Sen. Cur.: *Sally Main* •
Fine Arts Museum / University Museum – 1886
Art, Newcomb pottery, painting, sculpture, photography ... 51093

Pitot House Museum, 1440 Moss St, New Orleans, LA 70119 • T: +1 504 4820312 • F: +1 504 4820363 • smcclamroch@louisianalandmarks.org • www.pitothouse.org •
C.E.O. & Pres.: *Lyn Tomlison* •
Historical Museum – 1964
1799 French West Indies Plantation style house 51094

Tulane University Art Collection, Tulane University, 7001 Freret St, New Orleans, LA 70118 • T: +1 504 8655328 • F: +1 504 8655329 • www.newcombartgallery.com •
Cur.: *Joan G. Caldwell* •
Fine Arts Museum / University Museum – 1980 . 51095

New Paltz NY

College Art Gallery → Samuel Dorsky Museum of Art

Huguenot Historical Society Museum, 18 Broadhead Av, New Paltz, NY 12561-1403 • T: +1 845 2551660 • F: +1 845 2550376 • hhsoffice@hhs-newpaltz.org • www.hhs-newpaltz.org •
Dir.: *John Braunlein* • Asst. Dir.: *Stewart A. Crowell* •
Cur.: *Leslie LeFevre Stratton* • Sc. Staff: *Kenneth Shefsiek* (Education) • *Eric Roth* (Archives) • Admin. Asst.: *June L. Heneberry* •
Local Museum – 1894
History, architecture, decorative arts, historical houses – archives ... 51096

Samuel Dorsky Museum of Art, State University of New York New Paltz, 1 Hawk Drive, New Paltz, NY 12561 • T: +1 845 2573844 • F: +1 845 2573854 • sdma@newpaltz.edu • www.newpaltz.edu/museum •
Dir.: *Sara Pasti* • Chm.: *David A. Dorsky* • Cur.: *Brian Wallace* •
Fine Arts Museum / University Museum – 1963
Regional art, photographs, metals, Asian prints, 20th c works on paper ... 51097

New Philadelphia OH

Schoenbrunn Village State Memorial, 1984 E High Av, New Philadelphia, OH 44663 • T: +1 330 3393636 • F: +1 330 3394165 • schoenbrunn@tusco.net • www.ohiohistory.org/places/schoenbr/index.html •
C.E.O.: *Dr. Gary C. Ness* •
Open Air Museum / Local Museum – 1928
Hist of Ohio's first Christian settlement ... 51098

New Port Richey FL

West Pasco Historical Society Museum, 6431 Circle Blvd, New Port Richey, FL 34652 • T: +1 727 8470680 • wb2ium@att.net • www.rootsweb.com/~flwphs •
Pres.: *Leon Flood* •
Local Museum – 1983
Local history, photograph coll – library ... 51099

New Providence NJ

Cambria Historical Society Museum, 121 Chanlon Rd, New Providence, NJ 07974 • T: +1 908 6652846 •
Dir.: *Tamika Borden* • Cur.: *Donald Bruce* •
Local Museum / Decorative Arts Museum – 1950
China, dolls, furniture, glass, paintings ... 51100

New Roads LA

Pointe Coupee Museum, 8348 False River Rd, New Roads, LA 70760 • T: +1 225 6387788 • F: +1 504 6385555 •
Chm.: *Olinde S. Haag* •
Historical Museum – 1976
18th to early 19th c house furnishings ... 51101

New Rochelle NY

Castle Gallery, College of New Rochelle, 29 Castle Pl, New Rochelle, NY 10805 • T: +1 914 6545423 • F: +1 914 6545014 • www.cnr.edu/cg.htm •
Dir.: *Jennifer Zazo* •
University Museum / Fine Arts Museum
Art ... 51102

New Sweden ME

Lars Noak Blacksmith Shop, Larsson/Ostlund Log Home & One-Room Capitol School, Station Rd, New Sweden, ME 04762 • T: +1 207 8965624, 8963199 • F: +1 207 8965624 • rmhme@mfx.net • www.aroostook.me.us/newsweden/historical.html •
Dir.: *Allen Kampe* • *Linnea Helstrom* • *Ralph Ostlund* •
Local Museum / Open Air Museum – 1989
Historic village ... 51103

New Sweden Historical Museum, 116 Station Rd, New Sweden, ME 04762 • T: +1 207 8963018 • F: +1 207 8963120 •
C.E.O.: *Gary Dickinson* •
Local Museum – 1925
Historical replica of original colony capitel, photographs and portraits of immigrants ... 51104

New Ulm MN

Brown County Museum, 2 N Broadway, New Ulm, MN 56073-1714 • T: +1 507 2332616 • F: +1 507 3541068 • bchs@newulmtel.net • www.browncounty-historymnusa.org •
Exec. Dir.: *Robert Burgess* • Cur.: *Pam Krzmarzick* •
Local Museum – 1930
Dakota & Objiwe Indians, conflicts, regional art – library ... 51105

New Wilmington PA

Westminster College Art Gallery, Market St, New Wilmington, PA 16172 • T: +1 724 9467266 • F: +1 724 9467256 • www.westminster.edu •
Dir.: *Kathy Koop* •
Fine Arts Museum – 1854
Drawings, American paintings, prints ... 51106

New York NY

80 Washington Square East Gallery, New York University, 80 Washington Square East, New York, NY 10003-6697 • T: +1 212 9985747 • F: +1 212 9985752 • www.nyu.edu/pages/galleries •
Dir.: *M. Karp* • *R. D. Newman* •
University Museum
International arts ... 51107

Academy Gallery of Art, 2 E 63rd St, New York, NY 10021 • T: +1 212 8380230 • F: +1 212 8385226 • nyas@nyas.org •
Dir.: *Fred Moreno* •
Fine Arts Museum
Art ... 51108

Actual Art Foundation, 7 Worth St, New York, NY 10013 • T: +1 212 9660758 •
Dir.: *Valerie Shakespeare* •
Fine Arts Museum
Art ... 51109

Alternative Museum, 594 Broadway, Ste 402, New York, NY 10012 • T: +1 212 9664444 • F: +1 212 2262158 • altmuseum@aol.com • www.alternativemuseum.org •
Dir.: *Geno Rodriguez* • Asst. Dir.: *Jan Rooney* • Asst. Cur.: *Elisa White* •
Fine Arts Museum – 1975 ... 51110

American Academy of Arts and Letters Art Museum, 633 W 155th St, New York, NY 10032-7599 • T: +1 212 3685900 • F: +1 212 4914615 • artsandletters@mindspring.com •
Exec. Dir.: *Virginia Dajani* • Pres.: *Ned Rorem* • Exec. Asst.: *Rafael Díaz-Tusham* (Architecture, Literarture) • Sc. Staff: *Betsey Feeley* • *Ardith Holmgrain* (Music) • *Kathleen Kienholz* (Archivist) • *Jane Bolster* (Publications, Library) • *Souhad Rafey* (Exhibitions, Purchases, Art) • Coord.: *Richard Rogers* •
Fine Arts Museum – 1898
Coll of works by Childe Hassam and Eugene Speicher, art painting, sculpture, graphics, music – Archives, library ... 51111

American Craft Museum, 40 W 53rd St, New York, NY 10019 • T: +1 212 9563535 • F: +1 212 4590926 • ursula.neuman@americancraftmuseum.org • www.americancraftmuseum.org •
Dir.: *Holly Hotchner* • Pres.: *Nanette Laitman* • Vice Pres.: *Patricia Hynes* • Chm.: *Barbara Tober* • Chief.

U

Cur.: *David McFadden* • Cur.: *Ursula Ilse-Neuman* • Registrar: *Marsha Beitchman* •
Fine Arts Museum / Decorative Arts Museum – 1956
Crafts, design, decorative arts 51112

American Folk Art Museum, 45 W 53rd St, New York, NY 10019, mail addr: 1414 Av of the Americas, New York, NY 10019-2514 • T: +1 212 2651040, 9777170 ext 309 • F: +1 212 9778134 • info@folkartmuseum.org • www.folkartmuseum.org •
Dir.: *Gerard C. Wertkin* • Dir. & Sen.Cur.: *Stacy Hollander* (Exhibitions) • Dir./Cur.: *Brooke Davis Anderson* (Contemporary Center) • *Lee Kogan* (Folk Art Institute) •
Folklore Museum – 1961
American 18th-to 20th c folk sculpture and paintings, textile arts, decorative arts, hist 51113

American Irish Historical Society Museum, 991 Fifth Av, New York, NY 10028 • T: +1 212 2882263 • F: +1 212 6287927 • aihs@aihs.org • www.aihs.org •
Dir.: *William Cobert* • Pres.: *Kevin M. Cahill* • Chm.: *Donald R. Keough* •
Historical Museum – 1897
Hist of Irish in America – Library 51114

American Museum of Natural History, Central Park W at 79 St, New York, NY 10024 • T: +1 212 7695100 • F: +1 212 4965018 • postmaster@amnh.org • www.amnh.org •
Pres.: *Ellen V. Futter* • Cur.: *Dr. Enid Schildkrout* (Anthropology) • *Dr. Charles W. Myers* (Herpetology) • *Dr. Melanie L.J. Stiassny* (Ichthyology) • *Dr. Neil H. Landman* (Invertebrates) • *Dr. Ross D.E. MacPhee* (Mammalogy) • *Dr. Mark A. Norell* (Vertebrate Paleontology) • Ass. Cur.: *Dr. Edmond A. Mathez* (Earth and Planetary Sciences) • *Dr. David Grimaldi* (Entomology) • *Dr. George F. Barrowclough* (Ornithology) •
Natural History Museum / Historical Museum – 1869
Anthropology, hist buildings, street facades from New York 51115

American Numismatic Society Museum, 96 Fulton Str, New York, NY 10038 • T: +1 212 5714470 ext 1306 • F: +1 212 5714479 • info@numismatics.org • www.numismatics.org •
Dir.: *Dr. Ute Wartenberg* • Cur.: *Michael L. Bates* (Islamic Coins) • *Robert Hoge* (America) • *Peter van Alfen* (Greek Coins) • Libr.: *Francis D. Campbell* •
Special Museum – 1858 51116

Americas Society Art Gallery, 680 Park Av, New York, NY 10021 • T: +1 212 2498300 • F: +1 212 2495868 • exhibitions@as-coa.org • www.americas-society.org •
Pres.: *Myles R.R. Frechette* • Hon.Chm.: *David Rockefeller* • Chm.: *William Rhodes* • Dir.: *Marysol Nieves* (Visual Arts) • Assoc. Cur: *Sofia Hernandez* (Visual Arts) •
Fine Arts Museum – 1967
Temporary exhibitions, coll of Latin American paintings and drawings 51117

Amos Eno Gallery, 59 Franklin St, New York, NY 10012 • T: +1 212 2265342 •
Dir.: *Jane Harris* •
Fine Arts Museum
Contemporary art 51118

Anne Frank Center USA, 584 Broadway, Ste 408, New York, NY 10012 • T: +1 212 4317993 • F: +1 212 4318375 • director@annefrank.com • www.annefrank.com •
Historical Museum 51119

Anthology Film Archives, 32 Second Av, New York, NY 10003 • T: +1 212 5055181 • F: +1 212 4772714 •
Special Museum – 1970
Films, documentaries, photographs; changing exhibitions – library 51120

apexart, 291 Church St, New York, NY 10013 • T: +1 212 4315270 • F: +1 646 8272487 • info@apexart.org • www.apexart.org •
Public Gallery – 1994
Contemporary visual arts and culture 51121

Arsenal Gallery, Arsenal Bldg, Fifth Av at 64th St, New York, NY 10021 • T: +1 212 3608163 • F: +1 212 3601329 • Adian.Sas@parks.nyc.gov • www.nyc.gov/parks •
Cur.: *Adrian Sas* •
Natural History Museum
Mixed media-park and nature themes 51122

Art for Change, 1701 Lexington Av, New York, NY 10029 • T: +1 212 3487044 • info@artforchange.org • www.artforchange.org •
Dir.: *Eliana Godoy* •
Public Gallery 51123

Art Omi, 55 Fifth Av, New York, NY 10003 • T: +1 212 2065660 • artomi55@aol.com • www.artomi.org •
Fine Arts Museum
Visual arts 51124

Arthur A. Houghton Jr. Gallery and the Great Hall Gallery, 7 E 7th St, Foundation Bldg, New York, NY 10003-7120 • T: +1 212 3534155 •
Fine Arts Museum – 1859
Graphic design, fine arts, painting, sculpture, engineering 51125

The Artists Museum, 250 W 57th St, New York, NY 10107 • T: +1 914 8337864 • theartistsmuseum@usa.net •
Dir.: *Barbara Ingber* •
Fine Arts Museum
Sculpture, paintings 51126

Artist's Space, 38 Greene St, New York, NY 10013 • T: +1 212 2263970 • F: +1 212 9661434 • artspace@artistsspace.org • www.artistsspace.org •
Public Gallery 51127

Asia Society Museum, 725 Park Av, New York, NY 10021 • T: +1 212 2886400 • F: +1 212 5178315 • pr@asiasoc.org • www.asiasocietymuseum.org •
Pres.: *Vishakha N. Desai* • Dir.: *Melissa Chiu* • Cur.: *Adriana Poser* (Traditional Art) •
Fine Arts Museum – 1956
Asian art, Mr. and Mrs. John D. Rockefeller 3rd coll of Asian art 51128

Asian-American Arts Centre, 26 Bowery St, New York, NY 10013 • T: +1 212 2332154 • F: +1 212 7661287 • aaac@artspiral.org • www.artspiral.org •
C.E.O. & Cur.: *Robert Lee* • Chm.: *Sasha Hohri* •
Fine Arts Museum / Public Gallery – 1974
Folk art, contemporary visual art, June 4th Movement – research library 51129

Asian American Arts Centre, 26 Bowery, New York, NY 10013 • T: +1 212 2332154 • F: +1 212 7661287 • aaac@artspiral.org • www.artspiral.org •
Public Gallery 51130

The AXA Gallery, Equitable Tower, 787 Seventh Av, New York, NY 10019 • T: +1 212 5544731 • F: +1 212 5542456 • panistave@axa-financial.com • www.axa-financial.com/aboutus/gallery.html •
Dir.: *Pari Stave* •
Fine Arts Museum / Public Gallery – 1992
19th and 20th c paintings, sculpture and works on paper 51131

The Bard Graduate Center for Studies in the Decorative Arts, Design, and Culture, 18 W 86th St, New York, NY 10024 • T: +1 212 5013000 • F: +1 212 5013079 • generalinfo@BGC.bard.edu • www.bgc.bard.edu •
Dir.: *Dr. Susan Weber Soros* • Dir. Academy: *Dr. Andrew Morrall* • Sc. Staff: *Nina Stritzler-Levine* (Dir. Exhibitions) • *Elena Pinto Simon* (Academic Programming) • *Lorraine Bacalles* (Administration) • *Sarah B. Sherrill* (Decorative Arts) • *Lisa Beth Podos* (Public programs) •
Decorative Arts Museum / University Museum – 1992
Decorative arts – Library 51132

Bernard Judaica Museum, Congregation Emanuel of the City of New York, 1 E 65th St, New York, NY 10021 • T: +1 212 7441400 • F: +1 212 5700826 • info@emanuelnyc.org • www.emanuelnyc.org •
Sr.Cur.: *Elka Deitsch* •
Decorative Arts Museum / Folklore Museum / Religious Arts Museum – 1948
Judaica, liturgical items, hist documents and photos conc the Congregation, coll from Jewish communities 51133

Black Fashion Museum, 155 W 126th St, New York, NY 10027 •
Special Museum 51134

The Bohen Foundation, 415 W 13th St, New York, NY 10014 • T: +1 212 3342281, 4144575 • F: +1 212 3344178 • info@bohen.org •
Public Gallery 51135

Bowery Gallery, 530 W 25th St, New York, NY 10001 • T: +1 646 2306655 • boweryg@earthlink.net • www.bowerygallery.org •
Public Gallery – 1969
Paintings and sculpture 51136

Broadway Windows, New York University, 80 Washington Square East, New York, NY 10003-6697 • T: +1 212 9985747 • F: +1 212 9985752 • www.nyu.edu/pages/galleries •
Dir.: *M. Karp* • *R. D. Newman* •
University Museum
Art 51137

Burden Gallery, Aperture Foundation, 20 E 23 St, New York, NY 10010 • T: +1 212 5055555 ext 325 • F: +1 212 9797759 • gallery@aperture.org • www.aperture.org •
Exec. Dir.: *Michael E. Hoffman* • C.E.O.: *Paul Gottlieb* • Chm.: *Robert Anthoine* • Sc. Staff: *Sara Wolfe* (Gallery, Exhibitions) •
Fine Arts Museum – 1952
Prints by Robert Adams, Nick Waplington, Edward Steichen, Danny Lyon, Timothy O'Sullivan, Paul Strand archive, fine art 51138

Castle Clinton, Battery Park, New York, NY 10004 • T: +1 212 3447220 • F: +1 212 2856874 • joe_avery@nps.gov • www.nps.gov •
Historical Museum – 1950
Coll and archives conc the castle, immigration depot, miniature dioramas of 19th c Manhattan 51139

Cathedral Church of Saint John the Divine, 1047 Amsterdam Av (at 112th Street), New York, NY 10025 • T: +1 212 3167540 • F: +1 212 9327348 • info@stjohndivine.org • www.stjohndivine.org •
Dean of the Cathedral: *The Very Reverend James A. Kowalski* •
Religious Arts Museum – 1892
World's largest Cathedral, soaring Gothic & Romanesque architecture, magnificent stained glass windows, paintings, sculpture, tapestries, religious artefacts, Biblical Garden, wandering peacocks. – Textile Conservation Laboratory, Diocesan Archives 51140

Center for Book Arts, 28 W 27th St, New York, NY 10001 • T: +1 212 4810295 • F: +1 212 4819853 • info@centerforbookarts.org • www.centerforbookarts.org •
Dir.: *Rorey Golden* •
Public Gallery – 1974
Contemporary artists and hist books – Library 51141

Center for Figurative Painting, 115 W 30th St, New York, NY 10001 • T: +1 212 8683452 • F: +1 212 7361520 • cfpstudio@aol.com •
Dir.: *Jennifer Samet* •
Public Gallery
Paintings 51142

Center for Jewish History, American Jewish Historical Society, 15 W 16th St, New York, NY 10011 • T: +1 212 2946160 • F: +1 212 2946161 • ajhs@ajhs.org • www.ajhs.org •
Pres.: *Sydney Lapious* • Exec. Dir.: *Dr. Michael Feldberg* • Cur.: *Sarah Davis* •
Folklore Museum / Religious Arts Museum – 1892
American Jewish experience from the 16th c to present, Yiddish theatre coll of posters, music and actors, objects of daily life, portraits, photographs, documents – Library 51143

The Chaim Gross Studio Museum, 526 LaGuardia Pl, New York, NY 10012 • T: +1 212 5294906 • F: +1 212 5291966 • grossmuseum@earthlink.net •
Pres.: *Renee Gross* • Vice Pres.: *Miriam Gross* • Dir./Chief Cur.: *Dr. April Paul* • Sc. Staff: *Irwin Hersey* •
Fine Arts Museum – 1989
Sculpture, watercolours, drawings, prints 51144

Chancellor Robert R. Livingston Masonic Library and Museum, 71 W 23 St, New York, NY 10010-4171 • T: +1 212 3376620 • F: +1 212 6332639 • livmalib@pipeline.com • www.livmalib.org •
Pres.: *Richard H. Eberle* • Dir.: *Thomas M. Savini* •
Historical Museum / Library with Exhibitions – 1856
Masonic hist in NY State, hist of American fraternalism, Hermetic and esoteric thought 51145

The Chase Manhattan Bank Art Collections, 270 Park Av, New York, NY 10017 • T: +1 212 2700667 • F: +1 212 2700725 •
Dir.: *Manuel Gonzalez* • Asst. Cur.: *Stacey Gershon* •
Fine Arts Museum – 1959
17.000 largerly contemporary American works in all media 51146

Chelsea Art Museum, Miotte Foundation, 556 W 22nd St, New York, NY 10011 • T: +1 212 2550719 • F: +1 212 2552368 • contact@chelseaartmuseum.org • www.chelseaartmuseum.org •
Fine Arts Museum
European and American abstract artists, sculpture coll 51147

Children's Museum of Manhattan, 212 W 83rd St, The Tisch Bldg, New York, NY 10024 • T: +1 212 7211223 • F: +1 212 7211127 • mail@cmom.org • www.cmom.org •
Exec. Dir.: *Andrew S. Ackerman* • Assoc. Dir.: *Elizabeth London* •
Special Museum – 1973
Literacy, science & environment, arts, media & communications 51148

Children's Museum of the Arts, 182 Lafayette St, New York, NY 10013 • T: +1 212 2740986 • F: +1 212 2741776 • info@cmany.org • www.cmany.org •
Fine Arts Museum – 1988
Children's art from around the world 51149

China Institute Gallery, 125 E 65 St, New York, NY 10021 • T: +1 212 7448181 • F: +1 212 6284159 • gallery@chinainstitute.org • www.chinainstitute.org •
Dir.: *Hai Chang Willow* • Pres.: *Jack Maisano* • Chm.: *Virginia A. Kamsky* •
Public Gallery – 1926
Changing exhibitions of primarely Chinese art and culture 51150

City Arts, 525 Broadway, New York, NY 10012 • T: +1 212 9660377 • F: +1 212 9660551 • www.cityarts.org •
Public Gallery
7 51151

The Cloisters, The Metropolitan Museum of Art, Fort Tryon Park, 99 Margaret Corbin Dr, New York, NY 10040 • T: +1 212 9233700 • F: +1 212 9281146 • cloisters@metmuseum.org • www.metmuseum.org •
Cur. in Charge: *Peter Barnet* (Medieval Art, Cloisters) • Cur.: *Timothy Husband* • *Barbara Boehm* • Cons.: *Lucretia Kargere* • *Michele Marincola* • Sc. Staff: *Michael Carter* (Assistant Librarian) •
Fine Arts Museum / Decorative Arts Museum – 1938
Medieval art – garden, library 51152

The Collectors Club, 22 E 35th St, New York, NY 10016 • T: +1 212 6830559 • F: +1 212 4811269 • collectorsclub@nac.net • www.collectorsclub.org •
Pres.: *Thomas C. Mazza* •
Special Museum – 1896
Medals, stamps, awards, engravings prints, early postal equipment – library 51153

Con Edison Energy Museum (closed) 51154

Congregation Emanu-el Museum, 1 E 65th St, New York, NY 10021-6596 • T: +1 212 7441400 • F: +1 212 5700826 •
Dir.: *Reva G. Kirschberg-Grossman* • Cur.: *Cissy Grossman* •
Historical Museum
Congregational memorabilia, graphics, Judaica, Paintings 51155

Cooper-Hewitt National Design Museum, Smithsonian Institution, 2 E 91st St, New York, NY 10128 • T: +1 212 8498351 • F: +1 212 8498401 • www.si.edu/ndm •
Dir.: *Paul Warwick Thompson* • Asst. Dir.: *Susan Yelavich* (Public Projects) • Chm.: *Kay Allaire* • Cur.: *Marilyn Symmes* (Drawings and Prints) • *Ellen Lupton* • *Donald Albrecht* (Contemporary Design) • *Deborah Shinn* (Applied Arts, Industrial Design) • *Joanne Warner* (Wallcoverings) • *Matilda McQuaid* (Textiles) • Libr.: *Stephen H. Van Dyk* •
Decorative Arts Museum – 1897
300.000 works of design, decorative and applied arts representing a span of 300 years 51156

Cue Art Foundation, 511 W 25th St, New York, NY 10001 • T: +1 212 2063583 • F: +1 212 2060321 • info@cueartfoundation.org • www.cueartfoundation.org •
Dir.: *Jeremy Adams* •
Fine Arts Museum
Art 51157

Czech Center New York, 1109 Madison Av, New York, NY 10028 • T: +1 212 2880830 • F: +1 212 2880971 • info@czechcenter.com • www.czechcenter.com •
Dir.: *Iva Karolina Reisinger* •
Ethnology Museum / Folklore Museum – 1995
Civic art and culture 51158

Dactyl Foundation, 64 Grand St, New York, NY 10013 • T: +1 212 2192344 • info@dactyl.org • www.dactyl.org •
Dir.: *Neil Grayson* •
Fine Arts Museum
Art 51159

Dahesh Museum of Art, 580 Madison Av, New York, NY 10022 • T: +1 212 7590606 • F: +1 212 7591235 • information@daheshmuseum.org • www.daheshmuseum.org •
Dir.: *Peter Trippi* • Chief Cur.: *Stephen R. Edidin* • Assoc. Cur.: *Lisa Small* • Cur.: *Roger Diederen* • Sc. Staff: *Maria Celi* (Events) • *Jeremy Benjamin* • *Lisa Mansfield* (Registrar) •
Fine Arts Museum – 1995
European Art (19th-20th c) 51160

Dan Flavin Art Foundation Temporary Gallery, Dia Art Foundation, 5 Worth St, New York, NY 10013 • T: +1 212 4313033 •
Public Gallery 51161

Department of Art History & Archaeology Visual Resources Collection, Columbia University, 820-825 Schermerhorn Hall, New York, NY 10027 • T: +1 212 8543044 • F: +1 212 8547329 • www.mcah.columbia.edu •
Cur.: *Andrew Gessener* • Asst. Cur.: *Dorothy Krasowska* •
Fine Arts Museum
250.000 photographs, colls by Bartsch, Gaigleres, A. Kingsley Porter, Ware, Courtauld, Chatsworth and Millard Meiss; Marburger index, Windsor Castle – Berenson I-Tatti and Haseloff archives, Dial iconographic index 51162

Dia Art Foundation, 535 W 22nd St, New York, NY 10011 • T: +1 212 9895566 • F: +1 212 9894055 • info@diaart.org • www.diaart.org •
Fine Arts Museum – 1974 51163

Dia Center for the Arts, 141 Wooster St, New York, NY 10012 • T: +1 212 4738072 •
Fine Arts Museum
Exhibition of The New York Earth Room by Walter De Maria 51164

Dia Center for the Arts, 393 West Broadway, New York, NY 10012 • T: +1 212 9259397 •
Fine Arts Museum
Exhibition of The Broken Kilometer by Walter De Maria 51165

Dia:Chelsea, Dia Art Foundation, 548 W 22nd St, New York, NY 10011 • T: +1 212 9895566 • F: +1 212 9894055 • info@diaart.org • www.diaart.org •
Dir.: *Michael Govan* • Chm.: *Leonard Riggio* • Vize Chm.: *Ann Tenenbaum* • Cur.: *Lynne Cooke* • Sc. Staff: *Sara Tucker* (Digital Media) • *Karen Kelly* (Publications) • *Steven Evans* (Gallery) •
Fine Arts Museum – 1974
Contemporary American/European art – Library 51166

Drawing Center, 35 Wooster St, New York, NY 10013 • T: +1 212 2192166 • F: +1 212 9662976 • info@drawingcenter.org • www.drawingcenter.org •
C.E.O. & Exec.Dir.: *Catherine de Zegher* • Chm.: *Frances Beatty Adler* • Asst. Dir.: *Elizabeth Metcalf* • Cur.: *Luis Camnitzer* • Sc. Staff: *Lytle Shaw* • *Heidi O'Neill* (Asst. Cur.) • *Susette Min* (Contemporary Exhibitions) • *Adam Lehner* (Drawings) • *Meryl Zwanger* (Education) • *Tracey Fugami* (Asst. Cur., Registrar) • *Kathryn Thuma* (Historical Exhibitions) •
Fine Arts Museum – 1976
Art hist, painting and drawing 51167

Dyckman Farmhouse Museum, 4881 Broadway at 204 St, New York, NY 10034 • T: +1 212 3049422 • F: +1 212 3040635 • info@dyckman.org • www.dyckman.org •

Dir.: *Allyson Bowen* •
Historical Museum – 1916
18th/19th c furniture and furnishings, dec art, Revolutionary War uniforms, Hessian hut ... 51168

Eldridge Street Project, 12 Eldridge St, New York, NY 10002 • T: +1 212 2190888 • F: +1 212 9664782 • eldridge@eldridgestreet.org • www.eldridgestreet.org • Exec.Dir.: *Amy E. Waterman* • Chm.: *Michael Weinstein* • Pres.: *Lorinda Ash Ezersky* • Sc. Staff: *Hanna Griff* (Public Programs) • *Annie Polland* (Education) • *Erin Barlow* (Admin.) •
Folklore Museum / Religious Arts Museum / Historical Museum – 1986
Jewish culture and hist on the Lower East Side .. 51169

The Ernest W. Michel Historical Judaica Collection, c/o UJA-Federation of New York, 130 E 59 St, New York, NY 10022 • T: +1 212 8361720 • F: +1 212 7559183 •
Dir.: *Ernest W. Michel* • Cur.: *Paul Hunter* •
Ethnology Museum / Historical Museum – 1995
Material on Jewish and Non-Jewish personalities having been part of Jewish life in 19th/20th c, artifacts from the Holocaust, development of the State of Israel and of the Jewish community in NY ... 51170

Eva and Morris Feld Gallery, American Folk Art Museum, 2 Lincoln Sq, New York, NY 10023-6214 • T: +1 212 5959533 • F: +1 212 9778134 • www.folkartmuseum.org •
Dir.: *Gerard C. Wertkin* • Deputy Dir.: *Riccardo Salmona* • Cur.: *Stacy Hollander* • Sc. Staff: *Brooke Davis Anderson* (Contemporary Center) • *Elizabeth Warren* • *Lee Kogan* •
Folklore Museum – 1961 ... 51171

Eyebeam, 540 W 21st St, New York, NY 10011 • T: +1 212 9376580 • F: +1 212 9376582 • info@eyebeam.org • www.eyebeam.org •
Exec. Dir.: *Amanda McDonald Crowley* •
Public Gallery
Cultural dialogue at the intersection of the arts and sciences ... 51172

Federal Hall National Memorial, 26 Wall St, New York, NY 10005 • T: +1 212 8256888 • F: +1 212 8256874 • joe_avery@nps.gov • www.nps.gov/feha/ •
Historical Museum – 1939
Memorabilia of George Washington, evolution of the Bill of Rights, hist development of NY City, evolution of the Federal Hall ... 51173

Forbes Collection, 60 Fifth Av, New York, NY 10011 • T: +1 212 2065548 • jgstanley@jgstanley.com • www.forbes.com/forbescollection •
Decorative Arts Museum
Furnishings, decorative accessoires, jewelry, gifts ... 51174

Foundation for Hellenic Culture, 7 W 57th St, New York, NY 10019 • T: +1 212 3086908 • F: +1 212 3080919 •
Public Gallery
Art ... 51175

Fountain Gallery, 702 Ninth Av, New York, NY 10019 • T: +1 212 2622756 • info@fountaingallerynyc.com • www.fountaingallerynyc.com •
Public Gallery
Art by artists living with mental illness, their personal visions ... 51176

Fraunces Tavern Museum, 54 Pearl St, New York, NY 10004 • T: +1 212 4251778 • F: +1 212 5093467 • publicity@frauncestavernmuseum.org • www.frauncestavernmuseum.org •
Dir.: *Amy Northrop Adamo* • Chm.: *Kevin Hansley* • Cur.: *Nadezhda Williams* •
Historical Museum – 1907
Decorative arts, textiles, paintings, prints, war memorabilia ... 51177

Frick Collection, 1 E 70th St, New York, NY 10021 • T: +1 212 2880700 • F: +1 212 6284417 • info@frick.org • www.frick.org •
Dir.: *Anne L. Poulet* • Pres.: *Helen Clay Chace* • Chief Cur.: *Colin B. Bailey* • Cur.: *Susan Grace Galassi* • Deputy Dir.: *Robert B. Goldsmith* (Admin.) • Sc. Staff: *Patricia Barnett* (Librarian) • *Dennis F. Sweeney* (Operations) • *Diane Farynyk* (Registrar) •
Fine Arts Museum / Decorative Arts Museum – 1920
Paintings, sculptures, furniture, decorative art, prints ... 51178

Galerie at Lincoln Center, 136 W 65th St, New York, NY 10023 • T: +1 212 5804673 •
Fine Arts Museum
Performing arts ... 51179

Gallery Korea, 460 Park Av & 57th St, New York, NY 10022 • T: +1 212 7599550 • F: +1 212 6888640 • nyarts@koreanculture.org • www.koreanculture.org •
Fine Arts Museum – 1979
Art exhibits – Library ... 51180

Gallery of Henry Street Settlement, Abrons Arts Center, 466 Grand St, New York, NY 10002 • T: +1 212 5980400 • F: +1 212 5058329 • sfnarts@aol.com •
Dir.: *Jane Delgado* • Exec. Dir.: *Daniel Kronenfeld* • Admin. Asst.: *Amy Blitz* • Sc. Staff: *Josh Carr* • *Don Williams* • *Wanda Egipciaco* • *Karen Chait* (Registrar) •
Public Gallery – 1893
Contemporary Art – photo gallery ... 51181

Gallery of Prehistoric Paintings, 30 E 81st St, New York, NY 10028 • T: +1 212 8615152 •
Dir.: *Douglas Mazonowicz* •
Fine Arts Museum – 1975
Primitive art – Library ... 51182

General Grant National Memorial, Riverside Dr and W 122 St, New York, NY 10027 • T: +1 212 6661640 • F: +1 212 9329631 •
Dir.: *Joseph T. Avery* • Cur.: *Judith Muller* •
Historical Museum – 1897
History, sarcophagi of Gen. Grant and his wife, memorabilia of Pres. Ulysses S. Grant ... 51183

Glove Museum, 15 W 28th St, New York, NY 10001 • T: +1 212 8031600 • F: +1 212 6839099 • glovesla@aol.com •
Dir.: *Jay G. Ruckel* •
Special Museum – 1986
Veteran glove maker, antique gloves, vintage clothes, tools ... 51184

Goethe-Institut New York - Exhibitions, 1014 Fifth Av, New York, NY 10028 • T: +1 212 4398700 • F: +1 212 4398705 • program@goethe-newyork.org • www.goethe.de/uk/ney/enpausst.htm •
Public Gallery ... 51185

Gracie Mansion, East End Av at 88 St, New York, NY 10128 • T: +1 212 5704751 • F: +1 212 5704493 • dcarroll@cityhall.nyc.gov • www.nyc.gov •
Dir.: *Susan Danilow* • Cur.: *David L. Reese* •
Historical Museum / Decorative Arts Museum – 1981
Decorative arts of NY, paintings and prints by NY artists ... 51186

Grey Art Gallery, New York University Art Collection, 100 Washington Sq E, New York, NY 10003 • T: +1 212 9986780 • F: +1 212 9954024 • greygallery@nyu.edu • www.nyu.edu/greyart •
Dir.: *Lynn Gumpert* • Asst. Dir.: *Gwen Stolyarov* • Deputy Dir.: *Frank Poueymirou* • Asst. Deputy Dir.: *Jeniffer Bakal* • Sc. Staff: *Michele Wong* (Gallery, Registrar) • Preparator: *Christopher Skura* • Asst. Preparator: *David Colossi* •
Fine Arts Museum / University Museum – 1975
Paintings, sculpture, graphics, contemporary Asian and Middle East art ... 51187

Guggenheim Museum Soho, 575 Broadway, New York, NY 10012 • T: +1 212 4233500 • F: +1 212 4233787 • publicaffairs@guggenheim.org • www.guggenheim.org •
Dir.: *Thomas Krens* •
Fine Arts Museum / Public Gallery – 1992
Late 19th c and 20th c European and American art, esp. paintings, sculptures, works on paper ... 51188

Hamilton Grange National Memorial, 287 Convent Av, New York, NY 10031 • T: +1 212 2835154, 8258874 • F: +1 212 8256874 • joe_avery@nps.gov • www.nps.gov/hagr •
Cur.: *Judith Mueller* •
Historical Museum – 1924
Home of Alexander Hamilton (1757-1804), memorabilia, early 19th c furnishing, ... 51189

Hampden-Booth Theatre Library at the Players, 16 Gramercy Park, New York, NY 10003 • T: +1 212 2281861 • F: +1 212 2536473 • hampdenboo@aol.com •
Pres.: *Robert Winter-Berger* • Cur.: *Raymond Wemmlinger* •
Performing Arts Museum – 1888
American and English stage, manuscripts ... 51190

Hatch-Billops Collections, 491 Broadway, New York, NY 10012 • T: +1 212 9663231 • F: +1 212 9663231 • Hatch-Billops@worldnet.att.net •
Pres.: *Camille Billops* •
Fine Arts Museum – 1975
Collection of primary and secondary resource materials in the Black Cultural Arts – Owen & Edith Dodson Memorial Collection ... 51191

The Hispanic Society of America, 155th St and Broadway, New York, NY 10032 • T: +1 212 9262234 • F: +1 212 6900743 • info@hispanicsociety.org • www.hispanicsociety.org •
Dir.: *Mitchell A. Codding* • Pres.: *Theodore S. Beardsley* • Cur.: *Gerald J. MacDonald* (Modern Books) • *Patrick Lenaghan* (Iconography, Medals) • *Marcus B. Burke* (Paintings) • Asst. Cur.: *John O'Neill* (Manuscripts, Rare Books) •
Fine Arts Museum – 1904
Art, sculpture, furniture, ceramics, metalwork, jewellery, textiles, archaeology, costumes, glass ... 51192

Hunter College Art Galleries, 695 Park Av, New York, NY 10021 • T: +1 212 7724991 • F: +1 212 7724554 • www.hunter.cuny.edu/artgalleries •
Dir.: *Sanford Wurmfeld* • Cur.: *Tracy L. Adler* • Asst. Cur.: *Sarah Archino* • Preparator, Registrar: *Phi Nguyen* •
Fine Arts Museum / University Museum – 1984
American art since 1945 ... 51193

International Center of Photography, 1133 Ave of the Americans and 43rd Street, New York, NY 10036 • T: +1 212 8570000 • F: +1 212 8570090 • info@icp.org • www.icp.org •
Dir.: *Willis Hartshorn* • Deputy Dir.: *Phillip Block* (Programs) • *Brian Wallis* (Exhibitions & Collections) • *Colleen Criste* (External Affairs) • *Steve Rooney* (Administration) • Cur.: *Christopher Phillips* • *Carol Squiers* • *Edward Earle* (Collections, Digital Media) • Sc. Staff: *Phyllis Levine* (Communications) • *Barbara Woytowicz* (Registrar) • *Deidre Donohue* (Library) •
Special Museum – 1974
Vernacular photography, African American images from 1860 to 1940 – Library ... 51194

International Print Center New York, 526 W 26th St 824, New York, NY 1001 • T: +1 212 9895090 • F: +1 212 9896069 • contact@ipcny.org • www.icpny.org •
Dir.: *Anne Coffin* •
Fine Arts Museum
Contemporary and historical exhibitions of fine art prints ... 51195

Intrepid Sea-Air-Space Museum, W 46th St and 12th Av, New York, NY 10036 • T: +1 212 2450072 • www.intrepidmuseum.org •
Pres. & C.E.O.: *Martin Steel* •
Science&Tech Museum / Military Museum – 1982
900 ft long aircraft carrier intrepid, Vietnam era destroyer Edson, destroyer escort Slater, guided missile submarine Growler, MIG-21, Scimitar, Etendard, F-14 Super Tomcast, light ship Nantucket ... 51196

ISE Art Foundation, 555 Broadway, New York, NY 10012 • T: +1 212 9251649 • F: +1 212 2269362 • info@isefoundation.org • www.isefoundation.org •
Public Gallery – 1984 ... 51197

Italian American Museum, 28 W 44th St, New York, NY 10036 • T: +1 212 64220220 • F: +1 212 6422069 • info@italianamericanmuseum.org • www.italianamericanmuseum.org •
Pres.: *Dr. Joseph V. Scelsa* • Cur.: *Dr. Philip V. Cannistraro* •
Historical Museum – 2001
Italian artifacts, cultural heritage of Italy, contributions to American culture – Library ... 51198

Japan Society Gallery, 333 E 47th St, New York, NY 10017 • T: +1 212 7151233 • F: +1 212 7151262 • amunroe@japansociety.org • www.japansociety.org •
Dir.: *Alexandra Munroe* • Pres.: *Frank L. Ellsworth* • Man.: *Eleni Cocordas* (Exhibitions) •
Fine Arts Museum – 1907
Japanese art ... 51199

The Jewish Museum, 1109 Fifth Av, New York, NY 10128 • T: +1 212 4233200 • F: +1 212 4233232 • jewishmus@aol.com • www.thejewishmuseum.org •
Dir. & C.E.O.: *Joan Rosenbaum* • Chm.: *Susan Lytle Lipton* • Dep. Dir.: *Lynn Thommen* (External Affairs) • Cur.: *Fred Wassermann* • Sc. Staff: *John L. Vogelstein* (Press) • *Thomas A. Dougherty* (Admin.) • *Ruth Beesch* (Exhibitions, Programs) • *Mary Walling* • *Susan L. Braunstein* (Archaeology, Judaica) • *Susan Goodman* • *Norman Kleeblatt* (Fine Arts) • *Vivian Mann* (Judaica) • *Nancy McGary* (Collections and Exhebitions) • *Aviva Weintraub* (Media, Public Programms) • *Grace Rapkin* (Marketing) • *Anne Scher* (Communication) • *Debbie Schwab-Dorfman* (Marketing) • *Jane Rubin* (Registrar) • *Marcia Saft* (Visitor Service) •
Fine Arts Museum / Historical Museum / Religious Arts Museum – 1904
Judaica collection, ceremonial objects, paintings, drawings, sculpture, prints, textiles, antiquities, photographs, decorativ arts, coins, medals, historic manuscripts, artifacts, broadcast material ... 51200

John J. Harvey Fireboat Collection, 100 W 72nd St, New York, NY 10023 • www.fireboat.org •
Historical Museum ... 51201

Kiboko Projects, 44 Grand St, New York, NY 10013 • T: +1 212 3348014 • F: +1 212 2193588 • m.scheflen@verizon.net • www.kiboko.org •
Dir.: *Mark Scheflen* •
Fine Arts Museum
Fine art, video, photography ... 51202

Leslie-Lohman Gay Art Foundation, 127 B Prince St, New York, NY 10012 • T: +1 212 6737007 • F: +1 212 2600363 • LLDirector@earthlink.net • www.leslielohman.org •
Dir.: *Wayne Snellen* •
Public Gallery
Gay art world paintings and photographs ... 51203

Lladro Museum, 43 W 57th St, New York, NY 10019 • T: +1 212 8389352 •
Fine Arts Museum
Porcelain, leather ... 51204

The Lowe Gallery at Hudson Guild, 441 W 26 St, New York, NY 10001 • T: +1 212 7609800 •
Dir.: *James Furlong* •
Fine Arts Museum – 1948
Contemporary art ... 51205

Lower East Side Tenement Museum, 90 Orchard St, New York, NY 10002 • T: +1 212 4310233 • F: +1 212 4310402 • lestm@tenement.org • www.tenement.org •
C.E.O. & Pres.: *Ruth J. Abram* • Vice Pres.: *Renee Epps* (Admin.) • Chm.: *Raymond O'Keefe* • Cur.: *Steve Long* • Public Relations: *Katherine Snider* •
Historical Museum – 1988
Historical immigrant documents, artifacts ... 51206

Madame Tussaud's Wax Museum, 234 W 42nd St, New York, NY 10036 • T: +1 212 7199440 • sarah.sommerfield@madametussaudsny.com • www.madametussaudsny.com •
Fine Arts Museum ... 51207

Main Gallery of Henry Street Settlement, Abrons Arts Center, 466 Grand St, New York, NY 10002 • T: +1 212 5980400 • F: +1 212 5058329 •
Dir.: *Barbara L. Tate* •
Fine Arts Museum / Public Gallery – 1893 ... 51208

Marie Walsh Sharpe Art Foundation, Space-Program, 443 Greenwich St, New York, NY 10013 • T: +1 212 9253008 • F: +1 719 6353018 • shartpartfdn@qwest.net •
Public Gallery – 1997 ... 51209

Marymount Manhattan College Museum, 221 E 1st St, New York, NY 10021 • T: +1 212 5170692 • marymount.mmm.edu •
University Museum
Art – Library ... 51210

Merchant's House Museum, 29 E Fourth St, New York, NY 10003 • T: +1 212 7771089 • F: +1 212 7771104 • publicity@merchantshouse.org • merchantshouse.org •
Dir.: *Margaret Halsey Gardiner* • Pres.: *Anne Fairfax* • Cur.: *Mimi Sherman* •
Historical Museum – 1936
19th c furniture, decorative arts and textiles ... 51211

The Metropolitan Museum of Art, 1000 Fifth Av, New York, NY 10028-0198 • T: +1 212 8795500, 5357710 • F: +1 212 5703879 • webmaster@metmuseum.org • www.metmuseum.org •
Dir. & C.E.O.: *Philippe de Montebello* • Chm.: *Morrison H. Heckscher* (American Art) • *James C.Y. Watt* (Asian Art) • *George R. Goldner* (Drawings, Prints) • *Everett Fahy* (European Paintings) • *Ian Wardropper* (European Sculpture, Decorative Arts) • *William S. Lieberman* (Modern Art) • *Hubert von Sonnenburg* (Paintings Cons.) • Cur.: *Alice Cooney Frelinghuysen* (American Decorative Arts) • *Julie Jones* (Arts of Africa, Oceania and the Americas) • *H. Barbara Weinberg* (American Painting and Sculpture) • *Joan Aruz* (Ancient Near Eastern Art) • *Stuart Pyhrr* (Arms and Armor) • *Harold Koda* (Costume Institute) • *Dorothea Arnold* (Egyptian Art) • *Carlos Picon* (Greek and Roman Art) • *Dietrich von Bothmer* (Distinguished Research) • *Daniel Walker* (Islamic Art) • *Laurence B. Kanter* (Robert Lehman Coll) • *Peter Barnet* (Medieval Art and the Cloisters) • *J. Kenneth Moore* (Musical Instr) • *Maria Morris Hambourg* (Photographs) •
Fine Arts Museum – 1870
Ancient through modern art of Egypt, Greece, Rome, Near and Far East, Europe, Africa, Oceania, pre-Columbian cultures, USA ... 51212

The Mezzanine Gallery, The Metropolitan Museum of Art, 1000 Fifth Av, New York, NY 10028 • T: +1 212 5703767 • F: +1 212 3965049 • mezzgallery@metmuseum.org • www.metmuseum.org •
Dir.: *Michael Hladky* •
Fine Arts Museum
Printing ... 51213

Miriam and Ira D. Wallach Art Gallery, 116th St and Broadway, Schermerhorn Hall, New York, NY 10027 • T: +1 212 8547288, 8542877 • F: +1 212 8547800 • www.columbia.edu/cu/wallach •
Dir.: *Sarah Elliston Weiner* •
Fine Arts Museum / University Museum – 1986
Art coll of Columbia University ... 51214

MoMA, Museum of Modern Art, 11 W 53rd St, New York, NY 10019 • T: +1 212 7779400, 7089889 • info@moma.org • www.moma.org •
Dir.: *Glenn D. Lowry* • Chief Cur.: *Peter Galassi* (Photography) • *Deborah Wye* (Prints, Illustrated Books) • *John Elderfield* (Painting, Sculpture) • *Terence Riley* (Architecture, Design) • *Mary Lea Bandy* (Film, Video) • *Gary Garrels* (Drawings) •
Fine Arts Museum
Modern and contemporary art from 1880 to present ... 51215

The Morgan Library, 29 E 36th St, New York, NY 10016 • T: +1 212 6850008 • F: +1 212 4813484 • media@morganlibrary.org • www.morganlibrary.org •
Dir.: *Dr. Charles E. Pierce* • Dep. Dir.: *Brian Regan* • Cur.: *Robert E. Parks* (Literary, Historical Manuscripts) • *William M. Voelkle* (Medieval & Renaissance Manuscripts) • *John Bidwell* (Printed Books, Bindings) • *J. Rigbie Turner* (Music, Manuscripts and Books) • *Rhoda Eifel-Porter* (Drawings, Prints) • *Sidney H. Babcock* (Seals, Tablets) • Dir. Cons. Center: *Margaret Holben Ellis* • Cons.: *Patricia Reyes* • *Marie Fredericks* •
Fine Arts Museum / Library with Exhibitions – 1924
Paintings, art objects, manuscripts, drawings, prints – Gilbert & Sullivan coll ... 51216

Morris-Jumel Mansion, 65 Jumel Terrace, New York, NY 10032 • T: +1 212 9238008 • F: +1 212 9238947 • www.morisjumel.org •
Dir.: *Kenneth Moss* • Pres.: *Nancy Goshow* • Vice Pres.: *James Daly* • Cur.: *Joanna Pessa* •
Decorative Arts Museum / Historical Museum – 1904
Chippendale, Federal and Empire furniture, silver, china, crystal, prints and paintings from Colonial, Federal and Empire periods – archives ... 51217

Mount Vernon Hotel Museum, 421 E 61st St, New York, NY 10021 • T: +1 212 8386878 • F: +1 212 8387390 • info@mvhm.org • www.mvhm.org •
Dir.: *Minna Schneider* • Pres.: *Charlotte Armstrong* • Cur.: *Lisa Bedell* • Sc. Staff: *Lori Finkelstein* (Education) • *Rosalind Muggeridge* (Public Relations) •
Local Museum – 1939/2000
American decorative arts, 18th and 19th c documents and letters ... 51218

Municipal Art Society, 457 Madison Av, New York, NY 10022 • T: +1 212 9353960 • F: +1 212 7531816 • info@mas.org • www.mas.org •
Pres.: *Kent Barwick* • Vice Pres.: *Gloria Troy* • Chm.: *Frank Emile Sanchis* •
Association with Coll – 1893 ... 51219

El Museo del Barrio, 1230 Fifth Av, New York, NY 10029 • T: +1 212 8317272 • F: +1 212 8317927 • info@elmuseo.org • www.elmuseo.org •
Dir.: *Julian Zugazagoitia* • Chief Cur.: *Fatima Brecht* • Cur.: *Deborah Cullen* • Sc. Staff: *Noel Valentin* (Registrar) •

Fine Arts Museum – 1969
Works on paper, sculptures, prints, paintings, photography, 16 mm films on history, culture and art – junior museum, 510 seat theatre 51220

Museum at the Fashion Institute of Technology, Seventh Av at 27th St, New York, NY 10001-5992 • T: +1 212 2175970 • F: +1 212 2175978, 2175800 • steeleva@fitsuny.edu • www.fitnyc.suny.edu • Acting Dir./ Chief Cur.: *Dr. Valerie Steele* • Cur.: *Ellen Shanley* (Costume) • *Lynn Felsher* (Textiles) • Cons.: *Anahid Akasheh* • Asst. Cons.: *Glenn Peterson* • Registrar: *Deborah Nordon* •
Special Museum – 1967
Clothing, textiles 51221

Museum of American Finance, 48 Wall St, New York, NY 10005 • T: +1 212 9084110 • F: +1 212 9084601 • kaguilera@financialhistory.org • www.moaf.org •
Exec. Dir.: *N. N.* • Chm.: *John E. Herzog* • Asst. Dir.: *N. N.* • Sc. Staff: *Kristin Aguilera* (Communication) •
Historical Museum – 1988
American financial hist from the mid-18th c to present day, stock and bond certificates, books, periodicals 51222

The Museum of American Illustration at the Society of Illustrators, 128 E 63rd St, New York, NY 10021-7303 • T: +1 212 8382560 • F: +1 212 8382561 • society@societyillustrators.org • www.societyillustrators.org •
Dir.: *Terrence Brown* • Asst. Dir.: *Phyllis Harvey* •
Fine Arts Museum – 1901 51223

Museum of Arts and Design, 2 Columbus Circle, New York, NY 10019 • T: +1 212 2997777 • F: +1 212 4590926 • info@madmuseum.org • www.madmuseum.org •
Dir.: *Holly Hotchner* • Cur.: *David McFadden* • *Ursula Neumann* •
Fine Arts Museum
Contemporary craft, art and design from 20th c . 51224

Museum of Biblical Art, 1865 Broadway, New York, NY 10023 • T: +1 212 4081500 • F: +1 212 4081292 • info@mobia.org • www.mobia.org •
Exec. Dir.: *Dr. Erna Heller* • Chief Cur.: *Dr. Patricia Pongracz* • Sc. Staff: *Ute Schmid* (Exhibitions and Registration) •
Religious Arts Museum / Fine Arts Museum – 1816
Art with biblical themes, religious art 51225

Museum of Chinese in the Americas, 70 Mulberry St, New York, NY 10013 • T: +1 212 6194785 • F: +1 212 6194720 • info@moca-nyc.org • www.moca-nyc.org •
Pres.: *Fay Chew Matsuda* • C.E.O.: *Charles Lai* • Dep. Dir./Cur.: *Cynthia Lee* • Sc. Staff: *Micheal Hew Wing* (Program) • *Lamgen Leon* (Operations) • *Robert Meyer* (Collections) •
Fine Arts Museum / Historical Museum / Folklore Museum – 1980
Chinese history and culture in America, Contonese opera costumes, musical instruments, Chinatown store signs, laundry coll 51226

Museum of Jewish Heritage - A Living Memorial to the Holocaust, 36 Battery Pl, New York, NY 10280 • T: +1 646 4374200 • F: +1 646 4374311 • aspilka@mjhnyc.org • www.mjhnyc.org •
Dir.: *Dr. David G. Marwell* • Cur.: *Esther Brumberg* (Collections) • Dir.: *Dr. Louis Levine* (Collections) • Sc. Staff: *Anna Martin* (Registrar) • *Michael Minerva* (Operations) • *Abby R. Spilka* (Communication) •
Historical Museum / Folklore Museum – 1984
Jewish history, 20th-c memorial to the Holocaust 51227

Museum of Sex, 233 Fifth Av, New York, NY 10016 • T: +1 212 6896337, 9466323 • info@museumofsex.com • www.museumofsex.com •
Dir.: *Daniel Gluck* • Cur.: *Grady T. Turner* •
Special Museum – 2002 51228

The Museum of Television and Radio, 25 W 52nd St, New York, NY 10019 • T: +1 212 6216600 • F: +1 212 6216700 • www.mtr.org •
Cur.: *Ronald Simon* • *David Bushman* •
Science&Tech Museum – 1975 51229

Museum of the American Piano, 291 Broadway, New York, NY 10007-1814 • T: +1 212 2464646, 4066060 • F: +1 212 4065245 • pianomuseum@pianomuseum.com • www.museumforpianos.org •
Dir.: *Kalman Detrich* •
Music Museum – 1981 51230

Museum of the City of New York, 1220 Fifth Av at 103rd St, New York, NY 10029 • T: +1 212 5341672 ext 200 • F: +1 212 4230758 • mcny@mcny.org • www.mcny.org •
Dir.: *Susan Henshaw Jones* • Deputy Dir.: *Dr. Sarah Henry* (Programs) • Cur.: *Deborah D. Waters* (Decorative Arts) • *Bob Shamis* (Prints, Photographs) • *Phyllis Magidson* (Costumes) • *Marty Jacobs* (Theatre) • *Andrea Henderson Fahnestock* (Paintings, Sculpture) •
Local Museum – 1923
Social, economic, intellectual and political hist of NY, costumes. furniture, dec and fine arts, fire equip, theatrical and musical items, marine and military items 51231

National Academy Museum, 1083 Fifth Av, New York, NY 10128 • T: +1 212 3694880 • F: +1 212 3606795 • kbrady@nationalacademy.org • www.nationalacademy.com •
Dir.: *Dr. Annette Blaugrund* • Cur.: *Dr. David Dearinger* • Cons.: *Lucie Kinsolving* •
Fine Arts Museum – 1825
American painting, sculpture, graphic arts, drawings – Archives 51232

National Arts Club, 15 Gramercy Park S, New York, NY 10003 • T: +1 212 4753424 • F: +1 212 4753692 • www.nationalartsclub.org •
Pres.: *Alton James Jr.* • Chm.: *Cheryl Green* • Cur.: *Carol Lowrey* •
Fine Arts Museum – 1898
19th and 20th c American painting, sculpture, works on paper, decorative arts 51233

National Association of Women Artists Gallery, 80 Fifth Av, Ste 1405, New York, NY 10011 • T: +1 212 6751616 • F: +1 212 6751616 • office@nawanet.org • www.nawanet.org •
Dir.: *Marcelle Harwell* • *Jennifer Mahlman* •
Public Gallery / Fine Arts Museum
library 51234

National Museum of Catholic Art and History, 443 E 115th St, New York, NY 10029 • T: +1 212 8285209 • F: +1 212 8285208 • info@nmcah.org • www.nmcah.org •
Chief Cur.: *Peggy Hammerle-McGuire* •
Religious Arts Museum / Fine Arts Museum – 1995
Old master and contemporary religious paintings, sculpture, prints, liturgical vestments 51235

National Museum of the American Indian, Smithsonian Institution, George Gustav Heye Center, 1 Bowling Green, New York, NY 10004 • T: +1 212 5143700 • F: +1 212 5143800 • www.nmai.si.edu •
Dir.: *W. Richard West* • Dep. Dir.: *Douglas E. Evelyn* • Asst. Dir.: *Charlotte Heth* • *Donna A. Scott* • *Bruce Bernstein* • *James Volkert* • Chief Cur.: *Mary Jane Lenz* •
Ethnology Museum / Historical Museum – 1916
American Indian archaeology, ethnology, hist, art and culture, music, literature 51236

National Sculpture Society, 237 Park Av, New York, NY 10017 • T: +1 212 7645645 • www.nationalsculpture.org •
Association with Coll 51237

Neue Galerie New York, 1048 Fifth Av, New York, NY 10028 • T: +1 212 6286200 • F: +1 212 6288824 • museum@neuegalerie.org • www.neuegalerie.org •
Dir.: *Gerwald Sonnberger* • *Renée Price* •
Fine Arts Museum – 2001
German and Austrian fine and decorative art (20th c) 51238

Neustadt Museum of Tiffany Art, 5-26 46th Av, New York, NY 11101 • T: +1 718 3618489 • F: +1 718 3921420 • nmtamuseum@aol.com • www.neustadtmuseum.org •
Pres.: *Milton D. Hassol* • C.E.O.: *Nicholas Cass-Hassol* • Trustees: *Richard Hanna* • *Sheil Tabakoff* • *David Specter* • *Mary Alice McKay* •
Fine Arts Museum – 1969
Tiffany lamps, glass and jewels 51239

New Museum of Contemporary Art, 556 W 22nd St, New York, NY 10011 • T: +1 212 2191222 • F: +1 212 4315328 • newmu@newmuseum.org • www.newmuseum.org •
Dir.: *Lisa Phillips* • Deputy Dir.: *Lisa Roumell* • Cur.: *Dan Cameron* •
Fine Arts Museum – 1977 51240

New York City Fire Museum, 278 Spring St, New York, NY 10013 • T: +1 212 6911303 • F: +1 212 9240430 • www.nycfiremuseum.org •
Dir.: *Joann Kay* •
Historical Museum – 1987
18th c to present fire-related art and artifacts, horse and hand-drawn apparatus, fire buckets, trumpets, insurance marks 51241

New York City Police Museum, 100 Old Slip, New York, NY 10005 • T: +1 212 4803100 • F: +1 212 4809757 • info@nycpolicemuseum.org • www.nycpolicemuseum.org •
Dir.: *Ninfa Segarra* • Cur.: *Michael Cronin* •
Historical Museum – 1929 51242

The New York Historical Society Museum, 170 Central Park W, New York, NY 10024 • T: +1 212 8733400 • F: +1 212 8748706 • NYHS@interport.net • www.nyhistory.org •
Dir.: *Kenneth T. Jackson* • Pres. & C.E.O.: *Louise Mirror* • Chm.: *Nancy Newcomb* • Vice Pres.: *Jan Ramirez* • Sc. Staff: *Margaret Heilbrunn* (Library) • *Valerie Komor* (Print) • *Rober Del Bagno* (Head Exhibitions) • *Travis Steward* (Public Ralations) • *Nina Nazionale* (Public Service, Library) • *L.J. Krizner* (Education) • *Nicole Wells* (Reproductions) • *Ione Saroyan* (Visitor) • *Steven Jaffe* (Public Historien) • *Kathleen Hulser* (Public Historien) •
Historical Museum / Local Museum – 1804
Watercolors, paintings, portraits, silver, furniture, Tiffany lamps and glass, ceramics, glass, sculpture, toys, folk art, military and naval hist coll, prints, architectural drawings, rare books, documents, maps, manuscripts 51243

New York Public Library for the Performing Arts, 40 Lincoln Center Plaza, New York, NY 10023-7498 • T: +1 212 9701830 • F: +1 212 8701870 • bcohenstratyner@nypl.org • www.nypl.org •
Pres.: *Paul Leclerc* • Chm.: *Samuel Butler* • Exec. Dir.: *Jaqueline Z. Davis* (Performing Arts) • Sc. Staff: *Donald J. Vlack* (Design) • *Robert McGlynn* (Illustration) • *Barbara Cohen-Straytner* (Exhibiton) • *Robertine Taylor* (Theatre) • *Charles Eubanks* (Music) • *Donald McCormick* (Recording) • *Madeleine Nichols* (Dance) •
Performing Arts Museum / Library with Exhibitions – 1965
Prints, letters, manuscripts, photographs, posters, films, video tapes, memorabilia, dance, recordings 51244

New York Studio School of Drawing, Painting and Sculpture - Gallery, 8 W Eighth St, New York, NY 10011 • T: +1 212 6736466 • F: +1 212 7770996 • dcohen@nyss.org • www.nyss.org •
Dir.: *David Cohen* •
Public Gallery – 1964
Drawings, paintings, sculptures 51245

Nicholas Roerich Museum, 319 W 107th St, New York, NY 10025-2799 • T: +1 212 8647752 • F: +1 212 8647704 • director@roerich.org • www.roerich.org •
Dir.: *Daniel Entin* •
Fine Arts Museum – 1958
Paintings of Tibet, India, Himalaya area by N. Roerich 51246

The Painting Center, 52 Greene St, New York, NY 10013 • T: +1 212 3431060 • www.thepaintingcenter.com •
Dir.: *Christina Chow* •
Public Gallery
Paintings 51247

Pen and Brush Museum, 16 E 10 St, New York, NY 10003 • T: +1 212 4753669 • F: +1 212 4756018 • PenBrush99@aol.com • www.penandbrush.org •
Pres.: *Ivece Portiabohn* •
Fine Arts Museum – 1893
Oil, watercolours, pastel, graphics, sculpture, craft 51248

Pratt Manhattan Gallery, 114W 14th St, New York, NY 10011 • T: +1 718 6477778 • F: +1 718 6363785 • exhibits@pratt.edu • www.pratt.edu/exhibitions •
Dir.: *Loretta Yarlow* • Asst.Dir.: *Nicholas Battis* • Sc. Staff: *Katherine Davis* (Design) •
Fine Arts Museum / University Museum – 1975
European and American paintings, sculpture, prints, graphics, decorative art of 19th/20th c 51249

Rose Museum at Carnegie Hall, 154 W 57th St, New York, NY 10019 • T: +1 212 9039629 • F: +1 212 5825518 • gfrancesconi@carnegihall.org • www.carnegiehall.org •
Dir.: *Gino Francesconi* •
Performing Arts Museum – 1991
Hist and development of the Carnegie Hall, hist of the studios, hist events 51250

Salmagundi Museum of American Art, 47 Fifth Av, New York, NY 10003 • T: +1 212 2557740 • F: +1 212 2290172 • www.salmagundi.org •
Pres.: *Richard Pionk* • Chm.: *Edward Brennan* • Cur.: *Thomas Picard* • *Ruth Reininghaus* •
Fine Arts Museum / Association with Coll – 1871
Paintings, sculptures, photography – library 51251

The Salt Queen Foundation, 969 First Av, New York, NY 10022 • T: +1 212 7596561 • thesaltqueen-foundation@bettina-werner.com • www.thesaltqueen-foundation.org •
Dir.: *Bettina Werner* •
Fine Arts Museum 51252

Scandinavia House, 58 Park Av, New York, NY 10016 • T: +1 212 7793587 •
Fine Arts Museum 51253

Schomburg Center for Research in Black Culture, c/o New York Public Library, 515 Malcolm X. Blvd, New York, NY 10037-1801 • T: +1 212 4912200 • F: +1 212 4916760 • www.schomburgcenter.org •
Dir.: *Howard Dodson* • Head: *Genette McLaurin* (Reference) • *Victor Smyth* (Art, Artifacts) • *James Briggs Murray* (Moving Image, Recorded Sound) • *Diana Lachatenere* (Archives) •
Historical Museum / Ethnology Museum – 1925
Afro-American, African Diaspora and African life, culture and history 51254

Sidney Mishkin Gallery of Baruch College, 135 E 22nd St, New York, NY 10010 • T: +1 212 8022690 • F: +1 212 8022693 • www.baruch.cuny.edu/mishkin/gallery.html •
Dir.: *Sandra Kraskin* •
Fine Arts Museum / University Museum – 1981
20th c American and European drawings, paintings, photographs, prints and sculptures 51255

Skyscraper Museum, Southern Battery Park City, 39 Battery Pl, New York, NY 10280, mail addr: 55 Broad St, New York, NY 10004 • T: +1 212 9681961, 9456324 • F: +1 212 9681961 • admin@skyscraper.org • www.skyscraper.org •
Dir. & Pres.: *Carol Willis* • Chm.: *Jed Marcus* • Sc. Staff: *Kenneth Levine* • *Matt Pinto* (Public Relations) •
Special Museum – 1996/2003
Study oh high-rise buildings in past, present and future 51256

Soho Photo Gallery, 15 White St, New York, NY 10013 • T: +1 212 2268571 • larrydavis@asan.com • www.sohophoto.com •
Fine Arts Museum 51257

Solomon R. Guggenheim Museum, 1071 Fifth Av, New York, NY 10128 • T: +1 212 4233500 • F: +1 212 4233787 • visitorinfo@guggenheim.org • www.guggenheim.org •
Dir.: *Thomas Krens* • Deputy Dir. & Chief Cur.: *Lisa Dennison* • Cur.: *John Hanhardt* (Film, Media Arts) • *Jennifer Blessing* (Projects) • *Nancy Spector* • *Germano Celant* (Contemporary Art) • *Robert Rosenblum* • *Carmen Gimenez* (20th-C Art) • Cons.: *Paul Schwartzbaum* •
Fine Arts Museum – 1937
Modern and contemporary paintings, sculpture, video and works on paper of last 100 years 51258

Sony Wonder Technology Lab, 56th St at Madison Av, New York, NY 10022 • T: +1 212 8338100 • F: +1 212 8334445 • karen_kelso@sonyusa.com • www.sonywondertechlab.com •
Dir.: *Karen Kelso* •
Science&Tech Museum – 1994
Technology and science 51259

The South Street Seaport Museum, 12 Fulton St, New York, NY 10038 • T: +1 212 7488600 • F: +1 212 7488610 • richardstepler@southstseaport.org • www.southstseaport.org •
Cur.: *Sharon Holt* (Exhibitions) • *Norman Brouwer* (Ships) • *Barbara Henry* (Printing) •
Science&Tech Museum – 1967
Ship models, ship artifacts, navigational instruments, tools, fishmarket artifacts, historic ships, documents, photographs 51260

The Spanish Institute, 684 Park Av, New York, NY 10021 • T: +1 212 6280420 • F: +1 212 7344177 • www.spanishinstitute.org •
Public Gallery – 1954
Temporary exhibitions of Spanish art 51261

Statue of Liberty National Monument and Ellis Island Immigration Museum, Liberty Island, New York, NY 10004 • T: +1 212 3633200 • F: +1 212 3636302 • stli_museum@nps.gov • www.nps.gov/stli •
Dir.: *Diane H. Dayson* • Chief Cur.: *Diana R. Pardue* • Cur.: *Judith Giuricco* (Exhibitions, Media) • *Geraldine Santoro* (Collections) • Libr.: *Barry Moreno* • *Jeffrey Dosik* • Archivist: *George Tselos* • *Janet Levine* •
Historical Museum / Historic Site – 1924
History, folk art – library, archive 51262

The Studio Annex, 279 Fith Av, New York, NY 10016-6501 • T: +1 914 2731452 • F: +1 914 2736480 • thestudiony@optonline.net • www.thestudiony-alternative.com •
Dir.: *Katie Stratis* •
Fine Arts Museum 51263

The Studio Museum in Harlem, 144 W 125th St, New York, NY 10027 • T: +1 212 8644500 • F: +1 212 8644800 • smhny@aol.com • www.studiomuseuminharlem.org •
Dir.: *Lowery Stokes Sims* • Chm.: *Raymond J. McGuire* • Chief Cur.: *Thelma Golden* • Asst. Cur.: *Christine Y. Kim* •
Fine Arts Museum – 1967
19th and 20th c African American artists, traditional and contemporary African art, Caribbean art – Archives 51264

Swiss Institute - Contemporary Art, 495 Broadway, New York, NY 10012 • T: +1 212 9252035 • F: +1 212 9252040 • info@swissinstitute.net • www.swissinstitute.net •
Dir.: *Gianni Jetzer* •
Public Gallery 51265

Synagogue for the Arts, 49 White St, New York, NY 10013 • T: +1 212 9667141 • F: +1 212 9664968 • info@synagogueforthearts.org • www.synagogueforthearts.org •
Dir.: *Marilyn Sonntag* •
Fine Arts Museum
Paintings, drawings, relief sculpture, photography and prints 51266

T.F. Chen Cultural Center, New World Art Center, 250 Lafayette St, New York, NY 10012-4075 • T: +1 212 9664363 • F: +1 212 9665285 • chen@tfchen.org • www.tfchen.org •
Pres.: *Lucia Chen* • Cur.: *Julie Chen* •
Fine Arts Museum – 1996
International contemporary art, neo-iconographic painting 51267

Terrain Gallery, Aesthetic Realism Foundation, 141 Greene St, New York, NY 10012 • T: +1 212 7774490 • F: +1 212 7774426 • terraingallery@aestheticrealism.org • www.terraingallery.org •
Dir.: *Dorothy Koppelman* • *Carrie Wilson* • Coord.: *Marcia Rackow* •
Public Gallery – 1955
Abstract paintings, prints, drawings and photographs with commentary 51268

Theodore Roosevelt Birthplace, 28 E 20th St, New York, NY 10003 • T: +1 212 2601616 • F: +1 212 6773587 • masithrb@nps.gov • www.nps.gov/thrb/ •
Historical Museum / Historic Site – 1923
Theodore Roosevelt (1882-1945) memorabilia and historical items 51269

Tibet House Museum, 22 W 15th St, New York, NY 10011 • T: +1 212 8070563 • info@tibethouse.org • www.tibethouse.org •
Special Museum
Art and cultural hist 51270

Times Square Gallery, Hunter College City University of New York, 450 W 41st St, New York, NY 10036 • T: +1 212 7724991 • F: +1 212 7724554 •
University Museum
Modern and contemporary art 51271

Tresure Room Gallery of the Interchurch Center, 475 Riverside Dr, New York, NY 10115 • T: +1 212 8702200 • F: +1 212 8702440 • Dennis@interchurch-center.org • www.interchurch-center.org •
Dir.: *Mary E. McNamara* • C.E.O. & Pres.: *Sue M.*

U

Dennis • Chm.: *Joanne Fernandez* • Cur.: *Dorothy Cochran* • Sc. Staff: *Ernest Rubinstein* (Library) •
Library with Exhibitions – 1959
Bible exhibits; changing exhibitions of contemporary art ... 51272

Trinity Museum of the Parish of Trinity Church, Broadway and Wall St, New York, NY 10006 • T: +1 212 6020800 • F: +1 212 6029648 • djette@trinitywallstreet.org • www.trinitywallstreet.org •
C.E.O.: *Dr. James H. Cooper* • Cur.: *Gwynedd Cannan* • *Melissa Haley* •
Religious Arts Museum – 1966
Books, prints, photographs, paintings, artifacts and religious objects, documents – Archives ... 51273

The Ukrainian Museum, 203 Second Av, New York, NY 10003 • T: +1 212 2280110 • F: +1 212 2281947 • info@ukrainianmuseum.org • www.ukrainianmuseum.org •
Dir.: *Maria Shust* • Pres.: *Olha Hnateyko* • Admin. Dir.: *Daria Bajko* •
Fine Arts Museum / Folklore Museum – 1976
Fine art, folk art, photographs, documents, fkyers, posters, Ukrain art, culture, architecture and hist, hist of Ukrainian immigration, numismatics and philatelic coll ... 51274

Union of American Hebrew Congregation, 633 Third Av, New York, NY 10017 • T: +1 212 6504040 • F: +1 212 6504239 • synagoguemgmt@uahc.org • www.uahc.org •
Dir.: *Dale Glasser* •
Association with Coll / Library with Exhibitions
Books on Synagogue architecture and art, ceremonial objects and works of Jewish artists ... 51275

Visual Arts Gallery, 601 W 26th St, New York, NY 10001 • T: +1 212 5922010 • proffice@sva.edu •
Public Gallery ... 51276

Visual Arts Museum, 209 E 23rd St, New York, NY 10010 • T: +1 212 5922144 • F: +1 212 5922095 • fditommaso@adm.schoolofvisualarts.edu • www.schoolofvisualarts.edu •
Dir.: *Francis Di Tommaso* • Assoc. Dir.: *Rachel Gugelberger* • Asst. Dir.: *Geoffrey Detrani* (Exhibitions) • Sc. Staff: *Collin Mura-Smith* (Gallery) • *Laura Yeffeth* (Registrar) • *Michelle Meier* (Admin. Gallery and Museum) •
Fine Arts Museum – 1971
Photography, fine art, new media, digital graphic design ... 51277

Wally Findlay Galleries, 124 E 57th St, New York, NY 10022 • T: +1 212 4215390 • F: +1 212 8382460 •
Dir.: *Heather Parks* •
Fine Arts Museum
American and European arts in 19th-and 20th c 51278

Whitney Museum of American Art, 945 Madison Av, New York, NY 10021 • T: +1 212 5703600 • F: +1 212 5701807 • info@whitney.org • www.whitney.org •
Dir.: *Adam D. Weinberg* • *Alice Pratt Brown* • Assoc. Dir.: *Hillary Blass* • *Helena Rubinstein* • *Scott Elkins* • *Donna DeSalvo* (Cur. Permanent Collections) • *Carol Mancusi-Ungaro* (Conservation and Research) • *Christy Putnam* (Collections and Exhibitions) • *Jan Rothschild* (Communications) • Cur.: *Chissie Iles* • *Barbara Haskell* (Film, Video) • *Callie Angell* (Andy Warhol Film Project) • *David Kiehl* (Prints) • *Elisabeth Sussman* • *Sondra Gilman* • *Sylvia Wolf* (Photography) • Assoc. Cur.: *Dana Miller* • *Shamim Momin* • *Debra Singer* •
Fine Arts Museum – 1930
Paintings, sculpture, drawings, prints and photography, film and video ... 51279

Whitney Museum of American Art at Altria, 120 Park Av at 42 nd St, New York, NY 10017 • T: +1 917 6632453 •
Public Gallery ... 51280

Yeshiva University Museum, 15 W 16th St, New York, NY 10011 • T: +1 212 2948330 • F: +1 212 2948335 • sherskowitz@yum.cjh.edu • www.yumuseum.org •
Dir./C.E.O.: *Sylvia A. Herskowitz* • Dir.: *Gabriel M. Goldstein* (Programs and Exhibitions) • Cur.: *Rachelle Bradt* (Education) • *Bonni-Dara Michaels* (Collections) • *Reba Wulkan* (Contemporary Exhebitions) •
Historical Museum / University Museum / Ethnology Museum / Fine Arts Museum – 1973
Jewish ceremonial objects, Jewish hist, ethnography, art and culture, textile coll of ceremonial costumes and clothes ... 51281

Newark DE

University Gallery, University of Delaware, Main St at N College Av, Newark, DE 19716 • T: +1 302 8318242 • F: +1 302 8318251 • universitymuseums@udel.edu • www.udel.edu •
Dir.: *Belena S. Chapp* • Cur.: *Janet Gardner Broske* •
Fine Arts Museum / University Museum – 1978
Art ... 51282

University Museums, University of Delaware, 209 Mechanical Hall, Newark, DE 19716 • T: +1 302 8318037 • F: +1 302 8318057 • universitymuseums@udel.edu • www.udel.edu •
Dir.: *Janis A. Tomlinson* • Cur.: *Janet Gardner Broske* •
University Museum / Fine Arts Museum – 1978
19th-20th c American & European graphics, Pre-Columbian pottery, Gertrude Kasebier photographs, African art, Abraham Walkowitz drawings, Moholy-Nagy drawings, Lucy S. and Frederick Herman native American art ... 51283

Newark NJ

Aljira Center for Contemporary Art, 100 Washington St, Newark, NJ 07102 • T: +1 973 6436877 • F: +1 973 6433594 • info@aljira.org • www.aljira.org •
Exec. Dir.: *Victor Davson* •
Public Gallery – 1983 ... 51284

Junior Museum, Newark Museum, 49 Washington St, Newark, NJ 07101 • T: +1 973 5966605 • F: +1 973 6420459 •
Fine Arts Museum – 1926
Children's artwork ... 51285

New Jersey Historical Society Museum, 52 Park Pl, Newark, NJ 07102 • T: +1 973 5968500 • F: +1 973 5966957 • info@jerseyhistory.org • www.jerseyhistory.org •
Dir.: *Dr. Sally Yerkovich* • Deputy Dir.: *Janet Rassweiler* • Cur.: *Claudia Ocello* •
Historical Museum – 1845
New Jersy and American hist, transportation, industry, paintings, costumes – library ... 51286

The Newark Museum, 49 Washington St, Newark, NJ 07101-0540 • T: +1 973 5966550 • F: +1 973 6420459 • lmcconnell@newarkmuseum.org • www.newarkmuseum.org •
Dir.: *Mary Sue Sweeney Price* • Cur.: *Dr. Susan Auth* (Classical) • *Ulysses G. Dietz* (Decorative Arts) • *Valrae Reynolds* (Asian Coll.) • *Nii Quarcoopome* (Africa, The Americas, Pacific) • *Joseph Jacobs* (Painting, Sculpture) • *Dr. Sule Oygur* (Earth Sciences) •
Fine Arts Museum / Decorative Arts Museum / Archaeology Museum – 1909
Asian, American, African art, decorative art, Egypt, Greek and Roman Art – planetarium, zoo ... 51287

Robeson Center, Rutgers University, 350 Dr. Martin Luther King Jr. Blvd, Newark, NJ 07102 • T: +1 973 3531610 •
Dir.: *Corlis Cavalieri* •
University Museum ... 51288

Newark OH

Licking County Art Association Gallery, 50 S Secand St, Newark, OH 43055 • T: +1 740 3498031 • F: +1 740 3453787 • lcaa@msmisp.com •
Pres.: *Leah Mitchell* • Man.: *Vivian DuRant Smith* •
Fine Arts Museum – 1959
Prints and paaintings, ceramics, fiber pieces, glass, wood-sculpture ... 51289

Licking County Historical Society, Veterans Park, N 6th St, Newark, OH 43058-0785 • T: +1 740 3454898 • F: +1 740 3452983 • lchs@alltel.net • www.lchsohio.org •
Pres.: *Tim Riffle* •
Local Museum / Association with Coll – 1947
Regional history, hist buildings, ... 51290

Moundbuilders State Memorial and Museum, 99 Cooper Av, Newark, OH 43055 • ohiohistory.org •
Head: *James Kingery* •
Ethnology Museum / Archaeology Museum
Prehistoric Indian Art 1000 B.C. - 700 A.D. ... 51291

National Heisey Glass Museum, 169 W Church St, Newark, OH 43055 • T: +1 740 3452932 • F: +1 740 3459638 • curator@heiseymuseum.org • www.heiseymuseum.org •
Dir.: *Bill Douglas* • Cur.: *Walter Ludwig* •
Decorative Arts Museum – 1974
Heisey glassware, glassmakers' tools, factory hist ... 51292

Sherwood-Davidson House, Veterans Park, Sixth St, Newark, OH 43058-0785 • T: +1 740 3456525 • F: +1 740 3452983 • sherwooddavidson@yahoo.com • lchsohio.org •
Pres.: *Tim Riffle* •
Local Museum / Historical Museum – 1947
Furniture, hist costumes, toys, antique books, glass ... 51293

Webb House Museum, 303 Granville St, Newark, OH 43055 • T: +1 740 3458540 • webbhouse@nextek.net • lchsohio.org •
Cur.: *Mindy Honey Nelson* •
Historical Museum – 1976
Period furnishings, art, silver, china, glass ... 51294

The Works, Ohio Center for History, Art and Technology, 55 S First St, Newark, OH 43055 • T: +1 740 3499277 • F: +1 740 3457252 • www.attheworks.org •
C.E.O.: *Marcia W. Downes* •
Historical Museum / Fine Arts Museum / Science&Tech Museum – 1996
Hist coll, Licking County hist, glass blowing, letterpress ... 51295

Newark Valley NY

Bement-Billings Farmstead, 9142 Rte 38, Newark Valley, NY 13811, mail addr: POB 222, Newark Valley, NY 13811 • T: +1 607 6429516 • F: +1 607 6429516 • nvhistorical@juno.com • www.tier.net/nvhistory •
Dir.: *Harriet Miller* •
Historical Museum / Historic Site – 1977
1840's furniture, cookware, tools ... 51296

Newark Valley Depot Museum, Depot St, Newark Valley, NY 13811 • T: +1 607 6429516 • F: +1 607 6429516 •
Dir.: *Kim Mayhew* •
Science&Tech Museum – 1977
Railroad artifacts and memorabilia ... 51297

Newaygo MI

Newaygo County Museum, 85 Water St, Newaygo, MI 49337 • T: +1 231 6529281 • F: +1 231 6522461 • bbillerb@newaygo.net •
Pres.: *Allen Bradley* • C.E.O.: *Barb Billerbeck* •
Local Museum – 1968
Regional hist, military uniforms, civil war, costumes – library ... 51298

Newbern VA

Wilderness Road Regional Museum, State Rte 611, 5240 Wilderness Rd, Newbern, VA 24126-0373 • T: +1 540 6744835 • F: +1 540 6741266 • wrrm@psknet.com •
Dir.: *Ann S. Bailey* • Cur.: *Sara C. Zimmerman* •
Local Museum – 1980
Regional hist of New River Valley, dolls & Quilts, Civil War items, farm & hand tools ... 51299

Newburgh NY

David Crawford House, 189 Montgomery St, Newburgh, NY 12550 • T: +1 914 5612585 • historicalsocietynb@yahoo.com • www.newburghhistoricalsociety.com •
Pres.: *Carla Decker* •
Historical Museum – 1884
19th c furniture, dec arts, toys and dolls, ship models of Hudson River craft ... 51300

Washington's Headquarters, Corner of Liberty and Washington Sts, Newburgh, NY 12551-1476 • T: +1 914 5621195 • www.nysparks.state.ny.us •
Sc. Staff: *Thomas A. Hughes* •
Military Museum – 1850
American Revolution, documents and military artifacts, firearms, furnishings ... 51301

Newbury MA

Coffin House, 14 High Rd, Newbury, MA 01951 • T: +1 978 4622634 • F: +1 978 4624022 • CoffinHouse@HistoricNewEngland.org • www.historicnewengland.org/visit/homes/coffin.htm •
Historical Museum – 1929
Evolution of domestic life in rural New England ... 51302

Spencer-Peirce-Little Farm, 5 Little's Ln, Newbury, MA 01951 • T: +1 978 4622634 • F: +1 978 4624022 • SpencerPeirceLittleFarm@HistoricNewEngland.org • www.historicnewengland.org/visit/homes/little.htm •
Pres.: *Philip Zea* •
Folklore Museum – 1971
Life on the farm over the centuries, glass and ceramic vessels, decorative art ... 51303

Newbury Park CA

Stagecoach Inn Museum, 51 S Ventu Park Rd, Newbury Park, CA 91320 • T: +1 805 4989441 • stagecoachmuseum@msn.com • www.stagecoachmuseum.org •
Dir.: *Sandra Hildebrandt* • Cur.: *Jackie Pizitz* (Education) • *Miriam Sprankling* (History) • *Don Morris* (Archaeology) •
Local Museum – 1967
Local hisory, Chumash native people, Spanish-Mexican and Anglo-American, minerals, fossils – library, nature trails ... 51304

Newburyport MA

Cushing House Museum, 98 High St, Newburyport, MA 01950 • T: +1 978 4622681 • F: +1 978 4620134 • hson@greennet.net • www.newburyhist.com •
Pres.: *James Chanler* • Dir.: *Audrey Ladd* •
Local Museum – 1877
Local history, silver, glass and china, paintings, dolls and toys, paperweights, clocks, costumes, carriages, military – library ... 51305

Custom House Maritime Museum, 25 Water St, Newburyport, MA 01950 • T: +1 978 4628681 • F: +1 978 4628740 • info@themaritimesociety.org • www.themaritimesociety.org •
CEO: *Cari Conway* •
Historical Museum – 1969
Local maritime hist, trad. boat building ... 51306

Newcastle TX

Fort Belknap Museum, Farm to Market Rd 2 mi south of Newcastle, Newcastle, TX 76372 • T: +1 940 5491856 • www.forttours.com •
Dir.: *Dr. K.F. Neighbours* •
Military Museum – 1851
Indian artifacts, officer's quarters, army barracks, household items from 1850s-1890s ... 51307

Newcastle WY

Anna Miller Museum, 401 Delaware, Newcastle, WY 82701 • T: +1 307 7464188 • F: +1 307 7464629 • annamm@trib.com •
Dir.: *Bobbie Jo Tysdal* •
Local Museum – 1966
Local hist, Indian artifacts, natural hist, paleontology ... 51308

Newcomerstown OH

Temperance Tavern, 221 W Canal St, Newcomerstown, OH 43832 • T: +1 614 4987735 •
Dir.: *Barbara Scott* •
Historical Museum / Local Museum / Historic Site – 1923
Local hist, Indian artifacts, period furniture and costumes ... 51309

USS Radford National Naval Museum, 238 W Canal St, Newcomerstown, OH 43832 • T: +1 740 4984446 • F: +1 740 4988803 • vanescott@sbcglobal.net • www.ussradford446.org •
Pres.: *Vane S. Scott* •
Military Museum – 2000
USS Radford DD/DDE 446 - Destroyer, WWII (Korea, Vietnam), 3-D Diorama of Rescue at Kula Gulf ... 51310

Newell SD

Newell Museum, 108 Third St, Newell, SD 57760 • T: +1 605 4561310 • F: +1 605 4562587 • newellmuseum@sdplains.com • www.cityofnewell.com •
Cur.: *Linda Velder* •
Local Museum – 1983
Clothing, Native American artifacts and culture, fossils, antique toys and dolls, musical instruments ... 51311

Newfield ME

19th Century Willowbrook Village, Rte 11 and Elm St, Newfield, mail addr: PO Box 28, Newfield, ME 04056 • T: +1 207 7932784 • director@willowbrookmuseum.org • www.willowbrookmuseum.org •
C.E.O.: *Amelia E. Chamberlain* •
Historical Museum – 1970
Carriages and sleighs, 1894 carousel, historic trade shops, farm equipment, historic houses, antique engines ... 51312

Newfield NJ

Matchbox Road Museum, 17 Pearl St, Newfield, NJ 08344 • F: +1 856 6970762 • www.mbxroad.com •
Special Museum – 1992
27000 miniature Matchbox vehicles ... 51313

Newhall CA

William S. Hart Museum, Natural History Museum of Los Angeles County, 24151 Newhall Av, Newhall, CA 91321 • T: +1 661 2544584 • information@hartmuseum.org • www.hartmuseum.org •
Dir.: *Donna Chipperfield* •
Fine Arts Museum / Decorative Arts Museum – 1958
Paintings by Russell, De Yong, J.M. Flagg, Remington a.o., sculpture, woodcarvings, Navajo blankets, rugs, furniture, clothing ... 51314

Newland NC

Avery County Historical Museum, 1829 Schultz Circle, Newland, NC 28657 • T: +1 828 7337111 • averycm@intelink-cafe.com • averymuseum.com •
Local Museum / Folklore Museum – 1977
Local genealogy, local minerals ... 51315

Newport OR

Oregon Coast History Center, 545 SW Ninth, Newport, OR 97365 • T: +1 541 2657509 • F: +1 541 2653992 • coasthistory@newportnet.com • www.newportnet.com/coasthistory/home.htm •
C.E.O.: *Loretta Harrison* •
Historical Museum – 1948
Maritime objects, logging, coastal hist ... 51316

Newport RI

Artillery Company of Newport Military Museum, 23 Clarke St, Newport, RI 02840 • T: +1 401 8468488 • F: +1 401 8463311 • info@newportartillery.org • www.newportartillery.org •
Military Museum – 1959
Military hist ... 51317

Astors' Beechwood-Victorian Living History Museum, 580 Bellevue Av, Newport, RI 02840 • T: +1 401 8463772 • F: +1 401 8496998 • info@astorsbeechwood.com • www.astorsbeechwood.com •
C.E.O: *Robert B. Milligan* •
Historical Museum – 1983
Local hist ... 51318

Belcourt Castle, 657 Bellevue Av, Newport, RI 02840-4288 • T: +1 401 8491566, 8460669 • F: +1 401 8465345 • belcourt2001@aol.com • www.belcourtcastle.com •
Exec. Dir.: *Harle H. Tinney* •
Decorative Arts Museum – 1957
Romanesque, Gothic, Renaissance, Louis 14 and Louis 15 and classic revival style interiors, silks, embroideries, Oriental rugs, paintings, porcelains, ancient art ... 51319

Colony House, Washington Sq, Newport, RI 02840, mail addr: 82 Touro St, Newport, RI 02840 • T: +1 401 8460813 • F: +1 401 8461853 • www.newporthistorical.org •
Dir.: *Ruth Taylor* •
Historic Site – 1854
Antique furnishings 51320

International Tennis Hall of Fame Museum, 194 Bellevue Ave, Newport, RI 02840 • T: +1 401 8493990 • F: +1 401 8498780 • newport@tennisfame.com • www.tennisfame.com •
Dir.: *Douglas Stark* •
Special Museum – 1954
Sports, hist of tennis – Information Research Center 51321

The Museum of Newport History, 127 Thames St, Newport, RI 02840, mail addr: 82 Touro St, Newport, RI 02840 • T: +1 401 8460813, 8418770 • F: +1 401 8461853 • www.newporthistorical.org •
Dir.: *Ruth Taylor* • Cur.: *M. Joan Youngken* • Registrar: *Kim Hanrahan* •
Historical Museum – 1854
Exhibits, artifacts, paintings, photos, furniture related to Newport's hist 1600s to 1900s 51322

Museum of Yachting, Fort Adams State Park, Newport, RI 02840 • T: +1 401 8471018 • F: +1 401 8478320 • museum@moy.org • www.moy.org •
C.E.O: *David C. Brown* •
Science&Tech Museum – 1983
Yachting, sailing 51323

National Museum of American Illustration, Vernon Court, 492 Bellevue Av, Newport, RI 02840 • T: +1 401 8518949 • F: +1 401 8518974 • art@americanillustration.org • www.americanillustration.org •
Dir.: *Judy Goffman Cutler* •
Fine Arts Museum – 1998
Works of illustrators 51324

Naval War College Museum, 686 Cushing Rd, Newport, RI 02841-1207 • T: +1 401 8414052, 8411317 • F: +1 401 8417689 • cembrolr@nwc.navy.mil • www.nwc.navy.mil/museum •
Dir.: *Bob Cembrola* •
Military Museum / Historical Museum – 1978
Naval hist 51325

Newport Art Museum, 76 Bellevue Av, Newport, RI 02840 • T: +1 401 8488200 • F: +1 401 8488205 • info@newportartmuseum.com • www.newportartmuseum.com •
Dir.: *Christine Callahan* • Cur.: *Nancy Grinnell* •
Fine Arts Museum – 1912
Drawings, paintings, sculpture, prints, photography from New England and Newport 51326

Newport Mansions, 424 Bellevue Av, Newport, RI 02840 • T: +1 401 8471000 • F: +1 401 8479477 • jpaduette@newportmansions.org • www.NewportMansions.org •
C.E.O.: *Trudy Coxe* • Cur.: *Paul F. Miller* •
Local Museum – 1945
Group of 9 mansions 51327

Redwood Library and Athenaeum, 50 Bellevue Av, Newport, RI 02840 • T: +1 401 8470292 • F: +1 401 8415680 • redwood@edgenet.net • www.redwoodlibrary.org •
C.E.O.: *Cheryl V. Helms* • Pres.: *Stephen G.W. Walk* •
Fine Arts Museum – 1747
18th-19th c Anglo-American portraits 51328

Rough Point, Newport Restoration Foundation, 51 Touro St, Newport, RI 02840 • T: +1 401 8497300 • F: +1 401 8490125 • geninfo@newportrestoration.org • www.newportrestoration.org •
Exec. Dir.: *Bruce MacLeish* • Dir.: *Pieter N. Roos* (Collections) •
Historic Site / Decorative Arts Museum – 1968
European & Rhode Island furniture, decorative and fine arts, oriental porcelain, architecture, drawings, photos 51329

Royal Arts Foundation, Belcourt Castle, 657 Bellevue Av, Newport, RI 02840-4280 • T: +1 401 8460669 • F: +1 401 8465345 • kevint@belcourt.com • www.belcourtcastle.com •
Dir.: *Harle H. Tinney* •
Decorative Arts Museum / Local Museum – 1957
Historic house, paintings, sculpture, textiles, ceramics 51330

Thames Museum, 77 Long Wharf, Newport, RI 02840 • T: +1 401 8496966 • F: +1 401 8497144 • carol@thamesmuseum.org •
Pres.: *Jane A. Holdsworth* •
Natural History Museum – 1948 51331

Newport Beach CA

Orange County Museum of Art, 850 San Clemente Dr, Newport Beach, CA 92660 • T: +1 949 7591122 ext 242 • F: +1 949 7595623 • info@ocma.net • www.ocma.net •
Dir.: *Dennis Szakacs* • Dep. Dir./ Cur.: *Elizabeth Armstrong* (Programs) • Cur.: *Irene Hofmann* (Contemporary Art) • *Karen Moss* (Collections, Education) •
Fine Arts Museum – 1921
California art since the 19th c – library 51332

Newport News VA

Golf Museum, 1500 Country Club Rd, Newport News, VA 23606 • T: +1 757 5953327 • F: +1 757 5964807 •
Special Museum – 1932
International golf hist, clubs, balls, tees 51333

The Mariners' Museum, 100 Museum Dr, Newport News, VA 23606 • T: +1 757 5962222 • F: +1 757 5917311 • info@mariner.org • www.mariner.org •
Pres.: *John B. Hightower* •
Historical Museum / Science&Tech Museum – 1930
Maritime hist, international small craft, lighthouse & lifesaving equipment 51334

Newsome House Museum, 2803 Oak Av, Newport News, VA 23607 • T: +1 757 2472360 • F: +1 757 9286754 • mkayaselcuk@nngov.com • www.newsomehouse.org •
Man.: *Mary Kayaselcuk* •
Historical Museum – 1991
Personal papers of the Newsome family, local African-American hist & culture 51335

Peninsula Fine Arts Center, 101 Museum Dr, Newport News, VA 23606 • T: +1 757 5968175 • F: +1 757 5960807 • info@pfac-va.org • www.pfac-va.org •
Exec. Dir.: *Crystal C. Warlitner* •
Fine Arts Museum – 1962
Temporary exhib during the year 51336

Virginia Living Museum, 524 J. Clyde Morris Blvd, Newport News, VA 23601 • T: +1 757 5951900 • F: +1 757 5994897 • webmaster@valivingmuseum.org • www.valivingmuseum.org •
Dir.: *Gloria R. Lombardi* • *David C. Maness* (Astronomy) • *George K. Mathews* •
Natural History Museum – 1964
Live native animals, plants, birds, insects, mammals, reptiles 51337

The Virginia War Museum, 9285 Warwick Blvd, Huntington Park, Newport News, VA 23607 • T: +1 757 2478523 • F: +1 757 2478627 • info@warmuseum.org • www.warmuseum.org •
Military Museum – 1923
Weapons, uniforms, posters, prints, vehicles, tanks, insignia, artillery & accoutrements relating to U.S. military from 1776 to present 51338

Newton IL

Newton Museum, 100 S Vanburen, Newton, IL 62448 • T: +1 618 7838141, 7833860 • F: +1 618 7838141 • newtonpl@pfbnewton.com •
Local Museum – 1928
Military, Indian artifacts, farm implements, uniforms – library 51339

Newton KS

Harvey County Historical Society, 2nd & Main, Newton, KS 67114 • T: +1 316 2832221 • F: +1 316 2832221 • info@hchm.org • www.hchm.org •
Dir.: *Roger N. Wilson* • Cur.: *Jeannine Stults* •
Association with Coll – 1962
Local history, housed in a Carnegie library bldg 51340

Newton MA

Newton History Museum at Jackson Homestead, 527 Washington St, Newton, MA 02458 • T: +1 617 7961450 • F: +1 617 5527228 • dolson@ci.newton.ma.us • www.ci.newton.ma.us/jackson •
Dir.: *David A. Olson* • Cur.: *Susan Abele* • *Shelia M. Sibley* •
Historical Museum – 1950
Local history, costumes, toys, maps 51341

Newton NC

Catawba County Museum of History, 1 Court Sq, Newton, NC 28658 • T: +1 828 4650383 • F: +1 828 4659813 • ccha@gestaltmail.com • www.catawbahistory.org •
Dir.: *Sidney Halma* • Asst. Dir.: *Nathan Moehlmann* • Cur.: *Carroll Clark* •
Historical Museum – 1949
Upper Piedmont hist, historic buildings, textiles, folk art,household utensils, tools 51342

Newton NJ

Sussex County Historical Society Museum, 82 Main St, Newton, NJ 07860 • T: +1 973 3836010 • F: +1 973 3834911 • schs@tellurian.com • www.sussexcountyhistory.org •
Pres.: *Robert R. Longcore* •
Historical Museum – 1904
Local hist, American Indian artifacts, archaeology, genealogy 51343

Newtown PA

Artmobile, c/o Bucks County Community College, 275 Swamp Road, Newtown, PA 18940 • T: +1 215 9688432 • F: +1 215 5048530 • orlandof@bucks.edu • www.bucks.edu/artmobile •
Dir.: *Fran Orlando* •
Fine Arts Museum / University Museum – 1975
Temporary art exhibitions 51344

Hicks Art Center, c/o Bucks County Community College, 25 Swamp Rd, Newtown, PA 18940 • T: +1 215 9688425 • F: +1 215 5048530 •
Public Gallery – 1970 51345

Newtown Historic Museum, Court St and Centre Av, Newtown, PA 18940 • T: +1 215 9684004 • F: +1 215 9688925 • dcnhh@aol.com • www.newtownhistoric.org •
Pres.: *David Callahan* • Cur.: *Harriet Beckert* •
Historical Museum / Local Museum – 1965
Local history, early 1700s Court Inn 51346

Niagara Falls NY

Niagara Gorge Discovery Center, New York State Parks, Niagara Region, Robert Moses State Pkwy near Main St, Niagara Falls, NY 14303-0132, mail addr: Niagara State Parks, Western Dist, Niagara Region, Niagara Reservation State Park, POB 1132, Niagara Falls, NY 14303-0132 • T: +1 716 2781070 • F: +1 716 2785179 • www.nysparks.com •
Natural History Museum – 1971
Marine invertebrates from middle Silurian and Devonian periods, minerals 51347

Niagara University NY

Castellani Art Museum, Niagara University, Niagara University, NY 14109 • T: +1 716 2868200 • F: +1 716 2868289 • cam@niagara.edu • www.niagara.edu/~cam •
C.E.O. & Dir.: *Laurence Buckley* • Cur.: *Eric Jackson-Forsberg* (Collections) • *Kate Koperski* (Folk Arts) •
Fine Arts Museum / University Museum – 1973
19th and 20th c paintings, sculpture, prints and drawings, pre-Columbian pottery 51348

Niantic CT

Children's Museum of Southeastern Connecticut, 409 Main St, Niantic, CT 06357 • T: +1 860 6911111 • F: +1 860 6911194 • customerservice@childrensmuseum.org • www.childrensmuseum.org •
Exec. Dir.: *Mollica Tony* •
Historical Museum – 1992
History of Connecticut 51349

Niles IL

The Bradford Museum of Collector's Plates, 9333 Milwaukee Av, Niles, IL 60714 • T: +1 847 9662770 • F: +1 847 9663121 •
Decorative Arts Museum – 1978
Porcelain, plates 51350

Niles MI

Fort Saint Joseph Museum, 508 E Main St, Niles, MI 49120 • T: +1 269 6834702 • F: +1 269 6843930 • cbainbridge@qtm.net • www.ci.niles.mi.us •
Dir.: *Carol Bainbridge* •
Local Museum – 1932
Local history, H.A. Chapin carriage house, Sitting-Bull & Rain-in-the-Face pictographs, Sioux Indians, decorative arts – archives 51351

Niles OH

National McKinley Birthplace Memorial, 40 N Main St, Niles, OH 44446 • T: +1 330 6521704 • F: +1 330 6525788 • mckinley@oplin.lib.oh.us • www.mckinley.lib.oh.us •
Pres.: *William Jensen* •
Historic Site – 1911
Pres. Willam McKinley memorabilia and artifacts 51352

Nine Mile Falls WA

Spokane House Interpretive Center, 9711 W Charles, Nine Mile Falls, WA 99026 • T: +1 509 4655064 • F: +1 509 4655571 • manager@riversidestatepark.org • www.riversidestatepark.org •
Historical Museum – 1950
American Indian culture and artifacts, Tribes of the Plateau, archaelogy 51353

Ninety-Six SC

Ninety-Six National Historic Site, 1103 Hwy 248, Ninety-Six, SC 29666 • T: +1 864 5434068 • F: +1 864 5432058 • nisiadm@nps.gov •
Cur.: *Eric K. Williams* •
Local Museum / Open Air Museum – 1976
Local hist 51354

Noank CT

Noank Museum, 108 Main St, Noank, CT 06340 • T: +1 860 5363021 • noankhist@snet.net •
Cur.: *Kenneth O. Hodgson* •
Local Museum – 1966
Indian, local hist, geneaology 51355

Nobleboro ME

Nobleboro Historical Society Museum, 198 Center St, Nobleboro, ME 04555 •
Cur.: *George F. Dow* •
Local Museum – 1978
Agriculture, forestry, Indians, sailing vessels, industry 51356

Noblesville IN

Indiana Transportation Museum, 325 Cicero Rd, Noblesville, IN 46061-0083 • T: +1 317 7736000, 7767881 • F: +1 317 7735530 • npk587@iquest.net • www.itm.org •
C.E.O.: *David E. Witcox* •
Science&Tech Museum – 1960
Rail transportation and technology, 1930 railroad station from Hobbs 51357

Nogales AZ

Pimeria Alta Historical Society Museum, 136 N Grand Av, Nogales, AZ 85628-2281 • T: +1 520 2874621 •
Pres.: *Axel Holm* •
Local Museum – 1948
Prehistoric and historic Indians of Pimeria Alta, local hist and development 51358

Nome AK

Carrie M. McLain Memorial Museum, 223 Front St, Nome, AK 99762 • T: +1 907 4436630 • F: +1 907 4437955 • museum@ci.nome.ak.us • www.nomealaska.org/museum •
Dir.: *Laura Samuelson* • Asst. Dir.: *Bev Gelzer* •
Ethnology Museum / Fine Arts Museum – 1967
Local history, gold rush, Bering Strait Eskimo, aviation, ethnology, dolls, art from 1890-1998 51359

Norfolk CT

Norfolk Historical Museum, 13 Village Green, Norfolk, CT 06058 • T: +1 203 5425761 • norfolkhistoricalsociety@msn.com •
Pres.: *Barry Webber* •
Local Museum / Historical Museum – 1960 51360

Norfolk VA

The Chrysler Museum of Art, 245 West Olney Rd, Norfolk, VA 23510 • T: +1 757 6646200 • F: +1 757 6646201 • museum@chrysler.org • www.chrysler.org •
Dir.: *Dr. William Hennessey* • Deputy Dir.: *Catherine H. Jordan Wass* • Cur.: *Jefferson Harrison* (European Art) • *Gary E. Baker* (Glass) • *Brooks Johnson* (Photography) •
Fine Arts Museum – 1933
Arts from Egypt, Greece, rome, Near, Middle, & Far East, Africa etc. 51361

General Douglas MacArthur Memorial, MacArthur Sq, Norfolk, VA 23510 • T: +1 757 4412965 • F: +1 757 4415389 • macmem@norfolk.infi.net • www.macarthurmemorial.org •
Dir.: *William J. Davis* • Cur.: *Katherine A. Renfrew* •
Military Museum / Historical Museum – 1964
Military and personal artifacts 51362

Hampton Roads Naval Museum, One Waterside Dr, Ste 248, Norfolk, VA 23510-1607 • T: +1 757 3222987, 4448971 • F: +1 757 4451867 • bapoulliot@nsn.cmar.navy.mil • www.hrnm.navy.mil •
Dir.: *Elizabeth A. Poulliot* • Cur.: *Joseph M. Judge* •
Military Museum – 1979
Naval artifacts & artworks, ship & aircraft modls 51363

Hermitage Foundation Museum, 7637 N Shore Rd, Norfolk, VA 23505 • T: +1 757 4232052 • F: +1 757 4232410 • info@thfm.org • www.hermitagefoundation.org •
Dir.: *Karen E. Perry* •
Decorative Arts Museum – 1937
Woodcarvings, decorative arts, Chinese bronzes and ceramic tomb figures, lacquer ware, jades, Persian rugs, Spanish and English furniture, paintings, sculpture 51364

Lois E. Woods Museum, c/o Norfolk State University, 2401 Corfew Av, Norfolk, VA 23504 • T: +1 757 8232006 • F: +1 757 8232005 • jwoods@vger.nsu.edu • www.nsu.edu/resources/woods •
Dir.: *John S. Woods* •
Fine Arts Museum
African art coll, African-American art 51365

Moses Myers House, 331 Bank St, Norfolk, VA 23510 • T: +1 757 4411526 • F: +1 757 3331089 • mriley@chrysler.org • www.chrysler.org •
Dir.: *Dr. William Hennessey* •
Decorative Arts Museum / Local Museum – 1951
American & English furniture, glass, silver, china, textiles, costumes 51366

Nauticus, Hampton Roads Naval Museum, 1 Waterside Dr, Norfolk, VA 23510 • T: +1 757 6641000 • F: +1 757 6231287 • www.nauticus.org •
Exec. Dir.: *Richard C. Conti* •
Science&Tech Museum / Military Museum – 1994
Commerce and military, USS Wisconsin 51367

Norfolk Historical Society, 810 Front St, Norfolk, VA 23510 • T: +1 757 6251720 •
C.E.O.: *Amy Waters Yarsinske* •
Association with Coll – 1965 51368

Ohef Sholom Temple Archives, 530 Raleigh Av, Norfolk, VA 23507 • T: +1 757 6254295 • F: +1 757 6252775 • archives@ohefsholom.org • www.ohefsholom.org •
C.E.O.: *Amy Waters Yarsinske* •
Religious Arts Museum – 1992
Historical artifacts from our congregation and local Jewry including photographs, videos, records, military uniforms, marriage, naming, and birthing certificates, and temple records 51369

Old Dominion University Gallery, 350 W 21st ST, Norfolk, VA 23529 • T: +1 757 6836227 • F: +1 757 6835923 • rrlove@odu.edu • www.odu.edu •
Dir.: *Katherine Huntoon* •
Public Gallery – 1971
Paintigns, prints, sculpture 51370

Willoughby-Baylor House, 601 E Freemason St, Norfolk, VA 23510 • T: +1 757 6646283, 6646255 • F: +1 757 3331089 • mnley@chrysler.org • www.chrysler.org •
Dir.: *Dr. William Hennessey* •
Decorative Arts Museum / Local Museum – 1962
Historic house 51371

Normal IL

University Galleries, Illinois State University, 110 Center for the Visual Arts, Normal, IL 61761, mail addr: Campus Box 5600, Normal, IL 61790-5600 • T: +1 309 4385487 • F: +1 309 4385161 • gallery@ilstu.edu • www.cfa.ilstu.edu/galleries •
Dir.: *Barry Blinderman* •
Fine Arts Museum / University Museum – 1973
Contemporary paintings, sculptures 51372

Norman OK

Firehouse Art Center, 444 S Flood, Norman, OK 73069 • T: +1 405 3294523 • F: +1 405 2929763 • firehouse@telepath.com • www.normanfirehouse.com •
Dir.: *Danette Ward* •
Public Gallery – 1971 51373

Fred Jones Jr. Museum of Art, University of Oklahoma, 555 Elm Av, Norman, OK 73019 • T: +1 405 3253272 • F: +1 405 3257696 • www.ou.edu/fjima •
Dir.: *Eric M. Lee* • Chief Cur.: *Gail Kana Anderson* • Cur.: *Susan G. Baley* (Education) •
Fine Arts Museum / University Museum – 1936
French impressionism, European graphics 16th c to present, Native American art, American art, contemp art 51374

Norman Cleveland County Historical Museum, 508 N Peters, Norman, OK 73069 • T: +1 405 3210156 • cchs@coxinet.net • www.normanhistorichouse.com •
Dir.: *Jaime Evans* •
Historical Museum – 1973
Cleveland County hist; furniture, dec arts, textiles, toys 51375

Sam Noble Oklahoma Museum of Natural History, University of Oklahoma, 2401 Chautauqua, Norman, OK 73072-7029 • T: +1 405 3258978 • F: +1 405 3257699 • smomnh@ou.edu • www.snomnh.ou.edu •
Cur.: *Dr. Michael A. Mares* (Mammals) • *Dr. R.L. Cifelli* (Vertebrate Paleontology) • *Dr. L.J. Vitt* (Reptiles) • *Dr. J.P. Caldwell* (Amphibians) • *Dr. D. Wyckoff* (Archeology) • *Dr. S. Westrop* (Invertebrate Paleontology) • *Dr. B. Matthews* • *Dr. F. Marsh-Matthews* (Ichthyology) • *Dr. G. Schnell* (Ornithology) • *Dr. M. Linn* (Native American Languages) • *Dr. J.K. Braun* • *Dr. N.J. Czaplewski* •
University Museum / Natural History Museum – 1899
Mammals, birds, fish, amphibians, reptiles, insects, botany, anthropology, ethnology 51376

Norris TN

Museum of Appalachia, Hwy 61, Norris, TN 37828 • T: +1 865 4947680 • F: +1 865 4948957 • museum@museumofappalachia.org • www.museumofappalachia.com •
Dir.: *John Rice Irwin* • *Elaine Irwin Meyer* •
Local Museum / Folklore Museum – 1967
Local hist, folk art, over 200.000 early american & pioneer items of various types 51377

North Adams MA

Mass Moca, Massachusetts Museum of Contemporary Art, 87 Marshall St, North Adams, MA 01247 • T: +1 413 6644481 • F: +1 413 6638548 • info@massmoca.org • www.massmoca.org •
C.E.O.: *Joseph Thompson* • Cur.: *Laura Heon* •
Fine Arts Museum – 1988
Installations of sculpture and paintings – auditorium 51378

North Andover MA

North Andover Historical Museum, 153 Academy Rd, North Andover, MA 01845 • T: +1 978 6864035 • F: +1 978 6866616 • nahistory@juno.com • www.essexheritage.org/north_andover_hist_soc.htm •
Dir.: *Carol Majahad* •
Local Museum – 1913
Local history, lighting devices, tools 51379

Stevens-Coolidge Place, 137 Andover St, North Andover, MA 01845 • T: +1 978 6823580 •
Decorative Arts Museum – 1962
Chinese porcelain, decorative arts – gardens 51380

North Bend OR

Coos County Historical Society Museum, 1220 Sherman, North Bend, OR 97459 • T: +1 541 7566320 • F: +1 541 7566320 • museum@ucii.net •
Pres.: *Steve Greif* • Dir.: *Ann Koppy* •
Local Museum / Historical Museum – 1891 51381

North Bend WA

Snoqualmie Valley Historical Museum, 320 Bendigo Blvd S, North Bend, WA 98045 • T: +1 425 8883200 • F: +1 425 8883200 • snovalmuseum@isomedia.com • www.snoqualmievalleymuseum.org •
Pres.: *Kris Kirby* •
Local Museum – 1960
Local pioneer artifacts, local Indian hist 51382

North Bennington VT

The Park-McCullough House, Corner of Park & West Sts, North Bennington, VT 05257 • T: +1 802 4425441 • F: +1 802 4425442 • thehouse@sover.net • www.parkmccullough.org •
Exec. Dir.: *Matthew Schulte* •
Local Museum – 1968
Historic house with books, furniture, carriages, art, papers 51383

North Blenheim NY

Lansing Manor House Museum, NY State Rte 30, North Blenheim, NY 12131, mail addr: POB 898, North Blenheim, NY 12131 • F: +1 518 8276381 • steve.ramsey@nypa.gov •
Dir.: *Steve Ramsey* • Asst. Cur.: *Margaret Bailey* •
Historical Museum – 1977
Federal and Empire furniture, textiles, ceramics, manuscripts 51384

North Brunswick NJ

New Jersey Museum of Agriculture, College Farm Rd and Rte 1, North Brunswick, NJ 08902 • T: +1 732 2492077 • F: +1 732 2471035 • info@agriculturemuseum.org • www.agriculturemuseum.org •
Exec. Dir.: *Dana S. Vallely* • Cur.: *Coles Roberts* •
Agriculture Museum – 1984
Agriculture hist, household technology 51385

North Canton OH

Hoover Historical Center, 1875 E Maple St, North Canton, OH 44720-3331 • T: +1 330 4990287 • F: +1 330 4944725 • ahaines@hoover.com • www.hoover.com •
Dir.: *Jackie Love* •
Science&Tech Museum – 1978
Company hist, Hoover family home, evolution of vacuum cleaner, Hoover products for WW II 51386

North Chicago IL

Feet First: The Scholl Story, c/o Finch University, 3333 Green Bay Rd, North Chicago, IL 60064 • T: +1 312 2802904, +1 847 5788417 • F: +1 312 2802997 • kathy.peterson@rosalindfranklin.edu • www.rosalindfranklin.edu •
C.E.O.: *Terrence Albright* •
Historical Museum – 1993
Life of Dr. Scholl and his company, history of the institution 51387

North Conway NH

Conway Scenic Railroad, 38 Norcross Circle, North Conway, NH 03860 • T: +1 603 3565251 • F: +1 603 3567606 • info@conwayscenic.com • www.conwayscenic.com •
Pres.: *Russell G. Seybold* •
Science&Tech Museum – 1974
Railroad, wood frame railroad station 51388

Mount Washington Museum, 2936 White Mountain Hwy, North Conway, NH 03860 • T: +1 603 3562137 • F: +1 603 3560307 • observer@mountwashington.org • www.mountwashington.org •
Pres.: *Paul Fitzgerald* •
Natural History Museum – 1969
Natural hist artifacts, historical photos 51389

Weather Discovery Center, 2936 White Mountain Hwy, North Conway, NH 03860 • T: +1 603 3562137 • F: +1 603 3563070 • wdc@mountwashington.org • www.mountwashington.org •
Dir.: *Matthew White* •
Science&Tech Museum – 2000
Meterology, science, local history 51390

North East PA

Lake Shore Railway Museum, 31 Wall St, North East, PA 16428-0571 • T: +1 814 8252724 • F: +1 814 7251911 • lsrhs@velocity.net • www.lsrhs.railway.museum •
Pres.: *Charles W. Farrington* •
Science&Tech Museum – 1956
Railway 51391

North Easton MA

Children's Museum in Easton, Sullivan Av, North Easton, MA 02356 • T: +1 508 2303789 • F: +1 508 2307130 • cme417@yahoo.com • www.childrensmuseumineaston.org •
Dir.: *Paula J. Peterson* •
Historical Museum – 1988
Fire station in town's historical district 51392

North Freedom WI

Mid-Continent Railway Museum, E 8948 Diamond Hill Rd, North Freedom, WI 53951-0055 • T: +1 608 5224261 • F: +1 608 5224490 • info@midcontinent.org • www.midcontinent.org •
Cur.: *Donald W. Ginter* •
Science&Tech Museum – 1959
Railway equip, 1900 era steam locomotives 51393

North Haven CT

North Haven Historical Society Museum, 27 Broadway, North Haven, CT 06473 • T: +1 203 2397722 •
Cur.: *Barbara Pearsall* • *Gloria Furnival* •
Local Museum / Historical Museum – 1957 51394

North Hollywood CA

Portal of the Folded Wings, 10621 Victory Blvd, North Hollywood, CA 91606 • T: +1 818 7639121 • www.godickson.com/pfw.htm •
Dir.: *John Torres* •
Science&Tech Museum – 1996
Aviation 51395

North Little Rock AR

Arkansas National Guard Museum, Lloyd England Hall on Camp Robinson, North Little Rock, AR 72199-9600 • T: +1 501 2125215 • F: +1 501 2125228 • steve.rucker@ar.ngb.army.mil • www.arngmuseum.com •
Military Museum 51396

North Newton KS

Kauffman Museum, Bethel College, 2801 N Main, North Newton, KS 67117-0531 • T: +1 316 2831612 • kauffman@bethelks.edu • www.bethelks.edu/kauffman.html •
Dir.: *Rachel K. Pannabecker* • Cur.: *Charles Regier* •
University Museum / Local Museum / Natural History Museum – 1941
Cultural and natural history of Central Plains 51397

Mennonite Library and Archives Museum, Bethel College, 300 E 27th St, North Newton, KS 67117 • T: +1 316 2832500, 2845304 • F: +1 316 2845286 • mla@bethelks.edu • www.bethelks.edu/services/mla •
Sc. Staff: *John D. Thiesen* (Archivist) • *Barbara A. Thiesen* (Library) •
Religious Arts Museum – 1938
Anabaptist & Mennonite manuscripts, prints, paintings 51398

North Oxford MA

Clara Barton Birthplace, 68 Clara Barton Rd, North Oxford, MA 01537-0356 • T: +1 508 9875375 • F: +1 508 9872002 •
Dir.: *Shelley Yeager* •
Historical Museum – 1921
Historic house, American Red Cross – library 51399

North Plainfield NJ

North Plainfield Exempt Firemen's Museum, 300 Somerset St, North Plainfield, NJ 07060 • T: +1 908 7575720 • F: +1 908 7577383 • npexemt@juno.com •
Historical Museum – 1904
Antique fire apparatus, helmets 51400

North Platte NE

Buffalo Bill Ranch, 2921 Scouts Rest Ranch Rd, North Platte, NE 69101-8444 • T: +1 308 5358035 • F: +1 308 5358066 • bbranch@ns.nque.com • www.ngpc.state.ne.us •
Historical Museum – 1964
Photo coll, history 51401

Lincoln County Historical Society Museum, 2403 N Buffalo Bill Av, North Platte, NE 68101 • T: +1 308 5345640 •
Local Museum – 1974
County hist 51402

North Saint Paul MN

North Star Museum of Boy Scouting and Girl Scouting, 2640 E Seventh Av, North Saint Paul, MN 55109 • T: +1 651 7482880 • cnicholson@northstarmuseum.org • www.northstarmuseum.org •
Exec.Dir.: *Claudia J. Nicholson* •
Special Museum – 1976
Artifacts, films, photos, scouting publications, uniforms, equip, boy and girl scouting from arround the world and esp. the twin cities of Minneapolis ans Saint Paul – library 51403

North Salem NH

America's Stonehenge, 105 Haverhill Rd, North Salem, NH 03073, mail addr: POB 84, North Salem, NH 03073 • T: +1 603 8938300 • F: +1 603 8935889 • amstonehenge@worldnet.att.net • www.stonehengeusa.com •
C.E.O. & Pres.: *Robert E. Stone* •
Archaeology Museum – 1958
Archaeology on Mystery Hill, site of 4,000 yr old archaeo-astronomical site 51404

North Salem NY

Hammond Museum, 28 Deveau Rd, off Rte 124, North Salem, NY 10560 • T: +1 914 6695033 • F: +1 914 6698221 • gardenprogram@yahoo.com • www.hammondmuseum.org •
Dir.: *Lorraine Laken* •
Fine Arts Museum / Decorative Arts Museum – 1957
Asian art, decorative arts 51405

North Tonawanda NY

Herschell Carrousel Factory Museum, 180 Thompson St, North Tonawanda, NY 14120 • T: +1 716 6931885 • F: +1 716 7439018 • hcfm@carrouselmuseum.org • www.carrouselmuseum.org •
Dir.: *Elizabeth M. Brick* •
Science&Tech Museum – 1983
Company hist 51406

North Tonawanda History Museum, 314 Oliver St, North Tonawanda, NY 14120 • T: +1 716 6922681 • nthistory@aol.com •
Acting Dir.: *Donna Zellner* •
Historical Museum – 2004
Ethnic heritage, Erie Canal and Niagara River and development of the region in 19th and 20th c 51407

North Troy VT

Missisquoi Valley Historical Society, Main St, North Troy, VT 05859 • T: +1 802 9882397 • missisco@hotmail.com •
Pres.: *Nancy L. Allen* •
Association with Coll – 1976
General store memorabilia, photographs, household utensils, textiles, tools 51408

North Wildwood NJ

Hereford Inlet Lighthouse, 11 N Central Av, North Wildwood, NJ 08260 • T: +1 609 5224520 • F: +1 609 5238590 • herefordlh@yahoo.com • www.herefordlighthouse.org •
Head: *Aldo Palombo* •
Historical Museum – 1982
Maritime hist 51409

North Wilkesboro NC

Wilkes Art Gallery, 913 C St, North Wilkesboro, NC 28659 • T: +1 336 6672841 • F: +1 336 6679264 • wilkesartgal@wilkes.net • wilkesartgallery.org •
Dir.: *Kara Minton-Elmore* •
Fine Arts Museum / Public Gallery – 1962 51410

North Woburn MA

Rumford Historical Museum, 90 Elm St, North Woburn, MA 01801 • T: +1 617 9330781 • www.middlesexcanal.org/docs/rumford.htm •
Pres.: *Thomas Smith* •
Historical Museum – 1877
Count Rumford's birthplace, science and history 51411

Northampton MA

Calvin Coolidge Presidential Library and Museum, 20 West St, Northampton, MA 01060 • T: +1 413 5871011/14 • F: +1 413 5871015 • jbartlet@cwmars.org • www.forbeslibrary.org •
Cur.: *Lu Knox* •
Historical Museum / Library with Exhibitions – 1920
30th President of Mass. life and time, manuscripts 51412

Historic Northampton, 46 Bridge St, Northampton, MA 01060 • T: +1 413 5846011 • F: +1 413 5847956 • mailbox@historic-northampton.org • www.historic-northampton.org •
Dir.: *Kerry W. Buckley* •
Historical Museum – 1905
Local history, Parsons house, Isaac Damon house, Shepherd house & barn, decorative arts, regional textiles & costumes 51413

Smith College Museum of Art, Elm St at Bedford Tce, Northampton, MA 01063 • T: +1 413 5852760 • F: +1 413 5852782 • artmuseum@smith.edu • www.smith.edu/artmuseum •
Dir.: *Suzannah Fabing* • Cur.: *Linda Muehlig* (Paintings, Sculpture) • *Nancy Rich* (Education) • *Aprile Gallant* (Prints, Drawings, Photographs) •
Fine Arts Museum / University Museum – 1920
European and American paintings, sculptures, drawings, prints, photographs, decorative arts 17th-20th c – library 51414

Words and Pictures Museum, 140 Main St, Northampton, MA 01060 • T: +1 413 5868545 • F: +1 413 5869855 • curator@wordsandpictures.org • www.wordsandpictures.org •
Cur.: *J. Fiona Russell* •
Fine Arts Museum – 1991 51415

Northborough MA

Northborough Historical Museum, 50 Main St, Northborough, MA 01532 • T: +1 508 3936298 • letusgothen@juno.com • www.northboroughhistsoc.org •
Cur.: *Ellen Racine* •
Local Museum – 1906
Town history, lighting, apothecary and kitchen utensils, tools, clothing, portraits – library, archives 51416

Northeast Harbor ME

Great Harbor Maritime Museum, 125 Main St, Northeast Harbor, ME 04662 • T: +1 207 2765262 • F: +1 207 2765262 • ghmmuse@acadia.net •
Historical Museum – 1982
Regional boats, harbors, boat builders, arts 51417

U

Northfield MN

Carleton College Art Gallery, 1 N College St, Northfield, MN 55057 • T: +1 507 6464342, 6464469 • F: +1 507 6467042 • lbradley@carleton.edu • apps.carleton.edu/campus/gallery •
Dir.: *Laurel Bradley* •
Public Gallery
American paintings, photography, prints, woodcuts, Asian art, Greek and Roman antiquities ... 51418

Exhibition of the Norwegian-American Historical Association, 1510 Saint Olaf Av, Northfield, MN 55057-1097 • T: +1 507 6463221 • F: +1 507 6463734 • naha@stolaf.edu • www.naha.stolaf.edu •
Dir.: *Kim Holland* • Cur.: *Forrest Brown* •
Historical Museum – 1925
library, archives ... 51419

Flaten Art Museum, Saint Olaf College, 1520 Saint Olaf Av, Northfield, MN 55057 • T: +1 507 6463556 • F: +1 507 6463776 • ewald@stolaf.edu • www.stolaf.edu/depts/art/museum •
Dir.: *Jill Ewald* •
Fine Arts Museum / University Museum
American, European, Asian and African prints, paintings, photographs, sculpture ... 51420

Northfield Historical Society Museum, 408 Division St, Northfield, MN 55057 • T: +1 507 6459268 • nhsmuseum@rconnect.com • www.northfieldhistory.org •
Pres.: *Bruce Colwell* •
Local Museum – 1975
Whittier railroad, area items, bank raid of 1876 . 51421

Northfield OH

Palmer House, 9390 Olde Eight Rd, Northfield, OH 44067 • T: +1 330 4678538 • F: +1 330 4678322 • hson@worldnet.att.net • www.hson.info •
C.E.O.: *Arch Milani* •
Local Museum / Historical Museum – 1956
Filkore, Indian artifacts, agrculture, kitchen ware, costumes, tools ... 51422

Northfield VT

Norwich University Museum, White Chapel, Norwich University, Northfield, VT 05663 • T: +1 802 4852000, 4852360 • F: +1 802 4852580 • GLORD@Norwich.edu •
Cur.: *Gary T. Lord* •
University Museum / Historical Museum – 1819
Hist of Norwich exhib, personal memorabilia of the founder ... 51423

Northport AL

Kentuck Museum and Art Center, 503 Main Av, Northport, AL 35476-4483 • T: +1 205 7581257 • F: +1 205 7581258 • kentuck@kentuck.org • www.kentuck.org •
Dir.: *Miah Michaelsen* •
Local Museum / Fine Arts Museum – 1971
artist studios ... 51424

Northport NY

Northport Historical Society Museum, 215 Main St, Northport, NY 11768 • T: +1 631 7579859 • F: +1 631 7579398 • info@northporthistorical.org • www.northporthistorical.org •
Dir.: *Linda Furey* • Cur.: *George Wallace* •
Historical Museum – 1962
Ship building tools, domestic utensils ... 51425

Northridge CA

Art Galleries, California State University, Northridge, 18111 Nordhoff St, Northridge, CA 91330-8299 • T: +1 818 6772226, 6772156 • F: +1 818 6775910 • michelle.giacopuzzi@csun.edu • www.csun.edu/artgalleries •
Dir.: *Louise Lewis* •
Fine Arts Museum / University Museum – 1972 . 51426

Northumberland PA

Joseph Priestley House, 472 Priestley Av, Northumberland, PA 17857 • T: +1 570 4739474 • F: +1 570 4737901 • mbashore@state.pa.us • www.phmc.state.pa.us •
Pres.: *Hans Veening* •
Historical Museum – 1960
Historic house ... 51427

Northville MI

Mill Race Historical Village, Griswold St, Northville, MI 48167-0071 • T: +1 248 3481845 • F: +1 248 3480056 • history.northville.lib.mi.us •
Pres.: *Cheryl Gazlay* •
Local Museum – 1964
Victorian furniture, clothing, local hist – archives 51428

Norton KS

Station 15, Hway 36, Roadside Park, Norton, KS 67654 • T: +1 785 8772501 • F: +1 785 8773300 • nortoncc@ruraltel.net • us36.net/nortonkansas •
Exec.Dir.: *Karla Reed* •
Special Museum – 1961
Replica of stagecoach and wagontrain depot of 1859, costumes, Indian artifacts ... 51429

Norton MA

Beard Gallery, Wheaton College, 26 E Main St, Norton, MA 02766 • T: +1 508 2863578 • F: +1 508 2863565 • arts@wheatoncollege.edu • www.wheatoncollege.edu/acad/art/gallery •
Dir.: *Ann H. Murray* • Cur.: *Amy Friend* •
Fine Arts Museum / University Museum – 1960
American and European paintings, prints and drawings ... 51430

Norwalk CT

Connecticut Graphic Arts Center, 299 West Av, Norwalk, CT 06850 • T: +1 203 8997999 • F: +1 203 8997997 •
Dir.: *Grace Shanley* •
Fine Arts Museum
Prints, works on paper, photography ... 51431

Lockwood-Mathews Mansion Museum, 295 West Av, Norwalk, CT 06850 • T: +1 203 8389799 • F: +1 203 8381434 • director@lockwoodmathews.org • www.lockwoodmathews.org •
Dir.: *Zachary Studenroth* •
Local Museum / Music Museum – 1966 ... 51432

Stepping Stones Museum for Children, 303 West Av, Norwalk, CT 06850 • T: +1 203 8990606 • F: +1 203 8990530 • www.steppingstonesmuseum.org •
Exec. Dir.: *Rhonda Kiest* •
Special Museum – 1998
Hands on science, arts, culture and heritage – library ... 51433

Norwalk OH

Firelands Historical Society Museum, 4 Case Av, Norwalk, OH 44857 • T: +1 419 6686038 •
Cur.: *Marilou Creary* •
Historical Museum – 1857
Folkore, Indian artifacts, numismatic coll, primitive paintings, costumes ... 51434

Norwich CT

Faith Trumbull Chapter Museum, 42 Rockwell St, Norwich, CT 06360 • T: +1 860 8878737 •
Cur.: *Debra Graskoski* •
Local Museum / Historical Museum – 1893
American revolution, local history ... 51435

Leffingwell House Museum, 348 Washington St, Norwich, CT 06360 • T: +1 860 8899440 • F: +1 860 8874551 • www.visitnewengland.com •
Pres.: *Annetta Cannon* •
Decorative Arts Museum – 1901
Norwich silver, antique dolls, Indian artifacts ... 51436

Norwich Arts Council Gallery, 60 Broadway, Norwich, CT 06360 • T: +1 860 88706360 • administration@norwicharts.org • www.norwicharts.org •
Dir.: *Sheila K. Tabkoff* •
Fine Arts Museum
Art ... 51437

Slater Memorial Museum, 108 Crescent St, Norwich, CT 06360 • T: +1 860 8872506, 8850379 • F: +1 860 8896196 • zoev@norwichfreeacademy.com • www.norwichfreeacademy.com •
Dir.: *Vivian F. Zoe* •
Fine Arts Museum – 1888
American paintings, graphics, schulpture, Oriental art, costumes, marine, casts, decorative arts, textiles, guns ... 51438

Norwich NY

Chenango County Historical Society Museum, 45 Rexford St, Norwich, NY 13815 • T: +1 607 3349227 • F: +1 607 3347809 • drdave@ghufsc.net • www.chenangocounty.org/chencohistso •
Pres.: *Mary C. Weidman* • C.E.O. & Dir.: *R. David Drucker* •
Historical Museum – 1938
Folklore, folk art, costumes, Native American artifacts ... 51439

Northeast Classic Car Museum, 24 Rexford St, Norwich, NY 13815 • T: +1 607 3342886 • F: +1 607 3366745 • kay@classiccarmuseum.org • www.classiccarmuseum.org •
Dir.: *Kay Wells Zaia* •
Science&Tech Museum
Vintage and luxury cars ... 51440

Norwich OH

National Road/Zane Grey Museum, 8850 E Pike, Norwich, OH 43767 • T: +1 740 8723143 • F: +1 740 8723510 • zanegrey@globalco.net • www.ohiohistory.org •
Dir.: *Alan King* •
Science&Tech Museum – 1973
Vehicles, transportation items, Ohio art pottery .. 51441

Norwich VT

Montshire Museum of Science, 1 Montshire Rd, Norwich, VT 05055 • T: +1 802 6492200 • F: +1 802 6493637 • montshire@montshire.org • www.montshire.org •
Dir.: *David Goudy* • Assoc. Dir.: *Richard Stucker* • Cur.: *Joan Waltermire* •
Natural History Museum – 1975
Interactive science exhib ... 51442

Norwich Historical Society, 277 Main St, Norwich, VT 05055 • T: +1 802 6490124 • NHS@tpk.net •
Pres.: *William M. Aldrich* •
Association with Coll – 1951
Artifacts dating from 1769 to present ... 51443

Norwood NY

Norwood Historical Association Museum, POB 163, Norwood, NY 13668-0163 •
Dir.: *Gerald LaComb* •
Historical Museum – 1968
Railroad articles, folklore, local history ... 51444

Notre Dame IN

The Snite Museum of Art, University of Notre Dame, M. Krause Circle, Notre Dame, IN 46556-0368 • T: +1 219 6315466 • F: +1 219 6318501 • charles.r.loving.l@nd.edu • www.nd.edu/~sniteart •
Dir.: *Charles R. Loving* • Cur.: *Stephen B. Spiro* (Western Arts) •
Fine Arts Museum / University Museum – 1842
Ethnographic art, photography, design, native American art ... 51445

Novato CA

Marin Museum of the American Indian, 2200 Novato Blvd, Miwok Park, Novato, CA 94947 • T: +1 415 8974064 • F: +1 415 8927804 • www.marinindian.com •
Dir.: *Verna Smith* •
Ethnology Museum / Archaeology Museum – 1967
Archaeology, ethnography from Alaska to Peru, native Americans ... 51446

Novato History Museum, 815 De Long Av, Novato, CA 94945 • T: +1 415 8974320 • prcs@ci.novato.ca.us • www.ci.novato.ca.us •
Local Museum – 1976
Local hist – archive ... 51447

Nowata OK

Nowata County Historical Society Museum, 121 S Pine St, Nowata, OK 74048 • T: +1 918 2731191 •
Pres.: *Raymond Cline* •
Historical Museum – 1969
Cowboy room, dentist office, dining room, Indian suite, wood carvings, oil field equip ... 51448

Nyack NY

Edward Hopper House Art Center, 82 N Broadway, Nyack, NY 10960 • T: +1 845 3580774 • F: +1 845 3530774 • info@hopperhouse.org • www.hoppperhouse.org •
Exec. Dir.: *Carole Perry* •
Fine Arts Museum / Historic Site – 1971
House is on National Registry of Historic Places. Edward Hopper memorabilia; no art collection or Hopper artwork ... 51449

Oak Bluffs MA

Firehouse Gallery, 88 Dukes County Av, Oak Bluffs, MA 02557, mail addr: c/o MVCVA, POB 4377, Vineyard Haven, MA 02568 • T: +1 508 6939025 • dreyerc@earthlink.net • www.mvarts.org •
Pres.: *Chris Dreyer* •
Public Gallery – 1991 ... 51450

Oak Brook IL

Graue Mill and Museum, York and Spring Rds, Oak Brook, IL 60522-4533 • T: +1 630 6552090 • F: +1 630 9209721 • administrator@grauemill.org • www.grauemill.org •
Exec. Dir.: *Sandra Brubaker* •
Local Museum / Science&Tech Museum – 1950
Local history, housed in 1852 restored waterwheel gristmill ... 51451

Oak Creek WI

Oak Creek Pioneer Village, S 15th Av & E. Forest Hill Av, Oak Creek, WI 53154 • T: +1 414 7612572 •
Pres.: *Elroy Honadel* • Cur.: *Marge Berres* •
Local Museum – 1964
Local hist, farm equip ... 51452

Oak Park IL

Frank Lloyd Wright Home and Studio, 951 Chicago Av, Oak Park, IL 60302 • T: +1 708 8481976 • F: +1 708 8481248 • flwpr@wrightplus.org • www.wrightplus.org •
Pres.: *Joan B. Mercuri* •
Historical Museum – 1974
Residence and Studio of Frank Lloyd Wright ... 51453

Hemingway Museum, 200 N Oak Park Av, Oak Park, IL 60302 • T: +1 708 3864363 • F: +1 708 3868506 • ehfop@theramp.net • www.ehfop.org •
C.E.O: *Virginia Cassin* •
Special Museum – 1990
Birthplace and personal artifacts of Ernest Hemingway ... 51454

Historical Society of Oak Park and River Forest, 217 S Home Av, Oak Park, IL 60302-3101 • T: +1 708 8486755 • F: +1 708 8480246 • oprfhistorian@sbcglobal.net •
Dir.: *Frank Lipo* •
Public Gallery / Local Museum – 1968
Regional history, Prairie style mansion, photography ... 51455

Oak Ridge TN

American Museum of Science and Energy, 300 S Tulane Av, Oak Ridge, TN 37830 • T: +1 865 5763200 • F: +1 865 5766024 • jcomish@amse.org • www.amse.org •
Dir.: *Dr. Stephen Stow* •
Science&Tech Museum – 1949
Science, technology, hist of the Manhattan Project ... 51456

Children's Museum of Oak Ridge, 461 W Outer Dr, Oak Ridge, TN 37830 • T: +1 865 4821074 • F: +1 865 4814889 • chmor@bellsouth.net • www.childrensmuseumofoakridge.org •
Dir.: *Selma Shapiro* •
Special Museum – 1973
Coal mining artifacts, cherokee Indian artifacts .. 51457

Oak Ridge Art Center, 201 Badger Av, Oak Ridge, TN 37830 • T: +1 865 4821441 • F: +1 865 4821441 •
Dir.: *Leah Marcum-Estes* •
Fine Arts Museum – 1952
Mary & Alden coll, abstract expressionist art, contemporary regional art ... 51458

Oakdale CA

Oakdale Museum and History Center, 212 W F St, Oakdale, CA 95361 • T: +1 209 8479229, 8475822 • www.oakdalehistory.com •
Dir.: *Glenn Burghardt* •
Local Museum – 1985
Local history ... 51459

Oakdale NY

A. Giordano Gallery, Dowling College, Dept of Visual Art, Idle Hour Blvd, Oakdale, NY 11769 • T: +1 631 2443099 • F: +1 631 5896644 • brownp@dowling.edu •
Public Gallery ... 51460

Oakhurst CA

Fresno Flats Museum, School and Indian Springs Rds, Oakhurst, CA 93644, mail addr: POB 451, Oakhurst, CA 93644 • T: +1 559 6836570 • fresnoflatsmuseum@sti.net • www.fresnoflatsmuseum.org •
C.E.O.: *Sabine Cogdell* • Cur.: *Patricia McNichols* •
Historical Museum – 1968 ... 51461

Oakland CA

African American Museum at Oakland, 659 14th St, Oakland, CA 94612 • T: +1 510 6370200 • rmoss@oaklandlibrary.org • www.oaklandlibrary.org •
Cur.: *Rick Moss* •
Historical Museum – 1965
Western and Northern California Black history – library, archives ... 51462

Camron-Stanford House, 1418 Lakeside Dr, Oakland, CA 94612 • T: +1 510 8747802 • F: +1 510 8747803 • www.cshouse.org •
Cur.: *Wayne Mathes* •
Decorative Arts Museum – 1971
Decorative arts and funishings 1875-1885, photos – library ... 51463

Creative Growth Art Center, 355 24th St, Oakland, CA 94612 • T: +1 510 8362340 • gallery@creativegrowth.org • www.creativegrowth.org •
Dir.: *Tom DiMaria* •
Public Gallery – 1974
Art by physically, mentally and developmentally disabled adult artists ... 51464

June Steingart Art Gallery, Laney College, 900 Fallon St, Oakland, CA 94607 • T: +1 510 4643267 • pandreini@peralta.edu • www.laney.peralta.edu •
Public Gallery ... 51465

Merritt Museum of Anthropology, Merritt College, 12500 Campus Dr, Oakland, CA 94619 • T: +1 510 4362607, 5314911 • www.merritt.edu •
Dir.: *Dr. Barbara Joans* • Cur.: *Leslie Fleming* •
Ethnology Museum / University Museum – 1973
Ethnology of North and South America, Africa, Asia and Pacific, European coll, prehistory ... 51466

Mills College Art Museum, 5000 Mac Arthur Blvd, Oakland, CA 94613 • T: +1 510 4302164 • F: +1 510 4303168 • museum@mills.edu • www.mills.edu/museum •
Dir.: *Stephan Jost* •
Fine Arts Museum / University Museum – 1925
Asian and Latin American textiles, Japanese ceramics, prints and drawings, regionalist paintings ... 51467

Museum of Children's Art, 538 Ninth St, Oakland, CA 94607 • T: +1 510 4658770 • F: +1 510 4650772 • hello@mocha.org • www.mocha.org •
Dir.: *Karen Ransom* • *Rae Holzman* (Programs) •
Fine Arts Museum ... 51468

Oakland Aviation Museum, 8260 Boeing St, Oakland, CA 94614 • T: +1 510 6387100 • www.oaklandaviationmuseum.org •
Dir.: *Franz Steiner* •
Science&Tech Museum – 1980
Airplanes, aviation artifacts, photos, prints and films – library ... 51469

U

Oakland Museum of California, 1000 Oak St, Oakland, CA 94607 • T: +1 510 2382200 • F: +1 510 2382258 • www.museumca.org •
Dir.: *Dennis Power* • Chief Cur.: *Philip Linhares* (Art) • *Tom Steller* (Natural Sciences) • *Barbara Henry* (Education) • Cur.: *Harvey Jones* • *Christian Kleiger* (History) • *Karen Tsujimoto* (Art) • *Drew Johnson* (Art Photography) • *Inez Brooks-Myers* (Costumes and Textiles) • *Christopher Richard* (Natural Sciences) •
Fine Arts Museum / Historical Museum / Natural History Museum – 1969
California environment, culture, history and arts – library of natural sounds, archives, theater 51470

Pardee Home Museum, 672 11th St, Oakland, CA 94607 • T: +1 510 4442187 • F: +1 510 4447120 • info@pardeehome.org • www.pardeehome.org •
Dir.: *David Nicolai* •
Historical Museum / Historic Site – 1982
Former residence of Governor George Pardee, furniture, ethnography – archive 51471

Pro Arts at Oakland Art Gallery, 150 Frank H. Ogawa Plaza, Oakland, CA 94607 • T: +1 510 7634361 • F: +1 510 7639470 • info@proartsgallery.org • www.proartsgallery.org •
Dir.: *Margo Dunlap* •
Public Gallery – 1974 51472

Western Aerospace Museum → Oakland Aviation Museum

Oakland IA

Nishna Heritage Museum, 117 Main St, Oakland, IA 51560 • T: +1 712 4826802 • norm@nishnaheritagemuseum.com • www.nishnaheritagemuseum.com •
Dir.: *Wilson Pechacek* •
Fine Arts Museum / Local Museum – 1975
Heritage, local history 51473

Oakland MD

Garrett County Historical Museum, 107 S 2nd St, Oakland, MD 21550 • T: +1 301 3343226 •
Pres.: *Robert Boal* • Cur.: *Eleanor Callis* •
Local Museum – 1969
Local Indian artifacts, glassware, railroad exhibit, leather tools, coal mining exhibit and tools 51474

Oakley ID

Oakley Pioneer Museum, 108 W Main, Oakley, ID 83346 • T: +1 208 8623626 •
Dir.: *Becky Clark* •
Local Museum / Museum of Classical Antiquities – 1967
History, antiques 51475

Oakley KS

Fick Fossil and History Museum, 700 W 3rd St, Oakley, KS 67748 • T: +1 785 6724839 • F: +1 785 6723497 • fickmuseum@st-tel.net • www.discoveroakley.com •
Dir.: *Jeanet Bean* • Cur.: *Lucille Jennings* •
Historical Museum / Natural History Museum – 1972
Geology, paleontology, history 51476

Oberlin KS

Last Indian Raid Museum, 258 S Penn Av, Oberlin, KS 67749 • T: +1 785 4752712 • lirm@classicnet.net • skyways.lib.ks.us/museums/lirm •
Cur.: *Fonda Farr* •
Historic Site / Historical Museum – 1958
History, nr the site of 1878 last Indian raid on Kansas soil 51477

Oberlin OH

Allen Memorial Art Museum, Oberlin College, 87 N Main St, Oberlin, OH 44074 • T: +1 440 7758665 • F: +1 440 7756841 • leslie.miller@oberlin.edu • www.oberlin.edu/allenart •
Acting Dir.: *Katherine Solender* • Cur.: *Stephen Borys* (Western Art) • Sc. Staff: *Sara Hallberg* (Education) •
University Museum / Fine Arts Museum – 1917
Anvcient, African, Asian art, Italian Renaissance and Baroque, Dutch and Flemish art, Expressionism, contemporary art 51478

Oberlin Heritage Center, 73 1/2 S Professor St, Oberlin, OH 44074, mail addr: POB 0455, Oberlin, OH 44074 • T: +1 440 7741700 • F: +1 440 7748061 • members@oberlinheritage.org • www.oberlinheritage.org •
C.E.O.: *Patricia Murphy* •
Historical Museum – 1964
Hist buildings, Abolition, Underground Railroad, Women's history, reform movements 51479

Ocala FL

Appleton Museum of Art, 4333 NE Silver Springs Blvd, Ocala, FL 34470-5000 • T: +1 352 2367100 ext 0 • F: +1 352 2362621 • jorme@appleton.fsu.edu • www.appletonmuseum.org •
Dir.: *Jim Rosengren* •
Fine Arts Museum – 1986
European, Pre-Columbian, African, Asian and decorative arts, contemporary art, antiquities 51480

CFCC Webber Center, 3001 SW College Rd, Ocala, FL 34478, mail addr: POB 1388, Ocala, FL 34478-1388 • T: +1 352 2372111 ext 1552 • F: +1 352 8735886 • gonzalej@cf.edu • www.gocfcc.com/about/webber.htm •
Decorative Arts Museum – 1995
Folk culture, decorative arts, prints, sculpture 51481

Discovery Science Center of Central Florida, 1211 SE 22nd Rd, Ocala, FL 34471 • T: +1 352 4013900 • F: +1 352 4013939 • discover@ocalafl.org • www.ocalafl.org/recreationandp/dsoc.htm •
Historical Museum / Science&Tech Museum – 1993 51482

Don Garlits Museum of Drag Racing, 13700 SW 16th Av, I-75, Exit 67, Ocala, FL 34473 • T: +1 352 2458661 • F: +1 352 2456895 • garlits@pig.net • www.garlits.com •
C.E.O.: *Donald G. Garlits* •
Special Museum – 1976
Cars, pictures, model exhibitions 51483

Silver River Museum, 1445 N.E. 58th Av, Ocala, FL 34470 • T: +1 352 2365401 • F: +1 352 2367142 • Scott.Mitchell@marion.k12.fl.us • www.silverrivermuseum.com •
Dir./cur.: *Scott Mitchell* •
Natural History Museum / Archaeology Museum / Historical Museum – 1991 51484

Occoquan VA

Historic Occoquan, 413 Mill St, Occoquan, VA 22125 • T: +1 703 4917525 •
Cur.: *Nellie K. Curtis* •
Local Museum – 1969
Local history, Miller's cottage of Grist mill 51485

Ocean City NJ

Ocean City Art Center, 1735 Simpson Av, Ocean City, NJ 08226 • T: +1 609 3997628 • F: +1 609 3997089 • ocart@prousa.net •
Dir.: *Eunice Bell* •
Public Gallery – 1967
Paintings 51486

Ocean City Historical Museum, 1735 Simpson Av, Ocean City, NJ 08226 • T: +1 609 3991801 • F: +1 609 3990544 • ocnjhistmuseum@aol.com • www.ocnjmuseum.org •
Dir.: *Paul S. Anselm* •
Historical Museum – 1964
Victorian rooms, natural , local and military hist, Sindia shipwreck artifacts 51487

Ocean Grove NJ

Historical Society Museum of Ocean Grove, New Jersey, 50 Pittman Av, Ocean Grove, NJ 07756 • T: +1 732 7741869 • F: +1 732 7741684 • info@oceangrovehistory.org • www.oceangrovehistory.org •
Pres.: *Kevin Chambers* •
Local Museum – 1969
History of Ocean Grove, history of the camp meeting 51488

Ocean Springs MS

Walter Anderson Museum of Art, 510 Washington Av, Ocean Springs, MS 39564 • T: +1 228 8723164 • F: +1 228 8754494 • wama@walterandersonmuseum.org • www.walterandersonmuseum.org •
Dir.: *Marilyn Lyons* • Cur.: *Dr. Patricia Pinson* •
Fine Arts Museum – 1991
Murals, paintings, sculpture 51489

William M. Colmer Visitor Center, 3500 Park Rd, Davis Bayou Area, Ocean Springs, MS 39564 • T: +1 228 8750823 • F: +1 228 8722954 • www.nps.gov/guis/ •
Historical Museum / Natural History Museum – 1970
Pertaining to gulf ecosystem coastal fortification 51490

Oceanside CA

Mission San Luis Rey Museum, 4050 Mission Av, Oceanside, CA 92057 • T: +1 760 7573651 • www.sanluisrey.org •
Dir.: *Edward Gabarra* •
Religious Arts Museum – 1798
Local hist, decorative and fine arts 51491

Oceanside Museum of Art, 704 Pier View Way, Oceanside, CA 92054 • T: +1 760 4353720 • skip@oma-online.org • www.oma-online.org •
Exec. Dir.: *James R. Pahl* •
Fine Arts Museum – 1995
Art 51492

Oceanville NJ

The Noyes Museum of Art, Lily Lake Rd, Oceanville, NJ 08231 • T: +1 609 6528848 • F: +1 609 6526166 • info@noyesmuseum.org • www.noyesmuseum.org •
Exec. Dir.: *Lawrence R. Schmidt* •
Fine Arts Museum – 1983
20th c American paintings and sculpture, folk art, American arts and crafts 51493

Oconto WI

Oconto County Historical Society Museum, 917 Park Av, Oconto, WI 54153 • T: +1 920 8355733 • dscross@bayland.us • www.ocontocountyhistsoc.org •
Pres.: *Peter Stark* •
Local Museum – 1940
Local hist, Native American artifacts, fur trade, fishing and lumbering 51494

Odessa DE

Historic Houses of Odessa, Delaware, Main St, Odessa, DE 19730, mail addr: POB 507, Odessa, DE 19730 • T: +1 302 3784069 • F: +1 302 3784050 • www.winterthur.org •
Dir.: *Steven M. Pulinka* • Cur.: *Deborah Buckson* •
Local Museum – 1958
Historic house 51495

Odessa TX

The Ellen Noel Art Museum of the Permian Basin, 4909 E University Blvd, Odessa, TX 79762 • T: +1 432 5509696 • F: +1 432 5509226 • marilyn@noelartmuseum.org • noelartmuseum.org •
Dir.: *Marilyn Bassinger* •
Fine Arts Museum – 1985
Temporary exhib of art 51496

The Presidential Museum, 622 N Lee, Odessa, TX 79761 • T: +1 915 3327123 • F: +1 915 4984021 •
Dir.: *Carey F. Behrends* • Cur.: *Timothy M. Hewitt* •
Historical Museum – 1965
Highlights of all the US presidents 51497

Ogallala NE

Front Street Museum, 519 E First, Ogallala, NE 69153 • T: +1 308 2846000 • F: +1 308 2840865 •
C.E.O.: *Jan Nielsen* •
Local Museum – 1964
Cowboy items, saddles, farm tools, school house artifacts 51498

Ogden IA

Hickory Grove Rural School Museum, Don Williams Lake, Ogden, IA 50212 • T: +1 515 4321907 • bchs@opencominc.com •
Dir.: *Charles W. Irwin* •
Local Museum – 1972
Local hist 51499

Ogden UT

Eccles Community Arts Center, 2580 Jefferson Av, Ogden, UT 84401 • T: +1 801 3926935 • F: +1 801 3925295 • eccles@ogden4arts.org • www.ogden4arts.org •
Dir.: *Sandra H. Havas* •
Fine Arts Museum – 1957
Work of local artists 51500

Fort Buenaventura, 2450 A Av, Ogden, UT 84401 • T: +1 801 6214808, 3925581 • F: +1 801 3922431 • nrdpr.fbsp@state.ut.us •
Local Museum – 1980
Replica of 1848 fort located on site of first permanent white settlement in the Great Basin 51501

Myra Powell Art Gallery and Gallery at the Station, Ogden Union Station, Ogden, UT 84401 • T: +1 801 6298444 •
Exec. Dir.: *Bob Geier* •
Fine Arts Museum – 1979
Painting, sculpture 51502

Ogden Union Station Museums, 25th & Wall Av, Union Station, Ogden, UT 84401 • T: +1 801 3939886 • F: +1 801 6210230 • bobg@ci.ogden.ut.us • www.theunionstation.org •
Dir.: *Roberta Beverly* •
Local Museum – 1978
Art, railroads, cars, firearms, natural hist 51503

Weber State University Art Gallery, c/o Art Dept, Weber State University, 2001 University Circle, Ogden, UT 84408-2001 • T: +1 801 6267689 • F: +1 801 6266976 •
Dir.: *H. Barens* • Asst. Dir.: *Eden Betz* •
Public Gallery / University Museum – 1960
Contemporary prints, photos, drawings & ceramics 51504

Ogdensburg NJ

Sterling Hill Mining Museum, 30 Plant St, Ogdensburg, NJ 07439 • T: +1 973 2097212 • F: +1 973 2098505 • shm@tapnet.net • www.sterlinghill.org •
Pres.: *Richard Hauck* •
Science&Tech Museum – 1990
Mining 51505

Ogdensburg NY

Frederic Remington Art Museum, 303 Washington St, Ogdensburg, NY 13669 • T: +1 315 3932425 • F: +1 315 3934464 • info@fredericremington.org • www.fredericremington.org •
Dir.: *Lowell McAllister* • Cur.: *Laura Foster* •
Fine Arts Museum / Association with Coll – 1923
Frederic Remington's paintings, sculpture, drawings; decorative arts – archives 51506

Ogunquit ME

Ogunquit Museum of American Art, 543 Shore Rd, Ogunquit, ME 03907-0815 • T: +1 207 6464909 • F: +1 207 6466903 • ogunquitmuseum@aol.com • www.ogunquitmuseum.org •
Dir.: *Michael Culver* •
Fine Arts Museum – 1952
American art 51507

Oil City LA

Caddo-Pine Island Oil and Historical Society Museum, 200 S Land Av, Oil City, LA 71061 • T: +1 318 9956845 • F: +1 318 9956848 • oil@sos.louisiana.gov • www.sos.louisiana.gov •
Pres.: *H.D. Farrar* •
Local Museum – 1965 51508

Oil City PA

Venango Museum of Art, Science and Industry, 270 Seneca St, Oil City, PA 16301 • T: +1 814 6762007 • F: +1 814 6786719 • venangomuseum@venangomuseum.org •
Pres.: *Mary Balas* •
Fine Arts Museum / Science&Tech Museum – 1961
Art, science, industry 51509

Ojai CA

Ojai Center for the Arts, 113 S Montgomery St, Ojai, CA 93023 • T: +1 805 6460117 • info@ojaiartcenter.org • www.ojaiartcenter.org •
Dir.: *Teri Mettala* •
Public Gallery – 1939 51510

Ojai Valley Museum, 130 W Ojai Av, Ojai, CA 93023 • T: +1 805 6401390 • F: +1 805 6401342 • ojaimuseum@sbcglobal.net • www.ojaivalleymuseum.org •
Dir.: *Jane McClenahan* •
Local Museum – 1966
Local hist, art, cultural and natural hist 51511

Okemah OK

Territory Town USA, Old West Museum, 5 m W of Okemah, on I-40, exit 217, Okemah, OK 74859 • T: +1 918 6232599 •
Head: *Louise Parsons* •
Local Museum – 1967
Western relics, Civil War and Native American artefacts 51512

Oklahoma City OK

45th Infantry Division Museum, 2145 NE 36th St, Oklahoma City, OK 73111 • T: +1 405 4245313 • F: +1 405 4243748 • curator@45thdivisionmuseum.com • www.45thdivisionmuseum.com •
Cur.: *Michael E. Gonzales* •
Military Museum – 1976 51513

Harn Homestead and 1889er Museum, 313 NE 16th, Oklahoma City, OK 73104 • T: +1 405 2354058 • F: +1 405 2354041 • info@harnhomestead.com • harnhomestead.com •
Exec.Dir.: *Cher Lucewicz* •
Historical Museum – 1986
Hist bldgs, pre-1907 furnishings 51514

Hulsey Gallery, Oklahoma City University, Norick Art Center, 2501 N Blackwelder, Oklahoma City, OK 73106 • T: +1 405 5215226 • F: +1 405 5576029 • www.okcu.edu •
Dir.: *Kirk Niemeyer* •
Public Gallery 51515

Individual Artists of Oklahoma, 811 N Broadway, Oklahoma City, OK 73146, mail addr: POB 60824, Oklahoma City, OK 73146 • T: +1 405 2326060 • F: +1 405 2326061 • iao@telepath.com • www.iaogallery.org •
Dir.: *Suzanne Owens* •
Fine Arts Museum 51516

Kirkpatrick Science and Air Space Museum at Omniplex, 2100 NE 52nd St, Oklahoma City, OK 73111 • T: +1 405 6026664 • F: +1 405 6023767 • omnipr@omniplex.org • www.omniplex.org •
Dir.: *Eleanor Kirkpatrick* • *John Kirkpatrick* • Exec.Dir.: *Max L. Ary* •
Science&Tech Museum / Natural History Museum – 1958 51517

Melton Art Reference Library Museum, 4300 North Sewell, Oklahoma City, OK 73118 • T: +1 405 5253603 • F: +1 405 5250396 • meltonart@aol.com • www.marl-okc.org •
C.E.O. & Pres.: *Willard Johnson* • Dir.: *Suzanne Silvester* • Assistant: *Amena Butler* •
Library with Exhibitions – 1989
Art Books; Paintings; Decorative Arts; Auction Catalogues and Results; International Art Dictionaries; Art Reproductions Study Guides – Library, Adjunct Gallery 51518

National Cowboy and Western Heritage Museum, 1700 NE 63rd St, Oklahoma City, OK 73111 • T: +1 405 4782250 • F: +1 405 4784714 • info@nationalcowboymuseum.org • www.nationalcowboymuseum.org •
Dir.: *Chuck Schroeder* • Cur.: *Ed Muno* (Art) • *Richard Rattenbury* (History) • *Mike Leslie* (Ethnology) • *Don Reeves* (Colls) • *Steve Grafe* (Native American coll) •

U

Historical Museum – 1954
Contemporary Western and Native American fine arts, Western popular culture, cowboys and ranching, rodeo hist 51519

National Softball Hall of Fame and Museum Complex, 2801 NE 50, Oklahoma City, OK 73111 • T: +1 405 4245266 • F: +1 405 4243855 • info@softball.org • www.asasoftball.com •
Exec. Dir.: *Ron Radigonda* •
Special Museum – 1957 51520

Oklahoma City Museum of Art, 415 Couch Dr, Oklahoma City, OK 73102 • T: +1 405 2363100 • F: +1 405 2363122 • lspears@okcmoa.com • www.okcmoa.com •
Dir.: *Carolyn Hill* • Chief Cur.: *Hardy George* • Assoc.Cur.: *Alison Amick* • Cur.: *Brian Hearn* (Film) •
Fine Arts Museum – 1945
European, Asian and American art, German and American Expressionism, 1960's abstract art, figurative, conceptual and kinetic works 51521

Oklahoma Firefighters Museum, 2716 NE 50th, Oklahoma City, OK 73111 • T: +1 405 4243440 • F: +1 405 4251032 • osfa@osfa.info • www.osfa.info •
Dir.: *Jim Minx* • Cur.: *Sam Oruch* •
Special Museum – 1970
Fire-fighting equip, bldg of the first fire-station in Oklahoma (1869) 51522

Oklahoma Museum of African American Art, 3919 NW 10th St, Oklahoma City, OK 73107 • T: +1 405 9424896 •
Dir.: *Melvin R. Smith* • *Rose J. Smith* •
Fine Arts Museum 51523

Oklahoma Museum of History, 2100 N Lincoln Blvd, Oklahoma City, OK 73105 • T: +1 405 5225248 • F: +1 405 5215402 • www.ok-history.mus.ok.us •
C.E.O. & Dir.: *Dr. Bob Blackburn* • Dir.: *Dan Provo* • Cur.: *John Hill* (Exhibitions) •
Historical Museum – 1893
Historic site, state hist, dec arts, Native American coll – Library, Archives 51524

Overholser Mansion, Oklahoma Historical Society, 405 NW 15th, Oklahoma City, OK 73103 • T: +1 405 5288485 • overholser@ok-history.mus.ok.us •
Admin.: *Traci Lees* •
Historical Museum – 1972
1902-04 mansion, original furnishings, costumes, docs 51525

World of Wings Pigeon Center Museum, 2300 NE 63rd, Oklahoma City, OK 73111-8208 • T: +1 405 4785155 • F: +1 405 4784552 • jbrown@ionet.net • www.worldofwings.org •
Pres.: *Terry Cockrum* • Cur.: *Dr. Edgar Petty* •
Special Museum – 1973
Paintings, sketches, books, lithographics, WW I & WW II, European use of pigeons 51526

The World Organization of China Painters' Museum, 2641 NW 10th St, Oklahoma City, OK 73107 • T: +1 405 5211234 • F: +1 405 5211265 • wocporg@theshop.net • www.theshop.net/wocporg •
Dir.: *Patricia Dickerson* •
Fine Arts Museum / Decorative Arts Museum – 1967
19th/20th c hand painted porcelain 51527

Okmulgee OK

Creek Council House Museum, Town Square, 106 W Sixth, Okmulgee, OK 74447 • T: +1 918 7562324 • F: +1 918 7563671 • creekmuseum@sbcglobal.net •
Dir.: *David Anderson* •
Historical Museum / Ethnology Museum – 1923
Indian history, archaeology, hist of Muscogee Creek nation 51528

Okoboji IA

Higgins Museum, 1507 Sanborn Av, Okoboji, IA 51355 • T: +1 712 3325859 • F: +1 712 3325859 •
Cur.: *Glenn McConnell* •
Special Museum – 1978
National bank notes from 1863-1935 51529

Olathe KS

Mahaffie Stagecoach Stop and Farm, 1100 Kansas City Rd, Olathe, KS 66061 • T: +1 913 7826972 • F: +1 913 3975114 • ttalbott@olatheks.org •
C.E.O.: *Tim Talbott* •
Historical Museum – 1977
1865 Mahaffie house and stagecoach stop on Santa Fe Trail 51530

Old Bennington VT

Bennington Battle Monument, 15 Monument Circle, Old Bennington, VT 05201 • T: +1 802 8283051 • F: +1 802 4470550 • john.dumville@state.vt.us • www.historicvermont.org •
Historic Site – 1891
Historic site located near the site of the Bennington Battle of the Revolutionary War 51531

Old Bethpage NY

Old Bethpage Village Restoration, Round Swamp Rd, Old Bethpage, NY 11804 • T: +1 516 5728401 • F: +1 516 5728439 • www.oldbethpage.org •
Dir.: *James McKenna* •
Open Air Museum – 1970
American hist, tools, utensils, furnishings and dec arts 51532

Old Bridge Township NJ

Thomas Warne Historical Museum, 4216 Rte 516, Old Bridge Township, NJ 07747 • T: +1 732 5660348 • F: +1 732 5666943 •
Cur.: *Alvia D. Martin* •
Local Museum – 1964
Edison phonograph and records sheet music, pottery, carpenter tools, clothing, Indian artifacts, old school books – library 51533

Old Chatham NY

The Shaker Museum, 88 Shaker Museum Rd, Old Chatham, NY 12136 • T: +1 518 7949100 ext 100 • F: +1 518 7948621 • contact@shakermuseumandlibrary.org • www.smandl.org •
Exec. Dir.: *Lili R. Ott* • Cur.: *Starlyn D'Angelo* •
Folklore Museum / Historical Museum – 1950
Furniture, baskets, textiles, woodworking and metal working tools and machinery, craft industries 51534

Old Fort NC

Mountain Gateway Museum, Water and Catawba Sts, Old Fort, NC 28762 • T: +1 828 6689259 • F: +1 828 6680041 • gateway@wnclink.com • mcdowellcounty.org •
Dir.: *Sam Gray* •
Local Museum – 1971
Mountain life artifacts, pottery, farm implements, log cabins 51535

Old Lyme CT

Florence Griswold Museum, 96 Lyme St, Old Lyme, CT 06371 • T: +1 860 4345542 • F: +1 860 4346259, 4349778 • info@flogris.org • www.flogris.org •
Dir.: *Jeffrey W. Andersen* •
Fine Arts Museum / Historical Museum – 1936
American impressionist paintings, decorative arts, toys 51536

Old Shawneetown IL

Shawneetown Historic Site, 280 Washington St, Old Shawneetown, IL 62984-3401 • T: +1 618 2693303 •
Historical Museum / Open Air Museum – 1917
Historic bldg, 1839 4-story Greek Revival Bank 51537

Old Town ME

Old Town Museum, 153 S Main St, Old Town, ME 04468 • T: +1 207 8277256 • ozmaine@aol.com •
Local Museum – 1976
Antique telephones, costumes, glass ceramics 51538

Old Westbury NY

Old Westbury Gardens, 71 Old Westbury Rd, Old Westbury, NY 11568 • T: +1 516 3330048 • F: +1 516 3336807 • www.oldwestburygardens.org •
Pres. & C.E.O.: *Carol E. Large* •
Historical Museum / Decorative Arts Museum – 1959
English 18th c furniture, decorations and paintings, Chinese porcelain 51539

Olivet MI

Barker-Cawood Art Gallery, Olivet College, Art & Communication Dept., 320 S Main St, Mott Academic Center, Olivet, MI 49076 • T: +1 269 7497661 • F: +1 269 7497178 • dcrowe@olivetcollege.edu • www.olivetcollege.edu •
Dir.: *Donald Rowe* •
Fine Arts Museum
American Indian, Mesopotamian, Philippine and Thailand artifacts, primitive arts, sculpture, modern American prints – library, Armstrong coll 51540

Olustee FL

Olustee Battlefield, US 90, 2 m E, Olustee, FL 32072, mail addr: POB 40, Olustee, FL 32072 • T: +1 904 7580400 • www.floridastateparks.org/olustee/ •
Historic Site / Military Museum – 1909
State historic site, military 51541

Olympia WA

Bigelow House Museum, 918 Glass Av, Olympia, WA 98506 • T: +1 360 3576099 • annafitz@earthlink.net • www.bigelowhouse.org •
C.E.O.: *Annamary Fitzgerald* •
Local Museum – 1992
Local hist, 19th c furnishing, domestic art 51542

Evergreen Galleries, Evergreen State College, 2700 Evergreen Pkwy NW, Olympia, WA 98505-0002 • T: +1 360 8675125 • F: +1 360 8675430 • friedma@evergreen.edu • www.evergreen.edu/user/galleries •
Dir.: *Ann Friedman* •
Fine Arts Museum / University Museum – 1970
Photography, printmaking, functional and sculptural ceramics, Chicano posters 51543

Hands on Children's Museum, 106 11th Av SW, Olympia, WA 98501 • T: +1 360 9560818 • F: +1 360 7548626 • hocm@hocm.org • www.hocm.org •
Dir.: *Patty Belmonte* •
Science&Tech Museum / Natural History Museum – 1988 51544

Washington State Capital Museum, 211 W 21st Av, Olympia, WA 98501 • T: +1 360 7532580 • F: +1 360 5868322 • dvalley@wshs.wa.gov • www.wshs.com •
Dir.: *Derek R. Valley* • Cur.: *Susan Rohrer* • *Melissa Parr* •
Local Museum – 1941
Local hist, pioneer life, Indian artefacts 51545

Omaha NE

Artists' Cooperative Gallery, 405 S 11th St, Omaha, NE 68102 • T: +1 402 3429617 •
Pres.: *Nicholas W. Pella* •
Fine Arts Museum – 1975
American paintings 51546

Bemis Center for Contemporary Arts, 724 S 12th St, Omaha, NE 68102-3202 • T: +1 402 3417130 • F: +1 402 3419791 • bemis@novia.net • www.bemiscenter.org •
Dir.: *Ree Schonlau* •
Fine Arts Museum – 1985
Sculpture, ceramics 51547

Durham Western Heritage Museum, 801 S 10th St, Omaha, NE 68108 • T: +1 402 4445071 • F: +1 402 4445397 • info@dwhm.org • www.dwhm.org •
Dir.: *Randall Hayes* • Cur.: *Janelle Lindberg* •
Local Museum – 1975
Railroad memorabilia, furniture, arts, pioneering tools, photography coll 51548

Fine Arts Gallery, Creighton University, 2500 California Plaza, Omaha, NE 68178 • T: +1 402 2802509 ext 2831 • F: +1 402 2802320 • bohr@creightonuniversity.edu •
Dir.: *G. Ted Bohr* •
Fine Arts Museum – 1973
Drawings, graphics, paintings, photography, prints, sculpture 51549

General Crook House Museum, 5730 N 30 St, Omaha, NE 68111-1657 • T: +1 402 4559990 • F: +1 402 4539448 • director@omahahistory.org • www.omahahistory.org •
Dir.: *Betty J. Davis* • Cur.: *Patricia Pixley* •
Local Museum – 1956
Military hist, hist of Omaha and Douglas County 51550

Great Plains Black Museum, 2213 Lake St, Omaha, NE 68110 • T: +1 402 3452212 • F: +1 402 3452256 •
Pres.: *Joyce Young* • C.E.O.: *Jim Calloway* •
Historical Museum – 1975
Black hist, family hist 51551

Joslyn Art Museum, 2200 Dodge St, Omaha, NE 68102-1292 • T: +1 402 3423300 • F: +1 402 3422376 • info@joslyn.org • www.joslyn.org •
Dir.: *J. Brooks Joyner* • Cur.: *Marsha V. Gallagher* (Material, Culture) •
Fine Arts Museum – 1931
19th and 20th c European and American art 51552

El Museo Latino, 4701 S 25th St, Omaha, NE 68107-2728 • T: +1 402 7311137 • F: +1 402 7337012 • mgarcia@elmuseolatino.org • www.elmuseolatino.org •
Dir.: *Magdalena A. Garcia* •
Fine Arts Museum / Folklore Museum / Historical Museum – 1993
Latino art and hist 51553

Omaha Center for Contemporary Art, 1116 Jackson St, Omaha, NE 68102 • T: +1 402 3459711 •
Dir.: *Eadwierd Kroy* •
Fine Arts Museum – 2000 51554

Omaha Children's Museum, 500 S 20th St, Omaha, NE 68102-2508 • T: +1 402 3426164 ext 410 • F: +1 402 3426165 • discover@ocm.org • www.ocm.org •
Exec. Dir.: *Lindy Hoyer* •
Special Museum – 1977
Arts, science, humanities 51555

University of Nebraska Art Gallery at Omaha, Fine Arts Bldg, Rm 137, Omaha, NE 68182-0012 • T: +1 402 5542796 • F: +1 402 5543435 • n_kelly@unomaha.edu • www.unomaha.edu •
Cur.: *Nancy Kelly* •
Fine Arts Museum / University Museum – 1967
Prints 51556

Onancock VA

Kerr Place, 69 Market St, Onancock, VA 23417 • T: +1 757 7878012 • F: +1 757 7874271 • kerr@esva.net • www.kerrplace.org •
Dir.: *John H. Verrill* •
Local Museum – 1957
Glass & china, area reference books 51557

Onchiota NY

Six Nations Indian Museum, 1462 County Rte 60, Onchiota, NY 12989 • T: +1 518 8912299 • redmaple@northnet.org •
Dir.: *John Fadden* •
Ethnology Museum – 1954
Six Nations Indian artifacts, Iroquois culture and history, clothing, art work and dwellings 51558

Oneida NY

Cottage Lawn, Madison County Historical Society, 435 Main St, Oneida, NY 13421-0415 • T: +1 315 3634136 • F: +1 315 3634136 • mchs1900@dreamscape.com • www.dreamscape.com/mchs1900 •
Exec. Dir.: *Sydney L. Loftus* • Cur.: *Michael Martin* (Collections) •
Historical Museum – 1895
19th c furnishings, tools incl. agriculture, woodworking, black- and tinsmithing, carriages, carts and wagons 51559

Oneida Community Mansion House, 170 Kenwood Av, Oneida, NY 13421 • T: +1 315 3630745 • F: +1 315 3614580 • ocmh@dreamscape.com • www.oneidacommunity.org •
C.E.O.: *Bruce M. Moseley* •
Local Museum – 1987
Furniture, clothing, decorative arts, paintings, works of art on paper 51560

Shakowi Cultural Center, 5 Territory Rd, Oneida, NY 13421-9304 • T: +1 315 3631424 • F: +1 315 3631843 •
Dir.: *Birdy Birdick* •
Folklore Museum / Ethnology Museum – 1993
Native Americans, ethnography, beadwork 51561

Oneonta NY

Art Gallery, State University of New York College at Oneonta, Ravine Pkwy, Oneonta, NY 13820 • T: +1 607 4363717 • F: +1 607 4363715 • hallbc@oneonta.edu • www.oneonta.edu/academics/art/gallery.html •
Dir.: *Tim Sheesley* •
Public Gallery / University Museum 51562

Foreman Gallery, Hartwick College, Anderson Center for the Arts, Oneonta, NY 13820 • T: +1 607 4314825 • F: +1 607 4314191 • wilson@hartwick.edu • www.hartwick.edu •
Cur.: *Gloria Escobar* •
Public Gallery 51563

The National Soccer Hall of Fame, 18 Stadium Circle, Oneonta, NY 13820 • T: +1 607 4323351 • F: +1 607 4328429 • info@soccerhall.org • www.soccerhall.org •
Pres.: *Will Lunn* •
Special Museum – 1981
Sports hist 51564

Science Discovery Center of Oneonta, State University College, Oneonta, NY 13820-4015 • T: +1 607 4362011 • F: +1 607 4362654 • scdisc@oneonta.edu • www.oneonta.edu/academics/scdisc/index.html •
Dir.: *Albert J. Read* •
Science&Tech Museum / Natural History Museum – 1987
Physics and physical science, interactive exhibits 51565

The Yager Museum, Hartwick College, West St, Oneonta, NY 13820-4020 • T: +1 607 4314480 • F: +1 607 4314468 • museums@hartwick.edu • www.hartwick.edu/museum •
Interim Dir. & Cur.: *Dr. Fiona M. Dejardin* (Fine Arts) • Cur.: *Dr. David Anthony* (Anthropolgy Collection) •
Fine Arts Museum / University Museum / Archaeology Museum – 1797/1929
Upper Susquehanna Indian artifacts, basketry and pottery, Furman Coll of Mexican, Central & South American artifacts; Sandell Coll of Ecuadorian and Peruvian artifacts; Friess Coll of Mexican masks; Marks Coll of Pre-Columbian art; Vane Ess Fine Art Coll, Russian icons 51566

Onset MA

Porter Thermometer Museum, 49 Zarahemla Rd, Onset, MA 02558 • T: +1 508 2955504 • F: +1 508 2958323 • thermometerman@aol.com • members.aol.com/thermometerman •
Cur.: *Richard T. Porter* •
Science&Tech Museum – 1990
4,800 medical, maritime, antique, weather, souvenir thermometers 51567

Ontario CA

Museum of History and Art, 225 S Euclid Av, Ontario, CA 91762 • T: +1 909 3952510 • www.ci.ontario.ca.us/index.cfm/1605 •
Dir.: *Theresa Hanley* • Cur.: *Maricarmen Ruiz-Torres* • *Marcilyn Callejo* •
Local Museum / Fine Arts Museum – 1979
Local history, agriculture and industry, social and home life, viticulture, mining, paintings 51568

Ontario NY

Heritage Square Museum, 7147 Ontario Center Rd, Ontario, NY 14519-0462 • retoddl@usadata.net.net • heritagesquaremuseum.org •
Cur.: *Liz Albright* •
Local Museum – 1969
Iron ore mining, agriculture, farm machinery 51569

Ontario OR

Four Rivers Cultural Center, 676 SW 5th Av, Ontario, OR 97914 • T: +1 541 8898191 • F: +1 541 8897628 • cfugate@fmtc.com • www.4rcc.com •
Pres.: *Larry Sullivan* • Exec. Dir.: *Charlotte Fugate* •
Local Museum / Folklore Museum – 1987
Local hist, first settlement 51570

Ontonagon MI

Ontonagon County Historical Society Museum, 422 River St, Ontonagon, MI 49953-0092 • T: +1 906 8846165 • F: +1 906 8846094 • ochsmuse@up.net •
Pres.: *Bruce Johanson* •
Local Museum – 1957
Regional history, musical instr, pictures, copper mining, crafts, tools – archives 51571

Opelika AL

Mann Museum, 3161 Lee Rd 54, Opelika, AL 36803-2209, mail addr: POB 2209, Opelika AL 36804 • T: +1 334 7417776 •
C.E.O.: *George E. Mann* •
Natural History Museum
Natural hist 51572

Museum of East Alabama, 121 S 9th St, Opelika, AL 36801 • T: +1 334 7492751 • museum@eastalabama.org • www.eastalabama.org •
Dir./Cur.: *Barbara Barnhill-West* •
Historical Museum – 1989
Roanoke dolls, Camp Opelika (WW II) memorabilia, recording technology, agriculture implements, fire fighting tools and trucks, local hist items 51573

Opelousas LA

Opelousas Museum, 315 N Main St, Opelousas, LA 70570 • T: +1 337 9482589 • F: +1 337 9482592 • musdir@hotmail.com •
Dir.: *Sue DeVille* •
Local Museum – 1992
Hist and culture of the Opelousas area 51574

Opelousas Museum of Art, 106 N Union St, Opelousas, LA 70570 • T: +1 337 9424991 • F: +1 337 9424930 • omamuseum@aol.com •
Dir.: *Nan Wier* • Cur.: *Keith Guidry* •
Fine Arts Museum – 1997 51575

Oradell NJ

Hiram Blauvelt Art Museum, 705 Kinderkamack Rd, Oradell, NJ 07649 • T: +1 201 2610012 • F: +1 201 3916418 • maja218@att.net • www.blauveltmuseum.com •
Dir.: *Dr. Marijane Singer* •
Fine Arts Museum – 1940
Art 51576

Orange TX

Heritage House of Orange County Museum, 905 W Division, Orange, TX 77630 • T: +1 409 8865385 • F: +1 409 8860917 • hhmuseum@exp.net • www.heritagehouseoforange.org •
Pres.: *Sue Cowling* •
Historical Museum – 1977
Hist from early Indian settlements to present day, period rooms, decorative arts 51577

Stark Museum of Art, 712 Green Av, Orange, TX 77630 • T: +1 409 8836661 • F: +1 409 8836361 • starkmuseum@starkmuseum.org • www.starkmuseum.org •
C.E.O.: *Walter G. Riedel* • Dir.: *David Hunt* • Sc. Staff: *Richard Hunter* •
Fine Arts Museum – 1974
Western American art incl Paul Kane coll, porcelain & crystal 51578

The W.H. Stark House, 610 W Main St, Orange, TX 77630 • T: +1 409 8333513 • F: +1 409 8833530 • whstarkhouse@starkadmin.org • www.whstarkhouse.org •
C.E.O.: *Walter G. Riedel* •
Local Museum – 1981
Original furniture, rugs, family portraits, lace curtains 51579

Orange VA

The James Madison Museum, 129 Caroline St, Orange, VA 22960 • T: +1 540 6721776 • F: +1 540 6720231 • info@jamesmadisonmuseum.org • www.jamesmadisonmuseum.org •
Pres.: *Arthur H. Bryant* •
Historical Museum – 1976
Transportation, Madison artifacts, cube house 51580

Orange City IA

Thea G. Korver Visual Art Center, Northwestern College, 101 Seventh St SW, Orange City, IA 51041 • T: +1 712 7077000 • johnk@nwciowa.edu • www.nwciowa.edu •
Public Gallery – 2003 51581

Orangeburg SC

I.P. Stanback Museum, South Carolina State University, 300 College St, NE, Orangeburg, SC 29117 • T: +1 803 5367174, 5368119 • F: +1 803 5368309 • www.scsu.edu •
Pres.: *Dr. Leroy Davis* • Interim Dir.: *Frank Martin* •
Fine Arts Museum / University Museum – 1980
African and Afro-American art, Benin bronzes, photographyArt 51582

Orchard Lake MI

The Galeria, Saint Mary's College, 3535 Indian Trail, Orchard Lake, MI 48324 • T: +1 248 6830345 • F: +1 248 7386725 • dstearns@orchardlakeschools.com • www.orchardlakeschools.com/galeria/glripg.html •
Dir.: *Marian Owczarski* •
Fine Arts Museum – 1963
Former Michigan Military Academy, contemporary Polish paintings and sculpture, Polish printing, folk art, tapestries 51583

Orchard Park NY

Orchard Park Historical Society Museum, S-4287 S Buffalo St, Orchard Park, NY 14127 • T: +1 716 6622185 •
C.E.O. & Pres.: *Dennis J. Mill* • Cur.: *Yasabel Newton Gibson* •
Historical Museum – 1951
Agricultural and household artifacts, Indian artifacts, Quaker artifacts 51584

Oregon OH

Oregon-Jerusalem Historical Museum, 1133 Grasser St, Oregon, OH 43616 • T: +1 419 6917561 • F: +1 419 6937052 • ojhs@juno.com •
Dir.: *Kathy M. Clark* •
Local Museum – 1963
School records, maps, farm equip, musical instruments, office equip, household furniture, Indian artifacts, China, glass, costumes 51585

Oregonia OH

Fort Ancient Museum, 6123 State Rte 350, Exits 32 and 36 off I-71, Oregonia, OH 45054 • T: +1 513 9324421 • F: +1 513 9324843 • jblosser@ohiohistory.org • www.ohiohistory.org •
Man.: *Jack Blosser* •
Ethnology Museum / Natural History Museum – 1891
Prehist Indian life and culture 51586

Orient NY

Museum of Orient and East Marion History, Oysterponds Historical Society, Village Ln, Orient, NY 11957 • T: +1 631 3232480 • F: +1 631 3233719 • ohsorient@optonline.net •
C.E.O. & Dir.: *William MNaught* •
Local Museum – 1944
Indian artifacts, marine paintings, whaling and fishing, ship models, navigating instr, spinning and weaving tools, clothing, furniture, toys, dolls, agricultural tools 51587

Oriskany NY

Oriskany Battlefield, 7801 State Rte 69 W, Oriskany, NY 13424 • T: +1 315 7687224 • F: +1 315 3373081 • nancy.demyttenaere@oprhp.state.ny.us • www.nysparks.com •
Military Museum / Open Air Museum – 1927 51588

Orland ME

Orland Historical Museum, Castine Rd, Orland, ME 04472 • T: +1 207 4692476 •
Pres.: *Cindy Kimball* •
Local Museum – 1966
Local history 51589

Orlando FL

East Campus Galleries, Valencia Community College, 701 N Econlockhatchee Trail, Orlando, FL 32825 • T: +1 407 2995000 • F: +1 407 2493943 • east.valencia.cc.fl.us •
Dir.: *Jackie Otto-Miller* •
Fine Arts Museum – 1967
Contemporary and Florida art, small works – library 51590

Mennello Museum of American Folk Art, 900 E Princeton St, Orlando, FL 32803 • T: +1 407 2464278 • F: +1 407 2464329 • mennellomuseum@cityoforlando.net • www.mennellomuseum.com •
C.E.O.: *Frank Holt* •
Decorative Arts Museum – 1998
Paintings, folk art on American self-taught artists 51591

Orange County Regional History Center, 65 E Central Blv, Orlando, FL 32801 • T: +1 407 8368517 • F: +1 407 8368550 • hogla.reyes@ocfl.net • www.thehistorycenter.org •
Dir.: *Sara Van Arsdel* •
Local Museum – 1957
Local Central Florida history 51592

Orlando Museum of Art, 2416 N Mills Av, Orlando, FL 32803-1483 • T: +1 407 8964231 • F: +1 407 8969920, 8944314 • info@omart.org • www.omart.org •
Dir.: *Marena Grant Morrisey* • Cur.: *Susan Rosoff* •
Fine Arts Museum – 1924
American and African art, art of the ancient 51593

Orlando Science Center, 777 E Princeton St, Orlando, FL 32803 • T: +1 407 5142000, 5142012 • F: +1 407 5142001 • info@osc.org • www.osc.org •
Pres.: *Steven Goldman* •
Science&Tech Museum – 1959
Physical, natural, space and health sciences, applied technology-aerospace, lasers 51594

Terrace Gallery, 900 E Princeton St, Orlando, FL 32803 • T: +1 407 2464279 • F: +1 407 2464329 • public.art@cityoforlando.net •
Cur.: *Frank Holt* •
Fine Arts Museum – 1992
Florida art, temporary exhib 51595

Wordspring Discovery Center, 11221 John Wycliffe Blvd, Orlando, FL 32832 • T: +1 407 8523626 • F: +1 407 8523601 • www.wycliffe.org/wordspring •
Religious Arts Museum – 2002
Bible translations, ethnic costumes from around the world – library 51596

Orleans MA

French Cable Station Museum, 41 S Orleans Rd, Orleans, MA 02653 • T: +1 508 2550343 • F: +1 508 2406099 • miko@gis.net •
Special Museum – 1972
Atlantic Cable Terminal, communications 51597

Playhouse Gallery, Academy of Performing Arts, 120 Main St, Orleans, MA 02653-1843 • T: +1 508 2553075 • F: +1 508 2551963 • apa@cape.com • www.apa1.org/Playhouse-Gallery.htm •
Dir.: *Claire Scapellati* • *Ralph Bassett* •
Fine Arts Museum / University Museum
Contemporary art 51598

Ormond Beach FL

Fred Dana Marsh Museum, 2099 N Beach St, Tomoka State Park, Ormond Beach, FL 32174 • T: +1 904 6764045, 6764075 • F: +1 904 6764060 •
Open Air Museum / Natural History Museum – 1967
Village of Nocoroco of the 1605, site of British plantation, Mount Oswald, paintings, sculpture, Florida geology and history 51599

Ormond Memorial Art Museum, 78 E Granada Blvd, Ormond Beach, FL 32176 • T: +1 904 6763341 • F: +1 904 6763244 • omam78e@aol.com •
Dir.: *Ann Burt* •
Fine Arts Museum – 1946
Temporary exhibitions 51600

Orofino ID

Clearwater Historical Museum, 315 College Av, Orofino, ID 83544 • T: +1 208 4765033 • chmusuem@clearwater.net • www.clearwatermuseum.org •
Dir.: *Bernice Pullen* •
Local Museum – 1960
Local history, mining, logging, Nez Perce Indians, farming 51601

Orono ME

Hudson Museum, University of Maine, 5476 Maine Center for the Arts, Orono, ME 04469-5746 • T: +1 207 5811901 • F: +1 207 5811950 • hudsonmuseum@umit.maine.edu • www.umaine.edu/hudsonmuseum/ •
Pres.: *Ron Stegall* •
Ethnology Museum / Archaeology Museum – 1986
Ethnology coll, archaeological material 51602

Museum of Art, University of Maine, 109 Carnegie Hall, Orono, ME 04469 • T: +1 207 5813255 • F: +1 207 5813083 • umma@umit.maine.edu •
Dir.: *Wally Mason* •
Fine Arts Museum / University Museum – 1946
Art, housed in a library of Tusco-Doric (Palladian) design 51603

Oroville CA

Butte County Pioneer Memorial Museum, 2332 Montgomery St, Oroville, CA 95965 • T: +1 530 5382497 • www.cityoforoville.org •
Dir.: *Charles Miller* • Cur.: *David Dewey* •
Historical Museum – 1932
Gold rush era of California, artifacts 51604

Coyote Art Gallery, Butte College, 3536 Butte Campus Dr, Oroville, CA 95965 • T: +1 530 8952877 • F: +1 530 8952346 • www.butte.edu •
Dir.: *Alan Carrier* •
Public Gallery – 1981 51605

Oroville Chinese Temple, 1500 Broderick St, Oroville, CA 95965 • T: +1 530 5382415 • www.cityoforoville.org •
Dir.: *Charles Miller* • Cur.: *Vorin Dornan* •
Historical Museum / Religious Arts Museum
Restored temple, religious figures, screens, Chinese immigrants during the gold rush 51606

Osage IA

Cedar Valley Memories, 1 1/2 Mile W Hwy 9, Osage, IA 50461 • T: +1 641 7321269 •
Science&Tech Museum / Local Museum
Steam engines, gas running, agriculture 51607

Mitchell County Historical Museum, N 6th, Osage, IA 50461-8557 • T: +1 515 7321269 •
Pres.: *Dan Swann* •
Local Museum / Agriculture Museum – 1965
Local history, agriculture, Cedar Valley seminary 51608

Osawatomie KS

John Brown House, 10th and Main Sts, Osawatomie, KS 66064, mail addr: POB 37, Osawatomie, Ks 66064 • T: +1 913 7554384 • F: +1 913 7554164 • cityofoz2003@yahoo.com •
Military Museum – 1854
1856 Battle of Osawatomie, weapons, pictures, antiques, fireplace cooking equip 51609

Osceola NE

Polk County Historical Museum, 561 South St, Osceola, NE 68651 • T: +1 402 7477901 •
Pres./Dir.: *D. Ruth Lux* •
Local Museum – 1967 51610

Oshkosh NE

Garden County Museum, W First and Av E, Oshkosh, NE 69154-0913 • T: +1 308 7723115 •
Cur.: *Phyllis Chadwick* •
Local Museum – 1969
Mounted birds, Indian artifacts, pioneer machinery, fossils, local hist of schools 51611

Oshkosh WI

Allen Priebe Gallery, University of Wisconsin Oshkosh, 800 Algoma Blvd, Oshkosh, WI 54901 • T: +1 920 4242235 • F: +1 920 4240147 •
Dir.: *Jeff Lipschutz* •
Fine Arts Museum / University Museum
Art 51612

EAA AirVenture Museum, 3000 Poberezny Rd, Oshkosh, WI 54903-3065 • T: +1 920 4264800 • F: +1 920 4266765 • museum@eaa.org • www.airventuremuseum.com •
Dir.: *Adam. Smith* • Cur.: *Ron Twellman* •
Science&Tech Museum – 1963
Aviation, over 250 aircrafts, WW II aircrafts and memorabilia 51613

Military Veterans Museum, City Center, # 245, Oshkosh, WI 54901 • T: +1 920 4268615 • mvm@athenet.net • www.mvmwisconsin.com •
Pres.: *Dr. E.T. Sonnleitner* • Cur.: *Euclid Caron* •
Military Museum – 1985 51614

Oshkosh Public Museum, 1331 Algoma Blvd, Oshkosh, WI 54901 • T: +1 920 4244731 • F: +1 920 4244738 • info@publicmuseum.oshkosh.net • www.publicmuseum.oshkosh.net •
Dir.: *Bradley Larson* • Asst. Dir.: *Michael Breza* • Cur.: *Debra Daubert* •
Local Museum – 1924
Local hist 51615

Paine Art Center, 1410 Algoma Blvd, Oshkosh, WI 54901-2719 • T: +1 920 2356903 • F: +1 920 2356303 • info@paineartcenter.com • www.paineartcenter.com •
Dir.: *Aaron Sherer* •
Fine Arts Museum – 1947
American (19th-20th c) and European (18th - 20th c) paintings, sculpture and prints, dec art 51616

Oskaloosa IA

Nelson Pioneer Farm and Museum, 2294 Oxford Av, Oskaloosa, IA 52577 • T: +1 515 6722989 • www.nelsonpioneer.org •
Pres.: *Don Harrison* •
Historical Museum / Agriculture Museum – 1942
Daniel and Margaret Nelson Home, Nelson barn, local history, agriculture 51617

Oskaloosa KS

Old Jefferson Town, 703 Walnut, Hwy 59, Oskaloosa, KS 66066 • T: +1 785 8632070 •
Cur.: *Karen Heady* •
Local Museum / Open Air Museum – 1966
Seven c.1880 bldgs 51618

Osprey FL

Historic Spanish Point, 337 N Tamiami Trail, Osprey, FL 34229, mail addr: POB 846, Osprey, FL 34229 • T: +1 941 9665214 • F: +1 941 9661355 • diane@historicspanishpoint.org • www.historicspanishpoint.org •
Dir.: *Linda W. Mansperger* •
Historical Museum – 1980
Indian and Pioneer artifacts, shell tools, pottery, textiles 51619

Ossineke MI

Dinosaur Gardens Museum, 11160 US-23 S, Ossineke, MI 49766 • T: +1 989 4715477 • F: +1 989 4718032 •
Owner: *Frank A. McCourt* •
Natural History Museum – 1934
Paleontology, reproduction of 25 prehistoric animals and birds, artistic dinosaur sculpture 51620

Ossining NY

Ossining Historical Society Museum, 196 Croton Av, Ossining, NY 10562 • T: +1 914 9410001 • F: +1 914 9410001 • ohsm@bestweb.net • www.ossininghistorical.org •
Exec. Dir.: *Roberta Y. Arminio* •

U

Historical Museum / Fine Arts Museum – 1931
Culture of the 17th and 18th c residents, Indian artifacts; fine arts; doll coll, war relics, costumes, paintings 51621

Osterville MA

Osterville Historical Museum, 155 W Bay Rd, Osterville, MA 02655 • T: +1 508 4285861 • F: +1 508 4282241 • ohs@osterville.org • www.osterville.org •
Dir.: *Susan McGarry* •
Local Museum – 1931
Local history, furniture, textiles, folk and decorative arts, Crosby wooden catboats 51622

Oswego KS

Oswego Historical Museum, 410 Commercial St, Oswego, KS 67356 • T: +1 620 7954500 • history@oswego.net •
Historical Museum – 1967
Paintings, photos, furnishings, personal artifacts 51623

Oswego NY

Fort Ontario, 1 E Fourth St, Oswego, NY 13126 • T: +1 315 3434711 • F: +1 315 3431430 • paul.lear@oprhp.state.ny.us • www.nysparks.com •
Sc. Staff: *Paul Lear* •
Military Museum – 1949
Hist bldgs, military furnishings, firearms, equip, uniforms 51624

Richardson-Bates House Museum, 135 E Third St, Oswego, NY 13126 • T: +1 315 3431342 •
Dir.: *Terrence M. Prior* • Cur.: *Natalie Siembor* •
Historical Museum – 1896
Oswego County artifacts, documents; 19th c furnishings, paintings and dec arts 51625

Safe Haven Museum & Education Center, 2 E 7th St, Oswego, NY 13126 • T: +1 315 3423003 • F: +1 315 3421411 • safehaven@cnymail.com • www.oswegohaven.org •
Pres.: *Elizabeth Kahl* • Adm.: *Christine Sugrue* •
Historical Museum – 2002
History 51626

Tyler Art Gallery, State University of New York, College of Arts and Science, Oswego, NY 13126 • T: +1 315 3122113 • F: +1 315 3125642 • hzakin@oswego.edu •
Dir.: *Helen Zakin* • Asst. Dir.: *Mindy Ostrow* •
Fine Arts Museum / University Museum
Grant Arnold coll 51627

Ottawa IL

Ottawa Scouting Museum, 1100 Canal St, Ottawa, IL 61350 • T: +1 815 4319353 • scouter07@hotmail.com • www.ottawascoutingmuseum.org •
CEO: *Mollie Perrot* •
Special Museum – 1992
Camping gear, uniforms, badges, patches, awards, medals 51628

Ottawa KS

Old Depot Museum, 135 W Tecumseh, Ottawa, KS 66067 • T: +1 785 2421250 • F: +1 785 2421406 • history@ott.net • www.ott.net/~history •
Dir.: *Deborah Barker* •
Historical Museum – 1963
Local manufacturing, furniture, clothing 51629

Ottumwa IA

Airpower Museum, 22001 Bluegrass Rd, Ottumwa, IA 52501-8569 • T: +1 641 9382773 • F: +1 641 9382084 • antiqueairfield@sirisonline.com • www.aaa-apm.org •
C.E.O: *Robert L. Taylor* •
Science&Tech Museum – 1965
Aeronautics 51630

Wapello County Historical Museum, 210 W Main, Ottumwa, IA 52501 • T: +1 515 6828676 •
Pres.: *Edward Kuntz* •
Local Museum / Ethnology Museum – 1959
Regional history 51631

Overland Park KS

Gallery of Art, Johnson County Community College, 12345 College Blvd, Overland Park, KS 66210 • T: +1 913 4698500 • F: +1 913 4692348 • bhartman@jccc.net • www.jccc.net/gallery •
Pres.: *Dr. Charles Carlsen* • Dir.: *Bruce Hartman* •
Public Gallery / University Museum – 1969
Contemporary American paintings, ceramics, works on paper, Oppenheimer-Stein sculpture coll 51632

The Kansas City Jewish Museum, 5500 W 123rd, Overland Park, KS 66209 • T: +1 913 3122685 • F: +1 913 3452611 • www.kc-jewishmuseum.org •
Special Museum 51633

Overton NV

Lost City Museum, 721 S Hwy 169, Overton, NV 89040 • T: +1 702 3972193 • F: +1 702 3978987 • lostcity@comnett.net • www.comnett.net/~lostcity •
Cur.: *Kathryne Olson* •
Historical Museum / Local Museum – 1935 51634

Owatonna MN

Owatonna Arts Center, 435 Garden View Ln, Owatonna, MN 55060 • T: +1 507 4510533 • F: +1 507 4460198 • info@owatonnaartscenter.org • www.owatonnaartscenter.org •
Dir./Cur.: *Silvan Durben* •
Fine Arts Museum / Decorative Arts Museum – 1974
Marianne Young costume coll, prints and paintings, sculpture, oils by calendar illustrator Raymond James Stuart 51635

Owego NY

Tioga County Historical Society Museum, 110 Front St, Owego, NY 13827 • T: +1 607 6872460 • F: +1 607 6877788 • info@tiogahistory.org • www.tiogahistory.org •
Cur.: *Gwenyth Goodnight* •
Historical Museum / Local Museum – 1914
Indian artifacts, firearms, pioneer crafts; early commerce, industry, transportation and agriculture exhibits 51636

Owensboro KY

Anna Eaton Stout Memorial Art Gallery, c/o Brescia University, 717 Frederica, Owensboro, KY 42301 • T: +1 270 6853131 • maryt@brescia.edu •
Dir.: *Lance Hunter* •
Fine Arts Museum / University Museum – 1950 51637

International Bluegrass Music Museum, 207 E Second St, Owensboro, KY 42303 • T: +1 270 9267891 • F: +1 270 6899440 • gabriellegray@bluegrass-museum.org • www.bluegrass-museum.org •
Exec. Dir.: *Gabrielle Gray* •
Music Museum – 1992
Hist of bluegrass music 51638

Owensboro Area Museum of Science and History, 220 Daviess St, Owensboro, KY 42303 • T: +1 270 6872732 • F: +1 270 6872738 • owensboromuseum.com •
Dir.: *Jeff Jones* • Cur.: *Kathy Olson* •
Local Museum – 1966
Agricultural & industrial technology, social & natural hist, paleontology, geology, physical science 51639

Owensboro Museum of Fine Art, 901 Frederica St, Owensboro, KY 42301 • T: +1 270 6853181 • F: +1 270 6853181 • omfa@mindspring.com •
Dir.: *Mary Bryan Hood* • Asst. Dir.: *Jane Wilson* •
Fine Arts Museum – 1977
19th & 20th c American, English & French paintings 51640

Owls Head ME

Owls Head Transportation Museum, Rte 73, Owls Head, ME 04854 • T: +1 207 5944418 • F: +1 207 5944410 • info@ohtm.org • www.ohtm.org •
Dir.: *Charles Chiarchiaro* •
Science&Tech Museum – 1974
Pioneer air and ground vehicles 51641

Owosso MI

Curwood Castle, 224 Curwood Castle Dr, Owosso, MI 48867 • T: +1 989 7250597 • www.dupontcastle.com/castles/curwood.htm •
Cur.: *Ivan A. Conger* •
Special Museum – 1975
Studio of novelist James O. Curwood, furnishings, paintings 51642

The Movie Museum, 318 E Oliver St, Owosso, MI 48867-2351 • T: +1 989 7257621 • moviemuseum@aol.com •
Pres.: *Barbara Moore* • C.E.O.: *Don Schneider* •
Special Museum – 1972
Machinery, books, costumes, photos, records – library 51643

Steam Railroading Institute, 405 S Washington St, Owosso, MI 48867-0665 • T: +1 989 7259464 • F: +1 989 7231225 • info@mstrp.com • www.mstrp.com •
Exec. Dir.: *Dennis Braid* •
Science&Tech Museum – 1980
Railroad transportation & technology, steam loc from Pere Marquette Railway 51644

Oxford MA

Oxford Library Museum, 339 Main St, Oxford, MA 01540 • T: +1 508 9876003 • F: +1 508 9873896 • info@oxfordma.com •
Head: *Timothy Kelley* •
Library with Exhibitions – 1903
Local history, Nipmuc prehistoric Indians, colonial and civil war times – library 51645

Oxford MD

Oxford Museum, Morris and Market Sts, Oxford, MD 21654 • T: +1 410 2265331 • F: +1 410 2265670 • oxfmuse@dmv.com • www.oxfordmuseum.org •
Pres.: *Jeanne K. Foster* •
Local Museum – 1964
Local history, marine coll 51646

Oxford MI

Northeast Oakland Historical Museum, 1 N Washington St, Oxford, MI 48371 • T: +1 248 3911367 • mariee1@prodigy.net •
Cur.: *Marie English* •
Local Museum – 1971
Local history early farming, home & musical items, 2 WPA mural of Oxford – library 51647

Oxford MS

Rowan Oak, William Faulkner's Home, 916 Old Taylor Av, Oxford, MS 38655, mail addr: POB 1848University of Mississippi Museum, Oxford, MS 38677 • T: +1 662 9157073 • F: +1 662 9157035 • www.olemiss.edu •
Historic Site – 1977
1930-1962 home of the novelist and poet William Faulkner (1897-1962), personal artifacts, furniture 51648

University Gallery, c/o University of Mississippi, Bryandt Hall, Oxford, MS 38677 • T: +1 662 2327193 • F: +1 662 2325013 • art@sunset.backbone.oldmiss.edu •
Dir.: *Margaret Gorove* •
University Museum / Fine Arts Museum
Student art works 51649

University Museums, c/o University of Mississippi, 5th St & University Av, Oxford, MS 38677-1848 • T: +1 662 9157073 • F: +1 662 9157035 • museums@olemiss.edu • www.olemiss.edu/depts/u_museum •
Dir.: *Albert Sperath* •
University Museum / Fine Arts Museum – 1977
Local history, classical antiquities, folk art of the American South, antique scientific instruments 51650

Oxford NC

Granville County Historical Society Museum, 1 Museum Ln, Oxford, NC 27565 • T: +1 919 6939706 • F: +1 919 6939706 • gcmuseum@earthlink.net •
Dir.: *Pam Thornton* •
Local Museum – 1996
Regional hist, art, culture, science 51651

Harris Exhibit Hall, 1 Museum Ln, Oxford, NC 27565 • T: +1 919 6939706 • F: +1 919 6921030 • gcmuseum@gloryroad.net •
Dir.: *Pam Thornton* •
Public Gallery – 1996 51652

Oxford OH

Hefner Zoology Museum, Miami University, 100 Upham Hall, Oxford, OH 45056 • T: +1 513 5294617 • F: +1 513 5296900 • kaufmadg@muohio.edu • www.environmentaleducationohio.org •
C.E.O.: *Donald G. Kaufman* •
University Museum / Natural History Museum – 1951
Fauna of southwest Ohio 51653

Miami University Art Museum, Patterson Av, Oxford, OH 45056 • T: +1 513 5292232 • F: +1 513 5296555 • kretra@muohio.edu • www.muohio.edu/artmuseum •
Dir.: *Robert A. Kret* • Cur.: *Edna Carter Southard* • *Bonnie N. Mason* •
Fine Arts Museum / University Museum – 1978
Ancient, Islamic, Native American, pre-Columbian art, American and European painting and sculpture, and dec art, international folk art and textiles, contemporary Am folk art and outsider art 51654

Miami University Hiestand Galleries, Department of Art, Oxford, OH 45056 • T: +1 513 5291883 • F: +1 513 5291992 • sfagallery@muohio.edu • www.fna.muohio.edu/galleries •
Dir.: *Ann Taulbee* •
Public Gallery – 1978
Changing exhib of contemporary art 51655

William Holmes McGuffey Museum, 410 E Spring St, Oxford, OH 45056 • T: +1 513 5298380 • F: +1 513 5292637 • mcguffeymuseum@muohio.edu • www.muohio.edu/mcguffeymuseum •
Dir.: *Dr. Curtis W. Ellison* • Cur.: *Beverly Bach* •
Historical Museum – 1960
McGuffey memorabilia, furniture and decorative art of the 19th c 51656

Oxnard CA

Carnegie Art Museum, 424 S C St, Oxnard, CA 93030 • T: +1 805 3858157 • info@carnegieAM.org • www.carnegieam.org •
Dir.: *Suzanne Bellah* • Cur.: *Patricia Vincent* • *Jeanette LaVere* •
Fine Arts Museum – 1980
California art, Hollywood portrait photography, archarological and anthropological artifacts – Eastwood coll 51657

Ventura County Maritime Museum, 2731 S Victoria Av, Oxnard, CA 93035 • T: +1 805 9846260 • vcmminfo@aol.com • www.vcmm.org •
Dir.: *Mark S. Bacin* • Cur.: *Jackie Cavish* (Art) •
Historical Museum – 1991
Maritime county hist, ship models, POW bone models, maritime art – library 51658

Oxon Hill MD

Oxon Cove Park Museum, 6411 Oxon Hill Rd, Oxon Hill, MD 20745 • T: +1 301 8390211 • F: +1 301 7631066 • www.nps.gov/nace/oxhi •
Agriculture Museum – 1967
Agriculture, farm and farm outbuildings 51659

Oyster Bay NY

Coe Hall, Planting Fields Rd, Oyster Bay, NY 11771-0058 • T: +1 516 9229210 • F: +1 516 9229226 • coehall@worldnet.att.net • www.plantingfields.org •
Cur.: *Ellen Cone Busch* •
Historical Museum – 1979
Former residence of the Coe family, original furnishings, Baroque and Renaissance paintings, family papers and manuscripts 51660

Earle Wightman Museum, Oyster Bay Historical Society, 20 Summit St, Oyster Bay, NY 11771 • T: +1 516 9225032 • F: +1 516 9226892 • obhistory@aol.com • www.oysterbayhistory.org •
Dir.: *Thomas A. Kuehhas* • Cur.: *Stacie Hammond* •
Historical Museum – 1960
Reichman coll of early American tools and trades, Theodor Roosevelt coll, 18th/19th c household furnishings and dec art 51661

Raynham Hall Museum, 20 W Main St, Oyster Bay, NY 11771 • T: +1 516 9226808 • F: +1 516 9227640 • admin@raynhamhallmuseum.org • www.raynhamhallmuseum.org •
C.E.O. & Dir.: *Sarah K. Abruzzi* •
Decorative Arts Museum / Historical Museum – 1953
18th and 19th c household furnishings and dec arts, textiles, costumes, military coll 51662

Sagamore Hill National Historic Site, 20 Sagamore Hill Rd, Oyster Bay, NY 11771-1899 • T: +1 516 9224788 • F: +1 516 9224792 • sahi_superintendent@nps.gov • www.nps.gov/sahi •
Cur.: *Amy Verone* •
Historical Museum – 1963
Life, work and personal effects of Pres. Theodore Roosevelt (1858-1919) and his family 51663

Ozona TX

Crockett County Museum, 404 Eleventh St, Ozona, TX 76943 • T: +1 325 3922837 • F: +1 325 3925654 • ccmuseum@wcc.net •
Dir.: *Patsy White* •
Local Museum – 1939
Prehistoric &historic Indian tools, weapons & artifacts 51664

Pacific Grove CA

Pacific Grove Art Center, 568 Lighthouse Av, Pacific Grove, CA 93950 • T: +1 831 3752208 • F: +1 831 3752208 • pgart@mbay.net • www.pgartcenter.org •
Cur.: *Mark Davy* •
Fine Arts Museum – 1969
Paintings, sculpture, grapgics, photographs – artists' studios 51665

Pacific Grove Museum of Natural History, 165 Forest Av, Pacific Grove, CA 93950 • T: +1 831 6485716 • F: +1 831 3723256 • www.pgmuseum.org •
Dir.: *Paul M. Finnegan* •
Natural History Museum – 1881
Monterey County zoology, botany, geology and ethnology – herbarium, library 51666

Stowitts Museum, 591 Lighthouse Av, Pacific Grove, CA 93950 • T: +1 831 6554488 • F: +1 831 6554489 • stowitts@mbay.net • www.stowitts.org •
C.E.O.: *Robert W. Packard* • Cur.: *Anne Holliday* •
Fine Arts Museum – 1995
Paintings, prints, photographs, sculpture, theatrical costumes, performing arts books, films, Russian artifacts, letters and book manuscripts – archives 51667

Pacifica CA

Sanchez Adobe, 1000 Linda Mar Blvd, Pacifica, CA 94044 • T: +1 650 3591462 • F: +1 650 3591462 • www.cityofpacifica.org •
Dir.: *Mitchell P. Postel* •
Historical Museum / Local Museum – 1947
Archaeology, period furnishings 51668

Paducah KY

Museum of the American Quilter's Society, 215 Jefferson St, Paducah, KY 42001 • T: +1 270 4428856 • F: +1 270 4425448 • info@quiltmuseum.org • www.quiltmuseum.org •
Exec. Dir.: *May Louise Zumwalt* • Cur.: *Judy Schwender* •
Fine Arts Museum / Decorative Arts Museum – 1991
About 240 traditional, antique, contemporary & art quilts, bronze statues, pottery 51669

River Heritage Museum, 117 S Water St, Paducah, KY 42001 • T: +1 270 5759958 • F: +1 270 4449944 • www.riverheritagemuseum.org •
Exec.Dir.: *Julie Harris* • Chm.: *Ken Wheeler* •
Special Museum – 1990
River and nautical heritage, Civil War artifacts, boats and boat models 51670

U

William Clark Market House Museum, 121 S Second St, Market House Sq, Paducah, KY 42001 • T: +1 270 4437759 •
Exec. Dir.: *Penny Baucum-Fields* •
Local Museum – 1968
Hist, bldg was used as a farmers market 51671

Yeiser Art Center, 200 Broadway, Paducah, KY 42001-0732 • T: +1 270 4422453 • F: +1 270 4422243 • yacenter@paducah.com • www.yeiserartcenter.org •
Dir.: *Robert Durden* •
Fine Arts Museum – 1957
European, American, Asian & African 19th & 20th c works of art 51672

Page AZ

John Wesley Powell Memorial Museum, 6 N Lake Powell Blvd, Page, AZ 86040 • T: +1 928 6459496 • F: +1 928 6453412 • director@powellmuseum.org • www.powellmuseum.org •
Dir.: *Julia P. Betz* •
Special Museum – 1969
Anthropology, archaeology, ethnology, geology, manuscript coll – library 51673

Pagosa Springs CO

Fred Harman Art Museum, 2560 W Hwy 160, Pagosa Springs, CO 81147 • T: +1 970 7315785 • F: +1 970 7314832 • fharman@pagosa.net • www.harmanartmuseum.com •
C.E.O./Cur.: *Fred C. Harman* •
Fine Arts Museum – 1979 51674

Painesville OH

Indian Museum of Lake County-Ohio, c/o Lake Erie College, 391 W Washington, Painesville, OH 44077 • T: +1 440 3521911 •
Dir.: *Ann L. Dewald* •
Ethnology Museum / Archaeology Museum – 1980
Artifacts from 10.000 BC to 1650, present crafts and art of North American Native Americans 51675

Painted Post NY

Erwin Museum, Steuben and High St, Painted Post, NY 14870 • T: +1 607 9620249 • F: +1 607 9620249 • www.paintedpostny.com •
Local Museum – 1945
Indian lore; 1930's local hist, culture and industry 51676

Palm Beach FL

Flagler Museum, Cocoanut Row and Whitehall Way, Palm Beach, FL 33480 • T: +1 561 6552833 • F: +1 561 6552826 • mail@flaglermuseum.us • www.flaglermuseum.us •
Dir.: *John M. Blades* •
Historical Museum – 1959 51677

The Society of the Four Arts Gallery, 2 Four Arts Plaza, Palm Beach, FL 33480 • T: +1 561 6557227 • F: +1 561 6557233 • sfourarts@aol.com • www.fourarts.org •
Pres.: *Ervin S. Duggan* • Exec. Vice Pres.: *Nancy Mato* •
Fine Arts Museum – 1936
Art gallery, sculpture garden – library 51678

Palm Coast FL

Florida Agricultural Museum, 1850 Princess Place Rd, Palm Coast, FL 32137 • T: +1 904 4467630 • F: +1 904 4467631 • famuseum@pcfl.net •
C.E.O.: *Bruce J. Piatek* •
Agriculture Museum – 1983
Artifacts, Florida agriculture hist, railroads, historic domestic animals 51679

Palm Springs CA

Agua Caliente Cultural Museum, 219 S Palm Canyon Dr, Palm Springs, CA 92262 • T: +1 760 7781079 • F: +1 760 3227724 • mail@accmuseum.org • www.accmuseum.org •
Dir.: *Dr. Michael Hammond* • Cur.: *Ginger Ridgway* •
Fine Arts Museum – 1993
Art and culture of Agua Caliente Band of Cahuilla Indians – library 51680

Palm Springs Art Museum, 101 Museum Dr, Palm Springs, CA 92262-5659 • T: +1 760 3224800 • F: +1 760 3275069 • info@psmuseum.org • www.psmuseum.org •
Dir.: *Steven Nash* • Cur.: *Katherine Hough* •
Fine Arts Museum – 1938
Contemporary American, California, classic Western American and Native American art – theater, library, sculpture gardens 51681

Palmer Lake CO

Lucretia Vaile Museum, 66 Lower Glenway St, Palmer Lake, CO 80133, mail addr: POB 662, Palmer Lake, CO 80133 • T: +1 719 5590837 • history@ci.palmer-lake.co.us • www.ci.palmer-lake.co.us/plhs •
Dir.: *Roger Davis* • Cur.: *Susan Davis* •
Local Museum – 1981
Local relicts, photographs, manuscript coll, timeline 1873 to present 51682

Palmyra NY

Alling Coverlet Museum, 122 William St, Palmyra, NY 14522 • T: +1 315 5976981 • F: +1 315 5976981 • bjfhpinc@rochester.rr.com •
Exec. Dir.: *Bonnie J. Hays* •
Special Museum – 1976
Coverlets, quilts 51683

Hill Cumorah Visitors Center, 603 State Rte 21 S, Palmyra, NY 14522-9301 • T: +1 315 5975851 • F: +1 315 5970165 • www.hillcumorah.org •
Dir.: *Russell Homer* •
Religious Arts Museum – 1830
5 historical sites (farms and Book of Mormon Hist Publication Site) of significance in connection with the founding of the Church of Jesus Christ of Latter-day Saints 51684

Historic Palmyra, 132 Market St, Palmyra, NY 14522 • T: +1 315 5976981 • F: +1 315 5976981 • bjfhpinc@aol.com • www.palmyrany.com/historicpal/hpmuseum/hmgallery.htm •
Exec.Dir.: *Bonnie J. Hays* •
Local Museum / Historical Museum – 1967
Local history, military, textiles, Covelelt and quilts coll, naval admirals, W. Churchill's family, mormon hist 51685

Phelps Store Museum, 140 Market St, Palmyra, NY 14522 • T: +1 315 5976981 • F: +1 315 5974793 • bjfhpinc@rochester.rr.com • www.palmyrany.com/historicpal/phelps •
Exec.Dir.: *Bonnie J. Hays* •
Historical Museum – 1964 51686

Palmyra VA

The Old Stone Jail Museum, Court Sq, Palmyra, VA 22963 • T: +1 804 8423378 • F: +1 804 8423374 •
C.E.O.: *Ellen Miyagawa* •
Decorative Arts Museum / Local Museum – 1964
Firearms from Revolutionary War-WW I 51687

Palo Alto CA

Palo Alto Art Center, 1313 Newell Rd, Palo Alto, CA 94303 • T: +1 650 3292366 • F: +1 650 3266165 • ArtCenter@cityofpaloalto.org • www.cityofpaloalto.org/artcenter •
Dir.: *Linda Craighead* • Cur.: *Signe Mayfield* •
Public Gallery – 1971
Contemporary and historical fine arts – sculpture garden, studios 51688

Palo Alto Junior Museum, 1451 Middlefield Rd, Palo Alto, CA 94301 • T: +1 650 3292111 • www.cityofpaloalto.org •
Dir.: *Rachel Meyer* • Cur.: *Darin Wacs* •
Natural History Museum / Ethnology Museum – 1934
Natural hist, anthropology, historical living, ethnography – zoo 51689

Pampa TX

White Deer Land Museum, 112-116 S Cuyler, Pampa, TX 79065 • T: +1 806 6698041 • F: +1 806 6698030 • museum@pan-tex.net • www.pan-tex.net •
Cur.: *Anne Davidson* •
Local Museum – 1970 51690

Pamplin VA

Yesteryear Museum, Rte 1, Box 134 B, Pamplin, VA 23958-9465 • T: +1 804 2485357 • leemunsick@earthlink.net •
C.E.O.: *Lee R. Munsick* •
Music Museum – 1970
Mechanical and automatic music, sound recordings, Edisonia, ragtime, early broadcasting 51691

Panama City FL

The Junior Museum of Bay County, 1731 Jenks Av, Panama City, FL 32405 • T: +1 850 7696129 • F: +1 850 7696129 • jrmuseum@knology.net • www.jrmuseum.org •
Dir.: *William R. Barton* •
Ethnology Museum – 1969
Children education 51692

Museum of Man in the Sea, 17314 Panama City Beach Pkwy, Panama City, FL 32413-2020 • T: +1 904 2354101 • F: +1 904 2354101 • momits@bellsouth.net • museum-of-man-in-the-sea.panamacity-beachfanatic.com •
Exec. Dir.: *Douglas R. Hough* •
Historical Museum / Historic Site – 1981 51693

Visual Arts Center of Northwest Florida, 19 E Fourth St, Panama City, FL 32401 • T: +1 850 7694451 • F: +1 850 7859248 • visualartscenterofnwfla@comcast.net • www.visualartscenter.org •
Dir.: *Rebecca Walker* •
Fine Arts Museum / Public Gallery – 1988
Paintings in watercolor, oil, acrylics, drawings, photographs, sculpture 51694

Panhandle TX

Carson County Square House Museum, Texas Hwy 207 at Fifth St, Panhandle, TX 79068 • T: +1 806 5373524 • F: +1 806 5375628 • shm@squarehousemuseum.org • www.squarehousemuseum.org •
Dir.: *Dr. Viola Moore* •
Local Museum – 1965
Hist structures, mid-1880 Square House, caboose, reconstructed pionner dugout 51695

Paramus NJ

Bergen Museum of Art and Science, Bergen Mall, Rte 4 E, Paramus, NJ 07652 • T: +1 201 2918848 • F: +1 201 2652536 • nkolodzei@thebergenmuseum.com • www.thebergenmuseum.com •
Cur.: *Natalia Kolodzei* •
Fine Arts Museum / Natural History Museum / Science&Tech Museum – 1956
Paintings, prints, sculpture, Hackensack and Dwarskill mastodons, minerals, shell, fossils, aviation 51696

The Bergen Museum of Art and Science, Bergen Rte Mall 4 E, Paramus, NJ 07652 • T: +1 201 2918848 • F: +1 201 2652536 • BergenMuse@aol.com • www.thebergenmuseum.com •
Dir.: *Peter Knipe* •
Fine Arts Museum / Natural History Museum – 1956 51697

New Jersey Children's Museum, 599 Industrial Av, Valley Hills Pl, Paramus, NJ 07652 • T: +1 201 2622638 • F: +1 201 2620560 • esumers@aol.com • www.njcm.com •
Dir.: *Anne R. Sumers* •
Special Museum – 1992 51698

Paris AR

Arkansas Historic Wine Museum, 101 N Carbon City Rd, Paris, AR 72855 • T: +1 501 9633990 • cowie@cswnet.com • www.cowiewinecellars.com •
Dir.: *Robert G. Cowie* •
Agriculture Museum – 1967
History of wine making in Arkansas 51699

Logan County Museum, 202 N Vine St, Paris, AR 72855 • T: +1 501 9633936 • F: +1 501 9633936 • lcmuseum@apip.net •
Cur.: *Geneva Morton* •
Local Museum – 1975
Local history, former Logan County jail 51700

Paris IL

Bicentennial Art Center and Museum, 132 S Central Av, Paris, IL 61944 • T: +1 217 4668130 • F: +1 217 4668130 • bacm@1choice.net • www.parisartcenter.com •
Dir.: *Janet M. Messenger* •
Fine Arts Museum / Public Gallery – 1975
Paintings on oils and watercolors, drawings, sculpture and photography 51701

Edgar County Historical Museum, 408 N Main, Paris, IL 61944-1549 • T: +1 217 4635305 •
Pres.: *Mary Galway* •
Local Museum – 1971
Local history, housed in 1872 Arthur House 51702

Paris KY

Hopewell Museum, 800 Pleasant St, Paris, KY 40361 • T: +1 859 9877274 • F: +1 859 9877274 • museumgoddess@aol.com • www.hopewellmuseum.org •
Dir.: *Betsy Kephart* •
Local Museum – 1995
Local, regional & state art, artists, art hist 51703

Paris TX

Sam Bell Maxey House, 812 S Church St, Paris, TX 75460 • T: +1 903 7855716 • F: +1 903 7392924 • maxeyhouse@1starnet.com • www.maxeyhouse.com •
Dir.: *Robert Cook* •
Decorative Arts Museum / Local Museum – 1976
Household furnishings, 1830-1950 clothing 51704

Parishville NY

Parishville Museum, Main St, Parishville, NY 13672 • T: +1 315 2657619 •
Pres.: *Elton Tupper* •
Local Museum – 1964
Hist of early Parishville 51705

Park City UT

Kimball Art Center, 638 Park Av, Park City, UT 84060 • T: +1 435 6498882 • F: +1 435 6498889 • www.kimball-art.org •
Dir.: *Sarah Behrens* •
Public Gallery – 1976 51706

Park City Historical Society & Museum, 528 Main St, Park City, UT 84060 • T: +1 435 6497457 • F: +1 435 6497384 • smorrison@parkcityhistory.org • www.parkcityhistory.org •
Dir.: *Sandra Morrison* •
Local Museum – 1984
Hist artifacts of Western Summit County from 1860 to the present 51707

Park Hill OK

Murrell Home, Oklahoma Historical Society, 19479 E Murrell Home Rd, 3 miles S of Tahlequah, 1 mile E of SH 82, Park Hill, OK 74451 • T: +1 918 4562751 • F: +1 918 4562751 • murrellhome@intellex.com • www.ok-history.mus.ok.us •
Mgr.: *Shirley Pettengill* •
Historical Museum – 1948
Indian artifacts, costumes, pre-Civil War furnishings 51708

Park Rapids MN

North Country Museum of Arts, 301 Court Av Q Third St, Park Rapids, MN 56470 • T: +1 218 7325237 •
Cur.: *Johanna M. Verbrugghen* •
Fine Arts Museum – 1977
European paintings 15th-19th c, Native American pencil drawings, Nigerian arts & crafts (Yoruba, Ibo, Fulani & Hausa), contemporary art 51709

Park Ridge IL

Wood Library-Museum of Anesthesiology, 520 N Northwest Hwy, Park Ridge, IL 60068-2573 • T: +1 847 8255586 • F: +1 847 8251692 • wlm@asahq.org • www.asahq.org/wlm •
Cur.: *George S. Bause* •
Historical Museum – 1950
Medical history of anesthesiology, manuscripts, oral hist coll, books and journals 51710

Park Ridge NJ

Pascack Historical Society Museum, 19 Ridge Av, Park Ridge, NJ 07656 • T: +1 201 5730307 •
Pres.: *Richard Ross* •
Local Museum / Historical Museum – 1942
Local hist 51711

Parker AZ

Colorado River Indian Tribes Museum, Rte 1, Box 23-B, Parker, AZ 85344 • T: +1 928 6699211 ext 1335 • F: +1 928 6695675 •
Dir.: *Betty L. Cornelius* •
Ethnology Museum – 1970
American Indians, Mohave, Chemehuevi, Navajo, Hopi, Anasazi, Hohokam and Patayan tribes – library, archive 51712

Parkersburg WV

The Cultural Center of Fine Arts, 725 Market St, Parkersburg, WV 26101 • T: +1 304 4853859 • F: +1 304 4853850 •
Dir.: *Ed Pauley* •
Fine Arts Museum – 1938
American Realism, changing exhib 51713

Parris Island SC

Parris Island Museum, Bldg 111, MCRD, Parris Island, SC 29905-9001 • T: +1 843 2282951 • F: +1 843 2283065 • wiser@mcrdpi.usmc.mil •
Dir.: *Dr. Stephen R. Wise* •
Local Museum – 1976
Ceramics, metal work, uniforms, firearms 51714

Parrish FL

Florida Gulf Coast Railroad Museum, US Hwy 301, Parrish, FL 34219, mail addr: 2146 Ninth St, Sarasota, FL 34237-3404 • T: +1 941 7660906 • F: +1 941 9170081 • www.fgcrrm.org •
Pres.: *James Herron* •
Science&Tech Museum – 1983
Railroads 51715

Parsippany NJ

Craftsman Farms Foundation, 2352 Rte 10 W, Parsippany, NJ 07950 • T: +1 973 5401165 • F: +1 973 5401167 • craftsmanfarms@worldnet.att.net • www.strickleymuseum.org •
Dir.: *Tommy McPherson* • Cur.: *Beth Ann McPherson* •
Historical Museum / Open Air Museum / Agriculture Museum – 1989
Furniture, decorative arts and crafts 51716

Parsons KS

Parson Historical Commission Museum, 401 S 18th St, Parsons, KS 67357 • T: +1 620 4217000 •
Dir.: *Betty Olmsted* •
Local Museum – 1969 51717

Pasadena CA

Armory Center for the Arts, 145 N Raymond Av, Pasadena, CA 91103 • T: +1 626 7925101 • F: +1 626 4490139 • information@armoryarts.org • www.armoryarts.org •
Dir.: *Scott Ward* •
Public Gallery – 1974 51718

Kidspace Children's Museum, 480 N Arroyo Blvd, Pasadena, CA 91103 • T: +1 626 4499144 • info@kidspacemuseum.org • www.kidspacemuseum.org •
Special Museum – 1979
Hands-on, arts, science, humanities 51719

Norton Simon Museum of Art, 411 W Colorado Blvd, Pasadena, CA 91105 • T: +1 626 4496840 • F: +1 626 7964978 • www.nortonsimon.org •
Cur.: *Sara Campbell* (Art) •
Fine Arts Museum – 1924
European paintings and sculpture, Asian sculpture, American art, graphics, old masters paintings – theater 51720

Pacific Asia Museum, 46 N Los Robles Av, Pasadena, CA 91101 • T: +1 626 4492742 • info@pacificasiamuseum.org • www.pacificasiamuseum.org • Dir.: *Bruce Blomstrom* •
Fine Arts Museum / Ethnology Museum / Folklore Museum – 1971
Ceramics, sculpture, scrolls, screens, woodblock-prints, textiles, folk art, furniture, photos from Asia – library 51721

Pasadena City College Art Gallery, 1570 E Colorado Blvd, Pasadena, CA 91106 • T: +1 626 5857238 • bltucker@pasadena.edu • www.pasadena.edu/artgallery •
Dir.: *Brian Tucker* •
Public Gallery 51722

Pasadena Museum of California Art, 490 E Union St, Pasadena, CA 91101 • T: +1 626 5683665 • F: +1 626 5683674 • www.pmcaonline.org •
Dir.: *Brian Tucker* •
Fine Arts Museum – 2002
Contemporary and historic California art 51723

Pasadena Museum of History, 470 W Walnut St, Pasadena, CA 91103-3594 • T: +1 626 5771660 • F: +1 626 5771662 • info@pasadenahistory.org • www.pasadenahistory.org •
Dir.: *Jeanette O'Malley* •
Local Museum – 1924
Finnish folk art, costume, period furnishings, early California paintings, manuscripts – library 51724

Pascagoula MS

Old Spanish Fort Museum, 4602 Fort St, Pascagoula, MS 39567 • T: +1 228 7691505 • F: +1 228 7691432 • oldspanishfort@netscape.net •
Historical Museum – 1949 51725

Pasco WA

Franklin County Historical Museum, 305 N 4th Av, Pasco, WA 99301 • T: +1 509 5473714 • F: +1 509 5452168 • museums@franklincountyhistoricalsociety.org • www.franklincountyhistoricalsociety.org •
Pres.: *Marylin Krueger* •
Local Museum – 1982
Local hist, Indian culture, railroad hist, river transportation 51726

Pateros WA

Fort Okanogan, 1B Otto Rd, Pateros, WA 98846 • T: +1 509 9232473 • F: +1 509 9232980 • www.parks.wa.gov •
Dir.: *Rex Derr* •
Historical Museum – 1965
Indian and pioneer items, basketry, weapons 51727

Paterson NJ

Passaic County Community College Galleries, Broadway Gallery, 1 College Blvd, Paterson, NJ 07505-1179 • T: +1 973 6846555 • F: +1 973 5236085 • mgillan@pccc.edu • www.pccc.edu/art/gallery •
Dir.: *Maria Mazziotti Gillan* • Asst. Dir.: *Aline Papazian* • Cur.: *Jane Haw* •
Fine Arts Museum / University Museum – 1968
19th and earley 20th c paintings and sculpture coll 51728

Passaic County Historical Society Museum, Lambert Castle, 3 Valley Rd, Paterson, NJ 07503-2932 • T: +1 973 2470085 • F: +1 973 8819434 • lambertcastle@yahoo.com • www.lambertcastle.org •
Dir.: *Richard A. Sgritta* •
Local Museum – 1926
Local history, Koempel Spoon coll, textiles, paintings, decorative art, folk art – library 51729

Paterson Museum, 2 Market St, Paterson, NJ 07501 • T: +1 973 8813874 • F: +1 973 8813435 •
Dir.: *Giacomo R. DeStefano* • Cur.: *Bruce Balistrieri* (History) •
Historical Museum / Natural History Museum / Science&Tech Museum – 1925
Mineralogy, textiles, New Jersey Indian artifacts, local hist 51730

Patten ME

Lumberman's Museum, 25 Waters Rd, Patten, ME 04765 • T: +1 207 5282650 • curator@lumbermuseum.org • www.lumbermuseum.org •
Cur.: *Frank Rogers* •
Historical Museum / Ethnology Museum – 1959
Logging, lumbering 51731

Patterson LA

Louisiana State Museum, 394 Airport Circle, Patterson, LA 70392 • T: +1 504 3957067 • F: +1 504 3953179 •
Dir.: *Nicholas Neylon* •
Science&Tech Museum – 1975
Aviation 51732

Pawhuska OK

Osage County Historical Society Museum, 700 N Lynn Av, Pawhuska, OK 74056 • T: +1 918 2879924 •
Dir./Cur.: *J.B. Smith* •
Local Museum – 1964
Indian, pioneer, Western and military artifacts, early oil prod, early American Indian exhibits, cars 51733

Pawling NY

Gunnison Museum of Natural History, 397 Old Qaker Hill Rd, Pawling, NY 12564 • T: +1 845 8555099 • theakinlibrary@earthlink.net •
Pres.: *Thomas Schroth* •
Natural History Museum – 1960
Gunnison Coll of rocks, minerals, native birds and flora, agriculture, entomology, Indian artfacts, African and Asian musical instr 51734

Historical Society of Quaker Hill and Pawling, 126 E Main St, Pawling, NY 12564 • T: +1 914 8559316 • nadadavis@msn.com • www.pawlinghistory.org •
Pres.: *Nada Davies* •
Historical Museum – 1910
George Washington's headquarters in autumn 1778, Lowell Thomas, radio broadcaster – Hist houses 51735

Pawnee OK

Pawnee Bill Ranch, Oklahoma Historical Society, 1141 Pawnee Bill Rd, Pawnee, OK 74058, mail addr: POB 493, Pawnee, OK 74058 • T: +1 918 7622513 • F: +1 918 7622514 • pawneebill@okhistory.org • www.okhistory.org •
C.E.O.: *Dr. Bob Blackburn* • Mgr.: *Ron Brown* •
Historical Museum / Agriculture Museum – 1962
1910-1940 original furnishings of Pawnee Bill, exhibits of Wild West Shows, Indian artifacts, early days farming ans ranching equip 51736

Pawnee City NE

Pawnee City Historical Society Museum, Hwy 50/8 East, Pawnee City, NE 68420 • T: +1 402 8523131 •
Local Museum – 1968
Furniture, photographs, farm marchinery, local hist 51737

Pawtucket RI

Slater Mill Historic Site, 67 Roosevelt Av, Pawtucket, RI 02860 • T: +1 401 7258638 • F: +1 401 7223040 • info@slatermill.org • www.slatermill.org •
Dir.: *Jeanne L. Zavada* • Cur.: *Adrian Paquette* •
Science&Tech Museum / Historic Site – 1921
Hist of textile industry 51738

Paxton IL

Ford County Historical Society Museum, 255 West State St, Paxton, IL 60957 • T: +1 217 3794684 • www.rootsweb.com/~ilford •
Pres.: *Ed F. Karr* •
Local Museum – 1967
Local hist, Indian artifacts, Civil War memorabilia 51739

Paxton MA

Saint Luke's Gallery, Anna Maria College, Moll Art Center, 50 Sunset Ln, Paxton, MA 01612 • T: +1 508 8493300 ext 442 • F: +1 508 8493408 • admissions@annamaria.edu • www.annamaria.edu •
Dir.: *Elizabeth Killoran* •
Fine Arts Museum / University Museum – 1972 51740

Payson AZ

Rim Country Museum, 700 Green Valley Parkway, Payson, AZ 85547, mail addr: POB 2532, Payson, AZ 85547 • T: +1 928 4743483 • ngchs1@gmail.com • www.rimcountrymuseums.com •
Local Museum
Forest Ranger Station from 1907 51741

Peace Dale RI

Museum of Primitive Art and Culture, 1058 Kingstown Rd, Peace Dale, RI 02883 • T: +1 401 7835711 • www.primitiveartmuseum.org •
Pres.: *Wallace Campbell* • Dir.: *Sarah Peabody Turnbaugh* •
Ethnology Museum / Archaeology Museum – 1892
Anthropology 51742

Peaks Island ME

Fifth Maine Regiment Center, 45 Seashore Av, Peaks Island, ME 04108 • T: +1 207 7663330 • F: +1 207 7663083 • fifthmaine@juno.com • www.fifthmainemuseum.org •
Dir.: *Sharon L. McKenna* • Cur.: *Kim MacIsaac* •
Military Museum – 1954
Military hist 51743

Pearl River NY

Orangetown Historical Museum, 213 Blue Hill Rd, Pearl River, NY 10965 • T: +1 845 7350429 • F: +1 845 3598269 • otownmuseum@fcc.net • www.orangetown.com/parks •
C.E.O.: *Mary R. Cardenas* • Cur.: *Julie Trumpler* •
Local Museum – 1992
Town hist from Colonial era to present – archives 51744

Pecos NM

Pecos Park, State Rd 63, 2 mi S of Pecos, Pecos, NM 87552, mail addr: POB 418, Pecos, NM 87552 • T: +1 505 7576414 ext 221 • F: +1 505 7578460 • peco_visitor_information@nps.gov • www.nps.gov/peco •
Archaeology Museum – 1965
Archaeology, A.V. Kidder coll, Mission churches 51745

Pecos TX

West of the Pecos Museum, First at Cedar (U.S. 285), Pecos, TX 79772 • T: +1 432 4455076 • F: +1 432 4453149 • wpmuseum@nwol.net • www.westofthepecosmuseum.com •
Cur.: *Dorenda Zenegas-Millan* •
Local Museum – 1962
Local hist memorabilia, saloon, railroad & telegraph memorabilia, horse-drawn wagons 51746

Peebles OH

Serpent Mound Museum, 3850 State Rte 73, Peebles, OH 45660 • T: +1 937 5872796 • F: +1 937 5871116 • serpent@bright.net • www.ohiohistory.org •
Ethnology Museum – 1900
Adena Indian culture 51747

Peekskill NY

Peekskill Museum, 124 Union Av, Peekskill, NY 10566 • T: +1 914 7360473 • www.peekskillmuseum.com •
Pres.: *John Curran* •
Local Museum – 1946
Local memorabilia from Colonial era to present 51748

Pelham NY

Pelham Art Center, 155 Fifth Av, Pelham, NY 10803 • T: +1 914 7382525 • F: +1 914 7382686 • info@pelhamartcenter.org •
Pres.: *Paisley Kelling* • C.E.O.: *Lisa Robb* •
Fine Arts Museum / Folklore Museum – 1975
Annual visual art and craft exhibitions 51749

Pella IA

Pella Historical Village, 507 Franklin, Pella, IA 50219-0145 • T: +1 515 6282409 • F: +1 515 6289192 • pellatuliptime@iowatelecom.net • www.pellatuliptime.com •
Dir.: *Patsy Sadler* •
Ethnology Museum / Local Museum – 1965
Ethnography (Dutch), 20 historic buildings 51750

Scholte House Museum, 728 Washington, Pella, IA 50219 • T: +1 515 6283684 •
Cur.: *Shirley J. Rudd* •
Special Museum – 1982 51751

Pemaquid Point ME

Fishermen's Museum, Lighthouse Park, Pemaquid Point, ME 04554 • T: +1 207 6772494 •
Dir.: *Mary Norton Orrick* •
Historical Museum – 1972
Old pictures, models of fishing boats 51752

Pembina ND

Pembina State Museum, 805 State Hwy 59, Pembina, ND 58271-0456 • T: +1 701 8256840 • F: +1 701 8256383 • mbailey@state.nd.us •
Dir.: *Merlan E. Paaverud* •
Historical Museum – 1996
Fur trade, Meti's 51753

Pembroke MA

Pembroke Historical Society Museum, Center St, Pembroke, MA 02359 • T: +1 781 2939083 •
Cur.: *Lauren Richmond* •
Local Museum – 1950
Local history 51754

Pembroke NC

Native American Resource Center, University of North Carolina at Pembroke, Pembroke, NC 28372 • T: +1 910 5216282 • stan.knick@uncp.edu • www.uncp.edu/nativemuseum/ •
Dir.: *Dr. Stanley Knick* •
University Museum / Ethnology Museum – 1979
Indian artifacts, material on Native Americans and Lumbee Indians, handicrafts 51755

Pendleton OR

Tamastslikt Cultural Institute, 72789 Hwy 331, Pendleton, OR 97801 • T: +1 541 9669748 • F: +1 541 9669927 • www.tamastslikt.org •
C.E.O.: *Roberta Conner* •
Historical Museum – 1998
Hist and culture of American Indian tribes, esp. Cyuse, Utamilla and Walla Walla 51756

Umatilla County Historical Society Museum, 108 SW Frazer St, Pendleton, OR 97801 • T: +1 541 2760012 • F: +1 541 2767989 • uchs@oregontrail.net • www.umatillahistory.org •
Pres.: *Brenda Turner* • Exec. Dir.: *Julie Reese* •
Local Museum – 1974
Local hist 51757

Pendleton SC

Ashtabula Plantation, Hwy 88, Pendleton, SC 29670 • T: +1 864 6467249 • F: +1 864 6462506 • pendtour@innova.net • www.pendleton-district.org •
Decorative Arts Museum – 1960 51758

Pendleton District Agricultural Museum, 120 History Ln, Pendleton, SC 29670 • T: +1 864 6463782 • F: +1 864 6462506 • pendletontourism@bellsouth.net • www.pendleton-district.org •
Dir./Cur.: *Donna K. Roper* •
Agriculture Museum – 1976
Agriculture 51759

Woodburn Plantation, 130 History Ln, Pendleton, SC 29670 • T: +1 864 6467249 • phfdirector@aol.com •
Decorative Arts Museum – 1960 51760

Penland NC

Penland Gallery, Penland School of Crafts, Conley Ridge Rd, Penland, NC 28765-0037 • T: +1 828 7656211 • F: +1 828 7656211 • gallery@penland.org • www.penland.org •
Public Gallery 51761

Penn Valley CA

Museum of Ancient and Modern Art, Wildwood Business Center, 11392 Pleasant Valley Rd, Penn Valley, CA 95946-0975 • T: +1 530 4323080 • info@mama.org • www.mama.org •
Dir.: *Zoe Alowan* (Children's Art Academy) • Cur.: *Dr. Claude Needham* (Science, Technology) • *Jeff Spencer* (Antiquities) • *David Christie* (Mathematics, Science) • *Tim Elston* (Interactive Multimedia) • *Linda Corriveau* (Works on Paper) • Ass. Cur.: *Dr. James D. Wrey* (Stellar Astronomy) • Cons.: *Nancy Christie* (Ancient) • *Robbert Trice* (Modern) • Libr.: *Nancy Burns* •
Fine Arts Museum – 1981
Art, astromomy, prehistory – library 51762

Penn Yan NY

The Agricultural Memories Museum, 1110 Townline Rd, Penn Yan, NY 14527 • T: +1 315 5361206 •
Head: *Jennifer R. Jensen* •
Agriculture Museum – 1997
Hores-drawn carriages and sleighs, period tractors gasoline engines, toys, signs, pedal tractors 51763

Oliver House Museum, Yates County Genealogical and Historical Society, 200 Main St, Penn Yan, NY 14527 • T: +1 315 5367318 • F: +1 315 5360976 • ycghs@linkny.com • www.yatespast.com •
Exec. Dir.: *Idelle Dillon* • Cur.: *Charles Mitchell* •
Local Museum – 1860
19th c furnishings, costumes and photos, tools and agricultural implements, Indian artifacts, local hist and genealogy 51764

Pennsburg PA

Schwenkfelder Library and Heritage Center, 105 Seminary St, Pennsburg, PA 18073-1898 • T: +1 215 6793103 • F: +1 215 6798175 • info@schwenkfelder.com • www.schwenkfelder.com •
Exec. Dir.: *David W. Luz* •
Historical Museum – 1913
Local hist 51765

Pensacola FL

Historic Pensacola Village, 120 Church St, Pensacola, FL 32501 • T: +1 850 5955985 ext 103 • F: +1 850 5955989 • jdaniels@historicpensacola.org • www.historicpensacola.org •
Dir.: *John P. Daniels* • Chief Cur.: *Beatrice L. Robertson* •
Local Museum – 1967
Paintings, lithographs, photographs, weapons, hardware, marine artifacts, archaeology artifacts 51766

National Museum of Naval Aviation, 1750 Radford Blvd, Ste C, Pensacola, FL 32508-5402 • T: +1 904 4523604 ext 119 • F: +1 904 4523296 • naval.museum@cnet.navy.mil • naval.aviation.museum •
Dir.: *Robert Rasmussen* • Cur.: *Robert Macon* •
Military Museum / Science&Tech Museum – 1963
Naval aviation 51767

Pensacola Historical Museum, 115 E Zaragosa St, Pensacola, FL 32501 • T: +1 850 4341559 • F: +1 904 4331559 • phstaff@pcola.gulf.net •
Dir.: *Sandra L. Johnson* •
Historical Museum – 1960
Hist, historic Arbona bldg 51768

Pensacola Museum of Art, 407 S Jefferson St, Pensacola, FL 32501 • T: +1 850 4326247 • F: +1 850 4691532 • info@pensacolamuseumofart.org • www.pensacolamuseumofart.org •
Dir.: *Maria V. Butler* • Cur.: *Heather Roddenberry* •
Fine Arts Museum – 1954
Art Museum, housed in 1908 old city jail 51769

University of West Florida Art Gallery, 11000 University Pkwy, Pensacola, FL 32514 • T: +1 850 4742696, 4742045 • F: +1 850 4743247 • jmarkowitz@uwf.edu •
Dir.: *Debra Bond* • Ass. Dir.: *John Markowitz* •
Fine Arts Museum / Public Gallery / University Museum – 1970
Paintings, photos 51770

Visual Arts Gallery, 1000 College Blvd, Pensacola, FL 32504 • T: +1 850 4842554 • F: +1 850 4842564 • apeterson@pjc.cc.fl.us • www.pjc.edu/visarts •
Dir.: *Allan Peterson* •
Fine Arts Museum / Decorative Arts Museum / University Museum – 1970
Drawings, prints, paintings 51771

U

Peoria IL

Heuser Art Center, c/o Bradley University, 1501 W Bradley Av, Peoria, IL 61625 • T: +1 309 6772989 • F: +1 309 6773642 • jch@bradley.edu • www.bradley.edu •
Dir.: *John Heintzman* •
Public Gallery / University Museum ... 51772

Lakeview Museum of Arts and Sciences, 1125 W Lake Av, Peoria, IL 61614-5985 • T: +1 309 6867000 • F: +1 309 6860280 • kathleen@lakeview-museum.org • www.lakeview-museum.org •
Dir.: *James J. Richerson* •
Fine Arts Museum / Science&Tech Museum – 1965
18th -20th c European and American paintings, sculpture, works on paper, decorative arts, folk art, Western, African, Precolumbian, Native American, Asian art and artifacts ... 51773

Peoria Historical Society, 611 SW Washington, Peoria, IL 61602 • T: +1 309 6741921 • F: +1 309 6741882 • www.peoriahistoricalsociety.org •
Dir.: *Kathy Belsley* •
Historical Museum – 1934
Local history of industry, furniture, glassware, paintings, Oriental rugs, jewelry ... 51774

Wheels O' Time Museum, 11923 N Knoxville Av, Peoria, IL 61612 • T: +1 309 2439020 • F: +1 309 2435616 • wotmuseum@aol.com • www.wheelsotime.org •
Pres.: *Gary O. Bragg* •
Science&Tech Museum – 1977
Antique vehicles, clocks, musical devices, tools, toys, farm equip ... 51775

Perkasie PA

Pearl S. Buck House, Green Hills Farm, 520 Dublin Rd, Perkasie, PA 18944 • T: +1 215 2490100 • F: +1 215 2499657 • info@pearl-s-buck.org • www.pearl-s-buck.org •
Decorative Arts Museum / Historic Site – 1973
Stone farmhouse est. around 1820 with interior and paintings, coll of Pearl Buck documents – archive ... 51776

Perris CA

Lake Perris Regional Indian Museum, 17801 Lake Perris Dr, Perris, CA 92571 • T: +1 951 9405656 • LPSRA@parks.ca.gov • parks.ca.gov •
Historical Museum / Ethnology Museum – 1985
Southern California Native Americans culture and hist ... 51777

Orange Empire Railway Museum, 2201 S A St, Perris, CA 92572-0548 • T: +1 951 6572605 • info@oerm.org • www.oerm.org •
Pres./ CEO: *Thomas N. Jacobson* •
Science&Tech Museum – 1956
Railway, historical Santa Fe line from Chicago to San Diego, steam, diesel and electric locomotives ... 51778

Perry FL

Forest Capital State Museum, 204 Forest Park Dr, Perry, FL 32348 • T: +1 850 5843227 • F: +1 850 5843488 • forestcapital@perry.gulfnet.com •
Natural History Museum / Folklore Museum – 1973
Logging, lumber, timber industry, forestry ... 51779

Perry IA

Forest Park Museum, 1477 K Av, Perry, IA 50220 • T: +1 515 4653577 • F: +1 515 4653579 • info@dallas25.org • www.dallascountyconservation.org •
CEO: *Mike Wallace* •
Science&Tech Museum – 1953
Early transportation, farm machinery, tools, railroading, blacksmith shop ... 51780

Perry OK

Cherokee Strip Museum and Henry S. Johnston Library, Oklahoma Historical Society, 2617 W Fir St, Perry, OK 73077 • T: +1 580 3362405 • F: +1 580 3362064 • csmuseum@ionet.net • cherokee-strip-museum.org •
Cur.: *Kaye Bond* •
Historical Museum – 1965
Agricultural equip, medical and hist coll, Otoe-Missouri coll, costumes, glass ... 51781

Perrysburg OH

Fort Meigs State Memorial, State Rte 65, Perrysburg, OH 43551 • T: +1 419 8744121 • info@fortmeigs.org • www.ohiohistory.org •
Dir.: *Dr. Larry Nelson* •
Military Museum – 1975
Block houses, dioramas, weapons, cannon batteries ... 51782

Perryville KY

Perryville Battlefield Museum, 1825 Battlefield Rd, Perryville, KY 40468-0296 • T: +1 859 3328631 • F: +1 859 3322440 • kurt.holman@.ky.gov •
Head: *Kurt Holman* •
Military Museum – 1965
Civil War hist ... 51783

Perryville MD

Rodgers Tavern, 259 E Broad St, Perryville, MD 21903 • T: +1 410 6423703 • F: +1 410 6423641 • rtavern@iximd.com • www.Perryvillemd.org •
Local Museum – 1956 ... 51784

Perth Amboy NJ

Kearny Cottage, 63 Catalpa Av, Perth Amboy, NJ 08861 • T: +1 732 8261826 • writer@ptd.net •
Fine Arts Museum / Decorative Arts Museum – 1928
Roger Statuettes furniture, works by local artists 51785

The Royal Governor's Mansion, 149 Kearny Av, Perth Amboy, NJ 08861 • T: +1 732 8265527 • info@proprietaryhouse.org • www.proprietaryhouse.org •
C.E.O.: *Jack Stine* •
Local Museum – 1967
Historic house ... 51786

Peru IN

Circus City Festival Museum, 154 N Broadway, Peru, IN 46970 • T: +1 765 4723918 • F: +1 765 4722826 • perucirc@perucircus.com • www.perucircus.com •
Performing Arts Museum – 1959
Circus ... 51787

Grissom Air Museum, 6500 Hoosier Blvd, Peru, IN 46970 • T: +1 765 6898011 • F: +1 765 6882956 • info@grissomairmuseum.com • www.grissomairmuseum.com •
Cur.: *John S. Marsh* •
Military Museum / Science&Tech Museum – 1981
Military hist, aviation ... 51788

Miami County Museum, 51 N Broadway, Peru, IN 46970 • T: +1 765 4739183 • F: +1 317 4733880 • admin@miamicountymuseum.org • www.miamicountymuseum.org •
Cur.: *Mildred Kopis* •
Local Museum / Ethnology Museum – 1916
County history, circus, Miami Indians, transportation ... 51789

Peshtigo WI

Peshtigo Fire Museum, 400 Oconto Av, Peshtigo, WI 54157 • T: +1 715 5823244 •
C.E.O.: *Don Hansen* • Sc. Staff: *Ruth Erickson* • *Ruth Wiltzius* • *Arleen Behnke* • *Darlene Plump* • *Margaret Wood* • *Peggy Harrand* • *Marian Devory* • *Marilyn Olson* • *Mitzi Barclay* •
Local Museum – 1962
Local hist ... 51790

Petaluma CA

Petaluma Wildlife and Natural Science Museum, 201 Fair St, Petaluma, CA 94952 • T: +1 707 7784787 • F: +1 707 7784603 • www.petalumawildlifemuseum.org •
Pres.: *George Grossi* • Cur.: *Marsi H. Wier* •
Natural History Museum – 1990
Natural science, taxidermy specimens, minerals, fossils, African artifacts, dinosaur exhibit – library, zoo, aquarium ... 51791

Peterborough NH

Peterborough Historical Society Museum, 19 Grove St, Peterborough, NH 03458 • T: +1 603 9243235 • F: +1 603 9243200 • info@peterboroughhistory.org • www.peterboroughhistory.org •
Local Museum – 1902
Town artifacts, state hist ... 51792

Petersburg AK

Clausen Museum, 203 Fram St, Petersburg, AK 99833 • T: +1 907 7723598 • F: +1 907 7722698 • clausenmuseum@aptalaska.net • www.clausenmuseum.net •
C.E.O.: *Sue McCallum* •
Historical Museum / Local Museum – 1965
Fishing, fish, boats, local history ... 51793

Petersburg IL

Lincoln's New Salem Historic Site, R.R. 1, Petersburg, IL 62675 • T: +1 217 6324000 • F: +1 217 6324010 • newsalem@fgi.net • www.lincolnsnewsalem.com •
Pres.: *Carol Jenkins* •
Local Museum – 1917
New Salem Village, in the 1830s where Lincoln lived as a young man ... 51794

Petersburg KY

Creation Museum, 2800 Bullittsburg Church Rd, Petersburg, KY 41080, mail addr: POB 510, Hebron, KY 41048 • T: +1 859 7272222 • F: +1 859 7272299 • www.creationmuseum.org •
Religious Arts Museum – 2007
Pseudo scientific and religious creationism by bible fundamentalists – planetarium ... 51795

Petersburg VA

Archibald Graham McIlwaine House, The Petersburg Museums, 425 Cockade Alley, Petersburg, VA 23803 • T: +1 804 7332400 • F: +1 804 8630837 • petgtourism@earthlink.net • www.petersburg-va.org •
Dir.: *Suzanne T. Savery* •
Local Museum – 1984
19th c architecture ... 51796

Centre Hill Museum, The Petersburg Museums, Centre Hill Center, Petersburg, VA 23803 • T: +1 804 7332401 • F: +1 804 8630837 • petgcurator@earthlink.net • www.petersburg-va.org •
Dir.: *Suzanne T. Savery* • Cur.: *Laura Willoughby* •
Local Museum – 1976
Federal, Greek revival & colonial revival architecture, period furnishings, decorative arts ... 51797

Farmers Bank, The Petersburg Museums, 19 Bollingbrook St, Petersburg, VA 23803 • T: +1 804 7332400 • F: +1 804 8630837 • petgtourism@earthlink.net • www.petersburg-va.org •
Dir.: *Suzanne T. Savery* •
Special Museum – 1974
Pre-Civil War banking, furnishings used in banks of the periodgold storage area ... 51798

Pamplin Historical Park and the National Museum of the Civil War Soldier, 6125 Boydton Plank Rd, Petersburg, VA 23803 • T: +1 804 8612408 • F: +1 804 8612820 • generalmailbox@pamplinpark.org • www.pamplinpark.org •
Dir.: *A. Wilson Greene* •
Military Museum – 1994
Civil War military artifacts ... 51799

Petersburg Area Art League, 7 Olde St, Petersburg, VA 23803 • T: +1 804 8614611 • F: +1 804 8613962 •
Pres.: *George Tilarinos* •
Fine Arts Museum – 1932
Cvarious forms of rotating art of local artists ... 51800

The Petersburg Museums, 15 W Bank St, Petersburg, VA 23803 • T: +1 804 7332404 • F: +1 804 8630837 • petgtourism@earthlink.net • www.petersburg-va.com •
Dir.: *Suzanne T. Savery* • Cur.: *Roswitha M. Rash* •
Local Museum – 1972
Local hist ... 51801

Petersburg National Battlefield, 1539 Hickory Hill Rd, Petersburg, VA 23803-4721 • T: +1 804 7323531 • F: +1 804 7320835 • www.nps.gov/pete •
Historian: *James H. Blankenship jr* •
Historic Site / Military Museum – 1926
Military hist of campaign for Petersburg, artillery, maps ... 51802

Siege Museum, 15 W Bank St, Petersburg, VA 23803 • T: +1 804 7332404 • F: +1 804 8630837 • petgcurator@earthlink.net • www.petersburg-va.com •
Dir.: *Suzanne T. Savery* •
Military Museum – 1976
Civil war ... 51803

Trapezium House, The Petersburg Museums, Market St, Petersburg, VA 23803 • T: +1 804 7332400 • F: +1 804 7329212 • petgtourism@earthlink.net • www.petersburg-va.org •
Dir.: *Suzanne T. Savery* • Cur.: *Roswitha M. Rash* •
Local Museum – 1974
Historic house ... 51804

Petersham MA

Fisher Museum of Forestry, Harvard University, 326 N Main St, Petersham, MA 01366-0068 • T: +1 978 7243302 • F: +1 978 7243595 • jokeefe@fas.harvard.edu • www.harvardforest.fas.harvard.edu/mus.html •
Head: *John F. O'Keefe* •
Natural History Museum / University Museum – 1908
Forest history and ecology – library ... 51805

Petoskey MI

Crooked Tree Arts Center, 461 E Mitchell St, Petoskey, MI 49770 • T: +1 616 3474337 • F: +1 616 3475414 • dale@crookedtree.org • www.crookedtree.org •
Dir.: *Dale Hull* • Cur.: *Vivi Woodcock* •
Public Gallery – 1971 ... 51806

Little Traverse Historical Museum, 100 Depot St, Petoskey, MI 49770 • T: +1 231 3472620 • F: +1 616 3472875 • mce@freeway.net •
Exec. Dir.: *Mary Candace Eaton* •
Local Museum – 1971
Local hist, Hemingway artifacts ... 51807

Pharr TX

Old Clock Museum, 929 E Preston St, Pharr, TX 78577 • T: +1 956 7871923 •
Dir.: *Josephine Shawn* • *Barbara Shawn Barber* •
Science&Tech Museum – 1968
Horology, clocks ... 51808

Philadelphia PA

Academy of Natural Sciences of Philadelphia Museum, 1900 Ben Franklin Pkwy, Philadelphia, PA 19103-1195 • T: +1 215 2991000 • F: +1 215 2991028 • rosado@acnatsci.org • www.acnatsci.org •
Dir.: *James Baker* • Cur.: *Ruth Patrick* (Art, Artifacts) •
Natural History Museum – 1812
Natural science ... 51809

The African American Museum in Philadelphia, 701 Arch St, Philadelphia, PA 19106-1557 • T: +1 215 5740380 • F: +1 215 5743110 • trour@aampmuseum.org • www.aampmuseum.org •
Pres.: *Harry Harrison* •
Local Museum – 1976
Intellectual and material culture of African and American ... 51810

African Cultural Art Forum, 221 N 52nd St, Philadelphia, PA 19139 • T: +1 215 4760680 • F: +1 215 7480347 • snhewi@aol.com •
Fine Arts Museum / Ethnology Museum ... 51811

American Catholic Historical Society Museum, 263 S Fourth St, Philadelphia, PA 19106 • T: +1 215 9255752 • F: +1 215 2041663 • james.fitzsimmons@temple.edu • www.amchs.org •
Pres.: *Mary Louise Sullivan* • C.E.O.: *James P. McCoy* •
Historical Museum / Religious Arts Museum – 1884
Religious and American hist ... 51812

American Swedish Historical Museum, 1900 Pattison Av, Philadelphia, PA 19145 • T: +1 215 3891776 • F: +1 215 3897701 • info@americanswedish.org • www.americanswedish.org •
Exec. Dir.: *Richard Waldron* • Cur.: *Margaretha Talerman* •
Local Museum / Folklore Museum – 1926
Swedish-American history, art ... 51813

Artforms Gallery Manuyunk, 106 Levering St, Philadelphia, PA 19127 • T: +1 215 4833030 • F: +1 215 4838308 • artformsgallery@mac.com • www.artforms.org •
Dir.: *Ronna Cooper* •
Fine Arts Museum – 1994
Paintings, collage prints, ceramic sculpture, wood & stone sculpture & photography ... 51814

Arthur Ross Gallery, University of Pennsylvania, 220 S 34th St, Philadelphia, PA 19104-6303 • T: +1 215 8984401 • F: +1 215 5732045 • arg@pobox.upenn.edu • www.upenn.edu/ARG •
Dir.: *Dr. Dilys V. Winegrad* •
University Museum / Fine Arts Museum – 1983
Art ... 51815

Athenaeum of Philadelphia, 219 S 6th St, E Washington Sq, Philadelphia, PA 19106 • T: +1 215 9252688 • F: +1 215 9253755 • athena@philaathenaeum.org • www.philaathenaeum.org •
Dir.: *Dr. Roger W. Moss* • Ass. Dir.: *Eileen M. Magee* •
Cur.: *Bruce Laverty* (Architecture) •
Fine Arts Museum / Decorative Arts Museum – 1814
Art coll, manuscripts, architectural drawings, historic design ... 51816

Atwater Kent Museum - The History Museum of Philadelphia, 15 S 7th St, Philadelphia, PA 19106 • T: +1 215 6854830 • F: +1 215 6854837 • akmh@philadelphiahistory.org • www.philadelphiahistory.org •
Dir.: *Viki Sand* • Cur.: *Jeffrey Ray* •
Fine Arts Museum / Historical Museum – 1939
History ... 51817

The Balch Institute for Ethnic Studies, 18 S 7th St, Philadelphia, PA 19106 • T: +1 215 9258090 • F: +1 215 9258195 • balchlib@balchinstitute.org • www.balchinstitute.org •
Cur.: *Barbara Ward Grubb* •
Ethnology Museum – 1971
Ethnography ... 51818

Betsy Ross House, 239 Arch St, Philadelphia, PA 19106 • T: +1 215 6861252 • F: +1 215 6861256 • lori@betsyrosshouse.org • www.betsyrosshouse.org •
Exec. Dir.: *Lori Dillard Rech* •
Local Museum – 1898
Historic house, craftman's tools, furnishings – American Flag House ... 51819

Cedar Grove Mansion, Philadelphia Museum of Art, Fairmount Park, Lansdowne Dr, Philadelphia, PA 19101 • T: +1 215 7638100 ext 4013 • www.philamuseum.org •
Decorative Arts Museum ... 51820

Center for the Visual Arts, Brandywine Workshop, 730 S Broad St, Philadelphia, PA 19146 • T: +1 215 5463675 • F: +1 215 5462825 • brandwn@libertynet.org •
Dir.: *Allan L. Edmunds* •
Fine Arts Museum – 1972 ... 51821

Cigna Museum and Art Collection, 2 Liberty Pl, 1601 Chestnut St, Philadelphia, PA 19192-2078 • T: +1 215 7614907 • F: +1 215 7615596 • melissa.hough@cigna.com •
Dir.: *Melissa E. Hough* • Cur.: *Sue M. Levy* •
Fine Arts Museum / Historical Museum – 1925
Art, history ... 51822

Civil War Museum, 1805 Pine St, Philadelphia, PA 19103 • T: +1 215 7358196 • F: +1 215 7353812 • cwlm@netreach.net •
Dir.: *John J. Craft* •
Historical Museum – 1888
History ... 51823

Clay Studio, 139 N Second St, Philadelphia, PA 19106 • T: +1 215 9253453 • F: +1 215 9257774 • info@theclaystudio.org • www.theclaystudio.org •
Dir.: *Amy Sarner Williams* •
Decorative Arts Museum – 1974 ... 51824

Cliveden House, 6401 Germantown Av, Philadelphia, PA 19144 • T: +1 215 8481777 • F: +1 215 4382892 • chewhouse@aol.com • www.cliveden.org • Dir.: *Kris S. Kepford* • Cur.: *Phillip Sietz* •
Local Museum / Historic Site – 1972
Historic house, site of the Battle of Germantown 1777 ... 51825

Design Arts Gallery, Drexel University, College of Media Arts & Design, 33rd and Market Sts, Philadelphia, PA 19104 • T: +1 215 8952548 • F: +1 215 8956447 • gallery@drexel.edu •
Dir.: *Joseph Gregory* •
University Museum / Public Gallery
Changing exhibitions of contemporary and applied art, architecture, design, fashion, graphic, photography, painting and sculpture ... 51826

The Design Center at Philadelphia University, 4200 Henry Av, Philadelphia, PA 19144 • T: +1 215 9512860 • F: +1 215 9512662 • thedesigncenter@philau.edu • www.philau.edu/designcenter •
Dir.: *Hilary Jay* • Ass. Dir.: *Amanda Mullin* • Cur.: *Nancy Packer* •
Decorative Arts Museum – 1978
Textiles hist, costumes, design ... 51827

Eastern State Penitentiary Historic Site, 22nd St and Fairmount Av, Philadelphia, PA 19130 • T: +1 215 2363300 • F: +1 215 2365289 • e-state@liberty.net.org • www.easternstate.org •
Exec. Dir.: *Sara Jane Elk* •
Historical Museum ... 51828

Ebenezer Maxwell Mansion, 200 W Tulpehocken St, Philadelphia, PA 19144 • T: +1 215 4381861 • F: +1 215 4381861 • maxwellmansion@joimail.com • www.maxwellmansion.org •
Pres.: *Liza Hawley* •
Local Museum – 1975
Local hist ... 51829

Edgar Allan Poe House, 530-32 N 7th St, Philadelphia, PA 19123 • T: +1 215 5978780 • F: +1 215 5971901 • www.nps.gov/edal •
Man.: *Steve Sitarski* • Cur.: *Doris Fanelli* •
Special Museum – 1935
Edgar Allan'Poe brick house ... 51830

Elfreth's Alley Museum, 126 Elfreths Alley, Philadelphia, PA 19106 • T: +1 215 5740560 • F: +1 215 9227869 • elfreth@netzero.net • www.elfrethsalley.org •
Exec. Dir.: *Beth Richards* •
Historic Site – 1934
Historic bldgs ... 51831

Esther M. Klein Art Gallery, University City Science Center, 3600 Market St, Philadelphia, PA 19104 • T: +1 215 9666188 • F: +1 215 9666260 • kleinart@libertynet.org • www.kleinartgallery.org •
Dir./Cur.: *Dan Schimmel* •
University Museum / Fine Arts Museum – 1976
Paintings, sculpture, photography ... 51832

The Fabric Workshop and Museum, 1315 Cherry St, Philadelphia, PA 19107 • T: +1 215 5681111 • F: +1 215 5688211 • info@fabricworkshopandmuseum.org • www.fabricworkshopandmuseum.org •
Dir.: *Ellen B. Napier* •
Fine Arts Museum – 1977
Contemporary art, Studio focusing on fabric and new materials ... 51833

Fireman's Hall Museum, 147 N 2nd St, Philadelphia, PA 19106-2010 • T: +1 215 9231438 • F: +1 215 9230479 • firemusks@aol.com •
Pres.: *Charles D. Barber* • Cur.: *Henry J. Magee* •
Historical Museum – 1967
Fire fighting ... 51834

Fort Mifflin, Fort Mifflin Rd, Philadelphia, PA 19153 • T: +1 215 6854167 • F: +1 215 6854166 • lee_anderson@juno.com •
Exec. Dir.: *Eilee Young-Vignola* •
Historical Museum / Military Museum – 1962
Historic 1777 fort ... 51835

The Franklin Institute, 222 N 20th, Philadelphia, PA 19103-1194 • T: +1 215 4481208, 4481200 • F: +1 215 4481081 • kkirby@fi.edu • www.fi.edu •
Pres. & C.E.O.: *Dennis M. Wint* •
Science&Tech Museum – 1824
Science and technology ... 51836

Fred Wolf Jr. Gallery, 10100 Jamison Av, Philadelphia, PA 19116 • T: +1 215 6987300 • F: +1 215 6737447 • pactman@phillyjcc.com •
Dir.: *Phyllis Actman* •
Fine Arts Museum / Decorative Arts Museum – 1975
Judaic artifacts, prints, paintings, photographs, sculptures ... 51837

The Galleries at Moore, Moore College of Art and Design, 20th St and The Parkway, Philadelphia, PA 19103, mail addr: 1916 Race Street, Philadelphia, PA 19103 • T: +1 215 9654027 • F: +1 215 5685921 • galleries@moore.edu • www.thegalleriesatmoore.org •
Dir./ Chief Cur.: *Lorie Mertes* •
Fine Arts Museum / Public Gallery / University Museum – 1948
Non-collecting institution with 6 gallery spaces exhibiting contemporary art and design – All exhibitions feature programs & lectures ... 51838

Germantown Historical Society Museum, 5501 Germantown Av, Philadelphia, PA 19144-2291 • T: +1 215 8440514 • F: +1 215 8442831 • info@germantownhistory.org • www.germantownhistory.org •
Pres.: *Alex B. Humes* • C.E.O.: *Mary K. Dabney* •
Historical Museum – 1900
Local hist ... 51839

Grand Army of the Republic Civil War Museum, 4278 Griscom St, Philadelphia, PA 19124-3954 • T: +1 215 2896484, 6731688 • garmuslib@verizon.net • www.garmuslib.org •
Dir.: *Elmer F. Atkinson* •
Historical Museum – 1926
Hist of the Civil War ... 51840

Historical Society of Pennsylvania, 1300 Locust St, Philadelphia, PA 19107-5699 • T: +1 215 7326200 • F: +1 215 7322680 • hsp_lan@lsp.org • www.hsp.org •
Pres.: *David Moltke-Hansen* •
Local Museum / Library with Exhibitions – 1824
Historical Research Center ... 51841

Independence National Historical Park, 143 S 3rd St, Philadelphia, PA 19106 • T: +1 215 5978787 • F: +1 215 5975556 • www.nps.gov/inde •
Chief Cur.: *Karie Diethorn* •
Historical Museum / Historic Site – 1948
Local hist ... 51842

Independence Seaport Museum, Penn's Landing Waterfront, 211 S Columbus Blvd & Walnut St, Philadelphia, PA 19106 • T: +1 215 9255439 • F: +1 215 9256713 • mdigirolamo@phillyseaport.org • www.phillyseaport.org •
Dir.: *John S. Carter* • Cur.: *David Beard* •
Historical Museum – 1961
Maritime hist ... 51843

Institute of Contemporary Art, University of Pennsylvania, 118 S 36th St, Philadelphia, PA 19104-3289 • T: +1 215 8987108 • F: +1 215 8985050 • info@icaphila.org • www.icaphila.org •
Dir.: *Claudia Gould* • Cur.: *Ingrid Schaffner* •
Fine Arts Museum / University Museum – 1963
Contemporary visual culture ... 51844

John G. Johnson Collection, Philadelphia Museum of Art, Parkway and 26th St, Philadelphia, PA 19101-7646 • T: +1 215 6847646 • F: +1 215 6847616 •
Cur.: *Joseph Rishel* •
Fine Arts Museum
Italian renaissance paintings, French 19th c paintings ... 51845

La Salle University Art Museum, 20th St and Olney Av, Philadelphia, PA 19141 • T: +1 215 9511221, 9511000 • F: +1 215 9515096 • wistar@lasalle.edu • www.lasalle.edu/services/art-mus •
Dir.: *Daniel Burke* • Cur.: *Caroline Wistar* •
Fine Arts Museum / University Museum – 1975
16th-20th c European and American art ... 51846

Lemon Hill Mansion, Lemon Hill and Sedgelay Drs, East Fairmount Park, Philadelphia, PA 19130 • T: +1 215 6467084 • F: +1 215 6468472 • elliepen@msn.com • www.lemonhill.org •
Pres.: *H. Dawson Penniman* •
Decorative Arts Museum
Decorative arts of Philadelphia 1800 - 1836, local hist ... 51847

Levy Gallery, Moore College of Art and Design, 20th St and The Pkwy, Philadelphia, PA 19103-1179 • T: +1 215 9654027 • F: +1 215 5685921 • galleries@moore.edu • www.thegalleriesatmoore.org •
Dir.: *Molly Dougherty* • *Brian Wallace* (Exhibitions) •
Fine Arts Museum ... 51848

Loudoun Mansion, 4650 Germantown Av, Philadelphia, PA 19144 • T: +1 215 6862067, 6777830 •
Pres.: *Ursula Reed* •
Decorative Arts Museum
18th c furniture and paintings, costumes, glass, china ... 51849

Mario Lanza Museum, c/o Settlement Music School, 712 Montrose St, Philadelphia, PA 19148-0624 • T: +1 215 2389691 • F: +1 215 4681903 • mariolanzamuseum@aol.com • www.mario-lanza-institute.org •
Pres.: *Mary Galanti Papola* •
Music Museum – 1962
Music hist ... 51850

The Masonic Library and Museum of Pennsylvania, Masonic Temple, 1 N Broad St, Philadelphia, PA 19107-2520 • T: +1 215 9881932 • F: +1 215 9881972 • kwmccarty@pagrandlodge.org • www.pagrandlodge.org •
Dir.: *Kenneth W. McCarty* • Cur.: *Laura L. Libert* • Libr.: *Glenys A. Waldman* • *Catherine L. Giaimo* •
Local Museum
Local hist ... 51851

Minority Arts Resource Council Studio Art Museum, 1421 W Girard Av, Philadelphia, PA 19130 • T: +1 215 2362688, 2363977 • F: +1 215 2364255 •
C.E.O.: *Curtis E. Brown* •
Fine Arts Museum – 1978
Paintings, drawings, sculptures ... 51852

Mount Pleasant, Philadelphia Museum of Art, Fairmount Park, Philadelphia, PA 19101 • T: +1 215 7638100 ext 4014 • www.phila.museum.com •
Cur.: *Jack Lindsey* •
Decorative Arts Museum
Period furnishings ... 51853

Muse Art Gallery, 60 N Second St, Philadelphia, PA 19106 • T: +1 215 6275310 • membrane.com/philanet/muse_gallery •
Dir.: *Sissy Pizzollo* •
Fine Arts Museum – 1970 ... 51854

The Museum at Drexel University, Chestnut and 32nd Sts, Philadelphia, PA 19104 • T: +1 215 8951749 • F: +1 215 8956447 • www.drexel.edu •
Dir.: *Charles Morscheck* •
University Museum / Fine Arts Museum / Decorative Arts Museum – 1891
Sculpture, European painting, decorative arts and costumes, ceramics, hand printed India cottons, decorative arts of China, Europe, India and Japan ... 51855

The Museum of Nursing History, Pennsylvania Hospital, 8th and Spruce Sts, Philadelphia, PA 19107 • T: +1 610 7894277 • sdavis10@earthlink.net •
Cur.: *Joan T. Large* •
Historical Museum – 1976
Nursing hist ... 51856

Mutter Museum, College of Physicians of Philadelphia, 19 S 22nd St, Philadelphia, PA 19103-3097 • T: +1 215 5633737 ext 242 • F: +1 215 5616477 • worden@collphyphil.org • www.collphyphil.org •
Dir.: *Gretchen Worden* •
Historical Museum – 1863
Medicine ... 51857

Naomi Wood Collection at Woodford Mansion, 33rd and Dauphin Sts, E Fermont Park, Philadelphia, PA 19118 • T: +1 215 2296115 •
Historical Museum – 1926
Decorative arts, English Delftware, Colonial Philadelphia and American furnishings, pewter, silver, needlework, portraits ... 51858

National Exhibits by Blind Artists, 919 Walnut St, Philadelphia, PA 19107 • T: +1 215 6275930, 6833213 • info@nebaart.org • www.nebaart.org •
Pres.: *Mary Fediw* • Dir.: *Vickie Collins* •
Public Gallery – 1976
Art works by blind and physically handicapped artists ... 51859

National Liberty Museum, 321 Chestnut St, Philadelphia, PA 19106 • T: +1 215 9252800 • F: +1 215 9253800 • liberty@libertymuseum.org • www.libertymuseum.org •
Exec. Dir.: *Gwen Borowsky* •
Fine Arts Museum / Historical Museum – 2000
Glass sculptures, film, artwork ... 51860

National Museum of American Jewish History, 55 N 5th St, Independence Mall E, Philadelphia, PA 19106-2197 • T: +1 215 9233812 • F: +1 215 9230763 • nmajh@nmajh.org • www.nmajh.org •
Dir.: *Gwen Goodman* •
Historical Museum / Ethnology Museum / Religious Arts Museum – 1974
Social and ethnic hist ... 51861

Nexus Foundation for Today's Art, 137 N Second St, Philadelphia, PA 19106 • T: +1 215 6291103 • F: +1 215 6298683 • nexusartgallery@yhoo.com • www.nexusphiladelphia.com •
Dir.: *Chris Vecchio* • *Nick Cassway* •
Fine Arts Museum – 1975 ... 51862

Painted Bride Art Center Gallery, 230 Vine St, Philadelphia, PA 19106 • T: +1 215 9259914 • F: +1 215 9257402 • thebride@onix.com • www.paintedbride.org •
Dir.: *Gerry Givnish* • *Laurel Raczka* •
Fine Arts Museum – 1968
Contemporary art ... 51863

Pennsylvania Academy of the Fine Arts Gallery, 118 N Broad St, Philadelphia, PA 19102 • T: +1 215 9727600 • F: +1 215 5690153 • pafa@pafa.org • www.pafa.org •
Dir.: *Derek Gillman* • Cur.: *Alexander Baker* (Contemporary Art) • *Lynn Marsden-Atlass* • *Judith Ringold* (Education) • Cons.: *Aella Diamantopoulos* •
Fine Arts Museum – 1805
American art ... 51864

Philadelphia Art Alliance, 251 S 18th St, Philadelphia, PA 19103 • T: +1 215 5454305 • F: +1 215 5450767 • info@philartalliance.org • www.philartalliance.org •
Pres.: *Carole Price Shanis* • Exec. Dir.: *Sylvia Watts McKinney* •
Fine Arts Museum – 1915
Art ... 51865

Philadelphia Mummers Museum, 1100 S Second St, Philadelphia, PA 19147 • T: +1 215 3363050 • F: +1 215 3895630 • mummersmus@aol.com • www.mummers.com •
Dir.: *Palma B. Lucas* • Cur.: *Jack Cohen* •
Decorative Arts Museum / Performing Arts Museum – 1976
Costumes ... 51866

Philadelphia Museum of Art, 26th St and Benjamin Franklin Pkwy, Philadelphia, PA 19130 • T: +1 215 7638100 • F: +1 215 2364465 • pr@philamuseum.org • www.philamuseum.org •
Dir.: *Anne d' Harnoncourt* • Sen. Cur.: *Dean Walker* (European Decorative Arts and Sculpture) • *Innis H. Shoemaker* (Prints, Drawings and Photographs) • Cur.: *Joseph J. Rishel* (European Painting before 1900 and John G. Johnson Coll) • *Kathryn B. Hiesinger* (European Decorative Arts after 1700) • *Kathleen Foster* (American Art) • *Jack Lindsey* (American Decorative Arts) • *Ann Temkin* (20th c Art) • *Felice Fischer* (East Asian and Japanese Art) • *Darielle Mason* (Indian and Himalayan Art) • *Ann B. Percy* (Drawings) • *John Ittmann* (Prints) • *Dilys Blum* (Costume and Textiles) • *Katherine Ware* (Photographs) • Sen. Cons.: *P. Andrew Lins* (Decorative Arts and Sculpture) • *Mark S. Tucker* (Paintings) • Cons.: *David DeMuzio* (Furniture) • *Nancy Ash* (Works of Art on Paper) •
Fine Arts Museum – 1876
Art ... 51867

Philadelphia Museum of Judaica, Congregation Rodeph Shalom, 615 N Broad St, Philadelphia, PA 19123 • T: +1 215 6276747 • F: +1 215 6271313 • info@rodephshalom.org • www.rodephshalom.org •
Dir.: *Joan C. Sall* •
Historical Museum / Religious Arts Museum – 1975
Jewish art, religious ceremonials ... 51868

Philadelphia Society for the Preservation of Landmarks, 321 S Fourth St, Philadelphia, PA 19106 • T: +1 215 9252251 • F: +1 215 9257909 • phila.landworks@verizon.net • www.philalandmarks.org •
Association with Coll – 1931
Philadelphia and American-made furniture, decorative arts ... 51869

Plastic Club, 247 S Camac St, Philadelphia, PA 19107 • T: +1 215 5459324 • plasticclub@att.net • plasticc.libertynet.org •
Pres.: *Elizabeth H. MacDonald* •
Public Gallery – 1897
Paintings, drawings ... 51870

Please Touch Museum, 4231 Avenue of the Republic, Philadelphia, PA 19131 • T: +1 215 5813181 • F: +1 215 5813183 • ceo@pleasetouchmuseum.org • www.pleasetouchmuseum.org •
Pres.: *Nancy D. Kolb* •
Special Museum – 1976
Children's museum, visual and performing arts, sciences, humanities ... 51871

Presbyterian Historical Society Museum, 425 Lombard St, Philadelphia, PA 19147 • T: +1 215 6271852 • F: +1 215 6270509 • reference@history.pcusa.org • www.history.pcusa.org •
Dir.: *Frederick J. Heuser* •
Religious Arts Museum – 1852
Religious hist ... 51872

Print and Picture Collection, c/o Free Library of Philadelphia, 1901 Vine St, Philadelphia, PA 19103 • T: +1 215 6865405 • F: +1 215 5633628 • langw@library.phila.gov • www.library.phila.gov •
Dir.: *Elliot L. Shelkrot* •
Fine Arts Museum – 1954
Paintings, graphic arts ... 51873

The Print Center, 1614 Latimer St, Philadelphia, PA 19103 • T: +1 215 7356090 • F: +1 215 7355511 • info@printcenter.org • www.printcenter.org •
Cur.: *Jaqueline Van Rhyn* •
Public Gallery – 1915
Art ... 51874

Robert W. Ryerss Museum, Burholme Park, 7370 Central Av, Philadelphia, PA 19111 • T: +1 215 7453061 • ryerssmuseum@hotmail.com •
Dir.: *Theresa Stuhlman* •
Historical Museum – 1910
Asian collection ... 51875

Rodin Museum, Philadelphia Museum of Art, Benjamin Franklin Pkwy at 22nd St, Philadelphia, PA 19130 • T: +1 215 7638100 • F: +1 215 2350050 • www.rodinmuseum.org •
Dir.: *Anne d' Harnoncourt* • Cur.: *Joseph J. Rishel* (European Painting and Sculpture) •
Fine Arts Museum – 1926
Coll of works by Auguste Rodin ... 51876

Rosenbach Museum, 2010 Delancey Pl, Philadelphia, PA 19103 • T: +1 215 7321600 • F: +1 215 5457529 • info@rosenbach.org • www.rosenbach.org •
Dir.: *Derick Dreher* • Libr.: *Elizabeth E. Fuller* •
Special Museum – 1953
Rare books and manuscripts, antiques, prints, drawings and paintings ... 51877

Rosenwald-Wolf Gallery, University of the Arts, 333 S Broad St, Philadelphia, PA 19102 • T: +1 215 7176480 • F: +1 215 7176468 • sachs@uarts.edu • www.uarts.edu •
Pres.: *Miguel-Angel Corzo* • Dir.: *Sid Sachs* •
Fine Arts Museum / University Museum – 1876
Art ... 51878

Saint George's United Methodist Church Museum, 235 N Fourth St, Philadelphia, PA 19106 • T: +1 215 9257788 •
Religious Arts Museum – 1767 ... 51879

Saint Joseph's University Gallery, Boland Hall, 5600 City Av, Philadelphia, PA 19131-3195 • T: +1 610 6601840 • F: +1 610 6602278 • sfenton@sju.edu • www.sju.edu/gallery •
Dir.: *Dennis McNelly* •
Fine Arts Museum / University Museum – 1976
Contemporary photography, painting, printmaking, ceramics ... 51880

Samuel S. Fleisher Art Memorial, Philadelphia Museum of Art, 709-721 Catharine St, Philadelphia, PA 19147 • T: +1 215 9223456 • F: +1 215 9225327 • www.fleisher.org •
Dir.: *Thora E. Jacobson* •
Fine Arts Museum / Religious Arts Museum – 1898
Religious paintings and sculptures, Portuguese liturgical objects, Russian icons ... 51881

The Stephen Girard Collection, Girard College, Founder's Hall, 2101 S College Av, Philadelphia, PA 19121 • T: +1 215 7872680 • F: +1 215 7874404 • elaurent@girardrcollege.com • www.girardcollege.com •
Pres.: *Dominic M. Cermèle* • Cur.: *Elizabeth M. Laurent* •
Historical Museum – 1831
Furniture, decorative arts 51882

Taller Puertoriqueno, 2721 N Fifth St, Philadelphia, PA 19133 • T: +1 215 4263311 • F: +1 215 4265682 •
Dir.: *Annabella Roige* •
Fine Arts Museum – 1974 51883

Temple Gallery, Tyler School of Art, Temple University, 45 N Second St, Philadelphia, PA 19106 • T: +1 215 9257379 • F: +1 215 9257389 • tylerart@vm.temple.edu •
Dir.: *Kevin Melchionne* •
Fine Arts Museum / University Museum – 1985 51884

United States Mint-Philadelphia, 5th and Arch Sts, Philadelphia, PA 19106 • T: +1 215 4080114 • F: +1 215 4082700 • www.usmint.gov •
Head: *R.R. Robidoux* •
Special Museum – 1792
Numismatics 51885

University of Pennsylvania Museum of Archaeology and Anthropology, 3260 South St, Philadelphia, PA 19104-6324 • T: +1 215 8984000 • F: +1 215 8980657 • websiters@museum.upenn.edu • www.museum.upenn.edu •
Dir.: *Dr. Jeremy Sabloff* • Sc. Staff: *Gregory L. Possehl* (Asia) • *Robert Sharer* (America) • *Donald White* (Mediterranea) • *Janet Monge* (Physical Anthropology) • *Erle Leichty* (Babylonia) • *David Silverman* (Egypt) • *Richard Zettler* (Near East) • Sen. Cons.: *Virginia Greene* •
Ethnology Museum / Archaeology Museum – 1887
Archaeology, anthropology 51886

Vox Populi Gallery, 1315 Cherry St, Philadelphia, PA 19107 • T: +1 215 5685513 • F: +1 215 5685514 • vox@op.net • www.voxpopuligallery.org •
Public Gallery – 1988 51887

Wagner Free Institute of Science, 1700 W Montgomery Av, Philadelphia, PA 19121 • T: +1 215 7636529 • F: +1 215 7631299 • www.wagnerfreeinstitute.org •
Dir.: *Susan Glassman* •
Natural History Museum – 1855
Natural hist and science 51888

Woodmere Art Museum, 9201 Germantown Av, Philadelphia, PA 19118 • T: +1 215 2470476 • F: +1 215 2472387 • ngreene@woodmereartmuseum.org • www.woodmereartmuseum.org •
Pres.: *Joseph Nicholson* • Dir./Cur.: *Michael W. Schantz* •
Fine Arts Museum – 1940
Art 51889

Wyck Museum, 6026 Germantown Av, Philadelphia, PA 19144 • T: +1 215 8481690 • F: +1 215 8481612 • wyck@libertynet.org • www.wyck.org •
Dir.: *John M. Groff* • Cur.: *Elizabeth Solomon* •
Historical Museum – 1973
Local hist, furniture, glass, ceramics, metals, textiles 51890

Philip SD

Prairie Homestead, 131 off Interstate Hwy, Philip, SD 57567, mail addr: HCR01, Box 51, Philip, SD 57567 • T: +1 605 4335400 • F: +1 605 4335434 • klcrew@gwtc.net • www.prairiehomestead.com •
C.E.O.: *Keith Crew* •
Historic Site / Agriculture Museum / Open Air Museum – 1962
Historic house 51891

West River Museum, Center Av, Philip, SD 57567 •
Pres.: *Nancy Eksturm* • Ass. Dir.: *Shirley O'Connell* •
Local Museum – 1965
Local hist 51892

Phillips ME

Phillips Historical Society Museum, Pleasant St, Phillips, ME 04966 • T: +1 207 6395013 •
Pres.: *Ken Zieglar* •
Local Museum – 1959
Local history 51893

Phillipsburg KS

Old Fort Bissell, 501 Fort Bissell Av, Phillipsburg, KS 67661 • T: +1 913 5436212 • www.phillipsburgks.us •
Pres.: *Darlene Rumbaugh* • Cur.: *Eva Maria Hansen* •
Local Museum – 1961
Guns, doll coll, vrom 1740-present 51894

Philomath OR

Benton County Historical Museum, 1101 Main St, Philomath, OR 97370 • T: +1 541 9296230 • F: +1 541 9296261 • info@bentoncountymuseum.org • www.bentoncountymuseum.org •
Dir.: *William R. Lewis* •
Local Museum – 1980
Photograph coll, local hist and culture 51895

Gallery Gazelle, Oregon State University, 1136 Main St, Philomath, OR 97370 • T: +1 541 9296464 •
Public Gallery 51896

Phoenix AZ

Arizona Doll and Toy Museum, Heritage & Science Park, Historic Heritage Square, Phoenix, AZ 85004, mail addr: Heritage & Science Park, 115 N 6th Street, Phoenix, AZ 85004 • T: +1 602 2539337, 2625070 • rhhs@rossonhousemuseum.org • www.rossonhousemuseum.org •
Special Museum
Classroom scenario of dolls and toys from yesterday 51897

Arizona Jewish Historical Exhibitions, 4710 N 16th St, Phoenix, AZ 85016 • T: +1 602 2417870 • F: +1 602 2649773 • azjhs@aol.com • www.azjhs.com •
Exec. Dir.: *Risa Mallin* •
Historical Museum – 1981 51898

Arizona Mining and Mineral Museum, 1502 W Washington, Phoenix, AZ 85007 • T: +1 602 27711605 • dianebain@hotmail.com • www.admmr.state.az.us •
Dir.: *Madan M. Singh* • Cur.: *Jan Rasmussen* •
Natural History Museum / Science&Tech Museum – 1953
Minerals, mining, geology 51899

Arizona Science Center, 600 E Washington, Phoenix, AZ 85004-2394 • T: +1 602 7162000 • F: +1 602 7162099 • priscellak@azscience.org • www.azscience.org •
CEO: *Chevy Humphrey* •
Science&Tech Museum – 1984
Sciences, human body, psychology, weather, aerospace, geology – planetarium, giant screen theater 51900

Arizona State Capitol Museum, 1700 W Washington St, Phoenix, AZ 85007 • T: +1 602 9263620 • F: +1 602 2567985 • capmus@lib.az.us • www.lib.az.us/museum •
Dir.: *David Hoober* •
Historical Museum – 1974
Restored Capitol bldg 51901

Arizona Women Hall of Fame, Carnegie Center, 1101 W Washington, Phoenix, AZ 85007 • T: +1 602 2552110 • F: +1 602 2553314 • azwhof@lib.az.us • www.dlapr.lib.az.us/awhof •
Head: *Leslie J. Weaver* •
Local Museum – 1986
Women who have made important contributions to the hist of Arizona – archive 51902

Hall of Flame - Museum of Firefighting, 6101 E Van Buren St, Phoenix, AZ 85008-3421 • T: +1 602 2753473 • F: +1 602 2750896 • petermolloy@hallofflame.org • www.hallofflame.org •
Dir.: *Dr. Peter M. Molloy* • Cur.: *James Ward* • Rest.: *Donald G. Hale* •
Special Museum – 1961
History of firefighting, graphics, photos – library 51903

Heard Museum, 2301 N Central Av, Phoenix, AZ 85004-1323 • T: +1 602 2528840 • F: +1 602 2529757 • jmartin@heard.org • www.heard.org •
Dir.: *Frank H. Goodyear jr.* • Dept. Dir.: *Ann E. Marshall* (Research) • *Geri Wright* (Development) • *Juliet Martin* (Communication) • Cur.: *Diana Pardue* (Collections) •
Fine Arts Museum / Ethnology Museum – 1929
American Indian art and cultural materials, objects by Indigenous cultures from Africa, Asia, Oceania and upper Amazons – library 51904

Modified Art Gallery, 407 E Roosevelt St, Phoenix, AZ 85004 • T: +1 602 4625516 • kimber@modified.org • www.modified.org •
Public Gallery – 1999
Contemporary art 51905

Phoenix Art Museum, 1625 N Central Av, Phoenix, AZ 85004 • T: +1 602 2571880 • info@phxart.org • www.phxart.org •
Dir.: *James K. Ballinger* (American Art) • Cur.: *Dr. Claudia Brown* (Asian Art Research) • *Michael Komanecky* (European Art) • *Dr. Beverly Adams* (Latin American Art) • *Brady Roberts* (Modern and Contemporary Art) • *Dr. Janet Baker* (Asian Art) • *Dennita Sewell* (Fashion Design) •
Fine Arts Museum – 1959
American, European and Asian art, contemporary art, decorative arts, fashion design 18th-20th c – library 51906

Pioneer Arizona Living History Village, 3901 W Pioneer Rd, Phoenix, AZ 85086 • T: +1 623 4651052 • F: +1 623 4650683 • pioneerarizona@ • www.pioneer-arizona.com •
Dir.: *Becky Patton* • *Ron Heisner* •
Historical Museum – 1956
Living history complex, 19th c pioneer life, agriculture 51907

Pueblo Grande Museum, 4619 E Washington, Phoenix, AZ 85034-1909 • T: +1 602 4950901 • F: +1 602 4955645 • pueblo.grande.museum.pks@phoenix.gov • phoenix.gov/parks/pueblo •
Dir.: *Roger Lidman* • Cur.: *Glena Cain* (Exhibits) • *Holly Young* (Collections) •
Archaeology Museum – 1929
Archaeological finds from the Greater Southwest, regional ethnography 51908

Rosson House Museum, Heritage & Science Park, Historic Heritage Square, Phoenix, AZ 85004, mail addr: Heritage & Science Park, 115 N 6th St, Phoenix, AZ 85004 • T: +1 602 2625070 • www.rossonhousemuseum.org •
Local Museum
Victorian style house from 1895 51909

Shemer Art Center and Museum, 5005 E Camelbeck Rd, Phoenix, AZ 85018 • T: +1 602 2624727 • shemer@phoenix.gov • www.phoenix.gov/shemer •
Dir.: *Brian Flanigan* • Cur.: *Lindsay Palmer* •
Public Gallery – 1984
Contemporary art, changing exhibitions 51910

Wells Fargo History Museum, 145 W Adams, Phoenix, AZ 85003 • T: +1 602 3781578 • historicalservices@wellsfargo.com • www.wellsfargohistory.com •
Cur.: *Connie Whalen* •
Historical Museum / Fine Arts Museum – 2003
Concord Stagecoach, Wells Fargo banking and express hist, fine art, paintings by N.C. Wyeth, mining, staging, Phoenix hist 51911

Pickens SC

Pickens County Museum of Art History, 307 Johnson St, Pickens, SC 29671 • T: +1 864 8985963 • F: +1 803 8985947 • picmus@pickens.lib.sc.us • www.co.pickens.sc.us/culturalcommission •
Dir.: *C. Allen Coleman* • Cur.: *Ed L. Bolt* •
Fine Arts Museum / Local Museum / Historical Museum / Historic Site – 1976
Art, local hist 51912

Pickerington OH

Motorcycle Hall of Fame Museum, 13515 Yarmouth Dr, Pickerington, OH 43147 • T: +1 614 8562222 ext 1234 • F: +1 614 8562221 • info@motorcyclemuseum.org • www.motorcyclemuseum.org •
Dir.: *Mark Mederski* •
Science&Tech Museum – 1990
Motorcycles, riding clothing and accesories, related gear, racing 51913

Piedmont SD

Petrified Forest of the Black Hills, 8228 Elk Creek Rd, Piedmont, SD 57769-9520 • T: +1 605 7874560 • F: +1 605 7876477 • forest@elkcreek.org • www.elkcreek.org •
Dir./Cur.: *Timothy Scott* •
Natural History Museum – 1929
Geology 51914

Pierre SD

The Museum of the South Dakota State Historical Society, Cultural Heritage Center, 900 Governors Dr, Pierre, SD 57501-2217 • T: +1 605 7733458 • F: +1 605 7736041 • david.hartley@state.sd.us • www.sdhistory.org •
Dir.: *David B. Hartley* • Cur.: *Heather Bigeck* (Collections) •
Historical Museum – 1901
Local hist 51915

South Dakota Discovery Center, 805 W Sioux Av, Pierre, SD 57501 • T: +1 605 2248295 • F: +1 605 2242865 • kristiemaher@aol.com •
Pres.: *Tim Bjork* • C.E.O.: *Kristie Maher* •
Science&Tech Museum – 1989
Science 51916

Pilger NE

Stanton County Museum, 345 N Main St, Pilger, NE 68768-0213 • T: +1 402 4392952 • fullerel@stanton.net • www.stanton.net •
Local Museum – 1965
Indian artifacts, local hist 51917

Pilot Knob MO

Fort Davidson, 118 E Maple St, Pilot Knob, MO 63663 • T: +1 573 5463454 • F: +1 573 5462713 • fort.davidson.state.historic.site@dnr.mo.gov • www.mostateparks.com/ftdavidson.htm •
Head: *Walter Bush* • Cur.: *Brick Autry* •
Military Museum – 1969
Civil War artifacts 51918

Pima AZ

Eastern Arizona Museum, 2 N Main St, Pima, AZ 85543 • T: +1 928 4853032 • edres@zekes.com •
Cur.: *Edres Barney* •
Local Museum / Ethnology Museum – 1963
Local history – Old Cliff Hall 51919

Pine Bluff AR

Arts and Science Center for Southeast Arkansas, 701 Main St, Pine Bluff, AR 71601 • T: +1 870 5363375 • F: +1 870 5363380 • bhengel@artssciencecenter.org • www.artssciencecenter.org •
Exec. Dir.: *Mary Brock* •
Fine Arts Museum / Natural History Museum – 1968
Visual arts, physical and natural sciences – theater 51920

The Band Museum, 423 Main St, Pine Bluff, AR 71601 • T: +1 870 5344676 • F: +1 870 5344676 • bandmuseum@hotmail.com • bandmuseum.tripod.com •
C.E.O.: *Jerry G. Horne* •
Music Museum – 1994
Musical instruments 51921

Jefferson County Historical Museum, 201 E Fourth St, Pine Bluff, AR 71601 • T: +1 870 5415402 • marsha@jchswa.org •
Dir.: *Sue Trulock* •
Local Museum – 1980
History, culture and lifestyles of Jefferson County, farming, American Indiens, Civil war, doll coll 51922

Pine Ridge AR

Lum and Abner Museum, 4562 Hwy 88 W, Pine Ridge, AR 71966 • T: +1 870 3264442 • F: +1 870 3264442 •
Dir./ Owner: *Noah Lon Stucker* • Dir.: *Kathryn Moore Stucker* •
Science&Tech Museum / Historical Museum – 1971
Historic stores on which the Lum and Abner radio program 1931-55 was based, country store, rural life 51923

Pine Ridge SD

The Heritage Center, Red Cloud Indian School, 100 Mission Dr, Pine Ridge, SD 57770 • T: +1 605 8675491 • F: +1 605 8671209 • csimon@redcloudschool.org • www.redcloudschool.org •
Pres.: *Peter Klink* • Dir.: *C.M. Simon* •
Fine Arts Museum / Ethnology Museum – 1967
Native American art 51924

Pinedale WY

Museum of the Mountain Man, 700 E Hennick, Pinedale, WY 82941 • T: +1 307 3674101 • F: +1 307 3676768 • museummtman@wyoming.com • www.museumofthemountainman.com •
Exec. Dir.: *Laurie Hartwig* •
Historical Museum – 1990
Early settlement hist of Wyoming, fur trade, Plains Indians 51925

Pineville NC

James K. Polk Memorial, 308 S Polk St, Pineville, NC 28134 • T: +1 704 8897145 • F: +1 704 8893057 • polkmemorial@dasia.net • www.polk.nchistoricsites.org •
Pres.: *Don Woods* •
Historical Museum – 1961 51926

Piney Flats TN

Rocky Mount Museum, Rocky Mount Pkwy, 200 Hyder Hill Rd, Piney Flats, TN 37686 • T: +1 423 5387396 • F: +1 423 5381086 • info@rockymountmuseum.com • www.rockymountmuseum.com •
Pres.: *J. Paul Frye* • Exec. Dir.: *Gary Walrath* •
Open Air Museum / Local Museum – 1958
Historical village, separate museum of overmountain history 51927

Piney Woods MS

Laurence C. Jones Museum, Piney Woods Country Life School, Piney Woods, MS 39148 • T: +1 601 8452214 • F: +1 601 8452604 • astewart@pineywoods.org • www.pineywoods.org •
C.E.O.: *Charles H. Beady* •
Local Museum – 1986 51928

Pinson TN

Pinson Mounds State Archaeological Area, 460 Ozier Rd, Pinson, TN 38366 • T: +1 731 9885614 • F: +1 731 4243909 • www.tnstateparks.com/pinson •
Archaeology Museum – 1980
Regional & stie specific prehistory 51929

Pipestone MN

Pipestone County Historical Museum, 113 S Hiawatha Av, Pipestone, MN 56164 • T: +1 507 8252563 • F: +1 507 8252563 • pipctymu@rconnect.com • www.pipestoneminnesota.com/museum/index.htm •
Dir.: *Susan Hoskins* •
Historical Museum – 1880
Euro-American settlers, pipes, Dakota & Ojibwa tribes clothing, Indian saddles, American culture, prehistoric items 51930

Piqua OH

Piqua Historical Area State Memorial, 9845 N Hardin Rd, Piqua, OH 45356-9707 • T: +1 937 7732522 • F: +1 937 7734311 • johnstonfarm@mail2.wesnet.com • www.ohiohistory.org/places/piqua •
Head: *Andy Hite* •
Agriculture Museum – 1972
18th/19th c furnishings and tools, American Indian tools, weapons, costumes, canoes, trade items, hist of agriculture 51931

Piscataway NJ

Cornelius Low House/Middlesex County Museum, 1225 River Rd, Piscataway, NJ 08854 • T: +1 732 7454177, 7454489 • F: +1 732 7454507 • info@cultureheritage.org • www.cultureheritage.org •
Exec. Dir.: *Anna M. Aschkenes* •
Local Museum – 1979
Local hist 51932

East Jersey Olde Towne, 1050 River Rd and Old Hoes Ln, Piscataway, NJ 08855-0661 • T: +1 732 7453030, 7454489 • F: +1 732 4631086 • info@culturalheritage.org • www.culturalheritage.org •
Local Museum / Open Air Museum – 1971
Historic 18th c village ... 51933

Pittsburg KS

Crawford County Historical Museum, N of 69 Bypass and W 20th St, Pittsburg, KS 66762-8600 • T: +1 620 2311440 •
C.E.O.: *Alan Ross* •
Local Museum – 1968
Art, decorative arts, clothing, farming,goverment, household, industry ... 51934

Pittsburg State University Natural History Museum, c/o PSU Biology Dept, 1701 S Broadway, Heckert-Wells Hall, Pittsburg, KS 66762 • T: +1 620 2354732 • F: +1 620 2354194 • sford@pittstate.edu • www.pittstate.edu/biol •
Cur.: *Dr. Leon Dinkins* (Insects) • *Dr. S.D. Ford* (Mammals and Birds) •
Natural History Museum – 1903
Skin and skulls, alcohol specimens, insects ... 51935

Pittsburg PA

Associated American Jewish Museum, 4905 Fifth Av, Pittsburg, PA 15213 • T: +1 412 6216566 • F: +1 412 6215475 • jacob@rodefshalom.org •
Pres.: *Walter Jacob* •
Religious Arts Museum / Historical Museum – 1972
Historical and artistic Judaica, synagogue coll ... 51936

Pittsburg TX

Northeast Texas Rural Heritage Museum, 204 W Marshall, Pittsburg, TX 75686, mail addr: POB 157, Pittsburg, TX 75686 • T: +1 903 8561200 • F: +1 903 8561200 • CampCountyMuseum@aol.com • www.pittsburgtxmuseum.com •
Agriculture Museum – 1989
Artifacts & information related to the hist of NE Texas rural life & inhabitants from prehistoric times to 1950's ... 51937

Pittsburgh PA

The Andy Warhol Museum, 117 Sandusky St, Pittsburgh, PA 15212-5890 • T: +1 412 2378300 • F: +1 412 2378340 • warhol@alphaclp.clpgh.org • www.warhol.org •
Dir.: *Thomas Sokolowski* • Cur.: *Geralyn H. Huxley* (Film and Video) • *Jessica Arcand* (Education) • Ass. Cur.: *Margery King* •
Fine Arts Museum – 1994
Art, visual arts ... 51938

Art Institute of Pittsburgh, 420 Blvd of the Allies, Pittsburgh, PA 15233 • T: +1 412 4712473 •
Public Gallery ... 51939

Associated Artists of Pittsburgh Gallery, 937 Liberty Av, Pittsburgh, PA 15222 • T: +1 412 2632710 • F: +1 412 4711765 •
Pres.: *Rich Brown* • Exec. Dir.: *Frances Frederick* •
Fine Arts Museum – 1910
Art ... 51940

Car and Carriage Museum, 7227 Reynolds St, Pittsburgh, PA 15208-2923 • T: +1 412 3710600 • F: +1 412 7319415 • wsheerer@frickart.org • www.frickart.org •
Dir.: *William B. Bodine jr.* •
Science&Tech Museum – 1990 ... 51941

Carnegie Museum of Art, 4400 Forbes Av, Pittsburgh, PA 15213-4080 • T: +1 412 6223131 • F: +1 412 6223112 • www.cargiemuseum.org •
Dir.: *Richard Armstrong* • Ass. Dir.: *Maureen Rolla* • Chief Cur.: *Sarah Nichols* (Decorative Arts) • Cur.: *Louise Lippincott* (Fine Arts) • *Laura Hoptman* (Contemporary Art) • *Joseph Rosa* (Heinz Architectural Center) • *William D. Judson* (Film and Video) • *Marilyn M. Russell* (Education) • Chief Cons.: *Ellen Baxter* •
Fine Arts Museum – 1895 ... 51942

Carnegie Museum of Natural History, 4400 Forbes Av, Pittsburgh, PA 15213-4080 • T: +1 412 6223131 • F: +1 412 6228837 • www.carnegiemuseums.org/cmnh •
Dir.: *Billie R. DeWalt* • *Sylvia M. Keller* • Cur.: *Dr. Mary R. Dawson* (Vertebrate Paleontology) • *Dr. Bradley C. Livezey* (Birds) • *Dr. Frederick H. Utech* (Botany) • *Dr. John R. Wible* (Mammals) • *Dr. James B. Richardson* (Anthropology) • Ass. Cur.: *Dr. John J. Wiens* (Amphibians and Reptiles) •
Ethnology Museum / Natural History Museum – 1896
Natural hist, anthropology ... 51943

Carnegie Science Center, 1 Allegheny Av, Pittsburgh, PA 15212-5850 • T: +1 412 2373400 • F: +1 412 2373309 • www.carnegiesciencecenter.org •
Dir.: *Henry Buhl* •
Science&Tech Museum – 1991
Science and technology ... 51944

Center for American Music, Stephen Foster Memorial, University of Pittsburgh, 4301 Forbes Av, Pittsburgh, PA 15260 • T: +1 412 6244100 • F: +1 412 6247447 • amerimus@pitt.edu • www.library.pitt.edu/libraries/cam/cam.html •
Dir.: *Deane L. Root* •
Music Museum – 1937
American music hist, life and works of Stephen Collins Foster – Research library ... 51945

Chatham College Art Gallery, Woodland Rd, Pittsburgh, PA 15232 • T: +1 412 3651100 • F: +1 412 3651505 • www.chatham.edu •
Dir.: *Michael Pestel* •
Fine Arts Museum – 1960 ... 51946

Fort Pitt Museum, Point State Park, Pittsburgh, PA 15222, mail addr: 101 Commonwealth Pl, Pittsburgh, PA 15222 • T: +1 412 2819284 • F: +1 412 2811417 • www.fortpittmuseum.com •
Dir.: *Chuck Smith* •
Military Museum – 1964
Military and frontier objects, tools, muskets, cannon, furniture, maps, documents ... 51947

Frame Gallery, c/o Carnegie Mellon University, School of Fine Arts, Cnr Forbes Av and Margaret Morrison St, Pittsburgh, PA 15213-3890 • T: +1 412 2682409 • www.art.cfa.cmu.edu •
Dir.: *David Johnson* • *Kerry Schneider* •
Public Gallery / University Museum – 1969 ... 51948

The Frick Art Museum-Henry Clay Frick Estate, 7227 Reynolds St, Pittsburgh, PA 15208 • T: +1 412 3710600 • F: +1 412 3716140 • info@frickart.org • www.frickart.org •
Dir.: *William B. Bodine jr.* •
Fine Arts Museum – 1970
European paintings, sculpture and decorative art from 12th-18th c ... 51949

Hunt Institute for Botanical Documentation, c/o Carnegie Mellon University, 5000 Forbes Av, Pittsburgh, PA 15213-3890 • T: +1 412 2682434 • F: +1 412 2685677 • huntbot.andrew.cmu.edu •
Dir.: *Robert W. Kiger* • Cur.: *James J. White* •
Fine Arts Museum / University Museum – 1961
Botanical art and illustration ... 51950

John P. Barclay Memorial Gallery, c/o Art Institute of Pittsburgh, 420 Blvd of the Allies, Pittsburgh, PA 15219 • T: +1 412 2636600 •
Dir.: *Nancy Ruttner* •
Fine Arts Museum – 1921
Local artists, exhibits ... 51951

Manchester Craftsmen's Guild, 1815 Metropolitan St, Pittsburgh, PA 15233 • T: +1 412 3221773 • F: +1 412 3212120 •
C.E.O.: *William Strickland jr.* •
Decorative Arts Museum – 1968 ... 51952

Mattress Factory Museum, 500 Sampsonia Way, Pittsburgh, PA 15212 • T: +1 412 2313169 • F: +1 412 3222231 • info@mattress.org • www.mattress.org •
Dir.: *Barbara Luderowski* • Cur.: *Michael Olijnyk* •
Fine Arts Museum – 1977
Site-specific installations ... 51953

Mellon Financial Corporation's Collection, 1 Mellon Financial Ctr, Pittsburgh, PA 15258 • T: +1 412 2344775 • F: +1 412 2340831 •
Dir.: *Brian J. Lang* •
Fine Arts Museum
British drawings and paintings, American historical prints, contemporary works on paper, textiles, photography ... 51954

Photo Antiquities, Museum of Photographic History, 531 E Ohio St, Pittsburgh, PA 15212 • T: +1 412 2317881 • F: +1 412 2311217 • info-pa@photoantiquities.com • www.photoantiquities.com •
Dir.: *Frank X. Watters* •
Fine Arts Museum / Science&Tech Museum
Images, cameras and accessories that capture the earliest days of photography, Daguerreotype c. 1839 ... 51955

Pittsburgh Center for the Arts, 6300 Fifth Av, Pittsburgh, PA 15232 • T: +1 412 3610873, 3610455 • F: +1 412 3618338 • general@pittsburgharts.org • www.pittsburgharts.org •
Dir.: *Lourdes Karas* • Cur.: *Vicky A. Clark* •
Fine Arts Museum – 1945
Local art ... 51956

Pittsburgh Children's Museum, 10 Childrens Way, Pittsburgh, PA 15212 • T: +1 412 3225059 • F: +1 412 3224932 • info@pittsburghkids.org • www.pittsburghkids.org •
Pres.: *Judy Horgan* • Exec. Dir.: *Jane Werner* • Dir.: *Chris Siefert* (Project Mgr.) • *Penny Lodge* (Exhibits) •
Special Museum – 1980 ... 51957

Regina Gouger Miller Gallery, Carnegie Mellon University, 5000 Forbes Av, Pittsburgh, PA 15213-3890 • T: +1 412 2683877 • miller-gallery@andrew.cmu.edu • www.cmu.edu/millergallery •
Dir.: *Jenny Strayer* •
Public Gallery / University Museum ... 51958

Senator John Heinz Pittsburgh Regional History Center, Historical Society of Western Pennsylvania, 1212 Smallman St, Pittsburgh, PA 15222 • T: +1 412 4546000 • F: +1 412 4546031 • hswp@hswp.org • www.pghhistory.org •
Local Museum – 1879
Urban and regional hist, 20th c immigrant and ethnic experience, rural life ... 51959

Silver Eye-Center for Photography, 1015 E Carson St, Pittsburgh, PA 15203 • T: +1 412 4311810 • F: +1 412 4315777 • info@silvereye.org • www.silvereye.org •
Dir.: *Sam Berkovitz* •
Fine Arts Museum – 1979
Photography as an art ... 51960

Society for Contemporary Craft Museum, 2100 Smallman St, Pittsburgh, PA 15222 • T: +1 412 2617003 • F: +1 412 2611941 • info@contemporarycraft.org • www.contemporarycraft.org •
Dir.: *Janet L. McCall* • Ass. Dir.: *Kate Lydon* •
Fine Arts Museum / Decorative Arts Museum – 1972
Contemporary crafts in ceramics, glass, wood, metal and fiber ... 51961

Space 101 Gallery, Brew House Association, 2100 Mary St, Pittsburgh, PA 15203 • T: +1 412 3817767 • F: +1 412 3901977 • getinfo@brewhouse.org • www.brewhouse.org •
Public Gallery ... 51962

University Art Gallery, University of Pittsburgh, 104 Frick Fine Arts Bldg, Pittsburgh, PA 15260 • T: +1 412 6482400 • F: +1 412 6482792 • jpiller@pitt.edu •
Head: *Prof. David G. Wilkins* • Programming Dir.: *Josienne N. Piller* •
University Museum / Fine Arts Museum – 1966
Art ... 51963

Wood Street Galleries, 601 Wood St, Pittsburgh, PA 15222 • T: +1 412 4715605 • F: +1 412 2323262 •
Cur.: *Murray Horn* •
Public Gallery
Art ... 51964

Pittsfield MA

Arrowhead Museum, Berkshire County Historical Society, 780 Holmes Rd, Pittsfield, MA 01201 • T: +1 413 4421793 • F: +1 413 4431449 • info@berkshirehistory.org • www.berkshirehistory.org •
Cur.: *Catherine Reynolds* •
Local Museum – 1962
Regional history, Pittsfield home of author Herman Melville ... 51965

The Berkshire Artisans, Lichtenstein Center for the Arts, 28 Renne Av, Pittsfield, MA 01201 • T: +1 413 4999348 • F: +1 413 4428043 • berkart@taconic.net • www.berkshireweb.com/artisans •
Pres./Dir.: *Daniel M. O'Connell* •
Fine Arts Museum / Public Gallery – 1975
Public murals, current work of local and regional artists and craftspeople ... 51966

Berkshire Museum, 39 South St, Pittsfield, MA 01201 • T: +1 413 4437171 ext 10 • F: +1 413 4432135 • pr@berkshiremuseum.org • www.berkshiremuseum.org •
Dir.: *Michael Willson* • Cur.: *Thomas Smith* (Natural Sciences) •
Local Museum – 1903
Art, natural science, history – aquarium, library, auditorium ... 51967

Hancock Shaker Village, Rte 20, W Housatonic St, Pittsfield, MA 01202 • T: +1 413 4430188 • F: +1 413 4479357 • info@hancockshakervillage.org • www.hancockshakervillage.org •
Dir.: *Lawrence J. Yerdon* •
Historic Site / Open Air Museum – 1960
22 bldgs (dating back to 1790) – library, archive 51968

Pittsford NY

Historic Pittsford, 18 Monroe Av, Pittsford, NY 14534 • T: +1 716 3812941 •
C.E.O.: *Mary Menzie* •
Local Museum – 1966
Hist house, local architecture – Library ... 51969

Pittsford VT

New England Maple Museum, US Rte 7, Pittsford, VT 05763 • T: +1 802 4839414 • F: +1 802 7751650 • info@maplemuseum.com • www.maplemuseum.com •
Dir.: *Dona A. Olson* •
Special Museum – 1977
Maple sugaring, wooden sap buckets, antique sugar tubs, paintings by Vermont artitsts ... 51970

Pittsford Historical Society Museum, Main St, Ste 3399, Pittsford, VT 05763 • T: +1 802 4836040, 4836623 • www.pittsford-historical.org •
Pres.: *Ann Pelkey* •
Local Museum – 1980
Hist of the area from 17th c to present ... 51971

Placerville CA

El Dorado County Historical Museum, 104 Placerville Dr, Fairgrounds, Placerville, CA 95667 • T: +1 530 6215865 • museum@co.el-dorado.ca.us • www.co.el-dorado.ca.us/museum •
Local Museum – 1974
Local hist, Indians artifacts, mining, early transport – archives ... 51972

Plainfield NJ

Drake House Museum, 602 W Front St, Plainfield, NJ 07060 • T: +1 908 7555831 • F: +1 908 7550132 • cmerritt@thedrakehouse.org • www.drakehouse.org •
Pres.: *JoAnn Ball* •
Local Museum – 1921
Local hist, coll of American hist ... 51973

Plains TX

Tsa Mo Ga Memorial Museum, 1109 Av H, Plains, TX 79355 • T: +1 806 4568855 •
Dir.: *P.W. Saint Romain* •
Local Museum – 1959
Pioneer items, Civil War articles, two room bonus shack, early cowboy items depicting loc hist ... 51974

Plainview TX

Museum of the Llano Estacado, Wayland University, Plainview, TX 79072 • T: +1 806 2964735 • goffee@wbu.edu •
Dir.: *Eddie Guffee* • Cur.: *Patti Guffee* •
University Museum / Local Museum – 1976
Geological, archaeological development of the region, natural hist & hist ... 51975

Plano TX

Heritage Farmstead Museum, 1900 W 15 St, Plano, TX 75075 • T: +1 972 8810140 • F: +1 972 4226481 • museum@heritage.farmstead.museum • www.heritage.farmstead.museum •
C.E.O.: *Ted H. Peters* • Cur.: *Lolisa Moores-Franklin* •
Historical Museum – 1986
1891 Victorian farmhouse located on a 4-acre historic side ... 51976

Plant City FL

1914 Plant City High School Community Exhibition, 605 N Collins St, Plant City, FL 33566 • T: +1 813 7579226 •
Local Museum – 1974
Local pioneer family artifacts ... 51977

Plantation FL

Plantation Historical Museum, 511 N Fig Tree Ln, Plantation, FL 33317 • T: +1 954 7972722 • F: +1 954 7972717 • museum511@aol.com • www.plantation.org •
Dir./ Cur.: *Shirley Schuler* •
Local Museum – 1975
20th c artifacts, fire fighting equip ... 51978

Platteville CO

Fort Vasquez Museum, Colorado Historical Society Regional Museum, 13412 U.S. Hwy 85, Platteville, CO 80651 • T: +1 970 7852832 • F: +1 970 7859193 • susan.hoskinson@chs.state.co.us • www.coloradohistory.org •
Dir.: *Susan Hoskinson* •
Historic Site – 1958
Fur trade and Native American, reconstructed adobe fort (1835 trading post) ... 51979

Platteville WI

The Mining Museum, 405 E Main St, Platteville, WI 53818 • T: +1 608 3483301 • F: +1 608 3484640 • museums@platteville.org • www.platteville.org •
Dir.: *Stephen J. Kleefisch* • Cur.: *Stephanie Saager-Bourret* •
Science&Tech Museum – 1965
Mining, geology ... 51980

Rollo Jamison Museum, 405 E Main St, Platteville, WI 53818 • T: +1 608 3483301 • F: +1 608 3484640 • museums@platteville.org • www.platteville.org •
Dir.: *Stephen J. Kleefisch* • Cur.: *Stephanie Saager-Bourret* •
Historical Museum – 1981
Local hist ... 51981

Plattsburgh NY

Clinton County Historical Museum, 48 Court St, Plattsburgh, NY 12901 • T: +1 518 5610340 • F: +1 518 5610340 • clintoncohist@westelcom.com • www.clintoncountyhistorical.org •
Dir./Cur.: *John Tomkins* •
Local Museum – 1945
Local hist, paintings, maps, firearms, glass, ceramics, textiles, photos, furniture ... 51982

Kent-Delord House Museum, 17 Cumberland Av, Plattsburgh, NY 12901 • T: +1 518 5611035 • F: +1 518 5611035 • director@primelink1.net •
Dir.: *Gary Worthington* •
Local Museum – 1924
Period furnishings, paintings, 18th/19th c dec arts, manuscripts, photographs, textiles ... 51983

Plattsburgh Art Museum, State University of New York, 101 Broad St, Plattsburgh, NY 12901 • T: +1 518 5642474 • F: +1 518 5642473 • edward.brohel@plattsburgh.edu • www.plattsburgh.edu/museum •
Dir.: *Edward R. Brohel* •
Fine Arts Museum / University Museum – 1952
Rockwell Kent Coll, Nina Winkel coll of sculptures, Millet Asian Coll, Ackerman Coll of Modern Art, 19th c art ... 51984

Plattsmouth NE

Cass County Historical Society Museum, 646 Main St, Plattsmouth, NE 68048 • T: +1 402 2964770 • ccohsm@allnet.net • www.nebraskamuseums.org/casscountymuseum.htm •

U

Pres.: *George Miller* • Cur.: *Margo Prentiss* •
Local Museum – 1936
Minerals, fossils, tools, decorative arts, works of art 51985

Pleasanton CA

Museum on Main Street, 603 Main St, Pleasanton, CA 94566-6603 • T: +1 925 4622766 • info@museumonmain.org • www.museumonmain.org •
Local Museum – 1972
Amador-Livermore valley history, arcaeology and nature – library 51986

Pleasanton KS

Linn County Museum, Dunlap Park, Pleasanton, KS 66075 • T: +1 913 3528739 • F: +1 913 3528739 •
Pres.: *Ola May Earnest* •
Local Museum – 1973
County history, geneaology library 51987

Marais Des Cygnes Memorial Historic Site, Rte 2, Pleasanton, KS 66075 • T: +1 913 3528890 • minecreek@ckt.net • www.minecreek.org •
Cur.: *Russ Corder* •
Historical Museum
Home of Charles C. Hadsall, used as a fort by John Brown, nr massacre site 51988

Pleasanton TX

Longhorn Museum, 1959 Hwy 97 E, Pleasanton, TX 78064 • T: +1 830 5696313 •
Dir.: *Donna McCown* •
Historical Museum – 1976 51989

Pleasantville NY

Reader's Digest Art Gallery, Reader's Digest Rd, Pleasantville, NY 10570-7000 • T: +1 914 2445492 • F: +1 914 2445006 • www.rd.com •
Cur.: *Marianne Brunson Frisch* • Ass. Cur.: *Jill DeVonyar-Zansky* •
Fine Arts Museum 51990

Pleasantville PA

Historic Pithole City, Drake Well Museum, R.D. 1, Pleasantville, PA 16341 • T: +1 814 5897912, 8272797 • F: +1 814 8274888 • bzolli@state.pa.us • www.drakewell.org •
Dir.: *Barbara T. Zolli* •
Open Air Museum – 1975
Oil wells, model of city in 1865 51991

Plymouth IN

Marshall County Historical Museum, 123 N Michigan St, Plymouth, IN 46563 • T: +1 219 9362306 • F: +1 219 9369306 • mchistory@mchistoricalsociety.org • www.mchistoricalsociety.org •
Pres.: *Ronald Liechty* • Dir.: *Linda Rippy* •
Local Museum – 1957
Local history 51992

Plymouth MA

Howland House, 33 Sandwich St, Plymouth, MA 02360 • T: +1 508 7469590 • F: +1 508 8665056 • shawhardin@aol.com •
Decorative Arts Museum – 1897
Period furnishings, textiles, archaeological digs items 51993

Mayflower Society Museum, 4 Winslow St, Plymouth, MA 02360 • T: +1 508 7462590 • judybythebay@ •
Cur.: *Judith A. MacDonald* •
Local Museum – 1897
Local hist – archives, genealogical library 51994

Pilgrim Hall Museum, 75 Court St, Plymouth, MA 02360 • T: +1 508 7461620 • F: +1 508 7474228 • pegbaker@pilgrimhall.org • www.pilgrimhall.org •
C.E.O.: *Peggy M. Baker* • Cur.: *Jane L. Port* •
Local Museum – 1820
Local history, decorative and fine arts, archaeology – library 51995

Plymoth Plantation Museum, 137 Warren Av, Plymouth, MA 02362 • T: +1 508 7461622 ext 8601 • F: +1 508 7463407 • jmonac@plimoth.org • www.plimoth.org •
Exec. Dir.: *Nancy Brennan* •
Local Museum – 1947
Outdoor living history, archaeology, English & Native American artifacts, arms, Mayflower ship replica – library, theater 51996

Plymouth Antiquarian Society Museum, 126 Water St, Plymouth, MA 02360 • T: +1 508 7460012 • F: +1 508 7467908 • pasm@mindspring.com •
Dir.: *Donna D. Curtin* •
Local Museum – 1919
Local history, furnishings, dolls, quilts, china, costumes – garden 51997

Richard Sparrow House, 42 Summer St, Plymouth, MA 02360 • T: +1 508 7471240 • F: +1 508 7469521 • info@sparrowhouse.com • www.sparrowhouse.com •
Dir.: *Lois Atherton* •
Folklore Museum / Decorative Arts Museum – 1961
Local crafts, 17th c decorative arts 51998

Plymouth MI

Plymouth Historical Museum, 155 S Main St, Plymouth, MI 48170 • T: +1 734 4557797 • F: +1 734 4558571 • director@plymouthhistory.org • www.plymouthhistory.org •
Dir.: *Beth A. Stewart* •
Historical Museum – 1962
Local history, Petz Abraham Lincoln coll, prints, altar car, air rifles – library 51999

Plymouth NH

Karl Drerup Fine Arts Gallery, Dept of Fine Arts, Plymouth State College, 17 High St, Plymouth, NH 03264 • T: +1 603 5352614 • F: +1 603 5352938 • camidon@mail.plymouth.edu • www.plymouth.edu/psc/gallery/ •
Dir.: *Catherine Amidon* •
University Museum / Fine Arts Museum – 1969 52000

Plymouth WI

Bradley Gallery of Art, Lakeland College, W 3718 South Dr, Plymouth, WI 53073 • T: +1 920 5652111 • F: +1 920 5651206 • www.lakeland.edu •
Dir.: *Denise Presnell-Weidner* • *William Weidner* •
Public Gallery – 1988
Paintings and prints 52001

Gallery 110 North, Plymouth Arts Foundation, 520 E Mill St, Plymouth, WI 53073-0253, mail addr: POB 253, Plymouth, WI 53073-0253 • T: +1 920 8928409 • F: +1 920 8938473 • paf@excel.net • www.plymoutharts.org •
Dir.: *Donna Hahn* •
Public Gallery
Temporary exhibitions of art 52002

John G. Voigt House, W 5639 Anokijig Ln, Plymouth, WI 53073 • T: +1 920 8930782 • F: +1 920 8930873 •
Dir.: *Jim Scherer* •
Local Museum – 1850
Historic house, 1850 Log cabin with artifacts 52003

Plymouth Notch VT

Calvin Coolidge State Historic Site, Rte 100-A, Plymouth Notch, VT 05056 • T: +1 802 6723773 • F: +1 802 8283206 • john.dumville@state.vt.us • www.historicvermont.org •
Local Museum – 1951
Early 20th c village, Calvin Coolidge birthplace and homestead 52004

Pocatello ID

Bannock County Historical Museum, 3000 Alvord Loop, Pocatello, ID 83201 • T: +1 208 2330434 •
Dir./Cur.: *Mary Lien* •
Local Museum – 1963
Wrensted, Peake and Larsen colls, railroad, farm equip 52005

Idaho Museum of Natural History, 5th and Dillon, Pocatello, ID 83209 • T: +1 208 2823168 • F: +1 208 2825893 • imnh@isu.edu • imnh.isu.edu •
Dir.: *Linda T. Deck* • Cur.: *Dr. William A. Akersten* •
University Museum / Natural History Museum – 1934
Natural history 52006

John B. Davis Gallery of Fine Art, POB 8004, Pocatello, ID 83209 • T: +1 208 2362361 • F: +1 208 2824791 • dialgail@isu.edu • www.isu.edu/departments/art •
Dir.: *Amy Jo Johnson* •
Fine Arts Museum / University Museum – 1956 52007

Point Lookout MO

Ralph Foster Museum, College of the Ozarks, 1 Cultural Court, Point Lookout, MO 65726 • T: +1 417 3346411 ext 3407 • F: +1 417 3352618 • museum@cofo.edu • www.rfostermuseum.com •
Dir.: *Annette J. Sain* • Cur.: *Jeanelle Ash* • Cur.: *Thomas A. Debo* • *Gary Ponder* •
University Museum / Local Museum – 1930
Local history, art, firearms, natural history 52008

Point Marion PA

Friendship Hill, 223 New Geneva Rd, Point Marion, PA 15474 • T: +1 724 7259190 • F: 7242 7251999 • lawren-dunn@npg.gov • www.nps.gov/frhi •
Cur.: *Lawren Dunn* •
Historic Site – 1978
Local hist 52009

Point Pleasant OH

Grant's Birthplace State Memorial, US 52 and State Rte 232, Point Pleasant, OH 45153 • T: +1 513 5534911 •
Man.: *Loretta Fuhrman* •
Local Museum
Memorabilia of Ulysses S. Grant 52010

Point Pleasant WV

Tu-Endie-Wei State Park, 1 Main St, Point Pleasant, WV 25550 • T: +1 304 6750869 • F: +1 304 6746162 • www.tu-endie-weistatepark.-com •
Head: *Stephen Jones* •
Local Museum / Historic Site – 1901
Historic house and site, where the first battle of the Revolution was fought 52011

West Virginia State Farm Museum, Rte 1, Box 479, Point Pleasant, WV 25550 • T: +1 304 6755737 • F: +1 304 6755430 • wvsfm@ezwv.com • www.ezwv.com/~wvsfm •
Dir.: *Lloyd Akers* •
Agriculture Museum – 1974
Agriculture – 31 bldgs of agriculture museum and historic farm and village 52012

Point Richmond CA

Golden State Model Railroad Museum, 900 A Dornan Dr, Point Richmond, CA 94801, mail addr: POB 71244, Point Richmond, CA 94807-1244 • T: +1 510 2344884 • info@gsmrm.org • www.gsmrm.org •
C.E.O.: *John Morrison* •
Science&Tech Museum – 1985
Model railroads in 0, H0 and N scale with equipment, historic prototype railroad artifacts, railroad hist of California – library 52013

Polk City FL

Water Ski Hall of Fame, American Water Ski Educational Foundation, 1251 Holy Cow Rd, Polk City, FL 33868-8200 • T: +1 863 3242472 • F: +1 863 3243996 • awsefhalloffame@cs.com • usawaterski.org •
Pres.: *Dr. J.D. Morgan* • Exec. Dir.: *Carole Lowe* •
Special Museum – 1968
Show ski costumes, photos, water ski ropes and handles 52014

Polson MT

Miracle of America Museum, 36094 Memory Lane, Polson, MT 59860 • T: +1 406 8836804 • museum@cyberport.net, info@miracleofamericamuseum.org • www.cyberport.net/museum, www.miracleofamericamuseum.org •
Dir.: *Mel Adams* • *Cathleen Wilde* • C.E.O.: *W. Gilbert Mangels* •
Historical Museum / Science&Tech Museum – 1985
Ethnology, history, farm machinery, steam engine, fishing, cars, tractors, military vehicles, tranportation, decorative art 52015

Pomeroy OH

Meigs County Museum, 144 Butternut Av, Pomeroy, OH 45769 • T: +1 740 9923810, 9922264 • F: +1 740 9923810 • meigscountyhistoricalsociety@dragonlbs.com • www.meigscountyhistoricalsociety.com •
Pres.: *Margaret Parker* •
Local Museum – 1960
Hist of Meigs County 52016

Ponca City OK

Marland's Grand Home, 1000 E Grand, Ponca City, OK 74601 • T: +1 580 7670427 • www.marlandgrandhome.com •
Dir.: *Kathy Adams* •
Historical Museum / Historic Site – 1916
Hist and culture of the Ponca, Kaw, Otoe, Osage and Tonkawa tribes, ranch and Wild West Show memorabilia 52017

Pioneer Woman Statue and Museum, Oklahoma Historical Society, 701 Monument, Ponca City, OK 74604 • T: +1 580 7656108 • F: +1 580 7622498 • piown@ok-history.mus.ok.us • www.ok-history.mus.ok.us •
Dir.: *Valerie Haynes* •
Historical Museum – 1958
Relics and mementoes of Oklahoma's Pioneer Women, bronze statues, gawns coll 52018

Ponce Inlet FL

Ponce DeLeon Inlet Lighthouse, 4931 S Peninsula Dr, Ponce Inlet, FL 32127 • T: +1 386 7611821 • F: +1 386 7611321 • lighthouse@ponceinlet.org • www.ponceinlet.org •
Dir.: *Ann Caneer* (History) •
Historical Museum – 1972
Local hist and lighthouse keepers lives 52019

Pontiac MI

Creative Arts Center, 47 Williams St, Pontiac, MI 48341 • T: +1 248 3337849 • F: +1 248 3337841 • createpont@aol.com •
Dir.: *Carol Paster* • Cust.: *Leon Medlock* •
Fine Arts Museum – 1964
Regional, national and international art exhibits 52020

Museum of New Art, 7 N Saginaw St, Pontiac, MI 48342, mail addr: 327 W Second St, Rochester, MI 48307 • T: +1 248 2107560 • detroitmona@aol.com • www.detroitmona.com •
Dir.: *Jef Bourgeau* •
Fine Arts Museum
Experimental paintings and sculpture 52021

Pine Grove Historic Museum, 405 Cesar E. Chavez Av, Pontiac, MI 48342 • T: +1 248 3386732 • F: +1 248 3386731 • ocphs@wwnet.net •
Pres.: *Michael Willis* •
Historical Museum – 1874
Pine Grove, Gov. Moses Wisner house, tools and farm impl, vintage clothing coll – library, archives 52022

Pooler GA

Mighty Eighth Air Force Museum, 175 Bourne Av, Pooler, GA 31402 • T: +1 912 7488888 • F: +1 912 7480209 • www.mightyeighth.org •
Pres./Dir.: *C. J. Roberts* • Cur.: *Shasta Ireland* •
Military Museum – 1996
8th Air force hist (WW II to present), military and veterans artifacts, vehicles and aircraft – library 52023

Poplar Bluff MO

Margaret Harwell Art Museum, 421 N Main St, Poplar Bluff, MO 63901 • T: +1 573 6868002 • F: +1 573 6868017 • tina@mham.org • www.mham.org •
C.E.O.: *Tina M. Magill* •
Fine Arts Museum – 1981
Contemporary art media 52024

Port Allen LA

West Baton Rouge Museum, 845 N Jefferson Av, Port Allen, LA 70767 • T: +1 225 3362422 • F: +1 225 3362448 • contact_us@wbrmuseum.org • www.westbatonrougemuseum.com •
Dir.: *Caroline Kennedy* • Cur.: *Neal Roth* •
Historical Museum – 1968
Regional history 52025

Port Angeles WA

The Museum of Clallam County Historical Society, 1st St & Oak St, Federal Bldg, Port Angeles, WA 98362 • T: +1 360 4522662 • F: +1 360 4522662 • artifact@olypen.com •
Exec. Dir.: *Kathy Monds* •
Local Museum – 1948
Local hist, period furniture 52026

Olympic National Park Visitor Center, 3002 Mt. Angeles Rd, Port Angeles, WA 98362 • T: +1 360 5653130 • F: +1 360 5653147 • www.nps.gov.olymp •
Natural History Museum – 1957
Anthropology, botany, zoology, geology 52027

Port Angeles Fine Arts Center, 1203 E Lauridsen Blvd, Port Angeles, WA 98362 • T: +1 360 4573532, 4174590 • pafac@olypen.com • www.pafac.org •
Pres.: *Darlene Clemens* • Dir.: *Jake Seniuk* •
Fine Arts Museum / Open Air Museum – 1986
Art, coll of sculptures, contemporary art of the Northwest – art park 52028

Port Arthur TX

Museum of the Gulf Coast, 700 Procter St, Port Arthur, TX 77640 • T: +1 409 9827000 • F: +1 409 9829614 • shannon.hansen@lamarpa.edu • www.museum.lamarpa.edu •
Pres.: *Dr. Sam Monroe* • Dir.: *Shannon Hansen* •
Local Museum – 1964
Flora & fauna , paleo-Indian artifacts from McFadden Beach, pop culture memorabilia 52029

Port Austin MI

Huron City Museums, 7995 Pioneer Rd, Port Austin, MI 48467 • T: +1 989 4284123 • F: +1 989 4284123 • huroncity@centurytel.net • www.huroncitymuseums.org •
Local Museum – 1950
Local hist – library 52030

Port Clinton OH

Ottawa County Historical Museum, 126 W Third St, Port Clinton, OH 43452 • T: +1 419 7322237 • ochm@thirdplanet.net •
Cur.: *Vicki Ashton* •
Historical Museum – 1932
Indian relics, early household items, hand machine equip, clothing, guns, china, cameras 52031

Port Gamble WA

Port Gamble Historic Museum, 3 Rainier Av, Port Gamble, WA 98364 • T: +1 360 2978074 • F: +1 360 2977455 • ssmith@orminc.com • www.portgamble.com •
C.E.O.: *Dave Nunes* • Cur.: *Shana Smith* •
Historic Site – 1976
Local hist – several bldgs from 19th c 52032

U

Port Gibson MS

Grand Gulf Military State Park Museum, 12006 Grand Gulf Rd, Port Gibson, MS 39150 • T: +1 601 4375911 • F: +1 601 4372929 • park@grandgulf.state.ms.us • www.grandgulfpark.state.ms.us •
Chm.: *Dr. David Headley* •
Military Museum – 1958
Archaeology, Civil War relics, Indian arrowheads, period buggies & coaches, rifle pits 52033

Port Hueneme CA

Seabee Museum, US Navy Civil Engineer Corps, Bldg 99, NCBC, 1000 23rd Av, Port Hueneme, CA 93043-4301 • T: +1 805 9825167 • lara.godbille@nevy.mil • www.seabeehf.org •
Dir.: *Lara Godbille* •
Military Museum – 1947
Military hist, weapons, uniforms, culture and art, flags – archives 52034

Port Huron MI

Port Huron Museum, 1115 Sixth St, Port Huron, MI 48060 • T: +1 810 9820891 ext 13 • F: +1 810 9820053 • info@phmuseum.org • www.phmuseum.org •
Dir.: *Stephen R. Williams* • Cur.: *T.J. Gaffney* (Colls) • Sc. Staff: *Gloria Justice* (Education) •
Local Museum – 1967
Local hist, marine, natural hist, surgeon's instr, lumbering pictures & tools, prehistoric Indian artifacts 52035

Port Jefferson NY

Greater Port Jefferson Museum, Historical Society of Greater Port Jefferson, 115 Prospect St, Port Jefferson, NY 11777 • T: +1 631 4732665 • info@portjeffhistorical.org • www.portjeffhistorical.org •
Pres.: *Nick Acampora* •
Local Museum – 1967
Maritime coll, sailmakers' tools and loft, shipbuilding artifacts and tools, hist of Port Jefferson shipping industry, diorama of 1900 Port Jefferson Harbor 52036

Port Jervis NY

Minisink Valley Historical Society Museum, 125-133 W Main St, Port Jervis, NY 12771 • T: +1 914 8562375 • F: +1 914 8561049 • mvhs1889@magiccarpet.com • www.minisink.org •
Dir.: *Peter Osborne* •
Local Museum – 1889
Genealogical data, local hist, family files, photos 52037

Port Lavaca TX

Calhoun County Museum, 301 S Ann St, Port Lavaca, TX 77979 • T: +1 361 5534689 • F: +1 361 5534689 • director@calhouncauntymuseum.org • www.calhouncauntymuseum.org •
Cur.: *George Ann Cormier* •
Local Museum – 1964
Lens from the Matagorda Island Lighthouse 52038

Port Orange FL

Hungarian Folk-Art Museum, 546 Ruth St, Port Orange, FL 32127 • T: +1 904 7674292 • F: +1 904 7886785 • mhorvath@aol.com •
Dir.: *Margaret Pazmandy Horvath* •
Folklore Museum – 1979
Hungarian Folklore and folk art objects, ceramics, photos, videos 52039

Port Saint Joe FL

Constitution Convention Museum, 200 Allen Memorial Way, Port Saint Joe, FL 32456 • T: +1 850 2298029 • F: +1 850 2298029 • anne.harvey@dep.state.fl.us • www.floridastateparks.org/constitution-convention •
Local Museum – 1955
History 52040

Port Sanilac MI

Sanilac County Historical Museum, 228 S Ridge St, Port Sanilac, MI 48469 • T: +1 810 6229946 • F: +1 810 6229946 • museum@greatlakes.net •
Pres.: *Art Schlichting* •
Historical Museum – 1964
Local medical, marine, military & village hist, agriculture, Indians, firearms, glassware, dairy ind – library, archives 52041

Port Townsend WA

Centrum Arts and Creative Education, Fort Worden State Park, Port Townsend, WA 98368, mail addr: POB 1158, Port Townsend, WA 98368-0958 • T: +1 360 3853102 • F: +1 360 3852470 •
Exec. Dir.: *Carol Shiffman* •
Fine Arts Museum – 1973
Prints 52042

Jefferson County Historical Society Museum, 540 Water St, Port Townsend, WA 98368 • T: +1 360 3851003 • F: +1 360 3851042 • jchsmuseum@olympus.net • www.jchsmuseum.org •
Pres.: *Kevin Harris* • Dir.: *William Tennent* •
Historical Museum – 1951
Northwest Coast history and prehistory, maritime heritage, Indian and Alaskan baskets & artifacts 52043

Northwind Arts Center, 2409 Jefferson St, Port Townsend, WA 98368 • T: +1 360 3791086 • www.northwindartscenter.org •
Public Gallery
Local and regional American art 52044

Port Townsend Marine Science Center, Fort Worden State Park, 532 Battery Way, Port Townsend, WA 98368-3431 • T: +1 360 3855582 • F: +1 360 3857248 • info@ptmsc.org • www.ptmsc.org •
Dir.: *Anne Murphy* •
Military Museum – 1983
Marine hist 52045

Puget Sound Coast Artillery Museum at Fort Worden, Bldg 201, Fort Worden State Park, Port Townsend, WA 98368 • T: +1 360 3850373 • F: +1 360 3852328 • artymus@olypen.com • www.fortworden.org •
Military Museum – 1976
Military hist, seacoast artillery, harbor defenses, milit rifles 52046

Rothschild House, Franklin St, Port Townsend, WA 98368 • T: +1 360 3444401 • F: +1 360 3857248 • kate.burke@parks.wa.gov • www.fortworden.org •
Decorative Arts Museum – 1959
House from 1868, 19th c furnishing, artifacts of the Rothschild family 52047

Port Washington NY

Polish American Museum, 16 Belleview Av, Port Washington, NY 11050 • T: +1 516 8836542 • F: +1 516 7671936 • polishmuseum@aol.com •
Cur.: *Wilma Wierbicki* (History) •
Fine Arts Museum / Decorative Arts Museum / Historical Museum – 1977
Woodcarving, glass, art, china and tapestry from Poland, – Polish language library, Archacki archives 52048

Salgo Trust for Education, 95 Middle Neck Rd, Port Washington, NY 11050 • T: +1 516 7673654 • F: +1 516 7677881 • thesalgotrust@aol.com •
Cur.: *Nicolos Salgo* •
Fine Arts Museum / Decorative Arts Museum – 1994
Hungarian paintings, sculpture and silver 15th-18th c, saddles, saddle rugs, French and European furniture and decorations, Chinese art 52049

Sands-Willets House, Cow Neck Peninsula Historical Society, 336 Port Washington Blvd, Port Washington, NY 11050 • T: +1 516 3659074 • curator@cowneck.org • www.cowneck.org •
Pres.: *Joan Kent* • Cur.: *Kathleen Mullen* •
Local Museum – 1962
18th to early 20th c decorative art, 18th c costumes, 19th c quilts, furniture 52050

Portage WI

Fort Winnebago Surgeons Quarters, W 8687 St, Hwy 33 E, Portage, WI 53901 • T: +1 608 7422949 •
C.E.O.: *Shirley B. Dalton* • Cur.: *Rachel Wynn* •
Local Museum – 1938
Hist of the Fort and relics 52051

The Historic Indian Agency House, Agency House Rd, Portage, WI 53901-3116 • T: +1 608 7426362 • www.nscda.org/museums/wisconsin.htm •
Dir.: *Marilynn Baxter* • Pres.: *Sally Reuter* •
Decorative Arts Museum / Local Museum / Historic Site – 1932
Historic house and site, American Empire dec. art, Indian artifacts 52052

Portales NM

Blackwater Draw Museum, Eastern New Mexico University, 7 miles north of ENMU campus on Hwy 70, Portales, NM 88130 • T: +1 505 5622202 • F: +1 505 5622291 • matthew.hillsman@enmu.edu • www.enmu.edu •
Dir.: *Dr. John Montgomery* • Cur.: *Matthew Hillsman* • Sc. Staff: *Joanne Dickenson* (Archaeology) •
Archaeology Museum / University Museum – 1969
1932 America's first multi-cultural, paleoindian archaeological site 52053

Gallery of the Department of Art, Eastern New Mexico University, Portales, NM 88130 • T: +1 505 5622510 • F: +1 505 5622362 • www.enmu.edu •
Dir.: *Jim Bryant* •
Fine Arts Museum / University Museum – 1935
Students artworks 52054

Miles Mineral Museum, Eastern New Mexico University, ENMU Campus, Roosevelt Hall, Portales, NM 88130 • T: +1 505 5622651 • F: +1 505 5622192 • jim.constantopoulos@enmu.edu • w3a.enmu.edu/academics/excellence/museums/miles-mineral •
Dir.: *Dr. Jim Constantopoulos* •
University Museum / Natural History Museum – 1969
Minerals, rocks, fossils 52055

Natural History Museum, Eastern New Mexico University, ENMU, Station 33, Portales, NM 88130 • T: +1 505 5622862 • F: +1 505 5622192 • marv.lutnesky@enmu.edu • www.enmu.edu/academics/excellence/museums/natural-history •
Dir./Cur.: *Dr. Marvin Lutnesky* (Fishes, Reptiles and Amphibians) • *Dr. Dann Brown* (Plants) • *Dr. Gary Pfaffenberger* (Invertebrates) •
University Museum / Natural History Museum – 1968
Specimens of mammals, reptiles, fish, birds and amphibians 52056

Roosevelt County Museum, Eastern New Mexico University, ENMU, Station 39, Portales, NM 88130 • T: +1 505 5621011 • F: +1 505 5622578 • mark.romeo@emmu.edu •
Cur.: *Mark Romero* •
Local Museum – 1934
Technology, early settlers and native American artifacts, ethnology, folklore, costumes, numismatics, ranching artifacts, Roosevelt county historical artifacts – archive 52057

Runnels Gallery, Eastern New Mexico University, Art Department, Station 19, Portales, NM 88130 • T: +1 505 5622778 • F: +1 505 5622362 • www.enmu.edu/~bryant/runnels •
Dir.: *Jim Bryant* • *Mic Muhlbauer* •
Fine Arts Museum / University Museum
Art, painting, sculpture, photography, 52058

Porter ME

Parsonsfield-Porter History House, 92 Main St, Porter, ME 04068 • T: +1 207 6254667, 6257019 •
Local Museum – 1946
Bridge, local Baptist group 52059

Porterville CA

Porterville Historical Museum, 257 N D St, Porterville, CA 93257 • T: +1 559 7842053 • www.chamber.porterville.com •
Dir.: *Bill Scruggs* •
Historical Museum – 1965
Indian artifacts, farming, china, glass 52060

Zalud House, 393 N Hockett St, Porterville, CA 93257 • T: +1 559 7827548 • www.chamber.porterville.com •
C.E.O.: *Gil Meachum* • Cur.: *Lynn Shell* •
Local Museum – 1976
Hist house with interior, local history 52061

Portland ME

Children's Museum of Maine, 142 Free St, Portland, ME 04101 • T: +1 207 8281234 • F: +1 207 8285726 • suzanne@kitetails.com • www.kitetails.com •
Pres.: *Jim Brady* • Exec. Dir.: *Suzanne Olson* •
Special Museum – 1976 52062

Danforth Gallery, 34 Danforth St, Portland, ME 04101 • T: +1 207 7756245 • F: +1 207 7756550 • info@maineartistspace.org • www.maineartistspace.org •
Dir.: *Helen Rivas* •
Public Gallery – 1986 52063

Institute of Contemporary Art, Maine College of Art, 522 Congress St, Portland, ME 04101 • T: +1 207 8795742 ext 229 • F: +1 207 7800816 • ica@meca.edu • www.meca.edu/ica •
Dir.: *Cindy M. Foley* •
Fine Arts Museum – 1983
Art 52064

Maine Historical Society Museum, 489 Congress St, Portland, ME 04101 • T: +1 207 7741822 • F: +1 207 7754301 • info@mainehistory.org • www.mainehistory.com •
Dir.: *Richard d' Abate* •
Local Museum – 1822
Manuscripts, architectural and engeneering drawings, photographs, prints, paintings 52065

The Museum of African Tribal Art, 122 Spring St, Portland, ME 04101 • T: +1 207 8717188 • F: +1 207 7731197 • africart@maine.rr.com • www.tribalartmuseum.com •
Dir.: *Oscar O. Mokeme* •
Fine Arts Museum – 1998
African art, Sub-Saharan African tribal art 52066

Portland Museum of Art, 7 Congress Sq, Portland, ME 04101 • T: +1 207 7756148 • F: +1 207 7737324 • pma@maine.rr.com • www.portlandmuseum.org •
Dir.: *Daniel E. O'Leary* • Cur.: *Jessica Nicoll* •
Fine Arts Museum – 1882
Maine's largest art museum, prints, drawings, sculpture, photos 52067

Tate House, 1270 Westbrook St, Portland, ME 04102 • T: +1 207 7749781 • F: +1 207 7746177 • info@tatehouse.org • www.tatehouse.org •
Dir.: *Kristen Crean* •
Historical Museum – 1931
18th c furnishings, architecture, decorative arts 52068

Victoria Mansion, 109 Danforth St, Portland, ME 04101 • T: +1 207 7724841 • F: +1 207 7726290 • victoria@maine.rr.com • www.victoriamansion.org •
Pres.: *Wilmont M. Schwind* • Dir.: *Robert Wolterstorff* •
Local Museum / Decorative Arts Museum – 1943
Furniture, paintings, artworks, carpets, stained glass, porcelain, textiles and architecture 52069

Wadsworth-Longfellow House, 485 Congress St, Portland, ME 04101 • T: +1 207 7740427 • F: +1 207 7754301 • post@mainehistory.org • www.mainehistory.org •
Dir.: *Richard d' Abate* • Cur.: *Joyce Butler* •
Special Museum – 1822
Boyhood home of poet Henry Wadsworth Longfellow 52070

Portland OR

American Advertising Museum, 211 NW Fifth Av, Portland, OR 97209 • T: +1 503 AAM0000 • F: +1 503 2742576 • admuseum@aol.com • www.admuseum.org •
Pres.: *Steve Karakas* •
Special Museum – 1984
Hist of advertising in America, newspaper and magazine adv, posters, outdoor, radio and TV advertising 52071

Blue Sky, Oregon Center for the Photographic Arts, 1231 NW Hoyt St, Portland, OR 97209 • T: +1 503 2250210 • bluesky@blueskygallery.org • www.blueskygallery.org •
Dir.: *Chris Rauschenberg* •
Public Gallery – 1975 52072

Buckley Center Gallery, University of Portland, 5000 N Willamette Blvd, Portland, OR 97203 • T: +1 503 2837258 •
Dir.: *Michael Miller* • *Mary Margaret Dundore* •
University Museum / Fine Arts Museum – 1977 52073

Doll Gardner Art Gallery, 8470 SW Oleson Rd, Portland, OR 97223 • T: +1 503 2463351, 2441379 • www.whuuf.org •
Fine Arts Museum – 1970
Paintings, wall sculptures by local artists 52074

Douglas F. Cooley Memorial Art Gallery, Reed College, 3203 SE Woodstock Blvd, Portland, OR 97202-8199 • T: +1 503 7711112 • F: +1 503 7886691 • silas.b.cook@reed.edu • www.reed.edu •
Dir.: *Silas B. Cook* •
Fine Arts Museum – 1989
Prints, drawings, paintings, photographs and sculptures 52075

Feldman Gallery and Project Space, Pacific Northwest College of Art, 1241 NW Johnson St, Portland, OR 97209 • T: +1 503 8218969 • F: +1 503 2263587 • ncurtis@pnca.edu • www.pnca.edu •
Dir.: *Nan Curtis* •
Public Gallery 52076

Forest Discovery Center, 4033 SW Canyon Rd, Portland, OR 97221 • T: +1 503 2281367 • F: +1 503 2283624 • mail@worldforestry.org • www.worldforestry.org •
Pres. & C.E.O.: *Gary Hartshorn* • Dir.: *Mark Reed* •
Natural History Museum – 1964
Old growth forests and tropical rainforests, petrified woods, forest industry memorials 52077

Hoffman Gallery, Oregon College of Art and Craft, 8245 SW Barnes Rd, Portland, OR 97225 • T: +1 503 2975544 •
Dir.: *Arthur DeBow* •
Public Gallery 52078

Littman Gallery, Portland State University, 724 SW Harrison St, Portland, OR 97201 • T: +1 503 7255656 • F: +1 503 7255680 • www.pdx.edu •
Fine Arts Museum / University Museum – 1980 52079

Museum of Contemporary Craft, 724 NW Davis St, Portland, OR 97209 • T: +1 503 2232654 • F: +1 503 2230190 • info@MuseumofContemporaryCraft.org • www.MuseumofContemporaryCraft.org •
Cur.: *Namita Gupta Wiggers* •
Fine Arts Museum / Decorative Arts Museum – 1937
Coll of works by craft artists (incl. Betty Feves, Jack Lenor Larsen, Sam Maloof, John Mason and Peter Voulkos) and artists creating the Pacific Northwest's distinct craft identity (incl. Ray Grimm, Leroy Setziol, Ken Shores and Romona Solberg) 52080

North View Gallery, Portland Community College, 12000 SW 49th Av, Portland, OR 97219 • T: +1 503 9774264 • F: +1 503 9774874 •
Dir.: *Marie Watt* •
Fine Arts Museum – 1970
Contemporary Northwestern art 52081

Oregon Historical Society, 1200 SW Park Av, Portland, OR 97205 • T: +1 503 2221741 • F: +1 503 2212035 • orhist@ohs.org • www.ohs.org •
Dir.: *John C. Piers* • *Marsha Matthews* (Artifact Collections) •
Association with Coll / Historical Museum – 1898
Anthropological, ethnographic, archaeological artifacts, maritime coll, textiles, furnishings, tools, advertising, political memorabilia 52082

Oregon Museum of Science and Industry, 1945 SE Water Av, Portland, OR 97214 • T: +1 503 7974000 • F: +1 503 7974500 • clee@omsi.edu • www.omsi.edu •
Pres.: *Nancy Stueber* • Cur.: *Judy Margles* •
Natural History Museum / Science&Tech Museum – 1944
Mineralogy, paleontology, zoology, natural hist, arcaheology, health, fine art, 52083

Oregon Sports Hall of Fame and Museum, 321 SW Salmon St, Portland, OR 97204 • T: +1 503 2277466 • F: +1 503 2276925 • orsports@teleport.com • www.oregonsportshall.org •
Pres.: *Chuck Richards* •
Special Museum – 1978
Athletic memorabilia, implements, posters, photos 52084

Pittock Mansion, 3229 NW Pittock Dr, Portland, OR 97210 • T: +1 503 8233623 • F: +1 503 8233619 • pkdanc@ci.portland.or.us • www.pittockmansion.com •
Pres.: *Christine Bostick* • Dir. & C.E.O.: *Daniel T. Crandall* •
Local Museum / Historic Site – 1965
18th-early 20th c decorative arts, coll of Thomas Hill paintings 52085

Portland Art Museum, 1219 SW Park Av, Portland, OR 97205 • T: +1 503 2262811 • F: +1 503 2264842 • info@pam.org • www.portlandartmuseum.org •
Dir.: *John E. Buchanan jr.* • Cur.: *Bruce Guenther* (Contemporary and Modern Art) • *Donald Jenkins* (Asian Art) • *Margaret Bullock* (American Art) • *Bill Mercer* (Native American Art) • *Penelope Hunter-Stiebel* (European Art) • *Pamela Morris* (Prints and Drawings) • *Terry Toedtemeier* (Photography) • Sc. Staff: *Elizabeth Garrison* (Education) • Cons.: *Elizabeth Chambers* •
Fine Arts Museum – 1892
Asian, Northwest Coast native American, Cameroon West African art 52086

Portland Children's Museum, 4105 SW Canyon Rd, Portland, OR 97221 • T: +1 503 2236500 • F: +1 503 2236600 • cadams@portlandcm2.org • www.portlandchildrensmuseum.org •
Exec. Dir.: *Margaret Eickmann* •
Local Museum / Natural History Museum – 1949
Natural hist, paleontology, cultural hist, transportation, toys, dolls, stuffed animals of North America 52087

Portland State University Galleries, 725 SW Harrison, Portland, OR 97207, mail addr: POB 751/SD, Portland, OR 97207 • T: +1 503 7255656 • F: +1 503 7255680 •
C.E.O.: *Ann Amato* •
Fine Arts Museum / University Museum
Prints, paintings, sculpture 52088

Washington County Historical Society Museum, 17677 NW Springville Rd, Portland, OR 97229 • T: +1 503 6455353 • F: +1 503 6455650 • wchs@teleport.com • www.washingtoncountymuseum.org •
C.E.O.: *Mark E. Granlund* •
Local Museum – 1956 52089

Wells Fargo History Exhibits, 1300 SW Fifth Av, Portland, OR 97201 • T: +1 503 8861102 •
Special Museum – 2001
Wells fargo banking and express hist, mining, staging 52090

White Gallery, c/o Portland State University, POB 751, Portland, OR 97207 • T: +1 503 7255656 • F: +1 503 7255080 • www.pdx.edu •
Fine Arts Museum / University Museum – 1970 . 52091

Portola CA

Western Pacific Railroad Museum, 700 Western Pacific Way, Portola, CA 96122 • T: +1 530 8324131 • info@wplives.org • www.wplives.org •
Pres.: *Rod McClure* •
Science&Tech Museum – 1984
Western railroad hist, locomotives – archives 52092

Portsmouth NH

Children's Museum of Portsmouth, 280 Marcy St, Portsmouth, NH 03801 • T: +1 603 4363853 • F: +1 603 4367706 • staff@childrens-museum.org • www.childrens-museum.org •
Dir.: *Xanthi Gray* (Exhibits) •
Historical Museum – 1983
Children's literature, multicultural costumes, paleontology, space travel 52093

Governor John Langdon House, 143 Pleasant St, Portsmouth, NH 03801 • T: +1 603 4363205 • F: +1 617 2279204 • pgittlemann@spnea.org • www.historicnewengland.org •
Pres.: *Carl Nold* •
Local Museum – 1947
Artifacts from the 18th and 19th c 52094

John Paul Jones House Museum, 43 Middle St, Portsmouth, NH 03802-0728 • T: +1 603 4368420 • F: +1 603 4319973 • info@portsmouthhistory.org • www.portsmouthhistory.org •
Pres.: *Peter J. Narbonne* •
Local Museum – 1920
Guns, books, china, costumes, docs, furniture, glass, portraits and silver pertaining 52095

MacPheadris/Warner House, 150 Daniel St, Portsmouth, NH 03802-0895 • T: +1 603 4365909 • bb1798@aol.com • www.warnerhouse.org •
C.E.O.: *Dr. Robert L. Barth* •
Local Museum – 1931
Furniture and decorated accessoiries 52096

Moffatt-Ladd House, 154 Market St, Portsmouth, NH 03801 • T: +1 603 4368221 • F: +1 603 4319063 • moffatt-ladd@juno.com •
Pres.: *Dr. Barbara McLean Ward* •
Historic Site – 1911
Furniture, wallpaper, furnishings, American decorative arts, paintings, textiels 52097

Portsmouth Athenaeum, 6 - 8 Market Sq, Portsmouth, NH 03801 • T: +1 603 4312538 • F: +1 603 4317180 • athenaeum@gwi.net •
Pres.: *Michael Chubrich* • C.E.O.: *Tom Hardiman* •
Historical Museum – 1817
Maritinme history, ship models, paintings, portraits – library 52098

Rundlet-May House, 364 Middle St, Portsmouth, NH 03801 • T: +1 603 4363205 • F: +1 617 2279204 • pgittlemann@spnea.org • www.historicnewengland.org •
Pres.: *Carl Nold* •
Local Museum – 1971
Furnishings, Rumford roaster and range 52099

Strawbery Banke Museum, 454 Court St, Portsmouth, NH 03801, mail addr: POB 300, Portsmouth, NH 03802-0300 • T: +1 603 4331100 • F: +1 603 4331129 • www.strawberybanke.org •
Pres.: *Lawrence Yerdon* • Cur.: *Carolyn Roy* •
Local Museum – 1958
Decorative arts, tools, architectural, archaeological, maritime 52100

Wentworth-Coolidge Mansion, 375 Little Harbor Rd, Portsmouth, NH 03801 • T: +1 603 4366607 • F: +1 603 4369889 • wcc@wcmansion.org • www.nhstateparks.org •
Dir.: *Mark J. Sammons* •
Historical Museum – 1954
Home of first British Colonial Governor of New Hampshire 52101

Wentworth Gardner and Tobias Lear Houses, 50 Mechanic St, Portsmouth, NH 03801 • T: +1 603 4364406 • www.seacoastnh.com/wentworth •
Pres.: *John Baybut* •
Local Museum / Decorative Arts Museum – 1941
Historic House from 1760, early Georgian architecture 52102

Portsmouth OH

Southern Ohio Museum and Cultural Center, 825 Gallia St, Portsmouth, OH 45662, mail addr: POB 990, Portsmouth, OH 45662 • T: +1 740 3545629 • F: +1 740 3544090 • museum82@falcon1.net •
C.E.O. & Pres.: *C. Clayton Johnson* • Dir.: *Kay Bouyack* • *Sara Rardin Johnson* (Visual Arts) •
Fine Arts Museum / Folklore Museum – 1979
American art, folk art, decoratiove arts, prehistoric Native American artifacts 52103

Portsmouth RI

Portsmouth Museum, 870 E Main Rd and Union St, Portsmouth, RI 02871 • T: +1 401 6839178 •
Pres.: *Herbert Hall* •
Local Museum – 1938
Local hist 52104

Portsmouth VA

Children's Museum of Virginia, Portsmouth Museums, 420 High St, Portsmouth, VA 23704 • T: +1 757 3938393 • F: +1 757 3935228 • www.ci.portsmouth.va.us •
Special Museum
Educational & participatory exhib concentrating on science, art and the humanities 52105

Courthouse Gallery, Portsmouth Museums, 420 High St, Portsmouth, VA 23704 • T: +1 757 3938983 • F: +1 757 3935228 • perryn@ci.portsmouth.va.us • www.ci.portsmouth.va.us •
Dir.: *Nancy Perry* • Cur.: *Gayle Paul* •
Public Gallery – 1974 52106

The Hill House, Portsmouth Historical Association, 221 North St, Portsmouth, VA 23704 • T: +1 757 3930241 • F: +1 757 3935244 •
C.E.O. & Pres.: *Alice C. Hanes* •
Fine Arts Museum / Historical Museum / Museum of Classical Antiquities – 1957
Local hist, historic house, furnishings 52107

Lancaster Antique Toy and Train Collection, Portsmouth Museums, 420 High St, Portsmouth, VA 23704 • T: +1 757 3938393 • F: +1 757 3935228 • www.ci.portsmouth.va.us •
Special Museum 52108

Lightship Portsmouth Museum, Portsmouth Museums, Water and London Sts, Portsmouth, VA 23704 • T: +1 757 3938741 • F: +1 757 3935228 • navalmuseums@portsmouthva.gov • www.ci.portsmouth.va.us •
Dir.: *Michael Kerekesh* •
Science&Tech Museum 52109

Naval Shipyard Museum, Portsmouth Museums, 2 High St, Portsmouth, VA 23704 • T: +1 757 3938591 • F: +1 757 3935228 • navalmuseums@portsmouthva.gov • www.ci.portsmouth.va.us •
Dir.: *Alice C. Hanes* •
Military Museum / Historical Museum
Uniforms, cannon balls, models of ships, pieces of the CSS Virginia 52110

Portsmouth Museum of Military History, 701 Court St, Portsmouth, VA 23704 • T: +1 757 3932773 • F: +1 757 3932562 •
Military Museum – 1998
Military war artifacts 52111

Virginia Sports Hall of Fame and Museum, 420 High St, Portsmouth, VA 23704 • T: +1 757 3938031 • F: +1 757 3935228 • webbe@ci.portsmouth.va.us • wirginiasportshalloffamme.com •
Exec. Dir.: *Eddie Webb* •
Special Museum – 1971
Hist & contemporary Virginia sports figures 52112

Visual Arts Center, Tidewater Community College, 340 High St, Portsmouth, VA 23704 • T: +1 757 8226999 • F: +1 757 8226800 • tciotta@tcc.va.us • www.tcc.edu •
Performing Arts Museum 52113

Poteau OK

Robert S. Kerr Museum, 23009 Kerr Mansion Rd, Poteau, OK 74953 • T: +1 918 6479579 • F: +1 918 6473952 • kmansion@clnk.com • www.carlabert.edu/kerr_center/contact.htm •
Cur.: *Carol Spindle* •
Local Museum – 1968
Senator Kerr pictures and mementoes, Choctaw Indian artifacts, Viking runestones 52114

Potsdam NY

Potsdam Public Museum, Civic Center, Potsdam, NY 13676 • T: +1 315 2656910 • F: +1 315 2653149 • Museum@vi.potsdam.ny.us •
Dir./Cur.: *Betsy L. Travis* •
Local Museum / Decorative Arts Museum – 1940
Decorative arts, local hist, Burnap coll of English pottery, glass, American china 52115

Roland Gibson Gallery, State University of New York College at Potsdam, 44 Pierrepont Av, Potsdam, NY 13676 • T: +1 315 2673290 • F: +1 315 2674884 • www.potsdam.edu/gibson/gibson.html •
Dir.: *Romi Seald-Chudzinski* • *Margaret Price* •
Preparator: *Claudette Feffee* •
Fine Arts Museum / University Museum – 1968
Contemporary sculpture, prints, ceramics, drawings and paintings – Roland Gibson Coll of 20th c contemporary art 52116

Pottstown PA

Pottsgrove Manor, 100 West King St, Pottstown, PA 19464-6318 • T: +1 610 3264014 • F: +1 610 3269618 • pottsgrovemanor@mail.montcopa.org • www.montcopa.org •
Cur./Admin.: *William A. Brobst* •
Local Museum – 1988
Historic house 52117

Pottsville AR

Potts Inn Museum, 25 E Ash St, Pottsville, AR 72858, mail addr: Dept. of Parks & Tourism, 1 Capitol Mall, Little Rock, AR 72201 • T: +1 501 9681877 • info@arvtripeaks.com • www.arkansas.com •
Local Museum – 1978
Home and stage stop of Kirkbride Potts, ladies' hats 52118

Poughkeepsie NY

Art Gallery Marist College, 3399 North Rd, Poughkeepsie, NY 12601-1387 • T: +1 845 5753000 ext 2903 • F: +1 845 4716213 • donise.english@marist.edu •
Dir.: *Donise English* •
Fine Arts Museum – 1995
Paintings, sculptures, photos, contemporary works 52119

Dutchess County Historical Society, Glebe House & Clinton House, 549 Main Str, Poughkeepsie, NY 12601, mail addr: c/o Eileen Hayden, POB 88, Poughkeepsie, NY 12602 • T: +1 845 4711630 • F: +1 845 4718777 • dchistorical@earthlink.net • www.dutchesscountyhistoricalsociety.org •
Historical Museum – 1914 52120

The Frances Lehman Loeb Art Center, Vassar College, 124 Raymond Av, Poughkeepsie, NY 12604-0703 • T: +1 845 4375237 • F: +1 845 4377304 • jamundy@vassar.edu • departments.vassar.edu/~fllac/ •
Dir.: *James Mundy* • Cur.: *Patria Phagan* (Prints and Drawings) • *Joel Smith* (Collections) •
Fine Arts Museum – 1864
Medieval to modern American and European painting and sculpture, Greek and Roman antiquities, Asian art, drawings and watercolors 52121

Locust Grove, Samuel Morse Historic Site, 2683 South Rd, Poughkeepsie, NY 12601 • T: +1 845 4544500 ext 10 • F: +1 845 4857122 • locustgrovel@earthlink.net • www.morsehistoricsite.org •
Dir.: *Raymond J. Armater* •
Decorative Arts Museum – 1979
Furniture, China, costumes, dolls, docs, telegraph equip, life of Samuel F.B. Morse 52122

Mid-Hudson Children's Museum, 75 N Water St, Poughkeepsie, NY 12601 • T: +1 845 4710589 • F: +1 845 4710415 • info@mhcm.org • www.mhcm.org •
Exec. Dir.: *Diane Pedevillano* •
Historical Museum – 1989 52123

Poulsbo WA

Suquamish Museum, 15838 Sandy Hook Rd, Poulsbo, WA 98370 • T: +1 360 5983311, 3945275 • F: +1 360 5986295 • suqmuseum@hotmail.com • www.suquamish.nsn.us •
Dir.: *Marilyn Jones* •
Ethnology Museum – 1983
Culture of the Suquamish Tribe and other Puget Sound Tribes 52124

Pound Ridge NY

Pound Ridge Museum, 255 Westchester Av, Pound Ridge, NY 10576 • T: +1 914 7644333 • F: +1 914 7641778 •
Pres.: *Richard Major* •
Historical Museum / Local Museum – 1970
Prehistoric remains, hist of Pound Ridge, Native American Articles 52125

Powell WY

Northwest Gallery and Sinclair Gallery, Northwest College, 231 W Sixth St, Powell, WY 82435 • T: +1 307 7546111 • F: +1 307 7546245 • kelsayd@nwc.cc.wy.us • www.nwc.cc.wy.us/area/art/index.html •
Fine Arts Museum 52126

Powers Lake ND

Burke County Historical Powers Lake Complex, Powers Lake, ND 58773 • T: +1 701 5464491 •
C.E.O.: *Larry Tinjum* •
Local Museum – 1986
Hist bldgs 52127

Prairie du Chien WI

Fort Crawford Museum, 717 S Beaumont Rd, Prairie du Chien, WI 53821 • T: +1 608 3266960 • ftcrawmy@mhtc.net • www.fortcrawford.com •
Head: *Thomas L. Farrell* •
Local Museum – 1955
Military and medical artifacts, local hist 52128

Villa Louis Historic Site, 521 Villa Louis Rd, Saint Feriole Island, Prairie du Chien, WI 53821 • T: +1 608 3262721 • F: +1 608 3265507 • villalou@pearl.mhtc.net • www.wisconsinhistory.org/villalouis •
Dir.: *Michael Douglass* •
Local Museum – 1936
Victorian house, furnishing and dec. art, local hist 52129

Prairie du Rocher IL

Fort de Chartres Museum, 1350 State Rt 155, Prairie du Rocher, IL 62277 • T: +1 618 2847230 • F: +1 618 2847230 • ftdchart@htc.net •
Man.: *Darrell Duensing* •
Historic Site / Military Museum – 1913
Original site of the French Fort de Chartes built in 1753 52130

Prairie du Sac WI

Sauk Prairie Area Historical Society Museum, 565 Water St, Prairie du Sac, WI 53578 • T: +1 608 6436406 • F: +1 608 6448444 • spahs@chorus.net • www.saukprairiehistory.org •
Pres.: *Stephen Dittberner* • Cur.: *Orie Eilertson* •
Local Museum – 1961
Local hist, household items – Firehouse Museum, Old Hahn House, Tripp Memorial Museum 52131

Pratt KS

Pratt County Historical Society Museum, 208 S Ninnescah St, Pratt, KS 67124 • T: +1 316 6727874 • www.pratt.net •
Local Museum – 1967
Fossils, Plains Indian artifacts, pioneer period, military, jail 52132

Prattsburgh NY

Narcissa Prentiss House, 7226 Mill Pond Rd, Prattsburgh, NY 14873 • T: +1 607 5224537 •
C.E.O.: *William Willis* •
Local Museum – 1940
Birthplace of Narcissa Prentiss Whitman, missionary 52133

Prescott AZ

Phippen Museum of Western Art, 4701 Hwy 89 N, Prescott, AZ 86301 • T: +1 928 7781385 • F: +1 928 7784524 • phippen@phippenartmuseum.org • www.phippenartmuseum.org •
Dir.: *Kim Villalpando* • Cur.: *Deb Bentlage* •
Fine Arts Museum – 1974
Paintings and scultures of the American West – library 52134

Sharlot Hall Museum, 415 W Gurley St, Prescott, AZ 86301 • T: +1 928 4453122 • john@sharlot.org • www.sharlot.org •
Dir.: *John Langellier* • Cur.: *Mick Woodcock* (Colls) • *Scott Anderson* (Archivist) •
Local Museum – 1929
Regional hist, vehicles, medical and military hist – archives 52135

Smoki Museum, 147 N Arizona St, Prescott, AZ 86304 • T: +1 928 4451230 • info@smokimuseum.org • www.smokimuseum.org •
Dir.: *Ginny Chamberlin* •
Ethnology Museum – 1935
Anthropology, American Indian art and culture ... 52136

Presidio TX

Fort Leaton, River Rd FM 170, Presidio, TX 79845 • T: +1 432 2293613 • F: +1 432 2294814 • fortleatonshp@brooksdata.net •
C.E.O.: *Andrew Sansom* •
Local Museum – 1977
Local hist, birding, bats, adobe structure maintained, 2 wheel ox-cart 52137

Presque Isle ME

Reed Gallery, University of Main at Presque Isle, 181 Main St, Presque Isle, ME 04769 • T: +1 207 7689609 • F: +1 207 7689553 • gipson@umpi.maine.edu •
Dir.: *Shelley Gipson Adams* •
Fine Arts Museum / University Museum
Contemporary art 52138

Prestonsburg KY

East Kentucky Center for Scicece, Mathematics and Technology, 207 W Court St, Prestonsburg, KY 41653 • T: +1 606 8890303 • F: +1 606 8890306 • ekyscience@setel.com • www.wedoscience.org •
C.E.O.: *Raymond Shubinski* • Chm. & Pres.: *John Rosenberg* •
Natural History Museum
Rocks and minerals 52139

Price UT

College of Eastern Utah Prehistoric Museum, 155 E Main St, Price, UT 84501 • T: +1 435 6375060 • F: +1 435 6372514 • pam.miller@ceu.edu • www.ceu.edu •
Cur./Dir.: *Pamela Miller* •
University Museum / Ethnology Museum / Natural History Museum – 1961
Dinosaurs, minerals, fossils of Utah, paleontology, geology 52140

Gallery East, College of Eastern Utah, 451 E 400 N, Price, UT 84501 • T: +1 435 6135241 • F: +1 435 6134102 • noel.carmack@ceu.edu • www.ceu.edu •
Man.: *Brent Haddock* •
Public Gallery – 1937 52141

Princess Anne MD

Teackle Mansion, 11736 Mansion St, Princess Anne, MD 21853 • T: +1 410 6512238 • visitor@teacklemansion.org • www.teacklemansion.org •
Local Museum – 1958
Costumes, decoratives art, furniture, paintings 52142

Princeton IA

Buffalo Bill Cody Homestead, 28050 230th Av, Princeton, IA 52768 • T: +1 563 2252981 • F: +1 563 3812805 • www.scottcountyiowa.com •
C.E.O.: *Roger Kean* •
Local Museum – 1970
Historic house, boyhood home of Buffalo Bill Cody 52143

Princeton IL

Bureau County Historical Society Museum, 109 Park Av W, Princeton, IL 61356-1927 • T: +1 815 8752184 •
Dir.: *Barbara Hansen* •
Local Museum – 1911
Local history 52144

Princeton KY

Adsmore Museum, 304 N Jefferson St, Princeton, KY 42445 • T: +1 270 3653114 • F: +1 270 3653310 • ads@ziggycom.net • www.adsmore.org •
Cur.: *Ardell Jarratt* •
Historical Museum – 1986
Period furnishings, silver, China, linens, clothing, decorative accessories, photographs, letters, toys 52145

Princeton MO

Casteel-Linn House and Museum, 902 E Oak St, Princeton, MO 64673 • T: +1 816 7483905 •
Dir.: *Joe Dale Linn* • Cur.: *Nancy Paige Linn* •
Local Museum – 1982
Art & art hist, china & glass, railroad china & antiques, antique toys, portraits 52146

Mercer County Genealogical and Historical Society, 601 Grant, Princeton, MO 64673 • T: +1 660 7483725 • F: +1 660 7483723 •
Head: *Randy Ferguson* •
Association with Coll – 1965
Hist, antiques, costumes, furniture, china, geology 52147

Princeton NJ

Gallery at Bristol-Myers Squibb, POB 4000, Princeton, NJ 08543-4000 • T: +1 609 2524599 • F: +1 609 4972441 •
Dir.: *Pamela V. Sherin* •
Public Gallery – 1972 52148

Historic Morven, New Jersey State Museum, 55 Stockton St, Princeton, NJ 08540 • T: +1 609 6834495 • F: +1 609 4976390 • morvennj@aol.com • www.historicmorven.org •
Exec. Dir.: *Martha Leigh Wolf* • Cur.: *Pam Ruch* •
Archaeology Museum
Archaeology 52149

Princeton Museum, 158 Nassau St, Princeton, NJ 08542 • T: +1 609 9216748 • F: +1 609 9216939 • gailfstern@aol.com • www.princetonhistory.org •
C.E.O. & Pres.: *Dee Parberg* • Dir.: *Gail F. Stern* • Cur.: *Maureen M. Smith* •
Local Museum – 1938
Local hist artifacts, arts, photos, docs 52150

Princeton University Art Museum, Princeton University, Princeton, NJ 08544-1018 • T: +1 609 2583788 • F: +1 609 2585949 • artmuseum@princeton.edu • www.princetonartmuseum.org •
Dir.: *Susan M. Taylor* • Cur.: *Michael Padgett* (Ancient Art) •
Fine Arts Museum / University Museum – 1882
Ancient Mediterranean, British and American art, Chinese ritual bronze vessels, Far Eastern art, Pre-Columbian and African art – Index of Christian art, library 52151

Princeton University Museum of Natural History, Guyot and 9 Eno Hall, Princeton University, Princeton, NJ 08544-1003 • T: +1 609 2584102, 4661843 • F: +1 609 2581334 • hshorn@pnnuton.edu •
Cur.: *Elizabeth Horn* (Biological Collections) •
University Museum / Natural History Museum – 1805
Paleontology, geology, mineralogy, ornitology, ethnology, archaeology 52152

Rockingham, 108 CR 518, Princeton, NJ 08540 • T: +1 609 9218835 • F: +1 609 9218835 • peggi@superlink.net • www.rockingham.net •
Cur.: *Margaret Carlsen* •
Historic Site – 1896
1710 Berrien Mansion, Washington's headquarters while Continental Congress was in session in Princeton in 1738 52153

Thomas Clarke House-Princeton Battlefield, 500 Mercer Rd, Princeton, NJ 08540-4810 • T: +1 609 9210074 • F: +1 609 9210074 • pbsp@aol.com •
Cur.: *John K. Mills* •
Historical Museum / Military Museum – 1976 52154

Prineville OR

A.R. Bowman Memorial Museum, 246 N Main St, Prineville, OR 97754 • T: +1 541 4473715 • F: +1 541 4473715 • bowmuse@netscape.net • www.bowmanmuseum.org •
Pres.: *William Weberg* • Dir.: *Gordon Gillespie* •
Local Museum – 1971
American Indian artifacts, pioneer exhibits of the area, rocks 52155

Proctor VT

Vermont Marble Museum, 52 Main St, Proctor, VT 05765 • F: +1 802 4592948 • rpye@adelphia.net • www.vermont-marble.com •
Head: *Marsha Hemm* •
Natural History Museum / Science&Tech Museum – 1933
Mining, geology 52156

Wilson Castle, W Proctor Rd, Proctor, VT 05765 • T: +1 802 7733284 • F: +1 802 7733284 •
Dir.: *Blossom Wilson Davine* •
Decorative Arts Museum – 1961
European and Oriental objects d'art, stained glass, furniture, antiques 52157

Prospect Park PA

Morton Homestead, 100 Lincoln Av, Prospect Park, PA 19076 • T: +1 610 5837221 • F: +1 610 5832349 • erump@state.pa.us •
Dir.: *Elizabeth Rump* •
Local Museum – 1939
Local hist 52158

Prosser WA

Benton County Historical Museum, 1000 Paterson Rd, City Park, Prosser, WA 99350 • T: +1 509 7863842 •
Pres.: *Bob Elder* •
Local Museum – 1968
Indian artifacts, natural hist, guns, dolls, china, handwork, gown coll 52159

Providence RI

Annmary Brown Memorial, 21 Brown St, Providence, RI 02912 • T: +1 401 8631994 • carol_cramer@brown.edu • www.brown.edu •
Man.: *Carole Cramer* •
Fine Arts Museum / University Museum – 1905
Art 52160

Art and Art History Department, Providence College, River Ave and Eaton St, Providence, RI 02918 • T: +1 401 8652401 • F: +1 401 8652410 • art@providence.edu • www.providence.edu/art/art.htm •
Chm.: *James Janecek* •
Public Gallery 52161

Betsey Williams Cottage, Roger Williams Park, Providence, RI 02905 • T: +1 401 7859457 • F: +1 401 4615146 • museum@musnathist.com •
Dir.: *Roger Williams* • Cur.: *Marilyn Massaro* •
Historic Site – 1871
Historic house 52162

Culinary Archives and Museum, Johnson and Wales University, 315 Harborside Blvd, Providence, RI 02905 • T: +1 401 5982805 • F: +1 401 5982807 • bkuck@jwu.edu • www.culinary.org •
Exec. Dir.: *Morris Nathanson* •
University Museum / Special Museum – 1989
Culinary objects 52163

David Winton Bell Gallery, Brown University, List Art Center, 64 College St, Providence, RI 02912 • T: +1 401 8632932, 8632929 • F: +1 401 8639323 • www.brown.edu/bellgallery •
Dir.: *Jo-Ann Conklin* • Cur.: *Vesela Sretenovic* •
Fine Arts Museum / University Museum – 1971
20th c painting and sculpture, works on paper from the 15th c to the present 52164

Edward M. Bannister Gallery, Rhode Island College, 600 Mount Pleasant Av, Providence, RI 02908 • T: +1 401 4569765 • F: +1 401 4569718 • domalley@ric.edu • www.ric.edu/bannister •
Dir.: *Dennis M. O'Malley* •
Fine Arts Museum – 1978
Visual arts 52165

Governor Henry Lippitt House Museum, 199 Hope St, Providence, RI 02906 • T: +1 401 4530688 • F: +1 401 4538221 • sludin@ids.net •
C.E.O.: *Nicolas Brown* • Cur.: *Sally L. Barker* •
Local Museum – 1991
Local hist 52166

Governor Stephen Hopkins House, Hopkins and Benefit Sts, Providence, RI 02903 • T: +1 401 4210694 •
Pres.: *Joanne Dunlap* • Head: *Alice Walsh* •
Local Museum – 1707
Historic house, 1707 Governor Stephen Hopkins home 52167

John Brown House, 52 Power St, Providence, RI 02906 • T: +1 401 3318575 •
Cur.: *Linda Eppich* •
Decorative Arts Museum – 1942
Chinese export objects, antique dolls, furniture, portraits, China, glass, pewter and other decorative objects 52168

John Nicholas Brown Center for the Study of American Civilization, Brown University, 357 Benefit St, Providence, RI 02903 • T: +1 401 2720357 • F: +1 401 2721930 • joyce_botelho@brown.edu • www.brown.edu/jnbc •
Pres.: *Ruth Simmons* • Dir.: *Joce M. Botelho* •
University Museum / Decorative Arts Museum – 1983
Decorative arts of America and Europe, historic wallpaper – archives 52169

Museum of Art, Rhode Island School of Design, 224 Benefit St, Providence, RI 02903-2723 • T: +1 401 4546500 • F: +1 401 4546556 • museum@risd.edu • www.risdmuseum.org •
Dir.: *Lora Urbanelli* • Cur.: *Thomas Michie* (Decorative Arts) • *Maureen O'Brien* (Painting and Sculpture) • *Dr. Georgina Borromeo* (Antiquities) • *Susan Hay* (Costume and Textiles) • *Jan Howard* (Prints, Drawings and Photographs) • *Deborah del Gais Muller* (Asian Art) • *Judith Tannenbaum* (Contemporary Art) •
Fine Arts Museum / University Museum – 1877
European porcelain, Oriental textiles, ancient Oriental and ethnographic art, modern Latin American art, contempotary graphics 52170

Museum of Natural History, Roger Williams Park, Providence, RI 02905 • T: +1 401 7859457 • F: +1 401 4615146 • museum@musnathist.com • www.osfn.org/museum •
Dir.: *Tracey Keough* • Cur.: *Marilyn Massaro* •
Natural History Museum – 1896
Natural hist – Planetarium 52171

Providence Art Club, 11 Thomas St, Providence, RI 02903 • T: +1 401 3311114 •
Public Gallery 52172

Providence Athenaeum, 251 Benefit St, Providence, RI 02903 • T: +1 401 4216970 • F: +1 401 4212860 •
Public Gallery
library 52173

Providence Children's Museum, 100 South St, Providence, RI 02903 • T: +1 401 2735437 • F: +1 401 2731004 • provcm@childrenmuseum.org • www.childrenmuseum.org •
Pres.: *Jeffery R. Massotti* • Dir.: *Janice O'Donnell* •
Special Museum – 1976
Children's museum 52174

Providence Preservation Society, 21 Meeting St, Providence, RI 02903 • T: +1 401 8317440 • F: +1 401 8318283 • www.ppsri.org •
Exec.Dir.: *Jack Anthony Gold* •
Association with Coll – 1956
History, architecture – architecture library 52175

Rhode Island Black Heritage Society Museum, 65 Weyboset St, Providence, RI 02903, mail addr: POB 1656, Providence, RI 02901 • T: +1 401 7513490 • F: +1 401 7510040 • blkheritage@netzero.net •
Pres.: *Walter Stone* • Exec. Dir.: *Joaquina B. Teixeira* •
Fine Arts Museum / Historical Museum – 1975
Afro-American history and art 52176

Rhode Island Foundation Art Gallery, 1 Union Station, Providence, RI 02903 • T: +1 401 2744564 •
Fine Arts Museum 52177

Rhode Island Historical Society Exhibition, Aldrich House, 110 Benevolent St, Providence, RI 02906 • T: +1 401 3318575 • F: +1 401 3510127 • www.rihs.org •
Dir.: *Bernard B. Fishmann* • Chief Cur.: *Kirsten Hammerstrom* • Cur.: *Rick Stattler* (Manuscripts) •
Local Museum – 1822
Local hist – Historical library of 1874, John Brown House of 1786 52178

Roger Williams National Memorial, 282 N Main St, Providence, RI 02903 • T: +1 401 5217266 • F: +1 401 5217239 • rowi_interpretation@nps.gov • www.nps.gov/rowi •
Head: *Michael Creasey* •
Local Museum – 1965
Local hist 52179

Sol Koffler Graduate Gallery, Rhode Island School of Design, 169 Waybosset St, Providence, RI 02093 • T: +1 401 4546131 • F: +1 401 4546706 • www.risd.edu •
Public Gallery
Changing exhibitions 52180

Wheeler Gallery, 228 Angell St, Providence, RI 02906 • T: +1 401 4219230 • F: +1 401 7517674 • info@wheelergallery.org • www.wheelergallery.org •
Dir.: *Sue Carroll* •
Fine Arts Museum
Contemporary paintings, sculpture, photography, prints, ceramics, glass, installations 52181

Provincetown MA

Pilgrim Monument and Provincetown Museum, High Pole Hill, Provincetown, MA 02657 • T: +1 508 4871310 • F: +1 508 4874702 • cturley@pilgrim-monument.org • www.pilgrim-monument.org •
Exec. Dir.: *Chuck Turley* • Cur.: *Jeffory Morris* •
Local Museum – 1892
Local hist, Mayflower pilgrims, whaling and fishing, theater hist, pirate treasures, shipwrecks – archives, small library 52182

Provincetown Art Association and Museum, 460 Commercial St, Provincetown, MA 02657 • T: +1 508 4871750 • F: +1 508 4874372 • info@paam.org • www.paam.org •
Dir.: *Christine McCarthy* •
Fine Arts Museum – 1914
Art 52183

Provincetown Heritage Museum, 356 Commercial St, Provincetown, MA 02657 • T: +1 508 4877098 •
Dir.: *Dale A. Fanning* •
Local Museum – 1976
Local history, pilgrim artifacts, fishing schooner scale model, antique fire engine, fishing trade – art colony 52184

Provo UT

B.F. Larsen Gallery, c/o Birgham Young University, Harris Fine Arts Center, Provo, UT 84602 • T: +1 801 3782881, 4222881 • F: +1 801 3785964 • gallery303@byu.edu • cfac.byu.edu/va/Gallery/index.php •
Dir.: *Todd Frye* •
Public Gallery / University Museum – 1965 52185

Brigham Young University Earth Science Museum, N181 ESC, Provo, UT 84602-3300 • T: +1 801 3783939, 4222674 • F: +1 801 3787919 • klstadtman@geology.byu.edu • cpms.byu.edu •
Dir./Cur.: *Kenneth L. Stadtman* •
University Museum / Natural History Museum – 1987
Paleontology 52186

Monte L. Bean Life Science Museum, c/o Brigham Young University, 290 MLBM Bldg, Provo, UT 84602 • T: +1 801 422+5052 • F: +1 801 3783733 • office@museum.byu.edu • mlbean.byu.edu •
Dir.: *Dr. H. Duane Smith* • Ass. Dir.: *Dr. Douglas C. Cox* •
Natural History Museum / Science&Tech Museum – 1978
Entomology, herpetology, shells, ichthyology, mammalogy 52187

Museum of Art at Brigham Young University, North Campus Dr, Provo, UT 84602-1400 • T: +1 801 4228256, 8257 • F: +1 801 4220527 • moaadmin@byu.edu • www.byu.edu/moa •
Dir.: *Campbell B. Gray* • Ass. Dir.: *Ed Lind* • Cur.: *Dawn Pheysey* •
Fine Arts Museum / University Museum – 1993
19th & 20th c American coll 52188

Museum of Peoples and Cultures, c/o Brigham Young University, 700 N 100 E, Allen Hall, Provo, UT 84602 • T: +1 801 4220020 • F: +1 801 4220026 • mpc_programs@byu.edu • fhss.byu.edu/anthro/mopc/main.htm •
Dir.: *Dr. Marti L. Allen* •
Ethnology Museum – 1946
Anthropology, ethnology 52189

Pryor MT

Chief Plenty Coups Museum, Edgar Rd, Pryor, MT 59066 • T: +1 406 2521289 • F: +1 406 2526668 • plentycoups@plentycoups.org • www.plentycoups.org •
Dir.: *Rich Furber* •
Ethnology Museum – 1972
Ethnography of Crow Indians, paintings, drawings 52190

Pueblo CO

El Pueblo History Museum, 301 N Union St, Pueblo, CO 81003 • T: +1 719 5830453 • F: +1 719 5838214 • www.coloradohistory.org •
Dir.: *Deborah Espinosa* •
Local Museum – 1959
Local history 52191

Pueblo County Historical Society Museum, 217 S Grand Av, Pueblo, CO 81003 • T: +1 719 5421851 • pchs_h@yahoo.com • www.pueblohistory.org •
Local Museum – 1986
Local hist – Edward H. Broadhead Library 52192

Rosemount Museum, 419 W 14th St, Pueblo, CO 81003 • T: +1 719 5455290 • F: +1 719 5455291 • rosemnt@usa.net • www.rosemount.org •
Dir.: *Deborah Darrow* •
Local Museum – 1967
John A. Thatcher residence, Henry Hudson Holly 52193

Sangre de Cristo Arts Center and Buell Children's Museum, 210 N Santa Fe Av, Pueblo, CO 81003 • T: +1 719 5430130 • F: +1 719 5430134 • mail@sdc-arts.org • www.sdc-arts.org •
Dir.: *Maggie Divelbiss* •
Fine Arts Museum – 1972
Art education 52194

Pueblo of Acoma NM

Haakú Museum, Sky City Cultural Center, I-40 W, Exit 102, Pueblo of Acoma, NM 87034, mail addr: POB 310, Pueblo of Acoma, NM 87034 • T: +1 505 5527860 • F: +1 505 5527883 • www.skycity.com •
Dir.: *Brian D. Vallo* • Cur.: *Damian Garcia* •
Local Museum / Fine Arts Museum
Hist of Acoma, pottery, Native American art and artifacts, 16th c church – archives 52195

Pulaski VA

Fine Arts Center for New River Valley, 21 W Main St, Pulaski, VA 24301 • T: +1 540 9807363 • F: +1 540 9807363 • facnrv@adelphia.net •
Dir.: *Judy C. Ison* •
Fine Arts Museum / Public Gallery – 1978
Eclectic & contemporary works by local & regional artists 52196

Pullman WA

Charles R. Conner Museum, Washington State University, Pullman, WA 99164-4236 • T: +1 509 3353515 • F: +1 509 3353184 • connermuseum@wsu.edu • www.sci.wsu.edu/cm •
Dir.: *Dr. Richard E. Johnson* •
Natural History Museum – 1894
Coll of herpetology, ornithology and mammology 52197

Museum of Anthropology, Department of Anthropology, Washington State University, Pullman, WA 99164-4910 • T: +1 509 3353441 • F: +1 509 3353999 • collinsm@wsu.edu •
Dir.: *William Andrefsky* • Ass. Dir.: *Mary Collins* • Cur.: *Joy Mastrogiuseppe* •
University Museum / Ethnology Museum – 1966
Anthropology, archaeological and ethnographic coll 52198

Museum of Art, Washington State University, POB 647460, Pullman, WA 99164-7460 • T: +1 509 3351910 • F: +1 509 3351908 • artmuse@wsu.edu • www.wsu.edu/artmuse •
Dir.: *Ross Coates* • *Rob Snyder* • Cur.: *Roger H.D. Rowley* (Collections) • *Keith Wells* •
Fine Arts Museum / University Museum – 1973
Late 19th c, 20th c and contemporary American and Europen painting and graphic art 52199

Punta Gorda FL

Florida Adventure Museum, 260 W Retta Esplanade, Punta Gorda, FL 33950 • T: +1 941 6393777 • F: +1 941 6393505 • museum@sunline.net • www.charlotte-florida.com/museum •
Exec. Dir.: *Lori Tomlinson* •
Local Museum – 1969
Local and military history, science 52200

Purchase NY

Neuberger Museum of Art, Purchase College, State University of New York, 735 Anderson Hill Rd, Purchase, NY 10577-1400 • T: +1 914 2516100 • F: +1 914 2516101 • neuberger@purchase.edu • www.neuberger.org •
Interim Dir.: *Anne Bradner* • Cur.: *Dede Young* (Modern and Contemporary art) • *Marie Therese Brincard* (African art) •
Fine Arts Museum / University Museum – 1974
20th and 21st c art in all media, African art 52201

Put in Bay OH

Perry's Victory and International Peace Memorial, 93 Delaware Av, Put in Bay, OH 43456 • T: +1 419 2852184 • F: +1 419 2852516 • pevi_interpretation@nps.gov • www.nps.gov/pevi •
Historical Museum – 1936
Weapons and equip from the war of 1812, paintings, engravings, lithographs 52202

Putney VT

Putney Historical Society, Town Hall, Putney, VT 05346 • T: +1 802 3875862 • putneyhs@sover.net • www.putneyvt.org/history •
Cur.: *Laura Heller* • *Elaine Dixon* •
Local Museum / Folklore Museum – 1959
Artifacts relating to the town, Indian artifacts 52203

Putney School Gallery, Elm Lea Farm, Putney, VT 05346 • T: +1 802 3876241 • admission@putneyschool.org • www.putney.com •
Public Gallery
Painting, sculpture, prints, drawing, film and photography 52204

Puunene HI

Alexander and Baldwin Sugar Museum, 3957 Hansen Rd, Puunene, HI 96784 • T: +1 808 8718058 • F: +1 808 8717663 • sugarmus@maui.net • www.sugarmuseum.com •
Pres.: *Stephen Onaga* • Exec. Dir.: *Gaylord Kubota* •
Historical Museum / Agriculture Museum – 1980
Sugar industry and plantation life 52205

Puyallup WA

Ezra Meeker Mansion, 312 Spring St, Puyallup, WA 98372 • T: +1 253 8481770 • www.meekermansion.org •
Historic Site – 1970
Victorian house located at end of the Oregon Trail 52206

Paul H. Karshner Memorial Museum, 309 4th St, Puyallup, WA 98372 • T: +1 253 8418748 • F: +1 253 8408950 • reckerson@puyallup.k12.wa.us •
Dir.: *Rosemary Eckerson* • Cur.: *Stephen Crowell* • *Beth Bestrom* •
Historical Museum – 1930
History, Indian artifacts, American pioneer artifacts 52207

Quantico VA

United States Marine Corps Air-Ground Museum, 2014 Anderson Av, Quantico, VA 22134 • T: +1 703 7842606 • F: +1 703 7845856 •
Chief Cur.: *Charles A. Wood* • Cur.: *Kenneth L. Smith-Christmas* •
Military Museum – 1940
Military hist 52208

Quarryville PA

Robert Fulton Birthplace, 1932 Robert Fulton Hwy, Quarryville, PA 17566 • T: +1 717 5482679, 5482351 • slchs@aol.com • www.rootsweb.com/~paslchs •
Decorative Arts Museum – 1965
Steamboat Clermont builder 52209

Quemado NM

Dia Center for the Arts, Dia Art Foundation, Quemado, NM 87829, mail addr: POB 2993, Corrales, NM 87048 • T: +1 505 8983335 • F: +1 505 8983336 • www.diacenter.org •
Fine Arts Museum
Stainless steel poles, land art by Walter De Maria 52210

Quincy CA

Plumas County Museum, 500 Jackson St, Quincy, CA 95971 • T: +1 530 2836320 • F: +1 530 2836081 • pcmuseum@digitalpath.net • www.countyofplumas.com •
Dir.: *Scott J. Lawson* • Cur.: *Ilana Hirsch* •
Local Museum – 1964
Local hist, indian artifacts, geology, costume, Maidu baskets – library, archives 52211

Quincy FL

Gadsden Arts Center, 13 N Madison, Quincy, FL 32351, mail addr: POB 1945, Quincy, FL 32353 • T: +1 850 8754866, 8751606 • F: +1 850 6278606 • zoe@gadsdenarts.com • www.gadsdenarts.com •
C.E.O.: *Zoe C. Golloway* •
Fine Arts Museum – 1994
Fine arts 52212

Quincy IL

The Gardner Museum of Architecture and Design, 332 Maine St, Quincy, IL 62301 • T: +1 217 2246873 • F: +1 217 2240006 • gardnermuseu-marchitecture@adams.net • www.gardnermuseu-marchitecture.org •
Dir.: *Sherryl P. Lang* •
Historical Museum / Decorative Arts Museum – 1974
Architecture, design, housed in Richardsonian Romanesque Old Public Library 52213

Gray Gallery, c/o Quincy University, 1800 College Av, Quincy, IL 62301-2699 • T: +1 217 2285371 • www.quincy.edu •
Dir.: *Robert Lee Mejer* •
University Museum / Fine Arts Museum – 1968
Oriental and European prints, student and faculty works, American prints and drawings 52214

Quincy and Adams County Museum, 425 S 12th St, Quincy, IL 62301 • T: +1 217 2221835 • F: +1 217 2228212 • hspac@dstream.net •
Pres.: *Chuck Radel* •
Historical Museum – 1896
Local history, in home of former Governor of Illinois John Wood, also founder of Quincy 52215

Quincy Art Center, 1515 Jersey St, Quincy, IL 62301 • T: +1 217 2235900 • F: +1 217 2236950 • inelson@quincyartcenter.org • www.quincyartcenter.org •
Exec. Dir.: *Julie D. Nelson* •
Fine Arts Museum – 1923 52216

The Quincy Museum, 1601 Maine St, Quincy, IL 62301 • T: +1 217 2247669 • F: +1 217 2249323 • quinmu1@knsi.net • www.thequincymuseum.com •
Pres.: *Darin Prost* •
Local Museum / Natural History Museum – 1966
Fossils, minerals, oil paintings and furnishings from Victorian area, American Indian artifacts, natural hist items and dinosaurs 52217

Quincy MA

John Adams and John Quincy Adams Birthplaces, 133-141 Franklin St, Quincy, MA 02169, mail addr: 135 Adams St, Quincy, MA 02169 • T: +1 617 7731175/77 • F: +1 617 4719683 • kelly_cobble@nps.gov • www.nps.gov/adam •
Head: *Marianne Peak* • Cur.: *Kelly Cobble* •
Historical Museum – 1927
History 52218

Josiah Quincy House, 20 Muirhead St, Quincy, MA 02170 • T: +1 617 2273956 • F: +1 617 2279204 • QuincyHouse@HistoricNewEngland.org • www.historicnewengland.org •
Decorative Arts Museum – 1937
Family memorabila, period furnishings 52219

Quincy Historical Museum, Adams Academy Bldg, 8 Adams St, Quincy, MA 02169 • T: +1 617 7731144 • F: +1 617 4724990 • ghist@ci.quincy.ma.us • www.ci.quincy.ma.us/qhs.asp •
Dir./Cur.: *Dr. Edward Fitzgerald* •
Local Museum – 1893
Local history – library 52220

Rabun Gap GA

Hambidge Center for the Creative Arts and Sciences, Betty's Creek Rd, Rabun Gap, GA 30568 • T: +1 706 7465718 • F: +1 706 7469933 • hambidge@rabun.net • www.rabu.net/~hambidge •
Dir.: *Peggy McBride* •
Decorative Arts Museum / Folklore Museum – 1934
Crafts, folk art 52221

Racine WI

Charles A. Wustum Museum of Fine Arts, Racine Art Museum, 2519 N Western Av, Racine, WI 53401 • T: +1 262 6369177 • www.ramart.org •
Fine Arts Museum – 1941 52222

Racine Art Museum, 441 Main St, Racine, WI 53401 • T: +1 262 6388300 • F: +1 262 8981045 • www.ramart.org •
C.E.O.: *Bruce W. Pepich* • Dir. Exhibitions: *davira S. Taragin* •
Fine Arts Museum
20th c and contemporary ceramics, fibers, glass, jewelry and books 52223

Racine Heritage Museum, 701 S Main St, Racine, WI 53403-1211 • T: +1 414 6363926 • F: +1 414 6363940 • inquire@clmail.com • www.spiritofinnovation.org •
Dir.: *Christopher Paulson* • Cur.: *Karen Braun* •
Historical Museum / Science&Tech Museum – 1960
Southeast Wisconsin industrial and product hist 52224

Radford VA

Radford University Art Museum, 200 Powell Hall, Radford, VA 24142 • T: +1 540 8315475, 5475 • F: +1 540 8316799 • sarbury@radford.edu • www.radford.edu/~rumuseum •
Dir.: *Steve Arbury* •
Fine Arts Museum / University Museum – 1985
Late 20th c painting & sculpture, ancillary coll of Nigerian & Huichol art from Mexico 52225

Radium Springs NM

Fort Selden State Monument, Fort Selden Junction 185 and Interstate 25 on Exit 19, Radium Springs, NM 88054 • T: +1 505 5268911 • F: +1 505 6470421 • selden@zianet.com • www.nmculture.org •
Head: *Nathan Stone* •
Military Museum – 1973
Military artifacts, uniforms, military hist to the Indian Wars period 52226

Raleigh NC

Artspace, 201 E Davie St, Raleigh, NC 27601 • T: +1 919 8212787 • F: +1 919 8210383 • info@artspacenc.org • www.artspacenc.org •
Exec. Dir.: *Courtney Bailey* •
Fine Arts Museum – 1986
Fantasy art 52227

Contemporary Art Museum, 336 Fayetteville St, Raleigh, NC 27601 • T: +1 919 8360088 • F: +1 919 8362239 • nw@camnc.org • www.camnc.org •
Exec. Dir.: *Denise Dickens* •
Fine Arts Museum – 1983 52228

Exploris → Marbles Kids Museum

Frankie G. Weems Gallery & Rotunda Gallery, Meredith College, Gaddy-Hamrick ArtCenter, 3800 Hillsborough St, Raleigh, NC 27607-5298 • T: +1 919 7608465 • F: +1 919 7602347 • bankerm@meredith.edu •
Dir.: *Maureen Banker* •
Public Gallery
American art 52229

Marbles Kids Museum, 201 E Hargett St, Raleigh, NC 27601 • T: +1 919 8344040 • F: +1 919 8343516 • info@marbleskidsmuseum.org • www.marbleskidsmuseum.org •
Pres.: *Anne Bryan* • Cur.: *Linda Dallas* •
Special Museum – 1999 52230

Mordecai Historic Park, 1 Mimosa St, Raleigh, NC 27604 • T: +1 919 8344844 • F: +1 919 8347314 • CapPresInc@aol.com • capitalareapreservation.org •
Pres.: *Gary G. Roth* •
Open Air Museum / Historical Museum – 1972
Period furnishings and linnens 52231

North Carolina Museum of Art, 2110 Blue Ridge Rd, Raleigh, NC 27607-6494 • T: +1 919 8396262 • F: +1 919 7338034 • hmckinney@ncmamail.dcr.state.nc.us • www.ncartmuseum.org •
Dir.: *Lawrence J. Wheeler* • Assoc. Dir./ Cur.: *John W. Coffey* (American and Modern Art) • Cur.: *David H. Steel* (European Art) • *Huston Pascal* (Modern Art) • *Mary Ellen Soles* (Ancient Art) • *Dennis P. Weller* (European Art) • Chief Cons.: *Bill Brown Findley* •
Fine Arts Museum – 1956
European and American art, ancient art, African, Oceanic and pre-Columbian art, Judaic art 52232

North Carolina Museum of History, 5 E Edenton St, Raleigh, NC 27601-1011 • T: +1 919 8077900 • F: +1 919 7338655 • betsy.buford@ncmail.net • www.ncmuseumofhistory.org •
Dir.: *Elizabeth F. Buford* • *Dorothea Bitler* • Chief Cur.: *Dr. Joseph Porter* •
Historical Museum – 1902
Folklife, anthropology, military, medicine, transportation, furnishings, uniforms, tools and equip 52233

North Carolina Museum of Natural Sciences, 11 W Jones St, Raleigh, NC 27601-1029 • T: +1 919 7337450 • F: +1 919 7331573 • museum@naturalscienes.org • www.naturalsciences.org •
Dir.: *Dr. Betsy Bennett* • *Roy G. Campbell* (Exhibits) • *Dr. Stephen Busack* (Research and Collections) •
Natural History Museum – 1879
Zoology, geology, paleontology 52234

North Carolina State University Gallery of Art and Design, 2610 Cates Av, Raleigh, NC 27695-7306 • T: +1 919 5153503 • F: +1 919 5156163 • gallery@ncsu.edu • www.ncsu.edu/gad •
Dir.: *Dr. Charlotte V. Brown* •
University Museum / Fine Arts Museum – 1979
Coll of American, Indian, Asian & pre-Columbian textiles, ceramics, furniture, product design 52235

Raleigh City Museum, Briggs Bldg, 220 Fayetteville St Mall, Raleigh, NC 27601-1310 • T: +1 919 8323775 • F: +1 919 8323085 • dwescott@raleighcitymuseum.org • www.raleighcitymuseum.org •
C.E.O. & Dir.: *Dusty Wescott* •
Local Museum – 1993 52236

Ralls TX

Ralls Historical Museum, 801 Main St, Ralls, TX 79357 • T: +1 806 2532425 • rallshistoricalmuseum@nts-online.net •
Dir.: *Donna Harris* •
Local Museum – 1970
Art objects, hist, archaeology, military 52237

Ramah NM

El Morro National Monument, Hwy 53, 43 miles west of Grants, Ramah, NM 87321 • T: +1 505 7834226 • F: +1 505 7834689 •
Head: *Peri Eringen* •
Historical Museum / Archaeology Museum – 1906
Archaeological finds, pots, implements, handcrafts, weapons, religious items, prehistoric pottery, Spanish and South West exploration artifacts 52238

Rancho Cucamonga CA

Wignall Museum of Contemporary Art, Chaffey College, 5885 Haven Av, Rancho Cucamonga, CA 91737-3002 • T: +1 909 9412389 • wignall.staff@chaffey.edu • www.chaffey.edu/wignall •
Dir.: *Rebecca Trawick* •
Public Gallery / University Museum – 1972 52239

Rancho Mirage CA

Children's Discovery Museum of the Desert, 71-701 Gerald Ford Dr, Rancho Mirage, CA 92270 • T: +1 760 3210602 • F: +1 760 3211605 • cdminfo@cdmod.org • www.cdmod.org •
Exec. Dir.: *LeeAnne Vanderbeck* •
Special Museum – 1987
Hands-on museum, archaeological dig 52240

Heartland -California Museum of the Heart, 39600 Bob Hope Dr, Rancho Mirage, CA 92270 • T: +1 760 3243278 • F: +1 760 3461867 •
Dir./Cur.: *Adam Rubinstein* •
Historical Museum – 1988
Heart health 52241

Rancho Palos Verdes CA

Palos Verdes Art Center, 5504 W Crestridge Rd, Rancho Palos Verdes, CA 90275 • T: +1 310 5412479 • F: +1 310 5419520 • info@pvartcenter.org • www.pvartcenter.org •
Dir.: *Robert A. Yassin* • *Scott Canty* (Exhibits) • *Angela Hoffman* (Education) • *Gail Phinney* (Programs) •
Fine Arts Museum – 1931
Fine arts, ceramics, photos, print making 52242

Randleman NC

Saint Paul Museum, 411 High Point St, Randleman, NC 27317 • T: +1 336 4982447 •
Pres.: *Louise Hudson* •
Local Museum – 1966
Early domestic life 52243

Randolph NJ

Historical Museum of Old Randolph, 30 Carrel Rd, Randolph, NJ 07869 • T: +1 973 9897095 • hsor@juno.com •
Pres.: *Francis DeLucia* •
Local Museum – 1979
Local history 52244

Randolph VT

Chandler Gallery, Chandler Music Hall and Cultural Center, Main St, Randolph, VT 05060 • T: +1 802 7289878, 7283840 • F: +1 802 7284612 • chandler@quest.net •
Pres.: *Janet Watton* • Dir.: *Rebeeca B. McMeekin* •
Public Gallery – 1979
Rotating exhib ... 52245

Randolph Historical Society, Salisbury St, Randolph, VT 05060 • T: +1 802 72286677 • hatchasse@ earthlink.net •
Cur.: *Harriet Chase* •
Association with Coll – 1960
Costumes, pipe organ, early musical instr ... 52246

Rankin TX

Rankin Museum, 101 W Main St, Rankin, TX 79778 • T: +1 915 6932758, 6932422 • F: +1 915 6932303 • jcpg@yahoo.com • www.rootsweb.com/~txupton/ •
Pres.: *Judy Greer* •
Local Museum – 1974
Restored 1940 firetruck, fully operative, photo coll ... 52247

Rantoul IL

Korean War Veterans National Museum, 1007 Pacesetter Dr, Rantoul, IL 61866 • T: +1 217 8934111 • kwvm@kwvm.com • www.kwvmuseum.com •
C.E.O.: *Larry Sassorossi* •
Military Museum – 1997
Korean War memorabilia, photos, docs, books – library ... 52248

Octave Chanute Aerospace Museum, 1011 Pacesetter Dr, Rantoul, IL 61866- • T: +1 217 8931613 • F: +1 217 8925774 • director@ aeromuseum.org • www.aeromuseum.org •
Dir.: *James Snyder* •
Military Museum / Science&Tech Museum – 1992
Fighter, bomber and training aircraft used by USAAF and USAF ... 52249

Rapid City SD

Dahl Arts Center, 713 Seventh St, Rapid City, SD 57701 • T: +1 605 3944101 • F: +1 605 3946121 • contact@thedahl.org • www.thedahl.org •
Pres.: *Ron Johnsen* • Exec. Dir.: *Linda Anderson* •
Fine Arts Museum / Performing Arts Museum – 1974
Contemporary art ... 52250

Journey Museum, 222 New York St, Rapid City, SD 57701 • T: +1 605 3946923, 3942249 • F: +1 605 3946940 • journey@journeymuseum.org • www.journeymuseum.org •
Ethnology Museum / Archaeology Museum – 1997
Paleontology, geology, archaeology, history, Native American hist and culture ... 52251

Minnilusa Pioneer Museum, 222 New York St, Rapid City, SD 57701 • T: +1 605 3946099 • F: +1 605 3946940 •
Dir.: *Robert E. Preszler* • Cur.: *Eileen Howe* •
Local Museum – 1938
Local hist ... 52252

Museum of Geology, South Dakota School of Mines and Technology, 501 E St, Rapid City, SD 57701 • T: +1 605 3942467 • F: +1 605 3946131 • museum@sdsmt.edu • www.museum.sdsmt.edu •
Dir.: *Gale A. Bishop* • Cur.: *James E. Martin* (Paleontology) •
University Museum / Natural History Museum – 1885
Natural hist ... 52253

Sioux Indian Museum, 222 New York St, Rapid City, SD 57701 • T: +1 605 3942381 • F: +1 605 3486182 • montileaux@journeymuseum.org • www.journeymuseum.org •
Cur.: *Paulette Montileaux* •
Fine Arts Museum / Folklore Museum – 1939
Indian art ... 52254

Raton NM

Raton Museum, 216 S First St, Raton, NM 87740 • T: +1 505 4458979 •
C.E.O.: *Roger Sanchez* •
Local Museum – 1939
Railroading, mining, ranching, history of Raton ... 52255

Ravenna OH

Portage County Historical Society Museum, 6549 N Chestnut St, Ravenna, OH 44266 • T: +1 330 2963523 • history@config.com • www.history.portage.oh.us •
Pres.: *John Carson* •
Local Museum – 1951
Archaeological finds, hist, glassware ... 52256

Ravenswood WV

Washington's Lands Museum and Sayre Log House, Rte 68, Ravenswood, WV 26164 • F: +1 304 3727935 • edrauh@wirefire.com •
Local Museum – 1970
Local tools, clothing, Indian artifacts ... 52257

Rawlins WY

Carbon County Museum, 904 W Walnut St, Rawlins, WY 82301 • T: +1 307 3282740 • F: +1 307 3282666 • carbonc@wyoming.com •
Dir.: *Joyce J. Kelley* •
Local Museum – 1940 ... 52258

Raymond MS

Marie Hull Gallery, Hinds Community College, Raymond, MS 39154 • T: +1 601 8573275 • F: +1 601 8573392 •
Dir.: *Gayle McCarty* •
Fine Arts Museum
Prints, sculpture, paintings ... 52259

Reading MA

Parker Tavern Museum, 103 Washington St, Reading, MA 01867 • T: +1 781 9440885 •
Historical Museum – 1916
Headquarters for Scotch Highlanders prisoners of war during American revolution ... 52260

Reading OH

Reading Historical Society Museum, 22 W Benson St, Reading, OH 45215 • T: +1 513 7618535 •
Pres.: *James J. Lichtenberg* •
Local Museum – 1988
Local hist, furnishings, uniforms ... 52261

Reading PA

Freedman Gallery, Albright College, 1621 N 13th St, Reading, PA 19604 • T: +1 610 9217541, 9217715 • F: +1 610 9217768 • cyoungs@alb.deu • www.albright.edu •
Dir.: *Christopher Youngs* •
Fine Arts Museum / University Museum – 1976
Contemporary painting, sculpture, photography, prints, video, installation and performance art ... 52262

Historical Society of Berks County Museum, 940 Centre Av, Reading, PA 19601 • T: +1 610 3754375 • F: +1 610 3754376 • histsoc@berksweb.com • www.berksweb.com/histsoc •
Dir.: *Harold E. Yoder* •
Local Museum – 1869
Local hist ... 52263

Mid Atlantic Air Museum, 11 Museum Dr, Reading, PA 19605 • T: +1 610 3727333 • F: +1 610 3721702 • russ@maam.org • www.maam.org •
Cur.: *Linda T. Strine* •
Science&Tech Museum – 1980
54 aircrafts on aviation ... 52264

Reading Public Museum and Art Gallery, 500 Museum Rd, Reading, PA 19611-1425 • T: +1 610 3715850 • F: +1 610 3715632 • museum@ptd.net • www.readingpublicmuseum.org •
Dir./C.E.O.: *Ronald C. Roth* •
Fine Arts Museum / Natural History Museum / Science&Tech Museum – 1904
Art and science ... 52265

Reading Railroad Heritage Museum, Temple Station, Tuckerton Rd, Reading, PA 19605, mail addr: POB 15143, Reading, PA 19612-5143 • T: +1 610 9299902 • www.readingrailroad.org •
Pres.: *Duane E. Engle* •
Science&Tech Museum – 1976
Railroad hist and equip, artifacts, paperwork ... 52266

Reading VT

Reading Historical Society, Main St, Rte 106, Reading, VT 05062 • T: +1 802 4845005 •
Pres.: *Jonathan Springer* •
Association with Coll – 1953
Hist, military, medical, music ... 52267

Reads Landing MN

Wabasha County Museum, 871 Second St, Reads Landing, MN 55968-6813 • T: +1 507 2824027 • F: +1 507 9325204 • fpasse@rconnect.com • www.lakecitymn.org •
Pres.: *Eugene Passe* • Cur.: *Bessie Hoover* •
Historical Museum – 1965
Farming machinery, vehicles, school, country store, Laura Ingalls Wilder exhibit, Mississippi River artifacts ... 52268

Readsboro VT

Readsboro Historical Society, Main St, Readsboro, VT 05350 • T: +1 802 4235432 •
Pres.: *Betty Bolognani* •
Association with Coll – 1972
Day books of the area dating back to the 1700s, 19th & 20th c photographs, spinning wheels ... 52269

Red Bluff CA

Kelly-Griggs House Museum, 311 Washington St, Red Bluff, CA 96080 • T: +1 530 5271129 •
Dir.: *Jeraldine J. Wolfe* •
Decorative Arts Museum – 1965
Paintings, Indian artifacts, antique furniture, Victorian costumes ... 52270

Red Cloud NE

Webster County Historical Museum, 721 W Fourth Av, Red Cloud, NE 68970 • T: +1 402 7462444 •
Pres.: *Ron Gestring* • Dir.: *Joyce Terhune* •
Local Museum – 1965
Agriculture, period artifacts, natural hist, medical, military ... 52271

Willa Cather State Historic Site, 326 N Webster, Red Cloud, NE 68970 • T: +1 402 7462653 • F: +1 402 7462652 • wcpm@gpcom.net • www.willacather.org •
Pres.: *Mellanee Kuasnicka* • Exec. Dir.: *Betty Kort* •
Local Museum – 1978
Local hist, photographs, letters ... 52272

Red Wing MN

Goodhue County Historical Society Museum, 1166 Oak St, Red Wing, MN 55066 • T: +1 651 3886024 • F: +1 651 3883577 • goodhuecountyhis@qwest.net • www.goodhuehistory.mus.mn.us •
Exec. Dir.: *Char Henn* •
Historical Museum – 1869
Immigration & early settlement, pioneers, local government, military, industry, Dakota Indians, natural hist ... 52273

Redlands CA

Lincoln Memorial Shrine, 125 W Vine St, Redlands, CA 92373 • T: +1 909 7987632 • archives@akspl.org • www.lincolnshrine.org •
Pres.: *Jack D. Tompkins* • Cur.: *Donald McCue* •
Historical Museum – 1932
Rare manuscripts, soldiers and civilians into the Civil war – library ... 52274

Peppers Art Gallery, University of Redlands, 1200 E Colton Av, Redlands, CA 92373 • T: +1 909 7932121 ext 3660 • F: +1 909 7486293 • www.redlands.edu •
Dir.: *Barbara Thomason* •
Fine Arts Museum / University Museum
Ethnic art, graphics, paintings, sculpture, mixed media, drawings, photography, ceramics and design ... 52275

San Bernardino County Museum, 2024 Orange Tree Ln, Redlands, CA 92374 • T: +1 909 3072669 • F: +1 909 3070539 • rmckernan@sbcm.sbcounty.gov • www.sbcountymuseum.org •
Dir.: *Robert L. McKernan* • Sc. Staff: *J. Chris Sagebiel* (Geology) • *Kathleen Springer* (Geology) • *Michele Nielsen* (History) • *Dr. Adella Schroth* (Anthropology) • *Eric Scott* (Paleontology) • *Jolene Redvale* (Education) •
Natural History Museum – 1959
Local nature, history, archaeology, geology, art – library ... 52276

Redmond WA

Marymoor Museum of Eastside History, 6046 W Lake Sammamish Pkwy, Redmond, WA 98073 • T: +1 206 8853684 • F: +1 425 8853684 • eastsideheritagemuseum@msn.com •
Pres.: *Libby Walgamott* • Dir.: *Eden R. Toner* • Sc. Staff: *Barb Williams* (Education) • *Kathryn V. Innes* (Collections, Photos) •
Local Museum – 1965
Local hist ... 52277

Redwood City CA

San Mateo County History Museum, 2200 Broadway St, Redwood City, CA 94063 • T: +1 650 2990104 • info@historysmc.org • www.historysmc.org •
Dir.: *Mitchell P. Postel* • Cur.: *Karen Brey* •
Historical Museum – 1935
County hist, tools, horsedrawn vehicles, local culture – library ... 52278

Redwood Falls MN

Redwood County Museum, 915 W Bridge St, Redwood Falls, MN 56283 • T: +1 507 6373329 • dahunt@newulmtel.net •
Cur.: *Alice Hunt* •
Historical Museum – 1949
County hist, hand fans, farm tools, Indian artifacts, natural hist ... 52279

Reedsport OR

Umpqua Discovery Center, 409 Riverfront Way, Reedsport, OR 97467 • T: +1 541 2714816 • F: +1 541 2714816 • umpquadiscoverycenter@ charterinternet.com •
Dir.: *Diane Novak* •
Local Museum – 1993
Oregon coast hist, tides and their influence on coast towns, cultural and natural hist ... 52280

Reedville VA

Reedville Fishermen's Museum, 504 Main St, Reedville, VA 22539 • T: +1 804 4536529 • F: +1 804 4537159 • bunker@crosslink.net • www.rfmuseum.com •
Dir.: *Cara A. Sutherland* •
Historical Museum – 1986
Fishing industry ... 52281

Refugio TX

Refugio County Museum, 102 W West St, Refugio, TX 78377 • T: +1 512 5265555 • F: +1 512 5264943 •
C.E.O.: *Kathleen Campbell* •
Local Museum – 1983
Items preserving hist & culture of the County ... 52282

Regent ND

Hettinger County Historical Society Museum, Main St, Regent, ND 58650 • T: +1 701 5634543 •
Dir.: *Jess Kouba* •
Local Museum / Ethnology Museum – 1962 ... 52283

Rehoboth Beach DE

Rehoboth Art League, 12 Dodds Ln, Rehoboth Beach, DE 19971 • T: +1 302 2278408 • F: +1 302 2274121 • www.rehobothartleague.org •
Dir.: *Nancy Alexander* •
Fine Arts Museum – 1938
Art ... 52284

Reidsville NC

Chinqua-Penn Plantation, North Carolyna State University, 2138 Wentworth St, Reidsville, NC 27320 • T: +1 336 3494576 • F: +1 336 3424863 • charles_leffler@ncsu.edu • www.chinquapenn.com •
Cur.: *Carolyn Brady* •
Ethnology Museum / University Museum – 1966
Antique European furniture, oriental art objects, farm, garden and household tools and equip ... 52285

Remsen NY

Steuben Memorial State Historic Site, Star Hill Rd, Remsen, NY 13438, mail addr: Oriskany Battlefield, 7801 State Rte 69, Oriskany, NY 13424 • T: +1 315 7687224 • F: +1 315 3373081 • nancy.demyttenaere@oprhp.state.ny.us • www.nysparks.com •
Historical Museum / Historic Site – 1930
Burial Site of Baron Friedrich Wilhelm von Steuben (1730-1794), drillmaster of American Army, reprod log cabin, furniture, tomb – hist archives ... 52286

Reno NV

National Automobile Museum, Harrah Collection, 10 Lake St S, Reno, NV 89501 • T: +1 775 3339300 • F: +1 775 3339309 • info@automuseum.org • www.automuseum.org •
Exec. Dir.: *Jackie L. Frady* •
Science&Tech Museum – 1989
Classic and special vehicles ... 52287

Nevada Historical Society Museum, Nevada Department Museums and Library, 1650 N Virginia St, Reno, NV 89503 • T: +1 702 6881191 • F: +1 702 6882917 • www.nevadaculture.org •
Dir.: *Peter L. Bandurraga* • Cur.: *Eric N. Moody* (Manuscripts) •
Historical Museum – 1904
Library ... 52288

Nevada Museum of Art, 160 W Liberty St, Reno, NV 89501 • T: +1 775 3293333 • F: +1 775 3291541 • art@nma.reno.nv.us • www.nevadaart.org •
Dir.: *Steven S. High* • Cur.: *Diane Deming* •
Fine Arts Museum / Public Gallery – 1931
Temporary exhibitions – library ... 52289

Sheppard Fine Arts Gallery, University of Nevada, Reno, Church Fine Arts Complex, Mail Stop 224, Reno, NV 89557 • T: +1 775 7848015 • F: +1 775 7846655 • grays@unr.nevada.edu •
Dir.: *Sara Gray* •
Fine Arts Museum / University Museum – 1960
Print coll, drawings, sculpture, photos, paintings 52290

Sierra Arts Foundation, 200 Flint St, Reno, NV 89501 • T: +1 775 3292787 • F: +1 775 2391328 • jill@aierra-arts.org • www.sierra-arts.org •
Exec. Dir.: *Jill Berryman* •
Public Gallery – 1971
Contemporary works by emerging artists ... 52291

W.M. Keck Museum, Mackay School of Mines, University of Nevada, Reno, NV 89557 • T: +1 775 7846052 • F: +1 702 7841766 • rdolbier@unr.edu • www.mines.unr.edu/museum •
University Museum / Natural History Museum – 1908
Maetallurgical, paleontological, geological, fossils, mineralogical, hist of mining ... 52292

Wilbur D. May Museum, 1502 Washington St, Reno, NV 89503 • T: +1 775 7855961 • F: +1 775 7854707 •
Dir.: *Karen Mullen* • Pres.: *Melva Howell* •
Local Museum – 1985
Local hist ... 52293

Rensselaer NY

Crailo State Historic Site, 9 1/2 Riverside Av, Rensselaer, NY 12144 • T: +1 518 4638738 • F: +1 518 4331860 •
Head: *Donnarae Gordon* •
Local Museum – 1924
Hist building of the Van Rensselaer family, hist of Dutch culture in the upper Hudson Valley, Dutch colonial period artifacts ... 52294

Renton WA

Renton Museum, 235 Mill Av S, Renton, WA 98055-2133 • T: +1 425 2552330 • F: +1 425 2551570 • saanderson@ci.renton.wa.us • www.rentonhistory.org • Dir.: *Steve Anderson* • Cur.: *Steve Smith* •
Local Museum – 1966
Local hist, cultural and coal mining artifacts, clothing 52295

Republic KS

Pawnee Indian Village, 480 Pawnee Trail, Republic, KS 66964 • T: +1 785 3612255 • F: +1 785 3612255 • piv@kshs.org • www.kshs.org • Cur.: *Richard Gould* • Chm.: *Narveen Brzon* •
Ethnology Museum / Archaeology Museum / Historical Museum – 1967
Archaeology, preserved Pawnee site 52296

Republic MO

General Sweeney's Museum, 5228 S State Hwy ZZ, Republic, MO 65738 • T: +1 417 7321224 • F: +1 417 7321224 • tsweeney@alltel.net • www.civilwarmuseum.com • C.E.O.: *Karen Sweeney* •
Military Museum – 1992
Civil War in the Trans Mississippi 52297

Wilson's Creek National Battlefield, 6424 W Farm Rd 182, Republic, MO 65738 • T: +1 417 7322662 • F: +1 417 7321167 • www.nps.gov/wicr •
Military Museum – 1960 52298

Reston VA

The Greater Reston Arts Center, 11911 Freedom Dr, Ste 110, Reston, VA 20190 • T: +1 703 4719242 • F: +1 703 4710952 • graceart@restonart.com • Pres.: *James C. Cleveland* • C.E.O.: *Anne Brown* •
Fine Arts Museum / Folklore Museum – 1974
Civic art and culture 52299

Rexburg ID

Upper Snake River Valley Historical Museum, 51 N Center, Rexburg, ID 83440 • T: +1 208 3567030 • F: +1 208 3563379 • histsoc@onewest.net • Pres.: *David Morris* • Dir./Cur.: *Louis Clements* •
Local Museum – 1965
Agriculture, antiques, manuscripts, Idaho and local hist for teaching – library 52300

Rhinebeck NY

Rhinebeck Aerodrome Museum, Stone Church Rd and Norton Rd, Rhinebeck, NY 12572 • T: +1 845 7523200 • F: +1 914 7586481 • info@oldrhinebeck.org • www.oldrhinebeck.org • Pres.: *John Costa* • Dir.: *Tom Polapink* •
Science&Tech Museum – 1977
Aeronautics, airplanes of pioneer, WW I and Lindbergh era, Spirit of Saint Lewis memorabilia 52301

Rhinelander WI

Rhinelander Logging Museum, Pioneer Park, Martin Lynch Dr, Rhinelander, WI 54501 • T: +1 715 3622193, 3695004 • www.rhinelanderchamber.com • Pres.: *Walt Krause* •
Folklore Museum / Science&Tech Museum – 1932
Pioneer logging industry – School Musuem 52302

Rialto CA

Rialto Museum, 201-205 N Riverside Av, Rialto, CA 92376 • T: +1 909 8751750 •
Folklore Museum / Local Museum – 1971
Local hist, orange ind, Victorian furnishings – library 52303

Richey MT

Richey Historical Society, POB 218, Richey, MT 59259 • T: +1 406 7735615 • C.E.O.: *Jeff Brost* • Chief Cur.: *Valerie B. Baldwin Rehbein* •
Association with Coll – 1973
Farm tools, clocks, musical instr, furniture, clothing, machinery, local hist 52304

Richfield Springs NY

Petrified Creatures Museum of Natural History, Rte 20, Richfield Springs, NY 13439, mail addr: POB 751, Richfield Springs, NY 13439 • T: +1 315 8582868 • F: +1 315 8582868 • petrifiedcreaturesmuseum@yahoo.com • www.cooperstownchamber.org • Dir. & C.E.O.: *Stella C. Mlecz* • Ass. Dir.: *Sally E. Kennedy* • Cur.: *Richard S. Mlecz* •
Natural History Museum / Open Air Museum – 1934
Prehist animal life, petrified fossils, life size reproductions of dinosaurs 52305

Richland WA

Allied Arts Center and Gallery, 89 Lee Blvd, Richland, WA 99352 • T: +1 509 9439815 • F: +1 509 9434068 • Pres.: *Marion Goheen* •
Fine Arts Museum – 1947 52306

Columbia River Exhibition of History, Science and Technology, 95 Lee Blvd, Richland, WA 99352 • T: +1 509 9439000 • F: +1 509 9431770 • gwen@crehst.org • www.crehst.org • Dir.: *Gwen I. Leth* • Cur.: *Connie Estep* •
Science&Tech Museum – 1963
Science, nuclear hist 52307

Richland Center WI

Frank Lloyd Wright Museum, 300 S Church St, Richland Center, WI 53581 • T: +1 608 6472808 • Dir.: *Harvey W. Glanzer* • *Bethel I. Caulkins* •
Fine Arts Museum
Architecture by Frank Lloyd Wright, enginering/architectural models 52308

Richlands NC

Onslow County Museum, 301 S Wilmington St, Richlands, NC 28574 • T: +1 910 3245008 • F: +1 910 3242897 • museum@co.onslow.nc.us • www.co.onslow.nc.us/museum • Dir.: *Albert Potts* • Ass. Dir.: *Lisa Whitman-Grice* •
Local Museum – 1976
19th and early 20th c tools and farm implements, woodworking tools, 19th c costumes 52309

Richmond CA

Gallery of the National Institute of Art and Disabilities, 551 23rd St, Richmond, CA 94804 • T: +1 510 6200290 • F: +1 510 6200326 • admin@niadart.org • www.niadart.org • Dir.: *Allisen Asercion* • Cur.: *Denise M. Pranza* •
Public Gallery – 1984
Outsider art, painting and drawing, printmaking, ceramics, fabric arts 52310

Richmond Art Center, 2540 Barrett Av, Richmond, CA 94804 • T: +1 510 6206772 • F: +1 510 6206771 • admin@therichmondartcenter.org • www.therichmondartcenter.org • Dir.: *Amy Stimmel* •
Public Gallery – 1936 52311

Richmond Museum of History, 400 Nevin Av, Richmond, CA 94802 • T: +1 510 2357387 • www.richmondmuseumofhistory.org • Dir.: *Donald Bastin* •
Local Museum – 1952
Local hist, decorative arts, costume, WW II shipyard – archive, library 52312

Richmond IN

Joseph Moore Museum, Earlham College, Richmond, IN 47374 • T: +1 765 9831303 • F: +1 765 9831497 • johni@earlham.edu • www.earlham.edu • Dir.: *Dr. John B. Iverson* • Ass. Dir.: *Dr. William H. Buskirk* •
University Museum / Natural History Museum – 1887
Natural history 52313

Richmond Art Museum, 350 Hub Etchison Pkwy, Richmond, IN 47375 • T: +1 765 9660256 • F: +1 765 9733738 • shaund@rcs.k12.in.us • www.richmondartmuseum.org • C.E.O: *Shaun Dinwerth* •
Fine Arts Museum – 1898
American paintings, sculpture, graphics, decorative arts 52314

Wayne County Historical Museum, 1150 N A St, Richmond, IN 47374 • T: +1 765 9625756 • F: +1 765 9390909 • micheleb@infocom.com • www.wchm.org • Dir.: *Jan Livingston* •
Local Museum – 1930
General museum, housed in Hicksite Friends Meeting House 52315

Richmond KY

Fort Boonesborough Museum, 4375 Boonesboro Rd, Richmond, KY 40475 • T: +1 859 5273131 • F: +1 859 5273328 • phil.gray@mail.state.ky.us • www.kystateparks.com • Dir.: *Lilly Newman* (Craft) • Cur.: *Jerry Raisor* •
Historical Museum – 1974
Early Euro-American settlement of Kentucky, local Native American artifacts 52316

White Hall Historic Site, 500 White Hall Shrine Rd, Richmond, KY 40475 • T: +1 856 6239178 • F: +1 606 6268489 • Kathleen.white@ky.gov • www.state.ky.us/agencies/parks/whthall.htm • Dir.: *Kathleen White* • Cur.: *Lashe Mullins* •
Local Museum / Decorative Arts Museum – 1971
Historic house Georgian style bldg, added to in the Italianate style 52317

Richmond ME

CHTJ Southard House Museum, 75 Main St, Richmond, ME 04357 • T: +1 207 7378202 • F: +1 207 7378772 • fcmain@ctel.net • Pres./Cur.: *Carolyn Cooper Case* •
Local Museum – 1990
Toy and doll coll, textiles, local hist 52318

Richmond TX

Fort Bend Museum, 500 Houston, Richmond, TX 77469 • T: +1 281 3426478 • F: +1 281 3422439 • mmoore@georgeranch.org • www.fortbendmuseum.org • Dir.: *Michael R. Moore* •
Local Museum – 1967
Settlement of Stephen F. Austin's Colony 52319

Richmond VA

Agecroft Hall, 4305 Sulgrave Rd, Richmond, VA 23221 • T: +1 804 3534241 • F: +1 804 3532151 • www.agecrofthall.com • Dir.: *Richard W. Moxley* • Cur.: *Deborah Knott de Arechaga* •
Local Museum – 1967
Textiles, armor, musical instr 52320

The American Historical Foundation Museum, 1142 W Grace St, Richmond, VA 23220 • T: +1 804 3531812 • F: +1 804 3534895 • ahfoffice@aol.com • www.ahfrichmond.com • Pres./Cur.: *Robert A. Buerlein* •
Historical Museum / Military Museum – 1982
Military, hist, Marine Raider equip 52321

Anderson Gallery, c/o School of the Arts, Virginia Commonwealth University, 907 1/2 W Franklin St, Richmond, VA 23284-2514 • T: +1 804 8281522 • F: +1 804 8288585 • www.vcu.edu/artweb/gallery/ • Dir.: *Ted Potter* • Cur.: *Amy G. Moorefield* •
Public Gallery – 1969
Prints, photographs, contemporary painting & sculpture, folk art 52322

Artspace, Zero E 4th St, Richmond, VA 23224 • T: +1 804 2326464 • artspaceorg@gmail.com • www.artspacegallery.org • Dir.: *Christina Newton* •
Public Gallery
Changing exhibits 52323

Beth Ahabah Museum and Archives, 1109 W Franklin St, Richmond, VA 23220 • T: +1 804 3532668 • F: +1 804 3583451 • bama@bethahabah.org • www.bethahabah.org •
Historical Museum – 1977
Jewish hist 52324

Children's Museum of Richmond, 2626 W Broad St, Richmond, VA 23220-1904 • T: +1 804 4747000 • F: +1 804 4747099 • director@c-mor.org • www.c-mor.org •
Special Museum – 1977
Children's museum 52325

Crestar Bank Art Collection, 919 E Main St, Richmond, VA 23219 • T: +1 804 7825000 • F: +1 804 7825469 • www.suntrust.com •
Fine Arts Museum – 1970
Contemporary art 52326

Edgar Allan Poe Museum, 1914-16 E. Main St, Richmond, VA 23223 • T: +1 804 6485523 • F: +1 804 6488729 • info@poemuseum.org • www.poemuseum.org •
Local Museum – 1922
Poe books, manuscripts, memorabilia, illustrations, art galleries 52327

Hand Workshop Art Center, 1812 W Main St, Richmond, VA 23220 • T: +1 804 3530094 • F: +1 804 3538018 • Dir.: *Susan Glasser* •
Fine Arts Museum – 1963 52328

Joel and Lila Harnett Museum of Art, University of Richmond, 28 Westhampton Way, Richmond, VA 23173 • T: +1 804 2898276 • F: +1 804 2871894 • museums@richmond.edu • museums.richmond.edu/hma • Dir.: *Richard Waller* •
University Museum / Fine Arts Museum – 1968
All media of visual arts 52329

Joel and Lila Harnett Print Study Center, University of Richmond, 28 Westhampton Way, Richmond, VA 23173 • T: +1 804 2898276 • F: +1 804 2871894 • rwaller@richmond.edu • museums.richmond.edu/hpsc •
Fine Arts Museum / University Museum – 2001 52330

John Marshall House, 818 E Marshall St, Richmond, VA 23219 • T: +1 804 6487998 • F: +1 804 6485880 • johnmarshallhouse@apva.org • www.johnmarshall.org • Exec. Dir.: *Elizabeth Kostelny* •
Local Museum – 1790
Furnishings associated with the Chief Justice 18th to early 19th c glass 52331

Lora Robins Gallery of Design from Nature, University of Richmond, 28 Westhampton Way, Richmond, VA 23173 • T: +1 804 2898276 • F: +1 804 2871894 • museums@richmond.edu • museums.richmond.edu/lrg •
University Museum / Natural History Museum / Decorative Arts Museum – 1977
Jurassic dinosaur fossils, glasswork art by Dale Chihuly. Pre-columbian decorative vessels, Faberge jewels, rare gems, Oceanic art, African board games, prehistoric shells, florescent rocks 52332

Maggie L. Walker National Historic Site, 600 2nd St N, Richmond, VA 23219, mail addr: 3215 Broad St E, Richmond, VA 23223 • T: +1 804 7712017 • F: +1 804 7718522 • cynthiamacleod@nps.gov • www.nps.gov/mawa • Dir.: *Cynthia McLeod* • Cur.: *Thomas Klydie* •
Historical Museum – 1978
Hist house, original furnishings 1890 to 1930, African American history 52333

Maymont, 1700 Hampton St, Richmond, VA 23220 • T: +1 804 3587166 • F: +1 804 3589994 • info@maymont.org • www.maymont.org • Dir.: *Geoffrey Platt jr.* • Cur.: *Dot Rugus* (Carriages) •
Decorative Arts Museum / Local Museum / Agriculture Museum – 1925
19th c carriages 52334

The Museum of the Confederacy, 1201 E Clay St, Richmond, VA 23219-1615 • T: +1 804 6491861 • F: +1 804 6447150 • info@moc.org • www.moc.org • Exec. Dir.: *S. Waite Rawles III* • Cur.: *Robert Hancock* • Sc. Staff: *Melinda Gales* • *Dr. John M. Coski* (History) •
Historical Museum / Military Museum – 1896
Confederate military, artefacts, flags, docs – Eleanor Brockenbrough Library 52335

Old Dominion Railway Museum, 102 Hull St, Richmond, VA 23224 • T: +1 804 2236237 • F: +1 804 7454735 • www.odcnrhs.org • Cur.: *Greg A. Hodges* •
Science&Tech Museum – 1990
Railroading hist in Central Virginia, steam, caboose, passenger and freight equip, locomotives and rolling stock 52336

Richmond National Battlefield Park, 3215 E Broad St, Richmond, VA 23223 • T: +1 804 2261981 • F: +1 804 7718522 • www.nps.gov/rich • Head: *Cynthia MacLeod* •
Military Museum – 1936
Civil War hist 52337

Science Museum of Virginia, 2500 W Broad St, Richmond, VA 23220-2054 • T: +1 804 8641400 • F: +1 804 8641560 • smvfeedback@smv.mus.va.us • www.smv.org • Dir.: *Walter R.T. Witschey* • *Pat Fishback* •
Science&Tech Museum – 1970
Science, aviation 52338

Valentine Richmond History Center, 1015 E Clay St, Richmond, VA 23219-1590 • T: +1 804 6490711 • F: +1 804 6433510 • info@richmondhistorycenter.com • www.richmondhistorycenter.com • Dir.: *William J. Martin* • Cur.: *Colleen Callahan* (Costumes and Textiles) •
Historical Museum / Decorative Arts Museum – 1892
Urban hist 52339

Virginia Aviation Museum, International Airport, 5701 Huntsman Rd, Richmond, VA 23250-2416 • T: +1 804 2363620, 2363622 • F: +1 804 2363623 • mboehme@smv.org • www.vam.smv.org • Dir.: *Walter R.T. Witschey* •
Science&Tech Museum – 1987
Aircraft, aviation, WW 2 52340

Virginia Baptist Historical Society, c/o University of Richmond, Boatwright Library, Richmond, VA 23173 • T: +1 804 2898434 • F: +1 804 2898953 • www.baptistheritage.org • Pres.: *Thelma Hall Miller* •
Association with Coll – 1876
Baptist hist 52341

Virginia Historical Museum, Virginia Historical Society, 428 North Blvd, Richmond, VA 23220 • T: +1 804 3584901 • F: +1 804 3422399 • maribeth@vahistorical.org • www.vahistorical.org • Dir.: *Dr. Charles F. Bryan* • Cur.: *William Rasmussen* (Art) • Man.: *Tracey Bryan* •
Historical Museum – 1831
State hist 52342

Virginia Museum of Fine Arts, Blvd & Grove Av, Richmond, VA 23221-2466 • T: +1 804 3401400 • F: +1 804 3401548 • webmaster@vmfa.state.va.us • www.vmfa.state.va.us • Dir.: *Dr. Michael Brand* • Cur.: *Dr. Joseph M. Dye* (South Asian and Islamic Art) • *Dr. David Park Curry* (American Art) • *Richard Woodward* (African Art) • *Malcolm Cormack* (European Painting) • *Kathleen Schrader* (European Sculpture, Decorative Arts) • *Dr. Margaret Ellen Mayo* (Ancient Art) • *John Ravenal* (Art after 1900) • Cons.: *Carol Sawyer* (Paintings) • *Katharine Untch* (Objects) •
Fine Arts Museum / Folklore Museum – 1934
Ancient Egyptian, Greek, Etruscan, Roman 52343

Wilton House Museum, 215 S Wilton Rd, Richmond, VA 23226 • T: +1 804 2825936 • F: +1 804 2889805 • wiltonhouse@mindspring.com • www.wiltonhousemuseum.org • C.E.O.: *Dana Hand Evans* •
Decorative Arts Museum – 1934
Local hist, decorative arts 52344

Richmond Hill GA

Fort McAllister, 3894 Fort McAllister Rd, Richmond Hill, GA 31324 • T: +1 912 7272339 • F: +1 912 7273614 • ftmcallr@g-net.net • www.fortmcallister.org • Head: *Daniel Brown* •
Military Museum – 1958
1861 Confederate fort 52345

Ridgecrest CA

Maturango Museum of the Indian Wells Valley, 100 E Las Flores Av, Ridgecrest, CA 93555 • T: +1 760 3756900 • F: +1 760 3750479 • matmus6@maturango.org • www.maturango.org •

Dir.: *Harris Brokke* • Cur.: *Andrea Pelch* (Sylvia Winslow Art Gallery) • *Alexander K. Rogers* (Archaeology) • *Elizabeth Babcock* (History) • *Camille Anderson* (Natural History) •
Local Museum – 1962
Local hist, mining, Mojave desert flora and fauna, prehistoric rock art, regional artworks – library, archives ... 52346

Ridgefield CT

Aldrich Museum of Contemporary Art, 258 Main St, Ridgefield, CT 06877 • T: +1 203 4384519 • F: +1 203 4380198 • general@aldrichart.org • www.aldrichart.org •
Dir.: *Harry Philbrick* •
Fine Arts Museum – 1964 ... 52347

Keeler Tavern Museum, 132 Main St, Ridgefield, CT 06877 • T: +1 203 4385485 • F: +1 203 4389953 • keelertavernmuseum@earthlink.net • www.keelertavernmuseum.org •
Dir.: *Rosemary Morretta* •
Historical Museum – 1965
Local hist ... 52348

Ridgeland MS

Mississippi Crafts Center, Natchez Trace Pkwy, Ridgeland, MS 39157 • T: +1 601 8567546 • F: +1 601 8567546 •
Pres.: *Gayle Clark* • Dir.: *Martha Garrott* •
Decorative Arts Museum / Folklore Museum – 1975
Mississippi crafts, producer designed fine crafts from the southeastern US ... 52349

Ridgewood NJ

Schoolhouse Museum, 650 E Glen Av, Ridgewood, NJ 07450 • T: +1 201 6524584 •
Pres.: *Debbie Cofax* • Cur.: *D.A. Pangburn* •
Local Museum – 1949
New Jersey hist ... 52350

Rifle CO

Rifle Creek Museum, 337 East Av, Rifle, CO 81650 • T: +1 970 6254862 •
Dir.: *Kim Fazzi* •
Local Museum – 1967
Local and natural history, Indian artifacts, textiles 52351

Rincon GA

Georgia Salzburger Society Museum, 2980 Ebenezer Rd, Rincon, GA 31326 • T: +1 912 7547001 • F: +1 912 7547001 • info@GeorgiaSalzburgers.com • www.georgiasalzburgers.com •
Pres.: *Stuart Exley* •
Local Museum – 1925
Tools, furniture, letters, books ... 52352

Ringwood NJ

The Forges and Manor of Ringwood, 1304 Sloatsburg Rd, Ringwood, NJ 07456 • T: +1 973 9622240 • F: +1 973 9622247 • ringwood@warwick.net •
Cur.: *Elbertus Prol* •
Local Museum – 1935
Local history, industrial history ... 52353

Rio Hondo TX

Texas Air Museum, 1 Mile E Rio Hondo, Rio Hondo, TX 78583 • T: +1 956 7482112 • F: +1 956 7483500 • director@texasairmuseum.com • www.texasairmuseum.com •
Dir.: *John Houston* •
Science&Tech Museum – 1986
Concentration on aircraft used for warfare (1904-present) ... 52354

Rio Piedras RI

Museum of Antropology, History and Art, University of Puerto Rico, POB 21908, Rio Piedras, RI 00931 • T: +1 787 7642456 • F: +1 787 7634799 •
Dir.: *Luis Hernandez-Cruz* •
Historical Museum / Fine Arts Museum / University Museum
Puerto Rican and Latin American painting, graphics, photography and sculpture ... 52355

U

Ripley OH

Rankin House State Memorial, Rankin Hill Rd, Ripley, OH 45167 • T: +1 937 3921627 •
Local Museum – 1938
Home of abolitionist Rev. John Rankin ... 52356

Ripon WI

Little White Schoolhouse, 303 Blackburn St, Ripon, WI 54971 • T: +1 920 7486764 • F: +1 920 7486784 • exex@ripon-wi.com • www.ripon-wi.com •
Pres.: *Joan Karsten* • C.E.O.: *Paula T. Price* •
Historical Museum – 1951
Historic house, Birthplace of Republican Party, poltical hist, Republican presidential memorabilia ... 52357

Ripon College Art Gallery, 300 Seward St, Ripon, WI 54971 • T: +1 920 7488110, 7488115 • kaineu@ripon.edu •
Fine Arts Museum – 1965
Paintings, prints, sculpture, multi-media ... 52358

Ripon Historical Society Museum, 508 Watson St, Ripon, WI 54971 • T: +1 920 7485354 •
Historical Museum – 1899
Local hist ... 52359

Rittman OH

Rittman Historical Society Museum, 393 W Sunset Dr, Rittman, OH 44270 • T: +1 330 9257572 •
Pres.: *Jack Rice* •
Local Museum – 1960
Hist houses ... 52360

River Edge NJ

Bergen County Historical Society Museum, 1201 Main St, River Edge, NJ 07661 • T: +1 201 3439492 • www.carroll.com/bchs •
Cur.: *Matt Gebhardt* •
Local Museum – 1902
China, glass, farm and household tools, paintings, sculptures, toys, quilts, folk art, Wolfkiel pottery, clothing ... 52361

Steuben House Museum, 1201 Main St, River Edge, NJ 07661 • T: +1 201 3439492, 4871739 • F: +1 201 4981696 •
Cur.: *Matt Gebhardt* •
Local Museum – 1902
Artifacts of the Bergen Dutch 1680-1914 ... 52362

River Falls WI

Gallery 101, University of Wisconsin-River Falls, 410 S Third St, River Falls, WI 54022 • T: +1 715 4253266 • F: +1 715 4250657 • michael.a.padgett@uwrf.edu • www.uwrf.edu •
Head: *Michael Padgett* •
Public Gallery – 1973
Graphics, contemp. prints ... 52363

River Grove IL

Cernan Earth and Space Center, Triton College, 2000 N 5th Av, River Grove, IL 60171 • T: +1 708 4560300 ext 3372 • F: +1 708 5833153 • cernan@triton.edu • www.triton.edu/cernan/ •
Dir.: *Bart Benjamin* •
Natural History Museum – 1974
Meteorites, space hardware ... 52364

Riverdale Park MD

Riverdale Mansion, 4811 Riverdale Rd, Riverdale Park, MD 20737 • T: +1 301 8640420 • F: +1 301 9273498 • edward.day@pgparks.com • www.pgparks.com •
Dir.: *Edward Day* •
Local Museum – 1949
Personal and domestic artifacts, furnishing ... 52365

Riverhead NY

Hallockville Museum Farm, 6038 Sound Av, Riverhead, NY 11901 • T: +1 516 2985292 • F: +1 516 2985292 • hallockv@optonline.net • www.hallockville.com •
Pres.: *Richard Wines* • C.E.O.: *Jarod Kearney* •
Agriculture Museum – 1975
Rural and agricultural life 1880-1910 in the north fork of Long Island ... 52366

Railroad Museum of Long Island, 416 Griffling Av, Riverhead, NY 11901 • T: +1 631 7277920, 4770439 • F: +1 631 2616545 • secretaryrmli@aol.com • www.rmli.org •
Pres.: *Dennis Harrington* •
Science&Tech Museum – 1990
Railroading hist of Long Island, HO gauge ... 52367

Suffolk County Historical Society Museum, 300 W Main St, Riverhead, NY 11901 • T: +1 631 7272881 • F: +1 631 7273467 • schs@optonline.net • www.schs-museum.org •
Pres.: *John Sprague* • Dir.: *Wallace W. Broege* •
Local Museum – 1886
Indians, ceramics, decorative arts, whaling, early crafts, trade tools ... 52368

Riverside CA

California Museum of Photography, c/o University of California Riverside, 3824 Main St, UCR, Riverside, CA 92501 • T: +1 909 7874787 • F: +1 909 7874797 • www.cmp.ucr.edu •
Dir.: *Prof. Jonathan Green* • Cur.: *Georg Burwick* (Digital Media) • *Steve Thomas* (Collections) •
Fine Arts Museum / University Museum – 1973
Photography, stereoview coll, Bingham camera coll – library ... 52369

Entomology Research Museum, c/o University of California Riverside, Department of Entomology, Riverside, CA 92521-0314 • T: +1 951 8274315 • F: +1 951 8273086 • dyanega@ucr.edu • entmuseum.ucr.edu •
Dir.: *Dr. Serguei V. Triapitsyn* • Sc. Staff: *Dr. Doug A. Yanega* •
University Museum / Natural History Museum – 1962
Entomology ... 52370

March Field Air Museum, 22550 Van Buren Blvd, Riverside, CA 92518-6463, mail addr: POB 6463, Riverside, CA 92518 • T: +1 951 9025949 • info@marchfield.org • www.marchfield.org •
Dir.: *Patricia Korzec* •
Military Museum / Science&Tech Museum – 1979
Military and aviation items, uniforms, 69 aircraft – library, film theater ... 52371

Mission Inn Museum, 3696 Main St, Riverside, CA 92501 • T: +1 951 7889556 • F: +1 909 3416574 • www.missioninnmuseum.com •
Dir.: *John Worden* •
Decorative Arts Museum / Fine Arts Museum – 1976
Arts and crafts furniture, paintings, sculpture, Oriental art, dolls ... 52372

Riverside Art Museum, 3425 Mission Inn Av, Riverside, CA 92501 • T: +1 909 6847111 • F: +1 909 6847332 • www.riversideartmuseum.org •
Dir.: *Bobbie Powell* •
Fine Arts Museum – 1931
Paintings, sculpture, graphics, decorative arts ... 52373

Riverside Municipal Museum, 3580 Mission Inn Av, Riverside, CA 92501 • T: +1 909 8265273 • F: +1 909 3694970 • vmoses@riverside.ca.gov •
Dir.: *Vincent Moses* • Cur.: *Marjorie Mitchell* (Education) • *Brenda Buller-Focht* (Collections) • *Lynn Voorheis* (History) • *James Bryant* (Natural History) •
Local Museum – 1924
Local hist, ethnology and archaeology, decorative arts, local nature – archives, herbarium ... 52374

Sweeney Art Gallery, University of California, 3800 Main St, Riverside, CA 92501 • T: +1 951 8273755 • F: +1 951 8273798 • sweeney.ucr.edu •
Dir.: *John Divola* • Cur.: *Jennifer Frias* •
University Museum / Public Gallery – 1963 ... 52375

World Museum of Natural History, c/o La Sierra University, 4700 Pierce St, Riverside, CA 92515-8247 • T: +1 951 7852209 • www.lasierra.edu •
Natural History Museum / University Museum – 1971
Natural history, reptiles, SE Asian birds, Indian artifacts, meteorits, tektiles ... 52376

Riverton WY

Riverton Museum, 700 E Park Av, Riverton, WY 82501 • T: +1 307 8562665 •
Dir.: *Loren Jost* •
Local Museum – 1956
Local hist, Indian artifacts, crafts ... 52377

Robert A. Peck Gallery, Central Wyoming College, 2660 Peck Av, Riverton, WY 82501 • T: +1 307 8552211 • F: +1 307 8552090 • nkehoe@cwc.edu • www.cwc.edu •
Fine Arts Museum
Local art like paintings, drawings, sculpture ... 52378

Roanoke VA

Art Museum of Western Virginia, 1 Market Sq, Roanoke, VA 24011-1436 • T: +1 540 3425760 • F: +1 540 3425798 • info@artmuseumroanoke.org • artmuseumroanoke.org •
Dir.: *Dr. Judy L. Larson* • Cur.: *Susannah Koerber* •
Fine Arts Museum – 1951
19 th c to present American paintings, photograph, sculpture, prints & drawings, contemporary folk art, African art ... 52379

Catholic Historical Society of the Roanoke Valley, 400 W Campbell Av, Roanoke, VA 24016 • T: +1 540 9820152 • F: +1 540 9820152 •
Pres.: *Margaret M. Cochener* •
Association with Coll – 1983
Religious articles, photos ... 52380

History Museum & Historical Society of Western Virginia, 1 Market Sq, Roanoke, VA 24011 • T: +1 540 3425770 • F: +1 540 2241256 • history@roanoke.infionline.net • history-museum.org •
C.E.O.: *D. Kent Chrisman* •
Historical Museum – 1957
Archival & oral history repository, archeology, excavatioinal railroad coll ... 52381

Science Museum of Western Virginia, 1 Market Sq, Roanoke, VA 24011 • T: +1 540 3425710 • F: +1 540 2241240 • frontdesk@smwv.org • www.smwv.org •
Pres.: *Harry Nickens* •
Natural History Museum / Science&Tech Museum – 1970
Pertaining to earth science, health, energy, natural hist & physical science ... 52382

Virginia Museum of Transportation, 303 Norfolk Av, Roanoke, VA 24016 • T: +1 540 3425670 • F: +1 540 3426898 • info@vmt.org • www.vmt.org •
Science&Tech Museum – 1963
Transportation, trains, automobiles, carriages, aviation equip ... 52383

Rochester IN

Fulton County Historical Society Museum, 37 E 375 N, Rochester, IN 46975 • T: +1 219 2234436 • fchs@rtcol.com • www.icss.net/~fchs •
Dir.: *Melinda Clinger* •
Local Museum / Open Air Museum / Historical Museum – 1963
Round Barn, Potawatomi Trail of Death, Reference Room 17 rooms of permanent displays, village with 11 buildings ... 52384

Rochester MI

Meadow Brook Art Gallery, Oakland University, 209 Wilson Hall, Rochester, MI 48309-4401 • T: +1 248 3703005 • F: +1 248 3704208 • goody@oakland.edu • www.oakland.edu/mbag •
Dir.: *Dick Goody* •
Fine Arts Museum / University Museum – 1959
African, Pre-Columbian and Oceanian art, contemporary paintings, prints & graphics of America & Europe – sculpture park, theater ... 52385

Meadow Brook Hall, Oakland University, Rochester, MI 48309-4401 • T: +1 248 3703140 • F: +1 248 3704260 • stobersk@oakland.edu • www.meadowbrookhall.org •
Exec. Dir.: *Geoff Upward* • Cur.: *Brandy Hirschlieb* •
Decorative Arts Museum / University Museum – 1971
Paintings, sculpture, prints, drawings, furniture, glass, porcelain, silver, artifacts, photos, films ... 52386

Paint Creek Center for the Arts Exhibitions, 407 Pine St, Rochester, MI 48307 • T: +1 248 6514110 • F: +1 248 6514757 • mfortuna@pccart.org • www.pccart.org •
Dir.: *Mary Fortuna* •
Public Gallery – 1965
Contemporary art ... 52387

Rochester Hills Museum at Van Hoosen Farm, 1005 Van Hoosen Rd, Rochester, MI 48306 • T: +1 248 6564663 • F: +1 248 6088198 • rhmuseum@rochesterhills.org • www.rochesterhills.org/museum.htm •
Head: *Patrick J. McKay* •
Local Museum / Historical Museum – 1979
Local hist, household, agriculture, 20th c clothing ... 52388

Rochester MN

Olmsted County Museum, 1195 W Circle Dr SW, Rochester, MN 55903-6411 • T: +1 507 2829447 • F: +1 507 2895481 • businessoffice@olmstedhistory.com • www.olmstedhistory.com •
Exec. Dir.: *John Hunziker* • Cur.: *Margaret Ranweiller* •
Local Museum – 1926
Regional hist, farming machinery & implements, school – library ... 52389

Rochester Art Center, 40 Civic Center Dr SE, Rochester, MN 55904 • T: +1 507 2828629 • F: +1 507 2827737 • bjshigaki@juno.com • www.rochesterartcenter.org •
Exec. Dir.: *Betty Shigaki* •
Fine Arts Museum – 1946
Contemporary visual arts and crafts, prints, paintings, drawings ... 52390

Rochester NY

American Baptist - Samuel Colgate Historical Library, 1106 S Goodman St, Rochester, NY 14620 • T: +1 716 4731740 • F: +1 716 4733048 • abhs@crds.edu •
Library with Exhibitions – 1853
Paintings, books, records on Baptist hist and American religion ... 52391

Baker-Cederberg Museum, Rochester General Hospital, 1425 Portland Av, Rochester, NY 14621-3095 • T: +1 585 9223521 • F: +1 585 9225292 • phil.maples@viahealth.org •
Cur.: *Philip G. Maples* •
Science&Tech Museum – 1947
Local and hospital hist, Company museum ... 52392

Campbell Whittlesey House, The Landmark Society of Western New York, 123 Fitzhugh St, Rochester, NY 14608 • T: +1 585 5467029 • F: +1 585 5464788 • info@landmarksociety.org • www.landmarksociety.org •
Dir.: *Cindy Boyer* •
Decorative Arts Museum – 1937
Art, furnishings, decorative arts ... 52393

Dar-Hervey Ely House, 11 Livingston Park, Rochester, NY 14608 • T: +1 716 2324509 •
C.E.O.: *Alberta Greer* •
Decorative Arts Museum / Local Museum – 1894
Furniture of early 1800s, china glass, silver – Library ... 52394

International Museum of Photography and Film, George Eastman House, 900 East Av, Rochester, NY 14607 • T: +1 585 2713361 • F: +1 585 2713970 • info@geh.org • www.eastmanhouse.org •
Dir.: *Anthony Bannon* • Sen. Cur.: *Paolo Cherchi Usai* (Motion Picture Collection) • Cur.: *Alison Nordstrom* (Photography Collection) • *Kathy Connor* (George Eastman House) • *Deirdre Cunningham* (Landscape Collection) • *Todd Gustavson* (Technology) •
Fine Arts Museum – 1947
Photography, motion pictures ... 52395

Memorial Art Gallery of the University of Rochester, 500 University Av, Rochester, NY 14607 • T: +1 585 4737720 • F: +1 585 4736266 • maginfo@mag.rochester.edu • mag.rochester.edu •
Dir.: *Grant Holcomb* • Ass. Dir.: *Kim Hallatt* • Cur.: *Majorie Searl* (American Art) • *Nancy Norwood* (European Art) • *Marie Via* (Exhibitions) •
Fine Arts Museum / University Museum / Decorative Arts Museum – 1913 ... 52396

Mercer Gallery, Monroe Community College, 1000 E Henrietta Rd, Rochester, NY 14623 • T: +1 585 2922064 • F: +1 585 2923120 • kfarrell@monroecc.edu • www.monroecc.edu •
Dir.: *Kathleen Farrell* •
Fine Arts Museum / University Museum
Art ... 52397

Rochester Museum, 657 East Av, Rochester, NY 14607-2177 • T: +1 585 2714320 • F: +1 585 2715935 • katebennett@rmsc.org • www.rmsc.org •
C.E.O. & Pres.: *Kate Bennett* • Dir.: *Jennifer Pond* (Education) • *George MIntosh* (Collections) •
Historical Museum / Natural History Museum / Science&Tech Museum – 1912 ... 52398

Stone-Tolan House, The Landmark Society of Western New York, 2370 East Av, Rochester, NY 14608 • T: +1 585 5467029 • F: +1 585 5464788 • info@landmarksociety.org • www.landmarksociety.org •
Dir.: *Cindy Boyer* •
Decorative Arts Museum – 1937
Art, furnishings, decorative arts ... 52399

Strong National Museum of Play, 1 Manhattan Sq, Rochester, NY 14607 • T: +1 585 2632700 • F: +1 585 2632493 • info@strongmuseum.org • www.museumofplay.org •
Pres. & C.E.O.: *G. Rollie Adams* •
Special Museum – 1968
Cultural hist of American play, coll of toys, dolls, and play-related artifacts – library ... 52400

Susan B. Anthony House, 17 Madison St, Rochester, NY 14608 • T: +1 585 2356124 • F: +1 585 2356212 • susanbanthonyhouse.org •
Exec. Dir.: *Lorie Barnum* • Cur.: *Christin Grant* •
Local Museum – 1946
Home of S.B. Anthony (1820-1906), legendary American civil rights leader during the most politically active period of her life, and the site of her famous arrest for voting in 1872 ... 52401

Woodside House, Rochester Historical Society, 485 East Av, Rochester, NY 14607 • T: +1 716 2712705 • F: +1 716 2719089 • asalter@rochesterhistory.org • www.rochesterhistory.org •
Exec. Dir.: *Ann C. Salter* •
Association with Coll – 1861
Books and manuscripts, photos, paintings, costume coll, silver, furnishings, ceramics, firearms, toys ... 52402

Rock Hill SC

Museum of York County, 4621 Mount Gallant Rd, Rock Hill, SC 29732-9905 • T: +1 803 3292121 • F: +1 803 3295249 • myco@infoave.net • www.yorkcounty.org •
Dir.: *W. Van Shields* • Cur.: *Mary Lynn Norton* (Art) •
Local Museum – 1950
Local hist, African animals, art and ethnography, local art, archaeology and natural history ... 52403

Winthrop University Galleries, Rutledge Bldg, Rock Hill, SC 29733 • T: +1 803 3232126 • F: +1 803 3232333 • stanteyt@winthrop.edu • www.winthrop.edu/vpa •
Dir.: *Tom Stanley* •
Fine Arts Museum
Visual arts and design ... 52404

Rock Island IL

Augustana College Art Gallery, 7th Av and 38th St, Rock Island, IL 61201-2296 • T: +1 309 7947231, 7000 • F: +1 309 7947678 • armaurer@augustana.edu • www.augustana.edu/artmuseum •
Dir.: *Sherry C. Maurer* •
Fine Arts Museum – 1983
Swedish-American art, 19th - 20th c works, Schonstedt coll ... 52405

Black Hawk State Historic Site, Hauberg Indian Museum, Illinois Rte 5, Rock Island, IL 61201 • T: +1 309 7889536 • F: +1 309 7889865 • haubergmuseum@juno.com •
Dir.: *Elizabeth Carvey-Stewart* •
Historic Site / Historical Museum – 1927
1740-1831 site of the main villages of the Sauk and Fox Nations ... 52406

Colonel Davenport Historical Foundation, Hillman St, Rock Island Arsenal, Rock Island, IL 61201 • T: +1 309 7646471, 3889657 • coldav1833@yahoo.com • www.davenporthouse.org •
Pres.: *Helen Macalister* •
Local Museum – 1978
Furniture, personal artifacts ... 52407

Fryxell Geology Museum, c/o Augustana College, Swenson Hall of Science, Rock Island, IL 61201 • T: +1 309 7947318 • F: +1 309 7947564 • glhammer@augustana.edu • www.augustana.edu •
Dir.: *William R. Hammer* •
Natural History Museum – 1929
Paleontology, fossils, minerals, general geology ... 52408

Rock Island Arsenal Museum, Attn SMARI-CFS-M, 1 Rock Island Arsenal, Rock Island, IL 61299-5000 • T: +1 309 7825021 • F: +1 309 7823598 • leinickek@ria.army.mil • www.ria.army.mil/ •
Dir.: *Kris G. Leinicke* •
Military Museum – 1905
Military ... 52409

Rock Springs WY

Community Fine Arts Center, 400 C St, Rock Springs, WY 82901 • T: +1 307 3626212 • F: +1 307 3526657 • cfac@sweetwaterlibraries.com • www.cfac4art.com •
Dir.: *Debora Thaxton Soule* • Ass. Dir.: *Jennifer Messer* •
Fine Arts Museum – 1938
Contemp American art ... 52410

Rock Springs Historical Museum, 201 B St, Rock Springs, WY 82901 • T: +1 307 3623138 • F: +1 307 3521516 •
Pres.: *Cyndi Sullivan* • Dir.: *Bob Nelson* (Museum) •
Local Museum – 1988
Furnishings, tools and equip ... 52411

Western Wyoming College Art Gallery, 2500 College Dr, Rock Springs, WY 82902 • T: +1 307 3821600 • F: +1 307 3827665 • www.wwcc.wy.us •
Dir.: *Florence McEwin* •
Fine Arts Museum
Art ... 52412

Rockford IL

Burpee Museum of Natural History, 737 N Main St, Rockford, IL 61103 • T: +1 815 9653433 • F: +1 815 9652703 • info@burpee.org • www.burpee.org •
Pres.: *Lewis S.W. Crampton* •
Natural History Museum – 1942
Natural history, new Robert Solem wing ... 52413

Discovery Center Museum, 711 N Main St, Rockford, IL 61103 • T: +1 815 9636769 • F: +1 815 9680164 • webmaster@discoverycentermuseum.org • www.discoverycentermuseum.org •
Pres.: *David Thompson* • Exec. Dir.: *Sarah Wolf* •
Special Museum – 1981 ... 52414

Midway Village and Museum, 6799 Guilford Rd, Rockford, IL 61107 • T: +1 815 3979112 • F: +1 815 3979156 • lonna.converso@midwayvillage.com • www.midwayvillage.com •
Cur.: *Rosalyn Robertson* •
Open Air Museum – 1970
Hist Village, 24 bldg turn-of-the-c complex ... 52415

Rockford Art Museum, 711 N Main St, Rockford, IL 61103-6999 • T: +1 815 9682787 • F: +1 815 9680164 • staff@rockfordartmuseum.com • www.rockfordartmuseum.org •
Dir.: *Linda Dennis* • Cur.: *Scott Snyder* •
Fine Arts Museum – 1913
19th and 20th c paintings, sculpture, praphics and decorative arts, Arnold Gilbert coll of photographs, glass, African-American folk art ... 52416

Rockford College Art Gallery/Clark Arts Center, 5050 E State, Rockford, IL 61108 • T: +1 815 2264034 • F: +1 815 3945167 •
Dir.: *Philip Soosloff* •
Fine Arts Museum / University Museum – 1847
20th c paintings, prints, photography, ceramics, drawings, ethnographic art ... 52417

Tinker Swiss Cottage Museum, 411 Kent St, Rockford, IL 61102 • T: +1 815 9642424 • F: +1 815 9642466 • tinkercottage@sbcglobal.net • www.tinkercottage.com •
Pres.: *Robert Jakeway* • Exec. Dir.: *Laura Bachelder* •
Local Museum / Decorative Arts Museum – 1943
1865 Swiss-style home built by Robert H. Tinker 52418

Rockhill Furnace PA

Railways to Yesterday, Rockhill Trolley Museum, East Broad Top Railroad, State Rte 994, Rockhill Furnace, PA 17249 • T: +1 610 9659028, +1 814 4479576 • www.rockhilltrolley.org •
Pres.: *Joel Salomon* •
Science&Tech Museum – 1960
Transportation, specialization in collecting, restoring and maintaining cars that operated in Pennsylvania, trolley cars from Johnstown, Philadelphia, York, Scranton ... 52419

Rockhill Trolley Museum, Meadow St, Rockhill Furnace, PA 17249 • T: +1 610 4370448 • sgurley@prodigy.net • www.rockhilltrolley.org •
Science&Tech Museum – 1962
Trolleys from 1899-1947 ... 52420

Rockland ME

Shore Village Museum, 104 Limerock St, Rockland, ME 04841 • T: +1 207 5940311 • F: +1 207 5949581 • kenblack@midcoast.com • www.lighthouse.cc/shorevillage •
Dir.: *Kenneth N. Black* • Cur.: *Robert Nason Davis* (History) •
Historical Museum – 1977
Maritime, lighthouse, John W. Flint coll ... 52421

William A. Farnsworth Art Museum and Wyeth Center, 16 Museum St, Rockland, ME 04841 • T: +1 207 5966457 • F: +1 207 5960509 • farnsworth@midcoast.com • www.farnsworthmuseum.org •
Dir.: *Christopher B. Crosman* •
Fine Arts Museum – 1948
American art, Louise Nevelson coll – library, Wyeth study center ... 52422

Rockport MA

Rockport Art Association Galleries, 12 Main St, Rockport, MA 01966 • T: +1 978 5466604 • F: +1 978 5469767 • rockportart@sbcglobal.net • www.rockportartassn.org •
Exec. Dir.: *Carol Linsky* •
Fine Arts Museum – 1921
Paintings, graphics, photographs and sculpture by regional artists – small library ... 52423

Sandy Bay Museums, 40 King St, Rockport, MA 01966 • T: +1 978 5469533 •
C.E.O.: *Mary H. Sibbalds* • Cur.: *Cynthia A. Peckham* •
Local Museum – 1925
Local history, Sewall-Scripture house, old castle, Indian artifacts, granite quarry, local industry, Hannah Jumper, dolls, toys, glass ... 52424

Rockport ME

Center for Maine Contemporary Art, 162 Russell Av, Rockport, ME 04856, mail addr: POB 147, Rockport, ME 04856 • T: +1 207 2362875 • F: +1 207 2362490 • info@cmcanow.org • www.artsmaine.org •
Dir.: *Oliver L. Wilder* • Cur.: *Bruce Brown* •
Association with Coll – 1952
Contemporary art ... 52425

Rockport TX

Fulton Mansion, 317 Fulton Beach Rd, Rockport, TX 78382 • T: +1 361 7290386 ext 26 • F: +1 361 7296581 • fmdir@swbell.net • www.tpwd.state.tx.us/park/fulton •
Decorative Arts Museum – 1983
Period furniture, decorative arts ... 52426

Texas Maritime Museum, 1202 Navigation Circle, Rockport, TX 78382-2773 • T: +1 361 7291271 • F: +1 361 7299938 • klrd@pelicancoast.net • www.texasmaritimemuseum.org •
Pres.: *James Mayes* • Dir.: *Kathy Roberts-Douglass* •
Historical Museum – 1980
Nautical equip, commercial fishing equip, ship & small boat building tools ... 52427

Rockton IL

Macktown Living History Site and Whitman Trading Post, Macktown Forest Preserve, Hwy 75 W, Rockton, IL 61072 • T: +1 815 8776100 • F: +1 815 8776124 • wcfpd@wcfpd.org • www.wcfpd.org •
Pres.: *Ray Ferguson* • Dir.: *Marilyn Mohring* •
Local Museum – 1952
Two-story farm house, home to one of the first white settlers in Winnebago County, 1846 Whitman Trading Post ... 52428

Rockville MD

Jane L. and Robert H. Weiner Judaic Museum, 6125 Montrose Rd, Rockville, MD 20852 • T: +1 301 8810100, 2303711 • F: +1 301 8815512 • www.jccgw.org •
C.E.O.: *Arnie Sohinki* •
Fine Arts Museum / Religious Arts Museum – 1925
Judaica ... 52429

Latvian Museum, 400 Hurley Av, Rockville, MD 20850 • T: +1 301 3401914 • F: +1 301 3408732 • alanso@alausa.org • www.alausa.org •
Dir.: *Anna Graudina* • Cur.: *Lilita Bergs* •
Ethnology Museum / Folklore Museum – 1980
Latvian historic and cultural development from Ice Age to 20th c ... 52430

Montgomery County Historical Society Museum, 103 W Montgomery Av, Rockville, MD 20850-4212 • T: +1 301 3402825 • F: +1 301 3402871 • info@montgomeryhistory.org • www.montgomeryhistory.org •
Pres.: *John T. Beatyy* • Exec. Dir.: *Mary Kay Harper* •
Cur.: *Millicent Gay* •
Local Museum / Historical Museum – 1944
Beall-Dawson House, 1852 Stonestreet Medical Museum, local history ... 52431

Rockwell City IA

Calhoun County Museum, 150 E High St, Rockwell City, IA 50579 • T: +1 712 2978139, 2978585 •
Cur.: *Judy Webb* •
Local Museum – 1956
Regional history ... 52432

Rocky Ford CO

Rocky Ford Historical Museum, 1005 Sycamore Av, Rocky Ford, CO 81067 • T: +1 719 2546737 •
Cur.: *Diane Zimmerman* •
Local Museum – 1940
Local history, sugar produce, farming, military, 1st fire engine, ethnography, minerals and fossils – archives ... 52433

Rocky Hill CT

Academy Hall Museum of the Rocky Hill Historical Society, 785 Old Main St, Rocky Hill, CT 06067 • T: +1 860 5636704 • info@rockyhillhistory.org • www.rockyhillhistory.org •
Pres.: *Mike Martino* •
Local Museum – 1962
Local hist, Indian artifacts, clothing, domestics ... 52434

Rocky Mount NC

Four Sisters Gallery and the Robert Lynch Collection, North Carolina Wesleyan University, 3400 N Wesleyan Blvd, Rocky Mount, NC 27804-8630 • T: +1 252 9855100 • F: +1 252 9855319 • eadelman@ncwc.edu • www.ncwc.edu/arts/lynch •
Fine Arts Museum / University Museum
Self-taught visionary art, Lynch coll of outsider art ... 52435

Rocky Mount Arts Center, 1 Imperial Centre Plaza, Rocky Mount, NC 27803 • T: +1 252 9721163 • F: +1 252 9721563 • jackson@ci.rocky-mount.nc.us • www.ci.rocky-mount.nc.us/artscenter •
Dir.: *Jerry Jackson* • Cust.: *Raymond Draughn* •
Fine Arts Museum – 1956
NC art 1950 - present ... 52436

Rocky Mount Children's Museum, 1 Imperial Centre Plaza, Rocky Mount, NC 27803 • T: +1 252 9721167 • F: +1 252 9721535 • madrid@ci.rocky-mount.nc.us • www.ci.rocky-mount.nc.us/museum •
Dir.: *Candy L. Madrid* • Cur.: *Steve Armstrong* (Exhibits) • Education: *Leigh Lasher* •
Special Museum – 1952 ... 52437

Roebuck SC

Walnut Grove Plantation, 1200 Otts' Shoals Rd, Roebuck, SC 29376 • T: +1 864 5766546 • F: +1 864 5764058 • wannutgrove@mindspring.com • www.sparklenet.com/historicalassociation •
Pres.: *Susan Dunlap* • C.E.O.: *Susan Turpin* •
Historical Museum – 1961
Historic house ... 52438

Rogers AR

Daisy Airgun Museum, 202 W Walnut St, Rogers, AR 72756 • T: +1 501 9866873 • F: +1 501 9866875 • info@daisymuseum.com • www.daisymuseum.com •
Science&Tech Museum / Military Museum – 1966
Air guns ... 52439

Rogers Historical Museum, 322 S Second St, Rogers, AR 72756 • T: +1 501 6211154 • F: +1 501 6211155 • museum@rogersarkansas.com • www.rogersarkansas.com/museum •
Dir.: *Gaye K. Bland* •
Local Museum / Historical Museum – 1975
Hawkins house, local history, Frisco railroad ... 52440

Rogers MN

Ellingson Car Museum, 20950 Rogers Dr, Rogers, MN 55374 • T: +1 612 4287337 • F: +1 612 4284370 • ecmmuseum@mm.com • www.ellingsoncarmuseum.com •
Owner: *Eugene Ellingson* • Cur.: *Patty Zachman* •
Science&Tech Museum – 1994
Cars from 1900 to 1970 ... 52441

Rogers City MI

Presque Isle County Historical Museum, 176 W Michigan Av, Rogers City, MI 49779 • T: +1 517 7344121 •
Cur.: *Laural R. Maldonado* •
Local Museum – 1973
Area hist, maritime, ship model of USS "Carl D Bradley" ... 52442

Rogue River OR

Woodville Museum, First and Oak Sts, Rogue River, OR 97537 • T: +1 541 5823088 •
Public Gallery – 1986 ... 52443

Rohnert Park CA

University Art Gallery, Sonoma State University, 1801 E Cotati Av, Rohnert Park, CA 94928 • T: +1 707 6642295 • F: +1 707 6642054 • art.gallery@sonoma.edu • www.sonoma.edu/artgallery •
Dir.: *Michael Schwager* •
Public Gallery / University Museum – 1977
20th c American artworks on paper, Asian art, Garfield coll ... 52444

Rolla MO

Ed Clark Museum of Missouri Geology, 111 Fairgrounds Rd, Rolla, MO 65401 • T: +1 573 3682100 • F: +1 573 3682111 •
Natural History Museum – 1963
Geology, mineralogy, paleontology ... 52445

Memoryville USA, 2220 N Bishop Av, Rolla, MO 65401, mail addr: POB 569, Rolla, MO 65402-0569 • T: +1 573 3641810 • F: +1 573 3646975 • memoryvl@fidnet.com • www.memoryvilleusa.com •
Science&Tech Museum – 1969
Antique and classic automobiles, army truck with machine gun ... 52446

Rolla Minerals Museum, University of Missouri, 125 McNutt Hall, UMR Campus, Rolla, MO 65401 • T: +1 573 3414616 • F: +1 573 3416935 •
University Museum / Natural History Museum – 1870
Mineralogy, paleontology, geology ... 52447

Rome GA

Chieftains Museum, 501 Riverside Pkwy, Rome, GA 30161 • T: +1 706 2919494 • F: +1 706 2912410 • chmuseum@bellsouth.net • www.chieftainsmuseum.org •
Pres.: *Ansley Saville* •
Ethnology Museum / Historical Museum / Archaeology Museum – 1969
Plantation house belonging to Cherokee leader Major Ridge, native American archaeology 52448

Martha Berry Museum, 24 Veterans Memorial Hwy, Rome, GA 30161, mail addr: POB 490189, Mount Berry, GA 30149-0189 • T: +1 706 2911883 • F: +1 706 8020902 • oakhill@berry.edu • www.berry.edu/oakhill •
Local Museum – 1972
Furnishings, china. silver, artifacts, paintings, Italien and American art 52449

Rome NY

Erie Canal Village, 5789 New London Rd, RT46/49, Rome, NY 13440 • T: +1 888 3743226 • F: +1 315 3397595 • info@eriecanalvillage.net • www.eriecanalvillage.net •
Local Museum / Historic Site / Open Air Museum – 1973
Agricultural equip, domestic objects, textiles, costumes, canal hist and village life in 19th c, cheese manufacturing coll 52450

Fort Stanwix, 112 E Park St, Rome, NY 13440 • T: +1 315 3362090 • F: +1 315 3345051 • michael_caldwell@nps.gov • www.nps.gov/fost •
Local Museum – 1935
Archaeological coll, arms, clothing, glassware and pottery 52451

Rome Art and Community Center, 308 W Bloomfield St, Rome, NY 13440 • T: +1 315 3361040 • F: +1 315 3361090 • racc@borg.com • www.borg.com/~racc •
Pres.: *Ann Peach Lynch* • C.E.O.: *Deborah H. O'Shea* •
Fine Arts Museum – 1967
Changing exhibitions of art 52452

Rome Historical Society Museum, 200 Church St, Rome, NY 13440 • T: +1 315 3365870 • F: +1 315 3365912 • romehist@dreamscape.com •
Pres. & Dir.: *Merry Speicher* • Cur.: *Ann Swanson* •
Local Museum – 1936
!8th c to present hist of Rome, genealogy 52453

Rome City IN

Gene Stratton-Porter House, 1205 Pleasant Point, Rome City, IN 46784, mail addr: Box 639, Rome City, IN 46784 • T: +1 219 8543790 • F: +1 219 8549102 • gsporter@kuntrynet.com •
Cur.: *Martha Swartzlander* •
Historical Museum
Period furnishings, Gene Stratton Porter photographs and memorabilia 52454

Rosanky TX

Central Texas Museum of Automotive History, Hwy 304, Rosanky, TX 78953 • T: +1 512 2372635 • F: +1 512 7542424 • dburdick@thermon.com • www.tourtexas.com/rosanky •
Dir.: *Richard L. Burdick* • Cur.: *Kurt McCowan* • Sc. Staff: *Ray Terry* (Restoration) •
Science&Tech Museum – 1982
Emphasis on restoration & preservation of early automobiles 52455

Roseau MN

Roseau County Historical Museum, 110 Second Av NE, Roseau, MN 56751-1110 • T: +1 218 4631918 • rchsroseau@mncable.net • www.roseaucohistoricalsociety.org •
Dir.: *Charleen Haugen* •
Historical Museum – 1927 52456

Roseburg OR

Douglas County Museum of History and Natural History, 123 Museum Dr, Roseburg, OR 97470 • T: +1 541 9577007 • F: +1 541 9577017 • museum@co.douglas.or.us • www.co.douglas.or.us/museum •
Dir.: *Stacey B. McLaughlin* • Cur.: *Jena Mitchell* (History) • *Dennis Ruley* (Natural History) •
Local Museum / Natural History Museum – 1968
Native Indian artifacts, pioneer tools, agricultural equip, railroad equip, vehicles, machinery 52457

Lane House, 544 SE Douglas Av, Roseburg, OR 97470 • T: +1 541 4591393 •
Pres.: *Kenneth Shrum* •
Historical Museum – 1953
Period furnishing, hist of house and General J. Lane and his family 52458

Roselle IL

Law Enforcement Museum, POB 72835, Roselle, IL 60172 • T: +1 847 7951547 • F: +1 847 7952469 • forgottenheroes@aol.com • www.forgottenheroes-lema.org •
Pres.: *Ronald C. Van Raalte* •
Special Museum – 1989
Law, crime, police deaths 52459

Roselle NJ

Sons of the American Revolution, 101 W Ninth Av, Roselle, NJ 07203-1926 • T: +1 908 2451777 •
Military Museum – 1889 52460

Rosendale NY

Centura House Historical Society, A.J. Snyder Estate, 668 Rte 213, Rosendale, NY 12472-0150 • T: +1 845 6589900 • F: +1 845 6589277 • rosendalebuff@aol.com • www.centuryhouse.org •
Pres./Dir.: *Dietrich Werner* •
Historical Museum / Science&Tech Museum – 1988
Hist of cement manufacturing, barn, carriage house with 20 carriages, period furnishings, antiques 52461

Roseville OH

Ohio Ceramic Center, 7327 Ceramic Rd NE, Roseville, OH 43777 • T: +1 740 6977021 • F: +1 740 6970171 • webmaster@ceramic.22n.com • www.geocities.com/ceramiccenter •
Pres.: *James Swingle* •
Decorative Arts Museum / Science&Tech Museum – 1970
Develop from Ohio Bluebird potteries to modern ceramic industry 52462

Roslyn NY

Nassau County Museum of Art, 1 Museum Dr, Roslyn, NY 11576 • T: +1 516 4849338 • F: +1 516 4840710 • nassaumuseum@yahoo.com • www.nassaumuseum.com •
Dir.: *Constance Schwartz* • Ass. Dir.: *Fernanda Bennett* • Cur.: *Franklin Hill Perrell* •
Fine Arts Museum – 1989
19th and 20th c European and American art, changing exhibitions of modern and contemporary art 52463

Van Nostrand-Starkins House, 221 Main St, Roslyn, NY 11576 • T: +1 516 6254363 • F: +1 516 6254363 • roslynlandmarksociety@juno.com •
Pres.: *Harrison Hunt* • Dir.: *Catherine M. Tamis* •
Historical Museum – 1976
American decorative art, period room, architecture of 17th c house 52464

Rossville GA

Chief John Ross House, 200 E Lake Av, Rossville, GA 30741 • T: +1 706 8613954 •
Decorative Arts Museum – 1797
Cherokee alphabet, arrowheads, pictures, letters, furniture, rugs 52465

Roswell GA

Bulloch Hall, 180 Bulloch Av, Roswell, GA 30075 • T: +1 770 9921731, 9921951 • F: +1 770 5871840 •
Dir.: *Pam Billingsley* •
Folklore Museum – 1978
Historic c.1839 Antebellum Greek Revival House and Cottage 52466

Roswell NM

Anderson Museum of Contemporary Art, 409 E College Blvd, Roswell, NM 88202 • T: +1 505 6235600 • F: +1 505 6235603 •
Dir.: *Donald B. Anderson* •
Fine Arts Museum – 1994 52467

General Douglas L. McBride Museum, New Mexico Military Institute, 101 West College Blvd, Roswell, NM 88201-8107 • T: +1 505 6248220 • F: +1 505 6248258 • hardman@nmmi.edu • www.nmmi.cc.nm.us/museum •
Dir.: *Genee Hardman* •
Military Museum – 1983
Military artifacts from the 20th c conflicts 52468

Historical Center for Southeast New Mexico, 200 N Lea Av, Roswell, NM 88201 • T: +1 505 6228333 • F: +1 505 6228333 • hssnm@dfn.com • www.hssnm.net •
Dir.: *Roger K. Burnett* • Archivist: *Elvis Fleming* • Cur.: *Jean Rockhold* •
Historical Museum – 1976
1875-1936 period furniture and furnishings 52469

Roswell Museum and Art Center, 100 W 11, Roswell, NM 88201 • T: +1 505 6246744 ext 10 • F: +1 505 6246765 • rmac@roswellmuseum.org • www.roswellmuseum.org •
Dir.: *Jeannie Weiffenbach* • Cur.: *Ellen K. Moore* • *Wesley A. Rusnell* •
Fine Arts Museum / Local Museum – 1937
Regional and Native American fine arts and crafts, graphics, Western history, ethnology, folk art 52470

Round Top TX

Festival-Institute Museum, James Dick Foundation, State Hwy 237, Round Top, TX 78954 • T: +1 409 2493129 • F: +1 409 2495078 • lamarl@festivalhill.org • www.festivalhill.org •
Dir.: *James Dick* • Cur.: *Lamar Lentz* •
Fine Arts Museum / Decorative Arts Museum – 1971 52471

Winedale Historical Center, University of Texas, c/o Austin Center for American History, FM Road 2714, Round Top, TX 78954 • T: +1 979 2783530 • F: +1 979 2783531 • g.jaster@mail.utexas.edu • www.cah.utexas.edu •
Local Museum – 1967
Texas-German cultural hist, agricultural hist, American social hist, textiles 52472

Roundup MT

Musselshell Valley Historical Museum, 524 First W Roundup, Roundup, MT 59072 • T: +1 406 3231403 • F: +1 406 3231518 • dparrott@midrivers.com •
Dir.: *Bonnie DeMaio* • Ass. Dir.: *Shirley Parrott* •
Local Museum – 1972
Indian artifacts, fossils, crystals, coal mine, local wildlife 52473

Rowe MA

Kemp-McCarthy Memorial Museum, 288 Zoar Rd, Rowe, MA 01367-9774 • T: +1 413 3394729 • F: +1 413 3394700 • rowehistorical@netzero.net •
Dir./Cur.: *Alan Bjork* •
Local Museum – 1963
Furnishings of Revolutionary times, town hist 52474

Rowley MA

Rowley Historical Museum, 233 Main St, Rowley, MA 01969 • T: +1 508 9487483 •
C.E.O.: *Walter Farrell* •
Local Museum – 1918
Local history 52475

Roxbury NY

John Burroughs Memorial, Burroughs Memorial Rd, Roxbury, NY 12474, mail addr: Mine Kill State Park, POB 923 North Blenheim, NY 12131-0923 • T: +1 518 8276111 • F: +1 518 8276782 • laura.tully@oprhp.state.ny.us • www.nysparks.com •
Historical Museum – 1964
Burial Site of the writer John Burroughs 52476

Royalton VT

Royalton Historical Society, 4184 Rte 14, Royalton, VT 05068 • T: +1 802 8283051 • F: +1 802 8283206 • john.dumville@state.vt.us •
Pres.: *John P. Dumville* •
Association with Coll – 1967
Photographs, manuscripts, printed matter 52477

Rugby ND

Geographical Center Historical Museum, 1 Block E Hwy, Rugby, ND 58368 • T: +1 701 7766414 • www.artcom.com/museum/ •
Pres.: *Matt Mullally* •
Local Museum – 1964
Farm machinery, cars, buggies, wagons 52478

Rugby TN

Historic Rugby, State Hwy 52, Rugby, TN 37733 • T: +1 423 6282441 • F: +1 423 6282266 • rugbytn@highland.net • www.historicrugby.org •
Pres.: *Gerald Walker* • Exec. Dir.: *Barbara Stagg* •
Local Museum – 1966
Historic village 52479

Ruidoso Downs NM

Hubbard Museum of the American West, 841 Hwy 70 W, Ruidoso Downs, NM 88338 • T: +1 505 3784142 • F: +1 505 3784166 • info@hubbardmuseum.org • www.hubbardmuseum.org •
Dir.: *Greg Shuman* •
Local Museum – 1968
Local Hispanic and Native American culture, county war and Bill the Kid coll, medical instr, Buffalo Soldiers military 52480

Rushville IL

Schuyler Jail Museum, 200 S Congress St, Rushville, IL 62681 • T: +1 217 3226975 •
Cur.: *Nancy Stauffer* •
Local Museum – 1968
Local history, housed in Schuyler Jail 52481

Rushville IN

Rush County Historical Society Museum, 619 N Perkins, Rushville, IN 46173 • T: +1 765 9322492, 9322222 •
Local Museum – 1922 52482

Russell KS

Deines Cultural Center, 820 N Main St, Russell, KS 67665 • T: +1 785 4833742 • F: +1 785 4834397 • deinescenter@russellks.net •
Dir.: *Nance Selbe* •
Fine Arts Museum – 1990
E. Hubert Deines wood engravings, 20th c art 52483

Fossil Station Museum, 331 Kansas St, Russell, KS 67665 • T: +1 913 4833637 • rchs@russellks.net • www.rchs.russellks.net •
Pres.: *Connie Wagner* •
Local Museum / Science&Tech Museum – 1969
Native American artifacts, oil equip 52484

Gernon House and Blacksmith Shop, 808 Kansas St, Russell, KS 67665 • T: +1 913 4833637 • rchs@russellks.net • www.rchs.russellks.net •
Pres.: *Jeff McCoy* •
Local Museum – 1979
Historic bldgs 52485

Heym-Oliver House, 503 Kansas St, Russell, KS 67665 • T: +1 913 4833637 • rchs@russellks.net • www.rchs.russellks.net •
Cur.: *Ruby Blake* •
Ethnology Museum / Historical Museum – 1968 52486

Oil Patch Museum, Interstate 70 and Hwy 281, Russell, KS 67665 • T: +1 913 4836640, 4836960 • rchs@russellks.net • www.rchs.russellks.net •
Pres.: *Jeff McCoy* •
Science&Tech Museum – 1973
Industrial oil production 52487

Russell Springs KS

Butterfield Trail Historical Museum, Broadway and Hilts, Russell Springs, KS 67755 • T: +1 913 7514242 •
Pres.: *Jarett Hanemza* •
Historical Museum – 1964
Local history, housed in the Logan County Courthouse and Jail 52488

Russellville AR

Museum of Prehistory and History, Arkansas Tech University, Tucker Hall, 411 West N St, Russellville, AR 72801-2222, mail addr: ATV Box 8526, Tucker Hall Suite 12, Russellville, AR 72801-2222 • T: +1 479 9640826 • F: +1 479 9640872 • tech.museum@atu.edu •
Dir.: *Judith C. Stewart-Abernathy* •
Archaeology Museum / Historical Museum / University Museum – 1989
Archaeology, history 52489

Ruston LA

Lincoln Parish Museum, 609 N Vienna St, Ruston, LA 71270 • T: +1 318 2510018 • lpmuseum@tcainternet.com •
C.E.O. & Pres.: *William Davis Green* •
Local Museum – 1975
History and culture of North Central Louisiana 52490

Louisiana Tech Museum, Louisiana Tech University, 3175 Tech Station, Ruston, LA 71272 • T: +1 318 2572264, 2572737 • F: +1 318 2574735 • caesar@latech.edu • art.latech.edu •
Dir.: *C. Wade Meade* •
University Museum / Local Museum – 1982
Indian artifacts, paleontology and geologic coll, pre-WW II weapons, Roman and Egyptian archaeology 52491

Rutherford NJ

Meadowlands Museum, 91 Crane Av, Rutherford, NJ 07070 • T: +1 201 9351175 • F: +1 201 9359791 • meadowlandsmuseum@aol.com • www.meadowlandsmuseum.org •
Dir.: *Jackie Bunker-Lohrenz* •
Local Museum / Natural History Museum – 1961
Regional rocks and minerals, history 52492

Rutland VT

Chaffee Art Center, 16 S Main St, Rutland, VT 05701 • T: +1 801 7750356 • F: +1 802 7734401 • jdboughton@hotmail.com • www.chaffeeartcenter.org •
Exec. Dir.: *Jim Boughton* •
Public Gallery – 1961 52493

New England Maple Museum, POB 1615, Rutland, VT 05701 • T: +1 802 4839414 • F: +1 802 7751959 • info@maplemuseum.com • www.maplemuseum.com •
Pres.: *Thomas H. Olson* • Dir.: *Dona A. Olson* •
Agriculture Museum – 1977
Maple sugaring has been an early Spring tradition in Vermont ever since the Eastern Woodland Indians discovered that maple sap cooked over an open fire produces a sweet sugar 52494

Rye NY

The Rye Historical Society and Square House Museum, 1 Purchase St, Rye, NY 10580 • T: +1 914 9677588 • F: +1 914 9676253 • www.ryehistoricalsociety.org •
Dir.: *Kristina Bicher* •
Local Museum – 1964
Decorative arts, furniture, costumes and textiles, manuscripts 52495

Sabetha KS

Albany Historical Society, 415 Grant, Sabetha, KS 66534 • T: +1 785 2843446, 2843529 • www.albanydays.org •
Pres.: *Daryl Bechtelheimer* •
Association with Coll – 1965
Antique vehicles, agricultural machinery, train equip, military, two airplanes 52496

Sackets Harbor NY

Sackets Harbor Battlefield, 505 W Washington St, Sackets Harbor, NY 13685 • T: +1 315 6463634 • F: +1 315 6461203 •
Head: *Constance B. Barone* • Cur.: *Richard West* •

Military Museum / Open Air Museum – 1933
1812-1815 Fort Tompkins, 1812 War weapons, armamament accessoires and military clothing, 1814 artifacts from Brig Jefferson, 19th c household items 52497

Saco ME

Saco Museum, York Institute Museum, 371 Main St, Saco, ME 04072 • T: +1 207 2830684 • F: +1 207 2830754 • sacomuseum@maine.rv.com • www.sacomuseum.org •
Cur.: *Ryan Nutting* •
Fine Arts Museum – 1867
Colonial and Federal period fine and decorative arts, regional paintings and furniture, works by Brewster John, Bradbury Gibeon Elden, Granger Charles Henry 52498

Sacramento CA

California Automobile Museum, 2200 Front St, Sacramento, CA 95818 • T: +1 916 4426802 • F: +1 916 4422646 • director@calautomuseum.org • www.CalAutoMuseum.org •
Science&Tech Museum – 1982
Transportation, coll of cars 1900 - 1970s, Ford Motor Co. – library 52499

California Museum, 1020 O St, Sacramento, CA 95814 • T: +1 916 6537524 • F: +1 916 6530341 • info@californiamuseum.org • www.californiamuseum.org •
Cur.: *Larry Bishop* • *Kasey Schnormeier* •
Historical Museum – 1998
Artifacts, multimedia exhibit technology, state-of-the art technology – archives 52500

California State Capitol Museum, 10th and L Streets, Sacramento, CA 95814 • T: +1 916 3240333 • capitol@parks.ca.gov • www.statecapitolmuseum.ca.gov •
Dir.: *Todd Thames* • Cur.: *V. Joseph Sgromo* •
Historical Museum – 1981
State history 52501

California State History Museum → California Museum

California State Indian Museum, 2618 K Street, Sacramento, CA 95816 • T: +1 916 3240971 • www.parks.ca.gov •
Cur.: *Michael S. Tucker* •
Historical Museum – 1940
California Native American culture, contemporary artwork and hist 52502

California State Railroad Museum, 125 I St, Sacramento, CA 95814-2265 • T: +1 916 4456645 • www.csrmf.org •
Dir.: *Catherine A. Taylor* • Cur.: *Stephen E. Drew* •
Science&Tech Museum / Historical Museum – 1976
Railroads and railroading, 200 steam and diesel locomotives and cars – library 52503

Center for Sacramento History, 551 Sequoia Pacific Blvd, Sacramento, CA 95814 • T: +1 916 2647072 • F: +1 916 2647582 • csh@cityofsacramento.org • www.cityofsacramento.org/ccl/history •
Sc. Staff: *Patricia Johnson* • *Sally Stephenson* •
Historical Museum – 1953
Transportation, history, city and county records, photographs, printing 52504

Crocker Art Museum, 216 O St, Sacramento, CA 95814 • T: +1 916 8081179 • F: +1 916 8087372 • cam@cityofsacramento.org • www.crockerartmuseum.org •
Dir.: *Lial A. Jones* • Cur.: *Scott Shields* •
Fine Arts Museum – 1885
Drawings, European paintings, California art, photography, ceramics, Asian art – library 52505

Discovery Museum, Science and Space Center, 3615 Auburn Blvd, Sacramento, CA 95821 • T: +1 916 5753942 • info@PowerhouseScienceCenter.org • www.thediscovery.org •
Dir.: *Evangeline J. Higginbotham* • Cur.: *Milita Rios-Samaniego* •
Historical Museum / Science&Tech Museum – 1994
Hands-on history, science and technology, natural history – planetarium 52506

Governor's Mansion, 1526 H St, Sacramento, CA 95814 • T: +1 916 3233047 • www.parks.ca.gov •
Cur.: *Mike Tucker* •
Historic Site – 1967
Historic furnishings, California Govenor coll 1903-1967 52507

La Raza Galeria Posada, 1022-1024 22nd St, Sacramento, CA 95814 • T: +1 916 4465133 • larazagaleria@gmail.com • www.larazagaleriaposada.org •
Dir.: *Mario Gutierrez* • *Deborah V. Ortiz* • *Josh Pane* •
Exec. Dir.: *Stephanie Cornejo* •
Public Gallery / Performing Arts Museum – 1972
Chicano poster art, traditional folk and fine arts 52508

Matrixarts, 1713 25th St, Sacramento, CA 95816 • T: +1 916 4544988 • matrixarts@comcast.net • matrixarts.blogspot.com •
Public Gallery 52509

Sacramento Museum Collection Center → Center for Sacramento History

Towe Auto Museum → California Automobile Museum

Wells Fargo History Museum, 400 Capitol Mall, Sacramento, CA 95814 • T: +1 916 4404161 • www.wellsfargohistory.com •
Cur.: *Indulis Kalnins* •
Historical Museum – 1992
Banking hist, gold mining and specimens 52510

Wells Fargo History Museum - Old Sacramento, 1000 Second St, Sacramento, CA 95814 • T: +1 916 4404263 • www.wellsfargohistory.com •
Cur.: *Denise M. Pranza* •
Historical Museum – 1984
Company hist, gold scales 52511

Safety Harbor FL

Safety Harbor Museum of Regional History, 329 Bayshore Blvd S, Safety Harbor, FL 34695 • T: +1 727 7261668 • F: +1 727 7259938 • shmuseum@ij.net • www.safety-harbor-museum.org •
Historical Museum / Archaeology Museum – 1970
Fossils, artifacts pertaining to the pre-history of the Tocobaga Indians 52512

Safford AZ

Discovery Park, 1651 Discovery Park Blvd, Safford, AZ 85546 • T: +1 928 4286260 • F: +1 928 4288081 • discover@discoverypark.com • www.discoverypark.com •
Head: *Ed Sawyer* •
Science&Tech Museum – 1995
Science, agriculture, mining, history – planetarium, theater 52513

Sag Harbor NY

Sag Harbor Whaling and Historical Museum, 200 Main St, Sag Harbor, NY 11963 • T: +1 631 7250770 • F: +1 631 7255638 • www.sagharborwhalingmuseum.org •
C.E.O.: *Zachary N. Studenroth* •
Historical Museum – 1910
Whaling tools, scrimshaw, fishing rods, reels and lures, models, antiques 52514

Saginaw MI

Castle Museum of Saginaw County History, 500 Federal Av, Saginaw, MI 48607 • T: +1 989 7522861 ext 304 • F: +1 989 7521533 • info@castlemuseum.org • www.castlemuseum.org •
Dir.: *Irene D. Hensinger* • Cur.: *Sandy L. Schwan* • *Jeffrey Sommer* (Archaeology) •
Historical Museum – 1948
Regional history, costumes, logging tools, local industries 52515

Saginaw Art Museum, 1126 N Michigan Av, Saginaw, MI 48602 • T: +1 989 7542491 ext 201 • F: +1 989 7549387 • lreker@saginawartmuseum.org • www.saginawartmuseum.org •
Dir.: *Les Reker* • Cur.: *Michael S. Bell* (Exhibits, Colls) •
Sc. Staff: *Kara Harris* (Education) •
Fine Arts Museum – 1947
Former residence of Clark L. Ring family, African, Etruscan, Chinese, Japanes, European and American artworks on paper – library 52516

Saguache CO

Saguache County Museum, Hwy 285, Saguache, CO 81149 • T: +1 719 6552557, 2564272 • www.coloradotrails.com •
C.E.O.: *Evelyn Croft* •
Local Museum – 1958
Pioneer site, local history 52517

Sahuarita AZ

Titan Missile Museum, 1580 W Duval Mine Rd, Sahuarita, AZ 85629 • T: +1 520 6257736 • www.titanmissilemuseum.org •
Dir.: *Becky Roberts* •
Science&Tech Museum / Military Museum – 1986
Intercontinental ballistic missiles, Titan II 52518

Saint Albans VT

Saint Albans Historical Museum, 9 Church St, Saint Albans, VT 05478 • T: +1 802 5277933 • stamuseum.history@verizon.net • www.stalbansmuseum.org •
Dir.: *Dale Powers* •
Local Museum – 1971
China and glass, furniture, jewelry, quilts, linens, laces, paintings, Central Vermont railway, military, sports, toys, dolls, crafts, X-ray coll 52519

Saint Augustine FL

Castillo de San Marcos Museum, 1 S Castillo Dr, Saint Augustine, FL 32084 • T: +1 904 8296506 • F: +1 904 8239388 • casa_administration@nps.gov •
C.E.O: *Robert W. Harper* •
Historical Museum / Historic Site – 1935
Museum housed in 1672-95 restored Spanish Castillo de San Marcos 52520

Fort Matanzas, 8635 A1A S, Saint Augustine, FL 32080 • T: +1 904 4710116 • F: +1 904 4717605 • dave_parker@nps.gov •
Head: *Gordon Wilson* •
Historical Museum – 1935
Site of first European battle for control of New World 52521

Lightner Museum, 75 King St, Saint Augustine, FL 32085-0334 • T: +1 904 8242874 • F: +1 904 8242712 • lightner@aug.com • www.lightnermuseum.org •
Dir.: *Robert W. Harper* • Cur.: *Barry W. Myers Jr.* •
Local Museum – 1948
19th Century Fine and Decorative Art, Glass, Ceramics, Mechanical Music, 19th century Material Culture – library 52522

Oldest House Museum Complex, 271 Charlotte St, Saint Augustine, FL 32084 • T: +1 904 8242872 • F: +1 904 8242569 • sahsdirector@bellsouth.net • www.staugustinehistoricalsociety.org •
Dir.: *Dr. Overton G. Ganong* •
Historical Museum / Military Museum – 1918
Military, Gonzalez-Alvarez House, Tovar House, Webb Museum of Florida history 52523

Pena-Peck House, 143 St George St, Saint Augustine, FL 32084 • T: +1 904 8295064 • F: +1 904 8293898 • mpope@aug.com •
Decorative Arts Museum – 1932 52524

Saint Augustine Historical Society Museum, 271 Charlotte St, Saint Augustine, FL 32084 • T: +1 904 8242872 • F: +1 909 8242569 • oldhouse@aug.com • www.oldcity.com/ •
Dir.: *Taryn Rodriguez-Boette* •
Local Museum – 1883
Saint Augustin hist, artifacts, paintings – library 52525

Saint Augustine Lighthouse and Museum, 81 Lighthouse Av, Saint Augustine, FL 32080 • T: +1 904 8290745 • F: +1 904 8081248 • staugl@aug.com • www.staugustinelighthouse.com •
Dir.: *Kathy Allen Fleming* • Ass. Dir.: *Sue Van Vleet* •
Historical Museum / Historic Site – 1988
Lighthouse hist, maritim archaeology 52526

Saint Photios Greek Orthodox National Shrine, 41 St George St, Saint Augustine, FL 32085, mail addr: POB 1960, Saint Augustine, FL 32085 • T: +1 904 8298205 • F: +1 904 8298707 • info@stphotios.com • www.stphotios.com •
C.E.O.: *Nicholas Graff* •
Religious Arts Museum – 1982
1768 Greek landiing with Minorcans, Italians, Corsicans and Greeks, frescoes, icons 52527

The Spanish Quarter Museum, 29 Saint George St, Saint Augustine, FL 32084 • T: +1 904 8256830 • F: +1 904 8256874 • sqmuse@aug.com • www.historicstaugustine.com •
Man.: *Charles Dale* •
Folklore Museum – 1959
Restored village, traditional 18th c life skills 52528

World Golf Hall of Fame, 1 World Golf Pl, Saint Augustine, FL 32082 • T: +1 904 9404000 • F: +1 904 9404399 • www.wgv.com •
Dir.: *Jack Peter* •
Special Museum – 1974
Paintings, medals, sculpture, balls and other golf memorabilia 52529

Saint Bonaventure NY

The Regina A. Quick Center for the Arts, Saint Bonaventure University, Rte 417, Saint Bonaventure, NY 14778 • T: +1 716 3752494 • F: +1 716 3752690 • qac@sbu.edu • www.sbu.edu/go/arts-center/index.htm •
Exec.Dir.: *Joseph A. LoSchiavo* • Sr.Cur.: *Ruta Marino* •
Cur.: *Jason Trimmer* (Exhibitions, Prints and Drawings) •
Fine Arts Museum / University Museum – 1856
Paintings, sculpture, prints and drawings, Native American and pre-Columbian pottery, Asian porcelain 52530

Saint Charles IL

Saint Charles Heritage Center, 215 E Main St, Saint Charles, IL 60174 • T: +1 630 5846967 • F: +1 630 5846077 • info@stcmuseum.org • www.stchistory.org •
Pres.: *Cynthia Foster* •
Local Museum – 1940
Local hist 52531

Saint Charles MO

First Missouri State Capitol, 200-216 S Main St, Saint Charles, MO 63301 • T: +1 636 9403322 • F: +1 636 9403324 • dspfrst@mail.dnr.state.mo.us • www.dnr.state.mo.us •
Head: *David Klostermeier* •
Historical Museum – 1971
Period furnishings 52532

Harry D. Hendren Gallery, c/o Lindenwood University, 209 S Kings Hwy, Saint Charles, MO 63301 • T: +1 636 9494862 • F: +1 636 9494610 • etillinger@lindenwood.edu • www.lindenwood.edu •
Fine Arts Museum / University Museum – 1969
Contemporary American and European prints 52533

Lewis and Clark Boat House and Nature Center, 1050 Riverside Dr, Saint Charles, MO 63301 • T: +1 636 9473199 • F: +1 636 9160240 • www.lewisandclark.net •
Special Museum – 1985
Tools, fauna and flora, Indian artifacts 52534

Saint Charles County Historical Society, 101 S Main St, Saint Charles, MO 63301-2802 • T: +1 636 9469828 • scchs@mail.win.org • www.win.org/library/other/historical_society •
Association with Coll – 1956
American Indian artifacts, Tax records, naturalization, city directories & census 52535

Saint Clairsville OH

Ohio University Art Gallery, 209 Shannon Hall, Saint Clairsville, OH 43950 • T: +1 614 6951720 • www.eastern.ohiou.edu •
Public Gallery 52536

Saint Cloud MN

Atwood Memorial Center Art Collection, Saint Cloud State University, 720 Fourth Av S, Saint Cloud, MN 56301-4498 • T: +1 320 3084636 • atwood@stcloudstate.edu • www.stcloudstate.edu/atwood/art.asp •
Dir.: *Margaret Vos* •
Fine Arts Museum – 1967
Central Minnesota artists 52537

Evelyn Payne Hatcher Museum of Anthropology, 213A Stewart Hall, Saint Cloud State University, Saint Cloud, MN 56301 • T: +1 320 2552294 • F: +1 320 6545198 •
Cur.: *Richard B. Lane* •
Ethnology Museum / Archaeology Museum – 1972 52538

Kiehle Gallery, Saint Cloud State University, 102 Kiehle Visual Arts Center, 720 Fourth Av S, Saint Cloud, MN 56301-4498 • T: +1 320 3084717 • F: +1 320 3082232 • art@stcloudstate.edu •
Dir.: *Joseph Akin* •
Public Gallery / University Museum
Paintings, prints, sculpture, photographs 52539

Stearns History Museum, 235 S 33rd Av, Saint Cloud, MN 56301-3752 • T: +1 320 2538424 • F: +1 320 2532172 • info@stearns-museum.org • www.stearns.museum.org •
Dir.: *Anne Meline* • Cur.: *Steven Penick* •
Historical Museum – 1936
Local hist, miniature circus, dairy farming, regional nature, immigration, granite ind, Pan Car & Motor Co., local baseball – library 52540

Saint Croix VI

Christiansted National Historic Site, Danish Custom House, Kingswharf-Christiansted, Saint Croix, VI 00821 • T: +1 340 7731460 • F: +1 340 7732950 • chri_superintendent@nps.gov •
Head: *Anibal Colon* •
Local Museum – 1962
Hist houses, local hist, Danish Colonial development in the West Indies 52541

Saint Francis SD

Buechel Memorial Lakota Museum, Saint Francis Indian Mission, 350 S Oak St, Saint Francis, SD 57572 • T: +1 605 7472745 • F: +1 605 7475057 • www.museum.org •
Dir.: *J. Charmayne Young* •
Ethnology Museum / Folklore Museum – 1915
Lakota Indian culture 52542

Saint Francisville LA

Audubon State Historic Site, Louisiana State Hwy 965, Saint Francisville, LA 70775 • T: +1 225 6353739 • F: +1 225 7840578, +1 888 6772838 • audubon@crt.state.la.us • www.crt.state.la.us •
Fine Arts Museum – 1947
1st Audubon prints, early American lighting devices 52543

Saint George UT

Brigham Young's Winter Home, 67 W 200 N, Saint George, UT 84770 • T: +1 435 6732517 •
Dir.: *Tom Peterson* • Cur.: *Don Enders* (Exhibits and Historic Sites) • *Richard G. Oman* (Art and Artifacts) •
Local Museum – 1975
Historic house, Brigham Young winter home 52544

Rosenbruch Wildlife Museum, 1835 Convention Center Dr, Saint George, UT 84790 • T: +1 435 6560033 • F: 9866694 • www.rosenbruch.org •
Dir.: *Angie Rosenbruch-Hammer* •
Natural History Museum – 2001
Over 400 species of animals 52545

Saint George Art Museum, 47 E 200 North, Saint George, UT 84770 • T: +1 435 6345942 • F: +1 435 6345946 • museum@infowest.com • www.cist-george.ut.us •
Dir.: *David W. Pursley* •
Fine Arts Museum
Art 52546

Southwestern Utah Art Gallery, c/o Dixie College, 225 S 700th E, Saint George, UT 84770 • T: +1 435 6527500 • F: +1 435 6564000 • starkey@dixie.edu • www.dixie.edu •
Dir.: *Brent Hanson* •
Fine Arts Museum – 1960
Early and contemporary Utah painters 52547

Saint Helena CA

Robert Louis Stevenson Silverado Museum, 1490 Library Ln, Saint Helena, CA 94574 • T: +1 707 9633757 • dorothy@silveradomuseum.org • www.silveradomuseum.org •
Cur.: *Dorothy Mackay-Collins* •
Special Museum – 1969
First editions, Stevenson (1850-1894) memorabilia – library ... 52548

Saint Helens OR

Columbia County Historical Society Museum, Old County Courthouse, Saint Helens, OR 97051 • T: +1 503 3973868 • F: +1 503 3977257 •
Pres.: *R.J. Brown* •
Local Museum – 1969
Period furniture, Indian artfacts, local hist ... 52549

Saint Ignace MI

Father Marquette National Memorial and Museum, Straits State Park, 720 Church St, Saint Ignace, MI 49781 • T: +1 906 6438620 • F: +1 906 6439329 • www.nps.gov/fama •
Head: *Wayne Burnett* •
Religious Arts Museum – 1980
Michigan's earliest European settlement, French Jesuit missionary ... 52550

Saint Ignatius MT

Flathead Indian Museum, 1 Museum Ln, Saint Ignatius, MT 59865 • T: +1 406 7452951 • F: +1 406 7452961 •
Dir.: *Jeanine Allard* •
Ethnology Museum – 1974
American Indian arts & crafts, traditional dress, tools, buffalo exhib ... 52551

Saint James MO

Maramec Museum, The James Foundation, Maramec Spring Park, 21880 Maramec Spring Dr, Saint James, MO 65559 • T: +1 573 2657124 • F: +1 573 2658770 • tjf@socket.net •
Head: *Danny Marshall* •
Historical Museum / Folklore Museum / Science&Tech Museum – 1971
Model of the Village, grist mill, ron ore bank, geology of springs, sinkholes ... 52552

Saint Johns MI

Paine-Gillam-Scott Museum, 106 Maple Av, Saint Johns, MI 48879 • T: +1 517 2242894, 2247402 •
Dir.: *Catherine Rumbaugh* •
Local Museum – 1978
Clinton county hist ... 52553

Saint Johnsbury VT

Fairbanks Museum and Planetarium, 1302 Main St, Saint Johnsbury, VT 05819 • T: +1 802 7482372 • F: +1 802 7481893 • info@fairbanksmuseum.org • www.fairbanksmuseum.org •
C.E.O.: *Charles C. Browne* •
Local Museum / Natural History Museum – 1889
Local hist, ethnological coll ... 52554

Saint Johnsbury Athenaeum, 1171 Main St, Saint Johnsbury, VT 05819 • T: +1 802 7488291 • F: +1 802 7488086 • inform@stjathenaeum.org • www.stjathenaeum.org •
Dir.: *Lisa von Kann* •
Fine Arts Museum – 1873
American landscape paintings of the Hudson River School, sculpture ... 52555

Saint Johnsville NY

Fort Klock Historic Restoration, 7214 State Hwy 5, Saint Johnsville, NY 13452 • T: +1 518 5687779 • fortklock@hotmail.com • www.fortklock.com •
Pres.: *Cindy Sinchak* •
Local Museum – 1964
Hist Bldgs, early Dutch architecture, farmhouse furniture ... 52556

Saint Joseph MI

The Curious Kids' Museum, 415 Lake Blvd, Saint Joseph, MI 49085 • T: +1 616 9832543 • F: +1 616 9833317 • ckm@curiouskidsmuseum.org • www.curiouskidsmuseum.org •
Dir.: *Patricia Adams* • Cur.: *Gina Mason* (Programs) • Science&Tech Museum / Natural History Museum – 1988
Hands-on exhibits: science, technology, medical & natural hist, rain forest, volcano & earthquake ... 52557

Krasl Art Center, 707 Lake Blvd, Saint Joseph, MI 49085-1398 • T: +1 616 9830271 • F: +1 616 9830275 • info@krasl.org • www.krasl.org •
Exec. Dir.: *Darwin R. Davis* • Cur.: *Susan Wilczak* (Exhibits, Colls) • Sc. Staff: *Donna Metz* (Education) •
Fine Arts Museum – 1963
Sculpture ... 52558

Saint Joseph MN

Benedicta Arts Center, College of Saint Benedict, 37 S College Av, Saint Joseph, MN 56374 • T: +1 612 3635777 • F: +1 612 3636097 • jrule@csbsju.edu • www.csbsju.edu/finearts •
Dir.: *Anna M. Thompson* •
Fine Arts Museum / University Museum – 1963
Contemporary coll of crafts, drawings, paintings, prints and sculpture, East Asian and African colls – theater, concert hall ... 52559

Saint Joseph MO

Albrecht-Kemper Museum of Art, 2818 Frederick Av, Saint Joseph, MO 64506 • T: +1 816 2337003 • F: +1 816 2333413 • info@albrecht-kemper.org • www.albrecht-kemper.org •
Dir.: *Terry L. Oldham* •
Fine Arts Museum – 1914
Paintings, drawings, prints, sculpture ... 52560

Glore Psychiatric Museum, 3406 Frederick Av, Saint Joseph, MO 64506 • T: +1 816 3872300 • F: +1 816 3872170 • mcclars@mail.dmh.state.mo.us • www.gloremuseum.org •
C.E.O.: *Scott Clark* •
Special Museum – 1967 ... 52561

Jesse James Home Museum, 12th and Penn Sts, Saint Joseph, MO 64502 • T: +1 816 2328206 • F: +1 816 2323717 • patee@ponyexpress.net • www.stjoseph.net/ponyexpress •
Dir.: *Gary Chilcote* • Cur.: *Doug Chilcote* •
Historic Site / Historical Museum – 1939
1879 house where Jesse James was killed in 1882 ... 52562

Patee House Museum, 1202 Penn St, Saint Joseph, MO 64502 • T: +1 816 2328206 • F: +1 816 2328206 • patee@ponyexpress.net • www.stjoseph.net/ponyexpress •
Dir.: *Gary Chilcote* • Cur.: *Doug Chilcote* •
Science&Tech Museum / Historical Museum / Fine Arts Museum – 1964
Pony Express headquarters, trains, cars, fire trucks, antique town ... 52563

Pony Express Museum, 914 Penn St, Saint Joseph, MO 64503 • T: +1 816 2795059 • F: +1 816 2339370 • johnnyfry@ponyexpress.org • www.ponyexpress.org •
Pres.: *Richard N. DeShon* • Dir.: *F. Travis Boley* •
Local Museum – 1959 ... 52564

Saint Joseph Museum, 1100 Charles St, Saint Joseph, MO 64502 • T: +1 816 2328471 • F: +1 816 2328482 • sjm@stjosephmuseum.org • www.stjosephmuseum.org •
C.E.O. & Dir.: *Alberto C. Meloni* • Cur.: *Sarah M. Elder* •
Historical Museum / Ethnology Museum / Natural History Museum – 1927
American Indian art,anthropology, archaeology, ethnology, costumes, Eskimo art, native American art – library ... 52565

Saint Leonard MD

Jefferson Patterson Museum, 10515 Mackall Rd, Saint Leonard, MD 20685 • T: +1 410 5868500 • F: +1 410 5860080 • jppm@mdp.state.md.us • www.jefpat.org •
Pres.: *Victoria W. Childs* • Dir.: *Michael A. Smolek* •
Local Museum – 1983
Cultural heritage of people living in the Chesapeake Bay region, archaeology, agriculture, local history ... 52566

Saint Louis MO

Atrium Gallery, 7638 Forsyth Blvd, Saint Louis, MO 63105 • T: +1 314 7261066 • F: +1 314 7265444 • atrium@earthlink.net • www.atriumgallery.net •
Dir.: *Carolyn P. Miles* •
Public Gallery / Folklore Museum – 1986
Exhib of contemporary art, paintings, drawings & sculpture by living artists ... 52567

Boatmen's National Bank Art Collection, 1 Boatmen's Plaza, 800 Market St, Saint Louis, MO 63101 • T: +1 888 2793121 •
Fine Arts Museum
Political series by Caleb Bingham, transportation series by Oscar E. Berninghause, watercolours ... 52568

Campbell House Museum, 1508 Locust St, Saint Louis, MO 63103 • T: +1 314 4210325 • F: +1 314 4210113 • campbellhm@earthlink.net • stlouis.missouri.org/501c/chm •
C.E.O.: *John Dalzell* •
Decorative Arts Museum / Historical Museum – 1943
Decorative arts, costumes, letters, Lucas Place archives ... 52569

Chatillon-DeMenil Mansion, 3352 DeMenil Pl, Saint Louis, MO 63118 • T: +1 314 7715828 • F: +1 314 7713475 • demenil@stlouis.missouri.org • www.chatillondemenilhouse.com •
Pres.: *William Hart* • Dir.: *Paul R. Porter* •
Local Museum – 1965
Mid-Victorian furnishings, paintings, decorative arts ... 52570

City Museum, 701 N 15th St, Saint Louis, MO 63103 • T: +1 314 2312489 ext 110 • F: +1 314 2311009 • eparker@swbell.net • www.citymuseum.org •
Dir.: *Elizabeth Parker* •
Local Museum – 1997
Forest, architectural ornaments, glass blowers, weavers, potters and painters ... 52571

Concordia Historical Institute, 801 DeMun Av, Saint Louis, MO 63105 • T: +1 314 5057900 • F: +1 314 5057901 • chi@chi.lcms.org • chi.lcms.org •
Dir.: *Dr. Martin R. Noland* •
Religious Arts Museum – 1927
History of American Lutheranism ... 52572

Contemporary Art Museum of Saint Louis, Forum for Contemporary Art, 3540 Washington Av, Saint Louis, MO 63103 • T: +1 314 5354660 • F: +1 314 5351226 • info@contemporarystl.org • www.contemporarystl.org •
Dir.: *Paul Ha* • Cur.: *Shannon Fitzgerald* •
Fine Arts Museum – 1980
Photos, paintings ... 52573

Craft Alliance, 6640 Delmar Blvd, Saint Louis, MO 63130 • T: +1 314 7251151/1177 • F: +1 314 7252068 • info@craftalliance.org • www.craftalliance.org •
Public Gallery / Decorative Arts Museum – 1962
Craft arts, study coll ... 52574

Des Lee Gallery, Washington University School of Art, 1627 Washington Av, Saint Louis, MO 63103 • T: +1 314 6218735 • F: +1 314 6213703 • pslein@art.wustl.edu • www.wustl.edu •
University Museum / Fine Arts Museum
Temporary art ... 52575

Eugene Field House and Saint Louis Toy Museum, 634 S Broadway, Saint Louis, MO 63102 • T: +1 314 4214689 • F: +1 314 5889328 • fldhouse@swbell.org • www.eugenefieldhouse.org •
Pres.: *William Piper* • Dir.: *Julie Maio Kemper* •
Local Museum / Decorative Arts Museum – 1936 ... 52576

Gallery 210, University of Missouri Saint Louis, 210 Lucas Hall, Saint Louis, MO 63121 • T: +1 314 5165952 • F: +1 314 5165816 •
Dir.: *Terry Suhre* •
Fine Arts Museum – 1972
Contemporary art ... 52577

Gallery of Art, Washington university, 1 Brookings Dr, Saint Louis, MO 63130 • T: +1 314 9355490 •
Fine Arts Museum / University Museum
American and European paintings 19th-20th c, drawings and sculpture ... 52578

General Daniel Bissell House, 10225 Bellefontaine Rd, Saint Louis, MO 63137 • T: +1 314 8680973, 5545790 • F: +1 314 8688435 • jd_magurany@stlouisco.com • www.st-louiscountyparks.com •
Dir.: *J.D. Magurany* • Cur.: *Marc Kollbaum* •
Historic Site – 1960
Historic house ... 52579

International Bowling Museum and Hall of Fame, 111 Stadium Plaza, Saint Louis, MO 63102 • T: +1 314 2316340 • F: +1 314 2314054 • hofm@bowlingmuseum.com • www.bowlingmuseum.com •
Exec. Dir.: *Gerald W. Baltz* •
Special Museum – 1977
18th-20th c historical artifacts, equip, uniforms, graphics, trophies ... 52580

Jasper's Antique Radio Museum, 2022 Cherokee St, Saint Louis, MO 63118 • T: +1 314 4218313 • jaspers_stl_mo@yahoo.com • www.jaspers-stl-mo.com/jasanradmus.html •
Science&Tech Museum ... 52581

Jefferson Barracks, 533 Grant Rd, Saint Louis, MO 63125-4121 • T: +1 314 5445714, 5445790 • F: +1 314 6385009 • jd_magurany@stlouisco.com • www.st-louiscountyparks.com/jb.html •
Dir.: *J.D. Magurany* • Cur.: *Marc Kollbaum* •
Military Museum / Historic Site – 1826
Military post – Gen. Daniel Bissell house, Fort Belle ... 52582

Jefferson National Expansion Memorial, 11 N 4th St, Saint Louis, MO 63102 • T: +1 314 6551600 • F: +1 314 6551639 • jeff_superintendent@nps.gov • www.nps.gov/jeff •
Dir.: *David Grove* • Cur.: *Kathryn Thomas* •
Historical Museum – 1935
19th c Western American artifacts ... 52583

Laumeier Sculpture Park, 12580 Rott Rd, Saint Louis, MO 63127 • T: +1 314 8211209 • F: +1 314 8211248 • info@laumeier.org • www.laumeier.org •
Dir.: *Glen P. Gentele* • Cur.: *Clara Collins Coleman* •
Fine Arts Museum – 1976
Contemporary sculpture coll ... 52584

Magic House-Saint Louis Children's Museum, 516 S Kirkwood Rd, Saint Louis, MO 63122 • T: +1 314 8228900 • F: +1 314 8228930 • beth@magichouse.com • www.magichouse.com •
Pres.: *Elizabeth Fitzgerald* • Chm.: *Jerry Kent* •
Special Museum – 1979
Hands-on, participatory exhib ... 52585

Mildred Lane Kemper Art Museum, Forsyth and Skinker Blvds, Saint Louis, MO 63130 • T: +1 314 9355490 • F: +1 314 9357282 • kemperartmuseum@wustl.edu • www.kemperartmuseum.wustl.edu •
Dir.: *Dr. Mark S. Weil* • Cur.: *Sabine Eckmann* •
Fine Arts Museum / University Museum – 1881
Modern artists incl Miro, Ernst, Picasso, Leger, Moore, old masters, 19th-20th c paintings, sculpture, drawings, prints ... 52586

Missouri Historical Society, 5700 Lindell Blvd, Lindell and De Baliviere, Forest Park, Saint Louis, MO 63112-0040 • T: +1 314 7464599 • F: +1 314 4543162 • info@mohistory.org • www.mohistory.org •
Pres.: *Dr. Robert R. Archibald* • Dir.: *Becki Hartke* (Exhibition) •
Association with Coll – 1866
Exploration of the West, Lewis and Clark, artifacts ... 52587

Morton J. May Foundation Gallery, Maryville University, 13550 Conway Rd, Saint Louis, MO 63141 • T: +1 314 5769415 • F: +1 314 5299940 • nrice@maryville.edu •
Dir.: *Nancy N. Rice* •
Fine Arts Museum / University Museum
American paintings ... 52588

Museum of Art, Saint Louis University, 221 N Grand Blvd, Saint Louis, MO 63103 • T: +1 314 9773399 • F: +1 314 9776632 • sluma@slu.edu • sluma.slu.edu •
Dir.: *Nanette E. Boileau* •
Fine Arts Museum
Contemporary art ... 52589

Museum of Contemporary Religious Art, c/o Saint Louis University, 3700 John E Connelly Pedestrian Mall, Saint Louis, MO 63108 • T: +1 314 9777170 • F: +1 314 9772999 • mocra@slu.edu • mocra.slu.edu •
Cur.: *Terrénce E. Dempsey* •
Fine Arts Museum / Religious Arts Museum / University Museum – 1993
Art from all religious traditions ... 52590

Museum of Religious Art, Saint Louis University, 221 N Grand Blvd, Saint Louis, MO 63103 • T: +1 314 9777170 • F: +1 314 9772999 • mocra@slu.edu • mocra.slu.edu •
Dir.: *Terrénce E. Dempsey* •
Religious Arts Museum ... 52591

Museum of the Dog, 1721 S Mason Rd, Saint Louis, MO 63131 • T: +1 314 8213647 • F: +1 314 8217381 • dogarts@aol.com • www.akc.org •
Exec. Dir.: *Barbara Jedda McNab* •
Fine Arts Museum – 1981
Fine arts, pure bred dogs ... 52592

Museum of Transportation, 3015 Barrett Station Rd, Saint Louis, MO 63122 • T: +1 314 9658007 • F: +1 314 9650242 • barbara_bernsen@stlouisco.com • www.museumoftransport.org •
Pres.: *Lee Rottmann* • Dir.: *James Worton* •
Science&Tech Museum – 1944
Locomotives, railway cars, automoils, streetcars, buses trucks ... 52593

Old Cathedral Museum, 209 Walnut, Saint Louis, MO 63102 • T: +1 314 2313251 • F: +1 314 2314280 •
C.E.O.: *Jerome Billing* •
Religious Arts Museum – 1970
Religious hist items ... 52594

Saint Louis Art Museum, 1 Fine Arts Dr, Forest Park, Saint Louis, MO 63110-1380 • T: +1 314 6555406 • F: +1 314 7216172 • infotech@slam.org • www.slam.org •
Dir.: *Brent Benjamin* • Cur.: *Cornelia Homburg* (Modern Art) • *Steven D. Owyoung* (Asian Art) • *Cara McCarty* (Decorative and Design Arts) • *John Nunley* (Arts of Africa, Oceania and the Americas) • *Judith Mann* (Early European Art) • *Sid Goldstein* (Ancient, Islamic) • *Franesca Consagre* (Prints, Photographs) •
Fine Arts Museum – 1906
Prints, drawings, photographs, Oceanian, African, Pre-Columbian and American Indian objects, paintings, sculpture, European decorative art, Chinese bronzes and porcelain – Richardson Memorial Library ... 52595

Saint Louis Artists' Guild Museum, 2 Oak Knoll Park, Saint Louis, MO 63105 • T: +1 314 7276266 • F: +1 314 7279190 • âmurphy@stlouisartistsguild.org • www.stlouisartistsguild.org •
Pres.: *Joanne Stremsterfer* • C.E.O.: *Anne Murphy* •
Fine Arts Museum – 1886
Paintings, sculputre, graphics, photography, fine craft ... 52596

Saint Louis Car Museum, 1575 Woodson Rd, Saint Louis, MO 63114 • T: +1 314 9931330 • museumcars@waterwaysapartments.com • www.stlouiscarmuseum.com •
Science&Tech Museum – 1994 ... 52597

Saint Louis Science Center, 5050 Oakland Av, Saint Louis, MO 63110 • T: +1 314 2894400 • F: +1 314 2894420 • gjasper@slsc.org • www.slc.org •
Pres.: *Douglas R. King* •
Public Gallery / Science&Tech Museum – 1959 ... 52598

Samuel Cupples House, c/o Saint Louis University, 221 N Grand Blvd, Saint Louis, MO 63103 • T: +1 314 9773575 • F: +1 314 9773581 • meadows@slu.edu • www.slu.edu/the-arts/cupples •
Exec. Dir.: *Pamela E. Ambrose* •
Fine Arts Museum / University Museum – 1977
American impressionist paintings, Italian and northern renaissance paintings, American sculpture, American and European art glass and furniture ... 52599

The Sheldon Art Galleries, 3648 Washington Blvd, Saint Louis, MO 63108 • T: +1 314 5339900 • F: +1 314 5332958 • www.sheldonconcerthall.org •
Dir.: *Olivia Lahs-Gonzales* •
Fine Arts Museum
Art photography, architecture, Jazz hist ... 52600

Soldiers' Memorial Military Museum, 1315 Chestnut St, Saint Louis, MO 63103 • T: +1 314 6224550 • F: +1 314 6224237 •
Head: *Ralph D. Wiechert* •
Military Museum – 1938 ... 52601

Trova Foundation, 8112 Maryland Av, Saint Louis, MO 63105 • T: +1 314 7272444 • F: +1 314 7276084 • Dir.: *Clifford Samuels* • Pres.: *Philip Samuels* • Fine Arts Museum – 1988
Contemporary painting, collage, drawing and sculpture 52602

Saint Marks FL

San Marcos de Apalache Historic State Park, 148 Old Fort Rd, Saint Marks, FL 32355-0027 • T: +1 850 9226007 • F: +1 850 4880366 • sanmarcos@nettally.com • www.floridastateparks.org • Dir.: *Wendy Spencer* • Historical Museum / Historic Site – 1964
Second oldest fortification in Florida 52603

Saint Martinville LA

Longfellow-Evangeline State Historic Site, 1200 N Main St, Saint Martinville, LA 70582 • T: +1 337 3943754 • F: +1 337 3943553 • longfellow@crt.state.la.us • www.lastateparks.com • Cur.: *Suzanna Laviolette* • Historical Museum / Historic Site – 1934
Historic plantation house circa 1815.Portraits, textile arts, local craft and folk art, furniture, antiques, religious art, wood carvings, Acadian art 52604

Oliver House Museum, 1200 N Main St, Saint Martinville, LA 70582 • F: +1 318 3943754 • longfellow@crt.state.la.usa • Cur.: *Suzanna Laviolette* • Historical Museum – 1931
History, old plantation house of French and Carribean architecture 52605

Saint Marys OH

Auglaize County Historical Society Museum, 223 S Main St, Saint Marys, OH 45885 • Pres.: *Karen Dietz* • Local Museum – 1963
Fort St Marys, Fort Amanda and Fort Barbee artifacts, Indian artifacts, Miami Erie Canal, oil wells, early industry, Civil War 52606

Saint Marys PA

Saint Marys and Benzinger Township Museum, 99 Erie Av, Saint Marys, PA 15857 • T: +1 814 8346525 • Dir.: *Richard Dornisch* • Cur.: *Alice Beimel* • Local Museum – 1960
Local hist 52607

Saint Mary's City MD

Dwight Frederick Boyden Gallery, Saint Mary's College of Maryland, Montgomery Hall, 18952 Fisher Ln, Saint Mary's City, MD 20686 • T: +1 240 8954246 • F: +1 301 8620958 • smglasser@smcm.edu • www.smcm.edu/academics/gallery/ • C.E.O.: *Susan Glasser* • Fine Arts Museum / University Museum – 1839
20th c paintings, sculpture, prints, drawings 52608

Historic Saint Mary's City, Rte. 5, Saint Mary's City, MD 20686 • T: +1 240 8954960 • F: +1 301 8954968 • hsmc@smcm.edu • www.stmaryscity.org • C.E.O.: *Martin E. Sullivan* • Special Museum – 1966
Outdoor living history 52609

Saint Matthews SC

Calhoun County Museum, 303 Butler St, Saint Matthews, SC 29135 • T: +1 803 8743964 • F: +1 803 8744790 • calmus@oburg.net • Dir.: *Debbie U. Roland* • Local Museum – 1954
Local hist 52610

Saint Michaels MD

Chesapeake Bay Maritime Museum, Navy Point, Saint Michaels, MD 21663 • T: +1 410 7452916 • F: +1 410 7456088 • letters@cbmm.org • www.cbmm.org • Pres.: *John R. Valliant* • Cur.: *Ronald E. Lesher* • Historical Museum – 1965
Regional maritime history 52611

Saint Paul MN

Alexander Ramsey House, 265 S Exchange St, Saint Paul, MN 55102 • T: +1 612 2968760 • F: +1 612 2960100 • ramseyhouse@mnhs.org • www.mnhs.org/places/sites/arh • Dir.: *Craig Johnson* • Decorative Arts Museum – 1964
Late 19th c Victorian mansion, marble fireplaces, crystal chandeliers and furnishings 52612

American Museum of Asmat Art, 3510 Vivian Av, Saint Paul, MN 55126-3852 • T: +1 651 2871132 • F: +1 651 2871130 • museum@crosier.org • www.asmat.org • Dir.: *Mary E. Braun* • Fine Arts Museum / Ethnology Museum – 1995
Ethnological art by Western Papua & Indonesia Asmat people, culture 52613

College of Visual Arts Gallery, 173 Western Av, Saint Paul, MN 55102 • T: +1 651 2909379 • F: +1 651 2248854 • cvagallery@cva.edu • www.cva.edu • Cur.: *Leslie Mehren Gorev* • Public Gallery
Work of students, faculty, alumni and regional and nationally practicing artists 52614

Fort Snelling, 200 Tower Av, Saint Paul, MN 55111 • T: +1 612 7261171 • F: +1 612 7252429 • ftsnelling@mnhs.org • www.mnhs.org/fortsnelling • Military Museum – 1970
Military hist – theater 52615

The Goldstein Museum of Design, University of Minnesota, 244 McNeal Hall, 1985 Buford Av, Saint Paul, MN 55108-6236 • T: +1 612 6247434 • F: +1 612 6242750 • goldstein@che.umn.edu • goldstein.che.umn.edu • Dir.: *Dr. Lindsay Shen* • Cur.: *Dr. Marilyn DeLong* (Costumes) • *Dr. Gloria Williams* (Textiles) • *Dr. Stephen McCarthy* (Graphic Design) • Decorative Arts Museum / University Museum – 1976
Historic costumes, 20th c designer garments, decorative arts, furniture, glass, metal, ceramics, textiles – library 52616

Hamline University Galleries, Drew Fine Arts Center, 1536 Hewitt Av, Saint Paul, MN 55104-0161 • T: +1 651 5232386, 5232296 • F: +1 651 5233066 • llasansky@gw.hamline.edu • www.hamline.edu/art • Dir.: *Leonardo Lasansky* • Fine Arts Museum / University Museum – 1854
Modern sculpture, paintings, decorative arts 52617

James J. Hill House, 240 Summit Av, Saint Paul, MN 55102 • T: +1 651 2972555 • F: +1 651 2975655 • hillhouse@mnhs.org • www.mnhs.org/places/sites/jjhh • Head: *Craig Johnson* • Historic Site – 1978
Art exhibits, French landscape paintings, MN art hist, historic house museum 52618

Julian H. Sleeper House, 66 Saint Albans St S, Saint Paul, MN 55105 • T: +1 651 2251505 • C.E.O.: *Dr. Seth C. Hawkins* • Decorative Arts Museum – 1993
Eastlake-Vernacular house, Pres. James A. Garfield memorabilia, furnishings, Roseville pottery, postcards coll 52619

Macalester College Art Gallery, Janet Wallace Fine Arts Center, 1600 Grand Av, Saint Paul, MN 55105 • T: +1 651 6966416 • F: +1 651 6966266 • gallery@macalester.edu • www.macalester.edu/gallery • Cur.: *Devin A. Colman* • Public Gallery / University Museum – 1964
Regional, national & international contemporary art, Asian and British ceramics, African art 52620

Minnesota Air National Guard Museum, 670 General Miller Dr, Saint Paul, MN 55111-0598 • T: +1 612 7132523 • F: +1 612 7132525 • msp04332@isd.net • www.mnangmuseum.org • Chm.: *Bruce Graham* • Military Museum / Science&Tech Museum – 1980
Aircraft, hist of the MNG, 109th squadron – library 52621

Minnesota Children's Museum, 10 W Seventh St, Saint Paul, MN 55102 • T: +1 651 2256001 • F: +1 651 2256006 • mcm@mcm.org • www.mcm.org • Pres.: *Sarah Caruso* • Science&Tech Museum – 1979
Multidisciplinary hands-on exhibits, earth, art & nature 52622

Minnesota Historical Society Museum, 345 Kellogg Blvd W, Saint Paul, MN 55102-1906 • T: +1 651 2966126 • F: +1 651 2973343 • webmaster@mnhs.org • www.mnhs.org • Dir.: *Nina M. Archabal* • Cur.: *Dan Spock* • Historical Museum / Historic Site – 1849
Historic & prehistoric archaeological and cultural material, maps, paintings – state archives 52623

Minnesota Museum of American Art, 50 W Kellogg Blvd, Saint Paul, MN 55102 • T: +1 651 2661030 • F: +1 651 2912947 • blilly@mmaa.org • www.mmaa.org • Dir.: *Bruce Lilly* • Cur.: *Anita Gonzalez* • Fine Arts Museum – 1927
American art e.g. Paul Manship, contemporary art of Upper Midwest, 19th-20th c and native art – library, 2 studios 52624

Minnesota State Capitol Historic Site, 75 Rev. Dr Martin Luther King Jr. Blvd, Saint Paul, MN 55155 • T: +1 651 2962881 • F: +1 651 2971502 • www.mnhs.org/statecapitol • Historical Museum – 1969
Furniture, canvas murals, paintings, plaques, statues and busts, governors portraits, historic battle flags 52625

Minnesota Transportation Museum, 193 E Pennsylvania Av, Saint Paul, MN 55101-4319 • T: +1 651 2280263 • F: +1 651 2289412 • admin@mtmuseum.org • www.mtmuseum.org • Head: *Scott Reed* • Science&Tech Museum – 1962
Streetcars, locomotives, trucks, Mesaba Electric Railway 52626

Paul Whitney Larson Art Gallery, University of Minnesota, 2017 Buford Av, Saint Paul, MN 55108 • T: +1 612 6250214, 6265516 • F: +1 612 6248749 • larson@tcsu.umn.edu • www.sao.umn.edu/events/arts • Public Gallery / University Museum – 1979
Contemporary visual arts 52627

Schubert Club Museum of Musical Instruments, 75 W 5th St, Saint Paul, MN 55102 • T: +1 651 2923267 • F: +1 651 2924317 • schubert@schubert.org • www.schubert.org • Music Museum – 1972
Musical instr from around the world, keyboard instr, early phonographs, letters from composers and musicans – library 52628

Science Museum of Minnesota, 120 Kellogg Blvd W, Saint Paul, MN 55102 • T: +1 612 2219444 • F: +1 612 2214777 • postmaster@smm.org • www.smm.org • Dir.: *James L. Peterson* • Cur.: *Bruce R. Erickson* (Paleontology) • *Dr. Orrin Shane* (Anthropology for Archaeology) • Natural History Museum / Science&Tech Museum – 1907
Biological, anthropological, paleontological and geological specimens – research station 52629

University Saint Thomas Art Collection and Exhibitions, c/o Dept. of Art History, 2115 Summit Av, Ste 57P, Saint Paul, MN 55105 • T: +1 651 9625560 • F: +1 651 9625861 • manordtorpm@stthomas.edu • www.stthomas.edu/arthistory/exhibitions • Cur.: *Shelly Nordtorp-Madson* • Man.: *Sue Focke* (Exhibitions) • Fine Arts Museum – 1978
20th c American art, videos, oil paintings 52630

Saint Paul OR

Robert Newell House, DAR Museum, 8089 Champoeg Rd NE, Saint Paul, OR 97137 • T: +1 503 2663944 • DARcabin@stpaultel.com • www.ohwy.com/or/r/rnewellhm.htm • Local Museum / Historic Site – 1959
Indian artifacts, quilts and handwork, furniture and furnishings, Masonic artifacts 52631

Saint Peter MN

E. St. Julien Cox House, 500 S Washington, Saint Peter, MN 56082-1727 • T: +1 507 9312160 • F: +1 507 9310172 • nchs@hickorytech.net • Exec. Dir.: *Ben Leonard* • Historical Museum – 1971
Family items, furniture, glassware, linens, clothing, art 52632

Treaty Site History Center, 1851 N Minnesota Av, Saint Peter, MN 56082-1727 • T: +1 507 9342160 • F: +1 507 9348715 • museum@hickorytech.net • Exec. Dir.: *Ben Leonard* • Historical Museum – 1928
Nicollet couty hist, clothing, tools, military hist, Dakota, Native American & pioneer items – library, archives 52633

Saint Petersburg FL

The Arts Center, 719 Central Av, Saint Petersburg, FL 33701 • T: +1 727 8227872 • F: +1 727 8210516 • www.theaterscenter.org • Public Gallery
Art 52634

Children's Museum, 1925 Fourth St N, Saint Petersburg, FL 33704 • T: +1 727 8218992 • F: +1 727 8237287 • mbeairsto@greatex.org • www.greatexplorations.org • Exec. Dir.: *Murray Beairsto* • Special Museum – 1987
Art and history interactive exhibits 52635

Florida Craftsmen Gallery, 501 Central Av, Saint Petersburg, FL 33701 • T: +1 727 8217391 • F: +1 727 8224294 • info@floridacraftsmen.net • www.floridacraftsmen.net • Dir.: *Michele Tuegel* • Public Gallery
Temporary exhibitions 52636

Florida Holocaust Museum, 55 Fifth St, Saint Petersburg, FL 33701 • T: +1 813 8200100 ext 221 • F: +1 813 8218435 • nbrand@flholocaustmuseum.org • www.flholocaustmuseum.org • Dir.: *R. Andrew Maass* • Historical Museum – 1989
Artwork concerning the Holocaust genocide and the human condition, traveling art and cultural exhibitions 52637

Florida International Museum, 100 Second St N, Saint Petersburg, FL 33701 • T: +1 727 8223693, 8211448 • F: +1 727 8980248 • floridamuseum@mindspring.com • www.floridamuseum.org • C.E.O: *Kathleen C. Oathout* • Cur.: *Vera Espinola* • Historical Museum – 1994
Cuban Missile Crisis 52638

Great Explorations, The Pier, 800 2nd Av NE, Saint Petersburg, FL 33701 • T: +1 727 8218992 • F: +1 727 8237287 • greatest@greatexplorations.com • www.greatexplorations.org • Exec. Dir.: *Robert B. Patterson* • Science&Tech Museum – 1987 52639

Museum of Fine Arts, 255 Beach Dr, NE, Saint Petersburg, FL 33701 • T: +1 727 8962667 • F: +1 727 8944638 • jschloder@fine-arts.org • www.fine-arts.org • Dir.: *John E. Schloter* • Chief Cur.: *Jennifer Hardin* • Fine Arts Museum – 1961
Paintings, drawings, prints, sculpture and photographs, decorative arts, gallery of Steuben glass 52640

Saint Petersburg Museum of History, 335 Second Av NE, Saint Petersburg, FL 33701 • T: +1 727 8941052 • F: +1 727 8237276 • donna@museumofhistoryonline.org • www.saint-petersburg.com/museums/index.asp • Dir.: *Patricia Kelly* • Local Museum – 1920
Peninsula and Florida hist 52641

Salvador Dalí Museum, 1000 Third St S, Saint Petersburg, FL 33701 • T: +1 813 8233767 • F: +1 813 8946068 • info@salvadordalimuseum.org • www.salvadordalimuseum.org • Dir.: *Charles Henri Hine* • Fine Arts Museum – 1954
Salvador Dalí oils, drawings, watercolors, graphics and sculpture 52642

Science Center of Pinellas County, 7701 22nd Av N, Saint Petersburg, FL 33710 • T: +1 727 3840027 • F: +1 727 3435729 • scenter5@tampabay.rr.com • www.sciencecenterofpinellas.com • Natural History Museum – 1959
Anatomy, anthropology, archaeology, astronomy, botany, entomology, geology, herpetology, marine, medicine, natural hist, paleontology, zoology – aquarium 52643

Saint Simons Island GA

The Arthur J. Moore Methodist Museum, Arthur Moore Dr, Saint Simons Island, GA 31522 • T: +1 912 6384050 • F: +1 912 6389050 • methmuse@darientel.net • Dir.: *Mary L. Vice* • Religious Arts Museum – 1965
Religious, nr Oglethorpe, John and Charles Wesley's activities in 1736 52644

Coastal Center for the Arts, 2012 Demere Rd, Saint Simons Island, GA 31522 • T: +1 912 6340404 • F: +1 912 6340404 • coastalart@thebest.net • Dir.: *Mittie B. Hendrix* • Fine Arts Museum – 1947 52645

Fort Frederica, Rte 9, Box 286c, Saint Simons Island, GA 31522 • T: +1 912 6383639 • F: +1 912 6383639 • mike_tennent@nps.gov • www.nps.gov/fofr • Head: *Mike Tennent* • Historical Museum – 1936
Artifacts 52646

The Glynn Art Association, 319 Mallery St, Saint Simons Island, GA 31522 • T: +1 912 6388770 • F: +1 912 6342787 • glynnart@earthlink.net • www.glynnart.org • Dir.: *Pat Weaver* • Fine Arts Museum – 1948
Primitive painting, pottery 52647

Saint Simons Island Lighthouse Museum, 101 12th St, Saint Simons Island, GA 31522-0636 • T: +1 912 6384666 • F: +1 912 6386609 • ssi1872@bellsouth.net • www.saintsimonslighthouse.org • Exec. Dir.: *Patricia A. Morris* • Local Museum / Historical Museum – 1965
History, civil war 52648

Sainte Genevieve MO

Bolduc House Museum, 125 S Main St, Sainte Genevieve, MO 63670 • T: +1 573 8833105 • F: +1 573 8833359 • lstange@brick.net • Dir.: *Lorraine Stange* • Local Museum – 1770
Archaeoloty, preservation project 52649

Felix Valle State Historic Site, 198 Merchant St, Sainte Genevieve, MO 63670 • T: +1 573 8837102 • F: +1 573 8839630 • felix.valle.state.historic.site@dnr.mo.gov • www.mostateparks.com • Head: *James Baker* • Historic Site – 1970
American Empire furniture 52650

Sainte Genevieve Museum, Merchant and DuBourg Sts, Sainte Genevieve, MO 63670 • T: +1 573 8833461 • Pres.: *James Baker* • Dir.: *Ruby A. Stephens* • Local Museum – 1935
Folk art Indian artifacts, sketches, weapons 52651

Salado TX

Central Texas Area Museum, 1 Main St, Salado, TX 76571 • T: +1 254 947 5232 • F: +1 254 9475232 • webmaster@ctam-salado.org • www.ctam-salado.org • Local Museum – 1958
Genealogy, hist, arts and crafts, Scottish folklore, Tonkawa artifacts, antiques 52652

Salamanca NY

Salamanca Rail Museum, 170 Main St, Salamanca, NY 14779-1574 • T: +1 716 9453133 • F: +1 716 9452034 • C.E.O. & Cur.: *Gerald J. Fordham* • Science&Tech Museum – 1980
Telegraph keys, railroad memorabilia, train cars 52653

Seneca-Iroquois National Museum, 794-814 Broad St, Allegany Indian Reservation, Salamanca, NY 14779 • T: +1 716 9451738 • F: +1 716 9455397 • sue.grey@sni.org • www.senecamuseum.org •
Dir.: *Michele Dean Stock* •
Ethnology Museum – 1977
Culture and hist of the Seneca and other Iroquois Nations, archaeology, modern art 52654

Salem IN

Stevens Museum, 307 E Market St, Salem, IN 47167 • T: +1 812 8836495 • jhc@blueriver.net • www.stevensmuseum.com •
Pres.: *Willie Harlen* •
Local Museum / Open Air Museum – 1897
Local history, 1824 John Hay birthplace, pionner village – library 52655

Salem MA

The House of the Seven Gables, 54 Turner St, Salem, MA 01970 • T: +1 978 7440991 • F: +1 978 7414350 • postmaster@7gables.org • www.7gables.org •
Exec. Dir.: *Stanley G. Burchfield* •
Historical Museum / Open Air Museum – 1910
17th-19th c architecture, period furnishings, Nathaniel Hawthorne – colonial garden 52656

Peabody Essex Museum, E India Sq, Salem, MA 01970-3783 • T: +1 978 7459500 • F: +1 978 7403633 • pem@pem.org • www.pem.org •
Dir.: *Dan L. Monroe* • Sc. Staff: *Dr. Susan Bean* (Asian, Oceanic, African Art & Cultures) • *John Grimes* (Special Projects) • *William Sargent* (Asian Export Art) • *Janey Winchell* (Natural History) • *Dean Lahikainen* (American Decorative Arts) • *Dr. Daniel Finamore* (Maritime Arts and History) • *Kimberly Alexander* (Architecture) • *Andrew Maske* (Japanese Art) • *Nancy Berliner* (Chinese Art) • *Clark Worswick* (Photographic Collections) •
Fine Arts Museum / Decorative Arts Museum – 1799
Asian, Oceanic, African arts and cultures, native American art and archaeology, maritime arts and history, architecture, decorative arts, natural history – library, auditorium, lab, Gardner-Pingree House, Crowninshield-Bentley House, John Ward House, Andrew-Safford House, Peirce-Nichols House, Cotting-Smith-Assembly-House, Derby-Beebe Summer House, Lyle-Tapley Shoe Shop and Vaughn Doll House 52657

Salem Maritime Houses, Salem Maritime National Historic Site, 193 Derby St, Salem, MA 01970, mail addr: 160 Derby St, Salem, MA 01970 • T: +1 978 7401660 • F: +1 978 7401654 • sama_orientation_center@nps.gov • www.nps.gov/sama •
Cur.: *David Kayser* • Historian: *Emily Murphy* •
Historic Site / Archaeology Museum – 1938
Maritime, commerce and World trade, full-size replica of 1797 merchant ship – library 52658

Salem Pioneer Village, Forest River Park, West Av, Salem, MA 01970 • T: +1 978 7450525 • F: +1 978 7414350 • wwalker@7gables.org • www.7gables.org •
Exec. Dir.: *Stanley G. Burchfield* •
Historical Museum / Open Air Museum – 1930
Reproduction of a fishing village, English wigwam, furnishings 52659

Salem Witch Museum, Washington Sq N, Salem, MA 01970 • T: +1 978 7441692 • F: +1 978 7454414 • facts@salemwitchmuseum.com • www.salemwitchmuseum.com •
Dir.: *Patricia MacLeod* •
Historical Museum – 1971
Dramatic history, overview of the Witch Trials of 1692 – small theater 52660

Stephen Phillips Memorial House, 34 Chestnut St, Salem, MA 01970 • T: +1 978 7440440 • F: +1 978 7401086 • info@phillipsmuseum.org • www.phillipsmuseum.org •
Decorative Arts Museum – 1973
Travels, African wood carvings, pottery, porcelain, oriental carpets, cars and carriages 52661

U

Salem NJ

Salem County Historical Society Museum, 79-83 Market St, Salem, NJ 08079 • T: +1 856 9355004 • F: +1 856 9350728 • schs@snip.net • www.salemcounty.com/schs •
Pres.: *Donald L. Pierce* •
Local Museum – 1884
Archaeology, costumes, glass, furniture, Indian artifacts 52662

Salem OH

Salem Historical Society Museum, 208 S Broadway Av, Salem, OH 44460 • T: +1 330 3378514 • historicalsociety@salemohio.com • www.salemohio.com/historicalsociety/ •
Pres.: *George W.S. Hays* • Dir.: *David C. Stratton* •
Local Museum – 1971
Abolitiopnist artifacts and exhibits, Ohio Women's Right exhibits, Civail War, woodworking and coal mine exhibits, Quaker, Romanian and Saxony exhibits 52663

Salem OR

A.C. Gilbert's Discovery Village, 116 Marion St NE, Salem, OR 97301-3437 • T: +1 503 3713631 • F: +1 503 3163485 • info@acgilbert.org • www.acgilbert.org •
Exec. Dir.: *Pamela Vorachek* •
Special Museum – 1987
Toys and inventions by A.C. Gilbert 52664

Bush House Museum and Bush Barn Art Center, 600 Mission St SE, Salem, OR 97302 • T: +1 503 5812228 • F: +1 503 3713342 • info@salemart.org • www.salemart.org •
Dir.: *Julie Larson* • Cur.: *Paul Porter* •
Fine Arts Museum / Local Museum – 1919
Dec arts, original furnishing 52665

Hallie Ford Museum of Art, Willamette University, 700 State St, Salem, OR 97301 • T: +1 503 3706855 • F: +1 503 3755458 • museum-art@willamette.edu • www.willamette.edu/museum_of_art •
Dir.: *John Olbrantz* •
Fine Arts Museum / University Museum – 1970 52666

Historic Deepwood Estate, 1116 Mission St SE, Salem, OR 97302 • T: +1 503 3631825 • www.oregonlink.com/deepwood/ •
C.E.O. & Exec.Dir.: *Janice Palmquist* •
Historical Museum – 1919
Ornate, multi-gabled Queen Anne style home, furnishings, pesrsonal artifacts 52667

Marion County Historical Society Museum, 260 12th St SE, Salem, OR 97301-4101 • T: +1 503 3642128 • F: +1 503 3915356 • mchs@open.org • www.marionhistory.org •
C.E.O.: *Maureen Thomas* •
Local Museum – 1950 52668

Mission Mill Museum, 1313 Mill St, Salem, OR 97301 • T: +1 503 5857012 • F: +1 503 5889902 • info@missionmill.org • www.missionmill.org •
C.E.O.: *Maureen Thomas* •
Local Museum – 1964
Mill with equip, Methodist mission and settlers' houses 52669

Oregon Electric Railway Museum, 3395 Brooklake Rd, Salem, OR 97303 • T: +1 503 8884014 • www.oregonelectricrailway.org •
Dir.: *Greg Bonn* •
Special Museum – 1957
Light electric streetcars, intezrban cars, railroad lore, 1910 operating tram 52670

Salem VA

Olin Hall Galleries, Roanoke College, 221 College Ln, Salem, VA 24153 • T: +1 540 3752332 • F: +1 540 3752559 • reinhardt@roanoke.edu • www.roanoke.edu/finearts/galleries •
Dir.: *Beatrix R. Reinhardt* •
Public Gallery 52671

Salem Museum, 801 E Main St, Salem, VA 24153 • T: +1 540 3896760 • info@salemmuseum.org • www.salemmuseum.org •
Dir.: *John D. Long* •
Local Museum – 1992 52672

Salida CO

Salida Museum, 406 1/2 W Rainbow Blvd, Salida, CO 81201 • T: +1 719 5394602 •
Pres.: *Judy Micklich* •
Local Museum – 1954
Local history, Indian artifacts, textiles, mining 52673

Salina KS

Salina Art Center, 242 S Santa Fe, Salina, KS 67401 • T: +1 785 8271431 • F: +1 785 8270686 • info@salinaartcenter.org • www.salinaartcenter.org •
Dir.: *Saralyn Reece Hardy* •
Public Gallery – 1978
Tempoary exhib 52674

Smoky Hill Museum, 211 W Iron Av, Salina, KS 67401 • T: +1 785 3095776 • F: +1 785 8267414 • museum@salina.org • www.smokyhillmuseum.org •
Dir.: *Dee Harris* • Cur.: *Lisa Upshaw* (Collection) • Sc. Staff: *Dorothy Boyle* (Registration) • *Jenny Disney* (Education) • *Susan Hawksworth* (Exhibition) •
Historical Museum / Local Museum – 1983
History – Archives 52675

Salinas CA

Hartnell College Art Gallery, 156 Homestead Av, Salinas, CA 93901 • T: +1 831 7556791 •
Dir.: *Gary T. Smith* •
University Museum / Public Gallery – 1959
Works on paper, Virginia Bacher Haichol artifact coll, Leslie Fenton Netsuke coll 52676

National Steinbeck Center, 1 Main St, Salinas, CA 93901 • T: +1 831 7963833 • F: +1 831 7963828 • info@steinbeck.org • www.steinbeck.org •
Dir.: *Kim E. Greer* •
Special Museum – 1983
American literature, John Steinbeck memorabilia – library 52677

Salisbury MD

Salisbury State University Galleries, 1101 Camden Av, Salisbury State University, Salisbury, MD 21801 • T: +1 410 5436271, 5436000 • F: +1 410 5483002 • kabasile@ssu.edu • www.salisbury.edu •
Dir.: *Kenneth A. Basile* •
Fine Arts Museum / University Museum – 1962
Maryland regional artists, post-Impressionist prints and drawings, landscape photographs, 19th and 20th c American sculpture 52678

The Ward Museum of Wildfowl Art, Salisbury University, 909 S Schumaker Dr, Salisbury, MD 21804-8743 • T: +1 410 7424988 • F: +1 410 7423107 • ward@wardmuseum.org • www.wardmuseum.org •
Dir.: *Kenneth A. Basile* • Cur.: *Dan Brown* •
Fine Arts Museum – 1976
Art 52679

Salisbury NC

Horizons Unlimited Supplementary Educational Center, 1636 Parkview Circle, Salisbury, NC 28144 • T: +1 704 6393004 • F: +1 704 6393015 • www.rss.k12.nc.us •
Dir.: *Cynthia B. Osterhus* • Cur.: *Lisa Wear* (Natural Science) • *Susan R. Waller* (Local and Regional History) •
Local Museum / Natural History Museum – 1967
Natural science, astronomy, wildlife 52680

Rowan Museum, 202 N Main St, Salisbury, NC 28144 • T: +1 704 6335946 • F: +1 704 6339858 • rowanmuseum@vnet.net •
Pres.: *Edward Norvell* • Dir.: *Kaye Brown Hirst* •
Historical Museum – 1953
Hist bldgs and artifacts 52681

Waterworks Visual Arts Center, 123 E Liberty St, Salisbury, NC 28144 • T: +1 704 6361882 • F: +1 704 6361895 • info@waterworks.org • www.waterworks.org •
Pres.: *Jake F. Alexander* •
Public Gallery – 1959
Changing exhib of contemporary art – Art library, several galleries 52682

Salisbury NH

Salisbury Historical Society Museum, Salisbury Heights, Rte 4, Salisbury, NH 03268 • T: +1 603 6482774 •
Cur.: *Wendy Barrett* •
Local Museum – 1966
Tools, sylver, books, manuscripts, furniture 52683

Sallisaw OK

Sequoyah Cabin, Oklahoma Historical Society, Rte 1, Box 141, Sallisaw, OK 74955 • T: +1 918 7752413 • F: +1 918 7752413 • SeqCabin@ipa.net •
C.E.O.: *Dr. Bob Blackburn* • Cur.: *Jerry Dobbs* •
Local Museum – 1936
Hist house, furnishings, exhibits of Native Americans (prim Cherokee) and of the inventor of Chreokee Syllibary 52684

Salmon ID

Lemhi County Historical Museum, 210 Main, Salmon, ID 83467 • T: +1 208 7563342 • lemhimuseum@salmoninternet.com • www.sacajaweahome.com •
Pres.: *Hope Benedict* •
Local Museum – 1963
American Indian artifacts, Asian coll 52685

Salt Lake City UT

Atrium Gallery, 209 E 500 S, Salt Lake City, UT 84111 • T: +1 801 5248200 •
Public Gallery 52686

Beehive House, 67 E South Temple, Salt Lake City, UT 84111 • T: +1 801 2402681 • F: +1 801 2402695 •
Dir.: *McLea* •
Religious Arts Museum – 1961
Historic house, Brigham Young's residence and office 52687

Chase Home Museum of Utah Folk Art, Center of Liberty Park, Salt Lake City, UT 84105 • T: +1 801 5335760, 2367555 • F: +1 801 5334202 • cedison@utah.gov • www.folkartsmuseum.org •
Dir.: *Carol Edison* •
Folklore Museum – 1986
Folk & ethnic art by Utah residents, quilts, saddles, needlework 52688

The Children's Museum of Utah, 840 N 300 W, Salt Lake City, UT 84103 • T: +1 801 3283383 • F: +1 801 3283384 • mail@childmuseum.org • www.childmuseum.org •
Special Museum – 1979
Participatory exhib emphasizing science and technology, dolls, toys 52689

Classic Cars International, 355 W 700 S St, Salt Lake City, UT 84101 • T: +1 801 3225509 • F: +1 801 5826883 • classiccarsintl@hotmail.com • www.classiccarsmuseumsales.com •
C.E.O.: *Stacy Williams* •
Science&Tech Museum – 1975
Over 300 restored automobiles 1903-1970 52690

Daughters of Utah Pioneers Pioneer Memorial Museum, 300 N Main St, Salt Lake City, UT 84103-1699 • T: +1 801 5381050 • F: +1 801 5381119 • info@dupinternational.org • www.dupinternational.org •
Pres.: *Mary A. Johnson* • Cust.: *Edith Menna* •
Local Museum / Historical Museum – 1901
Pioneer vehicles, crafts, art, relics of the Uta pioneer period, manuscript coll 52691

Museum of Church History and Art, 45 N West Temple St, Salt Lake City, UT 84150-3470 • T: +1 801 2402299 • F: +1 801 2405342 • www.lds.org/churchhistory/museum •
Dir.: *Glen M. Leonard* • Sc. Staff: *Robert O. Davis* (Art) • *Donald L. Enders* (History) • Cur.: *Marjorie D. Conder* • *Mark L. Staker* • Cons.: *James L. Raines* • *Blanche Miles* •
Religious Arts Museum – 1869
Religious art and hist 52692

Salt Lake Art Center, 20 S West Temple, Salt Lake City, UT 84101 • T: +1 801 3284201 • F: +1 801 3224323 • allisons@slartcenter.org • www.slartcenter.org •
Dir.: *Ric Collier* • Ass. Dir.: *Allison South* •
Fine Arts Museum – 1931
Art 52693

This is the Place Heritage Park, 2601 Sunnyside Av, Salt Lake City, UT 84108 • T: +1 801 5821847 • www.thisistheplace.org •
Dir.: *Paul Williams* •
Open Air Museum – 1947 52694

Utah Museum of Fine Arts, University of Utah, 410 Campus Center Dr, Salt Lake City, UT 84112 • T: +1 801 5817332 • F: +1 801 5855198 • drobinson@umfa.utah.edu • www.utah.edu/umfa •
Dir.: *David L. Lee* • PR: *Isabelle Kalantzes* •
Fine Arts Museum / University Museum – 1951
19th c American paintings, 18th c French decorative art & tapestries, 17th-18th c English furniture 52695

Utah Museum of Natural History, c/o University of Utah, 1390 E President's Cir, Salt Lake City, UT 84112 • T: +1 801 5816927 • F: +1 801 5853684 • sgeorge@umnh.utah.edu • www.umnh.utah.edu •
Dir.: *Sarah B. George* • Cur.: *Michael Windham* (Herbarium) •
Natural History Museum – 1963
Utah geology, Jurassic & cretaceous dinosaurs, fossil mammals, Garrett herbarium 52696

Utah State Historical Society, 300 Rio Grande St, Salt Lake City, UT 84101-1182 • T: +1 801 5333500 • F: +1 801 5333503 • ushs@history.state.ut.us •
Dir.: *Wilson Martin* •
Association with Coll – 1897
Furnishings, hist prints, photographs, drawings, paintings, 20th c artifacts from Utah 52697

Wheeler Historic Farm, 6351 S 900 E, Salt Lake City, UT 84121 • T: +1 801 2642241 • F: +1 801 2642213 • www.wheelerfarm.com •
Dir.: *Vickie Rodman* • Cur.: *John Peterson* •
Local Museum / Agriculture Museum – 1976
Historic farm representing the initial statehood period and typical of Utah agriculture in 1898 52698

San Andreas CA

Calaveras County Museum, 30 N Main St, San Andreas, CA 95249 • T: +1 209 7541058 • F: +1 209 7541086 • cchs@goldrush.com • www.calaverascohistorical.com •
Pres.: *David Studley* •
Local Museum – 1936
County hist, Indian artifacts, gold rush, baskets – archives 52699

San Angelo TX

Children's Art Museum, 36 E Twohig St, San Angelo, TX 76903 • T: +1 325 6594391 • F: +1 325 6596800 • museum@samfa.org • www.samfa.org •
Dir.: *Howard Taylor* •
Fine Arts Museum – 1994 52700

Fort Concho, 630 S Oakes St, San Angelo, TX 76903 • T: +1 325 4812646, 6574444 • F: +1 325 6574540 • admin@fortconcho.com • www.fortconcho.com •
Dir.: *Robert Bluthardt* •
Historical Museum
Costumes, furniture, military artifacts, hist photographs hist bldg 52701

Helen King Kendall Memorial Art Gallery, c/o San Angelo Art Club, 119 W First St, San Angelo, TX 76903 • T: +1 915 6534405 •
Pres.: *Jean McFerrin* •
Fine Arts Museum – 1948 52702

Houston Harte University Center, c/o Angelo State University, POB 11027, San Angelo, TX 76909 • T: +1 915 9422062 • F: +1 915 9422354 •
Fine Arts Museum / Decorative Arts Museum / University Museum – 1970
Historical artifacts, modern drawings, photography, pottery, weaving 52703

San Angelo Museum of Fine Arts, One Love St, San Angelo, TX 76903 • T: +1 325 6533333 • F: +1 325 6596800 • museum@samfa.org • www.samfa.org •
Pres.: *Jan Duncan* • Dir.: *Howard Taylor* •
Fine Arts Museum – 1981
The work of Texas artists from 1945, contemporary American ceramics and religious Mexican folk art 52704

San Antonio TX

The Alamo, 300 Alamo Plaza, San Antonio, TX 78205 • T: +1 210 2251391 • F: +1 210 3543602 • dstewart@thealamo.org • www.thealamo.org •
C.E.O.: *David Stewart* • Cur.: *Dr. Bruce Winders* •
Historic Site / Military Museum / Historical Museum – 1905
Site of the Battle of the Alamo 52705

Art Gallery, University of Texas at San Antonio, 6900 N Loop, San Antonio, TX 78249-0641 • T: +1 210 4584352 • F: +1 210 4584356 • rboling@utsa.edu • altamira.arts.utsa.edu •
Dir.: *Ron Boling* •
Fine Arts Museum / University Museum – 1982
Portfolio of Elliott Erwitt & Manuel Alvarez Bravo 52706

Art Pace, 445 N Main Av, San Antonio, TX 78205-1441 • T: +1 210 2124900 • F: +1 210 2124990 • info@artpace.org • www.artpace.org •
Public Gallery
Art 52707

Blue Star Contemporary Art Center, 116 Blue Star, San Antonio, TX 78204 • T: +1 210 2276960 • F: +1 210 2299412 • info@bluestarartspace.org • www.bluestarartspace.org •
Dir.: *Carla Stellwes* •
Fine Arts Museum – 1986 52708

Bolivar Hall, 418 Villita, San Antonio, TX 78205 • T: +1 210 2245711 • F: +1 210 2246168 • conserve@saconservation.org • www.saconservation.org •
Exec. Dir.: *Bruce MacDougal* •
Local Museum – 1971 52709

The Bower Gallery, 1114 S Saint Mary's St, San Antonio, TX 78210 • T: +1 210 5353302 • thebowergallery@hotmail.com •
Dir.: *Joey Fauerso* • *Leslee Fraser* •
Public Gallery
Art 52710

Buckhorn Saloon & Museum, 318 E Houston St, San Antonio, TX 78205 • T: +1 210 2474002 • F: +1 210 2474020 • www.buckhornmuseum.com •
Dir.: *Dave George* •
Natural History Museum – 1881
Wildlife, marine, birds, cowboy coll & gear, cowboy art 52711

Casa Navarro, 228 S Laredo St, San Antonio, TX 78207 • T: +1 210 2264801 • F: +1 210 2264801 • navarro@ddc.net • www.tpwd.state.tx.us/park/jose/jose.htm •
Chief Cur.: *Joanne Avant* •
Historical Museum – 1964
Furnishings, Hispanic family 52712

Coppini Academy of Fine Arts Gallery and Museum, 115 Melrose Pl, San Antonio, TX 78212 • T: +1 210 8248502 • www.coppiniacademy.com •
Dir.: *Rhonda Coleman* •
Public Gallery / Local Museum – 1945
Oil paintings, sculpture 52713

Fine Art Gallery at Centro Cultural Aztlan, 803 Castroville Rd, San Antonio, TX 78237 • T: +1 210 4321896 • F: +1 210 4321899 •
Exec. Dir.: *Malena Gonzalez-Cid* •
Fine Arts Museum – 1977 52714

Guadalupe Cultural Arts Center, 1300 Guadalupe St, San Antonio, TX 78207 • T: +1 210 2710379 • F: +1 210 2713480 • info@guadalupeculturalarts.org • www.guadalupeculturalarts.org •
Chm.: *Noah Garcia* •
Fine Arts Museum – 1980 52715

Institute of Texan Cultures, c/o University of Texas, 801 S Bowie & Durango Blvd, San Antonio, TX 78205-3296 • T: +1 210 4583296 • F: +1 210 4582205 • www.texancultures.utsa.edu •
Exec. Dir.: *Rex Ball* •
Ethnology Museum / University Museum – 1968
Anthropology, ethnology, folk art, Afro-American art 52716

McNay Art Museum, 6000 N New Braunfels Av, San Antonio, TX 78209-0069, mail addr: P.O. Box 6069, San Antonio, TX 78209-0069 • T: +1 210 8245368 • F: +1 210 8240218 • info@mcnayart.org • www.mcnayart.org •
Dir.: *William J. Chiego* • Cur.: *Linda Hardberger* (Tobin Collection of Theater Arts) • *Lyle Williams* (Prints and Drawings) •
Fine Arts Museum – 1950
Medieval & Renaissance art, modern art, 19th-20th century European & American paintings, photographs & sculpture, arts & crafts of New Mexico – McNay Library and Archives, Valero Learning Centers, Chiego Lecture Hall 52717

Magic Lantern Castle Museum, 1419 Austin Hwy, San Antonio, TX 78209 • T: +1 210 8050011 • F: +1 210 8221226 • castle@magiclanterns.org • www.magiclanterns.org •
C.E.O./Cur.: *Jack Judson* •
Science&Tech Museum – 1991
Magic laterns, glass slides, related materials – library 52718

Pioneers, Trail and Texas Rangers Memorial Museum, 3805 Broadway, San Antonio, TX 78209 • T: +1 210 6614238 • F: +1 210 6665607 • gccarnes@ktc.com •
Cur.: *Charles Long* •
Historical Museum – 1936
Life in early Texas, gun coll, Texas Rangers, saddles, branding irons 52719

Russell Hill Rogers Gallery, Southwest School of Art and Craft, 300 Augusta St, Navarro Campus, San Antonio, TX 78205 • T: +1 210 2241848 • F: +1 210 2249337 • info@swschool.org • www.swschool.org/exhibitions.php •
Dir.: *Paula Owen* • Assoc. Cur.: *Kathy Armstrong-Gillis* •
Fine Arts Museum – 1965
Contemporary arts and crafts from prominent artists 52720

Saint Mary's University Art Center, 235 W Ligustrum, San Antonio, TX 78228 • T: +1 210 7718305 • F: +1 210 4315098 • cbehlmann@stmarytx.edu •
Dir.: *Cletus Behlmann* •
Public Gallery 52721

Sala Diaz, 517 Stieren, San Antonio, TX 78210 • T: +1 210 6955132 • F: +1 210 3893121 • saladiaz@safx.rr.com •
Dir.: *Hills Snyder* •
Fine Arts Museum – at 52722

San Antonio Children's Museum, 305 E Houston St, San Antonio, TX 78205 • T: +1 210 2124453 • F: +1 210 2421313 • frontdesk@sakids.org • www.sakids.org •
Exec. Dir.: *Chris Sinick* •
Special Museum – 1992 52723

San Antonio Conservation Society Museum, 107 King William St, San Antonio, TX 78204-1399 • T: +1 210 2246163 • F: +1 210 2246168 • conserve@saconservation.org • www.saconservation.org •
Exec. Dir.: *Bruce MacDougal* •
Historical Museum – 1924
Decorative arts, Victorian period furniture, family memorabilia & treasures 52724

San Antonio Missions Visitor Center, 6701 San Jose Dr, San Antonio, TX 78210 • T: +1 210 5348833 • F: +1 210 5341106 • saan_administration@nps.gov • www.nps.gov/saan •
Local Museum – 1978
Anthropology, archaeology, Indian artifacts, Spanish Colonial coll, technology 52725

San Antonio Museum of Art, 200 W Jones Av, San Antonio, TX 78215-1406 • T: +1 210 9788100 • F: +1 210 9788101 • info@samuseum.org • www.samuseum.org •
Dir.: *George W. Neubert* • Sen. Cur.: *Dr. Marion Oettinger* • Cur.: *Dr. Gerry Scott* (Ancient Art) • *Martha Blackwelder* (Asian Art) • *George W. Neubert* (Contemporary Art) •
Fine Arts Museum – 1981
Ancient Equptian, Greek & Roman art, Asian art, pre-Columbian art, American & contempoary art 52726

Spanish Governor's Palace, 105 Military Plaza, San Antonio, TX 78205 • T: +1 210 2240601 • F: +1 210 2240601 • nward@sanantonio.gov •
Dir.: *Malcom Matthew* •
Historical Museum – 1749
Historic bldg 52727

Steves Homestead Museum, 509 King William St, San Antonio, TX 78204 • T: +1 210 2255924 • F: +1 210 2239014 • dchenoweth@saconservation.org • www.saconservation.org •
Man.: *Diana Chenoweth* •
Historic Site – 1924
Historic house; decorative arts coll 52728

Ursuline Hallway Gallery, Southwest School of Art and Craft, 300 Augusta St, Ursuline Campus, San Antonio, TX 78205 • T: +1 210 2241848 • F: +1 210 2249337 • info@swschool.org • www.swschool.org/exhibitions.php •
Dir.: *Paula Owen* • Assoc. Cur.: *Kathy Armstrong-Gillis* •
Fine Arts Museum – 1965
Contemporary art by regional artists 52729

Visual Arts Annex Gallery, Guadalupe Cultural Arts Center, 325 S Salado St, San Antonio, TX 78207 • T: +1 210 2710379 • info@guadalupeculturalarts.org •
Public Gallery 52730

Witte Museum, 3801 Broadway, San Antonio, TX 78209 • T: +1 210 3571881 • F: +1 210 3571882 • witte@wittemuseum.org • www.wittemuseum.org •
Dir.: *Mimi Quintanilla* •
Historical Museum – 1926
Texana items, paintings, textiles & Costumes, Southwest Indian artifacts, dinosaurs 52731

Wooden Nickel Historical Museum, 345 Old Austin Rd, San Antonio, TX 78209 • T: +1 210 8291291 • F: +1 210 8328965 • museum@wooden-nickel.net • www.wooden-nickel.net •
C.E.O.: *Herb Hornung* •
Science&Tech Museum – 1998
Printing plates 52732

Yturri-Edmunds Historic Site, 107 King William St, San Antonio, TX 78204-1399 • T: +1 210 2246163 • F: +1 210 2246168 • conserve@saconservation.org • www.saconservation.org •
Exec. Dir.: *Bruce MacDougal* •
Historic Site – 1924
Historic houses 52733

San Bernardino CA

Robert V. Fullerton Art Museum, California State University, San Bernardino, 5500 University Pkwy, San Bernardino, CA 92407 • T: +1 909 8807373 • F: +1 909 8807068 • artmuseum@csusb.edu • rvf-artmuseum.csusb.edu •
Dir.: *Eva Kirsch* •
Fine Arts Museum / University Museum – 1965
Asian and Etruscan ceramics, African sculpture, Egyptian antiquities, contemporary prints, drawings, paintings, sculptures 52734

San Carlos CA

Hiller Aviation Institute, 601 Skyway Rd, San Carlos, CA 94070 • T: +1 650 6540200 • F: +1 650 6540220 • museum@hiller.org • www.hiller.org •
Pres./Dir.: *Jeffery Bass* •
Science&Tech Museum – 1995
Aviation artifacts, technology 52735

San Diego CA

African Arts Collection, San Diego Mesa College, 7250 Mesa College Dr, San Diego, CA 92111-0103 • T: +1 619 3882829 • F: +1 619 3885720 • www.sdmesa.edu/african-art/index.html •
Cur.: *Barbara W. Blackman* •
Fine Arts Museum / University Museum 52736

Centro Cultural de la Raza, 2125 Park Blvd, Balboa Park, San Diego, CA 92101 • T: +1 619 2356135 • F: +1 619 5950034 • centro@centroculturaldelaraza.org • www.centroculturaldelaraza.org •
Folklore Museum / Fine Arts Museum – 1970
Mexican and Indian culture, historic, folk and contemporary artworks by Chicano artists 52737

Marston House, 3525 Seventh Av, San Diego, CA 92103 • T: +1 619 2983142 • www.sandiegohistory.org •
Dir.: *John W. Wadas* •
Decorative Arts Museum – 1928
Arts and crafts movement, 19th c decorative art, furnishings 52738

Mingei International Museum, 1439 El Prado, Balboa Park, San Diego, CA 92101 • T: +1 619 2390003 • F: +1 619 2390605 • mingei@mingei.org • www.mingei.org •
Dir.: *Rob Sidner* •
Fine Arts Museum / Decorative Arts Museum – 1974
Folk art, craft, design from India, Japan, Indonesia, Mexico, Ethiopia and China, Palesinian costumes, Pre-Columbiana – library, theater 52739

Museum of Contemporary Art San Diego - Downtown, 1100 Kettner Blvd, San Diego, CA 92101 • T: +1 858 4543541 • info@mcasd.org • www.mcasandiego.org •
Pres.: *Dr. Peter C. Farrell* • Dir.: *Dr. Hugh M. Davies* •
Fine Arts Museum 52740

Museum of Photographic Arts, 1649 El Prado, Balboa Park, San Diego, CA 92101 • T: +1 619 2387559 ext 201 • F: +1 619 2388777 • info@mopa.org • www.mopa.org •
Dir.: *Deborah Klochko* • Cur.: *Carol McCusker* (Photography) • Dep. Dir.: *Vivian Kung Haga* •
Fine Arts Museum – 1983
History of photography, aesthetic movements and technological advancements, processes from Daguerreotypes, salt prints, Woodbury types, albumen prints, ambrotypes and tintypes – library, theater 52741

The New Children's Museum, 200 West Island Av, San Diego, CA 92101 • T: +1 619 2338792 ext 103 • F: +1 619 2338796 • experience@thinkplaycreate.org • www.thinkplaycreate.org •
Exec. Dir.: *Rachel Teagle* •
Special Museum – 1981
Arts, hands-on and minds-on experiences – theater 52742

Reuben H. Fleet Science Center, 1875 El Prado, Balboa Park, San Diego, CA 92101-1625 • T: +1 619 2381233 • F: +1 619 6855771 • www.rhfleet.org •
Exec. Dir.: *Dr. Jeffrey W. Kirsch* •
Science&Tech Museum – 1973
Interactive exhibits, simulator, virtual reality – planetarium, IMAX 52743

San Diego Air & Space Museum, 2001 Pan American Plaza, Balboa Park, San Diego, CA 92101 • T: +1 619 2348291 ext 100 • F: +1 619 2334526 • www.sandiegoairandspace.org •
Pres.: *Jim Kidrick* • Cur.: *Terry Brennan* •
Science&Tech Museum – 1961
Aviation, historical aircrafts, engines – library 52744

San Diego Art Institute, 1439 El Prado, Balboa Park, San Diego, CA 92101-1617 • T: +1 619 2360011 • F: +1 619 2361974 • director@sandiego-art.org • www.sandiego-art.org •
Fine Arts Museum 52745

San Diego Automotive Museum, 2080 Pan American Plaza, Balboa Park, San Diego, CA 92101-1636 • T: +1 619 2312886 • paula@sdautomuseum.org • www.sdautomuseum.org •
Exec. Dir.: *Paula Brandes* •
Science&Tech Museum – 1987
International cars, motorcycles, engines – libraries, auditorium 52746

San Diego Command Museum, c/o Marine Corps Recruit Depot, 1600 Henderson Av, Bldg 26, San Diego, CA 92140-5010 • T: +1 619 5248431 • F: +1 619 5240076 • barbara.mccurtis@usmc.mil • www.mcrdmuseumhistoricalsociety.org •
Dir.: *Barbara S. McCurtis* •
Military Museum – 1987
USMC military equip, battlefield souvenirs – library 52747

San Diego Hall of Champions Sports Museum, 2131 Pan America Plaza, Balboa Park, San Diego, CA 92101 • T: +1 619 2342544 • F: +1 619 2344543 • www.sdhoc.com •
Dir.: *Alan Kidd* •
Special Museum – 1961
Sports memorabila – library 52748

San Diego Historical Society Museum, 1649 El Prado, Casa de Balboa, Balboa Park, San Diego, CA 92101-1621 • T: +1 619 2326203 ext 107 • F: +1 619 2326297 • sdhsadmin@sandiegohistory.org • www.sandiegohistory.org •
Dir.: *John W. Wadas* • Cur.: *John Panter* (Programs, Collections) • *Kimberly Preciado* (Collections) •
Local Museum – 1928
Local hist and art since the 1840s – archives, library, theater 52749

San Diego Maritime Museum, 1492 N Harbor Dr, San Diego, CA 92101 • T: +1 619 2349153 • info@sdmaritime.com • www.sdmaritime.com •
Exec. Dir.: *Raymond E. Ashley* • Cur.: *Bob Crawford* • *Mark Montijo* •
Science&Tech Museum – 1948
3 historic ships (Star of India, ferryboat Berkeley, steam yacht Medea), ship models, navigation instr, tools, engineering – library 52750

San Diego Model Railroad Museum, 1649 El Prado, Balboa Park, San Diego, CA 92101 • T: +1 619 6960199 • info@sdmrm.org • www.sdmodelrailroadm.com •
Exec. Dir.: *John A. Rotsart* •
Science&Tech Museum – 1980
Railroad models of Southern California, San Diego county and Western roads – library 52751

San Diego Museum of Art, 1450 El Prado, Balboa Park, San Diego, CA 92101 • T: +1 619 2327931 • F: +1 619 2329367 • information@sdmart.org • www.sdmart.org •
Pres.: *Kenneth J. Widder* • Dep. Dir.: *Julia Marciari Alexander* • Cur.: *John Marciari* • *Sonia Quintanilla* • *Ami Galpin* •
Fine Arts Museum – 1925
European paintings sculpture and decorative arts, Asian art, ceramics, contemporary painting and sculpture, Indian paintings coll – library, auditorium 52752

San Diego Museum of Man, 1350 El Prado, Balboa Park, San Diego, CA 92101 • T: +1 619 2392001 • F: +1 619 2392749 • www.museumofman.org •
Interim Dir.: *Ned A. Smith* • Cur.: *Tori Randall* (Physical Anthropology) • *Phil Hoog* (Archaeology) • *Rose Tyson* (Physical Anthropology) • *Ken Hedges* (California coll.) • *Grace Johnson* (Latin America) •
Ethnology Museum / Archaeology Museum – 1915
Physical anthropology, peoples of the Western Americas, Egyptian antiquities – library 52753

San Diego Natural History Museum, 1788 El Prado, Balboa Park, San Diego, CA 92112-1390 • T: +1 619 2323821, 2550216 • F: +1 619 2320248 • mhager@sdnhm.org • www.sdnhm.org •
Dir.: *Dr. Michael Wall* (Director of Biodiversity Research Center of the Californias and Curator of Entomology) • Cur.: *Dr. Jon Rebman* (Botany) • *Dr. Thomas Demere* (Paleontology) • *Dr. Paisley Cato* (Collections) • *Dr. Brad Hollingworth* (Herpetology) • *Phil Unitt* (Ornitology, Mammalogy) •
Natural History Museum – 1874
Botany, mammalogy, ornitology, geology, herpetology, entomology, marine invertebrates, gems and minerals, rare books and paleontology – library, theater 52754

Serra Museum, 2727 Presidio Dr, Presidio Park, San Diego, CA 92103 • T: +1 619 2973258 • F: +1 619 2973281 • lemont@sandiegohistory.org • www.sandiegohistory.org •
Dir.: *John W. Wadas* • Cur.: *Dr. Therese Adams Muranaka* •
Local Museum – 1928
Period clothing, furniture and tools from the Native American 52755

Sushi Performance and Visual Art, 320 11th Av, San Diego, CA 92101 • T: +1 619 2358466 • F: +1 619 2358552 • info@sushiart.org • www.sushiart.org •
Dir.: *Vicki Wolf* •
Public Gallery – 1980
Interdisciplinary art 52756

Thomas Whaley House Museum, 2482 San Diego Av, San Diego, CA 92110 • T: +1 619 298248 •
C.E.O./Cur.: *Gary Beck* •
Historical Museum – 1960
1850-1880 furnishings 52757

Timken Museum of Art, 1500 El Prado, Balboa Park, San Diego, CA 92101 • T: +1 619 2395548 • F: +1 619 5319640 • info@timkenmuseum.org • www.timkenmuseum.org •
Dir.: *John Petersen* •
Fine Arts Museum – 1965
European paintings (13th-19th c), American paintings 19th c , gobbelin, tapestries, Russian icons (15th-19th c) 52758

University Art Gallery, San Diego State University, 5500 Campanile Dr, San Diego, CA 92182-4805 • T: +1 619 5945171 • artgallery@sdsu.edu • www.sdsu.edu/artgallery •
Dir.: *Tina Yapelli* •
Public Gallery / University Museum – 1977
Contemporary art, Asian sculpture, African and Mexican art 52759

Villa Montezuma, 1925 K St, San Diego, CA 92102, mail addr: 657 20th St, San Diego, CA 92102 • T: +1 619 2559367 • www.villamontezuma.org •
Decorative Arts Museum – 1928
Late 19th c furnishings and decorative arts 52760

Wells Fargo History Museum, 2733 San Diego Av, San Diego, CA 92110 • T: +1 619 2383929 • www.wellsfargohistory.com •
Cur.: *Allan E. Peterson* •
Special Museum – 1990
Banking company hist 52761

William Heath Davis House Museum, 410 Island Av, San Diego, CA 92101 • T: +1 619 2334692 • F: +1 619 2334148 • tracy@gaslampquarter.org • www.gaslampquarter.org •
Cur.: *Mary Joralmon* •
Historical Museum – 1984
Hist house with interior of different periods 52762

San Francisco CA

Adán E. Treganza Anthropology Museum, San Francisco State University, 1600 Holloway Av, San Francisco, CA 94132 • T: +1 415 3382046 • F: +1 415 3380530 • yamamoto@sfsu.edu • www.sfsu.edu/~treganza •
Dir.: *Yoshiko Yamamoto* •
University Museum / Archaeology Museum / Natural History Museum – 1958
African, Asian, Oceanian and New World ethnology and archaeology, ethnomusicology 52763

African American Historical and Cultural Society Museum, 762 Fulton St, 2nd Fl, San Francisco, CA 94102 • T: +1 415 2926172 • F: +1 415 4404231 • info@sfaahcs.org • www.sfaahcs.org •
Head: *Vandean Philpott* •
Folklore Museum / Historical Museum – 1955
African American culture, African and African American artists, incl Mary Ellen Pleasant, Alexander Leidesdorff and Sargent Claude Johnson, black businesses, Blacks in the west, Haitian art, protest movements 1960s-1970s – library 52764

American Indian Contemporary Arts, 23 Grant Av, San Francisco, CA 94108-5828 • T: +1 415 9897003 • F: +1 415 9897025 •
Dir.: *Janeen Antoime* •
Fine Arts Museum – 1983 52765

Asian Art Museum, Chong-Moon Lee Center, 200 Larkin St, San Francisco, CA 94102 • T: +1 415 5813500 • F: +1 415 5814700 • pr@asianart.org • www.asianart.org •
Dir.: *Dr. Emily J. Sano* • Chief Cur.: *Forrest McGill* (South and Southeast Asian Art) • Cur.: *Terese Tse Bartholomew* (Chinese and Himalayan Art, Decorative Art) • *Yoko Woodson* (Japanese Art) • *Kumja Paik Kim* (Korean Art) • *Michael Knight* (Chinese Art) •
Fine Arts Museum – 1966/ 2003
Cultural items from Asia last 6,000 yrs, ceramics, bronze, jade, lacquer, sculpture, metalwork, glass, textiles, applied art, paintings, screens – library 52766

Black Rock Arts Foundation, 1900 Third St, San Francisco, CA 94158 • T: +1 415 6261248 • info@blackrockarts.org • www.blackrockarts.org •
Exec. Dir.: *Tomas McCabe* •
Association with Coll – 1986 52767

California Academy of Sciences, 55 Music Concourse Drive, Golden Gate Park, San Francisco, CA 94118 • T: +1 415 3798000 • F: +1 415 3795720 • sholcomb@calacademy.org • www.calacademy.org •
Exec. Dir.: *Dr. Gregory C. Farrington* • Dean of Science and Research Collections: *Dr. David Mindell* •
Natural History Museum / Science&Tech Museum – 1853
Ichthyology, invertebrate zoology, ornithology, mammalogy, botany, entomology, herpetology, anthropology, geology, aquatic biology, comparative genomics – Natural history, planetarium, aquarium, scientific research, education 52768

Cartoon Art Museum, 655 Mission St, San Francisco, CA 94105 • T: +1 415 2278666 • www.cartoonart.org •
Dir.: *Rod Gilchrist* • Cur.: *Jenny Robb-Dietzen* •
Fine Arts Museum – 1984
Cartoons, flat graphics-animation 52769

Chinese Culture Center of San Francisco, 750 Kearny St, San Francisco, CA 94108 • T: +1 415 9861822 • F: +1 415 9862825 • info@c-c-c.org • www.c-c-c.org •
Fine Arts Museum / Folklore Museum – 1965
Chinese art, folk art and crafts, historical photographs 52770

Contemporary Jewish Museum, 736 Mission St, San Francisco, CA 94103 • T: +1 415 6557800 • F: +1 415 6557815 • info@thecjm.org • www.thecjm.org •
Dir.: *Connie Wolf* • Cur.: *Natasha Perlis* •
Fine Arts Museum / Historical Museum – 1984
Jewish art and culture 52771

De Young Museum, Fine Arts Museums of San Francisco, 50 Hagiwara Tea Garden Dr, Golden Gate Park, San Francisco, CA 94118 • T: +1 415 7503600 • F: +1 415 7507692 • contact@famsf.org • www.thinker.org •
Dir.: *Harry S. Parker* • Cur.: *Robert Flynn Johnson* (Prints and Drawings) • *Timothy Anglin Burgard* (American Arts) • *Kathleen Berrin* (Africa, Oceania and America) • *Dr. Lynn Federle Orr* (European Paintings) • *Diane Mott* (Textiles) • *Renee Dreyfus* (Ancient Art, Interpretation) • Education: *Sheila Pressley* (Education) •
Fine Arts Museum – 1894/2005
The Americas painting, sculpture and decorative art, Oceanic and African art, textiles, contemporary craft, new Guinea art 52772

The Exploratorium, 3601 Lyon St, San Francisco, CA 94123 • T: +1 415 5610360 • www.exploratorium.edu •
Exec. Dir.: *Goery Delacote* •
Fine Arts Museum / Science&Tech Museum / Ethnology Museum – 1969
Science, art, human perception 52773

Galeria de la Raza, 2857 24th St, San Francisco, CA 94110 • T: +1 415 8268009 • F: +1 415 8266235 • info@galeriadelaraza.org • www.galeriadelaraza.org •
Dir.: *Carolina Ponce de León* • Cur.: *Raquel de Anda* •
Public Gallery – 1970
Chicano/ Latino art and culture 52774

GLBT Historical Society Museum, Gay, Lesbian, Bisexual Transgender Historical Society, 657 Mission St, Ste 300, San Francisco, CA 94105 • T: +1 415 7775455 • F: +1 415 7775576 • info@glbthistory.org • www.glbthistory.org •
Dir.: *Terence Kissack* •
Historical Museum – 1996/ 2002
Social history of gay and lesbian life, gay community 52775

Haas-Lilienthal House, 2007 Franklin St, San Francisco, CA 94109 • T: +1 415 4413000 • F: +1 415 4413015 • info@sfheritage.org • www.sfheritage.org •
Dir.: *Jack A. Gold* •
Decorative Arts Museum / Fine Arts Museum – 1973
Historic house, furniture, paintings 52776

International Museum of Women, POB 190038, San Francisco, CA 94119-0038 • T: +1 415 5434669 • F: +1 415 5434668 • info@imow.org • www.imow.org •
Pres.: *Chris Yelton* •
Historical Museum – 1985
Lives of women around the world 52777

Intersection for the Arts Gallery, 446 Valencia St, San Francisco, CA 94103 • T: +1 415 6262787 • deborah@theintersection.org • www.theintersection.org •
Dir.: *Deborah Cullinan* • Cur.: *Kevin B. Chen* (Programs) • *Dan Hamaguchi* (Graphic Design) •
Public Gallery – 1965
Alternative art space 52778

Kent and Vicki Logan Galleries and Capp Street Project, Wattis Institute for Contemporary Arts, c/o California College of the Arts, 1111 Eighth St, San Francisco, CA 94107 • T: +1 415 5519210 • wattis@cca.edu • www.wattis.org •
Dir.: *Jens Hoffmann* •
Public Gallery – 1983
Installation art, national and international contemporary culture 52779

The Lab, 2948 16th St, San Francisco, CA 94103 • T: +1 415 8648855 • F: +1 415 8648860 • www.thelab.org •
Dir.: *Eilish Cullen* •
Public Gallery – 1983
Interdisciplinary artists' exhibits 52780

Legion of Honor, Fine Arts Museums of San Francisco, 100 34th Av, San Francisco, CA 94121 • T: +1 415 7503600 • contact@famsf.org • www.legionofhonor.org •
Dir.: *Harry S. Parker* • Sc. Staff: *Robert Flynn Johnson* (Prints and Drawings) • *Timothy Anglin Burgard* (American Arts) • *Kathleen Berrin* (Africa, Oceania and America) • *Dr. Lynn Federle Orr* (European Art) • *Diane Mott* (Textiles) • *Renee Dreyfus* (Ancient Art, Interpretation) • *Lee Hunt Miller* (European Decorative Arts and Sculpture) •
Fine Arts Museum – 1924
Ancient and European arts, sculptures, decorative arts – auditorium 52781

Luggage Store, 1007 Market St, San Francisco, CA 94103 • T: +1 415 2555971 • F: +1 415 8635509 • www.luggagestoregallery.org •
Cur.: *Darryl Smith* • *Laurie Lazer* •
Public Gallery 52782

The Mexican Museum, Fort Mason Center, Bldg D, Laguna and Marina Blvd, San Francisco, CA 94123 • T: +1 415 2029700 • F: +1 415 4417683 • info@mexicanmuseum.org • www.mexicanmuseum.org •
Dir.: *Lorraine Garcia-Nakata* • *William Moreno* • Cur.: *Tere Romo* •
Fine Arts Museum / Folklore Museum – 1975
Mexican and Chicani-Mexican arts and culture, folk art 52783

Mission Cultural Center for Latino Arts, 2868 Mission St, San Francisco, CA 94110 • T: +1 415 8211155 • info@missionculturalcenter.org • www.missionculturalcenter.org •
Dir.: *Jennie Rodriquez* •
Fine Arts Museum / Folklore Museum – 1977
Civic art, posters 52784

Museo Italoamericano, Fort Mason Center, Bldg C, Laguna and Marina Blvd, San Francisco, CA 94123 • T: +1 415 6732200 • F: +1 415 6732292 • sfmuseo@sbcglobal.net • www.museoitaloamericano.org •
Dir.: *Paola Bagnatori* • Cur.: *Jon Acquisti* •
Fine Arts Museum / Folklore Museum – 1978
Contemporary Italian and Italian-American art, Italian culture – library 52785

Museum of Craft and Folk Art, 51 Yerba Buena Lane, San Francisco, CA 94103 • T: +1 415 2274888 • F: +1 415 2274351 • jmccabe@mocfa.org • www.mocfa.org •
Dir.: *Jennifer McCabe* •
Decorative Arts Museum – 1983
Worldwide tribal, contemporary craft and folk art, outsider art – library 52786

Museum of Russian Culture, 2450 Sutter St, San Francisco, CA 94115 • T: +1 415 9217631 • F: +1 415 7718683 • program@russiancentersf.com • www.russiancentersf.com •
Dir.: *Dmitri G. Brauns* • Cur.: *Alex Karamzin* •
Ethnology Museum – 1948
Ethnic hist, military, numismatics – archives, library 52787

Museum of Vision, Foundation of the American Academy of Ophthalmology, 655 Beach St, San Francisco, CA 94109-1336 • T: +1 415 5618502 • F: +1 415 5618567 • faao@aao.org • www.aaofoundation.org •
Dir.: *Jenny E. Benjamin* •
Historical Museum – 1980
Medical hist, medical instr, ophthalmology, diagnostic and surgery, rare books – small library 52788

New Langton Arts, 1246 Folsom St, San Francisco, CA 94103 • T: +1 415 6265416 • F: +1 415 2551453 • nla@newlangtonarts.org • www.newlangtonarts.org •
Public Gallery – 1975 52789

Randall Museum, 199 Museum Way, San Francisco, CA 94114 • T: +1 415 5549600 • F: +1 415 5549609 • info@randallmuseum.org • www.randallmuseum.org •
Dir.: *Amy Dawson* • Cur.: *John Dillon* • *Carol Preston* (Science) • *Chris Boettcher* (Industrial Arts) •
Natural History Museum – 1937
Children's museum, rocks, minerals, shells, fossils, butterflies, model railroad 52790

San Francisco Airport Museums, San Francisco International Airport, San Francisco, CA 94128, mail addr: POB 8097, San Francisco, CA 94128 • T: +1 650 8216700 • curator@flysfo.com • www.sfoarts.org •
Dir./ Cur.: *Blake Summers* •
Science&Tech Museum – 1980
Commercial aviation in the Pacific region 52791

San Francisco Camerawork, 657 Mission St, 2nd Fl, San Francisco, CA 94105 • T: +1 415 5122020 • F: +1 415 5127109 • info@sfcamerawork.org • www.sfcamerawork.org •
Exec. Dir.: *Marnie Gillett* •
Public Gallery – 1974
Photography – library 52792

San Francisco Fire Department Museum, 655 Presidio Av, San Francisco, CA 94115 • T: +1 415 5583546, 5634630 • www.sffiremuseum.org •
Cur.: *Robert Kreuzberger* •
Historical Museum – 1964
Fire museum, specializing in SF history, vintage firehouses 52793

San Francisco Maritime National Historical Park, Hyde Street Pier, San Francisco, CA 94109, mail addr: Bldg E, Fort Mason Center, San Francisco, CA 94123 • T: +1 415 5617000 • F: +1 415 5561624 • www.nps.gov/safr •
Cur.: *Steve Canright* •
Historical Museum / Science&Tech Museum – 1951
Maritime hist, photographs, ship plans, log books, paintings and other fine arts, decorative arts – achives, libraray 52794

San Francisco Museum of Modern Art, 151 Third St, San Francisco, CA 94103 • T: +1 415 3574000 • F: +1 415 3574037 • commassistant@sfmoma.org • www.sfmoma.org •
Dir.: *Neal Benezra* • Cur.: *Janet C. Bishop* • *Madeleine Grynsztejn* (Painting and Sculpture) • *Dr. Sandra S. Phillips* (Photography) • *Benjamin E. Weil* (Media Arts) • *Joseph Rosa* (Architecture, Design) • Cons.: *Jill Sterrett* • Education, PR: *John Weber* •
Fine Arts Museum – 1935
Art s. early 20th c, international paintings, sculpture, grapgics, photos, design, media arts – library, theater 52795

Southern Exposure Gallery, 3030 20th St, San Francisco, CA 94110 • T: +1 415 8632141 • soex@soex.org • www.soex.org •
Dir.: *Courtney Fink* •
Public Gallery – 1974
Innovative, risk-taking contemporary art 52796

Visual Aid, 116 New Montgomery St, Ste 640, San Francisco, CA 94105 • T: +1 415 7778242 • F: +1 415 7778240 • julie@visualaid.org • www.visualaid.org •
Dir.: *Julie Blankenship* •
Special Museum 52797

The Walt Disney Family Museum, 104 Montgomery Street, The Presidio of San Francisco, San Francisco, CA 94129 • T: +1 415 3456800 • www.disney.go.com/disneyatoz/familymuseum •
Historical Museum – 2009 52798

Walter and McBean Galleries, San Francisco Art Institute, 800 Chestnut St, San Francisco, CA 94133 • T: +1 415 7494563 • exhibitions@sfai.edu • www.waltermcbean.com •
Pres.: *Larry Thomas* •
Public Gallery – 1871
Contemporary art – library 52799

Wells Fargo History Museum, 420 Montgomery St, San Francisco, CA 94163 • T: +1 415 3962619 • www.wellsfargohistory.com •
Cur.: *Anne M. Hall* •
Historical Museum – 1929
Company history, Wells Fargo Bank's headquarters 52800

Yerba Buena Center for the Arts, 701 Mission St, San Francisco, CA 94103 • T: +1 415 9782700, 9782787 • F: +1 415 9789635 • comments@YBCA.org • www.ybca.org •
Dir.: *Kenneth Foster* • Cur.: *Rene de Guzman* (Visual Arts) • *Joel Shepard* (Film Video) •
Fine Arts Museum – 1986
Changing exhibits – dance theater 52801

San Gabriel CA

San Gabriel Mission Museum, 428 S Mission Dr, San Gabriel, CA 91776 • T: +1 626 4573048 • www.sangabrielmission.org •
Dir.: *Ralph B. Berg* •
Religious Arts Museum
Religious hist, paintings 52802

San Jacinto CA

San Jacinto Museum, 695 Ash St, San Jacinto, CA 92583 • T: +1 951 6544952 • www.ci.san-jacinto.ca.us •
Local Museum – 1939
Regional hist, Indian archaeology, paleontology, mineralogy and geology 52803

San Jose CA

Art-Tech Exhibitions, Silicon Valley Institute of Art & Technology, 14300 Clayton Rd, San Jose, CA 95127-5201 • T: +1 408 9299969 • F: +1 408 9293330 • www.art-tech.org •
Head: *Meryle Holmes* •
Public Gallery – 1993
New art forms and media content 52804

Children's Discovery Museum of San Jose, 180 Woz Way, San Jose, CA 95110 • T: +1 408 2985437 • F: +1 408 2986826 • contactus@cdm.org • www.cdm.org •
Exec. Dir.: *Connie Martinez* • Dep. Dir.: *Jenni Martin* (Education, Programs) •
Special Museum – 1982
Regional hist, environmental sciences, cultural arts, health and physiognomy, banking 52805

Natalie and James Thompson Gallery, San Jose State University, 1 Washington Sq, San Jose, CA 95192-0089 • T: +1 408 9244320 • F: +1 408 9244326 • art.design@sjsu.edu • www.sjsu.edu •
Dir.: *John Loomis* •
Public Gallery – 1959 52806

Rosicrucian Egyptian Museum, 1664 Park Av, San Jose, CA 95191 • T: +1 408 9473636 • www.egyptianmuseum.org •
Dir.: *Julie Scott* • Cur.: *Lisa Schwappach* •
Historical Museum / Fine Arts Museum – 1929
Assyrian, Babylonian and Egyptian antiquities, archaeology, paintings, sculpture, graphics, Coptic textiles 52807

San Jose Institute of Contemporary Art, 560 S First St, San Jose, CA 95113 • T: +1 408 2838155 • F: +1 408 2838157 • info@sjica.org • www.sjica.org •
Dir.: *Cathy Kimball* •
Fine Arts Museum – 1980 52808

San Jose Museum of Art, 110 S Market St, San Jose, CA 95113 • T: +1 408 2942787, 2716840 • info@sjmusart.org • www.sanjosemuseumofart.org •
Dir.: *Daniel T. Keegan* • Dep. Dir.: *Deborah Norberg* • Chief Cur.: *Susan Landauer* •
Fine Arts Museum – 1969
20th-21st c art – library 52809

San Jose Museum of Quilts & Textiles, 520 South 1st St, San Jose, CA 95113-2806 • T: +1 408 9710323 ext 16 • F: +1 408 9717226 • jane@sjquiltmuseum.org • www.sjquiltmuseum.org •
Dir.: *Jane Przybysz* • *Deborah Corsini* (Exhibits) •
Fine Arts Museum – 1977
Quilts and textiles since 1850 – Quilts and textiles since 1850, ethnic textiles, contemporary fiber art 52810

The Tech Museum of Innovation, 201 S Market St, San Jose, CA 95113 • T: +1 408 2948324 • F: +1 408 2797167 • info@thetech.org • www.thetech.org •
Pres.: *Peter Friess* •
Science&Tech Museum – 1990 52811

San Juan Bautista CA

San Juan Bautista Mission Museum, Second and Mariposa Sts, San Juan Bautista, CA 95045 • T: +1 831 6232127 • F: +1 831 6232433 • ann@oldmissionsjb.org • www.oldmissionsjb.org •
Religious Arts Museum
Local Indian, Spanish, Mexican and religious artifacts, figures of saints ... 52812

San Juan Capistrano CA

Mission San Juan Capistrano Museum, 26801 Ortega Hwy, San Juan Capistrano, CA 92675 • T: +1 949 2341300 ext 320 • F: +1 949 4432061 • www.missionsjc.com •
C.E.O.: *Mechelle Lawrence-Adams* •
Local Museum / Historic Site – 1980
Local hist, archaeology, native american, period rooms – botanical garden ... 52813

San Luis Obispo CA

Mission San Luis Obispo de Tolosa, 751 Palm St, San Luis Obispo, CA 93401 • T: +1 805 7818220 • F: +1 805 7818214 • office@oldmissionslo.org •
Religious Arts Museum
Religious hist, Chumash Indian coll ... 52814

San Luis Obispo Art Center, 1010 Broad St, San Luis Obispo, CA 93401 • T: +1 805 5438562 • F: +1 805 5434518 • kkile@sloartcenter.org • www.sloartcenter.org •
Public Gallery – 1952 ... 52815

San Luis Obispo County Historical Museum, 696 Monterey St, San Luis Obispo, CA 93401 • T: +1 805 5430638 • F: +1 805 78322919 • www.slochs.org •
Pres.: *Randal Cruikshanks* •
Local Museum – 1956
Local hist, folklore, decorative arts, textiles – archives, small library ... 52816

San Marcos CA

Boehm Art Gallery, 1140 W Mission Rd, San Marcos, CA 92069 • T: +1 760 7441150 ext 2304 • www.palomar.edu/art/boehmgallery.html •
Dir.: *Vicki Cole* •
Fine Arts Museum – 1964
Art since 16th c ... 52817

San Marino CA

California Art Club Gallery at the Old Mill, 1120 Old Mill Rd, San Marino, CA 91108, mail addr: 75 S Grand Av, Pasadena, CA 91105 • T: +1 626 4495458 • F: +1 626 4491057 • beverly@brcart.com • www.californiaartclub.org •
Exec. Dir.: *Jack McQueen* • Cur.: *Jessica Eisenreich* •
Fine Arts Museum – 1999 ... 52818

The Huntington Art Collections, 1151 Oxford Rd, San Marino, CA 91108 • T: +1 626 4052140 • publicinformation@huntington.org • www.huntington.org •
Dir.: *John Murdoch* • Cur.: *Shelley Bennett* (British and Continental Art) • *Amy Meyers* (American Art) •
Fine Arts Museum – 1919
18th c British and European art, French descorative art, American paintings – library, botanical garden ... 52819

Old Mill Museum, El Molino Viejo Museo, 1120 Old Mill Rd, San Marino, CA 91108 • T: +1 626 4495458 • F: +1 626 4491057 • oldmill@sbcglobal.net • www.old-mill.org •
Dir.: *Jack McQueen* •
Local Museum
California hist, art – library ... 52820

San Martin CA

Wings of History Air Museum, 12777 Murphy Av, San Martin, CA 95046-0495 • T: +1 408 6832290 • F: +1 408 6832291 • www.wingsofhistory.org •
Dir.: *John McMains* • Rest.: *Vaughn Lamb* •
Science&Tech Museum – 1983
Homebuilt and antique aircraft – library ... 52821

San Mateo CA

Coyote Point Museum for Environmental Education, 1651 Coyote Point Dr, San Mateo, CA 94401 • T: +1 650 3427755 • F: +1 650 3427853 • info@coyoteptmuseum.org • www.coyoteptmuseum.org •
Dir.: *Nikii Finch-Morales* •
Natural History Museum – 1953
Natural science, birds, mammals, insects, plants, shells, reptiles from the area ... 52822

San Miguel CA

Mission San Miguel, 775 Mission St, San Miguel, CA 93451 • T: +1 805 4673256 • F: +1 805 4672141 • info@missionsanmiguel.org • www.missionsanmiguel.org •
Dir.: *William Short* •
Religious Arts Museum
Religious items, murals, paintings ... 52823

San Pedro CA

Fort MacArthur Museum, 3601 S Gaffey St, San Pedro, CA 90731 • T: +1 310 5482631 • F: +1 310 2410847 • director@ftmac.org • www.ftmac.org •
Dir./Cur.: *Stephen R. Nelson* •
Open Air Museum / Military Museum – 1985
Coast artillery, Nike missile, military vehicles ... 52824

Gallery A and Downstairs Gallery, c/o Angels Gate Cultural Center, 3601 S Gaffey St, San Pedro, CA 90731 • T: +1 310 5190936 • F: +1 310 5198698 • deborah@angelsgateart.org • www.angelsgateart.org •
Dir.: *Deborah Lewis* •
Public Gallery – 1981 ... 52825

Los Angeles Maritime Museum, Berth 84, Foot of 6th St, San Pedro, CA 90731 • T: +1 310 5487618 • F: +1 310 8326537 • museum@lamaritimemuseum.org • www.lamaritimemuseum.org •
Dir.: *Marifrances Trivelli* •
Historical Museum – 1980
Maritime, former Los Angeles municipal ferry bldg, ship building, US Navy – library ... 52826

San Rafael CA

Contemporary Art Gallery at Falkirk Cultural Center, 1408 Mission Av, San Rafael, CA 94901 • T: +1 415 4853328 • F: +1 415 4853404 • jane.lange@ci.san-rafael.ca.us • www.falkirkculturalcenter.org •
Dir.: *Jane Lange* • Cur.: *Beth Goldberg* •
Public Gallery – 1974
Contemporary arts ... 52827

Marin History Museum, 1125 B St, San Rafael, CA 94901 • T: +1 415 4548538 • F: +1 415 4546137 • info@marinhistory.org • www.marinhistory.org •
Dir.: *Merry Alberigi* •
Local Museum – 1935
Local hist, documents, clothing, San Quentin prison photos – library, photo archives ... 52828

San Marco Gallery, Dominican University of California, 50 Acacia A, Archbishop Alemany Library, San Rafael, CA 94901 • T: +1 415 4574440 • F: +1 415 4853262 • chilly@dominican.edu • www.dominican.edu •
Cur.: *Foad Satterfield* •
Public Gallery ... 52829

San Simeon CA

Hearst Castle, 750 Hearst Castle Rd, San Simeon, CA 93452 • T: +1 805 9272020 • curator@hearstcastle.com • www.hearstcastle.org •
Dir.: *Hoyt Fields* •
Decorative Arts Museum / Fine Arts Museum – 1957
Gothic and renaissance decorative and fine art from Spain and Italy, oriental rugs – garden ... 52830

Sand Springs OK

Sand Springs Cultural and Historical Museum, 6 E Broadway, Sand Springs, OK 74063 • T: +1 918 2462509 • F: +1 918 2457107 • swreaves@sandspringsok.org • www.sandspringsok.org •
Dir./Cur.: *Dr. Stacy Reaves* •
Local Museum – 1991 ... 52831

Sandersville GA

Brown House Museum, 268 N Harris St, Sandersville, GA 31082 • T: +1 478 5521965 • F: +1 478 5522963 • www.rootsweb.com •
Dir.: *Mary Alice Jordan* •
Historical Museum – 1999
Washington county hist ... 52832

Washington County Museum, 129 Jones St, Sandersville, GA 31082, mail addr: POB 6088, Sandersville, GA 31082 • T: +1 478 5526965 • wacogre@washemnc.net •
Dir.: *Mary Alice Jordan* •
Local Museum – 1978
Washington County hist ... 52833

Sandown NH

Sandown Historical Museum, Rte 121-A, Sandown, NH 03873 • T: +1 603 8874520 • history@sandownnh.com • www.sandown.us •
Dir.: *Paul M. Densen* • Cur.: *Bertha Deveau* •
Local Museum – 1977
Local history, railroad stn ... 52834

Sandpoint ID

Bonner County Historical Museum, 611 S Ella Av, Sandpoint, ID 83864 • T: +1 208 2632344 • F: +1 208 2632344 • bchsmuseum@nidlink.com • www.bonnercountyhistory.org •
Pres.: *Susan Kiebert* •
Historical Museum – 1972
American Indian artifacts, local hist, geology, sawmilling ... 52835

Sandusky OH

Follett House Museum, 404 Wayne St, Sandusky, OH 44870 • T: +1 419 6253834 • F: +1 419 6254574 • museumservices@sandusky.lib.oh.us • www.sandusky.lib.oh.us/ •
Local Museum – 1902
Householdd objects, furniture, toys ... 52836

Museum of Carousel Art and History, 301 Jackson St, Sandusky, OH 44870 • T: +1 419 6266111 • F: +1 419 6261297 • merrygor@aol.com • www.merrygoroundmuseum.org •
C.E.O.: *Veronica Vanden Bout* •
Fine Arts Museum / Science&Tech Museum – 1990
Carousel art, tools ... 52837

Sandwich IL

Sandwich Historical Society Museum, 315 E Railroad, Sandwich, IL 60548 • T: +1 815 7867936 •
Pres.: *Roger Peterson* •
Local Museum – 1969
Local history, housed in sturdy old structure locally known as the Old Stone Mill ... 52838

Sandwich MA

Heritage Museums, 67 Grove St, Sandwich, MA 02563 • T: +1 508 8883300 ext 100 • F: +1 508 8889535 • heritage@heritagemuseums.org • www.heritagemuseumsandgardens.org •
Exec. Dir.: *Glenn S. Pare* • Cur.: *Jean Gillis* • *Jennifer Madden* (Collection) •
Local Museum – 1969
American fine and folk arts, toys, military, cars – Cape Code Baseball League Hall of Fame, gardens, library, theater ... 52839

Old Hoxie House, Rte 130, Water St, Sandwich, MA 02563 • T: +1 508 8881173 •
Dir./Cur.: *Robert W. Singleton* •
Historical Museum – 1960
Oldest restored house on Cape Cod ... 52840

Sandwich Glass Museum, 129 Main St, Sandwich, MA 02563 • T: +1 508 8880251 • F: +1 508 8884941 • glass@sandwichglassmuseum.org • www.sandwichglassmuseum.org •
Dir.: *Bruce Courson* • Cur.: *Dorothy Schofield* •
Decorative Arts Museum – 1907
Local glass coll, history – library ... 52841

Thornton W. Burgess Museum, 4 Water St, Sandwich, MA 02563 • T: +1 508 8884668 • F: +1 508 8881919 • tburgess@capecod.net • www.thorntonburgess.org/Museum.htm •
Dir.: *Jeanne Johnson* • Historian: *Russell A. Lovell jr.* •
Cur.: *Bethany Rutledge* •
Natural History Museum – 1976
Life, works, and spirit of the children's author and naturalist Thornton Burgess – nature trails ... 52842

Yesteryears Doll and Toy Museum, Main and River Sts, Sandwich, MA 02563 • T: +1 508 8881711 •
Dir.: *Diane Costa* • Cur.: *Eileen Fair* •
Decorative Arts Museum – 1961
Dolls, toys ... 52843

Sandy Hook NJ

Fort Hancock Museum, Sandy Hook Unit, Gateway National Recreation Area, Sandy Hook, NJ 07732 • T: +1 732 8725970 • F: +1 732 8722256 • www.nps.gov./gate •
Sc. Staff: *Thomas Hoffman* (History) •
Natural History Museum – 1968
Natural, military, maritime and social hist ... 52844

Sandy Spring MD

Sandy Spring Museum, 17901 Bentley Rd, Sandy Spring, MD 20860 • T: +1 301 7740022 • F: +1 301 7748149 • dheibein@sandyspringmuseum.org • www.sandyspringmuseum.org •
Dir.: *Carol Perline* •
Local Museum / Historical Museum – 1980
Local history, quaker, farming ... 52845

Sanford FL

Museum of Seminole County History, 300 Bush Blvd, Sanford, FL 32773 • T: +1 407 3212489 • F: +1 407 6655220 • kjacobs@co.seminole.fl.us • www.co.seminole.fl.us •
Head: *Karen Jacobs* •
Local Museum – 1983
Artifacts from Civil War, modern times, documents and photos ... 52846

Sanford Museum, 520 E First St, Sanford, FL 32771 • T: +1 407 3021000 • F: +1 407 3305666 • clarkea@ci.sanford.fl.us • www.sanford.fl.us •
Cur.: *Alicia Clarke* •
Decorative Arts Museum / Historical Museum – 1957
Photos, culture, local baseball hist, 1820 -1890 portrait coll ... 52847

Sanford NC

House in the Horseshoe, 324 Alston House Rd, Sanford, NC 27330-8713 • T: +1 910 9472051 • F: +1 910 9472051 +1 call first • horseshoe@ncmail.net •
Head: *Elizabeth Faison* •
Local Museum – 1972
Furnished house from 1772 ... 52848

Railroad House Historical Museum, 110 Charlotte Av, Sanford, NC 27330 • T: +1 919 7767479 •
Pres.: *Jane S. Barringer* •
Local Museum – 1962
Local hist, railroad hist, Cole pottery ... 52849

Sanibel FL

Bailey-Matthews Shell Museum, 3075 Sanibel-Captiva Rd, Sanibel, FL 33957 • T: +1 941 3952233 • F: +1 941 3956706 • shell@shellmuseum.org • www.shellmuseum.org •
Dir.: *José H. Leal* •
Natural History Museum – 1986
Malacology ... 52850

Sankt Louis MO

Cecille R. Hunt Gallery, Webster University, 8342 Big Bend Blvd, Sankt Louis, MO 63119 • T: +1 314 9687171 • langtk@websteruniv.edu •
Chm.: *Tom Lang* •
Fine Arts Museum – 1950
Local, national and international art in all media ... 52851

Santa Ana CA

Bowers Museum of Cultural Art, 2002 N Main St, Santa Ana, CA 92706 • T: +1 714 5673600 • www.bowers.org •
Dir.: *Armand Labbe* (Collections, Research) • *Megan Shockro* (Education, Programs) • Cur.: *Janet Baker* (Asian Art) •
Fine Arts Museum – 1936
Ethnic art, decorative arts, ceramics ... 52852

Old Courthouse Museum, 211 W Santa Ana Blvd, Santa Ana, CA 92702-4048 • T: +1 714 8346605 • F: +1 714 8342280 • oldcourthouse@ocparks.com • www.ocparks.com •
Dir.: *Robert Selway* • Cur.: *Marshall Duell* •
Local Museum – 1987
County history, government and law ... 52853

Santa Ana College Art Galleries, 1530 W 17th St, Santa Ana, CA 92706 • T: +1 714 5645615 • www.sac.edu •
Dir.: *Mayde Herberg* •
Public Gallery / University Museum – 1970
Paintings, drawings – library ... 52854

Taco Bell Discovery Science Center, 2500 N Main St, Santa Ana, CA 92705 • T: +1 714 5422823 • contactus@discoverycube.org • www.discoverycube.org •
Science&Tech Museum – 1998
Hands-on science center for children ... 52855

Santa Barbara CA

Santa Babara Maritime Museum, 113 Harbor Way, Santa Barbara, CA 93109-2344 • T: +1 805 9628404 • F: +1 805 9627634 • museum@sbmm.org • www.sbmm.org •
Exec. Dir.: *Norman Bishop* • Cur.: *Angela Scott* (Collection) •
Historical Museum – 1994
Nautical instr and objects, hist, archaeology, photos, docs ... 52856

Santa Barbara Contemporary Arts Forum, Klausner Gallery and Norton Gallery, 653 Paseo Nuevo, Santa Barbara, CA 93101 • T: +1 805 9665373 • F: +1 805 9621421 • sbcaf@sbcaf.org • www.sbcaf.org •
Dir.: *Miki Garcia* •
Fine Arts Museum ... 52857

Santa Barbara Historical Museum, 136 E De la Guerra St, Santa Barbara, CA 93101 • T: +1 805 9661601 • F: 9661603 • www.santabarbarahistory.com •
Dir.: *George M. Anderjack* • Cur.: *David Bisol* • Ass. Cur.: *Douglas Diller* •
Local Museum – 1932
Local hist, Spanish, Mexican and American periods – library ... 52858

Santa Barbara Museum of Art, 1130 State St, Santa Barbara, CA 93101-2746 • T: +1 805 9634364 • F: +1 805 9666840 • info@sbma.net • www.sbmuseart.org •
Dir.: *Phillip M. Johnston* • Cur.: *Diana DuPont* (Contemporary Art) • *Susan Shin-tsu Tai* (Asian Art) • *Karen Sinsheimer* (Photography) •
Fine Arts Museum – 1941
Greek, Roman ans Egyptian antiquities, Asian art, European painting and sculpture, American paintings, decorative arts – library, auditorium ... 52859

Santa Barbara Museum of Natural History, 2559 Puesta del Sol Rd, Santa Barbara, CA 93105 • T: +1 805 6824711 • F: +1 805 5693170 • khutterer@sbnature2.org • www.sbnature.org •
Dir.: *Dr. Karl Hutterer* • Cur.: *Dr. John R. Johnson* (Anthropology) • *Dr. F.G. Hochberg* (Invertebrate Zoology) • *Paul W. Collins* (Vertebrate Zoology) •
Natural History Museum – 1916
Natural history, prehistory, anthropology, natural history art – library, archives, planetarium, auditoriums ... 52860

University Art Museum, University of California Santa Barbara, Santa Barbara, CA 93106-7130 • T: +1 805 8932951 • F: +1 805 8933013 • uam@uam.ucsb.edu • www.uam.ucsb.edu •
Dir.: *Dr. Bonnie G. Kelm* • Cur.: *Kurt Helfrich* (Architectural Drawing Collection) • Education: *Natalie Sanderson* •
Fine Arts Museum / University Museum – 1959
Paintings, renaissance medals and plaquettes, drawings, graphics, ceramics, sculpture ... 52861

U

Santa Clara CA

Intel Museum, Intel Corporation, 2200 Mission College Blvd, Santa Clara, CA 95054 • T: +1 408 7650503 • www.intel.com/go/museum •
Cur.: *Tracey Mazur* •
Science&Tech Museum – 1992
Chip development ... 52862

De Saisset Museum, Santa Clara University, 500 El Camino Real, Santa Clara, CA 95053-0550 • T: +1 408 5544528 • F: +1 408 5547840 • www.scu.edu/desaisset •
Dir.: *Rebecca M. Schapp* • Cur.: *Karen Kienzle* •
Fine Arts Museum / University Museum / Historical Museum – 1955
Fine and decorative arts, university history – library, auditorium ... 52863

Triton Museum of Art, 1505 Warburton Av, Santa Clara, CA 95050 • T: +1 408 2473754 • F: +1 408 2473796 • grivera@tritonmuseum.org • www.tritonmuseum.org •
Dir.: *George Rivera* • Cur.: *Susan Hillhouse* •
Fine Arts Museum – 1965
American paintings, contemporary drawings and prints – library ... 52864

Santa Claus IN

Wax Museum, 1 Holiday Sq, Santa Claus, IN 47579 • T: +1 812 9374401 • F: +1 812 9374405 • www.hollidayworld.com •
Pres.: *Will Koch* •
Special Museum
Dolls, toys, coll from 1950 ... 52865

Santa Cruz CA

Eloise Pickard Smith Gallery, University of California, Santa Cruz, c/o Cowell College, 1156 High St, Santa Cruz, CA 95064 • T: +1 831 4592953 • F: +1 831 4595348 • cowell11@ucsc.edu • www2.ucsc.edu/cowell/SmithGallery •
Dir.: *Linda Pope* •
Fine Arts Museum / University Museum ... 52866

Mary Porter Sesnon Art Gallery, University of California, Santa Cruz, c/o Porter College, 1156 High St, Santa Cruz, CA 95064 • T: +1 831 4593606, 4592314 • F: +1 831 4593535 • sgraham@ucsc.edu • arts.ucsc.edu/sesnon •
Dir.: *Shelby Graham* • Cur.: *Leslie Fellows* •
Public Gallery / University Museum – 1971
Drawings, paintings, video – archives ... 52867

The Museum of Art and History, McPherson Center, 705 Front St, Santa Cruz, CA 95060 • T: +1 831 4291964 ext 10 • F: +1 831 4291954 • admin@santacruzmah.org • www.santacruzmah.org •
Exec. Dir.: *Paul Figueroa* • Cur.: *Kathleen Moodie* (Exhibits) • *Marla Novo* (Collections) • *Anna Castillo* (Education) •
Fine Arts Museum / Historical Museum – 1981
Paintings, sculpture, local hist ... 52868

Santa Cruz Art League, 526 Broadway, Santa Cruz, CA 95060 • T: +1 831 4265787 • carol@scal.org • www.scal.org •
Pres.: *Stephanie Schriver* • Cur.: *Carol Jenied* •
Association with Coll – 1919
Local pioneer paintings, sculpture ... 52869

Santa Cruz Museum of Natural History, 1305 E Cliff Dr, Santa Cruz, CA 95062 • T: +1 831 4206115 • F: +1 831 4201137 • staff@santacruzmuseums.org • www.santacruzmuseums.org •
Dir.: *Carolyn O'Donnell* • Cur.: *Jennifer Lienau* •
Natural History Museum / Local Museum – 1904
Regional natural hist, surfing, ethnology, regional art – library ... 52870

Santa Fe NM

American Bicycle and Cycling Museum, POB 8533, Santa Fe, NM 87504 • T: +1 505 9897634 • F: +1 505 9897634 •
Dir.: *Sandra Vaillancourt* •
Science&Tech Museum – 1991 ... 52871

Center for Contemporary Arts of Santa Fe, 1050 Old Pecos Trail, Santa Fe, NM 87501 • T: +1 505 9821338 • F: +1 505 9829854 •
Dir.: *Guy Ambrosino* • *Jerry Barson* •
Fine Arts Museum – 1985 ... 52872

El Rancho de Las Golondrinas Museum, 334 Los Pinos Rd, Santa Fe, NM 87507 • T: +1 505 4712261 • F: +1 505 4715623 • mail@golondrinas.org • www.golondrinas.org •
Dir.: *George B. Paloheimo* • Sc. Staff: *Carla Gomez* (Textiles) • *Julie Anna Lopez* (Agriculture) • *Dr. Donna Pierce* (Collection) •
Open Air Museum – 1970
Spanish colonial lifestyle, agriculture, textiles ... 52873

Georgia O'Keeffe Museum, 217 Johnson St, Santa Fe, NM 87501 • T: +1 505 9461000 • F: +1 505 9461091 • main@okeeffemuseum.org • www.okeeffemuseum.org •
Dir.: *George G. King* •
Fine Arts Museum – 1997
Changing exhib of modern art, paintings ... 52874

Governor's Gallery, State Capitol Bldg, Rm 400, Santa Fe, NM 87503 • T: +1 505 8273089 • F: +1 505 8273026 • gov@gov.state.nm.us • www.governor.state.nm.us •
Dir.: *Dr. J. Edson Way* • Cur.: *Arif Khan* •
Public Gallery – 1973 ... 52875

Indian Arts Research Center, 660 Garcia St, Santa Fe, NM 87505, mail addr: POB 2188, Santa Fe, NM 87504 • T: +1 505 9547205 • F: +1 505 9547207 • iarc@sarsf.org • www.sarweb.org •
Fine Arts Museum / Decorative Arts Museum – 1907
Historic and contemporary SW Pueblo Indian pottery, Navajo and Pueblo Indian textiles and jewelry coll, Native American easel paintings, ethnography ... 52876

Institute of American Indian Arts Museum, 108 Cathedral Pl, Santa Fe, NM 87501 • T: +1 505 9838900, 9831777 • F: +1 505 9831222 • www.iaia.edu •
Dir.: *Charles Daily* • Cur.: *Linda Woody Cywink* • Ass. Cur.: *Tatiana Lomaheftewa-Slock* •
Fine Arts Museum – 1962
Contemporary native American fine art ... 52877

Liquid Paper Correction Fluid Museum, Rte 1, Santa Fe, NM 87501 • T: +1 505 4553848 •
Pres.: *Michael Nesmith* • Dir.: *Marcia J. Summers* •
Science&Tech Museum – 1980
Industrial hist ... 52878

Maria Martinez Museum, Rte 5, Box 315-A, Pueblo of San Ildefonso, Santa Fe, NM 87501 • T: +1 505 4552031, 4553549 •
Decorative Arts Museum
Arts and crafts, clothing, painting, pottery ... 52879

Museum of Fine Arts, 107 W Palace, Santa Fe, NM 87501 • T: +1 505 4765072 • F: +1 505 4765076 • mbol@mnn.state.nm.us •
Dir.: *Marsha C. Bol* • Cur.: *Joseph Traugott* (20th Century Art) • *Dr. Steven A. Yates* (Photography) • *Ellen Zieselman* (Education) • *Laura Addison* (Contemorary Art) • Library: *Mary Jebsen* •
Fine Arts Museum – 1917
Paintings, prints, drawings, photographs, sculpture, exhibitions on 20th c American art ... 52880

Museum of Indian Arts and Culture, 710 Camino Lejo, Santa Fe, NM 87505 • T: +1 505 4761250 • F: +1 505 4761330 • info@miaclab.org • www.miaclab.org •
Dir.: *Duane Anderson* • Ass. Dir.: *Chris Turnbow* • Cur.: *Antonio Chavarria* (Ethnology) • *Julia Clifton* (Archaeological research coll) • *John Torres* (Archaeology) • Ass. Cur.: *Dody Fugate* • *Tony Thibodeau* (Archaeology) • *Valerie Verzuh* (Collections) •
Fine Arts Museum / Ethnology Museum / Folklore Museum – 1909
Anthropology, ethnology and Indian art ... 52881

Museum of International Folk Art, 706 Camino Lejo, Santa Fe, NM 87505, mail addr: POB 2087, Santa Fe, NM 87505-2087 • T: +1 505 4761200 • F: +1 505 4761300 • info@moifa.org • www.moifa.org •
Dir.: *Dr. Joyce Ice* • Ass. Dir.: *Jacqueline Duke* • Sc. Staff: *Dr. Barbara Mauldin* (Latin American Folk Art) • *Dr. Tey Marianna Nunn* (Contemporary Hispano and Latino Collections) • *Dr. Barbara Sumberg* (Textiles) • Libr./Photo Archivist: *Ree Mobley* •
Folklore Museum – 1953
International folk art, costumes, ceramics, dolls, toys, Spanish colonial religious art, Southwest Hispanic folk art and furniture, jewelry, amulets, ceremonial and ritual objects ... 52882

Museum of New Mexico, 113 Lincoln Av, Santa Fe, NM 87501 • T: +1 505 4765060 • F: +1 505 4765088 • steve.cantrell@state.nm.us • www.museumofnewmexico.org •
Dir.: *Thomas A. Wilsony* • Dep. Dir.: *John J. McCarthy* • Assoc. Dir.: *Dr. Duane Anderson* (Anthropology) • Sen. Cons.: *Dale Kronkright* • Chief Cons.: *Claire Munzenrieder* • Cur.: *Dr. Joyce Ice* (Folk Art) • *Dr. Steve Yates* (Contemporary Photography) • *Joseph Traugott* (20th-Century Art) • *Charles Bennett* (History) • *Dr. Richard Rudisill* • *Art Olivas* (Historic Photography) • *John Torres* (Anthropological Collections) • *Robin Farwell-Gavin* (Spanish Colonial Folk Art) • *Barbara Mauldin* (Latin American Art) • *Renee Jolly* (Neutrogena) • *Dr. Tey Marianna Nunn* (Hispanic and Latino) • *Annie Carlano* (European and American) •
Local Museum – 1909
Anthropology, archaeology and ethnology of Southwest, fine arts, international folk art, Spanish colonial hist, hist of New Mexico ... 52883

Museum of Spanish Colonial Art, 750 Camino Lejo, Santa Fe, NM 87505 • T: +1 505 9822226 • F: +1 505 9824585 • museum@spanishcolonial.org • www.spanishcolonial.org •
C.E.O.: *William Field* •
Fine Arts Museum – 1925
Spanish colonial art, furniture, bultos, retablos, textiles, tinwork ... 52884

Palace of the Governors, Museum of New Mexico, One Plaza, Santa Fe, NM 87501, mail addr: POB 2087, Santa Fe, NM 87504-2087 • T: +1 505 4765094 • F: +1 505 4765104 •
Dir.: *Thomas Chavez* • Ass. Dir.: *Charles Bennett* • Cur.: *Diana Ortega DeSantis* • *Pamela Smith* • *Arthur Olivas* • *Dr. Richard Rudisill* • *David H. Snow* •
Local Museum – 1909
New Mexico and Southwest hist, religious and decorative arts, photos, silver, furniture ... 52885

Poeh Museum, Pueblo of Poloaque, 78 Cities of Gold Rd, Santa Fe, NM 87506 • T: +1 505 4553334 • F: +1 505 4550174 • www.poehmuseum.com •
Dir.: *Strum Christy* • Cur.: *Talachy Melissa* •
Archaeology Museum / Fine Arts Museum
Archaeological to contemporary art from villages of Mexico ... 52886

San Miguel Mission Church, 401 Old Santa Fe Trail, Santa Fe, NM 87501 • T: +1 505 9833974 • F: +1 505 9828722 • sanmiguel1610@yahoo.com •
Religious Arts Museum – 1610
Indian art, religious artifacts, pottery ... 52887

Santa Fe Children's Museum, 1050 Old Pecos Trail, Santa Fe, NM 87501 • T: +1 505 9898359 • F: +1 505 9897506 • children@santafechildrensmuseum.org • www.santafechildrensmuseum.org •
Pres.: *Peter Wirth* • Dir.: *Ellen Biderman* • C.E.O: *Londi Carbajal* •
Historical Museum – 1985 ... 52888

Santuario de Nuestra Senora de Guadalupe, 100 S Guadalupe St, Santa Fe, NM 87501 • T: +1 505 9882027 •
C.E.O.: *Leo Kahn* •
Religious Arts Museum / Fine Arts Museum – 1975
Spanish colonial mural, New Mexican Santero art ... 52889

Site Santa Fe, 1606 Paseo de Peralta, Santa Fe, NM 87501 • T: +1 505 9891199 • F: +1 505 9891188 • sitesantafe@sitesantafe.org • www.sitesantafe.org •
Public Gallery – 1995 ... 52890

Visual Art Gallery, Santa Fe Community College, 6401 Richards Av, Santa Fe, NM 87508 • T: +1 505 4281501 • cbaughan@sfccnm.edu • www.sfccnm.edu/gallery •
Dir.: *Clark Baughan* •
University Museum / Fine Arts Museum
Art ... 52891

Wheelwright Museum of the American Indian, 704 Camino Lejo, Santa Fe, NM 87505 • T: +1 505 9824636 • F: +1 505 9897386 • wheelwright@wheelwright.org • www.wheelwright.org •
Dir.: *Jonathan Batkin* • Cur.: *Cheri Falkenstien-Doyle* •
Folklore Museum / Ethnology Museum – 1937
Anthropology, archaeology, South Indian arts, folk hist ... 52892

Santa Fe Springs CA

Hathaway Ranch Museum, 11901 E Florence Av, Santa Fe Springs, CA 90670 • T: +1 562 7773444 • hrm@hathaworld.com • www.hathaworld.com •
Exec. Dir.: *Francine Rippy* •
Agriculture Museum – 1986
American furnishings 19th c, farming – archives 52893

Santa Maria CA

Natural History Museum, 412 S McClelland St, Santa Maria, CA 93454 • T: +1 805 6140806 • F: +1 805 6140806 • naturalhistory.santamaria @verizon.net • www.naturalhistorysantamaria.com •
Natural History Museum – 1996 ... 52894

Santa Maria Museum of Flight, 3015 Airpark Dr, Santa Maria, CA 93455 • T: +1 805 9228758 • F: +1 805 9228958 • smmof@msn.com • www.smmof.org •
Pres.: *Mike Geddry Sr.* •
Science&Tech Museum – 1984
Aeronautics, aircraft and aviation artifacts, space hist – library ... 52895

Santa Maria Valley Historical Society Museum, 616 S Broadway, Santa Maria, CA 93454 • T: +1 805 9223130 • richard@santamariahistory.org • www.santamariahistory.org •
Cur.: *Richard Chenoweth* •
Local Museum – 1955
Local hist, Indian artifacts, Rancho period – library ... 52896

Santa Monica CA

Angels Attic Museum, 516 Colorado Av, Santa Monica, CA 90410 • T: +1 310 3948331 • F: +1 310 6566865 • dollhses@earthlink.net • www.internetimpact.com/angelsattic •
Dir.: *Jackie McMahan* • *Eleanor La Vove* •
Special Museum – 1984
Dollhouse, miniatures, toys, dolls ... 52897

California Heritage Museum, 2612 Main St, Santa Monica, CA 90405 • T: +1 310 3928537/38 • F: +1 310 3960547 • calmuseum@earthlink.net • www.californiaheritagemuseum.org •
Dir.: *Tobi Smith* •
Historical Museum / Decorative Arts Museum – 1977
California culture and heritage, decorative and folk art, photography – photo archive ... 52898

Museum of Flying (closed temporarily), 2772 Donald Douglass Loop N, Santa Monica, CA 90405 • T: +1 310 3928822 • F: +1 310 3998783 • drodda@americanairports.net • www.museumofflying.com •
Exec. Dir.: *Dan Ryan* •
Military Museum / Science&Tech Museum – 1974
Military hist, aircrafts, WW II fighter planes ... 52899

Pete and Susan Barrett Art Gallery, Santa Monica College, Performing Arts Center, 11th St and Santa Monica Blvd, Santa Monica, CA 90405 • T: +1 310 4343434 • F: +1 310 4343646 • www.smcbarrettgallery.com •
Dir.: *Marian Winsryg* •
Public Gallery – 1973
Southern California prints and drawings ... 52900

Santa Monica Museum of Art, Bergamot Stn 61, 2525 Michigan Av, Santa Monica, CA 90404 • T: +1 310 5866488 • F: +1 310 5866487 • info@smmoa.org • www.smmoa.org •
Exec. Dir.: *Elsa Longhauser* • Dep. Dir.: *Lisa Melandri* (Exhibits, Program) • *Asuka Hisa* (Education) • *Christina Cassidy* (Development) •
Fine Arts Museum – 1985
Contemporary art ... 52901

Santa Paula CA

California Oil Museum, 1001 E Main St, Santa Paula, CA 93061 • T: +1 805 9330076 • F: +1 805 9330096 • info@oilmuseum.net • www.oilmuseum.net •
Dir.: *Mike Nelson* • Cur.: *John Nichols* •
Science&Tech Museum – 1950
Oil industry, photos, paintings ... 52902

Santa Rosa CA

The California Indian Museum, 5250 Aero Dr, Santa Rosa, CA 95403 • T: +1 707 5793004 • F: +1 707 5799019 • www.cimcc.org •
C.E.O.: *Leroy Lounibous* •
Historical Museum
Art, traveling exhibits photos display ... 52903

Charles M. Schulz Museum and Research Center, 2301 Hardies Ln, Santa Rosa, CA 95403 • T: +1 707 5794452 • F: +1 707 5794436 • inquiries@schulzmuseum.org • www.schulzmuseum.org •
Dir.: *Karen Johnson* •
Fine Arts Museum – 2002
Sculpture, paintings, drawings, photographs and memorabilia releated to life of Charles M. Schulz – library, research center ... 52904

Jesse Peter Museum, Santa Rosa Junior College, 1501 Mendocino Av, Santa Rosa, CA 95401 • T: +1 707 5274479 • www.santarosa.edu/museum •
Dir./Cur.: *Benjamin Foley Benson* •
Ethnology Museum – 1932
Native American artifacts, Latino, African and Asian cultures ... 52905

Luther Burbank Home, Santa Rosa and Sonoma Avs, Santa Rosa, CA 95402, mail addr: POB 1678, Santa Rosa, CA 95402 • T: +1 707 5245445 • F: +1 707 5245827 • lhall@srcity.org • www.lutherburbank.org •
Special Museum – 1979
Historic house, furnishings, tools of the gardener L. Burbank – gardens ... 52906

Pacific Coast Air Museum, 2230 Becker Blvd, Santa Rosa, CA 95403 • T: +1 707 5757900 • F: +1 707 5452813 • director@pacificcoastairmuseum.org • www.pacificcoastairmuseum.org •
Exec. Dir.: *Dave Pinsky* •
Science&Tech Museum – 1990
Aircrafts ... 52907

Santa Rosa Junior College Art Gallery, 1501 Mendocino Av, Santa Rosa, CA 95401 • T: +1 707 5274298 • F: +1 707 5274532 • mmcginnis@santarosa.edu • www.santarosa.edu/art/gallery •
Dir.: *Deborah Kirklin* • *Renata Breth* •
Public Gallery ... 52908

Sonoma County Museum, 425 Seventh St, Santa Rosa, CA 95401 • T: +1 707 5791500 • F: +1 707 5794849 • devans@sonomacountymuseum.org • www.sonomacountymuseum.org •
Dir.: *Natasha Boas* • Cur.: *Alexandra Quinn* (Education) • *Eric Stanley* (Exhibits) •
Local Museum – 1976
Local history and culture, paintings, Christo & Jeanne-Claude coll ... 52909

Sapulpa OK

Sapulpa Historical Museum, 100 E Lee, Sapulpa, OK 74066 • T: +1 918 2244871 • F: +1 918 2247765 • doris.yocham@juno.com •
Pres.: *Doris R. Yocham* •
Local Museum – 1968
Frisco railroad items, kitchen equip of early 1900's, glass industry items, Native American artifacts ... 52910

Saranac Lake NY

Robert Louis Stevenson Memorial Cottage, 44 Stevenson Ln, Saranac Lake, NY 12983 • T: +1 518 8911462 • pennypiper@capital.net • www.pennypiper.org •
Dir. & Pres.: *Susan Allen* • Cur.: *John M. Delahant* •
Special Museum – 1920
Personal mementos of the English writer (1850-1894), childhood photographs, original letters and articles ... 52911

Sarasota FL

Art Center Sarasota, 707 N Tamiami Trail, Sarasota, FL 34236-4050 • T: +1 941 3652032 • F: +1 941 3660585 • artsarasota@aol.com • www.artsarasota.org •
Pres.: *Alan P. Sloan* •
Fine Arts Museum / Public Gallery / Association with Coll – 1926
Paintings, sculpture, graphics, photography ... 52912

Crowley Museum, 16405 Myakka Rd, Sarasota, FL 34240 • T: +1 941 3221000 • F: +1 941 3221000 • crowleymuseum@aol.com • www.crowleymuseumnaturectr.org •
Exec. Dir.: *Debbie Dixon* •
Historical Museum / Natural History Museum – 1974
History, natural history 52913

The John and Mable Ringling Museum of Art, State Art Museum of Florida, 5401 Bay Shore Rd, Sarasota, FL 34243 • T: +1 941 3595700 • F: +1 941 3595745 • info@ringling.org • www.ringling.org •
Dir.: *John Wetenhall* • Cur.: *Aaron DeGroft* (Chief) •
Fine Arts Museum / University Museum – 1927
Archaeology, folk culture, costumes and textiles, decorative arts, paintings, photographs, prints, drawings, graphic arts, sculpture, furnishings, botany 52914

Sarasota Classic Car Museum, 5500 N Tamiami Trail, Sarasota, FL 34243 • T: +1 941 3556228 • F: +1 941 3588065 • info@sarasotacarmuseum.org • www.sarasotacarmuseum.org •
Pres.: *Martin Godbey* •
Science&Tech Museum – 1953
Automobile, horseless carriages, automotive hist, player pianos, organs, Edison's phonographs, music boxes 52915

Selby Gallery, Ringling School of Art and Design, 2700 N Tamiami Trail, Sarasota, FL 34234 • T: +1 941 3597563 • F: +1 941 3597517 • selby@ringling.edu • www.rsad.edu/ •
Dir.: *Kevin Dean* • Ass. Dir.: *Laura Avery* •
University Museum – 1986
20th c prints – Library 52916

The Turner Museum, POB 11073, Sarasota, FL 34278-1078 • T: +1 941 9245622, 3781885 • F: +1 941 9245622, 3781885 • turnermuseum@turnermuseum.org • www.turnermuseum.org •
Pres./Dir.: *Douglass J.M. Graham* •
Fine Arts Museum – 1973
Virtual museum of art of J.M.W. Turner 52917

Saratoga CA

Montalvo Arts Center, 15400 Montalvo Rd, Saratoga, CA 95070 • T: +1 408 9615800 • F: +1 408 9615850 • www.villamontalvo.org •
Dir.: *Elisbeth Challener* • *Michele Rowe-Shields* •
Public Gallery – 1930
Contemporary art, paintings, sculpture, decorative art – garden theater 52918

Saratoga WY

Saratoga Museum, 104 Constitution Av, Saratoga, WY 82331 • T: +1 307 3265511 • saratogamuseum@carbonpower.net •
Dir.: *Pat Bensen* •
Local Museum / Archaeology Museum – 1978
Local hist, Union Pacific memorabilia, archaeology 52919

Saratoga Springs NY

The Children's Museum at Saratoga, 69 Caroline Str, Saratoga Springs, NY 12866 • T: +1 518 5845540 • F: +1 518 5846049 • info@cmssny.org • www.childrensmuseumatsaratoga.org •
Exec. Dir.: *Barbara R. Milano* •
Special Museum – 1989 52920

National Museum of Dance and Hall of Fame, 99 S Broadway, Saratoga Springs, NY 12866 • T: +1 518 5842225 • F: +1 518 5844515 • info@dancemuseum.org • www.dancemuseum.org •
Dir.: *Garrett Smith* •
Performing Arts Museum – 1986
Dance and dance hist, photographs, personal items of dancers, costumes, videos 52921

National Museum of Racing and Hall of Fame, 191 Union Av, Saratoga Springs, NY 12866 • T: +1 518 5840400 • F: +1 518 5844574 • nmrinfo@racingmuseum.net • www.racingmuseum.org •
Dir.: *Peter Hammell* • Ass. Dir.: *Cathy Maguire* •
Special Museum – 1950
Sporting art, equine paintings, sculpture; thoroughbred racing trophies, racing colors, racing memorabilia 52922

New York State Military Museum, 61 Lake Av, Saratoga Springs, NY 12866 • T: +1 518 5815100 • F: +1 518 5815111 • www.dmna.state.ny.us/historic •
Dir.: *Michael Aikey* • Chief Cur.: *Courtney Burns* •
Military Museum – 1863 52923

The Saratoga Springs History Museum, Historical Society of saratoga Springs, Casino, Congress Park, Saratoga Springs, NY 12866 • T: +1 518 5846920 • F: +1 518 5811477 • historicalsociety@spa.net • www.saratogahistory.org •
Dir.: *James D. Parillo* • Ass. Dir.: *Maryann Fitzgerald* • Cur.: *Erin Doane* •
Historical Museum – 1883 52924

Schick Art Gallery, Skidmore College, Skidmore Campus, 815 N Broadway, Saratoga Springs, NY 12866-1632 • T: +1 518 5805049 • F: +1 516 5805029 • damiller@skidmore.edu • www.skidmore.edu •
Dir.: *David Miller* •
Fine Arts Museum / University Museum – 1926
Paintings and graphics, ceramics, sculpture 52925

Tang Teaching Museum and Art Gallery, Skidmore College, 815 N Broadway, Saratoga Springs, NY 12866 • T: +1 518 5808080 • F: +1 518 5805069 • tang@skidmore.edu • www.skidmore.edu/tang •
C.E.O.: *Charles Stainback* • Cur.: *Ian Berry* •
Fine Arts Museum
Drawings, paintings, prints, sculptures, video, installation art, ethnographic art 52926

Saugerties NY

Opus 40 and the Quarryman's Museum, 50 Fite Rd, Saugerties, NY 12477 • T: +1 845 2463400 • F: +1 845 2461997 • tad@opus40.org • www.opus40.org •
C.E.O.: *Pat Richards* •
Fine Arts Museum / Local Museum – 1978
Environmental bluestone sculpture, 19th c quarrymen's tools and artifacts 52927

Saugus MA

Saugus Iron Works, 244 Central St, Saugus, MA 01906 • T: +1 781 2330050 • F: +1 781 2317345 • sair_iron_works_house@nps.gov • www.nps.gov/sair •
Dir.: *Steve Kesselman* • Sc. Staff: *Carl Salmons-Perez* •
Science&Tech Museum – 1954
Blast furnace, forge and slitting mill, tools, domestic objects – library 52928

Sauk Centre MN

Sinclair Lewis Museum, 194 and US-71, Sauk Centre, MN 56378, mail addr: POB 25, Sauk Centre, MN 56378 • T: +1 320 3525201 •
Pres.: *Roberta Olson* •
Special Museum – 1960
Books by S. Lewis, his life, manuscripts – library 52929

Saukville WI

Ozaukee County Historical Society Pioneer Village, 4880 Hwy I, Saukville, WI 53080 • T: +1 262 3774510 • F: +1 262 3774510 • www.co.ozaukee.wi.us/ochs •
Pres.: *Mary Sayner* • Cur.: *Ruth Renz* •
Open Air Museum / Local Museum – 1960
Pioneer hist, folklore, agriculture, railroad artifacts 52930

Sault Sainte Marie MI

Le Sault de Sainte Marie Historical Sites, 501 E Water St, Sault Sainte Marie, MI 49783 • T: +1 906 6323658 • F: +1 906 6329344 • admin@thevalleycamp.com • www.thevalleycamp.com •
Dir.: *J.H. Hobaugh* •
Historical Museum – 1967
Maritime items, mid-19th to early 20th c items, ethnology 52931

Saum MN

First Consolidated School in Minnesota, 41982 Pioneer Rd NE, Saum, MN 56650 • T: +1 218 6478205, 6478531 •
C.E.O.: *Gari Krogseng* •
Historical Museum – 1962
1903 original log school, 1912 Saum School, desks, pictures, books, fixtures 52932

Saunderstown RI

Casey Farm, Route 1A, Saunderstown, RI 02874 • T: +1 617 2273956 • F: +1 617 2279204 • CaseyFarm@HistoricNewEngland.org • www.historicnewengland.org •
Pres.: *Carl Nold* •
Open Air Museum / Local Museum – 1955
Historic homestead 52933

The Gilbert Stuart Museum, 815 Gilbert Stuart Rd, Saunderstown, RI 02874 • T: +1 401 2943001 • F: +1 401 2943869 • info@gilbertstuartmuseum.org • www.gilbertstuartmuseum.com •
Cur.: *John Thompson* • *Deborah Thompson* •
Fine Arts Museum – 1930
Historic house, birthplace of artist Gilbert Stuart . 52934

Sausalito CA

Bay Area Discovery Museum, Fort Baker, 557 McReynolds Rd, Sausalito, CA 94965 • T: +1 415 3393900, 2897276 • contact@badm.org • www.baykidsmuseum.org •
Pres.: *Anne Dickerson Lind* • C.E.O.: *Lori Fogarty* •
Special Museum – 1984
Art, nature and science, culture 52935

Headlands Center for the Arts, 944 Fort Barry, Sausalito, CA 94965 • T: +1 415 3312787 ext 28 • F: +1 415 3313857 • info@headlands.org • www.headlands.org •
Dir./Cur.: *Tod Ruhstaller* • Cur.: *Susan Benedetti* (Education) • *Kylee Denning* (Collections) •
Public Gallery 52936

Sautee GA

Sautee-Nacoochee Museum, 283 Hwy 255 N, Sautee, GA 30571 • T: +1 706 8783300 • F: +1 706 8781395 • snca@alltel.net • www.snca.org •
Cur.: *Joanna Tinius* •
Decorative Arts Museum
Paintings, pottery, glass, jewelry 52937

Savannah GA

Archives Museum - Temple Mickve Israel, 20 E Gordon St, Savannah, GA 31401 • T: +1 912 2331547 • F: +1 912 2333086 • mickveisr@aol.com • www.mickveisrael.org •
Pres.: *Dr. David B. Byck* •
Religious Arts Museum – 1974
Congregation Mickve Israel Synagogue, oldest Jewish congregation in the South 52938

Davenport House Museum, 324 E State St, Savannah, GA 31401 • T: +1 912 2367938 • F: +1 912 2337706 • davenport@g-net.net •
Dir.: *Jamie Credle* •
Historical Museum / Historic Site – 1955
Historic House, court inventory, federal furniture, china 52939

Fort Pulaski, US-Hwy 80 E, Savannah, GA 31410 • T: +1 912 7865787 • F: +1 912 7866023 • june_devisfrutto@nps.gov • www.nps.gov/fopu •
Head: *John D. Breen* •
Military Museum – 1924
Uniform, bottle coll, personal effects of soldiers . 52940

Georgia Historical Society Museum, 501 Whitaker St, Savannah, GA 31401 • T: +1 912 6512125 • F: +1 912 6512831 • ghs@georgiahistory.com • www.georgiahistory.com •
Dir.: *Dr. W. Todd Groce* •
Local Museum – 1839
Books, manuscripts, maps, fotographs, prints, paintings, artifacts 52941

Juliette Gordon Low Birthplace, 10 E Oglethorpe Av, Savannah, GA 31401 • T: +1 912 2334501 • F: +1 912 2334659 • birthplace@girlscouts.org • www.juliettegordonlowbirthplace.org •
Dir.: *Fran Powell Harold* • Cur.: *Sherryl Lang* • Education/Archives: *Katherine Keena* •
Historical Museum – 1953
1818-1821 Wayne-Gordon House; 1912-1927 Early history of the Girl Scouts; 19th century furnishings, fine arts, decorative art; schulopture, paintings sketches by Juliette Gordon Low – Photographic collection; Family archives 52942

Old Fort Jackson, 1 Fort Jackson Rd, Savannah, GA 31404 • T: +1 912 2323945 • F: +1 912 2365126 •
Exec. Dir.: *Scott Smith* •
Military Museum – 1975
Artifacts to the military hist, American revolution, the War of 1812 and the Civil War 52943

Owens-Thomas House, 124 Abercorn, Savannah, GA 31401 • T: +1 912 2339743 • F: +1 912 2330102 • chamberlainc@telfair.org • www.telfair.org •
Dir.: *Diane B. Lesko* • Cur.: *Carola Hunt Chamberlain* •
Historical Museum – 1951
Original carriage house with slave quarters and walled garden 52944

SCAD Museum of Art, c/o Savannah College of Art and Design, 227 Martin Luther King Jr Blvd, Savannah, GA 31401 • T: +1 912 5257191 • F: +1 912 5257199 • info@scad.edu • www.scad.edu •
Dir.: *Kari Herrin* •
Fine Arts Museum – 1979
17th-20th c paintings, prints and photographs .. 52945

Ships of the Sea Maritime Museum, 41 MLK Blvd, Savannah, GA 31401 • T: +1 912 2321511 • F: +1 912 2347363 • shipssea@bellsouth.net • www.shipsofthesea.org •
Dir.: *Karl DeVries* •
Historical Museum – 1966
Ship models, paintings, maritime antiques, navigation instr, seafaring artifacts 52946

Telfair Museum of Art, 121 Barnard St, Savannah, GA 31401 • T: +1 912 2321177 • F: +1 912 2326954 • hadawys@telefair.org • www.telfair.org •
Dir.: *Diane B. Lesko* • Cur.: *Harry DeLorme* (Education) •
Fine Arts Museum / Science&Tech Museum – 1875
American and European paintings, prints, drawings, portraits, architecture 52947

William Scarbrough House, Ships of the Sea Museum, 41 Martin Luther King Blvd, Savannah, GA 31401 • T: +1 912 2321511 • F: +1 912 2347363 • shipssea@bellsouth.net • www.shipsofthesea.org •
Dir.: *Tony Pizzo* • Ass. Dir.: *Karl DeVries* •
Historical Museum – 1966
Maritime history, housed in 1819 Regency style house 52948

Wormsloe State Historic Site, 7601 Skidaway Rd, Savannah, GA 31406 • T: +1 912 3533023 • F: +1 912 3533023 • wormsloe@g-net.net • www.gastateparks.org •
Head: *Joe H. Thompson* •
Archaeology Museum – 1973
Ruins of 1739 fortified house 52949

Savannah MO

Andrew County Museum, 202 E Duncan Dr, Savannah, MO 64485-0012 • T: +1 816 3244720 • F: +1 816 3245271 • andcomus@ccp.com • www.artcom.com/museums.html •
Pres.: *Harold Johnson* • Dir./Cur.: *Patrick S. Clark* •
Local Museum – 1972
Early 19th c French and German dolls, old Hummel figurine coll 52950

Saxton PA

Captain Phillips' Rangers Memorial, Pennsylvania Hwy 26, Saxton, PA 16678 • T: +1 717 7873602 • F: +1 717 7831073 •
Historic Site – 1959
Site of July 16, 1780, Indian massacre of members of Capt. Phillips' militia 52951

Sayville NY

Sayville Historical Society Museum, Edwards St and Collins Av, Sayville, NY 11782 • T: +1 631 5630186, 5892609 •
Local Museum – 1944
Local hist 52952

Scarborough ME

Scarborough Historical Museum, 649A U.S. Rte 1, Dunstan, Scarborough, ME 04074 • T: +1 207 8833539 •
Pres.: *Rodney Laughton* •
Local Museum – 1961
Local hist 52953

Scarsdale NY

Scarsdale Historical Society Museum, 937 Post Rd, Scarsdale, NY 10583 • T: +1 914 7231744 • F: +1 914 7232185 • history@cloud9.net • www.scarsdalenet.com/historicalsociety •
Exec. Dir.: *Penny Brickman* • Cur.: *Karen Frederick* • *Kathy Craughwell-Varda* •
Local Museum – 1973
19th c textiles, costumes, furniture, rugs, farm equip, kitchen utensils 52954

Weinberg Nature Center, 455 Mamaroneck Rd, Scarsdale, NY 10583 • T: +1 914 7221289 • F: +1 914 7234784 • www.scarsdale.com/weinberg.asp •
Pres.: *Lois Weiss* • Dir.: *Walter D. Terrell* •
Natural History Museum – 1958
Rocks, shells, leaves 52955

Schaefferstown PA

Historic Schaefferstown, 106 N Market St, Schaefferstown, PA 17088 • T: +1 717 9492444 • info@hsimuseum.org • www.hsimuseum.org •
Pres.: *James Dibert* •
Agriculture Museum / Local Museum – 1966
Local hist 52956

Schaumburg IL

International Sculpture Park, The Chicago Athenaeum, c/o Robert O. Atcher Municipal Center, 190 S Roselle Rd, Schaumburg, IL 60193 • T: +1 847 8953950 • www.chi-athenaeum.org •
Dir.: *Christian K. Narkiewicz-Laine* •
Fine Arts Museum – 1988 52957

Motorola Museum, 1297 E Algonquin Rd, Schaumburg, IL 60196 • T: +1 847 5766400 • F: +1 847 5766401 • asd003@email.mot.com •
Dir.: *Sharon Darling* •
Science&Tech Museum – 1991 52958

Volkening Heritage Farm at Spring Valley, 1111 E Schaumburg Rd, Schaumburg, IL 60194 • T: +1 847 9852100 • F: +1 847 9859692 • www.parkfun.com •
C.E.O.: *Jean Schlinkmann* •
Natural History Museum / Agriculture Museum / Historical Museum – 1983
Living history farm, German-American farmlife of the 1880s 52959

Schenectady NY

Mandeville Gallery, Union College, Union St, Schenectady, NY 12308 • T: +1 518 3886729 • F: +1 518 3888340 • seligman@union.edu • www.union.edu/gallery •
Dir.: *Rachel Seligman* •
Public Gallery 52960

Schenectady County Historical Society Museum, 32 Washington Av, Schenectady, NY 12305 • T: +1 518 3740263 • F: +1 518 3740263 +1 call first • librarian@schist.org • www.schist.org •
Pres.: *Kim Mabee* •
Local Museum – 1905
County hist, Shaker items, Indian artifacts, dec arts 52961

Schenectady Museum, 15 Nott Terrace Heights, Schenectady, NY 12308 • T: +1 518 3827890 • F: +1 518 3827893 • schdymuse@schenectadymuseum.org • www.schenectadymuseum.org •
Exec. Dir.: *Lee Scott Theisen* • Dir.: *Stephanie Przybylek* (Collections) •
Local Museum – 1934
Technological and electrical industry, 19th and 20th c textiles and costumes, local hist, photograph coll 52962

Schoharie NY

Old Stone Fort Museum Complex, 145 Fort Rd, Schoharie, NY 12157 • T: +1 518 2957192 • F: +1 518 2957187 • office@schohariehistory.net • www.schohariehistory.net •
Dir.: *Carle J. Kopecky* • Cur.: *Daniel J. Beams* •
Local Museum / Historical Museum – 1889
Local history, American war for independence – Library, archives 52963

Schoharie Colonial Heritage Association Museum, 1743 Palatine House, Spring St, Schoharie, NY 12157 • T: +1 518 2957505 • F: +1 518 2956001 • scha@midtel.net • www.midtel.net/~scha •
C.E.O. & Pres.: *Judith Warner* •
Local Museum – 1963 ... 52964

Schwenksville PA

Pennypacker Mills, 5 Haldeman Rd, Schwenksville, PA 19473 • T: +1 610 2879349 • F: +1 610 2879657 • pennypackermills@mail.montcopa.org • www.montcopa.org/culture/history.htm •
Dir.: *Ella Aderman* •
Science&Tech Museum – 1981
Historic house ... 52965

Scituate MA

Scituate Historic Sites, 43 Cudworth Rd, Scituate, MA 02066 • T: +1 781 5451083 • F: +1 781 5441249 • director@scituatehistoricalsociety.org • www.scituatehistoricalsociety.org •
Pres.: *David Ball* •
Local Museum – 1916
Local history, farming, GAR Hall, lighthouses, mill ... 52966

Scituate Maritime and Irish Mossing Museum, The Driftway, Scituate, MA 02066 • T: +1 781 5452788, 5451083 • F: +1 781 5441249 • director@scituatehistoricalsociety.org • www.scituatehistoricalsociety.org/sites_maritime.html •
Cur.: *Eugene N. Trainor* •
Historical Museum – 1997
Maritime hist since the 17th c, North River shipbuilding, coast disasters, Irish mossing, lifesaving services ... 52967

Scobey MT

Daniels County Museum and Pioneer Town, 7 County Rd, Scobey, MT 59263 • T: +1 406 4875965 • F: +1 406 4875559 • papegg@nemontel.net • www.scobey.org •
Pres.: *Edgar Richardson* • Dir.: *Paul Richardson* • *Mike Thieven* •
Local Museum – 1965
House, bank, blacksmith shop, service station, sallon, shack, fire hall, funeral home, school ... 52968

Scotia CA

Scotia Museum, 125 Main St, Scotia, CA 95565 • T: +1 707 7642222 • F: +1 707 7644150 •
Dir.: *Robert Manne* •
Historical Museum – 1959
Logging and lumber industry, Pacific Lumber Company ... 52969

Scotia NY

Flint House, 421 Reynolds St, Scotia, NY 12302 • T: +1 518 3778799, 3741071 • F: +1 518 3740542 •
Local Museum – 1997
Village hist ... 52970

Scotia-Glenville Children's Museum, Traveling Museum, 303 Mohawk Av, Scotia, NY 12302 • T: +1 518 3461764 • F: +1 518 3776593 • dbennett@thetravelingmuseum.org • www.travelingmuseum.org •
Pres.: *Dr. Frank B. Strauss* • C.E.O.: *Nancy Lamb* •
Special Museum – 1978 ... 52971

Scotland SD

Scotland Heritage Chapel and Museum, 811 6th St, Scotland, SD 57059 • T: +1 605 5832344 •
Pres.: *Agnes Thum* • Cur.: *Marvin Thum* •
Local Museum / Religious Arts Museum – 1976
Local and religious hist ... 52972

Scott AR

Plantation Agriculture Museum, 4815 Hwy 161, Scott, AR 72142, mail addr: POB 87, Scott, AR 72142 • T: +1 501 9611409 • F: +1 501 9611579 • plantationagrimuseum@arkansas.com • www.arkansasstateparks.com/plantationagriculturemuseum •
Dir.: *Ben H. Swadley* • Cur.: *Randy Noah* •
Agriculture Museum / Local Museum – 1989
Cotton agriculture, plantation life, local hist ... 52973

Toltec Mounds Archeological State Park, 490 Toltec Mounds Rd, Scott, AR 72142 • T: +1 501 9619442 • F: +1 501 9619221 • toltecmounds@arkansas.com • www.ArkansasStateParks.com •
Park Sup.: *Stewart Carlton* •
Archaeology Museum – 1975
Human use of environment 1,000 yrs ago – lab ... 52974

Scottdale PA

West Overton Museums, West Overton Village, Scottdale, PA 15683 • T: +1 724 8877910 • F: +1 724 8825010 • womuseum@westol.com • www.westovertonmuseum.org •
Pres.: *Susan U. Endersbe* • Dir.: *Rodney Sturtz* •
Open Air Museum / Local Museum – 1928
Rural industrial village ... 52975

Scottsdale AZ

Rawhide Old West Museum, 23023 N Scottsdale Rd, Scottsdale, AZ 85255 • T: +1 602 5025600 • F: +1 602 5021301 • www.rawhide.com •
Dir.: *Victor Ostrow* •
Local Museum – 1972
Local hist ... 52976

Scottsdale Museum of Contemporary Art, 7374 E Second St, Scottsdale, AZ 85251 • T: +1 480 8744610 • F: +1 480 8744655 • smoca@scarts.org • www.smoca.org •
Dir.: *Susan Crane* • Cur.: *Marilu Knode* •
Fine Arts Museum – 1975/1999
Contemporary art, architecture, design – sculpture garden ... 52977

Sylvia Plotkin Judaica Museum, 10460 N 56th St, Scottsdale, AZ 85253 • T: +1 480 9510323 • museum@templebethisrael.org • www.spjm.org •
Dir.: *Pamela Levin* •
Religious Arts Museum – 1966
Religious antiques, housed in Temple belonging to oldest Jewish Congregation in the Phoenix area ... 52978

Scranton PA

The Catlin House Museum, 232 Monroe Av, Scranton, PA 18510 • T: +1 717 3443841 • F: +1 717 3443815 • lhs@albright.org • www.lackawannahistory.org •
Pres.: *Alan Sweeney* • Exec. Dir.: *Maryann Moran* •
Local Museum – 1886
Local hist ... 52979

Everhart Museum of Natural History, Science and Art, 1901 Mulberry St, Nay Aug Park, Scranton, PA 18510-2390 • T: +1 570 3467186 • F: +1 570 3460652 • exdir@everhart.museum.org • Everhart-museum.org •
Dir.: *Dr. Michael C. Illuzzi* • Ass. Dir.: *Julie Orloski* • Cur.: *Bruce Lanning* (Art) •
Fine Arts Museum / Natural History Museum / Science&Tech Museum – 1908
Art, science and natural hist ... 52980

Houdini Museum, 1433 N Main Av, Scranton, PA 18508 • T: +1 570 3425555 • magicusa@microserve.net • www.houdini.org •
C.E.O.: *John Bravo* •
Performing Arts Museum – 1992
Houdini's life theater, film and artifacts ... 52981

Marywood University Art Galleries, 2300 Adams Av, Scranton, PA 18509 • T: +1 717 3486278 • F: +1 717 3406023 • gallery@marywood.edu • www.marywood.edu/www2/galleries •
Pres.: *Mary Reap* • Dir.: *Sandra Ward Povse* •
Fine Arts Museum / University Museum – 1924
Art hist ... 52982

Pennsylvania Anthracite Heritage Museum, 22 Bald Mountain Rd, Scranton, PA 18504 • T: +1 570 9634804, 9634845 • F: +1 570 9634194 • www.anthracitemuseum.org •
Dir.: *Chester Kulesa* • Cur.: *Richard Stanislaus* •
Historical Museum – 1970
History ... 52983

Steamtown National Historic Site, 150 S Washington Av, Scranton, PA 18503 • T: +1 570 3405200 • F: +1 570 3405309 • stea_superintendent@nps.gov • www.nps.gov/stea •
Dir.: *Dr. Harold H. Hagen* • Cur.: *Ella S. Rayburn* •
Science&Tech Museum – 1986
Transportation and technology ... 52984

Sea Cliff NY

Sea Cliff Village Museum, 95 Tenth Av, Sea Cliff, NY 11579 • T: +1 516 6710090 • F: +1 516 6712530 • seacliffmuseum@aol.com •
Dir.: *Sara Reres* •
Historical Museum – 1979
Local hist ... 52985

Seabrook TX

Bay Area Museum, 5000 Nasa Rd I, Seabrook, TX 77586 • T: +1 281 3265950 • F: +1 281 3265950 •
Dir.: *Joy Smitherman* •
Local Museum – 1984
Hist of the area from late 19th c to the development of NASA ... 52986

Seaford DE

Governor Ross Plantation, 1101 N Pine St, Seaford, DE 19973 • T: +1 302 6289500 • F: +1 302 6282894 • www.seafordhistoricalsociety.com •
Pres./Dir.: *Earl B. Tull* •
Local Museum – 1977
Local hist, decorative arts ... 52987

Seaford NY

Seaford Historical Museum, 3890 Waverly Av, Seaford, NY 11783 • T: +1 516 8261150 •
Pres.: *Joshua Soren* •
Local Museum – 1968
Farm tools, local hist, maritime artifacts ... 52988

Seagraves TX

Loop Museum and Art Center, 201 Main, Seagraves, TX 79359 • T: +1 806 5462810 • F: +1 806 5462810 •
Dir.: *Bernell Thompson* •
Local Museum / Fine Arts Museum – 1974
Barbed wire coll, Indian artifacts, military ... 52989

Seagrove NC

North Carolina Pottery Center, 250 East Av, Seagrove, NC 27341-0531 • T: +1 336 8738430 • F: +1 336 8738530 • ncpc@atomic.net • www.ncpotterycenter.com •
C.E.O.: *Denny Hubbard Mecham* •
Decorative Arts Museum – 1991
North Carolina pottery, southern pottery, tools ... 52990

Searchlight NV

Searchlight Historic Museum and Mining Park, 200 Michael Wendell Way, Searchlight, NV 89046 • T: +1 702 2971642 •
Local Museum – 1989
Mining hist and artifacts, geological coll, furnishings, personal artifacts and photographs, Clara Bow hats ... 52991

Searsport ME

Penobscot Marine Museum, 5 Church St, Searsport, ME 04974-0498 • T: +1 207 5482529 • F: +1 207 5482520 • museumoffices@penobscotmarinemuseum.org • www.penobscotmarinemuseum.org •
Dir.: *T. M. Deford* • Cur.: *Benjamin A. G. Fuller* •
Historical Museum – 1936
Maritime hist, 3 sea captain's homes and Town Hall, paintings – Phillips library, Carver gallery ... 52992

Seattle WA

Blenko Museum of Seattle, 222 Westlake Av N, Seattle, WA 98109 • T: +1 206 6283117 • sherril2.user.msu.edu/blenko/blenko_museum.htm •
Fine Arts Museum
Hand-blown designer American glass from the Blenko Company of Milton ... 52993

Burke Museum of Natural History and Culture, 17th Av NE and NE 45th St, Campus, Seattle, WA 98105, mail addr: c/o University of Washington, POB 353010, Seattle, WA98195-3010 • T: +1 206 5437907 • F: +1 206 6853039 • recept@u.washington.edu • www.burkemuseum.org •
Dir.: *Dr. George MacDonald* • Sc. Staff: *Dr. Stevan Hawell* (Asian Ethnology) • *Dr. Sievert Rohwer* (Birds) • *Dr. Robin K. Wright* (Native American Art) • *Marvin Oliver* (Contemporary Native American Art) • *Dr. James Nason* (Pacific and American Ethnology) • *Dr. Elizabeth Nesbitt* (Invertebrate Paleontology) • *Dr. Ted Pietsch* (Fish) • *Dr. James Kenagy* (Mammals) • *Dr. Richard Olmstead* (Herbarium) • *Dr. Peter Lape* (Archaeology) •
Ethnology Museum / Natural History Museum / Folklore Museum – 1885
Pacific Rim anthropology, natural hist, Northwest Coast native art ... 52994

Center for Wooden Boats, 1010 Valley St, Seattle, WA 98109 • T: +1 206 3822628 • F: +1 206 3822699 • cwb@cwb.org • www.cwb.org •
Pres.: *Alex Bennet* • Exec. Dir.: *Elizabeth Davis* •
Science&Tech Museum – 1978
Boats ... 52995

Center on Contemporary Art, 410 Dexter Av, Seattle, WA 98109 • T: +1 206 7281980 • F: +1 206 7281980 • coca@cocaseattle.org • www.cocaseattle.org •
Dir.: *Don Hudgins* •
Fine Arts Museum
Contemporary art ... 52996

The Children's Museum, 305 Harrison St, Seattle, WA 98109 • T: +1 206 4411768 • F: +1 206 4480910 • tcm@thechildrensmuseum.org • www.thechildrensmuseum.org •
C.E.O.: *K.C. Gauldine* •
Special Museum – 1981 ... 52997

Experience Music Project, 325 5th Av N Seattle, WA 98109, mail addr: 2910 Third Av, Seattle, WA 98121 • T: +1 206 7702700 • F: +1 206 7702727 • www.emplive.com •
C.E.O.: *Bob Santelli* •
Music Museum – 1999
Jimi Hendrix, guitar, roots of rock'n'roll, hip-hop, punk, reggae ... 52998

Frye Art Museum, 704 Terry Av, Seattle, WA 98104 • T: +1 206 6229250 • F: +1 206 2231707 • info@fryemuseum.org • www.fryeart.org •
Pres.: *Richard L. Cleveland* • Interim Dir.: *Midge Bowman* • Sc. Staff: *Donna Kovalenko* (Collections) •
Fine Arts Museum – 1952
19th and 20th c European and American painting ... 52999

Henry Art Gallery, University of Washington, 15th Av NE and NE 41st St, Seattle, WA 98195, mail addr: Box 351410, Seattle, WA 98195-1410 • T: +1 206 5432280/81 • F: +1 206 6853123 • hartg@u.washington.edu • www.henryart.org •
Dir.: *Richard Andrews* • Cur.: *Elizabeth A. Brown* • *Judy Sourakli* (Collections) •
Fine Arts Museum / University Museum – 1927
19th and 20th c European and American paintings and prints, textiles and costumes ... 53000

King County Arts Commission Gallery, Smith Tower, Rm 1115, 506 Second Av, Seattle, WA 98104-2311 • T: +1 206 2968671 • F: +1 206 2968629 • jimkelly@metrokc.gov •
Dir.: *Jim Kelly* •
Public Gallery – 1967 ... 53001

Klondike Gold Rush National Historical Park, 117 S Main, Seattle, WA 98104 • T: +1 206 5537220 • F: +1 206 5530614 • klse_Ranger_Activities@nps.gov • www.nps.gov/klse •
Head: *Willie Russell* •
Historical Museum – 1979
History ... 53002

Memory Lane Museum at Seattle Goodwill, 1400 S Lane St, Seattle, WA 98144 • T: +1 206 3291000 • F: +1 206 7261502 • goodwill@seattlegoodwill.org •
Pres.: *Jill Jones* •
Local Museum – 1968
Local hist ... 53003

Museum of Flight, 9404 E Marginal Way S, Seattle, WA 98108 • T: +1 206 7645720 • F: +1 206 7645707 • info@museumofflight.org • www.museumofflight.org •
Pres.: *Ralph A. Bufano* • Cur.: *Dennis Parks* •
Science&Tech Museum / Military Museum / Historical Museum – 1964
Aeronautics, historic aircrafts, garments, flight test instr, model aircraft, space flight artifacts ... 53004

Museum of History and Industry, 2700 24th Av, Seattle, WA 98112 • T: +1 206 3241126 • F: +1 206 3241346 • information@seattlehistory.org • www.seattlehistory.org •
Dir.: *Leonard Garfield* • Cur.: *Elizabeth Furlow* (Collections) •
Historical Museum – 1914
Photos, mid 19th-20th c costumes, decorative arts, folk art, furniture, tools and equip of maritime, fishing, agriculture, logging, mining ... 53005

Nordic Heritage Museum, 3014 NW 67th St, Seattle, WA 98117 • T: +1 206 7895707 • F: +1 206 7893271 • nordic@intelistep.com • www.nordicmuseum.com •
Dir.: *Marianne Forssblad* • Cur.: *Lisa Hill-Festa* •
Special Museum – 1979
Local hist, nordic hist, nordic american art, folk costumes, woodcarvings, loggings, fishing and mining equip, furniture and household items ... 53006

NSCC Art Gallery, North Seattle Community College, 9600 College Way N, Instructional Bldg, Rm 1322, Seattle, WA 98103 • T: +1 206 5273709 • F: +1 206 5273784 • echriste@sccd.ctc.edu • www.northseattle.edu/services/art.htm •
Dir.: *Elroy Christenson* •
Public Gallery ... 53007

Odyssey Maritima Discovery Museum, 2205 Alaskan Way, Pier 66, Seattle, WA 98121 • T: +1 206 3744000 • F: +1 206 3744002 • info@ody.org • www.ody.org •
Science&Tech Museum – 1998 ... 53008

Pacific Arts Center, 1500 Lakeview Blvd E, Seattle, WA 98102 • T: +1 206 3292722 • F: +1 206 3297554 •
Pres.: *Marty Spiegel* •
Fine Arts Museum – 1983
Art ... 53009

Pacific Science Center, 200 2nd Av N, Seattle, WA 98109 • T: +1 206 4432001 • F: +1 206 4433631 • webmaster@pacsci.org • www.pacsci.org •
Pres. & C.E.O.: *R. Bryce Seidl* •
Science&Tech Museum – 1962
Science, technology ... 53010

Sacred Circle Gallery of American Indian Art, United Indians of All Tribes Foundation, Discovery Park, Daybreak Star Art and Cultural Center, Seattle, WA 98101, mail addr: POB 99100, Seattle, WA 98199 • T: +1 206 2854425 • F: +1 206 2823640 • info@unitedindians.com • www.unitedindians.com •
Dir.: *Steve Charles* • *Merlee Markishtum* •
Fine Arts Museum – 1977
Coll of Native American traditional and contemporary artwork and murals, incl works by Nathan Jackson, Robert Montoya, George Morrison, Jimmie Fife and T.C. Cannon, contemporary art ... 53011

The Science Fiction Museum and Hall of Fame, 325 5th Av, Seattle, WA 98109 • T: +1 206 7243428 • F: +1 206 7072727 • info@sfhomeworld.org • www.sfhomeworld.org •
Dir.: *Donna Shirley* •
Special Museum ... 53012

Seattle Art Museum, 100 University St, Seattle, WA 98101-2902 • T: +1 206 6258900 • F: +1 206 6543135 • webmaster@seattleartmuseum.org • www.seattleartmuseum.org •
Dir.: *Mimi Gardner Gates* • Cur.: *Lisa Corrin* (Modern Art) • *Barbara Brotherton* (Native American Art) • *Pamela McClusky* (Art of Africa and Oceania) • *Ruth J. Nutt* (Art of Africa and Oceania) • *Julie Emerson* (Decorative Arts) • *Chiyo Ishikawa* (European Painting) • *Yukiko Shirahara* (Asian Art) •
Fine Arts Museum – 1917

Asian art, Katherine White Collections of African Art, ancient American and Oceanic art, modern and contemporary art, ancient and middle age paintings 53013

Seattle Asian Art Museum, Volunteer Park, 1400 E Prospect, Seattle, WA 98112-3303 • T: +1 206 6258900 • F: +1 206 6543191 • webmaster@seattleartmuseum.org • www.seattleartmuseum.org • Dir.: *Mimi Gardner Gates* • Cur.: *Yukiko Shirahra* (Asian Art) • *Hsue-Man Shen* (Chinese Art) •
Fine Arts Museum – 1917 53014

Shoreline Historical Museum, 749 N 175th St, Seattle, WA 98133 • T: +1 206 5427111 • F: +1 206 5424645 • shorelinehistorical@juno.com • shorelinehistoricalmuseum.org •
Pres.: *Peggy Gerdes* • Exec.Dir. & C.E.O.: *Victoria Stiles* •
Local Museum – 1976
Local hist, memorabilia of old homes – school room, blacksmith shop, Old Richmond Beach post office, country store and farm 53015

Virginia Wright Fund, 407 Dexter Av N, Seattle, WA 98109 • T: +1 206 2648200 •
Fine Arts Museum 53016

Wing Luke Asian Museum, 407 Seventh Av, Seattle, WA 98104 • T: +1 206 6235124 • F: +1 206 6234559 • folks@wingluke.org • www.wingluke.org •
Dir.: *Ron Chew* •
Fine Arts Museum / Ethnology Museum / Historical Museum – 1967
Asian Pacific American hist, art and culture 53017

Sebago ME

The Jones Museum of Glass and Ceramics, 35 Douglas Mountain Rd, Sebago, ME 04029 • T: +1 207 7873370 • F: +1 207 7872800 •
Dir.: *John H. Holverson* •
Decorative Arts Museum – 1978
Glass, ceramics 53018

Sebewaing MI

Luckhard Museum, The Indian Mission, 612 E Bay St, Sebewaing, MI 48759 • T: +1 989 8832539, 8833730 •
Dir.: *Jim Bunke* •
Religious Arts Museum – 1957
Historic house, mission home, religious items – library 53019

Sebring FL

Civilian Conservation Corps Museum, Highlands Hammock State Park, 5931 Hammock Rd, Sebring, FL 33872 • T: +1 863 3866094 • F: +1 863 3866095 • www.floridastateparks.org •
Head: *Peter Anderson* •
Historical Museum / Natural History Museum – 1994
Housed in 1930s bldg constructed of heavy native timbers, lumber cut and fabricated 53020

Highlands Museum of the Arts, 351 W Center Av, Sebring, FL 33870 • T: +1 863 3855312 • F: +1 863 3855336 • haleague@strato.net • www.highlandsartleague.com •
Dir.: *Alice Stroppel* •
Fine Arts Museum – 1986
Art 53021

Second Mesa AZ

Hopi Cultural Center Museum, Rte 264, Second Mesa, AZ 86043 • T: +1 928 7342401 • hopi@psv.com • www.psv.com/hopi.html •
Folklore Museum – 1970
Silver, baskets, weaving, pottery, Kachina dolls 53022

Sedalia MO

Daum Museum of Contemporary Art, c/o State Fair Community College, 3201 W 16th St, Sedalia, MO 65301 • T: +1 660 5305888 • F: +1 660 5305890 • info@daummuseum.org • www.daummuseum.org •
Dir.: *Douglass Freed* •
Fine Arts Museum – 2001
Works from the last third on the 20th c 53023

Pettis County Historical Society, c/o Sedalia Public Library, 200 Dundee, Sedalia, MO 65301 • T: +1 660 8261314 • F: +1 660 8260396 •
Pres.: *Rhonda Chalfant* •
Association with Coll – 1943
Military, war agriculture, farm life, costumes 53024

Sedan KS

Emmett Kelly Historical Museum, 202 E Main, Sedan, KS 67361 • T: +1 316 7253470 • www.emmettkellymuseum.com •
Head: *Roger Floyd* •
Local Museum / Performing Arts Museum – 1967
Clowns, history, local history 53025

Sedgwick ME

Sedgwick-Brooklin Historical Museum, Rte 172, Sedgwick, ME 04676 • T: +1 207 3594447 • oblong@hypernet.com • www.mainemuseums.org •
Pres.: *John Bishof* •
Local Museum – 1963
Regional hist, tools, toys, photos 53026

Sedona AZ

Sedona Arts Center, Hwy 89a at Art Barn Rd, Sedona, AZ 86336, mail addr: POB 569, Sedona, AZ 86339 • T: +1 928 2823809 • F: +1 928 2821516 • sac@sedona.net • www.sedonaartscenter.com •
Exec. Dir.: *Michael McKitterick* •
Fine Arts Museum – 1961 53027

Seguin TX

Fiedler Memorial Museum, Texas Lutheran University, Seguin, TX 78155 • T: +1 830 3728038 • F: +1 830 3728188 •
Dir.: *Evelyn Fiedler Streng* •
Natural History Museum / Science&Tech Museum – 1973
Geology 53028

Los Nogales Museum, 415 S River, Seguin, TX 78155 • T: +1 830 3722649 • bj2bt@axs4u.net •
Dir.: *B.J. Comingore* •
Local Museum – 1952
Hist items, books, court minutes, early photos, doll house 53029

Selinsgrove PA

Lore Degenstein Gallery, Susquehanna University, 514 University Av, Selinsgrove, PA 17870-1001 • T: +1 570 3724059, 3740101 • F: +1 570 3722775 • livingst@susqu.edu • www.susqu.edu/art_gallery •
Dir./Cur.: *Dr. Valerie Livingston* •
Fine Arts Museum / University Museum – 1993
Art 53030

Selkirk NY

Bethlehem Historical Association Museum, 1003 River Rd, Selkirk, NY 12158 • T: +1 518 7679432 •
Pres.: *Parker D. Mathusa* •
Local Museum – 1965
Farming tools, ice-harvesting, fruit-growing tools, railroding 53031

Selma AL

Sturdivant Hall, 713 Mabry St, Selma, AL 36702 • T: +1 334 8725626 • info@sturdivanthall.com • www.sturdivanthall.com •
Dir.: *Manera S. Searcy* • Asst. Director: *Nancy M. Gantt* •
Local Museum – 1957
Historic house, antebellum furnishings, art collection 53032

Senatobia MS

Heritage Museum Foundation of Tate County, POB 375, Senatobia, MS 38668 • T: +1 622 5628715 • dperkins@gmi.net •
Pres.: *Deborah Perkins* •
Local Museum – 1977
History, agriculture, art, folklore 53033

Seneca SC

World of Energy at Keowee-Toxaway, 7812 Rochester Hwy, Seneca, SC 29672 • T: +1 864 8854600 • F: +1 864 8854605 •
Science&Tech Museum – 1969
Energy and electricity 53034

Seneca Falls NY

National Women's Hall of Fame, 76 Fall St, Seneca Falls, NY 13148 • T: +1 315 5688060 • F: +1 315 5682976 • womenshall@aol.com • www.greatwomen.org •
Head: *Billy Luisi-Potts* •
Local Museum – 1969 53035

Seneca Falls Historical Society Museum, 55 Cayuga St, Seneca Falls, NY 13148 • T: +1 315 5688412 • F: +1 315 5688426 • sfhs@flare.ner • www.sfhistoricalsociety.org •
C.E.O. & Exec.Dir.: *Philomena M. Cammuso* • Dir.: *Frances T. Barbieri* (Education) • Pres.: *Rebecca Holden* •
Local Museum – 1904
Local hist, 19th c American dec and fine art, pre-Industrial tools 53036

Sequim WA

Museum and Arts Center in the Sequim Dungeness Valley, 175 W Cedar, Sequim, WA 98382 • T: +1 360 6838110 • F: +1 360 6812353 • info@sequimmuseum.org • www.sequimmuseum.org •
Exec. Dir.: *Randy Sturgis* • Pres.: *Al Courtney* •
Fine Arts Museum / Decorative Arts Museum / Historical Museum – 1977 53037

Setauket NY

Gallery North, 90 N Country Rd, Setauket, NY 11733 • T: +1 631 7512676 • F: +1 631 7510180 • gallerynorth@aol.com • www.gallerynorth.org •
Pres.: *Elizabeth Goldberg* • Dir./Cur.: *Collen W. Hanson* •
Public Gallery – 1965
Contemporary works by Long Island artists and craftsmen 53038

Three Village Historical Society Museum, 93 N Country Rd, Setauket, NY 11733 • T: +1 631 7513730 • F: +1 631 7513936 • TVHistSoc@aol.com • www.tvhs.org •
Dir.: *Margaret Conover* •
Local Museum – 1964
Local hist 53039

Sevierville TN

Tennessee Museum of Aviation, 135 Air Museum Way, Sevierville, TN 37862 • T: +1 865 9089372, 9080171 ext 24 • F: +1 865 9088421 • bminter@tnairmuseum.com • www.tnairmuseum.com •
C.E.O. & Chm.: *R. Neal Melton* • Cur.& Asst.Dir.: *Tom Walker* •
Science&Tech Museum – 2000
Aviation hist from early flights to present, WW II combat aircraft, uniforms 53040

Sewanee TN

University Art Gallery of the University of the South, Guerry Hall, Georgia Av, Sewanee, TN 37383-1000 • T: +1 931 5981223 • F: +1 931 5983335 • jmgill@sewanne.edu • www.sewanee.edu/gallery •
Dir.: *Arlyn Ende* •
Fine Arts Museum / University Museum – 1965
Paintings, prints, drawings, graphics, sculpture, photos, furniture, silver, stained glass 53041

Seward AK

Seward Museum, 336 Third Av, Seward, AK 99664 • T: +1 907 2243902 • www.seward.com/directory •
Pres./Dir.: *Lee E. Poleske* •
Local Museum – 1967
Local photos and artifacts 53042

Seward NE

Marxhausen Art Gallery, Concordia University, 800 N Columbia Av, Seward, NE 68434 • T: +1 402 6433651 • F: +1 402 6434073 • jbockelman@seward.cune.edu • www.cune.edu •
Dir.: *James Bockelman* •
Fine Arts Museum / University Museum – 1951
Contemporary print multiples 53043

Sewickley PA

International Images, 514 Beaver St, Sewickley, PA 15143 • T: +1 412 7413036 • F: +1 412 7418606 • intimage@cobweb.net • www.intimage.com •
Pres.: *Elena Kornetchuk* • Dir.: *Charles M. Wiebe* •
Fine Arts Museum – 1978
Contemparary art from Russia, Georgia, Latvia, Lithuania, Estonia, Bulgaria, Cuba, US a.o. 53044

Seymour WI

Seymour Community Museum, Depot St, Seymour, WI 54165 • T: +1 414 8332868 •
Pres.: *Rita Gosse* •
Local Museum – 1976
Local hist 53045

Shady Side MD

Captain Salem Avery House Museum, 1418 EW Shadyside Rd, Shady Side, MD 20764 • T: +1 410 8674486 • F: +1 410 8674486 • captainavery@prodigy.net • www.averyhouse.org •
C.E.O.: *Janet Surrett* •
Historical Museum – 1988
Furniture, tools, 1860-1920 boats, artifacts, photographs, oral hist 53046

Shafter CA

Minter Field Air Museum, 401 Vultee St, Shafter, CA 93263, mail addr: PO Box 445, Shafter, CA 93263 • T: +1 661 3930291 • mfam@minterfieldairmuseum.com • www.minterfieldairmuseum.com •
C.E.O.: *James T. Whitehead* • Cur.: *Dr. David Day* •
Military Museum / Science&Tech Museum – 1981
WW II aircraft (BT-13, L-3B, Fokker DR.1 replica, Pietenpol Homebuild (1928 design); and vehicles (2 Jeeps, 1939 Staff car, 1945 Fire Truck); historical artifacts – Library of books and over 750 hrs of video on DVD 53047

Shaftsbury VT

Shaftsbury Historical Society, Historic 7a, Shaftsbury, VT 05262 • T: +1 802 4477488 •
C.E.O./Cur.: *Robert J. Williams* •
Association with Coll – 1967
Hist docs, decorative arts, crafts, industries 53048

Shaker Heights OH

Shaker Historical Society Museum, 16740 S Park Blvd, Shaker Heights, OH 44120 • T: +1 216 9211201 • F: +1 216 9212615 • shakhist@wviz.org • www.cwru.edu/affil/shakhist/shaker.htm •
Pres.: *Dr. Robert Spurney* • Dir.: *Cathie Winans* •
Local Museum – 1947
History of North Union and other Shaker communities 53049

Sharon MA

Kendall Whaling Museum, 27 Everett St, Sharon, MA 02067 • T: +1 781 7845642 • F: +1 781 7840451 • ehazen@kwm.org • www.kwm.org •
Dir.: *Stuart M. Frank* • Cur.: *Michael P. Dyer* •
Natural History Museum – 1956
History, maritime history, photos 53050

Sharon NH

Sharon Arts Center, 457 Rte 123, Sharon, NH 03458-9014 • T: +1 603 9247256 • F: +1 603 9246074 • sharonarts@sharonarts.org • www.sharonarts.org •
Pres.: *Elizabeth Rank-Beauchamp* • Cur.: *Randall Hoel* •
Fine Arts Museum / Public Gallery – 1947
Art hist and craft 53051

Sharonville OH

Heritage Village, 11450 Lebanon Pike, Rte 42, Sharonville, OH 45241 • T: +1 513 5639484 • F: +1 513 5630914 • bspitler@heritagevillagecincinnati.org • www.heritagevillagecincinnati.org •
Dir.: *Bing G. Spitler* •
Local Museum – 1964
Historic village, 19th c furniture and dec arts, farm equip and vehicles 53052

Sharpsburg MD

Antietam National Battlefield, Rte 65 N, Sharpsburg, MD 21782 • T: +1 301 4325124 • F: +1 301 4324590 • keith-snyder@nps.gov • www.nps.gov/anti •
Dir.: *Terry Carlstrom* •
Historic Site / Military Museum – 1890
Site of 1862 Civil War Maryland Campaign and battle of Antietam or Sharpsburg 53053

Chesapeake and Ohio Canal Tavern Museum, 16500 Shepherdstown Park, Sharpsburg, MD 21782 • T: +1 301 7394200 • F: +1 301 7142232 • doug.stover@nps.gov • www.nps.gov/choh •
Head: *Douglas D. Faris* •
Historical Museum – 1954
Great Falls Tavern, brick railroad station, with operating steam locomotive ride, Ferry Hill slave plantation 53054

Shaw Island WA

Shaw Island Historical Society Museum, Blind Bay Rd, Shaw Island, WA 98286 • T: +1 360 4684068 •
Cur.: *Sherie Christiansen* • *Marilyn Hoffman* •
Local Museum – 1966
Local hist 53055

Shawano WI

Shawano County Historical Society Museum, 524 N Franklin Str, Shawano, WI 54166, mail addr: POB 534, Shawano, WI 54166-0534 • T: +1 715 5269033 •
Local Museum – 1940 53056

Shawnee KS

Johnson County Museums, 6305 Lackman Rd, Shawnee, KS 66217 • T: +1 913 6316709 • F: +1 913 6316359 • jcmuseum@jocogov.org • www.jocomuseum.org •
Cur.: *Anne Marvin* •
Local Museum – 1967
Hist of 1950's suburbia 53057

Old Shawnee Town, 11501 W 57th St, Shawnee, KS 66203 • T: +1 913 2482360 • F: +1 913 2482363 • lcasey@cityofshawnee.org • www.cityofshawnee.org •
C.E.O.: *Laura Casey* •
Local Museum – 1966
Hist from 1840s to present 53058

Shawnee OK

Mabee-Gerrer Museum of Art, Saint Gregory's University, 1900 W MacArthur Dr, Shawnee, OK 74801-2499 • T: +1 405 8785300 • F: +1 405 8785198 • info@mabeegerrermuseum.org •
Dir.: *Debby Williams* •
Fine Arts Museum – 1917
Artifacts from ancient civilizations 53059

Sheboygan WI

Bradley Gallery of Art, Lakeland College, POB 359, Sheboygan, WI 53082-0359 • T: +1 920 5651280 •
Public Gallery / University Museum
Changing exhibitions of students works 53060

John Michael Kohler Arts Center, 608 New York Av, Sheboygan, WI 53081 • T: +1 920 4586144 • F: +1 920 4584473 • www.jmkac.org •
Dir.: *Ruth Kohler* • Cur.: *Leslie Umberger* •
Fine Arts Museum – 1967
Contemp ceramics, paintings, installations, American crafts and furniture 53061

Sheboygan County Historical Museum, 3110 Erie Av, Sheboygan, WI 53081 • T: +1 920 4581103 • F: +1 920 4585152 •
Pres.: *Ken Conger* • Exec. Dir.: *Robert Harker* •
Historical Museum – 1954
Local and local circus hist, farm implements, household articles 53062

Sheffield Lake OH

103rd Ohio Volunteer Infantry Civil War Museum, 5501 E Lake Rd, Sheffield Lake, OH 44054 • T: +1 440 9492790 • 103ovi.org •
Pres.: *Todd Bemis* • Cur.: *Olive Gerber* • *Ruth Wagner* •
Military Museum – 1972 ... 53063

Shelburne VT

National Museum of the Morgan Horse, 122 Bostwick Rd, Shelburne, VT 05482 • T: +1 802 9858665 • F: +1 802 9855242 • morgans@together.net • www.morganmuseum.org •
Cur.: *Kathlyn Furr* •
Special Museum – 1988
Concentration on the Morgan breed of horses & peoples ... 53064

Shelburne Museum, 5555 Shelburne Rd, Shelburne, VT 05482 • T: +1 802 9853346 • F: +1 802 9852331 • info@shelburnemuseum.org • www.shelburnemuseum.org •
Pres.: *Hope Alswang* •
Local Museum – 1947
Hist bldg, American folk art, objects of everyday life ... 53065

Shelby MT

Marias Museum of History and Art, 206 12th Av N, Shelby, MT 59474 • T: +1 406 4342551 • F: +1 406 4345422 • larrydar@northerntel.net •
C.E.O. & Pres.: *Larry Munson* •
Fine Arts Museum / Historical Museum – 1963
Rocks, fossils, American Indian artifacts, military guns, bottles, toys, clothing, glassware, silver ... 53066

Shelby NC

Cleveland County Historical Museum, Court Sq, Shelby, NC 28150 • T: +1 704 4828186 • F: +1 704 4828186 •
Dir./Cur.: *Lamar Wilson* •
Local Museum – 1976
Regional hist, clothing, textiles, houseware, furniture, tools ... 53067

Shelbyville IN

Louis H. and Lena Firn Grover Museum, 52 W Broadway, Shelbyville, IN 46176 • T: +1 317 3924634 • grover@lightbound.com • www.grovermuseum.org •
C.E.O.: *June Barnett* •
Local Museum – 1980
Local history ... 53068

Shell Lake WI

Museum of Woodcarving, 539 Hwy 63, Shell Lake, WI 54871 • T: +1 715 4687100 •
Cur.: *Maria McKay* •
Fine Arts Museum – 1950 ... 53069

Washburn County Historical Society Museum, 102 W Second Av, Shell Lake, WI 54871 • T: +1 715 4682982 • washburncohhistsoceby@hotmail.com •
C.E.O. & Pres.: *Rachel Gullickson* •
Local Museum – 1954
Crafts, costumes, logging tools, local hist ... 53070

Shelter Island NY

Shelter Island Historical Society Museum, 16 S Ferry Rd, Shelter Island, NY 11964 • T: +1 516 7490025 • F: +1 516 7491825 • sihissoc@hamptons.com • www.shelterislandhistsoc.org •
Pres.: *W.M. Anderson* • Dir.: *Louise Green* •
Historical Museum – 1965
Hist of the Town – archives ... 53071

Shepherdstown WV

Historic Shepherdstown Museum, 129 E German St, Shepherdstown, WV 25443, mail addr: POB 1786, Shepherdstown, WV 25443 • T: +1 304 8760910 • F: +1 304 8762679 • hsc86@citilink.net •
C.E.O.: *Susan Smith* •
Local Museum – 1983
Local hist ... 53072

Sheridan AR

Grant County Museum, 521 Shackleford Rd, Sheridan, AR 72150 • T: +1 870 9424496 • museum4@alltel.net •
Dir.: *Mary Beth Glover-Wilson* •
Local Museum – 1970
library ... 53073

Sheridan WY

Trail End State Historic Museum, Colorado-Wyoming Association of Museums, 400 Clarendon Av, Sheridan, WY 82801 • T: +1 307 6744589 • F: +1 307 6721720 • trailend@state.wy.us • www.trailend.org •
Cur.: *Nancy McClure* •
Historical Museum – 1982
local hist, esp 1913-1933 ... 53074

Sherman TX

Ida Green Gallery, Austin College, 900 N Grande Av, Sherman, TX 75090-4440 • T: +1 903 8132251 • F: +1 903 8132273 •
Public Gallery ... 53075

Sherwood WI

High Cliff General Store Museum, 7526 N Lower Cliff Rd, Sherwood, WI 54169-0001 • T: +1 920 9891636 • russjbishop@msn.com •
Pres.: *Russell J. Bishop* •
Local Museum – 1974
Local hist, household utensils, machinery and tools, wearing ... 53076

Shevlin MN

Clearwater County Museum, Hwy 2 W, Shevlin, MN 56676, mail addr: POB 241, Bagley, MN 56621 • T: +1 218 7852000 • F: +1 218 7852440 • cchshist@gvtel.com •
C.E.O.: *James Michel* •
Local Museum – 1967
Local hist, agriculture, timber, lumbering, military, Gran church, school ... 53077

Shiloh TN

Shiloh National Military Park and Cemetery, 1055 Pittsburg Landing Rd, Shiloh, TN 38376 • T: +1 731 6895275 • F: +1 731 6895450 • shil_administration@nps.gov •
Military Museum – 1894
Military hist, Civil War hist ... 53078

Shiner TX

Edwin Wolters Memorial Museum, 306 S Av, Shiner, TX 77984 • T: +1 512 5943774, 5943362 • F: +1 512 5943566 •
Cur.: *Bernard Siegel* •
Local Museum – 1963
Local hist, industry, Indian artifacts, archaeology 53079

Shippensburg PA

Kauffman Gallery, c/o Shippensburg University, 1871 Old Main Dr, Shippensburg, PA 17257 • T: +1 717 4771530 • F: +1 717 4774049 • vhmowe@ship.edu • www.ship.edu •
Dir.: *William Q. Hynes* •
Fine Arts Museum / University Museum – 1972 . 53080

Shippensburg Historical Society Museum, 52 W King St, Shippensburg, PA 17257 • T: +1 717 5326727 • www.ship.edu/~pegill/shs.html •
Cur.: *Ed Sheaffer* •
Local Museum – 1944
Local hist – Library ... 53081

Shoreview MN

Ramsey Center for Arts, 1071 W Hwy-96, Shoreview, MN 55126 • T: +1 651 4864883 • info@ramseycfa.org • www.ramseycfa.org •
Public Gallery
Community arts ... 53082

Shreveport LA

Ark-La-Tex Antique and Classic Vehicle Museum, 601 Spring St, Shreveport, LA 71101 •
Science&Tech Museum – 1993
Automotive art, antique and classic vehicles ... 53083

Louisiana State Exhibit Museum, 3015 Greenwood Rd, Shreveport, LA 71109 • T: +1 318 6322020 • F: +1 318 6322056 • lsem@sos.louisiana.gov • www.sos.louisiana.gov •
Dir.: *Forrest Dunn* •
Historical Museum – 1937
Murals, dioramas, historical relics, agriculture, industrial, archaeology, glass, china, Indian artifacts ... 53084

Meadows Museum of Art, Centenary College, 2911 Centenary Blvd, Shreveport, LA 71104 • T: +1 318 8695169 • F: +1 318 8695730 • ddufilho@centenary.edu • www.centenary.edu •
Dir.: *Diane Dufilho* • Cur.: *Bruce Allen* •
Fine Arts Museum / University Museum – 1976
Paintings by George Grosz, Alfred Maurer, William Glackens and Ernest Lawson, Latin American works on paper, African, American, Haitian, Inuit and tribial works ... 53085

Pioneer Heritage Center, LSU Shreveport, 1 University Pl, Shreveport, LA 71115 • T: +1 318 7975332 • F: +1 318 7975395 • mplummer@pilot.1sus.edu • www.lsus.edu •
Dir.: *Marguerite R. Plummer* • Asst. Dir.: *Marvin R. Young* •
Historical Museum – 1977
Nortwest Louisiana and 19th c Red River regional hist ... 53086

R.S. Barnwell Memorial Garden and Art Center, 601 Clyde Fant Pkwy, Shreveport, LA 71101 • T: +1 318 6737703 • F: +1 318 6737707 • jhirsch@ci.shreveport.la.us •
Pres.: *Barbara White* • C.E.O.: *Jan Hirsch* •
Fine Arts Museum / Public Gallery – 1970
Paintings ... 53087

The R.W. Norton Art Gallery, 4747 Creswell Av, Shreveport, LA 71106 • T: +1 318 8654201 • F: +1 318 8690435 • gallery@rwnaf.org • www.rwnaf.org •
Pres.: *Richard W. Norton jr.* •
Fine Arts Museum – 1946
Paintings, silver, glass, miniatures, tapestries ... 53088

Sci-Port Discovery Center, 820 Clyde Fant Pkwy, Shreveport, LA 71101 • T: +1 318 4243466 • F: +1 318 2225592 • apeek@sciport.org • www.sciport.org •
Pres.: *Andre Peek* •
Science&Tech Museum – 1994
Science ... 53089

Spring Street Historical Museum, 525 Spring St, Shreveport, LA 71101 • T: +1 318 4240964 • F: +1 318 4240964 • sshm@shreve.net • www.springstreetmuseum.com •
Dir.: *Debra M. Helton* •
Local Museum – 1977
Hist of Shreveport, maps, written docs, artifacts, newspaper, illustrations ... 53090

Stephens African-American Museum, 2810 Lindlohm, Shreveport, LA 71108 • T: +1 318 6352147 • F: +1 318 6360504 •
CEO: *Spencer Stephens* •
Fine Arts Museum
African-American art, paintings, sculptures, prints and drawings ... 53091

Shrewsbury NJ

Guild of Creative Art, 620 Broad St, Rte 35, Shrewsbury, NJ 07702 • T: +1 732 7411441 • guildofcreativeart@att.net • www.guildofcreativeart.com •
Public Gallery ... 53092

Shullsburg WI

Badger Mine and Museum, 279 Estey, Shullsburg, WI 53586 • T: +1 608 9654860 •
Dir.: *Tim Strang* •
Local Museum – 1964
Local hist, old mining tools, farm equip, Indian relics ... 53093

Sibley IA

McCallum Museum, 5th St and 8th Av, Sibley, IA 51249 • T: +1 712 7543882 • verstoff@heartlandtel.com • www.osceolacountyia.com/info/museums.htm •
Dir.: *Jan Stofferan* •
Local Museum – 1956
Local history ... 53094

Sibley MO

Fort Osage, 105 Osage St, Sibley, MO 64088 • T: +1 816 7958200 ext 1260 • F: +1 816 7957938 • juligor@gw.co.jackson.mo.us • www.co.jackson.mo.us •
Dir.: *Gary Salva* •
Local Museum – 1948
Hist, military, Indians, fur trade ... 53095

Sidney MT

Mondak Heritage Center, 120 Third Av SE, Sidney, MT 59270 • T: +1 406 4823500 • F: +1 406 4823500 • mondakheritagecenter@hotmail.com •
Local Museum – 1971
Historical artifacts, original art of the western US 53096

Sidney NE

Fort Sidney Museum and Post Commander's Home, 6th Av and Jackson St, Sidney, NE 69162 • T: +1 308 2542150 •
Pres.: *Thomas LaVerne* • Dir.: *Joan Olson* •
Military Museum – 1954
Military hist ... 53097

Sidney NY

Sidney Historical Association Museum, 21 Liberty St, Room 218, Sidney, NY 13838 • T: +1 607 5638787 • www.sidneyonline.com/sha.htm •
Cur.: *Neila C. Hayes* •
Local Museum – 1945
Hist, archaeology, genealogy, Indian artifacts – archives ... 53098

Siloam Springs AR

Siloam Springs Museum, 112 N Maxwell, Siloam Springs, AR 72761, mail addr: POB 1164, Siloam Springs, AR 72761 • T: +1 501 5244011 • ssmuseum@centurytel.net • www.siloamspringsmuseum.com •
Dir.: *Donald Warden* •
Local Museum – 1969
Local history, textiles, glass, decorative arts – archives ... 53099

Silver City NM

Arts Council Gallery, 1201 Pope St, Silver City, NM 88061 • T: +1 505 5382505 • F: +1 505 5382505 • arts@mrac.cc •
Dir.: *Faye McCalmot* •
Fine Arts Museum
Art ... 53100

Francis McCray Gallery, Western New Mexico University, 1000 College Av, Silver City, NM 88062 • T: +1 505 5386614, 5386515 • F: +1 505 5386619 • mccraygallery@wnmu.edu • www.wnmu.edu •
Dir.: *Mike Metcalf* •
Fine Arts Museum / University Museum – 1960
Art ... 53101

Gila Cliff Dwelling Visitor Center, Mile Post, 43 N Hwy 15, Silver City, NM 88061, mail addr: HC 68 Box 100, Silver City, NM 88061 • T: +1 505 5369461 • F: +1 505 5369461 • www.nps.gov/gicl •
Archaeology Museum / Natural History Museum
Archaeology, botany, ethnology, history, classic Mimbres, Tularosa Mogollon, Apache ... 53102

Silver City Museum, 312 W Broadway, Silver City, NM 88061 • T: +1 505 5385921 • F: +1 505 3881096 • scmuseum@zianet.com • www.silvercitymuseum.org •
Dir.: *Susan Berry* • Cur.: *Jim Carlson* •
Local Museum – 1967
Victorian objects and artifacts, native American artifacts, photos and docs ... 53103

Western New Mexico University Museum, Fleming Hall, 10 St, Silver City, NM 88062 • T: +1 505 5386386 • F: +1 505 5386178 •
Dir.: *Dr. Cynthia Ann Bettison* • Chm.: *Dr. Dale F. Giese* •
University Museum / Ethnology Museum – 1974 53104

Silver Cliff CO

Silver Cliff Museum, 606 Main St, Silver Cliff, CO 81252 • T: +1 719 7832615 • F: +1 719 7832615 • silverclifftown@piopc.net •
C.E.O.: *Susan Hutton* •
Local Museum – 1959
Local hist, textiles, silver, paintings, photos ... 53105

Silver Spring MD

George Meany Memorial Exhibition, 10000 New Hampshire Av, Silver Spring, MD 20903 • T: +1 301 4315451 • F: +1 301 4310385 • ldeloach@nlc.edu • www.georgemeany.org •
Dir.: *Michael Merrill* •
Historical Museum – 1980
American Federation of Labor-Congress of Industrial Organizations (AFL-CIO) ... 53106

National Capital Trolley Museum, 1313 Bonifant Rd, Silver Spring, MD 20905 • T: +1 301 3846352 • F: +1 301 3842865 • nctm@dotrolley.org • www.dctrolley.org •
Pres.: *Ken Rucker* •
Science&Tech Museum – 1959
Electric street cars, postal cards, demonstration railway ... 53107

Silverton CO

Mayflower Gold Mill, 1 Main St N, Silverton, CO 81433 • T: +1 970 3875838 •
Head: *Beverly Rich* •
Science&Tech Museum
Local history ... 53108

San Juan County Historical Museum, 1557 Greene St, Silverton, CO 81433 • T: +1 970 3875838 • F: +1 970 3875144 • silvertonhistoricalsociety.org •
Head: *Beverly Rich* •
Local Museum – 1964
Local history ... 53109

Simi Valley CA

Ronald Reagan Presidential Library and Museum, 40 Presidential Dr, Simi Valley, CA 93065 • T: +1 805 5222977 • F: +1 805 5209702 • info@reaganfoundation.org • www.reaganlibrary.com •
Dir.: *R. Duke Blackwood* •
Library with Exhibitions – 1991
Life, times and career of US-Pres. Ronald Reagan and his family ... 53110

Strathearn Museum, 137 Strathearn Pl, Simi Valley, CA 93063 • T: +1 805 5266453 • simimuseum@sbcglobal.net • www.simihistory.com •
Dir.: *Patricia Havens* •
Local Museum – 1970
Local hist, farm impl, furniture, household items 53111

Simsbury CT

Simsbury Historical Society Museum, 800 Hopmeadow St, Simsbury, CT 06070 • T: +1 860 6582500 • F: +1 860 6582500 • info@simburyhistory.org • www.simsburyhistory.org •
Pres.: *Dawn Hutchins Bobyrk* •
Local Museum – 1911
Local hist ... 53112

Sioux City IA

Sioux City Art Center, 225 Nebraska St, Sioux City, IA 51101-1712 • T: +1 712 2796272 ext 208 • F: +1 712 2552921 • jcollins@sioux-city.org • www.siouxcityartcenter.org •
Dir.: *Al Harris-Fernandez* •
Fine Arts Museum – 1914
Art ... 53113

Sioux City Public Museum, 2901 Jackson St, Sioux City, IA 51104-3697 • T: +1 712 2796174 • F: +1 712 2525615 • scpm@sioux-city.org • www.sioux-city.org/museum •

U

Dir.: *Dr. Steven D. Hansen* • Cur.: *Grace Linden* (History) •
Local Museum – 1886
General local history, Peirce mansion, Sergeant Floyd River Museum, Loren D. Callendar gallery 53114

Trinity Heights - Saint Joseph Center-Museum, 2509 33rd St, Sioux City, IA 51105 • T: +1 712 2398670 •
Exec. Dir.: *Bernard F. Cooper* •
Fine Arts Museum – 1992
Woodcarving 53115

Sioux Falls SD

Augustana College Center for Western Studies, Augustana College, Sioux Falls, SD 57197, mail addr: Box 7272, Sioux Falls, SD 57197 • T: +1 605 2744007 • F: +1 605 2744999 •
Dir.: *Arthur R. Huseboe* •
University Museum 53116

Battleship South Dakota Memorial, 12th and Kiwanis St, Sioux Falls, SD 57104 • T: +1 605 3677141, 3677060 • F: +1 605 3674326 • www.ohwy.com/sd/u/usssdbam.htm •
Pres.: *David Witte* •
Military Museum – 1968
Military hist 53117

Delbridge Museum, 805 S Kiwanis Av, Sioux Falls, SD 57104-3714 • T: +1 605 3677003 • F: +1 605 3678340 • info@gpzoo.org • gpzoo.org •
Dir.: *Christine Anderson* •
Natural History Museum – 1957
Natural hist 53118

Eide-Dalrymple Gallery, Augustana College, 29th St and S Grange Av, Sioux Falls, SD 57197 • T: +1 605 2744609 • F: +1 605 2744368 • hannus@inst.augie.edu •
Dir.: *L. Adrien Hannus* •
Fine Arts Museum – 1960
Primitive art, Japanese woodcuts, European prints, New Guinea masks 53119

Jim Savage Art Gallery and Museum, 3301 E 26th St, Village Sq, Sioux Falls, SD 57103 • T: +1 605 3327551 • F: +1 605 3327551 •
Dir.: *Barb Van Laar* •
Fine Arts Museum
Woodcarving 53120

Sioux Empire Medical Museum, 1305 W 18th St, Sioux Falls, SD 57117-5039 • T: +1 605 3336397 •
Head: *Thenetta Nield* •
Historical Museum – 1975
Medical hist, regional health care, uniforms, nursing school, hospital life, iron lung 53121

Siouxland Heritage Museums, 200 W 6th St, Sioux Falls, SD 57104-6001 • T: +1 605 3674210 • F: +1 605 3676004 • bhoskins@minnehahacounty.org • www.minnehahacounty.org •
Dir.: *William J. Hoskins* • Cur.: *Laura Hortz* • *April Woodside* • *Kevin Gansz* •
Local Museum – 1926
Local hist 53122

Washington Pavillion of Arts and Science, 301 S Main St, Sioux Falls, SD 57104 • T: +1 605 3677397 • F: +1 605 3677399 • info@washingtonpavilion.org • www.washingtonpavilion.org •
Dir.: *Steeven A. Hoffman* •
Fine Arts Museum / Folklore Museum / University Museum – 1961
Art 53123

Sisseton SD

Tekakwitha Fine Arts Center, 401 S 8th Av W, Sisseton, SD 57262-0208 • T: +1 605 6987058 • F: +1 605 6983801 •
Dir./Cur.: *Harold D. Moore* •
Fine Arts Museum – 1969
Dakotah Sioux art, former dormitory of the Tekakwitha Indian Childrens' home 53124

Sitka AK

Isabel Miller Museum → Sitka Historical Society Museum

Sheldon Jackson Museum, 104 College Dr, Sitka, AK 99835 • T: +1 907 7478981 • F: +1 907 7473004 • scott_mcadams@eed.state.ak.us • www.museums.state.ak.us •
Dir.: *Linda Thibodeau* • Cur.: *Bob Banghart* • *Rosemary Carlton* •
Ethnology Museum – 1888
Alaska native cultures and art, ivory and Haida argillite carvings, Eskimo implements 53125

Sitka Historical Society Museum, 330 Harbor Dr, Sitka, AK 99835 • T: +1 907 7476455 • F: +1 907 7476588 • sitka.history@yahoo.com • www.sitkahistory.org •
Dir.: *Bob Medinger* • Cur.: *Ashley Kircher* •
Local Museum – 1957
Local artifacts, paintings and docs 53126

Southeast Alaska Indian Cultural Center, Sitka National Historical Park, 106 Metlakatla St, Sitka, AK 99835 • T: +1 907 7478061, +1 907 7470110 • F: +1 907 7475938 • www.nps.gov •
Cur.: *Sue Thorsen* •
Historical Museum – 1910
Local hist, period furniture – totem walk 53127

Skagway AK

Skagway Museum, 700 Spring St, Skagway, AK 99840, mail addr: POB 521, Skagway, AK 99840 • T: +1 907 9832420 • F: +1 907 9833420 • info@skagwaymuseum.org • www.skagwaymuseum.org •
Cur.: *Judith Munns* •
Local Museum – 1961
Skagway history, gold rush 53128

Skowhegan ME

Skowhegan History House, 66 Elm St, Skowhegan, ME 04976 • T: +1 207 4746632 • skowheganhistoryhouse@hotmail.com • www.skowhegan.com/historyhouse •
Pres./Cur.: *Lee Z. Granville* •
Historical Museum – 1937
Furnishings, local hist 53129

Slidell LA

Slidell Art Center, 444 Erlanger St, Slidell, LA 70458 • T: +1 504 6464375 • F: +1 504 6464231 • kgustafson@cityofslidell.org • www.slidell.org •
Dir.: *Kelli Gustafson* •
Fine Arts Museum – 1989 53130

Slidell Museum, 2020 First St, Slidell, LA 70458, mail addr: POB 828, Slidell, LA 70459-0828 • T: +1 985 6464380 • F: +1 985 6464231 • www.slidell.la.us •
Head: *Priscilla Davis* •
Local Museum
History 53131

Smackover AR

Arkansas Museum of Natural Resources, 3853 Smackover Hwy, Smackover, AR 71762 • T: +1 870 7252877 • F: +1 870 7252161 • museum@amnr.org • www.amnr.org •
Dir.: *Pam Beasley* •
Historical Museum / Science&Tech Museum – 1977
Specimens, Arkansas history, petroleum technology and products 53132

Smithfield VA

Isle of Wight Courthouse, 130 Main St, Smithfield, VA 23430 • T: +1 804 3573502 • F: +1 804 7750802 • www.apva.org •
Exec. Dir.: *Elizabeth Kostelny* •
Historic Site – 1750
Historic bldg 53133

Smithtown NY

Smithtown Historical Society Museum, 5 N Country Rd, Smithtown, NY 11787 • T: +1 516 2656768 • F: +1 516 2656768 • historic@optonline.net • www.smithtownhistorical.org •
Dir.: *Carol Ghiorsi-Hart* •
Local Museum – 1955
Historic houses, local hist, dec arts 53134

Smithville TN

Appalachian Center for Crafts, 1560 Craft Center Dr, Smithville, TN 37166 • T: +1 615 5976801 • F: +1 615 5976803 • craftcenter@tntech.edu • www.tntech.edu/craftcenter •
Dir.: *Timothy Weber* • *Gail S. Looper* (Gallery) •
Ethnology Museum – 1979
Over 25 exhib annually of contemporary & traditional fine craft 53135

Snohomish WA

Blackman Museum, 118 Av B, Snohomish, WA 98290 • T: +1 360 5685235 •
Pres.: *Windsor Vest* •
Local Museum – 1969
Local hist, furniture and furnishings of the late 1800s, logging tools 53136

Snoqualmie WA

Northwest Railway Museum, 38625 SE King St, Snoqualmie, WA 98065 • T: +1 425 8883030 ext 201 • F: +1 425 8889311 • director@trainmuseum.org • www.trainmuseum.org •
Exec. Dir.: *Richard R. Anderson* •
Science&Tech Museum – 1957
Steam and diesel locomotives, passenger coaches and freight cars, rotary snow plow, cranes, observation cars, kitchen car 53137

Snow Hill MD

Julia A. Purnell Museum, 208 W Market St, Snow Hill, MD 21863 • T: +1 410 6320515 • F: +1 410 6320515 • purnellmuseum@dmv.com • www.purnellmuseum.com •
C.E.O.: *Mary Saint Hippolyte* •
Local Museum – 1942
History of American/ Worcester County, folk art, furniture, jewelry, machines, tools 53138

Snyder TX

Scurry County Museum, Western Texas College, 6200 College Av, Snyder, TX 79549 • T: +1 915 5736107 • F: +1 915 5739321 • scm@snydertex.com •
Dir.: *Charlene Akers* • Cur.: *Sue Goodwin* •
Local Museum – 1970
Farming & ranching equip, frontier furnishings, county records, contemp prints – archives 53139

Socorro NM

New Mexico Bureau of Geology Mineral Museum, New Mexico Tech, 801 Leroy Pl, Socorro, NM 87801 • T: +1 505 8355140 • F: +1 505 8356333 • vwlueth@nmt.edu • geoinfo.nmt.edu •
Dir./Cur.: *Virgil W. Lueth* • Asst. Cur.: *Patricia L. Frisch* •
Natural History Museum – 1926
Mineralogy, petrology, economic geology, mining hist, paleontology 53140

Sodus Point NY

Sodus Bay Historical Society Museum, 7606 N Ontario St, Sodus Point, NY 14555 • T: +1 315 4834936 • F: +1 315 4831398 • sodusbay@ix.netcom.com • www.peachey.com/soduslight •
C.E.O.: *Donna P. Jones* •
Local Museum – 1979
Local hist, Great Lakes maritime hist 53141

Solomons MD

Calvert Marine Museum, 14200 Solomons Island Rd, Solomons, MD 20688 • T: +1 410 3262042 • F: +1 410 3266691 • information@calvertmarinemuseum.com • www.calvertmarinemuseum.com •
Dir.: *C. Douglass Alves* • Cur.: *Karen E. Stone* •
Historical Museum / Natural History Museum – 1969
Marine, Drum Point Lighthouse on waterfront, seafood packing house 53142

Solvang CA

Old Mission Santa Ines Collection, 1760 Mission Dr, Solvang, CA 93464 • T: +1 805 6884815 • F: +1 805 6864468 • office@missionsantaines.org • www.missionsantaines.org •
Religious Arts Museum
Religious vestiments, Mexican and Spanish art 53143

Somers CT

Museum of Natural History and Primitive Technology, 332 Turnpike Rd, Somers, CT 06071 • T: +1 860 7494129 • somersmtn@aol.com • www.somersmountain.org •
Dir.: *Michael Stenz* •
Natural History Museum / Science&Tech Museum – 1936
Native American artifacts 53144

Somers NY

Somers Historical Society Museum and Museum of the Early American Circus, Elephant Hotel, Rtes 100 and 202, Somers, NY 10589 • T: +1 914 2774977 • F: +1 914 2774977 • oldbet@cloudnine.net • www.somersmuseum.org •
Cur.: *Terry Ariano* •
Historical Museum – 1956
Local hist, early American circus 53145

Somers Point NJ

Atlantic County Historical Society Museum, 907 Shore Rd, Somers Point, NJ 08244 • T: +1 609 9275218 • www.aclink.org/achs •
Cur.: *Allen Pergament* • Assist.Cur.: *Ruth C. Gold* •
Local Museum – 1913
Genealogy, local and U.S. history 53146

Somers Mansion, 1000 Shore Rd, Somers Point, NJ 08244 • T: +1 609 9272212 • F: +1 609 9271827 • history54@msn.com •
Sc.Staff: *Daniel T. Campbell* (Hist. Preservation) •
Local Museum – 1941
Furnishings 53147

Somerset PA

Somerset Historical Center, 10649 Somerset Pike, Somerset, PA 15501 • T: +1 814 4456077 • F: +1 814 4436621 • chfox@state.pa.us • www.somersethistoricalcenter.org •
Dir.: *Vernon Berkey* • *Charles Fox* • Cur.: *Barbara Black* • Educator: *Mark Ware* •
Local Museum – 1969
Local hist 53148

Somerville MA

Somerville Museum, 1 Westwood Rd, Somerville, MA 02143 • T: +1 617 6669810 •
Dir.: *Regina M. Pisa* •
Local Museum – 1897
Local hist, 18th-20th c American hist, furniture, fine arts, military, industry, science 53149

Somerville NJ

Old Dutch Parsonage, 71 Somerset St, Somerville, NJ 08876 • T: +1 908 7251015 •
Cur.: *Jim Kurzenberger* •
Local Museum – 1947
Dutch furnishings, replica of period pieces 53150

Printmaking Council of New Jersey, 440 River Rd, Somerville, NJ 08876 • T: +1 908 7252110 • F: +1 908 7252484 • info@printNJ.org • www.printNJ.org •
Dir.: *Elise Fuscarino* •
Public Gallery
Temporary Printing exhib 53151

Wallace House, 71 Somerset St, Somerville, NJ 08876 • T: +1 908 7251015 •
Cur.: *Jim Kurzenberger* •
Decorative Arts Museum / Local Museum – 1947
Revolutionary War, local hist 53152

Somesville ME

Mount Desert Island Historical Museum, 2 Oak Hill Rd, Somesville, ME 04660 • T: +1 207 2445043 • F: +1 207 2443991 • jroths@acadia.net •
Pres.: *Anne Mazlish* • Dir.: *Jaylene Roths* •
Local Museum – 1931
Local history 53153

Sonoma CA

Depot Park Museum, 270 First St W, Sonoma, CA 95476 • T: +1 707 9381762 • F: +1 707 9381762 • depot@vom.com • www.vom.com/~depot •
Dir.: *Diane Smith* •
Local Museum – 1979
Pioneers and settlers of the valley, native Americans, railroad equip 53154

Sonora TX

Miers Home Museum, Sutton county Historical Society, 307 Oak St, Sonora, TX 76950 • T: +1 915 3872855 • F: +1 915 3873303 •
Local Museum / Historical Museum – 1968
Furniture, branding irons of county, period ranching artifacts 53155

South Bend IN

College Football Hall of Fame, 111 S Saint Joseph St, South Bend, IN 46601 • T: +1 219 2359999 • F: +1 219 2355720 • Kent.Stephens@collegefootball.org • www.collegefootball.org •
Special Museum – 1978
Historical college football coll 53156

Copshaholm House Museum and Historic Oliver Gardens, Northern Indiana Center for History, 808 W Washington, South Bend, IN 46601 • T: +1 219 2359664 • F: +1 219 2359059 • director@centerforhistory.org • www.centerforhistory.org •
Cur.: *David S. Bainbridge* •
Local Museum – 1990
Furnishings, furniture, decorative arts, works of art, costumes 53157

Northern Indiana Center for History, 808 W Washington, South Bend, IN 46601 • T: +1 219 2359664 • F: +1 219 2359059 • www.centerforhistory.org •
Dir.: *Cheryl Taylor Bennett* • Cur.: *David S. Bainbridge* •
Ethnology Museum / Historical Museum – 1994
History, children's interactive museum, Voyages, history of the Saint Joseph River Valley Region, Changing gallery, Carroll gallery, Leighton gallery, traveling 53158

South Bend Regional Museum of Art, 120 S Saint Joseph St, South Bend, IN 46601 • T: +1 219 2359102 • F: +1 219 2355782 • info@southbendart.org • www.sbrma.org •
Dir.: *Susan R. Visser* • Cur.: *William Tourtillotte* •
Fine Arts Museum – 1947
American art 53159

Studebaker National Museum, 525 S Main St, South Bend, IN 46601 • T: +1 219 2359714, +1 888 3915600 • F: +1 219 2355522 • info@studebakermuseum.org • www.studebakermuseum.org •
C.E.O: *Rebecca Bonham* •
Science&Tech Museum – 1977
Industry history, vehicles, automobiles 53160

Worker's Home Museum, Northern Indiana Center for History, 808 W Washington St, South Bend, IN 46601 • T: +1 219 2359664 • F: +1 219 2359059 • director@centerforhistory.org • www.centerforhistory.org •
Cur.: *David S. Bainbridge* •
Historical Museum – 1994
History 53161

U

South Bend WA

Pacific County Historical Museum, 1008 W Robert Bush Dr, South Bend, WA 98586-0039 • T: +1 360 8755224 • F: +1 360 8755224 • museum@willapabay.org • www.pacificcohistory.org •
Pres.: *Vincent Shaudys* • Dir.: *Bruce Weilepp* •
Local Museum – 1970
Local hist, Indian artifacts and relics, pioneer relics, records 53162

South Berwick ME

Hamilton House, 40 Vaughan's Ln, South Berwick, ME 03908, mail addr: 141 Cambridge St, Boston, ME 03908 • T: +1 207 3842454, +1 617 2273956 • F: +1 617 2279204, +1 207 3848192 • HamiltonHouse@HistoricNewEngland.org • www.spnea.org •
Pres.: *Carl Nold* •
Special Museum – 1949
Personal and decorative objects from Mrs. Tyson 53163

Sarah Orne Jewett House, 5 Portland St, South Berwick, ME 03908 • T: +1 207 3842454, +1 617 2273956 • F: +1 617 2279204, +1 207 3848192 • JewettHouse@HistoricNewEngland.org • www.spnea.org •
Pres.: *Carl Nold* •
Special Museum – 1931
Residence of the author Sarah Orne Jewett 53164

South Boston VA

South Boston-Halifax County Museum of Fine Arts and History, 1540 Wilborn Av, South Boston, VA 24592 • T: +1 434 5729200 • F: +1 434 5728996 • sbhcm@halifax.com • sbhcmuseum.org •
Cur.: *Marge Holtman* •
Local Museum / Fine Arts Museum – 1981
Artifacts from South Boston, Halifax County & Southside Virginia 53165

South Charleston WV

South Charleston Museum, 312 Fourth Av, South Charleston, WV 25303 • T: +1 304 7449711 • F: +1 304 7448808 • SCMuseum@yahoo.com • www.geocities.com/scmuseum •
Dir.: *Teresa C. Whitt* •
Local Museum – 1989 53166

South Dartmouth MA

Children's Museum in Dartmouth, 276 Gulf Rd, South Dartmouth, MA 02748 • T: +1 508 9933361 • F: +1 508 9933332 •
Pres.: *Robert Howland* • Dir.: *Ronald Mayer* •
Ethnology Museum / Historical Museum – 1952 53167

South Deerfield MA

Yankee Candlemaking Museum, Rte 5, South Deerfield, MA 01373 • T: +1 413 6652020 • F: +1 413 6652399 • www.yankeecandle.com •
Head: *Steve Smith* •
Historical Museum 53168

South Elgin IL

Fox River Trolley Museum, 361 S La Fox St, South Elgin, IL 60177 • T: +1 630 6652581 • info@foxtrolley.org • www.foxtrolley.org •
Science&Tech Museum – 1958
Cars, diesel locomotives, trolley track 53169

South Hadley MA

Mount Holyoke College Art Museum, Lower Lake Rd, South Hadley, MA 01075-1499 • T: +1 413 5382245 • F: +1 413 5382144 • artmuseum@mtholyoke.edu • www.mtholyoke.edu/go/artmuseum •
Dir.: *Marianne Doezema* • Cur.: *Wendy M. Watson* •
Fine Arts Museum / University Museum – 1875
Ancient, medieval and Renaissance Asian and American art 53170

Skinner Museum, Mount Holyoke College, 35 Woodbridge St, South Hadley, MA 01075 • T: +1 413 5382245 • F: +1 413 5382144 •
Cur.: *Wendy Watson* •
Local Museum / University Museum – 1933
Agriculture, archaeology, geology, glass, hist, bird coll 53171

South Haven MI

Michigan Maritime Museum, 260 Dyckman Av, South Haven, MI 49090 • T: +1 269 6378078 • F: +1 269 6371594 • mmmuseum@accn.org • www.michiganmaritimemuseum.org •
Exec. Dir.: *Dr. Barbara K. Kreuzer* • Cur.: *Cobie Ball* •
Sc. Staff: *David Ludwig* (Boat Shed) •
Historical Museum – 1976
Great Lakes maritime history, historic vessels, tools & implements, models, marine art – library, boathouses 53172

South Hero VT

South Hero Bicentennial Museum, Rte 2, South Hero, VT 05486 • T: +1 802 3726615, 6725552 • lmjshvt@aol.com •
Pres./Cur.: *Barbara Winch* •
Local Museum – 1974
Bibles, local historic pieces 53173

South Holland IL

South Holland Historical Museum, 16250 Wausau Av, South Holland, IL 60473 • T: +1 708 5962722 •
Pres.: *Bill Paarlberg* •
Local Museum – 1969
Local history 53174

South Lake Tahoe CA

South Lake Tahoe Museum, 3058 Lake Tahoe Blvd, South Lake Tahoe, CA 96156 • T: +1 530 5415458 •
Pres.: *Richard Young* •
Historical Museum / Local Museum – 1968
Washo Indian, photos, South Lake Tahoe artifacts, industry, commerce and gaming – library, archives 53175

South Milwaukee WI

South Milwaukee Historical Society Museum, 717 Milwaukee Av, South Milwaukee, WI 53172 • T: +1 414 7688790, 7628852 • www.southmilwaukee.org •
C.E.O.: *Dean Marlowe* • Cur.: *Gertrude Endthoff* •
Local Museum – 1972
Local hist 53176

South Orange NJ

Seton Hall University Museum, 400 South Orange Av, South Orange, NJ 07079 • T: +1 973 7619459 • F: +1 973 2752368 • krafther@lanmail.shu.edu • www.shu.edu •
Dir.: *Charlotte Nichols* •
University Museum / Archaeology Museum – 1963
Archaeology 53177

South Pass City WY

South Pass City State Historic Site, 125 South Pass Main St, South Pass City, WY 82520 • T: +1 307 3326323 • F: +1 307 3323688 • southpass@onewest.net •
Cur.: *Scott Goetz* •
Local Museum – 1967
Gold mining memorabilia and artifacts, hist bldgs, furnishings 53178

South Portland ME

Museum of Glass and Ceramics, POB 2416, South Portland, ME 04116 • T: +1 207 5919035 • F: +1 207 5919037 • mglasceram@choiceonemail.com •
Dir.: *Holverson John* •
Decorative Arts Museum – 2002
Historic and contemporary glass and ceramics .. 53179

Portland Harbor Museum, Ford Rd, South Portland, ME 04106 • T: +1 207 7996337 • F: +1 207 7993862 • www.portlandharbormuseum.org •
Exec.Dir.: *Thompson Mark R.* •
Historical Museum – 1985
Portland Harbor and Casco Bay maritime hist and culture 53180

South Saint Paul MN

Dakota County Museum, 130 Third Av N, South Saint Paul, MN 55075 • T: +1 651 5527548 • F: +1 651 5527265 • dchs@mtn.org • www.dakotahistory.org •
Pres.: *Bill Wolston* • Exec. Dir.: *Chad Roberts* •
Historical Museum – 1939
Indian artifacts, agriculture, industry and trade, military, railroads – library 53181

South Sutton NH

South Sutton Old Store Museum, 12 Meeting House Hill Rd, South Sutton, NH 03273 • T: +1 603 9385843 •
Cur.: *Peggy Forand* •
Historical Museum – 1954
Tools, dolls, toys, photos, artifacts 53182

South Union KY

Shaker Museum at South Union, 850 Shaker Museum Rd, South Union, KY 42283 • T: +1 502 5424167, 5427734 • F: +1 502 5427558 • shakmus@logantele.com • www.shakermuseum.com •
Cur.: *Tommy Hines* •
Historical Museum – 1960
Site of 1807 South Union Shaker village 53183

South Williamsport PA

Peter J. McGovern Little League Baseball Museum, Rte 15, South Williamsport, PA 17701 • T: +1 570 3263607 • F: +1 570 3262267 • museum@littleleague.org • www.littleleague.org •
Dir./Cur.: *Michael Miller* •
Special Museum – 1982
Sports, Little League 53184

Southampton NY

The Parrish Art Museum, 25 Job's Ln, Southampton, NY 11968 • T: +1 631 2832118 ext 12 • F: +1 631 2837006 • fergusone@parrishart.org • thehamptons.com •
Dir.: *Trudy C. Kramer* • Chief Cur.: *Alicia Longwell* (Art and Education) • Cur.: *Katherine Crum* • *Lewis B. Cullman* • *Dorothy Cullmann* •
Fine Arts Museum – 1898
Italian Renaissance, 19th and early 20th c American paintings, graphic, sculpture, Japanese stencils – W.M. Chase Coll and Archives, Fairfield Porter Coll and Archives 53185

Southampton Historical Museum, 177 Meeting House Ln, Southampton, NY 11968 • T: +1 631 2832494 • F: +1 631 2834540 • hismusdir@hamptons.com • www.southamptonhistorymuseum.com •
C.E.O.: *Richard I. Barons* •
Local Museum – 1898
Local hist, South Fork culture, folk art, Native American, whaling, rural tools, area ethnic hist 53186

Southern Pines NC

Weymouth Woods-Sandhills Nature Preserve Museum, 1024 Fort Bragg Rd, Southern Pines, NC 28387 • T: +1 919 6922167 • F: +1 910 6928042 • weymouth@pinehurst.net •
Natural History Museum – 1969
Indian artifacts, turpentine industry artifacts, wildlife 53187

Southold NY

Southold Historical Society Museum, Main Rd and Maple Ln, Southold, NY 11971 • T: +1 631 7655500 • F: +1 631 7655500 • sohissoc@optonine.net • southoldhistoricalsociety.org •
Dir.: *Geoffrey Fleming* • Pres.: *Russell Thomes* •
Local Museum – 1960
Farm implements, agriculture hist, maritime coll, dec arts, historic houses 53188

Southold Indian Museum, Main Bayview Rd, Southold, NY 11971 • T: +1 631 7655577 • F: +1 631 7655577 • indianmuseum@aol.com • southoldindianmuseum.org •
Pres.: *Ellen Barcel* •
Ethnology Museum / Archaeology Museum – 1925
Hist, craft and culture 53189

Southwest Harbor ME

Wendell Gilley Museum, Main St and Herrick Rd, Southwest Harbor, ME 04679 • T: +1 207 2447555 • F: +1 207 2447555 • gilleymu@acadia.net • www.wendellgilleymuseum.org •
Pres.: *Kate Halle Briggs* • Dir.: *Nina Z. Gormley* •
Fine Arts Museum / Folklore Museum – 1979
Folk art, woodcarving 53190

Spalding ID

Nez Perce National Historical Park, 39063 US-Hwy 95, Spalding, ID 83540 • T: +1 208 8432261 • F: +1 208 8432001 • bob_chenoweth@nps.gov • www.nps.gov/nepe/ •
Cur.: *Bob Chenoweth* •
Natural History Museum / Ethnology Museum / Historical Museum – 1965
Prehistoric occupation, Indian ethnographic colls, photos, Dugout canoes, lawyer family coll, 1877 war – archives, research center 53191

Sparta WI

Monroe County Local History Room and Library, 200 W Main St, Sparta, WI 54656 • T: +1 608 2698680 • F: +1 608 2698921 • mclhr@centurytel.net • www.spartan.org/historyroom •
Local Museum – 1977 53192

Spartanburg SC

Milliken Gallery, Converse College, 580 E Main St, Spartanburg, SC 29302 • T: +1 864 5969181 • F: +1 864 5969606 • mac.boggs@converse.edu • www.converse.edu •
Dir.: *Mac Boggs* •
Fine Arts Museum / University Museum – 1971
Art 53193

Regional Museum of Spartanburg County, 100 E Main St, Spartanburg, SC 29306 • T: +1 864 5963501 • F: +1 864 5963501 • regionalmuseum@mindspring.com • www.spartanarts.org/history •
C.E.O.: *Susan Turpin* • Cur.: *Carolyn Creal* •
Local Museum – 1961
Local hist 53194

The Sandor Teszler Gallery, Wofford College, 429 N Church St, Spartanburg, SC 29303-3663 • T: +1 864 5974300 • F: +1 864 5974329 • coburnoh@wofford.edu • www.wofford.edu/sandorTeszlerLibrary/currentExhibit.asp •
Dir.: *Oakley H. Coburn* •
University Museum / Library with Exhibitions
Art, crafts, sciences, social sciences – library 53195

Spartanburg County Museum of Art, 385 S Spring St, Spartanburg, SC 29306 • T: +1 864 5827616 • F: +1 864 9485353 • museum@spartanarts.org • www.spartanburgartmuseumo.org •
Pres.: *Ray Eubanks* • Dir.: *Theresa Mann* •
Fine Arts Museum – 1969
Arts 53196

Spartanburg Science Center, 385 S Spring St, Spartanburg, SC 29306 • T: +1 864 5832777 • F: +1 864 9485353 •
Exec. Dir.: *John F. Green* •
Natural History Museum / Science&Tech Museum – 1978
Natural hist, science, exotic tortoises, reptiles, fossils, skulls, rocks, minerals, insects, shells, bird nests 53197

University of South Carolina at Spartanburg Art Gallery, 800 University Way, Spartanburg, SC 29303 • T: +1 864 5035838 • F: +1 864 5035825 •
Dir.: *Jane Nodine* •
University Museum / Fine Arts Museum – 1982
Contemporary painting, drawings, prints, photography and sculpture 53198

Spearfish SD

D.C. Booth Historic National Fish Hatchery, 423 Hatchery Circle, Spearfish, SD 57783 • T: +1 605 6427730 • F: +1 605 6422336 • dcbooth@fws.gov • www.fws.gov/dcbooth •
C.E.O.: *Steve Brimm* • Cur.: *Randi Smith* •
Special Museum – 1896
Historic fishery and fish culture artifacts, photos, fish car materials, U.S. fish and wildlife service hist, hist furnishings – archives 53199

Ruddell Gallery, Black Hills State University, 1200 University St, Spearfish, SD 57799 • T: +1 605 6426104 ext 6111 • F: +1 605 6426105 •
Dir.: *Jim Knutsen* •
Fine Arts Museum – 1936
Photography, regional, visual artists – library 53200

Spencer NC

North Carolina Transportation Museum, 411 S Salisbury Av, Spencer, NC 28159 • T: +1 704 6362889 • F: +1 704 6391881 • nctrans@vnet.net • www.nctrans.org •
Exec. Dir.: *Elizabeth Smith* •
Science&Tech Museum – 1977
1896 Southern Railway steam primary staging and repair facility complex containing 20 structures, 37 bay roundhouse, turntable and 90,000 feet back shop 53201

Spencerport NY

Ogden Historical Society Museum, 568 Colby St, Spencerport, NY 14559, mail addr: POB 777, Adams Basin, NY 14410 • T: +1 585 3520660 • F: +1 585 3525328 •
Pres.: *Ted Rogers* •
Local Museum – 1958
Early 19th c farmhouse implements, furnishings, costumes 53202

Spillville IA

Bily Clock Museum and Antonin Dvorak Exhibition, 323 S Main, Spillville, IA 52168 • T: +1 319 5623569 • F: +1 319 5624373 • bily@oneota.net • www.bilyclocks.org •
Special Museum – 1923
Clocks 53203

Spiro OK

Spiro Mounds Archaeological Center, Oklahoma Historical Society, 18154 First St, Spiro, OK 74959 • T: +1 918 9622062 • F: +1 918 9622062 • spiromds@ipa.net •
Head: *Dennis Peterson* •
Archaeology Museum – 1978 53204

Spokane WA

Children's Museum of Spokane, 110 N Post St, Spokane, WA 99201 • T: +1 509 6245437 • F: +1 509 6246453 • cms@childrensmuseum.net • www.childrensmuseum.net •
Exec. Dir.: *Don Kardong* •
Special Museum – 1995 53205

Jundt Art Museum, Gonzaga University, PearlSt & Cataldo, Spokane, WA 99258-0001, mail addr: 502 E Boone Av, Spokane, WA 99258-0001 • T: +1 509 3136611 • F: +1 509 3135525 • patnode@calvin.gonzaga.edu • www.gonzaga.edu •
Dir. & Cur.: *J. Scott Patnode* •
Fine Arts Museum / University Museum – 1995
Old Masters' print coll, photos, prints, Auguste Rodin sculpture coll, Chihuly glass 53206

Northwest Museum of Arts and Culture, 2316 W First Av W, Spokane, WA 99204 • T: +1 509 4563931 • F: +1 509 3635303 • brucee@northwestmuseum.org • www.northwestmuseum.org •
Dir.: *Bruce B. Eldredge* • Dep. Dir.: *Larry Schoonover* (Exhibits, Programs) • *Maurine Barrett* (Operations, Admin.) • *Joyce M. Cameron* (Development, Communication) •
Fine Arts Museum / Local Museum – 1916
Local hist, Indian culture coll of all Tribes of North America 53207

Spotsylvania VA

Spotsylvania Historical Association and Museum, 8956 Courthouse Rd, Spotsylvania, VA 22553 • T: +1 540 5827167 •
Pres.: *John E. Pruitt* • Dir./Cur.: *Martha C. Carter* •
Local Museum – 1962
Indian relics, early Colonial artifacts, china, tools, pottery, dolls 53208

Spring Green WI

The House on the Rock, 5754 Hwy 23, Spring Green, WI 53588 • T: +1 608 9353639 • F: +1 608 9359472 • www.thehouseontherock.com •
Pres.: *Susan Donaldson* •
Local Museum – 1961
Complex with historical houses, dollhouses, circus costumes, automatons, music machines and orchestrions, carousels 53209

Spring Valley MN

Spring Valley Community Museums, History Museum, Church Museum, Washburn-Zittleman House, Conley Camera Collection, 220 W Courtland St, Spring Valley, MN 55975 • T: +1 507 3467659 • F: +1 507 3467249 • www.ci.spring-valley.mn.us •
Dir.: *Rosalie Kruegel* •
Historical Museum – 1956 ... 53210

Springdale AR

Arts Center of the Ozarks-Gallery, 214 S Main, Springdale, AR 72764 • T: +1 501 7515441 • www.artscenteroftheozarks.org •
Dir.: *Stephanie Lewis* •
Public Gallery – 1948
Traditional handcrafts, contemporary arts and crafts ... 53211

Shiloh Museum of Ozark History, 118 W Johnson Av, Springdale, AR 72764 • T: +1 479 7508165 • shiloh@springdalear.org • www.springdaleark.org/shiloh •
Dir.: *Bob Besom* •
Local Museum – 1965
Native American artifacts, primitive paintings, pioneer life, poultry, fruit and timber industry, camera coll ... 53212

Springdale PA

Rachel Carson Homestead, 613 Marion Av, Springdale, PA 15144-1242 • T: +1 724 2745459 • F: +1 724 2751259 • www.rachelcarsonhomestead.org •
Exec. Dir.: *Vivienne Shaffer* •
Local Museum – 1975
Historic house ... 53213

Springdale UT

Zion Human History Museum, Zion National Park, Springdale, UT 84767-1099 • T: +1 435 7723256 • F: +1 435 7723426 • www.nps.gov/zion •
Cur.: *Leslie Newkirk* •
Historical Museum – 1919
Natural and human hist – archives ... 53214

Springfield IL

The Dana-Thomas House, 301 E Lawrence Av, Springfield, IL 62703 • T: +1 217 7826773 • F: +1 217 7889450 • dthf@warpnet.net • www.dana-thomas.org •
Dir.: *Dr. Donald P. Hallmark* •
Decorative Arts Museum – 1981
Frank Lloyd Wright prairie period house, art glass and furniture, decorative art, social history (20th c) ... 53215

Illinois State Museum, 502 S Spring St, Springfield, IL 62706-5000 • T: +1 217 7827387 • F: +1 217 7821254 • webmaster@museum.state.il.us • www.museum.state.il.us/ •
Dir.: *Dr. R. Bruce McMillan* • Cur.: *Robert Sill* •
Fine Arts Museum / Local Museum / Ethnology Museum / Natural History Museum – 1877
Natural history, anthropology, geology, botany, zoology, decorative art, art ... 53216

Lincoln Home, 413 S 8th St, Springfield, IL 62701-1905 • T: +1 217 4924241 • F: +1 217 4924673 • liho_superintendent@nps.gov • www.nps.gov/liho •
Cur.: *Susan M. Haake* •
Historical Museum / Historic Site – 1972
Home of Abraham Lincoln, 16th Pres. of the United States ... 53217

Lincoln Tomb, Oak Ridge Cemetery, Springfield, IL 62702 • T: +1 217 7822717 • F: +1 217 5243738 •
Historic Site – 1874
1874 the tomb of Abraham Lincoln ... 53218

Old State Capitol, Fifth at Adams Sts, 1 Old Capitol Pl, Springfield, IL 62701 • T: +1 217 7857960 • F: +1 217 5570282 • carol-andrews@ihpa.state.il.us •
Historical Museum – 1969
Period and original furnishings, Illinois hist ... 53219

The Pearson Museum, 801 N Rutledge, Springfield, IL 62794-9635 • T: +1 217 5458017, 5454261 • F: +1 217 5459605 • bmason@siumed.edu • www.siumed.edu/medhum/pearson •
Dir.: *Phillip V. Davis* •
University Museum / Historical Museum – 1974
19th and 20th c medical, dental and pharmaceutical artifacts, graphic art, photography ... 53220

Springfield Art Gallery, 700 N Fourth St, Springfield, IL 62702 • T: +1 217 5232631 • F: +1 217 5233866 • office@springfieldart.org • www.springfieldart.org •
Exec. Dir.: *Erika Fitzgerald* •
Fine Arts Museum / Public Gallery – 1913
19th-20th c paintings, African sculptures, Oriental artifacts, ceramic. bronze, glass, jewelry, eight rooms of 19th c American furniture ... 53221

Vachel Lindsay Home, 603 S Fifth St, Springfield, IL 62703 • T: +1 217 5240901 • F: +1 217 5570282 •
Special Museum – 1946
Drawings, letters, manuscripts, books, paintings, sculpture ... 53222

Springfield KY

Lincoln Homestead, 5079 Lincoln Park Rd, Springfield, KY 40069 • T: +1 859 3367461 • F: +1 859 3360659 • robert.bartholomai@mail.state.ky.us • www.state.ky.us/agencies/parks/linchome.htm •
Historic Site – 1936 ... 53223

Springfield MA

Connecticut Valley Historical Museum, 220 State St, Springfield, MA 01103 • T: +1 413 2636800 ext 304 • F: +1 413 2636898 • info@spfldlibmus.org • www.quadrangle.org •
Dir.: *Guy McLain* •
Local Museum – 1927
Regional history and art, maps – library, archives ... 53224

George Walter Vincent Smith Art Museum, 220 State St, Springfield, MA 01103 • T: +1 413 2636800 ext 302 • F: +1 413 2636897 • info@spfldlibmus.org • www.quadrangle.org •
Dir.: *Heather Haskell* •
Fine Arts Museum / Decorative Arts Museum – 1896
Japanese decorative arts, Chinese cloisonné, Middle Eastern rugs, American paintings of the 19th c – art studios ... 53225

Museum of Fine Arts, 220 State St, Springfield, MA 01103 • T: +1 413 2636800 ext 302 • F: +1 413 2636889 • info@spfldlibmus.org • www.quadrangle.org •
Dir.: *Heather Haskell* •
Fine Arts Museum – 1933
American and European painting, graphics, sculptures, decorative art – library, auditorium ... 53226

Naismith Memorial Basketball Hall of Fame, 1000 W Columbus Av, Springfield, MA 01105-2532 • T: +1 413 7816500 • F: +1 413 7811939 • robin@hoophall.com • www.hoophall.com •
Pres.: *John Doleva* • Cur.: *Michael Brooslin* •
Special Museum – 1959
library ... 53227

Springfield Armory Museum, 1 Armory Sq, Springfield, MA 01105-1299, mail addr: POB 4781, Springfield, MA 01101 • T: +1 413 7348551 • F: +1 413 7478062 • spar_interpretation@nps.gov • www.nps.gov •
Cur.: *James D. Roberts* •
Military Museum – 1870
Military ... 53228

Springfield Science Museum, 220 State St, Springfield, MA 01103 • T: +1 413 2636800 • F: +1 413 2636898 • info@spfldlibmus.org • www.quadrangle.org •
Dir.: *David Stier* • Cur.: *John P. Pretola* (Anthropology) •
Science&Tech Museum – 1859
African, Native American, minerals, dinosaur coll, Amazon rainforest & African savannah colls – Seymour planetarium ... 53229

William Blizard Gallery, Springfield College, 263 Alden St, Springfield, MA 01109 • T: +1 413 7473000, 7480204 • F: +1 413 7483580 • mhealey@spfldcol.edu • www.spfldcol.edu •
Dir.: *Mary Healey* •
Fine Arts Museum / University Museum – 1998 ... 53230

Springfield MO

Discovery Center of Springfield, 438 E St Louis, Springfield, MO 65806 • T: +1 417 8629910 ext 0 or ext 700 • F: +1 417 8626898 • efox@discoverycenter.org • discoverycenterofspringfield.org •
C.E.O.: *Emily Fox* •
Science&Tech Museum – 1991 ... 53231

History Museum for Springfield-Greene County, 830 Boonville, Springfield, MO 65802 • T: +1 417 8641976 • F: +1 417 8642019 • info@historymuseumsgc.org • www.historymuseumsgc.org •
Dir.: *Michal Dale* • Cur.: *Joan Hampton-Porter* •
Local Museum / Historical Museum – 1975
Costumes, textiles, decorative arts, archives, dolls, toys, Civil War, WW I & WW II ... 53232

Springfield Art Museum, 1111 E Brookside Dr, Springfield, MO 65807 • T: +1 417 8375700 • F: +1 417 8375704 • ArtMuseum@springfieldmo.gov • www.springfieldmogov.org •
Dir.: *Jerry A. Berger* • Cur.: *James M. Beasley* •
Fine Arts Museum – 1928
Decorative art, drawings, photography, paintings, prints, sculpture – library ... 53233

Springfield NJ

Springfield Historical Society Museum, 126 Morris Av, Springfield, NJ 07081 • T: +1 973 3764784 • www.springfield-nj.com/history.htm •
Pres.: *Margaret Bandrowski* •
Local Museum – 1954
Local hist, items, drawings, prints, decorative arts, costumes, textiles ... 53234

Springfield OH

Heritage Center of Clark County, Clark County Historical Society, 117 S Fountain Av, Springfield, OH 45502 • T: +1 937 3240657 • F: +1 937 3241992 • www.heritagecenter.us •
C.E.O.: *Robert Furman* • Cur.: *Virginia Weygandt* •
Local Museum – 1897
Newspaper files 1829-1970, early tools and farm equip, pioneer utensils ... 53235

Springfield Museum of Art, 107 Cliff Park Rd, Springfield, OH 45501 • T: +1 937 3254673 • F: +1 937 3254674 • www.spfld-museum-of-art.org •
Dir.: *Mark Chepp* • Cur.: *Dominique Vasseur* •
Fine Arts Museum – 1946
19th-20th c American art ... 53236

Springfield OR

Emerald Art Center, 500 Main St, Springfield, OR 97477 • T: +1 541 7268595 • F: +1 541 7262954 • emerald@epud.net • www.emeraldartcenter.org •
Dir.: *Cheryl Dorbolo* •
Fine Arts Museum / Public Gallery – 1957
Paintings, photographs and sculptures ... 53237

Springfield Museum, 590 Main St, Springfield, OR 97477 • T: +1 541 7263677 • F: +1 541 7263689 • kjensen@ci.springfield.or.us • www.springfieldmuseum.com •
Dir.: *Kathleen A. Jensen* • Cur.: *Estelle McCafferty* •
Local Museum – 1981
Local hist ... 53238

Springfield VT

Eureka School House, State Rte 11, Springfield, VT 05156 • T: +1 802 8283051 • F: +1 802 8283206 • john.dumville@state.vt.us • www.historicvermont.org •
Head: *John P. Dumville* •
Local Museum – 1968
Period furnishings, memorabilia, hist structures ... 53239

Miller Art Center, 9 Elm St, Springfield, VT 05156-0313 • T: +1 802 8852415 •
Dir.: *Robert McLaughlin* •
Local Museum / Fine Arts Museum – 1956
Portraits, pewter, Bennington pottery, paintings, toys, costumes, sculpture, crafts, machine tool industry ... 53240

Springs PA

Springs Museum, Rte 669, Springs, PA 15562 • T: +1 814 6622625 • F: +1 814 4452263 •
Pres.: *Joseph Bender* •
Local Museum – 1957
Local hist ... 53241

Springville NY

Warner Museum, Concord Historical Society, 98 E Main St, Springville, NY 14141 • T: +1 716 5920094 • www.townofconcordnyhistoricalsociety.org •
Cur.: *Donald Orton* •
Local Museum – 1953
Carpenter tools, Pop Warner's Indian artifact coll, local hist, furniture ... 53242

Springville UT

Springville Museum of Art, 126 E 400 S, Springville, UT 84663 • T: +1 801 4892727 • F: +1 801 4892739 • sharon@admn.shs.nebo.edu • www.sma.nebo.edu •
Dir.: *Dr. Vern G. Swanson* • Ass. Dir.: *Dr. Sharon R. Gray* •
Fine Arts Museum – 1903
Utah art ... 53243

Spruce Pine NC

Museum of North Carolina Minerals, Milepost 331, Blue Ridge Pkwy at Hwy 226, Spruce Pine, NC 28777 • T: +1 828 7652761 • F: +1 828 7650974 •
Head: *Tim Francis* •
Natural History Museum – 1955 ... 53244

Staatsburg NY

Staatsburgh State Historic Site, Old Post Rd, Staatsburg, NY 12580 • T: +1 845 8898851 • F: +1 845 8898321 • melodye.moore@oprhp.state.ny.us •
Pres.: *Eleanor Williamson* •
Local Museum – 1938
Mansion with 19th/20th c furniture of the Mills family, fine and dec arts ... 53245

Stafford County VA

George Washington's Ferry Farm, 268 Kings Hwy, Stafford County, VA 22405 • T: +1 540 3700732 • F: +1 540 3713398 • raudenbush@gwffoundation.org • www.kenmore.org •
Historical Museum
Archaeological, structures, hands-on exhib ... 53246

Stamford CT

International Museum of Cartoon Art, POB 16819, Stamford, CT 06905 • T: +1 203 9680748 • F: +1 203 9681863 • correspondence@cartoon.org • www.cartoon.org •
Fine Arts Museum
Cartoon drawings and illustrations ... 53247

Stamford Historical Society Museum, 1508 High Ridge Rd, Stamford, CT 06903 • T: +1 203 3221565 • F: +1 203 3221607 • shsadmin@flvax.ferg.lib.ct.us • www.stamfordhistory.org •
Pres.: *Thomas Zoubek* •
Local Museum / Decorative Arts Museum – 1901
History, decorative arts ... 53248

Stamford Museum and Nature Center, 39 Scofieldtown Rd, Stamford, CT 06903 • T: +1 203 3221646 • F: +1 203 3220408 • smnc@stamfordmuseum.org • www.stamfordmuseum.org •
Dir.: *Kenneth Marchione* • *Sharon Blume* • Cur.: *Rosa Portell* •
Fine Arts Museum / Natural History Museum – 1936
Natural hist, art ... 53249

Whitney Museum of American Art at Champion, 1 Champion Plaza, Stamford, CT 06921 • T: +1 203 3587652 • F: +1 203 3582975 •
Cur.: *Cynthia Roznoy* •
Fine Arts Museum – 1981
Art ... 53250

Standish ME

Marrett House, Rte 25, Standish, ME 04084 • T: +1 207 6423032, +1 207 3842454 • F: +1 207 3848192 • MarrettHouse@HistoricNewEngland.org • www.historicnewengland.org •
Historical Museum – 1944
Furnishings, local hist ... 53251

Stanfield NC

Reed Gold Mine State Historic Site, 9621 Reed Mine Rd, Stanfield, NC 28163 • T: +1 704 7214653 • F: +1 704 7214657 • reedmine@ctc.net • www.itpi.dpi.state.nc.us/reed •
Pres.: *Don McNeely* •
Science&Tech Museum – 1971 ... 53252

Stanford CA

Iris & B. Gerald Cantor Center for Visual Arts, Stanford University, Lomita Dr and Museum Way, Stanford, CA 94305-5060 • T: +1 650 7234177 • F: +1 650 7250464 • bbarryte@stanford.edu • www.museum.stanford.edu •
Dir.: *Thomas K. Seligman* • Chief Cur.: *Bernard Barryte* • Cur.: *Xiaoneng Yang* (Asian Art) • *Betsy Fryberger* (Prints and Drawings) • *Hilarie Faberman* (Modern and Contemporary Art) • *Barbara Thompson* (Art of Africa and the Americas) • *Patience Young* (Education) •
Fine Arts Museum / University Museum – 1891
Photos, paintings, sculpture, prints, drawings, decorative arts, European Oriental, pre-Columbian, American and African art – Rodin sculpture garden ... 53253

Stanford KY

William Whitley House, 625 William Whitley Rd, Stanford, KY 40484 • T: +1 606 3552881 • F: +1 606 3552778 •
Dir.: *Jack C. Bailey* •
Historical Museum – 1938
1792 William Whitley House, the first brick home W of the Alleghenies ... 53254

Stanford MT

Judith Basin County Museum, 19 Third S, Stanford, MT 59479 • T: +1 406 5662974, 5662277 ext.123 •
Pres. & Dir.: *Florence E. Harris* • Dir.: *Oliver Olson* • *Gene Ernst* • *Lorraine Boeck* • *Tess Brady* • Cur.: *Virginia B. Hayes* •
Local Museum – 1966
American Indian artifacts, period furnishings, Russel prints, period toys, clothes, canes, war items ... 53255

Stanton ND

Knife River Indian Villages National Historic Site, 564 County Rd 37, Stanton, ND 58571 • T: +1 701 7453309 • F: +1 701 7453708 • knri_information@nps.gov • www.nps.gov/knri •
Local Museum – 1974
Hist and culture of the Native Americans of the Northern Plains, Hidatsa and Mandan cultural artifacts ... 53256

Stanton TX

Martin County Historical Museum, 207 E Broadway, Stanton, TX 79782 • T: +1 915 7562722 •
Pres.: *Eugene Byrd* • Cur.: *Helen Thrailkill* •
Local Museum – 1969
County hist, farming, ranching & oil production ... 53257

Starkville MS

Oktibbeha County Heritage Museum, 206 Fellowship St, Starkville, MS 39759 • T: +1 601 3230211, 6511 •
Historical Museum – 1979
Old dairy equip, wedding dresses, military service uniforms ... 53258

State College PA

Centre County Historical Society Museum, 1001 E College Av, State College, PA 16801 • T: +1 814 2344779 • F: +1 814 2341694 • cchs@uplink.net • centrecountyhistory.org •
Local Museum – 1904
Victorian furnishings, clothing, tools, business, hist, Thompson family ... 53259

Staten Island NY

Alice Austen House Museum, 2 Hylan Blvd, Staten Island, NY 10305 • T: +1 718 8164506 • F: +1 718 8153959 • eaausten@aol.com • www.aliceausten.org •
Exec. Dir.: *Carl Rutberg* •
Local Museum – 1985
Historic house with Victorian furniture, local hist, photographs by Alice Auston ... 53260

The Conference House, 7455 Hylan Blvd, Staten Island, NY 10307 • T: +1 718 9846046 •
Pres.: *Madalen Bertolini* •
Local Museum – 1927
Hist house, site of peace conference during Revolutionary War, 18th c furnishings and accessoires ... 53261

Garibaldi-Meucci Museum, 420 Tompkins Av, Staten Island, NY 10305 • T: +1 718 4421608 • F: +1 718 4428635 • info@garibaldimeuccimuseum.org • garibaldimeuccimuseum.org •
Dir./Administrator: *Nicole Fenton* •
Local Museum / Historical Museum – 1956
Life and work of Antonio Meucci (1808-1889) and Giuseppe Garibaldi (1807-1882), Italian hist and culture, the Risorgimento ... 53262

Historic Richmond Town, 441 Clarke Av, Staten Island, NY 10306 • T: +1 718 3511611 • F: +1 718 3516057 • sihs-secretary@si.rr.com • www.historicrichmondtown.org •
Exec.Dir.: *John W. Guild* • Chief Cur.: *Maxine Friedman* •
Historical Museum – 1935
Historic village incl 28 historic bldgs from late 17th c to early 20th c ... 53263

Jacques Marchais Museum of Tibetan Art, 338 Lighthouse Av, Staten Island, NY 10306 • T: +1 718 9873500 • F: +1 718 3510402 • inquiries@tibetanmuseum.com • www.tibetanmuseum.com •
Exec. Dir.: *Jeanann Celli* •
Fine Arts Museum / Ethnology Museum / Folklore Museum – 1945
Tibetan and Buddhist art ... 53264

The Noble Maritime Collection, 1000 Richmond Tce, Staten Island, NY 10301 • T: +1 718 4476490 • F: +1 718 4476056 • erinurban@noblemaritime.org • www.noblemaritime.org •
C.E.O.: *Erin Urban* •
Fine Arts Museum – 1986
Lithographs, paintings, drawings, photos, marine artifacts ... 53265

Seguine-Burke Plantation, 440 Seguine Av, Staten Island, NY 10307, mail addr: 200 Nevada Av, Staten Island, NY 10306 • T: +1 718 9673542, 6676042 • F: +1 718 9675689 •
Decorative Arts Museum – 1989
Antique furnishings, stable, barn, carriage house 53266

Snug Harbor Cultural Center, Newhouse Center for Contemporary Art, 1000 Richmond Terrace, Staten Island, NY 10301 • T: +1 718 4482500 • F: +1 718 4428534 • newhouse@snug-harbor.org • snug-harbor.org •
C.E.O.: *Paul Goldberg* •
Historic Site / Public Gallery – 1976
Artifacts of Sailors Snug Harbor, glass, wood and iron works of nautical context, contemp art ... 53267

Staten Island Children's Museum, 1000 Richmond Terrace, Snug Harbor, Staten Island, NY 10301 • T: +1 718 2732060 • F: +1 718 2732836 • drosenthal@sichildrensmuseum.org •
Exec. Dir.: *Dina R. Rosenthal* •
Special Museum – 1974 ... 53268

Staten Island Historical Society Museum, 441 Clarke Av, Staten Island, NY 10306 • T: +1 718 3511611 • F: +1 718 3516057 • www.historicrichmondtown.org •
Dir.: *John Guild* • Chief Cur.: *Maxine Friedman* •
Local Museum – 1856
Folk art, furniture, china, glass, costumes, military, local hist ... 53269

Staten Island Institute of Arts and Sciences, 75 Stuyvesant Pl, Staten Island, NY 10301 • T: +1 718 7271135 • F: +1 718 2735683 • emailhenryk@aol.com •
Cur.: *Bart Bland* (Art) • *Edward Johnson* (Science) • *Patricia Salmon* (History) •
Fine Arts Museum / Historical Museum / Science&Tech Museum – 1881 ... 53270

Statesboro GA

Gallery 303, c/o Betty Foy Sanders Department of Art, Georgia Southern University, 224 Pittman Dr, POB 8032, Statesboro, GA 30460 • T: +1 912 6815358 • F: +1 912 6815104 • mdjones@georgiasouthern.edu • class.georgiasouthern.edu/art •
Public Gallery ... 53271

Georgia Southern University Museum, Rosenwald Bldg, Statesboro, GA 30460 • T: +1 912 6815444 • F: +1 912 6810729 • dharvey@gasou.edu • ceps.georgiasouthern.edu/museum/ •
Dir.: *Dr. Brent W. Tharp* • Cur.: *Dr. Jonathan Geisler* (Paleontology) •
University Museum / Natural History Museum – 1980
Natural and cultural hist of the Coastal Plain ... 53272

Statesville NC

Fort Dobbs State Historic Site, US 21, 1 mile off I-40 to SR 1930, West 1 1/2 miles on 1930, Statesville, NC 28677 • T: +1 704 8735866 • F: +1 704 8735866 • fortdobbs@statesville.net • www.fortdobbs.org •
C.E.O.: *Beth L. Carter* •
Military Museum / Open Air Museum – 1969
Mid-18th c fort and hist ... 53273

Iredell Museum of Arts & Heritage, 1335 Museum Rd, Statesville, NC 28625 • T: +1 704 8734734 • F: +1 704 8734407 • iredellmuseum@yahoo.com • www.iredellmuseum.com •
Exec. Dir.: *Henry Poore* •
Fine Arts Museum / Natural History Museum / Science&Tech Museum – 1956
Painting and sculpture, Indian artifacts and pioneer tools, 19th c glass, dolls and clothing ... 53274

Staunton VA

Frontier Culture Museum, 1290 Richmond Rd, Staunton, VA 24401 • T: +1 540 3327850 • F: +1 540 3329989 • info@frontiermuseum.state.va.us • www.frontiermuseum.org •
Folklore Museum – 1982
German, Scotch-Irish, English and American folklife and culture ... 53275

Hunt Gallery, Mary Baldwin College, Market and Vine Sts, Staunton, VA 24401 • T: +1 540 8877196 • F: +1 540 8877139 • pryan@mbc.edu • www.mbc.edu •
Dir.: *Paul Ryan* • Ass. Dir.: *Marlena Hobson* •
Public Gallery – 1842
20th c paintings, drawings, prints & photographs, contemporary works by regional & national artists ... 53276

Museum of American Frontier Culture, 1250 Richmond Rd, Staunton, VA 24401 • T: +1 540 3327850 • F: +1 540 3329989 • info@fcmv.virginia.gov • www.frontiermuseum.org •
Exec. Dir.: *John A. Walters* •
Historical Museum – 1982
European and American folklife and culture, focusing on Germany, England, Northern Ireland, America and immigration & culture ... 53277

Staunton Augusta Art Center, 1 Gyspy Hill Park, Staunton, VA 24401 • T: +1 540 8852028 • F: +1 540 8856000 • info@saartcenter.org • www.saartcenter.org •
Exec. Dir.: *Margo McGirr* •
Fine Arts Museum – 1961
Art ... 53278

Woodrow Wilson Presidential Library at his Birthplace, 20 N Coalter St, Staunton, VA 24401 • T: +1 540 8850897 • F: +1 540 8869874 • woodrow@cfw.com • www.woodrowwilson.org •
Cur.: *Edmund D. Potter* •
Historical Museum – 1938
Historic house, birthplace of Woodrow Wilson, academic and social life, presidential hist ... 53279

Steamboat Springs CO

Tread of Pioneers Museum, 8th and Oak Sts, Steamboat Springs, CO 80477 • T: +1 970 8792214 • F: +1 970 8796109 • topmuseum@springsips.com •
Dir.: *Candice Lombardo* • Cur.: *Katie Peck* •
Local Museum – 1959
Skiing hist, ranching, Routt County historical docs, native American coll, pioneer and turn-of-the c home furnishings, mining and agriculture hist ... 53280

Steeles Tavern VA

Cyrus H. McCormick Memorial Museum, 128 McCormick Farm Circle, Steeles Tavern, VA 24476 • T: +1 540 3772255 • F: +1 540 3775850 • dafiske@vt.edu •
Head: *David A. Fiske* •
Local Museum – 1956
Agriculture, replicas of first & subsequent reapers ... 53281

Steilacoom WA

Steilacoom Historical Museum, 112 Main St, Steilacoom, WA 98388-0016 • T: +1 253 5844133 •
Pres.: *Partricia J. Randall* • Sc. Staff: *Patricia J. Laughlin* (Registrar) •
Local Museum – 1970
Local hist, pioneer memorabilia ... 53282

Stephenville TX

Stephenville Museum, 525 E Washington St, Stephenville, TX 76401 • T: +1 254 9655880 • llohr@our-town.com •
Pres.: *Betty Heath* •
Local Museum – 1965
Pressed glass, rocks, eaarly day tools ... 53283

Sterling CO

Overland Trail Museum, Junction I-76 and Hwy-6 E, Sterling, CO 80751-0400 • T: +1 970 5223895 • F: +1 970 5210632 • www.sterlingcolo.com •
Cur.: *Anna Mae Hagemeier* • Ass. Cur.: *Cathy Asmus* • *Lana Tramp* • *Marilyn Hutt* •
Local Museum – 1936
Regional history ... 53284

Sterling IL

Sterling-Rock Falls Historical Society Museum, 1005 E 3rd St, Sterling, IL 61081 • T: +1 815 6226215 • srfhs@coiinc.com •
Pres.: *Tim Keller* • Cur.: *Terence Buckaloo* •
Local Museum – 1959
Local history, military, paintings, industrial – library, archives ... 53285

Sterling NY

Sterling Historical Society and Little Red Schoolhouse Museum, Rte 104A, Sterling, NY 13156, mail addr: 14352 Woods Rd, Sterling, NY 13156-4116 • T: +1 315 9476461 • sterlinghistory@lakeontario.net • www.lakeontario.net/sterlinghistory •
C.E.O.: *Don H. Richardson* •
Local Museum – 1976
Hist houses and rooms, Civil War and local hist, Indian artifacts, cameras and developing equip, glass negatives ... 53286

Sterling VA

Loudon Heritage Farm Museums, 21668 Heritage Farm Rd, Sterling, VA 20164 • T: +1 703 4215322 • F: +1 703 4339347 • www.loudonfarmmuseum.org •
Cur.: *Eric Larson* •
Agriculture Museum – 1999
Agricultural implements and machinery, horticulture ... 53287

Steubenville OH

Jefferson County Historical Association Museum, 426 Franklin Av, Steubenville, OH 43952 • T: +1 740 2831133 • www.rootsweb.com/~ohjcha •
Pres.: *Eleanor Naylor* •
Local Museum – 1976
Transportation exhibits, bridal gawns, presidents born in Ohio ... 53288

Stevens Point WI

The Museum of Natural History, 900 Reserve St, University of Wisconsin, Stevens Point, WI 54481 • T: +1 715 3464224, 3462858 • F: +1 715 3464213 • mark.williams@uwsp.edu • www.uwsp.edu/museum/ •
Cur.: *Edward Marks* •
Natural History Museum / University Museum – 1966
Mammals, birds, egg coll, reptiles, fossils ... 53289

Portage County Museum, 1475 Water St, Stevens Point, WI 54481 •
Pres.: *Tim Siebert* • Cur.: *Anton Anday* •
Local Museum – 1952
Area domestic liefe, military, local Jewish heritage ... 53290

Stevenson MD

Villa Julie College Gallery, 1525 Greenspring Valley Rd, Stevenson, MD 21153 • T: +1 410 4867000 • F: +1 410 4863552 •
University Museum / Fine Arts Museum
Contemporary and hist art exhibitions ... 53291

Stevenson WA

Columbia George Interpretive Center, 990 SW Rock Creek Dr, Stevenson, WA 98648 • T: +1 509 4278211 • F: +1 509 4277429 • info@columbiagorge.org • www.columbiagorge.org •
Dir.: *Sharon Tiffany* •
Local Museum – 1959
Indian artifacts, pioneer artifacts, Corliss steam engine, 1917 Curtiss JN-4 biplane ... 53292

Still River MA

Harvard Historical Society Museum, 215 Still River Rd, Still River, MA 01467 • T: +1 978 4568285 • curator@harvardhistory.org • www.harvardhistory.org •
Cur.: *Melanie Clifton-Harvey* •
Local Museum – 1897
Local hist, costumes, furniture, decorative arts ... 53293

Stillwater MN

Warden's House Museum, 602 N Main St, Stillwater, MN 55082 • T: +1 612 4395956 • brent@wchsmn.org • www.wchsmn.org •
Exec. Dir.: *Brent Peterson* •
Local Museum – 1941
Lumberjack tools, Indian artifacts, military & prison items, pictures, costumes, decorative arts ... 53294

Stillwater NY

Saratoga National Historical Park, 648 Rte 32, Stillwater, NY 12170 • T: +1 518 6649821 • F: +1 518 6643349 • sara_info@nps.gov • www.nps.gov/sara •
Dir.: *Gina Johnson* • Cur.: *Christine Robinson* •
Military Museum / Open Air Museum – 1938
Site of the Battles of Saratoga, Sept. 19 and Oct. 7, 1777 ... 53295

Stillwater OK

Gardiner Art Gallery, Oklahoma State University, 108 Bartlett Center, Stillwater, OK 74078 • T: +1 405 7446016 • F: +1 405 7445767 •
Dir.: *Mark A. White* •
Fine Arts Museum / University Museum – 1965
Contemporary art, graphics ... 53296

National Wrestling Hall of Fame, 405 W Hall of Fame Av, Stillwater, OK 74075 • T: +1 405 3775243 • F: +1 405 3775244 • info@wrestlinghalloffame.org • www.wrestlinghalloffame.org •
Pres.: *Myron Roderick* • Dir.: *Tony Linville* •
Special Museum – 1976 ... 53297

Stockbridge MA

Chesterwood Museum, 4 Williamsville Rd, Stockbridge, MA 01262-0827 • T: +1 413 2983579 • F: +1 413 2983973 • chesterwood@nthp.org • www.chesterwood.org •
Dir.: *Donna Hassler* • Cur. Asst.: *Anne Cathcart* •
Historic Site – 1955
Summer estate of Daniel Chester French, sculpture, paintings, pastels, drawings, photos – library ... 53298

Merwin House, 14 Main St, Stockbridge, MA 01262 • T: +1 617 2273956 • F: +1 617 2279204 • MerwinHouse@HistoricNewEngland.org • www.historicnewengland.org/visit/homes/merwin.htm •
Decorative Arts Museum / Local Museum – 1966
European and American decorative arts and furniture ... 53299

The Mission House, Main St, Stockbridge, MA 01262 • T: +1 413 2983239 • F: +1 413 2985239 • westregion@ttor.org • www.thetrustees.org/pages/324_mission_house.cfm •
Dir.: *Stephen McMahon* •
Religious Arts Museum – 1948
1739 home of Rev. John Sergeant, first missionary to the Stockbridge Indians ... 53300

Naumkeag House, Prospect Hill Rd, Stockbridge, MA 01262 • T: +1 413 2983239 • F: +1 413 2985239 • westregion@ttor.org • www.thetrustees.org/pages/335_naumkeag.cfm •
Dir.: *Stephen McMahon* •
Historical Museum – 1959
Shingle style house, Joseph Hodges Choate home, U.S. Ambassador to the Court of Saint James, decorative artsChinese porcelain ... 53301

Norman Rockwell Museum at Stockbridge, 9 Glendale Rd, Stockbridge, MA 01262 • T: +1 413 2984100 ext 221 • F: +1 413 2984142 • postmaster@nrm.org • www.nrm.org •
Dir.: *Laurie Norton Moffatt* • Chief Cur.: *Maureen Hart Hennessey* • Cur.: *Linda Szekely* (Norman Rockwell Collections) • Sc. Staff: *Stephanie Plunkett* (Illustration Art) • *Linda Pero* (Archive) • *Melinda Georgeson* (Education) •
Fine Arts Museum – 1967
Norman Rockwell original paintings, drawings and sketches, his studio, sculptures by Peter Rockwell – archive, library ... 53302

Stockbridge Library Historical Collection, 46 Main St, Stockbridge, MA 01262 • T: +1 413 2985501 • ballen@cwmars.org • www.masscat.org •
C.E.O.: *Rosemary Schneyer* • Cur.: *Babara Allen* •
Historical Museum – 1939
Local history, Indian artifacts, Cyrus W. Field – archives ... 53303

Stockholm ME

Stockholm Historical Society Museum, 280 Main St, Stockholm, ME 04783 • T: +1 207 8965759 • jhede@mfx.net •
Pres.: *Albertine Dufour* •
Local Museum – 1976
Local history ... 53304

Stockton CA

Children's Museum of Stockton, 402 W Weber Av, Stockton, CA 95203 • T: +1 209 4654386 • F: +1 209 4654394 • www.stocktongov.com/childrensmuseum •
Dir.: *Gina Delucchi* •
Special Museum – 1991 ... 53305

The Haggin Museum, 1201 N Pershing Av, Stockton, CA 95203-1699 • T: +1 209 9406300 • F: +1 209 4621404 • info@hagginmuseum.org • www.hagginmuseum.org •
Dir.: *Tod Ruhstaller* •
Fine Arts Museum / Local Museum – 1928
French and American art, American, European and Oriental decorative arts, glass, folk art, regional history, Indian artifacts incl baskets, agriculture – library, technology & shipbuilding archives ... 53306

Reynolds Gallery, University of the Pacific, Jeannette Powell Art Center, 1071 Mendecino Av, Stockton, CA 95211 • T: +1 209 9463096 • F: +1 209 9462518 • bschumacher1@pacific.edu • web.pacific.edu •
Dir.: *Bett Schumacher* •
Public Gallery / University Museum – 1975 ... 53307

U

Stockton Springs ME

Fort Knox State Historic Site, 711 Fort Knox Rd, Stockton Springs, ME 04981 • T: +1 207 4697719 • F: +1 207 4697719 •
Head: *Mike Wilusz* •
Military Museum / Historic Site – 1943 53308

Stone Mountain GA

Art Sation, 5384 Manor Dr, Stone Mountain, GA 30086 • T: +1 770 4691105 • F: +1 770 4690355 • info@artstation.org • www.artstation.org •
Dir.: *David T. Thomas* • *Jon Goldstein* •
Fine Arts Museum
Contemporary art 53309

Georgia's Stone Mountain Park, Hwy 78, Stone Mountain, GA 30086, mail addr: POB 689, Stone Mountain, GA 30086 • T: +1 770 4985690 • F: +1 770 4985607 • g.durham@stonemountainpark.org • www.stonemountainpark.org •
Head: *Curtis Branscome* •
Local Museum / Natural History Museum – 1958
Indian artifacts, geology, Civil War, automobiles, musical instr, English and American furniture 53310

Stonington CT

Captain Nathaniel B. Palmer House, 40 Palmer St, Stonington, CT 06378 • T: +1 860 5358445 • www.stoningtonhistory.org •
Pres.: *Michael Davis* • Cur.: *Constance B. Colon* •
Local Museum – 1996
Historic house, Antarctica discovery 53311

Old Lighthouse Museum, 7 Water St, Stonington, CT 06378 • T: +1 860 5351440 • director@stoningtonhistory.org • www.stoningtonhistory.org •
Pres.: *James Boylan* •
Local Museum – 1925
Local hist, paintings – archive 53312

Stony Brook NY

Long Island Museum of American Art, History and Carriages, 1200 Rte 25A, Stony Brook, NY 11790 • T: +1 631 7510066 • F: +1 631 7510353 • mail@longislandmuseum.com • www.longislandmuseum.org •
Pres. & C.E.O.: *Jacqueline Day* • Dep. Dir.: *William Ayres* (Collections, Interpretation) • Cur.: *Joshua Ruff* (History) • Asst. Cur.: *Eva Greguski* (Art) •
Fine Arts Museum / Historical Museum – 1935
American art, hist of transportation, carriages and accoutrements 53313

Museum of Long Island Natural Sciences, Earth and Space Sciences Bldg, State University of New York at Stony Brook, Stony Brook, NY 11794-2100 • T: +1 631 6328230 • F: +1 631 6328240 • Pamela.Stewart@sunysb.edu •
Dir.: *Pamela Stewart* • Cur.: *Steven E. Englebright* (Geology) •
University Museum / Natural History Museum – 1973
Natural hist coll of Long Island, insects and host plants, minerals and fossils, marine invertebrates 53314

University Art Gallery, Staller Center for the Arts, State University of New York at Stony Brook, Stony Brook, NY 11794-5425 • T: +1 516 6327240 • F: +1 516 6327354 • rcooper@notes.cc.sunysb.edu •
Dir.: *Rhonda Cooper* •
Fine Arts Museum / University Museum – 1975 53315

Stony Point NY

Stony Point Battlefield State Historic Site, Park Rd off US Rte 9W, Stony Point, NY 10980, mail addr: Box 182, Stony Point, NY 10980 • T: +1 914 7862521 • F: +1 914 7860463 • spbattle@ric.lhric.org • www.nysparks.com •
Head: *Julia M. Warger* •
Military Museum / Open Air Museum – 1897
Site of raid on British stronghold by Brigadier Gen. Anthony Wayne on July 15, 1779 53316

Storm Lake IA

Witter Gallery, 609 Cayuga St, Storm Lake, IA 50588 • T: +1 712 7323400 • wittergallery@yahoo.com • www.stormlake-ia.com/witter •
Dir.: *Jennifer Jochims* •
Fine Arts Museum – 1972
Art 53317

Storrs CT

Connecticut State Museum of Natural History, 2019 Hillside Rd, Unit 1023, Storrs, CT 06269-1023 • T: +1 860 4864460 • F: +1 860 4860827 • sue.broneill@uconn.edu • www.mnh.uconn.edu •
Dir.: *Leanne Kennedy Harty* •
University Museum / Natural History Museum – 1982
Natural hist 53318

Mansfield Historical Society Museum, 954 Storrs Rd, Storrs, CT 06268 • T: +1 860 4296575 • F: +1 860 4296575 • museum@mansfield-history.org • www.mansfield-history.org •
Dir.: *Ann Galonska* •
Local Museum – 1961
Local hist, farming, industry 53319

The William Benton Museum of Art, University of Connecticut, 245 Glenbrook Rd, Storrs, CT 06269-2140 • T: +1 860 4864520 • F: +1 860 4860234 • diane.lewis@uconn.edu • www.benton.uconn.edu •
Dir.: *Salvatore Scalora* • Cur.: *Thomas P. Bruhn* •
Fine Arts Museum / University Museum – 1966
Art 53320

Story WY

Fort Phil Kearny, 528 Wagon Box Rd, Story, WY 82842 • T: +1 307 6847629 • F: +1 307 6847967 • sreisc@missc.state.wy.us • www.philkearny.vcn.com •
Dir.: *John Keck* • Cur.: *Robert C. Wilson* •
Military Museum – 1913
Military hist, Native American hist 53321

Stoughton WI

Stoughton Historical Society Museum, 324 S Page Str, Stoughton, WI 53589, mail addr: 1547 Hwy 51 N, Stoughton, WI 53589 • T: +1 608 8733162 •
Local Museum – 1960 53322

Stoutsville MO

Mark Twain Birthplace, 37352 Shrine Rd, Stoutsville, MO 65283 • T: +1 573 5653449 • F: +1 573 5653718 • moparks@dnr.mo.gov • www.mostateparks.com •
Head: *John Huffman* •
Historical Museum – 1960
Paintings, manuscripts, Samuel Clemens memorabilia 53323

Stow MA

Stow West 1825 School Museum, Harvard Rd, Stow, MA 01775 • T: +1 978 8977417 • selectmen@stow-ma.gov •
Historical Museum – 1974
Local school hist, maps, furnishings 53324

Stowe VT

Helen Day Art Center, School St, Stowe, VT 05672 • T: +1 802 2538358 • F: +1 802 2532703 • mail@helenday.com • www.helenday.com •
C.E.O.: *Mickey Myers* •
Fine Arts Museum – 1981
Exhibit-related items for sale 53325

Strafford VT

Justin Smith Morrill Homestead, 214 Justin Morrill Memorial Hwy, Strafford, VT 05072 • T: +1 802 8283051 • F: +1 802 8283206 • john.dumville@state.vt.us • www.historicvermont.org •
Head: *John P. Dumville* •
Historic Site – 1969
Historic house, agricultural bldgs 53326

Strasburg CO

Comanche Crossing Museum, 56060 E Colfax Av, Strasburg, CO 80136 • T: +1 303 6224690 •
Cur.: *Sandy Miller* •
Local Museum / Science&Tech Museum – 1969
Regional history, site where the first continuous chain of rails was completed by the Kansas Pacific Railroad, Aug. 15, 1870 53327

Strasburg PA

The National Toy Train Museum, 300 Paradise Lane, Strasburg, PA 17579 • T: +1 717 6878976, 6878623 • F: +1 717 6870742 • toytrain@traincollectors.org • www.traincollectors.org •
Pres.: *Newton Derby* •
Special Museum – 1954
Toy and model trains 53328

Railroad Museum of Pennsylvania, 300 Gap Rd, Strasburg, PA 17579 • T: +1 717 6878629 • F: +1 717 6870876 • info@rrmuseumpa.org • www.rrmuseumpa.org •
Dir.: *David W. Dunn* • Cur.: *Bradley Smith* •
Science&Tech Museum / Historical Museum – 1963
Transportation, passanger and freight locomotives, railcars, PA railroad history – library, archives 53329

Strasburg VA

Crystal Caverns at Hupp's Hill, 33231 Old Valley Pike, Strasburg, VA 22657 • T: +1 540 4655884 • www.waysideofva.com •
C.E.O.: *Babs B. Melton* •
Special Museum – 1998
Hist cavern artifacts 53330

Jeane Dixon Museum and Library, 132 N Massanutten St, Strasburg, VA 22657 • T: +1 540 4655884 • wayside@shentel.net • www.waysideofva.com •
C.E.O.: *Babs B. Melton* •
Local Museum – 2002
Personal artifacts, furniture, artworks, Presidential studies 53331

Museum of American Presidents, 130 N Massanutten St, Strasburg, VA 22657 • T: +1 540 4655999 • F: +1 540 4658157 • wayside@shentel.net • www.waysideofva.com •
C.E.O. & Dir.: *Babs B. Melton* •
Historical Museum – 1996
Presidential portraits, signatures and memorabilia 53332

Stonewall Jackson Museum at Hupp's Hill, 33229 Old Valley Pike, Strasburg, VA 22657 • T: +1 540 4655884 • www.waysideofva.com •
C.E.O.: *Babs B. Melton* •
Special Museum – 1991
Civil War artifacts from 1862-1864, military and civilian artifacts, 1864 infantry and artillery trenches 53333

Strasburg Museum, 440 E King St, Strasburg, VA 22657 • T: +1 540 4653428, 4653175 • adamsons@shentel.net • csonner.net/museum.htm •
Pres.: *John Adamson* •
Folklore Museum – 1970
Strasburg pottery 1830-1910 53334

Stratford CT

The Stratford Historical Society and Catherine B. Mitchell Museum, 967 Academy Hill, Stratford, CT 06615 • T: +1 203 3780630 • F: +1 203 3782562 • judsonhousestfd@aol.com • www.stratfordhistoricalsociety.com •
Cur.: *Carol Lovell* •
Local Museum – 1925
Local hist 53335

Stratford VA

Robert E. Lee Memorial Association, 485 Great House Rd, Stratford, VA 22558 • T: +1 804 4938038 • F: +1 804 4930333 • shpedu@stratfordhall.org • www.stratfordhall.org •
Pres.: *Hugh G. Van der Veer* • C.E.O.: *Thomas C. Taylor* •
Association with Coll – 1929
18th & 19th c decorative arts, silver, archaeology artifacts – archives 53336

Strawberry Point IA

Wilder Memorial Museum, 123 W Mission, Strawberry Point, IA 52076 • T: +1 563 9334615 •
Pres.: *Larry DeShaw* •
Local Museum / Decorative Arts Museum – 1970
Local hist, dolls, antiques, art, tools 53337

Stroudsburg PA

Driebe Freight Station, 537 Ann St, Stroudsburg, PA 18360-2012 • T: +1 570 4241776 • F: +1 570 4219199 • mcha@ptd.net • www.mcha-pa.org •
Pres.: *William Ramsden* • Dir.: *Candace McGreevy* •
Local Museum – 1921
Local hist 53338

Elizabeth D. Walters Library, 900 Main St, Stroudsburg, PA 18360 • T: +1 570 4217703 • F: +1 570 4219199 •
Dir.: *Candace McGreevy* •
Decorative Arts Museum
Decorative arts, furniture, Indian artifacts, textiles 53339

Quiet Valley Living Historical Farm, 1000 Turkey Hill Rd, Stroudsburg, PA 18360 • T: +1 570 9926161 • F: +1 570 9929587 • farm@quietvalley.org • www.quietvalley.org •
Exec. Dir.: *Debra A. Seip* • Farm Dir.: *Milton Mosier* • Education Dir.: *Cheryl Statham* •
Agriculture Museum – 1963
Historic farm buildings, 19th century farming equipment, woodworking tools, blacksmithing tools, domestic rural collections 53340

Stroud Mansion, 900 Main St, Stroudsburg, PA 18360-1604 • T: +1 570 4217703 • F: +1 570 4219199 • mcha@ptd.net • mcha-pa.org/member.html •
Pres.: *William Ramsden* • Exec. Dir.: *Candace McGreevy* •
Local Museum – 1921
Historic house 53341

Stuart FL

Elliott Museum, 825 NE Ocean Blvd, Stuart, FL 34996 • T: +1 561 2251961 • F: +1 561 2252333 • reneeb@elliottmuseumfl.org • www.elliottmuseumfl.org •
C.E.O: *Robin Hicks-Connors* •
Local Museum – 1961
Furniture, antique cars, costumes, Indian artifacts 53342

Gilbert's Bar House of Refuge, 301 SE MacArthur Blvd, Stuart, FL 34996 • T: +1 561 2251961 • F: +1 561 2252333 • hsmc@bellsouth.net •
Historical Museum – 1875 53343

Sturbridge MA

Old Sturbridge Village, 1 Old Sturbridge Village Rd, Sturbridge, MA 01566 • T: +1 508 3473362 • F: +1 508 3470375 • osv@osv.org • www.osv.org •
Pres./ CEO: *Beverly K. Sheppard* •
Open Air Museum / Local Museum – 1946
Re-created rural New England village – library, archives 53344

Sturgeon Bay WI

Door County Maritime Museum, 120 N Madison Av, Sturgeon Bay, WI 54235 • T: +1 920 7435958 • F: +1 920 7439483 • dcmm@itol.com • www.dcmm.org •
Pres.: *Dan Austad* • Exec. Dir.: *Brian Kelsey* •
Science&Tech Museum – 1969
Maritime hist, shipbuilding, fishery artifacts 53345

Door County Museum, 18 N 4th & Michigan Av, Sturgeon Bay, WI 54235 • T: +1 414 7435809 • dcmuseum@co.door.wi.us •
Cur.: *Margaret S. Weir* •
Local Museum – 1939
Belgian and Scandinaavian settlers, farm and dairy, orchards 53346

Fairfield Center for Contemporary Art, 242 Michigan St, Sturgeon Bay, WI 54235 • T: +1 920 7460001 • info@fairfieldcenter.org • www.fairfieldcenter.org •
Fine Arts Museum
Works of Henry Moore, Kandinsky, Magritte, Giacometti 53347

The Farm, 4285 Hwy 57, Sturgeon Bay, WI 54235 • T: +1 920 7436666 • F: +1 262 3761551 • thetancks@aol.com •
Agriculture Museum – 1965
Agriculture, farm tools and implements 53348

Miller Art Museum, 107 S 4th Av, Sturgeon Bay, WI 54235 • T: +1 920 7460707 • F: +1 920 7460865 • bmam@dcwis.com • www.doorcountyarts.com •
Dir.: *Bonnie Hartmann* • Cur.: *Deborah Rosenthal* •
Public Gallery – 1975
Permanent coll of 20th c Wisconsin artists and changing exhib 53349

Sturgis SD

Bear Butte State Park Visitors Center, Hwy 79, Sturgis, SD 57785, mail addr: POB 688, Sturgis, SD 57785 • T: +1 605 3475240 • F: +1 605 3477627 • bearbutte@gfp.state.sd.us • www.sdgfp.info/Parks/Regions/WestRiver/BearButte.htm •
Local Museum – 1961
Native American clothing and religious artifacts, archaeological site, geology, military 53350

Stuttgart AR

Museum of the Arkansas Grand Prairie, 921 E Fourth St, Stuttgart, AR 72160 • T: +1 870 6737001 • ontheprairiebayou@yahoo.com • www.stuttgartmuseum.com •
Dir.: *Pat Peacock* • Cur.: *Sheila Stoner* (Collections) •
Folklore Museum / Agriculture Museum – 1972
Science of rice, hay and soybean industry, farming 53351

Sublette KS

The Haskell County Historical Museum, Haskell County Fairgrounds, Sublette, KS 67877 • T: +1 620 6758344 • museum@pld.com •
Dir.: *Rosa Kraber* •
Local Museum – 1983
Rocks and minerals, American Indian artifacts 53352

Sudbury MA

Longfellow's Wayside Inn Museum, 72 Wayside Inn Rd, Sudbury, MA 01776 • T: +1 978 4431776 • F: +1 978 4438041 • history@wayside.org • www.wayside.org •
Head: *Guy R. LeBlanc* •
Local Museum – 1716
Local history, period furnishings 53353

Suffolk VA

Ruddick's Folly, 510 N Main St, Suffolk, VA 23434 • T: +1 757 9341390 • F: +1 757 9340411 • riddicksfolly@prodigy.net • www.riddicksfolly.org •
Dir.: *Robin K. Rountree* •
Local Museum – 1978
Riddick family, Suffolk hist, furniture and decorative arts, peanut exhibit, Mills E. Godwin, jr 53354

Suffolk Museum, 118 Bosley Av, Suffolk, VA 23434 • T: +1 757 9232371 • F: +1 757 5380833 • nkinzinger@city.suffolk.va.us •
Dir.: *Nancy Kinzinger* •
Fine Arts Museum – 1986
Art 53355

Sugar Grove IL

Air Classic Museum of Aviation, 43 W 264 Rte 30, Sugar Grove, IL 60554 • T: +1 630 4660888 •
Pres.: *Robert Atac* •
Science&Tech Museum
Planes from the Korean War 2 to Desert Storm, automobiles 53356

Sugarland TX

The Museum of Southern History, 14080 SW Freeway, Sugarland, TX 77478-3553, mail addr: POB 2190, Sugar Land, TX 77487-2190 • T: +1 281 2697171 • F: +1 218 2697179 • snoddys@snbtx.com • www.museumofsouthernhistory.org •
Dir.: *Suzie Snoddy* •
Historical Museum / Military Museum / Museum of Classical Antiquities – 1978
Concentration on Civil War period 53357

Suisun City CA

The Western Railway Museum, 5848 State Hwy 12, Suisun City, CA 94585-9641 • T: +1 707 3742978 • F: +1 707 3746742 • www.wrm.org •
Dir.: *William C. Kluver* •
Science&Tech Museum – 1946
Electric railways, trolley cars, steam and diesel locomotives – library 53358

Suitland MD

Airmen Memorial Museum, 5211 Auth Rd, Suitland, MD 20746 • T: +1 301 8998386 • F: +1 301 8998136 • staff@assahq.org • www.assahq.org •
Exec. Dir.: *James D. Staton* •
Military Museum – 1986
Military ... 53359

Sullivan's Island SC

Fort Sumter National Monument, 1214 Middle St, Sullivan's Island, SC 29482 • T: +1 843 8833123 ext 22 • F: +1 843 8833910 • fosu_ranger_activitie@nps.gov • www.nps.gov/fosu •
Head: *Fran Norton* •
Military Museum – 1948
Military hist ... 53360

Sulphur LA

Brimstone Museum, 800 Picard Rd, Sulphur, LA 70663 • T: +1 337 5277142 • F: +1 337 5270860 • westcal@usunwired.net •
Dir.: *Glenda Vincent* •
Local Museum – 1975
Local history ... 53361

Summersville WV

Carnifex Ferry Battlefield State Park and Museum, State Hwy 129, Summersville, WV 26651 • T: +1 304 8720825 • F: +1 304 8723820 •
Head: *Mark Mengele* •
Local Museum
Local hist, Civil War relics ... 53362

Summerville SC

Old Dorchester State Historic Site, 300 State Park Rd, Summerville, SC 29485 • T: +1 843 8731740 • F: +1 843 8731740 • old_dorchester_sp@prt.state.sc.us • www.southcarolinaparks.com •
Dir.: *Larry Duncan* • Sc. Staff: *Ashley Chapman* (Archaeology) •
Local Museum – 1960
Archaeological site of 18th c Village of Dorchester ... 53363

Summit NJ

New Jersey Center for Visual Arts, 68 Elm St, Summit, NJ 07901 • T: +1 908 2739121 • F: +1 908 2731457 • mari@njcva.org • www.njcva.com •
Pres.: *Eric Pryor* • Chief Cur.: *Nancy Cohen* • Sc. Staff: *Alice Dillon* •
Fine Arts Museum – 1933
Temporary exhib ... 53364

Sumter SC

South Carolina National Guard Museum, 395 N Pike Rd, Sumter, SC 29150 • T: +1 803 8061107 • F: +1 803 8061121 • tng151fa@sc-ngnet.army.mil •
Dir.: *Roy Pipkin* • Cur.: *John S. Cato* •
Military Museum – 1982
Military hist ... 53365

The Sumter County Museum, 122 N Washington St, Sumter, SC 29150 • T: +1 803 7750908 • F: +1 803 4365820 • scmuseum@sumtercountymuseum.com • www.sumtercountymuseum.com •
Dir.: *Katherine Richardson* •
Local Museum – 1976
Local hist ... 53366

Sumter Gallery of Art, 200 Hassel St, Sumter, SC 29150 • T: +1 803 7750543 • F: +1 803 7782787 • sumtergallery@sumter.net •
Dir.: *Janice W. Williams* •
Fine Arts Museum – 1970
Paintings, etchings, drawings by Elizabeth White 53367

Sun Prairie WI

Sun Prairie Historical Museum, 115 E Main St, Sun Prairie, WI 53590 • T: +1 608 8372511 • F: +1 608 8256879 • museum@sun-prairie.com • www.sun-prairie.com •
Cur.: *Peter Klein* •
Local Museum – 1967
Local hist ... 53368

Sunbury PA

The Northumberland County Historical Society Museum, The Hunter House, 1150 N Front St, Sunbury, PA 17801 • T: +1 570 2864083 • nchsmusm@sunlink.net • www.northumberlandcountyhistoricalsociety.org •
Pres.: *Scott A. Heintzelman* •
Local Museum – 1925
Local hist ... 53369

Sundance WY

Crook County Museum and Art Gallery, 309 Cleveland St, Sundance, WY 82729 • T: +1 307 2833666 • F: +1 307 2831192 • crcogallery@vcn.com •
Dir.: *Linda Evans* •
Fine Arts Museum / Local Museum – 1971
Western hist, furniture, pictures ... 53370

Sunnyside WA

Sunnyside Historical Museum, 704 S 4th St, Sunnyside, WA 98944 • T: +1 509 8376010, 8372032, 8372105 • ssmuseum@bentonrea.com •
Cur.: *Don Wade* •
Local Museum – 1972
Indian hist, pioneer kitchen, dining and living room ... 53371

Sunnyvale CA

The Lace Museum, 552 S Murphy Av, Sunnyvale, CA 94086 • T: +1 408 7304695 • F: +1 408 7304695 • suzmeyer@sbcglobal.net • www.thelacemuseum.org •
C.E.O.: *Cherie Helm* • Dir.: *Eleanore Schwarts* •
Special Museum – 1980
Lace-trimmed clothing ... 53372

Sunnyvale Heritage Park Museum, Heritage Park, 570 E Remington Dr, Sunnyvale, CA 94087 • T: +1 408 7490220 • F: +1 408 7490220 • info@heritageparkmuseum.org • www.heritageparkmuseum.org •
Pres.: *Elizabeth Koning* •
Local Museum – 1973
Local hist ... 53373

Sunol CA

Golden Gate Railroad Museum, 5550 Niles Canyon Rd, Brightside Yard, Sunol, CA 92500, mail addr: 1755 East Bayshore Rd, Suite 19A, Redwood City, CA 94063 • T: +1 650 3652472 • www.ggrm.org •
Pres.: *James Bunger* •
Science&Tech Museum – 1991
Railroad equip and artifacts ... 53374

Sunset TX

Old West Museum, 11838 Hwy, 287 S Access Rd, Sunset, TX 76270 • T: +1 940 8729698 • F: +1 940 8728504 • cowboymuseum@earthlink.net • www.cowboymuseum.net •
Dir.: *Jack Glover* •
Local Museum / Historical Museum – 1956
Indian artifacts, agriculture, hist, early farming, paintings, sculpture, graphics ... 53375

Superior MT

Mineral County Museum, Second Av E, Superior, MT 59872 • T: +1 406 8224626 • mrshezzie@blackfoot.net •
Cur.: *Cathryn J. Strombo* •
Local Museum – 1975
Local and natural hist, geology, forest, mining, mullan ... 53376

Superior WI

Douglas County Historical Society Museum, 1401 Tower Av, Superior, WI 54880 • T: +1 715 3928449 • F: +1 715 3955639 • dchs@douglashistory.org • www.douglashistory.org •
C.E.O. & Pres.: *Robert LaBounty* • Dir.: *Kathy Laakso* •
Local Museum – 1854
Ojibwa Chippewa Indian crafts, Sioux Indian portraits, local hist ... 53377

Fairlawn Mansion and Museum, 906 E 2nd St, Superior, WI 54880 • T: +1 715 3945712 • F: +1 715 3942043 • www.visitsuperior.com •
Exec. Dir.: *Dr. Richard A. Sauers* •
Decorative Arts Museum – 1963
Pattison Family hist ... 53378

Old Fire House and Police Museum, 402 23 rd Av, Superior, WI 54880 • T: +1 715 3945712 • F: +1 715 3942043 •
Special Museum – 1983
Superior polic & fire dept. hist, Wisconsin fire & police hall of fame ... 53379

Suquamish WA

Suquamish Museum, Sandy Hook Rd, Suquamish, WA 98392 • T: +1 360 3945275 • F: +1 360 5986295 • www.suquamish.nsn.us •
Dir.: *Alexis Barry* • *Marilyn Jones* • Cur.: *Charles Sigo* •
Ethnology Museum – 1983
Hist of the Suquamish Native Indian tribe ... 53380

Surprise AZ

West Valley Art Museum, 17420 N Av of the Arts, Surprise, AZ 85374 • T: +1 602 9720635 • F: +1 602 9720456 • dtooker@wvam.org • www.wvam.org •
Dir.: *David P. Tooker* •
Fine Arts Museum – 1980
American art, emphasis on Henry Varnum Poor and George Resler ... 53381

Surry VA

Chippokes Farm and Forestry Museum, 695 Chippokes Park Rd, Surry, VA 23883 • T: +1 757 2943439 • F: +1 804 3718500 • cffmuseum@dcr.virginia.gov • www.dcr.virginia.gov •
Pres.: *Frederick M. Quayle* •
Agriculture Museum / Natural History Museum – 1990
Period farm & forestry equip, tools, housewares . 53382

Smith's Fort Plantation, 217 Smith's Fort Ln, Rte 31, Surry, VA 23883 • T: +1 757 2943872 •
C.E.O.: *Elizabeth Kustelny* •
Local Museum – 1925
Indian artifacts, 17th-18th c furnishings, formal boxwood garden ... 53383

Surry County VA

Bacon's Castle, 465 Bacon's Castle Trail, Surry County, VA 23883 • T: +1 757 3575976, +1 804 6481889 • F: +1 804 7750802 • www.apva.org •
Exec. Dir.: *Elizabeth Kostelny* •
Local Museum – 1665
Local hist, oldest brick house in North America .. 53384

Susanville CA

Lassen Historical Museum, 105 N Weatherlow St, Susanville, CA 96130 • T: +1 530 2574584 •
Pres.: *Tammi Brown* •
Local Museum – 1958 ... 53385

Sussex NJ

Space Farms Zoological Park and Museum, 218 Rt 519, Sussex, NJ 07461 • T: +1 973 8753223 • F: +1 973 8759397 • fpspace@warwick.net • www.spacefarms.com •
C.E.O.: *Parker Space* •
Local Museum / Natural History Museum – 1927
Mounted wildlife, live wild animals, farm machinery, Indian artifacts ... 53386

Swampscott MA

Atlantic 1 Museum, 1 Burrill St, Swampscott, MA 01907 • T: +1 781 5815833 •
Dir.: *Richard Maitland* •
Historical Museum – 1965
Fire-Fighting engines and canons ... 53387

John Humphrey House, 99 Paradise Rd, Swampscott, MA 01907 • T: +1 781 5991297 •
Local Museum – 1921
Local history ... 53388

Swarthmore PA

List Gallery, Swarthmore College, 500 College Av, Swarthmore, PA 19081-1397 • T: +1 610 3288488 • F: +1 610 3287793 • apackar1@swarthmore.edu •
Dir.: *Andrea Packard* •
Public Gallery ... 53389

Sweet Briar VA

Camp Gallery, Virginia Center for the Creative Arts, Mount San Angelo, Sweet Briar, VA 24595 • T: +1 804 9467236 • F: +1 804 9467239 • vcca@vcca.com • www.vcca.com •
Dir.: *Suny Monk* •
Fine Arts Museum – 1971 ... 53390

Sweet Briar College Art Gallery, Sweet Briar College, Sweet Briar, VA 24595 • T: +1 804 3816248 • F: +1 804 3816173 • rmlane@sbc.edu • www.artgallery.sbc.edu/ •
Dir.: *Rebecca Massie Lane* •
Fine Arts Museum / Public Gallery / University Museum – 1901
13th-20th c European & American paintings, drawings & prints, works on paper by contemp women 53391

Sweet Briar Museum, Sweet Briar College, Sweet Briar, VA 24595 • T: +1 434 3816248, 3816262 • F: +1 424 3816173 • ccarr@sbc.edu • www.sbc.edu •
Dir.: *Christian Carr* •
University Museum – 1985
Alumnae memorabilia ... 53392

Sweetwater TX

City County Pioneer Museum, 610 E Third St, Sweetwater, TX 79556 • T: +1 915 2358547 •
Dir.: *Franzas Cupp* •
Local Museum – 1968
Furniture, docs, artifacts, china, glassware, period gowns ... 53393

Sylacauga AL

Isabel Anderson Comer Museum and Arts Center, 711 N Broadway, Sylacauga, AL 35150 • T: +1 256 2454016 • comercenter@bellsouth.net • www.comermuseum.freeservers.com •
Exec.Dir.: *Donna Rentfrow* •
Historical Museum / Fine Arts Museum – 1982
Fine art, Indian artifacts, marble sculpture, furnishings, photographs, costumes, gold and platinum records ... 53394

Sylvania OH

American Gallery, 6600 Sylvana Av, Sylvania, OH 43560 • T: +1 419 8828949 •
Dir.: *Toni Andrews* •
Public Gallery
American art ... 53395

Syracuse NE

Otoe County Museum of Memories, 366 Poplar, Syracuse, NE 68446 • T: +1 402 2693482 •
Cur.: *Rose Garey* •
Local Museum – 1972
Local Indian artifacts, Thomas A. Edison exhibit . 53396

Syracuse NY

Erie Canal Museum, 318 Erie Blvd E, Syracuse, NY 13202 • T: +1 315 4710593 • F: +1 315 4717220 • www.eriecanalmuseum.org •
Exec. Dir.: *Nancy Furry* • Cur.: *Andrew Kitzmann* •
Local Museum – 1962 ... 53397

Everson Museum of Art, 401 Harrison St, Syracuse, NY 13202 • T: +1 315 4746064 • F: +1 315 4746943 • everson@everson.org • www.everson.org •
Dir.: *Sandra Trop* • Sen. Cur.: *Debora Ryan* •
Fine Arts Museum – 1896
Ceramics, American painting ... 53398

Joe and Emily Lowe Art Gallery, Syracuse University, Shaffer Art Bldg, Syracuse, NY 13244-1230 • T: +1 315 4434098 • F: +1 315 4431303 • jhart@vpa.syr.edu •
Dir.: *Dr. Edward A. Aiken* • *Dr. Alfred T. Collette* (Syracuse University Art Collection) • Cur.: *David Prince* •
Fine Arts Museum / University Museum – 1952
19th and 20th c American painting, sculpture and graphics, African and Indian art, photography 53399

Milton J. Rubenstein Museum of Science and Technology, 500 S Franklin St, Syracuse, NY 13202-1245 • T: +1 315 4259068 • F: +1 315 4259072 • director@most.org • www.most.org •
Exec. Dir. & Pres.: *Stephen A. Karon* •
Natural History Museum / Science&Tech Museum – 1978 ... 53400

Museum of Automobile Art and Design, 6710 Brooklawn Pkwy, Syracuse, NY 13211 • T: +1 315 4328282 • F: +1 315 4328256 • info@autolit.com •
C.E.O.: *Walter Miller* •
Science&Tech Museum – 1996
Advertising, styling artwork and memorabilia of the hist of trucks, automobiles and motorcycles ... 53401

Onondaga Historical Association Museum, 321 Montgomery St, Syracuse, NY 13202 • T: +1 315 4281864 • F: +1 315 4712133 • ohamail2002@yahoo.com • www.cnyhistory.org •
Exec.Dir.: *Paul H. Pflanz* • Asst.Dir. & Cur.: *Thomas Hunter* (Collections) • Cur.: *Dennis Connors* (History) •
Local Museum – 1862
Regional hist, Underground Railroad artifacts, New York State canal hist, art ... 53402

Syracuse University Art Collection, Sims Hall, Syracuse, NY 13244 • T: +1 315 4434097 • F: +1 315 4439225 • djiacono@syr.edu • sumweb.syr.edu/suart/ •
Dir.: *Dr. Alfred T. Collette* • Cur.: *David Prince* •
Fine Arts Museum – 1871
American art 1915-1965, prints, photography ... 53403

Table Rock NE

Table Rock Historical Society Museum, 414-416 Houston St, Table Rock, NE 68447 • T: +1 402 8394135 • F: +1 402 8394135 • fv64137@navix.net •
Pres.: *Floyd Vrtiska* •
Local Museum – 1965
Furnishing, antiques, paintings, Indian artifacts .. 53404

Tacoma WA

Camp 6 Logging Museum, 5 Mile Dr, Point Defiance Park, North End of Pearl St, Tacoma, WA 98407 • T: +1 253 7520047 • camp6museum@harbornet.com • www.camp-6-museum.org/c6.html •
Dir.: *Don Olson* •
Special Museum – 1964
Logging and lumber industry (bldgs, cars and operational logging railroad from logging camps) 53405

Commencement Art Gallery, 902 Commerce, Tacoma, WA 98402-4407 • T: +1 253 5915341 • F: +1 253 5912002 • director@commencementgallery.com • www.commencementgallery.org •
Fine Arts Museum – 1993
Drawings, paintings, photography, sculpture, ceramics, metalwork ... 53406

Fort Nisqually Living History Museum, 5400 N Pearl St, Tacoma, WA 98407 • T: +1 253 5915339 • F: +1 253 7596184 • fortnisqually@tacomaparks.com • www.fortnisqually.org •
Dir.: *Melissa S. McGinnis* • Cur.: *Doreen Beard-Simpkins* •
Local Museum / Historic Site – 1937
Local hist, agricultural and farming implements . 53407

Historic Fort Steilacoom, 9601 Steilacoom Blvd SW, Tacoma, WA 98498-7213 • T: +1 253 7563928 • adm@fortsteilacoom.com • www.fortsteilacoom.com •
Pres.: *Carol E. Neufeld* •
Military Museum – 1983
Bldgs of the Fort ... 53408

James R. Slater Museum of Natural History, University of Puget Sound, Tacoma, WA 98416 • T: +1 253 8793798 • F: +1 253 8793352 • dpaulson@ups.edu • www.ups.edu/biology/museum/museum.html •
Dir.: *Dr. Dennis R. Paulson* • Cur.: *Dr. Gary Shugart* (Collections) • *Dr. Kathy Ann Miller* (Botany) • *Dr. Peter Wimberger* (Ichthyology) •
University Museum / Natural History Museum – 1926
Natural hist, Northwest flora and fauna, mammal skins and skulls, bird skins, skeletons, wings, nests and eggs, reptiles and amphibians, invertebrates ... 53409

Karpeles Manuscript Museum, 407 South Gade St, Tacoma, WA 98405 • T: +1 253 3832575 • www.karpeles.com •
Dir.: *David Shawver* •
Fine Arts Museum
Contemporary art, manuscripts, illustrations and documents 53410

Kittredge Art Gallery, University of Puget Sound, 1500 N Lawrence, Tacoma, WA 98416 • T: +1 253 8792806 • F: +1 253 8793500 • gbell@ups.edu • www.ups.edu •
Dir.: *Greg Bell* •
University Museum / Fine Arts Museum – 1950
Paintings, drawings, contemporary American ceramics 53411

Museum of Glass, International Center for Contemporary Art, 1801 E Dock St, Tacoma, WA 98402 • T: +1 253 3961768 • F: +1 253 3961769 • info@museumofglass.org • www.museumofglass.org •
C.E.O.: *Josi Irene Callan* •
Fine Arts Museum – 1995
Changing exhib of internat glass art and of contemp art 53412

Pioneer Museum of Motorcycles, 1113 A St, Ste 207, Tacoma, WA 98402 • T: +1 253 7794800 • F: +1 253 5731506 • michele@museumofmotorcycles.com • www.museumofmotorcycles.com •
Science&Tech Museum – 2005
Motorcycles from the beginning up to now 53413

Tacoma Art Museum, 1701 Pacific Av, Tacoma, WA 98402 • T: +1 253 2724258 • F: +1 253 6271898 • info@tacomaartmuseum.org • www.tacomaartmuseum.org •
Dir.: *Janeanne A. Upp* • Chief Cur.: *Dr. Patricia McDonnell* •
Fine Arts Museum – 1891/1935
Northwest art from 1800 till present, Dale Chihuly installation, 19th c European paintings 53414

Thomas Handforth Gallery, 1102 Tacoma Av, Tacoma, WA 98402 • T: +1 253 5915666 • F: +1 253 5915470 • ddomkoski@tpl.lib.wa.us • www.tpl.lib.wa.us •
Asst. Dir.: *Darrell Matz* • Sc. Staff: *Gary Reese* (Special collections) •
Fine Arts Museum / Public Gallery – 1886
Art, Thomas Handforth coll, WW I poster coll, photo collections (A. & E. Curtis, Turner Richards, V. Haffer) 53415

Washington State History Museum, 1911 Pacific Av, Tacoma, WA 98402 • T: +1 253 2723500 • F: +1 253 2729518 • dnicandri@wshs.wa.gov • www.wshs.org •
Pres.: *David Edwards* • Dir.: *David L. Nicandri* • Sc. Staff: *Lynn Anderson* (Collections) • *Edward V. Nolan* (Special collections) •
Local Museum – 1891
Pacific Northwest hist, paintings, furnishings, textiles, ethnolgy 53416

Tahlequah OK

Cherokee National Museum, Willis Rd, Tahlequah, OK 74465 • T: +1 918 4566007 • F: +1 918 4566165 • info@cherokeeheritage.org • www.cherokeeheritage.org •
Dir.: *Mary Ellen Meredith* •
Historical Museum / Ethnology Museum / Folklore Museum – 1963
42-structure ancient village, 9-structure rural village, Cherokee pottery and heritage arts 53417

Tahoe City CA

Gatekeeper's Museum and Marion Steinbach Indian Basket Museum, 130 W Lake Blvd, Tahoe City, CA 96145 • T: +1 530 5831762 • F: +1 530 5838992 • info@northtahoemuseums.org • www.northtahoemuseums.org •
Dir.: *Stefanie Givens* •
Local Museum / Decorative Arts Museum – 1969
Local hist, American Indian basketry, archaeology, dolls – small library 53418

Tahoma CA

Ehrman Mansion, Sugarpine Point State Park, Hwy 89, Tahoma, CA 96142, mail addr: POB 266, Tahoma, CA 96142 • T: +1 530 5257892 • F: +1 530 5250138 •
Cur.: *Shirley J. Kendall* • *Judith K. Polanich* •
Natural History Museum – 1965
Natural hist 53419

Vikingsholm Museum, Emerald Bay State Park, Hwy 89, Tahoma, CA 96142, mail addr: POB 266, Tahoma, CA 95733 • T: +1 530 5413030 • F: +1 530 5256730 •
Cur.: *Judith K. Polanich* • *Shirley J. Kendall* •
Historic Site – 1953
Scaninavian furnishings and furniture 53420

Talkeetna AK

Talkeetna Historical Society Museum, First Alley and Village Airstrip, Talkeetna, AK 99676 • T: +1 907 7332487 • F: +1 907 7332484 • museum@talkeetnahistoricalsociety.org • www.talkeetnahistoricalsociety.org •
Cur.: *Rebecca Nelson* •
Local Museum – 1972
Local hist, mining and trapping equip, Alaskan art – library 53421

Tallahassee FL

Florida State University Museum of Fine Arts, Fine Arts Bldg, Rm 250, Tallahassee, FL 32306-1140 • T: +1 850 6446836, 6441254 • F: +1 850 6447229 • apcraig@mailer.fsu.edu • www.mofa.fsu.edu •
Dir.: *Allys Palladino-Craig* •
Fine Arts Museum – 1950
Decorative arts, paintings, photos, prints, drawings, graphics, sculpture 53422

Knott House Museum, 301 E Park Av, Tallahassee, FL 32301 • T: +1 850 9222459, 2456400 • F: +1 850 4137261 • bwcofellis@dos.state.fl.us • www.dos.state.fl.us •
Special Museum – 1992 53423

Lake Jackson Mounds Archaeological Park, 3600 Indian Mounds Rd, Tallahassee, FL 32303-2348 • T: +1 850 5620042, 9226007 • F: +1 850 4880366 • lakejackson@nettally.com • www.floridastateparks.org •
Dir.: *Wendy Spencer* •
Historic Site / Archaeology Museum – 1970
Archaeology, 1200-1500 ceremonial center 53424

LeMoyne Art Foundation, 125 N Gadsden St, Tallahassee, FL 32301 • T: +1 850 2228800 • F: +1 850 2242714 • kdozier@maddogweb.com • www.lemoyne.org •
Dir.: *Marybeth Foss* •
Fine Arts Museum – 1964
Contemporary art, sculpture 53425

Mary Brogan Museum of Art and Science, 350 S Duval St, Tallahassee, FL 32301 • T: +1 850 5130700 ext 221, 232 • F: +1 850 5130143 • cbarber@thebrogan.org • www.thebrogan.org •
Exec. Dir.: *Rena Minar* •
Natural History Museum / Fine Arts Museum – 1990
Visual arts, hands-on science 53426

Museum of Florida History, 500 S Bronough St, Tallahassee, FL 32399-0250 • T: +1 850 4881484 • F: +1 850 9212503 • jbrunson@dos.state.fl.us • www.flheritage.com/museum •
Dir.: *Jeana Brunson* •
Historical Museum / Archaeology Museum – 1967
Spanish numismatics and maritime objects, artifacts, archaeology 53427

Riley House Museum of African American History and Culture, 419 E Jefferson St, Tallahassee, FL 32311 • T: +1 850 6817881 • F: +1 850 3864368 • staff@rileymuseum.org • www.rileymuseum.org •
C.E.O.: *Althemese Barnes* •
Ethnology Museum – 1996
Hist of African Americans in Florida 53428

Tallahassee Museum of History and Natural Science, 3945 Museum Dr, Tallahassee, FL 32310-6325 • T: +1 850 5758684 • F: +1 850 5748243 • rdaws@tallahasseemuseum.org • www.tallahasseemuseum.org •
Dir.: *Russell S. Daws* • Cur.: *Linda Deaton* (Collections and Exhibits) •
Local Museum / Natural History Museum / Historical Museum – 1957
19th c regional and natural hist, artifacts, textiles, Florida wildlife 53429

Tampa FL

Children's Museum of Tampa - Kid City, 7550 North Blvd, Tampa, FL 33604 • T: +1 813 9358441 • F: +1 813 9150063 • childrens_museum@yahoo.com • www.flachildrensmuseum.com •
Dir.: *David R. Penin* •
Special Museum – 1987 53430

Contemporary Art Museum, University of South Florida, 4202 E Fowler Av, Tampa, FL 33620 • T: +1 813 9742849 • F: +1 813 9745130 • favata@arts.usf.edu • www.usfcam.usf.edu •
Dir.: *Margaret A. Miller* • Ass. Dir.: *Alexa Favata* • Cur.: *Peter Foe* •
Fine Arts Museum / University Museum – 1968
Graphics, public art 53431

Cracker Country Museum, 4800 N Hwy 301, Tampa, FL 33680 • T: +1 813 6217821 • F: +1 813 7403518 • stalver1@doacs.state.fl.us • www.crackercountry.org •
Dir.: *Rip Stalvey* •
Local Museum / Open Air Museum – 1979
Oil paintings, Lionell train coll 53432

Henry B. Plant Museum, 401 W Kennedy Blvd, Tampa, FL 33606 • T: +1 813 2541891 • F: +1 813 2587272 • cgandee@ut.edu • www.plantmuseum.com •
Dir.: *Cynthia Gandee* • Cur.: *Susan Carter* •
Historical Museum / Decorative Arts Museum – 1933
History and Decorative Arts Museum: housed in 1891 Tampa Bay Hotel 53433

Museum of Science and Industry, 4801 E Fowler Av, Tampa, FL 33617-2099 • T: +1 813 9876300 • F: +1 813 9876310 • blittlej@mosi.org • www.mosi.org •
Pres.: *Wit Ostrenko* •
Science&Tech Museum – 1962
Science, technology, industry, human hist 53434

Scarfone and Hartley Galleries, University of Tampa, 401 W Kennedy, Tampa, FL 33606 • T: +1 813 2533333, 2536217 • F: +1 813 2587211 • dcowden@ut.edu • www.ut.edu •
C.E.O: *Robert Blount* •
Fine Arts Museum / University Museum – 1977
Contemporary 3-D and 2-D sculpture, prints, paintings, drawings, ceramics 53435

Tampa Bay History Center, 225 S Franklin St, Tampa, FL 33602 • T: +1 813 2280097 • F: +1 813 2237021 • thistory@gte.net • www.tampabayhistorycenter.org •
Dir.: *Mark Gruetzmacher* • *Elizabeth L. Dunham* • Cur.: *Rodney Kite Powell* •
Historical Museum – 1989
Local hist, militaria, paleoindian archeology, seminole material culture 53436

Tampa Museum of Art, 600 N Ashley Dr, Tampa, FL 33602 • T: +1 813 2748130 • F: +1 813 2748732 • tm22@tampagov.net • www.tampamuseum.com •
Dir.: *Emily S. Kass* •
Fine Arts Museum – 1979
19th c paintings, sculpture and works on paper, Greek and Roman antiquities, C. Paul Jennewein coll of sculpture 53437

Ybor City State Museum, 1818 Ninth Av E, Tampa, FL 33605 • T: +1 813 2476323 • F: +1 813 2424010 • director@ybormuseum.org • www.ybormuseum.org •
Pres.: *Dan Haya* •
Historical Museum / Local Museum – 1980
Ethnically divere immigrants in Ybor City, cigar industry, mutual aid societies 53438

Tamworth NH

Remick Country Doctor Museum and Farm, 58 Cleveland Hill Rd, Tamworth, NH 03886 • T: +1 603 3237591 • F: +1 603 3238382 • pr@remickmuseum.org • www.remickmuseum.org •
Dir.: *Robert Cottrell* •
Local Museum – 1993
Local hist, medical hist 53439

Taos NM

Governor Bent Museum, 117 Bent St, Taos, NM 87571 • T: +1 505 7582376 • F: +1 505 7582376 • gnideon@laplaza.com •
C.E.O.: *Tom Noeding* •
Historical Museum / Natural History Museum – 1958
Indian artifacts, old Americana antiques, Eskimo coll, guns, spinning wheels, anthropology, mineralogy, archaeology, graphics 53440

Harwood Museum of Art of the University of New Mexico, 238 Ledoux St, Taos, NM 87571-6004 • T: +1 505 7589826 • F: +1 505 7581475 • harwood@unm.edu • www.harwoodmuseum.org •
Dir.: *Charles M. Lovell* • Cur.: *David Witt* •
Fine Arts Museum / University Museum – 1923
19th-20th c Santos and Spanish colonial furniture of Hispanic culture, paintings 53441

Millicent Rogers Museum of Northern New Mexico, 1504 Museum Rd, Taos, NM 87571 • T: +1 505 7582462 • F: +1 505 7585751 • mrm@newmex.com • www.millicentrogers.org •
Exec. Dir.: *Shelby J. Tisdale* • Cur.: *Steve Pettit* •
Fine Arts Museum / Ethnology Museum – 1953
Southwestern Native American jewelry, pottery, paintings, Southwestern Hispanic textiles, tinwork 53442

Stables Art Gallery of Taos Art Association, 133 N Pueblo Rd, Taos, NM 87571 • T: +1 505 7582036 • F: +1 505 7513305 • taa@taos.newmex.com • www.taosnet.com/taa/ •
Exec. Dir.: *Betsy Carey* •
Fine Arts Museum – 1953
Works by New Mexico artists 53443

Taos Center for the Arts, 133 Paseo del Pueblo N, Taos, NM 87571 • T: +1 505 7582052 • F: +1 505 7513305 • info@taoscenterforthearts.org • www.taoscenterforthearts.org •
Dir.: *Jean F. Marquardt* •
Public Gallery – 2001 53444

Taos Historic Museums, 222 Ledoux St, Taos, NM 87571 • T: +1 505 7580505 • F: +1 505 7580330 • thm@taoshistoricmuseums.com • www.taoshistoricmuseums.com •
Dir.: *Karen S. Young* •
Local Museum – 1949
Spanish colonial taos, furnishings, furniture – Ernest Blumenschein and Kit Carson home 53445

Tappan NY

Tappantown Historical Society Museum, Box 71, Tappan, NY 10983 • T: +1 845 3591149 •
Pres.: *Peter Schuerholz* •
Historical Museum – 1965
American architecture, 18th and 19th c homes 53446

Tarboro NC

Blount-Bridgers House/ Edgecombe County Art Museum, 130 Bridgers St, Tarboro, NC 27886 • T: +1 252 8234159 • F: +1 252 8236190 • edgecombearts@embarqmail.com • www.edgecombearts.org •
C.E.O.: *Jai Jordan* •
Fine Arts Museum / Local Museum – 1982
20th c art, 19th c silver, ceramics and American furniture 53447

Tarpon Springs FL

Leepa-Rattner Museum, POB 1545, Tarpon Springs, FL 34688 • T: +1 727 7125762 • F: +1 727 7125223 • LRMA@spcollege.edu • www.spcollege.edu/central/museum •
Dir.: *R. Lynn Whitelaw* •
Fine Arts Museum
Works by Abraham Rattner, 20th c art 53448

Tarpon Springs Cultural Center, 101 S Pinellas Av, Tarpon Springs, FL 34689 • T: +1 727 9425605 • F: +1 727 9382429 • bpoteat1@c1.tarpon-springs.fl.us •
Local Museum – 1987
Historical, artistic and ethnographic material 53449

Tarrytown NY

Historic Hudson Valley, 150 White Plains Rd, Tarrytown, NY 10591 • T: +1 914 6318200 • F: +1 914 6310089 • mail@hudsonvalley.org • www.hudsonvalley.org •
Cur.: *Kathleen E. Johnson* •
Local Museum – 1951
Historic houses, housing stained glass works by Matisse and Chagall 53450

Lyndhurst, 635 S Broadway, Tarrytown, NY 10591 • T: +1 914 6314481 • F: +1 914 6315634 • lyndhurst@nthp.org • www.lyndhurst.org •
Dir.: *Susanne Brendel-Pandich* • Cur.: *Judith Beil* •
Local Museum – 1964
19th c Gothic revival residence, Carriage house, 19th c dec and fine arts, textiles, furniture 53451

Tarrytowns Museum, The Historical Societey Sleepy Hollow and Tarrytown, 1 Grove St, Tarrytown, NY 10591 • T: +1 914 6318374 •
Cur.: *Sara Mascia* •
Local Museum – 1889
Native American artifacts, early Dutch hist of the region, local archaeology, American Revolution, WW I & WW II, art coll 53452

Taunton MA

Old Colony Historical Museum, 66 Church Green, Taunton, MA 02780 • T: +1 508 8221622 • F: +1 508 8806317 • oldcolony@oldcolonyhistoricalsociety.org • www.oldcolonyhistoricalsociety.org •
CEO: *Jane M. Hennedy* • Cur.: *Jane Emack Cambra* •
Local Museum – 1853
History, decorative arts, fire fighting, toys & dolls, military hist – library 53453

Tavares FL

Lake County Historical Museum, 317 W Main St, Tavares, FL 32778 • T: +1 352 3439600 • F: +1 352 3439696 • dkamp@co.lake.fl.us •
Dir.: *Dr. Diane D. Kamp* •
Local Museum – 1965
American artifacts, Spanish relics 53454

Taylors Falls MN

Folsom House Museum, 272 W Government Rd, Taylors Falls, MN 55084 • T: +1 612 4653125 •
Man.: *William W. Scott* •
Historical Museum – 1978
Folsom general store, family items 53455

Taylorsville MS

Watkins Museum, Eureka St, Taylorsville, MS 39168 • T: +1 601 7859816 • F: +1 601 7859816 • jrglenn@magamate.com •
Cur.: *Rosalyn Glenn* •
Local Museum – 1968
Hand printing press, early medical and farm equip, ancient bottle coll, clothing, typewriter 53456

Tazewell VA

Historic Crab Orchard Museum and Pioneer Park, Rte 19 and Rte 460, Tazewell, VA 24651 • T: +1 276 9886755 • F: +1 276 9889400 • info@craborchardmuseum.com • www.craborchardmuseum.com •
Dir.: *Randall A. Rose* • Cur.: *Corney A. Honaker* •
Folklore Museum / Historical Museum – 1978
Central Appalachian history 53457

Teague TX

Burlington-Rock Island Railroad and Historical Museum, 208 S 3rd Av, Teague, TX 75860 • T: +1 254 7393551 • www.therailroadmuseum.com •
Cur.: *Dorothy McVey* •
Science&Tech Museum / Historical Museum – 1969
Railroad memorabilia incl 1925 Baldwin locomotive, 1928 fire engine, grist mill, loom, farming 53458

Tecumseh NE

Johnson County Historical Society Museum, Third and Lincoln Sts, Tecumseh, NE 68450 • T: +1 402 3353258 •
Pres.: *John R. Fisher* • Dir.: *Boyd Mattox* •
Local Museum – 1962
Indian artifacts, silver, China, glass, toys, WW I objects 53459

Tekamah NE

Burt County Museum, 319 N 13th St, Tekamah, NE 68061 • T: +1 402 3741505 • burtcomuseum@huntel.net •
Pres.: *Rick Nelsen* •
Local Museum – 1967
Agriculture, furniture, photos, manuscript coll ... 53460

Telluride CO

Telluride Historical Museum, 201 W Gregory Av, Telluride, CO 81435, mail addr: POB 1597, Telluride, CO 81435 • T: +1 970 7283344 • F: +1 970 7286757 • info@telluridemuseum.org • www.telluridemuseum.org •
Exec. Dir.: *Lauren Bloemsma van Nes* •
Local Museum – 1964
Genealogy, mining, mineralogy, Indian artifacts, photographs, Rio Grande Southern railroad hist . 53461

Temecula CA

Temecula Valley Museum, 28314 Mercedes St, Temecula, CA 92592 • T: +1 951 6946450 • www.cityoftemecula.org •
Local Museum – 1985 ... 53462

Tempe AZ

American Museum of Nursing, 200 E Curry Rd, Tempe, AZ 85287 • T: +1 480 9652195 • rojann@asu.edu • www.asu.edu/museums/oc/nursing.htm •
Dir./Cur.: *Rojann R. Alpers* •
Historical Museum – 2002
Military, hospital missionary and other types of nursing memorabilia, history artifacts, photos, equip and uniforms – Library ... 53463

Arizona Historical Society Museum at Papago Park, 1300 N College Av, Tempe, AZ 85281 • T: +1 480 9290292 • F: +1 480 9675450 • AHSTempe@azhs.gov • www.arizonahistoricalsociety.org •
Dir.: *Vicki L. Berger* •
Local Museum – 1973
Regional hist, art and culture – archives ... 53464

Arizona State University Art Museum, 10th St and Mill Av, Nelson Fine Arts Center, Tempe, AZ 85287-2911 • T: +1 602 9652787 • F: +1 602 9655254 • asuartmuseum@asu.edu • asuartmuseum.asu.edu •
Dir.: *Marilyn Zeitlin* • Cur.: *Peter Held* • *Jean Makin* (Print coll) •
Fine Arts Museum / University Museum – 1950
Modern and contemporary American art, 15th c American and European print coll, Latin American art, paintings, prints, sculpture, ceramics, crafts, glass ... 53465

Computing Commons Gallery, c/o Arizona State University, 699 S Mill Av, Tempe, AZ 85287-8709 • T: +1 480 9653609 • F: +1 480 9658698 • ame.asu.edu/facilities/gallery.html •
Dir.: *Barbara Eschbach* •
Fine Arts Museum / University Museum
Arts, Media ... 53466

Halle Heart Center, 2929 S 48th St, Tempe, AZ 85282 • T: +1 602 2800 • F: +1 602 4145355 • halleheartcenter@heart.org • www.americanheart.org •
Dir.: *Jena Long* •
Science&Tech Museum – 1996
Medical equip, healthy living, heart exhibits – library ... 53467

Memorial Union Art Collection, c/o Arizona State University, 1290 S Normal Av, Tempe, AZ 85287-0901 • T: +1 480 9655728 • ehartley@mainex1.asu.edu • www.asu.edu/studentaffairs/mu/art_collection.htm •
Public Gallery
Paintings, photographs, sculptures and tapestries created by students ... 53468

Museum of Anthropology, Arizona State University, Anthropology Bldg, Tempe, AZ 85287-2402 • T: +1 480 9656213 • F: +1 480 9657671 • anthro.museum@asu.edu • www.asu.edu/anthropology/museum •
Dir.: *Gwyneira Isaac* • Cur.: *Emely Duwel* (Exhibits) • *C. Michael Barton* (Archaeology and Ethography) • *Diane Hawkey* (Physical Anthropology) •
Ethnology Museum / Archaeology Museum / University Museum – 1959
Physical anthropology, archaeology, ethnography – laboratories ... 53469

Museum of the Center for Meteorite Studies, Tempe Campus, Bateman Physical Sciences C-wing, Room 139, Tempe, AZ 85287-2504, mail addr: c/o Arizona State University, POB 872504, Tempe, AZ 85287-2504 • T: +1 480 9656511 • meteorites@asu.edu • meteorites.asu.edu •
Dir.: *Carleton B. Moore* •
University Museum / Natural History Museum – 1961
Specimens representing over 1635 separate meteorite falls ... 53470

Tempe Center for the Arts, 3340 S Rural Rd, Tempe, AZ 85282 • T: +1 480 3505183 • F: +1 480 3505161 • mary_fowler@tempe.gov • www.tempe.gov/TCA •
Dir.: *Claudia Anderson* • Cur.: *Patty Haberman* •
Fine Arts Museum – 1982/ 2003
Visual arts ... 53471

Tempe Historical Museum, 809 E Southern Av, Tempe, AZ 85282 • T: +1 480 3505100 • F: +1 480 3505150 • museum@tempe.gov • www.tempe.gov/museum •
Dir.: *Amy A. Douglass* • Cur.: *James Burns* (History) • *Richard Bauer* (Photographs, archive) • *Dan Miller* (Exhibits) • *Ann Poulos* (Colls) •
Local Museum – 1972
Local hist 19th-20th c – Petersen House Museum ... 53472

Temple TX

Czech Heritage Museum and Genealogy Center, 520 N Main St, Temple, TX 76501, mail addr: POB 434, Temple, TX 76503 • T: +1 254 7731575 • F: +1 254 7747447 • Ssandiwicker@aol.com • www.spjst.com •
Dir.: *Howard B. Leshikar* • Cur.: *Dorothy Pechal* •
Ethnology Museum / Historical Museum – 1971
Czech hist, culture and genealogy ... 53473

Railroad and Heritage Museum, 315 W Av B, Temple, TX 76501 • T: +1 254 2985172 • F: +1 254 2985171 • mirving@ci.temple.tx.us • www.rrdepot.org •
Dir.: *Mary L. Irving* •
Historical Museum / Science&Tech Museum – 1973
Transportation, history, Santa Fe, MKT and TX railroads – archives ... 53474

Spjst Library Archives & Museum → Czech Heritage Museum and Genealogy Center

Templeton MA

Narragansett Historical Society Museum, 1 Boynton Rd, Templeton, MA 01468 • T: +1 978 9392251 •
Local Museum – 1928
Regional history – library ... 53475

Tenafly NJ

African Art Museum of the S.M.A. Fathers, 23 Bliss Av, Tenafly, NJ 07670 • T: +1 201 8948611 • F: +1 201 5411280 • smausa-e@smafathers.org • www.smafathers.org •
Dir.: *Robert J. Koenig* •
Ethnology Museum – 1963
Sculpture, ritual objects, weapons, architectural elements, tools, furniture, textiels, jewelry ... 53476

Tenino WA

Tenino Depot Museum, 399 W Park, Tenino, WA 98589 • T: +1 360 2644321 • F: +1 360 8679138 •
Local Museum – 1974
Local hist, Tenino wooden money and money printing press, ... 53477

Tequesta FL

Lighthouse Gallery, 373 Tequesta Dr, Tequesta, FL 33469-3027 • T: +1 561 7463101 • F: +1 561 7463241 • info@lighthousearts.org • www.lighthousearts.org •
Exec. Dir.: *Margaret Inserra* •
Public Gallery – 1965
Drawings, painting-American, photography, sculpture, watercolors, bronzes, ceramics, collages, calligraphy ... 53478

Terra Alta WV

Americana Museum, 401 Aurora Av, Terra Alta, WV 26764 • T: +1 304 7892361 •
Dir.: *Ruth E. Teets* • *James W. Teets* • *Robert G. Teets* •
Local Museum – 1968
Local hist, artifacts, furniture, horse-drwan vehicles, antique cars ... 53479

Terre Haute IN

Eugene V. Debs Home, 451 N Eighth St, Terre Haute, IN 47807, mail addr: POB 843, Terre Haute, IN 47807 • T: +1 812 2322163, 2373443 • F: +1 812 2378072 • debsmuseum@verizon.net • www.eugenevdebs.com •
Dir.: *Karon Brown* •
Historical Museum – 1962
Furnishings, mementos of Deb's labor activism and his campaigns for the presidency, historical photographs ... 53480

Native American Museum, 5170 E Poplar St, Terre Haute, IN 47803 • T: +1 812 8776007 • F: +1 812 2327313 •
Cur.: *Amanda Smith* •
Historical Museum / Ethnology Museum – 1994
Indiana history ... 53481

Paul Dresser Memorial Birthplace, First and Farrington Sts, Terre Haute, IN 47802, mail addr: Terre Haute, IN 47802 • T: +1 812 2359717 • F: +1 812 2354998 • vchs@joink.com • web.indstate.edu/vchs/index.php •
Exec. Dir.: *Marylee Hagan* •
Historical Museum – 1967
Furniture, period artifacts ... 53482

Swope Art Museum, 25 S 7th St, Terre Haute, IN 47807-3692 • T: +1 812 2381676 • F: +1 812 2381677 • info@swope.org • www.swope.org •
Dir.: *David Vollmer* •
Fine Arts Museum – 1942
19th-20th c American art ... 53483

University Art Gallery, Indiana State University, Center for Performing and Fine Arts, 7th & Chestnut St, Terre Haute, IN 47809 • T: +1 812 2373720, 237 3697 • F: +1 812 2374369 • arkaz@isugw.indstate.edu • www.indstate.edu/artgallery •
Dir.: *Kaz McCue* •
Fine Arts Museum / University Museum – 1939
Paintings, sculpture, prints, ceramics ... 53484

Vigo County Historical Museum, Sage-Robison-Nagel house, 1411 S 6th St, Terre Haute, IN 47802 • T: +1 812 2359717 • F: +1 812 2354998 • vchs@joink.com • web.indstate.edu/vchs/index.php •
Dir.: *Marylee Hagan* • Ass. Dir.: *Barbara Carney* •
Local Museum – 1958 ... 53485

Terryville CT

Lock Museum of America, 230 Main St, Terryville, CT 06786 • T: +1 860 5896359 • F: +1 860 5896359 • thomasnsc@aol.com • www.lockmuseum.com/ •
Cur.: *Thomas F. Hennessy* •
Science&Tech Museum – 1972
Hist of locks ... 53486

Teutopolis IL

Teutopolis Monastery Museum, Rte 40 and S. Garrott St, Teutopolis, IL 62467, mail addr: 106 W Water St, Teutopolis, IL 62467 • T: +1 217 8573227 • F: +1 217 8573227 •
Pres.: *Ray Vahling* •
Local Museum / Religious Arts Museum – 1975
Pioneer artifacts, clocks, furniture, religious items and books ... 53487

Texarkana TX

Ace of Clubs House, 420 Pine St, Texarkana, TX 75501 • T: +1 903 7934831 • F: +1 903 7937108 • mnesbitt@cableone.net • www.texarkanamuseums.org •
Cur.: *Jamie Simmons* • *Melissa Nesbitt* •
Local Museum / Decorative Arts Museum – 1985
Bldg furnishings, personal artifacts – Wilbur Smith research library & archives ... 53488

Texarkana Museums System, 219 N State Line Av, Texarkana, TX 75501 • T: +1 903 7934831 • F: +1 903 7937108 • imcdowell@cableone.net • www.texarkanamuseums.org •
Exec. Dir.: *Ina McDowell* •
Local Museum – 1966
Caddo people, regional hist – Regional hist & Wilbur Smith Research Library and Archives, Discovery Place Children's Museum ... 53489

Texas City TX

Texas City Museum, 409 Sixth St, Texas City, TX 77592 • T: +1 409 6435799 • F: +1 409 9499972 • tcmuseum@prodigy.net •
Exec. Dir.: *Linda Turner* •
Historical Museum / Science&Tech Museum – 1991
Concentration on the Mainland area of Galveston County from 1900s to the present ... 53490

Thatcher AZ

Graham County Historical Museum, 3430 W Hwy 70, Thatcher, AZ 85546 • T: +1 928 3480470 •
Dir.: *Raydene Cluff* •
Local Museum – 1962
Local hist, pioneer and Indian artifacts ... 53491

Thermopolis WY

Hot Springs County Museum, 700 Broadway, Thermopolis, WY 82443 • T: +1 307 8645183 • F: +1 307 8642974 • history@trib.com • www.trib.com/~history •
Dir.: *Dr. Alexandra Service* •
Historical Museum – 1941
Regional history ... 53492

Old West Wax Museum, 119 S 6th, Thermopolis, WY 82443 • T: +1 307 8649396 • F: +1 307 8642657 • westwax@westwaxmuseum.com • www.westwaxmuseum.com •
Dir.: *Ellen Sue Blakey* •
Special Museum – 1999
American west figures, maps, clothing, folk hist . 53493

Wyoming Dinosaur Center, 110 Carter Ranch Rd, Thermopolis, WY 82443 • T: +1 307 8642997 • F: +1 307 8645762 • wdinoc@wyodino.org • www.wyodino.org •
C.E.O.: *Burkhart Pohl* •
Natural History Museum – 1995
Full size mounts of different dinosaurs, geology, prehistoric life on earth ... 53494

Wyoming Pioneer Home, 141 Pioneer Home Dr, Thermopolis, WY 82443 • T: +1 307 8643151 • F: +1 307 8642934 •
Dir.: *Sharon Skiver* •
Local Museum – 1950
Indian artifacts, pioneer relics ... 53495

Thetford VT

Thetford Historical Society Library and Museum, Bicentennial Bldg, 16 Library Rd, Thetford, VT 05074 • T: +1 802 7852068 •
Pres.: *Charles Latham* • Cur.: *Donald Fifield* •
Local Museum / Agriculture Museum – 1943
Clara Siprell photos, portraits, tools, agricultural implements – library ... 53496

Thomaston ME

General Henry Knox Museum, 30 High St, Thomaston, ME 04861 • T: +1 207 3548062 • F: +1 207 3543501 • genknox@midcoast.com • www.generalknoxmuseum.org •
Pres.: *Renny A. Stackpole* •
Historical Museum – 1931
Montpelier home of Major General Henry Knox .. 53497

Thomaston Historical Museum, 80 Knox St, Thomaston, ME 04861 • T: +1 207 3542295, 3548835 • katsmeow@adelphia.net • www.thomastonhistoricalsociety.com •
Dir.: *Eve Anderson* • Cur.: *Louise Crane* •
Local Museum – 1971
Local history, 1794 Henry Knox farmhouse ... 53498

Thomasville GA

Lapham-Patterson House, 626 N Dawson St, Thomasville, GA 31792 • T: +1 912 2254004 • F: +1 912 2272419 • lphouse@rose.net •
Cur.: *Cheryl Walters* •
Historical Museum – 1974
Historic Victorian House c.1884 ... 53499

Pebble Hill Plantation, 1251 US-319 S, Tallahassee Rd, Thomasville, GA 31792 • T: +1 912 2262344 • F: +1 912 2270095 • swhite@pebblehill.com • www.pebblehill.com •
Head: *Wallace Goodman* •
Local Museum – 1983
18th-19th c furniture and sporting art, photo coll, dec arts ... 53500

Thomasville Cultural Center, 600 E Washington St, Thomasville, GA 31792 • T: +1 912 2260588 • F: +1 912 2260589 • admin@tccarts.org • www.tccarts.org •
Cur.: *Gylbert Coker* •
Fine Arts Museum – 1978
American and European paintings, wildlife art, ceramics ... 53501

Thousand Oaks CA

Conejo Valley Art Museum, 193a N Moorpark, Thousand Oaks, CA 91358, mail addr: POB 1616, Thousand Oaks, CA 91358 • T: +1 805 3730054 • dessornes@earthlink.net • www.cvam.us •
Pres./ CEO: *Maria E. Dessornes* •
Fine Arts Museum – 1977 ... 53502

Three Forks MT

Headwaters Heritage Museum, 202 S Main Str, Three Forks, MT 59752, mail addr: Box 116, Three Forks, MT 59752 • T: +1 406 2854778 • phillips@holcim.com • www.threeforksmontana.com •
Dir. & Cur.: *Robin Cadby-Sorensen* •
Historical Museum – 1979
Natural and local hist, agricultural tools – coll of local newspapers on microfilm ... 53503

Three Oaks MI

Bicycle Museum, One Oak St, Three Oaks, MI 49128 • T: +1 269 7563361 • bryan@applecidercentury.com •
C.E.O.: *Bryan Volstorf* •
Special Museum – 1986
Bicycles, railroad tools & equip ... 53504

Tiburon CA

Schwartz Collection of Skiing Heritage, 9 Stephens Court, Tiburon, CA 94920 • T: +1 415 4351076 • F: +1 415 4355027 • info@picturesnow.com • www.picturesnow.com •
Dir.: *Gary Schwartz* • Cur.: *Rhoda Valowitz* •
Special Museum – 1986
Skiing sports, artworks ... 53505

Ticonderoga NY

Fort Ticonderoga, 30 Fort Rd, Ticonderoga, NY 12883 • T: +1 518 5852821 • F: +1 518 5852210 • fort@fort-ticonderoga.org • www.fort-ticonderoga.org •
C.E.O. & Dir.: *Nicholas Westbrook* • Cur.: *Christopher D. Fox* •
Military Museum – 1908
Military history (18th c), tourism (19th c) ... 53506

Tiffin OH

Seneca County Museum, 28 Clay St, Tiffin, OH 44883 • T: +1 419 4475955 • F: +1 419 4437940 •
Pres.: *Barry Porter* • Dir.: *Tonia Hoffert* •
Local Museum – 1942
Porcelain and glassware, costumes, pressed glass, art glass, costumes ... 53507

Tifton GA

Georgia Agrirama, 19th Century Living History Museum, I-75 Exit 20 at 8th St, Tifton, GA 31793 • T: +1 912 3863344 • F: +1 912 3863386 • market@ganet.org • www.agrirama.com •
Dir.: *James Higgins* •
Agriculture Museum / Ethnology Museum – 1972
Agricultural equip, printing and typesetting equip, medical and dentist equip, furnishings ... 53508

Tillamook OR

Tillamook County Pioneer Museum, 2106 Second St, Tillamook, OR 97141 • T: +1 503 8424553 • F: +1 503 8424553 • wjensen@tcpm.org •
Dir.: *M. Wayne Jensen* •
Local Museum – 1935
Pioneer artifacts, archaeology, natural hist, mineralogy, zoology, wildlife and natural hist ... 53509

Tillamook Naval Air Station Museum, 6030 Hangar Rd, Tillamook, OR 97141 • T: +1 503 8421130 • F: +1 503 8423054 • info@tillamookair.com • www.tillamookair.com •
C.E.O.: *Larry Schaible* •
Military Museum – 1994
US war birds, aircraft, WW II ... 53510

Tinley Park IL

Tinley Park Historical Society Museum, 6727 W 174 St, Tinley Park, IL 60477 • T: +1 708 4294210 • F: +1 708 4445099 • lrtphist@lincolnnet.net • www.lincolnnet.net/users/lrtphist •
Local Museum – 1974 ... 53511

Tishomingo OK

Chickasaw Council House Museum, Court House Sq, Tishomingo, OK 73460 • T: +1 580 3713351 • museum@chickasaw.net • www.chickasaw.net •
Mgr.: *Glenda Gavin* •
Ethnology Museum / Historic Site / Folklore Museum – 1970
Life and works of Chickasaws, first council house built and used by Chickasaws ... 53512

Margaret Lokey Gallery, Murray State College Campus Center Gallery, 1 Murray Campus, Suite FA 100, Tishomingo, OK 73460 • T: +1 580 3712371 ext. 232 •
Dir.: *Catherine Kinyon* •
Public Gallery – 1970 ... 53513

Titusville NJ

Howell Living History Farm, 101 Hunter Rd, Titusville, NJ 08560 • T: +1 609 7373299 • F: +1 609 7376524 • thefarm@bellatlantic.net • www.howellfarm.org •
Agriculture Museum – 1974
Farm life and farming ... 53514

Johnson Ferry House Museum, Washington Crossing State Park, 355 Washington Xing Penn Rd, Titusville, NJ 08560-1517 • T: +1 609 7372515 • F: +1 609 8189017 •
Local Museum – 1912
Colonial furniture, domestic and decorative items, American Revolutionary War artifacts ... 53515

Titusville PA

Drake Well Museum, 202 Museums Ln, Titusville, PA 16354-8902 • T: +1 814 8272797 • F: +1 814 8274888 • drakewell@usachoice.net • www.drakewell.org •
Pres.: *Glenn Cochran* • C.E.O.: *Brent D. Glass* •
Local Museum / Science&Tech Museum – 1934
Site of first commercially successful oil well ... 53516

Tobias NE

Tobias Community Historical Society Museum, Main St, Tobias, NE 68453 • T: +1 402 2432356 •
Pres.: *Judith K. Rada* •
Local Museum – 1968
Old bottles, kitchen wares, photos, military ... 53517

Toccoa GA

Traveler's Rest State Historic Site, 8162 Riverdale Rd, Toccoa, GA 30577 • T: +1 706 8862256 • travelersrest@alltel.net •
Man.: *Raymond L. Anderson* •
Historical Museum – 1955
1833 former stagecoach inn ... 53518

Toledo IA

Tama County Historical Museum, 200 N Broadway, Toledo, IA 52342 • T: +1 641 4846767 • tracers@pcpartner.net •
Pres.: *Joyce Wiese* •
Local Museum – 1942
Local history, housed in a former County Jail ... 53519

Toledo OH

Blair Museum of Lithophanes, 5403 Elmer Dr, Toledo, OH 43615 • T: +1 419 2451356 •
Special Museum ... 53520

Cosi Toledo, 1 Discovery Way, Toledo, OH 43604 • T: +1 419 2442674 • F: +1 419 2552674 • www.cosi.org •
C.E.O.: *William H. Booth* •
Science&Tech Museum / Natural History Museum – 1997 ... 53521

Museum of Science at the Toledo Zoo, 2700 Broadway, Toledo, OH 43609, mail addr: POB 140130, Toledo, OH 43614 • T: +1 419 3855721 • F: +1 419 3898670 • andi.norman@toledozoo.org • www.toledozoo.org •
Dir.: *William V.A. Dennler* • Cur.: *Jay F. Hemdal* (Fishes) • *R. Andrew Odum* (Herpetology) • *Robert Webster* (Birds) • *Randy Meyerson* (Mammals) • Sc. Staff: *Vanessa Neeb* (Interpretive Services) • *Bob Magdich* (Education) • *Bob Harden* (Operating) • *Peter Tolson* (Conservation Research) • *Tim Reichard* (Animal Health and Nutrition) •
Natural History Museum / Local Museum – 1899 53522

Toledo Museum of Art, 2445 Monroe St, Toledo, OH 43620 • T: +1 419 2558000 • F: +1 419 2555638 • info@toledomuseum.org • www.toledomuseum.org •
Dir.: *Don Bacigalupi* • Ass. Dir.: *Carol Bintz* • Chief Cur.: *Carolyn Putney* • Cur.: *Lawrence W. Nichols* (European Painting and Sculpture before 1900) • *Jutta Annette Page* (Glass) •
Fine Arts Museum – 1901
European and American art, esp glass, early glass, medieval art ... 53523

Tolland CT

The Benton Homestead, Metcalf Rd, Tolland, CT 06084 • T: +1 860 8728673 •
Dir.: *Gail W. White* •
Local Museum – 1969
Historic house ... 53524

Tolland County Jail and Warden's Home Museum, 52 Tolland Green, Tolland, CT 06084 • T: +1 860 8753544, 8709599 • tolland.historical@snet.net • pages.cthome.net/tollandhistorical •
Pres.: *Stewart Joslin* • Dir.: *Kathy Bach* •
Local Museum – 1856
Local hist, farming, Indians ... 53525

Tomball TX

Tomball Community Museum Center, 510 N Pine St, Tomball, TX 77375 • T: +1 281 4442449 •
Dir.: *Jean Alexander* •
Local Museum – 1961
19th c decorative art coll, pioneer country doctor's office, 1800s cotton gin ... 53526

Tome NM

Tome Parish Museum, 7 N Church Loop, Tome, NM 87060-0100 • T: +1 505 8657497 • F: +1 505 8657497 •
Dir.: *Fr. Carl Feil* •
Religious Arts Museum – 1966
Religion, paintings, santos statues ... 53527

Toms River NJ

Ocean County Historical Museum, 26 Hadley Av, Toms River, NJ 08754-2191 • T: +1 732 3411880 • F: +1 732 3414372 • oceancountyhistory@verizon.net • www.oceancountyhistory.org •
Cur.: *Barbara Rivolta* •
Local Museum – 1950
Indian artifacts, Victorian furniture and costumes 53528

Tonawanda NY

Tonawandas Museum, Historical Society of the Tonawandas, 113 Main St, Tonawanda, NY 14150-2129 • T: +1 716 6947406 •
Exec. Dir. & Cur.: *Jane Penvose* •
Historical Museum – 1961
Industry, transportation, military coll, Indian artifacts, Lumber, medical coll ... 53529

Tonkawa OK

A.D. Buck Museum of Natural History and Science, Northern Oklahoma College, 1220 E Grand, Tonkawa, OK 74653 • T: +1 405 6286200 • F: +1 405 6286209 •
Dir.: *Rex D. Ackerson* •
University Museum / Historical Museum / Natural History Museum – 1913
Indian artifacts, pioneer artifacts of local hist, mineralogy, archaeology, geology, zoology ... 53530

Tonopah NV

Central Nevada Museum, 1900 Logan Field Rd, Tonopah, NV 89049, mail addr: P.O. Box 326, Tonopah, NV 89049 • T: +1 702 4829676 • F: +1 702 4825423 • cnmuseum@citlink.net • www.tonopahnevada.com •
Asst. Curator: *Eva La Rue* • Researcher: *Angela Haag* •
Local Museum – 1981
History of central Nevada area, from prehistoric to present, indoor & outdoor displays – Research library ... 53531

Topeka KS

Alice C. Sabatini Gallery, Topeka and Shawnee County Public Library, 1515 W 10th, Topeka, KS 66604-1374 • T: +1 785 5804516, 5804400 • F: +1 785 5804496 • sbest@mail.tscpl.org • www.tscpl.org •
Dir.: *Sherry L. Best* (Gallery) • *David L. Leamon* • Dep. Dir.: *Robert Banks* •
Fine Arts Museum / Public Gallery – 1870
American ceramics, prints, paintings, antique/modern glass paperweights, West African cultural objects, wood carving, Chinese pewter, decorative objects – library ... 53532

Combat Air Museum, Forbes Field, Hangar 602, Topeka, KS 66619 • T: +1 785 8623303 • F: +1 785 8623304 • combatairmuseum@aol.com • www.combatairmuseum.org •
Head: *Adam Trupp* •
Military Museum / Science&Tech Museum – 1976
Topeka Army Air Field, later Forbes AFB, now Forbes Field ... 53533

Great Overland Station, 701 N Kansas Av, Topeka, KS 66608 • T: +1 785 2325533 • F: +1 785 2326259 • info@greatoverlandstation.com • greatoverlandstation.com •
Special Museum – 2000
Oregon Trail & Santa Fe Railway hist, hist railroad station ... 53534

Historic Ward-Meade Park Museum, 124 NW Fillmore, Topeka, KS 66606 • T: +1 785 3683888 • F: +1 785 3683890 • sleeth@topeka.org • www.topeka.org •
Exec.Dir.: *Sara Leeth* •
Special Museum
Clothing, toys, furniture, art works, hist houses .. 53535

Kansas Museum of History, 6425 SW 6th St, Topeka, KS 66615-1099 • T: +1 785 2728681 ext 401 • F: +1 785 2728682 • webmaster@kshs.org • www.kshs.org •
Dir.: *Robert J. Keckeisen* • Ass. Dir.: *Rebecca J. Martin* • Cur.: *Jill Keehner* (Exhibition) • *Blair D. Tarr* (Decorative Arts and Domestics) • Cons.: *Susanne Benda* •
Decorative Arts Museum / Historical Museum – 1875
Hist, decorative art ... 53536

Mulvane Art Museum, 17th and Jewell Sts, Topeka, KS 66621-1150 • T: +1 785 2311124 • F: +1 785 2311329 • mulvane.info@washburn.edu • www.washburn.edu/mulvane •
Dir.: *Edward Barr* •
Fine Arts Museum / University Museum – 1922
19th-20th c American art, 16th-20th c European prints, worlds arts ... 53537

Toppenish WA

Cultural Heritage Center Museum, Confederated Tribes and Bands of the Yakama Indian Nation, 100 Speelyi Loop, Toppenish, WA 98948 • T: +1 509 8652800 • F: +1 509 8655749 • heather@yakama.com • www.yakamamuseum.com •
Chm.: *Pamela K. Fabela* • Cur.: *Marilyn Malatare* •
Historical Museum / Ethnology Museum – 1980
Indian culture and hist ... 53538

Northern Pacific Railway Museum, 10 S Asotin Av, Toppenish, WA 98948, mail addr: POB 889, Toppenish, WA 98948 • T: +1 509 8651911 • lamarentr@aol.com • www.nprymuseum.org •
Dir. & Cur.: *Larry Rice* •
Science&Tech Museum – 1989
Railroad artifacts, 1911 depot ... 53539

Toppenish Museum, 1 S Elm St, Toppenish, WA 98948 • T: +1 509 8654510 • F: +1 509 8653864 •
Chm.: *Marian Ross* •
Local Museum – 1975
Antique artifacts, Indian baskets, regional contemp arts and crafts ... 53540

Topsfield MA

Topsfield Historical Museum, 1 Howlett St, Topsfield, MA 01983 • T: +1 978 8879724 • topshist@tiac.net • www.topsfieldhistory.org •
Cur.: *Jean Busch* •
Local Museum – 1894
Early American history – library ... 53541

Torrance CA

El Camino College Art Gallery, c/o El Camino College, 16007 Crenshaw Blvd, Torrance, CA 90506 • T: +1 310 6603010 • artgallery@elcamino.edu • www.elcamino.cc.ca.us •
Public Gallery
Prints, sculpture ... 53542

Torrey UT

Capitol Reef National Park Visitor Center, Capitol Reef National Park, Torrey, UT 84775 • T: +1 435 4253791 ext 111 • F: +1 435 4253026 • care_interpretation@nps.gov • www.nps.gov/care •
C.E.O.: *Albert J. Hendricks* •
Natural History Museum – 1968
Natural hist, Native American ethnology ... 53543

Torrington CT

Torrington Historical Society Museum, 192 Main St, Torrington, CT 06790 • T: +1 860 4828260 • torringtonhistorical@snet.net • www.torringtonhistoricalsociety.org •
Dir.: *Mark McEachern* • Cur.: *Gail Kruppa* •
Local Museum – 1944
Local hist ... 53544

Tougaloo MS

Tougaloo College Art Collection, Tougaloo, MS 39174 • T: +1 601 9777743 • F: +1 601 9777714 • art@tougaloo.edu • www.tougaloo.edu/artcolony •
Cur.: *Bruce O'Hara* (Photography) •
Fine Arts Museum / University Museum – 1963
Modern art, Afro-American & African art ... 53545

Towanda PA

Bradford County Historical Society Museum, 109 Pine St, Towanda, PA 18848 • T: +1 717 2652240 • bchs@cyber-quest.com •
C.E.O.: *Henry G. Farley* •
Local Museum / Historical Museum – 1870
Local hist – Library ... 53546

French Azilum, R.R. 2, Towanda, PA 18848 • T: +1 570 2653376 • www.frenchazilum.com •
Pres.: *Pam Emerson* • Dir.: *Thomas S. Owen* •
Local Museum – 1954
Historic house, 1793-1803 refuge of the French Royalists ... 53547

Townsend TN

Cades Cove Open-Air Museum, Great Smoky Mountains National Park, 10042 Campgrounds Dr, Townsend, TN 37882 • T: +1 865 4361256 • F: +1 865 4361220 • grsm_smokies_information@nps.gov • www.nps.gov/grsm •
Local Museum / Open Air Museum – 1951
Cable Mill Area and other historic structures typical of Southern Appalachia at turn of the 20th c ... 53548

Towson MD

Asian Arts and Culture Center, Towson University, 8000 York Rd, Towson, MD 21252 • T: +1 410 7042807 • F: +1 410 7044032 • sshieh@towson.edu • www.towson.edu/tu/asianarts •
Dir.: *Suewhei Shieh* •
Fine Arts Museum – 1971
Asian art from neolithic to modern times ... 53549

Hampton National Historic Site Museum, 535 Hampton Ln, Towson, MD 21286 • T: +1 410 8231309 • F: +1 410 8238394 • laurie_coughlan@nps.gov • www.nps.gov/hamp •
Cur.: *Lynne Dakin Hastings* •
Historical Museum – 1948
Late Georgian Mansion incl slave quarters, agricultural-industrial complex ... 53550

The Holtzman Art Gallery, c/o Towson University, 8000 York Rd, Towson, MD 21252-0001 • T: +1 410 7042808 •
Dir.: *Prof. Christopher Bartlett* •
Fine Arts Museum / University Museum – 1973
African & Asian arts, contemporary painting and sculpture, Maryland artists coll ... 53551

Trappe PA

Museum of the Historical Society of Trappe, Collegeville, Perkiomen Valley, 301 W Main St, Trappe, PA 19426 • T: +1 610 4897560 • F: +1 610 4897560 • ths708@aol.com •
Cur.: *John Shetler* •
Local Museum – 1964
Local hist ... 53552

Traverse City MI

Con Foster Museum → Grand Traverse Heritage Center

Dennos Museum Center of Northwestern Michigan College, 1701 E Front St, Traverse City, MI 49686 • T: +1 231 9951055 • F: +1 231 9951597 • dmc@nmc.edu • www.dennosmuseum.org •
Dir.: *Eugene A. Jenneman* • Cur.: *Kathleen Buday* •
Fine Arts Museum / University Museum – 1991
Contemporary prints & sculpture, Inuit art, Canadian Indian & Great Lakes Indian arts, Japanese prints – theater ... 53553

Grand Traverse Heritage Center, Museum and Con Foster Collection, 322 Sixth St, Traverse City, MI 49684 • T: +1 231 9950313 • F: +1 231 9466750 • info@gtheritagecenter.org • www.gtheritagecenter.org •
Dir./Cur.: *Lori Puckett* •
Local Museum – 1934
Local history, weapons, railroad history, logging, Native Americans, asylum & hospital, patient art, decorative arts, tools, toys – archives, library ... 53554

Trenton MI

Trenton Historical Museum, 306 Saint Joseph, Trenton, MI 48183 • T: +1 734 6752130 • F: +1 734 6754088 •
Dir.: *Nada Frost* •
Local Museum – 1962
Local hist, horse drawn carriages, Indian arrowheads ... 53555

Trenton NJ

Artworks, c/o Visual Art School of Trenton, 19 Everett Alley, Trenton, NJ 08611 • T: +1 609 3949436 •
Exec. Dir.: *Beth Daly* •
Fine Arts Museum – 1964 ... 53556

College of New Jersey Art Gallery, Holman Hall, Trenton, NJ 08650-4700 • T: +1 609 7712615 • F: +1 609 7712633 • masterjp@tcnj.edu • www.tcnj.edu/~tcag •
Dir.: *Prof. Dr. Lois Fichner-Rathus* •
Fine Arts Museum
Drawings, prints, photography, sculpture, African arts, crafts 53557

The Gallery, Mercer County Community College, 1200 Old Trenton Rd, Trenton, NJ 08690 • T: +1 609 5864800 ext 3589 •
Dir.: *Henry Hose* •
Fine Arts Museum – 1971
Cybis coll, paintings, sculpture, ceramics, Mexican art, folk art – library 53558

The Invention Factory, 650 S Broad St, Trenton, NJ 08611 • T: +1 609 3964214, 3962002 • F: +1 609 3960676 • dcarroll@inventionfactory.com • www.inventionfactory.com •
Exec. Dir.: *Daine L. Carroll* •
Science&Tech Museum
Science, Technology, Industry 53559

New Jersey State House, 125 W State St, Trenton, NJ 08625-0068 • T: +1 609 6332709 • F: +1 609 2921498 • dapril@njleg.org • www.njleg.state.nj.us •
Historical Museum – 1792
Period furnishings and equip, portraits of former governors and legislative figures 53560

New Jersey State Museum, 205 W State St, Trenton, NJ 08625, mail addr: POB 530, Trenton, NJ 08625-0530 • T: +1 609 2925420 • F: +1 609 5994098 • margaret.oreilly@sos.state.nj.us • www.newjerseystatemuseum.org •
Dir.: *Dr. Helen Shannon* • Sc. Staff: *Dr. Lorraine Williams* (Archaeology and Ethnology) • *Fran Mollet* (Archaeology, Ethnology) • *Jay Schwartz* (Planetarium) • *Margaret O'Reilly* (Art) • *Anthony Miskowski* (Natural history) • *Gregory Lattanzi* (Registrar, Archaeology, Ethnology) • *Shirley Albright* (Natural History) •
Local Museum / Historical Museum / Fine Arts Museum / Natural History Museum / Ethnology Museum / Archaeology Museum – 1895
Culture, archaeology, ethnology, science, fine and decorative arts 53561

Old Barracks Museum, Barrack St, Trenton, NJ 08608 • T: +1 609 3961776 • F: +1 609 7774000 • barracks@voicenet.com • www.barracks.org •
Dir.: *Richard Patterson* •
Military Museum – 1902
American decorative arts (1750-1820), archaeology, tools and household equipment, military artifacts 53562

Trenton City Museum, Cadwalader Park, 319 E State St, Trenton, NJ 08608 • T: +1 609 9893632 • F: +1 609 9893624 • brianohill@ellarslie.org • www.ellarslie.org •
Dir.: *Brian O. Hill* •
Local Museum – 1971
Ceramics, porcelain 53563

William Trent House, 15 Market St, Trenton, NJ 08611 • T: +1 609 9893027 • F: +1 609 2787890 • www.williamtrenthouse.org •
Dir.: *M.M. Pernot* •
Local Museum – 1939
Baroque furniture, decorative arts, ceramic, glass, local hist 53564

Trinidad CO

A.R. Mitchell Memorial Museum of Western Art, 150 E Main St, Trinidad, CO 81082 • T: +1 719 8464224 • F: +1 719 8464002 • themitch@rmi.net •
Dir.: *Pat Patrick* • Cur.: *Alan Peterson* •
Fine Arts Museum – 1979
Western art 53565

Baca House, Trinidad History Museum, 312 E Main St, Trinidad, CO 81082 • T: +1 719 8467217 • F: +1 719 8450117 • www.coloradohistory.org •
Pres.: *Georgianna Contiguglia* • Dir.: *Paula Manini* •
Historical Museum – 1870 53566

Bloom Mansion, Trinidad History Museum, 312 E Main St, Trinidad, CO 81082 • T: +1 719 8467217 • F: +1 719 8450117 • www.coloradohistory.org •
Pres.: *Georgianna Contiguglia* • Dir.: *Paula Manini* •
Local Museum / Decorative Arts Museum – 1955 53567

Louden-Henritze Archaeology Museum, Trinidad State Junior College, 600 Prospect St, Trinidad, CO 81082 • T: +1 719 8465508 • F: +1 719 8465667 • loretta.martin@trinidadstate.ed • www.trinidadstate.edu/museum •
Dir.: *Loretta Martin* •
University Museum / Archaeology Museum – 1955 53568

Santa Fe Trail Museum, Trinidad History Museum, 312 E Main St, Trinidad, CO 81082 • T: +1 719 8467217 • F: +1 719 8450117 •
Pres.: *Georgianna Contiguglia* • Dir.: *Paula Manini* •
Historical Museum – 1955
Hispanic hist 53569

Troutdale OR

Barn Museum and Rail Depot Museum, Troutdale Historical Society, 726 E Historic Columbia River Hwy, Troutdale, OR 97060 • T: +1 503 6612164 • F: +1 503 6742995 •
Cur.: *Mary Bryson* •
Local Museum / Science&Tech Museum – 1968 53570

Troy AL

Pioneer Museum of Alabama, 248 US Hwy 231 N, Troy, AL 36081 • T: +1 334 5663597 • F: +1 334 5663552 • jerry317@troycable.net • www.pioneer-museum.org •
Dir.: *Charlotte Gibson* •
Historical Museum / Agriculture Museum – 1969
Agriculture, local hist – amphitheatre 53571

Troy MI

Troy Museum and Historic Village, 60 W Wattles Rd, Troy, MI 48098 • T: +1 248 5243570 • F: +1 248 5243572 • museum@ci.troy.mi.us • www.ci.troy.mi.us/parks/museum •
Man.: *Loraine Campbell* •
Open Air Museum / Local Museum – 1927
Local agriculture, pioneer living, war memorial – archives 53572

Troy NY

The Arts Center of the Capital Region, 265 River St, Troy, NY 12180 • T: +1 518 2730552 • F: +1 518 2734591 • info@theartscenter.cc • www.artscenteronline.org •
Public Gallery
Changing exhib of contemp and folk (traditional) art 53573

The Junior Museum, 105 Eighth St, Troy, NY 12180 • T: +1 518 2352120 ext. 207 • F: +1 518 2356836 • info@juniormuseum.org • www.juniormuseum.org •
C.E.O. & Exec. Dir.: *Timothy S. Allen* •
Special Museum – 1954
Art, science and natural hist, shells, rocks, minerals, log cabin 53574

Rensselaer County Historical Society Museum, 57 Second St, Troy, NY 12180 • T: +1 518 2727232 • F: +1 518 2731264 • info@rchsonline.org • www.rchsonline.org •
Dir.: *Donna Hassler* • Cur.: *Stacy Pomeroy Draper* •
Local Museum – 1927
Hist of the area, genealogy, early 19th c furnishings, American dec arts, portraits, photos 53575

Troy OH

Overfield Tavern Museum, 121 E Water St, Troy, OH 45373 • T: +1 937 3354019 •
Dir.: *Robert Patton* • Cur.: *Kelly Smith* • Ass. Cur.: *Busser Howell* •
Local Museum – 1966
Pewter, redware, wrought-iron accessoires, brass, copper and tin 53576

Troy-Hayner Cultural Center, 301 W Main St, Troy, OH 45373 • T: +1 937 3390457 • F: +1 937 3356373 • hayner@tdnpublishing.com •
Dir.: *Linda Lee Jolly* •
Science&Tech Museum / Historical Museum – 1976
Hayner Distillery and Mary Jane Hayner family 53577

Truckee CA

Donner Memorial and Emigrant Trail Museum, 12593 Donner Pass Rd, Truckee, CA 96161 • T: +1 530 5827892 • F: +1 530 5502347 • ghack@parks.ca.gov • www.parks.ca.gov •
Cur.: *Judith K. Polanich* • *Shirley J. Kendell* •
Historic Site / Historical Museum – 1962
California emigrant trail-truckee rts, railroad, weapons, photos 53578

Truth or Consequences NM

Geronimo Springs Museum, 211 Main St, Truth or Consequences, NM 87901 • T: +1 505 8946600 • F: +1 505 8942888 •
Historical Museum – 1972
Local hist Indian artifacts, military hist 53579

Tubac AZ

Tubac Center of the Arts, 9 Plaza Rd, Tubac, AZ 85646, mail addr: POB 1911, Tubac, AZ 85646 • T: +1 520 3982371 • F: +1 520 3989511 • www.tubacarts.org •
Exec. Dir.: *Josie E. DeFalla* •
Public Gallery – 1964 53580

Tuckerton NJ

Barnegat Bay Decoy and Baymen's Museum, 137 W Main St, Tuckerton, NJ 08087 • T: +1 609 2968868 • F: +1 609 2965810 • tuckcport@aol.com • www.tuckertonseaport.org •
Pres.: *Malcolm J. Robinson* • Exec. Dir.: *John Gormley* •
Folklore Museum – 1989
Maritime culture 53581

Tucson AZ

Arizona Historical Museum, Arizona Historical Society Museum, 949 E Second St, Tucson, AZ 85719 • T: +1 520 6285774 • F: +1 520 6285695 • AHSTempe@azhs.gov • www.arizonahistoricalsociety.org •
Dir.: *Tom Peterson* •
Historical Museum – 1884
History, Spanish colonial time, Mexican culture, weapons, tranportation, ranching, mining – library, photo archive 53582

Arizona-Sonora Desert Museum, 2021 N Kinney Rd, Tucson, AZ 85743-8918 • T: +1 520 8832702 • info@desertmuseum.org • www.desertmuseum.org •
Dir.: *Robert D. Edison* • Cons.: *Richard C. Brusca* •
Natural History Museum – 1952
Natural history related to Sonoran desert region – botanical garden, zoo, aquarium 53583

Arizona State Museum, University of Arizona, 1013 E University Blvd, Tucson, AZ 85721-0026 • T: +1 520 6216302 • F: +1 520 6212976 • www.statemuseum.arizona.edu •
Dir.: *Beth Grindell* • Cur. Archaeology: *Chuck Adams* • *Suzanne Fish* • *Paul Fish* • *Mike Jacobs* • *Rich Lange* • Cur. Ethnohistory: *Diane Dittemore* • *Michael M. Brescia* • *Diana Hadley* • Cons.: *Nancy Odegaard* • Cur. Collections: *Davison Koenig* •
University Museum / Ethnology Museum / Archaeology Museum – 1893
Anthropology, archaeology, ethnology of the American Southwest and northern Mexico – library 53584

Center for Creative Photography, University of Arizona, 1030 N Olive Rd, Tucson, AZ 85721-0103 • T: +1 520 6217968 • F: +1 520 6219444 • www.creativephotography.org •
Dir.: *Douglas Nickel* • Ass. Dir.: *Barbara Allen* • Cur.: *Britt Salvesen* • Sc. Staff: *Cass Fey* (Education) •
Fine Arts Museum / Public Gallery / University Museum – 1975
archives, library 53585

DeGrazia Gallery in the Sun, 6300 N Swan Rd, Tucson, AZ 85718 • T: +1 520 2999191 • F: +1 520 2991381 • www.degrazia.org •
Dir.: *Kristine Peashock* •
Public Gallery – 1977
Southwest art e.g. by DeGrazia, oils, watercolours, pastels, lithographs, sculpture, enamels, jewelry 53586

Dinnerware Contemporary Art Gallery, 101 W Sixth St, Tucson, AZ 85701 • T: +1 520 7924503 • F: +1 520 7921282 • dinnerware@theriver.com • www.dinnerwarearts.com •
Dir.: *Sarah Hardesty* •
Public Gallery
Experimental work of art 53587

Downtown History Museum, Arizona Historical Society Museum, 140 N Stone Av, Tucson, AZ 85701 • T: +1 520 7701473 • F: +1 520 6298966 • AHSTucson@azhs.gov • www.arizonahistoricalsociety.org •
Cur.: *David T. Faust* •
Historical Museum – 2001
History of downtown Tucson, 18th c Spanish military, 19th c American military 53588

Flandrau Science Center, 1601 E University Blvd, Tucson, AZ 85721-0091, mail addr: c/o University of Arizona, POB 210091, Tucson, AZ 85721 • T: +1 520 6217827 • F: +1 520 6218451 • fsc@email.arizona.edu • www.flandrau.org •
Dir.: *Bill Buckingham* •
Science&Tech Museum – 1975
Science, astronomy exhibits – planetarium, observatory 53589

Fort Lowell Museum, Arizona Historical Society Museum, 2900 N Craycroft Rd, Tucson, AZ 85719, mail addr: 5230 E Fort Lowell Rd, Tucson, AZ 85712 • T: +1 520 3180219 • AHSTucson@azhs.gov • oflna.org/fort_lowell_museum/ftlowell.htm •
Pres.: *Bill Anderson* •
Military Museum – 1963
Period furnishings, costumes of frontier army life, uniforms, military artifacts 53590

The H. H. Franklin Museum, 1405 E Kleindale Rd, Tucson, AZ 85719 • T: +1 520 3268038 • FranklinMuseum@aol.com • franklinmuseum.org •
Dir.: *Bourke A. Runton* •
Science&Tech Museum – 1992
Franklin vehicles 1904-1934, 25 cars, air cooled motors 1938-1975, H.H Franklin company hist 1892-1936 53591

International Wildlife Museum, 4800 W Gates Pass Rd, Tucson, AZ 85745 • T: +1 520 6290100 • F: +1 520 6183561 • www.thewildlifemuseum.org •
Dir.: *Richard S. White* • Cur.: *Kristine Massey* •
Natural History Museum – 1988
Natural history dioramas, national heads and horns coll, Burnham-Eagle-Macomber coll, Wayde bird coll, insects, wildlife art 53592

Joseph Gross Gallery, 1013 N Olive Rd, Tucson, AZ 85721, mail addr: c/o University of Arizona, POB 210002, Tucson, AZ 85721-0002 • T: +1 520 6264215 • F: +1 520 6212353 • web.cfa.arizona.edu •
Dir.: *James Schaub* •
Fine Arts Museum / University Museum
Contemporary art 53593

Louis Bernal Gallery, c/o Pima College - Center for the Arts, West Campus, 2202 W Anklam Rd, Tucson, AZ 85709 • T: +1 520 2066986 • F: +1 520 2066670 • centerforthearts@pima.edu • www.pima.edu •
Dir.: *Katherine Monaghan* •
Public Gallery
Contemporary art 53594

Pima Air and Space Museum, 6000 E Valencia Rd, Tucson, AZ 85706 • T: +1 520 5740462 • F: +1 520 5749238 • tpate@pimaair.org • www.pimaair.org •
Exec. Dir.: *Daniel J. Ryan* • Rest.: *Scott Marchand* •
Historical Museum / Science&Tech Museum – 1966
Aeronautics, space, Arizona aviation – hall of fame 53595

Sosa-Carillo-Fremont House Museum, 151 S Granada, Tucson, AZ 85701 • T: +1 520 6220956, +1 520 6285774 • AHSTucson@azhs.gov • www.arizonahistoricalsociety.org •
Cur.: *Julia Arriola* •
Decorative Arts Museum – 1970
19th c furniture and decorative arts 53596

Tucson Children's Museum, 200 S Sixth Av, Tucson, AZ 85702-2609 • T: +1 520 7929985 • F: +1 520 7920639 • www.tucsonchildrensmuseum.org •
Dir.: *Michael Luria* •
Special Museum – 1986
Interactive exhibits 53597

Tucson Museum of Art and Historic Block, 140 N Main Av, Tucson, AZ 85701 • T: +1 520 6242333 • F: +1 520 6247202 • Info@TucsonMuseumofart.org • www.tucsonarts.com •
Dir.: *Laurie J. Rufe* • Cur.: *Susan Dolan* (Collections) • *Stephanie Coakley* (Education) • *Stephen Vollmer* (Art of The Americas) • *Julie Sasse* (Modern & Contemporary Art) •
Fine Arts Museum / Folklore Museum – 1924
Pre-Columbian, Spanish colonial, Latin American and Mexican folk art and contemporary art, American arts and crafts – library 53598

University of Arizona Mineral Museum, 1601 E University Blvd, Tucson, AZ 85721, mail addr: POB 210091Flandrau Science Center, Tucson, AZ 85721-0091 • T: +1 520 6214227 • F: +1 520 6218451 • www.uamineralmuseum.org •
Cur.: *Dr. Robert T. Downs* • *Sven Bailey* •
University Museum / Natural History Museum – 1891
Mineralogy from worldwide localities 53599

University of Arizona Museum of Art, 1031 N Olive Rd, Tucson, AZ 85721, mail addr: POB 210002, Tucson, AZ 85721-0002 • T: +1 520 6217567 • F: +1 520 6218770 • dhartman@u.arizona.edu • artmuseum.arizona.edu •
Dir.: *Charles A. Guerin* • Asst. Dir.: *Diane Hartman* • Cur.: *Loren Raab* • *John Kelly* (Exhibits) • *Carol Petrozzello* (Education) •
Fine Arts Museum / University Museum – 1955
Samuel H. Kress coll of 14th-19th c European art, contemporary art, C. Leonard Pfeiffer coll of American art – library 53600

University of Arizona Student Union Galleries, Student Union Rm 106B, Tucson, AZ 85721 • T: +1 520 6215853 • F: +1 520 6216930 • jsasse@u.arizona.edu • www.union.arizona.edu/csil/gallery •
Dir.: *Christina Lieberman* • Cur.: *Julie Sasse* •
Public Gallery / University Museum – 1971 53601

Western Archaeological and Conservation Center, 255 N Commerce Park Loop, Tucson, AZ 85745 • T: +1 520 7916406 •
Cur.: *Gloria J. Fenner* • Cons.: *Gretchen L. Voeks* •
Archaeology Museum – 1952
Archaeology and prehistory, ethnography, history, natural science of the Southwestern area – library, conservation & research labs 53602

Tucumcari NM

Dinosaur Museum, Mesalands Community College, 222 E Laughlin St, Tucumcari, NM 88401 • T: +1 505 4613466 • F: +1 505 4611901 • www.mesalands.edu •
Dir.: *Currell Craig* • Cur.: *Hungerbeuhler Axel* •
Natural History Museum / Archaeology Museum – 2000
Bronze prehistoric skeleton, paintings, sculpture, geological, paleontological 53603

Tucumcari Historical Research Institute Museum, 416 S Adams, Tucumcari, NM 88401 • T: +1 505 4614201 • F: +1 505 4612049 • museum@cityoftucumcari.com • www.cityoftucumcari.com •
Pres.: *Duane Moore* •
Local Museum – 1958
Local history, folk art, pictures 53604

Tujunga CA

McGroarty Cultural Art Center, 7570 McGroarty Terrace, Tujunga, CA 91042 • T: +1 818 3525285 • F: +1 818 9515348 • director@mcgroartyartscenter.org • www.mcgroartyartscenter.org •
Public Gallery 53605

Tulia TX

Swisher County Archives and Museum Association, 127 SW Second St, Tulia, TX 79088 • T: +1 806 9952819 •
Dir.: *Billie Sue Gayler* •
Association with Coll – 1965
Local memorabilia, furniture, clothing, stagecoach, needle art, blacksmith 53606

Tulsa OK

Alexandre Hogue Gallery, University of Tulsa, School of Art, 600 S College St, Tulsa, OK 74104 • T: +1 918 6312202 • F: +1 918 6313423 •
Public Gallery 53607

Gilcrease Museum, 1400 Gilcrease Museum Rd, Tulsa, OK 74127-2100 • T: +1 918 5962700 • F: +1 918 5962770 • gary-moore@utulsa.edu • www.gilcrease.org •
Dir.: *Hilary Kitz* • Ass. Dir.: *Gary F. Moore* • Sen. Cur.: *Daniel C. Swan* • *Sarah Erwin* • Cons.: *Gayle Clements* •
Fine Arts Museum / Historical Museum – 1949
American sculpture and painting, art of North American Indians, art and hist of Mexico ... 53608

Mac's Antique Car Museum, 1319 E Fourth St, Tulsa, OK 74120 • T: +1 918 5833101 • F: +1 918 5833108 • macselectric@sbcglobal.net •
Cur.: *D. McGlumphy* •
Science&Tech Museum – 1991
Antique and classic cars and trucks from 1910-1986 ... 53609

Philbrook Museum of Art, 2727 S Rockford Rd, Tulsa, OK 74114-4104 • T: +1 918 7485300 • F: +1 918 7434230 • mwintas@philbrook.org • www.philbrook.org •
Exec.Dir., Pres. & C.E.O.: *Brian Ferriso* • Dir.: *Christine Knop Kallenberger* (Exhibitions and Collections) • Dep. Dir.: *G. David Singleton* • Cur.: *James F. Peck* (European and American Art) •
Fine Arts Museum – 1938
Italian Renaissance paintings, native and contemp American art, oriental and African works ... 53610

Sherwin Miller Museum of Jewish Art, 2021 E 71st St, Tulsa, OK 74136 • T: +1 918 4921818 • F: +1 918 4921888 • info@jewishmuseum.net • www.jewishmuseum.net •
C.E.O.: *Arthur M. Feldman* •
Fine Arts Museum / Religious Arts Museum – 1966
Jewish hist, culture and art from biblical times to present, ritual objects, synagogue textiles, ethnographic artifacts, costumes, docs, prints, paintings, sculptures ... 53611

Tupelo MS

Bank of Mississippi Art Collection, 1 Mississippi Plaza, Tupelo, MS 38802 • T: +1 662 6802000 •
Pres.: *Aubrey B. Patterson jr.* •
Fine Arts Museum ... 53612

Gumtree Museum of Art, 211 W Main St, Tupelo, MS 38801 • T: +1 662 8442787 • F: +1 662 8449751 • info@gumtreemuseum.com • www.gumtreemuseum.com •
Exec. Dir.: *Tina Lutz* •
Fine Arts Museum – 1985
Art ... 53613

Natchez Trace Parkway, 2680 Natchez Trace Pkwy, Tupelo, MS 38801 • T: +1 662 6804004 • F: +1 662 6804036 • gretchen_ward@nps.gov • www.nps.gov/natr •
Head: *Wendell Simpson* •
Natural History Museum – 1938
Indian trails, 18th c trade rte, ethnology, geology 53614

Turlock CA

University Art Gallery, California State University Stanislaus, 801 W Monte Vista Av, Theatre Bldg, Turlock, CA 95382 • T: +1 209 6673186 • Art_Gallery@csustan.edu • www.csustan.edu/art/gallery •
Dir.: *Dean DeCocker* •
Public Gallery / University Museum – 1967 ... 53615

Tuscaloosa AL

Alabama Museum of Natural History, Campus, Smith Hall, Tuscaloosa, AL 35487-0340, mail addr: c/o University of Alabama, POB 870340, Tuscaloosa, AL 35487-0340 • T: +1 205 3487550 • F: +1 205 3489292 • museum.programs@ua.edu • www.museums.ua.edu/history •
C.E.O.: *Dr. Richard Diehl* •
University Museum / Natural History Museum – 1847
Natural hist ... 53616

Children's Hands-On Museum, 2213 University Blvd, Tuscaloosa, AL 35403 • T: +1 205 3494235 • F: +1 205 3494276 • info@chomonline.org • www.chomonline.org •
Dir.: *Miah Michaelsen* •
Special Museum – 1984
planetarium ... 53617

Gorgas House Museum, Capstone at McCorvy Dr, Tuscaloosa, AL 35487-0340, mail addr: c/o University of Alabama, POB 870340, Tuscaloosa, AL 35487 • T: +1 205 3487550 • F: +1 205 3489292 • ftucker@ua.edu • museums.ua.edu/gorgas •
Dir.: *Dr. Richard Diehl* • Cur.: *Marion Pearson* •
Decorative Arts Museum / University Museum – 1954
Spanish silver coll 18th-19th c ... 53618

Old Tavern Museum, 500 28th Av, Capitol Park, Tuscaloosa, AL 35401 • T: +1 205 7582238 • F: +1 205 7588163 • tcps@dbtech.net • www.historictuscaloosa.org •
Exec. Dir.: *Hannah Brown* •
Local Museum – 1965
Local hist, folklore ... 53619

Paul W. Bryant Museum, c/o University of Alabama, 300 Bryant Dr, Tuscaloosa, AL 35487-0385 • T: +1 205 3484668 • F: +1 205 3488883 • bryant-info@ua.edu • bryantmuseum.ua.edu •
Dir.: *Kenneth Gaddy* •
Special Museum / University Museum – 1985
Sports, memorabilia of university athletics ... 53620

Sarah Moody Gallery of Art, c/o University of Alabama, 103 Garland Hall, Tuscaloosa, AL 35487-0270 • T: +1 205 3485967 • wtdooley@bama.ua.edu • web.as.ua.edu/art •
Dir.: *Bill Dooley* •
Fine Arts Museum / University Museum – 1967
Primitive art, paintings, drawings, prints, photos, crafts, sculpture ... 53621

Tuscola IL

Korean War Veterans National Museum, C500 Tuscola Blvd, Tuscola, IL 61953 • T: +1 217 2535813 • F: +1 217 2539421 • kwmuseum@theforgottenvictory.org • www.theforgottenvictory.org •
Military Museum – 1997 ... 53622

Tuscumbia AL

Alabama Music Hall of Fame, 617 Hwy 72 W, Tuscumbia, AL 35674 • T: +1 256 3814417 • F: +1 256 3811031 • info@alamhof.org • www.alamhof.org •
Dir.: *David A. Johnson* • Ass. Dir.: *Marcia Weems* • Cur.: *George Lair* •
Music Museum – 1982
Musical instr, clothing, misic memorabilia – library, archives ... 53623

Ivy Green, Birthplace of Helen Keller, 300 North Commons St West, Tuscumbia, AL 35674 • T: +1 256 3834066 • www.helenkellerbirthplace.org •
Head: *Sue Pilkilton* •
Special Museum – 1952
Historic house, living and working conditions of blind people ... 53624

Tennessee Valley Art Center, 511 N Water St, Tuscumbia, AL 35674 • T: +1 256 3830533 • F: +1 256 3830535 • www.tvaa.net •
Dir.: *Mary Settle Cooney* • Asst. Dir.: *James C. Bennyman* •
Fine Arts Museum / Folklore Museum – 1962
Reynolds coll, H. Keller festival coll, fine art and crafts, coins, prints – theater ... 53625

Tuskegee Institute AL

George W. Carver Museum, c/o Tuskegee Institute, 1212 Old Montgomery Rd, Tuskegee Institute, AL 36088 • T: +1 334 7276390, 7273200 • F: +1 334 7271448 • www.nps.gov/tuin/ •
Head: *Brenda B. Caldwell* •
Natural History Museum – 1941
Natural history coll, science, paintings, needleworks – library ... 53626

Twin Falls ID

Herrett Center for Arts and Science, c/o College of Southern Idaho, 315 Falls Av, Twin Falls, ID 83301 • T: +1 208 7339554 ext 2655 • F: +1 208 7364712 • herritt@csi.edu • www.csi.edu/herrett •
Dir.: *James C. Woods* •
University Museum / Ethnology Museum / Natural History Museum – 1952
Anthropology, ethnology, Indian art – Faulkner Planetarium ... 53627

Twin Falls County Museum, Hwy 30, Filer, Twin Falls, ID 83301, mail addr: 214 Meadow Ln, Twin Falls, ID 83301 • T: +1 208 7364675 •
Pres.: *Chris Bolton* •
Local Museum – 1957
Period clothing, dishes, phonographs, musical instr, massacre site coll, sewing machines, office machines, farm machinery, gold washer, old picture gallery, mining ... 53628

Twinsburg OH

Twinsburg Historical Society Museum, 8996 Darrow Rd, Twinsburg, OH 44087 • T: +1 216 4875565 • twinsburghsoc@aol.com • www.twinsburg.com/historicalsociety •
Pres.: *Ken Roddie* •
Local Museum – 1963
Furniture, tools, clothes, toys, local artifacts ... 53629

Two Harbors MN

Lake County Museum, 520 South Av, Two Harbors, MN 55616-0128 • T: +1 218 8344898 • F: +1 218 8347198 • lakehist@lakenet.com • www.lakecountyhistoricalsociety.org •
C.E.O.: *Rachelle Malonly* •
Local Museum – 1926
Shipping & shipwrecks, lumbering, mineral coll, general local hist, mining, iron ore, steam engines, veterans ... 53630

Split Rock Lighthouse, 3713 Split Rock Lighthouse Rd, Two Harbors, MN 55616 • T: +1 218 2266372 • F: +1 218 2266373 • splitrock@mnhs.org • www.mnhs.org/places/sites/srl •
Dir.: *Lee Radzak* •
Historical Museum – 1976
Marine navigation ... 53631

Tybee Island GA

Tybee Museum and Tybee Island Light Station, 30 Meddin Dr, Tybee Island, GA 31328 • T: +1 912 7865801 • F: +1 912 7866538 • tybeelh@bellsouth.net • www.tybeelighthouse.org •
Dir.: *Cullen Chambers* •
Historical Museum – 1960
History, in old Spanish-American War Coastal Defense Battery, Tybee Island lighthouse and cottages ... 53632

Tyler TX

Goodman Museum, 624 N Broadway, Tyler, TX 75702 • T: +1 903 5311286 • F: +1 903 5311372 •
Cur.: *Boyd Sanders* •
Decorative Arts Museum – 1962
Cradles, china from England and France ... 53633

Smith County Historical Society, 125 S College Av, Tyler, TX 75702 • T: +1 903 5925561 • F: +1 903 5260924 • info@smithcountyhistory.org • www.smithcountyhistory.org •
Dir.: *Laura Petty* •
Association with Coll – 1959
Native American items, Civil War items, Gilded Age, games & toys, wireless telegraphic, equip, writing implements ... 53634

Tyler Museum of Art, 1300 S Mahon Av, Tyler, TX 75701 • T: +1 903 5951001 • F: +1 903 5951055 • info@tylermuseum.org • www.tylermuseum.org •
Dir.: *Kimberley Bush Tomio* •
Fine Arts Museum – 1969
Fine art from 19th c to the present ... 53635

The Wise Art Gallery, Tyler Junior College, 1400 E Fifth St, Tyler, TX 75711-9020 • T: +1 903 5102233 • F: +1 903 5102708 • dwhi@tjc.edu • www.tjc.edu •
University Museum / Fine Arts Museum
Art ... 53636

Tyringham MA

Santarella Museum, Tyringham Gingerbread House, 75 Main Rd, Tyringham, MA 01264 • T: +1 413 2432819 • info@santarella.us • www.santarella.us •
Dir.: *Denise Hoefer* • *Dennis Brandmeyer* •
Fine Arts Museum – 1953
Former art studio of Sir Henry Hudson Kitson – sculpture garden ... 53637

Ukiah CA

Grace Hudson Museum and The Sun House, 431 S Main St, Ukiah, CA 95482 • T: +1 707 4672836 • F: +1 707 4672835 • gracehudson@pacific.net • www.gracehudsonmuseum.org •
Dir.: *Sherrie Smith-Ferri* • Cur.: *Marvin Schenck* •
Local Museum / Fine Arts Museum – 1975
History, anthropology, Grace C. Hudson artworks, California Indians, Pomo Indians coll ... 53638

Held-Poage Memorial Home & Research Library, 603 W Perkins St, Ukiah, CA 95482-4726 • T: +1 707 4626969 • mchs@pacific.net • www.pacificsites.com/~mchs/ •
Dir.: *Paul Poulos* •
Historical Museum – 1970
Home of the mayor of Ukiah and county judge – library ... 53639

Ulster Park NY

Klyne Esopus Museum, 764 Rte 9 W, Ulster Park, NY 12487 • T: +1 845 2268221 • klynemuseum@aol.com •
Dir.: *Susan B. Wick* •
Local Museum – 1969
Local and regional hist ... 53640

Ulysses KS

Grant County Museum, Aka Historic Adobe Museum, 300 E Oklahoma, Ulysses, KS 67880 • T: +1 620 3563009 • F: +1 620 3565082 • ulyksmus@pld.com • www.historicadobemuseum.org •
Dir.: *Ginger Anthony* •
Local Museum – 1978
Indian artifacts, furnishings, toys, lothing, quilts, local and family hist ... 53641

Unalaska AK

Museum of the Aleutians, 314 Salmon Way, Unalaska, AK 99685, mail addr: POB 648, Unalaska, AK 99685 • T: +1 907 5815150 • F: +1 907 5816682 • Museum@akwisp.com • www.aleutians.org •
Dir.: *Dr. Richard Knecht* •
Local Museum – 1997
Archaeology, ethnology, historic items – archives 53642

Uncasville CT

Tantaquidgeon Indian Museum, Rte 32, 1819 Norwich-New London Rd, Uncasville, CT 06382 • T: +1 860 8489145 •
Cur.: *Gladys Tantaquidgeon* •
Historical Museum / Ethnology Museum – 1931
Indian hist and culture ... 53643

Union IL

Illinois Railway Museum, 7000 Olson Rd, Union, IL 60180 • T: +1 815 9234391 ext 404 • F: +1 815 9232006 • www.irm.org •
Pres.: *Barbara Lanphier* • C.E.O.: *Nick Kallas* •
Science&Tech Museum – 1953
Railways, housed in Marengo, Illinois rail depot ... 53644

McHenry County Historical Society Museum, 6422 Main St, Union, IL 60180 • T: +1 815 9232267 • F: +1 815 9232271 • info@mchsonline.org • www.mchsonline.org •
C.E.O.: *Nancy J. Fike* •
Local Museum – 1963
Local history and genealogy ... 53645

Union KY

Big Bone Lick State Park Museum, 3380 Beaver Rd, Union, KY 41091 • T: +1 859 3843522 • F: +1 859 3844775 • john.barker@mail.state.ky.us • www.state.ky.us/agencies/parks/bigbone.htm •
Archaeology Museum – 1971
Mastadon and bison bones ... 53646

Union ME

Matthews Museum of Maine Heritage, Union Fairgrounds, Union, ME 04862 • T: +1 207 7853321 • F: +1 207 7853321 • mitchell@tidewater.net •
Cur.: *Archie Mitchell* • *Irene Hawes* •
Local Museum – 1965
Local history ... 53647

Union NJ

James Howe Gallery, Kean University, Vaughn Eames Hall, Morris Av, Union, NJ 07083 • T: +1 908 7374000 • F: +1 908 5272804 •
Fine Arts Museum / University Museum – 1971
Paintings, prints, sculpture, photographs, design and furniture ... 53648

Liberty Hall Museum, 1003 Morris Av, Union, NJ 07083 • T: +1 908 5270400 • F: +1 908 3528915 • liberty-hall@juno.com • www.libertyhallnj.org •
C.E.O.: *John Kean* •
Historical Museum – 1968
Livingston and Kean family hist ... 53649

Union OR

Union County Museum, 311 S Main St, Union, OR 97883 • T: +1 541 5626003 • F: +1 541 5625196 • www.visitlagrande.com •
Dir.: *Val Stockhoff* • Cur.: *Kathleen Almquist* •
Local Museum – 1969
Artifacts, photos, manuscripts from 1830 – Library ... 53650

Union SC

Rose Hill Plantation State Historic Site, 2677 Sardis Rd, Union, SC 29379 • T: +1 864 4275966 • F: +1 864 4275966 • rose_hill_plantation_sp@scprt.com • www.southcarolinaparks.com •
Local Museum – 1960
Local hist ... 53651

Union County Historical Foundation Museum, American Federal Savings and Loan Bldg, Union, SC 29379 •
Cur.: *Tommy Bishop* •
Local Museum – 1960
Early American Revolutionary guns, currency, furniture, dolls, China and crystal, pottery, early tools, Confederate War ... 53652

Union City TN

Dixie Gun Works' Old Car Museum, 1412 W Reelfoot Av, Union City, TN 38261 • T: +1 731 8850561 • F: +1 731 8850440 • dixiegun@iswt.com • www.dixiegun.com •
Science&Tech Museum – 1954
American antique automotives, firearms ... 53653

U

Union Gap WA

Central Washington Agricultural Museum, 4508 Main St, Union Gap, WA 98903-0008 • T: +1 509 4578735, 2480432 •
Pres.: *Bob Herber* •
Agriculture Museum – 1979
Agricultural equip, tractor coll, blacksmith shop, horse drawn machinery, hand tools ... 53654

University Center MI

Marshall M. Fredericks Sculpture Museum, Saginaw Valley State University, Arbury Fine Arts Center, 7400 Bay Rd, University Center, MI 48710 • T: +1 989 9647125 • F: +1 989 9647221 • mfsm@svsu.edu • www.marshall fredericks.org •
Dir.: *Marilyn L. Wheaton* • Cur.: *Andrea Ondish* •
Fine Arts Museum / University Museum – 1988
Sculptures, plaster models, miniatures, drawings, tools – archives, education ... 53655

University Park IL

Nathan Manilow Sculpture Park, c/o Governors State University, Wagner House, University Park, IL 60466 • T: +1 708 5345000 • F: +1 708 5348959 • sculpture@govst.edu • www.govst.edu/sculpture • Dir.: *Beverly Goldberg* • University Museum / Open Air Museum – 1969 Sculpture 53656

University Park PA

The Frost Entomological Museum, c/o Department of Entomology, The Pennsylvania State University, 501 A.S.I. Bldg, University Park, PA 16802 • T: +1 814 8632863 • F: +1 814 8653048 • kck@psu.edu • www.ento.psu.edu/home/frost • Cur.: *Dr. Ke Chung Kim* • Natural History Museum – 1968 Natural hist 53657

HUB Robeson Galleries, c/o Pennsylvania State University, 241a HUB-Robeson, University Park, PA 16802 • T: +1 814 8652563 • F: +1 814 8630812 • galleries@sa.psu.edu • www.sa.psu.edu/galleries • Dir.: *Ann Shields* • University Museum / Public Gallery – 1976 53658

Palmer Museum of Art, The Pennsylvania State University, Curtin Rd, University Park, PA 16802-2507 • T: +1 814 8657672 • F: +1 814 8638608 • jlc5075@psu.edu • www.psu.edu/dept/palmermuseum/ • Dir.: *Jan Keene Muhlert* • Cur.: *Dr. Leo Mazow* (American Art) • *Charles V. Hallman* • *Dr. Patrick J. McGrady* • *Dr. Joyce Robinson* • Fine Arts Museum / University Museum – 1972 Art 53659

Upland CA

Cooper Regional History Museum, 217 E A St, Upland, CA 91786 • T: +1 909 9828010 • info@coopermuseum.org • www.coopermuseum.org • Local Museum – 1965 Local hist, domestic life, citrus industry 53660

Upland IN

Tyndale Galleria, Taylor University, Visual Art Department, c/o Modelle Metcalf Visual Arts Center, 236 W Reade Av, Upland, IN 46989-1001 • T: +1 317 9985322 • F: +1 765 9984680 • visualarts@tayloru.edu • www.taylor.edu/academics/acadDepts/art • Public Gallery – 2003 53661

Upland PA

Caleb Pusey House, 15 Race St, Upland, PA 19015 • T: +1 610 8745665 • www.delcohistory.org • Pres.: *Harold R. Peden* • Local Museum – 1960 Local hist 53662

Upper Marlboro MD

Billingsley House Museum, 6900 Green Landing Rd, Upper Marlboro, MD 20772 • T: +1 301 6270730 • F: +1 301 6277085 • billingsley@pgparks.com • www.pgparks.com • Dir.: *Greg Lewis* • Historic Site 53663

Darnall's Chance House Museum, 14800 Gov. Oden Bowie Dr, Upper Marlboro, MD 20772 • T: +1 301 9528010 • F: +1 301 9521773 • susan.reidy@pgparks.com • www.pgparks.com • Dir.: *Susan Reidy* • Local Museum / Museum of Classical Antiquities – 1988 Darnall and Carroll family home 53664

Upper Montclair NJ

Montclair State University Art Galleries, Valley Rd & Normal Av, Upper Montclair, NJ 07043 • T: +1 973 6555113 • F: +1 973 6557665 • pacel@montclair.edu • www.montclair.edu/pages/arts • Dir.: *Dr. Lorenzo Pace* • Ass. Dir.: *Teresa Rodriguez* • Fine Arts Museum / University Museum – 1973 Contemporary East Indian art, Japanese expressionism in paper – Calcia library 53665

Upper Sandusky OH

Indian Mill Museum State Memorial, 7417 Wyandot County Rd 47, Upper Sandusky, OH 43351 • T: +1 419 2943349, 2941491 • webmaster@ohiohistory.org • www.ohiohistory.org/places/indian • Historic Site – 1967 Grist mill 53666

Wyandot County Historical Society Museum, 130 S Seventh St, Upper Sandusky, OH 43351-0372 • T: +1 419 2943857 • wchs@udata.com • Cur.: *Kris Lininger* • Local Museum / Historical Museum – 1929 Agriculture, costomes, glass, Indian artifacts, musical instr 53667

Upton NY

Science Learning Center, Brookhaven National Laboratory, Upton, NY 11973-5000 • T: +1 631 3447098 • F: +1 631 3445832 • donoghue@bnl.gov • www.bnl.gov/slc • Dir.: *Gail Donoghue* • Science&Tech Museum – 1977 Children's Science exhibits and programs 53668

Urbana IL

Spurlock Museum, University of Illinois at Urbana-Champaign, 600 S Gregory St, Urbana, IL 61801 • T: +1 217 3332360 • F: +1 217 3449419 • ksheahan@uiuc.edu • www.spurlock.uiuc.edu • Dir.: *Dr. Douglas J. Brewer* • *Tandy Lacy* (Education) • Man.: *Christa Deacy-Quinn* (Collections) • Coord.: *Karen Flesher* (Program) • University Museum / Historical Museum / Ethnology Museum – 1911 Cultural, historical, and archaeological colls across the globe, e.g. Neolithic vessels, Egyptian mummy, Roman papyrus manuscripts, plaster casts, Merovingian artifacts, Ottomon coins, Japanese & Chinese carvings, bark cloth paintings from Columbia, Canelos Quichua ceramics, African masks, Inuit ivory carvings 53669

Urbana OH

Champaign County Historical Museum, 809 E Lawn Av, Urbana, OH 43078 • T: +1 937 6536721 • Pres.: *Chris Callison* • Local Museum – 1934 Civil War, local hist 53670

Urbandale IA

Living History Farms Museum, 2600 NW 111th St, Urbandale, IA 50322 • T: +1 515 2785286 • F: +1 515 2789808 • info@lhf.org • www.livinghistoryfarms.org • Dir.: *Rick Finch* • Agriculture Museum – 1967 Agriculture 53671

USAF Academy CO

Unites States Air Force Academy Visitor Center, 2346 Academy Dr, USAF Academy, CO 80840-9400 • T: +1 719 3332569 • F: +1 719 3334402 • Military Museum – 1961 Aviation and aerospace hist 53672

Utica IL

LaSalle County Historical Society Museum, Mill and Canal Sts, Utica, IL 61373 • T: +1 815 6674861 • F: +1 815 6675121 • museum@core.com • www.lasallecountymuseum.org • Dir.: *Mary C. Toraason* • Local Museum – 1967 Local history, housed in 1848 pre-Civil War stone warehouse 53673

Starved Rock State Park, POB 509, Utica, IL 61373 • T: +1 815 6674906, 6675356 • F: +1 815 6675354 • dnr.starvedrock@illinois.gov • Historical Museum – 1911 Former Indian village of Illinois Indians, 1673-1760 French occupation and 1683 French Fort Saint Louis 53674

Utica NY

Children's Museum of History, Natural History and Science at Utica, New York, 311 Main St, Utica, NY 13501 • T: +1 315 7246129 • F: +1 315 7246120 • www.museum4kids.net • Pres.: *Marlene Brown* • Natural History Museum – 1963 Indian artifacts, hist dioramas, rocks, minerals, shells, bird and animal mounts 53675

Munson-Williams-Proctor Arts Institute Museum of Art, 310 Genesee St, Utica, NY 13502 • T: +1 315 7970000 ext 2168 • F: +1 315 7975608 • pschweiz@mwpi.edu • www.mwpai.org • Dir.: *Dr. Paul D. Schweizer* • Cur.: *Mary Murray* (Modern & Contemporary Art) • *Anna T. D'Ambrosio* (Decorative Arts) • Fine Arts Museum – 1919 18th-20th c European and American fine art; prints, 19th c American dec art 53676

Oneida County Historical Society Museum, 1608 Genesee St, Utica, NY 13502-5425 • T: +1 315 7353642 • F: +1 315 7320806 • ochs@midyork.org • www.oneidacountyhistory.org • C.E.O. & Pres.: *Donald F. White* • Local Museum – 1876 Hist of the region, docs, manuscripts, maps 53677

Uvalde TX

Garner Memorial Museum, 333 N Park St, Uvalde, TX 78801 • T: +1 830 2785018 • F: +1 830 2790512 • m.lara@mail.utexas.edu • www.cah.utexas.edu • C.E.O.: *Dr. Don Carleton* • Dir.: *Dr. Patrick Cox* • Cur.: *Maria Lara* • University Museum / Historical Museum – 1960 Historic house, home of Vice President John N. Garner 53678

Vacaville CA

Vacaville Museum, 213 Buck Av, Vacaville, CA 95688 • T: +1 707 4474513 • vminfo@vacavillemuseum.org • www.vacavillemuseum.org • Dir.: *Shawn Lum* • Cur.: *Annie Farley* (Collections) • *Philip Nollar* (Exhibits) • Local Museum – 1981 Local hist, prehist 53679

Vail CO

Colorado Ski & Snowboard Museum and Hall of Fame, 231 S Frontage Rd, Vail, CO 81657, mail addr: P O Box 2975, Vail, CO 81658 • T: +1 970 4761876 • F: +1 970 4761879 • skimuseum@gmail.com • www.skimuseum.net • Dir.: *Susan Tjossem* • Cur.: *Justin Henderson* • Special Museum – 1976 Ski and Snowboard history, 10th Mountain Division, Hall of Fame, Tribute to Olympic Champions, Vail history 53680

Vails Gate NY

Edmonston House, c/o National Temple, Rte 94, Vails Gate, NY 12584 • T: +1 845 5615073 • F: +1 845 5615073 • swaddell@frontiernet.net • www.nationaltemplehill.org • Historical Museum – 1933 Period furnishings, American Revolutionary events 53681

Knox's Headquarters State Historic Site, Forge Hill Rd at Rte 94, Vails Gate, NY 12584, mail addr: PO Box 207, Vails Gate, NY 12584 • T: +1 845 5615498 • Michael.Clark@oprhp.state.ny.us • www.nysparks.com • C.E.O.: *Michael J. Clark* • Historic Site – 1922 18th century period furnishings, manuscripts 53682

New Windsor Cantonment State Historic Site and National Purple Heart Hall of Honor, 374 Rte 300 and Temple Hill Rd, Vails Gate, NY 12584 • T: +1 845 5611765 • F: +1 845 5616577 • www.friendsofpalisades.org • Head: *Michael J. Clark* • Local Museum – 1967 53683

Valdese NC

Museum of Waldensian History, c/o Waldensian Presbyterian Church, Roderet St, Valdese, NC 28690 • T: +1 828 8742531 • F: +1 704 8740880 • waldensian@nci.net • www.waldensianpresbyterian.com • Head: *Jewell P. Bounous* • Religious Arts Museum – 1955 Late 19th and early 2oth c clothing and household furnishings, implements and farming tools of the Waldensians, hist, furniture a.o. items of the Waldensian Presbyterian Church 53684

Valdez AK

Valdez Museum, 217 Egan Av, Valdez, AK 99686-0008 • T: +1 907 8352764 • F: +1 907 8355800 • info@valdezmuseum.org • www.valdezmuseum.org • Dir.: *Tabitha Gregory* • Cur.: *James Perry* • *Steve Richardson* • Local Museum – 1976 Local hist, gold and copper mining, lighthouse lens, transport and firefighting items – archives 53685

Valdosta GA

Lowndes County Museum, 305 W Central Av, Valdosta, GA 31601 • T: +1 229 2474780 • F: +1 229 2472840 • history@valdostamuseum.org • www.valdostamuseum.org • C.E.O.: *Renate Milner* • Local Museum – 1967 Artifacts, docs, photos 53686

Valdosta State University Fine Arts Gallery, College of the Arts, Valdosta, GA 31698-0110 • T: +1 912 3335835 • F: +1 912 2453799 • kgmurray@valdosta.edu • www.valdosta.edu/art • Dir.: *Karin G. Murray* • Fine Arts Museum / University Museum – 1906 Lamar Dodd Coll 53687

Vale OR

Stonehouse Museum, 255 Main St, Vale, OR 97918 • T: +1 541 4732070 • Pres.: *Gary Fugate* • Historical Museum / Local Museum – 1995 Oregon trail, Northern Piautes and Bannock Indians 53688

Valentine NE

Cherry County Historical Society Museum, Main St and Hwy 20, Valentine, NE 69201 • T: +1 402 3762015 • Local Museum – 1928 Indian artifacts, musical instr 53689

Sawyer's Sandhills Museum, 440 Valentine St, Valentine, NE 69201 • T: +1 402 3763293 • Science&Tech Museum / Local Museum – 1958 Antique autos, guns, Indian artifacts, musical instr, lamps 53690

Vallejo CA

Vallejo Naval and Historical Museum, 734 Marin St, Vallejo, CA 94590 • T: +1 707 6430077 • F: +1 707 6432443 • valmuse@pacbell.net • www.vallejomuseum.org • Dir.: *James E. Kern* • Local Museum – 1974 Local hist, Mare Island naval shipyard hist – library, auditorium 53691

Valley NE

Valley Community Historical Society Museum, 218 W Alexander St, Valley, NE 68064 • T: +1 402 3592678 • www.valleyne.org • Pres.: *Dick Allen* • Local Museum – 1966 53692

Valley City ND

Barnes County Historical Museum, 315 Central Av N, Valley City, ND 58072 • T: +1 701 8450966 • F: +1 701 8454755 • bchistoricalsociety@hotmail.com • C.E.O.: *Warren Ventsch* • Local Museum – 1930 Costumes, tools, dolls, military items, Far Eastern artifacts 53693

Valley Forge PA

National Center for the American Revolution, POB 122, Valley Forge, PA 19481-0122 • T: +1 610 9173451 • F: +1 610 9173188 • info@americanrevolutioncenter.org • www.americanrevolutioncenter.org • Dir.: *Thomas M. Daly* • Ass. Dir.: *Stacey A. Swigart* (Collections) • Historic Site / Historical Museum – 1918/2005 Hist, museum located on the site of Valley Forge encampment area 53694

Valley Forge National Historical Park, POB 953, Valley Forge, PA 19482-0953 • T: +1 610 7831000 • F: +1 610 7831053 • www.nps.gov/vafo • Head: *Arthur L. Stewart Swigart* • Historical Museum / Historic Site – 1893 1777-1778 site of Continental Army winter encampment 53695

Valley Glen CA

Los Angeles Valley College Art Gallery, 5800 Fulton Av, Valley Glen, CA 91401 • T: +1 818 7785536 • reeddj@lavc.edu • www.lavc.edu/arts/artgallery.html • Dir.: *Jim Marrin* • Public Gallery / University Museum – 1960 53696

Valparaiso FL

Heritage Museum of Northwest Florida, 115 Westview Av, Valparaiso, FL 32580 • T: +1 850 6782615 • F: +1 850 6784547 • bmoss@co.okaloosa.fl.us • www.heritage-museum.org • C.E.O: *Barbara Brundage* • Local Museum – 1970 Local history 53697

Valparaiso IN

Brauer Museum of Art, Valparaiso University Center for the Arts, Valparaiso, IN 46383-6349 • T: +1 219 4645365 • F: +1 219 4645244 • gregg.hertzlieb@valpo.edu • www.valpo.edu/artmuseum • Dir.: *Gregg Hertzlieb* • Fine Arts Museum / University Museum – 1953 Art 53698

Porter County Museum, Old Jail Bldg, 153 Franklin St, Valparaiso, IN 46383 • T: +1 219 4653595 • info@portercountymuseum.org • www.portercountymuseum.org • Dir.: *Teresa Schmidt* • Local Museum – 1916 Local history, criminology 53699

Porter County Old Jail Museum → Porter County Museum

Van Buren AR

Bob Burns Museum, 813 Main St, Van Buren, AR 72956 • T: +1 501 4746164 • vanburen@vanburen.org • www.vanburen.org • Local Museum / Historical Museum – 1994 Radio humorist Bob Burns 1930-40's, local history 53700

Van Horn TX

Clark Hotel Museum, 112 W Broadway, Van Horn, TX 79855 • T: +1 915 2838028 • oldwest@telstar1.com • Dir./Cur.: *Robert Stuckey* • Ass. Dir.: *Darice McVay* • Local Museum – 1975 Indian artifacts, pioneer articles, local art, antique bar, rocks & fossils 53701

Van Lear KY

Coal Miners' Museum, 78 Miller's Creek Rd, Van Lear, KY 41265 • T: +1 606 7899725 • coalcamp@yahoo.com • www.geocities.com/coalcamp • Science&Tech Museum – 1984 53702

U

Van Wert OH

Wassenberg Art Center, 643 S Washington St, Van Wert, OH 45891 • T: +1 419 2386837 • F: +1 419 2386828 • wassenberg@embarqmail.com • www.vanwert.com/wassenberg •
Dir.: *Michele L. Smith* •
Fine Arts Museum – 1954
Prints and original art ... 53703

Vancouver WA

Archer Gallery, Clark College, 1800 E McLoughlin Blvd, Gaiser Hall, Vancouver, WA 98663 • T: +1 360 9922246 • F: +1 360 9922828 • mhirsch@clark.edu • www.clark.edu/ •
Dir.: *Marjorie Hirsch* •
Fine Arts Museum – 1978
Installationa, fiber, ceramics ... 53704

Clark County Historical Museum, 1511 Main St, Vancouver, WA 98660 • T: +1 360 9935679 • F: +1 360 9935683 • cchm@pacifier.com •
Pres.: *Karleen Berlmont* • Dir.: *Susan Tissot* •
Historical Museum – 1917
Local hist the Northwest, pre-native culture ceramics, railway equip, household items ... 53705

Fort Vancouver, 612 E Reserve St, Vancouver, WA 98661 • T: +1 360 6967655 • F: +1 360 6967657 • FOVA_superintendent@nps.gov • www.nps.gov •
Cur.: *David K. Hansen* •
Historic Site – 1948
Local hist, archaeological and military coll, agriculture, manufacturing ... 53706

Pearson Air Museum, 1115 E Fifth St, Vancouver, WA 98661 • T: +1 360 6947026 • F: +1 360 6940824 • eliza.lane@vnhrt.org • www.pearsonairmuseum.org •
Dir.: *John Donnelly* • Cur.: *Tom Clark* •
Science&Tech Museum – 1987
Historic aircrafts, models, period furnishings ... 53707

Vandalia IL

The Little Brick House, 621 St Clair, Vandalia, IL 62471 • T: +1 618 2832371 •
Dir.: *Dale Timmermann* •
Historical Museum – 1960
1840-1860 James W. Berry property, furniture, glass, original bldgs from 1825-1985, manuscripts and autographed books by Vandalia authors, Lincol memorabilia ... 53708

Vandalia State House, 315 W Gallatin St, Vandalia, IL 62471 • T: +1 618 2831161 •
Head: *Robert Coomer* •
Historic Site – 1836
1836 Vandalia Statehouse is the oldest Capitol bldg in the state of Illinois ... 53709

Varnell GA

Praters Mill, c/o City Hall, 500 Praters Mill Rd, Varnell, GA 30756 • T: +1 706 6946455 • F: +1 706 6948413 • pratersmill@dalton.net • www.PratersMill.org •
Science&Tech Museum – 1970
Horse-drawn farm implements ... 53710

Vasa MN

Vasa Lutheran Church Museum, County Rd 7, Vasa, MN 55089 • T: +1 612 2584327 •
Pres.: *Everett Lindquist* •
Religious Arts Museum – 1930 ... 53711

Vashon WA

Vashon Maury Island Heritage Museum, 10105 SW Bank Rd, Vashon, WA 98070 • T: +1 206 4637808 •
Pres.: *David Swain* •
Local Museum – 1975
Pre-settlement, industry, agriculture ... 53712

Vaughan MS

Casey Jones Museum, Main St, Vaughan, MS 39179 • T: +1 601 6739864 • F: +1 601 6536693 • holmesp@ayrix.net • www.trainweb.org/caseyjones/home.html •
Science&Tech Museum – 1980
Railroad memorabilia ... 53713

Ventura CA

San Buenaventura Mission Museum, 211 E Main St, Ventura, CA 93001-2622 • T: +1 805 6434318 • F: +1 805 6437831 • mission@sanbuenaventuramission.org • www.sanbuenaventuramission.org •
Dir.: *Patrick J. O'Brien* •
Religious Arts Museum / Local Museum
Local hist, Indian artifacts, military, paintings, vestmentst, religious artifacts ... 53714

Ventura College Art Galleries, 4667 Telegraph Rd, Ventura, CA 93003 • T: +1 805 6546462, 6488974 • F: +1 805 6546466 • www.venturacollege.edu •
Dir.: *Sharon Coughran-Rayden* •
Public Gallery / University Museum – 1970 ... 53715

Ventura County Museum of History and Art, 100 E Main St, Ventura, CA 93001 • T: +1 805 6530323 ext 11 • F: +1 805 6535267 • director@VenturaMuseum.org • www.venturamuseum.org •
Dir.: *Tim Schiffer* •
Fine Arts Museum / Historical Museum – 1913
Native American, Hispanic and American Settlers periods, contemporary California art, farming – library ... 53716

Vergennes VT

The Lake Champlain Maritime Museum, 4472 Basin Harbor Rd, Vergennes, VT 05491 • T: +1 802 4752022 • F: +1 802 4752953 • info@lemm.org • www.lcmm.org •
Dir.: *Arthur B. Cohn* • Cons.: *David Robinson* •
Archaeology Museum – 1985
Maritime and nautical archaeology ... 53717

Vermilion OH

Inland Seas Maritime Museum, 480 Main St, Vermilion, OH 44089 • T: +1 440 9673467 • F: +1 440 9671519 • glhs1@inlandseas.org • www.inlandseas.org •
Pres.: *Anthony F. Fugaro* • Exec. Dir.: *Christopher H. Gillerist* •
Historical Museum – 1944
Great Lakes ship models, marine, racing and yachting artifacts, shipbuilding tools and equip ... 53718

Vermillion SD

America's National Music Museum, 414 E Clark St, Vermillion, SD 57069-2390 • T: +1 605 6775306 • F: +1 605 6776995 • aplarson@usd.edu • www.usd.edu/smm •
Dir.: *Dr. Andre P. Larson* • Cur.: *Dr. Margaret Downie Banks* • *Arian Sheets* • *Dr. Deborah Chek* • Cons.: *John Koster* •
Music Museum – 1973
Musical instr ... 53719

University Art Galleries, University of South Dakota, Warren M. Lee Center, 414 E Clark, Vermillion, SD 57069 • T: +1 605 6775481 • F: +1 605 6775988 • jday@usd.edu •
Dir.: *John A. Day* •
Fine Arts Museum / University Museum – 1976
Historic art of South Dakota, Asian, African and native American art ... 53720

W.H. Over Museum, 1110 Ratingen St, Vermillion, SD 57069, mail addr: 414 East Clark, Vermillion, SD 57069 • T: +1 605 6775228 • whover@usd.edu • www.usd.edu/whover •
Pres.: *Larry Bradley* • C.E.O.: *N. N.* •
Local Museum / Natural History Museum – 1883
Stanley J. Morrow Collection, Frontier photographs, Dakota Territory 1869-83 ... 53721

Vernal UT

Utah Field House of Natural History State Park, 235 E Main St, Vernal, UT 84078 • T: +1 435 7893799 • F: +1 435 7894883 •
Head: *Steven D. Sroka* •
Natural History Museum – 1946
Geology, fossils, archaeology, natural hist of the Uinta Basin ... 53722

Vernon TX

Red River Valley Museum, 4600 College Dr, Vernon, TX 76384 • T: +1 817 5531848 • F: +1 817 5531849 • rrvm1@yahoo.com • www.redrivervalleymuseum.org •
Exec. Dir.: *Mary Ann McCuistion* •
Historical Museum / Science&Tech Museum / Fine Arts Museum – 1963
History, science, fine arts ... 53723

Vernon VT

Vernon Historians, 4201 Fort Bridgman Rd, Vernon, VT 05354 • T: +1 802 2548015 • F: +1 802 2570292 •
Association with Coll – 1968
Books, bibles, postcards, photos ... 53724

Vernon Hills IL

Cuneo Museum, 1350 N Milwaukee, Vernon Hills, IL 60061 • T: +1 847 3623042, 3623054 • F: +1 847 3624130 • mcook@cuneomuseum.org • www.cuneomuseum.org •
Dir.: *James Bert* •
Fine Arts Museum / Historical Museum – 1991
Renaissance paintings, 17th c tapestries, period oriental rugs and furnishings ... 53725

Vernonia OR

Vernonia Historical Museum, 511 E Bridge St, Vernonia, OR 97064-1406 • T: +1 503 4293713 • VPMA@agalis.net •
Pres.: *Enid Parrow* • Vice. Pres.: *Lee Enneberg* •
Local Museum
Hist of Nehalem Valley pioneers, photos ... 53726

Vero Beach FL

McLarty Treasure Museum, 13180 Hwy A1A, Vero Beach, FL 32963-9400 • T: +1 561 5892147 • F: +1 321 9844854 • www.myflorida.com •
Dir.: *Ed Perry* •
Historical Museum – 1970
History, Film, Spanish Shipwreck ... 53727

Vero Beach Museum of Art, 3001 Riverside Park Dr, Vero Beach, FL 32963 • T: +1 772 2310707 • F: +1 772 2310938 • info@vbmuseum.org • www.vbmuseum.org •
Exec. Dir.: *Lucinda H. Gedeon* • Cur.: *Jeniffer A. Bailey* •
Fine Arts Museum – 1979
Art audiovisual and film, dec arts, paintings, photos, prints, drawings, graphic arts, sculpture ... 53728

Versailles IN

Ripley County Historical Society Museum, Water and Main, Versailles, IN 47023 •
Dir.: *Beatrice Boyd* • Cur.: *Cheryl Welch* •
Local Museum – 1930
Local history ... 53729

Vestal NY

Vestal Museum, 328 Vestal Pkwy E, Vestal, NY 13850 • T: +1 607 7481432 • vestalhistory@tier.net • www.tier.net/vestalhistory •
Dir.: *Jan Roosa* • Cur.: *Virginia Wood* •
Local Museum / Science&Tech Museum – 1976
Tools, railroad memorabilia, period clothing, hist life of the area ... 53730

Vevay IN

Switzerland County Historical Society Museum, Main and Market Sts, Vevay, IN 47043 • T: +1 812 4273560, 4273237 •
Pres.: *Martha Bladen* •
Local Museum – 1925
General county hist ... 53731

Vicksburg MS

Cairo Museum, 3201 Clay St, Vicksburg, MS 39180 • T: +1 601 6362199 • F: +1 601 6387329 • vick_interpretation@nps.gov • www.nps.gov/vick •
Cur.: *Elizabeth Joyner* •
Historical Museum / Military Museum / Natural History Museum – 1980
Official records of the Union and Confederate Armies ... 53732

Cedar Grove Mansion Inn, 2200 Oak St, Vicksburg, MS 39180 • T: +1 601 6361000 • F: +1 601 6346126 • info@cedargroveinn.com • www.cedargroveinn.com •
Dir.: *Colleen Small* •
Historical Museum – 1959
Regina music box, bohemian glass, original works of art, Aubusson rugs ... 53733

Old Court House Museum-Eva Whitaker Davis Memorial, 1008 Cherry St, Vicksburg, MS 39183 • T: +1 601 6360741 • www.oldcourthouse.org •
Dir.: *Gordon A. Cotton* • Ass. Dir.: *Blanche S. Terry* •
Local Museum – 1947
Civil War & Indian artifacts ... 53734

Vicksburg National Military Park, 3201 Clay St, Vicksburg, MS 39180 • T: +1 601 6360583 • F: +1 601 6369497 • www.nps.gov/vick •
Head: *William O. Nichols* • Cur.: *Elizabeth Joyner* •
Military Museum / Open Air Museum – 1899
Artifacts from ironclad gunboat ... 53735

Yesterday's Children Antique Doll & Toy Museum, 1104 Washington St, Vicksburg, MS 39183 • T: +1 601 6380650 • mbakarich@aol.com • www.yesterdayschildren museum.com •
Dir.: *Michael N. Bakarich* • Cur.: *Carolyn C. Bakarich* •
Special Museum – 1986
More than 1,000 dolls 1800s-1900s e.g. from China, France and Germany, contemporary dolls, boy's toys incl pedal cars, trucks, trains, guns, soldiers, cars, construction and farm equip ... 53736

Victor MT

Victor Heritage Museum, Blake & Main St, Victor, MT 59875 • T: +1 406 6423997 •
Pres.: *Mary K. Hafer* •
Historical Museum – 1989
Local hist, folk culture ... 53737

Victor NY

Ganondagan State Historic Site, 1488 Victor Bloomfield Rd at State Rte 444, Victor, NY 14564, mail addr: POB 113, Victor, NY 14564 • T: +1 585 9245848 • F: +1 585 7421732 • www.ganondagan.org •
Man.: *Gerald Peter Jemison* • Ass. Dir.: *John L. Clancy* •
Historical Museum – 1972
Seneca artifacts, European manufacture and trade items ... 53738

Valentown Museum, Victor Historical Society, Valentown Sq, Victor, NY 14564 • T: +1 585 9244170 • F: +1 585 9240523 • info@valentown.org • valentown.org •
Dir.: *Lilian Fisher* •
Local Museum – 1940
Archaeological Indian artifacts, local Civil War material, early Mormon artifacts ... 53739

Victoria TX

McNamara House Museum, 502 N Liberty, Victoria, TX 77901 • T: +1 361 5758227 • F: +1 361 5758228 • vrma@viptx.net • www.viptx.net/museum/ •
Historical Museum – 1959
Victorian era furniture, decorative arts & docs ... 53740

Nave Museum, 306 W Commercial, Victoria, TX 77901 • T: +1 361 5758227 • F: +1 361 5758228 • vrma@viptx.net • www.viptx.net/museum/ •
Dir.: *Dinah Mills* • Cur.: *Penelope Sherwood* •
Fine Arts Museum – 1976
Early 20th c paintings by Royston Nave ... 53741

Victorville CA

California Route 66 Museum, 16825 Rte 66 D St, Victorville, CA 92393 • T: +1 760 9510436 • F: +1 760 9510509 • cart66musm@aol.com • www.califrt66museum.org •
Local Museum – 1995
Hula Ville folk art, local hist ... 53742

Vienna GA

Georgia State Cotton Museum, 1321 E Union St, Vienna, GA 31092 • T: +1 229 2683663 • F: +1 229 2683664 • www.historicvienna.org •
Exec. Dir.: *Janet Joiner* • Cur.: *Margaret Hegidio* •
Special Museum – 1994
Period farm implements, cotton patch, history and production of cotton ... 53743

Walter F. George Law Office Museum, 106 Fourth St, Vienna, GA 31092 • T: +1 229 2683663 • F: +1 229 2683664 • www.historicvienna.org •
Exec. Dir.: *Janet Joiner* •
Historical Museum – 1991
U.S. Senator Walter F. George memorabilia, local hist ... 53744

Villanova PA

Villanova University Art Gallery, 800 Lancaster Av, Villanova, PA 19085 • T: +1 610 5194610 • F: +1 610 5196046 • maryanne.erwin@villanova.edu • www.artgallery.villanova.edu •
Dir.: *Richard G. Canulli* •
Public Gallery / University Museum ... 53745

Vinalhaven ME

The Vinalhaven Historical Society Museum, High St, Vinalhaven, ME 04863 • T: +1 207 8634410 • vhhissoc@midcoast.com • www.midcoast.com/~vhhissoc •
Pres.: *Wyman Philbrook* •
Historical Museum – 1963
Local history ... 53746

Vincennes IN

George Rogers Clark Park Museum, 401 S Second St, Vincennes, IN 47591 • T: +1 812 8821776 • F: +1 812 8827270 • www.nps.gov/gero •
Dir.: *Dale Phillips* •
Natural History Museum – 1967
American hist, American Revolution, military hist 18th c, frontier hist 18th c ... 53747

Indiana Territory Capitol, First and Harrison Sts, Vincennes, IN 47591 • T: +1 812 8827472 • F: +1 812 8820928 • www.state.in.us/sites/vincennes •
Cur.: *Bruce A. Beesley* •
Historical Museum – 1949 ... 53748

Michel Brouillet House and Museum, 509 N 1st St, Vincennes, IN 47591 • T: +1 812 8827422 •
Cur.: *Richard Day* •
Historical Museum – 1975
Furnishings, fur traders exhib, Indian weapons, domestic utensils ... 53749

Old State Bank, 114 N Second St, Vincennes, IN 47591 • T: +1 812 8827472 • F: +1 812 8820928 •
Cur.: *Bruce A. Beesley* •
Special Museum – 1838 ... 53750

William H. Harrison Museum/ Grouseland, 3 W Scott St, Vincennes, IN 47591 • T: +1 812 8822096 • F: +1 812 8827626 • grouseland@wvc.net •
Cur.: *Dr. E. Joseph Fabyan* •
Historical Museum / Historic Site – 1911
Local history, William Henry Harrison mansion, Grouseland ... 53751

Virginia Beach VA

Adam Thoroughgood House, 1636 Parish Rd, Virginia Beach, VA 23455 • T: +1 757 4314000 • F: +1 757 4313733 • mreed@vbgov.com •
Dir.: *Dr. William Hennessey* •
Decorative Arts Museum / Local Museum – 1961
Late 17th & early 18th c English artifacts incl furniture, ceramics & pewter ... 53752

Atlantic Wildfowl Heritage Museum, 1113 Atlantic Av, Virginia Beach, VA 23451 • T: +1 757 4378432 • F: +1 757 4379055 • atlanticwildfowl@rcn.com • www.awhm.org •
Dir.: *Thomas P. Beatty* •
Fine Arts Museum / Folklore Museum – 1995
Art and artifact concerning wildfowl ... 53753

Cape Henry Lighthouse, 583 Atlantic Av, Fort Story, Virginia Beach, VA 23459 • T: +1 757 4229421 • oldcapehenry@aol.com • www.apva.org • Exec. Dir.: *Elizabeth Kostelny* •
Historic Site – 1791 53754

Contemporary Art Center of Virginia, 2200 Parks Av, Virginia Beach, VA 23451 • T: +1 757 4250000 • F: +1 757 4258186 • KATE@CACV.ORG • www.cacv.org •
Dir.: *Cameron Kitchin* • Cur.: *Brenda LaBier* •
Fine Arts Museum – 1952
Rotating traveling exhib 53755

Francis Land House, 3131 Virginia Beach Blvd, Virginia Beach, VA 23452 • T: +1 757 4314000 • F: +1 757 4313733 • mreed@vbgov.com •
Local Museum – 1986
Period furnishing 53756

Lynnhaven House, Association for the Preservation of Virginia Antiquities, 4405 Wishart Rd, Virginia Beach, VA 23455 • T: +1 757 4560351, 4601688 • F: +1 757 4560997 • ahb@bueche.com • www.apva.org •
Exec. Dir.: *Elizabeth Kostelny* •
Historic Site – 1976
Household furnishings 53757

Virginia Beach Maritime Museum, 24th St and Atlantic Av, Virginia Beach, VA 23451 • T: +1 757 4221587 • F: +1 757 4918609 • FTylerVB2@aol.com • www.oldcoastguardstation.com •
Exec. Dir.: *Fielding L. Tyler* •
Science&Tech Museum – 1981
Maritime hist 53758

Virginia Marine Science Museum, 717 General Booth Blvd, Virginia Beach, VA 23451 • T: +1 757 4374949 • F: +1 757 4374976 • fish@vbgov.com • www.vmsm.com •
Dir.: *Lynn B. Clements* • Cur.: *Maylon White* • Ass. Cur.: *Mark Swingle* •
Natural History Museum – 1986
Marine science 53759

Virginia City MT

Virginia City Madison County Historical Museum, 221 Wallace St, Virginia City, MT 59755 • T: +1 406 8435500, 8435484 • F: +1 406 8435303 • madsnien@3rivers.net •
C.E.O.: *Daryl L. Tichenor* •
Local Museum – 1958
Mounted animals, historical western items, furniture 53760

Virginia City NV

Nevada State Fire Museum and Comstock Firemen's Museum, 117 S C St, Virginia City, NV 89440 • T: +1 775 8470701 •
Dir.: *Michael E. Nevin* •
Special Museum – 1979
Antique fire equipment 53761

Viroqua WI

Vernon County Historical Museum, 410 S Center St, Viroqua, WI 54665 • T: +1 608 6377396 • vcmuseum@frontiernet.net • www.frontiernet.net/~vcmuseum •
Dir.: *Marcia Andrew* • Cur.: *Judy K. Mathison* •
Local Museum – 1942
Local hist, hist bldgs, Indian artifacts, genealogy 53762

Visalia CA

Tulare County Museum, 27000 S Mooney Blvd, Visalia, CA 93277 • T: +1 559 7336616 • F: +1 559 7302653 • KMcGowan@co.tulare.ca.us • www.tularecountyhistoricalsociety.org/Museum.htm •
Cur.: *Kathy McCowan* •
Local Museum – 1948
Local hist, end of the trail, Indian relicts, agriculture 53763

Vista CA

Antique Gas and Steam Engine Museum, 2040 N Santa Fe Av, Vista, CA 92083 • T: +1 760 9411791 • F: +1 760 9410690 • rod_agsem@yahoo.com • www.agsem.com •
Dir.: *Rod Groenewold* •
Science&Tech Museum – 1976
Industrial hist, engines, horse drawn equipment 53764

Vista Historical Museum, 2317 Old Foothill Dr, Vista, CA 92084 • T: +1 760 6300444 • F: +1 760 2959993 • www.geocities.com/vistahistoricalmuseum •
Dir.: *Rosemary Conway* •
Local Museum – 1994
Artifacts from indigenous people of north County 53765

Volcano HI

Volcano Art Center, POB 129, Volcano, HI 96785 • T: +1 808 9677565 • F: +1 808 9678512 • director@volcanoartcenter.org • www.volcanoartcenter.org •
Dir.: *Marilyn L. Nicholson* •
Fine Arts Museum
Hawaiin fine art and crafts, literary, folk arts 53766

Volga SD

Brookings County Museum, 215 Samara Av, Volga, SD 57071 • T: +1 605 6279149 •
C.E.O. & Pres.: *Lawrence Barnett* •
Local Museum – 1968
Local hist 53767

Volo IL

Volo Antique Auto Museum, 27582 Volo Village Rd, Volo, IL 60073 • T: +1 815 3853644 • F: +1 815 3850703 • www.volocars.com •
C.E.O.: *Greg Grams* •
Science&Tech Museum / Folklore Museum – 1970
Automobile, village life 53768

Vulcan MI

Iron Mountain Iron Mine Museum, Hwy US-2, Vulcan, MI 49892, mail addr: POB 177, Iron Mountain, MI 49801 • T: +1 906 5638077, 7747914 • ironmine@uplogon.com • www.ironmountainironmine.com •
Dir.: *Albert H. Carollo* • *Eugene R. Carollo* •
Science&Tech Museum – 1956
Mining machinery, Menominee iron range, various drills, locs 53769

Wabasso MN

Wabasso Historical Society Museum, 564 South St, Wabasso, MN 56293-1600 • T: +1 507 3425367 •
Dir.: *Herman Melman* •
Local Museum – 1973
Farming, household items, guns, pictures, rocks 53770

Waco TX

Armstrong Browning Library, c/o Baylor University, 710 Speight St, Waco, TX 76798 • T: +1 254 7103566 • F: +1 254 7103552 • Cyndie_Burgess@baylor.edu • www.browninglibrary.org •
Dir.: *Mairi C. Rennie* • Cur.: *Rita Patteson* (Manuscripts) • *Cynthia Burgess* (Books and Printed Materials) •
University Museum / Library with Exhibitions – 1918
Library of Browningiana 53771

Art Center of Waco, 1300 College Dr, Waco, TX 76708 • T: +1 254 7524371 • F: +1 254 7523506 • info@artcenterwaco.org • www.artcenterwaco.org •
Dir.: *Jill Hoffman* •
Fine Arts Museum – 1972
Art works by Robert Wilson, Jack Boynton, Dick Wray, David Diao, Bob "Daddy-O" Wade, Arthur Tuner 53772

Dr. Pepper Museum and Free Enterprise Institute, 300 S Fifth St, Waco, TX 76701 • T: +1 254 7571024 • F: +1 254 7572221 • jackmck@drpeppermuseum.com • www.drpeppermuseum.com •
Dir.: *Jack N. McKinney* •
Science&Tech Museum – 1989
Artifacts dealing with the soft drink industry, machinery, emphasis on Dr. Pepper objects 53773

The Earle-Harrison House, 1901 N Fifth St, Waco, TX 76708 • T: +1 254 7532032 • F: +1 254 2353555 • info@earleharrison.com • www.earleharrison.com •
C.E.O.: *Stanley A. Latham* •
Local Museum – 1956 53774

Historic Waco Foundation, 810 S Fourth St, Waco, TX 76706 • T: +1 254 7535166 • F: +1 254 7141242 • hwf@hot.rr.com • www.historicwaco.org •
Exec. Dir.: *Pamela B. Crow* • Cur.: *Bruce Lipscombe* •
Association with Coll – 1967
Period furnishings & clothing 53775

Martin Museum of Art, c/o Baylor University, Waco, TX 76789-7263 • T: +1 254 7101867 • F: +1 254 7101566 • Art_Support@baylor.edu • www.baylor.edu/art •
Dir.: *Dr. Heidi Hornik* •
Fine Arts Museum / University Museum – 1967
Contemporary painting and sculpture, graphics, prints 53776

Masonic Grand Lodge Library and Museum, 715 Columbus Av, Waco, TX 76701, mail addr: POB 446, Waco, TX 76703 • T: +1 254 7537395 • F: +1 254 7532944 • gs@grandlodgeoftexas.org • www.grandlodgeoftexas.org •
Cur.: *Barbara Mechell* •
Historical Museum – 1936
Hist & biography of Masons & Texas 53777

Mayborn Museum Complex, 1300 S University Parks Dr, Waco, TX 76706, mail addr: c/o Baylor University, One Bear Place 97154, Waco, TX 76798 • T: +1 254 7101110 • F: +1 254 7101173 • Ellie_Caston@baylor.edu • www.baylor.edu •
Dir.: *Dr. Ellie Caston* •
Natural History Museum – 1893
Natural and cultural hist, science, zoology, herpetology, archaeology, antropology, botany, paleontology, botany, geology, mineralogy, numismatic 53778

Strecker Museum's Collection, Mayborn Museum Complex, 1300 S University Parks Dr, Waco, TX 76706, mail addr: c/o Baylor University, One Bear Place 97154, Waco, TX 76798 • T: +1 254 7101110 • F: +1 254 7101173 • Ellie_Caston@baylor.edu • www.baylor.edu •
Dir.: *Calvin B. Smith* • Cur.: *Melinda Herzog* •
Natural History Museum – 1893 53779

Texas Ranger Hall of Fame and Museum, 100 Texas Ranger Trail, Waco, TX 76706 • T: +1 254 7508631 • F: +1 254 7508629 • trhf@eramp.net • www.texasranger.org •
Dir.: *Byron A. Johnson* •
Historical Museum – 1968
Texas Ranger hist 53780

Texas Sports Hall of Fame, 1108 S University Parks Dr, Waco, TX 76706 • T: +1 254 7561633 • F: +1 254 7562384 • info@hallofame.org • www.tshof.org •
Dir.: *Steve Fallon* • Cur.: *Jay Black* •
Special Museum – 1989
Sports 53781

Waconia MN

Carver County Historical Society Museum, 555 W First St, Waconia, MN 55387-1203 • T: +1 952 4424234 • F: +1 952 4423025 • historical@co.carver.mn.us • www.carvercountyhistoricalsociety.org •
Exec. Dir.: *Leanne Brown* •
Local Museum – 1940
Pioneer agricultural tools, county business & hist 53782

Wadesboro NC

Anson County Historical Society Museum, 206 E Wade St, Wadesboro, NC 28170 • T: +1 704 6946694 • F: +1 704 6943763 • achs@vnet.net •
Dir.: *Paul W. Ricketts* • *Glenda A. Livingston* •
Local Museum – 1962 53783

Wahoo NE

Saunders County Historical Museum, 240 N Walnut, Wahoo, NE 68066 • T: +1 402 4433090 • www.visitsaunderscounty.org •
Pres.: *Kenneth Schoen* • Dir./Cur.: *Erin Hauser* •
Historical Museum – 1963
Agriculture, furniture, business machines, docorative arts, county hist 53784

Wahpeton ND

Richland County Historical Museum, Second St and Seventh Av N, Wahpeton, ND 58075 • T: +1 701 6423075 •
Pres.: *Reuben Brownlee* •
Local Museum – 1946
Indian artifacts, local hist 53785

Wailuku HI

Bailey House Museum, Mauri Historical Society, 2375a Main St, Wailuku, HI 96793 • T: +1 808 2443326 • F: +1 808 2423920 • info@mauimuseum.org • www.mauimuseum.org •
Historical Museum – 1951
Local hist, archaeology 53786

Waipahu HI

Hawaii's Plantation Village, 94-695 Waipahu St, Waipahu, HI 96797 • T: +1 808 6770110 • F: +1 808 6766727 • hpv.waipahu@verizon.net • www.hawaiiplantationvillage.org •
Ethnology Museum – 1973
Ethnic and cultural materials 53787

Waitsfield VT

General Wait House, Rte 100, Waitsfield, VT 05673 • T: +1 802 4963733 • F: +1 802 4963733 •
Historic Site – 1970 53788

Wake Forest NC

Wake Forest College Bithplace Society Museum, 414 N Main St, Wake Forest, NC 27587 • T: +1 919 5562911 • cappsgt@wfu.edu •
Exec.Dir.: *Gene Capps* •
Historical Museum – 1956
Memorabilia of Wake Forest college and town, period furniture 53789

Wakefield KS

Wakefield Museum Association, 604 6th St, Wakefield, KS 67487-0101 • T: +1 785 4615516 •
Cur.: *Lorraine Cowell* •
Local Museum / Association with Coll – 1973
Hist of Wakefield, music, textiles, hats, toys, furniture, governor's desk, books, pictures, newspapers 53790

Wakefield MA

Wakefield Historical Museum, Americal Civic Center, 467 Main St, Wakefield, MA 01880 • T: +1 781 2450549 • nbertran@concentric.net • www.wakefieldma.org/society.html •
Cur.: *Thomas Mahlstedt* •
Local Museum – 1890
Local history 53791

Wakefield RI

Hera Gallery, Hera Educational Foundation, 327 Main St, Wakefield, RI 02880-0336 • T: +1 401 7891488 •
Dir.: *Cynthia Farnell* •
Public Gallery – 1974 53792

Walden NY

Jacob Walden House, Historical Society of Walden and Wallkill Valley, 34 N Montgomery St, Walden, NY 12586 • T: +1 845 7781173 •
Pres.: *Patricia Eisley* •
Local Museum – 1958
Furnishings, local hist 53793

Walhalla ND

Gingras Trading Post, 10534 129 Av NE, Walhalla, ND 58282 • T: +1 701 3281476 • F: +1 701 3283710 • rcollin@state.nd.us • www.state.nd.us/HIST/gingras •
Special Museum – 1956
Fur trade, Antoine Gingras 53794

Walker MN

Cass County Museum, 201 Minnesota Av W, Walker, MN 56484 • T: +1 218 5477251 •
Pres.: *Phyllis Krueger* •
Natural History Museum / Ethnology Museum – 1937
Animals, Indian handicraft, mainly Ojibway/Chibbewa, butterflies, geology, hist, mineralogy 53795

Walla Walla WA

Carnegie Art Center, 109 S Palouse, Walla Walla, WA 99362 • T: +1 509 5254270 • cac@hscis.net •
Dir.: *Heidi Thomas* •
Public Gallery – 1970 53796

Fort Walla Walla Museum, 755 Myra Rd, Walla Walla, WA 99362 • T: +1 509 5257703 • F: +1 509 5257798 • info@fortwallawallamuseum.org • www.fortwallawallamuseum.org •
Exec. Dir.: *James Payne* •
Historical Museum – 1968
Pioneer village, agriculture, local hist 53797

Sheehan Gallery at Whitman College, Olin Hall, 814 Isaacs St, Walla Walla, WA 99362 • T: +1 509 5275249 • F: +1 509 5275039 • forbesdm@whitman.edu • www.whitman.edu/sheehan •
Dir.: *Ian Bayden* • Sc. Staff: *Thomas E. Cronin* (Collections) •
Fine Arts Museum / University Museum – 1972
Asian art coll, ceramics coll 53798

Wallace ID

Wallace District Mining Museum, 509 Bank St, Wallace, ID 83873 • T: +1 208 5561592 •
C.E.O.: *John Amonson* •
Science&Tech Museum – 1956
Mining exhib 53799

Wallingford CT

Wallingford Historical Society Museum, 180 S Main St, Wallingford, CT 06492 • T: +1 203 2941996 •
Pres.: *Robert N. Beaumont* •
Local Museum / Historical Museum – 1916
Local hist 53800

Walloomsac NY

Bennington Battlefield State Historic Site, Rte 67, Walloomsac, NY 12133, mail addr: POB 163, Grafton, NY 12082 • T: +1 518 2791155 • F: +1 518 2791902 • laura.conner@oprhp.state.ny.us • www.nysparks.state.ny.us •
Historical Museum – 1927
1777 Revolutionary War battle site, prints, sculptures 53801

Walnut Creek CA

Bedford Gallery, Dean Lesher Regional Center for the Arts, 1601 Civic Dr, Walnut Creek, CA 94596 • T: +1 925 2951417 • F: +1 925 2951486 • www.lesherartscenter.org •
Cur.: *Carrie Lederer* •
Fine Arts Museum / Public Gallery – 1963
library, theater 53802

Lindsay Wildlife Museum, 1931 First Av, Walnut Creek, CA 94597 • T: +1 925 9351978 • F: +1 925 9358015 • www.wildlife-museum.org •
Exec. Dir.: *Eunice E. Valentine* • Cur.: *Michele Setter* •
Natural History Museum – 1955
Natural hist, live native Califonia wildlife, Indian artifacts 53803

Walnut Grove MN

Laura Ingalls Wilder Museum, 330 Eighth St, Walnut Grove, MN 56180 • T: +1 507 8592358 • F: +1 507 8592933 • lauramuseum@walnutgrove.org • www.walnutgrove.org •
Exec. Dir.: *Stanley Gordon* •
Local Museum – 1975
Local hist, L. Ingalls Wilder memorabilia, author of the "Little House" books, toys 53804

Walsenburg CO

The Walsenburg Mining Museum and Fort Francisco Museum of La Veta, 112 W Fifth St, Walsenburg, CO 81089 • T: +1 719 7381992 • F: +1 719 7386218 •
Pres.: *Rudy Cordova* •
Science&Tech Museum / Historical Museum – 1987
Mining, history 53805

Walterboro SC

Colleton Museum, 239 N Jeffries Blvd, Walterboro, SC 29488 • T: +1 803 5492303 • F: +1 803 5497215 • museum@lowcountry.com • www.walterboro.org •
Dir.: *Martha Creighton* •
Local Museum – 1985
Local hist ... 53806

South Carolina Artisans Center, 334 Wichman St, Walterboro, SC 29488 • T: +1 843 5490011 • F: +1 843 5497433 • artisan@lowcountry.com • www.southcarolinaartisanscenter.org •
C.E.O: *Carol B. Privette* •
Decorative Arts Museum / Folklore Museum – 1994
Arts and crafts, drawings, ceramics, decorative and folk arts ... 53807

Waltham MA

Charles River Museum of Industry, 154 Moody St, Waltham, MA 02153 • T: +1 781 8935410 • F: +1 781 8914536 • charles_river@msn.com • www.crmi.org •
Dir.: *Dan Yaeger* •
Science&Tech Museum – 1980
Textiles, industry, innovation, watch coll, automobile and steam engine, tools – library ... 53808

Gore Place, 52 Gore St, Waltham, MA 02453-6866 • T: +1 781 8942798 • F: +1 781 8945745 • info@goreplace.org • www.goreplace.org •
Dir.: *Susan Robertson* •
Historical Museum – 1935
Family history, period tea room, farm house – library ... 53809

Rose Art Museum, Brandeis University, 415 South St, Waltham, MA 02254-9110 • T: +1 781 7363434 • F: +1 781 7363439 • jketner@brandeis.edu • www.brandeis.edu/rose •
Dir.: *Joseph D. Ketner* • Cur.: *Raphaela Platow* •
Fine Arts Museum / University Museum – 1961
International modern and contemporary art ... 53810

Waltham Historical Society Museum, 190 Moody St, Waltham, MA 02453-5300 • T: +1 781 8915815 • waynemccarthy@rcn.com • www.walthamhistoricalsociety.org •
Pres.: *Wayne McCarthy* • Dir.: *Sheila E. Fitzpatrick* •
Cur.: *Joan Sheridan* •
Local Museum – 1913
Local history, Francis Cabot Lowell Mill – small library ... 53811

The Waltham Museum, 196 Charles St, Waltham, MA 02543 • T: +1 781 8938017 • nancynatick@aol.com • www.walthammuseum.com •
Dir.: *Albert A. Arena* •
Local Museum – 1971
Local history, watches, cars, bicycles, steam engines, radios, tools, laterns – library ... 53812

Wamego KS

The Columbian Museum and Art Center, 521 Lincoln Av, Wamego, KS 66547 • T: +1 785 4562029 • F: +1 785 4569498 • ctheatre@kansas.net •
Fine Arts Museum / Decorative Arts Museum – 1990
Decorative arts ... 53813

Wantagh NY

Wantagh Historical & Preservation Society, 1700 Wantaugh Av, Wantagh, NY 11793 • T: +1 516 8268767 •
Local Museum – 1965
Hist bldgs ... 53814

Wapakoneta OH

Neil Armstrong Air and Space Museum, I-75 and Bellefontaine Rd, Wapakoneta, OH 45895 • T: +1 419 7388811 • F: +1 419 7383361 • namu@ohiohistory.org • ohiohistory.org •
Dir.: *Rebecca Moor* •
Science&Tech Museum – 1972
Aeronautical items, space flight suits, simulators, Neil Armstrong belongings ... 53815

Warm Springs GA

Little White House, 401 Little White House Rd, Warm Springs, GA 31830 • T: +1 706 6555870 • F: +1 706 6555872 • lwhs@peachnet.campus.mci.net • www.fdr-littlewhitehouse.org •
Head: *Frankie Mewborn* •
Historical Museum – 1946
Georgia home of Pres. Roosevelt, where he died April 12, 1945 ... 53816

Warm Springs OR

The Museum at Warm Springs, 2189 Hwy 26, Warm Springs, OR 97761 • T: +1 514 5533331 • F: +1 514 5533338 • info@warmsprings.com • www.warmsprings.com •
Exec. Dir.: *Carol Leone* • Cur.: *Natalie Kirk* • Archivist: *Alberta Comedown* •
Ethnology Museum – 1991
Tribal culture, ethnology, archaeology, history, folklore ... 53817

Warner OK

Wallis Museum at Connors State College, Rte 1, Box 1000, Warner, OK 74469 • T: +1 918 4632931 • F: +1 918 4636205 • jdigran@connors.cc.ok.us • www.connorsstate.edu •
C.E.O.: *Dr. Donnie Nero* • Cur.: *Dr. Forest Redding Jr.* •
University Museum / Local Museum – 1963
Indian artifacts, minerals, hist of the area ... 53818

Warner Robins GA

Museum of Aviation at Robins Air Force Base, Hwy 247 and Russell Pkwy, Warner Robins, GA 31099 • T: +1 478 9236600 • F: +1 478 9238807 • paul.hibbitts@robins.af.mil • www.museumofaviation.org •
Dir.: *Paul E. Hibbits* • Cur.: *Sarah Wolfe* •
Science&Tech Museum / Military Museum – 1984
Aviation s. WW II ... 53819

Warren ME

Warren Historical Museum, 225 Main St, Warren, ME 04864 • T: +1 207 2732726 •
Pres.: *Dick Ferren* • Cur.: *Barbara Larson* •
Local Museum – 1964
Regional history ... 53820

Warren MI

Ukrainian-American Museum, 26601 Ryan Rd, Warren, MI 48091 • T: +1 810 7578130, 9789239 • F: +1 810 7578684 •
Cur.: *Irene Zacharkiw* •
Military Museum – 1958
Ukrainian military in WW I, international military, art – archives, library ... 53821

Warren MN

Marshall County Historical Society Museum, POB 103, Warren, MN 56762 • T: +1 218 7454803 •
Pres.: *Delvin Potucek* • Dir./Cur.: *Ethel Thorlacius* •
Historical Museum – 1920
Antique farm machinery, business, household ... 53822

Marshall County Historical Society Museum, 808 E Johnson Av, Warren, MN 56762 • T: +1 218 4803 •
Dir./Cur.: *Ethel Thorlacius* •
Local Museum / Agriculture Museum – 1933
Local hist, agriculture, farm machinery, business and household items ... 53823

Warren OH

John Stark Edwards House, 303 Monroe St NW, Warren, OH 44483 • T: +1 330 3944653 •
Pres.: *David Ambrose* • Cur.: *Warren Hickman* •
Local Museum – 1938
1800s quilts and clothing, tools, musical instr ... 53824

National Packard Museum, 1899 Mahoning Av NW, Warren, OH 44483 • T: +1 330 3941899 • F: +1 330 3947796 • national@packardmuseum.org • www.packardmuseum.org •
C.E.O. & Cur.: *Mary Ann Porinchak* •
Science&Tech Museum – 1989
Packard motor car, Packard Electric, Ohio Lamp hist, Packard family hist ... 53825

Warren PA

Crary Art Gallery, 511 Market St, Warren, PA 16365 • T: +1 814 7234523 •
Pres./Cur.: *Ann Lesser* •
Fine Arts Museum – 1977
Oils and pastels drawings, photography, sculptures – Library ... 53826

Warren County Museum, 210 Fourth Av, Warren, PA 16365 • T: +1 814 7231795 • warrenhistory@kinzua.net • www.kinzua.net/warrenhistory/ •
Exec. Dir.: *Rhonda J. Hoover* •
Local Museum – 1900
Local hist ... 53827

Warrensburg MO

Art Center Gallery, Central Missouri State University, 217 Clark St, Warrensburg, MO 64093-5246 • T: +1 660 5434498 • F: +1 660 5438006 • www.ucmo.edu •
Dir.: *Morgan Dean Gallatin* •
Fine Arts Museum – 1984 ... 53828

Central Missouri State University Archives and Museum, James C. Kirkpatrick Library 1470, Warrensburg, MO 64093-5040 • T: +1 660 5434649 • admit@ucmo.edu • www.ucmo.edu •
Dir.: *John W. Sheets* • Ass. Dir.: *Vivian Richardson* •
University Museum / Local Museum / Natural History Museum – 1968
Archaeology, ethnology, hist, biology, university hist ... 53829

Johnson County Historical Society Museum, 302 N Main St, Warrensburg, MO 64093 • T: +1 660 7476480 •
Cur.: *Lisa Irle* •
Local Museum – 1920
Archives, folkore, general hist & cultural items ... 53830

Warrenton VA

The Old Jail Museum, Courthouse Sq, Warrenton, VA 20186 • T: +1 540 3475525 • oldjailmuseum@erols.com •
Dir.: *Jackie Lee* •
Local Museum – 1964
Civil War, WW I & WW II items, kitchen items ... 53831

Warsaw NY

Warsaw Historical Museum, 15 Perry Av, Warsaw, NY 14569 • T: +1 716 7865240 • mhconable@wycol.com •
Pres.: *Mary Conable* •
Local Museum – 1938
Period furniture, local hist, clothing, farm and carpentry tools ... 53832

Warwick MD

Old Bohemia Historical Museum, Bohemia Church Rd, Warwick, MD 21912 • T: +1 302 3785800 • F: +1 302 3785808 • office@stjosephmiddletown.com • www.stjosephmiddletown.com •
Pres.: *Margaret Matyniak* •
Local Museum / Religious Arts Museum – 1953
Religious artifacts (liturgical vessels, vestments, prayer books, devotional articles), farm conveyances and tools, historic cemetery ... 53833

Warwick NY

Pacem in Terris, 96 Covered Bridge Rd, Warwick, NY 10990 • T: +1 914 9864329 • F: +1 914 9864329 •
Dir.: *Dr. Frederick Franck* • Dep. Dir.: *Claske Berndes* •
Historic Site / Public Gallery – 1972
Temporary exhib, paintings, drawings, prints ... 53834

Town of Warwick Museum, The Historical Society of the Town of Warwick, Forester Av, Warwick, NY 10990, mail addr: POB 353, Warwick, NY 10990 • T: +1 845 9863236 • info@warwickhistoricalsociety.org • www.warwickhistoricalsociety.org •
Pres.: *Henry L. Neilsen* •
Historical Museum – 1906
Furniture from Queen to Duncan Phyfe periods, local cabinet works pf 1810-1830, hunting and trapping equip of the author Frank Forester (i.e. William Henry Herbert), old cariages, old farm tools, ice-cuttung equip ... 53835

Warwick RI

Art Department Gallery, Community College of Rhode Island, 400 East Av, Warwick, RI 02886 • T: +1 401 8252220 • F: +1 401 8251148 •
Chm.: *Nichola Sevigney* •
Public Gallery – 1972 ... 53836

Warwick Museum of Art, Kentish Artillery Armory, 3259 Post Rd, Warwick, RI 02886 • T: +1 401 7370010 • F: +1 401 7371796 • info@warwickmuseum.org • www.warwickmuseum.org •
Pres.: *Harvey Zimmermann* • C.E.O: *Damon A. Campagna* •
Fine Arts Museum – 1973
Art ... 53837

Waseca MN

Farmamerica, The Minnesota Agricultural Interpretive Center, County Rds 2 and 17, Waseca, MN 56093, mail addr: 7367 360th Av, Waseca, MN 56093 • T: +1 507 8352052 • F: +1 507 8352053 • farmamer@hickorytech.net • www.farmamerica.org •
Exec. Dir.: *Kathleen L. Backer* •
Agriculture Museum – 1978
Farm machines, tools, crops, early settlement, grain mill, dairy, decorative arts ... 53838

Waseca County Museum, 315 Second Av NE, Waseca, MN 56093 • T: +1 507 8357700 • director@historical.waseca.mn.us • www.historical.waseca.mn.us •
Exec. Dir.: *Margaret Sinn* •
Local Museum – 1938
Agriculture, decorative arts, costumes – archives 53839

Washburn ND

McLean County Historical Society Museum, 605 Main St, Washburn, ND 58577 • T: +1 701 4623744 • F: +1 701 4623744 • vmerkel@westriv.com •
Pres.: *Dan Wicklander* •
Local Museum – 1967
Pioneer and Indian artifacts, office and treching machines ... 53840

Washburn WI

Washburn Historical Museum, 1 E Bayfield St, Washburn, WI 54891 • T: +1 715 3735591 •
Local Museum – 1991
Area hist, industries and people, paintings ... 53841

Washington CT

Gunn Memorial Library and Museum, 5 Wykeham Rd, Washington, CT 06793 • T: +1 860 8687756 • F: +1 860 8687247 • gunnmus@biblio.org • www.biblio.org/gunn •
Dir.: *Jean Chapin* • Cur.: *Bruce Reinholdt* • Ass. Cur.: *Suzanne Fateh* •
Local Museum – 1899
Local hist ... 53842

Washington DC

AAF Museum, 1735 New York Av NW, Washington, DC 20006-5292 • T: +1 202 6267500 • F: +1 202 8797764 •
Archaeology Museum / Decorative Arts Museum
Archaeology, decorative arts ... 53843

American Red Cross Museum, 1730 E St NW, Washington, DC 20006 • T: +1 202 6393300 • F: +1 202 6281362 • CauseMarketing@usa.redcross.org • www.redcross.org •
Exec. Dir.: *Steven E. Shulman* •
Historical Museum – 1919
Paintings, decorative arts, uniforms, manuscripts, memorabilia – archives ... 53844

Anacostia Museum, 1901 Fort Pl SE, Washington, DC 20560-0004 • T: +1 202 2873306 • F: +1 202 2873183 • www.si.edu •
Dir.: *Steven C. Newsome* • Dep. Dir.: *Sharon Reinckens* •
Fine Arts Museum / Historical Museum – 1967
African American hist and culture ... 53845

Archives of American Art, Smithsonian Institution, 750 Ninth St NW, Ste 2200, Washington, DC 20560-0937 • T: +1 202 2751950 • F: +1 202 2751955 • aaaemref@aaa.si.edu • www.archivesofamericanart.si.edu •
Dir.: *Edwin Rifkin* • Ass. Dir.: *Jody Pettibone* •
Library with Exhibitions – 1954
Letters and diaries of artists and craft persons, critics, works on paper, photographs, recorded interviews ... 53846

Art in Embassies Program, U.S. Department of State, U.S. Depot of State, Washington, DC 20521-0611 • T: +1 703 8754202 • F: +1 703 8754182 • aip.state.gov •
Dir.: *Anne Johnson* • Cur.: *Virginia Shore* •
Fine Arts Museum – Paintings, sculpture, textiles 750730677 ... 53847

Art Museum of the Americas, 201 18th St NW, Washington, DC 20006 • T: +1 202 4586016/19 • F: +1 202 4586021 • artmus@oas.org • www.museum.oas.org •
Dir.: *Lydia Bendersky* • Cur.: *Maria Leyva* (Permanent Collection) • Sc. Staff: *Fabian Goncalves* (Temporary Exhibits) •
Fine Arts Museum – 1976
Latin American and Caribbean contemporary art 53848

Arthur M. Sackler Gallery, 1050 Independence Av SW, Washington, DC 20560 • T: +1 202 3574880 • F: +1 202 3574911 • www.si.edu/asia •
Dir.: *Dr. Julian Raby* • Cur.: *Louise Cort* (Ceramics) •
Fine Arts Museum – 1982
3500 objects of Asian art ... 53849

Articulate Gallery, WVSA Arts Connection, 1100 16th St NW, Washington, DC 20036 • T: +1 202 2969100 • www.wvsarts.org •
Public Gallery ... 53850

Arts Club of Washington, 2017 I St NW, Washington, DC 20006 • T: +1 202 3317282 • F: +1 202 8573678 • artsclub.membership@verizon.net • www.artsclubofwashington.org •
Dir.: *Charles Fritzell* •
Fine Arts Museum / Historical Museum – 1916
Washington art, local hist ... 53851

Arts in the Academy, c/o National Academy of Sciences, 2101 Constitution Av NW, Washington, DC 20418 • T: +1 202 3342436 • F: +1 202 3341210 • cpnas@nas.edu • www.nasonline.org •
Dir.: *Dr. Janis A. Tomlinson* •
Fine Arts Museum
Science related art ... 53852

B'nai B'rith Klutznick National Jewish Museum, 2020 K St NW, 7th Fl, Washington, DC 20006 • T: +1 202 8576647 • F: +1 202 8576601 • museum@bnaibrith.org • www.bnaibrith.org •
Cur.: *Cheryl A. Kempler* •
Religious Arts Museum – 1957
Judaica, fine arts – archives of B'nai B'rith ... 53853

Canadian Embassy Art Gallery, 501 Pennsylvania Av NW, Washington, DC 20001 • T: +1 202 6821740 • F: +1 202 6827791 • www.canadianembassy.org •
Public Gallery ... 53854

Capital Children's Museum → National Children's Museum

Communications and History Museum of Sutton, 17th St and New York Av, Washington, DC 20006 • T: +1 202 6391700 • F: +1 202 6391768 •
Pres./ Dir.: *David C. Levy* • Dep. Dir.: *Jack Cowart* • Cur.: *Terrie Sultan* (Contemporary Art) • *Philip Brookman* (Photo and Media Arts) • *Sarah Cash* (American Art) • *Susan Badder* (Education) •
Fine Arts Museum / Decorative Arts Museum – 1869 ... 53855

Corcoran Gallery of Art, 500 17th St NW, Washington, DC 20006 • T: +1 202 6391701/1770 • F: +1 202 6391768 • jcowan@corcoran.org • www.corcoran.org •
Dir.: *David C. Levy* • Chief Cur.: *Jacquelyn D. Serwer* •
Fine Arts Museum – 1869
American paintings and sculptures from 18th-20th c, European paintings, sculptures, drawings, prints, photos and dec arts ... 53856

Dadian Gallery, Wesley Theological Seminary Center for the Arts and Religion, 4500 Massachusetts Av NW, Washington, DC 20016 • T: +1 202 8858674 • F: +1 202 8858683 • www.wesleyseminary.edu •
Dir.: *Catherine Kapinan* • Cur.: *Deborah Soklove* •
Fine Arts Museum / Religious Arts Museum – 1989 ... 53857

Daughters of the American Revolution Museum, 1776 D St NW, Washington, DC 20006 • T: +1 202 8793241 • F: +1 202 6280820 • museum@dar.org • www.dar.org •
Dir.: *Diane L. Dunkley* • Cur.: *Olive Graffam* (Collections) •
Decorative Arts Museum / Historical Museum – 1890
Decorative arts, history ... 53858

Decatur House Museum, 748 Jackson Pl NW, Washington, DC 20006 • T: +1 202 8420920 • F: +1 202 8420030 • decatur_house@nthp.org • www.decaturhouse.org •
Cur.: *Richard Cote* •
Decorative Arts Museum – 1956
Decorative art, porcelain ... 53859

Department of the Treasury Museum, 15th and Pennsylvania NW, Washington, DC 20220 • T: +1 202 6221250 • F: +1 202 6222294 • richard.cote@do.treas.gov • www.ustreas.gov/curator •
Cur.: *Richard Cote* •
Fine Arts Museum
19th - 20th c furniture, artwork, portraits ... 53860

Dimock Gallery, c/o George Washington University, 730 21st St NW, Washington, DC 20052 • T: +1 202 9941525, 9947091 • F: +1 202 9941632 • ldmiller@gwu.edu • www.gwu.edu/~dimock •
Dir./Cur.: *Lenore D. Miller* •
Fine Arts Museum / University Museum – 1966
Art ... 53861

Discovery Creek Children's Museum of Washington DC, 2233 Wisconsins Av, Washington, DC 20007 • T: +1 202 3643111 • F: +1 202 3643114 • mail@discoverycreek.org • www.discoverycreek.org •
C.E.O.: *Susan M. Seligmann* •
Special Museum – 1991
100 yr old trees, windin creek, hidden rock, waterfalls, plant and animal life ... 53862

District of Columbia Arts Center, 2438 18th St NW, Washington, DC 20009 • T: +1 202 4627833 • F: +1 202 3287099 • info@dcartscenter.org • www.dcartscenter.org •
Dir.: *B. Stanley* •
Public Gallery – 1989 ... 53863

Dumbarton House, 2715 Que St NW, Washington, DC 20007-3071 • T: +1 202 3372288 • F: +1 202 3370348 • pr@dumbartonhouse.org • www.dumbartonhouse.org •
Dir.: *William S. Birdseye* •
Historical Museum – 1891
Furnishing and decorative arts ... 53864

Dumbarton Oaks Collections, Trustees for Harvard University, 1703 32nd St NW, Washington, DC 20007 • T: +1 202 3396400 • F: +1 202 3396419 • DumbartonOaks2009@doaks.org • www.doaks.org •
Dir.: *Jan Ziolkowski* • Museum Director: *Gudrun Bühl* •
Fine Arts Museum – 1940
Byzantine, Pre-Columbian, European and American paintings, sculpture, and decorative arts, books and manuscripts of landscape architecture ... 53865

Evans-Tibbs Collection, 1910 Vermont Av NW, Washington, DC 20001 • T: +1 202 2348164 •
Dir.: *E. Tibbs Thurlow jr.* •
Fine Arts Museum
19th-20th c African-American painting and sculpture ... 53866

Explorers Hall, National Geographic Society, 1145 17th St NW, Washington, DC 20036 • T: +1 202 8577588, 8577456 • F: +1 202 8575864 • www.nationalgeographic.com •
Pres.: *John M. Fahey* • Dir.: *Susan E.S. Norton* •
Natural History Museum / Science&Tech Museum – 1964
Science, geography, expedition equip ... 53867

Federal Reserve Board Art Gallery, 20 and C Sts NW, Washington, DC 20551 • T: +1 202 4523000 • F: +1 202 4525680 • www.federalreserve.gov •
Dir.: *Mary Anne Goley* •
Fine Arts Museum – 1975
19th and 20th c American and European pintings, prints and works on paper ... 53868

Folger Shakespeare Library, 201 E Capitol St, Washington, DC 20003 • T: +1 202 5444600 • F: +1 202 5444623 • webmaster@folger.edu • www.folger.edu •
Dir.: *Gail Kern Paster* •
Special Museum – 1932
Rare books and manuscripts of 15th-18th c on continental and English Renaissance, Shakespeare, theater hist, memorabilia ... 53869

Fondo Del Sol, Visual Art and Media Center, 2112 R St NW, Washington, DC 20008 • T: +1 202 4832777 • F: +1 202 6581078 •
Dir.: *W. Marc Zuver* •
Public Gallery – 1973
Pre- Columbian and Hispanic Santero coll; Latino, Chicano and Puerto Rican art ... 53870

Ford's Theatre, Lincoln Museum, 511 10th St NW, Washington, DC 20004 • T: +1 202 4266924 • F: +1 202 4261845 • ford's_theatre@nps.gov • www.nps.gov/foth/ •
Dir.: *Terry Carlstrom* •
Performing Arts Museum – 1933
History ... 53871

Frederick Douglass National Historic Site, 1411 W St SE, Washington, DC 20020 • T: +1 202 4265961, 4261452 • F: +1 202 4260880 • NACE_Frederick_Douglass_NHS@nps.gov •
Dir.: *Terry Carlstrom* • Cur.: *Cathy Ingram* •
Historic Site – 1916
Films, furnishings, docs and personal artifacts of Frederick Douglass ... 53872

Freer Gallery of Art, Jefferson Dr at 12th St SW, Washington, DC 20560 • T: +1 202 3574880 • F: +1 202 3574911 • publicaffairsAsia@si.edu • www.si.edu/asia •
Dir.: *Dr. Julian Raby* • Cur.: *Dr. James Ulak* (Japanese Art) • *Jan Stuart* (Chinese Art) • *Louise Cort* (Ceramics) • *Dr. Massumeh Farhad* (Islamic, Near Eastern) • *Kenneth Myers* (American Art) •
Fine Arts Museum – 1906
Chinese, Japanese, Korean, Islamic, Near Eastern, South and Southeast Asian bronze, jade sculpture, painting, lacquer, pottery, porcelain, manuscripts and metall works ... 53873

Georgetown University Art Collection, 3700 O St NW, Washington, DC 20057-1006 • T: +1 202 6871469 • F: +1 202 6877501 • llw@georgetown.edu • www.library.georgetown.edu/dept/speccoll/guac •
Coord.: *Lulen Walker* (Art collection) •
Fine Arts Museum / University Museum – 1789
Historical objects, works by Van Dyck and Gilbert Stuart, paintings, sculpture, graphics, American portraits, religious objects, decorative art ... 53874

Heurich House Museum, 1307 New Hampshire Av NW, Washington, DC 20036 • T: +1 202 7852068 • F: +1 202 8875785 • info@heurichhouse.org • www.brewmasterscastle.com •
Exec. Dir.: *Barbara Franco* •
Local Museum – 1894
Victorian furnishings, photographs and maps of the local hist of Washington DC ... 53875

Hillwood Museum, Hillwood Museum Foundation, 4155 Linnean Av NW, Washington, DC 20008 • T: +1 202 6868500 • F: +1 202 9667846 • info@hillwoodmuseum.org • www.hillwoodmuseum.org •
Exec. Dir.: *Frederick J. Fisher* •
Decorative Arts Museum – 1976
Russian decorative arts, paintings, sculptures and icons – Library, archives ... 53876

Hirshhorn Museum and Sculpture Garden, Smithsonian Institution, Seventh St and Independence Av SW, Washington, DC 20560 • T: +1 202 3573091 • F: +1 202 7862682 • lawrence_s@hmsg.si.edu • www.hirshhorn.si.edu •
Dir.: *Beverly Lang Pierce* • Chief Cons.: *Susan Lake* •
Fine Arts Museum – 1966
International modern and contemporary art, 19th and 20th c sculpture ... 53877

House of the Temple, 1733 16th St NW, Washington, DC 20009 • T: +1 202 2323579 • F: +1 202 3871843 • jsansbury@srmason-sj.org • www.srmason-sj.org •
Cur.: *Joan Sansbury* •
Historical Museum – 1801
Genealogy, history of the FBI and J. Edgar Hoover ... 53878

Howard University Gallery of Art, 2455 6th St NW, Washington, DC 20059 • T: +1 202 8067070 • F: +1 202 8066503 •
Dir.: *Dr. Tritobia H. Benjamin* • Asst. Dir.: *Scott Baker* • Registrar: *Eileen Johnston* •
Fine Arts Museum / University Museum – 1928
African-American and American paintings, sculptures, graphic arts, European graphic art, coll of Italien paintings and sculptures ... 53879

Howard University Museum, c/o Moorland Spingarn Research Center, 500 Howard Pl NW, Washington, DC 20059 • T: +1 202 8067239 • F: +1 202 8066405 • www.founders.howard.edu/moorland-spingarn •
Dir.: *Dr. Thomas C. Battle* •
University Museum / Historical Museum – 1914
Black hist ... 53880

IMF Center Exhibit, 700 19th St NW, Washington, DC 20431 • T: +1 202 6237000 • F: +1 202 6236562 • media@imf.org • www.imf.org/center •
C.E.O.: *Linda A. Kamel* •
Special Museum – 2000
Exhibit on the international monetary system, numismatics ... 53881

Indian Arts and Crafts Board, 1849 C St NW, Washington, DC 20240 • T: +1 202 2083773 • F: +1 202 2085196 • iacb@ios.doi.gov • www.iacb.doi.gov •
Dir.: *Meridith Stanton* •
Fine Arts Museum – 1935
Contemporary American Indian and Alaska native arts of the U.S ... 53882

International Spy Museum, 800 F St NW, Washington, DC 20004 • T: +1 202 3937798 • F: +1 202 3937797 • aabrell@spymuseum.org • www.spymuseum.org •
Exec. Dir.: *E. Peter Earnest* •
Historical Museum
Artifacts, tradecraft, hist and contemp role of spionage ... 53883

Jackson Art Center, Jackson School, 3048 1/2 R St NW, Washington, DC 20007 • T: +1 202 3429778 •
Public Gallery ... 53884

The Kreeger Museum, 2401 Foxhall Rd NW, Washington, DC 20007 • T: +1 202 3373050 • F: +1 202 3373051 • publicrelations@kreegermuseum.org • www.kreegermuseum.org •
Dir.: *Judy A. Greenberg* •
Fine Arts Museum – 1994
19th and 20th c paintings, sculptures, traditional African arts ... 53885

Library of Congress, 101 Independence Av SE, Washington, DC 20540 • T: +1 202 7074604 • F: +1 202 7075844 • www.loc.gov •
Library with Exhibitions – 1800
Worldwide newspaper, manuscripts, history and civilization, maps and views ... 53886

Lillian and Albert Small Jewish Museum, 600 I St NW, Washington, DC 20001 • T: +1 202 7890900 • F: +1 202 7890485 • info@jhsgw.org • www.jhsgw.org •
Exec. Dir.: *Laura C. Apelbaum* • Pres.: *Frank Gilbert* •
Religious Arts Museum / Historical Museum – 1975
Letters, photos, docs of Jewish hist ... 53887

Luther W. Brady Art Gallery, 805 21st St NW, Washington, DC 20052 • T: +1 202 9941525 • F: +1 202 9941632 • ldmiller@gwu.edu • www.gwu.edu/~dimock •
Fine Arts Museum – 1966
Paintings, sculpture, graphic arts and photos from 18th-20th c American art, Wright W. Lloyd coll of Washingtoniana, coll of photos, docs ... 53888

Marian Koshland Museum, 6th and E Sts NW, Washington, DC 20001 • T: +1 202 3341201 • F: +1 202 3341548 • ksm@nas.edu • www.koshlandsciencemuseum.org •
Dir.: *Patrice Legro* •
Historical Museum – 2004
Science, engineering, medicine ... 53889

Marine Corps Art Collection, Bldg 58 Washington Navy Yard, Washington, DC 20374 • T: +1 202 4333840 • F: +1 202 4337265 •
Dir.: *John W. Ripley* •
Fine Arts Museum – 1970
Art work by Marines, military music, combat art, historical illustrations, prints cartoons, recruiting posters, sculptures ... 53890

Mary McLeod Bethune House, 1318 Vermont Av NW, Washington, DC 20005 • T: +1 202 6732402 • F: +1 202 6732414 • www.nps.gov/mamc •
Dir.: *Terry Carlstrom* •
Local Museum – 1979
Photographs, communication artifacts – Archiv ... 53891

Meridian International Center - Cafritz Galleries, 1624-30 Crescent Pl NW, Washington, DC 20009 • T: +1 202 9395568 • F: +1 202 3191306 • nmatthew@meridian.org • www.meridian.org •
Pres.: *Walter L. Cutler* • Dir.: *Nancy Matthews* •
Fine Arts Museum / Folklore Museum – 1960
International arts and culture ... 53892

Millennium Arts Center International, 65 I St SW, Washington, DC 20024 • T: +1 202 4792572 • F: +1 202 4790946 • artatmac@aol.com •
Dir.: *William Wooby* •
Fine Arts Museum
Fine and peforming arts ... 53893

Museum of Contemporary Art, 1054 31st St NW, Washington, DC 20007 • T: +1 202 3426230 •
Fine Arts Museum ... 53894

National Air and Space Museum, Smithsonian Institution, Sixth St and Independence Av SW, Washington, DC 20560 • T: +1 202 3571745 • F: +1 202 3572426 • nasm@nasm.si.edu • www.nasm.si.edu •
Dir.: *John R. Dailey* • Dep. Dir.: *Donald Lopez* •
Science&Tech Museum – 1946
Aeronautical and astronautical items ... 53895

National Building Museum, 401 F St NW, Washington, DC 20001 • T: +1 202 2722448 • F: +1 202 2722564 • jneubauer@nbm.org • www.nbm.org •
Pres.: *Chase Rynd* •
Fine Arts Museum / Science&Tech Museum – 1980
Drawings, photos, models, artifacts ... 53896

National Children's Museum (opening 2013), 955 L'Enfant Plaza North, SW , Suite 5100, Washington, DC 20024 • T: +1 202 6754120 • F: +1 202 6754140 • info@ncm.museum • www.ncm.museum •
Pres.: *Kathy Dwyer Southern* •
Special Museum – 1974
Cultural exhibits on Mexico and Japan, animation and chemistry ... 53897

National Gallery of Art, 6 Constitution Av at Fourth St NW, Washington, DC 20565 • T: +1 202 7374215 • F: +1 202 7894976 • www.nga.gov •
Pres.: *Victoria P. Sant* • Dir.: *Earl A. Powell* • Dep. Dir./Chief Cur.: *Alan Shestack* •
Fine Arts Museum – 1937
European and American paintings, sculptures, decorative arts and graphics from 12th - 20th c, European old master paintings, French, Spanish, Italien, American and British 18th and 19th c paintings, Renaissance bronzes, Chinese porcelain ... 53898

National Museum of African Art, Smithsonian Institution, 950 Independence Av SW, Washington, DC 20560-0708 • T: +1 202 3574600, 3571300 • F: +1 202 3574879 • nmafaweb@nmafa.si.edu • www.si.edu/nmafa •
Asst. Dir.: *Alan Knezevich* • Chief Cur.: *David Binkley* •
Fine Arts Museum / Folklore Museum – 1964
7000 objects of African art, woods, metal, ceramics – Photographics archiv ... 53899

National Museum of American Art, Eighth and G Sts NW, Washington, DC 20001 • T: +1 202 6338998 • www.nmaa.si.edu •
Fine Arts Museum ... 53900

National Museum of American History, Smithsonian Institution, 14th St and Constitution Av NW, Washington, DC 20560 • T: +1 202 3572700 • F: +1 202 3571853 • americanhistory.si.edu •
Dir.: *Brent D. Glass* •
Historical Museum – 1846
American hist ... 53901

National Museum of American Jewish Military History, 1811 R St NW, Washington, DC 20009 • T: +1 202 2656280 • F: +1 202 4623192 • nmajmh@nmajmh.org • www.nmajmh.org •
Pres.: *Edwin Goldwasser* • Exec. Dir.: *Herb Rosenbleeth* •
Historical Museum / Military Museum – 1958
American Jewish military hist ... 53902

National Museum of Health and Medicine, Armed Forces Institute of Pathology, 6900 Georgia Av and Elder St NW, Washington, DC 20307 • T: +1 202 7822200 • F: +1 202 7823573 • nmhminfo@afip.osd.mil • www.nmhm.washingtondc.museum •
Dir.: *Adrianne Noe* • Cur.: *James T. H. Connor* (Historical coll) • *Archie Fobbs* (Neuropathology) • *Alan Hawk* (History) • *Michael Rhode* (Archives) •
Special Museum – 1862
Medical hist, human anatomy, neuroanatomy, civil war surgery, microscopes, AIDS ... 53903

National Museum of Natural History, 10th St and Constitution Av NW, Washington, DC 20560 • T: +1 202 3571300 • F: +1 202 3574779 • www.nmnh.si.edu •
Dir.: *Cristian Samper* •
Natural History Museum – 1846
Anthropology, entomology, botany, fossils, minerals ... 53904

National Museum of the American Indian, 4th St and Independence Av SW, Washington, DC 20560 • T: +1 202 6331000 • www.nmai.si.edu •
Dir.: *W. Richard West* •
Historical Museum – 2004
Archaeology objects, ceramics from Cost Rica, Central Mexico and Peru, textiles and gold from Andean cultures ... 53905

National Museum of Women in the Arts, 1250 New York Av NW, Washington, DC 20005 • T: +1 202 7835000 • F: +1 202 3933235 • info@mnwa.org • www.nmwa.org •
Dep. Dir.: *Dr. Susan Fisher Sterling* (Arts, Programs) •
Fine Arts Museum / Folklore Museum – 1981
Art ... 53906

National Portrait Gallery, 750 Ninth St NW, Ste 8300, Washington, DC 20560-0973 • T: +1 202 2752738 • F: +1 202 2751887 • npgweb@npg.si.edu • www.npg.si.edu •
Dir.: *Marc Pachter* • Dep. Dir.: *Carolyn Carr* • Cur.: *Ellen Miles* (Painting and Sculpture) • *Wendy Wick Reaves* (Prints) • *Ann Schumard* (Photographs) • *Beverly Jones Cox* (Exhibitions) • Cons.: *Cindy Lou Molnar* •
Fine Arts Museum / Historical Museum – 1962
Portraits of men and women ... 53907

National Postal Museum, Smithsonian Institution, 2 Massachusetts Av NE, Washington, DC 20560-0570 • T: +1 202 6332700 • F: +1 202 6339393 • npm@npm.si.edu • www.si.edu/postal •
Dir.: *Allen Kane* •
Special Museum – 1993
Postal hist ... 53908

National Society of the Children of the American Revolution Museum, 1776 D St NW, Washington, DC 20006 • T: +1 202 6383153 • F: +1 202 7373162 • hq@nscar.org • www.nscar.org •
Head: *Lawrence P. Goldschmidt* •
Historical Museum – 1895
Decorative and applied arts ... 53909

National Trust for Historic Preservation, 1785 Massachusetts Av NW, Washington, DC 20036 • T: +1 202 5886000 • F: +1 202 5886059 • feedback@nthp.org • www.nationaltrust.org •
Pres.: *James Vaughan* •
Historical Museum – 1949
Fine and decorative arts, furnishing ... 53910

The Navy Museum, 805 Kidder Breese St, Washington, DC 20374-5060 • T: +1 202 4334882 • F: +1 202 4338200 • www.history.navy.mil •
Dir.: *Kim Nielsen* • Cur.: *Dr. Edward Furgol* •
Military Museum – 1961
Docs, weapons, photos, medals, uniforms, 19th and 20th c guns ... 53911

The Octagon Museum, 1799 New York Av NW, Washington, DC 20006-5292 • T: +1 202 6383221 • F: +1 202 8797764 • info@archfoundation.org • www.archfoundation.org/octagon •
Cur.: *Sherry C. Birk* (Collections) •

U

Fine Arts Museum / Historic Site – 1942
Architectural drawings and photographs, archaeological artifacts, decorative arts of the peroid 1800-1828 53912

The Old Stone House, 3051 Main St NW, Washington, DC 20007 • T: +1 202 4266851 • F: +1 202 4260125 • old_stone_house@nps.gov • www.nps.gov/rocr •
Dir.: *Terry Carlstrom* •
Local Museum – 1950
18th c furniture, glassware, cooking ware 53913

The Phillips Collection, 1600 21st St NW, Washington, DC 20009-1090 • T: +1 202 3872151 • F: +1 202 3872436 • communications@phillipscollection.org • www.phillipscollection.org •
Dir.: *Jay Gates* • Cur.: *Eliza Rathbone* • *Elizabeth Hutton Turner* • *Elsa Mezvinski-Smithgall* •
Fine Arts Museum – 1918
19th-20th c European and American paintings and sculpture, 2332 works of art 53914

Renwick Gallery, Smithsonian American Art Museum, Pennsylvania Av NW at 17th St, Washington, DC 20006, mail addr: MRC 510 Box 37012, Washington, DC 20013-7012 • T: +1 202 3572700, 2531 • F: +1 202 7862810 • saaminfo@si.edu • www.americanart.si.edu •
Dir.: *Ellen M. Myette* •
Fine Arts Museum / Decorative Arts Museum – 1972
American crafts 53915

Sewall-Belmont House, 144 Constitution Av NE, Washington, DC 20002 • T: +1 202 5461210 • F: +1 202 5463997 • info@sewallbelmont.org • www.sewallbelmont.org •
Pres.: *Audrey Sheppard* • Exec. Dir.: *A. Page Harrington* •
Historical Museum – 1929
Women's History and Suffrage Gallery, residence of Albert Gallatin 1801-1813 and Alice Paul 1929-1972 – library 53916

Smithsonian American Art Museum, 8th and F St NW, Washington, DC 20006, mail addr: 750 9th St NW, Suite 3100, Washington, DC 20001 • T: +1 202 2751500 • F: +1 202 2751715 • saaminfo@si.edu • www.americanart.si.edu •
Dir.: *Dr. Elizabeth Broun* • Chief Cur.: *George Gurney* • Chief Cons.: *Stefano Scafetta* •
Fine Arts Museum – 1829/1846
American art from the clonial period, paintings, sculptures, graphics, photography, folk art 53917

The Society of the Cincinnati Museum, c/o Anderson House, 2118 Massachusetts Av, Washington, DC 20008-2810 • T: +1 202 7852040 • F: +1 202 2930729 • admin@societyofthecincinnati.org • www.societyofthecincinnati.org •
Exec. Dir.: *Jack D. Warren* • Cur.: *Emily L. Schulz* •
Fine Arts Museum / Historical Museum – 1783
Furniture, sculpture, paintings, textiels, ceramics, Asian and European decorative arts, artifacts, manuscripts, silver 53918

Studio Gallery, 2108 R St NW, Washington, DC 20008 • T: +1 202 2328734 • F: +1 202 2325894 • info@studiogallerydc.com • www.studiogallerydc.com •
Dir.: *Jana Lyons* •
Fine Arts Museum – 1964 53919

The Supreme Court of the United States Museum, One First St, NE, Washington, DC 20543 • T: +1 202 4793298 • F: +1 202 4792926 • mail76668@pop.net • www.supremecourtus.gov •
Cur.: *Catherine Fitts* •
Historical Museum – 1973
Historic agency and bldg 53920

The Textile Museum, 2320 S St NW, Washington, DC 20008 • T: +1 202 6670441 • F: +1 202 4830994 • info@textilemuseum.org • www.textilemuseum.org •
Pres.: *Bruce P. Baganz* • Cur.: *Ann P. Rowe* (Western Hemisphere) • *Sumru Krody* (Eastern Hemisphere) •
Special Museum – 1925
17000 historic and handmade textiles, Oriental carpets, pre-Columbian, Peruvian, Islamic, Coptic, Cuacasian, Southeast and central Asian, Latin and North American textiles 53921

Tudor Place Museum, 1644 31st St NW, Washington, DC 20007 • T: +1 202 9650400 ext 100 • F: +1 202 9650164 • info@tudorplace.org • www.tudorplace.org •
Exec. Dir.: *Leslie L. Buhler* •
Historical Museum – 1966
Washington-Custis-Peter memorabilia, furniture, silver, porcelain, sculpture, paintings, textiles, photos, manuscripts 53922

United States Capitol Visitor Center, Capitol Hill, Washington, DC 20515 • T: +1 202 2281222, 2256827 • F: +1 202 2281893 • www.aoc.gov •
Cur.: *Dr. Barbara A. Wolanin* •
Fine Arts Museum – 1793
Paintings, sculpture, decorative art, photography 53923

United States Department of the Interior Museum, 1849 C St NW, Washington, DC 20240 • T: +1 202 2084743 • F: +1 202 2081535 • www.doi.gov/museum •
Cur.: *Debra Berke* •
Historical Museum / Ethnology Museum – 1938
American Indian, Eskimo, Micronesia, Virgin Island and Guam handicraft and artifacts. paintings, sculptures, maps 53924

United States Holocaust Memorial Museum, 100 Raoul Wallenberg Pl SW, Washington, DC 20024-2150 • T: +1 202 4880400 • F: +1 202 4882690 • visitorsmail@ushmm.org • www.ushmm.org •
Dir.: *Sara J. Bloomfield* •
Historical Museum – 1980/1993
Reflecting the events of the Holocaust of 1933-1945 – archives 53925

United States Marine Corps Museum, Marine Corp Historical Center, 1254 Charles Morris St SE, Washington, DC 20374-5040 • T: +1 202 4333534, 4332484 • F: +1 202 4337265 •
Dir.: *Michael F. Monigan* • Cur.: *John T. Dyer* • *James A. Fairfax* •
Historical Museum / Military Museum – 1940
Military history 53926

United States Navy Art Gallery, Washington Navy Yard, 805 Kidder Breese SE, Washington, DC 20374-5060 • T: +1 202 4333815 • F: +1 202 4335635 • www.history.navy.mil •
Cur.: *Gale Munro* •
Fine Arts Museum
Graphic arts, paintings, sketches, sculptures 53927

United States Navy Memorial Foundation and Naval Heritage, 701 Pennsylvania Av NW, Washington, DC 20004-2608 • T: +1 202 7372300 • F: +1 202 7372308 • library@lonesailor.org • www.lonesailor.org •
Pres./C.E.O.: *Henry C. McKinney* • Cur.: *Mark T. Hacala* •
Military Museum – 1991
WW II veterans hist 53928

United States Senate Commission on Art Collection, Rm S-411, U.S. Capitol Bldg, Washington, DC 20510-7102 • T: +1 202 2242955 • F: +1 202 2248799 • curator@sec.senate.gov • www.senate.gov •
Cur.: *Diane K. Skvarla* •
Fine Arts Museum / Decorative Arts Museum – 1968
Paintings, sculpture, prints, furnishing 53929

Washington Center for Photography, 406 Seventh St NW, Washington, DC 20004 • T: +1 202 7370406 • F: +1 202 7370419 •
Pres.: *Jerry Smith* •
Fine Arts Museum – 1986 53930

Washington Dolls House and Toy Museum, 5236 44th St NW, Washington, DC 20015 • T: +1 202 2440024, 3636400 • F: +1 202 2371659 • dollhousemuseum@aol.com • www.dcnetworks.org •
Dir.: *Flora Gill Jacobs* •
Decorative Arts Museum – 1975
Antique Toy ans Dolls 53931

Washington National Cathedral, Cathedral Church of Saint Peter and Saint Paul, Massachusetts and Wisconsin Av, NW, Washington, DC 20016-5098 • T: +1 202 5378991, 5376200 • F: +1 202 3646611 • tours@cathedral.org • www.nationalcathedral.org •
C.E.O.: *Tina Mead* •
Religious Arts Museum – 1893
Gothic design Cathedral, Episcopal Cathedral, structure 53932

Watkins Gallery, American University, 4400 Massachusetts NW Av, Washington, DC 20016 • T: +1 202 8851300 • F: +1 202 8851140 • museum@american.edu • www.american.edu •
Dir.: *Ron Haynie* • Cur.: *Jonathan Bucci* •
Public Gallery / University Museum – 1945
American and European prints, drawings and paintings 53933

The White House, 1600 Pennsylvania Av NW, Washington, DC 20500 • T: +1 202 4562550 • F: +1 202 4566820 • www.whitehousehistory.org •
Cur.: *William G. Allman* • Ass. Cur.: *Lydia Tederick* •
Historical Museum – 1792
Hist, politics 53934

Woodrow Wilson House, 2340 S St NW, Washington, DC 20008 • T: +1 202 3874062 ext 14 • F: +1 202 4831466 • wilson_house@woodrowwilsonhouse.org • www.woodrowwilsonhouse.org •
Dir.: *Frank J. Aucella* • Cur.: *Meg Nowack* •
Historical Museum – 1963
Home of President Wilson 53935

Washington GA

Robert Toombs House, 216 E Robert Toombs Av, Washington, GA 30673 • T: +1 706 6782226 • F: +1 706 6787515 • toombs@nu-z.net • www.kudcom.com/www/att03.html •
Man.: *Marty Fleming* •
Natural History Museum – 1974
Furniture and furnishings from the period 1840-1900, Toombs family artifacts 53936

Washington Historical Museum, 308 E Robert Toombs Av, Washington, GA 30673-2038 • T: +1 706 6782105 • F: +1 706 6783752 • arikli@oconee.ga.us • www.visitoconeecounty.com •
Local Museum – 1959
1836 Barnett-Slaton House, local history 53937

Washington KS

Washington County Historical Society, 206-208 Ballard, Washington, KS 66968 • T: +1 785 3252198 •
Association with Coll – 1982
Artifacts from the 1860 - 1960s 53938

Washington LA

Washington Museum, 404 N Main St, Washington, LA 70589 • T: +1 318 8263627 •
Chief Cur.: *Melba Soileau* •
Local Museum – 1972
Artifacts, tools, 19th c docs, 19th c stethoscope, photos 53939

Washington MS

Historic Jefferson College, College and North St, Washington, MS 39190, mail addr: POB 700, Washington, MS 39190 • T: +1 601 4422901 • F: +1 601 4462912 • hjc@bkbank.com • mdah.state.ms.us •
Dir.: *Cheryl Munyer Branyan* •
Historical Museum – 1971
College uniforms, books, weapons 53940

Washington PA

David Bradford House, 175 S Main St, Washington, PA 15301 • T: +1 724 2223604 • mthart@pulsenet.com • www.bradfordhouse.org •
C.E.O. & Pres.: *Steven Tkach* •
Local Museum – 1960
Historic house 53941

Olin Fine Arts Center, Washington and Jefferson College, 285 E Wheeling St, Washington, PA 15301 • T: +1 724 2236546 • F: +1 724 2235271 • htaylor@washjeff.edu • www.washjeff.edu/ •
Fine Arts Museum / University Museum – 1980 53942

Pennsylvania Trolley Museum, 1 Museum Rd, Washington, PA 15301-6133 • T: +1 724 2289675 • F: +1 724 2289675 • ptm@pa-trolley.org • www.pa-trolley.org •
Exec. Dir.: *Scott R. Becker* •
Science&Tech Museum – 1949
Locomotives, cars 53943

Washington County Historical Society Museum, Le Moyne House, 49 E Maiden St, Washington, PA 15301 • T: +1 724 2256740 • F: +1 724 2258495 • info@wchspa.org • www.wchspa.org •
Cur.: *Rebecca Crum* •
Local Museum – 1900
Le Moyne House, a national historic landmark of the underground railroad incl historic house, doctor's office, gardens, crematory 53944

Washington TX

Barrington Living History Farm, 23100 Barrington Ln, Washington, TX 77880 • T: +1 936 8782214 • F: +1 936 8782810 • wilburt.scaggs@tpwd.state.tx.us • www.tpwd.state.tx.us/park/parks.htm •
Man.: *Tom Scaggs* •
Agriculture Museum – 1936
Historic house, home of Anson Jones, last president of the Republic of Texas 53945

Courtyard Gallery, Washington Studio School, 3232 P St NW, Washington, TX 78247 • T: +1 202 3332663 • F: +1 202 3334063 •
Fine Arts Museum 53946

Star of the Republic Museum, 23200 Park Rd 12, Washington, TX 77880 • T: +1 936 8782461 • F: +1 936 8782462 • star@blinncol.edu • www.starmuseum.org •
Dir.: *Houston McGaugh* • Cur.: *Dr. Shawn Carlson* •
Historical Museum – 1970
Texas hist 1836-1846 53947

Washington Court House OH

Fayette County Museum, 517 Columbus Av, Washington Court House, OH 43160 • T: +1 740 3352953 •
Pres.: *Roger Kirkpatrick* • Cur.: *Carol Karey* •
Local Museum – 1948
Period furnishings, hist and heritag of Fayette County 53948

Washington Crossing PA

Washington Crossing Historic Park, 1112 River Rd, Washington Crossing, PA 18977 • T: +1 215 4934076 • F: +1 215 4934820 • www.phmc.state.pa.us •
Head: *Michael Bertheaud* • Cur.: *Hilary Krueger* •
Historical Museum – 1917
Historic Site 53949

Washington Green CT

The Institute For American Indian Studies, 38 Curtis Rd, Washington Green, CT 06793-0260 • T: +1 860 8680518 • F: +1 860 8681649 • instituteamer.indian@snet.net • www.birdstone.org •
Exec. Dir.: *Alberto C. Meloni* • Sc. Staff: *Lucienne Lavin* (Archaeology) • *E. Barrie Kavasch* (Ethnobotany) •
Ethnology Museum / Archaeology Museum – 1975
American Indian culture and archaeology 53950

Washington's Birthplace VA

George Washington Birthplace National Monument, 1732 Popes Creek Rd, Washington's Birthplace, VA 22443 • T: +1 804 2241732 • F: +1 804 2242142 • gewa_park_information@nps.gov • www.nps.gov/gewa •
Local Museum – 1932
Birthplace of George Washington 53951

Washoe Valley NV

Bowers Mansion, 4005 US 395, N, Washoe Valley, NV 89704-9518 • T: +1 702 8490201 •
Cur.: *Betty Hood* •
Historical Museum – 1946
Restored 1846 mansion, furniture 53952

Wasilla AK

Dorothy G. Page Museum, 323 N Main St, Wasilla, AK 99654 • T: +1 907 3739071 • F: +1 907 3739072 • museum@ci.wasilla.ak.us • www.cityofwasilla.com •
Local Museum – 1966
Local hist, Willow Creek mining 53953

Museum of Alaska Transportation and Industry, 3800 W Neuser Dr, Wasilla, AK 99654, mail addr: POB 870646, Wasilla, AK 99687 • T: +1 907 3761211 • F: +1 907 3763082 • www.museumofalaska.org •
C.E.O.: *David Wharton* •
Science&Tech Museum / Historical Museum – 1967
Alaskan industrial hist and transportation 53954

Water Mill NY

Water Mill Museum, 41 Old Mill Rd, Water Mill, NY 11976 • T: +1 631 7264625 • www.watermillmuseum.org •
Pres.: *Gay Colina* •
Local Museum – 1942
Community life and milling in 19th c, tools, ice-harvesting, blacksmith, farmer, cooper tools, 53955

Waterbury CT

The Mattatuck Museum of the Mattatuck Historical Society, 144 W Main St, Waterbury, CT 06702 • T: +1 203 7530381 • F: +1 203 7566283 • svoghel@mattatuckmuseum.org • www.mattatuckmuseum.org •
Dir.: *Marie Galbraith* • Cur.: *Ann Smith* • Ass. Cur.: *Raechel Guest* •
Fine Arts Museum / Historical Museum – 1877
History and art Temple 53956

Timexpo Museum, 175 Union St, Brass Mill Commons, Waterbury, CT 06706 • T: +1 203 7558463 • F: +1 203 7558531 • crosa@timexpo.com • www.timexpo.com •
Dir.: *Geoff Giglierano* •
Special Museum – 2001
Watches and clocks 53957

Waterbury Center VT

Green Mountain Club, 4711 Waterbury Stowe Rd, Waterbury Center, VT 05677 • T: +1 802 2447037 • F: +1 802 2445867 • gmc@greenmountainclub.org • www.greenmountainclub.org •
Pres.: *Rolf Anderson* • Exec. Dir.: *Ben Rose* •
Association with Coll – 1910 53958

Waterford NY

Waterford Historical Museum, 2 Museum Ln, Waterford, NY 12188 • T: +1 518 2380809 • F: +1 518 2380809 • whmcc@nycap.rr.com • www.waterfordmuseum.com •
Pres.: *Dr. Brad L. Utter* •
Local Museum – 1964
Household items, local hist, Champlain and Erie Canal photos and artifacts 53959

Waterford PA

Fort Leboeuf, 31 High St, Waterford, PA 16441 • T: +1 814 7322575 • F: +1 814 7322629 • qolybwx@edinboro.edu •
Dir.: *Dr. Renata Wolynec* •
Historical Museum / University Museum – 1929
Period furniture, artifacts from French occupation 53960

Waterloo IA

Grout Museum of History and Science, 503 South St, Waterloo, IA 50701 • T: +1 319 2346357 • F: +1 319 2360500 • grout@forbin.net • www.groutmuseumdistrict.org •
Dir.: *Billie K. Bailey* •
Historical Museum / Natural History Museum – 1933
History, science 53961

Rensselaer Russell House Museum, 520 W 3rd St, Waterloo, IA 50701 • T: +1 319 2330262 • F: +1 319 2360500 • www.groutmuseumdistrict.org •
Dir.: *Billie K. Bailey* •
Historical Museum – 1861
Historic house built in Italianate architectural style 53962

Science Imaginarium, 322 Wahington St, Waterloo, IA 50701 • T: +1 319 2338708 • F: +1 319 2360500 • www.groutmuseumdistrict.org •
Dir.: *Billie K. Bailey* • Man.: *Alan Sweeney* •
Natural History Museum – 1993 53963

Snowden House, 306 Washington St, Waterloo, IA 50701 • T: +1 319 2346357 • F: +1 319 2360500 • www.groutmuseumdistrict.org •
Dir.: *Billie K. Bailey* •
Historical Museum – 1933
Summer house of William Snowden, pharmacy 53964

Waterloo Center for the Arts, 225 Commercial St, Waterloo, IA 50701 • T: +1 319 2914490 • F: +1 319 2914270 • museum@waterloo-ia.org • www.waterloo-ia.org/arts •
Dir.: *Cammie V. Scully* • Cur.: *Kent Shankle* •
Fine Arts Museum
Haitian art, Mid-West regional art 53965

Waterloo NY

Memorial Day Museum, 35 E Main St, Waterloo, NY 13165 • T: +1 315 5390533 • F: +1 315 5397798 • terwilliger@flare.net •
C.E.O. & Pres.: *Lynn Patti* • Dir.: *James T. Hughes* •
Military Museum – 1966 53966

Peter Whitmer Farm, 1451 Aunkst Rd, Waterloo, NY 13165 • T: +1 315 5392552 • F: +1 315 5970165 • www.hillcumorah.org •
Sc. Staff: *Calvin Christiansen* •
Historic Site – 1980 53967

Terwilliger Museum, 31 E Williams St, Waterloo, NY 13165 • T: +1 315 5390533 • F: +1 315 5397798 • terwilliger@elare.net • www.waterloony.com/library.html •
Dir.: *James T. Hughes* •
Local Museum / Historical Museum – 1960
History and development of Waterloo, genalogy 53968

Watertown CT

Watertown Historical Society Museum, 22 DeForest St, Watertown, CT 06795-2522 • T: +1 860 2741050 •
Cur.: *Stephen J. Bartkus* •
Local Museum – 1947
Local hist 53969

Watertown MA

Armenian Museum of America, 65 Main St, Watertown, MA 02472 • T: +1 617 9262562 ext 21 • F: +1 617 9260175 • gary@armenianlibraryandmuseum.org • www.armenianlibraryandmuseum.org •
Exec. Dir.: *Robert Parsekiau* • Cur.: *Susan Lind-Sinanian* • *Gary Lind-Sinanian* •
Ethnology Museum / Folklore Museum – 1971
Ethnic Armenian history, religious, folk and fine arts, photographs – library 53970

Perkins History Museum, c/o Perkins School for the Blind, 175 N Beacon St, Watertown, MA 02472 • T: +1 617 9727767, 9262027 • F: +1 617 9243434 • historymuseum@perkins.org • www.perkins.org/museum •
Cur.: *Betsy L. McGinnity* •
Special Museum – 1829
Picture coll, hist of blindness, Nella Braddy Henney coll, writing devices – library 53971

Watertown NY

Jefferson County Historical Society Museum, 228 Washington St, Watertown, NY 13601 • T: +1 315 7823491 • F: +1 315 7822913 •
Dir.: *Fred H. Rollins* • Cur.: *Elise Chan* (Collections) •
Education: *Melissa Widrick* •
Local Museum – 1886
19th c Americana, northern NY Indian artifacts, handmade agricultural and woodworking tools, early machinery 53972

Sci-Tech Center of Northern New York, 154 Stone St, Watertown, NY 13601 • T: +1 315 7881340 • F: +1 315 7882738 • scitech@imcnet.net • www.scitechcenter.org •
Exec. Dir.: *John Kunz* •
Science&Tech Museum – 1982
Science and technology 53973

Watertown SD

Codington County Heritage Museum, 27 First Av, Watertown, SD 57201 • T: +1 605 8867335 • F: +1 605 8824383 • director@cchsmuseum.org • www.cchsmuseum.org •
Exec. Dir.: *Tim Hoheisel* •
Local Museum – 1974
Local hist 53974

Mellette House, 421 Fifth Av, Watertown, SD 57201 • T: +1 605 8864730 •
Pres.: *Prudence K. Calvin* •
Local Museum – 1943
Historic house, home of Arthur Calvin Mellette, first Governor of South Dakota 53975

Watertown WI

Octagon House, 919 Charles St, Watertown, WI 53094 • T: +1 920 2612796 • www.watertownhistory.org •
Pres.: *William Jannke* • C.E.O.: *Linda Werth* •
Local Museum – 1933
Historic house, first Kindergarten in America 53976

Waterville ME

Colby College Museum of Art, 5600 Mayflower Hill, Waterville, ME 04901 • T: +1 207 8723000, 8723228 • F: +1 207 8723807 • museum@colby.edu • www.colby.edu/museum •
Dir.: *Daniel Rosenfeld* •
Fine Arts Museum / University Museum – 1959
Art, works of Alex Katz 53977

Redington Museum, 64 Silver St, Waterville, ME 04901 • T: +1 207 8729439 • www.redingtonmuseum.org •
Cur.: *Donnice Finnemore* • *Harry Finnemore* •
Historical Museum – 1903
Local history, residence of Asa Redington 53978

Watervliet NY

Watervliet Arsenal's Museum of the Big Guns, Watervliet Arsenal, Rte 32, 1 Buffington Str, Watervliet, NY 12189-4050 • T: +1 518 2665805 • F: +1 518 2665859 • swantek@wva.army.mill.us •
Dir./Cur.: *John E. Swantek* •
Military Museum – 1975
Military hist 53979

Watford City ND

Pioneer Museum, 104 Park Av W, South Main, Watford City, ND 58854 • T: +1 701 8422990 •
Cur.: *Sylvia Leiseth* •
Historical Museum – 1968
Pioneer furnishings 53980

Watkins Glen NY

International Motor Racing Research Center, 610 S Decatur St, Watkins Glen, NY 14891 • T: +1 607 5359044 • F: +1 607 5359039 • research@racingarchives.org • www.racingarchives.org •
Dir.: *Michael Rand* • Cur.: *Mark Steigerwald* •
Special Museum – 1998
Motorsport, paintings, photos 53981

Watkinsville GA

Eagle Tavern Museum, 26 N Main St, Watkinsville, GA 30677 • T: +1 706 7695197 • F: +1 706 7690705 • eagletavern@aol.com • www.oconeecounty.com •
Special Museum – 1956 53982

Watonga OK

T.B. Ferguson Home, Oklahoma Historical Society, 519 N Weigel, Watonga, OK 73772 • T: +1 580 6235069 •
Cur.: *Mildred Sanders* •
Local Museum – 1972
Period furnishings, 1892-1921 pioneer items, tax rolls 53983

Watrous NM

Fort Union, Rte 161, Watrous, NM 87753 • T: +1 505 4258025 • F: +1 505 4541155 • foun_administration@nps.gov • www.nps.gov/foun/ •
Military Museum – 1956
Military hist, Civil War 53984

Watseka IL

Iroquois County Historical Society Museum, 103 West Cherry St, Watseka, IL 60970 • T: +1 815 4322215 • iroqgene@yahoo.com • www.rootsweb.com/~ilicgs •
Pres.: *Wayne Hiles* •
Local Museum – 1967
Local history, housed in Old Iroquois County Courthouse 53985

Watsonville CA

Pajaro Valley Historical Museum, 332 E Beach St, Watsonville, CA 95076 • T: +1 831 7220305 • F: +1 831 7225501 •
Pres.: *James Dutra* •
Local Museum – 1940
Local hist, costumes – small library 53986

Wauconda IL

Lake County Discovery Museum, Rte 176, Fairfield Rd, Wauconda, IL 60084 • T: +1 847 9683400 • F: +1 847 5260024 • LCMuseum@LCFPD.org • www.lcfpd.org/discovery_museum •
Dir.: *Janet Gallimore* • Cur.: *Katherine Hamilton-Smith* •
Local Museum – 1976
Decorative arts, vehicles, regional hist, photos 53987

Wauconda Township Museum, 711 Old Rand Rd, Wauconda, IL 60084 • T: +1 847 5269303 •
Local Museum – 1973
Farm equip, toys, quilts, furnishings, Civil War cavalry equip, fans, cameras, china 53988

Waukegan IL

Haines Museum, 1917 N Sheridan Rd, Bowen Park, Waukegan, IL 60079 • T: +1 847 3361859, 3604772 • F: +1 847 6626190 • haines@waukeganparks.org • www.waukeganparks.org •
Pres.: *William Tyre* •
Local Museum – 1968
Civil War and Waukegan artifacts, furnishings, photos of local and county hist 53989

Waukesha WI

Waukesha County Museum, 101 W Main St, Waukesha, WI 53186-4811 • T: +1 262 5212859 • F: +1 262 5212865 • exdirector@voyager.net • www.waukeshacountymuseum.org •
Dir.: *Susan K. Baker* • Cur.: *Brook Swanson* (Collections) •
Local Museum – 1914
Local hist, military and Native American artifacts – research center 53990

Waupaca WI

Hutchinson House Museum, 321 S Main St, South Park, Waupaca, WI 54981 • T: +1 715 2585958, 2589980 • wauphistsoc@waupacaonline.net • www.waupacahistory.org •
Pres.: *Mark Ebner* • Cur.: *Linda Pope* • *Marion Snyder* •
Local Museum – 1956
Local hist, Victorian furnishing and clothing 53991

Waurika OK

Chisholm Trail Historical Museum, Oklahoma Historical Society, US 81 and State Rte 70, Waurika, OK 73573 • T: +1 580 2282166 • F: +1 580 2283290 •
Head: *Gin Dodson* •
Local Museum – 1965 53992

Wausau WI

Leigh Yawkey Woodson Art Museum, 700 N 12th St, Wausau, WI 54403-5007 • T: +1 715 8457010 • F: +1 715 8457103 • museum@lywam.org • www.lywam.org •
Dir.: *Kathy Kelsey Foley* • Cur.: *Andrew J. McGivern* • *Jane Weinke* •
Fine Arts Museum / Decorative Arts Museum – 1973
Art depicting natural world 53993

Marathon County Historical Society Museum, 410 McIndoe St, Wausau, WI 54403 • T: +1 715 8425750 • F: +1 715 8480576 • research@marathoncountyhistory.com • www.marathoncountyhistory.com •
Dir.: *Mary Jane Hettinga* •
Local Museum – 1952
Local hist, lumbering and pioneer artifacts 53994

Wauseon OH

Fulton County Historical Museum, 229 Monroe St, Wauseon, OH 43567 • T: +1 419 3377922 • museum@fulton-net.com • www.fultoncountyhs.org •
Pres.: *Marvin Leatherman* • Dir.: *Barbara Berry* •
Local Museum – 1883
Artifacts of Fulton County 1830-1976 53995

Wauwatosa WI

Lowell Damon House, 2107 Wauwatosa Av, Wauwatosa, WI 53213 • T: +1 414 2738288 • F: +1 414 2733268 • mchs@prodigy.net • www.milwaukeecountyhistsoc.org •
Pres.: *Robert B. Brumder* • Exec. Dir.: *Robert T. Teske* •
Local Museum – 1941
Historic house, 1840-1880 furniture 53996

Waverly TN

World O' Tools Museum, 2431 Hwy 13 S, Waverly, TN 37185 • T: +1 931 2963218 • hunterp@usit.net •
Cur.: *Hunter M. Pilkinton* •
Science&Tech Museum – 1973
25,000 early wood & metal working fraft tools 53997

Waverly VA

Miles B. Carpenter Museum, 201 Hunter St, Waverly, VA 23890 • T: +1 804 8342151, 8343327 •
Pres./Cur.: *Shirley S. Yancey* •
Folklore Museum / Agriculture Museum – 1986
Folk art, peanuts 53998

Waxahachie TX

Ellis County Museum, 201 S College St, Waxahachie, TX 75165 • T: +1 972 9370681 • F: +1 972 9373910 • ecmuseum@cnbcom.net • www.elliscountymuseum.org •
Dir./Cur.: *Shannon Simpson* •
Local Museum – 1967
Relics, photos, manuscripts 53999

Waxhaw NC

Museum of the Alphabet, Jaars Summer Institute of Linguistics, 6409 Davis Rd, Waxhaw, NC 28173 • T: +1 704 8436066 • F: +1 704 8436200 • info@jaars.org • www.jaars.org •
Dir.: *LaDonna Mann* •
Special Museum – 1990
Alphabet 54000

Museum of the Waxhaws and Andrew Jackson Memorial, Hwy 75 E, Waxhaw, NC 28173 • T: +1 704 8431832 • F: +1 704 8431767 • mwaxhaw@perigee.net • www.perigee.net/~mwaxhaw •
Dir.: *Jaiy Geno* •
Local Museum / Historical Museum – 1996
Civilian and military related artifacts, President Andrew Jackson 54001

Waycross GA

Okefenokee Heritage Center, 1460 N Augusta Av, Waycross, GA 31503 • T: +1 912 2854260 • F: +1 912 2832858 • ohc@acces.net • www.okeheritage.org •
Pres.: *T. Henry Clarke* • Dir.: *Catherine Larkins* •
Fine Arts Museum / Historical Museum – 1975
History, art 54002

Southern Forest World Museum, 1440 N Augusta Av, Waycross, GA 31503 • T: +1 912 2854056 • F: +1 912 2832858 • obediahs@okefenokeeswamp.com • www.okefenokeeswamp.com •
Pres.: *Earl Smith* •
Science&Tech Museum / Natural History Museum – 1981
Artifacts 54003

Wayland MA

Wayland Historical Museum, 12 Cochituate Rd, Wayland, MA 01778 • T: +1 508 3587959 • j.w.d.home.comcast.net/whs •
Cur.: *Joanne Davis* •
Historical Museum – 1954
Local hist, Indian artifacts, costumes 54004

Wayne MI

City of Wayne Historical Museum, 1 Town Sq, Wayne, MI 48184 • T: +1 734 7220113 • F: +1 734 7225052 • www.ci.wayne.mi.us/museum.php •
Dir.: *Virginia Presson* •
Local Museum – 1964
Local hist, maps & docs, carriage, General Wayne & the battles of Monmouth, Stony Point and Yorktown – small library 54005

Wayne NE

Nordstrand Visual Arts Gallery, Wayne State College, 1111 Main St, Wayne, NE 68787 • T: +1 402 3757000 • F: +1 402 3757204 •
Dir.: *Prof. Pearl Hansen* • *Prof. Marlene Mueller* •
Public Gallery – 1977 54006

Wayne NJ

Ben Shahn Galleries, William Paterson University, c/o William Paterson University of New Jersey, 300 Pompton Rd, Wayne, NJ 07470 • T: +1 973 7202654 • F: +1 973 7203290 • einreinhofern@wpunj.edu • www.wpunj.edu •
Dir./Cur.: *Nancy Einreinhofer* •
University Museum / Fine Arts Museum – 1969
19th c landscape paintings, sculptures, African and Oceanic, outdoor sculpture 54007

Dey Mansion, Washington's Headquarters Museum, 199 Totowa Rd, Wayne, NJ 07470 • T: +1 201 6961776 • F: +1 973 6961365 •
Dir./Cur.: *Raymond Wright* •
Military Museum – 1934 54008

Van Riper-Hopper House, 533 Berdan Av, Wayne, NJ 07470 • T: +1 973 6947192 • F: +1 973 8720586 • www.waynetownship.com/ •
Pres.: *Bob Brubaker* •
Historical Museum – 1964
Furnishing from Colonial period to 1860, Indian artifacts 54009

Wayne PA

The American Revolution Center, 435 Devon Park Dr, Suite 801, Wayne, PA 19087 • T: +1 610 9754948 • F: +1 610 2258420 • info@americanrevolutioncenter.org • /www.americanrevolutioncenter.org •
C.E.O.: *Thomas M. Daly* •
Historical Museum – 2000
Washington memorabilia and artifacts, 18th c firearms, decorative arts 54010

The Finley House, 113 W Beech Tree Ln, Wayne, PA 19087 • T: +1 610 6882668, 6884455 • www.radnorhistory.org •
Pres.: *Ted Pollard* •
Local Museum – 1948
Local hist 54011

Wayne Art Center, 413 Maplewood Ave, Wayne, PA 19087 • T: +1 610 6883553 • F: +1 610 9950478 • wayneart@worldnet.att.net • www.wayneart.com •
Dir.: *Nancy Campbell* •
Public Gallery – 1930 54012

Waynesboro PA

Renfrew Museum, 1010 E Main St, Waynesboro, PA 17268 • T: +1 717 7624723 • F: +1 717 7626384 • renfrew@innernet.net • www.waynesboro.org/renfrew •
Cur.: *Jeffrey Bliemeister* •
Decorative Arts Museum – 1973
Decorative arts 54013

Waynesboro VA

Humpback Rocks Mountain Farm Visitor Center, Blue Ridge Pkwy, Mile Post 5.9, Waynesboro, VA 24483 • T: +1 540 3772377 • F: +1 540 3776758 • www.nps.gov/blri •
Local Museum – 1939
Pioneer life, farm implements, tools, household items, toys, firearms 54014

Waynesburg PA

Greene County Historical Museum, 918 Rolling Meadows Rd, Waynesburg, PA 15370 • T: +1 724 6273204 • F: +1 724 6273204 • museum@greenepa.net • www.greenepa.net/~museum/ •
Pres.: *Paul Maytk* •
Local Museum – 1925
Local hist 54015

Weathersfield VT

Museum of the Weathersfield Historical Society, Reverend Dan Forster House, Weathersfield Center Rd, Weathersfield, VT 05156 • T: +1 802 2635230, 2635361 • F: +1 802 2639263 • ellen.clattenburg@dresden.us • www.weathersfield.net •
Pres.: *Willis Wood* •
Local Museum – 1951
1787 barn replica, local artifacts, old cobbler's tools, quilts 54016

Weaverville CA

Jack Jackson Memorial Museum, 508 Main St, Weaverville, CA 96093 • T: +1 530 6235211 • F: +1 530 6235053 • jake@trinitymuseum.org • www.trinitymuseum.org •
Dir.: *Bill McCoy* •
Local Museum – 1968
Local hist, mining, Chinese and Indian artifacts, carriages, antique firearms 54017

Weaverville NC

Zebulon B. Vance Birthplace, 911 Reems Creek Rd, Weaverville, NC 28787 • T: +1 828 6456706 • F: +1 828 6450936 • vance@brinet.com • www.ah.dcr.state.nc.us/sections/hs/vance/vance.htm •
Historical Museum – 1961
NC Governor Zebulon B. Vance 54018

Weedsport NY

Hall of Fame and Classic Car Museum, 1 Speedway Dr, Weedsport, NY 13166 • T: +1 315 8346606 • F: +1 315 8349734 • www.dirtmotorsports.com •
Cur.: *Romy Caruso* •
Science&Tech Museum – 1992
Race cars from 60s to present, classic cars 30s and 40s 54019

Old Brutus Historical Society Museum, 8943 N Seneca St, Weedsport, NY 13166 • T: +1 315 8349342 •
C.E.O. & Pres.: *Charles Kreplin* • Dir.: *Raymond Baker* • Sc. Staff: *Jeanne Baker* (History) •
Local Museum – 1967
Local hist, agricultural and household exhibits, furnishings, genealogy 54020

Weeping Water NE

Weeping Water Valley Historical Society Museum, 215 W Eldora Av, Weeping Water, NE 68463 • T: +1 402 2674925 • wd85407@navix.net •
Pres./Chm.: *Doris Duff* •
Local Museum – 1969
Archaeology, pioneer artifacts, Indian artifacts 54021

Weiser ID

Snake River Heritage Center, 2295 Paddock Av, Weiser, ID 83672 • T: +1 208 5490205, 5493834 • rain01@ruralnetwork.net •
Local Museum – 1962
Music, costumes, folklore, glass, geology 54022

Wellesley Hills MA

Davis Museum, Wellesley College, 106 Central St, Wellesley Hills, MA 02481-8203 • T: +1 781 2833386 • rdolloff@wellesley.edu • www.wellesley.edu/DavisMuseum •
Dir.: *David Mickenberg* • Cur.: *Anja Chavez* •
Fine Arts Museum / University Museum – 1889
European, Asian, African and American arts 54023

Wellesley Historical Society, 229 Washington St, Wellesley Hills, MA 02481 • T: +1 781 2356690 • F: +1 781 2390660 • www.wellesleyhsoc.com •
Exec. Dir.: *Laurel Sparks* •
Local Museum – 1925
Local hist, Dadmun-McNamara house, oldham lace coll, Denton Butterfly and moth coll – archives, small library 54024

Wellfleet MA

Cape Cod National Seashore Museum, 99 Marconi Site Rd, Wellfleet, MA 02667 • T: +1 508 2558925 • F: +1 508 3499052 • hope_morrilla@nps.gov • www.nps.gov/caco •
Cur.: *Hope Morrill* •
Historical Museum / Natural History Museum – 1961
Local hist, natural hist, archaeology – library 54025

Wellfleet Historical Society Museum, 266 Main St, Wellfleet, MA 02667 • T: +1 508 3492247 • museum@wellfleethistoricalsociety.com • www.wellfleethistoricalsociety.com •
Cur.: *Joan Hopkins Coughlin* •
Local Museum – 1953
Local history, whaling, shell-fishing, folk art paintings 54026

Wellington KS

Chisholm Trail Museum, 502 N Washington, Wellington, KS 67152 • T: +1 620 3263820, 3267466 •
Pres.: *Richard M. Gilfillan* •
Historical Museum – 1964
First hospital in Wellington, China, glass, silver, folk art, costumes, toys & dolls 54027

Wellington OH

Spirit of '76 Museum, 201 N Main St, Wellington, OH 44090 • T: +1 440 6474367 •
Pres.: *John Perry* •
Historical Museum – 1968
Paintings by A.M. Willard, hist of Southern Lorain County 54028

Wells ME

Historical Society of Wells and Ogunquit Museum, 938 Post Rd, Wells, ME 04090 • T: +1 207 6464775 • F: +1 207 6460832 • wohistory@gwi.net •
Local Museum – 1954
Costumes, paintings, crafts, decorative arts – archives 54029

Wells Auto Museum, 1181 Post Rd, Wells, ME 04090 • T: +1 207 6469064 •
Dir.: *Jonathan H. Gould* •
Science&Tech Museum – 1954
Automobile 54030

Wellston OH

Buckeye Furnace State Memorial, 123 Buckeye Park Rd, Wellston, OH 43692-9511 • T: +1 740 3843537 •
Science&Tech Museum – 1976
Restored 1851 Iron Furnace complex 54031

Wellsville NY

Mather Homestead, 343Main Str, Wellsville, NY 14895 • T: +1 585 5931636 •
Local Museum – 1981
Hist house, 1930s artifacts, music, books, toys, sound adventures – wildlife area 54032

Wellsville OH

River Museum, 1003 Riverside Av, Wellsville, OH 43968 •
Pres.: *Paul M. Bloor* •
Local Museum – 1955
Pennsylvania Railroad Cabin Car, Civil War, saber, Indian artifacts, pottery 54033

Wellsville UT

American West Heritage Center, 4025 S Hwy, 89-91, Wellsville, UT 84339 • T: +1 801 2456050 • F: +1 801 2456052 • awhc@cc.usu.edu • www.americanwestcenter.org •
C.E.O.: *Ronda Thompson* •
Agriculture Museum / Local Museum – 1979
Historic farm 54034

Wenatchee WA

Gallery '76, Wenatchee Valley College, 1300 5th St, Wenatchee, WA 98801 • T: +1 509 6826776 • F: +1 509 6642538 • gallery76@wvc.edu • www.wvc.edu •
Dir.: *Rae Dana* •
Fine Arts Museum – 1976
Changing art exhib 54035

North Central Washington Museum → Wenatchee Valley Museum

Rocky Reach Dam Museum, US-97a Chelau Hwy, Wenatchee, WA 98807, mail addr: POB 1231, Wenatchee, WA 98807-1231 • T: +1 509 6638121, 6614960 • F: +1 509 6673449 • gallaher@chelanpud.org • www.chelanpud.org •
Local Museum – 1961
History of electricity, Thomas Edison artifacts, geology, anthropology, local Indian history 54036

Wenatchee Valley Museum, 127 S Mission St, Wenatchee, WA 98801 • T: +1 509 8886240 • F: +1 509 8886256 • info@wvmcc.org • www.wenatcheevalleymuseum.com •
Dir.: *Dr. Keith Williams* • Cur.: *Mark Behler* •
Local Museum – 1939
Local hist, pioneer and farm equip, archaeological coll, fruit industry exhibits 54037

Wenham MA

Wenham Museum, 132 Main St, Wenham, MA 01984 • T: +1 978 4682377 • F: +1 978 4681763 • info@wenhammuseum.org • www.wenhammuseum.org •
Exec. Dir.: *Emily Stearns* •
Local Museum – 1921
Social history, dolls, toys, model trains, costumes, textiles, photography, ice industry – library 54038

Weslaco TX

Frontera Audubon Society, 1101 S Texas Blvd, Weslaco, TX 78596 • T: +1 956 9683275 • F: +1 956 9681388 • frontera@fronteraaudubon.org • www.fronteraaudubon.org •
C.E.O.: *Dr. Rey Ramirez* •
Association with Coll – 1974
Agricultural developments, human influences, organic orchard 54039

Weslaco Museum, 515 S Kansas, Weslaco, TX 78596-8062 • T: +1 956 9689142 • F: +1 956 9680955 • ahazlett@weslaco.l.b.tx.us • texasescapes.com •
Local Museum – 1971
Early 20th c local & border hist artifacts – Archives 54040

West Allis WI

International Clown Hall of Fame and Research Center, c/o Wisconsin State Fair Park, Tommy Thompson Youth Center, 640 S 84th St, West Allis, WI 53214 • T: +1 414 2900105 • contact@theclownmuseum.org • www.theclownmuseum.org •
Dir.: *Kathryn O'Dell* • Cur.: *Arthur Pedlar* •
Performing Arts Museum – 1986
History of clowns 54041

West Allis Historical Society Museum, 8405 West National Av, West Allis, WI 53227-1733 • T: +1 414 5416970 •
Pres.: *Devan Gracyalny* •
Local Museum – 1966
Local hist, toys, pioneer implements, hist dental and post office 54042

West Barnstable MA

Higgins Art Gallery, Cape Cod Community College, 2240 Iyanough Rd, West Barnstable, MA 02668-1599 • T: +1 508 3622131 ext 4381 • F: +1 508 3754020 • info@capecod.edu • www.capecod.mass.edu •
Dir.: *Sarah Ringler* •
Fine Arts Museum – 1989
Paintings, prints, drawings, graphics, sculpture, mixed media 54043

West Bend WI

Washington County Historical Society Museum, 320 S Fifth Av, West Bend, WI 53095 • T: +1 414 3354678 • F: +1 414 3354612 • wchs@historyisfun.com • www.historyisfun.com •
Dir.: *Chip Beckford* • Cur.: *Daniel J. Smith* •
Local Museum – 1937
Local hist 54044

West Bend Art Museum, 300 S 6th Av, West Bend, WI 53095 • T: +1 262 3349638 • F: +1 262 3348080 • officemanager@wbartmuseum.com • www.wbartmuseum.com •
Exec.Dir.: *Thomas D. Lidtke* •
Fine Arts Museum – 1961
Wisconsin Art from 1800 - 1950, Carl von Marr coll 54045

West Branch IA

Herbert Hoover National Historic Site, 110 Parkside Dr, West Branch, IA 52358 • T: +1 319 6432541 • F: +1 319 6435367 • dan_banta@nps.gov • www.nps.gov/heho •
Historic Site – 1965
Birthplace of Herbert Clark Hoover, graves of President and Mrs. Hoover 54046

Herbert Hoover Presidential Library-Museum, 210 Parkside Dr, West Branch, IA 52358 • T: +1 319 6435301 • F: +1 319 6435825 • library@nara.gov • www.hoover.archives.gov •
Dir.: *Timoth Walch* • Cur.: *Maureen Harding* • Ass. Cur.: *Christine Mouw* •
Library with Exhibitions – 1962
Presidential library – Archive 54047

West Chester PA

American Helicopter Museum & Education Center, 1220 American Blvd, West Chester, PA 19380 • T: +1 610 4369600 • F: +1 610 4368642 • info@helicoptermuseum.org • www.helicoptermuseum.org •
C.E.O: *Greg Kennedy* •
Science&Tech Museum – 1993
Aeronautics 54048

Chester County Historical Society Museum, 225 N High St, West Chester, PA 19380 • T: +1 610 6924800 • F: +1 610 6924357 • cchs@chesterhistorical.org • www.chestercohistorical.org •
Pres./Dir.: *Ann Mintz* • Cur.: *Ellen Endslow* •
Local Museum – 1893
Local hist 54049

West Chicago IL

Kruse House Museum, 527 Main St, West Chicago, IL 60185, mail addr: POB 246, West Chicago, IL 60186 • T: +1 630 2310564 • krusehouse@earthlink.net • www.krusehousemuseum.org •
Pres.: *Lance Conkright* •
Local Museum – 1976
Local hist, housed in 1917 Kruse House 54050

West Chicago City Museum, 132 Main St, West Chicago, IL 60185 • T: +1 630 2313376, 2932266 • F: +1 630 2933028 • museum@westchicago.org • www.westchicago.org •
Dir.: *LuAnn Bombard* •
Local Museum – 1976
Town and railroad hist, farm items, genealogy files 54051

West Columbia TX

Varner-Hogg Plantation, 1702 N 13th St, West Columbia, TX 77486 • T: +1 979 3454656 • F: +1 979 3454412 • kandy.taylor.hille@tpwd.state.tx.us • www.tpwd.state.tx.us •
Exec. Dir.: *Robert Cook* •
Decorative Arts Museum – 1958
Decorative arts, archaeology, Sam Houston, Gov James Stephen Hogg 54052

West Covina CA

Hurst Ranch, 1227 S Orange Av, West Covina, CA 91791 • T: +1 626 9191133 • F: +1 626 9191133 • hurstranch@yahoo.com • www.hurstranch.com •
Pres./C.E.O.: *Bob Bordon* •
Local Museum – 1996
19th c furniture, tools, farm implements, kitchen utensils, Coca-Cola memorabilia, antique automobiles 54053

West Fargo ND

Cass County History Museum at Bonanzaville, 1351 W Main Av, West Fargo, ND 58078 • T: +1 701 2822822 • F: +1 701 2827606 • info@bonanzaville.com • www.bonanzaville.com •
Pres.: *Howard Barlow* • Exec. Dir.: *Steve Stark* •
Historical Museum – 1954
Pioneer and Indian artifacts, crafts, textiles,period cars, farm machinery 54054

West Frankfort IL

Frankfort Area Historical Museum, 2000 E Saint Louis St, West Frankfort, IL 62896 • T: +1 618 9326159 •
Pres.: *Mary Burton* •
Local Museum – 1972
Southern Illinois hist, farm utensils, tools, furnishings, toys 54055

West Glacier MT

Glacier National Park Museum, Glacier National Park, West Glacier, MT 59936 • T: +1 406 8887936 • F: +1 406 8887808 • deirdre-shaw@nps.gov •
Cur.: *Deirdre Shaw* •
Natural History Museum / Historical Museum / Archaeology Museum – 1930
Geologic, bird, insect & animal specimens coll 54056

West Hartford CT

Joseloff Gallery, University of Hartford, Hartford Art School, Harry Jack Gray Center, 200 Bloomfield Av, West Hartford, CT 06117 • T: +1 860 7684090 • F: +1 860 7685159 • zdavis@mail.hartford.edu •
Dir.: *Zina Davis* •
Fine Arts Museum / University Museum – 1970 54057

Museum of American Political Life, University of Hartford (closed to the public), 200 Bloomfield Av, West Hartford, CT 06117 • T: +1 860 7684090 • F: +1 860 7685159 • pritting@hartford.edu •
Dir.: *Rand Ashton* •
Historical Museum – 1989 54058

Noah Webster House - Museum of West Hartford History, 227 S Main St, West Hartford, CT 06107-3430 • T: +1 860 5215362 • F: +1 860 5214036 • vfzoe@snet.net •
Dir.: *Vivian F. Zoe* •
Historical Museum – 1965
Local hist 54059

Saint Joseph College Art Gallery, 1678 Asylum Av, West Hartford, CT 06117 • T: +1 203 2324571 • F: +1 203 2315754 • artgallery@sjc.edu • www.sjc.edu/artgallery •
Pres.: *Winifred E. Coleman* • Dir.: *Vincenza Uccello* •
Fine Arts Museum / University Museum – 1932 54060

Science Center of Connecticut, 950 Trout Brook Dr, West Hartford, CT 06119 • T: +1 806 2312824 • F: +1 806 2320705 • dburke@sciencecenterct.org • www.sciencecenterct.org •
Dir.: *Edward J. Forand* •
Science&Tech Museum – 1927
Science and technology, natural history 54061

West Henrietta NY

New York Museum of Transportation, 6393 E River Rd, West Henrietta, NY 14586, mail addr: POB 136, West Henrietta, NY 14586 • T: +1 585 5331113 • info@nymtmuseum.org • www.nymtmuseum.org •
Pres.: *T.H. Strang* •
Science&Tech Museum – 1974
Trolleys, steam locomotive, horse-drawn and highway vehicles, local transportation hist, model railroads – archive, library 54062

West Hollywood CA

Gallery 825, Los Angeles Art Association, 825 N La Cienega Blvd, West Hollywood, CA 90069 • T: +1 310 6528272 • F: +1 310 6529251 • gallery825@laaa.org • www.laaa.org •
Dir.: *Ashley Emenegger* • Asst. Dir.: *Sinéad Finnerty* •
Public Gallery / Fine Arts Museum – 1925
Art of Southern California 54063

MAK Center for Art and Architecture, Schindler House, 835 N Kings Rd, West Hollywood, CA 90069-5409 • T: +1 323 6511510 • F: +1 323 6512340 • office@makcenter.org • www.makcenter.org •
Dir.: *Kimberli Meyer* •
Fine Arts Museum – 1994 54064

New Image Art, 7906 Santa Monica Blvd, West Hollywood, CA 90046 • T: +1 323 6542192 • info@newimageartgallery.com • www.newimageartgallery.com •
Dir.: *Marsea Goldberg* •
Public Gallery – 1995 54065

Pacific Design Center, Museum of Contemporary Art LA, 8687 Melrose Av, West Hollywood, CA 90069 • T: +1 310 657800 • F: +1 310 6528576 • marketing@pacificdesigncenter.com • www.pacificdesigncenter.com •
Fine Arts Museum
Architecture and design exhibits, contemporary art 54066

West Lafayette IN

Purdue University Galleries, Physics Bldg, 252 Northwestern Av, West Lafayette, IN 47907-2036 • T: +1 765 4943061 • F: +1 765 4962817 • galleries@purdue.edu • www.sla.purdue.edu/galleries • Dir.: *Martin Craig* •
Fine Arts Museum / University Museum – 1978
Art gallery 54067

West Liberty OH

Piatt Castles, 10051 Township Rd 47, West Liberty, OH 43357, mail addr: POB 497, West Liberty, OH 43357 • T: +1 937 4652821 • F: +1 937 4657774 • macochee@2access.net • www.piattcastles.org • Pres.: *Margaret Piatt* •
Local Museum / Folklore Museum – 1912
Hist houses, Piatt family furnishings and objects 54068

West Liberty WV

Women's History Museum, 108 Walnut St, West Liberty, WV 26074 • T: +1 304 3367159 • F: +1 304 3367893 • j_v_schramm@hotmail.com • Dir.: *Jeanne Schramm* • Cur.: *Robert W. Schramm* •
Historical Museum – 1990
Women's hist 54069

West Nyack NY

Rockland Center for the Arts, 27 S Greenbush Rd, West Nyack, NY 10994 • T: +1 845 3580877 • F: +1 845 3580971 • info@rocklandartcenter.org • www.rocklandartcenter.org •
Public Gallery
School for the Arts 54070

West Orange NJ

Edison National Historic Site, Main St at Lakeside Av, West Orange, NJ 07052 • T: +1 973 7360550 • F: +1 973 7368496 • edis_superintendent@nps.gov • www.nps.gov/edis •
Head: *Maryanne Gerbauckas* •
Historical Museum – 1956
Industrial and scientific machinery and equip, 19th c dec arts, photos 54071

West Palm Beach FL

Historical Museum of Palm Beach County, 400 N Dixie Hwy, West Palm Beach, FL 33401 • T: +1 561 8324164 • F: +1 561 8327965 •
Pres.: *Ken Brower* • C.E.O.: *Kristin H. Gaspari* •
Local Museum – 1937 54072

Norton Museum of Art, 1451 S Olive Av, West Palm Beach, FL 33401 • T: +1 561 8325196 • F: +1 561 6594689 • pangbornj@norton.org • www.norton.org • Dir.: *Christina Orr-Cahall* •
Fine Arts Museum – 1940
Chinese and contemporary art 54073

Robert and Mary Montgomery Armory Art Center, 1703 Lake Av, West Palm Beach, FL 33401 • T: +1 561 8321776 • F: +1 561 8320191 • garrett@armoryart.org • www.armoryart.org •
C.E.O.: *Amelia Ostrosky* •
Public Gallery – 1987
Library 54074

South Florida Science Museum, 4801 Dreher Trail N, West Palm Beach, FL 33405 • T: +1 561 8321988, 8335368 • F: +1 561 8330551 • info@sfsm.org • www.sfsm.org •
Dir.: *James R. Rollings* •
Science&Tech Museum – 1959
Natural hist, paleontology, archaeology, astronomy, geology, concology 54075

U

West Park NY

Slabsides, John Burroughs Association, Off John Burroughs Dr, West Park, NY 12493, mail addr: POB 439, West Park, NY 12493 • T: +1 212 7695169 • F: +1 212 3137182 • breslof@amnh.org • research.amnh.org/burroughs •
C.E.O. & Pres.: *Dave Liddell* •
Historical Museum – 1921
House of the author, naturalist and poet John Burroughs (1837-1921), original interior 54076

West Point NY

Constitution Island - Warner House, Constitution Island, Warner House NY 10996, mail addr: POB 41, West Point, NY 10996 • T: +1 914 4468676 • F: +1 208 2470138 • rdekoster@constitutionisland.org • www.constitutionisland.org •
Dir.: *Richard de Koster* •
Historical Museum – 1916
Historic Warner house, Victorian home of the writers Susan and Anna Warner, Fort Constitution documentation – library, archives 54077

West Point Museum, United States Military Academy, Bldg 2110, West Point, NY 10996 • T: +1 845 9382203 • F: +1 845 9387478 • ym5796@usma.edu • www.usma.edu •
Dir.: *Michael E. Moss* • Cur.: *David M. Reel* (Art) • *Leslie D. Jensen* (Arms) • *Michael J. McAfee* (Uniforms and Military History) • *Richard H. Clark* (Design) • Cons.: *Paul R. Ackermann* •
Military Museum – 1854
Hist of Military events and personalities, coll of weapons, military art, American and European uniforms 54078

West Salem WI

Hamlin Garland Homestead, 257 W Garland St, West Salem, WI 54669 • T: +1 608 7861399 •
C.E.O. & Pres.: *Errol Kindschy* •
Local Museum – 1976
Historic house, musical items, fans 54079

Palmer/Gullickson Octagon Home, 360 N Leonard, West Salem, WI 54669 • T: +1 608 7861399 •
Pres.: *Errol Kindschy* •
Local Museum – 1976
Hist building, local hist 54080

West Sayville NY

Long Island Maritime Museum, 86 West Av, West Sayville, NY 11796 • T: +1 631 8544974 • F: +1 631 8544979 • jab@limaritime.org • www.limaritime.org •
Head: *R. Douglas Shaw* •
Special Museum – 1966
Maritime hist, small craft, US Life Saving service equip, shipwreck artifacts 54081

West Springfield MA

Josiah Day House, 70 Park St, West Springfield, MA 01090 • T: +1 413 7348322, 7327230 • www.west-springfield.ma.us/Attractions/dayhouse_home.htm •
Cur.: *Raymond Wellspeake* •
Local Museum – 1903
18th c furniture and accesoires, original Day family possessions 54082

Storrowton Village Museum, 1305 Memorial Av, West Springfield, MA 01089 • T: +1 413 2055051 • F: +1 413 2055054 • storrow@thebige.com • www.thebige.com •
Dir.: *Dennis Picard* •
Open Air Museum / Historical Museum – 1929 54083

West Union IA

Fayette County Historical Museum, 100 N Walnut St, West Union, IA 52175 • T: +1 319 4225797 •
Pres.: *Ruth Brooks* • Dir.: *Laura Janssen* •
Local Museum – 1975
History, genealogy 54084

Westbrook CT

Military Historians Museum, N Main St, Westbrook, CT 06498 • T: +1 860 3999460 • military.historians@snet.net •
Cur.: *Jackie Reid* •
Military Museum – 1949/2000
Military uniforms 54085

Westbury NY

Town of North Hempstead Museum, 210 Plandome Rd, Westbury, NY 11590 • T: +1 516 8697757 • www.northhempstead.com •
Pres.: *Dr. George Williams* •
Historical Museum – 1963
Agriculture, botany, herbarium, Indian artifacts, hist, transportation 54086

Westerly RI

Hoxie Gallery, Westerly Public Library, 44 Broad St, Westerly, RI 02891 • T: +1 401 5962877 • F: +1 401 5965600 • reference@westerlylibrary.org • www.westerlylibrary.org •
Dir.: *Kathryn T. Taylor* •
Public Gallery
Local artists 54087

Western Springs IL

Western Springs Museum, 4211 Grand Av, Western Springs, IL 60558 • T: +1 708 2469230 •
Cur.: *Betty L. Howard* •
Local Museum – 1967
Local history, housed in a Old Water Tower 54088

Westerville OH

Anti-Saloon League Museum, 126 S State St, Westerville, OH 43081 • T: +1 614 8827277 ext 160 • F: +1 614 8825369 • bweinhar@wpl.lib.oh.us •
Special Museum – 1990
Publications, fliers, posters, song books, former League's headquarter 54089

Hanby House, Westerville Historical Society, 160 W Main St, Westerville, OH 43081, mail addr: POB 1063, Westerville, OH 43081 • T: +1 614 8956017 •
Pres.: *Marilyn Gale* •
Local Museum – 1937
Music, furniture, household items of the Hanby family 54090

The Ross C. Purdy Museum of Ceramics, 735 Ceramic Pl, Westerville, OH 43081 • T: +1 614 8904700 • F: +1 614 8996109 • customerservice@ceramics.org • www.acers.org •
Dir.: *Yvonne Manring* •
Decorative Arts Museum – 1977
Ceramics 54091

Westfield MA

Jasper Rand Art Museum, 6 Elm St, Westfield, MA 01085 • T: +1 413 5687833 • F: +1 413 5681558 • www.westath.org/art.html •
Dir.: *Christopher Lindauist* •
Public Gallery – 1927
Visual art, photography and crafts 54092

Westfield NJ

Miller-Cory House Museum, 614 Mountain Av, Westfield, NJ 07090 • T: +1 908 2321776 • F: +1 908 2321740 • mc@westfieldnj.com • www.westfieldnj.com •
Local Museum – 1972
Period furniture, farm implements, cooking and craft utensils 54093

Westfield NY

McClurg Mansion, Chautaqua County Historical Society, Main and Portage Sts, Westfield, NY 14787 • T: +1 716 3262977 •
Pres. & Dir.: *James O'Brien* •
Local Museum – 1883
Agriculture, hist, Native Americam artifacts, genealogy 54094

Westfield WI

Marquette County Museum, Marquette County Historical Society, 125 Lawrence St, Westfield, WI 53964 •
Dir.: *Esther Brancel* • Cur.: *D. Sprain* • *Joannie Ingraham* •
Local Museum – 1979
Historic house 54095

Westminster MD

Carroll County Farm Museum, 500 S Center St, Westminster, MD 21157 • T: +1 410 8487775 • F: +1 410 8768544 • ccfarm@ccg.carr.org • ccgov.carr.org/farm •
Head: *Dottie Freeman* • Cur.: *Victoria Fowler* •
Public Gallery / Historic Site / Agriculture Museum – 1965 54096

Esther Prangley Rice Gallery, c/o McDaniel College, 2 College Hill, Peterson Hall, Westminster, MD 21157 • T: +1 410 8572595 • pio@mcdaniel.edu • www.mcdaniel.edu •
Dir.: *Michael Losch* •
Ethnology Museum / Fine Arts Museum
Egyptian, African, native American, Asian prints 54097

Historical Society Museum of Carroll County, 210 E Main St, Westminster, MD 21157-5225 • T: +1 410 8486494 • F: +1 410 8483596 • hscc@carr.org • www.carr.org/hscc •
Cur.: *Catherine Baty* •
Local Museum – 1939
Hist and culture, clothing, ceramics and glass, furniture, clocks 54098

Union Mills Homestead and Grist Mill, 3311 Littles Town Pike, Westminster, MD 21158 • T: +1 410 8482288 • F: +1 410 8482288 • ejss61@aol.com • www.carr.org/tourism •
Pres.: *James M. Shriver* • Dir.: *Jane S. Sewell* •
Science&Tech Museum – 1797
18th c mill an homestad containing original household and agricultural items 54099

Westminster VT

Westminster Historical Society, Main St, Westminster, VT 05158 •
Pres.: *Virginia Lisai* • Dir.: *Karen Larsen* •
Association with Coll – 1966
Hist of the Abenaque Machine Works 54100

Weston CT

Coley Homestead and Barn Museum, 104 Weston Rd, Weston, CT 06883 • T: +1 203 2261804, 5449636 • F: +1 203 2274268 •
Pres.: *Paul Deysenroth* •
Local Museum / Agriculture Museum – 1961
Agriculture, handicraft, furnishings – herb garden 54101

Museum of Art and History, Weston Town Hall, N Field Rd, Weston, CT 06883 • T: +1 203 2273375 • www.weston-ct.com •
Cur.: *James Daniel* •
Local Museum
Sculpture, American art hist, design for the 1999 memorial coin, early iron/steel technology 54102

The Museum of Films for Children, Weston Woods Museum, 389 Newtown Tpke, Weston, CT 06883 • T: +1 203 2228000 • F: +1 203 2222263 • uryy@optonline.net • www.westonwoodinstitute.org •
Dir./ Founder: *Morton Schindel* •
Special Museum / Historic Site – 1983
Communication techniques anf films for children, animation cells, equipment (end of 19th c-end of 20th c) 54103

Weston MA

Carney Gallery, Regis College Fine Arts Center, 235 Wellesley St, Weston, MA 0293 • T: +1 781 7687032/34 • F: +1 781 7687030 • facbo@regiscollege.edu • www.regiscollege.edu •
Dir.: *Rosemary Noon* •
Fine Arts Museum – 1993
Works of contemporary women artists 54104

Golden Ball Tavern Museum, 662 Boston Post Rd, Weston, MA 02493 • T: +1 781 8941751 • F: +1 781 8629178 • joanb5@aol.com • www.goldenballtavern.org •
Dir.: *Dr. Joan P. Bines* •
Local Museum – 1964
Historic house, blue and white china, glass, silver, clothing, Jones family 54105

Museum of the Weston Historical Society, 358 Boston Post Rd, Weston, MA 02493 • T: +1 781 8938870 • F: +1 781 8995305 • verandy@aol.com •
Pres.: *William Martin* • Cur.: *Dr. Vera Laska* •
Historical Museum – 1963
Local history, town reports – archives 54106

Spellman Museum of Stamps and Postal History, c/o Regis College, 235 Wellesley St, Weston, MA 02493 • T: +1 781 7687331 • F: +1 781 7687332 • info@spellman.org • www.spellman.org •
Exec. Dir.: *David W. Gregg* • Cur.: *George Norton* •
Special Museum – 1960
Stamps and covers, postal services – library 54107

Weston VT

Farrar-Mansur House and Old Mill Museum, Main St, Weston, VT 05161 • T: +1 802 8246781 • F: +1 802 8244072 •
Dir.: *Jean Lindeman* •
Local Museum – 1933
Local period artifacts, murals, dolls, portraits, family bibles 54108

Weston WV

Jackson's Mill Historic Area, 160 Jackson Mill Rd, Weston, WV 26452 • T: +1 304 2878206 • F: +1 304 2693409 • jmill@wvu.edu • www.jacksonmill.com •
C.E.O.: *David Mann* •
Historic Site – 1968
Local hist, Boyhood home of "Stonewall" Jackson 54109

Westport CT

Westport Historical Museum, 25 Avery Pl, Westport, CT 06880 • T: +1 203 2221424 • F: +1 203 2210981 • info@westporthistory.org • www.westporthistory.org •
Pres.: *Wally Woods* • C.E.O.: *John M. Lupton* •
Local Museum – 1899
Diorama of historic downtown, costumes – archives 54110

Westport WA

Westport Maritime Museum, 2201 Westhaven Dr, Westport, WA 98595 • T: +1 360 2680078 • F: +1 360 2681278 • westport.maritime@verizon.net • www.westportwa.com/museum •
C.E.O.: *Robert W. Pitzer* •
Local Museum / Historical Museum – 1985
Local and coast guard hist, shipwrecks, marine mammals 54111

Wethersfield CT

The Webb-Deane-Stevens Museum, 211 Main St, Wethersfield, CT 06109 • T: +1 860 5290612 • F: +1 860 5718636 • info@webb-deane-stevens.org • www.webb-deane-stevens.org •
Dir.: *Jennifer S. Eifrig* • *Jane Peake* (Education) • Cur.: *Donna Baron* •
Local Museum – 1919
Local hist 54112

Wethersfield Museum, 150 Main St, Wethersfield, CT 06109 • T: +1 860 5297656 • F: +1 860 5632609 • director@wethhist.org • www.wethhist.org •
Dir.: *Douglas R. Shipman* •
Historical Museum – 1932
Local hist 54113

Wewoka OK

Seminole Nation Museum, 524 S Wewoka Av, Wewoka, OK 74884 • T: +1 405 2575580 • F: +1 405 2573184 • semuseum@okplus.com • www.wewoka.com/museum.htm •
C.E.O. & Pres.: *Stephen D. Walker* • Exec.Dir.: *Ricard Ellwanger* • Chief Cur.: *Margaret Jane Norman* •
Folklore Museum / Ethnology Museum / Historical Museum – 1974
Seminoles' hist and life, Oklahoma Oil Boom and railroad hist – art gallery 54114

Weyauwega WI

Little Red School House Museum, 411 W High St, Weyauwega, WI 54983 • T: +1 920 8672500 •
Historical Museum – 1970 54115

Wharton TX

Wharton County Historical Museum, 3615 N Richmond Rd, Wharton, TX 77488 • T: +1 979 5322600 • F: +1 979 5320871 • wchm@wcnet.net • www.whartoncountymuseum.com •
Pres.: *Linda Joy Stovall* • C.E.O.: *Marvin Albrecht* •
Local Museum – 1979
6,000 hist photos of the county 54116

Wheat Ridge CO

Wheat Ridge Sod House and Museum, Log Cabin, First Post Office and Tool Shed, 4610 Robb St, Wheat Ridge, CO 80033, mail addr: P.O Box 1833, Wheat Ridge, CO 80034 • T: +1 303 4219111 • F: +1 303 4672539 • wrhissoc@aol.com •
Pres.: *Claudia R. Worth* •
Historical Museum – 1970
Sod house furnished 1880, Log Cabin furnished 1860, Post Office 1913, Museum Bldg 1910 – Research Library, Collection of Jefferson County Schools history and Jefferson County History 54117

Wheaton IL

Billy Graham Center Museum, 500 E College Av, Wheaton, IL 60187 • T: +1 630 7525909 • F: +1 630 7525916 • bgcmus@wheaton.edu • bgc.gospelcom.net/museum •
Religious Arts Museum – 1975
History of evangelism, revival and missions in America, Christian art 54118

Du Page County Historical Museum, 102 E Wesley St, Wheaton, IL 60187 • T: +1 630 6826549 • museum@dupageco.org • www.dupageco.org/museum •
Dir.: *Jody A. Crago* • Sen. Cur.: *Steph McGrath* •
Historical Museum – 1967
County history exhibits with changing costumes and hands-on history for children 54119

The First Division Museum at Cantigny, 1 S 151 Winfield Rd, Wheaton, IL 60187-6097 • T: +1 630 2608185 • F: +1 630 2609298 • fdmuseum@tribune.com • www.rrmtf.org/firstdivision •
Dir.: *John F. Votaw* •
Military Museum – 1960
Story of first infantry division from WW I, American millitary hist, Chicago area hist 54120

Robert R. McCormick Museum at Cantigny, 1 S 151 Winfield Rd, Wheaton, IL 60187 • T: +1 630 2608159 • F: +1 630 2608160 • emanyon@tribone.com • www.cantignypark.com •
Dir.: *Ellen Manyon* •
Historical Museum – 1955
Joseph Medill, editor of the Chicago Tribune, Col. Robert R.McCormick (grandson of Medill) editor and publisher of the Chicago Tribune, hist of journalism 54121

Wheaton History Center, 606 N Main St, Wheaton, IL 60187-4167 • T: +1 630 6829472 • F: +1 630 6829913 • wtnhistory@aol.com • www.wheaton.lib.il.us/whc •
Dir.: *Alberta Adamson* •
Local Museum – 1986
Local hist, slavery, underground railroad 54122

Wheaton MN

Traverse County Historical Society Museum, 1201 Broadway, Wheaton, MN 56296 • T: +1 320 5634110 • F: +1 302 5634823 • www.cityofwheaton.com •
Pres.: *Clarence Juelich* •
Local Museum – 1977
Local history & businesses, farming, toys, mounted animals 54123

Wheeling WV

The Museums of Oglebay Institute, Oglebay Park, Wheeling, WV 26003 • T: +1 304 2427272 • F: +1 304 2424203 • www.oionline.com •
Dir.: *Dr. Frederick A. Lambert* • *Holly H. McCluskey* •
Asst. Cur.: *Travis Zeik* •
Local Museum – 1930
Local hist, decorative art, china, glass 54124

West Virginia Independence Hall, 1528 Market St, Wheeling, WV 26003 • T: +1 304 2381300 • F: +1 304 2381302 • wvindependencehall@wvculture.org • www.wvculture.org •
Dir.: *Gerry Reilly* •
Local Museum – 1964
Historic bldg, furnishings of Gevernor's office, Wheeling and Statehood Conventions 54125

West Virginia Northern Community College Alumni Association Museum, 1704 Market St, Wheeling, WV 26003-3699 • T: +1 304 2335900 ext 4268 • F: +1 304 2320965 •
Dir.: *Zak Witcherly* •
Local Museum – 1984
Local hist, glass and glass production, old railway passenger station 54126

White GA

Weinman Mineral Museum, 51 Mineral Museum Dr, White, GA 30184 • T: +1 770 3860579 • F: +1 770 3860600 • www.weinmanmuseum.org •
Dir./Cur.: *Jose Santamaria* •
Natural History Museum – 1983
Georgia minerals, geology and mining hist 54127

White Plains NY

Westchester Art Workshop Gallery, Westchester Community College, State University of New York, County Center Bldg 196, White Plains, NY 10606 • T: +1 914 6840094 • F: +1 914 6840608 • wayne.kartzinel@sunywcc.edu •
Public Gallery 54128

White Springs FL

Stephen Foster Folk Culture Center, US 41 N, White Springs, FL 32096, mail addr: POB 666, White Springs, FL 32096 • T: +1 386 3972733 • F: +1 386 397.4262 • floridastateparks.org •
Folklore Museum – 1939
Paintings, musical instr, minstrel show material, Stephen Foster's folk songs 54129

White Sulphur Springs MT

Meagher County Historical Association Castle Museum, 310 Second Av SE, White Sulphur Springs, MT 59645 • T: +1 406 5472324 •
Local Museum – 1967
Furniture & furnishings, Indian artifacts, carriages in new bldg 54130

White Sulphur Springs WV

President's Cottage Museum, 330 W Main St, White Sulphur Springs, WV 24986 • T: +1 304 5361110 ext 7314/7198 • F: +1 304 5367854 • www.greenbrier.com •
Dir.: *Ted J. Kleisner* • Cur.: *Dr. Robert S. Conte* •
Local Museum – 1932
Local hist 54131

White Swan WA

Fort Simcoe, 5150 Fort Simcoe Rd, White Swan, WA 98952 • T: +1 509 8742372 • F: +1 509 8742351 • www.fortsimcoestatepark.com •
Dir.: *Cleve Pinnix* •
Military Museum
Hist of life at the fort, military and Indian exhibits 54132

Whiteville NC

North Carolina Museum of Forestry, 415 S Madison St, Whiteville, NC 28472 • T: +1 910 9144185 • F: +1 910 6410385 • web.eenorthcarolina.org •
Dir.: *Harry Warren* •
Natural History Museum / Agriculture Museum – 2000
NC forest natural hist 54133

Whitewater WI

Crossman Art Gallery, University of Wisconsin Whitewater, Center of the Arts, 950 W Main St, Whitewater, WI 53190 • T: +1 262 4725708 • F: +1 262 4722808 • flanagam@uww.edu • www.uww.edu •
Dir.: *Michael Flanagan* •
Public Gallery – 1970
Contemp European and American art, outsider and folk art 54134

Whitewater Historical Museum, 275 N Tratt St, Whitewater, WI 53190 • T: +1 262 4736820 •
C.E.O.: *Dr. Alfred S. Kolmos* •
Local Museum – 1974
Local artifacts, clothing, tools, glass, china, furniture 54135

Whitingham VT

Whitingham Historical Museum, Stimpson Hill, Whitingham, VT 05361 • T: +1 802 3682448 •
Pres.: *Robert Coombs* •
Local Museum – 1973
Local costumes, hist papers, textiles 54136

Whittier CA

Rio Hondo College Art Gallery, 3600 Workman Mill Rd, Whittier, CA 90601-1699 • T: +1 562 9083471 • F: +1 562 9083446 • chansen@riohondo.edu • www.riohondo.edu •
Dir.: *Sheila Lynch* •
Fine Arts Museum – 1967
Contemporary paintings and graphics 54137

Whittington IL

Illinois Artisans and Visitors Centers, 14967 Gun Creek Trail, Whittington, IL 62897-1000 • T: +1 618 6292220 • F: +1 618 6292704 • www.museum.state.il.us •
Dir.: *Mary Lou Galloway* •
Decorative Arts Museum – 1990
Craft work of Illinois artisans 54138

Wichita KS

Decorative Arts Collection, Museum of Decorative Painting, 393 N McLean Blvd, Wichita, KS 67203-5968 • T: +1 316 2699300 ext 103 • F: +1 316 2699191 • sdp@decorativepainters.org • www.decorativepainters.com •
Chm.: *Andy B. Jones* • Sc. Staff: *Kay Blair* (Admin.) • *Jan Vavra* (Art collection) •
Fine Arts Museum – 1982
Decorative painting, Norwegian Rosemaling, early American Folk art 54139

Edwin A. Ulrich Museum of Art, Wichita State University, 1845 Fairmount, Wichita, KS 67260-0046 • T: +1 316 9783664 • F: +1 316 9783898 • ulrich@wichita.edu • www.ulrich.wichita.edu •
Dir.: *Dr. Patricia McDonnell* • Cur.: *Emily Stamey* •
Fine Arts Museum / University Museum – 1974
Outdoor sculpture coll, 20th-21st c art, paintings, works on paper & electronic art, contemp artist 54140

Exploration Place, 300 N McLean Blvd, Wichita, KS 67203 • T: +1 316 2633373 • F: +1 316 2634545 • www.exploration.org •
Dir.: *Marty Sigwing* •
Natural History Museum / Science&Tech Museum – 1984
Interactive exhib 54141

Gallery XII, 412 E Douglas Av, Wichita, KS 67202 • T: +1 316 2673607 •
Dir.: *Doug Billings* •
Fine Arts Museum – 1977 54142

Great Plains Transportation Museum, 700 E Douglas, Wichita, KS 67202 • T: +1 316 2630944 • www.gptm.org •
Science&Tech Museum – 1983 54143

Kansas African American Museum, 601 N Water St, Wichita, KS 67203 • T: +1 316 2627651 • F: +1 316 2656953 •
Exec. Dir.: *Eric Key* •
Fine Arts Museum – 1972
Visual art forms of African American life and culture – library 54144

Kansas Aviation Museum, 3350 S George Washington Blvd, Wichita, KS 67210 • T: +1 316 6839242 • F: +1 316 6830573 • ksaviation@kansasaviationmuseum.org • www.kansasaviationmuseum.org •
Exec. Dir.: *Teresa Day* •
Science&Tech Museum – 1991
Vintage airplanes manufactured in Kansas, photographic coll 54145

Mid-America All-Indian Center Museum, 650 N Seneca, Wichita, KS 67203 • T: +1 316 2625221 • F: +1 316 2624216 • cdendurent@wichita.gov • www.theindiancenter.com •
Pres.: *Dr. David Hughes* • C.E.O.: *Shelly Berger* •
Ethnology Museum / Folklore Museum – 1975
Native American life, art & religion, old Indian council grounds 54146

Old Cowtown Museum, 1865 W Museum Blvd Wichita, KS 67203, mail addr: 1871 Sim Park Dr, Wichita, KS 67203 • T: +1 316 2191871 • F: +1 316 2642937 • cowtown@southwind.net • www.oldcowtown.org •
Dir.: *JaLayne Wray* • Sc. Staff: *Vanya Scott* (Registrar, Collections) •
Local Museum / Open Air Museum – 1950
Open air living history, 1865-1880 era of Wichita and Sedgwick County Kansas 54147

Omnisphere and Science Center, 220 S Main, Wichita, KS 67202 • T: +1 316 3379174 •
Dir.: *Janice McKinney* •
Science&Tech Museum – 1976
Science 54148

Whittier Fine Arts Gallery, c/o Friends University, 2100 W University St, Wichita, KS 67213 • T: +1 316 2955877 • F: +1 316 2955656 •
Cur.: *Annie Lowrey* •
Fine Arts Museum / University Museum – 1963 54149

Wichita Art Museum, 1400 W Museum Blvd, Wichita, KS 67203 • T: +1 316 2684921 • F: +1 316 2684980 • poyntert@wichitaartmuseum.org • www.wichitaartmuseum.org •
Dir.: *Charles K. Steiner* • Chief Cur.: *Stephen Gleissner* •
Fine Arts Museum – 1935
American Art 54150

Wichita Center for the Arts, 9112 E Central, Wichita, KS 67206 • T: +1 316 6342787 • F: +1 316 6340593 • arts@wcfta.com • www.wcfta.com •
Exec. Dir.: *Howard W. Ellington* •
Fine Arts Museum – 1920
Paintings, drawings, fine prints, enamels, sculpture, pottery 54151

Wichita-Sedgwick County Historical Museum Association, 204 S Main, Wichita, KS 67202 • T: +1 316 2659314 • F: +1 316 2659319 • wschm@wichitahistory.org • www.wichitahistory.org •
Dir.: *Robert A. Puckett* •
Association with Coll – 1939
Local history, old City Hall, Indian artifacts, industrial 54152

Wichita Falls TX

Kell House Museum, 900 Bluff St, Wichita Falls, TX 76301 • T: +1 940 7232712 • F: +1 940 7675424 • kellhcurator@aol.com • www.wichitaheritage.org •
Cur.: *Ralph Gibson* •
Historical Museum – 1981
Hist from 1900-1980, fabrics, decorative arts 54153

Wichita Falls Museum and Art Center, 2 Eureka Circle, Wichita Falls, TX 76308 • T: +1 940 6920923 • F: +1 940 6965358 • contact@wfmac.org •
Dir.: *Jeff Desborough* • Cur.: *Danny Bills* •
Fine Arts Museum / Public Gallery – 1965
Contemporary American prints, art galleries 54154

Wickenburg AZ

Desert Caballeros Western Museum, 21 N Frontier St, Wickenburg, AZ 85390 • T: +1 520 6842272 • F: +1 520 6845794 • info@westernmuseum.org • www.westernmuseum.org •
Dir.: *Royce Kardinal* •
Historical Museum / Folklore Museum – 1960
Western art, Native American coll, mining, minerals 54155

Wickliffe KY

Wickliffe Mounds, Museum of Native American Village, 94 Green St, Wickliffe, KY 42087 • T: +1 270 3353681 • wmounds@brtc.net • campus.murraystate.edu/org/wmrc/wmrc.htm •
Archaeology Museum – 1932
Archaeology 54156

Wilber NE

Wilber Czech Museum, 102 W Third, Wilber, NE 68465 • T: +1 402 8212183 •
Dir.: *Irma Ourecky* •
Local Museum / Folklore Museum – 1962
Czechoslovakian cultures, dolls, clothing, pictures 54157

Wilberforce OH

National Afro-American Museum and Cultural Center, 1350 Brush Row Rd, Wilberforce, OH 45384-0578 • T: +1 937 3764944 ext. 110 • F: +1 937 3762007 • www.ohiohistory.org •
C.E.O.: *Vernon Courtney* • Cur.: *Dr. Floyd R. Thomas* •
Historical Museum / Folklore Museum – 1972
African American history 54158

Wild Rose WI

Pioneer Museum, Main St, Wild Rose, WI 54984 •
Pres.: *Pam Anderson* • Cur.: *Margaret Walters* •
Local Museum – 1964
Pioneer hist, costumes, agriculture, hist bldgs 54159

Wildwood MO

River Hills Park Museum, 800 Guy Park Dr, Wildwood, MO 63005 • T: +1 314 4583813 • F: +1 314 4589105 •
Head: *George Hosack* •
Natural History Museum – 1972 54160

Wildwood NJ

Boyer Museum, 3907 Pacific Av, Wildwood, NJ 08260 • T: +1 609 5230277 •
Dir.: *Robert J. Scully* •
Local Museum – 1963
Local, county and state hist 54161

Wilkes-Barre PA

Luzerne County Museum, 69 S Franklin St, Wilkes-Barre, PA 18701 • T: +1 570 8236244 • F: +1 570 8239011 • tonybrooks@luzernehistory.org • www.luzernehistory.org •
Pres.: *Frank E.P. Conyngham* • C.E.O.: *Anthony T.P. Brooks* •
Historical Museum / Natural History Museum / Local Museum – 1858
History, geology, archaeology, photography – Library and Manuscripts 54162

Sordoni Art Gallery, Wilkes University, 150 S River St, Wilkes-Barre, PA 18766 • T: +1 570 4084325 • F: +1 570 4087733 • bernier@wilkes.edu • www.wilkes.edu/arts/sordoni.asp •
C.E.O.: *Ronald R. Bernier* •
Fine Arts Museum – 1973
19th and 20th c American art 54163

Williamsburg VA

Abby Aldrich Rockefeller Folk Art Museum, 325 Francis St, Williamsburg, VA 23185 • T: +1 757 2291000 • F: +1 757 56658804 • museums@cwf.org • www.colonialwilliamsburg.org •
Dir.: *Carolyn Weekley* • Cur.: *Jan Gilliam* •
Fine Arts Museum – 1957 54164

Bassett Hall, 522 E Francis St, Williamsburg, VA 23185 • T: +1 757 2291000 • F: +1 757 2207173 • lkally@cwf.org • www.history.org •
Dir.: *Carolyn J. Weekley* •
Decorative Arts Museum – 1979
American folk art, ceramics, textiles, decorative arts 54165

Carter's Grove Visitor Center, 8797 Pocahontas Trail, Williamsburg, VA 23185 • T: +1 757 2291000 • F: +1 757 2207173 • lkelley@cwf.org • www.history.org •
Dir.: *Carolyn J. Weekley* •
Historical Museum – 1964
Archaeological coll 54166

Colonial Williamsburg, 134 N Henry St, Williamsburg, VA 23185 • T: +1 757 2291000 • F: +1 757 2207286 • shart@cwf.org • www.colonialwilliamsburg.org •
Pres.: *Colin Campbell* •
Local Museum – 1926
Historic town, archaeology, manuscripts, paintings, sculpture 54167

DeWitt Wallace Decorative Arts Museum, 325 Francis St, Williamsburg, VA 23185, mail addr: POB 1776, Williamsburg, VA 23187-1776 • T: +1 757 2207984 • F: +1 757 5658804 • mcotrill@cwf.org • www.history.org •
Decorative Arts Museum – 1985
English and American decorative arts from 1600-1830 ... 54168

Jamestown Settlement, Jamestown-Yorktown Foundation, Rte 31 S, Williamsburg, VA 23187, mail addr: POB 1607, Williamsburg, VA 23187-1607 • T: +1 757 8871776, 2534838 • F: +1 757 2535299 • www.historyisfun.org •
Dir.: *Philip G. Emerson* •
Historical Museum / Decorative Arts Museum / Fine Arts Museum – 1957/1976
Furniture, silver, paintings, prints, 18th c revolutionary war artifacts, Powhatan Indians America's first permanent English colony ... 54169

Muscarelle Museum of Art, College of William and Mary, Jamestown Rd, Williamsburg, VA 23185 • T: +1 757 2212710 • F: +1 757 2212711 • bgkelm@wm.edu • www.wm.edu/muscarelle •
Dir./Cur.: *Ann C. Madonia* •
Fine Arts Museum – 1982
16th-20th c Western paintings, works on Paper, sculpture & Asian, Africa, Islamic & American Contemp Art ... 54170

This Century Art Gallery, 219 North Boundary St, Williamsburg, VA 23185 • T: +1 757 2294949 • F: +1 757 2585624 • kirbyscorpio@aol.com • www.thiscenturyartgallery.org •
Pres.: *Michael Kirby* •
Fine Arts Museum – 1959
Art & crafts of contemporary artists ... 54171

Williamsburg Va

Presidential Pet Museum, 211 water country parkway, Williamsburg, Va 23185 • T: +1 757 2501121 • F: +1 757 220203 • info@presidentialpetmuseum.com • www.presidentialpetmuseum.com •
C.E.O.: *Claire McLean* •
Special Museum – 1999
Memorabilia and artifacts related to U.S. Presidents and their pets – library ... 54172

Williamsport PA

Thomas T. Taber Museum, 858 W 4th St, Williamsport, PA 17701-5824 • T: +1 570 3263326 • F: +1 570 3263689 • lchsmuseum@verizon.net • www.tabermuseum.org •
C.E.O., Dir. & Cur.: *Sandra B. Rife* • Cur.: *Scott Sagar* •
Local Museum / Historical Museum – 1895
Regional hist and archaeology ... 54173

Williamstown MA

Chapin Library of Rare Books, Williams College, Stetson Hall, 25 Hopkins Hall Dr, Williamstown, MA 01267 • T: +1 413 5972462 • F: +1 413 5972929 • chapin.library@williams.edu • www.williams.edu/resources/chapin •
Cust.: *Robert L. Volz* •
Library with Exhibitions – 1923
9th-20th c manuscripts, rare printed books, bookplates, 15th-20th c prints, original artworks for book illustrations ... 54174

Sterling and Francine Clark Art Institute, 225 South St, Williamstown, MA 01267 • T: +1 413 4589545 • F: +1 413 4582318 • info@clarkart.edu • www.clarkart.edu •
Dir.: *Michael Conforti* • Sen. Cur.: *Richard Rand* • Cur.: *Brian Allen* (American Paintings) • *James A. Ganz* (Prints, Drawings, Photographs) • Sc. Staff: *Michael Cassin* (Education) •
Fine Arts Museum – 1950
Decorative art, 19th c sculptures, photography, Old Master paintings, drawings and prints – library, auditorium ... 54175

Williams College Museum of Art, 15 Lawrence Hall Dr, Williamstown, MA 01267-2566 • T: +1 413 5972037, 5972429 • F: +1 413 4589017 • WCMA@williams.edu • www.williams.edu/WCMA/ •
Dir.: *Lisa Corrin* • Cur.: *Nancy Mowll Mathews* (19th-20th c Art) • *Deborah Menaker Rothschild* (Modern and Contemporary Art) • *Vivian Patterson* (Collections) •
Fine Arts Museum / University Museum – 1926
American and European art, modern and contemporary art, South Asian sculpture 9th-15th c, African art, Indian painting and sculpture – auditorium ... 54176

Williamstown WV

Fenton Glass Museum, 420 Caroline Av, Williamstown, WV 26187 • T: +1 304 3757772 • F: +1 304 3756459 • askfenton@fentonartglass.com • www.fentonartglass.com •
Dir.: *Dr. Frank M. Fenton* •
Decorative Arts Museum – 1977
Glass from upper Ohio valley 1880-1980, hist of Fenton Art Glass ... 54177

Willimantic CT

AKUS Gallery, Eastern Connecticut State University, 83 Windham St, Willimantic, CT 06226 • T: +1 860 4654659 • F: +1 860 4564647 • callism@easternct.edu • www.easternct.edu/depts/akus •
Dir.: *Marion M. Callis* •
Public Gallery ... 54178

Windham Textile and History Museum, 157 Union-Main St, Willimantic, CT 06226 • T: +1 860 4562178 • millmuseum@hotmail.com • www.millmuseum.org •
Dir.: *Beverly York* •
Historical Museum / Science&Tech Museum – 1985
Textile hist ... 54179

Williston ND

Fort Buford, 15349 39th Ln NW, Williston, ND 58801 • T: +1 701 5729034 • F: +1 701 5726509 • ghesser@state.nd.us • www.state.nd.us/hist/ •
Dir.: *Merl Paaverud* •
Military Museum – 1931 ... 54180

Fort Union Trading Post, 15550 Hwy 1804, Williston, ND 58801 • T: +1 701 5729083 • F: +1 701 5727321 • andrey_barnhart@nps.gov • www.nps.gov/fous •
Dir.: *Andrew Barta* • Cur.: *Audrey Barnhart* •
Historical Museum – 1966
Archaeological specimens, fur trading ... 54181

Frontier Museum, 6330 2nd Av W, Williston, ND 58801 • T: +1 701 5729751 •
Pres.: *Ruth Robinson* •
Local Museum – 1958
History, archaeology, decorative art, military ... 54182

Willits CA

Mendocino County Museum, 400 E Commercial St, Willits, CA 95490 • T: +1 707 4592736 • F: +1 707 4597836 • www.co.mendocino.ca.us/museum •
Dir.: *Herbert E. Pruett* • Cur.: *Elaine Hamby* •
Local Museum – 1972
Local hist, vehicles, tools, furnishings, clothing, logging and railroad equip – archives ... 54183

Willmar MN

Kandiyohi County Historical Society Museum, 610 Hwy 71 NE, Willmar, MN 56201 • T: +1 320 2351881 • F: +1 320 2351881 • kandhist@msn.com • www.kandimuseum.com •
Exec. Dir.: *Mona Nelson-Balcer* •
Local Museum – 1898
Local hist, transportation, Indian artifacts – archives, library ... 54184

Willow Street PA

Hans Herr House and Museum, 1849 Hans Herr Dr, Willow Street, PA 17584 • T: +1 717 4644438 • info@hansherr.org • www.hansherr.org •
Dir.: *Doug Nyce* •
Historical Museum – 1974
Mennonite rural life, agricultural items ... 54185

Wilmette IL

Wilmette Historical Museum, 609 Ridge Rd, Wilmette, IL 60091 • T: +1 847 8537666 • F: +1 847 8537706 • husseyk@wilmette.com • www.wilmette.com/ •
Dir.: *Kathy Hussey-Arntson* •
Local Museum – 1947
Local hist, costumes, photos ... 54186

Wilmington CA

The Banning Museum, 401 E Main St, Wilmington, CA 90744 • T: +1 310 5487777 • F: +1 310 5482644 • info@banningmuseum.org • www.banningmuseum.org •
Dir.: *Michael Sanborn* •
Decorative Arts Museum – 1974
Decorative arts, costume, furniture – library ... 54187

Drum Barracks Civil War Museum, 1052 Banning Blvd, Wilmington, CA 90744 • T: +1 310 5487509 • susan.ogle@lacity.org • www.drumbarracks.org •
Dir.: *Susan Ogle* •
Military Museum / Historical Museum – 1987
Military, civil war, weapons, Southern Califonia history – library ... 54188

Wilmington DE

Delaware Art Museum, 2301 Kentmere Pkwy, Wilmington, DE 19806 • T: +1 302 5719590 • F: +1 302 4281049 • info@delart.org • www.delart.org •
Exec. Dir.: *Danielle Rice* • Cur.: *Joyce K. Schiller* • Sc. Staff: *Mary F. Holahan* (Collections) • *Margaretta S. Frederick* (Bancroft Coll) • *Heather Campbell Coyle* •
Fine Arts Museum – 1912
American 19th-20th c art, English Pre-Raphaelite art ... 54189

Delaware Center for the Contemporary Arts, 200 S Madison St, Wilmington, DE 19801 • T: +1 302 6566466 • F: +1 302 6566944 • info@thedcca.org • www.thedcca.org •
Dir.: *Maxine Gaiber* •
Fine Arts Museum
National and regional arts ... 54190

Delaware Museum, 504 Market St, Wilmington, DE 19801 • T: +1 302 6560637 • hsd@hsd.org • www.hsd.org •
Dir.: *Barbara E. Benson* •
Local Museum – 1864
Local hist ... 54191

Delaware Museum of Natural History, 4840 Kennett Pike, Wilmington, DE 19807-0937 • T: +1 302 6589111 • F: +1 302 6582610 • ghalfpenny@delmnh.org • www.delmnh.org •
Dir.: *Geoff Halfpenny* •
Natural History Museum – 1957
Natural hist ... 54192

The Hagley Museum, 298 Buck Rd E, Wilmington, DE 19807-0630 • T: +1 302 6582400 • F: +1 302 6580568 • danmuir@udel.edu • www.hagley.org •
Dir.: *George L. Vogt* • Cur.: *Debra Hughes* •
Historical Museum / Science&Tech Museum – 1952
History, technology ... 54193

Hagley Museum and Library, 298 Bruck Rd E, Wilmington, DE 19807 • T: +1 302 6582400 • F: +1 302 6580568 • danmuir@udel.edu • www.hagley.org •
Cur.: *Jon Williams* (Photographs, prints) • *Debra Hughes* (Collection, Exhibits) • Cons.: *Ebenezer Kotei* •
Historical Museum / Science&Tech Museum – 1952
Industrial, economic, social and technology hist, archaeology ... 54194

Hendrickson House Museum and Old Swedes Church, 606 Church St, Wilmington, DE 19801 • T: +1 302 6525629 • F: +1 302 6528615 • oldswedes@aol.com • www.oldswedes.org •
Pres.: *Max Dooley* • C.E.O.: *Jo Thompson* •
Local Museum – 1947
Historic house and nation's oldest church ... 54195

Nemours Mansion and Gardens, 1600 Rockland Rd, Wilmington, DE 19803 • T: +1 302 6516912 • F: +1 302 6516933 • tours@nemours.org • www.nemoursmansion.org •
Exec. Dir.: *Grace Gary* •
Historical Museum / Fine Arts Museum – 1977
Historic house and art coll from the 14th-19th c 54196

Rockwood Museum, 610 Shipley Rd, Wilmington, DE 19809 • T: +1 302 7614340 • F: +1 302 7614345 • pnond@co.new-castle.de.us • www.rockwood.org •
Decorative Arts Museum / Local Museum – 1976
Local hist ... 54197

Wilmington NC

Battleship North Carolina, Battleship Dr, Eagles Island, Wilmington, NC 28402 • T: +1 910 2515797 • F: +1 910 2515807 • ncbb55@battleshipnc.com • www.battleshipnc.com •
Dir.: *David R. Scheu* •
Military Museum – 1961
Naval artifacts of WW II, U.S. Navy archival material ... 54198

Bellamy Mansion Museum of History and Design Arts, 503 Market St, Wilmington, NC 28401 • T: +1 910 2513700 • F: +1 910 7638154 • correspondence@bellamymansion.org • www.bellamymansion.org •
Exec.Dir.: *Gene Ayscue* •
Decorative Arts Museum / Historical Museum / Fine Arts Museum – 1993
Hist House Museum with original furniture, textiles and family material ... 54199

Burgwin-Wright Museum, 224 Market St, Wilmington, NC 28401 • T: +1 910 7620570 • F: +1 910 7628750 • burgwinw@bellsouth.net • www.geocities.com/PicketFence/Garden/4354 •
Dir.: *Susan E. Hart* •
Historical Museum – 1770
Colonial townhome, English and American antiques and works of art (18th-19th c) ... 54200

Cape Fear Museum, 814 Market St, Wilmington, NC 28401-4731 • T: +1 910 3414350 • F: +1 910 3414037 • ssullivan@nhcgov.com • www.capefearmuseum.com •
Dir.: *Ruth Haas* • Cur.: *Barbara Rowe* •
Historical Museum / Natural History Museum – 1898
History and culture of Cape Fear area, furniture, clothings, military material ... 54201

Louise Wells Cameron Art Museum, 3201 S 17th St, Wilmington, NC 28412 • T: +1 910 3955999 • F: +1 910 3955030 • www.cameronartmuseum.com •
Dir.: *C. Reynolds Brown* • Cur.: *Anne G. Brennan* •
Fine Arts Museum – 1962
NC paintings, sculpture, decorative arts and works on paper ... 54202

Lower Cape Fear Historical Society Museum, 126 S Third St, Wilmington, NC 28401 • T: +1 910 7620492 • F: +1 910 7635869 • info@latimerhouse.org • www.latimerhouse.org •
Pres.: *John Golden* • Exec. Dir.: *Candace McGreevy* •
Decorative Arts Museum / Local Museum – 1956
Original furnishings and decorative art ... 54203

Poplar Grove Historic Plantation, 10200 U.S. Hwy 17 N, Wilmington, NC 28411 • T: +1 910 6864868 • F: +1 910 6864309 • pgp@poplargrove.com • www.poplargrove.com •
Pres.: *Chris Wilcox* • Dir.: *Nancy Simon* •
Local Museum – 1980
Historic house, 19th c furnishings, agricultural implements ... 54204

Wilmington Railroad Museum, 501 Nutt St, Wilmington, NC 28401 • T: +1 910 7632634 • F: +1 910 7637655 • info@wrrm.org • www.wrrm.org •
Pres.: *Dr. Frank E. Funk* • Exec. Dir.: *Sadie Ann Hood* •
Special Museum – 1979
Railroad ... 54205

Wilmington OH

Clinton County Historical Society Museum, 149 E Locust St, Wilmington, OH 45177 • T: +1 937 3824684 • F: +1 937 3825634 • cchs@core.com • clintoncountyhistory.orgh •
Dir.: *Kay Fisher* •
Historical Museum – 1948
Quaker heritage and settlement of the county, sculpture, paintings ... 54206

Wilmore KY

Student Center Gallery, Asbury College, 1 Macklem Dr, Wilmore, KY 40390 • T: +1 859 8583511 ext 2242 • F: +1 606 8583921 • pr@asbury.edu • www.asbury.edu •
Chm.: *Kevin S. Sparks* •
Public Gallery ... 54207

Wilson AR

Hampson Archeological Museum, 2 Lake Dr, Wilson, AR 72395, mail addr: POB 156, Wilson, AR 72395 • T: +1 870 6558622 • hampsonarcheologicalmuseum@arkansas.com • www.arkansasstateparks.com •
Head: *Corinne Fletcher* •
Archaeology Museum – 1961
Late Mississippian culture, archaeology ... 54208

Wilson KS

House of Memories Museum, 415 27th St, Wilson, KS 67490-0271 • T: +1 785 6583505, 6583343 •
Dir.: *Jean T. Kingston* •
Historical Museum / Decorative Arts Museum – 1986
Czech heritage, trophies, hair wreath, tools, glassware for Czechoslovakia, costumes ... 54209

Wilson NC

Barton Museum, Whitehead and Gold St, Wilson, NC 27893, mail addr: c/o Barton College, POB 5000, Wilson, NC 27893 • T: +1 252 3996477 • F: +1 252 3996571 • www.barton.edu •
Dir.: *Chris Wilson* •
Fine Arts Museum / University Museum – 1960
Ceramics, recent drawings, painting, prints, sculpture, African masks ... 54210

Imagination Station Science Museum, 224 E Nash St, Wilson, NC 27893, mail addr: POB 2127, Wilson, NC 27894-2127 • T: +1 252 2915113 • F: +1 252 2912968 • mail@imaginescience.org • www.imaginescience.org •
C.E.O.: *Karl L. McKinnon* •
Natural History Museum / Science&Tech Museum – 1989 ... 54211

Wilson NY

Wilson Historical Museum, 645 Lake St, Wilson, NY 14172 • T: +1 716 7519886 • F: +1 716 7516141 • agaffiliat@aol.com •
Cur.: *Dorothy Maxfield* •
Local Museum – 1972
Woodworking tools, woodcarving, railroad artifacts, farm tools and implements ... 54212

Wilton CT

Weir Farm, 735 Nod Hill Rd, Wilton, CT 06897 • T: +1 203 8341896 • F: +1 203 8342421 • wefa_interpretation@nps.gov • www.nps.gov/wefa •
Local Museum – 1990
Art and history ... 54213

Wilton Historical Museums, 224 Danbury Rd, Wilton, CT 06897 • T: +1 203 7627257 • F: +1 203 7623297 • info@wiltonhistorical.org • www.wiltonhistorical.org •
Dir.: *Marilyn Gould* •
Local Museum – 1938
Local hist ... 54214

Wilton NH

Frye's Measure Mill, 12 Frye Mill Rd, Wilton, NH 03086 • T: +1 603 6546581, 6545345 • www.fryesmeasuremill.com •
Pres.: *Pamela Savage* • C.E.O.: *Harland Savage* •
Historical Museum / Science&Tech Museum – 1858
Industry, measure mill, shaker, colonial boxes ... 54215

Wilton NY

Ulysses S. Grant Cottage State Historic Site, Mount McGregor, Wilton, NY 12831, mail addr: POB 990, Saratoga Springs, NY 12866-0897 • T: +1 516 5878277 • grantcottage@townofwilton.com • www.museumsofsaratoga.com •
Historic Site – 1890
Cottage where Gen. Ulysses S. Grant spent the last weeks of his life, memorabilia, furnishings ... 54216

Wimberley TX

Pioneer Town, 333 Wayside Dr, Wimberley, TX 78676 • T: +1 512 8473289 • F: +1 512 8476705 • Dir.: *Raymond L. Czichos* • Local Museum – 1956 Reproduction of c.1880 old West Town 54217

Winchendon MA

Winchendon Historical Museum, 151 Front St, Winchendon, MA 01475 • T: +1 508 2972142 • historical@trysb.net • Cur.: *Shirley Parks* • *Julia White* • Local Museum – 1930 Local history, toys – library, archives 54218

Winchester IN

Randolph County Historical Museum, 416 S Meridian, Winchester, IN 47394 • T: +1 317 5841334 • Dir.: *Marjorie Birtwhistle* • Local Museum – 1968 Clothes, dishes, photos, tools, books, furniture and furnishings 54219

Winchester MA

Griffin Museum of Photography, 67 Shore Rd, Winchester, MA 01890 • T: +1 781 7291158 • F: +1 781 7212765 • photos@griffinmuseum.org • www.griffinmuseum.org • Exec. Dir.: *Blake Fitch* • Fine Arts Museum – 1992 Photographic art 54220

Winchester TN

Franklin County Old Jail Museum, 400 Dinah Shore Blvd, Winchester, TN 37398 • T: +1 931 9670524 • Head: *Nancy Hall* • Local Museum – 1973 Hist of Franklin County, tools, pictures, household articles, records 54221

Winchester VA

Winchester-Frederick County Historical Society, 1340 S Pleasant Valley Rd, Winchester, VA 22601 • T: +1 540 6626550 • F: +1 540 6626991 • wfchs@shentel.net • www.winchesterhistory.org • Pres.: *Robert F. Boxley* • Association with Coll – 1930 54222

Winder GA

Fort Yargo, 201 S Broad St, Winder, GA 30680 • T: +1 770 8673489 • F: +1 770 8677517 • fort_yargo_park@dnr.state.ga.us • www.gastateparks.org • Historical Museum / Historic Site – 1954 Restored blockhouse used during the Creek Indian Wars 54223

Windom MN

Cottonwood County Historical Society, 812 Fourth Av, Windom, MN 56101 • T: +1 507 8311134 • F: +1 507 8312665 • cchs@windomnet.com • www.rootsweb.ancestry.com • Dir.: *Linda Fransen* • Association with Coll – 1901 Guns, clothing, spinning wheels, artifacts pertaining to Cottonwood County, incl musical instr, early pioneer coll, post offiice 54224

Window Rock AZ

Navajo Nation Museum, Hwy 264 and PO Loop Rd, Window Rock, AZ 86515, mail addr: POB 1840, Window Rock, AZ 86515 • T: +1 928 8717941 • F: +1 928 8717942 • info@navajonationmuseum.org • www.navajonationmuseum.org • Dir.: *Geoffrey I. Brown* • Cur.: *Clarenda Begay* • Historical Museum / Fine Arts Museum – 1961 Anthropology, textiles, pottery, jewelry, Navajo Indian art and history – library 54225

Windsor NC

Historic Hope Plantation, 132 Hope House Rd, Windsor, NC 27983 • T: +1 252 7943140 • F: +1 252 7945583 • hopeplantation@coastalnet.com • www.hopeplantation.org • Pres.: *Dr. John Hill* • Cur.: *Gregory Tyler* • Decorative Arts Museum / Local Museum / Historical Museum / Historic Site – 1965 Hist bldgs, furniture, decorative arts – Reference library, geneology 54226

Windsor NY

Old Stone House Museum, 22 Chestnut St, Windsor, NY 13865 • T: +1 607 6551491 • Dir.: *Charles L. English* • Dep. Dir.: *Louella F. English* • Sc. Staff: *Bernard Osborne* (History) • Local Museum – 1970 Civil War artifacts, weapons, local hist 54227

Windsor VT

American Precision Museum, 196 Main St, Windsor, VT 05089 • T: +1 802 6745781 • F: +1 802 6742524 • info@americanprecision.org • www.americanprecision.org • C.E.O.: *Ann Lawless* • Science&Tech Museum – 1966 Industrial hist, engineering 54228

Old Constitution House, N Main St, Windsor, VT 05089 • T: +1 802 6723773 • F: +1 802 8283206 • john.dumville@state.vt.us • www.historicvermont.org • Head: *John P. Dumville* • Historical Museum / Decorative Arts Museum – 1961 Colonial and Civil War periods, pottery, furniture 54229

Windsor Locks CT

New England Air Museum of the Connecticut Aeronautical Historical Association, Bradley International Airport, Windsor Locks, CT 06096 • T: +1 860 6233305 • F: +1 860 6272820 • staff@neam.org • www.neam.org • C.E.O.: *Michael P. Speciale* • Science&Tech Museum – 1959 Aeronautics, propulsion systems 54230

Winfield IL

Kline Creek Farm, North and Geneva Rds, Winfield, IL 60190, mail addr: POB 5000, Wheaton, IL 60189-5000 • T: +1 630 8765900 • F: +1 630 2939421 • forest@dupageforest.com • www.dupageforest.com • Agriculture Museum – 1989 Farm equip, household furnishings 54231

Winfield KS

The Cowley County Historical Museum, 1011 Mansfield St, Winfield, KS 67156 • T: +1 316 2214811 • F: +1 316 2210793 • cchsm@iwinfield.com • www.cchsm.com • Pres.: *Brad Light* • Local Museum – 1931 Manuscripts, glassware, costumes, household equip from pioneer days 54232

Winlock WA

John R. Jackson House, Lewis and Clark State Park, Winlock, WA 98596 • T: +1 360 8642643 • Dir.: *Cleve Pinnix* • Local Museum – 1915 Historic house 54233

Winnabow NC

Brunswick Town/Fort Anderson State Historic Site, 8884 Saint Philips Rd SE, Winnabow, NC 28479 • T: +1 910 3716613 • F: +1 910 3833806 • brunswick@ncdcr.gov • www.nchistoricsites.org • Head: *James A. Bartley* • Local Museum – 1958 Excavations, 18th c English and Civil War artifacts 54234

Winnebago MN

Winnebago Area Museum, 18 First Av NE, Winnebago, MN 56098 • T: +1 507 8934660 • wmuseum@bevcomm.net • C.E.O.: *Pete Haight* • Historical Museum / Archaeology Museum – 1977 54235

Winnetka IL

Winnetka Historical Museum, 1140 Elm St, Winnetka, IL 60093 • T: +1 847 5016025 • F: +1 847 5013221 • www.winnetkahistory.org • Dir.: *Joan Evanich* • Cur.: *John Costersian* • Local Museum – 1988 Local hist, photos, costumes, paintings, docs, artifacts 54236

Winnsboro SC

Fairfield County Museum, 231 S Congress St, Winnsboro, SC 29180 • T: +1 803 6359811 • fairfieldmus@chestertel.com • www.fairfieldchamber.org • Dir.: *Pelham Lyles* • Local Museum – 1963 Local hist 54237

Winona MN

Arches Museum of Pioneer Life, Hwy 14, 9 mi W of Winona, Winona, MN 55987 • T: +1 507 4542723, 5232111 • F: +1 507 4540006 • wchs@luminet.net • Exec. Dir.: *Mark F. Peterson* • Cur.: *Jodi Brom* • Agriculture Museum / Local Museum – 1964 Windmill, farm and household, natural hist 54238

Winona County Historical Museum, 160 Johnson St, Winona, MN 55987 • T: +1 507 4542723 • F: +1 507 4540006 • info@winonahistory.org • www.winonahistory.org • Dir.: *Mark F. Peterson* • Cur.: *Jodi Brom* • Historical Museum – 1935 Artifacts from prehistoric to present times 54239

Winslow AZ

Homolovi Ruins, 87 North State Rte, Winslow, AZ 86047-9803 • T: +1 928 2894106 • F: +1 928 2892021 • homolovi@pr.state.az.us • www.pr.state.az.us • Archaeology Museum – 1986 Archaeological finds from Homolovi sites, natural hist 54240

Meteor Crater, Museum of Astrogeology, I-40, ext 233, Crater Rd 5.5 miles S, Winslow, AZ 86047, mail addr: POB 30940, Flagstaff, AZ 86003-0940 • T: +1 928 2895898, 2892362 • F: +1 928 2892598 • info@meteorcrater.com • www.meteorcrater.com • Pres.: *Brad Andes* • Special Museum – 1955 library 54241

Winston-Salem NC

Associated Artists of Winston-Salem, 301 W Fourth St, Winston-Salem, NC 27101 • T: +1 336 7220340 • staff@associatedartists.org • www.associatedartists.org • Public Gallery Temporary exhibits 54242

Charlotte and Philip Hanes Art Gallery, Wake Forest University, Art Dept, Winston-Salem, NC 27109 • T: +1 336 7585795, 7585310 • F: +1 336 7586014 • faccinto@wfu.edu • Dir.: *Victor Faccinto* • Ass. Dir.: *Paul Bright* • Fine Arts Museum / University Museum – 1976 54243

Diggs Gallery at Winston-Salem State University, 601 Martin Luther King Jr. Dr, Winston-Salem, NC 27110 • T: +1 336 7502458, 7502000 • F: +1 336 7502463 • Dir./Cur.: *Belinda Tate* • Fine Arts Museum / University Museum – 1990 African American art 54244

Historic Bethabara Park, 2147 Bethabara Rd, Winston-Salem, NC 27106 • T: +1 336 9248191 • F: +1 336 9240535 • ekutcher@triadbiz.rr.com • www.bethabarapark.org • Dir.: *Ellen M. Kutcher* • Historic Site / Open Air Museum / Natural History Museum / Religious Arts Museum – 1970 1753 site of first Moravian settlement in NC, hist bldgs, Moravian furniture and pottery 54245

Museum of Anthropology, Wingate Rd, Winston-Salem, NC 27109-7267 • T: +1 336 7585282 • F: +1 336 7585116 • moa@wfu.edu • www.wfu.edu/MOA • Dir.: *Dr. Stephen Wittington* • University Museum / Ethnology Museum – 1962 54246

Museum of Early Southern Decorative Arts, 924 S Main St, Winston-Salem, NC 27101 • T: +1 336 7217360 • F: +1 336 7217367 • research@oldsalem.org • www.mesda.org • C.E.O. & Pres.: *Paul C. Reber* • Dir.Coll. & Cur.: *Johanna Brown* • Decorative Arts Museum – 1965 54247

Old Salem, 600 S Main St, Winston-Salem, NC 27101 • T: +1 336 7217300 • F: +1 336 7217335 • webmaster@oldsalem.org • www.oldsalem.org • Pres.: *Paul Reber* • Open Air Museum / Local Museum – 1950 1766 Restoration village, hist bldgs, Moravian hist and culture – Toy and Children's Museum, library 54248

Reynolda House, Museum of American Art, 2250 Reynolda Rd, Winston-Salem, NC 27106-1765 • T: +1 336 7255325 • F: +1 336 7210991 • reynolda@reynoldahouse.org • www.reynoldahouse.org • C.E.O.: *Dr. John P. Anderson* • Fine Arts Museum – 1964 American paintings and sculpture, dec art 54249

Sciworks, 400 Hanes-Mill Rd, Winston-Salem, NC 27105 • T: +1 336 7676730 • F: +1 336 6611777 • bssanford@sciworks.org • www.sciworks.org • C.E.O.: *Dr. Beverly S. Sanford* • Dir.Exhibits: *Tom Wilson* • Natural History Museum / Science&Tech Museum – 1964 54250

Southeastern Center for Contemporary Art, 750 Marguerite Dr, Winston-Salem, NC 27106 • T: +1 336 7251904 • F: +1 336 7226059 • general@secca.org • www.secca.org • Dir.: *Vicki C. Kopf* • Cur.: *David J. Brown* • Fine Arts Museum – 1956 Changing exhib 54251

Wake Forest University Fine Arts Gallery, POB 7232, Winston-Salem, NC 27109 • T: +1 336 7585795 • F: +1 336 7586014 • faccinto@wfu.edu • Dir.: *Victor Faccinto* • Public Gallery / University Museum – 1976 54252

Winter Park FL

The Charles Hosmer Morse Museum of American Art, 445 N Park Av, Winter Park, FL 32789 • T: +1 407 6455311 • F: +1 407 6471284 • www.morsemuseum.org • Dir.: *Dr. Laurence J. Ruggiero* • Fine Arts Museum – 1942 American art 54253

The George D. and Harriet W. Cornell Fine Arts Museum, Rollins College, 1000 Holt Av, Winter Park, FL 32789-4499 • T: +1 407 6462526 • F: +1 407 6462524 • ablumenthal@rollins.edu • www.rollins.edu/cfam • Dir.: *Arthur R. Blumenthal* • Fine Arts Museum / University Museum – 1978 American and European paintings, sculptures, bronzes, prints, drawings, posters, photos, furniture, glass, ceramics 54254

Winterset IA

Birthplace of John Wayne, 224 S 2nd St, Winterset, IA 50273 • T: +1 515 4621044 • F: +1 515 4623289 • director@johnwaynebirthplace.org • www.johnwaynebirthplace.org • Pres.: *David Trask* • Dir.: *Carolyn Wilson* • Historical Museum – 1981 Historic house, birthplace of film star John Wayne 54255

Madison County Historical Society Museum, 815 S 2nd Av, Winterset, IA 50273 • T: +1 515 4622134 • F: +1 515 4624531 • mchistory@i-rule.net • www.madisoncountyhistoricalsociety.com • C.E.O.: *Carol Bass* • Local Museum – 1904 Local history 54256

Winterset Art Center, 216 S John Wayne Dr, Winterset, IA 50273 • T: +1 515 4623226, 4624847 • Dir.: *Don Thomas* • Fine Arts Museum – 1958 Art, housed in an Underground Railway stop during the Civil War 54257

Winterthur DE

Winterthur Museum, Rte 52, Winterthur, DE 19735 • T: +1 302 8884600 • F: +1 302 8884880 • www.winterthur.org • Dir.: *Leslie Green Bowman* • Chm.: *Bruce C. Perkins* • Dep. Dir.: *Gary Kulik* • Decorative Arts Museum / Historical Museum – 1951 Decorative arts, cultural history 54258

Winterville GA

Carter-Coile Country Doctors Museum, 111 Marigold Ln, Winterville, GA 30683, mail addr: POB 306, Winterville, GA 30683 • T: +1 706 7428600 • F: +1 706 7425476 • winterville@charter.net • www.cityofwinterville.com • Historical Museum – 1971 Medicine, office by Dr. Warren Carter and Dr. Frank Coile 54259

Wiscasset ME

The 1811 Old Lincoln County Jail and 1839 Jailer's House Museum, 133 Federal St, Wiscasset, ME 04578 • T: +1 207 8826817 • F: +1 207 8826817 • lcha@wiscasset.net • www.lincolncountyhistory.org • Exec. Dir.: *Margaret M. Shiels* • Local Museum – 1954 Regional hist, textiles, prints, photos, glass 54260

Castle Tucker, Lee St at High St, Wiscasset, ME 04578 • T: +1 207 8821769 • CastleTucker@HistoricNewEngland.org • www.historicnewengland.org • Historical Museum – 1974 Furnishings, local hist 54261

Maine Art Gallery, Warren St, Wiscasset, ME 04578 • T: +1 207 8827511 • Pres.: *Sally Loughridge* • Fine Arts Museum / Public Gallery – 1958 Art gallery 54262

Musical Wonder House, 18 High St, Wiscasset, ME 04578 • T: +1 207 8827163 • F: +1 207 8826373 • music@musicalwonderhouse.com • www.musicalwonderhouse.com/ • Pres.: *Danilo Konvalinka* • Music Museum – 1963 Mechanical musical instr 54263

Nickels-Sortwell House, 121 Main St, Rte US1, Wiscasset, ME 04578 • T: +1 207 8826218 • F: +1 617 2279204 • NickelsSortwellHouse@HistoricNewEngland.org • www.historicalnewengland.org • Historical Museum – 1958 54264

Wisconsin Rapids WI

South Wood County Historical Museum, 540 Third St, Wisconsin Rapids, WI 54494 • T: +1 715 4231580 • F: +1 715 4236369 • museum@wctc.net • www.swch-museum.com • Pres.: *J. Marshall Buehler* • Dir.: *Pam Walker* • Local Museum – 1955 Local hist, railway station, lumbering and blacksmith items 54265

Wisdom MT

Big Hole National Battlefield, 10 miles W of Wisdom Mount on Hwy 43, Wisdom, MT 59761 • T: +1 406 6893155 • F: +1 406 6893151 • www.nps.gov/biho • Head: *Jon G. James* • Military Museum / Open Air Museum / Historic Site – 1910 Nez Perce Indian culture, archaeology, 1970s amry accoutremtents, uniforms, firearms 54266

Wolf Point MT

Wolf Point Area Historical Society, 220 Second Av S, Wolf Point, MT 59201 • T: +1 406 6531912 • C.E.O. & Pres.: *Alma Hall* • Association with Coll – 1970 Early-day settlers, Assinaboine and Sioux culture, grain farming, livestock ranching 54267

Wolfeboro NH

Clark House Museum Complex, S Main St, Wolfeboro, NH 03894, mail addr: POB 1066, Wolfeboro, NH 03894 • T: +1 603 5694997 • Pres.: *Dianne Rogers* • Local Museum / Historical Museum – 1925 Schoolhouse, firehouse, pre-revolutionary farm house 54268

Libby Museum, 755 n Main St, Wolfeboro, NH 03894 • T: +1 603 5691035 • F: +1 603 5692246 • libbymus@metrocast.net • www.wolfeboroonline.com/libby • Dir.: *Patricia F. Smith* • Historical Museum / Natural History Museum – 1912 54269

Wright Museum, 77 Center St, Rte 28, Wolfeboro, NH 03894 • T: +1 603 5691212 • F: +1 603 5696326 • wrmuseum@aol.com • www.wrightmuseum.org • Pres. & Dir.: *David M. Wright* • Chm.: *Dennis I. Runey* • Historical Museum – 1984 WW II American homefront incl artiafcts, written materials, costumes 54270

Womelsdorf PA

Conrad Weiser Homestead, 28 Weiser Rd, Womelsdorf, PA 19567-9718 • T: +1 610 5892934 • F: +1 610 5899458 • sling@state.pa.us • Pres.: *David G. Sonnen* • Historic Site – 1928 Historic house 54271

Woodbine GA

Woodbine International Fire Museum, 110 Bedell Av, Woodbine, GA 31569-0058 • T: +1 912 5765351 • Dir.: *Jodie G. Briese* • Historical Museum – 1991 Fire fighting equip 54272

Woodbridge CT

Amity and Woodbridge Historical Society Museum, 1907 Litchfield Turnpike, Woodbridge, CT 06525 • T: +1 203 3872823 • woodbridgehistory.org • Dir.: *Donald Menzies* • Local Museum – 1936 Local hist 54273

Woodbridge NJ

Barron Arts Center and Museum, 582 Rahway Av, Woodbridge, NJ 07095 • T: +1 732 6340413 • F: +1 732 6348633 • barronarts@twp.woodbridge.nj.us • www.twp.woodbridge.nj.us • Dir.: *Cynthia Knight* • Fine Arts Museum – 1977 Arts, crafts and photograph 54274

Woodbury CT

The Glebe House Museum, Hollow Rd, Woodbury, CT 06798 • T: +1 203 2632855 • F: +1 203 2636726 • ghmgjg@snet.net • www.theglebehouse.org • Dir.: *Sarah Griswold* • Local Museum – 1923 Local hist, American episcopacy 54275

Woodbury KY

Green River Museum, Park St, Woodbury, KY 42261 • T: +1 270 5262133 • Cur.: *Jerry Martin* • Historical Museum – 1982 54276

Woodbury NJ

Gloucester County Historical Society Museum, 58 N Broad St, Woodbury, NJ 08096 • T: +1 856 8488531 • F: +1 856 8450131 • gchs@net-gate.com • www.rootsweb.com/~njglouce/gchs • Pres.: *Barbara L. Turner* • Local Museum – 1903 New Jersey and U.S. history, New Jersey decorative arts 54277

Woodland CA

Yolo County Historical Museum, 512 Gibson Rd, Woodland, CA 95695 • T: +1 530 6661045 • www.yolo.net/vme/ychm • Dir./Cur.: *Monika Stengert* • Local Museum – 1979 Local history, farming, furnishings – library 54278

Woodruff SC

Thomas Price House, 1200 Oak View Farms Rd, Woodruff, SC 29388 • T: +1 864 5766546, 4762483 • F: +1 864 5764058 • Pres.: *Glen Boggs* • C.E.O. & Exec. Dir.: *Susan Turpin* • Local Museum – 1972 Historic bldg 54279

Woods Hole MA

Woods Hole Oceanographic Institution Ocean Science Exhibit Center, 15 School St, Woods Hole, MA 02543 • T: +1 508 5481400, 2892663 • F: +1 508 4572034 • information@whoi.edu • www.whoi.edu • Dir.: *Susan Avery* • Natural History Museum – 1930 Oceanographic research photos, data, tools, instruments, vehicles, vessels 54280

Woodstock CT

Bowen House / Roseland Cottage, Rte 169, on the Common, Woodstock, CT 06281 • T: +1 860 9284074 • F: +1 860 9632208 • RoselandCottage@HistoricNewEngland.org • www.spnea.org • Head: *Pam Russo* • Local Museum – 1970 Historic houses, Gothic furniture 54281

Woodstock Historical Society Museum, 523 Rte 169, Woodstock, CT 06281 • T: +1 860 9281035 • Pres.: *Marilyn Pomeroy* • Local Museum – 1967 Local hist, agriculture 54282

Woodstock NY

Center for Photography at Woodstock, 59 Tinker St, Woodstock, NY 12498 • T: +1 845 6797747 • F: +1 845 6796337 • info@cpw.org • www.cpw.org • Dir.: *Colleen Kenyon* • *Kathleen Kenyon* • Ass. Dir.: *Kate Menconeri* • *Larry Lewis* • Fine Arts Museum – 1977 Fine art photography 54283

Woodstock Artists Gallery, 28 Tinker St, Woodstock, NY 12498 • T: +1 914 6792940 • F: +1 914 6792940 +1 call first • info@woodstockart.org • www.woodstockart.org • Dir.: *Linda Freaney* • Fine Arts Museum – 1920 Works of contemporary artists from the Woodstock art colony, Woodstock art from the past 54284

Woodstock School of Art Gallery, 2470 Rte 212, Woodstock, NY 12498 • T: +1 845 6792388 • F: +1 845 6792388 • wsart@earthlink.net • www.woodstockschoolofart.com • Dir.: *Karen White* • Public Gallery American art 54285

Woodstock VA

Woodstock Museum of Shenandoah County, 137 W Court St, Woodstock, VA 22664 • T: +1 540 4592542, 9848035 • ckn@shentel.com • Local Museum – 1969 Local hist, manuscript coll, genealogy 54286

Woodstock VT

Billings Farm and Museum, River Rd and Rte 12, Woodstock, VT 05091 • T: +1 802 4572355 • F: +1 802 4574663 • billings.farm@valley.net • www.billingsfarm.org • Dir.: *David A. Donath* • Ass. Dir.: *David A. Miles* • *David L. Yeatts* • *Darlyne S. Franzen* • Cur.: *Robert G. Benz* • Open Air Museum / Agriculture Museum / Historic Site – 1976 16,000 objects relating to agricultre and folklife 54287

Woodstock Historical Society, 26 Elm St, Woodstock, VT 05091 • T: +1 802 4571822 • F: +1 802 4572811 • whs@sover.net • www.woodstockhistsoc.org • Dir.: *Corwin Sharp* • Association with Coll – 1943 Portratis, toys, dolls, wintersports equip 54288

Woodville MS

Rosemont Plantation, Hwy 24 E, Woodville, MS 39669 • T: +1 601 8886809 • F: +1 601 8883327 • www.rosemontplantation.com • Decorative Arts Museum – 1971 Davis furniture, portraits 54289

Wilkinson County Museum, Courthouse Sq, Woodville, MS 39669 • T: +1 601 8883998 • F: +1 601 8883327 • wilkmuseum@aol.com • C.E.O.: *Ernesto Caldeira* • Local Museum / Decorative Arts Museum – 1971 Decorative arts from county plantations and homes, area hist 54290

Woodville TX

Allan Shivers Museum, 302 N Charlton, Woodville, TX 75979 • T: +1 409 2833709 • F: +1 409 2835258 • contact@allanshiverslibrary.com • www.allanshiverslibrary.com • Dir.: *Rosemary Bunch* • Local Museum – 1963 Historic house, mementos of the family 54291

Heritage Village Museum, Hwy 190 W, Woodville, TX 75979 • T: +1 409 2832272 • F: +1 409 2832194 • info@heritage-village.org • www.heritage-village.org • Dir.: *Ofiera Gazzaway* • Local Museum 36 bldg 1940-1900 early East Texas Pioneer farming village with replica and original bldg of the perod 54292

Woodward OK

Plains Indians and Pioneers Museum, 2009 Williams Av, Woodward, OK 73801 • T: +1 580 2566136 • F: +1 580 2562577 • pipm@swbell.net • Cur.: *Louise James* • Local Museum – 1966 Pioneer life, Indian artifacts, agricultural development 54293

Wooster OH

The College of Wooster Art Museum, c/o Ebert Art Center, 1220 Beall Av, Wooster, OH 44691 • T: +1 330 2632495 • F: +1 330 2632633 • jfuell@wooster.edu • www.wooster.edu/artmuseum • Dir.: *Kathleen McManus Zurko* • Fine Arts Museum / University Museum – 1930 Print coll, Chinese bronzes, African art, Cypriot pottery 54294

Wayne Center for the Arts, 237 S Walnut St, Wooster, OH 44691 • T: +1 330 2648596 • F: +1 330 2649314 • frontdesk@wayneartscenter.org • www.wayneartscenter.org • Exec. Dir.: *Lucy Spurgeon* • Public Gallery – 1973 54295

Wayne County Historical Society Museum, 546 E Bowman St, Wooster, OH 44691 • T: +1 330 2648856 • F: +1 330 2648823 • host@waynehistorical.org • www.waynehistorical.org/ • Pres.: *Greg Long* • Local Museum – 1954 Lustre ware, pressed glass, china, mounted birds and animals, militaria, Indian artifacts, hist bldgs 54296

Worcester MA

EcoTarium, Museum of Science and Nature, 222 Harrington Way, Worcester, MA 01604 • T: +1 508 9292700 • F: +1 508 9292701 • info@ecotarium.org • www.ecotarium.org • Exec. Dir.: *Stephen Pitcher* • Cur.: *Alexander Goldowsky* (Programs, Exhibits) • Natural History Museum – 1825 Environmental museum, birds, mammals, minerals, shells, interactive exhibits – library, zoo, planetarium, nature trail 54297

Higgins Armory Museum, 100 Barber Av, Worcester, MA 01606-2444 • T: +1 508 8536015 • F: +1 508 8527697 • higgins@higgins.org • www.higgins.org • Dir.: *Nikki Andersen* • Cur.: *Dr. Jeffrey Forgeng* • Historical Museum – 1928 Arms & armour, culture – library, auditorium 54298

Iris and B. Gerald Cantor Art Gallery, College of the Holy Cross, 1 College St, Worcester, MA 01610 • T: +1 508 7933356 • F: +1 508 7933030 • rhankins@holycross.edu • www.holycross.edu/departments/cantor/website/ • Dir.: *Roger Hankins* • University Museum / Fine Arts Museum – 1983 Rodin a.o. sculptures, contemporary photos 54299

Krikorian Gallery, Worcester Center for Crafts, 25 Sagamore Rd, Worcester, MA 01605 • T: +1 508 7538183 • F: +1 508 7975626 • wcc@worcestercraftcenter.org • www.worcestercraftcenter.org • Exec. Dir.: *Maryon Attwood* • Cur.: *Tom O'Malley* (Pottery, Ceramics) • Public Gallery – 1856 Arts, contemporary American craft 54300

Salisbury Mansion, 40 Highland St, Worcester, MA 01609 • T: +1 617 7538278 • F: +1 508 7538278 • worchistmu@aol.com • www.worcesterhistory.org • Exec. Dir.: *William D. Wallace* • Historical Museum – 1875 Stephen Salisbury & Worcester County's wealthiest families home 54301

Worcester Art Museum, 55 Salisbury St, Worcester, MA 01609-3196 • T: +1 508 7994406 • F: +1 508 7985646 • information@worcesterart.org • www.worcesterart.org • Dir./Cur.: *James A. Welu* (European Art) • Cur.: *David Brigham* (American Art) • *Louise Virgin* (Asian Art) • *David Acton* (Prints, Drawings, Photographs) • *Susan Stoops* (Contemporary Art) • Cons.: *Lawrence Becker* • Fine Arts Museum – 1896 American and Asian art, prints, drawings, 5,000 yrs art & culture – library, studios 54302

Worcester Historical Museum, 30 Elm St, Worcester, MA 01609 • T: +1 617 7538278 • F: +1 508 7539070 • worchistmu@aol.com • www.worcesterhistory.org • Pres.: *James J. Paugh* • Exec. Dir.: *William D. Wallace* • Local Museum – 1875 Local history, costumes, military, decorative and graphic arts – archives, library 54303

Worcester PA

Peter Wentz Farmstead, Shearer Rd, Worcester, PA 19490 • T: +1 610 5845104 • F: +1 610 5846860 • peterwentzfarmstead@mail.montcopa.org • www.montcopa.org • Pres.: *Dick Anderl* • Historic Site – 1976 Historic bldg 54304

Worland WY

Washakie Museum, 1115 Obie Sue, Worland, WY 82401 • T: +1 307 3474102 • F: +1 307 3474865 • washakiemuseum@rtconnect.net • www.washakiemuseum.com • Dir.: *Lisa Brahm-Lindberg* • Local Museum – 1986 Hist of Big Horn Basin, archaeology, geology, Native American art and culture 54305

Worthington MN

Nobles County Historical Society, 407 12 St, Ste 2, Worthington, MN 56187 • T: +1 507 3764431 • F: +1 507 3763005 • nchs@frontiernet.net • Pres.: *Alan Swanson* • Dir.: *Roxann L. Polzine* • Association with Coll – 1933 Harness, early power machinery, medical, mortuary 54306

Worthington OH

Ohio Railway Museum, 990 Proprietors Rd, Worthington, OH 43085 • T: +1 614 8857345 • info@ohiorailwaymuseum.net • www.ohiorailwaymuseum.net • Pres.: *Warren W. Hyer* • Science&Tech Museum – 1945 Hist steam and electric railway equip, electric streetcars and interurbans, engines and waggons 54307

Worthington Historical Society, 50 W New England Av, Worthington, OH 43085 • T: +1 614 8851247 • F: +1 614 8851040 • worthhsoc@aol.com • www.worthington.org • Cur.: *Jane Trucksis* • Historical Museum – 1955 Orange Johnson House and Old Rectory, Worthington Indian Mound, doll coll, community memorabilia 54308

Wrangell AK

Wrangell Museum, 296 Outer Dr, Wrangell, AK 99929, mail addr: POB 531, Wrangell, AK 99929 • T: +1 907 8743770 • museum@wrangell.com • www.wrangell.com • Dir./Cur.: *Dennis Chapman* • Local Museum – 1967 Southeast Alaska native cultures, Wrangell history and natural hist 54309

Wray CO

Wray Museum, 205 E Third St, Wray, CO 80758-0161 • T: +1 970 3325063 • Dir.: *Linda Marie Vermillion* • Local Museum / Archaeology Museum – 1969 Local history, archaeology, Paleo Indians, art 54310

Wright-Patterson Air Force Base OH

United States Air Force Museum, 1100 Spaatz St, Wright-Patterson Air Force Base, OH 45433-7102 • T: +1 937 2553286 ext. 302, 303 • F: +1 937 2553910 • usaf.museum@wpafb.af.mil • www.wpafb.af.mil/museum/ • Dir.: *Charles D. Metcalf* • Sen.Cur.: *Terry Aitken* • Military Museum – 1923 300 aircraft and missiles, aviation guns and instr, uniforms 54311

Wyandotte MI

Wyandotte Museum, 2610 Biddle Av, Wyandotte, MI 48192 • T: +1 734 3247297 • F: +1 734 3247283 • wymuseum@ili.net • www.angelfire.com/mi/WYMUSEUM • Dir.: *Marc Partin* • Local Museum – 1958 Local hist, Ford-MacNichol home, shipbuilding, chemical ind – library 54312

Wyckoff NJ

James A. McFaul Environmental Center of Bergen County, Crescent Av, Wyckoff, NJ 07481 • T: +1 201 8915571 • F: +1 201 3437249 • www.co.bergen.nj.us/parks • Dir.: *Frank DeBrari* • Natural History Museum – 1967 54313

Wyoming NY

Middlebury Historical Society Museum, 22 S Academy St, Wyoming, NY 14591-9801 • T: +1 716 4956582 • Cur.: *Karen Aman* • Local Museum – 1941 Military equip, musical instr, farm tools, photos 54314

Wytheville VA

Rock House Museum, 205 Tazwell St, Wytheville, VA 24382 • T: +1 540 2233330 • F: +1 540 2233315 • museum@wytheville.org • www.wytheville.org • Dir.: *Francis Emerson* • Local Museum – 1970 Furniture, artifacts, local memorabilia 54315

Thomas J. Boyd Museum, 295 Tazwell St, Wytheville, VA 24382 • T: +1 540 2233330 • F: +1 540 2233315 • Dir.: *Francis Emerson* • Local Museum Fire truck, farming and mining 54316

Xenia OH

Greene County Historical Society Museum, 74 W Church St, Xenia, OH 45385 • T: +1 513 3724606 • F: +1 372 3725660 • gchsxo@aol.com •
Pres.: *John Balmer* •
Local Museum – 1929
Military and Cosley coverlets, period furniture, agricultural implements, dolls 54317

Yacolt WA

Pomeroy Living History Farm, 20902 NE Lucia Falls Rd, Yacolt, WA 98675 • T: +1 360 6863537 • F: +1 360 6868111 • staff@pomeroyfarm.org • www.pomeroyfarm.org •
C.E.O.: *Bob Brink* •
Agriculture Museum
Pomeroy family coll, farm machinery 1920 to the present, pre-electrical farm life in SW Washington 54318

Yakima WA

Larson Gallery, Yakima Valley Community College, S 16th Ave and Nob Hill Blvd, Yakima, WA 98902, mail addr: PO Box 22520, Yakima, WA 98907-2520 • T: +1 509 5744875 • F: +1 509 5746826 • chahn@yvcc.edu • www.larsongallery.org •
Dir.: *Cheryl H. Hahn* •
Public Gallery – 1949
Fine arts, ethnic art 54319

Yakima Valley Museum, 2105 Tieton Dr, Yakima, WA 98902 • T: +1 509 2480747 • F: +1 509 4534890 • info@yakimavalleymuseum.org • www.yakimavalleymuseum.org •
Dir.: *John A. Baule* • Cur.: *Michael Siebol* • *Andrew Granitto* • *David Lynx* •
Local Museum – 1952
Native American artifacts, textile and costume coll, agricultural and houshold artifacts, mineral coll, horse-drawn vehicles 54320

Yale OK

Jim Thorpe Home, Oklahoma Historical Society, 706 E Boston, Yale, OK 74085 • T: +1 918 3872815 • jimthorpe@okhistory.org •
Cur.: *Alice Cussner* •
Local Museum – 1973
1920's furnishings and memorabilia 54321

Yankton SD

Bede Art Gallery, Mount Marty College, 1105 W 8th St, Yankton, SD 57078 • T: +1 605 6681011 • F: +1 605 6681607 • dkahle@mtmc.edu •
Public Gallery 54322

Dakota Territorial Museum, 610 Summit St, Yankton, SD 57078 • T: +1 605 6653898 • F: +1 605 6656314 • www.dakotaterritorialmuseum.org •
Pres.: *Fred Binder* •
Local Museum – 1936
Local hist, paintings, sculpture 54323

Yarmouth ME

Museum of Yarmouth History, Merrill Memorial Library, 215 Main St, Yarmouth, ME 04096 • T: +1 207 8466259 • yarmouth-history@inetmail.att.net • www.yarmouth.me.us •
Dir.: *Marilyn J. Hinkley* •
Local Museum – 1958
Local history 54324

Yarmouth Port MA

Historical Society of Old Yarmouth Museum, 11 Strawberry Ln, Yarmouth Port, MA 02675 • T: +1 508 3623021 • hsoy@comcast.net • www.hsoy.org •
Pres.: *Oliver Duncan* •
Local Museum – 1953
Local history, period furnishings, ship models and paintings – library, archives 54325

Winslow Crocker House, 250 Rte 6a, Old King's Hwy, Yarmouth Port, MA 02675 • T: +1 617 2273956 • F: +1 617 2279204 • WinslowCrockerHouse@HistoricNewEngland.org • www.historicnewengland.org/visit/homes/winslow.htm •
Local Museum / Decorative Arts Museum – 1935
Period furnishings, hooked & Oriental rugs, ceramics 54326

Yates Center KS

Woodson County Historical Society, 208 W Mary, Yates Center, KS 66783 • T: +1 620 6252371 • rcall1@cox.net •
Pres.: *Pete Watts* • Cur.: *Geri Town* • *Linda Call* •
Association with Coll – 1965
Historic church, country school, log cabin 54327

Yazoo City MS

Yazoo Historical Museum, 332 N Main St, Yazoo City, MS 39194 • T: +1 601 7461815 • yazoo@yazoo.org • www.yazoo.org •
Pres.: *Sam Olden* •
Historical Museum – 1977 54328

Yellow Springs OH

Glen Helen Ecology Institute Trailside Museum, 505 Corry St, Yellow Springs, OH 45387 • T: +1 937 7677375 • F: +1 937 7676659 • bkrisko@antioch-college.edu •
Ass. Dir.: *Rick Flood* •
Natural History Museum – 1951
Natural hist 54329

Noyes and Read Gallery and Herndon Gallery, Antioch College, 795 Livermore St, Yellow Springs, OH 45387 • T: +1 937 7691149 •
Dir.: *Nevin Mercede* •
Fine Arts Museum – 1972
Painting, photography and video 54330

Yonkers NY

The Hudson River Museum, 511 Warburton Av, Yonkers, NY 10701-1899 • T: +1 914 9634550 • F: +1 914 9638558 • info@hrm.org • www.hrm.org •
Dir.: *Michael Botwinick* • Ass. Dir.: *Diane Graf* • Cur.: *Laura L. Vookles* •
Fine Arts Museum / Local Museum / Folklore Museum – 1919
19th-21st c American art, local and regional memorabilia 54331

Philipse Manor Hall, 29 Warburton Av, Yonkers, NY 10701 • T: +1 914 9654027, 9650473 • F: +1 914 9656485 • heather.iannucci@oprhp.state.ny.us • nystaeparks.com •
Dir.: *Heather Iannucci* •
Decorative Arts Museum – 1911
Paintings, photographs, structures, architecture 54332

York ME

Elizabeth Perkins House, POB 312, York, ME 03909 • T: +1 207 3634974 • F: +1 207 3634021 •
Exec. Dir.: *Scott Stevens* •
Historical Museum 54333

Jefferds Tavern, Lindsey Rd, York, ME 03909 • T: +1 207 3634974 • F: +1 207 3634021 •
Exec. Dir.: *Scott Stevens* •
Historical Museum 54334

John Hancock Warehouse, Lindsey Rd, York, ME 03909 • T: +1 207 3634974 • F: +1 207 3634021 •
Exec. Dir.: *Scott Stevens* •
Historical Museum 54335

Old Gaol Museum, POB 312, York, ME 03909 • T: +1 207 3634974 • F: +1 207 3634021 •
Exec. Dir.: *Scott Stevens* • Cur.: *Tom Johnson* •
Historical Museum 54336

Old School House, Museums of Old York, York, ME 03909, mail addr: POB 312, York, ME 03909 • T: +1 207 3634974 • F: +1 207 3634021 • oyhs@oldyork.org • www.oldyork.org •
Exec. Dir.: *Scott Stevens* •
Historical Museum 54337

Old York Historical Museum, 207 York St, York, ME 03909, mail addr: POB 312, York, ME 03909-0312 • T: +1 207 3634974 • F: +1 207 3634021 • oyhs@oldyork.org • www.oldyork.org •
Dir.: *F. Scott Stevens* • Cur.: *Thomas B. Johnson* •
Local Museum – 1984
Historic bldgs, local hist 54338

York NE

Anna Bemis Palmer Museum, 211 E Seventh St, York, NE 68467 • T: +1 402 3632630 • F: +1 402 3632601 •
Dir.: *Jim Krejci* • Cur.: *Ginny L. Driewer* •
Historical Museum – 1967 54339

York PA

Fire Museum of York County, 757 W Market St, York, PA 17404 • T: +1 717 8455665 • jmummert@yorkheritage.org • www.yorkheritage.org •
C.E.O: *Delaine A. Toerper* •
Historical Museum – 1973
Fire fighting, Royal Fire Stn 6 54340

Industrial and Agricultural Museum, 217 W Princess St, York, PA 17403-2013 • T: +1 717 848 1587 • F: +1 717 8121204 • jmummert@yorkheritage.org • www.yorkheritage.org •
Cur.: *Richard Banz* •
Historical Museum – 1989
Local industry, wallpaper and wire cloth manufacturing, defense, physical fitness, refrigeration and air conditioning, automobile manufacturing 54341

York County Museum, 250 E Market St, York, PA 17403 • T: +1 717 8481587 • F: +1 717 8121204 • mprescott@yorkheritage.org • www.yorkheritage.org •
C.E.O. & Pres.: *Gayle Petty-Johnson* • Cur.: *Richard Banz* •
Local Museum – 1895
Local hist 54342

York Harbor ME

Sayward-Wheeler House, 9 Barrell Ln Extension, York Harbor, ME 03911 • T: +1 207 3842454, +1 617 2273956 • F: +1 617 2279204, +1 207 3848182 • SaywardWheelerHouse@HistoricNewEngland.org • www.historicalnewengland.org •
Decorative Arts Museum – 1977
Queen Anne and Chippendale furniture, portraits, china 54343

Yorktown TX

Yorktown Historical Museum, 144 W Main St, Yorktown, TX 78164 • T: +1 512 5642573 • F: +1 361 9385403 • muellers219@hotmail.com •
Pres.: *Shirley Mueller* •
Local Museum – 1978
Musical instr, antique tools & farm implements, school memorabilia 54344

Yorktown VA

On the Hill Cultural Arts Center, 121 Alexander Hamilton Blvd, Yorktown, VA 23690 • T: +1 757 8983076 •
Pres.: *Tom Coll* •
Fine Arts Museum – 1976
Art 54345

Watermen's Museum, 309 Water St, Yorktown, VA 23690, mail addr: POB 519, Yorktown, VA 23690 • T: +1 757 8872641 • F: +1 757 8882089 • admin@watermens.hrcoxmail.com • www.watermens.org •
Pres.: *John Hanna* •
Special Museum – 1980
Seafood industry 54346

Yorktown Battlefield Visitor Center Museum, Colonial National Historical Park, Old Rte 238 and Colonial Pkwy, Yorktown, VA 23690 • T: +1 757 8983400, 2534838 • F: +1 757 8986346, 2535299 • colo_interpretation@nps.gov • www.nps.gov/colo •
Dir.: *Alac Gould* • Cur.: *Richard Raymond* •
Historic Site / Historical Museum – 1930
Site of the last decisive battle of the American Revolution, America's struggle for independence 54347

Yorktown Colonial Park, Colonial Pkwy and Rte 238, Yorktown, VA 23690, mail addr: POB 210, Yorktown, VA 23690 • T: +1 757 8983400 • F: +1 757 8986346 • colo_interpretation@nps.gov • www.nps.gov/colo •
Historical Museum – 1930
17th and 18th-c arms and artifacts 54348

Yorktown Heights NY

Town of Yorktown Museum, 1974 Commerce St, Yorktown Heights, NY 10598 • T: +1 914 9622970 • F: +1 914 9624379 • museum@yorktownny.org • www.yorktownmuseum.org •
C.E.O.: *Linda G. Cooper* •
Local Museum – 1966
Agricultural and household tools and equip, textiles, costumes, toys, dolls, dollhouses, Indian artifacts, spinning and weaving equip 54349

Yosemite National Park CA

The Yosemite Museum, Museum Bldg, Yosemite National Park, CA 95389 • T: +1 209 3720283 • F: +1 209 3720255 • www.nps.gov/yose •
Cur.: *David M. Forgang* • *Barbara L. Beroza* (Collections) • *Craig D. Bates* (Ethnography) • Sc. Staff: *Jim Snyder* (History) •
Local Museum / Ethnology Museum / Natural History Museum – 1915
History, ethnology, natural science, anthropology, archaeology, geology, Indian culture – herbarium, library 54350

Youngstown NY

Old Fort Niagara, Fort Niagara State Park, Youngstown, NY 14174, mail addr: POB 169, Youngstown, NY 14174-0169 • T: +1 716 7457611 • F: +1 716 7459141 • ofn@oldfortniagara.org • www.oldfortniagara.org •
Pres.: *Ginger McNally* • C.E.O.: *Robert L. Emerson* •
Military Museum / Open Air Museum – 1927
Site related hist milit and archaeology 54351

Youngstown OH

Arms Family Museum of Local History, 648 Wick Av, Youngstown, OH 44502 • T: +1 330 7432589 • F: +1 330 7437210 • mvhs@mahoninghistory.org • www.mahoninghistory.org •
Dir.: *H. William Lawson* • Cur.: *Jessica D. Trickett* •
Local Museum – 1961
Native American hist, furnishings, regional media hist and exhibits 54352

Butler Institute of American Art, 524 Wick Av, Youngstown, OH 44502 • T: +1 330 7431711 ext.100 • F: +1 330 7439567 • k_earnhart@butlerart.com • www.butlerart.com •
Dir.: *Louis A. Zona* • Ass. Dir.: *M. Susan Carfano* •
Fine Arts Museum – 1919
American art, digital art, Western and marine coll 54353

John J. McDonough Museum of Art, 525 Wick Avenue, Youngstown, OH 44502, mail addr: YSU, 1 University Plaza, Youngstown, OH 44555 • T: +1 330 9411400 • F: +1 330 9411492 • labrothers@ysu.edu • http://mcdonoughmuseum.ysu.edu •
Dir.: *Leslie A. Brothers* •
University Museum / Fine Arts Museum – 1991 54354

Yountville CA

Napa Valley Museum, 55 Presidents Circle, Yountville, CA 94599 • T: +1 707 9440500 • F: +1 707 9450500 • tracy@napavalleymuseum.org • www.napavalleymuseum.org •
Dir.: *Eric Nelson* • Cur.: *Randy Murphy* •
Fine Arts Museum / Local Museum – 1973
Local History, art and natural science, Wappo Indiand, Mexicans and Chinese, viticulture 54355

Ypsilanti MI

Ford Gallery, Eastern Michigan University, Art Dept., EMU, 114 Ford Hall, Cross St, Ypsilanti, MI 48197 • T: +1 734 4871268 • larry.newhouse@emich.edu • www.emich.edu/fordgallery •
Dir.: *Larry Newhouse* •
Fine Arts Museum
Lithographs/ prints by Emil Weddige, students works 54356

Ypsilanti Historical Museum, 220 N Huron St, Ypsilanti, MI 48197 • T: +1 734 4824990 • F: +1 313 4837481 • jirc@aol.com • www.ypsilantihistoricalmuseum.org •
Pres.: *Joan Carpenter* •
Local Museum – 1960
City hist, Black hist, manuscript & photo colls – archives 54357

Yreka CA

Klamath National Forest Interpretive Museum, 1312 Fairlane Rd, Yreka, CA 96097 • T: +1 530 8426131 • F: +1 530 8426327 •
Head: *James T. Rock* (History) •
Natural History Museum – 1981
Mining, timber ind, geology, hist, natural hist – library, photo archive 54358

Siskiyou County Museum, 910 S Main St, Yreka, CA 96097 • T: +1 530 8423836 • F: +1 530 8423166 • hismus@att.net • www.co.siskiyou.ca.us/museum •
Dir./Cur.: *Michael Hendryx* •
Local Museum – 1950
Local hist, prehistoric marine, fossils, ethnographic Indian and Chinese artifacts – library, photo archive 54359

Yuba City CA

Community Memorial Museum of Sutter County, 1333 Butte House Rd, Yuba City, CA 95993 • T: +1 530 8227141 • museum@syix.com • www.syix.com/museum •
Dir.: *Julie Stark* •
Local Museum – 1975
Regional history 54360

Yucaipa CA

Mousley Museum of Natural History, San Bernardino County Museum, 35308 Panorama Dr, Yucaipa, CA 92399 • T: +1 909 7903163 • museum@sbcounty.gov • www.co.san-bernardino.ca.us •
Dir.: *Robert L. McKernan* •
Natural History Museum – 1970
Natural history, minerals, fossils 54361

Yucaipa Adobe, 32183 Kentucky St, Yucaipa, CA 92399 • T: +1 909 7953485 • museum@sbcounty.gov • www.co.san-bernardino.ca.us/museum/branches/yucadob.htm •
Dir.: *Robert L. McKernan* • Cur.: *Dr. Ann Deegan* •
Decorative Arts Museum / Agriculture Museum – 1958
Local hist, farm equip, decorative arts 54362

Yucca Valley CA

Hi-Desert Nature Museum, 57090 29 Palms Hwy, Yucca Valley, CA 92284 • T: +1 760 3697212 • F: +1 760 3691605 • museum@yucca-valley.org • www.hidesertnaturemuseum.org •
Dir./Cur.: *Jim Schooler* •
Natural History Museum – 1964
Natural history 54363

Yuma AZ

Sarguinetti House Museum, 240 S Madison Av, Yuma, AZ 85364 • T: +1 928 7821841 • F: +1 928 7830680 • AHSYuma@azhs.gov • www.arizonahistoricalsociety.org •
Dir.: *Megan Reid* • Cur.: *Carol Brooks* •
Historical Museum – 1963
History, housed in 1870s residences – historical garden, photo and manuscript archives 54364

Yuma Art Center and Galleries, 254 S Main St, Yuma, AZ 85364 • T: +1 928 3296607 • F: +1 928 3296616 • director@yumafinearts.org • www.yumafinearts.org •
Pres.: *Bill Butler* • Dir.: *Carolyn J. Bennett* •
Fine Arts Museum – 1962
Contemporary and historical Southwest art 54365

Yuma Territorial Prison Museum, 1 Prison Hill Rd, Yuma, AZ 85364-8792 • T: +1 928 7834771 • F: +1 928 7837442 • www.azstateparks.com •
Head: *Jesse Torres* •
Historical Museum – 1940
General hist, geology, numismatics, territorial prision time 54366

Zanesville OH

Dr. Increase Mathews House, 304 Woodlawn Av, Zanesville, OH 43701, mail addr: 115 Jefferson St • T: +1 740 4549500 •
Dir.: *Linda Smucker* •
Local Museum – 1970
Early settlers' exhibits, costumes, furniture, kitchen implements, art pottery ... 54367

Zanesville Art Center, 620 Military Rd, Zanesville, OH 43701 • T: +1 740 4520741 • F: +1 740 4520797 • info@zanesvilleartcenter.org • www.zanesvilleartcenter.org •
Pres.: *Charles Hunter* • Dir.: *Philip Alan LaDouceur* •
Fine Arts Museum – 1936
Fine art, ceramics, glass ... 54368

Zelienople PA

Zelienople Historical Museum, 243 S Main St, Zelienople, PA 16063 • T: +1 724 4529457 • zeliehistory@fyi.net • www.fyi.net/~zhs •
C.E.O.: *Joyce M. Bessor* •
Local Museum – 1975
Historic houses – library ... 54369

Zion IL

Zion Historical Museum, 1300 Shiloh Blvd, Zion, IL 60099 • T: +1 847 7462427, 8724566 •
Pres.: *Carol Ruesch* •
Local Museum – 1967
Shiloh House, residence of the founder of the city of Zion ... 54370

Zionsville IN

Munce Art Center, 205 W Hawthorne St, Zionsville, IN 46077 • T: +1 317 8736862 • amykindred@sbcglobal.net •
Dir.: *Amy Kindred* •
Public Gallery – 1981
Local and early Indian art ... 54371

P.H. Sullivan Museum and Genealogy Library, 225 W Hawthorn St, Zionsville, IN 46077 • T: +1 317 8734900 • F: +1 317 8732743 • www.sullivanmunce.org •
Exec. Dir.: *Andra Walters* •
Local Museum / Folklore Museum – 1973
Local history, woman's guild ... 54372

Zoar OH

Zoar State Memorial, 198 Main St, Zoar, OH 44697 • T: +1 330 8743011 • F: +1 330 8742936 • zoar@cannet.com • ohiohistory.org/places/zoar •
Local Museum / Historical Museum – 1930
Germanic-American folk arts and crafts, tools, furniture, textiles ... 54373

Zolfo Springs FL

Cracker Trail Museum, 2822 Museum Dr, Zolfo Springs, FL 33890 • T: +1 863 7350119 • F: +1 863 7730107 •
Cur.: *Areca Cotton* •
Natural History Museum – 1966
Aeronautics, agriculture, archaeology, anthropology ... 54374

Zuni NM

A:Shiwi A:Wan Museum, 02E Ojo Caliente Rd, Zuni, NM 87327 • T: +1 505 7824403 • F: +1 505 7824503 • aamhc@ashiwimuseum.org • www.ashiwimuseum.org •
Dir.: *Jamon Carlton* •
Folklore Museum / Fine Arts Museum – 1990
Photos, art, folk life ... 54375

Uruguay

Canelones

Museo Fernando García, Calle Camino Carrasco 7075, 90000 Canelones • T: +598 33 6019228, 618527 •
Natural History Museum ... 54376

La Barra

Museo del Mar, 20001 La Barra •
Historical Museum ... 54377

Madonado

Museo de Artes Plásticas, Casona de Don Antonio Lussich, 20000 Madonado •
Fine Arts Museum ... 54378

Maldonado

Museo de Arte Americano de Maldonado, Calle Treinta y Tres 823, esq Doders, 20000 Maldonado • T: +598 42 222276 • fondmaam@adinet.com.uy •
Fine Arts Museum ... 54379

Museo Didáctico Artiguista, Cuartel de Dragones, Calle 18 de Julio 871, 20000 Maldonado • T: +598 42 25378 •
Special Museum ... 54380

Museo Regional Francisco Mazzoni, Calle Ituzaingó 789, 20000 Maldonado • T: +598 42 221107 • F: 231786, 223781 • dgcultura@maldonado.gub.uy •
Dir.: *Mabel Plada* •
Local Museum ... 54381

Montevideo

Casa de Garibaldi, Museo Histórico Nacional, Calle 25 de Mayo 314, 11000 Montevideo • T: +598 2 9154257 • mhistoricnac@mixmail.com •
Historical Museum
Life of General José Garibaldi, Italian hist in Uruguay ... 54382

Casa de Giró, Museo Histórico Nacional, Calle Cerrito 586, 11000 Montevideo • T: +598 2 9157003 • mhistoricnac@mixmail.com •
Ethnology Museum
Uruguayian cultures ... 54383

Casa de Lavalleja, Museo Histórico Nacional, Calle Zabala 1469, 11000 Montevideo • T: +598 2 951028 • mhistoricnac@mixmail.com •
Historical Museum
Hist of Uruguay – library, historical archive ... 54384

Casa de Ximénez, Museo Histórico Nacional, Rambla 25 de Agosto de 1825 No 580, 11000 Montevideo • T: +598 2 9164984 • mhistoricnac@mixmail.com •
Historical Museum
Naval hist ... 54385

Casa Museo de Juan Zorrilla de San Martín, Museo Histórico Nacional, Calle J.L. Zorrilla de San Martín 96, 11300 Montevideo • T: +598 2 7101818 • mhistoricnac@mixmail.com •
Special Museum
Life and work of Juan Zorilla de San Martín ... 54386

Casa Quinta de José Battle y Ordóñez, Museo Histórico Nacional, Calle Teniente Rinaldi 3870, 11000 Montevideo • T: +598 2 2224933 • mhistoricnac@mixmail.com •
Historical Museum
Politican' life 20th c ... 54387

Casa Quinta de Luis Alberto de Herrera, Museo Histórico Nacional, Calle L.A. de Herrera 3760, Montevideo • T: +598 2 2008832 • mhistoricnac@mixmail.com •
Historical Museum
Hist of Uruguay 20th c ... 54388

Centro de Artistas Plásticos, Calle Charrúa 2009, 11200 Montevideo • T: +598 2 4080262 • F: 4087224 • agranese@adinet.com.uy •
Fine Arts Museum ... 54389

Centro Municipal de Exposiciones Subste, Av 18 de Julio y Julio Herrera y Obes, 11400 Montevideo • T: +598 2 9087643 • F: 4095701 • kikebada@adinet.com.uy •
Dir.: *Enrique Badaró Nadal* •
Public Gallery ... 54390

Galería de Cinemateca, Calle Dr. Lorenzo Carnelli 1311, 11200 Montevideo • T: +598 2 4189819 • F: 4194572 • cinemuy@chasque.net • www.cinemateca.org.uy •
Dir.: *Manuel Martínez Carril* •
Public Gallery
Fine arts, photography ... 54391

Galería Pocitos, Calle Alejandro Chucarro 1036, 11300 Montevideo • T: +598 2 7082957 •
Public Gallery ... 54392

Museo Aeronáutico, Calle Dámaso Antonio Larrañaga 4045, 12000 Montevideo • T: +598 2 2152039 •
Science&Tech Museum ... 54393

Museo Antártico, Calle 8 de Octubre 2958, Montevideo • T: +598 2 4878341 •
Natural History Museum ... 54394

Museo Botánico, Calle 19 de Abril 1181, 11000 Montevideo • T: +598 2 3364005 •
Natural History Museum ... 54395

Museo de Arte Contemporáneo, Calle 18 de Julio 965, 11000 Montevideo • T: +598 2 9006662, 9028915 • F: 9028916 •
Fine Arts Museum ... 54396

Museo de Descubrimiento, Museo Histórico Nacional, Calle Zabala 1583, 11000 Montevideo • T: +598 2 9159951 • mhistoricnac@mixmail.com •
Historical Museum
Evokes the journeys of Cristobal Colón, the meeting of the two worlds, maps, dioramas and photographs ... 54397

Museo del Automóvil, Calle Colonia y Yi, 11000 Montevideo • T: +598 2 9024792 • F: 9021406 • acu@acu.com.uy •
Dir.: *Alvaro Casal Tatlock* •
Science&Tech Museum ... 54398

Museo del Azulejo, Calle Cavia 3080, 11300 Montevideo • T: +598 2 7096352 • www.artemercosur.com.uy/azulejo •
Special Museum ... 54399

Museo del Círculo de Estudios Ferroviarios del Uruguay, Calle Paraguay 1763, Montevideo •
Science&Tech Museum ... 54400

Museo del Gaucho y de la Moneda, Calle 18 de Julio 998, 11000 Montevideo • T: +598 2 908764 •
Ethnology Museum ... 54401

Museo Egipcio, Calle 4 de Julio 3068, 11600 Montevideo • T: +598 2 6280743, 6225352 • juancast@yahoo.com • www.geocities.com/Paris/Musee/7650 •
Dir.: *Prof. Juan José Castillos* •
Museum of Classical Antiquities / Archaeology Museum ... 54402

Museo Geológico del Uruguay, Calle Hervidero 2861, 11800 Montevideo • T: +598 2 2001951/53 • F: 2094905 • secretaria@dinamige.gub.uy • www.dinamige.gub.uy •
Dir.: *Felipe Puig* •
Science&Tech Museum / Natural History Museum – 1934
Geology, mining, rock coll ... 54403

Museo Histórico Nacional - Casa Rivera (National Historical Museum), Calle Rincón 437, 11000 Montevideo • T: +598 2 9151051 • F: 9156863 • mhistoricnac@mixmail.com •
Dir.: *Prof. Enrique Mena Segarra* •
Historical Museum – 1901
Prehistoric and colonial Indian cultural hist, Uruguay political hist, military, armaments, music ... 54404

Museo Juan Manuel Blanes, Av Millán 4015, 11700 Montevideo • T: +598 2 362248 • F: 3367134 • www.montevideo.gub.uy/museoblanes •
Dir.: *María del Carmen Heguerte de Soria* • *Jorge Satut* •
Fine Arts Museum
Sculpture, paintings, drawings, graphic arts, woodcarvings ... 54405

Museo Martin Perez, Calle Solis esq Cerrito, Parroquia San Francisco, 11000 Montevideo •
Local Museum ... 54406

Museo Municipal Cabildo de Montevideo, Calle Juan Carlos Gómez 1362, 11000 Montevideo • T: +598 2 9159685 •
Dir.: *Jorge Rodriguez Delucchi* •
Local Museum – 1915
Paintings, jewelry, icons, furniture, documents, maps – library, archive ... 54407

Museo Municipal de Bellas Artes Juan Manuel Blanes, Calle Millán 1415, 11700 Montevideo • T: +598 2 3362248 • www.artemercosur.com.uy/museos/uruguay/bla.html •
Dir.: *Mario C. Tempone* •
Fine Arts Museum – 1928
Paintings, graphic arts, sculpture, woodcarvings 54408

Museo Municipal de Historia del Arte, Calle Ejido 1326, 11200 Montevideo • T: +598 2 9089252 int 457 • www.artemercosur.com.uy/museos/uruguay/his.html •
Fine Arts Museum ... 54409

Museo Municipal Precolombino y Colonial, Calle Ejido 1326, 11200 Montevideo • T: +598 2 9089252 • www.artemercosur.com.uy/museos/uruguay/pre.html •
Fine Arts Museum ... 54410

Museo Nacional de Artes Visuales, Calle Julio Herrera y Reissig s/n, Parque Rodó, 11300 Montevideo • T: +598 2 7116124/27 • F: 7116054 • mnav@mnav.gub.uy • www.zfm.com/mnav •
Dir.: *Ágel Kalenberg* •
Fine Arts Museum
library ... 54411

Museo Nacional de Bellas Artes, Av Tomás E. Giribaldi 2283, Parque Rodó, 11300 Montevideo • T: +598 2 43800 •
Dir.: *Angel Kalenberg* •
Fine Arts Museum
Paintings, sculptures, graphic arts of Uruguay and America, works by Joaquín Torres Carcía – library ... 54412

Museo Nacional de Historia Natural y Antropología, Calle Juan Carlos Gómez 1436, 11000 Montevideo, mail addr: CC 399, 11000 Montevideo • T: +598 2 9160908, 9167825 • F: 9170213 • mnhn@internet.com.uy • www.mec.gub.uy/natura •
Dir.: *Prof. Alvaro Mones* •
Natural History Museum – 1837
Botany, zoology, anthropology, geology, paleontology, history of science – library ... 54413

Museo Naval, Rambla Charles de Gaulle s/n, esq Luis Alberto de Herrera, Montevideo • T: +598 2 6221084 •
Historical Museum ... 54414

Museo Parque de Esculturas, Edificio Libertad, Calle Luis Alberto de Herrera 3350, Montevideo •
Ethnology Museum / Folklore Museum ... 54415

Museo Pedagógico José Pedro Varela, Pl de Cagancha 1175, 11100 Montevideo • T: +598 2 9004744, 9020915 • F: 9084131 • lemamc@adinet.com.uy • www.crnti.edu.uy/museo •
Dir.: *María del Carmen Lema Pensado* •
Special Museum – 1889
Educational hist ... 54416

Museo Romántico - Casa de Montero, Museo Histórico Nacional, Calle 25 de Mayo 428, 11000 Montevideo • T: +598 2 9155361 • mhistoricnac@mixmail.com •
Fine Arts Museum / Folklore Museum
Costume 19th c, music – hemerotheca ... 54417

Museo Torres García, Peatonal Sarandí 683, 11000 Montevideo • T: +598 2 9162663 • F: 9152635 • torresgarcia@montevideo.com.uy • www.torresgarcia.org.uy •
Dir.: *Dr. Jimena Perera Diaz Díaz* •
Fine Arts Museum ... 54418

Museo Virtual de Artes el Pais, Pl de Cagancha 1164, 11000 Montevideo • T: +598 2 9006662 • www.diarioelpais.com/muva •
Fine Arts Museum ... 54419

Museo y Centro Cultural AGADU, Calle Héctor Gutiérrez Ruiz 1181, Montevideo • T: +598 2 9012773 •
Folklore Museum ... 54420

Museo y Jardín Botánico Profesor Atilio Lombardo, Av 19 de Abril 1181, 11700 Montevideo • T: +598 2 3364005 • F: 3366488 • botanico@adinet.com.uy •
Dir.: *Carlos A. Brussa* •
Natural History Museum – 1902
Botany ... 54421

Museo Zoológico Dámaso A. Larrañaga, Rambla República de Chile 4215, 11400 Montevideo • T: +598 2 6220258 •
Dir.: *Juan Pablo Cuello* •
Natural History Museum – 1956
Zoology – library ... 54422

Museos del Gaucho y de la Moneda, c/o Banco de la República Oriental del Uruguay, Calle 18 de Julio 998, 11100 Montevideo • T: +598 2 9008764 • F: 9015686 •
Dir.: *Frederico Slinger* •
Special Museum
Banknotes, coins, history of banking ... 54423

Palacio Taranco, Museo de Artes Decorativas, Calle 25 de Mayo 376, Montevideo • T: +598 2 9156060, 9151101 • F: 9156060 • taranco@mec.gub.uy •
Decorative Arts Museum / Archaeology Museum
Archaeology, art ... 54424

Sala de Arte Carlos F. Sáez, Calle Rincón 575, 11000 Montevideo •
Fine Arts Museum ... 54425

Piriápolis

Museo de Piria, Ruta 37, 20200 Piriápolis • T: +598 43 3268 •
Local Museum ... 54426

Punta Ballena

Museo Antonio Lussich, Ruta 93, 20003 Punta Ballena • T: +598 42 578077 • F: 578077 • arboretumlussich@maldonado.gub.uy • www.artemercosur.com.uy/lussich •
Gen. Dir.: *Dr. Jorge Curbelo* • Dep. Dir.: *Dr. María Eloisa Rivero* •
Natural History Museum – 1997 ... 54427

Museo del Azulejo Francés, Casona de Don Antonio Lussich, Ruta 93, 20003 Punta Ballena • T: +598 42 578077 • F: 578077 • arboretumlussich@maldonado.gub.uy • www.maldonado.gub.uy •
Gen. Dir.: *Dr. Jorge Curbelo* • Dep. Dir.: *Dr. María Eloisa Rivero* •
Decorative Arts Museum – 1997 ... 54428

Punta del Este

Museo Ralli, Calle Los Arrayanes y Jacaranda, Barrio Beverly Hills, 20100 Punta del Este • T: +598 42 483476 • F: 487378 • museoralli_uruguay@hotmail.com •
Dir.: *Harry Recanati* •
Fine Arts Museum
Latin American ang European art (15th-18th c and post-impressionist period) ... 54429

Rivera

Museo Municipal de Artes Plásticas, Calle Artigas 1019, Rivera • T: +598 62 31900 • osantos@netgate.com.uy •
Dir.: *Osmar Santos* •
Fine Arts Museum – 1970 ... 54430

Museo Municipal de Historia y Arqueología, Calle Artigas 1019, Rivera • T: +598 62 31900 • osantos@netgate.com.uy •
Dir.: *Osmar Santos* •
Local Museum / Archaeology Museum – 1946 ... 54431

San Carlos

Museo Regional de San Carlos, Calle Carlos Reyles y Leonardo Olivera, 20400 San Carlos • T: +598 42 23781/82 •
Local Museum ... 54432

San Gregorio de Polanco

Museo San Gregorio de Polanco, 45200 San Gregorio de Polanco •
Fine Arts Museum ... 54433

San José de Mayo

Museo de Bellas Artes Departamental de San José, Calle Dr Julián Beccerro de Bengoa 493, 80000 San José de Mayo • T: 3642 •
Dir.: *César Bernesconi* •
Fine Arts Museum – 1947
Paintings, sculptures, graphic art, ceramics – library ... 54434

Tacuarembó

Museo del Indio y del Gaucho, Calle 25 de Mayo 315, 45000 Tacuarembó • Dir.: *Washington Escobar* • Ethnology Museum
Indian and Gaucho arts and crafts, armaments, implements 54435

Treinta y Tres

Museo de Bellas Artes Agustin Araujo, Pablo Zufriategui 1272, Treinta y Tres • Fine Arts Museum 54436

Uzbekistan

Afchona

Abu Ali Ibn Sina Yodgorlik Muzeyi (Abu Ali Ibn Sina Memorial Museum), Afchona settlement, 706450 Afchona • T: +998 65 3545142 • Dir.: *Mansur Nurullayev* • Special Museum 54437

Andijon

Andijon Viloyat Adabiyot va San'at Muzeyi (Andijan Regional Museum of Literature and Art), Bazaar küchasi 4, 710000 Andijon • T: +998 83742 257302, 251408 • F: 223087 • adsanmuz_1@rambler.ru • Dir.: *Tulkin Khuldashev* • Special Museum 54438

Andijon Viloyat Ülkashunoslik Muzeyi (Andijan Regional Museum of Local Lore), Fitrat küchasi 258, 710000 Andijon • T: +998 83742 251639, 251823 • Local Museum 54439

Babur Yodgorlik Muzeyi, Andijon Viloyat Adabiyot va San'at Muzeyi (Babur Memorial Museum - Dept. of Andijan Regional Museum of Literature and Art), Bazaar küchasi 21, Andijon • T: +998 83742 257302 • adsanmuz_1@rambler.ru • Dir.: *Tulkin Khuldashev* • Head of Dep.: *Abdullajon Djuraboyev* • Special Museum / Historic Site
Ark-ichi medresseh from 19th c, life and work of the Uzbek writer and statesman Zahiriddin Muhammad Babur and his dynasty 54440

Chulpan Yodgorlik Muzeyi, Andijon Viloyat Adabiyot va San'at Muzeyi (Chulpan Memorial Museum - Dept. of Andijan Regional Museum of Literature and Art), 1 Chulpan Av, Chulpan Park, Andijon • T: +998 83742 257302 • adsanmuz_1@rambler.ru • Dir.: *Tulkin Khuldashev* • Head of Dep.: *Muhammad Yusupov* • Special Museum
Life and work of the Uzbek writer and poet Abdulhamid Chulpan, memorabilia 54441

Tasviriy San'at Muzeyi - Andijon Qal'asi, Andijon Viloyat Adabiyot va San'at Muzeyi (Fine Art Museum - Andijan Fortress, Dept. of the Andijan Regional Museum of Literature and Art), 202 Bahovuddun Naqshband, Andijon • T: +998 83742 257302 • adsanmuz_1@rambler.ru • Dir.: *Tulkin Khuldashev* • Head of Dep.: *Avazbek Poziljonov* • Fine Arts Museum / Historic Site
Andijan region and Uzbek painting, fortress from 2nd half of 19th c 54442

Angren

Angren Viloyat Ülkashunoslik Muzeyi (Angren State Museum of Local Lore), 69, District 5-1a, 702500 Angren • T: +998 37166 33872 • Dir.: *L.L. Kerimova* • Local Museum 54443

Buxoro

Buxoro Qala Davlat Muzeyi Qo'riqxonasi (Buchara State Museum), Afrasiab küchasi 2, 705016 Buxoro • T: +998 65 2241349 • F: 2241349 • robert@bukhara.net • Dir.: *Robert V. Almeev* • Decorative Arts Museum / Historical Museum / Archaeology Museum / Fine Arts Museum / Natural History Museum – 1922
Textiles, carpets, gold embroidery, decorative arts, coins, seals, manuscripts, armour, photo coll (pre and post Soviet revolution, Soviet era documents, agricultural equipment, pre-revolutionary medical instruments, music, archeology, religion – 10 special museums 54444

Sitorai Mochi-Khosa, 705016 Buxoro • Decorative Arts Museum 54445

Farg'ona

Farg'ona Viloyat Ülkashunoslik Muzeyi (Fergana Regional Museum of Local Lore), B. Usmanchodjaeva küchasi 26, 712000 Farg'ona • T: +998 3732 243191, 243870 • F: 243191 • bahodir_hashimov@intal.uz • Dir.: *Bahodir Djuraevič Hashimov* • Local Museum 54446

Hamza Hakim-Jade Niazy Muzeyi (Hamza Hakim-Zadé Niazy Memorial Museum), Hamzaabad, 712000 Farg'ona der. Shohirmadon • T: +998 3732 26681 • Dir.: *A. Hamzaev* • Historical Museum 54447

Jizzax

Jizzax Viloyat Ülkashunoslik Muzeyi (Dzhizak State Museum of Local Lore), Rashidow küchasi 4, 708000 Jizzax • T: +998 37222 61721 • Local Museum – 1986 54448

Marg'ilon

Yoldosh Oxunboboev Yodgorlik Muzeyi (Memorial Museum of Yuldash Akhunbabayev), Ozodlik Maydoni 1-uy, Marg'ilon • T: +998 3732 333206, 724910 • muzey_yo@list.ru • Dir.: *Alijonov Ganijon* • Historical Museum – 1964 54449

Nukus

Natural History Museum, Karakalpakstan, Nukus • Natural History Museum
Fauna and flora of the Karakalpakstan region 54450

Qoraqalpoġiston Respublikasi Ülkashunoslik Muzeyi (Karakalpakstan State Regional Museum), pl Nezavisimosti, 742000 Nukus • T: +998 3612 227389 • hist-museum@mailgate.ru • Dir.: *Svetlana Nurabullaeva* • Local Museum – 1929 54451

Qoraqalpoġiston Tarikhi Muzeyi (Karakalpak Historical Museum), Rachmatov küchasi 3, 74200 Nukus • Historical Museum
Uzbek hist, military hist 54452

Savitsky nomidagi Qoraqalpoġiston Respublikasi Davlat San"at Muzeyi (Savitsky State Art Museum of Karakalpakstan), Rzaev 127, 742000 Nukus • T: +998 3612 222556 • Dir.: *Marinika Babanazarova* • Fine Arts Museum – 1966
Karakalpak applied art, art of ancient Khorezm, Russian avantgarde of 1910-1930 – library, archive 54453

Qarshi

Qashqadaryo Viloyat Ülkashunoslik Muzeyi (Kashkadarya Regional Museum of Local Lore), Husarskaya küchasi 15, Qarshi • T: +998 3752 231709 • Dir.: *B.R. Artykov* • Local Museum 54454

Qo'qon

Hamza Hakim-zade Niazy Yodgorlik Muzeyi, Qo'qon Viloyat Ülkashunoslik Muzeyi (Hamza Hakim-zade Niazy House Museum of the Kokand Regional Studies Museum), Matbuot küchasi 28, 713000 Qo'qon, mail addr: Istiklol küchasi 2, 713000 Qo'qon • T: +998 37355 36046, 26839 • F: 26839 • kokandmuseum@mail.ru • Dir.: *I. Rahmatulaev* • Special Museum
Life and work of the famous Uzbek poet 54455

K. Hajdarov Yodgorlik Muzeyi, Qo'qon Viloyat Ülkashunoslik Muzeyi (K. Hajdarov House-Museum of the Kokand Regional Studies Museum), 713000 Qo'qon, mail addr: Istiklol küchasi 2, 713000 Qo'qon • T: +998 37355 36046, 26839 • F: 26839 • kokandmuseum@mail.ru • Special Museum 54456

Mukimi- Hudjra Muzeyi, Qo'qon Viloyat Ülkashunoslik Muzeyi (Mukimi-Hudjta Study-room Museum of the Kokand Regional Studies Museum), 713000 Qo'qon, mail addr: Istiklol küchasi 2, 713000 Qo'qon • T: +998 37355 36046, 26839 • F: 26839 • kokandmuseum@mail.ru • Special Museum 54457

Qo'qon Viloyat Ülkashunoslik Muzeyi (Kokand Regional Studies Museum), Istiklol küchasi 2, 713000 Qo'qon • T: +998 37355 36046, 26839 • F: 26839 • kokandmuseum@mail.ru • Dir.: *I. Bekmuradov* • Fine Arts Museum / Historical Museum – 1925
Former palace of the Kokand khan Khudayarhan, Kokand khanat hist, 20th c fine painting and sculpture from Western Europe, Russia and Uzbekistan, numismatic coll, coins from Samanids and Karahanids reign period, coins of Kokand khanate, calligraphy fund – Architectural complex Djami 54458

Samarqand

Afrosiyob - Samarqand Thariki Muzeyi (Afrosyob - The Museum of the History of the Foundation of Samarkand), Toshkent küchasi 7, 703000 Samarqand • T: +998 3662 355336 • www.sammus.uz • Sc. Staff: *S. Zubajdullaeva* • Archaeology Museum / Historical Museum – 1970
Archaeology, items from the excavations at Afrasiab; paintings from late 7th/early 8th c from the palace of the Sogdiana khan 54459

Akmal Ikramov nomidagi Tarikhi Kültüri va San"atshunoslik Özbekiston Khalqlari Davlat Müzeyi (Akmal Ikramov State Museum of the History of Culture and Art of the Peoples of Uzbekistan), Pl Registan 1, 703001 Samarqand • T: +998 3662 353896, 355808 • F: 353896, 355808 • www.sammus.uz • Dir.: *N. Mahmudov* • Historical Museum / Fine Arts Museum / Historic Site – 1874
Cultural hist, fine art, ceramics, archeology, costumes, jewellery, manuscripts, coins, old photos – library, archive 54460

International Museum of Peace and Solidarity, 56 Mustaqillik küchasi, Central Recreation Park, Samarqand, mail addr: POB 76, 703000 Samarqand • T: +998 662 331753 • F: 331753 • imps@rol.uz • www.peace.museum.com • Dir.: *Anatoly Ionesov* • Special Museum 54461

Sadriddin Aini Yodgorlik Muzeyi (Sadriddin Aini House-Museum), Registan küchasi 7, 703000 Samarqand • T: +998 662 355153 • www.sammus.uz • Dir.: *A. Ganiev* • Historical Museum – 1964
Literature, hist, life and work of Sadriddin Aini (1878-1954), writer in Tajik and Uzbek languages 54462

Samarqand Ülkashunoslik Muzeyi (Samarkand Museum of Regional Studies), Rashidov küchasi 51, 703000 Samarqand • T: +998 662 330352 • www.sammus.uz • Archaeology Museum / Historical Museum – 1981 54463

Ulug-Beg Yodgorlik Muzeyi (Ulug-Beg Memorial Museum), c/o Ulug-Beg Observatory, Toshkent küchasi 1, 703000 Samarqand • T: +998 662 350345 • www.sammus.uz • Dir.: *N. Akramov* • Historical Museum – 1970
Hist of scientific thought in central Asia, life and work of the mathematician and astronomer Khorezmi (758-850), of the scientist-encyclopedist al-Biruni (973-1039) and of the scientist and astronomer Ulugbek (1394-1449), hist of Ulugbeks Star Atlas Ziddji-Guragoni from 1437 54464

Shakhrisabz

Amir Timur Museum, Ipak Yoli, Shakhrisabz • Historical Museum
Timur's history 54465

Termiz

Surxondaryo Vilojati Arxeologiya Muzeyi (Archeological Museum of Surkhandarya Region), At-Hakim Termizi 29, 732000 Termiz • T: +998 376 2275829 • Dir.: *I.T. Botirov* • Archaeology Museum – 2002
Archaeological coll from Bronce epoch, Early Iron Age, Hellenism, Kushan period, early and developed Midlle Ages, numismatical coll from 2nd c BC till 20th c AC – library, archives 54466

Toshkent

Alisher Navoii nomidagi Davlat Adabiyot Muzeyi (Alisher Navoi State Museum of Literature), Navoi küchasi 69, 700011 Toshkent • T: +998 71 1441268 • Dir.: *S.R. Khasanov* • Special Museum 54467

Centr Sovremennogo Iskusstva (Modern Art Center), Ahunbabaev küchasi 4, 700047 Toshkent • T: +998 71 1339743, 1338051 • F: 1336902 • Fine Arts Museum 54468

Centralnyj Muzej Vooružennych Sil Uzbekistana (Central Museum of the Army of Uzbekistan), Abdullaev küchasi 98, 700170 Toshkent • T: +998 71 1624175 • Military Museum 54469

Gelereja Iskusstva Uzbekistana (Art Gallery of Uzbekistan), Bujuk Turon küchasi 2, 700078 Toshkent • T: +998 71 1335674 • F: 1336902 • Public Gallery 54470

History Museum of the People of Uzbekistan, Sharaf Rashidova 30, 700047 Toshkent • Historical Museum
Hist of Turkestan from ancient times to present 54471

M.T. Oibek nomidagi Özbekiston Khalqlari Tarikhi Muzyi, Özbekiston Respublikasi Fanlar Akademiyasi (M.T. Aibek Museum of History of the Peoples of Uzbekistan at the Academy of Sciences of Uzbekistan), Babaev küchasi 15, 700047 Toshkent • T: +998 71 1335733 • F: 1391038 • Dir.: *K.Ch. Inojatov* • Historical Museum
library 54472

M.T. Oibek Yodgorlik Muzeyi (M.T. Aibek House-Museum), Tezetdinov küchasi 26, 700000 Toshkent • T: +998 71 3980900 • ainur2000@inbox.ru • Dir.: *Ainur Tashmuchamedova* • Special Museum
Life and work of the writer M.T. Aibek, manuscripts, first editions of Aibek's works; manuscripts of classic oriental literature 54473

Mouhtar Ashrafi Yodgorlik Muzeyi (Mouhtar Ashrafi Museum), C-1, 15-25, 700000 Toshkent • T: +998 71 1332384 • F: 322731 • Dir.: *G. Hamraeva* • Music Museum – 1982
Coll of classic music disks and music hall for concerts, coll of classic Uzbek music instr 54474

Moyle Mubarek Library Museum, Zarqaynar 114, Toshkent • T: +998 71 1600302 • Historical Museum
7th c Osman Quran, also 20000 additional books and rare manuscripts 54475

Muzej Geologii (Museum of Geology), Furkat küchasi 4, 700021 Toshkent • T: +998 71 451551, 451337 • Natural History Museum 54476

Muzej Kinoiskusstva Uzbekistana (Museum of Cinematography of Uzbekistan), Özbekiston pr 98, 700027 Toshkent • T: +998 71 458976 • Special Museum
library 54477

Muzej Olimpijskoj Slavy (Museum of Olympic Glory), Sharaf Rashidov pr 4a, 700017 Toshkent • T: +998 71 1447602, 1447603 • Special Museum 54478

Muzej Pamjati Žertv Repressij (Museum of Repressive Measures Victim Memory), Amir Temur küchasi, 700000 Toshkent • T: +998 71 1124246 • Historical Museum 54479

Muzej Prikladnogo Iskusstva Uzbekistana (Museum of Applied Art of Uzbekistan), Rakatboshi 15, 700031 Toshkent • T: +998 71 1367436 • Decorative Arts Museum – 1998
Hand-made embroidery, jewelry, ceramics, wood carving, carpet wearing, wood painting – library 54480

Muzej Zdravoochranenija (Museum of Health Services), Ahunbabaev küchasi 3, 700060 Toshkent • T: +998 71 1334053, 1334278 • Special Museum 54481

Muzej Železnodorožnoj Techniki (Museum of Railway Engineering), Turkiston küchasi 8, 700015 Toshkent • T: +998 71 596708, 1997040 • Science&Tech Museum 54482

Özbekiston Respublikasi San'at Akademiyasi (Uzbekistan Art Academy), Sharaf Rashidov Pr 40, 700029 Toshkent • T: +998 71 565046, 565047, 525625 • F: 525625 • acadartu@online.ru • www.arts-academy.uz • Public Gallery 54483

Özbekiston San'at Davlat Muzeyi (Fine Arts Museum Uzbekistan), Movarounnahr Küchasi 16, 700060 Toshkent • T: +998 71 1367436 • Dir.: *D.S. Rusibaev* • Fine Arts Museum – 1918
Art, architecture, sculpture, fine arts, graphics, music, theatre, ethnography – library 54484

Özbekiston Tarikhi Davlat Muzeyi (State Museum of History of Uzbekistan), Sharaf Rashidova küchasi 3, 700047 Toshkent • T: +998 71 1391778 • F: 1391038 • Historical Museum – 1922
Hist of the Central Asian area, exhibits on the life of Central Asiatic life ranging from primitive communal societies to the present time – library 54485

Sergej Borodin Muzeyi (Sergey Borodin House-Museum), Yukari-Yuz küchasi 18, 700000 Toshkent • T: +998 71 1330932 • F: 2406264 • Kalohc@aol.com • Dir.: *N. Chebanova* • Special Museum – 1978
Coins, books, life of Sergey Borodin 54486

Sergej Esenin Muzeyi (Literary Museum of Sergey Esenin), Tolstoy tup 20, 700000 Toshkent • T: +998 71 1371179 • esenin@tps.uz • Special Museum
Life of Sergey Esenin (1895-1925), Esenin and the Orient 54487

Tamara Hanum Yodgorlik Muzeyi (Tamara Khanum House-Museum), Tamara Hanum küchasi 1, 700000 Toshkent • T: +998 71 678690 • Special Museum 54488

Termuridi Tarikhi Muzeyi, Özbekiston Respublikasi Fanlar Akademiyasi (Museum of the History of Termurids at the Academy of Sciences of Uzbekistan), Amir Temur küchasi 1 (Yunusabadsk), 700000 Toshkent • T: +998 71 1336228 • Dir.: *Nozim Khabibullaev* • Ethnology Museum / Historical Museum 54489

Ural Tansykbaev Yodgorlik Muzeyi (Ural Tansykbaev House-Museum), Burchanov 1, 5, 700170 Toshkent • T: +998 71 1626230 • Special Museum 54490

Yuldosh Ahunbabaev Yodgorlik Muzeyi (Yuldosh Akhunbabaev House-Museum), Lashkarbegi küchasi 19, 700000 Toshkent • T: +998 71 1335131 • Special Museum 54491

Xiva

Ichan Qala Davlat Muzej Qöriqxonasi (Ichan-Kala National Reserve Museum), A. Baltaev 41, Xiva • T: +998 36237 53169 • F: 53169 • maqsud@tkt.uz • www.uztour.narod.ru • Dir.: *B. Davletov* • Local Museum
Local history, architecture 54492

Yangiyo'l

Usman Jusupov Yodgorlik Muzeyi (Usman Jusupov Memorial Museum), Dachnaya küchasi 6, Yangiyo'l • T: +998 37160 22519 • Special Museum 54493

U

Vanuatu

Port Vila

Michoutouchkine Pilioko Foundation, Road to Pango, Port Vila, mail addr: POB 224, Port Vila • T: +678 23053, 27753 • F: 24224 •
Dir.: *A. Pilioko* • *N. Michoutouchkine* •
Fine Arts Museum / Ethnology Museum 54494

Vanuatu Cultural Centre and National Museum, Rue Picardie, Port Vila, mail addr: POB 184, Port Vila • T: +678 22129 • F: 26590 • vks@vanuatu.com.vu •
Dir.: *Ralph Regenvanu* •
Ethnology Museum – 1959
Local art and ethnography 54495

Vatican City

Città del Vaticano

Archivio Segreto Vaticano, Cortile del Belvedere, 00120 Città del Vaticano • T: +39 0669883314 • asv@asv.va •
Head: *Sergio Pagano* •
Library with Exhibitions 54496

Cappella Sistina, Sala e Gallerie Affrescate, Musei Vaticani, Viale Vaticano, 00120 Città del Vaticano • T: +39 0669883332 •
Dir. Gen.: *Dr. Francesco Buranelli* • Dir.: *Dr. Fabrizio Mancinelli* •
Fine Arts Museum 54497

Collezione d'Arte Religiosa Moderna, Musei Vaticani, Palazzo Apostolico Vaticano, 00120 Città del Vaticano •
Cur.: *Dr. Arnold Nesselrath* •
Fine Arts Museum – 1973
Paintings, sculptures, drawings 54498

Galleria degli Arazzi, Musei Vaticani, Viale Vaticano, Palazzo Apostolico Vaticano, 00120 Città del Vaticano •
Fine Arts Museum
Tapestry coll 54499

Galleria Lapidaria, Musei Vaticani, Viale Vaticano, Palazzo Apostolico Vaticano, 00120 Città del Vaticano •
Museum of Classical Antiquities
Greek, Latin Cristian and Hebrew inscriptions 54500

Monumenti Musei e Gallerie Pontificie, Viale Vaticano, 00120 Città del Vaticano • T: +39 0669883333 • F: +39 0669885100 • musei@scv.va • www.vatican.va •
Dir. Gen.: *Dr. Francesco Buranelli* • Dep. Dir.: *Dr. Arnold Nesselrath* (Byzantine, Medieval and Modern Art) • *Roberto Zagnoli* (Ethnology) • Cur.: *Dr. Maurizio Sannibale* (Etruscian-Italian Antiquities) • *Dr. Paolo Liverani* (Classical Antiquities) • *Dr. Giandomenico Spinola* (Paleochristian Art) • *Dr. Giorgio Filippi* (Epigraphical Collection) • *Dr. Micol Forti* (18th c and Contemporary Art) • *Pietro Amato* (History) • *Dr. Anna Maria De Strobel* (Byzantine, Medieval and Modern Art) • *Dr. Ester Maria Console* (Ethnology) • Coord.: *Dr. Lorenzo Nigro* (Oriental Antiquities) • Photograph archive: *Dr. Guido Cornini* • Head: *Prof. Ulderico Santamaria* (Scientific Research) • Restorer: *Maurizio De Luca* (Painting) •
Early 16th c 54501

Museo Chiaramonti e Braccio Nuovo, Musei Vaticani, Palazzo Apostolico Vaticano, 00120 Città del Vaticano • T: +39 0669883333 •
Cur.: *Dr. Paolo Liverani* •
Fine Arts Museum
Statues of the Nile, of Demosthenes and of the Augustus 'of Prima Porta', Greek and Roman art 54502

Museo Gregoriano Egizio, Musei Vaticani, Palazzo Apostolico Vaticano, 00120 Città del Vaticano •
Museum of Classical Antiquities – 1839
Egyptian antiquities discovered in Rome, Roman imitations of Egyptian statues from Hadrian's villa in Hadrian's Tivoli 54503

Museo Gregoriano Etrusco, Musei Vaticani, Palazzo Apostolico Vaticano, 00120 Città del Vaticano •
Cur.: *Dr. Maurizio Sannibale* •
Decorative Arts Museum – 1837
Bronzes, terracottas, jewellery and Greek vases from Etruscan tombs 54504

Museo Gregoriano Profano, Musei Vaticani, Palazzo Apostolico Vaticano, 00120 Città del Vaticano •
Cur.: *Dr. Paolo Liverani* •
Fine Arts Museum – 1844
Roman sculptures 54505

Museo Missionario Etnologico, Musei Vaticani, Palazzo Apostolico Lateranese, Piazza San Giovanni 4, 00184 Città del Vaticano •
Cur.: *Roberto Zagnoli* •
Ethnology Museum – 1926
Ethnographical coll 54506

Museo Padiglione delle Carozza, Museo Storico (Carriage Museum), Vatican Gardens, 00120 Città del Vaticano •
Dir.: *Pietro Amato* •
Science&Tech Museum – 1973
Carriages, berlins and first cars used by the Popes 54507

Museo Pio Clementino, Musei Vaticani, Palazzo Apostolico Vaticano, 00120 Città del Vaticano •
Cur.: *Dr. Paolo Liverani* •
Fine Arts Museum
Greek and Roman art 54508

Museo Pio Cristiano, Musei Vaticani, Viale Vaticano, 00120 Città del Vaticano • T: +39 0669883041 • F: +39 0669885061 • secreteria.musei@scv.va •
Cur.: *Dr. Giandomenico Spinola* •
Museum of Classical Antiquities – 1854
Large coll of sarcophargi, Latin and Greek inscriptions from Christian cemeteries and basilicas, early Christian art 54509

Museo Profano, Biblioteca Apostolico Vaticano, 00120 Città del Vaticano • T: +39 066985051 •
Cur.: *Dr. Giovanni Morello* •
Fine Arts Museum / Decorative Arts Museum – 1767
Bronze sculptures and minor arts of the classical era 54510

Museo Sacro, Biblioteca Apostolico Vaticano, 00120 Città del Vaticano • T: +39 066985051 •
Dir.: *Dr. Giovanni Morello* •
Religious Arts Museum – 1756 54511

Museo Storico Artistico, Tesoro, Basilica di San Pietro, Capitolo di San Pietro, 00120 Città del Vaticano • T: +39 0669881840 • F: +39 0669883465 •
Dir.: *Ennio Francia* •
Religious Arts Museum
Sacred vestments, silver and goldsmith work 54512

Museo Storico Vaticano (Historical Museum), Palazzo Apostolico Lateranese, Piazza San Giovanni 4, 00184 Città del Vaticano •
Dir.: *Pietro Amato* •
Historical Museum – 1973
Arms, uniforms and armour of the pontificial court and Army Court 54513

Pinacoteca Vaticana, Musei Vaticani, Palazzo Apostolico Vaticano, 00120 Città del Vaticano •
Cur.: *Dr. Arnold Nesselrath* •
Fine Arts Museum – 1932
Paintings by Giotto, Fra Angelico, Raphael, Leonardo da Vinci, Titian and Caravaggio, the Raphael tapestries 54514

Venezuela

Acarigua

Museo José Antonio Páez, Páez en Curpa, Acarigua 3350 •
Historical Museum 54515

Aragua

Museo Ornitológico, Vía Ocumare de la Costa, Parque Nacional Rancho Grande, Aragua 2112 •
Natural History Museum
Ornithology 54516

Barcelona

Galería de Arte de la Escuela Armando Reverón, Calle Páez, Barcelona 6001 •
Fine Arts Museum 54517

Galería de la Asamblea Legislativa, Av Fuerzas Armadas, Barcelona 6001 •
Fine Arts Museum 54518

Galería del Museo Anzoátegui, Aeropuerto Internacional José Antonio Anzoátegui, Barcelona 6001 •
Fine Arts Museum 54519

Galería Municipal de Arte Moderno, Av Estadium, Barcelona 6001 •
Fine Arts Museum 54520

Museo de Anzoátegui, Calle Juncal 3-45, Barcelona 6001 •
Local Museum – 1981
17th-19th c painting and sculpture, regional hist, 19th c weapons, docs and furniture – library 54521

Museo El Sol de las Botellas, Peñalver, Barcelona 6001 •
Fine Arts Museum / Decorative Arts Museum
Painting, metal works, ceramics, graphic design 54522

Barinas

Museo Alberto Arvelo Torrealba, Av Medina Jiménez y Calle 5 de Julio, Barinas 5201 •
Special Museum – 1981 54523

Barquisimeto

Museo de Barquisimeto, Carrera 15 entre Calles 25 y 26, Barquisimeto 3001 • T: +58 251 310557, 317479 • F: 310889 • fundamuseo@cantv.net • www.geocities.com/Athens/Forum/4330/ •
Dir.: *Francisco Blavia* •
Local Museum 54524

Boca de Río

Museo del Mar, Universidad de Oriente, Península de Macanao, Boca de Río 6301 •
University Museum / Historical Museum 54525

Caracas

Casa Natal del Libertador Simón Bolívar (Simón Bolívar's Birthplace), Calle San Jacinto a Traposos, Pl El Venezolano, Caracas • T: +58 212 5412563 •
Cur.: *Josefina de Sandoval* •
Historical Museum
Murals by Tito Salas depicting the life of Simón Bolívar and the events of the Independence Movement 54526

Centro Cultural Corp Group, c/o Torre Corp Banca, Pl La Castellana, Caracas • T: +58 212 2061149 •
Public Gallery 54527

Centro de Arte La Cañuela, Av Principal, CC Alto Prado, 1080 Caracas • T: +58 212 9784624, 9781324 • mcanuela@telecel.net.ve •
Fine Arts Museum 54528

Centro de Arte La Estancia, Av Frederico de Miranda, La Floresta, 1060 Caracas • T: +58 212 2080412 •
Fine Arts Museum 54529

Colección Ornitológica W.H. Phelps, Blvd de Sabana Grande, Edificio Gran Sabana, Caracas 1050 • T: +58 212 7615631 • F: 7633695 • mlentino@reacciun.ve •
Dir.: *Kathleen D. Phelps* • Cur.: *Miguel Lentino* • Sc. Staff: *Robin Restall* (Research) • *Irving Carreño* • *Margarita Martínez* •
Natural History Museum
Ornithology 54530

Galeria de Arte de la UCV, Universidad Central de Venezuela, Adyacente al Aula Magna, Caracas • T: +58 212 619811 •
Fine Arts Museum 54531

Galería de Arte Nacional, Plaza de los Museos, Los Caobos, Caracas 1050 • T: +58 212 5781818 • F: 5781661 • fgan@infoline.wtfe.com • www.wtfe.com/gan •
Public Gallery – 1976
Venezuelan visual art from pre-Hispanic time to the present 54532

Museo Alejandro Otero, Complejo Cultural La Rinconada, Caracas 1010-A • T: +58 212 6820941, 6821841 • F: 6820023, 6820428 • museotero@cantv.net •
Pres.: *Marinelly Bello Morales* •
Fine Arts Museum 54533

Museo Armando Reverón, Callejón Colón, Av La Playa, Sec Las Quince Letras, Macuto, La Guaira, 1160 Caracas • T: +58 212 4611357 •
Special Museum 54534

Museo Arturo Michelena, Calle Urapal 82, Altagracia, Caracas 1010 • T: +58 212 8604802, 8623957 • museomichelena@cantv.net • www.museomichelena.arts.ve •
Fine Arts Museum
19th c home of the printer Arturo Michelena, memorabilia, docs 54535

Museo Audiovisual, Parque Central, Edificio Tacagua Nivel Bolívar, Caracas 1010, mail addr: c/o Academia Nacional de Ciencias y Artes del Cine y la Televisión, Apdo 17030, Caracas • T: +58 212 5721046, 5737757 •
Dir.: *Prof. Oscar Moraña* • *Sandra Ramón Vilarasau* •
Science&Tech Museum
Exhib of technical audiovisual instr and apparatus used in T.V. radio, and theatre – library, archives 54536

Museo Bolivariano (Bolívar Museum), Calle San Jacinto a Traposos, Pl El Venezolano, Caracas 1010 • T: +58 212 5459828 •
Dir.: *Flor Zambrano de Gentile* •
Historical Museum – 1911
Personal memorabilia of revolutionary hero Simón Bolívar (1783-1830), portraits, historical paintings of Bolívar contemporaries – library 54537

Museo Cuadra de Bolívar, Calle Piedras a Calle Bárcenas, Caracas 1010 •
Historical Museum
Murals by Titos Salas depicting the struggle for independence, correspondence and docs by Bolívar 54538

Museo de Arquitectura, Edificio Torre La Primera, Av Francisco de Miranda, Chacao, 1060 Caracas • T: +58 212 2620327, 2622171 •
Fine Arts Museum 54539

Museo de Arte Colonial, Quinta de Anauco, Av Panteón y Calle Gamboa, Caracas 1011 • T: +58 212 5518650, 5518190 • F: 5518517 • infoanauco@quintadeanauco.org.ve • www.quintadeanauco.org.ve •
Dir.: *Carlos F. Duarte* •
Fine Arts Museum / Decorative Arts Museum – 1942
Paintings, sculpture, furniture, china of the Venezuelar Colonial period – library 54540

Museo de Arte Contemporáneo de Caracas Sofía Imber, Parque Central, Edificio Anauco Nivel Lecuna, Caracas 1010, mail addr: Apdo 17093, Caracas 1010 • T: +58 212 5738289, 5734602 • F: 5771883 • info@maccsi.org.ve •
Dir.: *Sofía Imber* •
Fine Arts Museum / Decorative Arts Museum – 1973
Venezuelan and international works of art, sculpture, painting, arts and crafts, Calder's tapestries, works by Soto and Vasarely – library 54541

Museo de Arte Popular Petare, Calle Guánchez, CC Lino Clemente, Centro Histórico de Petare, Caracas 1073 • T: +58 212 218741, 218335 •
Folklore Museum 54542

Museo de Bellas Artes de Caracas, Plaza Morelos, Parque Los Caobos, Caracas 1050 • T: +58 212 5710169 • F: 5712119 • fmba@reacciun.ve •
Pres.: *María Elena Huizi* • Cur.: *Iris Peruga* • *Michaelle Ascencio* • *Tomás Rodríguez* • *Marco Rodríguez del Camino* • *Marta Liaño* • *Julieta González* • *Milagros González* • *Enrique Nóbrega* •
Fine Arts Museum – 1938
Latin American coll of paintings and sculptures, European paintings, coll of prints, drawings and photos, contemporary sculpture, Chinese ceramics, Egyptian art – library 54543

Museo de Caracas, Esq de Monjas, Alcaldía de Caracas Frente a la Pl Bolívar, Caracas • T: +58 212 5456706 •
Local Museum 54544

Museo de Ciencias, Pl de los Museos, Parque Los Caobos, Caracas 1010 • T: +58 212 5775094, 5775786 • F: 5749041, 5732368 • informacion@museo-de-ciencias.org.ve • www.museo-de-ciencias.org.ve •
Pres.: *Sergio Antillano Armas* • Dir.: *José Cesarino* • *Dr. Miguel Posani* • Sc. Staff: *Luis Galindo* (Conceptions) • *Sonia Anzola* (Environment) •
Natural History Museum / Science&Tech Museum – 1875
Archaeology, ethnography, herpetology, invertebrates, ichthyology, ornithology, paleontology, physical anthropology, mineralogy, scientific instr, theriology, toys – library, documentation center 54545

Museo de la Electricidad, Av Sanz, El Marqués, Caracas 1071 • T: +58 212 2088411 •
Science&Tech Museum 54546

Museo de la Estampa y del Diseño Carlos Cruz Diez, Av Bolívar, entre Calles Sur 11 y Este 8, Paseo Vargas, 1010 Caracas • T: +58 212 57231476, 5716910 • museocruz-diez@cantv.net •
Fine Arts Museum / Decorative Arts Museum 54547

Museo de la Fundación John Boulton, Av Universidad, Torre El Chorro, 1010 Caracas, mail addr: Apdo postal 929, 1010-A Caracas • T: +58 2 5644366 • F: 5631838 • fjb@reacciun.ve •
Fine Arts Museum
Paintings, prints, graphics, sculpture, numismatica, decorative arts, Bolivariana 54548

Museo de la Moneda, Edificio Sede del Banco Central de Venezuela, Mezzanina, 1030 Caracas • T: +58 212 8015111 •
Special Museum 54549

Museo de Los Niños de Caracas, Parque Central, Edificio Tacagua, Nivel Bolívar, mail addr: Apdo 14029, 1010-A Caracas • T: +58 212 5750295, 5750751 • F: 5754302 • info@maravillosarealidad.com • www.maravillosarealidad.com •
Dir.: *Alicia Pietri de Caldera* •
Science&Tech Museum – 1982
planetarium 54550

Museo del Beisbol (Museum of Baseball), Bello Campo, 1060 Caracas • T: +58 212 2632063 •
Special Museum 54551

Museo del Teclado, Parque Central, Edificio Tacagua, Nivel Mezzanina, Caracas 1010 • T: +58 212 5720713, 5729024 •
Local Museum 54552

Museo del Transporte Guillermo José Schael, Av Francisco de Miranda, Sec Santa Cecilia, Parque del Este, Caracas 1071-009 • T: +58 212 2342234, 2341621 • F: 2390652 • teremach@telcel.net.ve •
Dir.: *Alfredo Schael* •
Science&Tech Museum – 1970
Hist of transportation in Venezuela, early motor vehicles, railway hist – automobile saloon, aircrafts hangar, carriage saloon, two train stations, library and documental section 54553

Museo Jacobo Borges, Parque del Oeste, Av Sucre, Catia, Estación Gato Negro, Caracas 1030 • T: +58 212 8620427/8101 • F: 8623821, 8622989 • mujabo@hotmail.com • www.museojacoboborges.org.ve •
Dir.: *Adriana Meneses* •
Local Museum – 1993 54554

Museo Pedagógico de Historia del Arte, c/o Instituto del Arte, Facultad de Humanidades y Educación, Ciudad Universitaria, Caracas •
Fine Arts Museum / University Museum
Paintings, Latin American art 54555

Museo Sacro de Caracas, Calle Torre a Gradillas, La Catedral, Caracas 1010 • T: +58 212 8616562, 8615814 •
Religious Arts Museum 54556

Roberto Mata Taller de Fotografía, Av Trieste con Av Madrid, California Sur, Caracas • T: +58 212 2579745 • F: 2579745 • maryoriesanabria@robertomata.com • www.robertomata.com •
Public Gallery 54557

Sala Cadafe, Av Sanz, Edificio Cadafe, El Marqués, Caracas 1071 • T: +58 212 226256, 208500 •
Local Museum 54558

Sala Ipostel, Museo de Arte Contemporáneo de Caracas Sofía Imber, Av. José Angel Lamas, Edificio Sede Ipostel, San Martín, 1020 Caracas • T: +58 212 4512495 •
Fine Arts Museum 54559

Sala Mendoza, Av Andrés Bello, Edificio Las Fundaciones Planta Baja, 1020 Caracas • T: +58 212 5714731 • salamend@telcel.net.ve •
Public Gallery 54560

Catia la Mar

Museo Naval, Escuela Naval de Venezuela, Meseta de Mamo, Catia la Mar, Vargas •
Military Museum / Science&Tech Museum – 1965
History, ethnography, technology 54561

Ciudad Bolívar

Casa San Isidro, Av Táchira y calle 5 de Julio, Ciudad Bolívar 8001 •
Decorative Arts Museum
Local hist, first periodical of Venezuela 54562

Museo de Arte Moderno Jesús Soto, Av Germania, Ciudad Bolívar 8001, mail addr: Apdo 211, Ciudad Bolívar 8001-A • T: +58 285 24474, 20518 • F: 25854 • museosoto@cantv.net •
Pres.: *Ivanova Decán Gambús* •
Fine Arts Museum – 1973
20th c art, mainly by Venezuelan and European artists, Jesús Soto, Constructivism, kinetic art, new realism 54563

Museo de Ciudad Bolívar, Casa del Correo del Orinoco, Ciudad Bolívar 8001 •
Local Museum
Local hist, first periodical of Venezuela 54564

Colonia Tovar

Museo de Historia y Artesanía de la Colonial Tovar, Calle del Museo, Colonia Tovar 1030 •
Local Museum – 1970
Art and natural hist, anthropology, ethnography, archaeology, geology 54565

Cumaná

Casa de Andrés Eloy Blanco, Pl Bolívar, Cumaná 5001 •
Special Museum 54566

Casa Natal de José Antonio Ramos Sucre, Cumaná 5001 •
Special Museum 54567

Museo Gran Mariscal de Ayacucho, Av Humboldt, Parque Ayacucho, Cumaná 5001 •
Historical Museum 54568

El Tocuyo

Museo Colonial, El Tocuyo •
Local Museum – 1945
Relics relating mainly to the colonial period 54569

Falcón

Museo de Arte Coro y Museo Alberto Henríquez, Balcón Bolívar, Paseo Talavera, Falcón 4101 •
Fine Arts Museum 54570

Museo de Cerámica y Loza Popular, Balcón de los Arcaya, Calle Zamora, Zona Colonial, Falcón 4101 •
Decorative Arts Museum 54571

Museo Diocesano Lucas Guillermo Castillo, Calle Zamora, Zona Colonial, Falcón 4101 •
Religious Arts Museum / Fine Arts Museum 54572

Guanare

Museo de los Llanos, Parque José Antonio Páez, Guanare 3350 •
Local Museum – 1995
Local history and archaeology, colonial time, cultures – archive 54573

Museo Inés Mercedes Gómez Álvarez, Universidad Ezequiel Zamora, Casco Colonial, Guanare 3350 •
Historical Museum – 1984 54574

Guayana

Castillos de Guayana la Vieja, Av Manuel Piar, Guayana •
Historical Museum
Fortress San Francisco and San Diego de Alcalá 54575

Juangriego

Museo de Arte Popular Venezolano, Tacuantar, Juangriego 6301 •
Folklore Museum / Decorative Arts Museum 54576

La Asunción

Museo Nueva Cádiz, Calle Independencia, La Asunción 6311 •
Local Museum
Hist, ethnography, natural hist, archaeology 54577

La Guaira

Casa Natal de José María España, Calle San Francisco 9, La Guaira 1160 •
Historical Museum
Independence revolution of 1797 54578

Los Teques

Museo J.M. Cruxent, c/o Depto Antropología del Instituto Venezolano de Investigaciones Científicas, Carretera Panamericana, Los Teques 1201 • T: +58 212 5041227 • F: 5041085 • svidal@ivic.ivic.ve •
Dir.: *Silvia M. Vidal* •
Ethnology Museum / Archaeology Museum – 1959
Archaeology, ethnology 54579

Macuto

Museo Armando Reverón, Callejón Colon, Macuto, Vargas •
Historical Museum 54580

Maracaibo

Centro de Bellas Artes, Av 3F No 67-217, Colonia Bella Vista, Sector La Lago, Maracaibo • T: +58 261 7912950/2860 • F: 7912239 • centrobellasarte@iamnet.com •
Public Gallery – 1963 54581

Museo Urdaneta Histórico Militar (Museum of Military History), Maracaibo 4001 • T: +58 261 226778 •
Dir.: *Prof. J.C. Borges Rosales* •
Military Museum – 1936
Military hist 54582

Maracay

Museo Aeronáutico Colonel Luis Hernan Paredes, Av Las Delicias y Av 19 de Abril, Maracay 2101 • T: +58 243 333812 • F: 333812 •
Dir.: *Juan C. Flores* •
Science&Tech Museum – 1963
Aeronautical history 54583

Museo de Antropología e Historia, Fundación Lisandro Alvarado, Blvd La Alcaldía a la Plaza Girardot, Maracay 2101, mail addr: Apdo 4518, Maracay 2101-A • T: +58 243 2472521 •
Dir.: *Henriqueta Peñalver Gómez* • Pres. Honoraria: *Dr. Adelaida de Díaz Ungría* • Sc. Staff: *Mary Yamilet Bonilla* (Antropología Física) • *Prof. Delia García de Del Valle* (Información, Divulgación) •
Ethnology Museum / Historical Museum / Archaeology Museum – 1964
Pre-Columbian archeology, Venezuelan ethnology, anthropology, and history, vertebrate paleontology, religious art – library 54584

Museo Mario Abreu, Av 19 de Abril, Complejo Cultural Santos Michelena, Maracay 2107 • T: +58 243 336980 • F: 338534 •
Local Museum 54585

Mérida

Museo de Arte Colonial, Av 3, entre Calles 18 y 19, Mérida 5101 •
Fine Arts Museum 54586

Museo de Arte Moderno, Calle Flamboyán, Santa María, frente a la Plaza Beethoven, Mérida 5101 •
Fine Arts Museum – 1969
Modern Venezuelan art 54587

Museo del Estado de Mérida, c/o Universidad de Mérida, Mérida 5101 •
Local Museum
Regional ethnography and hist 54588

Nueva Cádiz

Museo Biblioteca Nueva Cádiz y Casa Natal de Juan Bautista Arismendi, Antigua Casa Capitular, Nueva Cádiz 6301 •
Historical Museum 54589

Pampatar

Museo Biblioteca Rosauro Rosa Acosta, Casa de la Aduana, Pampatar 6301 •
Decorative Arts Museum
Antiques, books, paintings, sculpture 54590

Píritu

Museo de la Ciudad de Píritu, Esteller, Píritu 3350 •
Local Museum 54591

Museo de las Muñecas (Doll Museum), Esteller, Píritu 3350 •
Decorative Arts Museum 54592

Porlamar

Museo de Arte Contemporáneo Francisco Narvárez, Complejo Rómulo Gallegos, Calle Igualdad, Porlamar 6301 •
Fine Arts Museum – 1981
Sculptures 54593

Puerto Ayacucho

Museo Etnológico de Amazonas, Iglesia, 7101 Puerto Ayacucho •
Ethnology Museum – 1984
Cultures of Piaroa, Guahiba, Yanomami, Arawak and Yekuana 54594

Trujillo

Museo Cristóbal Mendoza, Trujillo 3150 •
Historical Museum 54595

Valencia

Ateneo de Valencia, Av Bolívar Nort, Valencia 2001 • T: +58 241 8580046, 8581962 •
Fine Arts Museum 54596

Casa Museo Páez, Loc. 99-20, Av Boyacá, Valencia 2001 • T: +58 241 8571272 •
Local Museum 54597

Museo de Arte e Historia Casa de Los Celis, Av Soublette, Valencia 2001 • T: +58 241 8421245 •
Historical Museum / Archaeology Museum 54598

Valle del Espíritu Santo

Museo Diocesano, Santuario de Nuestra Señora del Valle, Valle del Espíritu Santo 6301 •
Religious Arts Museum 54599

Vietnam

Binh Minh

Chu Dong Tu Temple, Binh Minh, Khoai Chau •
Historical Museum
Rare and precious objects incl 2 Bach Tho vases 54600

Đà Lat

Bào Đai Palace Museum, Đà Lat •
Decorative Arts Museum / Historical Museum
Emperor Bào Đai seat 54601

Crazy House Museum, Crazy House Hotel, Đà Lat •
Head: *Dang Viet Nga* •
Special Museum – 1990 54602

Lâm Đông Museum, Đalat's Hilltop, Đà Lat •
Local Museum
Modern hist of Vietnam, Vietnamese popular arts, archaeology, costumes and musical instr of local ethnic minorities 54603

Da Nang

National Museum of Cham Sculpture, 2 Tiêu La St, Da Nang • T: +84 511 821951 • F: 821279 •
Archaeology Museum – 1915
Champá sculpture and architectural fragments from various periods and places, late 7th-15th c, in Central Vietnam, objects were collected from the temples y Hinduism and Buddhism of Champákinydom 54604

Ha Long

Quàng Ninh Museum, 165 Nguyen Van Cu, Ha Long • F: +84 33 845027 • tourismdepart@hn.vnn.vn •
Local Museum – 1964
Local hist, net-sinkers made of terracotta and hooks, needles made of bone from the Soi Nhu, Thoi Giêng, Ngoc Vùng cultures 54605

Ha Noi

Air Force Museum, Truong Chinh, Ha Noi •
Military Museum 54606

Anti-Aircraft Museum, Truong Chinh, Ha Noi • T: +84 4 8522658 •
Military Museum 54607

Bao Tàng Dân Tôc Hoc Viêt Nam (Vietnam Museum of Ethnology), Nguyen van Huyen St, Cau Giay, Ha Noi • T: +84 4 8360350 • F: 8360351 • vmel8@hn.vnn.vn •
Dir.: *Prof. Dr. Hguyen Van Huy* •
Ethnology Museum 54608

Bao Tàng Quan Doi (Army Museum), 28a Dien Bien Phu, Ha Noi • T: +84 4 7334682 •
Dir.: *Ma Luong Le* •
Military Museum
Military hist 54609

Bien Phong Museum, 2 Tran Hung Dao, Ha Noi • T: +84 4 8213835 •
Local Museum 54610

Exhibition Hall, c/o Ministry of Culture and Information, 29 Hang Bai St, Ha Noi •
Public Gallery 54611

Fine Art University Exhibition Hall, 42 Yet Kieu, Hoan Kiem, Ha Noi •
Public Gallery 54612

Geology Museum, 6 Pham Ngu Lao, Ha Noi • T: +84 4 8266802, 8249112 • F: 9331496 • danht-btdc@fpt.vn • www.idm.gov.vn •
Dir.: *Prof. Dr. Trinh Danh* •
Natural History Museum 54613

Ha Noi Museum, 5b Ham Long, Ha Noi • T: +84 4 8263982 •
Local Museum
Local history 54614

Ho Chi Minh Museum, 3 Ngoc Ha, Ba Dinh, Ha Noi • T: +84 4 8455525, 8463757 • F: 7439387 •
Dir.: *Cu Van Chuoc* •
Special Museum – 1977
Life and work of President Ho Chi Minh 54615

Vien Bao Tang Lich Sa Viet Nam (National Museum of Vietnamese History), 1 Pham Ngu Lao, Ha Noi • T: +84 4 8252853, 8242433 • F: 8252853 •
Dir.: *Phạm Quoc Quan* •
Historical Museum
Hist of Viet Nam from prehistoric times till 1945 54616

Viet Nam National Fine Arts Museum, 66 Nguyen Thai Hoc, Ha Noi • T: +84 4 8231085, 8233084 • F: 7341427 • binhtruong451@hn.vnn.vn •
Dir.: *Prof. Dr. Truong Quoc Binh* •
Fine Arts Museum – 1966
Folk drawings, modern paintings and sculptures, lacquers and silks, bronzes and stones, ancient ceramics, oil paintings, wood carvings – library, departments of research, propaganda and storage 54617

Viet Nam Revolution Museum, 25 Tong Dan, Ha Noi • T: +84 4 8254323, 8254151 • F: 9342064 • phammaihung@yahoo.com •
Dir.: *Prof. Dr. Pham Mai Hung* •
Military Museum / Historical Museum – 1959
Revolutionary hist of Viet Nam 54618

Viet Nam Women's Museum, 36 Ly Thuong Kiet, Hoan Kiem, Ha Noi • T: +84 4 8259935/38 •
Dir.: *Nguyen Thi Nghien* •
Ethnology Museum 54619

Hai Phong

Bao Tàng Hai Phong (Haiphong Museum), Hai Phong •
Local Museum – 1959
Local hist 54620

Ho Chi Minh City

Duc Minh Art Gallery, 31c Le Quy Don St, D3, Ho Chi Minh City • T: +84 8 9330498 • F: 9330497 • ducminh-art@hcm.vnn.vn •
Fine Arts Museum 54621

Geological Museum, 2 Nguyen Binh Khiem, Ho Chi Minh City • T: +84 8 8294821, 8221156 •
Dir.: *Trinh Van Hong* • Dep. Dir.: *Tran Huy An* •
Natural History Museum
Mineralogy, fossils 54622

Ho Chi Minh City Fine Arts Museum, 97a Pho Duc Chinh, Ho Chi Minh City • T: +84 8 8294441 • F: 8213508 • btmt@hcm.vnn.vn •
Dir.: *Nguyen Toan Thi* •
Fine Arts Museum 54623

Ho Chi Minh Museum, 1 Nguyen Tat Thanh, Ho Chi Minh City • T: +84 8 8255740, 9402060 •
Dir.: *Nguyen So* •
Historical Museum 54624

Revolutionary Museum Ho Chi Minh City, 114 Nam Ky Khoi Nghia, Dist 3, Ho Chi Minh City • T: +84 8 8298250, 8299743 •
Dir.: *Pham Van Cong* •
Historical Museum – 1978
History, revolutionary movement and War of Liberation – library 54625

Saigon Open City Studio Space, 3a Ton Duc Thang St, Ben Nghe, D1, Ho Chi Minh City • T: +84 8 9129337 • F: 9103277 • info@saigonopencity.org • www.saigonopencity.org •
Dir.: *Tran Luong* • *Le Quang Dinh* • *Do Thi Tuyet Mai* •
Public Gallery 54626

South-East Armed Force Museum, 247 Hoang Van Thu, Tan Binh, Ho Chi Minh City • T: +84 8 8421354, 8229357 •
Dir.: *Phan Oanh* •
Military Museum 54627

Thong Nhat Palace, 133 Nam Ky Khoi Nghia, Ho Chi Minh City • T: +84 8 294991 •
Fine Arts Museum 54628

Ton Duc Thang Museum, 5 Ton Duc Thang St, Ho Chi Minh City • T: +84 8 8295946, 8224887 •
Dir.: *Tran Thi Thuy Phuong* •
Local Museum 54629

Viet Nam History Museum Ho Chi Minh City, 2 Nguyen Binh Khiem, Dist 1, Ho Chi Minh City • T: +84 8 8298146, 8220743 •
Dir.: *Trung Le* •
Historical Museum 54630

War Remnants Museum, 28 Vo Van Tan St, Dist 3, Ho Chi Minh City • T: +84 8 9305153, 9306325, 9305587 • F: 9305153 • Warrm@cinet.vnnews.com •
Dir.: *Nguyên Khă Lân* •
Historical Museum
Requiem (photo coll), Vietnam 35 yrs war and peace, vestiges of crimes and aftermaths, weapons used in the Vietnam war, historical truths 54631

Women's Museum of South Vietnam, 202 Vo Thi Sau, Dist 3, Ho Chi Minh City • T: +84 8 9320322, 9325690 • F: 9327130 • baotangpnnb@hem.vnn.vn • www.womanmuseumsouth.edu.vn •
Dir.: *Tran Hong Anh* •
Ethnology Museum 54632

Hung-Yen

Hung-Yen Museum, Hung-Yen •
Local Museum
Local history and archaeology 54633

Nha Trang

Khanh Hoa Provincial Museum, 16 Tran Phu, Nha Trang • T: +84 58 82227, 813654 •
Local Museum – 1979
Ancient cultures of Khanh Hoa and Con Hamlet, stone axes and carvings, bronze drums,ceramics, numismatics, revolution tradition, weapons, president Ho Chi Minh 54634

Oceanographic Museum, Institute of Oceanography, 1 Cau Da, Nha Trang • T: +84 58 590032/36 • F: 590034 • haiduong@dng.vnn.vn •
Dir.: *Dr. Nguyen Tac An* •
Natural History Museum – 1923
Coll of marine fishes, sea weeds, coelenterates, sponges, polychaetes, crustacean, mollusks, echinoderms, reptiles, sea-birds, sea-mammals etc (about 20,000 specimens of 10,000 species) 54635

Thai Nguyen

Bao Tàng Văn Hóa Các Dân Tôc Viêt Nam (Museum of Cultures of Vietnam's Ethnic Groups), 359 Duong Minh St, Thai Nguyen • T: +84 280 855781, 852182 • F: 752940 • mangocdung@yahoo.com •
Dir.: *Ha Thi Nu* •
Ethnology Museum – 1960
Cultures of Vietnamese 54 ethnic groups in Viet Nam, clouthes, textiles, musical instr, tools, things of life, argiculture, houses 54636

Thua Thien Hue

Ho Chi Minh Memorial House, Duong No, Thua Thien Hue 47000 •
Special Museum 54637

Ho Chi Minh Museum, 06 Le Loi, Thua Thien Hue 47000 • T: +84 54 822152, 820250 •
Special Museum 54638

Thua Thien Hue Museum, 01-23 Thang Tam St, Thua Thien Hue 47000 • T: +84 54 822397, 823159 •
Dir.: *Pham Xuân Phuong* •
Archaeology Museum / Ethnology Museum – 1982
Cham sculptures (10th-11th c), ethnological objects, hist of the old capital 54639

Vinh

Nghe-Tinh Museum, Vinh •
Historical Museum
Hist of the Nghe-Tinh 'Soviet' uprising 1930-31 54640

Yemen

Aden

Ethnographical Museum, nr Tanks, Crater, Aden •
Ethnology Museum 54641

Military Museum, Sayla Rd, Crater, Aden •
Military Museum
Historic weapons 54642

National Museum, Crater, Aden •
Archaeology Museum
Archaeological find in an old sultan's palace 54643

The Tanks of Aden, On the Volcanic Slopes, Aden •
Science&Tech Museum
18 cisterns from 1st c AD 54644

Al-Mukalla

Al-Mukalla Museum, Former Sultans Palace, Riverside Promenade, Al-Mukalla •
Local Museum / Archaeology Museum
Folk costumes, traditional handicrafts, arcaeology 54645

Ataq

Ataq Museum, Wadi, Ataq, Shabwa •
Archaeology Museum
Relics from the archaeological sites of the ancient Qataban and Osan kingdom, ancient kingdom of Hadhramout capital 54646

Baihan

Baihan Museum, Wadi, Baihan, Shabwa •
Archaeology Museum
Examples of the Qataban archaeological sites dating back to pre-Islamic times 54647

Mahweet

Taweelah Museum of Mummies, Al-Taweelah, Mahweet •
Museum of Classical Antiquities – 2007
Embalmed mummies on the Arabian Peninsula from rocky cemeteries in Mahweet 54648

Sanaa

Military Museum, Tahrir Sq, Sanaa •
Military Museum 54649

National Museum, Dar as-Sa'd, nr Tahrir Sq, Sanaa, mail addr: POB 2606, Sanaa • T: +967 1 271648 •
Dir.: *Ahmed Naji Sari* •
Historical Museum / Archaeology Museum / Folklore Museum
South Arabian antiques of the pre-Islamic and Islamic periods, folklore, artefacts of ancient kingdoms of Saba, Ma'in, Ma'rib and Himyar 54650

Seiyun

Seiyun in Wadi Hadhramaut Museum, Seiyun • T: +967 5 5402258 •
Local Museum / Archaeology Museum / Folklore Museum
Housed in a former Sultans palace, handicrafts, folklore items, antiques found in the region 54651

Taizz

Taizz National Museum, Taizz •
Local Museum
Housed in a former Imams palace, items pertaining to the 1962 Revolution and the events that led up to it 54652

Zambia

Choma

Choma Museum and Crafts Centré, Great North Rd, between Lusaka and Livingstone Rds, Choma, mail addr: POB 630189, Choma • T: +260 32 20394 • F: 20394 • cmcc@coppernet.zm • www.catgen.net/cmcc •
Dir.: *Mwimanji Ndota Chellah* •
Local Museum – 1987
Material culture of the peoples inhabiting Southern Zambia, ethnographic hist of the Southern Province 54653

Limulunga

Nayuma Museum, POB 96, Limulunga • T: +260 7 221421 • F: 221351 •
Dir.: *Manyando Mukela* •
Local Museum 54654

Livingstone

Livingstone Museum, National Museum of Zambia, Mosi-oa-Tunya Rd, Livingstone, mail addr: POB 60498, Livingstone • T: +260 3 323566, 320495 • F: 320495 • livmus@zamnet.zm •
Dir.: *K.V. Katanekwa* •
Local Museum – 1934
Natural hist, exhibits, general hist, fine arts, relics of Dr Livingstone, archaeology, ethnography, zoology – library 54655

Railway Museum, 140 Chishimba Falls Rd, Livingstone, mail addr: POB 60124, Livingstone • T: +260 3 321820 • F: 324509 • nhcc@zamnet.zm •
Dir.: *N.M. Katanekwa* • *Patrick Lisina Wamulungwe* •
Science&Tech Museum – 1987
Vintage steam locomotives, wooden passenger carriages, and other memorabialia associated with the site 54656

Victoria Falls Information Centre, Mosi-Oa-Tunya Rd, Livingstone, mail addr: POB 60124, Livingstone • T: +260 3 323662 • F: 323653 • nhccsowe@zamnet.zm •
Dir.: *N.M. Katanekwa* •
Local Museum
Displays illustrating formation of the Victoria Falls, Stone Age implements 54657

Lusaka

Lusaka National Museum, Independence Av, Lusaka, mail addr: POB 50491, 15101 Lusaka • T: +260 1 228805/07 • F: 223788 • lusamus@zamnet.zm • www.museums.co.zm •
Dir.: *Flexon Moono Mizinga* •
Historical Museum
Cultural history, ethnography 54658

National Archives of Zambia, Ridgeway, Lusaka, mail addr: POB 50010, Lusaka • T: +260 1 250446 • F: 254080 • naz@zamnet.zm •
Dir.: *C. Hamooya* •
Historical Museum
Stamps and coins, local hist, photos 54659

University of Zambia Library Museums Collection, Great East Rd, Campus, Lusaka, mail addr: POB 32379, Lusaka • T: +260 1 250845 • F: 253952 • msimui@library.unza.zm •
Dir.: *Muyoyeta Simui* •
Library with Exhibitions – 1969 54660

Mbala

Moto Moto Museum, Lucheche Rd, Mbala, mail addr: POB 420230, Mbala • T: +260 4 450098, 450243 • F: 450243 • motomoto@zamtel.zm •
Dir.: *Flexon Moono Mizinga* •
Ethnology Museum / Historical Museum – 1974
Research in ethnography and cultural hist, traditional medicine and charms, musical instr, masks, Bemba girls initiation rite, prehistoric items – library 54661

Ndola

Copperbelt Museum, Buteko Av, Ndola, mail addr: POB 71444, Ndola • T: +260 2 617450, 613591 • F: 617450 • cbmus@zamnet.zm •
Dir.: *Stanford Mudenda Siachoono* •
Science&Tech Museum – 1962
Displays on the copper mining industry, nature and ecology, geology, ethnography 54662

Zimbabwe

Bulawayo

Khami Ruins Site Museum, POB 240, Bulawayo • T: +263 9 60045 • F: 64019 • monuments@telconet.co.zw •
Dir.: *Albert Kumirai* •
Open Air Museum / Archaeology Museum
Ruins of one of Mambo dynasty capitals (15th-19th c), archaeological finds from the area 54663

Matopos National Park Site Museums, Bulawayo •
Archaeology Museum
Prehistoric rock paintings, Stone Age tools, grave of Cecil Rhodes 54664

National Gallery Bulawayo, 75 Main St, Bulawayo, mail addr: POB 1993, Bulawayo • T: +263 9 70721 • F: 63343 • sabona@telcenet.co.zw •
Dir.: *Addelis Sibutha* •
Fine Arts Museum / Decorative Arts Museum – 1974
Works of contemp English and South African artists, loan exhib, Ndebele and Tonga baskets, Zimbabwean art and craft, works of Zimbabwean and Sadc artists 54665

Natural History Museum of Zimbabwe, Leopold Takawira Av and Park Rd, Bulawayo, mail addr: POB 240, Bulawayo • T: +263 9 60045/46 • F: 64019 • natmuse@telconet.co.zw •
Dir.: *Albert Kumirai* • Cur.: *F.D.P. Cotterill* (Mammals) • *R. Sithole* (Entomology) • *M. Fitzpatrick* (Arachnology) • *A. Msimanga* (Ornithology) • *R.P. Chidavaenzi* (Herpetology) • *D. Munyikwa* (Palaeontology) • *P. Makoni* (Ichthyology) •
Natural History Museum – 1901
Zoology coll covering Ethiopian region and Southern Africa – library, herbarium 54666

Gweru

Zimbabwe Military Museum, Lobengula Av, Gweru, mail addr: POB 1300, Gweru • T: +263 54 22816 • F: 20321 • gwerumuseum@comone.co.zw •
Dir.: *Tarisai Tsomondo* •
Military Museum – 1974
Weapons, military vehicles, uniforms, and equip, aircraft and aviation uniforms and equip, archaeology, mining, monuments 54667

Harare

Macgregor Museum, Causeway, Harare, mail addr: c/o Geological Survey Department, POB CY210, Harare • T: +263 4 726342/3 • F: 253626 • zgs@samara.co.zw •
Dir.: *T. Hawadi* •
Natural History Museum – 1940
Economic minerals, displays dealing with the Deweras and Umkondo groups, geology of Zimbabwe, great dyke, alkali ring complex rocks 54668

Museum of Human Sciences, Civic Centre, Rotten Row, Causeway, Harare, mail addr: POB CY33, Harare • T: +263 4 751797/98 • F: 774207 • nmmz@mweb.co.zw •
Dir.: *Pascall Taruvinska* •
Ethnology Museum / Archaeology Museum – 1902
National coll of Iron and Stone Age artifacts, rock paintings, ethnographic coll – library 54669

National Gallery of Zimbabwe, 20 Julius Nyerere Way, Harare • T: +263 4 704666/67 • F: 704668 • ngallery@mweb.co.zw •
Dir.: *D. Sibanda* •
Fine Arts Museum – 1957
Traditional African art, Zimbabwean stone sculpture (shona sculpture) – library 54670

Queen Victoria Museum, Rotten Row, Causeway, Harare, mail addr: POB 8006, Harare • T: +263 4 704831/32, 724915 • F: 77717 •
Dir.: *Lorraine Margaret Adams* •
Historical Museum 54671

Inyanga

Nyahokwe Ruins Site Museum, Inyanga •
Open Air Museum / Archaeology Museum
Relics of Iron Age, locally found relics of Ziwa culture 54672

Kwekwe

National Mining Museum, POB 512, Kwekwe • T: +263 55 23741 • F: (054) 20321 • minmus@globalzim.net •
Dir.: *Josiah Rungano Mhute* •
Science&Tech Museum – 1984
Mining antiquities, Globe and Phoenix Mining Company founded in 1895, machinery, tools, biographies 54673

Marondera

Children's Library Museum, The Green, Marondera • T: +263 3356 •
Special Museum
Objects illustrating man and his implements from the early Stone Age to pioneering days, rocks and minerals, local birds 54674

Masvingo

Great Zimbabwe Site Museum, POB 1060, Masvingo • T: +263 39 62080, 65084 • F: 63310 • greatzim@africaonline.co.zw •
Dir.: *Edward Matenga* •
Archaeology Museum
Zimbabwe Iron Age archaeology, Zimbabwe birds 54675

Mutare

Mutare Museum, Aerodrome Rd, Mutare • T: +263 20 63630, 63005 • F: 61100 • mutarmus@ecoweb.co.zw •
Dir.: *Traude Allison Rogers* •
Local Museum – 1959
Archaeology, hist, national coll of transport antiquites and firearms, local flora and fauna 54676

Raylton

National Railways of Zimbabwe Museum, Prospect Av and First St, Raylton, mail addr: POB 945, Bulawayo • T: +263 9 322507 •
Chm.: *G.W.T. Tyamzarne* •
Science&Tech Museum
Former Mashanaland and Bura Rhodesian Railways and National Railways of Zimbabwe, steam locomotives, coaches, and freight cars 54677

Museum Associations

Afghanistan

ICOM Afghanistan, c/o Moh. Yayha Mohibzada, National Museums, Unesco Office, Kabul • T: +93 0873-761 660769 • F: 660769 • jim.williams@undp.org • ... 54678

Algeria

Comitée National de ICOM de Algérie, c/o Musée des Arts Populaires et Traditions, Amamra Aicha, 69 Rue Mohamed Ahli Malek, Casbah, Alger 16000 • T: +213 21 713414 • ... 54679

Andorra

ICOM Andorra, c/o Marta Planas de la Maza, Patrimoni Artistic Nacional, Carretera de Bixessarri s/n, Aixovall • T: +376 844141 • F: 844343 • pca.gov@andorra.ad • ... 54680

Angola

ICOM Angola, Rua Major Kanyangulu 77, Luanda, mail addr: c/o M. Francisco Xavier Yambo, Instituto Nacional do Património Cultural, CP 1267, Luanda • T: +244 2 331139 • F: 332575 • ipc@snet.co.ao • ... 54681

Argentina

Centro de Museos de Buenos Aires, Av. de los Italianos 851, Buenos Aires • T: +54 11 45160941 • prensadgm@buenosaires.gov.ar • ... 54682

Comisión Nacional de Museos y de Monumentos, Av de Mayo 556, C1084AAN Buenos Aires • T: +54 11 43316151 • ... 54683

Federación Argentina de Amigos de Museos, Calle Bolívar 1131, C1066AAW Buenos Aires • T: +54 11 43070522 • F: 43070523 • fadam@fadam.org.ar • www.fadam.org.ar • ... 54684

ICOM Argentina, Calle Perú 272, Manzana de las Luces, C1067AAF Buenos Aires • T: +54 11 43426758 • F: 43426758 • info@icomargentina.org.ar • www.icomargentina.org.ar • ... 54685

Australia

Australian Federation of Friends of Museums, The Mint, 10 Macquarie St, Sydney, NSW 2000 • T: +61 2 92323466 • F: 92321699 • info@affm.org.au • www.affm.org.au • ... 54686

Collections Council of Australia, c/o SLSA, North Tce, Adelaide, SA 5006 • T: +61 8 82077287 • F: 82077207 • info@collectionscouncil.com.au • www.collectionscouncil.com.au • ... 54687

Council of Australian Art Museum Directors, c/o National Gallery of Victoria, POB 7259, Melbourne, Vic. 8004 • T: +61 3 86202276 • F: 86202520 • directorate@ngv.vic.gov.au • www.ngv.vic.gov.au • ... 54688

Council of Australian Museum Directors, c/o Powerhouse Museum, GPOB 346K, Haymarket, NSW 2000 • T: +61 2 92170363 • kevinf@phm.gov.au • ... 54689

ICOM Australia, c/o Dr. Ian Galloway, Queensland Museum, POB 3300, South Brisbane, Qld 4101 • T: +61 7 38407658 • F: 38429470 • icomaustralia@mac.com • australia.icom.museum • ... 54690

Museums Australia, POB 266, Civic Square, ACT 2608 • T: +61 2 62085044 • F: 62085015 • ma@museumsaustralia.org.au • www.museumsaustralia.org.au • ... 54691

Austria

ICDAD International Committee for Decorative Arts and Design Museums, c/o Dr. Rainald Franz, MAK Wien, Stubenring 5, 1010 Wien • T: +43 1 71136258 • Franz@MAK.at • kunstgeschichte.univie.ac.at • ... 54692

ICOM-Österreichisches Nationalkomitee, Burgring 5, 1010 Wien • T: +43 1 404206631 • F: 404206695 • armine.wehdorn@oenb.at • www.icom-oesterreich.at • 1978 ... 54693

Interessengemeinschaft niederösterreichische Museen und Sammlungen, c/o Museumsmanagement Niederösterreich, Donaulände 56, 3504 Krems -Stein • T: +43 2732 73999 • F: 7399933 • museen@volkskulturnoe.at • www.noemuseen.at • 1991 ... 54694

Landesverband Salzburger Volkskultur, Zugallistr 12 (Petersbrunnhof), 5020 Salzburg • T: +43 662 80422615 • F: 80422612 • volkskultur@salzburg.gv.at • www.salzburgervolkskultur.at • ... 54695

Museumsbund Österreich, Welserstr 20, 4060 Leonding • T: +43 732 674256182 • F: 674256185 • s.traxler@museumsbund.at • www.museumsbund.at • 1981 ... 54696

Museumsland Donauland Strudengau, Schlossberg 1/12, 4391 Waldhausen • T: +43 7260 45255 • F: 452554 • info@museumsland.at • www.museumsland.at • ... 54697

MUSIS, Verein zur Unterstützung der Museen und Sammlungen in der Steiermark, Strauchergasse 16, 8020 Graz • T: +43 316 738605 • F: 63860514 • office@musis.at • www.musis.at • ... 54698

Oberösterreichischer Musealverein, Gesellschaft für Landeskunde von Oberösterreich (Association of Heritage in Upper Austria), c/o Haus der Volkskultur, Promenade 33/103, 4020 Linz • T: +43 732 770218 • F: 770218 • office@ooelandeskunde.at • www.ooelandeskunde.at • 1833 ... 54699

Verbund Oberösterreichischer Museen, Welser Str 20, 4060 Leonding • T: +43 732 682616 • F: 682615 • office@ooemuseumsverbund.at • www.ooemuseumsverbund.at • ... 54700

Volkskultur Niederösterreich, Schlossplatz 1, 3452 Atzenbrugg • T: +43 2275 4660 • F: 466027 • office@volkskulturnoe.at • www.volkskulturnoe.at • 1998 ... 54701

Azerbaijan

ICOM Azerbaijan, c/o Roya Taghiyeva, State Museum of Azerbaijan Carpets and Applied Art Letif Kerimov, Neftçiler pr 123a, 1005 Bakı • T: +994 12 936685 • F: 930501 • azcarpetmuseum@azeurotel.com • 54702

Bangladesh

ICOM Bangladesh, Shahbagh, Dhaka, mail addr: c/o Bangladesh National Museum, POB 355, Dhaka 1000 • T: +880 2 9675601, 8625540 • F: 8615585 • jahangirhu@yahoo.com • www.bangladeshmuseum.org • ... 54703

Barbados

ICOM Barbados, c/o Barbados Museum, 11 Saint Ann's Fort, Garrison, 14038 Saint Michael • T: +1 246 427-0201 • F: (246) 429-5946 • museum@caribsurf.com • www.barbmuse.org.bb • ... 54704

Museums Association of the Caribbean, POB 112, Bridgetown • T: +1 246 228-2024 • F: (246) 228-2024 • macsecretariat@caribsurf.com • ... 54705

Belarus

Association of the State Literary Museums of the Republic of Belarus, Bogdanoviča ul 15, 220029 Minsk • T: +375 17 2347261 • F: 2347261 • nacbibl.org.by/litm/en/lithist.html • ... 54706

ICOM National Commitee of Belarus, c/o Museums Projects' Laboratory, Bogdanoviča ul 15, 220029 Minsk • T: +375 17 2347261 • F: 2347261 • icombelarus@tut.by • ... 54707

Belgium

L' Association Francophone des Musées de Belgique, c/o Musée des Arts Contemporains, 82 Rue Sainte-Louise, Site du Grand-Hornu, 7301 Hornu • T: +32 65613850 • F: +32 65613891 • info.macs@grand-hornu.be • www.mac-s.be • ... 54708

Association Internationale des Musées d'Armes et d'Histoire Militaire (International Association of Museums of Arms and Military History), c/o Musée d'Armes de Liège, Halles du Nord, 4 Rue de la Boucherie, 4000 Liège • T: +32 42219416/17 • F: +32 42219401 • claude.gaier@museedarmes.be • ... 54709

Conseil Bruxellois de Musées, 46 Rue des Bouchers, 1000 Bruxelles • T: +32 25127780 • F: +32 25122066 • info@brusselsmuseum.be • www.brusselsmuseums.be • ... 54710

European Collaborative for Science, Industry and Technology Exhibitions, 63 Blvd du Triomphe, 1160 Bruxelles • T: +32 26475098 • F: +32 26475098 • wststaveloz@ecsite.net • www.ecsite.net • ... 54711

Fédération des Amis des Musées de Belgique (Belgian Federation of Friends of Museums), 9 Rue du Musée, 1000 Bruxelles • ... 54712

ICOM - Belgian National Comittee, c/o Musée des Art Contemporains-MAC's, 82 Rue Sainte-Louise, 7301 Hornu • T: +32 65613851 • F: +32 65613891 • info.macs@grand-hornu.be • www.mac-s.be • ... 54713

International Association of Custom Museums, c/o Nationaal Museum, Kattendijkdok OK 22, 2000 Antwerpen • T: +32 32292260 • F: +32 32292261 • www.etat.lu/IACM • ... 54714

Musées et Société en Wallonie, 149 Rue des Brasseurs, 5000 Namur • T: +32 81229646 • F: +32 81229646 • maisondesmusees@msw.be • www.msw.be • ... 54715

Vlaamse Museumvereniging, Plaatsnijdersstr 2, 2000 Antwerpen • T: +32 32160360 • F: +32 32570861 • info@museumvereniging.be • www.museumvereniging.be • ... 54716

Benin

ICOM Bénin, c/o G.A. Hounkpevi, Musée Ethnographique, BP 299, Porto-Novo • T: +229 20213566 • F: 20212109 • toussyeb@altern.org • ... 54717

Bolivia

ICOM Bolivia, Casillo Correo 8083, La Paz • T: +591 2 2203683, 2201250 • F: 2201250 • icom@bolivia.com • ... 54718

Bosnia and Herzegovina

ICOM Bosnia-Hercegovina, c/o Zemaljski Musej, Zmaja od Bosne 3, 71000 Sarajevo • T: +387 33 668027, 262710 • F: 262 710 • icombih@gmail.com • www.zemaljskimuzej.ba • ... 54719

Botswana

ICOM Botswana, 331 Independence Av, 00114 Gaborone, mail addr: c/o Botswana National Museum, Private Bag 00114, 00267 Gaborone • T: +267 3610417 • F: 3902797 • matmmutle@yahoo.co.uk • ... 54720

Brazil

Comitê Brasileiro do ICOM, c/o Museu Lasar Segall, Rua Berta 111, São Paulo 04120-040 • T: +55 11 55747322 • F: 55723586 • icombr@terra.com.br • brazil.icom.museum • ... 54721

Federação de Amigos de Museus do Brasil (Brazilian Federation of Friends of Museums), Rua Horacio Lafer 702, São Paulo 04538-083 • msoalmeida@sti.com.br • ... 54722

Bulgaria

Associacija Muzej, c/o Peter Ivanov, 13a ul Gen. Skobelev, 9002 Varna • T: +359 52 258263 • F: 602079 • asociaciamusei@abv.bg • ... 54723

Bulgarian Museum Chamber, bul Vitoša 2, 1000 Sofia, mail addr: c/o Nacionalen istoričeski muzej, POB 1351, 1000 Sofia • T: +359 2 9802258, 9816600 • F: 980024 • nim@einet.bg • ... 54724

ICOM Bulgaria, c/o Nacionalen politechničeskij muzej, ul Opălčenska 66, 1303 Sofia • T: +359 2 9318018 • F: 9314036 • avalchev@mail.bg • ... 54725

Burkina Faso

ICOM Burkina Faso, c/o Patrimoine Culturel, Ministère des Arts et de la Culture, BP 7007, Ouagadougou 03 • T: +226 310927 • F: 316808 • patrimoine@mcc.gov.bf • ... 54726

Burundi

ICOM Burundi, c/o J. Mapfarakora, Musée National, BP 110, Gitega • T: +257 40 2359 • F: (02) 26231 • ... 54727

Cameroon

ICOM Cameroon, c/o R. Assombang Neba'ane, Ministère de la culture, BP 6544, Yaoundé • T: +237 2230614, 2224785 • F: 2221873/37 • ejm@camnet.cm • ... 54728

Canada

Alberta Museums Association, 9829 103 St, Edmonton, AB T5K 0X9 • T: +1 780 4242626 • F: +1 780 4251679 • info@museumsalberta.ab.ca • www.museumsalberta.ab.ca • ... 54729

Association of Manitoba Museums, 153 Lombard Av, Ste 206, Winnipeg, MB R3B 0T4 • T: +1 204 9471782 • F: +1 204 9423749 • info@museumsmanitoba.com • www.museumsmanitoba.com • ... 54730

British Columbia Museums Association, 204 26 Bastion Sq, Victoria, BC V8W 1H9 • T: +1 250 3565700 • F: +1 250 3871251 • bcma@museumsassn.bc.ca • www.museumsassn.bc.ca • ... 54731

Canadian Art Museum Directors Organization, 280 Metcalfe St, Ste 400, Ottawa, ON K2P 1R7 • T: +1 613 5670099 • F: +1 613 2335438 • info@museums.ca • ... 54732

Canadian Federation of Friends of Museums, c/o Art Gallery of Ontario, 317 Dundas St W, Toronto, ON M5T 1G4 • T: +1 416 9796650 • F: +1 416 9796674 • cffm_fcam@ago.net • ... 54733

Canadian Museums Association, 280 Metcalfe St, Ste 400, Ottawa, ON K2P 1R7 • T: +1 613 5670099 • F: +1 613 2335438 • info@museums.ca • www.museums.ca • ... 54734

Coalition of Canadian Healthcare Museums and Archives, 41 Melrose Av, Toronto, ON M5M 1Y6 • T: +1 416 4841878 • fpope@sympatico.ca • ... 54735

Commonwealth Association of Museums, POB 30192, Chinook Postal Outlet, Calgary, AB T2H 2V9 • T: +1 403 9383190 • F: +1 403 9383190 • irvinel@fclc.com • www.maltwood.uvic.ca/cam • ... 54736

Community Museums Association of Prince Edward Island, 440 University Av, Charlottetown, PE C1A 4N6, mail addr: POB 22002, Charlottetown, PE C1A 9J2 • T: +1 902 8928837 • F: +1 902 8921459 • info@museumspei.ca • www.museumspei.ca • ... 54737

ICOM Canada, c/o Canadian Museums Association, 280 Metcalfe St, Ste 400, Ottawa, ON K2P 1R7 • T: +1 613 5670099 ext 222 • F: +1 613 2335438 • fcaron@museums.ca • canada.icom.museum • ... 54738

International Committee for Exhibition Exchange, ICOM-ICEE, c/o Monique Horth, Canadian Museums Association, 280 Metcalfe Street, Suite 400, Ottawa, ON K2P 1R7 • T: +1 613 5670099 ext. 225 • F: 613 2335438 • mhorth@museums.ca • www.eastwestculture.org/icee/ • ... 54739

International Committee on Management, ICOM-INTERCOM, c/o John McAvity, Canadian Museums Association, 280 Metcalfe Street, Suite 400, Ottawa, ON K2P 1R7 • T: +1 613 5670099 ext. 225 • F: 613 2335438 • jmcavity@museums.ca • www.intercom.museum • ... 54740

Mississauga Heritage Foundation, 1921 Dundas St W, Mississauga, ON L5K 1R2 • T: +1 905 8288411 • F: +1 905 8288176 • info@heritagemississauga.org • www.heritagemississauga.com • ... 54741

Museum Association of Newfoundland and Labrador, One Springdale St, Saint John's, mail addr: POB 5785, Saint John's, NL A1C 5X3 • T: +1 709 7229034 • F: +1 709 7229035 • uokshevsky@nf.aibn.com • www.acfsj.ca • ... 54742

Museums Association of Saskatchewan, 422 McDonald St, Regina, SK S4N 6E1 • T: +1 306 7809279 • F: +1 306 7809463 • mas@saskmuseums.org • www.saskmuseums.org • 54743

Ontario Museum Association, 50 Baldwin St, Toronto, ON M5T 1L4 • T: +1 416 3488672 • F: +1 416 3480438 • www.museumsontario.com • 54744

Organization of Military Museums of Canada, POB 323, Gloucester, ON K1C 1S7 • T: +1 613 7373223, 9966799 • F: +1 613 7370821 • don.carrington@sympatico.ca • www.ommc.ca • 54745

Société des Musées Québécois, c/o UQAM, CP 8888, Succ Centre-Ville, Montréal, QC H3C 3P8 • T: +1 514 9873264 • F: +1 514 9873379 • info@smq.qc.ca • www.smq.qc.ca • 54746

Société Géographique Royale du Canada (Royal Canadian Geographical Society), 39 McArthur Av, Ottawa, ON K1L 8L7 • T: +1 613 745 4629 • F: +1 613 7440947 • rcgs@rcgs.org • www.rcgs.org • 54747

South Central Museum Association, POB 53, Wood Mountain, SK S0H 4L0 • 54748

Yukon Historical and Museums Association, 3126 3 Av, Whitehorse, YT Y1A 1E7 • T: +1 867 6674704 • F: +1 867 6674506 • yhma@yknet.yk.ca • www.yukonalaska.com/yhma • 54749

Central African Republic

Comitée National de ICOM Centralafrique, Rue des Industries et Av de France, Bangui, mail addr: c/o Musée National Barthélémy Boganda, BP 349, Bangui • T: +236 615367, 613533, 614568 • F: 615985, 614568 • 54750

Chad

ICOM du Tchad, c/o D. Bakary, Service du Patrimoine, BP 5394, N'Djamena • T: +235 524445 • F: 525538 • 54751

Chile

Federación Chilena de Amigos de los Museos (Chilean Federation of Friends of Museos), Calle Alameda 651, 8320255 Santiago de Chile • patrimon@ctcinternet.cl • 54752

ICOM Chile, c/o L. Nagel Vega, Centro de Documentación, Calle Tabare 654, Recoleta, 8420000 Santiago de Chile • T: +56 2 7321100, 6396488 • F: 7321133 • lnagel@cdbp.cl • 54753

China

Chinese Association of Natural Science Museums, c/o Beijing Natural History Museum, 126 Tien Chiao St, 100050 Beijing • T: +86 10 67053997 • F: 67011408 • cansm@vip.sina.com • www.cansm.org • 54754

Chinese Society of Museums, 29 Wusi St, Jing Shan Qian Jie, 100009 Beijing • T: +86 10 65551639 • F: 65551555, 65123119 • museums@public3.bta.net.cn • 54755

ICOM Chinese National Committee, c/o Numismatic Museum (Zhongguo Qianbi Bowuguan), 22 Xijiaominxiang St, 100031 Beijing • T: +86 10 66024179, 66058326 • F: 66071393 • chinumis@public2.bta.net.cn • 54756

Colombia

Asociación Colombiana de Museos, ACOM, c/o Institutos y Casas de Cultura, Calle 103a No 19-47, Bogotá • T: +57 1 6104235, 2367088 • F: 2186146 • 54757

Comité Nacional del Consejo Internacional de Museos, ICOM, c/o Casa Museo Quinta de Bolívar, Calle 20 No 2-91 Este, Bogotá • T: +57 1 3366410, 3432686 • F: 3366419 • icomcolombia@yahoo.com • 54758

Comoros

ICOM Comores, c/o Ali Mohamed Gou, BP 169, Moroni • T: +269 744187, 733980 • F: 744189, 732222 • cndrs@snpt.km • 54759

Congo, Democratic Republic

ICOM République Democratique du Congo, c/o S. Muteba Mwenema, Institut des Musées Nationaux, BP 13933, Kinshasa I • T: +243 12 583887, 546690 • matshung2002@yahoo.fr • 54760

ICOMAC Regional Organization for Central Africa, BP 13933, Kinshasa • T: +243 12 60263, 60008 • F: 43675 • 54761

Congo, Republic

ICOM Republique Congo, c/o Direction Générale de la Culture et des Arts, BP 459, Brazzaville • T: +242 5318652 • jpclabanz@yahoo.fr • 54762

Costa Rica

ICOM Costa Rica, c/o Museo Nacional de Costa Rica, Apdo 749, San José 1000 • T: +506 2568643 • nenamasis@yahoo.es • 54763

ICOM Latin America and the Caribbean Regional Organization, c/o Dirección General de Museos, Apdo 10277-1000, San Jose 1000 • T: +506 2553051 • F: 2552197 • lsanroma@terra.ecouncil.ac.cr • 54764

Côte d'Ivoire

ICOM Côté d'Ivoire, c/o Musée National du Costume, BP 311, Grand Bassam • T: +225 21301370, 21302329 • F: 20213359 • barroaminata@yahoo.fr • 54765

Croatia

Hrvatski Nacionalni Komitet ICOM, c/o Muzej Grada Zagreba, Opatička 20, 10000 Zagreb • T: +385 1 4851361, 4851359 • F: 4851359 • icom-croatia@mgz.htnet.hr • 54766

Hrvatsko Muzejsko Društvo, Habdelićeva 2, 10000 Zagreb • T: +385 1 14550424, 48511808 • F: 4851977 • hmd@hrmud.hr • www.hrmud.hr • 54767

Savez Muzejskih Društva Hrvatske, Mesnička 5, Muzejski Dokum Centar, 10000 Zagreb • T: +385 1 426534 • 54768

Cuba

ICOM Cuba, c/o Consejo Nacional de Patrimonio Cultural, Calle 4 No. 810, esq 13, Vedado, 10400 La Habana • T: +53 7 8334193 • F: 662106 • patrim@min.cult.cu • 54769

Cyprus

Cyprus Federation of Associations of Friends of Museums, 32a Heroes Av, 1105 Lefkosia • lana@spidernet.com.cy • 54770

ICOM Cyprus, c/o Dr. P. Flourentzos, Dept. of Antiquities, POB 2024, 1516 Lefkosia • T: +357 22302201, 22865800 • F: 22303148 • roctarch@cytanet.com.cy • 54771

Czech Republic

AMG - Association of Museums and Galleries of Czech Republic, Jindřišská 901/5, II. schodiště, 110 00 Praha 1 • T: +420 224210037 • F: 224210047 • amg@cz-museums.cz • www.cz-museums.cz • 54772

Czech Committee of ICOM, c/o Moravské zemské muzeum, Zelný trh 6, 659 37 Brno • T: +420 542210493 • F: 542210493 • icom@mzm.cz • www.cz-icom.cz • 54773

International Association of Agricultural Museums, c/o Dr. Vítězslav Koukal, Valašské Muzeum v Přírodě, Palackého 147, 756 61 Rožnov pod Radhoštěm • T: +420 571757101 • F: 654494 • kou@vmp.cz • 54774

Rada Galerii České Republiky (Council of Galleries of the Czech Republic), Husova 19-21, 110 00 Praha 1 • T: +420 222220218-19 • F: 222221190 • 54775

Societas Museologia, c/o Muzeum Těšínska, Hlavní 13, 737 01 Český Těšín • T: +420 65956795 • F: 65955060 • 54776

Denmark

CIMUSET - International Commitee of Museums of Science and Technology, c/o President of CIMUSET Jytte Thorndahl, Danish Museum of Energy, Bjerringbrovej 44, 8850 Bjerringbro • T: +45 87259744 • F: + 45 86680470 • jt@elmus.dk • www.cimuset.net •
Dir.: *Jytte Thorndahl* • 54777

Dansk ICOM, c/o Kongelige Danske Kunstakademi, Konservatorskolen, Esplanaden 34, 1263 København K • T: +45 33744700 • F: 33744777 • ms@kons.dk • denmark.icom.museum • 54778

Foreningen af Danske Kunstmuseer (Association of Danish Art Museums), c/o Nivågårds Malerisamling, Gammel Strandvej 2, 2990 Nivå • T: +45 49143966 • F: 49143967 • info@museek.dk • 54779

Foreningen af Danske Museumsmaend (Association of Danish Museum Curators), c/o Danske Kongers Kronologiske Samling, Rosenborg Slot, Øster Voldg 4a, 1350 København K • T: +45 33153286 • F: 33152046 • jh@dkks.dk • 54780

Foreningen af Danske Naturhistoriske Museer, c/o Naturhistorisk Museum, Wilhelm Meyers Allé 210, 8000 Århus C • T: +45 86129777 • F: 86130882 • nm@nathist.dk • 54781

Museumstjenesten (Danish Museum Service), Sjørupvej 1, Lysgaard, 8800 Viborg • T: +45 86666766 • F: 86667611 • mtj@museumstjenesten.com • www.museumstjenesten.com • 54782

Skandinavisk Museumsforbund, Danske Afdeling, c/o Jens Peter Munk, Reykjaviksg 2, 2300 København S • T: +45 32961171 • F: 33667191 • jenmun@tmf.kk.dk • 54783

Dominican Republic

ICOM República Dominicaine, c/o Dirección Nacional de Patrimonio Monumental, Hostos 154, Ciudad Colonial, Santo Domingo • T: 6883591 • maxcult@yahoo.com • 54784

Ecuador

ICOM Ecuador, c/o Museo Antropologico del Banco Central, Jose Antepara s/n y 9 de Octubre, Guayaquil • afreire@bceg.fin.ec • 54785

Egypt

ICOM Arab, Dept. of Egyptology, Faculty of Archaeology, Cairo University, Giza • T: +20 2 5678108 • F: 5675660 • a_nurelding@hotmail.com • 54786

ICOM Egypt, c/o Supreme Council of Antiquities, 4d Fakhri Abdel Nour St, Abbasiya, Cairo • T: +20 2 6843627 • F: 6831117 • hawass@sca.gov.eg • http://icom.museum/natcom/egypt.html • 54787

Estonia

Eesti Muuseumiühing Tallinn (Estonian Museum Society in Tallinn), Pikk 70, 10133 Tallinn • T: +372 6411410 • www.emy.kul.ee • 54788

Eesti Muuseumiühing Tartu (Estonian Museum Society in Tartu), Narva mnt 23, 51009 Tartu • T: +372 7461900 • F: 7461912 • heivi.pullerietś@katarina.ee • www.emy.kul.ee • 54789

Estonian Museums Association, c/o Tartu Linnamuuseum, Oru t 2, 2400 Tartu • T: +372 7422693 • muuseum@ekm.estnet.ee • 54790

ICOM Estonia, c/o Adamson-Eric Museum, Lühike jalg 3, 10130 Tallinn • T: +372 6445838 • F: 6445837 • kersti.koll@mail.ee • 54791

Fiji

Pacific Islands Museums Association, Government Bldgs, Suva, mail addr: POB 2023, Suva • T: +679 3306227 • F: 3306227 • pima@connect.com.fi • www.pacificislandsmuseum.org • 54792

Finland

ICOM Finland, c/o Jari Harju, Helsinki City Museum, PL 4300, 00099 Helsinki • T: +358 9 1693568 • F: 1693953 • jari.harju@hel.fi • finland.icom.museum • 54793

Skandinavisk Museumförbund - Skandinaavinen Museoliitto, Finnish Section, Department of Cultural Research, Unioninkatu 38 D 217, 2nd fl, 00014 Helsinki, mail addr: c/o Museology, PB 59, 00014 University of Helsinki • T: +358 9 23784, + 358 19 574 7525 • F: 22653 • marja-liisa.ronkko@helsinki.fi • www.helsinki.fi/museologia/english • 54794

Suomen Museoliitto (Finnish Museums Association), Annankatu 16 B 50, 00120 Helsinki • T: +358 9 58411700 • F: 58411750 • museoliitto@museoliitto.fi • www.museoliitto.fi, www.museot.fi, www.museums.fi •
1923 54795

France

Association des Conservateurs et du Personnel Scientifique des Musées de la Ville de Paris, c/o Musée d'Art Moderne, 9 Rue Gaston-de-Saint-Paul, 75116 Paris • T: +33 153674000 • F: +33 147233598 • 54796

Association des Musées Automobiles de France, 6 Pl de la Concorde, 75008 Paris • T: +33 143124324 • F: +33 143124343 • www.amaf.asso.fr • 54797

Association des Musées et Centres pour le Développement de la Culture Scientifique, Technique et Industrielle, AMCSTI, c/o Carole Grandgirard, 36 Rue Chabot Charny, 21000 Dijon • T: +33 380589875 • F: +33 380589858 • amcsti@u-bourgogne.fr • www.amcsti.org • 54798

Association Européenne des Musées d'Histoire des Sciences Médicales (AEMHSM), c/o Musée d'Histoire de la Médecine, 12 Rue de l'Ecole-de-Médecine, 75006 Paris • T: +33 140461693 • F: +33 140461892 • constructweb@yahoo.fr • www.aemhsm.org • 54799

Association Française des Directeurs de Centres d'Art, Av Conti, Parc Saint-Lèger CAC, 58320 Pougues-les-Eaux • T: +33 386909660 • F: +33 386909661 • pstleger@club-internet.fr • www.d-c-art.org • 54800

Association Française des Musées d'Agriculture et du Patrimoine Rural, AFMA, c/o Musée National des Arts et Traditions Populaires, 6 Av du Mahatma-Gandhi, 75116 Paris • T: +33 144176063 • F: +33 144176060 • jean-francois.charnier@culture.gouv.fr • www.afma.asso.fr/main.htm • 54801

Association Générale des Conservateurs des Collections Publiques de France, 6 Av du Mahatma Gandhi, 75116 Paris • T: +33 144176090 • F: +33 144176060 • nathalie.akrich@culture.gouv.fr • www.agccpf.com • 54802

Association Internationale des Musées d'Histoire (International Association of Museums of History), c/o Musée d'Histoire Contemporaine, Hôtel des Invalides, 75007 Paris • T: +33 144423150 • F: +33 144189384 • contact@euroclio.net • www.euroclio.net • 54803

Association Museum and Industries, 21 Rue Beaurepaire, 75010 Paris • T: +33 661260875 • F: +33 140226320 • museum-industries@club-internet.fr • www.museum-industries.com • 54804

Association pour les Musées de Marseille, c/o Musée des Beaux-Arts, Palais Longchamp, 13004 Marseille • T: +33 491145968 • F: +33 491145968 • ass.musees.marseille@wanadoo.fr • 54805

AVICOM - International Committee for the Audiovisual and Image and Sound New Technologies, c/o Marie-Françoise Delval, Ministère français de la Culture et de la Communication, Chargée de projet Internet, R de Valois, 75001 Paris • T: +33 147043847 • F: +33 153656965 • marie-francoise.delval@culture.gouv.fr • www.unesco.org/webworld/avicomfaimp/index.html • 54806

Comité National Français de l'ICOM, 13 Rue Molière, 75001 Paris • T: +33 142613202 • F: +33 142613202 • icomfrance@wanadoo.fr • france.icom.museum • 54807

Direction de Musées de France, 6 Rue des Pyramides, 75041 Paris Cedex 1 • T: +33 140153600 • F: +33 140153625 • mission-communication.dmf@culture.gouv.fr • www.culture.gouv.fr • 54808

Fédération des Ecomusées et des Musées de Société, 2 Av Arthur Gaulard, 25000 Besançon • T: +33 381832255 • F: +33 381810892 • info@fems.asso.fr • www.fems.asso.fr • 54809

Fédération Française des Sociétés d'Amis de Musées, 16-18 Rue de Cambrai, 75019 Paris • T: +33 142096610 • F: +33 142094471 • info@amis-musees.fr • www.amis-musees.fr • 54810

Fédération Internationale des Musées du Jouet et de l'Enfance, c/o Musée du Jouet, 5 Rue du Murgin, 39260 Moirans-en-Montagne • T: +33 384423864 • F: +33 384423897 • 54811

Fédération Mondiale des Sociétés d'Amis de Musées (World Federation of Friends of Museums), c/o FFSAM, Ellen Julia, 16-18 Rue de Cambrai, 75019 Paris • T: +33 142093891 • F: +33 142093891 • fmam.ellen@wanadoo.fr • www.amis-musees.fr/federation/fmam • 54812

Fédération nationale des Maisons d'Ecrivain et des Patrimoines Littéraires (Federation of Writers' Houses and Literary Collections), Place des Quatre-Piliers, 18001 Bourges Cedex, mail addr: c/o Bibliothèque patrimoniale, BP 18, 18001 Bourges Cedex • T: +33 248242916 • maisonsecrivain@yahoo.com • www.litterature-lieux.com •
Head: *Sophie Vannieuwenhuyze* •
1997
Aim of association is to federate all literary places and heritages; suggest and implement activities which aim at maintaining, protecting and developing the cultural influence of writers' houses, literary places or coll. 54813

Fédération Régionale des Amis des Musées du Nord-Pas de Calais, 1 Passage Pierre et Marie Curie, 59140 Dunkerque • T: +33 328650532 • 54814

Groupement des Amis de Musées de la Région Midi-Pyrénées, c/o Musée Ingres, 19 Rue de l'Hôtel de Ville, 82000 Montauban • T: +33 563201046 • F: +33 563664134 • www.amis-musees-castres.asso.fr/Arsam.htm • 54815

Groupement des Amis des Musées de la Région Centre, c/o Musée des Beaux-Arts, 1 Rue Fernand Rabier, 45000 Orléans • T: +33 238817279 • F: +33 238772206 • 54816

Groupement des Associations d'Amis de Musées de la Région Bretagne, 9 Lotissement des Dunes, 29920 Nevez • T: +33 299870639 • grsamb.rennes@wanadoo.fr • 54817

Groupement des Associations d'Amis de Musées de la Région Languedoc-Roussillon, c/o Amis du Musée Fabre, 7 Rue Verrerie Basse, 34000 Montpellier • T: +33 467925943 • 54818

Groupement des Associations d'Amis de Musées de la Région Limousin, c/o Musée Municipal de l'Evêché, Pl de la Cathédrale, 87000 Limoges • T: +33 555758099 • michele.bourzat@wanadoo.fr • 54819

Groupement des Associations des Amis de Musées Région Provence Alpes Côte d'Azur, c/o Claude Guieu, 73 Rue Jean Pezous, 83000 Toulon (Var) • T: +33 494928399 • claude.guieu@wanadoo.fr • 54820

Groupement Rhône-Alpes des Amis de Musée, c/o Palais Saint-Pierre, 17 Pl des Terreaux, 69001 Lyon • T: +33 476470928 • 54821

International Council of Museums ICOM, Maison de l'UNESCO, 1 Rue Miollis, 75732 Paris Cedex 15 • T: +33 147340500 • F: +33 143067862 • secretariat@icom.museum • icom.museum • 54822

International Council on Monuments and Sites, 49-51 Rue de la Fédération, 75015 Paris • T: +33 145676770 • F: +33 145660622 • secretariat@icomos.org • www.icomos.org • 54823

Réunion des Musées Nationaux, 49 Rue Etienne Marcel, 75039 Paris Cedex 01 • T: +33 140134903 • F: +33 140134973 • catherine.marquet@rmn.fr • www.rmn.fr • 54824

Réunion des Musées Nationaux - Nice, 6 Av Docteur Ménard, 06000 Nice • T: +33 493530546 • F: +33 493538739 • 54825

Germany

Arbeitsgemeinschaft Friedhof und Denkmal, Stiftung Zentralinstitut und Museum für Sepulkralkultur, Weinbergstr 25-27, 34117 Kassel • T: +49 561 918930 • F: 9189310 • info@sepulkralmuseum.de • www.sepulkralmuseum.de • 54826

Arbeitsgemeinschaft kirchlicher Museen und Schatzkammern, Domerschulstr 2, 97070 Würzburg • T: +49 931 386261 • F: 386262 • kunstreferat@bistum-wuerzburg.de • www.museum-am-dom.de •
1958 54827

Bundesverband Museumspädagogik e.V., Geschäftsstelle, c/o Landesstelle für nichtstaatliche Museen, Alter Hof 2, 80331 München • T: +49 89 18910761 • F: 18910761 • info@museumspaedagogik.org • www.museumspaedagogik.org • 54828

CIDOC International Committee for Documentation of the ICOM, c/o Prof. Monika Hagedorn-Saupe, Institut für Museumsforschung, In der Halde 1, 14195 Berlin • T: +49 30 8301460 • F: 8301504 • m.hagedorn@smb.spk-berlin.de • http:/cidoc.mediahost.org • 54829

Deutscher Museumsbund e.V., Büro Berlin, In der Halde 1, 14195 Berlin • T: +49 30 84109517 • F: 84109519 • office@museumsbund.de • www.museumsbund.de • 54830

Deutsches Nationalkomitee des Internationalen Museumsrates ICOM (German National Committee of the International Council of Museums ICOM), In der Halde 1, 14195 Berlin • T: +49 30 69504525 • F: 69504526 • icom@icom-deutschland.de • www.icom-deutschland.de • 54831

Düsseldorfer Museumsverein e.V., c/o Dr. Spennemann, Deutsche Bank 24, Königsallee 45-47, 40189 Düsseldorf • T: +49 211 8832458 • F: 883379 • 54832

Hessischer Museumsverband e.V., Kölnische Str 44-46, 34117 Kassel • T: +49 561 78896740 • F: 78896800 • iris.salomon@museumsverband-hessen.de • www.museumsverband-hessen.de • 54833

ICOM Committee for University Museums and Collections, c/o Dr. C. Weber, Humboldt-Universität, Helmholtz-Zentrum für Kulturtechnik, Unter den Linden 6, 10099 Berlin • T: +49 30 20932563 • F: 20931961 • weber@mathematik.hu-berlin.de, chair@umac.icom.museum • www.icom.museum/umac • 54834

Industriemuseen in der Euregio Maas-Rhein e.V., c/o Detlef Stender, Carl-Koenen-Str 25, 53881 Euskirchen • T: +49 2251 1488111 • F: 1488120 • detlef.stender@lvr.de • www.industriemuseen-emr.de • 54835

International Committee for Museums and Collections of Natural History, c/o Gerhard Winter, Naturmuseum Senckenberg, Senckenberganlage 25, 60325 Frankfurt am Main • T: +49 69 7542356 • F: 7542331 • erhard.winter@senckenberg.de • www.nathist.icom.museum • 54836

International Committee for Museums Security, ICMS, c/o H.-J. Harras, Staatliche Museen Berlin-PK, GD III 3, Ref. Sicherheit, Stauffenbergstr 42, 10785 Berlin • T: +49 30 2262900 • F: 2662969 • h.j.harras@smb.spk-berlin.de • icom.museum/international/icms.html • 54837

International Committee for the Training of Personnel, ICTOP, Eisenacher Str 76, 10823 Berlin, mail addr: c/o Prof. Dr. Angelika Ruge, FHTW Berlin, FB Museumskunde, Wilhelminenhofstr 76/77, 13129 Berlin • T: +49 30 78704673 • angelica.ruge@online.de • 54838

Kunst- und Kulturpädagogisches Zentrum der Museen in Nürnberg, Kartäusergasse 1, 90402 Nürnberg • T: +49 911 1331241, 1331238 • F: 1331318 • kpz@kubiss.de • www.kpz-nuernberg.de • 54839

Kunstsammlungen und Museen Augsburg, Maximilianstr 46, Schaezlerpalais, 86150 Augsburg • T: +49 821 3244102 • F: 3244105 • kunstsammlungen.stadt@augsburg.de • www.augsburg.de •
library 54840

Die Lübecker Museen, Kulturstiftung Hansestadt Lübeck (Museum Foundation), Schildstr 12, 23552 Lübeck • T: +49 451 1227560 • F: 1224106 • hansjoerg.wittern@luebeck.de • www.die-luebecker-museen.de • 54841

Museumslandschaft Hessen Kassel, Direktion und Verwaltung, Schloß Wilhelmshöhe, 34131 Kassel, mail addr: Postfach 410420, 34066 Kassel • T: +49 561 316800 • F: 31680111 • info@museum-kassel.de • www.museum-kassel.de •
Dir.: *Prof. Dr. Bernd Küster* • PR: *Judith Reitter* • Libr.: *Sabine Naumer* •
library 54842

Museumspädagogische Gesellschaft e.V., c/o Museumsdienst Köln, Richartzstr 2-4, 50667 Köln • T: +49 221 22126636 • F: 22127909 • museumsdienst@stadt-koeln.de • 54843

Museumsverband Baden-Württemberg e.V., c/o Städtische Museen Villingen-Schwenningen, Rietstr 37, 78050 Villingen-Schwenningen • T: +49 7721 822366 • F: 822357 • info@museumsverband-bw.de • www.museumsverband-bw.de • 54844

Museumsverband des Landes Brandenburg e.V., Am Bassin 3, 14467 Potsdam • T: +49 331 2327911 • F: 2327920 • museumsverband@t-online.de • www.museen-brandenburg.de • 54845

Museumsverband für Niedersachsen und Bremen e.V., Prinzenstraße 23, 30159 Hannover • T: +49 511 2144983 • F: 21449844 • info@mvnb.de • www.mvnb.de • 54846

Museumsverband Mecklenburg-Vorpommern e.V., Heidberg 15, 18273 Güstrow • T: +49 3843 344736 • F: 344743 • info@museumsverband-mv.de • www.museumsverband-mv.de •
1990 54847

Museumsverband Sachsen-Anhalt e.V., Käthe-Kollwitz-Str 11, 06406 Bernburg • T: +49 3471 628116 • F: 628983, 628116 • museumsverbandsachsen-anhalt@t-online.de • www.mv-sachsen-anhalt.de • 54848

Museumsverband Schleswig-Holstein e.V., c/o Museen im Kulturzentrum, Arsenalstr 2-10, 24768 Rendsburg • T: +49 4331 331338 • F: 27687 • museenrendsburg@web.de • www.museumsverband-sh.de • 54849

Museumsverband Thüringen e.V., Brühler Str 37, 99084 Erfurt • T: +49 361 5513865 • F: 5513879 • info@museumsverband.thueringen.de • www.thueringen.de/museen/ •
1990 54850

Museumsverbund Pankow - Prenzlauer Allee, Prenzlauer Allee 227/228, 10405 Berlin • T: +49 30 902953917 • F: 902953918 • museumsek@ba-pankow.verwalt-berlin.de • www.berlin.de/ba-pankow/museumsverbund •
Local Museum – 2001
History of Pankow – archive, library 54851

Museumsverbund Südniedersachsen e.V., Ritterplan 7-8, 37073 Göttingen • T: +49 551 4002883 • F: 4003135 • info@museumsverbund.de • www.museumsverbund.de • 54852

Museumsverein Mönchengladbach e.V, Städtisches Museum Abteiberg, Abteistr 27, 41061 Mönchengladbach • T: +49 2161 252647 • F: 252659 • museumsverein-moenchengladbach@t-online.de • www.museumsverein-moenchengladbach.de • 54853

Network of European Museum Organisations - NEMO, c/o German Museums Association, Büro Berlin, In der Halde 1, 14195 Berlin • T: +49 30 84109517 • F: 84109519 • office@ne-mo.org • www.ne-mo.org • 54854

Saarländischer Museumsverband e.V., Wilhelm-Heinrich-Straße 39, 66564 Ottweiler • T: +49 6824 8161 • F: 8161 • info@museumsverband-saarland.de • www.museumsverband-saarland.de • 54855

Sächsischer Museumsbund e.V., Wilsdruffer Str 2, 01067 Dresden • T: +49 351 4986635 • F: 8102239 • friedrich.reichert@museen-dresden.de • 54856

Schloss Glücksburg, Schloss, 24960 Glücksburg • T: +49 4631 442330 • F: 4423329 • info@schloss-gluecksburg.de • www.schloss-gluecksburg.de •
Dir.: *Elisabeth Prinzessin zu Ysenburg und Büdingen* •
Coll of arms and paintings 54857

Staatliche Museen Kassel → Museumslandschaft Hessen Kassel

Staatliche Museen zu Berlin - Stiftung Preußischer Kulturbesitz, Generaldirektion (National Museums in Berlin - Prussian Cultural Heritage Foundation - Directorate-general), Stauffenbergstr 41, 10785 Berlin • T: +49 30 2663231 • F: 2663025 • info@smb.spk-berlin.de • www.smb.museum • 54858

Stiftung Preußische Schlösser und Gärten Berlin-Brandenburg, Allee nach Sanssouci 5, 14471 Potsdam, mail addr: Postfach 601462, 14414 Potsdam • T: +49 331 96940 • F: 9694101 • info@spsg.de • www.spsg.de •
1995 54859

Vereinigung Westfälischer Museen e.V., Dompl 10, 48143 Münster • T: +49 251 5907256 • F: 5907210 • gerd.dethlefs@lwl.org • www.muenster.org/vereinigung-westfaelischer-museen •
1926 54860

Ghana

Ghana Museums and Monuments Board, c/o Patrick Essien, POB 281, Cape Coast • T: +233 42 32701 • ghana.icom.museum • 54861

ICOM Ghana, c/o Gramophone Records Museum and Research Centre of Ghana, POB UC 35, University of Cape Coast, Cape Coast • T: +233 24 568507 • sarpongkwame@yahoo.com • ghana.icom.museum • 54862

Organization of Museums, Monuments and Sites of Africa OMMSA, POB 3343, Accra • 54863

Greece

Archaeological Receipts Fund, Aghiou Panépistimiou 57, 105 72 Athinai • T: +30 2103253901 • F: 2103242254 • 54864

Association des Conservateurs d'Antiquités, Ermou 136, 105 53 Athinai • T: +30 2103252214 • F: 2103252214 • 54865

Department of Byzantine Museums, c/o Ministry of Culture, Bouboulinas Od 20-22, 106 82 Athinai • T: +30 2103304030 • F: 2103304009 • protocol@dbmm.culture.gr • www.culture.gr • 54866

Fédération Hellénique des Amis des Musées (Greek Federation of Associations of Friends of Museums), c/o Mouseio Mpenaki, 1 Odos Koumbaru, 106 74 Athinai • 54867

ICOM Grèce, Aghion Assomaton 15, 105 53 Athinai • T: +30 2103239414 • F: 2103239414 • icom@otenet.gr • greece.icom.museum • 54868

ICR International Committee for Regional Museums, c/o Yiannis Markakis, 29 Rodon Str, 71306 Heraklion • T: +30 2897023660, 2810242312 • F: 2897024674 • info@lychnostatis.gr • www.icr-icom.org • 54869

Greenland

Sammenslutningen af Grønlandske Lokalmuseer (Association of Local Museums in Greenland), Jukkorsuup Aqq 9, 3911 Sisimiut, mail addr: c/o Sisimiut Museum, Postboks 308, 3911 Sisimiut • T: +299 865087 • F: 864475 • sismus@greennet.gl • www.museum.gl • 54870

Guadeloupe

Musées Départementaux de la Guadeloupe, 24 Rue Peynier, 97110 Pointe-à-Pitre • T: +590 820804 • F: 837839 • 54871

Guatemala

Asociación de Museos de Guatemala, AMG, c/o Museo Universitario de San Carlos, 9a Av 9-79, Zona 1, 01009 Guatemala • T: +502 2327666, 2320721 • F: 2320721 • amg@hotmail.com • 54872

ICOM Guatemala, c/o A. García-Vassaux, Museo de Ciencia, Av Simeón Cañas Final, Hipodromo del Norte, 01002 Guatemala • T: +502 2541114, 2327666 • F: 2541114, 2320721 • agv@terra.com.gt • 54873

Haiti

ICOM Comité Haïtien, c/o H. Gaspard, Ministère de la Culture et de la Communication, Pl des Héros de l'Indépendance, Port-au-Prince • T: +509 2210004 • F: 2212318 • haroldgaspard@yahoo.ca • 54874

Hungary

European Association of Open Air Museums, c/o Dr. Miklós Cseri, Szabadtéri Népradzi Múzeum, Sztaravodai ú, 2000 Szentendre • T: +36 26 502500 • F: 502502 • sznm@sznm.hu • 54875

ICOM Magyar Nemzeti Bizottság (ICOM Hungarian National Comittee), Szent György tér 2, 1014 Budapest • T: +36 1 3559772 • F: 3559175 • cseri@sznm.hu • hungary.icom.museum • 54876

Pulszky Society - Hungarian Museums Association, c/o I. Matskási, Magyar Természettudományi Múzeum, Baross ú 13, 1088 Budapest • T: +36 1 2661481 • F: 3171669 • matskasi@zoo.zoo.nhmus.hu • www.museum.hu/pulszkytarsasag • 54877

Iceland

Association of Icelandic Curators, c/o National Museum of Iceland, Svourgotu 41, 101 Reykjavík • T: +354 5528967 • F: 5528967 • 54878

Icelandic Museums Association, c/o J. Asmundsson, Minjasafn Egilis Olafssonar, Hnjótur, 451 Patreksfjörður • T: +354 4561569 • F: 8676344 • museum@hnjotur.is • 54879

ICOM Iceland, c/o Þjóðminjasafn Islands, Suðurgata 41, 101 Reykjavík • T: +354 5302280 • F: 5302281 • lilja@natmus.is • www.icom.is • 54880

Scandinavian Museum Association, Icelandic Section, c/o National Museum of Iceland, Svourgotu 41, 101 Reykjavík • T: +354 5528967 • F: 5528967 • www.arbaejarsafn.is • 54881

India

Association of Indian Museums, c/o National Museum of Natural History, Barakhambra Rd, Delhi 110001 • T: +91 11 3319173 • 54882

ICOM India, c/o Indian Museum, 27 Jawaharlal Nehru Rd, Kolkata 700016 • T: +91 33 22499979, 22499902 • F: 22495669 • imbot@cal2.vsnl.net.in • india.icom.museum • 54883

Indian Association for the Study of Conservation of Cultural Property, c/o National Museum Institute, Janpath, Delhi 110011 • T: +91 11 3016098, 3792217 • F: 3019821, 3011901 • 54884

Museums Association of India, c/o National Museum, Janpath, Delhi 110011 • T: +91 11 383568 • 54885

National Council of Science Museums, Block GN, Sector V, Bidhan Nagar, Kolkata 700091 • T: +91 33 3579347/48 • F: 3576008 • ncsmin@giascl01.vsnl.net.in • 54886

Indonesia

ICOM Indonesia, c/o Dr. Hardiati, Pusat Penelitian Arkeologi Nasional, Jl Raya Condet, Pejaten 4, Pasar Minggu, Jakarta 13760 • T: +62 21 7988171 • F: 3811076 • munas@budpar.go.id • 54887

Iran

ICOM Iran, c/o E. Jalalian, Iranian Cultural Heritage Organization, 60 Larestan St, Motahhari Av, 15959-3481 Tehrān • T: +98 21 8898827 • F: 8898828 • iranicom_se@yahoo.com • 54888

Ireland

Geological Curators' Group, c/o Dr. Matthew A. Parkes, Natural History Museum, Merrion St, Dublin 2 • T: +353 87 1221967 • mparkes@museum.ie • www.geocurator.org • 54889

ICOM Ireland, c/o Dr. S. Mannion, Mayo County Museum, Arran Pl, Ballina, Co. Mayo • F: +353 44 62984 • smannion@mayo.coco.ie • 54890

Irish Museums Association, c/o Pearse Museum, Saint Enda's Park, Grange Rd, Rathfarnham, Dublin, 16 • F: +353 87 2790518 • office@irishmuseums.org • www.irishmuseums.org • 54891

Israel

Israel National Committee of ICOM, National Park, Ramat-Gan 52109, mail addr: c/o Man & the Living World Museum, POB 947, Ramat-Gan 52109 • T: +972 3 9565977 • F: 9565788 • icom98@netvision.net.il • www.icom.org.il • 54892

Museum Association of Israel, c/o Museum of Rishon Le-Zion, POB 7, Rishon Le-Zion 75100 • T: +972 3 9565977 • F: 9565788 • icom98@netvision.net.il • 54893

Italy

Associazione Musei Ecclesiastici Italiani, c/o Dr. Carlo Tatta, Via dei Pini 8, 05019 Orvieto • T: +39 0763305155 • 54894

Associazione Nazionale dei Musei dei Enti Locali ed Istituzionali, ANMLI, Via Tiziano 323, 25124 Brescia • T: +39 0302300307 • F: 0302300307 • 54895

Associazione Nazionale dei Musei Italiani, Piazza San Marco 49, 00186 Roma • T: +39 066791343 • F: 066791343 • 54896

Associazione Nazionale Musei Scientifici, ANMS, Via G. La Pira 4, 50121 Firenze • T: +39 0552757460 • F: 0552737373 • marcolin@mtsn.tn.it • www.anms.it • 54897

CIMCIM - International Committee of Musical Instrument Museums and Collections, Comité International des Musées et Collections d'Instruments de Musique, c/o Gabriele Rossi-Rognoni, Museo degli Strumenti Musicali, Galleria dell'Accademia, via Ricasoli 60, 50122 Firenze • secretary@cimcim.icom.museum • cimcim.icom.museum • 54898

DEMHIST-International Committee for Historic House Museums, c/o ICOM Italia - Museo della Scienza e della Tecnica, Via San Vittore 19-21, 20123 Milano • T: +39 0248555338 • F: 0243919840 • demhist@iol.it • 54899

Federazione Italiana delle Associazioni Amici dei Musei (Italian Federation of Associations of Friends of Museums), Via degli Alfani 39, 50121 Firenze • T: +39 055213558 • F: 055213558 • info.fidam@alice.it • www.fidam.org • 54900

ICOM Comitato Nazionale Italiano, c/o C. Cafaro, Via San Vittore 19-21, 20123 Milano • T: +39 0248555338, 024695693 • F: 0243919840 • icomit@iol.it • www.icom-italia.org • 54901

International Association of Libraries and Museums of Performing Arts (SIMBAS), c/o Maria Teresa Iovinelli-Biblioteca Burcardo, Via del Sudario 44, 00186 Roma • T: +39 066819471 • F: 0668194727 • biblioteca.burcardo@siae.it • www.theatrelibrary.org/sibmas/sibmas.html • 54902

International Centre for the Study of the Preservation and the Restoration of Cultural Property (ICCROM), Via di San Michele 13, 00153 Roma • T: +39 06585531 • F: 0658553349 • iccrom@iccrom.org • www.iccrom.org • 54903

International Committee for Conservation, c/o Joan Marie Reifsnyder, ICCROM, Via San Michele 13, 00153 Roma • T: +39 0658553410 • F: 0658553349 • secretariat@icom-cc.org • icom-cc.icom.museum • 54904

Istituto Italiano dei Castelli, Via G.A. Borgese 14, 20154 Milano • T: +39 02347237 • F: 02347237 • cast@cast.it • www.castit.it • 54905

Musei Civici Veneziani, San Marco 1, 30124 Venezia, mail addr: San Marco 52, 30124 Venezia • T: +39 0412405211, 0412715911 • F: 0415200935 • mtk.musei@comune.venezia.it • www.museiciviciveneziani.it • 54906

Jamaica

ICOM Jamaica, c/o Museums of History and Ethnography Institute of Jamaica, 10-16 East St, Kingston • T: +1 876 9220622 • F: 9223795 • museums@instituteofjamaica.org.jm • 54907

Japan

Japanese Council of Art Museums, c/o 21st Century Museum of Contemporary Art, 1-2-1 Hirosaka, Kanazawa 920-8509 • T: +81 76 2202828 • F: 2202829 • zenbikaigi@nifty.com • 54908

Japanese National Committee for ICOM, c/o Japanese Association of Museums, Shoyu Kaikan Bldg, 3-3-1 Kasumigaseki, Chiyoda-ku, Tokyo 100-8925 • T: +81 3 35917190 • F: 35917170 • icom@j-muse.or.jp • www.museum.or.jp/icom-japan • 54909

Nihon Hakubutsukan Kyokai (Japanese Association of Museums), Shoyu Kaikan Bldg, 3-3-1 Kasumigaseki, Chiyoda-ku, Tokyo 100-8925 • T: +81 3 35917190 • F: 35917170 • webmaster@j-muse.or.jp • www.j-muse.or.jp • 54910

Jordan

ICOM Jordan, Queen Misbah St, 11118 Amman, mail addr: c/o Dr. Al-Khrayshey, Department of Antiquities, POB 88, 11118 Amman • T: +962 6 4644336/320 • F: 4615848 • doa@nic.net.jo • 54911

Kazakhstan

ICOM Kazakhstan, c/o A. Kazteev State Museum of Arts, Satpaeva 30a, 480090 Almaty • T: +7 3272 478249 • F: 509567, 478669 • kazart@nursat.kz • 54912

Kenya

ICOM Kenya, c/o National Museum of Kenya, POB 40658, 00100 Nairobi • T: +254 20 3742131/34 • F: 3741427 • kitungulu@hotmail.com • kenya.icom.museum • 54913

International Council of African Museums (AFRICOM), Museum Hill Rd, Nairobi, mail addr: POB 38706 Ngara, 00600 Nairobi • T: +254 20 3748668 • F: 3748928 • secretariat@africom.museum • www.africom.museum • 54914

Kenya Museum Society, c/o National Museum, POB 40658, 00100 Nairobi • T: +254 20 3742131 ext 289, 3743808 • info@kenyamuseumsociety.org • www.kenyamuseumsociety.org • 54915

Programme for Museum Development in Africa, Old Law Courts Bldg, Nkrumah Rd, Mombasa, mail addr: POB 90010, Mombasa • T: +254 11 225114, 224846 • F: 227985 • pmda@africaonline.co.ke • www.heritageinafrica.org • 54916

Korea, Republic

ICOM Asia-Pacific, c/o Seyun Iron Museum, In-Kyung Chang, 97 Ohang-ri, Gamkok-myeon, Eumseong 369-854 • T: +82 43 8832321 • F: 8832321 • ladyiron@hanmail.net • www.asia-pacific.icom.museum • 54917

Korean Museum Association, c/o National Museum of Korea, 1-57 Sejong-no, Chongno-gu, Seoul 110-820 • T: +82 2 7350230, 3985190 • F: 7350231 • kormuseum@dreamwiz.com • www.museum.or.kr • 54918

Korean National Committee for the International Council of Museums (ICOM), c/o National Museum of Korea, 1-57 Sejong-no, Chongno-gu, Seoul 110-820 • T: +82 2 7350230, 3985190 • F: 7350231 • office@icomkorea.org • www.icomkorea.org • 54919

Kyrgyzstan

ICOM Kyrgyzstan, c/o Kyrgyz State Museum of Fine Arts, 196 Soviet st, 720000 Bishkek • T: +996 312 661623 • F: 228476 • erkina2002@mail.ru • 54920

Laos

ICOM Lao, Setthathirat Rd, Vientiane, mail addr: c/o Ministry of Information and Culture, Dept. of Museums and Archaeology, POB 122, Vientiane • T: +856 21 212423, 212895 • F: 212408 • 54921

Latvia

ICOM Latvijas Nacionāla Komiteja, c/o G. Gerharde-Upeniece, K. Valdemāra iela 10a, Riga, 1010 • T: +371 7324476 • F: 7325051 • vmm@latnet.lv • www.latvia.icom.museum.lv • 54922

Latvian Museum Council, c/o Valsts Makslas Muzejs, 10a K. Valdemara, Rīga, 1010 • T: +371 7325021 • F: 7325051 • vmm@latnet.lv • 54923

Latvijas Muzeju Asociācija (Latvian Museums Association), Skārņu iela 10/20, Rīga, 1050 • T: +371 7222235 • F: 7228083 • dlmm@apollo.lv • www.muzeji.lv • 54924

State Authority of Museums, Smilsu 20, Rīga, 1868 • T: +371 7228083 • F: 7227335 • muzasoc@acad.latnet.lv • 54925

Lebanon

ICOM Liban, Bliss St, Beirut, mail addr: c/o L. Badre, Musée de l'Université Americaine, BP 11-0236/9, Beirut • T: +961 1 340549 • F: 363235 • badre@aub.edu.lb • 54926

Libya

ICOM Libyan Arab Jamahiriya, c/o Jamahirya Museum, As-Saha al-Kradrah, Tripoli • T: +218 21 4449065 • F: 4440166 • 54927

Lithuania

ICOM Lietuva (ICOM Lithuania), c/o V. Balciunas, Vilniaus Paveikslu Galerija, Didžioji 4, Bokšto 5, 2001 Vilnius • T: +370 5 2120841 • F: 2120841 • galerija@ldm.lt • 54928

Lithuanian Museums Association, Saltoniskiu g 58, 2600 Vilnius • T: +370 2 790371, 790918 • F: 790213 • muzasoc@takas.lt • muziejai.mch.mii.lt • 54929

Luxembourg

ICMAH International Committee for Archaeology and History Museums, c/o Marie-Paule Jungblut, Musée d'Histoire de la Ville de Luxembourg, 2090 Luxembourg • T: +352 47964562 • F: 471707 • m.jungblut@musee-hist.lu • 54930

ICOM Luxembourg, c/o Paul Reiles, Musée National d'Histoire et d'Art, Marché-aux-Poissons, 2345 Luxembourg • T: +352 4793301 • F: 223760 • musee@mnha.etat.lu • www.mnha.lu • 54931

Macedonia

Association of Museum Professionals of the Republic of Macedonia, c/o Muzej na Sovremenata Umetnost, Samoilova bb, 1000 Skopje • T: +389 2 117734 • F: 110123 • 54932

ICOM F.Y.R. Macedonia, c/o Muzej na Sovremenata Umetnost Skopje, Samoilova b.b., Skopje 1000 • T: +389 91 117734/35 • F: 236372 • icom_mnc@yahoo.com • 54933

Muzejsko Društvo na Makedonija, c/o Muzej na Grad Skopje, Mito Hadzi Vasilev bb, 1000 Skopje • T: +389 2 54123, 9154123 • 54934

Madagascar

ICOM Madagascar, 17 Rue Dr. Villette, Isoraka, mail addr: c/o Musée d'Art et d'Archéologie, BP 564, Antananarivo 101 • T: +261 20 2225493 • F: 2225493 • chradimi@refer.mg • 54935

Malaysia

ICOM Malaysia, c/o National Museum, Dept. of Museums and Antiques, Jalan Damansara, 50566 Kuala Lumpur • T: +60 3 22826255, 3986321 • F: 22827294 • janetsm@jma.gov.my • 54936

Mali

ICOM Mali, Rue du Général Leclerc, Bamako, mail addr: c/o S. Sidibe, Musée National du Mali, BP 159, Bamako • T: +223 223486 • F: 231909 • musee@afribone.net.ml • 54937

ICOM Regional Organization for West Africa, BP 159, Bamako • T: +223 223489 • F: 231909 • musee@malinet.ml • 54938

Malta

Malta National Committee of ICOM, c/o T.M. Vella, National Museum of Fine Arts, Republic Street, Valletta VLT 11 • T: +356 21233034 • F: 21239915 • info@malta.icom.museum • 54939

Mauritania

ICOM Mauritanie, c/o Musée National, BP 5055, Nouakchott • T: +222 5253722, 5256017 • F: 5253244, 5255275 • ouldabidinesidi@hotmail.com • 54940

Mauritius

ICOM Mauritius, c/o Mauritius Museums Council, Mauritius Institute, POB 54, Port Louis • T: +230 2120639, 2122815 • F: 2125717 • mimuse@intnet.mu • 54941

Mexico

Federación Mexicana de Asociaciones de Amigos de los Museos (Mexican Federation of Associations of Friends of Museums), Calle General Léon 65bis, 11850 México • femam@prodigy.net.mx • 54942

ICOM México, Santa Veracruz 66, Col Guerrero, Cuauhtémoc, 06300 México • T: +52 55 55104377, 55182265 ext 246 • F: 55104377 • informes@icommexico.org • www.icommexico.org • 54943

Moldova

ICOM Moldova, c/o A. Cornetchi, Muzeul National de Istorie a Moldovei, Str 31 August 1989 No 121a, 2012 Chişinău • T: +373 2 2447031 • 54944

Monaco

ICOM - Comité National Monégasque du Conseil International des Musées, c/o Musée Océanographique de Monaco, Av Saint Martin, 98000 Monaco • T: +377 93153600 • F: 93505297 • nbarcoli@hotmail.com • 54945

Mongolia

ICOM Mongolia, c/o O. Norovtseren, Zanabazar Museum of Fine Arts, Barilgachidyn Talbai 46, Ulaanbaatar • T: +976 11 326060 • F: 323986 • tegshee5@yahoo.com • 54946

Montenegro

ICOM Crne Gore, c/o Narodni Muzej Crne Gore, Mirjana Dabovic, Novice Cerovića bb, 81250 Cetinje • T: +381 86 230310 • F: 230310 • dabcom@cg.yu • 54947

Muzejsko Društvo SR Crne Gore - Cetinje, N. Cerovica bb, 81250 Cetinje • T: +381 86 31682 • F: 31682 • museums@museumscg.org • 54948

Morocco

ICOM Maroc, c/o Direction du patrimoine culturel, Ministère de la Culture, 17 Rue Michlifen, Agdal, 10000 Rabat • T: +212 37 671385 • F: 671397, 760532 • nadiad@menara.ma • 54949

Namibia

ICOM Namibia, c/o Annaleen Eins, National Art Gallery of Namibia, POB 994, Windhoek • T: +264 61 231160 • F: 240930 • nagn@mweb.com.na • 54950

Museums Association of Namibia, MAN, POB 147, Windhoek • T: +264 61 22840 • F: 222544 • silvestj@iway.na • 54951

Nepal

ICOM Nepal, Patan Darbar Palace, Mangal Bazar, Lalitpur, mail addr: c/o Jal Krishna Shrestha, Patan Museum, GPOB 1040, Kathmandu • T: +977 1 5521492 • F: 4251479 • icomnepal@wlink.com.np • 54952

Netherlands

Association of European Open Air Museums, c/o Nederlands Openluchtmuseum, Postbus 649, 6800 AP Arnhem • T: +31 26 3576111 • F: 3576147 • info@openluchtmuseum.nl • 54953

CECA International Committee for Education and Cultural Action of Museums, c/o Arja van Veldhuizen, CODA, Vosselmanstraat 299, 7311 CL Apeldoorn • F: +31 55 5268499 • ceca.icom.museum • 54954

European Federation of Museum and Tourist Railways, De Akker 25, 7481 GA Haaksbergen • T: +31 53 5727357 • contact@fedecrail.org • www.fedecrail.org • 54955

ICAMT International Committee for Architecture and Museums Techniques, Prinsengracht 526, 1017 KJ Amsterdam, mail addr: c/o Architectenbureau Jowa, Postbus 15704, Jowa I. Kis-Jovak, Architectenbureau Jowa, 1001 Amsterdam • T: +31 20 6244380 • F: 6246469 • info@jowa.nl • www.icom.org/internationals.html#icamt • 54956

ICOM Nederland, Rapenburgerstr 123, 1011 VL Amsterdam, mail addr: c/o De Nederlandse Museumvereniging, Postbus 2975, 1000 CZ Amsterdam • T: +31 20 5512900 • F: 5512901 • info@museumvereniging.nl • www.icomnederland.nl • 54957

ICOMAM International Committee of Museums and Collections of Arms and Military History, c/o Jan Piet Puype, Koninklijk Nederlands Leger- en Wapenmuseum, Korte Geer 1, 2611 CA Delft • T: +31 15 2150500 • F: 2150544 • jppuype@hotmail.com • 54958

International Committee for Monetary Museums, c/o Stichting Het Geld- en Bankmuseum, Postbus 2407, 3500 GK Utrecht • T: +31 30 2910492 • F: 2910467 • info@sgbm.nl • www.geldmuseum.nl • 54959

International Confederation of Architectural Museums, Museumpark 25, 3015 CB Rotterdam, mail addr: c/o Nederlands Architektuurinstituut, Postbus 237, 3000 AE Rotterdam • service@nai.nl • www.icam-web.org • 54960

De Nederlandse Museumvereniging (Netherlands Museums Association), Rapenburgerstr 123, 1012 VM Amsterdam, mail addr: Postbus 2975, 1000 CZ Amsterdam • T: +31 20 5512900 • F: 5512901 • info@museumvereniging.nl • www.museumvereniging.nl • 54961

Vereniging van Rijksgesubsidieerde Musea (Association of State Subsidized Museums), Rapenburgerstr 123, 1011 VL Amsterdam, mail addr: c/o Museumhuis Bussenschut, Postbus 2975, 1000 CZ Amsterdam • T: +31 20 5512900 • F: 5512901 • vrm@netland.nl • 54962

New Zealand

ICOM New Zealand, c/o Greg McManus, Rotorua Museum, Government Gardens, Albert Dr, Rotorua 3220 • T: +64 7 3494350 • F: 3492819 • greg.mcmanus@rdc.govt.nz • 54963

Museums Aotearoa, Museum Association of New Zealand, 104 The Terrace, Wellington 6011, mail addr: POB 10928, Wellington 6143 • T: +64 4 4991313 • F: 4996313 • mail@museums-aotearoa.org.nz • www.museums-aotearoa.org.nz • 1947 54964

Niger

ICOM Comité National du Niger, c/o Musée National du Nieger, BP 10457, Niamey • T: +227 20 733727 • F: 733691 • saleytim@yahoo.fr • 54965

Nigeria

ICOM Nigeria, Old Residency, Calabar, mail addr: c/o National Commision for Museums and Monuments, PMB 1180, Calabar • T: +234 87 231776 • F: 230111 • nmuscal@skannet.com • 54966

Museums Association of Nigeria, King George V St, mail addr: c/o National Museum, PMB 12556, Lagos • T: +234 1 2636075 • 54967

National Commission for Museums and Monuments, Plot 2018, Cotonou Crescent, Wuse Zone 6, Abuja, mail addr: PMB 171, Garki, Abuja • T: +234 9 5230801, 5238253 • F: 5238254 • commissionmuseums@yahoo.com • www.ncfmm.org • 54968

Norway

ABM-Utvikling (Norwegian Archive, Library and Museum Authority), Kronprinsens G 9, 0033 Oslo, mail addr: Postboks 8145 DEP, 0033 Oslo • T: +47 23117500 • F: 23117501 • post@abm-utvikling.no • www.abm-utvikling.no • 54969

ICLM International Committee for Literary Museums, c/o E. Dahl, Edvard Grieg Museum, Troldhaugev 65, 5040 Paradis • T: +47 55910710 • F: 55911395 • edah@online.no • 54970

International Committee for Marketing and Public Relations, ICOM-MPR, c/o Paal Mork, Norsk Folkemuseum, Museumsv 10, 0287 Oslo • T: +47 22123700 • F: 22123777 • paal.mork@norskfolkemuseum.no • museumsnett.no/icommpr • 54971

Norges Museumsforbund, Ullevålsv 11, 0165 Oslo • T: +47 22201402 • F: 22112337 • museumsnytt@museumsforbundet.no • www.museumsforbundet.no • 54972

Norsk ICOM, Ullevålsv 11, 0165 Oslo • T: +47 22201402 • F: 22112337 • sekr@icom-norway.org • www.icom-norway.org • 54973

Norske Kunst- og Kulturhistoriske Museer, Ullevålsv 11, 0165 Oslo • T: +47 22201402 • F: 22112337 • 54974

Norske Museumspedagogiske Forening, c/o Nordlandsmuseet, Prinsensg 116, 8005 Bodø • T: +47 8121640 • 54975

Norwegian Federation of Friends of Museums, Postboks 95 Bygdoy, 0211 Oslo • 54976

Skandinavisk Museumsforbund, Norwegian Section, Haugar Vestfold Kunstmuseum, 3110 Tønsberg • T: +47 33307670 • F: 33307660 • einar.wexelsen@vfk.no • 54977

Oman

ICOM Oman, Al-Wazarat St, Muscat, mail addr: c/o Ministry of Heritage and Culture, POB 668, Muscat, 113 • T: +968 602225, 605400 • F: 602735, 604957 • mnhcgov@omantel.net.om • www.mnhc.gov.om • 54978

Pakistan

The Museums Association of Pakistan, c/o Bait al-Hikmah, Hamardy University, Sharae Madinat al-Hikmah, Karachi 74700 • T: +92 21 6996001/02 • F: 6350574 • huvc@cyber.net.pk • 54979

Panama

Comisión Nacional de Arqueología y Monumentos Historicos, Plaza 5 de Mayo, 662 Ciudad de Panamá • T: +507 621438 • 54980

ICOM Panamá, Entre Calle 5ta y 6ta, Ciudad de Panamá, mail addr: c/o Museo del Canal Interoceánico de Panamá, Apdo 1213, Zona 1, Ciudad de Panamá • T: +507 2111994 • F: 2111995 • museodelcanal@cwpanama.net • 54981

Paraguay

Dirección General de Museos de la Nación, Calle Humaita 673, Asunción • T: +595 21 440070 • 54982

ICOM Paraguay, c/o M.G. Gomzález Cáceres, Centro de Conservación, Calle J.E. O'Leary 1 y Barranco del Rio, Almacén Viola, Asunción • T: +595 21 446829 • F: 446829 • icom@patrimoniocultural-ccpc.org.py • 54983

Peru

Comité Peruano del Consejo Internacional de Museos, ICOM Perú, Jr Camaná 459, Lima 1 • T: +51 1 4602770 • F: 4602770 • acastel@pucp.edu.pe • icom.perucultural.org.pe • 54984

Philippines

ICOM Philippines, c/o National Museum of the Philippines, Padre Burgos St, Ermita, 1000 Manila • T: +63 2 5271215/242 • F: 5270306 • nmuseum@i-next.net • 54985

Museum Volunteers of the Philippines, 4114 Dasmariñas, mail addr: POB 8052, 1222 Makati • MVPhilippines@hotmail.com • mvphilippines.hypermart.net • 54986

Poland

ICOM Poland, c/o Muzeum Narodowe, Marcinkowskiego 9, Poznań • T: +48 61 8525969 • F: 8515898 • mnp@mnp.art.pl • 54987

Polish Association of Members of the Museum Professions, Wawel 5, 31-001 Kraków • T: +48 12 4225155 • F: 4215177 • 54988

Stowarzyszenie Muzeów na Wolnym Powietrzu (Association of Open Air Museums), Rynek Staromiejski 1, Ratusz, 87-100 Toruń • 54989

Stowarzyszenie Zwiazek Muzeów Polskich (Polish Museums Association), ul Krakowska 46, 31-066 Kraków • T: +48 12 4305575, 4305563 • F: 4306330 • mek@tele2.pl • 54990

Portugal

Associacão Portugesa de Museologia, c/o Panteão Nacional, Praça B à Travessa Sargento Abílio, Lote C1, 1500-567 Lisboa • T: +351 217780687 • F: 217780642 • apom@oninet.pt • www.museusportugal.org/apom • 54991

Federação de Amigos dos Museus de Portugal, Calçada do Combro 61, 1200-111 Lisboa • T: +351 213225435 • F: 213335437 • global@famportugal.pt • www.amigosdosmuseus.pt • 54992

ICOM Portugal, c/o Ana R. Godinho Carcoso, Museu Nacional de Arte Antiga, Rua das Janelas Verdes, 1249-017 Lisboa • T: +351 213912800/07 • F: 213973703 • mnaa@netcabo.pt • 54993

Instituto Português de Museus, Palacio Nacional da Ajuda Ala Sul, 1349-021 Lisboa • T: +351 213650800 • F: 213647821 • contactos@ipmuseus.pt • www.ipmuseus.pt • 54994

Qatar

ICOM Qatar, c/o National Council for Culture, Arts and Heritage, POB 2777, Doha • T: +974 4437993 • F: 4370572, 4437993 • my_taha@yahoo.co.uk • 54995

Romania

Asociatia Nationalá a Muzeelor in aer Liber - România, Piața Presei Libere 1, 013701 Bucureşti • T: +40 212244421 • F: 2244421 • manuela.tabacila@cultura.ro • 54996

ICOM Romania, c/o Muzeul Judetean de Istorie şi Arheologie Prahova, Str Toma Caragiu 10, 100042 Ploieşti • T: +40 2445144377 • F: 5144377 • cnricom@yahoo.com • romania.icom.museum • 54997

Russia

Associacija Muzeev Rossii (Russian Museums Association), Oktjabrskaja ul 14, 300002 Tula • T: +7 4872 779221, 776712 • F: 393196, 393599 • assoc@tsu.tula.ru • www.amr-museum.ru • 54998

CAMOC - International Committee for the Collections and Activities of Museums of Cities, c/o Galina Vedernikova, Pres., Moscow City Museum, Novaja pl 12, 103012 Moskva • migm@mail.ru • camoc.icom.museum • 54999

ICOFOM International Committee for Museology, c/o Dr. Olga Truevtseva, Altajskij Gosudarstvennyj Institut Iskusstv i Kultury, ul Jurina 277, 656055 Barnaul • T: +7 3852 445663 • F: 445709 • truevtseva@yandex.ru • www.lrz-muenchen.de/~iims/icofom • 55000

Museum Council, ul D. Uljanova 19, 117036 Moskva • T: +7 495 1251121 • F: 1260630 • i_zaitseva@hotmail.com • 55001

Rossijskij Komitet Meždunarodnogo Soveta Muzeev (IKOM Rossii), Volchonka ul 3-4, 119019 Moskva • T: +7 495 2036090 • F: 2038487 • icom@kremlin.museum.ru • www.icomrussia.ru • 55002

St. Lucia

INTERCOM International Committee for Museums Management, c/o L. Gill, Archaeological and Historical Society, 85 Chaussee Rd, Castries • T: +1 758 451-9251 • F: (758) 451-9324 • gilll@candw.lc • 55003

Senegal

ICOM Comité Sénégalais, c/o Musée du CRDS, BP 382, Saint-Louis • T: +221 9611050, 8251990 • F: 9611050 • signare@refer.sn • 55004

Serbia

ICOM Nacionalni Komitet Serbia, c/o Galerija Fresaka, Cara Urosa 20, 11000 Beograd • T: +381 11 182966 • F: 183655 • icom.serbia@gmail.com • serbia.icom.museum • 55005

Muzejsko Društvo Srbije, c/o Narodni Muzej, Trg Republike 1a, 11000 Beograd • T: +381 11 3671591 • F: 660170 • mdsrbije@infosky.net • www.mds.org.yu • 55006

Seychelles

Association of Museums of the Indian Ocean, c/o National Musuem of Seychelles, POB 720, Victoria • T: +248 321333 • F: 323183 • seymus@seychelles.net • 55007

ICOM Seychelles, c/o National Museum, POB 720, Victoria • T: +248 321333 • F: 323183 • seymus@seychelles.net • 55008

Singapore

ICOM Singapore, c/o National Heritage Board, Mita Bldg, 140 Hill St, Singapore 179369 • T: +65 63325593, 63323963 • F: 63391941 • Lim_Siok_Peng@nhb.gov.sg • 55009

Slovakia

ICOM Slovakia, c/o Slovenská Národná Galéria, Riečna 1, 815 13 Bratislava • T: +421 2 54430746 • F: 54433971 • uz@sng.sk • 55010

Zväz múzeí na Slovensku (Union of Museums in Slovakia), Kapitulská 23, 974 00 Banská Bystrica • T: +421 48 4123258 • F: 4123716 • www.zms.sk • 55011

Slovenia

ICOM Slovenia, c/o Slovenski Etnografski Muzej, Metelkova 2, 1000 Ljubljana • T: +386 1 3008700 • F: 3008700 • icom.slovenija@etno-muzej.si • slovenia.icom.museum • 55012

Skupnost Muzejev Slovenije (Museums Association of Slovenia), c/o Taja Cepić, Mestni Muzej, Gosposka 15, 1000 Ljubljana • T: +386 61 222930 • 55013

Slovensko Konservatorsko Društvo (Association of Conservators of Cultural Heritage of Slovenia), Trzaska 4, 1000 Ljubljana • T: +386 1 1256079 • F: 1254198 • 55014

Slovensko Muzejsko Društvo (Association of Museum Curators of Slovenia), Kongresni Trg 1, 1000 Ljubljana • T: +386 1 4213267 • F: 4213287 • 55015

South Africa

ICOM South Africa, c/o G. Balkwill, National Cultural History Museum, POB 28088, Sunnyside 0132 • T: +27 123346082 • F: 0123285173 • glyn@nfi.org.za • 55016

Iziko Museums of Cape Town, POB 61, Cape Town 8000 • T: +27 214813800 • F: 0214813993 • info@iziko.org.za • www.iziko.org.za •
Museum association 55017

South African Museums Association (SAMA), c/o SAMA National Office, POB 12413, Centrahil 6006 • T: +27 466243087 • F: 0466243276 • sama@futurserve.co.za • www.samaweb.org.za • 55018

Spain

Amics de los Museos de Cataluña, Palau de la Virreina, La Rambla 99, 08002 Barcelona • T: +34 933014379 • F: 933189421 • amics@amicsdelsmuseus.org • www.amicsdelsmuseus.org • 55019

Amigos de los Museos, c/o Museo de América, Av Reyes Católicos 6, 28040 Madrid • T: +34 913600057 • F: 915436106 • feamamigos@terra.es • www.amigosdemuseos.com • 55020

Asociacion de Amigos de los Museos Militares, c/o Museo del Ejercito, Calle Méndez Núñez 1, 28014 Madrid • F: +34 915218078 • 55021

Asociación Española de Archiveros, Bibliotecarios, Museólogos y Documentalistas, ANABAD, Calle Recoletos 5, 28001 Madrid • T: +34 915751727 • F: 915751727 • 55022

Asociación Española de Museólogos, Av Reyes Católicos 6, 28040 Madrid • T: +34 915430917 • F: 915440225 • aem@museologia.net • www.museologia.net • 55023

Associació del Museo de la Ciència i de la Tècnica i d'Arqueologia Industrial de Catalunya, Rambla d'Egara 270, 08221 Terrassa • T: +34 937803787 • F: 937806089 • associaciomct@eic.ictnet.es • . 55024

Comité Español del ICOM, c/o Museo de América, Av Reyes Católicos 6, 28040 Madrid • T: +34 915431820 • F: 915431820 • info@icom-ce.org • www.icom-ce.org • 55025

European Federation of Associations of Industrial and Technical Heritage, E-FAITH, c/o Museo de la Ciència i de la Tècnica, Rambla d'Egara 270, 08221 Terrassa • T: +34 937803787 • F: 937806089 • associaciomct@eic.ictnet.es • www.e-faith.org • 55026

Federación Española de Asociaciones de Amigos de los Museos, Av Reyes Católicos 6, 28040 Madrid • feamamigos@terra.es • 55027

International Committee for Museums and Collections of Modern Art, CIMAM, MACBA Pl des Àngels 1, 08001 Barcelona, mail addr: Pilar Cortada, Planella 43, 08001 Barcelona • T: +34 932036274 • F: 932036274 • info@cimam.org • www.cimam.org • 55028

Sri Lanka

ICOM Sri Lanka, c/o Dept. of National Museums, POB 854, Colombo 7 • T: +94 11 2694767/68 • F: 2695366 • nmdep@sltnet.lk • 55029

Swaziland

ICOM Swaziland, c/o Swaziland National Museum, POB 100, Lobamba H107 • T: +268 41 61178/79 • F: 61875 • staff@swazimus.org.sz • 55030

Sweden

ICOM Sweden, c/o M. Björkgren, Statens Maritima Museer, Box 27131, 102 52 Stockholm • T: +46 8 51954878 • F: 51954888 • margareta.bjorkgren@maritima.se • www.sweden.icom.org • 55031

International Association of Transport and Communication Museums, 262 52 Ängelholm, mail addr: c/o Swedish Railway Museum, Box 407, 801 05 Gävle • T: +46 26 144570 • F: 144598 • robert.sjoo@jarnvagsmuseum.sese • 55032

MPR International Committee for Museums Marketing and PR, Malmöhusvägen, 201 24 Malmö, mail addr: c/o Malmö Museer, Box 406, 201 24 Malmö • T: +46 40 344404 • F: 124097 • www.icom.museum/internationals.html • 55033

Skandinavisk Museumsforbund, Swedish Section, c/o Nordiska Museet, Djurgårdsvägen 6-16, 115 93 Stockholm • F: +46 8 51954580 • 55034

Sveriges Museimannaförbund (Swedish Association of Museum Curators), Planiavägen 13, 131 34 Nacka, mail addr: Box 760, 131 24 Nacka • T: +46 8 4662400 • F: 4662413 • kansli@dik.se • www.dik.se • 55035

Swedish Federation of Friends of Museums, c/o Groth, Box 1607, 150 23 Enhorna • goran@groth.com • www.museivanner.se • 55036

Switzerland

ICOM Schweiz, Internationaler Museumsrat, Museumstr 2, 8001 Zürich, mail addr: c/o Schweizerisches Landesmuseum, Postfach, 8021 Zürich • T: +41 442186588 • F: +41 442186589 • info@museums.ch • www.museums.ch •
1953 55037

Verband der Museen der Schweiz, Museumstr 2, 8001 Zürich, mail addr: c/o Schweizerisches Landesmuseum, Postfach, 8021 Zürich • T: +41 442186588 • F: +41 442186589 • info@museums.ch • www.museums.ch •
1966 55038

Taiwan

Chinese Association of Museums, 49 Nanhai Rd, Taipei • T: +886 2 23610270 ext 108 • F: 23890718 • cam@moe.nmh.gov.tw • www.cam.org.tw • 55039

Yilan Museums Association, 66 Jhioucheng W Rd, Yilan 260 • T: +886 3 9359748, 9357282 • F: 9358461 • ilan.ma@msa.hinet.net • www.ilma.org.tw •
2001 55040

Tanzania

ICOM Tanzania, c/o National Museums of Tanzania, POB 511, Dar-es-Salaam • T: +255 22 2700437 • F: 2112752 • nmt@museum.or.tz • 55041

Thailand

ICOM Thailand, c/o S. Charoenpot, Museum of H.M. The King's Golden Jubilee, Khlong 5, Pathumthani 12120 • T: +66 2 9029833, 9027568 • F: 9027833 • somlakc@hotmail.com • 55042

Togo

ICOM Togo, c/o Musée National des Sites et Monuments, BP 12156, Lomé • T: +228 2 217140 • F: 2221839 • musees.togo@rdd.tg • 55043

Tunisia

ICOM Tunisia, c/o Institut National du Patrimoine, 4 Pl du Château, 1008 Tunis • T: +216 71562622 • F: 71562452 • ali_drine@yahoo.fr • 55044

Turkey

ICOM Turkish National Committee, c/o Anitlar ve Müzeler Genel Müdürlügü, Meclis Binasi, Eski II, 06100 Ulus • T: +90 312 3109950 • F: 3116603 • anitlarmuzeler@kultur.gov.tr • 55045

Ukraine

ICOM Ukraine, c/o Lvivs'ka Galereja Mystectv, ul Stefanika 3, 79000 Lviv • T: +380 322 723009 • F: 723009 • dergach@city-adm.lviv.ua • 55046

United Kingdom

Alchemy: The Group for Collections Management in Yorkshire and Humberside, c/o Bankfield Museum, Boothtown Rd, Halifax HX3 6HG • T: +44 1422 354823 • F: 349020 • 55047

Association for Cultural Enterprises, c/o Everyevent, 41 Britannia Sq, Worcester WR1 3DN • T: +44 1905 724734 • F: 724744 • ACE@everyevent.co.uknt.co.uk • www.acenterprises.org.uk • 55048

Association of British Transport and Engineering Museums, c/o Heritage Motor Centre, Bunbury Rd, Gaydon CV35 0BJ • T: +44 1926 645105 • tbryan9@landrover.com • 55049

Association of Independent Museums, c/o Justeen Stone, AIM Administrator, 63 Wiston Avenue, Worthing BN14 7PX • www.aim-museums.co.uk • 55050

British Association of Friends of Museums, c/o Deborah Woodland, The Shrubbery, 14 Church St, Whitchurch RG28 7AB • T: +44 870 2248904 • admin@bafm.org.uk • www.bafm.org.uk •
1973 55051

British Aviation Preservation Council, c/o Centre for Lifelong Learning, University of Warwick, Coventry CV4 7AL • T: +44 1926 495191 • bapcchairman@tiscali.co.uk • 55052

Campaign for Museums, 35-37 Grosvenor Gardens, London SW1W 0BX • T: +44 20 72339796 • F: 72336770 • info@campaignformuseums.org.uk • www.campaignformuseums.org.uk • 55053

Care of Collections Forum, c/o Falkirk Museums Service, 7-11 Abbotsinch Rd, Grangemouth FK3 9UX • T: +44 1324 504689 • F: 503771 • carol.whittaker@falkirk.gov.uk • 55054

Dress and Textiles Specialists Group, c/o Ch. Stevens, Beamish Museum, Beamish NE12 8AP • T: +44 191 3704048 • F: 3704001 • christinestevens@beamish.org.uk •
1974 55055

Dualchas Heritage Service for Skye and Lochalsh, Tigh na Sgire, Park Ln, Portree IV51 9GP • T: +44 1478 613857 • F: 613855 • anne.macleod2@highland.gov.uk • www.highland.gov.uk • 55056

Dundee Art Galleries and Museum Association, c/o Frances Towns, 37 Ancrum Dr, Dundee, DD2 2JG • T: +44 1382 668720 • 55057

East Midlands Museums Service, POB 7221, Colston Bassett NG12 3WH • T: +44 1949 81734 • emms@emms.org.uk • www.emms.org.uk • 55058

English Heritage East Midlands Region, 44 Derngate, Northampton NN1 1UH • T: +44 1604 735400 • F: 735401 • www.english-heritage.org.uk • 55059

English Heritage North West Region, 3 Chepstow St, Manchester M1 5FW • T: +44 161 2421400 • F: 2421401 • www.english-heritage.org.uk • 55060

English Heritage South West Region, 29 Queen Sq, Bristol BS1 4ND • T: +44 117 9750700 • F: 9750701 • southwest@english-heritage.org.uk • www.english-heritage.org.uk • 55061

English Heritage West Midlands Region, 112 Colmore Row, Birmingham B3 3AG • T: +44 121 6256820 • F: 6256821 • www.english-heritage.org.uk • 55062

European Museum Forum, POB 913, Bristol BS99 5ST • T: +44 117 9238897 • F: 9732437 • EuropeanMuseumForum@compuserve.com • www.europeanmuseumforum.org • 55063

Federation of Museums and Galleries in Wales, c/o Ceredigion Museum, Coliseum, Terrace Rd, Aberystwyth SY23 2AQ • T: +44 1920 633088 • museum@ceredigion.gov.uk • 55064

Group for Directors of Museums, c/o Grosvenor Museum, 27 Grosvenor St, Chester CH1 2DD • T: +44 1244 402012 • F: 347587 • s.matthews@chestercc.gov.uk • 55065

Group for Education in Museums, Primrose House, 193 Gillingham Rd, Gillingham, Kent ME7 4EP • T: +44 1634 312409 • F: 312409 • gemso@blueyonder.co.uk • www.gem.org.uk • 55066

Heritage Railway Association, c/o HRA Travel, 8 Ffordd Dyfrig, Tywyn LL36 9EH • T: +44 1654 710344 • F: 712323 • hradw@globalnet.co.uk • www.hra.gb.com • 55067

The International Committee for Museums and Collections of Costume, c/o J. Marschner, Royal Ceremonial Dress Collection, Kensington Palace, London W8 4PX • T: +44 20 79379561 • F: 73760198 • joanna.marschner@hrp.org.uk • www.kent.edu/museum • 55068

Ironbridge Gorge Museum Trust, Coach Rd, Coalbrookdale TF8 7DQ • T: +44 1952 884391 • www.ironbridge.org.uk •
1967/1997
Items representing the hist of industry through paintings, engravings, books, decorative tiles, porcelain, iron (iron bridge over Severn) – archives, library.. 55069

Leicestershire County Council Museums Service, County Hall, Glenfield LE3 8RA • T: +44 116 2656781 • F: 2657370 • ycourtney@leics.gov.uk • www.leics.gov.uk •
1849 55070

London Federation of Museums and Art Galleries, Saint John's Gate, London EC1M 4DA • T: +44 20 72536644 • F: 74908835 • 55071

MDA, Museum Documentation Association, Spectrum Bldg, Michael Young Centre, Purbeck Rd, Cambridge CB2 2PD • T: +44 1223 415760 • F: 415960 • mda@mda.org.uk • www.mda.org.uk •
1979 55072

Midlands Federation of Museums and Art Galleries, Ironbridge Gorge Museum, Ironbridge, Telford TF8 7AW • T: +44 1952 432751 • F: 432237 • isabel_churcher@birmingham.gov.uk • www.bmag.org.uk • 55073

MLA West Midland, Regional Council for Museums, Libraries and Archives, Grosvenor House, 14 Bennetts Hill, Birmingham B2 5RS • T: +44 121 6315800 • F: 6315825 • info@mlawestmidlands.org.uk • www.mlawestmidlands.org.uk • 55074

Museum Development Service, Suffolk County Council, Endeavour House, Russell Road, Ipswich IP1 2BX • T: +44 1473 265241 • Lyn.gash@libher.suffolkcc.gov.uk • www.suffolkmuseums.org.uk • Off.: *Lyn Gash* •
Museum development service for the county – Library 55075

Museum Ethnographers Group, c/o Tabitha Cadbury, Royal Albert Memorial Museum, Queen St, Exeter EX4 3RX • T: +44 1271 342482 • 01271342482@talktalk.net • www.museumethnographersgroup.org.uk • 55076

Museum Professionals Group, c/o Nicola Hughes, Imperial War Museum, Duxford CB2 4QR • T: +44 1223 835000 • membership@museumprofessionalsgroup.org • www.museumprofessionalsgroup.org • 55077

Museums and Galleries Disability Association, c/o Hove Museum and Art Gallery, 19 New Church Rd, Hove BN3 4AB • T: +44 1273 292828 • F: 292827 • abigail.thomas@brighton-hove.gov.uk • www.magda.org.uk • 55078

Museums Association, 24 Calvin St, London E1 6NW • T: +44 20 74266950/970 • F: 74266961 • info@museumsassociation.org • www.museumsassociation.org • 55079

Museums in Essex Committee, Essex Record Office, Wharf Rd, Chelmsford CM2 6YT • T: +44 1245 244612, 244613 • F: 244655 • stephen.lowy@essexcc.gov.uk • 55080

Museums, Libraries and Archives Council, Victoria House, Southampton Row, London WC1B 4EA • T: +44 20 72731444 • F: 72731404 • info@mla.gov.uk • www.mla.gov.uk • 55081

Museums, Libraries and Archives North East, House of Recovery, Bath Ln, Newcastle-upon-Tyne NE4 5SQ • T: +44 191 2221661 • F: 2614725 • info@mlanortheast.org.uk • www.mlanortheast.org.uk • 55082

Museums North, Northern Federation of Museums and Art Galleries, Sir William Gray House, Clarence Rd, Hartlepool TS24 8BT • T: +44 1429 523443 • F: 523477 • karenteasdale@beamish.org.uk • 55083

Museums Weapons Group, Royal Armouries, Armouries Dr, Leeds LS10 1LT • T: +44 113 2201867 • F: 2201871 • grimer@armouries.org.uk • 55084

National Heritage, The Museums Action Movement, c/o Liz Moore, Rye Rd, Hawkhurst TN18 5DW • T: +44 1580 752052 • F: 755670 • www.nationalheritage.org.uk • 55085

National Museum Directors' Conference, c/o Imperial War Museum, Lambeth Rd, London SE1 6HZ • T: +44 20 74165202 • F: 74165485 • nmdc@iwm.org.uk • www.nationalmuseums.org.uk • 55086

National Museums and Galleries of Northern Ireland, c/o Ulster Museum, Botanic Gardens, Belfast BT9 5AB • T: +44 28 90383000 • F: 90383006 • www.magni.org.uk • 55087

North of England Museums Libraries and Archives Council, House of Recovery, Bath Lane, Newcastle-upon-Tyne NE4 5SQ • T: +44 191 2221661 • F: 2614725 • nemlac@nemlac.co.uk • www.nemlac.co.uk • 55088

North West Federation of Museums and Art Galleries, c/o Emma Varnam, Portland Basin Museum, Portland Pl, Ashton-under-Lyne OL7 0QA • emma.varnam@tameside.gov.uk • www.nwfed.org.uk • 55089

Northern Ireland Museums Council, 6 Crescent Gardens, Belfast BT7 1NS • T: +44 28 90550215 • F: 90550216 • info@nimc.co.uk • www.nimc.co.uk • 55090

Scottish Museums Council, County House, 20-22 Torphichen St, Edinburgh EH3 8JB • T: +44 131 2297465 • F: 2292728 • inform@scottishmuseums.org.uk • www.scottishmuseums.org.uk • 55091

Scottish Museums Federation, c/o Rowan Brown, Scottish Mining Museum, 55 Queen St, Edinburgh EH2 3PA • T: +44 131 6637519 • director@scottishminingmuseum.com • Dir.: *Rowan Brown* • 55092

Social History Curators Group, Katharine St, Croydon CR9 1ET, mail addr: c/o Room G11, Town Hall, Croydon Clocktower, Museum of Croydon • T: +44 208 2531026 • enquiry@shcg.org.uk • www.shcg.org.uk • Chair: *Victoria Rogers* • Treas.: *Michael Terwey* • 55093

Society of County Museum Officers, c/o Somerset Museums Service, Castle Green, Taunton TA1 4AA • T: +44 1823 320200 • F: 320229 • county-museum@somerset.gov.uk • www.somerset.gov.uk/museums • 55094

Society of Decorative Art Curators, c/o Decorative Art Department, National Museums Liverpool, 127 Dale St, Liverpool L2 2JH • T: +44 151 4784261/62 • F: 4784693 • pauline.rushton@liverpoolmuseums.org.uk • 55095

Society of Museum Archaeologists, c/o Museum of London, London Wall, London EC2Y 5HN • T: +44 20 76003699 • F: 76001058 • musmda@hants.gov.uk • www.socmusarch.org.uk • 55096

South Eastern Federation of Museums and Art Galleries, 2 Sackville St, Hayes, Bromley BR2 7JT • T: +44 20 84622228 • pwboreham@ukonline.co.uk • 55097

South Midlands Museums Federation, c/o Town House Museum, Queen St, King's Lynn PE30 5DQ • T: +44 1553 773450 • F: 773450 • robin.hanley@norfolk.gov.uk • 55098

South West Museums, Libraries and Archives Council, Creech Castle, Taunton TA1 2DX • T: +44 1823 259696 • F: 270933 • general@swmuseums.co.uk • www.swmlac.org.uk • 55099

South Western Federation of Museums and Art Galleries, c/o Alison Bevan, Penlee House Gallery & Museum, Penzance TR18 4HE • T: +44 1736 363625 • www.swfed.org.uk • Dir.: *Alison Bevan* • 55100

Standing Conference on Archives and Museums, c/o Ipswich Lelord Office, Gatacre Rd, Ipswich IP1 2LQ • T: +44 1473 584531 • jayne.austin@libher.suffolkcc.gov.uk • www.hmc.gov.uk/SCAM • 55101

Touring Exhibitions Group, 29 Point Hill, Greenwich, London SE10 8QW • T: +44 20 86912660 • F: 83331987 • admin@teg-net.org.uk • www.teg-net.org.uk • 55102

United Kingdom National Committee of the International Council of Museums, c/o British Library, Oliver Urquhart Irvine, 96 Euston Rd, London NW1 2DB • T: +44 20 78145501 • F: 76001058 • oliver.urquhart-irvine@bl.uk • www.mda.org.uk/icom-uk • 55103

University Museums Group, c/o Lapworth Museum of Geology, University of Birmingham, Edgbaston Park Rd, Birmingham B15 2TT • T: +44 121 4144173 • F: 4144942 • m.p.smith@bham.ac.uk • www.umg.org.uk • 55104

Visual Arts and Galleries Association, The Old School, High St, Witcham, Ely CB6 2LQ • T: +44 1353 776356 • F: 775411 • admin@vaga.co.uk • www.vaga.co.uk •
1978 55105

Yorkshire and Humberside Federation of Museums and Art Galleries, c/o Carolyn Dalton, Doncaster Museum Service, Chequer Rd, Doncaster DN1 2AE • T: +44 1302 734293 • F: 735409 • carolyn.dalton@doncaster.gov.uk • 55106

Yorkshire Museums Council, 1 Marshall St, Leeds LS11 9YP • T: +44 113 3944840 • F: 2439872 • info@ymlac.org.uk • www.ymlac.org.uk • 55107

United States of America

African American Museums Association, c/o Dr. M.Burroughs, Dusable Museum, 740 E 56th Pl, Chicago, IL 60637 • T: +1 312 9470600 • F: +1 312 9470677 • 55108

Alabama Museums Association, POB 2866, Birmingham, AL 35202-3802 • T: +1 205 9344475 • F: +1 205 3489292 • info@alabamamuseums.org • www.alabamamuseums.org • 55109

American Antiquarian Society, 185 Salisbury St, Worcester, MA 01609 • T: +1 508 7555221 • F: +1 508 7533311 • library@mwa.org • www.americanantiquarian.org • 55110

American Association for Museum Volunteers, c/o Denver Museum of Natural History, 2001 Colorado Blvd, Denver, CO 80205 • T: +1 303 3706419 • F: +1 303 3316492 • schristian@dmnh.org • 55111

American Association for State and Local History, 1717 Church St, Nashville, TN 37203-2991 • T: +1 615 3203203 • F: +1 615 3279013 • history@aaslh.org • www.aaslh.org • 55112

American Association of Museums, 1575 Eye St NW, Ste 400, Washington, DC 20005 • T: +1 202 2891818 • F: +1 202 2896578 • marketing@aam-us.org • www.aam-us.org • 55113

Arkansas Museums Association, c/o Arkansas National Guard Museum, Camp Robinson, North Little Rock, AR 72199-9600 • T: +1 501 2125215 • steve.rucker@ar.ngb.army.mil • www.armusa.org • 55114

Association for Living History, Farm and Agricultural Museums, c/o Judith Sheridan, 8774 Rte 45 NW, North Bloomfield, OH 44450-9701 • T: +1 440 6854410 • F: +1 440 6854410 • sheridan@orwell.net • www.alhfam.org • 55115

Association of Art Museum Directors, 41 E 65 St, New York, NY 10021 • T: +1 212 2494423 • F: +1 212 5355039 • aamd@amn.org • www.aamd.org • 55116

Association of Children's Museums, 1300 L St NW, Ste 975, Washington, DC 20005 • T: +1 202 8981080 • F: +1 202 8981086 • acm@childrensmuseums.org • www.childrensmuseums.org • 55117

Association of College and University Museums and Galleries, c/o Oklahoma Museum of Natural History, 1335 Asp Av, Norman, OK 73019-0606 • T: +1 405 3251009 • F: +1 405 3257699 • pbtirrell@ou.edu • 55118

Association of Indiana Museums, POB 24428, Indianapolis, IN 46224-0428 • T: +1 317 8825649 • F: +1 317 8657662 • wildcat@iquest.net • 55119

Association of Midwest Museums, POB 11940, Saint Louis, MO 63112 • T: +1 314 4543110 • F: +1 314 4543112 • mmcdirect2@aol.com • 55120

Association of Railway Museums, c/o Pennsylvania Trolley Museum, 725 W Chestnut St, Washington, PA 15301-4623 • T: +1 724 2289256 • F: +1 724 2289675 • 55121

Association of Science Museum Directors, c/o Santa Barbara Museum of Natural History, 2559 Puesta del Sol Rd, Santa Barbara, CA 93105 • T: +1 805 6824711 • khutterer@sbnature2.org • www.asmd.org • 55122

Association of Science-Technology Centers, 1025 Vermont Av NW, Ste 500, Washington, DC 20005-6310 • T: +1 202 7837200 • F: +1 202 7837207 • info@astc.org • www.astc.org • 55123

Association of South Dakota Museums, 200 W Sixth St, Sioux Falls, SD 57102 • T: +1 605 3674210 • F: +1 605 3676004 • 55124

Association of Systematics Collections, 1725 K St NW, Ste 601, Washington, DC 20006-1401 • T: +1 202 8359050 • F: +1 202 8357334 • general@nscalliance.org • www.ascoll.org • 55125

Brewery Art Association, 2100 N Main St, Los Angeles, CA 90031 • T: +1 323 3420717 • assoc@breweryart.org • www.the-brewery.net • Pres.: *Mat Gleason* • Dir.: *Stephanie Mercado* • 55126

California Association of Museums, POB 1455, Santa Cruz, CA 95061-1455 • T: +1 831 4719970 • F: +1 831 4719381 • cam@calmuseums.org • www.calmuseums.org • 55127

Colorado-Wyoming Association of Museums, 400 Clarendon Av, Sheridan, WY 82801 • T: +1 307 6744589 • F: +1 307 6721720 • admin@cwamit.org • www.cwamit.org • 55128

Connecticut Museum Association, POB 78, Roxbury, CT 06783 • T: +1 860 3543543 • 55129

Council of American Jewish Museums, 12 Eldridge St, New York, NY 10002 • T: +1 212 2190903 • F: +1 212 9664782 • aemsy@aol.com • 55130

Council of American Maritime Museums, c/o Maine Maritime Museum, 243 Washington St, Bath, ME 04530 • T: +1 207 4431316 • F: +1 207 4431665 • wilcox@bathmaine.com • 55131

Florida Art Museum Directors Association, 1990 Sand Dollar Ln, Vero Beach, FL 32963 • T: +1 561 3882428 • F: +1 561 3889313 • fmda@steds.org • 55132

Florida Association of Museums, 1114 Thomasville Rd, Tallahassee, FL 32303, mail addr: POB 10951, Tallahassee, FL 32302-2951 • T: +1 850 2226028 • F: +1 850 2226112 • fam@flamuseums.org • www.flamuseums.org • 55133

Folk Art Society of America, 1904 Byrd Av, Richmond, VA 23230 • T: +1 804 2854532 • F: +1 804 2854532 • fasa@folkart.org • www.folkart.org • Pres.: *Ann Oppenhimer* •
Folklore Museum – 1987
Folk, outsider & self taught art – library 55134

Georgia Association of Museums and Galleries, c/o Fernbank Science Center, 156 Heaton Park Dr, Atlanta, GA 30307 • T: +1 404 3784311 • F: +1 404 3701336 • gamg@gamg.org • www.gamg.org • 55135

Hawaii Museums Association, POB 4125, Honolulu, HI 96812-4125 • T: +1 808 2544292 • F: +1 808 2544153 • dpope@lava.net • 55136

Historic Columbus Foundation, 1440 Second Av, Columbus, GA 31901 • T: +1 706 3220756 • F: +1 706 5764760 • hcfinc@historiccolumbus.com • www.historiccolumbus.com •
1966
Five house museums incl c.1870 first brick house in original residential part of city 55137

History San José, 1650 Senter Rd, San Jose, CA 95112-2599 • T: +1 408 2872290 • F: +1 408 2872291 • spuckitt@historysanjose.org • www.historysanjose.org • Dir.: *David Crosson* • Cur.: *Alida Bray* (Colls, Exhibits) • *Sarah Puckitt* (Photography, Art) •
1971 55138

ICME - International Committee for Museums and Collections of Ethnography, c/o Annette B. Fromm, Ph.D. Director, 3060 Alton Road, Miami Beach, FL 33140 • T: +1 305 5323530 • F: +1 305 3482762 • president@icme.icom.museum • museumsnett.no/icme/ •
1948 55139

Idaho Association of Museums, c/o Idaho State Historical Museum, 7111 McMullen Rd, Boise, ID 83709 • T: +1 208 3342120 • F: +1 208 3342775 • 55140

Illinois Association of Museums, 1 Old State Capitol Plaza, Springfield, IL 62701-1507 • T: +1 217 5247080 • F: +1 217 7857937 • mturner@hpa084r1.state.il.us • 55141

Intermuseum Conservation Association, Allen Art Bldg, Oberlin, OH 44074 • T: +1 216 7757331 • 55142

International Association of Museum Facility Administrators, 11 W 53rd St, New York, NY 10019, mail addr: POB 1505, Washington, DC 20013-1505 • T: +1 212 7089413 • F: +1 212 3331169 • kowalczykr@si.edu • www.iamfa.org • 55143

International Committee for Egyptology in ICOM - CIPEG, c/o Dr. Regine Schulz, Walters Art Museum, 600 N Charles St, Baltimore, MD 21201-5185 • T: +1 410 5479000 ext 255 • F: +1 410 7524797 • rschulz@thewalters.org • 55144

International Committee for Museums and Collections of Glass, c/o Dr. J.A. Page, Toledo Museum of Art, 2445 Monroe St, Toledo, OH 43620 • T: +1 419 2558000 • F: +1 419 2555638 • museo@fcnv.es • 55145

International Congress of Maritime Museums, c/o Stuart Parnes, Connecticut River Museum, 67 Main St, Essex, CT 06426 • T: +1 860 7678269 • F: +1 860 7677028 • StuPames@aol.com • 55146

International Movement for a New Museology, c/o WAN Museum, POB 1009, Zuni, NM 87327 • T: +1 505 7824403 • F: +1 505 7822700 • aamhc@zuni.k12.nm.us • 55147

International Museum Theatre Alliance, c/o National Children's Museum, 800 Third St NE, Washington, DC 20002 • secretary@imtal.org • www.imtal.org • 55148

Iowa Museum Association, 225 Commercial St, Waterloo, IA 50701 • T: +1 319 2914490 • F: +1 319 2914270 • cammie.scully@waterloo-sa.org • www.iowamuseums.org • 55149

Kansas Museums Association, 6425 SW 6th Av, Topeka, KS 66615-1099 • T: +1 785 2728681 • F: +1 785 2728682 • starr@kshs.org • 55150

Kentucky Association of Museums, POB 9, Murray, KY 42071-0009 • T: +1 270 7623052 • F: +1 270 7623920 • 55151

Louisiana Association of Museums, 1609 Moreland Av, Baton Rouge, LA 70808-1173 • T: +1 225 3836800 • F: +1 225 3836880 • 55152

Maine Association of Museums, 37 Community Dr, Augusta, ME 04330 • T: +1 207 6238428 • F: +1 207 6265949 • 55153

Maryland Association of History Museums, POB 1806, Annapolis, MD 21404-1806 • T: +1 301 8099877 • F: +1 301 8099878 • info@mahm.org • www.mahm.org • 55154

Michigan Museums Association, POB 10067, Lansing, MI 48901-0067 • T: +1 517 4824055 • F: +1 517 4827997 • ashby.ann@acd.net • 55155

Mid-Atlantic Association of Museums, 800 E Lombard St, Baltimore, MD 21202 • T: +1 410 2231194 • F: +1 410 2232773 • info@midatlanticmuseums.org • www.midatlanticmuseums.org • 55156

Minnesota Association of Museums, POB 14825, Minneapolis, MN 55414-0825 • T: +1 612 6240089 • F: +1 612 6267704 • davis136@tc.umn.edu • 55157

Mississippi Museums Association, 640 S Canal St, Box E, Natchez, MS 39120 • T: +1 601 4466502 • F: +1 601 4466503 • 55158

Missouri Museums Association, 3218 Gladstone Blvd, Kansas City, MO 64123-1199 • T: +1 816 4838300 • 55159

Mountain-Plains Museums Association, 7110 W David Dr, Littleton, CO 80128-5405 • T: +1 303 9799358 • F: +1 303 9793553 • info@mountplainsmuseums.org • www.mountplainsmuseums.org • 55160

Museum Association of Arizona, POB 63902, Phoenix, AZ 85082-3902 • info@azmuseums.org • www.azmuseums.org • 55161

Museum Association of Montana, c/o Glacier National Park, West Glacier, MT 59936 • T: +1 406 8887936 • F: +1 406 8887808 • deirdre_shaw@nps.gov • 55162

Museum Association of New York, 265 River St, Troy, NY 12180 • T: +1 518 2733400 • F: +1 518 2733416 • info@manyonline.org • www.manyonline.org • 55163

Museum Education Roundtable, 621 Pennsylvania Av SE, Washington, DC 20003 • T: +1 202 5478378 • F: +1 202 5478344 • www.mer-online.org • 55164

Museum Trustee Association, 2025 M St NW, Ste 800, Washington, DC 20036-3309 • T: +1 202 3671180 • F: +1 202 3672180 • office@mta-hq.org • www.mta-hq.org • 55165

Museums Alaska, POB 242323, Anchorage, AK 99524 • T: +1 907 2434714, +1 907 3063409 • museums@alaska.net • www.museumsalaska.org • 55166

U

Natural Science Collections Alliance, 1725 K St, NW, Ste 601, Washington, DC 20006-1401 • T: +1 202 8359050 • F: +1 202 8357334 • general@nscalliance.org • www.nscalliance.org • 55167

Nebraska Museums Association, 208 16th St, Aurora, NE 68818-3009 • T: +1 402 6944032 • F: +1 402 6944035 • edgerton@hamilton.net • www.edgerton.org • 55168

New England Museum Association, 22 Mill St, Ste 409, Arlington, MA 02476 • T: +1 781 6410013 • F: +1 781 6410053 • nemadirector@tiac.net • www.nemanet.org • 55169

New Jersey Association of Museums, POB 877, Newark, NJ 07101-0877 • T: +1 888 3566526 • F: +1 973 7486607 • www.museumusa.org/sma/NJ • 55170

New Mexico Association of Museums, POB 5800, Albuquerque, NM 87185-1490 • T: +1 505 2843232 • F: +1 505 2843244 • jkwalth@sandia.gov • 55171

North Carolina Museums Council, POB 2603, Raleigh, NC 27602 • T: +1 336 7676730 • F: +1 336 6611777 • 55172

Northeast Mississippi Museums Association, POB 993, Corinth, MS 38835 • T: +1 662 2873120 • spardue@tsixroads.com • www2.dixie-net.com/nemma • 55173

Ohio Museums Association, 567 E Hudson St, Columbus, OH 43211 • T: +1 614 2982030 • F: +1 614 2982068 • ohmuseassn@juno.com • 55174

Oklahoma Museums Association, 2100 NE 52nd St, Oklahoma City, OK 73111 • T: +1 405 4247757 • F: +1 405 4245068 • info@okmuseums.org • www.okmuseums.org • 55175

Oregon Museums Association, POB 1718, Portland, OR 97207 • T: +1 503 8233623 • F: +1 503 8233619 • pkdanc@ci.portland.or.us • 55176

Pennsylvania Federation of Museums and Historical Organizations, POB 1026, Harrisburg, PA 17108-1026 • T: +1 717 7873253 • F: +1 717 7724698 • phmc@state.pa.us • 55177

Rhode Island Museum Network, 58 Walcott St, Pawtucket, RI 02860 • 55178

Small Museums Association, POB 1425, Clinton, MD 20735 • www.smallmuseum.org • 55179

Smithsonian Institution, 1000 Jefferson Dr SW, Washington, DC 20560, mail addr: POB 37012, Washington, DC 20013-7012 • T: +1 202 3572700, 6339126 • F: +1 202 7862515 • info@si.edu • www.smithsonian.org •
1846
Museum, education and research association with 19 museums, galleries, and institutes 55180

South Carolina Federation of Museums, POB 100107, Columbia, SC 29202 • T: +1 803 8984925 • F: +1 803 8984969 • scfm@museum.state.sc.us • 55181

Southeastern Museums Conference, 412 N 4th St, Ste 250, Baton Rouge, LA 70821, mail addr: POB 3494, Baton Rouge, LA 70821 • T: +1 225 3835042 • F: +1 225 3438669 • SEMCdirect@aol.com • www.SEMCdirect.net • 55182

Tennessee Association of Museums, 200 Hyder Hill, Piney Flats, TN 37686, mail addr: POB 330984, Nashville, TN 37203 • T: +1 866 3903638 • F: +1 615 4605688 • info@tnmuseums.org • www.tnmuseums.org • 55183

Texas Association of Museums, 3939 Bee Caves Rd, Bldg A, Ste 1-B, Austin, TX 78746 • T: +1 512 3286812 • F: +1 512 3279775 • tam@io.com • www.texasmuseums.org • 55184

United States Federation of Friends of Museums, 2 Gittings Av, Baltimore, MD 21212 • 55185

United States-International Council on Monuments and Sites, 401 F St, Ste 331, Washington, DC 20001 • T: +1 202 8421866 • F: +1 202 8421861 • edelage@usicomos.org • 55186

United States National Committee of the ICOM, ICOM-US, c/o American Association of Museums, 1575 Eye St NW, Ste 400, Washington, DC 20005 • T: +1 202 2899115 • F: +1 202 2896578 • icomus@aam-us.org • www.aam-us.org/icom •
1946 55187

Utah Museums Association, POB 2077, Salt Lake City, UT 84110-2077 • T: +1 435 6544092 • charlotte@utahmuseums.org • www.utahmuseums.org • 55188

Vermont Museum and Gallery Alliance, 4472 Basin Harbor Rd, Vergennes, VT 05491, mail addr: c/o Lake Champlain Maritime Museum, POB 489, Woodstock, VT 05091 • T: +1 802 4752022 • F: +1 802 4752953 • VCCP@sover.net • www.vmga.org • 55189

Virginia Association of Museums, 200 S Third St, Richmond, VA 23219 • T: +1 804 7885821 • mcarlock@vamuseums.org • www.vamuseums.org • 55190

Volunteer Committees of Art Museum, One Collins Diboll Circle, New Orleans, LA 70179, mail addr: c/o New Orleans Museum of Art, POB 19123, New Orleans, LA 70179-0123 • T: +1 504 7376301 • F: +1 504 7385103 • 55191

Washington Museum Association, POB 5817, Factoria Station, Bellevue, WA 98006 • T: +1 360 4663365 • F: +1 360 4661611 • wma@harbornet.com • 55192

Western Museums Association, 2960 San Pablo Av, Berkeley, CA 94702, mail addr: POB 8367, Emeryville, CA 94662 • T: +1 510 6650700 • F: +1 510 6659701 • info@westmuse.org • www.westmuse.org •
1935 55193

Wisconsin Federation of Museums, 700 N 12th St, Wausau, WI 54403 • T: +1 715 8457010 • F: +1 715 8457103 • 55194

Uruguay

ICOM Uruguay, c/o V. Serrana Prunell, Museo Ralli, Curupay y Los Arachanes, Barrio Beverly Hills, 20100 Punta del Este • T: +598 42 483478/79 • F: (02) 9085239 • icomuy@hotmail.com • 55195

Uzbekistan

ICOM Uzbekistan, c/o National Commission for UNESCO, 54 Buyuk Ipak Yuli, 700137 Toshkent • T: +998 71 21320211, 670542 • F: 670538, 670546 • unesco@natcom.org.uz • 55196

Venezuela

ICOM Venezuela, c/o M. Mago Tovar, Apdo 17376, Caracas 1015-A • T: +58 212 5741673 • F: 5741673 • artconserv@cantv.net • 55197

Vietnam

ICOM Viet Nam, c/o Dept. of National Cultural Heritage, Ministry of Culture and Information, 51-53 Ngo Quyen St, Ha Noi • T: +84 4 9437611 • F: 9439929 • nchdvn@hn.vnn.vn • 55198

Zambia

ICOM Zambia, Mosi-oa-Tunya Rd, Livingstone, mail addr: c/o Livingstone Museum, POB 60498, Livingstone • T: +260 3 320495 • F: 324509, 320495 • livmus@zamnet.zm • 55199

Zimbabwe

ICOM Zimbabwe, c/o A. Msimanga, POB AC192, Ascot • T: +263 9 230046 • F: 234019 • natmuse@telconet.co.zw • 55200

National Museums and Monuments of Zimbabwe, 107 Rotten Row, Harare, mail addr: POB CY 1485, Harare • T: +263 4 752876 • F: 753-085 • natmus@utande.co.zw • www.zimheritage.co.zw • 55201

Southern African Development Community Association of Museums and Monuments, c/o National Museums and Monuments of Zimbabwe, POB CY 1485, Harare • T: +263 4 752876 • F: 753-085 • natmus@baobab.cszim.co.zw • 55202

Index of Distinguished Persons

A

Aakjær, Jeppe 10251
Aalberg, Ida 11063
Aall, Jacob 33938
Aalto, Alvar 10849
Aaltonen, Wäinö 11093, 11408
Aas, Nils 33575
Aasen, Ivar 33568
Abbadie, Antoine d' 13200
Abbal, André 12279
Abbe, Ernst 19385
Abdessemed, Adel 11899
Abeghyan, Mher 00698
Abélard, Pierre 13628
Abovyan, Khachatour 00740
Abramov, Fëdor 36766, 37924
Accardi, Carla 26014
Achiţenie, Petre 36162
Achmatova, Anna Andreevna 37355, 37642, 37667
Adam, Robert 44331, 44542, 44994, 45120
Adams, Robert 51138
Adenauer, Konrad 16754
Ader, Clément 14335
Adimoolam, K.M. 24428
Adler, Friedrich 19859
Ady, Endre 23379, 23467, 36270, 36503
Adzak, Roy 14628
Affandi, Kusuma 24583
Affrey, Louis 41735
Agoncillo, Marcela M. 34508
Agricola, Georgius 18657
Aguéli, Ivan 40825
Aho, Juhani 10847
Aicard, Jean 13373, 15778
Aini, Sadriddin 54462
Aistis, Jonas 30348
Aizpiri, Paul 29119
Ajvazovskij, Ivan Konstantinovič 43039
Akersloot-Berg, Betzy R. 32899
Akiyev, Kalyk 29955
Aksakov, Sergej Timofeevič 36750
Akuraters, Jāņis 30070
Alain-Fournier, Henri 12841
Albers, Josef 17488, 21802, 29297
Albert, Eduard 09925
Albert, Prince of Saxe-Coburg and Gotha 44084
Alberti, Friedrich August von 21428
Alberti, Rafael 39978
Alcalá-Zamora y Torres, Niceto 39871
Alcott, Louisa May 47770
Aldridge, John 45539
Aldrovandi, Ulisse 25443
Alechinsky, Pierre 11549
Alecsandri, Vasile 36476
Aleichem, Sholom 25121
Aleš, Mikoláš 09538, 09539, 09656, 09766
Alesio, Perez de 34254
Alessandrini, Tommaso 27002
Alfano, Andrea 25770
Alfaro Siqueiros, David 31106
Alfelt, Else 10069
Alfieri, Vittorio 25313
Ali, Muhammad 50294
Alighieri, Dante 27371
Alkšņis, Ādams 30118
Allen, William Herbert 43292
Allende, Salvador 07457
Allilueva, Nadežda 37689
Allmayer, Josefine 02162
Allmers, Hermann 21484
Alt, Fritz 04515
Altamirano, Ignacio Manuel 31460
Alunāns, Ādolfs 30015
Alves, Castro 04791
Aman, Theodor 36230, 36525, 36646
Amaral, Tarsila do 04879
Amas, Ernest 13500
Amberes, Carlos de 39519
Ambrosi, Gustinus 02950
Ambrosino, Ferdinando 27800
Amerano, Giovanni Battista 26138
Amiet, Cuno 41080, 41707, 41885
Ammann, Othmar 41735
Ammannati, Bartolomeo 27042
Amorsolo, Fernando 34414
Ámos, Imre 23873
Ampère, André-Marie 14869
Amrhein, Willy 41352
Amundsen, Roald 33725, 33894, 33912, 33919
Ancher, Anna 10261
Ancher, Helga 10261
Ancher, Michael 10261
Andersen, Alfredo 04428
Andersen, Hans Christian 10215, 10216
Andersen, Magnus 33629
Anderson, Amos 10719
Andersson, Dan 40695
András, Jósa 23733
Andrée, Salomon 40562
Andreescu, Ion 36162, 36646
Andreev, Leonid Nikolaevič 37503, 37510
Andreotti, Libero 27173
Andres, Stefan 21680
Andri, Ferdinand 02686
Andrić, Ivo 38089
Anghel, Dimitrie 36477
Anich, Peter 02457
Anikušin, Michail 27800
Anipov, Konstantin 27800
Anjos Teixeira, Artur dos 35945
Anjos Teixeira, Pedro Augusto dos 35945
Anker, Albert 41885
Anker Aurdal, Synnøve 33925
Anna Amalia, Duchesse of Saxe-Weimar-Eisenach 22419, 22423
Anna, Margit 23873
Anner, Emil 41260
Anselmi, Michelangelo Buonarroti 27096
Anson, Peter 43664
Antes, Horst 08675, 17698
Anton Ulrich, Duke of Braunschweig-Lüneburg 17373
Antonelle, Patrick 46224
Antonello da Messina 27182, 28004
Antonioni, Michelangelo 26109
Aoki, Shigeru 29632
Apacible, Galiciano 34507
Apacible, Leon 34507
Apel, Heinrich 18884
Apollinaire, Guillaume 04049
Appel, Karel 32756, 50891
Appia, Adolphe 41216
Aquitaine, Aliénor d' 15462
Aragay, Josep 39057
Araki, Nobuyoshi 29395
Aralič, Stojan 08924
Arany, János 23727, 36580
Aratym, Hubert 02026
Aravantinos, Panos 23148
Arbaud, Paul 11519
Arbuzov, Aleksandr E. 36806, 37012, 37019
Arbuzov, Boris A. 37012, 37019
Archer, Colin 33629
Archipenko, Alexander 21459
Arden, Mary 45782
Arechavala, Joaquín 33287
Arghezi, Tudor 36190
Ariosto, Ludovico 26104
Arisman, Marshall 47782
Ariza, Alfonso 39889
Arkwright, Sir Richard 43932
Armand, Inessa 37584
Armando 31809, 32756
Armbruster, Ludwig 17086
Arndt, Ernst Moritz 17453, 18545, 21914
Arndt, Hermina 33154
Arnim, Achim von 20704
Arnim, Bettina von 20704
Arnim-Muskau, family 19700
Arntz, Gerd 21264
Arp, Hans 10103, 12537, 14909, 15252, 19736, 21259, 41570
Arramidis, M. 22934, 22935
Arrieta, Agustín 31267
Asahi, Shinpei 28399
Asakura, Fumio 29370
Asakura, Isokichi 29359
Asbjørnsen, Peter Christen 33559
Asch, Georg Thomas von 18690
Asch, Sholem 24934
Aso, Kaji 46930
Aspazija 30021, 30026, 30057, 30103, 41284
Assmann, Richard 22012
Atatürk, Kemal 42683, 42698, 42740, 42760, 42774, 42779, 42828, 42853, 42872, 42883, 42906, 42927, 42932, 42941
Auberjonois, René 41080
Aubry, Emile 00054
Auden, W.H. 02174
Audubon, John. J. 46465
Auerbach, Berthold 19271, 21961
Augundson, Thorgeir Øygarden 33777
August the younger, duke of Braunschweig-Wolfenbüttel 22653
Augustin, Jacques 15241
Augustinčić, Antun 08888
Auriol, Vincent 14335
Austen, Jane 43799
Austriņ, Antons 30141
Autengruber, Jan 09704
Avakian, Hrandt 36204
Avercamp, Barent 32440
Averkin, Aleksandr Petrovič 37755
Avery, Milton 50510
Axman, Emil 09484
Aytiyev, Gapar 29958
Azorín 39667
Azzinari, Franco 25210

B

Baar, Jindřich Šimon 09586
Babbage, Charles 45872
Babits, Mihály 23550, 23865
Bach, Johann Sebastian 16575, 18093, 19666, 19898, 20890
Bacher, Karl 03129
Bachmann, August 17576
Bachmann, Ingeborg 02191
Backer, Harriet 33419
Bacon, Francis 10103, 14722, 33717, 44496, 45285
Bacovia, George 36118, 36187
Baden-Powell, Olave 01643
Baden-Powell, Robert Stephenson Smyth 01643, 44787
Bähr, George 19850
Baekeland, Leo Hendrik 03636
Baer, Karl Ernst von 10650
Bärtschi, Arnold 41336
Baffier, Jean 15622
Bag, Nicholas 26342
Bahner, Hermann 19811
Baillie Scott, Mackay Hugh 43571
Bain, George 45513
Baird, John 43941
Bajcsy-Zsilinszky, Endre 23686
Bajén, Francisco 14147
Baker, Sir Herbert 38693
Baker, Josephine 12314
Bakunin, Michail 41127
Bakusen, Tsuchida 29237, 29238
Balaguer y Cirera, Víctor 40287
Balakirev, Milyj 37439
Bălcescu, Nicolae 36491
Balla, Giacomo 27800
Baltermants, Dmitrij 37254
Balzac, Honoré de 14618, 15145, 15742
Băncilă, Octav 36162
Bandaranaike, Sirimawo W.R.D. 40369
Bande, Francesco 27754
Banting, Frederick Grant 06181, 06404
Bara, Charlotte 41129
Barabino, Nicolo 26289
Baracca, Francesco 26543, 26544
Baranauskas, Antanas 30206
Barat, Madeleine-Sophie 13271
Barbella, Costantino 25873
Barbey d'Aurevilly, Jules-Amédée 15507
Barbieri, Olivo 25896
Barbosa, Rui 04705
Barbusse, Henri 11747
Barceló, Miquel 08675
Barcsay, Jenő 23872
Bardini, Stefano 26189
Bārdu, Anton 30030
Bārdu, Frič 30030
Bargheer, Eduard 18924
Barlach, Ernst 18827, 18828, 18926, 20414, 21196, 22353
Barnacle Joyce, Nora 24837
Barnard, Christian 38557
Barnes, William 43999
Barocci, Federico 22714
Barons, Krišjāņis 30073
Barrau y Buñol, Laureano 39999
Barrie, James M. 44617
Barrios, Ángel 39358
Bartels, Rudolf 21644
Bartholdi, Auguste 12582
Bartók, Béla 23387, 23494, 36586
Barton, Clara 48918
Baselitz, Georg 10103, 18024, 41763
Basho, Matsuo 29371
Basilico, Gabriele 25896
Battaglia, Lettizia 25896
Baud-Bovy, Auguste 41460
Bauer, Leopard 26327
Bauer, Wilhelm 17863
Bauernfeind, Gustav 21979
Bauernfeld, Eduard von 02956
Baumbach, Rudolf 20259
Baumeister, Willi 17698, 21948
Bawden, Edward 45539
Bažov, Pavel Petrovič 36890, 36891, 36897
The Beatles 18889
Beatrice, Princess of Battenberg 45243
Beauchesne, Gouin de 15390
Beaudoin, Eugene 15841
Beauharnais, Hortense de 41798
Bebel, August 18094
Becher, Bernd 19627, 21755
Becher, Hilla 19627, 21755
Bechstein, Ludwig 20259
Bechteler, Eduard 17619
Beck, Jules 41217
Beck, Richard 18421
Beckmann, Max 17924, 19005, 21960
Beckord, William 43402
Bede the Venerable 44545
Beecham, John 43845
Beerbohm Tree, Franko B. 43640
Beethoven, Ludwig van 01774, 02224, 02395, 02945, 02946, 02947, 02948, 17448, 23691, 38248
Behring, Emil von 20189
Beirão, Sarah 35958
Bejenaru, C. 36162
Bekker vom Rath, Hanna 19226
Belanger, François-Joseph 13925
Belinskij, Vissarion Grigorevič 36796, 36797
Beljutin, Ely Michailovič 27800
Bell, Alexander Graham 05369, 05466
Bell, Joseph 44160
Bellany, John 44037
Bellay, Joachim du 13775
Belli, Valerio 28252
Bellini, Emmanuel 12268
Bellini, Giovanni 25371, 25406, 27333, 28188, 28211, 43497
Bellini, Vincenzo 25772
Bellman, Carl Michael 40888
Bellmer, Hans 17244
Bellot, Léon 15622
Bellotto, Bernardo 27096
Bellwalder, Stefan 41434
Belmont, Alva 53916
Beloborodov, N.I. 37867
Beloff, Angelina 31111
Belting, Franz 19875
Belyj, Andrej 37227
Ben-Gurion, David 25100
Benczúr, Gyula 23733
Benda, Franz 09400
Benda, Georg 09400
Bendandi, Raffaele 26072
Benedito Vives, Manuel 39510
Beniuseviciute-Zymantiene, Julia 30218
Benjamin, Walter 17280
Benka, Martin 38295
Benlliure, José 40197
Benlliure, Mariano 39227
Benlliure, Peppino 40197
Bennàssar, Dionís 39850
Bennett, R. B. 05996
Benoit, Peter 03679
Benz, Carl 19761, 19762, 20174
Beobide-Kresala, Julio 40361
Berg, Werner 01788
Bergengruen, Werner 16884
Bering, Vitus 37554
Berliner, Emile 06361
Berlioz, Hector 13351
Bernadotte, Jean-Baptiste 14755
Bernadozzi, A. 31604
Bernard, Claude 15348
Bernard, Emile 31898
Bernegg, Theophil Sprecher von 41603
Bernhard, Thomas 02471, 02693
Berni, Antonio 00335, 00529
Bernier, Joseph-Elzéar 06175
Bernini, Gian Lorenzo 26741
Bernstein, F. W. 18357
Beron, Petăr 05077
Bert-Hubl 16546
Berta, Edoardo 41195
Bertrand, Henri Gratien 12459
Bertuch, Friedrich Justin 22421
Bērziņs, Ludis 30024
Berzsenyi, Dániel 23731
Bessenyei, György 23928
Best, Fritz 19724
Bethge, Hans 19549
Bethune, Norman 05907

Beuys, Joseph 04879, 14722, 16975, 17123, 17460, 17924, 17997, 19473, 19567, 21960, 29621, 42083, 51163
Beveridge Gray, Nelly 05948
Bewick, Thomas 45136
Bezruč, Petr 09504, 09597, 09694, 09701
Bialik, Nachman 25101
Bibiena, Antonio 26316
Bichsel, Peter 41215
Bickel, Karl 41997
Biegas, Boleslaw 14634
Bieruma Oosting, Jeanne 32522
Bieziņ, Aleksandr 30046
Bilderdijk, Willem 31835
Bílek, František 09488, 09735
Bilger, Margret 02707, 02796
Biliūnas, Jonas 30210
Bill, Max 04879, 17184
Billinger, Richard 02667
Birznieks-Upītis, Ernests 30151
Bismarck, Otto von 18479, 21610
Bissier, Julius 41129
Bjåland, Olav 33681
Bjarnadóttir, Halldóra 23988
Bjørnson, Bjørnstjerne 33502, 33620
Blackadder, Elizabeth 44037
Blackburne, Joseph 04200
Blaga, Lucian 36434
Blagoev, Dimităr 04977, 05175
Blaim, Adolf 02377
Blake, Robert 43606
Blake, William 45378
Blanche, Jacques-Emile 14497
Blank, Margarete 17487
Blankenstein, Hermann 17208
Blasius, Hueber 02457
Bláthy, Ottó Titusz 23434
Blaumaņis, Rūdolfs 30008, 30071
Blechen, Carl 17032, 17736
Bleken, Håkon 33537
Bleyl, Fritz 17058, 22833
Blicher, Steen Steensen 10068, 10119, 10314
Bligh, William 00912
Blixen, Karen 10253
Bljucher, Vasilij Konstantinovič 37850
Blok, Aleksandr A. 37786
Blossfeldt, Karl 19627
Blücher von Wahlstatt, Gebhard Leberecht 19482
Blüthgen, Victor 22816
Blum, Jeanne 13285
Blum, Léon 13285
Blumenbach, Johann Friedrich 18690
Bly, Robert 50382
Bobet, Louison 15413
Boccaccio, Giovanni 25814
Boccioni, Umberto 04879
Boch, Anna 20312
Boch, Eugène 20312
Bodoni, Giambattista 27100
Boeck, Félix de 03563
Boeckel, Lodewijk van 03820
Böckler, Hans 22080
Böcklin, Arnold 42083
Böckstiegel, Peter August 16438
Boehle, Fritz 18146
Böhlen, Max 19427
Böhme, Christa 17106
Böhme, Lothar 17106
Bössenroth, Carl 18053
Böttcher, Joachim 17106
Böttcher, Manfred 17106
Boghen, Felice 26052
Bol, Peter 32230
Boldini, Giovanni 26116
Bolívar, Simón 08546, 08591, 08747, 08754, 34066, 34285, 40358, 54526, 54537, 54538, 54548
Bollhagen, Hedwig 21303
Boltanski, Christian 12840, 14446
Bombardier, J. Armand 07196
Bonaiuto, Andrea di 26188
Bonevardi, Marcelo 00335
Bonnard, Pierre 15111, 42083
Bonnat, Léon 53253
Bonome, Rodrigo 00521
Bonsor, George Edward 39593
Bonzanigo, Giuseppe Maria 28004
Boon, Louis Paul 03259
Booth, Catherine 41206
Booth, Edward Thomas 43619
Booth, William 41206, 45301
Bora, Katharina von 20115, 22071
Boratynskij, Evgenij 37114
Bordalo Pinheiro, Columbano 35698
Bordalo Pinheiro, Rafael 35495, 35726
Bordes, Antoine 15390
Bores, Francisco 39565
Borgeaud, Georges 41215
Borlach, Johann Gottfried 16763
Borland, Christine 44281
Borodin, Sergej 54486
Bosch, Hieronymus 03865, 32375
Boscovits, Fritz 42062
Bose, Nataji Subhas 24279
Bose, Sarat Chandra 24279
Botero, Fernando 08591, 08675
Botev, Christo 05045
Bottema, Tjeerd 32443
Botticelli, Sandro 14714, 17172, 25406, 26157, 26160, 44138
Botticini, Francesco 26054
Boty, Pauline 45020
Bouchard, Henri 14635
Boucher, Alfred 14464
Boucher, François 11588, 14714, 15829
Boudou, Jean 15361
Boulle, André-Charles 14580
Bouman, Jan 21092
Boumeester, Christine 16193
Bourdelle, Emile-Antoine 14176, 14636
Bourdon, Sébastien 45378
Bourgault, Médard 06827
Bourgeoys, Marguerite 06374
Bourguiba, Habib 42663
Bovet, Claude 41310
Bowditch, Nathaniel 52658
Bowler, Jonathan B. 43406
Boyd, Arthur 00915
Boyle, Harry 07349
Boznanska, Olga 14634
Bozveli, Neofit 05077
Bragaglia, Alberto 27800
Brahe, Tycho 09400, 21387
Brahms, Johannes 01951, 02412, 16879, 18917, 19065
Braille, Louis 12648
Braith, Anton 17319
Brâncuşi, Constantin 36324, 36399
Brands, Eugène 32756
Brandsma, Titus 31981
Brangwyn, Frank 03406
Branly, Edouard 11587, 14706
Brant, Joseph 05500
Braque, Georges 11549, 12721, 14722, 16202, 17192, 27454, 41885
Braslavsky, Noam 17088
Braun-Blanquet, Josias 41300
Brayer, Yves 12625
Brecht, Bertolt 16613, 17055, 17620
Breck, John Leslie 13068
Brehm, Alfred Edmund 16498, 21279
Brehm, Christian Ludwig 21279
Breidvik, Mons 33704
Breinlinger, Hans 19678
Breithaupt, Friedrich August 18421
Breitner, Burghard 02367
Bremer, Frederick 45002
Brenders, Carl 43988
Brennand, Francisco 04667
Breuer, Marcel 17316, 17834
Breunig, Johann Michael 17639
Breuning, Olaf 12062
Breustedt, Hans Joachim 02796
Březina, Otokar 09554, 09722
Brigadere, Anna 30129
Brigaud, Florentin 13055
Brillat-Savarin, Jean-Anthelme 13795
Brinckman, John 18829
Bristowe, Ethel S.G. 43770
Brito, Jenny Mendes de 35559
Broch, Hermann 02800
Brochet, François 11773
Brodski, Isaak Izrailevič 37690
Bröker, Wilhelm 16763
Broniewski, Władysław 35278
Brontë, Anne 44430
Brontë, Charlotte 44326, 44430
Brontë, Emily Jane 44430
Broodthaers, Marcel 17998
Brookes, William Penny 45185
Brosch, Klemens 02317
Brossard, Georges 06341
Brouwer, Adriaen 03865
Brown, Alfred S. 06364
Brown, Jenny 41148
Brown, John 01207
Brown, , S. G. 45679
Brown, Sidney 41148
Brozincevic, Franz 42009
Brozzi, Renato 28051
Bruce, Michael 44606
Bruckner, Anton 01724, 02240
Bruegel, Pieter (the Elder) 03019, 03865, 27451
Brüggemann, Hans 22316
Brueghel, Jan (the Elder) 17833
Brueghel, Pieter (the Younger) 32375
Brüning, Peter 18024
Brunel, Isambard Kingdom 44809, 45819
Brunel, Sir Marc Isambard 44809
Brunelleschi, Filippo 26160, 26176
Brunovski, Albin 05016
Brus, Günter 03133
Brusau, René 00521
Bruzzone, Alberto 00443
Bu, Amund K. 33948
Buchan, John 43483
Buchser, Frank 41707, 41885
Buck, Pearl S. 51776
Büchler, Franz 16884
Büchner, Georg 21321
Bücker, Carl Clemens 19585
Bürgel, Ludwig 02735
Bürger, Gottfried August 20370, 22355
Bürkel, Heinrich 21041
Buffalo Bill 49223
Buffet, Bernard 14633, 29286
Buga, Kazimieras 30336
Bugatti, Carlo 26327
Buhe, Walter 16596
Buivydaitė, Bronė 30208
Bukovac, Vlaho 09003
Bulatov, Erik 17184
Bulgakov, Michail Afanasevič 43086
Bull, Jacob Breda 33778
Bulloch, Angela 12062
Bundtzen, Friedrich 22465
Bunin, Ivan Alekseevič 36909, 36910, 37508, 37510, 37972
Bunke, Franz 21644
Bunyan, John 43436, 43437
Burchfield, Charles E. 47119
Burduja, Aspazia 36162
Buren, Daniel 12062
Burgess, Thornton W. 52842
Burgkmair, Hans 44999
Burgos, José 34523
Burian, Zdeněk 09851
Burke, William 44160
Burliuk, David 50510
Burmeister, Hermann 21913
Burnand, Eugène 41650, 41651
Burne-Jones, Sir Edward 43884, 45327
Burnett, Calvin 46942
Burnot, C. Philippe 11894
Burns, Robert 43283, 43337, 44039, 44040, 44162, 44536, 44610, 45118, 45826
Burri, Alberto 25905
Burroughs, John 54076
Busch, Wilhelm 18038, 19008, 21725, 22540
Bush, George 47688
Busoni, Ferruccio 26052
Bustos, Hermengildo 30921
Buyx, Peter Michael 19524
Bykau, Vasil U. 03252
Byron, Gene 30911
Byron, George Gordon Noel Lord 23090, 27456, 45297

C

Caballero, José 40317
Caballero, Luis 08675
Cabanas, Manuel 36014
Cabiaca, Vincenzo 26289
Cabrini, Francesca Saverio 27722
Cadou, René-Louis 13836
Cahoon, Martha 47846
Cahoon, Ralph 47846
Cahun, Claude 45569
Caillebotte, Gustave 14908, 16306
Čajkovskij, Pëtr Il'ič 37052, 37682, 37977, 43056
Čaks, Aleksandrs 30060
Calder, Alexander 10103, 50224
Calderara, Antonio 25223
Callan, Nicholas 24875
Calle, Sophie 14446
Calloch, Jean-Pierre 13237
Calmette, Albert 13755
Calvin, Jean 14482, 41401
Calvo, Luis Antonio 08645
Camassei, Andrea 25413
Cambellotti, Duilio 26518
Cambellotti, Mario 26327
Cambiaso, Luca 26309
Cameron, Julia Margaret 44255
Camilar, Esebiu 36713
Camos, Honoré 11853
Campbell, Donald 43885
Campbell, Sir Malcolm 43885
Campendonk, Heinrich 21008
Campigli, Massimo 04879
Canaletto 13794, 15829, 43827, 45006
Cander, Fridrih 30068
Cankar, Ivan 38535
Cano, Francisco Antonio 08675
Canova, Antonio 26324, 27316
Canow, Carl 22613
Cantelon, Edgard 06979
Cantemir, Dimitrie 36346
Capa, Robert 51194
Čapaev, Vasilij I. 36840
Čapek, Josef 09646, 09845, 09890
Čapek, Karel 09646, 09845, 09890
Capella, Anna 39346
Capralos, Christos 22866, 22874
Carà, Ugo 26876
Caraffa, Emilio 00335
Caragiale, Ion Luca 36384, 36531
Caravaggio 14354, 26323, 27333, 27451, 27453
Carballo, Aida 00500
Cardini, Luigi 26138
Carducci, Giosuè 25446, 27205
Carême, Maurice 03491
Carigiet, Alois 41953
Carl-August, Duke of Saxe Weimar 22416
Carloni, Carlo Innocenzo 26316
Carlyle, Jane Welsh 44375, 44814
Carlyle, Thomas 44110, 44814
Carnacini, Ceferino 00671
Carnegie, Andrew 44054
Carneiro, António 35841
Carneiro, Carlos 35841
Carneo, Antonio 28110
Carnielo, Rinaldo 26162
Carolus-Duran, Emile 53253
Carpaccio, Vittore 26580, 28110, 28188
Carpeaux, Jean-Baptiste 16043
Carr, Emily 07226, 07246
Carracci, Annibale 12775, 27096, 27451, 27453
Carrington, Leonora 50952
Carrino, Nicola 26014
Carroll, Lewis 44371
Cartier-Bresson, Henri 28818, 51194
Cartier, Jacques 15389, 15391
Casals, Pablo 14936, 40260, 40263
Casanova, Giacomo 09498
Casanova, Rafael de 39656
Cascella, Michele 27004
Casorati, Felice 26289
Cassatt, Mary 13068, 47690
Castagnino, Juan Carlos 00529
Castagno, Andrea del 26147
Castel-Branco, Dorita 35943
Castellani, Enrico 10071
Castello, Valerio 26309
Castellucci, Marco 27800
Castres, Edouard 41589
Castro, Fernando de 35837
Castro, Ferreira de 35795, 35948
Castro, Rosalía de 39764
Catargi, Henri 36525, 36594
Catlett, Elizabeth 46942
Catlin, George 54162
Cattaneo, Carlo 41284
Cavour, Camillo Benso Conte di 27730
Cazimir, Otilia 36414
Cazin, Jean-Charles 15617
Cazneau, Harold 01120
Ceccobelli, Bruno 26014
Čech, Svatopluk 09628, 09688, 09702
Čechov, Anton Pavlovič 16892, 37129, 37161, 37195, 37830, 43188
Cedercreutz, Emil 10758
Čelakovský, František Ladislav 09849
Celaleddin, Mevlânâ 42889
Celommi, Pasquale 27559
Celso Barbosa, José 36028
Čepeláková, Zdenka 09703
Cercas, José 35406
Ceresa, Alice 41215
Cermignani, Vincenzo 26345
Cerna, Panait 36272
Cernătescu, Radu 36424
Černyševskij, Nikolaj Gavrilovič 36785, 37751

Cervantes Saavedra, Miguel de 38813, 39284, 40227
César, Arman 12754
Céspedes, Carlos Manuel de 09033
Cézanne, Paul 11513, 14674, 14719, 26580, 28628, 28718, 41763, 42083, 44937, 48935, 50224, 52717
Chabot, Hendrik 32703
Chadži, Dimităr 05166
Chagall, Marc 03257, 09700, 12636, 14436, 14722, 19204, 20414, 24956, 27800, 29062, 42083, 48662, 51258
Chaillioux, Julien 13010
Chamisso, Adelbert von 12365
Champaigne, Philippe de 15829
Champlain, Samuel de 06580
Champneys, Basil 45087
Champollion, Jean Francois 12932
Chandra, Prakash 24428
Chang Dai-Chien 42377
Chapais, Jean-Charles 06813
Chaplin, Charles Spencer 18358
Chappell, Reuben 44329
Chapu, Henri 13613, 14069
Charcot, Jean-Baptiste 15389
Charents, Yeghishe 00759
Charlemagne 16400, 19335, 41664
Charles VIII 11578
Charoux, Siegfried 02262
Chateaubriand, François-René de 12590, 15389
Chatterjee, Sarat Chandra 24342
Chavchavadze, Ilya 16366, 16367
Che Guevara, Ernesto 09297
Checa, Ulpiano 39199
Chen Bwo-Wen 42307
Chen Chung-Tsang 42180
Chéniers, André de 12285
Chénon, Emile 12445
Cherkassky, Zoya 17088
Chiattone, Mario 27239
Chiesa, Pietro 41195
Chifley, Ben 00839
Chillida, Eduardo 39231, 39387
Chinnaraj, T. 24428
Chiossone, Eduardo 26299
Chippendale, Thomas (the Younger) 45770
Chiro Iwasaki 29377
Chitov, Panaët 05167
Chittussi, Antonín 09819
Chlebnikov, Velimir Vladimirovič 36781
Chlebnikova, Vera Vladimirovna 36781
Chochrjakov, Nikolaj N. 37035
Choiseul, Étienne-François de 11578
Chopin, Frédéric 34529, 35037, 35117, 35216, 35242, 35288
Chopin, Nicolas 35288
Chrenikov, Tichon Nikolaevič 36907
Christie, Agatha 45866
Christoljubov, Nikolaj 27800
Christov, Ivan 05252
Chughtai, Muhammed Abdur Rahman 34018
Chung Li-Ho 42271
Churchill, Winston 48810
Cilea, Francisco 27086
Ciolkovskij, Konstantin Ėduardovič 36994, 36996, 37040
Čišinski, Jakub-Bart 20977
Čistjakov, P.P. 37578
Čiurlionis, Mikalojus Konstantinas 30224, 30262, 30338
Claes, Ernest 04004
Claesz, Pieter 32470
Claret, Sant Antoni María 39948
Clark, Jim 44065
Clarke, Peter E. 38576
Claudel, Camille 15111
Claudius, Hermann 19250
Claudius, Matthias 21253
Claver, Pedro 40265
Clayton, John 43839
Cleillida, Eduardo 40317
Clémenceau, Georges 14709, 15539
Clemens August, Duke of Bavaria 17607, 21791
Clercq, René de 03537
Cobbett, William 44218
Cocea, Nicolae D. 36616
Cockburn, Jack 06580
Cockburn, Sir John 01177
Cocteau, Jean 14083, 14122
Codesido, Julia 34208
Codreanu, Mihail 36411
Cody, William F. 49223
Coello, Alonso 34676
Coffermans, Marcellus 39638
Colbert, Edwin 46128
Colbert, Jean Baptiste 15688
Coleridge, Samuel Taylor 45201
Colladon, Jean-Daniel 41422
Collaert, Adrien 39638
Collingwood, William Gershom 43885
Collins, Thomas Francis 05996
Colombo, Cristoforo 26293
Colón, Cristóbal 39799, 39804, 39971, 54397
Columbus, Christopher 35870
Combas, Robert 12062
Comenius, Jan Amos 09406, 09415, 09506, 09525, 09602, 09788, 09804, 09888, 19133, 32563
Comte, Auguste 14617
Comyn, Christine 32230
Conca, Sebastiano 27182
Condorcet, Antoine Nicolas de Caritat 15034
Constable, John 17032, 44138, 44937, 44999, 45920
Constant, Benjamin 04704
Constantinescu, Paul 36535
Cook, Francis 43336
Cook, James 00912, 18690, 33200, 44352, 45139, 45994, 45998
Cooper, Thomas Sidney 43744
Čop, Matije 38543
Coper, Hans 45285
Copernicus, Nicolaus 34629, 35194
Čorbadžijski-Čudomir, Dimităr Christov 05057
Corbin Lukis, Frederick 45591
Corbineau, Fréres 13953
Corelli, Arcangelo 25826
Corelli, Mari 45781
Corinth, Lovis 21459
Corneille, Pierre 13631, 15120, 32756
Coronel, Pedro 31579
Coronel, Rafael 31579
Corot, Jean-Baptiste Camille 28754, 28880, 41460
Correggio, Antonio 26741, 27096, 27451
Corty, Jean 41195
Coşbuc, George 36314
Cosimo, Piero di 26160
Cossit, Ranna 07048
Costanzo, Guido 27004
Cotta, Bernhard von 18421
Coubertin, Pierre de 41550
Courbet, Gustave 12754, 14540, 17032, 27454, 28754
Courths-Mahler, Hedwig 20559
Cousin, Jean 15901
Couture, Thomas 15715
Cowper, William 45316
Coypel, Charles 12754
Cózar, José 38941
Craesbeeck, Joos Van 03865
Craig, Edward Gordon 45838
Craig-Martin, Michael 17673
Cranach, Lucas (the Elder) 16590, 16593, 17720, 19720, 20113, 20554, 21656, 22420, 22638
Cranach, Lucas (the Younger) 20113
Crary, Clare J. 53826
Creangă, Ion 36422
Crerar, H.D.G. 06090
Crişan, Eva 36285
Criswell, Warren 47782
Crite, Allan R. 46942
Crome, John 45276
Cromwell, Oliver 44500
Cross, Henri Edmond 12775
Cruz Fernández, Manuel 39396
Cruz, Juan Pantoja de la 39638
Cruz, Oswaldo 04691
Cruz, San Juan de la 40179
Csallány, Gábor 23884
Cuadrado, Juan 38861
Čudomir 05057
Čukovskij, Kornej Ivanovič 37195, 37539
Curie, Marie 14641, 35259
Curie, Pierre 14641
Curros Enríquez, Manuel 39160
Curry, John Steuart 50435
Curwood, James Oliver 51642
Custer, George Armstrong 49223
Cuyper, Herman De 03371
Cuypers, Pierre 32698
Cuza, Alexandru Ioan 36575
Cvetaeva, Marina I. 36753, 37059, 37162, 37476, 37837, 37911
Cygnaeus, Uno 10892
Czechowicz, Józef 34904
Czóbel, Béla 23871

D

Dąbrowska, Maria 35088
Dadd, Richard 43429
D'Addario, Salvatore 27800
Daens, Adolf 03259
Dahl, Johan Christian 33418
Daimler, Gottlieb 21934
Dal', Vladimir Ivanovič 37269, 43116
D'Albertis, Enrico Alberto 26304
D'Albertis, Luigi Maria 26304
Dalén, Gustaf 40881, 40882
Dalí, Salvador 05828, 14594, 17244, 19204, 28718, 39001, 39787, 40317, 44156, 44829, 52642
Dalinkevicius, Juozas Antanas 30397
Dalla Bella, Giovanni Antonio 35533
Dally, Fritz 16596
Dalyell, Thomas 44719
D'Amico, Victor 46248
D'Andrade, Alfredo 26289
Danicourt, Alfred 14785
Daniëls, René 32524
Danielson-Kalmari, Johan-Richard 11430
Danioth, Heinrich 41103
D'Annunzio, Gabriele 26272, 27162
Danov, Christo G. 05128
Dantan, Jean Pierre 09724
Dante Alighieri 26168
Darder i Llimona, Francesc 38948
Dargent, Yan 15511
Dariaus, Steponas 30220, 30449
Darío, Rubén 33275, 33284
Darvas, József 23738
Darwin, Charles 00509, 43735, 44018
Dass, Petter 33816
Dastrac, Raoul 11506
Daubigny, Charles-François 11770, 14908, 28880
Daudet, Alphonse 12977, 15155
Daumier, Honoré 15236, 22018, 28590
Daura, Pierre 12242, 14791, 50325
Davey, Alan 44037
David, Allen 00922
David, Gerard 31141
Davidis, Henriette 17918
Davies, John 45285
Davies, Stanley 44559
Davis, Jefferson 52335, 54289
Davison, Dorothy 45098
Davitt, Michael 24901
Davoine, René 12416
Dawley, Joseph 46224
De Bruyn, Pierre 03631
De Chirico, Giorgio 04879, 27444, 27454
De Ferrari, Gregorio 26324
De Kooning, Willem 50224
De Lorean, John 24851
De Louwe, Petrus 03289
De Maria, Walter 51163
De Merode 04148
De Pisis, Filippo 26289, 27800
De Smedt, Josepha-Marcella 03631
Deák, Ferenc 23658, 23805
Debeljanov, Dimcho 05070
Debussy, Claude 15289
Déchelette, Joseph 15059
Decurtins, Caspar 41953
Defesche, Pieter 32884
Degas, Edgar 13889, 14711, 14719, 43497
DeGrazia, Ted 53586
Deinhard, Johann Friedrich 19584
Deix, Manfred 02226
Dejneka, Aleksandr Aleksandrovič 37093
Del Piombo, Sebastiano 26321
Del Vaga, Perino 26321
Delacroix, Eugène 13889, 14354, 14732
Delauny, Sonia 26298
Delavigne, Casimir 14945
Delayrac, Nicolas 14335
Delderfield, R. F. 45679
Della Ragione, Alberto 26151
Delporte, Charles 03534
Delso, Pedro 38838
Delvaux, Paul 04016, 28512
Demard, Albert 12393
Demetrescu, Traian 36322
Demidov, Akinfy 36789
Demirchyan, Derenik 00731
Denchu, Hirakushi 28551
Denev, Boris 05252
Denis, Maurice 15290
Depero, Fortunato 26289, 27570, 28056
Depiau, Charles 13426
Derain, André 12380, 14020
Derby, Elias Hasket 52658
Déri, Miksa 23434
Deržavin, Gavrila Romanovič 37658
Descartes, René 12716
Deschamps, Paul 14728
Desiderio, Vincent 08675
Desmet, Jean 31841
Desmoulin, Fernand 12143
Despiau, Charles 14158
Desportes, Alexandre-François 13055
Destinnová, Ema 09565
Deubler, Konrad 01748
Deusser, August 41140
Dewantara, Ki Hadjar 24586
Dewey, Harold Nelson 45932
Di Bicci, Neri 26126
Di Blasio, Pier Giuseppe 27559
Di Cambio, Arnolfo 26176
Diabelli, Anton 02367
Dias, Cicero 04667
Diaz, Porfirio 06090
Dickens, Charles 32013, 43644, 43646, 44817, 45427, 45507
Dickhut, Willi 18580
Dickinson, Emily 46265
Dickinson, Frank R. 44902
Diefenbach, Karl Wilhelm 18850
Diefenbaker, John G. 06660, 07273
Diem, Peter 31837
Diesel, Rudolf 16628
Dietrich, Adolf 41201
Dietrich, Marlene 17069
Dietz, Günter 22343
Dill, Otto 20673
Dima, Gheorghe 36174
Dimitrescu, Ştefan 36646
Dimitrov, Georgi 05079, 05194
Dimitrov-Majstora, Vladimir 05067
Dimov, Dimităr 05176, 05192
Dine, Jim 19627
Dinis, Júlio 35800
Dios Mena, Juan de 00500, 00521
Disney, Walt 52798
Disraeli, Benjamin 44465
Disteli, Martin 41707
Diviš, Prokop 09925, 09948
Dix, Otto 02620, 16453, 17698, 18453, 18532, 18597, 21948
Djakonov, Nikolaj M. 37816
Dobeš, Milan 38224
Dobrian, Vasile 36566
Dobroljubov, Nikolaj Aleksandrovič 37451
Dobrovský, Josef 09487
Dobšinsky, Pavel E. 38352
Doderer, Heimito von 02954
Doerfler, Ludwig 21553
Doisneau, Robert 13044
Doll, Samie Keith 47153
Dollfuß, Engelbert 02806
Domela, César 41129
Domínguez, Lorenzo 00521
Domoto, Insho 28841
Domšaitis, Pranas 30277
Donatello 26176, 27042
Dongen, Kees van 15111, 32870
Doni, Adone 25413
Donizetti, Gaetano 25399, 25404
Dorbe, Herberts 30143
Doré, Gustave 12107
Doren, Emile van 03619
Doria, Andrea 26293
Doria, Giacomo 26297
Dospevski, Stanislav 05108
Dostoevskij, Fëdor Michajlovič 29728, 36873, 37195, 37316, 37480, 37656, 37658, 37796
Douglas, C. 24428
Douglass, Frederick 53872
Douwes-Dekker, E. 31860
Doyle, Sir Arthur Conan 41577, 41621, 44160, 44986
Drachmann, Holger 10259
Draewing, Paul Peter 21644
Drais, Karl von 20174
Drake, Sir Francis 39217, 46094
Drevon, Jean 15322

Droste-Hülshoff, Annette von 19056, 20247, 20248, 20508
Drummonds, James 01590
Duarte, António 35493
Duarte, Juan Pablo 10344
Dubček, Alexandr 38384
Dubois, Paul 14464
Dubuffet, Jean 14600, 14776, 15252
Duc, Viollet le 14728
Duca, I.G. 36401
Duchamp, Marcel 14719, 15252, 27454
Ducros, Abraham-Louis-Rodolphe 41544
Dürer, Albrecht 13122, 15004, 17172, 20755, 20923, 22714, 23823, 26323, 26580, 27171, 44999, 50737
Dürrenmatt, Friedrich 41215, 41678
Dufresne, Maurice 11811
Dufy, Raoul 12636, 14020, 14719, 14791, 15111
Duguay, Rodolphe 06456
Duguay-Trouin, René 15389
Duke, Doris 51329
Dumas, Alexandre (the elder) 13641, 16229
Dumas, Alexandre (the younger) 16229
Dumont, Louise 18010
Dunant, Henri 13795
Dunant, Henry 41423, 41475
Duncker, Johann Heinrich August 21190
Dunhill, Alfred 44780
Dunikowski, Xawery 35269
Dunman Ure, Annie 45472
Dunn, Harvey 47059
Dunn, John 38618
Dupain, Max 01120
Dupleix, Joseph-François 13500
Dupré, Amalia 26128
Dupré, Giovanni 26128
Dupré, Jules 13781
Durand, Marie 14940
Durand, Pierre 14940
Durant, Billy 48560
Durocher, Marie-Rose 06193
Durov, Anatolij Anatolevič 37824
Durov, Anatolij Leonidovič 37969
Durov, Vladimir Leonidovič 37377
Durova, Nadežda A. 36906
Dvořák, Antonín 09672, 09767, 09826, 09943
Dyck, Albert van 04005
Dyck, Anthonis van 14714, 15829, 17833, 26161, 26323, 26580, 27096, 28252, 32788, 43827, 44138, 45006, 45378, 53874
Džalil, Musa 37020
Dzierżon, Jan 34748

E

Eames, Charles 22379
Eames, Ray 22379
Eardley, Joan 44037
Ebert, Friedrich 19080
Ebihara, Kinosuke 28631
Eckenfelder, Friedrich 16899
Eckermann, Johann Peter 22606
Edelfelt, Albert 11093, 11225, 11226, 11280
Edelmann, Albert 41338
Edgeworth, David 01207
Edison, Thomas Alva 46675, 50298, 50681
Eertvet, Andries van 32788
Egger-Lienz, Albin 02301, 02302
Egk, Werner 17900
Ehmsen, Heinrich 19534
Ehrenberg, Christian Gottfried 17812
Ehrlich, Franz 17834
Eichendorff, Joseph von 21192, 22324
Eichler, Ernst 22485
Einstein, Albert 41205
Eisenhower, Dwight David 48056
Eisenstadt, Alfred 51194
Eistola, Pentti 11365
El-Kader, Abd 11578
Eldh, Carl 40889
Elek, Benedek 36144
Elgar, Edward 45038
Eliass, Ģederts 30017
Eliot, George 45305
Elisabeth, Empress of Austria 03009, 23565
Elisabeth of Thuringia 18459
Ellis, Florence Hawley 46128
Ellis, Richard 00790
Eluard, Paul 15236
Emerson, Ralph Waldo 47771
Eminescu, Mihai 36425, 36429
Emmler, Elisabeth 21041
Emre, Yunus 42782
Enckell, Magnus 11093
Ende, Hans am 22696, 22699
Enea, Nicu 36119
Enescu, George 36224, 36351, 36445, 36618
Engels, Friedrich 22742
Engert, Ernst Moritz 18850
Engström, Albert 40493, 40568, 40605
Enríquez, Carlos 09143
Ensor, James 03290, 03930, 28512, 28590
Epstein, Sir Jacob 45920
Erasmus, Desiderius 03450, 41157
Erben, Karel Jaromír 09654
Erdai, Zhao 42267
Erdös, Paul 36589
Erhard, Ludwig 18500
Erhardt, Alfred 17030
Erichsen, Thorvald 33555
Erikson, Gustaf 11099
Erixson, Sven X. 40871, 40986
Erkel, Ferenc 23593, 23594
Ermolova, Maria N. 37163
Erni, Hans 41591
Ernst August, Duke of Saxe-Weimar 17905
Ernst, Max 14722, 15252, 17244, 17606, 19005, 19204, 33420, 44156, 50224
Ernst, Paul 02642
Erošenko, Vasilij J. 37800
Ersoy, Mehmet Akif 42705
Erxleben, Dorothea Christiana 21150
Erzia, Stefan 00521
Escher, Maurits Cornelis 32071
Escher, Rolf 05016
Escutia, Juan 31440
Esenin, Sergej Aleksandrovič 37056, 37255, 54487
Eulenspiegel, Till 03536, 21618
Evans, Evan 18631
Evenepoel, Henri 04023
Everage, Dame Edna 01290
Exter, Julius 22147
Eyth, Max 19549

F

Fabro, Luciano 26881
Fader, Fernando 00389, 00438
Faidherbe, Négrier 13757
Fałat, Julian 34562
Falger, Josef Anton 01865
Falk-Breitenbach, Eugen 19047
Falk, Johannes Daniel 22405
Falkberget, Johan 33795
Falla, Manuel de 00121, 39359
Fallada, Hans 18301
Faraday, Michael 44917
Farago, Elena 36321
Farago, Odon 26327
Farey, Hélène 14481
Farina, Johann Maria 19600
Fath, Richard 13055
Faulkner, William 51011, 51648
Faust, Georg 19583
Faust, Heinrich 19583
Faust, Johann 19583
Faykod, Maria de 11748
Febus, Gaston 14050
Fedele, Ferdinando 26014
Fedkovič, Jurij 43011
Fehling, Heinz 21536
Fehr, Viktor 42001
Feininger, Lyonel 17041, 17834, 21151, 33420
Fekete, Esteban 21381
Fekete, István 23569
Felder, Franz Michael 02712
Felixmüller, Conrad 17698
Feluérez, Manuel 31579
Fenneker, Josef 17396
Feragutti, Adolfo 41195
Fergusson, John Duncan 45365
Fernández Alvarado, José 39396
Fernández de Latorre, Néstor Martín 39802
Ferrari, Defendente 28004
Ferry, Jules 15241
Festetics, George III. 23660
Feuerbach, Anselm 19448, 21836
Fičeto, Kolju 05024, 05025
Fichte, Johann Gottlieb 19388, 21175
Figl, Leopold 02603
Filion, Joseph 06901
Filla, Emil 09484, 09640
Filotti, Maria 36193
Firpo, Roberto 00561
Firth, Josiah 33145
Fischer, Jan Tomáš 09946
Fischer, Johann 17832
Fisher, Alvan 48023
Fitger, Arthur 17817
Fitzsimmons, Bob 44444
Fitzwilliam of Merrion, Richard VII Viscount 43727
Fjell, Kai 33752
Flacius Illyricus, Matthias 08906
Flagstad, Kirsten 33529
Flake, Otto 16884
Flath, Otto 16838
Flaubert, Gustave 12271, 15116
Flavin, Dan 12840, 46660, 51163, 51166
Flaxman, John 44999
Fleischmann, Arthur 38220
Fleming, Alexander 44779
Fleming, Ambrose 45679
Fleming, Paul 19018
Flinders, Matthew 01110, 01648
Flüe, Nikolaus von der 41786
Foch, Ferdinand 15861
Fojita, Tsuguji 14719
Fonck, René 15241
Fontana, Lucio 00521
Fontane, Theodor 17181, 20650, 21111, 22795
Forckenbeck, Oscar von 16399
Ford, Aleksander 34877
Ford, Henry 48008, 52499
Forel, Alexis 41643
Forster, Georg 18690
Forster, Reinhold 18690
Forster, Stephen 46607
Forte, Vicente 00521
Fortin, Marc-Aurèle 06373
Fortunato Bartolomeo, Félice de 42049
Foster, Con 53554
Foster, Norman 45285
Foujita, Leonard 29286
Foujita, Tsunguharu 29269
Fouqué, Caroline de la Motte 18404
Fouqué, Friedrich de la Motte 18404
Fowler, James 45037
Fra Angelico 26157, 26184, 27453
Fra Bartolomeo 26184
Fragonard, Jean-Honoré 13115, 13924
Fragoso, João 35494
Français, Louis 14845
Francis, Sam 08675, 28715, 50952
Francke, August Hermann 18891
Franco, Francisco 35616
Franco, Henrique 35616
François I 11578, 12966
Frangipane, Silvestro 28314
Frank, Anne 17034, 31828, 50256
Franke, Hanny 18233
Frankenthaler, Helen 08675, 50224
Franklin, Benjamin 44794
Franklin, John 43732
Franko, Ivan 07364, 43123
Franz Ferdinand, Archduke of Austria 01728
Franz Joseph I, Emperor of Austria 01753, 02639, 03009
Franzoni, Filippo 41195
Frappier, Armand 06145
Fraser, Simon 07078
Frasnedi, Alfonso 27800
Fraunhofer, Joseph von 16991, 20494
Fredericks, Marshall M. 53655
Freiligrath, Ferdinand 21961
French, Daniel Chester 53298
Frenette, Joseph 05579
Frenguelli, Joaquín 00633
Freud, Anna 03075, 44852
Freud, Lucian 44557
Freud, Sigmund 03075, 09805, 37709, 44852
Freundlich, Otto 14909
Freytag, Gustav 22325
Fried, Erich 02954
Friedel, Lutz 21096
Friedrich I Barbarossa 18570
Friedrich, Caspar David 17032
Friedrich the Great 21109, 21784
Friese, Georg 20087
Frind, August 09831
Frobisher, Marguerite 43699
Fröbel, Friedrich Wilhelm August 16698, 20849
Fröding, Gustaf 40638
Frölicher, Otto 41885
Frohner, Adolf 02225
Froment, Jean-Louis 12062
Frost, Robert 48094
Frunze, Michail V. 36949
Fuchs, Ernst 02981
Fürstenberg, Heinrich von 41333
Fugger, Jakob 16617
Fujii, Tatsuyoshi 29105
Fujiwara, Kei 28389
Fukuda, Heihachiro 29632
Fukuhara, Rintaro 28443
Fulla, Ludovít 38344
Fumbuka, David Clement 42518
Furneaux, Tobias 00912
Furphy, Joseph 01542

G

Gabin, Jean 14087
Gabo, Naum 45792
Gachnang, Johannes 41215
Gaddi, Taddeo 26126
Gadegård, Paul 10071
Gafuri, Mažit 37885
Gagarin, Jurij 36918, 37778
Gaidamavicius-Gele, Zigmas 30316
Gainsborough, Hals 45006
Gainsborough, Thomas 43405, 43497, 44840, 44894, 45797, 53253
Gaitán, Jorge Eliécer 08545
Gaj, Ljudevit 08901
Gajdar, Arkadij P. 36774, 37103, 37787
Galaction, Gala 36186
Galaune, Paulius 30238
Gale, Emile 29629
Galilei, Galileo 26192, 27237
Gallatin, Albert 41735, 53916
Gallatin Hoit, Albert 47314
Gallen-Kallela, Akseli 10728, 11093
Gallois, Genevieve 16084
Gambetta, Léon 12242, 15742
Gandersheim, Roswitha von 16738
Gandhi, Mahatma 24048, 24080, 24088, 24146, 24293, 24345, 24384, 38773
Gane, Nicolae 36413
Ganesh, Selva 24428
Ganghofer, Ludwig 02296, 19492
Gárate, Hermano Francisco 38929
Garborg, Arne 33573, 33945
García, Carlos P. 34434
García Lorca, Federico 39317, 39362
Gardel, Carlos 00170, 00264
Gárdonyi, Géza 23536, 23540
Gargallo, Pablo 40317
Garibaldi, Giuseppe 00381, 25405, 26429, 26654, 53262
Garinet, Jules 12363
Garland, Judy 49009, 49010
Garner, John N. 53678
Garnett, Annie 44559
Garret, Georges 16133
Gasparini, Osvaldo 00565
Gassendi, Pierre 12727
Gasteiger, Anna Sophie 22194
Gasteiger, Mathias 22194
Gaudí, Antoni 38953, 38986, 39001
Gauguin, Paul 14719, 14881, 16325, 30668, 31898, 43497
Gaul, August 18980
Gaulle, Charles de 11876, 12587, 13752
Gay-Lussac, Louis-Joseph 15369
Gaya, Ramón 39701
Gayarre, Julián 39913
Gazdov, Ivan 05230
Gebhardt, Eduard von 18666
Gebler, Friedrich A. von 36789
Geddes, Patrick 44052, 44115
Gegenbaur, Joseph Anton von 22328
Gehrcke, Hans-Otto 21707
Geier, Ernst 41260
Geitlinger, Ernst 20587
Gellée, Claude 12367, 12840
Gellert, Christian Fürchtegott 18873
Genichiro, Inokuma 28881
Gentil, Anton 16582
Gentileschi, Orazio 28004
Gentsch, Fritz 16568
Gentz, Wilhelm 20650
Genucchi, Giovanni 41195
Georganti, Loukia 22938
George, David Lloyd 43925
George II, Duke of Saxe-Meiningen 20262
George IV, King of England 43625
George, Stefan 17354
Georgi, Walter 19796
Gercen, Aleksandr 37195, 37231

Géricault, Théodore 13889, 14909
Gernhardt, Robert 18357
Gernsheim, Helmut 20163
Gerstäcker, Friedrich 17520
Gerstein, Noemí 00521
Gerstmayr, Hans 02372
Gerstner, Karl 41215
Gertsch, Franz 41275
Gerung, Mathis 19857
Gesellius, Herman 10963
Gevork, Grigorian 00736
Geyer zu Lauf, Hans von 18148
Gezelle, Guido 03415
Gfeller, Simon 41478
Ghiaţă, Dimitrie 36207, 36525, 36646
Ghiberti, Lorenzo 26176
Ghirlandaio, Domenico 26145, 26160, 26184
Ghirri, Luigi 25896
Giacometti, Alberto 10103, 17192, 27454, 41885, 41901, 42083, 45285, 50224
Giacometti, Augusto 41901
Giacometti, Giovanni 41080, 41901
Giallina, Angeliki 22973
Giallina, Angelos 22973
Giaquinto, Corrado 25371, 25413
Gibbons, Grinling 45378
Gibbons, Thomas 51222
Gibbs, May 01358
Gibran, Khalil 30166
Gide, André 16025
Giger, Hans Ruedi 41461
Gigli, Beniamino 27376
Gigli, Lorenzo 27378
Gil de Palacio, Léon 39565
Gilbert, Stuart 53874
Gill, Henry 17207
Gillray, James 28590
Giono, Jean 13940, 13942
Giordano, Luca 25371, 26200, 27182, 27453, 39565
Giorgione 27034, 28211
Giotto di Bondone 26157, 27032, 27034, 28250
Giovanoli, Samuele 41869
Giraudoux, Jean 11925
Girma, Lucien 15361
Gironas, Stasys 30220, 30449
Gironcoli, Bruno 02655
Gladel, Joseph 12668
Glauser, Friedrich 41215
Gleim, Johann Wilhelm Ludwig 18877
Gleizes, Albert 15492
Gleyre, Charles 41544
Gliczner, Erazm 35378
Glinka, Michail I. 37490, 37778
Glöckle, Fritz 21686
Gluck, Christoph Willibald von 16999
Gmunden, Johannes von 01951
Godefroy de Bouillon 03393
Gökalp, Ziya 42765
Görbing, Johannes 21039
Göschen, Georg Joachim 18760
Goethe, Cornelia 18146
Goethe, Johann Wolfgang von 12365, 15732, 16475, 16774, 17905, 17990, 18146, 18370, 18794, 19320, 19321, 19380, 20004, 20882, 21924, 22398, 22401, 22402, 22403, 22529, 27442
Goetz, Ferdinand 19923
Goetz, Henri 16193
Götz, K. O. 18024
Göz, Godefried Bernhard 19416
Goga, Octavian 36283, 36555
Gogh, Vincent van 03531, 11682, 11769, 15492, 27454, 31898, 32375, 32606, 32642, 33017, 42083, 44138, 44937, 45920, 48935, 52717
Gogol, Nikolaj Vasilevič 37164, 43040, 43201
Goitia, Francisco 31579
Goldblatt, Sarah 38696
Golden, Daan van 32756
Goldenvejzer, Aleksandr Borisovič 37317
Goldie, Charles Frederick 33211
Goldmark, Carl 01814
Goldoni, Carlo 28180
Goldsworthy, Andy 11899, 44557
Goldwater, Barry Sen. 51904
Goll, Claire 15241
Goll, Ivan 15241
Golovanov, N.S. 37315
Goltzius, Hendrick 17833, 19524
Golubkina, Anna Semënovna 37200, 37985
Gómez, Martin 00517
Gómez, Máximo 10330
Gomez, Pedro Nel 08685
Gončarov, Ivan Aleksandrovič 37658, 37898
Gončarova, Natalja 37765
González, Joaquín V. 00302
González, Julio 39001
Goodwin, Ginger 05664
Gordon, Charles George 43798
Gordon Low, Juliette 52942
Gore, Spencer 44695
Gorjuša, Dob 38416
Gorkij, Maksim 19144, 37011, 37279, 37318, 37452, 37454, 37460, 37616, 43202
Gorky, Arshile 50308
Gorsline, Douglas 12212
Gortani, Michele 27992
Gosselin, Joshua 45591
Goßmann, Gerhard 18495
Goto, Sumio 28646
Gotthelf, Jeremias 41277, 41578
Gottlob, Maurice 14299
Goudge, Edward 44344
Gounaropoulos, Giorgios 23247
Goupil, Adolphe 12072
Goya, Francisco de 11494, 12320, 15829, 26580, 31141, 39313, 39518, 39547, 39565, 40340, 40362
Goyen, Jan van 32489
Graber, Alfons 02763
Grade, Hans 17480
Graevenitz, Fritz von 21954
Graham, Benjamin 51222
Graham, Rodney 17123
Grainger, Percy Aldridge 01282
Granell, Eugenio 40028
Granello, Nicolosio 26317
Granovskij, Timofej Nikolajevič 37504, 37510
Grant, Ulysses S. 51183
Grard, George 03656
Grass, Günter 20075
Grassi, Nicola 28110
Grassi, Vittorio 26327
Grau, Miguel 34209, 34288
Graubner, Gotthard 17407
Graves, Abbott 49727
Graves, Morris 50838
Grebenščikov, G.D. 36791
El Greco 11588, 26741, 27096, 39001, 40127, 40362, 44138
Green, George 45292
Green, John 04200
Green, Thomas 44400
Gregorčič, Simon 38536
Grekov, M.B. 37477
Grela, Juan 00521
Grescny, Marie-Thérèse 13984
Gressly, Amanz 41333
Grétry, André Ernest Modeste 03814
Greuze, Jean-Baptiste 15957
Griboedov, Aleksandr Sergeevič 36863
Grieg, Edvard 33766
Grieshaber, HAP 17321
Grigorescu, Nicolae 36162, 36250, 36525, 36646
Grillparzer, Franz 02998
Grimdalen, Anne 33462
Grimm, Jacob 18883, 19459, 21877, 21878
Grimm, Ludwig Emil 19459, 21877
Grimm, Wilhelm 18883, 19459, 21877, 21878
Grimmelshausen, Johann Jakob Christoffel von 18571, 20812
Grin, Aleksandr S. 37036
Grinberg, Aleksandr 37254
Grincevičiūtė, Beatričė 30402
Grinkevičiūtė, Dalia 30298
Gris, Juan 40915
Grochtmann, Hermann 17800
Grock 21115
Groddeck, Georg 16884
Gröbli, Isaak 41700
Grohmann, Will 21960
Gropius, Walter 17041, 17316, 17834, 19536
Groth, Jan 33873
Groth, Klaus 19066
Grubenmann (builder family) 41932
Gruber, Franz Xaver 01690, 02047, 02253
Grubicy De Dragon, Vittore 41195
Grüger, Margit M. 17108
Gründgens, Gustaf 18010
Grünenwald, Jakob 18041
Grünewald, Matthias 12585, 19448
Grützke, Johannes 17698
Grundtvig, Nikolai Frederik Severin 10182
Gruodis, Juozas 30243, 30260
Grušas, Juozas 30244
Grybas, Vincas 30234
Guardi, Francesco 43405
Güiraldes, Ricardo 00566
Günther, Ignaz 16515
il Guercino 26323
Guericke, Otto von 20128
Guérin, Camille 13755
Guérin, Maurice de 11597
Guerrero, José 39360
Guerrero, Maria 00264
Guggenheim, Willy 41080
Guiguet, François 12619
Guillaumin, Émile 16307
Guimerà, Àngel 40259
Guisan, Henri 41737
Gulácsy, Lajos 23420, 23846
Gulbransson, Olaf 22018
Gulda, Friedrich 02234
Gulia, Dmitri 16350
Gully, John 33154
Gunarsa, Nyoman 24516
Gurney, Sir Goldsworthy 43667
Gursky, Andreas 19567
Gurvič, Josif 27800
Gustaitis, Motiejus 30301
Gustav II Adolf 20105, 22455, 40909
Gutenberg, Johannes 18131, 20143
Gutfreund, Otto 09764
Gutman, Nahum 25118
GutsMuths, Johann Christoph Friedrich 21150
Guttuso, Renato 26289
Guyer-Zeller, Adolf 41150
Gyarfas, Jenö 36599, 36601
Gyokudo, Kawai 29110

H

Haanpää, Pentti 11208
Haase, Hans 17086
Haberstock, Karl 16625
Hablik-Lindemann, Elisabeth 19370
Hablik, Wenzel 19370
Hacker, Dieter 08675
Hackert, Jakob Philipp 21118
Hadjikyriakos-Ghikas, Nikos 22941
Haeckel, Ernst 19377
Händel, Georg Friedrich 18902
Häusser, Robert 20163
Hagemeister, Karl 21707
Hager, Marie 17647
Hahn, Philipp Matthäus 16458
Haidar, Ali 24398
Haizmann, Richard 20699
Hajek, Otto Herbert 21948
Hale, Lilian 48023
Hale, Philip 48023
Halepas, Giannoulis 23162
Hall, Charles Martin 51479
Hall, John 45780
Haller, Anna 41812
Halonen, Eemil 11044
Hals, Frans 31141, 32295, 32849, 43497
Hamada, Shoji 15986, 28884, 28885, 29531
Hambleton, Richard 17184
Hamerling, Robert 02166
Hamilton, Alexander 51189, 51222
Hammar, Jörgen 11156
Hammerich, Gunnar 09959
Han, Oscar 36646
Hanada, Naojiro 29100
Hanak, Anton 02260
Handy, William Christopher 48566
Haniel, Franz 18018
Hanns, Georgi 21710
Hanrieder, Norbert 02543
Hans, Abraham 03712
Hansen, Helmer 33783
Hansjakob, Heinrich 19025
Harbutt, William 43406
Hardenberg, Georg Philip Friedrich von 22541
Hardman, Joseph 44559
Hardy, Adolphe 03560
Hardy, Oliver 21803, 45897
Hardy, Thomas 43999, 44467
Hare, William 44160
Haring, Keith 17184, 51140
Harlan, James 50882
Harrer, Heinrich 02103
Harris, Howell 43598
Harris, John 06183
Harrison, George 18889
Hart, Robert Edward 30685
Harth, Philipp 16967
Hartley, Marsden 50952, 52717
Hartmann, Oluf 10186
Harvard, John 45781
Has, Wojciech Jerzy 34877
Hasdeu, Bogdan Petriceicu 36249
Hasdeu, Julia 36249
Hasegawa, Machiko 29396
Hasenclever, Johann Peter 21264
Hassam, Childe 47690
Hasse, Sella 22613
Hathaway, Anne 45779
Hatmanu, Dan 36646
Hatry, Julius 20174
Hauff, Wilhelm 16896
Hauptmann, Gerhart 17181, 18007, 18215, 19574, 34703
Hauser, Erich 21426
Hauser, Kaspar 16560
Hausmann, Raoul 15067
Hausner, Rudolf 19736
Hauteclocque, Philippe François Marie de 14626
Haviland, Théodore 11564
Havlíček Borovský, Karel 09512
Hawkes, Samuel 52522
Hawthorne, Nathaniel 52658
Haydn, Johann Michael 02594, 02616
Haydn, Joseph 01860, 01861, 02594, 03000, 23557, 23558
Hayet, Louis 14908
Hazama, Inosuke 28624
He Xiangning 08220
Hearn, Lafcadio 28888
Hebbel, Friedrich 19535, 22511
Hebel, Johann Peter 19049, 20016, 21622
Hébert, Ernest 13453, 14731
Hechelmann, Friedrich 19366
Heckel, Erich 17058, 18196
Hedberg, Erik 40611
Heddon, James 48172
Heeg-Erasmus, Fritz 21622
Heel, Jan van 32522
Heesters, Jan 32757
Hegel, Georg Wilhelm Friedrich 21939
Hein, Piet 10170
Heine, Ferdinand 18851, 18878
Heine, Heinrich 17991
Heine, Salomon 18935
Heiner, Sebastian 21096
Heinrich, Hansjakob 18430
Heinsohn, Alfred 21644
Hellwag, Christoph Friedrich 18285
Helm, Walter 16584
Helmcken, J.S. 07250
Helmschmid, Desiderius 39586
Helmschmid, Kolman 39586
Helye, Helyas 41225
Hem, Louise de 03731
Hemingway, Ernest 09278, 49759, 51454, 51807
Henner, Jean-Jacques 14733
Henri, Robert 47690
Henrique Infante Dom 35862
Henry VIII 44968
Henry the Navigator 35862
Henry, O. 46502, 46517
Henţia, Sava 36525
Henze, Volker 17106
Hepworth, Barbara 32642, 43731, 45409, 45575, 45792, 45908
Heras, Henry 24317
Herbin, Auguste 13569
Herder, Johann Gottfried 22405, 34928
Heredia, José-Maria de 15035
Herites, František 09911
Herkomer, Hubert von 19794, 19796, 43699
Hernandez, Don Jaime 34467
Hernández, José 00672
Hernandez, Leon 34467
Hernández, Miguel 39740
Herrliberger, David 41616
Herschel, Caroline 43404
Herschel, William 43404
Hershey, General Lewis B. 46300
Hertervig, Lars 33873
Herwegh, Georg 41561
Herzen, Alexander 37195, 37231
Herzer, Klaus 20366
Herzog, Werner 17069

Hesse, Hermann 17678, 18531, 19549, 21961, 28531, 41127, 41215, 41633
Hester, Joy 00915
Hetzel, Pierre Jules 11584
Heuss, Theodor 17495, 21966
Heydt, Eduard v.d. 41127
Heyerdahl, Thor 33629, 33729
Heyn, Fritz 17219
Hickok, James Butler 49223
Hidalgo y Castulo, Miguel 30834
Hideyoshi, Toyotomi 29004
Hiemstra, Mary 06143
Hieronymus von Prag 19679
Higashiyama, Kaii 28954, 29201
Highsmith, Patricia 41215
Higueras, Jacinto 40039
Hilberseimer, Ludwig 17041
Hildegard von Bingen 17353
Hill, Derek 24861
Hill, James J. 52618
Hincks, William 44189
Hincz, Gyula 23944
Hippokrates of Kos 23064
Hirayama, Ikuo 28940
Hirche, Herbert 17280
Hirst, Damien 10108, 33717
His, Wilhelm 41153
Hitchens, Ivon 45409
Hjorth, Bror 40871, 41003
Ho Chi Minh 54615
Hobrecht, James 17207
Hochstetter, Ferdinand 02112
Hockaday, Hugh 49671
Hockney, David 29297
Hodgkins, Frances 33211
Hodler, Ferdinand 41080, 41885, 42083
Höch, Hannah 17184
Hoehme, Gerhard 18024
Hoel, Sigurd 33805
Hölderlin, Friedrich 16751, 22115
Hölzel, Adolf 17321, 21948
Höß, Crescentia 19484
Hoetger, Bernhard 17557
Hofer, Andreas 02134, 27654
Hoffmann, Ernst Theodor Amadeus 16907
Hoffmann von Fallersleben, August Heinrich 22663
Hoffmann, Heinrich 18397
Hofmann, Julius 21669
Hofner, Johann Baptist 21639
Hogarth, William 44876, 44937, 44987
Hogaş, Calistrat 36522
Hogg, James 45636
Hohenzollern-Veringen, Radu Prince of 36592
Hohmann, Walter 19011
Hokusai, Katsushika 29546
Holbein, Hans (the Younger) 41885
Holbein, Sigmund 44999
Holberg, Ludvig 09987
Holl, Moritz 02112
Hollar, Václav 09924
Hollenberg, Felix 17870
Hollenstein, Stephanie 02341
Holly, Ján 38202
Holm-Møller, Olivia 10092
Holman Hunt, William 43884, 45327
Holmes, Ralph 46405
Holst, Gustav 43811
Holub, Emil 09522
Holz, Paul 20987
Homar, Gaspar 26327
Honegger, Gottfried 14292
Hong Zhi-Sheng 42307
Honthorst, Gerard van 17833
Hoover, Herbert C. 01229, 54046
Hopper, Edward 51449, 52717
Horch, August 19585, 22831
Horn, Rebecca 18712, 26881
Hornel, Edward Atkinson 44610
Horov, Pavol 38182
Horstmann, Gustav 43406
Horstmann, Sidney 43406
Horta, Victor 03488
Horváth, Ödön von 20535
Houdin, Robert 12031
Houdon, Jean-Antoine 14734
Houssay, Bernardo A. 00171
Houston, Sam 49446
Hovhanisyan, Hovhannes 00696
Howe, Oscar 47059, 50764
Hrdlička, Aleš 09541
Hrebnicki, Adomas 30227
Hrozný, Bedřich 09644
Huber, Max 41298
Huber, Mida 02255
Hudetz, Karl Anton 22561
Hudson, Grace Carpenter 53638
Hudson Kitson, Henry 53637
Hübsch, Heinrich 22431
Hugo, Victor 11989, 14621, 16225, 30495, 45592
Huisman, Jopie 32964
Huldéns, Lars 11156
Huldéns, Mats 11156
Hulsdonck, Jacob van 14069
Humboldt, Alexander von 17140, 36789
Humboldt, Wilhelm von 17140
Hummel, Berta 20235
Hummel, Johann Nepomuk 38215
Humperdinck, Engelbert 17475
Humphrey, Hubert 50733
Hundertwasser, Friedensreich 03014
Hunt, Robert 45477
Hunt, William Henry 43699
Hunter, Adelaide 06817
Hunter, William 44285
Hus, Jan 09542, 19679
Hussein, Ibrahim 30612
Hutchinson, Peter Orlando 45679
Huygens, Constantijn 32905
Huys, Modest 04170
Hviezdoslav, Pavol O. 38251
Hydman-Vallien, Ulrica 40474
Hyrtl, Josef 02112, 02395

I

Ianchelevici, Idel 03768, 32272
Iancu, Avram 36116
Ibn Sina, Abu Ali 54437
Ibsen, Henrik 33519, 33727, 33834
Ibuse, Masuji 28443
Ignacio de Loyola 38928
Ikeda, Hosai 28525
Ikeda, Masuo 28767
Ikeda, Yoson 29066
Ikeda, Yushi 28525
Ikegami, Motoemon 29334
Iliescu-Călineşti, Gheorghe 36525
Imhoff, Berthold von 06886
Immendorff, Jörg 18712
Ingalls Wilder, Laura 53804
Ingres, Jean Auguste Dominique 14176, 14719
Ion Roată, Moş 36257
Ionescu, Mariana 36162
Iorga, Nicolae 36160, 36722
Iqbal, Allama Sheikh Muhammad 34021
Iramain, Juan Carlos 00600
Irimescu, Ion 36367
Irving, Henry 45838
Irving, Washington 47293
Isaacs, Sir Isaac 01674
Isabella of Castile 39589
Isac, Emil 36288
Isahakyan, Avetik 00705
Isembrant, Adrian 39638
Isepp, Sebastian 02451
Ishikawa, Takuboku 29350
Ishimoto, Tadashi 28915
Islandsmoen, Sigurd 33413
Israelsson, Fet-Mats 40516
Issahakian, Avetik 00728
Issel, Arturo 26138
Istrati, Panait 36168, 36170
Itaya, Hazan 28839, 29632
Ivanoff, Vassil 13318
Ivens, Joris 31841
Izquierdo, María 31070

J

Jabarly, Jafar 03151
Jabotinsky, Zeev 25110
Jackson, Andrew 49262
Jackson, Thomas J. "Stonewall" 52335
Jacobsen, Jens Peter 10296
Jacobsen, Robert 19736
Jäger, Ştefan 36432
Jahn, Friedrich Ludwig 18458
Jaindl, Othmar 02848
Jākobsons, Voldemārs 29990
Jakšić, Džuro 38154
James, Jesse 49713, 52562
Jammes, Francis 14542
Jamsz-Teles jr. 04667
Janáček, Leoš 09424, 09435, 09438, 09504, 09540
Janco, Marcel 24946
Janicki, Klemens 35378
Janků, Miloslav 09479
Jankus, Martynas 30217
Jannings, Emil 17069
Jans, Theo 32450
Janssen, Horst 20895, 22034
Jansson, Tove 11365
Japy, Frédéric 11890
Jarošenko, Nikola 37048
Jaroslavskij, Emeljan M. 36959
Jarquin, Tony 33276
Jászai, Mari 23337, 23788
Jaunius, Kazimieras 30303
Jaunsudrabiņ, Jāņis 30050
Jaurès, Jean 12318
Jauslin, Karl 41671
Javid, Huseyn 03157, 03184
Javornic'kij, Dmitrij Ivanovič 43035
Javorov, Pejo K. 05010, 05180, 05192
Jawlensky, Alexej von 17698, 17924, 22557, 41129, 41568
Jean, Émile 16245
Jean-Haffen, Yvonne 12747
Jeanne d'Arc 14537
Jedlik, Ányos 23434
Jefferies, Richard 43865
Jefferson, Thomas 47398
Jenewein, Felix 09613
Jenner, Edward 43454
Jennewein, C. Paul 53437
Jensens, Johannes Vilhelm 10017
Jerace, Francisco 25778
Jerome, Jerome K. 45919
Jesenský, Janko 38218
Jewett, Charles 51479
Jewett, Frank 51479
Jilemnický, Peter 38365
Jiménez, Juan Ramón 39653
Jindřich, Jindřich 09497, 09586
Jinnah, Mohammed Ali 34017
Jiquidi, Aurel 36646
Jirásek, Alois 09538, 09539, 09766
Jochumsson, Matthias 23987
Jodrell, Sir Alfred 44274
Johann, Archduke of Austria 02759, 02760
John, Augustus 45813, 45836
John, Gwen 45813, 45836
Johns, Jasper 26298, 28671, 49079, 50224
Johnson, Amy 44595
Johnson, Andrew 49050, 49052
Johnson, Anna C. 48998
Johnson, E. Pauline 06502
Johnson, Eyvind 40462
Johnson, Samuel 44706, 44839
Johnson, Sargent Claude 52764
Johnson, William H. 49079
Joja, Constantin 36185
Jókai, Mór 23352, 23467
Jomini, Antoine-Henri 41720
Jones, Anson 53945
Jones, Casey 49568
Jones, Inigo 44790
Jones, Lois Mailou 46942
Jones, Paul 01120
Jong, Adrianus Michael de 32573
Jonquieres, Eduardo 00521
Jónsson, Ásgrímur 24022, 24032
Jonsson, Einar 24021
Jónsson, Finnur 24022
Jonynas, Antanas 30205
Jonynas, Vytautas Kazimieras 30226
Jordaens, Jacob 17833
Jorn, Asger 10257
Josephsohn, Hans 41434
Jourdain, Francis 15236
Jourdan, Louis 13764, 15443
Jovellanos, Gaspar Melchor de 39335
Jovkov, Jordan 05020, 05269
Joyant, Jules-Romain 13781
Joyce, James 24803, 24804, 24837
József, Attila 23353, 23417, 23467, 23689, 23735
Juárez, Benito 31128, 31226, 31319
Jünger, Ernst 19828, 22570
Jünger, Friedrich Georg 22570
Jugan, Jeanne 12261
Jularbo, Carl 40449
Juliet, Charles 13795
Jun, Yong-Bok 28937
Jung-Dörfler, Karl 22589
Jungeblut, Nikolaus 19700
Juni, Juan de 39638
Junkers, Hugo 17835
Junqueiro, Guerra 35838
Jurkovič, Samuel 38354
Juvarra, Filippo 28004

K

Kabakov, Emilia 11899
Kabakov, Ilya 11899, 18712
Kableškov, Todor 05073
Kaempfer, Engelbert 19947
Kästner, Erich 17942
Kafenda, Frico 38310
Kafka, Franz 09745
Kahlen, Wolf 17290
Kahlo, Frida 31060, 31070, 31111, 31118
Kahn, Albert 12089
Kahnweiler, Daniel-Henry 11923, 21339
Kaim, Lorenz 19721
Kairis, Theophilos 22885
Kaiser, Georg 20125
Kalamaras, Dimitri 23002
Kalantiaw, Datu Benhadara 34344
Kallio, Kalervo 11144
Kallio, Kyösti 11144
Kallmann, Hans Jürgen 19364
Kálmán, Emmerich 23803
Kalpaks, Oskars 30152
Kaltenbach, Marianne 41940
Kalvoda, Alois 09839
Kamala, Šarif 37024
Kan, Kikuchi 29318
Kanakakis, Lefteris 23168
Kandinsky, Wassily 14719, 14722, 17041, 17834, 17997, 20534, 28671, 40915, 44830, 48191, 50224
Kandó, Kálmán 23434
Kaneshige, Toyo 29100
Kano, Hogai 29269
Kansetsu, Hashimoto 28820
Kantor, Tadeusz 34768, 34810
Kapoor, Anish 26881
Karalijčev, Angel 05173
Karavelov, Ljuben 05072
Karlstadt, Liesl 20493
Karsten, Ludvig 33752
Kašpar, Adolf 09639
Kaspar, Johann 20806
Kasprowicz, Jan 34690, 34691, 35356
Kass, János 23845
Kassák, Lajos 23418, 23467
Kassner, Rudolf 41866
Kastrioti, Gjergj 00026
Kastriotis, George 23145
Kat, Otto de 32522
Kato, Tokuro 29011
Katona, József 23648
Katsu, Kaishu 28679
Kaudzītes, Matiss 30031
Kaudzītes, Reinis 30031
Kauffmann, Angelika 02722
Kaufmann, Nikolaus 22488
Kaulbach, Friedrich 16670, 16671
Kaulbach, Friedrich August von 16671, 16670
Kaulbach, Hermann 16670, 16671
Kaulbach, Wilhelm von 16671, 20004, 16670
Kausopson, Kausop 38572
Kaván, František 09563
Kawaguchi, Kigai 29575
Kawai, Kanjiro 28821, 28839
Kawamura, Isamu 29630
Kawanabe, Kyosai 29576
Kazakov, Dimitar 05230
Kazantzakis, Nikos 23019, 23020
Kazinczy, Ferenc 23797, 23798
Kazuki, Yasuo 29269
Keats, John 27456, 44892
Kees, Peter 17088
Keith, William 50838
Kekkonen, Urho Kaleva 10834, 11204
Kekulé, Friedrich August 03636
Keller, Friedrich Gottlob 16825
Keller, Gottfried 41439
Keller, Helen 53624
Keller, Therese 41660
Kelley, Gary 47293
Kelley, Mike 17998
Kelly, Mary 43731
Kemal, Orhan 42833
Ķemers, Miervalž 30025
Kemp-Welch, Lucy 43699
Kempowski, Walter 22798
Kendall, Henry 01123
Kennedy, Edward M. 17216
Kennedy, Jacqueline 17216
Kennedy, John F. 17216, 47939
Kennedy, Robert F. 17216
Kenner, Rex 07199
Kenneth, King 24806
Kent, Corita 50218
Kepler, Johannes 19963, 21226, 22380
Kerényi, Jenő 23877
Kerg, Theo 21637
Kerner, Justinus 22432
Kernstock, Ottokar 01802

Kersting, Georg Friedrich 18829
Kette, Dragotin 38498
Khachaturyan, Aram 00726
Khare, G.H. 24355
Khnopff, Fernand 28512
Kiefer, Anselm 17123, 18024, 18712, 28671, 33717
Kienzl, Wilhelm 02479
Kierkegaard, Søren 10135
Kieślowski, Krzysztof 34877
Kießling, Franz 01830
Kilpi, Volker 11016
Kim, Jin-Hyuck 29796
Kim, N. 37638
Kim, Whanki 29933
King, Jessie M. 44037, 44611
King, William Lyon Mackenzie 06104
Kinoshita, Yuji 28443
Kinzing, Peter 20688
Kipling, Rudyard 38693, 43689, 44187
Kippenberger, Martin 17123
Kirchner, Ernst Ludwig 16592, 17058, 17172, 17319, 17698, 41321, 44156
Kirkland, Vance 48080
Kirov, Sergej Mironovič 37678
Kisfaludi Strobl, Zsigmond 23971
Kisfaludy, Károly 23825
Kisfaludy, Sándor 23344, 23825
Kishida, Ryusei 29269
Kita, Kazuaki 29261
Kitaoji, Rosanjin 28818, 28839, 29531
Kitchener, Horatio Herbert 43798
Kitching, James 38650
Kittelsen, Theodor 33608
Kiven, Aleksis 11315
Kivi, Aleksis 11185
Klee, Paul 17041, 17192, 17244, 17834, 17997, 19005, 28671, 28818, 33420, 41220, 41598
Kleist, Ewald Christian von 18404
Kleist, Franz Alexander von 18404
Kleist, Heinrich von 18404
Klenze, Leo von 21561
Klimt, Gustav 02620, 02951, 03003, 41885
Ključevskij, Vasilij Osipovič 37532
Ključnikov-Palantaj, Ivan 36977
Kløcker, Johannes Nicolai 33394
Klöpfer, Hans 01849
Klopstock, Friedrich Gottlieb 21150
Klotz, Matthias 20342
Klucis, Gustavs 30114
Kluk, Krzysztof 34608
Kmet, Andrej 38296
Kmetty, János 23877
Knapp, Franz 02503
Kneipp, Sebastian 16874
Knöpfer, Otto 18210
Knoll, Franz Xaver 22565
Knopp, Naum 00521
Knowles, Alison 21096
Knox, John 44129
Knox, Robert 44160
Kobayashi, Wasaku 28449
Kobell, Ferdinand 16590
Kobro, Katarzyna 34884
Koch, Alois 18850
Koch, Robert 17239, 35319
Kochanowski, Jan 34611
Kodály, Zoltán 23424
Kodishman, Menashe 29256
Koekkoek, Barend Cornelis 19568
Köleri, Johann 10668
König, Albert 22180
Koenig, Fritz 19805
Körner, Christian Gottfried 17950
Körner, Theodor 17950, 22637
Kogălniceanu, Mihail 36423
Kohán, György 23596
Koidulu, Lydia 10603
Kojima, Torajiro 28783, 28789
Kojoyan, Hakob 00725
Kokoschka, Oskar 02505, 02620, 02951, 42083
Kolbe, Georg 17117
Kolbenheyer, Erwin Guido 18602
Kolcov, Aleksej Vasiljevič 37972
Kolig, Anton 02451
Kollár, Ján 38308
Kollwitz, Käthe 17151, 19610, 20387, 28590
Kolstø, Fredrik 33537
Kommerell, Max 20500
Kondo, Yuzo 28828
Konenkov, Sergej Timofeevič 37246, 37771, 37778
Koniarek, Ján 38379
Kono, Takashi 41298
Konopnicka, Maria 34701, 35143
Konstantinov, Aleko 05221
Koons, Jeff 26881, 33717
Korin, Pavel Dmitrievič 37232
Korolenko, V.G. 36922
Korolev, S.P. 37229, 37337
Koschat, Thomas 02184
Kosciuszko, Tadeusz 41884
Kossuth, Joseph 21948
Kossuth, Lajos 23503, 23717
Kostede, Ulrich 19418
Kostka, Ferdiš 38363
Kostov, Dončo 05083
Koszta, József 23609, 23884
Kotopouli, Marika 23248
Kounellis, Jannis 26881
Kovács, Margit 23584, 23875
Kovalevskaja, Sofja Vassilievna 37570
Koyama, Yushi 28443
Kozin, Vadim Alekseevič 37120
Kozlov, Petr Kuzmič 37660
Kraljević, Miroslav 08930, 09003
Kramskoj, Ivan Nikolaevič 37519, 37520
Kranewetter, Franz Josef 02421
Krasilnikov, Gennadij Dmitrievič 36755, 36954
Krasiński, Zygmunt 34982
Kraszewski, Józef Ignacy 17949, 35086
Kraus, Joseph Martin 17615
Kreibig, Erwin von 20437
Kreutzer, Conradin 20307
Kreutzwald, Friedrich Reinhold 10676
Krévé-Mickevičius, Vincas 30447
Kricke, Norbert 17407
Kristiansen, Herleik 33448
Kristof, Agota 41215
Křižík, František 09715
Kroeyer, P.S. 10261
Krohg, Christian 33419
Kromer, Marcin 34554
Kropp, Georg 22728
Krúdy, Gyula 23733
Krüger, Franz 17032
Kruger, Paul 38748
Krumbacher, Karl 22565
Krupp, Alfred 18254
Krupp von Bohlen und Halbach, Alfried 18254
Krupp von Bohlen und Halbach, Gustav 18254
Krupp, Friedrich 18254
Krupp, Friedrich Alfred 18254
Kruse, Käthe 16763, 16776, 17897
Ku, Chieh Kuang 42417
Kuba, Ludvík 09418
Kubin, Alfred 02290, 02317, 02925
Kudirka, Vincas 30290
Kügelgen, Gerhard von 17950
Kues, Nikolaus von 17296
Kuindži, Archip Ivanovič 37693
Kuipers, Andre 32494
Kulisiewicz, Tadeusz 34707, 34710
Kullrich, Friedrich-Wilhelm 17774
Kumičić, Eugen 08849
Kunanbaev, Ibrahim Abaj 29669, 29729
Kuniyoshi, Yasuo 29100
Kunz, Emma 42047
Kupala, Janka 03251
Kupka, František 09764
Kuprin, Aleksandr Ivanovič 36919, 37445
Kuri Breña, José 31579
Kuroda, Kiyoteru 28628, 29632
Kuroda, Seiki 29426
Kurts, R. 31604
Kurz, Alfred 02297
Kurz, Hermann 19549, 21283
Kusama, Yayoi 29395
Kuster, Walter 41352
Kustodiev, Boris Michajlovič 37521
Kusube Yaichi 29066
Kuzmickis, Jonas 30212
Kvapil, Jaroslav 09487
Kwon, Jung-Ho 29796
Kyber, Manfred 20020
Kylych, Moldo 29955

L

La Fontaine, Jean de 12440
La Pérouse, Jean-François de Galaup de 11538
La Tour, Georges de 12840
La Villeglé, Jacques de 12840
Laban, Rudolf 41127
Labiş, Nicolae 36458
Labisse, Felix 13953
Labrador, José María 39396
Lachner, Franz 21168
Lachner, Ignaz 21168
Lachner, Vinzenz 21168
Laclos, Choderlos de 11587
Lacrocq, Louis 13159
Ladegast, Friedrich (Orgelbauer) 22459
Laemmle, Carl 19859
Lafage, Raimond 13786
Lafran, Paul 15200
Lagerlöf, Selma 40515, 40799
Laib, Wolfgang 21948
Lam, Wifredo 09176, 09292
Lamartine, Alphonse de 05129, 11522, 13908, 14815
Lamb, William 45174
Lambert, Margaret 45944
Lamennais, Felicité 15389
Lampi, Vilho 11073
Lamprecht, Hermann 19325
Landenberger, Christian 16453
Landowski, Paul 12091
Lang, Fritz 17069
Lange, Joseph 20661
Langenberg, Ferdinand 18666
Langenhoven, Cornelis Jacob 38696
Lansyer, Emmanuel 13794
Lanza, Mario 51850
Lao She 07545
Lapine, André 06293
Lara, Agustín 30737
Larbaud, Valéry 16144
Larese, Dino 41112
Largillière, Nicolas de 15829
Larionov, M. 37765
Laronze, Jean 12416
Larreta, Enrique 00181
Larrey, Dominique 11891
Larsen, Johannes 10116
Larsen, Thoger 10176
Larsson, Carl 40956
Lasker-Schüler, Else 21802
Lattement, Georges 16140
Lattre de Tassigny, Jean de 14300
Lauer, Nikolaus 21514
Lauge, Achille 13773
Laurel, José P. 34502
Laurel, Stan 45897
Laurel, Sten 21803
Laurence, Margaret 06421
Laurencin, Marie 04049, 28400
Laurens, Henri 16202
Laurier, Wilfrid 07261
Lauritzen, Jørgen 33977
Lavant, Christine 02191
Lavoisier, Antoine Laurent de 12073
Lawrence, D.H. 44108, 44109
Lawrence, Thomas 53253
Lawson, Henry 01133
Lazarević, Laze K. 38088
Lazaris, Theodoros 23079
Lázaro, Franco 30763
Lazykin, Aleksej 27800
Le Corbusier 14601, 14803, 15241, 21968, 41400, 42075
Le Fort, François 41735
Le Lievre, Peter 45591
Le Lorrain 12367, 12840
Le Maire, Jacob 15390
Le Mayeur, Adrien 24552
Leach, Bernard 15986
Leacock, Stephen 06515
Léandre, Charles 12758
Leber, Hermann 20326
Leblanc-Duvernoy, M. 11777
Lechuga, Ruth 31061
Leck, Bart van der 32642
Leclerc, Marc 13392
Leclerc, Philippe François Marie 14626
Lecoq, Henri 12552
Ledoux, Claude Nicolas 11661
Lee, Henry 54117
Lee, Robert E. 46357, 52335
Lee, Tze-Fan 42210
Lee, William 54117
Leger, Alexis 23277
Léger, Fernand 04879, 11998, 13122, 13256, 14722, 15252, 16202, 19005, 40915
Legge, Petrus 17499
Legler, Thomas 41326
Lehár, Franz 01754, 03021
Lehmann, Lotte 21013
Lehmann, Wilhelm 41685
Lehmbruck, Wilhelm 18027
Lehmden, Anton 01815
Leibl, Wilhelm 17032
Leibniz, Gottfried Wilhelm 22653
Leidesdorff, Alexander 52764
Leighton, Alfred Crocker 05522
Leighton, Barbara 05522
Leighton, Frederic Lord 44899
Leistikow, Walter 34578
Leitz, Ernst 22532
Leković, Veliš 31635
Lemire, Jules 16157
Lemonnier, Camille 03456
Lenau, Nikolaus 02782, 36438
Lenbach, Franz von 20491, 21639
Lenin, Vladimir Iljič 11353, 14717, 36926, 37005, 37327, 37614, 37897, 37903, 37906, 37907
Lenné, Peter Joseph 17249
Lennon, John 18889, 29187, 44731, 44737
Lenz, J.M.R. 18146
León, Luís de 39021
León y Castillo, Fernando de 40112
León y Castillo, Juan 40112
Leonardo da Vinci 04076, 11575, 11578, 26157, 26713, 27096, 28283, 28284, 28285, 34803, 44937, 48935
Leonhardi, Eduard 17957
Leopardi, Giacomo 27375
Lepe, Manuel 31297
Leppänen, Lauri 10842
Lerch, Maria 19168
Lermontov, Michail Jurjevič 37100, 37165, 37195, 37439, 37564, 37658
Leskov, Nikolaj Semenovič 37505, 37510
Lessing, Gotthold Ephraim 19420, 22653
Lesueur, Charles Alexandre 13599
Lettl, Wolfgang 16627
Levi, Carlo 25201
Levitan, Isaak Iljič 37566, 37808
Levski, Vasil 05044, 05050, 05087
Lewis, Sinclair 52929
Lewis, Wyndham 45021
Leyden, Lucas van 15829, 32489
Leyster, Judith 32295
Lhote, André 12069
Liapis, Kostas 22870
Lichtenstein, Roy 15252, 29297, 50224, 50891
Lichtner-Aix, Werner 15723
Liebermann, Max 17032, 17177, 17186, 21459
Liebig, Justus von 18639
Liebknecht, Karl 20040
Lightner, Otto C. 52522
Ligne, Charles-Joseph de 03358
Liipola, Yrjö 10985
Lilienthal, Otto 16551, 18705
Lima Couto, Vasco de 35549
Lin, Yutang 42401
Lincoln, Abraham 49191, 49307, 50427, 50882, 53217
Lindbergh, Charles A. 50133
Lindemann, Frederick 45679
Lindemann, Gustav 18010
Lindgren, Armas 10963
Liner, Carl August 41117, 41119
Liner, Carl Walter 41117, 41119
Liniers, Virrey 00120
Linley, Sambourne 44901
Linnankoski, Johannes 10716
Linné, Carl von 40487, 41007, 41008, 41009
Linnqvist, Hilding 40871
Lipatti, Dinu 36439
Lipchitz, Jacques 30223
Lippi, Filippo 26054, 26157, 26323, 26580, 27333, 27451
Lippmann, Johannes 18322
Lipsi, Morice 15106
List, Friedrich 21283
Lister, Joseph 44160
Liszt, Franz 01861, 02553, 16951, 22409, 23431
Litherland, Victor 01011
Livadićca, Ferde 08944
Livingstone, David 43542, 45491, 54655
Lo-Johansson, Ivar 40901
Lobo, Baltasar 40333
Locmanov, Andrej Vasilevič 36897
Löffler, Vojtech 38272
Löhe, Wilhelm 20607
Lönnrot, Elias 11298
Löns, Hermann 22316
Loetscher, Hugo 41215
Löwen, Axel von 21914
Logan, Andrew 45961
Loganathan, K. 24428

Lomonosov, Michail Vasilevič 37112, 37697
London, Jack 48919
Long, Richard 12062, 17123
Longo, Alessandro 25219
Longobardi, Nino 26014
Lonhy, Antoine de 28004
Loo, Jacob van 32788
Lope de Vega, Félix 39504
López de Legazpi, Miguel 40363
López y Planes, Vicente 00217
López Torres, Antonio 40143
López Velarde, Ramón 30973
Loppé, Gabriel 12381
Lorrain, Claude 14354, 27451, 45378
Loseleur, Hector 13953
Loti, Pierre 15073, 15462
Lotter, Hieronymus 18631
Lotto, Lorenzo 25406, 27379, 28188
Louis, Morris 50224
Louis XI 11578
Louis XIV 14160, 15688
Loukine, Rostilas 11704
Lowry, L.S. 45599, 45805
Lozano Sidro, Adolfo 39870
Lu Xun 07547, 07731, 08173, 08174, 08211
Lubbock, Sir John 45321
Lucaciu, Vasile 36624
Luchian, Ştefan 36162, 36525, 36646
Ludwig II, King of Bavaria 18279, 18540, 21669
Ludwig Leiner 19677
Ludwig, Otto 18108
Ludwig Rudolf, Duke of Braunschweig-Lüneburg 17373
Luebbert, Ernst 21870
Luise, Queen of Prussia 19523
Luna, Juan 34339, 34414
Luna, Pintor Antonio Rodríguez 39682
Lurçat, Jean 11904, 15241
Lutes Sikorski, Muriel 06318
Luther, Martin 18095, 18099, 20110, 20111, 20115, 20116, 20177, 20379
Luts, Oskar 10654
Lutz, Joseph Maria 21018
Luyten, Hendrik 32698
Lyautey, Louis Hubert Gonzalve 15899
Lyon, Danny 51138
Lyssenko, Andrej 27800

M

Mabille, Ernest 11578
Mabini, Apolinario 34429, 34515
McCahon, C. 33154
McCallum, Jane 46502
McCarthy, Paul 17123, 17998
McCartney, Paul 18889, 44731, 44739
McClung, Nellie 05872
McCouch, Gordon 41129
McCrae, Georgiana 01245
McCrae, John 05918
McCulloch, Thomas 05949, 06602
MacDiarmid, Hugh 43480
Macdonald, Sir John A. 06074, 06416
MacDonald, Wilson 06944
McDouall, John 44608
Macé, Jean 11915
Maceo, Antonio 09306
McGillivray, William 07078
Mácha, Karel Hynek 09494
Machado y Ruiz, Antonio 40049
Machar, Josef Svatopluk 09592
Macho, Victorio 40140
McIndoe, Ken 46224
Mačiulis, Jonas 30211, 30259
Mack, Alexander 16691
Macke, August 17447, 17460, 17924
Mackensen, Fritz 22696, 22699
MacKenzie, Sir Alexander 06575, 07078
McKenzie, Robert T. 05323
Mackenzie, William Lyon 07120
Mackillop, Sister Mary 01374
Mackintosh, Charles Rennie 44284, 44438
Mackintosh, Margaret MacDonald 44438
McLean Fraser, C. 07199
McMillan, Angus 01309
McNaughton, Daniel 43429
Macovei, Ligia 36162, 36646
Macpherson, Allan 06416
McTaggart-Cowan, Ian 07199
Maczek, Stanisław 32004
Madách, Imre 38249
Madigan, Elvira 10294
Madison, James 50827
Märkli, Peter 41434
Maeterlinck, Maurice 03633
Maeztu, Gustavo de 39286
Magalhães, João Jacinto de 35533
Magellan, Fernão de 15390
Magnasco, Alessandro 26316, 26323, 26324
Magnelli, Alberto 15252, 16050
Magritte, René 14722, 17244, 28512, 44156, 50224
Magsaysay, Ramón 34456
Mahaprabhu, Chaitanya 24430
Mahé de la Bourdonnais, Bertrand 15389
Mahen, Jiří 09451
Mahler, Gustav 02765, 09541, 09560
Mahringer, Anton 02451
Maillol, Aristide 14719, 14791
Mainssieux, Lucien 16267
Maiorano, Serafino 26014
Maironis 30211, 30259
Maistorov, Nikolai 05230
Majakovskij, Vladimir Vladimirovič 37199, 37355
Majorelle, Louis 14358
Makarenko, A.S. 37397
Makariopolski, Ilarion 05027
Makarova, N.V. 36982
Makart, Hans 03003
Makowski, Tadeusz 14634
Makuszyński, Kornel 35358
Malfatti, Anita 04879
Malhoa, José 35497, 35662
Mali, Christian 17319
Mallarmé, Stéphane 16280
Malraux, André 13598
Malura, Oswald 22178
Malyškin, A.G. 37138
Mameli, Goffredo 26301
Mamin-Sibirjak, Dmitrij N. 36887, 36891, 36897
Mammadguluzade, Jalil 03152
Manansala, Vicente 34346
Manby, George William 44360
Mandela, Nelson 38770
Mandelštam, Ossip 37355
Mané-Katz, Emmanuel 24964
Mánes, Josef 09452
Manet, Edouard 13043, 14354, 45920
Manfroce, Nicholas Antonio 27086
Manjarris, Michael 49533
Manley, Victor 45932
Mann, Golo 41215, 41503
Mann, Heinrich 09809, 20073
Mann, Thomas 09809, 20073, 41503, 42111
Mannerheim, Carl Gustaf Emil 10715, 10804, 11116
Manrique, César 40108
Mansart, François 13925
Mansfield, Katherine 33248
Manship, Paul 52624
Mantegna, Andrea 25406, 44999
Manzi, Homero 00250
Manzoni, Alessandro 26477, 26710
Manzoni, Piero 10071
Manzu, Giacomo 27454
Mao Zedong 08521
Mara, Pol 13085
Marc, Franz 19592
Marcello d'Chevrolet, Louis 41735
Marchese, Victor 00521
Marcks, Gerhard 17551
Marconi, Guglielmo 26436
Marey, Etienne-Jules 11905
Mariátegui, José Carlos 34207
Marini, Marino 26193, 27253
Marks, Sammy 38680
Marley, Bob 28324
Marmottan, Paul 12087
Marot, Clément 12242
Marquet, Albert 12069
Márquez Figueroa, José 31270
Marshall, Elisha 46502
Marsili, Luigi Ferdinando 25473
Marti, Hugo 41561
Martí, José 09126
Martin, Alexandre-Louis 03509
Martin-Bates, Pat 07253
Martin, Étienne 14478
Martin, Henri 12242
Martin, John 45229
Martin, Jonathan 43429
Martin, Sarah 44361
Martínez, Antonio 39906
Martínez Checa, Francisco 39396
Martínez, Efrain 08728
Martínez Murguía, Manuel 39764
Martini, Arturo 26289, 28069
Martinů, Bohuslav 09727
Maruja Díaz 40310
Marx, Enid 45944
Marx, Karl 22100
Mary, Queen of Scotland 44548
Masaccio 26157
Masaryk, Tomáš Garrigue 09453, 09519, 09618
Maside, Carlos 38888
Mason, Raymond 08675
Massari, Giuseppe 27004
Massier, Clément 13373
Massijs, Jan 26323
Masson, André 17184
Masson, Arthur 04094
Masurovsky, Gregory 14909
Mataré, Ewald 19567
Matejko, Jan 34769
Mathieu, Henri 12202
Matisse, Henri 13569, 14435, 14674, 14719, 14722, 14909, 15252, 16095, 17192, 40915, 41763, 41885, 45920, 50510, 52717
Matsumoto, Shunsuke 28703
Matsuo, Basho 28679
Matta, Roberto 08675
Matter, Herbert 41352
Matthiesen, Oscar 41079
Mattucci, Serafino 27559
Matulionis, Povilas 30293
Matutis, Anzelmas 30204
Mauersberger, Erhard 18804
Mauersberger, Rudolf 18804
Maura, Antonio 39559
Mauracher, Hans 01973
Maximilian I., Emperor 02124
Maximilian, Archduke of Austria 02115
Maximilian, Emperor of Mexico 02051
May, Karl 19244, 21160
Maydell, Ernst von 18734
Mayer-Franken, Georg 18337
Mayer, Tobias 20184
Mayrhuber, Sepp 01958
Mazies, Victor 13786
Mazoyer, Georges 11636
Mazzini, Giuseppe 26301, 27238, 41454
Mazzola, Francesco 26221, 27096
Meckenem, Israhel van 17396
Medina, Henrique 35480, 35567
Medrea, Cornel 36199, 36646
Medvecká, Mária 38383
Méheut, Mathurin 13492
Mehoffer, Józef von 34770
Meiche, Alfred 21710
Meier, Gerhard 41215
Meissonier, Ernst 53253
Meistermann, Georg 21802, 22623
Mejerchold, Vsevolod E. 37190, 37319, 37535
Melanchthon, Philipp 17584, 20116
Melantrich z Aventýna, Jiří 09823
Melba, Dame Nellie 01230, 01290
Melchior, Johann Peter 21191
Meletta, Carlo 41574
Melngail, Emil 30147
Melville, Herman 51965
Memling, Hans 03417, 28252
Mendel, Johann Gregor 09430, 09435
Mendeleev, Dmitri Ivanovič 36811, 37671
Mendelssohn Bartholdy, Felix 19924
Menéndez y Pelayo, Marcelino 40008
Menon, Vengalil Krishnan Krishna 24285
Menzel, Adolph von 17032, 17172
Mercator, Gerhard 03991, 04021, 18020
Mergenthaler, Ottmar 16784
Merkurov, Sergey Dmitrievich 00708
Merling, Hans 26323
Mernoz, Jean 11724
Merz, Mario 12062, 17123
Merz, Toni 21515
Mesdag, Hendrik Willem 32088, 32098, 32899
Mesonero Romanos, Ramón 39503
Messina, Francesco 26289
Messter, Oskar 17807
Meštrović, Ivan 08957, 08987
Metzkes, Harald 17106
Meucci, Antonio 53262
Meyer-Amden, Otto 41080
Meyer, Conrad Ferdinand 41503
Meyer, Hannes 17834
Meyer, Joseph 19179
Meyer, Traugott 41090
Meyerhold, Vsevolod E. 37190, 37319, 37535
Michael I, King of Romania 36592
Michelangelo Buonarroti 26141, 26143, 26159, 26176, 45327
Michelangelo Buonarroti (the Younger) 26143
Michelena, Arturo 54535
Michelet, Edmond 12188
Michelsen, Christian 33767
Michener, Roland 06125
Michoutouchkine, Nikolai 54494
Mickiewicz, Adam 14627, 30442, 35256, 35369
Micle, Veronica 36674
Miegel, Agnes 16793
Mieghem, Eugeen van 03285
Mies van der Rohe, Ludwig 09442, 17041, 17316
Migishi, Kotaro 29219
Mihelič, France 38397
Mikelson, Arnold 06996
Miki, Paul 28968
Miklos, Barabas 36601
Mikszáth, Kálmán 38351
Milcendeau, Charles 15799
Miles, Nelson 49223
Milev, Geo 05209
Millais, Sir John Everett 45569
Miller, Hugh 43928
Milles, Carl 40673
Millet, Jean-François 11588, 11846, 13145, 28754, 28880
Millian, Claudia 36188
Milton, John 43787
Miner, Jack 06093
Minguzzi, Luciano 26709
Minkowski, Maurice 00251
Minogue, Kylie 01290
Minulescu, Ion 36188
Mio, Vangjush 00024
Mirabeau, Honoré Gabriel Riqueti Comte de 11674
Miranda, Carmen 04702
Miró, Joan 11549, 15252, 16202, 19204, 27454, 38957, 39783, 39787, 40317, 44156, 48191, 50224
Mistral, Frédéric 13917
Mistral, Gabriela 07473
Mitchell, Joan 50952
Mitterhofer, Peter 27091
Mitterrand, François 14962
Miyagi, Michio 29438
Miyawaki, Aiko 29621
Miyonaki, Aiko 29256
Mkrtchyan, Mher 00706
Model, Lisette 51194
Modena, Barnaba da 26309
Modersohn-Becker, Paula 17557, 22691, 22696, 22699
Modersohn-Breling, Louise 22503
Modersohn, Otto 20954, 22503, 22696, 22699
Modigliani, Amedeo 04879, 14354, 14674, 16202, 27454, 45920
Modrzejewska, Helena 34810
Moe, Jørgen 33559
Möbius, Karl 19536
Mörike, Eduard 16783, 17366, 20611, 20713
Moffet, Joseph 07266
Moholy-Nagy, László 17834
Mohr, Joseph 02047, 02356
Mohs, Carl Friedrich 18421
Mokichi, Saito 28650
Moldagulova, Alia 29672
Moldo, Togolok 29955
Molengraaff, G.A.F. 32061
Molière, Jean-Baptiste 14808
Molina Campos, Florencio 00473
Molnár, Ferenc 23467
Moltke, Helmuth von 20982
Molwitz, Herbert 20344
Mommsen, Theodor 18539
Monaco, Lorenzo 26054, 26126
Mončys, Antanas 30327
Mondrian, Piet 31813, 32073
Monet, Claude 12073, 13067, 14354, 14674, 14720, 15252, 16123, 26580, 27454, 28628, 42083, 44138, 46309
Moniuszko, Stanisław 35216
Moniz, Egas 35570
Monkute-Marks, Janina 30270
Monmouth, James 43606
Monnet, Jean 11884
Monro, Thomas 43699
Monroe, James 48768

Montag, Karl 22355
Montandon, Paul 41217
Montecuccoli, Raimund von 02036
Monteiro, Vincente do Rego 04667
Montgolfier, Etienne de 11631
Montgolfier, Joseph de 11631
Montgomery, Lucy Maud 05593, 05594
Monţia, Emil 36623
Monticelli, Adolphe 31898
Montilla, Carmen 09116
Montoya, Baudilio 08626
Moon, Sarah 28818
Moore, Henry 10103, 10772, 18712, 28476, 32642, 45285, 45409, 45908
Móra, Ferenc 23666, 23667
Moral, Jean 13953
Morales, Tomás 39688
Moreau Braun, James 48842
Moreau, Gustave 14711
Moreau de Maupertuis, Pierre-Louis 15389, 41333
Morelli, Nicolò 26138
Morelos y Pavón, José María Teclo 30813
Moretti, Marino 25838
Morgan, William 43990
Morgner, Wilhelm 21799
Mori, Ogai 28886
Móricz, Zsigmond 23785, 23929
Moritz, Klaus 16596
Moroles, Jesus Bautista 49533
Morris, William 44554, 45015
Morse, Samuel F.B. 52122
Mosen, Julius 21055
Moser, Barry 47782
Mosset, Olivier 11899
Mossman, James 44129
Motherwell, W.R. 05308
Motzfeld, Benny 33925
Moulin, Jean 12065, 14626
Mourik, Jacques van 32561
Mozart, Anna Maria 02648
Mozart, Leopold 02618, 02619, 16630
Mozart, Marie Anna 02618, 02619, 02648
Mozart, Wolfgang Amadeus 02618, 02619, 02648, 03028, 09787, 16630
Mucha, Alphonse 09665, 09763, 32870
Muche, Georg 17041
Muchin, Igor 37254
Muchin, Nikolaj Semënovič 36976, 37494
Muehl, Otto 03133
Mühsam, Erich 41127
Müller-Brauel, Hans 22797
Müller, Carl Otto 18080
Müller, Elisabeth 41529
Müller, Gerhard Kurt 19910
Müller, Gottlieb 41260
Müller-Guttenbrunn, Adam 36740
Müller, Joachim Eugen 41352
Mueller, Otto 17058
Müller, Robert 41885
Müller-Scheeßel, Ernst 21536
Münchhausen, Hieronymus Karl Friedrich von 17414, 19754, 30006
Münster, Mia 21514
Münster, Sebastian 19335
Münter, Gabriele 17698, 20534, 20535
Müntzer, Thomas 16475, 19116, 20406, 21907
Münzner, Rolf 05016
Muir, Thomas 43521
Mukai, Junkichi 29440
Mukanov, Sabit 29681
Mukhtar, Mahmoud 10527
Mulders, Marc 32849
Multscher, Hans 28288
Munakata, Shiko 28361
Munch, Edvard 33403, 33419, 33420, 33636, 33732, 33735, 33752, 42083
Muñez Seca, Pedro 39879
Munirathinam, B. 24428
Munkacsi, Martin 51194
Munkácsy, Mihály 23367, 23370, 23371, 23467
Munnings, Sir Alfred 43961
Muñoz Marín, Luis 36063
Muñoz Rivera, Luis 36027
Munro, Alice 07349
Munthe, Axel 25224
Murakami, Kagaku 28818
Muraru, Nica 36162
Murat, Joachim 13462
Muratori, Ludovico A. 26755
Mureau, Marc 15021
Murer, Heinrich 42001
Murgaš, Jozef 38369
Murillo, Bartolomé Esteban 12320, 27453, 40065
Murnau, Friedrich Wilhelm 17069
Musäus, Johann Karl August 22405
Muschg, Adolf 41215
Music, Zoran 26351
Musil, Alois 09919
Musil, Robert 02191
Musorgskij, Modest Petrovič 37447
Musrepov, Gabit 29681
Muthesius, Hermann 17280
Myers, W.J. 44190
Mykolaitis-Putinas, Vincas 30310, 30371
Mylona, Alex 22928
Myung Hwan Kim 29781

N

Nabokov, Vladimir V. 37603, 37701
Nacura, Julian 28968
Näf, Matthias 41700
Näver, Pälle 41057
Nagaoka, Kounito 05016
Nagasawa, Hidetoshi 26014
Nagel, Gustav 16568
Nagy, István 23609
Nagy, Lajos 23336
Nagy, László 23621
Nakama, Satsuo 28631
Namora, Fernando 35547
Nandino, Elias 30779
Nansen, Fridtjof 33725, 33919
Napoléon I 01812, 03616, 03823, 11524, 11642, 11780, 12049, 12177, 12305, 12866, 12966, 13236, 13757, 15139, 19941, 27308, 27309, 27511, 27989, 41798
Napoléon III 12966, 41798
Napotnik, Ivan 38524
Narimanov, Nariman 03165
Nash, John 43625
Nash, Thomas 45783
Nast, Thomas 50891
Nasyri, Kajuma 37018
Natsagdorj, Dashdorjiyn 31629
Naumann, Bruce 17123
Naumann, J.F. 19667
Naumburger Meister 20554
Neamţu, Celia 36162
Necker, Jacques 41313
Negreiros, Almada 35698
Negri, Costache 36315
Negri, Pola 34877
Negroli, Filippo 39586
Negruzzi, Costache 36398
Nehru, Jawaharlal 24152
Neidhardt von Gneisenau, August Wilhelm Anton 21548
Nekrasov, Nikolaj Alekseevič 36746, 36871, 37002, 37661
Nelimarkka, Eero 10708
Nelkenmeister 42083
Nelson, George 22379
Nelson, Lord Horatio 45168, 45437
Němcová, Božena 09456, 09459, 09684
Němejc, Augustin 09673
Nemirovič-Dančenko, Vladimir I. 37320, 37336, 43152
Nerdrum, Odd 33636
Neris, Salomeja 30266, 30300
Nernst, Walther 18695
Neruda, Pablo 07470
Nervo, Amado 31439
Nesch, Rolf 33378, 33752
Netti, Francesco 25371
Netzer, Don 47782
Neuber, Friederike Caroline 21246
Neukams, Gottfried 19721
Neumann, Balthasar 22713, 22720
Neumayer, Georg von 16716
Neumeier, John 18919
Neuss, Wolfgang 17124
Neve, Hanneke de 46224
Neville Ure, Percy 45472
Newton, Helmut 17137, 17202
Newton, Isaac 44345
Newton, John 45316
Nezabitauskaite-Galauniene, Adele 30238
Niazy, Hamza Hakim-Zadé 54447, 54455
Nicholson, Ben 41129, 44557, 45792
Nicolaus de Cusa 17296
Niebuhr, Carsten 20269
Niekerk, Chris Van 38572
Niel, Adolphe 14335
Nielsen, Carl 09968, 10210
Nielsen Hauge, Hans 33792
Nielsen, Jens 10092
Niemeyer-Holstein, Otto 19692, 21394
Niemeyer, Oscar 04694
Nietlispach, Frieda 41940
Nietzsche, Friedrich 20555, 22412, 41870
Nightingale, Florence 44849
Nikitin, Ivan S. 37970
Nilsen, Torstein 33448
Nitsch, Hermann 02386, 03133
Nivelt, Roger 14481
Noakowski, Stanisław 34946
Nobel, Alfred 40632, 40921
Nobelius, Carl Axel 01070
Nobiling, Hans 04852
Noël, Ann 21096
Noguchi, Hideyo 28571, 50224
Noguchi, Yataro 28987
Nøis, Hilmar 33783
Nolan, Sidney 00915, 01581
Noland, Kenneth 50224
Nolde, Emil 17058, 17229, 19005, 20625
Nomellini, Plinio 26289
Nomura, Tokushichi 28857
Nonagase, Banka 29575
Nonomura, Ninsei 28367
Nordenskiöld, Adolf Erik 11151
Nordheim, Sondre 33682
Nordman, Maria 17407
Norton, Louis 49727
Nosilov, Konstantin Dmitrievič 36897
Nottara, Constantin C. 36206
Nottara, Constantin I. 36206
Novák, Vítězslav 09570
Novákova, Tereza 09809
Novalis 19388, 22541
Nowowiejski, Feliks 34535
Nüesch, Jacob 41151
Nunes, Clara 04379
Nunes Corrêa, Manuel 35812
Núñez del Prado, Marina 04209, 34261
Núñez, Rafael 08617
Nurmi, Paavo 10822
Nussbaum, Felix 20921
Nykänen, Matti 10822

O

Oates, Frank 45632
Oates, Lawrence 45632
Oberkampf, Christophe-Philippe 13286
Oberlin, Jean-Frédéric 16281
Oberth, Hermann 18310
Obregón, Alvaro 30942
Ocker, Dirk 32561
O'Connell, Daniel 24749
O'Connor, James 24911
Oeben, Jean François 14734
Oelze, Richard 22699
Oepts, Willem 33019
Ôgai, Mori 17190
Ogihara, Morie 28549
Ohara, Magosaburo 28789
Ohlmann, Dan 13897
Ohmeyer, Maria von 02518
Ohnesorge, Ilse 21710
Oka, Shikanosuke 29269
Okakura, Tenshin 28707
O'Keeffe, Georgia 52717
Okudžava, Bulat 37355, 37541
Ólafsson, Sigurjón 24023
Olbracht, Ivan 09836
Oldenburg, Claes 50224
Oldenburg, Ernst 22176
Oliveira Martins, Francisco Ernesto de 35437
Oller, Francisco 36029
Ollier, Léopold 13721
Olsson Hagalund, Olle 40871
O'Neill, Eugene 47959
O'Neill Verner, Elizabeth 47363
Onley, Toni 07253
Ono, Yoko 21096
Onofre, João 35698
Onufri 00021
Opitz, Theodor 41561
Oppenheim, Meret 41215, 41707, 41885
Opsomer, Isidore 03820
Al-Oraifi, Rashid 03213
Orff, Carl 17851
Orkan, Władisław 34944, 35070
Orozco, José Clemente 30885, 31070, 31106
Orthman, Thomas 08675
Ortiz Echagüe, Antonio 00295
Osaragi, Jiro 29595
Osborn Cunningham, Marion 46536
Ost, Alfred 04171, 32700
Osthaus, Karl Ernst 18857
Ostrovskij, Aleksandr Nikolaevič 37166, 37190, 37756
Ostrovskij, Nikolaj Alekseevič 37198, 37782, 43174
Ostwald, Wilhelm 18783
O'Sullivan, Timothy 51138
Oteiza, Jorge 40317
Otero Pedrayo, Ramón 38869
Othón, Manuel José 31352
Otto King of Greece 20957
Otto, Nicolaus August 19258
Oudry, Jean-Baptiste 14734
Overbeck, Fritz 17559, 22696, 22699
Overbeck-Rohte, Hermine 17559
Owen, Daniel 45163
Owen, Robert 45203, 45258

P

Pacheco, Francisco 12320
Paderewski, Ignace J. 41645
Padilla, José 39505
Padrón, Antonio 39319
Paegles, Leon 30148
Pagano, Giuseppe 26327
Pahl, Manfred 20138
Paik, Nam June 17998, 21096, 29621, 29945
Palaces, José Ángel 33299
Palacios, Alfredo L. 00158
Palacios, Pedro 00407
Palacký, František 09520
Paladino, Mimmo 26881
Palamas, Kostis 22948
Palitzsch, Johann Georg 17947
Palma, Jacopo 27034
Palma, Ricardo 34210
Palmié, Johann Charles 16596
Panaitescu-Perpessicius, Dumitru 36170
Pancera, Ella 01752
Pander, Pier 32477
Panetta, Pasquale 26506
Panger, Grete 02691
Panichi, Roberto 27800
Paniker, K.C.S. 24404
Pankok, Bernhard 20519
Pankok, Otto 16683, 19293, 21802
Pann, Anton 36550
Pannini, Giovanni Paolo 27182
Panton, Verner 22379
Paoli, Pascal 14285
Paolini, Giulio 26881
Papanastassiou, Alexandros 23076
Papineau, André-Benjamin 06147
Papio, Stanley 49758
Pappenheim, Gottfried Heinrich zu 20981
Papst, Georg Wilhelm 17069
Paracelsus, Theophrastus 41139
Paradise, Phil 46536
Parajanov, Sergei 00753
Pardo Bazán, Emilia 39214
Parks, Gordon 51194
Parmigianino 26221, 27096
Parnell, Charles Stewart 24889, 24890
Parrish, Clara Weaver 53032
Parry, Joseph 45131
Parthenis, K. 23228
Pascali, Pino 27271
Pascoli, G. 25359
Pasqualini, Alessandro 19405
Pass, Alfred A. de 38597
Pasternak, Boris L. 36866, 37195, 37355, 37540
Pasteur, Louis 11657, 12552, 12755, 13755, 14735
Pasteur, Vincent 13272
Patellani, Federico 25896
Paterson, Toby 44281
Patrick, Noel 01110
Patterson, Ben 21096
Pattison, Jim 06205
Patton, George S. 30474
Pátzay, Pál 23637
Paukštelis, Juozas 30271
Paul, Alice 53916
Paul, Jean 16954, 16955, 20259, 22731
Paule, François de 11578
Paulin Bertrand, André 13373
Paulin, Ida 16768
Paustovskij, Konstantin G. 37257, 37787, 43160
Pavelic, Myfanwy 07253
Pavlov, A.P. 37192
Pavlov, Ivan Petrovič 37594, 37662

Pavlović-Barilli, Milena 38131
Pavlovskij, Evgenij Nikanorovič 36817
Pawlowa, Maria 22398
Peace, Achille 27323
Pearsal, Vic 06205
Pearse, Patrick 24815
Pecharromán, Ricardo 39814
Pechstein, Max 16438, 17058, 22833
Peck, Sheldon 50165
Pedersen, Carl-Henning 10069
Pégoud, Adolphe 14207
Péguy, Charles 14534
Pelin, Elin 04973, 05192
Pellan, Jacques 13069
Pellicer Cámara, Carlos 31546
Peltonen, Vihtori 10716
Penev, Penjo 05013
Pengelly, William 45866
Perantinos, Nikolaos 22945
Perceval, John 00915
Pérez Galdós, Benito 39796
Pérez, Genaro 00335
Pergaud, Louis 11934
Perkins, Simeon 06179
Perlea, Ionel 36498
Permeke, Constant 03740
Perón, Evita 00231
Peroutka, Ferdinand 09845
Perpessicius, D.P. 36167
Perraudin, Jean-Pierre 41576
Perrier, Rosine 16167
Perrin, Léon 41647
Perrin de Puycousin, Maurice 15958
Perrone, Peter 27323
Perugino 13889, 14354
Peruzzini, Antonio Francesco 26316
Peské, Jean 12575
Pestalozzi, Johann Heinrich 41277, 42049
Peter I, the Great 10641, 32975, 37635, 37809, 37961
Petersson Döderhultarn, Axel 40802
Petiet, Marie 13773
Petkevičaitė-Bite, Gabrielė 30331, 30373
Petőfi, Sándor 23338, 23467, 23527, 23664, 23666, 23668, 23834, 36295
Petorutti, Emilio 00529
Petrarca, Francesco 12957, 13805, 25286
Petraşcu, Gheorghe 36162, 36646
Petraşcu, Mariana 36646
Petraska, Stasys 30206
Petrauskas, Kipras 30260
Petrauskas, Mykolas 30260
Petrescu, Cezar 36238
Petrov-Vodkin, Kuzma Sergeevič 36864
Pettoruti, Emilio 00521
Petzval, Johann 38355
Peukert, August 18980
Peynet, Raymond 12147
Peyrissac, Jean 12242
Pfahler, Georg Karl 21948
Pfanner, Franz 38686
Phleps, Roland 18438
Phomvihane, Kaysone 29972
Piaf, Edith 14705
Piatt, Donn 54068
Picabia, Francis 11549, 14020
Picard, Max 21622
Picasso, Pablo 10103, 11549, 11588, 11638, 11685, 12248, 13122, 13889, 14083, 14354, 14674, 14719, 14722, 14736, 14791, 15111, 15252, 16050, 16051, 16202, 16886, 17172, 17192, 17997, 19005, 19204, 19624, 20414, 20464, 20510, 21960, 28476, 28718, 29575, 33420, 38890, 39001, 39002, 39061, 39596, 39605, 39787, 40317, 40915, 41596, 41598, 41763, 42083, 44156, 46309, 48191, 48935, 50224, 51258, 52717
Piechowski, W. 34927
Pieck, Anton 32322
Piehl, Walter Jr. 50745
Piero della Francesca 27698
Pieterson, Hector 38656
Pietilä, Tuulikki 11365
Piette, Ludovic 14908
Piffetti, Pietro 28004
Pihkala, Lauri 10892
Pijnenburg, R. 32540
Pika, Heliodor 09531
Pilioko, Aloi 54494
Pilon, Veno 38397
Piłsudski, Józef 34788, 35262
Piñole, Nicanor 39340
Piola, Domenico 26316
Piotkowitz, Andreas 17088
Piotrowski, Maksymilian 34578
Pirandello, Luigi 25160, 27443
Piranese 13794
Piri, Markku 10897
Pirri, Alfredo 26014
Pisanello 25406
Pisani, Vettor 26014
Pisano, Giovanni 26309, 27032
Pisis, Filippo de 26109
Pissarro, Camille 14908, 45327, 45920
Pissarro, Lucien 14908
Pistoletto, Michelangelo Buonarroti 11899
Pitman, I. 43406
Pius X 27401
Pius XII 35482
Pjatnickij, Mitrofan E. 36754
Plančić, Juraj 08965
Plateau, Joseph 03636
Platter, Felix 41153
Plattner, Otto 41561
Pleasant, Mary Ellen 52764
Plechanov, Georgi Valentinovič 37104, 37105
Plessis, Jean de 15035
Pleyel, Ignaz Joseph 02012
Plūdoņs, Viljs 29995
Poe, Edgar Allen 46562, 47044, 51830, 52327
Poey, Felipe 09135
Pogor, Vasile 36415
Poincaré, Raymond 15619
Pol, Rein 32230
Pol, Wincenty 34896
Polanski, Roman 34877
Polenov, V.D. 37808
Polenz, Wilhelm von 17753
Polhem, Christopher 40516
Poliakoff, Serge 14719
Polke, Sigmar 10103, 14446, 15252
Pollock, Jackson 28671
Polychronopoulos, Costas 23089
Pommiès, André 12311
Pompidou, Georges 14187
Pompon, François 15656
Ponce de León, Juan 10328
Poni, Petru 36424
Popescu, Gabriel 36736
Popescu, Ştefan 36525
Popescu, Vasile 36525, 36646
Popov, Aleksandr Stepanovič 36898, 37080, 37081, 37085, 37664
Popp, Mişu 36646
Porras, Belisario 34075
Porsche, Ferdinand 01942, 21957
Porter, William Sydney 46502
Porumbescu, Ciprian 36642
Poška, Dionizas 30213
Post, Frans 04667
Poth, Chlodwig 18357
Pott, Karin 17088
Potter, Beatrix 43298, 44429, 45027
Potthast, Edward 48662
Pound, Ezra 27975
Poussin, Nicolas 13683, 27453, 44138, 44840
Povarova, V. 37638
Powell-Cotton, P.H.G. 43489
Požalostin, I.P. 37787
Prado, Leoncio 34134
Prassinos, Mario 15489
Prat de la Riba, Enric 39151
Pratt, Christopher 05402
Pratt, Katharine 48023
Prendergast, Maurice 47690
Preriati, Gaetano 26109
Presas, Leopoldo 00529
Prešeren, France 38543, 38544
Presley, Elvis 50573
Pretorius, M.W. 38715
Prieto, Gregorio 40189, 40191
Prišvin, Michail M. 36881, 37195
Prokofjev, Sergej Sergeevič 37350, 43074
Prorokov, Boris 36944
Proshyan, Perj 00692
Prost, Hippolyte 12416
Proust, Marcel 13246
Prus, Bolesław 34938
Przybyszewski, Stanisław 34691
Puccini, Giacomo 26532, 28037, 28245
Pückler-Muskau, Hermann Prince of 17736
Pueyrredón, Juan Martín de 00581
Pugačëv, Emeljan 37517
Puget, Pierre 26309
Pugh, Clifton 00922
Puig, Joseph 14792
Pułaski, Kazimir 35215
Pulvirenti, Giuseppe 26014
Pumpurs, Andrejs 30036
Pundzius, Bronius 30337
Purrmann, Hans 19816
Puškin, Aleksandr Sergeevič 36804, 36814, 37202, 37244, 37439, 37581, 37582, 37587, 37658, 37694, 37732, 37861, 37983, 38239, 43056, 43084, 43154
Puvis de Chavannes, Pierre Cécile 11588, 31141

Q

Qiu Jin 08211
Quaet-Faslem, Emanuel Bruno 20718
Queiroz, Eça de 35900
Queuille, Henri 14405
Quezon, Manuel L. 34493
Quinós, Bernardo Cesáreo de 00500
Quinquela Martín,Benito 00529
Quintruplets, Dione 05872

R

Raabe, Wilhelm 17528, 18235
Rabelais, François 15737
Rabinowitch, David 17407
Rabus, Carl 21802
Rachmaninov, Sergej Vasilevič 36943
Račić, Josip 09003
Racine, Jean 13362
Racioppi, Celso 04347
Racoviţă, Emil 36338
Rademacher, Heinrich 16596
Radetzky, Joseph Wenzel 02196
Radiščev, Aleksandr Nikolaevič 37589
Rădulescu-Motru, Gheorghe 36240
Radziwill, Franz 20897, 22203
Raevskij, Vladimir Fedoseevič 36813
Raffaello 15829, 26157, 26161, 27451, 28122, 44138, 45327, 48935
Ragana, Šatrijos 30452
Raidiger, Jerry 50402
Raiffeisen, Friedrich Wilhelm 20688
Raimu 12567
Raimund, Ferdinand 02025, 02515
Rainer, Arnulf 01772, 02317, 17407
Rainis 30004, 30021, 30026, 30057, 30102, 30103, 41284
Rais, Karel Václav 09620, 09708
Rakauskaite, Marijona 30258
Rakovski, Georgi S. 05077
Raku, Kichizaemon 28940
Raman, C.V. 24321
Ramón y Cajal, Santiago 39837
Ramoneda, Francisco 00386
Ramos, Mel 47782
Ran-In-Ting 42421
Ransome, Arthur 43885, 44559
Ranson, David E. 07351
Rapin, Aimée 41720
Raponi, Umberto 27323
Rasmussen, Jørgen Skafte 22825
Rasmussen, Knud 10104
Rasputin, Grigorij 37653
Ratcliffe, William 44695
Rathgeber, Valentin 20804
Rathinavelu, P.K. 24428
Rattner, Abraham 53448
Rauch, Christian Daniel 16668, 16669
Rauf, H. Abdul 24476
Rauschenberg, Robert 08675, 17123, 48191, 50224, 50952
Ravel, Maurice 14209
Ravilious, Eric 45539
Raynaud, Jean-Pierre 29395
Razin, Stepan Timofeevič 37478, 37798
Reagan, Ronald Wilson 53110
Rebeyrolle, Paul 12895
Rebreanu, Liviu 36189, 36487, 36634
Recchione, Oreste 27064
Redon, Odilon 12069, 14719, 17244
Rees, Lloyd 00838
Refling Hagen, Ingeborg 33901
Reger, Max 17503
Régio, José 35832, 35987
Rehmann, Erwin 41532
Reich, Ferdinand 18421
Reichardt, Margaretha 18197
Reiche, Maria 34281
Reichenbach, Georg von 20494
Reid, George A. 07349
Reiffenstuel, Hans 21024
Reimers, Lotte 17808
Reininger, Lotte 22124
Reis, Philipp 18476, 18571
Reist, Dölf 41217
Relp, Paula 43731
Remarque, Erich Maria 20920
Rembrandt 10229, 12840, 14714, 17172, 23823, 31861, 31862, 31883, 32085, 32489, 32849, 43497, 44138, 44840, 44894, 44937, 44999, 45006, 48935, 50737
Renan, Ernest 15983
Renda, Ascanio 26014
Rendl, Georg 02643
Reni, Guido 26323, 26741, 27445, 27453
Renner, Karl 01939
Renoir, Auguste 12239, 13770, 14674, 26580, 28628, 48935
Renoudot, Théophraste 13831
Renz, Ernst Jakob 21115
Repin, Ilja Efimovič 37590, 37808
Rerich, Elena Ivanovna 37249, 37679
Rerich, Jurij Nikolaevič 37249, 37679
Rerich, Nikolaj Konstantinovič 30114, 36791, 36958, 37249, 37486, 37679, 51246
Rerich, Svjatoslav Nikolaevič 37249, 37679
Rešetnikov, Fëdor M. 36888, 36891, 36897
Ressu, Camil 36525
Reteganul, Ion Pop 36564
Reuter, Fritz 17883, 18096, 21870
Reventlow, Christian Ditlev Frederik 10102
Reylaender, Ottilie 22699
Reynold, Gonzagne de 41215
Reynolds, Joshua 43497, 44894, 45006, 45378
Rhoades, Jason 17123
Rhodes, Cecil John 38693, 43524
Ribeiro, Aquilino 35822
Ribera, Jusepe de 11588, 27453
Ricci, Marco 26316
Ricci, Sebastiano 27096
Riccio, Andrea 27034
Richards, Ceri 45813
Richelieu, Armand 15035
Richmond, James. C. 33154
Richter, Gerhard 14446, 15252, 18024, 33717
Richter, Hans Werner 16925
Richter, Johann Paul Friedrich 16954, 16955, 20259, 22731
Rie, Lucie 45285
Riehl, Elli 01852
Riemenschneider, Till 22714
Riemenschneider, Tilman 19720, 22713
Ries, Adam 16552, 16843
Riesener, Jean Henri 14734
Rieuf, Elise 14028
Rigaud, Hyacinthe 14791, 26323
Riley, Bridget 44557, 45409
Rilke, Rainer Maria 41215, 41866
Rilski, Neofit 04978, 04979
Rimbaud, Arthur 12409
Rimoldi, Mario 25984
Rimski-Korsakov, Nikolaj Andreevič 37569, 37663, 37841
Rinckart, Martin 18089
Rincón de Gautier, Felisa 36072
Ringeisen, Dominikus 22188
Ringelnatz, Joachim 17755, 22751
Rio Branco, Baron of 04751
Riotte, Philipp Jakob 21514
Rippl-Rónai, József 23467, 23633
Riquet, Pierre-Paul 11798

Rist, Pipilotti 17123
Ristikivi, Karl 10651
Ritter, Carl 21150
Ritter, Johann Wilhelm 19388
Rivel, Charly 21115
Rivera, Diego 30912, 31060, 31070, 31106, 31118, 31127
Rizal, José 34359, 34382, 34441
Rizzarda, Carlo 26093
Robbi, Andrea 41869
Robbia, Luca della 26160, 26176
Robert, Hubert 16040
Roberts, Kenneth 49727
Robertson, David 45149
Robin, Mado 16315
Robinson, Joy 45941
Robinson, Theodore 13068, 47690
Roca, Julio A. 00272
Roch, André 41217
Rocha, Dardo 00415
Roché, Déodat 11695
Roche, Frank 01643
Rochow, Friedrich Eberhard von 19578
Rockox, Nicolaas 03313
Rockwell, Norman 53302
Rockwell, Peter 53302
Rodčenko, Aleksandr 37254
Rodgers, Jimmie 50599
Rodin, Auguste 11522, 12248, 12364, 14104, 14631, 15111, 15252, 28628, 42083, 52717, 53253
Rodrigues, Amalia 35709
Rodríguez, Mariano 09107
Röder, Paul 20220
Roelants, Joseph 03694
Römer, Willy 17202
Roentgen, Abraham 20688
Roentgen, David 20688
Röntgen, Wilhelm Conrad 21263, 22722
Roerich, Nicholas 30114, 36791, 36958, 37249, 37486, 37679, 51246
Roesch, Carl 41328
Rohlfs, Christian 16438, 18860
Rohrmann, Ludwig 19700
Rojas, Ricardo 00172
Rolland, Romain 41215
Roman, Notari 27323
Romanino, Girolamo 27034
Rømer, Ole 10295
Romero de Torres, Julio 39207
Romney, George 44557
Roosevelt, Franklin D. 07297, 49457, 49458, 49459, 53816
Rops, Félicien 03873, 03906
Roque Gameiro, Alfredo 35429, 35763
Rorschach, Hermann 41481
Rosa, João Guimarães 04417
Rosa, Salvator 26741
Rosegger, Peter 01706, 01707, 02236, 02237
Rosenquist, James 50224
Rosmini, A. 27568
Rossetti, Dante Gabriel 44995, 45327
Rossi, Gino 28069
Rossi, Luigi 41195
Rossini, Gioacchino 27155, 27158
Rostand, Edmond 12254
Rotella, Mimmo 26014, 27800
Roth, Dieter 17123, 18923, 21948, 21948, 41885
Róth, Miksa 23474
Rothenstein, Michael 45539
Rothko, Mark 28671, 29575, 49403, 50224
Roty, Oscar 13259
Rouault, Georges 28715, 50510
Rousseau, Jean-Jacques 12373, 12955, 14229, 14924, 41401, 41425, 41648, 41676
Rousseau, Théodore 11847, 28754
Roux, Arlette 16191
Rožek, Marcin 35318
Rozentāls, Jaņis 30071, 30123
Rubcov, Nikolaj 36766, 37353
Rubens, Pieter Paul 03024, 03290, 03315, 14354, 15829, 16043, 17833, 21756, 22639, 25229, 26161, 26323, 26580, 31141, 32788, 39001, 43497, 44138, 44790, 44840, 45006, 48935
Rubin, Reuven 25120
Rubinstejn, N.G. 37354, 37682
Rude, Olaf 10186
Rudnay, Gyula 23609
Rudolphi, Johann Georg 17499
Rückert, Friedrich 17715
Rückriem, Ulrich 21367
Ruelas, Julio 31579
Ruhlmann, Jacques-Emile 13088
Ruisdael, Salomon van 32085
Ruiz, José Martínez 39667
Runeberg, Johan Ludvig 11227
Runeberg, Walter 11231
Runge, Friedlieb Ferdinand 20909
Runge, Philipp Otto 18932, 21914, 22674
Rusanov, V.A. 37506
Rusiñol, Santiago 40083
Ruskin, John 43721, 43884, 43885, 44557, 44637
Russell, George 43317
Russell, John 44370
Rutherford, A.C. 05753
Ruzo, Victor 41641
Ryba, Jakub Jan 09824
Ryggen, Hannah 33925
Rylands, Enriqueta 45087
Rylands, John 45087
Ryti, Risto 10842
Ryushi, Kawabata 29461

S

Saar, Ferdinand von 02956
Saarinen, Eliel 10963
Sabā, Aboul Hassan 24693
Sabattier, Louis 11631
Sacharov, Andrej D. 37284, 37461
Sadoveanu, Mihail 36364, 36412, 36723
Saedeleer, Valerius de 03259
Saeki, Yuzo 29575
Saenredam, Pieter 32849
Sáenz, Manuelita 08546
Särev, Andres 10620
Šafárik, Pavel Jozef 38268
Saff, Don 48662
Sahia, Alexandru 36460
Sahut, Marcel 15759
Saint-Phalle, Niki de 19005, 29051, 41384
Saint-Saëns, Charles Camille 12721
Sajdašev, Salich 37023
Sakamoto, Ryoma 28746, 28747
Sakata, Kazuo 29100
Sakatsoume, Atsouo 05016
Saleški Finžgar, Fran 38545
Salis, Jean Rudolf von 41215
Šaljapin, Fëdor 36919, 37008, 37011, 37047, 37167, 37695
Salomons, David 45709
Salomons, David Lionel 45709
Saltykov-Ščedrin, Michail E. 37037, 37878
Salzillo, Francisco 39702
Sami Durai, K. 24428
Sammt, Albert 20709
Samojlov, Vasilij Vasiljevič 37659
San Martín, José de 00240
Sanati, Sayyed Ali Akbar 24619
Sanchez, Mario 49758
Sand, George 13035, 13340, 14465
Sand, Karl Ludwig 22731
Sandburg, Carl 48555
Sandby, Paul 01144
Sandell, Lina 40813
Sander, August 19627
Sandford, Marion 53826
Sandino, Augusto César 33305
Sandy, Isabelle 14136
Sandzén, Birger 50120, 50435
Sano, Ryuichi 28914
Santa María, Andrés de 08591, 08675
Santander, Nady 41140
Sante, Beato 26770
Santos, Bienvenido 34467
Sarasate Navascués, Pablo 39809
Sargent, John Singer 13068, 48948, 50308
Sargsyan, Ara 00725
Sarmento, Julião 35698
Sarmiento, Domingo Faustino 00151, 00243, 00585
Sarpaneva, Timo 10849
Saryan, Martiros 00732
Satie, Erik 13220, 14737
Sato, Churyo 28940
Sattler, Johann Michael 02621
Satylganov, Toktogul 29955
Saura, Antonio 40317
Sauser, Gustav 02112
Saussure, Horace Bénédict de 41422, 42052
Savery, Roeland 03759
Savignac, Raymond 29286
Savin, Risto 38540
Savio, John Andreas 33592
Savio, Manuel Nicolás Aristóbulo 00282
Ščerbov, P.E. 36919
Schad, Christian 16592
Schaible, Ivo 19859
Scharff, Edwin 20587
Schatzdorfer, Hans 02526
Schedoni, Bartolomeo 27096
Scheele, Carl-Wilhelm 21914
Scheffer, Ary 14665
Scheinpflugová, Olga 09845
Schelling, Friedrich 19388, 19963
Schenk, Carl 41529
Scherer, Joseph 17869
Scheving, Gunnlaugur 24022
Schiaffino, Francesco Maria 26309
Schiavo, Paolo 26147
Schiegg, Ulrich 20494
Schiele, Egon 02434, 02830, 02951, 09467
Schiering, Kurt 20210
Schikaneder, Emanuel 03021
Schill, Ferdinand von 21914
Schiller, Friedrich von 12365, 16774, 16935, 17968, 19390, 19932, 20055, 20169, 20181, 20182, 20259, 22398, 22402, 22417
Schilling, Alfons 03133
Schimansky, Hanns 17106
Schimmelmann, Heinrich Karl 16441
Schindler, Oskar 09857
Schinkel, Karl Friedrich 17032, 17172, 17227, 17249, 20650
Schippers, Arie 33019
Schirmer, Johann Wilhelm 19405
Schjerfbeck, Helene 10772, 11093
Schlegel, August Wilhelm 19388
Schlegel, Caroline 19388
Schlegel, Friedrich 19388
Schlemmer, Oskar 17834, 21956, 21960
Schliemann, Heinrich 16549, 20593
Schlosser, Cornelia 18146
Schmid, Mathias 02139
Schmidt, Franz 02486
Schmidt, Gustav 21001
Schmidt, Maximilian (gen. Waldschmidt) 18236
Schmidt-Rottluff, Karl 17058, 17697
Schnabel, Julian 08675
Schneider, Friedrich 18807
Schneider, Reinhold 16884
Schneider-Trnavsky, Mikulas 38378
Schnopfhagen, Hans 02456
Schöffer, Nicolas 14567, 23627
Schoemaker, Ruud 32230
Schönberg, Arnold 02397, 02934
Scholl, Hans 18339
Scholl, Sophie 18339
Scholte, Dominie 51751
Schoonbeke, Gilbert van 03291
Schoonhoven, Jan 32756, 32884
Schott, Otto 19385
Schouten, Cornelisz 15390
Schraemli, Harry 41940
Schreber, Daniel Paul 19931
Schreiber, Jakob Ferdinand 18275
Schreiner, Vinzenz 02367
Schreiter, Johannes 19812
Schreuder, Hans Paludin 38619
Schröder, Carl 22035
Schröter, Klaus 19796
Schubart, Christian Friedrich Daniel 17380
Schubert, Franz 01736, 03023, 03072, 03073, 38371
Schuch, Carl 21707
Schüppach, Michel 41529
Schütte, Thomas 17998
Schütz, Christian Georg 16590
Schütz, Heinrich 16765, 22456
Schütz, Willi 18040
Schuhmacher, Emil 18855
Schulenburg, Friedrich Werner von der 18289
Schulten, Ton 32632
Schulz, Charles M. 52904
Schulze-Delitzsch, Hermann 17812, 17813
Schumacher, Emil 17924
Schumann, Clara 17466, 19934, 22835
Schumann, Robert 15693, 17466, 19934, 22835
Schwarz, Joseph 17619
Schwarzenbach, Annemarie 41215
Schwarzkogler, Rudolf 03133
Schweig, Joseph 22465
Schweinfurth, Georg 45491
Schweitzer, Albert 13172, 13300, 18360, 19648, 22395
Schwind, Moritz von 19673
Schwitters, Kurt 14722, 15252, 19005, 44557, 45228
Scianna, Ferdinando 25896
Scorel, Jan van 32849
Scott, Robert Falcon 43732
Scott, Sir Walter 44162, 45124, 45638
Ščurovskij, G.E. 37192
Ščusev, Aleksej Viktorovič 31604
Sedlacek, Franz 02317
Seewald, Richard 41129
Segal, Arthur 41129
Segal, George 29575
Segantini, Giovanni 41604, 41813, 42083
Seghers, Anna 17033
Segovia, Andrés 39462
Segrelles, José 38805
Seguin, Marc 11631
Séguin, Robert-Lionel 07178
Seidel, Max 19700
Seiko, Sawada 28602
Sell Cotman, John 45276
Sella, Vittorio 25417
Sellaio, Jacopo del 26126
Semino, Ottavio 26317
Semmelweis, Ignác 23475
Semsey, Andor 23358
Sen, Sun Yat 07991
Senefelder, Alois 20494, 20882, 32870
Senlis, Séraphine de 15715
Šepka, Lionginas 30346
Šeremetev, V.P. 36982
Sergi, Sergio 00521
Serov, Valentin Aleksandrovič 37064
Serov, Vladimir Aleksandrovič 36914
Serpollet, Henri 13795
Serpollet, Léon 13795
Serra, Richard 17407, 26881
Serrano, Pablo 40348
Serruys, Yvonne 03868
Sertürner, Friedrich Wilhelm 18092
Setsuro, Takahashi 28387
Seume, Johann Gottfried 18760, 20106
Sevak, Paruyr 00763
Ševčenko, Taras Grigorevič 07141, 37696, 43060, 43081, 43087, 43107, 43182, 43183
Seward, Anna 44706
Shackleton, Ernest Henry 43732
Shahumya, Stepan 00718
Shaig, Abdulla 03162
Shakespeare, William 45779, 45780, 45782, 45783, 45784, 45785
Shakir, Ali 34028
Shaw, George Bernard 43343
Shawqi, Ahmed 10520
Shelley, Percy Bysshe 27456
Sheridan, Philip 49223
Sherman, Cindy 33717
Shigematsu, Hayashi 29571
Shiina, Rinzo 28513
Shiki, Masaoka 28906
Shimaki, Akahiko 29276
Shimizu, Takashi 28498
Shiraz, Hovhannes 00709
Shiro, Shitahata 29638
Shomu, Emperor 29038
Shortridge, G.C. 38675
Shukow, Georgi K. 17066
Shunzo, Kakite 28912
Siadristy, Nikolaï 00103
Sibelius, Jean 10748, 10863, 11397
Siebner, Herbert 07253
Siebold, Philipp Franz von 17475, 22724
Sieglbauer, Felix 02112
Siemens, Werner von 20483
Sienkiewicz, Henryk 34967, 35315
Signac, Paul 14020, 14908
Signer, Roman 32524, 41885
Sigurdson, Sigrid 18860
Sikorski, Władisław 35262, 35279
Silbermann, Gottfried 18413
Silcher, Friedrich 22441

Sillanpää, Frans Eemil 10740, 10741
Silone, Ignazio 27175
Silva, João da 35663
Silva, Mário Augusto da 35533
Silva Paranhos, José Maria da 04751
Silva, Vieira da 35666
Simberg, Hugo 11093
Šimkus, Stasys 30260
Simonaitytė, Ieva 30339
Simpson, Arthur 44559
Sinclair, Ross 44281
Singer, William H. 32465
Singh, S. Bhagat 24259
Siqueiros, David Alfaro 31070
Siret, Luis 38861
Sisakyan, Norayr 00691
Šiškin, Ivan Ivanovič 36904
Sison, Dolores 34467
Siswa, Taman 24586
Sitte, Willi 20295
Şiugariu, Ion 36132
Sivle, Per 33409
Sjollema, Joop 33019
Sjoo, Monica 45020
Skalbe, Kārlis 30142
Skanderbeg 00026
Skinner, J. Scott 43372
Skjoldborg, Johan 10312
Skovgaard, Joakim 10316
Skovgaard, Niels 10316
Skovgaard, P.C. 10316
Skrjabin, Aleksandr N. 37247
Sládek, Josef Václav 09929
Sladkevicius, Vincentas 30247
Slavejkov, Pencho 05181, 05192
Slavejkov, Petko Racho 05181, 05192
Slavici, Ioan 36623
Slevogt, Max 18056, 19887, 21459
Šliūpas, Jonas 30328
Sloan, John 52717
Slocum, Frances 54162
Slomšek, Anton Martin 38512
Sluyters, Jan 32870
Smet, Gust de 03547
Smet, Leon de 03545
Smetana, Bedřich 09400, 09544, 09565, 09687, 09768
Smidovič, Vikentij Vikentevič 37865
Smidt, Lajos 23903
Smirnenski, Christo 05174, 05192
Smith, Andre 50402
Smith, W. Eugene 51194
Smits, Jakob 03876
Smits, Ton 32195
Snellman, Johan Vilhelm 11006
Śniadecki, Jan 35378
Śniadecki, Jędrzej 35378
Snyder, Tom 52013
Soane, Sir John 44840, 44987
Soares, João 35552
Sobinov, Leonid V. 36970
Sobolewski, Klaus 16554
Šogencukov, A.A. 37441
Sohos, A. 23225
Soldan, Philipp 18343
Soldi, Raúl 00521
Soldner, Johann Georg von 20494
Solheiro Madureira, Marieta 35571
Solimena, Francesco 25371
Šolochov, Michail Aleksandrovič 37925
Solomos, Dionysios 22980, 23245
Sols, Alberto 40045
Somet, Jean-Marie 16171
Søndergaard, Jens 10175
Sørensen, Henrik 33555
Sorolla y Bastida, Joaquín 39581
Soto, Jesús 54563
Soulages, Pierre 15252
Soun, Tazaki 28374
Soundrajajan, N. 24428
Southey, Robert 44564
Soutter, Louis 41080, 41544
Souza-Cardoso, Amadeo de 35698
Sova, Antonín 09704
Soyer, Raphael 47782
Spacca, Ascensidonio 25413
Španiel, Otakar 09550
Spanzotti, Giovanni Martino 28004
Spencer, George J. 07199
Spencer, Stanley 43889
Spendiaryan, Alexander 00724
Spescha, Matias 41953
Spilimbergo, Lino Eneas 00335, 00529
Spinoza, Baruch de 32688
Spiridonov, Moisej S. 36839
Spitteler, Carl 41198, 41561
Spoerri, Daniel 41215
Spohr, Louis 19468
Spreitz, Karl 07253
Springs, Alice 17137
Spruce, Richard 45491
Spyri, Johanna 41484
Spyropoulos, Jannis 22990
Squires, Gerald 05451
Sruoga, Balys 30240
Staden, Hans 22655
Stäbli, Adolf 41080, 41260
Staël, Anne Louise Germaine de 41313
Šťáfl, Otakar 09513
Stahl, P.J. 11584
Stainer, Jakob 01688
Stalin, Jossif V. 16340, 37689
Stambolijski, Aleksandăr 05164, 05182
Stanczac, Julian 47782
Stănescu, Nichita 36528, 36684
Stanev, Emilijan 05192, 05254
Stanevicius, Simonas 30237
Stanislavskij, Konstantin S. 37168, 37190, 37336
Stanković, Borisav 38161
Stankowski, Anton 18577
Stanzione, Massimo 25371
Starobinski, Jean 41215
Starr, Ringo 18889
Stašek, Antal 09836
Staszic, Stanisław 34688, 35008
Stauffenberg, Claus Schenk Graf von 21963
Stażewski, Henryk 34884
Steele, Theodore Clement 50961
Steen, Jan 31883, 32085, 32489
Steeves, William Henry 05988
Ştefănescu, Eustaţiu 36162
Štefánik, Milan Rastislav 38270
Stefansson, David 23981
Steger, Milly 21802
Steichen, Edward 51138
Steiff, Margarete 18632
Steiger, Johann Ulrich 41140, 41142
Stein, Charlotte von 18794
Stein, Gertrude 13795
Steinbeck, John 52677
Steiner, Rudolf 01807
Steinien, Théophile-Alexandre 41544
Steinway, William 21724
Steisslinger, Fritz 17421
Stella, Frank 08675, 29297, 29575
Stelzhamer, Franz 02528
Sten, John 40604
Stendhal 13131
Stephansson, Stephan G. 05754
Stephenson, George 43949, 44263
Steriadi, Alexandru 36525
Sterl, Robert 21922
Stern, Carola 16925
Stern, Irma 38727
Sternbeck, Rudolf 16596
Sternberg, George 49225
Sterne, Laurence 43910
Sterzel, Johann Traugott 17700
Steuben, Friedrich Wilhelm von 52286, 52362
Stevenson, Robert 44254
Stevenson, Robert Louis 38001, 44162, 52548, 52911
Stewart, Hilda 06205
Stickel, Johann Gustav 19386
Stifter, Adalbert 02307, 02838, 03072, 09526
Still, Clyfford 50224
Stirner, Karl 18124
Stoddart, Margaret 33154
Stöhr, Ernst 02686
Stöhr, Gertrude 02874
Stöhrer, Walter 21948
Stojanov, Zachari 05090, 05152
Stokker, Lily van der 32524
Stollberg, Friedrich Leopold Graf zu 18285
Stoltze, Friedrich 18396
Stombuco, Andre 01142
Storck, Frederic 36191
Storm, P. 10023
Storm, Theodor 19108, 19302
Storno, Ferenc (jun) 23822
Storno, Ferenc (sen) 23822
Stoskopf, Gustave 12199
Stowe, Harriet Beecher 49194
Strádal, August 09831
Strand, Paul 51138
Straßner, Ernst 20326
Stratton, Mary Chase 48129
Straus, Oscar 01754
Strauß, Johann 03007
Stravinskij, Igor 43195
Streckenbach, Max 18053
Streuvels, Stijn 03733
Strich-Chapell, Walter 21748
Strieffler, Heinrich 19793
Strieffler, Marie 19793
Strigel, Bernhard 20280
Strindberg, August 02695, 40944
Stroux, Karl-Heinz 18010
Strozzi, Bernardo 26323
Strug, Andrzej 35235
Struth, Thomas 19567
Strzemiński, Władysław 34884
Stuart, J. E. B. 52335
Stuart, Raymond James 51635
Stubbs, George 43405, 43827
Stuck, Franz von 20469
Stuckenberg, Fritz 17817
Stüve, Gustav 20923
Stukenbrok, August 18092
Stumpf, Johannes 41264
Štúr, Ľudovít 38306, 38384
Sturt, Charles 01129
Sturtevant, Erich 19407
Stuyvaert, Victor 03633
Sucharda, Vojtěch 09677
Sudek, Josef 09748
Sudermann, Hermann 30304
Süßmuth, Richard 19325
Sugioka, Kason 29033
Suk, Josef 09607
Šukšin, Vasilij M. 36791, 37793
Suppé, Franz von 01924
Surcouf, Robert 15389
Susanin, Ivan 37810
Suter, August 41354
Sutter, Johann August 41277
Suvorov, Aleksandr Vasilevič 37055, 37643, 37920, 41847, 43045, 43048
Sveinsson, Jón 23985
Svensson, Uno 40815
Sverdrup, Otto 33725
Šverma, Jan 09660
Světlá, Karolina 09466
Svinhufvud, Pehr Evind 11086
Svoboda, Josef 09451
Svor, Anders 33561
Swagemakers, Theo 32303
Świętochowski, Aleksander 34674
Sypesteyn, C.H.C.A. van 32506
Szántó, Piroska 23476
Széchenyi, István 23723
Szenes, Arpad 35666
Szukalski, Stanislaw 35292
Szymanowski, Karol 35357

T

Tabarra, João 35698
Taeuber-Arp, Sophie 12537, 21259
Tagore, Rabindranath 24269, 24280, 24379
Tahon, Johan 03868
Taiga, Ike 28809, 28814
Tajiri, Shinkichi 32849
Tajovský, Jozef Gregor 38368
Takahama, Kyoshi 28679
Takamura, Kotaro 28494
Takashima, Hokkai 29269
Takehisa, Yumeji 28561
Talbot, William Henry Fox 44627
Talkov, Igor 37285
Talleyrand-Périgord, Charles-Maurice de 16033
Tallone, Guido 41195
Talvio, Maila 10760
Tamanyan, Alexander 00727
Tamási, Áron 36452
Tamayo, Rufino 31070, 31106, 31222, 50952
Tamil Pandian, M.K. 24428
Tammann, Gustav 18695
Tammsaare, Anton Hansen 10671
Tanaka, Kyokichi 29575
Taneev, Sergej Ivanovič 37439, 37990
Tanguy, Yves 15252
Tàpies, Antoni 08675, 19204, 39231, 39787
Tarancı, Cahit Sıtkı 42761
Tarkington, Booth 49727
Tarkovskij, Andrej 36984, 36985
Tarnowgrock, Otto 21644
Tartini, Giuseppe 38493
Tasaki, Hirosuke 28672
Tasso, Torquato 27474
Tattarascu, Gheorghe 36210, 36646
Tauto, Bruno 29605
Tăutu-Aloisie, Ludovic 36717
Tavarone, Lazzaro 26317
Tavčar, Ivan 38514
Tavel, Rudolf von 41498
Tavella, Carlo Antonio 26316
Taylor, Mary 44326
Tell, Wilhelm 41270
Tellgmann, Oscar 18237
Tempesta, Antonio 39638
Tennyson, Alfred Lord 44717
Tenshin, Okakura 28708
Teodoroiu, Ecaterina 36663
Tëplyj, N.I. 37638
Terborch, Gerard 32120, 32470
Terbrugghen, Hendrik 32120
Terešenko, Valentin 27800
Terrasse, Louis 16261
Terrus, Etienne 12818
Terry, Ellen 45838
Terry, José Antonio 00649
Tersteegen, Gerhard 20413
Tesla, Nikola 38071
Tessai, Tomioka 29329
Tétar van Elven, Paul 32065
Thaer, Albrecht Daniel 21248
Thalmann, Max 50242
Thatcher, Margaret 44345
Thater, Diana 17123
Theophilos 23232
Theotokopoulos, Domenikos 23019
Thiele, Johann Alexander 21817
Thiersch, Günther 21039
Thijssen, Theo 31895
Thoma, Hans 17291
Thoma, Ludwig 22016
Thomas, Steffen 47109
Thomayer, Josef 09586
Thomopoulos, Epaminondas 23139
Thompson, D'Arcy Wentworth 44052
Thoms, Ernst 20718
Thonet, Michael 17475, 18344, 19585
Thoreau, David 50093
Thorek, Max 47482
Thorma, János 23671
Thorvaldsen, Bertel 10155, 10321
Thyle von Frankenberg 18343
Ticho, Anna 25020
Tidemand, Adolph 33658
Tieck, Ludwig 19388
Tiepolo, Giovanni Battista 15829, 22713, 22714, 22720, 28110, 28113, 28211, 28252
Tietz, Ferdinand 22713
Tiffany, Louis Comfort 47690, 52522
Tilley, Leonard 05846
Timirjazev, Kliment Arkadevič 37324
Timlin, William 38674
Timmermans, Felix 03820
Tinel, Edgar 04012
Tinguely, Jean 41173, 41384, 41885
Tintoretto, Jacopo 13889, 14354, 25371, 26161, 26741, 27034, 28209, 28211, 39001
Tipu, Sultan 24398
Tischbein, Johann Heinrich (d.Ä.) 18871, 19473
Tischbein, Johann Heinrich (d.J.) 18871
Tischbein, Johann Heinrich Wilhelm 16673, 18285, 20897
Tito, Josip Broz 38066
Titov, German 36789
Titulescu, Nicolae 36492
Tiziano 26157, 26161, 26580, 27034, 27207, 27451, 28209, 28211, 39001, 44138, 45006, 45327
Tjutčev, Fëdor Ivanovič 37114, 37522
Tkačev, Aleksej Petrovič 36827
Tkačev, Sergej Petrovič 36827
Tobón Mejía, Marco 08675
Todo, Takeji 28628
Todorov, Petko J. 05027
Toe Boelop, Mechteld 32440
Togo, Seiji 29462
Toklas, Alice 13795
Tokugawa, Ieyasu 29284
Toledo, Francisco 31070
Tolstoj, Aleksej Konstantinovič 37083
Tolstoj, Aleksej Nikolaevič 37195, 37325, 37618, 37658
Tolstoj, Lev Nikolaevič 36975, 37204, 37205, 37222, 37380, 37449, 37658, 37794
Tomimoto, Kenkichi 28359, 28839, 29632
Tominz, Giuseppe 26351
Tomioka, Tessai 29066
Toombs, Robert 53936
Toorop, Jan 32849
Topârceanu, George 36410
Topelius, Sakari 11155
Toppel, Harald 17106
Torahiko, Terada 28748
Torner, Gustavo 39231
Tornyai, János 23609, 23614
Torre, Antonio de la 39396
Torrealba, Alberto Arvelo 54523
Torregrosa, Francisco 39708

S

T

Torrents Llado, Joaquin 39777
Torres Martínez, Manuel 39625
Torricelli, Evangelista 26078
Toscanini, Arturo 27092
Tosti, Francesco Paolo 27003
Tot, Amerigo 23757
Tóth, Menyhért 23609, 23644
Toulouse-Lautrec, Henri de 11540, 14384, 31898
Tour, Maurice-Quentin de la 15474
Tourlière, Michel 11904
Townsend, Steven 43988
Toyotomi, Hideyoshi 29140
Toyozou, Arakawa 28664
Trakl, Georg 02614
Traunfellner, Franz 02507
Travers, Mary 06440
Traxler, Hans 18357
Třebízský, Václav Beneš 09877
Treccani, Ernesto 27323
Trengrouse, Henry 44444
Trevithick, Richard 44569
Trifoglio, Antonio 26506
Trips, Wolfgang Graf Berghe von 19520
Trockel, Rosemarie 17998
Troelstra, Pieter Jelles 32479
Troll, Thaddäus 21961
Tropinin, Vassilij A. 37375
Trotsky, Leo 31059
Trotti, Sandro 27800
Trotzig, Ulf 40474
Trouvé Tatiana 12062
Troyon, Constant 28754
Truikys, Liudas 30258
Truman, Harry S. 49012, 49473, 49474, 49475, 49761, 49910
Tsarouchis, Yannis 23084
Tsouderos, E. 23019
Tsuguji, Fujita 28354
Tucholsky, Kurt 21304
Tucker, Albert 00915
Ţuculescu, Ion 36525
Tübke, Werner 16730
Tuerlinckx, Joëlle 32524
Tukaj, Gabdulla 37015, 37060, 37493
Tulloch, Bobby 46093
Tumanyan, Hovhannes 00739
Tumas-Vaižgantas, Juozas 30245
Tunnicliffe, C.F. 45067
Tupolev, A. 37337
Turgenev, Ivan Sergeevič 12085, 37507, 37510, 37658, 37792
Turgot, Anne Robert Jacques, baron de l'Aulne 13518
Turner, Joseph M. William 44138, 44557, 44886, 44937, 44987, 44995, 44999, 45327, 45378, 52917
Tvardovskij, A.T. 37770, 37984
Twachtman, John Henry 47690
Twain, Mark 49159, 49195, 53323
Twombly, Cy 17123, 20464, 42083, 49384
Twort, Flora 45373
Tycina, Anatolij 30229
Tyler, John 47351

U

U-fang, Lee 29395, 29621
Ubbelohde, Otto 19768
Uccello, Paolo 14714, 26188
Uecker, Günther 18712
Ugalde, Miguel Angel 00517
Uhde, Fritz von 17032
Ulmanis, Kārlis 29991
Ulrich, Christian 41529
Ulug-Beg 54464
Umehara, Ryuzaburo 29121
Unamuno y Jugo, Miguel de 39934
Uncini, Giuseppe 26014
Ungerer, Tomi 15828, 15833
Uniacke, Richard John 06401
Upītis, Pēteris 30001
Upītš, Andrej 30061, 30126
Urakami, Gyokudo 29100
Uriburu, José Evaristo 00559
Urquiza, Justo José de 00381
Urrutia, María Augusta 10425
Urteaga, Mario 34125
Uspenskij, Gleb Ivanovič 37767
Utrillo, Maurice 12636, 15624
Utzschneider, Joseph von 20494

V

Vacca, Antonio 27800
Vācietis, Ojārs 30100
Vaerten, Jan 03355
Vagis, Polygnotos 23048, 23204
Vainiūnas, Stasys 30433
Vajda, Lajos 23874
Valadon, Suzanne 13770, 14719
Valancius, Motiejus 30317
Valand, Norvald 33537
Valdés, Manolo 08675
Valenti, Italo 41129
Valentin, Karl 20493
Valera, Eamonn de 24747
Valéry, Paul 15735
Valle, Evaristo 39339
Vallgren, Ville 11226
Vallien, Bertil 40474
Vallotton, Félix 41544, 41885, 42083
Van Giffen, A.E. 32229
Van Kerckhove, Ann 03289
Vande Kerckhove, V. 03289
Vanderpoel, John H. 47484
VanDerZee, James 51194
Vaneev, A.A. 36916
Vangelli, Antonio 27800
Vanka, Maksimilijan 08896
Vanoni, Giovanni A. 41435
Vanorny, Otmar 09920
Vanvitelli, Luigi 25704
Vanzo, Julio 00521
Vapcarov, Nikola Ěnkov 04978, 04980, 05179, 05192
Vari, Sophia 08675
Varlin 41080, 41901
Varo, Remedios 31070
Vasarely, Victor 23490, 23780
Vasari, Giorgio 26183
Vasco y Vasco, Pedro 39919
Vasnecov, Apolinarij M. 37169, 37245, 37592
Vasnecov, Viktor M. 37169, 37245, 37592, 37808
Vassiliou, Spyros 22900
Vauban, Sébastien le Prestre de 11573, 11954
Vautier, Ben 21096
Vaylet, Joseph 12857
Vazov, Ivan 04987, 05178, 05192
Vega, Jurij 38414
Vega, Martine 14147
Veidenbaums, Eduards 30037
Veit, Dorothea 19388
Vela, Lorenzo 41566
Vela, Spartaco 41566
Vela, Vincenzo 41566
Velarde, Pedro 39705
Velazquez, Diego 12320, 26161, 26741, 27451
Velázquez, Diego 39001
Velazquez, Diego 44138, 45006
Velde, Henry van de 17638, 17696, 18860, 22404, 22412, 33925
Vélez, Eladio 08675
Veličkov, Konstantin 05106
Venclova, Antanas 30437
Venevitinov, Dmitrij Vasiljevič 37492
Venizelos, Eleftherios 22907, 22925, 22940
Vent, Hans 17106
Verbeek, Irene 32678
Verdaguer, Jacint 39301, 40251
Verdi, Giuseppe 25568
Verdier, Jean 13473
Veremans, Renaat 03820
Veres, Péter 23358, 23360
Veresaev, Vikentij Vikentevič 37865
Vereščagin, Vasilij V. 36851
Verhaeren, Emile 03710, 03980, 04014
Verlaine, Paul 13294
Vermeer, Jan 31883, 44138, 44894
Vermehren, Otto 18829
Verna, Claudio 26014
Vernadskij, V.T. 37192
Verne, Jules 11584, 11586, 11587, 14369
Vernet, Joseph 11789
Veronese, Paolo 12775, 13889, 14909, 23823, 25371, 26323, 26741, 27034, 28209, 28252, 43497
Verschaeve, Cyriel 03267
Vesalius, Andreas 41153
Viani, Domenico 27182
Vicente, Esteban 40052
Victoria, Queen of England 44084
Vieira Natividade, Joaquim 35391
Vieira Natividade, Manuel 35391
Vienuolis-Zukauskas, Antanas 30206
Vigeland, Emanuel 33721
Vigeland, Gustav 33735, 33758
Vigny, Alfred de 12385
Viita, Lauri 11352
Vijayvelu, T. 24428
Vila Gorgoll, Emilio 39469
Vila, Jean 11787
Viladomat, Josep 00094
Villa, Francisco 30935
Villa-Lobos, Heitor 04764
Villebois-Marenil 38572
Villeneuve, Arthur 05612
Villéon, Emmanuel de la 12987
Villon, Jacques 14719
Vinje, Å.O. 33983
Vinnen, Carl 22696
Vinogradov, A.P. 37234
Vinzons, Wenceslao Q. 34526
Viola, Bill 17998
Viollet-Le-Duc, Eugéne 14051
Virchow, Rudolf 17044
Vischer, Georg Matthäus 02545
Visdal, Jo 33954
Vissarion, I.C. 36317
Vitol, Jazep 30009
Vittoria, Alessandro 27042
Vivant Denon, Dominique 12360
Vivarini, Antonio 25371
Vivarini, Bartolomeo 25371
Vlahuţă, Alexandru 36098, 36361
Vlaminck, Maurice de 12425, 12636, 13781
Vleugels, Johannes 17615
Vörösmarty, Mihály 23629
Vogel, Alfred 41090
Vogel, Hermann 21055
Vogeler, Heinrich 22692, 22693, 22696, 22699
Voiculescu, Vasile 36511
Voigt, Lene 19931
Voivenel, Paul 12276
Volanakis, K. 23228
Volta, Alessandro 25956
Voltaire 41413
Voß, Johann Heinrich 18285
Vostell, Wolf 17184, 19204, 21096, 39609
Vouet, Simon 12754
Vovčok, Marko 37440
Vucetich, Juan 00414
Vuillard, Edouard 15111, 42083, 44156
Vulcan, Iosif 36504
Vydunas 30275
Vysockij, Vladimir S. 37224

W

Wacker, Hans 21707
Wada, Eisaku 28628
Wade, Charles Paget 43648
Waechter, F. K. 18357
Wagenbauer, Max Joseph 18734
Wagenfeld, Wilhelm 22465
Waggerl, Karl Heinrich 02882
Wagner, Christian 19961
Wagner, Ferdinand 21646
Wágner, Josef 09550
Wagner, Otto 02981, 03053
Wagner, Peter 22713
Wagner, Richard 16774, 16965, 18096, 21045, 41597
Waibl, Ignaz 17672
Wain, Louis 43429
Wajda, Andrzej 34810, 34877
Wakayama, Bokusui 29276
Walker, Gene Alden 53826
Wall, Jeff 17998
Wallace, Alfred Russel 45491
Wallace, Lew 47869
Wallach, Otto 18695
Wallenstein, Albrecht Wenzel Eusebius von 09473
Waller, Sam 07072
Wallis, Alfred 45792
Wallis, Thomas Wilkinson 45037
Walpole, Hugh 44564
Walser, Karl 41216, 41231
Walser, Robert 41231, 41481
Walton, Izaak 45646
Wandschneider, Wilhelm 21051
Wankel, Felix 20174
Waplington, Nick 51138
Warhol, Andy 12840, 15252, 17123, 17172, 20464, 29256, 29395, 33717, 38304, 41763, 50224, 51163, 51938
Warner, Anna 54077
Warner, Susan 54077
Washington, Booker T. 49167
Washington, George 46208, 50893, 53951
Watanabe, Kazan 28488
Waterhouse, E.G. 01120
Waterton, Charles 45914
Watsuji, Tetsuro 28513
Watt, James 44127, 44365, 44569
Watteau, Antoine 16043
Watts, George Frederick 43546, 43882
Wayne, John 54255
Weber, A. Paul 21195
Weber, Carl-Julius 19822
Weber, Carl Maria von 17940, 18285
Weber, Friedrich Wilhelm 16711
Weber-Lipsi, Hildegard 15106
Weber, Mili 41812
Weber, Rudolf 21001
Wech, Josephus 42001
Weckwerth, Werner 18473
Weddige, Emil 54356
Wegener, Alfred 21302
Weidemann, Jakob 33752
Weidemeyer, Carl 41129
Weigel, Helene 17055, 17620
Weinert, Erich 20125
Weinheber, Josef 02173
Weisbach, Albin 18421
Weise, Walter 16596
Weisgerber, Albert 21500
Weiß, Karl 17615
Weißkopf, Gustav 19969
Weizmann, Chaim 25090
Welk, Ehm 16548, 16709
Wellesley, Arthur 1st Duke of Wellington 04138, 44385, 44784
Welliver, Neil 08675
Wentworth, Sarah 01616
Wentworth, William C. 01616
Werefkin, Marianne von 41129
Werenskiold, Erik 33636
Werner, Abraham Gottlob 18421
Werner, Rolf 16926
Werth, Johann von 09400
Wesley, John 44891, 44929
Wessel, Johan Herman 10157
Wesselmann, Tom 15252
West, Franz 17123
Wet, C.R. de 38732
Wettstein, Johann Rudolf 41766
Weygang, August 20868
Weysser, Karl 19446
Wezel, Johann Carl 21817
Whistler, James McNeill 13068, 44284, 46309, 48935, 50308
Whistler, Rex 45381
White, Charles 46942
White, Elizabeth 53367
White, Gilbert 45632
Whiteley, Brett 01539
Whitman, Walt 47197, 49440
Whittle, Sir Frank 43351
Whyte, Catharine 05383
Whyte, Peter 05383
Wicksteed, Thomas 44569
Widmann, Josef Viktor 41561
Widukind 18173
Wiegele, Franz 02451
Wiegersma, Hendrik 32117
Wiegleb, Johann Christian 16773
Wieland, Christoph Martin 16424, 17322, 17325, 17326, 20928
Wierix, Johannes 39638
Wiertz, Antoine 03452
Wilberforce, William 44601
Wild Bill 49223
Wilder, Ingalls 50450
Wilder Lane, Rose 50450
Wildgans, Anton 02391
Wilhelm von Oranien-Nassau 17861
Wille, Fritz von 17367
Willette, Adolphe 32870
William II, Emperor of Germany 14541
William of Orange 32062
Williams, Emmett 21096
Willibrord, Saint 30469
Willumsen, J.F. 10028
Wilson, John 46942
Wilson, Woodrow 53279, 53935

Wimmer, Hans 20993
Winckelmann, Johann Joachim 17245, 21899, 27540
Windaus, Adolf 18695
Winkler, Clemens A. 18422
Winogrand, Garry 51194
Winter, Fritz 16436, 21948
Wirkkala, Tapio 10849
Wirth, Josef Alfons 20408
Wirth, Philipp 20327
Wislicenus, Hermann 18711
Withnell, Emma 01467
Withnell, John 01467
Witos, Wincenty 35302
Witten, Hans 17704
Wittgenstein, Ludwig 02167, 02822
Wlérick, Robert 14158
Wocher, Marquard 41941
Wöhler, Friedrich 18695
Wolf, Alexander 41200
Wolf, August 22431
Wolf, Caspar 41666
Wolf, Conny 41080
Wolf, Elise 41200
Wolf, Georg 22150
Wolf, Gustav 20875
Wolf, Hugo 02487
Wolfe, Thomas 46377
Wolfram von Eschenbach 22656
Wolker, Jiří 09810
Wols 17244
Wolseley, Frederick 24851
Wood, Grant 50891
Wood Perry, E. 47314
Woodall, George 47690
Woodall, Thomas 47690
Woodruff, Hale 46942
Woollaston, Mountford Tosswill 33154
Wordsworth, William 43300, 43869, 44348, 44564
Worthington, Alfred 43258
Wouters, Rik 03863
Wouw, Anton van 38574
Wright, Frank Lloyd 46204, 46619, 48080, 50690, 51453
Wright, Joseph 43965, 44999
Wright, Mary 48080
Wright, Russel 48080
Wüsten, Johannes 18679
Wunderlich, Fritz 19752
Wunderlich, Paul 05016
Wybicki, Józef 34954, 34955
Wyczółkowski, Leon 34578
Wyeth, Andrew 49079
Wyeth, N.C. 51911
Wynn, Steve 49981
Wyspiański, Stanisław 34809

X

Xavier, Francis 28968
Xu Beihong 07574

Y

Yaichi, Aizu 29058
Yamaga, Sokô 28517
Yamaguchi, Kaoru 28818
Yang, Yuyu 42441
Yasujiro, Ozu 28899
Yeats, William Butler 24840
Yokoyama, Ryuichi 28750
Yokoyama, Taikan 28707, 29509
Yole, Jean 15799
Yosano, Buson 28488
Young Simpson, James 44160
Yrurtia, Rogelio 00173

Z

Zabaleta, Rafael 39885
Zadkine, Ossip 14740
Zak, Igal 17088
Zakrevskaja-Benckendorff- Budberg, Maria 10575
Zambaccian, Krikor 36205
Zamora, José María 08675
Zapata, Melchor 38941
Zauli, Carlo 26073
Zdarsky, Mathias 02305
Zegadłowicz, Emil 35208
Zeh, Franz 02061
Zeisel, Eva 48080
Zeisig, Max 21013
Zeiss, Carl 19385
Zeißig, Johann Elias 18805
Zejtlin, Grigorij 27800
Zeppelin, Ferdinand von 20253, 20583
Żeromski, Stefan 34741, 34940
Zetkin, Clara 17360, 19653
Zeyer, Július 09911
Zichy, Mihály 23969
Zick, Januarius 19587
Ziem, F. 11903
Zikaras, Juozas 30246, 30330
Zille, Heinrich 17136, 17181, 20414
Zimmermann, Dominikus 17672
Zipernowsky, Károly 23434
Zittel, Andrea 12062
Zlatarov, Asen 05004
Zmuidzinavicius, Antanas 30239
Zóbel, Fernando 34414, 39231
Zola, Émile 14057
Zorilla de San Martín, Juan 54386
Zorn, Anders 40748
Zorrilla, José de 40229
Zoryan, Stepan 00720
Zoščenko, Michail Michajlovič 37657
Zox, Larry 47782
Zsigmondy, Richard 18695
Zubkov, G. 37638
Zuckmayer, Carl 41785
Zudermanis, Hermanis 30304
Zügel, Oscar 21802
Žukov, Georgij K. 31619
Žukovskij, Nikolaj Egorovič 37388, 37502, 37658
Zuloaga, Daniel 40056
Zuloaga, Ignacio 39253, 39821, 40056, 40362
Zurbar, Francisco de 34254
Zurbarán, Francisco de 31141, 39001, 39315, 39622, 40362
Zuse, Konrad 19290

Alphabetical Index to Museums

1. Deutsches Bockbier-Museum, Taucha ... 22011
I. Türkiye Büyük Millet Meclisi Binası, Ankara ... 42693
100 Mile & District Historical Society, 100 Mile House ... 05304
100th Bomb Group Memorial Museum, Diss ... 43984
The 100th Meridian Museum, Cozad ... 47864
103rd Ohio Volunteer Infantry Civil War Museum, Sheffield Lake ... 53063
106 South Division Gallery, Grand Rapids ... 49000
1078 Gallery, Chico ... 47524
12 Vancouver Service Battalion Museum, Richmond ... 06740
150 ans de Couture, Aix-les-Bains ... 11521
15th Field Artillery Regiment Museum, Vancouver ... 07197
17 Wing Heritage Collection, Winnipeg ... 07351
1796 Bantry French Armada Exhibition Centre, Bantry ... 24742
The 1811 Old Lincoln County Jail and 1839 Jailer's House Museum, Wiscasset ... 54260
1822-Forum Frankfurter Sparkasse, Frankfurt am Main ... 18352
1859 Jail-Marshal's Home and Museum, Independence ... 49471
1866 Battle of Chlum Memorial, Všestary ... 09917
1890 House-Museum and Center for Arts, Cortland ... 47828
19. Sajandi Tartu Linnakodaniku Muuseum, Tartu ... 10644
1910 Boomtown, Saskatoon ... 06909
1914 Plant City High School Community Exhibition, Plant City ... 51977
198 Gallery, London ... 44778
1997 Avrupa'da Yılın Müzesi, Atpazarı ... 42718
19th Century Willowbrook Village, Newfield ... 51312
2-28 Memorial Museum, Taipei ... 42371
II. Türkiye Büyük Millet Meclisi Binası, Ankara ... 42694
21-seiki Gendai Bijutsukan, Kanazawa ... 28654
21_21 Design Sight, Tokyo ... 29369
21st Century Museum of Contemporary Art, Kanazawa ... 28654
24HR Art, Parap ... 01404
26 Martyrs Museum, Nagasaki ... 28968
26th Field Regiment and XII Manitoba Dragoons Museum, Brandon ... 05457
3M Collection of Contemporary Western and Americana Art, Long Lake ... 50189
4-H Schoolhouse Museum, Clarion ... 47581
448th Bomb Group Memorial Museum, Seething ... 45631
45th Infantry Division Museum, Oklahoma City ... 51513
48th Highlands Museum, Toronto ... 07089
4th Infantry Division Museum, Fort Hood ... 48624
5 Elemente Museum, Waidhofen an der Ybbs ... 02887
50er Jahre Erlebniswelt, Zusmarshausen ... 22827
50er-Jahre-Museum, Büdingen ... 17626
600 Jahre Kunst - Studiensammlung der Universität Leipzig, Leipzig ... 19890
7th Ulans Regiment Museum, Mińsk Mazowiecki ... 34923
80 Washington Square East Gallery, New York ... 51107
82nd Airborne Division War Memorial Museum, Fort Bragg ... 48598
921 Earthquake Museum, Kengkou ... 42251

A

A.A. Chernov Geological Museum of the Institute of Geology of the Komi Scientific Center of Ural Branch of the Russian Academy of Sciences, Syktyvkar ... 37814
A.A. Kiselev Museum, Tuapse ... 37863
A.A. Shogentsukov Memorial Museum, Nalčik ... 37441
A.A. Vaneev Memorial House Museum, Ermakovskoe ... 36916
A.B. Goldenveyser Museum Apartment, Moskva ... 37317
A Badacsonyi Állami Gazdaság Üzemi Gyűjteménye, Badacsony ... 23342
A.C. Gilbert's Discovery Village, Salem ... 52664
A. Chaplygin Apartment Museum, Sankt-Peterburg ... 37691
A City Jinshangjing History Museum, Acheng 07476
A.D. Buck Museum of Natural History and Science, Tonkawa ... 53530
A.D. Sakharov Appartement Museum, Nižnij Novgorod ... 37461
A.E. Seaman Mineral Museum, Houghton ... 49377
A.G. Malyshkin Memorial Museum, Mokšan ... 37138
A. Gaydar Library-Museum, Kaniv ... 43057
A. Giordano Gallery, Oakdale ... 51460
A.H. Tammsaare Muuseum, Vetepere ... 10671
A.I. Koslov Museum of Geology and Mineralogy of the Far East State Technical University, Vladivostok ... 37937
A.K.A. Gallery, Saskatoon ... 06910
A. Kolcov Museum, Voronež ... 37972
A. Kotsky Memorial House-Museum, Užgorod ... 43196
A. Kuindshi Apartment Museum, Sankt-Peterburg ... 37693
A. Kusnetcov Regional Museum of Local History and Nature, Čita ... 36869
A.L. Durov House-Museum, Voronež ... 37969
A.L. Fetterman Educational Museum, Fort Valley ... 48694
A Logger's Life Provincial Museum, Grand Falls-Windsor ... 05890
A.M. de Jonghuis, Nieuw Vossemeer ... 32573
A.M. Gorki Appartment-Museum, Nižnij Novgorod ... 37460
A.M. Gorki House Museum, Moskva ... 37318
A.M. Gorki Museum, Moskva ... 37279
A.M. Vasnetsov Apartment Museum, Moskva 37245
The A Museum of Art, Takane ... 29323
A.N. Burdenko Memorial Museum, Moskva ... 37236
A.N Radishchev Memorial Museum of Literature, Radiščevo ... 37589
A.N. Skryabin Museum, Moskva ... 37247
A.P. Chekhov Apartment Museum, Moskva ... 37161
A.P. Chekhov Museum Taganrog, Taganrog ... 37830
A. Paul Weber-Museum, Ratzeburg ... 21195
A.R. Bowman Memorial Museum, Prineville ... 52155
A.R.C. Gallery, Chicago ... 47449
A.R. Mitchell Memorial Museum of Western Art, Trinidad ... 53565
A.S. Golubkina House-Museum, Zarajsk ... 37985
A.S. Grin House-Museum, Kirov ... 37036
A.S. Popov Central Museum of Communications, Sankt-Peterburg ... 37630
A.S. Popov Memorial Museum-Laboratory, Kronštadt ... 37085
A.S. Popov Radio Museum, Ekaterinburg ... 36898
A.S. Pushkin Memorial Apartment Museum, Moskva ... 37244
A. Schiiwesch Musée Rural, Binsfeld ... 30456
A.Shiwi A.Wan Museum, Zuni ... 54375
A. Sohos Museum, Tinos ... 23225
A Soldier's Life, Newcastle-upon-Tyne ... 45223
A.T. Bolotov Museum-Estate, Dvorjaninovo ... 36882
A Tegnap Autói, Tatabánya ... 23920
A.V. Nezhdanova House-Museum, Moskva ... 37321
A.V. Sidorov Mineralogical Museum, Irkutsk ... 36934
A.V. Suvorov Museum of Ismail, Izmaïl ... 43048
A.W.A. Electronic-Communication Museum, Bloomfield ... 46853
A.W. Campbell House Museum, Alvinston ... 05328
Aabenraa Museum, Aabenraa ... 09952
Aachener Printenmuseum, Aachen ... 16391
AAF Museum, Washington ... 53843
Aalborg Historiske Museum, Aalborg ... 09955
Aalholm Automobil Museum, Nysted ... 10207
Aaltense Oorlogs en Verzetscollectie 1940-1945, Aalten ... 31784
Aaly Tokombaev Memorial Museum, Biškek ... 29957
Aardewerkvermuseum Petrus Pegout, Noordbroek ... 32598
Aargauer Kunsthaus, Aarau ... 41080
Aaronson House, Zichron Yaakov ... 25134
Aasiaat Katersugaasiviat Museum, Aasiaat ... 23251
Aasland Museum Taxidermy, Cranbrook ... 05657
Aavik's Memorial Museum, Kuressaare ... 10586
Aavikute Majamuuseum, Kuressaare ... 10586
Abai Commemorative Literary Museum, Semey ... 29729
Abai Museum, Abaj ... 29669
Abashiri Kyodo Bijutsukan, Abashiri ... 28342
Abashiri Kyodo Hakubutsukan, Abashiri ... 28343
Abashiri Municipal Art Museum, Abashiri ... 28342
Abashiri Municipal Museum, Abashiri ... 28343
Abashiri Prison Museum, Abashiri ... 28344
Abbadia Castle, Hendaye ... 13200
The Abbasid Palace Museum, Baghdad ... 24710
Abbaye de Beaulieu en Rouergue, Ginals ... 13063
Abbaye de Bonnefont, Proupiary ... 14950
Abbaye de Flaran, Valence-sur-Baïse ... 16041
Abbaye de Saint-Amand de Coly, Saint-Amand-de-Coly ... 15158
Abbaye de Vauclair, Bouconville-Vauclair ... 12083
Abbaye du Val Richer, Saint-Ouen-le-Pin ... 15436
Abbaye Royale du Moncel, Pontpoint ... 14910
Abbaye Saint-André, Villeneuve-lès-Avignon ... 16213
Abbe Museum at Downtown, Bar Harbor ... 46600
Abbe Museum at Sieur de Monts Spring, Bar Harbor ... 46601
Abbey and Stones Museum, Margam ... 45104
Abbey House Museum, Leeds ... 44658
Abbey Museum of Art and Archaeology, Caboolture ... 00931
Abbey of Bellapais, Girne ... 33357
Abbey of Flaran, Valence-sur-Baïse ... 16041
Abbey of Pannonhalma Collection, Pannonhalma ... 23750
Abbey Pumping Station Museum, Leicester ... 44672
Abbeydale Industrial Hamlet, Sheffield ... 45648
Abbot Hall Art Gallery, Kendal ... 44557
Abbot's House, Arbroath ... 43311
Abbots Treasury, Korčula ... 08897
Abbotsford House, Melrose ... 45124
Abby Aldrich Rockefeller Folk Art Museum, Williamsburg ... 54164
ABC Architectuurcentrum Haarlem, Haarlem ... 32292
Abdeen Palace Museum, Cairo ... 10484
Abegg-Stiftung, Riggisberg ... 41767
Abel Tasman Kabinet, Lutjegast ... 32513
Abelardo Díaz Alfaro Pink House Museum, Caguas ... 36030
Abenteuermuseum Saarbrücken - Sammlung Heinz Rox-Schulz, Saarbrücken ... 21453
Aberconwy House, Conwy ... 43886
Aberdeen Art Gallery, Aberdeen ... 43233
Aberdeen Arts Centre, Aberdeen ... 43234
Aberdeen Maritime Museum, Aberdeen ... 43235
Aberdeen University Natural History Museum, Aberdeen ... 43236
Aberdeen University Natural Philosophy Museum, Aberdeen ... 43237
Aberdeen University Pathology and Forensic Medicine Collection, Aberdeen ... 43238
Aberdeenshire Farming Museum, Mintlaw ... 45159
Aberdona Gallery, Alloa ... 43281
Abergavenny Museum, Abergavenny ... 43252
Åbergs Museum, Bålsta ... 40451
Abernethy Museum, Abernethy ... 43254
Abernethy Nature Heritage Museum, Abernethy ... 05307
Aberseer Lipphaus, Strobl ... 02789
Abertillery Museum, Abertillery ... 43255
Aberystwyth Arts Centre, Aberystwyth ... 43256
Abgineh va Sofalineh, Tehrān ... 24654
Abguss Sammlung antiker Plastik, Berlin ... 17027
Abhai Smarak Panchal Sangrahalaya, Bareilly ... 24077
Abiko City Museum of Birds, Abiko ... 28346
Abiko-shiritsu Kinchou Hakubutsukan, Abiko ... 28346
Abingdon Museum, Abingdon ... 43260
Abington Art Center, Jenkintown ... 49619
Abington Park Museum, Northampton ... 45268
Abkar Museum, Tehrān ... 24655
ABM-Utvikling, Oslo ... 54969
Aboa Vetus - Arkeologis-historiallinen Museo, Turku ... 11387
Abony Lajos Falumúzeum, Abony ... 23334
Abony Lajos Village Museum, Abony ... 23334
Aboriginal Art Museum, Utrecht ... 32841
Abraham Lincoln Birthplace, Hodgenville ... 49307
Abraham Lincoln Museum, Harrogate ... 49191
Abraham Ojanperän Museo, Liminka ... 11072
Abraham's Mosterdmakerij, Eenrum ... 32176
Abramov Memorial and Literature Museum, Verkola ... 37924
Abramtsevo Estate Museum, Abramcevo ... 36747
Abri du Cap-Blanc, Marquay ... 13978
Abri Préhistorique de Laugerie Basse, Les Eyzies-de-Tayac-Sireuil ... 13697
ABSA Group Museum, Johannesburg ... 38645
Abtei Rommersdorf, Neuwied ... 20686
Abteilung Handschriften und Alte Kunstdrucke der Universitätsbibliothek, Würzburg ... 22706
Abteilung Handwerk und dörfliches Leben des Heimatmuseums Neu-Ulm, Neu-Ulm ... 20585
Abu Ali Ibn Sina Memorial Museum, Afchona ... 54437
Abu Ali Ibn Sina Yodgorlik Muzeyi, Afchona ... 54437
Abu Bakar Royal Museum, Johor Bahru ... 30558
Abuabdullo Roudaki Republican Historical and Regional Museum, Pandžakent ... 42505
Academia Brasileira de Arte, São Paulo ... 04847
Academia Brasileira de Letras, Rio de Janeiro 04679
Academic Museum of Medicine, Poznań ... 35029
Academician Doncho Kostov House-Museum, Lokorsko ... 05083
Academician F. Chernyshev Central Scientific-Researching Geological-Prospecting Museum, Sankt-Peterburg ... 37627
Academician K. Satpaev Memorial Museum, Almaty ... 29684
Academician Kurchatov Memorial House Museum, Moskva ... 37230
Academician Norayr Sisakyan House-Museum, Ashtarak ... 00691
Academie Salesienne Conservatoire d'Art et d'Histoire, Annecy ... 11624
Academisch Historisch Museum van de Universiteit Leiden, Leiden ... 32481
Academy Art Museum, Easton ... 48267
Academy Galleries, Wellington ... 33240
The Academy Gallery, Beverly Hills ... 46790
Academy Gallery, New Orleans ... 51074
Academy Gallery of Art, New York ... 51108
Academy Hall Museum of the Rocky Hill Historical Society, Rocky Hill ... 52434
Academy of Fine Arts Museum, Kolkata ... 24266
Academy of Natural Sciences of Philadelphia Museum, Philadelphia ... 51809
Academy of Sciences Museum of Zoology, Sankt-Peterburg ... 37738
Academy of Spherical Arts, Toronto ... 07090
Acadia University Art Gallery, Wolfville ... 07398
Acadian Historical Village, Caraquet ... 05555
Acadian Museum, Miscouche ... 06305
Acadian Museum and Archives of West-Pubnico, West Pubnico ... 07303
The Accokeek Foundation, Accokeek ... 46131
Accordeon en Harmonica Museum, Malden ... 32536
Accton Arts Foundation, Hsinchu ... 42204
Ace Art, Winnipeg ... 07352
Ace of Clubs House, Texarkana ... 53488
Acervo Artístico Cultural dos Palácios do Governo, São Paulo ... 04848
Acervo Didático de Invertebrados I e II, São Paulo ... 04849
Achamenian Museum, Shiraz ... 24646
Acharya Jogesh Chandra Purakirti Bhavan, Bishnupur ... 24102
Achenseer Museumswelt, Maurach am Achensee ... 02368
Achter de Zuilen, Overveen ... 32664
Achterhoeks Museum 1940-1945, Hengelo, Gelderland ... 32366
Achtertalbahn, Ottenhöfen im Schwarzwald ... 20945
Ackerbürgermuseum, Gartz ... 18543
Ackerbürgermuseum, Reichenbach, Oberlausitz ... 21247
Ackerbürgermuseum Haus Leck, Grebenstein 18741
Ackland Art Museum, Chapel Hill ... 47344
Acland Coal Mine Museum, Acland ... 00765
ACME Art Gallery, Columbus ... 47749
Acorn Hall House Museum, Morristown ... 50850
Acquario e Museo Etnografico del Po, Monticelli d'Ongina ... 26859
Acropolis Museum, Athinai ... 22917
Acton Scott Historic Working Farm, Acton Scott ... 43263
Actual Art Foundation, New York ... 51109
Açude Museum, Rio de Janeiro ... 04680
Adachi Bijutsukan, Yasugi ... 29588
Adachi Museum of Art, Yasugi ... 29588
Adair County Historical Society Museum, Kirksville ... 49799
Adalbert-Stifter-Haus, Linz ... 02307
Adam Art Gallery, Wellington ... 33241
Adam Chętnik Memorial Kurpic Region Open Air Museum, Nowogród ... 34953
Adam East Museum Art Center, Moses Lake ... 50866
Ádám Jenő Emlékház, Szigetszentmiklós ... 23889
Adam Mickiewicz Museum at Śmiełów, Żerków ... 35369
Adam Mickiewicz Museum of Literature, Warszawa ... 35256
Adam-Ries-Museum, Annaberg-Buchholz ... 16552
Adam Thoroughgood House, Virginia Beach ... 53752
Adam Wodziczko Natural History Museum of Wolin National Park, Międzyzdroje ... 34918
Ādama Alkšņa Memoriālais Muzejs, Rūjiena ... 30118
Adamo Petrausko Muziejus, Uoginiai ... 30389
Ādams Alksnis Memorial Museum, Rūjiena ... 30118
Adams County Historical Museum, Decatur ... 48018
Adams County Historical Society Museum, Gettysburg ... 48900
Adams County Historical Society Museum, Lind ... 50116
Adams County Museum, Brighton ... 47026
Adams Museum, Deadwood ... 48002
Adams Museum and Historic Adams House, Deadwood ... 48003
Adams Old Stone Grist Mill, Bellows Falls ... 46715
Adams State College Luther Bean Museum, Alamosa ... 46153
Adamson-Eric Museum, Tallinn ... 10619
Adamson House and Malibu Lagoon Museum, Malibu ... 50406
Adán E. Treganza Anthropology Museum, San Francisco ... 52763
Adana Archaeological Museum, Adana ... 42670
Adana Arkeoloji Müzesi, Adana ... 42670
Adana Atatürk Müzesi, Adana ... 42671
Adana Devlet Güzel Sanatlar Galerisi, Adana ... 42672
Adana Ethnographical Museum, Adana ... 42673
Adana Etnografya Müzesi, Adana ... 42673
Adana State Fine Art Gallery, Adana ... 42672
Adare Heritage Centre, Adare ... 24727
Addington - Ryde House of Heritage, Ryde ... 01475
Addison Gallery of American Art, Andover ... 46296
Adelaide Central Gallery, Norwood ... 01381
Adelaide Centre for the Arts, Adelaide ... 00777
Adelaide City Council Civic Collection, Adelaide ... 00770
Adelaide Gaol, Thebarton ... 01582
Adelaide House Museum, Alice Springs ... 00796
Adelaide Hunter Hoodless Homestead, Saint George ... 06817
Adelaide Lithuanian Museum, Norwood ... 01382
Adelaide Masonic Centre Museum, Adelaide ... 00771
Adelaide River Railway Heritage Precinct & Motorcycle Haven, Adelaide River ... 00784
Adelės ir Pauliaus Galaunių namai, Kaunas ... 30238
Adell McMillan Gallery, Eugene ... 48430
Adena State Memorial, Chillicothe ... 47533
Adirondack History Center, Elizabethtown ... 48340
Adirondack Museum, Blue Mountain Lake ... 46876
Adıyaman Müzesi, Adıyaman ... 42675

Adjara Arts Museum, Batumi 16333
Adjutant General's Corps Museum, Winchester 46026
Adler Museum of Medicine, Johannesburg 38646
Adlerskij Kraevedčeskij Muzej - Muzej Istorii Adlerskogo Rajona, Soči 37781
Admiral Digby Museum, Digby 05703
Admiral Nimitz National Museum of the Pacific War, Fredericksburg 48761
Admiralty House Museum and Archives, Mount Pearl 06400
Adolf-Bender-Zentrum, Sankt Wendel 21510
Adolf Dietrich-Haus, Berlingen 41201
Ādolfa Alunāna Memoriālais Muzejs, Jelgava 30015
Adsmore Museum, Princeton 52145
Ady Endre Emlékmúzeum, Budapest 23379
Adygean Republican Museum, Majkop 37123
Adygejskij Respublikanskij Kraevedčeskij Muzej, Majkop 37123
Äänekosken Kaupunginmuseo, Äänekoski 10705
Äänekoski Town Museum, Äänekoski 10705
Æbelholt Klostermuseum, Hillerød 10075
Aegean Maritime Museum, Mykonos 23102
Ägyptisches Museum der Universität Bonn, Bonn 17441
Ägyptisches Museum der Universität Leipzig - Georg Steindorff, Leipzig 19891
Ägyptisches Museum und Papyrussammlung, Berlin 17028
Ähtärin Kotiseutumuseo, Ähtäri 10706
Aera Memorial Museum, San Pablo 34499
Ærø Museum, Ærøskøbing 09957
Aero Nautical Museum-Armadillo, Mona Vale 01312
Aero Space Museum of Calgary, Calgary 05513
Aeronautical Museum, Telde 40113
Aeronautics Museum, Nanjing 08037
Aeronauticum - Deutsches Luftschiff- und Marinefliegermuseum Nordholz, Nordholz 20741
Aeroseum, Göteborg 40542
Aerospace and Technology Museum Park Merseburg, Merseburg 20293
Aérospatiale, Toulouse 15930
Afdeling Nuttige en Fraaie Handwerken, Uithuizermeeden 32837
Affiche Museum Hoorn, Hoorn 32408
Affresco di Piero della Francesca, Monterchi 26845
Afghanisches Museum, Hamburg 18913
Afif Abad Military Museum, Shiraz 24645
Africa Museum, Balatonederics 23351
Africa Museum, Seoul 29863
African American Historical and Cultural Society Museum, San Francisco 52764
African American Historical Museum, Cedar Rapids 47299
African American Museum, Cleveland 47621
African American Museum, Dallas 47924
African American Museum at Oakland, Oakland 51462
The African American Museum in Philadelphia, Philadelphia 51810
African American Museums Association, Chicago 55108
African Art Centre, Durban 38602
African Art Museum Atelier, San Juan 36075
African Art Museum of Maryland, Columbia 47712
African Art Museum of the S.M.A. Fathers, Tenafly 53476
African Arts Collection, San Diego 52736
African Cultural Art Forum, Philadelphia 51811
African Heritage Museum, San Juan 36069
Africana Museum, Mariannhill 38686
Africana Museum, Monrovia 30181
Afrika Centrum, Cadier en Keer 32032
Afrika Museum, Berg en Dal 31964
Afrika-Museum, Bergen, Kreis Celle 17008
Afrika-Museum, Zug 42122
Afrika Múzeum, Balatonederics 23351
Afrikaans Language Museum, Paarl 38699
Afrikaanse Taalmuseum, Paarl 38699
Afrikahaus Museum und Nold-Namibia-Bibliothek, Sebnitz 21709
Afrikamuseum, Gallneukirchen 01918
Afro-American Cultural Center, Charlotte 47384
Afro-American Historical Society Museum, Jersey City 49624
Afrosiyob - Samarqand Thariki Muzeyi, Samarqand 54459
Afrosyob - The Museum of the History of the Foundation of Samarkand, Samarqand 54459
Afval-Museum, Zwolle 33025
Afyon Devlet Güzel Sanatlar Galerisi, Afyon 42676
Afyon Etnografya Müzesi, Afyon 42677
Afyon Müzesi, Afyon 42678
Afyon State Gallery, Afyon 42676
Aga Khan Museum of Islamic Arts, Marawi 34459
Aga Village, Nå 33688
Agassiz-Harrison Historical Society Museum, Agassiz 05309
Agatunet, Nå 33688
Agder Naturmuseum, Kristiansand 33609
Age of Steam Railroad Museum, Dallas 47925
Agecroft Hall, Richmond 52320
Agensteinhaus und Simmentaler Hausweg, Erlenbach im Simmental 41356
Agharkar Museum, Allahabad 24058
Agia Triada Folklore Museum, Agia Triada 22867
Aginskij Okružnoj Kraevedčeskij Muzej im. G. Cybikova, Aginskoe 36748
Agios Andreas Convent, Kammeni Chora 23032
Agios Nikolaos Archaeological Museum, Agios Nikolaos 22872
Agios Stephanos Basilica, Kastoria 23042
Agnes Etherington Art Centre, Kingston 06073
Agnes Jamieson Gallery, Minden 06293
Agnes-Miegel-Haus, Bad Nenndorf 16793
Agora Açıkhava Müzesi, İzmir 42852
Agora Open Air Museum, İzmir 42852
Agrargeschichtliche Sammlung, Wels 02911
Agrarhistorisches Aktionsmuseum, Trendelburg 22087
Agrarhistorisches Museum, Domsühl 17891
Agrarhistorisches Museum, Schlepzig 21564
Agrarisch en Wagenmuseum, De Waal 32052
Agrarisch Museum Westerhem, Middenbeemster 32556
Agrarmuseum Wandlitz, Wandlitz 22321
Agrármúzeumi Emlékek Gyűjteménye, Zirc 23978
Agri Musée-Musée de l'Agriculture du Temps Passé, Rochehaut 03974
Agricultural Association Museum of Jhonghe, Jhonghe 42218
Agricultural Exhibition Hall, Taipei 42372
Agricultural Heritage Museum, Boerne 46888
Agricultural Machine Museum, Mezőkövesd 23699
The Agricultural Memories Museum, Penn Yan 51763
Agricultural Museum, Faisalabad 34001
Agricultural Museum, Jasin 30556
Agricultural Museum, Kangasala 10909
Agricultural Museum, Yangon 31731
Agricultural Museum and Cotton Museum, Cairo 10485
Agricultural Museum at Stone Mills, La Fargeville 49839
Agricultural Museum Brook, Brook 43657
Agricultural Museum, Miyazaki University, Miyazaki 28929
Agricultural Museum of Entre Douro and Minho, Vila do Conde 35988
Agricultural Museum of Jeollanam-do, Samho 29860
Agricultural Museum of New Brunswick, Sussex 07039
Agricultural Region House - Dutch-house, Dég 23523
Agricultural Tools Museum, Châteaurenard (Bouches-du-Rhône) 12455
L'Agriculture au Fil du Temps, Maniquerville 13937
L'Agriculture Bressane Musée, Saint-Germain-du-Bois 15288
Agriculture Museum, Imphal 24220
Agriculture Museum, Talsi 30127
Agriculture Museum Schloss Stainz, Stainz 02760
Agropolis Museum, Montpellier 14236
Agrotechnorama, Ettenhausen 41364
Agua Caliente Cultural Museum, Palm Springs 51680
Aguélimuseet, Sala 40825
Agung Rai Museum of Art, Ubud 24579
Ah-Tah-Thi-Ki Museum, Big Cypress 46794
Ahlat Müzesi, Ahlat 42680
Ahler Kråm - Volkskundliche Sammlung, Partenstein 20985
Ahmed Pervez Art Gallery, Karachi 34011
Ahola House Museum, Keitele 10936
Aholan Talomuseo, Keitele 10936
Ahsan Manzil Museum, Dhaka 03217
Åhus Museum, Åhus 40410
Aichi-ken Bijutsukan, Nagoya 28991
Aichi-ken Toji Siryokan, Seto 29250
Aichi Prefectural Ceramics Museum, Seto 29250
Aichi Prefectural Museum of Art, Nagoya 28991
Aigantighe Art Museum, Timaru 33211
Aikawa Archaeological Museum, Isesaki 28584
Aikawa Folk Museum and Exhibition Hall of Folk Crafts, Sado 29175
Aikawa Kokokan, Isesaki 28584
Aiken County Historical Museum, Aiken 46140
Aiken Thoroughbred Racing Hall of Fame and Museum, Aiken 46141
Ainaži Naval College Museum, Ainaži 29981
Ainažu Jūrskolas Memoriālais Muzejs, Ainaži 29981
Aine Art Museum, Tornio 11383
Aineen Taidemuseo, Tornio 11383
Aineh va Roshanaial Museum, Yazd 24703
Ainola, Järvenpää 10863
Ainola, The Home of Jean Sibelius, Järvenpää 10863
Ainsley House, Campbell 47207
Ainu Museum, Sapporo 29213
Air and Space museum, Le Bourget (Seine-Saint-Denis) 13564
Air Classic Museum of Aviation, Sugar Grove 53356
Air Defence Museum, Alouette 05326
Air Force Armament Museum, Eglin Air Force Base 48304
Air Force Flight Test Center Museum, Edwards 48299
Air Force Heritage Museum and Air Park, Winnipeg 07353
Air Force Museum, Christchurch 33067
Air Force Museum, Delhi 24133
Air Force Museum, Ha Noi 54606
Air Force Museum, Tehrān 24656
Air Mobility Command Museum, Dover, Air Force Base 48170
Air of Art Museum, Bologna 25437
Air Power Park and Museum, Hampton 49149
Air Victory Museum, Lumberton 50320
Air Victory Museum, Medford 50549
Airborne and Special Operations Museum, Fayetteville 48529
Airborne Forces Museum, Aldershot 43270
Airborne Museum Hartenstein, Oosterbeek 32624
Airmen Memorial Museum, Suitland 53359
Airport Gallery 1 & 2, Frankfurt am Main 18353
Airport Museum, Cairo 10486
Airpower Museum, Ottumwa 51630
Aitihasic Puratatva Sangrahalaya, Delhi 24134
Aitkin County Museum, Aitkin 46143
Aitonevan Turvemuseo, Kihniö 10954
Aître Saint-Saturnin, Blois 12030
Aiud Museum of History, Aiud 36103
Aiud Natural History Museum, Aiud 36102
Aiwen e Rifat Museum and Art Gallery, Karachi 34012
Aizkraukle History and Art Museum, Aizkraukle 29982
Aizkraukles Vēstures un Mākslas Muzejs, Aizkraukle 29982
Ajax Museum, Amsterdam 31824
Ajia-no Mingeikan, Noda 29079
Ájtte - Svenskt Fjäll- och Samemuseum, Jokkmokk 40618
Ájtte - Swedish Mountain and Sami Museum, Jokkmokk 40618
Aka-Renga Kokusaikan, Yokohama 29592
Akademie der Künste, Berlin 17029
Akademisches Kunstmuseum - Antikensammlung der Universität, Bonn 17442
Akademisches Münzkabinett, Jena 19375
Akadomari-mura Folk Museum, Sado 29176
Akama-jingu Homotsukan, Shimonoseki 29267
Akama Shrine Treasure House, Shimonoseki 29267
Akamon Bijutsukan, Sendai 29241
Akamon Museum, Sendai 29241
Akanes Folk Museum, Akranes 23980
Akaroa Museum, Akaroa 33033
Åkers Hembygdsmuseum, Strängnäs 40951
Akershus Fylkesmuseum, Strømmen 33887
Akershus Regional Museum, Strømmen 33887
Akesu Zone Museum, Akesu 07477
Akiba Anti-ku Okimono Bijutsukan, Nasu 29046
Akiba Museum of Antique Jewellry, Nasu 29046
Akita Daigaku Kogyo Hakubutsukan, Akita 28353
Akita-kenritsu Bijutsukan, Akita 28354
Akita-kenritsu Hakubutsukan, Akita 28355
Akita-kenritsu Kindai Bijutsukan, Yokote 29623
Akita Museum of Modern Art, Yokote 29623
Akita Prefectural Art Museum, Akita 28354
Akita Prefectural Museum, Akita 28355
Akita Senshu Bijutsukan, Akita 28356
Akita Senshu Museum of Art, Akita 28356
Akita University Mining Museum, Akita 28353
Akiyoshi-dai Kagaku Hakubutsukan, Shuuhou 29293
Akiyoshidai Museum of Natural History, Shuuhou 29293
Akizuki Hakubutsukan, Amagi 28358
Akizuki Museum, Amagi 28358
Akko Municipal Museum, Akko 24923
Akmal Ikramov nomidagi Tarikhi Kŭltŭri va San"atshunoslik Ŭzbekiston Khalqlari Davlat Mŭzeyi, Samarqand 54460
Akmal Ikramov State Museum of the History of Culture and Art of the Peoples of Uzbekistan, Samarqand 54460
Akpol Museum, Semarang 24554
Akron Art Museum, Akron 46146
Aksaray Müzesi ve Bağli Örenyerleri, Aksaray 42681
Akşehir Müzesi, Akşehir 42682
Aksel Bakuntz House-Museum, Goris 00702
Akshaya Kumar Maitreya Museum, Siliguri 24394
Aksum Archaeology Museum, Aksum 10683
Akta Lakota Museum, Chamberlain 47329
Aktives Museum Spiegelgasse für Deutsch-Jüdische Geschichte in Wiesbaden, Wiesbaden 22545
Aktives Museum Südwestfalen, Siegen 21753
Aktivmuseum Ländliches Brauchtum, Breitungen an der Werra 17545
Akureyri Museum, Akureyri 23983
Akuryri Art Museum, Akureyri 23982
AKUS Gallery, Willimantic 54178
Akwesasne Museum, Hogansburg 49308
Al-Ain Museum, Al-Ain 43217
Ål Bygdemuseum, Ål 33377
Al-Karak Museum for Archaeology and Folklore, Al-Karak 29641
Al Kassaba Museum Thafer al Masri Foundation, Nablus 34058
Al-Khor Museum, Al-Khor 36083
Al Mada'in Museum, Baghdad 24711
Al-Mahatah - Sharjah Aviation Museum, Sharjah 43223
Al-Mansora Museum Dar-ibn-Luqman, Daqahlia 10522
Al Manyal Palace and Nilometer, Cairo 10487
Al Mawsil Museum, Nineveh 24724
Al-Mukalla Museum, Al-Mukalla 54645
Al-Muntaza Palace Museum, Alexandria 10468
Al Shaheed Monument and Museum, Baghdad 24712
Al Sulaimaniya Museum, Sulaimaniya 24726
Al-Wakra Museum, Al-Wakra 36084
Al Zubara Fort, Doha 36085
Alabama Agricultural Museum, Dothan 48154
Alabama-Coushatta Indian Museum, Livingston 50150
Alabama Department of Archives and History Museum, Montgomery 50813
Alabama Mining Museum, Dora 48152
Alabama Museum of Natural History, Tuscaloosa 53616
Alabama Museum of the Health Science, Birmingham 46819
Alabama Museums Association, Birmingham 55109
Alabama Music Hall of Fame, Tuscumbia 53623
Alabama River Museum, Franklin 48745
Alacahöyük Müzesi, Alaca 42686
Alamance Battleground, Burlington 47138
Alamannenmuseum, Weingarten, Kreis Ravensburg 22425
Alamannenmuseum Ellwangen, Ellwangen 18120
The Alamo, San Antonio 52705
Alamo Township Museum - John E. Gray Memorial Museum, Kalamazoo 49663
Åland Art Museum, Mariehamn 11097
Aland Hunting and Fishing Museum, Eckerö 10720
Åland Maritime Museum, Mariehamn 11099
Ålands Jakt och Fiskemuseum, Eckerö 10720
Ålands Konstmuseum, Mariehamn 11097
Ålands Museum, Mariehamn 11098
Ålands Sjöfartsmuseum, Mariehamn 11099
Alanya Müzeleri ve Örenyerleri, Alanya 42687
Alanya Museum and Archaeological Sites, Alanya 42687
Alapítvány Érc és Ásványbányászati Múzeum, Rudabánya 23789
Alaska Aviation Museum, Anchorage 46285
Alaska Centennial Center for the Arts, Fairbanks 48469
Alaska Heritage Museum at Wells Fargo, Anchorage 46286
Alaska Indian Arts, Haines 49124
Alaska Museum of Natural History, Anchorage 46287
Alaska State Museum, Juneau 49656
Alatornion Pitäjämuseo, Tornio 11384
Albadia Museum, Osafia 25077
Albany Civil Rights Movement Museum, Albany 46154
Albany Historical Society, Sabetha 52496
Albany Institute of History and Art, Albany 46157
Albany Museum, Grahamstown 38629
Albany Museum of Art, Albany 46155
Albany Regional Museum, Albany 46166
Albany Residency Museum, Albany 00785
Albatros Flugmuseum, Stuttgart 21928
Albatrossmuseet, Romakloster 40814
Albergo-Museo Atelier sul Mare, Castel di Tusa 25718
Alberni Valley Museum, Port Alberni 06624
Albert County Museum, Hopewell Cape 05996
Albert Edelfeltin Ateljeemuseo, Porvoo 11225
Albert Edelfelt's Atelier Museum, Porvoo 11225
Albert Engström-Museerna, Grisslehamn 40568
Albert Kersten GeoCentre, Broken Hill 00902
Albert-König-Museum, Unterlüß 22180
Albert-Sammt-Zeppelin-Museum, Niederstetten 20709
Albert-Schweitzer-Gedenk- und Begegnungsstätte, Weimar 22395
Albert Schweitzer Haus, Königsfeld im Schwarzwald 19648
Alberta Art Foundation Collection, Edmonton 05738
Alberta Association of Registered Nurses Museum, Edmonton 05739
Alberta Aviation Museum, Edmonton 05740
Alberta Central Railway Museum, Wetaskiwin 07310
Alberta Forestry Service Museum, Hinton 05990
Alberta Hospital Museum, Ponoka 06620
Alberta Museums Association, Edmonton 54729
Alberta Railway Museum, Edmonton 05741
Alberta Sports Hall of Fame and Museum, Red Deer 06706

Albertfalva Regional History Collection and School Museum, Budapest 23380
Albertfalvai Helytörténeti Gyűjtemény és Iskolamúzeum, Budapest 23380
Albertina, Wien 02929
Albertland and Districts Museum, Wellsford 33262
Alberton, Auckland 33040
Alberton Museum, Alberton 05313
The Albin O. Kuhn Gallery, Baltimore 46544
Albion Academy Historical Museum, Edgerton 48287
Albrecht-Daniel-Thaer-Gedenkstätte-Landesausstellung, Reichenow-Möglin 21248
Albrecht-Dürer-Gesellschaft, Nürnberg 20754
Albrecht-Dürer-Haus, Nürnberg 20755
Albrecht-Kemper Museum of Art, Saint Joseph 52560
Albrechtsburg Meißen, Meißen 20263
Albright-Knox Art Gallery, Buffalo 47114
Albu Manor House, Albu 10559
Albu Mõisahoone, Albu 10559
Album-Het Nieuwe Reclamemuseum, Bruxelles 03430
The Albuquerque Museum of Art and History, Albuquerque 46175
Albury Regional Art Gallery, Albury 00792
Albury Regional Museum, Albury 00793
Alcan Museum and Archives, Montréal 06326
Alcázar de Segovia, Segovia 40048
Alchemy: The Group for Collections Management in Yorkshire and Humberside, Halifax 55047
Alchimistenmuseum, Kirchberg am Wagram 02165
Aldan Maadyr National Museum of the Republic of Tyva, Kyzyl 37099
Aldborough Roman Site, Aldborough 43266
Aldeburgh Museum, Aldeburgh 43267
Aldeia Museu de João Franco, Mafra 35748
Alden B. Dow Museum of Science and Art, Midland 50665
Alden Historical Society Museum, Alden 46192
Alden House, Duxbury 48226
Alderney Railway - Braye Road Station, Alderney 43268
Alderney Society Museum, Alderney 43269
Aldershot Military Museum and Rushmoor Local History Gallery, Aldershot 43271
Aldfaers Erf Route, Allingawier 31797
Aldheidskeamer Uldrik Bottema, Aldeboarn 31791
Aldrich Museum of Contemporary Art, Ridgefield 52347
Aleko Konstantinov Municipal Historical Museum, Svištov 05221
Aleksandar Rukavina Memorial Gallery, Brtonigla 08850
Aleksandr Caks Memorial Falt, Riga 30060
Aleksandra Čaka Memoriālais Dzīvoklis, Rīga 30060
Aleksandrov Literature and Art Museum of Marina and Anastasija Tsvetaeva, Aleksandrov 36753
Aleksandrovskaja Sloboda - Gosudarstvennyj Istoriko- architekturnyj i Chudožestvennyj Muzej-zapovednik, Aleksandrov 36751
Aleksandrovskij Chudožestvennyj Muzej, Aleksandrov 36752
Aleksandrovskij Dvorec - Gosudarstvennyj Muzej Carskoje Selo, Puškin 37577
Aleksandrovskij Literaturno-chudožestvennyj Muzej Mariny i Anastasii Cvetaevych, Aleksandrov 36753
Aleksandrovskij Narodnyj Muzej M.E. Pjatnickogo, Aleksandrovka 36754
Aleksis Kiven Syntymäkoti, Palojoki 11185
Aleksis Kivi's Birth-Place, Palojoki 11185
Ålen Bygdetun, Ålen 33379
Aleppo National Museum, Aleppo 42128
Alert Bay Museum, Alert Bay 05314
Aleš South Bohemian Gallery, Hluboká nad Vltavou 09515
Ålesunds Museum, Ålesund 33380
Alex Brown Cycle History Museum, Thornhill, Dumfriesshire 45854
Alex Youck School Museum, Regina 06714
Alexander and Baldwin Sugar Museum, Puunene 52205
Alexander Blok State Museum-Preserve of History and Literature, Solnečnogorsk 37786
Alexander Brest Museum, Jacksonville 49576
Alexander Fleming Laboratory Museum, London 44779
Alexander Graham Bell National Historic Site, Baddeck 05369
Alexander Griboedov State Museum-Preserve of History, Culture and Nature, Chmelita 36863
Alexander Kazbegi Museum, Kazbegi 16341
Alexander Keiller Museum, Avebury 43338
Alexander Kuprin Museum, Narovčat 37445
Alexander-Mack-Museum, Bad Berleburg 16691
Alexander McKay Geological Museum, Wellington 33242
Alexander Moutafov House-Museum, Sozopol 05208
Alexander Nevski Cathedral - Crypt - Museum of Old Bulgarian Art, Sofia 05184
Alexander Pushkin Literary Museum, Vilnius 30423
Alexander Ramsey House, Saint Paul 52612
Alexander Spendiaryan House Museum, Yerevan 00724
Alexander Stamboliiski House-Museum, Sofia 05182
Alexandra Museum, Alexandra 33034
Alexandra Timber Tramway and Museum, Alexandra 00794
Alexandre Hogue Gallery, Tulsa 53607
Alexandria Archaeology Museum, Alexandria 46198
Alexandria Black History Resource Center, Alexandria 46199
Alexandria Museum of Art, Alexandria 46195
Alexandria National Museum, Alexandria 10469
Alexandros Papanastassiou Museum, Levidion 23076
Alexandrovsk Art Museum, Aleksandrov 36752
Alexandrovsk Sloboda - Museum-Preserve of Art, History and Architecture, Aleksandrov 36751
Alexej K. Tolstoy Museum, Krasnyj Rog 37083
Alexej K. Tolstoy Park Museum, Brjansk 36829
Alexej N. Tolstoy Museum-Estate, Samara 37618
Alexej von Jawlensky-Archiv, Locarno 41568
Alexey N. Tolstoy Apartment Museum, Moskva 37325
Alföldfásítási Gyűjtemény, Bugac 23497
Alföldi Galéria, Hódmezővásárhely 23609
The Alfoldy Gallery, Erikson 05781
Alfons Graber-Museum, Steinach 02763
Alford Heritage Centre, Alford, Aberdeenshire 43276
Alford House-Anderson Fine Arts Center, Anderson 46291
Alford Manor House Museum, Alford, Lincolnshire 43278
Alfred Dunhill Museum, London 44780
Alfred East Art Gallery, Kettering 44566
Alfred-Erhardt-Stiftung, Berlin 17030
Alfred Mireka Museum of the Russian Harmonica, Moskva 37357
Alfred Ost Museum, Zwijndrecht 04171
Alfred-Vogel-Museum, Teufen 41931
Alfred Wegener Museum, Rheinsberg 21302
Alfredo Jonušo Žemaitiška-japoniška Sodyba-muziejus, Telšiai 30378
Alger County Heritage Center, Munising 50906
Algonquin College Museum, Nepean 06423
Algonquin Culture and Heritage Centre, Golden Lake 05881
Algonquin Logging Museum, Whitney 07332
Alice Austen House Museum, Staten Island 53260
Alice C. Sabatini Gallery, Topeka 53532
Alice Springs RSL War Museum, Alice Springs 00797
Alice T. Miner Colonial Collection, Chazy 47416
Aliceville Museum, Aliceville 46219
Alikartano, Numminen 11151
Alikartano Manor, Numminen 11151
Alimentarium, Vevey 41983
Alinari National Museum of Photography, Firenze 26195
Alingsås Kulturhus/Konsthall, Alingsås 40412
Alingsås Museum, Alingsås 40413
Alisgårdenin Kotiseutumuseo, Lapinkylä 11043
Alisher Navoi State Museum of Literature, Toshkent 54467
Alisher Navoii nomidagi Davlat Adabiyot Muzeyi, Toshkent 54467
Aliwal North Museum and Church Plein Museum, Aliwal North 38552
Alix Wagon Wheel Museum, Alix 05317
Aljira Center for Contemporary Art, Newark 51284
All Hallows Undercroft Museum, London 44781
All-Russian Museum of Decorative, Applied and Folk Art, Moskva 37412
Allahabad Museum, Allahabad 24059
Allan Macpherson House, Napanee 06416
Allan Shivers Museum, Woodville 54291
Allard Pierson Museum Amsterdam, Amsterdam 31825
Allegan County Historical and Old Jail Museum, Allegan 46222
Allegany County Historical Museum, Cumberland 47905
Allegany County Museum, Belmont 46719
Alleghany Highlands Arts and Crafts Center, Clifton Forge 47649
Allegheny Portage Railroad, Gallitzin 48841
Allen County-Fort Wayne Historical Society Museum, Fort Wayne 48703
Allen County Historical Society, Iola 49515
Allen County Museum, Lima 50084
Allen Gallery, Alton 43292
Allen Memorial Art Museum, Oberlin 51478
Allen Priebe Gallery, Oshkosh 51612
Allen R. Hite Art Institute, Louisville 50280
Allen Sapp Gallery, North Battleford 06465
Allentown Art Museum, Allentown 46225
Allerhande, Hippolytushoef 32392
Allgäu-Museum, Kempten 19510
Allgäuer Burgenmuseum, Kempten 19511
Allhallows Museum of Lace and Antiquities, Honiton 44483
Alliance and District Museum Society, Alliance 05318
Allie Griffin Art Gallery, Weyburn 07314
Allied Air Force Museum, Allentown 46226
Allied Arts Center and Gallery, Richland 52306
Alliierten-Museum Berlin, Berlin 17031
Allilujevy House-Museum, Sankt-Peterburg 37689
Alling Coverlet Museum, Palmyra 51683
Allison-Antrim Museum, Greencastle 49049
Allora Historical Museum, Allora 00803
Allschwiler Kunst-Zentrum, Allschwil 41100
Allwood House, Hurstbridge 01168
Alma Firehouse and Mining Museum, Alma 46234
Almaty City History Museum & Museum of Repression, Almaty 29674
Almaznyj Fond, Moskva 37140
Almond Valley Heritage Centre, Livingston 44751
Almonry Heritage Centre, Evesham 44191
Almviks Tegelbruksmuseum, Västervik 41036
Alor Gajah Museum, Alor Gajah 30549
Alp-Museum Fasons, Seewis Dorf 41858
Alpen-Adria-Galerie im Stadthaus, Klagenfurt 02180
Alpenländische Galerie, Kempten 19512
Alpenrail Swiss Model Village and Railway, Claremont, Tasmania 00975
Alpha Museum, Bangalore 24071
Alphons-Stübel-Sammlung früher Orientphotographien, Jena 19376
Alpin- und Heimatmuseum, Hohe Wand-Stollhof 02089
Alpinarium Galtür, Galtür 01920
alpinartgallery - Schatzhaus der Natur, Ebene Reichenau 01834
Alpine Curiosa, Stilfs 27905
Alpines Ballonsport-Museum, Mürren 41663
Alpines Museum des Deutschen Alpenvereins, München 20420
Alpineum, Hinterstoder 02079
Alpineum, Luzern 41586
Alpinmuseum, Kempten 19513
Alpinmuseum Dachstein, Ramsau am Dachstein 02557
Alpmuseum Riederalp, Riederalp 41762
Alpsennereimuseum, Hittisau 02083
Als Hjemstavnsmuseum, Hadsund 10052
Alšova Jihočeská Galerie, Hluboká nad Vltavou 09515
Alsters Herrgård, Karlstad 40638
Alstertal-Museum, Hamburg 18914
Alt-Arnstorf-Haus, Arnstorf 16576
Alt-Freden-Sammlung, Freden 18417
Alt-Rothenburger Handwerkerhaus, Rothenburg ob der Tauber 21405
Alt-Segeberger Bürgerhaus, Bad Segeberg 16836
Alt-Stade im Baumhaus, Stade 21847
Alta Museum, Alta 33384
Altai State Regional Studies Museum, Barnaul 36789
Altajskij Gosudarstvennyj Kraevedčeskij Muzej, Barnaul 36789
Altajskij Respublikanskij Kraevedčeskij Muzej im. A.V. Anochina, Gorno-Altajsk 36927
Altana, Bad Homburg v.d.Höhe 16749
Alte Apotheke, Ottweiler 20958
Alte Arche, Furtwangen im Schwarzwald 18519
Alte Börse, Leipzig 19892
Alte Cuesterey, Essen 18246
Alte Dorfmühle, Monzelfeld 20377
Alte Dorfschmiede, Waldbreitbach 22282
Alte Feste, Windhoek 31762
Alte Galerie am Landesmuseum Joanneum, Graz 01961
Alte Huf- und Wagenschmiede, Dietmanns 01818
Alte Lederfabrik Güttgemanns, Halle / Westfalen 18908
Alte Marktschmiede, Lasberg 02264
Alte Mühle Donsbrüggen, Kleve 19564
Alte Mühle Schrattenberg, Schrattenberg 02713
Alte Nationalgalerie, Berlin 17032
Alte Oberamtei, Gerabronn 18600
Alte Pinakothek, München 20421
Alte Saline Bad Reichenhall, Bad Reichenhall 16806
Alte Schmiede, Bedesbach 16976
Alte Schmiede, Bennhausen 16995
Alte Schmiede, Schwaikheim 21659
Alte Schmiede - Kunstverein Wien, Wien 02930
Alte Schule mit Clara-Zetkin-Gedächtnisstätte, Königshain-Wiederau 19653
Alte Synagoge, Essen 18247
Alte Synagoge, Hechingen 19060
Alte Wache, Groitzsch 18770
Alte Wassermühle, Splietsdorf 21841
Alte Welt Museum, Nußbach 20796
Das Alte Zollhaus Heimatmuseum, Hitzacker 19201
Alternative Museum, New York 51110
Altertumssammlung der Benediktinerabtei, Engelberg 41351
Altes Amtshaus, Werne 22493
Altes Bürgerhaus, Waidhofen an der Thaya 02883
Altes Hafenamt, Dortmund 17913
Altes Rathaus, Lüneburg 20094
Altes Rathaus - Potsdam Museum, Potsdam 21079
Altes Rathaus Wilster, Wilster 22591
Altes Schloss, Abteilung Gemäldegalerie und Kunsthandwerk, Gießen 18636
Altes und Neues Schloss Eremitage, Bayreuth 16944
Altfriesisches Haus, Sylt 21993
Althorp Museum, Althorp 43291
Altia Oyj, Rajamäki 11252
Altia's Rajamäki Industrial Museum, Rajamäki 11252
Altlandtechnik-Museum, Klein Upahl 19562
Altmärkisches Museum, Stendal 21897
Altmühltaler Mühlenmuseum, Dietfurt an der Altmühl 17854
Alton Museum of History and Art, Alton 46240
Altonaer Museum für Kunst und Kulturgeschichte, Hamburg 18915
Altstadtrathaus, Braunschweig 17517
Altun Ha Archaeological Site, Altun Ha 04172
Alūksne Museum of Local Studies and Art, Alūksne 29984
Alūksnes Novadpētniecības un Mākslas Muzejs, Alūksne 29984
Alumni Museum, Catonsville 47283
Alumny Creek Museum, Grafton 01126
Alupka Deržavnyj Palac ta Park Zapovidnyk, Alupka 42998
Alupka State Palace and Park Preserve, Alupka 42998
Alutiiq Museum and Archaeological Repository, Kodiak 49823
Alva DeMars Megan Chapel Art Center, Manchester 50420
Alvar Aalto Museum, Jyväskylä 10886
Alvdal Museum, Alvdal 33385
Alytaus Kraštotyros Muziejus, Alytus 30203
Alytaus Local Lore Museum, Alytus 30203
Alyth Museum, Alyth 43296
Amador County Museum, Jackson 49555
Amagermuseet, Dragør 09988
Amalienborg, København 10122
Åmåls Hembygdsmuseum, Åmål 40415
Åmåls Heritage Museum, Åmål 40415
Åmåls Järnvägsmuseum, Åmål 40416
Åmåls Konsthall, Åmål 40417
Amandine - d'Ostensjhe Ieslandvisjherie, Oostende 03927
Amarillo Museum of Art, Amarillo 46250
Amasra Historical Museum, Amasra 42688
Amasra Müzesi ve Tarihçesi, Amasra 42688
Amasya Devlet Güzel Sanatlar Galerisi, Amasya 42689
Amasya Müzesi, Amasya 42690
Amasya Museum, Amasya 42690
Amasya State Fine Art Gallery, Amasya 42689
Amathole Museum, King William's Town 38675
Ambalangoda Mask Museum, Ambalangoda 40365
Amber Inclusion Museum, Gdynia 34655
Amber World Museum, Santo Domingo 10356
Ambergris Museum and Cultural Centre, Ambergris Caye 04173
Amberley Museum and Heritage Centre, Amberley 43297
Amelia Earhart Birthplace Museum, Atchison 46406
Amelia Island Museum of History, Fernandina Beach 48535
Ameliasburgh Historical Museum, Ameliasburgh 05329
America-Goldbeach Museum, Ver-sur-Mer 16108
American Academy of Arts and Letters Art Museum, New York 51111
American Advertising Museum, Portland 52071
American Airlines C.R. Smith Museum, Fort Worth 48708
American Airpower Heritage Museum, Midland 50669
American Antiquarian Society, Worcester 55110
American Association for Museum Volunteers, Denver 55111
American Association for State and Local History, Nashville 55112
American Association of Museums, Washington 55113
American Baptist - Samuel Colgate Historical Library, Rochester 52391
American Bicycle and Cycling Museum, Santa Fe 52871
American Catholic Historical Society Museum, Philadelphia 51812
American Classical Music Hall of Fame and Museum, Cincinnati 47552
American Clock and Watch Museum, Bristol 47027
American Craft Museum, New York 51112
American Film Institute, Los Angeles 50208
American Folk Art Museum, New York 51113
American Gallery, Sylvania 53395
American Helicopter Museum & Education Center, West Chester 54048
American Heritage "Big Red" Fire Museum, Louisville 50300

The American Historical Foundation Museum, Richmond 52321
American Independence Museum, Exeter 48463
American Indian Contemporary Arts, San Francisco 52765
American Irish Historical Society Museum, New York 51114
American Jazz Museum, Kansas City 49682
American Labor Museum, Haledon 49127
American Legation Museum, Tanger 31694
American Maple Museum, Croghan 47885
American Merchant Marine Museum, Kings Point 49785
American Military Museum, Charleston 47356
American Motorcycle Museum, Raalte 32675
American Motorcycle Museum, Unterweitersdorf 02845
American Museum in Britain, Bath 43398
American Museum of Asmat Art, Saint Paul 52613
American Museum of Fire Fighting, Hudson 49410
The American Museum of Fly Fishing, Manchester 50426
American Museum of Magic, Marshall 50502
American Museum of Natural History, New York 51115
American Museum of Nursing, Tempe 53463
American Museum of Radio and Eletricity, Bellingham 46711
American Museum of Science and Energy, Oak Ridge 51456
American Museum of Straw Art, Long Beach 50172
American Museum of the Miniature Arts, Dallas 47926
American Museum of the Moving Image, Astoria 46399
American Numismatic Society Museum, New York 51116
American Police Center and Museum, Chicago 47450
American Precision Museum, Windsor 54228
American Printing House for the Blind, Callahan Museum, Louisville 50281
American Quarter Horse Heritage Center and Museum, Amarillo 46251
American Red Cross Museum, Washington 53844
The American Revolution Center, Wayne 54010
American Royal Museum, Kansas City 49683
American Saddlebred Horse Museum, Mexico 50617
American Saddlebred Museum, Lexington 50044
American Sport Art Museum, Daphne 47965
American Swedish Historical Museum, Philadelphia 51813
American Swedish Institute, Minneapolis 50727
American Textile History Museum, Lowell 50303
American Truck Historical Society, Kansas City 49684
American University of Beirut Archaeological Museum, Beirut 30157
American Visionary Art Museum, Baltimore 46545
American Watchmakers-Clockmakers Institute, Harrison 49184
American West Heritage Center, Wellsville 54034
American Work Horse Museum, Lexington 50068
Americana Manse, Whitney-Halsey Home, Belmont 46720
Americana Museum, El Paso 48315
Americana Museum, Terra Alta 53479
America's Black Holocaust Museum, Milwaukee 50704
America's Ice Cream and Dairy Museum, Medina 50556
America's National Music Museum, Vermillion 53719
Americas Society Art Gallery, New York 51117
America's Stonehenge, North Salem 51404
Amerigo Tot Múzeum, Pécs 23757
Amerika Museum Nederland, Cuijk 32040
Amerind Foundation, Dragoon 48181
Amerongs Historisch Museum/ Tabaksmuseum, Amerongen 31807
Amersham Museum, Amersham 43301
Amethystwelt Maissau, Maissau 02344
Ameziane Museum, Nador 31686
Amfiaraion Archaeological Museum, Kalamos 23026
AMG - Association of Museums and Galleries of Czech Republic, Praha 54772
Amgueddfa Cymru - National Museum of Wales, Cardiff 43748
Amgueddfa Pontypool Museums, Pontypool 45400
Amherst College Museum of Natural History, Amherst 46263
Amherst County Museum, Amherst 46273
Amherst History Museum, Amherst 46264
Amherst Museum, Amherst 46272
Ami Cultural Center, Chengkung 42161
Amics de los Museos de Cataluña, Barcelona 55019
Amigos de los Museos, Madrid 55020
Amir Timur Museum, Shakhrisabz 54465
Amity and Woodbridge Historical Society Museum, Woodbridge 54273
Ammerländer Schinken-Museum F. W. Meyer, Apen 16563
Amon Carter Museum, Fort Worth 48709
Amory Regional Museum, Amory 46275
Amos Anderson Art Museum, Helsinki 10767
Amos Andersonin Taidemuseo, Helsinki 10767
Amos Eno Gallery, New York 51118
Amos Herr House, Landisville 49935
Amos Seaman School Museum, Minudie 06300
Amparo Museum, Puebla 31265
Amr Ibrahim Palace, Cairo 10516
Amr Ibrahim Palace, Cairo 10515
Amreli Archaeological and Art Museum, Amreli 24064
Amrumer Museum und Windmühle, Nebel 20558
Amsterdam Public Transport Museum, Amsterdam 31827
Het Amsterdams Automuseum, Zwanenburg 33022
Amsterdams Historisch Museum, Amsterdam 31826
Amsterdams Openbaarvervoer Museum, Amsterdam 31827
Amtshofmuseum, Gernsbach 18606
Amtsrichterhaus, Schwarzenbek 21673
Amtsturm-Museum, Lüchow 20086
Amüseum - das Saarburger Museum am Wasserfall, Saarburg 21461
Amuri Historical Museum, Waihu 33219
Amuri Museum of Workers' Housing, Tampere 11348
Amurin Työläismuseokortteli, Tampere 11348
Amurskij Gorodskoj Kraevedčeskij Muzej, Amursk 36756
Amurskij Oblastnoj Kraevedčeskij Muzej im. G.S. Novikova-Daurskogo, Blagoveščensk 36809
L'Amusette, Mesvin 03872
AMVC-Letterenhuis, Antwerpen 03280
An Dun Transport and Heritage Museum, Ballinahown 24734
An Lanntair Art Centre, Stornoway 45766
Anacortes Museum, Anacortes 46278
Anacostia Museum, Washington 53845
Anadarko Philomathic Museum, Anadarko 46279
Anadolu Fortress Museum, İstanbul 42801
Anadolu Medeniyetleri Müzesi, Ankara 42695
Anadoluhisarı Müzesi, İstanbul 42801
The Anaesthesia Heritage Centre, London 44782
Anaesthetic Museum, London 44783
Anaheim Museum, Anaheim 46283
Anamur Culture Center, Anamur 42691
Anamur Kültür Merkezi, Anamur 42691
Anamur Müzesi ve Örenyerleri, Anamur 42692
Ananda Niketan Kirtishala, Bagnan 24069
Anapskij Archeologičeskij Muzej-zapovednik, Anapa 36758
Anasazi Heritage Center, Dolores 48150
Anasazi State Park Museum, Boulder 46962
Anatoliaus Tyčinos Paveikslų Ekspozicija, Gruzdžiai 30229
Anatomical Museum, Bangkok 42528
Anatomical Museum of the MUI, Innsbruck 02112
Anatomisch Museum, Groningen 32278
Anatomisch Museum, Utrecht 32842
Anatomisch Museum Nijmegen, Nijmegen 32587
Anatomisch Museum van de Rijksuniversiteit Leiden, Leiden 32482
Anatomische Sammlung, Erlangen 18218
Anatomische Sammlung, Greifswald 18747
Anatomische Sammlungen des Institut für Anatomie und Zellbiologie, Halle, Saale 18888
Anatomisches Museum, Innsbruck 02112
Anatomisches Museum Basel, Basel 41153
Anatomy Museum, Aberdeen 43239
Anatomy Museum, Chennai 24111
Anatomy Museum, Jakarta 24487
Anatomy Museum, Mumbai 24313
Anatomy Museum, Saint Lucia 01482
Anchor Graphics, Chicago 47451
Anchorage Museum at Rasmuson Center, Anchorage 46288
Ancien Couvent, Treigny 15985
Ancien Palais de Bruxelles, Bruxelles 03431
Ancien Palais de Justice - Centre d'Art et d'Histoire, Kamouraska 06040
Ancienne École de Médecine Navale, Rochefort (Charente-Maritime) 15069
Ancienne Mairie-Musée Annexe, Lillebonne 13760
Ancient Agora Museum, Athinai 22898
Ancient and Honorable Artillery Company Museum, Boston 46916
Ancient Architecture Museum, Beijing 07503
Ancient Belarusian Culture Museum, Minsk 03254
Ancient City, Samut Prakan 42597
Ancient Coins Museum, Beijing 07504
Ancient Foundry, Chlewiska 34593
Ancient High House, Stafford 45728
Ancient House Museum, Thetford 45846
Ancient Merv Historical Site, Mary 42979
Ancient Order of Foresters, Southampton 45701
Ancient Orient Museum, İstanbul 42817
Ancient Orient Museum, Tokyo 29413
Ancient Spanish Monastery of Saint Bernard de Clairvaux Cloisters, Miami 50619
Ancol Oceanarium Museum, Jakarta 24503
Andalusian Centre of Photography, Almería 38856
Anders Svors Museet, Hornindal 33561
Anderson/Abruzzo Albuquerque International Balloon Museum, Albuquerque 46176
Anderson County Arts Center, Anderson 46293
Anderson County Historical Museum, Garnett 48865
Anderson County Museum, Anderson 46294
Anderson Farm Museum, Lively 06176
Anderson Farm Museum, Madoc 06213
Anderson Gallery, Richmond 52322
Anderson Museum of Contemporary Art, Roswell 52467
Anderson Park Art Gallery, Invercargill 33128
Andersonville Prison, Andersonville 46295
Andhra Pradesh State Museum, Hyderabad 24210
Andhra Sahitya Parishat Government Museum, Kakinada 24249
Andijan Regional Museum of Literature and Art, Andijon 54438
Andijan Regional Museum of Local Lore, Andijon 54439
Andijon Viloyat Adabiyot va San'at Muzeyi, Andijon 54438
Andijon Viloyat Ülkashunoslik Muzeyi, Andijon 54439
Andong Folk Museum, Andong 29765
Andor Semsey Múzeum, Balmazújváros 23358
Andorra Romànica Centre d'Interpretació, La Massana 00095
Andover Historical Society Museum, Andover 46297
Andover Museum, Andover 43302
Andøymuseet, Risøyhamn 33783
Andrašiovská Obrazáreň, Krásnohorské Podhradie 38277
Andrassy Gallery, Krásnohorské Podhradie 38277
Andre Garitte Foundation, Berchem 03359
André Kertész Fotómúzeum, Szigetbecse 23888
André Kertész Photography Museum, Szigetbecse 23888
Andrea Robbi Museum, Sils Maria 41869
Andréemuseet, Gränna 40562
Andrej Upits' Memorial Apartment, Rīga 30061
Andreja Pumpura Muzejs, Lielvārde 30036
Andreja Upīša Memoriālmāja, Skrīveri 30126
Andreja Upītsa Memoriālais Muzejs, Rīga 30061
Andrejs Pumpurs Museum, Lielvārde 30036
Andres Pico Adobe, Mission Hills 50749
Andres Särev Memorial Museum of the Estonian Theatre and Music Museum, Tallinn 10620
Andres Särevi Kortermuuseum, Tallinn 10620
Andrew and District Local History Museum, Andrew 05340
Andrew Carnegie Birthplace Museum, Dunfermline 44054
Andrew County Museum, Savannah 52950
Andrew J. Blackbird Museum, Harbor Springs 49164
Andrew Johnson National Historic Site, Greeneville 49050
Andrew Logan Museum of Sculpture, Welshpool 45961
Andrew Ross Museum, Kangaroo Ground 01187
Andrey Rublev Museum of Ancient Russian Art, Moskva 37146
Andrey Sakharov Museum and Center World, Progress and Human Rights, Moskva 37284
Androscoggin Historical Society Museum, Auburn 46468
Andrøymuseet, Andenes 33392
The Andy Warhol Museum, Pittsburgh 51938
Andy Warhol Museum of the Modern Art, Medzilaborce 38304
Anekojikan - Kyoto Mangekyo Hakabutsukan, Kyoto 28802
Anfield Museum, Liverpool 44730
Angarskij Gorodskoj Vystavočnyj Zal, Angarsk 36759
Angarskij Muzej Časov, Angarsk 36760
Angarskij Muzej Mineralov, Angarsk 36761
Ånge Zoologiska Museum, Ånge 40424
Angel Kanchev House Museum, Trjavna 05228
Angel Mounds, Evansville 48455
Angel Row Gallery, Nottingham 45288
Angelika Kauffmann Museum - Gemeindemuseum, Schwarzenberg in Vorarlberg 02722
Angelos and Angeliki Giallina Gallery, Corfu 22973
Angels Attic Museum, Santa Monica 52897
Angermuseum, Erfurt 18196
Angermuseum - Margaretha-Reichardt-Haus, Erfurt 18197
Angkor Borei Museum, Angkor Borei 05282
The Angkor Conservation, Siem Reap 05290
Anglesey Heritage Gallery, Llangefni 44766
Anglican Cathedral Museum, Saint John's 06841
Angono National Museum, Angono 34333
Angren State Museum of Local Lore, Angren 54443
Angren Viloyat Ülkashunoslik Muzeyi, Angren 54443
Angus Folk Museum, Glamis 44272
Angyalföld Region History Collection and School Museum, Budapest 23381
Angyalföldi Helytörténeti Gyűjtemény, Budapest 23381
Anhaltische Gemäldegalerie, Dessau 17826
Anhui Agriculture Museum, Hefei 07774
Anhui Museum, Hefei 07775
Anhui Sheng Bo Wu Guan, Hefei 07776
Ani C, Longfosse 13807
Anicennes Mines de Grandfontaine, Grandfontaine 13102
Anich-Hueber-Museum, Oberperfuss 02457
Animal and Fish Museum, Evros 22996
Animal Ecology and Dinosaur Museum, Taipei 42373
Animal Museum, Amman 29643
Animalaine, Bastogne 03343
Animas Museum, Durango 48212
Anita Gallery, San Antonio 34497
Anıtkabir-Atatürk ve Kurtuluş Müzesi, Ankara 42696
Anjala Manor, Anjalankoski 10709
Anjalan Kartanomuseo, Anjalankoski 10709
Anji Museum, Anji 07478
Anka Gvozdanović Collection, Zagreb 09015
Ankang Zone Museum, Ankang 07479
Ankara Devlet Güzel Sanatlar Galerisi, Ankara 42697
Ankara State Fine Art Gallery, Ankara 42697
Anker's House Museum, Chester-le-Street 43819
Ankkapurha Industrial Museum, Anjalankoski 10710
Ankkapurhan Teollisuusmuseo, Anjalankoski 10710
Anli Art and Culture Museum, Shengang 42324
Anlu Museum, Anlu 07480
Ann Arbor Art Center, Ann Arbor 46302
Ann Arbor Hands-on Museum, Ann Arbor 46303
Ann Bryant Art Gallery, East London 38614
Anna Bemis Palmer Museum, York 54339
Anna Eaton Stout Memorial Art Gallery, Owensboro 51637
Anna Leonowens Gallery, Halifax 05929
Anna Miller Museum, Newcastle 51308
Anna-Seghers-Gedenkstätte, Berlin 17033
Annalan Kotiseutumuseo, Pyhäjoki 11244
Annandale, Tillsonburg 07083
Annapolis Valley Macdonald Museum, Middleton 06276
Annas Brigaderes Memoriālais Muzejs, Tērvete 30129
Anne Frank Center, Berlin 17034
Anne Frank Center USA, New York 51119
Anne Frank Huis, Amsterdam 31828
Anne Frank Zentrum, Berlin 17034
Anne & Gordon Samstag Museum of Art, Adelaide 00772
Anne Grimdalens Minne, Dalen i Telemark 33462
Anne Hathaway's Cottage, Stratford-upon-Avon 45779
Anne Hvides Gård, Svendborg 10288
Anne of Cleves House Museum, Lewes 44697
Anne of Green Gables Museum, Kensington 06059
The Anne Spencer Memorial Foundation, Lynchburg 50324
Anneberg-Samlingerne, Nykøbing Sjælland 10205
Annette Green Perfume Museum, Los Angeles 50209
Annie E. Woodman Institute, Dover 48166
Annie Riggs Memorial Museum, Fort Stockton 48688
Anniston Museum of Natural History, Anniston 46322
Annmary Brown Memorial, Providence 52160
Anno Art Museum, Tsuwano 29545
Anno Bijutsukan, Tsuwano 29545
L'Annonciade, Saint-Tropez 15525
Ano Moni Panagias Xenias, Almyros 22877
Anoka County Museum, Anoka 46323
Anola and District Museum, Anola 05346
Anqing Museum, Anqing 07481
Anqiu Museum, Anqiu 07482
Ansamblul Muzeal şi Monumental Dr. Vasile Lucaciu, Şişeşti 36624
L'Anse aux Meadows National Historic Park, Saint-Lunaire-Griquet 06865
Anse-Blanchette, Gaspé 05852
Anselmas Matutis Museum, Alytus 30204
Anselmo Matučio Memorialinis Muziejus, Alytus 30204
Anshan Museum, Anshan 07483
Anshun Zone Museum, Anshun 07484
Anson County Historical Society Museum, Wadesboro 53783
Antalya Devlet Güzel Sanatlar Galerisi, Antalya 42716
Antalya Müzesi, Antalya 42717
Antalya Museum, Antalya 42717
Antalya State Gallery, Antalya 42716
Antanas and Jonas Juskos Museum for Culture and Ethnography, Vilkija 30399
Antanas Jonynas Museum, Alytus 30205
Antanas Zmuidzinavicius Memorial Museum of the M.K. Čiurlionis National Art Museum, Kaunas 30239
Antano Baranausko ir Antano Vienuolio-Žukausko Memorialinis Muziejus, Anykščiai 30206

Antano ir Jono Juškų Etninėes Kultūros Muziejus, Vilkija 30399
Antano Jonyno Memorialinis Muziejus, Alytus 30205
Antano Mončio Namai-Muziejus, Palanga 30327
Antano Žmuidzinavičiaus Memorialinis Muziejus, Kaunas 30239
Antelope County Historical Museum, Neligh 51000
Antelope Valley Indian Museum, Lancaster 49914
Anten-Gräfsnäs Järnväg, Alingsås 40414
Antenne de l'Imprimerie de la Presse, Louhans 13832
Anthology Film Archives, New York 51120
Anthony Henday Museum, Delburne 05690
Anthony's Main Road Mining Museum, Idrija 38420
Anthropological Institute and Museum D.N. Anuchin, Moskva 37261
Anthropological Museum, Canchipur 24106
Anthropological Museum, Guwahati 24186
Anthropological Museum, Xiamen 08354
The Anthropological National Folklore Museum, Amman 29644
Anthropologische Sammlung der Universität Göttingen, Göttingen 18689
Anthropology Museum, DeKalb 48037
Anthropology Museum, Delhi 24135
Anthropology Museum, Johannesburg 38647
Anthropology Museum, Lucknow 24288
Anthropology Museum, Marshall 50504
Anthropology Museum, Ranchi 24371
Anthropology Museum, Saint Lucia 01483
Anthropology Museum, Winnipeg 07354
Anthropos Pavilion of the Moravian Museum, Brno 09439
Anti-Aircraft Defence Museum, Tuusula 11414
Anti-Aircraft Museum, Ha Noi 54607
Anti-Kriegs-Museum, Berlin 17035
Anti-Saloon League Museum, Westerville 54089
Anti War Museum, Berlin 17035
Antická Gerulata v Rusovciach, Bratislava 38203
Antico Frantoio, Massa Marittima 26625
Antietam National Battlefield, Sharpsburg 53053
Antifaschistische Mahn- und Gedenkstätte, Lieberose 19985
Antigonish Heritage Museum, Antigonish 05348
Antigua Cárcel de Jaruco, Jaruco 09188
Antigua Casa de los Marqueses de Campo Florido, San Antonio de los Baños 09272
Antiguo Ayuntamiento y Cárcel de Bejucal, Bejucal 09038
Antiken- und Abgußsammlung des Archäologischen Seminars, Marburg 20185
Antikenmuseum Basel und Sammlung Ludwig, Basel 41154
Antikenmuseum der Universität Leipzig, Leipzig 19893
Antikenmuseum und Abgusssammlung des Instituts für Klassische Archäologie Universität Heidelberg, Heidelberg 19069
Antikensammlung, Bern 41202
Antikensammlung, Kassel 19457
Antikensammlung, Kiel 19527
Antikensammlung der Friedrich-Alexander-Universität Erlangen-Nürnberg, Erlangen 18219
Antikensammlung im Pergamonmuseum, Berlin 17036
Antikes Wohnmuseum, Weinheim 22430
Antikmuseet, Göteborg 40543
Antikriegshaus Sieversshausen, Lehrte 19874
Den Antikvariske Samling, Ribe 10231
Antiquarian and Landmarks Society Museum, Hartford 49192
Antiquarium, Agropoli 25164
Antiquarium, Ariano Irpino 25274
Antiquarium, Avella 25330
Antiquarium, Borgia 25502
Antiquarium, Boscoreale 25517
Antiquarium, Brescello 25530
Antiquarium, Buccino 25558
Antiquarium, Caldarola 25596
Antiquarium, Cesenatico 25837
Antiquarium, Cimitile 25890
Antiquarium, Filadelfia 26133
Antiquarium, Fondi 26218
Antiquarium, Golasecca 26349
Antiquarium, Loreto Aprutino 26521
Antiquarium, Lugnano in Teverina 26542
Antiquarium, Manduria 26584
Antiquarium, Massarosa 26635
Antiquarium, Mergozzo 26665
Antiquarium, Minturno 26735
Antiquarium, Monasterace 26772
Antiquarium, Nonantola 26939
Antiquarium, Numana 26963
Antiquarium, Palazzolo Acreide 27058
Antiquarium, Palmi 27084
Antiquarium, Partinico 27111
Antiquarium, Patti 27114
Antiquarium, Porto Torres 27304
Antiquarium, Prato 27329
Antiquarium, Ravello 27367
Antiquarium, San Marzano sul Sarno 27665
Antiquarium, Santa Flavia 27700
Antiquarium, Santa Maria Capua Vetere 27704
Antiquarium, Sant'Antioco 27724
Antiquarium, Serravalle Scrivia 27817
Antiquarium, Tindari 27972
Antiquarium, Trieste 28075
Antiquarium Cantianense, San Canzian d'Isonzo 27620
Antiquarium Civico, Bagnolo San Vito 25349
Antiquarium Civico, Vazzano 28170
Antiquarium Civico, Vico del Gargano 28258
Antiquarium Comunale, Colleferro 25940
Antiquarium Comunale, Contessa Entellina 25963
Antiquarium Comunale, Milena 26729
Antiquarium Comunale, Monte Romano 26802
Antiquarium Comunale, Roma 27439
Antiquarium Comunale, Sezze 27831
Antiquarium Comunale, Tiriolo 27974
Antiquarium Comunale N. Pansoni, Cossignano 25994
Antiquarium del Castello Eurialo, Siracusa 27852
Antiquarium del Parco della Forza, Ispica 26418
Antiquarium del Seminario Vescovile, Nola 26935
Antiquarium del Serapeo, Tivoli 27977
Antiquarium del Teatro Greco-Romano, Taormina 27940
Antiquarium del Varignano, Portovenere 27313
Antiquarium delle Grotte di Catullo, Sirmione 27856
Antiquarium di Agrigento Paleocristiana Casa Pace, Agrigento 25159
Antiquarium di Canne, Barletta 25372
Antiquarium di Himera, Termini Imerese 27959
Antiquarium di Megara Hyblaea, Augusta 25327
Antiquarium di Monte Cronio, Sciacca 27794
Antiquarium di Nervia, Ventimiglia 28215
Antiquarium di Poggio Civitate, Murlo 26878
Antiquarium di Sant'Appiano, Barberino Val d'Elsa 25355
Antiquarium di Tesis, Vivaro 28301
Antiquarium di Torre Nao, Crotone 26011
Antiquarium di Villa Romana, Patti 27115
Antiquarium e Archeologica Museo, Cuglieri 26017
Antiquarium e Mosaico Romano, Bevagna 25412
Antiquarium e Zona Archeologica, Lugagnano Val d'Arda 26541
Antiquarium Forense, Roma 27440
Antiquarium Ipogeo dei Volumni, Perugia 27143
Antiquarium Jetino, San Cipirello 27624
Antiquarium Lucio Salvio Quintiano, Ossuccio 27015
Antiquarium Sebatum, Sankt Lorenzen 27693
Antiquarium Sestinale, Sestino 27821
Antiquarium Tellinum, Teglio 27951
Antiquarium Torre Cimalonga, Scalea 27781
Antiquarium - Villa Maritima, Minori 26734
Antique Auto and Race Car Museum, Bedford 46686
Antique Automobile Museum, Saint-Jacques 06825
Antique Boat Museum, Clayton 47600
Antique Gas and Steam Engine Museum, Vista 53764
Antique Tractor Museum and Frontier Village, Maple Creek 06231
Antiquités de Volubilis, Moulay-Idriss Zerhoun 31685
Antiquities Collection, University Museum of Cultural Heritage, Oslo 33757
Antiquities Museum, Alexandria 10470
Antiquities Museum, Saint Lucia 01484
Antiquities Museum of Tel Aviv-Yafo, Tel Aviv 25096
Antler River Historical Society Museum, Melita 06262
Anton Bruckner-Museum, Ansfelden 01724
Anton-Buchner-Heimatmuseum, Betzenstein 17311
Anton Chekhov House-Museum, Sumy 43188
Anton Hanak-Museum, Langenzersdorf 02260
Anton Pieck Museum, Hattem 32322
Anton van Wouw House, Brooklyn 38574
Anton Wildgans-Haus, Mödling 02391
Antona Austriņa Memoriālais Muzejs, Vecpiebalga 30141
Antonijev Rov - Rudnik Živega Srebra, Idrija 38420
Antonín Chittussi Gallery, Ronov nad Doubravou 09819
Antonín Dvořák Memorial, Nelahozeves 09672
Antonín Dvořak Memorial House, Rožmitál pod Třemšínem 09826
Antonín Dvořák Museum, Praha 09767
Antonopouleion Archaeological Museum, Pylos 23160
Antsla Gümnaasiumi Muuseum, Antsla 10560
Antwerp Fashion Museum, Antwerpen 03294
ANU School of Art Gallery, Acton 00766
Das Anwesen and Pharmacy Museum, New Braunfels 51021
Anxi Museum, Anxi, Fujian 07485
Anxi Museum, Anxi, Gansu 07486
Anyang Glaswegianjian Art Museum, Anyang 07487
Anyang Museum, Anyang 07488
Anyang Natural Impressions Garden Museum, Anyang 07489
Anyang Yinxu Bowu Yuan, Anyang 07489
Ányos Jedlik Memorial Exhibition, Győr 23583
Anyuan Museum, Anyuan 07490
ANZ Banking Museum, Melbourne 01273
Anza-Borrego Desert Park Museum, Borrego Springs 46915
Anzac Cottage, Mount Hawthorn 01328
ANZAC Memorial, Sydney 01543
Anzali Military Museum, Bandar-Anzali 24598
Anzio Beachhead Museum, Anzio 25245
Aohan Qi Museum, Aohan Qi 07491
Aomori-kenritsu Kyodokan, Aomori 28360
Aomori Prefectural Museum, Aomori 28360
Apartheid Museum, Johannesburg 38648
Apartment-Museum of G. Krasilnikov, Iževsk 36954
Apex Museum, Atlanta 46421
apexart, New York 51121
Aphrodisias Müzesi, Aydın 42719
Aphrodisias Museum, Aydın 42719
Apiculture Museum, Kalmthout 03747
Apolinario Mabini Birthplace Museum, Tanauan 34515
Apolinario Mabini Shrine and Museum, Manila 34429
Apostles Island National Lakeshore Maritime Museum, Bayfield 46656
Apotheken-Museum, Bad Münstereifel 16787
Apotheken-Museum, Schiltach 21555
Apothekenmuseum, Hofgeismar 19223
Apothicairerie, Gonesse 13080
Apothicairerie, Pont-de-Veyle 14884
Apothicairerie de l'Hôpital, Moutiers-Saint-Jean 14316
Apothicairerie de l'Hôtel Dieu-le-Comte, Troyes 15994
Appalachian Center for Crafts, Smithville 53135
Appaloosa Horse Club of Canada Museum, Claresholm 05618
The Appaloosa Museum and Heritage Center, Moscow 50863
Appartment-Museum of E.M. Yaroslavskyj, Jakutsk 36959
Appartment-Museum of Maksim Ammosov, Jakutsk 36960
Appenzell Cheese Dairy, Stein 41911
Appenzeller Brauchtumsmuseum, Urnäsch 41966
Appenzeller Schaukäserei, Stein 41911
Appenzeller Volkskunde-Museum, Stein 41912
The Apple Trees Museum, Burlington 47131
Appleton Art Center, Appleton 46329
Appleton Museum of Art, Ocala 51480
Applewood - Shaver Homestead, Etobicoke 05786
Applied Art Museum of the Glassworks in Bükk, Bükkszentkereszt 23496
Appomattox Court House, Appomattox 46333
Apriķu Novada Muzejs, Apriķi 29986
Apsheron Art Gallery, Baki 03140
Apskaitos Muziejus, Vilnius 30400
Apsley House, London 44784
Apteekkimuseo, Kouvola 10989
Apteekkimuseo ja Qwenselin Talo, Turku 11388
Aptucxet Trading Post Museum, Bourne 46966
Aqualeon Parc de la Natura, Albinyana 38810
Aquamarine Science Museum Fukushima, Iwaki 28605
Aquaria Vattenmuseum, Stockholm 40883
Aquarium-Musée de la Mer, Dinard 12750
Aquarium-Musée-Institut Paul Ricard, Six-Fours-les-Plages 15767
Aquarium-Muséum, Liège 03793
Aquarius Wassermuseum, Mülheim an der Ruhr 20409
Aquatic Hall of Fame and Museum of Canada, Winnipeg 07355
Aquatics Museum, Suez 10541
Aquazoo-Löbbecke Museum, Düsseldorf 17985
Aquilino Ribeiro House Museum, Peva 35822
Aquincumi Múzeum, Budapest 23382
Ara Sargsyan and Hakob Kojoyan House-Museum, Yerevan 00725
Arabia Museum-Gallery, Helsinki 10768
Arad Museum, Arad 24927
Aragon Museum, Ilupeju 33328
Arai Kinen Bijutsukan, Iwanai 28609
Arai Local History Museum, Arai 28362
Arai Memorial Museum of Art, Iwanai 28609
Arai-shiritsu Hakubutsukan, Arai 28362
Āraiši Museum Park, Lake Fortress, Drabeši 30002
Āraišu Muzejparks, Ezerpils, Drabeši 30002
Araki Shuseikan Museum, Nagoya 28992
Aram Khachaturyan Museum, Yerevan 00726
Arandjelovac Museum, Arandjelovac 38030
Arany Egyszarvú Patikamúzeum, Kőszeg 23683
Arany János Irodalmi Emlékház, Geszt 23564
Arany János Múzeum, Nagykőrös 23727
Arany Sas Patika - Könyvtár és Levéltár, Budapest 23383
Ararat Regional Art Gallery, Ararat 00809
Aratoi Wairarapa Museum of Art and History, Masterton 33140
Aravete Külamuuseum, Aravete 10561
Aravete Village Museum, Aravete 10561
Arbæjarsafn, Reykjavík 24014
Arbeia Roman Fort and Museum, South Shields 45697
Arbeidsgenot, Oudtshoorn 38696
Arbeitsgemeinschaft Friedhof und Denkmal, Kassel 54826
Arbeitsgemeinschaft "Historische Nahverkehrsmittel Leipzig" e.V., Leipzig 19894
Arbeitsgemeinschaft kirchlicher Museen und Schatzkammern, Würzburg 54827
Arbejdermuseet, København 10123
Arbetets Museum, Norrköping 40769
Arbil Museum, Arbil 24708
Arboga Bryggerimuseum, Arboga 40427
Arboga Museum, Arboga 40428
Årbols Skolmuseum, Åmål 40418
Arbor Lodge, Nebraska City 50995
Arbroath Art Gallery, Arbroath 43312
Arbroath Museum, Arbroath 43313
Arbuthnot Museum, Peterhead 45371
Arcade Historical Museum, Arcade 46335
Arcadia Historical Museum, Arcadia 46337
Arcadia University Art Gallery, Glenside 48936
ARCAM Architectuurcentrum Amsterdam, Amsterdam 31829
Archaelogical Museum Frankfurt, Frankfurt am Main 18354
Archaeolgical Museum of Catalonia-Girona, Girona 39343
Archaeological and Byzantine Museum Florina, Florina 22999
Archaeological and Ethnographic Museum, Edirne 42766
Archaeological and Ethnographic Museum, Łódź 34873
Archaeological and Ethnographical Museum, Corabia 36311
Archaeological and Ethnological Museum, Elazığ 42769
Archaeological and Historical Museum, Ramat Hashofet 25088
Archaeological and Land Study Collection, Alumot 24926
Archaeological Collection, Baltimore 46546
Archaeological Collection, Farsala 22997
Archaeological Collection, Filiatra 22998
Archaeological Collection, Kozani 23066
Archaeological Collection, Ljubuški 04268
Archaeological Collection, Loutra Aidipsou 23081
Archaeological Collection, Metković Split 08910
Archaeological Collection, Molyvos 23098
Archaeological Collection, Nin 08914
Archaeological Collection, Symi 23200
Archaeological Collection, Yehi'am 25130
Archaeological Collection and Lapidarium Dr. Grga Novak, Hvar 08878
Archaeological Collection of Agios Kirikos, Agios Kirikos 22871
Archaeological Collection of Amorgos, Amorgos 22879
Archaeological Collection of Pythagorion, Pythagorio 23164
Archaeological Collection of the University of Utrecht, Utrecht 32843
Archaeological Collection of the University Zürich, Zürich 42064
Archaeological Collection of Youth Village Nitzanim, Doar-Na Evtach 24943
Archaeological Collection Osor, Nerezine 08913
Archaeological Collection Panagia Kastrou, Platanos 23150
Archaeological Museum, Amman 29645
Archaeological Museum, Andros 22881
Archaeological Museum, Anuradhapura 40366
Archaeological Museum, Argostolion 22889
Archaeological Museum, Arhaia Olympia 22892
Archaeological Museum, Atalanti 22897
Archaeological Museum, Bijapur 24100
Archaeological Museum, Bišek 29951
Archaeological Museum, Bodhghaya 24103
Archaeological Museum, Břeclav 09416
Archaeological Museum, Budva 31638
Archaeological Museum, Chalkeio 22959
Archaeological Museum, Chalkida 22960
Archaeological Museum, Colombo 40368
Archaeological Museum, Corfu 22974
Archaeological Museum, Dedigama 40376
Archaeological Museum, Delhi 24136
Archaeological Museum, Delos 22981
Archaeological Museum, Gdańsk 34636
Archaeological Museum, Gianyar 24470
Archaeological Museum, Gorakhpur 24178
Archaeological Museum, Guntur 24183
Archaeological Museum, Guntur 24184
Archaeological Museum, Gwalior 24196
Archaeological Museum, Halebidu 24200
Archaeological Museum, Hampi 24201
Archaeological Museum, Hardwar 24203
Archaeological Museum, Iráklion 23018
Archaeological Museum, Jaunpur 24238
Archaeological Museum, Kamalapur 24251
Archaeological Museum, Karachi 34013
Archaeological Museum, Kavala 23046
Archaeological Museum, Khajuraho 24258
Archaeological Museum, Kibbutz Ein Dor 25036

A

Archaeological Museum, Kibbutz Gat 25040
Archaeological Museum, Kilkis 23053
Archaeological Museum, Konarak 24284
Archaeological Museum, Konya 42882
Archaeological Museum, Lamia 23070
Archaeological Museum, Lárissa 23072
Archaeological Museum, Lemesos 09384
Archaeological Museum, Mahastangarh 03231
Archaeological Museum, Mainamati 03232
Archaeological Museum, Merbok 30608
Archaeological Museum, Mykonos 23103
Archaeological Museum, Myohaung 31724
Archaeological Museum, Myrina 23105
Archaeological Museum, Nalanda 24334
Archaeological Museum, Néa Anchialos 23121
Archaeological Museum, Neishaboor 24637
Archaeological Museum, Nesebăr 05096
Archaeological Museum, Pagan 31725
Archaeological Museum, Paharpur 03233
Archaeological Museum, Paros 23135
Archaeological Museum, Poznań 35030
Archaeological Museum, Pune 24354
Archaeological Museum, Rethymnon 23165
Archaeological Museum, Rhodos 23170
Archaeological Museum, San Fernando del Valle de Catamarca 00571
Archaeological Museum, Sanchi 24375
Archaeological Museum, Sarnath 24380
Archaeological Museum, Sesimbra 35931
Archaeological Museum, Spanish Town 28340
Archaeological Museum, Swat 34043
Archaeological Museum, Taxila 34044
Archaeological Museum, Tbilisi 16351
Archaeological Museum, Trichur 24411
Archaeological Museum, Udaipur 24413
Archaeological Museum, Vaisali 24421
Archaeological Museum, Varanasi 24424
Archaeological Museum, Varna 05235
Archaeological Museum, Vila Viçosa 36018
Archaeological Museum, Vravrón 23241
Archaeological Museum, Wrocław 35326
Archaeological Museum and Portrait Gallery, Goa Velha 24177
Archaeological Museum at the A.Kh. Margulan Institut of Archaeology of the Academy of Sciences of the Republic of Kazakhstan, Almaty 29685
Archaeological Museum Blitar, Blitar 24455
Archaeological Museum Đerdap, Kladovo 38103
Archaeological Museum Eressos, Eressos 22993
Archaeological Museum Harappa, Harappa 34002
Archaeological Museum in Cracow, Kraków 34784
Archaeological Museum Kediri, Kediri 24512
Archaeological Museum Konarak, Puri 24363
Archaeological Museum Moen-Jo-Daro, Moen-Jo-Daro 34031
Archaeological Museum of Agrinion, Agrinion 22873
Archaeological Museum of Aiani, Aiani 22875
Archaeological Museum of Aigina, Aegina 22865
Archaeological Museum of Ancient Corinth, Ancient Corinth 22880
Archaeological Museum of Argos, Argos 22888
Archaeological Museum of Astros Kynourias, Astros 22895
Archaeological Museum of Astypalaia, Astypalaia 22896
Archaeological Museum of Baluchistan Quetta, Quetta 34038
Archaeological Museum of Chios, Chios 22967
Archaeological Museum of Chora, Messenia 23091
Archaeological Museum of Corinth, Kórinthias 23062
Archaeological Museum of Delphi, Delphi 22982
Archaeological Museum of Durrës, Durrës 00013
Archaeological Museum of Elefsina, Elefssina 22992
Archaeological Museum of Epidaurus, Asklepieion Lygourio 22894
Archaeological Museum of Eretria, Eretria 22994
Archaeological Museum of Galaxidi, Galaxidi 23004
Archaeological Museum of Gytheion, Gythio 23011
Archaeological Museum of Ioannina, Ioannina 23012
Archaeological Museum of Isthmia, Isthmia 23022
Archaeological Museum of Istria, Pula 08932
Archaeological Museum of Kampos Ikaria, Kampos 23033
Archaeological Museum of Karystos, Karystos 23039
Archaeological Museum of Kea, Kea 23049
Archaeological Museum of Kerameikos, Athinai 22923
The Archaeological Museum of Kfar Saba, Kfar Saba 25029
Archaeological Museum of Kimolos, Kimolos 23055
Archaeological Museum of Komotini, Komotini 23058
Archaeological Museum of Kos, Kos 23063
Archaeological Museum of Megara, Megara 23088
The Archaeological Museum of Meiji University, Tokyo 29432
Archaeological Museum of Mytilene, Mytilini 23109
Archaeological Museum of Nafplion, Nafplion 23115
Archaeological Museum of Naxos, Naxos 23119
Archaeological Museum of Nemea, Archaia Nemea 22887
Archaeological Museum of Nicopolis, Preveza 23158
Archaeological Museum of Patiala, Patiala 24343
Archaeological Museum of Patras, Patras 23138
Archaeological Museum of Pella, Pella 23141
Archaeological Museum of Piraeus, Piraeus 23143
Archaeological Museum of Plaka, Milos 23097
Archaeological Museum of Polygyros, Polygyros 23152
Archaeological Museum of Samos, Samos 23175
Archaeological Museum of Samothraki, Paleopoli 23130
Archaeological Museum of Shellmounds of Joinville, Joinville 04514
Archaeological Museum of Sifnos, Kastro 23044
Archaeological Museum of Siteia, Siteia 23188
Archaeological Museum of Spárti, Spárti 23194
Archaeological Museum of Stavros, Stavros Ithakis 23199
Archaeological Museum of Syros, Ermoupolis 22995
Archaeological Museum of Tegea, Tegea 23202
Archaeological Museum of Thassos, Thassos 23203
Archaeological Museum of the University of Münster, Münster 20503
Archaeological Museum of Thebes, Thebes 23205
Archaeological Museum of Thera, Thera 23206
Archaeological Museum of Thermos, Thermo 23208
Archaeological Museum of Thessaloniki, Thessaloniki 23209
Archaeological Museum of Tinos, Tinos 23226
Archaeological Museum of Veroia, Veria 23234
Archaeological Museum of Viuz-Faverges, Faverges 12908
Archaeological Museum Peshawar, Peshawar 34032
Archaeological Museum Poros, Poros 23156
Archaeological Museum Red Fort Delhi, Delhi 24137
Archaeological Museum Saidu Sharif, Saidu Sharif 34042
Archaeological Museum Serres, Serres 23179
Archaeological Museum Taj Mahal, Agra 24045
Archaeological Museum Umarkot, Umarkot 34045
Archaeological Museum Zadar, Zadar 08981
Archaeological Museums of Chania and Rethymnon, Chania 22961
Archaeological Park, Volterra 28311
Archaeological Preserve Nicopolis ad Istrum, Veliko Tărnovo 05249
Archaeological Receipts Fund, Athinai 54864
Archaeological Sit of Vitsa, Vitsa 23236
Archaeological Site, Paramythia 23134
Archaeological Site and Museum, Sicyon 23184
Archaeological Site Avedat, Avedat 24931
Archaeological Site Museum, Alampur 24056
Archaeological Site Nimrud (Calah), Nimrud 24723
Archaeological Site of Butrinti, Butrinti 00012
Archaeological Site of Edessa, Edessa 22989
Archaeological Site of Geraki, Geraki 23006
Archaeological Site of Lindos, Lindos 23078
Archaeological site of Megalopolis, Megalopolis 23087
Archaeological Site of Perachora, Perachora 23142
Archaeological Site Walraversijde 1465, Oostende 03928
Archaeological Survey of Alberta, Edmonton 05742
Archäologie am Landesmuseum Joanneum, Graz 01962
Archäologie-Museum Neuburg, Neuburg an der Donau 20597
Archäologiemuseum im Fränkischen Freilandmuseum, Bad Windsheim 16871
Archäologisch-Ethnographische Lehr- und Studiensammlung des Instituts für Altamerikanistik und Ethnologie der Universität Bonn, Bonn 17443
Archäologisch-Ökologisches Museum, Albersdorf 16451
Archäologische Ausgrabung Borg, Perl 21010
Archäologische Ausstellung in der Schule, Nassenfels 20548
Archäologische Gedenkstätte (Unterkirche), Klosterneuburg 02199
Archäologische Sammlung, Wels 02912
Archäologische Sammlung der Universität, Freiburg im Breisgau 18426
Archäologische Sammlung der Universität Fachbereich III, Trier 22095
Archäologische Sammlung der Universität Graz, Graz 01963
Archäologische Sammlung der Universität Rostock, Rostock 21386
Archäologische Sammlung der Universität Zürich, Zürich 42064
Archäologische Sammlung des Historischen Museums, Luzern 41587
Archäologische Sammlung Keltenhaus, Taufkirchen, Kreis München 22013
Archäologische Sammlung mit Heimatvertriebenenstube und Kruk-Sammlung, Königsbrunn 19644
Archäologische Staatssammlung München, München 20422
Archaeologische Verzameling van de Universiteit van Utrecht, Utrecht 32843
Archäologischer Park Cambodunum - APC, Kempten 19514
Archäologischer Park Carnuntum, Bad Deutsch-Altenburg 01741
Archäologischer Park Carnuntum, Petronell 02494
Archäologischer Park Magdalensberg, Pischeldorf 02500
Archäologischer Park mit Museums-Turm, Kellmünz an der Iller 19503
Archäologischer Park Ostkastell, Welzheim 22471
Archäologisches Freilichtmuseum Oerlinghausen, Oerlinghausen 20874
Archäologisches Landesmuseum, Schleswig 21565
Archäologisches Landesmuseum Baden-Württemberg, Konstanz 19676
Archäologisches Landesmuseum Brandenburg, Brandenburg an der Havel 17505
Archäologisches Landesmuseum Mecklenburg-Vorpommern, Schwerin 21691
Archäologisches Museum, Burgheim 17657
Archäologisches Museum, Donauwörth 17894
Archäologisches Museum, Oldendorf 20903
Archäologisches Museum, Rimpar 21333
Archäologisches Museum Amberg, Amberg 16532
Archäologisches Museum Colombischlössle, Freiburg im Breisgau 18427
Archäologisches Museum der Stadt Kelheim, Kelheim 19499
Archäologisches Museum der Stadt Weißenhorn, Weißenhorn 22460
Archäologisches Museum der Universität Münster, Münster 20503
Archäologisches Museum des Historischen Vereins für Oberfranken, Bayreuth 16945
Archäologisches Museum Essenbach, Essenbach 18271
Archäologisches Museum Frankfurt, Frankfurt am Main 18354
Archäologisches Museum Gablingen, Gablingen 18524
Archäologisches Museum in der Feldmühle, Rennertshofen 21276
Archäologisches Museum Innsbruck, Innsbruck 02113
Archäologisches Museum Lavant, Lavant 02271
Archäologisches Museum Neu-Ulm, Neu-Ulm 20586
Archäologisches Museum Renningen, Renningen 21277
Archäologisches Museum und Dokumentationszentrum Via Claudia, Fließ 01892
Archäologisches Pilgermuseum Hemmaberg-Juenna, Globasnitz 01935
Archäologisches Zentrum Hitzacker, Hitzacker 19202
Archaeology and History Museum, Głogów 34668
Archaeology and Nature Museum of Güzelyurt, Güzelyurt 33361
Archaeology Centre, Bagshot 43352
Archaeology Museum, Chennai 24112
Archaeology Museum, Korçë 00022
Archaeology Museum, Shamir 25094
Archaeology Museum Adamclisi, Adamclisi 36096
Archaeology Museum and Coin Cabinet, Graz 01985
Archaeology Museum Gozo, Victoria 30650
Archaeology Museum in Biskupin, Gąsawa 34630
Archaeology Museum in Cracow, Kraków 34785
Archaeology Museum of Lembah Bujang, Merbok 30609
Archangelos Mihail İkon Müzesi, Girne 33355
Archangelskij Gosudarstvennyj Muzej Zodčestva i Narodnogo Ikusstva - Malye Karely, Archangelsk 36765
Archangelskij Literaturnyj Muzej, Archangelsk 36766
Archangelskij Oblastnoj Kraevedčeskij Muzej, Archangelsk 36767
Archangelskij Oblastnoj Muzej Izobrazitelnych Iskusstv, Archangelsk 36768
Archangelskij Sobor, Moskva 37141
Archbishop's Palace, Trondheim 33923
Archdiocesan Library of Eger, Eger 23539
Archdiocesan Museum, Katowice 34724
Archdiocesan Museum, Poznań 35031
Archdiocesan Museum of Manila, Manila 34430
Arche Nebra, Wangen 22323
Archelaus Smith Museum, Cape Sable Island 05553
Archenhold Sternwarte, Berlin 17037
Archéo- et Géo-Espaces Moulin du Broukay, Eben-Emael 03567
Archéo-Musée du Kronos, Albens 11532
Archéo-Site Gaulois de Saint-Julien, Saint-Julien 15343
L'Archéodrome de Bourgogne, Meursault 14106
Archeologic Museum of Faro, Faro 35596
Archeologica di Sovana, Sorano 27878
Archeological and Natural History Museum at the National Park Brijuni, Brijuni 08848
Archeological and Nature Reserve, Krzemionki koło Ostrowca 34834
Archeological Collection of Montenegro, Podgorica 31653
Archeological Exhibition at Ahmed Bey Mosque, Kjustendil 05063
Archeological Museum, Bergama 42728
Archeological Museum, Burgas 04995
Archeological Museum, Čerepovec 36847
Archeological Museum, Plovdiv 05124
Archeological Museum, Radnevo 05142
Archeological Museum, Septemvri 05158
Archeological Museum, Sozopol 05206
Archeological Museum Dubrovnik, Dubrovnik 08860
Archeological Museum of Surkhandarya Region, Termiz 54466
Archeological Museum of the Kasan University, Kazan 37004
Archeological Museum-Preserve, Anapa 36758
Archeological Museums of Istanbul, İstanbul 42823
Archeological Open Air Museum in Březno at Louny, Louny 09642
Archeological Reserve, Giecz 34663
Archeological Reserve of Old Shumen and Shumen Fortress, Šumen 05212
Archeological Site Heraklea, Bitola 30499
Archeologičeska Ekspozicija v Džamia Achmed Bej, Kjustendil 05063
Archeologičeska Muzej, Silistra 05160
Archeologičeski Institut s Muzej, Sofia 05172
Archeologičeski Muzej, Burgas 04995
Archeologičeski Muzej, Plovdiv 05124
Archeologičeski Muzej, Radnevo 05142
Archeologičeski Muzej, Septemvri 05158
Archeologičeski Muzej, Sozopol 05206
Archeologičeski Muzej, Varna 05235
Archeologičeski Rezervat Starijat Grad Šumen i Šumenska Krepost, Šumen 05212
Archeologičeskij Muzej, Biškek 29951
Archeologičeskij Muzej, Čerepovec 36847
Archeologičeskij Muzej Kazanskogo Universiteta, Kazan 37004
Archeologické Múzeum, Bratislava 38204
Archeologické Muzeum, Břeclav 09416
Archeologiemuseum, Stein 32805
Archeologisch Centrum C.W. Bruinvis, Alkmaar 31792
Archeologisch en Paleontologisch Museum Hertogsgemaal, Gewande 32248
Archeologisch Museum, Harelbeke 03677
Archeologisch Museum, Mechelen 03856
Archeologisch Museum Abdij, Affligem 03262
Archeologisch Museum Haarlem, Haarlem 32293
Archeologisch Museum Van Bogaert-Wauters, Hamme 03672
Archeologisch Museum van de Universiteit, Gent 03622
Archeologisch Site Museum, Werken 04146
Archeologisch Streekmuseum Het Labyrint, Kemmel-Heuvelland 03751
Archeology Museum, Murol 14336
Archeology Museum of the Slovak National Museum, Bratislava 38204
Archeološki Grad Stobi, Negotino 30509
Archeološki Muzej Cherakleja, Bitola 30499
Archeoparc, Schnals 27793
Archéoscope, Le Mont-Saint-Michel 13620
Archéoscope du Pays de Salm, Vielsalm 04110
Archéoscope Godefroid de Bouillon, Bouillon 03391
Archéosite de l'Hospice des Pèlerins de Saint-Jacques-de-Compostelle, Pons 14876
Archéosite du Château et Village Médiéval, Montségur 14264
Archéosite Gallo-Romain des Flaviers, Mouzon 14318
Archéosite Préhistorique de Cambous, Saint-Martin-de-Londre 15399
Archer County Museum, Archer City 46341
Archer Gallery, Vancouver 53704
Arches Museum of Pioneer Life, Winona 54238
Archibald Graham McIlwaine House, Petersburg 51796
Archibald Historical Museum, La Rivière 06114
Archidiocesan Collections of Kalocsa - Library of the Cathedral, Kalocsa 23625
Archidiocesan Museum, Szczecin 35158
Archidiocesan Museum in Cracov, Kraków 34786
Archie Bray Foundation, Helena 49239
Archief en Museum van het Vlaams Leven te Brussel, Bruxelles 03432
Archief en Museum Zusters van Liefde van Jezus en Maria, Gent 03623
Archiepiscopal Castle, Fertőrákos 23561
Archiepiscopal Collection Center, Eger 23537
Archiepiscopal Palace, Sümeg 23828
Archiepiscopal Treasury of Kalocsa (Archidiocesan Museum of Kalocsa), Kalocsa 23624
Archipelagos Cultural Center, Athinai 22899

Architect Alexander Tamanyan Research Institute-Museum, Yerevan 00727
Architectur and Ethnographic Museum Talcy, Irkutsk 36931
Architectural Ensemble of the former Spas-Jakovlevsky Monastery, Rostov 37598
Architectural Museum Reserve Tsarevets, Veliko Tărnovo 05251
Architecture Foundation, London 44785
Architecture Museum, Vilnius 30401
Architecture Museum of Ljubljana, Ljubljana 38444
Architektur im Ringturm, Wien 02931
Architekturforum Zürich, Zürich 42065
Architekturgalerie Luzern, Ausstellungsraum Partikel, Luzern 41588
Architekturmuseum der Technischen Universität Berlin, Berlin 17038
Architekturmuseum der Technischen Universität München, München 20423
Architekturmuseum Schwaben, Augsburg 16611
Architekturno-chudožestvennyj Ansambl byvšego Spaso-Jakovleskogo Monastyrja, Rostov 37598
Architekturno-ètnografičeskij Muzej Angarskaja Derevnja, Bratsk 36820
Architekturno-ètnografičeskij Muzej Derevjannogo Zodčestva Chochlovka, Perm 37543
Architekturno-etnografski Muzej Etăr, Gabrovo 05029
Architekturno-istoričski Muzej-rezervat, Boženci 04993
Architekturno-muzeen Rezervat Arbanassi, Veliko Tărnovo 05250
Architekturno-muzeen Rezervat Carevec, Veliko Tărnovo 05251
Architektūros Muziejus, Vilnius 30401
Architekturzentrum Wien, Wien 02932
Archiv Bürgerbewegung Leipzig e.V., Leipzig 19895
Archiv der deutschen Jugendbewegung, Witzenhausen 22631
Archiv der Universität Wien, Schausammlung, Wien 02933
Archiv für Philatelie der Museumsstiftung Post und Telekommunikation, Bonn 17444
Archiv und Kollektion Paul Kaiser-Reka, Brandenburg an der Havel 17506
Archiv und Museum des Heimatkreises Leitmeritz, Fulda 18504
Archiv Werner Kittel, Hamburg 18916
Archives de la Construction Moderne, Lausanne 41534
Archives et Musée de la Littérature, Bruxelles 03433
Archives Museum - Temple Mickve Israel, Savannah 52938
Archives of American Art, Washington 53846
Archivio Segreto Vaticano, Città del Vaticano 54496
Archivio Storico Fiat, Torino 27996
Archivo-Museo Don Alvaro de Bazán, Viso del Marqués 40316
Archivo-Museo San Juan de Dios-Casa de los Pisas, Granada 39354
Arcola Museum, Arcola 05349
Arctic Corsair, Kingston-upon-Hull 44592
Ardennen Poteau '44 Museum, Sankt Vith 03998
Ardoisalle, Alle-sur-Semois 03263
Ardress Farmyard Museum, Annaghmore 43304
Ardrossan Historical Museum, Ardrossan 00811
Ards Art Centre, Newtownards 45260
Aremark Historielag, Ostfold 33760
Arend Dieperink Museum, Potgietersrus 38718
Arendal Bymuseum, Arendal 33394
Arents House, Brugge 03406
Arentshuis, Brugge 03406
Argauisch Kantonales Weinbaumuseum, Tegerfelden 41929
Argentine Museum of Sociology, Buenos Aires 00274
Argentine Notarial Museum, Buenos Aires 00266
Argeş County Museum, Piteşti 36526
Argo Gold Mine, Idaho Springs 49463
L'Argonaute, Paris 14565
The Argory, Dungannon 44057
Argyll and Sutherland Highlanders Regimental Museum, Stirling 45747
Arhaiologiko Mouseio Thessaloniki, Thessaloniki 23209
Arheološka Zbirka, Metković Split 08910
Arheološka Zbirka, Nin 08914
Arheološka Zbirka, Stari Grad 08964
Arheološka Zbirka Crne Gore, Podgorica 31653
Arheološka Zbirka i Lapidarij Dr. Grga Novak, Hvar 08878
Arheološka Zbirka Osor, Nerezine 08913
Arheološki i Prirodoslovni Muzej, Brijuni 08848
Arheološki Muzej, Budva 31638
Arheološki Muzej, Most na Soči 38484
Arheološki Muzej Đerdap, Kladovo 38103
Arheološki Muzej Dubrovnik, Dubrovnik 08860
Arheološki Muzej Istre, Pula 08932
Arheološki Muzej u Splitu, Split 08954
Arheološki Muzej Zadar, Zadar 08981
Arheološki Muzej Zagreb, Zagreb 08986
Arhitekturni Muzej Ljubljana, Ljubljana 38444
Århus Bymuseum, Århus 09961
Arimatsu Narumi Shibori Kaikan, Nagoya 28993
Arita-machi Minzoku Hakubutsukan, Arita 28363
Arita Porcelain Museum, Arita 28364
Arita Toji Bijutsukan, Arita 28364
Arithmeum, Bonn 17445
Arizona Doll and Toy Museum, Phoenix 51897
Arizona Historical Museum, Tucson 53582
Arizona Historical Society Museum at Papago Park, Tempe 53464
Arizona Historical Society Pioneer Museum, Flagstaff 48550
Arizona Jewish Historical Exhibitions, Phoenix 51898
Arizona Mining and Mineral Museum, Phoenix 51899
Arizona Museum for Youth, Mesa 50604
Arizona Museum of Natural History, Mesa 50605
Arizona Science Center, Phoenix 51900
Arizona-Sonora Desert Museum, Tucson 53583
Arizona State Capitol Museum, Phoenix 51901
Arizona State Museum, Tucson 53584
Arizona State University Art Museum, Tempe 53465
Arizona Wing Commemorative Air Force Museum, Mesa 50606
Arizona Women Hall of Fame, Phoenix 51902
Ark-La-Tex Antique and Classic Vehicle Museum, Shreveport 53083
ARKA Kulturwerkstatt e.V., Essen 18248
Arkadi Monastery Collection, Rethymnon 23166
Arkaeologisk Museum, Korsør 10167
Arkansas Air Museum, Fayetteville 48527
Arkansas Arts Center, Little Rock 50138
Arkansas County Doctor Museum, Lioncoln 50124
Arkansas Historic Wine Museum, Paris 51699
Arkansas Museum of Natural Resources, Smackover 53132
Arkansas Museums Association, North Little Rock 55114
Arkansas National Guard Museum, North Little Rock 51396
Arkansas Post Museum, Gillett 48908
Arkansas State University Art Gallery, Jonesboro 49646
Arkansas State University Museum, Jonesboro 49647
Arken Museum of Modern Art, Ishøj 10108
Arkeologisk Museum i Stavanger, Stavanger 33863
Arkeoloji Müzesi, Konya 42882
Arkeoloji Müzesi, Kütahya 42892
Arkeoloji ve Etnografya Müzesi, Elazığ 42769
Arkhangelsk Literary Museum, Archangelsk 36766
Arkhangelsk Regional Museum, Archangelsk 36767
Arkitekturmuseet, Oslo 33716
Arkitekturmuseet, Stockholm 40884
Arklio Muziejus, Anykščiai 30207
Arklow Maritime Museum, Arklow 24729
Arkoloji ve Etnografya Müzesi, Edirne 42766
Arkona Lion's Museum, Arkona 05353
Arktisk Museum Nanoq, Jakobstad 10864
The Arkwright Society, Cromford 43931
Arlington Antebellum Museum, Birmingham 46820
Arlington Arts Center, Arlington 46355
Arlington Court, Arlington 43315
Arlington Heights Historical Museum, Arlington Heights 46360
Arlington Historical Museum, Arlington 46356
Arlington Historical Society Collection, Arlington 46348
Arlington House - The Robert E. Lee Memorial, Arlington 46357
Arlington Mill Museum, Bibury 43474
Arlington Museum of Art, Arlington 46350
Armadale Community Museum, Armadale 43316
Armádní Museum Žižkov, Praha 09734
Armagh County Museum, Armagh 43317
Armand-Frappier Museum, Laval 06145
Armand Frappier Museum, Laval 06145
Armando Museum, Amersfoort 31809
Armed Forces Museum, Oslo 33724
Armed Forces Museum, Taipei 42374
Armed Forces of the Philippines Museum, Quezon City 34485
Armed Forces Survey Department Museum, Bangkok 42529
Armeemuseum 'Friedrich der Große', Kulmbach 19744
Armeepferdemuseum, Scherz 41833
Armémuseum, Stockholm 40885
Armenian Cathedral and Museum, Isfahan 24610
Armenian Museum, Jerusalem 34054
Armenian Museum of America, Watertown 53970
Armeria Reale, Torino 27997
Armidale Folk Museum, Armidale 00813
Armory Art Gallery, Blacksburg 46843
Armory at the Moscow Kremlin, Moskva 37392
Armory Center for the Arts, Pasadena 51718
Armoury, Dresden 17967
Arms Family Museum of Local History, Youngstown 54352
Arms Museum of Liège, Liège 03804
Armstrong Browning Library, Waco 53771
Armstrong-Spallumcheen Museum and Art Gallery, Armstrong 05354
Army History Museum and Park, Kecel 23642
(Army Horses Museum), Scherz 41833
Army Medical Corps Centre Museum, Lucknow 24289
Army Medical Services Museum, Mytchett 45187
Army Museum, Ha Noi 54609
Army Museum, Halifax 05930
Army Museum, Stockholm 40885
Army Museum and Fort Copacabana, Rio de Janeiro 04681
Army Museum Bandiana, Bandiana 00832
Army Museum of Infantry Center, Phachuab Khiri Khan 42591
Army Museum of New South Wales, Paddington, New South Wales 01398
Army Museum of South Australia, Keswick 01194
Army Museum of Western Australia, Fremantle 01091
Army Museum Žižkov, Praha 09734
Army Physical Training Corps Museum, Aldershot 43272
Arnhems Oorlogsmuseum 1940-45, Arnhem 31922
Arnol Blackhouse Museum, Arnol 43322
Arnold Schönberg Center, Wien 02934
Arnolfini, Bristol 43628
Arnot Art Museum, Elmira 48378
Arnprior and District Museum, Arnprior 05355
Arnulf Rainer Museum, Baden bei Wien 01772
Aron Museum, Westmount 07305
Aros Aarhus Kunstmuseum, Århus 09962
Aros Aarhus Museum, Århus 09962
Arp Museum Bahnhof Rolandseck, Remagen 21259
Arquebus Krigshistorisk Museum, Haugesund 33536
Arquivo Histórico e Museu de Canonas Dr. Sezefredo Azambuja Vieira, Canoas 04401
Arran Heritage Museum, Brodick 43650
Arrantzaleen Museoa, Bermeo 39030
Arras Fine Arts Museum, Arras 11698
Arrow Rock State Historic Site, Arrow Rock 46364
Arrowhead Museum, Pittsfield 51965
Arrowmont School of Arts and Crafts Collection, Gatlinburg 48873
Ars Aevi, Sarajevo 04272
Ars Electronica Center, Linz 02308
Ars Nova - Nykytaiteen Museo, Turku 11389
Arsbel Museum of Automatic Musical Instruments, Tendo, Tendo 29356
Arsbel Tendo O-tomachikku Gakki Hakubutsukan, Tendo 29356
Arsenal Gallery, New York 51122
Arsenal Museum, London 44786
Arsenal Museum, Lviv 43131
Arsenal - Museum of the Great Patriotic War, Voronež 37971
Arsenev State Museum of the Primorsky Region, Vladivostok 37949
Årsunda Viking Museum, Årsunda 40433
Art and Art History Department, Providence 52161
Art and Culture Center of Hollywood, Hollywood 49322
Art and Design Gallery, Hatfield 44418
Art and Exhibition Complex of the Museum of History and Culture of the Middle Prikamje, Sarapul 37744
Art and History Museum of Zarautz, Zarautz 40355
Art Association of Harrisburg, Harrisburg 49178
Art Cellar, Chabarovsk 36855
Art Center, Douliou 42183
Art Center, Flushing 48574
The Art Center, Mount Clemens 50871
Art Center, Tainan 42359
Art Center at Fuller Lodge, Los Alamos 50202
Art Center Berlin Friedrichstrasse, Berlin 17039
The Art Center Chagall, Ostrava 09700
Art Center Gallery, Warrensburg 53828
Art Center In Hargate, Concord 47773
Art Center of Battle Creek, Battle Creek 46643
Art Center of Corpus Christi, Corpus Christi 47817
Art Center of South Florida, Miami Beach 50640
Art Center of Waco, Waco 53772
Art Center Sarasota, Sarasota 52912
Art Centr Puškinskaja 10, Sankt-Peterburg 37624
Art Centre, Haifa 24952
Art Collection, Acton 00767
Art Collection, Stirling 45748
Art Collection of the Franciscan Monastery, Hvar 08881
Art Complex Museum, Duxbury 48227
Art Cru Muséum, Bordeaux 12061
Art Department Gallery, Warwick 53836
Art en Stalles, Pouzac 14932
Art Exhibition Hall of Bratsk, Bratsk 36821
Art Exhibition Room, Yilan 42464
Art for Change, New York 51123
Art Forum Würth, Capena 25655
Art Galleries, Northridge 51426
Art Galleries at UAH, Huntsville 49441
Art Galleries of Ramapo College, Mahwah 50400
Art Gallery, Atlanta 46422
Art Gallery, Balčik 04974
Art Gallery, Blagoevgrad 04990
Art Gallery, Borisoglebsk 36818
Art Gallery, Brooklyn 47066
Art Gallery, Burgas 04997
Art Gallery, Chaskovo 05003
Art Gallery, Čirpan 05008
The Art Gallery, College Park 47685
Art Gallery, Dimitrovgrad 05014
Art Gallery, Dobrič 05016
Art Gallery, Drjanovo 05022
Art Gallery, Gabrovo 05030
Art Gallery, Gălăbovo 05034
Art Gallery, Haifa 24953
Art Gallery, Havlíčkův Brod 09511
Art Gallery, Herceg-Novi 31645
The Art Gallery, Hillsborough 44468
Art Gallery, Jamestown 49595
Art Gallery, Karlovo 05048
Art Gallery, Kavarna 05053
Art Gallery, Kazanlăk 05056
Art Gallery, Kitahiroshima 28706
Art Gallery, Kruševac 38113
The Art Gallery, Kuala Lumpur 30571
Art Gallery, Levski 05082
Art Gallery, Loveč 05085
Art Gallery, Mezdra 05093
Art Gallery, Minneapolis 50728
Art Gallery, Novi Pazar 05098
Art Gallery, Oneonta 51562
Art Gallery, Provadia 05138
Art Gallery, Provadia 05139
Art Gallery, Radnevo 05143
Art Gallery, San Antonio 52706
Art Gallery, Silistra 05161
The Art Gallery, Singapore 38170
Art Gallery, Skopje 30525
Art Gallery, Sozopol 05207
Art Gallery, Velingrad 05259
Art Gallery, Vidin 05261
Art Gallery, Žeravna 05268
Art Gallery Daut Pasha Bath, Skopje 30515
Art Gallery Elena Karamikhailova, Šumen 05213
Art Gallery Georgi Papazov, Jambol 05041
Art Gallery - House of Puppets of Tatjana Kolinina, Petrozavodsk 37556
Art Gallery Kendall, Miami 50620
Art Gallery Kursumli An, Skopje 30516
Art Gallery Maribor, Maribor 38479
Art Gallery Marist College, Poughkeepsie 52119
Art Gallery New Zealand, Wellington 33243
Art Gallery Nikolaj Pavlovikh, Svištov 05219
Art Gallery of Algoma, Sault Sainte Marie 06928
Art Gallery of Bancroft, Bancroft 05378
Art Gallery of Calgary, Calgary 05514
Art Gallery of Greater Victoria, Victoria 07241
Art Gallery of Hamilton, Hamilton 05951
Art Gallery of Hungkuang University, Shalu 42319
Art Gallery of Mississauga, Mississauga 06308
Art Gallery of New South Wales, Sydney 01544
Art Gallery of Newfoundland and Labrador, Saint John's 06842
Art Gallery of North York, North York 06484
Art Gallery of Northumberland, Cobourg 05628
Art Gallery of Nova Scotia, Halifax 05931
Art Gallery of Ontario, Toronto 07091
Art Gallery of Peterborough, Peterborough 06589
Art Gallery of Prince Albert, Prince Albert 06659
Art Gallery of South Australia, Adelaide 00773
Art Gallery of Southwestern Manitoba, Brandon 05458
Art Gallery of Sudbury, Sudbury 07026
Art Gallery of Swift Current National Exhibition Centre, Swift Current 07044
Art Gallery of the College of Fine and Applied Arts, Khartoum 40386
Art Gallery of the Faculty of Painting, Sculpture and Graphic Art, Bangkok 42530
Art Gallery of the Republic of Srpska Banja Luka, Banja Luka 04256
Art Gallery of the Society for Macedonian Studies, Thessaloniki 23210
Art Gallery of Uzbekistan, Toshkent 54470
Art Gallery of Western Australia, Perth 01410
Art Gallery of Windsor, Windsor, Ontario 07344
Art Gallery of York University, North York 06485
Art Gallery on the Bastion, Pskov 37573
Art Gallery Ruse, Ruse 05151
Art Gallery Strumica, Strumica 30529
Art Gallery Varna, Varna 05236
Art Gallery Wozownia, Toruń 35191
Art Gluchowe, Glauchau 18656
Art Guild of Burlington Gallery, Burlington 47132
The Art Gym, Marylhurst 50522
Art Hall Miraku, Ikoma 28563
Art History and Archaeology Museum, Bellinzona 41193
Art House Project, Naoshima 29025
Art House - Vass collection, Veszprém 23961
Art in Embassies Program, U.S. Department of State, Washington 53847

A

The Art Institute of Boston Main Gallery, Boston 46917
The Art Institute of Chicago, Chicago 47452
Art Institute of Pittsburgh, Pittsburgh 51939
Art League of Houston, Houston 49380
Art League of Manatee County, Bradenton 46984
Art Library, Leeds 44659
Art Metropole, Toronto 07092
Art Museum, Denpasar 24467
Art Museum, Iaşi 36417
Art Museum, Ilulissat 23256
Art Museum, Jakarta 24499
Art Museum, Kokkola 10975
Art Museum, Los Gatos 50272
Art Museum, Medan 24528
Art Museum, Padang 24534
Art Museum, Staryj Oskol 37799
Art Museum, Târgu Jiu 36664
Art Museum, Thiruvananthapuram 24402
Art Museum, Tulcea 36703
Art Museum, Valujki 37915
The Art Museum at Florida International University, Miami 50621
Art Museum of Dazu Rock Carvings in Chongqing, Chongqing 07628
Art Museum of Estonia, Tallinn 10624
Art Museum of Greater Lafayette, Lafayette 49864
Art Museum of Hsinchu County, Jhubei 42223
Art Museum of Kouvola, Kouvola 10990
Art Museum of Northern Norway, Tromsø 33918
Art Museum of Sketches, Køge 10160
Art Museum of Sochi, Soči 37784
Art Museum of South Texas, Corpus Christi 47818
Art Museum of Southeast Texas, Beaumont 46670
Art Museum of Tartu University, Tartu 10661
Art Museum of the Americas, Washington 53848
Art Museum of the Chinese University of Hong Kong, Hong Kong 07792
Art Museum of The Republic Marij El, Joškar-Ola 36979
Art Museum of the University of Memphis, Memphis 50566
Art Museum of Western Virginia, Roanoke 52379
Art Nouveau and Art Déco Museum, Salamanca 39936
Art Nouveau House, Varde 10306
Art of Velvet Gallery De Stierenstal, Montfoort 32560
Art Omi, Ghent 48905
Art Omi, New York 51124
Art Pace, San Antonio 52707
Art Pavilion in Zagreb, Zagreb 09014
Art Pedagocical Museum of Toys, Sergiev Posad 37762
Art Podvalčik, Chabarovsk 36855
Art Retreat, Singapore 38171
Art Salon of the Museum of Contemporary Art, Beograd 38087
Art Sation, Stone Mountain 53309
Art Site of Chai-Yi Railway Warehouse, Chai-Yi 42149
The Art Studio, Beaumont 46671
Art-Tech Exhibitions, San Jose 52804
Art Tower Mito, Contemporary Art Gallery, Mito 28921
Art Via Mondsee, Mondsee 02403
Art West Gallery, Jackson 49570
Artarmon Galleries, Artarmon 00815
Artas Folklore Museum, Bethlehem 34048
Artbox, Mattersburg 02364
Artcentre Mältinranta, Tampere 11361
Artchivum, Budapest 23384
Artemis Archery Collection, Oldland 45315
Artemisia Gallery, Chicago 47453
Artesia Historical Museum and Art Center, Artesia 46365
Artforms Gallery Manuyunk, Philadelphia 51814
Artful Deposit Galleries, Allentown 46224
Arthe, Dinan 12745
Arthouse at the Jones Center, Austin 46500
Arthur A. Houghton Jr. Gallery and the Great Hall Gallery, New York 51125
Arthur Boskamp-Stiftung, Hohenlockstedt 19239
Arthur Cottage, Ballymena 43363
The Arthur J. Moore Methodist Museum, Saint Simons Island 52644
Arthur M. Sackler Gallery, Washington 53849
Arthur M. Sackler Museum, Cambridge 47172
Arthur M. Sackler Museum of Art and Archaeology, Beijing 07505
Arthur Ross Gallery, Philadelphia 51815
Arthurdale Heritage Museum, Arthurdale 46366
Arthur's Court - Dept of the Historical Museum Gdańsk, Gdańsk 34634
Arti et Amicitiae, Amsterdam 31830
Articulate Gallery, Washington 53850
Artificial Light in Art Centre, Eindhoven 32184
Artillery Company of Newport Military Museum, Newport 51317
Artillery Museum of Finland, Hämeenlinna 10742
Artis Geologisch Museum, Amsterdam 31831
Artistenmuseum, Marburg 20186
Artists Association of Nantucket Gallery, Nantucket 50942
Artists Center and Foundation, Makati 34412
Artists' Cooperative Gallery, Omaha 51546
Artists House Jerusalem, Jerusalem 24978
Artists House Tel Aviv, Tel Aviv 25097
Artists' Museum, Łódź 34875
The Artists Museum, New York 51126
Artists' Space, Makati 34413
Artist's Space, New York 51127
Artium - Centre Museum of Contemporary Art, Vitoria-Gasteiz 40317
Artium - Centro Museo Vasco de Arte Contemporáneo, Vitoria-Gasteiz 40317
Artjärven Kotiseutumuseo, Artjärvi 10713
Artjärvi Museum, Artjärvi 10713
Artmedia Art Gallery, Beograd 38034
Artmobile, Newtown 51344
Artonivo, Brugge 03407
artothek, Köln 19595
Artothek, Krems 02223
Artothek der Zentral- und Landesbibliothek Berlin, Berlin 17040
Artothek-Galerie, Wien 02935
Artothek im Bonner Kunstverein, Bonn 17446
Artothek Worpswede, Worpswede 22690
Artothèque, Angers 11601
Artothèque Antonin Artaud, Marseille 13988
Artothèque-Sud, Nîmes 14445
Artpoint Gallery and Studios Society, Calgary 05515
Artrain USA, Ann Arbor 46304
Artroom Würth Austria, Böheimkirchen 01792
Arts Alliance Center at Clear Lake, Nassau Bay 50983
Arts and Crafts Museum, Bhavnagar 24085
Arts and Science Center for Southeast Arkansas, Pine Bluff 51920
Arts Association Ibbenbueren, Ibbenbüren 19304
The Arts Center, Bangkok 42531
Arts Center, Jhongli 42219
The Arts Center, Saint Petersburg 52634
The Arts Center of the Capital Region, Troy 53573
Arts Center of the Ozarks-Gallery, Springdale 53211
Arts Centre, Washington 45946
Arts Centre of Christchurch, Christchurch 33068
The Arts Club of Chicago, Chicago 47454
Arts Club of Washington, Washington 53851
Arts Council Gallery, Silver City 53100
Arts Council of Fayetteville/Cumberland County, Fayetteville 48530
Arts Council of Wayne County - Museum, Goldsboro 48964
Arts etc., Baki 03141
Arts in the Academy, Washington 53852
Arts Midland Galleries, Midland 50666
Arts Project Australia, Northcote 01379
Artsakh State Museum, Stepanakert 03197
Artsonje Center, Seoul 29864
Artsonje Museum, Gyeongju 29825
Artspace, Auckland 33041
Artspace, Peterborough 06590
Artspace, Raleigh 52227
Artspace, Richmond 52323
Artspace, Woolloomooloo 01670
Artspace-Lima, Lima 50085
Artspace Mackay, Mackay 01246
Artspeak Gallery, Vancouver 07198
Artswatch, Louisville 50282
Artworks, Trenton 53556
Arundel Museum and Heritage Centre, Arundel 43323
Arundel Toy Museum Dolls House, Arundel 43324
Arusha Declaration Museum, Arusha 42507
Arvada Center for the Arts and Humanities, Arvada 46367
Arvidsjaurs Hembygdsmuseum, Arvidsjaur 40434
Arvika Fordonmuseum, Arvika 40436
Arvika Konsthall, Arvika 40437
Arxiu del Monestir, Sant Joan de les Abadess 39985
Arzamasskij Gosudarstvennyj Literaturno-Memorialnyj Muzej A.P. Gajdara, Arzamas 36774
Arzamasskij Istoriko-chudožestvennyj Muzej, Arzamas 36775
Arzemju Mákslas Muzejs, Riga 30062
Arzneipflanzenmuseum, Saarbrücken 21454
Arzney Küche, Bönnigheim 17425
Asa Packer Mansion, Jim Thorpe 49630
Asago Art Village, Asago 28366
Asahi-yaki Kiln Museum, Uji, Kyoto 29557
Asahikawa Bijutsukan, Asahikawa 28368
Asahikawa City Museum, Asahikawa 28369
Asahikawa Hakabutsukan, Asahikawa 28369
Asahikawa Museum of Art, Asahikawa 28368
Asahikawa Youth Science Museum, Asahikawa 28370
Asakura Isokichi Bijutsukan, Terai 29359
Asakura Sculpture Gallery, Tokyo 29370
Åsavallen, Edsbro 40491
Les Ascenceurs du Canal Historique du Centre, Houdeng-Goegnies 03718
Ascenseur Funiculaire pour Bateaux de Strépy-Thieu, Thieu 04061
Aschan Residence, Heinola 10766
Aschanska Gården, Eksjö 40492
Ascott House, Leighton Buzzard 44687
Asen Zlatarov House-Museum, Chaskovo 05004
Asesepän Paja, Ilomantsi 10852
Aşezământul de Artă şi Cultură Religioasă, Maia 36455
Ásgrimur Jónsson Gallery, Reykjavík 24032
Ash Lawn-Highland, Charlottesville 47395
Ashburton Aviation Museum, Ashburton 33036
Ashburton Museum, Ashburton 33037
Ashburton Museum, Ashburton 43325
Ashburton Vintage Car Club Museum, Tinwald 33213
Ashby-de-la-Zouch Museum, Ashby-de-la-Zouch 43326
Ashby-Hodge Gallery of American Art, Fayette 48525
Ashcroft Museum, Ashcroft 05357
Ashern Pioneer Museum, Ashern 05358
Asheville Art Museum, Asheville 46368
Ashford Gallery, Dublin 24777
Ashford Museum, Ashford 43327
Ashikaga Museum of Art, Ashikaga 28372
Ashikaga-shiritsu Bijutsukan, Ashikaga 28372
Ashiya City Museum of Art & History, Ashiya 28375
Ashiya-shiritsu Bijutsu oyo Rekishi Hakubutsukan, Ashiya 28375
Ashland - Henry Clay Estate, Lexington 50045
Ashland Historical Society Museum, Ashland 46381
Ashland Historical Society Museum, Ashland 46391
Ashland Logging Museum, Ashland 46382
Ashland Railroad Station Museum, Ashland 46384
Ashmolean Museum of Art and Archaeology, Oxford 45327
Ashtabula Arts Center, Ashtabula 46394
Ashtabula County Historical Society Museum, Geneva-on-the-Lake 48884
Ashtabula Plantation, Pendleton 51758
Ashton Court Visitor Centre, Bristol 43629
Ashton Villa, Galveston 48842
Ashwell Village Museum, Ashwell 43335
Ashworths Treasures of the Earth, Home Hill 01161
ASI Art Museum, Reykjavík 24020
Asia-Africa Museum, Genève 41398
Asia and Pacific Museum, Warszawa 35237
Asia and Pacific Museum - Asian Gallery, Warszawa 35226
Asia and Pacific Museum - Nusantara Gallery, Warszawa 35229
Asia Art Archive, Hong Kong 07793
Asia Society Museum, New York 51128
Asian-American Arts Centre, New York 51129
Asian American Arts Centre, New York 51130
Asian Art Museum, San Francisco 52766
Asian Art Museum, National Museums in Berlin - Prussian Cultural Heritage Foundation, Berlin 17201
Asian Arts and Culture Center, Towson 53549
Asian Center Museum, Quezon City 34486
Asian Civilisations Museum, Singapore 38172
Asian Cultures Museum, Corpus Christi 47819
Asian Folk Art Museum, Noda 29079
Asian Galleries, Sydney 01545
Asiatic Society Museum, Kolkata 24267
Asiatica Musée d'Art Oriental, Biarritz 11981
Asiatica Oriental Art Museum, Biarritz 11981
Asikkala Museum, Asikkala 10714
Asikkalan Kotiseutumuseo, Asikkala 10714
Aşiyan Müzesi, İstanbul 42802
Asiyan Museum, İstanbul 42802
Asker Museum, Hvalstad 33572
Askeri Müze ve Kültür Sitesi Komtuanlığı, İstanbul 42803
Askim Museum, Askim 33407
Askon Museo, Lahti 11025
Aslamazyan Sisters Museum, Gyumri 00704
Åsmundarsafn, Reykjavík 24015
Ásmundur Sveinsson Sculpture Museum, Reykjavík 24015
Aso-gun Kazan Hakubutsukan-Orugoru Kyouwakoku, Aso 28378
Aso Volcano Museum-Orgel Republic, Aso 28378
Asociación Colombiana de Museos, Bogotá 54757
Asociacion de Amigos de los Museos Militares, Madrid 55021
Asociación de Museos de Guatemala, Guatemala 54872
Asociación Española de Archiveros, Bibliotecarios, Museólogos y Documentalistas, Madrid 55022
Asociación Española de Museólogos, Madrid 55023
Asociatia Nationalá a Muzeelor in aer Liber - România, Bucureşti 54996
Aspazia House of the Jurmala City Museum, Jūrmala 30021
Aspazijas Māja, Jūrmala 30021
Aspen Art Museum, Aspen 46397
Aspen Historical Society Museum, Aspen 46398
Aspö-Tosterö Hembygdsmuseum, Strängnäs 40952
Asri Environmental Education Center, Bristol 47030
Assai al-Hamra (Red Castle), Tripoli 30185
Assam Forest Museum, Guwahati 24187
Assam State Museum, Guwahati 24188
Åssamuséet, Åtvidaberg 40446
Assiginack Historical Museum and Norisle Heritage Park, Manitowaning 06225
Assiniboia and District Museum, Assiniboia 05359
Associacão Portugesa de Museologia, Lisboa 54991
Associacija Muzeev Rossii, Tula 54998
Associacija Muzej, Varna 54723
Associació del Museo de la Ciència i de la Tècnica i d'Arqueologia Industrial de Catalunya, Terrassa 55024
Associated American Jewish Museum, Pittsburg 51936
Associated Artists of Pittsburgh Gallery, Pittsburgh 51940
Associated Artists of Winston-Salem, Winston-Salem 54242
Association des Artilleurs de la Garnison, Lévis 06165
Association des Conservateurs d'Antiquités, Athinai 54865
Association des Conservateurs et du Personnel Scientifique des Musées de la Ville de Paris, Paris 54796
Association des Musées Automobiles de France, Paris 54797
Association des Musées et Centres pour le Développement de la Culture Scientifique, Technique et Industrielle, Dijon 54798
Association des Musées Locaux d'Alsace, Wingen-sur-Moder 16297
Association Européenne des Musées d'Histoire des Sciences Médicales (AEMHSM), Paris 54799
Association for Cultural Enterprises, Worcester 55048
Association for Living History, Farm and Agricultural Museums, North Bloomfield 55115
Association Française des Directeurs de Centres d'Art, Pougues-les-Eaux 54800
Association Française des Musées d'Agriculture et du Patrimoine Rural, Paris 54801
L' Association Francophone des Musées de Belgique, Hornu 54708
Association Générale des Conservateurs des Collections Publiques de France, Paris 54802
Association Internationale des Amis de Nicolas Schöffer, Paris 14566
Association Internationale des Musées d'Armes et d'Histoire Militaire, Liège 54709
Association Internationale des Musées d'Histoire, Paris 54803
Association Mathieu Vivian, Saint-Pryvé Saint-Mesmin 15473
Association Museum and Industries, Paris 54804
Association of Art Museum Directors, New York 55116
Association of British Transport and Engineering Museums, Gaydon 55049
Association of Children's Museums, Washington 55117
Association of College and University Museums and Galleries, Norman 55118
Association of Conservators of Cultural Heritage of Slovenia, Ljubljana 55014
Association of Danish Art Museums, Nivå 54779
Association of Danish Museum Curators, København 54780
Association of European Open Air Museums, Arnhem 54953
Association of Heritage in Upper Austria, Linz 54699
Association of Icelandic Curators, Reykjavík 54878
Association of Independent Museums, Worthing 55050
Association of Indian Museums, Delhi 54882
Association of Indiana Museums, Indianapolis 55119
Association of Local Museums in Greenland, Sisimiut 54870
Association of Manitoba Museums, Winnipeg 54730
Association of Midwest Museums, Saint Louis 55120
Association of Museum Curators of Slovenia, Ljubljana 55015
Association of Museum Professionals of the Republic of Macedonia, Skopje 54932
Association of Museums of the Indian Ocean, Victoria 55007
Association of Open Air Museums, Toruń 54989
Association of Railway Museums, Washington 55121
Association of Science Museum Directors, Santa Barbara 55122
Association of Science-Technology Centers, Washington 55123
Association of South Dakota Museums, Sioux Falls 55124
Association of State Subsidized Museums, Amsterdam 54962
Association of Systematics Collections, Washington 55125

Association of the State Literary Museums of the Republic of Belarus, Minsk 54706
Association Photocenter of the Journalists' Association, Moskva 37390
Association pour les Musées de Marseille, Marseille 54805
Associazione Musei Ecclesiastici Italiani, Orvieto 54894
Associazione Nazionale dei Musei dei Enti Locali ed Istituzionali, Brescia 54895
Associazione Nazionale dei Musei Italiani, Roma 54896
Associazione Nazionale Musei Scientifici, Firenze 54897
Associón Museo del Automóvil, Lima 34205
Den Ast, Buggenhout 03505
Astan-e-Qods-e Razavi Museums, Mashad 24631
Astley-Cheetham Art Gallery, Stalybridge 45733
Astley Hall Museum and Art Gallery, Chorley 43840
Aston Hall, Birmingham 43495
Aston Manor Transport Museum, Birmingham 43496
Astor House Hotel Museum, Golden 48955
Astors' Beechwood-Victorian Living History Museum, Newport 51318
Astrachanskaja Gosudarstvennaja Kartinnaja Galereja im. B.M. Kustodieva, Astrachan 36777
Astrachanskij Gosudarstvennyj Obedinennyj Istoriko- architekturnyj Muzej-zapovednik, Astrachan 36778
Astrachanskij Kraevedčeskij Muzej, Astrachan 36779
Astrachanskij Kreml, Astrachan 36780
Astrakhan Kremlin, Astrachan 36780
Astrakhan Regional Museum at the Museum-Preserve, Astrachan 36779
Astrakhan State Gallery B.M. Kustodiev, Astrachan 36777
Astrakhan State Historical and Architectural United Museum- Preserve, Astrachan 36778
Astronautics Museum, Beijing 07506
Astronomisch Compensatieuurwerk Kamiel Festraets, Sint-Truiden 04026
Astronomisch-Physikalisches Kabinett, Kassel 19458
Astronomy Museum, Eger 23538
Astrup Fearnley Museet for Moderne Kunst, Oslo 33717
Astruptunet, Skei i Jølster 33831
Asturias Railway Museum, Gijón 39337
Asuka Historical Museum, Asuka 28379
Asuka Shiryokan, Asuka 28379
Asutosh Museum of Indian Art, Kolkata 24268
Ásványkiállítás, Kaposvár 23630
Ásványmúzeum, Siófok 23802
Asyl der Kunst, Häuslingen 18852
At-Bristol, Bristol 43630
Ataloa Lodge Museum, Muskogee 50927
Atanas Moncys House-Museum, Palanga 30327
Ataq Museum, Ataq 54646
Atascadero Historical Society Museum, Atascadero 46405
Atatürk and Culture Museum, Eskişehir 42779
Atatürk and Ethnographic Museum, Akşehir 42683
Atatürk Ethnography and Congress Museum, Sivas 42945
Atatürk Evi, Ankara 42698
Atatürk Evi Müzesi, Erzurum 42774
Atatürk Evi Müzesi, Mersin 42906
Atatürk Evi ve Müzesi, Sakarya 42927
Atatürk House, Ankara 42698
Atatürk House Havza, Havza 42795
Atatürk House Museum, Erzurum 42774
Atatürk House Museum, Mersin 42906
Atatürk House of Silifke, Silifke 42941
Atatürk Köşkü, Diyarbakır 42760
Atatürk Kongre ve Etnografya Müzesi, Sivas 42945
Atatürk Müzesi, Bursa 42740
Atatürk Müzesi, İzmir 42853
Atatürk Müzesi, Kayseri 42872
Atatürk Müzesi, Konya 42883
Atatürk Museum, Konya 42883
Atatürk ve Etnografya Müzesi, Akşehir 42683
Atatürk ve Etnografya Müzesi, Denizli 42757
Atatürk ve Etnografya Müzesi, Uşak 42959
Atatürk's Home Villa Museum, Ankara 42699
Atatürk's House, Diyarbakır 42760
Atatürk's Mausoleum and War of Independence Museum, Ankara 42696
Atatürkün Mekanı Müze Köşk, Ankara 42699
Atayal Museum, Ren-Ai 42302
Atchinson, Topeka and Santa Fe Depot, Alden 46191
Atchison County Historical Society, Atchison 46407
Atelier 13, Ellwangen 18121
L'Atelier de Nicolas Schöffer á la Villa des Arts, Paris 14567
Atelier des Bastides, Monpazier 14150
Atelier des Produits Résineux Vidal, Luxey 13874
Atelier d'Estampe Imago, Moncton 06314
Atelier Franz Knapp, Pöchlarn 02503
Atelier Galerie du Richard Rak, Coutures 12663
Atelier Galerie Gabriele Schurian, Villach 02846
Atelier Hoguet, Paris 14568
Atelier Ingeborg Chr. Berger, Villach 02847
Atelier Ismet Mujezinović, Tuzla 04288
Atelier Ivan Meštrović, Zagreb 08987
Atelier Jaindl, Villach 02848
Atelier Kea Homan, Assen 31935
L'Atelier Legault, Pouance 14925
Atelier Manoli-Musée et Jardin de Sculpture, La Richardais 13409
Atelier Manoli-Museum and Garden of Sculptures, La Richardais 13409
Atelier Memorial Paul Erdös, Satu Mare 36589
Atelier Municipal de Passementerie, Lyon 13879
Atelier Musée de la Soie, Montboucher-sur-Jabron 14186
Atelier-Musée de la Soie, Taulignan 15866
Atelier-Musée des Tisserands et de la Charentaise, Varaignes 16072
Atelier-Musée du Bâton Basque, Larressore 13530
Atelier-Musée du Chapeau, Chazelles-sur-Lyon 12494
Atelier-Musée du Sabotier, Lautrec 13538
Atelier Musée du Verre, Trélon 15988
Atelier-Musée du Vitrail et des Métiers d'Art du Verre, Gréoux-les-Bains 13140
Atelier-Musée Livre et Typographie, Grignan 13148
Atelier-Museu António Duarte, Caldas da Raínha 35493
Atelier-Museu João Fragoso, Caldas da Raínha 35494
Atelier Na-iem, Gees 32238
Atelier Otto Niemeyer-Holstein, Koserow 19692
Atelier Paul Cézanne, Aix-en-Provence 11513
Atelier Segantini, Maloja 41604
Atelier SoCa Sonja Capeller, Villach 02849
Atelier Spyros Vassiliou, Athinai 22900
Ateliergalerie Heinz Schweizer, Villach 02850
Atelierhaus Aachen, Aachen 16392
Atelierhaus der Akademie der bildenden Künste Wien, Wien 02936
Atelierhaus Philipp Harth, Bayrischzell 16967
Ateliermuseum und Skulpturengarten Erwin Rehmann, Laufenburg 41532
Ateliers de l'Abeille, Chavignon 12492
Ateliersammlung Herbert Hiesmayr, Sankt Thomas am Blasenstein 02689
Ateneo Art Gallery, Quezon City 34487
Ateneo de Valencia, Valencia 54596
Ateneo Puertorriqueño, San Juan 36058
Ateneum Art Museum of the Finnish National Gallery, Helsinki 10769
Ateneumin Taidemuseo, Helsinki 10769
Athanassakeion Archaeological Museum, Volos 23237
Athelhampton House, Dorchester 43997
Athelstan Museum, Malmesbury 45078
The Athenaeum, Alexandria 46200
Athenaeum of Philadelphia, Philadelphia 51816
Athens University Museum, Athinai 22901
Athlone Castle Museum, Athlone 24730
Athol Murray College of Notre Dame Museum, Wilcox 07334
Atholl Country Life Museum, Blair Atholl 43536
Athus et l'Acier, Athus 03332
Atikokan Centennial Museum, Atikokan 05360
Atkinson Art Gallery, Southport 45716
The Atlanta College of Art, Atlanta 46423
Atlanta Contemporary Art Center, Atlanta 46424
The Atlanta Cyclorama, Atlanta 46425
Atlanta History Museum, Atlanta 46426
Atlanta International Museum of Art and Design, Atlanta 46427
Atlanta Museum, Atlanta 46428
Atlantic 1 Museum, Swampscott 53387
Atlantic Canada Aviation Museum, Enfield 05777
Atlantic City Historical Museum, Atlantic City 46460
Atlantic County Historical Society Museum, Somers Point 53146
Atlantic Salmon Museum on the Miramichi River, Doaktown 05705
Atlantic Wall Museum, Oostende 03929
Atlantic Wildfowl Heritage Museum, Virginia Beach 53753
Atlantis Kindermuseum, Duisburg 18012
Atlas Coal Mine, East Coulee 05730
Atnabrufossen Vannbruksmuseum, Atna 33408
Atom Museum, Haigerloch 18868
Atomium, Bruxelles 03434
Atomium, Bruxelles 03434
Atrium an der Schleuse, Brunsbüttel 17610
Atrium Gallery, Fort Worth 48710
Atrium Gallery, Lewiston 50036
Atrium Gallery, Saint Louis 52567
Atrium Gallery, Salt Lake City 52686
Atsukeshi Folk Art Museum, Atsukeshi 28383
Atsukeshi-machi Mingeikan, Atsukeshi 28383
Atsuta-Jingu Museum, Nagoya 28994
Attila József Memorial House, Öcsöd 23735
Attila József Memorial Room, Budapest 23417
Åttingen Hantverk och Lantbruksmuseum, Aneby 40421
Attingham Park, Shrewsbury 45671
Attitudes, Genève 41399
Attleboro Area Industrial Museum, Attleboro 46462
Attleboro Museum-Cerner for the Arts, Attleboro 46463
Attmore-Oliver House, New Bern 51018
Åtvidabergs Bruks- och Facitmuseum, Åtvidaberg 40447
Atwater Kent Museum - The History Museum of Philadelphia, Philadelphia 51817
Atwell House Gallery, Alfred Cove 00795
Atwell-Wilson Motor Museum, Calne 43715
Atwood House Museum, Chatham 47404
Atwood Memorial Center Art Collection, Saint Cloud 52537
Atyrau HistoryMuseum, Atyrau 29695
Atyrau Museum of Decorative and Applied Art, Atyrau 29696
Atyrauskij Oblastnoj Istoriko-Kraevedčeskij Muzej, Atyrau 29695
Atyrauskij Oblastnoj Muzej Dekorativno-prikladnogo Iskusstva, Atyrau 29696
Au Coeur de l'Ardoise, Bertrix 03365
Au Glaize Village, Defiance 48036
Au Pays d'Ardenne - Original Museum, Bastogne 03344
Auberge Ganne, Barbizon 11845
Auberlehaus, Trossingen 22109
Aubonnière, Chaille-sous-les-Ormeaux 12355
Auburn-Cord-Duesenberg Museum, Auburn 46466
Auchindrain Museum, Inveraray 44518
Auckland Art Gallery, Auckland 33042
Auckland Museum and Auckland War Memorial Museum, Auckland 33043
Audi Fondation, Saïda 30172
Audi Museum Mobile, Ingolstadt 19337
Audie Murphy/American Cotton Museum, Greenville 49080
Audiovisuelles Zentrum Wien, Wien 02937
Audley End House, Saffron Walden 45538
Audorfer Museum im Burgtor, Oberaudorf 20801
Audrain Historical Museum, Graceland, Mexico 50618
Audubon House, Key West 49756
Audubon State Historic Site, Saint Francisville 52543
Auer von Welsbach Museum, Althofen 01715
The Auezov Museum of the Kazakh Dramatic Theatre, Almaty 29687
Auglaize County Historical Society Museum, Saint Marys 52606
Augsburger Puppentheatermuseum Die Kiste, Augsburg 16612
August-Becker-Museum, Klingenmünster 19571
August Deusser-Museum, Bad Zurzach 41140
August-Gottschalk-Haus, Esens 18238
August-Gottschalk-Haus, Esens 18239
August-Holländer-Museum, Emsdetten 18159
August Horch Museum Zwickau, Zwickau 22831
August Kitzbergi Tubamuuseum, Karksi 10581
August Macke Haus, Bonn 17447
August-Suter-Museum, Eptingen 41354
Augusta Historical Museum, Augusta 00819
Augusta Historical Museum, Augusta 46482
Augusta Museum of History, Augusta 46475
Augusta Raurica, Augst 41137
Augustana College Art Gallery, Rock Island 52405
Augustana College Center for Western Studies, Sioux Falls 53116
Augustiner Chorherrenstift Neustift, Varna 28167
Augustinerkloster und Kirche Hl. Augustinus, Wien 02938
Augustinermuseum, Freiburg im Breisgau 18428
Augustinermuseum, Rattenberg, Inn 02564
Augustines de la Miséricorde de Jésus de Roberval, Roberval 06763
Aula Didáctica de Prehistoria de Tito Bustillo, Ribadesella 39899
Aula Didáctica del Castro de Coaña, Coaña 39195
Auld Kirk Museum, Kirkintilloch 44613
Auld Sköll, Fair Isle 44202
Aule Mettmanner, Mettmann 20313
Aulestad, Follebu 33502
Auli Mølle, Årnes 33398
Aunt Margaret's Museum of Childhood, Dufresne 05719
Aurelio Sevilla Alvero Library and Museum, Pasay 34470
Aurland Bygdetun og Lensmannsstova, Aurland 33409
Aurnhammer-Sammlung, Treuchtlingen 22089
Aurora Art Museum, Taipei 42375
Aurora Historical Museum, Aurora 46489
Aurora Historical Museum, East Aurora 48233
Aurora Historical Society Museum, Aurora 46498
Aurora History Museum, Aurora 46488
Aurora Museum, Aurora 05363
Aurora Public Art Commission, Aurora 46490
Aurskog-Høland Bygdetun, Hemnes i Høland 33542
Ausgrabungen Michaelerplatz, Wien 02939
Ausgrabungen Wüstung Hard, Thaya 02809
Ausgrabungsdokumentation 6000 Jahre Wohnberg Oberleis, Ernstbrunn 01874
Ausgrabungsstätte Schanzberg, Thunau 02813
Aušros Alėjos Rūmai, Šiauliai 30352
Aušros Avenue Mansion of the Šiauliai Aušra Museum, Šiauliai 30352
Aussiedlermuseum im Schüttkasten, Allentsteig 01704
Ausstellung der Geigenbauschule Brienz, Brienz (Bern) 41254
Ausstellung der Kantonalen Schnitzerschule, Brienz (Bern) 41255
Ausstellung Fontane und Hankels Ablage, Zeuthen 22795
Ausstellung für Kraftwagen und Krafträder, Satteldorf 21522
Ausstellung Haus Wysburg und Ruine Wysburg, Remptendorf 21262
Ausstellung im Stadthaus, Zürich 42066
Ausstellung in der Universitätsbibliothek Basel, Basel 41155
Ausstellung 'Juden in Buttenhausen', Münsingen 20499
Ausstellung Kirche Zillis, Zillis 42056
Ausstellung Max-Kommerell, Münsingen 20500
Ausstellung zur Erd- und Vorgeschichte, Breitscheid, Lahn-Dill-Kreis 17544
Ausstellungräume Schloss Gloggnitz, Gloggnitz 01936
AusstellungsHalle, Frankfurt am Main 18355
Ausstellungshalle Neues Rathaus, Bayreuth 16946
Ausstellungshalle Zeitgenössische Kunst Münster, Münster 20504
Ausstellungsraum Klingental, Basel 41156
Ausstellungszentrum für die Archäologie des Emslandes, Meppen 20286
Ausstellungszentrum Heft, Hüttenberg 02101
Ausstellungszentrum Heiligenkreuzer Hof, Wien 02940
Ausstellungszentrum Kloster Cismar, Grömitz 18769
Ausstellungszentrum Kroch-Haus, Leipzig 19896
Ausstellungszentrum Lokschuppen, Rosenheim 21374
Aust-Agder Museet, Arendal 33395
Austefjord Museum, Austefjorden 33410
Austerlitz castle, Slavkov u Brna 09841
Austin Children's Museum, Austin 46501
Austin History Center, Austin 46502
Austin Museum of Art, Austin 46503
Austin Museum of Art Laguna Gloria, Austin 46504
Austin Nature and Science Center, Austin 46505
Australasian Golf Museum, Bothwell 00880
Australian Antarctic Division Display, Kingston 01203
Australian Centre for Contemporary Art, Southbank 01517
Australian Centre for Photography, Paddington, New South Wales 01399
Australian Children's Folklore Collection, Melbourne 01274
Australian Electric Transport Museum, Saint Kilda, South Australia 01477
Australian Exhibition Center, Chicago 47455
Australian Federation of Friends of Museums, Sydney 54686
Australian Flying Museum, Wallon 01625
Australian Freethought Heritage Library, Balwyn 00831
Australian Gallery of Sport and Olympic Museum, East Melbourne 01052
Australian Golf Museum, South Melbourne 01515
Australian Institute of Genealogical Studies, Blackburn 00873
Australian Museum, Sydney 01546
Australian National Maritime Museum, Darling Harbour 01025
Australian Pearling Exhibition, Darwin 01028
Australian Prospectors and Miners Hall of Fame, Kalgoorlie 01185
Australian Racing Museum and Hall of Fame, Melbourne 01275
Australian Stockman's Hall of Fame and Outback Heritage Centre, Longreach 01238
Australian Toy Museum, Collingwood 00994
Australian War Memorial, Campbell 00939
Austrått Fort, Brekstad 33454
Austråttborgen, Trondheim 33922
Austrian Army Museum, Wien 03001
Austrian Jewish Museum, Eisenstadt 01862
Austrian Museum of Paper Making and Printing, Steyrermühl 02779
Auswanderermuseum, Oberalben 20799
aut. architektur und tirol, Innsbruck 02114
Auto Museum, Escondido 48417
Auto Museum Dr. Carl Benz, Ladenburg 19761
Auto-Museum Fritz B. Busch, Wolfegg 22648
Auto Museum Moncopulli, Osorno 07442
Auto & Technik Museum Sinsheim e.V, Sinsheim 21783
Auto- und Motorrad-Museum, Öhringen 20863
Auto- und Motorradmuseum, Witzenhausen 22632
Automates Avenue, Falaise 12901
Automatik-Museum, Leipzig 19897
Automaton Museum, Barcelona 38987

Autómatuzsálem Gyűjtemény, Bordány 23377
Automobil-Museum, Asendorf 16598
Automobil Museum, Dortmund 17914
Automobil-Museum Siegfried Marcus, Stockerau 02781
Automobil- und Motorradmuseum Austria, Mitterndorf an der Fischa 02389
Automobile Driving Museum, Los Angeles 50210
Automobile Museum, Lima 34205
Automobilia Transport Museum, Hebden Bridge 44434
Automobilmuseum, Aspang 01732
Automobilmuseum, Stainz 02758
Automobilmuseum Stainz, Graz 01964
Automotive Hall of Fame, Dearborn 48005
Automusa, Bergeijk 31965
Automusée du Lyonnais, Seyssuel 15754
Automuseum Deventer, Deventer 32118
Automuseum Engstingen, Engstingen 18174
Automuseum Histo-Mobil, Giethoorn 32251
Automuseum Old Timer, Lo-Reninge 03825
Automuseum Oldie Monschau, Monschau 20372
Automuseum Praga, Zbuzany 09930
Automuseum Schagen, Schagen 32746
Automuseum Störy, Bockenem 17408
AutoMuseum Volkswagen, Wolfsburg 22658
Autosammlung, Linz 02309
AutoStadt ZeitHaus, Wolfsburg 22659
Autotron, Rosmalen 32701
Autoworld, Bruxelles 03435
Auwärter-Museum, Landau an der Isar 19785
Aux Portes de la Champagne, Bernex 41222
AV Gallery, Limerick 24863
Ava Gallery and Art Center, Lebanon 50010
Avampato Discovery Museum, Charleston 47378
Avanersuup Katersugaasivia, Qaanaaq 23264
Averkin Museum of the Russian Song, Sasovo 37755
Averoff Gallery, Athinai 22902
Averoff Gallery, Metsovo 23094
Averøy Skolemuseum, Averøy 33411
Avery County Historical Museum, Newland 51315
Avery Historical Museum, Smethwick 45691
Avery Research Center for African American History and Culture, Charleston 47357
Avetik Issahakyan House-Museum, Yerevan 00728
Aviation and Cosmonautical Museum K.E. Tsiolkovsky, Kirov 37040
Aviation Heritage Museum of Western Australia, Bull Creek 00914
Aviation Museum, București 36195
Aviation Museum, İstanbul 42822
Aviation Museum, Krumovo 05081
Aviation Museum, Praha 09759
Aviation Museum of Kentucky, Lexington 50046
AVICOM - International Committee for the Audiovisual and Image and Sound New Technologies, Paris 54806
Avijacijos Pradininko Aleksandro Griškevičiaus Muziejus, Viekšniai 30395
Avog's Crash Museum, Lievelde 32501
Avon Historical Society Museum, Avon 46526
Avon River Heritage Society Museum, Hants County 05968
Avoncroft Museum of Historic Buildings, Bromsgrove 43654
Avondale Discovery Farm, Beverley 00865
Avondale House, Rathdrum 24889
Avonlea and District Museum, Avonlea 05366
Ax Museum, Billnäs 10717
The AXA Gallery, New York 51131
Axe Néo-7 Art Contemporain, Gatineau 05855
Axel Ebbes Konsthall, Trelleborg 40980
Axel Stenross Maritime Museum, Port Lincoln 01431
Axiom Centre for Arts, Cheltenham 43807
Axminster Museum, Axminster 43341
Aya İrini Kilisesi Müzesi, İstanbul 42804
Ayabe City Museum, Ayabe 28384
Ayabe-shiritsu Hakubutsukan, Ayabe 28384
Ayala Museum, Makati 34414
Ayala Museum, Vigan 34521
Ayasofya Church Museum, Trabzon 42955
Ayasofya Müzesi, İstanbul 42805
Ayasofya Müzesi, Trabzon 42955
Aydın Devlet Güzel Sanatlar Galerisi, Aydın 42720
Aydın Müzesi, Aydın 42721
Aydın Müzesine Bağlı Örenyerleri, Aydın 42722
Aydin Museal Archaeology Institute, Aydın 42722
Aydin State Fine Art Gallery, Aydın 42720
Ayios Dimitrios Crypt Museum, Thessaloniki 23211
Aylmer and District Museum, Aylmer 05367
Aylmer Military Collection, Brading 43585
Aynalıkavak Kasrı, İstanbul 42806
Aynalıkavak Pavillion, İstanbul 42806
Ayrshire Yeomanry Museum, Ayr 43344
Ayscoughfee Hall Museum, Spalding 45721
Ayşe and Ercümend Kalmık Foundation- Museum, Gümüşsuyu 42791
Ayşe ve Ercümend Kalmık Vakfı-Müzesi, Gümüşsuyu 42791
Azadi Museum, Tehrān 24657
Azarbayezan Museum, Tabriz 24650
Azerbaijan State Historical and Architectural Museum Ichery Sheher, Bakı 03142
Azerbaijan State Museum of Independence, Bakı 03143
Azerbaycan Tarixi Muzeyi, Bakı 03144
The Azeri Miniature Art Center, Bakı 03145
Azovskij Kraevedčeskij Muzej, Azov 36786
Aztalan Museum, Jefferson 49610
Aztec Museum and Pioneer Village, Aztec 46527
Aztec Ruins, Aztec 46528

B

B. Bissoondoyal Memorial Centre, Tyack 30686
B.F. Larsen Gallery, Provo 52185
B.J. Hales Museum of Natural History, Brandon 05459
B.L. Pasternak House Museum, Peredelkino 37540
B.L. Pasternak Memorial Museum, Čistopol 36866
B.M. Birla Science and Technological Museum, Hyderabad 24211
B.M. Kustodiev Museum, Ostrovskoe 37521
The B & O Railroad Museum, Baltimore 46547
B & O Railroad Station Museum, Ellicott City 48364
Baarle-Hertog **31945**
Baba Nyonya Heritage Museum, Melaka 30593
Babamúzeum, Tihany 23924
Babe Didrikson Zaharias Museum, Beaumont 46672
Babe Ruth Birthplace & Museum, Baltimore 46548
Babits Mihály Emlékház, Szekszárd 23865
Bábkiállítás, Bosta 23378
Babls Uhrensammlung, Waffenbrunn 22268
Babol Museum, Babol 24597
Babur Memorial Museum - Dept. of Andijan Regional Museum of Literature and Art, Andijon 54440
Babur Yodgorlik Muzeyi, Andijon 54440
Babylon Museum, Babylon 24709
Babylonian Jewry Museum, Or Yehuda 25076
Baca House, Trinidad 53566
Bacău District Museum of History and Archaeology Iulian Anntonescu, Bacău 36122
Bach-Museum, Leipzig 19898
Bach-Stammhaus Wechmar, Günthersleben-Wechmar 18823
Bachčisarajs'kyj Palac Kryms'kych Chaniv, Bachčisaraj 42999
Bachelor's Club, Tarbolton 45826
Bachgau-Museum, Großostheim 18800
Bachhaus Eisenach, Eisenach 18093
Bachmann-Museum Bremervörde, Bremervörde 17576
Bachritterburg Kanzach, Kanzach 19428
Backnanger Stadtturm mit Gotischem Chor St. Michael, Backnang 16655
Backus Heritage Village, Simcoe 06978
Baclayon Church Museum, Baclayon 34336
Bacon's Castle, Surry County 53384
Bad Gasteiner Museum, Bad Gastein 01746
Bad Schwalbacher Kur-Stadt-Apotheken-Museum, Bad Schwalbach 16833
Badè Museum of Biblical Archaeology, Berkeley 46750
Baden-Powell House, London 44787
Badener Puppen- und Spielzeugmuseum, Baden bei Wien 01773
Badger Creek Museum, Cartwright 05571
Badger Mine and Museum, Shullsburg 53093
Badhaus Museum und Galerie, Kulmbach 19745
Badische Landesbibliothek, Karlsruhe 19435
Badischer Kunstverein, Karlsruhe 19436
Badisches Bäckereimuseum und Erstes Deutsches Zuckerbäckermuseum, Kraichtal 19695
Badisches Landesmuseum Karlsruhe, Karlsruhe 19437
Badong Museum, Badong 07493
Bäckerei- und Dorfgeschichtliches Museum, Bremervörde 17577
Bäckereimuseum, Rimpar 21334
Bäckereimuseum Poppinga's alte Bäckerei, Krummhörn 19730
Bäckereimuseum Vörstetten, Vörstetten 22250
Bäckermuseum, Wien 02941
Bärner Heimatstube, Langgöns 19838
Bärnstensmuseet, Höllviken 40595
Bäuerliches Gerätemuseum und Venezianer-Gatter, Innervillgraten 02111
Bäuerliches Heimatmuseum der Landtechnik, Geiselhöring 18556
Bäuerliches Museum im Achentaler Heimathaus, Rohrdorf 21366
Bäuerliches Wohnmuseum, Schellenberg, Schellenberg 30195
Bagamoyo Historical Museum, Bagamoyo 42510
Bagatti Valsecchi Museum, Milano 26690
Bağçasaray Palace of the Crimean Khans, Bachčisaraj 42999
Baghcheh-Jaq Palace, Urmieh 24699
Baghdad Museum, Baghdad 24713
Baghgheh Chogh Palace, Maku 24628
Bagn Bygdesamling, Bagn 33413
Bagpipe Museum, Gijón 39336
Bagpipe Music Museum, Ellicott City 48365
Bagpipes Museum, Scapoli 27785
Bagration Regional Museum of Kislyarsk, Kizljar 37051
Bagshaw Museum, Batley 43415
Baguio-Mountain Provinces Museum, Baguio 34340
Bahamas Historical Society Museum, Nassau 03203
Bahamas Museum, Nassau 03204
Bahawalpur Museum, Bhawalpur 33998
Bahay Laguna Museum, Liliw 34403
Bahay Tsinoy Museum, Manila 34431
Bahnzeit, Wels 02913
Bahrain Arts Society Gallery, Manama 03211
Bahrain National Museum, Manama 03212
Baicheng Museum, Baicheng 07494
Baierweinmuseum, Bach an der Donau 16652
Baihan Museum, Baihan 54647
Bailey House Museum, Wailuku 53786
Bailey-Matthews Shell Museum, Sanibel 52850
Bailian Dongdongkong Science Museum, Liuzhou 07965
Baimasi Han Wei Gucheng Display Center, Luoyang 07982
Baird Institute Museum, Cumnock 43941
Bairin Youqi Museum, Balin Youqi 07497
Bairin Zuoqi Museum, Bairin Zuoqi 07495
Bait Al Zubair, Muscat 33987
Bait Muzna Gallery, Muscat 33988
Baituna al Talhami Museum, Bethlehem 34049
Bajcsy-Zsilinszky Endre Emlékmúzeum, Kővágóörs-Pálköve 23686
Bajcsy-Zsilinszky Memorial Museum, Kővágóörs-Pálköve 23686
Bajor Gizi Emlékmúzeum, Budapest 23385
Bajuwarenmuseum Waging am See, Waging am See 22271
Bajza József Emlékszoba, Szűcsi 23907
Bak Negara Muzium Matawang, Kuala Lumpur 30572
Bakar Local History Museum, Bakar 08841
Bakelite Museum, Williton 46017
Baker-Cederberg Museum, Rochester 52392
Bakers' Museum, Sopron 23816
Bakersfield Museum of Art, Bakersfield 46536
Bakhruz Kangarli Museum, Naxçivan 03180
Baki Urmantshe Museum, Kazan 37013
Bakibey House, Burdur 42735
Bakibey Konagi, Burdur 42735
Bakkemuseet, Frederiksberg 10022
The Bakken, Minneapolis 50729
Bakkerij Museum, Oosterhout 32627
Bakkerijmuseum, Geel 03612
Bakkerijmuseum, Groot-Bijgaarden 03664
Bakkerijmuseum, Huizen 32421
Bakkerijmuseum De Grenswachter, Luijksgestel 32511
Bakkerijmuseum de Meelzolder, Almelo 31798
Bakkerijmuseum Walter Plaetinck, Veurne 04109
Bakony House, Veszprém 23958
Bakonybéli Erdok Háza, Bakonybél 23348
Bakonyi Ház, Veszprém 23958
Bakonyi Természettudományi Múzeum, Zirc 23979
Baktay Kortárs Képzőművészeti Múzeum, Dunaharaszti 23530
Baku Art Center, Bakı 03146
Baku Museum of Education, Bakı 03147
Bal Sangrahalaya, Lucknow 24290
Bala Lake Railway, Bala 43356
Balachninskij Kraevedčeskij Muzej, Balachna 36787
Balai Seni Lukis Negara, Kuala Lumpur 30573
Balaklava Museum - Centenary Hall, Balaklava 00822
Balaklava Museum - Urlwin Park, Balaklava 00823
Balance Museum of Heiner Lubja, Mustvee 10596
Balanghai Archaeological Site Museum, Butuan 34351
Balás Jenő Bauxitbányászati Múzeum, Gánt 23563
Balassa Bálint Múzeum, Esztergom 23549
Balassa Bálint Múzeum - Babits Mihály Emlékház, Esztergom 23550
Balatoni Múzeum, Keszthely 23659
Balay Negrense Museum, Silay 34505
The Balch Institute for Ethnic Studies, Philadelphia 51818
Balchašskij Gorodskoj Istoriko-Kraevedčeskij Muzej, Balchaš 29697
Balcony House, Nassau 03205
Baldomero Aguinaldo House, Kawit 34397
Baldwin County Heritage Museum, Elberta 48331
Baldwin Historical Society Museum, Baldwin 46539
Baldwin Photo Gallery, Murfreesboro 50913
Baldwin-Reynolds House Museum, Meadville 50545
Balfour Museum of Hampshire Red Cross History, Winchester 46027
Balıkesir Devlet Güzel Sanatlar Galerisi, Balıkesir 42723
Balıkesir State Fine Art Gallery, Balıkesir 42723
Balintescu Fine Art Collection, Costeşti 36316
Balio ir Vandos Sruogų Namai-Muziejus, Kaunas 30240
Balkh Provincial Museum, Mazar-i-Sharif 00008
Balkh Religious Museum, Mazar-i-Sharif 00009
Balkhash Historical and Regional Museum, Balchaš 29697
Ball Game Hall, Praha 09761
Ball State University Museum of Art, Muncie 50902
Ballan Shire Historical Museum, Ballan 00824
Ballance House, Glenavy 44309
Ballangen Bygdemuseet, Ballangen 33415
Ballarat Fine Art Gallery, Ballarat 00825
Ballarat Tramway Museum, Ballarat 00826
Ballerup Egnsmuseum, Ballerup 09975
Ballinamore Local Museum, Ballinamore 24735
Ballmuseum, Franeker 32231
Ballon- en Luchtvaartmuseum Zep/Allon, Lelystad 32492
Ballonmuseum Gersthofen, Gersthofen 18619
Ball's Falls Historical Park, Jordan 06032
Bally Schuhmuseum, Schönenwerd 41841
Ballybunion Heritage Museum, Ballybunion 24736
Ballycastle Museum, Ballycastle 43360
Ballyheigue Maritime Centre, Ballyheigue 24739
Ballymena Museum, Ballymena 43364
Ballymoney Museum, Ballymoney 43366
Balmoral Grist Mill Museum, Balmoral Mills 05376
Balneological Museum, Piešt'any 38327
Balneologické Múzeum, Piešt'any 38327
Baltic Art Centre, Visby 41067
Baltimore Civil War Museum, Baltimore 46549
Baltimore County Historical Museum, Cockeysville 47668
Baltimore Maritime Museum, Baltimore 46550
The Baltimore Museum of Art, Baltimore 46551
Baltimore Museum of Industry, Baltimore 46552
Baltimore Public Works Museum, Baltimore 46553
Baltimore Streetcar Museum, Baltimore 46554
Baltimore's Black American Museum, Baltimore 46555
Bałucki Ośrodek Kultury, Łódź 34859
Balvu Novada Muzejs, Balvi 29987
Balys and Vanda Sruogos Museum of the Maironis Lithuanian Literature Museum, Kaunas 30240
Balzekas Museum of Lithuanian Culture, Chicago 47456
BAM - Beaux Arts Mons, Mons 03879
Bamble Museum, Langesund 33624
Bamboo Doll Village Museum, Maruoka 28883
Bamboo Museum, Lugu 42259
Bamburgh Castle, Bamburgh 43368
Bamfield Community Museum, Bamfield 05377
Ban Kao Prehistoric Museum, Kanchanaburi 42573
Banbury Museum, Banbury 43370
Banchory Museum, Banchory 43372
Bancroft Historical Museum, Bancroft 05379
Bancroft Mill Engine Trust, Barnoldswick 43383
The Band Museum, Pine Bluff 51921
Bandaranaike Museum, Colombo 40369
Bandelier National Monument, Los Alamos 50203
Bandirma Arkeoloji Müzesi, Bandirma 42725
Bandit Museum, Ronda 39917
Bandwebermuseum, Wuppertal 22736
Banff Museum, Banff 43374
Banff Park Museum, Banff 05380
Bangabandhu Memorial Museum, Dhaka 03218
Bangabandhu Shek Mujibur Rahman Museum, Dhaka 03218
Bangbu Museum, Bangbu 07498
Bangiya Sahitya Parisad Museum, Kolkata 24269
Bangkok University Gallery, Bangkok 42532
Bangla Academy Folklore Museum, Dhaka 03219
Bangladesh National Museum, Dhaka 03220
Bangladesh Shamorik Jadughar, Dhaka 03221
Bangladesh Zoological Museum, Dhaka 03221
Bangor Historical Society Museum, Bangor 46595
Bangsbo Museum, Frederikshavn 10024
Bangunan Alat Kebesaran Diraja, Bandar Seri Begawan 04961
Banícke Múzeum, Gelnica 38256
Banícke Múzeum, Rožňava 38342
Banícky Dom, Vyšna Boca 38387
Bank Austria Kunstforum und Kunstsammlung, Wien 02942
Bank Leumi Museum, Tel Aviv 25098
Bank Museum, Meeteetse 50561
Bank Negara Money Museum, Kuala Lumpur 30572
Bank of England Museum, London 44788
Bank of Ireland Arts Centre, Dublin 24778
Bank of Mississippi Art Collection, Tupelo 53612
Bank of Montreal Museum, Montréal 06327
Bank of Valletta Museum, Sliema 30638
Bank of Victoria Museum, Yackandandah 01674
Bank One Fort Worth Collection, Fort Worth 48711
Bankfield Museum, Halifax 44382
Banking and Currency Museum, Kadina 01181
Bankmuseet, Stavanger 33864
Bankside Gallery, London 44789
Banner County Historical Museum, Harrisburg 49177
The Banning Museum, Wilmington 54187

Bannock County Historical Museum, Pocatello 52005
Bannockburn Heritage Centre, Stirling 45749
Banqueting House, London 44790
Banské Múzeum v Prírode, Banská Štiavnica 38188
Banting House, London 06181
Banting Interpretation Centre, Musgrave Harbour 06404
Bantock House, Wolverhampton 46057
Bantry House, Bantry 24743
Bányászati Kiállítás, Gyöngyös 23572
Bányászati Kiállítóhelye, Salgótarján 23790
Banys Àrabs, Palma de Mallorca 39775
Bào Đai Palace Museum, Đà Lat 54601
Bao Tàng Dân Tôc Hoc Viêt Nam, Ha Noi 54608
Bao Tàng Hai Phong, Hai Phong 54620
Bao Tàng Quan Doi, Ha Noi 54609
Bao Tàng Văn Hóa Các Dân Tôc Viêt Nam, Thai Nguyen 54636
Baoji Museum, Baoji 07500
Baoji Xiàn Museum, Baoji 07501
Baoji Zhouyuan Museum, Fufeng 07694
Baoshan Museum, Baoshan 07502
Bar David Museum, Kibbutz Bar Am 25033
Bar Harbor Historical Society Museum, Bar Harbor 46602
Barabinsk Regional Museum, Barabinsk 36788
Barabinskij Kraevedčeskij Muzej, Barabinsk 36788
De Baracquen Museum voor Bourtanger Bodemvondsten, Bourtange 31992
Baragova Rojstna Hiša, Dobrnič 38411
Barangay Museum, Mandaluyong 34424
Baranov Museum, Kodiak 49824
Barasoain Church Museum, Malolos 34420
Barbados Museum, Saint Michael 03242
Barbara and Steven Grossman Gallery, Boston 46918
Barbara Fritchie House Museum, Frederick 48754
Barbara Hepworth Museum and Sculpture Garden, Saint Ives, Cornwall 45575
Barbarlık Müzesi, Lefkoşa 33367
Barbed Wire Museum, La Crosse 49833
Barber Institute of Fine Arts, Birmingham 43497
Barberton Museum, Barberton 38555
Barbican Art Gallery, London 44791
Barbour's General Store Museum, Saint John 06834
Barcaldine and District Historical Museum, Barcaldine 00833
Barcelona Maritime Museum, Barcelona 38998
The Bard Graduate Center for Studies in the Decorative Arts, Design, and Culture, New York 51132
Bardelaere Museum, Lembeke 03783
Bardhaman Science Centre, Bardhaman 24076
Bardu Bygdetun, Salangsdalen 33806
Bārdu Dzimtas Muzejs Rumbiņi, Katvaru 30030
Bare Cove Fire Museum, Hingham 49297
Barfüsserkirche, Basel 41157
Barfüßerkirche, Erfurt 18198
Bargrave Collection, Canterbury 43740
Baripada Museum, Baripada 24078
Barker-Cawood Art Gallery, Olivet 51540
Barkerville Historic Town, Barkerville 05384
Barlanglakás Emlékkiállítás, Budapest 23386
Barleylands Farm Museum, Billericay 43485
Barlow Collection of Chinese Ceramics, Bronzes and Jades, Brighton 43618
Barmouth Sailors' Institute, Barmouth 43380
Barn Museum and Rail Depot Museum, Troutdale 53570
Barnacle Historic State Park, Coconut Grove 47670
Barnard's Mill Art Museum, Glen Rose 48922
Barnegat Bay Decoy and Baymen's Museum, Tuckerton 53581
Barnes County Historical Museum, Valley City 53693
Barnes Foundation, Merion 50601
Barnet Historical Society, Barnet 46611
Barnet Museum, London 44792
Barnsdall Art Center and Junior Arts Center, Los Angeles 50211
Barnum House Museum, Grafton 05886
Barnum Museum, Bridgeport 47016
Barnwell County Museum, Barnwell 46614
Barockjagdschloß Eckartsau, Eckartsau 01838
Barockschloss Mannheim, Mannheim 20162
Barockschloß Riegersburg, Riegersburg, Niederösterreich 02586
Baroda Museum and Picture Gallery, Vadodara 24418
Barometer World Museum, Merton 45132
Baroniet Rosendal, Rosendal 33799
Baroque Palace of Mannheim, Mannheim 20162
Barossa Valley Historical Museum, Tanunda 01570
Barr Colony Heritage Cultural Centre, Lloydminster 06180
Barre Historical Museum, Barre 46615
Barreau Le Maistre Art Gallery, Saint Helier 45569
Barrel Museum, Chécy 12495
Barrelorganmuseum Assen, Assen 31936
Barrett House, Forest Hall, New Ipswich 51065
Barrhead Centennial Museum, Barrhead 05385
Bārricdes 1991, Rīga 30063
Barrie's Birthplace & Camera Obscura, Kirriemuir 44617
Barrington Area Historical Museum, Barrington 46616
Barrington Living History Farm, Washington 53945
Barrington Woolen Mill Museum, Barrington 05389
Barron Arts Center and Museum, Woodbridge 54274
Barron County Historical Society's Pioneer Village Museum, Cameron 47204
Barthélemy Thimonnier Museum, Amplepuis 11593
Bartholomäus-Schmucker-Heimatmuseum, Ruhpolding 21445
Bartholomew County Historical Museum, Columbus 47741
Bartın Culture House, Bartın 42726
Bartın Kültür Evi, Bartın 42726
Bartlesville Area History Museum, Bartlesville 46617
Bartlett Museum, Amesbury 46260
Bartók Béla Emlékház, Budapest 23387
Bartolomeu Dias Museum Complex, Mossel Bay 38691
Barton County Historical Society Village and Museum, Great Bend 49027
Barton Museum, Bhavnagar 24086
Barton Museum, Wilson 54210
Bartow History Center, Cartersville 47263
Bartow-Pell Mansion Museum, Bronx 47039
Barunčina Škola, Česká Skalice 09458
Barunka's School, Česká Skalice 09458
Basaltpark Bad Marienberg, Bad Marienberg 16782
Base Borden Military Museum, Borden 05441
Baseball Hall of Fame and Museum, Tokyo 29506
Baseball Hall of Fame of Mexico, Monterrey 31190
Bashkirian State Art Museum, Ufa 37883
Basho Memorial Museum, Ueno 29554
Basho Museum Koto, Tokyo 29371
Basilian Fathers Museum, Mundare 06402
Basilica Collegiata di Santa Maria Assunta, San Gimignano 27635
Basílica de Nuestra Señora del Pilar, Zaragoza 40336
Basilica del Santo Niño Museum, Cebu 34364
Basilica Museum, Saint John's 06843
Basilica Museum, Schatkamer van de Onze-Lieve-Vrouwbasiliek, Tongeren 04074
Basilika Sipahi Ay. Trias and Kantara Castle, Kantara 33364
Basilique Royale Saint-Denis, Saint-Denis (Seine-Saint-Denis) 15234
Basin Head Fisheries Museum, Souris 06994
Basın Müzesi, İstanbul 42807
Basing House, Basingstoke 43394
Basingstoke Canal Exhibition, Mytchett 45188
Basis Wien, Wien 02943
Baškirskij Respublikanskij Chudožestvennyj Muzej im. M.V. Nesterova, Ufa 37883
Basler Papiermühle, Basel 41158
(Basque Gastronomic Museum), Llodio 39487
Basque Museum, Bilbao 39042
Basque Railway Museum, Azpeitia 38931
Basrah Museum, Basrah 24717
Bass Manor Museum, Malta 50409
Bass Museum of Art, Miami Beach 50641
Bassetlaw Museum and Percy Laws Memorial Gallery, Retford 45484
Bassett Hall, Williamsburg 54165
Bastion, Nanaimo 06409
Bastogne Historical Center, Bastogne 03345
Baszta Pranieńska, Chełmno 34589
Baszta Prochowa, Chełmno 34590
BAT Artventure Collectie, Amstelveen 31820
BAT Artventure Collectie, Zevenaar 32998
BAT CampusGalerie, Bayreuth 16947
Bata Shoe Museum, Toronto 07093
Batán Museo, Val de San Lorenzo 40185
Batavia Depot Museum, Batavia 46625
Bataviawerf-Nationaal Scheephistorisch Centrum, Lelystad 32493
Batchworth Lock Canal Centre, Rickmansworth 45494
Bate Collection of Musical Instruments, Oxford 45328
Bateman Historical Museum, Bateman 05395
Bateman's, Etchingham 44187
Bateman's Kipling's House, Burwash 43689
Bates College Museum of Art, Lewiston 50037
Bates-Scofield Homestead, Darien 47966
Bath Abbey Heritage Vaults, Bath 43399
Bath of Memi Pasha, Pécs 23766
Bath Postal Museum, Bath 43400
Bath Royal Literary and Scientific Institution, Bath 43401
Báthory Júlia Üveggyűjtemény, Dömsöd 23526
Bathurst Agricultural Museum, Bathurst 38556
Bathurst Heritage Museum Cultural Centre / Musée de Bathurst/Musée Centre Culturel, Bathurst 05397
Bathurst Museum, Bathurst 00837
Bathurst Regional Art Gallery, Bathurst 00838
Batik Gemi Müzesi, Girne 33356
Batley Art Gallery, Batley 43416
Batman Culture Center, Batman 42727
Batman Kültür Merkezi, Batman 42727
Batoche, Batoche 05399
Baton Rouge Gallery, Baton Rouge 46631
Båtsamlingarna på Bassholmen, Uddevalla 40991
Båtsmanskasernen, Karlskrona 40633
Batsto Village, Hammonton 49145
Battambang Museum, Battambang 05283
Battle Abbey, Battle 43417
Battle and Farm Museum, Naseby 45194
Battle Museum, Kibbutz Gesher 25041
Battle Museum of Local History, Battle 43418
Battle of Aljubarrota Interpretation Centre, Porto de Mós 35868
Battle of Lexington State Historic Site, Lexington 50062
Battle of Pákozd Memorial Site and Museum, Pákozd 23746
Battlefield House Museum, Stoney Creek 07011
Battles Museums of Rural Life, Girard 48914
Battlesbridge Motorcycle Museum, Battlesbridge 43420
Battleship Cove - Maritime Heritage Museums, Fall River 48498
Battleship North Carolina, Wilmington 54198
Battleship South Dakota Memorial, Sioux Falls 53117
Battleship Texas, La Porte 49854
Baturins'kyj Deržavnyj Istoričnyj Zapovidnyk Get'mans'ka Stolicja, Baturin 43000
Bauern-Technik-Museum, Steyr-Gleink 02774
Bauern- und Handwerker-Museum, Malgersdorf 20157
Bauern- und Heimatmuseum Haigermoos, Sankt Pantaleon 02675
Bauern- und Waldmuseum, Haidmühle 18865
Bauernfeind-Museum, Sulz am Neckar 21979
Bauerngerätemuseum Hundszell, Ingolstadt 19338
Bauernhaus Habach, Eppelborn 18178
Bauernhaus-Museum, Bielefeld 17337
Bauernhaus-Museum, Lindberg 20002
Bauernhaus-Museum, Muttenz 41669
Bauernhaus-Museum, Ruhmannsfelden 21444
Bauernhaus-Museum, Wolfegg 22649
Bauernhaus-Museum Bortfeld, Wendeburg 22474
Bauernhausmuseum, Grafenau, Kreis Freyung-Grafenau 18727
Bauernhausmuseum, Steinen 21886
Bauernhausmuseum Amerang des Bezirks Oberbayern, Amerang 16536
Bauernhausmuseum des Landkreises Erding, Erding 18193
Bauernhausmuseum Gebersheim, Leonberg 19960
Bauernhausmuseum Hattingen, Hattingen 19036
Bauernhausmuseum Hof Haina, Biebertal 17332
Bauernhausmuseum Hohenstein, Hohenstein 19243
Bauernhausmuseum Knechtenhofen, Oberstaufen 20840
Bauernhausmuseum Schniderlihof, Oberried 20828
Bauernhofmuseum, Wagrain 02881
Bauernhofmuseum Hof, Kirchanschöring 19542
Bauernhofmuseum Jexhof, Schöngeising 21609
Bauernhofmuseum Schreiner, Buchhofen 17617
Bauernkriegsmuseum, Weinstadt 22435
Bauernkriegsmuseum Kornmarktkirche, Mühlhausen/Thüringen 20401
Bauernmöbelmuseum Hirschbach, Hirschbach 02080
Bauernmöbelmuseum Propstkeusche, Malta 02345
Bauernmuseum, Dahlen 17770
Bauernmuseum, Kalladorf 02149
Bauernmuseum, Simmerath 21769
Bauernmuseum, Tuttlingen 22131
Bauernmuseum, Zabelitz 22764
Bauernmuseum, Zahna 22766
Bauernmuseum Althuus, Gurbrü 41463
Bauernmuseum Althuus, Gurbrü 41464
Bauernmuseum Blankensee, Trebbin 22081
Bauernmuseum Bundwerkstadel, Haigermoos 02037
Bauernmuseum Grottenhof, Kaindorf an der Sulm 02148
Bauernmuseum im Tschötscherhof, Castelrotto 25752
Bauernmuseum Inzigkofen, Inzigkofen 19349
Bauernmuseum Lanzenkirchen, Lanzenkirchen 02263
Bauernmuseum Liebenau, Geising 18559
Bauernmuseum Mühlhausen, Villingen-Schwenningen 22234
Bauernmuseum Ostdorf, Balingen 16898
Bauernmuseum Osternach, Ort im Innkreis 02472
Bauernmuseum Pfarrscheuer, Bösingen 17430
Bauernmuseum Rothenburg, Rothenburg/Oberlausitz 21413
Bauernmuseum Schwarzbach, Schwarzbach 21671
Bauernmuseum Selfkant, Selfkant 21736
Bauernmuseum Sollingerhof, Maria Schmolln 02353
Bauernmuseum Voadohuibn, Valle di Casies 28146
Bauernmuseum Wittstock, Nordwestuckermark 20746
Bauernmuseum Zabeltitz, Zabeltitz 22765
Bauernstube Pfahlheim, Ellwangen 18122
Baugeschichte der ehemaligen Fürstbischöflichen Residenz Schloß Neuhaus, Paderborn 20966
Baugeschichtliches Archiv der Stadt Zürich, Zürich 42067
Bauhaus-Archiv, Berlin 17041
Bauhaus-Museum, Weimar 22396
Bauhausbauten Dessau, Meisterhäuser, Dessau 17827
Baum Gallery, Conway 47782
Baumberger-Sandstein-Museum, Havixbeck 19055
Bauska Castle Museum, Bauska 29989
Bauska History and Art Museum, Bauska 29988
Bauskas Novadpētniecības un Mākslas Muzejs, Bauska 29988
Bauskas Pils Muzejs, Bauska 29989
Bauspar-Museum im Georg-Kropp-Haus, Wüstenrot 22728
Bautahaugen Samlinger, Hedalen 33541
Bavarian Army Museum, Ingolstadt 19339
Bavarian Textile and Industry Museum (tim), Augsburg 16638
BAWAG Foundation, Wien 02944
Baxter House Museum, Gorham 48975
Baxter Springs Heritage Center and Museum, Baxter Springs 46649
Bay Area Discovery Museum, Sausalito 52935
Bay Area Museum, Seabrook 52986
Bay Chaleur Military Museum, New Richmond 06433
Bay View Historical Museum, Bay View 46652
Bayanihan Folk Arts Museum, Manila 34432
Baycrafters, Bay Village 46653
Baycrest Heritage Museum, Toronto 07094
Bayerische Staatsbibliothek, München 20424
Bayerische Staatsgemäldesammlungen, München 20425
Bayerisches Armeemuseum, Ingolstadt 19339
Bayerisches Bienenmuseum, Illertissen 19315
Bayerisches Brauerei- und Bäckereimuseum Kulmbach, Kulmbach 19746
Bayerisches Eisenbahnmuseum, Nördlingen 20724
Bayerisches Moor- und Torfmuseum, Grassau 18739
Bayerisches Nationalmuseum, München 20426
Bayerisches Schulmuseum, Ichenhausen 19305
Bayerisches Strafvollzugsmuseum, Kaisheim 19415
Bayet Aisha Fahmi (Mogamaa el-Fonoon), Cairo 10488
Bayin Guoleng Zhou Museum, Kuerle 07915
Bayle Museum, Bridlington 43609
Bayou Bend Collection, Houston 49381
Bayreuther FeuerwehrMuseum, Bayreuth 16948
Baysgarth House Museum, Barton-upon-Humber 43391
Baznytinio Meno Skyrius, Kaunas 30241
Beaches Heritage Centre, Eastport 05733
Beachville District Museum, Beachville 05403
Beacon Historical Society Museum, Beacon 46659
The Beacon, Whitehaven, Whitehaven 46000
Beaconsfield Historic House, Charlottetown 05589
The Bead Museum, Glendale 48926
Beall Center for Art and Technology, Irvine 49529
Beaminster Museum, Beaminster 43421
Beamish, the North of England Open Air Museum, Beamish 43422
Bear Butte State Park Visitors Center, Sturgis 53350
Bear Mountain Trailside Museums Wildlife Center, Bear Mountain 46662
Bear Museum, Petersfield 45372
Beard Gallery, Norton 51430
Beate Uhse Erotik-Museum, Berlin 17042
Beatles Museum, Halle, Saale 18889
Beatles-Museum Little Cavern, Biebelnheim 17329
Beatles Museum Remember, Gemmerich 18581
The Beatles Story, Liverpool 44731
Beatrice Grinceviciute Museum - Beatrice's Home, Vilnius 30402
Beatrice M. Haggerty Gallery, Irving 49532
Beatričės Grincevičiūtės Memorialinis Muziejus, Beatričės Namai, Vilnius 30402
Beatrix Potter Gallery, Hawkshead 44429
Beatties Historic Photograph Museum, Dodges Ferry 01039
Beatty Museum, Newport Pagnell 45246
Beaty Biodiversity Museum, Vancouver 07199
Beau Fort Plantation Home, Bermuda 46765
Beau Village Museum, Saint Victor 06882
Beauchamp Newman Museum, Elizabeth 48337
Beaudesert Museum, Beaudesert 00845
Beaufort Historic Site, Beaufort 46665
Beaufort Museum, Beaufort 46667
Beaufort West Museum, Beaufort West 38557
Beaulieu Abbey Exhibition of Monastic Life, Brockenhurst 43649
Beaumaris Castle, Beaumaris 43425
Beaumaris Gaol and Courthouse, Beaumaris 43426
Beaumont Art League, Beaumont 46673
Beaumont Mill, Beaumont 05405
Beauport-Sleeper-McCann House, Gloucester 48944
Beauregard-Keyes House, New Orleans 51075
Beautiful Plains Museum, Neepawa 06419

Beauvais Heritage Center, Clark 47586
Beauvoir, Biloxi 46809
Beaver Island Historical Museum, Beaver Island 46682
Beaver River Museum, Beaverton 05409
Beaver Valley Military Museum, Clarksburg 05620
Beaverbrook Art Gallery, Fredericton 05828
Beaverhead County Museum, Dillon 48142
Beccles and District Museum, Beccles 43428
Beck Cultural Exchange Center, Knoxville 49811
Beck Isle Museum of Rural Life, Pickering 45381
Becker County Historical Museum, Detroit Lakes 48131
Beckford's Tower and Museum, Bath 43402
Beckoning Hills Museum, Boissevain 05433
Beck's Schreib- und Büromaschinenmuseum, Pfäffikon (Zürich) 41722
Bedale Museum, Bedale 43430
Bede Art Gallery, Yankton 54322
Bede's World, Jarrow 44545
Bedford Central Library Gallery, Bedford 43433
Bedford City/County Museum, Bedford 46691
Bedford Gallery, Walnut Creek 53802
Bedford Historical Society Museum, Bedford 46689
Bedford Museum, Bedford 43434
Bedfordshire and Hertfordshire Regiment Association Museum Collection, Luton 45046
Bedingfield Inn Museum, Lumpkin 50321
Bedřich Smetana Museum, Praha 09768
Bedrijfsmuseum NUON-ENW Amsterdam, Amsterdam 31832
Bee Family Museum, Yuanshan 42481
Bee-Keeping Museum, Moskva 37348
Beechworth Historic Court House, Beechworth 00847
Beecroft Art Gallery, Southend-on-Sea 45710
Beehive Collection, Gödöllő 23567
Beehive House, Salt Lake City 52687
Beekeeping Museum, Radovljica 38500
Beeldentuin Belling Garde, Bellingwolde 31958
Beenleigh and District Museum, Beenleigh 00852
Beerewein Museum, Eulenbis 18282
Be'eri Archaeological Collection, Be'eri 24935
Beeston Castle, Tarporley 45827
Beethoven Emlékmúzeum, Martonvásár 23691
Beethoven Eroicahaus, Wien 02945
Beethoven-Gedenkstätte Floridsdorf, Wien 02946
Beethoven-Haus Bonn, Bonn 17448
Beethoven Memorial, Dolná Krupá 38248
Beethoven-Memorial Museum, Martonvásár 23691
Beethoven Pasqualatihaus, Wien 02947
Beethoven Wohnung Heiligenstadt, Wien 02948
Beethovenhalle, Bonn 17449
Beethovenhaus, Krems 02224
Beethovenhaus "Haus der Neunten", Baden bei Wien 01774
Beeville Art Museum, Beeville 46692
Befreiungshalle Kelheim, Kelheim 19500
Le Befroi et le Musée des Frères Caudron, Rue 15136
Bega Valley Regional Gallery, Bega 00854
Begegnungsstätte Alte Synagoge, Wuppertal 22737
Begijnhofkerk, Sint-Truiden 04027
Begijnhofmuseum, Dendermonde 03540
Begijnhofmuseum, Herentals 03697
Begijnhofmuseum, Kortrijk 03758
Béguinage d'Anderlecht, Bruxelles 03436
Behnhaus/ Drägerhaus, Lübeck 20072
Het Behouden Blik, Uithuizermeeden 32838
Behringer-Crawford Museum, Covington 47858
Behzad Museum, Tehrān 24658
Behzod National Museum, Dušanbe 42494
Beiarn Bygdetun, Moldfjord 33680
Beijing Ancient Architecture Museum, Beijing 07507
Beijing Art Museum, Beijing 07508
Beijing Changcheng Hua Ren Huai, Beijing 07509
Beijing China Nationality Museum, Beijing 07510
Beijing Cultural Relic Protection Foundation, Beijing 07511
Beijing Dabao Taixi Western Han Tomb Museum, Beijing 07512
Beijing Folklore Museum, Beijing 07513
Beijing Gu Guanxiang Station, Beijing 07514
Beijing Katong Art Museum, Beijing 07515
Beijing Man Zhoukoudian Museum, Beijing 07516
Beijing Museum, Beijing 07517
Beijing Museum of Natural History, Beijing 07518
Beijing Police Museum, Beijing 07519
Beijing Telephone History Museum, Beijing 07520
Beijing Theater Museum, Beijing 07521
Beijing Wabgfujing Paleolithic Museum, Beijing 07522
Beijing World Art Museum, Beijing 07523
Beiliu Museum, Beiliu 07578
Beit Abraham Krinizi, Ramat Gan 25082
Beit al-Ajaib Museum with House of Wonders Museum, Zanzibar 42522
Beit-Alfa Ancient Synagogue, Kibbutz Chefzi-bah 25034
Beit Hagdudim, Avihail 24932
Beit-Hagefen Art Gallery, Haifa 24954
Beit Hameiri, Zefat 25132
Beit Hankin, Kfar Yehoshua 25031
Beit Ha'Omanim, Jerusalem 24978
Beit Hatefutsoth, Tel Aviv 25099
Beit Lohamei Haghetaot-Yad Layeled, Lohamei-Hagetaot 25062
Beit Miriam-Museum, Kibbutz Palmachim 25053
Beit-Shean Museum, Beit Shean 24937
Beit Shturman Museum, Kibbutz Ein Harod 25038
Beiteddine Museum, Shouf 30174
Beitou Hot Spring Museum, Taipei 42376
Beius Country Museum, Beiuş 36146
Békés County Museum Authority - Mihály Munkácsy Museum, Békéscsaba 23367
Békés Megyei Múzeumok Igazgatósága, Békéscsaba 23367
Bekko Crafts Museum, Nagasaki 28969
Béla Bartók Memorial Museum, Budapest 23387
Bela Krajina Museum - Metlika Castle, Metlika 38481
Béla Vidovszky History Collection - Town Gallery, Gyomaendrőd 23578
Belair Mansion, Bowie 46967
Belair Stable Museum, Bowie 46968
Belasting en Douane Museum, Rotterdam 32702
Belcarra Eviction Cottage, Castlebar 24755
Belcourt Castle, Newport 51319
Beldam Gallery, Uxbridge 45903
Belfast Historical Society, Belfast 05412
Belger Arts Center, Kansas City 49685
Belghazi-Musée, Fès 31669
Belgian Brewers Museum, Bruxelles 03467
Belgian Comic Strip Center, Bruxelles 03442
Belgian Federation of Friends of Museums, Bruxelles 54712
Belgian Museum of Radiology-Belgian Military Hospital, Bruxelles 03453
Belgorod History and Regional Museum, Belgorod 36794
Belgorodskij Gosudarstvennyj Chudožestvennyj Muzej, Belgorod 36793
Belgorodskij Gosudarstvennyj Istoriko-Kraevedčeskij Muzej, Belgorod 36794
Belgrade City Museum, Beograd 38065
Belgrave Hall Museums and Gardens, Leicester 44673
Belhaven Memorial Museum, Belhaven 46695
Believe It Or Not Curiosity Museum, Tamsui 42446
Belinskij Rajonnyj Kraevedčeskij Muzej, Belinskij 36796
Belinskyi Regional Museum, Belinskij 36796
Belize Audubon Society, Belize City 04174
Belize Maritime Terminal and Museum, Belize City 04175
Belkhandi Museum, Belkhandi 24081
The Belknap Mill Museum, Laconia 49861
Bell County Museum, Belton 46727
Bell Homestead National Historic Site, Brantford 05466
Bell Island Community Museum, Bell Island 05413
Bell Museum, Apolda 16565
Bell Pettigrew Museum, Saint Andrews 45551
Bell Rock Mill Museum, Bethany 05423
Bell Tower Museum, Lempäälä 11061
Bella Vista Historical Museum, Bella Vista 46696
Bellaghy Bawn, Magherafelt 45070
Bellamy Mansion Museum of History and Design Arts, Wilmington 54199
Bellapais Manastiri, Girne 33357
Belle Grove Plantation, Middletown 50663
Belle Meade Plantation, Nashville 50962
Belleville Area Museum, Belleville 46701
Belleville Public Library and Art Gallery, Belleville 05414
Bellevue Art Museum, Bellevue 46706
Bellevue House, Kingston 06074
Bellflower Genealogical and Historical Society Museum, Bellflower 46710
Bellport-Brookhaven Historical Society Museum, Bellport 46717
Bells and Carillons Museum, Tellin 04052
Belmont, Faversham 44220
Belmont and District Museum, Belmont 05418
Belmont County Victorian Mansion, Barnesville 46610
Belmont, Gari Melchers Estate and Memorial Gallery, Fredericksburg 48763
Beloborodov Memorial Museum, Tula 37867
Beloit Historical Society Museum, Beloit 46723
Belokranjski Muzej, Metlika 38481
Belorussian State Museum of the History of the Great Patriotic War, Minsk 03247
Belorusskij Gosudarstvennyj Muzej Istorii Velikoj Otečestvennoj Vojny, Minsk 03247
Beloveżskaja Pušča Muzej, Beloveżskaja Pušča 03245
Belozerskij Istoriko-chudožestvennyj Muzej, Belozersk 36798
Belozerskij Muzej Narodnogo Dekorativno-prikladnogo Iskusstva, Belozersk 36799
Belton House, Grantham 44344
Beltrami County Historical Center, Bemidji 46730
Belvedere - 20er Haus, Wien 02949
Belvedere auf dem Klausberg, Potsdam 21080
Belvedere - Augarten Contemporary, Wien 02950
Belvedere im Schlossgarten Charlottenburg, Berlin 17043
Belvedere - Oberes Belvedere, Wien 02951
Belvedere und Pomonatempel auf dem Pfingstberg, Potsdam 21081
Belvedere - Unteres Belvedere, Wien 02952
Bélyegmúzeum, Budapest 23388
Bembridge Windmill Museum, Mottistone 45180
Bement-Billings Farmstead, Newark Valley 51296
Bemis Center for Contemporary Arts, Omaha 51547
Ben-Gurion House, Tel Aviv 25100
Ben Maltz Gallery and Bolsky Gallery, Los Angeles 50212
Ben Shahn Galleries, Wayne 54007
Ben Uri Gallery, London 44793
Benakeion Archaeological Museum, Kalamata 23023
Benaki Museum, Athinai 22925
Benalla Art Gallery, Benalla 00857
Benalla Costume and Pioneer Museum, Benalla 00858
Benares Historic House, Mississauga 06309
Bencés Apátsági Múzeum, Tihany 23925
Benchamabophit National Museum, Bangkok 42533
Bendigo Art Gallery, Bendigo 00859
Benedek Péter Emlékház, Uszód 23942
Benedicta Arts Center, Saint Joseph 52559
Benedictan Abbey Museum, Tihany 23925
Benedictine Monastery - Church of the Multiplication of Loaves and Fish Tabgha, Tiberias 25128
Benedikt-Nimser-Haus, Wilhelmsdorf, Kreis Ravensburg 22572
Benediktinerstift und Stiftssammlungen, Altenburg 01710
The Benedykt Dybowsky Zoological Museum of the Ivan Franko National University of Lviv, Lviv 43143
Benesse House, Naoshima 29026
Benga Oral Historic Centre, Dandenong 01022
Bengal Natural History Museum, Darjeeling 24126
Bengough and District Museum, Bengough 05420
Benguet Museum, La Trinidad 34400
Benicia Historical Museum, Benicia 46734
Beningbrough Hall, York 46098
Benjamin and Dr. Edgar R. Cofeld Judaic Museum of Temple Beth Zion, Buffalo 47115
The Benjamin Banneker Museum, Baltimore 46556
Benjamin Franklin House, London 44794
Benjamin Patterson Inn Museum Complex, Corning 47809
Bennett Place, Durham 48215
Bennie Museum, Bathgate 43414
Bennington Battle Monument, Old Bennington 51531
Bennington Battlefield State Historic Site, Walloomsac 53801
The Bennington Museum, Bennington 46739
Benson Veteran Cycle Museum, Benson 43452
Bensusan Museum of Photography, Johannesburg 38649
Bentheim Castle, Bad Bentheim 16685
Benton County Historical Museum, Philomath 51895
Benton County Historical Museum, Prosser 52159
The Benton Homestead, Tolland 53524
Bentonville Battleground, Four Oaks 48726
Bent's Old Fort, La Junta 49849
Benxi Museum, Benxi 07579
Benzie Area Historical Museum, Benzonia 46745
Beobide Lantoki Museoa - Museo-Taller de Julio Beobide, Zumaia 40361
Beothuck Village, Grand Falls-Windsor 05891
Beothuk Interpretation Centre, Boyd's Cove 05451
Beppu Daigaku Fuzoku Hakubutsukan, Beppu 28388
Beppu University Museum, Beppu 28388
Berea College Burroughs Geological Museum, Berea 46746
Berea College Doris Ulmann Galleries, Berea 46747
Beredskapsmuseet, Viken 41060
Beregi Múzeum, Vásárosnamény 23955
Berend Lehmann Museum, Halberstadt 18874
Bereznikovskij Istoriko-kraevedčeskij Muzej, Berezniki 36803
Berg-Kragerø Museum, Kragerø 33607
Berg- und Stadtmuseum Obernkirchen, Obernkirchen 20822
Bergama Arkeoloji Müzesi, Bergama 42728
Bergbahnmuseum, Kitzbühel 02176
Bergbau- und Gotikmuseum Leogang, Leogang 02289
Bergbau- und Greifenstein-Museum, Ehrenfriedersdorf 18070
Bergbau- und Heimatmuseum, Gloggnitz 01937
Bergbau- und Heimatmuseum, Goldkronach 18704
Bergbau- und Heimatmuseum, Jochberg 02143
Bergbau- und Heimatmuseum im Paulushof, Essen 18249
Bergbau- und Industriemuseum Ostbayern, Kümmersbruck 19734
Bergbau- und Mineralienmuseum, Oberwolfach 20851
Bergbau- und Stadtmuseum Weilburg, Weilburg 22384
Bergbauernmuseum, Alpbach 01705
Bergbauernmuseum, Telfs 02801
Bergbauernmuseum Hausstein, Frankenfels 01898
Bergbauernmuseum z' Bach, Wildschönau 03104
Bergbaumuseum, Aldenhoven 16461
Bergbaumuseum, Altenberg 16493
Bergbaumuseum, Deuben 17843
Bergbaumuseum, Grünbach am Schneeberg 02013
Bergbaumuseum, Hall in Tirol 02042
Bergbaumuseum, Hausham 19052
Bergbaumuseum, Horgen 41487
Bergbaumuseum, Langelsheim 19808
Bergbaumuseum, Mechernich 20242
Bergbaumuseum, Mühlbach am Hochkönig 02408
Bergbaumuseum, Niedermoschel 20707
Bergbaumuseum, Oelsnitz 20870
Bergbaumuseum Achthal, Teisendorf 22019
Bergbaumuseum BUV Kleinzeche, Dortmund 17915
Bergbaumuseum Erzpoche, Hausach 19046
Bergbaumuseum Goberling, Stadtschlaining 02755
Bergbaumuseum Graubünden mit Schaubergwerk, Davos Platz 41320
Bergbaumuseum Grube Anna II, Alsdorf 16479
Bergbaumuseum Hückelhoven, Hückelhoven 19286
Bergbaumuseum im Besucherbergwerk Tiefer Stollen, Aalen 16410
Bergbaumuseum Klagenfurt, Klagenfurt 02181
Bergbaumuseum Knappenhaus, Altenmarkt bei Sankt Gallen 01712
Bergbaumuseum mit Mineralienschau und Schaubergwerk, Hüttenberg 02102
Bergbaumuseum Peißenberg, Peißenberg 21003
Bergbaumuseum Pöllau, Neumarkt in Steiermark 02438
Bergbaumuseum Schachtanlage Knesebeck, Bad Grund 16743
Bergbaumuseum und Schaubergwerk Röhrigschacht, Wettelrode 22519
Bergbaumuseum und Schaustollen, Fohnsdorf 01893
Bergbauschaustollen, Pölfing-Brunn 02512
Bergen Art Museum, Lysverket, Bergen 33418
Bergen Art Museum, Rasmus Meyer Collections, Bergen 33419
Bergen Art Museum, Stenersen Collection, Bergen 33420
Bergen-Belsen Memorial, Lohheide 20023
Bergen County Historical Society Museum, River Edge 52361
Bergen Kunsthall, Bergen 33417
Bergen Kunstmuseum - Lysverket, Bergen 33418
Bergen Kunstmuseum - Rasmus Meyers Samlinger, Bergen 33419
Bergen Kunstmuseum - Stenersens Samling, Bergen 33420
Bergen Maritime Museum, Bergen 33423
Bergen Museum, Bergen 33421
Bergen Museum of Art and Science, Paramus 51696
The Bergen Museum of Art and Science, Paramus 51697
Bergen Museum of Local History, Bergen 46749
Bergen Museum of Technology, Bergen 33424
Bergen Skolemuseum, Bergen 33422
Bergens Sjøfartsmuseum, Bergen 33423
Bergens Tekniske Museum, Bergen 33424
Bergerhaus, Gumpoldskirchen 02017
Bergführer-Museum, Sankt Niklaus 41814
Berggericht - Mineralogical Exposition, Banská Štiavnica 38189
Berggericht - Mineralogická Expozícia, Banská Štiavnica 38189
Bergische Museumsbahnen e.V, Wuppertal 22738
Bergisches Freilichtmuseum für Ökologie und bäuerlich-handwerkliche Kultur, Lindlar 20007
Bergisches Museum für Bergbau, Handwerk und Gewerbe, Bergisch Gladbach 17014
De Bergkerk, Deventer 32119
Bergmännisches Traditionskabinett, Breitenbrunn 17540
Bergmanns-Bauernmuseum, Breitenbach, Kreis Kusel 17538
Bergringmuseum, Teterow 22035
Bergsjö Konsthall, Bergsjö 40457
Bergstrom-Mahler Museum, Neenah 50997
Bergsturz-Museum Goldau, Goldau 41441
Bergtheil Museum, Westville 38777
Bergwerkmuseum Grube Glasebach, Straßberg 21917
Bergwerksmuseum, Penzberg 21007
Bergwerksmuseum Schacht Mehren, Käbschütztal 19410
Bergwinkelmuseum, Schlüchtern 21576
Berkeley Art Center, Berkeley 46751
Berkeley Art Museum and Pacific Film Archive, Berkeley 46752
Berkeley Castle, Berkeley 43453

The Berkeley Costume and Toy Museum, New Ross 24884
Berkeley Plantation, Charles City 47350
The Berkshire Artisans, Pittsfield 51966
Berkshire Mill Museum, Moora 01316
Berkshire Museum, Pittsfield 51967
Berkshire Scenic Railway Museum, Lenox 50024
Berkswell Village Museum, Berkswell 43456
Berlevåg Havnemuseum, Berlevåg 33435
Berlin airlift Museum Fassberg, Fassberg 18297
Berlin Art and Historical Collections, Berlin 46763
Berlin Historical Society Museum of Local History, Berlin 46764
Berlin State Library - Prussian Cultural Heritage, Berlin 17258
Berlin Subway Museum, Berlin 17046
Berliner Medizinhistorisches Museum der Charité, Berlin 17044
Berliner Planetarium und Sternwarte Wilhelm Foerster, Berlin 17045
Berliner S-Bahn-Museum, Potsdam 21082
Berliner U-Bahn-Museum, Berlin 17046
Berliner Unterwelten-Museum, Berlin 17047
Berliner Waldmuseum mit Waldschule Grunewald, Berlin 17048
Berlinische Galerie, Berlin 17049
Berlins Gem and Historical Museum, Kingaroy 01199
Berman Hall, Jerusalem 24979
Bermuda Historical Society Museum, Hamilton 04191
Bermuda Maritime Museum, Mangrove Bay 04194
Bermuda National Gallery, Hamilton 04192
Bermuda National Trust Museum, Saint George's 04196
Bermuda Natural History Museum, Flatts 04190
Bermuda Society of Arts, Hamilton 04193
Bernadottebiblioteket, Stockholm 40886
Bernandino Jalandoni Ancestral House, Silay 34506
Bernard Buffet Museum, Shizuoka 29286
Bernard Historical Museum, Delton 48051
Bernard Judaica Museum, New York 51133
Bernard Price Institute Paleontology Museum, Johannesburg 38650
Bernberg Fashion Museum, Johannesburg 38651
Berndt Museum of Anthropology, Crawley 01008
Bernera, Bernera 43457
Bernhard-Haus, Ohlsdorf 02471
Bernice Pauahi Bishop Museum, Honolulu 49335
Bernische Stiftung für angewandte Kunst und Gestaltung, Bern 41203
Bernstein Huus, Esens 18240
Bernstein Museum Bad Füssing, Bad Füssing 16736
Bernsteinmuseum, Sellin, Ostseebad 21740
Berrima District Art Society Gallery, Bowral 00883
Berrima Museum, Berrima 00863
Berry Museum, Berry 00864
Bersham Heritage Centre and Ironworks, Wrexham 46081
Das Berta-Hummel-Museum im Hummelhaus, Massing 20235
Berte Museum - Life in Countryside, Slöinge 40857
Berte Museum - Livet på Landet, Slöinge 40857
Bertha V.B. Lederer Arts Gallery, Geneseo 48876
Berthold-Auerbach-Museum, Horb am Neckar 19271
Bertram House Museum, Cape Town 38578
Berwick Barracks, Berwick-upon-Tweed 43458
Berwick Borough Museum and Art Gallery, Berwick-upon-Tweed 43459
Berwick Pakenham Historical Museum, Pakenham 01401
Bērzgales Pagasta Vēstures Muzejs, Bērzgale 29992
Berzsenyi Dániel Emlékmúzeum, Egyházashetye 23545
Berzsenyi Dániel Emlékmúzeum, Nikla 23731
Besættelsessamlingen 1940-45, Grindsted 10043
Besenmuseum & Galerie, Ehingen 18066
Besichtigungsstätte der Mumie des Ritter Kahlbutz, Neustadt, Dosse 20676
Bessemer Hall of History, Bessemer 46771
Bessenyei György Emlékszoba, Tiszabercel 23928
Bessie Surtees House, Newcastle-upon-Tyne 45224
Bestattungsmuseum der Bestattung Wien GmbH, Wien 02953
Besteckfabrik Hesse, Schmallenberg 21582
Besucher-Bergwerk Hüttenstollen, Salzhemmendorf 21478
Besucherbergwerk, Nothweiler 20749
Besucherbergwerk Bad Friedrichshall-Kochendorf, Heilbronn 19095
Besucherbergwerk "Drei Kronen und Ehrt", Elbingerode, Harz 18111
Besucherbergwerk Fell, Fell 18304
Besucherbergwerk Finstergrund, Wieden 22538
Besucherbergwerk Finstertal, Schmalkalden, Kurort 21579
Besucherbergwerk Frischglück, Neuenbürg 20604
Besucherbergwerk Grube Anna-Elisabeth, Schriesheim 21636
Besucherbergwerk Grube Bindweide, Steinebach 21884
Besucherbergwerk Grube Fortuna, Solms 21810
Besucherbergwerk Grube Gustav, Meißner 20268
Besucherbergwerk "Grube Silberhardt", Windeck 22593
Besucherbergwerk Grube Wohlfahrt, Hellenthal 19117
Besucherbergwerk "Mühlenberger Stollen", Bleialf 17386
Besucherbergwerk Rischbachstollen, Sankt Ingbert 21498
Besucherbergwerk Schiefergrube Christine, Willingen 22584
Besucherbergwerk und Ehemaliges KZ Kochendorf, Bad Friedrichshall 16735
Besucherbergwerk Vereinigt Zwitterfeld zu Zinnwald, Altenberg 16494
Bet Gordon, Deganya Aleph 24941
Bet Hashomer Museum, Kibbutz Kfar Giladi 25046
Bet Pinhas Museum of Nature, Haifa 24955
Beta Railway Station Historic Museum, Alpha 00804
Beth Ahabah Museum and Archives, Richmond 52324
Beth Tzedec Reuben and Helene Dennis Museum, Toronto 07095
Bethel Historical Society's Regional History Center, Bethel 46774
Bethlehem Historical Association Museum, Selkirk 53031
Bethlehem Museum, Bethlehem 38559
Bethlehem Museum, Karlštejn 09578
Bethlem Royal Hospital Archives and Museum, Beckenham 43429
Bethune Memorial House, Gravenhurst 05907
Betonzeitschiene, Dresden 17938
Betsey Williams Cottage, Providence 52162
Betsy Ross House, Philadelphia 51819
Bettina und Achim von Arnim Museum, Niederer Fläming 20704
Betts House Research Center, Cincinnati 47553
Betty Brinn Children's Museum, Milwaukee 50705
Betuws Fruitteelt Museum, Erichem 32220
Betws-y-coed Motor Museum, Betws-y-Coed 43466
Betygala Museum, Betygala 30212
Betygalos Muziejus, Betygala 30212
Beurs van Berlage Museum, Amsterdam 31833
Beverage Containers Museum, Millersville 50694
Beverley Art Gallery, Beverley 43468
Beverly Art Center, Chicago 47457
Beverly Historical Museum, Beverly 46785
Bevrijdingsmuseum van Boom, Boom 03381
Bewdley Museum, Bewdley 43469
Bewley's Café Museum, Dublin 24779
Bexhill Museum, Bexhill-on-Sea 43471
Bexhill Museum of Costume and Social History, Bexhill-on-Sea 43472
Bexley Historical Society Museum, Bexley 46793
Bexley Museum, London 44795
Beyer clock & watch museum, Zürich 42114
De Beyerd, Breda 32000
Beylerbeyi Palace, İstanbul 42808
Beylerbeyi Sarayı, İstanbul 42808
Beyoglu Sanat Merkezi, İstanbul 42809
Beypazari Culture and History Museum, Beypazarı 42729
Beypazarı Kültür ve Tarih Müzesi, Beypazarı 42729
Bezerédj Castle, Győr 23579
Bezerédj Kastély, Győr 23579
Ter Beziens, Kolhorn 32452
Bežigrajska Galerija, Ljubljana 38445
Bežigrajska Galerija 2, Ljubljana 38446
Bezirksheimatmuseum, Völkermarkt 02868
Bezirksheimatmuseum mit Zdarsky-Skimuseum, Lilienfeld 02305
Bezirksmuseum Alsergrund, Wien 02954
Bezirksmuseum Brigittenau, Wien 02955
Bezirksmuseum Buchen, Buchen 17615
Bezirksmuseum Dachau, Dachau 17763
Bezirksmuseum Döbling, Wien 02956
Bezirksmuseum Donaustadt, Wien 02957
Bezirksmuseum Favoriten, Wien 02958
Bezirksmuseum Floridsdorf, Wien 02959
Bezirksmuseum Hernals, Wien 02960
Bezirksmuseum Hietzing, Wien 02961
Bezirksmuseum Höfli, Bad Zurzach 41141
Bezirksmuseum Innere Stadt, Wien 02962
Bezirksmuseum Josefstadt, Wien 02963
Bezirksmuseum Landstraße, Wien 02964
Bezirksmuseum Leopoldstadt, Wien 02965
Bezirksmuseum Liesing, Wien 02966
Bezirksmuseum Margareten, Wien 02967
Bezirksmuseum Mariahilf, Wien 02968
Bezirksmuseum Marzahn-Hellersdorf, Berlin 17050
Bezirksmuseum Meidling, Wien 02969
Bezirksmuseum Neubau, Wien 02970
Bezirksmuseum Ottakring, Wien 02971
Bezirksmuseum Penzing, Wien 02972
Bezirksmuseum Rudolfsheim-Fünfhaus, Wien 02973
Bezirksmuseum Simmering, Wien 02974
Bezirksmuseum Stockerau, Stockerau 02782
Bezirksmuseum Währing, Wien 02975
Bezirksmuseum Wieden, Wien 02976
Bezoekerscentrum Binnenhof, Den Haag 32069
Bezoekerscentrum de Meinweg, Herkenbosch 32369
Bezoekerscentrum Mijl Op Zeven, Ospel 32637
Bezoekerscentrum Oortjespad, Kamerik 32434
BFI Southbank Gallery, London 44796
Bhagalpur Museum, Bhagalpur 24082
Bhagavan Mahavir Government Museum, Cuddapah 24123
Bharat Bhawan Museum, Bhopal 24089
Bharat Itihas Samshodhak Mandal, Pune 24355
Bharat Kala Bhavan, Varanasi 24425
Bharatiya Adim Jati Sevak Sangh Museum, Delhi 24138
Bhirasri Museum of Modern Art, Bangkok 42534
Bhuri Singh Museum, Chamba 24108
Bialik-Museum, Tel Aviv 25101
Białostockie Muzeum WSI, Wasilków 35293
Bibak Museum, Baguio 34341
Bibel Orient Museum, Fribourg 41383
Bībeles Muzejs, Alūksne 29985
Bibelgalerie Meersburg, Meersburg 20246
Bibelhaus Erlebnismuseum, Frankfurt am Main 18356
Bibelmuseum, Beuron 17313
Bibelmuseum, Münster 20505
Bibelmuseum Stuttgart, Stuttgart 21929
Bible Experience Museum, Frankfurt am Main 18356
Bible Lands Museum Jerusalem, Jerusalem 24980
Bible Museum, Tel Aviv 25102
Bible Museum, Trilofo 23230
Bible Orient Museum, Fribourg 41383
Biblical Arts Center, Dallas 47927
Biblical Museum of Canada, Vancouver 07200
Biblical Resource Center & Museum, Collierville 47692
Biblioludothèque, Paris 14569
Biblioteca Antônio Perdigão, Conseilheiro Lafaiete 04415
Biblioteca d'Arte, Milano 26672
Biblioteca Luis-Angel Arango, Bogotá 08543
Biblioteca Medicea Laurenziana, Firenze 26141
Biblioteca-Museo Alessandro Longo, Amantea 25219
Biblioteka Czartoryskich, Kraków 34764
Biblioteka-muzej A.P. Gajdara, Kaniv 43057
Biblioteka, Muzeum i Archiwum Warszawskiego Towarzystwa Muzycznego im. Stanisława Moniuszki, Warszawa 35216
Biblioteka Narodowa, Warszawa 35217
Biblioteka Sejmowa, Warszawa 35218
Biblioteksmuseet, Borås 40466
Bibliotheca Albertina, Leipzig 19899
Bibliotheca Wittockiana-Musée de la Reliure et du Hochet, Woluwe-Saint-Pierre 04155
Bibliothèque Centrale de Prêt du Saguenay Lac-Saint-Jean, Alma 05320
Bibliothèque d'Art ENSAV de la Cambre, Bruxelles 03437
Bibliothèque d'Art et d'Archéologie des Musées d'Art et d'Histoire de la Ville de Genève, Genève 41400
Bibliothèque d'Art et d'Archéologie Jacques Doucet, Paris 14570
Bibliothèque de Genève, Genève 41401
Bibliothèque de l'Agglomération, Saint-Omer 15430
Bibliothèque d'Histoire de l'Art, Genève 41402
Bibliothèque Marmottan, Boulogne-Billancourt 12087
Bibliothèque Mazarine, Paris 14571
Bibliothèque Musée, Clermont (Oise) 12554
Bibliothèque-Musée de la Comédie-Française, Paris 14572
Bibliothèque-Musée de l'Observatoire, Paris 14573
Bibliothèque-Musée de l'Opéra, Paris 14574
Bibliothèque-Musée Forney, Paris 14575
Bibliothèque-Musée Valéry Larbaud, Vichy 16144
Bibliothèque Publique et Universitaire, Salle Rousseau, Neuchâtel 41676
Bibliothèque Royale de Belgique- Koninklijke Bibliotheek van België, Bruxelles 03438
Bibracte Musée, Glux-en-Glenne 13076
Bicentennial Art Center and Museum, Paris 51701
Bicentennial Historical Museum, Cunnamulla 01017
Bickleigh Castle, Bickleigh 43476
Bicycle and Moped Museum, Zoutkamp 33010
Bicycle Culture Center, Tokyo 29403
Bicycle Culture Center, Tokyo 29372
Bicycle Museum, La Fresnaye-sur-Chédouet 13371
Bicycle Museum, Three Oaks 53504
Bicycle Museum Cycle Center, Sakai 29196
Bicycle Museum of America, New Bremen 51026
Bicycle Museum of the Šiauliai Aušra Museum, Šiauliai 30354
Bidwell House Museum, Monterey 50808
Bidwell Mansion, Chico 47525
Biebergrund-Museum, Biebergemünd 17330
Biedenharn Museum and Gardens, Monroe 50787
Bien Phong Museum, Ha Noi 54610
Bienen- und Wagnereimuseum, Tollet 02814
Bienenkundemuseum, Felsberg 18308
Bienenkundemuseum, Münstertal 20522
Bienenmuseum, Duisburg 18013
Bienenmuseum, Plau am See 21050
Bienenmuseum des VDRB, Alberswil 41098
Bienwaldmuseum, Kandel 19425
Bier- und Oktoberfestmuseum, München 20427
Bierarium, Oberwart 02464
Bierkrug-Museum, Kaiserslautern 19411
Bierreclamemuseum, Breda 32001
Biesbosch Museum, Werkendam 32937
Bieszczady National Park Museum, Ustrzyki Dolne 35204
Big Bear Trails Museum, Loon Lake 06194
Big Beaver Museum and Nature Centre, Big Beaver 05425
Big Bone Lick State Park Museum, Union 53646
Big Hole National Battlefield, Wisdom 54266
Big Horn County Historical Museum, Hardin 49165
Big Pit National Coal Museum of Wales, Blaenavon 43534
Big Red Barn Gallery, Clearmont 47608
Big River Memorial Museum, Big River 05426
Big Shell Museum, Tewantin 01577
Big Springs Museum, Caledonia 47165
Big Thunder Gold Mine, Keystone 49769
Big Timbers Museum, Lamar 49909
Big Well, Greensburg 49069
Bigelow House Museum, Olympia 51542
Biggar Gasworks Museum, Biggar 43479
Biggar Museum and Gallery, Biggar 05427
Biggenden Museum, Biggenden 00866
Biggs Museum of American Art, Dover 48161
Bihar Tribal Welfare Research Institute Museum, Ranchi 24372
Bihari Múzeum, Berettyóújfalu 23373
Bijbels Museum, Amsterdam 31834
Bijbels Openluchtmuseum, Heilig Landstichting 32345
Bijenteeltmuseum De Bankorf, Amen 31806
Bijenteeltmuseum Kalmthout, Kalmthout 03747
Bijskij Kraevedčeskij Muzej V.V. Bianki, Bijsk 36805
Bil & Teknikhistoriska Samlingarna, Köping 40645
Bilbao Fine Arts Museum, Bilbao 39038
Bilboko Arte Eder Museoa / Museo de Bellas Artes de Bilbao, Bilbao 39038
Bilboko Itsasadarra Itsas Museoa, Bilbao 39039
Bilboko Zezen Museoa, Bilbao 39040
Bild- und Tonarchiv, Graz 01965
Bildarchiv Foto Marburg, Marburg 20187
Bilderbogen-Galerie, Neuruppin 20648
Bilderdijkmuseum, Amsterdam 31835
Bildergalerie, Potsdam 21083
Bildhauermuseum Prof. Wandschneider, Plau am See 21051
Bildmuseet, Umeå 40997
Bilecik Devlet Güzel Sanatlar Galerisi, Bilecik 42730
Bilecik State Fine Art Gallery, Bilecik 42730
Bilger Breustedt Haus, Taufkirchen an der Pram 02796
Biljarsk Historical, Archaeological and Natural Museum Reserve, Biljarsk 36806
Biljarskij Gosudarstvennyj Istoriko-Archeologičeskij i Prirodnyj Muzej-Zapovednik, Biljarsk 36806
Bílkova vila, Praha 09735
Billingham Art Gallery, Billingham 43487
Billings Estate Museum, Ottawa 06529
Billings Farm and Museum, Woodstock 54287
Billingsley House Museum, Upper Marlboro 53663
Billnäsin Kirvesmuseo, Billnäs 10717
Billy Bishop Heritage Museum, Owen Sound 06560
Billy Graham Center Museum, Wheaton 54118
Billy The Kid Museum, Fort Sumner 48690
Bilmuseum, Vatnsoyrar 10702
Bilston Craft Gallery, Wolverhampton 46058
Biltmore Estate, Asheville 46369
Bily Clock Museum and Antonin Dvorak Exhibition, Spillville 53203
Binalong Motor Museum, Binalong 00867
Binchester Roman Fort, Bishop Auckland 43520
Bindereimuseum Hofbräu Kaltenhausen, Hallein 02045
Bingara Museum, Bingara 00868
Bingham Gallery, Columbia 47714
Binnenschifffahrts-Museum Oderberg, Oderberg 20858
Binscarth and District Gordon Orr Memorial Museum, Binscarth 05429
Biodiversity Museum, Dehradun 24129
Biodôme de Montréal, Montréal 06328
Biografmuséet, Säter 40822
Biohistoricum, Neuburg an der Donau 20598
Biola University Art Gallery, La Mirada 49851
Biological Museum, Imphal 24221
Biological Museum of the Al-Pharabi Kazakh National University, Almaty 29675
Biological Sciences Collection, Birmingham 43498
Biologičeskij Muzej, Almaty 29675
Biologiezentrum der Oberösterreichischen Landesmuseen, Linz 02310
Biologiska Museet, Örebro 40784

Biologiska Museet, Södertälje 40862
Biologiska Museet, Stockholm 40887
Biology Department Teaching Museum, Muncie 50903
Biology Museum, Amman 29646
Biology Museum GMU, Yogyakarta 24585
Biomedical Museum, Kuala Lumpur 30574
Birbal Sahni Institute of Palaeobotany Museum, Lucknow 24291
Birchip Local History Museum, Birchip 00869
Bird Museum, Jönköping 40612
Birds of Vermont Museum, Huntington 49433
Birdsville Working Museum, Birdsville 00870
Birendra Museum, Kathmandu 31773
Birger Sandzén Memorial Gallery, Lindsborg 50120
Birjand Museum, Birjand 24600
Birka Vikingastaden, Adelsö 40404
Birkenes Bygdemuseum, Birkeland 33437
Birkenhead Priory and Saint Mary's Tower, Birkenhead 43490
Birks Museum, Decatur 48014
Birla Academy of Art and Culture Museum, Kolkata 24270
Birla Industrial and Technological Museum, Kolkata 24271
Birla Museum, Bhopal 24090
Birla Museum, Pilani 24349
Birmingham Bloomfield Art Center, Birmingham 46828
Birmingham Civil Rights Institute Gallery, Birmingham 46821
Birmingham Historical Museum & Park, Birmingham 46829
Birmingham Institute of Art and Design, Birmingham 43499
Birmingham Museum and Art Gallery, Birmingham 43500
Birmingham Museum of Art, Birmingham 46822
Birsfelder Museum, Birsfelden 41238
Birštonas Local Lore Museum, Birštonas 30214
Birštonas Sacral Museum, Birštonas 30215
Birštono Muziejus, Birštonas 30214
Birštono Sakralinis Muziejus, Birštonas 30215
Birthplace of Beatified Anton Martin Slomšek, Šentjur 38512
Birthplace of Beatified Anton Martin Slomšek, Šentjur 38510
Birthplace of Friedrich Schiller, Marbach am Neckar 20182
Birthplace of John Wayne, Winterset 54255
Birthplace of Riste Savin, Žalec 38540
Birthplace of Sibelius, Hämeenlinna 10748
Birthplace of Simon Gregorcic, Vrsno 38536
Birthplace of Stasys Girenas, Vytogala 30449
Birthplace of Steponas Darius, Dariaus 30220
Birthplace of the Brothers-Musicians Motiejus, Bronius and Antanas Budriunai, Pabiržė 30324
Birzgale Regional Museum Dwarf, Birzgale 29993
Birzgales Novada Muzejs Rūķi, Birzgale 29993
Biržų Krašto Muziejus Sėla, Biržai 30216
Bisbee Mill Museum, Chesterfield 47435
Bisbee Mining and Historical Museum, Bisbee 46830
Bischöfliches Diözesanmuseum für christliche Kunst, Münster 20506
Bischöfliches Dom- und Diözesanmuseum, Trier 22096
Bischöfliches Dom- und Diözesanmuseum Mainz, Mainz 20141
Biscuit Tin Museum, Duns 44063
Bishnu Museum, Shillong 24385
Bishop Bonner's Cottage Museum, Dereham 43971
Bishop Gallery, Zacatecas 31576
Bishop Hill Colony, Bishop Hill 46832
Bishop Hill Heritage Museum, Bishop Hill 46833
Bishop Martin Museum, Natchitoches 50986
Bishop Motiejus Valancius Birthplace-Museum, Nasrėnų 30317
Bishop of Winchester's Palace Site, Witney 46051
Bishop's Courtyard - Moravian Museum, Brno 09420
Bishops' House Museum, Sheffield 45649
Bishop's Lodge Museum, Hay 01149
The Bishop's Palace, Galveston 48843
Bishop's Stortford Museum, Bishop's Stortford 43523
Bishops Waltham Museum, Bishop's Waltham 43525
Bisia Temple Museum, Yilan 42465
Biskupski Muzej i Katedralna Zbirka, Hvar 08879
Biskupský Dvůr, Brno 09420
Bismarck-Ausstellung, Friedrichsruh 18479
Bismarck-Museum, Schönhausen 21610
Bistrița-Nasaud District Museum, Bistrița, Năsăud 36152
Bitterne Local History Centre, Southampton 45702
Biwa-ko Bunkakan, Kusatsu 28796
Biwa-ko Sosui Kinenkan, Kyoto 28803
Bizarre Museum and Mouse Museum, Rieden bei Baden 41761
Bjärnums Museum, Bjärnum 40458
Bjarkøy Bygdemuseum, Bjarkøy 33438
Bjarne Ness Gallery, Fort Ransom 48672
Björklidens Gårdsmuseum, Ljusterö 40689
Bjørgan Prestegård, Kvikne 33620
Bjørn West-Museet, Matredal 33662
Bjugn Bygdatun, Bjugn 33442
BKW-Museum, Mühleberg 41654
Blaabjerg Egnsmuseum, Nørre Nebel 10193
Blaavandshuk Egnsmuseum, Oksbøl 10219
Black American West Museum and Heritage Center, Denver 48064
Black Arts National Diaspora, New Orleans 51076
Black Country Living Museum, Dudley 44029
Black Creek Pioneer Village, North York 06486
Black Cultural Archives, London 44797
Black Cultural Centre for Nova Scotia, Dartmouth 05674
Black Fashion Museum, New York 51134
Black Forest Open Air Museum, Gutach 18843
Black Hawk State Historic Site, Rock Island 52406
Black Heritage Museum, Miami 50622
Black Hills Mining Museum, Lead 50004
Black Hills Museum of Natural History, Hill City 49283
Black House, Szeged 23843
Black Kettle Museum, Cheyenne 47443
Black Legends of Professional Basketball Museum, Detroit 48106
Black Mountain College Museum and Arts Center, Asheville 46370
Black Nugget Museum, Ladysmith 06126
Black River Academy Museum, Ludlow 50317
Black River Historical Society of Lorain Museum, Lorain 50199
Black Rock Arts Foundation, San Francisco 52767
The Black Watch Castle & Museum, Perth 45364
Black Watch of Canada Regimental Memorial Museum, Montréal 06329
Black Woods Museum, Houbi 42203
Blackberry Farm-Pioneer Village, Aurora 46491
Blackburn Museum and Art Gallery, Blackburn, Lancashire 43526
Blackcurrant Museum, Nuits-Saint-Georges 14484
Blackgang Sawmill and Saint Catherine's Quay, Blackgang 43528
Blackhawk Museum, Danville 47958
Blackman Museum, Snohomish 53136
Blackpool Mill, Narberth 45192
Blackridge Community Museum, Blackridge 43531
Blacksmith Shop Museum, Dover-Foxcroft 48171
Blacksmith's Cottage, Bacchus Marsh 00820
Blackstone Valley Historical Society Museum, Lincoln 50111
Blackwater Draw Museum, Portales 52053
Blackwater Valley Museum, Benburb 43450
Blackwell Street Center for the Arts, Denville 48092
Blackwell - The Arts and Crafts House, Bowness-on-Windermere 43571
Blackwood Gallery, Mississauga 06310
Blagoveščenskij Sobor, Moskva 37142
Blaine County Historical Museum, Hailey 49123
Blaine County Museum, Chinook 47541
Blaine House, Augusta 46483
Blaine Lake Museum, Blaine Lake 05430
Blair Castle, Blair Atholl 43537
Blair House Museum, Kentville 06061
Blair Museum of Lithophanes, Toledo 53520
The Blairs Museum, Aberdeen 43240
Blaise Castle House Museum, Bristol 43631
Blake Museum, Bridgwater 43606
Blakesley Hall, Birmingham 43501
Blakstad Sykehus Museum, Asker 33405
Blanco Family Museum, Angono 34334
Bland District Historical Museum, West Wyalong 01644
Blanden Art Museum, Fort Dodge 48615
Blandford Forum Museum, Blandford Forum 43539
Blankschmiede Neimke, Dassel 17798
Blásastova, Norðagøta 10694
Blasinstrumenten-Sammlung, Zimmerwald 42058
Blaskovich Múzeum, Tápiószele 23911
Blasmusik-Museum, Ratten 02563
Blatské muzeum - Rožmberský Dům, Soběslav 09842
Blau-gelbe Viertelsgalerie und Kleine Schlossgalerie, Bad Fischau-Brunn 01744
Blau-gelbe Viertelsgalerie, Weinviertel, Mistelbach an der Zaya 02384
Blau-gelbe Viertelsgalerie Weistrach, Weistrach 02906
Blaues Schloss mit Museumspädagogischer Werkstatt Rotes Schloss, Obernzenn 20825
Blåvand Museum, Blåvand 09977
Blåvand Redningsbådsmuseum, Blåvand 09978
Bleak House, Broadstairs 43644
Bleikeller, Bremen 17546
Blekinge Museum, Karlskrona 40634
Bleloch Museum, Johannesburg 38652
Blenheim Palace, Woodstock 46066
Blenko Museum of Seattle, Seattle 52993
Bleschunov Municipal Museum of Personal Collections, Odesa 43156
Bletchley Park, Bletchley 43543
Blewitt-Harrison-Lee Museum, Columbus 47744
Blicheregnens Museum, Kjellerup 10119
Blichermuseet på Herningsholm, Herning 10068
Blickling Hall, Norwich 45274
Bligh Museum of Pacific Exploration, Bruny Island 00912
Blikmuseum De Blikvanger, Renesse 32681
Blindaž gde podpisan akt o kapituljacii, Kaliningrad 36986
Blindehistorisk Museum, Hellerup 10056
Blists Hill Victorian Town, Telford 45833
Blockhaus d'Éperlecques, Éperlecques 12828
Blockhouse Museum, Merrickville 06267
Blodveimuseet, Rognan 33788
Blokkodden Villmarksmuseum, Engerdal 33483
Blokwachterswoning Waterhuizen, Waterhuizen 32930
Bloodhound-Lenkwaffenstellung, Menzingen 41627
Bloom Mansion, Trinidad 53567
Bloomfield Academy Museum, Bloomfield 46854
Bloomfield Science Museum Jerusalem, Jerusalem 24981
Bloomington Art Center, Bloomington 46869
Bloomington Historical Museum, Bloomington 46870
Blount-Bridgers House/ Edgecombe County Art Museum, Tarboro 53447
Blount Mansion, Knoxville 49812
Bloxham Village Museum, Bloxham 43545
Blue Earth County Museums, Mankato 50446
Blue Hills Trailside Museum, Milton 50698
Blue Licks Battlefield Museum, Mount Olivet 50881
Blue Oyster Gallery, Dunedin 33088
Blue Ridge Institute and Museum, Ferrum 48538
Blue Sky, Portland 52072
Blue Star Contemporary Art Center, San Antonio 52708
Bluebell Railway, Uckfield 45892
Blüchermuseum, Kaub 19482
Bluestone Museum, Hinton 49301
Bluff Hall Museum, Demopolis 48054
Bluff Maritime Museum, Bluff 33064
Blumberger Mühle, Angermünde 16547
BMW Museum, München 20428
B'nai B'rith Klutznick National Jewish Museum, Washington 53853
Bo-Kaap Museum, Cape Town 38579
Bø Museum, Bø i Telemark 33443
Bo'ai Museum, Bo'ai 07581
Boal Mansion Museum, Boalsburg 46880
Boandik Cultural Museum, Beachport 00842
Boarman Arts Center, Martinsburg 50518
Boatmen's National Bank Art Collection, Saint Louis 52568
Boatmuseum Holmön, Holmön 40600
Bob Burns Museum, Van Buren 53700
Bob Campbell Geology Museum, Clemson 47613
Bob Jones University Museum & Gallery, Inc., Greenville 49078
Bob Marley Museum, Kingston 28324
Bobbitt Visual Arts Center, Albion 46172
Boca Raton Historical Society Museum, Boca Raton 46882
Boca Raton Museum of Art, Boca Raton 46883
Bocholter Brouwerij Museum, Bocholt 03375
Bocholter Handwerksmuseum, Bocholt 03376
Bockwindmühlenmuseum, Doberlug-Kirchhain 17879
Bocskai István Emlékkiállítás, Nagykereki 23726
Bodaskogs Affärs- och Postmuseum, Aneby 40422
Bode-Museum - Münzkabinett, Berlin 17051
Bode-Museum - Skulpturensammlung und Museum für Byzantinische Kunst, Berlin 17052
Bodelwyddan Castle Museum, Bodelwyddan 43546
Bodenheimer Heimatmuseum, Bodenheim 17412
Bodensee-Naturmuseum, Konstanz 19677
Bodensee-Obstmuseum, Frickingen 18463
Bodiam Castle, Robertsbridge 45503
Bodie Park Museum, Bridgeport 47015
Bodmin and Wenford Railway, Bodmin 43547
Bodmin Town Museum, Bodmin 43548
Bodrogköz Castle Museum, Pácin 23745
Bodrogközi Kastélymúzeum, Pácin 23745
Bodrum Museum of Underwater Archaeology, Bodrum 42731
Bodrum Sualtı Arkeolojisi Müzesi, Bodrum 42731
Böblingen German Butchers Museum, Böblingen 17417
Böcksteiner Montanmuseum Hohe Tauern, Böckstein 01791
Böd of Gremista Museum, Lerwick 44692
Boehm Art Gallery, San Marcos 52817
Böhmerwaldmuseum Wien und Erzgebirger Heimatstube, Wien 02977
Böke Museum, Leer 19868
Boer Tala Zhou Museum and Gallery, Bole 07582
Börde-Museum Burg Ummendorf, Ummendorf Bördekreis 22172
Das Bördemuseum, Lamstedt 19784
Bördenheimatmuseum, Heeslingen 19064
Boerderij Klein Hulze, Almen 31801
Boerderij Museum Duurswold, Slochteren 32783
Boerderijmuseum De Bovenstreek, Oldebroek 32614
Boerderijmuseum De Lebbenbrugge, Borculo 31982
Boerderijmuseum Zelhem, Zelhem 32996
Boerenbondsmuseum, Gemert 32242
Boerenkrijgmuseum, Overmere 03947
Boerenwagenmuseum, Buren, Gelderland 32024
Bofors Industrimuseum, Karlskoga 40629
Bogaev Museum, Pavlodar 29716
Boğazköy Müzesi, Boğazkale 42732
Bogd Khaan Winter Palace Museum, Ulaanbaatar 31617
Bogdan and Varvara Khanenko Museum of Arts, Kyïv 43097
Bognor Regis Museum, Bognor Regis 43552
Bogor Zoological Museum, Bogor 24457
Bogorodickij Dvorec-muzej i Park, Bogorodick 36812
Bogoroditsk Palace-Museum and Park, Bogorodick 36812
Bogstad Gård, Oslo 33718
Bogstad Manor, Oslo 33718
Bogtrykmuseet, Esbjerg 09999
Bogwood Sculpture Artists, Newtowncashel 24885
Bogyoke Aung San Museum, Yangon 31732
The Bohen Foundation, New York 51135
Bohol Museum, Tagbilaran 34513
Bohusläns Museum, Uddevalla 40992
Boiler House Steam and Engine Museum, Kurwongbah 01208
Boise Art Museum, Boise 46890
Bókasafn Menntaskólans á Ísafirði, Ísafirði 24007
Bokcheon Municipal Museum, Busan 29772
Bokn Bygdemuseet, Bokn 33447
Bolduc House Museum, Sainte Genevieve 52649
Bolesław Prus Museum, Nałęczów 34938
Bolin Wildlife Exhibit and Antique Collection, McKinney 50357
Bolinao National Museum, Bolinao 34347
Bolinas Museum, Bolinas 46894
Bolivar Hall, San Antonio 52709
Bolívar Museum, Caracas 54537
Bolling Hall Museum, Bradford 43575
Bollinger Mill, Burfordville 47129
Bollnäs Konsthallen, Bollnäs 40464
Bolšemuraškinskij Gosudarstvennyj Istoriko-chudožestvennyj Muzej, Bolšoe Muraškino 36815
Bolšoj Dvorec, Lomonosov 37106
Bolton Historical Museum, Bolton 46896
Bolton Library, Cashel 24754
Bolton Museum and Art Gallery, Bolton 43553
Bolton Steam Museum, Bolton 43554
Bolu Devlet Güzel Sanatlar Galerisi, Bolu 42733
Bolu Müzesi, Bolu 42734
Bolu State Fine Art Gallery, Bolu 42733
Bomann-Museum Celle, Celle 17683
Bomarsundmuseet i Prästö Lotsstuga, Sund 11332
De Bommelzolder, Zoeterwoude 33008
Bommen Elvemuseum, Vennesla 33970
Bomsholmen Museum, Arendal 33396
Bona Thompson Memorial Center, Indianapolis 49486
Bonavista Museum, Bonavista 05437
Bonavista North Regional Museum, Wesleyville 07300
Das Bonbon-Museum, Vaihingen an der Enz 22196
Bond Store Museum, Maryborough, Queensland 01266
Bo'ness Heritage Area, Bo'ness 43558
Bonnechere Museum, Eganville 05767
Bonnefantenmuseum, Maastricht 32524
Bonner County Historical Museum, Sandpoint 52835
Bonner Kunstverein, Bonn 17450
Bonnet House, Fort Lauderdale 48635
Bonsai Museum, Seeboden 02732
Bontoc Museum, Bontoc, Ifugao 34349
Bontoc Museum, Bontoc, Mountain Province 34350
Book Art Museum, Łódź 34878
Book Art Museum, Wrocław 35339
Book Cultural Museum, Qujiang 08134
Book Museum, Almaty 29688
Book Museum, Antigua 23292
Book Museum, Moskva 37308
Book Museum Prague, Žďár nad Sázavou 09933
Booker T. Washington National Monument, Hardy 49167
Bookmuseum of the Saxon State and University Library Dresden, Dresden 17939
Booleroo Steam and Traction Museum, Booleroo Centre 00875
Boomkwekerijmuseum, Boskoop 31990
Boondooma Homestead - Museum and Heritage Complex, Mundubbera 01335
Boone County Historical Center, Boone 46903
Boone County Historical Society Museum, Belvidere 46728
Boone's Lick Site, Boonesboro 46907
Boonsborough Museum of History, Boonsboro 46908
Boonshoft Museum of Discovery, Dayton 47982
The Boorman House, Mauston 50537
Boot Hill Museum, Dodge City 48147

Booth Museum of Natural History, Brighton ... 43619
Boothbay Railway Village, Boothbay ... 46910
Boothbay Region Art Foundation, Boothbay Harbor ... 46911
Borås Idrottsmuseum, Borås ... 40467
Borås Konstmuseum, Borås ... 40468
Borås Museum, Borås ... 40469
Bordentown Historical Society Museum, Bordentown ... 46913
Border Ethnographic and Regimental Museum, Caransebeş ... 36263
Border Museum, Imatra ... 10856
Border Regiment and King's Own Royal Border Regiment Museum, Carlisle ... 43757
Borg Verhildersum, Leens ... 32468
Borgarfjördur County Museum, Borgarnes ... 23989
Borgarsyssel Museum, Sarpsborg ... 33818
Borgazdasági Múzeum, Tolcsva ... 23935
Borgmuseum, Spøttrup ... 10283
Borgsjö Hembygdsgård, Erikslund ... 40500
Bori Kisjankó Memorial House, Mezőkövesd ... 23697
Boris-Becker-Zimmer, Leimen ... 19877
Boris Denev Art Gallery, Veliko Tărnovo ... 05252
Borisoglebskij Kraevedčeskij Muzej, Borisoglebsk ... 36816
Borlach-Museum, Bad Dürrenberg ... 16718
Bormshuis, Antwerpen ... 03281
Børnenens Museum, København ... 10124
Borner Forst- und Jagdmuseum, Born am Darß ... 17482
Bornholms Kunstmuseum, Gudhjem ... 10046
Bornholms Museum, Rønne ... 10241
Borno State Museum, Maiduguri ... 33334
Borodino State Museum of War and History Museum and Reserve, Borodino ... 36819
Borosov-Musatov Museum-Estate, Saratov ... 37750
Borough Museum and Art Gallery, Berwick-upon-Tweed ... 43460
Borsos Miklós Múzeum, Győr ... 23580
Borstei Museum, München ... 20429
Boryeong Coal Museum, Boryeong ... 29771
Boscobel House, Brewood ... 43601
Boscobel Restoration, Garrison ... 48868
Bosjökloster Slot & Trädgårdar, Höör ... 40597
Bosmuseum, Amsterdam ... 31836
Bosmuseum, Tessenderlo ... 04059
Bosmuseum Jan van Ruusbroec, Hoeilaart ... 03706
Bosmuseum Pijnven, Hechtel ... 03689
Bosmuseum Wildert, Essen ... 03585
Bosnian Region Historical and Ethnography Room, Áta ... 23339
Bosnyák Helytörténeti és Néprajzi Szoba, Áta ... 23339
Bosque de Ciência, Manaus ... 04551
Boston Athenæum, Boston ... 46919
Boston Children's Museum, Boston ... 46920
Boston Fire Museum, Boston ... 46921
Boston Guildhall Museum, Boston ... 43562
Boston National Historical Park, Boston ... 46922
Boston Public Library, Boston ... 46923
Boston Sculptors Gallery, Boston ... 46924
Boston University Art Gallery, Boston ... 46925
Boswell Museum of Music, East Springfield ... 48259
Bosworth Battlefield Heritage Centre and Country Park, Market Bosworth ... 45108
Botanic Garden Museum, Sapporo ... 29213
Botanic Gardens Museum, Southport ... 45717
Botanical Department of the Moravian Museum, Brno ... 09421
Botanical Museum, Helsinki ... 10790
Botanical Museum, Lund ... 40701
Botanical Museum, Rawalpindi ... 34040
Botanical Museum of Harvard University, Cambridge ... 47173
Botanical Museum of the Beja Agricultural Highschool, Beja ... 35462
Botanical Museum of the National Gardens, Athinai ... 22903
Botanical Museum of the Russian Academy of Sciences, Sankt-Peterburg ... 37625
Botanical Museum of the Universty of Oulu, Oulu ... 11175
Botanical Survey Museum of India, Pune ... 24356
Botanical Survey of India, Dehradun ... 24130
Botaničeskij Muzej, Sankt-Peterburg ... 37625
Botanické Oddělení, Brno ... 09421
Botanické Oddělení Národního Muzea, Průhonice ... 09812
Botanische Staatssammlung München, München ... 20430
Botanisches Museum Berlin-Dahlem, Berlin ... 17053
Botanisk Museum, København ... 10125
Botaniska Museet, Lund ... 40701
Botany Department of the National Museum, Průhonice ... 09812
Botany Museum, Faizabad ... 24171
Botany Museum, Gorakhpur ... 24179
Botany Museum, Jaunpur ... 24239
Botany Museum, Kanpur ... 24252
Botany Museum, Lucknow ... 24292
De Botermolen, Keerbergen ... 03750
Bothwell-Zone Oil Museum, Bothwell ... 05442
Botkyrka Konsthall och Xet-Museet, Tumba ... 40986
Bøtø Nor Gl. Pumpestation, Væggerløse ... 10303
Botswana National Museum and Art Gallery, Gaborone ... 04296
Bottleship Museum, Neuharlingersiel ... 20615
Botwood Heritage Centre, Botwood ... 05444
Boughton House, Kettering ... 44567
Boulder History Museum, Boulder ... 46955
Boulder Museum of Contemporary Art, Boulder ... 46956
Boulevard Arts Center, Chicago ... 47458
Boulia Stone House Museum, Boulia ... 00881
Boundary Museum, Grand Forks ... 05893
Bourbaki-Panorama Luzern, Luzern ... 41589
Bourgeneuf 1900, Saint-Julien-Chapteuil ... 15344
Bourne Hall Museum, Ewell ... 44192
Bournemouth Natural Science Society Museum, Bournemouth ... 43565
Bournemouth Transport Museum, Bournemouth ... 43566
Bournville Centre for Visual Arts, Birmingham 43502
La Bourrine à Rosalie, Sallertaine ... 15604
Bouvet-Ladubay Art Center, Saumur ... 15660
Bouzov castle, Bouzov ... 09412
Bov Museum, Padborg ... 10225
Bowden Pioneer Museum, Bowden ... 05447
Bowdoin College Museum of Art, Brunswick ... 47102
Bowen House / Roseland Cottage, Woodstock 54281
Bowen Island Museum, Bowen Island ... 05448
The Bower Gallery, San Antonio ... 52710
Bowers Mansion, Carson City ... 47259
Bowers Mansion, Washoe Valley ... 53952
Bowers Museum of Cultural Art, Santa Ana ... 52852
Bowers Science Museum, Cortland ... 47829
Bowery Gallery, New York ... 51136
The Bowes Museum, Barnard Castle ... 43381
Bowes Railway Heritage Museum, Gateshead 44263
Bowhill Collection, Selkirk ... 45634
Bowie Railroad Station and Huntington Museum, Bowie ... 46969
Bowie's Heritage and Children's Museum, Bowie ... 46970
Bowls New Zealand Museum, New Plymouth ... 33155
Bowman, Megahan and Penelec Galleries, Meadville ... 50546
Bowmanville Museum, Bowmanville ... 05449
Bowne House, Flushing ... 48575
Bowood House and Gardens, Calne ... 43716
Bowraville Folk and Military Museum, Bowraville ... 00884
La Box, Bourges ... 12114
Box Kraftverk, Bjärtrå ... 40459
Boxenstop Auto- und Spielzeugmuseum, Tübingen ... 22112
Boxfield Gallery, Stevenage ... 45741
Boyana Church National Museum, Sofia ... 05193
Boyang Museum, Boyang ... 07583
Boyer Museum, Wildwood ... 54161
Boyertown Museum of Historic Vehicles, Boyertown ... 46979
The Boyle House, Malvern ... 50410
Boyne Valley Historical Cottage, Ubobo ... 01608
Boys Town Hall of History & Father Flanagan House, Boys Town ... 46980
Bozhou Museum, Bozhou ... 07584
Bozsó Collection, Kecskemét ... 23643
Bozsó Gyűjtemény, Kecskemét ... 23643
BPS 22 - Espace de Creation Contemporaine, Charleroi ... 03513
Brabants Museum Oud Oosterhout, Oosterhout ... 32628
Brac Homeland Museum, Supetar ... 08966
Bracken Hall Countryside Centre, Baildon ... 43353
Brackendale Art Gallery Theatre Teahouse, Brackendale ... 05454
Bracki Muzej, Supetar ... 08966
Bradbury Science Museum, Los Alamos ... 50204
Bradford Brinton Memorial, Big Horn ... 46795
Bradford County Historical Society Museum, Towanda ... 53546
Bradford Historical Society, Bradford ... 46990
Bradford Industrial Museum and Horses at work, Bradford ... 43576
The Bradford Museum of Collector's Plates, Niles ... 51350
Bradford Ohio Railroad Museum, Bradford ... 46988
Bradford-on-Avon Museum, Bradford-on-Avon ... 43583
Brading Roman Villa, Brading ... 43586
Bradley Gallery of Art, Plymouth ... 52001
Bradley Gallery of Art, Sheboygan ... 53060
Bragegården, Vaasa ... 11422
Bragg-Mitchell Mansion, Mobile ... 50766
Brahms-Museum, Hamburg ... 18917
Brahmshaus, Baden-Baden ... 16879
Brahmshaus, Heide ... 19065
Braidwood Museum, Braidwood ... 00889
Brain-Watkins House, Tauranga ... 33205
Braintree District Museum, Braintree ... 43589
Braintree Historical Society Museum, Braintree ... 46992
Braith-Mali-Museum, Biberach an der Riß ... 17319
Braithwaite Fine Arts Gallery, Cedar City ... 47289
Brālu Amtmaņu Muzejs, Valle ... 30138
Brālu Jurjānu Memoriālais Muzejs Meņģeļi, Ērgļi ... 30007
Brālu Kaudzīšu Memoriālais Muzejs, Kaudzīši ... 30031
Brama Człuchowska, Chojnice ... 34596
Bramah Tea and Coffee Museum, London ... 44798
Bramall Hall, Bramhall ... 43590
Brambuk Living Cultural Centre, Halls Gap ... 01141
Bramsnæs Museum og Arkiv, Kirke Hyllinge ... 10118
Bramtidsmuseet, Borlänge ... 40475
Branch of the Regional Museum of South-Eastern Moravia in Zlín, Luhačovice ... 09643
Branchville Railroad Shrine and Museum, Branchville ... 46993
Brand Library Art Galleries, Glendale ... 48928
Brande Museum, Brande ... 09982
Brandeburgisches Apothkenmuseum, Cottbus ... 17729
Brandenburg-Preußen Museum, Wustrau ... 22755
Brandenburgischer Kunstverein Potsdam e.V., Potsdam ... 21084
Brandenburgisches Apothekenmuseum, Cottbus ... 17730
Brandenburgisches Eisenbahnmuseum Falkenberg (Elster) e.V., Falkenberg, Elster ... 18288
Brandenburgisches Forstmuseum Fürstenberg/Havel, Fürstenberg, Havel ... 18488
Brandenburgisches Freilichtmuseum Altranft, Bad Freienwalde ... 16732
Brandenburgisches Museum für Klein- und Privatbahnen, Gramzow ... 18735
Brandenburgisches Textilmuseum Forst, Forst, Lausitz ... 18341
Brander Museum, Huntly ... 44503
Brandkårsmuseet, Simonstorp ... 40837
Brandon Heritage Museum, Brandon ... 43591
Brandon Mental Health Centre Historical Society Museum and Archives, Brandon ... 05460
Brand's Health Museum, Lugang ... 42255
Brandval Museum, Brandval ... 33453
Brandweermuseum, Wouwse Plantage ... 32971
Brandweermuseum en Stormrampmuseum, Borculo ... 31983
Brandweermuseum Wassenaar, Wassenaar ... 32926
Brandýs Castle, Brandýs nad Labem ... 09414
Brandywine Battlefield, Chadds Ford ... 47321
Brandywine River Museum, Chadds Ford ... 47322
Brangakmenių Muziejus, Kaunas ... 30242
Branigan Cultural Center, Las Cruces ... 49957
Branko Deškovič Art Gallery, Bol ... 08846
Brant Museum, Brantford ... 05467
Brantwood, Coniston ... 43884
Brasilianisches Kulturinstitut in Deutschland - Galerie, Berlin ... 17054
Brasilienmuseum im Kloster Bardel, Bad Bentheim ... 16680
Braşov District Museum of History, Braşov ... 36180
Braşov Ethnographical Museum, Braşov ... 36179
Brass Rubbing Centre, Edinburgh ... 44114
Bratsk United Museum of the History of the Conquest of the Angara River, Bratsk ... 36822
Bratskij Chudožestvennyj Vystavočnyj Zal, Bratsk ... 36821
Bratskij Gorodskoj Objedinennyj Muzej Istorii Osvoenija Angary, Bratsk ... 36822
Brattle Farm Museum, Staplehurst ... 45740
Brattleboro Museum and Art Center, Brattleboro ... 46998
Brauchtums- und Trachtenpuppenmuseum der Banater Schwaben e.V., Würzburg ... 22707
Brauer Museum of Art, Valparaiso ... 53698
Brauerei-Kontor, Bochum ... 17397
Brauerei-Museum, Haßfurt ... 19030
Brauerei-Museum, Nesselwang ... 20576
Brauerei-Museum Dortmund, Dortmund ... 17916
Brauereichäller Kulturforum, Laufen ... 41530
Brauereikulturmuseum Gut Riedelsbach, Neureichenau ... 20645
Brauereimuseum, Aldersbach ... 16462
Brauereimuseum, Alpirsbach ... 16476
Brauereimuseum, Altomünster ... 16522
Brauereimuseum Barre's Brauwelt, Lübbecke ... 20066
Brauereimuseum Bräu im Moos, Tüßling ... 22130
Brauereimuseum im historischen Kronen-Brauhaus zu Lüneburg, Lüneburg ... 20095
Brauereimuseum Schneider, Weißenburg in Bayern ... 22449
Brauereimuseum Schöneck, Schöneck / Vogtland ... 21604
Brauhaus-Galerie, Freistadt ... 01903
Brauhausmuseum, Stadtlauringen ... 21858
Brauhinia Shire Bicentennial Art Gallery, Springsure ... 01523
Braumuseum, Leoben ... 02284
Braumuseum, Wieselburg an der Erlauf ... 03100
Braumuseum - Wolff Giersing's Brauerbe, Torgau ... 22069
Braunauer Heimatmuseum, Forchheim, Oberfranken ... 18336
Braunschweigisches Landesmuseum, Braunschweig ... 17518
Braunschweigisches Landesmuseum, Wolfenbüttel ... 22651
Braunton and District Museum, Braunton ... 43592
Bray Heritage Centre and Museum, Bray ... 24746
Brazilian Federation of Friends of Museums, São Paulo ... 54722
Brazoria County Historical Museum, Angleton ... 46299
Brazos Valley Museum of Natural History, Bryan ... 47105
Brazosport Museum of Natural Science, Brazosport ... 46999
Brazosport Museum of Natural Science, Clute ... 47666
Brdský Památník - Městské Muzeum, Rožmitál pod Třemšínem ... 09824
Bread & Puppet Museum, Glover ... 48950
Breadalbane Folklore Centre, Killin ... 44577
Breamore Countryside Museum, Breamore ... 43593
Breamore House Museum, Breamore ... 43594
Brebner Surgical Museum, Johannesburg ... 38653
Brechin Museum, Brechin ... 43595
Brecht-Weigel-Gedenkstätte, Berlin ... 17055
Brecht-Weigel-Museum, Berlin ... 17055
Brechthaus, Augsburg ... 16613
Brecknock Museum and Art Gallery, Brecon ... 43597
Brecon Mountain Railway, Merthyr Tydfil ... 45129
Breda's Begijnhof Museum, Breda ... 32002
Breda's Museum, Breda ... 32003
De Brede, Maasbree ... 32520
Brede Værk, Kongens Lyngby ... 10165
Bredovský Letohrádek na Lemberku, Jablonné v Podještědí ... 09546
Bredy Farm Old Farming Collection, Burton Bradstock ... 43687
Breginj Museum, Kobarid ... 38430
Breginjski Muzej, Kobarid ... 38430
Brehm-Gedenkstätte, Renthendorf ... 21279
Breidablikk, Stavanger ... 33865
Bremer Rundfunkmuseum, Bremen ... 17547
Bremerton Naval Museum, Bremerton ... 47004
Brenau University Galleries, Gainesville ... 48820
Brennan and Geraghty's Store Museum, Maryborough, Queensland ... 01267
Brennbergbányai Bányászati Emlékház, Sopron ... 23807
Brennereimuseum Haselünne, Haselünne ... 19023
Brensbach Museum und Galerie im alten Rathaus, Brensbach ... 17581
Brent Museum, London ... 44799
Brentwood Museum, Brentwood ... 43600
Brenzett Aeronautical Museum Trust, Romney Marsh ... 45511
Bresaylor Heritage Museum, Bresaylor ... 05472
Breslauer Sammlung Köln, Köln ... 19596
Bresse departmental Museum, Saint-Cyr-sur-Menton ... 15228
Bressingham Steam Museum, Diss ... 43985
Brett Whiteley Studio Museum, Surry Hills ... 01539
Breuberg-Museum, Breuberg ... 17587
Brevard Museum, Cocoa ... 47669
Brevard Museum of Art and Science, Melbourne ... 50564
Brevik Bymuseum, Brevik ... 33455
Brewarrina Aboriginal Cultural Museum, Brewarrina ... 00890
Brewery Art Association, Los Angeles ... 55126
Brewery Chapel Museum, Halstead ... 44390
Brewery Museum, Leoben ... 02284
Brewery Museum, Plzeň ... 09718
Brewery Museum Arboga, Arboga ... 40427
Brewery Museum Bocholt, Bocholt ... 03375
Brewhouse Yard Museum, Nottingham ... 45289
Brewing Greande Brasserie et Malterie Vosgienne, Ville-sur-Illon ... 16177
Briar Herb Factory Museum, Clyde ... 33080
Bribie Island Community Arts Centre, Bribie Island ... 00892
Brick House, Montgomery ... 50816
Brick Store Museum, Kennebunk ... 49727
Brickmuseum Vienna, Wien ... 03092
Bridewell Museum, Norwich ... 45275
Bridge Center for Contemporary Art, El Paso ... 48316
Bridgehampton Historical Society Museum, Bridgehampton ... 47012
Bridgestone Bijutsukan, Tokyo ... 29373
Bridgestone Museum of Art, Ishibashi Foundation, Tokyo ... 29373
Bridgetown Old Gaol Museum, Bridgetown ... 00893
Bridgnorth Northgate Museum, Bridgnorth ... 43603
Bridgton Historical Museum, Bridgton ... 47023
Bridlington Harbour Museum, Bridlington ... 43610
Bridport Museum, Bridport ... 43612
Briedes Krogs in the Burtnieki Parish, Burtnieki ... 29994
Briefmarkenmuseum im Kloster Bardel, Bad Bentheim ... 16681
Briercrest and District Museum, Briercrest ... 05476
Briged Pasukan Polis Hutan, Kuching ... 30590
Brigham City Museum-Gallery, Brigham City ... 47024

Brigham Young University Earth Science Museum, Provo 52186
Brigham Young's Winter Home, Saint George 52544
Brightlingsea Museum, Brightlingsea 43617
Brighton Fishing Museum, Brighton 43620
Brighton Museum and Art Gallery, Brighton 43621
Brikettfabrik Hermannschacht Zeitz, Zeitz 22775
Brillantes Ancestral House, Tayum 34517
Brillen-Museum, Homburg 19266
Brimstone Museum, Sulphur 53361
Brindley Bank Pumping Station and Museum, Rugeley 45529
Brink 7 Galerij, Yde 32972
Brinkens Museum, Malax 11095
Brisbane City Gallery, Brisbane 00895
Brisbane Tramway Museum, Ferny Grove 01079
Bristol Historical and Preservation Society Museum, Bristol 47031
Bristol's City Museum and Art Gallery, Bristol 43632
Bristol's Industrial Museum, Bristol 43633
Britain at War Experience, London 44800
Britannia Shipyard, Richmond 06741
British Association of Friends of Museums, Whitchurch 55051
British Aviation Preservation Council, Coventry 55052
British Balloon Museum, Newbury 45220
British Columbia Aviation Museum, Sidney 06970
British Columbia Congenital Heart Museum, Vancouver 07201
British Columbia Farm Machinery and Agricultural Museum, Fort Langley 05811
British Columbia Forest Discovery Centre, Duncan 05722
British Columbia Forest Service Museum, Victoria 07242
British Columbia Golf House Museum, Vancouver 07202
British Columbia Museum of Medicine, Vancouver 07203
British Columbia Museum of Mining, Britannia Beach 05482
British Columbia Museums Association, Victoria 54731
British Columbia Orchard Industry Museum, Kelowna 06050
British Columbia Regiment Museum, Vancouver 07204
British Columbia Sports Hall of Fame and Museum, Vancouver 07205
British Columbia Wine Museum, Kelowna 06051
British Commercial Vehicle Museum, Leyland 44701
British Cycling Museum, Camelford 43737
British Dental Association Museum, London 44801
British Empire and Commonwealth Museum, Bristol 43634
British Engineerium, Hove 44492
British Golf Museum, Saint Andrews 45552
British in India Museum, Nelson 45198
British Lawnmower Museum, Southport 45718
British Library, London 44802
The British Museum, London 44803
British Optical Association Museum, London 44804
British Photographic Museum, Totnes 45870
The British Postal Museum & Archive, London 44805
British Red Cross Museum and Archives, London 44806
British Schools Museums, Hitchin 44473
Brixham Heritage Museum, Brixham 43641
Brjanskij Gosudarstvennyj Obedinennyj Kraevedčeskij Muzej, Brjansk 36824
Brjanskij Literaturnyj Muzej, Brjansk 36825
Brjanskij Oblastnoj Chudožestvennyj Muzej, Brjansk 36826
Brno City Museum, Brno 09436
Brno Museum of Technology, Brno 09440
Broadcasting Museum of Sweden, Motala 40759
Broadfield House Glass Museum, Kingswinford 44604
Broadview Historical Museum, Broadview 05484
Broadway Magic Experience, Broadway 43647
Broadway Windows, New York 51137
Brockville Museum, Brockville 05486
Brodick Castle, Brodick 43651
Brodie Castle, Forres 44244
Brodsworth Hall & Gardens, Doncaster 43991
Bröhan-Museum, Berlin 17056
Bröhan-Museum, State Museum for Art Nouveau, Art Deco and Functionalism (1889-1939), Berlin 17056
Broelmuseum, Kortrijk 03759
Broken-Beau Pioneer Village Museum, Beauséjour 05406
Broken Hill Regional Art Gallery, Broken Hill 00903
Brome County Historical Society Museum, Knowlton 06108
Bromfield Art Gallery, Boston 46926
Bromfietsmuseum De Peel, Ospel 32638
Bromham Mill and Art Gallery, Bromham 43653
Bromsgrove Museum, Bromsgrove 43655
Bronck Museum, Coxsackie 47863
Bronės Buivydaitės Memorialinis Muziejus, Anykščiai 30208
Bronės ir Bronius Radeckų Keramikos Muziejus, Kuršėnai 30294
Brønshøj Museum, Brønshøj 09984
Brontë Parsonage Museum, Haworth 44430
Bronx County Historical Society, Bronx 47050
Bronx Museum of the Arts, Bronx 47040
Bronx River Art Center, Bronx 47041
Bronze and Brass Museum, Bhaktapur 31769
Brookfield Craft Center, Brookfield 47055
Brookgreen Gardens, Murrells Inlet 50920
Brookings Arts Council, Brookings 47058
Brookings County Museum, Volga 53767
Brooklands Museum, Weybridge 45985
Brookline Historical Society Museum, Brookline 47061
Brooklyn Arts Council Gallery, Bronx 47042
Brooklyn Children's Museum, Brooklyn 47067
Brooklyn Historical Society Museum, Brooklyn 47068
Brooklyn Historical Society Museum, Brooklyn 47086
Brooklyn Museum, Brooklyn 47069
Brooks Academy Museum, Harwich 49209
Brooks and District Museum, Brooks 05488
Brookside Museum, Ballston Spa 46542
Broome County Historical Society Museum, Binghamton 46814
Broome Museum, Broome 00909
Broomehill Museum, Broomehill 00910
Bror Hjorths Hus, Uppsala 41003
Brot- und Mühlen-Lehrmuseum, Gloggnitz 01938
Brotfabrik Galerie, Berlin 17057
Brother Kaudzītes Memorial Museum, Kaudzīši 30031
Broughton House and Garden, Kirkcudbright 44055
Broughty Castle Museum, Dundee 44045
Brouwershuis, Antwerpen 03282
Brouws Museum, Brouwershaven 32014
Broward County Historical Museum, Fort Lauderdale 48636
Brown County Art Gallery and Museum, Nashville 50959
Brown County Historical Society Pioneer Museum, Nashville 50960
Brown County Museum, New Ulm 51105
Brown House Museum, Sandersville 52832
Brown Mansion, Coffeyville 47673
Brownella Cottage, Galion 48837
Brownsbank Cottage, Biggar 43480
Brownsville Museum of Fine Art, Brownsville 47096
Browntown Museum, Lake City 49880
Brownvale North Peace Agricultural Museum, Brownvale 05490
Brownville Historical Society Museum, Brownville 47097
Brubacher House Museum, Waterloo 07284
Bruce Castle Museum, London 44807
Bruce County Museum, Southampton 06997
Bruce Gallery of Art, Edinboro 48290
Bruce Groner Museum, Groundbirch 05916
Bruce Mines Museum, Bruce Mines 05491
Bruce Museum of Arts and Science, Greenwich 49081
Brucemore, Cedar Rapids 47300
Brucknerzimmer, Kronstorf 02240
Brudavollen Bygdetun, Ørsta 33711
Bruder-Konrad-Museum, Altötting 16516
Brücke Museum, Berlin 17058
Brueckner Museum of Starr Commonwealth, Albion 46173
Brückturm-Museum über der 'Steinernen Brücke', Regensburg 21215
Brüder Grimm-Haus Steinau, Steinau an der Straße 21877
Brüder Grimm-Museum Kassel, Kassel 19459
Brüder-Grimm-Stube, Marburg 20188
Brünner Heimatmuseum, Schwäbisch Gmünd 21648
Brüxer und Komotauer Heimatstuben, Erlangen 18220
Bruggemuseum - Archaeology, Brugge 03408
Bruggemuseum - Archeologie, Brugge 03408
Bruggemuseum - Belfort, Brugge 03409
Bruggemuseum - Belfry, Brugge 03409
Bruggemuseum - Brugse Vrije, Brugge 03410
Bruggemuseum - Gruuthuse, Brugge 03411
Bruggemuseum - Gruuthuse, Brugge 03411
Bruggemuseum - Liberty of Bruges, Brugge 03410
Bruggemuseum - Onthaalkerk van Onze-Lieve-Vrouw, Brugge 03412
Bruggemuseum - Stadhuis, Brugge 03413
Bruggemuseum - Town Hall, Brugge 03413
Bruggemuseum - Welcome-Church of Our Lady, Brugge 03412
Brugwachterhuis, Dilsen-Stokkem 03556
Brukenthal National Museum, Sibiu 36613
Bruksmuseet Smedsgården, Sandviken 40830
Bruksmuseum, Hudiksvall 40603
Brunei Arts and Handicrafts Training Centre, Bandar Seri Begawan 04967
Brunei Forestry Museum, Sungai Liang 04970
Brunei Gallery, London 44808
Brunei Museum, Bandar Seri Begawan 04965
The Brunel Museum, London 44809
Brunn Museum, Nólsoy 10693
Brunnenmuseum, Bad Vilbel 16855
Brunner Heimathaus mit Rudolf-Steiner-Gedenkstätte, Brunn am Gebirge 01807
Brunner Mine Site, Dobson 33087
Brunnier Art Museum, Ames 46256
Bruno Weber Skulpturenpark, Dietikon 41329
Brunskogs Hembygdsgård, Brunskog 40480
Brunswick Railroad Museum, Brunswick 47101
Brunswick Town/Fort Anderson State Historic Site, Winnabow 54234
Brunswick Valley Museum, Mullumbimby 01334
Brush Art Gallery, Lowell 50304
Brush Country Museum, Cotulla 47847
Brussels Museum of Industry and Labor, Bruxelles 03455
Bruunshåb Gamle Papfabrik, Viborg 10313
Bruxella 1238, Bruxelles 03439
Bruzaholms Bruksmuseum, Bruzaholm 40481
Bry Gallery, Monroe 50788
Bryansk Museum of Fine Arts, Brjansk 36826
Bryansk Museum of Literature, Brjansk 36825
Bryansk State Museum of Local Lore, Brjansk 36824
Bryant Cottage, Bement 46729
Brydone Jack Observatory Museum, Fredericton 05829
Bryggens Museum, Bergen 33425
B'S Ballpark Museum, Centennial 47311
BSAT Gallery, Newmarket 45238
Bu Museum, Utne 33948
Buccleuch Mansion, New Brunswick 51031
Bucerius Kunst Forum, Hamburg 18918
Buchanan Center for the Arts, Monmouth 50782
Buchans Miners Museum, Buchans 05492
Buchar Zhyrau Museum of Literature and Art, Pavlodar 29717
Buchara State Museum, Buxoro 54444
Buchdruckerei Museum und Schauwerkstatt, Krakow am See 19698
Buchdruckmuseum, Rohrbach/OÖ 02596
Buchdruckmuseum Druckwerk, Dornbirn 01824
Buchheim Museum, Bernried 17301
s'Buchholze Hisli, Schuttertal 21643
Buchholzer Heimatmuseum, Buchholz in der Nordheide 17618
Buchmuseum, Solothurn 41880
Buchmuseum der Sächsischen Landesbibliothek - Staats- und Universitätsbibliothek Dresden, Dresden 17939
Buckeye Furnace State Memorial, Wellston 54031
Buckfast Abbey, Buckfastleigh 43659
Buckham Gallery, Flint 48558
Buckhaven Museum, Buckhaven 43662
Buckhorn Saloon & Museum, San Antonio 52711
Buckie Drifter Fishing Heritage Centre, Buckie 43663
Buckingham Movie Museum, Buckingham 43665
Buckingham Palace, London 44810
Buckinghamshire County Museum, Aylesbury 43342
Buckinghamshire Railway Centre, Quainton 45453
Buckland Abbey, Yelverton 46094
Buckler's Hard Village Maritime Museum, Beaulieu 43423
Buckley Center Gallery, Portland 52073
Buckleys Yesterday's World, Battle 43419
Bucksport Historical Museum, Bucksport 47110
Bucyrus Historical Society Museum, Bucyrus 47111
Budaörser Heimatmuseum, Bretzfeld 17586
Budapest Galéria Kiállítóháza, Budapest 23389
Budapest Galéria Kiállítóterem, Budapest 23390
Budapest Galéria Varga Imre Állandó Kiállítóház, Budapest 23391
Budapest Gallery, Budapest 23390
Budapest Gallery Exhibition Hall, Budapest 23389
Budapest Gallery Temporary Exhibition of Imre Varga, Budapest 23391
Budapest Historical Museum - Museum of Aquincum, Budapest 23382
Budapest History Museum, Budapest 23392
Budapest History Museum - Medieval Jewish Chapel, Budapest 23425
Budapesti Történeti Múzeum, Budapest 23392
Budavári Mátyás-templom Egyházművészeti Gyűjteménye, Budapest 23393
Buddelschiff-Museum, Neuharlingersiel 20615
Buddelschiff-Museum, Tangerhütte 22000
Buddelschiff-Museum, Wedel 22352
Buddenbrookhaus, Heinrich-und-Thomas-Mann-Zentrum, Lübeck 20073
The Buddhist Art Museum, Yangon 31733
Buddhist Art Photo Gallery Banri Namikawa, Hiroshima 28541
Buddhist Museum Koyasan, Koya 28773
Buddle Arts Centre, Wallsend 45916
Buddy Holly Center, Lubbock 50310
Bude-Stratton Museum, Bude 43667
Budenz-Ház - Ybl Gyűjtemény, Székesfehérvár 23851
Budenz House - The Ybl-Collection, Székesfehérvár 23851
Buderim Pioneer Cottage, Buderim 00913
Buderus Zentralheizungsmuseum, Lollar 20029
Budinok-muzej A. Čechova, Sumy 43188
Budinok-muzej Sevastopol's'ke Komunistične Pidpillja 1942-1944, Sevastopol 43175
Budyonny House-Museum, Proletarsk 37572
Buechel Memorial Lakota Museum, Saint Francis 52542
Büchnerhaus, Riedstadt 21321
Büchsenmacher- und Jagdmuseum, Ferlach 01887
Bücker-Luftfahrt- und Europäisches Eissegel-Museum, Rangsdorf 21178
Büecheler-Hus, Kloten 41506
Bükki Üveghuták Ipartörténeti Múzeuma, Bükkszentkereszt 23496
Buena Park Historical Society Museum, Buena Park 47112
Bündner Kunstmuseum, Chur 41299
Bündner Naturmuseum, Chur 41300
Buenting Teamuseum Leery/Eastfriesland, Leer 19869
Bünting Teemuseum Leer/Ostfriesland, Leer 19869
Bürger- und Bauernmuseum, Hilzingen 19191
Bürgermeister-Müller-Museum, Solnhofen 21813
Bürgermeister-Ringenhain-Haus, Torgau 22070
Bürgerstiftung Kunst für Güglingen, Güglingen 18821
Büro Otto Koch, Dessau 17828
Büromuseum, Mülheim an der Ruhr 20410
Bürsten- und Heimatmuseum, Schönheide 21611
Büyük Saray Mozaik Müzesi, İstanbul 42810
Buffalo and Erie County Historical Society Museum, Buffalo 47116
Buffalo and Erie County Naval and Military Park, Buffalo 47117
Buffalo Bill Cody Homestead, Princeton 52143
Buffalo Bill Historical Center, Cody 47671
Buffalo Bill Memorial Museum, Golden 48956
Buffalo Bill Museum of Le Claire, Iowa, Le Claire 50001
Buffalo Bill Ranch, North Platte 51401
Buffalo Museum of Science, Buffalo 47118
Buffalo Nations Luxton Museum, Banff 05381
Buffalo Trails Museum, Epping 48406
Buffs Museum, Canterbury 43741
Bugatti Trust, Cheltenham 43808
Bugey-Valmorey departmental Museum, Lochieu 13795
BuGok Railroad Exhibition Hall, Uiwang 29939
Buguruslanskij Istoriko-kraevedčeskij Muzej, Buguruslan 36830
Buguruslanskiy Historical-Regional Studies Museum, Buguruslan 36830
Building of Bath Museum, Bath 43403
Bujnakskij Istoriko-kraevedčeskij Muzej, Bujnaksk 36831
Bukyoung University Museum, Busan 29773
Bul-bul Memorial Museum, Bakı 03148
Bul-bul Museum-Flat, Bakı 03149
Bulakan Provincial Museum, Malolos 34422
Bulat Okudshava Memorial House-Museum, Peredelkino 37541
Bulgarian Museum Chamber, Sofia 54724
Bulkley Valley Museum, Smithers 06983
Bullfight Museum, Lima 34270
Bullfighting Museum, Alcochete 35400
Bullfighting Museum, Antequera 38876
Bullfighting Museum, Béjar 39015
Bullfighting Museum, Bogotá 08595
Bullfighting Museum, Cali 08615
Bullfighting Museum, Córdoba 39208
Bullfighting Museum, Hellín 39386
Bullfighting Museum, Huamantla 30939
Bullfighting Museum, La Línea de la Concepción 39466
Bullfighting Museum, Madrid 39582
Bullfighting Museum, Mijas 39649
Bullfighting Museum, Murcia 39704
Bullfighting Museum, Salamanca 39944
Bullfighting Museum, San Luis Potosí 31344
Bullfighting Museum, Valencia 40217
Bulloch Hall, Roswell 52466
Bullyworld Museum, Spraitbach 21842
Bulwagang Helena Benitez Gallery, Manila 34433
Bumbongan Dua Belas, Bandar Seri Begawan 04962
Bunbury Regional Art Galleries, Bunbury 00916
Bunbury Watermill, Bunbury 43671
Bund Museum, Shanghai 08155
Bundaberg Arts Centre, Bundaberg 00917
Bundaberg Historical Museum, Bundaberg 00918
Bundaberg Railway Museum, Bundaberg 00919
Bundelkhand Chhatrasal Museum, Banda 24070
Bundesarchiv, Rastatt 21183
Bundesbriefmuseum, Schwyz 41849
Bundesverband Museumspädagogik e.V., München 54828
Bungay Museum, Bungay 43672

H.K. Buniatyan Art Gallery, Gavar 00700
Bunin Memorial Literary Museum, Elec 36909
Bunin Museum, Orël 37508
Bunka Gakuen Costume Museum, Tokyo 29374
Bunka Gakuen Fukushoku Hakubutsukan, Tokyo 29374
Bunkamura Museum, Tokyo 29375
Bunker Hill Museum, Charlestown 47381
Bunker Museet, Frederikshavn 10025
Bunker Museum, Kaliningrad 36986
Bunker Museum 1940-1944, Miehikkälä 11110
Bunker Museum Wurzenpass, Riegersdorf 02588
Bunker of J.W. Stalin, Samara 37610
Bunker Stalina, Samara 37610
Das Bunkermuseum, Emden 18136
Bunkermuseum Oberhausen, Oberhausen 20807
Bunkermuseum Wurzenpass, Riegersdorf 02588
Bunnell Street Gallery, Homer 49329
Bunratty Castle and Folkpark, Bunratty 24748
Bunun Culture Exhibition Center, Taoyuan 42450
Buque Museo Corbeta Uruguay, Buenos Aires 00150
Buque Museo Ex BAP América, Iquitos 34191
Buque-Museo Fragata Presidente Sarmiento, Buenos Aires 00151
Burapha Museum of Art, Chon Buri 42571
Burbiškio Dvaro Istorijos Ekspozicija, Burbiškis 30219
La Burbuja-Museo del Niño, Hermosillo 30928
Burchfield-Penney Art Center, Buffalo 47119
Burden Gallery, New York 51138
Burdur Devlet Güzel Sanatlar Galerisi, Burdur 42736
Burdur Müzesi, Burdur 42737
Burdur Museum, Burdur 42737
Burdur State Fine Arts Gallery, Burdur 42736
Bureau County Historical Society Museum, Princeton 52144
Burg Berwartstein, Erlenbach bei Dahn 18229
Burg Eltz, Wierschem 22544
Burg Falkenberg, Falkenberg, Kreis Tirschenreuth 18289
Burg Gleichen, Wandersleben 22320
Burg Gnandstein, Kohren-Sahlis 19670
Burg Gößweinstein, Gößweinstein 18686
Burg Greifenstein, Greifenstein 01987
Burg Hagen, Hagen im Bremischen 18861
Burg Herzberg, Breitenbach am Herzberg 17537
Burg Hochosterwitz, Launsdorf 02268
Burg Hohenzollern, Burg Hohenzollern 17645
Burg Kreuzen, Bad Kreuzen 01758
Burg Kreuzenstein, Korneuburg 02217
Burg Kronberg, Kronberg im Taunus 19723
Burg Lauenstein, Ludwigsstadt 20063
Burg Meersburg GmbH, Meersburg 20247
Burg Mildenstein, Leisnig 19943
Burg-Museum, Parsberg 20984
Burg Neuhaus, Wolfsburg 22660
Burg Neustadt-Glewe, Neustadt-Glewe 20679
Burg Pappenheim mit Naturmuseum und Historischem Museum, Pappenheim 20981
Burg Prunn, Riedenburg 21314
Burg Reichenstein, Trechtingshausen 22083
Burg Sooneck, Niederheimbach 20706
Burg Stollberg, Stolberg, Rheinland 21909
Burg Stolpen, Stolpen 21911
Burg Taufers, Campo Tures 25633
Burg Thurant, Alken 16470
Burg Trausnitz, Landshut 19800
Burg Trifels, Annweiler am Trifels 16558
Burg- und Klosteranlage Oybin, Oybin 20965
Burg- und Mühlenmuseum Pewsum, Krummhörn 19731
Burg- und Schloßmuseum Allstedt, Allstedt 16475
Burg- und Schloßmuseum Jägersburg, Homburg 19267
Burg- und Stadtmuseum Königstein, Königstein im Taunus 19657
Burg Vischering, Lüdinghausen 20092
Burg zu Burghausen, Burghausen 17651
Burg Zug, Zug 42123
Burg Zwernitz, Wonsees 22679
Burgas Museum, Burgas 04996
Burgaski Muzej, Burgas 04996
Burgauer Tiermuseum, Burgau 17648
Burgenkundliche Sammlung im Burgenmuseum, Gossau (Sankt Gallen) 41443
Burgenkundliches Museum Schloss Alt-Kainach, Bärnbach 01777
Burgenländische Landesgalerie, Eisenstadt 01856
Burgenländisches Brotmuseum, Bad Tatzmannsdorf 01765
Burgenländisches Feuerwehrmuseum, Eisenstadt 01857
Burgenländisches Geschichte(n)haus, Bildein 01785
Burgenmuseum, Dahn 17775
Burgenmuseum Eisenberg, Eisenberg, Kreis Ostallgäu 18101
Burgeo Centennial Memorial Museum, Burgeo 05494
Burger Weeshuis, Zierikzee 33002
Burgerhuis, Stellenbosch 38734
Burgersdorp Cultural Historical Museum, Burgersdorp 38575
Burggalerie Laa, Laa an der Thaya 02246
Burghead Museum, Burghead 43674
Burgher's House - Helsinki City Museum, Helsinki 10814
Burghley House, Stamford 45735
Burghofmuseum, Soest 21795
Burgmuseum, Güssing 02015
Burgmuseum, Seebenstein 02731
Burgmuseum Altnußberg, Geiersthal 18553
Burgmuseum Archeo Norico, Deutschlandsberg 01816
Burgmuseum Bad Bodenteich, Bad Bodenteich 16702
Burgmuseum Burg Guttenberg, Haßmersheim 19032
Burgmuseum Clam, Klam 02193
Burgmuseum der Wasserburg Kapellendorf, Kapellendorf 19429
Burgmuseum Greifenstein, Greifenstein 18745
Burgmuseum Grünwald, Grünwald 18818
Burgmuseum Kirschau, Kirschau 19558
Burgmuseum Krautheim, Krautheim 19701
Burgmuseum Lenzen, Lenzen 19957
Burgmuseum Marksburg, Braubach 17513
Burgmuseum Nideggen, Nideggen 20695
Burgmuseum Ortenberg, Ortenberg, Hessen 20914
Burgmuseum Runneburg, Weißensee 22462
Burgmuseum Schlossfreiheit, Tangermünde 22001
Burgmuseum Schönfels, Schönfels 21608
Burgmuseum Sonnenberg, Wiesbaden 22546
Burgmuseum Sulzberg, Sulzberg 21986
Burgmuseum Waldaist-Stöckl, Tragwein 02816
Burgmuseum Waldeck, Waldeck 22285
Burgmuseum Wolfsegg, Wolfsegg 22669
Burgruine Montclair, Merzig 20298
Burgruine Münzenberg, Münzenberg 20525
Burgruine Teufelsberg, Überherrn 22144
Burgturm Davert, Ascheberg 16595
Burgwin-Wright Museum, Wilmington 54200
Burin Heritage House, Burin 05496
Burjatskij Respublikanskij Chudožestvennyj Muzej im. C.S. Sampilova, Ulan-Udė 37892
Burke County Historical Powers Lake Complex, Powers Lake 52127
Burke Museum of Natural History and Culture, Seattle 52994
Burk's Falls and District Museum, Burk's Falls 05497
Burleson County Historical Museum, Caldwell 47164
Burlington Art Centre, Burlington 05498
Burlington County Historical Society Museum, Burlington 47139
Burlington Historical Museum, Burlington 47136
Burlington-Rock Island Railroad and Historical Museum, Teague 53458
Burlövs Gamla Prästgård, Arlöv 40432
Burlovs Old Rectory, Arlöv 40432
Burnaby Art Gallery, Burnaby 05501
Burnaby Village Museum, Burnaby 05502
Burnett House and Myilly Point Heritage Precinct, Darwin 01029
Burnham on Crouch and District Museum, Burnham-on-Crouch 43675
Burnham Tavern, Machias 50349
Burnie Regional Art Gallery, Burnie 00923
Burns Cottage Museum, Alloway 43283
Burns House Museum, Mauchline 45118
Burnside Plantation, Bethlehem 46777
Burnt House Museum, Jerusalem 24982
Burntisland Edwardian Fair Museum, Burntisland 43681
Burøsund Bygdemuseum, Vannareid 33958
Burpee Museum of Natural History, Rockford 52413
Burra Mine Open Air Museum, Burra 00925
Burrell Collection, Glasgow 44275
Burren Centre, Kilfenora 24847
Burritt on the Mountain, Huntsville 49442
Burrough of the Gleann Museum, Glenboro 05872
Burrowes Mansion Museum, Matawan 50532
Bursa Archaeological Museum, Bursa 42741
Bursa Arkeoloji Müzesi, Bursa 42741
Bursa City Museum, Bursa 42743
Bursa Devlet Güzel Sanatlar Galerisi, Bursa 42742
Bursa Kent Müzesi, Bursa 42743
Bursa State Fine Art Gallery, Bursa 42742
Bursa Türk-İslâm Eserleri Müzesi, Bursa 42744
Bursa Turkish and Islamic Art Museum, Bursa 42744
Bursledon Windmill, Bursledon 43683
Burstarfell Local Museum, Vopnafjörður 24044
Burston Strike School, Burston 43685
Burt County Museum, Tekamah 53460
Burtnieku Pagasta Briedes Krogs, Burtnieki 29994
Burton Agnes Hall, Burton Agnes 43686
Burton Art Gallery and Museum, Bideford 43477
Burton Constable Hall, Kingston-upon-Hull 44593
Burton Court, Eardisland 44079
Burwell-Morgan Mill, Millwood 50697
Burwell Museum, Burwell 43690
Bury Art Gallery and Museum, Bury, Lancashire 43691
Bury Saint Edmunds Art Gallery, Bury Saint Edmunds 43694
Bury Transport Museum, Bury, Lancashire 43692
Busan Museum, Busan 29774
Busan University Museum, Busan 29775
Busan Woman's College Tea-ceremony Museum, Busan 29776
Buscot House, Buscot 43698
Buscot Park House, Faringdon 44213
Bush House Museum and Bush Barn Art Center, Salem 52665
Bushey Museum and Art Gallery, Bushey 43699
Bushwhacker Museum, Nevada 51006
Bushy Run Battlefield Museum, Jeannette 49604
Bussemakerhuis, Borne 31989
Butcher's Shop from 1920, Roskilde 10249
Bute Museum, Rothesay 45520
Butler County Historical Museum, Allison 46232
Butler County Museum, Hamilton 49138
Butler County Society Museum and Kansas Oil Museum, El Dorado 48311
Butler Gallery, Kilkenny 24848
Butler Institute of American Art, Youngstown 54353
Butte County Pioneer Memorial Museum, Oroville 51604
Butte Silver Bow Arts Chateau, Butte 47150
Butter Pat Museum, Bigfork 46802
Butterfield Trail Historical Museum, Russell Springs 52488
Butterfly and Bird Museum, Stockholm 40894
Butterfly House, La Queue-lèz-Yvelines 13404
Butterfly Museum, Saint-Quentin (Aisne) 15475
Butterfly Museum Foundation, Budapest 23488
The Buttonwoods Museum, Haverhill 49216
Butuan City Museum, Butuan 34352
Butuan Diocesan Museum, Butuan 34353
Butuan National Museum, Butuan 34354
Buxoro Qala Davlat Muzeyi Qo'riqxonasi, Buxoro 54444
Buxtehude Museum für Regionalgeschichte und Kunst, Buxtehude 17673
Buxton Museum, North Buxton 06473
Buxton Museum and Art Gallery, Buxton 43705
Buyeo National Museum, Buyeo 29783
Buzău District Museum, Buzău 36244
Buzulukskij Gorodskoj Istoriko-kraevedčeskij Muzej, Buzuluk 36832
Bydgoszcz Country Educational Museum, Bydgoszcz 34579
Byers-Evans House Museum, Denver 48065
Byfield and District Museum, Byfield 00929
Bygdarsavnið í Vestmanna, Vestmanna 10703
Bygdetunet Jutulheimen, Vågåmo 33954
Byggðasafn, Vestmannaeyjar 24042
Byggðasafn Akraness og Nærsveita, Akranes 23980
Byggðasafn Árnesinga, Eyrarbakki 23995
Byggðasafn Austur-Skaftafellssýslu, Höfn 24001
Byggðasafn Borgarfjardar, Borgarnes 23989
Byggðasafn Dalamanna, Búdardalur 23991
Byggðasafn Dalvíkurbyggdar, Dalvík 23992
Byggðasafn Hafnarfjarðar, Hafnarfjörður 23997
Byggðasafn Húnvetninga og Strandarmanna, Brú 23990
Byggðasafn Reykjanesbaejar, Reykjanesbaer 24012
Byggðasafn Skagfirdinga, Varmahlid 24041
Byggðasafn Snæfellinga og Hnappdæla, Stykkishólmur 24040
Byggðasafnið Grenjadarstad, Húsavík 24002
Byggðasafnið í Skógum, Hvolsvöllur 24006
Bygland Museum, Bygland 33459
Bygones Museum, Claydon 43850
Bygones Museum at Holkham, Wells-next-the-Sea 45959
Byland Abbey, Coxwold 43909
Bymodellen, Oslo 33719
Bymuseet, Oslo 33720
Bymuseet, Skjern 10267
Bymuseet Møntergården, Odense 10209
Byron Historical Museum, Byron 47154
Bytown Museum, Ottawa 06530
Byzantine and Christian Museum, Athinai 22904
Byzantine Museum, Mytilini 23110
Byzantine Museum, Paphos 09390
Byzantine Museum, Tinos 23227
Byzantine Museum and Art Galleries, Lefkosia 09372
Byzantine Museum of Antivouniotissa, Corfu 22975
Byzantine Museum of Ekatontapyliani, Paros 23136
Byzantine Museum of Naousa, Naousa 23118
Byzantine Prince Palace, İstanbul 42841

C

The C.A.S.E Museum, Jersey City 49625
C.E. Smith Museum of Anthropology, Hayward 49227
C. Grier Beam Truck Museum, Cherryville 47427
C.H. Moore Homestead, Clinton 47652
C.H. Nash Museum-Chucalissa Archaeological Museum, Memphis 50567
C.J. Bonnet Silkworks, Jujurieux 13292
C. Kermit Ewing Gallery of Art and Architecture, Knoxville 49813
C Konsthall, Upplands Väsby 41001
C.L. Alexander Museum, Tumby Bay 01604
C.M. Bellman Museum, Stockholm 40888
C.M. Booth Collection of Historic Vehicles, Rolvenden 45510
C.M. Russell Museum, Great Falls 49028
C/O Berlin, Berlin 17059
C.O. Card Home, Cardston 05560
C.O. Müller-Galerie, Eichstätt 18080
C.P. Nel Museum, Oudtshoorn 38697
C.S. Sampilov Art Museum of the Buryat Republic, Ulan-Udė 37892
C Y Tung Maritime Museum, Shanghai 08156
Ca' Bon Gallery, Motovun 08911
Cabanon de Le Corbusier, Roquebrune-Cap-Martin 15094
Cabinet d'Art Asiatique, Labarde 13458
Cabinet des Estampes du Musée d'Art et d'Histoire, Genève 41403
Cabinet des Estampes et des Dessins, Liège 03794
Cabinet des Estampes et des Dessins, Strasbourg 15809
Cabinet des Médailles, Bruxelles 03440
Cabinet des Médailles et Antiques, Paris 14576
Cabinet des Monnais et Médailles du Revest-les-Eaux, Le Revest-les-Eaux 13646
Cabinet of Prints and Drawings, Dresden 17954
Cable Natural History Museum, Cable 47155
Caboolture Historical Village and Museum, Caboolture 00932
Cabot Historical Society, Cabot 47156
Cabot's Pueblo Museum, Desert Hot Springs 48105
Cabri and District Museum, Cabri 05509
Caddo-Pine Island Oil and Historical Society Museum, Oil City 51508
Caddoan Mounds, Cherokee County 47424
Cades Cove Open-Air Museum, Townsend 53548
Cadillac Museum, Cadillac 05511
Caernarfon Maritime Museum, Caernarfon 43708
Caerphilly Castle, Caerphilly 43712
Caesarea Museum, Kibbutz Sdot Yam 25056
CAFAM, Los Angeles 50213
Cafe Gallery Projects, London 44811
Cafesjan Museum, Yerevan 00729
Cagayan Museum, Tuguegarao 34518
Caguana Indian Ceremonial Park and Museum, Utuado 36079
Cahit Sıtkı Tarancı House Museum, Diyarbakır 42761
Cahit Sıtkı Tarancı Müze-Evi, Diyarbakır 42761
Cahokia Courthouse, Cahokia 47158
Cahokia Mounds, Collinsville 47694
Cahoon Museum of American Art, Cotuit 47846
Caidian District Museum, Wuhan 08330
Caifeng Culture Museum, Danei 42175
Cailun Ciwen Guan (Museum), Yangxian 08425
Cairns Museum, Cairns 00933
Cairns Regional Gallery, Cairns 00934
Cairo Museum, Vicksburg 53732
Caixa Forum Madrid, Madrid 39501
CaixaForum Barcelona, Barcelona 38952
CaixaForum Palma de Mallorca, Palma de Mallorca 39776
Čajkovskaja Kartinnaja Galereja, Čajkovskij 36833
Čajkovskij Kraevedčeskij Muzej, Čajkovskij 36834
Cák Cellar Line - Open-Air Ethnographic Museum, Cák 23500
Cáki Pincesor - Szabadtéri Néprajzi Múzeum, Cák 23500
Cakiraga House, Ödemiş 42919
Çakirağa Konaği, Ödemiş 42919
Calala Cottage, Tamworth 01568
Calaveras County Museum, San Andreas 52699
Calbourne Water Mill and Rural Museum, Newport, Isle of Wight 45242
Calcografia, Roma 27441
Calcografía Nacional, Madrid 39502
Calderglen County Park, East Kilbride 44087
Caldicot Castle, Caldicot 43713
Caleb Pusey House, Upland 53662
Caledon Museum, Caledon 38576
Calgary Science Centre, Calgary 05516
Calhoun County Museum, Port Lavaca 52038
Calhoun County Museum, Rockwell City 52432
Calhoun County Museum, Saint Matthews 52610
Calico Museum of Textiles, Ahmedabad 24046
California Academy of Sciences, San Francisco 52768
California African American Museum, Los Angeles 50214
California Art Club Gallery at the Old Mill, San Marino 52818
California Association of Museums, Santa Cruz 55127
California Automobile Museum, Sacramento 52499
California Center for the Arts Museum, Escondido 48418

California Heritage Museum, Santa Monica 52898
The California Indian Museum, Santa Rosa 52903
California Museum, Sacramento 52500
California Museum of Ancient Art, Beverly Hills 46791
California Museum of Photography, Riverside 52369
California Oil Museum, Santa Paula 52902
California Route 66 Museum, Victorville 53742
California Science Center, Los Angeles 50215
California State Capitol Museum, Sacramento 52501
California State Indian Museum, Sacramento 52502
California State Mining and Mineral Museum, Mariposa 50491
California State Railroad Museum, Sacramento 52503
Calkins Field Museum, Iowa Falls 49523
Call of the Wild - Wildlife Museum, Gaylord 48874
Callahan County Pioneer Museum, Baird 46532
Callander Bay Heritage Museum, Callander 05535
Callanwolde Fine Arts Center, Atlanta 46429
Callendar House, Falkirk 44205
Calligraphy Museum, Tokyo 29472
Caloundra Regional Art Gallery, Caloundra 00935
Calumet County Historical Society Museum, New Holstein 51059
Calvert Marine Museum, Solomons 53142
Calverton Folk Museum, Calverton 43717
Calvin Coolidge Presidential Library and Museum, Northampton 51412
Calvin Coolidge State Historic Site, Plymouth Notch 52004
Calvinia Regional Museum, Calvinia 38577
Cámara Santa de la Catedral de Oviedo, Oviedo 39757
Camborne School of Mines Geological Museum and Art Gallery, Pool 45404
Cambria County Historical Society Museum, Ebensburg 48279
Cambria Historical Society Museum, New Providence 51100
Cambridge and County Folk Museum, Cambridge 43723
Cambridge Glass Museum, Cambridge 47189
Cambridge Historical Museum, Cambridge 47174
Cambridge Museum, Cambridge 33065
Cambridge Museum, Cambridge 47188
Cambridge Museum of Technology, Cambridge 43724
Cambridge University Collection of Air Photographs, Cambridge 43725
Camden Archives and Museum, Camden 47199
Camden Arts Centre, London 44812
Camden County Historical Society Museum, Camden 47195
Camden Haven Museum, Laurieton 01225
Camden Museum, Camden 00938
Camera Museum, Osmaniye 42923
Camera Obscura, Hainichen 18872
Camera Obscura, Mülheim an der Ruhr 20411
Camera Obscura and World of Illusions, Edinburgh 44115
Cameron Highlanders of Ottawa Regimental Museum, Ottawa 06531
Çamlı Köşk, İstanbul 42811
CAMOC - International Committee for the Collections and Activities of Museums of Cities, Moskva 54999
Camp 6 Logging Museum, Tacoma 53405
Camp Five Museum, Laona 49946
Camp Floyd, Fairfield 48486
Camp Gallery, Sweet Briar 53390
Camp Town Museum of the Budapest History Museum, Budapest 23479
Campaign for Museums, London 55053
Campbell Carriage Factory Museum, Middle Sackville 06275
Campbell Collections, Durban 38603
Campbell County Rockpile Museum, Gillette 48909
Campbell Hall Gallery, Monmouth 50784
Campbell Historical Museum, Campbell 47208
Campbell House Museum, Saint Louis 52569
Campbell House Museum, Toronto 07096
Campbell River and District Public Art Gallery, Campbell River 05538
Campbell River Museum, Campbell River 05539
Campbell River Optical Maritime Museum, Campbell River 05540
Campbell Whittlesey House, Rochester 52393
Campbellford-Seymour Heritage Centre, Campbellford 05542
Campbeltown Art Center, Campbelltown 00940
Campbeltown Museum, Campbeltown 43739
Camperdown and District Museum, Camperdown 00941
Camperdown Buggy Museum, Camperdown 00942
Campobello Island Museum, Welshpool 07297
Campus Gallery, Barrie 05386
Campus Martius Museum, Marietta 50472
Camron-Stanford House, Oakland 51463
Camrose and District Centennial Museum, Camrose 05544
Canaan Historical Society, Falls Village 48503
Canaan Historical Society Museum, Canaan 47210
Canada Agriculture Museum, Ottawa 06532
Canada Aviation Museum, Ottawa 06533
Canada House Gallery, London 44813
Canada Science and Technology Museum, Ottawa 06534
Canada War Museum, Adegem 03261
Canada's Aviation Hall of Fame, Wetaskiwin 07311
Canada's Penitentiary Museum, Kingston 06075
Canada's Sports Hall of Fame, Toronto 07097
Canadian Art Museum Directors Organization, Ottawa 54732
Canadian Automotive Museum, Oshawa 06521
Canadian Baseball Hall of Fame & Museum, Saint Marys 06868
Canadian Broadcast Museum Foundation, Toronto 07098
Canadian Bushplane Heritage Centre, Sault Sainte Marie 06929
Canadian Canoe Museum, Peterborough 06591
Canadian Centre for Architecture, Montréal 06330
Canadian Clay and Glass Gallery, Waterloo 07285
The Canadian Clock Museum, Deep River 05687
Canadian County Historical Museum, El Reno 48330
Canadian Embassy Art Gallery, Washington 53854
Canadian Federation of Friends of Museums, Toronto 54733
Canadian FireFighter's Museum, Port Hope 06642
Canadian Football Hall of Fame and Museum, Hamilton 05952
Canadian Force Base Gagetown Military Museum, Oromocto 06517
Canadian Golf Hall of Fame and Museum, Oakville 06494
Canadian Lacrosse Hall of Fame, New Westminster 06435
Canadian Medical Hall of Fame, London 06182
Canadian Military Engineers Museum, Oromocto 06518
Canadian Military Heritage Museum, Brantford 05468
Canadian Military Heritage Museum, Saskatoon 06911
Canadian Military Studies Museum, Limehouse 06171
Canadian Museum of Civilization, Gatineau 05862
Canadian Museum of Contemporary Photography, Ottawa 06535
Canadian Museum of Flight, Langley 06136
Canadian Museum of Hindu Civilization, Richmond Hills 06748
Canadian Museum of Nature, Gatineau 05860
Canadian Museum of Nature, Ottawa 06536
Canadian Museum of Rail Travel, Cranbrook 05658
Canadian Museums Association, Ottawa 54734
Canadian Postal Museum, Gatineau 05861
Canadian Scottish Regimental Museum, Victoria 07243
Canadian Scouting Museum, Ottawa 06537
Canadian Sculpture Centre, Toronto 07099
Canadian Ski Museum, Ottawa 06538
Canadian Society for Education Through Art, Toronto 07138
Canadian Society of Asian Arts, Vancouver 07206
Canadian Transportation Museum and Historic Village, Kingsville 06092
Canadian War Museum, Ottawa 06539
Canadian Warplane Heritage Museum, Mount Hope 06399
Canadian Wildlife and Wilderness Art Museum, Ottawa 06540
Canajoharie Library and Art Gallery, Canajoharie 47211
Çanakkale Anıtlari ve Savaş Müzesi, Çanakkale 42746
Çanakkale Devlet Güzel Sanatlar Galerisi, Çanakkale 42747
Çanakkale Martyrs Monument and War Museum, Çanakkale 42746
Çanakkale Müzesi, Çanakkale 42748
Çanakkale State Fine Art Gallery, Çanakkale 42747
Canal-de-Carillon, Saint-André-d'Argenteuil 06794
Canal-de-Lachine - Lieu Historique National du Canada, Lachine 06121
Canal Fulton Heritage Society Museum, Canal Fulton 47212
Canal Museum, Llangollen 44768
Canal Museum, Stoke Bruerne 45758
Canal Museum, Varkaus 11452
Canary Islands Regional Military Museum, Santa Cruz de Tenerife 39996
Canberra Bicycle Museum, Dickson 01038
Canberra Contemporary Art Space, Braddon 00888
Canberra Contemporary Art Space Manuka, Manuka 01259
Canberra Museum and Gallery, Canberra 00944
Canbulat Tomb and Museum, Gazimağusa 33350
Canbulat Türbe ve Müzesi, Gazimağusa 33350
Canby Depot Museum, Canby 47216
Caney Valley Historical Society, Caney 47217
Cango Caves Interpretive Centre, Oudtshoorn 38698
Cankarjev Dom, Ljubljana 38447
Çankırı Devlet Güzel Sanatlar Galerisi, Çankırı 42750
Çankırı Müzesi, Çankırı 42751
Çankırı State Fine Art Gallery, Çankırı 42750
Canmore Museum, Canmore 05545
Cannington Centennial Museum, Cannington 05546
Cannington Manor, Kenosee Lake 06058
Cannon Falls Area Historical Museum, Cannon Falls 47218
Cannon Hall Museum, Barnsley, South Yorkshire 43384
Cannon Village Visitor Center, Kannapolis 49677
Canolfan Y Plase, Bala 43357
Canolfan Y Plase, Y Bala 46090
Canon City Municipal Museum, Canon City 47219
Canso Museum, Canso 05547
Canterbury Heritage Museum, Canterbury 43742
Canterbury Museum, Christchurch 33069
Canterbury Shaker Village, Canterbury 47221
Canton Classic Car Museum, Canton 47226
Canton Historical Museum, Canton 47222
The Canton Museum of Art, Canton 47227
Canwood Museum, Canwood 05549
Canzani Center Galleries, Columbus 47750
CAP Art Center and Don Sergio Osmeña Memorabilia, Cebu 34365
Caparra Ruins Historical Museum, Guaynabo 36039
CAPC Bordaux Museum of Contemporary Art, Bordeaux 12062
CAPC Musée d'Art Contemporain, Bordeaux 12062
Cape Ann Historical Museum, Gloucester 48945
Cape Bonavista Lighthouse, Bonavista 05438
Cape Breton Centre for Heritage and Science, Sydney 07047
Cape Breton Miners' Museum, Glace Bay 05868
Cape Clear Heritage Centre, Skibbereen 24897
Cape Coast Castle Museum, Cape Coast 22850
Cape Cod Museum of Natural History, Brewster 47008
Cape Cod National Seashore Museum, Wellfleet 54025
Cape Fear Museum, Wilmington 54201
Cape Henry Lighthouse, Virginia Beach 53754
Cape May County Historical Museum, Cape May Court House 47236
Cape Museum of Fine Arts, Dennis 48057
Cape Ray Lightkeeper's House, Cape Ray 05552
Cape Sable Historical Society Centre, Barrington 05390
Cape Spear Lighthouse National Historic Site, Saint John's 06844
Cape Town Holocaust Centre, Cape Town 38580
Cape Vincent Historical Museum, Cape Vincent 47238
Y Capel, Llangollen 44769
Capella Brancacci, Firenze 26142
Ca'Pesaro - Galleria Internazionale d'Arte Moderna, Venezia 28178
Capharnaum Ancient Synagogue, Kfar Nahum 25028
Capital Museum, Beijing 07524
Capitol Reef National Park Visitor Center, Torrey 53543
Capitola Historical Museum, Capitola 47239
Capitolum and Diana Galleries at the City Palace, Pula 08934
Capodistria Museum, Koukouritsa, Evropouli 23065
Cappella degli Scrovegni, Padova 27032
Cappella Sistina, Sala e Gallerie Affrescate, Città del Vaticano 54497
Cappon House Museum, Holland 49313
Captain Cook Birthplace Museum, Middlesbrough 45139
Captain Cook Memorial Museum, Whitby 45994
Captain Cook Schoolroom Museum, Great Ayton 44352
Captain Forbes House Museum, Milton 50699
Captain George Flavel House Museum, Astoria 46400
Captain John Wilson House, Cohasset 47675
Captain Kennedy Museum and Tea House, Saint Andrews 06796
Captain Nathaniel B. Palmer House, Stonington 53311
Captain Phillips' Rangers Memorial, Saxton 52951
Captain Salem Avery House Museum, Shady Side 53046
Captain's Cottage Museum, Murray Bridge 01336
Car and Carriage Museum, Pittsburgh 51941
Car and Traffic Museum, Kangasala 10910
Car Life Museum, Charlottetown 05590
Car Museum, Leuze-en-Hainaut 03791
Car Museum, Vatnsoyrar 10702
Caracol Archaeological Site, Caracol 04177
Caramoor Center for Music and the Arts, Katonah 49707
Carberry Plains Museum, Carberry 05556
Carbethon Folk Museum and Pioneer Village, Crows Nest, Queensland 01013
Carbon County Museum, Rawlins 52258
Carbonear Museum, Carbonear 05558
Carcoar Historic Village, Carcoar 00954
Cardiff Castle, Cardiff 43749
Cardigan Heritage Centre, Cardigan 43756
Cardinal Gallery, Annapolis 46315
Cardinal Ó Fiaich Heritage Centre, Cullyhanna 43937
Cardinal Vincentas Sladkevicius Memorial Exposition of the Museum of the Kaunas Archdiocese, Kaunas 30247
Cardwell Shire Museum, Tully 01603
Care of Collections Forum, Grangemouth 55054
Carew Manor and Dovecote, Beddington 43432
Carey Museum, Serampore 24383
Ca'Rezzonico - Museo del Settecento Veneziano, Venezia 28179
Caribbean Primate Research Center Museum, San Juan 36059
Caricatura Galerie, Kassel 19460
Caricatura Museum Frankfurt, Frankfurt am Main 18357
Caricature Museum, Warszawa 35250
Caricyno - Gosudarstvennyj Istoriko-architekturnyj, Chudožestvennyj i Landšaftnyj Muzej-Zapovednik, Moskva 37143
Carillon Historical Park, Dayton 47983
Carisbrooke Castle Museum, Newport, Isle of Wight 45243
Carl-Benz-Haus, Ladenburg 19762
Carl Bosch Museum, Heidelberg 19070
Carl Eldhs Ateljémuseum, Stockholm 40889
Carl Goldmark-Gedenkhaus, Deutschkreutz 01814
Carl-Henning Pedersen og Else Alfelts Museum, Herning 10069
Carl-Hirnbein-Museum, Missen-Wilhams 20339
Carl Jularbo Museum, Avesta 40449
Carl-Julius-Weber-Gedenkstube, Langenburg 19822
Carl Larsson-Gården, Sundborn 40956
Carl-Lohse-Galerie - Stadtmuseum Bischofswerda, Bischofswerda 17364
Carl-Maria-von-Weber-Museum, Dresden 17940
Carl-Maria-von-Weber-Museum, Dresden 17940
Carl Nielsen Museet, Odense 10210
Carl Nielsens Barndomshjem, Årslev 09968
Carl Orff Museum, Dießen am Ammersee 17851
Carl Sandburg Home, Flat Rock 48555
Carl Sandburg State Historic Site, Galesburg 48832
Carl-Schweizer-Museum, Murrhardt 20536
Carl Zeiss Ausstellungszentrum, Oberkochen 20813
Carleton College Art Gallery, Northfield 51418
Carleton University Art Gallery, Ottawa 06541
Carlisle Collection of Miniature Rooms, Nunnington 45306
Carlos Gomes Museum, Campinas 04384
Carlos P. Garcia Memorabilia, Manila 34434
Carlos P. Romulo Collection, Manila 34435
Carlotte Weibulls Foklivscenter, Åkarp 40411
Carlow County Museum, Carlow 24751
Carlsbad Caverns National Park, Carlsbad 47248
Carlsbad Museum and Art Center, Carlsbad 47249
Carlsberg Museum Valby, Valby 10305
Carlson Tower Gallery, Chicago 47459
Carlsro Museum, Kristiinankaupunki 10995
Carlton County Historical Museum and Heritage Center, Cloquet 47663
Carlyle House, Alexandria 46201
Carlyle's House, London 44814
Carman House Museum, Iroquois 06022
Carmarthenshire County Museum, Carmarthen 43762
Carmen Museo Max Moreau, Granada 39355
Carmicheal-Stewart House, New Glasgow 06429
Carnaby Collection of Beetles and Butterflies, Boyup Brook 00887
Carnegie Art Center, Mankato 50447
Carnegie Art Center, Walla Walla 53796
Carnegie Art Museum, Oxnard 51657
Carnegie Center for Art & History, New Albany 51008
Carnegie Historical Museum, Fairfield 48483
Carnegie Museum, Inverurie 44527
Carnegie Museum of Art, Pittsburgh 51942
Carnegie Museum of Natural History, Pittsburgh 51943
Carnegie Science Center, Pittsburgh 51944
Carney Gallery, Weston 54104
Carnifex Ferry Battlefield State Park and Museum, Summersville 53362
Carnival Museum, Malmedy 03843
Carnton Plantation, Franklin 48750
Carolinas Aviation Museum, Charlotte 47385
Caroline Wheels of Time Museum, Caroline 05568
Carpenter Home Museum, Cle Elum 47603
Carpenter Museum of Antique Outboard Motors, Gilmanton 48911
Carpet Museum, Bakı 03150
Carpet Museum, Genemuiden 32245
Carpet Museum, Tehrān 24663
Carpetbagger Aviation Museum, Harrington 44398
Carpinteria Valley Museum of History, Carpinteria 47254
Carré d'Art, Nîmes 14446
Le Carré-Scéne Nationale, Château-Gontier 12434

Carrefour du Vitrail, Lavaudieu 13547
Carriage and Harness Museum & Murray Breweries Historic Museum, Beechworth 00848
Carriage House Museum, Camden 47198
Carriage House Museum, Colorado Springs 47698
Carriage Museum, Città del Vaticano 54507
Carriage Museum, Regensburg 21229
Carriage Museum, Villa Ciudad Parque 00674
Carriage Museum of Bermuda, Saint George's 04197
Carriage Museum of Seville, Sevilla 40070
Carriages Museum, Catanzaro 25781
Carrick Hill Museum, Springfield 01522
Carrie Chang Fine Arts Center, Tamsui 42447
Carrie M. McLain Memorial Museum, Nome 51359
Carrie Weedon Natural Science Museum, Galesville 48835
Carrière Souterraine d'Aubigny, Taingy 15845
Carroll County Farm Museum, Westminster 54096
Carroll Mansion Home - Leavenworth County Historical Society Museum, Leavenworth 50008
Carroll Reece Museum, Johnson City 49631
Carromato de Max, Mijas 39647
Carry A. Nation Home Memorial, Medicine Lodge 50554
Cars and Bikes Museum, Öhringen 20863
Cars of the Stars Motor Museum, Keswick 44561
Cars of Yesterday, Tatabánya 23920
Carscadden's Museum, Plenty 06612
Carson County Square House Museum, Panhandle 51695
Carson Valley Museum, Gardnerville 48863
Carss Cottage Museum, Blakehurst 00874
Cart Museum at the National Stud-farm of Bábolna, Bábolna 23340
Carter-Coile Country Doctors Museum, Winterville 54259
Carter County Museum, Ekalaka 48306
The Carter House, Franklin 48751
Carter's Grove Visitor Center, Williamsburg 54166
Cartier-Brébeuf, Québec 06676
Cartoon Art Museum, San Francisco 52769
Cartref Taid, Trevelin 00654
Cartuja de la Asunción, Granada 39356
Cartwright Hall Art Gallery, Bradford 43577
Carver County Historical Society Museum, Waconia 53782
Cary Cottage, Cincinnati 47554
Caryl House and Fisher Barn, Dover 48164
Casa Adobe de San Rafael, Glendale 48929
Casa Alegre de Sagrera, Terrassa 40114
Casa Anatta e Casa Selma, Ascona 41127
Casa-Atelier Gabriel Popescu, Vulcana Pandele 36736
Casa-Atelier Gheorghe Petraşcu, Târgovişte 36653
Casa Aymat, Sant Cugat del Vallès 39978
Casa Blanca Museum, San Juan 36065
Casa Bruzzone, Mar del Plata 00443
Casa Buonarroti, Firenze 26143
Casa Cercantes, Vélez Málaga 40256
Casa Contadina, Forenza 26222
Casa da Botica, Póvoa de Lanhoso 35871
Casa da Cerca, Almada 35412
Casa da Cultura, Machado 04549
Casa da Memória Arnaldo Estevão de Figueiredo, Campo Grande 04392
Casa Da Noal, Treviso 28068
Casa Daros, Rio de Janeiro 04682
Casa das Artes de Vigo, Vigo 40273
Casa de Andrés Eloy Blanco, Cumaná 54566
Casa de Cervantes, Esquivias 39284
Casa de Cora Coralina, Goiás 04477
Casa de Cultura Afrónio Peixoto, Lençóis 04536
Casa de Cultura João Ribeiro, Laranjeiras 04532
Casa de Deodoro, Rio de Janeiro 04683
Casa de Dulce María Loynaz, La Habana 09095
Casa de Garibaldi, Montevideo 54382
Casa de Giró, Montevideo 54383
Casa de Hidalgo, Dolores Hidalgo 30834
Casa de José Bonifácio, Rio de Janeiro 04684
Casa de José Jacinto Milanés, Matanzas 09211
Casa de la Constitución de 1814, Apatzingán 30717
Casa de la Cultura, Curuzú Cuatiá 00351
Casa de la Cultura "Augusto Rivera Garces", Bolívar 08597
Casa de la Cultura Benjamín Carrión, Cuenca 10380
Casa de la Panadería, Madrid 39503
Casa de la Ribera, Peñafiel 39825
Casa de la Vall, Andorra la Vella 00085
Casa de las Ciencias, A Coruña 39213
Casa de Lavalleja, Montevideo 54384
Casa de los Abuelos, Sonsón 08761
Casa de los Arabes, La Habana 09096
Casa de los Cinteros, Villar del Arzobispo 40306
Casa de los Morlanes, Zaragoza 40337
Casa de los Tiros, Granada 39357
La Casa de los Tratados, Girón 10397
Casa de Manuel Mujica Lainez, La Cumbre 00399
Casa de Marinos, Aranjuez 38878
Casa de Miguel Failde Pérez, Matanzas 09212
Casa de Oliveira Vianna, Niterói 04589
Casa de Osório, Rio de Janeiro 04685
Casa de Pilatos, Sevilla 40060
Casa de Unidad, Detroit 48107
Casa de Ximénez, Montevideo 54385
Casa del Águila y La Parra, Santillana del Mar 40032
Casa del Arte de la Universidad de Concepción, Concepción 07432
Casa del Boccaccio, Certaldo 25814
Casa del Cordón, Vitoria-Gasteiz 40318
Casa del Escribano Real Don Juan de Vargas, Tunja 08772
Casa della Nonna, Cairate 25590
Casa dell'Ariosto, Ferrara 26104
Casa di Carlo Goldoni, Venezia 28180
Casa di Giorgione, Castelfranco Veneto 25729
Casa di Goethe, Roma 27442
Casa di Lazzaro Spallanzani, Scandiano 27783
Casa di Luigi Pirandello, Roma 27443
Casa di Petrarca, Arquà Petrarca 25286
Casa di Pirandello, Agrigento 25160
Casa di Xilogravura, Campos de Jordão 04397
Casa do Corpo Santo, Setúbal 35934
Casa do Escultor Ervin Curt Teichmann, Pomerode 04638
Casa Domnească Brebu, Brebu 36183
Casa Dorita Castel-Branco, Sintra 35943
Casa dos Patudos - Museu de Alpiarça, Alpiarça 35423
Casa dos Sete Candeeiros, Salvador 04786
Casa Dr. José Celso Barbosa, Bayamón 36028
Casa Galería Marcos J. Alegría, Dorado 36037
Casa Gorordo Museum, Cebu 34366
Casa Goyeneche, Lima 34206
Casa Grande Ruins, Coolidge 47786
Casa Grande Valley History Museum, Casa Grande 47269
Casa Jesús T. Piñero, Canóvanas 36033
Casa João Turin, Curitiba 04426
Casa Leopardi, Recanati 27375
Casa Loma, Toronto 07100
Casa Luis Barragán, México 31036
Casa Manila Museum, Manila 34436
Casa Martirilor de la Moisei, Moisei 36480
Casa Memória d' Alte, Alte 35424
Casa Memorială a Poetului Ion Şiugariu, Băiţa 36132
Casa Memorială Alexandru Sahia, Mănăstirea 36460
Casa Memorială Alexandru Vlahuţă, Agapia 36098
Casa Memorială Aloisie Tăutu, Valea Vinului 36717
Casa Memorială Anton Pann, Râmnicu Vâlcea 36550
Casa Memorială Benedek Elek, Băţanii Mici 36144
Casa Memorială Cezar Petrescu, Buşteni 36238
Casa Memorială Constantin Joja, Bucureşti 36185
Casa Memorială Costache Negruzzi, Hermeziu 36398
Casa Memorială Crişan, Crişan 36330
Casa Memorială D.P. Perpessicius, Brăila 36167
Casa Memorială Dinu Lipatti, Leordeni 36439
Casa Memorială Dr. Vasile Lucaciu, Apa 36109
Casa Memorială Ecaterina Teodoroiu, Târgu Jiu 36663
Casa Memorială Elena Cuza, Soleşti 36633
Casa Memorială Elena Farago, Craiova 36321
Casa Memorială Emil Racoviţă, Dăneşti 36338
Casa Memorială Eusebiu Camilar, Udeşti 36713
Casa Memorială Gala Galaction, Bucureşti 36186
Casa Memorială George & Agatha Bacovia, Bucureşti 36187
Casa Memorială George Bacovia, Bacău 36118
Casa Memorială George Enescu, Liveni 36445
Casa Memorială George Enescu, Sinaia 36618
Casa Memorială George Pop de Băseşti, Băseşti 36143
Casa Memorială George Topârceanu, Iaşi 36410
Casa Memorială Gheorghe Dima, Braşov 36174
Casa Memorială Gheorghe Ionescu-Şişeşti, Şişeşti 36625
Casa Memorială Gheorghe Rădulescu-Motru, Butoieşti 36240
Casa Memorială Gyarfas Jenö, Sfântu Gheorghe 36599
Casa Memorială I.C. Vissarion, Costeştii din Vale 36317
Casa Memorială I.G. Duca, Horezu 36401
Casa Memorială Ion Creangă, Târgu Neamţ 36673
Casa Memorială Ion Minulescu & Claudia Millian, Bucureşti 36188
Casa Memorială Ion Popescu, Voiteşti 36734
Casa Memorială Ionel Perlea, Ograda 36498
Casa Memorială Liviu & Fanny Rebreanu, Bucureşti 36189
Casa Memorială Liviu Rebreanu, Ştefăneşti 36634
Casa Memorială Lucian Blaga, Lancrăm 36434
Casa Memorială Mihai Codreanu, Iaşi 36411
Casa Memorială Mihail Sadoveanu, Fălticeni 36364
Casa Memorială Mihail Sadoveanu, Iaşi 36412
Casa Memorială Mihail Sadoveanu, Vânători-Neamţ 36723
Casa Memorială Moş Dumitru, Baldovineşti 36133
Casa Memorială Moş Ion Roată, Câmpuri 36257
Casa Memorială N.D. Cocea, Sighişoara 36616
Casa Memorială Nichita Stănescu, Ploieşti 36528
Casa Memorială Nicolae Gane, Iaşi 36413
Casa Memorială Nicolae Iorga, Botoşani 36160
Casa Memorială Nicolae Labiş, Mălini 36458
Casa Memorială Nicu Enea, Bacău 36119
Casa Memorială Nikolaus Lenau, Lenauheim 36438
Casa Memorială Otilia Cazimir, Iaşi 36414
Casa Memorială Panait Cerna, Cerna 36272
Casa Memorială Panait Istrati, Brăila 36168
Casa Memorială Simion Florea Marian, Suceava 36640
Casa Memorială Tamasi Aron, Lupeni 36452
Casa Memorială Theodor Bălăşel, Ştefăneşti, Vâlcea 36636
Casa Memorială Traian Demetrescu, Craiova 36322
Casa Memorială Tudor Arghezi, Bucureşti 36190
Casa Memorială Tudor Vladimirescu, Vladimir 36733
Casa Memorială Vasile Alecsandri, Mirceşti 36476
Casa Memorială Vasile Pogor, Iaşi 36415
Casa Memorială Vasile Voiculescu, Pârscov 36511
Casa Memorială Veronica Micle, Târgu Neamţ 36674
Casa Mitropolitului Dosoftei, Iaşi 36416
Casa-Musei de Aquilino Ribeiro, Peva 35822
Casa Museo 20 de Julio de 1810, Bogotá 08544
Casa Museo A. Gramsci, Ghilarza 26333
Casa Museo A. Uccello, Palazzolo Acreide 27059
Casa Museo Alberto Sols, Sax 40045
Casa Museo Alfonso Ariza, La Rambla 39889
Casa Museo Almirante Miguel Grau, Piura 34288
Casa Museo Amado Nervo, Tepic 31439
Casa Museo Ángel Barrios, Granada 39358
Casa-Museo Antonio Machado, Segovia 40049
Casa Museo Antonio Nariñtildeo y Álvarez, Villa de Leyva 08780
Casa Museo Antonio Padrón, Gáldar 39319
Casa-Museo Azorín, Monóvar 39667
Casa Museo Blasco Ibáñez, Valencia 40195
Casa Museo Cadete Juan Escutia, Tepic 31440
Casa Museo Carlos Pellicer Cámara, Villahermosa 31546
Casa Museo Concha Piquer, Valencia 40196
Casa Museo Coronel Gregorio Méndez Magaña, Jalpa de Méndez 30968
Casa Museo Curros Enríquez, Celanova 39160
Casa Museo da Fundación Camilo Josè Cela, Padrón 39763
Casa Museo de Cervantes, Valladolid 40227
Casa Museo de Colón, Valladolid 40228
Casa Museo de Goya, Fuendetodos 39313
Casa Museo de Juan Zorrilla de San Martín, Montevideo 54386
Casa-Museo de Lope de Vega, Madrid 39504
Casa Museo de Rafael Alberti, El Puerto de Santa María 39878
Casa Museo de Unamuno, Salamanca 39934
Casa Museo de Zorrilla, Valladolid 40229
Casa-Museo de Zurbarán, Fuente de Cantos 39315
Casa Museo dei Puccini, Pescaglia 27160
Casa Museo del Banco Nacional, Ciudad de Panamá 34063
Casa Museo del Cima, Conegliano 25961
Casa Museo dell'Alta Valle del Cervo, Rosazza 27556
Casa Museo Depero, Rovereto 27567
Casa Museo di A. Oriani, Casola Valsenio 25707
Casa Museo di Monticello Amiata, Cinigiano 25895
Casa Museo Emilia Pardo Bazán, A Coruña 39214
Casa Museo F. Scaglione, Sciacca 27795
Casa Museo Fantoni, Rovetta 27572
Casa Museo Federico García Lorca, Fuente Vaqueros 39317
Casa Museo Fernando Fader, Ischilín 00389
Casa Museo Francesco Baracca, Lugo 26543
Casa Museo Francisco Torregrosa, Muro de Alcoy 39708
Casa Museo G. Mazzarino, Pescina 27174
Casa Museo Gene Byron, Guanajuato 30911
Casa Museo General San Martín, Cervatos de la Cueza 39164
Casa Museo Giorgio de Chirico, Roma 27444
Casa Museo Giovanni Verga, Catania 25771
Casa Museo Ignacio Zuloaga, Zumaia 40362
Casa Museo Ivan Bruschi, Arezzo 25264
Casa Museo Joaquin Torrents Llado, Palma de Mallorca 39777
Casa-Museo Jorge Eliécer Gaitán, Bogotá 08545
Casa-Museo José Benlliure, Valencia 40197
Casa Museo José Carlos Mariátegui, Lima 34207
Casa Museo José Clemente Orozco, Guadalajara 30885
Casa Museo José María de Cossío, Tudanca 40172
Casa Museo José Padilla, Madrid 39505
Casa Museo Julia Codesido, Lima 34208
Casa Museo Julián Gayarre, Roncal 39913
Casa Museo La Moreña, La Barca 30985
Casa-Museo La Sebastiana, Valparaiso 07470
Casa-Museo León y Castillo, Telde 40112
Casa Museo Leonardo Valdez Esquer, Etchojoa 30865
Casa Museo López Portillo, Guadalajara 30886
Casa Museo M. Moretti, Cesenatico 25838
Casa Museo Maestro Manuel Altamirano, Tixtla de Guerrero 31460
Casa-Museo Manuel de Falla, Granada 39359
Casa Museo María Augusta Urrutia, Quito 10425
Casa Museo Mario Urteaga, Cajamarca 34125
Casa-Museo Menéndez Pelayo, Santander 40008
Casa Museo Miguel Grau, Lima 34209
Casa Museo Miguel Hernández, Orihuela 39740
Casa Museo Modernista, Novelda 39717
Casa Museo Nuñez del Prado, La Paz 04209
Casa Museo Páez, Valencia 54597
Casa Museo Pavesiano, Santo Stefano Belbo 27732
Casa-Museo Pedralba 2000, Pedralba 39820
Casa Museo Pedro Muñez Seca, El Puerto de Santa María 39879
Casa Museo Pérez Galdós, Las Palmas de Gran Canaria 39796
Casa Museo Piccione d'Avola, Geraci Siculo 26329
Casa Museo Pinazo, Godella 39351
Casa Museo Posada del Moro, Torrecampo 40154
Casa Museo Quevedo Zornoza, Zipaquirá 08790
Casa Museo Quinta de Bolívar, Bogotá 08546
Casa Museo R. Bendandi, Faenza 26072
Casa Museo R. Siviero, Firenze 26144
Casa Museo Ricardo Gomez Campuzano, Bogotá 08547
Casa Museo Ricardo Palma, Lima 34210
Casa Museo Rosalía de Castro, Padrón 39764
Casa Museo Sa Domu Antiga, Santadi 27712
Casa Museo Signorini-Corsi, L'Aquila 26454
Casa Museo Silvestre Rodríguez, Nacozari de García 31208
Casa Museo Tiu Virgiliu, Suni 27923
Casa-Museo Tomás Morales, Moya 39688
Casa Museo V. Cermignani, Giulianova 26345
Casa Museo Valle Inclán, La Pobra do Caramiñal 39848
Casa Museo Vladimir Cora, Acaponeta 30689
Casa Museo Zenobia y Juan Ramón Jiménez, Moguer 39653
Casa Museu Abel Salazar da Universidade do Porto, São Mamede de Infesta 35922
Casa Museu Àngel Guimerà, El Vendrell 40259
Casa Museu Atelier Henrique Medina, Esposende 35567
Casa Museu Bissaya-Barreto, Coimbra 35526
Casa-Museu Castell Gala Dalí, La Pera 39835
Casa Museu Cristóvão Colombo, Porto Santo 35870
Casa Museu da Freira, Penacova 35815
Casa Museu da Fundação Dionísio Pinheiro e Alice Pinheiro, Águeda 35385
Casa-Museu da Fundação Medeiros e Almeida, Lisboa 35661
Casa Museu da Ordem Terceira de São Francisco, Ovar 35798
Casa-Museu d'Alfàbia, Bunyola-Mallorca 39063
Casa Museu de Aljustrel, Fátima 35600
Casa-Museu de Almeida Moreira, Viseu 36022
Casa-Museu de Camilo, Vila Nova de Famalicão 36003
Casa-Museu de Francisco Ernesto de Oliveira Martins, Angra do Heroísmo 35437
Casa Museu de Monsenhor Ribeiro de Almeida, Lamego 35655
Casa Museu de Penacova, Penacova 35816
Casa-Museu de Salvador Dalí de Cadaqués, Cadaqués 39085
Casa Museu de Sarah Beirão, Tábua 35958
Casa-Museu del Poeta Verdaguer, Folgueroles 39301
Casa Museu do Lorvão, Lorvão 35736
Casa Museu Dom Hugo, Porto 35836
Casa-Museu Dr. Anastácio Gonçalves, Lisboa 35662
Casa-Museu Egas Moniz, Estarreja 35570
Casa Museu Fernando de Castro, Porto 35837
Casa Museu Fernando Namora, Condeixa-a-Nova 35547
Casa-Museu Ferreira de Castro, Ossela 35795
Casa Museu Frederico de Freitas, Funchal 35609
Casa-Museu Gaudí, Barcelona 38953
Casa Museu Guerra Junqueiro, Porto 35838
Casa-Museu João Soares, Cortes 35552
Casa Museu José Furtado de Mendonça, Santo Estevão 35915
Casa-Museu José Régio, Portalegre 35832
Casa Museu José Régio, Vila do Conde 35987
Casa Museu Manuel Nunes Corrêa, Pedrógão Grande 35812
Casa-Museu Maria da Fontinha, Gafanhão 35624
Casa Museu Marieta Solheiro Madureira, Estarreja 35571
Casa Museu Marta Ortigão Sampaio, Porto 35839
Casa Museu Mestre João da Silva, Lisboa 35663
Casa Museu Pintor José Cercas, Aljezur 35406
Casa-Museu Prat de la Riba, Castellterçol 39151
Casa Museu Rafael Bordalo Pinheiro, Caldas da Rainha 35495
Casa Museu Roque Gameiro, Amadora 35429
Casa-Museu Sant Enric d'Ossó, Jesús 39430
Casa Museu Segrelles, Albaida 38805

Casa-Museu Teixeira Lopes, Vila Nova de Gaia 36009
Casa-Museu Torres Amat, Sallent 39947
Casa Museu Vasco Lima Couto, Constância 35549
Casa Museu Vieira Natividade, Alcobaça 35391
Casa Museu Vitorino Ribeiro, Porto 35840
Casa-Muzeu Constantin Brâncuşi, Hobiţa 36399
Casa-Muzeu Cristian, Cristian 36331
Casa-Muzeu Solca, Solca 36632
Casa Muzicii, Lugoj 36447
Casa Nacional de la Moneda, Potosí 04235
Casa Nadiua de Pau Casals, El Vendrell 40260
Casa Natal de Aquileo Parra, Barichara 08535
Casa Natal de Jesús Montané Oropesa, Nueva Gerona 09230
Casa Natal de José Antonio Ramos Sucre, Cumaná 54567
Casa Natal de José María España, La Guaira 54578
Casa Natal de Legazpi, Zumárraga 40363
Casa Natal de Regino Boti, Guantánamo 09090
Casa Natal de San Ignacio, Azpeitia 38928
Casa Natal de Sant Antoni María Claret, Sallent 39948
Casa Natal de Santa Teresa de Jesús, Ávila 38918
Casa Natal de Santiago Ramón y Cajal, Petilla de Aragón 39837
Casa Natal de Sarmiento, San Juan 00585
Casa Natal del Hermano Francisco Gárate, Azpeitia 38929
Casa Natal del Libertador Simón Bolívar, Caracas 54526
Casa Natal del Pintor Wifredo Lam, Santa Clara 09292
Casa Natale di A. Rosmini, Rovereto 27568
Casa Natale di Giosuè Carducci, Pietrasanta 27205
Casa Natale di Giuseppe Verdi, Busseto 25568
Casa Natale di Leonardo da Vinci, Vinci 28283
Casa Natale di Pascoli, San Mauro Pascoli 27668
Casa Natale di Pio V, Bosco Marengo 25515
Casa Natale di Raffaello, Urbino 28122
Casa Natale di Toscanini, Parma 27092
Casa Natale di V. Monti, Alfonsine 25196
Casa Natale Gaetano Donizetti, Bergamo 25399
Casa Navarro, San Antonio 52712
Casa-Oficina de António Carneiro, Porto 35841
Casa Pedro Domingo Murillo, La Paz 04210
Casa Personalităţilor, Lugoj 36448
Casa Quinta de José Battle y Ordóñez, Montevideo 54387
Casa Quinta de Luis Alberto de Herrera, Montevideo 54388
Casa Ramón Power y Giralt, San Juan 36060
Casa Real Shrine, Malolos 34421
Casa Rocca Piccola, Valletta 30640
Casa Roche Pombo, Morretes 04578
Casa Romei, Ferrara 26105
Casa San Isidro, Ciudad Bolívar 54562
Casa Santuario de San Pedro Claver, Verdú 40265
Casa Siglo XIX, Chihuahua 30764
Casa Tait, Porto 35842
Casa y Museo de El Greco, Toledo 40127
Casagwa National Museum, Daraga 34383
Casal Solleric, Palma de Mallorca 39778
La Casamaures, Saint-Martin-le-Vinoux 15406
Cascapedia River Museum, Cascapedia-Saint-Jules 05573
Cascina Museo della Civiltà Contadina, Cremona 26000
Casemate Museum, Fort Monroe 48658
Casey Farm, Saunderstown 52933
Casey Jones Home and Railroad Museum, Jackson 49568
Casey Jones Museum, Vaughan 53713
Cashmere Museum & Pioneer Village, Cashmere 47270
Casino dell'Aurora Pallavicini, Roma 27445
Casino Luxembourg - Forum d'Art Contemporain, Luxembourg 30476
Casita del Príncipe, Madrid 39506
Casón del Buen Retiro, Madrid 39507
Cass County Historical Society Museum, Logansport 50162
Cass County Historical Society Museum, Plattsmouth 51985
Cass County History Museum at Bonanzaville, West Fargo 54054
Cass County Museum, Walker 53795
Cass Sculpture Foundation, Goodwood 44328
Cassia County Museum, Burley 47130
Cassissium, Nuits-Saint-Georges 14484
Cassius Forum Bonn, Bonn 17451
Casteel-Linn House and Museum, Princeton 52146
Castel Beseno, Besenello 25410
Castel Prösels, Fié 26124
Castell Cartoixa Vallparadis, Terrassa 40115
Castell y Waun, Chirk 43837
Castellani Art Museum, Niagara University 51348
Castello dei Burattini - Museo Giordano Ferrari, Parma 27093
Castello dei Visconti di San Vito, Somma Lombardo 27866
Castello del Buonconsiglio, Trento 28054
Castello della Manta, Manta 26589
Castello di Costigliole d'Asti - Museo Nazionale del Territorio, Costigliole d'Asti 25995
Castello di Donnafugata, Ragusa 27358
Castello di Giulio II, Ostia Antica 27016
Castello di Masnago, Varese 28161
Castello di Montebello, Bellinzona 41190
Castello di Rivoli, Rivoli 27423
Castello di Sasso Corbaro, Bellinzona 41191
Castello di Stenico, Stenico 27901
Castello d'Issogne, Issogne 26419
Castello Ducale, Agliè 25155
Castello e Museo Civico F. Baldinucci, Lari 26461
Castello e Raccolta Archeologica, Montechiarugolo 26818
Castello Fénis, Fénis 26095
Castello Malaspina, Bobbio 25432
Castello Malaspina, Massa 26619
Castello Pallotta, Caldarola 25597
Castello Principesco, Merano 26655
Castello Reale, Racconigi 27353
Castello Reale, Venaria 28176
Castello Reggia di Venaria Reale, Torino 27998
Castelul Regal Săvârşin, Săvârşin 36592
Castillo de Javier, Javier 39424
Castillo de Loarre, Loarre 39490
Castillo de los Tres Reyes del Morro, La Habana 09097
Castillo de San Felipe del Morro, San Juan 36061
Castillo de San Marcos Museum, Saint Augustine 52520
Castillo de Santa Catalina, Cádiz 39087
Castillo de Villasobroso, Mondariz 39660
Castillos de Guayana la Vieja, Guayana 54575
Castine Scientific Society Museum, Castine 47278
The Castle, Marietta 50473
Castle Achtendries, Gent 03631
Castle Air Museum, Atwater 46464
Castle and Orthodox Church Museum, Olavinlinna 11160
Castle- and Playing Cards Museum, Altenburg 16499
Castle and Regimental Museum, Monmouth 45167
Castle Bogenšperk, Šmartno pri Litiji 38522
Castle Brand Visitor Centre, Nenagh 24882
Castle Budišov - Moravian Museum Brno, Budišov u Třebíče 09447
Castle Cary and District Museum, Castle Cary 43767
Castle Clinton, New York 51139
Castle Combe Museum, Castle Combe 43768
Castle Coole, Enniskillen 44177
Castle Cornet Military and Maritime Museums, Saint Peter Port 45589
Castle Douglas Art Gallery, Castle Douglas 43770
Castle Folk Museum, Fukuchiyama 28424
Castle Fraser, Sauchen 45619
Castle Gallery, New Rochelle 51102
Castle Grad, Grad 38419
Castle Hill National Historic Park, Placentia Bay 06609
Castle House Museum, Dunoon 44061
Castle Keep Museum, Newcastle-upon-Tyne 45225
Castle Menzies, Aberfeldy 43250
Castle Military Museum, Cape Town 38581
Castle Museum, Clitheroe 43854
Castle Museum, Colchester 43872
Castle Museum, Dunaföldvár 23529
Castle Museum, Gonder 10684
Castle Museum, Gyula 23601
Castle Museum, Kwidzyń 34840
Castle Museum, Niedzica 34943
Castle Museum, Pécsvárad 23784
Castle Museum, Pszczyna 35062
Castle Museum, Siklós 23800
Castle Museum, Simontornya 23801
Castle Museum, Szeged 23850
Castle Museum, Trakoščan 08967
Castle Museum, Veszprém 23963
Castle Museum, York 46099
Castle Museum and Art Gallery, Norwich 45276
Castle Museum in Bolków, Bolków 34569
Castle-Museum in Gołuchów, Gołuchów 34676
Castle Museum of Nagytétény, Budapest 23454
Castle Museum of Saginaw County History, Saginaw 52515
Castle Museum of the Hungarian National Museum, Esztergom 23555
Castle Museum of the Lithuania Minor History Museum, Klaipėda 30284
Castle Museum Sümeg, Sümeg 23829
Castle of Bouillon, Bouillon 03392
Castle of Hallwyl, Seengen 41856
Castle of Haut-Koenigsbourg, Orschwiller 14541
Castle of Saint John, Stranraer 45776
Castle of Valangin, Valangin 41971
Castle of Znojmo, South Moravian Museum, Znojmo 09949
Castle Park Arts Centre, Frodsham 44256
Castle Point Transport Museum, Canvey Island 43747
Castle Rajhenburg, Brestanica 38403
Castle Rushen, Castletown 43774
Castle Story, Newark-on-Trent 45209
Castle Tower of Gediminas Hill, Vilnius 30405
Castle Tucker, Wiscasset 54261
Castle Ward, Downpatrick 44019
Castle Zumberk, Žumberk 09951
Castle Zvolen of the Slovak National Gallery, Zvolen 38395
Castleford Museum Room, Castleford 43771
Castlemaine Art Gallery and Historical Museum, Castlemaine 00960
Castlemuseum Diosgyőr, Miskolc 23702
Castleton Historical Society, Castleton 47279
Castleton Village Museum, Castleton 43773
Castletown, Celbridge 24758
Castletown Museum, Gerston 44267
Častnyj Muzej Figurskoj Kultury, Almaty 29676
Častnyj Muzej Grammofonov i Fonografov, Sankt-Peterburg 37626
Castor and District Museum Society, Castor 05578
Cat Museum, Šiauliai 30358
Catacombes, Paris 14577
Catalina Island Museum, Avalon 46524
Catalyst, Widnes 46012
Catawba County Museum of History, Newton 51342
Catawba Science Center, Hickory 49266
Catawissa Railroad Museum, Catawissa 47281
Cater Museum, Billericay 43486
Cathedral Church of Saint John the Divine, New York 51140
Cathedral Museum, Angra do Heroísmo 35438
Cathedral Museum, Mdina 30630
Cathedral of the Annunciation at the Moscow Kremlin, Moskva 37142
Cathedral of the Archangel Michael at the Moscow Kremlin, Moskva 37141
Cathedral of the Dormition of the Virgin, also Cathedral of the Assumption at the Moscow Kremlin, Moskva 37410
Cathedral of the Holy Trinity, Sankt-Peterburg 37728
Cathedral Treasure Chamber, Esztergom 23552
Cathedral Treasury, Chichester 43824
Cathedral Treasury, Cathedral of Saint Duje, Split 08959
Cathedral Treasury in Dubrovnik, Dubrovnik 08873
Cathedral Treasury Museum, Carlisle 43758
Cathedral Treasury Museum, Viseu 36025
Cathédrale d'Images, Les Baux-de-Provence 13688
Catholic Historical Society of the Roanoke Valley, Roanoke 52380
Cathrinesminde Teglværk, Broager 09983
The Catlin House Museum, Scranton 52979
Catlins Historical Museum, Owaka 33167
Cats Museum, Moskva 37310
Catskill Fly Fishing Center and Museum, Livingston Manor 50152
Cattaraugus Area Historical Center, Cattaraugus 47284
Cattle Raisers Museum, Fort Worth 48712
Caudwell's Mill & Craft Centre, Matlock 45114
Causeway School Museum, Bushmills 43701
Cavalcade of Costume, Blandford Forum 43540
Cavalry Museum, Lappeenranta 11048
Cavalry Museum at the National Stud-farm of Bábolna, Bábolna 23341
Cavan County Museum, Ballyjamesduff 24740
La Cave aux Sculptures, Dénézé-sous-Doué 12713
Cave Dwelling Memorial Exhibition, Budapest 23386
Cave Museum of the Rhodopes, Čepelare 05001
Caveau Muséalisé, Cabrières 12220
Caveau-Musée Les Vignerons de Roueïre, Quarante 14963
Cavendish Laboratory, Cambridge 43726
Cavernes de l'Abîme à Couvin, Couvin 03528
Caves Muséalisées Byrrh, Thuir 15905
Les Caves Rupestres, Rochecorbon 15068
Cavite City Library and Museum, Cavite City 34361
Cawdor Castle, Cawdor 43781
Cayuga Museum, Auburn 46469
Cayuga-Owasco Lakes Historical Society Museum, Moravia 50839
CBC Museum, Toronto 07101
CBC Radio Museum, Saint John's 06845
CCA-Centre for Contemporary Arts, Glasgow 44276
CCC Weeks Gallery, Jamestown 49596
CCNOA, Bruxelles 03441
Čebelarski Muzej, Radovljica 38500
Čeboksarskij Vystavočnyj Zal, Čeboksary 36835
Cebu Archdiodesan Museum, Cebu 34367
Cebu City Museum, Cebu 34368
Cebu City State College Museum, Cebu 34369
CECA International Committee for Education and Cultural Action of Museums, Apeldoorn 54954
Cecelia Coker Bell Gallery, Hartsville 49202
Cecil Higgins Art Gallery, Bedford 43435
Cecille R. Hunt Gallery, Sankt Louis 52851
Cedar Falls Historical Museum, Cedar Falls 47291
Cedar Grove Historical Society Museum, Cedar Grove 47295
Cedar Grove Mansion, Philadelphia 51820
Cedar Grove Mansion Inn, Vicksburg 53733
Cedar Key Historical Society Museum, Cedar Key 47296
Cedar Key State Park Museum, Cedar Key 47297
Cedar Rapids Museum of Art, Cedar Rapids 47301
Cedar Ridge Creative Centre, Scarborough 06936
Cedar Valley Memories, Osage 51607
CEDIAS Musée Social, Paris 14578
Cefer Cabbarlının Ev-Muzeyi, Bakı 03151
Cefn Coed Colliery Museum, Neath 45195
Ceglédi Természetrajzi Vadászati Múzeum, Cegléd 23501
Ceiriog Memorial Institute, Glyn Ceiriog 44320
Celamuse - Porcelain Museum, Seoul 29865
Čelarevu Castle, Čelarevo 38097
Celbridge Motor Museum, Celbridge 24759
A Celebration of Immigration, London 44815
Celebration of Irish Museum, Carrickmacross 24752
Celestat, Kraków 34765
Celil Memmedquluzadenin Ev-Muzeyi, Bakı 03152
Čeljabinskaja Oblastnaja Kartinnaja Galereja, Čeljabinsk 36842
Čeljabinskij Oblastnoj Kraevedčeskij Muzej, Čeljabinsk 36843
Celler Garnison-Museum, Celle 17684
Celler Schützenmuseum, Celle 17685
Celtic Village Kulm, Kulm bei Weiz 02245
Cementmuseet Slemmestad, Slemmestad 33843
Cemetery Art Museum, Wrocław 35338
Cemetery - Museum, Sankt-Peterburg 37702
Cenacolo del Ghirlandaio, Firenze 26145
Cenacolo di Andrea del Sarto a San Salvi, Firenze 26146
Cenacolo di Sant'Apollonia, Firenze 26147
Cenacolo Vinciano, Milano 26673
Centar Savremene Umjetnosti Crne Gore, Podgorica 31654
Centar za Sovremeni Umetnosti Sarajevo, Sarajevo 04273
Centar za Sovremeni Umetnosti Skopje, Skopje 30514
The Centennial Museum at the University of Texas at El Paso, El Paso 48317
Centennial Museum of Sheguiandah, Sheguiandah 06952
Centennial Park 1910 Logging Museum, Thunder Bay 07076
Centennial Park Museum, Moose Factory 06384
Centennial Village Museum, Greeley 49032
Center Art Gallery, Athens 46409
Center Art Gallery, Grand Rapids 49001
Center for Agricultural Science and Heritage, Indianapolis 49487
Center for American Archeology, Kampsville 49673
Center for American History, Austin 46506
Center for American Music, Stephen Foster Memorial, Pittsburgh 51945
Center for Book Arts, New York 51141
Center for Contemporary Art, Kyïv 43076
Center for Contemporary Art, Rotterdam 32734
Center for Contemporary Art Kitakyushu, Kitakyushu 28711
Center for Contemporary Arts of Santa Fe, Santa Fe 52872
Center for Contemporary Graphic Art and Tyler Graphics Archive Collection, Sukagawa 29297
Center for Contemporary Non Objective Art, Bruxelles 03441
Center for Creative Photography, Tucson 53585
Center for Curatorial Studies, Annandale-on-Hudson 46314
Center for Exploratory and Perceptual Art, Buffalo 47120
Center for Figurative Painting, New York 51142
Center for Intuitive and Outsider Art, Chicago 47460
Center for Jewish History, New York 51143
Center for Maine Contemporary Art, Rockport 52425
Center for Photographic Art, Carmel 47250
Center for Photography at Woodstock, Woodstock 54283
Center for Puppetry Arts, Atlanta 46430
Center for Sacramento History, Sacramento 52504
Center for the Study of Political Graphics, Los Angeles 50216
Center for the Visual Arts, Denver 48066
Center for the Visual Arts, Denver 48067
Center for the Visual Arts, Philadelphia 51821
Center for Visual Communication, Coral Gables 47797
Center for Wooden Boats, Seattle 52995
Center Galleries, Detroit 48108
Center Gallery, Demarest 48052
Center of Nautical and Regional Archaeology, Nahshonim 25070
Center on Contemporary Art, Seattle 52996
Centerville Historical Museum, Centerville 47316
Centr Istorii Aviacionnych Dvigatelej im. N.D. Kuznecova, Samara 37611
Centr Sovremennogo Iskusstva, Toshkent 54468
Centr Sovremennogo Iskusstva Sorosa, Almaty 29677
Centraal Museum, Utrecht 32844

Central Archives for the History of the Jewish People, Jerusalem 24983
Central Art Gallery, Ashton-under-Lyne 43331
Central Australian Aviation Museum, Alice Springs 00798
Central Butte District Museum, Central Butte 05583
Central City Opera House Museum, Central City 47319
Central College Museum, Bangalore 24072
Central Exhibition Hall Manege, Sankt-Peterburg 37633
Central Exhibition Hall of the City, Perm 37544
Central Finland Aviation Museum, Tikkakoski 11376
Central Foundry Museum, Miskolc 23708
Central Hawke's Bay Settlers Museum, Waipawa 33223
Central House of Artists, Moskva 37144
Central Iowa Art Museum, Marshalltown 50511
Central Luzon State University Museum, Muñoz 34463
Central Missouri State University Archives and Museum, Warrensburg 53829
Central Montana Historical Association, Lewistown 50041
Central Museum, Bhopal 24091
Central Museum, Central 47318
Central Museum, Indore 24230
Central Museum, Nagpur 24332
Central Museum of Aviation and Cosmonautics, Moskva 37145
Central Museum of Frontier-Guard Regiments of Russia, Moskva 37147
Central Museum of Mining, Sopron 23812
Central Museum of Prisoners of War, Opole 34984
Central Museum of Railway Transport, Sankt-Peterburg 37631
Central Museum of Textiles, Łódź 34860
Central Museum of the Air Forces of the Russian Federation, Monino 37139
Central Museum of the Armed Forces, Moskva 37151
Central Museum of the Army of Uzbekistan, Toshkent 54469
Central Museum of the Border Protection Institute, Körmend 23675
Central Museum of the Internal Forces of the Ministry of Internal Affairs of Russia, Moskva 37150
Central Museum of the Ministry of Interior of Russia, Moskva 37148
Central Museum of the Russian Association of Blind People, Moskva 37152
Central Nation University Museum, Beijing 07525
Central National Herbarium, Haora 24202
Central Naval Museum, Sankt-Peterburg 37632
Central Nevada Museum, Tonopah 53531
Central New Brunswick Woodmen's Museum, Boiestown 05432
Central Sikh Museum, Amritsar 24066
Central State Museum of Kazakhstan, Almaty 29678
Central Texas Area Museum, Salado 52652
Central Texas Museum of Automotive History, Rosanky 52455
Central Washington Agricultural Museum, Union Gap 53654
Central Washington Gallery, Ellensburg 48356
Centrale Hydroélectrique du Val de Poix, Saint-Hubert 03994
Centralne Muzeum Jeńców Wojennych w Łambinowicach -Opolu, Opole 34984
Centralne Muzeum Morskie, Gdańsk 34631
Centralne Muzeum Pożarnictwa, Mysłowice 34936
Centralne Muzeum Włókiennictwa, Łódź 34860
Centralnyj Dom Chudožnikov, Moskva 37144
Centralnyj Geologorazvedočnyj Muzej im. Akademika F.N. Černyševa, Sankt-Peterburg 37627
Centralnyj Gosudarstvennyj Muzej Kazachstana, Almaty 29678
Centralnyj Muzej Aviacii i Kosmonavtiki, Moskva 37145
Centralnyj Muzej Drevnerusskoj Kultury i Isskustva im. Andreja Rubleva, Moskva 37146
Centralnyj Muzej Federalnoj Pograničnoj Služby RF, Moskva 37147
Centralnyj Muzej MVD Rossii, Moskva 37148
Centralnyj Muzej Oktjabrskoj Železnoj Dorogi, Sankt-Peterburg 37628
Centralnyj Muzej Počvovedenija im. V.V. Dokučaeva, Sankt-Peterburg 37629
Centralnyj Muzej Svjazi im. A.S. Popova, Sankt-Peterburg 37630
Centralnyj Muzej Velikoj Otečestvennoj Vojny, Moskva 37149
Centralnyj Muzej Vnutrennych Vojsk MVD Rossii, Moskva 37150
Centralnyj Muzej Voenno-Vozdušnych Sil Rossijskoj Federacii, Monino 37139
Centralnyj Muzej Vooružennych Sil, Moskva 37151
Centralnyj Muzej Vooružennych Sil Uzbekistana, Toshkent 54469
Centralnyj Muzej Vserossijskogo Obščestva Slepych, Moskva 37152
Centralnyj Muzej Železnodorožnogo Transporta, Sankt-Peterburg 37631
Centralnyj Voenno-morskoj Muzej, Sankt-Peterburg 37632
Centralnyj Vystavočnyj Zal g. Perm, Perm 37544
Centralnyj Vystavočnyj Zal Manež, Moskva 37153
Centralnyj Vystavočnyj Zal Manež g. Sankt-Peterburga, Sankt-Peterburg 37633
Centre Arqueològic de Sagunt, Sagunto 39931
Centre Artistique et Médiéval, Richerenches 15036
Centre Artistique Jean Baffier, Sancoins 15622
Centre Atlantique de la Photographie, Brest 12156
Centre Belge de la Bande Dessinée, Bruxelles 03442
Centre Commémoratif de l'Holocauste à Montréal, Montréal 06331
Centre County Historical Society Museum, State College 53259
Centre County Library Historical Museum, Bellefonte 46698
Centre Cultural de la Fundació La Caixa, Vic 40267
Centre Culturel de la Memoire Combattante, Carcassonne 12281
Centre Culturel de Lachute, Lachute 06124
Centre Culturel des Roches, Rochefort 03972
Centre Culturel et d'Interprétation de Havre Saint-Pierre, Havre-Saint-Pierre 05977
Centre Culturel et Patrimonial la Poudrière de Windsor, Windsor, Québec 07348
Centre Culturel Franco-Manitobain, Winnipeg 07356
Centre Culturel Ipousteguy, Doulcon 12781
Centre Culturel Marguerite Bourgeoys, Troyes 15995
Centre Culturel - Musée Montaut, Saint-Sever (Landes) 15513
Centre Culturel Ti Yarihuten, Wendake 07298
Centre d'Animation Permanente du Rail et de la Pierre, Jemelle 03744
Centre d'Archéologique et Historique, Neuville-aux-Bois 14408
Centre d'Art, Baie-Saint-Paul 05372
Centre d'Art, Port-au-Prince 23320
Centre d'Art Contemporain, Alençon 11543
Centre d'Art Contemporain, Bignan 11993
Centre d'Art Contemporain, Bruxelles 03443
Centre d'Art Contemporain, Dijon 12731
Centre d'Art Contemporain, Istres 13253
Centre d'Art Contemporain, Lacoux-Hauteville-Lompnes 13472
Centre d'Art Contemporain, Marne-la-Vallée 13977
Centre d'Art Contemporain, Meymac 14109
Centre d'Art Contemporain, Mont-de-Marsan 14157
Centre d'Art Contemporain, Pougues-les-Eaux 14928
Centre d'Art Contemporain, Revin 15032
Centre d'Art Contemporain, Rueil-Malmaison 15137
Centre d'Art Contemporain, Tanlay 15851
Centre d'Art Contemporain Bouvet-Ladubay, Saumur 15660
Centre d'Art Contemporain de Basse-Normandie, Hérouville-Saint-Clair 13207
Centre d'Art Contemporain de Brétigny, Brétigny-sur-Orge 12167
Centre d'Art Contemporain de Castres, Castres 12317
Centre d'Art Contemporain Genève, Genève 41404
Centre d'art contemporain Genève, Genève 41405
Centre d'Art Contemporain Georges Pompidou, Cajarc 12244
Centre d'Art Contemporain la Synagogue de Delme, Delme 12705
Centre d'Art Contemporain le Quartier, Quimper 14966
Centre d'Art de la Sarre, La Sarre 06115
Centre d'Art de Saint-Georges, Saint-Georges, Quebec 06819
Centre d'Art du Crestet, Crestet 12674
Centre d'Art et de Culture, Soignies 04034
Centre d'Art et de Plaisanterie l'Allan, Montbéliard 14182
Centre d'Art et d'Exposition Jacques-Henri Lartigue, L'Isle-Adam 13780
Centre d'Art et d'Histoire André Auclair, Cruas 12681
Centre d'Art Graphique, Parly 14751
Centre d'Art Neuchâtel, Neuchâtel 41677
Centre d'Art Nicolas de Staël, Braine-l'Alleud 03394
Centre d'Art Sacré, Ille-sur-Têt 13244
Centre d'Art Santa Mònica, Barcelona 38954
Centre d'Art Sébastien, Saint-Cyr-sur-Mer 15229
Centre d'Artistes Vaste et Vague, Carleton 05563
Centre de Création Contemporaine, Tours 15961
Centre de Cultura Contemporània de Barcelona, Barcelona 38955
Centre de Cultura Sa Nostra, Palma de Mallorca 39779
Centre de Culture Catalane, Perpignan 14787
Centre de Culture Scientifique la Rotonde Ecole des Mines, Saint-Étienne 15249
Centre de Découvert du Marais Breton, Le Barre-de-Monts 13555
Centre de Découverte de la Nature et du Patrimoine Cynégétique, Sault (Vaucluse) 15658
Centre de Découverte Martime, Penmarch 14772
Centre de Design de l'UQAM, Montréal 06332
Centre de Documentació i Museu Tèxtil, Terrassa 40116
Centre de la Gravure et de l'Image Imprimée de la Communauté Française de Belgique, La Louvière 03766
Centre de la Marionnette de la Communauté Française, Tournai 04081
Centre de la Mémoire d'Oradour, Oradour-sur-Glane 14515
Centre de la Mer, Paris 14579
Centre de la Nature Montagnarde, Sallanches 15600
Centre de la Photographie Genève, Genève 41406
Centre de la Vieille Charité, Marseille 13989
Centre de Natura de la Cortinada, Ordino 00098
Centre de Paléontologie et Archéologie, Valréas 16062
Centre de Patrimoine de Facture Instrumentale, Le Mans 13605
Centre de Photographie de Lectoure, Lectoure 13671
Centre de Recherches et d'Etudes Technologiques des Arts Plastiques, Bruxelles 03444
Centre de Recherches Historiques sur les Maîtres Ebénistes, Paris 14580
Centre de Secours Principal, Tourcoing 15950
Centre d'Edition Contemporaine, Genève 41407
Centre Départmental d'Art Contemporain Cimaise et Portique, Albi 11536
Centre des Dossiers, Québec 06677
Centre d'Étude du Masque et Mouvement, Walhain 04128
Centre d'Études Archéologiques, Eyguières 12892
Centre d'Études Edmond Michelet, Brive-la-Gaillarde 12188
Centre d'Études Historiques et Archéologiques, Le Barroux 13556
Centre d'Exposition Archéologique de l'Orangerie, Étiolles 12878
Centre d'Exposition Art-Image, Gatineau 05856
Centre d'Exposition d'Amos, Amos 05337
Centre d'Exposition de la Bibliothèque Gabrielle Roy, Québec 06678
Centre d'Exposition de Rouyon-Noranda, Rouyn-Noranda 06780
Centre d'Exposition de Saint-Hyacinthe, Saint-Hyacinthe 06820
Centre d'Exposition de Val d'Or, Val-d'Or 07193
Centre d'Exposition des Gouverneurs, Sorel 06992
Centre d'Exposition du Costume, Avallon 11782
Centre d'Exposition l'Imagier, Gatineau 05857
Centre d'Exposition Mont-Laurier, Mont-Laurier 06321
Centre d'Exposition Plein Sud, Longueuil 06192
Centre d'Exposition Raymond-Lasnier, Trois-Rivières 07170
Centre d'Exposition sur l'Industrie des Pâtes et Papiers, Trois-Rivières 07171
Centre d'Hébergement, d'Étude sur la Nature et l'Environnement, Allouville-Bellefosse 11558
Centre d'Histoire de la Résistance et de la Déportation, Lyon 13880
Centre d'Histoire de Montréal, Montréal 06333
Centre d'Histoire Locale, Tourcoing 15951
Centre d'Historie et d'Archéologie de la Métabetchouane, Desbiens 05697
Centre d'Iconographie Genevoise, Genève 41408
Centre Didactique de l'Eau, Seneffe 04007
Centre d'Interpretació del Castell de Gilida, Gelida 39330
Centre d'Interpretació p.n. d'Aigüestortes i Estany de Sant Maurici, Espot 39283
Centre d'Interprétation Archéo Topo, Bergeronnes 05421
Centre d'Interprétation de l'Agriculture et de la Ruralité, Métabetchouan-Lac-à-la-Croix 06270
Centre d'Interpretation de l'Ardoise, Richmond 06742
Centre d'Interprétation de l'Histoire de Sherbrooke, Sherbrooke, Québec 06958
Centre d'Interprétation de l'Outil, Braine-le-Château 03397
Centre d'Interprétation des Luys, Brassempouy 12148
Centre d'Interprétation des Mammifères Marins (CIMM), Tadoussac 07055
Centre d'Interprétation des Mines Minerais, Petit-Rocher 06597
Centre d'Interprétation du Bourg de Pabos, Pabos Mills 06566
Centre d'Interprétation du Moulin Bernier, Courcelles 05649
Centre d'Interprétation du Parc de l'Île-Bonaventure-et-du Rocher-Percé, Percé 06585
Centre d'Interprétation du Patrimoine de Plaisance, Plaisance 06610
Centre d'Interprétation du Patrimoine de Sorel, Sorel 06993
Centre d'interprétation du Phoque, Grande-Entrée 05899
Centre d'Interprétation et de la Recherche Philatélique du Canada, Terrebonne 07067
Centre d'Interprétation Faunique du Cap Tourmente, Saint-Joachim 06832
Centre du Livre d'Artiste Contemporain, Verderonne 16110
Centre Dürrenmatt, Neuchâtel 41678
Centre Européen d'Actions Artistiques Contemporaines, Strasbourg 15810
Centre Européen de la Paix, Souchez 15790
Centre Européen de l'Orgue, Marmoutier 13973
Centre Européen des Métiers d'Art, Givet 13071
Centre Européen du Résistant Déporté, Natzwiller 14382
Centre européen du résistant déporté, Natzwiller 14383
Centre for African Studies Collection, Rondebosch 38725
Centre for Art Tapes, Halifax 05932
Centre for Black and African Art and Civilization, Lagos 33333
Centre for Contemporary Art, Warszawa 35220
Centre for Contemporary Photography, Fitzroy 01080
Centre for Fine Art, Bruxelles 03498
Centre for Fine Arts, Bruxelles 03499
Centre for Rocks and Minerals, Jokkmokk 40619
Centre for the Magic Arts, London 44816
Centre for the Study of Greek Traditional Pottery, Athinai 22905
Centre for Visual Arts, Groningen 32279
Centre Franco-Ontarien de Folklore, Sudbury 07027
Centre Hill Museum, Petersburg 51797
Centre Historique Médiéval, Azincourt 11814
Centre Historique de la Résistance en Drôme et de la Déportation, Romans-sur-Isère 15087
Centre historique des Soeurs de Sainte-Anne, Lachine 06122
Centre Historique Minier, Lewarde 13735
Centre International d'Art Contemporain - CIAC, Port Aven 14912
Centre International d'Art Contemporain de Montréal, Montréal 06334
Centre International d'Art et du Paysage, Beaumont-du-Lac 11899
Centre International d'Art-Raymond du Puy, Le Poët-Laval 13636
Centre International du Vitrail, Chartres 12421
Centre Jean Giono, Manosque 13940
Centre Jean Rostand, Roquefort 15097
Centre Jeanne d'Arc, Orléans 14530
Centre Juno Beach, Courseulles-sur-Mer 12653
Centre Minier de Faymoreau, Faymoreau-les-Mines 12912
Centre Missionnaire Sainte-Therèse Musée, Montréal 06335
Centre National d'Artisanat d'Art, Ouagadougou 05277
Centre National de la Céramique d'Art de Sidi Kacem Jelizi, Tunis 42662
Centre National d'Exposition, Jonquière 06031
Centre National et Musée Jean Jaurès, Castres 12318
Centre of Contemporary Art, Christchurch 33070
Centre of Contemporary Art, Praha 09739
Centre of Polish Sculpture, Orońsko 34989
Centre Philatélique de Polynésie Française, Papeete 16323
Centre Photographique d'Ile-de-France, Pontault-Combault 14900
Centre Pierre-Paul Riquet, Avignonet-Lauragais 11798
Centre Pompidou Metz, Metz 14099
Centre Provincial d'Hébergement Le Caillou, Roisin 03980
Centre Régional d'Art Contemporain, Fontenoy-en-Puisaye 12972
Centre Régional d'Art Contemporain, Montbéliard 14183
Centre Régional d'Art Contemporain, Sète 15733
Centre Régional de la Photographie Nord-Pas-de-Calais, Douchy-les-Mines 12778
Centre Rhenan d'Art Contemporain, Altkirch 11561
Centre Social i Cultural de la Fundació La Caixa, Lleida 39472
Centre Social i Cultural de la Fundació La Caixa, Tarragona 40099
Centre Thématique Fossilifère, Notre-Dame-du-Nord 06492
Centre Touristique de la Laine & de la Mode, Verviers 04106
Centro Agost, Museo de Alfarería, Agost 38784
Centro Ambientale Archeologico, Legnano 26481
Centro Anadluz de Flamenco, Jerez de la Frontera 39425
Centro Andaluz de Arte Contemporáneo, Sevilla 40061
Centro Andaluz de la Fotografía, Almería 38856
Centro Andino de Tecnología Tradicional y Cultura de la Comunidades, Ollantaytambo 34282

Centro Atlántico de Arte Moderno, Las Palmas de Gran Canaria 39797
Centro Camuno di Studi Preistorici, Capo di Ponte 25657
Centro Ceremonial Indígena de Caguana, Utuado 36079
Centro Cultural Banco do Brasil, Rio de Janeiro 04686
Centro Cultural Conde Duque, Madrid 39508
Centro Cultural Corp Group, Caracas 54527
Centro Cultural de la Raza, San Diego 52737
Centro Cultural Eduardo León Jimenes, Santiago de los Caballeros 10338
Centro Cultural España Córdoba, Córdoba 00317
Centro Cultural Isidro Fabela, México 31037
Centro Cultural Paraguayo-Americano, Asunción 34088
Centro Cultural Paraguayo-Argentino, Concepción 34102
Centro Culturale Polivalente, Bagnacavallo 25342
Centro d'Arte Contemporanea Ticino, Bellinzona 41192
Centro d'Arte Verrocchio, Casole d'Elsa 25708
Centro de Arqueologia Annete Laminge Emperaire, Lagoa Santa 04527
Centro de Arte, Almería 38857
Centro de Arte, León 33282
Centro de Arte Caja de Burgos, Burgos 39065
Centro de Arte Contemporáneo de Málaga, Málaga 39594
Centro de Arte Contemporáneo Fundación Chirivella Soriano, Valencia 40198
Centro de Arte Contemporáneo la Fábrica, Abarca de Campos 38781
Centro de Arte Interpretación de las Hoces del Rio Duratón, Sepúlveda 40058
Centro de Arte José Guerrero de la Diputación de Granada, Granada 39360
Centro de Arte La Cañuela, Caracas 54528
Centro de Arte La Estancia, Caracas 54529
Centro de Arte Moderna José de Azevedo Perdigão, Lisboa 35664
Centro de Arte Moderno y Contemporáneo Daniel Vázquez Díaz, Nerva 39715
Centro de Arte Visuales, Lima 34211
Centro de Arte y Naturaleza. Fundación Beulas, Huesca 39397
Centro de Artes Tradicionais, Évora 35583
Centro de Artes Visuales, Asunción 34089
Centro de Artistas Plásticos, Montevideo 54389
Centro de Bellas Artes, Maracaibo 54581
Centro de Biologia Marinha, São Sebastião 04912
Centro de Ciencia y Tecnología, Guadalajara 30887
Centro de Desarrollo de las Artes Visuales, La Habana 09098
Centro de Diseño Ambiental, La Habana 09099
Centro de Documentación Musical, Linares 39462
Centro de Educação Ambiental Gralha Azul e Museu de História Natural, Cascavel 04406
Centro de Estudios Conservacionistas, Guatemala 23296
Centro de Estudios Folklóricos, Guatemala 23297
Centro de Estudos Murilo Mendes, Juiz de Fora 04519
Centro de Exhibiciones Marinas, Calzada de Amador 34061
Centro de Exposição Rural da Urzelina, Urzelina 35976
Centro de Exposiciónes de Geologia y Mineria, Buenos Aires 00152
Centro de Exposiciones La Casona de los Olivera, Buenos Aires 00153
Centro de Exposiciones-Ses Voltes Parc de la Mar, Palma de Mallorca 39780
Centro de Exposições, Arte e Multimédia, Sintra 35944
Centro de Fotografía, Santa Cruz de Tenerife 39993
Centro de Historia de Zaragoza, Zaragoza 40338
Centro de Información e Interpretación del Parque de Pagoeta, Aia 38796
Centro de Interpretação da Batalha de Aljubarrota, Porto de Mós 35868
Centro de Interpretación Baños de Montemayor, Baños de Montemayor 38944
Centro de Interpretación del Parque en San Martín, San Martín de Castañeda 39962
Centro de Investigaciones Culturales-Museo, Mexicali 31031
Centro de Museos de Buenos Aires, Buenos Aires 54682
Centro de Pesquisas Folclóricas, Rio de Janeiro 04687
Centro de Pesquisas Paleontológicas da Bacia do Araripe, Crato 04420
Centro del Diseño, Bilbao 39041
Centro di Documentazione delle Bonifiche Romane, Granze 26363
Centro di Documentazione Messapica, Oria 26989
Centro di Informazione e Documentazione Arti Visive, Prato 27330
Centro di Scienze Naturali e Raccolta Natura ed Arte, Prato 27331
Centro di Scultura Contemporanea, Cagli 25575
Centro Escolar University Museum and Archives, Manila 34437
Centro Etnográfico Joaquín Díaz, Urueña 40183
Centro Ferroviário de Cultura, Volta Grande 04960
Centro Fotográfico Álvarez Bravo, Oaxaca 31219
Centro Galego de Arte Contemporáneo, Santiago de Compostela 40014
Centro Histórico-Cultural de Enfermagem Ibero-americana, São Paulo 04850
Centro Maggiori, Monte San Giusto 26803
Centro Mundo Aborigen, Córdoba 00318
Centro Municipal de Exposiciones, Elche 39263
Centro Municipal de Exposiciones Subste, Montevideo 54390
Centro Museo Universitario di Storia Naturale e della Strumentazione Scientifica, Modena 26739
Centro Nacional de Artes, San Salvador 10548
Centro Nacional de las Artes, México 31038
Centro Ordóñez Falcón de Fotografía, Donostia-San Sebastián 39248
Centro Parque Nacional Picos de Europa - Casa Dago, Cangas de Onis 39122
Centro per l'Arte Contemporanea Luigi Pecci, Prato 27332
Centro Pró-Memória do Club Athlético Paulistano, São Paulo 04851
Centro Pró-Memória Hans Nobiling do Esporte Clube Pinheiros, São Paulo 04852
Centro Regional de Arqueologia Ambiental, Piraju 04631
Centro Regional de Pesquisas Archeológicas Mário Neme, Piraju 04632
Centro Studi I. Silone, Pescina 27175
Centro Studi Naturalistici del Pollino Il Nibbio, Morano Calabro 26866
Centro Visitantes Fabrica de Hielo, Sanlúcar de Barrameda 39974
Centrul de Cultură Palatele Brâncoveneşti, Mogoşoaia 36479
Centrum Arts and Creative Education, Port Townsend 52042
Centrum Beeldende Kunst, Groningen 32279
Centrum Beeldende Kunst Emmen, Emmen 32204
Centrum Beeldende Kunst Utrecht, Amersfoort 31810
Centrum Bohuslava Martinů v Poličce, Polička 09727
Centrum för Arbetarhistoria, Landskrona 40667
Centrum Hasičského Hnuti, Přibyslav 09807
Centrum Kultury Dwór Artusa, Toruń 35185
Centrum Kultury i Sztuki Afrykańskiej, Kraków 34766
Centrum Kultury Żydowskiej, Kraków 34767
Centrum Kunstlicht in de Kunst, Eindhoven 32184
Centrum Pamięci Gen. Józefa Hallera, Władysławowo 35305
Centrum Rysunku i Grafiki im. Tadeuza Kulisiewicza, Kalisz 34707
Centrum Rzeźby Polskiej, Orońsko 34989
Centrum Scenografii Polskiej - Galerija Anrzeja Majewskiego, Katowice 34719
Centrum Sztuki Studio im S.I. Witkiewicza, Warszawa 35219
Centrum Sztuki Współczesnej, Warszawa 35220
Centrum Sztuki Współczesnej Łaźnia, Gdańsk 34632
Centrum voor de Vlaamse Kunst van de 16de en de 17de Eeuw, Antwerpen 03283
Centrum voor Heiligenbeelden, Kranenburg 32459
Centrum voor Natuur en Landschap, Terschelling-West 32817
Centura House Historical Society, Rosendale 52461
Century Village Museum, Burton 47147
Ceol - Irish Traditional Music Centre, Dublin 24780
Ceramic Art Center, Mashiko 28884
Ceramic Centre and Pottery Museum, Tegelen 32814
Ceramic Museum, Béziers 11975
Ceramic Museum, İstanbul 42811
Ceramics Gallery, Aberystwyth 43257
Ceramics Museum, Holič 38259
Ceramics Museum, Kütahya 42893
Ceramics Museum, Tel Aviv 25103
Ceramics Museum, Týnec nad Sázavou 09886
Ceramics Museum of Brone and Bronius Radeckas, Kuršėnai 30294
Ceramics Museum of the M.K. Čiurlionis National Art Museum, Kaunas 30253
Čerdynskij Kraevedčeskij Muzej A.S. Puškina, Čerdyn 36846
Cereal Prairie Pioneer Museum, Cereal 05584
Ceredigion Museum, Aberystwyth 43258
Čerkas'kyj Chudožnij Muzej, Čerkasy 43001
Čerkas'kyj Oblasnyj Krajeznavčyj Muzej, Čerkasy 43002
Čerkechskij Muzej Jakutskaja Političeskaja Ssylka, Čerkech 36853
Cerkljanski Muzej, Cerkno 38408
Cerkno Museum, Cerkno 38408
Cerkov Ionna Predteči, Jaroslavl 36965
Cerkov Pokrova v Filjach, Moskva 37154
Cerkov Položenija riz Presvjatoj Bogorodicy, Moskva 37155
Cerkov Roždestva Christova, Jaroslavl 36966
Cerkov Troicy v Nikitnikach, Moskva 37156
CERN - Microcosm, Genève 41409
Cernan Earth and Space Center, River Grove 52364
Černigiv Mykhailo Kotsiubynsky Literary Museum, Černigiv 43005
Černigiv Regional Museum of Art, Černigiv 43004
Černigivs'kyj Oblasnyj Chudožnij Muzej, Černigiv 43004
Černigivs'kyj Literaturno-memorial'nyj Muzej-zapovidnik Mychajlo Kocjubynskogo, Černigiv 43005
Černivci Local Museum, Černivci 43010
Černivci Museum of Art, Černivci 43009
Černivci Museum of Folk Architecture and Folklore, Černivci 43014
Černivci Museum of the Bukovina Diaspora, Černivci 43013
Černivec'kyj Chudožnij Muzej, Černivci 43009
Černivec'kyj Krajeznavčyj Muzej, Černivci 43010
Černivec'kyj Literaturno-Memorial'nyj Muzej im. Ju.A. Fedkoviča, Černivci 43011
Černivec'kyj Literaturno-Memorial'nyj Muzej O. Kobyljans'koï, Černivci 43012
Černivec'kyj Muzej Bukovins'koï Diaspory, Černivci 43013
Černivec'kyj Muzej Narodnoï Architektury ta Pobutu, Černivci 43014
Cerny Inuit Collection, Bern 41204
Cēsis Exhibition Hall, Cēsis 29996
Cēsis History Museum, Cēsis 29997
České Muzeum Hudby - Oddělení Hudebních Nástrojů a Hudebně Historické Oddělení, Praha 09736
České Muzeum Hudby - Zámek Hořovice, Hořovice 09527
České muzeum stříbra - Hrádek, Kutná Hora 09612
České Muzeum Výtvarných Umění, Praha 09737
Çeşme Müzesi, Çeşme 42752
Cēsu Izstāžu Nams, Cēsis 29996
Cēsu Vēstures un Mākslas Muzejs, Cēsis 29997
Cetatea Braşov şi Fortificaţiile din Ţara Bârsei, Braşov 36175
Cetatea de Scaun a Sucevei, Suceava 36641
Cetatea Devei, Deva 36343
Cetatea Neamţului, Târgu Neamţ 36675
Cetatea Poienari, Căpăţânenii Pământeni 36258
Cetatea Sucidava, Corabia 36310
Cetatea Ţărănească Râşnov, Râşnov 36556
Cetinje Monastery Museum, Cetinje 31642
CFB Esquimalt Naval Museum and Military Museum, Victoria 07244
CFB Petawawa Military Museum, Petawawa 06588
CFCC Webber Center, Ocala 51481
Ch. Frenkelio Rūmai, Šiauliai 30353
Ch. Frenkel's Mansion of the Šiauliai Aušra Museum, Šiauliai 30353
Chaa Creek Natural History Centre, San Ignacio 04181
Chabarovskij Gosudarstvennyj Muzej Dalnego Vostoka im. N.I. Grodekova, Chabarovsk 36856
Chabarovskij Muzej Archeologii im. A.P. Okladnikova, Chabarovsk 36857
Chabot Museum, Rotterdam 32703
Chácara do Céu Museum, Rio de Janeiro 04688
Chaco Culture National Historical Park, Nageezi 50939
Chadron State College Main Gallery, Chadron 47324
Chaffee Art Center, Rutland 52493
Chagall House, Haifa 24956
Chagall Museum, Vitebsk 03257
Chagrin Falls Historical Society Museum, Chagrin Falls 47328
Chai-Yi 2-28 Memorial Park Museum, Chai-Yi 42150
Chai-Yi Museum of Stone and Art, Chai-Yi 42151
The Chaim Gross Studio Museum, New York 51144
Chaini-zu Toji Bijutsukan, Usuki 29565
Chaiya National Museum, Surat Thani 42606
Chakasskij Respublikanskij Kraevedčeskij Muzej, Abakan 36745
Chakdara Museum, Chakdara 33999
Chalet of the Golden Fleece, New Glarus 51044
Chamberlain Museum of Pathology, Birmingham 43503
Chamizal National Memorial, El Paso 48318
Champaign County Historical Museum, Champaign 47333
Champaign County Historical Museum, Urbana 53670
Champasak Historical Heritage Museum, Vientiane 29972
Champignonnière-Musée du Saut aux Loups, Montsoreau 14266
Champion Mill, Champion 47337
Champlain Trail Museum and Pioneer Village, Pembroke 06580
Champs Chapel Museum, East Hendred 44086
Chamsori Gramophone and Audio Science Museum, Gangneung 29806
Chancellor Robert R. Livingston Masonic Library and Museum, New York 51145
Chandler Gallery, Randolph 52245
Chandler Museum, Chandler 47338
Chandradhari Museum, Darbhanga 24125
Chang Da-Chien's Residence, Taipei 42377
Chang Foundation Museum, Taipei 42378
Chang Gung University Arts Center, Gueishan 42195
Changdao Museum, Changdao 07590
Changdao Voyage Museum, Changdao 07591
Changji Mulei County Museum, Changji 07592
Changjizhou Museum, Changji 07593
Changning District Revolutionary Display Center, Shanghai 08157
Changqing Museum, Changqing 07595
Changsha Museum, Changsha 07596
Changshu Beike Museum, Changshu 07599
Changshu Museum, Changshu 07600
Changtai Museum, Changtai 07601
Changting Museum, Changting 07602
Changwu Museum, Changwu 07603
Changxing Museum, Changxing 07604
Changyangtu Sept Self-Rule Museum, Yidu 08444
Changzhi Museum, Changzhi 07605
Changzhou Museum, Changzhou 07606
Chania Byzantine and Postbyzantine Collection, Chania 22962
Channel Folk Museum, Lower Snug 01243
Le Channel - Galerie de l'Ancienne Poste, Calais 12246
Channel Islands Military Museum, Saint Ouen 45587
Chantharakasem National Museum, Ayutthaya 42526
Chantier A.C. Davie, Lévis 06166
Chantilly Museum, Geraardsbergen 03649
Chanty-Mansijsk Regional Museum of Nature and Man, Chanty-Mansijsk 36862
Chaosamphraya National Museum, Ayutthaya 42527
Chaotiangong Palace, Nanjing 08038
Chaoyang Museum, Chaoyang 07608
Chaoyang Museum, Chaoyang, Liaoning 07610
Chaozhou Museum, Chaozhou 07611
La Chapeauthèque, Anduze 11598
Chapel Gallery, Bracebridge 05452
Chapel Gallery, North Battleford 06466
Chapel Hill Museum, Chapel Hill 47345
Chapel Hill Museum, Shag Harbour 06946
Chapel of the Holy Spirit of Seamen's Museologic Site, Sesimbra 35933
Chapelle de Bailly, Saint-Bris-le-Vineux 15192
Chapelle de Capucins, Aigues-Mortes 11500
Chapelle de Pénitents Blanc, Aigues-Mortes 11501
Chapelle des Indiens, Tadoussac 07056
Chapelle des Visitandines, Auxerre 11773
Chapelle Matisse, Vence 16095
Chapelle Saint-Calixte - Musée du Château Comtal, Mons 03880
Chapelle Sainte-Radegonde, Chinon 12513
Chapin Library of Rare Books, Williamstown 54174
Chapleau Centennial Museum, Chapleau 05586
Chaplin-Archiv, Frankfurt am Main 18358
Chapman Art Center Gallery, Cazenovia 47286
Chapman Gallery, Salford 45598
Chapman Historical Museum, Glens Falls 48934
Chapman Museum, Brandon 05461
Chappell Art Gallery, Chappell 47347
Chappell Hill Historical Society Museum, Chappell Hill 47348
Chard and District Museum, Chard 43790
Charkivs'kyj Chudožnij Muzej, Charkiv 43017
Charkivs'kyj Istoryčnyj Muzej, Charkiv 43018
Charkivs'kyj Literaturnyj Muzej, Charkiv 43019
Charlbury Museum, Charlbury 43793
Charles A. Grignon Mansion, Kaukauna 49711
Charles A. Lindbergh House, Little Falls 50133
Charles A. Weyerhaeuser Memorial Museum, Little Falls 50134
Charles A. Wustum Museum of Fine Arts, Racine 52222
Charles Allis Art Museum, Milwaukee 50706
Charles B. Aycock Birthplace, Fremont 48786
Charles B. Goddard Center for Visual and Performing Arts, Ardmore 46343
Charles Carroll House of Annapolis, Annapolis 46316
Charles Dickens Birthplace Museum, Portsmouth 45427
The Charles Dickens Museum, London 44817
Charles Gates Dawes House, Evanston 48446
Charles H. MacNider Museum, Mason City 50529
Charles H. Scott Gallery, Vancouver 07207
Charles H. Taylor Arts Center, Hampton 49150
Charles H. Wright Museum of African American History, Detroit 48109
The Charles Hosmer Morse Museum of American Art, Winter Park 54253
Charles J. Beldon Museum of Western Photography, Meeteetse 50562
Charles M. Schulz Museum and Research Center, Santa Rosa 52904

Charles R. Conner Museum, Pullman 52197
Charles River Museum of Industry, Waltham 53808
Charles Sturt Museum, Grange 01129
Charles Sturt University Art Collection, Wagga Wagga 01619
Charles Towne Landing 1670, Charleston 47358
Charleston Heights Art Center Gallery, Las Vegas 49965
The Charleston Museum, Charleston 47359
Charleston Trust, Firle 44229
Charleville Historic House Museum, Charlesville 00963
Charlie's Arthurs Seat Auto Museum, Arthurs Seat 00816
Charlotte and Philip Hanes Art Gallery, Winston-Salem 54243
Charlotte Berlins Museum, Ystad 41075
Charlotte County Historical Center, Charlotte Harbor 47394
Charlotte County Museum, Saint Stephen 06878
Charlotte Museum of History and Hezekiah Alexander Homesite, Charlotte 47386
Charlotte Nature Museum, Charlotte 47387
Charney Bassett Mill, Wantage 45928
Charnley-Persky House Museum, Chicago 47461
Charnwood Museum, Loughborough 45031
Chartered Insurance Institute's Museum, London 44818
Charterhouse School Museum, Godalming 44321
Chartwell, Westerham 45977
Chase and District Museum, Chase 05599
Chase County Historical Society, Cottonwood Falls 47844
Chase Home Museum of Utah Folk Art, Salt Lake City 52688
The Chase Manhattan Bank Art Collections, New York 51146
Châsse Sainte-Bernadette et Musée, Nevers 14415
The Chassidic Art Institute, Brooklyn 47070
Château, Arlay 11681
Château, Assas 11714
Château, Duilhac-sous-Peyrepertuse 12794
Château, Le Bourg-Saint-Léonard 13562
Château, Prudhomat 14952
Château, Rousson 15129
Château, La Sarraz 41822
Château, Vogue 16265
Château à Motte de la Haie Joulan, Saint-Sylvain-d'Anjou 15519
Château Beaufief, Mazeray 14049
Château d'Abbadie, Hendaye 13200
Château d'Aigle, Aigle 41093
Château d'Aigremont, Seraing 04009
Château d'Angers, Angers 11602
Château d'Arcelot, Arceau 11663
Château d'Assier, Assier 11715
Château d'Aubenas, Aubenas 11723
Château d'Auvers, Auvers-sur-Oise 11768
Château d'Avignon, Saintes-Maries-de-la-Mer 15586
Château d'Azay-le-Ferron, Azay-le-Ferron 11808
Château de Beloeil, Beloeil 03358
Château de Belvoir, Belvoir 11937
Château de Biron, Monpazier 14151
Château de Bonneval, Coussac-Bonneval 12660
Château de Bouges, Bouges-le-Château 12084
Château de Bourdeilles, Bourdeilles 12104
Château de Boury-en-Vexin, Boury-en-Vexin 12128
Château de Bruniquel Semateg, Bruniquel 12200
Château de Busca-Naniban, Mansencôme 13943
Château de Buzay, La Jarne 13387
Château de Cadillac, Cadillac 12224
Château de Castelnou, Castelnou 12316
Château de Cazeneuve, Préchac 14942
Château de Cénévières, Cénévières 12336
Château de Champchevrier, Cléré-les-Pins 12545
Château de Châteaudun, Châteaudun 12443
Château de Chevannes-Saint Racho, La Clayette 13345
Château de Chillon, Montreux 41637
Château de Coppet, Coppet 41313
Château de Courtanvaux, Bessé-sur-Braye 11965
Le Château de Craon, Craon 12667
Château de Crazannes, Crazannes 12669
Château de Deulin, Fronville 03610
Château de Fénelon, Sainte-Mondane 15576
Château de Fernelmont, Fernelmont 03598
Château de Fleurac, Nersac 14395
Château de Fleurigny, Thorigny-sur-Oreuse 15901
Château de Fléville, Fléville-devant-Nancy 12946
Château de Fougères-sur-Bièvre, Fougères-sur-Bièvre 12988
Château de Frolois, Frolois 13014
Château de Galard, Terraube 15872
Château de Germolles, Mellecey 14066
Château de Grandson, Grandson 41450
Château de Gratot, Gratot 13118
Château de Gruyères, Gruyères 41460
Château de Gy, Gy 13173
Château de Hautefort, Hautefort 13191
Château de Jannée, Pessoux 03950
Château de Kergrist, Ploubezre 14849
Château de la Bastie d'Urfé, Saint-Étienne-le-Molard 15260
Château de la Baume, Prinsuéjols-Marvejols 14948
Château de la Guerche, La Guerche 13382
Château de la Motte-Tilly, La Motte-Tilly 13396
Château de la Roche-Guyon, La Roche-Guyon 13415
Château de la Souche, Doyet 12787
Château de Lacypierre, Saint-Crépin-Carlucet 15222
Château de l'Alba-la-Romaine, Alba-la-Romaine 11529
Château de l'Angotière, Domjean 12759
Château de Larroque-Toirac, Larroque-Toirac 13531
Château de Léhélec, Béganne 11918
Chateau de l'Ermitage, Wavre 04140
Château de l'Hermitage, Condé-sur-l'Escaut 12608
Château de Lichtenberg, Lichtenberg 13742
Château de Losse, Thonac 15889
Château de Lourmarin, Lourmarin 13855
Château de Louvignies, Notre-Dame-de-Louvignies 03918
Château de Luynes, Luynes 13875
Château de Maintenon, Maintenon 13922
Château de Maisons-Laffitte, Maisons-Laffitte 13925
Château de Meillant, Meillant 14062
Château de Messilhac, Raulhac 14988
Château de Meux, Jonzac 13276
Château de Mogère, Montpellier 14237
Château de Montaigut, Camares 12250
Château de Montaner, Montaner 14168
Château de Montcornet, Montcornet-en-Ardenne 14193
Château de Monte-Cristo, Le Port-Marly 13641
Château de Montignies-sur-Roc, Montignies-sur-Roc 03891
Château de Montméry et son Parc, Ambazac 11564
Château de Montmuran, Les Iffs 13702
Chateau de Mores State Historic Site, Medora 50559
Château de Moulineaux Robert-le-Diable, Moulineaux 14302
Château de Parentignat, Parentignat 14564
Château de Poudenas, Poudenas 14927
Château de Prades, Lafox 13476
Château de Puymartin, Marquay 13979
Château de Ratilly, Treigny 15986
Château de Reinhardstein, Robertville 03970
Château de Rixensart, Rixensart 03969
Château de Rochambeau, Thoré-la-Rochette 15897
Château de Rochefort-en-Terre, Rochefort-en-Terre 15074
Château de Roquedols, Meyrueis 14112
Château de Rosanbo, Lanvellec 13519
Château de Saint-Aubin-sur-Loire, Saint-Aubin-sur-Loire 15178
Château de Saint-Jean-de-Beauregard, Saint-Jean-de-Beauregard 15321
Château de Solre-sur-Sambre, Solre-sur-Sambre 04037
Château de Sombreffe, Sombreffe 04038
Château de Sourdis, Gaujacq 13037
Château de Spontin, Yvoir 04160
Château de Sully-sur-Loire, Sully-sur-Loire 15839
Château de Talcy, Marchenoir 13950
Château de Tarascon, Tarascon (Bouches-du-Rhône) 15853
Château de Temniac, Sarlat-la-Canéda 15634
Château de Trazegnies, Trazegnies 04090
Château de Trevarez, Saint-Goazec 15303
Château de Vascœuil, Vascœuil 16078
Château de Vaurenard, Gleize 13075
Château de Véves, Celles 03511
Château de Villegongis, Villegongis 16197
Château de Villeneuve, Saint-Germain-Lembron 15295
Château de Villeneuve, Vence 16096
Château de Voltaire/Centre des Monuments Nationaux, Ferney-Voltaire 12922
Le Château d'Eau, Toulouse 15931
Le Château des Allymes, Ambérieu-en-Bugey 11566
Château des Pêcheurs, La Bussière 13322
Château des Princes de Chimay, Chimay 03523
Château des Rois Ducs, Sauveterre-la-Lémance 15670
Château des XIVe-XVIIe s, Époisses 12842
Château d'Estignols, Aurice 11752
Château d'Ételan, Saint-Maurice-d'Ételan 15411
Château d'Oiron, Oiron 14503
Château d'Oron, Oron-le-Châtel 41715
Château du Bousquet, Laguiole 13479
Château du Chassan, Faverolles (Cantal) 12909
Château du Clos de Vougeot, Vougeot 16273
Château du Clos-Lucé, Amboise 11575
Château du Fosteau, Leers-et-Fosteau 03782
Château du Haut-Koenigsbourg, Orschwiller 14541
Château du Lude, Le Lude 13604
Château du Riau, Villeneuve-sur-Allier 16219
Château du Rivau, Lemere 13680
Château du Val-Saint-Lambert, Seraing 04010
Château du XVIe s, Ancy-le-Franc 11595
Château d'Ussé, Rigny-Ussé 15041
Château et Musée de Filain-Fort, Filain 12934
Château et Musée de Valangin, Valangin 41971
Château Féodal de Beynac, Beynac-et-Cazenac 11971
Château feodal de Corroy-le-Château, Corroy-le-Château 03526
Château Fort, Bouillon 03392
Château Franc-Waret, Franc-Waret 03608
Château Guillaume le Conquérant, Falaise 12902
Château Histoire de Tours, Tours 15962
Château l'Hospitalet, Narbonne 14375
Château Logue, Maniwaki 06226
Château Lysice, Lysice 09645
Château Malromé, Saint-André-du-Bois 15168
Château Médiéval de Tancarville, Tancarville 15850
Château Médiéval et Musée Delporte, Thy-le-Château 04067
Château Médiéval et Renaissance, Châteaubriant 12441
Château-Musée, Aulan 11743
Château-Musée, Boën-sur-Lignon 12037
Château-Musée, Boulogne-sur-Mer 12092
Château-Musée, Brissac-Quincé 12187
Château-Musée, Carrouges 12300
Château-Musée, Collonges-lès-Bévy 12579
Château-Musée, Cons-la-Grandville 12616
Château-Musée, Cormatin 12627
Château-Musée, Davayat 12699
Château-Musée, Dinan 12746
Château-Musée, Ferrières-en-Brie 12925
Château-Musée, Haroué 13183
Château-Musée, Heudicourt 13214
Château-Musée, Lavardens 13546
Château-Musée, Lavoûte-sur-Loire 13553
Château-Musée, Le Plessis-Brion 13634
Château-Musée, Lenoncourt 13681
Château-Musée, Maillebois 13918
Château-Musée, Menthon-Saint-Bernard 14077
Château-Musée, Meung-sur-Loire 14105
Château-Musée, Montmirail 14228
Château-Musée, Pange 14559
Château-Musée, Pirou 14825
Château-Musée, Roche-la-Molière 15064
Château-Musée, Saumur 15661
Château-Musée, Tourrette-Levens 15960
Château-Musée, Valprivas 16061
Château-Musée, Vendeuvre 16098
Château-Musée, Villandry 16163
Château-Musée, Virieu-sur-Bourbre 16252
Château-Musée Anjony, Tournemire 15954
Château-Musée d'Andurain, Mauléon-Licharre 14036
Château-Musée d'Anet, Anet 11600
Château-Musée d'Art Populaire et Religieux, Clermont-en-Genevois 12546
Château-Musée d'Aulteribe, Sermentizon 15726
Château-Musée de Bois-Chevalier, Legé 13676
Château-Musée de Bonneville, Chamblac 12376
Château-Musée de Brie, Champagnac-la-Rivière 12383
Château-Musée de Bussy-Rabutin, Bussy-le-Grand 12211
Château-Musée de Cany, Cany-Barville 12272
Château-Musée de Château Dauphin, Pontgibaud 14904
Château-Musée de Chémery, Chémery 12498
Château-Musée de Chenonceau, Chenonceaux 12499
Château-Musée de Combourg, Combourg 12590
Château-Musée de Condé, Condé-en-Brie 12607
Château-Musée de Courson, Courson-Monteloup 12657
Château Musée de Cugney, Cugney 12685
Château-Musée de Culan, Culan 12687
Château-Musée de Dampierre-sur-Boutonne, Aulnay (Charente-Maritime) 11744
Château-Musée de Dieppe, Dieppe 12721
Château-Musée de Filières, Gommerville 13078
Château-Musée de Jussy, Jussy-Champagne 13296
Château-Musée de la Batisse, Chanonat 12398
Château-Musée de la Gataudière, Marennes 13965
Château-Musée de la Grange, Manom 13939
Château-Musée de la Nature, Bonnefamille 12053
Château-Musée de la Possonnière, Couture-sur-Loir 12662
Château-Musée de la Roche, Chaptuzat 12404
Château-Musée de la Roche-Racan, Saint-Paterne-Racan 15440
Château-Musée de la Rochelambert, Saint-Paulien 15451
Château-Musée de la Trémolière, Anglards-de-Salers 11614
Château-Musée de Lacassagne, Saint-Avit-Frandat 15179
Château-Musée de Lantheuil, Lantheuil 13518
Château-Musée de Maupas, Morogues 14284
Château-Musée de Miromesnil, Tourville-sur-Arques 15973
Château-Musée de Montaigu, Laneuveville-devant-Nancy 13502
Château-Musée de Montal, Saint-Jean-Lespinasse 15339
Château-Musée de Montfleury, Avressieux 11804
Château-Musée de Montgeoffroy, Mazé 14048
ChÂteau Musée de Montmorin, Beillom 11920
Château-Musée de Panloy, Port-d'Envaux 14915
Château-Musée de Pont-Chevron, Ouzouer-sur-Trézée 14548
Château-Musée de Préhistoire, Belesta-de-la-Frontière 11921
Château-Musée de Puyguilhem, Villars (Dordogne) 16170
Château-Musée de Ravignan, Perquie 14794
Château-Musée de Ripaille, Thonon-les-Bains 15894
Château-Musée de Serrant, Saint-Georges-sur-Loire 15284
Château-Musée de Teillan, Aimargues 11508
Château-Musée de Terre-Neuve, Fontenay-le-Comte 12968
Château-Musée de Vaulaville, Tour-en-Bessin 15947
Château-Musée de Veauce, Ébreuil 12805
Château-Musée de Villeprévost, Tillay-le-Péneux 15912
Château-Musée de Villersexel, Villersexel 16234
Château-Musée de Villesavin, Tour-en-Sologne 15948
Château-Musée de Vollore, Vollore-Ville 16268
Château-Musée d'Effiat, Aigueperse 11498
Château-Musée des Ducs, Duras 12800
Château-Musée des Martinanches, Saint-Dier-d'Auvergne 15242
Château-Musée des XVIe-XVIIe s, Tanlay 15852
Château-Musée du Fort La Latte, Rennes (Ille-et-Vilaine) 15011
Château-Musée du Grand-Saussay, Ballancourt-sur-Essonne 11831
Château-Musée du Maréchal Lyautey, Thorey-Lyautey 15899
Château-Musée du Moulin, Lassay-sur-Croisne 13533
Château-Musée du Plessis-Bouré, Ecuillé 12810
Château-Musée du Plessis-lès-Tours, La Riche 13410
Château-Musée du Rhône, Tournon-sur-Rhône 15956
Château-Musée du Touvet, Le Touvet 13656
Château-Musée du Val, Lanobre 13513
Château-Musée du XVe et XVIIIe Siècle, Commarin 12593
Château-Musée du XVIIe Siècle, Fontaine-Française 12961
Château-Musée d'Urtubie, Saint-Jean-de-Luz 15325
Château-Musée et Ecuries de Chaumont-sur-Loire, Chaumont-sur-Loire 12486
Château Musée Grimaldi, Cagnes-sur-Mer 12236
Château-Musée la Fayette, Chavaniac-Lafayette 12490
Château-Musée le Denacre, Wimille 16296
Château-Musée Listran, Jau-Dignac-et-Loirac 13264
Château-Musée Mansart, Sagonne 15147
Château-Musée Municipal, Nemours 14390
Château-Musée Renaissance, Azay-le-Rideau 11809
Château-Musée Turcan, Ansouis 11634
Château-Musée Verdus, Saint-Seurin-de-Cadourne 15512
Château of Monte-Cristo, Le Port-Marly 13641
Château-Promenade des Parfums, Chilleurs-aux-Bois 12512
Chateau Ramezay Museum, Montréal 06366
Château Royal, Collioure 12572
Château Royal d'Amboise, Amboise 11576
Château Saint-Hugues, Semur-en-Brionnais 15709
Chatelherault, Ferniegair 44223
Chatham College Art Gallery, Pittsburgh 51946
Chatham Islands Museum, Waitangi, Chatham Islands 33229
Chatham-Kent Museum, Chatham, Ontario 05602
Chatillon-DeMenil Mansion, Saint Louis 52570
Chatsworth House, Bakewell 43354
Chattahoochee Valley Art Museum, La Grange 49840
Chattanooga African Museum, Chattanooga 47406
Chattanooga Regional History Museum, Chattanooga 47407
Chattishgarh Coin and Papermoney Museum, Katora Talab 24256
Chattishgarh Mudra Sangrahalaya, Katora Talab 24256
Le Chaudron Magique, Brugnac 12194
Chauvin Trading Post, Tadoussac 07058
Chazen Museum of Art, Madison 50388
Cheb Castle, Cheb 09472
Cheb Museum, Cheb 09473
Chebský Hrad, Cheb 09472
Cheddar Man & the Cannibals, Cheddar 43801
Cheddleton Flint Mill, Cheddleton 43802
Chedworth Roman Villa, Yanworth 46091
Chedworth Roman Villa Museum, Cheltenham 43809
Cheekwood Museum of Art, Nashville 50963
cheese dairy museum, Lochen 02330
Chehel-Sotoon Museum, Qazvin 24638
Chehel-Sotoon Palace Museum, Isfahan 24611
Chehel-Stoun Museum, Ghazvin 24602
Chelmsford Historical Society Museum, Chelmsford 47418

C

C

Chelmsford Museum, Chelmsford 43803
Chelsea Art Museum, New York 51147
Chelsea Court House Museum, Chelsea 00965
Cheltenham Art Gallery and Museum, Cheltenham 43810
Chelyabinsk Regional Picture Gallery, Čeljabinsk 36842
Chemainus Valley Museum, Chemainus 05606
Chemin de Fer - Musée Blonay-Chamby, Blonay 41242
Chemin de la Soie, Molière-Cavaillac 14143
Le Chemin du Fauvisme, Collioure 12573
Chemung Valley History Museum, Elmira 48379
Chen Chung-Tsang Museum, Dongshan 42180
Chen Feng-Hsien Miniature Art Museum, Sindian 42333
Chen Jonghui Memorial Museum, Yuanli 42476
Chenango County Historical Society Museum, Norwich 51439
Cheney Homestead, Manchester 50416
Cheng Ta Art Gallery, Tainan 42360
Chengdu Museum, Chengdu 07614
Chenghai Museum, Chenghai 07624
Chengshiu University Art Center, Niaosong 42286
Cheongju Early Printing Museum, Cheongju 29787
Cheongju National Museum, Cheongju 29788
Cheongju University Museum, Cheongju 29789
Chepstow Museum, Chepstow 43812
Cheraw Lyceum Museum, Cheraw 47419
Cherokee County Historical Museum, Murphy 50917
Cherokee Courthouse Museum, Gore 48974
Cherokee National Museum, Tahlequah 53417
Cherokee Strip Land Rush Museum, Arkansas City 46347
Cherokee Strip Museum, Alva 46246
Cherokee Strip Museum and Henry S. Johnston Library, Perry 51781
Cherry County Historical Society Museum, Valentine 53689
Cherry Valley Museum, Cherry Valley 47426
Chertsey Museum, Chertsey 43814
Chesapeake and Ohio Canal Tavern Museum, Sharpsburg 53054
Chesapeake Bay Maritime Museum, Saint Michaels 52611
Chesapeake Beach Railway Museum, Chesapeake Beach 47428
Chesapeake Children's Museum, Annapolis 46317
Chesham Museum-The Stables, Chesham 43815
Cheshire County Museum, Keene 49718
Cheshire Military Museum, Chester 43816
Chess museum, Heves 23608
Chess Museum, Moskva 37359
Chester Art Guild, Chester 47434
The Chester Beatty Library, Dublin 24781
Chester County Historical Society Museum, West Chester 54049
Chester County Museum, Chester 47433
Chesterfield Historical Society of Virginia, Chesterfield 47438
Chesterfield Museum and Art Gallery, Chesterfield 43820
Chesterwood Museum, Stockbridge 53298
Chestico Museum, Port Hood 06641
Chestnut Museum, Joyeuse 13287
Chethams's Hospital and Library, Manchester 45081
Chetopa Historical Museum, Chetopa 47442
Cheviot Museum, Cheviot 33066
Cheyenne Frontier Days Old West Museum, Cheyenne 47444
Chi City Museum, Chicheng 07625
Chi Mei Museum, Rende 42304
Chia Chi Lan Wine Museum, Yilan 42466
Chia Hsien Fossil Museum, Jiasian 42229
Chia Nan Museum of Culture and Art, Rende 42305
Chiang Kai-shek's Shihlin Residence, Taipei 42379
Chiang Mai Contemporary Art Museum, Chiang Mai 42567
Chiang Mai National Museum, Chiang Mai 42568
Chiang Saen National Museum, Chiang Rai 42570
Chiayi Distillery Museum, Minsiong 42276
Chiayi Natural History Museum, Singang 42336
Chiba City Museum of Art, Chiba 28393
Chiba-kenritsu Awa Hakubutsukan, Tateyama, Chiba 29353
Chiba-kenritsu Bijutsukan, Chiba 28391
Chiba-kenritsu Boso Fudoki-No-Oka, Sakae 29193
Chiba-kenritsu Boso-no-Mura Hakubutsukan, Sakae 29194
Chiba-kenritsu Chuo Hakubutsukan, Chiba 28392
Chiba-kenritsu Indasutori oyo Kagakukan, Ichikawa 28553
Chiba-kenritsu Kazusa Hakubutsukan, Kisarazu 28704
Chiba-kenritsu Sonan Hakubutsukan, Isumi 28589
Chiba Museum of Science and Industry, Ichikawa 28553
Chiba Prefectional Awa Museum, Tateyama, Chiba 29353
Chiba Prefectural Boso Fudoki-No-Oka Museum, Sakae 29193
Chiba Prefectural Boso Vilage Museum, Sakae 29194
Chiba Prefectural Accident Museum, Isumi 28589
Chiba Prefectural Kazusa Museum, Kisarazu 28704
Chiba Prefectural Museum of Art, Chiba 28391
Chiba-shi Bijutsukan, Chiba 28393
Chicago Architecture Foundation, Chicago 47462
The Chicago Athenaeum, Chicago 47463
The Chicago Athenaeum - Europe, Néa Symirni 23125
The Chicago Athenaeum - Galena, Galena 48826
Chicago Children's Museum, Chicago 47464
Chicago Cultural Center, Chicago 47465
Chicago Historical Society, Chicago 47466
Chichester District Museum, Chichester 43825
Chichibunomiya Kinen Supohtsu Hakubutsukan, Tokyo 29376
Chichu Art Museum, Naoshima 29028
Chichu Art Museum, Naoshima 29027
Chickamauga-Chattanooga National Military Park, Fort Oglethorpe 48665
Chickasaw Council House Museum, Tishomingo 53512
Chickasaw County Historical Society Museum, Bradford Village, Nashua 50957
Chico Museum, Chico 47526
Chiddingstone Castle, Chiddingstone 43834
Chido Hakubutsukan, Tsuruoka 29542
Chido Museum, Tsuruoka 29542
Children's Museum House of Fairy Tales Once upon the Time, Moskva 37159
Chief John Ross House, Rossville 52465
Chief Oshkosh Native American Arts, Egg Harbor 48302
Chief Plenty Coups Museum, Pryor 52190
Chiefswood Museum, Ohsweken 06502
Chieftains Museum, Rome 52448
Chien Dai Sculpture Art Museum, Sanyi 42315
Ch'ien Mu House, Taipei 42380
Chiesa di Santa Maria Maddalena de Pazzi, Firenze 26148
Chifeng Museum, Chifeng 07626
Chifley Home, Bathurst 00839
Chigira Folklore Museum, Katashina 28679
Chigira Mingeikan, Katashina 28679
Chihaya-Akasaka Kyodo-shiryokan, Chihaya-Akasaka 28397
Chihiro Art Museum Azumino, Nagano 28953
Chihiro Art Museum Tokyo, Tokyo 29377
Chikatetsu Hakubutsukan, Tokyo 29378
Chikkyo Art Museum Kasaoka, Kasaoka 28676
Chikurin-ji Homotsukan, Kochi 28740
Chikurin-ji Temple Treasure House, Kochi 28740
The Child Museum, Cairo 10489
Childers Pharmaceutical Museum, Childers 00967
Childhood and Costume Museum, Bridgnorth 43604
Childhood Museum, Alger 00040
Children Art Palace, Islamabad 34006
Children Museum, Čerepovec 36848
Children Museum, Lisboa 35689
Children of Września Regional Museum, Września 35346
Childrenmuseum, Bruxelles 03476
Children's Art Gallery, Yerevan 00730
Children's Art Museum, San Angelo 52700
Children's Art Museum, Taipei 42381
Children's Art Museum, Warszawa 35271
Children's Creative Art Centre and Gallery, Toruń 35189
Children's Discovery Museum of Central Illinois, Bloomington 46858
Children's Discovery Museum of North San Diego, Carlsbad 47246
Children's Discovery Museum of San Jose, San Jose 52805
Children's Discovery Museum of the Desert, Rancho Mirage 52240
The Children's Discovery Museum of Vermont, Burlington 47142
Children's Folklore and Folkgame Museum, Sikou 42329
Children's Gallery, Taichung 42347
Children's Hands-On Museum, Tuscaloosa 53617
Children's Health Museum, Charlottesville 47396
Children's Heritage and Science Museum, Amman 29647
Children's Library Museum, Marondera 54674
The Children's Metamorphosis, Derry 48093
The Children's Metamorphosis, Londonderry 50169
Children's Museum, Alger 00042
Childrens' Museum, Almaty 29679
Children's Museum, Amreli 24065
Children's Museum, Bhavnagar 24087
Children's Museum, Detroit 48110
Children's Museum, Holyoke 49327
Childrens Museum, Imphal 24222
Children's Museum, København 10124
Children's Museum, La Habra 49845
Children's Museum, Lucknow 24290
Children's Museum, Quezon City 34488
Children's Museum, Qurm 33994
Children's Museum, Saint Petersburg 52635
Children's Museum, San Juan 36071
The Children's Museum, Seattle 52997
Children's Museum, Shanghai 08158
The Children's Museum at Saratoga, Saratoga Springs 52920
Children's Museum at Yunker Farm, Fargo 48508
Children's Museum Big Bang, Sakai 29195
children's museum frankfurt, Frankfurt am Main 18380
Children's Museum in Dartmouth, South Dartmouth 53167
Children's Museum in Easton, North Easton 51392
Children's Museum of Art and Creativity, Kaohsiung 42232
Children's Museum of Boca Raton, Boca Raton 46884
The Children's Museum of Cleveland, Cleveland 47622
Children's Museum of Denver, Denver 48068
The Children's Museum of Green Bay, Green Bay 49036
Children's Museum of Helsinki City Museum, Helsinki 10798
Children's Museum of History, Natural History and Science at Utica, New York, Utica 53675
The Children's Museum of Houston, Houston 49382
The Children's Museum of Indianapolis, Indianapolis 49488
The Children's Museum of Kansas City, Kansas City 49679
Children's Museum of Lake Charles, Lake Charles 49876
Children's Museum of Los Angeles, Los Angeles 50217
Children's Museum of Maine, Portland 52062
Children's Museum of Manhattan, New York 51148
The Children's Museum of Memphis, Memphis 50568
Children's Museum of Oak Ridge, Oak Ridge 51457
Children's Museum of Portsmouth, Portsmouth 52093
Children's Museum of Richmond, Richmond 52325
The Children's Museum of Rose Hill Manor Park, Frederick 48755
Children's Museum of South Carolina, Myrtle Beach 50932
Children's Museum of Southeastern Connecticut, Niantic 51349
Children's Museum of Spokane, Spokane 53205
Children's Museum of Stockton, Stockton 53305
Children's Museum of Tampa - Kid City, Tampa 53430
Children's Museum of the Arts, New York 51149
Children's Museum of Transportation and Communication, Taipei 42382
The Children's Museum of Utah, Salt Lake City 52689
Children's Museum of Virginia, Portsmouth 52105
Childrens Organisations Historical Museum, Zánka 23975
Children's Picture Gallery, Samara 37612
The Children's Science Center, Cape Coral 47232
Children's Science Center, Carbondale 47240
Childress County Heritage Museum, Childress 47530
Childventure Museum, Fort Washington 48699
Chilean Federation of Friends of Museos, Santiago de Chile 54752
Chili Pepper Museum, Maierà 26568
Chillagoe Historical Centre, Chillagoe 00969
Chilliwack Museum, Chilliwack 05613
Chiltern Athenaeum, Chiltern 00970
Chiltern Motor Museum, Chiltern 00971
Chiltern Open Air Museum, Chalfont Saint Giles 43786
Chimmey Point, Addison 46136
Chimneyville Crafts Gallery, Jackson 49557
China Agricultural Museum, Beijing 07526
China Art Museum, Beijing 07527
China Aviation Museum, Beijing 07528
China Caideng Museum, Zigong 08517
China Clay Country Park Mining and Heritage Centre, Saint Austell 45560
China Coal Museum, Taiyuan 08271
China First History Archive Museum, Beijing 07529
China Great Wall Museum, Badaling 07492
China Guche Museum, Linzi 07959
China Institute Gallery, New York 51150
China Jianian Warfare Museum, Weihai 08316
China Jiawu Zhan Zhen Guan, Weihai 08316
China Nationalities Museum of Inner Mongolia, Tongliao 08300
China Nationality Museum, Beijing 07530
China Palace Museum at the Museum-Preserve, Lomonosov 37109
China Penmanship Art Museum, Xian 08357
China People's Revolution Military Museum, Beijing 07531
China Revolution Museum, Beijing 07532
China Sex Culture Museum, Shanghai 08159
China Silk Museum, Hangzhou 07759
China Silk Museum, Shanghai 08160
China Sports Museum, Beijing 07533
China Stamp Museum, Beijing 07534
China Suxiu Art Museum, Suzhou 08253
China Tea Museum, Hangzhou 07760
China Tea Museum, Shanghai 08161
China Tradition Medicine Museum, Chongqing 07629
China Xiamen Huaqiao Museum, Xiamen 08355
China Yin Xue Museum, Hangzhou 07761
Chinati Foundation, Marfa 50466
Chinchilla and District Historical Society Museum, Chinchilla 00973
Chinchu Bijutsukan, Naoshima 29028
Chinese Ancient Architecture Museum, Yangmei 42463
Chinese Art and Ethnographic Museum, Parma 27102
Chinese Association of Museums, Taipei 55039
Chinese Association of Natural Science Museums, Beijing 54754
Chinese Catholic Museum and University History Exhibition Gallery, Sinjhuang 42338
Chinese Ceramics Museum, Usuki 29565
Chinese Characters Museum, Kyoto 28819
Chinese Cultural Centre of Vancouver, Vancouver 07208
Chinese Culture Center of San Francisco, San Francisco 52770
Chinese Furniture Museum, Taoyuan 42451
Chinese Historical Art Museum, Nagasaki 28970
Chinese House Museum, Echo 48280
Chinese Medicine Museum, Hangzhou 07762
Chinese Medicine Museum, Shanghai 08162
Chinese Museum, Ardebil 24596
Chinese Museum, Melbourne 01276
Chinese Museum of Military Weaponary, Tongzhou 08304
Chinese Opera Museum, Taipei 42383
Chinese Pavilion, Drottningholm 40489
Chinese Pavilion, Japanese Tower and Museum of Japanese Art, Bruxelles 03501
Chinese Sex Museum, Tóngli 08299
Chinese Society of Museums, Beijing 54755
Chinese University of Hong Kong Art Museum, Ma Liu Shui 07989
Chinesische Haus, Potsdam 21085
Chingsi Culture Museum, Kaohsiung 42233
Chinikhaneh Museum, Ardebil 24596
Chinqua-Penn Plantation, Reidsville 52285
CHIO Museum, Aachen 16393
Chios Maritime Museum, Chios 22970
Chios Museum of Byzantine and Postbyzantine Art, Chios 22968
Chiostro dello Scalzo, Firenze 26149
Chippenham Museum and Heritage Centre, Chippenham 43835
Chippewa County Museum, Montevideo 50811
Chippewa Valley Museum, Eau Claire 48276
Chipping Norton Museum of Local History, Chipping Norton 43836
Chippokes Farm and Forestry Museum, Surry 53382
Chiran Museum, Chiran 28403
Chiran Peace Museum for Kamikaze Pilots, Chiran 28404
Chiran Tokko Heiwa Kaikan, Chiran 28404
Chirk Castle, Chirk 43837
Chirurgiemuseum, Tuttlingen 22132
Chisenhale Gallery, London 44819
Chisholm Trail Heritage Center Museum, Duncan 48204
Chisholm Trail Historical Museum, Waurika 53992
Chisholm Trail Museum, Kingfisher 49781
Chisholm Trail Museum, Wellington 54027
Chiswick House, London 44820
Chita City Museum, Chita 28405
Chita-shi Rekishi Minzoku Hakubutsukan, Chita 28405
Sri Chitra Art Gallery, Thiruvananthapuram 24403
Chittagong University Museum, Chittagong 03214
Chittenango Landing Canal Boat Museum, Chittenango 47545
Chiverton House Museum, Northampton 01376
Chmel'nyc'kyj Chudožnij Muzej, Chmel'nyc'kyj 43026
Chmel'nyc'kyj Oblasnyj Krajeznavčyj Muzej, Chmel'nyc'kyj 43027
Choco-Musée Erico, Québec 06679
Chocolate Museum Alprose, Caslano 41282
Choctaw Museum, Hugo 49415
Chofu Museum, Shimonoseki 29268
Chohung Museum of Finance, Seoul 29866
Chojagahara Archaeological Museum, Itoigawa 28598
Chojagahara Kokokan, Itoigawa 28598
Chojin Lama Monastery Museum, Ulaanbaatar 31618
Chokaido Hakubutsukan, Yokkaichi 29591
Chokaido Museum, Yokkaichi 29591
Chokoku no Mori Bijutsukan, Hakone 28476

Choma Museum and Crafts Centré, Choma 54653
Chonbuk University Museum, Jeonju 29840
Chongming Museum, Shanghai 08163
Chongqing Museum, Chongqing 07630
Chongqing Natural History Museum, Chongqing 07631
Chonnam University Museum, Gwangju 29821
Chopin Family Drawing Room, Warszawa 35288
Chopin Memorial Room at the Radziwill Palace, Antonin 34529
Choraku Homotsukan, Shimoda 29263
Choraku Treasure House, Shimoda 29263
Choudomir House-Museum, Kazanlăk 05057
Christ Boelens Jukeboxen Museum, Sint Oedenrode 32775
Christ Church Cathedral Treasury, Oxford 45329
Christ Church Community Museum, Lakefield 06131
Christ Church Picture Gallery, Oxford 45330
Christ Museum, Borba 35472
Christ the King College Museum, Calbayog 34360
Christchurch Art Gallery Te Puna o Waiwhetu, Christchurch 33071
Christchurch Mansion, Ipswich 44528
Christian C. Sanderson Museum, Chadds Ford 47323
Christian Museum, Esztergom 23553
Christian-Wagner-Haus, Leonberg 19961
Christian-Wolff-Haus, Halle, Saale 18890
Christiansted National Historic Site, Saint Croix 52541
Christinehof Slott, Brösarp 40478
Christmas Museum Gallery, Horsley Park 01164
Christo Botev Memorial House, Kalofer 05045
Christo Danov House-Museum, Plovdiv 05128
Christophs Friseur-Museum, Leipheim 19888
Christos Capralos Museum, Aegina 22866
Christos Capralos Museum, Agrinion 22874
The Chrysler Museum of Art, Norfolk 51361
CHTJ Southard House Museum, Richmond 52318
Chu Dong Tu Temple, Binh Minh 54600
Chuan Cheng Art Center, Taipei 42384
Chudožestvena Galerija, Balčik 04974
Chudožestvena Galerija, Blagoevgrad 04990
Chudožestvena Galerija, Burgas 04997
Chudožestvena Galerija, Chaskovo 05003
Chudožestvena Galerija, Čirpan 05008
Chudožestvena Galerija, Drjanovo 05022
Chudožestvena Galerija, Gabrovo 05030
Chudožestvena Galerija, Gălăbovo 05034
Chudožestvena Galerija, Karlovo 05048
Chudožestvena Galerija, Kavarna 05053
Chudožestvena Galerija, Kazanlăk 05056
Chudožestvena Galerija, Levski 05082
Chudožestvena Galerija, Loveč 05085
Chudožestvena Galerija, Mezdra 05093
Chudožestvena Galerija, Pernik 05110
Chudožestvena Galerija, Plovdiv 05125
Chudožestvena Galerija, Provadia 05138
Chudožestvena Galerija, Provadia 05139
Chudožestvena Galerija, Radnevo 05143
Chudožestvena Galerija, Silistra 05161
Chudožestvena Galerija, Sozopol 05207
Chudožestvena Galerija, Velingrad 05259
Chudožestvena Galerija, Vidin 05261
Chudožestvena Galerija, Žeravna 05268
Chudožestvena Galerija Boris Denev, Veliko Tărnovo 05252
Chudožestvena Galerija Darenie Kollekcia Svetlin Russev, Pleven 05115
Chudožestvena Galerija Dobrič, Dobrič 05016
Chudožestvena Galerija Elena Karamichajlova, Šumen 05213
Chudožestvena Galerija Georgi Papazov, Jambol 05041
Chudožestvena Galerija Ilija Beškov, Pleven 05116
Chudožestvena Galerija Kiril Petrov, Montana 05094
Chudožestvena Galerija Nikola Marinov, Targovište 05224
Chudožestvena Galerija Nikolaj Pavlovič, Svištov 05219
Chudožestvena Galerija "Petar Persengiev", Novi Pazar 05098
Chudožestvena Galerija Prof. Ilija Petrov, Razgrad 05147
Chudožestvena Galerija Ruse, Ruse 05151
Chudožestvena Galerija Stanislav Dospevski, Pazardžik 05105
Chudožestvena Galerija Varna, Varna 05236
Chudožestvennaja Galereja - Dom Kukly Tatjany Kalininoj, Petrozavodsk 37556
Chudožestvenno-memorialnyj Muzej K.S. Petrova-Vodkina, Chvalynsk 36864
Chudožestvenno-pedagogičeskij Muzej Igruški - Kulturno-delovoj Centr, Sergiev Posad 37762
Chudožestvenno-vystavočnyj Kompleks, Sarapul 37744
Chudožestvennyj Muzej, Staryj Oskol 37799
Chudožestvennyj Muzej, Valujki 37915
Chudožestvennyj Muzej im. A.N. Radiščeva, Saratov 37747
Chudožestvennyj Muzej im. M.S. Tuganova, Vladikavkaz 37928
Chudožnij Muzej, Sevastopol 43176
Chudožnij Muzej, Žytomyr 43211
Chudožno-memorial'nyj Muzej Ivana Truša, Lviv 43119
Chudožno-memorial'nyj Muzej Leopol'da Levyc'kogo, Lviv 43120
Chudožno-memorial'nyj Muzej Oleksi Novakivs'kogo, Lviv 43121
Chudožno-memorial'nyj Muzej Oleni Kul'čic'koï, Lviv 43122
Chudožnyj Muzej im. O. Šovkunenka, Cherson 43022
Chughtai Museum Trust, Lahore 34018
Chulpan Memorial Museum - Dept. of Andijan Regional Museum of Literature and Art, Andijon 54441
Chulpan Yodgorlik Muzeyi, Andijon 54441
Chumbhot-Punthip Museum, Bangkok 42535
Chung Art Center, Seoul 29867
Chung-Cheng Art Gallery, Jamaica 49589
Chung Cheng Aviation Museum, Taoyuan 42452
Chung Hua University Arts Center, Hsinchu 42205
Chung Li-Ho Memorial Museum, Meinong 42271
Chung Yi Gallery, Danshuei 42177
Chungchern Art Museum and Jhongjheng Art Center, Pingtung 42290
Chungju Museum, Chungju 29790
Chungnam Telecommunication Museum, Daejeon 29797
Chungnam University Museum, Daejeon 29798
Churburg, Sluderno 27857
Church-Archaeological Museum, Nizná Mysla 38317
Church Art Exposition of the Museum of the Kaunas Archdiocese, Kaunas 30241
Church Farm Museum, Skegness 45685
Church Farmhouse Museum, London 44821
Church Museum, Dimitsana 22983
Church Museum, Längelmäki 11022
Church Museum, Laihia 11036
Church Museum, Marttila 11102
Church Museum, Nokia 11146
Church-Museum of the Holy Nicholas in Tolmachy, Moskva 37265
Church of the Deposition of the Robe at the Moscow Kremlin, Moskva 37155
Church-Waddel-Brumby House Museum, Athens 46410
Churchill County Museum and Archives, Fallon 48502
Churchill House and Marine Memorial Room Museum, Hantsport 05969
Churchill Island Agricultural Museum, Newhaven 01366
Churchill Museum and Cabinet War Rooms, London 44822
Chusonji Sankozo - Fujiwara Kinenkan, Hiraizumi 28520
Chusonji Temple, Fujiwara Memorial Hall, Hiraizumi 28520
Chuuko Tamba Yakimono Hakubutsukan, Sasayama 29230
Chuvash National Museum, Čeboksary 36837
Chuvash State Art Museum, Čeboksary 36836
Chuxiongzhou Museum, Chuxiong 07633
Chvalyn Museum of Local History, Nature and Culture, Chvalynsk 36865
Chvalynskij Kraevedčeskij Muzej, Chvalynsk 36865
CIA Museum, McLean 50361
Ciäsa Granda, Stampa 41901
Cider Museum and King Offa Distillery, Hereford 44453
CIDOC International Committee for Documentation of the ICOM, Berlin 54829
Ciências Educação e Artes Luiz de Queiráz, Piracibaba 04630
Ciencias Naturales P. Ignacio Sala, Valencia 40199
Cietumu Vēstures Muzejs, Jūrmala 30022
Cifrapalota, Kecskemét 23644
Çifte Medrese Müzesi, Kayseri 42873
Cigna Museum and Art Collection, Philadelphia 51822
Cilicia Museum, Antélias 30154
Cimarron Valley Railroad Museum, Cushing 47914
CIMCIM - International Committee of Musical Instrument Museums and Collections, Firenze 54898
CIMUSET - International Commitee of Museums of Science and Technology, Bjerringbro 54777
Cincinnati Art Museum, Cincinnati 47555
Cincinnati Fire Museum, Cincinnati 47556
Cincinnati Museum Center, Cincinnati 47557
Cinema Museum, London 44823
Cinema Museum, Moskva 37304
La Cinémathèque de Toulouse, Toulouse 15932
La Cinémathèque Française, Paris 14581
Cinémathèque Québécoise - Musée du Cinéma, Montréal 06336
Cinematographic Museum, Shkodër 00029
Cingyun Sword Craft Museum, Shalu 42320
Çini Müzesi, Kütahya 42893
Çinili Köşk Müzesi, İstanbul 42812
Circuit des Enclus Paroissiaux en Miniatures, Commana 12591
Circuit Géologique des Boves, Arras 11697
Circus City Festival Museum, Peru 51787
Circus Museum, Rapperswil 41742
Circus World Museum, Baraboo 46604
Circusmuseum "Bajasseum", Enkenbach-Alsenborn 18175
Cirencester Art & Local Studies Collection, Cirencester 43845
Cirencester Lock-Up, Cirencester 43846
Cirkevno-archeologické Múzeum, Nizná Mysla 38317
Ciruito Museale Urbano, Cascia 25703
Císařská Konírna, Praha 09738
Čišinski-Gedenkstätte, Panschwitz-Kuckau 20977
Cistercian Abbey of Bonnefont, Proupiary 14950
Cistercian Museum, Braunschweig 17531
Cistercian Museum, Hailes 44378
Cistercian Museum, Wąchock 35206
Cistersian Monastery Museum, Szczyrzyc 35167
The Citadel, Saint Helens 45566
The Citadel Archives and Museum, Charleston 47360
Cité de la Mer Estran, Dieppe 12722
Cité de l'Énergie, Shawinigan 06950
Cité des Métiers de Tradition, Saint-Laurent-de-la-Plaine 15357
Cité des Sciences et de l'Industrie, Paris 14582
Čitinskij Oblastnoj Chudožestvennyj Muzej, Čita 36868
Čitinskij Oblastnoj Kraevedčeskij Muzej im A.K. Kuznecova, Čita 36869
City and County Museum, Lincoln 44710
City Art Centre, Edinburgh 44116
City Art Gallery, Berkovica 04986
City Art Gallery, Leeds 44660
City Art Gallery, Pernik 05110
City Art Museum, Ljubljana 38452
City Art Museum, Wrocław 35341
City Arts, New York 51151
City County Pioneer Museum, Sweetwater 53393
City Exhibition Hall, Krasnojarsk 37070
City Exhibition Hall of Smolensk, Smolensk 37780
City Gallery, Chomutov 09477
The City Gallery, Leicester 44674
City Gallery at Chastain, Atlanta 46431
City Gallery East, Atlanta 46432
City Gallery Heerlen - Museum for Modern and Contemporary Art, Heerlen 32338
City Gallery of Beroun, Beroun 09403
City Gallery of Bratislava, Bratislava 38211
City Gallery of Fine Arts, Plovdiv 05127
City Gallery Wellington, Wellington 33244
City Hall Art Gallery, Ottawa 06542
City Hall Council Chamber Gallery, Charleston 47361
City Hall of Antwerpen, Antwerpen 03316
City History Museum, Szczecin 35161
City Island Nautical Museum, Bronx 47047
City Museum, Ahmedabad 24047
City Museum, Gent 03643
City Museum, Horažďovice 09523
City Museum, İstanbul 42839
City Museum, Klobouky u Brna 09588
City Museum, Kosovska Mitrovica 38105
City Museum, Križevci 08902
City Museum, Makale 10685
City Museum, Radnice u Rokycan 09813
City Museum, Rýmařov 09833
City Museum, Saint Louis 52571
City Museum, Yirgalem 10686
City Museum Aalst, Aalst 03259
City Museum and Art Gallery, Gloucester 44315
City Museum and Art Gallery, Plymouth 45390
City Museum and Gallery, Hořice 09524
City Museum and Museum of Textiles, Dvůr Králové nad Labem 09500
City Museum and Records Office, Portsmouth 45428
City Museum Bratislava, Bratislava 38219
City Museum Drenthe's Veste, Coevorden 32038
City Museum Erlangen, Erlangen 18225
City Museum in the Oslekov House, Koprivštica 05074
City Museum Innsbruck, Innsbruck 02131
City Museum of Göteborgs, Göteborg 40549
City Museum of Korčula, Korčula 08894
City Museum of Norrköping, Norrköping 40774
City Museum of Performing Arts, Omsk 37498
City Museum of Stuttgart-Bad Cannstatt, Stuttgart 21961
City Museum of Zagreb, Zagreb 09004
City Museum Pazin, Pazin 08927
City Museums of Pesaro, Pesaro 27153
City of Armadale History House Museum, Armadale, Western Australia 00812
City of Belmont Museum, Belmont 00856
City of Brea Gallery, Brea 47000
City of Calgary-Civic Art Collection, Calgary 05517
The City of Cologne's Documentation Centre on National Socialism, Köln 19626
City of Gilroy Museum, Gilroy 48912
City of Gosnells Museum, Gosnells 01124
City of Kingston Fire Department Museum, Kingston 06076
City of Las Vegas and Rough Riders Memorial Museum, Las Vegas 49963
City of Le Sueur Museum, Le Sueur 50003
City of London Police Museum, London 44824
City of Melbourne Collection, Melbourne 01277
City of Norwich Aviation Museum, Horsham Saint Faith 44490
City of Prague Museum, Praha 09769
City of Saint John Gallery, Saint John 06835
City of Unley Museum, Unley 01612
City of Wayne Historical Museum, Wayne 54005
City of York Museum, York 07413
City Park Radio Museum, Launceston 01220
City Wax Museum, México 31073
Civic Museum of the sea and the coast M. Zei, Sabaudia 27583
Civica Galleria d'Arte Contemporanea, Lissone 26494
Civica Galleria d'Arte Contemporanea F. Scroppo, Torre Pellice 28039
Civica Galleria d'Arte G. Sciortino, Monreale 26783
Civica Galleria d'Arte Moderna, Gallarate 26264
Civica Galleria d'Arte Moderna, Genova 26289
Civica Galleria d'Arte Moderna, Milano 26674
Civica Galleria di Palazzo Rocca, Chiavari 25857
Civica Pinacoteca V. Crivelli, Sant'Elpidio a Mare 27728
Civica Raccolta d'Arte, Medole 26648
Civica Raccolta d'Arte, Roseto degli Abruzzi 27559
Civica Raccolta d'Arte Applicata, Milano 26675
Civica Raccolta d'Arte B. Biancolini, Potenza Picena 27318
Civica Raccolta delle Stampe Achille Bertarelli, Milano 26676
Civica Raccolta di Terraglia-Museo Internazionale Design Cermico, Laveno Mombello 26471
Civica Raccolta e Ceramiche Rinascimentali, Camporgiano 25643
Civiche Raccolte d'Arte, Busto Arsizio 25571
Civici Musei di Villa Paolina, Viareggio 28243
Civici Musei e Gallerie di Storia ed Arte, Udine 28110
Civico Gabinetto dei Disegni, Milano 26677
Civico Museo Antropologico, Balestrate 25352
Civico Museo Archeologico, Acqui Terme 25149
Civico Museo Archeologico, Arsago Seprio 25287
Civico Museo Archeologico, Camogli 25626
Civico Museo Archeologico, Milano 26678
Civico Museo Archeologico, Ozieri 27025
Civico Museo Archeologico e di Scienze Naturali Federico Eusebio, Alba 25174
Civico Museo Archeologico L. Barni e Pinacoteca C. Ottone, Vigevano 28265
Civico Museo Bibliografico Musicale, Bologna 25438
Civico Museo Biblioteca dell'Attore del Teatro, Genova 26290
Civico Museo d'Arte Contemporanea, Milano 26679
Civico Museo d'Arte Moderne, Anticoli Corrado 25241
Civico Museo d'Arte Orientale, Trieste 28076
Civico Museo degli Strumenti Musicali, Milano 26680
Civico Museo del Castello di San Giusto, Trieste 28077
Civico Museo del Mare, Trieste 28078
Civico Museo del Merletto al Tombolo, Rapallo 27365
Civico Museo del Risorgimento e Sacrario Oberdan, Trieste 28079
Civico Museo della Risiera di San Sabba, Trieste 28080
Civico Museo di Guerra per la Pace Diego de Henriquez, Trieste 28081
Civico Museo di Scienze Naturali, Domodossola 26041
Civico Museo di Storia ed Arte e Orto Lapidario, Trieste 28082
Civico Museo di Storia Naturale della Lombardia, Jerago con Orago 26423
Civico Museo Dianese, Diano Marina 26038
Civico Museo e Antiquarium Comunale, Lanuvio 26451
Civico Museo Flaminio Massetano, Massa Martana 26634
Civico Museo Frignanese, Pavullo nel Frignano 27127
Civico Museo G. Galletti, Domodossola 26042
Civico Museo G.V. Parisi-Valle, Maccagno 26551
Civico Museo Ingauno, Albenga 25176
Civico Museo Insubrico di Storia Naturale, Induno Olona 26408
Civico Museo Morpurgo, Trieste 28083
Civico Museo Naturalistico, Ovada 27024
Civico Museo Naturalistico Ferruccio Lombardi, Stradella 27910
Civico Museo Revoltella e Galleria d'Arte Moderna, Trieste 28084
Civico Museo Sartorio, Trieste 28085

C

Civico Museo Setificio Monti, Abbadia Lariana 25137
Civico Museo Storico, Palmanova 27083
Civico Museo Storico Archeologico, Savona 27774
Civico Museo Teatrale di Fondazione Carlo Schmidl, Trieste 28086
Civico Studio Museo Francesco Messina, Milano 26681
Civil Guard Museum, Seinäjoki 11306
Civil War Museum, Carrollton 47256
Civil War Museum, Philadelphia 51823
Civil War Museum of Lone Jack, Jackson County, Lone Jack 50170
Civilian Conservation Corps Museum, Sebring 53020
Civilzation Gumelniţa Museum Olteniţa, Olteniţa 36499
Clackmannshire Council Museum, Alloa 43282
Clan Armstrong Museum, Langholm 44639
Clan Cameron Museum, Spean Bridge 45725
Clan Donnachaidh Museum, Pitlochry 45384
Clan Grant Museum, Alness 43285
Clan Gunn Museum, Latheron 44644
Clan Macpherson House and Museum, Newtonmore 45254
Clandestine Immigration and Naval Museum, Haifa 24957
Claphams Clocks-The National Clock Museum, Whangarei 33264
Clara Barton Birthplace, North Oxford 51399
Clara Barton National Historic Site, Glen Echo 48918
Clara-Zetkin-Museum, Birkenwerder 17360
Clare Ancient House Museum, Clare 43848
Clare Heritage Centre, Corofin 24772
Clare Museum, Ennis 24829
Clare National Trust Museum, Clare 00974
Claremont Museum, Claremont 47577
Claremont Museum, Claremont, Western Australia 00976
Clarencetown and District Historical Museum, Clarencetown 00978
Clarendon House, Evandale 01077
Claresholm Museum, Claresholm 05619
Clarin Ancestral House, Loay 34406
Clark Atlanta University Art Galleries, Atlanta 46433
Clark County Historical Museum, Vancouver 53705
Clark County Museum, Henderson 49249
Clark County Museum, Marshall 50501
Clark Hotel Museum, Van Horn 53701
Clark House Museum Complex, Wolfeboro 54268
Clark Humanities Museum, Claremont 47571
Clarke County Historical Museum, Berryville 46770
Clarke Hall, Wakefield 45907
Clarke Historical Museum, Eureka 48439
Clarke House Museum, Chicago 47467
Clarke Museum, Kirby 06096
Clarkson Historical Museum, Clarkson 47589
Clarksville-Montgomery County Museum-Customs House Museum, Clarksville 47592
Classic Cars International, Salt Lake City 52690
Classic Yuzen Art Museum, Kyoto 28826
Classical and Modern Art Museum Burganov House, Moskva 37307
Claude Crayston Museum, Glenora 05873
Clausen Museum, Petersburg 51793
Claustros del Pilar, Buenos Aires 00154
Clawson Historical Museum, Clawson 47596
Clay County Historical Museum, Liberty 50077
Clay County Museum, Moorhead 50834
Clay Studio, Philadelphia 51824
Clay Tobacco Pipe Museum, Broseley 43658
Claydon House, Middle Claydon 45137
Claypool-Young Art Gallery, Morehead 50840
Clayton Collection Museum, Chollerford 43839
Clayton McLain Memorial Museum, Cut Knife 05669
Cle Elum Telephone Museum, Cle Elum 47604
Clear Lake Area Historical Museum, Clear Lake 47606
Clearfield County Historical Society Museum, Clearfield 47607
Clearwater County Museum, Shevlin 53077
Clearwater Historical Museum, Orofino 51601
Cleethorpes Discovery Centre, Cleethorpes 43851
Clemenceau Heritage Museum, Cottonwood 47842
Clemens-Sels-Museum Neuss, Neuss 20654
Clément Ader Museum, Muret 14335
Clendening History of Medicine Library and Museum, Kansas City 49680
Cleo Redd Fisher Museum, Loudonville 50276
Clergy House, Alfriston 43279
Clermont State Historic Site, Clermont 47618
Cleve National Trust Museum, Cleve 00982
Clevedon Court, Clevedon 43852
Cleveland Artists Foundation at Beck Center for the Arts, Lakewood 49907
Cleveland County Historical Museum, Shelby 53067
Cleveland Institute of Art Gallery, Cleveland 47623
Cleveland Ironstone Mining Museum, Skinningrove 45687
Cleveland Living Arts Centre, Dunedin 33089
The Cleveland Museum of Art, Cleveland 47624
The Cleveland Museum of Natural History, Cleveland 47625
Cleveland Police Museum, Cleveland 47626
Cleveland State University Art Gallery, Cleveland 47627
Clewiston Museum, Clewiston 47646
Cliffe Castle, Keighley 44550
Clifton and District Historical Museum, Clifton 00985
Clifton Community Historical Society, Clifton 47647
Clifton Park Museum, Rotherham 45517
Climax Community Museum, Climax 05623
Clin d'Ailes, Payerne 41718
Clink Prison Museum, London 44825
Clinton Academy Museum, East Hampton 48243
Clinton County Historical Museum, Plattsburgh 51982
Clinton County Historical Society Museum, Wilmington 54206
Clinton County Museum, Clinton 47650
Clinton County Museum, Frankfort 48733
Clipper Ship Cutty Sark, London 44826
Clive House Museum, Shrewsbury 45672
Cliveden House, Philadelphia 51825
Cloches de Corneville, Pont-Audemer 14878
Clock Industry Museum, Villingen-Schwenningen 22241
Clock Museum, Angarsk 36760
Clock Museum, Ploieşti 36529
Clock Tower, Saint Albans 45541
Clock Tower Archive Centre, Youghal 24915
Clockmakers' Museum, London 44827
Clocktower Museum, Elsham 44170
The Cloisters, New York 51152
Cloître de Cadouin, Le Buisson-de-Cadouin 13567
Clonalis House, Castlerea 24757
Cloncurry and District Museum, Cloncurry 00986
Clonfert Museum of Ecclesiastical Art, Loughrea 24871
Clos Arqueologic Torre Llauder, Mataró 39631
Clos Arsene Lupin, Etretat 12882
Clothing Museum, La Rioja 00419
Clotworthy Arts Centre, Antrim 43307
Cloud County Historical Society Museum, Concordia 47777
Clough-Hanson Gallery, Memphis 50569
Clowes Fund Collection, Indianapolis 49489
Clun Local History Museum, Clun 43856
Clunes Museum, Clunes 00987
Clyde Historical Museum, Clyde 33081
Clydebank Museum, Clydebank 43857
Clydebuilt, Glasgow 44277
Clymer Museum of Art, Ellensburg 48357
CMC Art Museum, Taipei 42385
Co-operative Storage Museum, Albany 00786
Coach-House Gallery, Saint Peter Port 45590
Coach Museum of the National Ménes of Mezohegyes, Mezőhegyes 23696
Coachella Valley Museum, Indio 49508
Coal Creek Heritage Village, Korumburra 01206
Coal Mine Valkenburg, Valkenburg aan de Geul 32861
Coal Miners' Museum, Van Lear 53702
Coal Mining History Museum, Ube 29549
Coal Mining Museum, Zabrze 35349
Coalfields Museum, Collie 00993
Coalition of Canadian Healthcare Museums and Archives, Toronto 54735
Coalmine Fotogalerie, Winterthur 42024
Coalport China Museum, Coalport 43861
Coast Artillery Museum, Helsinki 10812
Coastal Center for the Arts, Saint Simons Island 52645
Coastal Discovery Museum, Hilton Head Island 49294
Coastal Heritage Museum, Crystal River 47896
Coastal Heritage Museum, Stadsbygd 33860
Coastal Museum of Godøy, Godøy 33514
Coasting Hill Pavilion, Lomonosov 37111
Coaticook Historical Society Museum, Coaticook 05625
Cobalt Mining Museum, Cobalt 05627
Cobb and Co. Museum, Toowoomba 01592
Cobb County Youth Museum, Marietta 50467
Cobb Institute of Archaeology Museum, Mississippi State 50750
Cobb Museum of Art, Marietta 50468
Cobbaton Combat Collection, Chittlehampton 43838
Cobblestones Museum, Greytown 33116
Cobdogla Irrigation and Steam Museum, Cobdogla 00989
Cobh Museum, Cobh 24764
Cobham Bus Museum, Cobham 43866
Cobra Museum of Modern Art, Amstelveen 31821
Cobra Museum voor Moderne Kunst, Amstelveen 31821
Coburger Puppen-Museum, Coburg 17714
The Coburn Gallery, Ashland 46387
Coca-Cola Memorabilia Museum of Elizabethtown, Elizabethtown 48339
The Coca-Cola Pavillion, Santa Cruz 34500
Coca Museum, Puno 34297
Cocalico Valley Museum, Ephrata 48403
Cochin Ceramics Museum, Chai-Yi 42152
Cockburn Museum, Edinburgh 44117
CODA Museum, Apeldoorn 31911
Codington County Heritage Museum, Watertown 53974
Codman Estate - The Grange, Lincoln 50089
Coe College Collection of Art, Cedar Rapids 47302
Coe Hall, Oyster Bay 51660
Coffin House, Newbury 51302
Coffin School Museum, Nantucket 50943
Cogges Manor Farm Museum, Witney 46052
Coggeshall Farm Museum, Bristol 47032
Coggeshall Heritage Centre and Museum, Coggeshall 43871
Cohuna Museum, Cohuna 00990
Coin and Medal Collection of the University of Helsinki, Helsinki 10810
Coin Cabinet, Dresden 17961
Coin Collection, University Museum of Cultural Heritage, Oslo 33734
Coin du Miroir, Dijon 12732
Coin Museum, Kerman 24620
Coin Museum, Tehrān 24695
Coins Museum, Amman 29648
Cokato Museum and Akerlund Photography Studio, Cokato 47679
Colac Otway Shire Hammerton Bottle Collection, Colac 00991
Colborne Lodge, Toronto 07102
Colburn Gem and Mineral Museum, Asheville 46371
Colby College Museum of Art, Waterville 53977
Colby-Curtis Museum, Stanstead 07005
Colchester Historical Museum, Truro 07180
Colcţia Muzeală a Mănăstirii Lainici, Bumbeşti-Jiu 36235
Cold Museum, Ilulissat 23252
Cold Spring Harbor Whaling Museum, Cold Spring Harbor 47682
Coldharbour Mill, Uffculme 45894
Coldstream Museum, Coldstream 43878
Coldwater Canadiana Heritage Museum, Coldwater 05629
Cole County Historical Museum, Jefferson City 49611
Cole Harbour Heritage Farm Museum, Cole Harbour 05630
Cole Land Transportation Museum, Bangor 46596
Cole Museum of Zoology, Reading 45467
Coleção Assis Pegado - Museu da Moto, Fortaleza 04461
Coleção de Artes Visuais, São Paulo 04853
Coleção do Palácio do Governo de Campos do Jordão, Campos de Jordão 04398
Coleção Entomológica de Referência, São Paulo 04854
Colecção Berardo, Lisboa 35665
Colecção Etnográfica da Câmara Municipal de Sobral de Monte Agraço, Sobral de Monte Agraço 35955
Colecção Etnográfica da Direcção-Geral Agrícola, Aljustrel 35410
Colecção Etnográfica da Paróquia de São Mamede da Ventosa, Torres Vedras 35970
Colección Anatomía, Madrid 39509
Colección Arqueológica, Génova 08646
Colección Arqueológica, Pijao 08721
Colección Benedito, Madrid 39510
Colección Carlos Alberto Pusineri Scala, Asunción 34090
Colección Confradía del Rollo, Jumilla 39432
Colección de Anestesioloxía e Reanimación, Santiago de Compostela 40015
Colección de Arqueología y Etnografía Americana, Madrid 39511
Colección de Arte y Museo Botero, Bogotá 08548
Colección de Escultura Contemporánea, Madrid 39512
Colección de la Biblioteca Musical, Madrid 39513
Colección de la Casa de Alba, Epila 39276
Colección de la Real Academia de la Historia, Madrid 39514
Colección de los Padres Escolapios, Madrid 39515
Colección de Minerales y Arqueología, Villanueva 08787
Colección de Mineralogía, Madrid 39516
Colección de Pintura de Olmeda de las Fuentes, Olmeda de las Fuentes 39726
Colección de Relojes Populares, Salamanca 39935
Colección de Roberto Montenegro, Guadalajara 30888
Colección del Banco de España, Madrid 39517
Colección del Castillo de Mazanares El Real, Manzanares el Real 39617
Colección del Poeta Baudilio Montoya, Clarcá 08626
Colecciòn dos Gabinetes de Física, Química e Historia Natural do Antigo, Santiago de Compostela 40016
Colección Etnográfica, Capela 39123
Colección Maruja Díaz, Villaviciosa de Odón 40310
Colección Monográfica Ecomuseo de Somiedo, Somiedo 40089
Colección Municipal, Granada 39361
Colección Municipal de Arte Contemporáneo de Alcorcón, Alcorcón 38824
Colección Museística de San Mateo, San Mateo 39965
Colección Museo Parroquial del Santuari de la Font de la Salut, Traiguera 40168
Colección Museográfica Casa de la Apicultura, Boal 39053
Colección Museográfica d'Cerámica de Potríes, Potríes 39862
Colección Museográfica de la Torre Musulmana, Silla 40080
Colección Museográfica d'Etnografia i Costums de la Mancomunitat de Municipis de la Vall d'Albaida, Benisoda 39027
Colección Museográfica Municipal de Borriol, Borriol 39056
Colección Museográfica Municipal de Castell de Castells, Castell de Castells 39144
Colección Museográfica Municipal de Cirat, Cirat 39185
Colección Museográfica Museo de la Baronía, Ribesalbes 39901
Colección Museográfica Permanente Arte Bizantino, Tángel 40096
Colección Museográfica Permanente de Biar, Biar 39036
Colección Museográfica Permanente de Buñol, Buñol 39062
Colección Museográfica Permanente de Cinctorres, Cinctorres 39184
Colección Museográfica Permanente de la Font de la Figuera, La Font de la Figuera 39303
Colección Museográfica Permanente de Municipal, Moncada 39659
Colección Museográfica Permanente Luís García de Fuentes, Caudete de las Fuentes 39156
Colección Museográfica Permanente Parroquial de Jérica, Jérica 39428
Colección Museográfica Permenent Municipal de Cocentaina, Cocentaina 39196
Colección Numismática, Bogotá 08549
Colección Ornitológica W.H. Phelps, Caracas 54530
Colección Parroquial, Liétor 39461
Colección Parroquial de Villahermosa del Río, Villahermosa del Río 40302
Colección Taurina, Gerena 39331
Colecţia Arheologică, Zlatna 36744
Colecţia Bibliotecii Seminarului Teologic Ortodox, Caransebeş 36261
Colecţia Boiţa, Tălmaciu 36651
Colecţia de Artă Plastică Al. Bălintescu, Costeşti 36316
Colecţia de Artă Plastică Fr. Storck şi Cecilia Cecilia Cuţescu Stock, Bucureşti 36191
Colecţia de Artă Veche Bisericească, Reşiţa 36561
Colecţia de Carte Veche şi Icoane Ştefan Crişan, Lipova 36442
Colecţia de Ceramică Populară de pe Valea Olteţului, Balş 36134
Colecţia de Etnografie şi Artă Populară, Buzău 36242
Colecţia de Etnografie şi Artă Populară, Chilia 36276
Colecţia de Istorie a Farmaciei Cluj-Napoca, Cluj-Napoca 36285
Colecţia de Istorie şi Arheologie, Piatra Olt 36523
Colecţia de Mineralogie Estetică a Fierului Constantin Gruescu, Ocna de Fier 36495
Colecţia de Obiecte Bisericeşti, Agnita 36100
Colecţia de Obiecte Bisericeşti, Movile 36484
Colecţia Etnografică, Bărbăteşti 36138
Colecţia Etnografică, Irimeşti 36430
Colecţia Etnografică, Poiana Sibiului 36538
Colecţia Etnografică, Răşinari 36554
Colecţia Etnografică Zahacinshi, Mihăileni 36474
Colecţia Eugenia şi Costache Burada, Dăneşti 36339
Colecţia Mănăstirii Zamfira, Lipăneşti 36441
Colecţia Maria şi Dr. George Severeanu, Bucureşti 36192
Colecţia Memorială Maria Filotti, Bucureşti 36193
Colecţia Memorială Octavian Goga, Răşinari 36555
Colecţia Mitropolit Antonie Plămădeală, Sâmbăta de Sus 36581
Colecţia Muzeală, Beliu 36147
Colecţia Muzeală, Colacu 36294
Colecţia Muzeală, Cotorca 36320
Colecţia Muzeală, Holod 36400
Colecţia Muzeală, Rimetea 36565
Colecţia Muzeală, Smeeni 36631
Colecţia Muzeală, Târcaia 36652
Colecţia Muzeală a Arhiepiscopiei Ortodoxe Cluj, Cluj-Napoca 36286
Colecţia Muzeală a Arhiepiscopiei Ortodoxe Galaţi, Galaţi 36377
Colecţia Muzeală a Bisericii Ortodoxe Sf. Nicolae, Bereşti 36148

Colecţia Muzeală a Episcopiei Armeano-Gregoriene, Bucureşti 36194
Colecţia Muzeală a Episcopiei Ortodoxe, Alba Iulia 36104
Colecţia Muzeală a Episcopiei Ortodoxe, Buzău 36243
Colecţia Muzeală a Episcopiei Ortodoxe, Oradea 36501
Colecţia Muzeală a Episcopiei Ortodoxe Sârbe, Timişoara 36687
Colecţia Muzeală a Mănăstirii Adormirea Maicii Domnului, Bodrogu Vechi 36156
Colecţia Muzeală a Mănăstirii Agapia, Agapia 36099
Colecţia Muzeală a Mănăstirii Bistriţa, Bistriţa, Vâlcea 36154
Colecţia Muzeală a Mănăstirii Cozia, Căciulata 36246
Colecţia Muzeală a Mănăstirii Dintr-un Lemn, Dezrobiţi 36345
Colecţia Muzeală a Mănăstirii Dragomirna, Dragomirna 36356
Colecţia Muzeală a Mănăstirii Govora, Mihăeşti 36473
Colecţia Muzeală a Mănăstirii Gura Motrului, Butoieşti 36241
Colecţia Muzeală a Mănăstirii Hurezi, Românii de Jos 36570
Colecţia Muzeală a Mănăstirii Jitianu, Podari 36537
Colecţia Muzeală a Mănăstirii Moldoviţa, Vatra Moldoviţei 36729
Colecţia Muzeală a Mănăstirii Neamţ, Vânători-Neamţ 36724
Colecţia Muzeală a Mănăstirii Nicula, Fizeşu Gherlii 36370
Colecţia Muzeală a Mănăstirii Ortodoxe Celic-Dere, Poşta 36540
Colecţia Muzeală a Mănăstirii Ortodoxe Cernica, Cernica 36275
Colecţia Muzeală a Mănăstirii Ortodoxe Ciolanu, Tisău 36695
Colecţia Muzeală a Mănăstirii Ortodoxe Cocoş, Niculiţel 36493
Colecţia Muzeală a Mănăstirii Ortodoxe Curtea de Argeş, Curtea de Argeş 36335
Colecţia Muzeală a Mănăstirii Ortodoxe Dealu, Aninoasa 36108
Colecţia Muzeală a Mănăstirii Ortodoxe Ghighiu, Bărcăneşti 36141
Colecţia Muzeală a Mănăstirii Ortodoxe Pasărea, Pasărea 36512
Colecţia Muzeală a Mănăstirii Ortodoxe Sinaia, Sinaia 36619
Colecţia Muzeală a Mănăstirii Ortodoxe Suzana, Măneciu 36462
Colecţia Muzeală a Mănăstirii Ortodoxe Ţigăneşti, Ciolpani 36279
Colecţia Muzeală a Mănăstirii Ortodoxe Viforâta, Viforâta 36730
Colecţia Muzeală a Mănăstirii Polovragi, Polovragi 36539
Colecţia Muzeală a Mănăstirii Putna, Putna 36544
Colecţia Muzeală a Mănăstirii Râmeţ, Râmeţ 36548
Colecţia Muzeală a Mănăstirii Rohia, Târgu Lăpuş 36668
Colecţia Muzeală a Mănăstirii Secu, Mănăstirea Neamţ 36461
Colecţia Muzeală a Mănăstirii Sf. Ilie, Topliţa 36699
Colecţia Muzeală a Mănăstirii Sf. Simion-Stâlpnicul, Arad 36110
Colecţia Muzeală a Mănăstirii Strehaia, Strehaia 36638
Colecţia Muzeală a Mănăstirii Suceviţa, Suceviţa 36648
Colecţia Muzeală a Mănăstirii Tismana, Tismana 36696
Colecţia Muzeală a Mănăstirii Văratec, Văratec 36725
Colecţia Muzeală a Mitropoliei Ortodoxe a Olteniei, Craiova 36323
Colecţia Muzeală a Protoieriei Ortodoxe Lugoj, Lugoj 36449
Colecţia Muzeală Batthyaneum, Alba Iulia 36105
Colecţia Muzeală Birchiş, Birchiş 36151
Colecţia Muzeală Colţi, Colţi 36296
Colecţia Muzeală de Argintărie, Braşov 36176
Colecţia Muzeală de Artă Bisericească Veche a Mitropoliei Banatului, Timişoara 36688
Colecţia Muzeală de Istorie, Săcueni 36578
Colecţia Muzeală Geamia Medgidia, Medgidia 36467
Colecţia Muzeală Sătească, Cârligele 36265
Colecţia Muzeală Sătească, Coteşti 36318
Colecţia Muzeală Sătească, Gugeşti 36393
Colecţia Muzeală Sătească, Paltin 36510
Colecţia Muzeală Sătească, Vulcăneasa 36737
Colecţia Oologică Kalaber Ladislau, Reghin 36558
Colecţia Sătească, Atid 36115
Colecţia Sătească, Breaza 36182
Colecţia Sătească, Cârţa 36266
Colecţia Sătească, Ciumani 36284
Colecţia Sătească, Corund 36313
Colecţia Sătească, Dăneşti 36340
Colecţia Sătească, Gura Râului 36395
Colecţia Sătească, Joseni 36433
Colecţia Sătească, Lăpuş 36435
Colecţia Sătească, Lăzarea 36436
Colecţia Sătească, Lunca de Sus 36451
Colecţia Sătească, Lupeni 36453
Colecţia Sătească, Mereşti 36471
Colecţia Sătească, Muşăteşti 36485
Colecţia Sătească, Ocland 36494
Colecţia Sătească, Plăieşii de Jos 36527
Colecţia Sătească, Praid 36541
Colecţia Sătească, Remetea 36560
Colecţia Sătească, Sâncrăieni 36582
Colecţia Sătească, Sândominic 36583
Colecţia Sătească, Sărmaş 36587
Colecţia Sătească, Subcetate 36639
Colecţia Sătească, Suseni 36649
Colecţia Sătească, Tulgheş 36708
Colecţia Sătească, Voşlăbeni 36735
Colecţia Sătească, Zetea 36743
Colecţia Vasile Heisu, Răcăciuni 36545
Colecţias Muzeală a Mănăstirii Ortodoxe Căldăruşani, Moara Vlăsiei 36478
Colegiata de Santa María la Mayor, Toro 40151
Colegio de Santa Isabel Museum, Naga 34465
Colegio del Arte Mayor de la Seda, Valencia 40200
Colegio del Rey, Alcalá de Henares 38811
Coleham Pumping Station, Shrewsbury 45673
Coleman House Museum, Ainsworth 46142
Coleraine Local History Museum, Coleraine 00992
Coleridge Cottage, Nether Stowey 45201
Colesberg/Kemper Museum, Colesberg 38596
Colesville and Windsor Museum at Saint Luke's Church, Harpursville 49176
Coley Homestead and Barn Museum, Weston 54101
Collecció la Porciuncula, L'Arenal 38884
Collecció Museogràfica d'Arqueologia de Pedreguer, Pedreguer 39822
Collecció Museogràfica d'Arqueologia Local, Riba-Roja de Turia 39896
Collecció Museogràfica Permanent de l'Almassara, Muro de Alcoy 39709
Colección Carlos Ferreyros Díaz, Bogotá 08550
Colleccion de Instrumentos Musicales "Monsenór José Ignacio Perdomo Escobar", Bogotá 08551
Colección Museográfica Permanente, Siete Aguas 40077
Colleció d'Autòbils de Salvador Claret Sargatal, Sils 40081
Collecíon de Antigüedades, San Andrés de Sotavento 08744
Collecíón Eugenio Fontaneda, Ampudia 38871
Collecions Municipals del Foment Arqueològic Excursionista Sallenti, Sallent 39949
Collectie A. Veltman, Bussum 32030
Collectie Anderson, Losser 32509
Collection Archéologique, Marmoutier 13974
Collection Archéologique, Villarzel-Cabardes 16172
Collection Archéologique de la Maison des Fouilles et Maison du Memontois, Malain 13927
Collection Archéologique Nationale, Tramont-Emy 15978
Collection Banco de la República, Bogotá 08552
Collection Charles de l'Escalopier, Amiens 11583
Collection d'Archéologie Biblique USH, Strasbourg 15811
Collection d'Art, Pauillac 14760
Collection d'Art Céramique, Longfosse 13808
Collection d'Art Religieux Ancien, Prunay-en-Yvelines 14953
Collection d'Arts Décoratifs, Cazaux-Saves 12332
Collection d'Arts Décoratifs, Villevielle 16240
Collection d'Automates, Gilley 13059
Collection de la Guerre de 1870, Le Bourget (Seine-Saint-Denis) 13563
Collection de l'Abbaye de l'Ile Chauvet, Bois-de-Céné 12047
Collection de l'Art Brut, Lausanne 41535
Collection de l'Aventure Automobile à Poissy, Carrieres-sous-Poissy 12299
Collection de l'Institut d'Égyptologie, Strasbourg 15812
Collection de Locomotives à Vapeur, Obermodern 14490
Collection de l'Outillage de Tonnellerie, Pauillac 14761
Collection de Matériel Ferroviaire, Burnhaupt-le-Haut 12209
Collection de Minéraux de l'Université Pierre et Marie Curie, Paris 14583
Collection de Peintures, Cavaillon 12328
Collection de Peintures, Issac 13249
Collection de Tapisseries d'Aubusson et des Flandres, eigene homepage, Saint-Chamant 15199
Collection de Téléphones Anciens, Saint-Jean-du-Gard 15336
Collection de Tire-Bouchons, Saint-Christophe-des-Bardes 15206
Collection de Tonnellerie, Boersch 12040
Collection de Vieux Pressoirs, Paulx 14765
Collection de Voitures Anciennes de S.A.S. Le Prince de Monaco, Monaco 31605
Collection d'Etains, Sierre 41865
Collection d'Orgues Mécaniques, Herzeele 13209
Collection d'Outillage Ancien de Tonnellerie, Cussac-Fort-Médoc 12689
Collection d'Outillage Viticole dans la Cave, Saint-Estèphe 15248
Collection du Château d'Attre, Attre 03334
Collection du Moulin du Cros, Montagnac-sur-Lède 14164
Collection du Musée Archéologique, Guise 13168
Collection Entomologique Ouellet-Robert, Outremont 06559
Collection Fondation Zoubov, Genève 41410
Collection for Organology, Tachikawa 29307
Collection Gallo-Romaine, La Roquebrussanne 13440
Collection Historial Cognacq-Jay, Paris 14584
Collection Historique, Douaumont 12777
Collection Historique de Fouets, Sorède 15783
Collection Historique de la Marqueterie d'Art Spindler, Boersch 12041
Collection Historique Maritime, Brest 12157
Collection in the Jamarski House - Krumperk Castle, Domžale 38416
Collection Internationale d'Épinettes, Dommartin-lès-Remiremont 12760
Collection Jules Verne, Amiens 11584
Collection Lambert en Avignon, Avignon 11786
Collection Lapidaire, Gimont 13060
Collection of Applied Folk Art, Kecskemét 23652
Collection of Archaeology and Land Study, Nahshonim 25071
Collection of Bratislav Petković - Automobile Museum, Beograd 38061
Collection of Bratislav Petković - Automobile Museum, Beograd 38062
Collection of Byzantine and Postbyzantine Art, Kythira 23069
Collection of Folk Ceramics of the Olt Valley, Balş 36134
Collection of Forestry, Timber Industry and Geodesy, Sopron 23808
Collection of Historical Agricultural Technique, Zirc 23978
Collection of historical biscuit tins, Läckerli-Huus Basel, Basel 41181
Collection of historical securities, Olten 41710
Collection of Icons and Church Heirlooms at Pyrgos, Thera 23207
Collection of Local History of Ferencváros, Budapest 23402
Collection of Martinware Pottery, Southall 45700
Collection of Physics, Chemistry and Natural History instruments from the old 2nd School, Santiago de Compostela 40016
Collection of Reitz Foundation, Helsinki 10813
Collection of Religious Art, Šibenik 08948
Collection of the Evangelismos tis Theotokou, Skiathos 23189
Collection of the Franciscan Monastery, Orebić 08921
Collection of the Franciscan Monastery, Sinj 08951
Collection of the Franciscan Monastery, Visovac 08979
Collection of the Mega Spileon Monastery, Kalavryta 23027
Collection of the Saint John the Divine Monastery, Patmos 23137
Collection Picasso Donation Rosengart, Luzern 41598
Collection Pierre Bouton, Berneuil 11949
Collection Préhistorique Pagès, Le Moustier 13625
Collection Rare de Tracteurs Viticoles, Savigny-les-Beaune 15680
Collection Républicaine, Urimenil 16019
Collection Roux, Villefranche-sur-Mer 16191
Collection Schlumpf, Mulhouse 14323
Collection Ski-Montagne, Briançon 12168
Collection Vini-Viticole et de Tonnellerie, Riquewihr 15050
Collection Viti-Vinicole, Banyuls-sur-Mer 11841
Collection Viticole, Mardeuil 13963
Collections Archéologiques, Parmain 14752
Collections Artistiques de l'Université de Liège, Liège 03795
Collections Baur-Musée des Arts d'Extrême-Orient, Genève 41411
Collections Brassicoles, Strasbourg 15813
Collections Claeyssens, Wambrechies 16282
Collections Council of Australia, Adelaide 54687
Collections d'Arts Populaires, Eschau 12850
Collections de la Bibliothèque, La Bernerie-en-Retz 13316
Collections de la Bibliothèque Municipale, Auxerre 11774
Collections de la Bibliothèque Sainte-Geneviève, Paris 14585
Collections de la Fondation de Coubertin, Saint-Rémy-lès-Chevreuse 15494
Collections de la Fondation Roger-van-Rogger, Bandol 11836
Collections de la Maison de la R.A.T.P., Paris 14586
Collections de la Maison Henri IV, Saint-Valery-en-Caux 15530
Collections de la Salle des Martyrs-Chapelle-Crypte, Paris 14587
Collections de la Serrurerie Dantin, Paris 14588
Collections de la Tour Eiffel, Paris 14589
Collections de l'Abbaye Bénédictine de Saint-Louis-du-Temple, Vauhallan 16084
Collections de l'Auberge du Cygne, Tôtes 15920
Collections de l'Ecole Nationale Supérieure des Beaux-Arts, Paris 14590
Collections du Château de Born, Saint-Europe-de-Born 15261
Collections du Château de Breil de Foin, Genneteil 13042
Collections du Château de Bridoré, Loches-en-Touraine 13793
Collections du Château de Courances, Courances 12649
Collections du Château de Gizeux, Gizeux 13074
Collections du Château de la Thibaudière, Juigné-Béné 13289
Collections du Château de Lantilly, Cervon 12351
Collections du Château de Sury-le-Comtal, Sury-le-Comtal 15844
Collections du Château des Rohan, Pontivy 14906
Collections du Château du Bouchet, Rosnay 15108
Collections du Château Médiéval des Cars, Les Cars 13696
Collections du Donjon, Bazoges-en-Pareds 11885
Collections du Haras, La Roche-sur-Yon 13418
Collections du Restaurant Procope, Paris 14591
Collections du Service Historique de l'Armée de l'Air, Armées 11688
Collections in Situ de l'Unesco, Paris 14592
Collections Internationales de Marionnettes, Charleville-Mézières 12408
Collections Littéraires, Kolbsheim 13305
Collections of Cypriot Painting, Sculpture and Graphics, Kolonaki 23057
Collections of the Coubertin Foundation, Saint-Rémy-lès-Chevreuse 15494
Collections of the Public Social Welfare Bureau, Bruxelles 03470
Collections of Winckelmann-Institute (Classical Archaeology), Berlin 17245
The Collectors Club, New York 51153
The College Art Gallery, Ewing 48461
College Football Hall of Fame, South Bend 53156
College Museum, Epsom 44183
College Museum, Maynooth 24875
College of Art Gallery, Kumasi 22855
College of Arts and Sciences Gallery, Coral Gables 47798
College of Eastern Utah Prehistoric Museum, Price 52140
College of Marin Art Gallery, Kentfield 49741
College of New Jersey Art Gallery, Trenton 53557
College of Visual Arts Gallery, Saint Paul 52614
The College of Wooster Art Museum, Wooster 54294
College Park Aviation Museum, College Park 47686
Collegium Hungaricum Berlin, Berlin 17060
Colleton Museum, Walterboro 53806
Collezione A. Checchi, Fucecchio 26254
Collezione Archeologica e Foro Romano, Assisi 25304
Collezione Borsari, Parma 27094
Collezione Calderara, Ameno 25223
Collezione Civica d'Arte, Pinerolo 27219
Collezione Contini-Bonacossi, Firenze 26150
Collezione d'Arte della Banca Carige, Genova 26291
Collezione d'Arte di Santa Maria di Piazza, Ostra 27020
Collezione d'Arte Religiosa Moderna, Città del Vaticano 54498
Collezione d'Arte Sacra, Saludecio 27601
Collezione di Carrozze, Maser 26616
Collezione di Ceramiche Mostra Permanente, Fiorano Modenese 26139
Collezione di Farfalle, Guardia Sanframondi 26383
Collezione di Fisica e Medicina, Pavia 27119
Collezione di Marionette Ferrari, Parma 27095
Collezione di Minerali e Fossili, Semione 41859
Collezione di Violini, Cremona 26001
Collezione Ducati Fratelli Saltarelli, Senigallia 27801
Collezione Etnografica, Forni Avoltri 26234
Collezione G. Lorenzi, Milano 26682
Collezione Indiana e dell' Estremo Oriente F. L. Pullè, Bologna 25439
Collezione Martini Carissimo, Oria 26990
Collezione Memorie della Comunità, Romano di Lombardia 27552
Collezione Minici Zotti-Un Museo di Magiche Visioni, Padova 27033
Collezione P. Mariani, Desio 26037
Collezione Permanente del Design Italiano '45-'90, Milano 26683

C

Collezione Permanente di Valentino Martinelli, Perugia 27144
Collezione Privata di Attrezzi ed Utensili Contadini, Camerota 25624
Collezione Sarda L. Piloni, Cagliari 25577
Collezione Scaglione, Locri 26504
Collezione Straka-Coppa, Spello 27885
Collezione Tessile della Tela Umbra, Città di Castello 25904
Collezione Titta Ruffo, Pisa 27235
Collezioni cartografica e topografica, Genova 26292
Collezioni Comunali d'Arte, Imola 26398
Collezioni d'Armi e Ceramiche della Rocca Sforzesca, Imola 26399
Collezioni d'Arte della Cassa di Risparmio, Savona 27775
Collezioni d'Arte e di Storia della Fondazione Cassa di Risparmio, Bologna 25440
Collezioni del '900, Firenze 26151
Collezioni del Dipartimento di Entomologia e Zoologia Agraria, Portici 27300
Collezioni dell'Accademia Nazionale Virgiliana, Mantova 26590
Collezioni di Presepi del Convento Muri-Gries, Bolzano 25483
Collezioni Egittologiche dell'Università di Pisa, Pisa 27236
Collier County Museum, Naples 50951
Collier Memorial State Park and Logging Museum, Chiloquin 47538
Collin County Farm Museum, McKinney 50358
Collingwood Library and Museum on Americanism, Alexandria 46202
Collingwood Museum, Collingwood 05632
Collins Gallery, Glasgow 44278
Collinsville Depot Museum, Collinsville 47695
Collotype Studio Leipzig, Leipzig 19917
Colne Valley Museum, Golcar 44323
Colombo National Museum, Colombo 40370
Colonel Ashley House, Ashley Falls 46393
Colonel Davenport Historical Foundation, Rock Island 52407
Colonel Oskars Kalpaks Museum in the Memorial Site Airītes, Zirņi 30152
Colonel Stephens Railway Museum, Tenterden 45837
Colonel William Jones House, Gentryville 48887
Colonial Burlington Foundation, Burlington 47140
Colonial Cottage Museum, Wellington 33245
Colonial Industrial Quarter, Bethlehem 46778
Colonial Pemaquid Historical Site, New Harbor 51047
Colonial Pennsylvania Plantation, Media 50553
Colonial Williamsburg, Williamsburg 54167
Colony House, Newport 51320
Colorado Gallery of the Arts, Littleton 50143
Colorado Historical Society Museum, Denver 48069
Colorado Photographic Arts Center, Denver 48070
Colorado Railroad Museum, Golden 48957
Colorado River Indian Tribes Museum, Parker 51712
Colorado School of Mines Geology Museum, Golden 48958
Colorado Ski & Snowboard Museum and Hall of Fame, Vail 53680
Colorado Springs Fine Arts Center, Colorado Springs 47699
Colorado Springs Museum, Colorado Springs 47700
Colorado University Art Galleries, Boulder 46957
Colorado University Heritage Center, Boulder 46958
Colorado-Wyoming Association of Museums, Sheridan 55128
Colosseum Art Gallery, Haifa 24958
Colour Museum, Bradford 43578
Coloured World of the Hungarian Calcit Crystals, Fertőrákos 23559
Colter Bay Indian Arts Museum, Colter Bay Village 47709
Colton Area Museum, Colton 47710
Colton Hall Museum, Monterey 50800
Columbia College Art Gallery, Chicago 47468
Columbia County Historical Museum, Lake City 49879
Columbia County Historical Society Museum, Kinderhook 49777
Columbia County Historical Society Museum, Saint Helens 52549
Columbia County Museum, Bloomsburg 46871
Columbia Fire Department Museum, Columbia 47722
Columbia George Interpretive Center, Stevenson 53292
Columbia Museum of Art, Columbia 47723
Columbia River Exhibition of History, Science and Technology, Richland 52307
Columbia River Maritime Museum, Astoria 46401
The Columbian Museum and Art Center, Wamego 53813
Columbus-Belmont Civil War Museum, Columbus 47743
Columbus Cultural Arts Center, Columbus 47751
The Columbus Museum, Columbus 47739
Columbus Museum of Art, Columbus 47752
Columbus Park Museum, Discovery Bay 28322
Colville Lake Museum, Colville Lake 05633
Colzium Museum, Kilsyth 44581
Comanche Crossing Museum, Strasburg 53327
Combat Air Museum, Topeka 53533
Combe Martin Museum, Combe Martin 43881
Combe Mill Beam Engine and Working Museum, Long Hanborough 45025
Comber and District Historical Society Museum, Comber 05634
Comenius Museum, Naarden 32563
Comenius Museum Přerov, Přerov 09804
Comer Museum, Morgantown 50844
Comisión Nacional de Arqueología y Monumentos Historicos, Ciudad de Panamá 54980
Comisión Nacional de Museos y de Monumentos, Buenos Aires 54683
Comitê Brasileiro do ICOM, São Paulo 54721
Comité Español del ICOM, Madrid 55025
Comité Nacional del Consejo Internacional de Museos, Bogotá 54758
Comité National Français de l'ICOM, Paris 54807
Comité Peruano del Consejo Internacional de Museos, Lima 54984
Comitée National de ICOM Centralafrique, Bangui 54750
Comitée National de ICOM de Algérie, Alger 54679
Commanda General Store Museum, Commanda 05636
Commandant's Quarters, Dearborn 48006
Commander Chr. Christensen's Whaling Museum, Sandefjord 33812
La Commanderie d'Arville, Arville 11711
Commandery Museum, Worcester 46070
Commencement Art Gallery, Tacoma 53406
Commerce Bancshares Fine Art Collection, Kansas City 49686
Commercial and Industrial Museum, Kanpur 24253
Commercial Museum, Guwahati 24189
Commissariat House, Saint John's 06846
Commissariat Store Museum, Brisbane 00896
The Commodity Museum of Meiji University, Tokyo 29433
Commonwealth Air Training Plan Museum, Brandon 05462
Commonwealth Association of Museums, Calgary 54736
Commonwealth Institute, London 44828
Commonwealth Museum, Boston 46927
Commonwealth of the Northern Mariana Islands Museum of History and Culture, Saipan 33374
Communication Development Museum, Kaunas 30265
Communications and History Museum of Sutton, Washington 53855
Communications Centre Lokki, Mikkeli 11118
Communications Monument, Seoul 29868
Communications Museum, Tbilisi 16352
Communications Museum, Tokyo 29483
Community Arts Centre, Wanganui 33233
Community Arts Council of the Alberni Valley, Port Alberni 06625
Community Arts Gallery, Detroit 48111
Community College Art Gallery, Baltimore 46557
Community Fine Arts Center, Rock Springs 52410
Community Gallery Spaces, Portland 01438
Community Historical Museum, Maxwell 50538
Community Memorial Museum of Sutter County, Yuba City 54360
Community Museums Association of Prince Edward Island, Charlottetown 54737
Community of Christ, Independence 49472
Comox Air Force Museum, Lazo 06151
Comox Archives & Museum, Comox 05637
Comox Valley Centre for the Arts, Courtenay 05650
Comox Valley Community Arts Council, Courtenay 05651
Complejo Arqueológico de Chavín, Huari 34183
Complejo Histórico Chivilcoy, Chivilcoy 00303
Complejo Museológico del Instituto Félix Bernasconi, Buenos Aires 00155
Complejo Museológico La Estación, Cruz Alta 00350
Complejo Museos Históricos Militares, La Habana 09100
Complesso Museale di Santa Chiara, Napoli 26879
Complesso Museale di Santa Maria della Scala, Siena 27836
Complex Muzeal de Ştiinţele Naturii, Galaţi 36378
Complexe de Betange, Florange 12948
Complexul Memorial Nicolae Titulescu, Nicolae Titulescu 36492
Complexul Monumental Curtea Domnească, Târgovişte 36654
Complexul Muzeal Arad, Arad 36111
Complexul Muzeal Bucovina, Suceava 36642
Complexul Muzeal de Stiinţe ale Naturii Constanţa, Constanţa 36300
Complexul Muzeal de Stiintele Naturii Ion Borcea, Bacău 36120
Complexul Muzeal Măldăreşti, Măldăreşti 36457
Complexul Muzeal Naţional Goleşti, Şefaneşti 36597
Complexul Muzeal Stelea, Târgovişte 36655
Complexul Muzeistic Ciprian Porumbescu, Ciprian Porumbescu 36280
Complexul Muzeistic Gheorghe Magheru, Râmnicu Vâlcea 36551
Complexul Naţional Muzeal Astra, Sibiu 36604
Complexul Naţional Muzeal Curtea Domnească, Târgovişte 36656
Compton Castle, Paignton 45342
Compton County Historical Museum, Eaton Corner 05734
Compton Verney, Warwickshire 45944
Computer Museum, Vilnius 30412
Computer- und Videospiele Museum, Berlin 17061
Computermuseum, Roeselare 03975
Computermuseum Aachen, Aachen 16394
Computing Commons Gallery, Tempe 53466
Comstock House, Moorhead 50835
Comtoise Uhren Museum, Düsseldorf 17986
Con Edison Energy Museum, New York 51154
Concentration Camp Memorial, Ulm 22159
Concentration Camp Memorial Ladelund, Ladelund 19760
Conception Bay Museum, Harbour Grace 05971
Concord Art Association, Concord 47766
Concord Heritage Museum, Concord 00995
Concord Museum, Concord 47767
Concordia Historical Institute, Saint Louis 52572
Concordia University-Wisconsin Art Gallery, Mequon 50593
Conejo Valley Art Museum, Thousand Oaks 53502
Confederate Memorial Hall-Bleak House, Knoxville 49814
Confederate Memorial State Historic Site, Higginsville 49273
Confederate Museum, Crawfordville 47872
Confederate Museum, New Orleans 51077
Confederation Centre Art Gallery, Charlottetown 05591
The Conference House, Staten Island 53261
Congdom Anatomical Museum, Bangkok 42536
Congregation Emanu-el Museum, New York 51155
Congregation Shaar Hashomayim Museum, Westmount 07306
Coninx Museum, Zürich 42068
Conjunt Monumental de la Esglésias de Sant Pere, Terrassa 40117
Conjunto Arqueológico de Baelo Claudia, Tarifa 40098
Conjunto Arqueológico de Carmona, Carmona 39131
Conjunto Arqueológico de Itálica, Santiponce 40037
Conjunto Arqueológico de Madinat Al Zahra, Córdoba 39202
Conjunto Cultural de Caixa, São Paulo 04855
Conjunto de la Ferrería y Molinos de Agorregi, Aia 38797
Conjunto Monumental Alcazaba de Almería, Almería 38858
Conklin Reed Organ and History Museum, Hanover 49161
Conneaut Railroad Museum, Conneaut 47779
Connecticut Audubon Birdcraft Museum, Fairfield 48480
Connecticut Fire Museum, East Windsor 48263
Connecticut Graphic Arts Center, Norwalk 51431
Connecticut Historical Society Museum, Hartford 49193
Connecticut Museum Association, Roxbury 55129
Connecticut River Museum, Essex 48421
Connecticut State Museum of Natural History, Storrs 53318
Connecticut Trolley Museum, East Windsor 48264
Connecticut Valley Historical Museum, Springfield 53224
Connemara History and Heritage Centre, Clifden 24760
Conner Prairie Living History Museum, Fishers 48546
Conrad Mansion Museum, Kalispell 49670
Conrad Weiser Homestead, Womelsdorf 54271
Conseil Bruxellois de Musées, Bruxelles 54710
Conservation Centre Presentation, Liverpool 44732
Conservation Départementale Musées des Pays de l'Ain, Bourg-en-Bresse 12106
Conservation Resource Centre, Glebe 01113
Conservatoire Botanique National de Brest, Brest 12158
Conservatoire de la Faune Sauvage de Sologne, Ménestreau-en-Villette 14074
Conservatoire de l'Air et de l'Espace d'Aquitaine, Mérignac (Gironde) 14088
Conservatoire de l'Estuaire de la Gironde, Blaye 12021
Conservatoire de l'Outil Maison de Pays des Crêtes, Évigny 12887
Conservatoire des Machines de Guerre Médiévales, Tiffauges 15908
Conservatoire du Machinisme et de Pratiques Agricoles, Vouillé 16274
Conservatoire du Patrimoine Religieux, Liessies 13743
Conservatoire Vinicole, Plassac 14832
Conservatoni du Musée, Gravelines 13121
Conservatorio di Santa Chiara, San Miniato 27670
Le Consortium, Dijon 12733
Constitution Convention Museum, Port Saint Joe 52040
Constitution Island - Warner House, West Point 54077
Constitution Museum, Tabriz 24651
Constitution Square, Danville 47961
Constitutional History Gallery, Bandar Seri Begawan 04963
Contemporary Art Center, Sarajevo 04273
Contemporary Art Center, Skopje 30514
Contemporary Art Center of Virginia, Virginia Beach 53755
Contemporary Art Centre of Montenegro, Podgorica 31654
Contemporary Art Centre of South Australia, Parkside 01405
Contemporary Art Gallery, Vancouver 07209
Contemporary Art Gallery at Falkirk Cultural Center, San Rafael 52827
Contemporary Art Museum, Raleigh 52228
Contemporary Art Museum, Tampa 53431
Contemporary Art Museum, Kumamoto, Kumamoto 28779
Contemporary Art Museum of Rome, Roma 27470
Contemporary Art Museum of Saint Louis, Saint Louis 52573
Contemporary Art Workshop, Chicago 47469
Contemporary Arts Center, Cincinnati 47558
Contemporary Arts Center, New Orleans 51078
Contemporary Arts Museum, Houston 49383
Contemporary Ceramic Museum, Santo Domingo 10346
Contemporary Fine Art Museum of Cerveira, Vila Nova de Cerveira 36002
Contemporary History Museum, Tehrān 24659
Contemporary Jewish Museum, San Francisco 52771
Contemporary Museum, Baltimore 46558
The Contemporary Museum, Honolulu 49336
The Contemporary Museum at First Hawaiian Center, Honolulu 49337
Contemporary Ramanian Painting Exhibition, Teremia Mare 36682
Contemporary Sculpture Museum, Tokyo 29393
Contemporary Spanish Engraving Museum, Marbella 39619
Convent museum Müstair, Müstair 41664
Convent of Saint Agnes of Bohemia, Praha 09755
Convent of Saint George, Praha 09756
Convento de San Antonio el Real, Segovia 40050
Convento de Santo Domingo el Antiguo, Toledo 40128
Conway Fire Museum Aka Vintage Fire Engines, New Albany 51009
Conway Scenic Railroad, North Conway 51388
Conwy Valley Railway Museum, Betws-y-Coed 43467
Conzen-Sammlung, Düsseldorf 17987
Cook County Historical Museum, Grand Marais 48997
Cookeville Depot Museum, Cookeville 47785
Cook's Creek Heritage Museum, Dugald 05720
Cooktown School of Art Gallery, Cooktown 00997
Cookworthy Museum of Rural Life in South Devon, Kingsbridge 44591
Coolspring Power Museum, Coolspring 47787
Cooma Cottage, Yass 01681
Cooper Gallery, Barnsley, South Yorkshire 43385
Cooper-Hewitt National Design Museum, New York 51156
Cooper Regional History Museum, Upland 53660
Coopersville Area Historical Society Museum, Coopersville 47793
Coop'ren Duikhelm, Giethoorn 32252
Coos Art Museum, Coos Bay 47794
Coos County Historical Society Museum, North Bend 51381
Copenhagen City Museum, København 10135
Copper Belt Museum, Whitehorse 07324
Copper Cliff Museum, Copper Cliff 05640
Copper King Mansion, Butte 47151
Copper Museum, Legnica 34848
Copper Village Museum and Arts Center, Anaconda 46277
Copperbelt Museum, Ndola 54662
Coppertown U.S.A. Mining Museum, Calumet 47169
Coppini Academy of Fine Arts Gallery and Museum, San Antonio 52713
Copshaholm House Museum and Historic Oliver Gardens, South Bend 53157
Coptic Museum, Cairo 10490
Coquille River Museum, Brandon 46994
Coquille River Museum, Bandon 46594
Coral Creation, the Cosmos of Buddha Coral Museum, Su-Ao 42343
Coral Gables Merrick House, Coral Gables 47799

C

Coral Springs Museum of Art, Coral Springs 47803
Corban Estate Arts Centre, Auckland 33044
Corbridge Roman Site and Museum, Corbridge 43891
Corcoran Gallery of Art, Washington 53856
Cordova Historical Museum, Cordova 47806
Core, Denver 48071
Corfe Castle Museum, Corfe Castle 43892
Corfu Byzantine Collection, Corfu 22976
Corfu Museum of Shells, Kastellanoi Mesis 23040
The Corgi Heritage Centre, Heywood 44464
Corgialenios Historical and Cultural Museum, Argostolion 22890
Corin Maman Ashdod Museum, Ashdod 24928
Corinium Museum, Cirencester 43847
Corita Art Center Gallery, Los Angeles 50218
Cork Butter Museum, Cork 24766
Cork Public Museum, Cork 24767
Cornelius Low House/Middlesex County Museum, Piscataway 51932
Cornell Museum, Delray Beach 48045
Corner Brook Museum & Archives, Corner Brook 05641
Cornice Museum of Ornamental Plasterwork, Peebles 45348
Corning Museum of Glass, Corning 47810
Cornish Mines, Engines and Cornwall Industrial Discovery Centre, Redruth 45476
Cornwall Geological Museum, Penzance 45359
Cornwall Historical Museum, Cornwall 47813
Cornwall Iron Furnace, Cornwall 47814
Cornwall Regional Art Gallery, Cornwall 05643
Coromandel School of Mines Museum, Coromandel 33082
Coronach District Museum, Coronach 05647
Coronado National Memorial Museum, Hereford 49256
Coronado-Quivira Museum, Lyons 50335
Corpsmuseum van het Utrechtsch Studenten Corps, Utrecht 32845
Corpus Christi Museum of Science and History, Corpus Christi 47820
Corredor, Gallery of the College of Fine Arts, Quezon City 34489
Corridor Gallery, Glenrothes 44313
Corrie ten Boomhuis, Haarlem 32294
Corrigin Pioneer Museum, Corrigin 01003
Corris Railway Museum, Machynlleth 45068
Corry Area Historical Society Museum, Corry 47824
Corsham Court, Corsham 43895
Cortijo Bacardi, Málaga 39595
Le Cortil, Falaën 03595
Cortland County Historical Society Museum, Cortland 47830
Çorum Archaeological Sites, Çorum 42756
Çorum Devlet Güzel Sanatlar Galerisi, Çorum 42754
Çorum Müzesi, Çorum 42755
Çorum Örenyerleri, Çorum 42756
Çorum State Fine Arts Gallery, Çorum 42754
Corvallis Art Center, Corvallis 47832
Corvin János Múzeum, Gyula 23596
Corvin János Múzeum, Gyula 23591
COSI Columbus, Columbus 47753
Cosi Toledo, Toledo 53521
CosmoCaixa Madrid, Alcobendas 38822
Cosmodôme, Laval 06146
Cossit House Museum, Sydney 07048
Costume Museum, Kyoto 28810
Costume Museum of Canada, Dugald 05721
Costume Museum of Shih Chien University, Taipei 42386
Cotabato National Museum, Cotabato 34381
Cotehele House, Saint Dominick, Cornwall 45562
Cothey Bottom Heritage Centre, Ryde, Isle of Wight 45534
Cotroceni National Museum, București 36218
Cotswold Motoring Museum and Toy Collection, Bourton-on-the-Water 43568
Cottage Blundell's, Canberra 00945
Cottage Grove Museum, Cottage Grove 47841
Cottage Lawn, Oneida 51559
Cottage Museum, Lancaster 44630
Cotton Spinning & Weaving Workshop, Shanghai 08164
Cottonlandia Museum, Greenwood 49085
Cottonwood County Historical Society, Windom 54224
Cottonwood House, Quesnel 06702
Coudray-Haus, Bad Berka 16689
Coughton Court, Alcester 43264
Coultershaw Beam Pump, Petworth 45376
Coumantaros Art Gallery of Sparta, Spárti 23195
Council of American Jewish Museums, New York 55130
Council of American Maritime Museums, Bath 55131
Council of Australian Art Museum Directors, Melbourne 54688
Council of Australian Museum Directors, Haymarket 54689
Council of Galleries of the Czech Republic, Praha 54775
Count Charles Esterházy Castle and Regional Museum, Pápa 23752
The Country Doctor Museum, Bailey 46530
Country Estate Museum Arkhangelskoe, Archangelskoe 36773
Country Heritage Park, Milton 06288
Country House and Open Air Ethnographic Museum, Bakonybél 23349
Country Museum, Vestmanna 10703
Country Museum, Vestmannaeyjar 24042
Country Museum of Borsod, Edelény 23534
Country Music Hall of Fame and Museum, Nashville 50964
Countryside Museum, East Budleigh 44082
Count's Castle, Gent 03627
County Donegal Railway Heritage Centre, Donegal 24775
County Gallery, Namsos 33692
County Hall, Eger 23541
County Hall Gallery, London 44829
County Museum, Sligo 24899
County Museum Dundalk, Dundalk 24827
County Museum of Somogy, Kaposvár 23635
County Museum Technical Centre, Halton 44391
County of Los Angeles Fire Museum, Los Angeles 50219
Court Hall Museum, Sittingbourne 45681
Court House and Museum, Pevensey 45379
Court House Museum, Cardston 05561
Court House Museum, Stroud 01534
Courtauld Institute Gallery, London 44830
Courtenay and District Museum and Paleontology Centre, Courtenay 05652
Courthouse Gallery, Portsmouth 52106
Courthouse Museum, Berrien Springs 46766
Courthouse Museum, Exira 48466
Courthouse Museum, Jacksonport 49574
Courthouse Square Museum, Charlotte 47383
Courtney and Lemmerman Galleries, Jersey City 49626
Courtyard Gallery, Washington 53946
Cousland Smiddy, Cousland 43898
Coutts Memorial Museum of Art, El Dorado 48312
Couven Museum, Aachen 16395
Coventry Toy Museum, Coventry 43899
Coventry Transport Museum, Coventry 43900
Coventry Watch Museum, Coventry 43901
Cowan Vertebrate Museum, Vancouver 07210
Cowboy-Museum Fatsy, Linz 02311
Cowbridge and District Museum, Cowbridge 43906
Cowes Maritime Museum, Cowes 43907
Cowichan Bay Maritime Centre, Cowichan Bay 05654
Cowichan Valley Museum, Duncan 05723
The Cowley County Historical Museum, Winfield 54232
Cowlitz County Historical Museum, Kelso 49723
Cowpens National Battlefield, Chesnee 47429
Cowper and Newton Museum, Olney 45316
Cowra Museums, Cowra 01006
Coyaba Gardens and Awarak Museum, Ocho Rios 28338
Coyote Art Gallery, Oroville 51605
Coyote Point Museum for Environmental Education, San Mateo 52822
Crabble Corn Mill, Dover 44011
Cracker Country Museum, Tampa 53432
Cracker Trail Museum, Zolfo Springs 54374
Cracow Saltworks Museum, Wieliczka 35300
Cradle of Aviation Museum, Garden City 48855
Craft Alliance, Saint Louis 52574
Craft Museum of Finland, Jyväskylä 10897
Crafts Council Gallery, London 44831
Crafts Galley of the National Museum of Modern Art Tokyo, Tokyo 29492
Crafts Museum, Delhi 24139
Crafts Museum, Mequon 50594
Craftselijke Zadelmakerij Museum, Bellingwolde 31959
Craftsman Farms Foundation, Parsippany 51716
Cragside House, Rothbury 45516
Craig Heritage Park Museum, Parksville 06568
Craigavon Museum, Craigavon 43912
Craigdarroch Castle, Victoria 07245
Craik Community Archives & Oral History Society, Craik 05655
Crail Museum, Crail 43913
Crailo State Historic Site, Rensselaer 52294
Craiova Art Museum, Craiova 36324
Crampton Tower Museum, Broadstairs 43645
Cranach-Haus, Lutherstadt Wittenberg 20113
Cranbrook Art Museum, Bloomfield Hills 46855
Cranbrook House, Bloomfield Hills 46856
Cranbrook Institute of Science - Natural History Collections, Bloomfield Hills 46857
Cranbrook Museum, Cranbrook 43914
Cranbury Historical and Preservation Society Museum, Cranbury 47866
Crane Museum, Izumi 28611
Crane Museum of Papermaking, Dalton 47949
Crane Point Hammock Museum, Marathon 50458
Crary Art Gallery, Warren 53826
Craters of the Moon, Arco 46342
Crathes Castle, Banchory 43373
Craven Museum and Gallery, Skipton 45688
Cravens House, Lookout Mountain 50197
Crawford Arts Centre, Saint Andrews 45553
Crawford Auto-Aviation Museum, Cleveland 47628
Crawford County Historical Museum, Pittsburg 51934
Crawford Municipal Art Gallery, Cork 24768
Crawford W. Long Museum, Jefferson 49605
Crawfordjohn Heritage Venture, Crawfordjohn 43915
Crawley Museum, Crawley 43916
Crazy House Museum, Đà Lat 54602
Crazy Mountain Museum, Big Timber 46801
Creamery Square Museum, Tatamagouche 07060
Creatabilitoys! - Museum of Advertising Icons, Coral Gables 47800
Creation Museum, Petersburg 51795
Creative Art Center, Jhongli 42220
Creative Arts Center, Kansas City 49687
Creative Arts Center, Pontiac 52020
Creative Arts Guild, Dalton 47947
Creative Arts Studio, Louisville 50283
Creative Center and Exhibition Hall Fyodor, Sankt-Peterburg 37729
Creative Discovery Museum, Chattanooga 47408
Creative Growth Art Center, Oakland 51464
Creative Spirit Art Centre, Toronto 07103
Creativiteitscentrum Gouden Handen, 's-Heerenberg 32330
Creek Council House Museum, Okmulgee 51528
Creetown Gem Rock Museum, Creetown 43918
Creetown Heritage Museum, Creetown 43919
Cregneash Folk Village, Cregneash 43920
Cregneash Village Folk Museum, Port Saint Mary 45413
Crepe Silk History Museum, Nodagawa 29080
Crescent Arts, Scarborough 45622
Crescent Bend/Armstrong-Lockett House and William P. Toms Memorial Gardens, Knoxville 49815
Crescentia-Gedenkstätte, Kaufbeuren 19484
Creski Muzej, Cres 08856
Cress Gallery of Art, Chattanooga 47409
Crestar Bank Art Collection, Richmond 52326
Crestet-Centre d'Art, Le Crestet 13575
Crestline Shunk Museum, Crestline 47877
Creston and District Museum, Creston 05659
Creswell Crags Museum and Education Centre, Welbeck 45954
Creswick Museum and Gold Battery, Creswick 01011
Le Creux de l'Enfer, Centre d'Art Contemporain, Thiers 15880
Crewkerne and District Museum and Heritage Centre, Crewkerne 43924
Criccieth Castle, Criccieth 43926
Crichton Museum, Dumfries 44034
Cricoteca-Galeria-Pracownia Tadeusza Kantora, Kraków 34768
Crime Museum, Vantaa 11447
Crimean Republics Regional Museum, Simferopol' 43185
Criminal Museum, Jakarta 24488
Criminological Museum of the Altay State University, Barnaul 36792
Criminology Museum, Abbasiya 10464
Cripple Creek District Museum, Cripple Creek 47882
Cripta e Museo di Sant'Anastasio, Asti 25312
Criş County Museum, Oradea 36505
Cristalleries Royale de Saint-Louis, Saint-Louis-les-Bitche 15378
Croatian Academy of Sciences and Arts - Glyptothèque, Zagreb 08993
Croatian Heritage Museum, Eastlake 48266
Croatian Historical Museum, Zagreb 08997
Croatian Museum of Architecture, Zagreb 08994
Croatian Museum of Literature and Theatre, Zagreb 09005
Croatian Museum of Naive Art, Zagreb 08995
Croatian Museum of Post and Telecommunications, Zagreb 08996
Croatian Natural History Museum, Zagreb 08998
Croatian School Museum, Zagreb 08999
Croatian Sports Museum, Zagreb 09000
Crocker Art Museum, Sacramento 52505
Crocker House, Mount Clemens 50872
Crockett County Museum, Ozona 51664
Crockett Tavern Museum, Morristown 50858
Croft Castle, Leominster 44690
Croft Museum, Luopioinen 11085
Crofter's House Museum, Somero 11324
Crofter's Museum of Laurinmäki, Janakkala 10870
Crofton Roman Villa, Orpington 45320
Croiseur Colbert, Bordeaux 12063
Crom Estate, Newtownbutler 45263
Cromarty Courthouse Museum, Cromarty 43927
Cromer Museum, Cromer 43929
Cromford Mill, Cromford 43932
Cromwell Museum, Huntingdon 44500
Cronomuseu de Queiriga, Queiriga 35877
Cronosaurio, Villa de Leyva 08781
Crook County Museum and Art Gallery, Sundance 53370
Crooked Tree Arts Center, Petoskey 51806
Crosby County Pioneer Memorial Museum, Crosbyton 47889
Crossman Art Gallery, Whitewater 54134
The Crossness Engines, London 44832
Crossroads Museum, Oyen 06565
Crossroads of America, Bethlehem 46776
Crow Canyon Archaeological Center, Cortez 47827
Crow Wing County Historical Museum, Brainerd 46991
Crowder College-Longwell Museum and Camp Crowder Collection, Neosho 51004
Crowley Art Association and Gallery, Crowley 47893
Crowley Museum, Sarasota 52913
Crown Gardens Museum, Dalton 47948
Crown Liquor Saloon, Belfast 43439
Crown Mill, Hämeenlinna 10747
Crown Point State Historic Site, Crown Point 47895
Crowsnest Museum, Coleman 05631
Crowsnest Pass Ecomuseum, Bellevue 05416
Croxteth Hall, Liverpool 44733
Croydon Museum, Croydon 43934
Croydon Natural History and Scientific Society, Croydon 43935
Cruizer Aurora Museum, Sankt-Peterburg 37688
Crusader Church, Abu Ghosh 24918
The Crypt - Town Hall Plate Room, Oxford 45331
Crypte Archéologique du Parvis de Notre-Dame, Paris 14593
Crypte Carolingienne de l'Ancienne Abbaye, Flavigny-sur-Ozerain 12943
Cryptoneum, Rostock 21387
Crystal Brook Heritage Centre, Crystal Brook 01015
Crystal Caverns at Hupp's Hill, Strasburg 53330
Crystal City Community Museum, Crystal City 05663
Crystal Kingdom, Coonabarabran 00999
Crystal Lake Falls Historical Association, Barton 46621
Crystal Mountain - Crystal & Mineral Gallery & Museum, Auckland 33045
Crystal Palace Museum, London 44833
Crystal River State Archaeological Site, Crystal River 47897
Csabai Tanya- és Gabonatörténeti Kiállítóhely, Békéscsaba 23368
Csavolyer Heimatstuben, Waiblingen 22275
Cselley Ház, Mosonmagyaróvár 23718
Cselley House of the Hanság Múzeum, Mosonmagyaróvár 23718
Csepel Galéria és Helytörténeti Gyűjtemény, Budapest 23394
Csepel Gallery and Region Historical Collection, Budapest 23394
Cserépfalvi Imre Emlékszoba, Cserépfalu 23506
Csohány Galerie, Pásztó 23754
Csók István Képtár, Székesfehérvár 23852
Csontváry Múzeum, Pécs 23758
Csúcsi Fazekasház, Hódmezővásárhely 23610
CTS Turner Museum, Elkhart 48345
Cu Heritage Center, Boulder 46959
Cuadros Vivos de la Sabana, Corozal 08628
Cuban Museum of Arts and Culture, Miami 50623
Cubus Kunsthalle, Duisburg 18014
Cuckfield Museum, Cuckfield 43936
Cudrio Kystmuseum, Langesund 33625
Cue Art Foundation, New York 51157
Cuivrerie de Cerdon, Cerdon 12337
Cukormúzeum, Szerencs 23886
Culbertson Mansion, New Albany 51010
Culbertson Museum, Culbertson 47900
Culinair Historisch Kookmuseum de Vleer, Appelscha 31916
Culinair Museum, Amersfoort 31811
Culinary Archives and Museum, Providence 52163
Cullman County Museum, Cullman 47902
Culloden Battlefield and Visitor Centre, Inverness 44523
Culross Palace, Culross 43938
Cultural Center and Museum, Wabag 34087
Cultural Center Hellenic Cosmos, Athinai 22906
The Cultural Center of Fine Arts, Parkersburg 51713
Cultural Center of the Philippines Museo, Pasay 34471
Cultural Centre de Dorval, Dorval 05709
Cultural Centre of Ronneby, Ronneby 40815
Cultural Heritage Center Museum, Toppenish 53538
Cultural Heritage Centre, Grand-Bassam 08835
Cultural Historical Museum, Dubrovnik 08865
Cultural History Museum, Búdardalur 23991
Cultural Museum, Kabacan 34394
Cultural Museum, Melaka 30594
Cultural Museum at Scarritt-Bennett Center's, Nashville 50965
Cultural Museum of Hemei, Hemei 42199
Cultural Research Institute, Kolkata 24272
Cultural Rights and Protection/Ute Indian Tribe, Fort Duchesne 48618
Culture and Exhibition Center Sokolniki, Moskva 37223

C

D

Culture and History Museum of Nobles Oginskiai, Rietavas 30345
Culture and Military Museum, Grafenwöhr 18733
Culture Exhibition Hall of Singang Fengtian Temple, Singang 42337
Culture Gallery of Naive Artists, Trebnje 38529
Culture House, National Centre for Cultural Heritage, Reykjavík 24029
Cultureel Centrum van Gogh, Zundert 33017
Cultureel Maçonniek Centrum Prins Frederik, Den Haag 32070
Cultuur-Historisch Museum Sorgdrager, Hollum 32399
Cultuurhistorisch Museum Valkenswaard, Valkenswaard 32868
Cultuurhistorisch Streek- en Handkarrenmuseum De Wemme, Zuidwolde, Drenthe 33014
Cultuurhistorisch Streekmuseum De Acht Zaligheden, Eersel 32178
Culzean Castle and Country Park, Maybole 45120
Cum Museum, Imphal 24223
Cumberland County Historical Society Museum, Greenwich 49084
Cumberland County Museum, Amherst 05331
Cumberland Gap National Historical Park, Middlesboro 50655
Cumberland Heritage Village Museum, Cumberland, Ontario 05665
Cumberland House, Regina 06715
Cumberland Museum and Archives, Cumberland 05664
Cumberland Pencil Museum, Keswick 44562
Cumberland Science Museum, Nashville 50966
Cumberland Theatre Arts Gallery, Cumberland 47906
Cumberland Township Historical Society, Ottawa 06543
Cumberland Toy and Model Museum, Cockermouth 43867
Cumbernauld Museum, Cumbernauld 43940
Cumhuriyet Devri Müzesi, Ankara 42700
Cumhuriyet Egitim Müzesi, İstanbul 42813
Cumhuriyet Eğitim Müzesi, Karataş 42865
Cuming Museum, London 44834
Cummer Museum of Art, Jacksonville 49577
Cunderdin Municipal Museum, Cunderdin 01016
Cuneo Museum, Vernon Hills 53725
Cunningham Dax Collection, Parkville, Victoria 01406
Cupar and District Heritage Museum, Cupar 05666
Cupertino Historical Museum, Cupertino 47910
Cupids Museum, Cupids 05667
Cupola House, Egg Harbor 48303
Curaçao Museum, Willemstad 09358
Curfman Gallery and Duhesa Lounge, Fort Collins 48605
Curiosamuseum Heraldiek en Folklore, Alveringem 03266
The Curious Kids' Museum, Saint Joseph 52557
Curonian Spit National Park Nature Museum, Klaipėda 30280
Currahee Military Museum, Virden 07268
Currency Museum, Institute for Monetary and Economics Studies, Bank of Japan, Tokyo 29406
Currency Museum of the Bank of Canada, Ottawa 06544
Currier Gallery of Art, Manchester 50421
Curry Historical Society Museum, Gold Beach 48954
Curtis Museum, Alton 43293
Curwood Castle, Owosso 51642
Cusanus-Geburtshaus, Bernkastel-Kues 17296
Cushing House Museum, Newburyport 51305
Cust Museum, Cust 33083
Custer Battlefield Museum, Garryowen 48869
Custer County 1881 Courthouse Museum, Custer 47915
Custer County Art Center, Miles City 50682
Custer County Historical Society Museum, Broken Bow 47037
Custer Museum, Clayton 47598
Custom House Maritime Museum, Newburyport 51306
Custom House Visitor Centre, Dublin 24782
Customs House, Oslo 33745
Customs House Nautical Museum, Robe 01462
Customs Museum, Helsinki 10831
Customs Museum, Lisboa 35687
Customs Museum, Stockholm 40948
Customs Museum, Taipei 42387
Customs National Museum, Bordeaux 12073
Čuvašskij Gosudarstvennyj Chudožestvennyj Muzej, Čeboksary 36836
Čuvašskij Nacionalnyj Muzej, Čeboksary 36837
Cuyler Manor Museum, Uitenhage 38767
Cuyuna Range Museum, Crosby 47887
CWU Anthropology Department Collection, Ellensburg 48358
Cy Twombly Gallery, Houston 49384
Cyfarthfa Castle Museum and Art Gallery, Merthyr Tydfil 45130
Cygnaeuksen Galleria, Helsinki 10770
Cylburn Nature Museum, Baltimore 46559
Cynon Valley Museum and Gallery, Aberdare 43232
Cypress College Fine Arts Gallery, Cypress 47918
Cyprus Federation of Associations of Friends of Museums, Lefkosia 54770
Cyprus Folk Art Museum, Lefkosia 09373
Cyprus Historical Museum, Lefkosia 09374
Cyprus Jewellers Museum, Laiki Geitonia 09366
Cyprus Medieval Museum, Lemesos 09385
Cyprus Museum, Lefkosia 09375
Cyprus National Struggle Museum, Lefkosia 09376
Cyprus Olympic Committee Museum, Lefkosia 09377
Cyriel Verschaeve Museum, Alveringem 03267
Cyrus H. McCormick Memorial Museum, Steeles Tavern 53281
Cz. Poborski Mineral Deposits Museum of the Silesian Technical University, Gliwice 34664
Czaar Peterhuisje, Zaandam 32975
Czartoryski Library of the National Museum in Cracow, Kraków 34764
Czech Center New York, New York 51158
Czech Committee of ICOM, Brno 54773
Czech Heritage Museum and Genealogy Center, Temple 53473
Czech Museum of Fine Arts, Praha 09737
Czech Museum of Music, Praha 09736
Czech Museum of Silver - Hrádek, Kutná Hora 09612
Czersk Castle, Czersk 34612
Czibor Zoltán Emlékszoba, Komárom 23676
Czibor Zoltán Memorial Room, Komárom 23676
Czóbel Múzeum, Szentendre 23871

D

D.A. Wurfel Grain Collection, Pinnaroo 01416
D. Baazov Georgian Jewish Historic and Ethnographic Museum, Tbilisi 16353
D.C. Booth Historic National Fish Hatchery, Spearfish 53199
D-Day Museum and Overlord Embroidery, Portsmouth 45429
D-Day Omaha Musée, Vierville-sur-Mer 16154
D.H. Lawrence Birthplace Museum, Eastwood 44108
D.I. Mendeleev Museum, Boblovo 36811
D.I. Mendeleev Museum and Archives, Sankt-Peterburg 37671
D.I. Nalbandyan Studio and Exhibition Hall, Moskva 37423
D.R. Barker Historical Museum, Fredonia 48776
D.R. Visual Art Center, Nakorn-Chaisri 42587
D. Ralph Hostetter Museum of Natural History, Harrisonburg 49186
D137, Sankt-Peterburg 37634
D137 The Gallery of Contemporary Art, Sankt-Peterburg 37634
DA, Hörstel 19217
Da Vinci-Museum, Tongerlo 04076
daadgalerie, Berlin 17062
Da'an Museum, Da'an 07634
DAAP Galleries, Cincinnati 47559
Dacorum Heritage, Berkhamsted 43455
Dacotah Prairie Museum and Lamont Art Gallery, Aberdeen 46116
Dactyl Foundation, New York 51159
Dadaocheng Art Museum, Taipei 42388
Dade Battlefield, Bushnell 47149
Dadian Gallery, Washington 53857
Dadiani Museum, Dadiani 16338
Daegu National Museum, Daegu 29792
Daegu National University of Education Museum, Daegu 29793
Daegwallyeong Museum, Gangneung 29807
Daejeon Museum of Art (DMA), Daejeon 29799
Daensmuseum en Archief van de Vlaamse Sociale Strijd, Aalst 03258
DAF Automobiel Museum, Eindhoven 32185
Dagali Museum, Geilo 33511
Dagbladmuseum Abraham Verhoevenhuis, Hoboken 03704
Dagestan Museum of Fine Arts, Machačkala 37116
Dagestan United Museum of History and Architecture, Machačkala 37115
Dagestanskij Gosudarstvennyj Objedinennyj Istoriko-architekturnyj Muzej, Machačkala 37115
Dagestanskij Muzej Izobrazitelnych Iskusstv, Machačkala 37116
Dagon Collection, Haifa 24959
Dagu Paotai Yi Zhi Museum, Tianjin 08281
Dahe Cun Yi Zhi Guan, Zhengzhou 08497
Dahe Primitive Village Ruins Museum, Zhengzhou 08497
Dahesh Museum of Art, New York 51160
Daheshite Museum, Beirut 30156
Dahl Arts Center, Rapid City 52250
Dahlonega Courthouse Gold Museum, Dahlonega 47920
Dai County Museum, Daixian 07635
Daigaku Bijutsukan, Kyoto 28804
Daigo Fukuryu Maru Exhibition Hall, Tokyo 29379
Daigoji Reihokan, Kyoto 28805
Daigoji Temple Treasure Hall, Kyoto 28805
Daihozoden Homotsukan, Ikoma 28564
Daihozoden Treasure House, Ikoma 28564
Daiichi Bijutsukan, Nagoya 28995
Daiichi Museum of Art, Nagoya 28995
Daikaku-ji Temple, Kyoto 28806
Daikan Yashiki and Setagaya Museum of History, Tokyo 29380
Daikan Yashiki oyo Setagaya Rekishi Hakubutsukan, Tokyo 29380
Dailininkės Veronikos Šleivytės Paveikslų Galerija, Viktariskes 30398
Dailininko Prano Domšaičio Galerija ir Kultūros Centras, Klaipėda 30277
Daimaru Museum Kobe, Kobe 28723
Daimaru Museum Kyoto, Kyoto 28807
Daimaru Museum Sapporo, Sapporo 29214
Daimaru Museum Shinsaibashi Osaka, Osaka 29123
Daimaru Museum Tokyo, Tokyo 29381
Daimaru Museum Umeda Osaka, Osaka 29124
Daimler Contemporary, Berlin 17063
Daimyo Clock Museum, Tokyo 29382
Daimyo Tokei Hakubutsukan, Tokyo 29382
Dairy-Museum, Saukkola 11300
Dairy Museum, Székesfehérvár 23862
Dairy Museum, Vaasa 11424
Daisy Airgun Museum, Rogers 52439
Dakota County Museum, South Saint Paul 53181
Dakota Dinosaur Museum, Dickinson 48138
Dakota Science Center, Grand Forks 48986
Dakota Territorial Museum, Yankton 54323
Dakotaland Museum, Huron 49450
Dakshina Chitra Museum, Muttukadu 24325
Dal-Paso Museum, Lamesa 49912
Dalälvarnas Flottningsmuseum, Sandviken 40831
Dalahästmuseum, Dala-Järna 40483
Dalane Folkemuseum, Egersund 33470
Dalarnas Idrottmuseum, Falun 40514
Dalarnas Indianmuseum, Dala Husby 40482
Dalarnas Museum, Falun 40515
Dalbeattie Museum, Dalbeattie 43944
Dalby Regional Gallery, Dalby 01020
Dalemain Historic House, Penrith 45356
Dalénium - Gustaf Daléns Upptäckarcentrum, Stenstorp 40881
Dalénmuseet, Stenstorp 40882
Dales Countryside Museum, Hawes 44424
Dalgarven Mill, Kilwinning 44583
Dalhousie Art Gallery, Halifax 05933
Dalí - Espace Montmartre, Paris 14594
Dali Museum, Dali 07636
Dalian Gallery, Dalian 07638
Dalian Natural History Museum, Dalian 07639
Dalizhou Museum, Dali 07637
Dalkeith Arts Centre, Dalkeith 43945
Dallam-Hartley XIT Museum, Dalhart 47922
The Dallas Center for Contemporary Art, Dallas 47928
Dallas Historical Society Museum, Dallas 47929
Dallas Holocaust Museum, Dallas 47930
Dallas Museum of Art, Dallas 47931
Dallas Museum of Natural History, Dallas 47932
The Dalles Art Association Gallery, The Dalles 47944
Dalmeny House, South Queensferry 45694
Dalnavert Museum, Winnipeg 07357
Dalnevostočnyj Chudožestvennyj Muzej, Chabarovsk 36858
Dals Ironworks Museum, Dalsbruk 10718
Dalslands konstmuseum, Åsensbruk 40443
Dalton Defenders Museum, Coffeyville 47674
Dalton Gallery, Decatur 48012
Dalvík Local Museum, Dalvík 23992
Daly House Museum, Brandon 05463
Dam Site Museum, Mangla 34030
Damerla Rama Rao Memorial Art Gallery, Rajahmundry 24368
Damjanich János Múzeum, Szolnok 23896
Dampffährschiff "Stralsund" Wolgast, Wolgast 22672
Dampfkornbranntweinbrennerei-Museum, Wildeshausen 22568
Dampflok-Museum, Hermeskeil 19147
Dampflok- und Modelleisenbahnmuseum, Tuttlingen/Möhringen 22139
Dampfmaschinenhaus-Moschee, Potsdam 21086
Dan Andersson Museum, Ludvika 40695
Dan City Museum, Dancheng 07645
Dan Estuary Museum, Shiyan 08237
Dan Flavin Art Foundation Temporary Gallery, New York 51161
Dan Flavin Art Institute, Bridgehampton 47013
Dan Kook University Museum, Seoul 29869
Dan O'Laurie Canyon Country Museum, Moab 50765
Dan Quayle Center and Museum, Huntington 49425
Dan Winters House - Ancestral Home, Loughgall 45036
Dan-yr-Ogof Showcaves Museum, Abercrave 43231
The Dana-Thomas House, Springfield 53215
Danbury Museum, Danbury 47952
Dance Collection - Stiftung John Neumeier, Hamburg 18919
Dance Cottage Museum, Ravensthorpe 01453
Dance Museum, La Habana 09140
Dance Museum, Stockholm 40890
Dandong Museum on the War to resist U.S. Aggression and Aid Korea, Dandong 07646
Dane G. Hansen Memorial Museum, Logan 50160
Danevirke Museum, Dannewerk 17782
Danford Collection of West African Art and Artefacts, Birmingham 43504
Danforth Gallery, Portland 52063
Danforth Museum of Art, Framingham 48729
Dangyang Museum, Dangyang 07647
Dániel Berzsenyi Memorial Museum, Egyházashetye 23545
Dániel Berzsenyi Memorial Museum, Nikla 23731
Daniel Boone Homestead, Birdsboro 46818
Daniel-Pöppelmann-Haus, Herford 19139
Daniel W. Tantoco jr. Collection, Manila 34438
Daniel Weetaluktuk Museum and Cultural Transmission Centre, Inukjuak 06015
Daniels County Museum and Pioneer Town, Scobey 52968
Danielson-Kalmari Villa, Vääksy 11430
Danielson-Kalmarin Huvila, Vääksy 11430
Danish Agricultural Museum, Auning 09973
Danish Amber Museum, Oksbøl 10220
Danish Canadian National Museum, Dickson 05700
The Danish Cultural Institute, Edinburgh 44118
Danish Film Institute, København 10128
Danish Immigrant Museum, Elk Horn 48343
Danish Museum of Books, København 10126
Danish Museum of Electricity, Bjerringbro 09976
Danish Museum of Hunting and Forestry, Hørsholm 10100
Danish Museum of the Medias, Odense 10212
Danish Museum Service, Viborg 54782
Danish Railway Museum, Odense 10211
Danmarks Bogmuseum, København 10126
Danmarks Cykelmuseum, Ålestrup 09960
Danmarks Fotomuseum, Herning 10070
Danmarks Jernbanemuseum, Odense 10211
Danmarks Keramikmuseum, Middelfart 10188
Danmarks Kloster Museum, Ry 10254
Danmarks Kunstbibliotek, København 10127
Danmarks Mediemuseum, Odense 10212
Danmarks Tekniske Museum, Helsingør 10063
Dansk ICOM, København 54778
Dansk Jagt- og Skovbrugsmuseum, Hørsholm 10100
Dansk Landbrugsmuseum, Auning 09973
Dansk Veteranflysamling, Skjern 10268
Det Danske Filminstitut, København 10128
Det Danske Kunstindustrimuseum, København 10129
Dansmuseet, Stockholm 40890
Danube Museum - Hungarian Museum of Water Administration and Environmental Protection, Esztergom 23551
Danubiana, Bratislava 38205
Danvers Historical Society Exhibition, Danvers 47956
Danville Museum of Fine Arts and History, Danville 47963
Danville Science Center, Danville 47964
Dapeng Gucheng Museum, Shenzhen 08219
Daphne du Maurier's Smugglers's at Jamaica Inn, Bolventor 43557
Dapu Museum, Dapu 07648
Daqing Dumeng Museum, Daqing 07649
Dar-es-Salaam National Museum, Dar-es-Salaam 42512
Dar-Hervey Ely House, Rochester 52394
Dar Maâkal Az-Zaïm, Tunis 42663
Dar Museum First Ladies of Texas Historic Costumes Collection, Denton 48060
Daraphirom Royal Mansion Museum, Bangkok 42537
Darbarhall Museum, Junagadh 24247
Darby Houses, Coalbrookdale 43858
Dargaville Museum, Dargaville 33085
Darl At Tifl Museum, Jerusalem 34055
Darlingford School Heritage Museum, Darlingford 05673
Darlington Arts Centre and Myles Meehan Gallery, Darlington 43948
Darlington Raceway Stock Car Museum, Darlington 47968
Darlington Railway Centre and Museum, Darlington 43949
Darnall's Chance House Museum, Upper Marlboro 53664
Daros Exhibitions, Zürich 42070
Daros Exhibitions, Zürich 42069
Daros-Latinamerica AG, Zürich 42070
Darß-Museum, Prerow, Ostseebad 21119
Darßer Arche, Wieck am Darß 22537
Dartford Borough Museum, Dartford 43951
Dartington Crystal Glass Museum, Torrington 45868
Dartmoor District Coach House Museum, Dartmoor 01027

Dartmouth Castle, Dartmouth 43952
Dartmouth Heritage Museum, Dartmouth 05675
Dartmouth Museum, Dartmouth 43953
Darvas József Irodalmi Emlékház és Kutatóépület, Orosháza 23738
Darvinovskij Muzej, Moskva 37157
Darwin Museum, Moskva 37157
Das Wattenmeerhaus, Wilhelmshaven 22574
DASA, Dortmund 17917
Dåsettunet, Lyngdal 33654
Dat lütt Museum, Vielank 22228
Datian Museum, Datian 07650
Datong Museum, Datong 07651
Dauerausstellung im Stadtarchiv, Heilbronn 19096
Dauerausstellung Württembergische Schwarzwaldbahn, Althengstett 16508
Daugavas Muzejs, Salaspils 30122
Daugavpils Museum of Local Studies and Art, Daugavpils 29998
Daugavpils Novadpētniecības un Mākslas Muzejs, Daugavpils 29998
Daugavpils Universitātes Muzejs, Daugavpils 29999
Daugavpils University Museum, Daugavpils 29999
Daughters of the American Revolution Museum, Washington 53858
Daughters of Utah Pioneers Pioneer Memorial Museum, Salt Lake City 52691
Daugyvene Museum-Reservation on History of Culture, Radviliškis 30341
Daugyvenės Kultūros Istorijos Muziejus-draustinis, Radviliškis 30341
Daum Museum of Contemporary Art, Sedalia 53023
Daura Gallery, Lynchburg 50325
Davao Museum, Davao 34385
Davenport House Museum, Savannah 52939
Davenport Museum of Art, Davenport 47969
Daventry Museum, Daventry 43956
David Bradford House, Washington 53941
David Collection, København 10130
David Conklin Farmhouse, Huntington 49428
David Crawford House, Newburgh 51300
The David Davis Mansion, Bloomington 46859
David L. Mason Children's Art Museum, Dunedin 48207
David Livingstone Centre, Blantyre 43542
David Palombo Museum, Jerusalem 24984
David Stefánsson Memorial Museum, Akureyri 23981
David Strawn Art Gallery, Jacksonville 49587
David Winton Bell Gallery, Providence 52164
Davids Samling, København 10130
Davidson County Historical Museum, Lexington 50063
Davidson County Museum of Art, Lexington 50064
Davis Art Center, Bountiful 46964
Davis Art Center, Davis 47975
Davis Art Gallery, Columbia 47715
Davis Museum, Wellesley Hills 54023
Davison Art Center, Middletown 50657
Dawes County Historical Society Museum, Chadron 47325
Dawlish Museum, Dawlish 43957
Dawson City Museum, Dawson City 05681
Dawson County Historical Museum, Lexington 50065
Dawson Creek Art Gallery, Dawson Creek 05683
Dawson Creek Station Museum, Dawson Creek 05684
Dawson Folk Museum, Theodore 01584
Dawson Springs Museum and Art Center, Dawson Springs 47981
Dawu Museum, Dawu 07652
Daxin Museum, Daxin 07653
Daye Museum, Daye 07654
Dayi Liu's Manor-House Museum, Chengdu 07615
Daylesford and District Historical Museum, Daylesford 01031
Dayton Art Institute, Dayton 47984
Dayton Historical Depot Society Museum, Dayton 47992
Dayton Visual Arts Center, Dayton 47985
Dazaifu Tenman-gu Hômotsuden, Dazaifu 28407
Dazaifu Tenman-gu Treasure House, Dazaifu 28407
Dazhong Siguzhong Museum, Beijing 07535
Dazibao, Montréal 06337
DB Museum, Nürnberg 20756
DDR-Motorradmuseum, Borna 17483
DDR Museum, Berlin 17064
DDR-Museum im Filmpalast, Malchow 20154
DDR-Museum Radebeul, Radebeul 21159
De Land Museum of Art, De Land 47996
De Locht - Streekmuseum, Nationaal Asperge- en Champignonmuseum, Horst 32414
De los Viejos Colonos, Bariloche 00143
De Mores State Historic Site, Bismarck 46835
De Morgan Centre, London 44835
De Pottenkarie, Dranouter 03561
De School van Toen, Gent 03624
De Smet Depot Museum, De Smet 48001
De Valera Library and Museum, Ennis 24830
De Velodroom, Snellegem 04033
De Young Museum, San Francisco 52772
Deaf Smith County Museum, Hereford 49257
Deák Ferenc Szülőháza, Söjtör 23805
Deák Kúria, Kehidakustány 23658
Deák Mansion, Kehidakustány 23658
Deakin University Art Gallery, Burwood 00927
Deal Castle, Deal 43958
Deal Maritime and Local History Museum, Deal 43959
Dean Castle, Kilmarnock 44578
The Dean Clough Galleries, Halifax 44383
Dean Forest Railway Museum, Lydney 45052
Dean Gallery, Edinburgh 44119
Dean Heritage Museum, Soudley 45692
De'an Museum, De'an 07657
Dearborn Historical Museum, Dearborn 48007
Death Valley National Park Museum, Death Valley 48010
Debao Museum, Guansi 42194
DeBolt and District Pioneer Museum, DeBolt 05686
Debrecen Literary Museum, Debrecen 23514
Debreceni Irodalmi Múzeum, Debrecen 23514
Decatur House Museum, Washington 53859
Decker, Meyerhoff and Pinkard Galleries, Baltimore 46560
Decorative Art Museum, Isfahan 24612
The Decorative Arts Center of Ohio, Lancaster 49919
Decorative Arts Collection, Wichita 54139
Decorative Arts Museum, Milano 26675
Decorative Arts Museum of Rhodos, Rhodos 23171
DeCordova Museum and Sculpture Park, Lincoln 50090
Dedham Historical Museum, Dedham 48023
Deep Cove Heritage Society, North Vancouver 06477
Deep Sea Adventure and Diving Museum, Weymouth 45987
Deep South Pioneer Museum, Ogema 06501
Deer Isle-Stonington Historical Society Museum, Deer Isle 48026
Deer Valley Rock Art Center, Glendale 48927
Deere Museum, Moline 50778
Deerfield Beach Historical Society Museum, Deerfield Beach 48034
Defence of Moscow Museum, Moskva 37201
Defence Services Museum, Adelaide 00774
Defence Services Museum, Yangon 31734
Degenhart Paperweight and Glass Museum, Cambridge 47190
Degerfors Fotbollsmuseum, Degerfors 40484
DeGrazia Gallery in the Sun, Tucson 53586
Dehua Ceramic Museum, Dehua 07658
Deià Museo Arqueológico, Deià 39241
Deichtorhallen Hamburg, Hamburg 18920
Deines Cultural Center, Russell 52483
Deinhard Cellar-Museum, Koblenz 19584
Deinhard-Kellermuseum, Koblenz 19584
Deir ez-Zor Museum, Deir ez-Zor 42134
Dejbjerg Jernalder, Skjern 10269
Dejima-machi Hakabutsukan, Nagasaki 28971
Dejima Museum of History, Nagasaki 28971
DeKalb Historical Society Museum, Decatur 48013
Dekanatsmuseum, Haus im Ennstal 02059
Dekema State, Jelsum 32431
Dekoratīvi Lietišķās Mākslas Muzejs, Rīga 30064
Delano Heritage Park Museum, Delano 48043
Delaware Agricultural Museum, Dover 48162
Delaware and Hudson Canal Museum, High Falls 49274
Delaware Art Museum, Wilmington 54189
Delaware Center for the Contemporary Arts, Wilmington 54190
Delaware Museum, Wilmington 54191
Delaware Museum of Natural History, Wilmington 54192
Delaware State Museums, Dover 48163
Delbridge Museum, Sioux Falls 53118
Đelekovec Folklore Collection, Đelekovec 08859
Delfina, London 44836
Delgatie Castle, Turriff 45882
Delhi Ontario Tobacco Museum & Heritage Centre, Delhi 05691
Delizsánsz Kiállítóterem, Debrecen 23515
Dell Mill, Isle-of-Lewis 44540
Deloraine Museum, Deloraine 01032
Delta Axis Contemporaray Arts Center, Memphis 50570
Delta Blues Museum, Clarksdale 47587
Delta County Historical Society Museum, Escanaba 48414
Delta County Museum, Delta 48048
Delta Cultural Center, Helena 49237
Delta Museum, Delta 05692
Demaison André, Restigné 15020
DEMHIST-International Committee for Historic House Museums, Milano 54899
Denbigh Castle, Denbigh 43963
Denbigh Museum and Gallery, Denbigh 43964
Dene Museum, Fort Good Hope 05810
Dengel-Galerie, Reutte 02582
Dengzhou Guchuan Museum, Penglai 08078
Denison University Art Gallery, Granville 49021
Deniz Müzesi, İstanbul 42814
Denizli Devlet Güzel Sanatlar Galerisi, Denizli 42758
Denizli State Fine Arts Gallery, Denizli 42758
Denki no Kagakukan, Nagoya 28996
Denkmal für die ermordeten Juden Europas, Berlin 17065
Denkmalhof Bechelsdorfer Schulzenhaus, Schönberg, Mecklenburg 21600
Denkmalhof Gererhof, Annaberg 01722
Denkmalhof Kösslerhäusl, Großarl 01999
Denkmalhof Maurerhaus, Zederhaus 03125
Denkmalhof Rauchhaus Möllin, Gadebusch 18525
Denkmalhof Retschow, Retschow 21281
DenkStätte Weiße Rose, Weiße Rose Stiftung, München 20431
Denman Island Museum, Denman Island 05695
Denmark Historical Museum, Denmark 01033
Dennison Railroad Depot Museum, Dennison 48058
Dennos Museum Center of Northwestern Michigan College, Traverse City 53553
Denny Ship Model Experiment Tank, Dumbarton 44033
Densheway National Museum, Densheway Village 10523
Dental Museum of Lyon, Lyon 13886
Dentalhistorisches Museum, Zschadraß 22822
Dentalmuseum, Zschadraß 22822
Denton County Historical Museum, Denton 48061
Denver Art Museum, Denver 48072
Denver Firefighters Museum, Denver 48073
Denver Museum of Miniatures, Dolls and Toys, Denver 48074
Denver Museum of Nature and Science, Denver 48075
Deokpojin Educational Museum, Gimpo 29815
Deosugung Art Museum, Seoul 29870
Department of Art History & Archaeology Visual Resources Collection, New York 51162
Department of Byzantine Museums, Athinai 54866
Department of Classical Archaeology and Ancient History, Göteborg University, Göteborg 40543
Department of Fine Arts Gallery, Decatur 48011
Department of the Treasury Museum, Washington 53860
Department of Zoology Museum, Kampala 42989
Departmental museum of History of Resistance and Deportation of Ain and Haut-Jura, Nantua 14374
DePaul University Art Gallery, Chicago 47470
DePauw University Anthropology Museum, Greencastle 49046
Deposit of Cars, La Habana 09101
Depósito del Automóvil, La Habana 09101
Dépôt Archéologique, Saint-Thibéry 15521
Dépôt Archéologique du Fort Vauban, Alès 11547
Dépôt de Fouilles Gallo-Romaines et Préhistoriques de Montmaurin, Montmaurin 14224
Dépôt-Musée Diocésain, Metz 14100
Depot Museum, Mammoth Spring 50413
Depot Museum and Fort Sedgwick Museum, Julesburg 49652
Depot Museum Complex, Condon 47778
The Depot Museum Complex, Henderson 49251
Depot Park Museum, Sonoma 53154
Depreciation Lands Museum, Allison Park 46233
DePree Art Center and Gallery, Holland 49314
Deqing Museum, Deqing, Guangdong 07660
Deqing Museum, Deqing, Zhejiang 07661
Derby Museum and Art Gallery, Derby 43965
Derenik Demirchyan House-Museum, Yerevan 00731
Déri Múzeum, Debrecen 23516
DeRicci Gallery, Madison 50389
Dernier Quartier Général de Napoléon, Genappe 03616
Derrymore House, Bessbrook 43464
Derrynane House, Caherdaniel 24749
Dervish Pasha Mansion Ethnographical Museum, Lefkoşa 33368
Deržavnyj Ajvazovskyj Kartynna Galereja, Feodosija 43039
Deržavnyj Istoriko-kul'turnyj Zapovidnyk, Kam'janec'-Podil's'kyj 43051
Deržavnyj Istoriko-kul'turnyj Zapovidnyk Bat'kivščyna Tarasa Ševčenka, Ševčenkove 43182
Deržavnyj Istoryko-kul'turnyj Zapovidnyk Pole Poltavskoï Bytvy, Poltava 43166
Deržavnyj Muzej Teatral'noho, Muzyčnoho ta Kinomystectva Ukraïny, Kyïv 43077
Deržavnyj Muzej Ukraïns'koho Narodnoho Dekorativnoho Mystectva, Kyïv 43078
Des Hommes dans la Guerre de 1870, Bitche 12005
Des Lee Gallery, Saint Louis 52575
Des Moines Art Center, Des Moines 48096
Des Plaines Historical Museum, Des Plaines 48104
DesBrisay Museum and Exhibition Centre, Bridgewater 05474
Deschutes County Historical Society Museum, Bend 46732
Descubre Museo Interactivo de Ciencia y Tecnología, Aguascalientes 30697
Desert Anthropology Museum, Nayin 24636
Desert Caballeros Western Museum, Wickenburg 54155
Desha County Museum, Dumas 48201
Deshine Bijutsukan, Nagoya 28997
Design Arts Gallery, Philadelphia 51826
The Design Center at Philadelphia University, Philadelphia 51827
Design Cerntre, Bilbao 39041
Design Exchange, Toronto 07104
Design Museum, London 44837
Design Museum, Nagoya 28997
Design Museum & Design Gallery, Nagoya 28998
Design Museum Gent, Gent 03625
Designmuseo, Helsinki 10771
Desing Artifact Museum, Sinpu 42339
Desmond Castle, Kinsale 24856
Desoto National Memorial, Bradenton 46985
Despića Kuća, Sarajevo 04274
Detlefsen-Museum, Glückstadt 18663
Detroit Artists Market, Detroit 48112
Detroit Historical Museum, Detroit 48113
The Detroit Institute of Arts, Detroit 48114
Detroit Repertory Theatre Gallery, Detroit 48115
Detskaja Kartinnaja Galereja, Samara 37612
Detskij Muzej, Čerepovec 36848
Detskij Muzej Buratino-Pinokkio, Moskva 37158
Detskij Muzej Dom-Skazok Žili-byli, Moskva 37159
Detskij Muzej Šantyrak, Almaty 29679
Deusserhaus, Monheim am Rhein 20371
Deutsch-Deutsches Museum Mödlareuth, Töpen 22068
Deutsch-Französisches Puppen und Spielzeugmuseum, Schweigen-Rechtenbach 21681
Deutsch-Russisches Museum Berlin-Karlshorst, Berlin 17066
Deutsche Bank Kunstsammlung, Frankfurt am Main 18359
Deutsche Barockgalerie, Augsburg 16614
Deutsche Guggenheim, Berlin 17067
Deutsche Kinemathek, Berlin 17068
Deutsche Kinemathek - Museum für Film und Fernsehen, Berlin 17069
Deutsche Raumfahrtausstellung e.V., Morgenröthe-Rautenkranz 20383
Das Deutsche Stickmuster-Museum Celle, Celle 17686
Deutscher Museumsbund e.V., Berlin 54830
Deutsches Albert-Schweitzer-Zentrum, Frankfurt am Main 18360
Deutsches Apotheken-Museum im Heidelberger Schloß, Heidelberg 19071
Deutsches Architekturmuseum, Frankfurt am Main 18361
Deutsches Auswandererhaus Bremerhaven, Bremerhaven 17568
Deutsches Automuseum Schloss Langenburg, Langenburg 19823
Deutsches Bananenmuseum, Sierksdorf 21760
Deutsches Bauernkriegsmuseum, Böblingen 17416
Deutsches Baumaschinen-Modellmuseum, Weilburg-Gaudernbach 22387
Deutsches Bergbau-Museum, Bochum 17398
Deutsches Bernsteinmuseum, Ribnitz-Damgarten 21309
Deutsches Bienenmuseum, Weimar 22397
Deutsches Blinden-Museum (DBM), Berlin 17070
Deutsches Boxsport-Museum, Sagard 21468
1. Deutsches Bratwurstmuseum, Wachsenburggemeinde 22260
Deutsches Buch- und Schriftmuseum der Deutschen Nationalbibliothek Leipzig, Leipzig 19900
Deutsches Buchbindermuseum, Mainz 20142
Deutsches Chemie-Museum, Merseburg 20291
Deutsches Currywurst Museum, Berlin 17071
Deutsches Damast- und Frottiermuseum, Großschönau 18805
Deutsches Dampflokomotiv-Museum, Neuenmarkt 20610
Deutsches Drachenmuseum und Stadtmuseum, Furth im Wald 18515
Deutsches Drahtmuseum, Altena 16486
Deutsches Edelsteinmuseum, Idar-Oberstein 19308
Deutsches Eichendorff-Museum, Wangen im Allgäu 22324
Deutsches Elfenbeinmuseum Erbach, Erbach, Odenwaldkreis 18190
Deutsches Epilepsiemuseum Kork, Kehl 19495
Deutsches Erdölmuseum Wietze, Wietze 22564
Deutsches Fächer Museum, Bielefeld 17338
Deutsches Fastnachtmuseum, Kitzingen 19560
Deutsches Feld- und Kleinbahnmuseum, Deinste 17810
Deutsches Feuerwehr-Museum, Fulda 18505
Deutsches Film- und Fototechnik Museum, Deidesheim 17807
Deutsches Filmmuseum, Frankfurt am Main 18362
Deutsches Fleischermuseum, Böblingen 17417
Deutsches Freimaurermuseum, Bayreuth 16949
Deutsches Gartenbaumuseum Erfurt, Erfurt 18199
Deutsches Glasmalerei-Museum Linnich, Linnich 20010

D

Deutsches Glockenmuseum, Greifenstein 18746
Deutsches Goldmuseum, Schalkau 21529
Deutsches Goldmuseum, Theuern 22051
Deutsches Goldschmiedehaus, Hanau 18975
Deutsches Harmonikamuseum, Trossingen 22110
Deutsches Hirtenmuseum, Hersbruck 19164
Deutsches Historisches Museum, Berlin 17072
Deutsches Hopfenmuseum, Wolnzach 22677
Deutsches Hugenotten-Museum, Bad Karlshafen 16757
Deutsches Hygiene-Museum Dresden, Dresden 17941
Deutsches Jagd- und Fischereimuseum, München 20432
Deutsches Kaltwalzmuseum, Hagen 18853
Deutsches Kartausen-Museum, Buxheim 17672
Deutsches Kartoffelmuseum Fußgönheim, Fußgönheim 18522
Deutsches Kinderwagenmuseum, Zeitz 22776
Deutsches Kleingärtnermuseum, Leipzig 19901
Deutsches Klingenmuseum Solingen, Solingen 21801
Deutsches Knopfmuseum, Bärnau 16895
Deutsches Kochbuchmuseum, Dortmund 17918
Deutsches Korbmuseum, Michelau in Oberfranken 20317
Deutsches Landwirtschaftsmuseum, Stuttgart 21930
Deutsches Landwirtschaftsmuseum Schloss Blankenhain, Crimmitschau 17749
Deutsches Märchen- und Wesersagenmuseum, Bad Oeynhausen 16798
Deutsches Maler- und Lackierer-Museum, Hamburg 18921
Deutsches Marinemuseum, Wilhelmshaven 22575
Deutsches Medizinhistorisches Museum, Ingolstadt 19340
Deutsches Meeresmuseum, Stralsund 21913
Deutsches Museum, München 20433
Deutsches Museum Bonn, Bonn 17452
Deutsches Museum Flugwerft Schleißheim, Oberschleißheim 20831
Deutsches Museum für Schulkunst, Hagen 18854
Deutsches Museum - Verkehrszentrum, München 20434
Deutsches Musikautomaten Museum, Bruchsal 17594
Deutsches Nationalkomitee des Internationalen Museumsrates ICOM, Berlin 54831
Deutsches Naturstein-Archiv (DNSA), Wunsiedel 22730
Deutsches Ofenmuseum, Burrweiler 17666
Deutsches Orthopädisches Geschichts- und Forschungsmuseum, Frankfurt am Main 18363
Deutsches Panzermuseum Munster, Munster 20530
Deutsches Peitschenmuseum, Burladingen-Killer 17665
Deutsches Pferdemuseum, Verden 22219
Deutsches Phonomuseum und Uhrensammlung, Sankt Georgen im Schwarzwald 21492
Deutsches Pinsel- und Bürstenmuseum, Bechhofen 16970
Deutsches Plakat Museum, Essen 18250
Deutsches Polizeimuseum, Salzkotten 21479
Deutsches Porzellanmuseum, Hohenberg an der Eger 19234
Deutsches Röntgen-Museum, Remscheid 21263
Deutsches Rundfunk-Museum e.V., Berlin 17073
Deutsches Salzmuseum - Industriedenkmal Saline Lüneburg, Lüneburg 20096
Deutsches Schaustellermuseum, Lambrecht 19782
Deutsches Schiefermuseum Steinach/Thür., Steinach, Kreis Sonneberg 21875
Deutsches Schiffahrtsmuseum, Bremerhaven 17569
Deutsches Schilder- und Lichtreklame-Museum, Lahr 19769
Deutsches Schloss-und Beschlägemuseum, Velbert 22210
Deutsches Schreibmaschinenmuseum, Bayreuth 16950
Deutsches Schuhmuseum Hauenstein, Hauenstein 19045
Deutsches Schustermuseum, Burgkunstadt 17659
Deutsches Schweinemuseum Ruhlsdorf, Teltow 22024
Deutsches Segelflugmuseum mit Modellflug, Gersfeld 18613
Deutsches Sielhafenmuseum in Carolinensiel, Wittmund 22625
Deutsches Skimuseum, Planegg 21048
Deutsches Spiele-Archiv, Nürnberg 20757
Deutsches SPIELEmuseum e.V, Chemnitz 17693
Deutsches Spielkartenmuseum, Leinfelden-Echterdingen 19881
Deutsches Spielzeugmuseum, Sonneberg 21818
Deutsches Sport- und Olympia Museum, Köln 19597
Deutsches Straßenmuseum, Germersheim 18604
Deutsches Tabak- und Zigarrenmuseum, Bünde 17634
Deutsches Tabakpfeifenmuseum, Oberelsbach 20804
Deutsches Tanzarchiv Köln, Köln 19598
Deutsches Tapetenmuseum, Kassel 19461
Deutsches Technikmuseum, Berlin 17074
Deutsches Telefon-Museum, Morbach 20381
Deutsches Teppichmuseum, Oelsnitz 20871
Deutsches Textilmuseum Krefeld, Krefeld 19703
Deutsches Theatermuseum, München 20435
Deutsches Traktoren und Modellauto Museum e.V., Paderborn 20967
Deutsches Uhrenmuseum, Furtwangen im Schwarzwald 18520
Deutsches Uhrenmuseum Glashütte, Glashütte 18653
Deutsches Verpackungs-Museum, Heidelberg 19072
Deutsches Vulkanmuseum, Mendig 20282
Deutsches Weinbaumuseum, Oppenheim 20906
Deutsches Werkzeugmuseum, Remscheid 21264
Deutsches Zeitungsmuseum, Wadgassen 22264
Deutsches Zementmuseum, Hemmoor 19127
Deutsches Zinnfigurenmuseum, Kulmbach 19747
Deutsches Zollmuseum, Hamburg 18922
Deutsches Zweirad- und NSU-Museum, Neckarsulm 20565
Deutschheim State Historic Site, Hermann 49260
Deutschordensmuseum Bad Mergentheim, Bad Mergentheim 16783
DeValera Museum and Bruree Heritage Centre, Bruree 24747
Devil's Coulee Dinosaur Heritage Museum, Warner 07278
Devils' Museum of the M.K. Čiurlionis National Art Museum, Kaunas 30268
Devils Tower Visitor Center, Devils Tower 48132
Devon and Cornwall Constabulary Museum, Exeter 44193
Devon Guild of Craftsmen Gallery, Bovey Tracey 43569
Devonport Gallery and Arts Centre, Devonport 01034
Devonport Maritime Museum, Devonport 01035
DeVos Art Museum, Marquette 50498
Dewa Roman Experience, Chester 43817
Dewa San-Zan History Museum, Tsuruoka 29543
Dewanoyuki Sake Museum, Tsuruoka 29544
Dewazakura Art Museum, Tendo 29357
Dewazakura Bijutsukan, Tendo 29357
Dewberry Valley Museum, Dewberry 05699
Dewey Hotel, Dewey 48133
Dewey Museum, Warminster 45932
DeWitt County Historical Museum, Cuero 47899
DeWitt Historical Society Museum, Ithaca 49546
DeWitt Stetten Jr., Bethesda 46775
DeWitt Wallace Decorative Arts Museum, Williamsburg 54168
Dewsbury Museum, Dewsbury 43980
Dexing Museum, Dexing 07662
Dexter Area Museum, Dexter 48136
Dexter Historical Society Museum, Dexter 48135
Dey Mansion, Wayne 54008
Deyang Museum, Deyang 07663
Dezign House, Jefferson 49607
DGEG Eisenbahnmuseum Würzburg, Würzburg 22708
Dhaka City Museum, Dhaka 03222
Dhenkanal Science Centre, Dhenkanal 24165
Dia Art Foundation, New York 51163
Dia:Beacon, Riggio Galleries, Beacon 46660
Dia Center for the Arts, New York 51164
Dia Center for the Arts, New York 51165
Dia Center for the Arts, Quemado 52210
Dia:Chelsea, New York 51166
Diamantmuseum, Antwerpen 03284
Diamantmuseum, Grobbendonk 03662
Diamantschleifermuseum, Brücken 17602
Dianbai Museum, Dianbai 07664
Dichter- und Stadtmuseum Liestal (DISTL)/ Herwegh-Archiv, Liestal 41561
Dichtermuseum Joseph Maria Lutz, Pfaffenhofen an der Ilm 21018
Dick Institute Museum and Art Gallery, Kilmarnock 44579
Dickens House Museum Broadstairs, Broadstairs 43646
Dickens Museumtheater, Bronkhorst 32013
Dickinson County Heritage Center, Abilene 46119
Dickson Mounds Museum, Lewistown 50040
Dickson Store Museum, Dickson 05701
Didaktische Ausstellung Urgeschichte, Chur 41301
Didcot Railway Centre, Didcot 43981
Didi Museum, Abuja 33312
Didrichsen Art Museum, Helsinki 10772
Didrichsenin Taidemuseo, Helsinki 10772
Didsbury and District Museum, Didsbury 05702
Didžiulių Muziejus, Anykščiai 30209
Die Neue Sammlung - The International Design Museum Munich, München 20436
Diecesan Museum, Włocławek 35308
Diecézní Muzeum, Brno 09422
Dief- en Duifhuisje, Capelle aan den IJssel 32035
Diefenbaker Canada Centre, Saskatoon 06912
Diefenbaker House Museum, Prince Albert 06660
Diefenbunker-Canada's Cold War Museum, Carp 05569
Diekirch Brewery Museum, Diekirch 30463
Diem Gallery, Amsterdam 31837
Dienvidu Elektrisko Tīlu Muzejs, Jelgava 30016
Dieselmuseum, Schramberg 21629
Dieter Roth-Museum, Hamburg 18923
Dietrichstein Palace of the The Moravian Museum, Brno 09423
Dietrichsteinský Palác, Brno 09423
DIG, York 46100
Diggs Gallery at Winston-Salem State University, Winston-Salem 54244
Digha Science Centre & National Science Camp, New Digha 24337
Digital Art Lab, Holon 24974
Digital Art Museum, Berlin 17075
Digterhuset, Farsø 10017
Dijecezanski Muzej Zagrebačke Nadbiskupije, Zagreb 08988
De Dijk Te Kijk, Petten 32668
Dikemark Sykehus Museum, Asker 33406
Diligence Exhibition Hall, Debrecen 23515
Dilijan Historic Center, Dilijan 00693
Dillard Mill State Historic Site, Davisville 47980
Dillaway-Thomas House, Boston 46928
Dimbola Lodge Museum, Freshwater Bay 44255
Dimcho Debelyanov Memorial House, Koprivština 05070
Dimităr Blagoev Memorial House, Bankja 04977
Dimităr Blagoev Memorial House, Sofia 05175
Dimitar Dimov House-Museum, Sofia 05176
Dimitar Peshev House-Museum, Kjustendil 05064
Dimitreion Oikima House, Agios Georgios Nilias 22870
Dimitrie Gusti National Village Museum, Bucureşti 36217
Dimitris Pierides Museum of Contemporary Art, Glyfada 23007
Dimock Gallery, Washington 53861
Dinajpur Museum, Dinajpur 03230
Dingles Steam Village - Fairground Museum, Lifton 44709
Dingling Museum, Beijing 07536
Dingwall Museum, Dingwall 43982
Dingxi Zone Museum, Dingxi 07665
Dingzhou Museum, Dingzhou 07666
Dinkelbergmuseum Minseln, Rheinfelden 21299
Dinnerware Contemporary Art Gallery, Tucson 53587
Dinosaur Gardens Museum, Ossineke 51620
Dinosaur Isle, Sandown 45613
Dinosaur Museum, Dorchester 43998
Dinosaur Museum, Malabon 34418
Dinosaur Museum, Savannakhet 29970
Dinosaur Museum, Tucumcari 53603
Dinosaur Provincial Park, Patricia 06574
Dinosaur Valley State Park, Glen Rose 48923
Dinosauria - Musée des Dinosaures, Espéraza 12860
Dinosaurier-Freilichtmuseum Münchehagen, Rehburg-Loccum 21240
Dinosaurland, Lyme Regis 45053
Dinsmore Homestead, Burlington 47135
Diocesan Museum, Brno 09422
Diocesan Museum, Opole 34985
Diocesan Museum, Płock 35013
Diocesan Museum, Sandomierz 35096
Diocesan Museum, Siedlce 35102
Diocesan Museum, Tarnów 35176
Diocesan Museum Pelplin, Pelplin 35005
Diocese Museum of Székesfehérvár, Székesfehérvár 23859
Diözesanmuseum, Bamberg 16906
Diözesanmuseum, Klagenfurt 02182
Diözesanmuseum Eisenstadt, Eisenstadt 01858
Diözesanmuseum Freising, Freising 18450
Diözesanmuseum Graz, Graz 01966
Diözesanmuseum Linz, Linz 02312
Diözesanmuseum Rottenburg, Rottenburg am Neckar 21420
Diözesanmuseum Sankt Afra, Augsburg 16615
Diözesanmuseum Sankt Pölten, Sankt Pölten 02681
Diözesanmuseum Sankt Ulrich, Regensburg 21216
Dion Archaeological Museum, Dion 22985
Dionizas Poška Hollowed Trunks of the Šiauliai Aušra Museum, Bijotai 30213
Dionizo Poškos Baubliai, Bijotai 30213
Dionne Quints Museum, North Bay 06469
Diorama Arts, London 44838
Diorama "Battle at the Dnepr", Dnipropetrovs'k 43031
Diorama Bethlehem, Einsiedeln 41342
Diorama "Bitva za Dnepr", Dnipropetrovs'k 43031
Diorama der Schlacht bei Rosbach, Reichardtswerben 21242
Diorama Kurskaja Bitva - Belgorodskoe Napravlenie, Belgorod 36795
Diorama Saint-Bénilde, Saugues-en-Gévaudan 15650
Diorama Storm of Ismail, Izmaïl 43045
Diorama Šturm Izmaïla, Izmaïl 43045
Dioramenschau Altötting, Altötting 16517
Diósgyőri Papírgyár Zrt. Papíripari Múzeuma, Miskolc 23701
Diósgyőri Vármúzeum, Miskolc 23702
Diplomatic History Museum, Seoul 29871
Diplomatic Record Office of the Ministry of Foreign Affairs, Tokyo 29387
Dirección General de Museos de la Nación, Asunción 54982
Direction de Musées de France, Paris 54808
Disagården, Uppsala 41004
Dischinger Heimatmuseum, Dischingen 17874
Disciples of Christ Historical Society Museum, Nashville 50967
Discover Houston County Visitors Center-Museum, Crockett 47884
Discovery Center, Fresno 48789
Discovery Center Museum, Rockford 52414
Discovery Center of Idaho, Boise 46891
Discovery Center of Springfield, Springfield 53231
Discovery Center of Taipei, Taipei 42389
Discovery Center of the Southern Tier, Binghamton 46815
Discovery Center Science Museum, Fort Collins 48606
Discovery Creek Children's Museum of Washington DC, Washington 53862
Discovery Harbour, Penetanguishene 06581
The Discovery Museum, Bridgeport 47017
Discovery Museum, Newcastle-upon-Tyne 45226
Discovery Museum, Sacramento 52506
The Discovery Museums, Acton 46133
Discovery North Bay Museum, North Bay 06470
Discovery Park, Safford 52513
Discovery Place, Charlotte 47388
Discovery Science Center of Central Florida, Ocala 51482
Discovery World - The James Lovell Museum of Science, Economics and Technology, Milwaukee 50707
Dishman Art Gallery, Beaumont 46674
Diss Museum, Diss 43986
Distillerie du Périgord, Sarlat-la-Canéda 15635
District Archaeological Museum, Dhar 24159
District Museum, Apriķi 29986
District Museum, Barpeta 24079
District Museum, Darrang 24128
District Museum, Gulbarga 24182
District Museum, Guntur 24185
District Museum, Jelenia Góra 34704
District Museum, Konin 34755
District Museum, Krosno 34830
District Museum, Kuldīga 30034
District Museum, Lublin 34905
District Museum, Náchod 09669
District Museum, Nowy Sącz 34959
District Museum, Ostrołęka 34993
District Museum, Pillalamari 24350
District Museum, Rzeszów 35094
District Museum, Sandomierz 35099
District Museum, Shimoga 24390
District Museum, Shivpuri 24391
District Museum, Żyradów 35381
District Museum in Chełm, Chełm 34587
District Museum in Leszno, Leszno 34850
District Museum in Leszno Judaistic Dept., Leszno 34851
District Museum in Toruń, Toruń 35195
District Museum of History, Galaţi 36380
District Museum Prague-East, Brandýs nad Labem 09413
District of Columbia Arts Center, Washington 53863
District Science Centre, Dharampur 24160
District Science Centre, Purulia 24364
District Science Centre, Tirunelveli 24407
District Six Museum, Cape Town 38582
Ditchling Museum, Ditchling 43987
Dithmarschen County Museum, Meldorf 20269
Dithmarscher Landesmuseum, Meldorf 20269
Dittmar Memorial Gallery, Evanston 48447
Dittrick Museum of Medical History, Cleveland 47629
Divan Edebiyatı Müzesi, İstanbul 42815
Diverse Works, Houston 49385
Divide County Historical Society Museum, Crosby 47888
Divine Word University Museum, Tacloban 34509
Divisão de Museus, Patrimônio e Arquivo Histórico, Taubaté 04922
División Museo e Investigaciones Históricas de la Policía Federal, Buenos Aires 00156
Divisional Kohima Museum, York 46101
Dix Art Mix, Chicago 47471
Dixie Gun Works' Old Car Museum, Union City 53653
The Dixon Gallery and Gardens, Memphis 50571
Diyabakir Archaeological Sites, Diyarbakır 42764
Diyarbakır Devlet Güzel Sanatlar Galerisi, Diyarbakır 42762
Diyarbakır Müzeleri ve Tarihi Kültürel Değerleri, Diyarbakır 42763

Diyarbakir Museums and Treasures of the Cultural History, Diyarbakır 42763
Diyarbakır Örenyerleri, Diyarbakır 42764
Diyarbakır State Fine Arts Gallery, Diyarbakır 42762
Djalil Mammadkulizade Home Museum, Naxçivan 03181
Djalil Mammadkulizade Memorial Museum, Nehram 03189
Djamshid Nakhchivansky Home Museum, Naxçivan 03182
Djanogly Art Gallery, Nottingham 45290
Djurslands Museum - Dansk Fiskerimuseum, Grenaa 10041
DLM - Deutsches Ledermuseum/ Schuhmuseum Offenbach, Offenbach am Main 20881
Dmanisi Museum-Reserve of History and Architecture, Dmanisi 16339
Dmitrov Kremlin - State Museum Reserve, Dmitrov 36876
Dneprodzerzhinsk History Museum, Dniprodzeržins'k 43030
Dnepropetrovsk State Art Museum, Dnipropetrovs'k 43032
Dnipropetrovskyj Chudožnij Muzej, Dnipropetrovs'k 43032
Doak House Museum, Greeneville 49051
Dobele Museum of Local Studies, Dobele 30000
Dobeles Novadpētniecības Muzejs, Dobele 30000
Dobergmuseum/Geologisches Museum Ostwestfalen-Lippe, Bünde 17635
Dobó István Vármúzeum, Eger 23536
Doboj Regional Museum, Doboj 04262
Dobrjanskij Istoriko-kraevedčeskij Muzej, Dobrjanka 36877
Dobrovo Castle, Dobrovo v Brdih 38413
Dock Museum, Barrow-in-Furness 43388
Dock och Leksaksmuseum, Landsbro 40666
Dock- och Textilmuseum, Katrineholm 40640
document Neupfarrplatz, Regensburg 21217
document Schnupftabakfabrik, Regensburg 21218
documenta Archiv für die Kunst des 20. und 21 Jahrhunderts, Kassel 19462
Documentary Centre of Czech Culture Martin, Martin 38293
Documentation and Memorial Place of the Federal Commissioner for State Security Service Records in the former STASI pre-trial prison, Rostock 21388
Documentation Center of Cambodia, Phnom Penh 05285
Documentation Centre Nazi Party Rally Grounds, Nürnberg 20758
Documentation centre of the state Mecklenburg-Vorpommern for the victims of German dictatorships, Schwerin 21692
Doddington Hall, Lincoln 44711
Dodengang en Oorlogsmuseum, Diksmuide 03553
Dodge County Historical Museum, Mantorville 50454
Dodge County Historical Society Museum, Beaver Dam 46681
Dodsland and District Museum, Dodsland 05706
Döderhultar Museum, Oskarshamn 40802
Döderhultarmuseet, Oskarshamn 40802
Dörflihaus-Museum, Spiringen 41894
Dörpmuseum Münkeboe, Südbrookmerland 21973
Doesburgsch Mosterd- en Azijnmuseum, Doesburg 32132
Dog Collar Museum, Maidstone 45072
Dog Mushing Museum, Fairbanks 48470
Dogana d'Arte, Atripalda 25324
Dogra Art Museum, Jammu 24236
Dokken, Hol i Hallingdal 33551
Dokuchaev Central Soil Museum, Sankt-Peterburg 37629
Dokumentačné Centrum Českej Kultúry Martin, Martin 38293
Dokumentation des ehemaligen Heilbades, Bad Pirawarth 01761
Dokumentation Obersalzberg - Orts- und Zeitgeschichte, Berchtesgaden 17000
Dokumentations- und Gedenkstätte der BStU in der ehemaligen U-Haft der STASI, Rostock 21388
Dokumentations- und Informationszentrum, Stadtallendorf 21855
Dokumentations- und Informationszentrum Emslandlager (DIZ), Papenburg 20979
Dokumentationsarchiv des österreichischen Widerstandes, Wien 02978
Dokumentationsraum für staufische Geschichte, Göppingen 18672
Dokumentationsraum Georg Matthäus Vischer, Raabs an der Thaya 02545
Dokumentationsstätte im Mesnerhaus, Scheiblingkirchen-Thernberg 02704
Dokumentationszentrum Alltagskultur der DDR, Eisenhüttenstadt 18105
Dokumentationszentrum des Landes für die Opfer der Diktaturen in Deutschland, Schwerin 21692
Dokumentationszentrum des Landes für die Opfer deutscher Diktaturen, Rostock 21389
Dokumentationszentrum ehemaliges KZ Osthofen, Osthofen 20941
Dokumentationszentrum NS-Zwangsarbeit Berlin-Schöneweide, Berlin 17076
Dokumentationszentrum Oberer Kuhberg-Ulm KZ Gedenkstätte, Ulm 22159
Dokumentationszentrum Prof. Sepp Mayrhuber, Golling an der Erlauf 01958
Dokumentationszentrum Reichsparteitagsgelände, Nürnberg 20758
Dolenjski Muzej Novo Mesto, Novo Mesto 38490
Doll and Bear Museum, Manunda 01260
Doll and Toy Museum, Landsbro 40666
The Doll and Toy Museum of NYC, Brooklyn 47071
Doll Gardner Art Gallery, Portland 52074
Doll Museum, Josselin 13279
Doll Museum, Píritu 54592
Doll Museum, Savonlinna 11302
Dollar Museum, Dollar 43989
Dolls and Toys Museum, Baden bei Wien 01773
Doll's House of Fehérvár, Székesfehérvár 23853
Doll's Museum, Tihany 23924
The Dolly Mixture, Langbank 44638
Dolmabahçe Sarayı Müzesi, İstanbul 42816
Dølmo Bygdetun, Tolga 33908
Dolnośląski Centrum Informacji Kulturalnej-Galeria, Wrocław 35322
Dolnośląskie Centrum Fotografii Domek Romański-Galeria, Wrocław 35323
Dolomythos, San Candido 27618
Dolphin Sailing Barge Museum, Sittingbourne 45682
Dom Artysta Plastyka, Warszawa 35221
Dom Eskenów, Toruń 35186
Dom Fotografie v Poprade, Poprad 38331
Dom Gerharta Hauptmanna, Jelenia Góra 34703
Dom Hudby Mikuláša Schneidera-Trnavského, Trnava 38378
Dom I.S. Ostrouchova, Moskva 37160
Dom Jana Matejki, Kraków 34769
Dom Józefa Mehoffera, Kraków 34770
Dom L'udového Bývania, Matúškovo 38303
Dom L'udového Bývania, Šala 38347
Dom Marina Držića, Dubrovnik 08861
Dom Mikuláša Galandu, Turčianske Teplice 38382
Dom-Museum, Fulda 18506
Dom-Museum Bremen, Bremen 17548
Dom-Museum Hildesheim, Hildesheim 19182
Dom-muzej A.P. Čechova, Jalta 43049
Dom-muzej Akademika E.N. Pavlovskogo, Borisoglebsk 36817
Dom-muzej L.N. Andreeva, Orël 37503
Dom-muzej M. Auėzova, Almaty 29680
Dom-muzej S.M. Budënnogo, Proletarsk 37572
Dom-muzej A.P. Čechova, Moskva 37161
Dom-muzej Velimira Chlebnikova, Astrachan 36781
Dom-muzej N.N. Chochrjakova, Kirov 37035
Dom-Muzej T.N. Chrenikova, Elec 36907
Dom-muzej G.V. Čičerina, Tambov 37831
Dom-muzej K.Ė. Ciolkovskogo, Kaluga 36994
Dom-muzej P.P. Čistjakova, Puškin 37578
Dom-muzej Čudomir, Kazanlăk 05057
Dom-muzej K.I. Čukovskogo, Peredelkino 37539
Dom-muzej M.I. Cvetaevoj, Moskva 37162
Dom-muzej Dekabristov, Kurgan 37090
Dom-muzej F.M. Dostoevskogo, Staraja Russa 37796
Dom-Muzej A.L. Durova, Voronež 37969
Dom-muzej A.A. Dydykina, Palech 37523
Dom-muzej Elin Pelin, Bajlovo 04973
Dom-muzej M.N. Ermolovoj, Moskva 37163
Dom-muzej M.V. Frunze, Samara 37613
Dom-muzej Gannibalov v Petrovskom, Puškinskie Gory 37585
Dom-muzej Generala I.A. Plieva, Vladikavkaz 37929
Dom-muzej N.V. Gogolja, Moskva 37164
Dom-muzej I.I. Golikova, Palech 37524
Dom-muzej M.B. Grekova, Novočerkassk 37477
Dom-muzej A.S. Grina, Kirov 37036
Dom-muzej im. T.N. Granovskogo, Orël 37504
Dom-muzej Istorii Molodëžnogo Dviženija, Rjazan 37593
Dom-muzej Ivan Vazov, Sopot 05205
Dom-muzej E.M. Jaroslavskogo, Jakutsk 36959
Dom-Muzej P.D. Korina, Palech 37525
Dom-muzej V.G. Korolenko, Gelendžik 36922
Dom-Muzej I.N. Kramskogo, Ostrogožsk 37519
Dom-muzej P.A. Krasikova, Krasnojarsk 37068
Dom-muzej Pavla Kuznecova, Saratov 37748
Dom-muzej V.I. Lenina, Kazan 37005
Dom-muzej V.I. Lenina, Samara 37614
Dom-muzej M.Ju. Lermontova, Moskva 37165
Dom-muzej N.S. Leskova, Orël 37505
Dom-muzej M.K. Ammosova, Jakutsk 36960
Dom Muzej M. Saryana, Yerevan 00732
Dom-muzej D.I. Mendeleeva, Boblovo 36811
Dom-muzej N.S. Muchina, Joškar-Ola 36976
Dom-muzej N.S. Muchina, Olikjal 37494
Dom-muzej Narodnogo Chudožnika V.A. Igoševa, Chanty-Mansijsk 36859
Dom-muzej N.A. Nekrasova, Čudovo 36871
Dom-muzej I.S. Nikitina, Voronež 37970
Dom-muzej Osipovych-Vulf v Trigorskom, Puškinskie Gory 37586
Dom-muzej A.N. Ostrovskogo, Moskva 37166
Dom-muzej B.L. Pasternaka, Peredelkino 37540
Dom-muzej Penjo Penev, Dimitrovgrad 05013
Dom-muzej Petra Pervogo (Petrovskij Domik), Vologda 37961
Dom-muzej M.M. Prišvina, Dunino 36881
Dom-muzej B.I. Prorokova, Ivanovo 36944
Dom-Muzej S.V. Rachmaninova, Ivanovka 36943
Dom-Muzej I.E. Repina, Žigulevsk 37987
Dom-muzej V.A. Rusanova, Orël 37506
Dom-muzej F.I. Šaljapina, Moskva 37167
Dom-muzej M.E. Saltykova-Ščedrina, Kirov 37037
Dom-muzej Semji Cvetaevych, Novo-Talicy 37476
Dom-muzej Semji Lopatinych, Stavropol 37802
Dom-muzej Sestër Nevzorovych, Nižnij Novgorod 37450
Dom-muzej Skulptora A.S. Golubkinoj, Zarajsk 37985
Dom-muzej K.S. Stanislavskogo, Moskva 37168
Dom-muzej Tukaevych, Košlauč 37060
Dom-muzej G.I. Uspenskogo, Sjabrenicy 37767
Dom-muzej V.M. Vasnecova, Moskva 37169
Dom-muzej Konstantin Veličkov, Pazardžik 05106
Dom-Muzej V.V. Veresaeva, Tula 37865
Dom-Muzej N.M. Zinovjeva, Djagilevo 36875
Dom-Muzej N.N. Žukova, Elec 36908
Dom na Humora i Satirata, Gabrovo 05031
Dom Naščokina, Moskva 37170
Dom Plastyka, Toruń 35187
Dom Poėta A.N. Širjaevca - Muzej Krestjanskogo Byta, Žigulevsk 37988
Dom Rodzinny Jana Kasprowicza, Inowrocław 34690
Dom Rodzinny Ojca Świętego Jana Pawła II, Wadowice 35207
Dom St. Peter, Fritzlar 18483
Dom Tkacza, Bielsko-Biała 34558
Dom- und Diözesanmuseum Wien, Wien 02979
Dom und Domschatz Halberstadt, Halberstadt 18875
Dom Uphagena, Gdańsk 34633
Dom Urodzenia Fryderyka Chopina, Sochaczew 35117
Domaine de l'Abbé Saunière, Rennes-le-Château 15017
Domaine National de Chambord, Chambord 12378
Domeček T.G.M-muzeum, Čejkovice 09453
Domein De Locht, Duffel 03564
Domein d'Oude Smelterij - Landbouwmuseum, Torhout 04077
Domek Boženy Němcové, Červený Kostelec 09456
Domek Żeromskiego, Gdynia 34654
Domgrabungsmuseum, Salzburg 02608
Domik Petra I, Sankt-Peterburg 37635
Dominicanian Amber Museum, San Felipe de Puerto Plata 10334
Dominiiklaste Kloostri Muuseum, Tallinn 10621
Dominikanerkloster Prenzlau, Prenzlau 21117
Dominikanermuseum Rottweil, Rottweil 21424
Dominus Flevit, Jerusalem 24985
Domkammer der Kathedralkirche Sankt Paulus zu Münster, Münster 20507
Dommuseum, Brandenburg an der Havel 17507
Dommuseum, Frankfurt am Main 18364
Dommuseum zu Salzburg, Salzburg 02609
Domnista Folklore and History Museum, Domnista 22986
Domonkos Kuny County Museum, Tata 23918
Domowniski muzej Dešno, Dissen-Striesow 17876
Domschatz, Trier 22097
Domschatz, Würzburg 22709
Domschatz der Stiftskirche St.Servatii, Quedlinburg 21145
Domschatz Sankt Ursen Kathedrale, Solothurn 41881
Domschatz- und Diözesanmuseum, Passau 20988
Domschatz- und Diözesanmuseum Eichstätt, Eichstätt 18081
Domschatzkammer, Aachen 16396
Domschatzkammer, Essen 18251
Domschatzkammer, Köln 19599
Domschatzkammer, Minden 20333
Domschatzkammer Sankt Petri, Bautzen 16939
Domschatzkammer und Diözesanmuseum, Osnabrück 20919
Domschatzmuseum, Chur 41302
Domschatzmuseum, Regensburg 21219
Domus - Casa del Hombre, A Coruña 39215
Domus Galilaeana, Pisa 27237
Domus Mazziniana, Pisa 27238
DoMuS - Museum und Galerie der Gemeinde Schaan, Schaan 30192
Domus Romana, Rabat 30636
Domus Romana Archéosite, Aubechies 03335
Domvorhalle, Goslar 18707
Don Bank Museum, North Sydney 01373
Don F. Pratt Memorial Museum, Fort Campbell 48603
Don Garlits Museum of Drag Racing, Ocala 51483
Don Harrington Discovery Center, Amarillo 46252
Don Murray's Museum of History, Neepawa 06420
Don River Railway Museum, Devonport 01036
Donacija Dr. Vladimir Malančec, Koprivnica 08890
Donald Agricultural Museum, Donald 01040
Donald G. Trayser Memorial Museum, Barnstable 46612
Donalda and District Museum, Donalda 05707
Donau-Schiffahrts-Museum-Regensburg, Regensburg 21220
Donauschwäbisches Zentralmuseum, Ulm 22160
Donazione Putti Biblioteca, Bologna 25441
Donazione Raffaele Vaccari, Brescello 25531
Donazione Sambo, Trieste 28087
Doncaster Museum and Art Gallery, Doncaster 43992
Donec'kij Oblasnyj Chudožnij Muzej, Donec'k 43037
Donec'kij Oblasnyj Krajeznavčyj Muzej, Donec'k 43038
Donegal County Museum, Letterkenny 24862
Donegal Historical Society Museum, Rossnowlagh 24895
Donetsk Museum of Art, Donec'k 43037
Dong-eui University Museum, Busan 29777
Dong Zuo Inkslab Art Museum, Ershuei 42185
Donga University Museum, Busan 29778
Donggang Museum, Donggou 07667
Dongguan Museum, Dongguan 07668
Dongguk University Museum, Seoul 29872
Dongha Museum, Dongen 32142
Dongjin Irrigation Folk Museum, Gimje 29814
Dongshan Kite Museum, Dongshan 42181
Dongshan Museum, Dongshan 07670
Dongtai Museum, Dongtai 07671
Dongtinghu Guan, Yuanjiang 08463
Dongying History Museum, Guangrao 07722
Donington Grand Prix Collection, Castle Donington 43769
Donjon, Ballon 11833
Donjon-Château Carondolet, Crupet 03529
Donjon Roman, Fondremand 12953
Donkey Milk House, Key West 49757
Donna Beam Fine Art Gallery, Las Vegas 49966
Dønna Bygdesamling, Dønna 33465
Donner Memorial and Emigrant Trail Museum, Truckee 53578
Donskoj Istoriko-kraevedčeskij Muzej, Donskoe 36879
Doon Heritage Crossroads, Kitchener 06100
Doon Valley Museum, Dalmellington 43946
Door County Maritime Museum, Gills Rock 48910
Door County Maritime Museum, Sturgeon Bay 53345
Door County Museum, Sturgeon Bay 53346
Doozoo Children's Museum, Grand Junction 48993
Doppelkapelle Sankt Crucis, Landsberg, Saalekreis 19797
Dorchester Abbey Museum, Dorchester-on-Thames 44005
Dorchester County Historical Society Museum, Cambridge 47186
Dordrechts Museum, Dordrecht 32149
Dorf- und Fahrradmuseum, Pflückuff 21025
Dorf- und Heimatmuseum, Winterbach, Rems-Murr-Kreis 22607
Dorf- und Kutschenmuseum, Eisgarn 01864
Dorf- und Uhrenmuseum, Gütenbach 18831
Dorfgemeinschaftshaus Heimatmuseum Dörrenbach, Sankt Wendel 21511
Dorfmuseum, Finkenstein 01889
Dorfmuseum, Günzach 18824
Dorfmuseum, Kritzendorf 02238
Dorfmuseum, Mertingen 20296
Dorfmuseum Ahnenhaus, Pliezhausen 21063
Dorfmuseum Alter Forsthof, Wetter 22521
Dorfmuseum Altkirchen, Altkirchen 16511
Dorfmuseum Bad Alexandersbad, Bad Alexandersbad 16667
Dorfmuseum Bellach, Bellach 41189
Dorfmuseum Bennwil, Bennwil 41198
Dorfmuseum Birmensdorf, Birmensdorf 41236
Dorfmuseum Birr, Birr 41237
Dorfmuseum Bönigen, Bönigen 41243
Dorfmuseum Bottmingen, Bottmingen 41247
Dorfmuseum Buchenberg, Königsfeld im Schwarzwald 19649
Dorfmuseum Daniel-Martin-Haus, Rauschenberg 21201
Dorfmuseum Deckenpfronn, Deckenpfronn 17802
Dorfmuseum Delligsen, Delligsen 17814
Dorfmuseum Dettingen und Illertal, Dettingen an der Iller 17842
Dorfmuseum Dierscheid, Dierscheid 17849
Dorfmuseum Dietersweiler, Freudenstadt 18455
Dorfmuseum Dobin am See, Dobin am See 17881
Dorfmuseum Düppenweiler, Beckingen 16971
Dorfmuseum Eichstetten am Kaiserstuhl, Eichstetten am Kaiserstuhl 18086
Dorfmuseum Emmingen-Liptingen, Emmingen-Liptingen 18156
Dorfmuseum Feldbrunnen, Feldbrunnen 41368
Dorfmuseum Gahlenz, Gahlenz 18529
Dorfmuseum Gebhardshain, Gebhardshain 18547
Dorfmuseum Georgenthal, Georgenthal 18589
Dorfmuseum Görzig, Görzig bei Köthen 18684

Dorfmuseum Gontenschwil, Gontenschwil 41442
Dorfmuseum Graberhaus, Strengelbach 41917
Dorfmuseum Großgörschen, Großgörschen 18791
Dorfmuseum Grube, Grube 18811
Dorfmuseum Hanweiler, Winnenden 22599
Dorfmuseum Hausen im Wiesental, Hausen im Wiesental 19049
Dorfmuseum Haynrode, Haynrode 19057
Dorfmuseum Herrnbaumgarten, Herrnbaumgarten 02073
Dorfmuseum Hoffeld, Hoffeld 19222
Dorfmuseum Holzburg, Schrecksbach 21635
Dorfmuseum Hüntwangen, Hüntwangen 41490
Dorfmuseum im Greifenhaus, Hausen, Oberfranken 19051
Dorfmuseum im Guggerhuus, Ettingen 41365
Dorfmuseum im Kulturzentrum, Fislisbach 41371
Dorfmuseum im Pulverturm, Schwanden 41846
Dorfmuseum Jockenhof, Simonswald 21772
Dorfmuseum Kelter, Marbach am Neckar 20178
Dorfmuseum Kienberg, Kienberg 19537
Dorfmuseum Kirchbözberg, Unterbözberg 41957
Dorfmuseum Konolfingen, Konolfingen 41509
Dorfmuseum Langendorf, Langendorf 41525
Dorfmuseum Langlingen, Langlingen 19839
Dorfmuseum Leo's Tenne, Schweisweiler 21689
Dorfmuseum Leutesdorf, Leutesdorf 19972
Dorfmuseum Linthal, Linthal 41567
Dorfmuseum Maschwanden, Maschwanden 41614
Dorfmuseum Melihus, Möhlin 41631
Dorfmuseum mit Pfarrer-Lehmann Stube im Bauernkrieghaus, Landau in der Pfalz 19788
Dorfmuseum Mönchhof, Mönchhof 02399
Dorfmuseum Niederbreitbach, Niederbreitbach 20702
Dorfmuseum Niederlinxweiler, Sankt Wendel 21512
Dorfmuseum Nikitsch, Nikitsch 02450
Dorfmuseum Nunningen, Nunningen 41689
Dorfmuseum Ostheim, Nidderau 20693
Dorfmuseum Pfaffenweiler, Pfaffenweiler 21022
Dorfmuseum Pöttelsdorf, Pöttelsdorf 02513
Dorfmuseum Poppenweiler, Ludwigsburg 20043
Dorfmuseum Ringenwalde, Temmen-Ringenwalde 22026
Dorfmuseum Rohrbach, Rohrbach, Pfalz 21365
Dorfmuseum Roiten, Rappottenstein 02560
Dorfmuseum Rupperswil, Rupperswil 41783
Dorfmuseum Schlossweid, Ringgenberg 41768
Dorfmuseum Schönbach, Großbothen 18782
Dorfmuseum Schönwalde, Schönwalde am Bungsberg 21616
Dorfmuseum Seon, Seon 41864
Dorfmuseum Stove, Boiensdorf 17433
Dorfmuseum Stricklerhuus, Hombrechtikon 41486
Dorfmuseum Therwil, Therwil 41935
Dorfmuseum Tremmen, Ketzin 19521
Dorfmuseum Tündern, Hameln 18966
Dorfmuseum Weckbach, Weilbach 22383
Dorfmuseum Weinburg, Weinburg 02901
Dorfmuseum Wiesen, Wiesen 42014
Dorfmuseum Zeihen, Zeihen 42051
Dorfmuseum Zeissholz, Bernsdorf 17305
Dorfmuseum Ziefen, Ziefen 42055
Dorfmuseum Zierow, Zierow 22801
Dorfmuseum Zwingendorf, Zwingendorf 03138
Dorfschmiede Neckargröningen, Remseck am Neckar 21267
Dorfschmiede- und Heimatmuseum, Empfingen 18157
Dorfschulmuseum, Ködnitz 19593
Dorfstube Münchholzhausen, Wetzlar 22523
Dorfstube Ötlingen, Weil am Rhein 22374
Doric House, Flemington 48557
Dorking and District Museum, Dorking 44006
Dorman Memorial Museum, Middlesbrough 45140
Dornburger Schlösser, Dornburg, Saale 17905
Dorney Court, Windsor 46040
Dornoch Heritage Museum, Dornoch 44009
Dorothea B. Hoover Historical Museum, Joplin 49649
Dorothea C. Witham Gallery, Lewiston 50038
Dorothy G. Page Museum, Wasilla 53953
Dorothy's House Museum, Port Hope 06643
Dorozsma Windmill, Szeged 23841
Dorozsmai Szélmalom, Szeged 23841
Dorpsmuseum de Kleuskes, Liempde 32500
Dorpsmuseum De Kluis, Eext 32179
Dorpsmuseum in de Drie Snoeken, Oud-Gastel 32646
Dorrigo Steam Railway and Museum, Dorrigo 01041
Dorset County Museum, Dorchester 43999
Dorset Teddy Bear Museum, Dorchester 44000
Dory Shop Museum, Shelburne 06953
Dossin Great Lakes Museum, Detroit 48116
Dostoevsky Literary-Memorial Museum, Sankt-Peterburg 37656
Dotzheimer Museum, Wiesbaden 22547
Doual'art, Douala 05294
Doucet House, Rustico 06783
Doug Kerr Vintage Museum, Oaklands 01387
Douglas County Historical Society Museum, Superior 53377
Douglas County Museum Complex, Armour 46362
Douglas County Museum of History and Natural History, Roseburg 52457
Douglas F. Cooley Memorial Art Gallery, Portland 52075
Douglas Heritage Museum, Douglas, Lanarkshire 44010
The Douglas Hyde Gallery, Dublin 24783
Douglass Historical Museum, Douglass 48160
Doukhobor Village Museum, Castlegar 05574
Doumen County Fine Art Calligraphy Society, Zhuhai 08512
Doumen Museum, Doumen 07673
Dove Cottage and the Wordsworth Museum, Grasmere 44348
Dover Castle, Dover 44012
Dover Museum, Dover 44013
Dover Transport Museum, Whitfield 46003
Dow Sheng Chinese Kuoshu Weapons Museum, Tamsui 42448
Dowd Fine Arts Gallery, Cortland 47831
Dowerin District Museum, Dowerin 01042
Down County Museum, Downpatrick 44020
The Downers Grove Park District Museum, Downers Grove 48174
Downey Museum of Art, Downey 48175
Downhill Castle, Castlerock 43772
Downieville Museum, Downieville 48176
Downpatrick Railway Museum, Downpatrick 44021
Downtown Gallery, Nanaimo 06410
Downtown History Museum, Tucson 53588
The Dowse, Hutt City 33124
DOX - Centrum současného umění, Praha 09739
Dr. Arratta Memorial Museum, Muttaburra 01343
Dr.-Arthur-Lindgens-Jagdmuseum, Schönsee 21613
Dr. Batthyány-Strattmann László Múzeum, Körmend 23674
Dr.-Bauer-Heimatmuseum, Bad Rothenfelde 16810
Dr. Berger Heimatmuseum, Forchtenberg 18338
Dr. Bhau Daji Lad Museum, Mumbai 24314
Dr. Bohuslav Horák Regional Museum, Rokycany 09818
Dr.-Carl-Haeberlin-Friesen-Museum, Wyk auf Föhr 22759
Dr. Daniel W. Kissam House, Huntington 49429
Dr. Eduard Benninger Heimathaus, Bad Wimsbach-Neydharting 01770
Dr. Eisenbarth- und Heimatmuseum, Oberviechtach 20848
Dr. Engelbert Dollfuß-Museum, Texing 02806
Dr. Fr.R. Kreutzwaldi Memoriaalmuuseum, Võru 10676
Dr. Francis Medical and Apothecary Museum, Jacksonville 49575
Dr. Giri's Museum, Chennai 24113
Dr. Heins Classic Car Collection, Christchurch 33072
Dr. Henry N. Payne Community Museum, Cow Head 05653
Dr. Increase Mathews House, Zanesville 54367
Dr. John Harris Dental Museum, Bainbridge 46531
Dr. Johnson's House, London 44839
Dr. Jono Šliūpo Memorialinė Sodyba, Palanga 30328
"Dr. José Evaristo Uriburu" Numismatic Museum, Buenos Aires 00245"Dr.
Dr. Juba Ferenc Magyar Tengerészettörténeti Gyűjteménye, Komáron 23677
Dr. Kui-Fang Lai Art Museum, Singapore 38173
Dr. Lin Sian Tang Gallery, Wufeng 42459
Dr. Park Collection and Gallery, Seoul 29873
Dr. Pepper Museum and Free Enterprise Institute, Waco 53773
Dr. Raj Bali Pandey Puratatva Sangrahalaya, Deoria 24157
Dr. Samuel D. Harris National Museum of Dentistry, Baltimore 46561
Dr. Sun Yat Sen Memorial Museum, Macau 07991
Dr. Vernon L. Watson Allied Arts Centre, Dauphin 05678
Dr. Vinco Kudirkos Muziejus, Kudirkos Naumiestis 30290
Dr. Vladimir Malančec Collection, Koprivnica 08890
Dr. William Robinson Plantation, Clark 47585
Dr. Woods House Museum, Leduc 06156
Draaiorgelmuseum, Assen 31936
Drachenhöhle, Syrau 21996
Drachenloch Museum, Vättis 41970
Drachmanns Hus, Skagen 10259
Dragon- og Frihedsmuseet, Holstebro 10089
Dragonermuseum, Wels 02914
Dragoon and Freedom Museum, Holstebro 10089
Dragør Museum, Dragør 09989
Drake House Museum, Plainfield 51973
Drake Well Museum, Titusville 53516
Drama Archaeological Museum, Drama 22987
Drammens Museum, Drammen 33466
Drammens Museum for Kunst og Kulturhistorie, Drammen 33467
Drangedal Bygdetun, Drangedal 33468
Drava Collection, Gola 08875
Drava Museum - Fire Extinguishing Historical Exhibition, Barcs 23362
Dráva Múzeum, Barcs 23361
Drawing Center, New York 51167
Drayton Hall, Charleston 47362
Drayton Valley Museum, Drayton Valley 05710
Dreamtime Cultural Centre, Rockhampton 01464
Drechsler- und Metalldrücker-Museum, Wendelstein 22475
Dreher Beer Museum, Budapest 23395
Dreher Sörmúzeum, Budapest 23395
Dreieich-Museum, Dreieich 17936
Drenewydd Museum, Bute Town 43704
Drents Museum, Assen 31937
De Drentse Glasbloazer, Dalen 32044
Dresden Transport Museum, Dresden 17976
Dresden University of Technology, Custody, Dresden 17955
Dress and Textiles Specialists Group, Beamish 55055
Drew County Historical Museum, Monticello 50818
Driebe Freight Station, Stroudsburg 53338
Driels Oudheidkundig Museum, Kerkdriel 32445
Drilandmuseum Gronau, Gronau, Westfalen 18773
Drill Hall Gallery, Acton 00768
Driminagh Castle, Dublin 24784
Drjanovski Manastir Muzej, Drjanovo 05023
Drogerie-Museum, Weißenstadt 22463
Droitwich Spa Heritage Centre, Droitwich 44024
Dromana and District Museum, Dromana 01043
Droogdok Jan Blanken, Hellevoetsluis 32359
Drostdy Museum, Swellendam 38744
Drostdy Museum, Uitenhage 38768
Droste-Museum, Havixbeck 19056
Droste-Museum, Münster 20508
Droste-Museum im Fürstenhäusle, Meersburg 20248
Drottningholm Castle, Drottningholm 40488
Drottningholms Slott, Drottningholm 40488
Druckereimuseum, Freienbach 41381
Druckereimuseum, Hatten 19034
Druckereimuseum, Kusel 19751
Druckereimuseum, Wildeshausen 22569
Druckereimuseum und Schaudepot der Museen der Stadt Erfurt im Benary-Speicher, Erfurt 18200
Druckhistorische Ausstellung, Kornwestheim 19689
Druckmuseum, Darmstadt 17786
Druckmuseum Graz, Graz 01967
Druivenmuseum, Overijse 03945
Drukkerijmuseum Meppel, Meppel 32549
Drum Barracks Civil War Museum, Wilmington 54188
Drum Castle, Drumoak 44028
Drum Museum Taikokan, Tokyo 29383
Drumlanrig Castle, Thornhill, Dumfriesshire 45855
Drumlanrig's Tower, Hawick 44425
Drumlin Farm, Lincoln 50091
Drummond Home, Hominy 49333
Drummond Island Historical Museum, Drummond Island 48183
Druskininkų Miesto Muziejus, Druskininkai 30222
Druskininkų Ž. Lipšico Žydų Muziejus, Druskininkai 30223
Druviena Old School, Druviena 30003
Druvienas Vecā Skola-Muzejs, Druviena 30003
Dry Falls Interpretive Center, Coulee City 47849
Dryanovo Monastery Museum, Drjanovo 05023
Dryden and District Museum, Dryden 05716
Državni Muzej, Cetinje 31640
Du Page County Historical Museum, Wheaton 54119
Du Tam-Tam au Satellite, Angers 11603
Dualchas Heritage Service for Skye and Lochalsh, Portree 55056
Dualchas-Museum Bharraigh Agus Bhatarsaidh, Isle-of-Barra 44537
Dubai Museum, Dubai 43218
De Dubbelde Palmboom, Rotterdam 32704
Dubbo Military Museum, Dubbo 01045
Dubbo Regional Gallery, Dubbo 01046
Dublin Castle, Dublin 24785
Dublin Civic Museum, Dublin 24786
Dublin-Laurens Museum, Dublin 48185
Dublin Print Museum, Dublin 24787
Dubois Museum, Dubois 48186
Dubrovin Farm Museum, Yisod Hama'ala 25131
Dubuque Museum of Art, Dubuque 48187
Duc Minh Art Gallery, Ho Chi Minh City 54621
Duchang Museum, Douchang 07672
Duché d'Uzès, Uzès 16023
Duck Lake Regional Interpretive Centre, Duck Lake 05717
Dudley and Mary Marks Lea Gallery, Findlay 48543
Dudley Museum and Art Gallery, Dudley 44030
Dudmaston House, Quatt 45454
Dülkener Narrenmuseum, Viersen 22231
Dümmer-Museum, Lembruch 19946
Dürer Hall, Gyula 23592
Dürer Terem, Gyula 23592
Dürstelerhaus, Ottikon 41717
Düsseldorfer Museumsverein e.V., Düsseldorf 54832
Duff Community Heritage Museum, Duff 05718
Dufferin County Museum, Rosemont 06774
Dufferin Historical Museum, Carman 05566
Dufftown Whisky Museum, Dufftown 44032
Duftmuseum im Farina Haus, Köln 19600
Dufu Caotang Museum, Chengdu 07616
Duggan-Cronin Gallery, Kimberley 38667
Dugger Coal Museum, Dugger 48190
Dugout Museum, Länkipohja 11023
Duke Homestead, Durham 48216
Duke University Union Museum and Brown Art Gallery, Durham 48217
Dukelské Múzeum - Vojenské Múzeum Svidník, Svidník 38366
Duksung Women's University Art Museum, Seoul 29874
Dukwon Museum of Art, Seoul 29875
Duluth Art Institute, Duluth 48193
Duluth Children's Museum, Duluth 48194
Dulwich Picture Gallery, London 44840
Dům Františka Bílka, Chýnov 09488
Dům Gustava Mahlera, Jihlava 09560
Dům Pánů z Kunštátu, Brno 09424
Dům u Černé Matky Boží - Muzeum kubismu, Praha 09740
Dům u Kamenného Zvonu, Praha 09741
Dům u Zlatého Prstenu, Praha 09742
Dům Umění, České Budějovice 09462
Dům Umění, Znojmo 09946
Dům Umění Města Brna, Brno 09425
Dumbarton House, Washington 53864
Dumbarton Oaks Collections, Washington 53865
Dumbo Arts Center, Brooklyn 47072
Dumfries and Galloway Aviation Museum, Dumfries 44035
Dumfries Museum & Camera Obscura, Dumfries 44036
Duna Múzeum - Magyar Környezetvédelmi és Vízügyi Múzeum, Esztergom 23551
Dunamelléki Református Egyházkerület Ráday, Budapest 23396
Dunamelléki Református Egyházkerület Ráday Múzeuma, Kecskemét 23645
Dunántúli Református Egyházkerület Tudományos Gyűjteményei Múzeum, Pápa 23751
Dunaskin Open Air Museum, Ayr 43345
Dunaújvárosi Papírgyár Üzemtörténeti Gyűjteménye, Dunaújváros 23532
Dunbar Town House Museum, Dunbar 44041
Dunblane Museum, Dunblane 44044
Duncan Cottage Museum, Metlakatla 50615
Duncan Gallery of Art, De Land 47997
Duncan-McAshan Visual Arts Center, Ingram 49511
Dundas Historical Society Museum, Dundas 05724
Dundas Valley School of Art Gallery, Dundas 05725
Dundee Art Galleries and Museum Association, Dundee 55057
Dundee Township Historical Society Museum, Dundee 48206
Dundurn Castle, Hamilton 05953
Dundy County Historical Society Museum, Benkelman 46736
Dunedin Fine Arts Center, Dunedin 48208
Dunedin Historical Society Museum, Dunedin 48209
Dunedin Public Art Gallery, Dunedin 33090
Dunfermline Museum, Dunfermline 44055
Dungarvan Museum, Dungarvan 24828
Dungeon and Museum of Namik Kemal, Gazimağusa 33352
Dungog Historical Museum, Dungog 01048
Dunguaire Castle, Kinvara 24858
Dunham Massey Hall, Altrincham 43294
Dunham Tavern Museum, Cleveland 47630
Dunhuang Folklorish Museum, Dunhuang 07676
Dunhuang Glasweglansu Guan, Dunhuang 07676
Dunhuang Museum, Dunhuang 07677
Dunkeld Cathedral Chapter House Museum, Dunkeld 44059
Dunkers Kulturhus, Helsingborg 40585
Dunklin County Museum, Kennett 49732
Dunlop Art Gallery, Regina 06716
Dunluce House Museum, Kimberley 38668
Dunn-Seiler Museum, Mississippi State 50751
Dunnet Pavilion, Castletown 43775
Dunolly Museum, Dunolly 01049
Dunree Military Museum, Inishowen 24842
Dunrobin Castle Museum, Golspie 44325
Duns Area Museum, Duns 44064
Dunster Castle, Dunster 44066
Dunvegan Castle, Dunvegan 44067
Dunwell and Community Museum, Weekes 07293
Dupont Gallery, Lexington 50069
DuPont Historical Museum, DuPont 48211
Durban Art Gallery, Durban 38604
Durban House Heritage Centre, Eastwood 44109
Durban Natural Science Museum, Durban 38605
Durbe Palace, Tukums 30130
Durbes Pils, Tukums 30130
Durham Art Gallery, Durham 05727
Durham Art Guild, Durham 48218
Durham Cathedral Treasures of Saint Cuthbert, Durham 44069
Durham Center Museum, East Durham 48238

Durham Heritage Centre and Museum, Durham 44070
Durham Historic Association Museum, Durham 48224
Durham Light Infantry Museum and Durham Art Gallery, Durham 44071
Durham Western Heritage Museum, Omaha 51548
Durjaven Kulturen Center Dvoreca, Balčik 04975
Durrell Museum, Durrell 05728
Durylin House-Museum, Korolev 37058
Dusable Museum of African-American History, Chicago 47472
Dutch Brick and Rooftile Museum, Zevenaar 33000
Dutch Dakota Association Exhibition, Schiphol 32759
Dutch Funeral Museum, Amsterdam 31874
Dutch Gold-, Silver- and Clockmuseum, Schoonhoven 32764
Dutch Leather and Shoe Museum, Waalwijk 32918
Dutch Museum of Ironing & Mangling Instruments, Noordbroek 32601
Dutch Museum of Lithography, Valkenswaard 32870
Dutch Period Museum, Colombo 40371
Dutch Poultry Museum, Barneveld 31950
Dutch Railway Museum, Utrecht 32852
Dutch Resistance Museum, Amsterdam 31899
Dutch Textile Museum, Tilburg 32829
Dutchess County Historical Society, Poughkeepsie 52120
Duwisib Castle, Maltahohe 31746
Duxbury Rural and Historical Society, Duxbury 48228
Dviračių Muziejus, Šiauliai 30354
Dvor Trakošćan, Trakošćan 08967
Dvorac u Čelarevu, Čelarevo 38097
Dvořákovo Muzeum Kolínska v Pravěku, Kolín 09589
Dvorec-Muzej Betnava, Maribor 38475
Dvorec-Muzej Petra I, Sankt-Peterburg 37636
Dvorec-muzej Petra III, Lomonosov 37107
Dvorec Petra I, Strelna 37809
Dvorec Štatenberg, Makole 38474
Dwight D. Eisenhower Library-Museum, Abilene 46120
Dwight Frederick Boyden Gallery, Saint Mary's City 52608
Dwór Artusa, Gdańsk 34634
Dwór w Dołędze, Zaborów 35348
Dworek Jana Matejki, Krzesławice 34835
Dworek Jana Matejki w Krzesławicach, Kraków 34771
Dworek Wincentego Pola, Lublin 34896
Dyckman Farmhouse Museum, New York 51168
Dydykin House-Museum, Palech 37523
Dyer Memorial Library, Abington 46127
Dylan Thomas Boathouse, Laugharne 44646
Dyrøy Bygdemuseum, Tennevoll 33902
Dyvelstens Flottningsmuseum, Forshaga 40525
Dzeržinsk City Museum, Dzeržinsk 36883
Dzeržinskij Gorodskoj Kraevedčeskij Muzej, Dzeržinsk 36883
Dzhizak State Museum of Local Lore, Jizzax 54448
Dział Archeologii i Etnografii, Bydgoszcz 34574
Dział Zbiorów Specjalnych - Muzeum, Warszawa 35222
Džuro Jakšić Memorial Museum, Srpska Crnja 38154

E

E.A. Boratynskyj Museum, Kazan 37014
E. Akhvlediani Children's Picture Gallery, Tbilisi 16354
E. Akhvlediani House Museum, Tbilisi 16355
E. Bindstouw, Viborg 10314
E.F. Gnesina Memorial Appartment Museum, Moskva 37243
E.M. Violette Museum, Kirksville 49800
E. St. Julien Cox House, Saint Peter 52632
E.T.A. Hoffmann-Haus, Bamberg 16907
E-Werk Freiburg, Freiburg im Breisgau 18429
EAA AirVenture Museum, Oshkosh 51613
Eagle Bluff Lighthouse, Fish Creek 48545
Eagle Harbor Lighthouse and Museums, Eagle Harbor 48232
Eagle Historical Society and Museums, Eagle City 48231
Eagle Tavern Museum, Watkinsville 53982
Eagle Valley Museum, Sicamous 06968
Ealing College Gallery, London 44841
EAM-live-Museum Wasserkraftwerk Merkenbach, Dillenburg 17859
EAM-live-Museum Wasserkraftwerk Wülmersen, Trendelburg 22088
Ear Falls District Museum, Ear Falls 05729
Earist Museum, Manila 34439
The Earle-Harrison House, Waco 53774
Earle Wightman Museum, Oyster Bay 51661
Earls Barton Museum of Local Life, Earls Barton 44080
Early American Museum, Mahomet 50399
Earlystreet, Norman Park 01370
Earlyworks Museums, Huntsville 49443
It Earmhus en Oud Friese Greidboerderij, Warten 32925
Earth and Man National Museum, Sofia 05196
Earth Science Museum, Rio de Janeiro 04725
Earth Sciences Museum, Waterloo 07286
Earthquake Museum, Aşgabat 42970
Easdale Island Folk Museum, Easdale Island 44081
East and Seaport Museum, Greenport 49058
East Anglia Transport Museum, Carlton Colville 43761
East Anglian Film Archive, Norwich 45277
East Anglian Railway Museum, Chappel 43789
East Bohemian Gallery, Pardubice 09706
East Bohemian Gallery, Pardubice 09705
East Bohemian Museum, Pardubice 09707
East Brunswick Museum, East Brunswick 48237
East Campus Galleries, Orlando 51590
East Carlton Steel Heritage Centre, East Carlton 44083
East Coast Museum of Technology, Gisborne 33108
East Coulee School Museum, East Coulee 05731
East Ely Railroad Depot Museum, Ely 48384
East Essex Aviation Society and Museum of the 40's, Saint Osyth 45586
East Gippsland Museum, Bairnsdale 00821
East Ham Nature Reserve, London 44842
East Hampton Historical Society Museum, East Hampton 48244
East Hants Historical Museum, Maitland 06219
East Iceland Heritage Museum, Egilsstadir 23993
East Jersey Olde Towne, Piscataway 51933
East Jordan Portside Art and Historical Museum, East Jordan 48252
East Kazakhstan Historical and Regional Museum, Özkemen 29712
East Kazakhstan Museum of Art, Özkemen 29713
East Kentucky Center for Sciece, Mathematics and Technology, Prestonsburg 52139
East London Museum, East London 38615
East Martello Museum, Key West 49758
East Midlands Museums Service, Colston Bassett 55058
East New Britain Historical & Cultural Center, Kokopo 34082
East Poultney Museum, East Poultney 48258
East Riddlesden Hall, Keighley 44551
East-Slovakian Gallery, Košice 38274
East Somerset Railway, Shepton Mallet 45657
East Surrey Museum, Caterham 43780
East Tennessee Discovery Center, Knoxville 49816
East Tennessee Historical Society Museum, Knoxville 49817
East Texas Oil Museum at Kilgore College, Kilgore 49774
East Timorese Cultural Centre Museum Collection, Dili 10364
East Troy Electric Railroad Museum, East Troy 48262
East-West Center, Honolulu 49338
East Wing Art Gallery, Gardner 48861
Eastbourne Heritage Centre, Eastbourne 44097
Eastchester Historical Society Museum, Bronxville 47054
Eastern Arizona Museum, Pima 51919
Eastern Azarbaijan Province History Museum, Tabriz 24652
Eastern California Museum, Independence 49469
Eastern Edge Art Gallery, Saint John's 06847
Eastern Häme Museum, Hartola 10760
Eastern Kazakhstan Regional Ethnographical Museum, Özkemen 29711
The Eastern Museum, Derby 43966
Eastern Oregon Museum on the Old Oregon Trail, Haines 49126
Eastern Shore Art Center, Fairhope 48488
Eastern Southland Gallery, Gore 33111
Eastern State Penitentiary Historic Site, Philadelphia 51828
Eastleigh Museum, Eastleigh 44104
Eastney Industrial Museum, Portsmouth 45430
Easton Farm Park, Easton 44106
Eastside Heritage Center, Bellevue 46707
Eastwood House, Glasgow 44279
Eaton-Buchan Gallery and Marvin Cone Gallery, Cedar Rapids 47303
Eavanston Ecology Center, Evanston 48448
Eaved House - Cultural and Art Centre, Lefkoşa 33372
Ebbamåla Bruk, Kyrkhult 40663
Ebbas Hus, Malmö 40718
Ebeltoft Museum, Ebeltoft 09994
Ebenezer Maxwell Mansion, Philadelphia 51829
Éber Emlékház, Baja 23345
Eber Memorial Hall, Baja 23345
Eberdt Museum of Communications, Sutton, Québec 07042
Ebersdorfer Schulmuseum, Chemnitz 17694
Ebetsu Ceramic Art Center, Ebetsu 28410
Ebetsu-shi Bijutsu Togeikan, Ebetsu 28410
Ebinger Heimatmuseum, Albstadt 16452
Eça de Queiroz Foundation, Santa Cruz do Douro 35900
Ecce Homo, Jerusalem 24986
Eccles Community Arts Center, Ogden 51500
Ecclesiastic Art Collection of the Matthias Church at Buda Castle, Budapest 23393
Ecclesiastic Museum and Treasury, Szeged 23842
Ecclesiastical Museum, Komotini 23059
Ecclesiastical Museum, Mullingar 24880
Ecclesiastical Museum, Sifnos 23185
Ecclesiastical Museum of Alexandroupolis, Alexandroupolis 22876
Echizen Crab Museum, Echizen 28411
Echizen Pottery Gallery, Echizen 28412
Echizen Take-ningyo-no-sato, Maruoka 28883
Echmiadzin Art Gallery, Echmiadzin 00694
Echmiadzin Cathedral Complex and Museum, Echmiadzin 00695
Echo Historical Museum, Echo 48281
Echuca Gem Club Collection, Echuca 01056
Echuca Museum, Echuca 01057
Eckley Miners' Village, Eckley 48283
Éco-Musée de la Blanchisserie, Grezieu-la-Varenne 13146
Eco-Musée de la Noix, Castelnaud-la-Chapelle 12312
Eco-Musée des Etouvières, Thônes 15891
Éco Musée Vivant de Provence, La Gaude 13377
Ecodrome Zwolle, Zwolle 33026
École Dentellière, Bailleul 11828
École du Rang II d'Authier, Authier 05365
Ecole-Musée de Champagny, Champagny 12388
Ecological Education Centre, Eindhoven 32190
Ecomare, De Koog 32047
Écomusée, Boigneville 12045
Écomusée, Saulge 15653
Ecomusée Agricole du Pays de Fayence, Fayence 12911
Ecomusée d'Alsace, Ungersheim 16017
Ecomusée d'Alzen, Alzen 11563
Écomusée d'Anticosti, Port-Menier 06645
Ecomusée d'Arts et Traditions Populaires, Ligny-le-Ribault 13749
Écomusée de Fresnes, Fresnes (Val-de-Marne) 13009
Ecomusée de Hauteluce, Hauteluce 13192
Écomusée de Hearst Ecomuseum, Hearst 05981
Écomusée de Hull, Gatineau 05858
Ecomusée de la Basse-Seine, Notre-Dame-de-Bliquetuit 14473
Ecomusée de la Batellerie et de la Voie d'Eau Le Thudo, Thuin 04064
Écomusée de la Bourrine du Bois Juquaud, Saint-Hilaire-de-Riez 15307
Ecomusée de la Brenne, Le Blanc 13558
Ecomusée de la Bresse Bourguignonne, Cuiseaux 12686
Ecomusée de la Bresse Bourguignonne, Pierre-de-Bresse 14812
Écomusée de la Cévenne, Saint-Laurent-de-Trèves 15358
Écomusée de la Combe de Savoie, Grésy-sur-Isère 13144
Ecomusée de la Communauté, Ciry-le-Noble 12530
Ecomusée de la Communauté Le Creusot Montceau-les-Mines, Le Creusot 13576
Écomusée de la Courneuve, La Courneuve 13354
Ecomusée de la Crau, Saint-Martin-de-Crau 15398
Ecomusée de la Ferme de Kervazegan, Pont-l'Abbé 14887
Écomusée de la Ferme de la Forêt, Saint-Trivier-de-Courtes 15523
Ecomusée de la Ferme du Chemin, Madré 13909
Ecomusée de la Ferme du Moulin Vanneau, Saints-en-Puisaye 15589
Ecomusée de la Ferme et des Vieux Métiers, Lizio 13791
Écomusée de la Forêt d'Orient, Brienne-la-Vieille 12175
Ecomusée de la Forêt Méditerranéenne, Gardanne 13032
Ecomusée de la Forge, Auchy-la-Montagne 11737
Écomusée de la Forge, Menton 14078
Ecomusée de la Grande Lande Marquèze, Sabres 15144
Ecomusée de la Haie, Châtillon-en-Dunois 12469
Ecomusée de la Lomagne, Miradoux 14128
Ecomusée de la Margeride, Loubaresse 13827
Ecomusée de la Margeride, Ruynes-en-Margeride 15142
Écomusée de la Métairie, Plouigneau 14860
Ecomusée de la Montagne des Reuns, Germaine 13048
Ecomusée de la Pêche, Thonon-les-Bains 15895
Ecomusée de la Pêche et de la Mer, Capbreton 12274
Ecomusée de la Pomme au Calvados, Le Sap 13649
Ecomusée de la Région de Fourmies Trélon en Avesnois, Trélon 15989
Ecomusée de la Région du Viroin, Treignes 04091
Ecomusée de la Sainte-Baume, Plan-d'Aups 14830
Ecomusée de la Tradition Basque Jean Vier, Saint-Jean-de-Luz 15326
Ecomusée de la Truffe, Sorges 15785
Ecomusée de la Vallee d'Aspe, Accous 11489
Ecomusée de la Vallée de l'Aigre, La Ferté-Villeneuil 13366
Ecomusée de la Vallée du Galeizon, Cendras 12335
Ecomusée de la Vigne et du Vin, Gradignan 13092
Ecomusée de l'Abeille, Auriac-du-Périgord 11751
Écomusée de l'Abeille, Grateloup 13117
Écomusée de l'Abeille, Le Faou 13581
Écomusée de l'Abeille et de la Flore Locale, Faugères 12907
Écomusée de l'Agriculture, Aigues-Vives 11505
Écomusée de l'Apiculture, Montargis 14170
Ecomusée de l'Armagnac, Labastide-d'Armagnac 13460
Ecomusée de l'Ile-de-Groix, Ile-de-Groix 13237
Écomusée de l'île d'Ouessant, Ile-d'Ouessant 13242
Ecomusée de Marais Breton Vendéen, La Barre-de-Monts 13311
Ecomusée de Marie Galante, Grand Bourg 23274
Ecomusée de Montjean, Montjean-sur-Loire 14218
Ecomusée de Plein Air du Quercy, Sauliac-sur-Célé 15654
Écomusée de Saint-Constant, Saint-Constant 06810
Ecomusée de Saint-Dégan, Brec'h 12151
Ecomusée de Saint-Joseph, Fargues-sur-Ourbise 12906
Écomusée de Saint-Léger, Saint-Léger-les-Mélèzes 15365
Ecomusée de Saint-Nazaire, Saint-Nazaire (Loire-Atlantique) 15423
Ecomusée de Salazie, Salazie 15592
Ecomusée de Savigny-le-Temple, Savigny-le-Temple 15679
Écomusée des Anciennes Carrières de Pierre "Les Capucins", Paris 14595
Ecomusée des Appeaux et de la Faune, Saint-Didier 15238
Écomusée des Boutères-Pilat, Riotord 15048
Ecomusée des Bruneaux, Firminy 12937
Ecomusée des Deux-Rives, Salaberry-de-Valleyfield 06903
Ecomusée des Goémoniers et de l'Algue, Plouguerneau 14855
Ecomusée des Monts d'Arrée, Commana 12592
Ecomusée des Monts d'Arrée, Saint-Rivoal 15499
Ecomusée des Pays de l'Oise, Beauvais 11908
Ecomusée des Pipes Ropp, Baume-les-Dames 11865
Ecomusée des Techniques, Thuin 04065
Ecomusée des Vallées du Paillon, Contes 12617
Ecomusée d'Hannonville, Hannonville-sous-les-Côtes 13179
Ecomusée d'Ichkeul, Bizerte 42618
Écomusée du Ancien Moulinage, Chirols 12518
Ecomusée du Bocage, Sains-du-Nord 15153
Ecomusée du Bommelaers Wall, Ghyvelde 13053
Écomusée du Carton, Mesnay 14095
Écomusée du Cheminot Veynois, Veynes 16134
Ecomusée du Cognac, Migron 14119
Ecomusée du Commerce et de l'Artisanat, Rochefort (Charente-Maritime) 15070
Ecomusée du Cuir de la Tannerie Nory, Sainghin-en-Weppes 15152
Écomusée du Domaine de la Petite Couère, Nyoiseau 14487
Écomusée du Fier Monde, Montréal 06338
Ecomusée du Haut Beaujolais, Marnand 13976
Ecomusée du Haut Pays de la Roya et de la Bevera, Breil-sur-Roya 12153
Ecomusée du Libournais, Montagne 14165
Écomusée du Liège, Gonfaron 13081
Écomusée du Linge, Montcy-Notre-Dame 14194
Ecomusée du Marais Salant, Loix-en-Ré 13804
Ecomusée du Mont-Lozère, Le Pont-de-Montvert 13640
Ecomusée du Montmorillon, Montmorillon 14230
Ecomusée du Moulin de Kerchuz, Saint-Thurien 15522
Écomusée du Moulin de la Sée, Brouains 12191
Écomusée du Moulin Saint-Martin, Cany-Barville 12273
Ecomusée du Pays de la Cerise de Fougerolles, Fougerolles 12989
Écomusée du Pays de la Roudoule, Puget-Rostang 14956
Ecomusée du Pays de Montfort, Montfort-sur-Meu 14210
Écomusée du Pays de Nied, Gomelange 13077
Ecomusée du Pays de Rennes, Rennes (Ille-et-Vilaine) 15012
Écomusée du Pays des Collines, Lahamaide 03773
Ecomusée du Paysan Gascon, Toujouse 15921
Écomusée du Perche de Saint-Cyr-la-Rosière, Bellême 11928
Écomusée du Phare de Trezien, Plouarzel 14847
Ecomusée du Port Ostréicole, La Teste-de-Buch 13447
Ecomusée du Roannais, Roanne 15058
Écomusée du Val de Saône, Seurre 15738
Ecomusée du Veron, Savigny-en-Veron 15678
Écomusée du Verre de Biot, Biot 11996
Ecomusée du Vignoble Drouais et des Artisans, Dreux 12791

D

E

Ecomusée du Village Breton de Poul-Fetan, Quistinic 14975
Ecomusée Haldensleben-Hundisburg, Haldensleben 18882
Ecomusée Industriel de Lochrist-Inzinzac, Lochrist-Inzinzac 13796
Ecomusée la Cité des Abeilles, Saint-Faust 15264
Écomusée la Ferme du Hameau, Bierre-les-Semur 11986
Écomusée La Vie d'Antan, Esparron-de-Verdon 12859
Ecomusée l'Abeille du Haut-Doubs, Aubonne 11733
Ecomusée Maison Michaud, Chapelle-des-Bois 12402
Ecomusée Picarvie, Saint-Valéry-sur-Somme 15531
Écomusée Rachana, Rachana 30171
Ecomusée Régional du Centre, Houdeng-Aimeries 03716
Écomusée Rural de Vigneux-La Pauelais, Vigneux-de-Bretagne 16160
Écomusée Sauvegarde du Vieil Auzon, Auzon 11781
Ecomuseo, Tordehumos 40145
Ecomuseo de Guinea, Frontera 39312
Ecomuseo del Caserio Vasco, Artea 38900
Ecomuseo del Habitat Subterráneo de Rojales, Rojales 39909
Ecomuseo della Resistenza, Coazze 25929
Ecomuseo della Val Nerina, Cerreto di Spoleto 25811
Ecomuseo dell'Archeologia Industriale, Schio 27792
Ecomuseo di Archeologia Industriale E. Crumi, Villar Pellice 28280
Ecomuseo Los Torres, Lerdo 31000
Ecomuseo-Luogo della Memoria, Lamezia Terme 26442
Ecomuseo - Molino de Zubieta, Zubieta 40359
Ecomuseo Regional Maipú, Maipú 00441
Ecomuseu, Rio de Janeiro 04689
Ecomuseu da Fazenda Boa Vista, Roseira 04781
Ecomuseu de Góis, Góis 35627
Ecomuseu de Itaipu, Foz do Iguaçu 04466
Ecomuseu de les Valls d'Aneu, Esterri d'Aneu 39290
Ecomuseu do Guadiana, Mértola 35759
Ecomuseu do Rabeirão do Ilha, Florianópolis 04452
Ecomuséum, Sainte-Anne-de-Bellvue 06889
Ecomuseum en Archief van de Boomse Baksteen, Boom 03382
Ecomuseum Simplon, Simplon Dorf 41871
Economic Botany Collection/Museum No. 1, Richmond, Surrey 45491
Economuseum voor Papier, Karton en Aandrijftechnieken, Alsemberg 03265
EcoTarium, Worcester 54297
Ecoteca-Museu João Lúcio, Olhão 35789
Ecovillage Centre Africain Reconstitué, Magny-Cours 13911
Écovillage de Lorraine, Ville-sur-Yron 16178
Écovillage Gaulois, Pleumeur-Bodou 14841
Ed Clark Museum of Missouri Geology, Rolla 52445
Eda Skans Museum, Åmotfors 40420
Edams Museum, Edam 32167
Edelfelt-Vallgren Museo, Porvoo 11226
Edelfelt-Vallgren Museum, Porvoo 11226
Edelsteenslijperij de Steenarend, Eindhoven 32186
Eden Camp, Malton 45079
Eden Court Art Gallery, Inverness 44524
Eden Killer Whale Museum, Eden 01060
Eden Valley Museum, Edenbridge 44111
Edgar Allan Poe Cottage, Bronx 47044
Edgar Allan Poe House, Philadelphia 51830
Edgar Allan Poe House and Museum, Baltimore 46562
Edgar Allan Poe Museum, Richmond 52327
Edgar County Historical Museum, Paris 51702
Edge of the Cedars State Park Museum, Blanding 46850
Edgerton and District Museum, Edgerton 05737
Edgerton Explorit Center, Aurora 46496
Edificiul Roman cu Mozaic, Constanța 36301
Edinburgh Printmakers Workshop and Gallery, Edinburgh 44120
Edinburgh Scout Museum, Edinburgh 44121
Edinburgh Square Heritage and Cultural Centre, Caledonia 05512
Edinburgh University Anatomy Museum, Edinburgh 44122
Edinburgh University Natural History Collections, Edinburgh 44123
Edirne Devlet Güzel Sanatlar Galerisi, Edirne 42767
Edirne State Fine Art Gallery, Edirne 42767
Edison and Ford Winter Estates, Fort Myers 48661
Edison Community College Gallery of Fine Art, Fort Myers 48662
Edison Museum, Beaumont 46675
Edison National Historic Site, West Orange 54071
Edisto Island Historic Preservation Society Museum, Edisto Island 48293
Edith-Ruß-Haus für Medienkunst, Oldenburg 20894
Edith Russ Site for Media Art, Oldenburg 20894
Edithburgh Museum, Edithburgh 01062
Edmond Historical Society Museum, Edmond 48294
Edmonds Arts Festival Museum, Edmonds 48295
Edmonds South Snohomish County Historical Society Museum, Edmonds 48296
Edmonston House, Vails Gate 53681
Edmonton Art Gallery, Edmonton 05743
Edmonton Public Schools Museum, Edmonton 05744
Edmonton Telephone Historical Centre, Edmonton 05745
Edmund Adler-Galerie, Mannersdorf 02347
Edna Historical Museum, Edna 48297
Edo-Tokyo Museum, Tokyo 29384
Edo-Tokyo Open Air Architectural Museum, Koganei 28757
Edo-Tokyo Tatemono-En, Koganei 28757
Edoardo Villa Museum, Tshwane 38745
Eds MC- och Motormuseum, Ed 40490
Edsel and Eleanor Ford House, Grosse Pointe 49094
Eduard Bargheer-Haus, Hamburg 18924
Eduarda Smiļga Teātra Muzejs, Rīga 30065
Eduarda Veidenbauma Memoriālais Muzejs, Liepa 30037
Eduards Smiļgis Theatre Museum, Rīga 30065
Educatieve Boerderij en Zijdemuseum de Schans, Oud-Gastel 32647
Education Museum, Hyderabad 34003
Educational Science Museum, Kuwait 29948
Eduskunnan Talvisotamuseo, Kauhajoki 10926
Edvard Grieg Museum, Paradis 33766
Edventure, Columbia 47724
Edward and Helen Mardigian Museum, Jerusalem 24987
Edward and Marthann Samek Art Gallery, Lewisburg 50029
Edward de Courcy Clarke Earth Science Museum, Crawley 01009
Edward-Dean Museum, Cherry Valley 47425
Edward H. White II Memorial Museum, Brooks Air Force Base 47089
Edward Hopper House Art Center, Nyack 51449
Edward M. Bannister Gallery, Providence 52165
Edward Williams Gallery, Hackensack 49110
Edwards County Historical Museum, Kinsley 49797
Edwards Memorial Museum, Chesterfield 47436
Edwin A. Ulrich Museum of Art, Wichita 54140
Edwin Scharff Museum, Neu-Ulm 20587
Edwin Wolters Memorial Museum, Shiner 53079
Edwin Young Collection, Salisbury 45602
Eells-Stow House, Milford 50683
Eemil Halonen Museum and Lapinlahti Art Museum, Lapinlahti 11044
Eemu Tulik, Linnuse 10593
Eemu Windmill, Linnuse 10593
Het Eerste Friese Schaatsmuseum, Hindeloopen 32390
Eerste Nederlandse Opel Automuseum, Tijnje 32826
Eesti Ajaloomuuseum, Tallinn 10622
Eesti Arhitektuurimuuseum, Tallinn 10623
Eesti Kirjandusmuuseum, Tartu 10645
Eesti Kunstimuuseum, Tallinn 10624
Eesti Loodusmuuseum, Tallinn 10625
Eesti Meremuuseum, Tallinn 10626
Eesti Muuseumiühing Tallinn, Tallinn 54788
Eesti Muuseumiühing Tartu, Tartu 54789
Eesti Muuseumraudtee, Lavassaare 10589
Eesti Piimandusmuuseum, Imavere 10574
Eesti Põllumajandusmuuseum, Tõrvandi 10664
Eesti Postimuuseum, Tartu 10646
Eesti Rahva Muuseum, Tartu 10647
Eesti Raudteemuuseum, Haapsalu 10563
Eesti Ringhäälingumuuseum, Türi 10665
Eesti Spordimuuseum, Tartu 10648
Eesti Tarbekunsti- ja Disainimuuseum, Tallinn 10627
Eesti Teatri- ja Muusikamuuseum, Tallinn 10628
Eesti Tervishoiu Muuseum, Tallinn 10629
Eesti Vabaõhumuuseum, Tallinn 10630
Eestimaa Vene Muuseum, Tallinn 10631
EFA Museum für Deutsche Automobilgeschichte, Amerang 16537
Efes Müzesi, Selçuk 42936
Effigy Mounds Museum, Harpers Ferry 49173
Egawa Bijutsukan, Nishinomiya 29072
Egawa Museum of Art, Nishinomiya 29072
Egerländer-Elbogner Heimatstuben, Illertissen 19316
Egerland-Museum, Marktredwitz 20222
Egersund Fayancemuseum, Egersund 33471
Egeskov Veteranmuseum, Kværndrup 10170
Egg Artistry Museum, Cyonglin 42170
Egge Museum, Steinkjer 33879
Egge-Museum Altenbeken, Altenbeken 16492
Egged Museum of Passengers Traffic, Holon 24975
Egg's Farm and Museum, Ciaotou 42164
Egham Museum, Egham 44163
Egill Olafssons Local Museum, Patreksfjörður 24011
Eğitim Özel Müzesi, Ankara 42701
Eğitim ve Kültür Vakfı Özel Müzesi, Malkara 42901
Eglise Saint-Rémy de Cuesmes, Cuesmes 03530
Egnsmuseet Jacob Michelsens Gård, Aabenraa 09953
Egon Schiele Art Centrum, Český Krumlov 09467
Egon Schiele-Museum, Tulln 02830
Egry József Emlékmúzeum, Badacsony 23343
Egyházmegyei Múzeum és Kincstár, Szeged 23842
Egypt Centre, Swansea 45812
Egypt Papyrus Museum, Haram 10532
The Egyptian Geological Museum, Cairo 10491
Egyptian Museum, Cairo 10492
Egyptian Museum and Papyrus Collection, Berlin State Museums - The Prussian Cultural Heritage Foundation, Berlin 17028
Egyptian Museum of Barcelona, Barcelona 38992
Ehagaki Shiryokan, Kobe 28724
Ehemalige Benediktinerabtei Seligenstadt, Seligenstadt 21737
Ehemalige Synagoge, Freudental 18457
Ehemalige Synagoge, Odenbach 20857
Ehemalige Synagoge Ahrweiler, Bad Neuenahr-Ahrweiler 16794
Ehemalige Synagoge Erfelden, Riedstadt 21322
Ehemalige Synagoge mit Ausstellung Juden auf dem Lande, Ichenhausen 19306
Ehemalige Synagoge Sankt Pölten, Sankt Pölten 02682
Ehemalige Synagoge Sulzburg, Sulzburg 21987
Ehemaligen Reichabtei Aachen - Kornelimünster, Aachen 16397
Ehemaliges Jagdschloß Bebenhausen, Tübingen 22113
Ehemaliges Postfuhramt, Berlin 17077
Ehime Bunkakan, Imabari 28567
Ehime Cultural Museum, Imabari 28567
Ehime-ken Bijutsukan, Matsuyama 28904
Ehime-kenritsu Hakubutsukan, Matsuyama 28905
Ehime-kenritsu Kagakukan, Niihama 29065
Ehime Mingeikan, Saijo 29185
Ehime Museum of Folk Crafts, Saijo 29185
Ehime Prefectural Museum, Matsuyama 28905
Ehime Prefectural Science Museum, Niihama 29065
Ehm Welk-Haus, Bad Doberan 16709
Ehm Welk- und Heimatmuseum, Angermünde 16548
Ehrensvärd-Museo, Helsinki 10773
Ehrensvärd Museum, Helsinki 10773
Ehrhart Museum, Antwerp 46326
Ehrman Mansion, Tahoma 53419
Ehrwalder Heimatmuseum, Ehrwald 01846
Eicha-Museum, Bergeijk 31966
Eichsfelder Heimatmuseum, Heiligenstadt, Heilbad 19107
Eide-Dalrymple Gallery, Sioux Falls 53119
Eidelstedter Heimatmuseum, Hamburg 18925
Eidskog Museum, Eidskog 33475
Eidsvold Historical Society Museum, Eidsvold 01063
Eidsvoll 1814 - Rikspolitisk Senter, Eidsvoll Verk 33476
Eidsvoll Bygdemuseet, Eidsvoll Verk 33477
Eidsvoll Rural Museum, Eidsvoll Verk 33477
Eifel-Heimatmuseum Bergweiler, Bergweiler 17025
Eifel-Vulkanmuseum, Daun 17801
Eifeler Bauernhausmuseum, Adenau 16430
Eifelmuseum Blankenheim, Blankenheim 17375
Eifelmuseum mit Deutschem Schieferbergwerk, Mayen 20241
Eikaasgalleriet, Skei i Jølster 33832
Eiktunet Kulturhistorisk Museum, Gjøvik 33513
Eiktunet Museum of Local History, Gjøvik 33513
Eildon Hall Sibbald Memorial Museum, Sutton, Ontario 07040
Eilean Donan Castle Museum, Dornie 44008
Einar Jónsson Museum, Reykjavík 24021
Einari Vuorelan Kirjailijakoti, Keuruu 10950
Einbecker Schreibmaschinenmuseum, Einbeck 18091
EinBlick-Dauerausstellung von Mission EineWelt, Neuendettelsau 20606
Einricher Heimatmuseum, Katzenelnbogen 19481
Einsiedelei Erzherzog Maximilians des Deutschmeisters, Innsbruck 02115
Einstein-Haus, Bern 41205
Eisei Bunko Museum, Tokyo 29385
Eisenbahn-Museum Lokschuppen Aumühle, Aumühle 16645
Eisenbahn- und Heimatmuseum Hochdahl, Erkrath 18217
Eisenbahn- und Technikmuseum Rügen, Prora 21129
Eisenbahnmuseum, Blumberg 17390
Eisenbahnmuseum, Deutsch Wagram 01811
Eisenbahnmuseum, Gerstetten 18616
Eisenbahnmuseum, Knittelfeld 02205
Eisenbahnmuseum, Losheim am See 20038
Eisenbahnmuseum Bayerischer Bahnhof zu Leipzig e.V., Leipzig 19902
Eisenbahnmuseum Bochum-Dahlhausen, Bochum 17399
Eisenbahnmuseum Darmstadt-Kranichstein, Darmstadt 17787
Eisenbahnmuseum der Köln-Bonner-Eisenbahn-Freunde, Wesseling 22512
Eisenbahnmuseum Historischer Wasserturm, Bebra 16968
Eisenbahnmuseum Lehmann, Hofheim in Unterfranken 19227
Eisenbahnmuseum Neustadt an der Weinstraße, Neustadt an der Weinstraße 20669
Eisenbahnmuseum - Sammlung Frey, Seifhennersdorf 21731
Eisenbahnmuseum Schwarzenberg, Schwarzenberg 21674
Eisenbahnmuseum Strasshof - Das Heizhaus, Strasshof 02788
Eisenbahnmuseum Vienenburg, Vienenburg 22229
Eisenbahnverein "Hei Na Ganzlin", Röbel/Müritz 21349
Eisenbergwerk Gonzen, Sargans 41816
Eisenerzer Krippenhaus, Eisenerz 01854
Eisenhammer Dorfchemnitz, Dorfchemnitz 17902
Eisenhower Birthplace, Denison 48056
Eisenhower National Historic Site, Gettysburg 48901
Eisenhüttenmuseum, Trippstadt 22106
Eisenkunstgussmuseum, Büdelsdorf 17625
Eisenkunstgußmuseum, Witzenhausen 22633
Eisenmuseum, Ainring 16449
Eisenmuseum Jünkerath, Jünkerath 19406
Eisenuhren- und Antikuhrenmuseum, Steyr 02768
Eisenverhüttungs-Schauraum, Oberpullendorf 02458
Eishockey-Museum, Augsburg 16616
Eisstockmuseum, Sankt Georgen im Attergau 02645
Eiteljorg Museum of American Indians and Western Art, Indianapolis 49490
Ekaterinburg Museum of Fine Art, Ekaterinburg 36884
Ekaterinburgskij Muzej Izobrazitelnych Iskusstv, Ekaterinburg 36884
Ekaterininskij Dvorec - Gosudarstvennyj Muzej Carskoje Selo, Puškin 37579
Ekehagens Forntidsby, Åsarp 40441
Ekenäs Museum, Ekenäs 10721
Ekenäs Slott, Linköping 40680
Eketahuna and Districts Museum, Eketahuna 33101
Eketorps Borg, Degerhamn 40485
Ekhof-Theater, Gotha 18717
Ekomuseum Bergslagen, Smedjebacken 40859
Ekomuseum Gränsland, Strømstad 40954
Ekotechnické Museum, Praha 09743
Ekron - The Philistine City and it's Culture, Shiqmim 25095
Eksi Şark Eserleri Müzesi, İstanbul 42817
Eksjö Museum, Eksjö 40493
Ekslibriso Muziejus, Šiauliai 30355
Ėkspozicija "Čelovek. Duša. Duchovnost.", Ivanovo 36945
Ėkspozicija "Ivanovskaja starina", Ivanovo 36946
Ekspozicija Kërpeeva Kašta, Kotel 05075
Ekspozycja Rzemiosło w Białym Spichrzu, Bydgoszcz 34575
EKZ Stromhaus Burenwisen, Glattfelden 41438
El-Barkai Museum, Khartoum 40387
El Camino College Art Gallery, Torrance 53542
El Campo Museum of Natural History, El Campo 48309
El Castell-Ecomuseu Urbà, Rubí 39921
El Dorado County Historical Museum, Placerville 51972
El Dorado Museum, Eldorado 01064
El-Ibied Regional Museum, Khartoum 40388
El Monte Historical Society Museum, El Monte 48313
El Morro National Monument, Ramah 52238
El Paso Holocaust Museum, El Paso 48319
El Paso Museum of Archaeology at Wilderness Park, El Paso 48320
El Paso Museum of Art, El Paso 48321
El Paso Museum of History, El Paso 48322
El Pueblo History Museum, Pueblo 52191
El Rancho de Las Golondrinas Museum, Santa Fe 52873
Elabuga Regional Museum, Elabuga 36905
Elabuga State Museum-Preserve of History, Architecture and Art, Elabuga 36903
Elabužskij Gosudarstvennyj Istoriko-architekturnyj i Chudožestvennyj Muzej-zapovednik, Elabuga 36903
Eläinmuseo, Helsinki 10774
Elagin Island Palace - Museum of Russian Decorative and Applied Art and Interior Design of the 18th-20th c., Sankt-Peterburg 37637
Elaginoostrovskij Dvorec-Muzej Russkogo Dekorativno-Prikladnogo Iskusstva i Interjera XVIII-XX vv., Sankt-Peterburg 37637
Elaine L. Jacob Gallery, Detroit 48117
Elazığ Archaeological Sites, Elazığ 42770
Elazığ Örenyerleri, Elazığ 42770
Elbert Hubbard-Roycroft Museum, East Aurora 48234
Elberton Granite Museum, Elberton 48332
Elbow Museum, Elbow 05769
Elbschiffahrtsmuseum mit stadtgeschichtlicher Sammlung, Lauenburg 19849
Elbschloss Bleckede, Bleckede 17384
Elbtal- Haus, Bleckede 17385
Elbtor Werben, Werben 22483
Elder Art Gallery, Lincoln 50094
Eldnäsets Flottningsmuseum, Ånge 40425

E

Eldon House, London 06183
Eldridge Street Project, New York 51169
Eleanor Barbour Cook Museum of Geology, Chadron 47326
Eleanor Roosevelt National Historic Site, Hyde Park 49457
Eleckij Literaturno-memorialnyj Muzej Pisatelja I.A. Bunina, Elec 36909
Electric Ladyland - The First Museum of Fluorescent Art, Amsterdam 31838
Electric Mountain - Museum of North Wales, Llanberis 44752
Electric Power Historical Museum, Yokohama 29608
Electrical Engineering Museum, Fredericton 05830
The Electrical Industry Museum, Tehrān 24660
Electricity Museum, Nagoya 28996
Électrium, Sainte-Julie 06896
Electrobroc, Broc-Fabrique 41259
Eleftherios K. Venizelos Museum, Athinai 22907
Elektriciteits- en Techniekmuseum, Hoenderloo 32396
Elektrizitätsmuseum, Münchenstein 41656
Elektrizitätszähler-Kabinett, Braunschweig 17519
Elektro Museum, Baden 41143
Elektro-Museum Rendsburg-E.ON Hanse, Rendsburg 21273
Elektromuseum Erfurt, Erfurt 18201
Elektronikmuseum Tettnang, Tettnang 22038
Elephantine Island's Museum, Aswan 10480
Élet az ártéren - kiállítás, Szekszárd 23866
Elfenbein-Museum, Michelstadt 20318
Elfenbeinmuseum, Walldürn 22309
Elfreth's Alley Museum, Philadelphia 51831
The Elgar Birthplace Museum, Lower Broadheath 45038
Elgin Apple Museum, Grabouw 38628
Elgin Area Museum, Elgin 48334
Elgin County Pioneer Museum, Saint Thomas 06879
Elgin Museum, Elgin 44166
Elgin Public Museum, Elgin 48335
Elham Valley Railway Museum, Folkestone 44236
Eli Whitney Museum, Hamden 49131
Eliasu Dzimtas Muzejs Zīlēni, Platone 30055
Elijah Clark Memorial Museum, Lincolnton 50114
Elimäen Kotiseutumuseo ja Koulumuseo, Elimäki 10724
Elimäen Koulumuseo, Elimäki 10725
Elimäki District Museum and Schoolmuseum, Elimäki 10724
Elimäki School Museum, Elimäki 10725
Elin Pelin House-Museum, Bajlovo 04973
Elingaard Museum, Gressvik 33518
Elisabet Ney Museum, Austin 46507
Elisarov Apartment Museum, Sankt-Peterburg 37692
Eliza Cruce Hall Doll Museum, Ardmore 46344
Elizabeth Bay House, Elizabeth Bay 01065
Elizabeth Castle, Saint Helier 45570
Elizabeth D. Walters Library, Stroudsburg 53339
Elizabeth Farm, Rosehill 01468
Elizabeth Myers Mitchell Art Gallery, Annapolis 46318
Elizabeth O'Neill Verner Studio Museum, Charleston 47363
Elizabeth P. Korn Gallery, Madison 50383
Elizabeth Perkins House, York 54333
Elizabeth Rozier Gallery, Jefferson City 49612
Elizabeth Sage Historic Costume Collection, Bloomington 46863
Elizabeth Slocumb Galleries, Johnson City 49632
Elizabethan House, Great Yarmouth 44357
Elizbarrutiko Arte Sakratuaren Museoa/ Museo Diocesano de Arte Sacro, Vitoria-Gasteiz 40319
Elk Grove Farmhouse Museum, Elk Grove Village 48342
Elkhart County Historical Museum, Bristol 47029
Ella Sharp Museum, Jackson 49556
The Ellen Noel Art Museum of the Permian Basin, Odessa 51496
Ellen Terry Memorial Museum, Tenterden 45838
Ellenroad Engine House, Milnrow 45151
Ellermeiers Burgmannshof, Hardegsen 19010
Elli Riehl Puppenwelt, Einöde 01852
Ellingson Car Museum, Rogers 52441
Ellington White Project-Art Exhibitions, Detroit 48118
Elliot Colliery Winding House, New Tredegar 45208
Elliot Lake Nuclear and Mining Museum, Elliot Lake 05773
Elliott Avedon Museum and Archives of Games, Waterloo 07287
Elliott Museum, Stuart 53342
Ellis County Historical Society Museum, Hays 49222
Ellis County Museum, Waxahachie 53999
Ellisland Farm, Auldgirth 43337
Ellwood House Museum, DeKalb 48038
Elm Museum, Yushu 08477
Elman W. Campbell Museum, Newmarket 06439
Elmbridge Museum, Weybridge 45986
Elmhurst Art Museum, Elmhurst 48374
Elmhurst Historical Museum, Elmhurst 48375
Elmina Castle Museum, Cape Coast 22851
Elmont Evergreen Firearms Museum, Belmont 05419
Elmore Progress Association Station Museum, Elmore 01066
Elmuseet, Bjerringbro 09976
Eloaitta-Kaavin Kotiseutumuseo, Kaavi 10900
Eloise Pickard Smith Gallery, Santa Cruz 52866
Elrose Heritage Museum, Elrose 05774
Elsa Forde Galleries, Bismarck 46836
Elsenzer Heimatstuben, Eppingen 18182
Elstow Moot Hall, Bedford 43436
Eltham Palace, London 44843
Elvarheim Museum, Åmli 33387
Elwood Haynes Museum, Kokomo 49825
Ely Museum, Ely 44172
Elztalmuseum - Regionalgeschichte und Orgelbau, Waldkirch 22297
Email-Museum in der Internationalen Begegnungsstätte Alte Mühle, Großlittgen 18798
Emailmuseum Gertrude Stöhr, Vorchdorf 02874
Emåns Ekomuseum, Bodafors 40461
Emanuel A Petersen Art Museum, Ilulissat 23253
Emanuel Vigeland Museum, Oslo 33721
Emba Museum of Chinese Modern Art, Ashiya 28376
Embom-Garpom Backstugusittarstuga, Liljendal 11069
Embom-Garpom Hill Croft, Liljendal 11069
Embroiderers Guild Museum, Mile End 01301
Embroiderers' Guild Museum Collection, East Molesey 44092
Embsay Bolton Abbey Steam Railway, Skipton 45689
Emeishan Museum, Emei 07678
Emerald Art Center, Springfield 53237
Emerald Art Gallery, Emerald 01069
Emerald Museum, Emerald 01070
Emerson Gallery, Clinton 47659
Emerson Gallery, McLean 50362
Emeryk Hutten-Czapski Museum of the National Museum in Cracow, Kraków 34791
Emigrantmuseet, Kisa 40644
Emil A. Blackmore Museum of the American Legion, Indianapolis 49491
Emil Cedercreutzin Museo, Harjavalta 10758
Emil Schumacher-Museum, Hagen 18855
Emil-von-Behring-Ausstellung, Marburg 20189
Emil Wikström Museo, Valkeakoski 11435
Emīla Dārziņa Muzejs, Jaunpiebalga 30011
Emīla Melngaiļa Memoriālais Muzejs, Vidriži 30147
Emile van Doren Museum, Genk 03619
Emilijos Digrevičienės Kūrybos Muziejus, Tytuvėnai 30383
Emilio Aguinaldo Shrine Museum, Kawit 34398
Emīls Dārziņš Museum, Jaunpiebalga 30011
Emily Carr House, Victoria 07246
Emily Dickinson Museum, Amherst 46265
Emirates Very Special Arts, Sharjah 43224
Emlékpont, Hódmezővásárhely 23611
Emma Kunz-Museum, Würenlos 42047
Emmanuel Gallery, Denver 48076
Emmaville Mining Museum, Emmaville 01071
Emmett Kelly Historical Museum, Sedan 53025
Emperor Maximilian of Mexico Museum, Hardegg 02051
Emperor Nicholas II Autograph Museum, Novosibirsk 37484
Emperors Chapel, Antwerpen 03289
Emphiedzhiev House-Museum, Kjustendil 05066
Empire Area Heritage Group Museum, Empire 48390
Empire State Aerosciences Museum, Glenville 48940
Emporia State University Geology Museum, Emporia 48391
Emser Bergbaumuseum, Bad Ems 16723
Emsland-Moormuseum, Geeste 18549
Emslandmuseum Lingen, Lingen 20008
Emslandmuseum Schloss Clemenswerth, Sögel 21791
Emsworth Museum, Emsworth 44175
Emu Park Historical Museum, Emu Park 01072
EnBW-Showroom Berlin, Berlin 17078
Enchanted Mansion Doll Museum, Baton Rouge 46632
Enchanted World Doll Museum, Mitchell 50762
Encsy György Tokaj-Hegyalja Földtani Gyűjteménye, Tállya 23910
End-O-Line Railroad Museum, Currie 47912
End of Steel Heritage Museum, Hines Creek 05989
Endeavour Museum, Wollombi 01661
Enderby and District Museum, Enderby 05776
Endicott College Museum, Beverly 46786
Endowment of King Petar I Karadjordjevic, Topola 38157
Endre Ady Memorial Museum, Budapest 23379
Endre Nemes Múzeum, Pécs 23759
Endrőd Region House, Gyomaendrőd 23576
Endrődi Tájház, Gyomaendrőd 23576
Enebakk Bygdetun, Trysil 33933
Enebakk Museet, Enebakk 33482
EnergeticA, Museum voor Energietechniek, Amsterdam 31839
Energie-Museum Berlin, Berlin 17079
Energiefabrik Knappenrode, Hoyerswerda 19281
Energiemuseum, Fürstenfeldbruck 18492
Energiemuseum, Rickenbach 21312
Enfield Historical Society Museum, Enfield 48397
Enfield Shaker Museum, Enfield 48398
Engadine Museum, Sankt Moritz 41811
Engadiner Museum, Sankt Moritz 41811
Engelenburg House Art Collection, Tshwane 38746
Engine House, Dunedin 33091
Engine House Project, Chelmsford 43804
The Engine Shed, Paekakariki 33169
Engineering Museum of Catalonia, Barcelona 38967
Englehart and Area Historical Museum, Englehart 05778
English Heritage East Midlands Region, Northampton 55059
English Heritage North West Region, Manchester 55060
English Heritage South East Region, Dover 44014
English Heritage South West Region, Bristol 55061
English Heritage West Midlands Region, Birmingham 55062
Englishman's Mill Cork Museum, Silves 35939
Engping Museum, Enping 07680
Engraving Print Village Art Museum, Sado 29177
Engsö Slott, Västerås 41030
Engströmsgården, Hult 40605
Enisejskij Kraevedčeskij Muzej, Enisejsk 36915
Enkhuizer Almanak Museum, Enkhuizen 32206
Enköpings Museum, Enköping 40497
Ennsmuseum Weyer, Weyer 02926
Enoagrarimuseum, Castelvetrano 25756
Enoch Turner Schoolhouse, Toronto 07105
Enoteca Regionale del Monferrato, Vignale Monferrato 28268
Enoteca Regionale Permanente la Serenissima, Gradisca d'Isonzo 26355
Ensival Historical Museum, Ensival 03583
Ensorhuis, Oostende 03930
Entlebucher Museum, Schüpfheim 41845
Entomological Department of the Moravian Museum, Brno 09426
Entomological Society Museum, Cairo 10493
Entomologické Oddělení, Brno 09426
Entomologické Oddělení, Praha 09744
Entomology Research Museum, Riverside 52370
Enviromental Museum, Azpeitia 38930
Environment Canada's Biosphère, Île-Saint-Hélène 06006
Environmental Education Center, Basking Ridge 46623
Enwald-Museo, Liperi 11075
Eötvös József Emlékmúzeum, Ercsi 23546
Eötvös Kúria, Vásárosnamény 23956
Eötvös Loránd Emlékkiállítás, Budapest 23397
Eötvös Mansion House, Vásárosnamény 23956
Epcot, Lake Buena Vista 49875
Epema-State, IJsbrechtum 32426
Ephesos-Museum, Wien 02980
Ephesus Museum, Selçuk 42936
Ephraim Foundation Museums, Ephraim 48402
Ephrata Cloister, Ephrata 48404
Epigraphical Museum, Athinai 22908
Epigraphy Museum, Tripoli 30186
Epping Forest District Museum, Waltham Abbey 45924
Eptek Art & Culture Centre, Summerside 07032
Epworth Old Rectory, Epworth 44184
Equine Museum, Gwacheon 29819
Equine Museum of Japan & JRA Racing Museum, Yokohama 29593
Erasmushouse, Bruxelles 03450
Ercsi Helytörténeti Gyűjtemény, Ercsi 23547
Ercsi Region Historical Collection, Ercsi 23547
Erdai Art Hall, Magong 42267
Erddig, Wrexham 46082
Erdefunkstelle Aflenz, Au bei Aflenz 01737
Erdészeti-, Faipari- és Földméréstörténeti Gyűjtemény, Sopron 23808
Erdészeti & Szabadtéri Erdészeti Múzeum, Szilvásvárad 23893
Erdgeschichtliches Museum, Lauffen am Neckar 19855
Erdős Kamill Cigány Múzeum, Pécs 23760
Erdos Renée-Ház - Kiállítóterem, Budapest 23398
Erdos Renée-House - Exhibition Hall, Budapest 23398
Erebouni Museum of the Establishment of Yerevan, Yerevan 00733
Ereğli Müzesi, Ereğli 42772
Eretz-Israel Museum Tel Aviv, Tel Aviv 25104
Erewash Museum, Ilkeston 44511
Erfahrungswelt Entfaltung der Sinne, Essen 18252
Erfatal-Museum Hardheim, Hardheim 19011
ErfinderZeiten, Schramberg 21630
Eric Carle Museum of Picture Book Art, Amherst 46266
Eric Thomas Galley Museum, Queenstown 01450
Erich Kästner Museum Dresden, Dresden 17942
Erich-Mäder-Glasmuseum, Grünenplan 18812
Erich Mäder-Glasmuseum, Grünenplan 18813
Erich Maria Remarque-Friedenszentrum, Osnabrück 20920
Erichsens Gaard, Rønne 10242
Erie Art Museum, Erie 48409
Erie Canal Museum, Syracuse 53397
Erie Canal Village, Rome 52450
Erie History Center, Erie 48410
Eriksbergs Museum, Tranås 40978
Eriksborg Carriage Museum, Ystad 41076
Eriksborg Vagnmuseum, Ystad 41076
Eriksdale Creamery Museum, Eriksdale 05780
Erinnerungsstätte an die Zwangsarbeit auf dem Gelände des Volkswagenwerk, Wolfsburg 22661
Erinnerungsstätte Baltringer Haufen - Bauernkrieg in Oberschwaben, Mietingen 20324
Erinnerungsstätte Luftbrücke Fliegerhorst Fassberg, Fassberg 18297
Erinnerungsstätte Notaufnahmelager Marienfelde, Berlin 17080
Erinnerungsstätte Ständehaus, Karlsruhe 19438
Erith Museum, Erith 44185
Eritrean Archaeological Museum, Asmara 10558
Erkebispegården, Trondheim 33923
Erkel Ferenc Emlékház, Gyula 23593
Erkel Ferenc Memorial House, Gyula 23593
Erkel Ferenc Múzeum, Gyula 23594
Erkenbert-Museum, Frankenthal, Pfalz 18351
Erland Lee Museum, Stoney Creek 07012
Erlauftaler Feuerwehrmuseum, Purgstall an der Erlauf 02539
Erlebnismuseum Anzenaumühle, Bad Goisern 01747
Erlebnismuseum "Mensch und Landschaft", Hermeskeil 19148
Erlebniswelt Blockflöte, Fulda 18507
Erlebniswelt Grundwasser, Langenau 19817
Erlebniswelt Renaissance - Schloß Bevern, Bevern, Kreis Holzminden 17314
Erlebniswelt Sipplingen, Sipplingen 21789
Erlebniswelt U-Boot, Sassnitz 21518
Erlianhaote Saurian Museum, Erlianhaote 07681
Ermatinger Old Stone House, Sault Sainte Marie 06930
Ermita de San Antonio de la Florida y Museo Panteón de Goya, Madrid 39518
Ermita de San Baudelio de Berlanga, Caltojar 39108
Ermita de Sant Feliu, Xàtiva 40326
Ermolova House Museum, Moskva 37163
Erna och Victor Hasselblads Fotografiska Centrum, Göteborg 40544
Ernest and Marion Davis Medical History Museum, Auckland 33046
Ernest Hemingway House Museum, Key West 49759
The Ernest W. Michel Historical Judaica Collection, New York 51170
Ernesta Birznieka-Upīša Muzejs Bisnieki, Zentene 30151
Ernests Birznieks-Upitis Museum, Zentene 30151
Ernie Pyle House, Dana 47951
Ernst Barlach Atelierhaus und Ausstellungsforum-Graphikkabinett, Güstrow 18827
Ernst Barlach Haus, Hamburg 18926
Ernst Barlach Museum, Ratzeburg 21196
Ernst Barlach Museum Wedel, Wedel 22353
Ernst-Bloch-Zentrum, Ludwigshafen am Rhein 20051
Ernst Fuchs Museum, Wien 02981
Ernst-Haeckel-Haus, Jena 19377
Ernst-Moritz-Arndt-Haus, Bonn 17453
Ernst-Moritz-Arndt-Museum, Garz 18545
Ernst Múzeum, Budapest 23399
Ernst Schering Foundation, Berlin 17247
Ernsts Samlinger, Assens 09971
Erotic Art Museum Hamburg, Hamburg 18927
Erotisch Museum, Amsterdam 31840
Errol Station Railway Heritage Centre, Blairgowrie 43538
Érseki Gyűjteményi Központ, Eger 23537
Ersta Diakonimuseum, Stockholm 40891
Erstes Allgäu-Schwäbisches Dorfschulmuseum, Erkheim 18214
Erstes Bayerisches Schulmuseum, Sulzbach-Rosenberg 21983
Erstes Circusmuseum in Deutschland, Preetz 21115
Erstes Deutsches Fliesenmuseum Boizenburg e.V, Boizenburg 17438
Erstes Deutsches Historic-Actien-Museum, Kürnbach 19739
Erstes Deutsches Motorroller-Museum, Kaub 19483
Erstes Deutsches Strumpfmuseum, Gelenau 18567
Erstes Deutsches Türmermuseum, Vilseck 22244
Erstes Europäisches Hundemuseum, Klostermarienberg 02197
Erstes Imaginäres Museum, Wasserburg am Inn 22343
Erstes Kärntner Handwerksmuseum, Baldramsdorf 01779
Erstes Niederbayerisches Automobil- und Motorrad-Museum, Adlkofen 16432
Erstes Niederrheinisches Karneval-Museum, Duisburg 18015

E

Erstes Oberösterreichisches Schnapsmuseum, Sankt Oswald bei Freistadt 02671
Erstes Österreichisches Biermuseum, Laa an der Thaya 02247
Erstes österreichisches Bodenschätzungsmuseum, Hollabrunn 02095
Erstes österreichisches Computermuseum, Bad Pirawarth 01762
Erstes Österreichisches Friedensmuseum und Heimatstube, Wolfsegg am Hausruck 03120
Erstes Österreichisches Funk- und Radiomuseum, Wien 02982
Erstes Österreichisches Museum für Alltagsgeschichte, Neupölla 02442
Erstes Österreichisches Rettungsmuseum, Hohenems 02091
Erstes Österreichisches Tischler-Museum, Pöchlarn 02504
Erstes Österreichisches Tonbandmuseum, Wien 02983
Erstes Rundfunkmuseum Rheinland-Pfalz, Münchweiler an der Alsenz 20497
Erstes Südburgenländisches Schnapsbrennereimuseum, Kukmirn 02244
Erstes Tiroler Feuerwehrmuseum, Schwaz 02723
Erstes Tiroler Holzmuseum, Wildschönau 03105
Erstes Urgeschichtliches Freilichtmuseum der Steiermark, Kulm bei Weiz 02245
Erstes Waldviertler Webereimuseum, Waidhofen an der Thaya 02884
Erstes Württembergisches Polizeimuseum, Deizisau 17811
Ervand Kochar Museum, Yerevan 00734
Erwin Museum, Painted Post 51676
Erwin von Kreibig-Museum, München 20437
Eryldene Heritage House, Gordon 01120
Erzbergbahn- und Wintersportmuseum, Vordernberg 02877
Erzbergbaumuseum und Besucherbergwerk, Bestwig 17309
Erzbischöfliches Diözesanmuseum und Domschatzkammer, Paderborn 20968
Erzgebirge Toys Museum & Museum of Picture Books, Karuizawa 28668
Erzgebirgische Volkskunst Ausstellung, Grünhainichen 18815
Erzgebirgischer Spielzeugwinkel, Eisenheim 18104
Erzgebirgisches Freilichtmuseum, Seiffen / Erzgebirge Kurort 21729
Erzgebirgisches Heimatmuseum Hospital Sankt Johannis, Sayda 21525
Erzgebirgisches Spielzeugmuseum, Seiffen / Erzgebirge Kurort 21730
Erzgebirgsmuseum mit Besucherbergwerk, Annaberg-Buchholz 16553
Erzherzog Franz Ferdinand Museum, Artstetten 01728
Erzherzog Johann-Dokumentation Thernberg, Thernberg 02811
Erzincan Culture Center, Erzincan 42773
Erzincan Kültür Merkezi, Erzincan 42773
Erzsébet Schaár Collection, Székesfehérvár 23858
Erzstollen Silbergründle, Seebach 21711
Erzurum Archeological Museum, Erzurum 42775
Erzurum Arkeoloji Müzesi, Erzurum 42775
Erzurum Culture Center, Erzurum 42776
Erzurum Kültür Merkezi, Erzurum 42776
Erzurum Resim ve Heykel Müzesi, Erzurum 42777
Erzurum State Painting and Sculpture Museum, Erzurum 42777
Es Baluard - Museu d'Art Modern i Contemporani de Palma, Palma de Mallorca 39781
Esbjerg Kunstmuseum, Esbjerg 10000
Esbjerg Museum, Esbjerg 10001
Escher in Het Paleis, Den Haag 32071
Escuela Museo de Bellas Artes General Urquiza, Buenos Aires 00157
Eshvin Literary Museum, Syktyvkar 37813
Esinemuseo ja Koulumuseo, Polvijärvi 11216
Eskbank House, Lithgow 01234
Eskens' House - District Museum in Toruń, Toruń 35186
Eskilstuna Konstmuseum, Eskilstuna 40501
Eskilstuna Stadsmuseum, Eskilstuna 40502
Eskimo Museum, Churchill 05614
Eskişehir Atatürk ve Kültür Müzesi, Eskişehir 42779
Eskişehir Devlet Güzel Sanatlar Galerisi, Eskişehir 42780
Eskişehir Müzeleri, Eskişehir 42781
Eskişehir Museums, Eskişehir 42781
Eskişehir State Fine Art Gallery, Eskişehir 42780
Eskoriatzako Museo Eskola, Eskoriatza 39280
Eslövs Stadsmuseum, Eslöv 40506
Espace Archéologique Forez-Jarez, Saint-Étienne 15250
Espace Architectural Normand, Crévecœur-en-Auge 12676
Espace Arlaud, Lausanne 41536
Espace Art Brenne, Concremiers 12606
Espace Arts Plastiques, Vénissieux 16102
Espace Automobiles Matra, Romorantin-Lanthenay 15090
Espace Baroque Maurienne, Lanslebourg-Mont-Cenis 13517
Espace Baroque Tarentaise, Séez 15698
Espace Cro-Magnon, Thonac 15890
Espace Culture Paul Doucet, Château-Thierry 12439
Espace Culturel Assens, Assens 41131
Espace Culturel La Douve, Langeais 13505
Espace Culturel Lucien Prigent, Landivisiau 13499
Espace d'Art Contemporain, Colmar 12580
Espace d'Art Contemporain, Marseille 13990
Espace d'Art Contemporain Camille-Lambert, Juvisy-sur-Orge 13299
Espace de Découvert du Cheval, Hennebont 13201
Espace de l'Art Concret, Mouans-Sartoux 14292
Espace des Arts, Le Boulou 13560
Espace des Inventions, Lausanne 41537
Espace des Saveurs, Herve 03702
Espace des Sciences, Rennes (Ille-et-Vilaine) 15013
Espace des Vieux Metiers, Ayen 11806
Espace d'Exposition Minal, Héricourt 13204
Espace du Cheminot, Nîmes 14447
Espace Exposition, Thionville 15883
Espace Gallo-Romain, Ath 03328
Espace Géologique "Les Frissons de la Terre", Château-Queyras 12437
Espace Giono, Lalley 13488
Espace Guisan, Pully 41737
Espace Historique de Sword Beach, Hermanville-sur-Mer 13206
Espace Horloger de la Vallée de Joux, Le Sentier 41863
Espace Jean de Joigny, Joigny (Yonne) 13270
Espace Jean Tinguely - Niki de Saint-Phalle, Fribourg 41384
Espace Joseph Besset, Vanosc 16070
Espace-Jouets, Firminy 12938
Espace Lausannoise d'Art Contemporain, Renens 41755
Espace Marais d'Anchais, Maillezais 13919
Espace Maritime et Portuaire des Docks Vauban, Le Havre 13594
Espace Minéralogique Minéraux Couleur Nature, Eymoutiers 12894
Espace Mira Phalaina, Montreuil-sous-Bois 14254
Espace Muséal, Rosheim 15107
Espace Muséal de la Maison du Berger, Châtillon-sur-Saône 12471
Espace Muséal de la Maison du Cordonnier, Châtillon-sur-Saône 12472
Espace Muséal de la Route Romane d'Alsace, Neuwiller-les-Saverne 14414
Espace Muséal Les Courtinals, Moureze 14310
Espace Musée, Freissinières 12999
Espace Musée Sainte-Madeleine, Besançon 11953
Espace Neige et Montagne, Bonneval-sur-Arc 12056
Espace Paroles d'Objets Nègres, Blangy-sur-Bresle 12011
Espace Patrimoine le Cœur, Tignes-le-Lac 15910
Espace Paul Rebeyrolle, Eymoutiers 12895
Espace Pierre Bonnard, Le Cannet 13568
Espace Pierres Folles, Saint-Jean-des-Vignes 15335
Espace Riquet, Béziers 11974
Espace Terre et Materiaux de la Faculté Polytechnique, Mons 03881
Espace Vauban, Besançon 11954
Espace Verre, Vannes-le-Châtel 16065
Espace Virtuel, Chicoutimi 05611
Espaço Cultural da Marinha, Rio de Janeiro 04690
Espaço dos Anjos, Leopoldina 04537
Espaço Lucio Costa, Brasília 04357
Espaço Museu da Vida, Rio de Janeiro 04691
Espaço UFF de Fotografia, Rio de Janeiro 04692
Esperance Municipal Museum, Esperance 01074
Esperanto-Museum der Österreichischen Nationalbibliothek und Sammlung für Plansprachen, Wien 02984
Espersit Ház, Makó 23689
Espersit House, Makó 23689
Espoo Car Museum, Espoo 10726
Espoo City Museum, Espoo 10727
Espoo City Museum, Glims Farmstead Museum, Espoo 10729
Espoon Automuseo, Espoo 10726
Espoon Kaupunginmuseo, Espoo 10727
Esposizione Forte Airolo, Airolo 41096
Èssentukskij Gorodskoj Kraevedčeskij Muzej im. V.P. Špakovskogo, Èssentuki 36917
Essex Historical Museum, Essex 48422
Essex Police Museum, Chelmsford 43805
Essex Regiment Museum, Chelmsford 43806
Essex Shipbuilding Museum, Essex 48423
Essig Museum of Entomology, Berkeley 46753
Essl Museum - Kunst der Gegenwart, Klosterneuburg 02200
Essley-Noble Museum, Aledo 46193
Estação Ciência / EST CIENC, São Paulo 04856
Estación Enológica de Haro y Museo del Vino, Haro 39384
Estación Experimental de Zonas Áridas, Almería 38859
Estación Ferrocarril Bejucal, Bejucal 09039
Esterházy-Ausstellung, Eisenstadt 01859
Esterhazy Community Museum, Esterhazy 05782
Esterházy Kastélymúzeum, Fertőd 23557
Esterhazy Palais of the Slovak National Gallery, Bratislava 38206
Esterházyo Palác, Bratislava 38206
Estes Park Area Historical Museum, Estes Park 48426
Estes-Winn Antique Automobile Museum, Asheville 46372
Estevan Art Gallery and Museum, Estevan 05784
Esther M. Klein Art Gallery, Philadelphia 51832
Esther Prangley Rice Gallery, Westminster 54097
The Esther Thomas Atkinson Museum, Hampden-Sydney 49146
Estonian Agricultural Museum, Tõrvandi 10664
Estonian Broadcasting Museum, Türi 10665
Estonian Dairy Museum, Imavere 10574
Estonian Health Care Museum, Tallinn 10629
Estonian History Museum, Tallinn 10622
Estonian History Museum, Maarjamäe Palace, Tallinn 10635
Estonian Literary Museum, Tartu 10645
Estonian Maritime Museum, Tallinn 10626
Estonian Museum of Applied Art and Design, Tallinn 10627
Estonian Museum of Natural History, Tallinn 10625
Estonian Museum of Postal Services, Tartu 10646
Estonian Museum Railway, Lavassaare 10589
Estonian Museum Society in Tallinn, Tallinn 54788
Estonian Museum Society in Tartu, Tartu 54789
Estonian Museums Association, Tartu 54790
Estonian National Defence College Museum, Tartu 10649
Estonian National Museum, Tartu 10647
Estonian Open Air Museum, Tallinn 10630
Estonian Railway Museum, Haapsalu 10563
Estonian Sports Museum, Tartu 10648
Estonian Theatre and Music Museum, Tallinn 10628
Estorick Collection of Modern Italian Art, London 44844
Estudio Museo Ramoneda, Humahuaca 00386
Eszterházy Károly Főiskola Csillagászati Múzeuma, Eger 23538
Etelä-Karjalan museo, Lappeenranta 11045
Etelä-Karjalan Taidemuseo, Lappeenranta 11046
Etelä-Pohjanmaan Maakuntamuseo, Seinäjoki 11304
Etele XI. Kerületi Helytörténeti Kör Kiállítóhelye, Budapest 23400
Eternity Museum, Wannian 08312
Ethnic Doll and Toy Museum, Whitstable 46006
Ethniki Pinakothiki & Mouseio Alexander Soytzoy, Athinai 22909
Ethnografic and Archeological Museum of the Ufa Scientific Section of the Russian Academy of Sciences, Ufa 37887
Ethnograghic Exhibition, Svištov 05220
Ethnographic and Archeological Collection, Visoko 04292
Ethnographic and Architectural Museum of Wood Arts and Crafts Hohlovka, Perm 37543
Ethnographic and Architectural Museum Village at Angara River, Bratsk 36820
Ethnographic and Folk Art Collection, Buzău 36242
Ethnographic and Folk Art Collection, Chilia 36276
Ethnographic Complex, Historical Museum, Vraca 05264
Ethnographic Exhibition of Šarišske Museum Bardejov, Bardejovské Kúpele 38196
Ethnographic Exhibition Stoyo House, Radomir 05144
Ethnographic Exposition, Pazardžik 05107
Ethnographic Exposition - Vajnory House, Bratislava 38235
Ethnographic Farmstead, Rusne 30349
Ethnographic Mining Museum, Oruro 04232
Ethnographic Museum, Addis Ababa 10678
Ethnographic Museum, Afyon 42677
Ethnographic Museum, Avgorou 09362
Ethnographic Museum, Berat 00010
Ethnographic Museum, Guwahati 24190
Ethnographic Museum, Konya 42884
Ethnographic Museum, Port Harcourt 33344
Ethnographic Museum, Smižany 38353
Ethnographic Museum, Ushguli 16388
Ethnographic Museum of Istria, Pazin 08926
Ethnographic Museum of Moldavia, Iaşi 36420
Ethnographic Museum of the Academy of Sciences of Tajikistan, Dušanbe 42497
Ethnographic Museum of the Kasan University, Kazan 37006
Ethnographic Museum of the National Museum in Wrocław, Wrocław 35329
Ethnographic Museum of the Peoples of the Baykal Region, Ulan-Udè 37893
Ethnographic Museum of the Slovak National Museum, Martin 38294
Ethnographic Museum of Transylvania, Cluj-Napoca 36287
Ethnographic Museum, University Museum of Cultural Heritage, Oslo 33722
Ethnographic Open Air Museum, Szombathely 23906
Ethnographic Open Air Museum of Ibresi, Ibresi 36929
Ethnographic Open-Air Museum of Latvia, Rīga 30079
Ethnographic Revival Complex, Kavarna 05054
Ethnographical Collection of Jr. József Lele, Szeged 23844
Ethnographical Collection of the Franciscan Monastery, Zaostrog 09016
Ethnographical Collection of the West Bohemian Museum in Pilsen, Plzeň 09717
Ethnographical Department of the Historical Museum, Praha 09782
Ethnographical Historical Museum of Lárissa, Lárissa 23073
Ethnographical House, Tápiószele 23912
Ethnographical House-Museum, Dobrič 05018
Ethnographical Institute of the Moravian Museum, Brno 09427
Ethnographical Museum, Aden 54641
Ethnographical Museum, Ankara 42702
Ethnographical Museum, Berkovica 04985
Ethnographical Museum, Burgas 04998
Ethnographical Museum, Doha 36086
Ethnographical Museum, Elbasan 00015
Ethnographical Museum, Elchovo 05026
Ethnographical Museum, Gjirokastër 00018
Ethnographical Museum, Khartoum 40389
Ethnographical Museum, Kraków 34792
Ethnographical Museum, Paphos 09391
Ethnographical Museum, Plovdiv 05126
Ethnographical Museum, Silistra 05162
Ethnographical Museum, Türkmenbaşi 42983
Ethnographical Museum, Varna 05237
Ethnographical Museum, Włocławek 35309
Ethnographical Museum, Zagreb 08989
Ethnographical Museum Dunabian Fishing and Boatbuilding, Tutrakan 05233
Ethnographical Museum of Liptov, Liptovský Hrádok 38285
Ethnographical Museum Tavrika, Simferopol' 43184
Ethnographical Museumof Serbia, Beograd 38035
Ethnographische Sammlung, Fribourg 41385
Ethnographisches Museum Pöykkölä, Rovaniemi 11275
Ethnographisches Museum Schloss Kittsee, Kittsee 02175
Ethnography and Folk Art Museum, Orăştie 36506
Ethnography Collection, Sopron 23814
Ethnography Exposition of Seduva, Seduva 30351
Ethnography Farmstead-Museum of Pranas Dzukas - Lazdijai Area Museum, Delnicai 30221
Ethnography Museum, Tripoli 30187
Ethnological Museum, Budva 31639
Ethnological Museum, Cairo 10494
Ethnological Museum, Chittagong 03215
Ethnological Museum, Pune 24357
Ethnological Museum, Ribnica na Pohorju 38506
Ethnological Museum, Tehrān 24661
Ethnological Museum, Tehrān 24662
Ethnological Museum of Pyrsogianni Stonemasons, Konitsa 23061
Ethnological Museum of the Archives and Museum Foundation of the UEM, Wuppertal 22747
Ethnologische Sammlung der Universität Göttingen, Göttingen 18690
Ethnologisches Museum, Berlin 17081
Ethnology Museum, Węgorzewo 35295
Ethnology Museum, Witzenhausen 22634
Etnografi Museoa, Arcienaga 38883
Ètnografičeskij Muzej Kazanskogo Universiteta, Kazan 37006
Ètnografičeskij Muzej Narodov Zabajkalja, Ulan-Udè 37893
Ètnografičeskij Muzej pod Otkrytym Nebom Torum Maa, Chanty-Mansijsk 36860
Etnografičeskij Muzej Tavrika, Simferopol' 43184
Etnografické Múzeum, Martin 38294
Etnografický Ústav, Brno 09427
Etnografiki Muzeý, Türkmenbaşi 42983
Etnografikoa Museoa Artziniega, Artziniega 38904
Etnografinė Prano Dzūko Sodyba-Muziejus, Delnicai 30221
Etnografische Verzamelingen van de Universiteit Gent, Gent 03626
Etnografisk Museum, Oslo 33722
Etnografiska Museet, Stockholm 40892
Etnografski Kompleks na Istoričeski Muzej, Vraca 05264
Etnografska Ekspozicija, Pazardžik 05107
Etnografska Ekspozicija, Svištov 05220
Etnografska Ekspozicija Stojova Kăšta, Radomir 05144
Etnografska Zbírka Františkánský Klášter, Zaostrog 09016
Etnografska Zbirka i Lapidarij Franjevačke Gimnazije, Visoko 04292
Etnografski Kompleks Starijat Dobrič, Dobrič 05017
Etnografski Muzej, Berkovica 04985
Etnografski muzej, Burgas 04998
Etnografski Muzej, Dobrič 05018
Etnografski Muzej, Elchovo 05026

E

Etnografski Muzej, Plovdiv 05126
Etnografski Muzej, Silistra 05162
Etnografski Muzej, Varna 05237
Etnografski Muzej, Zagreb 08989
Etnografski Muzej Dunavski Ribolov i Lodkostroenie, Tutrakan 05233
Etnografski Muzej Istre, Pazin 08926
Etnološki Muzej Rupe, Dubrovnik 08862
Etnografski Muzej Split, Split 08955
Etnografski Muzej Srbije, Beograd 38035
Etnografski-văzroždenski Kompleks, Kavarna 05054
Etnografya Müzesi, Ankara 42702
Etnografya Müzesi, Konya 42884
Etnološki Muzej, Budva 31639
Etnološki Muzej, Ribnica na Pohorju 38506
Etnomuseo dei Monti Lepini, Roccagorga 27430
Etobicoke's Museum - Montgomery's Inn, Etobicoke 05787
Eton College Natural History Museum, Eton 44189
Etowah Indian Mounds, Cartersville 47264
Etruria Industrial Museum, Stoke-on-Trent 45759
Etzel Museum, Tel Aviv 25105
Etzikom Museum, Etzikom 05788
Eubie Blake National Jazz Museum and Cultural Center, Baltimore 46563
Eugeen van Mieghem Museum, Antwerpen 03285
Eugene and Marilyn Glick Indiana History Center, Indianapolis 49492
Eugene Field House and Saint Louis Toy Museum, Saint Louis 52576
Eugene O'Neill Home, Danville 47959
Eugene V. Debs Home, Terre Haute 53480
Eugenijaus ir Leonardo Matuzevičių Memorialinis Muziejus, Krinčinas 30289
Eugenio Collectta Carrozze Serreria, Caronno Pertusella 25682
Die Eulenburg, Rinteln 21337
Eulenspiegel Museum, Mölln 20351
Eulenthurm-Museum, Weißenthurm 22464
Eumundi Museum, Eumundi 01075
Euphrat Museum of Art, Cupertino 47911
Eureka Fire Museum, Milltown 50695
Eureka - Parque de la Ciencia, Mendoza 00450
Eureka Pioneer Museum of McPherson County, Eureka 48442
Eureka School House, Springfield 53239
Eureka! The Museum for Children, Halifax 44384
Euro Tecnica Losheim, Hellenthal 19118
Europäische Glasmalerei- und Krippenausstellung, Schramberg 21631
Europäischer Skulpturenpark, Willebadessen 22582
Europäisches Brotmuseum, Ebergötzen 18037
Europäisches Industriemuseum für Porzellan und Europäisches Museum für Technische Keramik, Rosenthal Museum, Selb 21735
Europäisches Klempner- und Kupferschmiede-Museum, Karlstadt 19454
Europäisches Museum für Frieden, Stadtschlaining 02756
Europäisches Museum für Modernes Glas, Rödental 21351
Europäisches Spargelmuseum, Schrobenhausen 21638
European academy of fine arts, Trier 22099
European Association of Open Air Museums, Szentendre 54875
European Center on Resistance and Deportation, Natzwiller 14382
European Collaborative for Science, Industry and Technology Exhibitions, Bruxelles 54711
European Federation of Associations of Industrial and Technical Heritage, Terrassa 55026
European Federation of Museum and Tourist Railways, Haaksbergen 54955
European Folkcraft Museum, Teshikaga 29360
European Foundation for Drawing, Meina 26649
European Museum Forum, Bristol 55063
Europos Parkas, Vilnius 30403
Euskal Abertzaletasunaren Museoa, Artea 38901
Euskal Arkeologia, Etnografia eta Kondaira Museoa, Bilbao 39042
Euskal Gastronomi Museoa, Llodio 39487
Euskal Poliziaren Museoa, Arkaute 38891
Eustis Historical Museum, Eustis 48445
Euston Hall, Thetford 45847
Eva and Morris Feld Gallery, New York 51171
Eva Brook Donly Museum, Simcoe 06979
Eva Faschaunerin Heimatmuseum, Gmünd, Kärnten 01940
Evangelical National Museum - Lutheran Church Museum, Budapest 23401
Evangélikus Országos Múzeum, Budapest 23401
Evangelisches Bibelwerk im Rheinland - Bibelzentrum, Wuppertal 22739
Evangelisches Diözesanmuseum, Fresach 01905
Evangelisches Diözesanmuseum, Murau 02416
Evangelisches Kirchenmuseum Pleidelsheim, Pleidelsheim 21056
Evangelisches Museum in Wien, Wien 02985
Evangelisches Museum Oberösterreich, Rutzenmoos 02604
Evans-Tibbs Collection, Washington 53866
Evanston Art Center, Evanston 48449
Evansville Museum of Arts and Science, Evansville 48456
Evelyn Payne Hatcher Museum of Anthropology, Saint Cloud 52538
Ėvenkijskij Okružnoj Kraevedčeskij Muzej, Tura 37877
Evergreen Aviation Museum, McMinnville 50363
Evergreen Galleries, Olympia 51543
Evergreen House, Baltimore 46564
Evergreen House, Dartmouth 05676
Everhart Museum of Natural History, Science and Art, Scranton 52980
Everson Museum of Art, Syracuse 53398
Evje og Hornnes Museum, Evje 33485
Evolution of Education Museum, Prince Albert 06661
Evolutionsmuseet, Uppsala 41005
Evolutionsmuseet, Paleontologi, Uppsala 41006
Evraiko Mouseio tis Ellados, Athinai 22910
Ewelme Cottage, Auckland 33047
Ewha Womens University Museum, Seoul 29876
Ewha Yoja Taehakkyo Pakmulgwan, Seoul 29876
Ex Convento de Santo Domingo de Guzman, Oaxtepec 31228
Ex-Libris Museum of the Povilas Visinskis Public Library, Šiauliai 30355
Excell Blacksmith and Engineering Workshop Museum, Tumby Bay 01605
Excelsior House, Jefferson 49608
Excelsior-Lake Minnetonka Historical Museum, Excelsior 48462
Exchange Hotel Civil War Museum, Gordonsville 48973
The Executive Mansion, Frankfort 48734
Exhibit Museum of Natural History, Ann Arbor 46305
Exhibition and Gallery of Sanctuary of our Lady of Trsat, Rijeka 08941
Exhibition at the Shipyard, Västanfjärd 11434
Exhibition Center, Caotun 42148
Exhibition Center and Gallery, Iževsk 36957
Exhibition Center at the Artist Union, Bakı 03153
Exhibition Complex, Veliko Tărnovo 05253
Exhibition Gallery of the Islamic Dâwah Centre, Bandar Seri Begawan 04964
Exhibition Hall, Čeboksary 36835
Exhibition Hall, Čerepovec 36852
Exhibition Hall, Český Těšín 09471
Exhibition Hall, Ha Noi 54611
Exhibition Hall, Jablunkov 09548
Exhibition Hall, Krasnodar 37067
Exhibition Hall, Krasnoturinsk 37082
Exhibition Hall, Moskva 37418
Exhibition Hall, Moskva 37428
Exhibition Hall, Moskva 37417
Exhibition Hall Arsenals of the State Museum of Art, Riga 30115
Exhibition Hall at the House of Scientists of the Sibirian Section of the Russian Academy of Sciences, Novosibirsk 37489
Exhibition Hall Bogorodskoye, Moskva 37420
Exhibition Hall Folk Gallery, Moskva 37426
Exhibition Hall Ismailovo, Moskva 37422
Exhibition Hall of Agricultural Implements, Neipu 42284
Exhibition Hall of Geology, Beijing 07537
Exhibition Hall of the Artists' Union of Russia, Sankt-Peterburg 37735
Exhibition Hall of the House of Friendship with Foreign Peoples, Moskva 37421
Exhibition Hall of the Karelian State Philharmonic, Petrozavodsk 37563
Exhibition Hall of the Kirov Regional Museum, Kirov 37045
Exhibition Hall of the Madona Museum of Local Studies and Art, Madona 30044
Exhibition Hall of the Moscow Artists' Union, Moskva 37425
Exhibition Hall of the Moscow Fund of Culture, Moskva 37424
Exhibition Hall of the Regional Museum, Perm 37552
Exhibition Hall of the Republic Marij El, Joškar-Ola 36981
Exhibition Hall of the Russian State Library of Art, Moskva 37427
Exhibition Hall of the Russian State Library of Art, Moskva 37419
Exhibition Hall of the Saint-Petersburg Section of the Artists' Union of Russia, Sankt-Peterburg 37733
Exhibition Hall of the State Museum Reserve, Vologda 37965
Exhibition Hall of the Town, Angarsk 36759
Exhibition Hall of the Tver Regional Art Gallery, Tver 37882
Exhibition Hall of the Volgograd Art Museum, Volgograd 37960
Exhibition Hall of Victor and Apollinari Vasnetsov Art Museum, Kirov 37044
Exhibition Hall Rainbow, Čeboksary 36841
Exhibition Hall Rainbow, Samara 37615
Exhibition Hall - Russian Artists of the Government Samara, Žigulevsk 37989
Exhibition Hall Sirak Skitnik, Sliven 05165
Exhibition Hall Spasskyi Church, Irkutsk 36940
Exhibition Halls at the Old Arbat Street, Moskva 37414
Exhibition Halls in Aksakov's House, Moskva 37416
Exhibition Halls of Liepāja Museum, Liepāja 30039
Exhibition Halls of the Russian Academy of Arts, Moskva 37415
Exhibition Halls of the Russian Fund of Culture, Moskva 37429
Exhibition Ivanovo's Ancient Times, Ivanovo 36946
Exhibition Man-Soul-Mind, Ivanovo 36945
Exhibition of Greek-Roman Statue Copies, Tata 23917
Exhibition of Mining of Brennbergbánya, Sopron 23807
Exhibition of Tapestries, Felletin 12919
Exhibition of the Norwegian-American Historical Association, Northfield 51419
Exhibition Ward of the Csabai Homestead and the History of Grain, Békéscsaba 23368
Exhibitions @ Soap Factory, Minneapolis 50730
Exhibiton Hall of the Ekaterinburg Culture Association, Ekaterinburg 36902
Exhibtion Centre Ancient Tula of the Kulikovo Field Museum- preserve, Tula 37872
Exhibtion Hall in Rotermann Salt Storage, Tallinn 10637
Exhibtion Hall of the Tula Museum of Fine Arts, Tula 37876
Exmouth Museum, Exmouth 44197
Expeditionsmuseum Werner Freund, Merzig 20299
Experience Children's Museum, Erie 48411
Experience Music Project, Seattle 52998
Experimental Art Foundation, Adelaide 00775
Experimentarium, Hellerup 10057
Explora Museum, Frankfurt am Main 18365
Explora Science Center and Children's Museum, Albuquerque 46177
Exploration Place, Wichita 54141
Exploration Station, Bourbonnais 46965
Exploratorio, San Isidro 00580
The Exploratorium, San Francisco 52773
Exploratorium-Kindermuseum Stuttgart und Region, Stuttgart 21931
Explore & More ... a Children's Museum, East Aurora 48235
Explorers Hall, Washington 53867
Explorers' Museum of the District Museum in Toruń, Toruń 35196
Explorion, Heerlen 32336
Explorium of Lexington, Lexington 50047
Explosion! The Museum of Naval Firepower, Gosport 44332
Expo Haringvliet, Stellendam 32806
Expo-Musée Renault, Boulogne-Billancourt 12088
Expopetro - Museo del Petróleo, Lima 34212
Exporail, Saint-Constant 06811
Exposición de Gregorio de Laferrere, Gregorio de Laferrere 00380
Exposición Permanente Castro Celtibéricon de el Ceremeño, Herrería 39388
Exposición Permanente de Cerámica Indígena, Manizales 08666
Exposición Roberto J. Mouras, Carlos Casares 00293
Expositie 40-45, Blitterswijk 31974
Expositie Versteend Leven, Drouwen 32159
Exposition Archéologique de la Maison de la Cour, Givrand 13073
Exposition Archéologique Gallo-Romaine, Mandeure 13935
Exposition Artistique et Viticole, Aigues-Mortes 11502
Exposition au Château, Les Ternes 13718
Exposition Bernardini, Obernai 14491
Exposition Caillebotte, Yerres 16306
Exposition d'Archéologie Médiévale, Antignac 11644
Exposition d'Art et d'Histoire, Pionsat 14823
Exposition de Coiffes et Broderies Bretonnes, Guiclan 13164
Exposition de la Medievale, Evol 12888
Exposition de la Préhistoire, Mazières-en-Mauges 14055
Exposition de la Tour de Castellane, Épernay 12829
Exposition de Patrimoine Meylanais, Meylan 14108
Exposition de Tapisseries, Felletin 12919
Exposition des Éditions Tarmeye, Mazet-Saint-Voy 14054
Exposition d'Saint-Romain de la Préhistoire, Saint-Romain 15500
Exposition du Château, Cadillac-en-Fronsadais 12226
Exposition du Château Fort du Fleckenstein, Lembach 13678
Exposition du Cloître des Cordeliers, Tarascon (Bouches-du-Rhône) 15854
Exposition du Fort de l'Estissac, Port-Cros 14913
Exposition du Fort Sainte-Agathe, Porquerolles 14911
Exposition du Jouet en Bois, Arvieux-en-Queyras 11708
Exposition du Mémorial Capétien, Bernières-sur-Mer 11950
Exposition du Moulin à Huile, La Cadière-d'Azur 13323
Exposition du Petit Monde de Pagnol, Aubagne 11717
Exposition Historique, Cahors 12240
Exposition Historique du Château, La Roche-Maurice 13416
Exposition Historique du Monastère, Sauxillanges 15673
Exposition Jadis, La Roche-Derrien 13414
Exposition Kyorpeev's House, Kotel 05075
Exposition La Guerre de Cent Ans, Saint-Mesmin 15414
Exposition of Folk Architecture and Dwelling, Martovce 38302
Exposition of Historic Clocks Chateaux Budimir, Budimír 38240
Exposition of Holocaust, Vilnius 30410
Exposition of Home Décor at the Kežmarok Museum, Kežmarok 38265
Exposition of Old Bulgarian Carpets, Kotel 05076
Exposition on History of the Burbiskis Manor, Burbiškis 30219
Exposition Permanente, Aigueperse 11499
Exposition Permanente, Thanvillé 15876
Exposition Permanente d'Archéologie, Ainvelle 11510
Exposition Permanente d'Art, Brasparts 12145
Exposition Permanente d'Art, Éguilly 12813
Exposition Permanente d'Art Contemporain, Bonnemazon 12054
Exposition Permanente d'Arts Plastiques, Brioude 12183
Exposition Permanente Pierre Miquel, Maurs 14042
Exposition Permanente sur François Pétrarque, Lombez 13805
Exposition Permanente sur la Déportation, Charmes 12414
Exposition Permanente sur les Églises Fortifiées de la Thiérache, Vervins 16130
Exposition Permanente "Valence à la Croisée des Chemirs", Valence (Drôme) 16038
Exposition Préhistorique, Choranche 12525
Exposition Ruchon Morin, Le Merlerault 13614
Exposition sur la Parfumerie, Eze 12897
Exposition sur l'Art Cistercien, Bétête 11966
Exposition Universelle des Vins et Spiritueux, Bandol 11837
Expositions de Chapelle Sainte-Thècle, Seguret 15701
Expositions Pipe et Diamant, Saint-Claude 15215
Expostion Grotte Chauvet Pont d'Arc, Vallon-Pont-d'Arc 16054
Expozice Čelákovického Košíkářství, Čelákovice 09454
Expozice Dominikánský Klášter, Jablonné v Podještědí 09547
Expozice Městského Vlastivědného Mudruňkovo Muzeuma na Městském úřadu, Říčany u Prahy 09816
Expozice Oddělení Starších Českých Dějin, Vrchotovy Janovice 09916
Expozice Řemesel, Moravské Budějovice 09663
Expozícia Čičmany, Čičmany 38246
Expozícia Historických Hodín, Bratislava 38207
Expozícia Historických Hodín Kaštiel' v Budimíre, Budimír 38240
Expozícia L'udovej Architektúry a Bývania (Skanzen), Humenné 38261
Expozícia L'udovej Archtektúry a Bývania, Martovce 38302
Expozícia Meštianskej Bytovej Kultúry, Kežmarok 38265
Expozícia Zbraní a Stredovekého Mestského Opevnenia, Bratislava 38208
Expoziţia Arheologică Tibiscum, Caransebeş 36262
Expoziţia Curtea Domnească, Piatra-Neamţ 36516
Expoziţia de Arheologie şi Istorie Locală, Deta 36342
Expoziţia de Artă Feudală, Domneşti 36348
Expoziţia de Artă Populară Iulia Folea Troceanu, Buziaş 36245
Expoziţia de Etnografie şi Artă Populară, Ciacova 36277
Expoziţia de Etnografie şi Artă Populară, Pucioasa 36543
Expoziţia de Istoria Farmaciei, Sibiu 36605
Expoziţia de Istorie a Culturii Nămăeşti, Câmpulung 36251
Expoziţia de Istorie Locală, Ciacova 36278
Expoziţia de Pictură Românească Contemporană, Teremia Mare 36682
Expoziţia de Viticultură, Teremia Mare 36683
Expoziţia Etnografică, Horezu 36402
Expoziţia Foto-documentară Ion Ionescu de la Brad, Negri 36490

E

Expoziţia Foto-documentară Vasile Pârvan, Huruieşti 36406
Expoziţia Hidrocentrală Porţile de Fier, Drobeta-Turnu Severin 36358
Expoziţia Memorială Adam Müller-Guttenbrunn, Zăbrani 36740
Expoziţia Memorială Béla Bartók, Sânnicolau Mare 36586
Expoziţia Memorială Dimitrie Cantemir, Dimitrie Cantemir 36346
Expoziţia Memorială Ion Borcea, Racova 36546
Expoziţia Memorială Silviu Dragomir, Gurasada 36396
Expoziţia Memorială Ştefan Jäger, Jimbolia 36432
Expoziţia Memorială Traian Vuia, Traian Vuia 36702
Expoziţia Muzeală, Turnu Roşu 36712
Expoziţia Muzeală de Etnografie, Cisnădie 36281
Expoziţia Muzeală I.L. Caragiale, Ghirdoveni 36384
Expoziţia Muzeală Istoricul Industriei Textile, Cisnădie 36282
Expoziţia Muzeală Vasile Dobrian, Rod 36566
Expoziţia Nichita Stănescu, Teremia Mare 36684
Expoziţia Permanentă de Artă Plastică Contemporană, Ştefăneşti, Botosani 36635
Expoziţia Permanentă de Etnografie, Pietroşiţa 36524
Expoziţia Permanentă de Istoria Farmaciei Montanistice Bănăţene, Oraviţa 36507
Expoziţia Sătească, Stejaru 36637
Expoziţia Teatrului, Arad 36112
Expoziţie Sătească, Lisa 36444
Expoziţie Sătească, Sericu 36598
Extraordinary Museum, Ansouis 11636
Eyam Museum, Eyam 44199
Eye Level Gallery, Halifax 05934
Eyebeam, New York 51172
Eyemouth Museum, Eyemouth 44200
Eyvind Johnson-Stugan, Boden 40462
Ezekiel Harris House, Augusta 46476
Ezhou Museum, Ezhou 07682
Ezra and Cecile Zilkha Gallery, Middletown 50658
Ezra Meeker Mansion, Puyallup 52206
Ėžvinskij Literaturnyj Muzej, Syktyvkar 37813

F

F.I. Tyutchev Literary Museum, Ovstug 37522
F.M. Dostoevsky Apartment Museum, Moskva 37316
F.M. Dostoevsky Commemorative Literary Museum, Semey 29728
F.M. Dostoevsky Memorial and Literature Museum, Novokuzneck 37480
F.M. Dostoevsky Museum-Estate in Darovoe, Darovoe 36873
F. Manayla Memorial House-Museum, Užgorod 43197
F.R.I. Museums, Dehradun 24131
F.W.Raiffeisen Museum, Hamm, Sieg 18968
Faaborg Arrest, Faaborg 10006
Faaborg Kulturhistoriske Museer, Faaborg 10007
Faaborg Museum for Fynsk Malerkunst, Faaborg 10008
Fabbrica Casa Museo G. Mazzotti 1903, Albissola 25183
Fabriano Paper and Watermark Museum, Fabriano 26069
The Fabric Workshop and Museum, Philadelphia 51833
Fabrica, Brighton 43622
Fabricius-Ház, Sopron 23809
Fabricius House, Sopron 23809
Fabrikmuseum der Leonischen Industrie, Roth 21401
Fabyan Villa Museum and Dutch Windmill, Geneva 48878
Face Painting Culture Museum, Sanyi 42316
Fachausstellung Arbeitsschutz/ Sicherheitstechnik, München 20438
Das Fachwerk, Bad Salzuflen 16821
Fachwerkbaumuseum im Ständerbau, Quedlinburg 21146
Factory Collection, Badacsony 23342
Factory Museum, Eskilstuna 40502
Fädernegården i Rejpelt, Vörå 11474
Fälschermuseum, Binz, Ostseebad 17355
Fälschermuseum, Wien 02986
Fængselshistoriske Museum, Horsens 10095
Färbermuseum, Gutau 02023
Färgargården, Norrköping 40770
Färgfabriken, Stockholm 40893
Fästningsmuseet Karlsborg, Karlsborg 40627
Fågelmuseet, Jönköping 40612
Fågelsjö Gammelgård Bortom Åa, Los 40694
Fagus-Gropius-Ausstellung, Alfeld 16464
Fahl Kro, Hemmet 10067
Fahrngruber Hammer mit Köhlereimuseum, Ybbsitz 03123
Fahrradmuseum, Ybbs 03122
Fahrradmuseum Zumhaus und Brauereimuseum, Feuchtwangen 18312
Fahrzeugmuseum, Frankenberg, Sachsen 18345
Fahrzeugmuseum Marxzell, Marxzell 20231
Fahrzeugmuseum Suhl, Suhl 21976
Fahua Ta Tayuan Museum, Shanghai 08165
Faïences et Emaux, Longwy 13812
Fairbanks Community Museum, Fairbanks 48471
Fairbanks Gallery, Corvallis 47833
Fairbanks House, Dedham 48024
Fairbanks Museum and Planetarium, Saint Johnsbury 52554
Fairbridge Chapel Museum, Mill Bay 06284
Fairfax House Museum, York 46102
Fairfax Museum and Visitor Center, Fairfax 48477
Fairfield Art Museum, Fairfield 48484
Fairfield Center for Contemporary Art, Sturgeon Bay 53347
Fairfield City Museum and Gallery, Smithfield 01508
Fairfield County Museum, Winnsboro 54237
Fairfield Museum, Bothwell 05443
Fairlawn Mansion and Museum, Superior 53378
Fairlynch Museum, Budleigh Salterton 43668
Fairport Historical Museum, Fairport 48492
Fairport Marine Museum, Fairport Harbor 48493
Fairview Museum of History and Art, Fairview 48494
Fairview R.C.M.P. Centennial Museum, Fairview 05791
Fairy Tale House, Békéscsaba 23369
Fairymead House Sugar Museum, Bundaberg 00920
Faith Trumbull Chapter Museum, Norwich 51435
Fake Art Museum, Vledder 32896
Fakenham Museum of Gas and Local History, Fakenham 44203
Falbygdens Museum, Falköping 40512
Falconer Museum, Forres 44245
Faleide Skulemuseum, Stryn 33889
Falger-Museum, Elbigenalp 01865
Falkenbergs Museum, Falkenberg 40508
Falkland Palace and Garden, Falkland 44207
Falknereimuseum Burg Hohenwerten, Werfen 02922
Fall River County Historical Museum, Hot Springs 49374
Fall River Historical Society Museum, Fall River 48499
Fallingwater - Western Pennsylvania Conservancy, Mill Run 50690
Falmouth Art Gallery, Falmouth 44208
Falmouth Historical Museum, Falmouth 48505
Falstadminnet, Levanger 33634
Falsterbo Museum, Falsterbo 40513
Falu Gruva Gruvmuseum, Falun 40516
Falumúzeum, Hollókő 23616
Falumúzeum, Törökbálint 23933
Familie Vrouwenhofje - Hofje van Aerden, Leerdam 32470
Familienmuseum, Schwerin 21693
Familienmuseum Bad Homburg, Bad Homburg v.d.Höhe 16750
Familienmuseum im Altbauernhaus, Göstling an der Ybbs 01955
Familienmuseum Schloss Braunfels, Braunfels 17514
Familistère de Guise, Guise 13169
Family Museum of Arts and Science, Bettendorf 46784
Family Museum of Postcards - House of the Red Chair, Praha 09792
Family Vault of the Dukes of Courland Exhibition, Jelgava 30018
Famine Museum, Strokestown 24902
Fan Museum, London 44845
Fan Tower Museum, Fanshi 07685
Fanchang Museum, Fanchang 07683
Fang County Museum, Fangxian 07684
Faninadahen Kosas Guahan, Adelup 23284
Fanø Kunstmuseum, Fanø 10013
Fanø Museum, Fanø 10014
Fanø Skibsfarts- og Dragtsamling, Fanø 10015
Fanshawe Pioneer Village, London 06184
Faqir Khana Museum, Lahore 34019
F'ar, Lausanne 41538
Far Eastern Art Museum, Chabarovsk 36858
Far Eastern Art Museum, Toruń 35197
Far North Regional Museum, Kaitaia 33135
Farga Rossell Centre d'Interpretació del Ferro, La Massana 00096
Fargo Air Museum, Fargo 48509
Farg'ona Viloyat Ülkashunoslik Muzeyi, Farg'ona 54446
Farleigh Hungerford Castle, Farleigh Hungerford 44214
The Farm, Sturgeon Bay 53348
Farm House Museum, Ames 46257
Farm Life Museum, Tréflaouenan 15980
Farm Museum, Szabadkígyós 23831
Farm Museum at Kopáncs, Hódmezővásárhely 23612
Farm-Museum El Abra, Siguanea 09332
Farm Shed Museum, Kadina 01182
Farmaceutická Expozícia, Bratislava 38209
Farmacevtsko-Medicinska Zbirka Bohuslava Lavičke, Ljubljana 38448
Farmacia Museo Taquechel, La Habana 09102
Farmācijas Muzejs, Rīga 30066
Farmamerica, Waseca 53838
Farmamuseo Sa Potecaria, Villacidro 28275
Farmer Museum, Novellara 26960
Farmer Museum, Xochitepec 31569
Farmer's Arms Hotel Museum, Euroa 01076
Farmers Bank, Petersburg 51798
Farmers' Bank of Rustico, Hunter River 06002
Farmer's House, Metulla 25064
The Farmers' Museum, Cooperstown 47788
Farmers' Museum Rothenburg/OL, Rothenburg/Oberlausitz 21413
Farmhouse, Cserépfalu 23507
Farmhouse Museum, Ylijärvi 11480
Farming and Fishing Museum of Sihu, Sihu 42328
Farming Museum, Tótkomlós 23938
Farming Museum of the Hungarian Agricultural Museum, Lajosmizse - Bene 23688
Farming Through the Ages, Whyte Yarcowie 01650
Farming Village Museum, Cingshuei 42167
Farmington Historic Home, Louisville 50284
Farmington Museum, Farmington 48517
Farmland Museum and Denny Abbey, Waterbeach 45949
Farmleigh House, Dublin 24788
Farmlife Centre, Thornhill, Central 45853
Farnham Galleries, Indianola 49507
Farnhem Maltings Gallery, Farnham 44216
Farní muzeum Zábřeh, Zábřeh 09923
Faroese Museum of Natural History, Tórshavn 10698
Farouq Corner Museum, Helwan 10534
Farrar-Mansur House and Old Mill Museum, Weston 54108
Farsh Museum, Tehrān 24663
Farshchian Museum, Tehrān 24664
Farvergården, Ebeltoft 09995
Farvergården, Kerteminde 10115
Faschings- und Brauchtumsmuseum, Knittelfeld 02206
Fashion and Textile Museum, London 44846
Fashion-Costume Museum, Digoin 12730
Fashion History Museum, Buenos Aires 00260
Fashion Museum, Hasselt 03682
Fasnacht- und Heimatmuseum Noaflhaus, Telfs 02802
Fasnachtsmuseum, Imst 02107
Fasnets-Museum am Vorderochsen, Weingarten, Kreis Ravensburg 22426
Fassbinder- und Weinbaumuseum, Straß im Straßertale 02785
Fastnachtmuseum, Nassereith 02421
Fastnachtmuseum Schloß Langenstein, Orsingen-Nenzingen 20912
Fastnachtsmuseum, Bräunlingen 17496
Fastnachtsmuseum, Herbstein 19136
Fastnachtsmuseum, Sankt Ingbert 21499
Fastnachtsmuseum Fasenickl, Kipfenberg 19540
Faszination Zugspitze, Ehrwald 01847
Faszinosum, Borna 17484
Father Marquette National Memorial and Museum, Saint Ignace 52550
Father Pandosy Mission, Kelowna 06052
Father Weyland Gallery, Epworth 48407
Faulconer Gallery at Grinnell College, Grinnell 49090
Faulkner County Museum, Conway 47783
Fauske Museum, Fauske 33490
Faust-Museum, Knittlingen 19583
Favell Museum of Western Art and Indian Artifacts, Klamath Falls 49806
Fawick Art Gallery, Berea 46748
Fayette Art Museum, Fayette 48524
Fayette County Historical Museum, West Union 54084
Fayette County Museum, Washington Court House 53948
Fayette Heritage Museum, La Grange 49843
Fayetteville Museum of Art, Fayetteville 48531
Fazekasház, Tihany 23926
FBW-Museum, Wetzikon 42009
Featherston Heritage Museum, Featherston 33102
Featherston Memorabilia Museum, Featherston 33103
Federação de Amigos de Museus do Brasil, São Paulo 54722
Federação de Amigos dos Museus de Portugal, Lisboa 54992
Federación Argentina de Amigos de Museos, Buenos Aires 54684
Federación Chilena de Amigos de los Museos, Santiago de Chile 54752
Federación Española de Asociaciones de Amigos de los Museos, Madrid 55027
Federación Mexicana de Asociaciones de Amigos de los Museos, México 54942
Federal Hall National Memorial, New York 51173
Federal Reserve Bank of Boston Collection, Boston 46929
Federal Reserve Board Art Gallery, Washington 53868
Fédération des Amis des Musées de Belgique, Bruxelles 54712
Fédération des Ecomusées et des Musées de Société, Besançon 54809
Fédération Française des Sociétés d'Amis de Musées, Paris 54810
Fédération Hellénique des Amis des Musées, Athinai 54867
Fédération Internationale des Musées du Jouet et de l'Enfance, Moirans-en-Montagne 54811
Fédération Mondiale des Sociétés d'Amis de Musées, Paris 54812
Fédération nationale des Maisons d'Ecrivain et des Patrimoines Littéraires, Bourges 54813
Federation of Museums and Galleries in Wales, Aberystwyth 55064
Federation of Writers' Houses and Literary Collections, Bourges 54813
Fédération Régionale des Amis des Musées du Nord-Pas de Calais, Dunkerque 54814
Federazione Italiana delle Associazioni Amici dei Musei, Firenze 54900
Federseemuseum, Bad Buchau 16706
Fedrenes Minne, Hidrasund 33547
Feering and Kelvedon Local History Museum, Kelevedon 44553
Feet First: The Scholl Story, North Chicago 51387
Fehér Miklós Emlékmúzeum, Győr 23581
Fehér Miklós Memorial Museum, Győr 23581
Fehérvári Babaház, Székesfehérvár 23853
Fehn- und Schiffahrtsmuseum, Rhauderfehn 21291
Fehnmuseum Eiland, Großefehn 18786
Feiligrath Museum, Rüdesheim am Rhein 21434
Fekete-Ház, Szeged 23843
Fekete István Emlékház, Gölle 23569
Fekete Sas Patikamúzeum, Székesfehérvár 23854
Felberturmmuseum, Mittersill 02390
Felbrigg Hall, Norwich 45278
Feld-Haus - Museum für populäre Druckgrafik, Neuss 20655
Feld- und Grubenbahnmuseum, Solms 21811
Feld- und Industriebahnmuseum (FIM), Freiland 01901
Feldherrnhalle, München 20439
Feldman Gallery and Project Space, Portland 52076
Feledy-Gyűjtemény - Feledy Ház, Miskolc 23703
Feledy House and Feledy Collection of the Miskolc Gallery of Fine Arts, Miskolc 23703
Feliks Nowowiejski Museum, Barczewo 34535
Felix Jenewein Gallery, Kutná Hora 09613
Felix-Müller-Museum im Zehntspeicher, Neunkirchen am Brand 20638
Felix-Nussbaum-Haus Osnabrück, Osnabrück 20921
Felix Valle State Historic Site, Sainte Genevieve 52650
Fell Locomotive Museum, Featherston 33104
Fellenbergmühle Feinmechanisches Museum, Merzig 20300
Felsenburg Neurathen, Lohmen, Sachsen 20025
Felsendome Rabenstein, Chemnitz 17695
Felsengarten Sanspareil mit Morgenländischem Bau, Wonsees 22680
Felsenkeller Brauerei-Museum, Monschau 20373
Felsenkeller-Labyrinth im Hirschberg - Brauereimuseum, Beilngries 16983
Felsenmuseum, Bernstein 01782
Fembohaus City Museum, Nürnberg 20791
Femörefortet, Oxelösund 40805
Fender Museum of Music and the Arts, Corona 47816
Fenelon Falls Museum, Fenelon Falls 05792
Feng Chia University Art Center, Taichung 42348
Feng City Museum, Fengcheng 07686
Fenghua Xikou Museum, Fenghua 07687
Fengkai Museum, Fengkai 07688
Fengshan Dr. Sun Yat-sen Memorial Hall, Fengshan 42187
Fengshun Museum, Fengshun 07689
Fengxian Museum, Shanghai 08166
Fengxiang Museum, Fengxiang 07690
Fengyuan Lacquer Art Museum, Fengyuan 42188
Fenimore Art Museum, Cooperstown 47789
Fenland and West Norfolk Aviation Museum, West Walton 45973
Fenomenalen, Visby 41068
Fenton Glass Museum, Williamstown 54177
Fenton History Center-Museum and Library, Jamestown 49597
Fenton House, London 44847
Fenxi Museum, Linfen 07940
Fenyang Museum, Fenyang 07691
Fenyi Museum, Fenyi 07692
Ferdinand Domela Nieuwenhuis Museum, Heerenveen 32332
Ferenc Deák Birthhouse, Söjtör 23805
Ferenc Hopp Museum of Eastern Asiatic Arts, Budapest 23413
Ferenc Kölcsey Memorial House, Szatmárcseke 23837
Ferenc Lajos and Contemporary Graphic Collection, Gyula 23598

Ferenc Medgyessy Memorial Musuem, Debrecen 23518
Ferenc Móra Memorial House, Kiskunfélegyháza 23667
Ferenc Nagy Gallery, Tab 23908
Ferencvárosi Helytörténeti Gyűjtemény, Budapest 23402
Ferenczy Museum - Barcsay Collection, Szentendre 23872
Ferenczy Museum - Imre Ámos - Margit Anna Memorial Museum, Szentendre 23873
Ferenczy Museum - Lajos Vajda Memorial Exhibition, Szentendre 23874
Ferenczy Museum - Margit Kovács Ceramics Exhibition, Szentendre 23875
Ferenczy Múzeum, Szentendre 23876
Ferens Art Gallery, Kingston-upon-Hull 44594
Feres Archaeological Site, Velestino 23233
Fergana Regional Museum of Local Lore, Farg'ona 54446
Fergusson Gallery, Perth 45365
Ferihegy Aviation Memorial Park, Budapest 23403
Ferihegy Repülogép Emlékpark, Budapest 23403
Feringa Sach - Ortsgeschichte und heimatkundliche Sammlung, Unterföhring 22179
Fermanagh County Museum at Enniskillen Castle, Enniskillen 44178
Ferme aux Abeilles, Saint-Brévin-les-Pins 15189
La Ferme Caussenarde d'Autrefois, Hures-la-Parade 13230
Ferme d'Antan, Plédéliac 14837
Ferme de la Claison, Saint-Étienne-du-Bois 15258
Ferme de Paris, Paris 14596
Ferme des Bois Nois, Saint-Eloy La-Glaciere 15245
La Ferme des Étoiles, Mauroux 14041
Ferme des Mangettes, Saint-Étienne-du-Bois 15259
Ferme d'Hier et d'Aujourd'hui, Aviré 11800
Ferme du Château, Saint-Fargeau 15262
La Ferme Marine-Musée de l'Huître et du Coquillage, Cancale 12260
Ferme-Musée à Pans de Bois de Sougey, Montrevel-en-Bresse 14257
Ferme-Musée de Bray, Sommery 15781
Ferme Musée de la Soyotte, Sainte-Marguerite 15564
Ferme-Musée de la Tonnelière, Courcerault 12651
Ferme-Musée des Frères Perrel, Moudeyres 14294
Ferme-Musée du Contentin, Sainte-Mère-Eglise 15573
Ferme-Musée du Léon, Tréflaouenan 15980
Ferme-Musée du Montagnon, Fournets-Luisans 12993
Fermes-Musée du Pays Horloger, Grand-Combe-Châteleu 13097
Fermstead Vlkolínec, Ružomberok 38345
Fernbank Museum of Natural History, Atlanta 46434
Fernbank Science Center, Atlanta 46435
Fernhill House, Belfast 43440
Fernie and District Historical Society Museum, Fernie 05795
Fernmeldemuseum Aachen, Aachen 16398
Ferrari Museum of Art, Gotemba 28458
Ferreria de Mirandaola, Legazpia 39449
Ferrum - Museum Kremayr, Ybbsitz 03124
Ferrymead Heritage Park, Christchurch 33073
Fersman Mineralogical Museum of the Russian Academy of Sciences, Moskva 37250
Festetics Castle, Dég 23522
Festetics-Kastély, Dég 23522
Festetics-Kastély, Keszthely 23660
Festival-Institute Museum, Round Top 52471
Festivity of Remedios Museum, Remedios 09265
Festung Kniepass, Unken 02839
Festung Königstein, Königstein 19656
Festung Marienberg mit Fürstenbaumuseum, Würzburg 22710
Die Festung Rosenberg - Deutsches Festungsmuseum, Kronach 19719
Festung Wilhelmstein im Steinhuder Meer, Wunstorf 22732
Festungs- und Heimatmuseum, Kufstein 02243
Festungsmuseum Crestawald, Filisur 41370
Festungsmuseum Heldsberg, St. Margrethen 41897
Festungsmuseum Reuenthal, Reuenthal 41756
Festungsmuseum Salzburg, Salzburg 02610
Fethard Park and Folk Museum, Fethard 24833
Fethiye Cami Müzesi, İstanbul 42818
Fethiye Culture Center, Fethiye 42783
Fethiye Kültür Merkezi, Fethiye 42783
Fethiye Mosque Museum, İstanbul 42818
Fethiye Müzesi, Fethiye 42784
Fetlar Interpretive Centre, Fetlar 44225
Fetsund Lenser Fløtingsmuseum, Fetsund 33491
Feuchtmayer Museum, Salem, Baden 21470
Feudal Castle of Corroy-le-Château, Corroy-le-Château 03526
Feuerschiff, Wilhelmshaven 22576
Feuerwehr-Helm-Museum, Stuttgart 21932
Feuerwehr Museum, Lahr 19770
Feuerwehr-Museum, Stuttgart 21933
Feuerwehr-Museum Schloß Waldmannshofen, Creglingen 17739
Feuerwehr- und Handwerkermuseum, Endingen 41350
Feuerwehr- und Technikmuseum, Eisenhüttenstadt 18106
Feuerwehrhistorische Ausstellung, Dahlen 17771
Feuerwehrmuseum, Adligenswil 41089
Feuerwehrmuseum, Altenkirchen, Westerwald 16501
Feuerwehrmuseum, Asperg 16600
Feuerwehrmuseum, Dobersberg 01819
Feuerwehrmuseum, Hermeskeil 19149
Feuerwehrmuseum, Kaufbeuren 19485
Feuerwehrmuseum, Kreuzlingen 41512
Feuerwehrmuseum, Lengenfeld 19952
Feuerwehrmuseum, Niederwiesa 20715
Feuerwehrmuseum, Pegau 20998
Feuerwehrmuseum, Perchtoldsdorf 02485
Feuerwehrmuseum, Rehau 21237
Feuerwehrmuseum, Roding 21347
Feuerwehrmuseum, Sachsenheim 21465
Feuerwehrmuseum, Salem, Baden 21471
Feuerwehrmuseum, Sankt Leonhard am Forst 02660
Feuerwehrmuseum, Sayda 21526
Feuerwehrmuseum, Spalt 21827
Feuerwehrmuseum, Stadtprozelten 21860
Feuerwehrmuseum, Stetten am kalten Markt 21903
Feuerwehrmuseum, Steyrermühl 02778
Feuerwehrmuseum, Türnitz 02828
Feuerwehrmuseum, Tuttlingen 22133
Feuerwehrmuseum, Wasserburg am Inn 22344
Feuerwehrmuseum Berlin, Berlin 17082
Feuerwehrmuseum des Landkreises Harburg, Marxen 20229
Feuerwehrmuseum Ellricher Spritzenhaus, Ellrich 18118
Feuerwehrmuseum Gahlenz, Gahlenz 18530
Feuerwehrmuseum Grethen, Parthenstein 20986
Feuerwehrmuseum Hannover, Hannover 18989
Feuerwehrmuseum Hordorf, Cremlingen 17745
Feuerwehrmuseum Kalmbach, Riedbach 21313
Feuerwehrmuseum Kradolf-Schönenberg, Kradolf 41510
Feuerwehrmuseum Marxen, Marxen 20230
Feuerwehrmuseum Musberg, Leinfelden-Echterdingen 19882
Feuerwehrmuseum Nürnberg, Nürnberg 20759
Feuerwehrmuseum Ravensburg, Ravensburg 21203
Feuerwehrmuseum Salzbergen, Salzbergen 21473
Feuerwehrmuseum Schleswig-Holstein, Norderstedt 20734
Feuerwehrmuseum Schröttinghausen, Preußisch Oldendorf 21124
Feuerwehrmuseum Winnenden, Winnenden 22600
Feuerwehrmuseum Zeven, Zeven 22796
Fiakermuseum, Wien 02987
Fichtelgebirgsmuseum, Wunsiedel 22731
Fick Fossil and History Museum, Oakley 51476
Fideicomisio de Conservación e Historia de Vieques, Vieques 36081
FIDM Museum and Galleries, Los Angeles 50220
Fiedler Memorial Museum, Seguin 53028
Field Marshal's Hall, München 20439
Field Museum of Natural History, Chicago 47473
Fielding L. Wright Art Center, Cleveland 47620
Fier Historical Museum, Fier 00016
Fife Folk Museum, Ceres 43785
Fifth Maine Regiment Center, Peaks Island 51743
Fifth Meeting House, Lancaster 49917
Fighter World, Williamtown 01653
Figurencabinett Madame Lucrezia, Rosegg 02598
Fiji Museum, Suva 10704
Filey Museum, Filey 44227
Fillmore County Historical Center, Fountain 48724
Fillmore Historical Museum, Fillmore 48541
Film Forum - Film Archives, Collingswood 47693
Film Museum in Łódź, Łódź 34877
Film Museum of Iran, Tehrān 24665
Film- und Fernsehmuseum Hamburg e.V., Hamburg 18928
Film- und Kinomuseum Baden-Württemberg, Biberach an der Riß 17320
Filmmuseet, Kristianstad 40653
Filmmuseet, Oslo 33723
Filmmuseum, Antwerpen 03286
Filmmuseum, Bendestorf 16987
Filmmuseum, Bruxelles 03481
Filmmuseum, München 20440
Filmmuseum Foundation Nederlands, Amsterdam 31841
Filmmuseum Landeshauptstadt Düsseldorf, Düsseldorf 17988
Filmmuseum Potsdam, Potsdam 21087
The Filson Historical Society, Louisville 50285
Finanzgeschichtliche Sammlung der Bundesfinanzakademie, Brühl 17604
Finch Foundry Working Museum, Okehampton 45311
Finchcocks, Goudhurst 44339
Findlater Museum, Dublin 24789
Findorffhof, Grasberg 18738
Fine Art Collection & Gallery 1C03, Winnipeg 07358
Fine Art Gallery, Hodonín 09518
Fine Art Gallery, Roudnice nad Labem 09820
Fine Art Gallery at Centro Cultural Aztlan, San Antonio 52714
Fine Art Museum - Andijan Fortress, Dept. of the Andijan Regional Museum of Literature and Art, Andijon 54442
Fine Art Museum at the Ervin Baktay Secondary School, Dunaharaszti 23530
Fine Art Museum Rolf Werner, Bansin, Seebad 16926
Fine Art University Exhibition Hall, Ha Noi 54612
Fine Arts Center for New River Valley, Pulaski 52196
Fine Arts Center Galleries, Bowling Green 46977
Fine Arts Center Gallery, Maryville 50525
Fine Arts Center of Hot Springs, Hot Springs 49372
Fine Arts Center of Kershaw County, Camden 47200
Fine Arts Department Gallery, Cincinnati 47560
Fine Arts Exhibitions, Austin 46508
Fine Arts Gallery, Breckenridge 47001
Fine Arts Gallery, Columbus 47745
Fine Arts Gallery, Great Falls 49029
Fine Arts Gallery, Los Angeles 50221
Fine Arts Gallery, Luc'k 43111
Fine Arts Gallery, Náchod 09668
Fine Arts Gallery, New Orleans 51079
Fine Arts Gallery, Omaha 51549
Fine Arts Gallery, Ostrava 09696
Fine Arts Gallery, Tiranë 00030
Fine Arts Museum, Blois 12035
Fine Arts Museum, Girne 33359
Fine Arts Museum, Narbonne 14377
Fine Arts Museum, Sevastopol 43176
Fine Arts Museum, Tbilisi 16356
Fine Arts Museum, Tehrān 24666
Fine Arts Museum, Włocławek 35312
Fine Arts Museum of Córdoba, Córdoba 39205
Fine Arts Museum of Santander, Santander 40009
Fine Arts Museum Uzbekistan, Toshkent 54484
Fine Arts School Museum, Buenos Aires 00157
Fine Piece Museum, Lintong 07953
Fingerhutmuseum, Creglingen 17740
Fini's Hoeve, Vriezenveen 32910
Finish Forestareas Museum, Viksjöfors 41063
Finlaggan Centre, Ballygrant 43362
Finland Swede's School Museum, Munsala 11124
Finlands Svenska Skolmuseum, Munsala 11124
Finlandssvenskt Konstcentrum, Ekenäs 10722
The Finley House, Wayne 54011
Finnetunet, Grue Finnskog 33523
Finney County Kansas Historical Museum, Garden City 48854
Finnish Aviation Museum, Vantaa 11448
Finnish Boxing Museum, Tampere 11360
Finnish Dataprocessing Museum, Jyväskylä 10898
Finnish Folk Instrument Museum, Kaustinen 10933
Finnish Football Museum, Valkeakoski 11438
Finnish Foundry Museum and Högfors Blast Furnace, Karkkila 10919
Finnish Glass Museum, Riihimäki 11270
Finnish Ice Hockey Museum, Tampere 11358
Finnish Labour Museum Werstas, Tampere 11367
Finnish Museum of Art and Design, Helsinki 10771
Finnish Museum of Horology, Espoo 10733
Finnish Museum of Photography, Helsinki 10823
Finnish Museums Association, Helsinki 54795
Finnish National Gallery, Helsinki 10835
Finnish Paint Museum, Kemi 10942
Finnish Railway Museum, Hyvinkää 10846
Finnish School Museum, Tampere 11359
Finnkroken Bygdemuseum, Tromsdalen 33917
Finnskogsmuseet, Viksjöfors 41063
Finspångs Slott, Finspång 40521
Finta Múzeum, Túrkeve 23939
Fire Brigade Museum, İstanbul 42829
Fire Brigade Museum Schleswig-Holstein, Norderstedt 20734
Fire Brigade Museum Waldmannshofen, Creglingen 17739
Fire Brigades Museum, Reigate 45480
Fire Brigadesmen's Museum, Wellington 33246
Fire Fighters Museum of Winnipeg, Winnipeg 07359
Fire Fighting Museum, Sairaalamäki 11294
Fire-Fighting Museum of Latvia, Riga 30086
Fire Fighting Museum of the Retro Technology Club "Retrom", Kärevere 10579
Fire Fly Museum of Dongshi Forest Farm, Dongshi 42182
Fire Museum, Helsinki 10807
Fire Museum, Minsk 03248
Fire Museum of Maryland, Lutherville 50323
Fire Museum of Memphis, Memphis 50572
Fire Museum of Texas, Beaumont 46676
Fire Museum of York County, York 54340
Fire Police Museum, Sheffield 45650
Fire Protection Museum, Šiauliai 30360
Fire Safety Museum of Taipei Fire Department, Taipei 42390
Fire Services Museum of Victoria, East Melbourne 01053
Fire Technical Exhibition, Sankt-Peterburg 37721
Firearms Technology Museum, Orange 01394
Firefighter Museum, Bonyhád 23375
Firefighter Museum, Matanzas 09214
Firefighter Museum, Vörs 23967
Firefighters Museum of Nova Scotia, Yarmouth 07409
Firefighting Museum, Lisboa 35696
Firefighting Museum, Vila do Conde 35992
Firefighting Museum of the Dravinjska Valley, Slovenske Konjice 38520
Firehouse Art Center, Norman 51373
Firehouse Art Gallery, Garden City 48856
Firehouse Gallery, Oak Bluffs 51450
Firehouse Museum, Ellicott City 48366
Firehouse No. 1 Museum, Nevada City 51007
Firelands Historical Society Museum, Norwalk 51434
Firelight Museum, Leongatha 01228
Fireman's Hall Museum, Philadelphia 51834
Firemen's House, La Habana 09177
Firepower - Royal Artillery Museum, London 44848
Fireworks and Regional Memorial House, Vál 23949
Firle Place, Firle 44230
Firmenmuseum Novatech, Reutlingen 21282
First Austrian Scotch Whisky Museum, Steyregg 02776
First Cavalry Division Museum, Fort Hood 48625
First Consolidated School in Minnesota, Saum 52932
The First Division Museum at Cantigny, Wheaton 54120
First Garden City Heritage Museum, Letchworth 44694
First German Bockbeer-Museum, Taucha 22011
First Grocery Shop Albert Heijn, Zaandam 32979
First Hussars Citizen Soldiers Museum, London 06185
First Michigan Museum of Military History, Holland 49315
First Missouri State Capitol, Saint Charles 52532
First National Bank of Chicago Art Collection, Chicago 47474
First Penny Farthing Museum, Knutsford 44622
First Raadsaal Museum, Bloemfontein 38561
First Territorial Capitol of Kansas, Fort Riley 48675
First Turkish Grand National Assembly Museum, Ankara 42693
First White House of the Confederacy, Montgomery 50814
First Wing Historical Center (1WHC) The Golden Falcon, Beauvechain 03351
Firstsite at the Minories Art Gallery, Colchester 43873
Firth Tower Historical Reserve, Matamata 33145
Fischer- und Webermuseum Steinhude, Wunstorf 22733
FischerBauKunst, Polling 21073
Fischerei-Museum, Zug 42124
Fischereimuseum, Wassertrüdingen 22348
Fischerstube, Mönkebude 20359
Fischinger Heimathaus mit Schimuseum, Fischen im Allgäu 18323
Fischlandhaus, Wustrow, Ostseebad 22756
Fish Museum, Pescara 27165
Fish Museum Golden Angel, Damascus 42130
Fishbourne Roman Palace and Museum, Chichester 43826
Fisher Gallery, Los Angeles 50222
Fisher Grove Country School, Frankfort 48744
Fisher Museum of Forestry, Petersham 51805
Fisheries Museum of the Atlantic, Lunenburg 06202
Fisherman's Cabin, Kauhava 10929
Fisherman's Cottage Museum, Tooradin 01591
Fishermen's Life Museum, Jeddore Oyster Ponds 06028
Fishermen's Museum, Bermeo 39030
Fishermen's Museum, Hastings 44414
Fishermen's Museum, Pemaquid Point 51752
Fishermen's Museum, Port-de-Grave 06635
Fishermen's Museum, Saint Vincent's 06884
Fishery and Harbor Museum, Sassnitz 21521
Fishery Exhibition Hall, Lugang 42256
Fishery Museum, Hel 34687
Fishery Museum, Jastarnia 34698
Fishing and Maritime Museum, Esbjerg 10002
Fishing Museum, Kaskinen 10924
Fishing Museum, Merikarvia 11109
Fishing Museum, Palamós 39768
Fisk University Galleries, Nashville 50968
Fiska- og Nåttúrngripasafn Vestmannaeyja, Vestmannaeyjar 24043
Fiskars Museum, Fiskars 10734
Fiskemuséet i Kaskö, Kaskinen 10924
Fiskeri- og Søfartsmuseet, Esbjerg 10002
Fiskeriets Hus - Vestkyst Akvariet, Hvide Sande 10107
Fiskerimuseet på Hjertøya, Molde 33676
Fiskerimuseum i Måløy, Måløy 33657
Fitchburg Art Museum, Fitchburg 48548
Fitchburg Historical Society Museum, Fitchburg 48549
The Fitzwilliam Museum, Cambridge 43727
Five Civilized Tribes Museum, Muskogee 50928

Fizikos Muziejus, Vilnius 30404
Fjärilshuset i Hagaparken, Stockholm 40894
Fjerritslev Bryggeri- og Egnsmuseum, Fjerritslev 10020
Fjordane Forsvarsmuseum, Nordfjordeid 33700
Fjordmuseet, Jyllinge 10111
Fjøsangersamlingene, Fjøsanger 33495
Flachsbrechhütte, Creglingen 17741
Flachsmuseum Wegberg-Beeck, Wegberg 22359
Flagler Museum, Palm Beach 51677
Flagstaff House Museum of Tea Ware, Hong Kong 07794
Flakkebjerg Skolemuseum, Slagelse 10274
Flambards Experience, Helston 44442
Flambards Victorian Village and Gardens, Helston 44443
Flambro Head Heritage Museum, Job's Cove 06029
Flame!, Carrickfergus 43766
Flamingo Gallery, Las Vegas 49967
Flamish Heritage, Groede 32274
Flanagan Valley Campus Art Gallery, Lincoln 50112
Flanders Center, Osaka 29125
Flandrau Science Center, Tucson 53589
Flaske-Peters-Samling, Ærøskøbing 09958
Flatanger Bygdemuseum, Flatanger 33496
Flaten Art Museum, Northfield 51420
Flathead Indian Museum, Saint Ignatius 52551
Flatowturm, Potsdam 21088
Flatrock Museum, Flatrock 05798
Flaxbourne Settlers Museum, Ward 33237
Fleet Air Arm Museum - Concorde, Yeovilton 46097
Fleet Air Arm Museum of New Zealand, Auckland 33048
Fleetwood Museum, Fleetwood 44232
Flekkefjord Museum, Flekkefjord 33497
Fleming Historical Museum, Fleming 48556
Flensburger Schiffahrtsmuseum, Flensburg 18327
Flessenscheepjesmuseum, Enkhuizen 32207
Fleur de Lis Heritage Centre, Faversham 44221
Flieger Flab Museum, Dübendorf 41335
Fliegermuseum Altenrhein, Altenrhein 41107
Flin Flon Museum, Flin Flon 05800
Flinders University Art Museum, Bedford Park 00846
Flinders University City Gallery, Adelaide 00776
Flint House, Scotia 52970
Flint Institute of Arts, Flint 48559
Flint Ridge State Memorial Museum, Glenford 48932
Flintham Museum, Flintham 44233
Flipje en Jam Museum Tiel, Tiel 32823
Flößer- und Heimatmuseum, Wolfach 22645
Flößer- und Schiffermuseum, Kamp-Bornhofen 19422
Flößerei- und Verkehrsmuseum, Gengenbach 18582
Flößereimuseum, Lychen 20118
Flößermuseum Unterrodach, Marktrodach 20223
Flood Museum, Johnstown 49640
Floors Castle, Kelso 44555
Flora Twort Gallery, Petersfield 45373
Floral Museum, Chiba 28394
Florence Court House, Enniskillen 44179
Florence Griswold Museum, Old Lyme 51536
Florence Hawley Ellis Museum of Anthropology, Abiquiu 46128
Florence McLeod Hazard Museum, Columbus 47746
Florence Mine Heritage Centre, Egremont 44165
Florence Museum of Art, Science and History, Florence 48570
Florence Nightingale Museum, London 44849
Florence Price Pioneer Museum, Florence 48568
Florence Ranch Homestead, Mesquite 50612
Florewood State Park, Greenwood 49086
Florianerbahn, Sankt Florian 02636
Florida Adventure Museum, Punta Gorda 52200
Florida Agricultural Museum, Palm Coast 51679
Florida Art Museum Directors Association, Vero Beach 55132
Florida Association of Museums, Tallahassee 55133
Florida Craftsmen Gallery, Saint Petersburg 52636
Florida Gulf Coast Railroad Museum, Parrish 51715
Florida History Museum, Jupiter 49659
Florida Holocaust Museum, Saint Petersburg 52637
Florida International Museum, Saint Petersburg 52638
The Florida Museum of Hispanic and Latin American Art, Coral Gables 47801
Florida Museum of Natural History, Gainesville 48813
Florida State University Museum of Fine Arts, Tallahassee 53422
Florina Museum of Modern Art, Florina 23000
Florissant Valley Historical Society, Florissant 48571
Flottmann-Hallen, Herne 19152
Flour Mill Museum, Sudbury 07028
Flowerdew Hundred Foundation, Hopewell 49363
Floyd County Historical Museum, Charles City 47349
Flugausstellung Hermeskeil, Hermeskeil 19150
Flughafen Modellschau, Hamburg 18929
Flugmuseum Aviaticum, Wiener Neustadt 03095
Flugpionier-Gustav-Weißkopf-Museum, Leutershausen-Wiedersbach 19969
Flur-Galerie im Kulturamt, Berlin 17083
Fluweelengrot Valkenburg, Valkenburg, Limburg 32862
Flyhistorisk Museum, Sola 33848
Flynderupgård Museet, Espergærde 10005
Flypast Museum of Australian Army Flying, Oakey 01386
Flyvapenmuseum, Malmslätt 40728
FN Museet, Padborg 10226
Fo Guang Yuan Art Gallery and Museum of Treasures, Dashu 42178
Foam Fotografiemuseum Amsterdam, Amsterdam 31842
Foam Lake Museum, Foam Lake 05801
Foam Photography Museum, Amsterdam 31842
Focal Point Gallery, Southend-on-Sea 45711
Fochabers Folk Museum, Fochabers 44235
Focke-Museum, Bremen 17549
focusTerra, Zürich 42071
Föglö Museet, Föglö 10735
Főegyházmegyei Könyvtár, Eger 23539
Förderkreis Industrie- und Technikgeschichte, Frankfurt am Main 18366
Főszékesegyházi Kincstár és Könyvtár, Esztergom 23552
Fogelsangh State, Veenklooster 32877
El Fogon de los Arrieros, Resistencia 00521
Fojnica Monastery Museum Collection, Fojnica 04263
Foldalbruket, Kjøllefjord 33594
Folger Shakespeare Library, Washington 53869
Folk and Applied Arts Museum, Tbilisi 16357
Folk and Costume Museum, Câtel Guernsey 43779
Folk Art and Photography Galleries, Atlanta 46436
Folk Art House, Heves 23607
Folk Art Institute, Lefkoşa 33369
Folk Art Museum, Constanţa 36303
Folk Art Museum, Girne 33360
Folk Art Museum, Krosno 34831
Folk Art Museum, Lemesos 09386
Folk Art Museum, Moskva 37338
Folk Art Museum, Osiek nad Notecią 34990
Folk Art Museum, Pyrgi 23161
Folk Art Museum, Yeroskipos 09393
Folk Art Society of America, Richmond 55134
Folk Cultural Collection Museum, Jhongpu 42222
Folk Custom Museum, Xiangfen 08369
Folk Customs Museums of Bucovina, Gura Humorului 36394
Folk Experience Museum, Dranouter 03562
Folk Heritage Museum, Thimphu 04202
Folk Historical and Regional Museum of Tatarsk, Tatarsk 37838
Folk Museum, Anuradhapura 40367
Folk Museum, Eyrarbakki 23995
Folk Museum, Kolbuszowa 34749
Folk Museum, Mbarara 42995
Folk Museum, Mevagissey 45135
Folk Museum, Millom 45147
Folk Museum, Miyama 28924
Folk Museum, Ohrid 30512
Folk Museum, Soroti 42997
Folk Museum Naujene, Naujene 30049
Folk Museum of Higashimurayama, Higashimurayama 28507
Folk Museum of Indian Immigration, Moka 30678
Folk Museum of Ota District, Tokyo 29458
Folk Museum of Stranda, Mönsterås 40744
Folk Music Museum, Genova 26305
Folkenborg Museum, Mysen 33687
Folkestone Museum and Art Gallery, Folkestone 44237
Folklife Display, Lisnaskea 44726
Folklore and Ethnological Museum of Macedonia and Thrace, Thessaloniki 23212
Folklore Museum, Cairo 10495
Folklore Museum, Mysore 24328
Folklore Museum, Pogoniani 23151
Folklore Museum, Salzburg 02629
Folklore Museum, Thessaloniki 23213
Folklore Museum, Victoria 30651
Folklore Museum, Xanthi 23242
Folklore Museum and Municipal Art Gallery, Karditsa 23034
Folklore Museum at Drama, Drama 22988
Folklore Museum of Arita, Arita 28363
Folklore Museum of Folegandros, Folegandros 23003
Folklore Museum of Polygyros, Polygyros 23153
Folklore Museum of Seto, Seto 29251
Folklore Museum of the Culture Club Florina, Florina 23001
Folkloremuseum Florishof, Oostduinkerke 03924
Folkloristisch Museum, Swalmen 32813
Folldal Bygdetun, Folldal 33500
Folldal Gruver, Folldal 33501
Follett House Museum, Sandusky 52836
Follo Museum, Drøbak 33469
Folsom History Museum, Folsom 48581
Folsom House Museum, Taylors Falls 53455
Folsom Museum, Folsom 48582
Foltermuseum, Burghausen 17652
Foltermuseum, Seeboden 02733
Fond d'art contemporain de la Ville de Genève (Fmac), Genève 41412
Fond Régional d'Art Contemporain, Angoulême 11616
Fondaçao-Museu Chissano, Matola 31713
Fondation Beyeler, Riehen 41763
Fondation Cartier pour l'Art Contemporain, Paris 14597
Fondation Carzou, Manosque 13941
Fondation Charles de Gaulle, Colombey-les-Deux-Eglises 12587
Fondation Custodia, Collection Frits Lugt, Paris 14598
Fondation Cziffra, Senlis 15714
Fondation de Jau, Cases-de-Pène 12303
Fondation de l'Automobile Marius-Berliet, Lyon 13881
Fondation de l'Hermitage, Lausanne 41539
Fondation Dosne-Thiers, Paris 14599
Fondation Dubuffet, Paris 14600
Fondation Dubuffet, Périgny-sur-Yerres 14776
Fondation et Musée René Carcan, Bruxelles 03445
Fondation Félicien Rops, Mettet 03873
Fondation Folon, La Hulpe 03765
Fondation Herzog, Basel 41159
Fondation Jean Paul Blachère, Abidjan 08829
Fondation Jean Paul Blachère, Apt 11652
Fondation La Tène, Hauterive 41473
Fondation Le Corbusier, Paris 14601
Fondation L'Estrée, Ropraz 41775
Fondation Louis Moret, Martigny 41607
Fondation Maeght, Saint-Paul-de-Vence 15444
Fondation Marius Vazeilles, Meymac 14110
Fondation Martin Bodmer, Cologny 41308
Fondation Musée-Volti, Villefranche-sur-Mer 16192
Fondation Oscar et Janine Wiggli, Muriaux 41667
Fondation Pierre Gianadda, Martigny 41608
Fondation pour l'Architecture, Bruxelles 03446
Fondation pour l'Art Contemporain Claudine et Jean-Marc Salomon, Alex 11552
Fondation Prouvost, Marcq-en-Barœul 13959
Fondation Rainer Maria Rilke, Sierre 41866
Fondation René Pous, Auterive 11760
Fondation Saner, Studen 41918
Fondation Van-Gogh Arles, Arles 11682
Fondation Vasarely, Aix-en-Provence 11514
Fondazione Antonio Mazzotta, Milano 26684
Fondazione Bandera per l'Arte, Busto Arsizio 25572
Fondazione Dazio Grande, Rodi-Fiesso 41770
Fondazione Europea del Disegno, Meina 26649
Fondazione Galleria Gottardo, Lugano 41580
Fondazione M. Minucci, Vittorio Veneto 28297
Fondazione Magnani Rocca, Mamiano di Traversetolo 26580
Fondazione Merz, Torino 27999
Fondazione Palazzo Albizzini Collezione Burri, Città di Castello 25905
Fondazione Palazzo Coronini Cronberg, Gorizia 26350
Fondazione Prada, Milano 26685
Fondazione Rolf Gérard, Ascona 41128
Fondazione Romano nel Cenacolo di Santo Spirito, Firenze 26152
Fondazione Sandretto Re Rebaudengo, Torino 28000
Fondazione T. Balestra, Longiano 26515
Fonderie de Cloches, Robecourt 15060
Fondo Del Sol, Washington 53870
Fonds Ancien, Avranches 11801
Fonds ancien de la Bibliothèque cantonale jurassienne, Porrentruy 41727
Fonds Régional d'Art Contemporain, Saint-Paul 36094
Fonds Régionaux d'Art Contemporain Bourgogne, Dijon 12734
Fonthill Museum of the Bucks County Historical Society, Doylestown 48177
Foot-Binding Culture Museum, Tucheng 42457
Foothills Art Center, Golden 48959
Footscray Community Arts Centre, Footscray 01083
Forbes Collection, New York 51174
Forbes Museum, Forbes 01085
Forbes Museum, Tanger 31695
The Forbidden City Museum, Lujhou 42262
Ford County Historical Society Museum, Paxton 51739
Ford End Watermill, Ivinghoe 44543
Ford Gallery, Ypsilanti 54356
Ford Green Hall, Stoke-on-Trent 45760
Ford Museum, Dumalag 34387
Forde Abbey, Chard 43791
Ford's Theatre, Washington 53871
Fordyce Joiner's Visitor Centre, Fordyce 44240
Foredown Tower, Portslade-by-Sea 45426
Foreman Art Gallery of Bishop's University, Lennoxville 06157
Foreman Gallery, Oneonta 51563
Foreningen af Danske Kunstmuseer, Nivå 54779
Foreningen af Danske Museumsmaend, København 54780
Foreningen af Danske Naturhistoriske Museer, Århus 54781
Forensic Medicine Museum, Bangkok 42538
Forest and Heritage Centre, Geeveston 01105
Forest Capital State Museum, Perry 51779
Forest Department Utilisation Division Museum, Kampala 42990
Forest Discovery Center, Portland 52077
Forest Entomology Museum, Bangkok 42539
Forest History Center, Grand Rapids 49008
Forest-Lambton Museum, Forest 05802
Forest Lawn Museum, Glendale 48930
Forest Museum, Gyeonggi-do 29824
Forest Museum, Lycksele 40710
Forest Museum, Peshawar 34033
Forest Museum, Taungdwingyi 31729
Forest Museum in Bojas Palace, Aizpute 29983
Forest Museum of Tytuvenai, Tytuvėnai 30385
Forest Park Museum, Perry 51780
Forestburg and District Museum, Forestburg 05803
Forestry Exhibition Hall, Fengyuan 42189
Forestry Museum, Viitasaari 11466
Forestry Museum at Stajer-House, Kőszeg 23684
Forestry Museum Siljansfors, Mora 40746
Forestry of Nyírség and the forestry of Debrecen - The Country House of Bánk, Debrecen 23519
Forestry & Open Air Forestry Museum, Szilvásvárad 23893
Forestry Research Institute Malaysia Museum, Kepong 30562
Forge-Menuiserie Cauchon, Rivière-Malbaie 06762
Forge Mill Needle Museum and Bordesley Abbey Visitor Centre, Redditch 45475
Forge-Musée d'Etueffont, Etueffont 12884
Forge Musée d'Ittre, Ittre 03734
Forge Museum, Much Hadham 45182
Forge Museum in Warsaw, Warszawa 35254
The Forges and Manor of Ringwood, Ringwood 52353
Forma-Centro Internazionale di Fotografia, Milano 26686
Former Governors' Mansion, Bismarck 46837
Former Hong Kong and Shanghai Bank Museum, Nagasaki 28972
Former Jewish local house of Esens, Memorial and museum, Esens 18238
Former Palace of Brussels, Archaeological Site and Museum of the Coudenberg, Bruxelles 03431
Forncett Industrial Steam Museum, Forncett Saint Mary 44243
Forney Transportation Museum, Denver 48077
Fornminnesgården, Eksjö 40494
Føroya Fornminnissavn, Hoyvík 10687
Føroya Náttúrugripasavn, Tórshavn 10698
Forrester Gallery, Oamaru 33161
Forsand Bygdemuseum, Forsand 33504
Forssa Natural History Museum, Forssa 10736
Forssan Luonnonhistoriallinen Museo, Forssa 10736
Forst- und Köhlerhof, Rostock 21390
Forstbotanisches Museum, Tharandt 22049
Forstmuseum, Reichraming 02573
Forstmuseum Heringen, Heringen (Werra) 19142
Forstmuseum im Hochwildpark Karlsberg, Weikersheim 22371
Forsvarsmuseet, Oslo 33724
Forsvarsmuseet på Bornholm, Rønne 10243
Forsviks Bruk, Forsvik 40526
Forsvik's Industrial Heritage, Forsvik 40526
Fort Abercrombie Historic Site, Abercrombie 46115
Fort Amherst Heritage Museum, Chatham 43794
Fort Ancient Museum, Oregonia 51586
Fort Anne, Annapolis Royal 05342
Fort Atkinson State Historical Park, Fort Calhoun 48601
Fort Beaufort Museum, Fort Beaufort 38620
Fort Beauséjour, Aulac 05362
Fort Bedford Museum, Bedford 46690
Fort Belknap Museum, Newcastle 51307
Fort Bend Museum, Richmond 52319
Fort Benton Museum of the Upper Missouri, Fort Benton 48592
Fort Bliss Museum, Fort Bliss 48594
Fort Boonesborough Museum, Richmond 52316
Fort Bridger State Museum, Fort Bridger 48600
Fort Brockhurst, Gosport 44333
Fort Buenaventura, Ogden 51501
Fort Buford, Williston 54180
Fort Calgary, Calgary 05518
Fort Carlton, Regina 06717
Fort Caroline Memorial Museum, Jacksonville 49578
Fort Carré, Antibes-Juan-les-Pins 11639
Fort Casey Interpretive Center, Coupeville 47853
Fort Caspar Museum, Casper 47271
Fort-Chambly, Chambly 05585
Fort Charles Maritime Museum, Kingston 28325
Fort Charlotte, Nassau 03206
Fort Chipewyan Bicentennial Museum, Fort Chipewyan 05806
Fort Christmas, Christmas 47548

F

Fort Clark Trading Post, Center 47313
Fort Clatsop National Memorial, Astoria 46402
Fort Clinch, Fernandina Beach 48536
Fort Collins Museum, Fort Collins 48607
Fort Columbia House Museum, Chinook 47542
Fort Concho, San Angelo 52701
Fort Crawford Museum, Prairie du Chien 52128
Fort Croghan Museum, Burnet 47145
Fort Crook Museum, Fall River Mills 48501
Fort Dalles Museum, The Dalles 47945
Fort d'Artillerie A46, Champex 41290
Fort d'Aubin-Neufchâteau et son Musée, Neufchâteau 03911
Fort Dauphin Museum, Dauphin 05679
Fort Davidson, Pilot Knob 51918
Fort Davis Museum, Fort Davis 48611
Fort de Chartres Museum, Prairie du Rocher 52130
Fort de Seyne, Seyne-les-Alpes 15746
Fort de Vallorbe Pré-Giroud, Vallorbe 41974
Fort de Verny, Verny 16124
Fort Découvertes, Martigny 41609
Fort Delaware, Narrowsburg 50956
Fort Delaware Society Museum, Delaware City 48044
Fort Discovery, Augusta 46477
Fort Dix Museum, Fort Dix 48614
Fort Dobbs State Historic Site, Statesville 53273
Fort Dodge Historical Museum, Fort Dodge 48616
Fort Donelson National Battlefield Museum, Dover 48169
Fort Dorsner, Giromagny 13064
Fort Douglas Military Museum, Fort Douglas 48617
Fort Edmonton Park, Edmonton 05746
Fort Erie Historical Museum, Ridgeway 06750
Fort Erie Railroad Museum, Fort Erie 05807
Le Fort et son Musée, Flémalle 03601
Fort Fetterman, Douglas 48158
Fort Fincastle, Nassau 03207
Fort Fisher, Kure Beach 49828
Fort Fleur d'Epée, Le Gosier 23275
Fort Frances Museum, Fort Frances 05809
Fort Frederica, Saint Simons Island 52646
Fort Frederick, Big Pool 46796
Fort Garry Horse Regimental Museum, Winnipeg 07360
Fort George, Niagara-on-the-Lake 06451
Fort George G. Meade Museum, Fort Meade 48653
Fort Gibson Historic Site, Fort Gibson 48622
Fort Glanville, Semaphore Park 01499
Fort Grey and Shipwreck Museum, Saint Pierre du Bois 45595
Fort Hancock Museum, Sandy Hook 52844
Fort Harker Museum Complex, Kanopolis 49678
Fort Hartsuff, Burwell 47148
Fort Hays State Historic Site, Hays 49223
Fort Henrietta, Echo 48282
Fort Henry, Kingston 06077
Fort Hill Museum, Hillsboro 49286
Fort Hill - The John C. Calhoun House, Clemson 47614
Fort Huachuca Museum, Fort Huachuca 48626
Fort Hunter Mansion, Harrisburg 49179
Fort Ingall, Cabano 05508
Fort Inglish, Bonham 46898
Fort Jackson Museum, Fort Jackson 48628
Fort Jesus Museum, Mombasa 29746
Fort Jones Museum, Fort Jones 48632
Fort Kearney Museum, Kearney 49714
Fort Kearny, Kearney 49715
Fort Kijkduin, Castricum 32037
Fort King George, Darien 47967
Fort Klock Historic Restoration, Saint Johnsville 52556
Fort Knox State Historic Site, Stockton Springs 53308
Fort La Reine Museum and Pioneer Village, Portage-la-Prairie 06651
Fort Langley, Fort Langley 05812
Fort Laramie, Fort Laramie 48634
Fort Larned, Larned 49954
Fort Laurens State Memorial, Bolivar 46895
Fort Leaton, Presidio 52137
Fort Leboeuf, Waterford 53960
Fort l'Ecluse, Leaz 13669
Fort Lee Historic Park and Museum, Fort Lee 48644
Fort Lennox, Saint-Paul-de-l'Île-aux-Noix 06872
Fort Lewis Military Museum, Fort Lewis 48649
Fort Ligonier Museum, Ligonier 50082
Fort Loudoun, Chambersburg 47331
Fort Lowell Museum, Tucson 53590
Fort McAllister, Richmond Hill 52345
Fort MacArthur Museum, San Pedro 52824
Fort McHenry, Baltimore 46565
Fort McKavett State Historic Park, Fort McKavett 48650
Fort Mackinac, Mackinac Island 50351
Fort Macon, Atlantic Beach 46459
Fort Malden, Amherstburg 05332
Fort Massac, Metropolis 50616
Fort Matanzas, Saint Augustine 52521
Fort Meigs State Memorial, Perrysburg 51782
Fort Mifflin, Philadelphia 51835
Fort Monostor, Komárom 23680
Fort Montagu, Nassau 03208
Fort Morgan Museum, Fort Morgan 48659
Fort Morris, Midway 50676
Fort Mountain, Chattsworth 47415
Fort Muséalisé d'Uxegney, Uxegney 16022
Fort Museumof the North West Mounted Police, Fort Macleod 05814
Fort N5 - Muzej Istorii Velikoj Otečestvennoj Vojny, Kaliningrad 36987
Fort Namutoni, Tsumeb 31758
Fort Napoleon, Oostende 03931
Fort Nashborough, Nashville 50969
Fort Necessity National Battlefield, Farmington 48518
Fort Nelson Heritage Museum, Fort Nelson 05818
Fort Nisqually Living History Museum, Tacoma 53407
Fort No 5 - Museum of the Great Patriotic War, Kaliningrad 36987
Fort Normandeau, Red Deer 06707
Fort-Numéro-Un de la Pointe-de-Lévy, Lévis 06167
Fort Okanogan, Pateros 51727
Fort Ontario, Oswego 51624
Fort Osage, Sibley 53095
Fort Ostell Museum, Ponoka 06621
Fort Peck Museum, Fort Peck 48666
Fort Pelley Livingston Museum, Pelly 06578
Fort Phil Kearny, Story 53321
Fort Pike, New Orleans 51080
Fort Pitt Museum, Pittsburgh 51947
Fort Plain Museum, Fort Plain 48670
Fort Point Museum, LaHave 06128
Fort Polk Military Museum, Fort Polk 48671
Fort Pulaski, Savannah 52940
Fort Qu'appelle Museum, Fort Qu'appelle 05819
Fort Queenscliff Museum, Queenscliff 01447
Fort Raleigh, Manteo 50452
Fort Rammekens, Ritthem 32691
Fort Recovery Museum, Fort Recovery 48674
Fort Richardson, Jacksboro 49554
Fort Robinson Museum, Crawford 47868
Fort Rodd Hill, Victoria 07247
Fort Sackville Foundation, Bedford 05411
Fort Saint George Museum, Chennai 24114
Fort Saint James, Fort Saint James 05820
Fort Saint Jean Baptiste, Natchitoches 50987
Fort Saint John-North Peace Museum, Fort Saint John 05821
Fort Saint Joseph, Richard's Landing 06737
Fort Saint Joseph Museum, Niles 51351
Fort Sam Houston Museum, Fort Sam Houston 48678
Fort San Pedro National Museum, Cebu 34370
Fort Saskatchewan Museum, Fort Saskatchewan 05822
Fort Savannah Museum, Lewisburg 50033
Fort Scott National Historic Site, Fort Scott 48680
Fort Scratchley, Newcastle 01361
Fort Selden State Monument, Radium Springs 52226
Fort Selkirk, Yukon 07418
Fort Sidney Museum and Post Commander's Home, Sidney 53097
Fort Sill Museum, Fort Sill 48682
Fort Siloso, Singapore 38174
Fort Simcoe, White Swan 54132
Fort Sisseton, Lake City 49881
The Fort Smith, Fort Smith 48683
Fort Smith Art Center, Fort Smith 48684
Fort Smith Museum of History, Fort Smith 48685
Fort Snelling, Saint Paul 52615
Fort Spokane Visitor Center and Museum, Davenport 47971
Fort Stanwix, Rome 52451
Fort Steele Heritage Town, Fort Steele 05824
Fort Stewart Museum, Fort Stewart 48687
Fort Sumner, Fort Sumner 48691
Fort Sumter National Monument, Sullivan's Island 53360
Fort Tejon, Lebec 50014
Fort Témiscamingue, Ville-Marie 07265
Fort Ticonderoga, Ticonderoga 53506
Fort Totten State Historic Museum, Fort Totten 48692
Fort Towson Military Park, Fort Towson 48693
Fort Uncompahgre History Museum, Delta 48049
Fort Union, Watrous 53984
Fort Union Trading Post, Williston 54181
Fort Vancouver, Vancouver 53706
Fort Vasquez Museum, Platteville 51979
Fort Verde, Camp Verde 47206
Fort Walla Walla Museum, Walla Walla 53797
Fort Walsh, Maple Creek 06232
Fort Ward Museum and Historic Site, Alexandria 46203
Fort Washington, Fort Washington 48698
Fort Washington Museum, Fort Washington 48700
Fort Washita, Durant 48214
Fort Wayne Firefighters Museum, Fort Wayne 48704
Fort Wayne Museum of Art, Fort Wayne 48705
Fort Wellington, Prescott 06656
Fort Whoop-Up Centre, Lethbridge 06161
Fort Widley, Portsmouth 45431
Fort Wilkins Historic Complex, Copper Harbor 47796
Fort William Henry Museum, Lake George 49883
Fort Winnebago Surgeons Quarters, Portage 52051
Fort Worth Museum of Science and History, Fort Worth 48713
Fort Worth Public Library Arts and Humanities, Fort Worth 48714
Fort Yargo, Winder 54223
Fort York Armoury, Toronto 07106
Fort Zachary Taylor, Key West 49760
Fortaleza de Maputo, Maputo 31705
Fortaleza de Nuestra Señora de los Angeles de Jagua, Jagua 09186
Fortaleza de Sagres, Sagres 35894
Fortaleza San Cristóbal, San Juan 36062
Forte Rinascimentale, San Leo 27652
Forteresse de Mornas, Mornas 14283
Forteresse de Salses, Salses-le-Château 15614
Forteresse Médiévale de Ducs, Bourbon-l'Archambault 12100
Fortifications-de-Québec, Québec 06680
Fortress Museum, Cesena 25830
Fortress Museum Karlsborg, Karlsborg 40627
Fortress of Louisbourg, Louisbourg 06197
Forts Folle Avoine Historic Park, Danbury 47954
Forty Hall Museum, Enfield 44176
Forum Art Gallery, Sankt-Peterburg 37638
Forum Artis Museum, Montese 26851
Forum der Schweizer Geschichte, Schwyz, Schwyz 41850
Forum Flüh, Flüh 41376
Forum Fränkischer Hof, Bad Rappenau 16805
Forum Frohner, Krems 02225
Forum für Angewandte Kunst Nürnberg e.V, Nürnberg 20760
Forum für Fotografie, Köln 19601
Forum Hall, Bad Hall 01751
Forum Internationale Photographie, Mannheim 20163
Forum Konkrete Kunst Erfurt, Erfurt 18202
Forum Marinum, Turku 11390
Forum Museo, Omegna 26982
Forum of Swiss History Schwyz, Swiss National Museums, Schwyz 41850
Forum Schloßplatz, Aarau 41081
Forum Stadtscheune-Ortsmuseum und Raum für Kunst, Mellingen 41623
Forum Würth, Arlesheim 41124
Forwarders' Museum, Prescott 06657
Foshan Museum, Foshan 07693
Fosnes Bygdemuseum, Jøa 33583
Fossa Magna Museum, Itoigawa 28599
Les Fosses d'Enfer, Saint-Rémy-sur-Orne 15495
Fossesholm Herregård, Vestfossen 33975
Fossil Country Museum, Kemmerer 49724
Fossil Grove, Glasgow 44280
Fossil Museum, Fossil 48722
Fossil Station Museum, Russell 52484
Fossilien-Museum, Fürth 18497
Fossilien- und Heimatmuseum, Messel 20304
Fossilien- und Steindruck-Museum, Gunzenhausen 18838
Fossilienmuseum im Werkforum Rohrbach Zement, Dotternhausen 17933
Fossilienschauraum, Ernstbrunn 01875
Foster and District Historical Society Museum, Foster 01089
Foster Gallery, Eau Claire 48277
Fosterfields Living Historical Farm, Morristown 50851
Fostoria Area Historical Museum, Fostoria 48723
Fota House, Carrigtwohill 24753
Fotevikens Museum, Höllviken 40596
Fotland Bygdemølle, Bryne 33456
Fotland Kraftverk, Bryne 33457
Fotland Power Plant, Bryne 33457
Fotland Village Mill, Bryne 33456
Foto Museet Sundvall - Bildens Hus, Sundsvall 40958
Fotoarchiv Sojuza Fotochudožnikov Rossii, Moskva 37171
Fotoforum West, Innsbruck 02116
Fotogalereja im. Bratjev Ljumjer, Moskva 37172
Fotogaleria 1/3, Toruń 35188
Fotogalerie Friedrichshain, Berlin 17084
Fotogalerie Wien, Wien 02988
Fotografičeskij Muzej Dom Metenkova, Ekaterinburg 36885
Fotograficzny Muzeum, Mielec 34919
Fotografie Forum International, Frankfurt am Main 18367
Fotografijos Muziejus, Šiauliai 30356
Fotoğrafın Tarihi, Osmaniye 42923
Fotohistorische Sammlung G. und W. Pabst, Uhingen 22156
Fotomuséet i Osby, Osby 40801
Fotomuseet Olympia, Falkenberg 40509
Fotomuseum, Motala 40750
Fotomuseum Den Haag, Den Haag 32072
Fotomuseum Provincie Antwerpen, Antwerpen 03287
Fotomuseum Winterthur, Winterthur 42025
Fotomuzej Vlastja, Gorenja Vas 38418
Fotoplastikon, Warszawa 35223
Fotostiftung Schweiz, Winterthur 42026
Fototeca de Cuba, La Habana 09103
Fototeca Municipal de Reus, Reus 39893
Fouilles Archéologiques, Jonzac 13277
Foundation for Hellenic Culture, New York 51175
Foundation for Historic Christ Church, Irvington 49537
Foundation for Women's Art, London 44850
Founders' Hall, Charlottetown 05592
Founders Historic Park, Nelson 33152
Founders Room, Raanana 25080
Foundling Museum, London 44851
The Foundry, Design Museum, Cape Town 38583
Foundry Museum Komarov, Komárov 09593
Foundry Museum of the National Museum Science and Technology, Budapest 23459
Fountain Gallery, New York 51176
Four Rivers Cultural Center, Ontario 51570
Four Sisters Gallery and the Robert Lynch Collection, Rocky Mount 52435
Four Solaire, Mont-Louis 14159
Four Wheel Drive Foundation, Clintonville 47662
Fours à Chaux, Antoing 03277
Fővárosi Képtár, Budapest 23404
Fowler Museum of Cultural History, Los Angeles 50223
Fox Harbour Community Museum, Fox Harbour 05826
Fox Island Historical Society Museum, Fox Island 48727
Fox Lake Historical Museum, Fox Lake 48728
Fox River Trolley Museum, South Elgin 53169
Fox Talbot Museum, Lacock 44627
Foxton Canal Museum, Foxton 44250
Foxton Flax Stripper Museum, Foxton 33106
Foxton Museum, Foxton 33107
Foyer / James Hockey Gallery, Farnham 44217
Foyer Rural, Jupilles 13295
Foynes Flying Boat Museum, Foynes 24834
FRAC du Centre, Orléans 14531
FRAC Limousin, Limoges 13763
Fränkische Galerie, Kronach 19720
Fränkische Hopfenscheune mit heimatkundlicher Sammlung, Neunkirchen am Sand 20640
Fränkische Schweiz Museum, Pottenstein 21113
Fränkisches Bauern- und Handwerkermuseum, Iphofen 19350
Fränkisches Brauereimuseum, Bamberg 16908
Fränkisches Freilandmuseum, Bad Windsheim 16872
Fränkisches Freilandmuseum Fladungen mit dem Rhön-Zügle, Fladungen 18324
Fränkisches Museum, Feuchtwangen 18313
Fränkisches Spielzeugmuseum, Gößweinstein 18687
Fränkisches Spitalmuseum Aub, Aub 16604
Fränkisches Turmuhrenmuseum, Mistelbach 20340
Fraeylemaborg, Slochteren 32784
Fragrance Museum Farina House, Köln 19600
Frame Gallery, Pittsburgh 51948
Framgi Dadabhoy Alpaiwalle Museum, Mumbai 24315
Framingham Historical Museum, Framingham 48730
Frammuseet, Oslo 33725
Franča Trasuna Muzejs Kolnasāta, Sakstagals 30120
Frances Burke Textile Resource Centre-Collections, Melbourne 01278
The Frances Lehman Loeb Art Center, Poughkeepsie 52121
Frances Willard House Museum, Evanston 48450
Francis Colburn Gallery, Burlington 47143
Francis Land House, Virginia Beach 53756
Francis Ledwidge Cottage and Museum, Slane 24898
Francis McCray Gallery, Silver City 53101
Francis Trasuns Museum Kolnasāta, Sakstagals 30120
Franciscan Monastery, Ystad 41077
Franciscan Monastery Museum Gorica, Livno 04267
Franciscan Monastery Museum Jajce, Jajce 04264
Franciscan Monastery Museum Kraljeva Sutjeska, Kraljeva Sutjeska 04265
Franciscan Monastery Museum Krešovo, Krešovo 04266
Francisco Fort Museum, La Veta 49856
Francke Foundations, Halle, Saale 18891
Die Franckeschen Stiftungen zu Halle, Halle, Saale 18891
François Duesberg Decorative Arts Museum, Mons 03888
Franjevački Samostan Gorica, Livno 04267
Franjevački Samostan Jajce, Jajce 04264
Franjevački Samostan Kraljeva Sutjeska, Kraljeva Sutjeska 04265
Franjevački Samostan Krešovo, Krešovo 04266

Frank Cameron Museum of Wynyard, Wynyard 07407
Frank H. McClung Museum, Knoxville 49818
Frank Lloyd Wright Home and Studio, Oak Park 51453
Frank Lloyd Wright Museum, Richland Center 52308
Frank Lloyd Wright's Pope-Leighey House, Alexandria 46204
Frank-Loebsches Haus, Landau in der Pfalz 19789
Frank Partridge V.C. Military Museum, Bowraville 00885
Frank Phillips Home, Bartlesville 46618
Frank Slide Interpretive Centre, Crowsnest Pass 05661
Frankenberger Art Gallery, Charleston 47379
Frankenburger Heimatstube, Frankenburg 01896
Frankenmuth Historical Museum, Frankenmuth 48732
Frankenwaldmuseum, Kronach 19721
Frankfort Area Historical Museum, West Frankfort 54055
Frankfurter Äpfelwein-Museum, Frankfurt am Main 18368
Frankfurter Feldbahnmuseum, Frankfurt am Main 18369
Frankfurter Goethe-Haus, Frankfurt am Main 18370
Frankfurter Sportmuseum, Frankfurt am Main 18371
Frankie G. Weems Gallery & Rotunda Gallery, Raleigh 52229
Franklin Arts and Cultural Centre, Pukekohe 33190
Franklin County Historical Museum, Pasco 51726
Franklin County Museum, Brookville 47090
Franklin County Old Jail Museum, Winchester 54221
Franklin D. Roosevelt Presidential Library-Museum, Hyde Park 49458
Franklin Furnace Archive, Brooklyn 47073
Franklin G. Burroughs-Simeon B. Chapin Art Museum, Myrtle Beach 50933
Franklin Historical Society Museum, Franklin 48747
Franklin House, Launceston 01221
The Franklin Institute, Philadelphia 51836
Franklin Mineral Museum, Franklin 48748
Franklin Mint Museum, Franklin Center 48752
Franklin Pierce Homestead, Hillsborough 49292
Franks Antique Doll Museum, Marshall 50508
Frankston Primary Old School Museum, Frankston 01090
Frans Hals Museum, Haarlem 32295
Frans Walkate Archief/ SNS Historisch Archief, Kampen 32435
Fransu Cottage, Himanka 10839
Fransu-tupa, Himanka 10839
Franz Binder Museum of Universal Ethnography, Sibiu 36607
Franz Daurach-Sammlung, Hadres 02033
Franz Kafka museum, Praha 09745
Franz-Liszt-Museum der Stadt Bayreuth, Bayreuth 16951
Franz Liszt Zentrum, Raiding 02553
Franz Marc Museum, Kochel am See 19592
Franz Michael Felder-Stube, Schoppernau 02712
Franz Radziwill Haus, Varel 22203
Franz Schmidt Gedenkraum, Perchtoldsdorf 02486
Franz Schubert Memorial Site, Želiezovce 38371
Franz-Stock-Museum, Arnsberg 16571
Franz Traunfellner-Dokumentation, Pöggstall 02507
Franz Xaver Gruber-Gedächtnishaus, Ach 01690
Franz Xaver Gruber-Museum, Lamprechtshausen 02253
Franzensburg, Laxenburg 02272
Franzensburg, Laxenburg 02272
Franziskanermuseum, Villingen-Schwenningen 22235
Franztaler Heimatstube, Mondsee 02404
Frasassi-Le Grotte, Genga 26286
Fraser-Fort George Regional Museum, Prince George 06666
Fraserburgh Museum, Fraserburgh 44253
Frau Holle Expreß, Hessisch Lichtenau 19170
Frauen Museum, Bonn 17454
Frauen Museum Wiesbaden, Wiesbaden 22548
Frauenkultur, Leipzig 19903
Frauenmuseum, Hittisau 02084
Fraunces Tavern Museum, New York 51177
Frazer's Museum, Beauval 05407
Fred Dana Marsh Museum, Ormond Beach 51599
Fred Harman Art Museum, Pagosa Springs 51674
Fred J. Johnston House Museum, Kingston 49789
Fred Jones Jr. Museum of Art, Norman 51374
Fred Light Museum, Battleford 05400
Fred Wolf Jr. Gallery, Philadelphia 51837
Frederic Remington Art Museum, Ogdensburg 51506
Frederick C. Robie House, Chicago 47475
Frederick Douglass National Historic Site, Washington 53872
Frederick Law Olmsted Collection, Brookline 47062
Frederick R. Weisman Art Foundation, Los Angeles 50224
Frederick R. Weisman Art Museum, Minneapolis 50731
Frederick R. Weisman Museum of Art, Malibu 50407
Fredericksburg and Spotsylvania National Military Park, Fredericksburg 48764
Fredericksburg Area Museum and Cultural Center, Fredericksburg 48765
Frederikshavn Kunstmuseum- og Exlibrissamling, Frederikshavn 10026
Frederiksværk Bymuseum, Frederiksværk 10029
Fredriksdal Museer och Trädgårdar, Helsingborg 40586
Fredrikstad Museum, Fredrikstad 33505
Fredy's Mechanisches Musikmuseum, Lichtensteig 41558
Free Public Library Special Collections, New Bedford 51012
Free State Voortrekker Museum, Winburg 38778
Freeborn County Museum, Albert Lea 46170
Freedman Gallery, Reading 52262
Freeport Arts Center, Freeport 48781
Freeport Historical Society Museum, Freeport 48784
Freer Gallery of Art, Washington 53873
Freestone County Historical Museum, Fairfield 48485
Fregatten Jylland, Ebeltoft 09996
Freiämter Strohmuseum, Wohlen 42044
Freiburger Photohistorische Sammlung, Schallstadt 21532
Freies Museum Berlin, Berlin 17085
Freiland-Grenzmuseum Sorge, Sorge 21825
Freilandmuseum Ammerländer Bauernhaus, Bad Zwischenahn 16876
Freilandmuseum Grassemann Naturpark-Infostelle, Warmensteinach 22338
Freilandmuseum Lehde, Lübbenau 20070
Freilicht- und Heimatmuseum Donaumoos, Karlshuld 19434
Freilichtanlage Gschlößl, Leithaprodersdorf 02281
Freilichtmuseum, Bad Tatzmannsdorf 01766
Freilichtmuseum, Diesdorf 17850
Freilichtmuseum, Herbertingen 19131
Freilichtmuseum, Wolfach 22646
Freilichtmuseum Alt Schwerin, Alt Schwerin 16483
Freilichtmuseum am Kiekeberg, Rosengarten, Kreis Harburg 21373
Freilichtmuseum am Rätischen Limes, Rainau 21170
Freilichtmuseum Apriacher Stockmühlen, Heiligenblut 02065
Freilichtmuseum Beuren, Beuren 17312
Freilichtmuseum des Gebirgskrieges, Kötschach-Mauthen 02214
Freilichtmuseum Domäne Dahlem, Berlin 17086
Freilichtmuseum Ensemble Gerersdorf, Gerersdorf 01930
Freilichtmuseum Erdöl- und Erdgaslehrpfad, Prottes 02534
Freilichtmuseum Finsterau, Mauth 20240
Freilichtmuseum "Frelsdorfer Brink", Frelsdorf 18454
Freilichtmuseum Fürstenhammer, Lasberg 02265
Freilichtmuseum Glentleiten, Großweil 18809
Freilichtmuseum Groß Raden, Sternberg 21900
Freilichtmuseum Hackenschmiede, Bad Wimsbach-Neydharting 01771
Freilichtmuseum Handwerkerhaus Stegwagner, Windhaag bei Freistadt 03109
Freilichtmuseum Hayrl, Reichenthal 02571
Freilichtmuseum Hessenpark, Neu-Anspach 20581
Freilichtmuseum Kalte Kuchl, Rohr im Gebirge 02592
Freilichtmuseum Katzensteiner Mühle, Weyer 02927
Freilichtmuseum Klausenhof, Herrischried 19161
Freilichtmuseum Klockenhagen, Ribnitz-Damgarten 21310
Freilichtmuseum Ledermühle, Sankt Oswald bei Freistadt 02672
Freilichtmuseum Massing, Massing 20236
Freilichtmuseum "Mittelalterliche Bergstadt Bleiberg", Frankenberg, Sachsen 18346
Freilichtmuseum Mondseer Rauchhaus, Mondsee 02405
Freilichtmuseum Neuhausen ob Eck, Neuhausen ob Eck 20621
Freilichtmuseum Oberlienz, Lienz 02300
Freilichtmuseum Ostenfelder Bauernhaus, Husum 19298
Freilichtmuseum Pelmberg, Hellmonsödt 02069
Freilichtmuseum Petronell, Petronell 02495
Freilichtmuseum Rätersiedlung Himmelreich, Volders 02871
Freilichtmuseum Rhöner Museumsdorf, Tann 22004
Freilichtmuseum Römerbad, Jagsthausen 19372
Freilichtmuseum Römersteinbruch, Sankt Margarethen 02664
Freilichtmuseum Römervilla in Brederis, Rankweil 02558
Freilichtmuseum Säge Buch, Buch 41265
Freilichtmuseum Scherzenmühle Weidenberg, Weidenberg 22368
Freilichtmuseum Schwebsingen, Bech-Kleinmacher 30453
Freilichtmuseum spätbronzezeitlicher Hügelgräber, Siegendorf 02740
Freilichtmuseum Stade, Stade 21848
Freilichtmuseum Stehrerhof - Dreschmaschinen-museum, Neukirchen an der Vöckla 02432
Freilichtmuseum Sumerauerhof, Sankt Florian 02637
Freilichtmuseum Tiroler Bauernhöfe, Kramsach 02221
Freilichtmuseum Venetianersäge, Windhaag bei Freistadt 03110
Freilichtmuseum Vorau, Vorau 02873
Freimaurermuseum der Großen Landesloge der Freimaurer von Deutschland, Sankt Michaelisdonn 21506
Frejamuseét, Kil 40642
Fremantle Arts Centre, Fremantle 01092
Fremantle History Museum, Fremantle 01093
Fremantle Prison Precinct, Fremantle 01094
Fremont County Pioneer Museum, Lander 49933
French Algeria Museum, Perpignan 14789
French Art Colony, Gallipolis 48839
French Azilum, Towanda 53547
French Cable Station Museum, Orleans 51597
French Foreign Legion Museum, Aubagne 11719
French Foreign Legion Uniforms Museum, Puyloubier 14961
The French Legation Museum, Austin 46509
French Natural History National Museum, Paris 14608
Frenchman Butte Museum, Frenchman Butte 05843
Freshford House Museum, Bloemfontein 38562
Fresno Art Museum, Fresno 48790
Fresno Flats Museum, Oakhurst 51461
Fresno Metropolitan Museum, Fresno 48791
Freud Museum, London 44852
Fri-Art, Fribourg 41386
Friary Art Gallery, Lichfield 44703
The Frick Art Museum-Henry Clay Frick Estate, Pittsburgh 51949
Frick Collection, New York 51178
Fricktaler Museum, Rheinfelden 41757
Frico Kafenda Memorial Site, Mošovce 38310
Fričovo Muzeum, Lázně Bělohrad 09619
Frida & Fred - Grazer Kindermuseum, Graz 01968
Friderik Baraga Birth House, Dobrnič 38411
Fridge Gallery, Petone 33181
Fridrich János Fényírdája, Szentes 23883
Frīdriha Candera Kosmosa Muzejs, Rīga 30067
Frīdriha Candera Memoriālā Māja, Rīga 30068
friedens räume, mehr als ein museum, Lindau 19999
Friedenshistorisches Museum, Bad Hindelang 16748
Friedensmuseum, Meeder 20244
Friedensmuseum, Remagen 21260
Friedenszimmer im Schloß Altranstädt, Großlehna 18797
Friedrich-der-Große-Museum im Museumshof Lerchenest, Sinsheim 21784
Friedrich-Eckenfelder-Galerie, Balingen 16899
Friedrich-Fröbel-Museum, Bad Blankenburg 16698
Friedrich-Gerstäcker-Museum, Braunschweig 17520
Friedrich-Gottlob-Keller Museum Krippen, Bad Schandau 16825
Friedrich-Ludwig-Jahn-Gedenkstätte, Lanz 19840
Friedrich-Ludwig-Jahn-Museum, Freyburg (Unstrut) 18458
Friedrich-Rückert-Gedächtnisstätte, Coburg 17715
Friedrich-Wilhelm-Weber-Museum, Bad Driburg 16711
Friedrich Zander Space Museum of the University of Latvia, Rīga 30067
Friends of Firemen's Museum, New Bern 51019
Friendship Firehouse, Alexandria 46205
Friendship Hill, Point Marion 52009
Friendship Museum, Darhan 31616
Fries Landbouw Museum, Exmorra 32228
Fries Museum, Leeuwarden 32473
Fries Natuurmuseum, Leeuwarden 32474
Fries Scheepvaart Museum, Sneek 32789
Friese-Galerie, Lüchow 20087
Friesenstube Honkenswarf, Langeneß 19830
Friesisches Heimatmuseum, Niebüll 20697
Frietkotmuseum, Antwerpen 03288
Frigate Suomen Joutsen, Turku 11399
Frihedsmuseet, København 10131
Frilandsmuseet, Kongens Lyngby 10166
Frilandsmuseet i Maribo, Maribo 10184
Friluftmuseet Hallandsgården, Halmstad 40577
Friluftmuseet Skansen, Stockholm 40895
Friluftsmuseet, Jönköping 40613
Friluftsmuseet Gammelgården, Köping 40646
Friluftsmuseet i Apladalen, Värnamo 41028
Frisco Historical Society Museum, Frisco 48797
Frisco Native American Museum and Natural History Center, Frisco 48798
Frist Center for the Visual Arts, Nashville 50970
Fritz-Best-Museum, Kronberg im Taunus 19724
Fritz-Reuter-Literaturmuseum, Stavenhagen 21870
Fritz-von-Wille-Museum im Haus Beda, Bitburg 17367
Fritz-Winter-Haus, Ahlen 16436
Fritzøe Museum, Larvik 33627
Frobisher Thresherman's Museum, Frobisher 05844
Frösö Hembygdsmuseum, Frösön 40528
Frövifors Pappersbruksmuseum, Frövi 40529
Frog Hollow Vermont State Craft Center, Middlebury 50651
Frog Museum, Den Haag 32077
Frog Museum, Estavayer-le-Lac 41362
Frogmore House, Windsor 46041
From Holocaust to Revival Museum, Chof Ashkelon 24940
Frome Museum, Frome 44257
Front Street Museum, Ogallala 51498
Frontera Audubon Society, Weslaco 54039
Frontier Army Museum, Fort Leavenworth 48642
Frontier Culture Museum, Staunton 53275
Frontier Gateway Museum, Glendive 48931
Frontier Museum, Williston 54182
Frontier Times Museum, Bandera 46593
Frontier Troops Museum, Vilnius 30429
Frontier Village Museum, Jamestown 49593
Frontiers of Flight Museum, Dallas 47933
Frontmuseet - Rintamamuseo, Karis 10917
Frontmuseum, Karis 10917
Froschmuseum, Münchenstein 41657
Frøslevlejrens Museum, Padborg 10227
The Frost Entomological Museum, University Park 53657
Frosta Bygdemuseet, Frosta 33507
Frøya Bygdemuseum, Sistranda 33827
Fruchtkasten, Tuttlingen 22134
Frühchristliche Kirchenanlage, Feistritz an der Drau 01877
Frühmittelaltermuseum Carantana, Rothenthurn 02600
Fruitlands Museums, Harvard 49207
Fruitmarket Gallery, Edinburgh 44124
Fruitteeltmuseum, Kapelle 32441
Frunse House-Museum, Samara 37613
Frunze House Museum, Biškek 29952
Fry Art Gallery, Saffron Walden 45539
Fry Model Railway, Demesne 24773
Fry Model Railway Museum, Malahide 24873
Fryderyk Chopin Birth House, Sochaczew 35117
Fryderyk Chopin Museum, Warszawa 35242
Frye Art Museum, Seattle 52999
The Fryeburg Fair Farm Museum, Fryeburg 48804
Fryer Memorial Museum, Munnsville 50908
Frye's Measure Mill, Wilton 54215
Fryxell Geology Museum, Rock Island 52408
Fu Hakubutsukan, Hijikawa 28508
Fuchu City Field Museum, Fuchu 28418
Fuchu-shi Kyodo-no-mori Ajisai Matsuri, Fuchu 28418
Führichův Dům, Chrastava 09481
Fünf-Giebel-Haus, Friedrichstadt 18480
Fürdő Múzeum, Budapest 23405
Fürst Thurn und Taxis Zentralarchiv und Hofbibliothek, Regensburg 21221
Fürstenberger Hof, Zell am Harmersbach 22778
Fürstengruft mit russisch-orthodoxer Kirche, Weimar 22398
Fürstlich Esterházy'sche Sammlungen Burg Forchtenstein, Forchtenstein 01894
Fürstlich Fürstenbergische Sammlungen, Donaueschingen 17892
Fürstlich Hohenzollernsche Sammlungen, Sigmaringen 21763
Fürstlich Leiningensche Sammlungen - Heimatmuseum Amorbach, Amorbach 16541
Fürstlich Ysenburg- und Büdingensches Schloßmuseum, Büdingen 17627
Fürstliches Residenzschloß, Detmold 17837
Fürstliches Schloß Mausoleum, Bückeburg 17621
Fürstliches Schloß und Kreuzgang Sankt Emmeram, Regensburg 21222
Fuerte de la Punta del Sauce, La Carlota 00398
Fufeng Museum, Fufeng 07695
Fugger-Museum, Babenhausen, Kreis Unterallgäu 16651
Fuggerei-Museum, Augsburg 16617
Fugitive Art Center, Nashville 50971
Fuglsøcentret, Knebel 10120
Fugou Museum, Fugou 07698
Fuhlrott-Museum, Wuppertal 22740
Fuhrwerkswaage Kunstraum, Köln 19602
Fujairah Museum, Fujairah 43221
Fuji Art Museum, Fujinomiya 28423
Fuji Visitor Center, Kawaguchiko 28687
Fujian Museum, Fuzhou 07706
Fujii Bijutsu Mingeikan, Takayama 29335
Fujii Folkart Craft Gallery, Takayama 29335
Fujii Saisei-kai Yurinkan, Kyoto 28808
Fujimi Historical Museum, Fujimi 28421
Fujimi-machi Rekishi Hakubutsukan, Fujimi 28421
Fujita Art Museum, Fukui 28426
Fujita Bijutsukan, Fukui 28426

Fujita Bijutsukan, Osaka 29126
Fujita Museum of Art, Osaka 29126
Fujiyama Museum, Fuji-Yoshida 28420
Fujiyama Museum of Historical Data, Oonishi 29122
Fukagawa Edo Museum of Television, Tokyo 29420
Fukahori Kaizukai Iseki Museum, Nagasaki 28973
Fukuchiyama City Clothes Museum, Fukuchiyama 28425
Fukuchiyama Shato-Mingeikan, Fukuchiyama 28424
Fukuchiyama-shi Tanba Seikatsui-kan, Fukuchiyama 28425
Fukui City Art Museum, Fukui 28430
Fukui City Museum of Natural History, Fukui 28427
Fukui-ken Togeikan, Echizen 28413
Fukui-kenritsu Bijutsukan, Fukui 28428
Fukui-kenritsu Bunka Hakubutsukan, Fukui 28429
Fukui-kenritsu Kyoryu Hakabutsukan, Katsuyama 28681
Fukui Prefectural Ceramics Museum, Echizen 28413
Fukui Prefectural Dinosaur Museum, Katsuyama 28681
Fukui Prefectural Fine Arts Museum, Fukui 28428
Fukui Prefectural Museum of Cultural History, Fukui 28429
Fukui-shiritsu Bijutsukan, Fukui 28430
Fukuoka Ajian Bijutsukan, Fukuoka 28431
Fukuoka Art Museum, Fukuoka 28433
Fukuoka Camera Museum, Fukuoka, Toyoma 28438
Fukuoka City Museum, Fukuoka 28434
Fukuoka-kenritsu Bijutsukan, Fukuoka 28432
Fukuoka-machi Shashinki Hakubutsukan, Fukuoka, Toyoma 28438
Fukuoka Prefectural Museum of Art, Fukuoka 28432
Fukuoka-shi Bijutsukan, Fukuoka 28433
Fukuoka-shi Hakubutsukan, Fukuoka 28434
Fukushima Kaiyo Kagakukan, Iwaki 28605
Fukushima-ken Bunka Centre, Fukushima 28439
Fukushima-kenritsu Bijutsukan, Fukushima 28440
Fukushima-kenritsu Hakubutsukan, Aizuwakamatsu 28350
Fukushima Museum, Aizuwakamatsu 28350
Fukushima Prefectural Museum of Art, Fukushima 28440
Fukushima Prefecture Cultural Center, Fukushima 28439
Fukuyama Auto and Clock Museum, Fukuyama 28441
Fukuyama Bijutsukan, Fukuyama 28442
Fukuyama Bungakukan, Fukuyama 28443
Fukuyama Castle Museum, Fukuyama 28444
Fukuyama City Human Rights and Peace Museum, Fukuyama 28445
Fukuyama Museum of Art, Fukuyama 28442
Fukuyama Museum of Literature, Fukuyama 28443
Fukuyama Prefectural Museum of History, Fukuyama 28446
Fukuyama-renritsu Rekishi Hakubutsukan, Fukuyama 28446
Fukuzawa Ichiro Memorial Museum, Nakatsu 29022
Fuling Museum, Fuling 07699
Fuller Craft Museum, Brockton 47036
Fullerton Museum Center, Fullerton 48805
Fullyjie Zone Museum, Bijie 07580
Fulton County Historical Museum, Wauseon 53995
Fulton County Historical Society Museum, Rochester 52384
Fulton County Museum, Groversville 49100
Fulton Mansion, Rockport 52426
Fultz House Museum, Lower Sackville 06200
Fundação Arpad Szenes e Vieira da Silva, Lisboa 35666
Fundação Casa de Jorge Amado, Salvador 04787
Fundação Clóuis Salgado - Palácio das Artes, Belo Horizonte 04342
Fundação Cultural Ema Gordon Klabin, São Paulo 04857
Fundação Eça de Queiroz, Santa Cruz do Douro 35900
Fundação Eva Klabin Rappaport, Rio de Janeiro 04693
Fundação Hansen Bahia, Cachoeira 04375
Fundação Maria Luisa e Oscar Americano, São Paulo 04858
Fundação Memorial da América Latina, São Paulo 04859
Fundação Museu do Zebu Edilson Lamartine Mendes, Uberaba 04940
Fundaçao Oscar Niemeyer, Rio de Janeiro 04694
Fundação Ricardo do Espírito Santo Silva, Lisboa 35667
Fundació Antoni Tàpies, Barcelona 38956
Fundació Barceló, Palma de Mallorca 39782
Fundació Casa Museu Llorenç Villalonga, Binissalem 39049
Fundacio Espais, Girona 39341
Fundacio Fita, Girona 39342
Fundació Joan Miró, Barcelona 38957
Fundació Pilar i Joan Miró, Palma de Mallorca 39783
Fundació Pública Institut d'Estudis Ilerdencs, Lleida 39473
Fundación Alfredo L. Palacios, Buenos Aires 00158
Fundación Archivo Gráfico y Museo Histórico de la Ciudad de San Francisco y la Región, San Francisco 00576
Fundación Banco Francés, Buenos Aires 00159
Fundación Carlos de Amberes, Madrid 39519
Fundación César Manrique, Teguise 40108
Fundación Cultural Otero Pedrayo, Amoeiro 38869
Fundación de Fomento a las Artes, La Paz 04211
Fundación Federico Jorge Klemm, Buenos Aires 00160
Fundación Folch, Barcelona 38958
Fundación Juan March, Madrid 39520
Fundación Luis Muñoz Marín, San Juan 36063
Fundación Luis Seoane, A Coruña 39216
Fundación Museo de las Ferias, Medina del Campo 39638
Fundación NMAC Montenmedio de Arte Contemporáneo, Vejer de la Frontera 40255
Fundación Pablo o Ruiz Picasso Museo Casa Natal, Málaga 39596
Fundacion Privada Espai Guinovart, Agramunt 38785
Fundación Vela Zanetti, León 39452
Fundación Yannick y Ben Jakober, Alcúdia (Baleares) 38831
Fundaziun Planta Samedan Wohnmuseum, Samedan 41800
Fundy Geological Museum, Parrsboro 06569
Fune no Kagakukan, Tokyo 29386
Funen Village, Odense 10214
Funeral Collection, Budapest 23419
Funnefoss Industriarbeidermuseum, Årnes 33399
Fur Museum, Fur 10032
Fureai Minatokan, Osaka 29127
Fureaikan - Kyoto Museum of Traditional Crafts, Kyoto 28809
Furesø Museer, Farum 10019
The Furlong Gallery, Menomonie 50591
Furnas-Gosper Historical Society and Museum, Arapahoe 46334
Furneaux Museum, Flinders Island 01082
Furness Abbey, Barrow-in-Furness 43389
Furniture Industry Museum Virserum, Virserum 41066
Furniture Manufacturing Museum in Tainan, Rende 42306
Furniture Museum, Markušovce 38292
Furniture Museum, Tokyo 29405
Furniture Museum in Moscow, Moskva 37331
Fursdon House, Cadbury 43706
Furthmühle, Egenhofen 18060
Furukawa Art Museum, Nagoya 28999
Furukawa Bijutsukan, Nagoya 28999
Fushun Museum, Fushun 07700
Fusiliers London Volunteers' Museum, London 44853
Fusiliers' Museum Lancashire, Bury, Lancashire 43693
Fusiliers Museum of Northumberland, Alnwick 43286
Fusions Gallery, Fortitude Valley 01086
Futagawa Syuku Fortress Museum, Toyohashi 29527
Fuxin Chahai Ruins Museum, Fuxin 07701
Fuxin Museum, Fuxin 07702
Fuyang Museum, Fuyang, Anhui 07703
Fuyang Museum, Fuyang, Zhejiang 07704
Fuyu Museum, Fuyu 07705
Fuzhou Cangshan Museum, Fuzhou 07707
Fuzhou City Hualin Si Museum, Fuzhou 07708
Fuzoku Hakubutsukan, Kyoto 28810
Fyns Kunstmuseum, Odense 10213
Den Fynske Landsby, Odense 10214
Fyresdal Bygdemuseet, Fyresdal 33508
Fyresdal Open Air Museum, Fyresdal 33508
Fyvie Castle, Fyvie 44258

G

G.I. Katsigras Museum, Lárissa 23074
G.I. Uspensky House-Museum, Sjabrenicy 37767
G.K. Zhukov Museum, Ulaanbaatar 31619
G. Leonidze State Literature Museum of Georgia, Tbilisi 16358
G.N. Jha Kendriya Sanskrit Vidyapeetha, Allahabad 24060
G.N. Prosritelev and G.K. Prave Regional Museum, Stavropol 37805
G.S. Novikov-Daurskiy Regional Museum of the Amur Region, Blagoveščensk 36809
G. Tabidze House Museum, Tbilisi 16359
G. Tsybikov Regional Museum of Aginskoe, Aginskoe 36748
G.V. Tchitcherin House-Museum, Tambov 37831
GAA Museum, Dublin 24790
Gaál Imre Galéria, Budapest 23406
Gabdulla Tukaj Museum, Kazan 37015
Gabinet de les Arts Gràfiques, Barcelona 38959
Gabinet Numizmatyczny Mennicy Państwowej, Warszawa 35224
Gabinet Pisarza Pawła Hulki-Lasowskiego, Żyradów 35380
Gabinet Prof. Kazimierza Sosnowskiego, Kraków 34772
Gabinet Rycin Biblioteki Uniwersyteckiej w Warszawie, Warszawa 35225
Gabinet Rycin Polskiej Akademii Umiejętności w Krakowie, Kraków 34773
Gabinete de Antigüedades, Madrid 39521
Gabinete Numismático de Cataluña, Barcelona 38960
Gabinetto Comunale delle Stampe, Roma 27446
Gabinetto dei Disegni e delle Stampe, Bologna 25442
Gabinetto Disegni e Stampe, Pisa 27239
Gabinetto Disegni e Stampe degli Uffizi, Firenze 26153
Gabinetto Fotografico, Firenze 26154
Gabinetto Geologico e Botanico Piccinini, Pergola 27138
Gabinetto Nazionale delle Stampe, Roma 27447
Gablonzer Archiv Museum, Kaufbeuren 19486
Gablonzer Schmuckmuseum, Enns 01869
Gabriele Petkevicaite-Bite Literature Museum, Smilgiai 30373
Gabriele Petkevicaite-Bite's Apartment and Museum, Panevėžys 30331
Gabrielės Petkevičaitės-Bitės Literatūrinis Muziejus, Smilgiai 30373
Gabrielės Petkevičaitės Memorialinė Ekspozicija, Panevėžys 30331
Gabriola Museum, Gabriola 05845
Gacheon Museum, Incheon 29832
Gadsby's Tavern Museum, Alexandria 46206
Gadsden Arts Center, Quincy 52212
Gadsden Museum, Mesilla 50610
Gadsden Museum of Fine Arts, Gadsden 48811
Gällersta Forngård, Örebro 40785
Gärtner- und Häckermuseum, Bamberg 16909
Gærup Skolemuseum og Brahetrolleborg Sogns Folkemindesamling, Faaborg 10009
Gäubodenmuseum, Straubing 21918
Gävle Konstcentrum, Gävle 40531
Gage County Historical Museum, Beatrice 46663
Gahlberg Gallery, Glen Ellyn 48920
Gaidu un Skautu Muzejs, Ogre 30051
Gailtaler Heimatmuseum-Sammlung Georg Essl, Hermagor 02071
Gaimusho, Gaiko Shiryokan, Tokyo 29387
Gaineswood Mansion, Demopolis 48055
Gainsborough Old Hall, Gainsborough 44259
Gainsborough's House, Sudbury, Suffolk 45797
Gairloch Heritage Museum, Gairloch 44260
Gakki Hakubutsukan, Tokyo 29388
Gakkigaku Shiryôkan, Tachikawa 29307
Gakumon Kagaku Hakubutsukan, Tokyo 29389
Galapagos Art Space, Brooklyn 47074
Galata - Museo del Mare e della Navigazione, Genova 26293
Galaune Family Museum of the M.K. Čiurlionis National Art Museum, Kaunas 30238
Galchon Mask Museum, Goseong 29818
Galena-Jo Daviess County Historical Museum, Galena 48827
Galena Mining and Historical Museum, Galena 48831
Galera Kombëtare e Arteve Figurative, Tiranë 00030
Galereja Forum, Sankt-Peterburg 37638
Galereja Gerceva, Moskva 37173
Galereja Iskusstv Zuraba Cereteli, Moskva 37174
Galereja-klub v Saburovo, Moskva 37175
Galereja-masterskaja Chudožnika G.S. Rajševa, Chanty-Mansijsk 36861
Galereja Meždunarodnogo Universiteta, Moskva 37176
Galereja na Bastionnoj, Pskov 37573
Galereja Odoevskogo, Ekaterinburg 36886
Galereja Sacha-Art, Moskva 37177
Galereja Sem Bruk, Moskva 37178
Galereja Sojuza Tvorčestvo, Moskva 37179
Galereja Sovremennogo Iskusstva Arka, Vladivostok 37935
Galereja Sovremennogo Iskusstva Artėtaž, Vladivostok 37936
Galereja Zarubežnogo Iskusstva im. Professora M.F. Gabyševa, Jakutsk 36961
Galeri Suav ve Küsav, İstanbul 42819
Galeri Valfisken, Simrishamn 40838
Galeria, Lublin 34897
Galeria, Nowy Sącz 34956
The Galeria, Orchard Lake 51583
Galeria 2piR, Poznań 35022
Galeria 3-2-0, Bielsko-Biała 34559
Galeria 5, Kalisz 34708
Galeria 5, Katowice 34720
Galeria Akademii Sztuk Pięknych im. Władysława Strzemińskiego, Łódź 34861
Galeria Amfilada, Szczecin 35156
Galeria Architektury SARP, Katowice 34721
Galeria Architektury SARP, Kraków 34774
Galeria Arsenal, Białystok 34545
Galeria Art-Foto, Częstochowa 34613
Galeria Azjatycka, Warszawa 35226
Galeria Bałucka, Łódź 34862
Galeria Biała, Lublin 34898
Galeria Białostockiego Towarzystwa Przyjaciół Pięknych, Białystok 34546
Galeria Biblioteka, Łódź 34863
Galería Bienes Culturales, Victoria de las Tunas 09344
Galería Cultural de la Escuela Nacional de Bellas Artes, Lima 34213
Galeria d'Art Cal Talaveró, Verdú 40266
Galeria d'Art Mar, Barcelona 38961
Galeria d'Art Terra Ferma, Lleida 39474
Galeria de Artă, Pitești 36525
Galeria de Artă Ioan Sima, Zalău 36741
Galeria de Artă Lascăr Viorel, Piatra-Neamț 36517
La Galería de Arte, La Habana 09104
Galería de Arte de la Escuela Armando Reverón, Barcelona 54517
Galería de Arte de la UCV, Caracas 54531
Galería de Arte de Tabasco, Villahermosa 31547
Galería de Arte del Aeropuerto, México 31039
Galeria de Arte Dos V's, Lima 34214
Galería de Arte Iztapalapa, México 31040
Galería de Arte Jaguar Despertado, Villahermosa 31548
Galería de Arte Nacional, Caracas 54532
Galeria de Arte UFF, Rio de Janeiro 04695
Galeria de Arte UNICAMP, Campinas 04385
Galería de Arte UNIFE, Lima 34215
Galería de Arte Universal, Santiago de Cuba 09303
Galería de Arte Valerio Calles, Oruro 04229
Galería de Artes Visuales, Lima 34216
Galería de Catalanes Ilustres, Barcelona 38962
Galería de Cinemateca, Montevideo 54391
Galería de Escuela Nacional de Artes Plásticas, México 31041
Galeria de Exposições Temporárias da CMMP, Macau 07992
Galería de Historia o Museo del Caracol, México 31042
Galería de la Asamblea Legislativa, Barcelona 54518
Galería de la Escuela de Artes Plásticas, San José 08811
Galería de la Municipalidad de Miraflores, Lima 34217
Galería de la Raza, San Francisco 52774
Galería del Banco Continental, Lima 34218
Galería del Instituto Cultural Peruano-Juan Pardo Heeren, Lima 34219
Galería del Instituto Cultural Peruano-Norteamericano, Lima 34220
Galería del Museo Anzoátegui, Barcelona 54519
Galería del Sur, México 31043
Galeria Domek Ogrodnika, Łódź 34864
Galeria Domu Kultury Włókniarzy, Bielsko-Biała 34560
Galería Dr. Miguel Angel Gómez Ventura, Villahermosa 31549
Galeria Engram, Katowice 34722
Galería Episcopal, Zacatecas 31576
Galeria Fabryka Kestenberga, Łódź 34865
Galeria Foksal, Warszawa 35227
Galeria Forum Fotografii, Łódź 34866
Galeria Fotografii Instytutu Sztuki Akademii Pedagogicznej, Kraków 34775
Galeria i Ośrodek Plastycznej Twórczości Dziecka, Toruń 35189
Galería IFAL, México 31044
Galeria Instituto d'Estudis Fotografics da Catalunya, Barcelona 38963
Galéria Jána Koniarka v Trnave - Kopplov Kaštieľ, Trnava 38379
Galeria Jatki, Nowy Targ 34962
Galería José María Velasco, México 31045
Galéria Jozefa Kollára, Banská Štiavnica 38190
Galéria KOMART, Bratislava 38210
Galeria Krytyków Pokaz, Warszawa 35228
Galeria la Capella de l'Antic, Barcelona 38964
Galería Las Pintadas, Estelí 33276
Galería Los Oficios, La Habana 09105
Galéria M.A. Bazovského, Trenčín 38375
Galeria Malarstwa Akademii Sztuk Pięknych im. Jana Matejki, Kraków 34776
Galéria Márie Medveckej, Tvrdošín 38383
Galeria Marii Ritter, Nowy Sącz 34957
Galéria Mesta Bratislavy, Bratislava 38211
Galería Mexicana de Diseño, México 31046
Galeria Międzynarodowego Centrum Kultury, Kraków 34777
Galeria Miejsko, Stary Sącz 35131
Galería Monumento Histórico Nacional a la Bandera, Rosario 00541
Galeria Municipal de Arte, Almada 35413
Galería Municipal de Arte Moderno, Barcelona 54520
Galeria Municipal Trem, Faro 35594
Galeria Muz, Toruń 35190
Galería Nacional, San Juan 36064

Galería Nacional de Arte, Tegucigalpa 23331
Galería Nacional de Arte Contemporáneo, San José 08812
Galería Nacional y Museo de los Niños, San José 08813
Galeria Nowa, Łódź 34867
Galeria Nowa Przestrzeń, Łódź 34868
Galeria Nusantara, Warszawa 35229
Galeria Oamenilor de Seamă, Fălticeni 36365
Galería Oriente, Santiago de Cuba 09304
Galeria Osiedlowego Domu Kultury Pod Lipami, Poznań 35023
Galeria Ostrołęka, Ostrołęka 34991
Galeria Patio, Łódź 34869
Galéria Petra Michala Bohúňa, Liptovský Mikuláš 38286
Galería Piurarte, Piura 34289
Galería Pocitos, Montevideo 54392
Galeria Podlaska, Biała Podlaska 34541
Galeria Prof. Sylvio Vasconcellos, Belo Horizonte 04343
Galeria Rynek, Olsztyn 34977
Galeria SARP, Wrocław 35324
Galeria Sleńdzińskich, Białystok 34547
Galéria Slovenskej Sporitel'ne, Bratislava 38212
Galeria Słowiańska przy Parafii Prawosławnej, Kraków 34778
Galeria Stara, Łódź 34870
Galeria Sztuki Dawna Synagoga, Nowy Sącz 34958
Galeria Sztuki im. Włodzimierca i Jerzego Kulczyckich, Zakopane 35354
Galeria Sztuki Polskiej XIX w. w Sukiennicach, Kraków 34779
Galeria Sztuki Polskiej XIX wieku w Sukiennicach, Kraków 34780
Galeria Sztuki Wozownia, Toruń 35191
Galeria Sztuki Współczesnej, Katowice 34723
Galeria Sztuki Współczesnej, Myślenice 34933
Galeria Sztuki Współczesnej, Szczecin 35157
Galeria Sztuki Współczesnej Profil & Galeria Fotografii pf, Poznań 35024
Galeria Sztuki Współczesnej Winda, Kielce 34740
Galeria Sztuki Współczesnj, Łomża 34889
Galeria Teatro Nacional Enrique Echandi y Joaquín García Monge, San José 08814
Galeria Test, Warszawa 35230
Galéria Umelcov Spiša, Spišská Nová Ves 38358
Galéria Umenia, Nové Zámky 38321
Galería Universitaria Artistos, México 31047
Galeria Virtual da Casa das Rosas, São Paulo 04860
Galeria w Hallu, Kalisz 34709
Galeria Willa, Łódź 34871
Galeria Władysława Hasiora, Zakopane 35355
Galeria XXI, Warszawa 35231
Galeria Zamek w Reszlu, Reszel 35080
Galeria Ziemi Pałuckiej Mariusza Niewolskiego, Żnin 35376
Galerías Pacífico de Centro Cultural Borges, Buenos Aires 00161
Galerie 100, Berlin 17087
Galerie 101, Ottawa 06545
Galerie 5020, Salzburg 02611
Galerie a Muzeum Litoměřické Dieceźe, Litoměřice 09629
Galerie Adlergasse, Dresden 17943
Galerie Adolf Blaim, Messern 02377
Galerie Albstadt, Albstadt 16453
Galerie "Alte Brennerei", Ebersberg 18044
Galerie Alte Reichsvogtei, Schweinfurt 21682
Galerie Alte Schule Wittstedt, Bramstedt 17502
Galerie am Alten Markt, Rostock 21391
Galerie am Dom, Wetzlar 22524
Galerie am Domhof, Zwickau 22832
Galerie am Domplatz, Halle, Saale 18892
Galerie am Graben, Augsburg 16618
Galerie am Kamp, Teterow 22036
Galerie Am Markt, Bad Saulgau 16823
Galerie am Ratswall, Bitterfeld 17369
Galerie an der Bleiche, Ludwigslust 20059
Galerie Anna Paulownahuis, Soest 32792
Galerie Antenna, Dakar 38019
Galerie Antického Umění, Hostinné 09530
Galerie Antonína Chittussiho, Ronov nad Doubravou 09819
Galerie Architektur und Arbeit, Gelsenkirchen 18572
Galerie Architektury Brno, Brno 09428
Galerie art one, Zürich 42072
Galerie at Lincoln Center, New York 51179
Galerie B des "Frankfurter Kunstverein e.V.", Frankfurt (Oder) 18401
Galerie Bab El Kébir, Rabat 31687
Galerie Beaubourg, Vence 16097
Galerie Benedikta Rejta v Lounech, Louny 09640
Galerie Bernau des Förderkreises Bildende Kunst e.V., Bernau bei Berlin 17287
Galerie Biebertal, Biebertal 17333
Galerie Blaugelbe Zwettl, Zwettl 03134
Galerie Bodenseekreis, Meersburg 20249
Galerie Bovary, Ry 15143
Galerie Bratří Špillarů, Domažlice 09495
galerie.bruehl, Brühl 17605
Galerie Capazza, Nançay 14347
Galerie Colline, Edmundston 05763
Galerie Contact, Böblingen 17418
Galerie Coup d'Oeuil, Rimouski 06752
Galerie Daguerre, Paris 14602
Galerie d'Art, Moncton 06315
Galerie d'Art Contemporain, Chamalières 12368
Galerie d'Art Contemporain, La Seyne-sur-Mer 13442
Galerie d'Art Contemporain, Menton 14079
Galerie d'Art Contemporain, Rablay-sur-Layon 14977
Galerie d'Art Contemporain Am Tunnel, Luxembourg 30477
Galerie d'Art de Créteil, Créteil 12675
Galerie d'Art de Matane, Matane 06247
Galerie d'Art d'Esch, Esch-sur-Alzette 30472
Galerie d'Art du Centre Culturel, Sherbrooke, Québec 06959
Galerie d'Art du Château, Bettembourg 30455
Galerie d'Art du Parc, Trois-Rivières 07172
Galerie d'Art du Vieux Presbytère, Sainte-Flavie 06891
Galerie d'Art l'Union-Vie, Drummondville 05714
Galerie d'Art Moderne, Passavant-la-Rochère 14754
Galerie d'Art Municipale, Oberkorn 30489
Galerie d'Art Municipale, Rumelange 30490
Galerie d'Art Stewart Hall, Pointe-Claire 06617
Galerie d'Art Stewart Hall, Pointe-Claire 06618
Galerie de Bourglinster, Bourglinster 30457
Galerie de Chai du Terral, Saint-Jean-de-Védas 15332
Galerie de la Marine, Nice 14423
Galerie de la Tour, Bazens 11883
Galerie de l'Ancienne Douane, Strasbourg 15814
Galerie de Lange, Emmen 32205
Galerie de l'École d'Art, Marseille 13991
Galerie de l'École la Chaufferie, Strasbourg 15815
Galerie de l'École Nationale des Beaux-Arts, Nancy 14350
Galerie de l'École Régionale des Beaux-Arts, Nantes 14360
Galerie de l'UQAM, Montréal 06339
Galerie de Minéralogie, Paris 14603
Galerie de Nesle - Exposition, Paris 14604
La Galerie de Noisy, Noisy-le-Sec 14470
Galerie de Paléobotanique, Paris 14605
Galerie de Pouzauges, Pouzauges 14933
Galerie de Prêt Ville d'Angers, Angers 11604
Galerie der 3 Generationen, Villach 02851
Galerie der Bezirksbibliothek Rheinhausen, Duisburg 18016
Galerie der Fekete-Stiftung, Roßdorf 21381
Galerie der Hochschule für Graphik und Buchkunst Leipzig, Leipzig 19904
Galerie der Jenaoptik AG, Jena 19378
Galerie der Kreissparkasse Esslingen-Nürtingen, Esslingen 18272
Galerie der Künste, Berlin 17088
Galerie der Kunststudenten, Wien 02989
Galerie der Kunstuniversität Linz, Linz 02313
Galerie der Mitte, Linz 02314
Galerie der Stadt, Bad Wimpfen 16865
Galerie der Stadt Backnang, Backnang 16656
Galerie der Stadt Fellbach, Fellbach 18305
Galerie der Stadt Plochingen, Plochingen 21064
Galerie der Stadt Remscheid, Remscheid 21265
Galerie der Stadt Salzburg im Mirabellgarten, Salzburg 02612
Galerie der Stadt Sindelfingen, Sindelfingen 21774
Galerie der Stadt Traun, Traun 02823
Galerie der Stadt Tuttlingen, Tuttlingen 22135
Galerie der Stadt Vöcklabruck, Vöcklabruck 02864
Galerie der Stadt Wels, Wels 02915
Galerie der Stadt Wendlingen am Neckar, Wendlingen am Neckar 22477
Galerie des Bildungshauses Sodalitas, Tainach 02792
Galerie des Dames, Chenonceaux 12500
Galerie des Estampes, Rouen 15112
Galerie des Expositions Pablo Picasso, Denain 12707
Galerie des Jenaer Kunstvereins im Optischen Museum, Jena 19379
Galerie des Kultur-Forum-Amthof, Feldkirchen in Kärnten 01884
Galerie des Kultur- und Festspielhauses Wittenberge, Wittenberge 22620
Galerie des Kunstvereins Aalen e.V., Aalen 16411
Galerie des Kunstvereins Neustadt an der Weinstraße, Neustadt an der Weinstraße 20670
Galerie des Landkreises Rügen, Putbus 21140
Galerie des Polnischen Instituts Düsseldorf, Düsseldorf 17989
Galerie des Ponchettes, Nice 14424
Galerie des Transports, Marseille 13992
Galerie des Verbandes Bildender Künstler Thüringen e.V., Erfurt 18203
Galerie d'Exposition d'Art et d'Histoire, Villaries 16168
Galerie d'Exposition sur la Bourse, Paris 14606
Galerie DKM, Duisburg 18017
Galerie Dorée du Comte de Toulouse, Paris 14607
Galerie du Château, Nice 14425
Galerie du Conseil Général du Bas-Rhin, Strasbourg 15816
Galerie du Musée Montebello, Trouville-sur-Mer 15992
Galerie du Vitrail, Amiens 11585
Galerie Eurêka-Espace Montagne, Chambéry 12371
Galerie Europeenne de la Forêt et du Bois, Dompierre (Orne) 12761
Galerie Felixe Jeneweina, Kutná Hora 09613
Galerie Fernand Léger - Le Crédac, Ivry-sur-Seine 13256
Galerie Freihausgasse, Villach 02852
Galerie für Zeitgenössische Kunst Leipzig, Leipzig 19905
Galerie Georges-Goguen, Moncton 06316
Galerie Gersag Emmen, Emmenbrücke 41348
Galerie Glockengasse, Köln 19603
Galerie Gluri-Suter-Huus, Wettingen 42008
Galerie Gmünd, Gmünd, Kärnten 01941
Galerie Goethe 53, München 20441
Galerie Grünstraße, Berlin 17089
Galerie Handwerk, München 20442
Galerie 'Hans Steger', Zeulenroda-Triebes 22793
Galerie Haus 23, Cottbus 17731
Galerie Heinz Teufel Multimedia, Zingst, Ostseebad 22804
Galerie Heinz Teufel Schule des Sehens, Zingst, Ostseebad 22805
Galerie Heinz Teufel Villa Ruh, Zingst, Ostseebad 22806
Galerie Heitz, Strasbourg 15817
Galerie Hinter dem Rathaus, Wismar 22612
Galerie Hlavního Města Prahy - Městská knihovna, Praha 09746
Galerie Hotel Leipziger Hof, Leipzig 19906
Galerie im 44er Haus, Leonding 02292
Galerie im Alten Rathaus, Prien am Chiemsee 21126
Galerie im Amtshaus, Bad Wurzach 16875
Galerie im Angelburger Kunst- und Kulturhaus, Angelburg 16546
Galerie im Bildhauerhaus und Skulpturenwanderweg, Einöde 01853
Galerie im Bürger- und Gemeindezentrum, Hofstetten-Grünau 02088
Galerie im Fontanehaus, Berlin 17090
Galerie im Gemeinschaftshaus, Berlin 17091
Galerie im Helferhaus, Backnang 16657
Galerie im Kehrwiederturm, Hildesheim 19183
Galerie im Körnerpark, Berlin 17092
Galerie im Kornhauskeller der Ulmer Kunststiftung Pro Arte, Ulm 22161
Galerie im Kulturhaus Spandau, Berlin 17093
Galerie im Kunsthoken, Quedlinburg 21147
Galerie im Malzhaus, Plauen 21053
Galerie im Polnischen Institut, Leipzig 19907
Galerie im Prater, Berlin 17094
Galerie im Rathaus, Bad Aussee 01738
Galerie im Rathaus Köpenick, Berlin 17095
Galerie im Saalbau, Berlin 17096
Galerie im Schloss Porcia, Spittal an der Drau 02751
Galerie im Taxispalais, Innsbruck 02117
Galerie im Teisenhoferhof der Malschule Motiv Wachau, Weißenkirchen in der Wachau 02904
Galerie im Theater am Saumarkt, Feldkirch 01880
Galerie im Traklhaus, Salzburg 02613
Galerie im Troadkasten, Pram 02522
Galerie im Turm, Berlin 17097
Galerie im Volkspark, Halle, Saale 18893
Galerie im Willy-Brandt-Haus, Berlin 17098
Galerie in der Alten Post, Eislingen 18109
Galerie in der Brasilianischen Botschaft, Berlin 17099
Galerie in der Hauptschule, Ulrichsberg 02837
Galerie in der Klosterschule, Glonn 18661
Galerie in der Ungarischen Botschaft, Berlin 17100
Galerie Ingeborg Assam, Villach 02853
Galerie Inkatt, Bremen 17550
Galerie Jamborův Dům, Tišnov 09870
Galerie Jaroslava Fragnera, Praha 09747
Galerie Jiřího Trnky, Plzeň 09716
Galerie Josefa Sudka, Praha 09748
Galerie Kulturzentrum Lerchhaus, Eibiswald 01848
Galerie Kunst-Lücke, Villach 02854
Galerie Kunstbrücke Osteuropa, Berlin 17101
Galerie Kunstlade, Zittau 22811
Galerie Kunstverein Mistelbach, Mistelbach an der Zaya 02385
Galerie Kunstverein Wilhelm Kimmich, Lauterbach, Schwarzwald 19863
Galerie L'Aquarium, Valenciennes 16042
Galerie Le Carré, Bayonne 11877
Galerie Litvinov, Litvínov 09636
Galerie Ludvika Kuby a Městské Muzeum, Březnice 09418
Galerie Ludwig Bürgel, Seekirchen am Wallersee 02735
Galerie M, Berlin 17102
Galerie Markt Bruckmühl, Bruckmühl 17600
Galerie Marschall, Bernried 17302
Galerie Melantrich, Rožďalovice 09822
Galerie Města Blanska, Blansko 09408
Galerie Městského Muzea Čáslav, Čáslav 09450
Galerie Mitte, Berlin 17103
Galerie Moderniho Umění, Hradec Králové 09533
Galerie Montcalm, Gatineau 05859
Galerie Münsterland, Emsdetten 18160
Galerie Municipale, Montesquieu 14202
Galerie Municipale, Vitry-sur-Seine 16258
Galerie Municipale d'Art Contemporain Julio-Gonzalez, Arcueil 11666
Galerie Municipale Edouard Manet, Gennevilliers 13043
Galerie-Musée Le Petit Chapitre, Fosses-la-Ville 03607
Galerie Museum Rabalderhaus, Schwaz 02724
Galerie N, Dahn 17776
Galerie Nationale d'Art, Dakar 38020
Galerie Nationale de la Tapisserie et de l'Art Textile, Beauvais 11909
Galerie Neue Meister, Dresden 17944
Galerie Nikolai - die andere Galerie, Villach 02855
Galerie Nord-Kunstverein Tiergarten, Berlin 17104
Galerie Numero 16, Noordbroek 32599
Galerie Offenes Atelier D.U.Design, Villach 02856
Galerie Palais Walderdorff, Trier 22098
Galerie Pankow, Berlin 17105
Galerie Parterre, Berlin 17106
Galerie pod Zámkem, Rožmitál pod Třemšínem 09825
Galerie Port-Maurice, Saint-Léonard 06862
Galerie Portyč, Písek 09712
Galerie Pravěk Očima Zdeňka Buriana, Dvůr Králové nad Labem 09499
Galerie Privée Hôtel les Templiers, Collioure 12574
Galerie Rainmühle, Emmenbrücke 41349
Galerie Remise, Halle / Westfalen 18909
Galerie Renée-Blain, Brossard 05489
Galerie Rudolfinum, Praha 09749
Galerie Sala Terrena, Mödling 02392
Galerie Sans Nom, Moncton 06317
Galerie Schloß Albeck, Sirnitz 02744
Galerie Schloss Borbeck, Essen 18253
Galerie Schloß Rimsingen, Breisach am Rhein 17535
Galerie Schloß Weyer, Gmunden 01948
Galerie Sonnensegel, Wien 02990
Galerie Stephanie Hollenstein, Lustenau 02341
Galerie Šternberk, Šternberk 09847
Galerie Stihl Waiblingen, Waiblingen 22276
Galerie Taisei, Tokyo 29390
Galerie Terra Viva, Saint-Quentin-la-Poterie 15478
Galerie Theodor von Hörmann, Imst 02108
Galerie Umění Karlovy Vary, Karlovy Vary 09573
Galerie und Museum im Franck-Haus, Marktheidenfeld 20216
Galerie und Oberösterreichisches Literaturhaus im Stifterhaus, Linz 02315
Galerie Unie Výtvarný Umělců Teplice, Teplice 09862
Galerie Volkskundesammlung Piaty, Waidhofen an der Ybbs 02888
Galerie Vyšehrad, Praha 09750
Galerie Výtvarného Umění, Havlíčkův Brod 09511
Galerie Výtvarného Umění, Hodonín 09518
Galerie Výtvarného Umění, Most 09666
Galerie Výtvarného Umění, Náchod 09668
Galerie Výtvarného Umění, Ostrava 09696
Galerie Výtvarného Umění, Roudnice nad Labem 09820
Galerie Waldhof in Birkenwerder, Birkenwerder 17361
Galerie Weberhaus, Weiz 02910
Galerie Weib-art, Berlin 17107
Galerie Weisses Häusle, Hechingen 19061
Galerie Zázvorka, Nové Město nad Metují 09680
Galerie zum Alten Rathaus, Stockerau 02783
Galeriehaus, Nürnberg 20761
Les Galeries de l'Historie, Neuchâtel 41679
Galeries de Paléontologie et d'Anatomie Comparée, Paris 14608
Galeries du Panthéon Bouddhique du Japon et de la Chine, Paris 14609
Galeries Nationales du Grand Palais, Paris 14610
Galerieverein Leonberg e.V., Leonberg 19962
Galeriile de Artă Emilia Dumitrescu, Brăila 36169
Galeriile de Artă Plastică Ştefan Luchian, Botoşani 36161
Galerija Akademije Likovnih Umetnosti Sarajevo, Sarajevo 04275
Galerija Antuna Augustinčića, Klanjec 08888
Galerija Biblioteke grada Beograda, Beograd 38036
Galerija Božidar Jakac, Kostanjevica na Krki 38435
Galerija Brčko, Brčko 04261
Galerija Bukovac, Cavtat 08853
Galerija Cate Dujšin Ribar, Trogir 08968
Galerija Centăr za Săvremenno Izkustvo, Varna 05238
Galerija Centralnog Kluba Vojske Srbije i Crne Gore, Beograd 38037
Galerija Daut Pašin Amam, Skopje 30515
Galerija Doma kulture Studentski Grad, Beograd 38038
Galerija Doma Omladine, Beograd 38039
Galerija Fresaka, Beograd 38040
Galerija Grada Krapine, Krapina 08899
Galerija Izvorne Umjetnosti, Zlatar 09018

Galerija Juraj Plančić, Stari Grad 08965
Galerija Klovićevi Dvori, Zagreb 08990
Galerija Koprivnica, Koprivnica 08891
Galerija Kulturnog Centra Politehničke Akademije, Beograd 38041
Galerija Kuršumli An, Skopje 30516
Galerija Lazar Vozarević, Sremska Mitrovica 38150
Galerija Legat Milice Zorić i Radoljuba Čolakovića, Beograd 38042
Galerija Likovni Susret, Subotica 38155
Galerija Likovnih Samorastnikov, Trebnje 38529
Galerija Likovnih Umetnosti, Slovenj Gradec 38517
Galerija Likovnih Umjetnosti, Osijek 08922
Galerija Likovnih Umjetnosti Republike Srpske Banja Luka, Banja Luka 04256
Galerija Matice Srpske, Novi Sad 38120
Galerija Mestne Hiše, Kranj 38436
Galerija Milan Konjović, Sombor 38146
Galerija Milene Pavlović Barili, Požarevac 38131
Galerija Miroslav Kraljević, Zagreb 08991
Galerija Murska Sobota, Murska Sobota 38486
Galerija-Muzej Lendava, Lendava 38443
Galerija na Centar za Kultura, Mostar 04269
Galerija na Ikoni, Skopje 30517
Galerija Naivne Umetnosti - Donacija Ivana Generalića, Hlebine 08877
Galerija Naivnih Slikara i Etnološka Zbirka, Kovačica 38106
Galerija Portreta Tuzla, Tuzla 04289
Galerija Remont, Beograd 38043
Galerija Rigo, Novigrad 08918
Galerija Savremene Likovne Umetnosti, Niš 38117
Galerija Savremene Umetnosti Likovna Jesen, Sombor 38147
Galerija Sebastian, Dubrovnik 08863
Galerija Singidunum, Beograd 38044
Galerija Skulptura Ivan Sabolić, Peteranec 08928
Galerija Slika Sava Šumanović, Šid 38140
Galerija Sodobne Umetnosti, Celje 38405
Galerija Srpske Akademije Nauka i Umetnosti, Beograd 38045
Galerija ULUS, Beograd 38046
Galerija ULUS, Beograd 38047
Galerija Umetničkih Dela, Beograd 38048
Galerija Umjetnina, Split 08956
Galerija Umjetnina Narodnog Muzeja Zadar, Zadar 08982
Galerija Vincent iz Kastva, Pula 08933
Galerija Vjekoslav Karas, Karlovac 08885
Galerija Zadužbine Ilije M. Kolarca, Beograd 38049
Galerija Zavjetnih Slika Brodova, Dubrovnik 08864
Galerije Capitolum i Diana, Pula 08934
Galerije Sivčeva Hiša, Radovljica 38501
Galesburg Civic Art Center, Galesburg 48833
Galicja Muzeum, Kraków 34781
Galion Historical Museum, Galion 48838
Gallagher Gallery, Dublin 24791
Galle National Museum, Galle 40377
Gallen-Kallela Museum, Espoo 10728
Gallen-Kallelan Museo, Espoo 10728
Galleri 21, Malmö 40719
Galleri 54, Göteborg 40545
Galleri Espolin, Kabelvåg 33584
Galleri Svalbard, Longyearbyen 33649
Galleri Würth, Hagan 33524
Galleria Alberoni, Piacenza 27182
Galleria Aurora Pallavicini, Roma 27448
Galleria Baumgartner, Mendrisio 41624
Galleria Civica, Valdagno 28136
Galleria Civica Anna e Luigi Parmiggiani, Reggio Emilia 27386
Galleria Civica Brunico, Brunico 25554
Galleria Civica d'Arte Contemporanea, Caltagirone 25603
Galleria Civica d'Arte Contemporanea, Suzzara 27933
Galleria Civica d'Arte Contemporanea, Termoli 27961
Galleria Civica d'Arte Contemporanea, Trento 28055
Galleria Civica d'Arte Moderna, Santhià 27731
Galleria Civica d'Arte Moderna, Spoleto 27891
Galleria Civica d'Arte Moderna e Contemporanea, Torino 28001
Galleria Civico, Modena 26740
Galleria Colonna, Roma 27449
Galleria Comunale d'Arte, Cagliari 25578
Galleria Comunale d'Arte, Lecco 26476
Galleria Comunale d'Arte Moderna e Contemporanea, Arezzo 25265
Galleria Comunale Santa Croce, Cattolica 25784
Galleria Corsini, Firenze 26155
Galleria Credito Siciliano, Acireale 25142
Galleria Credito Valtellinese, Sondrio 27871
Galleria d'Arte Contemporanea, Ascoli Piceno 25293
Galleria d'Arte Contemporanea delle Pro Civitate Christiane, Assisi 25305
Galleria d'Arte Contemporanea L. Spazzapan, Gradisca d'Isonzo 26356
Galleria d'Arte Contemporanea V. Stoppioni, Santa Sofia 27711
Galleria d'Arte Moderna, Firenze 26156
Galleria d'Arte Moderna, Udine 28111
Galleria d'Arte Moderna A. Discovolo, Bonassola 25493
Galleria d'Arte Moderna Arnoldo Bonzagni, Cento 25800
Galleria d'Arte Moderna Carlo Rizzarda, Feltre 26093
Galleria d'Arte Moderna Comunale, San Severino Marche 27683
Galleria d'Arte Moderna e Contemporanea R. Guttuso, Bagheria 25340
Galleria d'Arte Moderna F. Montanari, Moncalvo 26773
Galleria d'Arte Moderna Giannoni, Novara 26949
Galleria d'Arte Moderna Moretti, Civitanova Marche 25921
Galleria d'Arte Moderna O. Marchesi, Copparo 25967
Galleria d'Arte Moderna Ricci-Oddi, Piacenza 27183
Galleria d'Arte Sacra dei Contemporanei, Milano 26687
Galleria degli Arazzi, Città del Vaticano 54499
Galleria degli ex-voto del Santuario di Montenero, Livorno 26496
Galleria degli Uffizi, Firenze 26157
Galleria del Costume, Firenze 26158
Galleria del Maggio, Villa Minozzo 28273
Galleria dell'Accademia, Firenze 26159
Galleria dell'Accademia di Belle Arti, Napoli 26880
Galleria dell'Accademia Nazionale di San Luca, Roma 27450
Galleria dell'Accademia Tadini, Lovere 26528
Galleria dello Spedale degli Innocenti, Firenze 26160
Galleria di Arte Moderna Palazzo Forti, Verona 28228
Galleria di Palazzo degli Alberti, Prato 27333
Galleria di Piazza San Marco, Venezia 28181
Galleria Doria Pamphilj, Roma 27451
Galleria Estense, Modena 26741
Galleria Ferrari, Maranello 26600
Galleria Giorgio Franchetti alla Ca' d'Oro, Venezia 28182
Galleria Gruppo Credito Valtellinese, Milano 26688
Galleria Lapidaria, Città del Vaticano 54500
Galleria M. Rizzi, Sestri Levante 27829
Galleria Matasci, Tenero 41930
Galleria Municipale, Francavilla al Mare 26243
Galleria Nazionale, Parma 27096
Galleria Nazionale d'Arte Antica, Trieste 28088
Galleria Nazionale d'Arte Antica Palazzo Barberini, Roma 27452
Galleria Nazionale d'Arte Antica Palazzo Corsini, Roma 27453
Galleria Nazionale d'Arte Moderna e Contemporanea, Roma 27454
Galleria Nazionale delle Marche, Urbino 28123
Galleria Nazionale dell'Umbria, Perugia 27145
Galleria Nazionale di Arte Moderna e Contemporanea, San Marino 38002
Galleria Nazionale di Cosenza, Cosenza 25990
Galleria Nazionale di Palazzo Spinola, Genova 26294
Galleria Palatina e Appartamenti Reali, Firenze 26161
Galleria Palazzo Cini, Venezia 28183
Galleria Pinacotheca, Jyväskylä 10887
Galleria Regionale d'Arte Moderna e Contemporanea, Sassoferrato 27760
Galleria Regionale della Sicilia, Palermo 27065
Galleria Regionale di Palazzo Bellomo, Siracusa 27853
Galleria Rinaldo Carnielo, Firenze 26162
Galleria Sabauda, Torino 28002
Galleria Spada, Roma 27455
Gallerie Arts Graphiques, Villars-Colmar 16169
Gallerie dell'Accademia, Venezia 28184
Gallerie di Palazzo Leoni Montanari, Vicenza 28251
Gallerie Restigouche, Campbellton 05543
The Galleries at Moore, Philadelphia 51838
Galleries of Justice, Nottingham 45291
The Galleries of Texas Tech, Lubbock 50311
The Gallery, Gateshead 44264
The Gallery, La Romana 10331
The Gallery, Trenton 53558
Gallery 101, Ottawa 06545
Gallery 101, River Falls 52363
Gallery 110 North, Plymouth 52002
Gallery 181, Ames 46258
Gallery 2, Chicago 47476
Gallery 210, Saint Louis 52577
Gallery 218, Milwaukee 50708
Gallery 303, Statesboro 53271
Gallery 400 at the University of Illinois at Chicago, Chicago 47477
Gallery 44, Toronto 07107
Gallery 5 at the Center of Culture and Art, Kalisz 34708
Gallery 57, Cambridge 47175
Gallery '76, Wenatchee 54035
Gallery 825, West Hollywood 54063
Gallery A and Downstairs Gallery, San Pedro 52825
Gallery Abarth, Yamanakako 29582
Gallery and Museum of the Diocese of Litomerice, Litoměřice 09629
Gallery Arcturus, Toronto 07108
Gallery at Bristol-Myers Squibb, Princeton 52148
Gallery at the Wharf, Coupeville 47854
The Gallery at UTA, Arlington 46351
The Gallery & Beyond Words Gallery, Detroit 48119
Gallery Center for Contemporary Art, Varna 05238
Gallery Connexion, Fredericton 05831
Gallery East, Price 52141
Gallery Gachet, Vancouver 07211
Gallery Gazelle, Philomath 51896
Gallery II, Bradford 43579
Gallery II, Kalamazoo 49664
Gallery in the Hall at the Center of Culture and Art, Kalisz 34709
Gallery Istvan Csók, Székesfehérvár 23852
Gallery Korea, New York 51180
Gallery Lambton, Sarnia 06907
Gallery Loft of the Vydunas Culture Centre in Kintai, Kintai 30276
Gallery Lolani, Kaneohe 49675
Gallery M, Oaklands Park 01388
Gallery-Museum in the Castle Lendava, Lendava 38443
Gallery Nature and Temptation, Boston 46930
Gallery North, Setauket 53038
Gallery of Antique Art, Hostinné 09530
Gallery of Architecture, Brno 09428
Gallery of Architecture of the Association of Polish Architects, Katowice 34721
Gallery of Art, Cedar Falls 47292
Gallery of Art, Dickinson 48139
Gallery of Art, Fort Valley 48695
Gallery of Art, Overland Park 51632
Gallery of Art, Saint Louis 52578
Gallery of Arts, Nové Zámky 38321
Gallery of Artworks, Beograd 38048
Gallery of British Columbia Ceramics, Granville Island 05905
Gallery of Contemporary Art, Colorado Springs 47701
The Gallery of Contemporary Art, Fairfield 48479
Gallery of Contemporary Art Arka, Vladivostok 37935
Gallery of Contemporary Art Artetage, Vladivostok 37936
Gallery of Contemporary Art Celje, Celje 38405
Gallery of Contemporary Painting, Niš 38117
Gallery of Costume, Manchester 45082
Gallery of Early Canadian Flight, Wiaton 07333
Gallery of Fine Arts, Osijek 08922
Gallery of Fine Arts, Plovdiv 05125
Gallery of Fine Arts, Slovenj Gradec 38517
Gallery of Fine Arts, Split 08956
Gallery of Florina Artists, Florina 23002
Gallery of Frescoes of the National Museum, Beograd 38040
Gallery of Henry Street Settlement, New York 51181
Gallery of Horyu-ji Treasures, Tokyo 29399
Gallery of Hualien County, Hualien 42214
Gallery of Labin, Labin 08905
Gallery of Modern Art, Glasgow 44281
Gallery of Modern Art, Pančevo 38129
Gallery of Modern Art, South Brisbane 01511
Gallery of Modern Art, Zagreb 09003
Gallery of Modern Greek Art of Kalamata, Kalamata 23024
Gallery of Murska Sobota, Murska Sobota 38486
Gallery of Musical Instruments, Delhi 24140
Gallery of Naive Art and Ethnological Collection, Kovačica 38106
Gallery of National Taiwan Normal University, Taipei 42391
Gallery of Notable People, Fălticeni 36365
Gallery of Original Art, Zlatar 09018
Gallery of Prehistoric Paintings, New York 51182
Gallery of Sailor's Votive Paintings, Dubrovnik 08864
Gallery of Šariš, Prešov 38337
Gallery of Slovenská Sporiteľna, Bratislava 38212
Gallery of Spiš Artists, Spišská Nová Ves 38358
Gallery of Szombathely, Szombathely 23904
Gallery of the Association of Fine Artists of Serbia, Beograd 38046
Gallery of the Department of Art, Portales 52054
Gallery of the Department of Art and Art History, Lincoln 50095
Gallery of the Faculty of Architecture, Bangkok 42540
Gallery of the Ilija Kolarac Foundation, Beograd 38049
Gallery of the Jan Matejko Academy of Fine Arts, Kraków 34776
Gallery of the Municipality of Patras, Patras 23139
Gallery of the National Institute of Art and Disabilities, Richmond 52310
Gallery of the Serbian Academy of Sciences and Arts, Beograd 38045
Gallery of the Society for Epirot Studies, Ioannina 23013
Gallery of the State University College at Fredonia, Fredonia 48777
Gallery of Traditional Ornamental Art, Minsk 03249
Gallery of Visual Arts, Missoula 50752
Gallery Oldham, Oldham 45314
Gallery One, Ellensburg 48359
Gallery One One One, Winnipeg 07361
Gallery Pinacotheca of the Jyväskylä University Museum, Section of Cultural History, Jyväskylä 10887
Gallery Primeval Ages by Zdeněk Burian's Eyes, Dvůr Králové nad Labem 09499
Gallery Sculpturama, Bornem 03384
Gallery Slav Epoch of Alfons Mucha on Castle Moravsky Krumlov, Moravský Krumlov 09665
Gallery Stratford, Stratford 07019
Gallery-Studio of the Painter G. Rayshev, Chanty-Mansijsk 36861
Gallery TPW (Toronto Photographers Workshop), Toronto 07109
Gallery Tresco, Tresco 45874
Gallery - University of Toronto, Scarborough 06937
Gallery Vincent of Kastav at the Istra Folk Theatre, Pula 08933
Gallery West, Alexandria 46207
Gallery XII, Wichita 54142
Gallier House, New Orleans 51081
Gallo-Roman Mosaics, Amou 11592
Gallo-Roman Museum, Martigny 41612
Gallo-Romeins Museum, Tongeren 04075
Galloway Deer Museum, Clatteringshaws 43849
Galloway House and Village-Blakely Museum, Fond du Lac 48583
Galloway Station Museum, Edson 05765
Galtebosamlingen, Spydeberg 33858
Galtze Mühle, Seelbach 21718
Galveston Arts Center, Galveston 48844
Galveston County Historical Museum, Galveston 48845
Galway Arts Centre, Galway 24835
Gambia National Museum, Banjul 16331
Gambrinus Drivers Museum, Romedenne 03983
Game and Fisheries Museum, Entebbe 42986
GAMeC - Galleria d'Arte Moderna e Contemporanea, Bergamo 25400
Gamla Bankgården, Vrigstad 41073
Gamla Brukets Muséer, Munkfors 40761
Gamla Krukmakarverkstaden, Arvika 40438
Gamla Linköping, Linköping 40681
Gamla Skeninge, Skänninge 40842
Gamle Bergen Museum, Bergen 33426
Den Gamle By, Århus 09963
Det Gamle Fiskerhus og Skibshallerne, Gilleleje 10035
Den Gamle Gård, Faaborg 10010
Gamle Hvam Museum, Årnes 33400
Gamle Kvernes Bygdemuseum, Averøy 33412
Gamlehaugen, Paradis 33767
Gamli Prestagarðurin og Bygdarsavnið, Nes, Eysturoy 10692
Gammel Estrup - Herregårdsmuseet, Auning 09974
Gammel Estrup - The Manor Museum, Auning 09974
Gammel Havn, Fur 10033
Gammelgården Friluftsmuseet, Bengtsfors 40455
Gammelgruva, Løkken Verk 33643
Gamvik Museum, Gamvik 33509
Gan County Museum, Ganxian 07709
Gan-Song Art Museum, Seoul 29877
Gananoque Historical Museum, Gananoque 05847
Gandahus, Vals 41978
Gandhi Centenary Museum, Karimnagar 24255
Gandhi Memorial Museum, Madurai 24303
Gandhi Museum, Bhavnagar 24088
Gandhi Museum, Lucknow 24293
Gandhi Museum, Sevagram 24384
Gandhi National Memorial, Pune 24358
Gandhi Sangrahalaya, Patna 24345
Gandhi Smarak Sangrahalaya, Ahmedabad 24048
Gandhi Smarak Sangrahalaya, Barrackpore 24080
Gandhi Smarak Sangrahalaya, Sevagram 24384
Ganga Government Museum, Bikaner 24101
Ganghofer-Museum, Leutasch 02296
Gangidori Art Gallery, Joetsu 28618
Gangidori Bijutsukan, Joetsu 28618
Gangneung Municipal Museum, Gangneung 29808
Gangwon Folk Museum, Cheongil 29786
Gannan Storezu Zone Museum, Xiahe 08352
Ganondagan State Historic Site, Victor 53738
Gansu Museum, Lanzhou 07921
Ganyu Museum, Ganyu 07710
Ganzhou Museum, Ganzhou 07711
Ganzhou Zone Museum, Ganzhou 07712
Ganzi Storezu Museum, Kangding 07913
Gao'an Museum, Gao'an 07713
Gaomi Museum, Gaomi 07714
Gaoming Museum, Gaoming 07715
Gaowish Museum, Zhaoqing 08493
Gaozhou Museum, Gaozhou 07716
Gapar Aytiyev Studio-Museum, Biškek 29958
Garda Museum, Dublin 24792
Garden County Museum, Oshkosh 51611

G

Garden Museum, Przemyśl 35054
Garden of Fine Art Kyoto, Kyoto 28867
Garden of Roman Ruins, Szentendre 23879
Garden of the Gulf Museum, Montague 06323
Garden Park Farm Museum, Alberta Beach 05312
Gardiner Art Gallery, Stillwater 53296
Gardiner Museum of Ceramic Art, Toronto 07110
Gardner House Museum, Albion 46174
Gardner Museum, Gardner 48862
The Gardner Museum of Architecture and Design, Quincy 52213
Gárdonyi Géza Emlékmúzeum, Eger 23540
La Gare de Rivière Blanche, Saint-Ulric-de-Rivière-Blanche 06881
Garfield Farm Museum, Lafox 49869
Garibaldi-Meucci Museum, Staten Island 53262
Garlogie Mill Power House Museum, Garlogie 44262
Garnavillo Historical Museum, Garnavillo 49864
Garner Memorial Museum, Uvalde 53678
Garnisonmuseum Nürnberg, Nürnberg 20762
Garnisonsmuseet, Boden 40463
Garnisonsmuseet i Skövde, Skövde 40850
Garnisonsmuseum, Zossen 22819
Garnisonsmuseum Graz-Kanonenbastei, Graz 01969
Gåröströms Industrimuseum, Gnosjö 40539
Garrett County Historical Museum, Oakland 51474
Garrett Historical Museum, Garrett 48867
Garrison Gallery Museum, Millers Point 01304
Garst Museum, Greenville 49076
The Garstang Museum of Archaeology, Liverpool 44734
De Garstkamp, Overasselt 32662
Gartenatelier-Galerie-Berlin, Berlin 17108
Gartenkunst-Museum Schloss Fantasie, Eckersdorf 18054
Garter Lane Arts Centre, Waterford 24907
Gas Museum, Kodaira 28751
Gas Science Museum, Takaishi 29315
Gas Science Museum, Tokyo 29391
Gascoigne Gallery, Harrogate 44399
Gascoyne Junction Museum, Gascoyne Junction 01098
Gaselec Musée, Melilla 31682
Gaslaternen-Freilichtmuseum Berlin, Berlin 17109
Gasmuseum, Budapest 23407
Gáspár Károli Museum and Bible Exhibition, Gönc 23571
Gaspar Museum, Arlon 03326
Gaspesian British Heritage Village, New Richmond 06432
Gass Forest Museum, Coimbatore 24121
Gasthuismuseum Sint-Dimpna, Geel 03613
Gaston County Museum of Art and History, Dallas 47923
Gastronomiaren Euskal Museoa, Laudio 39446
Gasu no Kagakukan, Takaishi 29315
Gasu no Kagakukan, Tokyo 29391
Gatčinskij Literaturno-memorialnyj Muzej-usadba P.E. Ščerbova, Gatčina 36919
Gate P, Berlin 17110
Gatekeeper's Museum and Marion Steinbach Indian Basket Museum, Tahoe City 53418
Gately House, East London 38616
Gates House, Machiasport 50350
Gates Mills Historical Society Museum, Gates Mills 48872
Gateway to Science, Bismarck 46838
Gateway to the Panhandle, Gate 48871
Gatineau Valley Historical Society, Chelsea 05604
Gatov Gallery, Long Beach 50173
Gatzaren Ekomuseoa, Leintz-Gatzaga 39450
Gaudete, Museum voor Volksdevotie, Boxtel 31995
Gaudnek-Museum, Altomünster 16523
Gauermann-Museum, Miesenbach 02381
Gauselmann Collection - German museum of coin-operated machines, Espelkamp 18245
Gavar Local Tradition Museum, Gavar 00701
Gavilan College Art Gallery, Gilroy 48913
Gawharah Museum, Cairo 10511
Gawharah Palace Museum, Cairo 10514
Gawler Museum, Gawler 01100
Gawsworth Hall, Macclesfield 45063
Gawthorpe Hall, Padiham 45340
Gay Nineties Button and Doll Museum, Eureka Springs 48444
Gaya Museum, Gaya 24176
Gayer Anderson Museum, Cairo 10496
Gayndah Museum, Gayndah 01101
Gayo Museum, Takengon 24573
Gaytal-Park, Körperich 19664
Gaz Factory Museum, Nižnij Novgorod 37457
Gazdaház, Cserépfalu 23507
Gazi Müzesi, Samsun 42930
Gaziantep Archaeological Museum, Gaziantep 42785
Gaziantep Arkeoloji Müzesi, Gaziantep 42785
Gaziantep Devlet Güzel Sanatlar Galerisi, Gaziantep 42786
Gaziantep State Fine Art Gallery, Gaziantep 42786
Gazirah Museum, Cairo 10497
Gázmúzeum, Budapest 23407
GDR-Museum Radebeul, Radebeul 21159
Geamia Esman Sultan, Mangalia 36464
Gebirgsjägermuseum, Sonthofen 21822
Gebrüder-Lachner-Museum, Rain 21168
Geburtshaus des heiligen Konrad von Parzham, Bad Griesbach 16741
Geburtshaus Levi Strauss, Buttenheim 17670
Gedächtnisstätte & Galerie Maria v. Ohmeyer, Poysdorf 02518
Gedächtnisstätte Mida Huber, Landsee 02255
Gedächtnisstätte von Scheidt-Saalfeld, Bad Bergzabern 16686
Geddes Historic District Village, Geddes 48875
Gede Museum and National Monument, Watamu 29752
Gedenk- und Dokumentationsstätte KZ Drütte, Salzgitter 21474
Gedenk- und Dokumentationsstätte Opfer politischer Gewaltherrschaft, Frankfurt (Oder) 18402
Gedenkräume Schloss Hartheim, Alkoven 01703
Gedenkraum 1945, Hochwolkersdorf 02087
Gedenkstätte Abraham a Sancta Clara, Leibertingen 19876
Gedenkstätte Bergen-Belsen, Lohheide 20023
Gedenkstätte Berlin-Hohenschönhausen, Berlin 17111
Gedenkstätte Berliner Mauer, Berlin 17112
Gedenkstätte Breitenau, Guxhagen 18844
Gedenkstätte Buchenwald, Weimar 22399
Gedenkstätte Deutsche Teilung Marienborn, Marienborn 20199
Gedenkstätte Deutscher Widerstand, Berlin 17113
Gedenkstätte Ehemalige Synagoge, Adelsheim 16428
Gedenkstätte ehemaliges SS-Sonderlager / KZ Hinzert, Hinzert-Pölert 19197
Gedenkstätte "Erinnern für die Zukunft", Leipzig 19908
Gedenkstätte für die Opfer der NS-"Euthanasie" Bernburg, Bernburg 17294
Gedenkstätte Goldener Löwe, Eisenach 18094
Gedenkstätte Hadamar, Hadamar 18848
Gedenkstätte in der JVA Wolfenbüttel, Wolfenbüttel 22652
Gedenkstätte Köpenicker Blutwoche Juni 1933, Berlin 17114
Gedenkstätte Langenstein-Zwieberge, Halberstadt 18876
Gedenkstätte Moritzplatz Magdeburg für die Opfer politischer Gewaltherrschaft 1945-1989, Magdeburg 20122
Gedenkstätte Münchner Platz Dresden, Dresden 17945
Gedenkstätte Museum in der "Runden Ecke", Leipzig 19909
Gedenkstätte/Museum Seelower Höhen, Seelow 21720
Gedenkstätte Plötzensee für die Opfer des Nationalsozialismus, Berlin 17115
Gedenkstätte "Roter Ochse", Halle, Saale 18894
Gedenkstätte Synagoge, Dornum 17908
Gedenkstätte und Museum Sachsenhausen, Oranienburg 20908
Gedenkstätte und Museum Trutzhain, Schwalmstadt 21661
Gedenkstätte zum Konzentrationslager im Schloss Lichtenburg, Prettin 21121
Gedenkturm Großbeeren 1813, Großbeeren 18780
Ģederta Eliasa Jelgavas Vēstures un Mākslas Muzejs, Jelgava 30017
Ģederts Eliass Jelgava History and Art Museum, Jelgava 30017
Gedimino Pilies Bokštas, Vilnius 30405
Het Gedistilleerd Museum, Schiedam 32752
GEDOK Künstlerinnenforum, Karlsruhe 19439
Geelong Gallery, Geelong 01102
Geelong Heritage Centre, Geelong 01103
Geelong Naval and Maritime Museum, North Geelong 01372
Geels & Co. Koffie- en Theemuseum, Amsterdam 31843
Geeson Brothers Motor Cycle Museum, South Witham 45699
Geevor Tin Mining Museum, Pendeen 45353
Geffrye Museum, London 44854
Geidai Bijutsukan, Tokyo 29392
Geigenbau-Museum, Bubenreuth 17613
Geigenbaumuseum, Mittenwald 20342
Geiseltalmuseum, Halle, Saale 18895
Gekkeikan Okura Kinenkan, Kyoto 28811
Das Gelbe Haus, Flims Dorf 41374
Het Geld-en Bankmuseum, Utrecht 32846
Gelders Geologisch Museum, Velp 32881
Gelders Schuttersmuseum, Didam 32123
Gelderse Smalspoor Museum, Erlecom 32221
Geldgeschichtliches Museum, Köln 19604
Geldmuseum der Deutschen Bundesbank, Frankfurt am Main 18372
Geldmuseum der Oesterreichischen Nationalbank, Wien 02991
Gelendžikskij Istoriko-kraevedčeskij Muzej, Gelendžik 36923
Gelereja Iskusstva Uzbekistana, Toshkent 54470
Geležinkelio Muziejus, Šiauliai 30357
Geležinkelių Muziejus, Vilnius 30406
Gellert-Museum Hainichen, Hainichen 18873
Gelsenkirchen im Nationalsozialismus, Gelsenkirchen 18573
Gem Museum, Ylämaa 11477
Gemälde- und Skulpturensammlung, Nürnberg 20763
Gemäldegalerie, Berlin 17116
Gemäldegalerie Alte Meister, Dresden 17946
Gemäldegalerie Alte Meister, Kassel 19463
Gemäldegalerie Dachau, Dachau 17764
Gemäldegalerie der Akademie der Bildenden Künste, Wien 02992
Gemeentegrot Valkenburg, Valkenburg, Limburg 32863
Gemeentelijk Archaeologisch Museum, Grobbendonk 03663
Gemeentelijk Archeologisch Museum, Aardenburg 31787
Gemeentelijk Heemkundig Museum, Hoeilaart 03707
Gemeentelijk Heemkundig Museum, Rijkevorsel 03967
Gemeentelijk Historisch Museum Ouder-Amstel, Ouderkerk aan de Amstel 32655
Gemeentelijk Museum, Melle 03866
Gemeentelijk Museum Gevaert-Minne, Sint-Martens-Latem 04018
Gemeentelijk Museum 't Oude Slot, Veldhoven 32880
Gemeentelijk Museum van Evere, Bruxelles 03447
Gemeentelijk Museum voor Geschiedenis en Heemkunde, Destelbergen 03544
Gemeentelijk Streekmuseum De Veste, Sint-Stevens-Woluwe 04025
Gemeentelijke Expositieruimte Kampen, Kampen 32436
Gemeentelijke Oudheidkamer, Ermelo 32222
Gemeentelijke Oudheidkamer, Geldrop 32239
Gemeentemuseum, Maassluis 32522
Gemeentemuseum, Temse 04053
Gemeentemuseum, Weesp 32934
Gemeentemuseum Den Haag, Den Haag 32073
Gemeentemuseum Elburg, Elburg 32197
Gemeentemuseum Helmond, Helmond 32363
Gemeentemuseum Het Hannemahuis, Harlingen 32316
Gemeentemuseum Het Markiezenhof, Bergen op Zoom 31969
Gemeentemuseum 't Behouden Huijs, Terschelling-West 32818
Gemeentemuseum 't Sterkenhuis, Bergen 31967
Gemeentemuseum Weert - De Tiendschuur, Weert 32932
Gemeentemuseum Weert - Jacob van Horne, Weert 32933
Gemeinde-Heimatmuseum Harsum, Harsum 19017
Gemeinde- und Forstmuseum, Oftersheim 20889
Gemeindegalerie der Evangelischen Brüdergemeinde Korntal, Korntal-Münchingen 19687
Gemeindegalerie im Alten Rathaus, Wiener Neudorf 03094
Gemeindemuseum, Bad Krozingen 16769
Gemeindemuseum, Hittisau 02085
Gemeindemuseum, Krauchthal 41511
Gemeindemuseum, Regensdorf 41750
Gemeindemuseum, Schweiggers 02730
Gemeindemuseum Absam, Absam 01688
Gemeindemuseum im Historischen Rathaus, Seeheim-Jugenheim 21717
Gemeindemuseum Rothus, Oberriet 41699
Gemersko-Malohontské Múzeum, Rimavská Sobota 38340
Gemossenschaft Pro Sagi Samstagern, Samstagern 41802
Gemstone Gallery, Kemi 10938
Genadendal Museum, Genadendal 38623
Genç Culture Center, Genç 42789
Genç Kültür Merkezi, Genç 42789
Gendai Chokoku Bijutsukan, Tokyo 29393
Gendai Gangu Hakubutsukan, Mimasaka 28909
Gene Fornby, Örnsköldsvik 40793
Gene Stratton-Porter House, Rome City 52454
Generaal Maczek Museum, Breda 32004
General Adam Stephen House, Martinsburg 50519
General Crook House Museum, Omaha 51550
General Daniel Bissell House, Saint Louis 52579
General Douglas L. McBride Museum, Roswell 52468
General Douglas MacArthur Memorial, Norfolk 51362
General George C. Marshall Museum, Zwijndrecht 33024
General Grant National Memorial, New York 51183
General Henry Knox Museum, Thomaston 53497
General Jacob Brown Historical Society Museum, Brownville 47099
General John J. Pershing Boyhood Home, Laclede 49860
General Lew Wallace Museum, Crawfordsville 47869
General Lewis B. Hershey Museum, Angola 46300
General Patton Memorial Museum, Chiriaco Summit 47543
General Patton Memorial Museum, Ettelbruck 30474
General Pliyev House Museum, Vladikavkaz 37929
General Sterling Price Museum, Keytesville 49773
General Sweeney's Museum, Republic 52297
General Wait House, Waitsfield 53788
Generali Foundation, Wien 02993
Genesee Country Village and Museum, Mumford 50901
Genia Schreiber University Art Gallery, Ramat Aviv 25081
Genichiro-Inokuma Museum of Contemporary Art, Marugame 28881
Genoa Courthouse Museum, Genoa 48885
Genocide Victims Museum, Beograd 38081
Genocido Aukų Muziejus, Vilnius 30407
Gentil-Haus, Aschaffenburg 16582
Geo Inn, Oosterbeek 32625
Geo Milev Memorial House, Stara Zagora 05209
Geochang Museum, Geochang 29811
Geoffrey Kaye Museum of Anaesthetic History, Melbourne 01279
Geographical Center Historical Museum, Rugby 52478
Geographical Collection of Hungary, Érd 23548
Geographical Museum of Slovenia, Ljubljana 38470
Geography Museum, Faizabad 24172
Geoje Museum, Geoje 29812
Geológiai Múzeum, Tata 23916
Geologian museo, Helsinki 10775
Geological and Mineralogical Museum, Biškek 29953
Geological Curators' Group, Dublin 54889
Geological-mineralogical Museum, Moskva 37182
Geological Museum, Almaty 29686
Geological Museum, Bandung 24442
Geological Museum, Bangkok 42541
Geological Museum, Čeljabinsk 36845
Geological Museum, Dublin 24793
Geological Museum, Guwahati 24191
Geological Museum, Ho Chi Minh City 54622
Geological Museum, Johannesburg 38654
Geological Museum, Khartoum 40390
Geological Museum, København 10153
The Geological Museum, Laramie 49947
Geological Museum, Lisboa 35710
Geological Museum, Lucknow 24294
Geological Museum, Mizpe Ramon 25068
Geological Museum, Oulu 11174
Geological Museum, Riyadh 38015
Geological Museum, Saskatoon 06913
Geological Museum, Tata 23916
Geological Museum, Varanasi 24426
Geological Museum, Yerevan 00735
Geological Museum and Art Gallery, Redruth 45477
Geological Museum of Central Siberia, Krasnojarsk 37075
Geological Museum of China, Beijing 07538
Geological Museum of Hungary, Budapest 23461
Geological Museum of Iran, Tehrān 24667
Geological Museum of the Finnish Museum of Natural History, Helsinki 10775
Geological Museum of the Joint Stock Company Polar Uralgeology, Vorkuta 37967
Geological Museum of the N.M. Fedorovsky All-Russian Scientific-Research Institute of Mineral Resources, Moskva 37181
Geological Museum of the Polish Geological Institute, Warszawa 35244
Geological Museum of University of Valencia, Burjassot 39074
Geological Museum, University of Wrocław, Wrocław 35330
Geological Museum Volgageology, Nižnij Novgorod 37466
Geological Survey and Mines Museum, Entebbe 42987
Geological Survey Department Museum, Accra 22845
Geological Survey Museum, Dhaka 03223
Geological Survey Museum, Ipoh 30554
Geological Survey Museum, Windhoek 31763
Geologičeskij i Mineralogičeskij Muzej, Biškek 29953
Geologičeskij Muzej, Vorkuta 37967
Geologičeskij Muzej im. A.A. Černova, Syktyvkar 37814
Geologičeskij Muzej im. Professora V.E. Eršova MGGU, Moskva 37180
Geologičeskij Muzej Vserossijskogo NII Mineralnogo Syrja im. N.M. Fedorovskogo, Moskva 37181
Geologijos ir Mineralogijos Muziejus, Vilnius 30408
Geologisch-Mineralogische Sammlung, Hillesheim 19187
Geologisch Museum Hofland, Laren, Noord-Holland 32464
Geologisch-Paläontologisches Museum der Universität Hamburg, Hamburg 18930

Geologisch Streek-Museum de IJsselvallei, Olst 32618
Geologische Landessammlung von Vorpommern, Greifswald 18748
Geologische Sammlung Schloß Kapfenstein, Kapfenstein 02152
Geologische Schausammlung Eselsmühle, Leinfelden-Echterdingen 19883
Geologisches Freilichtmuseum, Bad Bentheim 16682
Geologisches Freilichtmuseum, Gmünd, Niederösterreich 01944
Geologisches Freimuseum, Bettendorf 17310
Geologisches Museum, Kamp-Lintfort 19423
Geologisches Museum München, München 20443
Geologisches und Mineralogisches Museum der Christian-Albrechts-Universität, Kiel 19528
Geologo-mineralogičeskij Muzej, Kazan 37007
Geologo-mineralogičeskij Muzej, Moskva 37182
Geologo-mineralogičeskij Muzej im. A.I. Kozlova, Vladivostok 37937
Geology and Geophysics Museum, Roorkee 24373
Geology Collections, London 44855
Geology Department Museum, Aberdeen 43241
Geology Education Museum, Durban 38606
Geology Museum, Amman 29649
Geology Museum, Auckland 33049
Geology Museum, Ha Noi 54613
Geology Museum, Huntington 49434
Geology Museum, Kampala 42991
Geology Museum, Kingston 28326
Geology Museum, Legon 22860
Geology Museum, Lucknow 24295
Geology Museum, Tainan 42361
Geology Museum of the University of Otago, Dunedin 33092
Geology Museum of the University of Stellenbosch, Stellenbosch 38735
Geomuseum der Universität Münster, Münster 20509
GeoMuseum der Universität zu Köln, Köln 19605
Georg Elser Gedenkstätte, Königsbronn 19642
Georg Jensen Museum, København 10132
Georg-Kolbe-Museum, Berlin 17117
Georg-Meistermann-Museum, Wittlich 22623
Georg Papendicks Faschings- und Karnevalsordenmuseum, Bad Reichenhall 16807
Georg Rendl-Museum, Sankt Georgen bei Salzburg 02643
Georg Trakl Forschungs- und Gedenkstätte, Salzburg 02614
George A. Spiva Center for the Arts, Joplin 49650
The George Adams Gallery, Melbourne 01280
George B. Dorr Museum of Natural History, Bar Harbor 46603
George Bush Presidential Library and Museum, College Station 47688
George C. Marshall Museum, Lexington 50070
The George D. and Harriet W. Cornell Fine Arts Museum, Winter Park 54254
George Eliot Hospital Museum, Nuneaton 45304
George Fraser Gallery, Auckland 33050
George Hooey Wildlife Collection, North Battleford 06467
George I. Ashby Memorial Museum, Copper Center 47795
George J. Doizaki Gallery, Los Angeles 50225
George Johnston Tlingit Indian Museum, Teslin 07069
George L. Sturman Museum, Las Vegas 49968
George Meany Memorial Exhibition, Silver Spring 53106
George Museum, George 38624
George Paton Gallery, Victoria 01618
George Rogers Clark Park Museum, Vincennes 53747
George Taylor House, Catasauqua 47280
George W. Brown jr. Ojibwe Museum, Lac du Flambeau 49857
George W. Carver Museum, Tuskegee Institute 53626
George W. Somerville Historical Library, Chillicothe 47531
George Walter Vincent Smith Art Museum, Springfield 53225
George Washington Birthplace National Monument, Washington's Birthplace 53951
George Washington Carver House, Diamond 48137
George Washington Masonic National Memorial, Alexandria 46208
George Washington's Ferry Farm, Stafford County 53246
George Washington's Headquarters, Cumberland 47907
George Washington's Mount Vernon, Mount Vernon 50893
Georgetown College Gallery, Georgetown 48891
Georgetown Museum, Columbus 47748
Georgetown University Art Collection, Washington 53874
Georgi Benkovski Memorial House, Koprivštica 05071
Georgi Dimitrov Memorial House, Kovačevci 05079
Georgi Dimitrov National Museum, Sofia 05194
Georgi Velchev Art Museum, Varna 05240
Georgia Agrirama, Tifton 53508
Georgia Association of Museums and Galleries, Atlanta 55135
Georgia Capitol Museum, Atlanta 46437
Georgia Historical Society Museum, Savannah 52941
Georgia Mountains History Museum, Gainesville 48821
Georgia Museum of Art, Athens 46411
Georgia Museum of Natural History, Athens 46412
Georgia Music Hall of Fame, Macon 50365
Georgia O'Keeffe Museum, Santa Fe 52874
Georgia Salzburger Society Museum, Rincon 52352
Georgia Southern University Museum, Statesboro 53272
Georgia State Cotton Museum, Vienna 53743
Georgia State University Art Gallery, Atlanta 46438
Georgia Veterans Memorial Museum, Cordele 47805
The Georgian, Lancaster 49920
Georgian House, Bristol 43635
The Georgian House, Edinburgh 44125
Georgian National Gallery, Tbilisi 16360
Georgian Silk Museum, Tbilisi 16361
Georgian State Museum of Oriental Art, Tbilisi 16362
Georgian Theatre Royal Museum, Richmond, North Yorkshire 45488
Georgia's Stone Mountain Park, Stone Mountain 53310
Georgikon Farm Museum, Keszthely 23661
Georgikon Majormúzeum, Keszthely 23661
Georgina Pioneer Village, Keswick 06066
Geosammlung der Technischen Universität Clausthal, Clausthal-Zellerfeld 17708
Geoscience Museum, Tshwane 38747
Geoscience Museum (University of Münster), Münster 20509
Geoskop-Urweltmuseum, Thallichtenberg 22044
Géothermia-Musée de la Géothermie et du Thermalisme, Chaudes-Aigues 12478
Geowissenschaftliche Sammlungen, Freiberg, Sachsen 18421
Geowissenschaftliche Sammlungen und Geologischer Garten, Halle, Saale 18896
Geowissenschaftliches Museum am Geowissenschaftlichen Zentrum der Universität Göttingen, Göttingen 18691
GeoZentrum Gams, Gams bei Hieflau 01923
Gerätemuseum des Coburger Landes, Ahorn 16440
Gerätesammlung Koch, Königsberg in Bayern 19641
Gerald R. Ford Museum, Grand Rapids 49002
Geraldton Regional Art Gallery, Geraldton 01106
Geralka Rural Farm, Spalding 01520
Gerbermuseum, Bretten 17583
Gerbermuseum Lohmühle, Frickingen 18464
Gerdarsafn, Reykjavík 24016
Gerðuberg Menningarmiðstöð, Reykjavík 24017
Gereedschap Museum Mensert, Delft 32056
Gerhard-Kurt-Müller-Stiftung, Leipzig 19910
Gerhard-Marcks-Haus, Bremen 17551
Gerhart-Hauptmann-Haus, Kloster 19574
Gerhart-Hauptmann-Museum, Erkner 18215
Gerhart Hauptmann's House, Jelenia Góra 34703
Gerichtsmuseum Bad Fredeburg, Schmallenberg 21583
German Clock Museum, Furtwangen im Schwarzwald 18520
German Emigration Center Bremerhaven, Bremerhaven 17568
German Fairytale Museum, Bad Oeynhausen 16798
German Fan Museum, Bielefeld 17338
German Folk Collection, Gyula 23599
German Gemstone Museum, Idar-Oberstein 19308
German Horticultural Museum of Erfurt, Erfurt 18199
German Museum for the History of Medicine, Ingolstadt 19340
German Museum of Agriculture, Rural Heritage Open Air Museum of Central Germany, Crimmitschau 17749
German Museum of Chemistry, Merseburg 20291
German National Committee of the International Council of Museums ICOM, Berlin 54831
German Newspaper Museum, Wadgassen 22264
German Occupation Museum, Forest 44241
German Pharmacy Museum, Heidelberg 19071
German Plastics Museum Society, Düsseldorf 17999
The German Sampler Museum, Celle 17686
German Toy Museum, Sonneberg 21818
German Wipe Museum, Burladingen-Killer 17665
Germanisches Nationalmuseum, Nürnberg 20764
Germantown Historical Society Museum, Philadelphia 51839
Gernika Peace Museum, Gernika-Lumo 39332
Gernon House and Blacksmith Shop, Russell 52485
Geroldsecker-Museum im Storchenturm, Lahr 19771
Geroldsecker Waffenschmiede, Seelbach 21719
Geronimo Berenguer de los Reyes jr. Museum, Cavite City 34362
Geronimo Springs Museum, Truth or Consequences 53579
Gerringong and District Museum, Gerringong 01108
Gerrit Valk's Bakkerij- en Ijsmuseum, Hellendoorn 32354
Gertrud Theiler Tick Museum, Onderstepoort 38695
Gertrude Contemporary Art Spaces, Fitzroy 01081
Gertrude Herbert Institute of Art, Augusta 46478
Gertrude Posel Gallery, Johannesburg 38655
Gertrude Smith House, Mount Airy 50869
Gertrudenkapelle, Güstrow 18828
Gertsev Gallery, Moskva 37173
Gervais Wheels Museum, Alida 05316
Geschichte auf Rädern, Melle 20271
Geschichtlich-heimatkundliche Sammlung, Aschheim 16597
Geschichtliche Begegnungsstätte Hachborn, Ebsdorfergrund 18049
Geschichtliche Sammlung der Schule-Ausbildung für Blinde, Hannover 18990
Geschichtliches Museum Enns-Donauwinkel, Sankt Valentin 02690
Geschichtsmuseum Lüdenscheid, Lüdenscheid 20088
Geschichtswerkstatt Herrenwyk, Lübeck 20074
Geschiedkundig Museum Mesen, Mesen 03871
Geschirr-Museum, Wilhelmsburg an der Traisen 03108
Gesellschaft der Musikfreunde in Wien, Wien 02994
Gesellschaft für Aktuelle Kunst GAK, Bremen 17552
Gesellschaft für das Weinbaumuseum am Zürichsee, Au 41133
Gesenkschmiede Lubenbach, Zella-Mehlis 22784
Gesterby Manor, Kirkkonummi 10962
Gesterbyn Kartano - Kirkkonummen Kunnan Museoalue, Kirkkonummi 10962
Gestütsmuseum Offenhausen, Gomadingen 18706
Gettysburg National Military Park, Gettysburg 48902
Gevorg Grigoryan - Giotto - Studio-Museum, Yerevan 00736
Gewerbe-Museum, Spaichingen 21826
Gewerbemuseum, Winterthur 42027
Gewerbemuseum im Germanischen Nationalmuseum, Nürnberg 20765
Gewölbe der Geheimnisse, Viechtach 22224
Geymüller Schlössel, Wien 02995
Géza Gárdonyi Memorial Museum, Eger 23540
Ggantija Prehistoric Temples, Xaghra 30658
Ghalib Museum, Delhi 24141
Ghana Armed Forces Museum, Kumasi 22856
Ghana Museums and Monuments Board, Cape Coast 54861
Ghar Dalam Cave and Museum, Birzebbuga 30627
Gharb Folklore Museum, Gharb 30628
Ghazni Provincial Museum, Ghazni 00001
Gheorghe Dima Memorial House, Braşov 36174
Gherla History Museum, Gherla 36383
Ghetto Fighters' House-Children's Memorial Museum, Lohamei-Hagetaot 25062
Ghoochan Museum, Ghoochan 24604
GianFu Classic Art Museum, Beijing 07539
Giant's Castle, Marazion 45102
Giant's Causeway and Bushmills Railway, Bushmills 43702
Giant's Causeway Visitor Centre, Bushmills 43703
Giardino di Boboli, Firenze 26163
Giardino Storico e Villa Garzoni, Pescia 27170
Gibbes Museum of Art, Charleston 47364
Gibbs Farm Museum, Falcon Heights 48496
Gibraltar Museum, Gibraltar 22864
Gibran Museum, Bsharri 30166
Gibson Barham Gallery, Lake Charles 49877
Gibson Gallery, Amherstburg 05333
Gibson House Museum, Boston 46931
Gibson House Museum, North York 06487
Gifu-ken Bijutsukan, Gifu 28457
Gifu-ken Gendai Togei Bijutsukan, Tajimi 29311
Gifu-ken Hakubutsukan, Seki 29240
Gifu-ken Toji Shiryokan, Tajimi 29312
Gifu Prefectural Ceramics Museum, Tajimi 29312
Gifu Prefectural Museum, Seki 29240
Gig Harbor Peninsula Historical Society Museum, Gig Harbor 48907
Gila Cliff Dwelling Visitor Center, Silver City 53102
Gila County Historical Museum, Globe 48943
Gilbert Collection, London 44856
Gilbert House, Atlanta 46439
The Gilbert Stuart Museum, Saunderstown 52934
Gilbert White's House and the Oates Museum, Selborne 45632
Gilbert's Bar House of Refuge, Stuart 53343
Gilcrease Museum, Tulsa 53608
Gilde-Museum, Oldenburg in Holstein 20901
Gildehaus, Bardowick 16927
Het Gildenhuys, Blokzijl 31975
Gildeskål Bygdesamling, Gildeskål 33512
Gilgandra Museum, Gilgandra 01109
Gillam Community Museum, Gillam 05865
Gilleleje Museum, Gilleleje 10036
Gillesgården, Pohja 11214
Gillespie Museum of Minerals, De Land 47998
Gillingham Museum, Gillingham, Dorset 44268
Gilman Garrison House, Exeter 48464
Gilman Museum, Hellertown 49243
Gilman Town Hall Museum, Issaquah 49544
Gilmore Car Museum, Hickory Corners 49269
Gilpin History Museum, Central City 47320
Gilstrap Heritage Centre, Newark-on-Trent 45210
Gingras Trading Post, Walhalla 53794
Ginkgo Museum, Weimar 22400
Ginny Williams Family Foundation, Denver 48078
Gintaro Galerija-Muziejus, Nida 30319
Gintaro Galerija-Muziejus, Vilnius 30409
Ginučių Vandens Malūno Ekspozicija, Ginučiai 30228
Giorgi Chitaia Ethnography Museum, Tbilisi 16363
Giorgi Chitaia Museum of Architecture, Tbilisi 16364
Gippsland Art Gallery, Sale 01488
Gippsland Heritage Park, Moe 01309
Gipsformerei, Berlin 17118
Gipsmuseum Oberwiesen, Schleitheim 41835
Gipsoteca, Lucca 26530
Gipsoteca Canoviana e Casa del Canova, Possagno 27316
Gipsoteca del Castello Normanno Svevo, Bari 25361
Gipsoteca F. Jerace, Catanzaro 25778
Gipsoteca Istituto d'Arte, Firenze 26164
Giraffen-Museum Dortmund, Dortmund 17919
Giresun Müzesi, Giresun 42790
Girne Kalesi, Girne 33358
Gironcoli Museum, Sankt Johann bei Herberstein 02655
Girouxville Museum, Girouxville 05867
Gisborne Steam Park, New Gisborne 01359
Gislaved Industry Museum, Gislaved 40535
Gislaveds Industrimuseum, Gislaved 40535
Gislöfs Smidesmuseum, Simrishamn 40839
Gitstappermolen, Vlodrop 32901
Gitxsan Treaty Office, Hazelton 05978
Giuseppe Tartini Memorial Room, Piran 38493
Giustina Gallery, Corvallis 47834
Giustiniani Palace, Chios 22969
Give-Egnens Museum, Give 10037
Giza Zoological Museum, Giza 10526
Gizi Bajor Memorial and Theatre Museum, Budapest 23385
Gjenreisningsmuseet, Hammerfest 33531
Gjesdal Bygdemuseum, Ålgård 33382
Glace Bay Heritage Museum, Glace Bay 05869
Glacier National Park Museum, West Glacier 54056
Gladstone and District Museum, Gladstone 05870
Gladstone Court Museum, Biggar 43481
Gladstone Gaol, Gladstone, South Australia 01112
Gladstone Maritime Museum, Gladstone, Queensland 01110
Gladstone Pottery Museum, Stoke-on-Trent 45761
Gladstone Regional Art Gallery and Museum, Gladstone, Queensland 01111
Gläsermuseum, Gerersdorf 01931
Glamis Castle, Glamis 44273
Glandford Shell Museum, Glandford 44274
Glanmore National Historic Site of Canada, Belleville 05415
Glanum, Saint-Rémy-de-Provence 15485
Glarner Naturmuseum, Engi 41353
Glas- & Heimat-Museum Warndt, Völklingen 22248
Glas- und Keramikmuseum, Großalmerode 18779
Glas- und Steinmuseum, Gmünd, Niederösterreich 01945
Glasatelier Alois Hechl, Villach 02857
Glasatelier Heimo Setten, Villach 02858
Glasbruksmuseet i Surte, Surte 40965
Glasgow Print Studio, Glasgow 44282
Glasgow School of Art - Mackintosh Collection, Glasgow 44283
Glashaus-Derneburg, Holle 19251
Glashütte Annenwalde, Templin 22027
Glashüttenmuseum des Erzgebirges, Neuhausen / Erzgebirge 20619
Glasmuseet Ebeltoft, Ebeltoft 09997
Glasmuseum, Frauenau 18410
Glasmuseum, Gehlberg 18551
Glasmuseum, Hergiswil 41480
Glasmuseum, Immenhausen 19325
Glasmuseum, Warmensteinach 22339
Glasmuseum Alter Hof Herding, Coesfeld 17721
Glasmuseum Boffzen, Boffzen 17431
Glasmuseum der Dorotheenhütte, Wolfach 22647
Glasmuseum Grünenplan, Grünenplan 18814
Glasmuseum Hadamar, Hadamar 18849
Glasmuseum in der Glashütte, Monschau 20374
Glasmuseum Passau, Passau 20989
Glasmuseum Rheinbach, Rheinbach 21297
Glasmuseum Steina, Bad Sachsa 16811
Glasmuseum Veenhuizen, Veenhuizen 32875
Glasmuseum Weisswasser, Weißwasser 22465
Glasmuseum Wertheim, Wertheim 22502
Glasmuseum Zalto, Nagelberg 02420
Glass Gallery and Main Gallery, El Paso 48323

The Glass Museum, Dunkirk 48210
Glass Museum, Harrachov 09507
Glass Museum, Kamenický Šenov 09571
Glass Museum, Marinha Grande 35756
Glass Museum, Nový Bor 09683
Glass Museum, San Ildefonso 39957
Glass Museum of Hsinchu City, Hsinchu 42206
Glass Museum Železný Brod, Železný Brod 09938
Glasschmelzofenbau-Hütte mit Glasausstellung, Plößberg 21066
Glasses Museum, Morez 14272
Glassware and Ceramic Museum of Iran, Tehrān 24654
Glastonbury Abbey Museum, Glastonbury 44306
Glastonbury Gild Arts Center, New Canaan 51036
Glastonbury Lake Village Museum, Glastonbury 44307
Glastonbury Museum, Glastonbury 48916
Glauberg-Museum, Glauburg 18655
Glazovskij Kraevedčeskij Muzej, Glazov 36924
GLBT Historical Society Museum, San Francisco 52775
Glebe House and Gallery, Letterkenny 24861
The Glebe House Museum, Woodbury 54275
Gledswood Farm Museum and Homestead, Catherine Fields 00961
Das Gleimhaus, Halberstadt 18877
Glen Eira City Council Gallery, Caulfield South 00962
Glen Ewen Community Antique Centre, Glen Ewen 05871
Glen Helen Ecology Institute Trailside Museum, Yellow Springs 54329
Glen Innes Museum, Glen Innes 01114
Glenbow Museum, Calgary 05519
Glencairn Museum, Bryn Athyn 47107
Glencoe and North Lorn Folk Museum, Glencoe 44310
Glencolmcille Folk Village Museum, Glencolmcille 24839
Glendon Gallery, Toronto 07111
Glendower State Memorial, Lebanon 50011
Glenesk Folk Museum, Brechin 43596
Glenesk Folk Museum, Glenesk 44311
Glenfarclas Distillery Museum, Ballindalloch 43359
Glenfinnan Station Museum, Fort William 44247
Glengarry Pioneer Museum, Dunvegan 05726
Glengarry Sports Hall of Fame, Maxville 06253
Glenhyrst Art Gallery of Brant, Brantford 05469
Glenluce Motor Museum, Glenluce 44312
Glenn H. Curtiss Museum, Hammondsport 49143
Glenn House, Cape Girardeau 47233
Glenreagh Memorial Museum, Glenreagh 01116
Glensheen-Historic Congdon Estate, Duluth 48195
Glentworth Museum, Glentworth 05874
Glenview Area Historical Museum, Glenview 48937
Glessner House Museum, Chicago 47478
Gletschergarten Luzern, Luzern 41590
Glimmingehus, Hammenhög 40581
Glims Talomuseo, Espoo 10729
Global Arts Link, Ipswich 01176
Global Health Odyssey, Atlanta 46440
Globenmuseum der Österreichischen Nationalbibliothek, Wien 02996
Glocken-Museum, Siegen 21754
Glockenmuseum, Apolda 16565
Glockenmuseum, Laucha an der Unstrut 19843
Glockenmuseum der Glockengiesserei, Innsbruck 02118
Glockenschmiede, Ruhpolding 21446
Gloddfa Ganol Slate Mine, Blaenau Ffestiniog 43532
Glomdalsmuseet, Elverum 33480
Glommersträsks Hembygdsmuseum, Glommersträsk 40538
Glor na nGael Museum, Carrickmacross 24752
Glore Psychiatric Museum, Saint Joseph 52561
Gloria Maris Schelpengalerie, Giethoorn 32253
Glossop Heritage Centre, Glossop 44314
Gloucester County Historical Society Museum, Woodbury 54277
Gloucester Folk Museum, Gloucester 44316
Gloucester Historic Museum, Gloucester 01117
Gloucester Museum, Gloucester 05875
Gloucestershire Warwickshire Railway, Toddington 45862
Glove Museum, New York 51184
Glovertown Heritage Museum, Glovertown 05876
Glud Museum, Horsens 10096
Glyndor Gallery and Wave Hill House Gallery, Bronx 47045
The Glynn Art Association, Saint Simons Island 52647
Glynn Vivian Art Gallery, Swansea 45813
Gnadenhutten Historical Park and Museum, Gnadenhutten 48951
Gneisenau Museum, Schildau, Gneisenaustadt 21548
Goa Science Centre, Panaji 24340
Gobabis Museum, Gobabis 31739
Goclette La Recouvrance, Brest 12159
Godalming Museum, Godalming 44322
Godegårds Bruks- och Porslinsmuseum, Motala 40751
Godfrey Dean Art Gallery, Yorkton 07414
Godøy Kystmuseum, Godøy 33514
Godwin-Ternbach Museum, Flushing 48576
Göcseji Falumúzeum, Zalaegerszeg 23970
Göcseji Múzeum, Zalaegerszeg 23971
Gödöllői Királyi Kastélymúzeum, Gödöllő 23565
Gödöllői Városi Múzeum, Gödöllő 23566
Göltzschtalgalerie-Nicolaikirche, Auerbach, Vogtland 16607
Goemanszorg - Streek- en Landbouwmuseum, Dreischor 32156
Gömöri Múzeum, Putnok 23786
Göreme Açıkhava Müzesi, Nevşehir 42915
Göreme Open Air Museum, Nevşehir 42915
Görög-Római Szobormásolatok Kiállítása, Tata 23917
Görög Templom Kiállítóhely, Vác 23943
Görögkatolikus Egyházművészeti Gyűjtemény, Nyíregyháza 23732
Göschenhaus / Seume-Gedenkstätte, Grimma 18760
Gösta Serlachiuksen Taidemuseo, Mänttä 11093
Gösta Serlachius Museum of Fine Arts, Mänttä 11093
Göta Kanalutställning, Motala 40752
Göteborg Museum of Art, Göteborg 40547
Göteborg Natural History Museum, Göteborg 40548
Göteborgs Konsthall, Göteborg 40546
Göteborgs Konstmuseum, Göteborg 40547
Göteborgs Naturhistoriska Museum, Göteborg 40548
Göteborgs Stadsmuseum, Göteborg 40549
Goethe and Schiller Archive, Weimar 22402
Goethe-Gedenkstätte, Jena 19380
Goethe-Institut New York - Exhibitions, New York 51185
Goethe-Institut Sydney, Woollahra 01669
Goethe-Museum Düsseldorf, Düsseldorf 17990
Goethe-Museum Stützerbach mit Museum zur Geschichte des technischen Glases, Stützerbach 21924
Goethe National Museum with Goethe's Home, Weimar 22401
Goethe-Nationalmuseum mit Goethes Wohnhaus, Weimar 22401
Goethe- und Schiller-Archiv, Weimar 22402
Goethe's Garden House, Weimar 22403
Goethes Gartenhaus, Weimar 22403
Goethes Gartenhaus 2, Bad Sulza 16848
GoetheStadtMuseum Ilmenau, Ilmenau 19320
Goetz Collection, München 20478
Goetz/Fleischack Museum, Potchefstroom 38713
Gohltalmuseum, Neu-Moresnet 03910
Gol Bygdemuseet, Gol 33515
Golan Archaeological Museum, Qatzrin 25079
Gold Coast City Art Gallery, Surfers Paradise 01537
Gold Coast Railroad Museum, Miami 50624
Gold Discovery Museum, Coloma 47696
Gold Museum, Ballarat 00827
Gold of Africa Museum, Cape Town 38584
Gold Prospector Museum, Tankavaara 11370
Gold Treasury Museum, Melbourne 01281
Gold- und Silberschmiedemuseum, Mülsen 20419
Gold- und Silberschmiedemuseum, Wien 02997
Goldberg-Museum, Riesbürg 21331
Goldborgsund Museum, Nykøbing Falster 10200
Golden and District Museum, Golden 05880
Golden Ball Tavern Museum, Weston 54105
Golden Bay Museum and Gallery, Golden Bay 33110
Golden City Paddle Steamer Museum, Ballarat 00828
Golden Dragon Museum, Bendigo 00860
Golden Gate Railroad Museum, Sunol 53374
Golden Hind Museum, Brixham 43642
Golden Hinde Educational Trust, London 44857
Golden Key Museum, Karlovy Vary 09576
Golden Pioneer Museum, Golden 48960
Golden Spike National Historic Site, Brigham City 47025
Golden State Model Railroad Museum, Point Richmond 52013
Golden Unicorn Pharmacy Museum, Kőszeg 23683
Goldsmiths' Hall, London 44858
Goldsmith's House Museum, La Habana 09118
The Goldstein Museum of Design, Saint Paul 52616
Goldstream Region Museum, Victoria 07248
Golestan Palace Museum, Tehrān 24668
Goleşti National Museum, Ştefaneşti 36597
Golf Museum, Far Hills 48506
Golf Museum, Newport News 51333
Golfmuseum, Regensburg 21223
Goliad State Historical Park, Goliad 48967
Golikov House-Museum, Palech 37524
Het Gols-Station, Winterswijk 32956
Gomez Foundation for Mill House, Marlboro 50494
Gomshall Gallery, Gomshall 44327
Goncharov Museum, Uljanovsk 37898
Gongju Folk Drama Museum, Gongju 29816
Gongju National Museum, Gongju 29817
Gongpyeong Art Center, Seoul 29878
Gonzales Memorial Museum, Gonzales 48969
Goochland County Museum, Goochland 48970
Goodhue County Historical Society Museum, Red Wing 52273
Goodman Museum, Tyler 53633
Goodnow Museum, Manhattan 50431
Goodsoil Historical Museum, Goodsoil 05882
Goodwood House, Chichester 43827
Goois Museum, Hilversum 32388
Goold Catholic Museum, East Melbourne 01054
Goole Museum and Art Gallery, Goole 44329
Goolwa National Trust Museum, Goolwa 01118
Goors Historisch Museum, Goor 32261
Goppert Gallery, Leavenworth 50009
Gorcums Museum, Gorinchem 32263
Gordion Müzesi, Polatlı 42925
Gordion Museum, Polatlı 42925
Gordon Brown Collection, Ravenshead 45463
Gordon Highlanders Museum, Aberdeen 43242
Gordon Museum, London 44859
Gordon Snelgrove Art Gallery, Saskatoon 06914
Gore Airforce Museum, Gore 33112
Gore Historical Museum and Hokonui Heritage Research Centre, Gore 33113
Gore Place, Waltham 53809
Gorenjski Muzej Kranj, Kranj 38437
Gorgan Museum, Gorgan 24605
Gorgas House Museum, Tuscaloosa 53618
Goris Local Tradition Museum, Goris 00703
Goriški Muzej, Nova Gorica 38489
Gorj District Museum, Târgu Jiu 36666
Gorj Museum of Folk Architecture, Bumbeşti-Jiu 36236
Gorj Museum of Folk Architecture, Urlaţi 36715
Gorki Memorial Museum, Kazan 37011
Gornjesavski Muzej Jesenice, Jesenice 38425
Górnośląski Park Etnograficzny, Chorzów 34603
Gornyj Muzej, Sankt-Peterburg 37639
Gorodec Regional Museum, Gorodec 36928
Gorodeckij Kraevedčeskij Muzej, Gorodec 36928
Gorodka Galeries, Sarlat-la-Canéda 15636
Gorodskoj Istoriko-Kraevedčeskij Muzej, Ridder 29724
Gorodskoj Kraevedčeskij Muzej, Elec 36910
Gorodskoj Muzej Boevoj Slavy, Kaluga 36995
Gorodskoj Muzej Iskusstvo Omska, Omsk 37497
Gorodskoj Muzej Teatralnogo Iskusstva, Omsk 37498
Gorodskoj Vystavočnyj Zal, Petrozavodsk 37557
Gorrotxategi Konfiteri Museoa, Tolosa 40141
Gorsium Roman Ruin, Tác 23909
Gorsium Szabadtéri Múzeum, Tác 23909
Gorsline Museum, Bussy-le-Grand 12212
Goschenhoppen Folklife Museum, Green Lane 49042
Gosford Regional Gallery and Arts Centre, Gosford 01122
Goshen Historical Society Museum, Goshen 48977
Goslarer Museum, Goslar 18708
Gosport Museum, Gosport 44334
Gosudarstvennaja Kollekcija Unikalnych Muzykalnych Instrumentov, Moskva 37183
Gosudarstvennaja OK Istoriko-ètnografičeskij i Architekturnyj Muzej-zapovednik Staraja Sarepta, Volgograd 37954
Gosudarstvennaja Tretjakovskaja Galerja, Moskva 37184
Gosudarstvennaja Tretjakovskaja Galerja na Krymskom Valu, Moskva 37185
Gosudarstvennoe Muzejnoe Objedinenie Chudožestvennaja Kultura Rossijskogo Severa, Archangelsk 36769
Gosudarstvennyj Biologičeskij Muzej im. K.A. Timirjazeva, Moskva 37186
Gosudarstvennyj Borodinskij Voenno-istoričeskij Muzej-zapovednik, Borodino 36819
Gosudarstvennyj Centr Sovremennogo Iskusstva, Moskva 37187
Gosudarstvennyj Centralnyj Muzej Muzykalnoj Kultury im. M.I. Glinki, Moskva 37188
Gosudarstvennyj Centralnyj Muzej Sovremennoj Istorii Rossii, Moskva 37189
Gosudarstvennyj Centralnyj Teatralnyj Muzej im. Bachrušina, Moskva 37190
Gosudarstvennyj Chudožestvenno-architekturnyj Dvorcovo-parkovyj Muzej-zapovednik Oranienbaum, Lomonosov 37108
Gosudarstvennyj Chudožestvenno-architekturnyj Dvorcovo-parkovyj Muzej-zapovednik Pavlovsk, Pavlovsk 37528
Gosudarstvennyj Chudožestvennyj Istoriko-architekturnyj i prirodno-landšaftnyj Muzej-zapovednik Kolomenskoe, Moskva 37191
Gosudarstvennyj Chudožestvennyj Muzej Altajskogo Kraja, Barnaul 36790
Gosudarstvennyj Dom-muzej P.I. Čajkovskogo, Klin 37052
Gosudarstvennyj Dom-muzej I.V. Stalina, Gori 16340
Gosudarstvennyj Ėrmitaž, Sankt-Peterburg 37640
Gosudarstvennyj Geologičeskij Muzej im V.I. Vernadskogo, Moskva 37192
Gosudarstvennyj Istoričeskij Muzej, Moskva 37193
Gosudarstvennyj Istoričeskij Muzej Kyrgyzskoj Respubliki, Biškek 29954
Gosudarstvennyj Istoričeskij Zapovednik Gorki Leninskie, Gorki Leninskie 36926
Gosudarstvennyj Istoriko-architekturnyj i Chudožestvennyj Muzej-zapovednik Drevnij Derbent, Derbent 36874
Gosudarstvennyj Istoriko-architekturnyj i Ėtnografičeskij Muzej-zapovednik Kiži, Kiži 37050
Gosudarstvennyj Istoriko-architekturnyj i Prirodno-landšaftnyj Muzej Zapovednik Izborsk, Izborsk 36953
Gosudarstvennyj Istoriko-architekturnyj i Prirodnyj Muzej-Zapovednik Monrepo, Vyborg 37981
Gosudarstvennyj Istoriko-chudožestvennyj Dvorcovo-parkovyj Muzej-zapovednik Gatčina, Gatčina 36920
Gosudarstvennyj Istoriko-ètnografičeskij Muzej-zapovednik Šušenskoe, Šušenskoe 37811
Gosudarstvennyj Istoriko-kulturnyj i Prirodnyj Muzej-zapovednik A.S. Griboedova - Chmelita, Chmelita 36863
Gosudarstvennyj Istoriko-kulturnyj Muzej-zapovednik "Moskovskij Kreml", Moskva 37194
Gosudarstvennyj Istoriko-literaturnyj i Prirodnyj Muzej-zapovednik A.A. Bloka, Solnečnogorsk 37786
Gosudarstvennyj Istoriko-literaturnyj Muzej-Zapovednik A.S. Puškina s usadbami Vjazemy i Zacharovo, Bolšie Vjazemy 36814
Gosudarstvennyj Istoriko-memorialnyj Sankt-Peterburgskij Muzej Smolnyj, Sankt-Peterburg 37641
Gosudarstvennyj Istoriko-memorialnyj Zapovednik Rodina V.I. Lenina, Uljanovsk 37897
Gosudarstvennyj Literaturno-memorialnyj Muzej Anny Achmatovoj v Fontannom Dome, Sankt-Peterburg 37642
Gosudarstvennyj Literaturno-memorialnyj Muzej N.A. Dobroljubova, Nižnij Novgorod 37451
Gosudarstvennyj Literaturno-memorialnyj Muzej im. M.Ju. Lermontova, Pjatigorsk 37564
Gosudarstvennyj Literaturno-memorialnyj Muzej N. Ostrovskogo, Soči 37782
Gosudarstvennyj Literaturno-memorialnyj Muzej A.N. Radiščeva, Radiščevo 37589
Gosudarstvennyj Literaturno-memorialnyj Muzej-zapovednik A.P. Čechova, Melichovo 37129
Gosudarstvennyj Literaturno-memorialnyj Muzej-zapovednik N.A. Nekrasova "Karabicha", Karabicha 37002
Gosudarstvennyj Literaturno-memorialnyj Muzejnyj Kompleks Sabita Mukanova i Gabita Musrepova, Almaty 29681
Gosudarstvennyj Literaturnyj Muzej, Moskva 37195
Gosudarstvennyj Literaturnyj Muzej A.M. Gorkogo, Nižnij Novgorod 37452
Gosudarstvennyj Literaturnyj Muzej im. I.S. Turgeneva, Orël 37507
Gosudarstvennyj Memorialnyj Dom-muzej Bulata Okudžavy, Peredelkino 37541
Gosudarstvennyj Memorialnyj Dom-muzej N.A. Rimskogo-Korsakova, Tichvin 37841
Gosudarstvennyj Memorialnyj i Prirodnyj Muzej-zapovednik A.N. Ostrovskogo - Ščelykovo, Ščelykovo 37756
Gosudarstvennyj Memorialnyj i Prirodnyj Muzej-zapovednik I.S. Turgeneva, Spasskoe-Lutovinovo 37792
Gosudarstvennyj Memorialnyj Istoriko-chudožestvennyj i Prirodnyj Muzej-zapovednik V.D. Polenova, Strachovo 37808
Gosudarstvennyj Memorialnyj Muzej im. A.V. Suvorova, Sankt-Peterburg 37643
Gosudarstvennyj Memorialnyj Muzej Oborony i Blokady Leningrada, Sankt-Peterburg 37644
Gosudarstvennyj Memorialnyj Muzej-usadba V.G. Belinskogo, Belinskij 36797
Gosudarstvennyj Moskovskij Muzej Naivnogo Iskusstva, Moskva 37196
Gosudarstvennyj Muzej Detskich Teatrov, Moskva 37197
Gosudarstvennyj Muzej K.A. Fedina, Saratov 37749
Gosudarstvennyj Muzej - Gumanitarnyj Centr 'Preodolenie' im. N.A. Ostrovskogo, Moskva 37198
Gosudarstvennyj Muzej Istorii Aviacii, Sankt-Peterburg 37645
Gosudarstvennyj Muzej Istorii i Kultury Belarusa, Minsk 03250
Gosudarstvennyj Muzej Istorii Kosmonavtiki im. K.Ė. Ciolkovskogo, Kaluga 36996
Gosudarstvennyj Muzej Istorii Religii, Sankt-Peterburg 37646
Gosudarstvennyj Muzej Istorii Sankt-Peterburga, Sankt-Peterburg 37647
Gosudarstvennyj Muzej Izobrazitelnych Iskusstv Respubliki Kalmykija, Ėlista 36911
Gosudarstvennyj Muzej A.I. Kuprina, Narovčat 37445
Gosudarstvennyj Muzej V.V. Majakovskogo, Moskva 37199

Gosudarstvennyj Muzej-masterskaja Skulptora A.S. Golubkinoj, Moskva 37200
Gosudarstvennyj Muzej Oborony Moskvy, Moskva 37201
Gosudarstvennyj Muzej Palechskogo Iskusstva, Palech 37526
Gosudarstvennyj Muzej-panorama Stalingradskaja Bitva, Volgograd 37955
Gosudarstvennyj Muzej Političeskoj Istorii Rossii, Sankt-Peterburg 37648
Gosudarstvennyj Muzej A.S. Puškina, Moskva 37202
Gosudarstvennyj Muzej-Reservat Pamjatniki Drevnego Taraza, Taraz 29732
Gosudarstvennyj Muzej Vadima Sidura, Moskva 37203
Gosudarstvennyj Muzej L.N. Tolstogo, Moskva 37204
Gosudarstvennyj Muzej L.N. Tolstogo, Moskva 37205
Gosudarstvennyj Muzej L.N. Tolstogo "Jasnaja Poljana", Jasnaja Poljana (Tula) 36975
Gosudarstvennyj Muzej-usadba Archangelskoe, Archangelskoe 36773
Gosudarstvennyj Muzej-usadba P.I. Čajkovskogo, Votkinsk 37977
Gosudarstvennyj Muzej Vyborgskij Zamok, Vyborg 37982
Gosudarstvennyj Muzej-zapovednik S.A. Esenina, Konstantinovo 37056
Gosudarstvennyj Muzej-zapovednik M.A. Šolochova, Vešenskaja 37925
Gosudarstvennyj Muzej Životnovodstva im. Akademika E.F. Liskuna, Moskva 37206
Gosudarstvennyj Nacionalnyj Muzej Ust-Ordynskogo Burjatskogo Avtonomnogo Okruga, Ust-Ordynsk 37913
Gosudarstvennyj Naučno-issledovatelskij Muzej Architektury im. A.V. Ščuseva, Moskva 37207
Gosudarstvennyj Okružnoj Muzej Prirody i Čeloveka Goroda Chanty-Mansijsk, Chanty-Mansijsk 36862
Gosudarstvennyj Russkij Muzej, Sankt-Peterburg 37649
Gosudarstvennyj Russkij Muzej, Sankt-Peterburg 37650
Gosudarstvennyj Severnyj Morskoj Muzej, Archangelsk 36770
Gosudarstvennyj Vladimiro-Suzdalskij Istoriko-architekturnyj i Chudožestvennyj Muzej-zapovednik, Vladimir 37934
Gosudarstvennyj Voenno-istoričeskij i Prirodnyj Muzej-zapovednik Kulikovo Pole, Tula 37866
Gosudarstvennyj Voenno-istoričeskij Muzej-zapovednik Prochorovskoe Pole, Prochorovka 37571
Gosudarstvennyj Vystavočnyj Zal Beljaevo, Moskva 37208
Gosudarstvennyj Vystavočnyj Zal Chodynka, Moskva 37209
Gosudarstvennyj Vystavočnyj Zal Kovčeg, Moskva 37210
Gosudarstvennyj Vystavočnyj Zal-muzej Nasledie, Moskva 37211
Gosudarstvennyj Vystavočnyj Zal Na Kaširke, Moskva 37212
Gosudarstvennyj Vystavočnyj Zal Na Soljanke, Moskva 37213
Gosudarstvennyj Vystavočnyj Zal Peresvetov Pereulok, Moskva 37214
Gosudarstvennyj Vystavočnyj Zal Solncevo, Moskva 37215
Gosudarstvennyj Vystavočnyj Zal Zamoskvorečie, Moskva 37216
Gothenburg Radio Museum, Göteborg 40554
Gotisches Haus, Wörlitz 22638
Gotlands Konstmuseum, Visby 41069
Gotland's Museum of Natural History, Visby 41070
Gotlands Naturmuseum, Visby 41070
Goto Sumio Bijutsukan, Kamifurano 28646
Goto Sumio Museum of Art, Kamifurano 28646
Gotoh Bijutsukan, Tokyo 29394
The Gotoh Museum, Tokyo 29394
Gottfried-August-Bürger-Museum, Molmerswende 20370
Gottfried-Keller-Zentrum, Glattfelden 41439
Gotthelf-Stube, Lützelflüh 41578
Gottlieb-Daimler-Geburtshaus, Schorndorf 21625
Gottlieb Daimler-Gedächtnisstätte, Stuttgart 21934
Gottlieb-Häußler-Heimatmuseum, Filderstadt 18317
Gottliebin-Dittus-Haus, Bad Liebenzell 16779
Gottlob Auwärter Museum, Stuttgart 21935
Gottscheer Stube, Bad Wildbad 16858
Goudreau Museum of Mathematics in Art and Science, New Hyde Park 51063
Goulandris Collection, Athinai 22911
Goulandris Natural History Museum, Kifissia 23051
Goulburn Regional Art Gallery, Goulburn 01125
Gounaropoulos Museum of Zografou, Zografou 23247
Gourd Sculpture Museum, Shanlin 42323
Gouverneur Historical Association Museum, Gouverneur 48980
Het Gouverneurshuis, Heusden 32383
Government Art Collection, London 44860
Government Central Museum, Jaipur, Rajasthan 24233
Government Educational Museum, Bulandshahar 24104
Government Educational Museum, Deoria 24158
Government Educational Museum, Etawah 24170
Government House, Sydney 01547
Government House Museum and Heritage Property, Regina 06718
Government Industrial and Commercial Museum, Kolkata 24273
Government J.T.C. Museum, Faizabad 24173
Government Museum, Ajmer 24054
Government Museum, Alwar 24063
Government Museum, Bharatpur 24084
Government Museum, Chennai 24115
Government Museum, Hassan 24206
Government Museum, Jodhpur 24245
Government Museum, Kodagu 24263
Government Museum, Mathura 24309
Government Museum, Mount Abu 24312
Government Museum, Pudukottai 24353
Government Museum, Udaipur 24414
Government Museum and Art Gallery, Chandigarh 24109
Government Museum Jhalawar, Jhalawar 24242
Government Museum Vellore, Vellore 24428
Governmental College of Engineering and Textile Technology Museum, Hooghly 24208
Governmental Educational Museum, Muzaffarnagar 24326
Governor Bent Museum, Taos 53440
Governor General's Foot Guards Museum, Ottawa 06546
Governor Henry Lippitt House Museum, Providence 52166
Governor John Langdon House, Portsmouth 52094
Governor Printz Park, Essington 48424
Governor Ross Plantation, Seaford 52987
Governor Seay Mansion, Kingfisher 49782
Governor Stephen Hopkins House, Providence 52167
Governor's Gallery, Santa Fe 52875
Governor's Mansion, Columbia 47725
The Governor's Mansion, Frankfort 48735
Governor's Mansion, Sacramento 52507
Govett-Brewster Art Gallery, New Plymouth 33156
Gower Heritage Centre, Gower 44340
Gowganda and Area Museum, Gowganda 05885
Gozo Cathedral Museum, Victoria 30652
Gozo Heritage, Gozo 30629
Graaf-Reinet Museum, Graaff-Reinet 38626
Grabkapelle auf dem Württemberg, Stuttgart 21936
Gråbo Minnenas Museum, Karlskoga 40630
Grabungsmuseum Kirchhof, Coburg 17716
Grabungsmuseum Langenburg, Langenburg 19824
Grace Campbell Gallery, Prince Albert 06662
Grace Darling Museum, Bamburgh 43369
Grace Hudson Museum and The Sun House, Ukiah 53638
The Grace Museum, Abilene 46124
Gracefield Arts Centre, Dumfries 44037
Graceland, Memphis 50573
Gracie Mansion, New York 51186
Grad Bogenšperk, Šmartno pri Litiji 38522
Grad Dobrovo, Dobrovo v Brdih 38413
Grad Podsreda, Podsreda 38494
Grad pri Gradu, Grad 38419
Gradska Chudožestvena Galerija, Berkovica 04986
Gradska Chudožestvena Galerija, Plovdiv 05127
Gradska Chudožestvena Galerija Petko Čurčuliev, Dimitrovgrad 05014
Gradska Galerija Bihać, Bihać 04258
Gradska Galerija Labin, Labin 08905
Gradski Archeologičeski Muzej, Chisar 05006
Gradski Archeologičeski Muzej, Nesebăr 05096
Gradski Archeologičeski Muzej, Sandanski 05157
Gradski Istoričeski Muzej, Asenovgrad 04971
Gradski Istoričeski Muzej, Balčik 04976
Gradski Istoričeski Muzej, Botevgrad 04992
Gradski Istoričeski Muzej, Čirpan 05009
Gradski Istoričeski Muzej, Dalgopol 05011
Gradski Istoričeski Muzej, Karnobat 05051
Gradski Istoričeski Muzej, Kavarna 05055
Gradski Istoričeski Muzej, Klisura 05069
Gradski Istoričeski Muzej, Lom 05084
Gradski Istoričeski Muzej, Melnik 05091
Gradski Istoričeski Muzej, Nova Zagora 05097
Gradski Istoričeski Muzej, Orjachovo 05101
Gradski Istoričeski Muzej, Panagjurište 05102
Gradski Istoričeski Muzej, Peruštica 05112
Gradski Istoričeski Muzej, Samakov 05156
Gradski Istoričeski Muzej, Sevlievo 05159
Gradski Istoričeski Muzej, Teteven 05226
Gradski Istoričeski Muzej Aleko Konstantinov, Svištov 05221
Gradski Istoričeski Muzej Iskra, Kazanlăk 05058
Gradski Muzej, Bečej 38032
Gradski Muzej, Bjelovar 08845
Gradski Muzej, Kosovska Mitrovica 38105
Gradski Muzej, Križevci 08902
Gradski Muzej, Požega 08930
Gradski Muzej, Senj 08945
Gradski Muzej, Sremski Karlovci 38152
Gradski Muzej, Umag 08974
Gradski Muzej Bakar, Bakar 08841
Gradski muzej Čazma, Čazma 08855
Gradski Muzej Jastrebarsko i Gradska Galerija, Jastrebarsko 08884
Gradski Muzej Karlovac, Karlovac 08886
Gradski Muzej Korčula, Korčula 08894
Gradski Muzej Kruševac, Kruševac 38112
Gradski Muzej Makarska, Makarska 08907
Gradski muzej Nova Gradiška, Nova Gradiška 08916
Gradski Muzej Sisak, Sisak 08952
Gradski Muzej Sombor, Sombor 38148
Gradski Muzej Varaždin, Varaždin 08975
Gradski Muzej - Városi Muzeum, Subotica 38156
Gradski Muzej Virovitica, Virovitica 08978
Gradski Muzej Vršac, Vršac 38163
Graeco-Roman Museum, Alexandria 10471
Gräfliche Sammlungen Schloss Erbach, Erbach, Odenwaldkreis 18191
Graeme Park/Keith Mansion, Horsham 49371
Grängesbergsbanornas Järnvägsmuseum, Grängesberg 40560
Graf Harrach'sche Familiensammlung, Rohrau 02593
Graf-Luxburg-Museum, Bad Bocklet 16699
De Grafelijke Torenmolen, Zeddam 32990
Grafička Zbirka Nacionalne i Sveučilišne Knjižnice, Zagreb 08992
Grafični Muzej, Rogaška Slatina 38507
Grafični Muzej Rogoška Slatina, Rogoška Slatina 38509
Grafik Museum Stiftung Schreiner, Bad Steben 16845
Grafikens Hus, Mariefred 40730
Grafisch Historisch Centrum - Drukkerijmuseum, Etten Leur 32224
Grafisch Musem-Atelier in den Groenen Zonck, Wouw 32970
Grafisch Museum Groningen, Groningen 32280
Grafisch Museum Zutphen, Zutphen 33018
Grafische Sammlung, Augsburg 16619
Grafische Sammlung, Nürnberg 20766
Grafschafter Museum im Moerser Schloss, Moers 20362
Grafschaftsmuseum, Wertheim 22503
Grafton Historical Society, Grafton 48982
Grafton Regional Gallery, Grafton 01127
Graham County Historical Museum, Thatcher 53491
Grain Store Museum, Vihti 11464
Grainger Museum, Melbourne 01282
Grains Industry Museum, Cabanatuan 34357
Gralglasmuseum, Dürnau 17984
Gramophone Records Museum, Cape Coast 22852
Grampian Transport Museum, Alford, Aberdeenshire 43277
Gramzower Heimatstuben, Gramzow 18736
Granary Road Museum, Renko 11266
Granatium, Radenthein 02549
Grand Army of the Republic Civil War Museum, Philadelphia 51840
Grand Army of the Republic Memorial and Veteran's Military Museum, Aurora 46492
Grand Canyon Museum Collection, Grand Canyon 48985
Grand Coteau Heritage and Cultural Centre, Shaunavon 06949
Grand County Museum, Hot Sulphur Springs 49376
Grand Curtius, Liège 03796
Grand Encampment Museum, Encampment 48395
Grand Falls Museum, Grand Falls 05889
Grand Forks Art Gallery, Grand Forks 05894
Grand Forks County Historical Society Museum, Grand Forks 48987
Grand Gulf Military State Park Museum, Port Gibson 52033
Grand-Hornu, Hornu 03713
Grand Island Trader's Cabin, Munising 50907
Grand Lake Area Historical Museum, Grand Lake 48996
The Grand Lodge of Scotland Museum, Edinburgh 44126
Grand Manan Museum and Walter B. McLaughlin Marine Gallery, Grand Manan 05896
Grand Portage National Monument, Grand Portage 48999
Grand-Pré National Historic Site, Grand-Pré 05897
Grand Prix Museum, Macau 07993
Grand Quartier Général Allemand, Brûly-de-Pesche 03429
Grand Rapids Art Museum, Grand Rapids 49003
Grand Rapids Children's Museum, Grand Rapids 49004
Grand River Historical Society & Museum, Chillicothe 47532
Grand Shaft, Dover 44015
Grand Traverse Heritage Center, Traverse City 53554
Grand Village of the Natchez Indians, Natchez 50984
La Grande Ferme, Saint-Joachim 06833
Grande Museo, Fabriano 26068
Grande Prairie Museum, Grande Prairie 05900
Grangärde Hembygdsgården, Grangärde 40565
The Grange, Toronto 07112
La Grange aux Oies, Collonges-la-Rouge 12577
La Grange Béranger, Beblenheim 11914
Grange des Artisans, Sainte-Suzanne 15580
Grangemouth Museum, Grangemouth 44342
Granger Homestead Society Museum, Canandaigua 47213
Granger House, Marion 50480
Granges de Servette, Chens-sur-Léman 12502
Granisle Museum, Granisle 05903
Granitabbaumuseum, Königshain 19652
Granitsa Folklore Museum, Granitsa 23009
Granma Memorial, La Habana 09106
Granösunds Fiskeläge, Södra Vallgrund 11318
Grans Lantbruksmuseum, Piteå 40808
Grant County Historical Museum, Canyon City 47231
Grant County Historical Museum, Elbow Lake 48333
Grant County Historical Museum, Ephrata 48405
Grant County Museum, Elgin 48336
Grant County Museum, Hyannis 49456
Grant County Museum, Medford 50550
Grant County Museum, Sheridan 53073
Grant County Museum, Ulysses 53641
Grant-Humphreys Mansion, Denver 48079
Grant-Kohrs Ranch, Deer Lodge 48028
Grant Medical College Museum, Mumbai 24316
Grant Museum of Zoology and Comparative Anatomy, London 44861
Grantham Museum, Grantham 44345
Grantown Museum, Grantown-on-Spey 44347
Grant's Birthplace State Memorial, Point Pleasant 52010
Grants Old Mill, Winnipeg 07362
Grants Pass Museum of Art, Grants Pass 49016
Granville County Historical Society Museum, Oxford 51651
Granville Historical Museum, Granville 49022
Granvin Bygdamuseum, Granvin 33516
Grapevine Museum, Chiltern 00972
Graphic and Cartographical Collections of the Jagiellonian Library, Kraków 34815
Graphic Art Collection, Nürnberg 20766
Graphic Museum, Rogaska Slatina 38507
Graphic Museum Rogaska Slatina, Rogoška Slatina 38509
Graphik-Kabinett der Stadt Backnang, Backnang 16658
Graphikmuseum Pablo Picasso Münster, Münster 20510
Graphische Sammlung, Kassel 19464
Graphische Sammlung am Kunsthistorischen Institut der Universität Tübingen, Tübingen 22114
Graphische Sammlung der Eidgenössischen Technischen Hochschule Zürich, Zürich 42073
Graphische Sammlung der Universität, Erlangen 18221
Graphothek Berlin, Berlin 17119
Graslitzer Gedenk- und Informationsraum, Aschaffenburg 16583
Grassagården Miljömuseum, Strängnäs 40953
Grassi-Museum für Angewandte Kunst, Leipzig 19911
Grassi-Museum für Musikinstrumente der Universität Leipzig, Leipzig 19912
GRASSI-Museum für Völkerkunde zu Leipzig, Leipzig 19913
Grassi-Museum of Applied Arts, Leipzig 19911
Graue Mill and Museum, Oak Brook 51451
Grave Creek Mound Archaeology Complex, Moundsville 50868
Gravelbourg and District Museum, Gravelbourg 05906
Gravensteen Museum voor Gerechtsvoorwerpen, Gent 03627
Graves Art Gallery, Sheffield 45651
Gravesham Museum, Gravesend, Kent 44350
Gravestone Museum, İstanbul 42844
Gray Gallery, Quincy 52214
Gray's Printing Press, Strabane 45774
Grays School of Art Gallery and Museum, Aberdeen 43243
Grazer Kunstverein, Graz 01970
Great Bardfield Cage, Great Bardfield 44353
Great Bardfield Cottage Museum, Great Bardfield 44354
Great Basin Historical Society & Museum, Delta 48050
Great Blacks in Wax Museum, Baltimore 46566
Great Brak River Museum, Great Brak River 38635
Great Central Railway Museum, Loughborough 45032
Great Cobar Heritage Centre, Cobar 00988
Great Explorations, Saint Petersburg 52639
Great Fish River Museum, Cradock 38599
Great Hall, Conwy 43887
Great Hall of the Clans Museum, English Town 05779

The Great Hanshin-Awaji Earthquake Memorial, Kobe 28725
Great Harbor Maritime Museum, Northeast Harbor 51417
Great Lakes Historical Museum, Tuncurry 01607
Great Lakes Lighthouse Museum, Mackinaw City 50353
Great Lakes Naval Memorial and Museum, Muskegon 50923
Great Lakes Science Center, Cleveland 47631
Great Laxey Wheel and Mines Trail, Laxey 44650
Great Orme Tramway, Llandudno 44758
Great Ormond Street Hospital for Children Museum, London 44862
Great Overland Station, Topeka 53534
Great Palace at the Museum-Preserve, Lomonosov 37106
Great Palace Mosaic Museum, İstanbul 42810
Great Patriotic War Museum, Moskva 37149
Great Plains Art Collection in the Christlieb Gallery, Lincoln 50096
Great Plains Black Museum, Omaha 51551
Great Plains Transportation Museum, Wichita 54143
Great Sandhills Museum, Sceptre 06939
Great Valley Museum of Natural History, Modesto 50774
Great War Flying Museum, Cheltenham 05605
Great Yarmouth Museums, Great Yarmouth 44358
Great Zimbabwe Site Museum, Masvingo 54675
Greater Harvard Area Historical Society, Harvard 49206
Greater Hazleton Historical Society Museum, Hazleton 49233
Greater Loveland Historical Society Museum, Loveland 50302
Greater Manchester Police Museum, Manchester 45083
Greater Port Jefferson Museum, Port Jefferson 52036
The Greater Reston Arts Center, Reston 52299
Greater Vernon Museum, Vernon, British Colombia 07236
Greek Catholic Art Collection, Nyíregyháza 23732
Greek Church Exhibition Space, Vác 23943
Greek Federation of Associations of Friends of Museums, Athinai 54867
Greek Orthodox Museum, Jerusalem 24988
Greek Shadow Theatre Museum, Zografou 23248
Greek Women's Association Collection, Rethymnon 23169
Green Bay Packers Hall of Fame, Green Bay 49037
Green Dragon Museum, Stockton-on-Tees 45756
Green Family Forge - A Blacksmith Museum, Trinity 07163
Green Gables House, Charlottetown 05593
Green Hill Center for North Carolina Art, Greensboro 49060
Green Howards Regimental Museum, Richmond, North Yorkshire 45489
Green Mountain Club, Waterbury Center 53958
Green Palace Museum, Tehrān 24669
Green Park Shipbuilding Museum and Yeo House, Port Hill 06640
Green River Museum, Woodbury 54276
Greenan Farm Museums, Greenan 24841
Greenaway Gallery, Camberwell 00937
Greenbelt Museum, Greenbelt 49044
Greene County Historical Museum, Waynesburg 54015
Greene County Historical Society Museum, Xenia 54317
Greene Memorial Museum, Milwaukee 50709
Greenfield Historical Society Museum, Greenfield 49057
Greenfield Museum, Greenfield 49056
Greenfield Valley Museum, Greenfield 44362
Greenhill Covenanter's House, Biggar 43482
Greenland National Museum and Archives, Nuuk 23262
Green's Windmill and Science Centre, Nottingham 45292
Greensboro Artists' League Gallery, Greensboro 49061
Greensboro Historical Museum, Greensboro 49062
Greensward Gallery, Corvallis 47835
Greentown Glass Museum, Greentown 49071
Greenville County Museum of Art, Greenville 49079
Greenville Museum of Art, Greenville 49074
Greenwich Heritage Centre, London 44863
Greenwich Local History Centre, London 44864
Greenwood Centre for Living History, Hudson 05998
Greenwood County Historical Society, Eureka 48441
Greenwood Great House Museum, Greenwood 28323
Greenwood Military Aviation Museum, Greenwood, Nova Scotia 05910
Greenwood Museum, Greenwood, British Columbia 05909
Gregg County Historical Museum, Longview 50194
Gregor Mendel Memorial, Brno 09430
Greinburg Castle, Ducal art collection Saxe-Coburg and Gotha & Upper Austrian Nautical Museum, Grein an der Donau 01989
Grekiskt Kulturcentrum, Stockholm 40896
Grenada National Museum, Saint George's 23273
Grenen Kunstmuseum, Skagen 10260
Grenfell and District History Museum, Grenfell 01130
Grenfell Community Museum, Grenfell 05911
Grenfell House Museum, Saint Anthony 06802
Grenjadarstadur Folk Museum, Húsavík 24002
Grenna Museum, Gränna 40563
Grenslandmuseum, Dinxperlo 32131
Grenzhuus, Schlagsdorf 21562
Grenzland-Museum Bad Sachsa, Bad Sachsa 16812
Grenzland- und Trenckmuseum, Waldmünchen 22302
Grenzlandheimatstuben des Heimatkreises Marienbad, Neualbenreuth 20590
Grenzlandmuseum, Schnackenburg 21588
Grenzlandmuseum Eichsfeld, Teistungen 22020
Grenzlandmuseum Raabs an der Thaya, Raabs an der Thaya 02546
Grenzlandmuseum Swinmark, Schnega 21592
Grenzmuseum Philippsthal (Werra), Philippsthal 21038
Grenzmuseum Schifflersgrund, Asbach-Sickenberg 16581
Grenzsteinmuseum Ostrach, Ostrach 20942
Grenzwald-Destillation Museum, Crottendorf 17752
Gresham History Museum, Gresham 49088
Greve Museum, Greve 10042
Grevenbroek Museum, Hamont-Achel 03673
Grevesmühlener Mühle, Grevesmühlen 18757
Grey Art Gallery, New York 51187
Grey Roots Heritage, Owen Sound 06561
Greybull Museum, Greybull 49089
Greyhound Hall of Fame, Abilene 46121
Greytown Museum, Greytown 38637
Grier Musser Museum, Los Angeles 50226
Grietje Tump Museum, Landsmeer 32462
Griffin Museum of Photography, Winchester 54220
Griffith Artworks, Nathan 01355
Griffith Pioneer Park Museum, Griffith 01131
Griffith Regional Art Gallery, Griffith 01132
Griffiths Sea Shell Museum and Marine Display, Lakes Entrance 01214
Grigore Antipa National Museum of Natural History, Bucureşti 36222
Grillparzer-Gedenkzimmer, Wien 02998
Grimsby Museum, Grimsby 05912
Grimsby Public Art Gallery, Grimsby 05913
Grindheim Bygdemuseum, Kollungtveit 33597
Grindsted Museum, Grindsted 10044
Grinnell College Art Gallery, Grinnell 49091
Grinnell Historical Museum, Grinnell 49092
Grinter Place, Kansas City 49681
Grinter Place State Historical Site, Fairway 48495
Gripes Modelltheatermuseum, Nyköping 40781
Gripsholm Castle, Mariefred 40731
Gripsholms Slott, Mariefred 40731
Grissom Air Museum, Peru 51788
Grist Mill, Keremeos 06063
Grizzly Bear Prairie Museum, Wanham 07277
Groam House Museum, Rosemarkie 45513
Grocer's Shop of Roskilde Museum, Roskilde 10248
Gród Piastowski w Gieczu, Giecz 34663
Grødaland Bygdetun, Nærbø 33689
Grodnenski Dziarzhauny Hystorika-Archealagičny Muziej, Grodno 03246
Grodno State Historical-Archaeological Museum, Grodno 03246
De Groene Schuur, Culemborg 32042
Grönegau-Museum, Melle 20272
Groeninge Museum, Brugge 03414
Groeningemuseum, Brugge 03414
Grönsöö Slott, Enköping 40498
Gróf Esterházy Károly Kastély- és Tájmúzeum, Pápa 23752
Grohar House, Spodnja Sorica 38525
Groharjeva Hiša v Sorici, Spodnja Sorica 38525
Grong Bygdamuseum, Grong 33521
Groninger Museum, Groningen 32281
Groninger Schaatsmuseum, Sappemeer 32744
Groot Constantia Manor House and Wine Museum, Constantia 38597
Groot Constantin Homestead and Wine Museum, Constantia 38598
Groot Kasteel van Loppem, Loppem 03832
Grootfontein Museum Das Alte Fort, Grootfontein 31740
Grootmoeders Keukenmuseum, Nijmegen 32588
Gropius House, Lincoln 50092
Grosse Ile Historical Museum, Grosse Ile 49093
Große Kunstschau im Roselius Museum, Worpswede 22691
Großer Wappensaal im Landhaus, Klagenfurt 02183
Großherzoglich-Hessische Porzellansammlung, Darmstadt 17788
Großmitterberger Troadkasten, Laussa 02269
Grosvenor Museum, Chester 43818
Groton Historical Museum, Groton 49096
Grotte de Bédeilhac, Tarascon-sur-Ariège 15857
Grotte de Blanot, Blanot 12017
Grotte de la Madeleine, Saint-Remèze 15481
Grotte de la Vache, Alliat 11557
Grotte de Pair non Pair, Prignac-et-Marcamps 14947
Grotte de Rouffignac, Rouffignac-Saint-Cernin 15124
Grotte d'Osselle, Saint-Vit 15540
Grotte et Parc Préhistoriques, Foissac 12950
Grotte Préhistorique Ornée, Miers 14117
Grotte Scladina et Centre Archéologique, Sclayn 04006
Grottes d'Arcy-sur-Cure, Arcy-sur-Cure 11667
Grottes de Vallorbe, Vallorbe 41975
Grottes Préhistoriques de Sare-Lezea, Sare 15630
Groudle Glen Railway, Onchan 45319
Grounds for Sculpture, Hamilton 49133
Group for Directors of Museums, Chester 55065
Group for Education in Museums, Gillingham, Kent 55066
Groupement des Amis de Musées de la Région Midi-Pyrénées, Montauban 54815
Groupement des Amis des Musées de la Région Centre, Orléans 54816
Groupement des Associations d'Amis de Musées de la Région Bretagne, Nevez 54817
Groupement des Associations d'Amis de Musées de la Région Languedoc-Roussillon, Montpellier 54818
Groupement des Associations d'Amis de Musées de la Région Limousin, Limoges 54819
Groupement des Associations des Amis de Musées Région Provence Alpes Côte d'Azur, Toulon (Var) 54820
Groupement Rhône-Alpes des Amis de Musée, Lyon 54821
Grout Museum of History and Science, Waterloo 53961
Grove House Museum, Ramsey 24917
Groveland Yosemite Gateway Museum, Groveland 49098
Grover Cleveland Birthplace, Caldwell 47162
Grover M. Hermann Fine Arts Center, Marietta 50474
Grovewood Gallery, Asheville 46373
Grubb Shaft Gold and Heritage Museum, Beaconsfield 00844
Grube Alte Hoffnung Erbstolln, Mittweida 20345
Grubenmann-Sammlung, Teufen 41932
Grünauer Wassersportmuseum, Berlin 17120
Gründerzeitmuseum im Gutshof Mahlsdorf, Berlin 17121
Grüttert Uhrenmuseum Bremen, Bremen 17553
Gruetunet Museum, Kirkenær i Solør 33591
Grumman Memorial Park, Calverton 47170
Grundtvigs Mindestuer i Udby, Lundby 10182
Grundy Art Gallery, Blackpool 43529
Grunt Gallery, Vancouver 07212
Grunwald Center for the Graphic Arts, Los Angeles 50227
Gruvmuseet, Sala 40826
Grytøy Bygdetun, Lundenes 33653
Gsellmanns Weltmaschine, Edelsbach bei Feldbach 01839
GSIS Gallery, Pasay 34472
Gu City Museum, Gucheng, Hubei 07739
Gu City Qi Ruins Museum, Linzi 07960
Guadalupe Cultural Arts Center, San Antonio 52715
Guam Museum, Agana 23285
Guan Return to the Ancient Dian Art Museum, Beijing 07540
Guandu Museum, Guandu 07719
Guangdong Beginxing Museum, Shixing 08236
Guangdong Folk Arts and Crafts Museum, Guangzhou 07723
Guangdong Museum, Guangzhou 07724
Guangdong Museum of Art, Guangzhou 07725
Guangdong Revolutionary History Museum, Guangzhou 07726
Guangning Museum, Guangning 07721
Guangshui Museum, Xiaogan 08377
Guangxi Natural Museum, Nanning 08054
Guangxi Zhuang Museum, Nanning 08055
Guangzhou Art Gallery, Guangzhou 07727
Guangzhou Fine Arts Museum, Guangzhou 07728
Guangzhou Huangpu War Exhibition Center, Guangzhou 07729
Guangzhou Liwan District Museum, Guangzhou 07730
Guangzhou Lu Xun Museum, Guangzhou 07731
Guangzhou Museum, Guangzhou 07732
Guard House and Soldiers Barracks, Fredericton 05832
Guard Ship 'Red Pennant' of 1911, Vladivostok 37950
Guardhouse Museum, Tattershall 45828
The Guards Museum, London 44865
Gucity Museum, Gucheng, Jiangsu 07740
Gudbrandsdal Krigsminnesamling, Kvam 33618
Guelph Civic Museum, Guelph 05917
Günter Grass Haus, Forum für Literatur und bildende Kunst, Lübeck 20075
Guernsey County Museum, Cambridge 47191
Guernsey Museum and Art Gallery, Saint Peter Port 45591
Guest House Museum, Fort Bragg 48596
Güzel Sanatlar Müzesi, Girne 33359
Güzelyurt Müzesi, Güzelyurt 33361
Guggenheim Bilbao Museoa, Bilbao 39043
Guggenheim Hermitage Museum, Las Vegas 49969
Guggenheim Museum Bilbao, Bilbao 39043
Guggenheim Museum Soho, New York 51188
Guido Gezelle Museum, Brugge 03415
Guido Gezellemuseum, Brugge 03415
Guifei Muwen Guan (Museum), Xinping 08401
Guild Gallery, Bristol 43636
Guild Hall Museum, East Hampton 48245
Guild Museum, Tîrgu Secuiesc 36694
Guild of Creative Art, Shrewsbury 53092
Guildford Cathedral Treasury, Guildford 44369
Guildford House Gallery, Guildford 44370
Guildford Museum, Guildford 44371
Guildhall, Exeter 44194
Guildhall, Leicester 44675
Guildhall, Thaxted 45845
Guildhall Art Gallery, London 44866
Guildhall Collection, Kingston-upon-Hull 44595
Guildhall Gallery, Winchester 46028
Guildhall Museum, Carlisle 43759
Guildhall Museum, Chichester 43828
Guildhall Museum, Rochester 45507
Guildhall Museum, Sandwich 45615
Guilford College Art Gallery, Greensboro 49063
Guilford Courthouse National Military Park, Greensboro 49064
Guilin Museum, Guilin 07741
Guilin Stone Inscription Museum, Guilin 07742
Guimet National Museum of Asian Arts, Paris 14725
Guinness Storehouse, Dublin 24794
Guinness World of Records Museum, Las Vegas 49970
Guiping Museum, Guiping 07745
Guitar Museo Ex Teresa Arte Actual, México 31048
Guizhou Provincial Museum, Guiyang 07746
Gukgiwon Taekwondo Museum, Seoul 29879
Gulbenes Vēstures un Mākslas Muzejs, Gulbene 30010
Gulf Beach Art Center, Indian Rocks Beach 49481
Gulf Coast Exploreum, Mobile 50767
Gulf Museum, Port-aux-Basque 06630
Gulf of Georgia Cannery, Richmond 06743
Gulf Shores Museum, Gulf Shores 49105
Gulf Station Historic Farm, Yarra Glen 01678
Gulli Skoletun, Andebu 33391
Gumanitarnyj Centr-muzej Tvorčestva Autsajderov, Moskva 37217
Gumbez Manace Literary Ethnographic Museum, Taš-Aryk 29965
Gumeracha and District History Centre, Gumeracha 01134
Gumtree Museum of Art, Tupelo 53613
Gun Museum, Nishinomote 29076
Gunarsa Museum, Klungkung 24516
Gundagai Historical Museum, Gundagai 01135
Gundelsheimer Oldtimermuseum, Gundelsheim, Württemberg 18835
Gunderson House, Kenyon 49743
Gunma-kenritsu Kindai Bijutsukan, Takasaki 29330
Gunma-kenritsu Rekishi Hakubutsukan, Takasaki 29331
Gunma-kenritsu Shizenshi Hakubutsukan, Tomioka 29512
Gunma Museum of Natural History, Tomioka 29512
Gunma Prefectural Museum of History, Takasaki 29331
Gunn Memorial Library and Museum, Washington 53842
Gunnebo House and Gardens, Mölndal 40742
Gunnebo Slott och Trädgårdar, Mölndal 40742
Gunnersbury Park Museum, London 44867
Gunnes gård, Upplands Väsby 41002
Gunnison County Pioneer and Historical Museum, Gunnison 49106
Gunnison Museum of Natural History, Pawling 51734
Gunsmith's Shop, Ilomantsi 10852
Gunston Hall Plantation, Mason Neck 50530
Gunze Museum, Ayabe 28385
Guovdageainnu Gilisillju, Kautokeino 33590
Gurkha Museum, Winchester 46029
Gurusaday Museum, Kolkata 24274
Gush Ezyon Museum, Gush Ezyon 24950
Gustaf-Dalman-Institut für biblische Landes- und Altertumskunde, Greifswald 18749
Gustafson Gallery, Fort Collins 48608
Gustav-Adolf-Gedenkstätte, Lützen 20105
Gustav-Adolf-Museum, Geleitshaus, Weißenfels 22455
Gustav-Freytag-Archiv und Museum, Wangen im Allgäu 22325
Gustav III:s Antikmuseum, Stockholm 40897
Gustav III:s Paviljong, Solna 40870
Gustav III's Museum of Antiquities, Stockholm 40897

Gustav III's Pavilion, Solna 40870
Gustav Jeeninga Museum of Bible and Near Eastern Studies, Anderson 46292
Gustav-Lübcke Museum, Hamm, Westfalen 18969
Gustav Wolf Kunstgalerie, Östringen 20875
Gustava Skiltera Memoriālais Muzejs, Rīga 30069
Gustavsbergs Porslinmuseum, Gustavsberg 40569
Gut of Canso Museum, Port Hastings 06639
Gut Seekamp der Hans Kock-Stiftung, Kiel 19529
Gut Zernikow, Großwoltersdorf 18810
Gutao Ure Museum, Beijing 07541
Gutenberg-Gedenkstätte, Eltville am Rhein 18131
Gutenberg-Museum, Mainz 20143
Gutian Museum, Gutian 07748
Gutshaus Steglitz, Berlin 17122
Gutshof Pinnow, Pinnow 21040
Guwahati Medical Museum, Guwahati 24192
Guyana Museum, Georgetown 23318
Guyuan Museum, Guyuan 07749
Gwangju Municipal Folk Museum, Gwangju 29822
Gwangju National Museum, Gwangju 29823
Gwili Steam Railway, Bronwydd 43656
Gwinnett History Museum, Lawrenceville 49996
Gwynedd Education and Culture, Caernarfon 43709
Gwynedd Museum and Art Gallery, Bangor, Gwynedd 43376
Gyemyung University Museum, Daegu 29794
Gyeongbo Museum of Paleontology, Namjeong 29855
Gyeongbuk National University Museum, Daegu 29795
Gyeonggi Provincial Museum, Yongin 29941
Gyeongju National Museum, Gyeongju 29826
Gyeongju University Museum, Gyeongju 29827
Gyermek- és ifjúságszervezet-történeti Múzeum, Zánka 23975
Gympie and District Historical and Gold Mining Museum, Gympie 01138
Gyöngyös Monument Library of the Hungarian Franciscan Library, Gyöngyös 23573
Győr Diocese Library and Treasure Chamber, Győr 23582
Győr Egyházmegyei Könyvtár es Kincstár, Győr 23582
Györffy István Nagykun Múzeum, Karcag 23638
György Bessenyei Memorial Room, Tiszabercel 23928
Gyokudo Art Museum, Ome 29110
Gyokudo Bijutsukan, Ome 29110
Gyor-Moson-Sopron County Fire Extinguisher Exhibition, Mosonmagyaróvár 23719
Gyor-Moson-Sopron Megyei Tűzoltó Kiállítás, Mosonmagyaróvár 23719
It Gysbert Japicxhûs Museum, Bolsward 31979
Gyűjtemények Háza, Kiskunhalas 23669
Gyula Hincz Memorial Museum and Exhibition Building, Vác 23944

H

H.C.Andersen-koen Kodomotachi Hakubutsukan, Funabashi 28452
H.C.Andersen Park Childrens Museum, Funabashi 28452
H. Earl Clack Museum, Havre 49218
H.G. Albee Memorial Museum, Estherville 48428
The H. H. Franklin Museum, Tucson 53591
H.H. Maharaja Jayaji Rao Scindia Museum, Gwalior 24197
H.J. van de Kamp Museum, Sint Oedenrode 32776
H.M. Frigate Unicorn, Dundee 44046
H.R. MacMillan Space Centre, Vancouver 07213
H2 - Zentrum für Gegenwartskunst im Glaspalast, Augsburg 16620
Hå Bygdemuseum, Varhaug 33962
Hå Gamle Prestegard, Varhaug 33963
Ha Noi Museum, Ha Noi 54614
Haager Heimatstuben, Haag am Hausruck 02028
Haags Historisch Museum, Den Haag 32074
Haags Openbaar Vervoer Museum, Den Haag 32075
Haakú Museum, Pueblo of Acoma 52195
Haardplatenmuseum Klarenbeek, Klarenbeek 32448
Haarener Mühle, Waldfeucht 22292
Haas Gallery of Art, Bloomsburg 46872
Haas-Lilienthal House, San Francisco 52776
Haastrup Folkemindesamling, Faaborg 10011
Hachiga Folk Art Gallery, Takayama 29336
Hachiga Minzoku Bijutsukan, Takayama 29336
Hachinohe City Museum, Hachinohe 28462
Hachinohe City Museum of Art, Hachinohe 28461
Hachinohe Historical Museum, Hachinohe 28463
Hachinohe-shi Bijutsukan, Hachinohe 28461
Hachinohe-shi Hakubutsukan, Hachinohe 28462
Hachinohe-shiritsu Rekishi Hakubutsukan, Hachinohe 28463
Hachioji City Historical Museum, Hachioji 28465
Hachioji-shi Rekishi Hakubutsukan, Hachioji 28465
Hacı Bayram Camii, Ankara 42703
Hacı Ömer Sabancı Culture Center, Adana 42674
Hacı Ömer Sabancı Kültür Merkezi, Adana 42674
Hacıbektaş Müzesi, Hacibektaş 42792
Hacienda Buena Vista, Ponce 36045
Hacienda El Paraíso, El Cerrito 08635
Hacken- und Holzmuseum, Köflach 02210
Hackenberger Mühle, Saarburg 21462
Hackettstown Historical Society Museum, Hackettstown 49112
Hackley & Hume Houses, Muskegon 50924
Hackney Museum, London 44868
Hadamar Memorial, Hadamar 18848
Haddo House, Methlick 45134
Haddonfield Museum, Haddonfield 49114
Hadeland Bergverksmuseum, Grua 33522
Hadeland Folkemuseum, Jaren 33579
Hadeland Glassverks Museum, Jevnaker 33581
Haden Hall House, Cradley Heath 43911
Haderslev Museum, Haderslev 10049
Hadiqat al-Asmak Museum, Cairo 10498
Hadley Farm Museum, Hadley 49117
Hadsund Egns Museum, Hadsund 10053
Hadtörténeti Aero Múzeum, Szolnok 23897
Hadtörténeti Múzeum, Budapest 23408
Haegang Ceramics Museum, Icheon 29829
Hægebostad Bygdemuseum, Tingvatn 33905
Hägnan Friluftsmuseet i Gammelstad, Gammelstad 40534
Hägnan Open Air Museum, Gammelstad 40534
Haeju Historical Museum, Haeju 29753
Häkkilän Museo, Toholampi 11380
Hälleforsnäs Gjuterimuseum, Hälleforsnäs 40571
Hälleviks Fiskemuseum, Sölvesborg 40866
Hällisch-Fränkisches Museum, Schwäbisch Hall 21652
Hällristningsmuseet, Norrköping 40771
Hälsinglands Museum, Hudiksvall 40604
Häme Castle, Hämeenlinna 10743
Häme Museum, Tampere 11349
Hämeen Linna, Hämeenlinna 10743
Hämeen Museo, Tampere 11349
Hämeenkyrö Museum, Hämeenkyrö 10739
Hämeenkyrön Kotiseutumuseo, Hämeenkyrö 10739
Hämeenlinna Art Museum, Hämeenlinna 10746
Hämeenlinna Historical Museum, Hämeenlinna 10744
Hämeenlinna Telephone Museum, Hämeenlinna 10745
Hämeenlinnan Kaupungin Historiallinen Museo, Hämeenlinna 10744
Hämeenlinnan Puhelinmuseo, Hämeenlinna 10745
Hämeenlinnan Taidemuseo, Hämeenlinna 10746
Haenel-Pancera-Familienmuseum, Bad Ischl 01752
Härjedalens Fjällmuseum, Funäsdalen 40530
Härkätien Museo, Renko 11266
Härtsfeld-Museum, Neresheim 20571
Härtsfeldbahn-Museum, Neresheim 20572
Hässleholms Museum, Hässleholm 40574
Haffenreffer Museum of Anthropology, Bristol 47033
Haffmuseum Ueckermünde, Ueckermünde 22149
Hafnarborg Art Gallery, Hafnarfjörður 23998
Hafnarfjördur Museum, Hafnarfjörður 23997
Hafnarhúsið, Reykjavík 24018
Haft Tappeh Museum, Haft Tappeh 24607
Hagadorn House Museum, Almond 46236
Hagan, Eggedal 33472
The Haggin Museum, Stockton 53306
Haghna Museum, Tel Aviv 25106
Hagi Hakubutsukan, Hagi 28468
Hagi Museum, Hagi 28468
Hagi-shi Uragami Bijutsukan, Hagi 28469
Hagi Uragami Museum, Hagi 28469
Hagia Sophia Museum, İstanbul 42805
Hagley Hall, Hagley 44377
The Hagley Museum, Wilmington 54193
Hagley Museum and Library, Wilmington 54194
Hahndorf Academy Public Art Gallery and Museum, Hahndorf 01139
Hahns Peak Area Historical Museum, Clark 47584
Hahoedong Mask Museum, Andong 29766
HAHS Museum, Hattiesburg 49212
Haian Museum, Haian 07750
Haida Gwaii Museum at Qay'llnagaay, Skidegate 06982
Haidhausen-Museum, München 20444
Haifa City Museum, Haifa 24960
Haifa Museum of Art, Haifa 24961
Haig-Brown House, Campbell River 05541
Haig Colliery Mining Museum, Whitehaven 46001
Haikou Museum, Haikou 07751
Haileybury Heritage Museum, Haileybury 05925
Hailsham Heritage Centre, Hailsham 44379
Hailuodon Kotiseutumuseo Kniivilä, Hailuoto 10751
Hailuoto Museum Kniivilä, Hailuoto 10751
Haina Psychiatry Museum, Haina (Kloster) 18870
Hainan Museum, Haikou 07752
Hainan Nationality Museum, Tongshi 08301
Hainanzhou Nationality Museum, Gonghe 07718
Haines Museum, Waukegan 53989
Haining Museum, Haining 07753
Haiphong Museum, Hai Phong 54620
Haixizhou Nationality Museum, Delingha 07659
Haiyan Museum, Haiyan 07754
Haiyang Museum, Haiyang 07755
Hajdusági Múzeum, Hajdúböszörmény 23603
Haji Bayram Mosque, Ankara 42703
Hajózástörténeti Múzeum, Zebegény 23976
Hakasalmen Huvila, Helsinki 10776
Hakasalmi Villa - Helsinki City Museum, Helsinki 10776
Hakel Fossils Museum, Byblos 30167
Hakhamaneshi Museum, Shiraz 24646
Hakka Cultural Museum of Shigang, Shigang 42325
Hakka Culture and History Museum, Kaohsiung 42234
Hakka Relic and Folk Custom Museum, Meinong 42272
Hakodate Bijutsukan, Hakodate 28472
Hakodate City Museum, Hakodate 28474
Hakodate Museum of Art, Hakodate 28472
Hakodate Musical Boxes Hall, Hakodate 28473
Hakodate-shi Meijikan, Hakodate 28473
Hakodate-shiritsu Hakubutsukan, Hakodate 28474
Hakone Bijutsukan, Hakone 28477
Hakone-Jinja Homotsuden, Hakone 28478
Hakone Museum of Art, Hakone 28477
Hakone Open-Air Museum, Hakone 28476
Hakone Shrine Treasure House, Hakone 28478
Håkon's Hall and The Rosenkrantz Tower, Bergen 33427
Håkonshallen og Rosenkrantztårnet, Bergen 33427
Hakubutsukan Meiji-Mura, Inuyama 28573
Hakutsuru Fine Art Museum, Kobe 28726
Hakutsuru Sake Brewery Museum, Kobe 28727
Halasi Csipke Közalapítvány - Csipkeház, Kiskunhalas 23670
Halden Historiske Samlinger, Halden 33525
Haldensvassdragets Kanalmuseum, Ørje 33709
Haldimand County Museum, Cayuga 05581
Hale Farm and Village, Bath 46630
Hale School Museum, Wembley Downs 01641
Halesworth and District Museum, Halesworth 44381
Halfa Museum, Wadi Halfa 40398
Halı, Kilim ve Düz Dokuma Yaygılar Müzesi, İstanbul 42820
Hali Logging Camp Museum, Virrat 11470
Haliburton Highlands Museum, Haliburton 05927
Haliburton House Museum, Windsor, Nova Scotia 07342
Halifax Citadel, Halifax 05935
Halifax Historical Museum, Daytona Beach 47993
Halifax Police Museum, Halifax 05936
Halikko Local Museum, Halikko 10752
Halikkon Museo, Halikko 10752
Halin Metsäkämppämuseo, Virrat 11470
Halk Sanatları Müzesi, Girne 33360
Hall i'th' Wood Museum, Bolton 43555
Hall Mint and Minttower, Hall in Tirol 02043
Hall Museum, Meeteetse 50563
Hall of Art, Budapest 23453
Hall of Fame, Zouk Mikayel 30178
Hall of Fame and Classic Car Museum, Weedsport 54019
Hall of Fame for Great Americans, Bronx 47046
Hall of Flame - Museum of Firefighting, Phoenix 51903
Hall of Health, Berkeley 46754
Hall Place, Bexley 43473
Hallaton Museum, Hallaton 44389
La Halle aux Oiseaux, Marseillan 13987
Halle für Kunst Lüneburg e.V., Lüneburg 20097
Halle Heart Center, Tempe 53467
Halle Saint-Pierre Museum, Paris 14611
Hallen für neue Kunst, Schaffhausen 41828
De Hallen Haarlem, Haarlem 32296
Haller Feuerwehrmuseum, Schwäbisch Hall 21653
Hallerbosmuseum, Halle 03667
Hallertauer Heimat- und Hopfenmuseum, Mainburg 20136
Hallertauer Hopfen- und Heimatmuseum, Geisenfeld 18558
Hallie Ford Museum of Art, Salem 52666
Hallingdal Folkemuseum, Nesbyen 33698
Halliwell's House Museum, Selkirk 45635
Hallmark Society, Victoria 07249
Hallockville Museum Farm, Riverhead 52366
Hall's Croft, Stratford-upon-Avon 45780
Hallwalls Arts Center, Buffalo 47121
Hallwylska Museet, Stockholm 40898
Halmens Hus, Bengtsfors 40456
Halosenniemi Museo, Tuusula 11413
Halsey Gallery, Charleston 47365
Halton Region Museum, Milton 06289
Ham House, Richmond, Surrey 45492
Hama National Museum, Hama 42135
Hamaland-Museum, Vreden 22258
Hamamatsu City Museum of Art, Hamamatsu 28485
Hamamatsu Museum of Musical Instruments, Hamamatsu 28483
Hamamatsu Science Museum, Hamamatsu 28486
Hamamatsu-shi Gakki Hakubutsukan, Hamamatsu 28483
Hamamatsu-shi Hakubutsukan, Hamamatsu 28484
Hamamatsu-shiritsu Bijutsukan, Hamamatsu 28485
Hamamatsu-shiritsu Kagakukan, Hamamatsu 28486
Hamana-ko Orugorukan, Hamamatsu 28487
Hamana-ko Tateyama-ji Bijutsukan, Hamamatsu 28488
Hamarøy Bygdetun, Harmarøy 33532
Hambacher Schloss, Neustadt an der Weinstraße 20671
Hambidge Center for the Creative Arts and Sciences, Rabun Gap 52221
Hamburg Dungeon, Hamburg 18931
Hamburg School Museum, Hamburg 18933
Hamburg State Park Museum, Mitchell 50759
Hamburger Bahnhof - Museum für Gegenwart - Berlin, Berlin 17123
Hamburger Kunsthalle, Hamburg 18932
Hamburger Schulmuseum, Hamburg 18933
Hameau du Fromage, Cléron 12556
Hameau en Beaujolais, Romanèche-Thorins 15084
Hameau Troglodytique de la Fosse, Dénezé-sous-Doué 12714
Hamhung Historical Museum, Hamhung 29754
Hamilton Art Gallery, Hamilton, Victoria 01144
Hamilton Artists, Hamilton 05954
The Hamilton Children's Museum, Hamilton 05955
Hamilton Grange National Memorial, New York 51189
Hamilton Heritage Museum, Hamilton, Tasmania 01143
Hamilton House, South Berwick 53163
Hamilton Hume Museum, Yass 01682
Hamilton Library and Two Mile House, Carlisle 47243
Hamilton Low Parks Museum, Hamilton 44392
Hamilton Military Museum, Hamilton 05956
Hamilton Museum of Steam and Technology, Hamilton 05957
Hamilton Psychiatric Hospital Museum, Hamilton 05958
Hamilton van Wogener Museum, Clifton 47648
Hamilton's Gallery, London 44869
Hamina Town Museum, Hamina 10753
Haminan Kaupunginmuseo, Hamina 10753
Hamiota Pioneer Club Museum, Hamiota 05965
Hamizgaga Museum, Kibbutz Nachsholim 25051
Hamlin Garland Homestead, West Salem 54079
Hamline University Galleries, Saint Paul 52617
Hammaslääketieteen Museo, Helsinki 10777
Hammer Museum, Los Angeles 50228
Hammer- und Waffenschmiede-Museum Hexenagger, Altmannstein 16513
Hammerichs Hus, Ærøskøbing 09959
Hammermuseum zu Hannover, Hannover 18991
Hammerschmiede Gröningen, Satteldorf 21523
Hammerschmiede mit Heimatmuseum und Bienengarten, Lahr 19772
Hammerschmiede und Stockerhof, Neuburg an der Kammel 20602
Hammond Castle Museum, Gloucester 48946
Hammond-Harwood House Museum, Annapolis 46319
Hammond Historical Museum, Hammond 49141
Hammond Museum, North Salem 51405
Hamnmuseét, Varberg 41047
Hampden-Booth Theatre Library at the Players, New York 51190
Hampshire Farm Museum, Botley, Hampshire 43563
Hampson Archeological Museum, Wilson 54208
Hampstead Museum, London 44870
Hampton County Historical Museum, Hampton 49148
Hampton Court Palace, East Molesey 44093
Hampton History Museum, Hampton 49151
Hampton National Historic Site Museum, Towson 53550
Hampton Plantation, McClellanville 50341
Hampton-Preston Mansion, Columbia 47726
Hampton Roads Naval Museum, Norfolk 51363
Hampton University Museum, Hampton 49152
Hamptonne Country Life Museum, Saint Lawrence 45580
Hamura Museum, Hamura 28489
Hamura-shiritsu Hakubutsukan, Hamura 28489
Hamza Collection and Jász Gallery, Jászberény 23622
Hamza Gyűjtemény és Jász Galéria, Jászberény 23622
Hamza Hakim-Jade Niazy Muzeyi, Farg'ona 54447
Hamza Hakim-zade Niazy House Museum of the Kokand Regional Studies Museum, Qo'qon 54455
Hamza Hakim-Zadé Niazy Memorial Museum, Farg'ona 54447
Hamza Hakim-zade Niazy Yodgorlik Muzeyi, Qo'qon 54455
Han Kuk Art Museum, Yongin 29942
Han Military Forces Yong Museum, Xuzhou 08417
Han Tomb Cultural Relics Exhibition Hall, Mawangdui 08010
Hana Cultural Center, Hana, Maui 49156
Hana Senesh House, Kibbutz Sdot Yam 25057
Hanamaki City Museum, Hanamaki 28490
Hanamaki-shiritsu Hakubutsukan, Hanamaki 28490
Hanbat Education Museum, Daejeon 29800

Hanbisuan Mikado Pheasant Ecology Museum, Su-Ao 42344
Hanbury Hall, Droitwich 44025
Hanby House, Westerville 54090
Hanch Hall, Lichfield 44704
Hancheng Museum, Hancheng 07756
Hanchuan Museum, Hanchuan 07757
Hancock County Historical Museum, Carthage 47267
Hancock Historical Museum, Findlay 48544
Hancock Historical Society Museum, Hancock 49157
Hancock House, Hancock's Bridge 49158
Hancock Museum, Newcastle-upon-Tyne 45227
Hancock Shaker Village, Pittsfield 51968
Hand Workshop Art Center, Richmond 52328
Handa Museum & Niimi Nankichi Memorial Museum, Handa 28495
Handa-shiritsu Hakubutsukan & Niimi Nankichi Kinenkan, Handa 28495
Handan Museum, Handan 07758
Handarbeitsmuseum, Traunkirchen 02826
Handel House Museum, London 44871
Handels- og Søfartsmuseet paa Kronborg, Helsingør 10064
Handmade Fabrics Museum, Kyoto 28860
Handover Gifts Museum, Macau 07994
Hands on Children's Museum, Olympia 51544
Hands on History, Kingston-upon-Hull 44596
Hands-on House, Lancaster 49922
Hands On! Regional Museum, Johnson City 49633
Handsworth Saint Mary's Museum, Handsworth 44393
Handweaving Museum and Arts Center, Clayton 47601
Handweberei Henni Jaensch-Zeymer, Schwielowsee 21706
Handwebmuseum Rupperath, Bad Münstereifel 16788
Handwerk & Gewerbe Museum Sagar, Krauschwitz 19700
Handwerker Gallery of Art, Ithaca 49547
Handwerker- und Heimatstube Gernsbach, Gernsbach 18607
Handwerkermuseum, Freinsheim 18446
Handwerkermuseum, Schönecken 21606
Handwerkermuseum Laufenbach, Taufkirchen an der Pram 02797
Handwerkerstuben, Feuchtwangen 18314
Handwerkskunst-Museum, Rattenberg, Inn 02565
Handwerksmuseum, Deggendorf 17805
Handwerksmuseum, Horneburg 19277
Handwerksmuseum, Sankt Leonhard am Hornerwald 02662
Handwerksmuseum am Mühlenberg, Suhlendorf 21978
Handwerksmuseum Groß-Gerau, Groß-Gerau 18775
Handwerksmuseum Kork, Kehl 19496
Handwerksmuseum Ovelgönne, Ovelgönne 20964
Haney House Museum, Maple Ridge 06235
Hanf-Museum, Berlin 17124
Hanford Mills Museum, East Meredith 48257
Hangar Bicocca - spazio d'arte contemporanea, Milano 26689
Hangguk Minsok-chon, Kihung 29845
Hangzhou City Construction Exhibition Hall, Shanghai 08167
Haniel Museum, Duisburg 18018
Hanila Muuseum, Hanila 10568
Hanita Museum, Kibbutz Hanita 25043
Hank Snow Country Music Centre, Liverpool 06177
Hankasalmen Kotiseutumuseo, Hankasalmi 10755
Hankasalmi Local History Museum, Hankasalmi 10755
Hanko Front Museum, Hanko 10756
Hanko Museo, Hanko 10757
Hanley House, Clayton 47599
Hanna Pioneer Museum, Hanna 05967
Hannah Lindahl Children's Museum, Mishawaka 50748
Hannes Hus og Fonden Gamle Sønderho, Fanø 10016
Hannoversches Straßenbahn-Museum, Sehnde 21728
Hanns Schell Collection, Graz 01971
Hanover Historical Society Museum, Hanover 49163
Hanover House, Clemson 47615
Hanover Museum, Lucea, Hanover 28337
Hanrieder-Gedenkraum, Putzleinsdorf 02543
Hans-Böckler-Geburtshaus, Trautskirchen 22080
Hans Breinlinger Museum, Konstanz 19678
Hans Christian Andersen Museum, Odense 10215
Hans Christian Andersens Barndomshjem, Odense 10216
Hans Erni Museum, Luzern 41591
Hans-Fallada-Haus, Feldberger Seenlandschaft 18301
Hans-Grade-Museum, Borkheide 17480
Hans Gross Kriminalmuseum, Graz 01972
Hans Herr House and Museum, Willow Street 54185
Hans Mauracher-Museum, Graz 01973
Hans Nielsens Hauges Minne, Rolvsøy 33792
Hans Schatzdorfer Haus, Pramet 02526
Hans-Thoma-Gedächtnisstätte und Vortaunusmuseum, Oberursel 20847
Hans-Thoma-Museum, Bernau im Schwarzwald 17291
Hans Trudel-Haus Stiftung, Baden 41144
Hans van Riessen Museum, Vledderveen 32897
Hans Werner Richter Haus, Bansin, Seebad 16925
Hanśági Múzeum, Mosonmagyaróvár 23720
Hanseatic Booth or The Pier House, Symbister 45821
Det Hanseatiske Museum og Schøtstuene, Bergen 33428
Hansjakobmuseum im Freihof, Haslach im Kinzigtal 19025
Hansong Womens College Museum, Seoul 29880
Hantverks- och Sjörfartsmuseét på Norra Berget, Sundsvall 40959
Hantverksloftet, Simrishamn 40840
Hantverksmuseet, Ängelholm 40408
Hanul Domnesc, Suceava 36643
Hanwon Art Museum, Seoul 29881
Har-Ber Village, Grove 49097
Hara Gendai Bijutsukan, Tokyo 29395
Hara Museum ARC, Shibukawa 29256
Hara Museum of Contemporary Art, Tokyo 29395
Harbin Museum, Harbin 07769
Harbin Museum, Harbin 07770
Harbor Branch Oceanographic Institution, Fort Pierce 48667
Harbor Defense Museum of New York City, Brooklyn 47075
Harborough Museum, Market Harborough 45109
Harbour Life Exhibition, Bridport 43613
Harbour Museum, Derry 43973
Harbourmuseum, Rotterdam 32705
Hardanger Folkemuseum, Utne 33949
Hardanger Folkemuseum, Utne 33950
Hardin County Historical Museum, Kenton 49742
Harding Home and Museum, Marion 50487
Harding Museum, Franklin 48749
Hardtmuseum, Durmersheim 18031
Hardwick Hall, Chesterfield 43821
Hardwick Historical Museum, Hardwick 49166
Hardy's Cottage, Higher Bockhampton 44467
Haremere Hall, Etchingham 44188
Harestanes Countryside Visitor Centre, Jedburgh 44546
Harewood House, Harewood 44394
Hariphunchai National Museum, Lampun 42578
Hariya Razna and Ahmed Urabi Museum, Zaqaziq 10543
Harjamäen Sairaalamuseo, Siilinjärvi 11310
Harjukosken Vesimylly, Mikkeli 11112
Harjumaa Muuseum, Keila 10583
Harlan-Lincoln House, Mount Pleasant 50882
Harlech Castle, Harlech 44395
Harley Gallery, Worksop 46076
Harlinger Aardewerk Museum, Harlingen 32317
Harlow Carr Museum of Gardening, Harrogate 44400
Harlow Ceramics Collection & Kleven Print Collection, Bemidji 46731
Harlow Gallery, Hallowell 49129
Harmar Station, Marietta 50475
Harmonikamuseum Zwota, Zwota 22844
Harmonikazentrum, Graz 01974
Harmonium-Art-Museum, Schelle 04001
Harmonium-Museum, Liestal 41562
Harn Homestead and 1889er Museum, Oklahoma City 51514
Harness Racing Museum and Hall of Fame, Goshen 48978
Harney County Historical Museum, Burns 47146
Harold M. Freund American Museum of Baking, Manhattan 50432
Harold Warp Pioneer Village Foundation, Minden 50723
Harper City Historical Museum, Harper 49172
Harpers Ferry National Historical Park, Harpers Ferry 49174
Harput Müzesi, Elazığ 42771
Harriet Beecher Stowe Center, Hartford 49194
Harris Art Gallery, James M. Lykes Maritime Gallery and Hutchings Gallery, Galveston 48846
Harris Exhibit Hall, Oxford 51652
Harris Heritage and Museum, Harris 05973
Harris Museum and Art Gallery, Preston 45444
Harrison County Historical Museum, Marshall 50509
Harrison House, Branford 46995
Harrow Museum, Harrow 44404
Harrströmin Myllymäkimuseo, Harrström 10759
Harry D. Hendren Gallery, Saint Charles 52533
Harry J. Music Museum, Kingston 28327
Harry Oppenheimer Diamond Museum, Ramat Gan 25083
Harry Ransom Humanities Research Center, Austin 46510
Harry S. Truman Birthplace, Lamar 49910
Harry S. Truman Home, Independence 49473
Harry S. Truman Little White House Museum, Key West 49761
Harry S. Truman Office and Courtroom, Independence 49474
Harry S. Truman Presidential Library and Museum, Independence 49475
Harry S. Washbrook Museum, Edam 05735
Harry's Truman Farm Home, Grandview 49012
Harsefelder Museumszug, Harsefeld 19015
Hart-Cam Museum, Hartney 05976
Hart House, Toronto 07113
Hartenbos Museum, Hartenbos 38639
Hartford House Museum, Manhattan 50433
Hartland Quay Museum, Bideford 43478
Hartland Quay Museum, Hartland 44405
Hartlepool Art Gallery, Hartlepool 44406
Hartmanin Mökki, Oripää 11166
Hartman's Cabin, Oripää 11166
Hartnell College Art Gallery, Salinas 52676
Den Hartogh Ford Automobielmuseum, Hillegom 32384
Hartsville Museum, Hartsville 49203
Hartung's Auto and License Plate Museum, Glenview 48938
Harvard Historical Society Collection, Harvard 49208
Harvard Historical Society Museum, Still River 53293
Harvard House, Stratford-upon-Avon 45781
Harvard Museum of Natural History, Cambridge 47176
Harvard University Art Museums, Cambridge 47177
Harvard University Semitic Museum, Cambridge 47178
Harvery Art Gallery, Harvey 01146
Harvey County Historical Society, Newton 51340
Harvey Grant Heritage Centre, Springdale 07001
Harvey House Museum, Florence 48569
Harwich Maritime Museum and Harwich Lifeboat Museum, Harwich 44410
Harwich Redoubt Fort, Harwich 44411
Harwood Museum of Art of the University of New Mexico, Taos 53441
Haryana Prantiya Puratatva Sangrahalaya, Jhajjar 24241
Harzer Roller-Kanarien-Museum, Sankt Andreasberg 21486
Harzmuseum, Wernigerode 22495
Hasan Süzer Etnografya Müzesi, Gaziantep 42787
Hasegawa Machiko Kinen Bijutsukan, Tokyo 29396
Hasegawa Machiko Memorial Museum of Art, Tokyo 29396
Haselünner Heimathäuser, Haselünne 19024
Hasenmuseum, Bubikon 41263
The Hash Marihuana Hemp Museum, Amsterdam 31844
Hashimoto Kansetsu Kinenkan, Kyoto 28812
Hashimoto Kansetsu Museum, Kyoto 28820
Hasičské Muzeum, Bechyně 09396
The Haskell County Historical Museum, Sublette 53352
Haslemere Educational Museum, Haslemere 44413
Haslev Museum, Haslev 10055
Hassanbey Zardabi Museum of Natural History, Bakı 03154
Hassett's Uralla Military Museum, Uralla 01613
Hastings Community Arts Centre, Hastings 33118
Hastings Museum, Hastings 49211
Hastings Museum and Art Gallery, Hastings 44415
Hastings-Western Port Historical Museum, Hastings 01147
Hat Making Museum, Chazelles-sur-Lyon 12494
Hat Sanatları Müzesi, İstanbul 42821
Hat Works, Museum of Hatting, Stockport 45753
Hatakeyama Memorial Museum of Fine Art, Tokyo 29397
Határőrség Központi Múzeuma, Körmend 23675
Hatay Devlet Güzel Sanatlar Galerisi, Hatay 42793
Hatay Müzesi, Hatay 42794
Hatay Museum, Hatay 42794
Hatay State Fine Art Gallery, Hatay 42793
Hatch-Billops Collections, New York 51191
Hathaway Ranch Museum, Santa Fe Springs 52893
Hatton-Eielson Museum, Hatton 49214
Hatton Gallery, Fort Collins 48609
Hatton Gallery, Newcastle-upon-Tyne 45228
Hauchs Physiske Cabinet, Sorø 10280
Hauffs Märchen-Museum, Baiersbronn 16896
Haugar Art Museum, Tønsberg 33910
Haugar Vestfold Kunstmuseum, Tønsberg 33910
Haugesund Art Gallery, Haugesund 33537
Haugesund Billedgalleri, Haugesund 33537
Hauho Museum, Hauho 10761
Hauhon Esinemuseo, Hauho 10761
Haukipudas School Museum, Haukipudas 10762
Haukiputaan Koulumuseo, Haukipudas 10762
Hauptstaatsarchiv Stuttgart, Stuttgart 21937
Haus am Kleistpark, Berlin 17125
Haus am Lützowplatz, Berlin 17126
Haus am Waldsee, Berlin 17127
Haus der badisch-pfälzischen Fasnacht, Speyer 21833
Haus der Bayerischen Geschichte, Augsburg 16621
Haus der Brandenburgisch-Preußischen Geschichte, Potsdam 21089
Haus der Donauschwaben, Sindelfingen 21775
Haus der Dorfgeschichte Bliesmengen-Bolchen, Mandelbachtal 20160
Haus der erzgebirgischen Tradition, Deutschneudorf 17844
Haus der Fotografie - Dr. Robert-Gerlich-Museum, Burghausen 17653
Haus der Geschichte Baden-Württemberg, Stuttgart 21938
Haus der Geschichte der Bundesrepublik Deutschland, Bonn 17455
Haus der Geschichte Wittenberg, Lutherstadt Wittenberg 20114
Haus der Heimat, Olbernhau 20892
Haus der Kulturen der Welt, Berlin 17128
Haus der Kunst, München 20445
Haus der Kunst, Nümbrecht 20752
Haus der Natur, Eichendorf 18077
Haus der Natur, Salzburg 02615
Haus der Natur - Cismar, Cismar 17707
Haus der Naturfreunde, Duisburg 18019
Haus der Ortsgeschichte, Schwäbisch Hall 21654
Haus der Ost- und Westpreußen, Oberschleißheim 20832
Haus der Seidenkultur Paramentenweberei Hubert Gotzes e.V., Krefeld 19704
Haus der Stadtgeschichte, Donauwörth 17895
Haus der Stadtgeschichte, Offenbach am Main 20882
Haus der Völker, Schwaz 02725
Haus der Wannsee-Konferenz, Berlin 17129
Haus des Apfels, Puch bei Weiz 02535
Haus des Gastes mit heimatkundlicher Sammlung, Gößweinstein 18688
Haus des Moores, Heidenreichstein 02060
Haus des Waldes, Köln 19606
Haus Fischerzunft, Bad Säckingen 16813
Haus für Kunst Uri, Altdorf 41102
Haus für Kunst Uri, Altdorf 41103
Haus Hohe Pappeln, Weimar 22404
Haus Kickelhain, Mosbach 20388
Haus Konstruktiv, Zürich 42074
Haus Mährisch-Schönberg, Bad Hersfeld 16746
Haus Opherdicke, Holzwickede 19263
Haus Papst Benedikt XVI. - neue Schatzkammer und Wallfahrtsmuseum, Altötting 16518
Haus Peters, Tetenbüll 22034
Haus Rottels, Neuss 20656
Haus Ruhrnatur, Mülheim an der Ruhr 20412
Haus Stöcken, Haan 18846
Haus Tannenbusch, Dormagen 17903
Haus Völker und Kulturen - Völkerkundliches Museum, Sankt Augustin 21489
Haus Wittgenstein, Wien 02999
Haus zum Dolder, Beromünster 41223
Haus zum Kirschgarten, Basel 41160
Haus zum Stockfisch - Stadtmuseum, Erfurt 18204
Haus zum Torggel, Berneck 41221
Haus zur Geschichte der IBM Datenverarbeitung, Sindelfingen 21776
Hausenhäusl, Reit im Winkl 21256
Hausjärven Kotiseutumuseo, Hausjärvi 10763
Hausjärvi Museum, Hausjärvi 10763
Hausmuseum Kloster Kreuzlingen, Kreuzlingen 41513
Hausmuseum Sankt Katharinental, Diessenhofen 41327
Les Hautes-Mynes, Le Thillot 13653
Hauteville House, Saint Peter Port 45592
Havacılık Müzesi, İstanbul 42822
Havant Museum, Havant 44420
Havayeda Museum, Holon 24976
Havelock Museum, Havelock 33119
Het Havenmuseum, Rotterdam 32705
Haverford Township Historical Society Museum, Havertown 49217
Haverhill and District Local History Centre, Haverhill 44423
Havesdonckhoeve, Bornem 03385
Havets Hus, Lysekil 40711
De Havilland Aircraft Heritage Centre, Saint Albans 45542
Havråtunet, Lonevåg 33647
Havre de Grace Decoy Museum, Havre de Grace 49219
Havza Atatürk Evi, Havza 42795
Haw Pha Kaeo, Vientiane 29973
Hawaii Children's Discovery Center, Honolulu 49339
Hawaii Maritime Center, Honolulu 49340
Hawaii Museums Association, Honolulu 55136
Hawaii Pacific University Gallery, Honolulu 49341
Hawaii State Foundation on Culture and the Arts, Honolulu 49342
Hawaii's Plantation Village, Waipahu 53787
Hawick Museum, Hawick 44426
Hawke House, Bordertown 00878
Hawker Museum, Hawker, South Australia 01148
Hawke's Bay Museum, Napier 33150

Hawkesbury Regional Museum, Windsor, New South Wales 01656
Hawkeye Log Cabin, Burlington 47133
Hawks Inn, Delafield 48041
Hawks Nest State Park, Ansted 46324
Haworth Art Gallery, Accrington 43262
Hawthorne Cottage, Brigus 05479
Hawthorns Urban Wildlife Centre, Southampton 45703
Hay Gaol Museum, Hay 01150
Hay House Museum, Macon 50366
Hayart Cultural Center, Yerevan 00737
Hayashibara Bijutsukan, Okayama 29098
Hayashibara Museum of Art, Okayama 29098
Hayashibara Museum of Natural Sciences, Okayama 29099
Hayashibara Shizenkagaku Hakubutsukan, Okayama 29099
Haydn Birthplace Museum, Rohrau 02594
Haydn-Geburtshaus, Rohrau 02594
Haydn-Haus Eisenstadt, Eisenstadt 01860
Haydnhaus, Wien 03000
Haydon Gallery, Lincoln 50097
Hayloft Gallery, Armagh 43318
Haynes International Motor Museum, Sparkford 45724
Haystack Mountain School of Crafts Gallery, Deer Isle 48027
Hayward Area Museum, Hayward 49228
Hayward China Museum, Saint John 06836
Hayward Gallery, London 44872
Hazama Inosuke Bijutsukan, Kaga 28624
Hazan Itaya Memorial Museum, Shimodate 29266
Hazelhurst Regional Gallery and Arts Centre, Gymea 01137
Hazor Museum, Ayelet-Hashahar 24933
Hazrati Museum, Semnan 24643
HAZU Print Room, Zagreb 09001
He County Museum, Hexian 07785
He Xiang Ning Art Museum, Shenzhen 08220
Head-Smashed-In Buffalo Jump, Fort Macleod 05815
Headland Historical Museum, Nambucca Heads 01345
Headlands Center for the Arts, Sausalito 52936
Headley-Whitney Museum, Lexington 50048
Headlund Museum, Hugo 49414
Headquarters Museum, Mikkeli 11116
Headwaters Heritage Museum, Three Forks 53503
Healdton Oil Museum, Healdton 49235
Health Adventure, Asheville 46374
Health Museum, Hyderabad 24212
Health Museum, Yangon 31735
Health Science Museum, Hiroshima 28539
Health Sciences Museum, Praha 09800
Healthspace Cleveland, Cleveland 47632
Healy House and Dexter Cabin, Leadville 50005
Heaman's Antique Autorama, Carman 05567
Heard Museum, Phoenix 51904
Heard Natural Science Museum, McKinney 50359
Hearing aid museum, Buenos Aires 00221
Hearst Art Gallery, Moraga 50838
Hearst Castle, San Simeon 52830
Heart of Dixie Railroad Museum, Calera 47166
Heart of West Texas Museum, Colorado City 47697
Hearthstone Historic House Museum, Appleton 46330
Heartland -California Museum of the Heart, Rancho Mirage 52241
Heart's Content Cable Station, Heart's Content 05982
Heatherslaw Mill, Cornhill-on-Tweed 43893
Heaton Hall, Manchester 45084
Heaven Farm Museum, Uckfield 45893
Hebbel-Museum, Wesselburen 22511
Hebei Museum, Shijiazhuang 08231
Hebei Provincial Museum, Shijiazhuang 08232
Hebei Provincial Museum, Wuchang 08328
Hebezeug-Museum, Witten 22617
Hebi Museum, Hebi 07773
Hechter-Schulz Afro-Culture Museum, Boksburg 38571
Heckington Windmill, Heckington 44435
Heckscher Museum of Art, Huntington 49430
Hector Exhibit Centre, Pictou 06601
Hector Pieterson Museum and Memorial, Johannesburg 38656
Heddal Bygdetun, Notodden 33705
Heddon Museum, Dowagiac 48172
Hedmarksmuseet og Domkirkeodden, Hamar 33528
Hedon Museum, Hedon 44437
Hedrum Bygdetun, Kvelde 33619
Heemerf De Schutsboom, Goirle 32260
Heemhuis en Sigarenmuseum, Arendonk 03322
Heemkamer Barthold Van Heessel, Aarle Rixtel 31788
Heemkundemuseum Beek, Beek 31952
Heemkundemuseum Paulus Van Daesdonck, Ulvenhout 32839
Heemkundig Hagelands Museum, Tielt-Winge 04068
Heemkundig Museum, Asse 03327
Heemkundig Museum, Borgerhout 03383
Heemkundig Museum, Bree 03402
Heemkundig Museum, Moen 03875
Heemkundig Museum, Ommeren 32621
Heemkundig Museum, Tervuren 04057
Heemkundig Museum, Waasmunster 04122
Heemkundig Museum, Wilrijk 04154
Heemkundig Museum, Zolder 04164
Heemkundig Museum Boekhoute, Boekhoute 03378
Heemkundig Museum De Drie Rozen, Gravenwezel 03660
Heemkundig Museum Gerard Meeusen, Essen 03586
Heemkundig Museum Sincfala, Knokke 03753
Heemkundig Museum Slag van Lafelt, Vlijtingen 04121
Heemkundig Museum 't Smiske, Rijmenam 03968
Heemkundig Museum Tempelhof, Beerse 03354
Heemkundig Museum Wetteren, Wetteren 04149
Heemkundig Museum Wissekerke, Kruibeke 03763
Heemkundig Streekmuseum Jan Uten Houte, Etten Leur 32225
Heemmuseum Bystervelt, Schelle 04002
Heemmuseum De Kaeck, Wommelgem 04156
Heemmuseum De Pastorij, Edegem 03573
Heemmuseum Die Swane, Heist-op-den-Berg 03691
Heemmuseum Eeklo, Eeklo 03575
Heemmuseum van de Zuiderkempen, Westerlo 04148
Heemschuur Maasland, Heesch 32340
Heeresgeschichtliches Museum HGM, Wien 03001
Heesberg Museum, Gevensleben 18629
Heeswijk Castle, Heeswijk-Dinther 32342
Hefeng Museum, Hefeng 07777
Hefner Zoology Museum, Oxford 51653
Hegau-Museum, Singen 21780
Hegel-Museum, Stuttgart 21939
Hegelhaus, Stuttgart 21939
Hegmataneh Museum, Samedan 24641
Hegyvidék Galéria, Budapest 23409
Hegyvidék Region Historical Collection, Budapest 23410
Hegyvidéki Helytörténeti Gyűjtemény, Budapest 23410
De Heibergske Samlinger - Sogn Folkemuseum, Kaupanger 33589
Heide Museum of Modern Art, Bulleen 00915
Heidelberger Kunstverein, Heidelberg 19073
Heidemuseum Dat ole Huus, Bispingen 17365
Heidemuseum Rischmannshof Walsrode, Walsrode 22316
Heidi Weber-Museum, Zürich 42075
Heihe Museum, Heihe 07778
Heijo Palace Site Museum, Nara 29029
Heijokyuseki Shiryokan, Nara 29029
Heikintupa, Pohjaslahti 11215
Heikki's Cabin, Pohjaslahti 11215
Heiliggeistkirche, Landshut 19801
Heilongjiang Nationality Museum, Harbin 07771
Heilsarmee - Museum und Archiv, Bern 41206
Heimat-Haus Großderschau, Großderschau 18785
Heimat-Museum Maintal, Maintal 20140
Heimat- und Bauernmuseum Bruck, Bruck in der Oberpfalz 17599
Heimat- und Bauernmuseum Sulz, Sulz am Neckar 21980
Heimat- und Bergbaumuseum Lugau, Lugau 20107
Heimat- und Bergbaumuseum Nentershausen, Nentershausen 20570
Heimat- und Bergbaumuseum Reinsdorf, Reinsdorf bei Zwickau 21255
Heimat und Bildhauer Kern Museum, Forchtenberg 18339
Heimat- und Brauereimuseum Pleinfeld, Pleinfeld 21058
Heimat- und Braunkohlemuseum Steinberg, Steinberg, Kreis Schwandorf 21883
Heimat- und Buddelmuseum Osten, Osten 20930
Heimat- und Bürstenbindermuseum, Ramberg 21173
Heimat- und Burgmuseum, Kirkel 19557
Heimat- und Feuerwehrmuseum, Schauenstein 21534
Heimat- und Flößermuseum Calmbach, Bad Wildbad 16859
Heimat- und Grimmelshausenmuseum, Oberkirch 20812
Heimat- und Hafnermuseum, Heidenheim 19089
Heimat- und Handfeuerwaffenmuseum, Kemnath 19507
Heimat-und Handwerkermuseum, Leutershausen-Wiedersbach 19970
Heimat- und Heringsfängermuseum Heimsen, Petershagen 21015
Heimat- und Hugenotten-Museum, Stutensee 21926
Heimat- und Humboldtmuseum, Eibau 18072
Heimat- und Industriemuseum Kolbermoor, Kolbermoor 19674
Heimat- und Industriemuseum Wackersdorf, Wackersdorf 22261
Heimat- und Keramikmuseum, Kandern 19426
Heimat- und Kleinbrennermuseum, Steinach, Ortenaukreis 21876
Heimat- und Krippenmuseum, Zirl 03131
Heimat- und Landlermuseum, Bad Goisern 01748
Heimat- und Miedermuseum Heubach, Heubach 19174
Heimat- und Montanmuseum in Schloß Oberkindberg, Kindberg 02163
Heimat- und Naturkunde-Museum Wanne-Eickel, Herne 19153
Heimat- und Ortsmuseum, Zürich 42076
Heimat- und Palitzsch-Museum Prohlis, Dresden 17947
Heimat- und Pfarrmuseum, Wildalpen 03102
Heimat- und Rebbaumuseum, Spiez 41892
Heimat- und Schiffahrtsmuseum, Heinsen 19113
Heimat- und Schmiedemuseum, Widdern 22536
Heimat- und Ski-Museum, Braunlage 17516
Heimat- und Ständemuseum, Tragöß 02815
Heimat- und Torfmuseum, Gröbenzell 18767
Heimat- und Uhrenmuseum, Villingen-Schwenningen 22236
Heimat- und Waldmuseum, Eppenbrunn 18181
Heimat- und Weinbaumuseum, Graach an der Mosel 18721
Heimat- und Zunftmuseum, Adenau 16431
Heimatarchiv Zwittau, Esslingen 18273
Heimatdiele Kutenholz, Kutenholz 19753
Heimatecke am Seifenbach, Beierfeld 16981
Heimatgeschichtliche Sammlung, Bodelshausen 17410
Heimatgeschichtliche Sammlung des Kirchspiels, Schwabstedt 21647
Heimathaus, Königswiesen 02212
Heimathaus, Litzelsdorf 02329
Heimathaus, Neufelden 02427
Heimathaus, Obernberg am Inn 02454
Heimathaus, Pregarten 02530
Heimathaus, Raab 02544
Heimathaus, Sankt Georgen an der Gusen 02641
Heimathaus, Sankt Peter im Sulmtal 02680
Heimathaus, Schörfling am Attersee 02711
Heimathaus, Steinbach am Attersee 02764
Heimathaus, Ulrichsberg 02838
Heimathaus Alte Mühle, Schladen 21561
Heimathaus Aying, Aying 16650
Heimathaus Beandhaus, Sankt Johann am Walde 02653
Heimathaus Bevergern, Hörstel 19218
Heimathaus De Theeshof, Schneverdingen 21593
Heimathaus der Stadt Lauingen, Lauingen 19857
Heimathaus der Stadt Warendorf, Warendorf 22337
Heimathaus des Rupertiwinkels und Heimatstube der Sudetendeutschen, Tittmoning 22066
Heimathaus Dingden, Hamminkeln 18974
Heimathaus Ebensee, Ebensee 01835
Heimathaus Gallneukirchen, Gallneukirchen 01919
Heimathaus Greven Worth, Selsingen 21742
Heimathaus-Handwerkermuseum, Scheidegg 21539
Heimathaus im Alten Turm, Haslach 02054
Heimathaus Julbach, Julbach 02147
Heimathaus Mehedorf, Bremervörde 17578
Heimathaus mit Photographiemuseum, Mariazell 02357
Heimathaus Mörbisch, Mörbisch 02401
Heimathaus Nesselwang, Nesselwang 20577
Heimathaus Neuchl-Anwesen, Hohenschäftlarn 19242
Heimathaus Neudorf - Seoski muzej, Neudorf bei Parndorf 02426
Heimathaus Ollershof, Munster 20531
Heimathaus Pfarrkirchen, Pfarrkirchen 21023
Heimathaus Plettenberg, Plettenberg 21060
Heimathaus Richard Eichinger und Steinmuseum, Enzenkirchen 01872
Heimathaus Sittensen, Sittensen 21790
Heimathaus Sonthofen, Sonthofen 21823
Heimathaus-Stadtmuseum Vöcklabruck, Vöcklabruck 02865
Heimathaus und Schimuseum, Saalbach 02605
Heimathaus und Stadtmuseum, Perg 02490
Heimathaus Wartberg an der Krems, Wartberg an der Krems 02894
Heimathaus Welver, Welver 22470
Heimathaus Wenigzell, Wenigzell 02921
Heimathausanlage Schafstall, Bremervörde 17579
Heimathof Emsbüren, Emsbüren 18158
Heimathof Hüll, Drochtersen 17978
Heimatkabinett Westerholt, Herten 19166
Heimatkundehaus und Münzkabinett, Sankt Marien 02665
Heimatkundesammlung Inzersdorf, Herzogenburg 02075
Heimatkundliche Sammlung, Richterswil 41758
Heimatkundliche Sammlung, Uznach 41969
Heimatkundliche Sammlung Bergendorf, Holzheim 19259
Heimatkundliche Sammlung Blaustein, Blaustein 17382
Heimatkundliche Sammlung Burg Rheinfels, Sankt Goar 21496
Heimatkundliche Sammlung Elzach, Elzach 18133
Heimatkundliche Sammlung Heideck, Heideck 19068
Heimatkundliche Sammlung Heroldsbach, Heroldsbach 19158
Heimatkundliche Sammlung Isen, Isen 19356
Heimatkundliche Sammlung, Mühle Oberkleinich, Kleinich 19563
Heimatkundliche Sammlung Pfronten, Pfronten 21032
Heimatkundliche Sammlung Strick, Bad Mitterndorf 01760
Heimatkundliche Sammlung und Ausstellung, Hermannsburg 19145
Heimatkundliches Museum, Friedeburg 18470
Heimatkundliches Museum Medaria, Matrei in Osttirol 02363
Heimatkundliches Museum Wetzlhäusl, Sankt Gilgen 02647
Heimatmuseum, Abstatt 16420
Heimatmuseum, Adelsdorf 16426
Heimatmuseum, Aesch 41090
Heimatmuseum, Affalterbach 16434
Heimatmuseum, Ahlen 16437
Heimatmuseum, Aken 16450
Heimatmuseum, Alfdorf 16463
Heimatmuseum, Algermissen 16469
Heimatmuseum, Allendorf 16472
Heimatmuseum, Allensbach 16473
Heimatmuseum, Altenkirchen, Kreis Kusel 16500
Heimatmuseum, Althütte 16509
Heimatmuseum, Altlichtenwarth 01716
Heimatmuseum, Amöneburg 16540
Heimatmuseum, Annaburg 16557
Heimatmuseum, Arendsee 16568
Heimatmuseum, Arneburg 16570
Heimatmuseum, Aßlar 16602
Heimatmuseum, Attiswil 41132
Heimatmuseum, Auetal 16609
Heimatmuseum, Bad Aibling 16666
Heimatmuseum, Bad Dürrheim 16719
Heimatmuseum, Bad Eilsen 16721
Heimatmuseum, Bad Endbach 16726
Heimatmuseum, Bad König 16760
Heimatmuseum, Bad Laer 16772
Heimatmuseum, Bad Lauterberg 16775
Heimatmuseum, Bad Oldesloe 16801
Heimatmuseum, Bad Rodach 16809
Heimatmuseum, Bad Schandau 16826
Heimatmuseum, Bad Teinach-Zavelstein 16850
Heimatmuseum, Bad Überkingen 16852
Heimatmuseum, Bad Vilbel 16856
Heimatmuseum, Balingen 16900
Heimatmuseum, Bammental 16922
Heimatmuseum, Baunach 16936
Heimatmuseum, Beelitz 16977
Heimatmuseum, Belgern 16985
Heimatmuseum, Bergneustadt 17024
Heimatmuseum, Berkatal 17026
Heimatmuseum, Bermel 17286
Heimatmuseum, Bernau bei Berlin 17288
Heimatmuseum, Biebesheim am Rhein 17334
Heimatmuseum, Blaubeuren 17377
Heimatmuseum, Bleicherode 17387
Heimatmuseum, Blindheim 17389
Heimatmuseum, Bobenheim-Roxheim 17392
Heimatmuseum, Bodenfelde 17411
Heimatmuseum, Brackenheim 17494
Heimatmuseum, Breckerfeld 17533
Heimatmuseum, Brigachtal 17589
Heimatmuseum, Brüssow 17609
Heimatmuseum, Brugg 41260
Heimatmuseum, Brunsbüttel 17611
Heimatmuseum, Buchenberg 17616
Heimatmuseum, Buchloe 17619
Heimatmuseum, Büddenstedt 17624
Heimatmuseum, Burg-Reuland 03506
Heimatmuseum, Burg Stargard 17646
Heimatmuseum, Calau 17674
Heimatmuseum, Camburg 17681
Heimatmuseum, Crivitz 17751
Heimatmuseum, Dachwig 17768
Heimatmuseum, Dahlen 17772
Heimatmuseum, Dahme 17774
Heimatmuseum, Darlingerode 17785
Heimatmuseum, Denkendorf 17819
Heimatmuseum, Dienheim 17847
Heimatmuseum, Dietenhofen 17853
Heimatmuseum, Dietzenbach 17857
Heimatmuseum, Dirlewang 17873
Heimatmuseum, Dömitz 17883
Heimatmuseum, Dohna 17887
Heimatmuseum, Domersleben 17888
Heimatmuseum, Donzhausen 41332
Heimatmuseum, Dornstetten 17906
Heimatmuseum, Dossenheim 17932
Heimatmuseum, Duderstadt 17981
Heimatmuseum, Dunningen 18029
Heimatmuseum, Ebermannstadt 18039
Heimatmuseum, Ebern 18040
Heimatmuseum, Echzell 18051
Heimatmuseum, Eggenstein-Leopoldshafen 18064
Heimatmuseum, Eibelstadt 18073

H

Heimatmuseum, Eich 18075
Heimatmuseum, Eichenzell 18078
Heimatmuseum, Eisenbach 18100
Heimatmuseum, Ellrich 18119
Heimatmuseum, Elsenfeld 18128
Heimatmuseum, Eltmann 18130
Heimatmuseum, Elze 18134
Heimatmuseum, Emskirchen 18162
Heimatmuseum, Epfenbach 18177
Heimatmuseum, Eppelborn 18179
Heimatmuseum, Ergoldsbach 18211
Heimatmuseum, Eriskirch 18212
Heimatmuseum, Erlach 01873
Heimatmuseum, Eschenburg 18234
Heimatmuseum, Eschwege 18237
Heimatmuseum, Falkensee 18291
Heimatmuseum, Falkenstein, Vogtland 18296
Heimatmuseum, Finsterbergen 18320
Heimatmuseum, Föhren 18333
Heimatmuseum, Forchheim, Breisgau 18335
Heimatmuseum, Freiamt 18418
Heimatmuseum, Freigericht 18443
Heimatmuseum, Freudenstadt 18456
Heimatmuseum, Friedewald 18471
Heimatmuseum, Friedland 18472
Heimatmuseum, Friesenheim 18481
Heimatmuseum, Fußgönheim 18523
Heimatmuseum, Gadebusch 18526
Heimatmuseum, Gaildorf 18533
Heimatmuseum, Garbsen 18535
Heimatmuseum, Geisa 18555
Heimatmuseum, Geislingen an der Steige 18561
Heimatmuseum, Geretsried 18601
Heimatmuseum, Gersfeld 18614
Heimatmuseum, Gerstungen 18620
Heimatmuseum, Geversdorf 18630
Heimatmuseum, Giesen 18634
Heimatmuseum, Gimbsheim 18645
Heimatmuseum, Ginsheim-Gustavsburg 18648
Heimatmuseum, Glonn 18662
Heimatmuseum, Görwihl 18683
Heimatmuseum, Gößnitz 18685
Heimatmuseum, Graben-Neudorf 18723
Heimatmuseum, Grafenau, Kreis Böblingen 18726
Heimatmuseum, Grafing bei München 18734
Heimatmuseum, Gransee 18737
Heimatmuseum, Greußen 18755
Heimatmuseum, Grindelwald 41455
Heimatmuseum, Großenlüder 18790
Heimatmuseum, Großröhrsdorf 18802
Heimatmuseum, Grüningen 41456
Heimatmuseum, Grünstadt 18817
Heimatmuseum, Gschwend 18819
Heimatmuseum, Günzburg 18825
Heimatmuseum, Gundheim 18837
Heimatmuseum, Haiger 18866
Heimatmuseum, Haimhausen 18869
Heimatmuseum, Hallau 41468
Heimatmuseum, Hallstadt 18911
Heimatmuseum, Haßloch 19031
Heimatmuseum, Haynsburg 19058
Heimatmuseum, Hebertsfelden 19059
Heimatmuseum, Heiligenhafen 19104
Heimatmuseum, Herbrechtingen 19134
Heimatmuseum, Hergensweiler 19141
Heimatmuseum, Hettenleidelheim 19172
Heimatmuseum, Heuchelheim bei Frankenthal 19177
Heimatmuseum, Heusenstamm 19178
Heimatmuseum, Hillscheid 19188
Heimatmuseum, Himmelpforten 19193
Heimatmuseum, Hohberg 19229
Heimatmuseum, Hohenroda 19241
Heimatmuseum, Hohenwestedt 19247
Heimatmuseum, Holle 19252
Heimatmuseum, Holzgerlingen 19257
Heimatmuseum, Hornburg 19276
Heimatmuseum, Hoya / Weser 19280
Heimatmuseum, Hüttenberg 19295
Heimatmuseum, Igel 19312
Heimatmuseum, Ihringen 19313
Heimatmuseum, Ilvesheim 19323
Heimatmuseum, Immendingen 19324
Heimatmuseum, Immenstaad am Bodensee 19326
Heimatmuseum, Kaprun 02153
Heimatmuseum, Karlstein am Main 19456
Heimatmuseum, Kastl, Kreis Amberg-Sulzbach 19480
Heimatmuseum, Kefenrod 19494
Heimatmuseum, Kehl 19497
Heimatmuseum, Keltern 19506
Heimatmuseum, Kenn 19517
Heimatmuseum, Kirchen 19545
Heimatmuseum, Kirchhain 19547
Heimatmuseum, Klipphausen 19573
Heimatmuseum, Kloster 19575
Heimatmuseum, Kölleda 19594
Heimatmuseum, Königs Wusterhausen 19637
Heimatmuseum, Königsdorf 19647
Heimatmuseum, Kraiburg am Inn 19694
Heimatmuseum, Kroppenstedt 19728
Heimatmuseum, Küssnacht am Rigi 41520
Heimatmuseum, Kuhardt 19742
Heimatmuseum, Ladbergen 19757
Heimatmuseum, Landau an der Isar 19786
Heimatmuseum, Langenzenn 19835
Heimatmuseum, Leimersheim 19879
Heimatmuseum, Leipheim 19889
Heimatmuseum, Lengenfeld 19953
Heimatmuseum, Lenggries 19954
Heimatmuseum, Leutenbach 19968
Heimatmuseum, Lilienthal 19988
Heimatmuseum, Lommatzsch 20030
Heimatmuseum, Lorch, Württemberg 20034
Heimatmuseum, Luckenwalde 20041
Heimatmuseum, Lügde 20093
Heimatmuseum, Magstadt 20132
Heimatmuseum, Mainz-Kostheim 20153
Heimatmuseum, March 20196
Heimatmuseum, Markgröningen 20204
Heimatmuseum, Markranstädt 20210
Heimatmuseum, Marktl 20217
Heimatmuseum, Marne 20227
Heimatmuseum, Maßbach 20234
Heimatmuseum, Meerane 20245
Heimatmuseum, Mehrstetten 20257
Heimatmuseum, Meinhard 20258
Heimatmuseum, Mellrichstadt 20274
Heimatmuseum, Melsungen 20276
Heimatmuseum, Merchweiler 20288
Heimatmuseum, Mering 20289
Heimatmuseum, Merkendorf 20290
Heimatmuseum, Merzenich 20297
Heimatmuseum, Miesbach 20323
Heimatmuseum, Millstatt 02382
Heimatmuseum, Mindelheim 20328
Heimatmuseum, Mitterfels 20343
Heimatmuseum, Möckmühl 20350
Heimatmuseum, Moosburg an der Isar 20380
Heimatmuseum, Moringen 20384
Heimatmuseum, Mügeln 20395
Heimatmuseum, Mülheim an der Ruhr 20413
Heimatmuseum, Müllrose 20418
Heimatmuseum, Murg 20533
Heimatmuseum, Nauheim 20552
Heimatmuseum, Naumburg, Hessen 20553
Heimatmuseum, Neckargerach 20564
Heimatmuseum, Neidenstein 20567
Heimatmuseum, Nellingen 20568
Heimatmuseum, Neu-Ulm 20588
Heimatmuseum, Neukirch/Lausitz 20623
Heimatmuseum, Neukirchen, Schwalm-Eder-Kreis 20626
Heimatmuseum, Neumagen-Dhron 20631
Heimatmuseum, Neunkirchen, Neckar-Odenwald-Kreis 20641
Heimatmuseum, Neupotz 20644
Heimatmuseum, Neureichenau 20646
Heimatmuseum, Neuried, Ortenaukreis 20647
Heimatmuseum, Neustadt in Sachsen 20682
Heimatmuseum, Neustadt, Wied 20683
Heimatmuseum, Nidda 20692
Heimatmuseum, Nideggen 20696
Heimatmuseum, Nieder-Olm 20700
Heimatmuseum, Niederaichbach 20701
Heimatmuseum, Niedernhall 20708
Heimatmuseum, Niederstetten 20710
Heimatmuseum, Northeim 20747
Heimatmuseum, Nüdlingen 20751
Heimatmuseum, Ober-Flörsheim 20797
Heimatmuseum, Oberderdingen 20803
Heimatmuseum, Oberndorf am Lech 20818
Heimatmuseum, Oberndorf am Neckar 20819
Heimatmuseum, Obernfeld 20821
Heimatmuseum, Oberstdorf 20842
Heimatmuseum, Oberweningen 41701
Heimatmuseum, Ockenheim 20856
Heimatmuseum, Oederan 20859
Heimatmuseum, Oettingen in Bayern 20878
Heimatmuseum, Osterhofen 20934
Heimatmuseum, Osterwieck 20938
Heimatmuseum, Ostrach 20943
Heimatmuseum, Otterberg 20949
Heimatmuseum, Papenburg 20980
Heimatmuseum, Perl 21011
Heimatmuseum, Perleberg 21013
Heimatmuseum, Pfaffenhausen 21017
Heimatmuseum, Pfullendorf 21033
Heimatmuseum, Philippsburg 21037
Heimatmuseum, Plankstadt 21049
Heimatmuseum, Polch 21072
Heimatmuseum, Polling 21074
Heimatmuseum, Postbauer-Heng 21077
Heimatmuseum, Preetz 21116
Heimatmuseum, Pressath 21120
Heimatmuseum, Pretzsch 21123
Heimatmuseum, Prien am Chiemsee 21127
Heimatmuseum, Queidersbach 21153
Heimatmuseum, Quierschied 21156
Heimatmuseum, Radeburg 21162
Heimatmuseum, Radevormwald 21164
Heimatmuseum, Radstadt 02551
Heimatmuseum, Rattenberg 21194
Heimatmuseum, Raunheim 21200
Heimatmuseum, Rauschenberg 21202
Heimatmuseum, Rehlingen-Siersburg 21241
Heimatmuseum, Reilingen 21250
Heimatmuseum, Reinach (Basel-Land) 41754
Heimatmuseum, Reinheim 21254
Heimatmuseum, Renningen 21278
Heimatmuseum, Rheda-Wiedenbrück 21292
Heimatmuseum, Rieneck 21328
Heimatmuseum, Rietz 02589
Heimatmuseum, Rodenberg 21344
Heimatmuseum, Rodewald 21345
Heimatmuseum, Rötha 21358
Heimatmuseum, Rötz 21360
Heimatmuseum, Ronnenberg 21370
Heimatmuseum, Roßwein 21385
Heimatmuseum, Rotenburg, Wümme 21400
Heimatmuseum, Rothrist 41777
Heimatmuseum, Sandhausen 21483
Heimatmuseum, Sankt Aegyd am Neuwalde 02632
Heimatmuseum, Sankt Egidien 21491
Heimatmuseum, Sankt Martin am Wöllmißberg 02668
Heimatmuseum, Schalkau 21530
Heimatmuseum, Scheeßel 21536
Heimatmuseum, Schermbeck 21544
Heimatmuseum, Schifferstadt 21547
Heimatmuseum, Schliersee 21574
Heimatmuseum, Schlitz 21575
Heimatmuseum, Schnaittach 21589
Heimatmuseum, Schöneck / Vogtland 21605
Heimatmuseum, Schöneiche bei Berlin 21607
Heimatmuseum, Schöningen 21612
Heimatmuseum, Schwalmtal 21663
Heimatmuseum, Schwanenstadt 02717
Heimatmuseum, Schwanfeld 21666
Heimatmuseum, Seeg 21716
Heimatmuseum, Seelze 21721
Heimatmuseum, Sersheim 21746
Heimatmuseum, Seßlach 21749
Heimatmuseum, Siegbach 21750
Heimatmuseum, Simbach am Inn 21767
Heimatmuseum, Sinzig 21787
Heimatmuseum, Sohland an der Spree 21800
Heimatmuseum, Spangenberg 21829
Heimatmuseum, Speicher 21831
Heimatmuseum, Sprendlingen 21844
Heimatmuseum, Stade 21849
Heimatmuseum, Stadt Wehlen 21854
Heimatmuseum, Stadtilm 21857
Heimatmuseum, Stadtsteinach 21861
Heimatmuseum, Staufenberg 21869
Heimatmuseum, Steinbach, Taunus 21882
Heimatmuseum, Steinwiesen 21894
Heimatmuseum, Sternberg 21901
Heimatmuseum, Stockstadt am Main 21905
Heimatmuseum, Strasburg 21916
Heimatmuseum, Strausberg 21919
Heimatmuseum, Stromberg 21921
Heimatmuseum, Talheim 21998
Heimatmuseum, Tambach-Dietharz 21999
Heimatmuseum, Todtmoos 22067
Heimatmuseum, Trebur 22082
Heimatmuseum, Trechtingshausen 22084
Heimatmuseum, Treffurt 22086
Heimatmuseum, Triptis 22107
Heimatmuseum, Undenheim 22173
Heimatmuseum, Unterschneidheim 22184
Heimatmuseum, Vallendar 22201
Heimatmuseum, Varel 22204
Heimatmuseum, Velburg 22213
Heimatmuseum, Venningen 22218
Heimatmuseum, Viernheim 22230
Heimatmuseum, Wadern 22262
Heimatmuseum, Wächtersbach 22266
Heimatmuseum, Wald 41996
Heimatmuseum, Waldaschaff 22281
Heimatmuseum, Waldfischbach-Burgalben 22294
Heimatmuseum, Waldheim 22295
Heimatmuseum, Waldsee 22304
Heimatmuseum, Waldthurn 22305
Heimatmuseum, Waltensburg 42000
Heimatmuseum, Wanna 22331
Heimatmuseum, Wedel 22354
Heimatmuseum, Weener 22356
Heimatmuseum, Weisendorf 22442
Heimatmuseum, Weissach im Tal 22446
Heimatmuseum, Weißenborn 22448
Heimatmuseum, Wemding 22473
Heimatmuseum, Wennigsen 22480
Heimatmuseum, Wertach 22501
Heimatmuseum, Wertingen 22505
Heimatmuseum, Westerhausen 22514
Heimatmuseum, Westheim 22516
Heimatmuseum, Wilfersdorf 03106
Heimatmuseum, Windeck 22594
Heimatmuseum, Winden 22595
Heimatmuseum, Wippra 22610
Heimatmuseum, Wolfratshausen 22657
Heimatmuseum, Wolfstein 22671
Heimatmuseum, Wusterhausen 22753
Heimatmuseum, Zarrentin am Schaalsee 22767
Heimatmuseum, Zemmer 22787
Heimatmuseum, Zerf 22789
Heimatmuseum, Zingst, Ostseebad 22807
Heimatmuseum, Zöblitz 22814
Heimatmuseum, Zorge 22818
Heimatmuseum, Zusmarshausen 22828
Heimatmuseum Aarburg, Aarburg 41086
Heimatmuseum Achental, Achenkirch 01692
Heimatmuseum Adelboden, Adelboden 41088
Heimatmuseum Adlhoch-Haus, Altdorf, Kreis Landshut 16485
Heimatmuseum Adolphshöhe, Wallerfangen 22313
Heimatmuseum Ahlem, Hannover 18992
Heimatmuseum Ahrbergen, Giesen 18635
Heimatmuseum Allschwil, Allschwil 41101
Heimatmuseum Alt-Falkenstein, Klus 41507
Heimatmuseum Alte Weberstube, Niedercunnersdorf 20703
Heimatmuseum Alte Windmühle Groß Reken, Reken 21257
Heimatmuseum Alter Pfarrhof, Raisting 21171
Heimatmuseum Altes Rathaus, Böhl-Iggelheim 17423
Heimatmuseum Altes Schulhaus, Böhl-Iggelheim 17424
Heimatmuseum am Kobernaußerwald, Lohnsburg am Kobernaußerwald 02334
Heimatmuseum am Pfäffikersee, Pfäffikon (Zürich) 41723
Heimatmuseum Amt Blankenstein, Gladenbach 18652
Heimatmuseum Arnold Bärtschi, Dulliken 41336
Heimatmuseum Aschen, Diepholz 17848
Heimatmuseum Assel, Drochtersen 17979
Heimatmuseum auf dem Klosterhof, Steinheim am Albuch 21889
Heimatmuseum Bad Großpertholz, Bad Großpertholz 01750
Heimatmuseum Barbaramühle, Longkamp 20031
Heimatmuseum Berching mit Ritter-v.-Gluck-Archiv, Berching 16999
Heimatmuseum Bergen-Enkheim, Frankfurt am Main 18373
Heimatmuseum Bernhardsthal, Bernhardsthal 01781
Heimatmuseum Beuel, Bonn 17456
Heimatmuseum Beutelsbach, Weinstadt 22436
Heimatmuseum Bezau, Bezau 01784
Heimatmuseum Bierstadt, Wiesbaden 22549
Heimatmuseum Bislich, Wesel 22506
Heimatmuseum Blankenau, Hosenfeld 19279
Heimatmuseum Blankenloch, Stutensee 21927
Heimatmuseum Bockwindmühle, Lebusa 19866
Heimatmuseum Börgerende, Börgerende-Rethwisch 17429
Heimatmuseum Boizenburg, Boizenburg 17439
Heimatmuseum Boizenburg, Außenstelle am Elbberg, Boizenburg 17440
Heimatmuseum Borgloh, Hilter am Teutoburger Wald 19190
Heimatmuseum Boxberg, Boxberg 17491
Heimatmuseum Brueckenhof, Königswinter 19658
Heimatmuseum Bucheggberg, Buchegg 41266
Heimatmuseum Bühl, Baden, Bühl 17630
Heimatmuseum Carl Swoboda, Schirgiswalde 21558
Heimatmuseum Crumstadt, Riedstadt 21323
Heimatmuseum Dachtel, Aidlingen 16447
Heimatmuseum Davos, Davos Dorf 41319
Heimatmuseum Debstedt, Langen, Kreis Cuxhaven 19813
Heimatmuseum Delkenheim, Wiesbaden 22550
Heimatmuseum der Albert-Edelmann-Stiftung, Ebnat-Kappel 41338
Heimatmuseum der Deutschen aus Bessarabien, Stuttgart 21940
Heimatmuseum der Gemeinde Weißenkirchen, Perschling 02491
Heimatmuseum der Hauptschule Taufkirchen, Taufkirchen an der Pram 02798
Heimatmuseum der Parabutscher Donauschwaben, Bad Schönborn 16828
Heimatmuseum der Stadt Bad Tölz, Bad Tölz 16851
Heimatmuseum der Stadt Gernsheim, Gernsheim 18610
Heimatmuseum der Stadt Herrnhut, Herrnhut 19162
Heimatmuseum der Stadt Ketzin, Ketzin 19522
Heimatmuseum der Stadt Rerik, Rerik, Ostseebad 21280
Heimatmuseum der Stadt Rheinau in Freistett, Rheinau 21296
Heimatmuseum der Stadt Wilsdruff, Wilsdruff 22590
Heimatmuseum der Talschaft Lauterbrunnen, Lauterbrunnen 41552
Heimatmuseum der Verbandsgemeinde Wallhalben, Schmitshausen 21587
Heimatmuseum des Daadener Landes, Daaden 17761
Heimatmuseum des Heimatbundes Jauernig, Vaihingen an der Enz 22197
Heimatmuseum des Landkreises Regensburg, Altenthann 16504
Heimatmuseum des Nordböhmischen Niederlandes, Böblingen 17419
Heimatmuseum Dietingen, Dietingen 17856
Heimatmuseum Dykhus, Borkum 17481
Heimatmuseum Ebenhausen, Oerlenbach 20873

Heimatmuseum Ebersbach in der Humboldt-Baude, Ebersbach, Sachsen 18043
Heimatmuseum Egg, Egg 01842
Heimatmuseum Eisenbach, Obernburg am Main 20816
Heimatmuseum Elgg, Elgg 41347
Heimatmuseum Elsbethen, Elsbethen 01866
Heimatmuseum Erbenheim, Wiesbaden 22551
Heimatmuseum Esterau, Holzappel 19256
Heimatmuseum Eversberg, Meschede 20302
Heimatmuseum Flacht, Weissach 22445
Heimatmuseum Frankenberg, Frankenberg, Sachsen 18347
Heimatmuseum Freudenthal, Memmingen 20279
Heimatmuseum Friedrichshain, Berlin 17130
Heimatmuseum Fronfeste, Limbach-Oberfrohna 19992
Heimatmuseum Fügen, Fügen 01909
Heimatmuseum für Stadt und Kreis Gelnhausen, Gelnhausen 18568
Heimatmuseum für Stadt- und Landkreis, Kirchheimbolanden 19553
Heimatmuseum Gaaden, Gaaden 01915
Heimatmuseum Gablitz, Gablitz 01917
Heimatmuseum Gadernheim, Lautertal 19864
Heimatmuseum Garbenheim, Wetzlar 22525
Heimatmuseum Gersau, Gersau 41431
Heimatmuseum Gnas, Gnas 01953
Heimatmuseum Grabow, Grabow 18724
Heimatmuseum Grafenwörth, Grafenwörth 01960
Heimatmuseum Grafling, Oberschneiding 20839
Heimatmuseum "Grafschaft Molbach", Kreuzau 19713
Heimatmuseum Greene, Kreiensen 19709
Heimatmuseum Grimmen "Im Mühlentor", Grimmen 18766
Heimatmuseum Gröbming, Gröbming 01992
Heimatmuseum Groß-Enzersdorf, Groß-Enzersdorf 01995
Heimatmuseum Großengersdorf, Großengersdorf 02000
Heimatmuseum Großes Walsertal, Sonntag 02747
Heimatmuseum Großschönau, Großschönau 02011
Heimatmuseum Großseelheim, Kirchhain 19548
Heimatmuseum Großzschepa, Hohburg 19231
Heimatmuseum Guntramsdorf, Guntramsdorf 02018
Heimatmuseum Hadersdorf, Hadersdorf 02031
Heimatmuseum Harthausen, Winterlingen 22608
Heimatmuseum Haueneberstein, Baden-Baden 16880
Heimatmuseum Haus der Südostdeutschen, Böchingen 17422
Heimatmuseum Haus Horn, Niederzier 20716
Heimatmuseum Heidelsheim, Bruchsal 17595
Heimatmuseum Heinrich Zoller, Wittelshofen 22616
Heimatmuseum Helfs Hof, Bochum 17400
Heimatmuseum Herzogenweiler, Villingen-Schwenningen 22237
Heimatmuseum Heuchelheim-Kinzenbach, Heuchelheim 19176
Heimatmuseum Hochzeitshaus, Homberg (Efze) 19264
Heimatmuseum Höchstädt, Höchstädt an der Donau 19210
Heimatmuseum Hoheneggelsen, Söhlde 21792
Heimatmuseum Holzgau, Holzgau 02098
Heimatmuseum Hopfenhaus, Aidlingen 16448
Heimatmuseum Hossingen, Meßstetten 20308
Heimatmuseum Hus tu Löwenberg, Löwenberger Land 20017
Heimatmuseum Ilz, Ilz 02106
Heimatmuseum im Ackerbürgerhaus, Vohenstrauß 22253
Heimatmuseum im alten Bruderhaus und in der Dechanathofscheune, Altenmarkt im Pongau 01713
Heimatmuseum im Alten Kloster, Erbendorf 18192
Heimatmuseum im Alten Leuchtturm, Wangerooge, Nordseebad 22330
Heimatmuseum im alten Marktturm, Fischamend 01890
Heimatmuseum im alten Pfarrhaus, Schwieberdingen 21705
Heimatmuseum im Alten Rathaus, Bruchköbel 17593
Heimatmuseum im alten Zollhaus, Kobersdorf 02208
Heimatmuseum im Amonhaus, Lunz 02340
Heimatmuseum im Berchtoldshof, Uhingen 22157
Heimatmuseum im Einlegerhaus, Obertrum am See 02463
Heimatmuseum im Fünfeckigen Turm, Neckarbischofsheim 20560
Heimatmuseum im Hartmannhaus, Marktoberdorf 20218
Heimatmuseum im Haus Catoir, Bad Dürkheim 16714
Heimatmuseum im Herrenhaus, Burghaun 17650
Heimatmuseum im Hintermeierhaus, Donauwörth 17896
Heimatmuseum im Kolb'schen Haus, Bickenbach 17328
Heimatmuseum im Oberen Torturm, Lauchheim 19846
Heimatmuseum im Schelfenhaus, Volkach 22255
Heimatmuseum im Schillerhaus, Oberkochen 20814
Heimatmuseum im Schlösschen, Ochsenfurt 20852
Heimatmuseum im Schloß, Zörbig 22816
Heimatmuseum im Studierstüberl, Krieglach 02236
Heimatmuseum im Tuchmacherhaus, Thannhausen 22048
Heimatmuseum im Unteren Turm, Leutershausen-Wiedersbach 19971
Heimatmuseum im Vogteigebäude, Hüttlingen 19296
Heimatmuseum im Wachtturm, Geyer 18631
Heimatmuseum im Wiggensbacher Informationszentrum, Wiggensbach 22565
Heimatmuseum im Wolfschneider-Hof, Taufkirchen, Kreis München 22014
Heimatmuseum in der Mühle Gettorf, Gettorf 18628
Heimatmuseum in der Scheune, Mössingen 20365
Heimatmuseum Ingersleben/Neudietendorf, Ingersleben 19336
Heimatmuseum Ittersbach, Karlsbad 19432
Heimatmuseum Kalchofengut, Unken 02840
Heimatmuseum Kals, Kals am Großglockner 02150
Heimatmuseum Kappe, Zehdenick 22768
Heimatmuseum Karlsdorf, Karlsdorf-Neuthard 19433
Heimatmuseum Katsdorf, Katsdorf 02156
Heimatmuseum Kaumberg, Kaumberg 02158
Heimatmuseum Kaunertal, Kaunertal 02159
Heimatmuseum Kautzen, Kautzen 02160
Heimatmuseum Kematen, Kematen an der Ybbs 02161
Heimatmuseum Kiechlinsbergen, Endingen am Kaiserstuhl 18164
Heimatmuseum Kirberg, Hünfelden 19291
Heimatmuseum Kirchdorf, Insel Poel 19348
Heimatmuseum Kirchham, Kirchham 02171
Heimatmuseum Kirchheim, Heidelberg 19074
Heimatmuseum Kirchschlag, Kirchschlag 02172
Heimatmuseum Klösterle, Schönau im Schwarzwald 21597
Heimatmuseum Klosterlangheim, Lichtenfels 19979
Heimatmuseum Köpenick, Berlin 17131
Heimatmuseum Kohbauernhaus, Köstendorf 02213
Heimatmuseum Kornburg, Nürnberg 20767
Heimatmuseum Kuchl, Kuchl 02242
Heimatmuseum Lahnau-Waldgirmes, Lahnau 19765
Heimatmuseum Laisa, Battenberg 16933
Heimatmuseum Lamspringe, Lamspringe 19783
Heimatmuseum Langelsheim, Langelsheim 19809
Heimatmuseum Langenau, Langenau 19818
Heimatmuseum Langenlois, Langenlois 02256
Heimatmuseum Langenseifen, Bad Schwalbach 16834
Heimatmuseum Langenselbold, Langenselbold 19834
Heimatmuseum Langenzersdorf, Langenzersdorf 02261
Heimatmuseum Laubach, Laubach 19842
Heimatmuseum Lauchhammer, Lauchhammer 19844
Heimatmuseum Lauda-Königshofen, Lauda-Königshofen 19847
Heimatmuseum Lauenau, Lauenau 19848
Heimatmuseum Leeheim, Riedstadt 21324
Heimatmuseum Leer, Leer 19870
Heimatmuseum Lembach, Lembach 02282
Heimatmuseum Lengsdorf, Bonn 17457
Heimatmuseum Lette, Coesfeld 17722
Heimatmuseum Leuk Stadt, Leuk Stadt 41557
Heimatmuseum Lich, Lich 19977
Heimatmuseum Lichtenberg, Berlin 17132
Heimatmuseum Liebenwalde im historischen Stadtgefängnis, Liebenwalde 19984
Heimatmuseum Litschau, Litschau 02328
Heimatmuseum Losenstein, Losenstein 02337
Heimatmuseum Lübbecke und Kunstgalerie im Speicher, Lübbecke 20067
Heimatmuseum Lütgendortmund, Dortmund 17920
Heimatmuseum Mank, Mank 02346
Heimatmuseum Marchegg, Marchegg 02349
Heimatmuseum Markt Piesting, Markt Piesting 02361
Heimatmuseum Mautern, Mautern in Steiermark 02370
Heimatmuseum Medenbach, Wiesbaden 22552
Heimatmuseum Meerholz, Gelnhausen 18569
Heimatmuseum-Menton, Teningen 22031
Heimatmuseum Metzgerhaus, Kirchdorf 02170
Heimatmuseum Michelbacher Schlößchen, Alzenau in Unterfranken 16529
Heimatmuseum Michelhausen, Michelhausen 02380
Heimatmuseum mit Folterkammer, Pöggstall 02508
Heimatmuseum mit Franz Jonas-Gedenkraum, Teesdorf 02799
Heimatmuseum mit ostdeutscher Heimatstube, Hanerau-Hademarschen 18985
Heimatmuseum mit Raimund von Montecuccoli-Gedenkraum, Hafnerbach 02036
Heimatmuseum Mittelbuchen, Hanau 18976
Heimatmuseum Möhringen, Stuttgart 21941
Heimatmuseum Münchingen, Korntal-Münchingen 19688
Heimatmuseum Mureck, Mureck 02419
Heimatmuseum Naunheim, Wetzlar 22526
Heimatmuseum Naurod, Wiesbaden 22553
Heimatmuseum Neckarau, Mannheim 20164
Heimatmuseum Neuenwalde, Langen, Kreis Cuxhaven 19814
Heimatmuseum Neunkirchen, Neunkirchen 02440
Heimatmuseum Neustadtgödens, Sande 21482
Heimatmuseum Nied, Frankfurt am Main 18374
Heimatmuseum Niederleis, Niederleis 02445
Heimatmuseum Niederndorf, Niederndorf 02447
Heimatmuseum Niemes und Prachatitz, Ingolstadt 19341
Heimatmuseum Nürnberg, Nürnberg 20768
Heimatmuseum Nutli-Hüschi, Klosters 41505
Heimatmuseum Obbornhofen, Hungen 19297
Heimatmuseum Oberbichl, Prägraten 02521
Heimatmuseum Obere Mühle, Loßburg 20039
Heimatmuseum Obergünzburg mit Südseesammlung, Obergünzburg 20806
Heimatmuseum Oberhofen, Oberhofen im Inntal 02453
Heimatmuseum Obermühle, Ach 01691
Heimatmuseum Oberwölz und Österreichisches Blasmusikmuseum, Oberwölz 02466
Heimatmuseum Ole Schüne, Drakenburg 17934
Heimatmuseum Oltingen-Wenslingen-Anwil, Oltingen 41711
Heimatmuseum Pabneukirchen, Pabneukirchen 02478
Heimatmuseum Peldemühle, Wittmund 22626
Heimatmuseum Persenbeug-Gottsdorf, Persenbeug 02492
Heimatmuseum Pfarrscheuer, Römerstein 21355
Heimatmuseum Pfunds, Pfunds 02498
Heimatmuseum Plieningen, Stuttgart 21942
Heimatmuseum Prättigau, Grüsch 41459
Heimatmuseum Pressbaum, Pressbaum 02532
Heimatmuseum Prestegg, Altstätten 41108
Heimatmuseum Prieros, Heidesee 19094
Heimatmuseum Pulkau, Pulkau 02537
Heimatmuseum Rabenstein, Rabenstein an der Pielach 02548
Heimatmuseum Rain am Lech, Rain 21169
Heimatmuseum Rapperswil, Rapperswil 41743
Heimatmuseum Ratzenried, Argenbühl 16569
Heimatmuseum Rauchkate, Zetel 22790
Heimatmuseum Rauchstubenhaus Edelschachen, Anger 01721
Heimatmuseum Reinickendorf, Berlin 17133
Heimatmuseum Reischenau, Dinkelscherben 17869
Heimatmuseum Reiterhaus, Neusalza-Spremberg 20653
Heimatmuseum Reith, Reith im Alpbachtal 02577
Heimatmuseum Reutlingen, Reutlingen 21283
Heimatmuseum Rheiderland, Weener 22357
Heimatmuseum Rheinwald, Splügen 41895
Heimatmuseum Rhüden, Seesen 21723
Heimatmuseum Rodenstein, Fränkisch-Crumbach 18342
Heimatmuseum Römstedthaus, Bergen, Kreis Celle 17009
Heimatmuseum Rohrbach, Heidelberg 19075
Heimatmuseum Rohrhirschmühle, Bühl 17631
Heimatmuseum Rother Hof, Pottendorf 02514
Heimatmuseum Runder Turm, Sigmaringen 21764
Heimatmuseum Sammlung Fuchs, Kirchbach in Steiermark 02164
Heimatmuseum Sandweier, Baden-Baden 16881
Heimatmuseum Sanhofen, Mannheim 20165
Heimatmuseum Sankt Laurentius, Dahlenburg 17773
Heimatmuseum Schanfigg, Arosa 41126
Heimatmuseum Schierstein, Wiesbaden 22554
Heimatmuseum Schinznach-Dorf, Schinznach Dorf 41834
Heimatmuseum Schloß Brenz, Sontheim an der Brenz 21821
Heimatmuseum Schloß Fels, Fels am Wagram 01885
Heimatmuseum Schloß Hochhaus, Vorchdorf 02875
Heimatmuseum Schloß Schönebeck, Bremen 17554
Heimatmuseum Schloß Wolkenstein, Wolkenstein 22675
Heimatmuseum Schönau, Günselsdorf 02014
Heimatmuseum Schrems, Schrems 02714
Heimatmuseum Schulte-Wessels, Neuenhaus 20608
Heimatmuseum Schwanheim, Frankfurt am Main 18375
Heimatmuseum Schwarzbubenland, Dornach 41333
Heimatmuseum Schwarzes Tor, Sankt Georgen im Schwarzwald 21493
Heimatmuseum Seehof, Buonas 41273
Heimatmuseum Seulberg, Friedrichsdorf 18475
Heimatmuseum Sissach, Sissach 41879
Heimatmuseum Spittel, Büren an der Aare 41269
Heimatmuseum Sprucker Mühle, Guben 18820
Heimatmuseum Spycher, Rickenbach (Luzern) 41760
Heimatmuseum Stadt Bernstadt, Bernstadt auf dem Eigen 17307
Heimatmuseum Stadt Nagold, Nagold 20545
Heimatmuseum Stadt Teltow, Teltow 22025
Heimatmuseum Stadt und Landkreis Neudek im Erzgebirge, Augsburg 16622
Heimatmuseum Stammheim, Stuttgart 21943
Heimatmuseum Stein, Stein am Rhein 41913
Heimatmuseum Steinbach, Ottweiler 20959
Heimatmuseum Steinfischbach, Waldems 22286
Heimatmuseum Steyregg, Steyregg 02777
Heimatmuseum Strümpfelbach, Weinstadt 22437
Heimatmuseum Suhr, Suhr 41920
Heimatmuseum Tannheimer Tal, Tannheim 02794
Heimatmuseum Tarrenz, Tarrenz 02795
Heimatmuseum Thaya, Thaya 02810
Heimatmuseum Tieringen, Meßstetten 20309
Heimatmuseum - Traktoren- und Landmaschinenmuseum, Großengottern 18788
Heimatmuseum Traun, Traun 02824
Heimatmuseum Treptow, Berlin 17134
Heimatmuseum Trimmis, Trimmis 41950
Heimatmuseum Trubschachen, Trubschachen 41952
Heimatmuseum Uelzen, Uelzen 22150
Heimatmuseum Uhl'sches Haus, Göllheim 18671
Heimatmuseum und Apothekenmuseum, Mauthausen 02372
Heimatmuseum und Dokumentationszentrum zur Deutschen Nachkriegsgeschichte, Wanfried 22322
Heimatmuseum und Galerie Neue Diele, Jork 19402
Heimatmuseum und Gedenkstätte Bützow, Bützow 17641
Heimatmuseum und Hedwig Courths-Mahler Archiv, Nebra 20559
Heimatmuseum und Keltenhaus, Ligist 02304
Heimatmuseum und Klöpferhaus, Eibiswald 01849
Heimatmuseum und landwirtschaftliche Sammlung, Alberschwende 01702
Heimatmuseum und Naturalienkabinett, Waldenburg, Kreis Chemnitzer Land 22290
Heimatmuseum und Schaubergwerk, Arzberg 01729
Heimatmuseum und Schiele Museum, Neulengbach 02434
Heimatmuseum und Textil-Schauwerkstatt, Greiz 18752
Heimatmuseum Untergrombach, Bruchsal 17596
Heimatmuseum Usseln, Willingen 22585
Heimatmuseum Vietze, Höhbeck 19213
Heimatmuseum Vilsbiburg - Kröninger Hafnermuseum, Vilsbiburg 22243
Heimatmuseum Vollmers Mühle, Seebach 21712
Heimatmuseum Walldorf, Mörfelden-Walldorf 20360
Heimatmuseum Waltersweier, Offenburg 20885
Heimatmuseum Wandsbek, Hamburg 18934
Heimatmuseum Warnemünde, Rostock 21392
Heimatmuseum Wartberg, Wartberg im Mürztal 02895
Heimatmuseum Weesen, Weesen 42003
Heimatmuseum Weiler, Schorndorf 21626
Heimatmuseum Weitenung, Bühl 17632
Heimatmuseum Weng, Weng im Innkreis 02920
Heimatmuseum Wiesenbach, Wiesenbach 22559
Heimatmuseum Windischgarsten, Windischgarsten 03116
Heimatmuseum Winterberg im Böhmerwald, Freyung 18460
Heimatmuseum Wißmar, Wettenberg 22520
Heimatmuseum Wörgl, Wörgl 03117
Heimatmuseum Wolfskehlen, Riedstadt 21325
Heimatmuseum Wommelshausen, Bad Endbach 16727
Heimatmuseum Worben, Worben 42046
Heimatmuseum Worms Abenheim, Worms 22681
Heimatmuseum Worms Horchheim, Worms 22682
Heimatmuseum Worms-Weinsheim, Worms 22683
Heimatmuseum Zehlendorf, Berlin 17135
Heimatmuseum Zwischen Venn und Schneifel, Sankt Vith 03999
Heimatscheune, Zschepplin 22824
Heimatstätte der Kuhländer, Ludwigsburg 20044
Heimatstube, Altenbrak 16496
Heimatstube, Altenstadt an der Waldnaab 16502
Heimatstube, Benneckenstein 16993
Heimatstube, Brüel 17603
Heimatstube, Calbe, Saale 17676
Heimatstube, Dabel 17762
Heimatstube, Derenburg 17824
Heimatstube, Ditfurt 17877
Heimatstube, Eggesin 18065
Heimatstube, Elsteraue 18129
Heimatstube, Enkirch 18176
Heimatstube, Esslingen 18274
Heimatstube, Fehrbellin 18300
Heimatstube, Gittelde 18649
Heimatstube, Goslar 18709
Heimatstube, Großraming 02007
Heimatstube, Hasselfelde 19027
Heimatstube, Hasselfelde 19028
Heimatstube, Hinternah 19194
Heimatstube, Hohenzieritz 19248
Heimatstube, Lahstedt 19774
Heimatstube, Lenzkirch 19958
Heimatstube, Liepgarten 19987

H

Heimatstube, Marlow 20226
Heimatstube, Matzlow-Garwitz 20237
Heimatstube, Neulußheim 20629
Heimatstube, Neuruppin 20649
Heimatstube, Quellendorf 21154
Heimatstube, Remseck am Neckar 21268
Heimatstube, Schäplitz 21528
Heimatstube, Schochwitz 21595
Heimatstube, Schönwalde, Altmark 21615
Heimatstube, Schonach im Schwarzwald 21619
Heimatstube, Schwarzenberg am Böhmerwald 02720
Heimatstube, Schwedt / Oder 21677
Heimatstube, Stapelburg 21863
Heimatstube, Tanne 22007
Heimatstube, Wettin 22522
Heimatstube, Zörbig 22817
Heimatstube 1813 Möckern, Möckern 20349
Heimatstube Abbenrode, Abbenrode 16416
Heimatstube Adlergebirge, Waldkraiburg 22299
Heimatstube Altenau, Altenau 16490
Heimatstube Arenborn, Oberweser 20850
Heimatstube Batzsch-Brestowatz, Königsbach-Stein 19640
Heimatstube Bliedersdorf, Bliedersdorf 17388
Heimatstube Börnecke, Blankenburg 17371
Heimatstube Bredenbeck, Wennigsen 22481
Heimatstube Brunkensen, Alfeld 16465
Heimatstube Bühlau, Großharthau 18792
Heimatstube der Gemeinde und des Harzklub Zweigvereins Bad Suderode, Bad Suderode 16846
Heimatstube "Der Große Hof", Wienrode 22543
Heimatstube der Stadt Saaz, Roth 21402
Heimatstube des Heimatkreises Reichenberg, Augsburg 16623
Heimatstube Dessau-Alten, Dessau 17829
Heimatstube Dreetz, Dreetz 17935
Heimatstube Eime, Eime 18090
Heimatstube Elbingerode, Elbingerode, Harz 18112
Heimatstube Eltze, Uetze 22153
Heimatstube Endersbach, Weinstadt 22438
Heimatstube Feldberg, Feldberger Seenlandschaft 18302
Heimatstube Fischbach, Niedereschach 20705
Heimatstube Frauenwald und Bunkermuseum, Frauenwald 18414
Heimatstube Freest, Kröslin 19718
Heimatstube für Volkskunst, Zschadraß 22823
Heimatstube Gadenstedt, Lahstedt 19775
Heimatstube Gräfenhain/ Nauendorf, Georgenthal 18590
Heimatstube Groß Denkte, Denkte 17822
Heimatstube Groß Gleidingen, Vechelde 22206
Heimatstube Groß Lafferde, Lahstedt 19776
Heimatstube Groß Nemerow, Groß Nemerow 18777
Heimatstube Großbodungen, Großbodungen 18781
Heimatstube Gutenberg, Oberostendorf 20826
Heimatstube Hänigsen, Uetze 22154
Heimatstube Hetzdorf, Uckerland 22143
Heimatstube Hohenhameln, Hohenhameln 19237
Heimatstube im alten Rathaus, Hessigheim 19169
Heimatstube im Burgturm, Plau am See 21052
Heimatstube im Elbtor, Werben 22484
Heimatstube im "Spieker Anno 1754", Tarmstedt 22008
Heimatstube in der Wasserburg, Gerswalde 18621
Heimatstube Kappel, Lenzkirch 19959
Heimatstube Lebus, Lebus 19865
Heimatstube Lichtenstadt, Zirndorf 22809
Heimatstube Mederitzsch, Pflückuff 21026
Heimatstube Mühlhausen, Villingen-Schwenningen 22238
Heimatstube Münstedt, Lahstedt 19777
Heimatstube Neckarrems, Remseck am Neckar 21269
Heimatstube Nendingen, Tuttlingen 22136
Heimatstube Neuhaus-Schierschnitz, Neuhaus-Schierschnitz 20618
Heimatstube Oberg, Lahstedt 19778
Heimatstube Petznick, Templin 22028
Heimatstube Podersam-Jechnitz, Kronach 19722
Heimatstube Radegast, Radegast 21163
Heimatstube Rautheim, Braunschweig 17521
Heimatstube Röckwitz, Röckwitz 21350
Heimatstube Röhrnbach-Kaltenbach, Röhrnbach 21354
Heimatstube Rothenklempenow, Rothenklempenow 21416
Heimatstube Ruhland, Ruhland 21443
Heimatstube Schandelah, Cremlingen 17746
Heimatstube Schlesien, Kaufbeuren 19487
Heimatstube Seeheilbad Graal-Müritz, Graal-Müritz, Seeheilbad 18722
Heimatstube Sitzenroda, Schildau, Gneisenaustadt 21549
Heimatstube Sperenberg, Am Mellensee 16531
Heimatstube Stadt und Landkreis Saaz, Georgensgmünd 18587
Heimatstube Thal, Ruhla 21440
Heimatstube Tribsees, Tribsees 22093
Heimatstube Trochtelfingen, Bopfingen 17471
Heimatstube Untere Baranya, Gingen an der Fils 18646
Heimatstube Usedom, Usedom 22190
Heimatstube Vietmannsdorf, Templin 22029
Heimatstube Wahrenbrück, Uebigau-Wahrenbrück 22148
Heimatstube Warthe, Boitzenburger Land 17435
Heimatstube Wendhausen, Schellerten 21542
Heimatstube Wiesental, Waghäusel 22269
Heimatstube Wilthen, Wilthen 22592
Heimatstuben, Langelsheim 19810
Heimatstuben, Waldbronn 22283
Heimatstuben der Stadt, Titisee-Neustadt 22064
Heimatstuben im Arresturm, Spalt 21828
Heimatstuben im Oberen Tor, Scheinfeld 21540
Heimatstuben Weipert und Erzgebirgsschau, Gunzenhausen 18839
Heimatsube, Uchtspringe 22142
Heimatsube Klaushagen, Boitzenburger Land 17436
Heimatsube Staupitz, Pflückuff 21027
Heimattreff Gersweiler, Saarbrücken 21455
Heimatverein Räntzelstecher Güsten Osmarsleben e.V, Güsten 18826
Heimatvertriebenen-Stuben, Traun 02825
Heimethüs, Denzlingen 17823
Heimilisidnardarsafnid Halldórustofa, Blönduós 23988
Heimschneidermuseum, Großwallstadt 18808
Heimtali Koduloomuuseum, Heimtali 10569
Heimweberei-Museum, Schalkenmehren 21531
Heine-Haus, Hamburg 18935
Heineken Experience, Amsterdam 31845
Heino Lubja Kaalumuuseum, Mustvee 10596
Heinola Town Museum, Heinola 10765
Heinolan Kaupunginmuseo, Heinola 10765
Heinrich-Blickle-Museum, Sammlung Gußeiserner Ofenplatten, Rosenfeld 21372
Heinrich-Büssing-Haus, Wolfsburg 22662
Heinrich-Hansjakob-Gedenkstätte, Freiburg im Breisgau 18430
Heinrich Harrer-Museum, Hüttenberg 02103
Heinrich-Heine-Institut-Museum, Düsseldorf 17991
Heinrich-Schliemann-Gedenkstätte, Neubukow 20593
Heinrich-Schliemann-Museum, Ankershagen 16549
Heinrich-Schütz-Haus, Bad Köstritz 16765
Heinrich-Schütz-Haus Weißenfels, Weißenfels 22456
Heinrich-Sohnrey-Archiv und Gedächtnisstätte, Mautturm, Jühnder Schloß, Jühnde 19404
Heinrich-Vogeler Museum, Worpswede 22692
Heinrich Vogeler Stiftung Haus im Schluh, Worpswede 22693
Heinrich-von-Zügel-Gedächtnisgalerie, Wörth am Rhein 22641
Heinrich Zille Museum, Berlin 17136
Hein's Mühle, Bendorf 16988
Heinz Nixdorf MuseumsForum, Paderborn 20969
Heisey Museum, Lock Haven 50154
Heizungs- und Schulmuseum Wien, Wien 03002
De Heksenwaag, Oudewater 32658
Held Joan Museum, Carlton South 00955
Held-Poage Memorial Home & Research Library, Ukiah 53639
Helen Day Art Center, Stowe 53325
Helen E. Copeland Gallery, Bozeman 46981
Helen King Kendall Memorial Art Gallery, San Angelo 52702
Helen Louise Allen Textile Collection, Madison 50390
Helena Gallery, Helena 49240
Helena Modjeska Historic House, Modjeska Canyon 50776
Helena Thompson Museum, Workington 46075
Helensville and District Pioneer Museum, Helensville 33122
Helfštýn castle, Týn nad Bečvou 09883
Helgenæs Præstegård, Knebel 10121
L'Hélice Terrestre, Saint-Georges-des-Sept-Voies 15281
The Helicopter Museum, Weston-super-Mare 45980
Helikon Castle Museum, Keszthely 23660
Heliloojate Kappide Majamuuseum, Suure-Jaani 10617
Helinä Rautavaaran museo, Espoo 10730
Hellenic Antiquities Museum, Melbourne 01283
Hellenic Children's Museum-The Station, Athinai 22912
Hellenic Folklore Research Centre, Athinai 22913
Hellenic Maritime Museum, Piraeus 23147
Hellenic Museum and Cultural Center, Chicago 47479
Hellenic Theatre Museum and Study Center, Athinai 22914
Hellens Manor, Much Marcle 45184
Hellweg-Museum Unna, Unna 22175
Helmcken House, Victoria 07250
Helme Kihelkonnamuuseum, Helme 10570
Helmeringhausen Museum, Helmeringhausen 31741
Helmhaus Zürich, Zürich 42077
Helms-Museum, Hamburg 18936
Helmshore Mills Textile Museum, Helmshore 44441
Helmstedter Freundeskreis für Paramentik und christliche Kunst, Helmstedt 19121
Helmut Newton Stiftung im Museum für Fotografie, Berlin 17137
Helsingborgs Idrottsmuseum, Helsingborg 40587
Helsingborgs Skolmuseum, Helsingborg 40588
Helsinge Hembygdsmuseum, Vantaa 11445
Helsinge Hestevogns Museum, Helsinge 10060
Helsingin Automuseo, Helsinki 10778
Helsingin Diakonissalaitoksen Museo, Helsinki 10779
Helsingin Kaupungin Taidemuseo, Meilahti, Helsinki 10780
Helsingin Kaupungin Taidemuseo, Tennispalatsi, Helsinki 10781
Helsingin Kaupunginmuseo, Helsinki 10782
Helsingin Taidehalli, Helsinki 10783
Helsingin Väestönsuojelumuseo, Helsinki 10784
Helsingin Yliopistomuseo Arppeanum, Helsinki 10785
Helsingin Yliopiston Maatalousmuseo, Helsinki 10786
Helsingør Bymuseum, Helsingør 10065
Helsinki Car Museum, Helsinki 10778
Helsinki City Art Museum, Art Museum Meilahti, Helsinki 10780
Helsinki City Art Museum, Tennis Palace, Helsinki 10781
Helsinki City Museum, Helsinki 10782
Helsinki Deaconess Institute Museum, Helsinki 10779
Helsinki Parish Museum, Vantaa 11445
Helsinki University Museum Arppeanum, Helsinki 10785
Helston Folk Museum, Helston 44444
Heltborg Museum, Hurup Thy 10105
Helvetia Museum, Helvetia 49244
Helvetisches Goldmuseum, Burgdorf 41274
Helwan Corner Museum, Helwan 10534
Helytörténeti Emléktár, Dunaharaszti 23531
Helytörténeti Gyu#5jtemény, Isaszeg 23620
Helytörténeti Gyűjtemény, Tiszaújváros 23932
Helytörténeti Gyűjtemény és Tájház, Várpalota 23951
Helytörténeti Kiállítás és J. Haydn Emlékszoba, Fertőd 23558
Helytörténeti Múzeum, Csempeszkopács 23505
Hembygdsgården, Bjurholm 40460
Hembygdsgården i Kristvalla, Nybro 40779
Hembygdsgården Labbas, Storby 11329
Hembygdsgården Rots Skans, Älvdalen 40406
Hembygdsmuseet i Korsberga, Korsberga 40652
Hemet Museum, Hemet 49245
Hemingway Museum, Oak Park 51454
Hemne Bygdemuseum, Kyrksæterøra 33622
Hemnes Museum, Bjerka 33439
Hemsedal Bygdatun, Hemsedal 33543
Hemudu Ruins Museum, Yuyao 08478
Henan Museum, Zhengzhou 08498
Hendrickson House Museum and Old Swedes Church, Wilmington 54195
Henfield Museum, Henfield 44448
Hengchuen Peninsula Aboriginal Arts and Crafts Museum, Mudan 42278
Hengelo History Museum, Hengelo, Overijssel 32367
Hengfeng Museum, Hengfeng 07780
Hengistburgmuseum & Volkskundliche Sammlung, Hengsberg 02070
Hengyang Museum, Hengyang 07781
Henie Onstad Kunstsenter, Høvikodden 33569
Henkerhaus, Bernau bei Berlin 17289
Henneberg-Museum, Münnerstadt 20498
Hennebergisches Museum, Kloster Veßra 19579
Hennepin History Museum, Minneapolis 50732
Henri-Arnaud-Haus - Waldensermuseum, Ötisheim 20877
Henrik Ibsen Museum, Skien 33834
Henry Art Gallery, Seattle 53000
Henry B. Plant Museum, Tampa 53433
Henry Blogg Lifeboat Museum, Cromer 43930
Henry County Historical Museum, New Castle 51040
Henry County Museum and Cultural Arts Center, Clinton 47655
Henry-Dunant-Museum, Heiden 41475
Henry Ford Estate, Dearborn 48008
Henry Ford Museum, Dearborn 48009
Henry Guest House, New Brunswick 51032
Henry H. Blommel Historic Automotive Data Collection, Connersville 47780
Henry Kendall Cottage and Museum, Gosford 01123
Henry Lawson Centre, Gulgong 01133
Henry Moore Foundation, Much Hadham 45183
Henry Moore Institute, Leeds 44661
Henry S. Reuss Ice Age Visitor Center, Campbellsport 47209
The Henry Sheldon Museum of Vermont Histoy, Middlebury 50652
Henry-Stuart House, Québec 06681
Henry van de Velde Museum, Chemnitz 17696
Henry Whitfield State Historical Museum, Guilford 49102
Henryk Sienkiewicz Museum, Oblęgorek 34967
Henryk Tomaszewski Toy Museum, Karpacz 34716
Heptonstall Museum, Heptonstall 44452
Hepu Museum, Hepu 07783
The Hepworth Wakefield, Wakefield 45908
Her Majesty Customs and Excise National Museum, Liverpool 44735
Her Majesty Tower of London, London 44873
Hera Gallery, Wakefield 53792
Heraldic Museum, Dublin 24795
Heras Institute of Indian History and Culture, Mumbai 24317
Herat Provincial Military Museum, Herat 00002
Herat Provincial Museum, Herat 00003
Herbarium, University of Turku, Turku 11391
Herbert Art Gallery and Museum, Coventry 43902
Herbert CPR Train Station Museum, Herbert 05983
Herbert E. Clark Collection of Near Eastern Antiquities, Jerusalem 24989
Herbert F. Johnson Museum of Art, Ithaca 49548
Herbert-Gerisch-Stiftung, Neumünster 20635
Herbert Hoover National Historic Site, West Branch 54046
Herbert Hoover Presidential Library-Museum, West Branch 54047
Herberts Dorbe Memorial Museum Senču Putekļi (The Dust of Ancestors), Ventspils 30143
Herdenkingsbordenmuseum, Lochem 32504
Herdla Museum, Herdla 33544
Hereford Inlet Lighthouse, North Wildwood 51409
Hereford Museum and Art Gallery, Hereford 44454
Herefordshire Regiment Museum, Hereford 44455
Herendi Porcelánművészeti, Herend 23605
Herinneringscentrum Kamp Westerbork, Hooghalen 32406
Heriot-Watt University Museum, Edinburgh 44127
Heritage Area of Kuokkala, Lempäälä 11059
Heritage Art Center, Quezon City 34490
Heritage Center, Dayton 47991
The Heritage Center, Pine Ridge 51924
Heritage Center Museum, Berryville 46768
Heritage Center of Clark County, Springfield 53235
Heritage Center of Lancaster County, Lancaster 49923
Heritage Centre Canada - HVACR, Mississauga 06311
Heritage Centre of Saint Louis Convent, Monaghan 24877
Heritage Collection Catherina Brand, Ladybrand 38681
Heritage Conservancy, Doylestown 48178
Heritage Farm & Village, North Battleford 06468
Heritage Farmstead Museum, Plano 51976
Heritage Gallery, Budapest 23460
Heritage Hall Museum and Archives, Freeman 48780
Heritage Hazenmore Museum, Hazenmore 05980
Heritage Hill Museum, Dandenong 01023
Heritage Hill State Park, Green Bay 49038
Heritage Hjemkomst Interpretive Center, Moorhead 50836
Heritage House Museum, Smiths Falls 06986
Heritage House Museum of Montgomery County, Mount Ida 50880
Heritage House of Orange County Museum, Orange 51577
Heritage Models Centre, River Herbert 06756
Heritage Motor Centre, Gaydon 44266
The Heritage Museum, Astoria 46403
Heritage Museum, Baltimore 46567
Heritage Museum, Deer Lake 05689
Heritage Museum, Idaho Springs 49464
Heritage Museum, Kuah 30567
Heritage Museum, Saint-Albert 06790
Heritage Museum and Potton House, Big Spring 46799
The Heritage Museum at Falfurrias, Falfurrias 48497
Heritage Museum Foundation of Tate County, Senatobia 53033
Heritage Museum of Fine Arts for Youth, Detroit 48120
Heritage Museum of Northwest Florida, Valparaiso 53697
Heritage Museums, Sandwich 52839
Heritage North Museum, Thompson 07074
Heritage of the Americas Museum, El Cajon 48307
Heritage Park, Fort McMurray 05816
Heritage Park Folk Museum, Saint Phillip 03243
Heritage Park - Historical Village, Calgary 05520
Heritage Park Museum, Terrace 07066
Heritage Railway Association, Tywyn 55067
The Heritage Society Museum, Houston 49386
Heritage Square Museum, Los Angeles 50229
Heritage Square Museum, Ontario 51569
Heritage Sutton, Sutton, Québec 07041
Heritage Village, Mountain Lake 50894
Heritage Village, Sharonville 53052
Heritage Village Museum, Woodville 54292
Heritage Village - Pinellas County Historical Museum, Largo 49953
Heritage Walk Museum, Escondido 48419
Herkimer County Historical Society Museum, Herkimer 49259
Herkimer Home, Little Falls 50136

Herkomer-Museum am Mutterturm, Landsberg am Lech 19794
Herkules Villa, Budapest 23411
Herkules Villa of the Budapest History Museum, Budapest 23411
Herman Ottó Múzeum, Miskolc 23704
Herman Ottó Múzeum Ásványtára, Miskolc 23705
Herman Ottó Múzeum, Department of Mineralogy, Miskolc 23705
Herman Ottó Múzeum - Exhibition Hall, Miskolc 23707
Herman Ottó Múzeum - Herman Ottó Emlékház, Miskolc 23706
Herman Ottó Múzeum - Herman Ottó Memorial House, Miskolc 23706
Herman Ottó Múzeum - Kiállítóhelye, Miskolc 23707
Herman Teirlinckmuseum, Beersel 03356
Hermann-Allmers-Haus, Sandstedt 21484
Hermann Broch-Museum, Teesdorf 02800
Hermann-Grima House, New Orleans 51082
Hermann-Grochtmann-Museum Datteln, Datteln 17800
Hermann-Hesse-Höri-Museum, Gaienhofen 18531
Hermann-Hesse-Museum, Calw 17678
Hermann Hesse Museum, Hiroshima 28531
Hermann-Köhl-Museum, Pfaffenhofen an der Roth 21021
Hermann-Oberth-Raumfahrt-Museum e.V., Feucht 18310
Hermann Sudermann Museum-Silute Museum, Macikai 30304
Hermano Blodės Muziejus, Nida 30320
Hermano Zudermano Memorialinis Muziejus, Macikai 30304
Hermesvilla, Wien 03003
The Hermitage, Ho-Ho-Kus 49302
Hermitage And Gatehouse Museum, Ancaster 05338
Hermitage Foundation Museum, Norfolk 51364
The Hermitage - Home of President Andrew Jackson, Hermitage 49262
Hermitage of San Antonio de la Florida, Madrid 39518
Hermitage Rooms, London 44874
Herne Bay Museum and Gallery, Herne Bay 44461
Herning Kunstmuseum, Herning 10071
Herning Museum, Herning 10072
Heron Corn Mill and Museum of Papermaking, Beetham 43438
Herons Reef Historic Gold Diggings, Fryerstown 01097
Herøy Bygdesamling, Herøy 33545
Herøy Kystmuseum, Herøy 33546
Herregården Museum, Larvik 33628
Herreshoff Marine Museum/America's Cup Hall of Fame, Bristol 47034
Herrett Center for Arts and Science, Twin Falls 53627
Herring Era Museum of Iceland, Siglufjörður 24039
Herrliberger-Sammlung, Maur 41616
Herron Gallery, Indianapolis 49493
The Herschel Museum of Astronomy, Bath 43404
Herschell Carrousel Factory Museum, North Tonawanda 51406
Hershey Museum, Hershey 49263
Herter Art Gallery, Amherst 46267
Hertford Museum, Hertford 44462
Herttoniemen Museo, Helsinki 10787
Herttoniemi Museum, Helsinki 10787
Hervey Bay Historical Village and Museum, Scarness 01495
Hervey Bay Regional Art Gallery, Pialba 01413
Herzen Memorial House Museum, Moskva 37231
Herzliya Museum of Contemporary Art, Herzliya 24972
Herzog Anton Ulrich-Museum, Braunschweig 17522
Herzog August Bibliothek, Wolfenbüttel 22653
Herzogkasten, Abensberg 16419
Herzogliches Museum Gotha, Gotha 18718
Herz'sche Heimatstiftung Hilgerhof, Pittenhart 21047
Heshan Museum, Heshan 07784
Hessaby Museum, Tehrān 24670
Hessischer Museumsverband e.V., Kassel 54833
Hessisches Braunkohle Bergbaumuseum, Borken, Hessen 17478
Hessisches Forst-Kulturhistorisches Museum, Biebergemünd 17331
Hessisches Forstmuseum Alte Fasanerie, Hanau 18977
Hessisches Kutschen- und Wagenmuseum, Lohfelden 20022
Hessisches Landesmuseum Darmstadt, Darmstadt 17789
Hessisches Puppenmuseum, Hanau 18978
Hester Rupert Art Museum, Graaff-Reinet 38627
Het Cuypershuis, Roermond 32698
Hetjens-Museum, Düsseldorf 17992
Hetman Museum, Kyïv 43088
Hettinger County Historical Society Museum, Regent 52283
Heuneburgmuseum, Herbertingen 19132
Heureka - The Finnish Science Centre, Vantaa 11446
Heurich House Museum, Washington 53875
Heuser Art Center, Peoria 51772
Heuson-Museum im Rathaus, Büdingen 17628
Heussenstamm Stiftung, Frankfurt am Main 18376
Hever Castle and Gardens, Edenbridge 44112
Heves Museum Exhibition Hall, Heves 23606
Hevesi Múzeumi Kiállítóhely, Heves 23606
Hexenbürgermeisterhaus, Lemgo 19947
Hexenmuseum, Riegersburg, Steiermark 02587
Hexenmuseum, Ringelai 21336
Hexham Old Gaol, Hexham 44463
Heydar Aliyev Museum, Naxçivan 03183
Heym-Oliver House, Russell 52486
Heyuan Museum, Heyuan 07786
Heyward-Washington House, Charleston 47366
Heze Museum, Heze 07787
Heze Zone Museum, Heze 07788
Hezlett House, Coleraine 43880
HfG-Archiv, Ulm 22162
Hi-Desert Nature Museum, Yucca Valley 54363
Hibbing Historical Museum, Hibbing 49265
Hibel Museum of Art, Jupiter 49660
Hickories Museum, Elyria 48386
Hickory Grove Rural School Museum, Ogden 51499
Hickory Museum of Art, Hickory 49267
Hicks Art Center, Newtown 51345
Hicksville Gregory Museum, Hicksville 49270
Hida Folk Village, Takayama 29338
Hida Folklore and Archaeological Museum, Takayama 29337
Hida Minzoku Kokokan, Takayama 29337
Hida Minzoku-mura, Takayama 29338
Hida Takayama Bijutsukan, Takayama 29339
Hida Takayama Laquerware Museum, Takayama 29340
Hida Takayama Museum of Art, Takayama 29339
Hida Takayama Shunkei Kaikan, Takayama 29340
Hidalgo Pumphouse Heritage, Hidalgo 49271
Hiddenite Center, Hiddenite 49272
Hideyo Noguchi Memorial Hall, Inawashiro 28571
Hið Íslenzka Reðasafn, Húsavík 24003
Hiekan Taidemuseo, Tampere 11350
Hiekka Art Museum, Tampere 11350
Hierapolis Archeological Museum, Denizli 42759
Hierapolis Arkeoloji Müzesi, Denizli 42759
Higashi-Hiroshima City Museum of Art, Higashi-Hiroshima 28502
Higashi-Hiroshima-shiritsu Bijutsukan, Higashi-Hiroshima 28502
Higashi-Osaka Art Center, Higashi-Osaka 28505
Higashi-Osaka Municipal Local Museum, Higashi-Osaka 28506
Higashi-Osaka-shiritsu Kyodo Hakubutsukan, Higashi-Osaka 28506
Higashi-yamate District Historic Preservation Center, Nagasaki 28974
Higashimurayama Mingeikan, Higashimurayama 28507
Higashiyama Kaii Gallery, Nagano 28954
Higgins Armory Museum, Worcester 54298
Higgins Art Gallery, West Barnstable 54043
Higgins Museum, Okoboji 51529
High Cliff General Store Museum, Sherwood 53076
High Desert Museum, Bend 46733
High Museum of Art, Atlanta 46441
High Plains Heritage Center, Great Falls 49030
High Plains Museum, Goodland 48971
High Plains Museum, McCook 50343
High Point Museum, High Point 49275
High Prairie and District Museum, High Prairie 05985
Highland Cultural Arts Gallery, Highland 49278
Highland House Museum, Hillsboro 49287
Highland Maple Museum, Monterey 50809
Highland Museum of Childhood, Strathpeffer 45788
Highland Park Historical Museum, Highland Park 49281
Highland Pioneers Museum, Baddeck 05370
Highland Village Museum, Iona 06019
Highlands, Fort Washington 48701
Highlands Museum and Discovery Center, Ashland 46380
Highlands Museum of the Arts, Sebring 53021
Highwic, Auckland 33051
Hiihtomuseo, Lahti 11026
Hiiumaa Muuseum, Kärdla 10578
Hikarinotani Metaru Bijutsukan, Imba 28570
Hikone Castle Museum, Hikone 28509
Hildene, Manchester 50427
Hill Aerospace Museum, Hill Air Force Base 49282
Hill Country Museum, Kerrville 49749
Hill Cumorah Visitors Center, Palmyra 51684
Hill-Hold Museum, Montgomery 50817
Hill House, Helensburgh 44438
The Hill House, Portsmouth 52107
Hill of Tarvit Mansion House, Cupar 43942
Hill-Stead Museum, Farmington 48513
Hill Top, Ambleside 43298
Hill Tribes Museum, Bangkok 42542
Hillcrest Museum, Souris 06995
Hillel Jewish Student Center Gallery, Cincinnati 47561
Hiller Aviation Institute, San Carlos 52735
Hilleshuis Museum, Rotterdam 32706
Hillforest House Museum, Aurora 46495
Hillman strongpoint, Colleville-Montgomery 12571
Hillsboro Area Historical Society Museum, Hillsboro 49289
Hillsboro Museums, Hillsboro 49284
Hillsborough Railway Museum, Hillsborough 05987
Hillview Museum, Bagenalstown 24732
Hillwood Art Museum, Brookville 47091
Hillwood Museum, Washington 53876
Hilprecht-Sammlung Vorderasiatischer Altertümer, Jena 19381
Himachal State Museum, Shimla 24389
Himangan Kotiseutumuseo, Himanka 10840
Himanka Museum, Himanka 10840
Himeji Castle, Himeji 28510
Himeji City Museum of Art, Himeji 28512
Himeji City Museum of Literature, Himeji 28513
Himeji-jo, Himeji 28510
Himeji-kenritsu Budokan, Himeji 28511
Himeji-shiritsu Bijutsukan, Himeji 28512
Himeji-shiritsu Bungaku Hakubutsukan, Himeji 28513
Himezaki Lighthouse Museum, Ryotsu 29172
Himley Hall, Himley 44470
Himmelkeller, Kronberg im Weinviertel 02239
Himmelsberga Museum - Ölands Museum för Konst och Kulturhistoria, Borgholm 40472
Hinchingbrooke House, Huntingdon 44501
Hinchman House Natural Science Museum, Nantucket 50944
Hinckley and District Museum, Hinckley 44471
Hinckley Fire Museum, Hinckley 49296
Hinckley Foundation Museum, Ithaca 49549
Hincz Gyula Állandó Gyűjtemény, Vác 23944
Hindi Sangrahalaya, Allahabad 24061
Hindley Museum, Hindley 44472
Hinkler House Museum, Bundaberg 00921
Hino City Historical Museum, Hino 28516
Hino-shiritsu Shinsengumino Furusato Hakubutsukan, Hino 28516
Hinsdale Historical Society Museum, Hinsdale 49299
Hinterglasmuseum, Sandl 02631
Hinterlandmuseum Schloss Biedenkopf, Biedenkopf 17335
Hipologické Múzeum, Topol'čianky 38373
Hipp-Halle, Gmunden 01949
Hippo Regius Museum, Annaba 00050
Hippological Museum, Slatiňany 09840
Hippologické Muzeum, Slatiňany 09840
Hippology Museum, Topol'čianky 38373
Hirado Kanko Historical Museum, Hirado 28517
Hirado Kanko Shiryokan, Hirado 28517
Hiraide Archaeological Museum, Shiojiri 29275
Hiraide Gatokan, Shiojiri 29275
Hiraizumi Historical Artifacts Museum, Hiraizumi 28521
Hirakata Municipal Gotenyama Art Center, Hirakata 28524
Hirakawa-go Open Air Museum, Shirakawa 29281
Hiram Blauvelt Art Museum, Oradell 51576
Hiram Smith Octagon House, Neenah 50998
Hirano Image Library, Osaka 29128
Hirano Soundscape Museum, Yamato-koriyama 29585
Hirata Folk Art Museum, Takayama 29341
Hirata Kinenkan, Takayama 29341
Hiratsuka Bijutsukan, Hiratsuka 28527
Hiratsuka City Museum, Hiratsuka 28528
Hiratsuka Museum of Art, Hiratsuka 28527
Hiratsuka-shi Hakubutsukan, Hiratsuka 28528
Hirosaki City Museum, Hirosaki 28529
Hirosaki-shiritsu Hakubutsukan, Hirosaki 28529
Hirose-machi Rekishi oyo Minzoku Hakubutsukan, Hirose 28530
Hirose Museum of History and Folklore, Hirose 28530
Hiroshima Bijutsukan, Hiroshima 28532
Hiroshima Castle, Hiroshima 28533
Hiroshima Children's Museum, Hiroshima 28534
Hiroshima City Museum of Contemporary Art, Hiroshima 28538
Hiroshima City Museum of History and Traditional Ctrafts, Hiroshima 28535
Hiroshima Heiwa Kinen Shiryokan, Hiroshima 28536
Hiroshima-kenritsu Bijutsukan, Hiroshima 28537
Hiroshima Museum of Art, Hiroshima 28532
Hiroshima Peace Memorial Museum, Hiroshima 28536
Hiroshima Prefectural Art Museum, Hiroshima 28537
Hiroshima-shi Gendai Bijutsukan, Hiroshima 28538
Hirschberger Heimatstuben, Alfeld 16466
Den Hirschsprungske Samling, København 10133
Hirschwirtscheuer-Museum für die Künstlerfamilie Sommer, Künzelsau 19735
Hirsel Homestead Museum, Coldstream 43879
Hirshhorn Museum and Sculpture Garden, Washington 53877
Hirsmäen Museoalue, Valkeala 11439
Hiscock House, Trinity 07164
The Hispanic Society of America, New York 51192
Histo-Rail Limousin du Chemin de Fer, Saint-Léonard-de-Noblat 15368
Historama Kloster Hornbach, Hornbach 19274
Historama - Museum für Technik und Verkehr, Ferlach 01888
Historia Elstal, Wustermark 22754
Historial de la Grande Guerre, Péronne 14784
Historial du Gouffre, Padirac 14551
Historial/Musée de Cire, La Chaise-Dieu 13327
Historial of the Great War, Péronne 14784
Historiallinen Museo, Kokkola 10971
Historic Allaire Village, Allaire 46221
Historic and Ethnic Instrument Collection, Nedlands 01356
Historic Annapolis Foundation, Annapolis 46320
Historic Babcock Mill, Odessa 06500
Historic Bath, Bath 46629
Historic Bethabara Park, Winston-Salem 54245
Historic Bethany, Bethany 46772
Historic Bowens Mills and Pioneer Park, Middleville 50664
Historic Brattonsville, McConnells 50342
Historic Burlington County Prison Museum, Mount Holly 50878
Historic Camden Revolutionary War Site, Camden 47201
Historic Carson House, Marion 50486
Historic Charleston Foundation, Charleston 47367
Historic Charlton Park Museum, Hastings 49210
Historic Cherry Hill, Albany 46158
Historic Columbia Foundation, Columbia 47727
Historic Columbus Foundation, Columbus 55137
Historic Commercial Vehicle Museum, Hartmannsdorf 19020
Historic Costume & Textiles Collection, Columbus 47754
Historic Courthouse Museum, Lakeport 49903
Historic Crab Orchard Museum and Pioneer Park, Tazewell 53457
Historic Cragfont, Castalian Springs 47276
Historic Daniel Boone Home and Boonesfield Village, Defiance 48035
Historic Deepwood Estate, Salem 52667
Historic Deerfield, Deerfield 48032
Historic Dockyard Chatham, Chatham 43795
Historic Edenton, Edenton 48284
Historic-Ethnographical Collection, Oświęcim 34998
Historic Fallsington, Fallsington 48504
Historic Ferryland Museum, Ferryland 05797
Historic Fort Steilacoom, Tacoma 53408
Historic Fort Stockton, Fort Stockton 48689
Historic Fort York, Toronto 07114
Historic Gardner's Basin, Atlantic City 46461
Historic General Dodge House, Council Bluffs 47850
Historic Georgetown, Georgetown 48888
Historic Governors' Mansion, Cheyenne 47445
Historic Halifax, Halifax 49128
Historic Hat Creek Ranch, Cache Creek 05510
Historic Hermann Museum, Hermann 49261
Historic Hope Plantation, Windsor 54226
Historic Houses of Odessa, Delaware, Odessa 51495
Historic Hudson Valley, Tarrytown 53450
The Historic Indian Agency House, Portage 52052
Historic Jefferson College, Washington 53940
Historic Kenmore, Fredericksburg 48766
Historic Landmarks Foundation of Indiana, Indianapolis 49494
Historic Langhorne Museum, Langhorne 49936
Historic Latta Plantation, Huntersville 49420
Historic Liberty Jail Visitors Center, Liberty 50078
Historic Lincoln, Lincoln 50109
Historic London Town, Edgewater 48288
Historic Lyme Village, Bellevue 46704
Historic Madison House, Madison 50378
Historic Mann House, Concord 47772
Historic Markerville Creamery, Markerville 06238
Historic Mill Creek, Mackinaw City 50354
Historic Mining Center, Lewarde 13735
Historic Morven, Princeton 52149
Historic Museum of Arkansas, Little Rock 50139
Historic Museum of Obwald, Sarnen 41820
Historic New England, Boston 46932
Historic New Harmony, New Harmony 51048
The Historic New Orleans Collection, New Orleans 51083
Historic Newton Home, Decatur 48019
Historic Northampton, Northampton 51413
Historic Occoquan, Occoquan 51485
Historic O'Keefe Ranch, Vernon, British Columbia 07237
Historic Palmyra, Palmyra 51685
Historic Pasha House, Mudanya 42911
Historic Pensacola Village, Pensacola 51766
Historic Pithole City, Pleasantville 51991
Historic Pittsford, Pittsford 51969

H

Historic Preservation Association of Bourbon County, Fort Scott 48681
Historic Resources Centre, Annan 43305
Historic Richmond Town, Staten Island 53263
Historic Rock Ford, Lancaster 49924
Historic Rosedale, Charlotte 47389
Historic Rugby, Rugby 52479
Historic Saint Mary's City, Saint Mary's City 52609
Historic Sauder Village, Archbold 46340
Historic Schaefferstown, Schaefferstown 52956
Historic Scotland, Edinburgh 44128
Historic Shepherdstown Museum, Shepherdstown 53072
Historic Smithfield, Blacksburg 46844
Historic Spa and Goethe-Theatre, Bad Lauchstädt 16774
Historic Spanish Point, Osprey 51619
Historic Speedwell, Morristown 50852
Historic Stagville, Bahama 46529
Historic Stagville, Durham 48219
Historic Waco Foundation, Waco 53775
Historic Ward-Meade Park Museum, Topeka 53535
Historic White Pine Village, Ludington 50316
Historic Yale Museum, Yale 07408
Historic Zion Schoolhouse, Toronto 07115
Historical and Archaeological Museum, Montségur 14265
Historical and Architectural Museum Kiev Fortress, Kyïv 43079
Historical and Architectural Museum Red Hill, Kemerovo 37030
Historical and Art Museum, Uglič 37891
Historical and Cultural Museum and Nature Park of the Republic Karatchaevo-Tcherkessiya, Čerkessk 36854
Historical and Cultural Preserve Field of the Battle of Poltava, Poltava 43166
Historical and Cultural Preserve Kamyanka, Kam'janka 43054
Historical and Cultural Preserve The Batkyvshchyna of Taras Shevchenko, Ševčenkove 43182
Historical and Culture Museum-Preserve on the Poem of Igor's Struggle, Novgorod-Sivers'kyj 43153
Historical and Ethnographical Museum, Mary 42978
Historical and Ethnographical Museum, Türkmenabat 42982
Historical and Ethnological Museum, Sankt Gallen 41803
Historical and Ethnological Museum of Cappadocia, Kavala 23047
Historical and Ethnological Museum of Patras, Patras 23140
Historical and Ethnological Museum of the Mani, Kranae 23068
Historical and Etnographic Museum of Želovce, Želovce 38389
Historical and Folk Art Museum of Gavalohori, Chania 22963
Historical and Folklore Museum of Rethymnon, Rethymnon 23167
Historical and Genealogical Society of Indiana County, Indiana 49482
Historical and Memorial Museum Complex Bobriki, Donskij 36878
Historical and Military Museum of Kamtchatka Region, Petropavlovsk-Kamčatskij 37554
Historical and Natural History Museum of Natick, Natick 50989
Historical and Regional Museum, Merke 29710
Historical and Regional Museum, Ridder 29724
Historical and Regional Museum, Rudnyj 29725
Historical and Regional Museum, Uro-Teppa 42506
Historical and Regional Museum Donskoe, Donskoe 36879
Historical and Regional Museum of Bujnaksk, Bujnaksk 36831
Historical and Regional Museum of Gelendzhik, Gelendžik 36923
Historical, Architectural and Archeological Museum-Reserve, Staraja Ladoga 37795
Historical, Architectural and Artistic Museum-Reserve, Velikij Ustjug 37923
Historical, Architectural and Artistic Museum Reserve of Ples, Ples 37568
Historical, Architectural and Nature Museum-Reserve Monrepo, Vyborg 37981
Historical, Architecture and Art Museum of Zvenigorod, Zvenigorod 37991
Historical Archive and University Museum of Ljubljana, Ljubljana 38471
Historical Capuchin Museum, Camerino 25622
Historical Center for Southeast New Mexico, Roswell 52469
Historical Collection Electronical Engineering, Delft 32057
Historical Collection of the Dunaújváros Paper Factory, Dunaújváros 23532
Historical demonstrational weaving mill Braunsdorf, Braunsdorf 17532
Historical Electronics Museum, Linthicum 50123
Historical Estate Boat of Peter I, the Great, Veskovo 37926
Historical-Ethnographical Museum, Chojnice 34596
Historical-Ethnographical Museum, Chojnice 34599
Historical-Ethnographical Museum - Gallery of Polish Art, Chojnice 34598
Historical-ethnographical Museum House of Galsky, Čerepovec 36849
Historical Exhibition and J. Haydn Memorial Hall, Fertőd 23558
Historical, Folk Art and Natural History Museum of Kozani, Kozani 23067
Historical Goverment House, Takayama 29344
Historical Heritage Complex in Old Harry - CAMI, Old Harry 06506
Historical Memorial Museum of the Family Fadeevy-Hahn-Blavatsky, Dnipropetrovs'k 43034
Historical Memorial State Preserve Homeplace of V.I. Lenin, Uljanovsk 37897
Historical Museum, Białystok 34549
Historical Museum, Białystok 34551
Historical Museum, Blagoevgrad 04991
Historical Museum, Bracigovo 04994
Historical Museum, Città del Vaticano 54513
Historical Museum, Dnipropetrovs'k 43033
Historical Museum, Drjanovo 05024
Historical Museum, Gabrovo 05032
Historical Museum, Jerusalem 24990
Historical Museum, Kruševo 30505
Historical Museum, Lund 40702
Historical Museum, Mahébourg 30677
Historical Museum, Razgrad 05148
Historical Museum, Tutrakan 05234
Historical Museum, Vlorë 00035
Historical Museum and Pioneer Village, Sundre 07035
Historical Museum at Fort Missoula, Missoula 50753
The Historical Museum at Saint Gertrude, Cottonwood 47843
Historical Museum Batak, Batak 04981
Historical Museum Chokokan, Saga 29178
Historical Museum House at the Embankment, Moskva 37218
Historical Museum in Wrocław, Wrocław 35331
Historical Museum Kamyanka, Kam'janka 43055
Historical Museum Memorial of Victory, Krasnojarsk 37071
Historical Museum of Anthiros, Karditsa 23035
Historical Museum of Arta, Arta 22893
Historical Museum of Azogire, Azogire 22957
Historical Museum of Bay County, Bay City 46650
Historical Museum of Bosnia and Hercegovina, Sarajevo 04276
Historical Museum of Cecil County, Elkton 48355
Historical Museum of Crete, Iráklion 23019
Historical Museum of Frederick County, Frederick 48756
Historical Museum of Hokkaido, Sapporo 29216
Historical Museum of Old Randolph, Randolph 52244
Historical Museum of Palm Beach County, West Palm Beach 54072
Historical Museum of Portitsa, Karditsa 23036
Historical Museum of Private Schools at the Foreign Settlement, Nagasaki 28975
Historical Museum of Saint James-Assiniboia, Winnipeg 07363
Historical Museum of Sansia Town, Sansia 42311
Historical Museum of Sciences Carlos J. Finlay, La Habana 09146
Historical Museum of Serbia, Beograd 38050
Historical Museum of Sofia, Sofia 05204
Historical Museum of Southern Florida, Miami 50625
Historical Museum of Sughd Province, Khojand 42501
Historical Museum of Talbot County, Easton 48268
Historical Museum of Tarnobrzeg, Tarnobrzeg 35175
Historical Museum of the Agricultural Department of the West-Hungarian University, Mosonmagyaróvár 23721
Historical Museum of the Babes-Bolyai University, Cluj-Napoca 36291
Historical Museum of the City of Cracow, Kraków 34798
Historical Museum of the city of Cracow - Pomorska Branch, Kraków 34799
Historical Museum of the Moscow Metro, Moskva 37292
Historical Museum of the National Museum in Prague, Praha 09751
Historical Museum of the National Resistance of Rentina, Agrapha, Karditsa 23037
Historical Museum of the Old City of Warsaw, Warszawa 35247
Historical Museum of the Principality of Stavelot-Malmedy, Stavelot 04050
Historical Museum of Tracadie, Tracadie-Sheila 07158
Historical Museum Thurgau, Frauenfeld 41377
Historical Museum, University Museum of Cultural History, Oslo 33726
Historical Site Abiverd, Kaka 42976
Historical Site Geok-Depe Fortress, Gökdepe 42975
Historical Site Sarakhs, Saragt 42981
Historical Society Museum, Hudson 49412
Historical Society Museum of Carroll County, Westminster 54098
Historical Society Museum of Ocean Grove, New Jersey, Ocean Grove 51488
Historical Society Museum of Saint Kilda, Saint Kilda, Victoria 01478
Historical Society Museum of the Town of Greenwich, Greenwich 49082
Historical Society of Berks County Museum, Reading 52263
Historical Society of Bloomfield Museum, Bloomfield 46851
Historical Society of Brookfield, Brookfield 47056
Historical Society of Kent County Museum, Chestertown 47439
Historical Society of Middletown and the Wallkill Precinct, Middletown 50661
Historical Society of Oak Park and River Forest, Oak Park 51455
Historical Society of Old Yarmouth Museum, Yarmouth Port 54325
Historical Society of Pennsylvania, Philadelphia 51841
Historical Society of Quaker Hill and Pawling, Pawling 51735
Historical Society of Wells and Ogunquit Museum, Wells 54029
Historical State Preserve The Hetman Capital, Baturin 43000
Historical Woolscour Blackall, Blackall 00872
Historické a Národopisné Múzeum Želovce, Želovce 38389
Historické Múzeum, Bratislava 38213
Historické Muzeum, Praha 09751
Historicum, Sargans 41817
Historicum - Sammlung für Bergbaukunde, Modellsammlung, Winkler-Gedenkstätte und Karzer, Freiberg, Sachsen 18422
Historiekamer Hardenberg, Hardenberg 32311
Historiengewölbe, Rothenburg ob der Tauber 21406
Historijski Muzej Bosne i Hercegovine, Sarajevo 04276
Historisch-Archäologisches Museum, Markgrafneusiedl 02360
Historisch Documentatiecentrum voor het Nederlands Protestantisme 1800-Heden, Amsterdam 31846
Historisch Museum, Capelle aan den IJssel 32036
Historisch Museum Arnhem, Arnhem 31923
Historisch Museum De Bevelanden, Goes 32258
Historisch Museum de Scheper, Eibergen 32183
Historisch Museum De Tien Malen, Putten 32673
Historisch Museum Den Briel, Brielle 32011
Historisch Museum Deventer, Deventer 32120
Historisch Museum Ede, Ede 32168
Historisch Museum Grenadiers en Jagers, Arnhem 31924
Historisch Museum Haarlemmermeer, Hoofddorp 32404
Historisch Museum Hedel, Hedel 32327
Historisch Museum Hengelo, Hengelo, Overijssel 32367
Historisch Museum Het Kleine Veenlo, Veenendaal 32873
Historisch Museum Het Palthe Huis, Oldenzaal 32615
Historisch Museum Piet Dorenbosch, Boxtel 31996
Historisch Museum Rotterdam Het Schielandshuis, Rotterdam 32707
Historisch Museum Ter Aar, Ter Aar 31786
Historisch Museum Tweestromenland, Beneden Leeuwen 31961
Historisch Museum Warsenhoeck, Nieuwegein 32577
Historisch Museum Wolfheze, Wolfheze 32960
Historisch Museum Zuid-Kennemerland, Haarlem 32297
Historisch Muzeum Hazerswoude, Hazerswoude Dorp 32326
Historisch Openluchtmuseum Eindhoven, Eindhoven 32187
Historisch Pijp en Tabaksmuseum, Sint-Niklaas 04019
Historisch-Technisches Informationszentrum, Peenemünde 20996
Historische Arztpraxis, Burgstädt 17663
Historische Ausstellung Krupp, Essen 18254
Historische Bibliothek, Quedlinburg 21148
Historische Collectie UMCG, Groningen 32282
Historische Dauerausstellung Schloss Borbeck, Essen 18255
Historische Einungsmeisterstube, Dogern 17886
Historische Fahrzeugsammlung, Simmelsdorf 21768
Historische Fraunhofer-Glashütte, Benediktbeuern 16991
Historische Genie Verzameling, Vught 32914
Historische Gerberei, Burg bei Magdeburg 17643
Historische Gesellenherberge, Blankenburg 17372
Historische Grünes Gewölbe und Neues Grünes Gewölbe, Dresden 17948
Historische Hammerschmiede, Blaubeuren 17378
Historische Handwerkerstuben, Gingst 18647
Historische Höhler, Gera 18593
Historische Käsküche, Wiggensbach 22566
Historische Kuranlagen und Goethe-Theater Bad Lauchstädt, Bad Lauchstädt 16774
Historische Lehrsammlung der Feuerwehr Köln, Köln 19607
Historische Mühle, Potsdam 21090
Historische Ochsentretanlage im Brunnenhausmuseum, Schillingsfürst 21552
Historische Ölmühle, Simonswald 21773
Historische Ortssammlung, Reigoldswil 41752
Historische Räume des Stadtschlosses, Fulda 18508
Historische Sammlung der von Bodelschwingh'schen Anstalten, Bielefeld 17339
Historische Sammlung Eltville, Eltville am Rhein 18132
Historische Sammlungen im Gesundheitspark, Bad Gottleuba 16740
Historische Schauweberei, Braunsdorf 17532
Historische Schlosserei Pelzer, Jena 19382
Historische Senfmühle, Monschau 20375
Historische Spinnerei Gartetal, Gleichen 18659
Historische Turmstuben, Winnenden 22601
Historische Verzameling Aan- en Afvoertroepen, Stroe 32810
Historische Verzameling Intendance, Bussum 32031
Historische Verzameling Korps Nationale Reserve, Harskamp 32318
Historische Verzameling Luchtdoelartillerie, Ede 32169
Historische Verzameling Militair Geneeskundige Dienst, Loosdrecht 32505
Historische Verzameling Regiment van Heutsz, Arnhem 31925
Historische Wassermühle Dickendorfer Mühle, Dickendorf 17845
Historische Wehranlage, Mühlhausen/Thüringen 20402
Historische Werkstatt Nordenstadt, Wiesbaden 22555
Historischen Fabrikanlage Maste-Barendorf mit Nadelmuseum und Gelbgießerei, Iserlohn 19357
Historischer Eisenhammer, Roth 21403
Historischer Hafen Berlin, Berlin 17138
Historischer Kunstbunker, Nürnberg 20769
Historischer Saal der Fischerzunft, Würzburg 22711
Historischer Schieferbergbau Lehesten, Lehesten 19873
Historisches Archiv der Stadt Köln, Köln 19608
Historisches Bergamt Bad Wildungen-Bergfreiheit, Bad Wildungen 16860
Historisches Binz Museum, Binz, Ostseebad 17356
Historisches Centrum Hagen, Hagen 18856
Historisches Erzbergwerk im Silberbergwerk Bodenmais, Bodenmais 17413
Historisches Feuerwehrmuseum Lüchow-Dannenberg, Dannenberg 17779
Historisches Glasapparate-Museum, Cursdorf 17754
Historisches Heimatmuseum Östringen, Östringen 20876
Historisches Kabinett Burgruine Eckartsburg, Eckartsberga 18052
Historisches Käsereimuseum, Altusried 16527
Historisches Museum, Bischofszell 41239
Historisches Museum am Hohen Ufer, Hannover 18993
Historisches Museum am Strom - Hildegard von Bingen, Bingen am Rhein 17353
Historisches Museum Aurich, Aurich 16647
Historisches Museum Bamberg, Bamberg 16910
Historisches Museum Basel, Basel 41161
Historisches Museum Bayreuth, Bayreuth 16952
Historisches Museum Bern, Bern 41207
Historisches Museum Bielefeld, Bielefeld 17340
Historisches Museum Blumenstein, Solothurn 41882
Historisches Museum Bremerhaven, Bremerhaven 17570
Historisches Museum der Pfalz Speyer, Speyer 21834
Historisches Museum der Stadt Baden, Baden 41145
Historisches Museum Frankfurt am Main, Frankfurt am Main 18377
Historisches Museum für Mittelanhalt und Bach-Gedenkstätte im Schloss Köthen, Köthen 19666
Historisches Museum Hanau - Schloß Philipsruhe, Hanau 18979
Historisches Museum Heiden, Heiden 41476
Historisches Museum im Marstall, Paderborn 20970
Historisches Museum im Spital, Dinkelsbühl 17866
Historisches Museum Luzern, Luzern 41592
Historisches Museum Obwalden, Sarnen 41820

Historisches Museum Olten, Olten 41706
Historisches Museum Regensburg, Regensburg 21224
Historisches Museum Saar, Saarbrücken 21456
Historisches Museum Schloss Arbon, Arbon 41121
Historisches Museum Schloß Gifhorn, Gifhorn 18642
Historisches Museum Schloß Urach, Bad Urach 16853
Historisches Museum Thurgau, Frauenfeld 41377
Historisches Museum Uri, Altdorf 41104
Historisches Museum Verden, Verden 22220
Historisches Museum Wiedlisbach, Wiedlisbach 42013
Historisches Puppen-, Spielzeug- und Ikonenmuseum, Bernkastel-Kues 17297
Historisches Schmucksteinbergwerk Silberschacht, Bach an der Donau 16653
Historisches Schuhmuseum, Landsberg am Lech 19795
Historisches Silberbergwerk "Hella-Glücksstollen", Neubulach 20594
Historisches Silbererzbergwerk Grube Samson und Heimatmuseum Sankt Andreasberg, Sankt Andreasberg 21487
Historisches Straßenbahn-Depot Sankt-Peter, Nürnberg 20770
Historisches Straßenbahndepot, Halle, Saale 18897
Historisches und Völkerkundemuseum, Sankt Gallen 41803
Historisches Zentrum, Wuppertal 22741
Historisches Zentrum Engelshaus, Wuppertal 22742
Historisk Museum, Oslo 33726
Historisk Museum, Slangerup 10277
Historiska Museet, Lund 40702
Historiska Museet, Stockholm 40899
Historium de Sedan, Sedan 15696
History and Archaeology Museum, Marsilly 14015
History and Archaeology Museum, Tulcea 36705
History and Architecture Museum Fortress of Kuznetsk, Novokuzneck 37479
History and Art Museum, Gulbene 30010
History and Art Museum, Krāslava 30033
History and Art Museum Jurino, Jurino 36982
History and Folklore Museum, Shimosuwa 29271
History and Legendary Museum, Kuah 30568
History and Natural History Museum, Alger 00038
History and Regional Museum, Vsevoložsk 37980
History and Regional Museum of Kronstadt, Kronštadt 37084
History and Regional Museum of the Komi Republic, Syktyvkar 37815
History and Tambow Museum, Amnéville-les-Thermes 11590
History and Traditions Museum, Lackland Air Force Base 49859
History Center of the United Methodist Church, Madison 50384
History & Local Studies Museum, Semey 29727
History Museum, Ahmednagar 24053
History Museum, Bansko 04978
History Museum, Bratislava 38213
History Museum, Burgas 04999
History Museum, Čiprovci 05007
History Museum, Dimitrovgrad 05015
History Museum, Goce Delčev 05035
History Museum, Gorna Orjachovica 05036
History Museum, Grahamstown 38630
History Museum, Ichtiman 05039
History Museum, Isperich 05040
History Museum, Jēkabpils 30014
History Museum, Los Gatos 50273
History Museum, Martakert 03179
History Museum, Omurtag 05099
History Museum, Petrič 05113
History Museum, Pomorie 05133
History Museum, Popovo 05134
History Museum, Pravec 05137
History Museum, Provadia 05140
History Museum, Razlog 05149
History Museum, Svištov 05222
History Museum, Târgu Jiu 36665
History Museum, Velingrad 05260
History Museum for Springfield-Greene County, Springfield 53232
History Museum & Historical Society of Western Virginia, Roanoke 52381
History Museum Landing of the Russian Army at Svishtov, Svištov 05223
History Museum of Armenia, Yerevan 00738
History Museum of Aspe, Aspe 38907
History Museum of Joshkar-Ola, Joškar-Ola 36978
History Museum of Latvia, Rīga 30093
History Museum of Moldavia, Iaşi 36418
History Museum of Neamţ County, Piatra-Neamţ 36521
History Museum of the Academy of Mining and Metallurgy, Kraków 34795
History Museum of the People of Uzbekistan, Toshkent 54471
History Museum of the Romanian Jews, Bucureşti 36203
History Museum of the Šiauliai University, Šiauliai 30364
History Museum of the Taiping Heavenly Kingdom in Nanjing, Nanjing 08039
History Museum of Tsuyama Region, Tsuyama 29548
History Museum-Palace in Dukla, Dukla 34626
History Museum Pavlikeni, Pavlikeni 05104
History of Grudziądz, Grudziądz 34683
History of Medicine Collections, Durham 48220
History of Transportation, Moose Jaw 06385
The History Place, Morehead City 50842
History San José, San Jose 55138
History Shop, Wigan 46013
Historyczny Oddział Muzeum w Grudziądzu, Grudziądz 34683
Hitachi Kyodo Hakubutsukan, Hitachi 28543
Hitachi Local Museum, Hitachi 28543
Hitchin Museum and Art Gallery, Hitchin 44474
Hitomi Bijutsukan, Eigenji 28415
Hitomi Museum, Eigenji 28415
Hiwa Kagaku Hakubutsukan, Hiwa 28544
Hiwa Museum for Natural History, Hiwa 28544
Hiwan Homestead Museum, Evergreen 48460
Hiwwelhaus, Marpingen 20228
Hixon House, La Crosse 49835
Hiyas ng Bulakan Museum, Malolos 34422
Hjedding Mejerimuseum, Ølgod 10221
Hjemmevaernsmuseet, Padborg 10228
Hjemsted Oldtidspark, Skærbæk 10258
Hjerl Hedes Frilandsmuseum, Vinderup 10318
HMAS Castlemaine - Museum Ship, Williamstown 01651
HMCS Haida, Hamilton 05959
HMCS Sackville, Halifax 05937
HMS Belfast, London 44875
HMS Ganges Museum, Ipswich 44529
HMS Trincomalee, Hartlepool 44407
HMS Victory, Portsmouth 45432
HMS Warrior 1860, Portsmouth 45433
Ho-Am Art Museum, Yongin 29943
Ho Chi Minh City Fine Arts Museum, Ho Chi Minh City 54623
Ho Chi Minh Memorial House, Thua Thien Hue 54637
Ho Chi Minh Museum, Ha Noi 54615
Ho Chi Minh Museum, Ho Chi Minh City 54624
Ho Chi Minh Museum, Thua Thien Hue 54638
Hoard Historical Museum, Fort Atkinson 48590
Hobart Heritage Museum, Hobart 01153
Hobart Historical Society Museum, Hobart 49303
Hobbies Museum of Fretwork, Dereham 43972
Hobby- och Leksaksmuseum, Stockholm 40900
Hobel- und Werkzeugmuseum, Laxenburg 02273
Hobelmuseum Langenfeld, Langenfeld 19832
Hobro Museum, Hobro 10082
Hochburg-Museum, Emmendingen 18144
Hochheimer Kunstsammlung, Hochheim am Main 19204
Hochofen-Museum Bundschuh, Thomatal 02812
Hochofenmuseum Radwerk IV, Vordernberg 02878
Hochrheinmuseum, Bad Säckingen 16814
Hochschul-und Landesbibliothek, Fulda 18509
Hochschwabmuseum, Sankt Ilgen 02651
Hochstift Meissen, Meißen 20264
Hochwaldmuseum, Hermeskeil 19151
Hockaday Museum of Arts, Kalispell 49671
Hockey Hall of Fame, Toronto 07116
Hocking Valley Museum of Theatrical History, Nelsonville 51002
Hodgen House Museum Complex, Ellsworth 48370
Hodgeville Community Museum, Hodgeville 05991
Höbarth- und Madermuseum der Stadt Horn, Horn 02099
Höganäs Museum och Konsthall, Höganäs 40594
Högarps Bymuseum, Vetlanda 41055
Höhlenkundemuseum Dechenhöhle, Iserlohn 19358
Höhlenmuseum, Frasdorf 18409
Höhlenmuseum, Obertraun 02462
Höhlenmuseum, Sundlauenen 41922
Höhlenmuseum Eisensteinhöhle, Bad Fischau-Brunn 01745
Höhlenmuseum in der Lurgrotte, Peggau 02482
Höhlenmuseum Kubacher Kristallhöhle und Freilicht-Steinemuseum, Weilburg 22385
Hölderlinturm, Tübingen 22115
Hörby Museum, Hörby 40598
Hörselbergmuseum, Wutha-Farnroda 22758
Hof van Hessen, Huissen 32420
Hofburg Innsbruck, Innsbruck 02119
Hoffman Gallery, Portland 52078
Hoffmann Collection, Berlin 17242
Hoffmann-von-Fallersleben-Museum, Wolfsburg 22663
Hofgarten an der Residenz, München 20446
Hofjagd- und Rüstkammer, Wien 03004
Hofje Van Gratie, Delft 32058
Hofmarkmuseum, Eggenfelden 18061
Hofmarkmuseum Schloß Eggersberg, Riedenburg 21315
Hofmobiliendepot, Wien 03005
Hofstra Museum, Hempstead 49246
Hofu Tenmangu History Museum, Hofu 28545
Hofwyl-Broadfield Plantation, Brunswick 47100
Hogarth's House, London 44876
Hohberger Bienenmuseum, Hohberg 19230
Hohenhof, die Jugendstil-Villa des Osthaus Museums Hagen, Hagen 18857
Hohenlohe-Museum, Neuenstein, Württemberg 20613
Hohenloher Freilandmuseum, Schwäbisch Hall 21655
Hohenloher Kunstverein e.V., Langenburg 19825
Hohenloher Urweltmuseum, Waldenburg, Hohenlohekreis 22289
Hohenzollerisches Landesmuseum, Hechingen 19062
Hohhaus-Museum, Lauterbach, Hessen 19862
Hohhot City Museum, Hohhot 07789
Hoiles-Davis Museum, Greenville 49072
Højer Mølle- og Marskmuseum, Højer 10086
Hojing Cultural Museum, Kaohsiung 42235
Hokkaido Daigaku Nogakubu Hakubutsukan, Sapporo 29215
Hokkaido Home-Making School Museum, Tsugaru 29539
Hokkaido Kaitaku Kinenkan, Sapporo 29216
Hokkaido-kenritsu Kita-hitobito Hakubutsukan, Abashiri 28345
Hokkaido Kotaro Migishi Museum of Art, Sapporo 29219
Hokkaido Museum of Literature, Sapporo 29217
Hokkaido Museum of Modern Art, Sapporo 29218
Hokkaido Museum of Northern Peoples, Abashiri 28345
Hokkaido Obihiro Museum of Art, Obihiro 29085
Hokkaido-ritsu Bungakukan, Sapporo 29217
Hokkaidoritsu Kindai Bijutsukan, Sapporo 29218
Hokkaidoritsu Migishi Kotaro Bijutsukan, Sapporo 29219
Hokkaidoritsu Obihiro Bijutsukan, Obihiro 29085
Hokuetsu Art Museum, Nakajo 29016
Hokuetsu Bijutsukan, Nakajo 29016
Hol Bygdemuseum, Hol i Hallingdal 33552
Holarium - 3D-Museum und Digital Art Gallery, Esens 18241
Holburne Museum of Art, Bath 43405
Holden Gallery, Manchester 45085
Holden Historical Society Museum, Holden 05992
Holdhus Skulemuseum, Eikelandsosen 33479
Holländer-Windmühle, Boiensdorf 17434
Holland Area Arts Council, Holland 49316
Holland Experience, Amsterdam 31847
Holland Historical Society Museum, Holland 49319
Holland Land Office Museum, Batavia 46626
Holland Museum, Holland 49317
Holland Tunnel Art, Brooklyn 47076
Het Hollands Kaasmuseum, Alkmaar 31793
De Hollandsche Schouwburg, Amsterdam 31848
Hollenberg Pony Express Station, Hanover 49160
Holló László Emlékmúzeum, Debrecen 23517
Holló László Galéria, Putnok 23787
Hollycombe Steam Collection, Liphook 44724
The Hollycroft Foundation, Ivoryton 49552
Hollytrees Museum, Colchester 43874
Hollywood Bowl Museum, Los Angeles 50230
Hollywood Entertainment Museum, Los Angeles 50231
Hollywood Guinness World of Records Museum, Los Angeles 50232
Hollywood Heritage Museum, Los Angeles 50233
Hollywood Wax Museum, Branson 46996
Hollywood Wax Museum, Los Angeles 50234
Holman Museum, Holman 05993
Holmes à Court Gallery, East Perth 01055
Holmes County Historical Society Museum, Millersburg 50692
Holmestova, Frekhaug 33506
Holmestrand Museum, Holmestrand 33553
Holmgrens Volkswagenmuseum, Pålsboda 40806
Holmöns Båtmuseum, Holmön 40600
Holmsbu Billedgalleri, Holmsbu 33555
Holocaust Documentation and Education Center, Miami 50626
Holocaust Memorial Center, Budapest 23412
Holocaust Memorial Center of Central Florida, Maitland 50401
Holocaust Museum, Chełmno nad Nerem 34592
Holocaust Museum Houston, Houston 49387
Holocaust Museum of Treblinka, Kosów Lacki 34760
Holokausto Ekspozicija, Vilnius 30410
Holokauszt Emlékközpont, Budapest 23412
Hølonda Skimuseum of Bygdasamling, Gåsbakken 33510
Holowood - Holographiemuseum Bamberg, Bamberg 16911
Holst Birthplace Museum, Cheltenham 43811
Holstebro Kunstmuseum, Holstebro 10090
Holstebro Museum, Holstebro 10091
Holsworthy Museum, Holsworthy 44478
Holt Skolemuseum, Tvedestrand 33937
Holter Museum of Art, Helena 49241
The Holtzman Art Gallery, Towson 53551
Holy Defense Museum, Kerman 24617
Holyhead Maritime Museum, Holyhead 44481
Holyland Arts Museum, Bethlehem 34050
Holyland Exhibition, Los Angeles 50235
Holzgerätemuseum, Schlüchtern 21577
Holzkanumuseum, Stuttgart 21944
Holzknechtmuseum im Salzkammergut, Bad Goisern 01749
Holzknechtmuseum Ruhpolding, Ruhpolding 21447
Holzschnitt-Museum Klaus Herzer, Mössingen 20366
Holztechnisches Museum, Rosenheim 21375
Holzwurm-Museum, Quedlinburg 21149
Home-Museum of Azim Azimzadeh, Bakı 03155
Home-Museum of Venclova Family at the Vilnius Ethnic Activity Centre, Vilnius 30437
Home of Chales Darwin - Down House, Downe 44018
Home of Franklin D. Roosevelt, Hyde Park 49459
Home of History, Rotterdam 32708
Home of Signatories, Vilnius 30432
Home of Stone - a Ford County Museum, Dodge City 48148
Home Sweet Home Museum, East Hampton 48246
Homeland Museum, Hanušovce nad Topl'ou 38257
Homeland Museum, Nové Zámky 38322
Homeland Museum, Trebišov 38374
Homeopathy Works, Berkeley Springs 46761
Homer Watson House and Gallery, Kitchener 06101
Homestead Antique Museum, Drumheller 05712
Homestead Museum, Biggar 05428
Homestead National Monument of America, Beatrice 46664
Homestead Park Pioneer Museum, Port Augusta 01428
Homeville Antique Fire Department, Homer 49331
Homeville Museum, Homer 49332
Homewood House Museum, Baltimore 46568
Homewood Museum, Prescott 06658
Homolovi Ruins, Winslow 54240
Homs Museum, Homs 42136
Honarhaye Moaser Tehran, Tehrān 24671
Honarhaye Tazini - Rakibkhaneh, Isfahan 24612
Honda Collection Hall, Motegi 28941
Honda Museum, Kanazawa 28655
Hondenmuseum, Eindhoven 32188
Hondo City Amakusa Christian Hall, Hondo 28547
Hondo City Folk Museum, Hondo 28548
Hondo-shi Amakusa Kirishitankan, Hondo 28547
Hondo-shiritsu Mingeikan, Hondo 28548
Honey Cake and Candle Making Museum, Szekszárd 23868
Honeybee Science Center, Tokyo 29398
Honeywood Heritage Centre, London 44877
Hong-Gah Museum, Taipei 42392
Hong-ik University Museum, Seoul 29882
Hong Kong Correctional Services Museum, Hong Kong 07795
Hong Kong Heritage Museum, Hong Kong 07796
Hong Kong Maritime Museum, Hong Kong 07797
Hong Kong Museum of Art, Hong Kong 07798
Hong Kong Museum of Coastal Defence, Hong Kong 07799
Hong Kong Museum of History, Hong Kong 07800
Hong Kong Museum of Medical Sciences, Hong Kong 07801
Hong Kong Racing Museum, Hong Kong 07802
Hong Kong Railway Museum, Hong Kong 07803
Hong Kong Science Museum, Hong Kong 07804
Hong Kong Space Museum, Hong Kong 07805
Hong Kong Visual Arts Centre, Hong Kong 07806
Hongdong Museum, Hongdong 07819
Honghu Revolution History Museum, Honghu 07820
Hongsan Museum, Seoul 29883
Honjin Hakubutsukan Futagawa Syuku, Toyohashi 29527
Honma Bijutsukan, Sakata 29204
Honma Museum of Art, Sakata 29204
Honolulu Academy of Arts Museum, Honolulu 49343
Honolulu House Museum, Marshall 50503
Honouring Hall of Juárez, Guelatao de Juárez 30924
Hontza Natur Zientzien Museoa, Mañaria 39611
Hoo-Hoo International Forestry Museum, Gurdon 49107
Hood Museum of Art, Hanover 49162
Hood River County Historical Museum, Hood River 49359
Hooge Crater Museum, Zillebeke 04161
Hoosier Salon Gallery, Indianapolis 49495
Hooton Park Exhibition Centre, Ellesmere Port 44167
Hoover Dam Museum, Boulder City 46963
Hoover Historical Center, North Canton 51386
Hooy Kaye Museum, Bruxelles 03448
Hop Museum, Poperinge 03953
Hope Cottage, Kingscote 01202
Hope Historical Society Museum, Hope 49361
Hope Lodge and Mather Mill, Fort Washington 48702
Hope Museum, Hope 05994
Hopetoun House, South Queensferry 45695
Hopewell Culture National Historic Park, Chillicothe 47534
Hopewell Furnace Hopewell Furnace National Historic Site, Elverson 48383

Hopewell Museum, Hopewell 49362
Hopewell Museum, Paris 51703
Hopfen Erlebnis Hof, Altmannstein 16514
Hopfenmuseum, Tettnang 22039
Hopi Cultural Center Museum, Second Mesa 53022
Hôpital Notre-Dame de la Rose, Lessines 03786
Hopkins House Gallery, Haddon Township 49113
Hopmuseum De Stadsschaal, Poperinge 03953
Hopp Ferenc Kelet-Ázsiai Művészeti Múzeum, Budapest 23413
Hopsewee Plantation, Georgetown 48893
Horace Greeley House, Chappaqua 47346
Horácká Galerie, Nové Město na Moravě 09678
Horácké Muzeum, Nové Město na Moravě 09679
Horaiji-san Shizenkshi Hakubutsukan, Hourai 28550
Horatio Colony House Museum, Keene 49719
Hordamuseet, Fana 33488
Horehronské Múzeum, Brezno 38238
Horie Orgel Museum, Nishinomiya 29073
Horie Orugorukan, Nishinomiya 29073
Horim Museum, Seoul 29884
Horizons Unlimited Supplementary Educational Center, Salisbury 52680
Horlogeriemuseum, Mechelen 03857
Hornby Museum, Hornby 49368
Hornické Muzeum OKD, Ostrava 09697
Hornické Muzeum Příbram, Příbram 09806
Hornický Skanzen důl Mayrau, Vinařice u Kladna 09908
Horniman Museum, London 44878
Hornsbury Mill, Chard 43792
Hornsea Museum of Village Life, Hornsea 44484
Horology Museum, Šternberk 09848
Hororata Historic Museum, Darfield 33084
Horry County Museum, Conway 47784
Horse-Breeding Museum, Moskva 37309
Horse Museum, Cussac-Fort-Médoc 12692
Horse Museum, Sümeg 23826
Horsens Kunstmuseum, Horsens 10097
Horsens Museum, Horsens 10098
Horsens Museum of Modern Art, Horsens 10097
Horseshoe Bend National Military Park, Daviston 47979
Horsforth Village Museum, Horsforth 44486
Horsham Arts Centre, Horsham 44488
Horsham Museum, Horsham 01162
Horsham Museum, Horsham 44489
Horsham Regional Art Gallery, Horsham 01163
Hørsholm Egns Museum, Hørsholm 10101
Horst-Janssen-Museum, Oldenburg 20895
Horster Motorrad-Museum, Gelsenkirchen 18574
Horten Bilmuseum, Horten 33562
Hortobágyi Pásztormúzeum, Hortobágy 23619
Horváth József Festőművész Állandó Kiállítása, Sopron 23810
Horyu-ji Homotsukan, Tokyo 29399
Horyuji Kondo Hekiga Mosha Tenjikan, Nagakute 28951
Hosanna Museum, Chai-Yi 42153
Hoskins House, Burlington 47141
Hosomi Bijutsukan, Kyoto 28813
Hosomi Museum, Kyoto 28813
Hospitaalmuseum of Our Lady of the Potterie, Brugge 03416
Hospitaalmuseum Onze-Lieve-Vrouw ter Potterie, Brugge 03416
Hospitaalmuseum - Saint John's Hospital, Brugge 03417
Hospitaalmuseum Sint-Janshospitaal, Brugge 03417
Hospital and Medical Care Museum, Deventer 32122
Hospital de los Venerables, Sevilla 40062
Hospital de San Juan Bautista, Toledo 40129
Hospital Museum, Jyväskylä 10895
Hospital Museum, Kellokoski 10937
Hospital Museum, Lahti 11034
Hospital Museum, Turku 11393
Hospitalmuseum Sint-Dimpna, Geel 03613
Hospitals and Clinics Medical Museum, Iowa City 49517
Hospitalsmuseet, Vadstena 41019
Hosyokan Byodoin Museum, Uji, Kyoto 29558
Hot Poppentheatermuseum, Gouda 32265
Hot Springs County Museum, Thermopolis 53492
Hotel and Restaurant Museum, Helsinki 10788
Hotel de Paris Museum, Georgetown 48889
Hôtel Dieu - Musée Greuze, Tournus 15957
Hôtel Groslot, Orléans 14532
Hotel-Museum Arthur Merghelynck, Ieper 03727
Hotelli- ja Ravintolamuseo, Helsinki 10788
Hotelmuseum, Flims Waldhaus 41375
Hou Wang Chinese Temple and Museum, Atherton 00817
Houdini Museum, Scranton 52981
Houghton County Historical Museum, Lake Linden 49887
Houghton Hall Soldier Museum, King's Lynn 44585
Housatonic Museum of Art, Bridgeport 47018
House for the History of IBM Data Processing, Sindelfingen 21776
House "Hohe Pappeln", Weimar 22404
House in the Horseshoe, Sanford 52848
House Mill, London 44879
House Museum, Polvijärvi 11217
House Museum and Memorial House of Yordan Yovkov, Dobrič 05020
House-Museum of F.M. Dostoevsky, Staraja Russa 37796
House-Museum of Jafar Jabarly, Bakı 03151
House-Museum of Jalil Mammadguluzade, Bakı 03152
House Museum of Karl Ristikivi, Tartu 10651
House Museum of Oskar Luts, Tartu 10654
House-Museum of Samad Vurgun, Bakı 03171
House-Museum of the Academician E.N. Pavlovsky, Borisoglebsk 36817
House-Museum of the conductor N.S. Golovanov, Moskva 37315
House-Museum of the Decembrists, Kurgan 37090
House-Museum of the Hannibal Family, Puškinskie Gory 37585
House Museum of the History of Youth Movement, Rjazan 37593
House-Museum of the Lopatin Family, Stavropol 37802
House-Museum of the Nevsorovy Sisters, Nižnij Novgorod 37450
House-Museum of the Painter V.A. Igoshev, Chanty-Mansijsk 36859
House-Museum of the Poet Shirjaevts - Museum of Farmers' Life, Žigulevsk 37988
House-Museum of the Sevastopol Communist Underground 1942-1944, Sevastopol 43175
House-Museum of the Tsvetaeva Family, Novo-Talicy 37476
House-Museum of Uzeyir Hajibeyov, Bakı 03156
House-Museum of V. Veresaev, Tula 37865
House-Museum Peter the Great, Vologda 37961
House-Museum Rayna Knyaginya, Panagjurište 05103
House of a Thousand Dolls, Loma 50163
House of Art, Brno 09425
House of Arts, České Budějovice 09462
House of Arts, South Moravian Museum, Znojmo 09946
House of Bottles, Kinglake 01201
House of Bottles and Bottle Museum, Tewantin 01578
House of Commons, Ottawa 06547
House of Dun, Montrose 45171
The House of Emigrants, Växjö 41043
House of Folk Dwelling, Matúškovo 38303
House of Folk Dwelling, Šala 38347
House of Forests at the Balaton National Park, Bakonybél 23348
House of Hadjigeorgakis Kornesios, Lefkosia 09378
House of Happy Walls, Glen Ellen 48919
House of History, Malone 50408
House of Humour and Satire, Gabrovo 05031
House of Literature - Miklos Mészöly Museum, Szekszárd 23867
The House of Lords, Dublin 24796
House of Manannan, Peel 45350
House of Memories, Latchford 06144
House of Memories Museum, Wilson 54209
House of Music Mikulas Schneider Trnavsky, Trnava 38378
House of Music Parma, Parma 27097
House of Negritude and Human Rights, Champagney 12386
House of Nobility, Stockholm 40929
House of Pataphony, Dinant 03557
House of Photography Poprad, Poprad 38331
House of Popular Arts, Szentendre 23878
House of Potery, Bouffioulx 03389
House of Roses, Deadwood 48004
House of Straw, Bengtsfors 40456
House of Technology, Luleå 40700
House of Terror, Budapest 23482
House of the Anchors, Kibbutz Ein Gev 25037
House of the Beguine, Brugge 03420
House of the Binns, Linlithgow 44719
House of the History of Larra, Larra 13528
House of the Martyrs Museum, La Habana 09120
House of the Rurality, Thuillies 04063
House of the Scribe, Almog 24925
The House of the Seven Gables, Salem 52656
House of the Templar Knights, Caudebec-en-Caux 12322
House of the Temple, Washington 53878
House of Twelve Roofs, Bandar Seri Begawan 04962
House of Wickersham, Juneau 49657
House of World Cultures, Berlin 17128
House on Crutches Museum, Bishop's Castle 43522
House on Feet, Võrs 23968
House on the Hill Museums Adventure, Stansted Mountfitchet 45738
The House on the Rock, Spring Green 53209
House under the Star, Toruń 35192
Houseboatmuseum, Amsterdam 31903
Household Cavalry Museum, Windsor 46042
Household Museum of Skriaudziai, Skriaudžiai 30370
Housesteads Roman Fort and Museum, Housesteads 44491
Houston Center for Photography, Houston 49388
Houston Fire Museum, Houston 49389
Houston Harte University Center, San Angelo 52703
Houston Museum of Decorative Arts, Chattanooga 47410
Houston Museum of Natural Science, Houston 49390
Houston Police Museum, Houston 49391
Høvåg Museum, Høvåg 33567
Hovden Jernvinnemuseum, Bykle 33460
Hove Museum and Art Gallery, Hove 44493
Hovercraft Museum, Gosport 44335
Hovhannes Hovhanisyan House-Museum, Echmiadzin 00696
Hovhannes Tumanyan Museum, Yerevan 00739
How we lived then - Museum of Shops, Eastbourne 44098
Howard B. Owens Science Center, Lanham-Seabrook 49939
Howard County Center of African American Culture, Columbia 47713
Howard County Historical Museum, Kokomo 49826
Howard County Historical Society Museum, Ellicott City 48367
Howard-Dickinson House Museum, Henderson 49252
Howard Gardens Gallery, Cardiff 43750
Howard House of Artifacts, Old Perlican 06508
Howard Steamboat Museum, Jeffersonville 49616
Howard University Gallery of Art, Washington 53879
Howard University Museum, Washington 53880
Howard W. Cannon Aviation Museum, Henderson 49250
Howell Harris Museum, Brecon 43598
Howell Historical Society & Committee Museum, Howell 49407
Howell Living History Farm, Titusville 53514
Howick Historical Village, Pakuranga 33173
Howick Museum, Howick 38643
Howland House, Plymouth 51993
Hoxie Gallery, Westerly 54087
Hoyt Institute of Fine Arts, New Castle 51042
Hoyt Sherman Place, Des Moines 48097
Hozat Culture Center, Hozat 42796
Hozat Kültür Merkezi, Hozat 42796
Hrad a Muzeum Hodin, Šternberk 09848
Hrad Devín, Bratislava 38214
Hrad Helfštýn, Týn nad Bečvou 09883
Hrad Krásna Hôrka, Krásnohorské Podhradie 38278
Hrad Roštejn, Telč 09860
Hrdlička's Museum of Man, Praha 09752
Hrdličkovo Muzeum Člověka, Praha 09752
Hristo Smirenski Memorial House, Sofia 05174
Hrvatska Akademija Znanosti i Umjetnosti - Gliptoteka, Zagreb 08993
Hrvatski Muzej Arhitekture, Zagreb 08994
Hrvatski Muzej Naivne Umjetnosti, Zagreb 08995
Hrvatski Nacionalni Komitet ICOM, Zagreb 54766
Hrvatski Poštansko-telekomunikacijski Muzej, Zagreb 08996
Hrvatski Povijesni Muzej, Zagreb 08997
Hrvatski Prirodoslovni Muzej, Zagreb 08998
Hrvatski Školski Muzej, Zagreb 08999
Hrvatski Sportski Muzej, Zagreb 09000
Hrvatsko Muzejsko Društvo, Zagreb 54767
Hsin Wang Ceramics Memorial Gallery, Yingge 42472
Hsinchu City Fire Museum, Hsinchu 42207
HTS-Museum, Rødovre 10239
Hu Shih Memorial Hall, Taipei 42393
Huafan Cultural Gallery, Shihtin 42326
Huai'an Museum, Huai'an 07821
Huaibei Museum, Huaibei 07824
Huaihua Zone Museum, Huaihua 07825
Huaiji Museum, Huaiji 07826
Huainan Museum, Huainan 07827
Huaiyang Museum, Huaiyang 07828
Huaiyin Museum, Huai'an 07822
Hualien County Stone Sculpture Museum, Hualien 42215
Hualien Railway Museum, Hualien 42216
Huanaki Cultural Centre, Alofi 33348
Huanduoci Curiosity Museum, Hengshan 42202
Huang Guei-li Memorial Hall, Banciao 42144
Huang Mun Art Center, Tongsiao 42454
Huanggang Museum, Huanggang 07830
Huanghe Museum, Zhengzhou 08499
Huangmei Museum, Huangmei 07831
Huangping Nationality Holiday Museum, Huangping 07832
Huangshan Beauty Spot Museum, Huangshan 07833
Huangshan Museum, Huangshan 07834
Huangshi Museum, Huangshi 07838
Huangsi Art Exhibition, Changhua 42155
Huangyan Museum, Huangyan 07839
Huangzhong Museum, Huangzhong 07840
Huangzhou Museum, Huangzhou 07841
Huating Museum, Huating 07842
Huazhou Museum, Huazhou 07843
Hub Local History Collection, Aberfoyle Park 00764
HUB Robeson Galleries, University Park 53658
Hubbard Museum of the American West, Ruidoso Downs 52480
Hubbardton Battlefield Museum, Hubbardton 49409
Hubbell Trading Post, Ganado 48853
Huberhus Museum, Lech 02277
Huble Homestead, Prince George 06667
Hubschraubermuseum, Bückeburg 17622
Huddersfield Art Gallery, Huddersfield 44496
Huddleston Farmhouse Inn Museum, Cambridge City 47192
Hudetz-Turm, Wiesent 22561
Hudobná Expozícia - Rodný Dom J.N. Hummela, Bratislava 38215
Hudobné Múzeum, Bratislava 38216
Hudson Bay Museum, Hudson Bay 05999
Hudson Museum, Orono 51602
Hudson River Maritime Museum, Kingston 49790
The Hudson River Museum, Yonkers 54331
Hudson Valley Children's Museum, Nanuet 50949
Hudson's Hope Histrical Society Museum, Hudson's Hope 06000
Huebmer-Gedächtnisstätte, Naßwald 02422
Huerta de San Vicente, Granada 39362
Hürten-Heimatmuseum, Bad Münstereifel 16789
Hüseyn Cavidin Ev-Muzeyi, Bakı 03157
Hüseyn Cavidin Ev-Muzeyi, Naxçivan 03184
Hütten- und Technikmuseum Ilsenburg, Ilsenburg 19322
Hüttenberger Heimatmuseum, Linden 20003
Hüttenmuseum Thale, Thale (Harz) 22042
Huf - und Hackenschmiede Lindermayr, Spital am Pyhrn 02749
Hufschmiede Museum Frehrking, Neustadt am Rübenberge 20663
Hufschmiedemuseum, Engelhartszell 01867
Hufthamartunet, Storebø 33886
Hugenotten-Museum, Berlin 17139
The Hugh Lane Municipal Gallery of Modern Art, Dublin 24797
Hugh Mercer Apothecary Shop, Fredericksburg 48767
Hugh Miller's Cottage, Cromarty 43928
Hughenden Manor, High Wycombe 44465
Hughes Fine Arts Center, Grand Forks 48988
Hugo Wolf-Haus, Perchtoldsdorf 02487
Huguan Museum, Huguan 07844
Huguenot Historical Society Museum, New Paltz 51096
Huguenot Memorial Museum, Franschhoek 38621
Hui No'eau Visual Arts Center, Makawao 50404
Hui'an Museum, Hui'an 07845
Huichang Museum, Huichang 07846
Huidong Museum, Huidong 07847
Huilai Museum, Huilai 07848
Huis de Lalaing, Oudenaarde 03940
Huis Doorn, Doorn 32143
Huis Hellemans, Edegem 03574
Huis Marseille, Amsterdam 31849
Huis Singraven, Denekamp 32114
Huis Ten Bosch Museums, Sasebo 29234
Het Huis van Alijn, Gent 03628
Huis Verwolde, Laren, Gelderland 32463
Huis Zypendaal, Arnhem 31926
Huittisten Museo, Huittinen 10842
Huize Betje Wolff, Middenbeemster 32557
Huize Keizer, Denekamp 32115
Huize Nijenstede, Hardenberg 32312
Huizer Museum Het Schoutenhuis, Huizen 32422
Huizhou Museum, Huizhou 07849
Huldigungssaal und Rathaus, Goslar 18710
Huldreheimen, Bykle 33461
Huleh Valley Regional Prehistoric Museum, Kibbutz Maayan Baruch 25050
Hulihee Palace, Kailua-Kona 49662
Hull and East Riding Museum, Kingston-upon-Hull 44597
Hull Lifesaving Museum, Hull 49417
Hull Maritime Museum, Hull 44499
Hull University Art Collection, Kingston-upon-Hull 44598
Hulsey Gallery, Oklahoma City 51515
Huludao Museum, Huludao 07850
Humačka Arheološko Zbirka, Ljubuški 04268
Humanitarian Centre - Museum of Outsider Art, Moskva 37217
Humboldt and District Museum and Gallery, Humboldt 06001
Humboldt County Historical Association Museum, Dakota City 47921
Humboldt County Historical Association Museum, Humboldt 49418
Humboldt County Old Settler's Museum, Livermore 50145
Humboldt Historical Museum, Humboldt 49419
Humboldt-Museum Schloß Tegel, Berlin 17140
Humpback Rocks Mountain Farm Visitor Center, Waynesboro 54014
Humphrey Forum, Minneapolis 50733

H

Hunan Provincial Museum, Changsha 07597
Hunan Revolutionary Cemetery, Changsha 07598
Hundsmarktmühle, Thalgau 02808
Hunebedcentrum, Borger 31987
Hung-Yen Museum, Hung-Yen 54633
Hungarian Building Industry Museum, Veszprém 23960
Hungarian Building Industry Museum - Brick Museum, Veszprém 23962
Hungarian Chemical Museum, Várpalota 23952
Hungarian Cultural Center, Moskva 37411
Hungarian Film Historical Collection, Budapest 23436
Hungarian Firefighter Museum, Budapest 23486
Hungarian Folk-Art Museum, Port Orange 52039
Hungarian Jewish Museum and Archives, Budapest 23447
Hungarian Marine Collection of Dr. Ferenc Juba, Komáron 23677
Hungarian Museum, Bogdand 36157
Hungarian Museum for Science, Technology and Transport - Industrial Collection, Budapest 23414
Hungarian Museum of Architecture, Budapest 23435
Hungarian Museum of Electrical Engineering, Budapest 23434
Hungarian Museum of Naive Arts, Kecskemét 23651
Hungarian Museum of Trade and Tourism, Budapest 23438
Hungarian National Gallery, Budapest 23441
Hungarian National Museum, Budapest 23442
Hungarian National Railway Museum, Budapest 23446
Hungarian Natural History Museum, Budapest 23445
Hungarian Open Air Museum, Szentendre 23880
Hungarian Ortodox Clerical Art Museum, Miskolc 23709
Hungarian Photography Museum, Kecskemét 23650
Hungarian Plains Forestration Collection, Bugac 23497
Hungarian Red Pepper Museum, Kalocsa 23626
Hungarian Technical Museum, Budapest 23443
Hungarian Theatre Museum and Institute, Budapest 23464
Hunsrück-Museum, Simmern 21771
Hunsrücker Holzmuseum, Morbach 20382
Hunsu Nationality Museum, Xingyi 08395
Hunt Gallery, Staunton 53276
Hunt Institute for Botanical Documentation, Pittsburgh 51950
Hunt-Morgan House, Lexington 50049
The Hunt Museum, Limerick 24864
Hunter College Art Galleries, New York 51193
Hunter-Dawson State Historic Site, New Madrid 51071
Hunter House, East Kilbride 44088
Hunter Museum of American Art, Chattanooga 47411
Hunter Valley Museum of Rural Life, Scone 01497
Hunterdon Museum of Art, Clinton 47657
Hunterian Art Gallery, Glasgow 44284
Hunterian Museum, Glasgow 44285
Hunterian Museum, London 44880
Hunterian Museum, Zoology Section, Glasgow 44286
Hunters Hill Historical Museum, Hunters Hill 01165
Hunting and Fishing Museum of Southern Tyrol, Racines 27355
Hunting & Fishing Museum, Ilulissat 23254
Hunting Museum of Finland, Riihimäki 11271
Hunting Museum Schloss Stainz, Stainz 02759
Huntingdon County Museum, Huntingdon 49421
The Huntington Art Collections, San Marino 52819
Huntington Beach Art Center, Huntington Beach 49438
Huntington Beach International Surfing Museum, Huntington Beach 49439
Huntington County Historical Society Museum, Huntington 49426
Huntington Museum of Art, Huntington 49435
Huntington Railroad Museum, Huntington 49436
Huntington Trade School, Huntington 49431
Huntly and Districts Historical Museum, Huntly 01166
Huntsman Marine Science Centre Aquarium & Museum, Saint Andrews 06797
Huntsville Museum of Art, Huntsville 49444
Huon Valley Apple and Heritage Museum, Huonville 01167
Hurda Farmstead Museum, Hurda 10571
Hurda Talumuuseum, Hurda 10571
Hurdal Bygdetun, Hurdal 33571
Hurley Patentee Manor, Hurley 49447
Huron City Museums, Port Austin 52030
Huron County Museum, Goderich 05879
Huronia Museum, Midland 06277
Hurricane Valley Heritage Park Museum, Hurricane 49451
Hurst Ranch, West Covina 54053
Hurtigrutemuseet, Stokmarknes 33883
Hurworth Cottage, New Plymouth 33157
Hus-Museum, Konstanz 19679
Húsavík Museum, Húsavík 24004
Huseby Bruk, Grimslöv 40567
Huseyn Javid Home Museum, Naxçivan 03184
Huseyn Javid Memorial Museum, Bakı 03157
Húsipari Múzeum, Budapest 23415
Húsipari Üzemtörténeti Gyűjtemény, Gyula 23595
Husitské Muzeum v Táboře, Tábor 09858
Huskvarna Stadsmuseum, Huskvarna 40607
Husqvarna Fabriksmuseum, Huskvarna 40608
Hussar Museum, Balatonszentgyörgy 23356
Hussite Museum in Tabor, Tábor 09858
Hussite Museum - Weis' House, Veselí nad Lužnicí 09904
Husum Castle, Husum 19301
Huszármúzeum, Balatonszentgyörgy 23356
Huszita-Ház - Helytörténeti Gyűjtemény, Gönc 23570
Huszita-House - Region Historical Collection, Gönc 23570
Hutchinson Art Center, Hutchinson 49453
Hutchinson County Museum, Borger 46914
Hutchinson House Museum, Waupaca 53991
Hutchison House Museum, Peterborough 06592
Hutnícke Múzeum Železiarne Podbrezová, Podbrezová 38329
Huttermuseum, Erdweg 18195
Huygensmuseum Hofwijck, Voorburg 32905
Huzhou Museum, Huzhou 07851
Hvaler Coastal Museum, Vesterøy 33974
Hvar Heritage Museum, Hvar 08880
Hvitträsk, Kirkkonummi 10963
Hwa Kang Museum, Taipei 42394
Hwajeong Museum, Seoul 29885
Hwataoyao - Kiln Exhibition Gallery, Yuanli 42477
Hyde Art Gallery, El Cajon 48308
The Hyde Collection, Glens Falls 48935
Hyde Hall, Cooperstown 47790
Hyde Hall, East Springfield 48260
Hyde Park Art Center, Chicago 47480
Hyde Park Barracks Museum, Sydney 01548
Hyderabad Museum, Hyderabad 34004
Hydro Industripark-Museum, Porsgrunn 33769
Hydrobiological Museum, Alexandria 10472
Hyeonchungsa Shrine, Asan 29769
Hyland House, Guilford 49103
Hylténs Industrimuseum, Gnosjö 40540
Hyman and Sons General Store, Gaspé 05853
Hymers Museum, Hymers 06004
Hyogo-ken Togeikan, Sasayamashi 29233
Hyogo-kenritsu Bijutsukan, Kobe 28728
Hyogo kenritsu Hito to Shizen no Hakubutsukan, Sanda 29212
Hyogo-kenritsu Rekishi Hakubutsukan, Himeji 28514
Hyogo Prefectural Martial Art Museum, Himeji 28511
Hyogo Prefectural Museum of Art, Kobe 28728
Hyogo Prefectural Museum of History, Himeji 28514
Hypocaust, Saint Albans 45543
Hyrax Hill Museum, Hyrax Hill 29739
Hythe Local History Room, Hythe, Kent 44507

I

I-5 Gallery, Los Angeles 50236
I.D. Sytin Museum-Appartment, Moskva 37323
I.E. Repin House-Museum, Žigulevsk 37987
I.E. Repin Museum-Estate, Repino 37590
I.J. Padarowski and Polish Americans Museum, Warszawa 35249
I. Karpenko-Karo Museum, Kirovograd 43069
I.N. Kramskoy House-Museum, Ostrogožsk 37519
I. Nikoladze House Museum, Tbilisi 16365
I.P. Pavlov Apartment Museum at the I.P. Pavlov Physiological Institut of the Russian Academy of Sciences, Sankt-Peterburg 37662
I.P. Stanback Museum, Orangeburg 51582
I.S. Klyuchnikov-Palantai Memorial Museum, Joškar-Ola 36977
I.S. Nikitin House-Museum, Voronež 37970
I.S. Shemanovsky Museum of the Jamalo-Nenetsk Autonomous Region, Salechard 37608
I.S. Turgenev State Literary Museum, Orël 37507
I.S. Turgenev State Memory Museum-Estate, Spasskoe-Lutovinovo 37792
I Space, Champaign 47334
I.V.K. Rajwade Sanshodhan Mandal Museum, Dhule 24166
I.V. Tsvetaev Art Museum of Copies, Moskva 37409
The Ian Potter Museum of Art, Melbourne 01284
Iaşi Natural History Museum, Iaşi 36419
Ibara Kyodo Denchu Bijutsukan, Ibara 28551
Ibara Municipal Denchu Art Museum, Ibara 28551
Ibaraki Ceramic Art Museum, Kasama 28673
Ibaraki Daigaku Izura Bijutsu Bunka Kenkyujo, Kitaibaraki 28707
Ibaraki-ken Kindai Bijutsukan, Mito 28918
Ibaraki-ken Tenshin Kinen Izura Bijutsukan, Kitaibaraki 28708
Ibaraki-kenritsu Rekishikan, Mito 28919
Ibaraki-kenritsu Shizenshi Hakubutsukan, Iwai 28604
Ibaraki-kenritsu Tougeikan, Kasama 28673
Ibaraki Nature Museum, Iwai 28604
Ibaraki Prefectural Museum of History, Mito 28919
Ibaraki University Izura Institut of Arts and Culture, Kitaibaraki 28707
IBRA Collection, Cardiff 43751
Ibrahim Hussein Foundation Museum, Pulau Langkawi 30612
Ibresinskij Ětnografičeskij Muzej pod Otkrytym Nebom, Ibresi 36929
Ibsen-museet, Oslo 33727
Ibsenhuset og Grimstad Bymuseet, Grimstad 33519
ICAMT International Committee for Architecture and Museums Techniques, Amsterdam 54956
ICDAD International Committee for Decorative Arts and Design Museums, Wien 54692
Ice Museum, Kamikawa 28647
Icebreaker Angara, Irkutsk 36938
Icebreaker Baykal, Angarsk 36762
The Icelandic Folk and Outsider Art Museum, Akureyri 23986
Icelandic Institute of Natural History, Reykjavík 24026
Icelandic Institute of Natural History, Akureyri Division, Akureyri 23984
Icelandic Maritime Museum, Reykjavík 24036
Icelandic Museums Association, Patreksfjörður 54879
Icelandic Phallological Museum, Húsavík 24003
Icelandic Postal Museum, Hafnarfjörður 24000
Iceni Village and Museums, Cockley Cley 43870
Ichan-Kala National Reserve Museum, Xiva 54492
Ichan Qala Davlat Muzej Qŏriqxonasi, Xiva 54492
Ichikawa City Museum of Archaeology, Ichikawa 28554
Ichikawa-shiritsu Koko Hakubutsukan, Ichikawa 28554
Ichikawa-shiritsu Shizenshi Hakubutsukan, Ichikawa 28555
Ichinomiya City Museum, Ichinomiya 28557
Ichinomiya-shiritsu Hakubutsukan, Ichinomiya 28557
Ichinoseki City Museum, Ichinoseki 28558
Ichinoseki-shiritsu Hakubutsukan, Ichinoseki 28558
Ickworth House, Horringer 44485
ICLM International Committee for Literary Museums, Paradis 54970
ICMAH International Committee for Archaeology and History Museums, Luxembourg 54930
ICME - International Committee for Museums and Collections of Ethnography, Miami Beach 55139
ICOFOM International Committee for Museology, Barnaul 55000
ICOM Afghanistan, Kabul 54678
ICOM Andorra, Aixovall 54680
ICOM Angola, Luanda 54681
ICOM Arab, Giza 54786
ICOM Argentina, Buenos Aires 54685
ICOM Asia-Pacific, Eumseong 54917
ICOM Australia, South Brisbane 54690
ICOM Azerbaijan, Bakı 54702
ICOM Bangladesh, Dhaka 54703
ICOM Barbados, Saint Michael 54704
ICOM - Belgian National Comittee, Hornu 54713
ICOM Bénin, Porto-Novo 54717
ICOM Bolivia, La Paz 54718
ICOM Bosnia-Hercegovina, Sarajevo 54719
ICOM Botswana, Gaborone 54720
ICOM Bulgaria, Sofia 54725
ICOM Burkina Faso, Ouagadougou 54726
ICOM Burundi, Gitega 54727
ICOM Cameroon, Yaoundé 54728
ICOM Canada, Ottawa 54738
ICOM Chile, Santiago de Chile 54753
ICOM Chinese National Committee, Beijing 54756
ICOM Comitato Nazionale Italiano, Milano 54901
ICOM Comité Haïtien, Port-au-Prince 54874
ICOM Comité National du Niger, Niamey 54965
ICOM - Comité National Monégasque du Conseil International des Musées, Monaco 54945
ICOM Comité Sénégalais, Saint-Louis 55004
ICOM Committee for University Museums and Collections, Berlin 54834
ICOM Comores, Moroni 54759
ICOM Costa Rica, San José 54763
ICOM Côté d'Ivoire, Grand Bassam 54765
ICOM Crne Gore, Cetinje 54947
ICOM Cuba, La Habana 54769
ICOM Cyprus, Lefkosia 54771
ICOM du Tchad, N'Djamena 54751
ICOM Ecuador, Guayaquil 54785
ICOM Egypt, Cairo 54787
ICOM Estonia, Tallinn 54791
ICOM F.Y.R. Macedonia, Skopje 54933
ICOM Finland, Helsinki 54793
ICOM Ghana, Cape Coast 54862
ICOM Grèce, Athinai 54868
ICOM Guatemala, Guatemala 54873
ICOM Hungarian National Comittee, Budapest 54876
ICOM Iceland, Reykjavík 54880
ICOM India, Kolkata 54883
ICOM Indonesia, Jakarta 54887
ICOM Iran, Tehrān 54888
ICOM Ireland, Ballina 54890
ICOM Jamaica, Kingston 54907
ICOM Jordan, Amman 54911
ICOM Kazakhstan, Almaty 54912
ICOM Kenya, Nairobi 54913
ICOM Kyrgyzstan, Biškek 54920
ICOM Lao, Vientiane 54921
ICOM Latin America and the Caribbean Regional Organization, San Jose 54764
ICOM Latvijas Nacionāla Komiteja, Riga 54922
ICOM Liban, Beirut 54926
ICOM Libyan Arab Jamahiriya, Tripoli 54927
ICOM Lietuva, Vilnius 54928
ICOM Lithuania, Vilnius 54928
ICOM Luxembourg, Luxembourg 54931
ICOM Madagascar, Isoraka 54935
ICOM Magyar Nemzeti Bizottság, Budapest 54876
ICOM Malaysia, Kuala Lumpur 54936
ICOM Mali, Bamako 54937
ICOM Maroc, Rabat 54949
ICOM Mauritanie, Nouakchott 54940
ICOM Mauritius, Port Louis 54941
ICOM México, México 54943
ICOM Moldova, Chişinău 54944
ICOM Mongolia, Ulaanbaatar 54946
ICOM Nacionalni Komitet Serbia, Beograd 55005
ICOM Namibia, Windhoek 54950
ICOM National Commitee of Belarus, Minsk 54707
ICOM Nederland, Amsterdam 54957
ICOM Nepal, Lalitpur 54952
ICOM New Zealand, Rotorua 54963
ICOM Nigeria, Calabar 54966
ICOM-Österreichisches Nationalkomitee, Wien 54693
ICOM Oman, Muscat 54978
ICOM Panamá, Ciudad de Panamá 54981
ICOM Paraguay, Asunción 54983
ICOM Philippines, Manila 54985
ICOM Poland, Poznań 54987
ICOM Portugal, Lisboa 54993
ICOM Qatar, Doha 54995
ICOM Regional Organization for West Africa, Bamako 54938
ICOM República Dominicaine, Santo Domingo 54784
ICOM Republique Congo, Brazzaville 54762
ICOM République Democratique du Congo, Kinshasa 54760
ICOM Romania, Ploieşti 54997
ICOM Schweiz, Zürich 55037
ICOM Seychelles, Victoria 55008
ICOM Singapore, Singapore 55009
ICOM Slovakia, Bratislava 55010
ICOM Slovenia, Ljubljana 55012
ICOM South Africa, Sunnyside 55016
ICOM Sri Lanka, Colombo 55029
ICOM Swaziland, Lobamba 55030
ICOM Sweden, Stockholm 55031
ICOM Tanzania, Dar-es-Salaam 55041
ICOM Thailand, Pathumthani 55042
ICOM Togo, Lomé 55043
ICOM Tunisia, Tunis 55044
ICOM Turkish National Committee, Ulus 55045
ICOM Ukraine, Lviv 55046
ICOM Uruguay, Punta del Este 55195
ICOM Uzbekistan, Toshkent 55196
ICOM Venezuela, Caracas 55197
ICOM Viet Nam, Ha Noi 55198
ICOM Zambia, Livingstone 55199
ICOM Zimbabwe, Ascot 55200
ICOMAC Regional Organization for Central Africa, Kinshasa 54761
ICOMAM International Committee of Museums and Collections of Arms and Military History, Delft 54958
Icon Museum, Girne 33355
Icon Museum, Korčula 08895
Icon Museum of Iskele, İskele 33363
Icon Museum of the City of Frankfurt on Main, Foundation Dr. Schmidt-Voigt, Frankfurt am Main 18378
Iconografisch Bureau, Den Haag 32076
Iconographic Museum Onufri, Berat 00011
ICR International Committee for Regional Museums, Heraklion 54869
ICU Hachiro Yuasa Memorial Museum, Mitaka 28916
ID Galerie, Helmond 32364
Ida Aalbergin Lapsuudenkoti, Leppäkoski 11063
Ida Aalberg's Childhood Home, Leppäkoski 11063
Ida Green Gallery, Sherman 53075
Idaho Association of Museums, Boise 55140
Idaho Museum of Mining and Geology, Boise 46892
Idaho Museum of Natural History, Pocatello 52006
Idaho State Historical Museum, Boise 46893
Idea Art and Design Center, Hsinchu 42208
Idemitsu Bijutsukan, Tokyo 29400
Idemitsu Museum of Arts, Tokyo 29400
Idemitsu Museum of Arts, Moji, Kitakyushu 28715

Idnakar - Historical and Cultural Museum-Preserve, Glazov 36925
Idojiri Archaeological Museum, Fujimi 28422
Idojiri Kokokan, Fujimi 28422
Idrija Municipal Museum at Castle Gewerkenegg, Idrija 38421
Idrottsmuseet, Malmö 40720
Idstedt-Gedächtnishalle, Idstedt 19310
Ieva Simonaitytė Memorial Museum, Priekulė 30339
Ievos Simonaitytės Memorialinis Muziejus, Priekulė 30339
If Museum Inner Spaces, Poznań 35025
ifa-Galerie Berlin, Berlin 17141
ifa-Galerie Stuttgart, Stuttgart 21945
Ifield Watermill, Crawley 43917
Ifj. Lele J. Néprajzi Gyűjteménye, Szeged 23844
Iföverkens Industrimuseum, Bromölla 40479
IG Halle, Rapperswil 41744
Iga and Shigaraki Pottery Museum, Ueno 29555
Iga-ryu Ninja Yashiki, Ueno 29556
Igarskij Kraevedčeskij Kompleks Muzej Večnoj Merzloty, Igarka 36930
Igartzako Museoa, Beasain 39011
Iglesia ni Cristo Museum and Gallery, Quezon City 34491
Ignaz Günther Museum, Altmannstein 16515
Ignaz J. Pleyel Museum, Großweikersdorf 02012
Igor Talkov Museum, Moskva 37285
II. Yesemek Heykel Atölyesi ve Açikhava Müzesi, İslahiye 42798
Iida City and Art Museum, Iida 28559
Iida Municipal Archaeological Museum, Iida 28560
Iida-shi Bijutsu Hakubutsukan, Iida 28559
Iida-shiritsu Kokokan, Iida 28560
Iisakin Jussin Tupa, Kauhava 10928
Iisakin Jussi's Cabin, Kauhava 10928
Iisaku Muuseum, Iisaku 10573
Iittala Glass Museum, Iittala 10849
Iittalan lasimuseo, Iittala 10849
Ijevan Local Lore Museum, Ijevan 00710
Ike Taiga Art Museum, Kyoto 28814
Ike Taiga Bijutsukan, Kyoto 28814
Ikebana Bijutsukan, Kyoto 28815
Ikeda 20-seiki Bijutsukan, Ito 28593
Ikeda Museum of 20th Century Art, Ito 28593
Ikegami Museum of Folk, Takatomachi Inashi 29334
Ikon Gallery, Birmingham 43505
Ikonen-Museum, Recklinghausen 21208
Ikonen-Museum der Stadt Frankfurt am Main, Frankfurt am Main 18378
Ikonenmuseum, Autenried 16649
Ikonenmuseum, Bad Breisig 16703
Ikonenmuseum, Potzneusiedl 02516
Ikonenmuseum Schloß Autenried, Ichenhausen 19307
Ikonenzentrum Alexej Saweljew, Traben-Trarbach 22074
Ikuno Kobutsukan no Kogyo Hakubutsukan, Ikuno 28566
Ikuno Minerals and Mining Museum, Ikuno 28566
Ikuo Hirayama Bijutsukan, Onomichi 29118
Ikuo Hirayama Museum of Art, Onomichi 29118
Il Mauriziano, Reggio Emilia 27387
Ilana Goor Museum, Tel Aviv 25107
Ilchester Museum, Ilchester 44508
Ilfracombe Machinery and Heritage Museum, Ilfracombe 01171
Ilfracombe Museum, Ilfracombe 44510
Ilia Chavchavadze's house & museum, Kvareli 16346
Ilias Lalaounis Jewelry Museum, Athinai 22915
Iligan Museum, Iligan 34388
Ilijanum Museum of Naive Art, Šid 38141
Ilja Chavchavadze House Museum, Tbilisi 16366
Ilja Chavchavadze Literature-Memorial Museum, Tbilisi 16367
Illawarra Museum, Wollongong 01662
Illingworth Kerr Gallery, Calgary 05521
Illinois and Michigan Canal Museum, Lockport 50156
Illinois Art Gallery, Chicago 47481
Illinois Artisans and Visitors Centers, Whittington 54138
Illinois Association of Museums, Springfield 55141
Illinois Citizen Soldier Museum, Galesburg 48834
Illinois Railway Museum, Union 53644
Illinois State Museum, Springfield 53216
Het Illuseum, Gent 03629
Illusoria-Land, Ittigen 41497
Ilmajoen Museo, Ilmajoki 10850
Ilmajoki Museum, Ilmajoki 10850
Ilmatorjuntamuseo, Tuusula 11414
Ilmavoimien Viestikoulun Perinnehuone, Tikkakoski 11375
Ilmin Museum of Art, Seoul 29886
Ilocaniana Cultural Museum, Magsingal 34410
Iloniemen Pajamuseo, Raatala 11249
Iloniemi Smithy Museum, Raatala 11249
Îlot des Palais, Québec 06682
Ilulissani Katersugaasivik, Ilulissat 23255
Ilulissat Town Museum, Ilulissat 23255
Ilwaco Heritage Museum, Ilwaco 49468
Ilyo Voivoda House-Museum, Kjustendil 05065
Imabari City Kono Art Museum, Imabari 28568
Imabari-shiritsu Kono Bijutsukan, Imabari 28568
Image Museum of Hsinchu City, Hsinchu 42209
Imagina-Museo Interactivo Puebla, Puebla 31264
Imaginarium Discovery Center, Anchorage 46289
Imaginarium Hands-On Museum, Fort Myers 48663
Imagination Station Science Museum, Wilson 54211
Imaizumi Hakubutsukan, Shiozawa 29277
Imaizumi Museum, Shiozawa 29277
Imari Municipal Folk History Museum, Imari 28569
Imari-shi Rekishi Minzoku Shiryokan, Imari 28569
Imatra Art Museum, Imatra 10854
Imatra Museum of Cultural History, Imatra 10853
Imatran Kulttuurihistoriallinen Kaupunginmuseo, Imatra 10853
Imatran Taidemuseo, Imatra 10854
IMF Center Exhibit, Washington 53881
Imhoff Art Gallery, Saint Walburg 06886
Imker- und Heimatmuseum Wassertrüdingen, Wassertrüdingen 22349
Imkereigeschichtliche und bienenkundliche Sammlungen, Celle 17687
Imkereimuseum, Grüningen 41457
Imkereimuseum, Pöggstall 02509
Imkereimuseum, Pramet 02527
Imkereimuseum Grüningen, Wetzikon 42010
Imkerijmuseum Poppendamme, Grijpskerke 32273
Imkermuseum, Sitzenberg-Reidling 02745
Immigrant City Museum, Lawrence 49995
Immigrant-institutets Museum, Borås 40470
Immigration Museum, Melbourne 01285
Immingham Museum, Immingham 44514
Imogene Herbert Historical Museum, Manistique 50440
Les Impatients, Montréal 06340
Imperato Collection of West African Artifacts, Chanute 47340
Imperial and District Museum, Imperial 06009
Imperial Calcasieu Museum, Lake Charles 49878
Imperial Palace, Innsbruck 02119
Imperial Stables, Praha 09738
Imperial War Museum, London 44881
Imperial War Museum Duxford, Duxford 44075
Imperial War Museum North, Manchester 45086
Impression 5 Science Center, Lansing 49941
Impressions Gallery of Photography, Bradford 43580
Imre Cserépfalvi Memorial Room, Cserépfalu 23506
Imre Kálmán Memorial Museum, Siófok 23803
Imre Madach Manor House of the Bratislava Museum of Hungarians Culture in Slovakia, Dolná Strehová 38249
Imre Madách Memorial Museum, Csesztve 23510
Imre Patkó Collection, Győr 23585
Imura Art Museum, Kyoto 28816
In Buri National Museum, Sing Buri 42600
In de Gecroonde Duijvekater, Zaandam 32976
In de Zevende Hemel, Apeldoorn 31912
In Flanders Fields Museum, Ieper 03728
Inari Museum of Art, Kasama, Kasama 28674
Inatura - Erlebnis Naturschau Dornbirn, Dornbirn 01825
Inax Tile Museum, Tokoname 29364
Inazawa City Oguiss Memorial Art Museum, Inazawa 28572
İnce Minare Müzesi, Konya 42885
Incheon Metroploitan City Museum, Incheon 29833
Independence Hall, Tel Aviv 25108
Independence Hall of Korea, Cheonan 29784
Independence Historical Museum, Independence 49470
Independence National Historical Park, Philadelphia 51842
Independence Seaport Museum, Philadelphia 51843
Independence Visitors Center, Independence 49476
Inderøy Museums- og Historielag, Inderøy 33574
India Arts Museum, Chennai 24116
Indian Academy of Fine Arts Gallery, Amritsar 24067
Indian Arts and Crafts Board, Washington 53882
Indian Arts Research Center, Santa Fe 52876
Indian Association for the Study of Conservation of Cultural Property, Delhi 54884
Indian City U.S.A., Anadarko 46280
Indian Head Museum, Indian Head 06011
Indian Hill Historical Society Museum, Cincinnati 47562
Indian King Tavern Museum, Haddonfield 49115
Indian Land Museum, Gossau (Zürich) 41445
Indian Mill Museum State Memorial, Upper Sandusky 53666
Indian Museum, Kolkata 24275
Indian Museum, Lemmer 32497
Indian Museum of Lake County-Ohio, Painesville 51675
Indian Museum of North America, Crazy Horse 47873
Indian Museum of the Carolinas, Laurinburg 49988
Indian Pueblo Cultural Center, Albuquerque 46178
Indian Springs State Park Museum, Flovilla 48573
Indian Temple Mound Museum, Fort Walton Beach 48696
Indian War Memorial Museum, Delhi 24142
Indiana Basketball Hall of Fame, New Castle 51041
Indiana Medical History Museum, Indianapolis 49496
Indiana State Museum, Indianapolis 49497
Indiana Territory Capitol, Vincennes 53748
Indiana Transportation Museum, Noblesville 51357
Indiana University Art Museum, Bloomington 46864
Indiana War Memorials Museum, Indianapolis 49498
Indianapolis Art Center, Indianapolis 49499
Indianapolis Motor Speedway Hall of Fame Museum, Indianapolis 49500
Indianapolis Museum of Art, Indianapolis 49501
Indianapolis Museum of Art - Columbus Gallery, Columbus 47742
Indira Gandhi National Centre for the Arts, Delhi 24143
Individual Artists of Oklahoma, Oklahoma City 51516
Indo-Cuban Bani Museum, Banes 09029
Indre Sør-Troms Distriktsmuseum, Tennevoll 33903
Industrial and Agricultural Museum, York 54341
Industrial and Commercial Museum, Lahore 34020
Industrial Heritage Complex - Merrickville Lockstation, Merrickville 06268
Industrial Museum, Lübeck 20074
Industrial Museum, Nottingham 45293
Industrie- und Filmmuseum Wolfen, Wolfen 22650
Industrieel Smalspoor Museum, Erica 32219
Industriemuseen in der Euregio Maas-Rhein e.V., Euskirchen 54835
Industriemuseum, Limbach-Oberfrohna 19993
Industriemuseum Brandenburg, Brandenburg an der Havel 17508
Industriemuseum Elmshorn, Elmshorn 18126
Industriemuseum Freudenthaler Sensenhammer, Leverkusen 19974
Industriemuseum Hartha, Hartha 19019
Industriemuseum Lauf, Lauf an der Pegnitz 19851
Industriemuseum Lohne, Lohne 20026
Industrieviertel-Museum, Wiener Neustadt 03096
Industrimuseet, Horsens 10099
Industrimuseet, Kopperå 33606
Industrimuséet C.W. Thorstensons Mekaniska Verkstad, Åmål 40419
Industrimuseet Tollereds Öfvre Kraftstation, Tollered 40973
Industrion Museum for Industry and Society, Kerkrade 32446
Infantrie Museum, Harskamp 32319
Infantry and Small Arms School Corps Weapons Collection, Warminster 45933
Infantry Museum, Mikkeli 11113
Infantry Museum, Palhais 35805
Informationsstelle des Naturparks Fichtelgebirge, Zell, Oberfranken 22782
Informationszentrum am Wasserstraßenkreuz Minden, Minden 20334
Informationszentrum Donnersberghaus, Dannenfels 17781
Informationszentrum Schloss Paffendorf, Bergheim (Erft) 17012
Ingatestone Hall, Ingatestone 44515
Ingeborgmuseet Fredheim, Tangen 33901
Ingeborrarps Friluftsmuseum, Örkelljunga 40792
Ingebrigt Vik's Museum, Øystese 33765
Ingersoll Cheese Factory Museum and Sports Hall of Fame, Ingersoll 06012
Ingledale, Stoney Creek 07013
Inglewood District Historical Museum, Inglewood 01173
Ingrow Loco Museum, Keighley 44552
Ingurugiro Etxea, Azpeitia 38930
Ingušskij Gosudarstvennyj Muzej Kraevedenija im. T.Ch. Malsagova, Nazran 37448
Inland Seas Maritime Museum, Vermilion 53718
Inland-Water Ecological Laboratory Collection, Jerusalem 24991
Inn-Museum, Rosenheim 21376
Inner Mongolia Museum, Hohhot 07790
Inner Mongolia Painting And Calligraphy House, Hohhot 07791
Innisfail and District Historical Museum, Innisfail 01174
Innisfail and District Historical Village, Innisfail 06013
Inniskeen Folk Museum, Inniskeen 24844
Innisville and District Museum, Innisville 06014
Innovatum Kunskapens Hus, Trollhättan 40982
Inns of Court and City Yeomanry Museum, London 44882
Innviertler Freilichtmuseum Brunnbauerhof, Andorf 01720
İnönü Evi Müzesi, İzmir 42854
Inönü House Museum, İzmir 42854
Inosuke Hazama Museum of Art, Kaga 28624
Inpa Coleções Zoológicas, Manaus 04552
Inquisitor's Palace, Vittoriosa 30655
Inro Bijitsukan, Takayama 29342
Inro Museum, Takayama 29342
Insect Collections at the Entomology Department, Taipei 42395
Insectarium de Montréal, Montréal 06341
Insects Museum, Amman 29650
Insects World, Wunstorf 22734
Inselmuseum, Pellworm 21006
Inselmuseum Spiekeroog, Spiekeroog 21839
Insights - El Paso Science Center, El Paso 48324
Inspire - Hands-on Science Centre, Norwich 45279
Institut Culturel Avataq, Westmount 07307
Institut d'Art Contemporain, Villeurbanne 16236
Institut de France, Saint-Jean-Cap-Ferrat 15316
Institut et Musée Voltaire, Genève 41413
Institut für Auslandsbeziehungen e.V. - Abt. Kunst, Stuttgart 21946
Institut für Auslandsbeziehungen e.V.- Büro Berlin, Berlin 17142
Institut für Geschichte der Medizin der Robert Bosch Stiftung, Stuttgart 21947
Institut für moderne Kunst, Nürnberg 20771
Institut für moderne Kunst im Studio des Zumikon Nürnberg, Nürnberg 20772
Institut für Museumsforschung, Berlin 17143
Institut Mathildenhöhe, Darmstadt 17790
Institut Tessin, Paris 14612
The Institute For American Indian Studies, Washington Green 53950
Institute for Jewish Studies Gallery, Jerusalem 24992
Institute for Protection of Cultural Monuments and Museum - Strumica, Strumica 30530
Institute Menezes Braganza, Panaji 24341
Institute of American Indian Arts Museum, Santa Fe 52877
Institute of Archaeology and Museum Studies, Jos 33329
Institute of Archaeology with Museum, Sofia 05172
Institute of Arts and Crafts Gallery, Dhaka 03224
Institute of Contemporary Art, Boston 46933
Institute of Contemporary Art, Philadelphia 51844
Institute of Contemporary Art, Portland 52064
Institute of Contemporary Arts, London 44883
Institute of Geological and Nuclear Sciences, Hutt City 33125
Institute of Jamaica, Kingston 28328
Institute of Meteoritics Meteorite Museum, Albuquerque 46179
Institute Of Modern Art, Fortitude Valley 01087
Institute of Nature Study, Tokyo 29471
Institute of Social History, Gent 03630
Institute of Texan Cultures, San Antonio 52716
Institute of Visual Arts, Milwaukee 50710
Instituto Amatller de Arte Hispánico, Barcelona 38965
Instituto Cultural Cabañas, Guadalajara 30889
Instituto de Artes Gráficas de Oaxaca, Oaxaca 31220
Instituto de Investigaciones Arqueológicas y Museo Prof. Mariano Gambier, Albardón 00118
Instituto de Investigaciones Históricas de la Manzana de las Luces, Buenos Aires 00162
Instituto de Valencia de Don Juan, Madrid 39522
Instituto Gómez-Moreno de la Fundación Rodríguez-Acosta, Granada 39363
Instituto Mexicano Norteamericano de Jalisco, Guadalajara 30890
Instituto Moreira Salles, São Paulo 04861
Instituto Moreira Salles, Belo Horizonte 04344
Instituto Moreira Salles, Poços de Caldas 04636
Instituto Moreira Salles, Rio de Janeiro 04696
Instituto Português de Museus, Lisboa 54994
Instituts- und Studiensammlung des Instituts für Ur- und Frühgeschichte, Wien 03006
Instituut voor Sociale Geschiedenis, Gent 03630
Instytut Sztuki Wyspa, Gdańsk 34635
Insurance Museum, Kraków 34812
Intel Museum, Santa Clara 52862
Inter Art - Sonsbeek Art and Design, Arnhem 31927
Inter Oceans Museum, Saint-Barthélemy 23279
Intercisa Múzeum, Dunaújváros 23533
INTERCOM International Committee for Museums Management, Castries 55003
Intercultural Museum and Art Gallery, Baltimore 46569
Interessengemeinschaft niederösterreichische Museen und Sammlungen, Krems 54694
Interlaken Historical Society Museum, Interlaken 49512
Intermedia Arts Minnesota-Galleries, Minneapolis 50734
Intermuseum Conservation Association, Oberlin 55142
Internasjonale Barnekunstmuseet, Oslo 33728
Internationaal Exlibriscentrum, Sint-Niklaas 04020
International Art Center, Miami 50627
International Association of Agricultural Museums, Rožnov pod Radhoštěm 54774
International Association of Custom Museums, Antwerpen 54714

International Association of Libraries and Museums of Performing Arts (SIMBAS), Roma 54902
International Association of Museum Facility Administrators, New York 55143
International Association of Museums of Arms and Military History, Liège 54709
International Association of Museums of History, Paris 54803
International Association of Transport and Communication Museums, Ängelholm 55032
International Bluegrass Music Museum, Owensboro 51638
International Bottle Museum, Sonkajärvi 11325
International Bowling Museum and Hall of Fame, Saint Louis 52580
International Center of Art in Łódź, Łódź 34872
International Center of Graphic Arts, Ljubljana 38451
International Center of Photography, New York 51194
International Centre for the Study of the Preservation and the Restoration of Cultural Property (ICCROM), Roma 54903
International Centre - Museum of N.K. Roerich, Moskva 37249
International Ceramics Museum, Bechyně 09398
International Clown Hall of Fame and Research Center, West Allis 54041
International Coffee Cup Museum, Posio 11233
International Committee for Conservation, Roma 54904
International Committee for Egyptology in ICOM - CIPEG, Baltimore 55144
International Committee for Exhibition Exchange, Ottawa 54739
International Committee for Marketing and Public Relations, Oslo 54971
International Committee for Monetary Museums, Utrecht 54959
The International Committee for Museums and Collections of Costume, London 55068
International Committee for Museums and Collections of Glass, Toledo 55145
International Committee for Museums and Collections of Modern Art, Barcelona 55028
International Committee for Museums and Collections of Natural History, Frankfurt am Main 54836
International Committee for Museums Security, Berlin 54837
International Committee for the Training of Personnel, Berlin 54838
International Committee on Management, Ottawa 54740
International Confederation of Architectural Museums, Rotterdam 54960
International Congress of Maritime Museums, Essex 55146
International Council of African Museums (AFRICOM), Nairobi 54914
International Council of Museums ICOM, Paris 54822
International Council on Monuments and Sites, Paris 54823
International Crane Foundation Museum, Baraboo 46605
International Cultural Center for Youth in Jerusalem, Jerusalem 24993
International Fox Museum, Summerside 07033
International Gospel Music Hall of Fame and Museum, Detroit 48121
International Hippocratic Foundation of Kos, Kos 23064
International Ice Hockey Federation Museum, Kingston 06078
International Images, Sewickley 53044
International Institute of Metropolitan Detroit, Detroit 48122
International Jewish Sports Hall of Fame, Netanya 25073
International Motor Racing Research Center, Watkins Glen 53981
International Movement for a New Museology, Zuni 55147
International Museum Centre of the Arsenev State Museum of the Primorsky Region, Vladivostok 37941
International Museum of Art, El Paso 48325
International Museum of Cartoon Art, Boca Raton 46885
International Museum of Cartoon Art, Stamford 53247
International Museum of Children's Art, Oslo 33728
International Museum of Children's Art, Vårby 41049
International Museum of Cultures, Dallas 47934
International Museum of Miniature, Lyon 13897
International Museum of Naive Art of Brazil, Rio de Janeiro 04754
International Museum of Peace and Solidarity, Samarqand 54461
International Museum of Photography and Film, Rochester 52395
International Museum of Surgical Science, Chicago 47482
International Museum of the Horse, Lexington 50050
International Museum of Women, San Francisco 52777
International Museum Theatre Alliance, Washington 55148
International Print Center New York, New York 51195
International Printing Museum, Carson 47257
International Puppet Theatre Museum, Stockholm 40912
International Red Cross and Red Crescent Museum, Genève 41423
International Sculpture Center, Hamilton 49134
International Sculpture Park, Schaumburg 52957
International Sport Aviation Museum, Lakeland 49900
International Spy Museum, Washington 53883
International Swimming Hall of Fame, Fort Lauderdale 48637
International Tennis Hall of Fame Museum, Newport 51321
International Wildlife Museum, Tucson 53592
International Women's Air and Space Museum, Cleveland 47633
Internationale Jugendbibliothek, München 20447
Internationale Spitzensammlung, Nordhalben 20736
Internationale Tage - Boehringer Ingelheim, Ingelheim am Rhein 19334
Internationales Artistenmuseum in Deutschland, Klosterfelde 19581
Internationales Keramik-Museum, Weiden in der Oberpfalz 22366
Internationales Luftfahrtmuseum, Villingen-Schwenningen 22239
Internationales Maritimes Museum Hamburg, Hamburg 18937
Internationales Muschelmuseum, Wangerland 22329
Internationales Radiomuseum, Bad Laasphe 16771
Internationales Wind- und Wassermühlenmuseum, Gifhorn 18643
Internationales Zeitungsmuseum der Stadt Aachen, Aachen 16399
Internationella Barnkonstmuseet, Vårby 41049
Interoceanic Canal Museum of Panamá, Ciudad de Panamá 34071
Intersection for the Arts Gallery, San Francisco 52778
Intrepid Sea-Air-Space Museum, New York 51196
Inuit Heritage Center, Baker Lake 05375
Inuyama Castle, Inuyama 28574
Inuyama Kenbunkan, Inuyama 28574
The Invention Factory, Trenton 53559
Inveraray Castle, Inveraray 44519
Inveraray Jail, Inveraray 44520
Inveraray Maritime Museum, Inveraray 44521
Inverkeithing Museum, Inverkeithing 44522
Inverness Miners Museum, Inverness 06017
Inverness Museum and Art Gallery, Inverness 44525
Inženernyj (Michailovskij) Zamok, Sankt-Peterburg 37651
Ioannina Museum, Ioannina 23014
Iolani Palace, Honolulu 49344
Ion Luca Caragiale Museum, Ploiești 36531
Iona Abbey and Nunnery, Isle-of-Iona 44538
Iona Heritage Centre, Isle-of-Iona 44539
Ionad Arann Heritage Centre, Aran Islands 24728
Ionnina Byzantine Museum, Ioannina 23015
Iosco County Historical Museum, East Tawas 48261
Ioseb Grishashvili Tbilisi History Museum, Tbilisi 16368
Iowa Aviation Museum, Greenfield 49053
Iowa Great Lakes Maritime Museum, Arnolds Park 46363
Iowa Masonic Library and Museum, Cedar Rapids 47304
Iowa Museum Association, Waterloo 55149
Iowa Railroad Historical Museum, Boone 46904
Iparművészeti Múzeum, Budapest 23416
Ipogeo dei Volumni, Ponte San Giovanni 27286
Ippuku Bijutsukan, Asahi 28367
Ippuku Museum of Art, Asahi 28367
Ipswich Historical Society, Ipswich 49524
Ipswich Museum, Ipswich 44530
Ipswich Transport Museum, Ipswich 44531
Iqbal Museum, Lahore 34021
Iraan Museum, Iraan 49526
Iran Bastan Museum, Tehrān 24672
Iranian National Museum of Water, Tehrān 24673
Iraq Natural History Research Centre and Museum, Baghdad 24714
Iraqi National Museum, Baghdad 24715
Irbid Archaeological Museum, Irbid 29662
Irchester Narrow Gauge Railway Museum, Wellingborough 45956
Iredell Museum of Arts & Heritage, Statesville 53274
Ireland House at Oakridge Farm, Burlington 05499
Ireland's Historic Science Centre, Birr 24745
Irene Cullis Gallery, Greensboro 49065
Iris and B. Gerald Cantor Art Gallery, Worcester 54299
Iris & B. Gerald Cantor Center for Visual Arts, Stanford 53253
Irish Agricultural Museum, Wexford 24912
Irish American Heritage Museum, East Durham 48239
Irish American Museum, Albany 46159
Irish Architectural Archive, Dublin 24798
Irish Horse Museum, Kildare 24846
Irish Jewish Museum, Dublin 24799
Irish Labour History Society Museum, Dublin 24800
Irish Linen Centre and Lisburn Museum, Lisburn 44725
Irish Moss Interpretive Centre and Museum, Miminegash 06292
Irish Museum of Modern Art, Dublin 24801
Irish Museums Association, Dublin 54891
Irish Music Hall of Fame, Dublin 24802
Irish Palatine Association Museum, Rathkeale 24891
Irkutskij Architekturno-ėtnografičeskij Muzej Talcy, Irkutsk 36931
Irkutskij Oblastnoj Chudožestvennyj Muzej im. V.P. Sukačeva, Irkutsk 36932
Irkutskij Oblastnoj Kraevedčeskij Muzej, Irkutsk 36933
Irma Dahn-Anderssons Nävermuseum, Bollnäs 40465
Irma Stern Museum of the University of Cape Town, Rosebank 38727
Irodalom Háza - Mészöly Miklós Múzeum, Szekszárd 23867
Iron and Steel Museum of Alabama, McCalla 50340
Iron Bridge Historical Museum, Iron Bridge 06021
Iron County Museum, Caspian 47274
Iron Creek Museum, Lougheed 06196
Iron Forging Museum, Kropa 38441
Iron Garden, Bridgetown 03238
Iron Mill Complex, Aia 38797
Iron Mission Museum, Cedar City 47290
Iron Mountain Iron Mine Museum, Vulcan 53769
Iron Museum, Saint-Hubert 03997
Iron Museum and Moncorvo Regional Museum, Torre de Moncorvo 35967
Ironbridge Gorge Museum Trust, Coalbrookdale 55069
Ironbridge Tollhouse, Ironbridge 44532
Ironwood Area Historical Museum, Ironwood 49528
Ironworld Discovery Center, Chisholm 47544
Iroquois County Historical Society Museum, Watseka 53985
Iroquois Falls Pioneer Museum, Iroquois Falls 06023
Iroquois Indian Museum, Howes Cave 49408
Irrseer Heimathaus, Zell am Moos 03126
Iruma City Museum Alit, Iruma 28579
Iruma-shi Hakubutsukan Alit, Iruma 28579
Irvin Goodon International Wildlife Museum, Boissevain 05434
Irvine Burns Club Museum, Irvine 44534
Irvine Fine Arts Center, Irvine 49530
Irvine Museum, Irvine 49531
Irving Arts Center, Irving 49533
Irving House Historic Centre and New Westminster Museum, New Westminster 06436
Isaac Farrar Mansion, Bangor 46597
Isaak Brodsky Apartment Museum, Sankt-Peterburg 37690
Isaak Kaplan Old Yishuv Coourt Museum, Jerusalem 24994
Isaak Pasha Palace, Ağri 42679
Isaakevskij Sobor - Memorialnyj Muzej-pamjatnik, Sankt-Peterburg 37652
Isabel Anderson Comer Museum and Arts Center, Sylacauga 53394
Isabella Stewart Gardner Museum, Boston 46934
Isahakyan House Museum, Gyumri 00705
Isamu Noguchi Garden Museum, Long Island City 50182
Isanti County Museums, Cambridge 47187
Ischua Valley Historical Society, Franklinville 48753
ISE Art Foundation, New York 51197
Isenbaye Art Gallery and Cultural Troupe, Oshogbo 33339
Isergebirgs Museum Neugablonz, Kaufbeuren 19488
Iseum Romkert, Szombathely 23900
Iseum Ruin Garden, Szombathely 23900
Isfara Historical and Regional Museum, Isfara 42499
İshak Paşa Sarayi, Ağri 42679
Ishavsmuseet Aarvak, Brandal 33452
Ishibashi Museum of Art and Asian Gallery, Kurume 28795
Ishii Chawan Bijutsukan, Hagi 28470
Ishii Tea Bowl Museum, Hagi 28470
Ishikawa-kenritsu Bijutsukan, Kanazawa 28656
Ishikawa-kenritsu Dento Sangyo Kogeikan, Kanazawa 28657
Ishikawa-kenritsu Rekishi Hakubutsukan, Kanazawa 28658
Ishikawa -kenritsu Shizenshi Shiryokan, Kanazawa 28659
Ishikawa Museum of Natural History, Kanazawa 28659
Ishikawa Prefectural Arts and Crafts Museum, Kanazawa 28657
Ishikawa Prefectural History Museum, Kanazawa 28658
Ishikawa Prefectural Museum of Art, Kanazawa 28656
Ishikawa Takuboku Memorial Hall, Tamayama 29350
Ishinomaki Bunkakan, Ishinomaki 28587
Ishinomaki Culture Center, Ishinomaki 28587
Isinger Dorfmuseum - Alte Kelter, Tübingen 22116
Isis District Historical Complex, Childers 00968
İskele İon Müzesi, İskele 33363
Iskitimskij Kraevedčeskij Muzej, Iskitim 36941
Iskolamúzeum, Békés 23363
Iskolatörténeti és Egyházművészeti Múzeum, Cegléd 23502
Iskolatörténeti Kiállítás, Kecskemét 23646
Isla Center for the Arts at the University of Guam, Mangilao 23287
Islamabad Museum, Islamabad 34007
Islamia College Museum, Peshawar 34034
Islamic Museum, Abbasiya 10465
Islamic Museum, Amman 29651
Islamic Museum, Jerusalem 24995
Islamic Museum, Jerusalem 34056
Islamic Museum, Tehrān 24674
Islamic Museum, Tripoli 30188
Island County Historical Society Museum, Coupeville 47855
Island Fishery of Oostende, Oostende 03927
Island Fortress, Helsinki 10825
Island Mountain Arts Public Gallery, Wells 07295
Islands Museum, Tiverton 07085
Isle La Motte Historical Society, Isle La Motte 49540
Isle of Wight Courthouse, Smithfield 53133
Isle of Wight Steam Railway and Isle of Wight Railway Heritage Museum, Havenstreet 44421
Isle of Wight Wax Works, Brading 43587
Isleford Historical Museum, Isleford 49541
Isles of Scilly Museum, Saint Mary's 45584
Islesboro Historical Museum, Islesboro 49542
Islington Education Artefacts Library, London 44884
Islington Museum, London 44885
Islip Art Museum, East Islip 48251
Ismail Local Lore and History Museum of the Danube Region, Izmail 43047
Ismail Picture Gallery, Izmail 43046
Ismailia Museum, Ismailia 10535
Isokichi Asakura Museum of Art, Terai 29359
Isokyrö Local History Museum, Isokyrö 10862
Isonkyrön Kotiseutumuseo, Isokyrö 10862
İsparta Devlet Güzel Sanatlar Galerisi, İsparta 42799
İsparta Müzesi, İsparta 42800
Isparta State Fine Art Gallery, İsparta 42799
Israel Bible Museum, Zefat 25133
Israel Children's Museum, Holon 24977
Israel Defense Forces History Museum, Tel Aviv 25109
Israel Goor Theater Archive and Museum, Jerusalem 24996
Israel Museum, Jerusalem, Jerusalem 24997
Israel National Committee of ICOM, Ramat-Gan 54892
Israel Oil Industry Museum, Haifa 24962
Israel Railway Museum, Haifa 24963
Israeli Air Force Museum, Hatzerim 24971
Israeli Museum of Photography, Tel Hai 25125
Issaquah Depot Museum, Issaquah 49545
Issyk-Kul State Historical Cultural Museum-Preserve, Čolpan-Ata 29961
Issyk-Kulskij Gosudarstvennyj Istoriko-Kulturnyj Muzej-Zapovednik, Čolpan-Ata 29961
Istana Batu, Kota Bharu 30563
İstanbul Arkeoloji Müzeleri, İstanbul 42823
İstanbul Büyükşehir Belediyesi Karikatür ve Mizah Müzesi, İstanbul 42824
İstanbul Devlet Güzel Sanatlar Galerisi, İstanbul 42825
İstanbul Modern Sanat Müzesi, İstanbul 42826
İstanbul Museum of Caricature and Humour, İstanbul 42824
İstanbul Sanat Müzesi, İstanbul 42827
İstanbul Şişli'deki Ataürk Evi, İstanbul 42828
İstanbul State Fine Art Gallery, İstanbul 42825
Istituto della Cultura Arbereshe Giuseppe Gangale, Caraffa 25671
Istituto di Fotografia Alpina Vittorio Sella, Biella 25417
Istituto Italiano dei Castelli, Milano 54905
Istituto per la Storia della Resistenza e della Società Contemporanea nelle Province di Biella e Vercelli, Varallo 28155
Istoki, Belozersk 36800

I

Istoričeskaja Usadba Botik Petra I, Veskovo 37926
Istoričeski Muzej, Bansko 04978
Istoričeski Muzej, Blagoevgrad 04991
Istoričeski Muzej, Bracigovo 04994
Istoričeski Muzej, Burgas 04999
Istoričeski Muzej, Čiprovci 05007
Istoričeski Muzej, Dimitrovgrad 05015
Istoričeski Muzej, Dobrič 05019
Istoričeski Muzej, Drjanovo 05024
Istoričeski Muzej, Gabrovo 05032
Istoričeski Muzej, Goce Delčev 05035
Istoričeski Muzej, Gorna Orjachovica 05036
Istoričeski Muzej, Ichtiman 05039
Istoričeski Muzej, Isperich 05040
Istoričeski Muzej, Karlovo 05049
Istoričeski Muzej, Loveč 05086
Istoričeski Muzej, Omurtag 05099
Istoričeski Muzej, Petrič 05113
Istoričeski Muzej, Pomorie 05133
Istoričeski Muzej, Popovo 05134
Istoričeski Muzej, Pravec 05137
Istoričeski Muzej, Provadia 05140
Istoričeski Muzej, Razgrad 05148
Istoričeski Muzej, Razlog 05149
Istoričeski Muzej, Svištov 05222
Istoričeski Muzej, Velingrad 05260
Istoričeski Muzej Batak, Batak 04981
Istoričeski Muzej Belogradčik, Belogradčik 04982
Istoričeski Muzej Pavlikeni, Pavlikeni 05104
Istoričeski Muzej - Preminavane na Ruskite Vojski pri Svištov, Svištov 05223
Istoričeski Muzej Silistra, Silistra 05163
Istoričeski Muzej Targovište, Targovište 05225
Istoričeskij Muzej, Tutrakan 05234
Istoričeskij Muzej Dom na Nabereznoj, Moskva 37218
Istoričnij Muzej, Dnipropetrovs'k 43033
Istorijski Muzej Crne Gore, Cetinje 31641
Istorijski Muzej Srbije, Beograd 38050
Istoriko-Architekturnyj, Chudožestvennyj i Archeologičeskij Muzej "Zarajskij Kreml", Zarajsk 37986
Istoriko-Architekturnyj i Chudožestvennyj Muzej Ivangorodskaja Krepost, Ivangorod 36942
Istoriko-architekturnyj Muzej Kuzneckaja Krepost', Novokuzneck 37479
Istoriko-chudožestvennyj Muzej, Uglič 37891
Istoriko-ėtnografičeskij Muzej Usadba Galskich, Čerepovec 36849
Istoriko-Kraevedčeskij Muzej, Ivanovo 36947
Istoriko-Kraevedčeskij Muzej, Merke 29710
Istoriko-Kraevedčeskij Muzej, Rudnyj 29725
Istoriko-Kraevedceskij Muzej, Tokmok 29966
Istoriko-kraevedčeskij Muzej g. Kronštadta, Kronštadt 37084
Istoriko-kulturnyj Centr Krasnye Palaty XVII Veka, Moskva 37219
Istoriko-Kulturnyj Muzej-zapovednik Azret Sultan, Turkestan 29734
Istoriko-kulturnyj Muzej-zapovednik Idnakar, Glazov 36925
Istoriko-literaturnyj Muzej I.A. Gončarova, Uljanovsk 37898
Istoriko-memorialnyj i Architekturno-chudožestvennyj Muzej, Tichvin 37842
Istoriko-memorialnyj Kompleks Gerojam Stalingradskoj Bitvy na Mamaevom Kurgane, Volgograd 37956
Istoriko-memorialnyj Muzej M.V. Lomonosova, Lomonosovo 37112
Istoriko-memorial'nyj Muzej Polityčnych V'jazniv, Ternopil' 43192
Istoriko-memorialnyj Muzej Semji Fadeevyh-Chan-Blavackij, Dnipropetrovs'k 43034
Istoriko-memorialnyj Muzejnyj Kompleks Bobriki, Donskij 36878
Istoriko-sportivnyj Muzej, Moskva 37220
Istoriski Muzej, Kruševo 30505
Istoryko-architekturna Pam'jatka-muzej Kyïvska Fortecja, Kyïv 43079
István Bocskai Memorial Exhibition - Bocskai Castle Museum, Nagykereki 23726
István Dobó Castle Museum, Eger 23536
István Fekete Memorial House, Gölle 23569
István Járdányi Paulovics Ruin Garden, Szombathely 23901
István Nagy Gallery, Baja 23346
István Sinka Memorial House - Region Historical Exhibition, Vészto 23964
István Széchenyi Memorial Museum, Nagycenk 23723
István Szőnyi Memorial Museum, Zebegény 23977
Itabashi Art Museum, Tokyo 29401
Itabashi Kuritsu Bijutsukan, Tokyo 29401
Itä-Hämeen Museo, Hartola 10760
Ital Reding-Haus, Schwyz 41851
Italian American Museum, New York 51198
Italian Federation of Associations of Friends of Museums, Firenze 54900
Italian Gold History Museum, Predosa 27342
Italian Museum of Cast Iron, Longiano 26518
Italienisches Kulturinstitut, Berlin 17144
Itami City Museum, Itami 28591
Itami City Museum of Art, Itami 28590
Itami-shiritsu Bijutsukan, Itami 28590
Itami-shiritsu Hakubutsukan, Itami 28591
Itasca Heritage Center Museum, Grand Rapids 49009
Itaya Hazan Kinenkan, Shimodate 29266
Itayanagi Folk Art Museum, Itayanagi 28592
Itayanagi-machi Mingeikan, Itayanagi 28592
İtfaiye Müzesi, İstanbul 42829
Itimuseum, Buenos Aires 00163
Itsas Begia, Ciboure 12528
Itsukaichi Folklore Museum, Akiruno 28352
Itsukushima-jinja Homotsukan, Hatsukaichi 28499
Itsukushima Shrine Treasury, Hatsukaichi 28499
Itsuo Bijutsukan, Ikeda 28562
Itsuoh Art Museum, Ikeda 28562
Itsutsubashikan, Sendai 29242
Ittinger Museum, Warth 42001
Ituna Cultural and Historical Museum, Ituna 06026
Iupui Cultural Arts Gallery, Indianapolis 49502
IVAM Centre Julio González, Valencia 40201
Ivan Dougherty Gallery, Paddington, New South Wales 01400
Ivan Franko Museum, Winnipeg 07364
Ivan Gonchar Museum, Kyïv 43091
Ivan Kotliarevsky Literary and Memorial Museum, Poltava 43169
Ivan Meštrović Gallery, Split 08957
Ivan Sabolić Sculpture Gallery, Peteranec 08928
Ivan Vazov Memorial House, Berkovica 04987
Ivan Vazov Memorial House, Sofia 05178
Ivan Vazov Museum, Sopot 05205
Ivane Javakhishvili Samtskhe-Javaxeti History Museum, Akhaltsikhe 16332
Ivano-Frankivs'kyj Chudožnij Muzej, Ivano-Frankivs'k 43041
Ivano-Frankivs'kyj Krajeznavčyj Muzej, Ivano-Frankivs'k 43042
Ivano-Frankovsk Fine Art Museum, Ivano-Frankivs'k 43041
Ivano-Frankovsk Local Museum, Ivano-Frankivs'k 43042
Ivanovo Art Museum, Ivanovo 36948
Ivanovskij Oblastnoj Chudožestvennyj Muzej, Ivanovo 36948
Ivar Aasen-tunet, Hovdebygda 33568
Ivar Lo-Museet, Stockholm 40901
Iveland og Vegusdal Bygdemuseum, Vatnestrom 33966
Ivy Green, Tuscumbia 53624
Iwaki City Art Museum, Iwaki 28606
Iwaki-shi Bijutsukan, Iwaki 28606
Iwakuni Art Museum and Kashiwara Collections, Iwakuni 28608
Iwakuni Historical Museum & Kikkawa Family Museum, Iwakuni 28607
Iwakuni-shi Bijutsukan no Kashiwara Shuu, Iwakuni 28608
Iwalewa Haus, Bayreuth 16953
Iwalewa House, Bayreuth 16953
Iwami Adachi Bijutsukan, Hamada 28482
Iwami Adachi Museum of Art, Hamada 28482
Iwami Art Museum, Masuda 28886
Iwami Bijutsukan, Masuda 28886
Iwasaki Bijutsukan, Ibusuki 28552
Iwasaki Castle Museum, Nisshin-city, Aichi Prefecture 29078
Iwasaki Kinen Kenbunkan, Nisshin-city, Aichi Prefecture 29078
Iwasaki Museum, Yokohama 29594
Iwasaki Museum of Art, Ibusuki 28552
Iwata Collectable Art Museum, Inuyama 28575
Iwata-gun Kaori Hakubutsukan, Toyoda 29526
Iwata Senshinkan, Inuyama 28575
Iwata-shi Kyu Mitsukegakko, Iwata, Shizuoka 28610
Iwate Handicraft Museum, Morioka 28936
Iwate-kenritsu Bijutsukan, Morioka 28934
Iwate-kenritsu Hakubutsukan, Morioka 28935
Iwate-kenritsu Meisho Hakubutsukan, Morioka 28936
Iwate Museum of Art, Morioka 28934
Iwate Prefectural Museum, Morioka 28935
Iwayama Urushi Art Museum, Morioka 28937
Iwayama Urushi Bijutsukan, Morioka 28937
Ixchel Museum of Indigenous Textiles and Clothing, Guatemala 23300
Iyoboya Salmon Museum, Murakami 28944
Izaak Walton's Cottage, Shallowford 45646
Izba Muzealna im. Antoniego Krajewskiego, Lanckorona 34841
Izba Pamięci Żydów Lublina, Lublin 34899
Iziko Museums of Cape Town, Cape Town 55017
Izložba na Stari Kotlenski Kilimi, Kotel 05076
Izložben Kompleks, Veliko Tărnovo 05253
Izložbena Zala Sirak Skitnik, Sliven 05165
Izmaïl's'ka Kartynna Galereja, Izmaïl 43046
Izmaïl's'kyj Krajeznavčyj Muzej Pridunav'ja, Izmaïl 43047
Izmaïl's'kyj Muzej O.V. Suvorova, Izmaïl 43048
Izmir Archaeological Museum, İzmir 42855
İzmir Arkeoloji Müzesi, İzmir 42855
İzmir Etnografya Müzesi, İzmir 42856
İznik Müzeleri, İznik 42860
Iznik Museums, İznik 42860
Izstāžu Zāle, Madona 30044
Izu Orugorukan, Ito 28594
Izukogen Kirie Bijutsukan, Ito 28595
Izumi-mura Rekishi no Sato, Izumi, Fukui 28612
Izumi Village History Park - Flute, Folklore and Local Museums, Izumi, Fukui 28612
Izumo Folklore Museum, Izumo 28616
Izumo Mingeikan, Izumo 28616
Izumo Taisha Homutsuden, Izumo 28617
Izumo Taisha Treasure House, Izumo 28617

J

J.A.V. David Museum, Killarney 06067
J.B. Wallis Museum of Entomology, Winnipeg 07365
J.E. Reeves Home and Museum, Dover 48167
J. Evetts Haley History Center, Midland 50670
J. F. Schreiber-Museum, Esslingen 18275
J.F. Willumsens Museum, Frederikssund 10028
J.G. Tajovský Memorial House, Tajov 38368
J.K. MacCarthy Museum, Goroka 34081
J.L. Runebergin Koti, Porvoo 11227
J.L. Runeberg's Home, Porvoo 11227
J.L. Shellshear Museum of Physical Anthropology and Comparative Anatomy, Sydney 01549
J. Łukasiewicz Regional Museum, Gorlice 34679
J.M. Davis Arms and Historical Museum, Claremore 47578
J.M.W. Turner House, London 44886
J. Paul Getty Museum, Los Angeles 50237
J. Stalin Staten Museum, Gori 16340
J.T. Wilson Museum of Human Anatomy, Sydney 01550
J.U. Polimski Muzej, Berane 31637
J.V. Sládek Museum, Zbiroh 09929
J.V. Snellman Home Museum, Kuopio 11006
J.V. Snellmanin Kotimuseo, Kuopio 11006
J.W. Parmely Historical Home Museum, Ipswich 49525
J. Wayne Stark University Center Galleries, College Station 47689
Jabatan Muzium Sabah, Kota Kinabalu 30565
Jabonga Parish Museum, Jabonga 34391
Jabotinsky Museum, Tel Aviv 25110
Jacek Malczewski Museum, Radom 35073
Jachthuis Sint-Hubertus, Hoenderloo 32397
Jacinto Courthouse, Corinth 47807
Jack Heath Art Gallery, Pietermaritzburg 38704
Jack Jackson Memorial Museum, Weaverville 54017
Jack Lynn Memorial Museum, Horsefly 05997
Jack Miner Museum, Kingsville 06093
Jack S. Blanton Museum of Art, Austin 46511
Jackfield Tile Museum, Jackfield 44544
Jackson Art Center, Washington 53884
Jackson County Historical Museum, Lakefield 49899
Jackson County Historical Museum, Maquoketa 50457
Jackson County Historical Society Museum, Black River Falls 46841
Jackson Hall Gallery, Frankfort 48736
Jackson Hole Historical Museum, Jackson 49571
Jackson-Washabaugh Historical Museum, Kadoka 49661
Jackson's Mill Historic Area, Weston 54109
Jacksonville Fire Museum, Jacksonville 49579
Jacksonville Museum of Modern Art, Jacksonville 49580
Jacob Breda Bullmuseet, Rendalen 33778
Jacob Kelley House Museum, Hartsville 49204
Jacob Walden House, Walden 53793
Jacqueline Casey Hudgens Center for the Arts, Duluth 48191
Jacques Marchais Museum of Tibetan Art, Staten Island 53264
Jacques van Mourik Museum, Mook 32561
Jägerndorfer Heimatstuben, Ansbach 16559
Jægerspris Slot, Jægerspris 10109
Jael Stichting Mystic Museum, Dronten 32158
Jämtlands Flyg och Lottamuseum, Östersund 40795
Jämtlands Läns Museum, Östersund 40796
Jäneda Muuseum, Jäneda 10575
Jærmuseet, Nærbø 33690
Järnvägs- och Industrimuseum, Hagfors 40575
Järnvägsmuseet, Kristianstad 40654
Järva-Jaani Fire-fighters Asociation Museum, Järva-Jaani 10576
Järva-Jaani Tultõrjeseltsi Muuseum, Järva-Jaani 10576
Järvamaa Muuseum, Paide 10605
Jafar Jabarly State Museum of Theatre, Baki 03158
Jaffna National Museum, Jaffna 40379
Jaffrey Civic Center, Jaffrey 49588
Jagd- und Falknereimuseum, Riedenburg 21316
Jagd- und Fischereimuseum, Adelsdorf 16427
Jagd- und Fischereimuseum Schloss Tambach, Weitramsdorf 22468
Jagd- und Fischereimuseum Schloß Wolfstein und Galerie Wolfstein, Freyung 18461
Jagd- und Naturkunde Museum, Niederstetten 20711
Jagd und Naturkundemuseum, Hahn am See 18864
Jagd- und Schloßmuseum Spangenberg, Spangenberg 21830
Jagdhaus Gabelbach, Ilmenau 19321
Jagdmuseum Gußwerk, Gußwerk 02021
Jagdmuseum Pfahlheim, Ellwangen 18123
Jagdmuseum Schloss Stainz, Stainz 02759
Jagdmuseum Wulff, Dedelstorf 17804
Jagdschau im Jagdschloss Springe, Springe 21845
Jagdschloß, Gelbensande 18565
Jagdschloß Granitz, Binz, Ostseebad 17357
Jagdschloss Grunewald, Berlin 17145
Jagdschloß Paulinzella, Rottenbach 21419
Jagdschloss Stern, Potsdam 21091
Jagdschoss Letzlingen, Letzlingen 19966
Jagiellonian University Museum, Kraków 34813
Jaguar Daimler Heritage Trust, Coventry 43903
Jail Museum, Horsens 10095
Jail Training School Museum, Lucknow 24296
Jailhouse Galleries, Morganton 50843
Jakarta History Museum & Museum Batavia, Jakarta 24483
Jakob-Grünenwald-Gedächtnisstätte, Ebersbach an der Fils 18041
Jakober Foundation, Alcúdia (Baleares) 38831
Jakobstads Museum, Jakobstad 10865
Jakopič Gallery of the Architecture Museum of Ljubljana, Ljubljana 38449
Jakopičeva Galerija, Ljubljana 38449
Jákovali Hasszán Múzeum, Pécs 23761
Jaktvårdsmuseum, Eksjö 40495
Jakutskij Gosudarstvennyj Muzej Istorii i Kultury Narodov Severa im. Jaroslavskogo, Jakutsk 36962
Jalasjärven Museo, Jalasjarvi 10869
Jalkaväkimuseo, Mikkeli 11113
Jalokivi Galleria, Kemi 10938
Jamahiriya Museum, Tripoli 30189
Jamaica Center for Arts, Jamaica 49590
Jamaica Folk Museum, Kingston 28329
Jamaican People's Museum of Craft and Technology, Spanish Town 28341
Jamalo-Neneckij Okružnoj Kraevedčeskij Muzej im. I.S. Šemanovskogo, Salechard 37608
James A. Garfield National Historic Site, Mentor 50592
James A. McFaul Environmental Center of Bergen County, Wyckoff 54313
James A. Michener Art Museum, Doylestown 48179
James and Meryl Hearst Center for the Arts, Cedar Falls 47293
James Bond 007 Museum, Nybro 40780
James Buchanan Foundation for the Preservation of Wheatland, Lancaster 49925
James Cook Museum, Cooktown 00998
James Dunklin House, Laurens 49987
James E. Lewis Museum of Art, Baltimore 46570
James Ford Bell Museum of Natural History, Minneapolis 50735
James Hall Museum of Transport, Johannesburg 38657
James Hogg Exhibition, Selkirk 45636
James House Museum, Bridgetown 05473
James Howe Gallery, Union 53648
James J. Hill House, Saint Paul 52618
James J. O'Mara Pharmacy Museum, Saint John's 06848
James Joyce Centre, Dublin 24803
James Joyce Museum, Dublin 24804
James K. Polk Ancestral Home, Columbia 47735
James K. Polk Memorial, Pineville 51926
The James Madison Museum, Orange 51580
James Madison's Montpelier, Montpelier Station 50827
James Mitchell Museum, Galway 24836
James Mitchell Varnum House and Museum, East Greenwich 48240
James Monroe Museum and Memorial Library, Fredericksburg 48768
James Paterson Museum, Moniaive 45166
James R. Slater Museum of Natural History, Tacoma 53409
James W. Dillon House Museum, Dillon 48144
The James Watrous Gallery, Madison 50391
James Whitcomb Riley Birthplace and Museum, Greenfield 49054
James Whitcomb Riley Museum, Indianapolis 49503
Jamestown Museum, Jamestown 01177
Jamestown Museum, Jamestown 49599
Jamestown Settlement, Williamsburg 54169
Jamestown Visitor Center & Museum, Jamestown 49601
Jamieson Museum, Moosomin 06389
Jamtli, Östersund 40797
Jan Amos Comenius Pedagogical Museum, Praha 09788
Jan Becher Muzeum, Karlovy Vary 09574
Jan Bouman Haus, Potsdam 21092
Jan Dzierżon Museum, Kluczbork 34748

Jan Holly's Birth House, Borský Mikuláš 38202
Jan Kasprowicz Birthplace, Inowrocław 34690
Jan Kochanowski Muzeum, Czarnolas 34611
Jan Kollar Museum, Mošovce 38308
Jan Koniarek Gallery in Trnava - Koppel Manor, Trnava 38379
Jan Koniarek Gallery - Synagogue - Centre of Contemporary Art, Trnava 38380
Jan Matejko House of the National Museum in Cracow, Kraków 34769
Jan Rosental Museum of History and Art, Saldus 30123
Jan Vissermuseum, Helmond 32365
Jāņa Akuratera Muzejs, Rīga 30070
Jāņa Jaunsudrabiņa Muzejs Riekstiņi, Nereta 30050
Jaņa Rozentāla Saldus Vēstures un Mākslas Muzejs, Saldus 30123
Jaņa Rozentāla un Rūdolfa Blaumaņa Muzejs, Rīga 30071
Jāņa Zabera Memoriālais Muzejs, Meirani 30048
Janaček Museum, Brno 09438
Janco-Dada Museum, Ein Hod 24946
Jandovo Muzeum - Vodní hrad, Budyně nad Ohří 09448
Jane Addams' Hull-House Museum, Chicago 47483
Jane Austen's House, Chawton 43799
Jane L. and Robert H. Weiner Judaic Museum, Rockville 52429
Jane Neville-Rolfe Gallery, Alpha 00805
Jane Voorhees Zimmerli Art Museum, New Brunswick 51033
Jane Welsh Carlyle's House, Haddington 44375
Janet Turner Print Museum, Chico 47527
Janfusun Museum, Gukeng 42198
Janggigot Lighthouse Museum, Pohang 29858
Janina Monkute-Marks Museum-Gallery, Kėdainiai 30270
Janinos Monkutės-Marks Muziejus-Galerija, Kėdainiai 30270
Jāņis Akuraters Museum, Rīga 30070
Janis Jaunsudrabins Museum Riekstini, Nereta 30050
Janis Rozentals and Rudolfs Blaumanis Museum, Rīga 30071
Janko Kral Museum, Liptovský Mikuláš 38287
Jannis Spyropoulos Museum, Ekali 22990
János Arany Literary Memorial House, Geszt 23564
János Fridrich Photographer's Museum, Szentes 23883
János Kmetty Memorial Museum - Kerényi Collection, Szentendre 23877
János Szántó Kovács János Regional Museum, Orosháza 23740
János Thorma Museum - House of Collections, Kiskunhalas 23669
János Thorma Museum - Végh Cottage, Kiskunhalas 23672
Jantyik Mátyás Museum - Békés Region House, Békés 23366
Jantyik Mátyás Múzeum, Békés 23364
Jantyik Mátyás Múzeum - Békési Galéria, Békés 23365
Jantyik Mátyás Múzeum - Békési Tájház, Békés 23366
Janus Pannonius Museum - Amerigo Tot Museum, Pécs 23757
Janus Pannonius Museum - Archeological Department and Permanent Archeological Exhibition, Pécs 23772
Janus Pannonius Museum - Csontváry Museum, Pécs 23758
Janus Pannonius Museum - Early Christian Mausoleum, Pécs 23770
Janus Pannonius Museum - Endre Nemes Museum, Pécs 23759
Janus Pannonius Museum - Erzsébet Schaár Exhibition Utca, Pécs 23775
Janus Pannonius Museum - Ethnographic Department, Pécs 23769
Janus Pannonius Museum - Martyn Ferenc Museum, Pécs 23763
Janus Pannonius Museum - Modern Hungarian Gallery I, Pécs 23767
Janus Pannonius Museum - Modern Hungarian Gallery II, Pécs 23768
Janus Pannonius Museum - Museum of Town History, Pécs 23779
Janus Pannonius Museum - Natural History Department, Pécs 23778
Janus Pannonius Museum - Renaissance-Lapidarium, Pécs 23773
Janus Pannonius Museum - Victor Vasarely Museum, Pécs 23780
Janus Pannonius Museum - Zsolnay Ceramic and Porcelain Exhibition, Pécs 23781
Janus Pannonius Múzeum, Pécs 23762
Japan Amateur Art Museum, Ogi 29090
Japan Calligraphy Art Museum, Tokyo 29451
Japan Clay Doll Museum, Nakano 29020
Japan Folk Art Museum, Osaka 29139
Japan Folk Crafts Museum, Tokyo 29450
The Japan Folk Crafts Museum, Osaka, Suita 29296
Japan Football Mueum, Tokyo 29402
Japan Footwear Museum & Japan Folk Toy and Doll Museum, Fukuyama 28447
Japan Kokeshi Museum, Naruko 29042
Japan-Mongolia Folk Museum, Tantou 29351
Japan Museum of Contemporary Toy, Mimasaka 28909
Japan Society Gallery, New York 51199
Japan Toy Museum, Kodera 28753
Japan Ukiyo-e Museum, Matsumoto 28896
Japane Open Air Folk House Museum, Kawasaki 28694
Japanese American National Museum, Los Angeles 50238
Japanese Association of Museums, Tokyo 54910
The Japanese Canadian Museum, Kaslo 06045
Japanese Canadian National Museum, Burnaby 05503
Japanese Coiffure Museum, Kyoto 28855
Japanese Council of Art Museums, Kanazawa 54908
Japanese Cultural Center of Hawaii, Honolulu 49345
Japanese Demon Museum, Oe 29089
Japanese Earthen Bells Museum, Shirotori 29282
Japanese National Committee for ICOM, Tokyo 54909
Japanese Overseas Migration Museum, Yokohama 29592
Japanese Rural Toy Museum, Kurashiki 28788
Japanese Ski Memorial Museum, Joetsu 28619
Japanese Sword Museum, Tokyo 29487
Japanisches Kulturinstitut, Köln 19609
Jarash Archaeological Museum, Jarash 29664
Járdányi Paulovics István Romkert, Szombathely 23901
Jardim-Museu Agrícola Tropical, Lisboa 35668
Le Jardin aux Papillons, Vannes (Morbihan) 16066
Jardin de Sculptures et d'Agrumes du Palais, Menton 14080
Le Jardin des Papillons, Paris 14613
Le Jardin Ferroviaire, Chatte 12477
Jardin Géologique d'Obourg, Obourg 03919
Jardins del Princep, Tortosa 40163
Jaroslavl Fine Art Museum - Governor's House, Jaroslavl 36967
Jaroslavl Historical and Architectural Museum-Preserve, Jaroslavl 36968
Jaroslavl History Museum, Jaroslavl 36972
Jaroslavl Museum of Combat Fame, Jaroslavl 36969
Jaroslavskij Chudožestvennyj Muzej - Gubernatorskij Dom, Jaroslavl 36967
Jaroslavskij Istoriko-architekturnyj i Chudožestvennyj Muzej-zapovednik, Jaroslavl 36968
Jaroslavskij Muzej Boevoj Slavy, Jaroslavl 36969
Jarrell Plantation Georgia, Juliette 49653
Jartunet, Åsgreina 33404
Jasin Museum, Jasin 30557
Jasper Cultural and Historical Centre, Maple Creek 06233
Jasper Rand Art Museum, Westfield 54092
Jasper-Yellowhead Museum and Archives, Jasper 06027
Jasper's Antique Radio Museum, Saint Louis 52581
Jász Múzeum, Jászberény 23623
Jászai Mari Emlékház, Ászár 23337
Jászai Mari Emlékszoba, Rábatamási 23788
Jauering Museum of Local History, Vaihingen an der Enz 22197
Jaunmokas Palace Museum, Tukums 30131
Jaunmoku Pils Muzejs, Tukums 30131
Jaunpils Museum, Jaunpils 30012
Jaunpils Muzejs, Jaunpils 30012
Javaherat, Tehrān 24675
Jawaharlal Nehru State Museum, Itanagar 24231
Jay I. Kislak Foundation, Miami Lakes 50645
Sri Jayachamarajendra Art Gallery, Mysore 24329
Jazep Vitol Memorial Museum, Gaujiena 30009
Jāzepa Vītola Memoriālais Muzejs, Gaujiena 30009
Jazz Museum, Kolbäck 40650
Jazzens Museum, Kolbäck 40650
Jazzothèque, Nice 14426
JCII Camera Museum, Tokyo 29447
Jean Giraudoux Birthplace, Bellac 11925
Jean Lafitte National Historical Park and Preserve, New Orleans 51084
Jean P. Haydon Museum, Pago Pago 00084
Jean-Paul-Museum der Stadt Bayreuth, Bayreuth 16954
Jean Paul Slusser and Warren Robbins Galleries, Ann Arbor 46306
Jean-Paul-Zimmer in der Rollwenzelei, Bayreuth 16955
Jeane Dixon Museum and Library, Strasburg 53331
Jedburgh Castle Jail Museum, Jedburgh 44547
Jedlik Ányos Emlékkiállítás, Győr 23583
Jeff Matthews Memorial Museum, Galax 48825
Jefferds Tavern, York 54334
Jeffers Petroglyphs Historic Site, Comfrey 47762
Jefferson Barracks, Saint Louis 52582
Jefferson County Historical Association Museum, Steubenville 53288
Jefferson County Historical Museum, Madison 50379
Jefferson County Historical Museum, Pine Bluff 51922
Jefferson County Historical Society Museum, Port Townsend 52043
Jefferson County Historical Society Museum, Watertown 53972
Jefferson Historical Museum, Jefferson 49609
Jefferson Landing State Historic Site, Jefferson City 49613
Jefferson National Expansion Memorial, Saint Louis 52583
Jefferson Patterson Museum, Saint Leonard 52566
Jefferson Telephone Museum, Jefferson 49606
Jehangir Art Gallery, Mumbai 24318
Jeju-do Folklore and Natural History Museum, Jeju 29836
Jeju Education Museum, Jeju 29837
Jeju Folklore Museum, Jeju 29838
Jeju National University Museum, Jeju 29839
Jēkaba Prīaņa Anatomijas Muzejs, Rīga 30072
Jēkabpils Vēstures Muzeja Brīvdabas Nodaļa, Jēkabpils 30013
Jēkabpils Vēstures Muzejs, Jēkabpils 30014
Jekabs Primanis Anatomy Museum, Rīga 30072
Jekyll Island Museum, Jekyll Island 49617
Jelly van den Bosch Museum, Roden 32694
Jemez State Monument, Jemez Springs 49618
Jenbacher Museum, Jenbach 02141
Jeneverhuis Van Hoorebeke, Eeklo 03576
Jenisch Haus, Hamburg 18938
Jenko Barracks Museum, Zgornje Jezersko 38542
Jenkova Kasarna, Zgornje Jezersko 38542
Jenle Museum, Roslev 10251
Jenner Museum, Berkeley 43454
Jennings-Brown House Female Academy, Bennettsville 46737
Jenő Ádám Memorial House, Szigetszentmiklós 23889
Jenő Balás Bauxite Mining Museum, Gánt 23563
Jens Nielsens & Olivia Holm-Møller Museet, Holstebro 10092
Jens Søndergaard Museet, Lemvig 10175
Jensen Arctic Museum, Monmouth 50785
Jeonju Municipal Museum, Jeonju 29841
Jeonju National Museum, Jeonju 29842
Jeoufen Art Gallery, Rueifang 42307
Jeremiah Lee Mansion, Marblehead 50460
Jericho Historical Society, Jericho Corners 49622
Jerilderie Doll World, Jerilderie 01179
Jermuk Art Gallery, Jermuk 00711
Jerome County Historical Society, Jerome 49623
Jerome K. Jerome Birthplace Museum, Walsall 45919
Jersey Battle of Flowers Museum, Saint Ouen 45588
Jersey City Museum, Jersey City 49627
Jersey Heritage Trust, Saint Dominick, Jersey 45564
Jersey Museum, Saint Helier 45571
Jersey Photographic Museum, Saint Helier 45572
Jersey War Tunnels, Saint Lawrence 45581
Jerusalemhaus mit Verwaltung der Städtischen Sammlungen Wetzlar, Wetzlar 22527
Jerzy Dunin-Borkowski Museum, Krośniewice 34828
Jesse Besser Museum, Alpena 46237
Jesse James Bank Museum, Liberty 50079
Jesse James Farm and Museum, Kearney 49713
Jesse James Home Museum, Saint Joseph 52562
Jesse Peter Museum, Santa Rosa 52905
Jeu de Paume, Paris 14614
Jeu de Paume, Paris 14615
Jevrejski Istorijski Muzej, Beograd 38051
Jewel Tower, London 44887
Jewell County Historical Museum, Mankato 50445
Jewelry Fond at the Moscow Kremlin, Moskva 37140
Jewett Hall Gallery, Augusta 46484
Jewish Battalions Museum, Ahivil 24922
Jewish Culture Museum Augsburg-Swabia, Augsburg 16624
Jewish Historical Museum, Amsterdam 31850
Jewish Historical Museum, Beograd 38051
Jewish Holocaust Centre, Elsternwick 01067
Jewish Institute for the Arts, Boca Raton 46886
Jewish Legions Museum WW I, Avihail 24932
Jewish Military Museum, London 44888
Jewish Museum, London 44889
Jewish Museum, Merano 26659
The Jewish Museum, New York 51200
Jewish Museum, Prešov 38338
Jewish Museum, Sarajevo 04278
Jewish Museum Berlin, Berlin 17147
Jewish Museum Finchley, London 44890
Jewish Museum in Prague, Praha 09801
Jewish Museum in Stockholm, Stockholm 40902
Jewish Museum Munich, München 20448
Jewish Museum of Australia, Saint Kilda, Victoria 01479
Jewish Museum of Belgium, Bruxelles 03490
Jewish Museum of Buenos Aires Dr. Salvador Kibrick, Buenos Aires 00249
Jewish Museum of Deportation and Resistance, Mechelen 03858
Jewish Museum of Florida, Miami Beach 50642
The Jewish Museum of Franconia in Fürth, Fürth 18498
Jewish Museum of Greece, Athinai 22910
Jewish Museum of Maryland, Baltimore 46571
Jewish Museum of Thessaloniki, Thessaloniki 23214
Jewish National Fund House, Tel Aviv 25111
Jewry Wall Museum, Leicester 44676
Jeypore Branch Museum, Jaipur, Orissa 24232
JFK Special Warfare Museum, Fort Bragg 48599
Jhalawar Archaeology Museum, Jhalawar 24243
Jheng Cheng-Gong Memorial Hall, Sikou 42330
Jhenlangong Mazu Exhibition Hall, Dajia 42173
Jiading District Museum, Shanghai 08168
Jian City Museum, Jian, Jilin 07854
Jian Museum, Jian 42227
Jian Museum, Jian, Jiangxi 07853
Jiangdu Museum, Jiangdu 07857
Jiangmen Museum, Jiangmen 07860
Jiangnan Gongyuan History Display Center, Nanjing 08040
Jiangning Museum, Jiangning 07861
Jiangshan Museum, Jiangshan 07862
Jiangxi Provincial Museum, Nanchang 08028
Jiangxian Museum, Jiangxian 07863
Jiangyin Museum, Jiangyin 07864
Jian'ou Museum, Jian'ou 07866
Jianyang Museum, Jianyang 07867
Jiaoling Museum, Jiaoling 07868
Jiaonan Museum, Jiaonan 07869
Jiaoshan Beike Stone Museum, Zhenjiang 08501
Jiaosi Museum, Jiaosi 42228
Jiaozhou Museum, Jiaozhou 07870
Jiaozhuang Hu Didao Station, Beijing 07542
Jiaozuo Museum, Jiaozuo 07871
Jiashan Museum, Jiashan 07872
Jiaxing Museum, Jiaxing 07873
Jiayuguan Changcity Museum, Jiayuguan 07874
Jibei Cultural Relics Exhibition Hall, Baisha 42141
Jierjia Sweet Potato Museum, Taichung 42349
Jiexi Museum, Jiexi 07875
Jiexiu Museum, Jiexiu 07876
Jieyang Museum, Jieyang 07877
Jigsaw Gallery and Bridgetown Heritage Museum, Bridgetown 00894
Jihočeské Motocyklové Muzeum, České Budějovice 09463
Jihočeské Muzeum v Českých Budějovicích, České Budějovice 09464
Jihomoravské Muzeum ve Znojmě, Znojmo 09947
Jiji Railway Museum, Jiji 42230
Jikjiseongbo Museum, Gimcheon 29813
Jilin City Museum, Jilin 07878
Jilin Provincial Museum, Changchun 07585
Jilin Provincial Museum of Natural History, Changchun 07586
Jilin Provincial Museum of Revolution, Changchun 07587
Jilin University Museum, Changchun 07588
Jilin Wenmiao Museum, Jilin 07879
Jim Clark Room, Duns 44065
Jim Crow Museum, Big Rapids 46797
Jim Gatchell Memorial Museum, Buffalo 47127
Jim Savage Art Gallery and Museum, Sioux Falls 53120
The Jim Thompson House, Bangkok 42543
Jim Thorpe Home, Yale 54321
Jimmie Rodgers Museum, Meridian 50599
The Jimmy Stewart Museum, Indiana 49483
Jimo Museum, Jimo 07880
Jin Ancestral Hall Museum, Taiyuan 08272
Jinan Museum, Jinan 07881
Jinan Revolutionary Mausoleum of Fallen Heroes, Jinan 07882
Jincheng-gu Construction Art Museum, Jincheng 07887
Jingchuan Museum, Jingchuan 07888
Jingdezhen Ceramic History Museum, Jingdezhen 07889
Jinggangshan Revolution Museum, Jinggangshan 07890
Jingmen Museum, Jingmen 07891
Jingsha Art Museum, Jingsha 07892
Jingsha Jingzhou Museum, Jingsha 07893
Jingtong Mining Museum, Pingsi 42288
Jingu Agricultural Museum, Ise 28582
Jingu Art Museum, Ise 28580
Jingu Bijutsukan, Ise 28580
Jingu Hakubutsukan, Ise 28581
Jingu History Museum, Ise 28581
Jingu Nogyokan, Ise 28582
Jingxi Chuang Museum, Jingxi 07894
Jingyang Museum, Jingyang 07895
Jining Museum, Jining 07896

Jinišk9is Museum of History and Culture, Joniškis 30231
Jinjiang Museum, Jinjiang 07897
Jinju National Museum, Jinju 29844
Jinshan Museum, Shanghai 08169
Jintai Art Museum, Beijing 07543
Jinze Xuan Art Center, Shenzhen 08221
Jinzhou District Museum, Dalian 07640
Jinzhou Museum, Jinzhou 07899
Jioufen Kite Museum, Rueifang 42308
Jiro Osaragi Kinenkan, Yokohama 29595
Jiro Osaragi Memorial Museum, Yokohama 29595
Jishan Museum, Jishan 07900
Jishui Museum, Jishui 07902
Jitensha Bunka, Tokyo 29403
Jitensha Hakubutsukan Saikuru Senta, Sakai 29196
Jiujiang Museum, Jiujiang 07903
Jiuquan Museum, Jiuquan 07904
Jixian Display Center, Jixian 07905
Jiyang Museum, Jiyang 07906
Jiyuan Museum, Jiyuan 07907
Jizdárna Pražského Hradu, Praha 09753
Jizzax Viloyat Ülkashunoslik Muzeyi, Jizzax 54448
Joachim Ringelnatz Museum, Cuxhaven 17755
Jødisk Museum, Trondheim 33924
Jodrell Bank Visitor Centre, Macclesfield 45064
Joe and Emily Lowe Art Gallery, Syracuse 53399
Joel and Lila Harnett Museum of Art, Richmond 52329
Joel and Lila Harnett Print Study Center, Richmond 52330
Jönköping County Museum, Jönköping 40614
Jönköpings Läns Museum, Jönköping 40614
Joensuu Art Museum, Joensuu 10872
Joensuun Taidemuseo, Joensuu 10872
Jörg-Metzler-Stube, Ravenstein 21206
Joetsu Municipal Museum, Joetsu 28620
Joetsu-shiritsu Hakubutsukan, Joetsu 28620
Jõevere Farmstead Museum, Lindora 10592
Jõevere Talumuuseum, Lindora 10592
Johan Nygaardsvold Museum, Hommelvik 33557
Johann-Adams-Mühle, Tholey 22054
Johann-Baptist-Graser-Schulmuseum, Bayreuth 16956
Johann Friedrich Böttger und die Geschichte der Porzellanmanufaktur Meissen, Schleiz 21563
Johann-Friedrich-Danneil-Museum, Salzwedel 21480
Johann Jacobs Museum, Zürich 42078
Johann Köleri Muuseum, Vastemõisa 10668
Johann Michael Haydn Museum, Salzburg 02616
Johann Michael Haydn-Museum, Salzburg 02616
Johann-Peter-Frank Museum, Rodalben 21343
Johann Strauß Wohnung, Wien 03007
Johann Ulrich Steiger-Freilichtmuseum, Bad Zurzach 41142
Johann Wülfing & Sohn Museum, Radevormwald 21165
Johann-Wunter-Museum Andernach, Andernach 16544
Johanna Museet, Skurup 40855
Johanna-Spyri-Museum im Alten Schulhaus, Hirzel 41484
Johannes a Lasco Bibliothek, Emden 18137
Johannes Brahms-Museum, Mürzzuschlag 02412
Johannes Jørgensens Mindestuer, Svendborg 10289
Johannes Larsen Museet, Kerteminde 10116
Johannes Linnankosken Nuoruudenkoti, Askola 10716
Johannes Linnankoski's Childhood Home, Askola 10716
Johannes Lohilammen Museo, Sammatti 11297
Johannes Stegmann Art Gallery, Bloemfontein 38563
Johannesburg Art Gallery, Johannesburg 38658
Johanniter-Museum, Krautheim 19702
Johanniter- und Maltesermuseum, Heitersheim 19114
Johanniterhalle, Schwäbisch Hall 21656
John A. Hermann jr. Memorial Art Museum, Bellevue 46705
John A. Logan College Museum, Carterville 47266
John Abbott II House, Hamilton 49135
John Adams and John Quincy Adams Birthplaces, Quincy 52218
The John and Mable Ringling Museum of Art, Sarasota 52914
John B. Davis Gallery of Fine Art, Pocatello 52007
John Brown Farm, Lake Placid 49890
John Brown House, Osawatomie 51609
John Brown House, Providence 52168
John Brown Wax Museum, Harpers Ferry 49175
John Buchan Centre, Biggar 43483
John Bunyan Museum, Bedford 43437
John Burroughs Memorial, Roxbury 52476
John Creasey Museum, Salisbury 45603
John Curtin Gallery, Bentley 00862
John Day Fossil Beds National Monument, Kimberly 49776
John Deere House, Dixon 48145
John Dony Field Centre, Bushmead 43700
John E. and Walter D. Webb Museum of Vintage Fashion, Island Falls 49539
John E. Conner Museum, Kingsville 49796
John E. Weatherhead Gallery, Fort Wayne 48706
John Elliott Classics Museum, Sandy Bay 01494
John F. Kennedy Presidential Library-Museum, Boston 46935
John Fitzgerald Kennedy House, Brookline 47063
John G. Diefenbaker Replica Law Office, Wakaw 07273
John G. Johnson Collection, Philadelphia 51845
John G. Neihardt Center, Bancroft 46592
John G. Voigt House, Plymouth 52003
John Gorrie Museum, Apalachicola 46327
John Greenleaf Whittier Home, Amesbury 46261
John H. Vanderpoel Art Gallery, Chicago 47484
John Hancock Warehouse, York 54335
John Hansard Gallery, Southampton 45704
John Harris/Simon Cameron Mansion, Harrisburg 49180
John Hastie Museum, Strathaven 45787
John Humphrey House, Swampscott 53388
John Hutchings Museum of Natural History, Lehi 50021
John J. Harvey Fireboat Collection, New York 51201
John J. McDonough Museum of Art, Youngstown 54354
John James Audubon Museum, Henderson 49247
John Jarrold Printing Museum, Norwich 45280
John Jay French House, Beaumont 46677
John Jay Homestead, Katonah 49708
John King Workshop Museum, Pinxton 45383
John Knox's House, Edinburgh 44129
John Lennon Museum, Saitama 29187
John McDouall Stuart Museum, Kirkcaldy 44608
John Mariani Art Gallery, Greeley 49033
John Marshall House, Richmond 52331
John Michael Kohler Arts Center, Sheboygan 53061
John Moore Countryside Museum, Tewkesbury 45842
John Muir's Birthplace, Dunbar 44042
John N. Leamon Museum, Brigus 05480
John Nicholas Brown Center for the Study of American Civilization, Providence 52169
John P. Barclay Memorial Gallery, Pittsburgh 51951
John P. Fisher Memorial Museum, Kingston 06079
John P. McGovern Museum of Health and Medical Science, Houston 49392
John Paul Jones Birthplace Museum, Arbigland 43309
John Paul Jones House Museum, Portsmouth 52095
The John Q. Adams Center for the History of Otolaryngology - Head and Neck Surgery, Alexandria 46209
John R. Jackson House, Winlock 54233
John R. Park Homestead, Harrow 05975
John Ralston Museum, Mendham 50585
John Rivers Communications Museum, Charleston 47368
John Rylands University Library-Special Collections, Manchester 45087
John S. Barry Historical Society Museum, Constantine 47781
John Slade Ely House, New Haven 51051
John Smart House, Medina 50557
The John Southern Gallery, Dobwalls 43988
John Stark Edwards House, Warren 53824
John Strong Mansion, Addison 46137
John Sydnor's 1847 Powhatan House, Galveston 48847
John Taylor Bellfoundry Museum, Loughborough 45033
John-Wagener-Haus Sievern, Langen, Kreis Cuxhaven 19815
John Walter Museum, Edmonton 05747
John Weaver Sculpture Museum, Hope 05995
The John Wesley House and The Museum of Methodism, London 44891
John Wesley Powell Memorial Museum, Page 51673
John Woolman Memorial, Mount Holly 50879
John Wornall House Museum, Independence 49477
The Johns Hopkins University Archaeological Collection, Baltimore 46572
Johnson Atelier, Mercerville 50597
Johnson Collection of Photographs, Movies and Memorabilia, Chanute 47341
Johnson County Historical Society Museum, Coralville 47804
Johnson County Historical Society Museum, Tecumseh 53459
Johnson County Historical Society Museum, Warrensburg 53830
Johnson County History Museum, Franklin 48746
Johnson County Museums, Shawnee 53057
Johnson Ferry House Museum, Titusville 53515
Johnson Hall, Johnstown 49638
Johnson Heritage Post Art Gallery, Grand Marais 48998
Johnson-Humrickhouse Museum, Coshocton 47838
Johnstown Area Heritage Association, Johnstown 49641
Johnstown Historical Society Museum, Johnstown 49639
Jókai Mór Emlékmúzeum, Balatonfüred 23352
Jókai Mór Emlékszoba és Tájház, Tardona 23915
Jokioinen Parsonage Museum, Jokioinen 10876
Jokioinen Railway Museum and Narrow Gauge Museum, Jokioinen 10874
Jokioisten Museorautatie ja Kapearaidemuseo, Jokioinen 10874
Jokioisten Naulamuseo, Jokioinen 10875
Jokioisten Pappilamuseo, Jokioinen 10876
Jokkmokks Stencenter, Jokkmokk 40619
Joliet Area Historical Museum, Joliet 49643
Jølstramuseet og Kunstgalleriet, Vassenden 33965
Jomon Bijutsukan, Nakaniida 29018
Jomon Clay Museum, Kizukuri 28722
Jomon-Kan, Kizukuri 28722
Jonas Aistis Museum, Rumšiškės 30348
Jonathan Hager House and Museum, Hagerstown 49120
Jonava Museum, Jonava 30230
Jonavos Krašto Muziejus, Jonava 30230
The Jones Museum of Glass and Ceramics, Sebago 53018
Jonesborough-Washington County History Museum, Jonesborough 49648
Joniškio Istorijos ir Kultūros Muziejus, Joniškis 30231
Jono Biliūno Memorialinis Muziejus, Anykščiai 30210
Jonson Gallery, Albuquerque 46180
Joods Historisch Museum, Amsterdam 31850
Joods Museum van Deportatie en Verzet, Mechelen 03858
Joodse Schooltje, Leek 32466
Joongang University Museum, Seoul 29887
Jopie Huisman Museum, Workum 32964
Jordan Archaeological Museum, Amman 29652
Jordan Folklore Museum, Amman 29653
Jordan Historical Museum of the Twenty, Jordan 06033
Jordan Museum of Popular Tradition, Amman 29654
Jordan National Gallery of Fine Arts, Amman 29655
Jordan Natural History Museum, Irbid 29663
Jordan Schnitzer Museum of Art at the University of Oregon, Eugene 48431
Jorge B. Vargas Museum and Filipiana Research Center, Quezon City 34492
Jorge Barlin National Monument, Baao 34335
Joroinen Local Arts and Crafts Museum, Joroinen 10877
Joroisten Kotiseutumuseo, Joroinen 10877
Jorvik Museum, York 46103
Jos National Museum and Museum of Traditional Nigerian Architecture, Jos 33330
Jósa András Múzeum, Nyíregyháza 23733
José Drudis-Biada Art Gallery, Los Angeles 50239
Jose P. Laurel Memorial Museum, Manila 34440
José P. Laurel Monument Museum, Santo Tomas 34502
José Rizal Shrine Museum, Calamba 34359
José Rizal Shrine Museum, Dapitan 34382
José Rizal Shrine Museum, Manila 34441
Josef Albers Museum Quadrat Bottrop, Bottrop 17488
Josef Hoffmann House of Birth, Brtnice 09444
Josef Sudek Gallery, Praha 09748
Josef Suk Memorial, Křečovice 09607
Josef Weinheber-Museum, Kirchstetten 02173
Josefine-Weihrauch-Heimatmuseum Neudenau, Neudenau 20603
Joseloff Gallery, West Hartford 54057
Joseph A. Cain Memorial Art Gallery, Corpus Christi 47821
The Joseph A. Callaway Archaeological Museum, Louisville 50286
Joseph A. Tallman Museum, Cherokee 47420
Joseph Brant Museum, Burlington 05500
Joseph Gross Gallery, Tucson 53593
Joseph Manigault House, Charleston 47369
Joseph Misson-Gedenkstätte, Mühlbach am Mannhartsberg 02409
Joseph Moore Museum, Richmond 52313
Joseph Parry's Ironworkers Cottage, Merthyr Tydfil 45131
Joseph Priestley House, Northumberland 51427
Joseph Schneider Haus, Kitchener 06102
Joseph Smith Historic Center, Nauvoo 50991
Joshibi Art Museum, Sagamihara 29181
Joshibi Bijutsukan, Sagamihara 29181
Josiah Day House, West Springfield 54082
Josiah Quincy House, Quincy 52219
Joslowitzer Heimatstube, Zwingendorf 03139
Joslyn Art Museum, Omaha 51552
Jost House Museum, Sydney 07049
Jotenkaku Hakubutsukan, Kyoto 28817
Jotenkaku Museum, Kyoto 28817
Jouëts d'Autrefois Rêves d'Aujourd'hui, Azay-le-Rideau 11810
Jourdan-Bachman Pioneer Farm, Austin 46512
Journey Museum, Rapid City 52251
Journey's End Cottage and Laishley House, Auckland 33052
Joutsa Museum, Joutsa 10878
Joutsan Kotiseutumuseo, Joutsa 10878
Jovan Cvijic Memorial Museum, Beograd 38067
The Joyo Geibun Center, Mito 28920
Józef Czechowicz Literature Museum, Lublin 34904
Jozef Gregor Tajovský Pamätný Dom, Tajov 38368
Jozef I. Kraszewski Museum, Poznań 35047
Jozef Kollar Art Gallery, Banská Štiavnica 38190
Józef Kraszewski Museum, Romanów 35086
Józef Mehoffer House of the National Museum in Cracow, Kraków 34770
Jozef Murgas Memorial Site, Tajov 38369
Józef Wybicki Memorial Museum, Nowy Karczma 34955
József Attila Emlékház, Öcsöd 23735
József Attila Emlékmúzeum, Balatonszárszó 23353
József Attila Emlékszoba, Budapest 23417
József Attila Múzeum, Makó 23690
József Bajza Memorial Room, Szűcsi 23907
József Darvas Literary Memorial House, Orosháza 23738
József Egry Memorial Museum, Badacsony 23343
József Eötvös Memorial Museum, Ercsi 23546
József Katona Memorial House, Kecskemét 23647
Juan Luna Shrine Museum, Badoc 34339
Judah L. Magnes Museum, Berkeley 46755
The Judah P. Benjamin Confederate Memorial at Gamble Plantation, Ellenton 48362
The Judaica Museum, Bronx 47047
Judaica Museum of Central Synagogue, Irvington 49536
Judenbad, Friedberg, Hessen 18467
Judge Roy Bean Visitor Center, Langtry 49938
Judge's Lodging, Presteigne 45443
Judges' Lodgings, Lancaster 44631
Judiska Museet i Stockholm, Stockholm 40902
Judith Basin County Museum, Stanford 53255
Judy Garland Museum, Grand Rapids 49010
Jüdische Galerie Berlin, Berlin 17146
Jüdische Gedenkstätte und ehemalige Synagoge, Wallhausen 22315
Jüdisches Kulturmuseum Augsburg-Schwaben, Augsburg 16624
Jüdisches Kulturmuseum und Synagoge, Veitshöchheim 22208
Jüdisches Kulturzentrum, Fulda 18510
Jüdisches Museum, Creglingen 17742
Jüdisches Museum, Frankfurt am Main 18379
Jüdisches Museum, Göppingen 18673
Jüdisches Museum, München 20448
Jüdisches Museum, Steinbach am Glan 21880
Jüdisches Museum Berlin, Berlin 17147
Jüdisches Museum der Nordpfalz, Winnweiler 22603
Jüdisches Museum der Schweiz, Basel 41162
Jüdisches Museum der Stadt Wien, Wien 03008
Jüdisches Museum Emmendingen, Emmendingen 18145
Jüdisches Museum Franken in Fürth, Fürth 18498
Jüdisches Museum Franken in Schnaittach, Schnaittach 21590
Jüdisches Museum Georgensgmünd, Georgensgmünd 18588
Jüdisches Museum Hohenems, Hohenems 02092
Jüdisches Museum im Raschi-Haus, Worms 22684
Jüdisches Museum Rendsburg, Rendsburg 21274
Jüdisches Museum Shalom Europa, Würzburg 22712
Jüdisches Museum Westfalen, Dorsten 17910
Jünger Haus, Langenenslingen 19828
Jünger-Haus Wilflingen, Wilfingen 22570
Jugend Museum Schöneberg, Berlin 17148
Jugendhuset, Varde 10306
Jugendkunstschule Pankow, Berlin 17149
Jugoslovenska Galerija Umetnickih Dela, Beograd 38052
Juhani Ahon Museo, Iisalmi 10847
Juice and Vinegar Museum, Doesburg 32132
Jukeboxmuseum, Menen 03867
Jukkasjärvi Museet, Jukkasjärvi 40620
Jule Collins Smith Museum of Fine Arts, Auburn 46465
Julfa Regional History Museum, Julfa 03178
Julia A. Purnell Museum, Snow Hill 53138
Júlia Báthory Glass Collection, Dömsöd 23526
Julia C. Butridge Gallery, Austin 46513
Julian H. Sleeper House, Saint Paul 52619
Juliette Gordon Low Birthplace, Savannah 52942
Julistemuseo, Lahti 11027
Julita Sveriges Lantbruksmuseums, Julita 40621
Julius Gordon Africana Centre, Riversdale 38723
Jumalon Museum and Art Gallery, Cebu 34371
Juming Museum, Chin-shan 42163
Jun Pottery Shu House, Tongxiang 08302
Junagadh Museum, Junagadh 24248
Junction City Historical Society Museum, Junction City 49655
Junction Cottage, Pontypool 45401

Jundt Art Museum, Spokane 53206
June Steingart Art Gallery, Oakland 51465
Juneau-Douglas City Museum, Juneau 49658
Junee Historical Museum, Junee 01180
Junge Kunst e.V., Wolfsburg 22664
Junge Kunsthalle, Karlsruhe 19440
das junge museum, Bottrop 17489
Jungmun Folklore Museum, Seogwipo 29861
Juniata College Museum of Art, Huntingdon 49422
Junior Museum, Newark 51285
The Junior Museum, Troy 53574
The Junior Museum of Bay County, Panama City 51692
JuniorMuseum im Ethnologischen Museum, Berlin 17150
Junkichi Mukai Annex Art Museum, Tokyo 29440
Juodkrante Exhibition Hall of the Lithuanian Art Museum, Juodkrantė 30232
Juodkrantės Parodų Salėa, Juodkrantė 30232
Juozas Gruodis Memorial Museum, Kaunas 30243
Juozas Grusas Memorial Museum of the Maironis Lithuanian Literature Museum, Kaunas 30244
Juozas Tumas-Vaizgantas Museum of the Maironis Lithuanian Literature Museum, Kaunas 30245
Juozas Zikaras' House and Museum, Paliukai 30330
Juozas Zikaras Memorial Museum of the M.K. Čiurlionis National Art Museum, Kaunas 30246
Juozo Gruodžio Memorialinis Muziejus, Kaunas 30243
Juozo Grušo Memorialinis Muziejus, Kaunas 30244
Juozo Naujalio Memorial Museum, Raudondvaris 30343
Juozo Naujalio Memorialinis Muziejus, Raudondvaris 30343
Juozo Paukštelio Memorial Museum, Kėdainiai 30271
Juozo Paukštelio Memorialinis Muziejus, Kėdainiai 30271
Juozo Tumo-Vaižganto Memorialinis Muziejus, Kaunas 30245
Juozo Zikaro Memorialinis Muziejus, Kaunas 30246
Juozo Zikaro Memorialinis Muziejus, Paliukai 30330
Jura-Bauernhof-Museum, Hitzhofen 19203
Jura-Museum Eichstätt, Eichstätt 18082
Jurbarkas Area History Museum, Jurbarkas 30235
Jurbarko Krašto Istorijos Muziejus, Jurbarkas 30235
Jurica Nature Museum, Lisle 50126
Jurinskij Istoriko-chudožestvennyj Muzej, Jurino 36982
Jurjeveckij Kraevedčeskij Muzej, Jurjevec 36983
Jurkovič Museum at Nyaryi Manor-House, Sobotište 38354
Jurmala City Museum, Jūrmala 30023
Jūrmalas Pilsētas Muzejs, Jūrmala 30023
Jussi Björlingmuseet, Borlänge 40476
Just for Laughs Museum, Montréal 06370
Justice and Police Museum, Sydney 01551
Justice Brothers Racing Museum, Duarte 48184
Justin Smith Morrill Homestead, Strafford 53326
Justina M. Barnicke Gallery, Hart House, Toronto 07117
Justinus-Kerner-Haus, Weinsberg 22432
Jusupovskij Dvorec, Sankt-Peterburg 37653
Jute Museum, Kolkata 24276
Jutulheimen Bygdemuseum, Vaagaa 33952
Juuan Pitäjänmuseo, Juuka 10881
Juvan Museo, Juva 10884
Juxian Museum, Juxian 07908
Južno-Kazachstanskij Istoriko-Kraevedčeskij Muzej, Čimkent 29698
Jysk Automobilmuseum, Gjern 10038
Jyväskylä Art Museum, Jyväskylä 10890
Jyväskylä School Museum, Jyväskylä 10888
Jyväskylä University Museum - Section of Natural History, Jyväskylä 10891
Jyväskylän Lyseon Museo, Jyväskylä 10888
Jyväskylän Näkövammaisten Koulun Museo, Jyväskylä 10889
Jyväskylän Taidemuseo, Jyväskylä 10890
Jyväskylän Yliopiston Museo - Keski-Suomen Luontomuseo, Jyväskylä 10891
Jyväskylän Yliopiston Museo - Kulttuurihistoriallinen Osasto, Jyväskylä 10892

K

K.A. Timiryazev Apartment Museum, Moskva 37324
K.A. Timiryazev State Museum of Biology, Moskva 37186
K. Banio Etnografinė Žvejo Sodyba-Muziejus, Rusne 30349
K.C.S. Paniker Gallery, Thiruvananthapuram 24404
K.E. Kivirikon Lintu- ja Nisäkäskokoelma, Helsinki 10789
K.H. Renlund Museum and Camera Collection, Kokkola 10972
K.H. Renlundin Museo ja Kamerakokoelma, Kokkola 10972
K. Hajdarov House-Museum of the Kokand Regional Studies Museum, Qo'qon 54456
K. Hajdarov Yodgorlik Muzeyi, Qo'qon 54456
K-HOF-Galerie, Gmunden 01950
K-HOF Kammerhof Museen Gmunden, Gmunden 01951
K.I. Tchukovsky House-Museum, Peredelkino 37539
K&K Keramikmuseum, Bad Radkersburg 01763& K-Museum, Türnitz 02829 **Vladikavkaz** 37933
K Museum, Shimizu 29261
K.S. Stanislavsky House Museum, Moskva 37168
K.u.k. Hausmuseum, Weyregg am Attersee 02928
K.V. Ivanov Literary Museum, Čeboksary 36838
K. Vasilyev Art Gallery, Kazan 37009
Ka-Do-Ha Indian Village Museum, Murfreesboro 50911
Kaarinan Koulumuseo, Littoinen 11076
Kaarlela Local History Museum, Kokkola 10973
Kaarlelan Kotiseutumuseo, Kokkola 10973
Kaarlo Kangasniemen Urheilumuseo, Kullaa 11003
Kaasboerderijmuseum de Weistaar, Maarsbergen 32515
Kaasmuseum Bodegraven, Bodegraven 31976
Kaaswaag Gouda, Gouda 32266
Kaatsmuseum, Franeker 32231
Kaatza Station Museum, Lake Cowichan 06130
Kaban ng Hiyas ng Lungsod ng Mandaluyong, Mandaluyong 34425
Kabardino-Balkarian Art Museum, Nalčik 37438
Kabardino-Balkarskij Gosudarstvennyj Muzej Izobrazitelnych Iskusstv, Nalčik 37438
Kabarnet Museum, Kabarnet 29740
Kabayan National Museum, Kabayan 34395
Kabinet Grafike, Zagreb 09001
De Kabinetten van de Vleeshal, Middelburg 32551
Kabusa Konsthall, Köpingebro 40649
Kachchh Museum, Bhuj 24099
Kachelmuseum De Drie Kronen, Boekel 31977
Kadman Numismatic Pavilion, Tel Aviv 25112
Kadrior Palace - The Kadriorg Art Museum of the Estonian Art Museum, Tallinn 10632
Kadrioru Loss Väliskunsti Muuseum, Tallinn 10632
Kärnan, Helsingborg 40589
Kärntner Fischereimuseum, Seeboden 02734
Kärntner Freilichtmuseum mit Bauernmöbelsammlung, Maria Saal 02352
Kärsämäen Kotiseutumuseo, Kärsämäki 10902
Käseglocke, Museum für Worpsweder Möbel und Kunsthandwerk, Worpswede 22694
Käserei-Museum, Endingen am Kaiserstuhl 18165
Käsereimuseum, Lochen 02330
Käthe-Kollwitz-Museum Berlin, Berlin 17151
Käthe Kollwitz Museum Köln, Köln 19610
Käthe Kruse-Poppenmuseum, Den Helder 32109
Käthe-Kruse-Puppen-Museum, Donauwörth 17897
Kaffeekannenmuseum, Schöppenstedt 21617
Kaga Art Gallery, Kaga 28625
Kaga-shi Bijutsukan, Kaga 28625
Kagaku Gijutsukan, Tokyo 29404
Kagawa History Museum, Takamatsu 29317
Kagawa-ken Bunka Kaikan, Takamatsu 29316
Kagawa-kenritsu Higashiyama Kaii Setouchi Bijutsukan, Sakaide 29201
Kagawa-kenritsu Rekishi Hakubutsukan, Takamatsu 29317
Kagawa Prefectural Higashiyama Kaii Setouchi Art Museum, Sakaide 29201
Kagawa Prefecture Cultural Center, Takamatsu 29316
Kagoshima City Modern Literature Museum & Marchen Fairy Tale Museum, Kagoshima 28626
Kagoshima City Museum of Archaeology, Kagoshima 28629
Kagoshima City Museum of Art, Kagoshima 28628
Kagoshima-kenritsu Hakubutsukan, Kagoshima 28627
Kagoshima Prefectural Museum, Kagoshima 28627
Kagoshima-shiritsu Bijutsukan, Kagoshima 28628
Kagoshima-shiritsu Kokokan, Kagoshima 28629
Kagu no Hakubutsukan, Tokyo 29405
Kahari Local Arts and Crafts Museum, Noormarkku 11148
Kahei Hakubutsukan, Tokyo 29406
Kahitsukan - Kyoto Museum of Contemporary Art, Kyoto 28818
Kahnweilerhaus, Rockenhausen 21339
Kahramanmaraş Devlet Güzel Sanatlar Galerisi, Kahramanmaraş 42861
Kahramanmaraş Müzesi, Kahramanmaraş 42862
Kahramanmaraş Museum, Kahramanmaraş 42862
Kahramanmaraş State Fine Art Gallery, Kahramanmaraş 42861
Kahutara Canoes and Taxidermy Museum, Featherston 33105
Kai 10, Düsseldorf 17993
Kaieji Buddhist Temple Site Museum, Sennan 29249
Kaifeng Museum, Kaifeng 07909
Kaigan Bijutsukan, Chikura 28399
Kaigan Museum of Art, Chikura 28399
Kaikohe Pioneer Village and Museum, Kaikohe 33133
Kaikoura District Museum, Kaikoura 33134
Kainuun Museo, Kajaani 10903
Kaiping Huaqiao Museum, Kaiping 07912
Kaisendo Museum, Kaminoyama 28649
Kaiser Franz-Josef-Museum für Handwerk und Volkskunst, Baden bei Wien 01775
Kaiser Maximilian von Mexiko Museum, Hardegg 02051
Kaiserappartements, Sisi Museum, Silberkammer, Wien 03009
Kaiserburg Nürnberg, Nürnberg 20773
Kaisergruft, Wien 03010
Kaiserpfalz Gelnhausen, Gelnhausen 18570
Kaiserpfalz mit Sankt Ulrichskapelle, Goslar 18711
Kaisersaal Schloß Schwarzburg, Schwarzburg 21672
Kaiserstühler Heimatmuseum, Endingen am Kaiserstuhl 18166
Kaiserstühler Weinbaumuseum, Vogtsburg im Kaiserstuhl 22251
Kaiservilla, Bad Ischl 01753
Kaishu Ichizo Museum, Aizuwakamatsu 28351
Kaišiadorių Muziejus, Kaišiadorys 30236
Kaitseväe Ühendatud Õppeasutuste Muuseum, Tartu 10649
Kaivolan Museo, Laitila 11038
Kajaani Art Museum, Kajaani 10904
Kajaanin Taidemuseo, Kajaani 10904
Kajuma Nasyri Museum, Kazan 37018
Kakabdze Art Gallery, Kutaisi 16342
Kakamigahara Aerospace Science Museum, Kakamigahara 28636
Kakamigahara-shiritsu Uchuu Kagakukan, Kakamigahara 28636
Kake Bijutsukan, Kurashiki 28782
Kake Museum of Art, Kurashiki 28782
Kakegawa-shi Ninomaru Bijutsukan, Kakegawa 28637
Kakh-e-Rejat va Ebrat & Mellat, Tehrān 24676
Kakunodate Museum of Traditional Crafts and Industry, Kakunodate 28639
Kakurin-ji Treasure House, Katsuura 28680
Kala Gallery, Berkeley 46756
Kalahari-Oranje Museum, Upington 38771
Kalajoen Kalastusmuseo, Kalajoki 10905
Kalajoen Kotiseutumuseo, Kalajoki 10906
Kalajoki Fishing Museum, Kalajoki 10905
Kalamazoo Aviation History Museum, Kalamazoo 49665
Kalamazoo Institute of Arts, Kalamazoo 49666
Kalamazoo Valley Museum, Kalamazoo 49667
Kalamunda History Village, Kalamunda 01183
Kalan Torppa, Kauhava 10929
Kalanag-Museum, Backnang 16659
Kalannin Kotiseutumuseo, Kalanti 10907
Kalanti Local History Museum, Kalanti 10907
Kalantiaw Shrine Museum, Batan 34344
Kalaupapa Historical Park, Kalaupapa 49668
Kalbininko Kazimiero Būgos Memorialinis Muziejus, Paziegė 30336
Kalbininko, Kunigo Kazimiero Jauniaus Kletele-muziejus, Lembo 30303
Kaleidoscope, Kansas City 49688
Kaleidoscope Museum of Kyoto, Kyoto 28802
Kaleko Mill, Faaborg 10012
Kaleko Mølle, Faaborg 10012
Kali-Bergbaumuseum Volpriehausen, Uslar 22192
Kalimuseum, Buggingen 17642
Kalinčeva Kăšta, Trjavna 05227
Kalincheva House, Trjavna 05227
Kaliningrad Amber Museum, Kaliningrad 36991
Kaliningrad Art Gallery, Kaliningrad 36988
Kaliningradskaja Chudožestvennaja Galereja, Kaliningrad 36988
Kaliningradskij Oblastnoj Istoriko-chudožestvennyj Muzej, Kaliningrad 36989
Kaliningradskij Oblastnoj Kraevedčeskij Muzej, Kaliningrad 36990
Kalix Flottningsmuseum, Kalix 40622
Kalkaska County Historical Museum, Kalkaska 49672
Kalkofenmuseum, Steinbach am Ziehberg 02766
Kalkofenmuseum, Untermarchtal 22181
Kállay Collection, Nyíregyháza 23734
Kállay Gyűjtemény, Nyíregyháza 23734
Kallenaution Kievari ja Kotiseutumuseo, Juupajoki 10883
Kallioniemi, Taivalkoski 11343
Kallmann-Museum, Ismaning 19364
Kálmán Imre Emlékmúzeum, Siófok 23803
Kálmán Mikszáth Memorial House, Horpács 23618
Kalman Mikszath Memorial Site, Sklabiná 38351
Kalmar Art Museum, Kalmar 40623
Kalmar Castle, Kalmar 40626
Kalmar Konstmuseum, Kalmar 40623
Kalmar Läns Museum, Kalmar 40624
Kalmar Maritime Museum, Kalmar 40625
Kalmar Sjöfartsmuseum, Kalmar 40625
Kalmar Slott, Kalmar 40626
Kalmyckaja Gosudarstvennaja Kartinnaja Galereja, Elista 36912
Kalmyckij Respublikanskij Kraevedčeskij Muzej im. Prof. N.N. Palmova, Elista 36913
Kalmyk State Picture Gallery, Elista 36912
Kalncempji Parish Museum of Local Studies - Ates Mill, Kalncempji 30028
Kalncempju Pagasta Novadpētniecības Muzejs Ates Dzirnavās, Kalncempji 30028
Kalocsai Érseki Kincstár, Kalocsa 23624
Kalocsai Foegyházmegyei Gyűjtemények - Foszékesegyházi Könyvtár, Kalocsa 23625
Kaluga Regional Art Museum, Kaluga 36997
Kaluga Regional Museum, Kaluga 36998
Kalundborg og Omegns Museum, Kalundborg 10113
Kalužskij Oblastnoj Chudožestvennyj Muzej, Kaluga 36997
Kalužskij Oblastnoj Kraevedčeskij Muzej, Kaluga 36998
Kalvystės Muziejus, Klaipėda 30278
Kamakura Kindai Bijutsukan, Kamakura 28640
Kamakura Kokuhokan, Kamakura 28641
Kamakura Museum of Literature, Kamakura 28642
Kamakura Museum of Modern Art, Kamakura 28640
Kamakura-shi Bungaku, Kamakura 28642
Kamakura Treasure Museum, Kamakura 28641
Kamakuragu Homotsukan, Kamakura 28643
Kamakuragu Shrine Treasure Hall, Kamakura 28643
Kamčatskij Oblastnoj Kraevedčeskij Muzej, Petropavlovsk-Kamčatskij 37553
Kamčatskij Voenno-istoričeskij Muzej, Petropavlovsk-Kamčatskij 37554
Kamenný Dům, Kutná Hora 09614
Kamera- und Fotomuseum Leipzig, Leipzig 19914
Kamernyj Šaljapinskij Zal, Kazan 37008
Kami no Hakubutsukan, Tokyo 29407
Kamienica Celejowska, Kazimierz Dolny 34732
Kamienica Hipolitów, Kraków 34782
Kamienica pod Gwiazdą, Toruń 35192
Kamigata Ukiyo-e Bijutsukan, Osaka 29129
Kamill Erdős Gipsy Museum, Pécs 23760
Kaminokuni Folk Art Museum, Kaminokumi 28648
Kaminokuni Mingeikan, Kaminokumi 28648
Kaminski House Museum, Georgetown 48894
Kam'jans'ka Kartynna Galereja, Kam'janka 43053
Kam'jans'kyj Deržavnyj Istoriko-kul'turnyj Zapovidnyk, Kam'janka 43054
Kam'jans'kyj Istoryčnyj Muzej, Kam'janka 43055
Kamloops Art Gallery, Kamloops 06035
Kamloops Museum & Archives, Kamloops 06036
Kammerhof - Baníctvo na Slovensku, Banská Štiavnica 38191
Kammerhof - Mining in Slovakia, Banská Štiavnica 38191
Kammerhofmuseum, Bad Aussee 01739
Kamnik Museum, Kamnik 38427
Kamniški Muzej, Kamnik 38427
Kamogawa Ceramic Art Museum, Kamogawa 28651
Kamogawa-shiritsu Tougeikan, Kamogawa 28651
Kamoliddin Behzad State Museum of History, Regional Studies and Art of the Republic of Tajikistan, Dušanbe 42498
Kamper Tabaksmuseum, Kampen 32437
Kampffmeyersche Holzlöffelsammlung, Bad Dürkheim 16715
Kamphaeng Phet National Museum, Kamphaeng Phet 42572
Kamsack and District Museum, Kamsack 06042
Kamthieng House, Bangkok 42544
Kamyanka Picture Gallery, Kam'janka 43053
Kan Kikuchi Kinenkan, Takamatsu 29318
Kanabec History Center, Mora 50837
Kanagawa-kenritsu Kanazawabunko Museum, Yokohama 29596
Kanagawa Kenritsu Kindai Bijutsukan, Hayama 28500
Kanagawa-kenritsu Rekishi Hakubutsukan, Yokohama 29597
Kanagawa-kenritsu Seimei-no-Hoshi Chikyu Hakubutsukan, Odawara 29087
Kanagawa Kindai Bungakukan, Yokohama 29598
Kanagawa Museum of Modern Literature, Yokohama 29598
Kanagawa Prefectural Gallery, Yokohama 29599
Kanagawa Prefectural Kanazawa-Bunko Museum, Yokohama 29596
Kanagawa Prefectural Museum of Cultural History, Yokohama 29597
Kanagawa Prefectural Museum of Natural History, Odawara 29087
Kanai History and Folk Museum, Kanai 28652
Kanai-machi Hakubutsukan, Kanai 28652
Kanalmuseet, Trollhättan 40983
Kanalmuseum Skantzen, Hallstahammar 40576
Kanavamuseo, Varkaus 11452
Kanazawa Daigaku Hakubutsukan, Kanazawa 28660
Kanazawa University Museum, Kanazawa 28660
Kandahar Provincial Museum, Kandahar 00006
Kandava Regional Museum, Kandava 30029
Kandavas Novada Muzejs, Kandava 30029
Kandeloos Museum, Kojoor 24626

Kandelous Museum, Kandelous 24614
Kandiyohi County Historical Society Museum, Willmar 54184
Kandos Bicentennial Industrial Museum, Kandos 01186
Kandy National Museum, Kandy 40380
Kanemori Museum with Baccarat Collection, Hakodate 28475
Kang Xicao Museum, Beijing 07544
Kangasalan Kotiseutumuseo, Kangasala 10908
Kangaslammin Kotiseutumuseo, Kangaslampi 10911
Kangasniemen Kotiseutumuseo, Kangasniemi 10912
Kaniv Historical Museum, Kaniv 43058
Kaniv Museum of Folk Decorative Art, Kaniv 43059
Kanivs'kyj Istoryčnyj Muzej, Kaniv 43058
Kanivs'kyj Muzej Narodnogo Dekoratyvnogo Mystectva, Kaniv 43059
Kanizsai Dorottya Múzeum, Mohács 23716
Kankaanpään Kaupunginmuseo, Niinisalo 11141
Kankakee County Historical Society Museum, Kankakee 49676
Kanken Kanji Hakubutsukan, Kyoto 28819
Kannada Research Institute Museum, Dharwad 24163
Kannonkosken Kotiseutukokoelma, Kannonkoski 10914
Kannuksen Museo, Kannus 10915
Kanra Ancient Art Museum, Kanra 28665
Kansainvälinen Kahvikuppimuseo, Posio 11233
Kansainvälinen Pullomuseo, Sonkajärvi 11325
Kansallispukukeskus, Jyväskylä 10893
Kansas African American Museum, Wichita 54144
Kansas Aviation Museum, Wichita 54145
Kansas City Art Institute Gallery, Kansas City 49689
The Kansas City Jewish Museum, Overland Park 51633
Kansas City Museum, Kansas City 49690
Kansas Cosmosphere and Space Center, Hutchinson 49454
Kansas Health Museum, Halstead 49130
Kansas Museum of History, Topeka 53536
Kansas Museums Association, Topeka 55150
The Kansas Teachers' Hall of Fame, Dodge City 48149
Kansetsu Hashimoto Bijutsukan, Kyoto 28820
Kansetsu Hashimoto Memorial Museum, Kyoto 28812
Kant- en Textielcentrum, Diest 03551
Kanta Museum Argungu, Argungu 33315
Kantcentrum, Brugge 03418
Kantmuseum, Uithuizen 32835
Kantmuseum Olsene, Olsene 03922
Kantonales Fischereimuseum Mettlen, Netstal 41674
Kantonales Museum Altes Zeughaus, Solothurn 41883
Kántor Sándor Fazekasház, Karcag 23639
Kanuma Municipal Art Museum of Kawakami Sumio, Kanuma 28666
Kanzeon-ji Temple Treasure House, Dazaifu 28408
Kao Chong, Trang 42608
Kao Luang, Nakhon Si Thammarat 42584
Kao Ta Phet, Sukhothai 42602
Kao Yai, Saraburi 42599
Kaohsiung Astronomical Museum, Kaohsiung 42236
Kaohsiung County Museum of Natural History, Lujhu 42263
Kaohsiung Museum of Fine Arts, Kaohsiung 42237
Kaohsiung Museum of History, Kaohsiung 42238
Kapan Local Tradition Museum, Kapan 00712
Kapan Museum of Contemporary Art, Kapan 00713
Kapenguria Museum, Kapenguria 29741
Kaperdalen Samemuseum, Senja 33825
Kapilavastu Museum, Lumbani 31783
Kapisa Provincial Museum and Archeological Site, Mahmud-i-Raqi 00007
Kapitän Tadsen Museum, Langeneß 19831
Kapiteinshuis Pekela, Nieuwe Pekela 32576
Kapiti Coast Museum Waikanae, Wellington 33247
Kaplanstöckl, Hohenzell 02094
Káplár Miklós Gyűjtemény, Hajdúböszörmény 23604
Kaple Sv. Kříže, Praha 09754
Kaposvar Historic Site, Esterhazy 05783
Kappan Hakubutsukan, Tokyo 29408
Kapucijnenmuseum, 's-Hertogenbosch 32371
Kapunda Museum, Kapunda 01188
Karačaevo-Čerkesskij Istoriko-kulturnyj i Prirodnyj Muzej-Zapovednik, Čerkessk 36854
Karacolskij Istoričeskij Kraevedčeskij Muzej, Karacol 29962
Karaganda Ecological Museum, Karaganda 29702
Karaganda Historical and Regional Museum, Qaraghandy 29722
Karaganda Museum of Fine Arts, Qaraghandy 29723
Karagandinskij Oblastnoj Istoriko-Kraevedčeskij Muzej, Qaraghandy 29722
Karagandinskij Oblastnoj Muzej Izobrazitelnogo Iskusstva, Qaraghandy 29723
Karaimų Etnografinė Paroda, Trakai 30381
Karaite Ethnographic Exhibition of the Trakai Historical Museum, Trakai 30381
Karakalpak Historical Museum, Nukus 54452
Karakalpakstan State Regional Museum, Nukus 54451
Karaliichev House-Museum, Sofia 05173
Karaman Müzesi, Karaman 42864
Karaman Museum, Karaman 42864
Karamzin Public Library Museum, Uljanovsk 37901
Karasek-Museum, Seifhennersdorf 21732
Karatay Çini Eserler Müzesi, Konya 42886
Karatepe-Aslantaş Açıkhava Müzesi, Osmaniye 42924
Karatepe-Aslantas Open Air Museum, Osmaniye 42924
Kardemir Demir Çelik Müzesi, Karabük 42863
Kardemir Iron and Steel Museum, Karabük 42863
Kardinal-Bea-Museum, Blumberg 17391
Kardinolo Vincento Sladkevičiaus Memorialinis Butas-Muziejus, Kaunas 30247
Karelian Farmhouse, Imatra 10855
Karelian Museum of Fine Arts, Petrozavodsk 37561
Karelian State Regional Museum, Petrozavodsk 37558
Karelskij Gosudarstvennyj Kraevedčeskij Muzej, Petrozavodsk 37558
Karen Blixen Museet, Rungsted Kyst 10253
Karen Blixen Museum, Nairobi 29747
Karepa Kalame Talumuuseum, Vihula 10672
Karepe Kalame Fishermen and Farm Museum, Vihula 10672
Kareum - Gamla Bilsalongen, Sparreholm 40877
Kargopol State Museum of History, Architecture and Art, Kargopol 37003
Kargopolskij Gosudarstvennyj Istoriko-Architekturnyj i Chudožestvennyj Muzej, Kargopol 37003
Kariandusi Prehistoric Site Museum, Gilgil 29738
Karijoen Kotiseutumuseo, Karijoki 10916
Karikatür ve Mizah Müzesi, İstanbul 42830
Karikatur und Cartoon Museum Basel, Basel 41163
Karikaturmuseum Krems, Krems 02226
Kariya City Art Museum, Kariya 28667
Kariya-shiritsu Bijutsukan, Kariya 28667
Kariye Müzesi, İstanbul 42831
Karjalainen Kotitalo, Imatra 10855
Karjalan Kotiseutumuseo, Mynämäki 11130
Karkaralinsk Historical and Regional Museum, Karkaralinsk 29703
Karkaralinskij Rajonnyj Istoriko-Kraevedčeskij Muzej, Karkaralinsk 29703
Karkkila Ironworks Museum, Karkkila 10918
Karkkilan Ruukkimuseo, Karkkila 10918
Karl Bacher-Stüberl, Zellerndorf 03129
Karl Drerup Fine Arts Gallery, Plymouth 52000
Karl Ernst von Baeri Muuseum, Tartu 10650
Karl Heinrich Waggerl-Haus, Wagrain 02882
Karl-May-Haus Hohenstein-Ernstthal, Hohenstein-Ernstthal 19244
Karl-May-Museum, Radebeul 21160
Karl-Otto-Braun-Museum, Ludwigshafen am Rhein 20052
Karl Ristikivi Majamuuseum, Tartu 10651
Karl und Magdalene Haberstock Stiftung, Augsburg 16625
Karl-Wagenplast-Museum, Schwaigern 21658
Karl-Wörn-Haus, Schwetzingen 21702
Kārļa Skalbes Memoriālais Muzejs Saulrieti, Vecpiebalga 30142
Kārļa Ulmaņa Piemiņas Muzejs Pikšas, Bērze 29991
Karlebo Museum, Kokkedal 10161
Karlinger-Schmiede, Unterweissenbach 02844
Karlis Skalbe Memorial Museum Saulrieti, Vecpiebalga 30142
Karlovac Municipal Museum, Karlovac 08886
Karlovarské Muzeum, Karlovy Vary 09575
Karlovské Muzeum, Velké Karlovice 09900
Karlovy Vary Art Gallery, Karlovy Vary 09573
Karlshamns Museum, Karlshamn 40628
Karlskoga Konsthall, Karlskoga 40631
Karlsruhe Palace, Karlsruhe 19437
Karmøy Fiskerimuseum, Vedavågen 33967
Karmsund Folkemuseum, Haugesund 33538
Karnataka Government Museum and Venkatappa Art Gallery, Bangalore 24073
Karnevalmuseum, Mönchengladbach 20356
Karnevalsmuseum, Blankenheim 17376
Karol Plicka Múzeum, Blatnica 38200
Karol Szymanovsky Museum of Music Culture, Kirovograd 43066
Karol Szymanowski Museum in the Atma Chalet in Zakopane, Zakopane 35357
Károli Gáspár Múzeum és Bibliakiállítás, Gönc 23571
Károly Kisfaludy Memorial Exhibition, Tét 23923
Károlyi Castle, Fehérvárcsurgó 23556
Károlyi Castle, Füzérradvány 23562
Károlyi-Kastély, Fehérvárcsurgó 23556
Károlyi-Kastély, Füzérradvány 23562
Karottenmuseum-Musée de la Carotte, Raeren 03959
Karpatendeutsches Museum, Karlsruhe 19441
Karpeles Manuscript Museum, Buffalo 47122
Karpeles Manuscript Museum, Charleston 47370
Karpeles Manuscript Museum, Duluth 48196
Karpeles Manuscript Museum, Jacksonville 49581
Karpeles Manuscript Museum, Tacoma 53410
Karrasburg Museum Coswig, Coswig 17727
Karrenmuseum, Essen 03587
Kars Devlet Güzel Sanatlar Galerisi, Kars 42866
Kars Müzesi, Kars 42867
Kars Museum, Kars 42867
Kars State Fine Art Gallery, Kars 42866
Karsh-Masson Gallery, Ottawa 06548
Karstulan Kotiseutumuseo, Karstula 10920
Kartäusermuseum Tückelhausen, Ochsenfurt 20853
Kartause Aggsbach, Aggsbach Dorf 01697
Kartause Gaming, Gaming 01921
Karthaus-Prüll, Regensburg 21225
Kartinnaja Galereja A.A. Plastova, Uljanovsk 37899
Kartinnaja Galereja K. Vasiljeva, Kazan 37009
Kartinnaja Galereja Ministerstva Kultury Respubliki Adygeja, Majkop 37124
Kartinnaja Galereja Pejzažej P.M. Grečiškina, Stavropol 37803
Kartinnaja Galerija im. P.I. Šolochova, Borisoglebsk 36818
Kartinnaya Gallery, Stepanakert 03198
Das Kartoffelmuseum, München 20449
Karttulan Kotiseutumuseo, Karttula 10921
Kartynna Galereja, Kam'janec'-Podil's'kyj 43052
Kartynna Galereja, Luc'k 43111
Karuizawa Wild Flower Museum of Art, Karuizawa 28669
Karvian Museo ja Ulkomuseoalue, Karvia 10923
Karyes-Protaton, Agion Oros 22868
Kasama Inari Bijutsukan, Kasama 28674
Kasama Nichido Bijutsukan, Kasama 28675
Kasama Nichido Museum of Art, Kasama 28675
Kashi Zone Silu Museum, Kashi 07914
Kashima-jingu Homotsukan, Kashima, Ibaraki 28678
Kashima Shrine Treasure House, Kashima, Ibaraki 28678
Kashkadarya Regional Museum of Local Lore, Qarshi 54454
Kashubian Museum, Kartuzy 34718
Kaskade Infohaus Wasser, Jade 19371
Kaskö Hembygdsmuseum, Kaskinen 10925
Kasori Kaizuka Hakubutsukan, Chiba 28395
Kasori Shellmounds Site Museum, Chiba 28395
Kass Galéria, Szeged 23845
Kassák Múzeum, Budapest 23418
Kăšta-muzej Akademik Dončo Kostov, Lokorsko 05083
Kăšta-muzej Aleksandăr Mutafov, Sozopol 05208
Kăšta-muzej Angel Kănčev, Trjavna 05228
Kăšta-muzej Angel Karalijčev, Sofia 05173
Kăšta-muzej Asen Razcvetnikov, Draganovo 05021
Kăšta-muzej Asen Zlatarov, Chaskovo 05004
Kăšta-muzej Canko Cerkovski, Bjala Čerkva 04989
Kăšta-muzej Chadži Dimităr, Sliven 05166
Kăšta-muzej Christo Botev, Kalofer 05045
Kăšta-muzej Christo G. Danov, Plovdiv 05128
Kăšta-muzej Christo Smirnenski, Sofia 05174
Kăšta-muzej Dimcho Debeljanov, Koprivštica 05070
Kăšta-muzej Dimităr Blagoev, Bankja 04977
Kăšta-muzej Dimităr Blagoev, Sofia 05175
Kăšta-muzej Dimităr Dimov, Sofia 05176
Kăšta-muzej Dimităr Pešev, Kjustendil 05064
Kăšta-muzej Dimităr Poljanov, Karnobat 05052
Kăšta-muzej Dobri Voinikov, Šumen 05214
Kăšta-muzej Emilijan Stanev, Veliko Tărnovo 05254
Kăšta-muzej Geo Milev, Stara Zagora 05209
Kăšta-muzej Georgi Benkovski, Koprivštica 05071
Kăšta-muzej Georgi Dimitrov, Kovačevci 05079
Kăšta-muzej Iljo Vojvoda, Kjustendil 05065
Kăšta-muzej Ivan Lazarov, Sofia 05177
Kăšta-muzej Ivan Vazov, Berkovica 04987
Kăšta-muzej Ivan Vazov, Sofia 05178
Kăšta-muzej Jordan, Jovkov, Žeravna 05269
Kăšta-muzej Lamartine, Plovdiv 05129
Kăšta-muzej Ljuben Karavelov, Koprivštica 05072
Kăšta-muzej Neofit Rilski, Bansko 04979
Kăšta-muzej Nikola Ěnkov Vapcarov, Bansko 04980
Kăšta-muzej Nikola Vapcarov, Sofia 05179
Kăšta-muzej Panaët Chitov, Sliven 05167
Kăšta-muzej Panaët Volov, Šumen 05215
Kăšta-muzej Pančo Vladigerov, Šumen 05216
Kăšta-muzej Pejo Javorov, Sofia 05180
Kăšta-muzej Pejo K. Javorov, Čirpan 05010
Kăšta-muzej Petko i Pencho Slavejkovi, Sofia 05181
Kăšta-muzej Rajna Knjaginja, Panagjurište 05103
Kăšta-muzej Rusi Čorbadži, Žeravna 05270
Kăšta-muzej s Dom-pametnik Jordan Jovkov, Dobrič 05020
Kăšta-muzej Sava Filaretov, Žeravna 05271
Kăšta-muzej Slavejkovi, Trjavna 05229
Kăšta-muzej Stanislav Dospevski, Pazardžik 05108
Kăšta-muzej Stojan i Vladimir Zaimovi, Pleven 05117
Kăšta-muzej Svetoslav Obretenov, Provadia 05141
Kăšta-muzej Todor Kableškov, Koprivštica 05073
Kăšta-muzej Zachari Stojanov, Medven 05090
Kăšta-muzej Zachari Stojanov, Ruse 05152
Kastamonu Arkeoloji Müzesi, Kastamonu 42868
Kastamonu Devlet Güzel Sanatlar Galerisi, Kastamonu 42869
Kastamonu State Fine Art Gallery, Kastamonu 42869
Kasteel Achtendries, Gent 03631
Kasteel Amerongen, Amerongen 31808
Kasteel Ammersoyen, Ammerzoden 31819
Kasteel Beauvoorde, Wulveringem 04158
Kasteel Cannenburch, Vaassen 32860
Kasteel De Doornenburg, Doornenburg 32145
Kasteel de Haar, Haarzuilens 32304
Kasteel Doorwerth, Doorwerth 32146
Kasteel Duivenvoorde, Voorschoten 32907
Kasteel Groeneveld, Baarn 31946
Kasteel Heeswijk, Heeswijk-Dinther 32342
Kasteel Heeze - Collectie H.N.C. Baron Van Tuyll Van Serooskerken Van Heeze en Leende, Heeze 32344
Kasteel Hernen, Hernen 32370
Kasteel Het Nijenhuis, Heino 32352
Kasteel Hoensbroek, Hoensbroek 32398
Kasteel Middachten, De Steeg 32051
Kasteel-Museum Sypesteyn, Loosdrecht 32506
Kasteel Ooidonk, Deinze 03538
Kasteel Radboud, Medemblik 32542
Kasteel Rosendael, Rozendaal 32737
Kasteel van Beersel, Beersel 03357
Kasteel van Bornem, Bornem 03386
Kasteel van Gaasbeek, Gaasbeek 03611
Kasteel van Horst, Sint-Pieters-Rode 04024
Kasteel van Leeuwergem, Zottegem 04167
Kasteel van Wijnendale, Torhout 04078
Kasteelmuseum Slot van Laarne, Laarne 03772
Kasteelruïne Valkenburg, Valkenburg, Limburg 32864
Kasteeltoren IJsselstein, IJsselstein 32427
Kaštel Town Museum, Kaštel Lukšić 08887
Kaštela Gradski Muzej, Kaštel Lukšić 08887
Kaštelet Crikvine, Split 08958
Kastelholm Castle, Sund 11333
Kastelholms Slott, Sund 11333
The Kasteyev State Museum of Arts of the Republic of Kazakhstan, Almaty 29683
Kaštiel a Kysuce Galéria, Oščadnica 38323
Kaštiel' Imre Madácha a Literárne osobnosti, Dolná Strehová 38249
Kastoria Byzantine Museum, Kastoria 23043
Kastrupgårdsamlingen, Kastrup 10114
Kasuga Shrine Treasure Hall, Nara 29030
Kasuga Taisha Homostuden, Nara 29030
Kasvimuseo, Helsinki 10790
Kasvimuseo, Turku 11391
Katamatite Museum, Katamatite 01189
Katanning Historical Museum, Katanning 01190
Kataray Ceramic Art Museum, Konya 42886
The Kate Chopin House and Bayou Folk Museum, Cloutierville 47665
Kate Shelley Railroad Museum, Boone 46905
Katedralna Riznica, Split 08959
Kateri Galleries, Auriesville 46487
Katharina-Luther-Stube, Torgau 22071
Katharinas Puppenhaus, Pleystein 21061
Katherine Mansfield Birthplace, Wellington 33248
Katherine Museum, Katherine 01191
Katherine Nash Gallery, Minneapolis 50736
Kathmandu - Hot Spring in the Art Museum, Yuzawa 29634
Katina Paxinou Museum, Athinai 22916
Katinų Muziejus, Šiauliai 30358
Katona József Emlékház, Kecskemét 23647
Katona József Múzeum, Kecskemét 23648
Katonah Museum of Art, Katonah 49709
Katori-jingu Homotsukan, Sawara 29236
Katori Shrine Treasure House, Sawara 29236
Katrineholms Hembygdsmuseet, Kristineholm 40657
Katschtaler Heimatmuseum, Rennweg 02578
Katsushika Hokusai Art Museum, Tsuwano 29546
Katsushika Hokusai Bijutsukan, Tsuwano 29546
Katsushika-ku Kyoudo to Tenmon no Hakubutsukan, Tokyo 29409
Katsuyama Castle Museum, Katsuyama 28682
Katsuyama Kenbunkan, Katsuyama 28682
Katten Kabinet, Amsterdam 31851
Kattlunds Museigård, Havdhem 40583
Katumuseo, Helsinki 10791
Katwijks Museum, Katwijk aan Zee 32443
Katzenmuseum, Traunstein 22076
Kau Rainforest Museum & Conservation Area, Madang 34083
Kauai Children's Discovery Museum, Kapaa 49706
Kaua'i Museum, Lihu'e 50083
Kauffman Gallery, Shippensburg 53080
Kauffman Museum, North Newton 51397
Kaufmannsmuseum, Haslach 02055
Kauhajoen Museo, Kauhajoki 10927
Kauhavanmuseo, Kauhava 10930
Kaunas' 9th Fort Museum, Kaunas 30249
Kaunas Picture Gallery of the M.K. Čiurlionis National Art Museum, Kaunas 30251
Kaunislehdon Talomuseo, Hyrynsalmi 10845

K

Kaunislehto Farmhouse, Hyrynsalmi 10845
Kauno Arkivyskupijos Muziejus, Kaunas 30248
Kauno IX Forto Muziejus, Kaunas 30249
Kauno Lėlių Muziejus, Kaunas 30250
Kauno Paveikslų Galerija, Kaunas 30251
Kauno Tado Ivanausko Zoologijos Muziejus, Kaunas 30252
Kauppila House, Humppila 10843
Kauppilan perinnetalo, Humppila 10843
Kauppilan Umpipiha, Laitila 11039
Kauppilanmäen Museo, Valkeakoski 11436
The Kauri Museum, Matakohe 33144
Kaurila School Museum, Tohmajärvi 11378
Kaurilan Koulumuseo, Tohmajärvi 11378
Kaustisen Kotiseutumuseo, Kaustinen 10932
Kauttua Factory Museum, Kauttua 10934
Kauttuan Tehtaan Museo, Kauttua 10934
Kavalierhaus Gifhorn - Museum für bürgerliche Wohnkultur, Gifhorn 18644
Kaw Mission, Council Grove 47852
Kawagoe City Museum, Kawagoe 28684
Kawagoe Museum of Art, Kawagoe 28683
Kawagoe-shi Bijutsukan, Kawagoe 28683
Kawagoe-shiritsu Hakubutsukan, Kawagoe 28684
Kawaguchi-ko Motor Museum, Narusawa 29043
Kawaguchi-ko Music Forest Automatic Musical Instruments, Kawaguchi-ko 28686
Kawaguchi Science Museum, Kawaguchi 28685
Kawaguchi-shiritsu Kagakukan, Kawaguchi 28685
Kawaguchiko Bijutsukan, Kawaguchiko 28688
Kawaguchiko Museum of Art, Kawaguchiko 28688
Kawai Kanjiro's House, Kyoto 28821
Kawamura Art Museum, Yuda 29630
Kawamura Bijutsukan, Yuda 29630
Kawamura Memorial Museum of Art, Sakura 29209
Kawanabe Kyosai Memorial Museum, Warabi 29576
Kawara Museum of Takahama City, Takahama 29314
Kawasaki City Museum, Kawasaki 28693
Kawasaki Juvenile Science Museum, Kawasaki 28691
Kawasaki Peace Museum, Kawasaki 28692
Kawasaki-shi Hakubutsukan, Kawasaki 28693
Kawasaki-shiritsu Nihon Minkaen, Kawasaki 28694
Kawhia Regional Museum Gallery, Kawhia 33136
Kaya Steam Locomotive Square, Kaya 28698
Kayseri Archaeological Museum, Kayseri 42874
Kayseri Arkeoloji Müzesi, Kayseri 42874
Kayseri Devlet Güzel Sanatlar Galerisi, Kayseri 42875
Kayseri Etnografya Müzesi, Kayseri 42876
Kayseri State Fine Art Gallery, Kayseri 42875
Kazakhstan Paleolithic Museum of the Al-Pharabi Kazakh National University, Almaty 29682
Kazakstan Paleolit Muzeji, Almaty 29682
Ķazaķstan Respublikasynyņ Memlekettīk Ävilchan Ķasteev Atyndaghy Öner Muzejī, Almaty 29683
Kazanlăška Roza, Kazanlăk 05059
Kazantzakis Museum, Iráklion 23020
Kazemattenmuseum, Oosterend 32626
Kazemattenmuseum Kornwerderzand, Kornwerderzand 32457
Kazimierz Pułaski Museum, Warka-Winiary 35215
Kazinczy Ferenc Emlékcsarnok, Sátoraljaújhely 23797
Kazinczy Ferenc Memorial Hall, Sátoraljaújhely 23797
Kazinczy Ferenc Múzeum, Sátoraljaújhely 23798
Kazio Varnelio Namai-Muziejus, Vilnius 30411
Kazuaki Iwasaki Space Art Gallery, Ito 28596
Kazuno Cultural Property Preservation Center, Kazuno 28700
Kazuo Museum, Chaoyang 07609
Kazys Varnelis House-Museum of the National Museum of Lithuania, Vilnius 30411
KdF-Museum in der Kulturkunststatt Prora, Prora 21130
Kealley's Gemstone Museum, Nannup 01346
Kearney Area Children's Museum, Kearney 49716
Kearney County Historical Museum, Minden 50724
Kearney Mansion Museum, Fresno 48792
Kearny Cottage, Perth Amboy 51785
Keats House, London 44892
Keats-Shelley House, Roma 27456
Kebon Binatang Bandung, Bandung 24440
Kedah Royal Museum, Alor Setar 30550
Kedah State Art Gallery, Alor Setar 30551
Kedah State Museum, Alor Setar 30552
Kėdainiai Area Museum, Kėdainiai 30272
Kėdainių Krašto Muziejus, Kėdainiai 30272
Keeler Tavern Museum, Ridgefield 52348
Keelung Mid-Summer Ghost Festival Museum, Keelung 42247
Keep, Helsingborg 40589
Keetmanshoop Museum, Keetmanshoop 31742
Keewatin Maritime Museum, Douglas 48157
Kegworth Museum, Kegworth 44549
Kegyeleti Szakgyűjtemény, Budapest 23419
Kehdinger Küstenschiffahrts-Museum, Wischhafen 22611
Kei Fujiwara Art Museum, Bizen 28389
Kei Fujiwara Bijutsukan, Bizen 28389
Keighley and Worth Valley Railway Museum, Haworth 44431
Keiko Art Museum, Hirara 28526
Keiko Bijutsukan, Hirara 28526
Keikyän Kotiseutumuseo, Keikyä 10935
Keillor House Museum, Dorchester 05708
Keio Daigaku Bijutsu Senta, Tokyo 29410
Keio University Arts Center, Tokyo 29410
Keir Memorial Museum, Malpeque 06224
Keitele-Museo, Suolahti 11337
Keith Harding's World of Mechanical Music, Northleach 45271
Keith Museum, Keith 01192
Keizerskapel, Antwerpen 03289
Keizo Koyama Art Museum, Komoro 28769
Kekava Museum, Kekava 30032
Kekavas Novadpētniecības Muzejs, Kekava 30032
Kékfestő Múzeum, Pápa 23753
Keladi Museum, Keladi 24257
Kelantan State Museum, Kota Bharu 30564
Kelham Island Museum, Sheffield 45652
Kell House Museum, Wichita Falls 54153
Kellenberger Clock and Watch Collection, Winterthur 42040
Kellermuseum Preßhaus, Großkrut 02006
Kellerviertel Heiligenbrunn, Heiligenbrunn 02067
Kelley House Museum, Mendocino 50586
Kellie Castle, Pittenweem 45388
Kelliher and District Heritage Museum, Kelliher 06049
Kellogg Historical Society Museum, Kellogg 49721
Kellokosken Sairaalan Museo, Kellokoski 10937
Kellross Heritage Museum, Leross 06159
Kelly-Griggs House Museum, Red Bluff 52270
Kelmės Krašto Muziejus, Kelmė 30273
Kelmscott Manor, Kelmscott 44554
Kelnhof-Museum, Bräunlingen 17497
Kelowna Art Gallery, Kelowna 06053
Kelowna Museum, Kelowna 06054
Kelsey Museum of Archaeology, Ann Arbor 46307
The Kelso House Museum, Kent 49738
Kelsterbacher Museum für Porzellan, Kelsterbach 19504
Kelten Römer Museum Manching, Manching 20159
Keltendorf Mitterkirchen, Mitterkirchen 02388
Keltenmuseum, Hallein 02046
Keltenmuseum Gracarca, Sankt Kanzian 02658
Keltenmuseum Hochdorf/Enz, Eberdingen 18035
Keltenpark Schwarzenbach, Schwarzenbach 02719
Kelter- und Weinbaumuseum, Niederstetten 20712
Kelton House Museum, Columbus 47755
Kelvedon Hatch Secret Nuclear Bunker, Kelvedon Hatch 44556
Kelvingrove Art Gallery and Museum, Glasgow 44287
Kemajuan Kraftangan Malaysia, Rawang 30615
Kemerer Museum of Decorative Arts, Bethlehem 46779
Kemerovo Museum of Fine Art, Kemerovo 37032
Kemerovskij Istoriko-architekturnyj Muzej Krasnaja Gorka, Kemerovo 37030
Kemerovskij Oblastnoj Kraevedčeskij Muzej, Kemerovo 37031
Kemerovskij Oblastnoj Muzej Izobrazitelnych Iskusstv, Kemerovo 37032
Kemi Art Museum, Kemi 10940
Kemi Culture Historical Museum, Kemi 10939
Kemijärven Kotiseutumuseo, Kemijärvi 10943
Kemin Kulttuurihistoriallinen Museo, Kemi 10939
Kemin Taidemuseo, Kemi 10940
Kemin Työläismuseo ja Kähertäjämuseo, Kemi 10941
Keminmaan Kotiseutumuseo, Keminmaa 10944
Kemp-McCarthy Memorial Museum, Rowe 52474
Kempeleen Kotiseutumuseo, Kempele 10945
Kemper Museum of Contemporary Art, Kansas City 49691
Kempisch Museum, Brecht 03400
Kempsey Museum, Kempsey 01193
Ken Domon Museum of Photography, Sakata 29205
Kendal Museum, Kendal 44558
Kendall Campus Art Gallery, Miami 50628
The Kendall College Mitchell Museum of the American Indian, Evanston 48451
Kendall Gallery, Grand Rapids 49005
Kendall Whaling Museum, Sharon 53050
Kenderdine Art Gallery, Saskatoon 06915
Kenilworth Historical Society Museum, Kenilworth 49725
Kenji Igarashi Memorial Museum of Laundry, Tokyo 29411
Kenji Miyazawa Memorial Museum, Hanamaki 28491
Kenkichi Tomimoto Memorial Museum, Ando 28359
Kenkoukanri Hakubutsukan, Hiroshima 28539
Kenmin Hall Art Museum Toyama, Toyama 29521
Kennebunkport Historical Museum, Kennebunkport 49728
Kennedy-Douglass Center for the Arts, Florence 48563
Kennedy Museum of Art, Athens 46417
Kennesaw Civil War Museum, Kennesaw 49730
Kennesaw Mountain National Battlefield Park, Kennesaw 49731
Kennet and Avon Canal Trust Museum, Devizes 43978
Kenneth G. Fiske Museum, Claremont 47572
Keno City Mining Museum, Keno City 06056
Kenosha County Museum, Kenosha 49733
Kenosha Public Museum, Kenosha 49734
Kenpu Hakubutsukan, Komoro 28768
Kenpu Hakubutsukan, Yokohama 29600
Kenritsu Kagakukan Joetsu, Joetsu 28621
Kensei Kinenkan, Niigata 29054
Kensington Palace State Apartments, London 44893
Kent and East Sussex Railway, Tenterden 45839
Kent and Medway Museum, West Malling 45971
Kent and Sharpshooters Yeomanry Museum, Edenbridge 44113
Kent and Vicki Logan Galleries and Capp Street Project, San Francisco 52779
Kent Art Association, Kent 49736
Kent Battle of Britain Museum, Hawkinge 44428
Kent Campus Museum and Gallery, Jacksonville 49582
Kent-Delord House Museum, Plattsburgh 51983
Kent Police Museum, Chatham 43796
Kent State University Art Galleries, Kent 49739
Kent State University Museum, Kent 49740
Kentuck Museum and Art Center, Northport 51424
Kentucky Association of Museums, Murray 55151
Kentucky Derby Museum, Louisville 50287
Kentucky Folk Art Center, Morehead 50841
Kentucky Historical Society, Frankfort 48737
Kentucky Military History Museum, Frankfort 48738
The Kentucky Museum, Bowling Green 46973
Kentucky Museum of Arts and Design, Louisville 50288
Kentucky Railway Museum, New Haven 51058
Kentucky State Capitol, Frankfort 48739
Kenwood House, London 44894
Kenya Museum Society, Nairobi 54915
Keokuk Art Center, Keokuk 49744
Keokuk River Museum, Keokuk 49745
Képcsarnok - Gulácsy Lajos Terem, Szeged 23846
Kepler-Gedächtnishaus, Regensburg 21226
Kepler-Museum, Weil der Stadt 22380
Képzőművészek Gulácsy Lajos, Budapest 23420
Keramiekcentrum Tiendschuur, Tegelen 32814
Keramik-Museum, Triengen 41949
Keramik-Museum Berlin, Berlin 17152
Keramik-Museum Bürgel, Bürgel 17638
Keramik Museum Mettlach, Mettlach 20312
Keramikmuseum, Rheinsberg 21303
Keramikmuseum Ludwigsburg, Ludwigsburg 20045
Keramikmuseum Scheibbs, Scheibbs 02702
Keramikmuseum Schloss Obernzell, Obernzell 20824
Keramikmuseum Staufen, Staufen im Breisgau 21867
Keramikmuseum und Maria-Felchin-Sammlung, Matzendorf 41615
Keramikmuseum Westerwald, Höhr-Grenzhausen 19214
Keramikos Muziejus, Kaunas 30253
Keramisch Museum Goedewaagen, Nieuw Buinen 32571
Kerava Art Museum, Kerava 10947
Keravan Museo, Kerava 10946
Keravan Taidemuseo, Kerava 10947
Kerbyville Museum, Kerby 49748
Kerch Museum-Preserve of Culture and History, Kerč 43062
Kerčs'kyj Deržavnyj Istoriko-kul'turnyj Zapovidnyk, Kerč 43062
Keresztény Múzeum, Esztergom 23553
Kerkmuseum Elst, Elst 32203
Kerkmuseum Janum, Janum 32430
Kermis- en Circusmuseum Steenwijk, Steenwijk 32802
Kern County Museum, Bakersfield 46537
Kernave Archaeological and Historical Museum-Reserve, Kernavė 30274
Kernavės Archeologijos ir Istorijos Muziejus, Kernavė 30274
Kernstock-Museum, Bruck an der Lafnitz 01802
Kerr Place, Onancock 51557
Kerrobert and District Museum, Kerrobert 06065
The Kerry Bog Village Museum, Glenbeigh 24838
Kerry County Museum, Tralee 24904
Kershaw County Historical Society Museum, Camden 47202
Kesälahden Museo, Kesälahti 10949
Keshike Tengqi Museum, Chifeng 07627
Keski-Suomen Ilmailumuseo, Tikkakoski 11376
Keski-Suomen Museo, Jyväskylä 10894
Keski-Suomen Sairaanhoitopiirin Sairaalamuseo, Jyväskylä 10895
Keski-Suomen Tieliikennemuseo, Kintaus 10961
Keskisuomalainen Museo, Jyväskylä 10896
Kestnergesellschaft, Hannover 18994
Keswick Mining Museum, Keswick 44563
Keswick Museum and Art Gallery, Keswick 44564
Kettenschmiedemuseum, Fröndenberg 18486
Ketterer-Haus-Museum, Biberach, Ortenaukreis 17327
Kettle River Museum, Midway 06281
Kettle's Yard, Cambridge 43728
Keumho Museum of Art, Seoul 29888
Keuruu Open Air Museum, Keuruu 10951
Keuruun Kotiseutumuseo, Keuruu 10951
Keuruun Museo, Keuruu 10952
Kew Bridge Steam Museum, Kew 44569
Kew Historical Museum, Kew 01195
Kew Palace Museum & Queen Charlotte's Cottage, Kew 44570
Kewaunee County Historical Museum, Kewaunee 49755
Key West Lighthouse Museum, Key West 49762
Key West Museum of Art and History, Key West 49763
Keystone Area Museum, Keystone 49770
Keystone Pioneer Museum, Roblin 06764
KGB Cells Museum of Tartu City-Museum, Tartu 10652
KGB Kongide Muuseum, Tartu 10652
Kgosi Bathoen II (Segopotso) Museum, Kanye 04298
Khabarovsk State Museum of the Far East N.I. Grodekov, Chabarovsk 36856
Khachatour Abovyan House-Museum, Yerevan 00740
Khadzhi Dimitär Memorial House, Sliven 05166
Khajana Buildings Museum, Hyderabad 24213
Khalifa's House Museum, Omdurman 40396
Khalif's House Museum Abdallah at-Ta'ayushi, Khartoum 40391
Khalkhal Museum, Khalkhal 24622
Khama III Memorial Museum, Serowe 04301
Khami Ruins Site Museum, Bulawayo 54663
The Khan Museum, Hadera 24951
Khanh Hoa Provincial Museum, Nha Trang 54634
Khariton Akhvlediani State Museum, Batumi 16334
Kharkiv Art Museum, Charkiv 43017
Kharkiv Historical Museum, Charkiv 43018
Kharkiv Literary Museum, Charkiv 43019
Kherson Local Museum, Cherson 43024
Khiching Museum, Khiching 24260
Khmelnitski Museum of Art, Chmel'nyc'kyj 43026
Khmelnitski Regional Museum, Chmel'nyc'kyj 43027
Khodshent Museum of Archaeology and Fortification, Chudžand 42492
Khodshents Historical and Regional Museum, Chudžand 42493
Khokhryakov House-Museum, Kirov 37035
Khon Kaen National Museum, Khon Kaen 42574
Khorassan Natural History Museum, Mashad 24632
Khoren Ter-Harutyan Museum, Echmiadzin 00697
Khorug Historical and Regional Museum, Chorog 42491
Khoy Museum, Khoy 24625
Khyber Digital Media Centre, Halifax 05938
Kiangan National Museum, Kiangan 34399
Kiasma Nykytaiteen Museo, Helsinki 10792
Kiato Archaeological Museum, Kiato 23050
Kibbutz Art Gallery, Tel Aviv 25113
Kibbutz Negba Museum, Kibbutz Negba 25052
Kibi Archaeological Collection, Yamateson 29584
Kibi Kokoshizou, Yamateson 29584
Kibiji Kyodo Hakubutsukan, Soja 29294
Kibiji Local Museum, Soja 29294
Kibo Art Gallery, Marangu 42516
Kiboko Projects, New York 51202
Kidderminster Railway Museum, Kidderminster 44572
Kids' Science Museum of Photons, Kizu 28721
Kidspace Children's Museum, Pasadena 51719
Kidsquest Children's Museum, Bellevue 46708
Kidwelly Industrial Museum, Kidwelly 44574
Kiehle Gallery, Saint Cloud 52539
Kiek in de Kök, Tallinn 10633
't Kiekhuus, Wolvega 32961
Kieler Stadt- und Schiffahrtsmuseum, Kiel 19530
Kienzler-Schmiede, Kirchzarten 19555
Kieppi-Luontomuseo ja Mineraalikokoelma, Kokkola 10974
Kieppi Nature Museum and Minerals Collection, Kokkola 10974
Kiev Museum of Alexander Pushkin, Kyïv 43084
Kiev Museum of Russian Art, Kyïv 43085
Kihniön Museo, Kihniö 10955
Kii-Fudoki-no-oka Museum of History, Wakayama 29572
Kiihtelysvaaran Museo, Heinävaara 10764
Kiikalan Kotiseutumuseo, Kiikala 10956
Kiikoisten Kotiseutumuseo, Kiikoinen 10957
Kiimingin Museo, Kiiminki 10958
Kijk en Luister Museum, Bennekom 31962
Kijk- en Luistermuseum, Bant 31948
KiK - Kultur im Kloster, Klostermarienberg 02198
Kikinda National Museum, Kikinda 38102
Kikkermuseum, Den Haag 32077
Kiko's Fossil and Mineral Museum, Banciao 42145
Kikuchi Kan Memorial Museum, Takamatsu 29318

Kikusui Handicraft Museum, Kawanishi 28690
Kilby Store and Farm, Harrison Mills 05974
Kildonan Museum, Kildonan, Isle of South Uist 44576
Kilkivan Historical Museum, Kilkivan 01197
Killarney Centennial Museum, Killarney 06068
Killhope, the North of England Lead Mining Museum, Weardale 45951
Kilmainham Gaol and Museum, Dublin 24805
Kilmallock Museum, Kilmallock 24853
Kilmore Quay Maritime Museum, Kilmore Quay 24854
Kilmore Quay Maritime Museum, Wexford 24913
Kilsyth's Heritage, Kilsyth 44582
Kilwinning Abbey Tower, Kilwinning 44584
Kimba and Gawler Ranges Historical Society Museum, Kimba 01198
Kimball Art Center, Park City 51706
Kimball House Museum, Battle Creek 46644
Kimbell Art Museum, Fort Worth 48715
Kimberley District Heritage Museum, Kimberley 06069
Kimberley Mine Museum, Kimberley 38669
Kimble County Historical Museum, Junction 49654
Kimchi Field Museum, Seoul 29889
Kimo Bruks Museum and Gallery, Oravais 11162
Kina Slott, Drottningholm 40489
Kincaid Museum, Kincaid 06070
Kincsem Múzeum, Tápiószentmárton 23913
Kinder-Akademie Fulda Werkraummuseum, Fulda 18511
Kinder & JugendMuseum im Prenzlauer Berg, Berlin 17153
kinder museum frankfurt, Frankfurt am Main 18380
Kinder- und Jugendgalerie Sonnensegel, Brandenburg an der Havel 17509
Kinder- und Jugendmuseum, Nürnberg 20774
Kinder- und Jugendmuseum im G.-Lübcke-Museum, Hamm, Westfalen 18970
Kinder- und Jugendmuseum München, München 20450
Kindergarten Museum, Helsinki 10799
Kindergarten Museum, Potsdam 21093
Kindergartenmuseum, Bruchsal 17597
Kindergartenmuseum Nordrhein-Westfalen, Bergisch Gladbach 17015
Kindermuseum, Aflenz 01696
Kindermuseum, Wuppertal 22743
Kindermuseum im FEZ-Berlin, Berlin 17154
Kindersley and District Plains Museum, Kindersley 06071
Kinderwagenmuseum, Nieuwolda 32583
Kinderwagens van toen, Dwingeloo 32161
Kinderwelt Walchen, Vöcklamarkt 02867
Kindheitsmuseum, Marburg 20190
Kindheitsmuseum, Schönberg, Holstein 21598
Kinematographisches Museum, Wuppertal 22744
Kinenkan Mikasa, Yokosuka 29620
King County Arts Commission Gallery, Seattle 53001
King Hooper Mansion, Marblehead 50461
King Island Historical Museum, Currie 01019
King John's Castle, Limerick 24865
King John's Hunting Lodge, Axbridge 43340
King Kamehameha V - Judiciary History Center, Honolulu 49346
King Manor, Jamaica 49591
King Matthias Palace of the Hungarian National Museum, Visegrád 23966
King Saint Stephen Museum, Székesfehérvár 23860
King Saint Stephen Museum - János Vajda Memorial House, Vál 23950
King Saint Stephen Museum - Monastery, Székesfehérvár 23861
King Saud University Museum, Riyadh 38016
King Sejong The Great Memorial Exhibition, Seoul 29890
King Township Historical Museum, King City 06072
Kingaroy Shire Art Gallery, Kingaroy 01200
Kingman County Historical Museum, Kingman 49784
Kingman Museum of Natural History, Battle Creek 46645
Kingman Tavern Historical Museum, Cummington 47908
The King's College, Alcalá de Henares 38811
Kings County Museum, Hampton 05966
Kings County Museum, Kentville 06062
Kings Landing Historical Settlement, Prince William 06673
King's Lynn Arts Centre, King's Lynn 44586
King's Lynn Museum, King's Lynn 44587
King's Mill Visitor Centre, Wrexham 46083
Kings Mountain National Military Park, Blacksburg 46842
King's Own Royal Regiment Museum, Lancaster 44632
King's Own Scottish Borderers Regimental Museum, Berwick-upon-Tweed 43461
King's Own Yorkshire Light Infantry Regimental Museum, Doncaster 43993
The King's Royal Hussars Museum, Winchester 46030
Kings Weston Roman Villa, Bristol 43637
Kingsbury Watermill Museum, Saint Albans 45544
Kingsgate Gallery, London 44895
Kingsland Homestead, Flushing 48577
Kingsley Plantation, Jacksonville 49583
Kingston Archaeological Centre, Kingston 06080
Kingston Mills Blockhouse, Elgin 05770
Kingston Museum, Kingston-upon-Thames 44602
Kingston Pioneer Museum, Kingston 01204
Kingussie Museum, Kingussie 44605
Kinistino and District Pioneer Museum, Kinistino 06094
Kinizsi Pál Vármúzeum, Nagyvázsony 23728
Kinloch Historical museum, Laxay 44649
Kinlough Folk Museum, Kinlough 24855
Kinneil Museum and Roman Fortlet, Bo'ness 43559
Kinney Pioneer Museum, Clear Lake 47605
Kinnulan Kotiseutumuseo, Kinnula 10960
Kinpo-cho Rekishi Akusha, Kinpo 28702
Kinpo History Material Pavilion, Kinpo 28702
Kinross Museum, Kinross 44607
Kinsale Regional Museum, Kinsale 24857
Kioskshop Berlin, Berlin 17155
Kipling District Museum, Kipling 06095
Kipp Gallery, Indiana 49484
Ķirbiži Forest Museum, Vitrupe 30149
Ķirbižu Meža Muzejs, Vitrupe 30149
Kirche am Hohenzollernplatz, Berlin 17156
Kirchengeschichtliches Museum in der Pfalzkapelle, Bad Wimpfen 16866
Kirchenhäusl, Sankt Oswald bei Freistadt 02673
Kirchenmuseum, Ernen 41357
Kirchenmuseum, Marienhafe 20200
Kirchenschatz, Bremgarten (Aargau) 41253
Kirchenschatz, Glarus 41436
Kirchenschatz, Sursee 41923
Kirchenschatz des Chorherrenstiftes, Beromünster 41224
Kirchenschatz im Sankt Fridolinsmünster, Bad Säckingen 16815
Kirchenschatz-Museum, Baden 41146
Kirchenschatz-Museum Sankt Martin, Altdorf 41105
Kirchenschatz Sankt Pelagius Kirche, Bischofszell 41240
Kirchliches Heimatmuseum Bärenthal, Bärenthal 16894
Kirchner Museum Davos, Davos Platz 41321
Kirie Museum of Art, Izukogen, Ito 28595
Kırıkkale Culture Center, Kırıkkale 42878
Kırıkkale Kültür Merkezi, Kırıkkale 42878
Kiril Petrov Art Gallery, Montana 05094
Kirillo-Belosersky Historical, Architectural and Artistic Museum-Reserve, Kirillov 37034
Kirillo-Belozerskij Istoriko-Architekturnyj i Chudožestvennyj Muzej-Zapovednik, Kirillov 37034
Kirishitan Shiryokan, Hirado 28518
Kirkaldy Testing Museum, London 44896
Kirkcaldy Museum and Art Gallery, Kirkcaldy 44609
Kirkham House, Paignton 45343
Kirkkomuseo, Nokia 11146
Kirkland Art Center, Clinton 47660
Kirkland Fine Arts Center-Perkinson Gallery, Decatur 48015
Kirkland Museum of Fine and Decorative Art, Denver 48080
Kırklareli Müzesi, Kırklareli 42879
Kirkleatham Museum, Redcar 45473
Kirkpatrick Science and Air Space Museum at Omniplex, Oklahoma City 51517
Kirkuk Museum, Kirkuk 24719
Kirms-Krackow-Haus, Weimar 22405
Kirov Regional Museum, Kirov 37039
Kirovograd Literary and Memorial Museum to M. Kropyvnytskyj, Kirovograd 43064
Kirovograd Museum of Literature, Kirovograd 43065
Kirovograd O. Osmorkin Memorial Museum of Fine Art, Kirovograd 43063
Kirovograd Regional Museum of Fine Art, Kirovograd 43067
Kirovograd Regional Museum of Local History, Art and Nature, Kirovograd 43068
Kirovograds'kyj Chudožno-memorial'nyj Muzej O. Osmorkina, Kirovograd 43063
Kirovograds'kyj Literaturno-memorial'nyj Muzej M.L. Kropyvnyc'kogo, Kirovograd 43064
Kirovograds'kyj Literaturnyj Muzej, Kirovograd 43065
Kirovograds'kyj Muzej Musyčnoï Kul'tury im. Karolja Šymanovs'kogo, Kirovograd 43066
Kirovograds'kyj Oblasnyj Chudožnij Muzej, Kirovograd 43067
Kirovograds'kyj Oblasnyj Krajeznavčyj Muzej, Kirovograd 43068
Kirovskij Oblastnoj Chudožestvennyj Muzej im. Viktora i Apollinarija Vasnecovych, Kirov 37038
Kirovskij Oblastnoj Kraevedčeskij Muzej, Kirov 37039
Kirpilä Art Collection, Helsinki 10827
Kirriemuir Gateway to the Glens Museum, Kirriemuir 44618
Kirschenmuseum, Endingen am Kaiserstuhl 18167
Kırşehir Culture Center, Kırşehir 42881
Kırşehir Devlet Güzel Sanatlar Galerisi, Kırşehir 42880
Kırşehir Müzesi, Kırşehir 42881
Kırşehir State Fine Art Gallery, Kırşehir 42880
Kirsten Flagstad Museum, Hamar 33529
Kirsten Kjærs Museum, Frøstrup 10031
Kirsti, Home of a Seaman Museum, Rauma 11257
Kirsti Rauman Museo, Rauma 11257
Kirtland Temple Historic Center, Kirtland 49802
Kis Géza Ormánság Múzeum és Skanzen, Sellye 23799
Kisbey Museum, Kisbey 06099
Kiscelli Múzeum, Budapest 23421
Kisfaludy Károly Emlékkiállítás, Tét 23923
Kisfaludy Sándor Emlékház, Sümeg 23825
Kishu-furitsu Hakubutsukan, Shirahama 29280
Kishu Museum, Shirahama 29280
Kisjankó Bori Emlékház, Mezőkövesd 23697
Kiskastély - Art House, Nagykanizsa 23724
Kiskastély - Művészetek Háza, Nagykanizsa 23724
Kiskon Kotiseutumuseo, Kisko 10964
Kiskun Múzeum, Kiskunfélegyháza 23666
Kiskunsag National Park - The House of Nature, Kecskemét 23656
Kislovodsk Historical Museum Fortress, Kislovodsk 37046
Kislovodskij Istoriko-kraevedčeskij Muzej Krepost', Kislovodsk 37046
Kiso-ji Bijutsukan, Agematsu 28347
Kiso-ji Museum of Art, Agematsu 28347
Kispest Region Historical Collection, Budapest 23422
Kispesti Helytörténeti Gyűjtemény, Budapest 23422
Kiss Pál Múzeum, Tiszafüred 23930
KISS - Temporäres Museum zeitgenössischer Kunst, Abtsgmünd 16421
Kistefos-museet, Jevnaker 33582
Kisumu Museum, Kisumu 29742
Kit Carson Historical Society Museum, Kit Carson 49804
Kit Carson Museum, Las Animas 49956
KIT Tropenmuseum, Amsterdam 31852
Kita-kamakura Museum, Kamakura 28644
Kitai Hakubutsukan, Kodaira 28751
Kitajskij Dvorec-muzej, Lomonosov 37109
Kitakami City Museum, Kitakami 28709
Kitakami-shiritsu Hakubutsukan, Kitakami 28709
Kitakyushu Municipal Museum of Art, Kitakyushu 28712
Kitakyushu Museum of Natural History & Human History, Kitakyushu 28713
Kitakyushu-shiritsu Bijutsukan, Kitakyushu 28712
Kitakyushu-shiritsu Shizenshi Hakubutsukan oyo Ningenteki Rekishi, Kitakyushu 28713
Kitale Museum, Kitale 29743
Kitamae-senshu-no-yakata Ukon-ke, Echizen 28414
Kitami-Abashiri-furitsu Bunkakan, Kitami 28716
Kitami-Abashiri Regional Cultural Centre, Kitami 28716
Kitami Peppermint Museum, Kitami 28717
Kitamura Bijutsukan, Kyoto 28822
Kitamura Museum, Kyoto 28822
Kitano Bijutsukan, Nagano 28955
Kitano Bunkakan, Nagano 28956
Kitano Cultural Center, Nagano 28956
Kitano Museum of Art, Nagano 28955
Kitano Temman-gu Treasure House, Kyoto 28823
Kitayama Cedar Museum, Kyoto 28824
Kitazawa Bijutsukan, Suwa 29301
Kitazawa Museum of Art, Suwa 29301
Kitazawa Museum of Art, Makuhari, Chiba 28396
Kitchener-Waterloo Art Gallery, Kitchener 06103
Kite Museum, Melaka 30599
Kite Museum, Tokyo 29481
Kiteen Kotiseutumuseo, Kitee 10965
Kithur Rani Channamma Memorial Museum, Kittur 24261
Kitikmeot Heritage Society, Cambridge Bay 05537
Kitimat Centennial Museum, Kitimat 06105
Kitsap County Historical Museum, Bremerton 47005
Kitsos Makris Folk Art Center, Volos 23238
Kitte no Hakubutsukan, Tokyo 29412
Kittelsenhuset, Kragerø 33608
Kittery Historical and Naval Museum, Kittery 49805
Kittilän Kotiseutumuseo, Kittilä 10966
Kittilbu Utmarksmuseum, Østre Gausdal 33761
Kittredge Art Gallery, Tacoma 53411
Kittson County History Center Museum, Lake Bronson 49874
Kiuchi Gallery, Ichikawa 28556
Kiuruveden Kotiseutumuseo, Kiuruvesi 10967
Kivikauden Kylä, Saarijärvi 11285
Kivimuseo, Tampere 11351
Kivirannan Kotimuseo, Säkylä 11289
Kivisakasti, Mikkeli 11114
Kiwanis Van Slyke Museum Foundation, Caldwell 47160
Kiyoharu Shirakaba Museum, Nagasaka 28967
Kiyomizu Sannenzaka Museum, Kyoto 28825
Kiyosato Kitazawa Bijutsukan, Takane 29324
Kiyosato Museum of Photographic Arts, Takane 29325
Kiyose Local Museum, Kiyose 28719
Kiyose-shi Kyodo Hakubutsukan, Kiyose 28719
Kız Teknik Öğretim Müzesi, Ankara 42704
Kizhi - State Open Air Museum of Architecture, History and Ethnography, Petrozavodsk 37559
Kiži - Gosudarstvennyj Istoriko-architekturnyj i Ėtnografičeskij Muzej-zapovednik, Petrozavodsk 37559
Kizljarskij Kraevedčeskij Muzej im. P.I. Bagrationa, Kizljar 37051
Kjarvalsstaðir, Reykjavík 24019
Kjerringøy Gamle Handelssted, Bodø 33444
Klaavola Museo, Tuusula 11415
Klæbu Bygdemuseum, Klæbu 33595
Klaipeda Clock Museum of the Lithuanian Art Museum, Klaipėda 30281
Klaipeda Picture Gallery of the Lithuanian Art Museum, Klaipėda 30279
Klaipėdos Paveikslų Galerija, Klaipėda 30279
Klamath County Museum, Klamath Falls 49807
Klamath National Forest Interpretive Museum, Yreka 54358
Klapka György Múzeum, Komáron 23678
Klappzahlenwecker Museum, Leipzig 19915
Das Klassenzimmer, Tauberbischofsheim 22009
Klassik Stiftung Weimar, Weimar 22406
Klášter Sv. Anežky České, Praha 09755
Klašter Sv. Jiří, Praha 09756
Klasztorna, Grudziądz 34684
Klaus-Groth-Museum und Neue Museumsinsel Lüttenheid, Heide 19066
Klausner Stadtgalerie, Chiusa 25879
Kleboniskes Village Household Exposition, Kleboniškes 30286
Kleboniškių Kaimo Buities Ekspozicija, Kleboniškes 30286
Klederdracht en Visserijmuseum, Bunschoten Spakenburg 32020
Klein Museum, Mobridge 50773
Klein Saab Museum d'Oude Bolneus, Woerden 32958
Kleine Galerie, Wien 03011
Kleine Humboldt-Galerie, Berlin 17157
Das Kleine Museum, Gelsenkirchen 18575
Das Kleine Museum, Hagnau am Bodensee 18863
Das kleine Spielzeugmuseum, Baden-Baden 16882
Das kleine Wäschemuseum, Otter 20947
Kleine Welt, Freiburg im Breisgau 18431
Kleines Heimatmuseum Windmühle, Werdum 22489
Kleines Kirchengeschichtsmuseum Sankt Gumbertus, Illesheim 19317
Kleines Militärmuseum Burg Laurenburg, Laurenburg 19860
Kleines Museum am Hafen, Romanshorn 41772
Kleines Museum - Kinder der Welt, Frankfurt (Oder) 18403
Kleines Plakatmuseum, Bayreuth 16957
Kleist-Museum, Frankfurt (Oder) 18404
Klemettimuseo, Kuortane 11012
Klemm Gallery, Adrian 46138
Klepp Bygdemuseum, Kleppe 33596
Klepper Museum, Rosenheim 21377
Klerksdorp Museum, Klerksdorp 38677
Klever Schwanenturm, Kleve 19565
Klevfos Industrimuseum, Ådalsbruk 33375
Klick Kindermuseum Hamburg, Hamburg 18939
Klickitat County Historical Society Museum, Goldendale 48962
Kline Creek Farm, Winfield 54231
Klingspor-Museum, Offenbach am Main 20883
Klinskoe Muzejnoe Obedinenie, Klin 37053
Klöppelmuseum, Abenberg 16417
Klokkengieterij Museum, Heiligerlee 32346
Klokkengieterijmuseum, Vries 32909
Klokkenmuseum, Frederiksoord 32233
Klokkenmuseum, Hechtel 03690
Klokkenmuseum van de Sint-Martinusbasiliek, Halle 03668
Klokkergården Bygdetun Lyngdal Misjonsmuseum, Lyngdal 33655
Klompenmuseum De Platijn, Best 31970
Klompenmuseum Gebr. Wietzes, Eelde 32173
Klompenmuseum 't Oale Ambacht, Goor 32262
Klompenmuseum 't Schöpke Enter, Enter 32215
De Klomphoek, Beilen 31956
Klondike Gold Rush National Historical Park, Seattle 53002
Klondike National Historic Sites, Dawson City 05682
Klooster Ter Apel, Ter Apel 31909
Klopstock house, Quedlinburg 21150
Klopstockhaus, Quedlinburg 21150
Kloster Alpirsbach, Alpirsbach 16477
Kloster Chorin, Chorin 17706
Kloster Ebstorf, Ebstorf 18050
Kloster Frauenchiemsee, Frauenchiemsee 18411
Kloster Hadmersleben, Hadmersleben 18851
Kloster Irsee, Irsee 19354
Kloster Isenhagen, Hankensbüttel 18986
Kloster Lüne mit Textilmuseum, Lüneburg 20098
Kloster Mariawald, Heimbach 19109

K

Kloster Mariensee, Neustadt am Rübenberge 20664
Kloster Medingen, Bad Bevensen 16695
Kloster-Museum, Sankt Märgen 21505
Kloster Sankt Marienberg, Helmstedt 19122
Kloster Schöntal, Schöntal 21614
Kloster Seeon, Seeon 21722
Kloster Walsrode, Walsrode 22317
Kloster Wiblingen, Ulm 22163
Kloster Wienhausen, Wienhausen 22542
Kloster Zinna, Kloster Zinna 19580
Klosteranlage mit Klostermuseum, Maulbronn 20239
Klosterbräu Brauereimuseum, Irsee 19355
Klostergalerie, Zehdenick 22769
Klosterkirche Blaubeuren, Blaubeuren 17379
Klosterkirche Himmelpfort, Fürstenberg, Havel 18489
Klosterlund Museum og Naturcenter, Engesvang 09998
Klostermühle, Boitzenburger Land 17437
Klostermühlenmuseum, Thierhaupten 22052
Klostermuseum, Roggenburg 21362
Klostermuseum der Benediktiner-Abtei, Ottobeuren 20955
Klostermuseum der Sankt Josefskongregation, Ursberg 22188
Klostermuseum Disentis, Disentis 41331
Klostermuseum Heggbach, Maselheim 20233
Klostermuseum Hirsau, Calw 17679
Klostermuseum in der Pfarrkirche Sankt Magnus, Bad Schussenried 16830
Klostermuseum Jerichow, Jerichow 19394
Klostermuseum Müstair, Müstair 41664
Klostermuseum St. Georgen, Stein am Rhein 41914
Klostermuseum und Caspar Wolf-Kabinett, Muri 41666
Klostermuseum Wittichen, Schenkenzell 21543
Klosterruine Memleben, Memleben 20277
Klostersammlung der Benediktinerabtei, Michaelbeuern 02378
Klostertal Museum, Dalaas 01810
Klostertunet, Halsnøy Kloster 33526
Klostervilla Adelberg, Adelberg 16425
Kluane Museum of Natural History, Burwash Landing 05507
Klub Galeria Kisielice, Poznań 35026
Kluuvin Galleria - Glogalleriet, Helsinki 10793
Klyne Esopus Museum, Ulster Park 53640
Klyuchevsky Museum, Penza 37532
Knappogue Castle, Quin 24888
Knaresborough Castle and Old Courthouse Museum, Knaresborough 44619
Knauf-Museum, Iphofen 19351
Knaut Rhuland House Museum, Lunenburg 06203
Knebworth House, Knebworth 44621
Knee Hill Historical Society Museum, Three Hills 07075
Kneipp-Museum, Bad Endbach 16728
Kner Nyomdaipari Múzeum, Gyomaendrőd 23577
Kner Printing Industry Museum, Gyomaendrőd 23577
Knife River Indian Villages National Historic Site, Stanton 53256
Knight Museum, Alliance 46230
Knights of Columbus Museum, New Haven 51052
Kninski Muzej, Knin 08889
Knock Museum, Knock 24860
Knole, Sevenoaks 45641
Knott House Museum, Tallahassee 53423
Knox County Historical Society Museum, Edina 48289
Knox County Museum, Benjamin 46735
Knox County Museum, Knoxville 49810
Knox's Headquarters State Historic Site, Vails Gate 53682
Knoxville Museum of Art, Knoxville 49819
Knud Rasmussens Hus, Hundested 10104
Knudaheio - Garborgheimen, Undheim 33945
Knutsford Heritage Centre, Knutsford 44623
Knysna Museum, Knysna 38678
KOA Art Gallery, Honolulu 49347
Kőbánya Region Historical Collection, Budapest 23423
Kőbányai Helytörténeti Gyűjtemény, Budapest 23423
Kobarid Museum, Kobarid 38431
Kobariški Muzej, Kobarid 38431
Kobe City Museum, Kobe 28735
Kobe Doll Museum, Kobe 28729
Kobe Fashion Museum, Kobe 28733
Kobe Lamp Museum, Kobe 28730
Kobe Lampwork Glass Museum, Kobe 28731
Kobe Maritime Museum, Kobe 28732
Kobe Science Museum, Kobe 28734
Kobe-shi Fukushoku Hakubutsukan, Kobe 28733
Kobe-shi Kagakukan, Kobe 28734
Kobe-shiritsu Hakubutsukan, Kobe 28735
København Museum for Moderne Kunst, København 10134
Københavns Bymuseum, København 10135
Kobisenmühle, Sankt Georgen im Schwarzwald 21494
Kocaeli Devlet Güzel Sanatlar Galerisi, İzmit 42858
Kocaeli Müzeleri ve Örenyerleri, İzmit 42859
Kocaeli Museums and Archaeological Sites, İzmit 42859
Kocaeli State Fine Art Gallery, İzmit 42858
Kochberg Palace, Park and Amateur Theatre, Großkochberg 18794
Kochi Bungakukan, Kochi 28741
Kochi-ken Kaitokukan, Kochi 28742
Kochi-kenritsu Bijutsukan, Kochi 28743
Kochi Liberty and People's Rights Museum, Kochi 28744
Kochi Literary Museum, Kochi 28741
Kochi Prefectural Kaito House, Kochi 28742
Kocsimúzeum, Bábolna 23340
Kodai Orient Hakubutsukan, Tokyo 29413
Kodai Yuzen-en Bijutsukan, Kyoto 28826
Kodály Zoltán Emlékmúzeum és Archívum, Budapest 23424
Kodama Museum, Kagoshima 28630
Kodisjoen Kotiseutumuseo, Kodisjoki 10968
Kodolányi Emlékmúzeum, Vajszló 23948
Kodolányi Memorial Museum, Vajszló 23948
Köhlereimuseum, Hasselfelde 19029
Kölcsey Ferenc Emlékszoba, Szatmárcseke 23837
Kölner Weinmuseum, Köln 19611
Kölnisches Stadtmuseum, Köln 19612
Königliches Schloss Berchtesgaden und Rehmuseum, Berchtesgaden 17001
Königshaus am Schachen, Garmisch-Partenkirchen 18540
Königspesel, Hooge 19269
Königspfalz Freilichtmuseum, Tilleda 22061
Königstein Fortress, Königstein 19656
Koenraad Bosman Museum, Rees 21212
Könyvtár és Múzeum - Borsodi Tájház, Edelény 23534
Köpings Museum, Köping 40647
Köyliö Croft-Museum, Tuiskula 11386
Köyliön Torpparimuseo, Tuiskula 11386
Középkori O-Zsinagóga, Sopron 23811
Középkori Romkert - Nemzeti Emlékhely, Székesfehérvár 23855
Középkori Zsidó Imaház, Budapest 23425
Közlekedési Múzeum, Budapest 23426
Közlekedési Múzeum - Kiskörösi Közúti Múzeum, Kiskőrös 23663
Központi Bányászati Múzeum, Sopron 23812
Központi Kohászati Múzeum, Miskolc 23708
Koffie- en Theemuseum, Amsterdam 31853
Koffie- en Winkelmuseum, Pieterburen 32669
Koffiebanderij De Gulden Tas, Bree 03403
The Koffler Gallery, Toronto 07118
Kofukuji Kokuhokan, Nara 29031
Kofukuji National Treasure House, Nara 29031
Koga Masao Museum of Music, Tokyo 29414
Koganei City Hakenomori Art Museum, Koganei 28758
Køge Museum, Køge 10159
Kogen Art Museum, Komagane, Komagane 28763
Kohán György Múzeum, Gyula 23596
Kohl Children's Museum of Greater Chicago, Glenview 48939
Kohtla Kaevanduspark-muuseum, Ida 10572
Kohtla Oil Shale Mining Museum, Ida 10572
Koishibara-mura Yakimono Bijutsukan, Koishibara 28760
Koishibara Pottery Museum, Koishibara 28760
Koizumi Yakumo Kinenkan, Matsue 28888
Koji Pottery Museum, Sindian 42334
Kojima Torajiro Kinenkan, Kurashiki 28783
Kokand Regional Studies Museum, Qo'qon 54458
Koke'e Natural History Museum, Kauai 49710
Kokemäen Maatalousmuseo ja Sotilaspoikamuseo, Kokemäki 10969
Kokemäen Ulkomuseo, Kokemäki 10970
Kokemäki Agricultural Museum and Military Museum, Kokemäki 10969
Kokemäki Open Air Museum, Kokemäki 10970
Kokerei Zollverein, Essen 18256
Kokoschka-Haus Pöchlarn, Pöchlarn 02505
Kokubunji Municipal Archaeological Gallery, Kokubunji 28761
Kokubunji-shi Bunkazai Hozonkan, Kokubunji 28761
Kokugakuin University Archaeology Museum, Tokyo 29415
Kokuritsu Kagaku Hakubutsukan, Tokyo 29416
Kokuritsu Kagaku Hakubutsukan , Shinju-ku, Tokyo 29417
Kokuritsu Kokusai Bijutsukan, Osaka 29130
Kokuritsu Minzokugaku Hakubutsukan, Minpaku, Suita 29295
Kokuritsu Miraikan, Tokyo 29418
Kokuritsu Rekishi Minzoku Hakubutsukan, Sakura 29210
Kokuritsu Seiyo Bijutsukan, Tokyo 29419
Kolarin Kunnan Kotiseutumuseo, Sieppijärvi 11307
Kolbenheyer-Archiv und Gedenkstätte, Geretsried 18602
Koldo Mitxelena Kulturunea Erakustaret, Donostia-San Sebastián 39249
Kolekcija Historyczno-regionalna Rodziny Makowskich, Chojnice 34597
Kolhapur Museum, Kolhapur 24265
Kolín Regional Museum, Kolín 09591
Kolju Ficheto Museum, Drjanovo 05025
Kolmanskop Museum, Kolmanskop 31743
Kolomoki Mounds State Park Museum, Blakely 46849
Kolomya State Museum of Folk Art, Kolomija 43072
Kolpings- und Handwerksmuseum im Faltertor, Dettelbach 17841
Columba, Köln 19613
Komagane Kogen Bijutsukan, Komagane 28763
Komagata Jukichi Bijutsu Kinenkan, Nagaoka 28958
Komagata Jukichi Memorial Art Museum, Nagaoka 28958
Komatsu City Museum, Komatsu 28765
Komatsu Hitoshi Art Museum, Kyoto 28827
Komatsu Shiritsu Hakubutsukan, Komatsu 28765
Komboloi Museum, Nafplion 23116
Komi-Permjackij Okružnoj Kraevedčeskij Muzej im. Subbotina-Permjaka, Kudymkar 37088
Komi-Permyatsk Regional Museum, Kudymkar 37088
Komi Respublikanskij Istoriko-Kraevedčeskij Muzej, Syktyvkar 37815
Komiszkenyér Kiállítás, Komáron 23679
Komlói Természettudományi Gyűjtemény, Komló 23681
Kommandør Chr. Christensen's Hvalfangst Museum, Sandefjord 33812
Kommunale Galerie Berlin, Berlin 17158
Kommunale Galerie Gelsenkirchen, Gelsenkirchen 18576
Komperon Kotiseutumuseo, Viinijärvi 11465
Kompiuterijos Muziejus, Vilnius 30412
Kompleksi Monumental i Lidhjes Shqiptare të Prizrenit, Prizren 38134
Komponierstube Gustav Mahlers, Steinbach am Attersee 02765
Kompozitoriaus Antano Budriūno Kambarys, Vilnius 30413
Kon-Tiki Museum, Oslo 33729
Konak Knegine Ljubice, Beograd 38053
Konak Kneza Miloša, Beograd 38054
Konan-ji Homotsukan, Setoda 29252
Konceptkonstmuseum, Rydboholm 40820
Kondo Yuzo Kinenkan, Kyoto 28828
Konfeksjonsmuseet, Molde 33677
Den Kongelige Mønt- og Medaillesamling, København 10136
Den Kongelige Mynts Museum, Kongsberg 33599
Kongenhus Mindeparks Museum, Viborg 10315
Konginkankaan Kotiseutumuseo, Konginkangas 10977
Kongosho-ji Treasure House, Ise 28583
Kongsberg Arms and Industry Collection, Kongsberg 33601
Kongsberg Ski Museum, Kongsberg 33600
Kongsberg Våpenfabrikks Museum, Kongsberg 33601
Kongsvinger Festningsmuseum, Kongsvinger 33604
Koninklijk Militair Historisch Museum, Delft 32059
Koninklijk Museum voor Midden-Afrika, Tervuren 04058
Koninklijk Museum voor Schone Kunsten Antwerpen, Antwerpen 03290
Koninklijk Oudheidkundig Genootschap, Amsterdam 31854
Koninklijk Paleis te Amsterdam, Amsterdam 31855
Koninklijk Tehuis voor Oud-Militairen en Museum Bronbeek, Arnhem 31928
Koninklijke Hoofdgilde Sint-Sebastiaan, Brugge 03419
Koninklijke Manufactuur De Wit, Mechelen 03859
Konjušenny Muzej-Muzej Avtomobilnogo Transporta, Sankt-Peterburg 37654
Konneveden Kotiseutumuseo, Konnevesi 10978
Konoki one Artist Museum, Nasu 29047
Konstantin Fedin Museum, Saratov 37749
Konstantin Palace, Sankt-Peterburg 37655
Konstantin Vasilyev Memorial Museum of the National Museum of the Tatarian Republic, Vasiljevo 37916
Konstantin Velichkov House-Museum, Pazardžik 05106
Konstantinas Sankenis Memorialinis Muziejus, Veleniskiai 30392
Konstantino Glinskio Sodyba-Muziejus, Jurbarka 30233
Konstantinovskij Dvorec, Sankt-Peterburg 37655
Konsthallen, Borgholm 40473
Konsthallen Sandviken, Sandviken 40832
Konsthallen Trollhättan, Trollhättan 40984
Kontiolahden Museo, Kontiolahti 10979
Konventmuseum der Barmherzigen Brüder, Wien 03012
Konya Devlet Güzel Sanatlar Galerisi, Konya 42887
Konya State Fine Art Gallery, Konya 42887
Koochiching County Historical Museum & Bronko Nagurski Museum, International Falls 49513
Kook Museum, Jerusalem 24998
Kookmin College Museum, Seoul 29891
Koopmans-De Wet House Museum, Cape Town 38585
Koorda Museum, Koorda 01205
Koorie Heritage Trust Cultural Centre-Galleries, Melbourne 01286
Koornmarktspoort, Kampen 32438
Kootenay Gallery of Art, History and Science, Castlegar 05575
Kopáncsi Tanyamúzeum, Hódmezővásárhely 23612
Kópavogur Art Museum, Kópavogur 24008
Koper Regional Museum, Koper 38434
Kopermolen, Vaals 32859
Koperslagersmuseum Van der Beele, Horst 32415
Kopiergerätemuseum, Wien 03013
Koprivnica Art Gallery, Koprivnica 08891
Koprivnica Municipal Museum, Koprivnica 08892
Koptisches Museum, Borgentreich 17476
Korablinskij Kraevedčeskij Muzej, Korablino 37057
Korallen- und Heimatmuseum, Nattheim 20550
Korbinian-Brodmann-Museum, Hohenfels 19236
Korbmacher-Museum, Beverungen 17315
Korbmacher-Museum, Malsfeld 20158
Kordegarda, Warszawa 35232
Korea Maritime University Maritime Museum, Busan 29779
Korea University Museum, Seoul 29899
Korea University Museum, Seoul 29892
Korea War Memorial, Seoul 29893
Korean Agricultural Museum, Seoul 29894
Korean American Museum, Los Angeles 50240
Korean Art Gallery, Pyongyang 29756
Korean Central Ethnographic Museum, Pyongyang 29757
Korean Central Historical Museum, Pyongyang 29758
Korean Christian Museum, Seoul 29895
Korean Cultural Center Art Gallery, Los Angeles 50241
Korean Folk Village, Kihung 29845
Korean Folklore Museum, Longjing 07970
Korean Furniture Museum, Seoul 29896
Korean Magazine Museum, Seoul 29897
Korean Museum Association, Seoul 54918
Korean Museum of Contemporary Clothing, Seoul 29898
Korean National Committee for the International Council of Museums (ICOM), Seoul 54919
Korean Revolutionary Museum, Pyongyang 29759
Korean Ski Museum, Ganseong 29810
Korean War Veterans National Museum, Rantoul 52248
Korean War Veterans National Museum, Tuscola 53622
Korekawa Archaeological Museum, Hachinohe 28464
Koren en Pelmolen De Noordstar, Noordbroek 32600
Korenmolen de Hoop, Hellendoorn 32355
Korenmolen de Hoop, Klarenbeek 32449
Korenmolen de Phenix, Nes 32568
Korenmolen de Zandhaas, Santpoort 32742
Koreshan Historic Site, Estero 48425
Korin Memorial House Museum, Moskva 37232
Koriyama City Hands-on Science Center, Koriyama 28771
Koriyama City Museum of Art, Koriyama 28770
Koriyama-shi Fureai Kagakukan, Koriyama 28771
Korkenzieher Museum Kaiserstuhl, Vogtsburg im Kaiserstuhl 22252
Kornbrennerei-Museum, Telgte 22021
Kornel Makuszyński Museum, Zakopane 35358
Kornhaus Georgenthal-Klosterruinen, Georgenthal 18591
Kornhaus-Museum, Weiler-Simmerberg 22388
Kornhausforum, Bern 41208
Korom Gyula Fodrászati Eszközök Gyűjtemény, Budapest 23427
Korom Gyula Hair Dresser Tools Collection, Budapest 23427
Korosk County Museum, Slovenj Gradec 38518
Koroški Muzej Ravne na Koroškem, Ravne na Koroškem 38503
Koroški Pokrajinski Muzej, Slovenj Gradec 38518
Korpilahden Kotiseutumuseo, Korpilahti 10980
Korppoon Kotiseutumuseo, Korppoo 10981
Korsnäs Hembygdsmuseum, Korsnäs 10982
Korsør By- og Overfartsmuseum, Korsør 10168
Kortedala Museum, Göteborg 40550
Kortesjärven Luontomuseo, Kortesjärvi 10983
Kortrijk 1302, Kortrijk 03760
Korvenkylän Nuorisoseuran Museo, Muhniemi 11120
Korvensuun Voimalaitos- ja Konepajamuseo, Mynämäki 11131
Koryo Museum of Art, Kyoto 28829
Koryo Taehakyo Pakmulgwan, Seoul 29899
Koryu-ji Reihoden, Kyoto 28830
Koryu-ji Temple Treasury, Kyoto 28830
Kos Manor House, Jesenice 38426
Kosanji Temple Museum, Setoda 29253
Koschatmuseum, Klagenfurt 02184

K

Kosciuszko-Museum, Solothurn 41884
Kosetsu Bijutsukan, Kobe 28736
Kosetsu Museum of Art, Kobe 28736
Koshare Indian Museum, La Junta 49850
Kosken Kotiseutumuseo, Koski 10984
Koskenniskan Mylly- ja Kievarimuseo, Tuupovaara 11411
Kosova Graščina, Jesenice 38426
Kosovo Museum, Priština 38133
Kossuth Lajos Emlékmúzeum, Monok 23717
Kossuth Müzesi, Kütahya 42894
Kossuth Museum Ship of the Transport Museum, Budapest 23428
Kossuth Múzeum, Cegléd 23503
Kossuth Múzeumhajó, Budapest 23428
Kostanaj Historical and Regional Museum, Kostanaj 29706
Kostanaj Memorial Museum of I. Altynsarin, Kostanaj 29705
Kostanajskij Memorialnyj Muzej im. I. Altynsarina, Kostanaj 29705
Kostanajskij Oblastnoj Istoriko-Kraevedčeskij Muzej, Kostanaj 29706
Kostede Museum für Überlebenskunst, Kamen 19418
Kostens hus, Luleå 40697
Kostroma City Gallery of Arts, Kostroma 37063
Kostroma Historical and Architectural Museum-Preserve, Kostroma 37061
Kostroma Museum of Fine Arts, Kostroma 37062
Kostromskoj Gosudarstvennyj Istoriko-architekturnyj Muzej-zapovednik, Kostroma 37061
Kostromskoj Gosudarstvennyj Obedinennyj Chudožestvennyj Muzej, Kostroma 37062
Kostuummuseum De Gouden Leeuw, Noordhorn 32602
Koszta József Múzeum, Szentes 23884
Kotel Enlighteners Museum and Rakovski Pantheon, Kotel 05077
Kotiseutumuseo, Jurva 10880
Kotiseutumuseo Lukkarin Puustelli, Loppi 11081
Kotiseutumuseo Muina, Rauma 11258
Kotiseutumuseo Pellava ja Sahti, Alfred Kordelinin Ajokalumuseo, Lammi 11042
Kotiseututalo Kahari, Noormarkku 11148
Kotkaniemi - Presidentti P.E. Svinhufvudin koti, Luumäki 11086
Kotkaniemi - The Home of President P.E. Svinhufvud, Luumäki 11086
Koto-ku Fukagawa Edo Siryokan, Tokyo 29420
Koto-ku Nakagawa Funabansho Museum, Tokyo 29421
Kotohira-gu Museum, Kotohira 28772
Kotor Maritime Museum, Kotor 31647
Kotsu Kagaku Hakubutsukan, Osaka 29131
Kotula's Timber House, Havířov 09508
Kotulova Dřevěnka, Havířov 09508
Koulumuseo, Helsinki 10794
Koulumuseo, Somero 11320
Kourion Museum, Episkopi 09363
Kouseiroudoushou Sangyo Anzen Gijutsukan, Tokyo 29422
Kouvolan Taidemuseo, Kouvola 10990
Kovács Margit Állandó Kerámia Kiállítás, Győr 23584
Kovalenko Art Museum, Krasnodar 37065
Kovaski Muzej Kropa, Kropa 38441
Kovjoki Museo, Kovjoki 10993
Koxinga Museum, Tainan 42362
Koyama Keizo Bijutsukan, Komoro 28769
Koyasan Reihokan, Koya 28773
Koyunoglu Metropolit Museum, Konya 42888
Koyunoğlu Şehir Müzesi, Konya 42888
Kozanji Temple Museum, Setoda 29252
Kozu Kobunka Museum, Kyoto 28831
KPM Welt, Berlin 17159
Kraevaja Detskaja Kartinnaja Galereja, Vladivostok 37938
Kraevedčeskij Muzej Evrejskoj Avtonomnoj Oblasti, Birobidžan 36807
Kraevedčeskij Muzej g. Lomonosova, Lomonosov 37110
Kraevedčeskij Muzej g. Puškina, Puškin 37580
Kraevedčeskij Muzej g. Puškino, Puškino 37584
Kraevedčeskij Muzej Novgorodskogo Rajona, Velikij Novgorod 37919
Kraevoj Muzej Istorii Literatury, Iskusstva i Kultury Altaja, Barnaul 36791
Kraftwerkmuseum Löntsch, Netstal 41675
Krahuletz-Museum, Eggenburg 01843
Krajanské Múzeum MS Bratislava, Bratislava 38217
Krajská Galerie Výtvarného Umění ve Zlíně, Zlín 09940
Krajské Múzeum, Prešov 38335
Krajské Muzeum Cheb, Cheb 09473
Krajské Muzeum Sokolov, Sokolov 09844
Kråkö Båtbyggarmuseum, Kråkö 10994
Kråkö Boatbuilding Museum, Kråkö 10994
Kråksundet notnaust Sjøbruksmuseum, Tustna 33936
Kralingsmuseum, Rotterdam 32709
Kraljevo Public Museum, Kraljevo 38111
Královská Mincovna v Jáchymove, Jáchymov 09549
Královský Letohrádek, Praha 09757
Krambuvika Bygdemuseum, Tennevoll 33904
Das Kranichhaus, Otterndorf 20950
Krankenhaus-Museum am Klinikum Bremen-Ost, Bremen 17555
Krankenhausmuseum, Nürnberg 20775
Krannert Art Museum, Champaign 47335
Krasl Art Center, Saint Joseph 52558
Krāslavas Vēstures un Mākslas Muzejs, Krāslava 30033
Krasna Horka Castle, Krásnohorské Podhradie 38278
Krasnodarskij Chudožestvennyj Muzej im. F.A. Kovalenko, Krasnodar 37065
Krasnodarskij Gosudarstvennyj Istoriko-archeologičeskij Muzej-zapovednik, Krasnodar 37066
Krasnodarskij Vystavočnyj Zal, Krasnodar 37067
Krasnojarsk Regional Museum, Krasnojarsk 37072
Krasnojarskij Chudožestvennyj Muzej im. V.I. Surikova, Krasnojarsk 37069
Krasnojarskij Gorodskoj Vystavočnyj Zal, Krasnojarsk 37070
Krasnojarskij Istoričeskij Muzej Memorial Pobedy, Krasnojarsk 37071
Krasnojarskij Kraevoj Kraevedčeskij Muzej, Krasnojarsk 37072
Krasnojarskij Kulturno-istoričeskij Muzejnyj Kompleks, Krasnojarsk 37073
Krasnokamsk Picture Gallery, Krasnokamsk 37078
Krasnokamskaja Kartinnaja Galereja, Krasnokamsk 37078
Krasnokamskij Kraevedčeskij Muzej, Krasnokamsk 37079
Krasnoturinsk Regional Museum, Krasnoturinsk 37080
Krasnoturinskij Kraevedčeskij Muzej, Krasnoturinsk 37080
Kraszewski-Museum, Dresden 17949
Kražiai Ethnographic Museum, Kražiai 30287
Kražių Kraštotyros Muziejus, Kražiai 30287
The Kreeger Museum, Washington 53885
Kreidemuseum Gummanz, Sagard 21469
Kreis Feuerwehrmuseum Lohmen, Lohmen, Kreis Güstrow 20024
Kreis- und Heimatmuseum, Wittstock/Dosse 22628
Kreis- und Heimatmuseum Bogen, Bogen 17432
Kreis- und Stadtmuseum Dieburg, Dieburg 17846
Kreis- und Universitätsmuseum Helmstedt, Helmstedt 19123
Kreisagrarmuseum, Dorf Mecklenburg 17901
Kreisbergbaumuseum Sassenroth, Herdorf 19138
Kreisgalerie, Dahn 17777
Kreisgalerie, Mellrichstadt 20275
Kreisheimatmuseum, Grimma 18761
Kreisheimatmuseum Demmin, Demmin 17818
Kreisheimatmuseum Frankenberg, Frankenberg, Eder 18343
Kreisheimatmuseum Striedieck Hof, Bünde 17636
Kreisheimatmuseum Weißes Haus, Rotenburg an der Fulda 21399
Kreisheimatstube, Ellzee 18125
Kreismuseum, Osterburg 20931
Kreismuseum Bad Liebenwerda, Bad Liebenwerda 16778
Kreismuseum Bitburg-Prüm, Bitburg 17368
Kreismuseum Bitterfeld, Bitterfeld 17370
Kreismuseum der Heimatvertriebenen, Seligenstadt 21738
Kreismuseum Finsterwalde, Finsterwalde 18321
Kreismuseum Geilenkirchen, Geilenkirchen 18554
Kreismuseum Heinsberg, Heinsberg 19112
Kreismuseum Herzogtum Lauenburg, Ratzeburg 21197
Kreismuseum Jerichower Land, Genthin 18586
Kreismuseum Mühldorf, Mühldorf am Inn 20399
Kreismuseum Oberhavel, Oranienburg 20909
Kreismuseum Peine, Peine 21001
Kreismuseum Prinzeßhof, Itzehoe 19369
Kreismuseum Schönebeck, Schönebeck 21602
Kreismuseum Syke, Syke 21991
Kreismuseum Walderbach, Walderbach 22291
Kreismuseum Wewelsburg, Büren 17637
Kreismuseum Zons, Dormagen 17904
Krell'sche Schmiede, Wernigerode 22496
Kresge Art Museum, East Lansing 48253
Kresz Géza Ambulance Museum, Budapest 23429
Kresz Géza Mentomúzeum, Budapest 23429
Kretinga Museum, Kretinga 30288
Kretingos Muziejus, Kretinga 30288
Kreuzberg Museum, Berlin 17160
Kreuzstadelmuseum, Mogersdorf 02402
Krikorian Gallery, Worcester 54300
Kriminalmuseum der Kantonspolizei Zürich, Zürich 42079
Krippana - Exposition Internationale de Crèches d'Église, Büllingen 03504
Krippen- und Spielzeugmuseum, Bubendorf 41262
Krippenmuseum, Dornbirn 01826
Krippenmuseum, Fulpmes 01911
Krippenmuseum, Glattbach 18654
Krippenmuseum, Telgte 22022
Krippenmuseum, Vösendorf 02870
Krippensammlung, Großraming 02008
Krippenstube-Heimatstube, Plößberg 21067
Krishna Menon Museum and Art Gallery, Kozhikode 24285
Krišjāņa Barona Muzejs, Rīga 30073
Krišjānis Barons Museum, Rīga 30073
Kristall-Museum, Obergesteln 41693
Kristallgrotte, Guttannen 41465
Kristallmuseum, Guttannen 41466
Kristallmuseum, Viechtach 22225
Kristallmuseum Riedenburg, Riedenburg 21317
Kristalmuseum 't Los Hoes, Borculo 31984
Kristianhus Motormuseum, Vik i Sogn 33978
Kristiansand Kanonmuseum, Kristiansand 33610
Kristinehamn Museum of Art, Kristinehamn 40656
Kristinehamns Konstmuseum, Kristinehamn 40656
Kristjan Raua Majamuuseum, Tallinn 10634
Kristjan Raud House-Museum of the Estonian Art Museum, Tallinn 10634
Krivoriz'kyj Krajeznavčyj Muzej, Kryvyj Rig 43075
Krkonošské Muzeum, Jilemnice 09563
Krkonošské Muzeum - Augustiniánský Klášter, Vrchlabí 09914
Krkonošské Muzeum - Tři historické domy, Vrchlabí 09915
Krøderbanen, Krøderen 33617
Kröller-Müller Museum, Otterlo 32642
Krönungssaal des Aachener Rathauses, Aachen 16400
Kronbloms Kök, Örebro 40786
Kronborg Castle, Helsingør 10066
Kronborg Slot, Helsingør 10066
Kronobergs Lantbruksmuseum, VÄXJÖ 41038
Krookila Local History Museum, Raisio 11251
Krookilan Kotiseutukeskus, Raisio 11251
Krshishevskij and Starkov Memorial Museum, Minusinsk 37134
Krudttaarnsmuseet, Frederikshavn 10027
Krudtværksmuseet, Frederiksværk 10030
Krügemuseum, Creußen 17747
Krügersches Haus, Geesthacht 18550
Kruger Museum, Tshwane 38748
Kruidentuin de Groene Kruidhof, Elburg 32198
Kruiersmuseum, Balen 03339
krupp stadt museum, Berndorf 01780
Kruse House Museum, West Chicago 54050
Kruševac City Museum, Kruševac 38112
Krylov Museum of the Tula Museum of Fine Arts, Tula 37868
Krymski Respublikanskij Kraevedčeskij Muzej, Simferopol' 43185
Krypta 182, Bergisch Gladbach 17016
Krypte en Schatkamer van de O.L.V. Basiliek, Halle 03669
Kryptemuseum onder de Sint-Hermeskerk, Ronse 03985
Ksan Historical Village and Museum, Hazelton 05979
Kuala Lumpur Craft Museum, Kuala Lumpur 30575
Kuandu Museum of Fine Arts, Taipei 42396
Kubalu School Museum, Dundaga 30005
Kubalu Skolas Muzejs, Dundaga 30005
Kubin-Haus, Wernstein am Inn 02925
Kubiner Heimatstube, Weinsberg 22433
Kubinkabinett im Badhaus, Leogang 02290
Kubinyi Ferenc Múzeum, Szécsény 23840
Kuboso Memorial Museum of Arts, Izumi, Izumi, Osaka 28613
Kubus Hannover, Hannover 18995
Kuća Cveća, Beograd 38055
Kuddnäs Museum, Nykarleby 11155
Küfer-Museum, Langenau 19819
Küferei-Museum, Eberbach 18032
Kügelgenhaus, Dresden 17950
Kuenburg Kunststätte, Payerbach 02480
Künstlerbahnhof Ebernburg, Bad Münster am Stein-Ebernburg 16786
Künstlerbunker, Leverkusen 19975
Künstlerforum Bonn, Bonn 17458
Künstlergarten Wietzen - Alfred Kurz, Lichtenau 02297
Künstlerhaus Büchsenhausen, Innsbruck 02120
Künstlerhaus Exter, Übersee 22147
Künstlerhaus Graz am Landesmuseum Joanneum, Graz 01975
Künstlerhaus Hooksiel, Hooksiel 19270
Künstlerhaus Marktoberdorf, Marktoberdorf 20219
Künstlerhaus mit Galerie, Göttingen 18692
Künstlerhaus Schloß Balmoral, Bad Ems 16724
Künstlerhaus Walter Helm, Aschaffenburg 16584
Künstlerzeche Unser Fritz 2/3, Herne 19154
Kürbismühlenmuseum, Preding 02529
Küstenmuseum, Juist 19408
Küstenmuseum Wilhelmshaven, Wilhelmshaven 22577
Kütahya Devlet Güzel Sanatlar Galerisi, Kütahya 42895
Kütahya Müzeleri ve Örenyerleri, Kütahya 42896
Kütahya Museums and Archaeological Sites, Kütahya 42896
Kütahya State Fine Art Gallery, Kütahya 42895
Kufa Gallery, London 44897
Kuhlmann King Historical House and Museum, Boerne 46889
Kuhmalahden Kotiseutumuseo, Kuhmalahti 10998
Kuhmoisten Kotiseutumuseo, Kuhmoinen 11001
Kuivaniemen Kotiseutumuseo, Kuivaniemi 11002
Kuḱa-Muzej na Narodniot Cheroj Christijan Todorovski Karpoš, Kumanovo 30506
Kukkola Fiskemuseum, Haparanda 40582
Kuknip Minsok Pakmulgwan, Seoul 29900
Kukolnaja Galereja Vachtanov, Moskva 37221
Kuldīgas Novada Muzejs, Kuldīga 30034
Kulikovo Field - Museum-preserve of Military History and Nature, Tula 37866
Kulla Gunnarstorps Mölla, Helsingborg 40590
Kullaan Kotiseutumuseo, Kullaa 11004
Kullängsstugan, Askersund 40444
Kultamuseo Tankavaara, Tankavaara 11370
Kulttuurien Museo, Helsinki 10795
Kultur Bahnhof Eller, Düsseldorf 17994
Kultur-Gut-Museum, Aigen im Mühlkreis 01699
Kultur-Historisches Museum Grenchen, Grenchen 41452
Kultur- und Heimattreff, Tholey 22055
Kultur- und Kommunikationszentrum Lagerhalle Osnabrück, Osnabrück 20922
Kultur- und Militärmuseum, Grafenwöhr 18733
Kultur- und Museumszentrum Schloss Glatt, Sulz am Neckar 21981
Kultur- und Orgelzentrum, Valley 22202
Kultur- und Stadthistorisches Museum Duisburg, Duisburg 18020
Kulturama, Zürich 42080
Kulturamt Mitte, Berlin 17161
Kulturamt Neukölln, Berlin 17162
Kulturbunker Mülheim, Köln 19614
Kulturcentrum i Ronneby, Ronneby 40815
Kulturelles Forum im Freiherr-vom-Stein-Haus, Langenfeld 19833
Kulturfabrik Hainburg, Hainburg an der Donau 02038
Kulturforum Alte Post, Neuss 20657
Kulturforum Burgkloster und Museum für Archäologie, Lübeck 20076
Kulturforum Würth, Kolding 10162
Kulturforum Würth Chur, Chur 41303
Kulturgården på Lassfolk, Lillby 11071
Kulturgeschichtliches Museum Bügeleisenhaus, Hattingen 19037
Kulturgeschichtliches Museum Osnabrück, Osnabrück 20923
Kulturgut Höribach, Sankt Lorenz 02663
Kulturhalle Tübingen, Tübingen 22117
Kulturhaus der Otto-Hellmeier-Stiftung, Raisting 21172
Kulturhistorie Sønderborg Slot, Sønderborg 10279
KulturhistorieTønder, Tønder 10298
Kulturhistorische Sammlung, Iphofen 19352
Kulturhistorisches Karnevalsmuseum des Regionalverband Düren, Niederzier 20717
Kulturhistorisches Museum, Görlitz 18679
Kulturhistorisches Museum, Magdeburg 20123
Kulturhistorisches Museum der Hansestadt Stralsund, Stralsund 21914
Kulturhistorisches Museum Franziskanerkloster, Zittau 22812
Kulturhistorisches Museum Prenzlau im Dominikanerkloster, Prenzlau 21118
Kulturhistorisches Museum Rostock, Rostock 21393
Kulturhistorisches Museum Schloß Merseburg, Merseburg 20292
Kulturhistorisk Museum, Randers 10229
Kulturhistoriska Museet i Bunge, Fårösund 40517
Kulturhistoriska Museet 'Kulturen i Lund', Lund 40703
De Kulturhistoriske Samlinger, Bergen 33429
Kulturhuset, Stockholm 40903
Kulturkunststatt Prora, Prora 21131
Kulturmühle Lützelflüh, Lützelflüh 41579
Kulturno-obrazovatelnyj centr Škola L.N. Tolstogo, Moskva 37222
Kulturno Povijesni Muzej, Dubrovnik 08865
Kulturno-vystavočnyj Centr Raduga, Samara 37615
Kulturno-vystavočnyj Centr Raduga, Uljanovsk 37900
Kulturno-Vystavočnyj Kompleks Sokolniki, Moskva 37223
Kulturnyj Centr Kazan, Kazan 37010
Kulturnyj Centr-muzej V.S. Vysockogo, Moskva 37224
Kulturpfad Root, Root 41774
Kulturschloß Reichenau, Reichenau an der Rax 02569
Kulturspeicher im Schamuhn Museum, Uelzen 22151
Kulturstiftung Schloss Britz, Berlin 17163
Kulturzentrum Hans-Reiffenstuel-Haus, Pfarrkirchen 21024
Kulturzentrum Kolvenburg, Billerbeck 17352
Kulturzentrum Ostpreußen, Ellingen 18116
Kulturzentrum Schloß Bonndorf, Bonndorf im Schwarzwald 17470

Kulturzentrum Sinsteden des Rhein-Kreises Neuss, Rommerskirchen ... 21367
Kulunqi Religion Museum, Kailu ... 07911
Kuma Bijutsukan, Kuma ... 28774
Kuma Museum of Art, Kuma ... 28774
Kumagai Morikazu Museum, Tokyo ... 29423
Kumamoto Castle, Kumamoto ... 28775
Kumamoto City Museum, Kumamoto ... 28780
Kumamoto International Folklore Museum, Kumamoto ... 28778
Kumamoto-kenritsu Bijutsukan, Kumamoto ... 28776
Kumamoto-kenritsu Gigei Bijutsukan, Kumamoto ... 28777
Kumamoto Kokusai Mingeikan, Kumamoto ... 28778
Kumamoto Prefectural Arts and Crafts Museum, Kumamoto ... 28777
Kumamoto Prefectural Museum of Art, Kumamoto ... 28776
Kumamoto-shiritsu Gendai Bijutsukan, Kumamoto ... 28779
Kumamoto-shiritsu Hakubutsukan, Kumamoto ... 28780
Kumaya Art Museum, Hagi ... 28471
Kumaya Bijutsukan, Hagi ... 28471
Kume Bijutsukan, Tokyo ... 29424
Kume Museum of Art, Tokyo ... 29424
Kun-Shan University Art Center, Yongkang ... 42474
Kun Zsigmond Lakásmúzeuma, Budapest ... 23430
Kunaicho Sannomaru Shozokan, Tokyo ... 29425
Kundencenter der Stadtwerke Strausberg, Strausberg ... 21920
Kunffy Lajos Emlékmúzeum, Somogytúr ... 23806
Kung Long Mineral Museum, Sincheng ... 42331
Kungajaktmuseet, Vargön ... 41053
Kungl. Myntkabinettet, Stockholm ... 40904
Kungliga Akademien för de Fria Konsterna, Stockholm ... 40905
Kungliga Slottet i Stockholm, Stockholm ... 40906
Kungsstugan och Cajsa Wargs Hus, Örebro ... 40787
Kungsudden Local Museum, Kungsör ... 40661
Kungurskij Kraevedčeskij Muzej, Kungur ... 37089
Kunigo Stanislovo Rimkaus Muziejus, Laukuva ... 30297
Kunisaki History and Folklore Museum, Kunisaki ... 28781
Kunisaki-machi Rekishi Minzoku Shiryokan, Kunisaki ... 28781
Kunizakari Sake Museum, Handa ... 28496
Kunja-Urgench Historical Site Museum, Könürgenç ... 42977
Kunozan-Toshogu-ji Homotsukan, Shizuoka ... 29284
Kunozan-Toshogu Shrine Treasure Museum, Shizuoka ... 29284
Kûnqû Bówùguăn, Shanghai ... 08170
Kunst Archiv Darmstadt, Darmstadt ... 17791
Kunst aus Nordrhein-Westfalen, Aachen ... 16401
Kunst Haus Wien, Wien ... 03014
Kunst im Alten Schützenhaus, Zofingen ... 42060
Kunst im Rohnerhaus, Lauterach ... 02270
Kunst im Turm, Sömmerda ... 21793
Kunst Pavillon, München ... 20451
Kunst Raum Goethestraße, Linz ... 02316
Kunst Raum Riehen, Riehen ... 41764
Kunst:raum Sylt Quelle, Rantum ... 21181
Kunst- und Ausstellungshalle der Bundesrepublik Deutschland, Bonn ... 17459
Kunst- und Gewerbeverein Regensburg, Regensburg ... 21227
Kunst- und Heimathaus, Paderborn ... 20971
Kunst- und Kulturpädagogisches Zentrum der Museen in Nürnberg, Nürnberg ... 54839
Kunst und Museum, Hollfeld ... 19253
Kunst- und Museumsbibliothek, Köln ... 19615
Kunst- und Museumsverein Wuppertal e.V., Wuppertal ... 22745
Kunst- und Skulpturenmuseum, Heilbronn ... 19097
Kunst- und Wunderkammer Burg Trausnitz, Landshut ... 19802
Kunstamt Tempelhof-Schöneberg, Berlin ... 17164
Kunstausstellung des Rothenburger Künstlerbundes, Rothenburg ob der Tauber ... 21407
Kunstbibliothek, Berlin ... 17165
Kunstbunker Tumulka, München ... 20452
Kunstcentrum De Waagh, Oldenzaal ... 32616
Kunstcentrum Pand Paulus, Schiedam ... 32753
Kunstenaarscentrum De Fabriek, Eindhoven ... 32189
Kunstfabrik am Flutgraben, Berlin ... 17166
Kunstforeningen GL Strand, København ... 10137
Kunstforum Arabellapark, München ... 20453
Kunstforum Bâloise, Basel ... 41164
Kunstforum der Berliner Volksbank, Berlin ... 17167
Kunstforum Ebendorf Raumacht, Wien ... 03015
Kunstforum Ostdeutsche Galerie, Regensburg ... 21228
Kunstforum Würth, Turnhout ... 04098
Kunstgewerbemuseum, Berlin ... 17168
Kunstgewerbemuseum, Dresden ... 17951
Kunstgewerbemuseum im Schloss Köpenick, Berlin ... 17169
Kunstgewerbesammlung, Zürich ... 42081
Kunstgußmuseum Lauchhammer, Lauchhammer ... 19845
Kunsthal Rotterdam, Rotterdam ... 32710
Kunsthal Sint-Pietersabdij, Gent ... 03632
Kunsthalle, Bremerhaven ... 17571
Kunsthalle Arbon, Arbon ... 41122
Kunsthalle Barmen, Wuppertal ... 22746
Kunsthalle Basel, Basel ... 41165
Kunsthalle Bern, Bern ... 41209
Kunsthalle Bielefeld, Bielefeld ... 17341
Kunsthalle Bremen, Bremen ... 17556
Kunsthalle Darmstadt, Darmstadt ... 17792
Kunsthalle der Europäischen Kunstakademie, Trier ... 22099
Kunsthalle der Hypo-Kulturstiftung, München ... 20454
Kunsthalle der Sparkasse Leipzig, Leipzig ... 19916
Kunsthalle Dominikanerkirche, Osnabrück ... 20924
Kunsthalle Dresden im Art'otel Dresden, Dresden ... 17952
Kunsthalle Düsseldorf, Düsseldorf ... 17995
Kunsthalle Emden, Emden ... 18138
Kunsthalle Erfurt im Haus zum Roten Ochsen, Erfurt ... 18205
Kunsthalle Exnergasse, Wien ... 03016
Kunsthalle Faust, Hannover ... 18996
Kunsthalle Fridericianum, Kassel ... 19465
Kunsthalle Gießen, Gießen ... 18637
Kunsthalle Göppingen, Göppingen ... 18674
Kunsthalle Harry-Graf-Kessler Weimar, Weimar ... 22407
Kunsthalle Helsinki, Helsinki ... 10783
Kunsthalle im Schloss, Isny im Allgäu ... 19366
Kunsthalle Jesuitenkirche, Aschaffenburg ... 16585
Kunsthalle Krems, Krems ... 02227
Kunsthalle Leoben, Leoben ... 02285
Kunsthalle Lingen, Lingen ... 20009
Kunsthalle Loppen, Loppem ... 03833
Kunsthalle Mainz, Mainz ... 20144
Kunsthalle Mannheim, Mannheim ... 20166
Kunsthalle Nürnberg, Nürnberg ... 20776
Kunsthalle Palazzo, Liestal ... 41563
Kunsthalle Rostock, Rostock ... 21394
Kunsthalle Sankt-Annen und Sankt-Annen-Museum, Lübeck ... 20077
Kunsthalle Sankt Gallen, Sankt Gallen ... 41804
Kunsthalle Schweinfurt, Schweinfurt ... 21683
Kunsthalle Tübingen, Tübingen ... 22118
Kunsthalle Vierseithof, Luckenwalde ... 20042
Kunsthalle Villa Kobe, Halle, Saale ... 18898
Kunsthalle Weishaupt, Ulm ... 22164
Kunsthalle Wien, Wien ... 03017
Kunsthalle Wien project space, Wien ... 03018
Kunsthalle Wil, Wil ... 42016
Kunsthalle Wilhelmshaven, Wilhelmshaven ... 22578
Kunsthalle Winterthur, Winterthur ... 42028
Kunsthalle Würth, Schwäbisch Hall ... 21657
Kunsthalle Ziegelhütte, Appenzell ... 41117
Kunsthalle zu Kiel, Christian-Albrechts-Universität, Kiel ... 19531
Kunsthalle Zürich, Zürich ... 42082
Kunsthallen Brandts, Odense ... 10217
Kunsthaus Apolda Avantgarde, Apolda ... 16566
Kunsthaus Baselland, Muttenz ... 41670
Kunsthaus Bregenz, Bregenz ... 01795
Kunsthaus Centre PasquArt, Biel ... 41228
Kunsthaus Dresden, Dresden ... 17953
Kunsthaus Erfurt, Erfurt ... 18206
Kunsthaus Essen, Essen ... 18257
Kunsthaus Glarus, Glarus ... 41437
Kunsthaus Graz am Landesmuseum Joanneum, Graz ... 01976
Kunsthaus Grenchen, Grenchen ... 41453
Kunsthaus Hamburg, Hamburg ... 18940
Kunsthaus im KunstkulturQuartier, Nürnberg ... 20777
Kunsthaus Kannen, Münster ... 20511
Kunsthaus Kaufbeuren, Kaufbeuren ... 19489
Kunsthaus Köflach, Köflach ... 02211
Kunsthaus Langenthal, Langenthal ... 41526
Kunsthaus Meyenburg, Nordhausen ... 20737
Kunsthaus Mürzzuschlag, Mürzzuschlag ... 02413
Kunsthaus Rehau und Archiv Eugen Gomringer, Rehau ... 21238
Kunsthaus Stade, Stade ... 21850
Kunsthaus Stadt Bocholt, Bocholt ... 17393
Kunsthaus Zürich, Zürich ... 42083
Kunsthaus Zug, Zug ... 42125
Kunsthistorisches Museum Wien, Wien ... 03019
Kunsthof Bad Salzelmen, Schönebeck ... 21603
Kunsthof Zürich, Zürich ... 42084
Kunsthuis Secretarie, Meppel ... 32550
Kunstigalerii, Narva ... 10597
Kunstindustrimuseet, Oslo ... 33730
Kunstkammer der Pfarrkirche Sankt Georg, Bocholt ... 17394
Kunstkammer des Innsbrucker Servitenklosters, Innsbruck ... 02121
Kunstkaten Ahrenshoop, Ahrenshoop, Ostseebad ... 16442
Kunstkeller Annaberg, Annaberg-Buchholz ... 16554
KunstLANDing, Aschaffenburg ... 16586
Kunstlocatie Würth, 's-Hertogenbosch ... 32372
Kunstmühle Schwaan, Schwaan ... 21644
Kunstmuseen Krefeld, Krefeld ... 19705
Kunstmuseet Brundlund Slot, Aabenraa ... 09954
Kunstmuseet i Tønder, Tønder ... 10299
Kunstmuseet Køge Skitsesamling, Køge ... 10160
Kunstmuseum, Sankt Gallen ... 41805
Kunstmuseum Ahlen, Ahlen ... 16438
Kunstmuseum Basel, Basel ... 41166
Kunstmuseum Bayreuth mit Tabakhistorischer Sammlung der British American Tobacco, Bayreuth ... 16958
Kunstmuseum Bern, Bern ... 41210
Kunstmuseum Bochum, Bochum ... 17401
Kunstmuseum Bonn, Bonn ... 17460
Kunstmuseum Celle mit Sammlung Robert Simon, Celle ... 17688
Kunstmuseum des Kantons Thurgau, Warth ... 42002
Kunstmuseum Dieselkraftwerk Cottbus, Cottbus ... 17732
Kunstmuseum Dr. Krupp, Weinbach ... 22424
Kunstmuseum Gehrke-Remund, Baden-Baden ... 16883
Kunstmuseum Gelsenkirchen, Gelsenkirchen ... 18577
Kunstmuseum Heidenheim, Heidenheim ... 19090
Kunstmuseum Hohenkarpfen, Hausen ob Verena ... 19050
Kunstmuseum Kloster Unser Lieben Frauen, Magdeburg ... 20124
Kunstmuseum Liechtenstein, Vaduz ... 30197
Kunstmuseum Luzern, Luzern ... 41593
Kunstmuseum Mülheim an der Ruhr, Mülheim an der Ruhr ... 20414
Kunstmuseum Nörvenich, Nörvenich ... 20728
Kunstmuseum Olten, Olten ... 41707
Kunstmuseum Poschiavo, Poschiavo ... 41730
Kunstmuseum Solingen, Solingen ... 21802
Kunstmuseum Solothurn, Solothurn ... 41885
Kunstmuseum Stuttgart, Stuttgart ... 21948
Kunstmuseum Thun, Thun ... 41936
Kunstmuseum Thun, Thun ... 41937
Kunstmuseum Walter, Augsburg ... 16626
Kunstmuseum Winterthur, Winterthur ... 42029
Kunstmuseum Wolfsburg, Wolfsburg ... 22665
KunstNatturLandschaft, Heideblick ... 19067
Kunstpalais, Erlangen ... 18222
Kunstpanorama Luzern, Luzern ... 41594
Kunstpaviljoen, Nieuw Roden ... 32572
Kunstraum Aarau, Aarau ... 41082
Kunstraum Baden, Baden ... 41147
Kunstraum Bruxelles, Bruxelles ... 03449
Kunstraum Dornbirn, Dornbirn ... 01827
Kunstraum Düsseldorf, Düsseldorf ... 17996
Kunstraum Farmsen, Hamburg ... 18941
Kunstraum Innsbruck, Innsbruck ... 02122
Kunstraum Klosterkirche, Traunstein ... 22077
Kunstraum Kreuzberg/Bethanien, Berlin ... 17170
Kunstraum Leuphana, Lüneburg ... 20099
Kunstraum Notkirche, Essen ... 18258
Kunstraum Sankt Virgil, Salzburg ... 02617
Kunstraum Stein, Krems ... 02228
Kunstsammlung der Universität Göttingen, Göttingen ... 18693
Kunstsammlung des Herzoglichen Georgianums, München ... 20455
Kunstsammlung Hans und Marlis Suter, Oberhofen am Thunersee ... 41695
Kunstsammlung im Schloß Rheinstein, Trechtingshausen ... 22085
Kunstsammlung im Stadtmuseum, Jena ... 19383
Kunstsammlung Lausitz, Senftenberg ... 21743
Kunstsammlung Lorenzkapelle Rottweil, Rottweil ... 21425
Kunstsammlung Neubrandenburg, Neubrandenburg ... 20591
Kunstsammlung Nordrhein-Westfalen, Düsseldorf ... 17997
Kunstsammlung Nordrhein-Westfalen, Düsseldorf ... 17998
Kunstsammlung Robert Spreng, Reiden ... 41751
Kunstsammlung Villa Schüpbach, Steffisburg ... 41910
Kunstsammlungen, Schwerin ... 21694
Kunstsammlungen Böttcherstraße, Bremen ... 17557
Kunstsammlungen Chemnitz, Chemnitz ... 17697
Kunstsammlungen Chemnitz - Museum Gunzenhauser, Chemnitz ... 17698
Kunstsammlungen der Ruhr-Universität Bochum, Bochum ... 17402
Kunstsammlungen der Stadt Limburg, Limburg an der Lahn ... 19995
Kunstsammlungen der Veste Coburg, Coburg ... 17717
Kunstsammlungen des Stiftes Kremsmünster, Kremsmünster ... 02233
Kunstsammlungen und Graphische Sammlung, Furth bei Göttweig ... 01913
Kunstsammlungen und Museen Augsburg, Augsburg ... 54840
Kunstsammlungen Zwickau, Zwickau ... 22833
Kunstschacht Katernberg, Essen ... 18259
Kunststätte Bossard, Jesteburg ... 19395
Kunststiftung Baden-Württemberg, Stuttgart ... 21949
Kunststiftung Erich Hauser, Rottweil ... 21426
Kunststiftung Poll, Berlin ... 17171
Kunststoff-Museums-Verein e.V., Düsseldorf ... 17999
't Kunstuus, Heinkenszand ... 32350
Kunstverein Bad Salzdetfurth, Bad Salzdetfurth ... 16819
Kunstverein Braunschweig e.V., Braunschweig ... 17523
Kunstverein für die Rheinlande und Westfalen, Düsseldorf ... 18000
Kunstverein Gelderland, Geldern ... 18566
Kunstverein Hannover, Hannover ... 18997
Kunstverein Heilbronn, Heilbronn ... 19098
Kunstverein in Hamburg e.V, Hamburg ... 18942
Kunstverein Ingolstadt e.V., Ingolstadt ... 19342
Kunstverein Kärnten, Klagenfurt ... 02185
Kunstverein Köln rechtsrheinisch e.V., Köln ... 19616
Kunstverein Konstanz e.V., Konstanz ... 19680
Kunstverein Passau - Sankt-Anna-Kapelle, Passau ... 20990
Kunstverein Reutlingen, Reutlingen ... 21284
Kunstverein Ruhr e.V., Essen ... 18260
Kunstverein Schwerte, Schwerte ... 21699
Kunstverein Speyer, Speyer ... 21835
Kunstverein Springhornhof e.V., Neuenkirchen, Kreis Soltau-Fallingbostel ... 20609
Kunstverein "Talstrasse" e.V, Halle, Saale ... 18899
Kunstverein Ulm, Ulm ... 22165
Kunstverein Villa Wessel, Iserlohn ... 19359
Kunstverein Weiden, Weiden in der Oberpfalz ... 22367
Kunstverein zu Frechen e.V., Frechen ... 18415
Kunstvereniging Diepenheim, Diepenheim ... 32124
Kunstwerk, Köln ... 19617
Kunst(Zeug)Haus, Rapperswil ... 41745
Kuntsi Art Collection, Vaasa ... 11423
Kuntsin Taidekokoelma, Vaasa ... 11423
Kuny Domokos Megyei Múzeum, Tata ... 23918
Kunying Art and Culture Museum, Beimen ... 42146
Kuo Chang-Hsi Knife and Sword Museum, Jiading ... 42226
Kuo Yuan Ye Museum of Cake and Pastry, Taipei ... 42397
Kuokkalan Museoraitti, Lempäälä ... 11059
Kuopio Art Museum, Kuopio ... 11009
Kuopio Natural History Museum, Kuopio ... 11008
Kuopion Korttelimuseo, Kuopio ... 11007
Kuopion Luonnontieteellinen Museo, Kuopio ... 11008
Kuopion taidemuseo, Kuopio ... 11009
Kupferberg-Museum, Mainz ... 20145
Kupferbergwerk, Beckingen ... 16972
Kupferschmiede-Museum, Tangermünde ... 22002
Kupferstich-Kabinett, Dresden ... 17954
Kupferstichkabinett, Basel ... 41167
Kupferstichkabinett, Wien ... 03020
Kupferstichkabinett - Museum of Prints and Drawings, Berlin ... 17172
Kupferstichkabinett - Sammlung der Zeichnungen und Druckgraphik, Berlin ... 17172
Kupiškio Etnografijos Muziejus, Kupiškis ... 30292
Kupiškis Ethnographic Museum, Kupiškis ... 30292
Kur-und Stadtmuseum Bad Ems, Bad Ems ... 16725
Kura Gallery, Wellington ... 33249
Kuralan Kylämäki, Turku ... 11392
Kuramayama Museum, Kyoto ... 28832
Kurashiki Archaeological Museum, Kurashiki ... 28787
Kurashiki City Art Museum, Kurashiki ... 28785
Kurashiki Japanese Art Swords Museum, Kurashiki ... 28786
Kurashiki Mingeikan, Kurashiki ... 28784
Kurashiki Museum of Folkcraft, Kurashiki ... 28784
Kurashiki-shi Bijutsukan, Kurashiki ... 28785
Kurashiki-shi Nihon Token Bijutsukan, Kurashiki ... 28786
Kurashiki-shiritsu Kokokan, Kurashiki ... 28787
Kurayoshi Hakubutsukan, Kurayoshi ... 28792
Kurayoshi Museum, Kurayoshi ... 28792
Kurchatov Museum, Kurchatov ... 29707
The Kurdish Museum, Brooklyn ... 47077
Kure Municipal Museum of Art, Kure ... 28793
Kure-n Hakubutsukan, Izumi ... 28611
Kure-shiritsu Bijutsukan, Kure ... 28793
Kurfürstliche Kanzlei, Torgau ... 22072
Kurfürstliches Gärtnerhaus im Baumschulwäldchen, Bonn ... 17461
Kurgan Regional Art Museum, Kurgan ... 37091
Kurgan Regional Museum, Kurgan ... 37092
Kurganskij Oblastnoj Chudožestvennyj Muzej, Kurgan ... 37091
Kurganskij Oblastnoj Kraevedčeskij Muzej, Kurgan ... 37092
Kurikan Museo, Kurikka ... 11014
Kurikka Museum, Kurikka ... 11014
Kurioses Muschel-Museum, Nordholz ... 20742
Kurioses Muschel-Museum, Spiekeroog ... 21840
Kurioses Muschel-Museum Wremen, Wremen ... 22701
Kurita Hakabutsukan, Ashikaga ... 28373
Kurita Museum, Ashikaga ... 28373
Kurmuseum, Bad Tatzmannsdorf ... 01767
Kurnik House, Tržič ... 38530
Kurnikova Hiša, Tržič ... 38530
Kuroda Kinenshitsu, Tokyo ... 29426
Kuroda Memorial Hall, Tokyo ... 29426
Kurpfälzisches Museum der Stadt Heidelberg, Heidelberg ... 19076
Kursenai Calendars' Museum, Kuršėnai ... 30295

K

Kuršėnų Kalendorių Muziejus, Kuršėnai 30295
Kuršių Nerijos Nacionalino Parko Gamtos Muziejus, Klaipėda 30280
Kursk Art Gallery Alexander Dejneka, Kursk 37093
Kursk State Museum of Archeology, Kursk 37095
Kursk State Regional Museum, Kursk 37094
Kurskaja Kartinnaja Galereja im. A.A. Dejneki, Kursk 37093
Kurskij Gosudarstvennyj Oblastnoj Kraevedčeskij Muzej, Kursk 37094
Kurskij Gosudarstvennyj Oblastnoj Muzej Archeologii, Kursk 37095
Kurt-Mühlenhaupt-Museum, Zehdenick 22770
Kurt Tucholsky Literaturmuseum, Rheinsberg 21304
Kuru Open-Air Museum, Kuru 11015
Kurukshetra Panorama & Science Centre, Kurukshetra 24287
Kurun Ulkomuseo, Kuru 11015
Kurza Stopa - Galeria Współczesnej Sztuki Polskiej, Chojnice 34598
Kurzemes Cietokšņa Muzejs, Zante 30150
Kurzemes Hercogu Kapenes, Jelgava 30018
Kuser Farm Mansion, Hamilton 49136
Kushiro Art Museum, Kushiro 28798
Kushiro City Museum, Kushiro 28799
Kustavin Kotiseutumuseo, Kustavi 11016
Kustodie, Dresden 17955
Kutaisi History Museum, Kutaisi 16343
Kutaisi State Museum of History and Ethnography, Kutaisi 16344
Kutaisskij Gosudarstvennyj Muzej Istorii i Etnografii, Kutaisi 16344
Kutschen-, Schlitten- und Wagenmuseum Rottach-Egern, Rottach-Egern 21417
Kutschen- und Heimatmuseum, Obertilliach 02461
Kutschen- und Schlittenmuseum, Großraming 02009
Kutschenkabinett, Landau in der Pfalz 19790
Kutschenmuseum, Augustusburg 16639
Kutschenmuseum, Basel 41168
Kutschenmuseum, Laa an der Thaya 02248
Kutschenmuseum, Ludwigslust 20060
Kutschenmuseum, Neustadt, Dosse 20677
Kutschenmuseum Hessisches Landgestüt, Dillenburg 17860
Kutschensammlung, Eigeltingen 18088
Kutschensammlung, Weimar 22408
Kutschensammlung Robert Sallmann, Amriswil 41111
Kutuzov's Hut, Moskva 37313
Kutuzov's Hut and Battle of Borodino Panorama, Moskva 37345
Kuurojen Museo, Helsinki 10796
Kuusamon Kotiseutumuseo, Kuusamo 11017
Kuusisto Manor, Kuusisto 11019
Kuusiston Kartano, Kuusisto 11019
Kuva-yi Milliye Müzesi, Balıkesir 42724
Kuva-yi National Museum, Balıkesir 42724
Kuwait National Museum, Kuwait 29949
Kuwana City Museum, Kuwana 28800
Kuwana-shiritsu Hakubutsukan, Kuwana 28800
Kuwayama Art Museum, Nagoya 29000
Kuwayama Bijutsukan, Nagoya 29000
Kuzminki - Museum of the Russian Estate Culture, Moskva 37358
Kuzuu-machi Yoshizawa Kinen Bijutsukan, Kuzuu 28801
Kvæfjord Bygdetun, Harstad 33533
Kvæfjord Open Air Museum, Harstad 33533
Kvæven Bygdetun, Tjørhom 33907
Kvam Bygdemuseum, Norheimsund 33703
Kvarken Boat Museum, Malax 11096
Kvarkens Båtmuseum, Malax 11096
Kvarnbacken Prylmuseum, Sandviken 40833
Kvevlax Hembygdsmuseum, Kvevlax 11020
Kvikmyndasafn Islands, Hafnarfjörður 23999
Kvindemuseet i Danmark, Århus 09964
Kviteseid Bygdetun, Kviteseid 33621
KW Institute for Contemporary Art, Berlin 17173
Kwa Muhle Museum, Durban 38607
Kwame Nkrumah University of Science and Technology Museum, Kumasi 22857
Kwang-Ya Art Center, Zaociao 42486
Kwangdong College Museum, Gangneung 29809
Kwinitsa Railway Station Museum, Prince Rupert 06670
Kyffhäuser-Denkmal, Bad Frankenhausen 16729
Kyïvs'kyj Literaturno-memorial'nyj Budinok-muzej M. Ryl's'kogo, Kyïv 43080
Kyïvs'kyj Literaturno-memorial'nyj Budinok-muzej Tarasa Ševčenka, Kyïv 43081
Kyïvs'kyj Literaturno-memorial'nyj Muzej-kvartyra P. Tyčyny, Kyïv 43082
Kyïvs'kyj Memorial'nyj Budinok-muzej M. Zan'kovec'koï, Kyïv 43083
Kyïvskyj Muzej Puškina, Kyïv 43084
Kyïvs'kyj Muzej Rosijs'koho Mystectva, Kyïv 43085
Kyle and Carrick District Library and Museum, Ayr 43346
Kymenlaakson maakuntamuseo, Kotka 10986
Kyneton Museum, Kyneton 01210
Kyngston House Museum, Saint Albans 45545
Kyocera Museum of Fine Ceramics, Kyoto 28833
Kyösti- ja Kalervo Kallion Museo, Nivala 11144
Kyongju University Museum, Kyongju 29849
Kyongnam College Museum, Masan 29850
Kyoto Arashiyama Orgel Museum & Guido Reuge Museum, Kyoto 28834
Kyoto Arashiyama Orugoru Hakubutsukan oyo Guido Reuge Hakubutsukan, Kyoto 28834
Kyoto Arts and Crafts Museum, Kyoto 28839
Kyoto Asahi Beer Oyamazaki Villa Museum, Oyamazaki 29169
Kyoto Chopstick Culture Museum, Kyoto 28835
Kyoto City Archaeological Museum, Kyoto 28848
Kyoto Daigaku Bungakukan, Kyoto 28836
Kyoto Daigaku Hakubutsukan, Kyoto 28837
Kyoto-fu Kyoto Bunka Hakubutsukan, Kyoto 28838
Kyoto Gigei Bijutsukan, Kyoto 28839
Kyoto Gion Oil Lamp Museum, Kyoto 28840
Kyoto-kenritsu Insho-Domoto Bijutsukan, Kyoto 28841
Kyoto-kenritsu Tango Minzoku Shiryokan, Miyau 28926
Kyoto-kenritsu Yamashiro Minzoku Shiryokan, Yamashiro 29583
Kyoto Kokuritsu Hakubutsukan, Kyoto 28842
Kyoto Kokuritsu Kindai Bijutsukan, Kyoto 28843
Kyoto Municipal Museum of Art, Kyoto 28845
Kyoto Municipal Museum of School History, Kyoto 28847
Kyoto Municipal Science Center for Youth, Kyoto 28846
Kyoto Museum for World Peace, Kyoto 28844
Kyoto National Museum, Kyoto 28842
Kyoto Prefectural Insho-Domoto Museum of Fine Arts, Kyoto 28841
Kyoto Prefectural Tango Folk Museum, Miyau 28926
Kyoto Prefectural Yamashiro Folk Museum, Yamashiro 29583
Kyoto-shi Bijutsukan, Kyoto 28845
Kyoto-shi Seishonen Kagaku, Kyoto 28846
Kyoto-shiritsu Gakkou Rekishi Hakubutsukan, Kyoto 28847
Kyoto-shiritsu Kokokan, Kyoto 28848
Kyoto University Literature Museum, Kyoto 28836
Kyoto University Museum, Kyoto 28837
Kyoto Yuzen Bunka Kaikan, Kyoto 28849
Kyoto Yuzen Culture Hall, Kyoto 28849
Kyrenia Castle, Girne 33358
Kyrgyz State Museum of Fine Arts Gapar Aitiyev, Biškek 29956
Kyrgyz State Toktogul Satylganov Museum of Literature and Fine Arts, Biškek 29955
Kyrgyzskij Gosudarstvennyj Muzej Literatury i Iskusstva im. Toktogula Satylganova, Biškek 29955
Kyrgyzskij Nacionalnyj Muzej Izobrazitelnych Iskusstv im. Gapara Ajtijeva, Biškek 29956
Kystmuseet Hvaler, Vesterøy 33974
Kystmuseet i Nord-Trøndelag, Rørvik 33797
Kystmuseet i Øygarden, Rong 33794
Kystmuseet i Sogn og Fjordane, Florø 33499
Kystmuseet i Sør-Trøndelag, Hitra 33548
Kysucka Galéria, Čadca 38242
Kysucke Múzeum, Čadca 38243
Kyu Ikegamike, Takatomachi Inashi 29334
Kyunghee University Museum, Seoul 29901
Kyungin Museum of Fine Art, Seoul 29902
Kyungsung Museum, Busan 29780
Kyushu Ceramic Museum, Saga, Arita 28365
Kyushu Historical Museum, Dazaifu 28409
Kyushu Rekishi Shiryokan, Dazaifu 28409
Kyushu Sangyo Daigaku Hakubutsukan, Fukuoka 28435
Kyyjärven Kotiseutumuseo, Kyyjärvi 11021
Kyzylorda Historical and Regional Museum, Kyzylorda 29708
Kyzylordynskij Istoriko-Kraevedčeskij Muzej, Kyzylorda 29708
KZ-Gedenk- und Begegnungsstätte, Ladelund 19760
KZ-Gedenkstätte Dachau, Dachau 17765
KZ-Gedenkstätte Flossenbürg, Flossenbürg 18332
KZ-Gedenkstätte Husum-Schwesing, Schwesing 21701
KZ-Gedenkstätte Mauthausen, Mauthausen 02373
KZ-Gedenkstätte Mittelbau-Dora, Nordhausen 20738
KZ Gedenkstätte Moringen, Moringen 20385
KZ-Gedenkstätte Neckarelz, Mosbach 20389
KZ-Gedenkstätte Neuengamme, Hamburg 18943
KZ-Gedenkstätte Sandhofen, Mannheim 20167
KZ-Gedenkstätte und Zeitgeschichte-Museum Ebensee, Ebensee 01836
KZNSA Gallery, Durban 38608

L

L.A.C. Lieu d'Art Contemporain, Sigean 15756
L.A. Mayer Museum for Islamic Art, Jerusalem 24999
L.C. Bates Museum, Hinckley 49295
L.D. Brinkman Art Foundation, Kerrville 49750
L. Kuindzhi Exhibition Hall of the Museum of Local Lore, Mariupol' 43147
L. Lange's Ovnmuseum, Svendborg 10290
L. Laptsy Museum-Appartment, Salechard 37609
L.N. Andreev House Museum, Orël 37503
L.N. Tolstoy Country Estate Museum, Moskva 37380
L.N. Tolstoy School - Cultural Center, Moskva 37222
L.N. Tolstoy State Museum, Moskva 37204
L.N. Tolstoy State Museum, Moskva 37205
L.N. Tolstoy State Museum, Stancija Lev Tolstoj 37794
L. Štur and A. Dubček House, Uhrovec 38384
LÁ Art Museum, Hveragerði 24005
La Chaumière de Granièr, Thiézac 15882
La Criée Centre d'Art Contemporain, Rennes (Ille-et-Vilaine) 15014
La Ferme Musée de Marcq, Enghien 03580
La Grande Vigne - Artists house, Dinan 12747
La Grenette, Sion 41872
La Hougue Bie Museum, Grouville 44368
La Maison du Gruyère, Pringy 41736
La Maison du Gruyère - demonstration cheese-dairy, Pringy 41736
La Para Municipal Historical Museum, La Para 00404
La Piscine - Musée d'Art et d'Industrie André Diligent, Roubaix 15111
La Raza Galeria Posada, Sacramento 52508
La Remise Musée de la Vie Rurale, Offaing 03920
La Rioja Museum of Sacred Art, La Rioja 00417
La Salle University Art Museum, Philadelphia 51846
La Serre aux Papillons, La Queue-lèz-Yvelines 13404
La Trobe University Art Museum, Bundoora 00922
The Lab, San Francisco 52780
Laband Art Gallery, Los Angeles 50242
Lábasház, Sopron 23813
Laboratário de Ensino de Ciências, Ribeirão Preto 04675
Laboratório de Anatomia e Identificação de Madeiras, Manaus 04553
Laboratório de Demonstração do Instituto de Fisica, São Paulo 04862
Labour Union Museum, Amsterdam 31872
Labrador Interpretation Centre, North West River 06483
Labrador Straits Museum, L'Anse-au-Loup 06139
Labråten, Hvalstad 33573
L'Abri Pataud, Les Eyzies-de-Tayac-Sireuil 13698
Labyrinth Kindermuseum Berlin, Berlin 17174
Lac Cardinal Regional Pioneer Village Museum, Grimshaw 05914
Lac-Du-Bonnet and District Historical Society Museum, Lac-Du-Bonnet 06118
Lac-La-Hache Museum, Lac-La-Hache 06119
Lac Qui Parle County Historical Museum, Madison 50382
Lac Sainte-Anne Historical Society Pioneer Museum, Rochfort Bridge 06766
Lace House, Kiskunhalas 23670
Lace House Museum, Black Hawk 46840
LACE - Los Angeles Contemporary Exhibitions, Los Angeles 50243
Lace Museum, Prachatice 09732
The Lace Museum, Sunnyvale 53372
Lace museum, Vamberk 09898
Lacey Historical Society Museum, Forked River 48586
Lackham Museum of Agriculture and Rural Life, Lacock 44628
Lacquerware Museum, Chiapa de Corzo 30762
Lacquerware Museum, Pagan 31726
The Lacrosse Museum and National Hall of Fame, Baltimore 46573
Laczkó Dezső Múzeum, Veszprém 23959
Ladbyskibsmuseet, Kerteminde 10117
Ladew Manor House, Monkton 50781
Ladics-ház, Gyula 23597
Ladics House, Gyula 23597
Lady Denman Maritime Museum, Huskisson 01170
Lady Franklin Gallery, Lenah Valley 01227
Lady Lever Art Gallery, Port Sunlight 45415
Lady Waterford Hall, Berwick-upon-Tweed 43462
Lady Wilson Museum, Dharampur 24222
Ladysmith Railway Museum, Ladysmith 06127
Läänemaa Muuseum, Haapsalu 10564
Lääninkivalteri Aschanin Talo, Heinola 10766
Läckö Slott, Lidköping 40674
Lähetysmuseo, Helsinki 10797
Lände, Kressbronn am Bodensee 19711
Ländliche Bildergalerie, Dischingen 17875
Längelmäen Kirkkomuseo, Längelmäki 11022
Längelmäen Korsumuseo, Länkipohja 11023
Längelmäen Pellavamuseo, Länkipohja 11024
Länsmuseet Gävleborg, Gävle 40532
Länsmuseet Halmstad, Halmstad 40578
Länsmuseet Västernorrland, Härnösand 40572
Länsmuseet Varberg, Varberg 41048
Læsø Museum, Læsø 10171
Lafayette College Art Gallery, Easton 48269
Lafayette Miners Museum, Lafayette 49863
Lafayette Museum - Alexandre Mouton House, Lafayette 49866
Lafayette Natural History Museum, Lafayette 49867
Lafcadio Hearn Memorial Museum, Matsue 28888
Laforet Hakuba Bijutsukan, Hakuba 28481
Lafortet Hakuba Museum of Art, Hakuba 28481
Lagan Lookout Centre, Belfast 43441
Laganland Bilmuseum, Lagan 40664
Lågdalsmuseet, Kongsberg 33602
LaGuardia and Wagner Archives, Long Island City 50183
Laguna Art Museum, Laguna Beach 49871
Lahaina Restoration Foundation, Lahaina 49872
LaHave Islands Marine Museum, LaHave 06129
Lahden Historiallinen Museo, Lahti 11028
Lahden Kaupunginmuseo, Lahti 11029
Lahden Taidemuseo, Lahti 11030
Lahn-Marmor-Museum, Villmar 22242
Lahnsteiner Fastnachtsmuseum, Lahnstein 19766
Lahore Fort Museum, Lahore 34022
Lahore Museum, Lahore 34023
Lahti Art Museum, Lahti 11030
Lahti City Museum, Lahti 11029
Lahti Historical Museum, Lahti 11028
Lahti Poster Museum, Lahti 11027
Laidhay Croft Museum, Dunbeath 44043
Laidley District Historical Village, Laidley 01211
Laihian Kirkkomuseo, Laihia 11036
Laihian Museo, Laihia 11037
Laiho Museum, Changhua 42156
Laikrodžių Muziejus, Klaipėda 30281
Laing Art Gallery, Newcastle-upon-Tyne 45229
Laing Museum, Newburgh 45218
Laipa - Galerija Valmiera, Valmiera 30139
Laisvės Kovų Muziejus Krauji Takais, Lazdijai 30299
Laiyang Museum, Laiyang 07918
Laizhou Museum, Laizhou 07919
Lajos Ferenc és Kortársai Grafikai Gyűjtemény, Gyula 23598
Lajos Gulácsy House, Szeged 23846
Lajos Kossuth Memorial Museum, Monok 23717
Lajos Kunffy Memorial Museum, Somogytúr 23806
Lajos Nagy Memorial Exhibition, Apostag 23336
Lake Biwa Canal Museum of Kyoto, Kyoto 28803
Lake Biwa Museum, Kusatsu 28796
Lake Broadwater Natural History Museum, Dalby 01021
The Lake Champlain Maritime Museum, Vergennes 53717
Lake Country Museum, Okanagan 06504
Lake County Discovery Museum, Wauconda 53987
Lake County Historical Museum, Tavares 53454
Lake County Historical Society Museum, Kirtland Hills 49803
Lake County Museum, Two Harbors 53630
Lake District Museum of South Ostrobothnia - Väinö's House, Vasikka-Aho 11457
Lake Dongting Museum, Yuanjiang 08463
Lake Erie Nature and Science Center, Bay Village 46654
Lake Forest-Lake Bluff Historical Society Museum, Lake Forest 49882
Lake George Historical Association Museum, Lake George 49884
Lake Guernsey Museum, Guernsey 49101
Lake Hamana Museum of Music Boxes, Hamamatsu 28487
Lake Hamana Tateyama Temple Art Museum, Hamamatsu 28488
Lake Hopatcong Historical Museum, Landing 49934
Lake House Arts Centre, Auckland 33053
Lake Jackson Historical Museum, Lake Jackson 49885
Lake Jackson Mounds Archaeological Park, Tallahassee 53424
Lake Macquarie City Art Gallery, Booragul 00877
Lake Malawi Museum, Mangochi 30545
Lake Meredith Aquatic and Wildlife Museum, Fritch 48799
Lake of the Red Cedars Museum, Cedar Lake 47298
Lake of the Woods County Museum, Baudette 46648
Lake of the Woods Museum, Kenora 06057
Lake Perris Regional Indian Museum, Perris 51777
Lake Placid-North Elba Historical Society Museum, Lake Placid 49891
Lake Shore Railway Museum, North East 51391
Lake Superior Maritime Visitors Center, Duluth 48197
Lake Superior Railroad Museum, Duluth 48198
Lake Tabourie Museum, Lake Tabourie 01213
Lake Taupo Museum and Art Gallery, Taupo 33204
Lake Waccamaw Depot Museum, Lake Waccamaw 49894
Lake Wales Depot Museum, Lake Wales 49895
Lakeland Motor Museum, Grange-over-Sands 44341
Lakes District Museum, Arrowtown 33035
Lakes District Museum Society, Burns Lake 05506
Lakeside Museum, Bulyea 05493
Lakeview Museum of Arts and Sciences, Peoria 51773

Lakewood's Heritage Culture and the Arts Galleries, Lakewood 49905
Lalbagh Fort Museum, Dhaka 03225
Lâm Dông Museum, Đà Lat 54603
Lamanai Archaeological Site, Lamanai 04178
Lamar Dodd Art Center, La Grange 49841
Lamartine Memorial House, Plovdiv 05129
Lambertsmühle, Burscheid 17667
Lambing Flat Museum, Young 01686
Lambropoulos Archaeological Collection, Polygyros 23154
Lambton Heritage Museum, Grand Bend 05888
The Lamont Gallery, Exeter 48465
Lampenmuseum, Geel 03614
Lampenmuseum, Wezemaal 04151
Lampi-Museo, Liminka 11073
Lampi Museum, Liminka 11073
Lamu Museum, Lamu 29744
L'Amusant Musée, Juigne-sur-Sarthe 13290
Lan Sang, Tak 42607
Lanark and District Museum, Lanark 06132
Lanark Museum, Lanark 44629
Lancaster Antique Toy and Train Collection, Portsmouth 52108
Lancaster City Museum, Lancaster 44633
Lancaster County Museum, Lancaster 49926
Lancaster Historical Society Museum, Lancaster 49918
Lancaster Maritime Museum, Lancaster 44634
Lancaster Museum / Art Gallery, Lancaster 49915
Lancaster Museum of Art, Lancaster 49927
Lancefield Courthouse Museum, Lancefield 01215
Lancer Centennial Museum, Lancer 06134
Land- en Tuinbouwmuseum, Etten Leur 32226
Land Fossils and Minerals Museum, Yongkang 42475
Land Mine Museum, Siem Reap 05291
Land of the Yankee Fork Historical Museum, Nampa 50941
Land Reform Museum, Taipei 42398
Land Surveying Museum, Helsinki 10802
Het Land van Strijen, Strijen 32809
Het Land van Thorn/ Panorama Thorn, Thorn 32820
Landbouw- en Juttersmuseum Swartwoude, Buren, Friesland 32023
Landbouwmuseum Erve Niehof, Diepenheim 32125
Landbouwmuseum Leiedal, Bissegem 03370
Landbrugs- og Interiørmuseet, Farsø 10018
Landbrugsmuseet Melstedgård, Gudhjem 10047
Landbrugsmuseet Skarregaard, Nykøbing Mors 10202
Landbruksmuseet for Møre og Romsdal, Vikebukt 33979
Landcommanderij Alden Biesen, Bilzen 03367
Landes-Feuerwehrmuseum, Stendal 21898
Landesbergbaumuseum Baden-Württemberg, Sulzburg 21988
Landesbibliothek Oldenburg, Oldenburg 20896
Landesbibliothek und Murhardsche Bibliothek der Stadt Kassel, Kassel 19466
Landesfeuerwehrmuseum Meetzen Mecklenburg-Vorpommern, Meetzen 20254
Landesgalerie Linz, Linz 02317
Landesgeschichtliche Sammlung der Schleswig-Holsteinischen Landesbibliothek, Kiel 19532
Landesmuseum Birkenfeld, Birkenfeld 17359
Landesmuseum Burgenland, Eisenstadt 01861
Landesmuseum für Kunst und Kulturgeschichte, Schleswig 21566
Landesmuseum für Kunst und Kulturgeschichte Oldenburg, Oldenburg 20897
Landesmuseum für Natur und Mensch, Oldenburg 20898
Landesmuseum für schaumburg-lippische Geschichte, Landes- und Volkskunde, Bückeburg 17623
Landesmuseum für Vorgeschichte, Dresden 17956
Landesmuseum für Vorgeschichte Sachsen-Anhalt, Halle, Saale 18900
Landesmuseum Kärnten, Klagenfurt 02186
Landesmuseum Koblenz, Koblenz 19585
Landesmuseum Mainz, Mainz 20146
Landesmuseum Württemberg, Stuttgart 21950
Landesrabbiner Dr. I.E. Lichtigfeld-Museum, Michelstadt 20319
Landesverband Salzburger Volkskultur, Salzburg 54695
Landeszeughaus am Landesmuseum Joanneum, Graz 01977
Landhausgalerie Ausstellungsbrücke, Sankt Pölten 02683
Landings Gallery, Edinburgh 44130
Landis Valley Museum, Lancaster 49928
Landlmuseum Sulzbürg, Mühlhausen, Oberpfalz 20400
Landlord's Manor House Museum, Dayi 07656
Landmark Forest Heritage Park, Carrbridge 43765
Landmaschinenmuseum - Sammlung Speer, Rimbach 21332
Lands Museum, Dokka 33464
Landsborough Shire Historical Museum, Landsborough 01216
Landscapes of the Holy Land Park, Tel Aviv 25114
Landschaftsinformationszentrum (LIZ) Hessisches Kegelspiel Rasdorf, Rasdorf 21182
Landschaftsmuseum, Seligenstadt 21739
Landschaftsmuseum Angeln, Langballig 19807
Landschaftsmuseum der Dübener Heide, Bad Düben 16713
Landschaftsmuseum der Kulmregion, Pischelsdorf 02501
Landschaftsmuseum im Schloß Trautenfels, Trautenfels 02827
Landschaftsmuseum Obermain, Kulmbach 19748
Landschaftsmuseum Schönhengstgau, Göppingen 18675
Landschaftsmuseum Westerwald, Hachenburg 18847
Landschloss Pirna-Zuschendorf - Botanische Sammlungen der TU Dresden, Pirna 21044
Landschulmuseum Göldenitz, Dummerstorf 18028
Landskrona Konsthall, Landskrona 40668
Landskrona Museum, Landskrona 40669
Landsratamt Rems-Murr-Kreis, Waiblingen 22277
Landtechnik-Museum Braunschweig Gut Steinhof, Braunschweig 17524
Landtechnikmuseum, Leiben 02279
Landtechnisches Museum Burgenland, Sankt Michael im Burgenland 02670
Landwirtschaftliches Museum und Pfarrer Jungblut-Museum, Prinzendorf 02533
Landwirtschaftliches Museum Wetzlar, Wetzlar 22528
Landwirtschafts- und Heimatmuseum, Karben 19431
Landwirtschaftsmuseum, Rhede 21295
Landwirtschaftsmuseum, Weil am Rhein 22375
Landwirtschaftsmuseum Lüneburger Heide, Suderburg 21972
Landwirtschaftsmuseum Reitscheid, Freisen 18447
Landwirtschaftsmuseum Schloss Ehrental, Klagenfurt 02187
Landwirtschaftsmuseum Schloss Stainz, Stainz 02760
Lane Community College Art Gallery, Eugene 48432
Lane County Historical Museum, Dighton 48140
Lane County Historical Museum, Eugene 48433
Lane House, Roseburg 52458
The Lane Place, Crawfordsville 47870
Lanesfield School, Edgerton 48286
Lang Pioneer Village, Keene 06048
Lang Water Powered Grist Mill, Peterborough 06593
Langage Plus, Alma 05321
Långbans Gruvby, Filipstad 40519
Langbeinmuseum, Hirschhorn 19199
Långelanda Tingshus, Årjäng 40429
Langelands Museum, Rudkøbing 10252
Langen Foundation, Neuss 20658
Langham and District Heritage Village Museum, Langham 06135
Langham Cultural Centre Galleries, Kaslo 06046
Långholmen Prison Museum, Stockholm 40907
Långholmens Fängelsemuseum, Stockholm 40907
Langi Morgala Museum, Ararat 00810
Langingkoski Imperial Fishing Lodge, Kotka 10987
Langlade County Historical Society Museum, Antigo 46325
Langley Centennial Museum and National Exhibition Centre, Fort Langley 05813
Langobardenwerkstatt Zethlingen, Zethlingen 22792
Langsdorfer Heimatmuseum, Lich 19978
Langton Matravers Museum, Langton Matravers 44640
Lanhydrock, Bodmin 43549
Lanier Mansion, Madison 50380
Lanier Museum of Natural History, Buford 47128
Lanigan and District Heritage Centre, Lanigan 06137
Lanman Museum, Framlingham 44251
Lanna Museum, Lampang 42575
Lansing Art Gallery, Lansing 49942
Lansing Manor House Museum, North Blenheim 51384
Lansing Veterans Memorial Museum, Lansing 49940
Lantbruksmuseum, Ingatorp 40609
Lanxi Museum, Lanxi 07920
Lanyang Museum of Yilan County, Yilan 42467
Lanzhou Museum, Lanzhou 07922
Lao National Museum, Vientiane 29974
Lao She Museum, Beijing 07545
Laoag Museum, Vigan 34522
Laohekou Museum, Laohekou 07923
Lapham-Patterson House, Thomasville 53499
Lapidari Museum, Lefkoşa 33373
Lapidarij Muzeja Grada Trogira, Trogir 08969
Lapidario Civico, Gradisca d'Isonzo 26357
Lapidario Comunale, Osimo 27010
Lapidarium, Berlin 17175
Lapidarium, Neustadt am Main 20662
Lapidarium der Stiftsbibliotheh St. Gallen, Sankt Gallen 41806
Lapidarium of Roman Age Cathedral Museum, Pécs 23774
Lapidarium of the City of Stuttgart, Stuttgart 21962
Lapin Maakuntamuseo, Rovaniemi 11276
Lapin Metsämuseo, Rovaniemi 11277
Lapin Tl. Kotiseutumuseo, Lappi 11051
Lapinlahden Taidemuseo ja Eemil Halosen Museo, Lapinlahti 11044
Lapland Forestry Museum, Rovaniemi 11277
Lappa Valley Steam Railway, Newquay 45248
Lappstaden, Arvidsjaur 40435
Lapu-Lapu City Science Museum, Lapu-Lapu 34401
Lapworth Museum of Geology, Birmingham 43506
Laramie Plains Museum, Laramie 49948
Larchmont Historical Society Museum, Mamaroneck 50412
Larco Archaeological Museum, Lima 34225
Lårdal Bygdemuseum, Høydalsmo 33570
Lårdal Rural Museum, Høydalsmo 33570
Laredo Children's Museum, Laredo 49952
Largs Museum, Largs 44642
Lark Lane Motor Museum, Liverpool 44736
Larnaca District Archaeological Museum, Larnaka 09367
Larnaca District Museum, Larnaka 09368
Larne Museum, Larne 44643
Larraulgo Museoa, Larraul 39443
Larrybane and Carrick-A-Rede, Ballycastle 43361
Lars Noak Blacksmith Shop, Larsson/Ostlund Log Home & One-Room Capitol School, New Sweden 51103
Larsmo Hembygdsmuseum, Larsmo 11054
Larson Gallery, Yakima 54319
Larvik Maritime Museum, Larvik 33629
Larvik Sjøfartsmuseum, Larvik 33629
Larz Anderson Auto Museum, Brookline 47064
Las Cruces Museum of Natural History, Las Cruces 49958
Las Vegas Art Museum, Las Vegas 49971
Las Vegas International Scouting Museum, Las Vegas 49972
Las Vegas Natural History Museum, Las Vegas 49973
LaSalle County Historical Society Museum, Utica 53673
Lasarettimuseo, Turku 11393
Lascaux II, Montignac 14214
Lashburn Centennial Museum, Lashburn 06143
Lashenden Air Warfare Museum, Headcorn 44432
Laško Museum Collection, Laško 38442
Lassaner Mühle, Lassan 19841
Lassen Historical Museum, Susanville 53385
Lassen Museum Waikiki, Honolulu 49348
Last Indian Raid Museum, Oberlin 51477
Lastenmuseo, Helsinki 10798
Lastentarhamuseo, Helsinki 10799
László Holló Memorial Museum, Debrecen 23517
László Nagy Memorial House, Iszkáz 23621
László Németh Memorial Exhibition, Hódmezővásárhely 23613
Latcham Gallery, Stouffville 07016
Laténium, Hauterive 41474
Latgale Culture and History Museum, Rēzekne 30059
Latgales Kultūrvēstures Muzejs, Rēzekne 30059
Latin American Art Museum, Miami 50629
Latin American Cultural Center Museum, Koyang 29847
Latinovits Zoltán Emlékmúzeum, Balatonszemes 23354
Latinovits Zoltán Memorial Museum, Balatonszemes 23354
Latitude 53 Contemporary Visual Culture, Edmonton 05748
Latrobe Bicycle Race Club Museum, Latrobe 01218
Latrobe Court House Museum, Latrobe 01219
Latrobe Regional Gallery, Morwell 01325
Lattelekom Telekomunikciju Muzejs, Rīga 30074
Latvia Museum of the History of Chemistry, Rīga 30082
Latvia University Museum of Botany, Rīga 30087
Latvia University Museum of Education, Rīga 30089
Latvia University Museum of Geology, Rīga 30088
Latvian Architecture Museum, Rīga 30076
Latvian Centre for Contemporary Art, Rīga 30075
Latvian History Museum, Toronto 07119
Latvian Museum, Rockville 52430
Latvian Museum, Wayville 01639
Latvian Museum Council, Rīga 54923
Latvian Museum of Culture Dauderi, Rīga 30083
Latvian Museums Association, Rīga 54924
Latvian Photography Museum, Rīga 30080
Latvian Railway History Museum, Jelgava Branch, Jelgava 30019
Latvian Railway History Museum, Riga Branch, Rīga 30078
Latvian War Museum, Rīga 30081
Latvijas Arhitektūras Muzejs, Rīga 30076
Latvijas Dabas Muzejs, Rīga 30077
Latvijas Dzelzceļa Vēstures Muzeja Jelgavas Nodaļa, Jelgava 30019
Latvijas Dzelzceļa Vēstures Muzeja Rīgas Nodaļa, Rīga 30078
Latvijas Etnogrāfiskais Brīvdabas Muzejs, Rīga 30079
Latvijas Fotogrāfijas Muzejs, Rīga 30080
Latvijas Kara Muzejs, Rīga 30081
Latvijas Ķīmijas Vēstures Muzejs, Rīga 30082
Latvijas Kultūras Muzejs Dauderi, Rīga 30083
Latvijas Lauksaimniecības Muzejs, Talsi 30127
Latvijas Lauksaimniecības Universitātes Muzejs, Jelgava 30020
Latvijas Muzeju Asociācija, Rīga 54924
Latvijas Okupācijas Muzejs, Rīga 30084
Latvijas Republikas Meliorācijas un Zemkopības Muzejs, Mālpils 30047
Latvijas Sporta Muzejs, Rīga 30085
Latvijas Ugunsdzēsības Muzejs, Rīga 30086
Latvijas Universitātes Botānikas Muzejs, Rīga 30087
Latvijas Universitātes Ģeoloģijas Muzejs, Rīga 30088
Latvijas Universitātes Pedagoģijas Muzejs, Rīga 30089
Latvijas Universitātes Skaitļošanas Tehnikas un Informatikās Muzejs, Rīga 30090
Latvijas Universitātes Vēstures Muzejs, Rīga 30091
Latvijas Universitātes Zooloģijas Muzejs, Rīga 30092
Latvijas Vēstures Muzejs, Rīga 30093
Laudal Museum, Øyslebø 33763
Lauder Museum, Amityville 46274
Lauderdale House, London 44898
Laufener Stiftsschatz, Laufen 19854
Laugø Smedie Museum, Helsinge 10061
Laukminiškių Kaimo Muziejus, Laukminiškes 30296
Laumeier Sculpture Park, Saint Louis 52584
Launceston Steam Railway, Launceston 44647
Laura and Charles Ferguson Museum, Mentone 01293
Laura and Paul Mesaros Galleries, Morgantown 50845
Laura Ingalls Wilder Museum, Walnut Grove 53804
Laura Ingalls Wilder-Rose Wilder Lane Historic Home and Museum, Mansfield 50450
Laura Secord Homestead, Niagara Falls 06441
Laure A. Sprague Art Gallery, Joliet 49644
Laurel and Hardy Museum, Ulverston 45897
The Laurel Museum, Laurel 49982
Laurel und Hardy Museum, Solingen 21803
Lauren Rogers Museum of Art, Laurel 49985
Laurence C. Jones Museum, Piney Woods 51928
Laurens Van der Post Museum, Philippolis 38702
Laurent Hubert Musée, Orbey 14521
Lauri Viidan Museo, Tampere 11352
Laurier House, Ottawa 06549
Laurinmäen Torpparimuseo, Janakkala 10870
Lauriston Castle, Edinburgh 44131
Lavanttaler Heimatmuseum, Wolfsberg 03118
Lavanttaler Obstbaumuseum, Sankt Paul 02677
Lavian Kotiseutumuseo, Lavia 11056
Le Lavoir, Mougins 14295
Law Enforcement Museum, Roselle 52459
Law Uk Folk Museum, Hong Kong 07807
Lawn Heritage Museum, Lawn 06150
Lawndale Art Center, Houston 49393
Lawrence County Gray House Museum, Ironton 49527
Lawrence County Historical Museum, Bedford 46687
Lawrence Hall of Science, Berkeley 46757
Lawrence House Museum, Launceston 44648
Lawrence House Museum, Maitland 06220
Lawrence L. Lee Scouting Museum, Manchester 50422
Lawrence Wilson Art Gallery, Crawley 01010
Laws Railroad Museum, Bishop 46831
Lawton Gallery, Green Bay 49039
Laxfield and District Museum, Laxfield 44652
Layland Museum, Cleburne 47612
Lazdijai Area Museum, Lazdijai 30300
Lazdijų Krašto Muziejus, Lazdijai 30300
Lazdynų Pelėdos Muziejus, Paragiai 30334
Laznia Centre for Contemporary Art, Gdańsk 34632
Le Carmel, Tarbes 15860
Le Creuset, Waterloo 04135
Le Grand Filon, Saint-Georges-d'Hurtieres 15282
Le Musée Archéologique de Montaigle, Falaën 03596
Le Petit Musée, Wavre 04141
Le Petit Musée du Pain, Chievres 03521
Le Point Fortifié Hillmann, Colleville-Montgomery 12571
Le Puits-Marie, Blegny 03373
Le Roy House and Jell-o Gallery, Le Roy 50002
Le Sault de Sainte Marie Historical Sites, Sault Sainte Marie 52931
Le Sueur County Historical Museum, Elysian 48388
Le Village des Automates, Saint-Cannat 15196
Lea County Cowboy Hall of Fame and Western Heritage Center, Hobbs 49304
Leadhills Miners' Library Museum, Leadhills 44653

L

Leading Tickles Heritage Centre, Leading Tickles 06152
Leadshan Museum, Qianshan, Jiangxi 08102
Leaf Rapids National Exhibition Centre, Leaf Rapids 06153
Leaming's Run Garden and Colonial Farm, Cape May Court House 47237
Leamington Art Gallery, Leamington 06154
Leamington Spa Art Gallery and Museum, Leamington Spa 44654
Le'an Museum, Le'an 07924
Leanin' Tree Museum of Western Art, Boulder 46960
Learmonth Museum, Learmonth 01226
Leaseholdermuseum, Kiikoinen 10957
Leather Museum, Chicago 47485
Leather Museum, Haram 10533
Leather Museum, Swakopmund 31756
Leatherhead Museum of Local History, Leatherhead 44655
Lebell Residence, Kristiinankaupunki 10996
Lebendes Textilmuseum, Groß-Siegharts 01998
Lebendiges Museum, Bad Wildungen 16861
Lebzelterei- und Wachsziehereimuseum, Pfaffenhofen an der Ilm 21019
Lechang Museum, Lechang 07925
Lechfeldmuseum, Königsbrunn 19645
Lechflößermuseum, Lechbruck am See 19867
Lechner Museum, Ingolstadt 19343
Ledaal Museum, Stavanger 33866
Ledbury Heritage Centre, Ledbury 44657
Leder- und Gerbermuseum, Mülheim an der Ruhr 20415
Lee Chapel and Museum, Lexington 50071
Lee County Historical Society Museum, Loachapoka 50153
Lee-Fendall House Museum, Alexandria 46210
Lee Tze-Fan Memorial Art Gallery, Hsinchu 42210
Leeds Industrial Museum, Leeds 44662
Leeds Metropolitan University Gallery, Leeds 44663
Leeds Museum, Leeds 44664
Leedy-Voulke's Art Center, Kansas City 49692
Leelanau Historical Museum, Leland 50022
Leepa-Rattner Museum, Tarpon Springs 53448
Lee's Horsey Fine Art Studio, Kaohsiung 42239
Lefferts Historic House, Brooklyn 47078
Leffingwell House Museum, Norwich 51436
Lefortovo Museum, Moskva 37326
Left Bank Art Gallery, Greymouth 33114
Lefteris Kanakakis Gallery, Rethymnon 23168
Legacy of Nedeljko Gvozdenovic, Beograd 38056
Legado Valeriano Salas, Béjar 39013
Legaspi City Museum, Legaspi 34402
Legat Nedeljka Gvozdenovića, Beograd 38056
Legends of New Mexico Museum, Algodones 46217
Legends of the Game Baseball Museum, Arlington 46352
Legged House, Sopron 23813
Legion of Honor, San Francisco 52781
Legislative Building Art Galleries, Regina 06719
Lehap Regional Museum, Turkmenabat 42985
Lehár-Schikaneder-Schlößl, Wien 03021
Lehár-Villa, Bad Ischl 01754
Lehigh County Historical Society Museum, Allentown 46227
Lehigh University Art Galleries/Museum, Bethlehem 46780
Lehman College Art Gallery, Bronx 47048
Lehmmuseum, Buchberg 17614
Lehr- und Schaubergwerk Frisch Glück, Johanngeorgenstadt 19400
Lehr- und Schaubergwerk Herkules Frisch Glück, Beierfeld 16982
Lehrerhaus-Friemersheim, Duisburg 18021
Lehtimäen Kotiseutumuseo, Lehtimäki 11057
Lehtse Country History Museum, Lehtse 10590
Lehtse Koduloomuuseum, Lehtse 10590
Lei Cheng Uk Han Tomb Museum, Hong Kong 07808
Leib'sches Haus, Abteilung Stadtgeschichte und Volkskunde, Gießen 18638
Leica Gallery Prague, Praha 09758
Leica Stammbaum und Museum, Solms 21812
Leicestershire CCC Museum, Leicester 44677
Leicestershire County Council Museums Service, Glenfield 55070
Leigh Heritage Centre, Leigh-on-Sea 44686
Leigh Yawkey Woodson Art Museum, Wausau 53993
Leighton Art Centre, Calgary 05522
Leighton Buzzard Railway, Leighton Buzzard 44688
Leighton Hall, Carnforth 43763
Leighton House Museum, London 44899
Leimarel Museum and Research Centre, Imphal 24224
Leinen- und Spitzenmuseum Haigerseelbach, Haiger 18867
Leinfelder Haus, Leinfelden-Echterdingen 19884
Leipzig Forum of Contemporary History, Leipzig 19942
Leirfjord Bygdesamlinger, Leirfjord 33630
Leitch Collieries, Crowsnest Pass 05662
Leith Hall, Huntly 44504
Leitheimer Schloß-Museum, Kaisheim 19416
Leithenmühle, Windhaag bei Freistadt 03111
Lejre Forsøgscenter, Lejre 10174
Leka Bygdemuseum, Leka 33631
Leksands Museum, Leksand 40670
Leksbergs Skolmuseum, Mariestad 40736
Leksvik Bygdesamling, Leksvik 33632
Leīlu Mākslas muzeja, Rīga 30094
Lembaga Muzium Negeri, Pulau Pinang 30613
Lemee House, Natchitoches 50988
Lemhi County Historical Museum, Salmon 52685
Lemin Kotiseutumuseo, Lemi 11058
Lemon Hill Mansion, Philadelphia 51847
LeMoyne Art Foundation, Tallahassee 53425
Lemun Kotiseutumuseo, Lemu 11062
Lemvig Museum Vesterhus, Lemvig 10176
Lenbachmuseum, Schrobenhausen 21639
Lenin House-Museum, Kazan 37005
Lenin-Museo, Tampere 11353
Lenin Museum, Moskva 37327
Lenin Museum, Tampere 11353
Lennon House, Liverpool 44737
Lennox and Addington County Museum, Napanee 06417
Lennoxlove House, Haddington 44376
Lennoxville Ascot Historical and Museum Society, Lennoxville 06158
Lenoir Forge Museum, Arichat 05352
L'Enrajolada Casa-Museu Santacana, Martorell 39626
Lentos Kunstmuseum Linz, Linz 02318
The Lentz Center for Asian Culture, Lincoln 50098
Lenvik Bygdemuseet, Finnsnes 33492
Leo Baeck Institute New York, Berlin 17176
Leominster Folk Museum, Leominster 44691
Leon Apacible Historical Landmark Museum, Taal 34507
Leon Wyczółkowski Regional Museum, Bydgoszcz 34578
Leona Paegles Memoriālais Muzejs, Vidriži 30148
Leonard & Bina Ellen Art Gallery, Montréal 06342
Leonhardi-Museum, Dresden 17957
Leonhardi-Museum Aigen, Bad Füssing 16737
Leonora Gwalia Historical Museum, Leonora 01229
Leopold Figl-Museum, Rust im Tullnerfeld 02603
Leopold-Hoesch-Museum Düren, Düren 17982
Leopold Museum, Wien 03022
Lepikko Cottage, Pielavesi 11204
Lepikon Torppa, Pielavesi 11204
Leping Museum, Leping 07926
Leppävirran Kotiseutumuseo, Leppävirta 11064
Leppävirta Local History Museum, Leppävirta 11064
Lepramuseet Sant Jørgens Hospital, Bergen 33430
Leptis Magna Museum, Leptis Magna 30183
Lermontov Apartment Museum, Moskva 37165
Lermontov State Literary Memorial Museum, Pjatigorsk 37564
Lermontov State Museum-Estate Tarkhany, Lermontovo 37100
Lermontovskij Gosudarstvennyj Muzej-zapovednik Tarchany, Lermontovo 37100
Leroy and District Heritage Museum, Leroy 06160
Les Ex-Voto, Monthermé 14213
Les Grottes de Gargas, Aventignan 11784
Les Jacobins, Morlaix 14275
Les Roues de L'Areuse, Saint-Sulpice (Neuchâtel) 41794
Les Territoires de la Mémoire, Liège 03797
Lesbian Herstory Educational Foundation, Brooklyn 47079
Lesja Bygdatun, Lesja 33633
Lesjöforsmuseum, Lesjöfors 40671
Leskowsky Collection of Musical Instruments, Kecskemét 23649
Leskowsky Hangszergyűjtemény Közalapítvány, Kecskemét 23649
Leslie Hill Open Farm, Ballymoney 43367
Leslie-Lohman Gay Art Foundation, New York 51203
Leslie M. Frost Natural Resources Centre, Minden 06294
Lesnícke a Drevárske Múzeum, Zvolen 38394
Lesnické, Myslivecké a Rybářské Muzeum, Hluboká nad Vltavou 09516
Lesotho National Museum, Maseru 30179
Lessing-Museum, Kamenz 19420
Lester-Garland Premises, Trinity 07165
Lestijärven Kotiseutumuseo, Lestijärvi 11065
L'Etang d'Art, Bages (Aude) 11817
Letchworth Museum and Art Gallery, Letchworth 44695
Letecké Muzeum VHÚ Praha, Praha 09759
Letif Kerimov ad. Azerbaycan Xalçasi ve Xalq Tetbiqi Seneti Dövlet Muzeyi, Bakı 03159
Letnisko, Milanówek 34922
Letohrádek Ostrov, Ostrov nad Ohří 09703
Letschiner Heimatstuben, Letschin 19965
Letterkundig Museum en Kinderboekenmuseum, Den Haag 32078
Lettl-Atrium, Augsburg 16627
Lev Tolstoy Museum-Estate Nikolskoe-Vjasemskoe, Nikolskoe-Vjazemskoe 37449
Levanger Museum, Levanger 33635
Leventis Municipal Museum of Nicosia, Lefkosia 09379
Levere Memorial Temple, Evanston 48452
Leverett Historical Museum, Leverett 50027
Levi Coffin House, Fountain City 48725
Levine Museum of the New South, Charlotte 47390
Levitan Memorial House-Museum, Ples 37566
Levy Gallery, Philadelphia 51848
Lewes Castle and Barbican House Museum, Lewes 44698
Lewes Historical Society Museum, Lewes 50028
Lewis and Clark Boat House and Nature Center, Saint Charles 52534
Lewis Collection, Cambridge 43729
Lewis County Historical Museum, Chehalis 47417
Lewis County Historical Society Museum, Lowville 50309
Lewis County Historical Society Museum, Lyons Falls 50337
Lewis Elton Gallery, Guildford 44372
Lewis, James and Nellie Stratton Gallery, Columbia 47716
Lewis Textile Museum, Blackburn, Lancashire 43527
Lewisham Local Studies Centre, London 44900
Lewistown Art Center, Lewistown 50042
Lexington County Museum, Lexington 50067
Lexington Historical Society, Lexington 50060
Lexington History Museum, Lexington 50051
Leye Museum, Leye 07927
Lgovskij Kraevedčeskij Muzej, Lgov 37101
Lgovskij Literaturno-memorialnyj Muzej N.N. Aseeva, Lgov 37102
Lgovskij Literaturno-memorialnyj Muzej A.P. Gajdara, Lgov 37103
Lhasa Museum, Lhasa 07928
LHU Arts Center, Gueishan 42196
L'Huillier-Coburg Castle, Edelény 23535
L'Huillier-Coburg-Kastély, Edelény 23535
Li County Museum, Lixian 07967
Li Mei-Shu Memorial Gallery, Sansia 42312
Li Tien-lu Hand Puppet Historical Museum, Sanjhi 42310
Li Xian Jun Former Residence, Nanjing 08041
Li Zong Ren Display Center, Guilin 07743
Liane and Danny Taran Gallery, Montréal 06343
Liangshan Archaeological Museum, Xichang 08383
Liangshan Yizu Thrall Society Museum, Xichang 08382
Liangshanzhou Guan, Xichang 08383
Liangzhu Culture Museum, Hangzhou 07763
Lianjiang Museum, Lianjiang, Fujian 07930
Lianjiang Museum, Lianjiang, Guangdong 07931
Liannan Nationality Museum, Qingyuan 08115
Lianping Museum, Lianping 07932
Lianshan Chuang Yaozu Self-Rule Nationality Museum, Qingyuan 08116
Lianyungang Museum, Lianyungang 07933
Lianzhou Museum, Qingyuan 08117
Liaocheng Zone Museum, Liaocheng 07934
LiaoJin Cheng Yuan, Beijing 07546
LiaoJin Dynasty Citywall Museum, Beijing 07546
Liaoning Art Gallery, Shenyang 08214
Liaoning Jin Modern City Museum, Shenyang 08215
Liaoning Provincial Museum, Shenyang 08216
Liaoyang Museum, Liaoyang 07935
Libby Museum, Wolfeboro 54269
Liberace Museum, Las Vegas 49974
Liberty County Museum, Chester 47430
Liberty Hall, Frankfort 48740
Liberty Hall Historic Center, Lamoni 49913
Liberty Hall Museum, Union 53649
Liberty Memorial Museum of World War One, Kansas City 49693
Liberty Village Arts Center and Gallery, Chester 47431
Libertyville-Mundelein Historical Society Museum, Libertyville 50080
Library & Gallery, Cambridge 05536
Library Museum, Borås 40466
Library, Museum and Archives of the Stanisław Moniuszko Warsaw Musical Society, Warszawa 35216
Library of Art and Archaeology of the Museum of Art and History of the City of Geneva, Genève 41400
Library of Congress, Washington 53886
Library of the Greek Orthodox Patriarchate of Alexandria and All Africa, Alexandria 10473
Licheng District Museum, Licheng 07936
Lichfield Heritage Centre, Lichfield 44705
Lichtdruck, Dresden 17958
Lichtdruck-Kunst Leipzig e.V., Leipzig 19917
Lichtenburg Museums and Art Gallery, Lichtenburg 38682
Lichtentaler Pfarrmuseum, Wien 03023
Licking County Art Association Gallery, Newark 51289
Licking County Historical Society, Newark 51290
Lidköpings Konsthall, Lidköping 40675
Liebenberger Geschichtsmuseum, Löwenberger Land 20018
Liebermann House, Nahariya 25069
Liebermann-Villa am Wannsee, Berlin 17177
Liebig-Museum, Gießen 18639
Lieblinger Heimatstube, Willstätt 22588
Liechtenstein Museum, Wien 03024
Liechtenstein National Library, Vaduz 30198
Liechtenstein National Museum, Vaduz 30199
Liechtenstein Schloss Wilfersdorf, Wilfersdorf 03107
Liechtensteinische Landesbibliothek, Vaduz 30198
Liechtensteinisches Landesmuseum, Vaduz 30199
Lied Discovery Children's Museum, Las Vegas 49975
Liedenpohjan Kylämuseo, Liedenpohja 11066
Liedon Vanhalinna, Vanhalinna 11444
Liemers Museum, Zevenaar 32999
Liepāja Okupaciju Rezimos, Liepāja 30038
Liepājas Muzeja Izstāžu Zāles, Liepāja 30039
Liepājas Muzejs, Liepāja 30040
Lier Bygdetun, Tranby 33914
Lierne museer, Nordli 33701
Lietuvos Aviacijos Muziejus, Kaunas 30254
Lietuvos Banko Muziejus, Vilnius 30414
Lietuvos Dailės Muziejus, Vilnius 30415
Lietuvos Energetikos Muziejus, Vilnius 30416
Lietuvos Etnokosmologijos Muziejus, Kulionių 30291
Lietuvos Geologijos Muziejus, Vievis 30397
Lietuvos Jūrų Muziejus, Klaipėda 30282
Lietuvos Laisvės Kovų ir Kančių Muziejus, Merkinė 30312
Lietuvos Liaudis Buities Muziejus, Rumšiškės 30347
Lietuvos Matematikų Muziejus, Vilnius 30417
Lietuvos Medicinos ir Farmacijos Istorijos Muziejus, Kaunas 30255
Lietuvos Nacionaline Dailės Galerija, Vilnius 30418
Lietuvos Nacionalinis Muziejus, Vilnius 30419
Lietuvos Prezidento K. Griniaus Memorialinis Muziejus, Marijampolė 30307
Lietuvos Radijo ir Televizijos Muziejus, Vilnius 30420
Lietuvos Šaulių Sąjungos Muziejus, Kaunas 30256
Lietuvos Savanorių Muziejus, Vilnius 30421
Lietuvos Sporto Istorijos Ekspozicija, Kaunas 30257
Lietuvos Teatro, Muzikos ir Kino Muziejus, Vilnius 30422
Lieu d'Art Contemporain, Colomiers 12588
Lieu Historique National de Côteau-du-Lac, Côteau-du-Lac 05648
Lieu Historique National des Forges du Saint-Maurice, Trois-Rivières 07173
Lieu Historique National du Canada de la Bataille-de-la-Châteauguay, Ormstown 06516
Lieu Historique National du Canada de la Bataille-de-la-Ristigouche, Pointe-à-la-Croix 06616
Lieu Historique National du Manoir Papineau, Montebello 06324
Lieu Historique National Sir Wilfrid Laurier, Ville-des-Laurentides 07264
Lieutenant General Ashton Armoury Museum, Victoria 07251
Life in the inundation area - exhibition, Szekszárd 23866
Life Sciences Museum, Johannesburg 38659
Lifeboat House, Margate 45105
Lifeboat Museum, Eastbourne 44099
Lifu Museum of Chinese Medicine, Taichung 42350
Light and Greenery Art Museum, Sagamihara 29182
The Light Factory, Charlotte 47391
Light Infantry Museum, Winchester 46031
Light Square Gallery, Adelaide 00777
The Lightbox, Woking 46055
Lighthouse Gallery, Poole 45405
Lighthouse Gallery, Tequesta 53478
Lighthouse Keepers Cottage, Carnarvon 00959
Lighthouse Museum, La Martre 06112
Lightner Museum, Saint Augustine 52522
Lightship Portsmouth Museum, Portsmouth 52109
Lignorama - Holz- und Werkzeugmuseum, Riedau 02585
Lihula Muuseum, Lihula 10591
Lijiang Museum, Lijiang 07937
Lijiashan Bronze Museum, Jiangchuan 07856
Likeio Ton Ellinidon Collection, Rethymnon 23169
Lilienhof, Lilienthal 19989
Liling Museum, Liling 07938
Liljendal Hembygdsmuseum, Liljendal 11070
Liljevalch's Art Gallery, Stockholm 40908
Liljevalchs Konsthall, Stockholm 40908
Lille Heddinge Rytterskole, Rødvig Stevns 10240
Lille Kunterbunt, Minden 20335
Lille Metropolitan Museum of Modern Art, Villeneuve-d'Ascq 16202
Lillehammer Art Museum, Lillehammer 33636
Lillehammer Kunstmuseum, Lillehammer 33636
Lillesand By-og Sjøfartsmuseum, Lillesand 33640
Lillesand Town and Maritime Museum, Lillesand 33640
Lillian and Albert Small Jewish Museum, Washington 53887
Lillian and Coleman Taube Museum of Art, Minot 50745
Lillian Stevenson Nursing Archives Museum, Saint John's 06849

Lillie Art Gallery, Milngavie 45150
Lilliput Antique Doll and Toy Museum, Brading 43588
Lillooet District Museum, Lillooet 06170
Lima Art Museum, Lima 34232
Lima Hembygdsgård, Lima 40677
Limassol District Museum, Lemesos 09387
Limassol Municipal Art Gallery, Lemesos 09388
Limbaži Museum, Limbaži 30042
Limbažu Muzejs, Limbaži 30042
Limberlost State Historic Site, Geneva 48879
Limburgs Miniatuurmuseum, Sint Geertruid 32772
Limburgs Museum, Venlo 32883
Limburgs Openluchtmuseum Eynderhoof, Nederweert Eind 32566
Limerick City Gallery of Art, Limerick 24866
Limerick Museum, Limerick 24867
Limesmuseum Aalen, Aalen 16412
Limfjordsmuseet, Løgstør 10179
Limhamns Museum, Malmö 40721
Limingan Kotiseutumuseo, Liminka 11074
Limon Heritage Museum, Limon 50086
Lin An-Tai Historical Home, Taipei 42399
Lin Liu-Hsin Puppet Theatre Museum, Taipei 42400
Lin San Zhi Art Display Center, Zhujiang 08514
Lin Yuan Simplicity Art Museum, Yuchi 42482
Lin Yutang House, Taipei 42401
Linan Museum, Hangzhou 07764
Linchuan Museum, Linchuan 07939
Lincoln Cathedral Treasury, Lincoln 44712
Lincoln Children's Museum, Lincoln 50099
Lincoln County Historical Museum, Davenport 47972
Lincoln County Historical Society, Lincoln 50088
Lincoln County Historical Society Museum, North Platte 51402
Lincoln County Historical Society Museum of Pioneer History, Chandler 47339
Lincoln County Pioneer Museum, Hendricks 49254
Lincoln Home, Springfield 53217
Lincoln Homestead, Springfield 53223
Lincoln Log Cabin, Lerna 50025
Lincoln Memorial Shrine, Redlands 52274
The Lincoln Museum, Fort Wayne 48707
Lincoln Parish Museum, Ruston 52490
Lincoln Park Historical Museum, Lincoln Park 50113
The Lincoln-Tallman Restorations, Janesville 49602
Lincoln Tomb, Springfield 53218
Lincoln Train Museum, Gettysburg 48903
Lincoln's New Salem Historic Site, Petersburg 51794
Lincolnshire Road Transport Museum, Lincoln 44713
Lindås Skulemuseum, Isdalstø 33577
Linden-Museum Stuttgart, Stuttgart 21951
Linden-Museum Stuttgart, State Museum of Ethnology, Stuttgart 21951
Linden Railroad Museum, Linden 50117
Linden - Saint Kilda Centre for Contemporary Arts, Saint Kilda, Victoria 01480
Lindenau-Museum, Altenburg 16497
Lindenfelser Museum, Lindenfels 20006
Lindesbergs Museum, Lindesberg 40679
Lindesnes Bygdemuseum, Sør-Audnedal 33851
Lindholm Høje Museet, Nørresundby 10196
Lindigtmühle am Lindenvorwerk, Kohren-Sahlis 19671
Lindisfarne Priory, Holy Island 44480
Lindleinturm-Museum, Creglingen 17743
Lindsay Gallery, Lindsay 06172
Lindsay Wildlife Museum, Walnut Creek 53803
Linen and Home-brew Museum, Alfred Kordelin Vehicle Museum, Lammi 11042
Linen Museum, Länkipohja 11024
Linfen Museum, Linfen 07941
Linfen Zone Museum, Linfen 07942
Lingao Museum, Lingao 07943
Lingnan Fine Arts Museum, Taipei 42402
Lingnan Hua Pai Memorial Hall, Guangzhou 07733
Lingshan Museum, Qinzhou 08122
Lingtai Museum, Lingtai 07944
Lingtong Exhibition, Changzhou 07607
Lingyuan Museum, Lingyuan 07946
Lingyun Museum, Lingyun 07947
Linhai Museum, Linhai 07948
Linköpings Konsthall, Linköping 40682
Linköpings Slotts- och Domkyrkomuseum, Linköping 40683
Linley Sambourne House, London 44901
Linli Museum, Linli 07949
Linlithgow Palace, Linlithgow 44720
Linlithgow Story, Linlithgow 44721
Linlithgow Union Canal Museum, Linlithgow 44722
Linn County Historical Museum and Moyer House, Brownsville 47095
Linn County Historical Society Museum, Cedar Rapids 47305
Linn County Museum, Pleasanton 51987
Linnæus Museum, Uppsala 41007
Linnan Museo ja Ortodoksinen Kirkkomuseo, Olavinlinna 11160
Linnanmäen Museo, Helsinki 10800
Linnanmäki Toy and Game Museum, Helsinki 10800
Linnémuseet, Uppsala 41007
Linnenschmidt'sche Vormals Landesherrliche Mühle zu Venne, Ostercappeln 20933
Linnés Hammarby, Uppsala 41008
Linnés Råshult, Diö 40487
Linnoitus- ja Kotiseutumuseo, Luumäki 11087
Linotype Museum, Schopfheim 21621
Linqing Museum, Linqing 07950
Linquan Museum, Linquan 07951
Linton and District Museum, Linton 01231
Lintong County Museum, Lintong 07952
Linxiazhou Museum, Tiyuchang 08295
Linyi Museum, Linyi, Shandong 07955
Linyi Museum, Linyi, Shanxi 07957
Linyou Museum, Linyou 07958
LinzGenesis, Linz 02319
Liohuo Aboriginal Stone-roofed Hut Museum, Neipu 42285
Lion Dance Ceremony Exhibition Hall, Takayama 29343
Lion Salt Works, Northwich 45272
Lionel Wendt Art Gallery, Colombo 40372
Lion's Kingdom Museum, Toucheng 42455
Lipeckij Oblastnoj Kraevedčeskij Muzej, Lipeck 37104
Lipetsk Regional Museum, Lipeck 37104
Lipica Museum, Szilvásvárad 23894
Lipicai Múzeum, Szilvásvárad 23894
Lippische Landesbibliothek, Detmold 17838
Lippisches Landesmuseum, Detmold 17839
LIPS Slotenmuseum, Dordrecht 32150
LIPSIKUS Kindermuseum, Leipzig 19918
Liptovské Múzeum, Ružomberok 38343
Liquid Paper Correction Fluid Museum, Santa Fe 52878
Liquid Paper Museum, Nambe 50940
Lisakovsk Museum of History and Culture Verhnego Pritobolya, Lisakovsk 29709
Lisakovskij Muzej Istorii i Kultury Verchnego Pritobolja, Lisakovsk 29709
Lisbon Historical Society Museum, Lisbon 50125
Liselund Gamle Slot, Borre 09981
Lishui Museum, Lishui, Jiangsu 07962
Lishui Museum, Lishui, Zhejiang 07963
Liskun State Museum of Cattle Breeding, Moskva 37206
Lisle Station Park, Lisle 50127
Lismore Regional Gallery, Lismore 01232
List Gallery, Swarthmore 53389
List Visual Arts Center, Cambridge 47179
Lista Museum, Vanse 33959
Listasafn Árnesinga, Hveragerði 24005
Listasafn Así, Ásmundarsalur, Reykjavík 24020
Listasafn Einars Jónssonar, Reykjavík 24021
Listasafn Íslands, Reykjavík 24022
Listasafn Kópavogs - Gerðasafn, Kópavogur 24008
Listasafn Sigurjóns Ólafssonar, Reykjavík 24023
Listasafnið á Akureyri, Akureyri 23982
Listasafnið Reykjanesbæjar, Reykjanesbær 24013
Listasavn Føroya, Tórshavn 10699
Liszt Ferenc Emlékmúzeum és Kutatóközpont, Budapest 23431
Liszt Ferenc Memorial Museum and Research Centre, Budapest 23431
Liszt-Haus, Weimar 22409
Liszt's House, Weimar 22409
Lit Sing Kwang Chinese Temple, Innisfail 01175
Litaraturny Muzej Janki Kupaly, Minsk 03251
Litcham Village Museum, Litcham 44727
Litchfield Historical Society Museum, Litchfield 50128
Literair Museum, Hasselt 03683
Literarisches Museum, Badenweiler 16892
Literarisches Museum im Max-Eyth-Haus, Kirchheim unter Teck 19549
Literárna Expozícia - Múzeum Janka Jesenského, Bratislava 38218
Literárne a Hudobné Múzeum, Banská Bystrica 38183
Literárne a Vlastivedné Múzeum, Svätý Jur 38365
Literárne Múzeum Pavla Horova, Bánovce nad Ondavou 38182
Literární Oddělení, Brno 09429
Literary and Memorial House-Museum of Maxim Rylsky, Kyïv 43080
Literary and Memorial House-Museum of Mikhail Bulgakov, Kyïv 43086
Literary and Memorial Museum-Appartment of Pavlo Tychyna, Kyïv 43082
Literary and Memorial Museum of Mikhailo Kotsiubynsky, Vinnycja 43203
Literary and Memorial Museum of Taras Shevchenko, Ševčenkove 43183
Literary Department of the Moravian Museum, Brno 09429
Literary Documentation Centre H.B. Yasin, Jakarta 24475
Literary History Museum, Minsk 03252
Literary Lviv in the First Half of the 20th c, Lviv 43124
Literary Memorial Museum of A.S. Pushkin and P.I. Chaikovski, Kam'janka 43056
Literary-Memorial Museum of Anna Achmatova, Slobidka Šelechivs'ka 43187
Literary Memorial Museum of Ivan Franko, Lviv 43123
Literary-Memorial Museum of Lesya Ukrayinka, Kolodjažne 43070
Literary Memorial Museum of N.A. Ostrovski, Šepetovka 43174
Literary Museum, Abakumcevo 36746
Literary Museum, Ulaanbaatar 31629
Literary Museum of Sergey Esenin, Toshkent 54487
Literary Museum of the Institute of Russian Literature of the Russian Academy of Sciences, Sankt-Peterburg 37658
Literaturarchiv Sulzbach-Rosenberg, Sulzbach-Rosenberg 21984
Literature and Art Museum of Marina Tsvetaeva, Usen-Ivanovskoe 37911
Literature and Memorial Museum N.N. Aseev, Lgov 37102
Literature and Memorial Museum of A.P. Gaydar, Lgov 37103
Literature and Music Museum, Rīga 30104
Literature and Music Museum Dacha Shalyapina, Kislovodsk 37047
Literature Museum of Literary Kherson, Cherson 43023
Literature Museum of Naltchik, Nalčik 37439
Literaturhaus, Magdeburg 20125
Literaturhaus München, München 20456
Literatūrinis Aleksandro Puškino Muziejus, Vilnius 30423
Literaturmuseum Altaussee, Altaussee 01708
Literaturmuseum Baumbachhaus, Meiningen 20259
Literaturmuseum Brecht-Weigel-Haus, Buckow 17620
Literaturmuseum der Moderne, Marbach am Neckar 20179
Literaturmuseum im Gartenhaus der Stadtbibliothek Baden-Baden, Baden-Baden 16884
Literaturmuseum Theodor Storm, Heiligenstadt, Heilbad 19108
Literaturno-chudožestvennyj Muzej M. Cvetaevoj, Usen-Ivanovskoe 37911
Literaturno-chudožestvennyj Muzej-usadba Prijutino, Vsevoložsk 37979
Literaturno-Memoral'nyj Muzej N.A Ostrovskogo, Šepetovka 43174
Literaturno-Memorialny Muzej A.A. Achmatovoï, Slobidka Šelechivs'ka 43187
Literaturno-memorial'nyj Budinok-muzej M. Bulgakova, Kyïv 43086
Literaturno-memorialnyj Dom-Muzej D.N. Mamina-Sibirjaka, Ekaterinburg 36887
Literaturno-memorialnyj Dom-Muzej F.M. Rešetnikova, Ekaterinburg 36888
Literaturno-memorialnyj Dom-muzej Sulejmana Stalskogo, Ašaga-Stalsk 36776
Literaturno-memorialnyj Muzei O.S. Puškina ta P.I. Čajkovskogo, Kam'janka 43056
Literaturno-memorialnyj Muzej F.A. Abramova, Verkola 37924
Literaturno-memorialnyj Muzej F.M. Dostoevskogo, Novokuzneck 37480
Literaturno-memorialnyj Muzej F.M. Dostoevskogo, Sankt-Peterburg 37656
Literaturno-memorialnyj Muzej F.M. Dostoevskogo, Semey 29728
Literaturno-memorialnyj Muzej Vasilija Erošenko, Staryj Oskol 37800
Literaturno-memorialnyj Muzej A.M. Gorkogo, Kazan 37011
Literaturno-memorialnyj Muzej M. Gorkogo, Samara 37616
Literaturno-memorialnyj Muzej Ivana Franko, Lviv 43123
Literaturno-Memorial'nyj Muzej Lesi Ukrayinky, Kolodjažne 43070
Literaturno-Memorial'nyj Muzej M. Eminesku, Černivci 43015
Literaturno-memorial'nyj Muzej Puškina, Odesa 43154
Literaturno-memorialnyj Muzej Stancionnyj Smotritel, Vyra 37983
Literaturno-memorial'nyj Muzej T.G. Ševčenka, Ševčenkove 43183
Literaturno-memorialnyj Muzej M.M. Zoščenko, Sankt-Peterburg 37657
Literaturno-muzykalnyj Muzej Dača Šaljapina, Kislovodsk 37047
Literaturno-mystec'kyj Muzej, Chmel'nyc'kyj 43028
Literaturno-teatralnyj Muzej N.M. Djakonova, Syktyvkar 37816
Literaturnyj Memorialnyj Dom-muzej Abaja, Semey 29729
Literaturnyj Lviv Peršoï Polovyny XXst, Lviv 43124
Literaturnyj Muzej, Krasnojarsk 37074
Literaturnyj Muzej, Penza 37529
Literaturnyj Muzej Abakumcevo, Abakumcevo 36746
Literaturnyj Muzej Goroda Nalčik, Nalčik 37439
Literaturnyj Muzej im. K.V. Ivanova, Čeboksary 36838
Literaturnyj Muzej Instituta Russkoj Literatury, Sankt-Peterburg 37658
Literaturnyj Muzej Literaturna Chersonščina, Cherson 43023
Literaturnyj Muzej Maksima Bogdanoviča, Minsk 03253
Literaturnyj Muzej F.I. Tjutčeva, Ovstug 37522
Lithuania Minor History Museum, Klaipėda 30283
Lithuanian Army Volunteers Museum, Vilnius 30421
Lithuanian Art Museum, Vilnius 30415
Lithuanian Aviation Museum, Kaunas 30254
Lithuanian Energy Museum, Vilnius 30416
Lithuanian Ethno-Cosmology Museum, Kulionių 30291
The Lithuanian Museum, Chicago 47486
Lithuanian Museum-Archives of Canada, Mississauga 06312
Lithuanian Museums Association, Vilnius 54929
Lithuanian National Gallery, Vilnius 30418
Lithuanian Riflemen's Union Museum, Kaunas 30256
Lithuanian Sea Museum, Klaipėda 30282
Lithuanian Theater, Music and Film Museum, Vilnius 30422
Little Bighorn Battlefield Museum, Crow Agency 47892
The Little Brick House, Vandalia 53708
Little-Carpathian Museum in Pezinok, Pezinok 38324
Little Claybelt Homesteaders Museum, New Liskeard 06431
Little Compton Historical Society Museum, Little Compton 50132
Little Cowboy Bar and Museum, Fromberg 48800
Little Falls Historical Museum, Little Falls 50137
Little Holland House, London 44902
Little Houses, Dunkeld 44060
Little Moreton Hall, Congleton 43883
Little Museum, Tewkesbury 45843
Little Norway, Blue Mounds 46875
Little Prairie Heritage Society, Chetwynd 05610
Little Red School House Museum, Weyauwega 54115
Little Red Schoolhouse-Living Library, Beloit 46722
Little Stone School, Saskatoon 06916
Little Sweets Museum, Osaka 29132
Little Traverse Historical Museum, Petoskey 51807
Little White House, Warm Springs 53816
Little White Schoolhouse, Ripon 52357
Little White Schoolhouse, Truro 07181
Little World Museum of Man, Inuyama 28576
Little World - Museum of Miniature Puppets, Semipalatinsk 37760
Littlehampton Museum, Littlehampton 44729
Littleton Historical Museum, Littleton 50144
Littman Gallery, Portland 52079
Liudas Truikys and Marijona Rakauskaite Memorial Museum of the M.K. Čiurlionis National Art Museum, Kaunas 30258
Liudo Truikio ir Marijonos Rakauskaitės Memorialinis Muziejus, Kaunas 30258
Liuhe County Display Center, Liucheng 07964
Liuksialan Maatalousmuseo, Kangasala 10909
Liuligongfang Museum, Shanghai 08171
Liuzhou Museum, Liuzhou 07966
Livadijskij Dvorec, Jalta 43050
Livadiya Palace, Jalta 43050
Livapaşa Konağı Etnografya Müzesi, Kastamonu 42870
Liverpool Scottish Regimental Museum, Liverpool 44738
Livesey Museum for Children, London 44903
Living Art Museum LAM, Reykjavík 24028
The Living Arts and Science Center, Lexington 50052
Living History Farms Museum, Urbandale 53671
Living Prairie Museum, Winnipeg 07366
Living Word National Bible Museum, Aledo 46194
Livingston County Historical Society Museum, Geneseo 48877
Livingston Depot Center, Livingston 50148
Livingstone and Stanley Memorial, Tabora 42521
Livingstone Museum, Livingstone 54655
Livonijas Ordeņa Pils Tornis, Tukums 30132
Livrustkammaren, Stockholm 40909
Lizauli Cultural Village, Lizauli 31744
Liznjekova Hiša, Kranjska Gora 38439
Lizzadro Museum of Lapidary Art, Elmhurst 48376
Ljósmyndasafn Reykjavíkur Grófarhús, Reykjavík 24024
Ljungbergmuseet, Ljungby 40686
Ljungby Museum, Ljungby 40687
Ljusdalsbygdens Museum, Ljusdal 40688
Ljusterö Hembygdsmuseum, Ljusterö 40690
LKAB Gruvmuseum, Kiruna 40643
Lladro Museum, New York 51204
Llanberis Lake Railway, Llanberis 44753
Llancaiach Fawr Living History Museum, Nelson 45199

Llandudno Museum and Art Gallery, Llandudno 44759
Llangollen Motor Museum, Llangollen 44770
Llangollen Railway, Llangollen 44771
Llanidloes Museum, Llanidloes 44773
Llechwedd Slate Caverns, Blaenau Ffestiniog 43533
Lleyn Historical and Maritime Museum, Nefyn 45197
La Llonja, Palma de Mallorca 39784
Lloyd George Museum, Cricciert 43925
Llywernog Silver-Lead Mine Museum, Ponterwyd 45399
Lobaumuseum, Wien 03025
Lobdengau-Museum, Ladenburg 19763
Lobkovicky Palác, Praha 09760
Lobkovicz Palace, Praha 09760
Local and Maritime Museum, Newhaven 45235
Local Antiquities Museum, Chitradurga 24120
Local Art Museum, Pingtung 42291
Local Arts and Crafts Museum, Tenala 11373
Local Cultural Museum, Keelung 42248
Local Cultural Museum, Lujhu 42264
Local Historic Museum and Art Gallery, Vassenden 33965
Local History and Ethnography Museum, Brad 36166
Local History Collection, Szigetszentmiklós 23890
Local History Museum, Buzsák 23499
Local History Museum, Csempeszkopács 23505
Local History Museum, Kryvyj Rig 43075
Local History Museum in Pöytyä, Pöytyä 11212
Local History Museum of Kassel, Kassel 19478
Local Hsitory Museum, Leer 19870
Local Museum, Aleksinac 38029
Local Museum, Ashkelon 24930
Local Museum, Bechyně 09397
Local Museum, Bhanpura 24083
Local Museum, Brú 23990
Local Museum, Buzet 08851
Local Museum, Chrastava 09482
Local Museum, Danilovgrad 31644
Local Museum, Fort Shevchenko 29700
Local Museum, Höfn 24001
Local Museum, Kibbutz Maabarot 25049
Local Museum, Kibbutz Sasa 25055
Local Museum, Našice 08912
Local Museum, Nikšić 31650
Local Museum, Ogulin 08919
Local Museum, Pljevlja 31652
Local Museum, Polvijärvi 11216
Local Museum, Považská Bystrica 38334
Local Museum, Ruma 38137
Local Museum, Skiptvet 33838
Local Museum, Stykkishólmur 24040
Local Museum, Törökbálint 23933
Local Museum, Travnik 04286
Local Museum, Ulcinj 31659
Local Museum and Jan Hala Gallery, Važec 38385
Local Museum and Rindell Cabin Museum, Rusko 11283
Local Museum Collection, Čajetina 38096
Local Museum of Amursk, Amursk 36756
Local Museum of Helsingør, Helsingør 10065
Local Museum of History and Culture, Lugans'k 43117
Local Museum of Perniö, Perniö 11200
Local Museum of Stuttgart-Möhringen, Stuttgart 21941
Local Museum Ozalj, Ozalj 08925
Local Museum Poreč, Poreč 08929
Local Museum Zemun, Zemun 38165
Local Studies Collection @ Library, Marrickville 01265
Localbahnmuseum, Bayerisch Eisenstein 16943
Localbahnmuseum, Innsbruck 02123
Loch Ness 2000 Exhibition, Drumnadrochit 44026
Lock Museum of America, Terryville 53486
Lock-up Museum and John Paynter Gallery, Newcastle 01362
Lockehaven Schoolhouse Museum, Enfield 48399
Locke's Distillery Museum, Kilbeggan 24845
Lockhouse-Friends of the Delaware Canal, New Hope 51061
Lockmaster's House Museum, Elgin 05771
Lockport Gallery, Lockport 50157
Locksmith's House, Willenhall 46015
Lockwood-Mathews Mansion Museum, Norwalk 51432
Locomotion - The National Railway Museum at Shildon, Shildon 45665
Locust Grove, Louisville 50289
Locust Grove, Poughkeepsie 52122
Locust Lawn and Terwilliger House, Gardiner 48859
Łódź Archidiocesan Museum, Łódź 34874
Lödöse Museum, Lödöse 40691
Löfstad Slott Museum, Norrköping 40772
Löhe-Zeit-Museum, Neuendettelsau 20607
Lönnström Art Museum, Rauma 11259
Lönnströmin Taidemuseo, Rauma 11259
Lötschentaler Museum, Kippel 41504
Lövångers Sockenmuseum, Lövånger 40692
Löwen-Drogerie und Museum, Oelsnitz 20872
Löwenburg, Kassel 19467
Löwenpalais, Berlin 17178
Lofoten Krigsminnemuseum, Svolvær 33896
Lofotmuseet, Kabelvåg 33585
Log Cabin Museum, Murray Harbour 06403
Log Cabin Museum, Saanichton 06784
Log Cabin Village, Fort Worth 48716
The Log Farm, Nepean 06424
Log House Museum, Columbiana 47738
Logan Art Gallery, Logan 01237
Logan County Historical Society Museum, Bellefontaine 46697
Logan County Museum, Paris 51700
Logan Hall, Ottawa 06550
Logan Museum of Anthropology, Beloit 46724
Logging Museum, Peachland 06576
Lohgerber-, Stadt- und Kreismuseum, Dippoldiswalde 17871
Lohjan Museo, Lohja 11077
Lohrbacher Heimatstuben, Mosbach 20390
Lohtajan Kotiseutumuseo, Lohtaja 11079
Loimaan Kotiseutumuseo, Loimaa 11080
Lois E. Woods Museum, Norfolk 51365
Loisium, Langenlois 02257
Lokalhistorisches Pulkautalmuseum, Hadres 02034
Løkken Museum, Løkken 10181
Lokpark Ampflwang, Ampflwang im Hausruckwald 01718
Lokpark Brosigstraße, Braunschweig 17525
Lokstallet Strömstadt Konsthall, Grebbestad 40566
Lokwelt Freilassing, Freilassing 18444
Lokwelt Freilassing - The World of Railways, Freilassing 18444
Lolland-Falsters Stiftsmuseum, Maribo 10185
Lolland-Falsters Traktor- og Motormuseum, Eskildstrup 10004
Lom Bygdamuseum, Lom 33645
Lomakov Museum of Antique Cars and Motorcycles, Moskva 37225
Lomakovskij Muzej Starinnych Avtomobilej i Motociklov, Moskva 37225
Lombard Historical Museum, Lombard 50164
Lommedalsbanen, Rykkinn 33802
Lompoc Museum, Lompoc 50166
Lon C. Hill Home, Harlingen 49169
London Borough of Bromley Museum, Orpington 45321
London Brass Rubbing Centre, London 44904
The London Brass Rubbing Centre in Washington D.C., Gaithersburg 48824
London Canal Museum, London 44905
London Federation of Museums and Art Galleries, London 55071
London Fire Brigade Museum, London 44906
London Motorcycle Museum, Greenford 44363
London Museum of Archaeology, London 06186
London Regional Children's Museum, London 06187
London Scottish Regimental Museum, London 44907
London Sewing Machine Museum, London 44908
Londonderry Mines Museum, Londonderry 06191
London's Transport Museum, London 44909
Lone Star Flight Museum/Texas Aviation Hall of Fame, Galveston 48848
Long Beach Museum of Art, Long Beach 50174
Long Branch Historical Museum, Long Branch 50181
Long Eaton Town Hall, Long Eaton 45024
Long Island Children's Museum, Garden City 48857
Long Island Culture History Museum, Commack 47763
Long Island Maritime Museum, West Sayville 54081
Long Island Museum, Long Island 03202
Long Island Museum of American Art, History and Carriages, Stony Brook 53313
Long Island Museum of Science and Technology, Garden City 48858
Long-Nan Museum of Natural Lacquer Ware, Puli 42294
Long Shop Museum, Leiston 44689
Long Warehouse, Coalbrookdale 43859
Longboat Key Center for the Arts, Longboat Key 50191
Longchuan Museum, Longchuan 07968
Longdale Craft Centre and Museum, Ravenshead 45464
Longfellow-Evangeline State Historic Site, Saint Martinville 52604
Longfellow House, Cambridge 47180
Longfellow's Wayside Inn Museum, Sudbury 53353
Longgang Ke Jia Folklorish Museum, Shenzhen 08222
Longhorn Museum, Pleasanton 51989
Longhua Mausoleum of Fallen Heroes, Shanghai 08172
Longhua Museum, Longhua 07969
Longjing Koreazu Glaswegiansu Guan, Longjing 07970
Longkou Museum, Longkou 07971
Longleat House, Warminster 45934
Longmont Museum, Longmont 50193
Longquan Museum, Longquan 07972
Longquanyi District Museum, Chengdu 07617
Longquanyi Mingshu Wangling Museum, Chengdu 07618
Longreach Power House Museum, Longreach 01239
Longrock Museum, Longyan 07974
Longstreet Farm, Holmdel 49324
Longue Vue House, New Orleans 51085
Longview Museum of Fine Art, Longview 50195
Longwood Center for the Visual Arts, Farmville 48523
Longyear Museum, Chestnut Hill 47440
Longyou Museum, Longyou 07975
La Lonja, Zaragoza 40339
Lónyai Castle, Tuzsér 23940
Lónyai-Kastély, Tuzsér 23940
Look Out Discovery Centre, Bracknell 43573
Loop Museum and Art Center, Seagraves 52989
Loos Koboltgruva, Loos 40693
Loosduins Museum De Korenschuur, Den Haag 32079
Lopdell House Gallery, Waitakere 33227
Lopez Island Historical Museum, Lopez Island 50198
Lopez Memorial Museum, Pasig 34478
Loquat Museum of Sinshe, Sinshe 42340
Lora Robins Gallery of Design from Nature, Richmond 52332
Loránd Eötvös Collection, Budapest 23397
Lord Howe Island Museum, Lord Howe Island 01241
Lord Strathcona's Horse Museum, Calgary 05523
Lore Degenstein Gallery, Selinsgrove 53030
Loreley-Museum im historischen Stadtturm, Sankt Goarshausen 21497
Lørenskog Bygdemuseum, Skårer 33829
Lorenzo State Historic Site, Cazenovia 47287
Lori Local Tradition Museum, Vanadzor 00719
Loringiano Museum, Málaga 39603
Lorn Museum, Ledaig 44656
Lorne Scots Regimental Museum, Brampton 05455
Los Alamos County Historical Museum, Los Alamos 50205
Los Altos History Museum, Los Altos 50207
Los Angeles County Museum of Art, Los Angeles 50244
Los Angeles Maritime Museum, San Pedro 52826
Los Angeles Municipal Art Gallery, Los Angeles 50245
Los Angeles Museum of the Holocaust, Los Angeles 50246
Los Angeles Police Museum, Los Angeles 50247
Los Angeles Valley College Art Gallery, Valley Glen 53696
Los Nogales Museum, Seguin 53029
Loseley House, Guildford 44373
Loshi Stone Art Museum, Ershuei 42186
Loški Muzej, Škofja Loka 38514
Lossiemouth Fisheries and Community Museum, Lossiemouth 45029
Lost City Museum, Overton 51634
Lost River State Park Museum, Mathias 50534
Lostwithiel Museum, Lostwithiel 45030
Lószerszám Múzeum, Sümeg 23826
Løten Bankmuseum, Løten 33651
Lotherton Hall, Aberford 43251
Lothringer 13, München 20457
Lotschnauer Heimatstube, Eislingen 18110
Lotte Reimers-Stiftung, Deidesheim 17808
Lotte Reiniger Museum, Tübingen 22119
Lottehaus, Wetzlar 22529
Lotus Explorer Museum, Baihe 42140
Lotus Image Museum, Luodong 42265
Lou Holtz and Upper Ohio Valley Hall of Fame, East Liverpool 48255
Louden-Henritze Archaeology Museum, Trinidad 53568
Loudon Heritage Farm Museums, Sterling 53287
Loudoun Mansion, Philadelphia 51849
Loudoun Museum, Leesburg 50018
Lougheed House Conservation Society, Calgary 05524
Louhisaaren Kartanolinna, Askainen 10715
Louhisaari Manor, Askainen 10715
Louis Bernal Gallery, Tucson 53594
Louis Couperus Museum, Den Haag 32080
Louis E. May Museum, Fremont 48787
Louis H. and Lena Firn Grover Museum, Shelbyville 53068
Louis-S.-Saint-Laurent, Compton 05638
Louis Spohr-Gedenk- und Forschungsstätte, Kassel 19468
Louis Tussaud Wax Museum, København 10138
Louis Tussaud's Waxworks Museum, Niagara Falls 06442
Louisa County Historical Society, Louisa 50277
Louisbourg Marine Museum, Louisbourg 06198
Louisburg College Art Gallery, Louisburg 50278
Louise-Henriette-Stift, Kloster Lehnin 19576
Louise Wells Cameron Art Museum, Wilmington 54202
Louisiana Arts and Science Center, Baton Rouge 46633
Louisiana Association of Museums, Baton Rouge 55152
Louisiana Children's Museum, New Orleans 51086
Louisiana Country Music Museum, Marthaville 50515
Louisiana Museum of Modern Art, Humlebæk 10103
Louisiana Naval War Memorial U.S.S. Kidd, Baton Rouge 46634
Louisiana Old State Capitol, Center for Political and Governmental History, Baton Rouge 46635
Louisiana State Exhibit Museum, Shreveport 53084
Louisiana State Museum, New Orleans 51087
Louisiana State Museum, Patterson 51732
Louisiana State University Museum of Art, Baton Rouge 46636
Louisiana Tech Museum, Ruston 52491
Louisville Science Center, Louisville 50290
Louisville Slugger Museum, Louisville 50291
Louisville Visual Art Museum, Louisville 50292
Lounais-Hämeen Museo, Forssa 10737
Louth Museum, Louth 45037
Louwman Collection - Het National Automobilmuseum, Raamsdonksveer 32677
Lovački Muzej, Zagreb 09002
Lovasmúzeum, Bábolna 23341
Lovecko Lesnicke Muzeum, Úsov 09891
Loveland Museum and Gallery, Loveland 50301
Lovely Lane Museum, Baltimore 46574
Loviisa Maritime Museum, Loviisa 11083
Loviisa Town Museum, Loviisa 11082
Loviisan Kaupunginmuseo, Loviisa 11082
Loviisan Merenkulkumuseo, Loviisa 11083
Lovriner Stube, Donauwörth 17898
Lowe Art Museum, Coral Gables 47802
The Lowe Gallery at Hudson Guild, New York 51205
Lowell Damon House, Wauwatosa 53996
Lower Cape Fear Historical Society Museum, Wilmington 54203
Lower Columbia College, Longview 50196
Lower Danube Museum, Călărași 36248
Lower East Side Tenement Museum, New York 51206
Lower Fort Garry, Saint Andrews 06798
Lower Methil Heritage Centre, Lower Methil 45039
Lower Saxony Asparagus Museum Nienburg, Nienburg 20719
Lower Saxony State Museum Hanover, Hannover 19002
Lower White River Museum, Des Arc 48095
Lowestoft and East Suffolk Maritime Museum, Lowestoft 45041
Lowestoft Museum, Lowestoft 45042
Lowewood Museum, Hoddesdon 44476
Lowndes County Museum, Valdosta 53686
The Lowry Gallery, Salford 45599
Loxton Historical Village, Loxton 01244
Loyal Edmonton Regiment Military Museum, Edmonton 05749
Loyalist Cultural Centre, Bath 05396
Loyalist House Museum, Saint John 06837
Lu Xun and Qiu Jin Museum, Shaoxing 08211
Lu Xun Former Residence, Shanghai 08173
Lu Xun Memorial Hall, Shanghai 08174
Lu Xun Museum, Beijing 07547
Luanping Museum, Luanping 07977
Lubelskie Towarzystwo Zachęty Sztuk Pięknych, Lublin 34900
Lubeznik Center for the Arts, Michigan City 50646
Lublin Archidiocesan Museum, Lublin 34901
Lubovňa Castle Museum, Stará Lubovňa 38361
Lubovnianske Múzeum - Hrad, Stará Lubovňa 38361
Lubovniansky Skanzen, Stará Lubovňa 38362
Lubuskie Muzeum Wojskowe w Zielonej Górze z/s w Drzonowie, Letnica 34852
Luchtvaart Museum Texel, De Cocksdorp 32046
Luchtvaart Museum Twenthe, Enschede 32210
Lucian Blaga Memorial House, Lancrăm 36434
Lucifersmuseum, Geraardsbergen 03650
Luckhard Museum, Sebewaing 53019
Luckman Gallery, Los Angeles 50248
Lucretia Vaile Museum, Palmer Lake 51682
Lucy Craft Laney Museum of Black History, Augusta 46479
Lucy Hayes Heritage Center, Chillicothe 47535
Lucy Maud Montgomery Birthplace, Charlottetown 05594
Luda Bērziņa Memoriālais Muzejs, Jūrmala 30024
Ludis Berzins Memorial Museum, Jūrmala 30024
Ludlow Museum, Ludlow 45045
L'udovit Fulla Galéria, Ružomberok 38344
Ludovit Fulla Gallery of the Slovak National Gallery, Ružomberok 38344
Ludvika Gammelgård och Gruvmuseum, Ludvika 40696
Ludvika Old Homestead and Museum of Mining, Ludvika 40696
Ludwig-Doerfler-Galerie, Schillingsfürst 21553
Ludwig Forum für Internationale Kunst, Aachen 16402
Ludwig Galerie Schloss Oberhausen, Oberhausen 20808

L

Ludwig-Gebhard-Museum, Tiefenbach, Oberpfalz 22059
Ludwig-Harms-Haus, Hermannsburg 19146
Ludwig Museum im Deutschherrenhaus, Koblenz 19586
Ludwig Museum - Museum of Contemporary Art, Budapest 23432
Ludwig Múzeum Budapest - Kortárs Művészeti Múzeum, Budapest 23432
Ludwig-Thoma-Haus, Tegernsee 22016
Ludwig Wittgenstein - Gedenkraum, Trattenbach 02822
Ludwig Wittgenstein - Wirklichkeit und Mythos, Kirchberg am Wechsel 02167
Ludza Museum of Local Studies, Ludza 30043
Ludzas Novadpētniecības Muzejs, Ludza 30043
Die Lübecker Museen, Lübeck 54841
Lüderitz Museum, Lüderitz 31745
Lügenmuseum, Kyritz 19754
Lüzinyan Evi, Lefkoşa 33370
Lufeng Saurian Museum, Lufeng 07978
Luftfahrt-Museum Laatzen-Hannover e.V., Laatzen 19755
Luftfahrt- und Technik- Museumspark Merseburg, Merseburg 20293
Luftfahrtausstellung im Alten Straßenbahndepot, Brandenburg an der Havel 17510
Luftfahrthistorische Sammlung Flugplatz Finow, Schorfheide 21623
Luftfahrttechnisches Museum, Rechlin 21207
Luftfahrttechnisches Museum, Rothenburg/Oberlausitz 21414
Luftwaffenmuseum der Bundeswehr, Berlin 17179
Lugang Folk Arts Museum, Lugang 42257
Lugansk Museum of Fine Art, Lugans'k 43114
Lugansk Regional Museum of Local Lore, Lugans'k 43115
Lugans'kyj Chudožnij Muzej, Lugans'k 43114
Lugans'kyj Oblasnyj Krajeznavčyj Muzej, Lugans'k 43115
Luggage Store, San Francisco 52782
Lugoj Orthodox District Micro-Museum, Lugoj 36449
Lugouqiao Display Center, Beijing 07548
Lugu Museum, Lugu 42260
Lugu Tea Museum, Lugu 42261
Luigi Cattaneo Anatomical Wax Model Museum, Bologna 25456
Luis Muñoz Rivera Museum, Barranquitas 36027
Luisenhütte, Balve 16905
Lullingstone Roman Villa, Eynsford 44201
Lum and Abner Museum, Pine Ridge 51923
Lumberman's Museum, Patten 51731
Lummis Home 'El Alisal', Los Angeles 50249
Luna Mimbres Museum, Deming 48053
Lund Bygdemuseum, Moi 33675
Lund University Art Collection, Lund 40704
Lundar Museum, Lundar 06201
Lunds Konsthall, Lund 40705
Lunds Universitets Antikmuseum, Lund 40706
Lundströmska Gården, Sigtuna 40834
Lundy's Lane Historical Museum, Niagara Falls 06443
Lunenburg Art Gallery, Lunenburg 06204
Luneville, castle's museum, Lunéville 13865
Lungauer Heimatmuseum Tamsweg, Tamsweg 02793
Lungauer Landschaftsmuseum, Mauterndorf 02371
Lunt Roman Fort, Baginton 43350
Luochuan Museum, Luochuan 07979
Luoding Museum, Luoding 07980
Luohe Wieldyang Museum, Wuyang 08345
Luonan Museum, Luonan 07981
Luonnonhistoriallinen Museo, Porvoo 11228
Luonnontieteellinen museo, Helsinki 10801
Luontotalo Arkki, Pori 11219
Luopioinen Museum, Luopioinen 11084
Luopioisten Museo, Luopioinen 11084
Luopioisten Torpparimuseo, Luopioinen 11085
Luostarinmäen Käsityöläismuseo, Turku 11394
Luostarinmäki Handicraft Museum, Turku 11394
Luotsitupa, Uusikaupunki 11419
Luoyang Capitol Museum, Luoyang 07983
Luoyang Ducheng Guan, Luoyang 07983
Luoyang Folklorish Museum, Luoyang 07984
Luoyang Glaswegiansu Guan, Luoyang 07984
Luoyang Museum, Luoyang 07985
Luoyang Tumulus Museum, Luoyang 07986
Lura Watkins Museum, Middleton 50656
Lusaka National Museum, Lusaka 54658
Lusatian Mining Museum, Hoyerswerda 19281
Luseland and Districts Museum, Luseland 06205
Lushan Museum, Lushan 07987
Lushun Museum, Dalian 07641
Lushun Snakes Dao Natural Museum, Dalian 07642
Lushun Su Jun Mausoleum of Fallen Heroes, Dalian 07643
Lusignan House, Lefkoşa 33370
Lusto-Suomen Metsämuseo, Punkaharju 11238
Lusto-The Finnish Forest Museum, Punkaharju 11238
Luther Burbank Home, Santa Rosa 52906
Luther College Fine Arts Collection, Decorah 48021
Luther-Stube Mansfeld, Mansfeld 20177
Luther W. Brady Art Gallery, Washington 53888
Lutheran Brotherhood Gallery, Minneapolis 50737
Lutherhaus mit Bibel-Café, Eisenach 18095
Lutherhaus Wittenberg, Lutherstadt Wittenberg 20115
Lutherkirche und Lutherzimmer im Dorfgemeinschaftshaus, Moorgrund 20379
Lutz Children's Museum, Manchester 50417
Lutz Mountain Meeting House and Heritage Museum, Moncton 06318
Luusua Village Museum, Luusua 11088
Luusuan Kylämuseo, Luusua 11088
Luvian Kotiseutumuseo, Luvia 11089
Luweizhao Painting and Calligraphy House, Pinghu 08083
Luxe Art Museum, Singapore 38175
Luxembourg City History Museum, Luxembourg 30482
Luxor Museum of Ancient Art, Al-Uqsur 10467
Luza Acht, Lübben 20068
Luzerne County Museum, Wilkes-Barre 54162
Luzhou Bowu Guan, Luzhou 07988
Lviv Historical Museum, Lviv 43126
Lviv Memorial Museum of Mykhailo Hrushevskyi, Lviv 43129
Lviv Museum of Folk Architecture and Foklore, Lviv 43134
Lviv Museum of the History of Religion, Lviv 43127
Lviv Picture Gallery, Lviv 43125
Lvivs'ka Galereja Mystectv, Lviv 43125
Lvivs'kyj Istoryčnyj Muzej, Lviv 43126
Lvivs'kyj Muzej Istorïi Religiï, Lviv 43127
LVR - Archäologischer Park Xanten, Xanten 22760
LVR-Industriemuseum, Bergisch Gladbach 17017
LVR-Industriemuseum, Engelskirchen 18171
LVR-Industriemuseum, Rheinisches Landesmuseum für Industrie- und Sozialgeschichte, Oberhausen 20809
LVR-Industry Museum, Oberhausen 20809
LVR-Landesmuseum Bonn, Bonn 17462
LWL-Freilichtmuseum Detmold, Detmold 17840
LWL-Freilichtmuseum Hagen, Hagen 18858
LWL-Industriemuseum, Glashütte Gernheim, Petershagen 21016
LWL-Industriemuseum, Henrichshütte Hattingen, Hattingen 19038
LWL-Industriemuseum, Schiffshebewerk Henrichenburg, Waltrop 22319
LWL-Industriemuseum, Textilmuseum in Bocholt, Bocholt 17395
LWL-Industriemuseum, Westfälisches Landesmuseum für Industriekultur, Dortmund 17921
LWL-Industriemuseum, Zeche Hannover, Bochum 17403
LWL-Industriemuseum, Zeche Nachtigall, Witten 22618
LWL-Industriemuseum, Ziegeleimuseum in Lage, Lage 19764
LWL-Landesmuseum für Kunst und Kulturgeschichte, Münster 20512
LWL-Museum für Archäologie, Herne 19155
LWL-Museum für Naturkunde, Münster 20513
LWL-Römermuseum, Haltern am See 18912
The Lyceum - Alexandria's History Museum, Alexandria 46211
Lyceum Museum, Puškin 37582
Lydd Town Museum, Lydd 45051
Lydia Koidulu Memoriaalmuuseum, Pärnu 10603
Lydiard House, Swindon 45817
Lyly Signal Depot Museum, Lyly 11090
Lylyn Viestivarikon Museo, Lyly 11090
Lyman Allyn Museum of Art, New London 51066
Lyman House Memorial Museum, Hilo 49293
Lyme Hall, Disley 43983
Lyme Regis Philpot Museum, Lyme Regis 45054
Lyn and Exmoor Museum, Lynton 45057
Lynchburg Fine Arts Center, Lynchburg 50326
Lynchburg Museum System, Lynchburg 50327
Lynden Pioneer Museum, Lynden 50331
Lyndhurst, Tarrytown 53451
Lyndon B. Johnson National Historical Park, Johnson City 49637
Lyndon Baines Johnson Museum, Austin 46514
Lyndon House Art Center, Athens 46413
Lynn Canyon Ecology Centre, North Vancouver 06478
Lynn Lake Mining Town Museum, Lynn Lake 06206
Lynn Museum, Lynn 50333
Lynnhaven House, Virginia Beach 53757
Lyon County Historical Society and Museum, Emporia 48392
Lyon County Historical Society Museum, Marshall 50505
Lyonel-Feininger-Galerie, Quedlinburg 21151
The Lyons Redstone Museum, Lyons 50334
Lyth Arts Centre, Lyth 45058
Lytham Hall, Lytham Saint Anne's 45059
Lytham Heritage Centre, Lytham Saint Anne's 45060
Lytham Lifeboat Museum, Lytham Saint Anne's 45061
Lytham Windmill Museum, Lytham Saint Anne's 45062
Lyttelton Historical Museum, Christchurch 33074
Lytton Museum, Lytton 06207
Lyuben Karavelov Memorial House, Koprivštica 05072

M

m. a. x. museo, Chiasso 41298
M. Auezov House Museum, Almaty 29680
M.B. Grekov House-Museum, Novočerkassk 37477
M.C. Mehta Gallery, Ahmedabad 24049
M. Christina Geis Gallery, Lakewood 49906
M.E. Pjatnicky Folk Museum, Aleksandrovka 36754
M. Eminesku Literary Memorial Museum, Černivci 43015
M.F. Gabyshin Gallery of Foreign Art, Jakutsk 36961
M.F. Husain's Museum of Art and Cinema, Hyderabad 24214
M. Gorky Memorial Museum, Samara 37616
M.I. Glinka Museum Estate, Novospasskoe 37490
M.I. Glinka State Central Museum of Music, Moskva 37188
M. Javakhishvili Museum, Tbilisi 16369
M. Kostava Memorial House Museum, Tbilisi 16370
M.M. Prishvin House-Museum, Dunino 36881
M.M. Soshchenko Literary Memorial Museum, Sankt-Peterburg 37657
M - Museum Leuven, Leuven 03787
M. Onatsky Sumy State Art Museum, Sumy 43190
M.R. Stefanik Museum, Košariská 38270
M.S. Spiridonov Memorial Museum-Apartment, Čeboksary 36839
M.S. Tuganov Museum of Fine Art, Vladikavkaz 37928
M.T. Aibek House-Museum, Toshkent 54473
M.T. Aibek Museum of History of the Peoples of Uzbekistan at the Academy of Sciences of Uzbekistan, Toshkent 54472
M.T. Oibek nomidagi Özbekiston Khalqlari Tarikhi Muzyi, Toshkent 54472
M.T. Oibek Yodgorlik Muzeyi, Toshkent 54473
M.V. Lomonosov Memorial Museum, Lomonosovo 37112
M.V. Lomonosov Museum, Sankt-Peterburg 37697
M.W. Williams Geological Museum, Vancouver 07214
Ma City Revolution Museum, Jiangling 07859
Maagdenhuismuseum, Antwerpen 03291
Maanmittausmuseo, Helsinki 10802
Ma'anshan Museum, Ma'anshan 07990
Maarjamäe Loss, Tallinn 10635
Maarten van Rossum Museum, Zaltbommel 32987
Maas- en Scheepvaartmuseum, Maasbracht 32519
Maas en Waals Museum 1939-1945, Winssen 32954
Mabee-Gerrer Museum of Art, Shawnee 53059
Mabel and Victor D'Amico Studio and Archive, Amagansett 46248
Mabel Hartzell Historical Home, Alliance 46231
Mabel Larson Fine Arts Gallery, Clarksville 47593
Måbødalen Kulturlandskapsmuseum, Eidfjord 33474
Mabry-Hazen House, Knoxville 49820
MAC, Birmingham 43507
MAC/VAL, Musée d'art contemporain du Val-de-Marne, Vitry-sur-Seine 16259
Macalester College Art Gallery, Saint Paul 52620
McAllen International Museum, McAllen 50339
McAllister House Museum, Colorado Springs 47702
Macao Museum, Macau 07995
Macao Tea Culture House, Macau 07996
Macau Museum of Art, Macau 07997
Macau Security Forces Museum, Macau 07998
Macau Wine Museum, Macau 07999
Macaulay Heritage Park, Picton 06600
Macaulay Museum of Dental History, Charleston 47371
MacBride Museum, Whitehorse 07325
MacCallum More Museum, Chase City 47403
McCallum Museum, Sibley 53094
McCartney House, Liverpool 44739
McClelland Gallery, Langwarrin 01217
Macclesfield Silk Museum, Macclesfield 45065
McClurg Mansion, Westfield 54094
McCollum-Chidester House, Camden 47193
McConaghy House, Hayward 49229
McCone County Museum, Circle 47568
McConnell Mansion, Moscow 50864
McCord and District Museum, McCord 06209
McCord Museum, Montréal 06371
McCormick Gallery, Midland 50671
McCoy House, Lewistown 50043
McCrae Homestead Museum, McCrae 01245
McCrae House, Guelph 05918
McCrossin's Mill Museum, Uralla 01614
Macculloch Hall Historical Museum, Morristown 50853
McCulloch House Museum and Hector Exhibit Centre, Pictou 06602
McCutchen Overland Inn, McCutchenville 50345
McDade Museum, McDade 50346
MacDonald House Museum, East Lake Ainslie 05732
Macdonald Stewart Art Centre, Guelph 05919
McDougall Mill Museum, Renfrew 06733
McDowell House Museum, Danville 47962
Macedon Historical Society Museum, Macedon 50347
Macedonian Museum of Contemporary Art, Thessaloniki 23215
McFaddin-Ward House, Beaumont 46678
McFarland Historical Society Museum, McFarland 50348
McFarland House, Niagara Falls 06444
McGillis Pioneer Home, Saint Victor 06883
Macgregor Museum, Harare 54668
McGregor Museum and Duggan Cronin Gallery, Kimberley 38670
McGroarty Cultural Art Center, Tujunga 53605
Machan Museum, Lagrange 49870
McHenry County Historical Society Museum, Union 53645
McHenry Museum, Modesto 50775
Machida City Museum, Machida 28874
Machida City Museum of Graphic Arts, Machida 28875
Machida City Museum of Photography, Machida 28876
Machida-shiritsu Hakubutsukan, Machida 28874
Machida-shiritsu Kokusai Hanga Bijutsukan, Machida 28875
Machida-shiritsu Shashin Bijutsukan, Machida 28876
MACHmit! Museum für Kinder, Berlin 17180
Macikai Concentration Camp Museum-Silute Museum, Macikai 30305
Macikų Lagerio Karceris-muziejus, Macikai 30305
McIntosh County Historical Society Museum, Ashley 46392
McIntosh Gallery, London 06188
McKechnie Institute, Girvan 44270
McKenna Children's Museum, New Braunfels 51022
Mackenzie and District Museum, Mackenzie 06210
MacKenzie Art Gallery, Regina 06720
Mackenzie Crossroads Museum, High Level 05984
Mackenzie House, Toronto 07120
Mackenzie Printery and Newspaper Museum, Queenston 06700
McKissick Museum, Columbia 47728
Macklin and District Museum, Macklin 06211
Macktown Living History Site and Whitman Trading Post, Rockton 52428
MacLachlan Woodworking Museum, Kingston 06081
McLane Center at Silk Farm, Concord 47774
MacLaren Art Centre, Barrie 05387
McLarty Treasure Museum, Vero Beach 53727
Maclaurin Art Gallery, Ayr 43347
McLean-Alanreed Area Museum, McLean 50360
McLean County Arts Center, Bloomington 46860
McLean County Historical Society Museum, Washburn 53840
McLean County Museum of History, Bloomington 46861
McLean Mill, Port Alberni 06626
McLean Museum and Art Gallery, Greenock 44365
Macleay Museum, Sydney 01552
McLellan Galleries, Glasgow 44288
McLurg Museum, Wilkie 07335
McManus Galleries, Dundee 44047
McMaster Museum of Art, Hamilton 05960
McMichael Canadian Art Collection, Kleinburg 06107
McMinn County Living Heritage Museum, Athens 46420
McMullen Museum of Art, Chestnut Hill 47441
McNamara House Museum, Victoria 53740
McNay Art Museum, San Antonio 52717
Macon County Museum Complex, Decatur 48016
MacPheadris/Warner House, Portsmouth 52096
McPherson College Gallery, McPherson 50371
McPherson County Old Mill Museum, Lindsborg 50121
McPherson Library Special Collections, Victoria 07252
McPherson Museum, McPherson 50372
MacPherson's Mill and Farm Homestead, New Glasgow 06430
Macroom Museum, Macroom 24872
Macrorie House Museum, Pietermaritzburg 38705
Macrorie Museum, Macrorie 06212
Mac's Antique Car Museum, Tulsa 53609
MAC's - Musée des Arts Contemporains, Hornu 03714
McWane Science Center, Birmingham 46823
Mad Musée, Liège 03798
Madaba Archaeology Museum, Ajloun 29640
Madách Imre Emlékmúzeum, Csesztve 23510
The Madam Brett Homestead, Beacon 46661
Madame Tussaud's, London 44910

Madame Tussaud's Amsterdam, Amsterdam 31856
Madame Tussaud's Wax Museum, New York 51207
Madang Museum, Madang 34084
Madara Nacionalen Istoriko-archeologičeski Rezervat, Madara 05089
Madeline Island Historical Museum, La Pointe 49852
Madera County Museum, Madera 50374
Madhavan Nair Foundation, Edapally 24167
Madhya Pradesh Tribal Research and Development Institute, Bhopal 24092
Madison Children's Museum, Madison 50392
Madison County Historical Museum, Edwardsville 48300
Madison County Historical Society Museum, Winterset 54256
Madison Historical Society Museum, Madison 50375
Madison-Morgan Cultural Center Collection, Madison 50377
Madison Museum of Contemporary Art, Madison 50393
Madona Museum of Local Studies and Art, Madona 30045
Madonas Novadpētniecības un Mākslas Muzejs, Madona 30045
Madonna House Museum, Combermere 05635
Madonna's House, Selçuk 42937
Madras Christian College Museum, Chennai 24117
MADRE - Museo d'Arte Donnaregina, Napoli 26881
Madres Dominicanas - Monasterio Sancti Spiritus, Toro 40152
Madrid Railway Museum, Madrid 39556
Madurodam, Den Haag 32081
Maeght Musée, Paris 14616
Mährisch-Schlesisches Heimatmuseum, Klosterneuburg 02201
Maejima Memorial Museum, Joetsu 28622
Mälsåkers Slott, Stallarholmen 40879
Die Männer von Brettheim, Rot am See 21397
Mäntsälän Kotiseutumuseo, Mäntsälä 11091
Mäntyharju Museum, Mäntyharju 11094
Mäntyharjun Museo, Mäntyharju 11094
Märchenhain Gföhl, Gföhl 01932
Märchenhaus und Heimatmuseum Ottenschlag, Ottenschlag 02476
Märchenofen, Neu-Ulm 20589
Märkisches Glasmuseum, Stechlin 21871
Märkisches Museum, Berlin 17181
Märkisches Museum, Witten 22619
Märkisches Ziegelei-Museum Glindow, Werder 22486
Märklin-Museum, Göppingen 18676
Maffeischächte der Grube Auerbach-Nitzlbuch, Auerbach, Kreis Deggendorf 16606
Maffra Sugarbeet Museum, Maffra 01249
Mafikeng Museum, Mafikeng 38685
Magadanskij Oblastnoj Kraevedčeskij Muzej, Magadan 37119
Magan Sangrahalaya Samiti, Wardha 24431
Magasin 3 Stockholm Konsthall, Stockholm 40910
Magasin - Centre National d'Art Contemporain, Grenoble 13130
Magdalen College Museum, Wainfleet All Saints 45906
Magdeburger Museen, Magdeburg 20126
Magersfontein Battlefields Museum, Kimberley 38671
Magevney House, Memphis 50574
Maggie L. Walker National Historic Site, Richmond 52333
Magic House-Saint Louis Children's Museum, Saint Louis 52585
Magic Lantern Castle Museum, San Antonio 52718
La Magnanerie du Coudray, Le Coudray-Macquard 13573
Magnanerie Muséalisée de la Roque, Molezon 14142
Magnesitmuseum, Radenthein 02550
Magnetawan Historical Museum, Magnetawan 06214
Magnitogorsk Picture Gallery, Magnitogorsk 37121
Magnitogorsk Regional Museum, Magnitogorsk 37122
Magnitogorskaja Kartinnaja Galereja, Magnitogorsk 37121
Magnitogorskij Kraevedčeskij Muzej, Magnitogorsk 37122
Magnolia Grove House Museum, Greensboro 49059
Magnolia Manor Museum, Cairo 47159
Magnolia Mound Plantation, Baton Rouge 46637
Magnolia Plantation, Charleston 47372
Magnus Dagestad Museet, Voss 33985
Magoffin Home, El Paso 48326
Magsingal National Museum, Magsingal 34411
Magyar Állami Operaház Múzeuma, Budapest 23433
Magyar Alumíniumipari Múzeum, Székesfehérvár 23856
Magyar Elektrotechnikai Múzeum, Budapest 23434
Magyar Építészeti Múzeum, Budapest 23435
Magyar Építőipari Múzeum, Veszprém 23960
Magyar Ferences Könyvtár Gyöngyösi Muemlék Könyvtára, Gyöngyös 23573
Magyar Filmtörténeti Gyűjtemény, Budapest 23436
Magyar Földrajzi Gyűjtemény, Érd 23548
Magyar Fotográfiai Múzeum, Kecskemét 23650
Magyar Fuszerpaprika Múzeum, Kalocsa 23626
Magyar Képzőművészek és Iparművészek Szövetsége, Budapest 23437
Magyar Kereskedelmi és Vendéglátóipari Múzeum, Budapest 23438
Magyar Mezőgazdasági Múzeum, Budapest 23439
Magyar Mezogazdasági Múzeum Tanyamúzeuma, Lajosmizse - Bene 23688
Magyar Motorok Múzeuma, Bodrogkeresztúr 23374
Magyar Naiv Művészek Múzeuma, Kecskemét 23651
Magyar Nemzeti Bank Látogatóközpont, Budapest 23440
Magyar Nemzeti Galéria, Budapest 23441
Magyar Nemzeti Múzeum, Budapest 23442
Magyar Olajipari Múzeum, Zalaegerszeg 23972
Magyar Ortodox Egyházművészeti Múzeum, Miskolc 23709
Magyar Technikai Múzeum Alapítvány, Budapest 23443
Magyar Térképészet Szakmatörténeti Múzeum, Budapest 23444
Magyar Természettudományi Múzeum, Budapest 23445
Magyar Vasúttörténeti Park, Budapest 23446
Magyar Vegyészeti Múzeum, Várpalota 23952
Magyar Zsidó Múzeum és Levéltár, Budapest 23447
Magyarországi Kalcitkristályok Színes Világa, Fertőrákos 23559
Maha Al Saqqa Center - Palestinian Folk Museum, Bethlehem 34051
Maha Wirawong National Museum, Nakhon Ratchasima 42583
Mahaffie Stagecoach Stop and Farm, Olathe 51530
Mahakoshal Art Gallery, Raipur 24366
Mahant Ghasidas Memorial Museum, Raipur 24367
Maharaja Banaras Vidya Mandir Museum, Varanasi 24427
Maharaja Fate Singh Museum, Vadodara 24419
Maharaja Sawai Man Singh II Museum, Jaipur, Rajasthan 24234
Mahatma Gandhi Hindi Sangrahalaya, Kalpi 24250
Mahatma Gandhi Museum, Mangalore 24307
Mahatma Gandhi Museum, Verulam 38773
Mahatma Phule Vastu Sangrahalaya, Pune 24359
Mahendra Museum, Kathmandu 31774
Mahn- und Gedenkstätte Düsseldorf, Düsseldorf 18001
Mahn- und Gedenkstätte Isenschnibber-Feldscheune, Gardelegen 18537
Mahn- und Gedenkstätte Ravensbrück, Fürstenberg, Havel 18490
Mahn- und Gedenkstätte Steinwache, Dortmund 17922
Mahn- und Gedenkstätte Wernigerode, Wernigerode 22497
Mahn- und Gedenkstätten Wöbbelin, Wöbbelin 22637
Mahone Bay Settlers Museum, Mahone Bay 06216
Mahtra Talurahvamuuseum, Juuru 10577
Mahymobiles - Musée Automobile, Leuze-en-Hainaut 03791
Maiden Tower Studio & Gallery, Bakı 03167
Maidenhair Tree Museum, Weimar 22400
Maidstone and District Museum, Maidstone 06217
Maidstone Museum and Bentlif Art Gallery, Maidstone 45073
Maier Museum of Art, Lynchburg 50328
Maigne Arsene, Gramat 13093
Maihaugen, Lillehammer 33637
Mail and Telecommunication Museum, Gdańsk 34645
Main Art Gallery, Fullerton 48806
Main Centre Heritage Museum, Main Centre 06218
Main Gallery of Henry Street Settlement, New York 51208
Main Line Art Center, Haverford 49215
Main Museum, Crescent City 47874
Maine Art Gallery, Wiscasset 54262
Maine Association of Museums, Augusta 55153
Maine Historical Society Museum, Portland 52065
Maine Maritime Museum, Bath 46628
Maine State Museum, Augusta 46485
Mainfränkisches Museum Würzburg, Würzburg 22713
Mainichi Art Collection, Nagoya 29001
Mairangi Arts Centre, Auckland 33054
Maironio Lietuvių Literatūros Muziejus, Kaunas 30259
Maironis Lithuanian Literature Museum, Kaunas 30259
Maisel's Brauerei- und Büttnerei-Museum, Bayreuth 16959
Maison à Pondalez, Morlaix 14276
Maison A. Ravier, Morestel 14268
Maison Adolphe Hardy, Dison 03560
Maison André-Benjamin-Papineau, Laval 06147
Maison André Derain, Chambourcy 12380
Maison-Atelier Foujita, Villiers-le-Bacle 16242
Maison Buttin-de Loës, Grandvaux 41451
Maison Centrale des Filles de la Charité du Sacré-Coeur de Jésus, Sherbrooke, Québec 06960
Maison Champenoise, Oeuilly 14495
Maison Chevalier, Québec 06683
Maison Communale Gérard Philippe, La Garde 13372
Maison Creole Eureka, Moka 30679
Maison Cyr - Musée de Chapelle, Saint-Basile 06803
Maison d'Abraham Mazel à Falguières, Mialet 14115
Maison d'Artiste de la Grande Vigne, Dinan 12747
Maison d'Auguste Comte, Paris 14617
La Maison d'Aure, Sailhan 15150
Maison d'Autrefois, Chasseneuil-du-Poitou 12429
Maison de Balzac, Paris 14618
Maison de Géologie et de l'Environnement de Haute-Alsace, Sentheim 15721
Maison de George Sand, Nohant-Vic 14465
Maison de Gergovie, La Roche-Blanche 13413
Maison de Jean Monnet, Bazoches-sur-Guyonne 11884
Maison de Jules Verne, Amiens 11586
Maison de Kerner, Riantec 15033
Maison de la Baie, Courtils 12659
Maison de la Baie du Mont Saint-Michel, Vains 16029
Maison de la Beauce, Orgeres-en-Beauce 14525
Maison de la Broderie, Villers-Outréaux 16232
Maison de la Canne, Les Trois-Ilets 30669
Maison de la Chataigne, Mourjou 14311
Maison de la Châtaigne et du Châtaignier, Saint-Martin-de-Boubaux 15396
La Maison de la Chicorée, Orchies 14523
Maison de la Culture Ahuntsic-Cartierville, Montréal 06344
Maison de la Culture Côte-des-Neiges, Montréal 06345
Maison de la Culture Mercier, Montréal 06346
Maison de la Culture Plateau-Mont-Royal, Montréal 06347
Maison de la Culture Pointe-aux-Trembles, Montréal 06348
Maison de la Culture Rosemont-Petite-Patrie, Montréal 06349
Maison de la Dentelle, Urçay 16018
Maison de la Dernière Cartouche, Bazeilles 11882
La Maison de la Dordogne Quercynoise, Souillac 15793
Maison de la Faïence, Badonviller 11815
Maison de la Faïence, Desvres 12717
Maison de la Faune, Murat (Cantal) 14333
Maison de la Faune, Vaujany 16085
Maison de la Flore, Longchaumois 13806
Maison de la Fonderie, Vrigne-aux-Bois 16277
Maison de la Forêt, Raismes 14979
La Maison de la Forêt et du Bois, Saint-Martin-en-Bresse 15404
Maison de la Fourdonne, Saint-Cirq-Lapopie 15210
Maison de la Lentille, Chouday 12526
Maison de la Loire, Montlouis-sur-Loire 14220
Maison de la Magie Robert Houdin, Blois 12031
Maison de la Manufacture d'Armes Blanches, Klingenthal 13304
Maison de la Mariée, Saint-Joachim 15341
Maison de la Mer, Courseulles-sur-Mer 12654
Maison de la Metallurgie et de l'Industrie de Liège, Liège 03799
Maison de la Meunerie, Nieul-sur-l'Autise 14444
Maison de la Musique Mécanique, Mirecourt 14133
Maison de la Nature, Eze 12898
Maison de la Nature, Paris 14619
Maison de la Nature et de l'Oiseau, Neuville-sur-Ailette 14410
Maison de la Négritude et des Droits de l'Homme, Champagney 12386
Maison de la Pataphonie, Dinant 03557
Maison de la Pétanque, Vallauris 16048
Maison de la Pierre de Volvic, Volvic 16269
Maison de la Pierre et du Ciment, Montalieu-Vercieu 14167
Maison de la Pisciculture, Sizun 15769
Maison de la Pomme et de la Poire, Barenton 11851
Maison de la Pomme et de la Poire, Carrouges 12301
Maison de la Pomme et Forge, Sainte-Opportune-la-Mare 15577
Maison de la Poterie, Bouffioulx 03389
Maison de la Réserve Naturelle du Lac de Remoray, Labergement-Sainte-Marie 13464
Maison de la Résistance, Chasseneuil-sur-Bonnieure 12430
Maison de la Rivière, Châteauneuf-sur-Sarthe 12451
Maison de la Rivière, Chinon 12514
Maison de la Rivière, de l'Eau et de la Pêche, Sizun 15770
Maison de la Rivière et du Pêcheur, Saint-Georges-de-Montaigu 15279
Maison de la Ruralité, Thuillies 04063
Maison de la Sainte-Victoire, Saint-Antonin-sur-Bayon 15173
Maison de la Science, Liège 03800
Maison de la Sirène, Collonges-la-Rouge 12578
Maison de la Thiérache, Liart 13740
Maison de la Vallée d'Ossau, Laruns 13532
Maison de la Vie Paysanne, Han-sur-Lesse 03674
Maison de la Vie Rurale, Massiac 14027
Maison de la Vigne, Essoyes 12867
Maison de l'Abbé Grégoire, Embermenil 12820
Maison de l'Abeille et du Miel, Riez 15039
Maison de l'Agoût, Fraisse-sur-Agout 12994
Maison de l'Air, Paris 14620
Maison de l'Amandier, Saint-Rémy-de-Provence 15486
Maison de l'Amiral, Pouldreuzic 14929
Maison de l'Âne, Beauvoir-sur-Mer 11913
Maison de l'Araire, Yzeron 16313
Maison de l'Archéologie des Vosges du Nord, Niederbronn-les-Bains 14443
Maison de l'Ardoise, Rimogne 15043
Maison de l'Argile, Aubagne 11718
Maison de l'Automate, Tournemire 15955
Maison de l'Eau, Auxerre 11775
La Maison de l'Eau, Mallièvre 13933
Maison de l'Eau, Neuvy-sur-Barangeon 14413
Maison de l'Eau et de la Pêche, Besse-en-Chandesse 11961
Maison de l'Eau et de la Pêche, Neuvic (Corrèze) 14404
Maison de l'Eclusier, Saint-Malo-de-Guersac 15386
Maison de l'Émail, Morez 14271
Maison de l'Espace Vert, Nancy 14351
Maison de l'Etain, Villedieu-les-Poêles 16181
Maison de l'Éventail, Bèze 11973
Maison de l'Histoire, Larra 13528
Maison de l'Histoire de la Terre de Gorze, Gorze 13087
Maison de l'Industrie, Saint-Berthevin 15184
Maison de Loire, Belleville-sur-Loire 11932
Maison de l'Oiseau, Lancheres 13497
Maison de l'Oiseau et du Poisson, Outines 14547
Maison de l'Olivier, Le Val 13659
Maison de Louis Pasteur, Arbois 11657
La Maison de l'Ours, Saint-Lary-Soulan 15355
Maison de l'Outil et de la Pensée Ouvrière, Troyes 15996
Maison de Magayant, Sainte-Marie-de-Re 15567
Maison de Nos Aïeux, Sainte-Famille 06890
Maison de Nostradamus, Salon-de-Provence 15609
Maison de Pauline, Gréoux-les-Bains 13141
Maison de Pays, Bage-le-Chatel 11816
Maison de Pays, Hede 13198
Maison de Pays, Melle 14065
Maison de Pays, Mornant 14282
Maison de Pays, Pont-sur-Sambre 14896
Maison de Pays, Sainte-Marie-aux-Mines 15565
Maison de Pays, Seuilly 15736
Maison de Pays, Tourzel-Ronzieres 15974
Maison de Robert Schumann, Scy-Chazelles 15693
Maison de Sainte Marie-Madeleine-Postel, Barfleur 11852
Maison de Sainte-Thérèse, Lisieux 13776
Maison de Troglodytique Seigneuriale, Turquant 16011
La Maison de Van Gogh, Auvers-sur-Oise 11769
Maison de Victor Hugo, Paris 14621
Maison d'Elsa Triolet et Louis Aragon, Saint-Arnoult-en-Yvelines 15176
Maison d'Émile Zola, Medan 14057
Maison Départementale de l'Innovation, Clermont-Ferrand 12547
Maison d'Erasme, Bruxelles 03450
Maison des Abeilles, Brignoles 12179
Maison des Alpages, Besse-en-Oisans 11963
Maison des Amis du Blanc, Le Blanc 13559
Maison des Ancêtres de Bossuet, Seurre 15739
Maison des Archers, Quimperlé 14972
Maison des Artisans, Angers 11605
Maison des Artisans & Créateurs, Séez 15699
La Maison des Arts, Bages (Aude) 11818
Maison des Arts, Malakoff 13928
Maison des Arts du Belvédère Dar El-Founoun, Tunis 42664
Maison des Arts et du Vin, La Réole 13405
Maison des Arts et Tradition Populaires de la Landes Médocaine, Carcans 12280
Maison des Arts et Traditions Populaires, Châteauroux 12458
Maison des Arts Ruraux, Saint-Germain-de-Marencennes 15287
Maison des Chapais, Saint-Denis 06812
Maison des Chapais, Saint-Denis-de-la-Bouteillerie 06813
Maison des Châteaux Forts, Obersteinbach 14494
Maison des Cinq Sens, Paris 14622
Maison des Compagnons du Devoir du Tour de France, Strasbourg 15818
Maison des Deux Marines, Briare 12171
Maison des Eaux Minérales, Vic-sur-Cère 16142
Maison des Esclaves, Dakar 38021
Maison des Etangs, Saint-Viâtre 15535

M

Maison des Fortifications, Maubeuge 14031
Maison des Fromages, Les Moussières 13707
Maison des Fromages d'Auvergne, Egliseneuve-d'Entraigues 12812
Maison des Géants, Ath 03329
Maison des Insectes, Charlesbourg 05587
Maison des Jardies-Musée Léon Gambetta, Sèvres 15742
Maison des Jésuites de Sillery, Sillery 06976
Maison des Johnnies, Roscoff 15102
Maison des Liqueurs de l'Auzonnet, Saint-Florent-sur-Auzonnet 15268
Maison des Marais, Marchesieux 13951
Maison des Marais Mouillés, Coulon 12646
Maison des Métiers du Cuir, Graulhet 13120
Maison des Métiers du Monde Rural, Tamnay-en-Bazois 15848
Maison des Minéraux, Crozon 12680
Maison des Paludiers, Saille 15151
Maison des Papillons, Saint-Tropez 15526
Maison des Rochers, Graufthal 13119
Maison des Sirènes et des Siréniens, Castellane 12307
Maison des Sports Bretons, Berrien 11951
Maison des Traditions, Chassignolles 12431
Maison des Traditions, La Chapelle-des-Marais 13333
Maison des Tresses et des Lacets, La Terrasse-sur-Dorlay 13446
Maison d'Histoire Locale, Cocheren 12564
Maison Dieu, Faversham 44222
Maison Dieu, Ospringe 45322
Maison dite de François 1er, Aubigny-sur-Nère 11729
Maison dite de la Duchesse Anne, Morlaix 14277
Maison Docteur Joseph-Frenette, Causapscal 05579
Maison d'Offwiller, Offwiller 14498
Maison d'Ossau, Arudy 11707
Maison du Bâti, Sansac-de-Marmiesse 15625
Maison du Bison d'Europe, Sainte-Eulalie-en-Margeride 15558
La Maison du Blé et du Pain, Echallens 41339
Maison du Boutis Les Cordelles, Calvisson 12249
Maison du Buronnier, Laveissière 13550
La Maison du Cerf, Villeny 16224
Maison du Cheval, Corlay 12626
Maison du Cidre et du Terroir, Le Hezo 13600
Maison du Colon, Ville-Marie 07266
Maison du Comté, Poligny 14870
Maison du Fier, Les Portes-en-Ré 13709
Maison du Fleuve Rhône, Givors 13072
Maison du Forgeron et Maréchal Ferrand, Carnusse 12293
Maison du Fromages et de la Vache Salers, Salers 15594
Maison du Gabardan, Gabarret 13023
Maison du Granit, Lac-Drolet 06117
Maison du Haut-Rhône, Seyssel 15752
Maison du Jazz de Liège & de la Communauté Française, Liège 03801
Maison du Journalier, Pleaux 14835
Maison du Kochersberg, Truchtersheim 16004
La Maison du Lac, Les Salles-sur-Verdon 13717
Maison du Lauzeron, Champclause 12389
Maison du Lin, Routot 15131
Maison du Littoral, Perros-Guirec 14797
Maison du Livre, de l'Image et du Son, Villeurbanne 16237
Maison du Livre et de l'Affiche, Chaumont (Haute-Marne) 12482
Maison du Luthier-Musée, Jenzat 13269
Maison du Marbre, Rinxent 15044
Maison du Mont Aiguille Vercors-Trièves, Chichilianne 12510
Maison du Notaire, Trois-Pistoles 07169
Maison du Pain d'Alsace, Sélestat 15703
Maison du Parc et des 4 Montagnes, Autrans 11762
Maison du Parc National des Ecrins, Vallouise 16056
La Maison du Parc Naturel des Plaines de l'Escaut, Bon-Secours 03379
Maison du Parc Naturel Régional de Camargue, Saintes-Maries-de-la-Mer 15587
Maison du Parc Naturel Régional de Lorraine, Bisping 12003
Maison du Parc Naturel Régional des Ballons des Vosges, Munster 14332
Maison du Parc Naturel Régional des Vosges du Nord, La Petite-Pierre 13401
Maison du Parc Naturel Régional du Luberon, Apt 11653
Maison du Parc naturel régional Normandie-Maine, Carrouges 12302
Maison du Patrimoine, Argentat 11673
Maison du Patrimoine, Beauchastel 11889
Maison du Patrimoine, Blanquefort 12018
Maison du Patrimoine, Bourguignon-les-Morey 12127
Maison du Patrimoine, Grande-Synthe 13100
Maison du Patrimoine, Lampaul-Guimiliau 13495
Maison du Patrimoine, Le Crès 13574
Maison du Patrimoine, Lesterps 13726
Maison du Patrimoine, Marsannay-la-Côte 13986
Maison du Patrimoine, Matour 14029
Maison du Patrimoine, Pleyben 14843
Maison du Patrimoine, Sarcelles 15628
Maison du Patrimoine, Villard-de-Lans 16166
Maison du Patrimoine Archéologique, Voingt 16266
Maison du Patrimoine Culturel, Grand-Bassam 08835
Maison du Patrimoine de Saint-Chef, Saint-Chef 15201
Maison du Patrimoine des Kerhorres, Le Relecq-Kerhuon 13645
Maison du Patrimoine d'Oloron et du Haut Béarn, Oloron-Sainte-Marie 14511
Maison du Patrimoine et Ilsle, Crémieu 12672
Maison du Patrimoine Historique du Donezan, Rouze 15132
Maison du Patrimoine Maritime, Camaret-sur-Mer 12251
Maison du Pays des Étangs, Tarquimpol 15865
Maison du Pays Roussillonnais, Roussillon (Isère) 15128
La Maison du Père Mousset, Vrigny 16278
Maison du Platin, La Flotte 13370
Maison du Portal, Levens 13729
Maison du Protestantisme, Nîmes 14448
Maison du Protestantisme Poitevin Beaussais, La Couarde 13353
Maison du Recteur et de la Paroisse, Loqueffret 13818
Maison du Recteur et des Pilhaoueriens, Loqueffret 13819
Maison du Roi-Musée de la Ville, Bruxelles 03451
Maison du Rugby, Avignonet-Lauragais 11799
Maison du Sabot Neufchâtelois, Neufchâtel-en-Saosnois 14399
Maison du Sabotier, La Chapelle-des-Marais 13334
Maison du Saint-Curé d'Ars, Ars-sur-Formans 11703
Maison du Saint-Nectaire, Saint-Nectaire 15424
Maison du Sauvetage, Grand-Fort-Philippe 13098
Maison du Seigle, Menessaire 14073
Maison du Sucre, Frasnes-lez-Anvaing 03609
Maison du Terril, Rieulay 15037
Maison du Terroir, Saint-Martin-Lacaussade 15405
Maison du Tisserand, Quintin 14974
Maison du Val, Abondance 11488
Maison du Val d'Azun, Arrens-Marsous 11700
Maison du Val de Villé, Albé 11531
Maison du Verre, Puy-Guillaume 14960
Maison du Vieill Alby, Albi 11537
Maison du Vieux-Zinal, Zinal 42059
Maison du Village de Val-David, Val-David 07192
Maison Dumulon, Rouyn-Noranda 06781
Maison et Atelier de Jean-François Millet, Barbizon 11846
Maison et la Tour de Mélusine, Vouvant 16275
Maison et Sculptures de Jean Cacoud, Fransèches 12998
Maison Européenne de la Photo, Paris 14623
Maison Francis Jammes, Orthez 14542
Maison Guillaume de Rubrouck, Rubrouck 15135
Maison Hamel-Bruneau, Sainte-Foy 06892
Maison Henri IV, Cahors 12241
Maison Historique de Léon et Jeanne Blum, Jouy-en-Josas 13285
Maison J.A. Vachon, Sainte-Marie-de-Beauce 06897
Maison Jacques Prévert, Omonville-la-Petite 14513
Maison Jean-François Millet, Gréville-Hague 13145
Maison Jean Giono, Manosque 13942
Maison Jean-Vilar, Avignon 11787
Maison Jeanne d'Arc, Orléans 14533
Maison Jonathas, Enghien 03581
Maison Lamontagne, Rimouski 06753
Maison Lansyer, Loches-en-Touraine 13794
Maison Léon-Provancher, Cap-Rouge 05550
Maison LePailleur, Châteauguay 05601
Maison Levanneur, Chatou 12475
Maison Littéraire de Victor Hugo, Bièvres 11989
Maison Lorraine de la Polyculture, Lucey 13864
Maison Louis Bertrand, L'Isle-Verte 06174
Maison Louis-Hippolyte Lafontaine, Boucherville 05445
Maison Marie Henry, Clohars-Carnoët 12561
Maison Médard-Bourgault, Saint-Jean-Port-Joli 06827
Maison Michel, Coppet 41314
Maison-Musée de Jeanne Jugan, Cancale 12261
Maison-Musée de Kerlard, Ile-de-Groix 13238
Maison-Musée des Charcot, Neuilly-sur-Seine 14402
Maison-Musée des Sciences, Lettres et Arts de Cholet et de sa Région, Cholet 12521
Maison-Musée du Parc Naturel Régional du Haut-Jura, Lajoux 13487
Maison Musée René Descartes, Descartes 12716
Maison Natale de Claude Gellée, Chamagne 12367
Maison Natale de Jean Giraudoux, Bellac 11925
Maison Natale de Jeanne d'Arc, Domremy-la-Pucelle 12765
Maison Natale de Jeanne d'Arc et Centre d'Interprètation, Domremy-la-Pucelle 12766
Maison Natale de la Religieuse Sophie Barat, Joigny (Yonne) 13271
Maison Natale de Saint-Vincent-de-Paul, Saint-Vincent-de-Paul 15537
Maison Natale du Brave Crillon, Murs 14339
Maison Natale du Maréchal Foch, Tarbes 15861
Maison Natale du Maréchal Jourdan, Limoges 13764
Maison Natale Sainte Therese de l'Enfant-Jésus, Alençon 11544
Maison Nationale Bonaparte, Ajaccio 11524
Maison Nationale de la Pêche et de l'Eau, Ornans 14538
Maison Nationale des Patriotes, Saint-Denis-sur-Richelieu 06814
Maison Peyne, Laurède 13536
Maison Pierre-Lacroix, Sainte-Marie-de-Beauce 06898
Maison Rodolphe-Duguay, Nicolet 06456
Maison Rouge, Paris 14624
Maison Rurale de l'Outre-Forêt, Kutzenhausen 13306
Maison Saint-Dominique, Fanjeaux 12904
Maison Satie, Honfleur 13220
Maison Stendhal, Grenoble 13131
Maison Tavel, Genève 41414
La Maison Tournaisienne - Musée de Folklore, Tournai 04082
Maison Traditionnelle de la Boucherie, Limoges 13765
Maison Trestler, Vaudreuil-Dorion 07233
Maison Tsawenhohi, Wendake 07299
Maison Vincent Van Gogh, Cuesmes 03531
Maison Vinette, Malmedy 03842
Maison Visinand, Montreux 41638
Maisons-Ateliers de Jean Arp et Sophie Taeuber, Clamart 12537
Maitland Art Center, Maitland 50402
Maitland City Art Gallery, Maitland, New South Wales 01250
Maitland Historical Museums, Maitland 50403
Maitland Museum, Maitland, South Australia 01251
Maizuru Museum of Salvage, Maizuru 28878
Maizuru-shiritsu Sekei Renga Hakubutsukan, Maizuru 28879
Maizuru World Brick Museum, Maizuru 28879
Majolika Galerie, Karlsruhe 19442
The Major General Frederick Funston Boyhood Home and Museum, Iola 49516
Major John Bradford House, Kingston 49787
MAK - Austrian Museum of Applied Arts / Contemporary Arts, Wien 03026
MAK Center for Art and Architecture, Los Angeles 50250
MAK Center for Art and Architecture, West Hollywood 54064
MAK - Österreichisches Museum für angewandte Kunst / Gegenwartskunst, Wien 03026
Makati Museum, Makati 34415
Makedonsko Archeološko Društvo - Zavod i Muzej Prilep, Prilep 30513
Makerere Art Gallery, Kampala 42992
Makina Erremintaren Museoa, Elgoibar 39273
Makonde Art Museum, Bammental 16923
Makonde Art Museum, Futami 28454
Makonde Bijutsukan, Futami 28454
Maksim Bogdanovich Literature Museum, Minsk 03253
Makuhari Kitazawa Bijutsukan, Chiba 28396
Mała Galeria Białostockiego Ośrodka Kultury w Famie, Białystok 34548
Mala Galerija, Ljubljana 38450
Mala Galerija Udruženja Likovnih Umetnika Primenjenih Umetnosti i Dizajnera Srbije, Beograd 38057
Malabar Farm State Park, Lucas 50315
Malacañan Palace Presidential Museum, Manila 34442
Malacological Museum, Makarska 08908
Malagash Area Heritage Museum, Malagash 06222
Malahide Castle, Malahide 24874
Malakhov Bastion, Sevastopol 43179
Malakološki Muzej, Makarska 08908
Malatya Devlet Güzel Sanatlar Galerisi, Malatya 42897
Malatya Müzesi, Malatya 42898
Malatya Museum, Malatya 42898
Malatya Sabancı Culture Center, Malatya 42899
Malatya Sabancı Kültür Merkezi, Malatya 42899
Malatya State Fine Art Gallery, Malatya 42897
Malay Ethnographic Museum, Kuala Lumpur 30576
Malay Technology Museum, Bandar Seri Begawan 04966
Malaysian Aboriginal Museum, Melaka 30606
Malaysian Youth Museum, Melaka 30595
Malazgirt Culture Center, Malazgirt 42900
Malazgirt Kültür Merkezi, Malazgirt 42900
Malba - Fundación Costantini, Buenos Aires 00164
Malbork Castle Museum, Malbork 34914
Malda Museum, Malda 24305
Malden Public Library Art Collection, Malden 50405
Maldon and District Agricultural and Domestic Museum, Goldhanger 44324
Maldon District Museum, Maldon 45076
Maldon Museum, Maldon 01252
Malé Máslovické Muzeum Másla, Vodochody 09912
Malek Museum, Tehrān 24677
Måleriyrkets Museum, Stockholm 40911
Malermuseum Wetzikon, Wetzikon 42011
Malerstübchen Willingshausen, Willingshausen 22587
Målilla Hembygdspark, Målilla 40713
Målilla Motormuseum, Målilla 40714
Målilla Sanatoriums Museum, Målilla 40715
Málinga- og Bátasavnið, Leirvík 10691
Malki Museum, Banning 46599
Mall Galleries, London 44911
Mallaig Heritage Centre, Mallaig 45077
Mallala and Districts Historical Museum, Mallala 01253
Mallalieu Motor Museum, Christ Church 03239
Mallawy Museum, Mallawy 10536
Mallory-Neely House, Memphis 50575
La Malmaison, Cannes 12267
Malmö Konsthall, Malmö 40722
Malmö Konstmuseum, Malmö 40723
Malmö Museer, Malmö 40724
Malmö Museums, Malmö 40724
Malmska Gården, Jakobstad 10866
Malojaroslaveckaja Gorodskaja Kartinnaja Galereja, Malojaroslavec 37125
Malojaroslaveckij Istoriko-kraevedčeskij Muzej, Malojaroslavec 37126
Malojaroslaveckij Muzej Voennoj Istorii 1812 Goda, Malojaroslavec 37127
Malojaroslaveckij Muzejno-vystavočnyj Centr, Malojaroslavec 37128
Malokarpatské Múzeum v Pezinku, Pezinok 38324
Malomipari Múzeum, Budapest 23448
Malouinière-Musée, Saint-Malo (Ille-et-Vilaine) 15387
Maloyaroslavets City Art Gallery, Malojaroslavec 37125
Maloyaroslavets Exhibition Center, Malojaroslavec 37128
Maloyaroslavets Historical Museum, Malojaroslavec 37126
Maloyaroslavets Museum of History of the War of 1812, Malojaroslavec 37127
Målselv Bygdemuseum, Moen 33671
The Malta at War Museum, Vittoriosa 30656
Malta Aviation Museum, Ta'Qali 30639
Malta Classic Car Collection, Qawra 30635
Malta Maritime Museum, Vittoriosa 30657
Malta National Committee of ICOM, Valletta 54939
Malteser-Museum, Mailberg 02343
Malton Museum, Malton 45080
Maltwood Art Museum and Gallery, Victoria 07253
Malungs Hembygdsgård, Malung 40729
Malura Museum, Unterdießen 22178
Malvern Museum, Great Malvern 44356
Mamco, Genève 41415
Mamie Doud Eisenhower Birthplace, Boone 46906
Mammad Seid Ordubad Home Museum, Ordubad 03190
Mammoth Site of Hot Springs, Hot Springs 49375
Mammutmuseum, Niederweningen 41687
Mamnu"gohi Bostonshinosii Panjakent, Pandžakent 42504
Mamoa de Lamas, Lamas 35654
Man and His Health Museum, Varna 05239
Man and His Work Center in Memory of S. Avitsur, Tel Aviv 25115
Man and the Living World Museum, Ramat Gan 25084
Man From Snowy River Folk Museum, Corryong 01004
Man in the Galilee Museum, Kibbutz Ginosar 25042
MAN-Museum, Augsburg 16628
Manakova Kuća, Beograd 38058
Manak's Mansion of the Ethnographical Museum of Serbia, Beograd 38058
Manassas Museum System, Manassas 50414
Manassas National Battlefield Park, Manassas 50415
Manatee Village Historical Park Museum, Bradenton 46986
Manchester Art Gallery, Manchester 45088
Manchester Craftsmen's Guild, Pittsburgh 51952
Manchester Historic Association Museum, Manchester 50423
Manchester Historical Society Museum, Manchester 50419
Manchester Jewish Museum, Manchester 45089
The Manchester Museum, Manchester 45090
Manchester United Museum and Tour Centre, Manchester 45091
Mandal Bymuseum, Mandal 33658
Mandal og Opplands Folkemuseum, Øyslebø 33764
Mandalay Palace, Mandalay 31718
Mander and Mitchenson Theatre Collection, London 44912
Mandeville Gallery, Schenectady 52960
Mandurah Community Museum, Mandurah 01255
Mané-Katz Museum, Haifa 24964
Maneesi Sotamuseo, Helsinki 10803
Manezh Central Exhibition Hall, Moskva 37153

M

Manfreds Haus- und Hofmuseum, Weisweil ... 22466
Mangapps Railway Museum, Burnham-on-Crouch ... 43676
Manggha Centrum Sztuki i Techniki Japońskiej, Kraków ... 34783
Mångkulturellt Centrum, Tumba ... 40987
Mango Explorer Museum, Yujing ... 42483
Mangystau Historical and Regional Museum, Aktau ... 29670
Mangystauskij Oblastnoj Istoriko-Kraevedčeskij Muzej, Aktau ... 29670
Manhattan Arts Center, Manhattan ... 50434
Manhyia Palace Museum, Kumasi ... 22858
Manial Palace Museum, Cairo ... 10499
Maniitsoq Katersugaasivia, Maniitsoq ... 23257
Manila Police Department Museum and Gallery, Manila ... 34443
Manilla Historical Royce Cottage Museum, Manilla ... 01256
Maniototo Early Settlers Museum, Naseby ... 33151
Manipur State Museum, Imphal ... 24225
Manipur University Museum, Canchipur ... 24107
Manisa Devlet Güzel Sanatlar Galerisi, Manisa ... 42902
Manisa Müzesi, Manisa ... 42903
Manisa Museum, Manisa ... 42903
Manisa State Art Gallery, Manisa ... 42902
Manistee County Historical Museum, Manistee ... 50439
Manitoba Agricultural Hall of Fame Inc., Brandon ... 05464
Manitoba Agricultural Museum, Austin ... 05364
Manitoba Amateur Radio Museum, Brandon ... 05465
Manitoba Automobile Museum, Elkhorn ... 05772
Manitoba Baseball Hall of Fame & Museum, Morden ... 06390
Manitoba Children's Museum, Winnipeg ... 07367
Manitoba Crafts Museum and Library, Winnipeg ... 07368
Manitoba Electrical Museum & Education Centre, Winnipeg ... 07369
Manitoba Museum, Winnipeg ... 07370
Manitoba Sports Hall of Fame and Museum, Winnipeg ... 07371
Manitowoc County History Museum, Manitowoc ... 50442
Manjimup Timber Museum, Manjimup ... 01257
Manjusha Museum, Dharmasthala ... 24162
Mankota Museum, Mankota ... 06227
Manly Art Gallery and Museum, Manly ... 01258
Mann Museum, Opelika ... 51572
Mann-Simons Cottage, Columbia ... 47729
Mannaminne, Nordingrå ... 40766
Manneken-Pis-Museum, Geraardsbergen ... 03651
Mannenzaal van het Sint Pieters en Bloklands Gasthuis, Amersfoort ... 31812
Mannerheim museo, Helsinki ... 10804
The Mannerheim Museum, Helsinki ... 10804
Manning Park Visitor Centre, Manning Park ... 06228
Manning Regional Art Gallery, Taree ... 01573
Manoel Theatre and Museum, Valletta ... 30641
Manoir Automobile, Lohéac ... 13802
Manoir de Kernault, Mellac ... 14064
Manoir de la Ville de Martigny, Martigny ... 41610
Manoir de Veygoux, Scénomusée des Combrailles, Charbonnières-les-Varennes ... 12407
Manoir de Villers, Saint-Pierre-de-Manneville ... 15460
Manoir du Huis Bois-Maison P.N.R. du Boulonnais, Le Wast ... 13668
Manoir Leboutillier, Anse-au-Griffon ... 05347
Manoir Mauvide-Jenest, Saint-Jean ... 06826
Manor Cottage Heritage Centre, Southwick ... 45719
Manor Farm, Bursledon ... 43684
Manor House, Donington-le-Heath ... 43996
Manor House, Sandford Orcas ... 45610
Manor House and Kysuce Gallery, Oščadnica ... 38323
Manor House Art Gallery and Museum, Ilkley ... 44512
The Manor House in Łopuszna, Łopuszna ... 34891
Manor House Museum, Kettering ... 44568
Manor Museum, Manor ... 06229
Manos Faltaits Museum, Skyros ... 23192
Manro-Classic - Auto & Musik Museum, Koppl ... 02216
Mansfeld-Museum Hettstedt, Hettstedt ... 19173
Mansfelder Bergwerksbahn e.V., Benndorf ... 16992
Mansfield Art Center, Mansfield ... 50451
Mansfield Historical Society Museum, Storrs ... 53319
Mansfield Museum and Art Gallery, Mansfield ... 45101
Mansfield State Historc Site, Mansfield ... 50449
Manship House Museum, Jackson ... 49558
Mansion House, Warkworth ... 33238
Manuel L. Quezon Memorial Shrine Museum, Quezon City ... 34493
Manuel Lepe Museum Gallery, Puerto Vallarta ... 31297
Manufaktur der Träume, Annaberg-Buchholz ... 16555
Manujlivs'kyj Literaturno-memorialnyj Muzej O.M. Gor'kogo, Verchnja Manujlivka ... 43202
Manurewa Community Arts Centre, Auckland ... 33055
Manuscript Museum, Alexandria ... 10474
Manx Museum, Douglas ... 24916
Manyara Museum, Manyara ... 42515
Man'yo Bunkakan, Wakayama ... 29573
Man'yo-Museum, Wakayama ... 29573
Manzoni Museum, Lesa ... 26487
Manzoni Villa, Lecco ... 26477
Mao Dun Former Residence, Beijing ... 07549
Maoling Museum, Xingping ... 08394
Maoming Museum, Maoming ... 08009
Maple Ridge Art Gallery, Maple Ridge ... 06236
Maple Ridge Museum, Maple Ridge ... 06237
Maple Syrup Museum, Saint Jacobs ... 06823
Maple Valley Historical Museum, Maple Valley ... 50456
Maplewood Farm, North Vancouver ... 06479
Mappa Mundi and Chained Library, Hereford ... 44456
Mapungubwe Museum, Pretoria ... 38719
Maragheh Museum, Maragheh ... 24629
Marais Des Cygnes Memorial Historic Site, Pleasanton ... 51988
Maramec Museum, Saint James ... 52552
Maramureş County Museum, Baia Mare ... 36128
Maras Icon Museum, Gazimağusa ... 33351
Maraş Ikon Müzesi, Gazimağusa ... 33351
Maratha History Museum, Pune ... 24360
Marathon County Historical Society Museum, Wausau ... 53994
Mårbacka Manor, Östra Ämtervik ... 40799
Marble Hill House, Twickenham ... 45886
Marble Historical Museum, Marble ... 50459
Marble Mountain Library and Museum, West Bay ... 07301
Marble Palace Art Gallery and Zoo, Kolkata ... 24277
Marble Springs State Historic Farmstead Governor John Sevier Memorial, Knoxville ... 49821
Marblehead Museum and J.O.J. Frost Folk Art Gallery, Marblehead ... 50462
Marbles Kids Museum, Raleigh ... 52230
Marburger Universitätsmuseum für Kunst und Kulturgeschichte, Marburg ... 20191
Marc Chagall Kunsthuis, Amsterdam ... 31857
Marcela M. Agoncillo Historical Museum, Taal ... 34508
Marcella Sembrich Opera Museum, Bolton Landing ... 46897
Marcellus Historical Society Museum, Marcellus ... 50465
Marcelo H. del Pilar Museum, Malolos ... 34423
March and District Museum, March ... 45103
March Field Air Museum, Riverside ... 52371
March-Museum, Vorderthal ... 41993
Marchfeldmuseum, Weikendorf ... 02899
Marcos Museum, Batac ... 34343
Mardi Gras Museum, Biloxi ... 46810
Mardin Müzesi, Mardin ... 42904
Mardin Museum, Mardin ... 42904
Mardom-shenasi-ye Kavir, Nayin ... 24636
Mareeba District Rodeo Museum, Mareeba ... 01262
Mareeba Heritage Museum, Mareeba ... 01263
Marès Lac Museum, Arenys de Mar ... 38886
Margam County Park, Port Talbot ... 45417
Margaree Salmon Museum, North East Margaree ... 06474
Margaret Fort Trahern Gallery, Clarksville ... 47594
Margaret Harvey Gallery, Saint Albans ... 45546
Margaret Harwell Art Museum, Poplar Bluff ... 52024
Margaret Hutchinson Compton Gallery, Cambridge ... 47181
Margaret Laurence Home, Neepawa ... 06421
Margaret Lokey Gallery, Tishomingo ... 53513
Margaret Mitchell House and Museum, Atlanta ... 46442
Margarete Blank Gedenkstätte, Borsdorf ... 17487
Margarete-Steiff-Museum, Giengen an der Brenz ... 18632
Margaretha Museum, Laag Keppel ... 32461
Margate Old Town Hall Local History Museum, Margate ... 45106
Margret-Knoop-Schellbach-Museum, Körle ... 19663
Margrove - South Cleveland Heritage Centre, Boosbeck ... 43561
Mari Jászai Memorial House, Ászár ... 23337
Mari Jászai Memorial Room, Rábatamási ... 23788
Maria de Lisitzin Fine Art and Historical Museum, Porvoo ... 11229
Maria Martinez Museum, Santa Fe ... 52879
Maria Medvecka Gallery, Tvrdošín ... 38383
Maria Ritter Gallery, Nowy Sącz ... 34957
Mária-ruhák Múzeum, Andocs ... 23335
Mariager Museum, Mariager ... 10183
Marian Koshland Museum, Washington ... 53889
Marianna Kistler Beach Museum of Art, Manhattan ... 50435
Mariano Rodríguez Gallery, La Habana ... 09107
Mariánské Lázně Municipal Museum, Mariánské Lázně ... 09649
Marias Museum of History and Art, Shelby ... 53066
Maribor National Liberation Museum, Maribor ... 38476
Marie Curie Museum, Warszawa ... 35259
Marie de Museum of World Clock and Orgel, Inuyama ... 28577
Marie-Elisabeth-Lüders-Haus, Berlin ... 17182
Marie-Hager-Haus, Burg Stargard ... 17647
Marie Hull Gallery, Raymond ... 52259
Marie Rawdon Museum, Matjiesfontein ... 38687
Marie Sekei Tokei oyo Orugoru Hakubutsukan, Inuyama ... 28577
Marie Tak van Poortvliet Museum, Domburg ... 32141
Marie Walsh Sharpe Art Foundation, New York ... 51209
Mariefreds Hembygdsmuseum, Mariefred ... 40732
Marienfelde Refugee Centre Memorial, Berlin 17080
Marienglashöhle Friedrichroda, Friedrichroda 18474
Mariestads Industrimuseum, Mariestad ... 40737
Marietta House Museum, Glenn Dale ... 48933
Marietta Museum of History, Marietta ... 50469
Marija and Jurgis Slapeliai House and Museum, Vilnius ... 30424
Marijampole Local Lore Museum, Marijampolė ... 30308
Marijampole Local Lore Museum - Tauras District Partisans and Deportation Museum, Marijampolė ... 30309
Marijampole Local Lore Museum - Vincas Mykolaitis-Putinas Memorial Museum, Marijampolė ... 30310
Marijampolės Kraštotyros Muziejus, Marijampolė ... 30308
Marijos ir Jurgio Šlapelių Namas-muziejus, Vilnius ... 30424
Marijos Savickienės Gamtos Išdaigų Muziejus, Tytuvėnai ... 30384
Marika Kotopouli Museum, Zografou ... 23248
Marilla Historical Society Museum, Marilla ... 50477
Marin Drzic House, Dubrovnik ... 08861
Marin History Museum, San Rafael ... 52828
Marin Museum of the American Indian, Novato ... 51446
Marine Biological Museumn, Nov Dojran ... 30511
Marine Corps Art Collection, Washington ... 53890
Marine-Ehrenmal und Technisches Museum U 995, Laboe ... 19756
Marine Museum, Toronto ... 07121
Marine Museum at Fall River, Fall River ... 48500
Marine-Museum Limburg, Limburg an der Lahn ... 19996
Marine Museum of Manitoba, Selkirk ... 06943
Marine Museum of the Great Lakes, Kingston 06082
Marine Science Museum, Shizuoka ... 29285
Marine Science Museum, Tokai University, Shizuoka ... 29291
Marine Sciences Museum, Mayagüez ... 36043
Marine Transportation Museum, Ogi ... 29091
Marineland Ocean Museum, Marineland ... 50478
Marinemuseet, Horten ... 33563
Marinemuseum, Den Helder ... 32110
Marinemuseum Stralsund, Stralsund ... 21915
The Mariners' Museum, Newport News ... 51334
Mariners' Park Museum, Milford ... 06283
Marinette County Historical Museum, Marinette ... 50479
Mariniersmuseum, Rotterdam ... 32711
Marinmuseum, Karlskrona ... 40635
Mario Lanza Museum, Philadelphia ... 51850
Marion Art Center, Marion ... 50485
Marion County Historical Society Museum, Marion ... 50488
Marion County Historical Society Museum, Salem ... 52668
Marion Heritage Center, Marion ... 50481
Marion Public Library Museum, Marion ... 50483
Marion Scott Gallery, Vancouver ... 07215
Marionettmuseet, Stockholm ... 40912
Marischal Museum, Aberdeen ... 43244
Maritiem en Jutters Museum, Oudeschild ... 32657
Maritiem Museum Rotterdam, Rotterdam ... 32712
Maritiman, Göteborg ... 40551
Maritime and Historical Museum of Croatian Littoral, Rijeka ... 08939
Maritime and Metalwork Museum, Reykjavík ... 24035
Maritime and Seafood Industry Museum, Biloxi ... 46811
Maritime Archaeology Museum, Melaka ... 30596
Maritime Cameroon Museum, Douala ... 05295
Maritime Centre, Cape Town ... 38586
Maritime Centre Newcastle, Newcastle ... 01363
Maritime Collections at Troense, Svendborg ... 10292
Maritime Command Museum, Halifax ... 05939
Maritime Heritage Society of Vancouver, Britannia Beach ... 05483
Maritime Industry Museum at Fort Schuyler, Bronx ... 47049
Maritime Institute of the Åbo Academy, Turku ... 11398
Maritime Museum, Cohasset ... 47676
The Maritime Museum, Dublin ... 24806
Maritime Museum, Dubrovnik ... 08872
Maritime Museum, Eyrarbakki ... 23996
Maritime Museum, Jakarta ... 24482
Maritime Museum, Kristiinankaupunki ... 10997
Maritime Museum, Low Head ... 01242
Maritime Museum, Melaka ... 30601
Maritime Museum, Nadur, Gozo ... 30633
Maritime Museum, Oskarshamn ... 40804
Maritime Museum, Piran ... 38492
Maritime Museum, Singapore ... 38176
Maritime Museum, Szczecin ... 35162
Maritime Museum, Vaasa ... 11429
Maritime Museum, Zadar ... 08984
The Maritime Museum and Aquarium, Göteborg ... 40556
Maritime Museum of Andros, Andros ... 22882
Maritime Museum of British Columbia, Victoria ... 07254
Maritime Museum of Crete, Chania ... 22964
Maritime Museum of Finland, Kotka ... 10988
Maritime Museum of Macau, Macau ... 08000
Maritime Museum of Porsgrunn, Porsgrunn ... 33771
Maritime Museum of Tasmania, Hobart ... 01154
Maritime Museum of the Atlantic, Halifax ... 05940
Maritime Museum of Thera (Santorini), Oia ... 23126
Maritime Museum of Townsville, Townsville ... 01597
Maritime Museum Orebić, Orebić ... 08920
Maritime Museum Rostock, Rostock ... 21395
Maritime Museum Zierikzee, Zierikzee ... 33003
Maritime Self-Defense Force Sasebo Museum, Sasebo ... 29235
Maritime Tradition Museum, Piraeus ... 23144
Mariupol's'kij Krajeznavčyj Muzej, Mariupol' ... 43144
Marjorie Kinnan Rawlings Historic State Park, Cross Creek ... 47890
Mark Twain Birthplace, Stoutsville ... 53323
Mark Twain Home and Museum, Hannibal ... 49159
Mark Twain House, Hartford ... 49195
Markarydsortens Hembygdsmuseum, Markaryd ... 40739
Markeim Art Center, Haddonfield ... 49116
Marker Museum, Marken ... 32540
Market Gallery, Toronto ... 07122
Market Lavington Village Museum, Market Lavington ... 45111
Markfield Beam Engine and Museum, London ... 44913
Markgräfler Museum Müllheim, Müllheim ... 20417
Markgräfliches Opernhaus, Bayreuth ... 16960
Markgrafenmuseum Ansbach, Ansbach ... 16560
Markham Museum and Historic Village, Markham ... 06239
Marko Vovchok Memorial Museum, Nalčik ... 37440
Marks-Hirshfeld Museum of Medical History, Herston ... 01151
Markt Einersheimer Heimatstuben, Markt Einersheim ... 20211
Markt- und Schaustellermuseum, Essen ... 18261
Marland's Grand Home, Ponca City ... 52017
Marlboro County Historical Museum, Bennettsville ... 46738
Marlboro Historical Society, Marlboro ... 50495
Marlborough Provincial Museum, Blenheim ... 33063
Marlipins Museum, Shoreham-by-Sea ... 45668
Marmaris Church and Archaeological Museum, Marmaris ... 42905
Marmaris Kalesi ve Arkeoloji Müzesi, Marmaris ... 42905
Marmorbad, Kassel ... 19469
Marmormuseum, Adnet ... 01695
Marmorpalais, Potsdam ... 21094
Marnardal Museum, Marnardal ... 33661
Maroc Telecom Museum, Rabat ... 31689
Marquerite-Bourgeoys Museum, Montréal ... 06374
Marquette County History Museum, Marquette ... 50499
Marquette County Museum, Westfield ... 54095
Marquisat de Pont-Croix, Pont-Croix ... 14882
Marrett House, Standish ... 53251
Marringhurst Pioneer Park Museum, Pilot Mound ... 06605
Marsh Collection Society, Amherstburg ... 05334
Marshall County Historical Museum, Holly Springs ... 49321
Marshall County Historical Museum, Plymouth ... 51992
Marshall County Historical Society Museum, Warren ... 53822
Marshall County Historical Society Museum, Warren ... 53823
Marshall County Museum, Marshalltown ... 50512
Marshall M. Fredericks Sculpture Museum, University Center ... 53655
Marsil Museum, Saint-Lambert ... 06861
Marstal Maritime Museum, Marstal ... 10187
Marstal Søfartsmuseum, Marstal ... 10187
Marstallmuseum, München ... 20458
Marstallmuseum, Regensburg ... 21229
Marston House, San Diego ... 52738
Mart Saare Majamuuseum, Suure-Jaani ... 10618
Marta Herford, Herford ... 19140
Marten River Logging Museum, Marten River ... 06241
Martha Berry Museum, Rome ... 52449
Martha's Vineyard Exhibition, Edgartown ... 48285
Martin and Osa Johnson Safari Museum, Chanute ... 47342
Martin Art Gallery, Allentown ... 46228
Martin Benka Múzeum, Martin ... 38295
Martin County Historical Museum, Stanton ... 53257
Martin County Pioneer Museum, Fairmont ... 48489
The Martin D'Arcy Museum of Art, Chicago ... 47487

M

Martin-Gropius-Bau, Berlin 17183
Martin-Heidegger-Museum, Meßkirch 20305
Martin-Lauterburg-Stiftung, Bern 41211
Martin Luther King jr. Center and Preservation District, Atlanta 46443
Martin Luthers Geburtshaus, Lutherstadt Eisleben 20110
Martin Luthers Sterbehaus, Lutherstadt Eisleben 20111
Martin Museum of Art, Waco 53776
Martin Van Buren Historic Site, Kinderhook 49778
Martin-von-Wagner-Museum der Universität Würzburg, Würzburg 22714
Martinez Museum, Martinez 50517
Martinustoren, Losser 32510
Martiros Sarian Museum, Yerevan 00741
Marton Historic Village Museum, Marton 33139
Marttilan Kirkkomuseo, Marttila 11102
Marttilan Kotiseutumuseo, Marttila 11103
Martuni Art Gallery, Martuni 00714
Martyn Ferenc Múzeum, Pécs 23763
Martyno Jankaus Muziejus, Bitenai 30217
Martyr's Memorial at Sports City, Amman 29656
Martyrs' Shrine, Midland 06278
Marugame Bijutsukan, Marugame 28880
Marugame Genichiro-Inokuma Gendai Bijutsukan, Marugame 28881
Marugame Hirai Museum, Marugame 28882
Marugame Museum of Art, Marugame 28880
Marusu Museum, Ishinomaki 28588
Maruzian Museum, Szczytno 35168
Marvin Newton House, Brookfield 47057
Marvol Museum, Kuilsriver 38679
Marwen Foundation, Chicago 47488
Marxhausen Art Gallery, Seward 53043
Mary and Leigh Block Museum of Art, Evanston 48453
Mary Arden's House and the Countryside Museum, Stratford-upon-Avon 45782
Mary Baker Eddy Library for the Betterment of Humanity, Boston 46936
Mary Ball Washington Museum, Lancaster 49932
Mary Brogan Museum of Art and Science, Tallahassee 53426
Mary E. Black Gallery, Halifax 05941
Mary Leu Fine Carving Gallery and Meili Museum, Sansing 42314
Mary Mackillop Place Museum, North Sydney 01374
Mary McLeod Bethune House, Washington 53891
Mary March Regional Museum and Logging Exhibition, Grand Falls-Windsor 05892
Mary Moffat Museum, Griquatown 38638
Mary Porter Sesnon Art Gallery, Santa Cruz 52867
Mary Queen of Scots' House, Jedburgh 44548
Mary Rose Museum, Portsmouth 45434
Mary Todd Lincoln House, Lexington 50053
Mary Washington College Galleries, Fredericksburg 48769
Mary Washington House, Fredericksburg 48770
Maryfield Museum, Maryfield 06243
Maryhill Museum of Art, Goldendale 48963
Maryland Art Place, Baltimore 46575
Maryland Association of History Museums, Annapolis 55154
Maryland Historical Society Museum, Baltimore 46576
Maryland Institute Museum, Baltimore 46577
Maryland Science Center, Baltimore 46578
Marymoor Museum of Eastside History, Redmond 52277
Marymount Manhattan College Museum, New York 51210
Maryport Maritime Museum, Maryport 45112
Marystown Heritage Museum, Marystown 06244
Marywood University Art Galleries, Scranton 52982
Mas de la Pyramide, Saint-Rémy-de-Provence 15487
Masaki Art Museum, Osaka 29133
Masaryk-Museum, Hodonín 09519
Masarykovo Muzeum, Hodonín 09519
Maschenmuseum, Albstadt 16454
Maschinen-u.Heimatmuseum, Eslohe 18244
Maschinenmuseum-Musée de la Machine-Centre Müller, Biel 41229
Mashantucket Pequot Museum, Mashantucket 50527
Mashiko Reference Collection Museum, Mashiko 28885
Mashiko Sankokan, Mashiko 28885
Masia Museu Can Magarola, Alella 38837
Maskenmuseum, Denkendorf 17820
Maskun Museo, Masku 11104
Mason County Museum, Mason 50528
Mason Gross Art Galleries, New Brunswick 51034
Masonic Grand Lodge Library and Museum, Waco 53777
The Masonic Library and Museum of Pennsylvania, Philadelphia 51851
Måsøy Museum, Havøysund 33540
Mass Moca, North Adams 51378
Massa Museum of the National Technical Museum, Miskolc 23710
Massa Múzeum, Miskolc 23710
Massachusetts Historical Society, Boston 46937
Massacre Rocks State Park Museum, American Falls 46255
Massey Area Museum, Massey 06246
Massillon Museum, Massillon 50531
Massnes Villmarksmuseum, Bjordal 33440
Massua-Educational Museum on the Holocaust, Tel Yitzhak 25127
The Masterworks Foundation Collection, Paget 04195
Masur Museum of Art, Monroe 50789
Matagorda County Museum, Bay City 46651
Match Museum, Jönköping 40616
Match Museum, Tomar 35964
Matchbox Road Museum, Newfield 51313
Matchboxmuseum Latent, Prinsenbeek 32671
Matches Museum, Bystrzyca Kłodzka 34583
Matchimawas National Museum, Songkhla 42601
Matenadaran Manuscript Museum, Yerevan 00742
Mateureka - Museo del Calcolo, Pennabilli 27131
Mathematikum, Gießen 18640
Mathematisch-Physikalischer Salon, Dresden 17959
Mather Homestead, Wellsville 54032
Mather Post Office Museum, Cartwright 05572
Matheson Museum, Gainesville 48814
Mathias Ham House, Dubuque 48188
Mathias Schmid-Museum, Ischgl 02139
Matilda Roslin-Kalliolan Kirjailijakoti, Merikarvia 11108
Matopos National Park Site Museums, Bulawayo 54664
Mátra Múzeum, Gyöngyös 23574
Matrica Museum and Archeological Park, Százhalombatta 23839
Matrica Múzeum, Százhalombatta 23839
Matrioshka Museum, Escaldes-Engordany 00092
Matrixarts, Sacramento 52509
Matschgerermuseum, Absam 01689
Matsqui Sumas Abbotsford Museum, Abbotsford 05305
Matsudo Museum, Matsodo 28887
Matsumae-cho Bunkakan, Matsumae 28892
Matsumae Culture Museum, Matsumae 28892
Matsumoto City Museum, Matsumoto 28895
Matsumoto City Museum of Art, Matsumoto 28894
Matsumoto Museum of Folk and Arts, Matsumoto 28893
Matsumoto-shiritsu Bijutsukan, Matsumoto 28894
Matsumoto-shiritsu Hakubutsukan, Matsumoto 28895
Matsuo Museum, Takigawa 29347
Matsuoka Museum of Art, Tokyo 29427
Matsura History Museum, Hirado 28519
Matsura Shiryo Hakubutsukan, Hirado 28519
Matsusaka Oroshi Kinenkan, Matsusaka 28897
Matsushima Historical Museum, Matsushima 28901
Matsushima Organ Museum, Matsushima 28902
Matsushima Orugoru Hakubutsukan, Matsushima 28902
Matsushita Art Museum, Fukuyama, Aira-gun 28451
Matsushita Bijutsukan, Fukuyama, Aira-gun 28451
Matsushita Hakubutsukan, Aira 28349
Matsushita Museum, Aira 28349
Matsuyama Municipal Shiki-Kinen Museum, Matsuyama 28906
Mattancherry Palace Museum, Kochi 24262
Mattapoisett Museum and Carriage House, Mattapoisett 50535
The Mattatuck Museum of the Mattatuck Historical Society, Waterbury 53956
Mattawa and District Museum, Mattawa 06250
Matterhorn Museum, Zermatt 42052
Matthews Museum of Maine Heritage, Union 53647
Mattisrud Småbruksmuseum, Løten 33652
Mattress Factory Museum, Pittsburgh 51953
Matt's Gallery, London 44914
Matt's Museum, Lawtell 49999
Mattye Reed African Heritage Center, Greensboro 49066
Maturango Museum of the Indian Wells Valley, Ridgecrest 52346
Matvej Khangalov History Museum of Buryatia, Ulan-Udė 37894
Mátyás Király Palota Múzeum, Visegrád 23966
Mátyás Varga's Theatre Historical Exhibition, Szeged 23849
Matyó Múzeum, Mezőkövesd 23698
Maud Railway Museum, Maud 45119
Maude I. Kerns Art Center, Eugene 48434
Maud's Place - The Putnam House, Brookfield 05487
Mauermuseum - Museum Haus am Checkpoint Charlie, Berlin 17184
Mauersberger-Museum, Großrückerswalde 18804
Maurermuseum und Dachdeckermuseum, Langenlois 02258
Mauritianum, Altenburg 16498
Mauritianum, Natural History Museum Altenburg, Altenburg 16498
Mauritius Postal Museum, Port Louis 30682
Mauritshuis, Den Haag 32082
Mausefallen-Museum, Neroth 20573
Mausefallen- und Kuriositätenmuseum mit Galerie der Stillen Örtchen, Güntersberge 18822
Mauseum, Bergheim (Erft) 17013
Mausoleum im Schlossgarten Charlottenburg, Berlin 17185
Mauthausen Memorial, Mauthausen 02373
Max Böhlen Museum, Kandern 19427
Max Ernst Museum Brühl des LVR, Brühl 17606
Max Euwe Centrum, Amsterdam 31858
Max Liebermann Haus, Berlin 17186
Max-Reger-Gedächtniszimmer, Brand 17503
Maxhütten-Museum, Bergen, Kreis Traunstein 17010
Maxim-Gorki-Gedächtnisstätte, Heringsdorf, Seebad 19144
Maxim Gorki Literary Memorial Museum, Verchnja Manujlivka 43202
Maximilianmuseum, Augsburg 16629
Maxwell Museum of Anthropology, Albuquerque 46181
Maxxi - Museo delle Arti dell XXI secolo, Roma 27457
May and Herbert Gibbs Art Collection, South Perth 01516
May Museum and Park, Farmville 48522
May Natural History Museum and Museum of Space Exploration, Colorado Springs 47703
Mayang Miaozu Self-Rule Museum, Mayang 08011
Mayborn Museum Complex, Waco 53778
Mayflower Gold Mill, Silverton 53108
Mayflower Society Museum, Plymouth 51994
Mayibuye Centre Collection, Bellville 38558
Maymont, Richmond 52334
Mayne Island Museum, Mayne Island 06254
Mayo County Museum, Ballina 24733
The Mayor's Residence Arts Salon, Taipei 42403
Mayrau Mine Open Air Museum of Mining, Vinařice u Kladna 09908
Mayville Area Museum of History and Genealogy, Mayville 50540
Mayville Historical Society Museum, Mayville 50542
Mazar Islamic Museum, Al-Mazar 29642
Mazda Motor Museum, Hiroshima 28540
Mazeikiai Museum, Mažeikiai 30311
Mažeikių Muziejus, Mažeikiai 30311
Mazomanie Historical Society Museum, Mazomanie 50543
Mažosios Lietuvos Istorijos Muzieus, Klaipėda 30283
Mazot Musée, Troistorrents 41951
Mazovian Museum in Płock, Płock 35014
Mazu Folklore Museum of Tianhou Temple, Lugang 42258
Mazzini-Gedenkstätte, Grenchen 41454
MC-Museum Gyllene Hjulet, Surahammar 40963
MCAD Gallery, Minneapolis 50738
MCC Museum, London 44915
MDA, Cambridge 55072
Mead Art Museum, Amherst 46268
Mead Gallery, Coventry 43904
Meade County Historical Society Museum, Meade 50544
Meadow Brook Art Gallery, Rochester 52385
Meadow Brook Hall, Rochester 52386
Meadow Farm Museum, Glen Allen 48917
Meadow Garden Museum, Augusta 46480
Meadow Lake Museum, Meadow Lake 06255
Meadowcroft Museum of Rural Life, Avella 46525
Meadowlands Museum, Rutherford 52492
Meadows Museum, Dallas 47935
Meadows Museum of Art, Shreveport 53085
Meaford Museum, Meaford 06256
Meagher County Historical Association Castle Museum, White Sulphur Springs 54130
Meanskinisht Museum, Kitwanga 06106
Measham Museum, Measham 45121
Meat Factory Collection, Gyula 23595
Mechanical Music and Doll Collection, Chichester 43829
Mechanical Music Museum and Bygones, Cotton 43897
Mechanical Organ Museum-Pension Humpty Dumpty, Nikko 29067
Mechanischer Musiksalon, Weiskirchen 22443
Mechanisches Musikmuseum, Sursee 41924
Mecklenburgisches Orgelmuseum, Malchow 20155
Mecklenburgisches Volkskundemuseum Schwerin, Schwerin 21695
Mecosta County Historical Museum, Big Rapids 46798
Mecseki Banyászati Múzeum, Pécs 23764
Mecseki Banyászati Múzeum, Pécs 23765
Mecseki Bányászati Múzeum - Uránérc Bányászati Kiállítás, Kővágószőlős 23687
Médaillier Municipal, Bordeaux 12064
Medal Art Museum, Wrocław 35340
Medelhavsmuseet, Stockholm 40913
Medgyessy Ferenc Emlékmúzeum, Debrecen 23518
Medical Historical and Pharmacy Exhibition, Karcag 23641
Medical Historical Exhibition of the József Katona Museum, Kecskemét 23654
Medical History Museum, Göteborg 40552
Medical History Museum, Melbourne 01287
Medical History Museum of Berlin, Berlin 17044
Medical Museum, Amman 29657
Medical Museum, Beograd 38059
Medical Museum of hte Russian Academy of Medical Sciences, Moskva 37226
Medicine Hat Clay Industries National Historic District, Medicine Hat 06257
Medicine Hat Museum and Art Gallery, Medicine Hat 06258
Medicinhistoriska Museet, Göteborg 40552
Medicinhistoriska Museet, Lund 40707
Medicinhistoriska Museet, Vänersborg 41024
Medicinhistoriska Museet Eugenia, Stockholm 40914
Medicinsk-Historisk Museum, København 10139
Medicinski Muzej, Beograd 38059
Medicinskij Muzej, Moskva 37226
Medieval Dungeons, Nürnberg 20779
Medieval Old Synagogue, Sopron 23811
Medieval Pilgrims Hospice, Pons 14876
Medieval Ruin Garden, Székesfehérvár 23855
Medina Railroad Museum, Medina 50555
Medisch Farmaceutisch Museum de Griffioen, Delft 32060
Medizin- und Pharmaziehistorische Sammlung Kiel der Christian-Albrechts-Universität, Kiel 19533
Medizinhistorische Sammlung der Ruhr-Universität Bochum, Bochum 17404
Medizinhistorische Sammlungen, Chemnitz 17699
Medizinhistorisches Museum, Wien 03027
Medizinhistorisches Museum, Zürich 42085
Mednjarodni Grafični Likovni Centar, Ljubljana 38451
Medrese Müzesi, Akşehir 42684
Medway Heritage Centre, Chatham 43797
Medway Towns Gallery, Rochester 45508
Meeker County Museum and G.A.R. Hall, Litchfield 50130
Meeker Home Museum, Greeley 49034
Sri Meenakshi Sundaresvara Temple Museum, Madurai 24304
Meeresmuseum, Öhringen 20864
Meerrettichmuseum, Baiersdorf 16897
Meezkeret-Batia Museum, Meezkeret-Batia 25063
Megalo Chorio Folklore Museum, Megalo Chorio 23085
Megginch Castle, Errol 44186
Meghalaya State Museum, Shillong 24386
Megisti Museum, Kastellorizo 23041
Meguro Gajoen Bijutsukan, Tokyo 29428
Meguro Gajoen Museum of Art, Tokyo 29428
Meguro Kiseichukan, Tokyo 29429
Meguro Museum of Art, Tokyo 29430
Meguro Parasitological Museum, Tokyo 29429
Megyeháza Börtönépület, Eger 23541
Méhészeti Gyűjtemény, Gödöllő 23567
Mehmet Akif Ersoy Evi, Ankara 42705
Mehrangarh Museum, Jodhpur 24246
Mei Ling Arts Museum, Puzi 42300
Mei Museum, Meixian 08013
Mei Museum, Shueili 42327
Mei-Nong Country Gallery, Qishan 42301
Meidiyas Museum of Art and Papertole, Linkou 42253
Meierimuseet, Ås 33401
Meierische Museumboerderij, Heeswijk-Dinther 32343
Meigle Museum of Sculptured Stones, Meigle 45122
Meigs County Museum, Pomeroy 52016
Meijerimuseo, Saukkola 11300
Meiji-Daigaku Hakubutsukan, Tokyo 29431
Meiji-Daigaku Kouko Hakubutsukan, Tokyo 29432
Meiji-Daigaku Shouhin Hakubutsukan, Tokyo 29433
Meiji Jingu Homotsuden, Tokyo 29434
Meiji Memorial Picture Gallery, Tokyo 29435
Meiji Shrine Treasure Museum, Tokyo 29434
Meiji University Museum, Tokyo 29431
Meinong Hakka Museum, Meinong 42273
Meir Dizengoff Museum, Tel Aviv 25116
Meishan Farming Village Museum, Meishan 42274
Meissener Porzellan-Sammlung Sigrid Pless, Bad Soden am Taunus 16840
Meißener Porzellan-Sammlung Stiftung Ernst Schneider, Oberschleißheim 20833
Meißener-Porzellan- und Christofle Tafelsilber-Ausstellung- Schloß Weyer, Gmunden 01952
Mejerimuseet vid Korsholms Lantbruksskolor, Vaasa 11424
Mekaanisen Musiikin Museo, Varkaus 11453
Mel Fisher Maritime Heritage Museum, Key West 49764
Melaka Govenor's Museum, Melaka 30605
Melaka Museum Corporation, Melaka 30607
Melanchthonhaus Bretten, Bretten 17584
Melanchthonhaus Wittenberg, Lutherstadt Wittenberg 20116
Melbourne Clocks Museum, Hampton 01145
Melbourne Hall, Melbourne 45123

M

Melbourne Maritime Museum, Southbank 01518
Melbourne Museum, Carlton South 00956
Melbourne Museum of Printing, Footscray 01084
Melbourne Tank Museum, Narre Warren North 01352
Melbourne's Living Museum of the West, Maribyrnong 01264
Meldal Bygdemuseum, Meldal 33665
Melford Hall, Long Melford, Sudbury, Suffolk 45027
Melfort and District Museum, Melfort 06261
Melicharovo Městské Muzeum, Unhošt 09889
Mellat Museum, Tehrān 24678
Mellemværftet, Kristiansund 33614
Mellerstain House, Gordon 44331
Mellette House, Watertown 53975
Melli Museum, Tehrān 24679
Mellilä Radio and Telephone Museum, Mellilä 11106
Mellilän Museoriihi, Mellilä 11105
Mellilän Radio- ja Puhelinmuseo, Mellilä 11106
Mellon Financial Corporation's Collection, Pittsburgh 51954
Melnikov Memorial House-Museum, Sarapul 37745
Meløy Bygdemuseum, Ørnes 33710
Meløy Local Museum, Ørnes 33710
Melrose Abbey Museum, Melrose 45125
Melrose Courthouse Museum, Melrose 01292
Melrose House, Tshwane 38749
Melrose Plantation Home Complex, Melrose 50565
Melton Art Reference Library Museum, Oklahoma City 51518
Melton Carnegie Museum, Melton Mowbray 45126
Melville Discovery Centre, Applecross 00807
Melville Heritage Museum, Melville 06264
Melville Railway Museum, Melville 06265
Melvin Art Gallery, Lakeland 49901
Melvin B. Tolson Black Heritage Center, Langston 49937
Mem-Erie Historical Museum, Erie 48408
Memento Park Budapest, Budapest 23449
Memento Park Budapest - Remains of communist dictatorship, Budapest 23449
Memi Pasa Fürdoje, Pécs 23766
Memme-Taadi Kamber, Karksi 10582
Memmed Seid Ordubadinin Xatire Muzeyi, Bakı 03160
Mémoire de la Nationale 7, Piolenc 14821
Mémoire de la Photo, Donzère 12768
Mémoire de Saumon, Saint-Louis-la-Chaussée 15375
Mémoire du Rhin, Saint-Louis-la-Chaussée 15376
Les Mémoires d'Obersoultzbach, Obersoultzbach 14493
Mémorial 1939-1945, Saint-Malo (Ille-et-Vilaine) 15388
Memorial Alexander Pushkin Museum of the Odessa Museum of Literature, Odesa 43154
Memorial and Literature Museum of A.P. Bondin, Ekaterinburg 36889
Memorial and Museum Auschwitz-Birkenau, Oświęcim 34997
Memorial and Museum Sachsenhausen, Oranienburg 20908
Memorial Appartment of Andrej Belyj, Moskva 37227
Memorial Appartment of G.M. Krshishanovskiy, Moskva 37228
Memorial Art Gallery of the University of Rochester, Rochester 52396
Memorial Center for Liberation Movements in German History, Rastatt 21183
Memorial Centre of Jozef Haller, Władysławowo 35305
Memorial Coplex Voroshilov Battery, Vladivostok 37940
Memorial Day Museum, Waterloo 53966
Le Mémorial de Caen, Caen 12229
Mémorial de la Shoah, Paris 14625
Memorial de Sergipe, Aracaju 04312
Mémorial de Vendée, Saint-Sulpice-le-Verdon 15518
Mémorial des Guerres en Indochine, Fréjus 13000
Mémorial des Soldats de La Nouvelle France, Le Château-d'Oléron 13570
Le Mémorial d'Essigny le Grand, Essigny-le-Grand 12865
Mémorial d'Évocation Vaudois et Huguenot, La Roque-d'Anthéron 13438
Memorial di Imigração Ucraniana, Curitiba 04427
Memorial do Imigrante, São Paulo 04863
Mémorial du Maréchal Leclerc de Hauteclocque et de la Libération de Paris et Musée Jean Moulin, Paris 14626
Memorial Exhibition of Deportee Dr. Dalia Grinkeviciute, Laukuva 30298
Memorial Farmstead of Dr. Jonas Sliupas - National Museum of Lithuania, Palanga 30328
Memorial for former political Prisoners 1945-1989, Magdeburg 20122
Mémorial Fort Montbarey, Brest 12160
Memorial Gallery of Tako Child Atelier, Tako 29348
Memorial Geroïčnoï Oboroni Odesi - 411 Beregova Linija, Odesa 43155
Memorial Guard Submarine S-56 from 1939, Vladivostok 37939
Memorial Hall Museum, Deerfield 48033
Memorial Hall of the Zunyi Conference, Zunyi 08521
Memorial Historic Site Danica, Koprivnica 08893
Memorial Historical Complex to the Heroes of the Stalingrad Battle on Mamaev Mount, Volgograd 37956
Memorial House Moseum of S.T. Aksakov, Ufa 37884
Memorial House-Museum of A. Uchtomskij, Rybinsk 37604
Memorial House-Museum of D. Mamin-Sibiryak, Ekaterinburg 36887
Memorial House-Museum of F. Reshetnikov, Ekaterinburg 36888
Memorial House-Museum of G. Krasilnikov, Alnaši 36755
Memorial House-Museum of G.V. Plekhanov, Lipeck 37105
Memorial House-Museum of I.I. Shishkin, Elabuga 36904
Memorial House-Museum of I.M. Vinogradov, Velikie Luki 37917
Memorial House-Museum of L.V. Sobinov, Jaroslavl 36970
Memorial House-Museum of M. Zankovetska, Kyiv 43083
Memorial House Museum of Mashit Gafuri, Ufa 37885
Memorial House-Museum of N.G. Slavyanov, Perm 37546
Memorial House-Museum of P. Bazhov, Ekaterinburg 36890
Memorial House-Museum of Taras Shevchenko, Kyiv 43081
Memorial House-Museum of Taras Shevchenko, Kyiv 43087
Memorial House-Museum of the Ulyanov Family, Astrachan 36782
Memorial House of Adalbert Stifter, Horní Planá 09526
Memorial House of Bedřich Smetana, Jabkenice 09544
Memorial House of Božena Němcová, Červený Kostelec 09456
Memorial House of D. Javornickij, Dnipropetrovs'k 43035
Memorial House of Dimitar Polyanov, Karnobat 05052
Memorial House of František Palacký, Hodslavice 09520
Memorial House of Friedrich Zander of the University of Latvia, Rīga 30068
Memorial House of Jan Amos Comenius, Brandýs nad Orlicí 09415
Memorial House of Jan Hus, Husinec 09542
Memorial House of Josef Mánes - Castle, Čechy pod Košířem 09452
Memorial House of Karel Čapek, Stará Hut u Dobříše 09845
Memorial House of Karel J. Erben, Miletin 09654
Memorial House of Karel V. Rais, Lázně Bělohrad 09620
Memorial House of King Jiří z Poděbrad and Lapidarium, Poděbrady 09723
Memorial House of Laza K. Lazarević, Beograd 38088
Memorial House of Leoš Janáček, Hukvaldy 09540
Memorial House of Mikuláš Aleš and Matej Kopecky, Mirotice 09656
Memorial House of Petr Bezruč, Museum Prostejov, Kostelec na Hané 09597
Memorial House of Rippl-Rónai, Kaposvár 23633
Memorial House of Svatopluk Čech, Ostředek 09702
Memorial House of the Jews of Lublin, Lublin 34899
Memorial House of Václav Beneš Třebízský and of Svatopluk Čech, Liteň 09628
Memorial JK, Brasília 04358
Memorial Karel Havlíček Borovský, Havlíčkův Brod 09512
Memorial Konstantin Paustovsky Museum of the Odessa Museum of Literature, Odesa 43160
Memorial Library and Art Gallery, Saskatoon 06917
Memorial Literature Museum Estate A.P. Chekhov, Melichovo 37129
Memorial Literature Museum of N.A. Dobrolyubov, Nižnij Novgorod 37451
Mémorial Musée du Régiment, Spa 04040
Memorial Museum, Naxçıvan 03185
Memorial Museum, Yogyakarta 24586
Memorial Museum Apartment of Rimsky-Korsakov, Sankt-Peterburg 37663
Memorial Museum B.P. Hasdeu, Câmpina 36249
Memorial Museum Illegal Printery of the Perm Committee of the Russian Social Democratic Labour Party of 1906, Perm 37545
Memorial Museum Ion Creangă's Cottage, Iași 36422
Memorial Museum M.V. Frunse, Ivanovo 36949
Memorial Museum of A.S. Popov, Sankt-Peterburg 37664
Memorial Museum of A.S. Popov - the Inventor of the Radio, Krasnoturinsk 37081
Memorial Museum of Alia Moldagulova, Aktöbe 29672
Memorial Museum of Avram Iancu, Avram Iancu 36116
Memorial Museum of Cosmonautics, Moskva 37242
Memorial Museum of Eugenijus and Leonardas Matuzeviciai, Krinčinas 30289
Memorial Museum of France Prešeren, Kranj 38438
Memorial Museum of Holy Maximilian Kolbe, Zduńska Wola 35367
Memorial Museum of Hostages, Begunje na Gorenjskem 38398
Memorial Museum of Ivo Andrić, Beograd 38089
Memorial Museum of Karim Khakimov, Ufa 37886
Memorial Museum of M. Saryan, Yerevan 00732
Memorial Museum of Matsusaka Merchant, Matsusaka 28897
Memorial Museum of Paneriai, Vilnius 30428
Memorial Museum of Professor Povilas Matulionis, Kupiškis 30293
Memorial Museum of the Academician V.I. Vernadskyj, Moskva 37235
Memorial Museum of the Academy Members A.E. and B.A. Arbusov, Kazan 37012
Memorial Museum of the Kanto Earthquake Disaster, Tokyo 29436
Memorial Museum of the Mexican Academy of Language, México 31107
Memorial Museum of the Sculptor Bronius Pundzius - Mazeikiai Museum, Pieveniai 30337
Memorial Museum of the War of Liberation, Pyongyang 29760
Memorial Museum of the Writer Satrijos Ragana - Mazeikiai Museum, Židikai 30452
Memorial Museum of Yuldash Akhunbabayev, Marg'ilon 54449
Memorial Museum Passchendaele 1917, Zonnebeke 04166
Memorial Museum & Pioneer Museum, Ada 46134
Memorial Museum-preserve of V.D. Polenov, Strachovo 37808
Memorial Museum Saint Isaacs Cathedral, Sankt-Peterburg 37652
Memorial Museum The Grave of T.G. Shevchenko, Kaniv 43060
Memorial Museum The October 21st, Kragujevac 38110
Memorial Museum-Workshop of Theodisius Brish, Lviv 43128
Memorial N.I. Vavilov Museum of the Vavilov-Institute of General Genetics of the Russian Academy of Sciences, Moskva 37248
Memorial of Jan Amos Comenius, Bílá Třemešná 09406
Memorial of Otakar Štáfl, Havlíčkův Brod 09513
Memorial of the Kralice Bible - Moravian Museum Brno, Kralice nad Oslavou 09602
Memorial Pengisytiharan Kemerdekaan, Melaka 30597
Memorial Piskarevskoye Cemetery - Museum, Sankt-Peterburg 37720
Memorial Pontes de Miranda da Justiça do Trabalho em Alagoas, Maceió 04541
Memorial Prins Karel, Oostende 03932
Memorial Room of Professor Stasys Usinkas, Pakruojis 30326
Memorial Room of the Composer A. Budriunas, Vilnius 30413
Memorial Ship Mikasa, Yokosuka 29620
Memorial Studio Museum of the Sculptor S.T. Konenkov, Moskva 37246
Memorial Sultan Tidore Museum, Tidore 10367
Memorial Tancredo Neves, São João Del Rei 04833
Memorial to the Murdered Jews of Europe, Berlin 17065
Memorial Union Art Collection, Tempe 53468
Memorial Union Art Gallery, Corvallis 47836
Memorial Union Gallery, Fargo 48510
Memorial Workroom-Museum of the Academician A.N. Nesmejanov, Moskva 37240
Memorial Workroom-Museum of the Academician A.P. Vinogradov, Moskva 37234
Memorial Workroom-Museum of the Academician M.V. Keldysh, Moskva 37239
Memorial Workroom-Museum of the Academician N.D. Zelinskiy, Moskva 37241
Memorial Workroom-Museum of the Academician N.V. Belov, Moskva 37237
Memorial Workroom-Museum of the Academician P.L. Kapitsa, Moskva 37238
Memorial Workroom of A.V. Lunacharsky, Moskva 37233
Memorialinė Tremtines, Gydytojos Dalios Grinkevičiūtės Ekspozicija, Laukuva 30298
Memorialmuseum Friedrich Fröbel, Oberweißbach 20849
Memorialnaja Gvardejskaja Krasnoznamënnaja Podvodnaja Lodka S-56 1939 g., Vladivostok 37939
Memorialnaja Kvartira Andreja Belogo, Moskva 37227
Memorialnaja Kvartira G.M. Kržižanovskogo, Moskva 37228
Memorialno-chudožestvennyj Dom-muzej Vladimira Aleksandroviča Serova, Ėmmaus 36914
Memorialno-literaturnyj Muzej A.P. Bondina, Ekaterinburg 36889
Memorial'nyj Budinok-muzej V. Dalja, Lugans'k 43116
Memorial'nyj Budinok-muzej A.A. Kocky, Užgorod 43196
Memorial'nyj Budinok-muzej F.F. Manajla, Užgorod 43197
Memorial'nyj Budinok-muzej Tarasa Ševčenka, Kyiv 43087
Memorialnyj Chudožestvennyj Muzej Valentina Aleksandroviča Serova v Domotkanovo, Krasnaja Nov 37064
Memorialnyj Dom-Muzej Aaly Tokombaeva, Biškek 29957
Memorialnyj Dom-muzej Akademika S.P. Koroleva, Moskva 37229
Memorialnyj Dom-muzej Akademika I.V. Kurčatova, Moskva 37230
Memorialnyj Dom-muzej Akademika A.A. Uchtomskogo, Rybinsk 37604
Memorialnyj Dom-muzej Akademika I.M. Vinogradova, Velikie Luki 37917
Memorialnyj Dom-muzej S.T. Aksakova, Ufa 37884
Memorialnyj Dom-Muzej P.P. Bažova, Ekaterinburg 36890
Memorialnyj Dom-muzej Gercena, Moskva 37231
Memorialnyj Dom-muzej P.D. Korina, Moskva 37232
Memorialnyj Dom-muzej G.D. Krasilnikova, Alnaši 36755
Memorialnyj Dom-muzej I.I. Levitana, Ples 37566
Memorialnyj Dom-muzej Mažita Gafuri, Ufa 37885
Memorialnyj Dom-muzej G.V. Plechanova, Lipeck 37105
Memorialnyj Dom-muzej Podpolnaja Tipografija Permskogo Komiteta RSDRP 1906 g., Perm 37545
Memorialnyj Dom-muzej I.P. Požalostina, Solotča 37787
Memorialnyj Dom-Muzej I.I. Šiškina, Elabuga 36904
Memorialnyj Dom-muzej im. N.G. Slavjanova, Perm 37546
Memorialnyj Dom-muzej L.V. Sobinova, Jaroslavl 36970
Memorialnyj Dom-muzej Uljanovych, Astrachan 36782
Memorialnyj Dom-muzej A.A. Vaneeva, Ermakovskoe 36916
Memorialnyj Dom-muzej Viktora i Apolinarija Vasnecovych, Rjabovo 37592
Memorialnyj Dom-muzej Marko Vovčok, Nalčik 37440
Memorialnyj Dom-muzej N.E. Žukovskogo, Orechovo 37502
Memorialnyj Istoriko-biografičeskij Dom-muzej Akademika N.V. Melnikova, Sarapul 37745
Memorialnyj Kabinet A.V. Lunačarskogo, Moskva 37233
Memorialnyj Kabinet-muzej Akademika A.P. Vinogradova, Moskva 37234
Memorialnyj Kompleks G. Tukaja, Novyj Kyrlaj 37493
Memorialnyj Kompleks 'Vorošilovskaja Batareja', Vladivostok 37940
Memorialnyj Muzej Akademika K.I. Satpaeva, Almaty 29684
Memorialnyj Muzej Akademika V.I. Vernadskogo, Moskva 37235
Memorialnyj Muzej Akademikov A.E. i B.A. Arbuzovych, Kazan 37012
Memorialnyj Muzej N.I. Beloborodova, Tula 37867
Memorialnyj Muzej A.N. Burdenko, Moskva 37236
Memorialnyj Muzej-dača A.S. Puškina, Puškin 37581
Memorialnyj Muzej Jurija Gagarina, Gagarin 36918
Memorial'nyj Muzej V. Ivasjuka, Černivci 43016
Memorialnyj Muzej-kabinet Akademika N.V. Belova, Moskva 37237
Memorialnyj Muzej-kabinet Akademika P.L. Kapicy, Moskva 37238
Memorialnyj Muzej-kabinet Akademika M.V. Keldyša, Moskva 37239
Memorialnyj Muzej-kabinet Akademika A.N. Nesmejanova, Moskva 37240
Memorialnyj Muzej-kabinet Akademika N.D. Zelinskogo, Moskva 37241
Memorialnyj Muzej-kabinet M.V. Frunze, Ivanovo 36949
Memorialnyj Muzej Karima Chakimova, Ufa 37886
Memorialnyj Muzej I.S. Ključnikova-Palantaja, Joškar-Ola 36977
Memorialnyj Muzej Kosmonavtiki, Moskva 37242
Memorialnyj Muzej-kvartira Aktërov Samojlovych, Sankt-Peterburg 37659
Memorialnyj Muzej-kvartira E.F. Gnesinoj, Moskva 37243
Memorialnyj Muzej-Kvartira S.M. Kirova, Vladikavkaz 37930

M

Memorialnyj Muzej-kvartira Vadima Alekseeviča Kozina, Magadan 37120
Memorialnyj Muzej-kvartira P.K. Kozlova, Sankt-Peterburg 37660
Memorialnyj Muzej-kvartira G.M. Kržižanovskogo i V.V. Starkova, Minusinsk 37134
Memorialnyj Muzej-kvartira N.A. Nekrasova, Sankt-Peterburg 37661
Memorialnyj Muzej-kvartira I.P. Pavlova, Sankt-Peterburg 37662
Memorialnyj Muzej-kvartira A.S. Puškina, Moskva 37244
Memorialnyj Muzej-kvartira Rimskogo-Korsakova, Sankt-Peterburg 37663
Memorialnyj Muzej-kvartira A.A. Šogencukova, Nalčik 37441
Memorialnyj Muzej-kvartira M.S. Spiridonova, Čeboksary 36839
Memorialnyj Muzej-kvartira A.M. Vasnecova, Moskva 37245
Memorialnyj Muzej-laboratorija A.S. Popova, Kronštadt 37085
Memorialnyj Muzej Licej, Puškin 37582
Memorial'nyj Muzej-majsternja Teodizii Briž, Lviv 43128
Memorialnyj Muzej A.G. Malyškina, Mokšan 37138
Memorialnyj Muzej-Masterskaja Skulptora S.T. Konenkova, Moskva 37246
Memorialnyj Muzej Alii Moldagulovoj, Aktöbe 29672
Memorial'nyj Muzej Mychajla Hruševskoho u Lvovi, Lviv 43129
Memorial'nyj Muzej V.I. Nemiroviča-Dančenka, Neskučne 43152
Memorialnyj Muzej B.L. Pasternaka, Čistopol 36866
Memorialnyj Muzej A.S. Popova, Sankt-Peterburg 37664
Memorialnyj Muzej A.S. Popova - Izobretatelja Radio, Krasnoturinsk 37081
Memorialnyj Muzej A.N. Skrjabina, Moskva 37247
Memorialnyj Muzej-usadba Akademika I.P. Pavlova, Rjazan 37594
Memorialnyj Muzej-usadba Chudožnika N.A. Jarošenko, Kislovodsk 37048
Memorialnyj Muzej-usadba M.P. Musorgskogo, Naumovo 37447
Memorialnyj Muzej Konstantina Vasiljeva, Vasiljevo 37916
Memorialnyj Muzej N.I. Vavilova, Moskva 37248
Memorialnyj Muzej-zapovednik N.A. Rimskogo-Korsakova Ljubensk-Večaša, Pljussa 37569
Memorialul Ipoteşti-Centrul Naţional de Studii Mihai Eminescu, Ipoteşti 36429
Memorialul Victimelor Comunismului şi al Rezistenţei, Sighetu Marmaţiei 36614
Memorijalna Galerija Aleksandar Rukavina, Brtonigla 08850
Memorijalna Zbirka Eugena Kumičića, Brseč 08849
Memorijalni Muzej Džuro Jakšić, Srpska Crnja 38154
Memorijalnyj Budinok-muzej D. Javornic'kogo, Dnipropetrovs'k 43035
Memorium Nürnberger Prozesse, Nürnberg 20778
Memorium Nuremberg Trials, Nürnberg 20778
Memory Hall of the Military School of Montaña, Jaca 39419
Memory Lane Museum at Seattle Goodwill, Seattle 53003
Memory Museum, Kyoto 28858
Memoryville USA, Rolla 52446
Memphis Belle B17 Flying Fortress, Memphis 50576
Memphis Brooks Museum of Art, Memphis 50577
Memphis College of Art Gallery, Memphis 50578
Memphis Pink Palace Museum, Memphis 50579
Men in the french-prussian war of 1870, Bitche 12005
Menard Art Museum, Komaki 28764
Mencendorfa Nams - Rīdzinieku Māja-Muzejs, Rīga 30095
Menczer Museum of Medicine and Dentistry, Hartford 49196
Mendel Art Gallery, Saskatoon 06918
Mendelianum - Památník Gregora Mendela, Brno 09430
Mendelssohn House Museum, Leipzig 19924
Mendocino Art Center Gallery, Mendocino 50587
Mendocino County Museum, Willits 54183
Mengeli - Memorial Museum of the Jurjanu Brothers, Ērgļi 30007
Mengzhou Museum, Mengzhou 08014
The Menil Collection, Houston 49394
Menin Gate of the Brno City Museum, Brno 09431
Měnínská Brána, Brno 09431
Mennello Museum of American Folk Art, Orlando 51591
Mennonite Heritage Center, Harleysville 49168
Mennonite Heritage Museum, Goessel 48953
Mennonite Heritage Museum, Rosthern 06778
Mennonite Heritage Village, Steinbach 07007
Mennonite Historical Society of British Columbia, Abbotsford 05306
Mennonite Library and Archives Museum, North Newton 51398
Menominee County Heritage Museum, Menominee 50589
Mentalvårdsmuseet Säter, Säter 40823
Mentlhof-Bergbauernmuseum, Heiligenblut 02066
Mentor Graham Museum, Blunt 46879
MenTouGou Museum, Beijing 07550
Mentzendorff's House - City-dwellers' House-Museum, Rīga 30095
Menzer Heimatstube, Stechlin 21872
Meopham Windmill, Meopham 45127
Meråker Bygdemuseum, Meråker 33666
Merano Arte - Kunst Meran, Merano 26656
Mercantile Arts Centre Kintyre, Tarbert 45825
Mercator Museum, Sint-Niklaas 04021
Mercators Schrijn Graventoren, Rupelmonde 03991
Merced County Courthouse Museum, Merced 50595
Mercedes-Benz Museum, Stuttgart 21952
Mercer Art Gallery, Harrogate 44401
Mercer County Genealogical and Historical Society, Princeton 52147
Mercer County Historical Museum, Celina 47310
Mercer County Museum, Mercer 50596
Mercer Gallery, Rochester 52397
Mercer Museum of the Bucks County Historical Society, Doylestown 48180
Mercer Union - Centre for Contemporary Visual Art, Toronto 07123
Merchant Adventurers' Hall, York 46104
Merchant's House Museum, New York 51211
Merchant's House Museum, Plymouth 45391
Mercosur Itinerant Contemporary Art Museum, Buenos Aires 00163
Mercury Bay and District Museum, Whitianga 33267
Merdøgård, Arendal 33397
Mere Museum, Mere 45128
Meredith Gallery, Baltimore 46579
Meriden Historical Society Museum, Meriden 50598
Meridian International Center - Cafritz Galleries, Washington 53892
Meridian Museum of Art, Meridian 50600
Merijärven Kotiseutumuseo, Merijärvi 11107
Merikarvian Kalastusmuseo, Merikarvia 11109
Merimiehen Koti, Uusikaupunki 11420
Merimiehenkotimuseo, Oulu 11170
Meritalo Museum, Salo 11296
Meritalon Museo, Salo 11296
Meriwether Lewis Dredge Museum, Brownville 47098
Meriwether Lewis National Monument, Hohenwald 49309
Merkine Local Lore Museum, Merkinė 30313
Merkinės Kraštotyros Muziejus, Merkinė 30313
Merklappenmuseum, Dieteren 32128
Meroogal House, Nowra 01383
Merowe Museum, Merowe 40395
Merredin Military Museum, Merredin 01296
Merredin Museum, Merredin 01297
The Merrick Art Gallery, New Brighton 51027
Merrill Historical Museum, Merrill 50602
Merrill Museum, Carrollton 47255
Merritt Museum of Anthropology, Oakland 51466
Merriwa Colonial Museum, Merriwa 01298
Mersea Island Museum, West Mersea 45972
Merseburg Treasury in the cloister and chapter house, Merseburg 20294
Merseburger Domschatz in der Südklausur und Kapitelhaus, Merseburg 20294
Merseyside Maritime Museum, Liverpool 44740
Mersin Müzesi, Mersin 42907
Mersin Museum, Mersin 42907
Meru Museum, Meru 29745
Les Merveilles de la Mer, Jdeidet El-Metn 30170
Mervyn Quinlan Museum, Benenden 43451
Merwin and Wakeley Galleries, Bloomington 46862
Merwin House, Stockbridge 53299
Meryem Ana Evi, Selçuk 42937
Mesa Contemporary Arts, Mesa 50607
Mesa Historical Museum, Mesa 50608
Mesa Verde National Park Museum, Mesa Verde 50609
Meseház, Békéscsaba 23369
Meshkinshahr Museum, Meshkinshahr 24634
Mesnil aux Abeille, Beautheil 11907
Mesolithic Museum, Abinger Common 43261
Mesquite Heritage Museum, Mesquite 50611
Messerermuseum, Steinbach an der Steyr 02767
Messina Museum, Messina 38688
Messner Mountain Museum - Dolomites, Cibiana di Cadore 25885
Messner Mountain Museum - Firmian, Bolzano 25484
Messner Mountain Museum - Juval, Castelbello 25721
Messner Mountain Museum - Ortles, Solda 27864
Messner Mountain Museum - Ripa, Brunico 25555
Mestna Galerija, Šoštanj 38523
Mestna Galerija Ljubljana, Ljubljana 38452
Mestni Muzej Idrija, Idrija 38421
Mestni Muzej Ljubljana, Ljubljana 38453
Mestni Muzej Radovljica, Radovljica 38502
Mestská Galéria, Rimavská Sobota 38341
Městská Galerie, Polička 09728
Městská Galerie Art Club, Týn nad Vltavou 09884
Městská Galerie Beroun, Beroun 09403
Městská Galerie Josefa Lieslerova, Kadaň 09567
Městská Galerie Karla Havlíčka, Kadaň 09568
Městská Galerie Špejchar, Chomutov 09477
Městská Galerie v Zámeckém Parku, Veselí nad Moravou 09905
Mestská Výstavná Sieň Barónka Kežmarok, Kežmarok 38266
Městské Muzeum, Aš 09395
Městské Muzeum, Bechyně 09397
Městské Muzeum, Bílovec 09407
Městské Muzeum, Bystřice nad Pernštejnem 09449
Městské Muzeum, Čáslav 09451
Městské Muzeum, Čelákovice 09455
Městské Muzeum, Česká Třebová 09461
Městské Muzeum, Chotěboř 09479
Městské Muzeum, Chrast u Chrudimě 09480
Městské Muzeum, Chrastava 09482
Městské Muzeum, Dobruška 09493
Městské Muzeum, Františkovy Lázně 09502
Městské Muzeum, Frenštát pod Radhoštěm 09503
Městské Muzeum, Holešov 09521
Městské Muzeum, Horažďovice 09523
Městské Múzeum, Ilava 38263
Městské Muzeum, Jesenice 09556
Městské muzeum, Kamenice nad Lipou 09570
Městské Muzeum, Klobouky u Brna 09588
Městské Muzeum, Králíky 09603
Městské Muzeum, Krnov 09609
Městské Muzeum, Lanškroun 09617
Městské Muzeum, Ledeč nad Sázavou 09621
Městské Muzeum, Letohrad 09623
Městské Muzeum, Mnichovo Hradiště 09660
Městské Muzeum, Moravská Třebová 09662
Městské Muzeum, Moravský Krumlov 09664
Městské Muzeum, Napajedla 09670
Městské Muzeum, Nepomuk 09673
Městské Muzeum, Netvořice 09675
Městské Muzeum, Nové Město nad Metují 09681
Městské Muzeum, Nové Strašecí 09682
Městské Muzeum, Nový Bydžov 09684
Městské Muzeum, Počátky 09721
Městské Muzeum, Přelouč 09803
Městské Muzeum, Příbor 09805
Městské Muzeum, Přibyslav 09808
Městské Muzeum, Protivín 09811
Městské Muzeum, Radnice u Rokycan 09813
Městské Muzeum, Rtyně v Podkrkonoší 09829
Městské Muzeum, Rýmařov 09833
Městské Múzeum, Sabinov 38346
Městské Muzeum, Sadská 09834
Městské Muzeum, Sedlčany 09835
Městské Muzeum, Skuteč 09837
Městské Muzeum, Strážnice 09852
Městské Muzeum, Stříbro 09853
Městské Muzeum, Valašské Klobouky 09895
Městské Muzeum, Velká Bíteš 09899
Městské Muzeum, Veselí nad Moravou 09906
Městské Muzeum, Vimperk 09907
Městské Muzeum, Volyně 09913
Městské Muzeum, Žamberk 09925
Městské Muzeum, Zbiroh 09928
Městské Múzeum, Zlaté Moravce 38392
Městské Múzeum a Galéria, Vel'ký Šariš 38386
Městské Muzeum a Galerie, Břeclav 09417
Městské Muzeum a Galerie, Dačice 09489
Městské Muzeum a Galerie, Dvůr Králové nad Labem 09500
Městské Muzeum a Galerie, Hlinsko v Čechách 09514
Městské Muzeum a Galerie, Hořice 09524
Městské Muzeum a Galerie, Hranice 09536
Městské Muzeum a Galerie, Lomnice nad Popelkou 09637
Městské Muzeum a Galerie, Svitavy 09857
Městské Muzeum a Galerie, Vodňany 09911
Městské Muzeum a Galerie J. Karse, Velvary 09903
Městské Muzeum a Galerie Julie W. Mezerové, Úpice 09890
Městské Muzeum a Galerie Otakara Spaniela a Josefa Wágnera, Jaroměř 09550
Městské Muzeum Antonína Sovy, Pacov 09704
Městské Muzeum Dr. E. Beneše, Kožlany 09601
Městské Múzeum Fil'akovo, Fil'akovo 38254
Městské Muzeum - Františkánský klášter, Kadaň 09569
Městské Muzeum Klenotnice, Nová Paka 09676
Městské Múzeum Michala Tillnera, Malacky 38291
Městské Múzeum Poltár, Poltár 38330
Městské Múzeum Rajec, Rajec 38339
Městské Múzeum v Bratislave, Bratislava 38219
Městské Muzeum v Kralupech nad Vltavou, Kralupy nad Vltavou 09604
Městské Muzeum v Mariánských Lázních, Mariánské Lázně 09649
Městské Múzeum v Pezinku, Pezinok 38325
Městské Vlastivědné Mudruňkovo Muzeum, Říčany u Prahy 09817
Městské Vrbasovo Muzeum, Ždánice 09932
Mestský Vlastivedný Dom, Šamorín 38348
Metal Art Museum Hikarinotani, Imba 28570
Metallhandwerksmuseum, Steinbach-Hallenberg 21881
Metallmuseum, Tullnerbach-Lawies 02835
Metallurgical Museum, Podbrezová 38329
Metamora Courthouse, Metamora 50613
Metamorphosis Church Museum, Meteora 23092
Metchosin School Museum, Victoria 07255
Metenkov House - Museum of Photography, Ekaterinburg 36885
Meteor Crater, Winslow 54241
Meteorkrater-Museum, Steinheim am Albuch 21890
Meteorological Museum of the Main Observatory, Sankt-Peterburg 37698
Methil Heritage Centre, Methil 45133
Metiers d'Art Sant Roch, Céret 12339
Les Métiers du Champagne, Épernay 12830
Metro Arts, Brisbane 00897
Metrological Coll of the Central Office of Measures, Warszawa 35260
Metrònom, Barcelona 38966
Metropole Gallery, Folkestone 44238
Metropolit Chambers - Museum of Ancient Russian Art, Jaroslavl 36971
The Metropolitan Museum of Art, New York 51212
Metropolitan Museum of Manila, Manila 34444
Metropolitan Police Historical Collection, London 44916
Metsamor Museum, Metsamor 16348
Metz von Tessin'sche Kutschensammlung, Starzach 21865
Metzgermuseum Zunft zum Widder, Zürich 42086
Meux Home Museum, Fresno 48793
Mevlânâ Müzesi, Konya 42889
Mevlânâ Rumi Museum, Konya 42889
Mevlevi Tekke Müzesi, Lefkoşa 33371
Mexic-Arte Museum, Austin 46515
Mexican Federation of Associations of Friends of Museums, México 54942
Mexican Fine Arts Center Museum, Chicago 47489
The Mexican Museum, San Francisco 52783
Meyerhof Scheeßel, Scheeßel 21537
Meyerhold Museum of Scenic Performance, Penza 37535
Meža Muzejs Boju Pilī, Aizpute 29983
Meždunarodnyj Centr-Muzej im. N.K. Rericha, Moskva 37249
Meždunarodnyj Muzejno-vystavočnyj Centr, Vladivostok 37941
Mézédes emlékeink - Mézeskalácsos, Cukorka- és Gyertyakészítő Múzeum, Szekszárd 23868
Mezinárodní Muzeum Keramiky, Bechyně 09398
Mezőgazdasági Tájmúzeum - Holland-ház, Dég 23523
Mezőföld Regional Peasant House Museum, Dég 23524
Mezőföldi Tájház, Dég 23524
Mezőgazdasági és Élelmiszeripari Szakképző Intézet, Szabadkígyós 23830
Mezőgazdasági Eszköz- és Gépfejlődéstörténeti Szakmúzeum, Gödöllő 23568
Mezőgazdasági Gépmúzeum, Mezőkövesd 23699
Mezőgazdaságtudományi Kar Történeti Múzeum, Mosonmagyaróvár 23721
Mezőhegyesi Állami Ménes Kocsimúzeuma, Mezőhegyes 23696
Mezquita Catedral de Córdoba, Córdoba 39203
The Mezzanine Gallery, New York 51213
Mher Abeghyan Museum, Echmiadzin 00698
Mher Mkrtchyan House Museum, Gyumri 00706
MIAAO - Museo Internazionale delle Arti Applicate oggi, Torino 28003
Miami Art Central, Miami 50630
Miami Art Museum, Miami 50631
Miami Children's Museum, Miami 50632
Miami County Museum, Peru 51789
Miami Museum, Miami 06271
Miami Museum of Science, Miami 50633
Miami Station Museum, Miami 06272
Miami University Art Museum, Oxford 51654
Miami University Hiestand Galleries, Oxford 51655
Mian County Museum, Mianxian 08015
Miandoab Museum, Urmieh 24700
Mianposhteh Palace Museum, Bandar-Anzali 24599
Mianyang Museum, Mianyang 08016
Miaoli Ceramics Museum, Gongguan 42192
Miaoli Railway Museum, Miaoli 42275
Miaoli Wood Sculpture Museum, Sanyi 42317
Miasskij Kraevedčeskij Muzej, Miass 37130
Michael Bruce Cottage Museum, Kinnesswood 44606
Michael C. Carlos Museum, Atlanta 46444
Michael C. Rockefeller Arts Center Gallery, Fredonia 48778
Michael Davitt National Memorial Museum, Straide 24901
Michael Faraday's Laboratory and Museum, London 44917
Michael og Anna Anchers Hus og Saxilds Gaard, Skagen 10261
Michaelis Collection, Cape Town 38587

Michaels Privatmuseum, Nellingen 20569
Michajlovskij Istoričeskij Muzej, Michajlov 37131
Michajlovskoe - Gosudarstvennyj Memorialnyj Istoriko-literaturnyj i Prirodno-landšaftnyj Muzej-Zapovednik A.S. Puškina, Puškinskie Gory 37587
Michel Brouillet House and Museum, Vincennes 53749
Michel-Buck-Gedenkstätte im Kaplaneihaus, Ertingen 18231
Michelham Priory, Hailsham 44380
Michelson Museum of Art, Marshall 50510
Michelstettner Schule, Asparn an der Zaya 01733
Michener House Museum, Lacombe 06125
Michie Tavern Ca. 1784, Charlottesville 47397
Michif Historical and Cultural Preservation Society Museum, Surrey 07036
Michigan Chapter Gallery, Detroit 48123
Michigan Historical Museum, Lansing 49943
Michigan Maritime Museum, South Haven 53172
Michigan Museums Association, Lansing 55155
Michigan Sports Hall of Fame, Detroit 48124
Michigan State University Museum, East Lansing 48254
Michigan Transit Museum, Mount Clemens 50873
Michio Miyagi Memorial Hall, Tokyo 29438
Michizo Tachihara Memorial Museum, Tokyo 29480
Michoutouchkine Pilioko Foundation, Port Vila 54494
Mico INAFCA Museum, Kingston 28330
Mičovna, Praha 09761
Micro-Musée, La Trinité 30667
Micro-Musée de la Lessive, Seyne-les-Alpes 15747
Micro-Musée de l'École, Seyne-les-Alpes 15748
Micro-Musée de l'Ex-Colonie Pénitentiaire, Mettray 14098
Micro-Musée du Tailleur, Seyne-les-Alpes 15749
Micromusée de la Légion Etrangère, Douzillac 12785
Micromusée de la Maison de Retraîte Jeanne Pierretée Carnot, Nolay 14471
Mid-America Air Museum, Liberal 50076
Mid-America All-Indian Center Museum, Wichita 54146
Mid-America Science Museum, Hot Springs 49373
Mid Atlantic Air Museum, Reading 52264
Mid-Atlantic Association of Museums, Baltimore 55156
Mid-Atlantic Center for the Arts, Cape May 47235
Mid-Continent Railway Museum, North Freedom 51393
Mid Hants Railway 'Watercress Line', Alresford 43289
Mid-Hudson Children's Museum, Poughkeepsie 52123
Mid-Richmond Historical Museum, Coraki 01000
Middelfart Museum, Middelfart 10189
Middelheimmuseum, Antwerpen 03292
Middendorpshuis, Den Ham 32107
Middle Border Museum of American Indian and Pioneer Life, Mitchell 50763
Middle East Museum, Yerevan 00743
Middle East Technical University Art Gallery, Ankara 42708
Middle East University Museum, Ankara 42707
Middle Eastern Culture Centre in Japan, Mitaka 28917
Middle Lake Museum, Middle Lake 06273
Middleborough Historical Museum, Middleborough 50648
The Middlebury College Museum of Art, Middlebury 50653
Middlebury Historical Society Museum, Wyoming 54314
Middlesbrough Institute of Modern Art, Middlesbrough 45141
Middlesex Canal Collection, Lowell 50305
Middlesex County Historical Society Museum, Middletown 50659
Middleton Place Foundation, Charleston 47373
Middleton Top Engine House, Middleton-by-Wirksworth 45143
Middletown Fine Arts Center, Middletown 50662
Middletown Valley Historical Society Museum, Middletown 50660
Middleville and District Museum, Lanark 06133
Midland Air Museum, Baginton 43351
Midland County Historical Museum, Midland 50667
Midland County Historical Museum, Midland 50672
Midland Motor Museum, Bridgnorth 43605
Midland Railway Butterley, Ripley, Derbyshire 45497
Midlands Federation of Museums and Art Galleries, Telford 55073
Midtown Art Center, Houston 49395
Midtsønderjyllands Museum, Gram 10040
Midttunet, Skei i Jølster 33833
Midway Museum, Midway 50677
Midway Village and Museum, Rockford 52415
Midwest Museum of American Art, Elkhart 48346
Midwest Old Settlers and Threshers Association Museum, Mount Pleasant 50883
Mie-kenritsu Bijutsukan, Tsu 29536
Mie-kenritsu Hakubutsukan, Tsu 29537
Mie Prefectural Art Museum, Tsu 29536
Mie Prefectural Museum, Tsu 29537
Międzynarodowe Centrum Sztuki w Łódźi, Łódź 34872
Miegunyah Pioneer Women's Memorial House Museum, Bowen Hills 00882
Miejsce Pamięci i Muzeum Auschwitz-Birkenau, Oświęcim 34997
Miejska Galeria Arsenał, Poznań 35027
Miejska Galeria Sztuki, Częstochowa 34614
Miele-Museum, Gütersloh 18832
Miellerie, Villeneuve-de-Berg 16208
Miellerie des Gorges de la Loire, Roche-la-Molière 15065
Miers Home Museum, Sonora 53155
Miervalža Ķemera Memoriālais Muzejs, Jūrmala 30025
Mies-Pilsner-Heimatmuseum, Dinkelsbühl 17867
Mies van der Rohe Haus, Berlin 17187
Mieszkanie-Pracownia Kazimiery Iłłakowiczówny, Poznań 35028
Mifflinburg Buggy Museum, Mifflinburg 50679
Mighty Eighth Air Force Museum, Pooler 52023
Migration Museum, Adelaide 00778
Migros Museum für Gegenwartskunst, Zürich 42087
Miguel Malvar Museum, Santo Tomas 34503
Mihály Babits Memorial House of the Bálint Balassa Museum, Esztergom 23550
Mihály Babits Memorial Museum, Szekszárd 23865
Mihály Vörösmarty Memorial Museum, Kapolnásnyék 23629
Mihály Zichy Memorial Museum, Zala 23969
Miharu Folk Doll Museum, Miharu 28908
Miharu-machi Ningyo Hakubutsukan, Miharu 28908
Mihkli Talumuuseum, Kihelkonna 10584
Miho Museum, Shigaraki 29258
Mikailov Historical Museum, Michajlov 37131
Mikalojaus Konstantino Čiurlionio Memorialinis Muziejus, Druskininkai 30224
Mikalojus Konstantinas Čiurlionis Memorial Museum of the M.K. Čiurlionis National Art Museum, Druskininkai 30224
Mikalojus Konstantinas Čiurlionis National Art Museum, Kaunas 30262
Mikas and Kipras Petrauskas Museum of the Lithuanian Music, Kaunas 30260
Mikhailovskoye - State Memorial Historical-Literary and Natural-Landscape Museum-Reserve of A.S. Pushkin, Puškinskie Gory 37587
Mikkel Museum of the Estonian Art Museum, Tallinn 10636
Mikkeli Art Museum, Mikkeli 11115
Mikkeli Muuseum, Tallinn 10636
Mikkelin Taidemuseo, Mikkeli 11115
Miklós Káplár Memorial Exhibition, Hajdúböszörmény 23604
Miklós Müller Exhibition Ward, Orosháza 23739
Miklova Hiša, Ribnica 38505
Miko ir Kipro Petrauskų Lietuvių Muzikos Muziejos, Kaunas 30260
Miksa Róth Memorial House, Budapest 23474
Mikszáth Kálmán Emlékház, Horpács 23618
Mikulas Galanda's House, Turčianske Teplice 38382
Mikumi Museum, Mikumi 42517
Milam County Museum and Annex, Cameron 47203
Milan Dobeš Museum, Bratislava 38224
Milan Historical Museum, Milan 50680
Milano Model and Toy Museum, Elmhurst 48377
Milas Culture Center, Milas 42908
Milas Kültür Merkezi, Milas 42908
Milas Müzesi, Milas 42909
Milas Museum, Milas 42909
MilbenkäseMuseum Würchwitz, Würchwitz 22705
Milchpintli, Rubigen 41779
Milchwirtschaftliches Museum, Kiesen 41502
Milden Community Museum, Milden 06282
Mildenhall and District Museum, Mildenhall, Suffolk 45144
Mildred Lane Kemper Art Museum, Saint Louis 52586
Mildred M. Mahoney Silver Jubilee Dolls House Gallery, Fort Erie 05808
Mildura Arts Centre, Mildura 01300
Miles and District Historical Village Museum, Miles 01302
Miles B. Carpenter Museum, Waverly 53998
Miles Mineral Museum, Portales 52055
Milestones: Hampshire's Living History Museum, Basingstoke 43395
Milet Müzesi, Söke 42948
Miletitscher Heimatstube, Schönaich 21596
Milevské Muzeum, Milevsko 09655
Milford Haven Museum, Milford Haven 45145
Milford Historical Museum, Milford 50686
Milford Historical Society Museum, Milford 50684
Milford Museum, Milford 50685
Mili Weber-Haus, Sankt Moritz 41812
Mili-Weber-House, Sankt Moritz 41812
Milieu Educatie Centrum, Eindhoven 32190
Milieucentrum De Grote Rivieren, Heerewaarden 32334
Militär-Museum Kossa, Kossa 19693
Militär- und Jagdgeschichtliche Sammlung, Bad Wildungen 16862
Militärgeschichtliche Ortssammlung, Schnifis 02710
Militärhistorisches Museum der Bundeswehr, Dresden 17960
Militärmuseum Luzern, Kriens 41517
Militärmuseum Rosenau, Sonntagberg 02748
Militärmuseum St. Luzisteig, Maienfeld 41603
Militair Historisch Museum De Veteraan, Eefde 32172
Militaire Luchtvaart Museum, Soesterberg 32794
Militarial Musée Mémorial pour la Paix, Boissezon 12048
Military Academy Museum, Seoul 29903
Military Aviation Museum, Tangmere 45824
Military Communications and Electronics Museum, Kingston 06083
Military Dependants Village Museum, Hsinchu 42211
Military Heritage Museum, Lewes 44699
Military Heritage Tours, Longford 24869
Military Historians Museum, Westbrook 54085
Military Historical Aero Museum, Szolnok 23897
Military-Historical Museum of Artillery, Engineer and Signal Corps, Sankt-Peterburg 37730
Military Historical Museum of the Pacific Fleat, Vladivostok 37952
Military History Museum War of Liberation 1877-1878, Bjala 04988
Military Medical Museum, Sankt-Peterburg 37731
Military Museum, Aden 54642
Military Museum, Barcelona 38999
Military Museum, Białystok 34553
Military Museum, Bodmin 43550
Military Museum, Cairo 10500
Military Museum, El-Alamein 10524
Military Museum, Kingston 28331
Military Museum, Logatec 38472
Military Museum, Piešťany 38328
Military Museum, Sanaa 54649
Military Museum, Tehrān 24680
Military Museum, Ulaanbaatar 31620
Military Museum, Yerevan 00744
Military Museum and Cultural Centre Command, İstanbul 42803
Military Museum - City Armoury, Wrocław 35333
Military Museum Fort, Bloemfontein 38564
Military Museum Manege, Helsinki 10803
Military Museum of Belgrade, Beograd 38091
Military Museum of Devon and Dorset, Dorchester 44001
Military Museum of Finland, Helsinki 10818
Military Museum of Southern New England, Danbury 47953
Military Museum Svidnik, Svidník 38366
Military Museum Trenčín, Trenčín 38377
Military Technical Museum Lešany, Krhanice 09608
Military Vehicle Museum, Sunderland 45798
Military Veterans Museum, Oshkosh 51614
Militia History Museum, Sankt-Peterburg 37680
Miljømuseet Dokken, Haugesund 33539
Milkyuan Yaozu Self-Rule Nationality Museum, Ruyuan 08146
Mill Green Museum and Mill, Hatfield 44419
Mill House Cider Museum and Dorset Collection of Clocks, Owermoigne 45326
Mill Industry Museum, Budapest 23448
Mill Museum, Juuka 10882
Mill Museum, Sappee 11299
Mill Museum, Värtsilä 11433
Mill of Kintail Conservation Area, Almonte 05323
Mill of the Castle, Amsterdam 31859
Mill Race Historical Village, Northville 51428
Mill Trail Visitor Centre, Alva 43295
Millard Fillmore House, East Aurora 48236
Millenaris- House os Future exhibition, Palace of Wonders, Budapest 23450
Millenaris- Jövő Háza Kiállítás, Csodák Palotája, Budapest 23450
Millenium Underground Museum, Budapest 23451
Millennium Arts Center International, Washington 53893
Millennium Galleries, Sheffield 45653
Millenniumi Földalatti Vasúti Múzeum, Budapest 23451
Miller Art Center, Springfield 53240
Miller Art Museum, Sturgeon Bay 53349
Miller Bakehouse Museum, Palmyra 01402
Miller-Cory House Museum, Westfield 54093
Miller House Museum, Hagerstown 49121
Miller Museum of Geology, Kingston 06084
Millertown Museum, Millertown 06285
Millesgården, Lidingö 40673
Millet and District Museum, Millet 06286
Millicent National Trust Museum, Millicent 01305
Millicent Rogers Museum of Northern New Mexico, Taos 53442
Milliken Gallery, Spartanburg 53193
Millmount Museum and Tower, Drogheda 24776
Millpetersch Haus Güdesweiler, Oberthal 20845
Mills College Art Museum, Oakland 51467
Mills County Museum, Glenwood 48941
Mills Observatory Museum, Dundee 44048
Millstreet Local Museum, Millstreet 24876
Milltown-Head Bay d'Espoir Museum, Milltown 06287
Millwall FC Museum, London 44918
Milne Bay Military Museum, Toowoomba 01593
Milos Crnjanski Emlékszoba, Csongrád 23511
Milos Crnjanski Memorial Hall, Csongrád 23511
Milton Blacksmith Shop Museum, Milton 06290
Milton Historical Museum, Milton 50702
Milton House Museum, Milton 50703
Milton J. Rubenstein Museum of Science and Technology, Syracuse 53400
Milton Keynes Gallery, Milton Keynes 45153
Milton Keynes Museum, Milton Keynes 45154
Milton's Cottage, Chalfont Saint Giles 43787
Milwaukee Art Museum, Milwaukee 50711
Milwaukee County Historical Society Museum, Milwaukee 50712
Milwaukee Public Museum, Milwaukee 50713
Milwaukee Very Special Art Center, Milwaukee 50714
Minas Avetisyan Museum, Yerevan 00745
Mind and Matter Gallery, South Surrey 06996
Minden Hills Museum, Minden 06295
Mindener Museum für Geschichte, Landes- und Volkskunde, Minden 20336
Mindszenty Emlékhely, Esztergom 23554
Mindszenty Memorial Exhibition of the Christian Museum, Esztergom 23554
Mine d'Argent Saint-Barthélemy, Sainte-Marie-aux-Mines 15566
La Mine-Image, La Motte-d'Aveillans 13395
Mine Témoin d'Alès, Alès 11548
Minera Lead Mines, Minera 45157
Mineral and Gem Geological Museum, Parrsboro 06570
Mineral and Lapidary Museum of Henderson County, Hendersonville 49253
Mineral County Museum, Superior 53376
Mineral Exhibition, Kaposvár 23630
Mineral Museum, Butte 47152
Mineral- och Gruvmuseet, Holsbybrunn 40602
Mineral Resources Divisional Museum, Dodoma 42514
Mineralien-Fossilien-Bergbau, Langenlois 02259
Mineralien-Kabinett des Mineralogisch-Petrografischen Museums der TU Braunschweig, Braunschweig 17526
Mineralien Museum, Seedorf 41855
Mineralien-Museum Andreas Gabrys, Lam 19781
Mineralien- und Heimatmuseum Pleystein, Pleystein 21062
Mineralienmuseum, Bad Säckingen 16816
Mineralienmuseum, Einsiedeln 41343
Mineralienmuseum, Freisen 18448
Mineralienmuseum, Neubulach 20595
Mineralienmuseum Essen, Essen 18262
Mineralienmuseum Pforzheim-Dillweißenstein, Pforzheim 21028
Mineraliensammlung Lauingen, Lauingen 19858
Mineraliensammlung Steger, Wiener Neustadt 03097
Mineralienschau, Vohenstrauß 22254
Mineralienzentrum Steinstadel, Aggsbach Dorf 01698
Mineraligical Museum Wiener Neustadt, Wiener Neustadt 03097
Mineralogical-geological Rock collection, Schwarzenfeld 21676
Mineralogical Museum, Merksem 03869
Mineralogical Museum of École des Mines d'Alès, Alès 11551
Mineralogical Museum of Harvard University, Cambridge 47182
Mineralogical Museum of Kamarzia, Attica 22955
Mineralogical Museum of Lavrion, Lavrion 23075
Mineralogical Museum of the Institute of Geological Sciences, Wrocław 35334
Mineralogical Museum of the Moscow Geological Academy, Moskva 37251
Mineralogical Museum, University of Hamburg, Hamburg 18944
Mineralogičeskij Muzej, Apatity 36763
Mineralogičeskij Muzej im. A.E. Fersmana Rossijskoj Akademii Nauk, Moskva 37250
Mineralogičeskij Muzej im. A.V. Sidorova, Irkutsk 36934
Mineralogičeskij Muzej Moskovskoj Geologorazvedyvatelnoj Akademii, Moskva 37251
Mineralogické muzeum - PřF-UK, Praha 09762
Mineralogisch-Geologisch Museum, Delft 32061
Mineralogisch-Geologische Gesteins-Sammlung, Schwarzenfeld 21676
Mineralogisch Museum, Grou 32289
Mineralogisch Museum, Merksem 03869
Mineralogisch-petrographische Sammlung der Universität Leipzig, Leipzig 19919
Mineralogische Sammlung, Oberthal 20846

M

Mineralogische Schau- und Lehrsammlung, Tübingen 22120
Mineralogisches Museum der Philipps-Universität Marburg, Marburg 20192
Mineralogisches Museum der Universität, Würzburg 22715
Mineralogisches Museum der Universität Bonn, Bonn 17463
Mineralogisches Museum der Universität Hamburg, Hamburg 18944
Mineralogisches Museum (Kristalle und Gesteine), Münster 20514
Mineralogisches Museum Oberkirchen, Freisen 18449
Mineralogy Museum, Bangkok 42545
Mineralogy Museum Baia Mare, Baia Mare 36127
Mineralogy Museum of the Geological Institute of the Kola Scientific Center of the Russian Academy of Sciences, Apatity 36763
Mineralovodskij Kraevedčeskij Muzej, Mineralnye Vody 37132
Minerals Museum, Siófok 23802
Mineralų Muziejus, Vilnius 30425
Miner's House, Vyšna Boca 38387
Mines d'Asphalte de la Presta, Travers 41947
Mines de Sel de Bex, Bex 41226
Mingcheng Yuan Shi Museum, Nanjing 08042
Mingei International Museum, Escondido 48420
Mingei International Museum, San Diego 52739
Mingenback Art Center Gallery, Lindsborg 50122
Mingenew Museum, Mingenew 01306
Mingshan Chashi Museum, Mingshan 08018
Mingxi Museum, Mingxi 08019
Minhauzena Muzejs, Dunte 30006
Minhe Hui Tuzu Self-Rule Museum, Minhe 08020
Mini-Château, Amboise 11577
Mini-Musée, Habsheim 13175
Mini-Musée Architectural, Charnat 12415
Mini Musée des Miniatures, Sospel 15786
Miniatur Wunderland Hamburg, Hamburg 18945
Miniature Books Museum, Bakı 03161
Miniatures Museum of Taiwan, Taipei 42404
Miniaturmuseum Arikalex, Berlin 17188
Miniaturpark Klein-Erzgebirge, Oederan 20860
Miniatuur Walcheren, Middelburg 32552
Miniatuurstad, Antwerpen 03293
Minimotor Museum, Rīga 30096
Minimotor Muzejs, Rīga 30096
Minimundus, Klagenfurt 02188
Minimuseum, Groningen 32283
Mining and Iron Works Museum, Złoty Stok 35375
Mining and Town Museum Weilburg, Weilburg 22384
Mining Exhibition, Gyöngyös 23572
Mining Exhibition, Salgótarján 23790
Mining Museum, Bogatynia 34566
Mining Museum, Gelnica 38256
Mining Museum, Nowa Ruda 34947
Mining Museum, Ostrava 09697
Mining Museum, Outokumpu 11180
Mining Museum, Petroşani 36515
The Mining Museum, Platteville 51980
Mining Museum, Příbram 09806
Mining Museum, Rožňava 38342
Mining Museum, Sala 40826
Mining Museum, Sankt-Peterburg 37639
Mining Museum, São Pedro da Cova 35924
Mining Museum Altenberg, Altenberg 16493
Mining Museum of Mecsek - Exhibition on Mining Uranium, Kővágószőlős 23687
Mining Museum of Mecsek - Exhibition on the History of Mining and Library, Pécs 23764
Mining Museum of Slovenia, Velenje 38533
Mining Museum of Slovenia, Železniki 38541
Mining Museum Roşia Montană, Roşia Montană 36572
Mining Museum - Underground Exhibition, Pécs 23765
Mining Skansen Museum Queen Louise, Zabrze 35351
Miniota Municipal Museum, Miniota 06297
Minischeepvaartmuseum, Ouwerkerk 32660
Minisink Valley Historical Society Museum, Port Jervis 52037
Minjasafn Austurlands, Egilsstadir 23993
Minjasafn Egils Ólafssonar, Patreksfjörður 24011
Minjasafnid Burstarfelli, Vopnafjörður 24044
Minjasafnida á Akureyri, Akureyri 23983
Minlaton National Trust Museum, Minlaton 01307
Minle Museum, Minle 08021
Minneapolis Institute of Arts, Minneapolis 50739
Minnedosa and District Museum, Minnedosa 06298
Minnesota Air National Guard Museum, Saint Paul 52621
Minnesota Association of Museums, Minneapolis 55157
Minnesota Children's Museum, Saint Paul 52622
Minnesota Historical Society Museum, Saint Paul 52623
Minnesota Museum of American Art, Saint Paul 52624
Minnesota Pioneer Park Museum, Annandale 46313
Minnesota State Capitol Historic Site, Saint Paul 52625
Minnesota Transportation Museum, Saint Paul 52626
Minnilusa Pioneer Museum, Rapid City 52252
Minokamo City Museum, Minokamo 28911
Minokamo-shiritsu Hakubutsukan, Minokamo 28911
Minority Arts Resource Council Studio Art Museum, Philadelphia 51852
Minqin Museum, Minqin 08022
Minqing Museum, Minqing 08023
Minsk Mazowiecki Regional Museum, Mińsk Mazowiecki 34924
Minster Abbey Gate House Museum, Minster-in-Sheppey 45158
Minster Abbey Gatehouse Museum, Sheerness 45647
Mint Museum, Osaka 29134
Mint Museum of Art, Charlotte 47392
Mint Museum of Craft and Design, Charlotte 47393
Mintai Tieshi Museum, Quanzhou 08129
Minter Field Air Museum, Shafter 53047
Minto Museum, Minto 06299
Minusinskaja Chudožestvennaja Kartinnaja Galereja, Minusinsk 37135
Minusinskij Regionalnyj Kraevedčeskij Muzej im. N.M. Martjanova, Minusinsk 37136
Minute Man, Concord 47768
Minxi Revolution History Museum, Minxi 08024
Minzoku Shiryokan, Miyama 28924
Mir-Emad Calligraphy Museum, Tehrān 24681
Miracle of America Museum, Polson 52015
Miramar Zeemuseum, Vledder 32893
Miramichi Natural History Museum, Miramichi 06301
Miramón Kutxa Espacio de la Ciencia, Donostia-San Sebastián 39250
Miramont Castle Museum, Manitou Springs 50441
Mirasaka Peace Museum of Art, Mirasaka 28912
Mirehouse, Keswick 44565
Mireuksa Temple Museum, Iksan 29830
Miriam and Ira D. Wallach Art Gallery, New York 51214
Mirouge Museum, Mizhi 08025
Mirror and District Museum, Mirror 06304
Miryang Municipal Museum, Miryang 29852
Mirza Fatali Akhundov House Museum, Tbilisi 16371
Mishima Taisha Treasure House, Mishima 28913
Mishkan Le'Omanut, Ein Harod 24945
Misijné Múzeum Nitra, Nitra 38313
Misis Mozaik Müzesi, Yakapınar 42964
Misjonsmuseet, Stavanger 33867
Misjonsmuseet på Fjellhaug, Oslo 33731
Mis'ka Chudožna Galereja im. S.I. Vasyl'kivs'kogo, Charkiv 43020
Miskolc Gallery of Fine Arts - Rákóczi House, Miskolc 23712
Miskolc University Historical Collection, Miskolc 23711
Miskolci Egyetemtörténeti Gyűjtemények, Miskolc 23711
Miskolci Galéria - Városi Művészeti Múzeum - Rákóczi-ház, Miskolc 23712
Misora Hibari Kinenkan, Kyoto 28850
Misora Hibari Memorial Hall, Kyoto 28850
Missiemuseum Steijl, Steijl 32804
Mission and Folklore Museum, Polanica Zdrój 35020
Mission Bell Museum, Coweta 47862
Mission Cultural Center for Latino Arts, San Francisco 52784
Mission Gallery, Swansea 45814
The Mission House, Stockbridge 53300
Mission Houses Museum, Honolulu 49349
Mission Inn Museum, Riverside 52372
Mission Mill Museum, Salem 52669
Mission Museum, Helsinki 10797
Mission Museum, Krosno 34829
Mission Museum, Mission 06306
Mission San Carlos Borromeo del Rio Carmelo, Carmel 47251
Mission San Juan Capistrano Museum, San Juan Capistrano 52813
Mission San Luis Obispo de Tolosa, San Luis Obispo 52814
Mission San Luis Rey Museum, Oceanside 51491
Mission San Miguel, San Miguel 52823
Missionary Museum, King William's Town 38676
Missionary Museum of the N.G. Kerk, Stellenbosch 38736
Missions-Ethnographisches Museum St. Gabriel, Mödling 02393
Missionsmuséet Nårunga, Vårgårda 41050
Missionsmuseum, Sankt Ottilien 21508
Missionsmuseum, Schwarzach am Main 21670
Missionsmuseum Bug, Bamberg 16912
Missionsmuseum Dachsberg, Prambachkirchen 02525
Missionsmuseum der Liebenzeller Mission, Bad Liebenzell 16780
Missionsmuseum der Mariannhiller Missionare, Würzburg 22716
Missionsmuseum der Pallottiner Limburg, Limburg an der Lahn 19997
Missionsmuseum des päpstlichen Missionswerks der Kinder in Deutschland, Aachen 16403
Missisquoi Museum, Stanbridge East 07004
Missisquoi Valley Historical Society, North Troy 51408
Mississauga Heritage Foundation, Mississauga 54741
Mississippi Baptist Historical Commission, Clinton 47656
Mississippi County Historical Society, Charleston 47355
Mississippi Crafts Center, Ridgeland 52349
Mississippi Governor's Mansion, Jackson 49559
Mississippi Museum of Art, Jackson 49560
Mississippi Museum of Natural Science, Jackson 49561
Mississippi Museums Association, Natchez 55158
Mississippi Petrified Forest, Flora 48561
Mississippi River Museum, Dubuque 48189
Mississippi River Museum at Mud Island River Park, Memphis 50580
Mississippi University for Women Museum, Columbus 47747
Mississippi Valley Textile Museum, Almonte 05324
Mississipppi Agriculture and Forestry/National Agricultural Aviation Museum, Jackson 49562
Missoula Art Museum, Missoula 50754
Missouri Historical Society, Saint Louis 52587
Missouri Museums Association, Kansas City 55159
Missouri State Museum, Jefferson City 49614
Missouri Town 1855, Lees Summit 50017
Mistasinihk Place Interpretive Centre, LaRonge 06140
Místní Muzeum, Mladá Vožice 09659
Místní Muzeum, Netolice 09674
Mistni Vlastivědné Muzeum, Železnice 09936
Místní Vlastividni Muzeum, Kopidlno 09594
MIT Museum, Cambridge 47183
Mitchell County Historical Museum, Osage 51608
Mitchell Gallery of Flight, Milwaukee 50715
Mitchell House, Nantucket 50945
Mitchell Library Exhibition, Glasgow 44289
Mitchell Museum at Cedarhurst, Mount Vernon 50892
Mitchell Museum of the American Indian, Evanston 48454
Mithrasheiligtum, Mundelsheim 20526
Mithrász-szentély, Fertőrákos 23560
Mito Geijutsu-kan, Gendai Bijutsu Gallery, Mito 28921
Mitropoličji Palaty - Muzej Drevnerusskogo Iskusstva, Jaroslavl 36971
Mitspielzeugmuseum, Leipzig 19920
Mitsubishi Auto Gallery, Okazaki 29104
Mitsubishi Minatomirai Indasutoriaru Hakubutsukan, Yokohama 29601
Mitsubishi Minatomirai Industrial Museum, Yokohama 29601
Mitsui Kinen Bijutsukan, Tokyo 29437
Mitsui Memorial Museum, Tokyo 29437
Mitsuke Historical School, Iwate, Iwata, Shizuoka 28610
Mitte Museum, Berlin 17189
Mittelalterliche Lochgefängnisse unter dem Alten Rathaus, Nürnberg 20779
Mittelalterliches Foltermuseum, Rüdesheim am Rhein 21435
Mittelalterliches Kriminalmuseum, Rothenburg ob der Tauber 21408
Mittelbau-Dora Concentration Camp Memorial, Nordhausen 20738
Mittelmoselmuseum im Barockhaus Böcking, Traben-Trarbach 22075
Mittelrhein-Museum Koblenz, Koblenz 19587
Mittelschwäbisches Heimatmuseum Krumbach, Krumbach 19729
Mitzpe Revivim Museum, Kibbutz Revivim 25054
Miyagi-ken Bijutsukan, Sendai 29243
Miyagi Michio Kinenkan, Tokyo 29438
Miyagi Museum of Art, Sendai 29243
MIYAI Furoshiki and Fukusa Gallery, Kyoto 28851
Miyake Bijutsukan, Kagoshima 28631
Miyake Museum of Art, Kagoshima 28631
Miyakonojo City Museum of Art, Miyakonojo 28923
Miyakonojo-shi Bijutsukan, Miyakonojo 28923
Miyama-cho Kayabuki Bijutsukan oyo Kyodo Hakubutsukan, Miyama 28925
Miyama Kayabuki Art Museum and Local Museum, Miyama 28925
Miyandoab Museum, Miyandoab 24635
Miyazaki-ken Sogo Hakubutsukan, Miyazaki 28927
Miyazaki-kenritsu Bijutsukan, Miyazaki 28928
Miyazaki Prefectural Art Museum, Miyazaki 28928
Miyazaki Prefectural Museum of Nature and History, Miyazaki 28927
Miyazawa Kenji Kinenkan, Hanamaki 28491
Mizel Museum, Denver 48081
Mizgiris' Amber Gallery-Museum, Nida 30319
Mizgiris' Amber Gallery-Museum, Vilnius 30409
Mizu-to-Midori-no Fureaikan, Okutama 29108
Mizuhomachi Kyodo Hakubutsukan, Mizuho 28931
Mizunami Ceramic Museum, Mizunami 28933
Mizunami Fossil Museum, Mizunami 28932
Mizunami-shi Kaseki Hakubutsukan, Mizunami 28932
Mizunami Toji Shiryokan, Mizunami 28933
Mjellby Konstmuseum, Halmstad 40579
Mjölby Hembygdsgård, Mjölby 40740
MLA West Midland, Birmingham 55074
Mo Tu Uv - Museum of Mechanical Toys, Ulaanbaatar 31621
MOA Bijutsukan, Atami 28382
MOA Museum of Art, Atami 28382
Moanalua Gardens Foundation, Honolulu 49350
Moat Park Heritage Centre, Biggar 43484
Moaven Al- Molk Museum, Kermanshah 24621
Mobach Keramiek Museum, Utrecht 32847
Mobile Medical Museum, Mobile 50768
Mobile Museum of Art, Mobile 50769
Mobiles Kindermuseum im Freizeitheim Vahrenwald, Hannover 18998
Mobilia-Auto- ja Tiemuseo, Kangasala 10910
Mobius Gallery, Boston 46938
Mockbeggar Plantation, Bonavista 05439
Modalen Skulemuseum, Modalen 33669
Model House Gallery, Llantrisant 44774
Modelbouwmuseum, Wassenaar 32927
Modellbahnmuseum-Muggendorf, Wiesenttal 22562
Modellbaumuseum, Büdingen 17629
Modellpark Mönichkirchen, Mönichkirchen 02400
Modemuseum Feigel, Ehingen 18067
Modemuseum im Schloss Ludwigsburg, Ludwigsburg 20046
Modemuseum Provincie Antwerpen, Antwerpen 03294
Modemuseum Schloss Meyenburg, Meyenburg 20316
Modern Art Center, Toshkent 54468
Modern Art Museum of Bucaramanga, Bucaramanga 08599
Modern Art Museum of Fort Worth, Fort Worth 48717
Modern Art Oxford, Oxford 45332
Modern Fuel Artist - Run Centre, Kingston 06085
Modern Gallery, Valjevo 38159
Modern History Museum, Dubrovnik 08871
Modern Magyar Képtár I, Pécs 23767
Modern Magyar Képtár II, Pécs 23768
Modern Transportation Museum, Osaka 29131
Moderna Galerija, Ljubljana 38454
Moderna Galerija, Podgorica 31655
Moderna Galerija, Valjevo 38159
Moderna Galerija, Zagreb 09003
Moderna Museet, Stockholm 40915
Moderne Galerie im Weinstadtmuseum Krems, Krems 02229
Modest P. Musorgski Memorial Estate Museum, Naumovo 37447
Modified Art Gallery, Phoenix 51905
Modoc County Historical Museum, Alturas 46244
Moe Historical Museum, Moe 01310
Mödlinger Stadtverkehrsmuseum, Mödling 02394
Möhkö Ironworks Museum, Möhkö 11119
Möhkön Ruukki, Möhkö 11119
Möllner Museum Historisches Rathaus, Mölln 20352
Mölndals museum, Mölndal 40743
Mönchehaus Museum Goslar, Goslar 18712
Mönchguter Museen im Ostseebad Göhren, Göhren, Ostseebad 18670
Möniste Muuseum, Möniste 10595
Mörfelden Museum, Mörfelden-Walldorf 20361
Mörike Museum im alten Schulhaus, Neuenstadt am Kocher 20611
Mörike Zimmer Wermutshausen, Niederstetten 20713
Mörikehaus Ochsenwang, Bissingen an der Teck 17366
Mörsbacher Museum, Mörsbach 20364
Moervaartstede, Wachtebeke 04124
Moesgård Museum, Højbjerg 10084
Möwe Bay Museum, Möwe Bay 31747
Moffat Museum, Moffat 45161
Moffatt-Ladd House, Portsmouth 52097
Mohave Museum of History and Arts, Kingman 49783
Mohawk Trail School Museum, Hamilton 05961
Moheda Skolmuseum, Moheda 40745
Møhlman Art Gallery, Upplands Väsby 41001
Mohnmuseum, Ottenschlag 02477
Mohonk Preserve, Gardiner 48860
Mohri Hakubutsukan, Hofu 28546
Moira Furnace, Moira 45162
Mõisaküla Muuseum, Mõisaküla 10594
Moisburger Mühlenmuseum, Moisburg 20367
Moji-ko Orugorukan, Kitakyushu 28714
Moji-kuritsu Idemitsu Bijutsukan, Kitakyushu 28715
Mok Am Museum of Art, Koyang 29848
Moka Buddhist Museum, Gangcheon 29804
Mokichi Saito Museum, Kaminoyama 28650

Mokolo City Museum, Mokolo 05300
Mola e Serradora de Cal Pal, Ordino 00099
Mola- het Provinciaal Molenmuseum, Wachtebeke 04125
Mold Museum, Mold 45163
Moldovan State Art Museum, Chişinău 31600
Molen-, Bakkerij- en Stoommuseum, Essen 03588
Molen De Vriendschap, Weesp 32935
Molen van Frans Expositie, Mander (Tubbergen) 32537
Molen van Sloten, Amsterdam 31859
Molenmuseum, Koog aan de Zaan 32455
Molenmuseum, Lommel 03828
Molenmuseum, Sint-Amands 04013
Molenmuseum de Assumburg, Nieuw Vossemeer 32574
Molenmuseum De Valk, Leiden 32483
Molenmuseum De Wachter, Zuidlaren 33013
Molerhiisli, Hausach 19047
Molermuseet, Nykøbing Mors 10203
Molėtai Area Museum, Molėtai 30314
Molėtų Krašto Muziejus, Molėtai 30314
Moli d'en Gaspar, Lluchmajor 39489
Molkerei-Museum, Bernbeuren 17293
Mollfulleda Mineralogy Museum, Arenys de Mar 38887
Molly Brown House Museum, Denver 48082
Molnár C. Pál Műterem-múzeum, Budapest 23452
Molong Historical Museum, Molong 01311
Mølsteds Museum, Dragør 09990
Moltke-Gedächtnisstätte, Parchim 20982
Moluks Historisch Museum, Utrecht 32848
MoMA, Museum of Modern Art, New York 51215
MOMA Wales, Machynlleth 45069
Mompesson House, Salisbury 45604
Mon State Museum, Mawlamyine 31722
Monacensia, München 20459
Monaco Modern Art Museum, Monaco 31606
Monaghan County Museum, Monaghan 24878
Monarch Historical Museum, Williamtown 01654
Monash Gallery of Art, Wheelers Hill 01647
Monash University Museum of Art, Clayton 00980
Monastère de Mauriac, Mauriac 14038
Monasterie Royal de Brou, Bourg-en-Bresse 12107
Monasterio Cisterciense de Santa María, Sobrado de los Monjes 40086
Monasterio de la Cartuja, Sevilla 40063
Monasterio de la Encarnación, Madrid 39523
Monasterio de la Encarnación, Osuna 39751
Monasterio de las Descalzas Reales, Madrid 39524
Monasterio de San Juan de los Reyes, Toledo 40130
Monasterio de San Lorenzo de El Escorial, San Lorenzo de El Escorial 39961
Monasterio de San Pedro de Cardeña, Castrillo del Val 39152
Monasterio de Santa Catalina, Arequipa 34108
Monasterio de Santa Clara de Tordesillas, Tordesillas 40146
Monasterio de Santa María de la Rábida, Palos de la Frontera 39804
Monasterio de Santa María San Salvador, Cañas 39115
Monasterio San Juan de Duero, Soria 40091
Monasterio Santa María la Real de las Huelgas, Burgos 39066
Monasterium De Wijngaard, Brugge 03420
Monastery of Aghiou Tessarakonta, Spárti 23196
Monastery of Evangelistria Daponte, Skopelos 23190
Monastery of Leimon Collection, Kalloni 23030
Monastery of Petra, Kataphygi 23045
Monastery of Saint Catherine, Gebel Katherîna 10525
Monastery of Saint Francis' Art Collections, Zadar 08985
Monastery of Santa Clara-a-Velha, Coimbra 35527
Monastery of the Archangel Michael at Panormitis, Panormitis 23132
Monastery of the Taxiarchs, Megalo Livadi 23086
Monastery of the Virgin Xenia, Almyros 22877
Monastery of the Zoodochos Pighi, Poros 23157
Monastic Museum of Denmark, Ry 10254
Moncton Museum, Moncton 06319
Moncur Gallery, Boissevain 05435
Mondak Heritage Center, Sidney 53096
Le Monde de Belibaste, Villerouge-Termenès 16227
Mondriaanhuis - Museum voor Constructieve en Concrete Kunst, Amersfoort 31813
Monemvasia Archaeological Collection, Monemvasia 23100
Monesterio de Santes Creus, Santes Creus 40013
Monestir de Cura, Randa 39890
The Money Museum, Kansas City 49694
Money Museum, Kingston 28332
Money Museum, Tehrān 24682
Money Museum of the UFJ Bank, Nagoya 29002
MoneyMuseum, Zürich 42088
Mongolian and Tibetan Cultural Center, Taipei 42405
Mongolian Art Museum, Ulaanbaatar 31622
Mongolian National Modern Art Gallery, Ulaanbaatar 31623
Mongolian Theatre Museum, Ulaanbaatar 31624
The Monhegan Museum, Monhegan 50780
Mõniste Muuseum, Kuutsi 10588
Monks' Dormitory Durham Cathedral, Durham 44072
monks house museum Goslar, Goslar 18712
Monkwearmouth Station Museum, Sunderland 45799
Monmouth County Historical Museum, Freehold 48779
Monmouth Museum, Lincroft 50115
Monocacy National Battlefield, Frederick 48757
Monostori Erőd, Komárom 23680
Monroe County Heritage Museum, Monroeville 50795
Monroe County Historical Museum, Monroe 50791
Monroe County Historical Society, Bloomington 46865
Monroe County Local History Room and Library, Sparta 53192
Monroe Historical Society Museum, Monroe 50786
Mons Breidvik Galleries, Norheimsund 33704
Møns Museum, Stege 10284
Monsignor Hawes Priest House Museum, Mullewa 01332
Monsignor Yatco Ecclesiastical Museum, Batangas 34345
Mont De Lancey Historical Museum, Wandin 01626
Mont Orgueil Castle, Saint Martin 45583
Montafoner Bergbaumuseum Silbertal, Silbertal 02743
Montafoner Heimatmuseum Schruns, Schruns 02716
Montafoner Tourismusmuseum Gaschurn, Gaschurn 01927
Montagu Museum, Montagu 38690
Montague Museum, Montague 50796
Montalvo Arts Center, Saratoga 52918
Montana Auto Museum, Deer Lodge 48029
Montana Historical Society, Helena 49242
Montana Museum of Art and Culture, Missoula 50755
Montanmuseum, Gußwerk 02022
Montanmuseum, Hieflau 02077
Montauk, Clermont 47617
Montauk Point Lighthouse Museum, Montauk 50797
Montclair Art Museum, Montclair 50798
Montclair Historical Society Museum, Montclair 50799
Montclair State University Art Galleries, Upper Montclair 53665
Monte L. Bean Life Science Museum, Provo 52187
Montecatini, Concepción 07433
Monterey History and Art Association, Monterey 50801
Monterey Maritime and History Museum, Monterey 50802
Monterey Museum of Art, Monterey 50803
Monterey Museum of Art at La Mirada, Monterey 50804
Monteverdi Car Collection, Binningen 41234
Montfort-Museum, Tettnang 22040
Montgomery County Historical Society Center, Dayton 47986
Montgomery County Historical Society Museum, Rockville 52431
Montgomery House Museum, Frankville 05827
Montgomery Museum and Lewis Miller Regional Art Center, Christiansburg 47547
Montgomery Museum of Fine Arts, Montgomery 50815
Monticello, Home of Thomas Jefferson, Charlottesville 47398
Monticello Railway Museum, Monticello 50819
Montpelier Cultur Arts Center, Laurel 49983
Montpellier Mansion, Laurel 49984
Montréal Insectarium (Montréal's Nature Museums), Montréal 06341
Montréal Museum of Archaeology and History, Montréal 06378
Montréal Museum of Fine Arts, Montréal 06357
Montreal Science Centre, Montréal 06350
Montrose Air Station Museum, Montrose 45172
Montrose County Historical Museum, Montrose 50829
Montrose Museum and Art Gallery, Montrose 45173
Montshire Museum of Science, Norwich 51442
Montville Township Historical Museum, Montville 50832
Monumen Nasional, Jakarta 24476
Monument Lefebvre, Memramcook 06266
Monument of Životice's Tragedy, Havířov 09509
Monumental Complex of the Prizren Albanian League, Prizren 38134
Monumental Rooms of the Marciana Library, Venezia 28209
Monumenti Musei e Gallerie Pontificie, Città del Vaticano 54501
Monumento a Cristo Rey, Belalcázar 08541
Moody County Museum, Flandreau 48554
The Moody Mansion Museum, Galveston 48849
Sree Moolam Shastyabdapurti Memorial Institute, Thiruvananthapuram 24405
Moonah Arts Centre, Moonah 01313
Moonta Mines Museum, Moonta 01314
Moonta National Trust Museum, Moonta 01315
Moor und Bauer-Museum, Wardenburg 22333
Moor- und Fehnmuseum, Barßel 16930
Moorabbin Air Museum, Cheltenham 00966
Moore County Historical Museum, Dumas 48202
Moore Home, Lerna 50026
Moore Museum, Mooretown 06383
Moores Creek National Battlefield, Currie 47913
Moormuseum Moordorf, Südbrookmerland 21974
Mooroopna Hospital and District Museum, Mooroopna 01317
Moose Factory Centennial Museum, Kirkland Lake 06097
Moose Jaw Museum and Art Gallery, Moose Jaw 06386
Moose River Gold Mines Museum, Middle Musquodoboit 06274
Moosehorn Heritage Museum, Moosehorn 06388
Mór Jókai Memorial Museum, Balatonfüred 23352
Mór Jókai Memorial Room and Region House, Tardona 23915
Mór Wosinsky County Museum, Szekszárd 23869
Móra Ferenc Emlékház, Kiskunfélegyháza 23667
Móra Ferenc Múzeum, Szeged 23847
Moran Open-air Museum, Namyangju 29857
Morat-Institut für Kunst und Kunstwissenschaft, Freiburg im Breisgau 18432
Moravian Gallery Brno, Brno 09433
Moravian Gallery Brno, Brno 09434
Moravian Gallery Brno, Brno 09432
Moravian Historical Society Museum, Nazareth 50993
Moravian Museum, Brno 09435
Moravian Museum, Pudsey 45450
Moravian Museum of Bethlehem, Bethlehem 46781
Moravian Museum - The House of the Lords of Kunštát, Brno 09424
Moravská Galerie Brno, Brno 09432
Moravská Galerie Brno - Pražákův Palác, Brno 09433
Moravská Galerie Brno - Uměleckoprůmyslové Muzeum, Brno 09434
Moravské Zemské Muzeum, Brno 09435
Morawa Old Police Station Museum, Morawa 01318
Moray Firth Wildlife Centre, Spey Bay 45726
Mordecai Historic Park, Raleigh 52231
Morden and District Museum, Morden 06391
Mordialloc Historical Society Museum, Mentone 01294
Mordovian Museum of Fine Arts, Saransk 37739
Mordovskij Muzej Izobrazitelnych Isskustv im. S.D. Ėrzi, Saransk 37739
Mordovskij Respublikanskij Kraevedčeskij Muzej, Saransk 37740
Moree Plains Gallery, Moree 01319
Moreton's Harbour Museum, Moreton's Harbour 06392
The Morgan Library, New York 51216
Morgan Row Museum, Harrodsburg 49188
Mori Art Museum, Tokyo 29439
Mori Bijutsukan, Tokyo 29439
Mori Museum, Hofu 28546
Mori-no-Naka no Chiisana Ongakukan, Nikko 29067
Mori-Ôgai-Gedenkstätte, Berlin 17190
Moriarty Historical Museum, Moriarty 50847
Móricz Zsigmond Emlékház, Prügy 23785
Móricz Zsigmond Emlékház, Tiszacsécse 23929
Morija Museum, Morija 30180
The Morikami Museum and Japanese Gardens, Delray Beach 48046
Morioka City Local Hall, Morioka 28938
Morishita Art Museum, Hinase 28515
Morishita Bijutsukan, Hinase 28515
Morita Historical Folk Art Museum, Morita 28939
Morita-mura Rekishi Mingeikan, Morita 28939
Moritzburg Castle & Little Pheasant Castle, Moritzburg 20386
Morlan Gallery, Lexington 50054
Morley Gallery, London 44919
Mormon Station Museum, Genoa 48886
Mormon Visitors Center, Independence 49478
Morningside Nature Center, Gainesville 48815
Mornington Peninsula Regional Gallery, Mornington 01322
Morohashi Kindai Bijutsukan, Kitashiobara 28718
Morohashi Museum of Modern Art, Kitashiobara 28718
Moromi Folklore Museum, Okinawa 29106
Moromi Mingeikan, Okinawa 29106
Morpeth Chantry Bagpipe Museum, Morpeth 45177
Morphett's Enginehouse Museum, Burra 00926
Morrin Centre, Québec 06684
Morrin Museum, Morrinsville 33147
Morris and District Centennial Museum, Morris 06394
Morris and Helen Belkin Art Gallery, Vancouver 07216
Morris Belknap Gallery, Dario Covi Gallery and SAL Gallery, Louisville 50293
Morris-Butler House Museum, Indianapolis 49504
Morris Graves Museum of Art, Eureka 48440
Morris-Jumel Mansion, New York 51217
The Morris Museum, Morristown 50854
Morris Museum of Art, Augusta 46481
Morrison Museum of the Country School, Islay 06025
Morristown National Historical Park, Morristown 50855
Morro Bay State Park Museum of Natural History, Morro Bay 50861
Morrow Mountain State Park Museum, Albemarle 46168
Morse Museum, Morse 06396
Morskoj Muzej, Petrozavodsk 37560
Morskoj Muzej-Akvarium, Vladivostok 37942
Morslands Historiske Museum, Nykøbing Mors 10204
Morton Grove Historical Museum, Morton Grove 50862
Morton Homestead, Prospect Park 52158
Morton House Museum, Benton Harbor 46743
Morton J. May Foundation Gallery, Saint Louis 52588
Morton Museum of Cooke County, Gainesville 48823
Morven Historical Museum, Morven 01324
Morven Park Museum, Leesburg 50019
Morwellham Quay Historic Port & Copper Mine, Morwellham 45178
Mosaic Gallery, Amman 29658
Mosaïque Gallo-Romaine, Amou 11592
Moschakeion, Moschato 23101
Moscow Archaeology Museum, Moskva 37262
Moscow House of Photography, Moskva 37254
Moscow Kremlin State Museum-Preserve of History and Culture, Moskva 37194
Moscow Museum-Estate Ostankino, Moskva 37259
Moscow Museum of Modern Art, Moskva 37258
Moscow State Esenin Museum, Moskva 37255
Moscow State Museum of Naive Art, Moskva 37196
Mose Toidze House Museum, Tbilisi 16372
Mosel-Weinmuseum, Bernkastel-Kues 17298
Moser Glass Museum, Karlovy Vary 09577
Moses Lake Museum and Art Center, Moses Lake 50867
Moses Myers House, Norfolk 51366
Moskovskaja Gosudarstvennaja Kartinnaja Galereja A. Šilova, Moskva 37252
Moskovskij Centr Iskusstva, Moskva 37253
Moskovskij Dom Fotografii, Moskva 37254
Moskovskij Gosudarstvennyj Muzej S.A. Esenina, Moskva 37255
Moskovskij Gosudarstvennyj Vystavočnyj Zal Novyi Manež, Moskva 37256
Moskovskij Literaturnyj Muzej-centr K.G. Paustovskogo, Moskva 37257
Moskovskij Muzej Sovremennogo Iskusstva, Moskva 37258
Moskovskij Muzej-usad'ba Ostankino, Moskva 37259
Moskvitch Factory Museum, Moskva 37385
Moss Mansion Museum, Billings 46804
Moss-Thorns Gallery of Arts, Hays 49224
Mossbank and District Museum, Mossbank 06397
Most čerez Stiks - Pons per Styx, Sankt-Peterburg 37665
MostBirnHaus Stift Ardagger, Ardagger 01725
Mosteiro de S. Martinho de Tibães, Mire de Tibães 35766
Mosteiro de Santa Clara-a-Velha, Coimbra 35527
Mosteiro de São João de Tarouca, São João de Tarouca 35921
Mostmuseum, Neumarkt im Mühlkreis 02437
Mostmuseum, Sankt Leonhard am Forst 02661
Mostmuseum und Heimathaus, Sankt Marienkirchen an der Polsenz 02666
Mostra Archeologica G. Venturini, San Felice sul Panaro 27632
Mostra Cabriniana, Codogno 25932
Mostra Cartografica, Mendatica 26652
Mostra degli Antichi Mestieri di Calabria, Tropea 28104
Mostra dei Pupi Siciliani, Caltagirone 25604
Mostra della Civiltà Contadina, Lavello 26470
Mostra della Civiltà Contadina, Massa Marittima 26626
Mostra della Civiltà Contadina e Pastorale, Avezzano 25334
Mostra di Cimeli del Risorgimento, Salemi 27592
Mostra di Palazzo Farnese, Piacenza 27184
Mostra Etnografica Museo Contadino della Piana, Capannori 25653
Mostra Mineraria Permanente, Sutri 27929
Mostra Nazionale di Pittura Contemporanea, Marsala 26610
Mostra Permanente del Costume Arbereshe, Vaccarizzo Albanese 28130
Mostra Permanente del Presepio, Muggia 26874

Mostra Permanente della Biblioteca Estense, Modena 26742
Mostra Permanente della Ceramica, San Lorenzello 27655
Mostra Permanente della Civiltà Contadina, Montefoscoli 26832
Mostra Permanente della Cultura Materiale, Levanto 26489
Mostra Permanente della Giudaica e Raccolta di Minerali, Laino Borgo 26439
Mostra Permanente della Resistenza, Lugo 26544
Mostra Permanente della Resistenzo, Massa Marittima 26627
Mostra Permanente della Tradizione Mineraria, Tarvisio 27944
Mostra Permanente di Archeologia, Rivello 27422
Mostra Permanente di Paleontologia, Terni 27962
Mostra Permanente di Xilografie di Pietro Parigi, Firenze 26165
Mostra Permanente P. Mariani, Bordighera 25497
Mostviertelmuseum Haag, Haag, Niederösterreich 02029
Mostviertler Bauernmuseum, Amstetten 01719
Mosul Museum, Mosul 24720
Mosvik Museum, Mosvik 33686
Motala Brandförsvarsmuseum, Motala 40753
Motala Industrimuseum, Motala 40754
Motala Motormuseum, Motala 40755
Motala Museum - Charlottenborgs Slott, Motala 40756
Mother Armenia Military Museum, Yerevan 00746
Mother of Pearl Museum, Méru 14091
Motherwell Heritage Centre, Motherwell 45179
Motherwell Homestead, Abernethy 05308
Motiejaus Gustaičio Memorialinis Muziejus, Lazdijai 30301
Motiejus Gustaitis Memorial Museum - Lazdijai Area Museum, Lazdijai 30301
Motley Palace of the József Katona Museum, Kecskemét 23644
Moto Moto Museum, Mbala 54661
Motobu-cho Hakubutsukan, Motobu 28942
Motobu Museum, Motobu 28942
Motoori Norinaga Kinenkan, Matsusaka 28898
Motor Car Museum of Japan, Futatsunashi 28455
Motor Museum, Filching 44226
Motor & Nostalgimuséet, Grängesberg 40561
Motor-Sport-Museum Hockenheimring, Hockenheim 19206
Motor Technica Museum, Bad Oeynhausen 16799
Motorbåtmuseet Museihuset, Vreta Kloster 41072
Motorbike Museum, Lesná u Znojma 09622
The Motorboat Museum, Basildon 43392
Motorcycle Hall of Fame Museum, Pickerington 51913
Motorcycle Museum, Canillo 00087
Motormuseet, Strømmen 33888
Motorola Museum, Schaumburg 52958
Motorrad-Museum, Augustusburg 16640
Motorrad Museum, Ibbenbüren 19303
Motorrad-Museum, Otterbach 20948
Motorrad-Museum Krems-Egelsee, Krems 02230
Motorrad-Träume, Zschopau 22825
Motorrad-Veteranen- und Technikmuseum, Großschönau 18806
Motorradmuseum, Gossau (Sankt Gallen) 41444
Motorradmuseum, Neunkirchen 02441
Motorradmuseum, Sulz 02791
Motorradwelt der DDR, Prora 21132
Motown Historical Museum, Detroit 48125
Motsuji Homotsukan, Hiraizumi 28522
Motsuji Treasure Museum, Hiraizumi 28522
Motts Military Museum, Groveport 49099
Motueka District Museum, Motueka 33148
Mouhtar Ashrafi Museum, Toshkent 54474
Mouhtar Ashrafi Yodgorlik Muzeyi, Toshkent 54474
Moulagensammlung des Universitätsspitals und der Universität Zürich, Zürich 42089
Mouldsworth Motor Museum, Mouldsworth 45181
Moulin à Eau Maître Marcel, Sainte-Agathe-en-Donzy 15545
Moulin à Musique Mécanique, Mormoiron 14279
Le Moulin à Papier de Brousses, Brousses-et-Villaret 12192
Moulin à Papier Vallis-Clausa, Fontaine-de-Vaucluse 12956
Moulin à Vent Gaillardin, Chapelon 12403
Moulin de Beaumont, Beaumont 05405
Moulin de Boschepe, Boeschepe 12042
Moulin de la Chevrotière, Deschambault 05698
Moulin de la Herpinière, Turquant 16012
Moulin de Malet, Clermont-Soubiran 12555
Moulin de Pierre, Artenay 11705
Moulin de Pierre, Hauville 13196
Moulin de Rotrou et son Musée, Vaas 16026
Moulin de Traou-Meur-Écomusée, Pleudaniel 14838
Moulin des Grands Vignes, Périssac 14781
Moulin des Jésuites, Charlesbourg 05588
Moulin d'Eschviller, Eschviller 12851
Moulin du Bois-Landon, Beaumont-sur-Sarthe 11901
Moulin Fleming, LaSalle 06141
Moulin Légaré, Saint-Eustache 06815
Moulin Michel de Gentilly, Bécancour 05410
Moulin-Musée de la Brosserie, Saint-Félix 15265
Les Moulins, Sagy 15148
Les Moulins de l'isle-aux-Coudres, Saint-Louis-de-île-aux-Coudres 06864
Moulins Souterrains, Le Locle 41571
Moundbuilders State Memorial and Museum, Newark 51291
Mount Airy Museum of Regional History, Mount Airy 50870
Mount Bruce Pioneer Museum, Masterton 33141
Mount Clare Museum House, Baltimore 46580
Mount Desert Island Historical Museum, Somesville 53153
Mount Dora Center for the Arts, Mount Dora 50876
Mount Edgcumbe House, Cremyll 43921
Mount Holly Community Historical Museum, Belmont 46721
Mount Holyoke College Art Museum, South Hadley 53170
Mount Horaiji Natural History Museum, Hourai 28550
Mount Laura Homestead Museum, Whyalla 01648
Mount Mary College Costume Museum, Milwaukee 50716
Mount Morgan Historical Museum, Mount Morgan 01329
Mount Pleasant, Philadelphia 51853
Mount Pleasant Historical Society Museum, Mount Pleasant 50886
Mount Prospect Historical Society Museums, Mount Prospect 50889
Mount Pulaski Courthouse, Mount Pulaski 50890
Mount Rainier National Park Museum, Ashford 46378
Mount Royal College Gallery, Calgary 05525
Mount Rushmore National Memorial, Keystone 49771
Mount Saint Vincent University Art Gallery, Halifax 05942
Mount Stewart House, Newtownards 45261
Mount Vernon Hotel Museum, New York 51218
Mount Vernon Museum of Incandescent Lighting, Baltimore 46581
Mount Victoria and District Historical Museum, Mount Victoria 01330
Mount Washington Museum, North Conway 51389
Mountain Farm Museum, Cherokee 47422
Mountain Gateway Museum, Old Fort 51535
Mountain Heritage Center, Cullowhee 47903
Mountain Life Museum, London 50168
Mountain Mills Museum, Saint Catharines 06806
Mountain-Plains Museums Association, Littleton 55160
Mountain Resort Museum, Chengde 07612
Mountain View Doukhobor Museum, Grand Forks 05895
Mountain View Museum, Olds 06509
Mountaineering Museum, Darjeeling 24127
Mountaineering Museum, Hanamaki 28492
Mountbatten Exhibition, Romsey 45512
Mounted Branch Museum, East Molesey 44094
Mouseio Akropoleos, Athinai 22917
Mouseio Ellinikis Laïkis Technis, Athinai 22918
Mouseio Ellinikis Paidikis Tehnis, Athinai 22919
Mouseio Ethnikon Archaiologikon, Athinai 22920
Mouseio Isrorias tis Ellinikis Endymasias, Athinai 22921
Mouseio Istorias Laografias Aharnon, Athinai 22922
Mouseio Kerameikos, Athinai 22923
Mouseio Kompologiou, Nafplion 23116
Mouseio Laikon Organon, Athinai 22924
Mouseio Mpenaki, Athinai 22925
Mouseio Trenon, Athinai 22926
Mouseio Vorre, Paiania 23129
Mouseio Zakynthou, Zakynthos 23244
Mouseion Goulandri Fysikis Istorias, Kifissia 23051
Mousley Museum of Natural History, Yucaipa 54361
Moussa Castle Museum, Beiteddine 30163
Mout- en Brouwhuis De Snoek, Alveringem 03268
Mouvement Socio-Culturel de la Tuque, La Tuque 06116
The Movie Museum, Owosso 51643
Movieland Wax Museum of the Stars, Niagara Falls 06445
Moyle Mubarek Library Museum, Toshkent 54475
Moyse's Hall Museum, Bury Saint Edmunds 43695
Mozaik Müzesi, İstanbul 42832
Mozart Wohnhaus, Salzburg 02618
Mozarthaus, Augsburg 16630
Mozarthaus St.Gilgen, Sankt Gilgen 02648
Mozarthaus Vienna, Wien 03028
Mozart's Birthplace, Salzburg 02619
Mozarts Geburtshaus, Salzburg 02619
MPG Town Gallery, Odiongan 34468
MPR International Committee for Museums Marketing and PR, Malmö 55033
Mramornyj Dvorec, Sankt-Peterburg 37666
Mrštík Brothers Memorial House, Diváky 09492
MSCC Forsyth Center Galleries, College Station 47690
MTA Natural History Museum, Ankara 42706
MTA Tabiat Tarihi Müzesi, Ankara 42706
Mtengatenga Postal Museum, Namaka 30547
Mtskheta Museum, Mtskheta 16349
Mu.ZEE - Kunstmuseum aan Zee, Oostende 03933
Mu-Zee-Um, Oostende 03934
Muang Sing Exhibitions Museum, Muang Sing 29969
Muaseum of Nature and Agriculture, Macau 08001
muba - Museum für Baukultur, Neutal 02444
Much Wenlock Museum, Much Wenlock 45185
Mucha Museum, Praha 09763
The Muchnic Gallery, Atchison 46408
Muchovo muzeum, Praha 09763
Muckenthaler Gallery and Mansion Museum, Fullerton 48807
Muckleburgh Collection, Holt 44479
Muckross House Gardens and Traditional Farms, Killarney 24850
Műcsarnok, Budapest 23453
mudac - Musée de Design et d'Arts Appliqués Contemporains, Lausanne 41540
Mudanya Armistice House Museum, Mudanya 42910
Mudanya Mutareke Evi Müzesi, Mudanya 42910
Mudgee Colonial Inn Museum, Mudgee 01331
Mudrány Kúria, Szabolcs 23832
Mudrány Mansion House, Szabolcs 23832
Die Mühle, Eberswalde 18046
Mühle Laar, Zierenberg 22800
Mühle Oberneuland, Bremen 17558
Mühlen- und Landwirtschaftsmuseum, Fehmarn 18298
Mühlen- und Landwirtschaftsmuseum, Jever 19396
Mühlen- und Mennonitenmuseum Bettinger Mühle, Schmelz 21586
Das Mühlendorf, Ottenhöfen im Schwarzwald 20946
Mühlenfachmuseum, Aurich 16648
Mühlenhof Altkalen, Altkalen 16510
Mühlenhof-Freilichtmuseum Münster, Münster 20515
Mühlenmuseum, Haren (Ems) 19012
Mühlenmuseum, Oberkrämer-Vehlefanz 20815
Mühlenmuseum, Rheinsberg 21305
Mühlenmuseum Brüglingen, Münchenstein 41658
Mühlenmuseum Katzbrui, Apfeltrach 16564
Mühlenmuseum Mitling-Mark mit Sammlung Omas Küche, Westoverledingen 22517
Mühlenmuseum Moisburg, Moisburg 20368
Mühlenmuseum Teichmühle, Steinwiesen 21895
Mühlenmuseum und Mühlenweg, Maria Luggau 02351
Mühlerama, Zürich 42090
Mühlviertler Keramikwerkstätte Hafnerhaus, Leopoldschlag 02294
Mühlviertler Kulturgütermuseum, Sankt Johann am Wimberg 02654
Mühlviertler Schlossmuseum Freistadt, Freistadt 01904
Mühlviertler Vogelkundeweg, Gutau 02024
Mühlviertler Waldhaus, Windhaag bei Freistadt 03112
Müller Miklós Kiállítóterem, Orosháza 23739
Müllmuseum, Bad Säckingen 16817
Muenchausen's Museum, Dunte 30006
Münchhausen-Museum, Bodenwerder 17414
Münchner Feuerwehrmuseum, München 20460
Münchner Puppenmuseum, München 20461
Münchner Stadtmuseum, München 20462
Münchner Tiermuseum, München 20463
Münter-Haus, Murnau am Staffelsee 20534
Münze Hall und Münzerturm, Hall in Tirol 02043
Münzkabinett, Dresden 17961
Münzkabinett am Landesmuseum Joanneum, Graz 01978
Münzkabinett der TUI AG, Hannover 18999
Münzkabinett im Pergamonmuseum, Berlin 17191
Münzkabinett im Stadtarchiv Mainz, Mainz 20147
Münzkabinett und Antikensammlung der Stadt Winterthur, Winterthur 42030
Müritzeum, Waren (Müritz) 22335
Muestra Arqueológica, Armenia 08532
Muestra Arqueológica, Montenegro 08691
Muestra Arqueológica, Quimbaya 08735
Muestra Arqueológica Colección Rosmary, Colosó 08627
Muestrario Arqueológico de Amazonas, Chachapoyas 34144
Mughal Museum, Lahore 34024
Muğla Culture Center, Muğla 42913
Muğla Devlet Güzel Sanatlar Galerisi, Muğla 42912
Muğla Kültür Merkezi, Muğla 42913
Muğla Müzesi, Muğla 42914
Muğla State Fine Art Gallery, Muğla 42912
Muh Sheng Museum of Entomology, Puli 42295
Muhammad Ali Center, Louisville 50294
Muhoksen Kotiseutumuseo, Muhos 11121
Muiderslot, Muiden 32562
Mukai Junkichi Fuzoku Bijutsukan, Tokyo 29440
Mukhtar Museum, Giza 10527
Mukimi- Hudjra Muzeyi, Qo'qon 54457
Mukimi-Hudjta Study-room Museum of the Kokand Regional Studies Museum, Qo'qon 54457
Mukti Juddha Museum, Dhaka 03226
Mulberry Phospate Museum, Mulberry 50899
Mulford House and Farm, East Hampton 48247
Mulgi Külamuuseum, Halliste 10566
Mulgrave Settlers Museum, Gordonvale 01121
Mull Museum, Tobermory 45861
Mullach Ban Folk Museum, Mullach Ban 45186
Mullingar Military Museum, Mullingar 24881
Multatuli Museum, Amsterdam 31860
Multian Kotiseutumuseo, Multia 11122
Multicultural Heritage Centre, Stony Plain 07014
Multivision "Worpswede und das Teufelsmoor", Worpswede 22695
Mulvane Art Museum, Topeka 53537
Munakata Prints Museum, Kamakura 28645
Munakata Shiko Kinenkan, Aomori 28361
Munakata Shikou Art Station, Muika 28943
Munakata Taisha Shinpōkan, Genkai 28456
Muncaster Castle, Ravenglass 45460
Muncaster Watermill, Ravenglass 45461
Munce Art Center, Zionsville 54371
Munch-museet, Oslo 33732
Munch Museum, Oslo 33732
Munch's house, Åsgårdstrand 33403
Munchs Hus, Åsgårdstrand 33403
Muncie Children's Museum, Muncie 50904
Mundaneum - Centre d'Archives de la Communauté Française, Mons 03882
El Mundo de Muñecas, Icod de los Vinos 39406
Munich Residenz, München 20476
Munich State Coin Collection, München 20488
Municipal and Open Air Museum, Reykjavik 24014
Municipal Archaeological Museum, Mojokerto 24529
Municipal Archaeological Museum, Sandanski 05157
Municipal Archeological Museum, Chisar 05006
Municipal Art Gallery, Molyvos 23099
Municipal Art Gallery, Thessaloniki 23216
Municipal Art Gallery of Xanthi, Xanthi 23243
Municipal Art Society, New York 51219
Municipal Cartoon Art Museum Saitama, Saitama 29191
Municipal d'Arts Plastiques, Choisy-le-Roi 12520
Municipal Gallery, Orosháza 23741
Municipal Gallery Arsenal, Poznań 35027
Municipal Gallery Gladbeck, Gladbeck 18651
Municipal Gallery of Athens, Athinai 22927
Municipal Gallery of Corfu, Corfu 22977
Municipal Gallery of Herakleion, Iráklion 23021
Municipal Gallery of Lamia, Lamia 23071
Municipal Gallery of Livadia, Livadia 23079
Municipal Gallery of Piraeus, Piraeus 23145
Municipal Gallery of Rhodes, Rhodos 23172
Municipal Gallery of Samos, Samos 23176
Municipal Gallery of Volos, Volos 23239
Municipal Historical Museum, Asenovgrad 04971
Municipal Historical Museum, Balčik 04976
Municipal Historical Museum, Belogradčik 04982
Municipal Historical Museum, Dalgopol 05011
Municipal Historical Museum, Karlovo 05049
Municipal Historical Museum, Karnobat 05051
Municipal Historical Museum, Kavarna 05055
Municipal Historical Museum, Klisura 05069
Municipal Historical Museum, Lom 05084
Municipal Historical Museum, Melnik 05091
Municipal Historical Museum, Orjachovo 05101
Municipal Historical Museum, Panagjurište 05102
Municipal Historical Museum, Peruštica 05112
Municipal Historical Museum, Samakov 05156
Municipal Historical Museum, Sevlievo 05159
Municipal Historical Museum, Teteven 05226
Municipal History Museum, Botevgrad 04992
Municipal History Museum, Čirpan 05009
Municipal History Museum, Etropole 05028
Municipal History Museum, Kazanlăk 05058
Municipal History Museum, Nova Zagora 05097
Municipal History Museum of Écija, Écija 39257
Municipal Homeland House, Šamorín 38348
Municipal Museum, Bečej 38032
Municipal Museum, Chojnów 34600
Municipal Museum, Chotěboř 09479
Municipal Museum, Dimona 24942
Municipal Museum, Kangasniemi 10912
Municipal Museum, Kőszeg 23685
Municipal Museum, Letohrad 09623
Municipal Museum, Sremski Karlovci 38152
Municipal Museum, Subotica 38156
Municipal Museum, Žamberk 09925
Municipal Museum and Cheesemaking Museum, Nakkila 11136
Municipal Museum and Gallery, Vel'ký Šariš 38386
Municipal Museum and Gallery Breclav, Břeclav 09417
Municipal Museum Anna Akhmatova - Silver Century, Sankt-Peterburg 37667
Municipal Museum Bergkamen - Art Gallery "sohle 1", Bergkamen 17021
Municipal Museum De Lakenhal, Leiden 32489

M

Municipal Museum Den Bosch (SM's), 's-Hertogenbosch 32379
Municipal Museum Gwalior, Gwalior 24198
Municipal Museum in Pezinok, Pezinok 38325
Municipal Museum Nova Gradiška, Nova Gradiška 08916
Municipal Museum of Amsterdam, Amsterdam 31888
Municipal Museum of Art, Győr 23589
Municipal Museum of Art - Margit Kovács Permanent Ceramics Exhibition, Győr 23584
Municipal Museum of Art - Miklós Borsos Museum, Győr 23580
Municipal Museum of Art - Napoleon House, Győr 23588
Municipal Museum of Art - Péter Váczy Collection, Győr 23587
Municipal Museum of Guantanamo - Old Prison, Guantánamo 09093
Municipal Museum of Kavala, Kavala 23048
Municipal Museum of Montijo, Montijo 35778
Municipal Museum of Natural History in Ichikawa, Ichikawa 28555
Municipal Museum of Pásztó, Pásztó 23755
Municipal Museum of Suzu Ware, Suzu 29306
Municipal Museum of Železný Brod, Železný Brod 09937
Municipality Museum, Hebron 34053
Municipalnaja Chudožestvennaja Galereja g. Kostromy, Kostroma 37063
Municipalnyj Objedinennyj Muzej Pisatelej Urala, Ekaterinburg 36891
Municipialnyj Muzej Anna Achmatova - Serebrjannyj vek, Sankt-Peterburg 37667
Municypal'nyj Muzej Pryvatnych Kollekcij im. O. Bleščunova, Odesa 43156
Munkácsy Memorial House, Békéscsaba 23370
Munkácsy Mihály Emlékház, Békéscsaba 23370
Munkácsy Mihály Múzeum, Békéscsaba 23371
Munsala Hembygdsmuseum, Munsala 11125
Munsin Museum, Masan 29851
Munson-Williams-Proctor Arts Institute Museum of Art, Utica 53676
Munster Literature Centre, Cork 24769
Munt- en Penningkabinet van de Spaar- en Voorschotbank, Surhuisterveen 32811
Munting Museo ng Namayan, Mandaluyong 34426
Muonion Kotiseutumuseo, Muonio 11126
Murakami Folk Museum, Murakami 28945
Murauchi Art Museum, Hachioji 28466
Murauchi Bijutsukan, Hachioji 28466
Murchison Falls National Park Museum, Murchison Falls 42996
Murchison Museum, Murchison 33149
Murchison Settlement Museum, Mullewa 01333
Mures District Art Museum, Târgu Mureş 36670
Mures District Museum of Ethnography and Folk Art, Târgu Mureş 36671
Mures District Museum of Natural History, Târgu Mureş 36672
Murgtal-Museum, Forbach 18334
Murie Museum, Kelly 49722
Murmansk Regional Art Museum, Murmansk 37433
Murmansk Regional Museum, Murmansk 37434
Murmanskij Oblastnoj Chudožestvennyj Muzej, Murmansk 37433
Murmanskij Oblastnoj Kraevedčeskij Muzej, Murmansk 37434
Murney Tower Museum, Kingston 06086
Murols School's Painters Museum, Murol 14337
Muroran Folklore and Historic Museum, Muroran 28948
Muroran-shi Seishonen Kagakukan, Muroran 28947
Muroran-shi Tontenkan, Muroran 28948
Muroran Youth Science Museum, Muroran 28947
Murray-Lindsay Mansion, Lindsay 50119
Murray's Motorcycle Museum, Laxey 44651
Murrell Home, Park Hill 51708
Murrumburrah-Harden Historical Museum, Murrumburrah 01337
Murrurundi Museum, Murrurundi 01338
Murtoa Water Tower Museum, Murtoa 01339
Murtovaara Farmhouse, Valtimo 11441
Murtovaaran Talomuseo, Valtimo 11441
Mus-Olles Museum, Nälden 40764
Musa Dshalil Museum-Appartment, Kazan 37020
Musa Heritage Gallery, Kumbo-Nso 05298
MUSA - Museum - Startgalerie - Artothek, Wien 03029
MUSAC, León 39453
Musaem Chontae Dhún na nGall, Letterkenny 24862
Músaem Corca Dhuibhne, Ballyferriter 24738
Musafir-Khana Palace, Cairo 10501
Musashi-murayama City Museum of History, Folklore and Archaeology, Musashi-murayama 28949
Musashi-murayama-shi Rekishi Minzoku Shiryokan, Musashi-murayama 28949
Musashino Art University Museum and Library, Kodaira 28752
Muscarelle Museum of Art, Williamsburg 54170
Muscat Gate Museum, Muscat 33989
Muscatine Art Center, Muscatine 50921
Muschel- und Schneckenmuseum, Norden 20729
Muschelkalkmuseum Hagdorn, Ingelfingen 19333
Muse Art Gallery, Philadelphia 51854
Muse Gallery, Jackson 49572
MUSE-O, Stuttgart 21953
Museactron, Maaseik 03837
Muséal Adolphe Pégoud, Montferrat 14207
Museal Complex of Natural Sciences, Bacău 36120
Musée, Aigues-Mortes 11503
Musée, Bages (Pyrénées-Orientales) 11819
Musée, Bischwiller 12002
Musée, Brignon 12181
Musée, Chaillac 12353
Musée, Chazé-sur-Argos 12493
Musée, Cransac 12666
Musée, Cruzy 12683
Musée, Evolène 41367
Musée, Frontignan-la-Peyrade 13016
Musée, Gramat 13094
Musée, Hautvillers 13195
Musée, Joinville (Haute-Marne) 13274
Musée, Kampong Thom 05284
Musée, Le Landeron 41523
Musée, Laragne-Montéglin 13524
Musée, Lisieux 13777
Musée, Livry 13789
Musée, Lupiac 13867
Musée, Moyen 14320
Musée, Orainville 14516
Musée, Plouay 14848
Musée, Quiberon 14965
Musée, Saint-Hilaire-en-Lignière 15309
Musée, Saint-Julien-Molhesabate 15349
Musée, Tauriers 15867
Musée, Ville-d'Avray 16173
Musée 1900, Arpaillargues-et-Aureillac 11692
Musée 1900, Cauterets 12327
Musée 1900 Arts et Techniques, Champlitte 12391
Musée 2ème Guerre Mondiale, Ambleteuse 11572
Musée 39-45, Roscoff 15103
Musée 39-45 du Grand Blockhaus, Batz-sur-Mer 11861
Musée 40-44, Monceau-Imbrechies 03878
Musée à Flot de l'Escorteur d'Escadre, Nantes 14361
Musée à Flot du Galion, Plouharnel 14858
Musée à la Ferme de Rome, Bézu-la-Forêt 11980
Musée à la Mémoire des Combattants et de Victimes de Guerre, Mâcon 13906
Musée a la Recherche du Temps Perdu, Bricquebec 12173
Musée A la Rencontre des Vieux Métiers, Bouray-sur-Juine 12096
Musée Aal Braanebourg, Brandenbourg 30458
Musée Abbaye d'Airvault, Airvault 11512
Musée-Abbaye de Charroux, Charroux 12419
Musée-Abbaye Saint-Germain, Auxerre 11776
Musée Abbé Deletoille, Fruges 13017
Musée Abbé Jules Lemire, Vieux-Berquin 16157
Musée Acadien, Archigny 11664
Le Musée Acadien, Caraquet 05554
Musée Acadien, Cheticamp 05607
Musée Acadien, Miscouche 06305
Musée Acadien, Moncton 06320
Musée Acadien de Pubnico-Ouest, West Pubnico 07303
Musée Acadien du Québec, Bonaventure 05436
Musée Accous-Fermiers Basco-Béarnais, Accous 11490
Musée Adam Mickiewicz, Paris 14627
Musée Adja Swa, Yamoussoukro 08840
Musée Adjaman Mihiin Raphaël, Vavoua 08839
Musée Adrien Mentienne, Bry-sur-Marne 12203
Musée Adzak, Paris 14628
Musée Aéronautique, La Baule-Escoublac 13315
Musée Aérorétro, Albon 11541
Musée Africain, Ile-d'Aix 13235
Musée Africain, Lyon 13882
Musée Africain de Namur, Namur 03898
Musée Agathois, Agde 11491
Musée Agricole, Coffrane 41307
Musée Agricole, Sainte-Geneviève (Meurthe-et-Moselle) 15561
Musée Agricole, Villy-le-Maréchal 16246
Musée Agricole Bras de Brosne, Marles-sur-Canche 13971
Musée Agricole de Brullioles ou la Cadole, Brullioles 12195
Musée Agricole de la Haute Hesbaye, Liernu 03822
Musée Agricole Départemental Marcel-Mouillseaux, Botans 12080
Musée Agricole des Ruralies, Prahecq 14939
Musée Agricole du Château de Didonne, Semussac 15710
Musée Agricole du Parlement, Gap 13030
Musée Agricole Vivant, Bissey-la-Pierre 12004
Musée Agro-Pastoral d'Aussois, Aussois 11759
Musée Airborne, Sainte-Mère-Eglise 15574
Musée Al Mathaf El Lubnani, Beiteddine 30164
Musée Alain Fournier et Jacques Rivière, La Chapelle-d'Angillon 13332
Musée Albert André, Bagnols-sur-Cèze 11825
Musée Albert Schweitzer, Gunsbach Village 13172
Musée Albert Schweitzer, Kaysersberg 13300
Musée Alésia, Alise-Sainte-Reine 11554
Musée Alexandre Dumas, Villers-Cotterêts 16229
Musée Alexandre-Louis Martin, Carnières 03509
Musée Alexis Forel, Morges 41643
Musée Alfred Bonnot, Chelles 12497
Musée Alfred de Vigny, Champagne-Vigny 12385
Musée Alice Taverne, Ambierle 11571
Musée Alphonse Daudet, Fontvieille 12977
Musée Alphonse Daudet, Saint-Alban-Auriolles 15155
Musée Alpin, Chamonix-Mont-Blanc 12381
Musée Alpin, Verbier 41980
Musée Alsacien, Haguenau 13176
Musée Alsacien, Strasbourg 15819
Musée America-Gold Beach, Ver-sur-Mer 16108
Musée Amérindien de Mashteuiatsh, Mashteuiatsh 06245
Musée Amérindien et Inuit de Godbout, Godbout 05878
Musée Amora, Dijon 12735
Musée Amphoralis, Sallèles-d'Aude 15601
Musée André Abbal, Carbonne 12279
Musée André Dunoyer-de-Ségonzac, Boussy-Saint-Antoine 12131
Musée André-Marie Ampère, Poleymieux-au-Mont-d'Or 14869
Musée Angladon, Avignon 11788
Musée Animalier, Ville-sous-Anjou 16175
Musée Animalier à Jougne, Jougne 13282
Musée Animé des Arts et Traditions Populaires, Serralongue 15727
Musée-Animé du Jouet et des Petits Trains, Colmar 12581
Musée Animé du Vin et de la Tonnellerie, Chinon 12515
Musée Anne de Beaujeu, Moulins (Allier) 14303
Musée Antiquités Gallo-Romain, Aoste 11648
Musée Antoine Brun, Sainte-Consorce 15554
Musée Antoine Lécuyer, Saint-Quentin (Aisne) 15474
Musée Antoine Vivenel, Compiègne 12597
Musée Antoine Wiertz, Bruxelles 03452
Musée Août 1944, Falaise 12903
Musée Apicole, Nedde 14389
Musée-Aquariophile, Dunkerque 12796
Musée Aquarium, Bergerac 11942
Musée-Aquarium d'Arcachon, Arcachon 11662
Musée Aquarium de la Rivière Dordogne, Creysse 12678
Musée-Aquarium du Laboratoire Arago, Banyuls-sur-Mer 11842
Musée Archéologique, Antigny 11645
Musée Archéologique, Arlon 03323
Musée Archeologique, Azé 11812
Musée Archéologique, Bagnols-en-Forêt 11824
Musée Archéologique, Banassac 11835
Musée Archéologique, Bezouce 11979
Musée Archéologique, Blaye 12022
Musée Archéologique, Brumath 12197
Musée Archéologique, Champagnole 12387
Musée Archéologique, Charleroi 03514
Musée Archéologique, Corseul 12633
Musée Archéologique, Cutry 12694
Musée Archéologique, Delme 12706
Musée Archéologique, Dijon 12736
Musée Archéologique, Eauze 12804
Musée Archéologique, Ensisheim 12821
Musée Archéologique, Entrains-sur-Nohain 12824
Musée Archéologique, Escolives-Sainte-Camille 12852
Musée Archéologique, Gafsa 42625
Musée Archéologique, Gouzeaucourt 13090
Musée Archéologique, L'Aigle 13482
Musée Archéologique, Laissac 13486
Musée Archéologique, Larache 31676
Musée Archéologique, Lectoure 13672
Musée Archéologique, Magnac-Laval 13910
Musée Archéologique, Martizay 14021
Musée Archéologique, Marvejols 14023
Musée Archéologique, Monségur 14153
Musée Archéologique, Mormoiron 14280
Musée Archéologique, Murol 14336
Musée Archéologique, Namur 03899
Musée Archéologique, Nîmes 14449
Musée Archéologique, Peyriac-de-Mer 14806
Musée Archéologique, Pithiviers-le-Vieil 14826
Musée Archéologique, Rabat 31688
Musée Archéologique, Saint-Gilles-du-Gard 15300
Musée Archéologique, Saint-Martin-de-Bromes 15397
Musée Archéologique, Saint-Pal-de-Mons 15438
Musée Archéologique, Saint-Paulien 15452
Musée Archéologique, Sainte-Agnès 15546
Musée Archéologique, Sainte-Bazeille 15551
Musée Archéologique, Saintes 15582
Musée Archéologique, Saverne 15674
Musée Archéologique, Simiane-Collongue 15758
Musée Archéologique, Sollières-Sardières 15776
Musée Archéologique, Soulosse-sous-Saint-Élophe 15800
Musée Archéologique, Soyons 15805
Musée Archéologique, Strasbourg 15820
Musée Archéologique, Suippes 15838
Musée Archéologique, Tétouan 31700
Musée Archéologique, Thérouanne 15877
Musée Archéologique, Toulouse 15933
Musée Archéologique, Vachères 16027
Musée Archéologique, Vaison-la-Romaine 16031
Musée Archéologique, Vienne-en-Val 16153
Musée Archéologique, Villemagne-l'Argentière 16198
Musée Archéologique, Villeneuve-d'Ascq 16201
Musée Archéologique Armand Viré, Luzech 13877
Musée Archéologique Blasimon, Blasimon 12020
Musée Archéologique d'Argentomagus, Saint-Marcel (Indre) 15393
Musée Archéologique d'Autelbas, Bastogne 03346
Musée Archéologique de Chimtou, Jendouba 42629
Musée Archéologique de Djemila, Djemila 00059
Musée Archéologique de la Basse-Meuse, Oupeye 03944
Musée Archéologique de la Porte du Croux, Nevers 14416
Musée Archéologique de la Préhistoire à l'Époque Médiévale, Nyons 14488
Musée Archéologique de la Région de Breteuil, Breteuil 12166
Musée Archéologique de Lamta, Lamta 42635
Musée Archéologique de l'Université Americaine, Beirut 30157
Musée Archéologique de Makthar, Makthar 42638
Musée Archéologique de Rom-Sainte-Soline, Rom 15082
Musée Archéologique de Saint-Pierre, Vienne 16149
Musée Archéologique de Sbeïtla, Sbeïtla 42649
Musée Archéologique de Sfax, Sfax 42650
Musée Archéologique de Site, Liffol-le-Grand 13744
Musée Archéologique de Sousse, Sousse 42656
Musée Archéologique de Thésée-la-Romaine, Thésée-la-Romaine 15878
Musée Archéologique de Touraine, Tours 15963
Musée Archéologique de Viuz-Faverges, Faverges 12908
Musée Archéologique Départemental, Jublains 13288
Musée Archéologique Départemental, Saint-Bertrand-de-Comminges 15185
Musée Archéologique Départemental du Val-d'Oise, Guiry-en-Vexin 13167
Musée Archéologique des Vestiges et Musée Agricole, Jonvelle 13275
Musée Archeologique du Cap Bon, Nabeul 42644
Musée Archéologique du Château Féodal, Bressieux 12154
Musée Archéologique du Gâtinais, Montargis 14171
Musée Archéologique du Minervois, Olonzac 14510
Musée Archéologique du Théâtre de Guelma, Guelma 00065
Musée Archéologique d'Uxellodunun, Vayrac 16089
Musée Archéologique - Eglise Saint-Laurent, Grenoble 13132
Musée Archéologique et de Paléontologie, Minerve 14125
Musée Archéologique et d'Histoire Locale, Saint-Prix 15472
Musée Archéologique et Historique Cantonal de Gimont, Gimont 13061
Musée Archéologique et Historique du Comte de Logne, Vieuxville 04113
Musée Archéologique et Historique Municipal, Ermont 12844
Musée Archéologique et Lapidaire, Blois 12032
Musée Archéologique et Lapidaire de Montauban, Montauban-sous-Buzenol 03890
Musée Archéologique et Missionnaire, Saxon-Sion 15687
Musée Archéologique Gallo-Romain, Revel-Tourdan 15029
Musée Archéologique Henri Prades, Lattes 13534
Musée Archéologique - Hôtel de Sade, Saint-Rémy-de-Provence 15488
Musée Archéologique Hôtel Dieu, Cavaillon 12329
Musée Archéologique Jean Régnier, Mont-Saint-Vincent 14163
Musée Archéologique Le Cloître, Elne 12816
Musée Archéologique Léon Alègre, Bagnols-sur-Cèze 11826
Musée Archéologique Ludna, Saint-Georges-de-Reneins 15280
Musée Archéologique Municipal, Martres-Tolosane 14022
Musée Archéologique Municipal, Pélissanne 14769
Musée Archéologique Municipal de Fréjus, Fréjus 13001
Musée Archéologique Régional d'Orp-le-Grand, Orp-le-Grand 03939
Musée Archéologique Régional d'Oupeye-(M.A.R.O.), Vivegnis 04120
Musée Archeologique René Galloux, Montrichard 14259
Musée Archéologique Géologique et Ethnologique des Vans, Les Vans 13720

Musée Archipélitude, Saint-Pierre 37997
Musée-Archives de Saint-Pierre, Saint-Pierre 37998
Musée Ardeche d'Autrefois, Thueyts 15904
Musée Ardoisier, Saint-Julien-Mont-Denis 15351
Musée Ariana, Genève 41416
Musée Arménien de France, Paris 14629
Musée Art du Chocolat, Lisle-sur-Tarn 13785
Musée Art et Culture, Lescar 13722
Musée Art Paysan, Villeneuve-de-Berg 16209
Musée Arteum, Châteauneuf-le-Rouge 12448
Musée Arthur Batut, Labruguiere 13467
Musée Arthur Le-Duc, Torigni-sur-Vire 15918
Musée Arthur Rimbaud, Charleville-Mézières 12409
Musée Artisanal et Rural, Clion-sur-Seugne 12560
Musée Artisanal et Rural d'Arts, Ars 11702
Musée Arts d'Afrique et d'Asie, Vichy 16145
Musée Arts et Histoire, Bormes-les-Mimosas 12077
Musée Arts et Traditions Populaires, Firmi 12936
Musée Atelier d'Automates, Lusigny-sur-Barse 13869
Musée Atelier de Poterie, Neuvic-sur-l'Isle 14407
Musée-Atelier de Poupées Francépoque, Nances-Lac d'Aiguebelette 14348
Musée-Atelier du Cuivre, Villedieu-les-Poêles 16182
Musée-Atelier du Cuivre et de l'Argent, Paris 14630
Musée Atelier du Trompe l'Oeil et du Décor Peint, Périgueux 14777
Musée-Atelier Werner Lichtner-Aix, Sérignan-du-Comtat 15723
Musée Atger, Montpellier 14238
Musée Athois, Ath 03330
Musée au Fil de la Pierre, Antoing 03278
Musée Au fil du papier, Pont-à-Mousson 14877
Musée Aubois d'Histoire de l'Éducation, Troyes 15997
Musée Audemars Piguet, Le Brassus 41251
Musée Auguste Jacquet de Beaucaire, Beaucaire 11887
Musée Auguste Rodin, Paris 14631
Musée Augustin Bernard, Bourbon-l'Archambault 12101
Musée Auto Moto Vélo, Châtellerault 12462
Musée Auto-Retro, Trooz 04096
Musée Automobile, La Clayette 13346
Musée Automobile, Saint-Germain-Laval 15292
Musée Automobile, Saint-Jacques 06825
Musée Automobile Bellenavois, Bellenaves 11929
Musée Automobile de la Sarthe, Le Mans 13606
Musée Automobile des Voitures des Chéfs d'État, Sauvigny-Le-Bois 15672
Musée Automobile Espera Sbarro, Pontarlier 14897
Musée aux Dominicaines, Pont-l'Évêque 14890
Musée Baccarat, Paris 14632
Musée Baillet-Latour, Latour 03779
Musee Bajén-Vega, Monesties 14147
Musée Balzac, Saché 15145
Musée Barbey d'Aurévilly, Saint-Sauveur-le-Vicomte 15507
Musée Barbier-Mueller, Genève 41417
Musée Baron Gérard, Bayeux 11873
Musée Baron Martin, Gray 13126
Musée Baroncelli, Saintes-Maries-de-la-Mer 15588
Musée Barrois, Bar-le-Duc 11844
Musée Barthélemy Thimonnier de la Machine à Coudre et du Cycle, Amplepuis 11593
Musée Bartholdi, Colmar 12582
Musée-Base V 3 de Mimoyecques, Landrethun-le-Nord 13501
Musée Basque et de l'Histoire de Bayonne, Bayonne 11878
Musée Bastien Lepage, Montmédy 14225
Musée Baud, L'Auberson 41134
Musée Baudy, Giverny 13066
Musée Bayard et de la Chevalerie, Pontcharra 14901
Musée Beaubois, Carcassonne 12282
Musée Beauceron du Grand Breau, Tivernon 15916
Musée Beaulne, Coaticook 05626
Musée Belge de Radiologie-Belgisch Museum voor Radiologie, Bruxelles 03453
Musée Benoist, Le Gavre 13586
Musée Benoît de Puydt, Bailleul 11829
Musée Benoît XIII, Châteaurenard (Bouches-du-Rhône) 12454
Musée Bernadette, Lourdes 13838
Musée Bernadotte, Pau 14755
Musée Bernard Buffet, Paris 14633
Musée Bernard Buffet, Shizuoka 29286
Musée Bernard d'Agesci, Niort 14455
Musée Bernard Palissy, Lacapelle-Biron 13468
Musée Bibliothèque, Anost 11632
Musée-Bibliothèque de l'Histoire du Protestantisme, Joigny (Yonne) 13272
Musée-Bibliothèque François Pétrarque, Fontaine-de-Vaucluse 12957
Musée-Bibliothèque Humaniste, Sélestat 15704
Musée-Bibliothèque Pierre-André Benoît, Alès 11549
Musée Bigouden, Pont-l'Abbé 14888
Musée Bijoux de Braque, Saint-Dié-des-Vosges 15240
Musée Binger de Zaranou, Abengourou 08827
Musée Biochet-Brechot, Caudebec-en-Caux 12322
Musée Bizarre und Mäusemuseum, Rieden bei Baden 41761
Musée Boilhet-Christofle, Saint-Denis (Seine-Saint-Denis) 15235
Musée Boleslaw Biegas, Paris 14634
Musée Bon-Pasteur, Québec 06685
Musée Bonaparte, Auxonne 11780
Musée Bonaventure Fieullien, Regniowez 14994
Musée Bonnat, Bayonne 11879
Musée Bonnevallais, Bonneval 12055
Musée Bossuet, Meaux 14056
Musée Botanique, Angers 11606
Musée Bouchard, Paris 14635
Musée Boucher-de-Perthes, Abbeville 11486
Musée Bourbonnais, Bellerive-sur-Allier 11930
Musée Bourbonnais, Moulins (Allier) 14304
Musée Bourdelle, Paris 14636
Musée Bourgeon-Jallieu, Bourgoin-Jallieu 12123
Musée Bourguiba de Art Islamique, Monastir 42643
Musée Bourguignon Perrin de Puycousin, Tournus 15958
Musée Boutdeville, Langueux 13511
Musée Brasserie des Fagnes, Mariembourg 03854
Musée Bres Marcel, Saint-Léger-près-Troyes 15366
Musée Breton, Quimper 14967
Musée Briard et de la Crypte Mérovingienne, Jouarre 13280
Musée Bruno Danvin, Saint-Pol-sur-Ternoise 15465
Musée Bruxellois de la Gueuze, Bruxelles 03454
Musée Bruxellois de l'Industrie et du Travail, Bruxelles 03455
Musée Buffon, Buffon 12204
Musée C.C. Olsommer, Veyras 41988
Musée Cadou, Louisfert 13836
Musée Calbet, Grisolles 13151
Musée Calvet, Avignon 11789
Musée Calvin, Noyon 14482
Musée Camille Lemonnier, Bruxelles 03456
Musée Camille Pissarro, Pontoise 14908
Musée Campagnard Agricole, Fillières 12935
Musée Campagnard du Haut-Beaujolais, Saint-Christophe 15205
Musée Campanaire, L'Isle-Jourdain 13782
Musée Campanaire Bollée, Saint-Jean-de-Braye 15323
Musée Campredon, L'Isle-sur-la-Sorgue 13783
Musée Canadien de la Nature, Gatineau 05860
Musée Canadien de la Poste, Gatineau 05861
Musée Canadien des Civilisations, Gatineau 05862
Musée Cantini, Marseille 13993
Musée Cantonal d'Archéologie, Sion 41873
Musée Cantonal d'Archéologie et d'Histoire, Lausanne 41541
Musée Cantonal de Géologie, Lausanne 41542
Musée Cantonal de Zoologie, Lausanne 41543
Musée Cantonal des Beaux-Arts, Lausanne 41544
Musée Cantonal d'Histoire, Sion 41874
Musée Cantonal d'Histoire Naturelle, Sion 41875
Musée Cappon, Marans 13945
Musée Cappuis-Fähndrich, Develier 41325
Musée Carnavalet - Histoire de Paris, Paris 14637
Musée Casa De Les Alberes, Argelès-sur-Mer 11670
Musée Casimir Delavigne, Pressagny-l'Orgueilleux 14945
Musée Casino de Sa, Saxon 41825
Musée Castral, Wintzenheim 16298
Musée Catalan des Arts et Traditions Populaires, Perpignan 14788
Musée Caverne du Dragon, Oulches-la-Vallée-Foulon 14546
Musée Cavernicole de la Grotte de Balme, La Balme-les-Grottes 13308
Musée Cavernicole de la Grotte de Thaïs, Saint-Nazaire-en-Royans 15422
Musée Célestin Freinet, Bonnut 12059
Musée Celte - Gallo Romain, Seltz 15706
Musée-Centre Charles Péguy, Orléans 14534
Musée Centre Culturel, Vichy 16146
Musée Centre d'Art et de Littérature, L'Echelle 13670
Musée-Centre d'Arts Plastiques Chanot, Clamart 12538
Musée-Centre de Documentation Alfred Desmasures, Hirson 13218
Musée Centre de l'Alchimie, Tiffauges 15909
Musée-Centre de l'Imaginaire Arthurien, Concoret 12605
Musée Centre Historique des Sapeurs-Pompiers, Firminy 12939
Musée Centre National Jean Moulin, Bordeaux 12065
Musée Centre Nature de Botrange, Robertville 03971
Musée Cernuschi, Paris 14638
Musée Césaire Phisalix, Mouthier-Haute-Pierre 14315
Musée Cévenol, Le Vigan (Gard) 13666
Musée-Chalutier Le Kifanlo, Les Sables-d'Olonne 13713
Musée Champenois de l'Imprimerie, Épernay 12831
Musée Champollion, Figeac 12932
Musée-Chapelle Bellini, Cannes 12268
Musée-Chapelle Saint-Blaise-des-Simples, Milly-la-Forêt 14122
Musée Charbonneau-Lassay, Loudun 13830
Musée Charles Cros, Fabrezan 12900
Musée Charles de Bruyères, Remiremont 15008
Musée Charles de Gaulle, Lille 13752
Musée Charles et Marguerithe Bieth, Abengourou 08828
Musée Charles Léandre, Domfront 12758
Musée Charles-Louis Philippe, Cérilly 12342
Musée Charles Milcendeau-Jean Yole, Soullans 15799
Musée Charles-Portal, Cordes-sur-Ciel 12623
Musée Charles VII, Mehun-sur-Yèvre 14061
Musée Charlier, Bruxelles 03457
Musée Charolais du Machinisme Agricole, Neuvy-Grandchamp 14412
Musée-Château, Grignan 13149
Musée-Château, La Ferté-Saint-Aubin 13364
Musée-Château, Montreuil-Bellay 14253
Musée-Château, Senonches 15718
Musée-Château, Ville-d'Avray 16174
Musée Château Abbaye de Cassan, Roujan 15126
Musée Château d'Aigremont, Awirs 03336
Musée Château d'Annecy et Observatoire Régional des Lacs Alpins, Annecy 11625
Musée-Château de Camponac, Pessac 14802
Musée-Château de Cheverny, Contres 12618
Musée-Château de Fontaine-Henry, Fontaine-Henry 12962
Musée-Château de Goulaine, Haute-Goulaine 13189
Musée-Château de La Barben, La Barben 13309
Musée-Château de la Garnache, La Garnache 13375
Musée-Château de La Verrerie, Oizon 14505
Musée-Château de Pommard, Pommard 14873
Musée-Château de Pyrénée, Lourdes 13839
Musée-Château de Rochebrune, Etagnac 12870
Musée-Château de Saint-Germain-de-Livet, Saint-Germain-de-Livet 15286
Musée-Château de Saint-Ouen, Saint-Ouen 15433
Musée-Château des Adhémar, Montélimar 14197
Musée Château des Princes de Ligne et Musée Lapidaire, Antoing 03279
Musée-Château des Roures, La Bastide-de-Virac 13312
Musée-Château d'Espeyran, Saint-Gilles-du-Gard 15301
Musée-Château d'Etoges, Etoges 12881
Musée Château du Bucheneck, Soultz-Haut-Rhin 15801
Musée-Château du Moine Sacristain, Aurec-sur-Loire 11750
Musée Château-Ferme de Falaën, Falaën 03597
Musée-Château Saint-Brisson, Saint-Brisson-sur-Loire 15194
Musée Chez Manuel, Migne-Auxances 14118
Musée Chintreuil, Pont-de-Vaux 14883
Musée Christhi, Lourdes 13840
Musée Christian-Dior, Granville 13103
Musée Ciotaden, La Ciotat 13344
Musée Claude Bernard, Saint-Julien-en-Beaujolais 15348
Musée Claude Chappe, Brûlon 12196
Musée Claude Debussy, Saint-Germain-en-Laye 15289
Musée Claude Monet, Giverny 13067
Musée Claude Nicolas Ledoux Saline Royale, Arc-et-Senans 11661
Musée Clément Ader, Muret 14335
Musée Cloître Roman Saint-André-le-Bas, Vienne 16150
Musée Cognacq-Jay, Paris 14639
Musée Cointreau, Saint-Barthélemy-d'Anjou 15181
Musée Colby-Curtis, Stanstead 07005
Musée Colette, Saint-Sauveur-en-Puisaye 15506
Musée Collection Eugène Pesch, Differdange 30466
Musée Collection Ferroviaire, Miramas 14129
Musée Colombophile du Mont-Valérien, Suresnes 15840
Musée Communal, Clérac 12544
Musée Communal, Pierre-Perthuis 14813
Musée Communal, Xhoris 04159
Musée Communal d'Archéologie, d'Art et d'Histoire, Nivelles 03917
Musée Communal d'Archéologie et d'Art Religieux, Amay 03270
Musée Communal d'Archéologie Hesbignonne, Saint-Georges-sur-Meuse 03992
Musée Communal de Braine-l'Alleud, Braine-l'Alleud 03395
Musée Communal de Folklore Léon Maes, Mouscron 03895
Musée Communal de Huy, Huy 03723
Musée Communal de la Pierre de Sprimont, Sprimont 04045
Musée Communal de la Résistance et de la Déportation, Vénissieux 16103
Musée Communal de la Ville de Braine-le-Comte, Braine-le-Comte 03399
Musée Communal de Woluwe-Saint-Lambert, Bruxelles 03458
Musée Communal du Comte de Jette, Bruxelles 03459
Musée Communal Georges Mulpas, Elouges 03577
Musée Communal Herstalien d'Archéologie et de Folklore, Herstal 03701
Musée Communal Robert Tatin, Cossé-le-Vivien 12637
Musée Communautaire des Soeurs de Sainte-Croix, Montréal 06351
Musée Comtadin, Carpentras 12295
Musée Comtadin-Duplessis, Carpentras 12296
Musée Comtois, Besançon 11955
Musée Condé, Chantilly 12399
Musée Condorcet, Ribemont 15034
Musée Conrad-et-Marcel-Schlumberger, Crèvecœur-en-Auge 12677
Musée Conservation, Saint-Paul-Trois-Châteaux 15448
Musée-Conservatoire de la Dentelle, Luxeuil-les-Bains 13871
Musée Conservatoire de la Vie Agricole et Rurale, Hétomesnil 13212
Musée Conservatoire de la Vie Agricole et Rurale d'Autrefois, Gimont 13062
Musée Conservatoire de l'Agriculture, Chartres 12422
Musée-Conservatoire de l'Outil, Brecy-Brières 12152
Musée Conservatoire des Cépages, Saint-Ilpize 15314
Musée Conservatoire des Productions Fruitières, Saint-Jean-d'Arvey 15320
Musée Conservatoire des Traditions Rurales, Mauriac 14039
Musée-Conservatoire du Patrimoine, La Garde-Freinet 13374
Musée Conservatoire Ethnologique de Haute-Provence, Mane 13936
Musée Conservatoire Maritime et Rural, Narbonne 14376
Musée-Conservatoire Nationale de la Typographie, Paris 14640
Musée-Conservatoire Rural d'Arts et Traditions Populaires, Villefagnan 16186
Musée Constantin Meunier, Bruxelles 03460
Musée Costi, Bergerac 11943
Musée Crozatier, Le Puy-en-Velay 13643
Musée Curie, Paris 14641
Musée da Guine Portugesa, Bissau 23316
Musée da Silva des Arts et de la Culture, Porto-Novo 04187
Musée d'Abéché, Abéché 07423
Musée d'Allard, Montbrison 14188
Musée d'Allauch, Allauch 11555
Musée Damien Bec, Laval-Roquecezière 13544
Musée d'Anatomie Delmas-Orfila-Rouvière, Paris 14642
Musée d'Anatomie Lesbres-Tagand de l'Ecole Vétérinaire de Lyon, Marcy-l'Étoile 13961
Musée dans la Mairie Vitrine, Panossas 14561
Musée d'Ansembourg, Liège 03802
Musée d'Antan, Aigremont 11497
Musée d'Antan, La Léchère 13389
Musée d'Anthropologie, Kinshasa 08795
Musée d'Anthropologie, Sainte-Foy 06893
Musée d'Anthropologie Préhistorique, Monaco 31607
Musée Dapper, Paris 14643
Musée d'Aquitaine, Bordeaux 12066
Musée Dar Belghazi, Km 17, Route de Kénitra 31673
Musée Dar Cheraït, Tozeur 42661
Musée Dar-El-Annabi, Sidi Bou Saïd 42653
Musée Dar Hammamet, Hammamet 42627
Musée Dar Sghir, Mahdia 42636
Musée Dar-Si-Said, Marrakech 31677
Musée Dar Taieb, Nabeul 42645
Musée d'Archéologie, Antibes-Juan-les-Pins 11640
Musée d'Archéologie, Capdenac-le-Haut 12275
Musée d'Archéologie, Izernore 13257
Musée d'Archéologie, Le Pègue 13630
Musée d'Archéologie, Séverac-le-Château 15741
Musée d'Archéologie, Tournai 04083
Musée d'Archéologie Bargoin, Clermont-Ferrand 12548
Musée d'Archéologie, du Costume, d'Art et Traditions Populaires, Saint-Jean-de-Maurienne 15329
Musée d'Archéologie du Jura, Lons-le-Saunier 13815
Musée d'Archéologie et de Sciences Naturelles, Douai 12774
Musée d'Archéologie et d'Ethnologie, Villeneuve (Alpes-de-Haute-Provence) 16200
Musée d'Archéologie et d'Histoire Locale, Denain 12708
Musée d'Archéologie et Folklore, Verviers 04107
Musée d'Archéologie Industrielle, Vaulx 04103
Musée d'Archéologie Méditerranéenne, Marseille 13994
Musée d'Archéologie Précolombienne et de Préhistoire de la Martinique, Fort-de-France 30665

M

Musée d'Archéologie Préhistorique de l'Université de Liège, Liège 03803
Musée d'Archéologie Sublacustre, Sanguinet 15623
Musée d'Archéologie Tricastin, Saint-Paul-Trois-Châteaux 15449
Musée d'Archéologieque, Saint-Ciers-sur-Gironde 15208
Musée d'Archéologique, Saint-Raphaël 15480
Musée d'Architecture - La Loge, Bruxelles 03461
Musée d'Arenberg, Rebecq 03962
Musée d'Argonne, Varennes-en-Argonne 16075
Musée d'Armes Anciennes, La Cluse-et-Mijoux 13347
Musée d'Armes-Citadelle de Dinant, Dinant 03558
Musée d'Armes de Liège, Liège 03804
Musée d'Armes du Bordj Nord, Fès 31670
Musée d'Armes et d'Historie Militaire, Tournai 04084
Musée d'Arromanches, Arromanches-les-Bains 11701
Musée d'Art, Toulon (Var) 15923
Musée d'Art Africain, Dakar 38022
Musée d'Art Africain, Piré-sur-Seiche 14824
Musée d'Art Africain Religieux, Gentinnes 03647
Musée d'Art Américain, Giverny 13068
Musée d'Art Ancien, Bruxelles 03462
Musée d'Art Breton, Saint-Vougay 15541
Musée d'Art Contemporain, Dunkerque 12797
Musée d'Art Contemporain, Marseille 13995
Musée d'Art Contemporain, Noyers-sur-Serein 14480
Musée d'Art Contemporain de Lyon, Lyon 13883
Musée d'Art Contemporain de Montréal, Montréal 06352
Musée d'Art Contemporain de Tanger, Tanger 31696
Musée d'Art Contemporain en Plein Air le Cyclop, Milly-la-Forêt 14123
Musée d'Art Contemporain Fernet Branca, Saint-Louis-la-Chaussée 15377
Musée d'Art Contemporarin Africain à Cotonou, Cotonou 04183
Musée d'Art Contemporarin des Laurentides, Saint-Jérôme 06830
Musée d'Art de Joliette, Joliette 06030
Musée d'Art de Mont-Saint-Hilaire, Mont-Saint-Hilaire 06322
Musée d'Art du Valais, Sion 41876
Musée d'Art et d'Archéologie, Aurillac 11754
Musée d'Art et d'Archéologie, Hyères 13233
Musée d'Art et d'Archéologie, Laon 13521
Musée d'Art et d'Archéologie, Senlis 15715
Musée d'Art et d'Archéologie, Soulac-sur-Mer 15797
Musée d'Art et d'Archéologie Cluny, Cluny 12562
Musée d'Art et d'Archéologie de l'Université d'Antananarivo, Antananarivo 30534
Musée d'Art et d'Archéologie de Périgord, Périgueux 14778
Musée d'Art et d'Artisanat, Corsavy 12632
Musée d'Art et de Folklore Régional, Fessy 12930
Musée d'Art et de Tradition Populaire, Bazoges-en-Pareds 11886
Musée d'Art et de Traditions Populaires André Abet, Saint-Laurent-de-Cerdans 15356
Musée d'Art et des Traditions Populaires, Sizun 15771
Musée d'Art et d'Histoire, Belfort 11922
Musée d'Art et d'Histoire, Bulgnéville 12206
Musée d'Art et d'Histoire, Chaumont (Haute-Marne) 12483
Musée d'Art et d'Histoire, Cholet 12522
Musée d'art et d'histoire, Fribourg 41387
Musée d'Art et d'Histoire, Genève 41418
Musée d'Art et d'Histoire, Kaysersberg 13301
Musée d'Art et d'Histoire, Livry-Gargan 13790
Musée d'Art et d'Histoire, Meudon 14103
Musée d'Art et d'Histoire, Narbonne 14377
Musée d'Art et d'Histoire, Nice 14427
Musée d'Art et d'Histoire, Rochefort (Charente-Maritime) 15071
Musée d'Art et d'Histoire, Rueil-Malmaison 15138
Musée d'Art et d'Histoire, Saint-Denis (Seine-Saint-Denis) 15236
Musée d'Art et d'Histoire, Saint-Julien-du-Sault 15347
Musée d'Art et d'Histoire, Thorens-Glières 15898
Musée d'Art et d'Histoire Alfred Douët, Saint-Flour (Cantal) 15270
Musée d'Art et d'Histoire de Baugé, Baugé 11863
Musée d'Art et d'Histoire de Cognac, Cognac 12565
Musée d'Art et d'Histoire de la Chapelle Notre-Dame, Luz-Saint-Sauveur 13876
Musée d'Art et d'Histoire de Langres, Langres 13509
Musée d'Art et d'Histoire de Lisieux, Lisieux 13778
Musée d'Art et d'Histoire de Provence, Grasse 13109
Musée d'Art et d'Histoire de Puisaye, Villiers-Saint-Benoît 16244
Musée d'Art et d'Histoire de Toul, Toul 15922
Musée d'Art et d'Histoire des Côtes d'Armor, Saint-Brieuc 15191
Musée d'Art et d'Histoire du Judaïsme, Paris 14644
Musée d'Art et d'Histoire du Pays Blayais, Blaye 12023
Musée d'Art et d'Histoire Local, Pezens 14809
Musée d'Art et d'Histoire Louis Senlecq, L'Isle-Adam 13781
Musée d'Art et d'Histoire Marcel Dessal, Dreux 12792
Musée d'Art et d'Histoire Neuchâtel, Neuchâtel 41680
Musée d'Art et d'Histoire Roger Rodière, Montreuil-sur-Mer 14256
Musée d'Art et d'Histoire Romain Rolland, Clamecy 12539
Musée d'Art et d'Industrie, Saint-Étienne 15251
Musée d'Art et Historique, Montbéliard 14184
Musée d'Art et Tradition Populaire, Lamballe 13491
Musée d'Art et Traditions Populaires, Binic 11995
Musée d'Art et Traditions Populaires, Nancy 14352
Musée d'Art et Traditions Populaires, Tunis 42665
Musée d'Art Funeraire de Marville, Marville 14024
Musée d'Art Haïtien, Port-au-Prince 23321
Musée d'Art Inuit Brousseau, Québec 06686
Musée d'Art Islamique, Marrakech 31678
Musée d'Art Islamique de Rakkada, Rakkada 42646
Musée d'Art Islamique Raqqada, Kairouan 42631
Musée d'Art Juif, Paris 14645
Musée d'Art Juif J.C. Katz, Colmar 12583
Musée d'Art Local, Oisemont 14504
Musee d'Art Mercian Karuizawa, Miyota 28930
Musée d'Art Militaire, Vincey 16249
Musée d'Art Moderne, Belfort 11923
Musée d'Art Moderne, Céret 12340
Musée d'Art Moderne, Troyes 15998
Musée d'Art Moderne de la Ville de Paris, Paris 14646
Musée d'Art Moderne de Saint-Étienne Métropol, Saint-Étienne 15252
Musée d'Art Moderne et Contemporain, Strasbourg 15821
Musée d'Art Moderne et Contemporain, Toulouse 15934
Musée d'Art Moderne et d'Art Contemporain, Liège 03805
Musée d'Art Moderne et d'Art Contemporain, Nice 14428
Musée d'Art Moderne Grand-Duc Jean, Luxembourg 30478
Musée d'Art Moderne Jean Peské, Collioure 12575
Musée d'Art Moderne Lille Métropole, Villeneuve-d'Ascq 16202
Musée d'Art Moderne - Museum voor Moderne Kunst, Bruxelles 03463
Musée d'Art Moderne Richard Anacréon, Granville 13104
Musée d'Art Naïf, Lasne-Chapelle-Saint-Lambert 03777
Musée d'Art Naïf, Nice 14429
Musée d'Art Nègre, Yaoundé 05301
Musée d'Art Populaire, Le Monastier-sur-Gazeille 13617
Museé d'Art Populaire du Hauran, Bosra 42129
Musée d'Art Populaire Religieux, Saint-Julien-Molhesabate 15350
Musée d'Art Religieux, Blois 12033
Musée d'Art Religieux, Castelnau-Magnoac 12310
Musée d'Art Religieux, Fréland 13005
Musée d'Art Religieux, Marsal 13984
Musée d'Art Religieux, Vermand 16117
Musée d'Art Religieux et d'Art Mosan, Liège 03806
Musée d'Art Roger Quilliot, Clermont-Ferrand 12549
Musée d'Art Rustique, Vianden 30493
Musée d'Art Sacré, Dijon 12737
Musée d'Art Sacré, Hambye 13178
Musée d'Art Sacré, Le Monêtier-les-Bains 13619
Musée d'Art Sacré, Le Val 13660
Musée d'Art Sacré, Lectoure 13673
Musée d'Art Sacré, Lorgues 13820
Musée d'Art Sacré, Paris 14647
Musée d'Art Sacré, Saint-Béat 15182
Musée d'Art Sacré, Saint-Cirq-Lapopie 15211
Musée d'Art Sacré, Senez 15713
Musée d'Art Sacré, Viviers-en-Ardèche 16262
Musée d'Art Sacré Copte et Byzantin Scete, Le Revest-les-Eaux 13647
Musée d'Art Sacré de Fourvière, Lyon 13884
Musée d'Art Sacré de Rostilas Loukine, Arsonval 11704
Musée d'Art Sacré du Gard, Pont-Saint-Esprit 14893
Musée d'Art Sacré - Francis Poulenc, Rocamadour 15061
Musée d'Art Sacré Occidental, Le Revest-les-Eaux 13648
Musée d'Art Singulièr Raymond Reynaud, Sénas 15712
Musée d'Art Urbain, Montréal 06353
Musée d'Artisanat Monuments en Allumettes et Sciences Naturelles, Fontet 12976
Musée d'Arts Africains, Langonnet 13508
Musée d'Arts Décoratifs, Saint-Géoire-en-Valdaine 15278
Musée d'Arts et de Traditions Populaires, Amélie-les-Bains-Palalda 11581
Musée d'Arts et de Traditions Populaires, Cervione 12350
Musée d'Arts et de Traditions Populaires, Mussidan 14341
Musée d'Arts et de Traditions Populaires, Palinges 14558
Musée d'Arts et de Traditions Populaires, Poncé-sur-le-Loir 14875
Musée d'Arts et de Traditions Populaires, Saint-Paul-de-Fenouillet 15441
Musée d'Arts et d'Histoire du Donjon, Niort 14456
Musée d'Arts et Traditions Populaires, Cancale 12262
Musée d'Arts et Traditions Populaires, La Guérinière 13383
Musée d'Arts et Traditions Populaires, Marmoutier 13975
Musée d'Arts et Traditions Populaires, Yssingeaux 16308
Musée d'Arts et Traditions Populaires de Castellane et du Moyen Verdon, Castellane 12308
Musée d'Arts et Traditions Rurales, Saint-Alban-Auriolles 15156
Musée d'Arts Populaires, Laroque-des-Albères 13527
Musée d'Arts Populaires de Lille-Sud, Lille 13753
Musée d'Arts Reliqieux, Verdelais 16109
Musée d'Assier, Feurs 12931
Musée Daubigny, Auvers-sur-Oise 11770
Musée Dauphinois, Grenoble 13133
Musée David, Angers 11607
Musée David et Alice Van Buuren, Bruxelles 03464
Musée d'Avions Montélimar, Montélimar 14198
Musée de Baalbeck, Baalbeck 30155
Musée de Bagnes, Le Châble 41286
Musée de Basse-Navarre et des Chemins de Saint-Jacques, Saint-Palais (Pyrénées-Atlantiques) 15439
Musée de Bastia, Bastia 11859
Musée de Bataille, Mutzig 14343
Musée de Béjaïa, Béjaïa 00052
Musée de Beni Abbès, Bechar 00051
Musée de Bible et Terre Sainte, Paris 14648
Musée de Bibracte, Saint-Léger-sous-Beuvray 15367
Musée de Biot, Biot 11997
Musée de Boinchoux, Besse-et-Saint-Anastaise 11964
Musée de Boîtes à Musique et Automates, Sainte-Croix 41796
Musée de Bonoua, Bonoua 08832
Musée de Borda, Dax 12701
Musée de Bordj Moussa, Béjaïa 00053
Musée de Botanique, Lausanne 41545
Musée de Bouxwiller et du Pays de Hanau, Bouxwiller 12134
Musée de Bretagne, Rennes (Ille-et-Vilaine) 15015
Musée de Bulla Regia, Jendouba 42630
Musée de Cahors Henri-Martin, Cahors 12242
Musée de Calès, Lamanon 13490
Musée de Calligraphie et Epigraphie Arabe, Damascus 42131
Musée de Carouge, Carouge 41280
Musée de Castel-Merle, Sergeac 15722
Musée de Centre Social, Luzy 13878
Musée de Cerdagne, Sainte-Léocadie 15562
Musée de Cerdagne d'Eyne, Eyne 12896
Musée de Cerisiers La Grange a Janou, Cerisiers 12343
Musée de Cervières, Cervières 12349
Musée de Charlevoix, La Malbaie 06111
Musée de Charroux et de son Canton, Charroux-d'Allier 12420
Musée de Château, Durtal 12802
Musée de Château Guillaume, Lignac 13745
Musée de Châtillon-Coligny, Châtillon-Coligny 12468
Musée de Chauny, Chauny 12487
Musée de Cherchell, Cherchell 00055
Musée de Chine, Bruxelles 03465
Musée de Circonscription Archéologique, Sétif 00073
Musée de Cire de l'Historial, Sainte-Anne-d'Auray 15549
Musée de Cire et Salle de Documents Historiques, Perros-Guirec 14798
Musée de Cire-Historial de la Haute-Auvergne, Aurillac 11755
Musée de Cire Historique, Byblos 30168
Musée de Cires, Monaco 31608
Musée de Citadelle, Sisteron 15762
Musée de Clichy et Société Historique, Clichy 12559
Musée de Cohennoz, Cohennoz 12569
Musée de Cox, Cox 12665
Musée de Dar Jamaï, Meknès 31681
Musée de Diamare, Maroua 05299
Musée de Djelfa, Djelfa 00058
Musée de Dol, Dol-de-Bretagne 12753
Musée de Domaine de Lacroix-Laval, Marcy-l'Étoile 13962
Musée de Drummond Castle, Ile-de-Molène 13239
Musée de Enchanteurs, Antrain-sur-Couesnon 11646
Musée de Faïences Christine Viennet, Béziers 11975
Musée de Faykod, Aups 11748
Musée de Ferroviaire à Orient-Express, Sospel 15787
Musée de Figurines Végétales, Basse-Pointe 30664
Musée de Flandre, Bailleul 11830
Musée de Folklore, Nismes 03913
Musée de Folklore de la Chée, Heiltz-le-Maurupt 13199
Musée de Folklore et d'Histoire Armand Pellegrin, Opheylissem 03937
Musée de Folklore Flamand Jeanne Devos, Wormhout 16302
Musée de France de Berck-sur-Mer, Berck-sur-Mer 11941
Musée de Fresques, Blain 12007
Musée de Gajac, Villeneuve-sur-Lot 16220
Musée de Géologie, Lozanne 13861
Musée de Géologie Botanique - Espace Geologique, La Bourboule 13320
Musée de géologie René-Bureau, Sainte-Foy 06894
Musée de Germigny et Grange Germignonne, Germigny-des-Prés 13049
Musée de Glozel, Ferrières-sur-Sichon 12926
Musée de Grenoble, Grenoble 13134
Musée de Groesbeeck de Croix, Namur 03900
Musée de Guérin, Guérin 05920
Musée de Hottemme, Barvaux 03341
Musée de Jean Chouan, Saint-Ouen-des-Toits 15434
Musée de Jouet et da la Poupée Ancienne, L'Isle-sur-la-Sorgue 13784
Musée de Jouques, Jouques 13283
Musée de Jules Gounon-Loubens, Loubens-Lauragais 13828
Musée de Kabgayi, Kabgayi 37993
Musée de Kamouraska, Kamouraska 06041
Musée de Kent, Bouctouche 05446
Musée de Kerhinet, Saint-Lyphard 15380
Musée de la Baleine, Port-la-Nouvelle 14917
Musée de la Bande Dessinée et Musée Municipal, Angoulême 11617
Musée de la Banderette, Travers 41948
Musée de la Banque Nationale de Belgique, Bruxelles 03466
Musée de la Basilique et des Thermes Romains, Arlon 03324
Musée de la Bataille, Fontenoy-en-Puisaye 12973
Musée de la Bataille, Rocroi 15077
Musée de la Bataille de Tilly sur Seulles 1944, Tilly-sur-Seulles 15913
Musée de la Bataille des Ardennes, La Roche-en-Ardenne 03769
Musée de la Bataille des Ardennes, Wiltz 30497
Musée de la Bataille du 6 Août 1870, Woerth 16301
Musée de la Bataille Mai-Juin 1940, Semuy 15711
Musée de la Batellerie, Briennon 12178
Musée de la Batellerie, Conflans-Sainte-Honorine 12612
Musée de la Batellerie de l'Ouest, Redon 14991
Musée de la Batellerie d'Offendorf, Offendorf 14496
Musée de la Batellerie et du Plan Incliné, Saint-Louis-Arzviller 15374
Musée de la Batterie de Merville, Merville-Franceville-Plage 14094
Musée de la Bécane à Grand-Père, Sault-lès-Rethel 15657
Musée de la Bendrologie, Manéga 05275
Musée de la Bière, Armentières 11689
Musée de la Bière, Sankt Vith 04000
Musée de la Bière, Stenay 15808
Musée de la Blanchisserie Artisanale Joseph Gladel, Craponne (Rhône) 12668
Musée de la Bohème, Yviers 16311
Musée de la Boisselerie, Bois-d'Amont 12046
Musée de la Boite à Biscuits, Opheylissem 03938
Musée de la Boîte en Fer Blanc Lithographiée, Grand-Hallet 03659
Musée de la Bonneterie, Troyes 15999
Musée de la Bonneterie et du Négoce de la Toile, Quevaucamps 03957
Musée de la Boulangerie, Bonnieux 12058
Musée de la Boulangerie Rurale, La Haye-de-Routot 13384
Musée de la Bourrellerie et de la Ferronnerie, Lahamaide 03774
Musée de La Bouteille, Saint-Émilion 15246
Musée de la Brasserie-Museum van de Belgische Brouwers, Bruxelles 03467
Musée de la Broderie, Fontenoy-le-Château 12975
Musée de la Cabane des Bangards, Thann 15873
Musée de la Cadillac, Saint-Michel-sur-Loire 15420
Musée de la Calligraphie et de l'Imagerie Ancienne, Capvern-les-Bains 12277
Musée de la Camargue, Arles 11683
Musée de la Caricature, La Cassagne 13324
Musée de la Carte Postale, Mortagne-sur-Gironde 14288
Musée de la Casse, Orcières 14524
Musée de la Castre, Cannes 12269
Musée de la Cavalerie, Saumur 15662
Musée de la Cave, Damery 12697

Musée de la Cave des Champagnes de Castellane, Épernay 12832
Musée de la Céramique, Digoin 12729
Musée de la Céramique, Lezoux 13737
Musée de la Céramique, Rouen 15113
Musée de la Céramique d'Andenne, Andenne 03274
Musée de la Céramique et de l'Ivoire, Commercy 12595
Musée de la Céramique Saint-Jean-l'Aigle, Longwy 13813
Musée de la Cervoise, du Gruyt et des Bières Mediévales, Anthisnes 03276
Musée de la Chalosse, Montfort-en-Chalosse 14208
Musée de la Chapellerie, Espéraza 12861
Musée de la Chapellerie, Le Somail 13651
Musée de la Charronnerie, Brienne-la-Vieille 12176
Musée de la Chartreuse et Fondation Bugatti, Molsheim 14144
Musée de la Chasse, Boutenac 12132
Musée de la Chasse, de la Vénerie et de la Protection de la Nature, Lavaux-Sainte-Anne 03780
Musée de la Chasse et de la Nature, Baurech 11869
Musée de la Chasse et de la Nature, Chambord 12379
Musée de la Chasse et de la Nature, Foix 12951
Musée de la Chasse et de la Nature, Paris 14649
Musée de la Chasse et de la Venerie, Villiers-le-Duc 16243
Musée de la Châtaigneraie, Joyeuse 13287
Musée de la Chemiserie et de l'Elégance Masculine, Argenton-sur-Creuse 11676
Musée de la Chevalerie, Armes et Archeries, Carcassonne 12283
Musée de la Chimie, Jarrie 13261
Musée de la Chouannerie et des Guerres de l'Ouest, Plouharnel 14859
Musée de la Citadelle Vauban, Belle-île-en-Mer 11926
Musée de la Civilisation, Québec 06687
Musée de la Cloche, Sévrier 15744
Musée de la Cloche et du Carillon, Tellin 04052
Musée de la Cloches et de Sonnailles, Hérépian 13203
Musée de la Cochonnaille, Tourzel-Ronzieres 15975
Musée de la Cohue, Vannes (Morbihan) 16067
Musée de la Coiffe, Blesle 12026
Musée de la Coiffe et du Costume Oléronnais, Le Grand-Village-Plage 13588
Musée de la Coiffe et du Pays Pelebois le Prieuré, Souvigne (Deux-Sèvres) 15803
Musée de la Colombophilie, Bouvignies 12133
Musée de la Comédie, Preignac 14943
Musée de la Commanderie, Viaprès-le-Petit 16139
Musée de la Communication en Alsace, Riquewihr 15051
Musée de la Compagnie des Indes, Port-Louis 14918
Musée de la Compagnie Royale des Anciens Arquebusiers de Visé, Visé 04117
Musée de la Comtesse de Ségur, Aube 11721
Musée de la Conciergerie, Paris 14650
Musée de la Confiture, Pau 14756
Musée de la Confrérie des Vignerons, Vevey 41984
Musée de la Conscription, Strasbourg 15822
Musée de la Conserverie Le Gall, Loctudy 13798
Musée de la Construction Navale, Noirmoutier-en-l'Ile 14466
Musée de la Contrefaçon, Paris 14651
Musée de la Corbillière, Mer 14084
Musée de la Cordonnerie, Alby-sur-Chéran 11542
Musée de la Correrie, Saint-Pierre-de-Chartreuse 15456
Musée de la Corse, Corte 12635
Musée de la Course Landaise, Bascons 11856
Musée de la Coutellerie, Nogent-en-Bassigny 14461
Musée de la Coutellerie, Thiers 15881
Musée de la Création Franche, Bègles 11919
Musée de la Crèche, Chaumont (Haute-Marne) 12484
Musée de la Crèche Provençale, Cavaillon 12330
Musée de la Cuisine, Saint-Charles-de-Drummond 06809
Musée de la Curiosité et de la Magie, Paris 14652
Musée de la Dame aux Camélias, Gacé 13024
Musée de la Défense Aérienne, Alouette 05326
Musée de la Dentelle, Caudry 12324
Musée de la Dentelle, Chamalières-sur-Loire 12370
Musée de la Dentelle, Marche-en-Famenne 03846
Musée de la Dentelle, Mirecourt 14134
Musée de la Dentelle à la Main, Arlanc-en-Livradois 11680
Musée de la Dentelle du Puy-la-Galerie, Le Puy-en-Velay 13644
Musée de la Déportation et de la Résistance, Tarbes 15862
Musée de la Distillerie, Pontarlier 14898
Musée de la Distillerie Combier, Saumur 15663
Musée de la Douane et des Frontières, Hestrud 13211
Musée de la Draperie, Vienne 16151
Musée de la Fabuloserie, Dicy 12719
Musée de la Faculté de Médecine, Strasbourg 15823
Musée de la Faïence, Forges-les-Eaux 12979
Musée de la Faïence, Marseille 13996
Musée de la Faïence, Montereau-Fault-Yonne 14201
Musée de la Faïence de la Manufacture Royal Boch, La Louvière 03767
Musée de la Faïence et de la Poterie, Ferrière-la-Petite 12924
Musée de la Faïence et des Arts de la Table, Samadet 15615
Musée de la Faïence Jules Verlingue, Quimper 14968
Musée de la Faïencerie de Gien, Gien 13054
Musée de la Famille, Chasselay 12428
Musée de la Faucillonnaie, Vitré 16253
Musée de la Faune, Bellevaux 11931
Musée de la Faune, Lanhélin 13512
Musée de la Faune Alpine, Servoz 15731
Musée de la Faune et de la Nature, Séez 15700
Musée de la Faune et la Flore, Etaples 12873
Musée de la Fausse Monnaie, Saillon 41789
Musée de la Faux, Pont-Salomon 14895
Musée de la Fédération Nationale de Spéléologie, Revel 15028
Musée de la Femme Henriette Bathily, Gorée 38024
Musée de La Ferblanterie, La Tour-Blanche 13448
Musée de la Ferme, Blangy-sur-Bresle 12012
Musée de la Ferme d'Antan, Ploudiry 14850
Musée de la Ferme et des Vieux Métiers, Bosquentin 12079
Musée de la Ferronnerie, Francheville (Eure) 12996
Musée de la Fête Foraine, Plancher-les-Mines 14831
Musée de la Fève et de la Crèche, Blain 12008
Musée de la Figurine, La Grande-Fosse 13380
Musée de la Figurine, Toulon (Var) 15924
Musée de la Figurine, Tulette 16005
Musée de la Figurine Historique, Charre 12418
Musée de la Figurine Historique, Compiègne 12598
Musée de la Figurine Historique et du Jouet Ancien, Le Val 13661
Musée de la Filature, Angoustrine-Villeneuve 11623
Musée de la Flore et de la Faune du Haut Bugey, Hotonnes 13225
Musée de la Foire et du Théâtre Itinérant, Saint-Ghislain 03993
Musée de la Folie Marco, Barr 11854
Musée de la Fondation Lorin, Tanger 31697
Musée de la Fontaine et de l'Eau, Genval 03648
Musée de la Forêt, Saugues-en-Gévaudan 15651
Musée de la Forêt, Spa 04041
Musée de la Forêt Ardennaise, Renwez 15019
Musée de la Forge, Denazé 12711
Musée de la Forge, Romedenne 03982
Musée de la Forteresse, Luxembourg 30479
Musée de la Fortification Cuirassée, Villey-le-Sec 16241
Musée de la Foudre, Marcenat 13948
Musée de la Fourche et de la Vie Rurale, Mortier 03894
Musée de la Fourme et du Sabot, Sauvain 15669
Musée de la Fraise, Wépion 04144
Musée de la Fraise et du Patrimoine, Plougastel-Daoulas 14852
Musée de la Franc-Maçonnerie, Paris 14653
Musée de la France Protestante de l'Ouest, Monsireigne 14154
Musée de la Gare, Temiscaming 07065
Musée de la Gaspésie, Gaspé 05854
Musée de la Gendarmerie, Melun 14068
Musée de la Gendarmerie, Toamasina 30541
Musée de la Géologie et de l'Archéologie, Blangy-sur-Bresle 12013
Musée de la Glace, Mazaugues 14047
Musée de La Glacerie, La Glacerie 13378
Musée de la Goubaudière, Cholet 12523
Musée de la Gourmandise, Hermalle-sous-Huy 03700
Musée de la Grenouillère, Croissy-sur-Seine 12679
Musée de la Grosse Forge, Aube 11722
Musée de la Grotte de Cougnac, Payrignac 14767
Musée de la Grotte des Huguenots, Vallon-Pont-d'Arc 16055
Musée de la Guerre, Calais 12247
Musée de la Guerre, Clervaux 30459
Musée de la Guerre, Fromelles 13015
Musée de la Guerre 1939-45, Pourrain 14931
Musée de la Guerre au Moyen Âge, Castelnaud-la-Chapelle 12313
Musée de la Guerre de 1870, Gravelotte 13123
Musée de la Halte du Pèlerin, Larressingle 13529
Musée de la Hardt, Fessenheim 12928
Musée de la Haute-Auvergne, Saint-Flour (Cantal) 15271
Musée de la Haute Haine, Carnières 03510
Musée de la Haute Sûre, Martelange 03855
Musée de la Hesbaye, Remicourt 03965
Musée de la Houille Blanche et de ses Industries, Lancey 13496
Musée de la la Faune des Pyrénées, Nébias 14388
Musée de la Laine, du Vêtement de Laine et du Tissage, Louhossoa 13835
Musée de la Lauze, Annoisin-Châtelans 11629
Musée de la Lavande, Cabrières-d'Avignon 12221
Musée de la Lavande, Saint-Remèze 15482
Musée de la Légion Étrangère, Aubagne 11719
Musée de la Lessive, Spa 04042
Musée de la Liberation, Aywaille 03337
Musée de la Libération, Cherbourg-Octeville 12503
Musée de la Libération 15 Août 1944, Le Muy 13626
Musée de la Libération de la Normandie, Surrain 15843
Musée de la Ligne Koningshooikt-Wavre, Chaumont 03520
Musée de la Ligne Maginot, Bambiderstroff 11834
Musée de la Ligne Maginot, Hatten 13185
Musée de la Ligne Maginot, Hunspach 13228
Musée de la Ligne Maginot, Lembach 13679
Musée de la Ligne Maginot, Longuyon 13811
Musée de la Ligne Maginot Alpine, Sainte-Agnès 15547
Musée de la Ligne Maginot Immerhof, Hettange-Grande 13213
Musée de la Locomotion et de l'Attelage de l'île de France, Coignières 12570
Musée de la Loire, Cosne-Cours-sur-Loire 12636
Musée de la Lunette, Morez 14272
Musée de la Lutherie et de l'Archèterie Françaises, Mirecourt 14135
Musée de la Machine à Vapeur, Arc-en-Barrois 11660
Musée de la Machine à Vapeur, Oignies 14501
Musée de la Machine Agricole, Vroncourt 16279
Musée de la Machine Agricole Ancienne, Saint-Loup 15379
Musée de la Machine Agricole et à Vapeur, Ambert 11568
Musée de la Machine de Guerre, Espalion 12854
Musée de la Maille, Riorges 15047
Musée de la Main, Lausanne 41546
Musée de la Mairie de Crécy, Crécy-la-Chapelle 12670
Musée de la Maison Carrée, Nay 14386
Musée de la Maison de Tartarin, Tarascon (Bouches-du-Rhône) 15855
Musée de la Maison s'Kleenderfel, Weyersheim 16295
Musée de la Maquette, Clervaux 30460
Musée de la Marine, Etaples 12874
Musée de la Marine, Grasse 13110
Musée de la Marine, Honfleur 13221
Musée de la Marine, Saint-Brévin-les-Pins 15190
Musée de la Marine de la Seine, Caudebec-en-Caux 12323
Musée de la Marine de Loire, Châteauneuf-sur-Loire 12450
Musée de la Marine et de l'Économie de Marseille, Marseille 13997
Musée de la Médaille Miraculeuse, Lourdes 13841
Musée de la Médecine, Bruxelles 03468
Musée de la Mémoire de la Vigne et du Vin, Saint-Pierre-de-Mézoargues 15461
Musée de la Mémoire des Murs et d'Archéologie Locale, Verneuil-en-Halatte 16121
Musée de la Mémoire du Royans, Rochechinard 15066
Musée de la Mémoire et de la Paix, Clerval 12558
Musée de la Mémoire Paysanne, Estissac 12868
Musée de la Mer, Biarritz 11982
Musée de la Mer, Cannes 12270
Musée de la Mer, Gorée 38025
Musée de la Mer, Hane 16320
Musée de la Mer, îles-de-la-Madeleine 06008
Musée de la Mer, Le Grau-du-Roi 13590
Musée de la Mer, Nosy-Bé 30540
Musée de la Mer, Paimpol 14553
Musée de la Mer de Pointe-au-Père, Rimouski 06754
Musée de la Mer des Faluns, Doué-la-Fontaine 12779
Musée de la Mer OCEAM, Les Sables-d'Olonne 13714
Musée de la Métallurgie, Bogny-sur-Meuse 12043
Musée de la Métallurgie, Saint-Chély-d'Apcher 15202
Musée de la Météorite, L'Aigle 13483
Musée de la Météorologie, Valleraugues 16052
Musée de la Meunerie, Bannegon 11840
Musée de la Meunerie, Braine-le-Château 03398
Musée de la Meunerie, Challain-la-Potherie 12357
Musée de la Meunerie, Courtelevant 12658
Musée de la Meunerie, Le Breuil 13565
Musée de la Meunerie et de la Boulangerie, Harzé 03681
Musée de la Meunerie-Moulin de la Petite Strument, La Roche-en-Ardenne 03770
Musée de la Mine, Brassac-les-Mines 12146
Musée de la Mine, Bruay-la-Buissière 12193
Musée de la Mine, Champagnac-les-Mines 12384
Musée de la Mine, Grande-Synthe 13101
Musée de la Mine, Gréasque 13129
Musée de la Mine, La Machine 13391
Musée de la Mine, La Ricamaire 13408
Musée de la Mine, Nœux-les-Mines 14486
Musée de la Mine, Saint-Étienne 15253
Musée de la Mine, Sainte-Maime 15563
Musée de la Mine de Bois-du-Luc, Houdeng-Aimeries 03717
Musée de la Mine de Cap Garonne, Le Pradet 13642
Musée de la Mine de Cuivre et de Soufre, Saint-Pierre-de-la-Palud 15458
Musée de la Mine, de la Verrerie et du Chemin de Fer, Epinac 12838
Musée de la Mine et de l'École, Harnes 13181
Musée de la Mine et de Meunerie-Moulin de Marcy, Le Molay-Littry 13616
Musée de la Mine et des Hommes, Blanzy 12019
Musée de la Mine et des Techniques Minières, Giromagny 13065
Musée de la Mine Lucien-Mazars, Aubin 11732
Musée de la Mine Marcel Maulini, Ronchamp 15093
Musée de la Mine-Mémoire Ouvrière, Harchies 03676
Musée de la Mine, Musée de la Clouterie, Musée du Gazomètre, Fontaine-l'Evêque 03606
Musée de la Miniature, Montélimar 14199
Musée de la Miniature, Saint-Sauves-d'Auvergne 15504
Musée de la Mode, Marseille 13998
Musée de la Mode et du Textile, Paris 14654
Musée de la Mode Retrouvée, Digoin 12730
Musée de la Moineaudière, Xonrupt-Longemer 16305
Musée de la Moisson, Sagy (Magny-en-Vexin) 15149
Musée de la Monnaie de Paris, Paris 14655
Musée de la Montagne, Le Haut-du-Them 13592
Musée de la Montagne Saint-Pierre, Lanaye 03775
Musée de la Montre Droz-Flores, Villers-le-Lac 16230
Musée de la Mosaïque et des Émaux, Briare 12172
Musée de la Mosaique Gallo-Romaine, Sorde-l'Abbaye 15782
Musée de la Moto, Marseille 13999
Musée de la Moto, Savigny-les-Beaune 15681
Musée de la Motocyclette, Entrevaux 12827
Musée de la Musique, Anduze 11599
Musée de la Musique, Paris 14656
Musée de la Musique de Ouagadougou, Ouagadougou 05278
Musée de la Musique et des Instruments, Blangy-sur-Bresle 12014
Musée de la Musique Mécanique, Haubourdin 13187
Musée de la Musique Mécanique, Les Gets 13700
Musée de la Musique Mécanique, Mormoiron 14281
Musée de la Mytiliculture, Esnandes 12853
Musée de la Mytiliculture, Pénestin 14771
Musée de la Nacre et de la Tabletterie, Méru 14091
Musée de la Naissance de l'Aviation Suisse, Payerne 41719
Musée de la Natation, Chissey-les-Mâcon 12519
Musée de la Nativité, Lourdes 13842
Musée de la Nature, Ettelbruck 30475
Musée de la Nature, Sivry 04032
Musée de la Nature et de la Chasse, Thuir 15906
Musée de la Nature et des Sciences, Sherbrooke, Québec 06961
Musée de la Navigation Fluviale/ Ascenseur à Bateaux, Arques (Pas-de-Calais) 11696
Musée de la Pailletterie, Amilly 11589
Musée de la Paléontologie et de l'Évolution, Montréal 06354
Musée de la Papeterie-Canson et Montgolfier, Davézieux 12700
Musée de la Parfumerie, Paris 14657
Musée de la Parfumerie Cyrnarom, Bastia 11860
Musée de la Parfumerie Fragonard, Grasse 13111
Musée de la Parfumerie Fragonard, Paris 14658
Musée de la Parfumerie Galimard, Grasse 13112
Musée de la Parfumerie Molinard, Grasse 13113
Musée de la Parfumerie Molinard, Nice 14430
Musée de la Paysannerie, Baguer-Morvan 11827
Musée de la Paysannerie et des Vieux Métiers, Valentigney 16045
Musée de la Pêche, Concarneau 12602
Musée de la Pêche et de la Nature, Levallois-Perret 13728
Musée de la Pêche et des Poissons, Amberloup 03271
Musée de la Pelote Basque, Sare 15631
Musée de la Perle, Papeete 16324
Musée de la Petite Gare, Olonne-sur-Mer 14508
Musée de la Pétrification, Savonnières 15686
Musée de la Pharmacie, Tence 15869
Musée de la Pharmacie Albert Ciurana, Montpellier 14239
Musée de la Photo, Gracay 13091
Musée de la Photographie, Charleroi 03515
Musée de la Photographie, Mougins 14296
Musée de la Photographie, Port Louis 30683
Musée de la Pierre, Trambly 15977
Musée de la Pierre à Fusil, Luçay-le-Male 13863
Musée de la Pierre à Fusil, Meusnes 14107
Musée de la Pierre Écrite, Vézelay 16136
Musée de la Pierre et du Marbre, Basècles 03342
Musée de la Pierre et Site des Carrières, Maffle 03840
Musée de la Pierre Philosophale, Issoire 13250
Musée de la Pince, Montécheroux 14196

Musée de la Pointe Saint-Mathieu, Plougonvelin 14853
Musée de la Poire Tapée, Rivarennes 15055
Musée de la Police de l'île de Montréal et de l'île Bizard, Montréal 06355
Musée de la Police et de la Gendarmerie, Floreffe 03603
Musée de la Pomme, Bourg-Saint-Maurice 12109
Musée de la Pomme et du Cidre, Pleudihen-sur-Rance 14839
Musée de la Pomme et du Cidre, Vaudeurs 16083
Musée de la Pomme et du Cidre et des Métiers Traditionnels, Rosay 15101
Musée de la Porcelaine, Couleuvre 12643
Musée de la Porcelaine, Saint-Yrieix-la-Perche 15543
Musée de la Porte, Tubize 04097
Musée de la Porte de Hal-Hallepoortmuseum, Bruxelles 03469
Musée de la Porte de Thann, Cernay 12346
Musée de la Poste, Amélie-les-Bains-Palalda 11582
Musée de la Poste, Nantes 14362
Musée de la Poste, Paris 14659
Musée de la Poste, Prunelli-di-Fium'Orbu 14954
Musée de la Poste, Saint-Flour (Cantal) 15272
Musée de la Poste et des Techniques de Communication de Basse-Normandie, Caen 12230
Musée de la Poterie, La Borne 13317
Musée de la Poterie, Nospelt 30488
Musée de la Poterie, Vallauris 16049
Musée de la Poterie de Grès, Betschdorf 11968
Musée de la Poterie du Sartre, Saint-Alban-Auriolles 15157
Musée de la Poterie et de la Céramique, Sadirac 15146
Musée de la Poterie et Musée du Jambon d'Ardenne, La Roche-en-Ardenne 03771
Musée de la Poterne, Preuilly-sur-Claise 14946
Musée de la Poudrerie Nationale, Vaujours 16086
Musée de la Poupée, Aubeterre-sur-Dronne 11725
Musée de la Poupée, La Côte-d'Aime 13349
Musée de la Poupée, Le Beausset 13557
Musée de la Poupée, Montbrison 14189
Musée de la Poupée, Paris 14660
Musée de la Poupée et des Jouets Anciens, Guérande 13156
Musée de la Poupée et du Jouet Ancien, Wambrechies 16283
Musée de la Préfecture de Police, Paris 14661
Musée de la Préhistoire, Goyet-Gesves 03658
Musée de la Préhistoire, Jemeppe-sur-Sambre 03745
Musée de la Préhistoire, Le Mas-d'Azil 13611
Musée de la Prehistoire, Rânes 14984
Musée de la Préhistoire, Saint-Flovier 15273
Musée de la Préhistoire, Saint-Porchaire 15467
Musée de la Préhistoire, Solesmes 15775
Musée de la Préhistoire Charentaise, La Rochefoucauld 13421
Musée de la Préhistoire du Vercors, Vassieux-en-Vercors 16079
Musée de la Préhistoire en Wallonie, Flémalle 03602
Musée de la Préhistoire-Le Cairn, Saint-Hilaire-la-Forêt 15310
Musée de la Préhistoire/Paléontologie, Peyzac-le-Moustier 14807
Musée de la Préhistorique, Mozet 03896
Musée de la Première Guerre Mondiale, Villeneuve-les-Convers 16215
Musée de la Princerie, Verdun (Meuse) 16112
Musée de la Protohistoire, Beynac-et-Cazenac 11972
Musée de la Prune et du Pruneau, Lafitte-sur-Lot 13475
Musée de la Publicité, Paris 14662
Musée de la Radio, Mons 03883
Musée de la Radio, Villers-le-Lac 16231
Musée de la Radio et du Phonographe, Lanobre 13514
Musée de la Radio Galletti, Saint-Maurice-de-Rotherens 15410
Musée de la Radio Les Années 30, Saint-Victor-sur-Arlanc 15536
Musée de la Reddition, Reims 14998
Musée de la Régence, Ensisheim 12822
Musée de la Reine Bérengère, Le Mans 13607
Musée de la Reliure, Beaumesnil 11898
Musée de La Remise, Gex 13052
Musée de la Réparation Navale, Marseille 14000
Musée de la Reproduction du Son, Saint-Fargeau 15263
Musée de la Résistance, Bondues 12051
Musée de la Résistance, Cabanès (Aveyron) 12216
Musée de la Résistance, Clavières 12540
Musée de la Résistance, Fussy 13020
Musée de la Résistance, La Balme-de-Thuy 13307
Musée de la Résistance, Limoges 13766
Musée de la Résistance, Mussy-sur-Seine 14342
Musée de la Résistance, Sospel 15788
Musée de la Résistance, Varennes-Vauzelles 16076
Musée de la Résistance Bretonne, Saint-Marcel (Morbihan) 15394
Musée de la Résistance d'Anterrieux de la 7ème Cie et des Maquis du Réduit de la Truyère, Anterrieux 11637
Musée de la Résistance, de l'Internement et de la Déportation, Chamalières 12369
Musée de la Résistance du Mont Mouchet, Auvers (Haute-Loire) 11767
Musée de la Résistance en Morvan, Saint-Brisson (Nièvre) 15193
Musée de la Résistance en Zone Interdite, Denain 12709
Musée de la Résistance et de la Déportation, Agen 11493
Musée de la Résistance et de la Déportation, Angoulême 11618
Musée de la Résistance et de la Déportation, Besançon 11956
Musée de la Résistance et de la Déportation, Castelnau-le-Lez 12309
Musée de la Résistance et de la Déportation, Grenoble 13135
Musée de la Résistance et de la Déportation, Montauban 14174
Musée de la Résistance et de la Déportation en Picardie, Fargniers 12905
Musée de la Résistance et Déportation, Blois 12034
Musée de la Résistance et des Anciens Combattants, Floing 12947
Musée de la Résistance Henri Queuille, Neuvic (Corrèze) 14405
Musée de la Résistance Nationale, Champigny-sur-Marne 12390
Musée de la Révolution Française, Vizille 16264
Musée de la Rose et des Arts, Les Rosiers-sur-Loire 13711
Musée de la Roue Tourne, La Mothe-Achard 13394
Musée de la Route, Henri-Chapelle 03695
Musée de la Route, Mons 03884
Musée de la Rubanerie Cominoise, Comines 03525
Musée de la Ruralité, Donzac 12767
Musée de la Sablière, Châteaubriant 12442
Musée de la Sahla, Audun-le-Tiche 11742
Musée de la Schlitte et des Métiers du Bois, Muhlbach-sur-Munster 14322
Musée de la Science-Fiction, de l'Utopie et des Voyages Extraordinaires, Yverdon-les-Bains 42048
Musée de la Scierie Ariel-Vincent, Sainte-Croix-aux-Mines 15555
Musée de la Seconde Guerre Mondiale, Digne-les-Bains 12724
Musée de la Seconde Guerre Mondiale, Le Val-Saint-Père 13663
Musée de la Seconde Guerre Mondiale, Tourcoing 15952
Musée de la Seconde Guerre Mondiale Roger Bellon, Conlie 12613
Musée de la Sénatorie, Guéret 13158
Musée de la SNSM, Ile-de-Molène 13240
Musée de la Société Archéologique de Corseul, Corseul 12634
Musée de la Société Archéologique et Historique de la Charente, Angoulême 11619
Musée de la Société de l'Histoire du Protestantisme Français, Paris 14663
Musée de la Société d'Histoire et d'Archéologie, Briord 12182
Musée de la Soie, Monoblet 14149
Musée de la Soierie, Charlieu 12411
Musée de la Sorcellerie, Blancafort 12010
Musée de La Source, Valignat 16046
Musée de la Station Biologique, Roscoff 15104
Musée de la Statue Religieuse, Lourdes 13843
Musée de la Straize, Gy-en-Sologne 13174
Musée de la Taillanderie, Nans-sous-Sainte-Anne 14359
Musée de la Tannerie Granval, Maringues 13969
Musée de la Tapiserie et des Arts du Tissu, Tournai 04085
Musée de la Tapisserie Contemporaine, Angers 11608
Musée de la Tapisserie de Bayeux, Bayeux 11874
Musée de la Targette, Neuville-Saint-Vaast 14409
Musée de la Terre, Rambervillers 14980
Musée de la Thiérache, Vervins 16131
Musée de la Toinette, Murat-le-Quaire 14334
Musée de la Tonnellerie, Cadillac 12225
Musée de la Tonnellerie, Chécy 12495
Musée de la Tonnellerie, Generac 13039
Musée de la Tonnellerie, Listrac-Médoc 13787
Musée de la Tonnellerie Animée, Vergèze 16115
Musée de la Tonnellerie et du Vin, Aloxe-Corton 11560
Musée de la Tour, Coucy-le-Château-Auffrique 12639
Musée de la Tour aux Puces, Thionville 15884
Musée de la Tour Carrée, Rosans 15100
Musée de la Tour de l'Abbaye, Jouarre 13281
Musée de la Tour de l'Orle d'Or, Semur-en-Auxois 15707
Musée de la Tour des Echevins, Luxeuil-les-Bains 13872
Musée de la Tour d'Honneur, Lesparre-Médoc 13724
Musée de la Tour du Moulin, Marcigny 13957
Musée de la Tour, Histoire et Traditions Locales, Antibes-Juan-les-Pins 11641
Musée de la Tour Romaine, Arlon 03325
Musée de la Tour Salamandre, Beaumont 03349
Musée de la Tournerie, Lavans-lès-Saint-Claude 13545
Musée de la Tradition Champenoise, Épernay 12833
Musée de la Tradition de la Brigade des Sapeurs-Pompiers de Paris, Paris 14664
Musée de la Tuile, Blajan 12009
Musée de la Tuilerie, Malbrans 13930
Musée de la Vache et de Alpages, Frangy 12997
Musée de la Vallée à Barcelonnette, Barcelonnette 11848
Musée de la Vallée à Jausiers, Jausiers 13266
Musée de la Vallée à Lauzet, La Lauzet 13388
Musée de la Vallée à Meyronnes, Meyronnes 14111
Musée de la Vallée à Pontis, Pontis 14905
Musée de la Vallée à Saint-Paul-sur-Ubaye, Saint-Paul-sur-Ubaye 15447
Musée de la Vallée d'Aure, Ancizan 11594
Musée de la Vallée de la Creuse, Eguzon 12814
Musée de la Vannerie, Cadenet 12223
Musée de la Vannerie, Remilly-sur-Lozon 15007
Musée de la Vannerie des Compagnons du Terroir, Villaines-les-Rochers 16161
Musée de la Vannerie et de l'Artisant, Vallabrègues 16047
Musée de la Vannerie L'Osarium, Le Mayet-en-Montagne 13612
Musée de la Vénerie et des Spahis, Senlis 15716
Musée de la Verrerie, Blangy-sur-Bresle 12015
Musée de la Verrerie, Saint-Galmier 15275
Musée de la Verrerie, Veauche 16090
Musée de la Vie, Bidon 11985
Musée de la Vie Artisanale, Ham-sur-Heure 03671
Musée de la Vie Bourguignonne Perrin-de-Puycousin, Dijon 12738
Musée de la Vie d'Autrefois, Marchin 03849
Musée de la Vie d'Autrefois, Vivoin 16263
Musée de la Vie Lensoise, Lens 03784
Musée de la Vie Locale de Walhain, Walhain 04129
Musée de la Vie Locale et de la Coutellerie, Gembloux 03615
Musée de la Vie Monastique, Saint-Hilaire-du-Harcouët 15308
Musée de la Vie Paysanne, Montquintin 03892
Musée de la Vie Populaire, Soumagne 04039
Musée de la Vie Régionale, Cerfontaine 03512
Musée de la Vie Régionale des Rièzes et des Sarts, Cul-des-Sarts 03532
Musée de la Vie Romantique, Paris 14665
Musée de la Vie Rurale, Desaignes 12715
Musée de la Vie Rurale, Mont-Saint-Jean 14161
Musée de La Vie Rurale, Saint-Quentin-Fallavier 15476
Musée de la Vie Rurale, Steenwerck 15807
Musée de la Vie Rurale, Tusson 16013
Musée de la Vie Rurale Condruze, Nandrin 03908
Musée de la Vie Rurale d'Huissignies, Huissignies 03722
Musée de la Vie Rurale en Wallonie, Saint-Hubert 03995
Musée de la Vie Rurale et Forestière, Saint-Michel-en-Thiérache 15419
Musée de la Vie Rurale Li Vîle Grègne, Erezée 03584
Musée de la Vie Tihangeoise, Tihange 04071
Musée de la Vie Wallonne, Liège 03807
Musée de La Vieille École, Les Valagans 13719
Musée de la Vieille Prune, Souillac 15794
Musée de la Vigne, La Roche-Vineuse 13420
Musée de la Vigne, Verzenay 16132
Musée de la Vigne Benoit-Raclet, Romanèche-Thorins 15085
Musée de la Vigne et de la Vie Rurale, Plauzat 14834
Musée de la Vigne et du Patrimoine, Brioude 12184
Musée de la Vigne et du Terroir, Saint-Pourçain-sur-Sioule 15468
Musée de la Vigne et du Vin, Aigle 41094
Musée de la Vigne et du Vin, Ansouis 11635
Musée de la Vigne et du Vin, Aubière 11728
Musée de la Vigne et du Vin, Boën-sur-Lignon 12038
Musée de la Vigne et du Vin, Boudry 41248
Musée de la Vigne et du Vin, Buzet-sur-Baise 12215
Musée de la Vigne et du Vin, Cabara 12217
Musée de la Vigne et du Vin, Cérons 12348
Musée de la Vigne et du Vin, Gabarnac 13022
Musée de la Vigne et du Vin, Gigondas 13057
Musée de la Vigne et du Vin, Gruissan 13153
Musée de la Vigne et du Vin, Le Mesnil-sur-Oger 13615
Musée de la Vigne et du Vin, Lézignan-Corbières 13736
Musée de la Vigne et du Vin, Lhomme 13739
Musée de la Vigne et du Vin, Mareau-aux-Près 13964
Musée de la Vigne et du Vin, Onzain 14514
Musée de la Vigne et du Vin, Pierreclos 14814
Musée de la Vigne et du Vin, Portets 14923
Musée de la Vigne et du Vin, Vaux-en-Beaujolais 16088
Musée de la Vigne et du Vin d'Anjou, Saint-Lambert-du-Lattay 15354
Musée de la Vigne et du Vin de Franche-Comté, Arbois 11658
Musée de la Vigne et du Vin-Galerie d'Art, Lods 13801
Musée de la Vigne-Moulin du Haut-Benauge, Gornac 13086
Musée de la Vilaine Maritime, La Roche-Bernard 13412
Musée de la Villa Romaine de Pully, Pully 41738
Musée de la Ville de Saint-Quentin-en-Yvelines, Montigny-le-Bretonneux 14217
Musée de la Ville de Saverne, Saverne 15675
Musée de la Ville d'Eaux, Spa 04043
Musée de la Viscose, Échirolles 12807
Musée de la Viticulture, Santenay 15626
Musée de la Voie Sacrée, Souilly 15796
Musée de la Voiture de Course, Savigny-les-Beaune 15682
Musée de la Voiture Hippomobile et de la Citadelle de Bourg, Bourg-sur-Gironde 12112
Musée de l'Abbaye, Echternach 30469
Musée de l'Abbaye, Gaillac 13025
Musée de l'Abbaye de Bourgueil, Bourgueil 12124
Musée de l'Abbaye de Clairvaux, Ville-sous-la-Ferté 16176
Musée de l'Abbaye de Font-Caude, Cazedarnes 12333
Musée de l'Abbaye de Port-Royal, Magny-lès-Hameaux 13914
Musée de l'Abbaye de Saint-Mathieu, Plougonvelin 14854
Musée de l'Abbaye d'Orval, Villers-devant-Orval 04114
Musée de l'Abbaye Romane, La Sauve-Majeure 13441
Musée de l'Abbaye Saint-Vigor, Cérisy-la-Forêt 12344
Musée de l'Abbaye Sainte-Croix, Les Sables-d'Olonne 13715
Musée de l'Abeille, Château Richer 05600
Musée de l'Abeille, Le Tronchet 13658
Musée de l'Abeille, Tilff 04072
Musée de l'Abeille et du Braconnage, Salbris 15593
Musée de l'Abeille Vivante, Vitré 16254
Musée de l'Abri, Hatten 13186
Musée de l'Absinthe, Auvers-sur-Oise 11771
Musée de l'Académie Saint-Anselme, Aosta 25246
Musée de l'Accordéon, Montmagny 06325
Musée de l'Accordéon, Siran 15761
Musée de Lachine, Lachine 06123
Musée de l'Affiche de la Carte Postale et de l'Art Graphique, Toulouse 15935
Musée de Laghouat, Laghouat 00068
Musée de l'Agriculture Auvergnate, Coltines 12589
Musée de l'Agriculture Catalane, Saint-Michel-de-Llotes 15415
Musée de l'Agriculture et de la Vie Rurale, Labrousse 13466
Musée de l'Aile Delta, Athis-Mons 11716
Musée de l'Air et de l'Espace, Le Bourget (Seine-Saint-Denis) 13564
Musée de l'Alambic, Quimper 14969
Musée de l'Alambic, Saint-Desirat 15237
Musée de l'Albanais, Rumilly 15140
Musée de l'Algérie Française, Perpignan 14789
Musée de l'Alta Rocca, Lévie 13732
Musée de l'Amande et du Macaron, Montmorillon 14231
Musée de l'Amérique Française, Québec 06688
Musée de l'Amour et des ses Traditions, Oger 14499
Musée de l'Ancien Collège des Jésuites, Reims 14999
Musée de l'Ancien Évêché, Grenoble 13136
Musée de l'Ancien Havre, Le Havre 13595
Musée de l'Ancienne Abbaye de Landévennec, Landévennec 13498
Musée de l'Ancienne Chartreuse de la Verne, Collobrières 12576
Musée de L'Ancienne Forge, Pesmes 14800
Musée de l'Ancienne Orangerie-Site Buffon, Montbard 14179
Musée de l'Âne, Venosc 16104
Musée de l'Arc de Triomphe, Paris 14666
Musée de l'Archéologie, Céret 12341
Musée de l'Archerie Traditionnelle, Ayssenes 11807
Musée de l'Architecture Traditionnelle, Sfax 42651
Musée de l'Ardenne, Charleville-Mézières 12410
Musée de l'Ardoise, Fumay 13019
Musée de l'Ardoise, Renazé 15010
Musée de l'Ardoise, Trélazé 15987
Musée de l'Areuse, Boudry 41249
Musée de l'Argonne, Sainte-Menehould 15572
Musée de l'Arles et Provence Antiques, Arles 11684
Musée de l'Armagnac, Condom 12609
Musée de l'Armée, Paris 14667
Musée de l'Armée des Alpes, Sospel 15789
Musée de l'Armée Nationale Populaire, Alger 00039

Musée de l'Art Campanaire, Labergement-Sainte-Marie 13465
Musée de l'Art Culinaire, Villeneuve-Loubet 16217
Musée de l'Art du Sucre, Cordes-sur-Ciel 12624
Musée de l'Art Forain et de la Musique Mécanique, Conflans-en-Jarnisy 12611
Musée de l'Art Wallon, Liège 03808
Musée de l'Artillerie, Draguignan 12788
Musée de l'Artisanat et des Traditions Rurales, Brebotte 12150
Musée de l'Artisanat Rural Ancien, Tigy 15911
Musée de l'Assistance Publique, Bruxelles 03470
Musée de l'Assistance Publique-Hôpitaux de Paris, Paris 14668
Musée de l'Association des Marches Folkloriques de l'Entre-Sambre-et-Meuse, Gerpinnes 03655
Musée de l'Atelier de Rosa Bonheur, Thomery 15888
Musée de l'Atelier de Xaintonge, Pont-l'Abbé-d'Arnoult 14889
Musée de l'Atlantique, La Rochelle 13423
Musée de l'Attelage, Marquillies 13981
Musée de l'Attelage et du Cheval, Sérignan (Hérault) 15725
Musée de l'Auberge Symmes, Gatineau 05863
Musée de l'Audiovisuel, Montarcher 14169
Musée de l'Auditoire, Sainte-Suzanne 15581
Musée de l'Auld Alliance Franco-Écossaise, Aubigny-sur-Nère 11730
Musée de l'Auto, Moto et Cycle Jean Tua, Genève 41419
Musée de l'Automate et de la Robotique, Souillac 15795
Musée de l'Automobile, Martigny 41611
Musée de l'Automobile, Orgon 14527
Musée de l'Automobile, Plan-les-Ouates 41724
Musée de l'Automobile, Valençay 16036
Musée de l'Automobile, Velaine-en-Haye 16092
Musée de l'Automobile de Vendée, Talmont-Saint-Hilaire 15846
Musée de l'Automobile Française, Reims 15000
Musée de l'Automobile La Belle Époque, Pont-l'Évêque 14891
Musée de l'Automobile Miniature, Fontenay-sur-Mer 12971
Musée de l'Automobile Sportive, Sanary-sur-Mer 15620
Musée de l'Automobiliste, Mougins 14297
Musée de l'Avallonnais, Avallon 11783
Musée de l'Aventure Industrielle du Pays d'Apt, Apt 11654
Musée de l'Aviation, Perpignan 14790
Musée de l'Aviation, Savigny-les-Beaune 15683
Musée de l'Aviation, Warluis 16286
Musée de l'Aviation Légère de l'Armée de Terre, Dax 12702
Musée de l'Eau, Pont-en-Royans 14886
Musée de l'Eau de Vie et des Vieux Métiers, Valognes 16057
Musée de l'Échevinage, Saintes 15583
Musée de l'Eclairage, Dinant 03559
Musée de l'École, Bourges 12115
Musée de l'Ecole, Bruxelles 03471
Musée de l'École, Carcassonne 12284
Musée de l'Ecole, Chartres 12423
Musée de l'École, Châtelus (Allier) 12464
Musée de l'Ecole, Echery 12806
Musée de l'École, Saint-Étienne-de-Tinée 15255
Musée de l'École, Saint-Remy 15484
Musée de l'Ecole, Villeneuve-d'Ascq 16203
Musée de l'Ecole 1900, Saint-Martin-des-Olmes 15402
Musée de l'Ecole Biblique et Archéologique Française, Jerusalem 34057
Musée de l'Ecole de Nancy, Nancy 14353
Musée de l'École Jules Ferry, Fraize 12995
Musée de l'Ecole Navale, Lanvéoc-Poulmic 13520
Musée de l'Ecole Rurale, Trégarvan 15981
Musée de l'École Rurale d'Auvergne, Messeix 14097
Musée de l'Ecorché d'Anatomie, Le Neubourg 13627
Musée de Leers, Leers 13675
Musée de l'Eglise Abbatiale de Floreffe, Floreffe 03604
Musée de l'Église de Lorris, Lorris 13825
Musée de l'Électricité, Dijon 12739
Musée de l'Électricité, Saint-Aignan 15154
Musée de l'Elysée, Lausanne 41547
Musée de l'Empéri, Salon-de-Provence 15610
Musée de l'Empéri de Salon et de la Crau, Salon-de-Provence 15611
Musée de l'Enfance, Alger 00040
Musée de l'Entité de Walcourt, Walcourt 04126
Musée de l'Ephèbe, Agde 11492
Musée de l'Epicerie et des Commerces, Lignerolles 13747
Musée de l'Erotisme, Paris 14669
Musée de l'Escrime Charles Debeur, Bruxelles 03472
Musée de l'Espace, Kourou 16318
Musée de l'Estérel, Les Adrets-de-l'Estérel 13682
Musée de l'Etang de Thau, Bouzigues 12136
Musée de l'Etiquette, Langon 13507
Musée de l'Evéché, Sion 41877
Musée de l'Eventail, Paris 14670
Musée de l'Habitat Ancien, Bessans 11960
Musée de l'Habitat Rural en Pays d'Aigues, La Tour-d'Aigues 13450
Musée de l'Habitat Rural Passé et des Traditions, La Garnache 13376
Musée de l'Heure et du Feu, Villerest 16226
Musée de l'Histoire, Sartrouville 15648
Musée de l'Histoire de France, Paris 14671
Musée de l'Histoire de France à travers son Armée, Montet et Bouxal 14204
Musée de l'Histoire de la Médecine, Vandœuvre-les-Nancy 16064
Musée de l'Histoire de la Médecine de Toulouse, Toulouse 15936
Musée de l'Histoire de la Poterie, La Chapelle-des-Pots 13335
Musée de l'Histoire de l'Abbaye, Saint-Sauveur-le-Vicomte 15508
Musée de l'Histoire de l'Emigration Percheronne au Canada, Tourouvre 15959
Musée de l'Histoire des Bastides et de l'Evolution de l'Habitat, Bassoues 11858
Musée de l'Histoire du Costume, Scey-sur-Saône-et-Saint-Albin 15690
Musée de l'Histoire du Fer, Jarville 13262
Musée de l'Histoire du Feu, Campuzan 12259
Musée de l'Histoire et de la Vie Salmiennes, Vielsalm 04111
Musée de l'Histoire Paysanne, Maraye-en-Othe 13946
Musée de l'Histoire Vivante Parc Montreau, Montreuil-sous-Bois 14255
Musée de l'Historie de Montpellier, Montpellier 14240
Musée de l'Histoirie de Rosny, Rosny-sous-Bois 15109
Musée de l'Holographie, Paris 14672
Musée de l'Homme de Néandertal, La Chapelle-aux-Saints 13331
Musée de l'Homme et de l'Industrie, Le Creusot 13577
Musée de l'Hôpital, Yssingeaux 16309
Musée de l'Hôpital Général de Dijon, Dijon 12740
Musée de l'Hôpital Psychiatrique, Brumath 12198
Musée de l'Horlogerie, Pesmes 14801
Musée de l'Horlogerie, Saint-Nicolas-d'Aliermont 15426
Musée de l'Horlogerie Ancienne, Fougères (Ille-et-Vilaine) 12985
Musée de l'Horlogerie du Haut-Doubs, Morteau 14289
Musée de l'Horlogerie et de l'Emaillerie, Genève 41420
Musée de l'Horlogerie et du Décolletage, Cluses 12563
Musée de l'Hospice Comtesse, Lille 13754
Musée de l'Hospice Saint-Roch, Issoudun 13251
Musée de l'Hôspital, La Rochefoucauld 13422
Musée de l'Hôtel de Barral, Soissons 15773
Musée de l'hôtel de Berny, Amiens 11587
Musée de l'Hôtel de Vermandois, Senlis 15717
Musée de l'Hôtel de Ville, Amboise 11578
Musée de l'Hôtel de Ville, Hondschoote 13219
Musée de l'Hôtel-Dieu, Beaune 11902
Musée de l'Hôtel-Dieu, Mantes-La Jolie 13944
Musée de l'Hôtel-Dieu Porrentruy (MHDP), Porrentruy 41728
Musée de l'Hôtel Sandelin, Saint-Omer 15431
Musée de l'Hydraviation, Biscarrosse 11999
Musée de Lignan et du Canton de Créon, Lignan-de-Bordeaux 13746
Musée de l'Iguanodon, Bernissart 03364
Musée de l'Île de France, Sceaux 15688
Musée de l'Île d'Oleron Aliénor d'Aquitaine, Saint-Pierre-d'Oléron 15462
Musée de l'Image, Épinal 12839
Musée de l'Image Populaire, Pfaffenhoffen 14810
Musée de l'Impression sur Etoffes, Mulhouse 14324
Musée de l'Imprimerie, Bordeaux 12067
Musée de l'Imprimerie, Nantes 14363
Musée de l'Imprimerie de Lyon, Lyon 13885
Musée de l'Imprimerie-Museum van de Boekdrukkunst, Bruxelles 03473
Musée de l'Industrie Régionale de la Chaussure, Saint-André-de-la-Marche 15166
Musée de l'Infanterie, Montpellier 14241
Musée de l'Insolite, Cabrerets 12218
Musée de l'Insolite, Loriol-sur-Drôme 13822
Musée de l'Institut du Monde Arabe, Paris 14673
Musée de l'Institut Pasteur de Lille, Lille 13755
Musée de l'Invasion Aéroportée, Bénouville 11939
Musée de l'Océanie, Saint-Symphorien-sur-Coise 15520
Musée de Lodève, Lodève 13800
Musée de l'Œuf et de l'Icône, Banne 11839
Musée de l'Œuvre, Avignon 11790
Musée de l'Œuvre Notre-Dame, Strasbourg 15824
Musée de l'Œuvre Viollet-le-Duc, Vézelay 16137
Musée de l'Olivette, Sorède 15784
Musée de l'Olivier, Les Baux-de-Provence 13689
Musée de l'Olivier, Nyons 14489
Musée de l'Olivier des Mées, Mées 14058
Musée de l'Opéra, Vichy 16147
Musée de l'Opinel, Saint-Jean-de-Maurienne 15330
Musée de l'Optique, Biesheim 11987
Musée de l'Optométrie, Bures-sur-Yvette 12208
Musée de l'Or, Jumilhac-le-Grand 13293
Musée de l'Orangerie, Paris 14674
Musée de l'Oratoire Saint-Joseph, Montréal 06356
Musée de l'Orce, Saint-Georges-sur-la-Prée 15283
Musée de l'Ordre de la Libération, Paris 14675
Musée de l'Ordre de Malte, Bardonnex 41152
Musée de l'Orfevrerie de la Communaute Française, Seneffe 04008
Musée de l'Ostréiculture, Marennes 13966
Musée de l'Ostrévant, Bouchain 12081
Musée de Lourdes, Lourdes 13844
Musée de l'Ours des Cavernes, Entremont-le-Vieux 12826
Musée de l'Outil, Magny-en-Vexin 13913
Musée de l'Outil des Métiers du Bois, Montjoie-Saint-Martin 14219
Musée de l'Outil et de la Vie au Village, Saint-Privat-des-Près 15470
Musée de l'Outil et des Métiers, Tinténiac 15915
Musée de l'Outil et des Vieux Métiers, La Celle-Guenand 13325
Musée de l'Outil, la Forge Médiévale, Wy-Dit-Joli-Village 16304
Musée de l'Outillage Agricole et Viticole du Château Aney, Cussac-Fort-Médoc 12690
Musée de l'Outillage Ancien et des Traditions, Briançonnet 12170
Musée de l'Outillage Artisanal Rural et Bourguignon, Cruzille-en-Mâconnais 12682
Musée de l'Outillage et des Techniques Traditionnelles de la Taille de Pierre, Coucouron 12638
Musée de Louvain-la-Neuve, Louvain-la-Neuve 03834
Musée de l'Uniforme Légionnaire, Puyloubier 14961
Musée de l'Union Compagnonnique des Devoirs Unis, Nantes 14364
Musée de l'Université de Tuléar, Tuléar 30543
Musée de l'Uranium, Lachaux 13471
Musée de Machines Agricoles Miniatures, Guerlesquin 13161
Musée de Maison-Alfort, Maisons-Alfort 13923
Musée de Maquettes Animées, Vallon-en-Sully 16053
Musée de Marine, Faranui 16319
Musée de Maroc Telecom, Rabat 31689
Musée de Marrakech, Marrakech 31679
Musée de Matériel Agricole, Saint-Laurent-des-Autels 15359
Musée de Matériel Viticole, Saint-Laurent-des-Autels 15360
Musée de Matière Médicale, Paris 14676
Musée de Mayombe, Kinshasa 08796
Musée de Mbandaka, Mbandaka 08802
Musée de Megève, Megève 14059
Musée de Melun, Melun 14069
Musée de Miellerie, Saint-Saturnin-les-Avignon 15503
Musée de Minéralogie, Strasbourg 15825
Musée de Minéralogie de l'Ecole des Mines de Paris, Paris 14677
Musée de Minéralogie et de Géologie, Bruxelles 03474
Musée de Minéralogie et du Pétrographie, Ambazac 11565
Musée de Miquelon, Miquelon 37996
Musée de Modèles Réduits, La Rochelle 13424
Musée de Modélisme, Savigny-les-Beaune 15684
Musée de Montaut-Centre Culturel, Montaut 14178
Musée de Montmartre, Paris 14678
Musée de Moulages, Strasbourg 15826
Musée de Musiques Méchanique, L'Auberson 41135
Musée de Natitingou, Natitingou 04184
Musée de Nevers Magny-Cours, Magny-Cours 13912
Musée de Nogent, Nogent-sur-Marne 14463
Musée de Normandie, Caen 12231
Musée de Notre-Dame-de-Paris, Paris 14679
Musée de Ouadane, Nouakchott 30674
Musée de Paléonthologie et Préhistoire, Palaja 14556
Musée de Paléontologie, La Voulte-sur-Rhône 13457
Musée de Paléontologie, Menat 14070
Musée de Paléontologie, Villers-sur-Mer 16233
Musée de Paléontologie, Villeurbanne 16238
Musée de Paléontologie Christian Guth, Chilhac 12511
Musée de Paléontologie Humaine de Terra Amata, Nice 14431
Musée de Palmyra, Palmyra 42137
Musée de Payerne et Abbatiale, Payerne 41720
Musée de Peinture J. Drevon, Saint-Jean-de-Bournay 15322
Musée de Peinture Louis Français, Plombières-les-Bains 14845
Musée de Peinture Mario Prassinos, Saint-Rémy-de-Provence 15489
Musée de Peintures Contemporaines Hors du Temps, Pierrefeu (Alpes-Maritimes) 14816
Musée de Personnages de Cire, Waterloo 04136
Musée de Picardie, Amiens 11588
Musée de Pierre Corneille, Le Petit-Couronne 13631
Musée de Plein Air, Gueugnon 13163
Musée de Plein Air, Parakou 04186
Musée de Plein Air, Saint-Vallier-de-Thiey 15532
Musée de Plein Air, Seurre 15740
Musée de Plein-Air de Cherchell, Cherchell 00056
Musée de Plein Air des Maisons Comtoises, Nancray 14349
Musée de Plein Air du Quercy de Cuzals, Sauliac-sur-Célé 15655
Musée de Pont-Aven, Pont-Aven 14880
Musée de Pont-Saint-Esprit, Saint-Paulet-de-Caisson 15450
Musée de Pontarlier, Pontarlier 14899
Musée de Poteries de Terre Vernissée, Saint-Jean-de-Fos 15324
Musée de Préhistoire, Aurignac 11753
Musée de Préhistoire, Coulmiers 12644
Musée de Préhistoire, Echternach 30470
Musée de Préhistoire, Kinshasa 08797
Musée de Préhistoire Amédée Lemozi, Cabrerets 12219
Musée de Préhistoire des Gorges du Verdon, Quinson 14973
Musée de Préhistoire d'Île-de-France, Nemours 14391
Musée de Préhistoire James Miln-Zacharie le Rouzic, Carnac 12292
Musée de Préhistoire Libanaise, Beirut 30158
Musée de Préhistoire Mésolitique L. Coulonges, Sauveterre-la-Lémance 15671
Musée de Préhistoire Régionale, Menton 14081
Musée de Provins et du Provinois, Provins 14951
Musée de Pully, Pully 41739
Musée de Racour, Racour 03958
Musée de Radio France, Paris 14680
Musée de Rauranum, Rom 15083
Musée de Région Auguste Chabaud, Graveson 13124
Musée de Rome, Caen 12232
Musée de Rouget de Lisle, Lons-le-Saunier 13816
Musée de Royan, Royan 15133
Musée de Saint-Boniface, Saint-Boniface 06804
Musée de Saint-Eustache et de ses Patriotes, Saint-Eustache 06816
Musée de Saint-Gengoux-le-Royal, Saint-Gengoux-le-National 15277
Musée de Saint-Georges, Saint-Georges, Manitoba 06818
Musée de Saint-Germain-Laval, Saint-Germain-Laval 15293
Musée de Saint-Gervais, Saint-Gervais-sur-Mare 15296
Musée de Saint-Martin, Saint-Martin 23281
Musée de Saint-Maurin, Saint-Maurin 15412
Musée de Saint-Paul, Saint-Paul-de-Vence 15445
Musée de Saint-Pierre-Jolys, Saint-Pierre-Jolys 06875
Musée de Sainte-Anne-de-Beaupré, Sainte-Anne-de-Beaupré 06888
Musée de Sainte-Sigolène, Sainte-Sigolène 15579
Musée de Salers, Salers 15595
Musée de Sarreguemines, Sarreguemines 15644
Musée de Schirmeck, Schirmeck 15691
Musée de Sétif, Sétif 00074
Musée de Sismologie et Magnétisme Terrestre, Strasbourg 15827
Musée de Site, Sare 15632
Musée de Site Galloromain Villa Loupian, Loupian 13837
Musée de Site Le Bazacle, Toulouse 15937
Musée de Skikda, Skikda 00076
Musée de Sologne, Cheverny 12508
Musée de Sologne, Romorantin-Lanthenay 15091
Musée de Sonis, Bataille du 2 Décembre 1870, Loigny-la-Bataille 13803
Musée de Sonvilier, Sonvilier 41890
Musée de Soueida, Soueida 42138
Musée de Souvigny, Souvigny 15804
Musée de Spéléologie, Fontaine-de-Vaucluse 12958
Musée de Statues d'Enfants, Les Mages 13704
Musée de Sucy, Sucy-en-Brie 15836
Musée de Suresnes René Sordes, Suresnes 15841
Musée de Tabarka, Tabarka 42658
Musée de Tahiti et des Iles - Te Fare Manaha, Punaauia 16327
Musée de Tauroentum, Saint-Cyr-les-Lecques 15227
Musée de Tautavel-Centre Européen de la Préhistoire, Tautavel 15868
Musée de Tazoult, Tazoult 00078
Musée de Téboursouk, Téboursouk 42660
Musée de Temouchent, Ain Temouchent 00037
Musée de Tessé, Le Mans 13608
Musée de Tigran, Champvoux 12395

M

Musée de Timgad, Timgad-Batna 00080
Musée de Tipasa, Tipasa 00081
Musée de Tissage, Chauffailles 12479
Musée de Tlemcen, Tlemcen 00082
Musée de Tlemcen, Tlemcen 00083
Musée de Tourisme et d'Artisanat, Lys-Saint-Georges 13905
Musée de Traditions Populaires et d'Archéologie, Chauvigny 12488
Musée de Ua Huka, Vaipaee 16328
Musée de Ventabren, Ventabren 16106
Musée de Verrières-le-Buisson, Verrières-le-Buisson 16125
Musée de Vieux Cormeilles, Cormeilles-en-Parisis 12629
Musée de Vieux-la-Romaine, Vieux-la-Romaine 16158
Musée de Vieux-Lavaux, Cully 41317
Musée de Vieux Logis, Sainte-Enimie 15557
Musée de Vieux Sallèles, Sallèles-d'Aude 15602
Musée de Voitures Anciennes et Sellerie, Menetou-Salon 14075
Musée de Voitures Hippomobiles Anciennes, Tour-en-Sologne 15949
Musée de Vulliod Saint-Germain, Pézenas 14808
Musée de Wanne, Trois-Ponts 04095
Musée de Warcoing, Warcoing 04130
Musée de Waterloo, Waterloo 04137
Musée de Zarzis, Zarzis 42669
Musée d'Eben, Eben-Emael 03568
Musée d'Egyptologie, Villeneuve-d'Ascq 16204
Musée d'El-Oued, El-Oued 00062
Musée Delphinal, Beauvoir-en-Royans 11912
Musée d'Enfidha, Enfidaville 42624
Musée d'Ennery, Paris 14681
Musée Dentaire de Lyon, Lyon 13886
Musée d'Entomologique, Chaveyriat 12491
Musée d'Entrecasteaux, Entrecasteaux 12825
Musée Départemantal des Sapeurs-Pompiers de l'Orne, Bagnoles-de-l'Orne 11823
Musée Départemental, Baume-les-Messieurs 11866
Musée Départemental Albert Kahn, Boulogne-Billancourt 12089
Musée Départemental Alexandre Franconie, Cayenne 16316
Musée Départemental Breton, Quimper 14970
Musée Départemental d'Archéologie, Hastingues 13184
Musée Départemental d'Archéologie, Montrozier 14263
Musée Départemental d'Art Ancien et Contemporain, Épinal 12840
Musée Départemental d'Art Contemporain de Rochechouart, Rochechouart 15067
Musée Départemental d'Art Religieux, Digne-les-Bains 12725
Musée Départemental d'Art Religieux, Sées 15697
Musée Départemental d'Art Sacré Contemporain, Saint-Pierre-de-Chartreuse 15457
Musée Départemental de Flandre, Cassel 12304
Musée départemental de la Bresse - Domaine des Planons, Saint-Cyr-sur-Menton 15228
Musée Départemental de la Préhistoire, Le Grand-Pressigny 13587
Musée Départemental de la Résistance, Le Teil 13652
Musée Départemental de la Résistance en Vercors, Vassieux-en-Vercors 16080
Musée Departemental de la Résistance et de la Déportation, Forges-les-Eaux 12980
Musée Départemental de la Résistance et de la Déportation, Lorris 13826
Musée Départemental de la Résistance et de la Déportation, Manneville-sur-Risle 13938
Musée Départemental de la Résistance et de la Déportation, Tulle 16006
Musée Départemental de la Résistance et de la Déportation Jean Philippe, Toulouse 15938
Musée Départemental de la Tapisserie, Aubusson 11734
Musée Départemental de la Vigne et des Pressoirs, Champlitte 12392
Musée Départemental de l'Abbaye de Saint-Riquier, Saint-Riquier 15497
Musée Départemental de l'École Publique, Chevregny 12509
Musée Départemental de l'École Publique, Saint-Clar-de-Lomagne 15214
Musée Départemental de l'Ecole Publique, Vergne 16116
Musée Départemental de l'Éducation, Saint-Ouen-l'Aumône 15435
Musée Départemental de l'Oise, Beauvais 11910
Musée Départemental de Préhistoire, Solutré-Pouilly 15779
Musée Départemental de Préhistoire Corse, Sartène 15647
Musée Départemental de Saint-Antoine, Saint-Antoine-l'Abbaye 15170
Musée Départemental del'Ariège, Foix 12952
Musée Départemental des Antiquités de la Seine-Maritime, Rouen 15114
Musée Départemental des Arts et Traditions Populaires de la Haute-Loire et du Massif Central, Saint-Didier-en-Velay 15239
Musée Départemental des Arts et Traditions Populaires du Perche, Saint-Cyr-la-Rosière 15225
Musée Départemental des Hautes-Alpes, Gap 13031
Musée Départemental des Pays de Seine-et-Marne, Saint-Cyr-sur-Morin 15231
Musée Départemental des Sapeurs-Pompiers, Sciez 15692
Musée Départemental des Sapeurs-Pompiers du Val-d'Oise, Osny 14543
Musée départemental du Bugey-Valmorey, Lochieu 13795
Musée départemental du Revermont, Treffort 15979
Musée Départemental du Sel, Marsal 13985
Musée Departemental du Textile, Lavelanet 13551
Musée Départemental Georges de La-Tour, Vic-sur-Seille 16143
Musée Départemental Hébert, La Tronche 13453
Musée Départemental Ignon-Fabre, Mende 14071
Musée Départemental Jérôme-Carcopino, Aléria 11546
Musée Départemental Maurice Denis, Saint-Germain-en-Laye 15290
Musée Départemental Pascal Paoli, Morosaglia 14285
Musée Départemental Stéphane Mallarmé, Vulaines-sur-Seine 16280
Musée des 18 Jours-Mémoires de Guerre, Pipaix 03951
Musée des 4 Saisons, Generagues 13040
Musée des Abénakis, Odanak 06499
Musée des Affiches Léonetto-Cappiello, Épernay 12834
Musée des Alambics, Saint-Romain-de-Benet 15501
Musée des Allues, Meribel 14086
Musée des Alpilles, Saint-Rémy-de-Provence 15490
Musée des Amis de Castrum Vetus, Châteauneuf-les-Martigues 12449
Musée des Amis de Montmélian, Montmélian 14226
Musée des Amis de Thann, Thann 15874
Musée des Amis du Vieux Corbie, Corbie 12621
Musée des Amis du Vieux Donzère, Donzère 12769
Musée des Amis du Vieux Guérigny, Guérigny 13160
Musée des Amis du Vieux Lormont, Lormont 13823
Musée des Amis du Vieux Montrichard, Montrichard 14260
Musée des Amis du Vieux Nolay, Nolay 14472
Musée des Amis du Vieux Saint-Étienne, Saint-Étienne 15254
Musée des Amoureux, Werentzhouse 16293
Musée des Anciens Enfants de Troupe, Autun 11763
Musée des Anciens Outils Agricoles, Saint-Mathurin 15408
Musée des Années 30, Boulogne-Billancourt 12090
Musée des Antiquités Nationales, Saint-Germain-en-Laye 15291
Musée des Appareils Photographiques et de Cinéma d'Amateurs, Marsilly 14014
Musée des Archives, Saint-Hubert 03996
Musée des Archives Départementales de l'Aveyron, Rodez 15078
Musée des Armes, Tulle 16007
Musée des Arômes de Provence, Saint-Rémy-de-Provence 15491
Musée des Arômes et du Parfum, Graveson 13125
Musée des Art et Tradition, Reulle-Vergy 15027
Musée des Arts Africains, Océaniens, Amérindiens, Marseille 14001
Musée des Arts Anciens du Namurois, Namur 03901
Musée des Arts Asiatiques, Nice 14432
Musée des Arts Décoratifs, Bordeaux 12068
Musée des Arts Décoratifs, Lyon 13887
Musée des Arts Décoratifs, Paris 14682
Musée des Arts Décoratifs, Strasbourg 15828
Musée des Arts Décoratifs et de la Modernité, Gourdon (Alpes-Maritimes) 13088
Musée des Arts Décoratifs Sabatier d'Espeyran, Montpellier 14242
Musée des Arts du Cognac, Cognac 12566
Musée des Arts et de l'Enfance, Fécamp 12913
Musée des Arts et des Meubles, Poilley 14865
Musée des Arts et des Sciences, Sainte-Croix 41797
Musée des Arts et d'Histoire, Locronan 13797
Musée des Arts et Metiers, Paris 14683
Musée des Arts et Métiers, Saint-Gildas-de-Rhuys 15298
Musée des Arts et Métiers de la Vigne et du Vin, Moulis-en-Médoc 14309
Musée des Arts et Tradition Apicoles, Fontan 12967
Musée des Arts et Traditions Populaires, Bressuire 12155
Musée des Arts et Traditions Populaires, Djerba 42622
Musée des Arts et Traditions Populaires, Ermont 12845
Musée des Arts et Traditions Populaires, Esse 12863
Musée des Arts et Traditions Populaires, La Petite-Pierre 13402
Musée des Arts et Traditions Populaires, Le Sel-de-Bretagne 13650
Musée des Arts et Traditions Populaires, Saint-Arnoult-en-Yvelines 15177
Musée des Arts et Traditions Populaires, Saint-Esprit 30670
Musée des Arts et Traditions Populaires, Treignac 15984
Musée des Arts et Traditions Populaires, Valbonne 16034
Musée des Arts et Traditions Populaires, Wattrelos 16291
Musée des Arts et Traditions Populaires 'Dar Ben Abdallah' de Tunis, Tunis 42666
Musée des Arts et Traditions Populaires de la Haute-Loire, Lavaudieu 13548
Musée des Arts et Traditions Populaires de la Vallée de la Cère, Arpajon-sur-Cère 11694
Musée des Arts et Traditions Populaires de Moyenne Provence, Draguignan 12789
Musée des Arts Forains, Paris 14684
Musée des Arts Patrimoine et des Traditions, Nanteuil-en-Vallée 14373
Musée des Arts Populaires, Noyant 14479
Musée des Arts Populaires, Reillanne 14997
Musée des Arts Populaires et Traditions, Alger 00041
Musée des Arts Populaires et Traditions Dar Jallouli, Sfax 42652
Musée des Arts Sahariens de Laâyoune, Laâyoune 31674
Musée des Arts Traditionnels, Cussac-Fort-Médoc 12691
Musée des Arts Traditionnels et de la Culture, Foumban 05297
Musée des Attelages de la Belle Epoque, Saint-Martin-du-Lac 15403
Musée des Attelages, de la Carrosserie et du Charronnage, Vonnas 16272
Musée des Augustines de l'Hôpital Général de Québec, Québec 06689
Musée des Augustines de l'Hôtel Dieu de Québec, Québec 06690
Musée des Augustins, Toulouse 15939
Musée des Automates, La Rochelle 13425
Musée des Automates, Neuilly-sur-Seine 14403
Musée des Automates, Sauclières 15649
Musée des Automates de Grenoble, Grenoble 13137
Musée des Automates Éma, Lyon 13888
Musée des Ballons, Balleroy 11832
Musée des Bastides, Monflanquin 14148
Musée des Batailles, Montesquieu-des-Albères 14203
Musée des Bateliers de la Dordogne, Port-Sainte-Foy 14920
Musée des Beaux-Arts, Agen 11494
Musée des Beaux-Arts, Angers 11609
Musée des Beaux-Arts, Angoulême 11620
Musée des Beaux-Arts, Arras 11698
Musée des Beaux-Arts, Beaune 11903
Musée des Beaux-Arts, Beirut 30159
Musée des Beaux-Arts, Béziers 11976
Musée des Beaux-Arts, Blois 12035
Musée des Beaux-Arts, Brest 12161
Musée des Beaux-Arts, Caen 12233
Musée des Beaux-Arts, Carcassonne 12285
Musée des Beaux-Arts, Chambéry 12372
Musée des Beaux-Arts, La Chaux-de-Fonds 41293
Musée des Beaux-Arts, Dijon 12741
Musée des Beaux-Arts, Dole 12754
Musée des Beaux-Arts, Gaillac 13026
Musée des Beaux-Arts, Kinshasa 08798
Musée des Beaux-Arts, La Ferté-Macé 13359
Musée des Beaux-Arts, La Rochelle 13426
Musée des Beaux-Arts, Libourne 13741
Musée des Beaux-Arts, Lille 13756
Musée des Beaux-Arts, Le Locle 41572
Musée des Beaux Arts, Lons-le-Saunier 13817
Musée des Beaux-Arts, Lyon 13889
Musée des Beaux-Arts, Marseille 14002
Musée des Beaux-Arts, Menton 14082
Musée des Beaux-Arts, Mirande 14130
Musée des Beaux-Arts, Montbard 14180
Musée des Beaux-Arts, Mulhouse 14325
Musée des Beaux-Arts, Nancy 14354
Musée des Beaux-Arts, Nantes 14365
Musée des Beaux-Arts, Nîmes 14450
Musée des Beaux-Arts, Orléans 14535
Musée des Beaux-Arts, Pau 14757
Musée des Beaux-Arts, Quimper 14971
Musée des Beaux-Arts, Reims 15001
Musée des Beaux-Arts, Rennes (Ille-et-Vilaine) 15016
Musée des Beaux-Arts, Rouen 15115
Musée des Beaux-Arts, Saintes 15584
Musée des Beaux-Arts, Strasbourg 15829
Musée des Beaux-Arts, Tourcoing 15953
Musée des Beaux-Arts, Tournai 04086
Musée des Beaux-Arts, Tours 15964
Musée des Beaux-Arts, Valenciennes 16043
Musée des Beaux-Arts, Vannes (Morbihan) 16068
Musée des Beaux-Arts, Villeneuve-sur-Yonne 16222
Musée des Beaux-Arts Charleroi, Charleroi 03516
Musée des Beaux-Arts de Bordeaux, Bordeaux 12069
Musée des Beaux-Arts de Dunkerque, Dunkerque 12798
Musée des Beaux-Arts de Montréal, Montréal 06357
Musée des Beaux-Arts de Sherbrooke, Sherbrooke, Québec 06962
Musée des Beaux-Arts Denys Puech, Rodez 15079
Musée des Beaux-Arts et Arts Décoratifs Raymond Ritter, Morlanne 14278
Musée des Beaux-Arts et Céramique, Verviers 04108
Musée des Beaux-Arts et d'Archéologie, Besançon 11957
Musée des Beaux-Arts et d'Archéologie, Vienne 16152
Musée des Beaux-Arts et d'Archeologie Joseph Déchelette, Roanne 15059
Musée des Beaux-Arts et de la Dentelle, Alençon 11545
Musée des Beaux-Arts et de la Dentelle, Calais 12248
Musée des Beaux-Arts et d'Histoire, Saint-Lô 15372
Musée des Beaux-Arts Hyacinthe Rigaud, Perpignan 14791
Musée des Beaux-Arts Jules Chéret, Nice 14433
Musée des Bêtises de Cambrai, Cambrai 12255
Musée des Bières Belges, Lustin 03836
Musée des Blasons, Saint-Jean-de-Veleriscle 15333
Musée des Blindés, Saumur 15664
Musée des Bois Jolis, Felleries 12918
Musée des Bois Sculptes, Cancale 12263
Musée des Bories, Gordes 13082
Musée des Bujoliers, Saint-Cezaire 15198
Musée des Calèches, Apremont-sur-Allier 11651
Musée des Camions de Brasserie, Romedenne 03983
Musée des Canadiens Nord-Bourguignons, Sainte-Colombe-sur-Seine 15553
Musée des Canonniers Sédentaires de Lille, Lille 13757
Musée des Carrosses, Versailles 16126
Musée des Caves de la Cantrie, Saint-Fiacre-sur-Maine 15266
Musée des Celtes, Libramont 03792
Musée des Champs, Saint-Ségal 15509
Musée des Chantiers de Jeunesse, Châtel-Guyon 12460
Musée des Charmettes, Chambéry 12373
Musée des Chartrons, Bordeaux 12070
Musée des Chasseurs à Pied, Charleroi 03517
Musée des Chasseurs Ardennais, Marche-en-Famenne 03847
Musée des Chemins de Fer Belges-Museum van de Belgische Spoorwegen, Bruxelles 03475
Musée des Chemins de Fer Militaires et Industriels, La Neuville-les-Bray 13398
Musée des Civilisations, Saint-Just-Saint-Rambert 15353
Musée des Civilisations de Côte d'Ivoire, Abidjan 08830
Musée des Civilisations de l'Europe et de la Méditerranée, Marseille 14003
Musée des Coiffes, Fresnay-sur-Sarthe 13008
Musée des Coiffes et des Traditions, Les Ponts-de-Cé 13708
Musée des Commandos, Flawinne 03600
Musée des Commerces Anciens, Doué-la-Fontaine 12780
Musée des Contenants Vinaires, Pouzols-Minérvois 14934
Musée des Corbières, Sigean 15757
Musée des Cordeliers, Saint-Jean-d'Angély 15318
Musée des Cultures Guyanaises, Cayenne 16317
Musée des Cultures Légumières et ses Réserves, La Courneuve 13355
Musée des Cultures Taurines, Nîmes 14451
Musée des Débuts de l'Aviation, Douzy 12786
Musée des Distilleries Limougeaudes, Limoges 13767
Musée des Docks Romains, Marseille 14004
Musée des Eaux de Vie, Lapoutroie 13523
Musée des Egouts de Paris, Paris 14685
Le Musée des Enfants-Het Kindermuseum, Bruxelles 03476
Musée des Engins de Travaux Publics, Mallemort 13932
Musée des Épaves Sous-Marines, Commes 12596
Musée des Équipages, Maincy 13921
Musée des Évolutions, Bousies 12129
Musée des Fagnes, Roly 03981
Musée des Faïences Anciennes de la Tour d'Aigues, La Tour-d'Aigues 13451
Musée des Faïences de Thiviers, Thiviers 15886
Musée des Faïences de Varages, Varages 16071
Musée des Feuillardiers, Chalus 12366
Musée des Filles de Jésus, Trois-Rivières 07174
Musée des Flûtes du Monde, Bollène 12050
Musée des Fontaines Salées, Saint-Père-sous-Vezelay 15454

Musée des Forges, Brocas-les-Forges 12190
Musée des Fossiles, Montceau-les-Mines 14191
Musée des Francs et de la Famenne, Marche-en-Famenne 03848
Musée des Fusiliers Mont-Royal, Montréal 06358
Musée des Galvachers, Anost 11633
Musée des Glaciers, Lourtier 41576
Musée des Goemoniers, Plouguerneau 14856
Musée des Grenouilles, Estavayer-le-Lac 41362
Musée des Guerres de Vendée, Le Pin-en-Mauges 13633
Musée des Guerres de Vendée, Saint-Florent-le-Vieil 15267
Musée des Hospices Civils de Lyon, Lyon 13890
Musée des Hospitalières de l'Hôtel-Dieu de Montréal, Montréal 06359
Musée des Imprimés en Luxembourg, Redu 03963
Musée des Industries du Viemeu, Friville-Escarbotin 13013
Musée des Instruments de Musique, L'Aigle 13484
Musée des Instruments de Musique à Vent, La Couture-Boussey 13356
Musée des Instruments de Musique/Muziekinstrumentenmuseum, Bruxelles 03477
Musée des Ivoires, Yvetot 16310
Musée des Jacobins, Auch 11735
Musée des Jacobins, Saint-Sever (Landes) 15514
Musée des Jasseries des Monts du Forez, Saint-Anthème 15169
Musée des Lettres et Manuscrits, Paris 14686
Musée des Liqueurs Cherry-Rocher, La Côte-Saint-André 13350
Musée des Maîtres et Artisans, Montréal 06360
Musée des Majorettes, Saint-Quentin-Fallavier 15477
Musée des Maladies de la Peau, Paris 14687
Musée des Manufactures de Dentelles, Retournac 15023
Musée des Maquettes, Forges-les-Eaux 12981
Musée des Maquettes, Saint-Privat-des-Près 15471
Musée des Marais Salants, Batz-sur-Mer 11862
Musée des Mariniers, Chouzé-sur-Loire 12527
Musée des Mariniers du Rhône, Serrières 15729
Musée des Marionnettes, Aubeterre-sur-Dronne 11726
Musée des Médailles et Vieux Courseulles, Courseulles-sur-Mer 12655
Musée des Mégalithes et Centre d'Exposition de Wéris, Wéris 04145
Musée des Meilleurs Ouvriers de France, Bourges 12116
Musée des Merveilles, Tende 15870
Musée des Métiers Anciens et de la Tradition Rurale, Estinnes-au-Mont 03590
Musée des Métiers d'Autrefois, Pierrefort 14819
Musée des Métiers de France, Argent-sur-Sauldre 11671
Musée des Métiers de la Pierre, Saint-Beauzély 15183
Musée des Métiers du Bois, Labaroche 13459
Musée des Métiers et des Légendes de la Forêt d'Orléans, Loury 13856
Musée des Métiers et Traditions Paysannes, Argy 11679
Musée des Métiers Ruraux, Eternoz 12877
Musée des Mille et une Racines, Cornimont 12630
Musée des Minéralogie, Strasbourg 15830
Musée des Minéraux et de la Faune des Alpes, Le Bourg-d'Oisans 13561
Musée des Minéraux et des Fossiles, Rougemont 41778
Musée des Minéraux et Fossiles des Ardennes, Bogny-sur-Meuse 12044
Musée des Mines d'Argent des Gorges du Fournel, L'Argentière-la-Bessée 13526
Musée des Mines d'Aumetz, Aumetz 11745
Musée des Mines de Fer, Aumetz 11746
Musée des Mines de Fer de Lorraine, Neufchef 14400
Musée des Missions, Plouguerneau 14857
Musée des Monnais et Médailles J. Puig, Perpignan 14792
Musée des Moulages, Lyon 13891
Musée des Moulages, Montpellier 14243
Musée des Moulins, Villeneuve-d'Ascq 16205
Musée des Musiques Arabes et Méditerranéennes Ennejma Ezzahra, Sidi Bou Saïd 42654
Musée des Musiques Populaires, Montluçon 14221
Musée des Oiseaux, Trignac 15990
Musée des Ondes Emile Berliner, Montréal 06361
Musée des Ostensions et de la Religon Populaire, Esse 12864
Musée des Outil d'Hier, Chauvirey-le-Vieil 12489
Musée des Outils Agraires, Châteaurenard (Bouches-du-Rhône) 12455
Musée des Outils Agraires, Châteaurenard (Bouches-du-Rhône) 12456
Musée des Outils du Bois et de la Vie Paysanne, Bligny-sur-Ouche 12028
Musée des Outils et Appareils Ménagers, Merceuil 14085
Musée des Papeteries Canson et Montgolfier, Annonay 11630
Musée des Papillons, Fuveau 13021
Musée des Papillons, Saint-Chély-d'Apcher 15203
Musée des Papillons de Liège, Liège 03809
Musée des Parachutistes, Pau 14758
Musée des Patenôtriers, Brain-sur-Allonnes 12140
Musée des Peintres de l'Ecole de Murol, Murol 14337
Musée des Peintres du Pays, Saint-Cyprien 15223
Musée des Pénitents Blancs, Marsac-en-Livradois 13983
Musée des Petits Hommes en Pyrénées, Arreau 11699
Musée des Petits Métiers de la Ferme, Wervicq-Sud 16294
Musée des Petits Meubles Bretons, Plozevet 14863
Musée des Phares, La Martre 06112
Musée des Pioneers, Saint-André-Avellin 06793
Musée des Pionniers, Grande-Digue 05898
Musée des Plans-Reliefs, Paris 14688
Musée des Plantes Médicinales et de la Pharmacie, Bruxelles 03478
Musée des Pompiers, Fontainebleau 12963
Musée des Pompiers, Wasquehal 16287
Musée des Pompiers Auxiliaires de Mont-Royal, Montréal 06362
Musée des Postes et Telecommunications, Luxembourg 30480
Musée des Poupées, Josselin 13279
Musée des Poupées d'Antan, Malicorne-sur-Sarthe 13931
Musée des Poupées d'Antan et de la Tonnellerie, Maisons-lès-Chaource 13926
Musée des Processions, Hyon-Ciply 03726
Musée des Racines et d'Histoire régionale à Amel, Amel 03272
Musée des Rangers, Grandcamp-Maisy 13099
Musée des Records, La Tour-Blanche 13449
Musée des Religions, Nicolet 06457
Musée des Rochers Sévigné, Vitré 16255
Musée des Santons Animés, Maussane-les-Alpilles 14043
Musée des Santons des Baux, Les Baux-de-Provence 13690
Musée des Santons Di-Landro, Aubagne 11720
Musée des Sapeurs Pompiers, Bièvres 11990
Musée des Sapeurs-Pompiers, Fessenheim 12929
Musée des Sapeurs-Pompiers, Thônes 15892
Musée des Sapeurs-Pompiers de France, Montville 14267
Musée des Sapeurs-Pompiers du Grand Lyon, Lyon 13892
Musée des Sapeurs Pompiers-L'Univers des Pompiers, Athus 03333
Musée des Sciences et Techniques, Laval (Mayenne) 13540
Musée des Sciences Naturelles, Orléans 14536
Musée des Soeurs de Miséricorde, Montréal 06363
Musée des Soldats du Feu, Wambrechies 16284
Musée des Sous-Stations R.A.T.P., Paris 14689
Musée des Souvenirs Napoléoniens et des Archives du Palais, Monaco 31609
Musée des Spahis, La Horgne 13386
Musée des Spahis, Valence (Drôme) 16039
Musée des Suisses dans le Monde, Pregny-Chambésy 41735
Musée des Tanneurs, Montargis 14172
Musée des Tapisseries, Aix-en-Provence 11515
Musée des Techniques Faïencières, Sarreguemines 15645
Musée des Techniques Fromagères, Saint-Pierre-sur-Dives 15463
Musée des Technologies Ferroviaires, La Seyne-sur-Mer 13443
Musée des Télécommunications, Perros-Guirec 14799
Musée des Télécoms, Pleumeur-Bodou 14842
Musée des Téléphones de Genève, Plan-les-Ouates 41725
Musée des Temps Barbares, Marle 13970
Musée des Terre-Neuvas et de la Pêche, Fécamp 12914
Musée des Thoniers, Etel 12876
Musée des Timbres et des Monnaies, Monaco 31610
Musée des Tisserands Mayennais, Ambrières-les-Vallées 11580
Musée des Tissus, Lyon 13893
Musée des Toiles Peintes, Colombier (Neuchâtel) 41310
Musée des Tourneurs, Aiguines 11507
Musée des Tradition de l'Île, Noirmoutier-en-l'Île 14467
Musée des Traditions, Les Avanchers-Valmorel 13687
Musée des Traditions Agricoles du Ségala, Pradinas 14938
Musée des Traditions Béarnaises, Salies-de-Béarn 15596
Musée des Traditions Bugistes, Saint-Rambert-en-Bugey 15479
Musée des Traditions Comtadines, Pernes-les-Fontaines 14782
Musée des Traditions du Ban-de-la-Roche, Neuviller-la-Roche 14411
Musée des Traditions du Vieux Dirinon, Dirinon 12752
Musée des Traditions et Arts Normands, Martainville-Epreville 14017
Musée des Traditions et de l'Histoire de Biscarrosse, Biscarrosse 12000
Musée des Traditions et des Barques du Léman, Saint-Gingolph 41790
Musée des Traditions Locales, Plounéour-Trez 14861
Musée des Traditions Locales, Sainte-Maxime 15570
Musée des Traditions Namuroises, Namur 03902
Musée des Traditions Populaires, Escaudain 12849
Musée des Traditions Populaires, Olonne-sur-Mer 14509
Musée des Traditions Populaires, Saint-Gilles-du-Gard 15302
Musée des Traditions Populaires Cap Al Campestre, Lherm 13738
Musée des Traditions Populaires Marius Audin, Beaujeu 11894
Musée des Traditions Rurales et Artisanales, Revel-Tourdan 15030
Musée des Traditions Stéphanoises, Saint-Étienne-de-Tinée 15256
Musée des Traditions Vésubiennes, Saint-Martin-Vésubie 15407
Musée des Traditions Vigneronnes, Vongnes 16271
Musée des Tramways à Vapeur et des Chemins de Fer Secondaires Français, Butry-sur-Oise 12213
Musée des Tramways et Autobus, Luxembourg 30481
Musée des Transmissions, Suresnes 15842
Musée des Transports - Amitram, Marquette-Lez-Lille 13980
Musée des Transports de Pithiviers, Bouzonville-en-Beauce 12138
Musée des Transports, du Tourisme et des Communications, Toulouse 15940
Musée des Transports en Commun du Pays de Liège, Liège 03810
Musée des Trophées de Chasse, La Chapelle-Glain 13336
Musée des Troupes de Marine, Fréjus 13002
Musée des Troupes de Montagne, Grenoble 13138
Musée des Tumulus de Bougon, Bougon 12086
Musée des Ursulines, Mâcon 13907
Musée des Ursulines, Québec 06691
Musée des Ursulines, Trois-Rivières 07175
Musée des Ustensiles de Cuisine Anciens, Saint-Denis-la-Chevasse 15233
Musée des Vallées Cévenoles, Saint-Jean-du-Gard 15337
Musée des Vieilles Automobiles, Canet-en-Roussillon 12264
Musée des Vieilles Landes, Lit-et-Mixe 13788
Musée des Vieux Métiers, Ardes-sur-Couze 11668
Musée des Vieux Métiers, Saissac 15591
Musée des Vieux Métiers et des Eaux-de-Vie, Dabo 12695
Musée des Vieux Métiers et des Traditions Populaires, Sainte-Croix-Volvestre 15556
Musée des Vieux Outils, Auterive 11761
Musée des Vieux Outils, Fourcés 12991
Musée des Vieux Outils Noël-Morard, Caromb 12294
Musée des Vieux Outils Vignerons, Aragon 11656
Musée des Vieux Tacots, Les Riceys 13710
Musée des Vins de Touraine, Tours 15965
Musée des XVIes Jeux Olympiques d'Hiver, Albertville 11534
Musée Despiau Wlérick, Mont-de-Marsan 14158
Musée d'Estavayer, Estavayer-le-Lac 41363
Musée d'Ethnographie, Genève 41421
Musée d'Ethnographie, Neuchâtel 41681
Musée d'Ethnographie de Genève - Annexe de Conches, Conches 41312
Musée d'Ethnographie de l'Université de Bordeaux II, Bordeaux 12071
Musée d'Ethnographie et d'Art Populaire, Honfleur 13222
Musée d'Ethnographie et d'Histoire de Conflans, Albertville 11535
Musée d'Evreux, Evreux 12889
Musée d'Exposition des Fouilles Archéologiques, Javols 13267
Musée d'Eysses, Villeneuve-sur-Lot 16221
Musée d'Héritage de Notukeu, Ponteix 06622
Musée d'Hippone, Annaba 00050
Musée d'Histoire, La Chaux-de-Fonds 41294
Musée d'Histoire, Elne 12817
Musée d'Histoire, La Neuveville 41684
Musée d'Histoire, Obernai 14492
Musée d'Histoire, Ouidah 04185
Musée d'Histoire Contemporaine, Paris 14690
Musée d'Histoire de la Brasserie de Diekirch, Diekirch 30463
Musée d'Histoire de la Médecine, Paris 14691
Musée d'Histoire de la Résistance et de la Déportation de l'Ain et du Haut-Jura, Nantua 14374
Musée d'Histoire de la Ville de Luxembourg, Luxembourg 30482
Musée d'Histoire de la Ville et d'Ethnographie du Pays Malouin, Saint-Malo (Ille-et-Vilaine) 15389
Musée d'Histoire de Marseille, Marseille 14005
Musée d'Histoire des Sciences, Genève 41422
Musée d'Histoire du 20e Siècle, Estivareilles 12869
Musée d'Histoire et d'Archéologie, Les Baux-de-Provence 13691
Musée d'Histoire et d'Archéologie, Longjumeau 13809
Musée d'Histoire et d'Archéologie, Marcinelle 03850
Musée d'Histoire et d'Archéologie, Moutiers (Savoie) 14317
Musée d'Histoire et d'Archéologie, Vannes (Morbihan) 16069
Musée d'Histoire et d'Archéologie, Warneton 04133
Musée d'Histoire et d'Archéologie de Die et du Diois, Die 12720
Musée d'Histoire et d'Archéologie de la Chartreuse, Molsheim 14145
Musée d'Histoire et d'Archéologie du Pays d'Apt, Apt 11655
Musée d'Histoire et d'Archéologie René Bauberot, Châteauponsac 12452
Musée d'Histoire et d'Art, Mirande 14131
Musée d'Histoire et d'Art, Villeneuve-Loubet 16218
Musée d'Histoire et d'Arts Populaires, Pellafol-le-Vieux 14770
Musée d'Histoire et de l'Éclairage, Liège 03811
Musée d'Histoire et de Tambow, Amnéville-les-Thermes 11590
Musée d'Histoire et de Vie Locale, Genappe 03617
Musée d'Histoire et des Arts Décoratifs, Tournai 04087
Musée d'Histoire et d'Ethnographie, Aneho 42610
Musée d'Histoire et du Patrimoine de Dorval, Montréal 06364
Musée d'Histoire et Traditions Populaires, L'Oudon-Notre-Dame-de-Fresnay 13829
Musée d'Histoire Jean Garcin 1939-1945, Fontaine-de-Vaucluse 12959
Musée d'Histoire Local, Billom 11994
Musée d'Histoire Local, Maubourguet 14033
Musée d'Histoire Locale, Belin-Beliet 11924
Musée d'Histoire Locale, Cagnes-sur-Mer 12237
Musée d'Histoire Locale, Deuil-la-Barre 12718
Musée d'Histoire Locale, Ecaussinnes 03570
Musée d'Histoire Locale, Mougins 14298
Musée d'Histoire Locale, Saint-Paul-de-Vence 15446
Musée d'Histoire Locale, Saint-Renan 15496
Musée d'Histoire Locale et des Traditions, Fréjus 13003
Musée d'Histoire Locale et des Vieux Mètiers, Saint-Ciers-sur-Gironde 15209
Musée d'Histoire Locale et Traditions Populaires, Olargues 14506
Musée d'Histoire Locale Pierre Salvi, Viarmes 16140
Musée d'Histoire Médiévale, Châtel-sur-Moselle 12461
Musée d'Histoire Militaire à Bandera, Ajaccio 11525
Musée d'Histoire Naturelle, Aix-en-Provence 11516
Musée d'Histoire Naturelle, Amnéville-les-Thermes 11591
Musée d'Histoire Naturelle, Angers 11610
Musée d'Histoire Naturelle, Baume-les-Messieurs 11867
Musée d'Histoire Naturelle, Besançon 11958
Musée d'Histoire Naturelle, Chambéry 12374
Musée d'Histoire Naturelle, La Chaux-de-Fonds 41295
Musée d'Histoire Naturelle, Eloyes 12819
Musée d'Histoire Naturelle, Fribourg 41388
Musée d'Histoire Naturelle, Gray 13127
Musée d'Histoire Naturelle, La Rochelle 13427
Musée d'Histoire Naturelle, Nantes 14366
Musée d'Histoire Naturelle de Mons, Mons 03885
Musée d'Histoire Naturelle de Toulon et du Var, Toulon (Var) 15925
Musée d'Histoire Naturelle du Parc de Miguasha, Nouvelle 06493
Musée d'Histoire Naturelle et de Géologie, Lille 13758
Musée d'Histoire Naturelle et de la Préhistoire, Nîmes 14452
Musée d'Histoire Naturelle et d'Ethnographie, Colmar 12584
Musée d'Histoire Naturelle et d'Histoire Locale, Elbeuf-sur-Seine 12815
Musée d'Histoire Naturelle et Régionale, Gruffy 13152
Musée d'Histoire Naturelle et Vivarium, Tournai 04088
Musée d'Histoire Naturelle Gabriel Foucher, Bourges 12117
Musée d'Histoire Naturelle Philadelphe Thomas, Gaillac 13027
Musée d'Histoire Urbaine de Bergerac, Bergerac 11944
Musée d'Histoires Naturelles, Boulogne-sur-Mer 12093

Musée d'Histore Naturelle Victor-Brun, Montauban 14175
Musée d'Homme Palais Royaux de Porto-Novo, Porto-Novo 04188
Musée d'Horlogerie du Locle, Le Locle 41573
Musée d'Huez et de l'Oisans, L'Alpe-d'Huez 13489
Musée Dicks, Vianden 30494
Musée d'Iconographie Historique, Onoz 03923
Musée d'Ifri, Ifri 00066
Musée d'Initiation à la Nature de Normandie, Caen 12234
Musée d'Initiation à la Préhistoire, Tamnies 15849
Musée d'Institut National de Archéologique d'El-Jem, El-Jem 42623
Musée d'Intérêt National du Tabac, Bergerac 11945
Musée Diocésain d'Art Sacré, Moulins (Allier) 14305
Musée Diocésain et Trésor de la Cathédrale Saint-Aubain, Namur 03903
Musée d'Istres, Istres 13254
Musée d'Ixelles-Museum van Elsene, Bruxelles 03479
Musée d'Izieu Mémorial des Enfants Juifs Exterminés, Izieu 13258
Musée d'Métiers et Traditions du Pays d'Olliergues, Olliergues 14507
Musée Dobrée, Nantes 14367
Musée Docteur Joseph Fau, Conques 12614
Musée Documentaire Dar Ayed, Ksar Hellal 42633
Musée Domaine de Gourjade, Castres 12319
Musée Domaine de Samara, La Chaussée-Tirancourt 13341
Musée Don Bosco, Doékoué 08834
Musée d'Orainville, Orainville 14517
Musée d'Orange, Orange 14518
Musée d'Orbe, Orbe 41713
Musée d'Orbigny-Bernon, La Rochelle 13428
Musée d'Orgues de Kermesse, Bruxelles 03480
Musée Dorothée Salhab Kazemi, Broumana 30165
Musée d'Orsay, Paris 14692
Musée Doubrère, Marciac 13954
Musée d'Ourthe-Amblève, Comblain-au-Pont 03524
Musée d'Sacré, Mours-Saint-Eusèbe 14312
Musée du 11 Septembre 1709, Bavay 11870
Musée du 3è Lanciers, Namur 03904
Musée du Bailles, Nismes 03914
Musée du Bain Rituel Juif, Bischheim 12001
Musée du Barreau de Paris, Paris 14693
Musée du Bas Saint-Laurent, Rivière-du-Loup 06761
Musée du Bassin Houiller Lorrain, Petite-Rosselle 14805
Musée du Bateau-Lavoir Le Saint-Julien, Laval (Mayenne) 13541
Musée du Batha, Fès 31671
Musée du Batiment, Moulins (Allier) 14306
Musée du Béret, Nay 14387
Musée du Berry, Bourges 12118
Musée du Biterrois, Béziers 11977
Musée du Blé et du Pain, Verdun-sur-le-Doubs 16113
Musée du Bocage Normand, Saint-Lô 15373
Musée du Bois, Aubonne 41136
Musée du Bois, Papuakeikaha 16326
Musée du Bois, Seyssel 15753
Musée du Bois de Maredret-Galerie A la Fontaine, Maredret 03853
Musée du Bois et de la Forêt, La Chaise-Dieu 13328
Musée du Bois et de l'Outil, Montgobert 14212
Musée du Bonbon Haribo, Uzès 16024
Musée du Bonsaï, Châtenay-Malabry 12466
Musée du Bouchardais, L'Ile-Bouchard 13750
Musée du Bouton-Château d'Ambleville, Ambleville 11574
Musée du Breuil, Bourbon-Lancy 12097
Musée du Bronze d'Inverness, Inverness 06018
Musée du Buis, Durbuy 03565
Musée du Café Gondrée, Bénouville 11940
Musée du Calvados et des Métiers Anciens de Pont-l'Évêque, Pont-l'Évêque 14892
Musée du Calvaire de Pontchateau, Pontchâteau 14902
Musée du Camembert, Vimoutiers 16247
Musée du Camp d'Internement du Vernet d'Ariège, Le Vernet d'Ariège 13665
Musée du Camp Romain, Saint-Jean-le-Vieux 15338
Musée du Canal, Cleden-Poher 12543
Musée du Canal, Ecuisses 12811
Musée du Canal de Berry, Audes-Reugny 11739
Musée du Canal d'Orléans, Fay-aux-Loges 12910
Musée du Canal du Duc de Berry, Dun-sur-Auron 12795
Musée du Capitellu, Ajaccio 11526
Musée du Cardinal Verdier, Lacroix-Barrez 13473
Musée du Cartonnage et de l'Imprimerie, Valréas 16063
Musée du Casque de Combat, Saint-Vincent-sur-Jabron 15538
Musée du Centre Acadien de l'Université Sainte-Anne, Pointe de l'Eglise 06619
Musée du Centre de Recherches et de Documentation du Sénégal, Saint-Louis 38027
Musée du Centre de Recherches sur les Monuments Historiques, Paris 14694
Musée du Centre Élisabeth Bergeron, Saint-Hyacinthe 06821
Musée du Cerf-Volant, Perrigny-lès-Dijon 14796
Musée du Chablais, Thonon-les-Bains 15896
Musée du Champignon, Saumur 15665
Musée du Champignon, Tillet-Saint-Ode 04073
Musée du Chanvre, Bligny-sur-Ouche 12029
Musée du Chapitre, Soignies 04035
Musée du Charroi Rural, Salmiech 15608
Musée du Châtaignier, Marrons et Champignons, Villefranche-du-Périgord 16190
Musée du Château, Baugé 11864
Musée du Château, Flers 12944
Musée du Château, Fougères (Ille-et-Vilaine) 12986
Musée du Château, Lunéville 13865
Musée du Château, Margaux 13967
Musée du Château, Montbard 14181
Musée du Château, Montmaur 14223
Musée du Château, Noirmoutier-en-l'Ile 14468
Musée du Château, Pauillac 14762
Musée du Château, Roquebrune-Cap-Martin 15095
Musée du Château, Siorac-en-Périgord 15760
Musée du Château Comtal, Rochefort 03973
Museè du Château de Bourbilly, Vic-de-Chassenay 16141
Musée du Château de Dourdan, Dourdan 12783
Musée du Château de Fagnolles, Fagnolles 03594
Musée du Château de Flaugergues, Montpellier 14244
Musée du Château de Franchimont, Theux 04060
Musée du Château de la Guignardière, Avrillé 11805
Musée du Château de la Louviere, Montluçon 14222
Musée du Château de Mascarras, Garlin 13036
Musée du Château de Mesnières-en-Bray, Neufchâtel-en-Bray 14397
Musée du Château de Monbazillac, Monbazillac 14146
Musée du Château de Montrond-les-Bains, Montrond-les-Bains 14262
Musée du Château de Pontécoulant, Pontécoulant 14903
Musée du Château de Ravel, Ravel 14989
Musée du Château de Vincennes, Vincennes 16248
Musée du Château de Vitré, Vitré 16256
Musée du Château des Baux, Les Baux-de-Provence 13692
Musée du Château des Comtes de Marchin, Modave 03874
Musée du Château des Ducs de Bretagne, Nantes 14368
Musée du Château des Ducs de Wurtemberg, Montbéliard 14185
Musée du Château d'Henri IV, Nérac 14392
Musée du Château d'Oigny-en-Valois, Oigny-en-Valois 14502
Musée du Château du Marais, Val Saint-Germain 16033
Musée du Château Dufresne, Montréal 06365
Musée du Château-Fort, Ecaussinnes 03571
Musée du Château Fort-Liberia, Villefranche-de-Conflent 16187
Musée du Château Paul Pastre, Marsillargues 14013
Musée du Château Ramezay, Montréal 06366
Musée du Château Saint-Jean, Nogent-le-Rotrou 14462
Musée du Châtillonnais, Châtillon-sur-Seine 12474
Musée du Chemin de Fer, Apremont-la-Forêt 11650
Musée du Chemin de Fer Miniature, Clécy 12541
Musée du Cheminot, Ambérieu-en-Bugey 11567
Musée du Cheval, La Sarraz 41823
Musée du Cheval en 1900, Cussac-Fort-Médoc 12692
Musée du Chien, Villeneuve-les-Convers 16216
Musée du Chocolat, Barcelonnette 11849
Musée du Cidre, Domagné 12756
Musée du Cidre, Montaure 14177
Musée du Cidre, Saint-Jean-des-Champs 15334
Musée du Cidre Breton, Argol 11677
Musée du Cidre du Pays d'Othe, Eaux-Puiseaux 12803
Musée du Cimetière Américain Henri-Chapelle, Hombourg 03709
Musée du Cinéma, Bruxelles 03481
Musée du Cinéma, Nismes 03915
Musée du Cinquantenaire-Jubelparkmuseum, Bruxelles 03482
Musée du Circuit de Spa-Francorchamps, Stavelot 04048
Musée du Cirque, Vatan 16081
Musée du Cirque, Wasquehal 16288
Musée du Clocher, Haute-Rivoire 13190
Musée du Cloître, Tulle 16008
Musée du Cloître Notre-Dame en Vaux, Châlons-en-Champagne 12361
Musée du Clou, Saint-Cornier-des-Landes 15221
Musée du Collège de Lévis, Lévis 06168
Musée du Colombier, Alès 11550
Musée du Colombier, Sarniguet 15637
Musée du Commissariat de l'Armée de Terre, Montpellier 14245
Musée du Compagnonnage, Paris 14695
Musée du Compagnonnage, Tours 15966
Musée du Coquillage, Le Lavandou 13601
Musée du Coquillage, Papara 16322
Musée du Corp des Magasins Militaires, Montréal 06367
Musée du Corps de la Cavalerie Française 1940, Jandrain-Jandrenouille 03742
Musée du Corps Franc Pommiès, Castelnau-Magnoac 12311
Musée du Costume, Château-Chinon 12432
Musée du Costume Breton, Paimpol 14554
Musée du Costume Breton, Sainte-Anne-d'Auray 15550
Musée du Costume Comtadin, Pernes-les-Fontaines 14783
Musée du Costume Comtois, Ornans 14539
Musée du Costume et de la Dentelle, Bruxelles 03483
Musée du Costume Militaire, Sainte-Mesme 15575
Musée du Costume Provençal, Solliès-Ville 15777
Musée du Costume Savogard, Sévrier 15745
Musée du Costume Savoyard, Bourg-Saint-Maurice 12110
Musée du Coticule, Vielsalm 04112
Musée du Couteau de Laguiole, Laguiole 13480
Musée du Crest-Cherel, Ugine 16015
Musée du Cristal, Bayel 11872
Musée du Crozet, Le Crozet 13579
Musée du Cuir et de la Tannerie, Bort-les-Orgues 12078
Musée du Cuir et de la Tannerie, Château-Renault 12438
Musée du Cuir et des Industries du Péruwelzis, Péruwelz 03949
Musée du Cuir et du Parchemin, Levroux 13734
Musée du Cuivre à Stolzembourg, Stolzembourg 30492
Musée du Cwarmê, Malmedy 03843
Musée du Cycle, Huccorgne 03721
Musée du Cycle et de la Moto, Lunéville 13866
Musée du Cyclisme, Lavaux-Sainte-Anne 03781
Musée du Débarquement à Utah Beach, Sainte-Marie-du-Mont 15568
Musée du Désert, Mialet 14116
Musée du Dessin et de l'Estampe Originale, Gravelines 13122
Musée du Diamant, Durbuy 03566
Musée du Docteur Faure, Aix-les-Bains 11522
Musée du Dolder, Riquewihr 15052
Musée du Dolon, Chanas 12396
Musée du Donjon de la Toque, Huriel 13231
Musée du Faouët, Le Faouët 13582
Musée du Felicien, Argentine 11675
Musée du Fer, Arles-sur-Tech 11687
Musée du Fer, Reichshoffen 14996
Musée du Fer à Repasser, Longwy 13814
Musée du Fer Blanc Ancien et Moderne, Saint-Arcons-d'Allier 15175
Musée du Fer et de la Métallurgie Ancienne, Saint-Hubert 03997
Musée du Fer et du Chemin de Fer, Vallorbe 41976
Musée du Fer et du Fil, Dompierre (Orne) 12762
Musée du Feu Adjudant-Chef Gérard, Châlons-en-Champagne 12362
Musée du Feutre, Mouzon 14319
Musée du Fjord, Ville de Saguenay, La Baie 07263
Musée du Flacon à Parfum, La Rochelle 13429
Musée du Florentinois, Saint-Florentin 15269
Musée du Florival, Guebwiller 13154
Musée du Foie Gras, Frespech 13011
Musée du Foie Gras, Roquettes 15099
Musée du Foie Gras, Thiviers 15887
Musée du Foie Gras et des Traditions Populaires, Samatan 15616
Musée du Folklore, Ittre 03735
Musée du Folklore, Moulins (Allier) 14307
Musée du Folklore et de la Vie Montoise, Mons 03886
Musée du Football en France - Musée du Sport, Paris 14696
Musée du Fort de Battice, Battice 03348
Musée du Fort de Huy, Huy 03724
Musée du Fort de la Pompelle, Puisieulx 14957
Musée du Fort de la Pompelle, Reims 15002
Musée du Fort de Vaux, Vaux-devant-Damloup 16087
Musée du Fort d'Embourg, Embourg 03579
Musée du Fort Saint-Jean, Saint-Jean-sur-Richelieu 06828
Musée du Fort Vauban, Ambleteuse 11573
Musée du Fossile, Lion-sur-Mer 13774
Musée du Frère Mutien-Marie, Malonne 03845
Musée du Fromage Cheddar, Saint-Prime 06876
Musée du Fromage de la Chèvrerie, Noirmoutier-en-l'Ile 14469
Musée du Fromage et de la Vache Salers, Saint-Paul-de-Salers 15442
Musée du Fusil Rouillé, Bioul 03369
Musée du Gabarier, Saint-Simon (Charente) 15515
Musée du Gâteau Basque, Sare 15633
Musée du Gemmail, Lourdes 13845
Musée du Gemmail, Tours 15967
Musée du Général Thys, Dalhem 03533
Musée du Grand-Lens, Lens 41553
Musée du Grand Saint-Bernard, Le Grand-Saint-Bernard 41449
Musée du Granit, Bécon-les-Granits 11916
Musée du Granit, Saint-Michel-de-Montjoie 15416
Musée du Grenier du Roy Louis XIII, Villemur-sur-Tarn 16199
Musée du Grès, Saint-Amand-en-Puisaye 15159
Musée du Groupe Jovinien de Résistance Bayard, Joigny (Yonne) 13273
Musée du Guesclin, Châteauneuf-de-Randon 12446
Musée du Hard Rock Café, Paris 14697
Musée du Haut-Poitou, Mirebeau 14132
Musée du Haut-Richelieu, Saint-Jean-sur-Richelieu 06829
Musée du Haut-Rouergue, Laguiole 13481
Musée du Haut-Val-d'Arly, Megève 14060
Musée du Huchier, Poucharramet 14926
Musée du Jamais Vu, Puymaurin 14962
Musée du Jeu Ancien, Saint-Hilaire-de-Brethmas 15305
Musée du Jeu de l'Oie, Rambouillet 14981
Musée du Jeu de Paume, Versailles 16127
Musée du Jouet, Bruxelles 03484
Musée du Jouet, Clervaux 30461
Musée du Jouet, Moirans-en-Montagne 14138
Musée du Jouet, Poissy 14866
Musée du Jouet Ancien "Atlantrain", Saint-Just-Luzac 15352
Musée du Jouet Ancien Automobile, Rocamadour 15062
Musée du Jouet et de l'Enfant, Ferrières 03599
Musée du Jus de Fruit, Saint-Andiol 15165
Musée du Lac, Nernier 14394
Musée du Lac de Paladru, Charavines 12405
Musée du Lait, Belvedere 11935
Musée du Lait, Montebourg 14195
Musée du Lait et des Traditions, Saint-Étienne-de-Tinée 15257
Musée du Lapidaire, Lamoura 13494
Musée du Lavage et du Repassage, Verneuil-en-Bourbonnais 16120
Musée du Lavandin, Carennac 12289
Musée du Léman, Nyon 41690
Musée du Léon, Lesneven 13723
Musée du Libertaire, Fontenay-les-Briis 12970
Musée du Liège, Maureillas-las-Illas 14037
Musée du Liège et du Bouchon, Mézin 14114
Musée du Littoral et de des Fours à Chaux, Regnéville-sur-Mer 14992
Musée du Livre, Langres 13510
Musée du Livre, de la Pierre et du Verre, Conches-en-Ouche 12603
Musée du Livre et Cabinets de Donations de la Bibliothèque Royale de Belgique, Bruxelles 03485
Musée du Loup, Le Cloître-Saint-Thégonnec 13572
Musée du Louvre, Paris 14698
Musée du Luxembourg, Paris 14699
Musée du Lycée Militaire de Saint-Cyr, Saint-Cyr-l'École 15226
Musée du Machinisme Agricole, Bourbon-Lancy 12098
Musée du Machinisme Agricole, Neuvic (Corrèze) 14406
Musée du Machinisme Agricole, Prissac 14949
Musée du Machinisme Agricole, Revest-Saint-Martin 15031
Musée du Machinisme Agricole, Tullins-Fures 16009
Musée du Machinisme Agricole et du Cheval de Trait de 1830 à nos Jours, Juvigny-sous-Andaine 13298
Musée du Maïs, Morlaàs 14273
Musée du Maison Forte, Rarécourt 14986
Musée du Malgré-Tout, Treignes 04092
Musée du Manoir, Saussey 15668
Musée du Manoir d'Argentelles, Villebadin 16179
Musée du Maquettisme, de la Fonderie et de la Moulerie, Blangy-sur-Bresle 12016
Musée du Marbre, Rance 03961
Musée du Marbre et de la Pierre Bleue, Bellignies 11933
Musée du Maréchal Murat, Labastide-Murat 13462
Musée du Matériel Agricole, La Villeneuve-les-Convers 13456
Musée du Memoire de la Terre, Tataouine 42659
Musée du Mémorial de Sidi Ferruch, Port-Vendres 14921
Musée du Meuble Normand, Villedieu-les-Poêles 16183
Musée du Miel, Lobbes 03826
Musée du Miel et de l'Abeille, Gramont 13095
Musée du Mineur, Ensisheim 12823
Musée du Mineur, La Grand-Combe 13379
Musée du Modèle Réduit, Varces 16074
Musée du Monde Rural, Rettel 15024
Musée du Monde Souterrain, Han-sur-Lesse 03675
Musée du Monde Souterrain et Historique de la Spéléologie Française, Saint-Remèze 15483
Musée du Mont Corbier, Saint-Jean-de-Maurienne 15331
Musée du Mont-de-Piété et de la Condition des Soies, Avignon 11791
Musée du Mont-Repais, Montmelon 41636

Musée du Mont Riant, Alger 00042
Musée du Montreux, Montreux 41639
Musée du Moteur, Saumur 15666
Musée du Moteur et de l'Outil à la Ferme, Juvigné 13297
Musée du Moudjahid, Skikda 00077
Musée du Moulin, Saint-Jean-Saint-Nicolas 15340
Musée du Moulin, Wissant 16299
Musée du Moulin à Huile, Rustrel 15141
Musée du Moulin à Marée, Trégastel 15982
Musée du Moulin de Boly, Lourdes 13846
Musée du Moulins des Bouillons, Gordes 13083
Musée du Mouton et de la Bergerie Nationale, Rambouillet 14982
Musée du Mur de l'Atlantique, Audinghen 11741
Musée du Mur de l'Atlantique, Ouistreham 14544
Musée du Nord de la Vendée, Montaigu (Vendée) 14166
Musée du Nougat Arnaud Soubeyran, Montélimar 14200
Musée du Nouveau-Brunswick, Saint John 06838
Musée du Nouveau Monde, La Rochelle 13430
Musée du Noyonnais, Noyon 14483
Musée du Pain, Uhlwiller 16016
Musée du Pain Bouilli et du Four Communal, Villar-d'Arène 16164
Musée du Pain d'Épices d'Alsace et des Douceurs, Gertwiller 13050
Musée du Palais de la Reine, Antananarivo 30535
Musée du Palais des Evêques, Saint-Lizier 15371
Musée du Palais du Dey, Casbah, Alger 00043
Musée du Panache Gascon, Termes-d'Armagnac 15871
Musée du Panorama de la Vie de Sainte-Bernadette, Lourdes 13847
Musée du Panthéon National Haïtien, Port-au-Prince 23322
Musée du Papier Chiffon, Muzillac 14345
Musée du Papier Le Nil, Angoulême 11621
Musée du Papier Peint, Rixheim 15057
Musée du Papillon et des Arts Africains, Aubeterre-sur-Dronne 11727
Musée du Papogay, Rieux-Volvestre 15038
Musée du Parc M'Ploussoue, Bonoua 08833
Musée du Parchemin et de l'Enluminure, Duras 12801
Musée du Passé Agricole de Beauvechain, Beauvechain 03352
Musée du Pastel, Magrin 13916
Musée du Patrimoine, Guellala 42626
Musée du Patrimoine, Néris-les-Bains 14393
Musée du Patrimoine, Roquebrune-sur-Argens 15096
Musée du Patrimoine, Six-Fours-les-Plages 15768
Musée du Patrimoine Amazigh, Agadir 31660
Musée du Patrimoine Berlugan André Cane, Beaulieu-sur-Mer 11895
Musée du Patrimoine de l'Isle Crémieu, Hières-sur-Amby 13215
Musée du Patrimoine des Pères de Bétharram, Lestelle-Bétharram 13725
Musée du Patrimoine et de la Dentelle, Chantilly 12400
Musée du Patrimoine Montagnard, Nages 14346
Musée du Patrimoine Religieux et des Croyances Populaires, Moustey 14313
Musée du Patrimoine Rural, Dompierre (Orne) 12763
Musée du Paye de Guérande, Guérande 13157
Musée du Pays, Ussel 16020
Musée du Pays Brignolais, Brignoles 12180
Musée du Pays d'Albe, Sarralbe 15638
Musée du Pays de Luchon, Bagnères-de-Luchon 11822
Musée du Pays de Retz, Bourgneuf-en-Retz 12122
Musée du Pays de Sarrebourg, Sarrebourg 15643
Musée du Pays de Thônes, Thônes 15893
Musée du Pays du Der, Sainte-Rémy-en-Bouzemont 15578
Musée du Pays et Val de Charmey, Charmey 41291
Musée du Pays Rabastinois, Rabastens 14976
Musée du Pays Vaurais, Lavaur 13549
Musée du Pays Welche, Fréland 13006
Musée du Peigne et de la Plasturgie, Oyonnax 14550
Musée du Père Gèdèon, Sainte-Marie-de-Beauce 06899
Musée du Père Jean-Marie de la Mennais, Ploërmen 14844
Musée du Père Noël, Chailly-en-Bière 12356
Musée du Petit Bateau et des Aventures de la Mer, Canet-en-Roussillon 12265
Musée du Petit Format, Nismes 03916
Musée du Petit Lourdes, Lourdes 13848
Musée du Petit-Palais, Avignon 11792
Musée du Pétrogliphe, Hokatu 16321
Musée du Pétrole, Merkwiller-Pechelbronn 14090
Musée du Peuple Haïtien, Fermath 23319
Musée du Phare de Cordouan, Le Verdon-sur-Mer 13664
Musée du Phonographe et de la Musique Mécanique, Sainte-Maxime 15571
Musée du Phonographe et de Musique Mécanique, Limoges 13768
Musée du Pilori, Niort 14457
Musée du Plan Relief, Bitche 12006
Musée du Poète Ferrailleur, Lizio 13792
Musée du Premier Empire, Fixin 12941
Musée du Préservatif, Condom 12610
Musée du Président Jacques Chirac, Sarran 15640
Musée du Présidial, Castelnaudary 12315
Musée du Présidial Louis Lacrocq, Guéret 13159
Musée du Prieuré, Charolles 12416
Musée du Prieuré, Harfleur 13180
Musée du Prieuré, Perrecy-les-Forges 14795
Musée du Prieuré d'Amilhac, Servian 15730
Musée du Prieuré de Graville, Le Havre 13596
Musée du Protestantisme Dauphinois, Le Poët-Laval 13637
Musée du Protestantisme en Haut Languedoc, Ferrières (Tarn) 12927
Musée du Quai Branly, Paris 14700
Musée du Quercorb, Puivert 14958
Musée du Radar Allemand, Douvres-la-Délivrande 12784
Musée du Rail, Cajarc 12245
Musée du Rail, Dinan 12748
Musée du Ranquet, Clermont-Ferrand 12550
Musée du Rhum, Sainte-Marie 30673
Musée du Rhum, Sainte-Rose 23282
Musée du Rocher des Fourches, Cornus 12631
Musée du Roi de Cuba, Mushenge 08803
Musée du Rouergue, Espalion 12855
Musée du Rouergue Arts Métiers, Salles-la-Source 15607
Musée du Roy Soleil, Mont-Louis 14160
Musée du Royal 22e Régiment, Québec 06692
Musée du Sable, Le Château-d'Olonne 13571
Musée du Sabot, Buironfosse 12205
Musée du Sabot, Etang-sur-Arroux 12872
Musée du Sabot, La Haye-de-Routot 13385
Musée du Sabot, Porcheresse 03955
Musée du Sabotier, Soucht 15792
Musée du Safran, Boynes 12139
Musée du Santon, Generagues 13041
Musée du Santon, Le Val 13662
Musée du Santon, Marseille 14006
Musée du Santon, Pierrefeu-du-Var 14817
Musée du Santon et des Traditions de Provence, Fontaine-de-Vaucluse 12960
Musée du Santons, Gréoux-les-Bains 13142
Musée du Sapeur-Pompier, Bouzigues 12137
Musée du Sapeur Pompier, Saint-Hippolyte-du-Fort 15311
Musée du Sapeurs-Pompiers, Ille-sur-Têt 13245
Musée du Saut du Tarn, Saint-Juéry 15342
Musée du Savignéen, Savigné-sur-Lathan 15677
Musée du Savon de Marseille Marius Fabre, Salon-de-Provence 15612
Musée du Scaphandre, Espalion 12856
Musée du Sceau Alsacien, La Petite-Pierre 13403
Musée du Scribe, Saint-Christol-lès-Alès 15204
Musée du Sel, Salies-de-Béarn 15597
Musée du Séminaire de Saint-Hyacinthe, Saint-Hyacinthe 06822
Musée du Septennat, Château-Chinon 12433
Musée du Service de Santé des Armées, Paris 14701
Musée du Site Balnéaire, Dinard 12751
Musée du Site de l'Outil au Moutier-d'Ahun, Ahun 11496
Musée du Ski, Besse-en-Chandesse 11962
Musée du Ski, Briançon 12169
Musée du Ski des Laurentides, Saint-Sauveur 06877
Musée du Ski et de la Tradition Rousselande, Les Rousses 13712
Musée du Sous-Officier, Saint-Maixent-l'École 15385
Musée du Souvenier, Seillans 15702
Musée du Souvenir, Guer 13155
Musée du Souvenir, Soulac-sur-Mer 15798
Musée du Souvenir 1914-1918, Ablain-Saint-Nazaire 11487
Musée du Souvenir des Guerres 1914-18 et 1940-45, Ecaussinnes 03572
Musée du Souvenir du Combattant, Chalon-sur-Saône 12358
Musée du Souvenir du Génie Militaire, Angers 11611
Musée du Souvenir Militaire de Lyon, Lyon 13894
Musée du Souvenir Militaire de Thiérache, Martigny 14019
Musée du Squelette, île-Verte 06007
Musée du Suaire, Cadouin 12227
Musée du Sucre d'Orge, Moret-sur-Loing 14269
Musée du Tacot, Cléron 12557
Musée du Tapis et des Arts Textiles, Clermont-Ferrand 12551
Musée du Téléphone, Nancy 14355
Musée du Téléphone, Warneton 04134
Musée du Téléphone Narbonne, Narbonne 14378
Musée du Témiscouata, Notre-Dame-du-Lac 06491
Musée du Temple de Minerve, Tebessa 00079
Musée du Temps, Besançon 11959
Musée du Temps, Ceillac 12334
Musée du Temps des Baigneurs, Cordemais 12622
Musée du Temps Passé, Sainte-Agnès 15548
Musée du Terrain d'Aviation de Conde-Vraux, Vraux 16276
Musée du Terroir, Sisteron 15763
Musée du Terroir, Villeneuve-d'Ascq 16206
Musée du Terroir de Peille, Peille 14768
Musée du Terroir des Clots, Mizoën-en-Oisans 14137
Musée du Terroir et Musée de la Volaille, Romenay 15089
Musée du Terroir Hérissonnais, Hérisson 13205
Musée du Terroir Marseillais, Marseille 14007
Musée du Terroir Normand, Caugé 12325
Musée du Terroir Saint-Dalmas, Valdeblore 16035
Musée du Terroire, Conches-en-Ouche 12604
Musée du Textile, Cholet 12524
Musée du Textile des Vosges, Ventron 16107
Musée du Textile et de la Vie Sociale, Fourmies 12992
Musée du Textilen, Labastide-Rouairoux 13463
Musée du Thé, Paris 14702
Musée du Théâtre, Couilly-Pont-aux-Dames 12641
Musée du Théâtre Forain et Musée Archéologique, Artenay 11706
Musée du Théâtre Royal de Toone, Bruxelles 03486
Musée du Tire-Bouchon, Ménerbes 14072
Musée du Tissage et de la Soierie, Bussières 12210
Le Musée du Tisserand Dauphinois, La Batie-Montgascon 13313
Musée du Touage et Canal de Saint-Quentin, Riqueval 15049
Musée du Touquet, Le Touquet-Paris-Plage 13655
Musée du Tour Automatique et d'Histoire de Moutier, Moutier 41652
Musée du Tour de France, Arvieux-en-Queyras 11709
Musée du Tourneur, Cinquétral 12529
Musée du Train, Clairac 12534
Musée du Train à Vapeur, Sainte-Foy-l'Argentière 15560
Musée du Train et de la Miniature, Saint-Sauves-d'Auvergne 15505
Musée du Train et des Equipages Militaires, Tours 15968
Musée du Train et du Jouet, Arpaillargues-et-Aureillac 11693
Musée du Transmanche, Escalles 12848
Musée du Transport Urbain Bruxellois, Bruxelles 03487
Musée du Travail, Montfermeil 14205
Musée du Travail Charles Peyre, Montfermeil 14206
Musée du Trégor, Guimaec 13165
Musée du Tribunal Révolutionnaire, Tinchebray 15914
Musée du Trièves, Mens 14076
Musée du Trophée des Alpes, La Turbie 13454
Musée du Val-de-Cher, Selles-sur-Cher 15705
Musée du Valois, Vez 16135
Musée du Valois et de l'Archerie, Crépy-en-Valois 12673
Musée du Veinazès, Lacapelle-del-Fraisse 13469
Musée du Vélo, Cormatin 12628
Musée du Vélo, Fenneviller 12921
Musée du Vélo, Periers 14775
Musée du Vélo et de la Moto, Domazan 12757
Musée du Vélo Le Belle Échappée, La Fresnaye-sur-Chédouet 13371
Musée du Vélocipède, Cadouin 12228
Musée du Veneur, Céré-la-Ronde 12338
Musée du Verre, Blaye-les-Mines 12024
Musée du Verre, Liège 03812
Musée du Verre, Marcinelle 03851
Musée du Verre, Sars-Poteries 15646
Musée du Verrier, Saint-Prex 41793
Musée du Vieaux Château, Bricquebec 12174
Musée du Vieil Aix, Aix-en-Provence 11517
Musée du Vieil Argenteuil, Argenteuil 11674
Musée du Vieil Auvillar, Auvillar 11772
Musée du Vieil Eygalières, Eygalières 12891
Musée du Vieil Outil Hastière, Hastière 03688
Musée du Vieux-Baulmes, Baulmes 41187
Musée du Vieux Biarritz, Biarritz 11983
Musée du Vieux Bourg, Mimizan 14124
Musée du Vieux Brest, Brest 12162
Musée du Vieux-Château, Laval (Mayenne) 13542
Musée du Vieux-Cimetière, Soignies 04036
Musée du Vieux Clairac, Clairac 12535
Musée du Vieux Courseulles, Courseulles-sur-Mer 12656
Musée du Vieux Figeac, Figeac 12933
Musée du Vieux Granville, Granville 13105
Musée du Vieux Lambesc, Lambesc 13493
Musée du Vieux L'Arbresle, L'Arbresle 13525
Musée du Vieux Marseille, Marseille 14008
Musée du Vieux Mazères, Mazères (Ariège) 14050
Musée du Vieux Merville, Merville 14093
Musée du Vieux-Monthey, Monthey 41635
Musée du Vieux Montpellier, Montpellier 14246
Musée du Vieux-Moudon, Moudon 41650
Musée du Vieux Moulin, Villar-Loubière 16165
Musée du Vieux Moulin à Vent, Vensac 16105
Musée du Vieux Nîmes, Nîmes 14453
Musée du Vieux Nimy, Mons 03887
Musée du Vieux-Pays, Le Châble 41287
Musée du Vieux Pays, Dieulouard 12723
Musée du Vieux-Pays d'Enhaut, Château-d'Oex 41292
Musée du Vieux Pérouges, Pérouges 14786
Musée du Vieux-Phare, Matane 06248
Musée du Vieux Plassac, Plassac 14833
Musée du Vieux Pommiers, Pommiers-en-Forez 14874
Musée du Vieux Pressoir et de la Vigne, Coulanges-la-Vineuse 12642
Musée du Vieux-Saxon, Saxon 41826
Musée du Vieux Sisteron, Sisteron 15764
Musée du Vieux Toulon, Toulon (Var) 15926
Musée du Vieux-Toulouse, Toulouse 15941
Musée du Vieux-Tréport, Le Tréport 13657
Musée du Vieux Warcq, Warcq 16285
Musée du Vigneron, Buxy 12214
Musée du Vigneron, Chaumont-le-Bois 12485
Musée du Vigneron, Les Marches 13705
Musée du Vigneron, Verdigny 16111
Musée du Vigneron, Wuenheim 16303
Musée du Vigneron Paul Coulon et Fils, Rasteau 14987
Musée du Vignoble et des Vins d'Alsace, Kientzheim 13302
Musée du Vignoble Nantais, Le Pallet 13628
Musée du Village, Moknine 42642
Musée du Village, Saint-Sorlin-d'Arves 15516
Musée du Vin, Châteauneuf-du-Pape 12447
Musée du Vin, Cheignieu-la-Balme 12496
Musée du Vin, Cissac-Médoc 12531
Musée du Vin, Ehnen 30471
Musée du Vin, Nissan-lez-Ensérune 14459
Musée du Vin, Paris 14703
Musée du Vin, Saint-Cirq-Lapopie 15212
Musée du Vin, Saint-Hilaire-de-Loulay 15306
Musée du Vin dans l'Art, Pauillac 14763
Musée du Vin de Bourgogne, Beaune 11904
Musée du Vin de Cahors et de la Table Lotoise, Cahors 12243
Musée du Vin de Champagne Piper-Heidsieck, Reims 15003
Musée du Vin de la Batellerie et de la Tonnellerie, Bergerac 11946
Musée du Vin et de Saint-Jacques de Compostelle, Pomerol 14872
Musée du Vin-Mimoria di Vino, Luri 13868
Musée du Vitrail, Curzac-sur-Vonne 12688
Musée du Vitrail et du Verre, Gordes 13084
Musée du Vivarais Protestant, Pranles 14940
Musée du Zinc, Montpellier 14247
Musée Ducal, Bouillon 03393
Musée Duhamel du Monceau, Pithiviers-le-Vieil 14827
Musée d'Unterlinden, Colmar 12585
Musée Dupleix et Amas, Landrecies 13500
Musée Dupuytren, Paris 14704
Musée d'Uxellodunum et Musée de Martel, Martel 14018
Musée d'Yverdon-les-Bains et sa Région, Yverdon-les-Bains 42049
Musée E. Chenon, Château-Meillant 12436
Musée Eclaté d'Arts et de Traditions Populaires, Cardaillac 12288
Musée-École, Saint-Christophe-en-Brionnais 15207
Musée-Ecole de Broderie, Cilaos 36089
Musée École de la Perrine, Laval (Mayenne) 13543
Musée-École du Grand Meaulnes, Épineuil-le-Fleuriel 12841
Musée Ecole Publique en Charente, Saint-Fraigne 15274
Musée Écologique Vanier, Laval 06148
Musée EDF Electropolis, Mulhouse 14326
Musée Edgar Melik, Cabriès 12222
Musée Edgard Clerc, Le Moule 23276
Musée Edith Piaf, Paris 14705
Musée Edmond Rostand, Cambo-les-Bains 12254
Musée Edouard Branly, Paris 14706
Musée Edouard Mahé, Retiers 15022
Musée Éducatif de Préhistoire, Villeneuve-la-Comtesse 16211
Musée Éducatif d'Entomologie de Levens, Levens 13730
Musée Église Catholique de l'Assomption, Rosenwiller 15105
Musée El Kobba des Arts et Traditions, Sousse 42657
Musée Elise Rieuf, Massiac 14028
Musée Emile Aubry, Béjaïa 00054
Musée Emile Chénon, Châteaumeillant 12445
Musée Émile Guillaumin, Ygrande 16307
Musée Émile Jean, Villiers-sur-Marne 16245
Musée Émile Verhaeren, Honnelles 03710
Musée Emir Abdelkader, Miliana 00069
Musée Emmanuel de la Villéon, Fougères (Ille-et-Vilaine) 12987
Musée en Herbe, Paris 14707
Musée en Piconrue, Bastogne 03347
Musée en Plein Air de Sculptures Contemporaines, Étiolles 12879
Musée en Plein Air du Sart-Tilman, Liège 03813
Musée Entomologique, Saint-Quentin (Aisne) 15475

Musée Ermitage John lo Scorpione, Saint-Ame 15163
Musée Ernest Cognacq, Saint-Martin-de-Ré 15400
Musée Ernest Renan, Tréguier 15983
Musée Espace Bérenger Saunière, Rennes-le-Château 15018
Musée Espéranto a Gray, Gray 13128
Musée Estève, Bourges 12119
Musée Estrine, Saint-Rémy-de-Provence 15492
Musée et Atelier d'Art Animalier, La Chapelle-Thouarault 13338
Musée et Centre de Transmission de la Culture Daniel Weetaluktuk, Inukjuak 06015
Musée et du Vieux Chinon, Chinon 12516
Musée et Maison des Vins, Bandol 11838
Musée et Parc Archéologiques de Montauban, Buzenol 03508
Musée et Plaisir du Chocolat, Biarritz 11984
Musée et Salon Permanent de l'Automobile de Sport Prestige et Collection, Chauffailles 12480
Musée et Site Archéologiques de Nice-Cimiez, Nice 14434
Musée et Site Gallo-Romains de Saint-Romain-en-Gal - Vienne, Saint-Romain-en-Gal 15502
Musée et Temple des Bastides, Labastide-d'Armagnac 13461
Musée et Trésor de l'Hospice, Bourg-Saint-Pierre 41250
Musée Ethnographique, Chefchaouen 31666
Musée Ethnographique, Mouans-Sartoux 14293
Musée Ethnographique, Tétouan 31701
Musée Ethnographique, Tétouan 31702
Musée Ethnographique Alexandre Senou Adande, Porto-Novo 04189
Musée Ethnographique de l'Olivier, Cagnes-sur-Mer 12238
Musée Ethnographique des Oudaïa, Rabat 31690
Musée Ethnographique du Donjon, Niort 14458
Musée Ethnographique du Peuple Basque, Isturitz 13255
Musée Ethnographique Regional, Bouar 07422
Musée Ethnologique Intercommunal du Vermandois, Vermand 16118
Musée Ethnolopique Provincial, Tshikappa 08804
Musée Etienne-Jules Marey, Beaune 11905
Musée Étienne Martin, Noyal-sur-Vilaine 14478
Musée Etival dans le Temps, Etival-Clairefontaine 12880
Musée Eudore-Dubeau, Montréal 06368
Musée Eugène Boudin, Honfleur 13223
Musée Eugène Burnand, Moudon 41651
Musée Eugène Farcot, Sainville 15590
Musée Eugène le Roy et Vieux Métiers, Montignac 14215
Musée Européen de Sculptures Contemporaines en Plein Air, Launstroff 13535
Musée-Expo de la Grange aux Abeilles, Giffaumont-Champaubert 13056
Musée Expo Forêt, Gérardmer 13046
Musée Extraordinaire Georges-Mazoyer, Ansouis 11636
Musée Fabre, Montpellier 14248
Musée Faniahy, Fianarantsoa 30539
Musée Fantastique de la Bête du Gévaudan, Saugues-en-Gévaudan 15652
Musée Fayet, Béziers 11978
Musée Fenaille, Rodez 15080
Musée-Ferme Conservatoire Bigourdane, Péré 14774
Musée Ferme d'Autrefois et Matériel Artisanal, Thourie 15903
Musée Ferme de la Patte d'Oie, Le Plessis-Patte-d'Oie 13635
Musée-Ferme des Castors du Chili, Arzacq-Arraziguet 11712
Musée Ferroviaire, Longueville 13810
Musée Ferroviaire, Saint-Sulpice-de-Pommeray 15517
Musée Ferroviaire de Saint-Géry, Saint-Géry 15297
Musée Ferroviaire de Treignes, Treignes 04093
Musée Ferroviaire et des Vieilles Gares, Guîtres 13170
Musée Fesch, Ajaccio 11527
Musée Flaubert et d'Histoire de la Médecine, Rouen 15116
Musée Flottant-Architecture Navale, Audierne 11740
Musée Folklorique de Ghardaïa, Ghardaïa 00063
Musée Folklorique et Archéologique, Châtelus-le-Marcheix 12465
Musée Folklorique et Viticole à Possen, Bech-Kleinmacher 30454
Musée-Fondation Alexandra David Néel, Digne-les-Bains 12726
Musée-Fondation Bemberg, Toulouse 15942
Musée Fondation Deutsch, Belmont-sur-Lausanne 41196
Musée Fondation Léon Perrin, Môtiers 41647
Musée-Fondation Louis Jou, Les Baux-de-Provence 13693
Musée for Fossils & Minerals, Erfoud 31667
Musée-Fort de Leveau, Feignies 12917
Musée Fort de Sucy, Sucy-en-Brie 15837
Musée Fort Lagarde, Prats-de-Mollo-la-Preste 14941
Musée Fougou, Montpellier 14249
Musée Fournaise, Chatou 12476
Musée Fragonard de l'Ecole Nationale Vétérinaire d'Alfort, Maisons-Alfort 13924
Musée Français - 1e Armée Française Mai 1940, Cortil-Noirmont 03527
Musée Français de la Brasserie, Saint-Nicolas-de-Port 15428
Musée Français de la Carte à Jouer, Issy-les-Moulineaux 13252
Musée Français de la Photographie, Bièvres 11991
Musée Français de la Spéléologie, Courniou-les-Grottes 12652
Musée Français des Phares et Balises, Ile-d'Ouessant 13243
Musée Français du Chemin de Fer, Mulhouse 14327
Musée Francisque Mandet, Riom 15045
Musée Franco-Australien, Villers-Bretonneux 16228
Musée Franco-Tchécoslosvaque, Darney 12698
Musée François Desnoyer, Saint-Cyprien 15224
Musée François Duesberg des Arts Décoratifs, Mons 03888
Musée François Mauriac, Saint-Maixant (Gironde) 15384
Musée François-Mauriac, Vemars 16094
Musée François Mitterrand, Jarnac 13260
Musée François Pilote, La Pocatière 06113
Musée François Pompon, Saulieu 15656
Musée Frédéric Japy, Beaucourt 11890
Musée Frédéric Mistral, Maillane 13917
Musée Friry, Remiremont 15009
Musée Fromages et Patrimoine, Ambert 11569
Musée Frugès - Le Corbusier, Pessac 14803
Musée Gadagne, Lyon 13895
Musée Galerie Honoré Camos, Bargemon 11853
Musée Gallé-Juillet, Creil 12671
Musée Galliera, Paris 14708
Musée Gallo-Romain, Berneau 03363
Musée Gallo-Romain, Biesheim 11988
Musée Gallo-Romain, Blicquy 03374
Musée Gallo-Romain, Chanaz 12397
Musée Gallo-Romain, Périgueux 14779
Musée Gallo-Romain, Petit-Bersac 14804
Musée Gallo-Romain, Rumes 03989
Musée Gallo-Romain d'Aoste, Aoste 11649
Musée Gallo-Romain de Lyon-Fourvière, Lyon 13896
Musée Gallo-Romain de Tauroentum, Saint-Cyr-sur-Mer 15230
Musée Gallo-Romain de Waudrez, Waudrez 04139
Musée Gallo-Romain d'Octodure, Martigny 41612
Musée Gantner, La Chapelle-sous-Chaux 13337
Musée Gardanne Autrefois, Gardanne 13033
Musée Gaspar, Arlon 03326
Musée Gassendi, Digne-les-Bains 12727
Musée Gaston Fébus, Mauvezin (Hautes-Pyrénées) 14044
Musée Gaston Grégor, Salles-d'Angles 15606
Musée Gatien-Bonnet, Lagny-sur-Marne 13477
Musée Gaumais, Virton 04116
Musée Gay-Lussac, Saint-Léonard-de-Noblat 15369
Musée Général Estienne, Berry-au-Bac 11952
Musée Géo Charles, Échirolles 12808
Musée Géologique, Conakry 23309
Musée Géologique, Dakar 38023
Musée Géologique, La Charguia 42634
Musée Geologique, Vernet-les-Bains 16119
Musée Géologique du Rwanda, Kigali 37994
Musée Géologique et Paléontologique, Langé 13503
Musée Géologique et Salle Archéologique, Saint-Avit-Senieur 15180
Musée Géologique Fossiles Préhistoire, Neuil 14401
Musée Geologique Sengier-Cousin, Jadotville 08793
Musée George Sand et de la Vallée Noire, La Châtre 13340
Musée Georges Borias, Uzès 16025
Musée Georges Clemenceau, Paris 14709
Musée Georges Garret, Vesoul 16133
Musée Georges Labit, Toulouse 15943
Musée Georges Perraudin-Résistance en Morvan, Saint-Honoré-les-Bains 15313
Musée Georges Pompidou, Montboudif 14187
Musée Georgette Lemaire, Angoulême 11622
Musée Gilles-Villeneuve, Berthierville 05422
Musée Girodet, Montargis 14173
Musée Girouxville, Girouxville 05867
Musée Goethe, Sessenheim 15732
Musée Goetz-Boumeester, Villefranche-sur-Mer 16193
Musée Gorsline, Bussy-le-Grand 12212
Musée Goupil, Bordeaux 12072
Musée Goya, Castres 12320
Musée Granet, Aix-en-Provence 11518
Musée Grétry, Liège 03814
Musée Grévin, La Rochelle 13431
Musée Grévin, Le Mont-Saint-Michel 13621
Musée Grévin, Paris 14710
Musée Grévin, Saint-Jean-de-Luz 15327
Musée Grévin de la Provence, Salon-de-Provence 15613
Musée Grévin de Lourdes, Lourdes 13849
Musée Grobet-Labadié, Marseille 14009
Musée Grotte de Limousis, Limousis 13772
Musée Gruérien, Bulle 41272
Musée Guerre et Paix en Ardennes, Novion-Porcien 14477
Musée Guillaume Apollinaire, Stavelot 04049
Musée Guillion, Romanèche-Thorins 15086
Musée Gustave Moreau, Paris 14711
Musée Gustave Stoskopf, Brumath 12199
Le Musée Gutenberg des Arts Graphiques et de la Communication, Fribourg 41389
Musée Gyger, Vallorbe 41977
Musée Hansi, Riquewihr 15053
Musée Hardy, Clécy 12542
Musée Harmas Jean-Henri Fabre, Sérignan-du-Comtat 15724
Musée Haut Savoyard de la Résistance, Bonneville 12057
Musée Haviland, Limoges 13769
Musée Hector Berlioz, La Côte-Saint-André 13351
Musée Helmut Warzecha, Saint-Philbert-de-Grand-Lieu 15455
Musée Henri Barbusse, Aumont-en-Halatte 11747
Musée Henri-Barre, Thouars 15902
Musée Henri Boez, Maubeuge 14032
Musée Henri Chapu, Le Mée-sur-Seine 13613
Musée Henri Dupuis, Saint-Omer 15432
Musée Henri Giron, Le Vigan (Lot) 13667
Musée Henri Malartre, Rochetaillée-sur-Saône 15075
Musée Henri Mathieu, Bruyères 12202
Musée Henry Clews, Mandelieu-La Napoule 13934
Musée Henry de Monfreid, Ingrandes (Indre) 13247
Musée Héritage, Saint-Albert 06790
Musée Héritage, Saint-Pierre 37999
Musée Hermès, Paris 14712
Musée Hippomobile, Saint-Michel-des-Andaines 15417
Musée Historial, Grézolles 13147
Musée Historial, Mazères (Gironde) 14051
Musée Historie Locale, Marchiennes 13953
Musée Historique, Antananarivo 30536
Musée Historique, Basse-Rentgen 11857
Musée Historique, Bourail 33031
Musée Historique, Carhaix-Plouguer 12290
Musée Historique, Cérisy-la-Forêt 12345
Musée Historique, Châtillon-sur-Saône 12473
Musée Historique, Clervaux 30462
Musée Historique, Fénétrange 12920
Musée Historique, Gourdon (Alpes-Maritimes) 13089
Musée Historique, Haguenau 13177
Musée Historique, La Couvertoirade 13357
Musée Historique, Mont-Saint-Michel 14162
Musée Historique, Mulhouse 14328
Musée Historique, Paris 14713
Musée Historique d'Abomey, Abomey 04182
Musée Historique de Gorée, Gorée 38026
Musée Historique de Graffiti Anciens et d'Archéologie, Marsilly 14015
Musée Historique de la Faïence, Moustiers-Sainte-Marie 14314
Musée Historique de la Ligne Maginot Aquatique, Sarralbe 15639
Musée Historique de la Principauté de Stavelot-Malmedy, Stavelot 04050
Musée Historique de la Tour de la Chaîne, La Rochelle 13432
Musée Historique de la Ville de Strasbourg, Strasbourg 15831
Musée Historique de Lausanne, Lausanne 41548
Musée Historique de l'Ile-de-Sein, Ile-de-Sein 13241
Musée Historique de Saint-Gilles-les-Hauts, Saint-Paul 36095
Musée Historique de Saint-Paul, Saint Paul 06870
Musée Historique de Saint-Pierre, Saint-Pierre 30671
Le Musée Historique de Tracadie, Tracadie-Sheila 07158
Musée Historique de Troyes et de la Champagne, Troyes 16000
Musée Historique de Verfeil et des Petites Filles Modèles, Verfeil 16114
Musée Historique de Vevey, Vevey 41985
Musée Historique de Villèle, Saint-Gilles-les-Hauts 36093
Musée Historique des Communications Eberdt, Sutton, Québec 07042
Musée Historique des Soeurs de l'Assomption, Nicolet 06458
Musée Historique du Bailliage de Rouffach, Rouffach 15123
Musée Historique du Centre-Var, Le Luc 13602
Musée Historique du Château, Langoiran 13506
Musée Historique du Cidre, Melleray-la-Vallée 14067
Musée Historique du Domaine National, Saint-Cloud 15219
Musée Historique du Madawaska, Edmundston 05764
Musée Historique du Papier, Ambert 11570
Musée Historique du Papier, Couze-et-Saint-Front 12664
Musée Historique du Port, La Rochelle 13433
Musée Historique et Archéologique, Civaux 12532
Musée Historique et Archéologique, Montségur 14265
Musée Historique et Archéologique de l'Orléanais, Orléans 14537
Musée Historique et de la Pêche, Talmont-sur-Gironde 15847
Musée Historique et des Porcelaines, Nyon 41691
Musée Historique et Diorama 1914-1918, Souchez 15791
Musée Historique et Erckmann-Chatrian, Phalsbourg 14811
Musée Historique et Lapidaire, Isle-Aumont 13248
Musée Historique et Logis Thiphaine, Le Mont-Saint-Michel 13622
Musée Historique et Militaire, Brest 12163
Musée Historique et Militaire, Freyming-Merlebach 13012
Musée Historique et Militaire, Huningue 13227
Musée Historique et Préhistorique Van den Steen, Jehay-Amay 03743
Musée Historique Valdec-de Lessart, Portets 14924
Musée Hofer-Bury, Lavérune 13552
Musée Honoré Mercier, Sabrevois 06786
Musée Horlogerie Automates Yves Cupillard, Morteau 14290
Musée Horreum Romain, Narbonne 14379
Musée Horta, Bruxelles 03488
Musée Hospitalier, Charlieu 12412
Musée Hospitalier - Apothicairerie, Bourg-en-Bresse 12108
Musée Hospitalier et Apothicairerie, Louhans 13833
Musée-Hôtel Bertrand, Châteauroux 12459
Musée-Hôtel de Beaumont, Valognes 16058
Musée Hôtel de la Dentelle, Brioude 12185
Musée Hôtel de Ville, Thann 15875
Musée-Hôtel Granval-Caligny, Valognes 16059
Musée Hôtel Lallemand, Bourges 12120
Musée-Hôtel Le Vergeur, Reims 15004
Musée Humoristique Albert Dubout, Palavas-les-Flots 14557
Musée Huperel de l'Histoire Cathare, Minerve 14126
Musée Huppy Autrefois, Huppy 13229
Musée Hydrogéologique, Chaudfontaine 03519
Musée Ianchelevici, La Louvière 03768
Musée Impérial de la Miniature, Verneuil-sur-Avre 16122
Musée in Situ des Bains Romains, Aix-les-Bains 11523
Musée Industriel Corderie Vallois, Notre-Dame-de-Bondeville 14474
Musée Industriel du Val-de-Travers, Fleurier 41373
Musée Industriel et d'Ethnologie, Lille 13759
Musée Ingres, Montauban 14176
Musée Intercommunal d'Histoire et d'Archéologie, Louvres 13859
Musée International d'Art Naïf, Vicq 16148
Musée International d'Art Naïf Yvon-M. Daigle, Magog 06215
Musée International d'Arts Modestes, Sète 15734
Musée International de Akum, Bamenda 05293
Musée International de la Base Aérienne de Chièvres, Chievres 03522
Musée International de la Chasse, Gien 13055
Musée International de la Chaussure, Romans-sur-Isère 15088
Musée International de la Croix-Rouge et du Croissant-Rouge, Genève 41423
Musée International de la Faune, Argelès-Gazost 11669
Musée International de la Faune, Québriac 14964
Musée International de la Marionnette, Bruxelles 03489
Musée International de la Miniature, Lyon 13897
Musée International de la Parfumerie, Grasse 13114
Musée International de la Reforme, Genève 41424
Musée International de l'Automobile, Le Grand-Saconnex 41448
Musée International de Mobilier Miniature, Vendeuvre 16099
Musée International des Hussards, Tarbes 15863
Musée International des Polices et des Gendarmeries, Charvieu-Chavagneux 12427
Musée International d'Horlogerie, La Chaux-de-Fonds 41296
Musée International du Carnaval et du Masque, Binche 03368
Musée International du Coquillage, Saint-Barthélemy 23279
Musée International du Long Cours Cap Hornier, Saint-Malo (Ille-et-Vilaine) 15390
Musée International, Pétanque et Boules, Saint-Bonnet-le-Château 15187
Musée Isérables, Isérables 41494
Musée Itinérant d'Art Byzantin, Montréal 06369
Musée Ivan Tourguéniev, Bougival 12085
Musée Ivanoff, La Borne 13318

Musée J. Armand Bombardier, Valcourt 07196
Musée J.B. Mathon et André Durand, Neufchâtel-en-Bray 14398
Musée J.-E. Blanche, Offranville 14497
Musée J.-F. Leyraud, La Roche-sur-le-Buis 13417
Musée Jacquemart-André, Fontaine-Chaalis 12955
Musée Jacquemart-André, Paris 14714
Musée Jacques Cartier à Saint-Malo, Saint-Malo (Ille-et-Vilaine) 15391
Musée Jacques Pellan, Giverny 13069
Musée Jadis Allevard, Allevard-les-Bains 11556
Musée-Jardin Paul Landowski, Boulogne-Billancourt 12091
Musée Jean Aicard, Solliès-Ville 15778
Musée Jean Aicard - Paulin Bertrand, La Garde 13373
Musée Jean-Charles Cazin, Samer 15617
Musée Jean Cocteau, Menton 14083
Musée Jean de La Fontaine, Château-Thierry 12440
Musée Jean Gabin, Mériel 14087
Musée Jean-Henri Fabre, Saint-Léons 15370
Musée Jean-Honoré Fragonard, Grasse 13115
Musée Jean-Jacques Rousseau, Genève 41425
Musée Jean-Jacques Rousseau, Montmorency 14229
Musée Jean-Jacques Rousseau, Môtiers 41648
Musée Jean-Lurçat, Angers 11612
Musée Jean Lurçat, Saint-Laurent-les-Tours 15363
Musée Jean Macé, Beblenheim 11915
Musée Jean-Marie Somet, Villars (Loire) 16171
Musée Jean-Marie Souplet, Engis 03582
Musée Jean-Mermoz, Aubenton 11724
Musée Jean-Moulin, Paris 14715
Musée Jean Racine, La Ferté-Milon 13362
Musée Jeanne d'Aboville, La Fère 13358
Musée Jeanne d'Arc, Chinon 12517
Musée Jeanne d'Arc, Rouen 15117
Musée Jeanne d'Arc, Sainte-Cathérine-de-Fierbois 15552
Musée Jenisch, Vevey 41986
Musée Joachim du Bellay, Liré 13775
Musée Johannique, Vaucouleurs 16082
Musée Joseph Abeilhe, Marciac 13955
Musée Joseph Denais, Beaufort-en-Vallée 11892
Musée Joseph Durand, Saint-Clément-des-Baleines 15216
Musée Joseph Filion, Sainte-Thérèse 06901
Musée Joseph Lhoménède, Frugières-le-Pin 13018
Musée Joseph Vaylet, Espalion 12857
Musée Jouenne, Saint-Rémy-de-Provence 15493
Musée Judéo-Alsacien, Bouxwiller 12135
Musée Juif Comtadin, Cavaillon 12331
Musée Juif de Belgique-Joods Museum van België, Bruxelles 03490
Musée Juin 1944, L'Aigle 13485
Musée Jules Desbois, Parçay-les-Pins 14563
Musée Jules Destrée, Charleroi 03518
Musée Jules Garinet, Châlons-en-Champagne 12363
Musée Jules Romains, Saint-Julien-Chapteuil 15345
Musée Jules Verne, Nantes 14369
Musée Jurassien d'Art et d'Histoire, Delémont 41324
Musée Jurassien des Arts, Moutier 41653
Musée Jurassien des Sciences Naturelles et Jardin botanique, Porrentruy 41729
Musée Juste pour Rire, Montréal 06370
Musée Kasba, Tanger 31698
Musée Kateri Tekakwitha, Kahnawake 06034
Musée Kaysone Phomvihane, Vientiane 29975
Musée Kio-Warini, Loretteville 06195
Musée Kotama, Jijel 00067
Musée Kwok-On, Paris 14716
Musée La Batterie d'Azeville, Azeville 11813
Musée La Diana, Montbrison 14190
Musée la Faïencerie, Ancy-le-Franc 11596
Musée La Ferme Forézienne, Saint-Bonnet-les-Oules 15188
Musée la Magie des Automates, Lans-en-Vercors 13515
Musée la Main et l'Outil, Pannes 14560
Musée La Maison du Bucheron, Pourcy 14930
Musée La Maison du Passé, Royat-les-Bains 15134
Musée la Maison du Pêcheur et "Le Hope", Saint-Gilles-Croix-de-Vie 15299
Musée la Muse, Mérindol 14089
Musée la Percée du Bocage 1944, Saint-Martin-des-Besaces 15401
Musée La Pouponnière, Bourg-Charente 12105
Musée la Taille de Pierre, Saint-Antoine-l'Abbaye 15171
Musée la Vie des Jouets, Mauleon 14035
Musée la Villa Aurélienne, Fréjus 13004
Musée Labasso, Bangassou 07420
Musée l'Abeille Vivante, Le Faouët 13583
Musée Labenche d'Art et d'Histoire, Brive-la-Gaillarde 12189
Musée Lacoune, Lacaune 13470
Musée Laennec, Nantes 14370
Musée Lamartine, Mâcon 13908
Musée Lamartine, Pierreclos 14815
Musée Lambinet, Versailles 16128
Musée Lamothe-Cadillac, Saint-Nicolas-de-la-Grave 15427
Musée Langlois, Beaumont-en-Auge 11900
Musée Languédocien, Montpellier 14250
Musée Lapérouse, Albi 11538
Musée Lapidaire, Aire-sur-l'Adour 11511
Musée Lapidaire, Avignon 11793
Musée Lapidaire, Baume-les-Messieurs 11868
Musée Lapidaire, Carpentras 12297
Musée Lapidaire, Morlaàs 14274
Musée Lapidaire, Narbonne 14380
Musée Lapidaire, Saint-Germain-Laval 15294
Musée Lapidaire, Saint-Guilhem-le-Désert 15304
Musée Lapidaire, Saint-Ursanne 41795
Musée Lapidaire, Villeveyrac 16239
Musée Lapidaire de l'Abbaye, Savigny (Rhône) 15685
Musée Lapidaire de l'Abbaye Saint-Pierre, Maillezais 13920
Musée Lapidaire de Mozac, Mozac 14321
Musée Lapidaire du Cloître de l'Ancienne Cathédrale, Vaison-la-Romaine 16032
Musée Lapidaire et d'Art Religieux, Charlieu 12413
Musée Lapidaire et Exposition Permanente sur la Restauration, Carcassonne 12286
Musée Larrey, Beaudean 11891
Musée L'Art en Marche Hauterives Art Brut, Neuve Invention, Hauterives 13193
Musée L'Art en Marche Lapalisse Art Brut, Lapalisse 13522
Musée Laumonier de la Locomotive à Vapeur, Vierzon 16155
Musée Laurier, Victoriaville 07261
Musée Lautrecois, Lautrec 13539
Musée Le Grenier de l'Histoire, Lèves 13731
Musée Le Jacquemart, Langeac 13504
Musée Le Masque en Liberté, Nibelle 14421
Musée le Mémoire d'Oyé, Oyé 14549
Musée Le Minerve, Yzeures-sur-Creuse 16314
Musée Le Mont-Saint-Michel, Le Mont-Saint-Michel 13623
Musée Le Patrimoine du Sapeur-Pompier, Beaune 11906
Musée Le Petit Louvre, La Pacaudière 13399
Musée le Petit Mont-Saint-Michel, Saint-Marcan 15392
Musée Le Pionnier, Saint-Malo 06866
Musée Le Secq des Tournelles - Musée de la Ferronnerie, Rouen 15118
Musée le Soum, Saint-Véran 15533
Musée le Temple du Soleil, Saint-Michel-en-l'Herm 15418
Musée Le Vieux Cormenier, Champniers-en-Vienne 12394
Musée Leblanc-Duvernoy, Auxerre 11777
Musée Leclerc Briant, Épernay 12835
Musée Lella Hadria, Midoun 42641
Musée Lénine, Paris 14717
Musée Leon de Smet, Deurle 03545
Musée Léon Dierx, Saint-Denis 36091
Musée Léon Marès, Lovagny 13860
Musée Les 3 Salles, Lectoure 13674
Musée Les Amis de l'Outil, Bièvres 11992
Musée Les Ateliers du Bois, Vadencourt 16028
Musée Les Flocons, Ittre 03736
Musée Les Sanglots Longs, Reguiny 14995
Musée les Sources d'Hercule, Deneuvre 12712
Musée les Thermes d'Antonin, Carthage 42619
Musée Les Trésors de Lisette, Archingeay 11665
Musée les Trois Pignons, Cheticamp 05608
Musée les Voltigeurs de Québec Manège Militaire, Québec 06693
Musée Libertador du Général San Martin, Boulogne-sur-Mer 12094
Musée Ligier Richier, Saint-Mihiel 15421
Musée Lillet, Podensac 14864
Musée Littéraire Victor Hugo, Vianden 30495
Musée Local, Boën-sur-Lignon 12039
Musée Local de Préhistoire, Sceaux-du-Gatinais 15689
Musée Local des Arts et Traditions Populaires, Auxi-le-Château 11779
Musée Loire Métiers, Saint-Clement-des-Levées 15218
Musée Lombard, Doullens 12782
Musée Longines, Saint-Imier 41791
Musée Lorrain, Nancy 14356
Musée Lorrain des Cheminots, Rettel 15025
Musée Lorrain du Cheval, Gondrecourt-le-Château 13079
Musée Lou Férouil, Gilette 13058
Musée Louis Braille, Coupvray 12648
Musée Louis Cornu, Wassigny 16289
Musée Louis-Hémon, Péribonka 06586
Musée Louis Jourdan, Saint-Paul-de-Varax 15443
Musée Louis Lutaud, Ferrière-la-Grande 12923
Musée Louis Pasteur, Strasbourg 15832
Musée Louis Pergaud, Belmont 11934
Musée Louis Philippe, Eu 12885
Musée Louis-Vouland, Avignon 11794
Musée Louise Cottin, Vigneulles-lès-Hattonchâtel 16159
Musée Louise Weiss, Saverne 15676
Musée Louison Bobet, Saint-Méen-le-Grande 15413
Musée Lucien Roy, Beure 11969
Musée McCord, Montréal 06371
Musée Maçonnique de la Grande Loge de France, Paris 14718
Musée Mado Robin, Yzeures-sur-Creuse 16315
Musée Magic de la Musique Mécanique et de la Machine Parlante, Mesnil-Raoul 14096
Musée Magie des Abeilles, Villechauve 16180
Musée Maginot de Rohrbach-Fort Casso, Rohrbach-les-Bitche 15081
Musée Magnanerie, Lagorce 13478
Musée Magnelli, Vallauris 16050
Musée Magnin, Dijon 12742
Musée Mai Juin 1940, Margut 13968
Musée Maillol, Banyuls-sur-Mer 11843
Musée Maillol, Paris 14719
Musée Mainssieux, Voiron 16267
Musée Mairie, Le Pont-de-Beauvoisin 13638
Musée-Maison d'Art Contemporain Chaillioux, Fresnes (Val-de-Marne) 13010
Musée-Maison de Chateaubriand, Châtenay-Malabry 12467
Musée Maison de Clemenceau, Saint-Vincent-sur-Jard 15539
Musée Maison de George Sand, Gargilesse-Dampierre 13035
Musée Maison de la Baleine, Luc-sur-Mer 13862
Musée Maison de la Brique, Saint-Martin-d'Aubigny 15395
Musée Maison de la Dame de Brassempouy, Brassempouy 12149
Musée-Maison de la Nature, Samer 15618
Musée-Maison de la Nature et de l'Estuaire de l'Orne, Sallenelles 15603
Musée Maison de la Radio TSF, Saint-Appolinard 15174
Musée Maison de la Soie, Saint-Hippolyte-du-Fort 15312
Musée-Maison de la Vie Rurale, La Flocellière 13368
Musée Maison de l'Armateur, Le Havre 13597
Musée-Maison de l'Aspre, Thuir 15907
Musée-Maison de l'Erdre, Nantes 14371
Musée Maison de l'Huître, Gujan-Mestras 13171
Musée-Maison d'Ecole, Montceau-les-Mines 14192
Musée Maison Déodat Roché, Arques (Aude) 11695
Musée-Maison des Amis de la Forêt, Mervent 14092
Musée-Maison des Arts des Hauts-Cantons, Bédarieux 11917
Musée-Maison des Canuts, Lyon 13898
Musée-Maison des Dentelles et du Point d'Argentan, Argentan 11672
Musée-Maison des Milelli, Ajaccio 11528
Musée-Maison des Pierreux, Massangis 14026
Musée-Maison du Café, Vieux-Habitants 23283
Musée-Maison du Canal, Fégréac 12916
Musée-Maison du Jambon de Bayonne, Arzacq-Arraziguet 11713
Musée-Maison du Jardinier, Gétigné 13051
Musée-Maison du Pain, Commeny 12594
Musée Maison du Pays, Moisdon-la-Rivière 14139
Musée-Maison du Paysan, Touvois 15976
Musée-Maison du Pêcheur, La Chevrolière 13342
Musée-Maison du Petit Poitou, Chaillé-les-Marais 12354
Musée Maison du Potier, Le Fuilet 13585
Musée-Maison du Saumon et de la Rivière, Brioude 12186
Musée-Maison du Terroir, Châteauponsac 12453
Musée-Maison du Verre, Meisenthal 14063
Musée-Maison Lohobiague dite Maison Louis XIV, Saint-Jean-de-Luz 15328
Musée-Maison Marcel Proust, Illiers-Combray 13246
Musée-Maison Natale de Jean de Lattre de Tassigny, Mouilleron-en-Pareds 14300
Musée-Maison Natale Gustave Courbet, Ornans 14540
Musée-Maison Paternelle de Bernadette, Lourdes 13850
Musée Maison Picassiette, Chartres 12424
Musée Maison Robert Doisneau, Gentilly 13044
Musée Maison Saint-Gabriel, Montréal 06372
Musée Malraux, Le Havre 13598
Musée M'Am Jeanne, Fontenoy-en-Puisaye 12974
Musée Manoir de Kerazan, Loctudy 13799
Musée-Manoir de la Touche, Nantes 14372
Musée Manoir de Vains, Vains 16030
Musée-Manoir des Sciences, Réaumur 14990
Musée Marc-Aurèle Fortin, Montréal 06373
Musée Marc Deydier, Cucuron 12684
Musée Marc Leclerc, La Ménitré 13392
Musée Marc Mureau, Restigné 15021
Musée Marcel Collet, Loverval 03835
Musée Marcel de Marcheville, Romorantin-Lanthenay 15092
Musée Marcel Lenoir, Montricoux 14261
Musée Marcel Sahut, Volvic 16270
Musée Maréchal Joffre, Rivesaltes 15056
Musée Marguerite Audoux, Aubigny-sur-Nère 11731
Musée Marguerite Bourgeoys, Montréal 06374
Musée Marguerite de Bourgogne, Tonnerre 15917
Musée Marguerite Vidal des Arts et Traditions Populaires, Moissac 14140
Musée Marguerite Yourcenar, Saint-Jans-Cappel 15315
Musée Marial de Beauraing, Beauraing 03350
Musée Marie Baz, Deir-el-Qamar 30169
Musée Marie Laurencin, Chino 28400
Musée Marie-Rose Durocher, Longueuil 06193
Musée-Marinarium du Haut Lavédan, Pierrefitte-Nestalas 14818
Musée Marine et Montagne, Mons (Var) 14152
Musée Maritime, Camaret-sur-Mer 12252
Musée Maritime, Carantec 12278
Musée Maritime, Le Mont-Saint-Michel 13624
Musée Maritime Cameroun, Douala 05295
Musée Maritime Chantereyne, Cherbourg-Octeville 12504
Musée Maritime de Charlevoix, Saint-Joseph-de-la-Rive 06860
Musée Maritime de l'Ile de Tatihou, Saint-Vaast-la-Hougue 15527
Musée Maritime de l'île de Tatihou, Saint-Vaast-la-Hougue 15528
Musée Maritime de Regnéville, Regnéville-sur-Mer 14993
Musée Maritime de Tadoussac, Tadoussac 07057
Musée Maritime du Québec, L'Islet 06175
Musée Maritime, Fluvial et Portuaire, Rouen 15119
Musée Maritime Islais, Port-Joinville 14916
Musée Maritime La Rochelle, La Rochelle 13434
Musée Maritime l'Esvale et Ostréicole, La Tremblade 13452
Musée Marius Barbeau, Saint-Joseph-de-Beauce 06859
Musée Marmottan Monet, Paris 14720
Musée Marsa El-Kharez, El-Kala 00060
Musée Marsil - Musée du Costume et du Textile du Québec, Saint-Lambert 06861
Musée Martin Duby, Auriol 11757
Musée Massey, Tarbes 15864
Musée Matheysin, La Mure-d'Isère 13397
Musée Mathurin Méheut, Lamballe 13492
Musée Matisse, Le Cateau-Cambrésis 13569
Musée Matisse, Nice 14435
Musée Maurice Beaumont, Luxeuil-les-Bains 13873
Musée Maurice Carême, Bruxelles 03491
Musée Maurice Dufresne, Azay-le-Rideau 11811
Musée Maurice et Eugénie de Guérin, Andillac 11597
Musée Maurice Genevoix, Saint-Denis-de-l'Hôtel 15232
Musée Maurice Gottlob, Mougins 14299
Musée Maurice Poignant, Bourg-sur-Gironde 12113
Musée Maurice Ravel, Montfort-l'Amaury 14209
Musée Maxime Mabilleau, Saumur 15667
Musée Mazot de Plan-Cerisier, Martigny 41613
Musée Médiéval, Brain-sur-Allonnes 12141
Musée Medieval, Serralongue 15728
Musée Médiumnique Fernand Desmoulin, Brantôme 12143
Musée Mémoires de la Terre et Horizons Cathares, Mazamet 14045
Musée Mémoires du Moyen Âge, Carcassonne 12287
Musée Mémorial 1914-1918, Wattwiller 16292
Musée Memorial Alain, Mortagne-au-Perche 14286
Musée-Mémorial de la Bataille de l'Atlantique, Camaret-sur-Mer 12253
Musée Mémorial de la Bataille de Normandie, Bayeux 11875
Musée Mémorial de la Bataille de Verdun, Fleury-devant-Douaumont 12945
Musée Mémorial de la Ligne Maginot du Rhin, Marckolsheim 13958
Musée Mémorial des Combat de la Poche de Colmar, Turckheim 16010
Musée Mémorial du Front d'Orient 1915-1918, Troissereux 15991
Musée Mémorial du Général de Gaulle, Bayeux 11876
Musée-Mémorial du Linge, Orbey 14522
Musée Mémorial Maginot de Haute-Alsace, Uffheim 16014
Musée-Mémorial National du Débarquement en Provence (1944), Toulon (Var) 15927
Musée Memorial Omaha Beach 6 Juin 1944, Saint-Laurent-sur-Mer 15364
Musée Mémorial Pegasus, Ranville 14985
Musée Mer et Désert, Villerville 16235
Musée Metlili, Ghardaïa 00064
Musée Michel Aboujaoudé, Zalka 30176
Musée Michel Braibant, Montolieu 14235
Musée Militaire, Cambrai 12256
Musée Militaire, Colombier (Neuchâtel) 41311
Musée Militaire, Sonvilier 41891
Musée Militaire, Veckring 16091
Musée Militaire Baie-des-Chaleurs, New Richmond 06433
Musée Militaire de Bure, Bure-Tressange 12207
Musée Militaire de la Guerre 1914 - 1918, Thiaucourt 15879

Musée Militaire des Gloires et Souvenirs du Périgord, Périgueux 14780
Musée Militaire du 12e Régiment Blindé du Canada, Trois-Rivières 07176
Musée Militaire Faller, Mars-la-Tour 13982
Musée Militaire Genevois, Cologny 41309
Musée Militaire Ligne Mareth, Mareth 42640
Musée Militaire National, Manouba 42639
Musée Militaire Vaudois, Morges 41644
Musée Mine, Cagnac-les-Mines 12235
Musée Mineralexpo, La Bourboule 13321
Musée Minéralogique, Nîmes 14454
Musée Minéralogique d'Asbestos, Asbestos 05356
Musée Minéralogique de la Société Industrielle, Mulhouse 14329
Musée Minéralogique de l'École des Mines d'Alès, Alès 11551
Musée Minéralogique et Minier de Thetford Mines, Thetford Mines 07073
Musée Minéralogique et Paléontologique, Saint-Pierre-de-la-Palud 15459
Musée Mini-Trains-Lyonnais, Lyon 13899
Musée Miniatures et Poupées, Gréoux-les-Bains 13143
Musée Minier d'Halanzy, Halanzy 03665
Musée Missionnaire, Pontmain 14907
Musée Mistorique, Jaulny 13265
Musée Mnemosina, Prunelli-di-Fium'Orbu 14955
Musée Moissan, Paris 14721
Musée Monétaire Cantonal, Lausanne 41549
Musée Monopoli, Barsy-Flostoy 03340
Musée Monsieur Pigneau de Behaine, Origny-en-Thiérache 14529
Musée Montagnard, Les Houches 13701
Musée Montagnard du Lavedan, Aucun 11738
Musée Montagnard et Pictural, Les Haudères 41471
Musée Montebello, Trouville-sur-Mer 15993
Musée Montenot Agricole et Archéologique, Genay 13038
Musée Moralès, Port-de-Bouc 14914
Musée Morice Lipsi, Rosey 15106
Musée Moto du Douhet, Le Douhet 13580
Musée-Moulin à Vent de Rairé, Sallertaine 15605
Musée Moulin Cézanne, Le Tholonet 13654
Musée-Moulin de Guerin, Saint-Julien-Chapteuil 15346
Musée-Moulin de l'Espine, Fougax-et-Barrineuf 12984
Musée Moulin de Maupertuis, Donzy 12771
Musée-Moulin du Papé Malaval, Tornac 15919
Musée Moulin Faber, Hotton 03715
Musée Municipal, Barentin 11850
Musée Municipal, Bergues 11947
Musée Municipal, Bernay 11948
Musée Municipal, Bourbonne-les-Bains 12102
Musée Municipal, Châlons-en-Champagne 12364
Musée Municipal, Chartres 12425
Musée Municipal, Châtellerault 12463
Musée Municipal, Coutances 12661
Musée Municipal, Cusset 12693
Musée Municipal, Dudelange 30467
Musée Municipal, Épernay 12836
Musée Municipal, Forcalquier 12978
Musée Municipal, Gerberoy 13047
Musée Municipal, Hazebrouck 13197
Musée Municipal, Hennezel-Clairey 13202
Musée Municipal, Hesdin 13210
Musée Municipal, La Charité-sur-Loire 13339
Musée Municipal, La Réole 13406
Musée Municipal, La Roche-sur-Yon 13419
Musée Municipal, Le Monastier-sur-Gazeille 13618
Musée Municipal, Lillebonne 13761
Musée Municipal, Louhans 13834
Musée Municipal, Mauriac 14040
Musée Municipal, Mazan 14046
Musée Municipal, Nuits-Saint-Georges 14485
Musée Municipal, Richelieu 15035
Musée Municipal, Saint-Amand-les-Eaux 15160
Musée Municipal, Saint-Antonin-Noble-Val 15172
Musée Municipal, Saint-Calais 15195
Musée Municipal, Saint-Dizier 15244
Musée Municipal, Saint-Jean-d'Angély 15319
Musée Municipal, Sainte-Maure-de-Touraine 15569
Musée Municipal, Semur-en-Auxois 15708
Musée Municipal, Villandraut 16162
Musée Municipal, Vire 16251
Musée Municipal A. Danicourt, Péronne 14785
Musée Municipal A.G. Poulain, Vernon 16123
Musée Municipal Albert Marzelles, Marmande 13972
Musée Municipal Alfred Canel, Pont-Audemer 14879
Musée Municipal Ancienne Abbaye Saint-Léger, Soissons 15774
Musée Municipal Auguste Grasset, Varzy 16077
Musée Municipal Bourbon-Lancy, Bourbon-Lancy 12099
Musée Municipal d'Archéologie, Roquefort-sur-Soulzon 15098
Musée Municipal d'Archéologique, Murviel-les-Montpellier 14340
Musée Municipal d'Art et d'Histoire, Colombes 12586
Musée Municipal d'Art Naïf, Noyers-sur-Serein 14481
Musée Municipal d'Arts Asiatiques, Toulon (Var) 15928
Musée Municipal d'Avranches, Avranches 11802
Musée Municipal de Cambrai, Cambrai 12257
Musée Municipal de Civray, Civray 12533
Musée Municipal de Draguignan, Draguignan 12790
Musée Municipal de la Chartreuse, Douai 12775
Musée Municipal de la Poterie, La Chapelle-aux-Pots 13330
Musée Municipal de la Toile de Jouy, Jouy-en-Josas 13286
Musée Municipal de la Vie Sauvage, Le Bugue 13566
Musée Municipal de l'Évêché, Limoges 13770
Musée Municipal de Louviers, Louviers 13858
Musée Municipal de Millau, Millau 14120
Musée Municipal de Mokolo, Mokolo 05300
Musée Municipal de Montmorillon, Montmorillon 14232
Musée Municipal de Moret, Moret-sur-Loing 14270
Musée Municipal de Pithiviers, Pithiviers-le-Vieil 14828
Musée Municipal de Préhistoire, Fismes 12940
Musée Municipal de Préhistoire, Les Matelles 13706
Musée Municipal de Préhistoire des Matelles, Viols-le-Fort 16250
Musée Municipal de Préhistoire Régionale, Saint-Pons-de-Thomières 15466
Musée Municipal de Saint-Barthélemy, Saint-Barthélemy 23280
Musée Municipal de Saint-Cloud, Saint-Cloud 15220
Musée Municipal de Saint-Gaudens, Saint-Gaudens 15276
Musée Municipal de Saint-Maur, La Varenne-Saint-Hilaire 13455
Musée Municipal de Sault, Sault (Vaucluse) 15659
Musée Municipal de Vendôme, Vendôme 16100
Musée Municipal Département Jouet, La Ferté-Macé 13360
Musée Municipal d'Ernée, Ernée 12846
Musée Municipal des Beaux-Arts et d'Archéologie, Valence (Drôme) 16040
Musée Municipal des Beaux Arts et d'Histoire Naturelle, Châteaudun 12444
Musée Municipal des Capucins, Coulommiers 12645
Musée Municipal des Fossiles, Tourtour 15972
Musée Municipal des Fours Banaux, Vierzon 16156
Musée Municipal des Mosaïques Romaines, Diekirch 30464
Musée Municipal d'Étampes, Etampes 12871
Musée Municipal d'Histoire et d'Archéologie, Harnes 13182
Musée Municipal d'Orbec, Orbec 14520
Musée Municipal du Château, Gannat 13028
Musée Municipal du Peigne, Ezy-sur-Eure 12899
Musée Municipal du Présidial, Bazas 11881
Musée Municipal Emile Villez, Guînes 13166
Musée Municipal François Guiguet, Corbelin 12619
Musée Municipal Frédéric Blandin, Nevers 14417
Musée Municipal Gautron du Coudray, Marzy 14025
Musée Municipal Georges Turpin, Parthenay 14753
Musée Municipal Joseph Jacquiot, Montgeron 14211
Musée Municipal Max Claudet, Salins-les-Bains 15598
Musée Municipal Méditerranéen de Cassis, Cassis 12305
Musée Municipal-Musée d'Art et d'Archéologie, Château-Gontier 12435
Musée Municipal Nicolas Poussin, Les Andelys 13683
Musée Municipal Paul Lafran, Saint-Chamas 15200
Musée Municipal Raymond Touchard, Lussac-les-Châteaux 13870
Musée Municipal Robert Dubois-Corneau, Brunoy 12201
Musée Municipal Saint-John-Perse, Pointe-à-Pitre 23277
Musée Municipal Saraleguinea, Guéthary 13162
Musée Municipal Urbain-Cabrol, Villefranche-de-Rouergue 16189
Musée Municipal Vendôme, Naveil 14385
Musée Municipal Yvon Gueret, Pierrelatte 14820
Musée Municipale, Lay-Saint-Christophe 13554
Musée Municipale, Mont-de-Lans 14156
Musée Municipale, Saint-Laurent-d'Olt 15361
Musée Municipale, Villefranche-de-Lonchat 16188
Musée Nader, Port-au-Prince 23323
Musée Namesokanjic, Lac-Mégantic 06120
Musée Napoléon, Ligny 03823
Musée Napoléon Ier et Tresors des Eglises, Brienne-le-Château 12177
Musée Napoléonien, Antibes-Juan-les-Pins 11642
Musée Napoléonien, Ile-d'Aix 13236
Musée Napoléonien d'Art et d'Histoire Militaire, Fontainebleau 12964
Musée Napoléonien du Château de Grosbois, Boissy-Saint-Léger 12049
Musée Napoléonien-Mémoire 1814, Essises 12866
Musée Nasréddine Dinet de Bou-Sâala, M'Sila 00070
Musée Natale de Pierre Corneille, Rouen 15120
Musée National, Beirut 30160
Musée National, Moroni 08792
Musée National, N'Djamena 07424
Musée National Adrien Dubouché, Limoges 13771
Musée National Ahmed Zabana - Demaeght Museum, Oran 00071
Musée National Auguste Rodin, Meudon 14104
Musée National Barthélémy Boganda, Bangui 07421
Musée National Cirta de Constantine, Constantine 00057
Musée National d'Archéologie de Sétif, Sétif 00075
Musée National d'Art Brassicole et Musée de la Tannerie, Wiltz 30498
Musée National d'Art Moderne, Paris 14722
Musée National de Burkina Faso, Ouagadougou 05279
Musée National de Carthage, Carthage 42620
Musée National de Céramique, Sèvres 15743
Musée National de Conakry, Conakry 23310
Musée National de Géologie, Antananarivo 30537
Musée National de Gitega, Gitega 05281
Musée National de Guinée, Conakry 23311
Musée National de Kananga, Kananga 08794
Musée National de la Céramique, Safi 31693
Musée National de la Coopération Franco-Américaine, Blérancourt 12025
Musée National de la Légion d'Honneur et des Ordres de Chevalerie, Paris 14723
Musée National de la Marine, Brest 12164
Musée National de la Marine, Paris 14724
Musée National de la Marine, Port-Louis 14919
Musée National de la Marine, Rochefort (Charente-Maritime) 15072
Musée National de la Marine, Toulon (Var) 15929
Musée National de la Renaissance, Ecouen 12809
Musée National de la Résistance, Bruxelles 03492
Musée National de la Résistance, Esch-sur-Alzette 30473
Musée National de la Voiture et du Tourisme, Compiègne 12599
Musée National de l'Assurance Maladie, Lormont 13824
Musée National de l'Éducation, Montréal 06375
Musée National de l'Éducation, Rouen 15121
Musée National de Lubumbashi, Lumumbashi 08801
Musée National de Monaco, Monaco 31611
Musée National de Nouakchott, Nouakchott 30675
Musée National de Phnom Penh, Phnom Penh 05286
Musée National de Préhistoire, Les Eyzies-de-Tayac-Sireuil 13699
Musée National de Préhistoire et d'Ethnographie du Bardo, Alger 00044
Musée National de Sarh, Sarh 07426
Musée National de Yaoundé, Yaoundé 05302
Musée National d'Enéerune, Nissan-lez-Ensérune 14460
Musée National des Antiquités, Alger 00045
Musée National des Arts Asiatiques Guimet, Paris 14725
Musée National des Arts d'Afrique et d'Océanie, Paris 14726
Musée National des Arts et Traditions, Libreville 16329
Musée National des Arts et Traditions Populaires, Paris 14727
Musée National des Beaux-Arts d'Alger, Alger 00046
Musée National des Beaux-Arts du Québec, Québec 06694
Musée National des Châteaux de Malmaison et de Bois-Préau, Rueil-Malmaison 15139
Musée National des Châteaux de Versailles et de Trianon, Versailles 16129
Musée National des Deux Victoires, Mouilleron-en-Pareds 14301
Musée National des Douanes, Bordeaux 12073
Musée National des Granges de Port Royal, Magny-lès-Hameaux 13915
Musée National des Jeux de Paume, Ath 03331
Musée National des Mines de Fer Luxembourgeoises, Rumelange 30491
Musée National des Monuments Français, Paris 14728
Musée National des Prisons, Fontainebleau 12965
Musée National d'Histoire et d'Art (MNHA), Luxembourg 30483
Musée National d'Histoire Militaire, Diekirch 30465
Musée National d'Histoire Naturelle, Luxembourg 30484
Musée National du Bardo, Alger 00047
Musée National du Bardo, Le Bardo 42616
Musée National du Château de Compiègne - Musée du Second Empire, Compiègne 12600
Musée National du Château de Fontainebleau, Fontainebleau 12966
Musée National du Château de Pau, Pau 14759
Musée National du Congo, Brazzaville 08805
Musée National du Costume, Grand-Bassam 08836
Musée National du Gabon, Libreville 16330
Musée National du Jihad, Alger 00048
Musée National du Mali, Bamako 30624
Musée National du Moudjahid, Alger 00049
Musée National du Moudjahid, El Madania 00061
Musée National du Moyen Age, Paris 14729
Musée National du Niger, Niamey 33308
Musée National du Papier, Malmedy 03844
Musée National du Rwanda, Butare 37992
Musée National du Scoutisme en France, Thorey-Lyautey 15900
Musée National du Sport, Paris 14730
Musée National du Togo, Lomé 42612
Musée National d'Utique, Utique 42668
Musée National Ernest Hébert, Paris 14731
Musée National Eugène Delacroix, Paris 14732
Musée National Fernand Léger, Biot 11998
Musée National Jean-Jacques Henner, Paris 14733
Musée National Message Biblique Marc Chagall, Nice 14436
Musée National N'Djamena, N'Djamena 07425
Musée National Picasso La Guerre et la Paix, Vallauris 16051
Musée National Suisse - Château de Prangins, Prangins 41732
Musée National Suisse de l'Audiovisuel - Audiorama, Montreux 41640
Musée Nature en Provence, Riez 15040
Musée Naturel du Sud, Chenini Gabes 42621
Musée Naval, Monaco 31612
Musée Naval de Nice, Nice 14437
Musée Naval de Sarcelles, Sarcelles 15629
Musée Naval Fort Balaguier, La Seyne-sur-Mer 13444
Musée Navale du Québec, Québec 06695
Musée Nicéphore Niépce, Chalon-sur-Saône 12359
Musée Nicholas Ibrahim Sursock, Beirut 30161
Musée Nissim de Camondo, Paris 14734
Musée Nivernais de l'Education, Nevers 14418
Musée No.4 Commando, Ouistreham 14545
Musée Normandie-Niemen, Les Andelys 13684
Musée Notre-Dame, Le Folgoët 13584
Musée Notre-Dame, Lourdes 13851
Musée Notre Dame de la Pierre, Sarrance 15641
Musée Nungesser et Coli, Etretat 12883
Musée Oberlin, Waldersbach 16281
Musée-Observatoire de la Vallée d'Anjou, Saint-Mathurin-sur-Loire 15409
Musée Océanien de Cuet, Montrevel-en-Bresse 14258
Musée Océanographique, Bonifacio 12052
Musée Océanographique, Kingersheim 13303
Musée Océanographique, L'Ile-Rousse 13751
Musée Océanographique, Saint-Macaire 15381
Musée Océanographique Dar-El-Hout de Salammbô, Salammbô 42647
Musée Océanographique de l'Odet, Ergué-Gabéric 12843
Musée Océanographique de Monaco, Monaco 31613
Musée Olivier-de-Serres, Mirabel (Ardèche) 14127
Musée Ollier, Les Vans 13721
Le Musée Olympique Lausanne, Lausanne 41550
Musée Omega, Biel 41230
Musée Organistrum et des Vielles à Roues du Périgord Noir, Belvès 11936
Musée Ornithologique Charles Payraudeau, La Chaize-le-Vicomte 13329
Musée Oscar Roty, Jargeau 13259
Musée Ostréicole, Bourcefranc-le-Chapus 12103
Musée Ostréicole, Saint-Trojan-les-Bains 15524
Musée P. Dubois-A. Boucher, Nogent-sur-Seine 14464
Musée Pablo Casals, Prades 14936
Musée Paderewski, Morges 41645
Musée Palais el Bahia, Marrakech 31680
Musée Palais Lascaris, Nice 14438
Musée Palaisien du Hurepoix, Palaiseau 14555
Musée Paléo-Écologique, Marchamp 13949
Musée Paléontologique, Aguessac 11495
Musée Paléontologique et Minéraux, Rougemont 15125
Musée-Palombière, Le Mas-d'Agenais 13610
Musée Parc de la Drosera, Jeansagnière 13268
Musée Passé Simple, Pleudihen-sur-Rance 14840
Musée Pasteur, Dole 12755
Musée Pasteur, Paris 14735
Musée Pastoral le Monde Paysan d'Autrefois, Saint-Pé-de-Bigorre 15453
Musée Patrimoine et Traditions, Mirepoix 14136
Musée Paul Arbaud, Aix-en-Provence 11519
Musée Paul Charnoz, Paray-le-Monial 14562
Musée Paul Dini, Villefranche-sur-Saône 16196
Musée Paul Dupuy, Toulouse 15944
Musée Paul-Emile Victor, Prémanon 14944
Musée Paul Gauguin, Le Carbet 30668
Musée Paul Gauguin, Papeete 16325
Musée Paul Gauguin, Pont-Aven 14881
Musée Paul Géradin, Jodoigne 03746
Musée Paul-José Gosselin, Saint-Vaast-la-Hougue 15529

Musée Paul Raymond, Pont-Saint-Esprit 14894
Musée Paul Surtel, Reuilly (Indre) 15026
Musée Paul Valéry, Sète 15735
Musée Paul Voivenel, Capoulet-Junac 12276
Musée Paysan, Espezel 12862
Musée Paysan, Le Grand-Village-Plage 13589
Musée Paysan, Viuz-en-Sallaz 16260
Musée Paysan de Bourgogne Nivernaise, La Celle-sur-Loire 13326
Musée Paysan de la Save, Espaon 12858
Musée Paysan du Moulin Neuf, Saint-Diéry 15243
Musée Paysan du Sundgau, Oltingue 14512
Musée Paysan et Artisanal, La Chaux-de-Fonds 41297
Musée Peleforo Gbon Coulibaly, Korhogo 08837
Musée Percheron, Mortagne-au-Perche 14287
Musée Petiet, Limoux 13773
Musée Peugeot, Sochaux 15772
Musée Peynet et du Dessin Humoristique, Antibes-Juan-les-Pins 11643
Musée Picasso, Antibes 11638
Musée Picasso, Paris 14736
Musée Pierre Bayle, Carla-Bayle 12291
Musée Pierre Borrione, Aime-en-Tarentaise 11509
Musée Pierre Boucher, Trois-Rivières 07177
Musée Pierre de Luxembourg, Villeneuve-lès-Avignon 16214
Musée Pierre Gaudin, Puteaux 14959
Musée Pierre-Loti, Rochefort (Charente-Maritime) 15073
Musée Pierre-Mondanel, Pont-du-Château 14885
Musée Pierre Noël, Saint-Dié-des-Vosges 15241
Musée Pierre Vigne, Boucieu-le-Roi 12082
Musée Pillon, Chaumont-en-Vexin 12481
Musée Pincé, Angers 11613
Musée Piperie Léonard, Andenelle 03273
Musée-Placard d'Erik Satie, Paris 14737
Musée Plain Air Régional, Guérin 05921
Musée Pobe Mengao, Mengao 05276
Musée Poche de Royan, Le Gua 13591
Musée Pol Mara, Gordes 13085
Musée Pontois, Le Pont-de-Beauvoisin 13639
Musée Portuaire, Dunkerque 12799
Musée Postal, Rabat 31691
Musée Postal d'Aquitaine, Saint-Macaire 15382
Musée Postal de Tunis, Tunis 42667
Musée-Préau des Accoules, Marseille 14010
Musée Préfectoral de Boké, Boké 23308
Musée Préfectoral de Kissidougou, Kissidougou 23313
Musée Préfectoral de Koundara, Koundara 23314
Musée Préfectoral de N'Zerekore, N'Zerekore 23315
Musée Préhistorique, Eymet 12893
Musée Préhistorique, Marsoulas 14016
Musée Préhistorique Charles Nardin, Sainte-Foy-la-Grande 15559
Musée Préhistorique du Tumulus de Barnenez, Plouézoch 14851
Musée Préhistorique Finistérien, Penmarch 14773
Musée-Privé de Cire du Vieil, Albi 11539
Musée Privé de Documentation Ferroviaire-Mupdofer, Bruxelles 03493
Musée Privé Saint-Jean l'Aigle, Herserange 13208
Musée Privee de N'Douci, N'Douci 08838
Musée-Promenade Marly-le-Roi, Louveciennes 13857
Musée Protestant de la Grange de Wassy, Wassy 16290
Musée Provençal des Arts et Traditions Poplaires, Donzère 12770
Musée Provençal des Transports Urbains et Régionaux, La Barque 13310
Musée Provençal du Costume et du Bijou, Grasse 13116
Musée Provençal et Folklorique Taradeau, Les Arcs 13685
Musée Provincial de Douala, Douala 05296
Musée Provincial de la Forêt, Namur 03905
Musée Provincial du Houet, Bobo-Dioulasso 05272
Musée Provincial du Poni, Gaoua 05273
Musée Provincial du Sanmatenga, Kaya 05274
Musée Provincial Félicien Rops, Namur 03906
Musée Prytanée Militaire, La Flèche 13367
Musée Pyrénéen, Lourdes 13852
Musée Pyrénéen de Niaux, Niaux 14420
Musée Québécois de Culture Populaire, Trois-Rivières 07178
Musée Quentovic, Etaples 12875
Musée Rabelais, Seuilly 15737
Musée Raimu, Cogolin 12567
Musée Rambolitrain, Rambouillet 14983
Musée Raoul Dastrac, Aiguillon 11506
Musée Rath, Genève 41426
Musée Raymond Lafage, Lisle-sur-Tarn 13786
Musée Raymond Peynet, Brassac-les-Mines 12147
Musée Raymond Poincaré, Sampigny 15619
Musée Raymond Rochette, Saint-Sernin-du-Bois 15510
Musée Réattu, Arles 11685
Musée Reflet Brionnais, Iguerande 13234
Musée Regards sur le Passé, Hourtin 13226
Musée Régimentaire les Fusiliers de Sherbrooke, Sherbrooke, Québec 06963
Musée Regional, Kisangani 08800
Musée Régional André Grenard Matsoua, Kinkala 08806
Musée Regional d'Argenteuil, Saint-André-d'Argenteuil 06795
Musée Régional d'Armes, Mutzig 14344
Musée Régional d'Arts et de Traditions Populaires, Grimaud 13150
Musée Régional d'Auvergne, Riom 15046
Musée Régional de Fouras, Fouras 12990
Musée Régional de Géologie Pierre Vetter, Decazeville 12704
Musée Régional de la Côte-Nord, Sept-Îles 06945
Musée Régional de la Poterie bas Normande, Ger 13045
Musée Régional de la Résistance et de la Déportation, Thionville 15885
Musée Régional de la Sagne, La Sagne 41788
Musée Regional de la Vigne et du Vin, Montmélian 14227
Musée Régional de l'Air, Marcé 13947
Musée Régional de l'Alsace Bossue, Sarre-Union 15642
Musée Régional de l'Orléanais, Beaugency 11893
Musée Regional de Mahdia, Mahdia 42637
Musée Régional de Préhistoire, Orgnac-l'Aven 14526
Musée Régional de Rimouski, Rimouski 06755
Musée Régional de Sikasso, Sikasso 30626
Musée Régional de Sokadé, Sokadé 42613
Musée Régional de Vaudreuil-Soulanges, Vaudreuil-Dorion 07234
Musée Régional de Zinder, Zinder 33309
Musée Régional des Arts de la Table, Arnay-le-Duc 11690
Musée Régional des Enrôles de Force, Dudelange 30468
Musée Régional des Métiers, La Cheze 13343
Musée Régional des Mines de Malartic, Malartic 06223
Musée Régional des Savanes, Dapaong 42611
Musée Régional des Télécommunications en Flandres, Marcq-en-Barœul 13960
Musée Régional d'Ethnologie, Béthune 11967
Musée Régional d'Histoire et d'Archéologie, Visé 04118
Musée Régional d'Histoire et d'Artisanat du Val-de-Travers, Môtiers 41649
Musée Régional d'Histoire et d'Ethnographie, Fort-de-France 30666
Musée Régional du Cerel, Toamasina 30542
Musée Régional du Chemin de Fer, Rosny-sous-Bois 15110
Musée Régional du Chemin de Fer Nord-Pas-de-Calais, Denain 12710
Musée Régional du Cidre et du Calvados, Valognes 16060
Musée Régional du Machinisme Agricole, La Ferté-Milon 13363
Musée Régional du Sahel, Gao 30625
Musée Régional du Timbre et de la Philatélie, Le Luc 13603
Musée Régional Dupuy-Méstreau, Saintes 15585
Musée Régional Ex-Dixième Riaom, Thiès 38028
Musée Régional Ma-Loango Diosso, Pointe-Noire 08807
Musée Relais du Cheval de Trait Comtois, Levier 13733
Musée Religieux d'Art Franciscain, Nice 14439
Musée Religieux Franco-Canadien, Tours 15969
Musée Remember 39-45, Léhon 13677
Musée Remise aux Outils Champenois, Dormans-en-Champagne 12772
Musée Renaudot, Loudun 13831
Musée René Davoine, Charolles 12417
Musée Renoir, Cagnes-sur-Mer 12239
Musée Requien, Avignon 11795
Musée Rêve Auto-Jeunesse, Mosnac 14291
Musée Rêve et Miniatures, Brantôme 12144
Musée Revoir le Passé, Les Loges-Marchis 13703
Musée Ribauri, Lasne-Chapelle-Saint-Lambert 03778
Musée Rignault, Saint-Cirq-Lapopie 15213
Musée Rizla Croix, Mazères-sur-Salat 14053
Musée Rochelais d'Histoire Protestante, La Rochelle 13435
Musée Rodava, Rêves 03966
Musée Rodriguez Huguette, Saint-Macaire 15383
Musée Roger Thières, Payrac 14766
Musée Rolin, Autun 11764
Musée Romain, Avenches 41138
Musée Romain, Nyon 41692
Musée Romain de Lausanne-Vidy, Lausanne 41551
Musée Romain de Vallon, Vallon 41973
Musée Romand de la Machine Agricole, Gingins 41432
Musée Ronsard, La Riche 13411
Musée Rosine Perrier, Villargondran 16167
Musée Royal de l'Armée et d'Histoire Militaire, Bruxelles 03494
Musée Royal de Mariemont, Morlanwelz-Mariemont 03893
Musée Royaume de l'Horloge, Villedieu-les-Poêles 16184
Musée Roybet Fould, Courbevoie 12650
Musée Rude, Dijon 12743
Musée Rue Etienne Lenoir, Mussy-la-Ville 03897
Musée Rupert de Chièvres, Poitiers 14867
Musée Rural, Foucherans 12983
Musée Rural, Saint-André-d'Embrun 15167
Musée Rural, Usson-en-Forez 16021
Musée Rural Berry-Marche, Boussac 12130
Musée Rural d'Arts et Traditions Populaires La Combà, Autrafé, La Combe-de-Lancey 13348
Musée Rural d'Auris-en-Oisans des Choses d'Autrefois, Auris-en-Oisans 11758
Musée Rural de la Sologne Bourbonnaise, Beaulon 11897
Musée Rural de l'Educations dans les Côtes d'Armor, Saint-Nicolas-du-Pelem 15429
Musée Rural des Art Populaires, Laduz 13474
Musée Rural du Bois, Coupiac 12647
Musée Rural du Champsaur, Bénévent-et-Charbillac 11938
Musée Rural du Porzay, Plomodiern 14846
Musée Rural du Porzay, Saint-Nic-Pontrez 15425
Musée Rural Jurassien, Les Genevez 41430
Musée Ruzo, Montreux 41641
Musée Sabatier, Saint-Bonnet-de-Joux 15186
Musée Sabatier d'Eyspéran, Montpellier 14251
Musée Saharien de Ouargla, Ouargla 00072
Musée Sahut, Simiane-Collongue 15759
Musée Saint-Brieux, Saint-Brieux 06805
Musée Saint-Étienne, Honfleur 13224
Musée Saint-Euxpéry, Laâyoune 31675
Musée Saint-François-Régis, La Louvesc 13390
Musée Saint-Jacques, Moissac 14141
Musée Saint-Jean, Anneyron 11628
Musée Saint-Jean, La Rochette (Savoie) 13437
Musée Saint-Joachim, La Broquerie 06109
Musée Saint-Joseph, Saint-Joseph 06858
Musée Saint-Loup, Troyes 16001
Musée Saint-Martin, Tours 15970
Musée Saint-Nicolas, Vitré 16257
Musée Saint-Pierre Fourier, Mattaincourt 14030
Musée Saint-Raymond, Toulouse 15945
Musée Saint-Remi, Reims 15005
Musée Saint-Sauveur Nibelle Autrefois, Nibelle 14422
Musée Saint-Simon, La Ferté-Vidame 13365
Musée Saint-Vic, Saint-Amand-Montrond 15161
Musée Sainte-Croix, Poitiers 14868
Musée Sainte-Jeanne-Antide-Thouret, Sancey-le-Long 15621
Musée Sainte-Martine, Sainte-Martine 06900
Musée Salies, Bagnères-de-Bigorre 11820
Musée-Salle Communale, Yvoire 16312
Musée Salle Documentaire, Coucy-le-Château-Auffrique 12640
Musée Santons du Monde, Clermont-l'Hérault 12553
Musée Sarret de Grozon, Arbois 11659
Musée Saturnin Garimond, Fons 12954
Musée Savoisien, Chambéry 12375
Musée Schiller et Goethe, Châlons-en-Champagne 12365
Musée Schœlcher, Pointe-à-Pitre 23278
Musée Scout Baden-Powell, Sisteron 15765
Musée Sculpture sur Bois, Lans-en-Vercors 13516
Musée Sellier, Cogolin 12568
Musée Serret, Saint-Amarin 15162
Musée Servie de Santé des Armées au Val-du-Grâce, Paris 14738
Musée Sherlock Holmes, Lucens 41577
Musée Sidi Ben-Aissa, Le Kef 42632
Musée Sidi Bou Saïd, Sidi Bou Saïd 42655
Musée Sidi Mohamed Ben Abdellah, Essaouira 31668
Musée Sidi Zitouni, Houmt Souk 42628
Musée Simon-Segal, Aups 11749
Musée Site Départemental d'Archéologie de Bavay, Bavay 11871
Musée Skolig al Louarn, Plouvien 14862
Musée Sobirats, Carpentras 12298
Musée Somme 1916, Albert 11533
Musée Souleïado Charles Demery, Tarascon (Bouches-du-Rhône) 15856
Musée Sous-Glaciaire, La Grave 13381
Musée Sous Marin Pays Lorient, Lorient 13821
Musée souterrain de la Poterie, Saint-Émilion 15247
Musée Souterrain de l'Outil, Chacé 12352
Musée Souvenir du Dirigeable R101, Beauvais 11911
Musée Spadois du Cheval, Spa 04044
Musée Spectacle de Musiques du Monde, Montoire-sur-le-Loir 14233
Musée Spitfire, Florennes 03605
Musée Sterna de l'Armée et des Trois Guerres, Seynod 15751
Musée Stewart au Fort de île Sainte-Hélène, Montréal 06376
Musée Suffren et du Vieux Saint-Cannat, Saint-Cannat 15197
Musée Suisse de la Figurine Historique, Morges 41646
Musée Suisse de la Machine à Coudre, Fribourg 41390
Musée Suisse de la Marionnette, Fribourg 41391
Musée suisse de la Mode, Yverdon-les-Bains 42050
Musée Suisse de l'Appareil Photographique, Vevey 41987
Musée Suisse de l'Orgue, Roche 41769
Musée Suisse de Spéléologie, Chamoson 41289
Musée Suisse du Jeu, La Tour-de-Peilz 41945
Musée Suisse du Sel, Aigle 41095
Musée Sundgauvien, Altkirch 11562
Musée sur l'Artillerie, la Cavalerie et l'Infanterie, Seclin 15695
Musée Tancrède, Hauteville-la-Guicharde 13194
Musée Tauromachique, Pomarez 14871
Musée Tavet-Delacour, Pontoise 14909
Musée Tchantchès - Musée de la République Libre d'Outre-Meuse, Liège 03815
Musée Terbol, Zahlé 30175
Musée Terrasse, Viverols 16261
Musée Territorial de Nouvelle-Calédonien, Nouméa 33032
Musée Terrus, Elne 12818
Musée Testut-Latarjet de Médecine et d'Anatomie, Lyon 13900
Musée Textile de Haute-Alsace, Husseren-Wesserling 13232
Musée Théâtre Ki-Yi, Abidjan 08831
Musée Théodore Rousseau, Barbizon 11847
Musée Théophile Jouglet, Anzin 11647
Musée Thimonnier Doyen, Saint-Germain-au-Mont-d'Or 15285
Musée Thomas-Henry, Cherbourg-Octeville 12505
Musée Tomi Ungerer - Centre International de l'illustration, Strasbourg 15833
Musée Toulouse-Lautrec, Albi 11540
Musée Toulouse-Lautrec, Naucelle 14384
Musée Tour de la Lanterne, La Rochelle 13436
Musée Tour d'Eben-Ezer, Eben-Emael 03569
Musée Traditions et Vie, Châtillon-sur-Chalaronne 12470
Musée Traditions Verrieres, Eu 12886
Musée Trésor d'Église, Metz 14101
Musée Troglodytique, Louresse 13854
Musée Truschbaum, Elsenborn 03578
Musée Ukraina, Saskatoon 06919
Musée Une Halte sur le Chemin de Saint-Jacques, Borce 12060
Musée Urbain Tony Garnier, Lyon 13901
Musée Utrillo-Valadon, Sannois 15624
Musée Valaisan de la Vigne et du Vin, Sierre 41867
Musée Valentin Haüy, Paris 14739
Musée Valery Stuyver, Seraing-le-Château 04011
Musée Van Œveren, Bourgueil 12125
Musée Vauban, Neuf-Brisach 14396
Musée Vauban et de la Fortification dans les Alpes, Mont-Dauphin 14155
Musée Vendéen, Fontenay-le-Comte 12969
Musée Verger Tarin, Autun 11765
Musée Verlaine, Juniville 13294
Musée Vert, muséum d'histoire naturelle du Mans, Le Mans 13609
Musée Victor Aubert, Maule 14034
Musée Victor Hugo, Villequier 16225
Musée Vie et Métiers d'Autrefois dans le Haut-Jura, La Pesse 13400
Musée Vigneron, Vendres 16101
Musée Villa Tamaris Pacha, La Seyne-sur-Mer 13445
Musée Villageois d'Orgon, Orgon 14528
Musée Villeneuvien, Villeneuve-sur-Yonne 16223
Musée Villien, Avesnes-sur-Helpe 11785
Musée Vini-Viticole Gallo-Romain, Beaucaire 11888
Musée Vissoie, Vissoie 41990
Musée Viticole, Saint-Yzans-de-Médoc 15544
Musée Viticole Charles Bernard, Fixin 12942
Musée Vivant d'Archéologie Industrielle Ferroviaire, Sprimont 04046
Musée Vivant de Bujumbura, Bujumbura 05280
Musée Vivant de la Brasserie à Vapeur, Pipaix 03952
Musée Vivant de la Faisanderie, Milly 14121
Musée Vivant de la Forge, Seyne-les-Alpes 15750
Musée Vivant de la Laine et du Mouton, Saint-Pierreville 15464
Musée Vivant de la Métairie de Crampet, Mazères (Gironde) 14052
Musée Vivant de la Mine, Auchel 11736
Musée Vivant de la Passementerie, Jonzieux 13278
Musée Vivant de la Plante Aquatique, Limal 03824
Musée Vivant de la Pomme de Terre, Genappe 03618
Musée Vivant de la Porcelaine, Foëcy 12949
Musée Vivant de l'Abeille, Corbeny 12620
Musée Vivant de l'Abeille, Valensole 16044
Musée Vivant de l'Apiculture Gâtinaise, Châteaurenard (Loiret) 12457
Musée Vivant de l'Egypte pour Enfants, Jambes 03741
Musée Vivant Denon, Chalon-sur-Saône 12360
Musée Vivant des Vieux Métiers, Argol 11678

M

Musée Vivant du Chemin de Fer Vicinal, Thuin 04066
Musée Vivant du Cheval, Chantilly 12401
Musée Vivant du Cheval de Trait, Waret-l'Evêque 04132
Musée Vivant du Cheval de Trait et de l'Attelage, Pradelles 14935
Musée Vivant du Cinéma, Lyon 13902
Musée Vivant du Cochon, Chambonas 12377
Musée Vivant du Jambon - La Ferme des Sanglochons, Verlaine 04105
Musée Vivant du Roman d'Aventures, Jassans-Riottier 13263
Musée Vivant et Historique du Train Historique du Lac de Rillé, Rillé 15042
Musée Vivant Européen du Cirque, Piolenc 14822
Musée Vivant Médiéval, Murol 14338
Musée Vivarois César Filhol, Annonay 11631
Musée Voland, Cerny 12347
Musée Vosgien de la Brasserie, Ville-sur-Illon 16177
Musée Vulcanologique Franck-Arnold Perret, Saint-Pierre 30672
Musée Wagon de l'Armistice, Compiègne 12601
Musée Wellington, Waterloo 04138
Musée Westercamp, Wissembourg 16300
Musée Würth, Erstein 12847
Musée Yan Dagent, Saint-Servais 15511
Musée Yani Faux, Orbais-l'Abbaye 14519
Musée Yves Brayer, Cordes-sur-Ciel 12625
Musée Yves Brayer, Les Baux-de-Provence 13694
Musée Zadkine, Les Arques 13686
Musée Zadkine, Paris 14740
Musée Zervos, Vézelay 16138
Musée Ziem, Martigues 14020
Musée Zoologique de l'Université Louis Pasteur et de la Ville de Strasbourg, Strasbourg 15834
Museen Alte Bischofsburg, Wittstock/Dosse 22629
Museen der Stadt Bad Kösen, Bad Kösen 16763
Museen der Stadt Landshut in der Stadtresidenz, Landshut 19803
Museen für Kunst und Kulturgeschichte der Hansestadt Lübeck - Verwaltung, Lübeck 20078
Museen im Alten Schloss, Neustadt an der Aisch 20666
Museen im Kulturzentrum, Rendsburg 21275
Museen im Markgrafen-Schloss, Emmendingen 18146
Museen im Rathaus, Perchtoldsdorf 02488
Museen im Wehrturm, Perchtoldsdorf 02489
Museerne i Fredericia, Fredericia 10021
Musées d'Archéologie et d'Arts Décoratifs, Liège 03816
Musées de la Citadelle, Villefranche-sur-Mer 16194
Musées de la Cour d'Or, Metz 14102
Musées de La Poëslerie et de la Dentelle, Villedieu-les-Poëles 16185
Musées de la Réole, La Réole 13407
Musées de l'Horlogerie Pater, Longueville 03831
Musées de Sens, Sens 15719
Musées de Vieille Forge, de la Traversée du Siècle et des Automates, Fréland 13007
Musées Départementaux Albert Demard, Champlitte 12393
Musées Départementaux de la Guadeloupe, Pointe-à-Pitre 54871
Musées des Colombiers Cauchois, Oherville 14500
Musées des Technique et Cultures Comtoises, Salins-les-Bains 15599
Musées et Société en Wallonie, Namur 54715
Musées Royaux d'Art et d'Histoire, Bruxelles 03495
Musées Universitaire, Kinshasa 08799
Museet Ett Hem, Turku 11395
Museet Færgegården, Jægerspris 10110
Museet for Dansk Bladtegning, København 10140
Museet for Fotokunst, Odense 10218
Museet for Holbæk og Omegn, Holbæk 10087
Museet for Religiøs Kunst, Lemvig 10177
Museet for Samtidskunst, Oslo 33733
Museet for Samtidskunst, Roskilde 10244
Museet for Thy og Vester Hanherred, Thisted 10296
Museet for Varde By og Omegn, Varde 10307
Museet Holmen, Løgumkloster 10180
Museet i Mælandsgården, Skudeneshavn 33841
Museet Kvarnen, Filipstad 40520
Museet Kystens Arv, Stadsbygd 33860
Museet Leksaksland, Malmö 40725
Museet Näktergalen, Vimmerby 41064
Museet på Gudhjem Station, Gudhjem 10048
Museet på Koldinghus, Kolding 10163
Museet på Sønderskov, Brørup 09985
Museet Psykiatrisk Hospital i Århus, Risskov 10238
Museet Tikøb Frysehus, Tikøb 10297
Museet ved Sct. Ansgar Kirke, København 10141
Museet ved Trelleborg, Slagelse 10275
Musei Anatomici, Modena 26743
Musei Capitolini, Roma 27458
Musei Civici, Lecco 26478
Musei Civici, Pavia 27120
Musei Civici, Reggio Emilia 27388
Musei Civici di Padova, Padova 27034
Musei Civici di Palazzo d'Avalos, Vasto 28169
Musei Civici di Pesaro, Pesaro 27153
Musei Civici e Pinacoteca L. Sturzo, Caltagirone 25605
Musei Civici Veneziani, Venezia 54906
Musei del Castello Scaligero, Malcesine 26574
Musei della Civiltà Contadina del Friuli Imperiale, Aiello del Friuli 25166
Musei della Cultura Popolare Radana, San Benedetto Po 27613
Musei Provinciali di Gorizia, Gorizia 26351
Musei Sistini del Piceno, San Benedetto del Tronto 27608
Museifartyget Pommern, Mariehamn 11100
Museion - Museo d'Arte Moderna e Contemporanea, Bolzano 25485
Museisällskapet Jädraås-Tallås Järnväg, Jädraås 40610
Museispårvägen Malmköping, Malmköping 40716
Museo 13 de Marzo, Pinar del Río 09246
Museo 9 de Abril, Santa Clara 09293
Museo A. Binda, Cittiglio 25914
Museo A. Cassioli, Asciano 25290
Museo A. Colocci, Jesi 26424
Museo A. e C. Gaffoglio, Rapallo 27366
Museo a la Bandera y Santuario de la Patria, Iguala de la Independencia 30955
Museo A. Lia, La Spezia 26432
Museo A. Murer, Falcade 26081
Museo A.R. Giorgi, Reggiolo 27393
Museo A. Ricci dei Minerali Elbani, Rio nell'Elba 27413
Museo A. Versace, Bagnara Calabra 25343
Museo Abarth, Marietta 50470
Museo Abel Santamaría, La Habana 09108
Museo Abel Santamaria, Santa Clara 09294
Museo Abraham González, Vicente Guerrero 31540
Museo Acacia, La Habana 09109
Museo Academia de Historia Leonardo Tascón, Buga 08602
Museo Activo del Pago de los Arroyos, Rosario 00542
Museo Ada e Giuseppe Marchetti, Fumone 26257
Museo ADIAC, León 33283
Museo Adolfo López Mateos, Atizapán de Zaragoza 30725
Museo Adolfo Lozano Sidro, Priego de Córdoba 39870
Museo Aeronáutico, Baradero 00141
Museo Aeronáutico, Montevideo 54393
Museo Aeronautico Caproni di Taliedo, Roma 27459
Museo Aeronautico Caproni-Taliedo, Vizzola Ticino 28302
Museo Aeronáutico Colonel Luis Hernan Paredes, Maracay 54583
Museo Aeronáutico del Perú, Lima 34221
Museo Aeronáutico Municipalo, Los Alcázares 38821
Museo Aeronáutico Torreón de Gando, Telde 40113
Museo Aeronautico y del Espacio, Quito 10426
Museo Aerospaziale Monte di Apollo, Perugia 27146
Museo Africano, Roma 27460
Museo Africano, Verona 28229
Museo Africano Mundo Negro, Madrid 39525
Museo Afro Antillano de Panamá, Ciudad de Panamá 34064
Museo Agrario Tropicale, Firenze 26166
Museo Agrarista de Tzurúmutaro, Pátzcuaro 31254
Museo Agrícola El Patio, Teguise 40109
Museo Agricolo, Ronco Briantino 27554
Museo Agricolo Brunnenburg, Tirolo di Merano 27975
Museo Agricolo della Civiltà Contadina L. Carminati, San Giuliano Milanese 27649
Museo Agricolo e del Vino Ricci Curbastro, Capriolo 25668
Museo Agricolo e Museo dell'Arte Conciaria, Castelfranco Veneto 25730
Museo Agro-Forestale San Matteo, Erice 26061
Museo Agro-Pastorale dell'Area Ellenofona, Bova Marina 25520
Museo Agrumario, Reggio Calabria 27380
Museo Aguilar y Eslava, Cabra 39077
Museo Alberto Arvelo Torrealba, Barinas 54523
Museo Alcamese d'Arte Sacra, Alcamo 25188
Museo Alcázar de Colón, Santo Domingo 10341
Museo Aldrovandiano, Bologna 25443
Museo Alejandro Otero, Caracas 54533
Museo Alejo Carpentier, Camagüey 09048
Museo Alessi, Enna 26055
Museo Alfar, Santa Elena de Jamuz 39998
Museo Alfieriano, Asti 25313
Museo Alfonsiano, Pagani 27055
Museo Alfredo Ramírez de Arellano & Rosello, San Germán 36056
Museo All'Aperto di Storia dell'Agricultura, Sassari 27752
Museo Almacén El Recreo, Chivilcoy 00304
Museo Almafuerte, La Plata 00407
Museo Alpino Duca degli Abruzzi, Courmayeur 25997
Museo Alpujarreño de Costumbre Populares, Capileira 39125
Museo Alta Valle Scrivia, Valbrevenna 28135
Museo Altar de la Patria, Ixcateopan de Cuauhtémoc 30958
Museo Alternativo Remo Brindisi, Comacchio 25945
Museo Alto Bierzo, Bembibre 39022
Museo Amano, Lima 34222
Museo Amazigh, Melilla 39641
Museo Ambientale del W.W.F., Orbetello 26986
Museo Ambientalistico Madonita, Polizzi Generoso 27273
Museo Ambiente Cubano, Gibara 09084
Museo Amedeo Bocchi, Parma 27098
Museo Americanista, Lomas de Zamora 00430
Museo Americanistico Federico Lunardi, Genova 26295
Museo Amparo, Puebla 31265
Museo Anatomia Veterinaria, Napoli 26882
Museo Anatomico, Siena 27837
Museo Anatómico, Valladolid 40230
Museo Anatomico G. Tumiati, Ferrara 26106
Museo Anatómico Pedro Ara, Córdoba 00319
Museo Anatomico Veterinario, Pisa 27240
Museo Anconetani del Acordeón, Buenos Aires 00165
Museo Andrés Avelino Cáceres, Lima 34223
Museo Andrés Cué, Amancio 09024
Museo Anenecuilco, Ayala 30729
Museo Anestesiológico y Biblioteca Histórica de la Asociación de Anestesiología de Buenos Aires, Buenos Aires 00166
Museo Angel Orensanz y Artes de Serralbo, Sabiñánigo 39929
Museo Antártico, Montevideo 54394
Museo Antica Miniera di Talco Brunetta, Cantoira 25649
Museo Antiguo Colegio de San Ildefonso, México 31049
Museo Antiguo Cuartel de Caballería del Ejército Español, Camagüey 09049
Museo Antiguo Palacio de la Medicina Mexicana, México 31050
Museo Antimperialista, Nueva Gerona 09231
Museo Antiquarium Archeologico, Falerone 26082
Museo Antón, Candás 39116
Museo Antoniano, Padova 27035
Museo Antonio Lussich, Punta Ballena 54427
Museo Antonio Maceo, San Pedro 09282
Museo Antonio Manuel Campoy, Cuevas del Almanzora 39237
Museo Antonio Raimondi, La Merced 34201
Museo Antonio Raimondi, Lima 34224
Museo Antonio Ricaurte, Villa de Leyva 08782
Museo Antrológico, Ceretè 08623
Museo Antropologia e Historia, Victoria 31543
Museo Antropologia e Historia Valle de Sula, San Pedro Sula 23330
Museo Antropológico, Ibagué 08658
Museo Antropológico, Sucre 04243
Museo Antropológico de Antofagasta, Antofagasta 07428
Museo Antropológico de Iquique, Iquique 07438
Museo Antropológico de Tunja, Tunja 08773
Museo Antropológico del Banco Central, Guayaquil 10400
Museo Antropológico Montané, La Habana 09110
Museo Antropológico Reina Torres de Araúz, Ciudad de Panamá 34065
Museo Antropológico Shuar, Quito 10427
Museo Antropológico Tairona, Santa Marta 08745
Museo Antzetik Ta Jtelum, Zinacantán 31598
Museo Anzoátegui, Pamplona 08703
Museo Aperos de Labranza, Lerdo 31001
Museo Apistico, Bologna 25444
Museo Apistico Didattico, Bregnano 25527
Museo Archaeologico di Santa Restituta, Lacco Ameno 26437
Museo Archeologica, Savignone 27770
Museo Archeologico, Acqualagna 25144
Museo Archeologico, Amelia 25221
Museo Archeologico, Arborea 25257
Museo Archeologico, Artena 25288
Museo Archeologico, Atri 25321
Museo Archeologico, Bonorva 25495
Museo Archeologico, Caltanissetta 25607
Museo Archeologico, Camaiore 25612
Museo Archeologico, Castello di Godego 25741
Museo Archeologico, Cavriana 25791
Museo Archeologico, Cecina e Marina 25792
Museo Archeologico, Cesena 25825
Museo Archeologico, Colle di Val d'Elsa 25936
Museo Archeologico, Dorgali 26044
Museo Archeologico, Enna 26056
Museo Archeologico, Fermo 26098
Museo Archeologico, Frosinone 26253
Museo Archeologico, Isola della Scala 26417
Museo Archeologico, Lentini 26484
Museo Archeologico, Marianopoli 26606
Museo Archeologico, Molfetta 26763
Museo Archeologico, Montelupo Fiorentino 26837
Museo Archeologico, Ostiglia 27019
Museo Archeologico, Palazzuolo sul Senio 27061
Museo Archeologico, Povegliano Veronese 27320
Museo Archeologico, Priverno 27345
Museo Archeologico, Pula 27347
Museo Archeologico, Rocca di Mezzo 27427
Museo Archeologico, San Gimignano 27636
Museo Archeologico, San Miniato 27671
Museo Archeologico, Santadi 27713
Museo Archeologico, Santorso 27734
Museo Archeologico, Scansano 27784
Museo Archeologico, Siena 27838
Museo Archeologico, Teramo 27955
Museo Archeologico, Varallo Pombia 28156
Museo Archeologico, Venafro 28175
Museo Archeologico, Ventotene 28218
Museo Archeologico, Veroli 28225
Museo Archeologico, Vetulonia 28241
Museo Archeologico, Villasimius 28281
Museo Archeologico al Teatro Romano, Verona 28230
Museo Archeologico Antiquarium Arborense, Oristano 26996
Museo Archeologico Antonio Santarelli, Forlì 26223
Museo Archeologico Aquaria, Soncino 27868
Museo Archeologico Attrezzato e Cattedrale di San Zeno, Pistoia 27247
Museo Archeologico Baglio Anselmi, Marsala 26611
Museo Archeologico Civico, Carassai 25673
Museo Archeologico Civico, Forlimpopoli 26231
Museo Archeologico Civico-Diocesano G. B. Leopardi, Penne 27134
Museo Archeologico Comprensoriale, Teti 27970
Museo Archeologico Comunale, Vicchio 28248
Museo Archeologico Comunale di Artimino, Carmignano 25680
Museo Archeologico dei Campi Flegrei, Bacoli 25338
Museo Archeologico del Finale, Finale Ligure 26138
Museo Archeologico del Territorio, Cupra Marittima 26022
Museo Archeologico del Territorio, San Lorenzo in Campo 27656
Museo Archeologico del Territorio di Populonia, Piombino 27228
Museo Archeologico della Badia, Licata 26490
Museo Archeologico della Lucania Occidentale, Padula 27054
Museo Archeologico della Valle Sabbia, Gavardo 26276
Museo Archeologico dell'Abruzzo, Crecchio 25998
Museo Archeologico dell'Agro Nocerino, Nocera Inferiore 26932
Museo Archeologico dell'Alto Adige, Bolzano 25486
Museo Archeologico dell'Alto Canavese, Cuorgnè 26021
Museo Archeologico dell'Antica, Alife 25202
Museo Archeologico dell'Antica Capua, Santa Maria Capua Vetere 27705
Museo Archeologico dell'Antica Corfinio, Corfinio 25974
Museo Archeologico dell'Arte della Lana, Arpino 25283
Museo Archeologico di Calatia, Maddaloni 26561
Museo Archeologico di Casteggio, Casteggio 25714
Museo Archeologico di Morgantina, Aidone 25165
Museo Archeologico di Naxos, Giardini Naxos 26334
Museo Archeologico di Sepino, Sepino 27808
Museo Archeologico di Vallecamonica, Cividate Camuno 25918
Museo Archeologico di Villa Sulcis, Carbonia 25675
Museo Archeologico e Antiquarium, Sala Consilina 27591
Museo Archeologico e Antiquarium Nazionale, Formia 26233
Museo Archeologico e d'Arte della Maremma, Grosseto 26371
Museo Archeologico e della Collegiata, Casole d'Elsa 25709
Museo Archeologico e della via Flaminia, Cagli 25576
Museo Archeologico e Paleontologico, Asti 25314
Museo Archeologico e Paleontologico di Farneta, Cortona 25987
Museo Archeologico e Pinacoteca, Fano 26084
Museo Archeologico e Pinacoteca, Piacenza 27185
Museo Archeologico e Pinacoteca Comunale, Foligno 26214
Museo Archeologico e Storico Aristico, Bra 25523
Museo Archeologico ed Etnografico, Ittireddu 26421
Museo Archeologico Eoliano, Lipari 26493
Museo Archeologico Etneo, Adrano 25152
Museo Archeologico Etnografico, Paulilatino 27118
Museo Archeologico F. Milizia, Oria 26991
Museo Archeologico G. Allevi, Offida 26973
Museo Archeologico G. Cilnio Mecenate, Arezzo 25266
Museo Archeologico G. Moretti, San Severino Marche 27684
Museo Archeologico G. Rambotti, Desenzano del Garda 26034
Museo Archeologico Girolamo Rossi, Ventimiglia 28216
Museo Archeologico Ibleo, Ragusa 27359

M

Museo Archeologico Iulium Carnicum, Zuglio 28320
Museo Archeologico Lametino, Lamezia Terme 26443
Museo Archeologico Liutprando, San Giovanni in Persiceto 27645
Museo Archeologico Lomellino, Gamboló 26269
Museo Archeologico Lucus Feroniae, Capena 25656
Museo Archeologico Luigi Donini, San Lazzaro di Savena 27650
Museo Archeologico Medievale, Attimis 25325
Museo Archeologico Michele Janora, Irsina 26409
Museo Archeologico Nazionale, Adria 25153
Museo Archeologico Nazionale, Aquileia 25253
Museo Archeologico Nazionale, Capaccio 25652
Museo Archeologico Nazionale, Cassino 25711
Museo Archeologico Nazionale, Cividale del Friuli 25916
Museo Archeologico Nazionale, Crotone 26012
Museo Archeologico Nazionale, Eboli 26050
Museo Archeologico Nazionale, Firenze 26167
Museo Archeologico Nazionale, Gioia del Colle 26341
Museo Archeologico Nazionale, Manfredonia 26587
Museo Archeologico Nazionale, Mantova 26591
Museo Archeologico Nazionale, Napoli 26883
Museo Archeologico Nazionale, Parma 27099
Museo Archeologico Nazionale, Quarto d'Altino 27350
Museo Archeologico Nazionale, Sperlonga 27887
Museo Archeologico Nazionale, Spoleto 27892
Museo Archeologico Nazionale, Taranto 27941
Museo Archeologico Nazionale, Venezia 28185
Museo Archeologico Nazionale, Viterbo 28291
Museo Archeologico Nazionale d'Abruzzo, Campli 25631
Museo Archeologico Nazionale del Melfese, Melfi 26651
Museo Archeologico Nazionale della Sibaritide, Cassano allo Jonio 25710
Museo Archeologico Nazionale dell'Abruzzo, Chieti 25871
Museo Archeologico Nazionale dell'Abruzzo, Chieti 25872
Museo Archeologico Nazionale delle Marche, Ancona 25227
Museo Archeologico Nazionale dell'Umbria, Perugia 27147
Museo Archeologico Nazionale di Cagliari, Cagliari 25579
Museo Archeologico Nazionale di Cosa, Orbetello 26987
Museo Archeologico Nazionale di Locri Epizefiri, Locri 26505
Museo Archeologico Nazionale di Luni, Ortonovo 27005
Museo Archeologico Nazionale di Metaponto, Bernalda 25407
Museo Archeologico Nazionale di Palestrina, Palestrina 27081
Museo Archeologico Nazionale di Venosa, Venosa 28214
Museo Archeologico Nazionale Etrusco, Cerveteri 25819
Museo Archeologico Nazionale Etrusco, Tuscania 28109
Museo Archeologico Oliveriano, Pesaro 27154
Museo Archeologico per la Preistoria e Protostoria del Tigullio, Chiavari 25858
Museo Archeologico Provinciale, Bari 25362
Museo Archeologico Provinciale, Potenza 27317
Museo Archeologico Provinciale, Salerno 27594
Museo Archeologico Provinciale F. Ribezzo, Brindisi 25545
Museo Archeologico Regionale, Agrigento 25161
Museo Archeologico Regionale, Aosta 25247
Museo Archeologico Regionale, Gela 26282
Museo Archeologico Regionale A. Salinas, Palermo 27066
Museo Archeologico Regionale Paolo Orsi, Siracusa 27854
Museo Archeologico Remo Fumagalli, Broletto di Novara 25550
Museo Archeologico San Giovanni in Compito, Savignano sul Rubicone 27769
Museo Archeologico Sentinate, Sassoferrato 27761
Museo Archeologico Statale, Arcevia 25259
Museo Archeologico Statale, Ascoli Piceno 25294
Museo Archeologico Statale, Cingoli 25891
Museo Archeologico Statale, Teano 27947
Museo Archeologico Statale, Urbisaglia 28128
Museo Archeologico Statale V. Capialbi, Vibo Valentia 28246
Museo Archeologico Statale V. Laviola, Amendolara 25222
Museo Archeologico Tifernum Mataurense, Sant'Angelo in Vado 27718
Museo Archeologico Villa Abbas, Sardara 27740
Museo Archiconfradia de la Sangre, Murcia 39691
Museo-Archivio di Fotografia Storica, Roma 27461
Museo-Archivo de Falset y Comarca, Falset 39293
Museo Archivo Rubén Darío, León 33284
Museo Arcivescovile, Ravenna 27370
Museo Argentino de Caza Mayor, Buenos Aires 00167
Museo Argentino de Ciencias Naturales Bernardino Rivadavia, Buenos Aires 00168
Museo Argentino de Motos Antiguas, Mendoza 00451
Museo Argentino del Títere, Buenos Aires 00169
Museo Armando Reverón, Caracas 54534
Museo Armando Reverón, Macuto 54580
Museo Armería 9 de Abril, La Habana 09111
Museo Arqueología Regional, La Romana 10332
Museo Arqueològic, Gandía 39324
Museo Arqueològic, Llíria 39484
Museo Arqueològic Municipal, Vallada 40226
Museo Arqueológica de Mongua, Mongua 08689
Museo Arqueológical Rafael Larco Herrera, Lima 34225
Museo Arqueológico, Aguilares 00117
Museo Arqueológico, Almuñécar 38864
Museo Arqueológico, Aranzazu 08530
Museo Arqueológico, Cacabelos 39081
Museo Arqueológico, La Carolina 39133
Museo Arqueológico, Cartago 08621
Museo Arqueológico, Cochabamba 04206
Museo Arqueológico, Cuenca 10381
Museo Arqueológico, Curumaní 08632
Museo Arqueológico, Gata de Gorgos 39328
Museo Arqueológico, Gualeguaychú 00382
Museo Arqueológico, Inzá 08661
Museo Arqueológico, Manizales 08667
Museo Arqueológico, Nerja 39714
Museo Arqueológico, Novelda 39718
Museo Arqueológico, Oiba 08700
Museo Arqueológico, Paéz 08701
Museo Arqueológico, Pasca 08707
Museo Arqueológico, Puerto Berrío 08731
Museo Arqueológico, Puerto de la Cruz 39876
Museo Arqueológico, Sagunto 39932
Museo Arqueológico, Salamina 08742
Museo Arqueológico, Santa Cruz de Tenerife 39994
Museo Arqueológico, Sogamoso 08760
Museo Arqueológico, Tafí del Valle 00638
Museo Arqueológico, La Tebaida 08767
Museo Arqueológico, Trinidad 04253
Museo Arqueológico Adàn Quiroga, San Fernando del Valle de Catamarca 00571
Museo Arqueológico Alejandro Pezzia Aseretto, La Angostura 34198
Museo Arqueológico Alto Maestrazgo, Benasal 39026
Museo Arqueológico Atahualpa, Ibarra 10411
Museo Arqueológico Calima, Calima 08616
Museo Arqueológico Carlos Aiiercado, Esmeraldas 10395
Museo Arqueológico Casa de Madame Augustín, Santa Marta 08746
Museo Arqueologico Casa del Marqués de San Jorge, Bogotá 08553
Museo Arqueológico Chiametlán, Rosario 31307
Museo Arqueológico Chimica, El Copey 08638
Museo Arqueológico Comarcal, Orihuela 39741
Museo Arqueológico Comunal, Carquín 34138
Museo Arqueológico Cuicuilco, México 31051
Museo Arqueológico de Ancash, Huaraz 34181
Museo Arqueológico de Apaxco, Apaxco 30719
Museo Arqueológico de Asturias, Oviedo 39758
Museo Arqueológico de Benalmádena, Benalmádena 39025
Museo Arqueológico de Cajamarca y Museo Médico, Cajamarca 34126
Museo Arqueológico de Campeche, Campeche 30739
Museo Arqueológico de Cancún, Cancún 30747
Museo Arqueológico de Chasicó, Chasicó 00301
Museo Arqueológico de Chusís, Sechura 34306
Museo Arqueológico de Ciudad Guzmán, Guzmán 30925
Museo Arqueológico de Comayagua, Comayagua 23325
Museo Arqueológico de Comitán, Comitán de Domínguez 30792
Museo Arqueológico de Esteli, San Juan de Limay 33304
Museo Arqueológico de Guacarí, Guacarí 08648
Museo Arqueológico de La Caleta, La Caleta 10325
Museo Arqueológico de la Ciudad, Emiliano Zapata 30851
Museo Arqueológico de la Costa Grande, Zihuatanejo 31596
Museo Arqueológico de La Serena, La Serena 07440
Museo Arqueológico de la Universidad Nacional del Altiplano, Pucara 34295
Museo Arqueológico de Lagos de Moreno, Lagos de Moreno 30991
Museo Arqueológico de Lumbrales, Lumbrales 39500
Museo Arqueológico de Martins Sarmento, Guimarães 35637
Museo Arqueológico de Mazatlán, Mazatlán 31020
Museo Arqueológico de Muzo, Muzo 08693
Museo Arqueológico de Nextlalpan, Nextlalpan 31214
Museo Arqueológico de Puerto Esperanza, Puerto Esperanza 00506
Museo Arqueológico de Saipina, Santa Cruz 04239
Museo Arqueológico de Santa Trega, A Guardia 39380
Museo Arqueológico de Solentiname, Solentiname 33307
Museo Arqueológico de Tepeapulco, Tepeapulco 31432
Museo Arqueológico de Tilcara, Tilcara 00647
Museo Arqueológico de Tizatlán, Tizatlán 31462
Museo Arqueológico de Tonalá, Tonalá, Chiapas 31495
Museo Arqueológico de Tula Jorge R. Acosta, Tula de Allende 31508
Museo Arqueológico de Úbeda, Úbeda 40178
Museo Arqueológico de Venustiano Carranza, Venustiano Carranza 31534
Museo Arqueológico de Xochimilco, México 31052
Museo Arqueológico d'Eivissa i Formentera, Eivissa 39259
Museo Arqueologico del Banco Central de Bahia, Bahia de Carácluez 10372
Museo Arqueológico del Centro de Estudios Histórico Sociales Julio Espejo Nuñez, Jauja 34196
Museo Arqueológico del Cerro de la Estrella, México 31053
Museo Arqueológico del Cesar, Valledupar 08779
Museo Arqueológico del Colegio Juan XXIII, Lima 34226
Museo Arqueológico del Huila, Neiva 08695
Museo Arqueológico del Instituto de Arqueología y Etnología, Mendoza 00452
Museo Arqueológico del Sinú, Tierralta 08769
Museo Arqueológico del Soconusco, Tapachula 31401
Museo Arqueológico Dr. Román Piña Chan, Tenango del Valle 31422
Museo Arqueológico e Histórico de Andahuaylas, Andahuaylas 34107
Museo Arqueológico Etnográfico, San Pedro del Pinatar 39968
Museo Arqueológico-Etnológico Gratiniano Baches, Pilar de la Horadada 39840
Museo Arqueológico Etnológico Municipal, Guardamar de Segura 39379
Museo Arqueológico Frederico Galvez Durand, Huancayo 34176
Museo Arqueológico Gregorio Aguilar Barea, Juigalpa 33280
Museo Arqueológico Guamuhaya, Trinidad 09334
Museo Arqueológico Horacio Urteaga, Cajamarca 34127
Museo Arqueológico José Casinelli, Santa Inés 34305
Museo Arqueologico José María Morante de la Universidad Nacional de San Augustín de Arequipa, Arequipa 34109
Museo Arqueologico Juan Sepúlveda y Museo Eduardo Ruiz, Uruapán 31520
Museo Arqueológico Julio César Cubillos, Cali 08606
Museo Arqueológico Julio César Cubillos, San Agustín 08743
Museo Arqueológico La Bagatela, Villa del Rosario 08785
Museo Arqueològico La Encomienda, Calasparra 39096
Museo Arqueológico La Merced, Cali 08607
Museo Arqueológico Luis G. Urbina, México 31054
Museo Arqueológico Marco Fidel Micolta, Guamo 08653
Museo Arqueológico Mesoamericano, Tijuana 31455
Museo Arqueológico Municipal, Alcúdia (Baleares) 38832
Museo Arqueológico Municipal, Burriana 39075
Museo Arqueológico Municipal, Caravaca de la Cruz 39126
Museo Arqueológico Municipal, Cartagena 39136
Museo Arqueológico Municipal, Cehegín 39158
Museo Arqueológico Municipal, Crevillente 39226
Museo Arqueológico Municipal, Elda 39271
Museo Arqueológico Municipal, Jerez de la Frontera 39426
Museo Arqueológico Municipal, Lorca 39492
Museo Arqueológico Municipal, Marchena 39621
Museo Arqueológico Municipal, Montoro 39681
Museo Arqueológico Municipal, San Fulgencio 39956
Museo Arqueológico Municipal, Santaella 40007
Museo Arqueológico Municipal Alejandro Ramos Folqués, Elche 39264
Museo Arqueológico Municipal Antonio Ballester, Callosa de Segura 39103
Museo Arqueológico Municipal Camilo Visedo Molto, Alcoy 38827
Museo Arqueológico Municipal Cayetano Mergelina, Yecla 40330
Museo Arqueológico Municipal de Alhambra, Alhambra 38842
Museo Arqueológico Municipal de Obulco, Porcuna 39859
Museo Arqueológico Municipal José María Soler, Villena 40311
Museo Arqueológico Municipal y Centro de Interpretación del Mar, Águilas 38793
Museo Arqueológico Municipal y Museo Municipal Miguel Justino Ramírez, Piura 34290
Museo Arqueológico Nacional, Madrid 39526
Museo Arqueológico Nacional Brüning, Lambayeque 34202
Museo Arqueológico Padre Martín Recio, Estepa 39287
Museo Arqueológico-Paleontológico Municipal de Rojales, Rojales 39910
Museo Arqueológico Pío P. Díaz, Cachi 00285
Museo Arqueológico Provincial, Alicante 38843
Museo Arqueológico Provincial, Badajoz 38933
Museo Arqueológico Provincial, Huesca 39398
Museo Arqueológico Provincial, Ourense 39754
Museo Arqueológico Provincial, San Salvador de Jujuy 00607
Museo Arqueológico Provincial, Sevilla 40064
Museo Arqueológico Provincial Andalgalá, Andalgalá 00123
Museo Arqueológico Provincial Condor Huasi (interim), Belén 00146
Museo Arqueológico R.P. Gustavo Le Paige S.J., San Pedro de Atacama 07449
Museo Arqueológico Regional, Alcalá de Henares 38812
Museo Arqueológico Regional Anibal Montes, Río Segundo 00538
Museo Arqueológico Regional Guane, Floridablanca 08643
Museo Arqueológico Regional Inca Huasi, La Rioja 00416
Museo Arqueológico San Miguel de Azapa, Arica 07430
Museo Arqueológico-Torre del Agua, Osuna 39752
Museo Arqueológico y Colonial, Fuerte Quemado 00374
Museo Arqueológico y Etnográfico, Betancuria 39031
Museo Arqueológico y Etnológico Córdoba, Córdoba 39204
Museo Arqueológico y Etnológico de Granada, Granada 39364
Museo Arqueológico y Etnológico de Lucena, Lucena 39496
Museo Arqueologico y Etnologico Municipal Dámaso Navarro, Petrer 39839
Museo Arqueológico y Paleontológico de Guane, Barichara 08536
Museo Arqueoloógico y Etnográfico, Ibarra 10412
Museo Arqueolóxico e Histórico, A Coruña 39217
Museo Arquidiocesano, Santiago de Cuba 09305
Museo Arquidiocesano de Arte Religioso, Pamplona 08704
Museo Arquidiocesano de Arte Religioso de Popayán, Popayán 08723
Museo Arquitectura de Trinidad, Trinidad 09335
Museo Arrese, Corella 39211
Museo Art Nouveau-Art Déco, Salamanca 39936
Museo Artequin, Santiago de Chile 07450
Museo Artes Decorativo, Gibara 09085
Museo Artes Plásticas, Pinar del Río 09247
Museo Artes y Costumbres Populares Alto Guadalquivir, Cazorla 39157
Museo Artesanal de Gualaceo, Gualaceo 10398
Museo Arti e Mestieri di un Tempo, Cisterna d'Asti 25901
Museo Arti e Tradizioni Popolari del Gargano Giovanni Tancredi, Monte Sant'Angelo 26806
Museo Artístico de Roncesvalles, Roncesvalles 39914
Museo Artistico della Bambola, Suvereto 27931
Museo Artistico Industriale F. Palizzi, Napoli 26884
Museo Arturo Michelena, Caracas 54535
Museo Arxiu Capitular, Tortosa 40164
Museo Astronomico e Copernicano, Roma 27462
Museo Astronomico e Geofisico, Modena 26744
Museo Atelier Antonio Ortiz Echagüe, Carro Quemado 00295
Museo Audiovisual, Caracas 54536
Museo Augustín Landívar, Cuenca 10382
Museo Augusto Righi, Montese 26852
Museo Aurelio Castelli, Siena 27839
Museo Aurelio Espinosa Polit, Cotocollao 10379
Museo Aurelio Espinoza Pólit, Quito 10428
Museo Aurelio Marena, Bitonto 25429
Museo Austral de Pintura Primitiva Moderna Naif, Esquel 00369
Museo Auto d'Ecopa, San Marino 38003
Museo Azuayo del Folklore, Cuenca 10383
Museo B. Gigli, Recanati 27376
Museo Badia di San Gemolo, Valganna 28141
Museo Bagatti Valsecchi, Milano 26690
Museo Bahía de los Angeles, Ensenada 30855

Museo Balaa Xte Guech Gulal, Teotitlán del Valle 31429
Museo Balear Ciencies Naturals, Sóller 40087
Museo Balseros del Mar del Sur, Salango 10458
Museo Baluarte de Santiago, Veracruz 31535
Museo Banco Central de Reserva del Perú, Lima 34227
Museo Banco del Pacifico, Guayaquil 10401
Museo Bandini, Fiesole 26126
Museo Barjola, Gijón 39334
Museo Baroffio e del Santuario del Sacro Monte sopra Varese, Varese 28162
Museo Barovier e Toso, Venezia 28186
Museo Barracco, Roma 27463
Museo Beata Camilla Battista Varano, Camerino 25616
Museo Beata P. Morosini, Albino 25180
Museo Beato Angelico, Vicchio 28249
Museo Beato Angelo, Acri 25150
Museo Belisario Porras, Las Tablas 34075
Museo Bellapart, Santo Domingo 10342
Museo Bello Piñeio, Ferrol 39296
Museo Benalauría, Benalauría 39024
Museo Benedettino Nonantolano e Diocesano di Arte Sacra, Nonantola 26940
Museo Benigno Malo, Cuenca 10384
Museo Benito Hortiz, Sancti-Spíritus 09283
Museo Benito Juárez, La Habana 09112
Museo Benito Juárez, José Azueta 30978
Museo Berenziano, Cremona 26002
Museo Bernabé de las Casas, Mina 31166
Museo Bernardino Caballero, Asunción 34091
Museo Bernardo de Muro, Tempio Pausania 27952
Museo Bernardo Martínez Villegas, Bácum 30733
Museo Bernardo Samper Sordo, Bogotá 08554
Museo Berrocal, Vilanueva de Algaidas 40288
Museo Bersano delle Contadinerie e Stampe antiche del Vino, Nizza Monferrato 26929
Museo Biblioteca Comunale, Etroubles 26067
Museo Biblioteca Municipal Cesar Fernández Navarro, Santa Fé 00613
Museo Biblioteca Nueva Cádiz y Casa Natal de Juan Bautista Arismendi, Nueva Cádiz 54589
Museo Biblioteca Palacio Postal, México 31055
Museo Biblioteca Palafoxiana, Puebla 31266
Museo Biblioteca Pape, Monclova 31172
Museo Biblioteca Rosauro Rosa Acosta, Pampatar 54590
Museo Bicknell, Bordighera 25498
Museo Bicocchi, Pomarance 27276
Museo Bodoniano, Parma 27100
Museo Bolivar, Ciudad de Panamá 34066
Museo Bolívar Periodista, Piura 34291
Museo Bolivariano, Caracas 54537
Museo Bolivariano, Pativilca 34285
Museo Bolivariano de Arte Contemporáneo, Santa Marta 08747
Museo Bologna-Buonsignori, Siena 27840
Museo Boncompagni Ludovisi, Roma 27464
Museo Bonsor, Mairena del Alcor 39593
Museo Borromeo, Isola Bella 26415
Museo Botánico, Córdoba 00320
Museo Botánico, Montevideo 54395
Museo Botánico, Tuxtla Gutiérrez 31511
Museo Botanico Cesare Bicchi, Lucca 26531
Museo Botanico della Lessinia, Fumane 26256
Museo Bottacin, Padova 27036
Museo Bottega della Tarsia Lignea, Sorrento 27880
Museo Brandolini e Giol, Oderzo 26968
Museo Brembano di Scienze Naturali, San Pellegrino Terme 27677
Museo C. Nivola, Orani 26985
Museo Ca' la Ghironda, Bologna 25445
Museo Cabrera, Ica 34187
Museo Cacique Balata, La Cumbre 00400
Museo Calabozo de Hidalgo, Chihuahua 30765
Museo Calabrese di Etnografia e Folklore Raffaele Corso, Palmi 27085
Museo Camillo Leone, Vercelli 28220
Museo Camilo Egas del Banco Central, Quito 10429
Museo Camilo Ortega de la Ciudad Masaya, Masaya 33293
Museo Camón Aznar de Ibercaja, Zaragoza 40340
Museo Campagna G. Bruno, Campagna 25628
Museo Campamento Irritila, San Pedro 31365
Museo Camuno e Biblioteca Civica, Breno 25528
Museo Canario, Las Palmas de Gran Canaria 39798
Museo Cándido Bido, Bonao 10324
Museo Canonicale, Verona 28231
Museo Cantonale d'Arte, Lugano 41581
Museo Cantonale di Storia Naturale, Lugano 41582
Museo Capilla Alfonsina, México 31056
Museo Capitán Juan de Zevallos, Valle Hermoso 00665
Museo Capitolare, Assisi 25306
Museo Capitolare, Atri 25322
Museo Capitolare, Cagliari 25580
Museo Capitolare, Ortona 27002
Museo Capitolare di Arte Sacra, Gravina di Puglia 26366
Museo Cappella San Severo, Napoli 26885
Museo Cappuccini Emiliani, Reggio Emilia 27389
Museo Capuaniano, Mineo 26733
Museo Carducci - Biblioteca e Casa, Bologna 25446
Museo Carlo Venturini, Massa Lombarda 26623
Museo Carlo Zauli, Faenza 26073
Museo Carlos Dreyer, Puno 34298
Museo Carlos Emilio Grijalva, Tulcán 10461
Museo Carlos Torres Barraza, Valentín Gómez Farías 31522
Museo Carnico delle Arti Popolari Michele Gortani, Tolmezzo 27992
Museo Carozzi, Seriate 27810
Museo Le Carrozze d'Epoca, Roma 27465
Museo Casa, Valverde del Camino 40253
Museo Casa Alejandro de Humboldt, La Habana 09113
Museo Casa Blanca, San Juan 36065
Museo Casa Busoni, Empoli 26052
Museo Casa Carlos Gardel, Buenos Aires 00170
Museo Casa Cautiño, Guayama 36038
Museo Casa Clautana E. Borsatti, Claut 25927
Museo Casa Colonial, Pamplona 08705
Museo Casa Cultural "Alfonso López Pumarejo", Honda 08656
Museo Casa de Africa, La Habana 09114
Museo Casa de Agustín Arrieta, Puebla 31267
Museo Casa de Asia, La Habana 09115
Museo Casa de Benalcazar, Quito 10430
Museo Casa de Bernardo A. Houssay, Buenos Aires 00171
Museo Casa de Bolívar, Bucaramanga 08598
Museo Casa de Carmen Montilla, La Habana 09116
Museo Casa de Carranza, Cuatrocienegas 30811
Museo Casa de Carranza, México 31057
Museo Casa de Colón, Las Palmas de Gran Canaria 39799
Museo Casa de Colón, San Sebastián de La Gomera 39971
Museo Casa de Cultura de Ixtlán, Ixtlán del Río 30963
Museo Casa de Dulcinea, El Toboso 40125
Museo Casa de Hidalgo la Francia Chiquita, San Felipe 31330
Museo Casa de Isla Negra, San Felipe 07448
Museo Casa de la Bola, México 31058
Museo Casa de la Cultura, Montemorelos 31176
Museo Casa de la Cultura de Barichara, Barichara 08537
Museo Casa de la Cultura de Cuacutecuta, Cuacutecuta 08629
Museo Casa de la Cultura de Pensilvania, Pensilvania 08717
Museo Casa de la Cultura "Luis Camacho Rueda", Socorro 08759
Museo Casa de la Cultura Simón Patiño, Oruro 04230
Museo Casa de la Libertad, Sucre 04244
Museo Casa de la Masacre de Ponce, Ponce 36046
Museo Casa de la Moneda, Mariquita 08672
Museo Casa de la Obrapía, La Habana 09117
Museo Casa de la Orfebrería, La Habana 09118
Museo Casa de la Plomada, Vallecillo 31528
Museo Casa de la Poesía, La Habana 09119
Museo Casa de la Tecnología, Monterrey 31177
Museo Casa de la Troya, Santiago de Compostela 40017
Museo Casa de León Trotsky, México 31059
Museo Casa de les Rocas, Valencia 40202
Museo Casa de los Mártires, La Habana 09120
Museo Casa de Morelos, Ecatepec de Morelos 30845
Museo Casa de Morelos, Morelia 31191
Museo Casa de Policarpa Salavarrieta, Guaduas 08650
Museo Casa de Puerto Rico y Sala del Tabaco, La Habana 09121
Museo Casa de Ramón López Velarde, Jerez de García Salinas 30973
Museo Casa de Ricardo Rojas, Buenos Aires 00172
Museo Casa de Sais Montes de Oca Hermanos, Pinar del Río 09248
Museo Casa de Sucre, Quito 10431
Museo Casa de Yrurtia, Buenos Aires 00173
Museo Casa de Zela, Tacna 34310
Museo Casa del Agua, Tehuacán 31417
Museo Casa del Alfeñique, Puebla 31268
Museo Casa del Deán, Puebla 31269
Museo Casa del Diego, Celaya 30752
Museo Casa del Dr. Mora, Comonfort 30795
Museo Casa del Duca, Venezia 28187
Museo Casa del Fundador Ponzalo Suárez Rendón, Tunja 08774
Museo Casa del General Alvaro Obregón, Huatabampo 30942
Museo Casa del Libertador Simón Bolívar, La Habana 09122
Museo Casa del Maestro José Márquez Figueroa, Puebla 31270
Museo Casa del Marques de Aguayo y Museo Municipal, Mazapil Zacatecas 31019
Museo Casa del Padre, Orselina 41716
Museo Casa del Teatro, Buenos Aires 00174
Museo Casa di Dante, Firenze 26168
Museo Casa Diego Rivera, Guanajuato 30912
Museo Casa Dorada-Maison d'Or, Tarija 04250
Museo Casa Dr. Belisario Domínguez, Comitán de Domínguez 30793
Museo Casa Estudio Diego Rivera y Frida Kahlo, México 31060
Museo Casa Francisco Villa, Hidalgo del Parral 30935
Museo Casa-Fuerte Juan de Ponce de León, Higüey 10328
Museo Casa Fundación Oswaldo Guayasamín, La Habana 09123
Museo Casa Galimberti, Cuneo 26018
Museo Casa Generalísimo Máximo Gómez, Montecristi 10330
Museo Casa Hacienda San Jacinto, San Jacinto 33303
Museo Casa Ibáñez, Olula del Río 39730
Museo Casa José Lezama Lima, La Habana 09124
Museo Casa Mata, Matamoros 31017
Museo Casa Maya, Cancún 30748
Museo Casa Mosquera, Popayán 08724
Museo Casa Murillo, Sevilla 40065
Museo Casa Natal de Antonio Maceo, Santiago de Cuba 09306
Museo Casa Natal de Calixto Garcia, Holguín 09178
Museo Casa Natal de Camilo Cienfuegos, La Habana 09125
Museo Casa Natal de Carlos Manuel de Céspedes, Bayamo 09033
Museo Casa Natal de Celia Sanchez Manduley, Media Luna 09221
Museo Casa Natal de Cervantes, Alcalá de Henares 38813
Museo Casa Natal de Frank País, Santiago de Cuba 09307
Museo Casa Natal de Hermanos Ameijeiras, Pueblo Viejo 09258
Museo Casa Natal de Ignacio Agramonte, Camagüey 09050
Museo Casa Natal de Isabel la Católica, Madrigal de las Altas Torres 39589
Museo Casa Natal de José María Heredia, Santiago de Cuba 09308
Museo Casa Natal de José Martí, La Habana 09126
Museo Casa Natal de Jovellanos, Gijón 39335
Museo Casa Natal de Morelos, Morelia 31192
Museo Casa Natal de Pedro Martínez Brito, Ciego de Avila 09061
Museo Casa Natal de Serafín Sánchez, Sancti-Spíritus 09284
Museo Casa Natal General Santander, Villa del Rosario 08786
Museo Casa Natal Juan Manuel Márquez, La Habana 09127
Museo Casa Natal Rubén Darío, Darío 33275
Museo Casa Natal y Museo de Niceto Alcalá-Zamora y Torres, Priego de Córdoba 39871
Museo Casa Natale di Giotto, Vicchio 28250
Museo Casa Natale di San Pio X, Riese Pio X 27401
Museo Casa Natale di Tiziano Vecellio, Pieve di Cadore 27207
Museo Casa Natale G. D'Annunzio, Pescara 27162
Museo Casa Natale Giacomo Puccini, Lucca 26532
Museo Casa Negret, Popayán 08725
Museo Casa Paoli, Ponce 36047
Museo Casa Pascoli, Barga 25359
Museo Casa Rafael Núñez, Cartagena 08617
Museo Casa Roig, Humacao 36041
Museo Casa Rosada Abelardo Díaz Alfaro, Caguas 36030
Museo Casa Rossini, Pesaro 27155
Museo Casa Ruth Lechuga, México 31061
Museo Casa Studio F. Melani, Pistoia 27248
Museo Casa Taller Celia Chevalier, Buenos Aires 00175
Museo Casa Walser, Macugnaga 26558
Museo Casabianca, Malo 26577
Museo Casita Blanca de Agustín Lara, Boca del Río 30737
Museo Cassinelli, Trujillo 34317
Museo Castellano-Manchego del Juguete, Mohedas de la Jara 39655
Museo Castello Masegra, Sondrio 27872
Museo Castillo de la Fortaleza, Santa Lucía 40001
Museo Castillo del Macho, Peñíscola 39833
Museo Castillo Serrallés, Ponce 36048
Museo Catcco, Ollantaytambo 34283
Museo Catedral, Palma de Mallorca 39785
Museo Catedral de La Plata, La Plata 00408
Museo Catedralicio, Almería 38860
Museo Catedralicio, Badajoz 38934
Museo Catedralicio, Baeza 38940
Museo Catedralicio, Burgo de Osma 39064
Museo Catedralicio, Burgos 39067
Museo Catedralicio, Cádiz 39088
Museo Catedralicio, Ciudad Real 39186
Museo Catedralicio, Cuenca 39230
Museo Catedralicio, Guadix 39378
Museo Catedralicio, Jaén 39420
Museo Catedralicio, Málaga 39597
Museo Catedralicio, Ourense 39755
Museo Catedralicio, Palencia 39769
Museo Catedralicio, Santo Domingo de la Calzada 40040
Museo Catedralicio, Segovia 40051
Museo Catedralicio, Sigüenza 40078
Museo Catedralicio, Zamora 40332
Museo Catedralicio de La Seo de Zaragoza, Zaragoza 40341
Museo Catedralicio de Tui, Tui 40175
Museo Catedralicio Diocesano, Ciudad Rodrigo 39192
Museo Catedralicio-Diocesano, León 39454
Museo Catedralicio Diocesano, Pamplona 39806
Museo Catedralicio-Diocesano, Valencia 40203
Museo Catedralico y Giralda, Sevilla 40066
Museo Católica de Santa María, Cercado 34141
Museo Católico Peninsular, Mérida 31023
Museo Cavour, Santena 27730
Museo Cemento Rezola, Donostia-San Sebastián 39251
Museo Cemí, Jayuya 36042
Museo Centrale del Risorgimento, Roma 27466
Museo Centrale dell'Istituto Storico della Resistenza, Imperia 26404
Museo Centrale Montemartini, Roma 27467
Museo Centro Cultural Pachamama, Amaicha del Valle 00122
Museo Centro de Cultura Mayo de el Júpare Blas Mazo, Huatabampo 30943
Museo Centro de Interpretacion de Monesterio, Monesterio 39662
Museo Centro de la Imagen, México 31062
Museo Centro Studi A.G. Barrili, Carcare 25677
Museo Centro Taurino Potosino, San Luis Potosí 31344
Museo Ceramoteca, Salamanca 31310
Museo Cerezo Moreno, Villatorres 40309
Museo Cerlogne, Saint Nicolas 27587
Museo Cerralbo, Madrid 39527
Museo Cerro de la Campana, Santiago Suchilquitongo 31385
Museo Cerro de los Huizaches, San Pablo Huixtepec 31364
Museo Cervantino, El Toboso 40126
Museo Cervi, Gattatico 26275
Museo Chairama, Santa Marta 08748
Museo Chiaramonti e Braccio Nuovo, Città del Vaticano 54502
Museo Chihuahuense de Arte Contemporáneo, Chihuahua 30766
Museo Chileno de Arte Precolombino, Santiago de Chile 07451
Museo Chillida Leku, Hernani 39387
Museo Chorotega-Nicarao llamado "Enrique Berio Mántica Deshon", Chinandega 33273
Museo Chorro de Maita, Yaguajay 09353
Museo Christiano, Cividale del Friuli 25917
Museo Ciénaga de San Pedro, Madera 31009
Museo Ciencias Naturales, Nueva Gerona 09232
Museo Ciencias Naturales Carlos de la Torre, Sancti-Spíritus 09285
Museo Ciudad de Baeza, Baeza 38941
Museo Ciudades del Mundo, La Habana 09128
Museo Civico, Aci Castello 25141
Museo Civico, Agrigento 25162
Museo Civico, Airola 25167
Museo Civico, Alatri 25172
Museo Civico, Amalfi 25214
Museo Civico, Andreis 25233
Museo Civico, Apice 25250
Museo Civico, Argenta 25271
Museo Civico, Ariano Irpino 25275
Museo Civico, Asolo 25302
Museo Civico, Atina 25319
Museo Civico, Avola 25337
Museo Civico, Baranello 25353
Museo Civico, Bardonecchia 25357
Museo Civico, Barga 25360
Museo Civico, Bassano del Grappa 25378
Museo Civico, Belluno 25389
Museo Civico, Bisacquino 25423
Museo Civico, Bojano 25435
Museo Civico, Bojano 25436
Museo Civico, Bolzano 25487
Museo Civico, Borgo Velino 25506
Museo Civico, Bormio 25510
Museo Civico, Busseto 25569
Museo Civico, Cabras 25574
Museo Civico, Caldarola 25598
Museo Civico, Campagnano di Roma 25629
Museo Civico, Campochiaro 25638
Museo Civico, Canneto sull'Oglio 25646
Museo Civico, Caprino Veronese 25667
Museo Civico, Casale Monferrato 25695
Museo Civico, Casamicciola Terme 25701
Museo Civico, Castel Bolognese 25715
Museo Civico, Castelleone 25738
Museo Civico, Castelnovo Bariano 25744
Museo Civico, Castelnuovo Scrivia 25750
Museo Civico, Castelraimondo 25751
Museo Civico, Castroreale 25766
Museo Civico, Centuripe 25803
Museo Civico, Cerchio 25806

M

Museo Civico, Champdepraz 25845
Museo Civico, Cittadella 25911
Museo Civico, Conversano BA 25965
Museo Civico, Cori 25975
Museo Civico, Crema 25999
Museo Civico, Crotone 26013
Museo Civico, Cuneo 26019
Museo Civico, Erba 26058
Museo Civico, Feltre 26094
Museo Civico, Ferentino 26097
Museo Civico, Finale Emilia 26136
Museo Civico, Foggia 26209
Museo Civico, Fondi 26219
Museo Civico, Fornovo San Giovanni 26236
Museo Civico, Fratta Polesine 26249
Museo Civico, Fucecchio 26255
Museo Civico, Gangi 26271
Museo Civico, Garessio 26273
Museo Civico, Gibellina 26337
Museo Civico, Guardiagrele 26384
Museo Civico, Iesi 26394
Museo Civico, Larciano 26460
Museo Civico, Larino 26462
Museo Civico, Lodi 26507
Museo Civico, Lucignano 26539
Museo Civico, Luino 26546
Museo Civico, Maddaloni 26562
Museo Civico, Magliano Sabina 26566
Museo Civico, Manerbio 26586
Museo Civico, Mattinata 26642
Museo Civico, Mazara del Vallo 26643
Museo Civico, Medicina 26647
Museo Civico, Mirandola 26737
Museo Civico, Mondavio 26775
Museo Civico, Monsampolo del Tronto 26786
Museo Civico, Montelparo 26836
Museo Civico, Nocera Umbra 26933
Museo Civico, Noto 26945
Museo Civico, Osimo 27011
Museo Civico, Patrica 27113
Museo Civico, Pegognaga 27130
Museo Civico, Pescia 27171
Museo Civico, Piedimonte Matese 27200
Museo Civico, Pieve di Cento 27212
Museo Civico, Pistoia 27249
Museo Civico, Pizzighettone 27260
Museo Civico, Pontecorvo 27289
Museo Civico, Putignano 27348
Museo Civico, Ragogna 27357
Museo Civico, Recanati 27377
Museo Civico, Rende 27395
Museo Civico, Rieti 27402
Museo Civico, Riva del Garda 27421
Museo Civico, Rovereto 27569
Museo Civico, Salemi 27593
Museo Civico, San Bonifacio 27616
Museo Civico, San Cesario di Lecce 27622
Museo Civico, San Ferdinando di Puglia 27633
Museo Civico, San Gimignano 27637
Museo Civico, San Severo 27687
Museo Civico, Sanremo 27696
Museo Civico, Sansepolcro 27698
Museo Civico, Santa Giusta 27701
Museo Civico, Sesto Calende 27823
Museo Civico, Sulmona 27917
Museo Civico, Susa 27925
Museo Civico, Taverna 27946
Museo Civico, Terrasini 27968
Museo Civico, Tolfa 27991
Museo Civico, Trevignano Romano 28067
Museo Civico, Trinitapoli 28098
Museo Civico, Troia 28102
Museo Civico, Vasanello 28168
Museo Civico, Viterbo 28292
Museo Civico A.E. Baruffaldi, Badia Polesine 25339
Museo Civico A. Giacomelli, Montagnana 26793
Museo Civico A. Klitsche de la Grange, Allumiere 25203
Museo Civico A. Parazzi, Viadana 28242
Museo Civico A. Tubino, Masone 26618
Museo Civico Ala Ponzone, Cremona 26003
Museo Civico Albano, Albano Laziale 25175
Museo Civico Alta Val Brembana, Valtorta 28153
Museo Civico Antiquarium, Nettuno 26921
Museo Civico Antiquarium Platina, Piadena 27190
Museo Civico Antonino Olmo e Gipsoteca, Savigliano 27766
Museo Civico Antonio Cordici, Erice 26062
Museo Civico Archeologica e Quadreria, Monterubbiano 26850
Museo Civico Archeologico, Angera 25235
Museo Civico Archeologico, Anzio 25244
Museo Civico Archeologico, Arona 25281
Museo Civico Archeologico, Barbarano Romano 25354
Museo Civico Archeologico, Bene Vagienna 25393
Museo Civico Archeologico, Bergamo 25401
Museo Civico Archeologico, Bologna 25447
Museo Civico Archeologico, Camerino 25617
Museo Civico Archeologico, Canosa di Puglia 25647
Museo Civico Archeologico, Castelfranco Emilia 25727
Museo Civico Archeologico, Castelleone di Suasa 25739
Museo Civico Archeologico, Castro dei Volsci 25765
Museo Civico Archeologico, Castrovillari 25768
Museo Civico Archeologico, Cavaion Veronese 25787
Museo Civico Archeologico, Chiomonte 25877
Museo Civico Archeologico, Ciró Marina 25899
Museo Civico Archeologico, Cologna Veneta 25943
Museo Civico Archeologico, Concordia Sagittaria 25960
Museo Civico Archeologico, Fiesole 26127
Museo Civico Archeologico, Gazzo Veronese 26281
Museo Civico Archeologico, Gottolengo 26354
Museo Civico Archeologico, Grotte di Castro 26377
Museo Civico Archeologico, Latronico 26468
Museo Civico Archeologico, Marciana 26602
Museo Civico Archeologico, Massa Marittima 26628
Museo Civico Archeologico, Mel 26650
Museo Civico Archeologico, Mondragone 26778
Museo Civico Archeologico, Montalto Marche 26797
Museo Civico Archeologico, Nepi 26919
Museo Civico Archeologico, Nicotera 26925
Museo Civico Archeologico, Norma 26942
Museo Civico Archeologico, Padria 27053
Museo Civico Archeologico, Pitigliano 27255
Museo Civico Archeologico, Portoferraio 27306
Museo Civico Archeologico, Ramacca 27361
Museo Civico Archeologico, Remedello 27394
Museo Civico Archeologico, Roccagloriosa 27429
Museo Civico Archeologico, Rosignano Marittimo 27561
Museo Civico Archeologico, Rutigliano 27581
Museo Civico Archeologico, Salò 27599
Museo Civico Archeologico, Santa Marinella 27708
Museo Civico Archeologico, Sarteano 27749
Museo Civico Archeologico, Tolentino 27984
Museo Civico Archeologico, Treia 28052
Museo Civico Archeologico, Verucchio 28238
Museo Civico Archeologico Arsenio Crespellani, Bazzano 25385
Museo Civico Archeologico C. Cellini, Ripatransone 27414
Museo Civico Archeologico della Val Tenesi, Manerba del Garda 26585
Museo Civico Archeologico di Fregellae, Ceprano 25805
Museo Civico Archeologico e Paleobotanico, Perfugas 27137
Museo Civico Archeologico Eno Bellis, Oderzo 26969
Museo Civico Archeologico Etnologico, Modena 26745
Museo Civico Archeologico F. Saverio Majellaro, Bisceglie 25424
Museo Civico Archeologico Genna Maria, Villanovaforru 28279
Museo Civico Archeologico L. Fantini, Monterenzio 26847
Museo Civico Archeologico M. Petrone, Vieste 28263
Museo Civico Archeologico O. Nardini, Velletri 28171
Museo Civico Archeologico P. Giovio, Como 25951
Museo Civico Archeologico Paleoambientale, Budrio 25559
Museo Civico Archeologico 'Romualdi', Notaresco 26944
Museo Civico Archeologico Sa Domu Nosta, Senorbì 27807
Museo Civico Archeologico Trebula Mutesca, Monteleone Sabino 26834
Museo Civico Archeologico U. Formentini, La Spezia 26433
Museo Civico Archeologico U. Granafei, Mesagne 26666
Museo Civico Arecheologico delle Acque, Chianciano Terme 25850
Museo Civico Aufidenate, Castel di Sangro 25717
Museo Civico Aufidenate De Nino, Alfedena 25195
Museo Civico B. Romano, Termini Imerese 27960
Museo Civico Belliniano, Catania 25772
Museo Civico C. G. Nicastro, Bovino 25522
Museo Civico C. Verri, Biassono 25416
Museo Civico Casa Cavassa, Saluzzo 27602
Museo Civico Castel Nuovo, Napoli 26886
Museo Civico Castello Ursino, Catania 25773
Museo Civico Chiusa, Chiusa 25880
Museo Civico Craveri di Storia Naturale, Bra 25524
Museo Civico D. dal Lago, Valdagno 28137
Museo Civico d'Arte, Modena 26746
Museo Civico d'Arte, Pordenone 27297
Museo Civico d'Arte Antica e Palazzo Madama, Torino 28004
Museo Civico d'Arte Contemporanea, Albissola 25184
Museo Civico d'Arte Contemporanea U. Apollonio, San Martino di Lupari 27662
Museo Civico dei Fossili di Besano, Besano 25409
Museo Civico dei Villaggi Scomparsi, Villa Estense 28272
Museo Civico del Castello, Conegliano 25962
Museo Civico del Marmo, Carrara 25689
Museo Civico del Risorgimento, Bologna 25448
Museo Civico del Risorgimento, Modena 26747
Museo Civico del Sigillo, La Spezia 26434
Museo Civico del Territorio, San Daniele del Friuli 27630
Museo Civico della Ceramica, Nove 26958
Museo Civico della Cultura Materiale, Roseto degli Abruzzi 27560
Museo Civico della Laguna Sud San Francesco Fuori le Mura, Chioggia 25876
Museo Civico della Media Valle del Liri, Sora 27876
Museo Civico della Navigazione Fluviale, Battaglia Terme 25384
Museo Civico della Val Fiorentina Vittorino Cazzetta, Selva di Cadore 27798
Museo Civico delle Armi L. Marzoli, Brescia 25534
Museo Civico delle Arti e Tradizioni Popolari, Micigliano 26671
Museo Civico delle Centuriazioni Romane Padovana, Borgoricco 25507
Museo Civico delle Ceramiche, Santo Stefano di Camastra 27733
Museo Civico delle Scienze, Pordenone 27298
Museo Civico di Archeologia Ligure, Genova 26296
Museo Civico di Arte Contadina, Romagnese 27550
Museo Civico di Arte Moderna, Mombercelli 26771
Museo Civico di Belriguardo, Voghiera 28305
Museo Civico di Merano - Stadtmuseum Meran, Merano 26657
Museo Civico di Palazzo Te, Mantova 26592
Museo Civico di Paleontologia, Empoli 26053
Museo Civico di Paleontologia e Archeologia, Ugento 28115
Museo Civico di Paleontologia e Paletnologia, Borgosesia 25508
Museo Civico di San Francesco, Montefalco 26821
Museo Civico di San Francesco, Montone 26862
Museo Civico di Santa Caterina, Treviso 28069
Museo Civico di Santa Maria della Consolazione, Altomonte 25208
Museo Civico di Schifanoia, Ferrara 26107
Museo Civico di Scienze Naturali, Faenza 26074
Museo Civico di Scienze Naturali, Lovere 26529
Museo Civico di Scienze Naturali, Voghera 28303
Museo Civico di Scienze Naturali E. Caffi, Bergamo 25402
Museo Civico di Scienze Naturali ed Archeologia della Valdinievole, Pescia 27172
Museo Civico di Storia Naturale, Carmagnola 25679
Museo Civico di Storia Naturale, Casalnuovo Monterotaro 25698
Museo Civico di Storia Naturale, Chies d'Alpago 25869
Museo Civico di Storia Naturale, Cittanova 25913
Museo Civico di Storia Naturale, Crocetta del Montello 26009
Museo Civico di Storia Naturale, Ferrara 26108
Museo Civico di Storia Naturale, Milano 26691
Museo Civico di Storia Naturale, Morbegno 26868
Museo Civico di Storia Naturale, Trieste 28089
Museo Civico di Storia Naturale, Verona 28232
Museo Civico di Storia Naturale del Liceo Classico N. Machiavelli, Lucca 26533
Museo Civico di Storia Naturale del Salento, Calimera 25602
Museo Civico di Storia Naturale Giacomo Doria, Genova 26297
Museo Civico di Vignola, Vignola 28269
Museo Civico di Villa Rathgeb, Abano Terme 25135
Museo Civico di Zoologia, Roma 27468
Museo Civico Didattico di Scienze Naturali, Pinerolo 27220
Museo Civico Didattico Gian Andrea Irico, Trino 28099
Museo Civico Diocesano La Castellina, Norcia 26941
Museo Civico Domenico Mambrini, Galeata 26263
Museo Civico Don Queirolo, Vado Ligure 28131
Museo Civico e Antiquarium, Latina 26465
Museo Civico e Archeologico, Locarno 41569
Museo Civico E. Barba, Gallipoli 26268
Museo Civico e d'Arte Sacra, Colle di Val d'Elsa 25937
Museo Civico e della Ceramica, Cerreto Sannita 25813
Museo Civico e Diocesano d' Arte Sacra, Montalcino 26795
Museo Civico E. Nardi, Poggio Mirteto 27269
Museo Civico e Pinacoteca, Asti 25315
Museo Civico e Pinacoteca, Sassoferrato 27762
Museo Civico e Pinacoteca, Urbania 28119
Museo Civico e Pinacoteca A. Vernarecci, Fossombrone 26241
Museo Civico e Pinacoteca E. Lancerotto, Noale 26930
Museo Civico e Pinacoteca P.F. Crociani, Montepulciano 26844
Museo Civico e Risorgimentale Don Giovanni Verità, Modigliana 26760
Museo Civico e Teatro e Serbatoio Romano, Falerone 26083
Museo Civico Ernesto e Teresa della Torre, Treviglio 28066
Museo Civico Etno-Antropologico, Petralia Sottana 27178
Museo Civico Etnografico, Atri 25323
Museo Civico Etnografico, Casale di Scodosia 25694
Museo Civico Etnografico, Stanghella 27898
Museo Civico Etnografico C.G. Franchini, Oleggio 26978
Museo Civico Etnografico La Lessinia, Bosco Chiesanuova 25514
Museo Civico Etnografico M.G. Fontana, Sappada 27737
Museo Civico F. Bodini, Gemonio 26285
Museo Civico F. de Rocco, San Vito al Tagliamento 27688
Museo Civico F. L. Belgiorno, Modica 26758
Museo Civico F. Rittatore Von Willer, Farnese 26087
Museo Civico Fedrigo, Sona 27867
Museo Civico Filippo Meli, Ciminna 25889
Museo Civico G.B. Adriani, Cherasco 25848
Museo Civico G. Sutermeister, Legnano 26482
Museo Civico G. Ugonia, Brisighella 25547
Museo Civico G. Zannato, Montecchio Maggiore 26816
Museo Civico Gaetano Filangieri, Napoli 26887
Museo Civico-Galleria di Palazzo Pretorio, Prato 27334
Museo Civico Geologia e Etnografia, Predazzo 27341
Museo Civico Giacomo Rodolfo, Carignano 25678
Museo Civico Giovanni Fattori, Livorno 26497
Museo Civico Giulio Ferrari, Carpi 25685
Museo Civico Goffredo Bellini, Asola 25301
Museo Civico Gonzaga, Novellara 26959
Museo Civico Il Correggio, Correggio 25980
Museo Civico L. Mallé, Dronero 26048
Museo Civico Luigi Bailo, Treviso 28070
Museo Civico Luigi Dalla Laita, Ala 25168
Museo Civico M. Trucco, Albissola 25185
Museo Civico Marinaro Gio-Bono Ferrari, Camogli 25627
Museo Civico Marinaro U. Mursia, Milano 26692
Museo Civico Medievale, Bologna 25449
Museo Civico Messapico, Alezio 25194
Museo Civico Naturalistico, Maserada sul Piave 26617
Museo Civico Naturalistico dei Monti Prenestini, Capranica Prenestina 25662
Museo Civico Naturalistico Severo Sini, Villa d'Almè 28271
Museo Civico Orientale Santa Cardu, Cagliari 25581
Museo Civico Ornitologico, Lonato 26512
Museo Civico Ornitologico e di Scienze Naturali Luigi Scanagatta, Varenna 28160
Museo Civico P. A. Garda, Ivrea 26422
Museo Civico P. Rosario, Ascoli Satriano 25299
Museo Civico Padre Michele Jacobelli, Casalvieri 25699
Museo Civico Paleontologico, Macerata Feltria 26557
Museo Civico per la Preistoria del Monte Cetona, Cetona 25842
Museo Civico Pietro Lotti, Ischia di Castro 26410
Museo Civico-Pinacoteca, Vicenza 28252
Museo Civico Pio Capponi, Terracina 27967
Museo Civico Polivalente, Mistretta 26738
Museo Civico Preistorico, Pofi 27264
Museo Civico Rocca Flea, Gualdo Tadino 26380
Museo Civico Rogadeo, Bitonto 25430
Museo Civico Speleo-Archeologico, Nuoro 26964
Museo Civico Storico Archeologico, Leno 26483
Museo Civico Storico Artistico, Antrodoco 25242
Museo Civico Storico ed Etnografico, Primaluna 27344
Museo Civico Storico-Etnografico, Ripatransone 27415
Museo Civicodi Scienze Naturali, Randazzo 27363
Museo Civicodi Storia Naturale, Stazzano 27899
Museo Clarisse Coulombie de Goyaud, Ituzaingó 00391
Museo Clemente Rospigliosi, Pistoia 27250
Museo Coahuila y Texas, Monclova 31173
Museo Colección de Veterinaria Militar, Madrid 39528
Museo-Colección del Observatorio Astronómico Nacional, Madrid 39529
Museo-Colección Específico del Suboficial, Tremp 40169
Museo-Colección Específico Militar El Desnarigado, Ceuta 39170
Museo Colecciones ICO, Madrid 39530
Museo Colombiano de Informática, Bogotá 08555
Museo Colonial, El Tocuyo 54569
Museo Colonial Charcas, Sucre 04245
Museo Colonial de Acolman, Acolman 30695
Museo Colonial del Conde de Sierra Gorda, Santander Jiménez 31382
Museo Colonial e Histórico, Luján 00433
Museo Coltzin, Culiacán 30824
Museo Comandancia La Plata, Santo Domingo 09327

M

Museo Comarcal, Zalduondo 40331
Museo Comarcal da Fonsagrada, A Fonsagrada 39302
Museo Comarcal de Arte Sacro de Peñafiel, Peñafiel 39826
Museo Comarcal Velezano Miguel Guirao, Vélez Rubio 40257
Museo Come Eravamo, Borgamanero 25499
Museo Comercial, Hellín 39385
Museo Comico Verscio, Verscio 41981
Museo Comité Olímpico, México 31063
Museo Communale Antiquarium, Atena Lucana 25318
Museo Comparativo de Biología Marina, Viña del Mar 07474
Museo Complesso della Collegiata, Castiglione Olona 25763
Museo Comunal de Chañar Ladeado, Chañar Ladeado 00299
Museo Comunal Pedro Bargero, Emilio Bunge 00364
Museo Comunal Regional de San Guillermo, San Guillermo 00579
Museo Comunale, Montaione 26794
Museo Comunale, Monte San Savino 26805
Museo Comunale, Nova Ponente 26946
Museo Comunale, Praia a Mare 27323
Museo Comunale, Roggiano Gravina 27436
Museo Comunale A. Mendola, Favara 26092
Museo Comunale Archeologico, Viddalba 28262
Museo Comunale d'Arte Contemporanea, Polignano a Mare 27271
Museo Comunale d'Arte Moderna, Ascona 41129
Museo Comunale della Ceramica, Cutrofiano . 26028
Museo Comunale della Manifattura Chini, Borgo San Lorenzo 25504
Museo Comunale della Valle dei Nuraghi, Torralba 28033
Museo Comunale della Valle del Sarno, Sarno 27744
Museo Comunale E. Durio, Civiasco 25915
Museo Comunale La Malaria e la sua Storia, Pontinia 27292
Museo Comunale Storico, Acquaviva delle Fonti 25147
Museo Comunale Trappeto Maratea, Vico del Gargano 28259
Museo Comunidad Las Terrazas, San Cristobal 09276
Museo Comunitario, Mineral del Monte 31168
Museo Comunitario, Puente de Ixtla 31293
Museo Comunitario Altagracia de Arauz, Ensenada 30856
Museo Comunitario Amuzgo de Xochistlahuaca, Xochistlahuaca 31567
Museo Comunitario Colonia Vicente Guerrero, Colonia Vicente Guerrero 30787
Museo Comunitario Concuemitl, Cuencamé ... 30815
Museo Comunitario Coronela Amelia La Güera Robles, Xochipala 31566
Museo Comunitario Crónica y Arte Agropecuario de la Sierra Gorda, Landa de Matamoros 30994
Museo Comunitario Cuarenta Casas, Valentín Gómez Farías 31523
Museo Comunitario de Antropología e Historia, Apatzingán 30718
Museo Comunitario de Atoyac, Atoyac 30727
Museo Comunitario de Carrillo Puerto, Alvaro Obregón 30713
Museo Comunitario de Ciudad Lerdo, Lerdo .. 31002
Museo Comunitario de Estación Coahuila, Mexicali 31032
Museo Comunitario de Historia de San José de Gracia, San José de Gracia 31333
Museo Comunitario de Huitzapula, Huitzapula 30951
Museo Comunitario de Ixtlahuacán, Ixtlahuacán 30962
Museo Comunitario de la Ballena, La Paz 30986
Museo Comunitario de la Sal, Armería 30724
Museo Comunitario de las Varas, Compostela 30796
Museo Comunitario de Pejelagartero, Huimanguillo 30949
Museo Comunitario de Pipillola, Españita 30864
Museo Comunitario de San Andrés Mixquic, México 31064
Museo Comunitario de San Nicolás Zoyatlán, San Nicolás Zoyatlán 31362
Museo Comunitario de San Rafael del Sur, San Rafael del Sur 33306
Museo Comunitario de Santa Eulalia, Santa Eulalia 31379
Museo Comunitario de Santa Martha, Mulegé 31205
Museo Comunitario de Tototepec, Tototepec .. 31507
Museo Comunitario de Villa El Salvador, Villa El Salvador 34325
Museo Comunitario de Villa Regina, Villa Regina 00684
Museo Comunitario de Xalisco, Xalisco 31561
Museo Comunitario de Xolalpan, Ixtacamaxtitlán 30960
Museo Comunitario del Valle de Guadalupe, Ensenada 30857
Museo Comunitario el Asalto a las Tierras, Mexicali 31033
Museo Comunitario El Rosario, El Rosario 30850
Museo Comunitario Felipe Carrillo Puerto, Motul 31203
Museo Comunitario Francisco I. Madero, Francisco I. Madero, Coahuila 30867
Museo Comunitario General Francisco Villa, Durango 30837
Museo Comunitario Hicupa, Tehuacán 31418
Museo Comunitario Hitalulu, San Martín Huamelulpan 31354
Museo Comunitario Huitza Chilin, San Pedro . 31366
Museo Comunitario Iluikatlachiyalistli, Yahualica 31570
Museo Comunitario Inti Raymi, Lima 34228
Museo Comunitario Ismael Girón González, Pánuco, Zacatecas 31250
Museo Comunitario Itzmal Kauil, Izamal 30964
Museo Comunitario Ji, Ocosingo 31232
Museo Comunitario Jna Niingui, San Miguel Tequixtepec 31359
Museo Comunitario Joyonaque, Tuxtla Gutiérrez 31512
Museo Comunitario Juan García Aldama de la Comunidad Cucapá, El Mayor Indígena 30848
Museo Comunitario Kumkuy Ys Untzi, Copainalá 30799
Museo Comunitario Maika, Poanas 31263
Museo Comunitario Note Ujia, Santa María Asunción Tlaxiaco 31380
Museo Comunitario Nu-kuiñe, Cuquila 30830
Museo Comunitario Raramuri, Guachochi 30882
Museo Comunitario Rubén Jaramillo, Tlaquiltenango 31475
Museo Comunitario Serafín Olarte, Papantla .. 31251
Museo Comunitario Shan-Dany, Santa Ana del Valle 31372
Museo Comunitario sin Paredes Colonia Orizaba, Mexicali 31034
Museo Comunitario Snuuvico, San Juan Mixtepec Juxtlahuaca 31342
Museo Comunitario Tejamen, Nuevo Ideal 31218
Museo Comunitario Tzakualli, Zacualpan 31587
Museo Comunitario Unidad Indígena Emiliano Zapata, Hueyapan de Ocampo 30947
Museo Comunitario Unión y Progreso, Cuatrocienegas 30812
Museo Comunitario Xiximes, Gómez Palacio .. 30878
Museo Comunitario Xolalpan Calli, Hueyapan 30946
Museo Comunitario Ya Nfädi Yu Nohño Los Conocimientos de los Otomíes, Tolimán 31483
Museo Comunitario Yucu-Iti, Santa María Yucu-Iti 31381
Museo Comunitario Ze Acatl Topiltzin Quetzalcoatl, Tepoztlán 31446
Museo Concha Ferrant, La Habana 09129
Museo Conchiliologico e della Carta Moneta, Bellaria Igea Marina 25388
Museo Contadino, Montalto Pavese 26799
Museo Contadino, Varese Ligure 28166
Museo Contadino della Bassa Pavese, San Cristina e Bissone 27629
Museo Contadino Foiano - Bauernmuseum Völlan, Lana 26445
Museo Contalpa de Nombre de Dios, Nombre de Dios 31217
Museo Contemporaneo, Montelupo Fiorentino . 26838
Museo Contisuyo, Moquegua 34278
Museo Convento de San Diego, Quito 10432
Museo Convento de Santa Clara, Salamanca . 39937
Museo Convento del Desierto de la Candelaria, Ráquira 08736
Museo Convento Santo Ecce Homo, Sutamarchán 08766
Museo Conventual de las Carmelitas Descalzas de Antequera, Antequera 38874
Museo Conventual de Santa Paula, Sevilla 40067
Museo Cooperativo Eldorado, Eldorado 00361
Museo Coronel Félix Luque Plata, Guayaquil . 10402
Museo Correale di Terranova, Sorrento 27881
Museo Correr e Biblioteca d'Arte e Storia Veneziana, Venezia 28188
Museo Costumbrista de Sonora, Alamos 30709
Museo Costumbristá Juan de Vargas, La Paz 04212
Museo Coyug-Curá, Pigüé 00492
Museo Criollo de los Corrales, Buenos Aires .. 00176
Museo Cristero Ing. Efrén Quezada, Encarnación de Díaz, Jalisco 30854
Museo Cristero Señor Cura Cristóbal Magallanes, Totatiche 31504
Museo Cristóbal Mendoza, Trujillo 54595
Museo Cruz Herrera, La Línea de la Concepción 39464
Museo Cuadra de Bolívar, Caracas 54538
Museo Cuartel de Emiliano Zapata, Tlaltizapán 31470
Museo Cuartel Zapatista, México 31065
Museo Cuevas de Las Maravillas, La Romana 10333
Museo Cultura Antigua Ribera del Bernesga, Cuadros 39229
Museo Cultura e Musica Popolare dei Peloritani, Messina 26667
Museo Cultura y Tradición de Jalcomulco, Jalcomulco 30966
Museo Cultural del Instituto Geografico Militar, Quito 10433
Museo Cultural Iijdio Guaranga, Guaranga 10399
Museo D. Agostinelli, Roma 27469
Museo d' Arte e Scienza, Milano 26693
Museo D. Chalonge, Erice 26063
Museo da Terra de Melide, Melide 39640
Museo Dantesco, Ravenna 27371
Museo d'Art Contemporani, Elche 39265
Museo d'Arte Antica, Milano 26694
Museo d'Arte Contemporanea D. Formaggio, Teolo 27954
Museo d'Arte Contemporanea di Roma - MACRO, Roma 27470
Museo d'Arte Contemporanea di Villa Croce, Genova 26298
Museo d'Arte Costantino Barbella, Chieti 25873
Museo d'Arte d'Ammobigliamento, Stupinigi . 27915
Museo d'Arte della Città di Ravenna, Ravenna 27372
Museo d'Arte dello Splendore, Giulianova 26346
Museo d'Arte e Ammobiliamento, Nichelino ... 26924
Museo d'Arte e Archeologia I. Mormino, Palermo 27067
Museo d'Arte e Storia Antica Ebraica, Casale Monferrato 25696
Museo d'Arte G. Bargellini, Pieve di Cento 27213
Museo d'Arte Mendrisio, Mendrisio 41625
Museo d'Arte Moderna, Brescia 25535
Museo d'Arte Moderna, Lugano 41583
Museo d'Arte Moderna, Tertenia 27969
Museo d'Arte Moderna di Bologna - MAMbo, Bologna 25450
Museo d'Arte Moderna e Contemporanea, Ferrara 26109
Museo d'Arte Moderna e Contemporanea di Trento e Rovereto, Rovereto 27570
Museo d'Arte Moderna e Contemporanea di Trento e Rovereto, Trento 28056
Museo d'Arte Moderna e Contemporanea - MAM, Gazoldo degli Ippoliti 26279
Museo d'Arte Moderna Mario Rimoldi, Cortina d'Ampezzo 25984
Museo d'Arte Moderna Pagani, Castellanza 25734
Museo d'Arte Naturale, Gioiosa Ionica 26342
Museo d'Arte Nuoro, Nuoro 26965
Museo d'Arte Orientale, Venezia 28189
Museo d'Arte Orientale Edoardo Chiossone, Genova 26299
Museo d'Arte P. Pini, Milano 26695
Museo d'Arte Pietro Cavoti, Galatina 26262
Museo d'Arte Preistorica, Pinerolo 27221
Museo d'Arte S. Pertini, Savona 27776
Museo d'Arte Sacra, Asciano 25291
Museo d'Arte Sacra, Baiardo 25350
Museo d'Arte Sacra, Camaiore 25613
Museo d'Arte Sacra, Chiaramonte Gulfi 25852
Museo d'Arte Sacra, Gradoli 26360
Museo d'Arte Sacra, Montespertoli 26855
Museo d'Arte Sacra, Ponzone 27294
Museo d'Arte Sacra, San Casciano in Val di Pesa 27621
Museo d'Arte Sacra, San Gimignano 27638
Museo d'Arte Sacra, San Giovanni in Persiceto 27646
Museo d'Arte Sacra, San Leo 27653
Museo d'Arte Sacra, Tavarnelle Val di Pesa 27945
Museo d'Arte Sacra, Viterbo 28293
Museo d'Arte Sacra Cardinal Agnifili, Rocca di Mezzo 27428
Museo d'Arte Sacra della Insigne Collegiata di San Lorenzo, Montevarchi 26856
Museo d'Arte Sacra della Marsica, Celano 25797
Museo d'Arte Sacra della Santa Maria Assunta, Gallarate 26265
Museo d'Arte Sacra della Val d'Arbia, Buonconvento 25564
Museo d'Arte Sacra L. Acquarone, Lucinasco. 26540
Museo d'Arte Sacra San Marco, Mombaroccio 26768
Museo d'Arte Sacra Silvestro Frangipane, Zagarise 28314
Museo d'Arte Sacro San Martino, Alzano Lombardo 25213
Museo d'Arte Siamese Stefano Gardu, Cagliari 25582
Museo d'Arti e Mestieri Antichi, Montelupone 26840
Museo das Mariñas, Betanzos 39033
Museo das Peregrinacións, Santiago de Compostela 40018
Museo Davia Bargellini, Bologna 25451
Museo David Paltán, San Gabriel 10460
Museo David Paltan, Tulcán 10462
Museo David Ramírez Lavoignet, Misantla 31169
Museo de Abancay, Abancay 34105
Museo de Academia de San Carlos, México ... 31066
Museo de Acaponeta, Acaponeta 30690
Museo de Aduanas y Puerto, Buenos Aires 00177
Museo de Aeronáutica y Astronáutica, Madrid 39531
Museo de Aguascalientes, Aguascalientes 30698
Museo de Ahumada, Villa Ahumada y Anexas .. 31545
Museo de Akil Uyotoch Cah, Akil 30708
Museo de Albacete, Albacete 38803
Museo de Albarracín, Albarracín 38807
Museo de Alcalá la Real, Alcalá la Real 38816
Museo de Alfarería Tradicional Vasca, Elosu . 39274
Museo de Almería, Almería 38861
Museo de Ámbar Dominicano, San Felipe de Puerto Plata 10334
Museo de Ambiente Histórico Cubano, Santiago de Cuba 09309
Museo de Ameca, Vetagrande 31538
Museo de América, Madrid 39532
Museo de Anatomía, México 31067
Museo de Animales Venenosos, Buenos Aires 00178
Museo de Antioquia, Medellín 08675
Museo de Antofagasta, Antofagasta 07429
Museo de Antropolocía Forense, Paleopatología y Criminalística Profesor Reverte, Madrid 39533
Museo de Antropología, Barranquilla 08539
Museo de Antropología, Córdoba 00321
Museo de Antropología, Ibarra 10413
Museo de Antropología, Tecolotlán 31413
Museo de Antropología David J. Guzmán, San Salvador 10549
Museo de Antropología de Tenerife, La Laguna de Tenerife 39438
Museo de Antropología de Xalapa, Xalapa 31557
Museo de Antropología e Historia, Maracay ... 54584
Museo de Antropología e Historia de Allende, Allende 30711
Museo de Antropología e Historia de Rivas, Rivas 33301
Museo de Antropología e Historia de Santiago Papasquiaro, Santiago Papasquiaro 31384
Museo de Antropología e Historia Natural Los Desmochados, Casilda 00296
Museo de Antropología y Ciencias Naturales, Concordia 00315
Museo de Antropología y Municipal de Boaco, Juigalpa 33281
Museo de Anzoátegui, Barcelona 54521
Museo de Armas de la Nación, Buenos Aires .. 00179
Museo de Armería de Álava, Vitoria-Gasteiz ... 40320
Museo de Arqueología, Acapulco 30691
Museo de Arqueología, La Habana 09130
Museo de Arqueología, Ixtepec 30961
Museo de Arqueología Colonial Rodolfo I. Bravo, Cafayate 00286
Museo de Arqueología de Álava, Vitoria-Gasteiz 40321
Museo de Arqueología de Córdoba, Córdoba . 30800
Museo de Arqueología de la Universidad Nacional de Trujillo, Trujillo 34318
Museo de Arqueología de la Universidad Nacional de Tucumán, San Miguel de Tucumán 00595
Museo de Arqueología de León, León 30997
Museo de Arqueología de Occidente de México, Guadalajara 30891
Museo de Arqueología de Salta, Salta 00554
Museo de Arqueología e Historia, Melilla 31683
Museo de Arqueología e Historia Natural de Yungay, Yungay 34330
Museo de Arqueología el Chamizal, Juárez ... 30980
Museo de Arqueología Josefina Ramos de Cox, Lima 34229
Museo de Arqueología Samuel Humberto Espinoza Lozano, Huaytará 34185
Museo de Arqueología y Antropología, Lima . 34230
Museo de Arqueometalurgia de Cerro Muriano, Obejo 39721
Museo de Arquitectura, Caracas 54539
Museo de Arquitectura Leopoldo Rother, Bogotá 08556
Museo de Arte, Santa Marta 08749
Museo de Arte Abstracto del Sur, Buenos Aires 00180
Museo de Arte Abstracto Español, Cuenca 39231
Museo de Arte Abstracto Manuel Felguérez, Zacatecas 31577
Museo de Arte Alberto Mena Caamaño, Quito 10434
Museo de Arte Americano de Maldonado, Maldonado 54379
Museo de Arte Carrillo Gil, México 31068
Museo de Arte Colonial, Antigua 23289
Museo de Arte Colonial, Bogotá 08557
Museo de Arte Colonial, Caracas 54540
Museo de Arte Colonial, Mérida 54586
Museo de Arte Colonial, Morelia 31193
Museo de Arte Colonial, Quito 10435
Museo de Arte Colonial de la Habana, La Habana 09131
Museo de Arte Colonial de Palacio Valle Iznaga, Sancti-Spíritus 09286
Museo de Arte Colonial de San Francisco, Santiago de Chile 07452
Museo de Arte Colonial Pedro de Osma, Lima 34231

M

Museo de Arte Contemporánea de Vigo, Vigo 40274
Museo de Arte Contemporáneo, Bahía Blanca 00133
Museo de Arte Contemporáneo, Bogotá 08558
Museo de Arte Contemporáneo, Ciudad de Panamá 34067
Museo de Arte Contemporáneo, Córdoba 00322
Museo de Arte Contemporáneo, Cusco 34156
Museo de Arte Contemporáneo, Montevideo 54396
Museo de Arte Contemporáneo, San Salvador 10550
Museo de Arte Contemporáneo, Santiago de Chile 07453
Museo de Arte Contemporáneo, Toledo 40131
Museo de Arte Contemporaneo, Valdivia 07468
Museo de Arte Contemporáneo al Aire Libre, Aracena 38877
Museo de Arte Contemporáneo Alfredo Zalce, Morelia 31194
Museo de Arte Contemporáneo Ángel Miguel de Arce, Sasamón 40043
Museo de Arte Contemporáneo Ángel Zárraga, Durango 30838
Museo de Arte Contemporáneo Ateneo de Yucatán, Mérida 31024
Museo de Arte Contemporáneo Beato Angélico, La Plata 00409
Museo de Arte Contemporáneo de Caracas Sofía Imber, Caracas 54541
Museo de Arte Contemporáneo de Chihuahua, Chihuahua 30767
Museo de Arte Contemporáneo de Monterrey, Monterrey 31178
Museo de Arte Contemporáneo de Oaxaca, Oaxaca 31221
Museo de Arte Contemporáneo de Puerto Rico, San Juan 36066
Museo de Arte Contemporáneo Esteban Vicente, Segovia 40052
Museo de Arte Contemporáneo Francisco Narváez, Porlamar 54593
Museo de Arte Contemporáneo José María Moreno Galván, La Puebla de Cazalla 39874
Museo de Arte Contemporáneo Moderno de Ibagué, Ibagué 08659
Museo de Arte Contemporáneo Número 8, Aguascalientes 30699
Museo de Arte Contemporáneo Pajar de Agustín, Valle de Hecho 40247
Museo de Arte Contemporáneo Union Fenosa - MACUF, A Coruña 39218
Museo de Arte Coro y Museo Alberto Henríquez, Falcón 54570
Museo de Arte Costarricense, San José 08815
Museo de Arte de Caguas, Caguas 36031
Museo de Arte de Lima, Lima 34232
Museo de Arte de Matanzas, Matanzas 09213
Museo de Arte de Mazatlán, Mazatlán 31021
Museo de Arte de Orizaba y del Estado de Veracruz, Orizaba 31238
Museo de Arte de Pereira, Pereira 08719
Museo de Arte de Ponce, Ponce 36049
Museo de Arte de Puerto Rico, San Juan 36067
Museo de Arte de Querétaro, Querétaro 31299
Museo de Arte de Sinaloa, Culiacán 30825
Museo de Arte e Historia, Durango 39256
Museo de Arte e Historia, Lima 34233
Museo de Arte e Historia Alberto Mena Caamaño, Quito 10436
Museo de Arte e Historia Casa de Los Celis, Valencia 54598
Museo de Arte e Historia de la Ciudad "Jorge y Elisa", Cuacutecuta 08630
Museo de Arte e História de Los Toldos, Los Toldos 00431
Museo de Arte E. Minnicelli, Río Gallegos 00530
Museo de Arte en Ciudad Juárez, Juárez 30981
Museo de Arte Español Enrique Larreta, Buenos Aires 00181
Museo de Arte Francisco Oller, Bayamón 36029
Museo de Arte Gauchesco La Recova, San Antonio de Areco 00564
Museo de Arte Gráficas Juan Pablos, México 31069
Museo de Arte Guillermo Ceniceros, Durango 30839
Museo de Arte Habana, La Habana 09132
Museo de Arte Hispanoamericano Isaac Fernández Blanco, Buenos Aires 00182
Museo de Arte, Historia y Cultura Casa Alonso, Vega Baja 36080
Museo de Arte Italiano, Lima 34234
Museo de Arte José Luis Bello y González, Puebla 31271
Museo de Arte José Luis Bello y Zetina, Puebla 31272
Museo de Arte Moderno, Cuenca 10385
Museo de Arte Moderno, Mérida 54587
Museo de Arte Moderno, México 31070
Museo de Arte Moderno, Santiago de Chile 07454
Museo de Arte Moderno, Santo Domingo 10343
Museo de Arte Moderno de Barranquilla, Barranquilla 08540
Museo de Arte Moderno de Bogotá, Bogotá 08559
Museo de Arte Moderno de Bogotá II, Bogotá 08560
Museo de Arte Moderno de Bucaramanga, Bucaramanga 08599
Museo de Arte Moderno de Buenos Aires, Buenos Aires 00183
Museo de Arte Moderno de Cartagena, Cartagena 08618
Museo de Arte Moderno de Durango, Gómez Palacio 30879
Museo de Arte Moderno de la Casa de la Cultura Ecuatoriana, Quito 10437
Museo de Arte Moderno de Medellín, Medellín 08676
Museo de Arte Moderno del Estado de México, Toluca 31484
Museo de Arte Moderno Jesús Soto, Ciudad Bolívar 54563
Museo de Arte Moderno La Tertulia, Cali 08608
Museo de Arte Moderno Ramírez Villamizar, Pamplona 08706
Museo de Arte Olga Costa, Guanajuato 30913
Museo de Arte Oriental, Ávila 38919
Museo de Arte Polular, México 31071
Museo de Arte Popular Americano, Santiago de Chile 07455
Museo de Arte Popular de la Universidad, Puebla 31273
Museo de Arte Popular de la Universidad Nacional San Cristóbal de Huamanga, Ayacucho 34119
Museo de Arte Popular de Quinua, Quinua 34300
Museo de Arte Popular de San Bartolo de Coyotepec, San Baltazar Chichicapan 31322
Museo de Arte Popular José Hernández, Buenos Aires 00184
Museo de Arte Popular Petare, Caracas 54542
Museo de Arte Popular Poblano y Museo de Artesanías, Puebla 31274
Museo de Arte Popular Tlajomulco, Tlajomulco de Zúñiga 31467
Museo de Arte Popular Venezolano, Juangriego 54576
Museo de Arte Precolombino, Cusco 34157
Museo de Arte Prehispánico, Mascota 31015
Museo de Arte Prehispánico Carlos Pellicer, Tepoztlán 31447
Museo de Arte Prehispánico de México Rufino Tamayo, Oaxaca 31222
Museo de Arte Religioso, La Habana 09133
Museo de Arte Religioso, Huaquechula 30941
Museo de Arte Religioso Cardenal Luque, Tenjo 08768
Museo de Arte Religioso Colonial, Ciudad de Panamá 34068
Museo de Arte Religioso Colonial, Comayagua 23326
Museo de Arte Religioso Concepción, Riobamba 10454
Museo de Arte Religioso Convento de San Francisco, Cajamarca 34128
Museo de Arte Religioso de Girón, Girón 08647
Museo de Arte Religioso de la Catedral, Lima 34235
Museo de Arte Religioso de la Catedral de San Nicolás, Rionegro 08738
Museo de Arte Religioso de Monguí, Monguí 08690
Museo de Arte Religioso de Piura, Piura 34292
Museo de Arte Religioso de San Gil, Atienza 38915
Museo de Arte Religioso de Santa Mónica, Puebla 31275
Museo de Arte Religioso Ex Convento Franciscano, Huejotzingo 30944
Museo de Arte Religioso i Palenteológico de San Bartolomé, Atienza 38916
Museo de Arte Religioso Juan de Tejeda, Córdoba 00323
Museo de Arte Religioso Julio Franco Arango, Duitama 08633
Museo de Arte Religioso la Concepción, Baños 10374
Museo de Arte Religioso Presbítero Antonio María Franco, Supía 08765
Museo de Arte Religioso Santo Domingo de Porta Coeli, San Germán 36057
Museo de Arte Religioso Tiberio de J. Salazar y Herrera, Sonsón 08762
Museo de Arte Religioso y Colonial, Cali 08609
Museo de Arte Religoso, Cusco 34158
Museo de Arte Sacra das Clarisas, Monforte de Lemos 39664
Museo de Arte Sacra de Iria Flavia, Padrón 39765
Museo de Arte Sacra de Santa Clara, Allariz 38848
Museo de Arte Sacra de Santa María do Campo, A Coruña 39219
Museo de Arte Sacra de Vilanova, Lourenzá 39495
Museo de Arte Sacro, Betancuria 39032
Museo de Arte Sacro, Castrojeriz 39154
Museo de Arte Sacro, Chihuahua 30768
Museo de Arte Sacro, Huete 39402
Museo de Arte Sacro, León 33285
Museo de Arte Sacro, Osuna 39753
Museo de Arte Sacro, Santiago de Compostela 40019
Museo de Arte Sacro de La Rioja, La Rioja 00417
Museo de Arte Sacro de Tineo, Tineo 40123
Museo de Arte Sacro San Francisco, Salta 00555
Museo de Arte Sacro San Miguel, Oruro 04231
Museo de Arte Sacro Santiago de los Caballeros, Gáldar 39320
Museo de Arte San Miguel Arcángel, Mula 39689
Museo de Arte Taino, San Felipe de Puerto Plata 10335
Museo de Arte Universidad Nacional de Colombia, Bogotá 08561
Museo de Arte Virreinal de Santa Teresa, Arequipa 34110
Museo de Arte Virreinal de Taxco, Taxco de Alarcón 31403
Museo de Arte y Artesanía de Linares, Linares 07441
Museo de Arte y Diseño Contemporáneo, San José 08816
Museo de Arte y Tradiciones Populares, Bogotá 08562
Museo de Artes Decorativas de Palacio Salcines, Guantánamo 09091
Museo de Artes e Industrias Populares, Pátzcuaro 31255
Museo de Artes Gráficas, Bogotá 08563
Museo de Artes Plásticas, Madonado 54378
Museo de Artes Plásticas Eduardo Sívori, Buenos Aires 00185
Museo de Artes Populares, Tepic 31441
Museo de Artes Visuales Aramara, Tepic 31442
Museo de Artes y Artesanías Enrique Estrada Bello, Santo Tomé 00632
Museo de Artes y Costumbres Populares, Málaga 39598
Museo de Artes y Costumbres Populares, Pedrafita do Cebrero 39819
Museo de Artes y Costumbres Populares, Sevilla 40068
Museo de Artes y Costumbres Populares e internacional de Arte Naïf, Jaén 39421
Museo de Artes y Oficios, Monóvar 39668
Museo de Artes y Tradiciones Populares, Espinal 08639
Museo de Artes y Tradiciones Populares, Lima 34236
Museo de Artes y Tradiciones Populares, Madrid 39534
Museo de Artes y Tradiciones Populares, San Sebastián de los Reyes 39973
Museo de Artes y Tradiciones Populares de Ráquira, Ráquira 08737
Museo de Artes y Tradiciones Populares de Tlaxcala, Tlaxcala 31476
Museo de Artesanía Iberoamericana, La Orotava 39749
Museo de Artesanias Cefa, Quito 10438
Museo de Artesanías del Mundo, Silvia 08755
Museo de Artesanías Tradicionales Folklóricas, Corrientes 00346
Museo de Astronomía y Geodesia, Madrid 39535
Museo de Atoyac, Atoyac de Alvarez 30728
Museo de Autos y del Transporte Humberto Lobo Villarreal, Monterrey 31179
Museo de Ávila, Ávila 38920
Museo de Aviones Históricos en Vuelo, Madrid 39536
Museo de Ayutla, Ayutla de los Libres 30730
Museo de Bahía de Cochinos, Playa Girón 09256
Museo de Baltasar Lobo, Zamora 40333
Museo de Baracoa, Baracoa 09030
Museo de Bardados del Paso Azul, Lorca 39493
Museo de Barquisimeto, Barquisimeto 54524
Museo de Batopilas, Francisco I. Madero, Coahuila 30868
Museo de Belas Artes da Coruña, A Coruña 39220
Museo de Bellas Artes, Badajoz 38935
Museo de Bellas Artes, Comayagüela 23327
Museo de Bellas Artes, Córdoba 39205
Museo de Bellas Artes, Valencia 40204
Museo de Bellas Artes, Viña del Mar 07475
Museo de Bellas Artes Agustin Araujo, Treinta y Tres 54436
Museo de Bellas Artes al Aire Libre Caminito, Buenos Aires 00186
Museo de Bellas Artes Bonaerense, La Plata 00410
Museo de Bellas Artes Ceferino Carnacini, Villa Ballester 00671
Museo de Bellas Artes Claudio León Sempere, Burzaco 00284
Museo de Bellas Artes de Álava, Vitoria-Gasteiz 40322
Museo de Bellas Artes de Asturias, Oviedo 39759
Museo de Bellas Artes de Caracas, Caracas 54543
Museo de Bellas Artes de Castellón, Castellón de la Plana 39148
Museo de Bellas Artes de Granada, Granada 39365
Museo de Bellas Artes de Gravina, Alicante 38844
Museo de Bellas Artes de la Boca Quinquela Martín, Buenos Aires 00187
Museo de Bellas Artes de Murcia, Murcia 39692
Museo de Bellas Artes de Santander, Santander 40009
Museo de Bellas Artes de Sevilla, Sevilla 40069
Museo de Bellas Artes de Toluca, Toluca 31485
Museo de Bellas Artes Departamental de San José, San José de Mayo 54434
Museo de Bellas Artes Domingo Faustino Sarmiento, Mercedes 00460
Museo de Bellas Artes Franklin Rawson, San Juan 00586
Museo de Bellas Artes Juan Yapari, Posadas 00497
Museo de Bellas Artes Laureano Brizuela, San Fernando del Valle de Catamarca 00572
Museo de Bellas Artes Octavio de la Colina, La Rioja 00418
Museo de Bermejillo, Bermejillo 30736
Museo de Bomberos, Madrid 39537
Museo de Bordados Paso Blanco, Lorca 39494
Museo de Buenos Aires, Tacotalpa 31394
Museo de Burgos, Burgos 39068
Museo de Cáceres, Cáceres 39082
Museo de Cádiz, Cádiz 39089
Museo de Calcos y Esculturas Comparda Ernesto de La Cárcova, Buenos Aires 00188
Museo de Calzado, Arnedo 38892
Museo de Campeche, Campeche 30740
Museo de Caracas, Caracas 54544
Museo de Carmen Conde y Antonio Oliver, Cartagena 39137
Museo de Carruajes de Sevilla, Sevilla 40070
Museo de Casa Popenoe, Antigua 23290
Museo de Caza Benito Albarrán, Zapopan 31589
Museo de Centro Cultural Recoleta, Buenos Aires 00189
Museo de Cera, Guadalajara 30892
Museo de Cera, Madrid 39538
Museo de Cera de la Ciudad de México, México 31072
Museo de Cera de la Villa, México 31073
Museo de Cera de Tijuana, Tijuana 31456
Museo de Cerámica de la Cartuja-Pickman, Salteras 39950
Museo de Cerámica La Peñuela, Chinchilla de Monte-Aragón 39176
Museo de Cerámica Ruíz de Luna, Talavera de la Reina 40095
Museo de Cerámica y Bellas Artes Julián de la Herreria, Asunción 34092
Museo de Cerámica y Loza Popular, Falcón 54571
Museo de Ceuta, Ceuta 31662
Museo de Chilecito, Chilecito 00302
Museo de Ciencia, Cuernavaca 30816
Museo de Ciencia y Juego, Santa Marta 08750
Museo de Ciencia y Técnica, Buenos Aires 00190
Museo de Ciencia y Tecnología del Estado de Veracruz, Xalapa 31558
Museo de Ciencias, Caracas 54545
Museo de Ciencias, Huánuco 34179
Museo de Ciencias Antropológicas y Naturales, Santiago del Estero 00628
Museo de Ciencias de Ensenada, Ensenada 30858
Museo de Ciencias Explora, León 30998
Museo de Ciencias Morfológicas, Rosario 00543
Museo de Ciencias Naturales, Arnedo 38893
Museo de Ciencias Naturales, Asunción 34093
Museo de Ciencias Naturales, Coronel Pringles 00343
Museo de Ciencias Naturales, Cusco 34159
Museo de Ciencias Naturales, Managua 33288
Museo de Ciencias Naturales, Pamplona 39807
Museo de Ciencias Naturales, Salta 00556
Museo de Ciencias Naturales, Santa Cruz de Tenerife 39995
Museo de Ciencias Naturales, Toluca 31486
Museo de Ciencias Naturales, Valverde de la Virgen 40252
Museo de Ciencias Naturales, Villafranca del Bierzo 40300
Museo de Ciencias Naturales Augusto G. Schulz, Resistencia 00522
Museo de Ciencias Naturales Bartolomé Mitre, Córdoba 00324
Museo de Ciencias Naturales Carlos de la Torre, Holguín 09179
Museo de Ciencias Naturales de Álava, Vitoria-Gasteiz 40323
Museo de Ciencias Naturales de Atarfe, Atarfe 38913
Museo de Ciencias Naturales de Especialidad Zoológica Médica, Buenos Aires 00191
Museo de Ciencias Naturales de Oberá, Oberá 00479
Museo de Ciencias Naturales del Colegio de San José de La Salle, Medellín 08677
Museo de Ciencias Naturales del Departamento San Cristóbal, San Cristóbal 00570
Museo de Ciencias Naturales Dr. Amado Bonpland, Corrientes 00347
Museo de Ciencias Naturales e Historia, Posadas 00498
Museo de Ciencias Naturales Federico Carlos Lehmann, Cali 08610
Museo de Ciencias Naturales Florentino Ameghino, Las Flores 00424
Museo de Ciencias Naturales Luma, Tanti 00643

Museo de Ciencias Naturales Rvdo. P. Antonio Scasso, San Nicolás de los Arroyos 00603
Museo de Ciencias Naturales Tomás Romay, Santiago de Cuba 09310
Museo de Ciencias Naturales y Antropológicas Juan Cornelio Moyano, Mendoza 00453
Museo de Ciencias Naturales y Antropológicas Prof. Antonio Serrano, Paraná 00484
Museo de Ciencias Naturales y Misional del Colegio San José, Esperanza 00366
Museo de Ciencias y Antropología, Cartago 08622
Museo de Ciudad Bolívar, Ciudad Bolívar 54564
Museo de Ciudad Real, Ciudad Real 39187
Museo de Ciudades Hermanas, Guadalajara 30893
Museo de Coamo, Coamo 36036
Museo de Cochoapa, Ometepec 31237
Museo de Cocula es el Mariachi, Cocula 30778
Museo de Colexiata do Sar, Santiago de Compostela 40020
Museo de Colón, Colón 30785
Museo de Condega, Condega 33274
Museo de Conkal, Conkal 30798
Museo de Cosautlán de Carvajal, Cosautlán de Carvajal 30804
Museo de Coscomatepec, Coscomatepec de Bravo 30805
Museo de Costumbres y Artes Populares de los Montes, Guadamur 39377
Museo de Covadonga, Santuario de Covadonga 40042
Museo de Criminalística de la Policía Nacional del Perú, Lima 34237
Museo de Cuajinicuilapa, Cuajinicuilapa 30809
Museo de Cuenca, Cuenca 39232
Museo de Cultura Popular Angel Gil Hermidas, Villahermosa 31550
Museo de Cultura Popular Costa Rica, Heredia 08809
Museo de Cultura Populares del Estado de México, Toluca 31487
Museo de Culturas Populares, Durango 30840
Museo de Culturas Populares de Chiapas, San Cristóbal de las Casas 31324
Museo de Culturas Populares de San Luis Potosí, San Luis Potosí 31345
Museo de Culturas Populares e Indígenas, Obregón 31229
Museo de Culturas Populares e Indígenas de Sonora, Hermosillo 30929
Museo de Descubrimiento, Montevideo 54397
Museo de Dibujo "Julio Gavín-Castillo de Larrés", Larrés (Huesca) 39444
Museo de Dinosaurios, Salas de los Infantes 39945
Museo de Duarte, Santo Domingo 10344
Museo de Ejército Oriental, Holguín 09180
Museo de Emiliano Zapata, Emiliano Zapata 30852
Museo de Escritores Bolivianos, La Paz 04213
Museo de Escuela de Educación Técnica Número 1, Dolores 00354
Museo de Escultura al Aire Libre, Madrid 39539
Museo de Escultura al Aire Libre de Leganés, Leganés 39448
Museo de Escultura Jacinto Higueras, Santisteban del Puerto 40039
Museo de Escultura Novarro Santafé, Villena 40312
Museo de Esculturas Luis Perlotti, Buenos Aires 00192
Museo de Etnografía, Quito 10439
Museo de Etnohistoria de Suyo, Suyo 34309
Museo de Farmacia, Burgos 39069
Museo de Farmacia, Llívia 39485
Museo de Farmacia, Peñaranda de Duero 39832
Museo de Farmacia Militar, Madrid 39540
Museo de Ferrocarriles en Miniatura Werner Dura, La Falda 00401
Museo de Filatelia de Oaxaca, Oaxaca 31223
Museo de Filigrana de Cobre, La Esperanza 34199
Museo de Firmat, Firmat 00372
Museo de Física de la Universidad Nacional de La Plata, La Plata 00411
Museo de Física Recreativa, Puebla 31276
Museo de Fósiles, Igea 39407
Museo de Frank País, Holguín 09181
Museo de Galera, Galera 39322
Museo de Gendarmería Nacional, Buenos Aires 00193
Museo de General Zaragoza, General Zaragoza 30876
Museo de Geología, Managua 33289
Museo de Geología, Manizales 08668
Museo de Geología de la UNAM, México 31074
Museo de Geología del IPN, México 31075
Museo de Geología Dr. Jenaro González Reyna, Morelia 31195
Museo de Geologia, Paleontologia i Prehistòria del Bruc, El Bruc 39058
Museo de Geología y Paleontología, Neuquén 00477
Museo de Geología y Suelos, Armenia 08533
Museo de Gomellano, Gumiel de Hizán 39383
Museo de Granjita Siboney, Santiago de Cuba 09311
Museo de Guadalajara, Guadalajara 39375
Museo de Guadalupe, Guadalupe 30906
Museo de Guamúchil de Salvador Alvarado, Guamúchil 30910
Museo de Heráldica Alavesa, Mendoza 39644
Museo de Herrera, Chitré-Herrera 34062
Museo de Historia de Agüimes, Agüimes 38789
Museo de Historia de Ciencias Naturales, Chuquimarca 34152
Museo de Historia de Ensenada, Ensenada 30859
Museo de Historia de Higueras, Higueras 30936
Museo de Historia de la Automoción, Salamanca 39938
Museo de Historia de la Medicina, Bogotá 08564
Museo de Historia de la Medicina y de la Cirugía Vicente A. Risolía, Buenos Aires 00194
Museo de Historia de la Minería de Santa Rosalía, Mulegé 31206
Museo de Historia de Lampazos de Naranjo y las Armas Nacionales, Lampazos de Naranjo 30993
Museo de Historia de Linares, Linares 31006
Museo de Historia de Panamá, Ciudad de Panamá 34069
Museo de Historia de Tabasco, Villahermosa 31551
Museo de Historia de Tenerife, La Laguna de Tenerife 39439
Museo de História del Deporte, La Habana 09134
Museo de Historia del Transporte, Guadalupe 30907
Museo de Historia Económica, Bogotá 08565
Museo de Historia Mexicana, Monterrey 31180
Museo de Historia Natural, Alta Suiza 08527
Museo de Historia Natural, Bogotá 08566
Museo de Historia Natural, Bucaramanga 08600
Museo de Historia Natural, Ciudad de Panamá 34070
Museo de Historia Natural, Guatemala 23298
Museo de Historia Natural, La Paz 30987
Museo de Historia Natural, Manizales 08669
Museo de Historia Natural, Mérida 31025
Museo de Historia Natural, México 31076
Museo de Historia Natural, Pereira 08720
Museo de Historia Natural, Popayán 08726
Museo de Historia Natural, Santiago de Compostela 40021
Museo de Historia Natural, Sucre 04246
Museo de História Natural, Tunja 08775
Museo de Historia Natural, Villahermosa 31552
Museo de Historia Natural, Zapotlán del Rey 31593
Museo de Historia Natural Alcide D'Orbigni, Cochabamba 04207
Museo de Historia Natural Alfredo Dugès, Guanajuato 30914
Museo de Historia Natural de Concepción, Concepción 07434
Museo de Historia Natural de El Salvador, San Salvador 10551
Museo de Historia Natural de la Universidad Antenor Orrego, Trujillo 34319
Museo de Historia Natural de la Universidad Ricardo Palma, Lima 34238
Museo de Historia Natural de San Pedro Nolasco, Santiago de Chile 07456
Museo de Historia Natural de Valparaíso, Valparaíso 07471
Museo de Historia Natural del Ateneo Fuente, Saltillo 31312
Museo de Historia Natural del Paraquay, Asunción 34094
Museo de Historia Natural Dr. Manuel M. Villada, Toluca 31488
Museo de Historia Natural Dr. Manuel Martínez Solórzano, Morelia 31196
Museo de Historia Natural Dr. Ricardo S. Vadell, Suipacha 00636
Museo de Historia Natural Felipe Poey, La Habana 09135
Museo de Historia Natural Javier Prado, Lima 34239
Museo de Historia Natural José Vilet Brullet, Mexquitic 31162
Museo de Historia Natural Noel Kempf Mercado, Santa Cruz 04240
Museo de Historia Natural Pedro Villalobos, Cobija 04205
Museo de Historia Natural Regional, Lerdo 31003
Museo de Historia Natural Valle de la Prehistoria, Santiago de Cuba 09312
Museo de Historia Natural Víctor Vaca Aguinaga, Lambayeque 34203
Museo de Historia Reginal San Lorenzo y Galería del Libertador General San Martín, San Lorenzo 00590
Museo de Historia Regional de Pánuco, Pánuco, Veracruz 31249
Museo de Historia Regional de Sabinas Hidalgo, Sabinas Hidalgo 31309
Museo de Historia Regional de Tristán Suárez, Trisán Suárez 00657
Museo de Historia Republicana, Tegucigalpa 23332
Museo de Historia y Artesanía de la Colonial Tovar, Colonia Tovar 54565
Museo de História y Ciencias Naturales, Lobería 00429
Museo de Historia y Cultura Casa Pedrilla, Cáceres 39083
Museo de Historia y Geografía, Santo Domingo 10345
Museo de Histórico, Tunja 08776
Museo de Hostotipaquillo, Hostotipaquillo 30937
Museo de Hualpén, Concepción 07435
Museo de Huelva, Huelva 39396
Museo de Iconos - Casa Grande, Torrejón de Ardoz 40156
Museo de Imáges de la Semana Santa en Popayán, Popayán 08727
Museo de Inmigrantes de Alemania, Mataglpa 33295
Museo de Insectos, San José 08817
Museo de Instituto Magnasco, Gualeguaychú 00383
Museo de Instituto Nacional de Estudios de Teatro, Buenos Aires 00195
Museo de Instrumentos Musicales, Buenos Aires 00196
Museo de Instrumentos Musicales de Bolivia, La Paz 04214
Museo de Instrumentos Musicales Pedro Pablo Traversari, Quito 10440
Museo de Intendencia, Madrid 39541
Museo de Investigaciones de Zonas Aridas, Lima 34240
Museo de Jaén, Jaén 39422
Museo de Jala, Jala 30965
Museo de Jamapa, Jamapa 30970
Museo de Jérica, Jérica 39429
Museo de la Abadía del Sacro-Monte, Granada 39366
Museo de la Academia de Infantería, Toledo 40132
Museo de la Acuarela del Estado de México, Toluca 31489
Museo de la Administración Federal de Ingresos Públicos, Buenos Aires 00197
Museo de la Agricultura, Torremocha de Jarama 40158
Museo de la Alhambra, Granada 39367
Museo de la Amistad México-Cuba, Tuxpan de Rodríguez Cano 31509
Museo de la Antigua Estación Ferroviaría, Teocelo 31425
Museo de la Arriería Maragata, Santiago Millas 40030
Museo de la Artesanía Tradicional Juan Alfonso Carrizo, La Plata 00412
Museo de la Asegurada, Alicante 38845
Museo de la Aviación Naval Argentina, Bahía Blanca 00134
Museo de la Balanza, Buenos Aires 00198
Museo de la Basílica de Guadalupe, México 31077
Museo de la Basílica de Santa María, Elche 39266
Museo de la Basílica Nuestra Señora de la Candelaria, Candelaria 39117
Museo de la Basilica Tardorromana, Ceuta 31663
Museo de la Biblia, Lima 34241
Museo de la Biblioteca Nacional, Madrid 39542
Museo de la Cabrera, Encinedo 39275
Museo de la Caña de Azúcar, El Cerrito 08636
Museo de la Canción Yucateca, Mérida 31026
Museo de la Capilla Real, Granada 39368
Museo de la Caricatura, México 31078
Museo de la Caricatura Severo Vaccaro, Buenos Aires 00199
Museo de la Casa de Acuerdo de San Nicolás, San Nicolás de los Arroyos 00604
Museo de la Casa de Cultura de Tepeapulco, Tepeapulco Hidalgo 31434
Museo de la Casa de Cultura de Zacapú, Zacapu 31575
Museo de la Casa de Gobierno, San Miguel de Tucumán 00596
Museo de la Casa de la Cultura, Pinchote 08722
Museo de la Casa de la Cultura, San Luis Potosí 31346
Museo de la Casa de la Independencia, Asunción 34095
Museo de la Casa de la Música Mexicana, México 31079
Museo de la Casa de las Artesanías del Estado de Chihuahua, Bocoyna 30738
Museo de la Casa del Risco, México 31080
Museo de la Casa Histórica de la Independencia Nacional, San Miguel de Tucumán 00597
Museo de la Casa Rosada, Buenos Aires 00200
Museo de la Catedral, Astorga 38908
Museo de la Catedral, Cusco 34160
Museo de la Catedral, Granada 39369
Museo de la Catedral, Murcia 39693
Museo de la Catedral, Trujillo 34320
Museo de la Catedral de Santiago de Compostela, Santiago de Compostela 40022
Museo de la Cerámica Contemporánea, Santo Domingo 10346
Museo de La Cerería, Tlayacapan 31480
Museo de la Charrería, México 31081
Museo de la Ciencia de Valladolid, Valladolid 40231
Museo de la Ciencia y del Agua, Murcia 39694
Museo de la Ciencia y el Cosmos, La Laguna de Tenerife 39440
Museo de la Ciencia y el Juego, Bogotá 08567
Museo de la Ciudad, Buenos Aires 00201
Museo de la Ciudad, Campeche 30741
Museo de la Ciudad, Guadalajara 30894
Museo de la Ciudad, Madrid 39543
Museo de la Ciudad, Melilla 39642
Museo de la Ciudad, Mérida 31027
Museo de la Ciudad, Murcia 39695
Museo de la Ciudad, Paraná 00485
Museo de la Ciudad, Quito 10441
Museo de la Ciudad, Salta 00557
Museo de la Ciudad, Tecate 31412
Museo de la Ciudad, Victoria 00668
Museo de la Ciudad, Vinaròs 40315
Museo de la Ciudad de Chetumal, Chetumal 30760
Museo de la Ciudad de Colón, Colón 00308
Museo de la Ciudad de Córdoba, Córdoba 30801
Museo de la Ciudad de Irapuato, Irapuato 30957
Museo de la Ciudad de La Habana, La Habana 09136
Museo de la Ciudad de León, León 30999
Museo de la Ciudad de México, México 31082
Museo de la Ciudad de Píritu, Píritu 54591
Museo de la Ciudad de Querétaro, Querétaro 31300
Museo de la Ciudad de San Francisco, San Francisco 00577
Museo de la Ciudad de Sombrerete, Sombrerete 31392
Museo de la Ciudad de Tepatitlan de Morelos, Tepatitlán de Morelos 31430
Museo de la Ciudad de Veracruz, Veracruz 31536
Museo de la Ciudad Tuxtlán, Tuxtla Gutiérrez 31513
Museo de la Clandestinidad, Nueva Gerona 09233
Museo de la Coca, La Paz 04215
Museo de la Cofradía del Espíritu Santo de los Congos, Villa Mella 10362
Museo de la Colegiata, Belmonte 39020
Museo de la Colegiata, Medinaceli 39639
Museo de la Colegiata de Santa Juliana y Claustro, Santillana del Mar 40033
Museo de la Colegiata San Luis, Villagarcía de Campos 40301
Museo de la Colonia Siete de Marzo, Juan Aldama 30979
Museo de la Colonización, Esperanza 00367
Museo de la Comisión Provincial de Monumentos, León 39455
Museo de la Compañía de Bomberos Italia No. 5, Bellavista 34123
Museo de la Comunidad de Agua Blanca, Puerto López 10424
Museo de la Comunidad de Chordeleg, Chordeleg 10377
Museo de la Concatedral de San Pedro, Soria 40092
Museo de la Constitución, Guáimaro 09087
Museo de la Coria, Trujillo 40171
Museo de la Cueva del Señor del Calvario, México 31083
Museo de la Cultura de Albán, Albán 08525
Museo de la Cultura Huasteca, Tampico 31399
Museo de la Cultura José Arens Berg, Sullana 34308
Museo de la Cultura Martínez de Compañon, Tambogrande 34315
Museo de la Cultura Maya, Chetumal 30761
Museo de la Cultura Popular, Cañete 39120
Museo de la Cultura Potosina, San Luis Potosí 31347
Museo de la Diócesis de Matagalpa, Mataglpa 33296
Museo de la Dirección Nacional del Antártico, Buenos Aires 00202
Museo de la Diversidad Biológica y Cultural del Pacífico, Quibdó 08734
Museo de la E.E.M. Número 3, Dolores 00355
Museo de la Electricidad, Caracas 54546
Museo de la Electricidad, Lima 34242
Museo de la Encarnación, Corella 39212
Museo de la Energía, Embalse 00363
Museo de la Escuela Militar de Montaña y Operaciones Especiales, Huesca 39399
Museo de la Escultura Mexica Eusebio Dávalos, Acatitlán 30693
Museo de la Estampa, Cuitzeo 30823
Museo de la Estampa del Estado de México, Toluca 31490
Museo de la Estampa y del Diseño Carlos Cruz Diez, Caracas 54547
Museo de la Evangelización, Huejotzingo 30945
Museo de la Fabrica Nacional de Moneda y Timbre, Madrid 39544
Museo de la Facultad de Ciencias de la Salud, Sucre 04247
Museo de la Facultad de Ciencias Naturales y del Instituto Miguel Lillo, San Miguel de Tucumán 00598
Museo de la Familia Dominicana del Siglo XIX, Santo Domingo 10347
Museo de la Farmacia Dra. D'Alessio Bonino, Buenos Aires 00203
Museo de la Farmacia Hispana, Madrid 39545
Museo de la Fauna, Aguascalientes 30700
Museo de la Fauna de Veracruz, Xalapa 31559

Museo de la Fauna y Ciencias Naturales, Monterrey 31181
Museo de la Ferrería de La Pradera, Subachoque 08764
Museo de la Fiesta, Caravaca de la Cruz 39127
Museo de la Forja, El Franco 39309
Museo de la Fotografía, Pachuca 31241
Museo de la Fotografía, Rafaela 00513
Museo de la Fuerza de Aviación Naval, Callao 34131
Museo de la Fuerza de Submarinos, Callao 34132
Museo de la Fundación Guayasamin, Quito 10442
Museo de la Fundación John Boulton, Caracas 54548
Museo de la Gaita, Gijón 39336
Museo de la Gráfica de la Historia Social de Taxco Siglo XX, Taxco de Alarcón 31404
Museo de la Guardia Civil, Madrid 39546
Museo de la Guerra de Castas, Felipe Carrillo Puerto 30866
Museo de la Guerra Hispano-Cubano-Norteamericana, Santiago de Cuba 09313
Museo de la Historia de Ponce, Ponce 36050
Museo de la Hoya, Laguardia 39436
Museo de la Huaxteca, Amatlán de los Reyes 30714
Museo de la Huerta, Almoradi 38863
Museo de la Iglesia, Oviedo 39760
Museo de la Iglesia de la Compañía de Jesús, Ica 34188
Museo de la Iglesia de San Pedro, Chucuito 34148
Museo de la Iglesia de Santa María la Real, Sasamón 40044
Museo de la Iglesia Parroquial San Pedro González Telmo y su Claustro Histórico, Buenos Aires 00204
Museo de La Imagen, Vista Alegre 09352
Museo de la Independencia, Dolores Hidalgo 30835
Museo de la Indumentaria Mexicana Luis Márquez Romay, México 31084
Museo de la Industria Azucarera, Abel Santamaría 09020
Museo de la Industria Azucarera Obispo Colombres, San Miguel de Tucumán 00599
Museo de la Inmaculada Concepción, Cuenca 10386
Museo de la Inmigración Japonesa, Lima 34243
Museo de la Insurgencia, Rincón de Romos 31305
Museo de la Isla de Cozumel, Cozumel 30808
Museo de la Labranza, Francisco I. Madero, Hidalgo 30871
Museo de la Lealtad Republicana, Chihuahua 30769
Museo de la Legión, Ceuta 31664
Museo de la Lucha Clandestina, Nueva Gerona 09234
Museo de la Lucha Clandestina, Santiago de Cuba 09314
Museo de la Lucha Estudiantil Jesús Suárez Gayol, Camagüey 09051
Museo de la Lucha Obrera, Cananea 30745
Museo de la Luz, México 31085
Museo de la Máquina Agrícola, Esperanza 00368
Museo de la Máscara, Morelia 31197
Museo de la Medicine, Cuenca 10387
Museo de la Memoria, Cuacutecuta 08631
Museo de la Memoria, Tlaxcala 31477
Museo de la Miel, Peñalver 39830
Museo de la Minería, San Juan de Guadalupe 31337
Museo de la Minería del Estado de México, El Oro 30849
Museo de la Minería del País Vasco, Gallarta 39323
Museo de la Minería José de la Borda, Taxco de Alarcón 31405
Museo de la Minería y de la Industria de Asturias, San Martín del Rey 39963
Museo de la Moneda, Caracas 54549
Museo de la Moto Clásica de Hervás, Hervás 39389
Museo de la Muerte, San Juan del Río 31338
Museo de la Música Alejandro García Caturla, Santa Clara 09295
Museo de la Música Grabada, Zapotlanejo 31594
Museo de la Música Puertorriqueña, Ponce 36051
Museo de la Música Rodrigo Prats, Santa Clara 09296
Museo de la Nación, Lima 34244
Museo de la Nacionalidad, La Villa de Los Santos 34074
Museo de la Naturaleza de Cantabria, Carrejo-Cantabria 39134
Museo de la No Intervención Fuerte de Loreto, Puebla 31277
Museo de la Paeria, Lleida 39475
Museo de la Palabra, Villa Adelina 00670
Museo de La Palabra y La Imagen, San Salvador 10552
Museo de la Parroquia de Gámbita, Gámbita 08645
Museo de la Partitura Historica, Rosario 00544
Museo de la Pasión Boquense, Buenos Aires 00205
Museo de la Patagonia Dr. Francisco P. Moreno, Bariloche 00144
Museo de la Paz de Gernika, Gernika-Lumo 39332
Museo de la Pintura Mural y Museo Manuel Gamio, Teotihuacán 31426
Museo de la Piratería, Santiago de Cuba 09315
Museo de La Plata, La Plata 00413
Museo de la Platería, Taxco de Alarcón 31406
Museo de la Pluma, México 31086
Museo de la Policía del Chaco, Resistencia 00523
Museo de la Policía Federal Argentina, Buenos Aires 00206
Museo de la Policía Preventiva de la Ciudad de México, México 31087
Museo de la Porcelana, Santo Domingo 10348
Museo de la Provincia de Granma, Bayamo 09034
Museo de la Provincia El Dorado, El Dorado 34172
Museo de la Psicología Experimental Argentina Horacio G. Piñero, Buenos Aires 00207
Museo de la Purísima Concepción, Celaya 30753
Museo de la Quemada, Guadalupe 30908
Museo de la Radio y TV - Lulio Moscoso, Santo Domingo 10349
Museo de la Radiodifusión, Puebla 31278
Museo de la Real Academia de Bellas Artes de la Purísima Concepción, Valladolid 40232
Museo de la Real Academia de Bellas Artes de San Fernando, Madrid 39547
Museo de la Real Academia Nacional de Farmacia, Madrid 39548
Museo de la Real Capilla, Arenas de San Pedro 38885
Museo de la Real Colegiata, Roncesvalles 39915
Museo de la Real Maestranza de Caballería de Ronda, Ronda 39916
Museo de la Real Plaza de Toros de Sevilla, Sevilla 40071
Museo de la Recoleta, Arequipa 34111
Museo de la Recoleta, Sucre 04248
Museo de la Reconquista, Orihuela 39742
Museo de la Reconquista, Tigre 00644
Museo de la Región de Ansenuza Aníbal Montes, Miramar 00467
Museo de la Reincorporación, Tacna 34311
Museo de la Resistencia Indígena, Ixcateopan de Cuauhtémoc 30959
Museo de la Revolución, La Habana 09137
Museo de la Revolución, Nueva Gerona 09235
Museo de la Revolución, Puebla 31279
Museo de la Revolución de San Pedro, San Pedro 31367
Museo de la Revolución Salvadoreña, Perquín 10546
Museo de la Rioja, Logroño 39491
Museo de la Sal, Zipaquirá 08791
Museo de la Sal y el Azúcar, Aguadulce 34059
Museo de la Salle, Dasmariñas 34384
Museo de la Santa Casa de Ejercicios Espirituales Sor María Antonia de la Paz y Figueroa, Buenos Aires 00208
Museo de la Santa Iglesia Catedral, Ávila 38921
Museo de la Secretaria de Hacienda y Credito Publico Antiguo Palacio del Arzobispado, México 31088
Museo de la Selva Juan Forster, Miramar 00468
Museo de la Semana Santa de Cuenca, Cuenca 39233
Museo de la Semana Santa Marinera de Valencia Salvador Caurín Alarcón, Valencia 40205
Museo de la Sep, México 31089
Museo de la Shoá- Fundación Memoria del Holocausto, Buenos Aires 00209
Museo de la Sidra Copa de Oro, San Andrés Cholula 31321
Museo de la Sidra de Asturias, Nava 39711
Museo de la Sociedad Amigos de Laguardia, Laguardia 39437
Museo de la Soledad, Oaxaca 31224
Museo de la Solidaridad Salvador Allende, Santiago de Chile 07457
Museo de la Tortuga, Aquila 30722
Museo de la Universidad, Salamanca 39939
Museo de la Universidad de Valladolid, Valladolid 40233
Museo de la Universidad Nacional Hermilio Valdizán, Huánuco 34180
Museo de la Vera Cruz, Caravaca de la Cruz 39128
Museo de la Vida Animal Herpetario, Puebla 31280
Museo de la Vida Rural, L'Espluga de Francolí 39282
Museo de la Villa de Ahome, Ahome 30704
Museo de la Villa de Almonte, Almonte 38862
Museo de Labranza de la Torre Vieja, San Clemente 39952
Museo de Laca, Chiapa de Corzo 30762
Museo de Lagos de Moreno, Lagos de Moreno 30992
Museo de lan deuda externa, Buenos Aires 00210
Museo de Las Americas, Denver 48083
Museo de Las Américas, San Juan 36068
Museo de las Artes, Guadalajara 30895
Museo de las Artes Populares Cidap, Cuenca 10388
Museo de las Artes Populares de Jalisco, Guadalajara 30896
Museo de las Atarazanas Reales, Santo Domingo 10350
Museo de las Aves de México, Saltillo 31313
Museo de las Californias, Los Cabos 31008
Museo de las Californias, Tijuana 31457
Museo de las Campanas, Mina Clavero 00466
Museo de las Carmelitas, Peñaranda de Bracamonte 39831
Museo de las Casas Reales, Santo Domingo 10351
Museo de las Ciencias de Castilla La Mancha, Cuenca 39234
Muséo de Las Conceptas, Cuenca 10389
Museo de las Cortes de Cádiz, Cádiz 39090
Museo de las Culturas, Cotacahi 10378
Museo de las Culturas, Saltillo 31314
Museo de las Culturas Aborígenes, Cuenca 10390
Museo de las Culturas Afromestizas, Cuajinicuilapa 30810
Museo de las Culturas de la Huasteca Potosina Tamuantzán, Valles 31529
Museo de las Culturas de Oaxaca, Oaxaca 31225
Museo de las Culturas de Occidente, Colima 30780
Museo de las Culturas del Norte, Monte Verde 31175
Museo de las Culturas Prehispánicas, Puerto Vallarta 31295
Museo de las Culturas Prehispánicas de Durango, Durango 30841
Museo de las Encartaciones, Sopuerta 40090
Museo de las Escuelas, Buenos Aires 00211
Museo de las Esculturas de Copán, Copán 23328
Museo de las Estelas Mayas, Campeche 30742
Museo de las Leyendas, Guanajuato 30915
Museo de las Luchas Obreras, Manzanillo 09207
Museo de las Máscaras Ceremoniales, Chichicastenango 23293
Museo de las Miniaturas Poblanas, Palmar de Bravo 31247
Museo de las Misiones Jesuitas, Loreto 31007
Museo de las Momias, Guanajuato 30916
Museo de las Muñecas, Píritu 54592
Museo de las Parrandas de Remedios, Remedios 09265
Museo de las Pinturas Rupestres de San Ignacio, Pueblo de San Ignacio 31292
Museo de las Revoluciones Mariano Jiménez, San Luis Potosí 31348
Museo de las Telecomunicaciones, Madrid 39549
Museo de las Telecomunicationes, México 31090
Museo de las Termas Públicas de Caesaraugusta, Zaragoza 40342
Museo de las Tradiciones Potosinas, San Luis Potosí 31349
Museo de las Villas Pasiegas, Vega de Pas 40254
Museo de l'Enginyeria de Catalunya, Barcelona 38967
Museo de León, León 39456
Museo de León - Edificio Pallares, León 39457
Museo de Lluc, Escorca 39279
Museo de los Altos de Chiapas, San Cristóbal de las Casas 31325
Museo de los Andes, Socha 08758
Museo de los Asentamientos, Federación 00371
Museo de los Bomberos, Guayaquil 10403
Museo de los Bomberos, Matanzas 09214
Museo de los Caminos, Astorga 38909
Museo de los Combates contra los Bandidos, Trinidad 09336
Museo de los Combatientes del Morro de Arica, Lima 34245
Museo de los Concilios y de la Cultura Visigoda, Toledo 40133
Museo de los Condestables de Castilla, Medina de Pomar 39634
Museo de los Corporales, Daroca 39240
Museo de los Corrales Viejos de Parque de los Patricios, Buenos Aires 00212
Museo de Los Ferrocarrileros, México 31091
Museo de los Héroes de la Restauración, Santiago de los Caballeros 10339
Museo de los Instrumentos del Teatro Colón, Buenos Aires 00213
Museo de los Llanos, Guanare 54573
Museo de los Metales, Cuenca 10391
Museo de los Minerales y Rocas, Porriño 39861
Museo de los Molinos de Gregorio Prieto, Valdepeñas 40189
Museo de los Molinos de Viento, Campo de Criptana 39112
Museo de los Muralles Reales, Ceuta 31665
Museo de los Niños, Bogotá 08568
Museo de los Niños Abasto, Buenos Aires 00214
Museo de los Niños Barrilete, Córdoba 00325
Museo de Los Niños de Caracas, Caracas 54550
Museo de los Niños Tin Marin, San Salvador 10553
Museo de los Origenes, Madrid 39550
Museo de los Orishas, La Habana 09138
Museo de los Patricios de Buenos Ayres, Buenos Aires 00215
Museo de los Pioneros, Río Gallegos 00531
Museo de Málaga, Málaga 39599
Museo de Málaga, Málaga 39600
Museo de Maní y Museo del Convento San Miguel Arcángel, Maní 31011
Museo de Marionetas, Bogotá 08569
Museo de Mas de las Matas, Mas de las Matas 39628
Museo de Medicina Legal y Ciencas Forenses, Bogotá 08570
Museo de Medicina Maya, San Cristóbal de las Casas 31326
Museo de Medicina Tradicional y Herbolaria, Cuernavaca 30817
Museo de Medicina Tropical Carlos J. Finlay, La Habana 09139
Museo de Medios de Comunicación, Resistencia 00524
Museo de Menorca, Mahón 39591
Museo de Metales Preciosos Precolombinos o Museo del Oro, La Paz 04216
Museo de Metrología - Museo de Pesas y Medidas, Tres Cantos 40170
Museo de Mexticacan, Mexticacán 31163
Museo de Mier y Noriega, Mier y Noriega 31165
Museo de Minerales del Instituto Geofísico, Bogotá 08571
Museo de Minerales y Fósiles, Zumárraga 40364
Museo de Mineralogía, Guanajuato 30917
Museo de Mineralogía, Pachuca 31242
Museo de Mineralogía Dr Miguel Romero Sánchez, Tehuacán 31419
Museo de Mineralogía y Geología Dr. Alfredo Stelzner, Córdoba 00326
Museo de Mineralogía y Geología Ing. Luis Silva Ruelas, Morelia 31198
Museo de Minería de la Compañía Real del Monte, Pachuca 31243
Museo de Minería e Historia de Cosalá, Cosalá 30803
Museo de Miniaturas, Huaraz 34182
Museo de Miniaturas Militares Ciudadela de Jaca, Jaca 39416
Museo de Mitos y Leyendas, Tafí del Valle 00639
Museo de Momias de Celaya, Celaya 30754
Museo de Monseñor Ramón Ibarra y González, Puebla 31281
Museo de Moyogalpa, Moyogalpa 33298
Museo de Municipal, Bielsa 39037
Museo de Murcia, Murcia 39696
Museo de Museos Colsubsidio, Bogotá 08572
Museo de Navajas y Curiosidades del Mundo, Puebla 31282
Museo de Navarra, Pamplona 39808
Museo de Nosa Señora da Antiga, Monforte de Lemos 39665
Museo de Nuestra Raíz Africana, San Juan 36069
Museo de Nuestra Señora del Rosario, Talpa de Allende 31398
Museo de Numismática, San José 08818
Museo de Numismática del Estado de México, Toluca 31491
Museo de Odontología, Madrid 39551
Museo de Odontología, México 31092
Museo de Oftalmología Doctor Saracíbar, Valladolid 40234
Museo de Oro Precolombino, San José 08819
Museo de Oro Vicús, Piura 34293
Museo de Orozko, Orozko 39750
Museo de Oxolotán, Tacotalpa 31395
Museo de Palencia, Palencia 39770
Museo de Paleontología, Delicias 30832
Museo de Paleontología de Guadalajara, Guadalajara 30897
Museo de Paleontología de la Preparatoria, Apaxtla de Castrejón 30720
Museo de Paleontología de Rincón Gallardo, General Cepeda 30875
Museo de Paleontología Eliseo Palacios Aguilera, Tuxtla Gutiérrez 31514
Museo de Paracho, Paracho 31253
Museo de Patología, Buenos Aires 00216
Museo de Patones, Patones 39817
Museo de Penonomé, Penonomé 34078
Museo de Pericos, Culiacán 30826
Museo de Piedra y Artesanía Canaria, Ingenio 39411
Museo de Piria, Piriápolis 54426
Museo de Pontevedra, Pontevedra 39857
Museo de Prehistoria e Arqueoloxía de Vilalba, Vilalba 40282
Museo de Prehistoria y Arqueología de Cantabria, Santander 40010
Museo de Prehistoria y Paleontología, Orce 39739
Museo de Prehistoria y Servicio de Investigación Prehistórica, Valencia 40206
Museo de Pungarabato, Pungarabato 31298
Museo de Quejana, Valdegovia 40186
Museo de Relojes, A Coruña 39221
Museo de Reproducciones Artísticas, Bilbao 39044
Museo de Rocas, Yaiza 40328
Museo de Roda de Isábena, Roda de Isábena 39907
Museo de Ruiz, Ruiz 31308
Museo de SADAIC Vincente López y Planes, Buenos Aires 00217
Museo de Salamanca, Salamanca 39940
Museo de San Antolín, Tordesillas 40147
Museo de San Esteban, Salamanca 39941
Museo de San Esteban Tetelpa, Zacatepec de Hidalgo 31585

Museo de San Francisco, Potosí 04236
Museo de San Jerónimo, Cusco 34161
Museo de San José María de Yermo y Parres, Puebla 31283
Museo de San Juan, San Juan 36070
Museo de San Juan Raya, Zapotitlán Salinas 31592
Museo de San Juan Tlacotenco, San Juan Tlacotenco 31343
Museo de San Lorenzo Tenochtitlán, Texistepec 31454
Museo de San Miguel, Tlaxiaco 31479
Museo de San Pablo, San Pablo Huitzo 31363
Museo de San Pedro, Navolato 31213
Museo de San Pedro, San Pedro Lagunillas 31369
Museo de San Salvador, Toro 40153
Museo de Sanidad e Higiene Pública, Madrid 39552
Museo de Santa Clara, Bogotá 08573
Museo de Santa Cruz, Toledo 40134
Museo de Santa Monica, Panay 34469
Museo de Santiago de los Caballeros, Antigua 23291
Museo de Santuario di Montevergine, Santuario di Montevergine 27736
Museo de Sayula Juan Rulfo, Sayula 31389
Museo de Segóbriga, Saelices 39930
Museo de Segovia, Segovia 40053
Museo de Semana Santa, Medina de Rioseco 39635
Museo de Semana Santa, Orihuela 39743
Museo de Semana Santa, Zamora 40334
Museo de Sitio Arqueologico el Mongote Real Alto, Guayaquil 10404
Museo de Sitio Arqueológico Kuntur Wasi, San Pablo 34304
Museo de Sitio Boca de Potrerillos, Mina 31167
Museo de Sitio Casa de Juárez, Oaxaca 31226
Museo de Sitio Castillo de Teayo, Castillo de Teayo 30751
Museo de Sitio Complejo Arqueológico de Sillustani, Atuncolla 34117
Museo de Sitio de Cacaxtla, Nativitas 31210
Museo de Sitio de Cempoala, Cempoala 30755
Museo de Sitio de Chalcatzingo, Chalcatzingo 30757
Museo de Sitio de Chan Chan, Trujillo 34321
Museo de Sitio de Chichen Itza, Timun 31459
Museo de Sitio de Chinchero, Urubamba 34324
Museo de Sitio de Cholula, San Pedro Cholula 31368
Museo de Sitio de Coatetelco, Coatetelco 30777
Museo de Sitio de Comalcalco, Comalcalco 30791
Museo de Sitio de el Tajín, Papantla 31252
Museo de Sitio de el Zapotal, Ignacio de la Llave 30954
Museo de Sitio de Higueras, Vega de Alatorre 31532
Museo de Sitio de Huallamarca, Lima 34246
Museo de Sitio de Ingapirca, Azogues 10371
Museo de Sitio de La Isabela, Luperón 10329
Museo de Sitio de la Pampa Galeras, Quinua 34301
Museo de Sitio de la Vega Vieja, La Vega Vieja 10361
Museo de Sitio de La Venta, Huimanguillo 30950
Museo de Sitio de la Zona Arqueológica de Tzintzuntzan, Tzintzuntzan 31519
Museo de Sitio de Monte Albán, Oaxaca 31227
Museo de Sitio de Narihualá, Catacaos 34140
Museo de Sitio de Ocuilan, Ocuilan 31236
Museo de Sitio de Pachacamac, Lima 34247
Museo de Sitio de Palenque Alberto Ruz Lhuillier, Palenque 31246
Museo de Sitio de Paracas Julio C. Tello, Pisco 34287
Museo de Sitio de Pomona, Tenosique de Pino Suárez 31424
Museo de Sitio de Puruchuco, Lima 34248
Museo de Sitio de Querétaro, Querétaro 31301
Museo de Sitio de San Isidro, Tembladera 34316
Museo de Sitio de San Lorenzo Tenochtitlán, Tenochtitlán 31423
Museo de Sitio de San Miguel Ixtapan, San Miguel Ixtapan 31358
Museo de Sitio de Sipán, Zaña 34331
Museo de Sitio de Tenayuca, Tlalnepantla 31468
Museo de Sitio de Teotihuacán, Teotihuacán 31427
Museo de Sitio de Toniná, Ocosingo 31233
Museo de Sitio de Tres Zapotes, Santiago Tuxtla 31386
Museo de Sitio de Túcume, Campo 34136
Museo de Sitio de Uxmal, Santa Elena 31378
Museo de Sitio de Wari, Wari 34326
Museo de Sitio de Xochicalco, Xochicalco 31565
Museo de Sitio de Xochitecatl, Nativitas 31211
Museo de Sitio Degollado Jalisco, Degollado 30831
Museo de Sitio del Alto de la Alianza, Tacna 34312
Museo de Sitio del Claustro de Sor Juana, México 31093
Museo de Sitio del Mirador del Cerro San Cristóbal, Lima 34249
Museo de Sitio del Rumicucho, San Antonio 10459
Museo de Sitio El Algarrobal, El Algarrobal 34171
Museo de Sitio El Torreoncito, Torreón 31498
Museo de Sitio Huaca El Dragón, La Esperanza 34200
Museo de Sitio Huaca Pucllana, Lima 34250
Museo de Sitio Manuel Chávez Ballón, Aguas Calientes 34106
Museo de Sitio Ocotelulco, Totolac 31505
Museo de Sitio Parque Reducto 2, Lima 34251
Museo de Sitio Polyforum Siqueiros, México 31094
Museo de Sitio San Agustín, Juárez 30982
Museo de Sitio Talavera, México 31095
Museo de Sitio Tecpan, México 31096
Museo de Sitio Wari, Ayacucho 34120
Museo de Sitio Wari Willca, Huancayo 34177
Museo de Siyasa, Cieza 39180
Museo de Sor de María Romero, Granada 33278
Museo de Suelos, Iquitos 34192
Museo de Suelos de Colombia, Bogotá 08574
Museo de Susticacán, Susticacán 31393
Museo de Tabar, Navarra 39712
Museo de Tabar, Tabar 40094
Museo de Tacuichamona, Culiacán 30827
Museo de Tahdziú, Tahdziú 31396
Museo de Tapalpa, Tapalpa 31402
Museo de Tapices, Oncala 39732
Museo de Tarlac, Tarlac 34516
Museo de Taximaroa, Hidalgo 30934
Museo de Tecnología Aeronáutica y Espacial, Córdoba 00327
Museo de Telecomunicaciones, Buenos Aires 00218
Museo de Telefonía, Delicias 30833
Museo de Tepeticpa, Tepeticpa 31436
Museo de Terra Santa, Santiago de Compostela 40023
Museo de Texcalyacac, Texcalyacac 31452
Museo de Textiles Andinos Bolivianos, La Paz 04217
Museo de Tlahualilo, Tlahualilo de Zaragoza 31465
Museo de Tlalancaleca, San Matías Tlalancaleca 31355
Museo de Tlapa, Tlapa de Comonfort 31471
Museo de Tlayacapan, Tlayacapan 31481
Museo de Todos los Santos, La Paz 30988
Museo de Totolapan, Totolapan 31506
Museo de Tradiciones Populares de Aguadas, Aguadas 08523
Museo de Trajes Regionales de Colombia, Bogotá 08575
Museo de Transportes, Luján 00434
Museo de Tudela, Tudela 40173
Museo de Ulía, Montemayor 39675
Museo de Unidades Acorazadas, Madrid 39553
Museo de Urología de Buenos Aires, Buenos Aires 00219
Museo de Usos y Costumbres Etnológicas, Paracuellos de Jarama 39810
Museo de Valladolid, Valladolid 40235
Museo de Velardeña, Velardeña 31533
Museo de Xiutetelco, Xiutetelco 31564
Museo de Xochitepec, Xochitepec 31568
Museo de Xolalpancalli, Zacapoaxtla 31574
Museo de Yautepec, Yautepec 31571
Museo de Zamora, Zamora 40335
Museo de Zaragoza, Zaragoza 40343
Museo de Zaragoza, Zaragoza 40344
Museo de Zoología, Córdoba 00328
Museo de Zoología Alfonso L. Herrera, México 31097
Museo de Zoología Comparada de Vertebrados, Madrid 39554
Museo de Zoología Juan Ormea, Trujillo 34322
Museo de Zurbarán, Marchena 39622
Museo degli Affreschi G.B. Cavalcaselle, Verona 28233
Museo degli Alpini, Bassano del Grappa 25379
Museo degli Alpini, Savignone 27771
Museo degli Arazzi delle Madrice, Marsala 26612
Museo degli Argenti, Ariano Irpino 25276
Museo degli Argenti, Firenze 26169
Museo degli Argenti Contemporanei, Sartirana Lomellina 27750
Museo degli Arredi Sacri, Castel Sant'Elia 25719
Museo degli Attrezzi Agricoli, Suno 27924
Museo degli Attrezzi Agricoli Ferretti Florindo, Sefro 27796
Museo degli Ex Voto, Viterbo 28294
Museo degli Indios dell'Amazonia, Assisi 25307
Museo degli Insetti, Bisegna 25426
Museo degli Scavi, Piuro 27257
Museo degli Strumenti della Riproduzione del Suono, Roma 27471
Museo degli Strumenti Musica Populare Sardi, Tadasuni 27934
Museo degli Strumenti Musicali, Cesena 25826
Museo degli Strumenti Musicali Meccanici, Savio 27772
Museo degli Strumenti Musicali Meccanici, Sestola 27827
Museo degli Strumenti per il Calcolo, Pisa 27241
Museo degli Usi e Costumi della Gente di Romagna, Santarcangelo di Romagna 27725
Museo degli Usi e Costumi della Gente Trentina, San Michele all'Adige 27669
Museo degli Usi e Costumi della Valle di Goima, Zoldo Alto 28319
Museo degli Usi e Costumi delle Genti dell'Etna, Giarre 26335
Museo dei Bambini Explora, Roma 27472
Museo dei Benedettini, Pontida 27291
Museo dei Bozzetti, Pietrasanta 27206
Museo dei Bronzi Dorati, Pergola 27139
Museo dei Burattini, Mantova 26593
Museo dei Burattini e delle Figure, Cervia 25820
Museo dei Campionissimi, Novi Ligure 26961
Museo dei Cappuccini, Bassano del Grappa 25380
Museo dei Combattenti, Soncino 27869
Museo dei Costumi della Partita a Scacchi, Marostica 26609
Museo dei Cuchi, Roana 27425
Museo dei Damaschi, Lorsica 26526
Museo dei Ferri Chirurgici, Pistoia 27251
Museo dei Ferri Taglienti, Scarperia 27787
Museo dei Fossili, Lugo di Vicenza 26545
Museo dei Fossili, Meride 41629
Museo dei Fossili, Vestenanova 28240
Museo dei Fossili, Zogno 28316
Museo dei Fossili Don Giuseppe Mattiacci, Serra San Quirico 27812
Museo dei Fossili e dei Minerali, Smerillo 27859
Museo dei Fossili e di Storia Naturale, Montefalcone Appennino 26822
Museo dei Fossili e Minerali del Monte Nerone, Apecchio 25249
Museo dei Fossili e Mostra dei Funghi, Pioraco 27232
Museo dei Grandi Fiumi, Rovigo 27574
Museo dei Legni Processionali, Petriolo 27180
Museo dei Magli, Ponte Nossa 27285
Museo dei Minerali e della Miniera, Oltre il Colle 26981
Museo dei Minerali Elbani, Rio Marina 27412
Museo dei Navigli, Milano 26696
Museo dei Parati Sacri, Montemarano 26842
Museo dei Reperti Bellici e Storici della Guerra 1915-1918, Capovalle 25661
Museo dei Reperti della Civiltà Contadina, Alberobello 25179
Museo dei Soldati del Montello, Nervesa della Battaglia 26920
Museo dei Soldatino e della Figurina Storica, Calenzano 25599
Museo dei Tasso e della Storia Postale, Camerata Cornello 25615
Museo dei Trasporti, Ranco 27362
Museo dei Trombini di Lessinia, San Bortolo 27617
Museo del Aceite de Oliva, Cabra 39078
Museo del Agua y del Suelo, Viedma 00669
Museo del Alcázar, Sevilla 40072
Museo del Alcázar, Toledo 40135
Museo del Almirez, Les Coves de Vinromà 39225
Museo del Almudín, Xàtiva 40327
Museo del Ámbar de Chiapas, San Cristóbal de las Casas 31327
Museo del Ambiente Histórico Cubano, Santiago de Cuba 09316
Museo del Anís de Rute, Rute 39923
Museo del Antiguo Recinto del Congreso Nacional, Buenos Aires 00220
Museo del Antiquo Parque de Bombas, Ponce 36052
Museo del Arbol El Genízaro, Nagarote 33299
Museo del Archivo Histórico de Granada, Granada 33279
Museo del Area Fundacional, Mendoza 00454
Museo del Arroz, Valencia 40207
Museo del audifono, Buenos Aires 00221
Museo del Automóvil, Buenos Aires 00222
Museo del Automóvil, Luján 00435
Museo del Automóvil, México 31098
Museo del Automóvil, Montevideo 54398
Museo del Automóvil Club Argentino, Buenos Aires 00223
Museo del Automóvil de Santa Teresita, Santa Teresita 00627
Museo del Automóvil Puebla, Puebla 31284
Museo del Automovilismo Juan Manuel Fangio, Balcarce 00138
Museo del Azulejo, Montevideo 54399
Museo del Azulejo, Onda 39733
Museo del Azulejo Francés, Punta Ballena 54428
Museo del Baile Flamenco, Sevilla 40073
Museo del Banco Central, Cuenca 10392
Museo del Banco Central, Loja 10417
Museo del Banco Central, Riobamba 10455
Museo del Banco Central de Manta, Manta 10420
Museo del Banco Provincia de Córdoba, Córdoba 00329
Museo del Bandolero, Ronda 39917
El Museo del Barrio, New York 51220
Museo del Barro, Asunción 34096
Museo del Beisbol, Caracas 54551
Museo del Belén, San Javier 39959
Museo del Bigallo, Firenze 26170
Museo del Bijou, Casalmaggiore 25697
Museo del Biroccio Marchigiano e Museo Beltrami, Filottrano 26134
Museo del Borgo, Amato 25220
Museo del Bosco, Sovicille 27884
Museo del Botijo, Toral de los Guzmanes 40144
Museo del Botijo, Villena 40313
Museo del Brigante, Sonnino 27874
Museo del Café, Mataglpa 33297
Museo del Café, San Cristóbal de las Casas 31328
Museo del Calzado, Elda 39272
Museo del Calzado El Borceguí, México 31099
Museo del Camino Real de Hecelchakan, Hecelchakán 30927
Museo del Camoscio, Opi 26984
Museo del Campesino, Xochitepec 31569
Museo del Camposanto Vecchio e dell'Opera, Pisa 27242
Museo del Canal Interoceanico de Panamá, Ciudad de Panamá 34071
Museo del Cañamo, Callosa de Segura 39104
Museo del Canasto Cafetero y Muestra Arqueológica de la Cultura Quimbaya, Filandia 08640
Museo del Capitolo di San Lorenzo, Perugia 27148
Museo del Capoluogo, Castelbellino 25720
Museo del Cappello, Alessandria 25191
Museo del Cappello, Montappone 26800
Museo del Capriolo, Bisegna 25427
Museo del Carajo, Cali 08611
Museo del Carbón, Lenguazaque 08665
Museo del Carmen, México 31100
Museo del Carmen, Villa de Leyva 08783
Museo del Carmen de Maipú, Santiago de Chile 07458
Museo del Carnaval, Santiago de Cuba 09317
Museo del Carro, Buendía 39059
Museo del Carro y Aperos de Labranza, Tomelloso 40142
Museo del Carruaje "El Tacú", Villa Ciudad Parque 00674
Museo del Cartel José P.H. Hernández, Río Grande 36054
Museo del Casin, Bagni di Lucca 25345
Museo del Castagno, Pescaglia 27161
Museo del Castello, Fumone 26258
Museo del Castello, Santa Severina 27709
Museo del Castello Colleoni Porto Thiene, Thiene 27971
Museo del Castello di Porciano, Stia 27902
Museo del Castello di Sarre, Sarre 27745
Museo del Castello e della Città, Piombino 27229
Museo del Castello Scaligero, Torri del Benaco 28041
Museo del Castillo de Peralada, Peralada 39836
Museo del Cattedrale, Ferrara 26110
Museo del Cavallo Giocattolo, Grandate 26362
Museo del Cenedese, Vittorio Veneto 28298
Museo del Centro Caprense Ignazio Cerio, Capri 25665
Museo del Centro Cultural Joaquín Arcadio Pagaza, Valle de Bravo 31525
Museo del Centro Cultural La Cantera, Jonacatepec 30976
Museo del Centro Cultural Patio de los Ángeles, Guadalajara 30898
Museo del Centro Cultural Sor Juana Inés de la Cruz, Tepetlixpa 31437
Museo del Centro Cultural Universitario Quinta Gameros, Chihuahua 30770
Museo del Centro Cultural Zacualpan, Comala 30788
Museo del Centro de Estudios de la Revolución Mexicana, Jiquilpan 30975
Museo del Centro de Historia de Tuluá, Tuluá 08770
Museo del Centro de Investigación Arqueológica de Ancón, Lima 34252
Museo del Centro Histórico, San Juan del Río 31339
Museo del Centro Studi, Albissola 25186
Museo del Cervo Sardo, Sinnai 27851
Museo del Chiodo, Forno di Zoldo 26235
Museo del Chocolate, Astorga 38910
Museo del Cielo, Villa Carlos Paz 00673
Museo del Cine, Durango 30842
Museo del Cine Pablo C. Ducrós Hicken, Buenos Aires 00224
Museo del Cine y la Fotografía, Puebla 31285
Museo del Cinema, Milano 26697
Museo del Cioccolato, Norma 26943
Museo del Círculo de Bellas Artes, Pozoblanco 39863
Museo del Círculo de Estudios Ferroviarios del Uruguay, Montevideo 54400
Museo del Cobre, Bogotá 08576
Museo del Colegio De La Salle de Buenos Aires Hermano Esteban, Buenos Aires 00225
Museo del Colegio Médico, Santiago de Compostela 40024
Museo del Colegio Nacional Amauta, Chupaca 34150
Museo del Colegio Nacional Pedro Vicente Maldonado, Riobamba 10456
Museo del Colegio Padres Franciscanos, Ontinyent 39737
Museo del Colegio San José, Buenos Aires 00226
Museo del Colegio Stella Maris, Atacames 10370

M

Museo del Coltello Sardo, Arbus 25258
Museo del Combattente, Modena 26748
Museo del Conservatorio die Napoli, Napoli 26888
Museo del Convento de La Merced, Mantas 34277
Museo del Convento de los Descalzos, Lima 34253
Museo del Convento de San Francisco de Asís, Lima 34254
Museo del Convento de Santa Catalina, Cusco 34162
Museo del Convento de Santa Clara, Carrión de Los Condes 39135
Museo del Convento de Santa Úrsula, Salamanca 39942
Museo del Convento di San Francesco, Susa 27926
Museo del Corallo, Ravello 27368
Museo del Corallo, Torre del Greco 28036
Museo del CosmoCaixa Barcelona, Barcelona 38968
Museo del Costume Arberesh, Frascineto 26247
Museo del Costume e della Tradizione Nostra Gente, Guardiagrele 26385
Museo del Costume Farnesiano, Gradoli 26361
Museo del Costume R. Piraino, Palermo 27068
Museo del Costume Tradizionale, Grosio 26370
Museo del Cristallo, Colle di Val d'Elsa 25938
Museo del Cristo Negro de Portobelo, Portobelo 34079
Museo del Cuale, Puerto Vallarta 31296
Museo del Danza, La Habana 09140
Museo del Deporte, Cienfuegos 09065
Museo del Deporte, Holguín 09182
Museo del Deporte, Pabellón de Arteaga 31240
Museo del Deporte Pierre de Coubertin, La Falda 00402
Museo del Desierto, Saltillo 31315
Museo del Disco, Zipacón 08789
Museo del Duomo, Città di Castello 25906
Museo del Duomo, Fidenza 26122
Museo del Duomo, Guardiagrele 26386
Museo del Duomo, Massa 26620
Museo del Duomo, Milano 26698
Museo del Duomo, Ravello 27369
Museo del Duomo, Udine 28112
Museo del Duomo, Valenza 28139
Museo del Ejército, Callao 34133
Museo del Ejército, Madrid 39555
Museo del Ejército y Fuerza Aérea Bethlemitas, México 31101
Museo del Ejército y Fuerza Aérea Mexicanos Cuartel Colorado, Guadalajara 30899
Museo del Ejidatario, Guasave 30923
Museo del Emigrante Canario, Teguise 40110
Museo del Esparto, Cieza 39181
Museo del Estado, Morelia 31199
Museo del Estado de Mérida, Mérida 54588
Museo del Este, Monte Caseros 00471
Museo del Ex Convento de Culhuacán, México 31102
Museo del Falegname T. Sana-Onlus, Almenno San Bartolomeo 25204
Museo del Ferro e della Ghisa, Follonica 26215
Museo del Ferrocarril, Aguascalientes 30701
Museo del Ferrocarril, Águilas 38794
Museo del Ferrocarril, Castejón 39142
Museo del Ferrocarril, Guadalajara 30900
Museo del Ferrocarril, Mérida 39645
Museo del Ferrocarril, Monterrey 31182
Museo del Ferrocarril, Ponferrada 39852
Museo del Ferrocarril, Torreón 31499
Museo del Ferrocarril de Asturias, Gijón 39337
Museo del Ferrocarril de Madrid, Madrid 39556
Museo del Festero, Villena 40314
Museo del Fin del Mundo, Ushuaia 00660
Museo del Fiore, Acquapendente 25145
Museo del Folklore e delle Tradizioni Popolari, Caltanissetta 25608
Museo del Foro de Caesaraugusta, Zaragoza 40345
Museo del Fossile del Monte Baldo, Brentonico 25529
Museo del Foto - Gabinetto di Fisica, Urbino 28124
Museo del Frer, Chiaverano 25865
Museo del Fuerte de San Felipe, Bacalar 30732
Museo del Fuerte de San Juan de Ulúa, Veracruz 31537
Museo del Fungo e di Scienze Naturali, Boves 25521
Museo del Fútbol Aguileño, Águilas 38795
Museo del Gaucho y de la Moneda, Montevideo 54401
Museo del General Toribio Ortega, Coyame 30806
Museo del Giocattolo, Palermo 27069
Museo del Giocattolo, Sezze 27832
Museo del Gioco del Pallone a Bracciale e Tamburello, Santarcangelo di Romagna 27726
Museo del Grabado a la Estampa Digital Artes, Ribeira 39900
Museo del Grabado Español Contemporáneo, Marbella 39619
Museo del Grano, Cerignola 25809
Museo del Hombre, Tegucigalpa 23333
Museo del Hombre Dominicano, Santo Domingo 10352
Museo del Hombre en el Universo, Sopó 08763
Museo del Hombre y su Entorno, Caleta Olivia 00287
Museo del Hospital Santo Toribio de Mogrovejo, Lima 34255
Museo del Humor, San Antonio de los Baños 09273
Museo del Indio, Los Toldos 00432
Museo del Indio y del Gaucho, Tacuarembó 54435
Museo del Instituto Americano del Arte, Cusco 34163
Museo del Instituto de Investigaciones Antropológicas, México 31103
Museo del Instituto Nacional de Cultura, Arequipa 34112
Museo del Int. Presepio in Miniatura Guido Colitti, Campobasso 25635
Museo del Istmo, La Línea de la Concepción 39465
Museo del Jade, San José 08820
Museo del Jazz G. Dagnino, Genova 26300
Museo del Juguete Tradicional Mexicano, San Francisco de los Romo 31331
Museo del Labrador, Humanes de Madrid 39404
Museo del Lago Gutierrez Dr. Rosendo Pascual, Bariloche 00145
Museo del Larimar, Santo Domingo 10353
Museo del las Comunicaciones de Mar del Plata, Mar del Plata 00444
Museo del Lavoro, Castelmagno 25742
Museo del Lavoro Contadino, Brisighella 25548
Museo del Lavoro Contadino, Campobello di Mazara 25637
Museo del Lavoro e della Civiltà Rurale, Palaia 27056
Museo del Lavoro e della Tradizioni Popolari della Versilia Storica, Seravezza 27809
Museo del Lavoro Povero e della Civiltà Contadina, Livraga 26502
Museo del Lazio Meridionale, Anagni 25225
Museo del Legno, Cantù 25650
Museo del Libro, San Javier 00582
Museo del Libro Antiguo, Antigua 23292
Museo del Limarí, Ovalle 07444
Museo del Lino, Pescarolo 27167
Museo del Litoral Boliviano, La Paz 04218
Museo del Lupo Appenninico, Civitella Alfedena 25925
Museo del Maglio, Breganze 25526
Museo del Malcantone, Curio 41318
Museo del Mantecado, Estepa 39288
Museo del Mar, Boca de Río 54525
Museo del Mar, Cádiz 39091
Museo del Mar, La Barra 54377
Museo del Mar, Mazatlán 31022
Museo del Mar, San Pedro del Pinatar 39969
Museo del Mar, Santa Pola 40006
Museo del Mar El Rodadero, Santa Marta 08751
Museo del Mar Mundo Marino, Santa Marta 08752
Museo del Mar y de la Sal, Torrevieja 40160
Museo del Marco, Madrid 39557
Museo del Mare, Furci Siculo 26259
Museo del Mare, Ischia Ponte 26411
Museo del Mare, Nardò 26914
Museo del Mare, Pesaro 27156
Museo del Mare, Piombino 27230
Museo del Mare, Pizzo 27261
Museo del Mare, Pollica 27275
Museo del Mare e della Costa M. Zei, Sabaudia 27583
Museo del Médico, Santa Fé 00614
Museo del Medioevo e del Rinascimento, Sorano 27879
Museo del Merletto, Burano 25565
Museo del Merletto, Offida 26974
Museo del Merletto a Tombolo, Pescocostanzo 27176
Museo del Milagroso, Buga 08603
Museo del Modellismo Storico, Voghiera 28306
Museo del Molino de Papel, México 31104
Museo del Monasterio de San Joaquín y Santa Ana, Valladolid 40236
Museo del Monasterio de Santa Isabel, Valladolid 40237
Museo del Monasterio de Santa María de la Vid, La Vid 40271
Museo del Monasterio de Tulebras, Tulebras 40177
Museo del Monastero della Beata Filippa Mareri, Petrella Salto 27179
Museo del Mondo Contadino, Savoca 27773
Museo del Motociclo, Rimini 27406
Museo del Motociclo Moto Guzzi, Mondello del Lario 26777
Museo del Mundo, Buenos Aires 00227
Museo del Niño, Albacete 38804
Museo del Niño, San Juan 36071
Museo del Niño, Santa Ana del Valle 31373
Museo del Niño Globo mi Ciudad Pequeña, Guadalajara 30901
Museo del Niño Papalote, México 31105
Museo del Niño - Trampolín, Santo Domingo 10354
Museo del Olivar y del Aceite, Baena 38938
Museo del Opera Omnia R. Musa e Pinacoteca Parmigiani, Bedonia 25386
Museo del Oppidum de Iruña, Iruña de Oca 39414
Museo del Origen, Santiago Ixcuintla 31383
Museo del Oro, Bogotá 08577
Museo del Oro, Ipiales 08663
Museo del Oro Calima, Cali 08612
Museo del Oro Nariño, Pasto 08708
Museo del Oro Quimbaya, Armenia 08534
Museo del Oro Quimbaya, Manizales 08670
Museo del Paesaggio, Castelnuovo Berardenga 25746
Museo del Paesaggio e Museo Storico Artistico, Verbania Pallanza 28219
Museo del Paesaggio Storico dell'Appennino, Firenzuola 26204
Museo del Palacio de Bellas Artes, México 31106
Museo del Palacio Nacional, Guatemala 23299
Museo del Palazzo del Podestà e Gipsoteca Libero Andreotti, Pescia 27173
Museo del Pane, Sant'Angelo Lodigiano 27719
Museo del Papiro, Siracusa 27855
Museo del Parco Nazionale Gran Paradiso, Noasca 26931
Museo del Parque Abel Santamaría Cuadrado, Santiago de Cuba 09318
Museo del Parque Arqueológico El Caño, Natá 34077
Museo del Parque Nacional Los Alerces, Villa Futalaufquen 00679
Museo del Pasado Cuyano, Mendoza 00455
Museo del Pastor, Los Barrios de Luna 39009
Museo del Pastor, Masegoso de Tajuña 39629
Museo del Patriarca, Valencia 40208
Museo del Patrimonio Histórico - Palacio de Aguas Corrientes, Buenos Aires 00228
Museo del Patrimonio Industriale, Bologna 25452
Museo del Patrimonium, Sutri 27930
Museo del Peperoncino, Maierà 26568
Museo del Periodismo Bonaerense, Capilla del Señor 00290
Museo del Periodismo y las Artes Gráficas, Guadalajara 30902
Museo del Pianoforte Antico, Ala 25169
Museo del Po e Young Museum, Revere 27397
Museo del Policía, Monterrey 31183
Museo del Presepe, Modena 26749
Museo del Presepio, Dalmine 26031
Museo del Presepio Tipologico Internazionale, Roma 27473
Museo del Primer Supremo Tribunal de Justicia de la Nación, Ario de Rosales 30723
Museo del Profumo, Milano 26699
Museo del Pueblo, Guanajuato 30918
Museo del Pueblo, Pirané 00494
Museo del Pueblo de Asturias, Gijón 39338
Museo del Pueblo-Hospital, Quiroga 31303
Museo del Pueblo Maya Dzibilchaltún, Chablekal 30756
Museo del Pueblo y Museo Pre-Colombino de Chagüitillo, Chagüitillo 33272
Museo del Puerto de Ingeniero White, Ingeniero White 00388
Museo del Queso, Villarejo-Periesteban 40307
Museo del Quijote, Ciudad Real 39188
Museo del Real Monasterio de San Juan de la Peña, Jaca 39417
Museo del Real Seminario de Bergara, Bergara 39029
Museo del Recuerdo, México 31107
Museo del Reggimento "Giovani Fascisti", Ponti sul Mincio 27290
Museo del Retablo, Burgos 39070
Museo del Retrato Hablado, México 31108
Museo del Río Magdalena, Honda 08657
Museo del Risorgimento, Brescia 25536
Museo del Risorgimento, Castelfidardo 25723
Museo del Risorgimento, Catanzaro 25779
Museo del Risorgimento, Genova 26301
Museo del Risorgimento, Imola 26400
Museo del Risorgimento, Lucca 26534
Museo del Risorgimento, Massa Marittima 26629
Museo del Risorgimento, Milano 26700
Museo del Risorgimento, Piacenza 27186
Museo del Risorgimento, Solferino 27865
Museo del Risorgimento, Villafranca di Verona 28276
Museo del Risorgimento Aurelio Saffi, Forlì 26224
Museo del Risorgimento Duca d'Aosta, Sanluri 27694
Museo del Risorgimento e della Resistenza, Ferrara 26111
Museo del Risorgimento e della Resistenza, Vicenza 28253
Museo del Risorgimento F. Tanara, Langhirano 26449
Museo del Risorgimento L. Musini, Fidenza 26123
Museo del Risorgimento Vittorio Emanuele Orlando, Palermo 27070
Museo del Risorgmento e Resistenza R. Giusti, Mantova 26594
Museo del Románico, Aguilar de Campoo 38791
Museo del Ron, Santiago de Cuba 09319
Museo del Ron Havana Club, La Habana 09141
Museo del Rubinetto e della sua Tecnologia, San Maurizio d'Opaglio 27666
Museo del Sacrario, Asiago 25300
Museo del Sacro dell'Alta Val Trebbia, Montebruno 26810
Museo del Sale, Paceco 27029
Museo del Samán, Guacarí 08649
Museo del Sannio, Benevento 25394
Museo del Santo Rosario, Aroche 38894
Museo del Santuario de la Virgen de Agua Santa, Baños 10375
Museo del Santuario di Crea, Serralunga di Crea 27813
Museo del Santuario di Montevergine, Mercogliano 26664
Museo del Santuario di Santa Maria Santissima della Fontenuova, Monsummano Terme 26789
Museo del Seminario Conciliar de Nueva Caceres, Naga 34466
Museo del Sidecar, Cingoli 25892
Museo del Siglo XIX, Bogotá 08578
Museo del Sitio de Qorikancha, Cusco 34164
Museo del Sud-Ovest Americano, Cuveglio 26030
Museo del Tabaco Herminio Torres Grillo, Caguas 36032
Museo del Tamburo Parlante, Montone 26863
Museo del Tango Roberto Firpo, Salto 00561
Museo del Tasso, Roma 27474
Museo del Teatro, Cesena 25827
Museo del Teatro, Spoleto 27893
Museo del Teatro Argentina, Roma 27475
Museo del Teatro dell'Aquila, Fermo 26099
Museo del Teatro G. Borgatti, Cento 25801
Museo del Teatro Municipale, Piacenza 27187
Museo del Teclado, Caracas 54552
Museo del Tejido Sedero Valenciano, Valencia 40209
Museo del Templo Mayor, México 31109
Museo del Tequila, Tequila 31449
Museo del Territorio, Finale Emilia 26137
Museo del Territorio, Foggia 26210
Museo del Territorio, Longiano 26516
Museo del Territorio, Riccione 27400
Museo del Territorio, San Severino Marche 27685
Museo del Territorio Biellese, Biella 25418
Museo del Territorio dell'Alta Valle dell'Aulella, Casola in Lunigiana 25706
Museo del Territorio F.R. Fazio, Roccapalumba 27433
Museo del Tesoro, Lodi 26508
Museo del Tesoro del Duomo, Berceto 25397
Museo del Tesoro del Duomo, Monza 26864
Museo del Tesoro del Duomo, Rieti 27403
Museo del Tesoro del Duomo, Veroli 28226
Museo del Tesoro della Cattedrale, Anagni 25226
Museo del Tesoro della Cattedrale, Aosta 25248
Museo del Tesoro della Cattedrale, Savona 27777
Museo del Tesoro della Cattedrale di San Lorenzo, Genova 26302
Museo del Tesoro Santuario della Misericordia, Savona 27778
Museo del Tessile, Chieri 25866
Museo del Tessile e della Tradizione Industriale, Busto Arsizio 25573
Museo del Tessuto, Prato 27335
Museo del Traje, La Rioja 00419
Museo del Traje, Madrid 39558
Museo del Traje en Tabasco, Villahermosa 31553
Museo del Transporte, Garza García 30872
Museo del Transporte Guillermo José Schael, Caracas 54553
Museo del Transporte Terrestre, Santiago de Cuba 09320
Museo del Trattore Agricolo, Senago 27799
Museo del Tren, Murcia 39697
Museo del Tricolore, Reggio Emilia 27390
Museo del Turismo dell'Alta Pusteria, Villabassa 28274
Museo del Vajont, Longarone 26514
Museo del Valle de Tehuacán, Tehuacán 31420
Museo del Valle de Xico, Valle de Chalco 31526
Museo del Vetro, Murano 26877
Museo del Vicino Oriente, Roma 27476
Museo del Vidrio, Monterrey 31184
Museo del Vidrio, San Ildefonso 39957
Museo del Viento, L'Aldea 38835
Museo del Villaggio, Chiusa 25881
Museo del Vino, Caldaro 25595
Museo del Vino, Peñafiel 39827
Museo del Vino, Torgiano 27993
Museo del Vino, Villafamés 40298
Museo del Vino de la D. O. Campo de Barja, Vera de Moncayo 40264
Museo del Vino de Montilla, Montilla 39678
Museo del Vino e della Moda, Cilivergho di Mazzano 25888
Museo del Vino Juan Carcelén, Jumilla 39433
Museo del Vino Valdepeñas, Valdepeñas 40190
Museo del Virreinato, Lima 34256
Museo del Virrey Ezpeleta, Guaduas 08651
Museo del Volcán Masaya, Masaya 33294
Museo Deleddiano, Nuoro 26966
Museo dell' Antica Civiltà Locale, Bronte 25551
Museo della Badia Benedettina, Cava dei Tirreni 25786

M

Museo della Badia di San Salvatore, Vaiano 28133
Museo della Barca Lariana, Pianello del Lario 27193
Museo della Basilica, Gandino 26270
Museo della Basilica, Grotte di Castro 26378
Museo della Basilica di San Michele Arcangelo, Monte Sant'Angelo 26807
Museo della Basilica di San Nicola, Bari 25363
Museo della Basilica di San Pancrazio, Roma 27477
Museo della Basilica di San Venanzio, Camerino 25618
Museo della Basilica di Santa Maria della Passione, Milano 26701
Museo della Basilica Patriarcale di Santa Maria degli Angeli, Porziuncola 27315
Museo della Basilica Patriarcale Santa Maria degli Angeli, Assisi 25308
Museo della Battaglia, San Martino della Battaglia 27661
Museo della Battaglia, Vittorio Veneto 28299
Museo della Battaglia del Senio, Alfonsine 25197
Museo della Battaglia di Marengo, Alessandria 25192
Museo della Battaglia di San Martino, Desenzano del Garda 26035
Museo della Biblioteca Maldotti, Guastalla 26388
Museo della Bilancia, Campogalliano 25639
Museo della Bonifica, Argenta 25272
Museo della Bonifica, San Donà di Piave 27631
Museo della Calzatura, Sant'Elpidio a Mare 27729
Museo della Calzatura d'Autore di Rossimoda, Strà 27908
Museo della Calzatura P. Bertolini, Vigevano 28266
Museo della Canzone e della Riproduzione Sonora, Vallecrosia 28147
Museo della Carrozza, Macerata 26552
Museo della Carta, Amalfi 25215
Museo della Carta e della Filigrana, Fabriano 26069
Museo della Carta e della Filigrana, Pioraco 27233
Museo della Cartolina S. Nuvoli, Isera 26413
Museo della Casa Carnica, Prato Carnico 27340
Museo della Casa Carsica, Monrupino 26785
Museo della Casa Contadina, Bolognetta 25480
Museo della Casa Contadina, Corciano 25970
Museo della Casa Trevigiana, Treviso 28071
Museo della Casina delle Civette, Roma 27478
Museo della Cattedrale, Bari 25364
Museo della Cattedrale, Barletta 25373
Museo della Cattedrale, Chiusi Città 25883
Museo della Cattedrale, Fossano 26239
Museo della Centuriazione, Cesena 25828
Museo della Ceramica, Bassano del Grappa 25381
Museo della Ceramica, Castellamonte 25732
Museo della Ceramica, Fiorano Modenese 26140
Museo della Ceramica, Forlì 26225
Museo della Ceramica, Montelupo Fiorentino 26839
Museo della Ceramica, Patti 27116
Museo della Ceramica del Castello, Salerno 27595
Museo della Ceramica di Castelli, Castelli 25740
Museo della Certosa, Certosa di Pavia 25816
Museo della Certosa, Serra San Bruno 27811
Museo della Chiesa di San Francesco, Mercatello sul Metauro 26662
Museo della Chiesa di San Lorenzo, Portovenere 27314
Museo della Città, Ancona 25228
Museo della Città, Aquino 25256
Museo della Città, Rimini 27407
Museo della Città e del Territorio, Monsummano Terme 26790
Museo della Civiltà Contadina, Acri 25151
Museo della Civiltà Contadina, Aliano 25200
Museo della Civiltà Contadina, Amandola 25217
Museo della Civiltà Contadina, Andrate 25232
Museo della Civiltà Contadina, Ateleta 25317
Museo della Civiltà Contadina, Barolo 25375
Museo della Civiltà Contadina, Bastiglia 25383
Museo della Civiltà Contadina, Bentivoglio 25396
Museo della Civiltà Contadina, Borghi 25500
Museo della Civiltà Contadina, Borgo San Lorenzo 25505
Museo della Civiltà Contadina, Calvisano 25611
Museo della Civiltà Contadina, Castelfranco Emilia 25728
Museo della Civiltà Contadina, Castelnovo no' Monti 25745
Museo della Civiltà Contadina, Cavenago d'Adda 25790
Museo della Civiltà Contadina, Cervia 25821
Museo della Civiltà Contadina, Copparo 25968
Museo della Civiltà Contadina, Curtatone 26026
Museo della Civiltà Contadina, Fabriano 26070
Museo della Civiltà Contadina, Figline Valdarno 26131
Museo della Civiltà Contadina, Filottrano 26135
Museo della Civiltà Contadina, Guardia Piemontese 26382
Museo della Civiltà Contadina, Loreto Aprutino 26522
Museo della Civiltà Contadina, Maierato 26569
Museo della Civiltà Contadina, Mairano 26572
Museo della Civiltà Contadina, Moio della Civitella 26762
Museo della Civiltà Contadina, Mombaroccio 26769
Museo della Civiltà Contadina, Montechiaro d'Acqui 26817
Museo della Civiltà Contadina, Montefiore dell'Aso 26827
Museo della Civiltà Contadina, Novellara 26960
Museo della Civiltà Contadina, Perdifumo 27136
Museo della Civiltà Contadina, Roscigno 27557
Museo della Civiltà Contadina, San Nicola La Strada 27674
Museo della Civiltà Contadina, Sappada 27738
Museo della Civiltà Contadina, Sassocorvaro 27758
Museo della Civiltà Contadina, Serrastretta 27816
Museo della Civiltà Contadina, Sesto Fiorentino 27824
Museo della Civiltà Contadina, Seui 27830
Museo della Civiltà Contadina, Spinetoli 27890
Museo della Civiltà Contadina, Teggiano 27948
Museo della Civiltà Contadina, Todi 27980
Museo della Civiltà Contadina, Torre Pallavicina 28038
Museo della Civiltà Contadina, Verdello 28222
Museo della Civiltà Contadina, Zibello 28315
Museo della Civiltà Contadina A. Marsiano, Niscemi 26927
Museo della Civiltà Contadina D. Bianco, Sammichele di Bari 27606
Museo della Civiltà Contadina del Cilento, Montecorice 26820
Museo della Civiltà Contadina del Mendrisiotto, Stabio 41898
Museo della Civiltà Contadina dell'Alto Monferrato e della Bassa Langa, Castagnole delle Lanze 25713
Museo della Civiltà Contadina e Artigianale d'Enza, Montecchio Emilia 26815
Museo della Civilta Contadina e della Canapa G. Romagnoli, Argelato 25270
Museo della Civiltà Contadina e Tradizioni Popolari, Tuglie 28107
Museo della Civiltà Contadina ed Artigiana, Monterosso Calabro 26848
Museo della Civiltà Contadina ed Artigiana, Ripatransone 27416
Museo della Civilta Contadina Friulana, Farra d'Isonzo 26088
Museo della Civiltà Contadina in Val Vibrata, Controguerra 25964
Museo della Civiltà Contadina L. Rubat, Piscina 27246
Museo della Civiltà Contadina Masseria Lupoli, Crispiano 26008
Museo della Civiltà Contadina nell'Area, Montefalcone di Val Fortore 26823
Museo della Civiltà Contadina Valle dell'Aniene, Roviano 27573
Museo della Civiltà del Lavoro, Campiglia Marittima 25630
Museo della Civiltà Locale, Buseto Palizzolo 25567
Museo della Civiltà Locale - Raccolta d'Arte di San Francesco, Trevi 28064
Museo della Civiltà Messapica, Poggiardo 27265
Museo della Civiltà Montanara, Sestola 27828
Museo della Civiltà Romana, Roma 27479
Museo della Civiltà Rupestre, Zungri 28321
Museo della Civiltà Rurale, Altamura 25205
Museo della Civiltà Rurale del Vicentino, Malo 26578
Museo della Civiltà Salinara, Cervia 25822
Museo della Civiltà Solandra, Malé 26575
Museo della Civiltà Valligiana, Bardi 25356
Museo della Civitella, Chieti 25874
Museo della Collegiata, Castell'Arquato 25735
Museo della Collegiata, Chianciano Terme 25851
Museo della Collegiata, San Candido 27619
Museo della Collegiata di Santa Maria a Mare, Maiori 26571
Museo della Collegiata di Sant'Andrea, Empoli 26054
Museo della Collezione Gambarotta, Piozzo 27234
Museo della Comunità Ebraica, Venezia 28190
Museo della Comunità Ebraica Carlo e Vera Wagner, Trieste 28090
Museo della Confraternita della Misericordia, Anghiari 25237
Museo della Congregazione Mechitarista dei Padri Armeni, Venezia 28191
Museo della Corte, Porto Viro 27305
Museo della Croce Rossa Italiana, Campomorone 25640
Museo della Cultura Alpina del Comelico, Comelico Superiore 25949
Museo della Cultura Arbereshe, San Paolo Albanese 27675
Museo della Cultura Marinara, Tortoreto Lido 28045
Museo della Cultura Mezzadrile, Morro d'Alba 26872
Museo della Cultura Popolare e Contadina, Carrega Ligure 25692
Museo della Donna, Angrogna 25239
Museo della Donna E. Ortner, Merano 26658
Museo della Esposizione Permanente, Milano 26702
Museo della Farmacia Picciola, Trieste 28091
Museo della Figurina, Modena 26750
Museo della Figurina di Gesso e dell'Emigrazione, Coreglia Antelminelli 25973
Museo della Filigrana, Campo Ligure 25632
Museo della Flora, Fauna e Mineralogia, Auronzo di Cadore 25329
Museo della Fondazione Giovanni Scaramangà di Altomonte, Trieste 28092
Museo della Fondazione Horne, Firenze 26171
Museo della Fondazione Mandralisca, Cefalù 25794
Museo della Fondazione Querini Stampalia, Venezia 28192
Museo della Fondazione Ugo da Como, Lonato 26513
Museo della Fotografia Storica e Contemporanea, Torino 28005
Museo della Frutticoltura Adofo Bonvicini, Massa Lombarda 26624
Museo della Gente dell'Appennino Pistoiese, Cutigliano 26027
Museo della Gente Senza Storia, Altavilla Irpina 25207
Museo della Geotermia, Pomarance 27277
Museo della Grancia, Serre di Rapolano 27819
Museo della Guerra, Castel del Rio 25716
Museo della Guerra, Gardone Riviera 26272
Museo della Guerra Adamellina, Spiazzo 27888
Museo della Guerra Bianca, Temù 27953
Museo della Guerra Bianca, Vermiglio 28223
Museo della Linea Gotica, Montefiore Conca 26825
Museo della Liquirizia, Rossano 27564
Museo della Liuteria, Arpino 25284
Museo della Macchina da Scrivere P. Mitterhofer, Parcines 27091
Museo della Magnifica Comunità di Cadore, Pieve di Cadore 27208
Museo della Mail Art, Montecarotto 26812
Museo della Marineria dell'Alto e Medio Adriatico, Cesenatico 25839
Museo della Marionetta, Torino 28006
Museo della Memoria, Sant'Andrea Apostolo 27717
Museo della Miniera, Cesena 25829
Museo della Miniera, Massa Marittima 26630
Museo della Miniera, Prata di Pordenone 27328
Museo della Miniera, Schilpario 27790
Museo della Miniera Aurifera della Guia, Macugnaga 26559
Museo della Miniera di Zolfo di Cabernardi, Sassoferrato 27763
Museo della Misericordia, San Miniato 27672
Museo della Montagna, Macugnaga 26560
Museo della Nave Romana, Comacchio 25946
Museo della Nostra Terra, Pieve Torina 27216
Museo della Paglia e dell'Intreccio D. Michalacci, Signa 27850
Museo della Pesca, Caslano 41281
Museo della Pesca, Magione 26565
Museo della Pesca e delle Civiltà Marinara, San Benedetto del Tronto 27609
Museo della Pietà Popolare, Molfetta 26764
Museo della Pietra Serena, Firenzuola 26205
Museo della Pievania, Corciano 25971
Museo della Pieve, Sant'Ambrogio di Valpolicella 27716
Museo della Pieve di San Pietro, Prato 27336
Museo della Pieve di Staggia, Poggibonsi 27267
Museo della Pipa, Gavirate 26278
Museo della Plastica, Pont Canavese 27281
Museo della Preistoria, Celano 25798
Museo della Preistoria della Tuscia, Valentano 28138
Museo della Preistoria e Museo Egizio, Milano 26703
Museo della Propositura di San Pietro, Montecatini Val di Nievole 26814
Museo della Regina, Cattolica 25785
Museo della Repubblica Partigiana, Montefiorino 26829
Museo della Resistenza e della Civiltà Contadina, Pietrabruna 27203
Museo della Resistenza Cà Malanca, Brisighella 25549
Museo della Resistenza e del Folclore Valsabbino, Pertica Bassa 27141
Museo della Rocca, Dozza 26046
Museo della Rocca Malatestiana, Cesena 25830
Museo della Sanità e dell'Assistenza, Bologna 25453
Museo della Santa, Bologna 25454
Museo della SAT, Trento 28057
Museo della Scuola, Bolzano 25488
Museo della Scuola, Pramollo 27326
Museo della Scuola Grande dei Carmini, Venezia 28193
Museo della Seta Abegg, Garlate 26274
Museo della Sindone, Torino 28007
Museo della Società di Esecutori di Pie Disposizioni, Siena 27841
Museo della Società di Studi Patri, Gallarate 26266
Museo della Specola, Bologna 25455
Museo della Stampa, Jesi 26425
Museo della Stampa, Soncino 27870
Museo della Stampa e Museum Graphia, Urbino 28125
Museo della Tarsia, Rolo 27438
Museo della Tecnica e del Lavoro MV Agusta, Samarate 27604
Museo della Tecnica e del Lavoro MV Augusta, Trequanda 28063
Museo della Tecnologia Contadina, Santu Lussurgiu 27735
Museo della Terza Armata, Padova 27037
Museo della Tipografia, Città di Castello 25907
Museo della Tonnara, Milazzo 26727
Museo della Tonnara, Stintino 27907
Museo della Transumanza e del Costume Abruzzese Molisano, Sulmona 27918
Museo della Val Codera, Novate Mezzola 26957
Museo della Val Gardena, Ortisei 27001
Museo della Valchiavenna, Chiavenna 25863
Museo della Valle, Cavargna 25789
Museo della Valle Cannobina, Gurro 26393
Museo della Valle Intelvi, Lanzo d'Intelvi 26452
Museo della Via Ostiense, Roma 27480
Museo della Vicaria di San Lorenzo, Zogno 28317
Museo della Villa Bettoni, Bogliaco 25434
Museo della Villa San Carlo Borromeo, Senago 27800
Museo della Vita, Castelmagno 25743
Museo della Vita Contadina, Castelnuovo Don Bosco 25748
Museo della Vita Contadina in Romagnolo, Russi 27579
Museo della Vita e del Lavoro della Maremma Settentrionale, Cecina e Marina 25793
Museo della Vita e delle Tradizioni Popolari Sarde, Nuoro 26967
Museo della Vite e del Vino, Carmignano 25681
Museo della Vite e del Vino, Rufina 27578
Museo della Viticoltura, Prarostino 27327
Museo della Xilografia, Carpi 25686
Museo della Zampogna, Scapoli 27785
Museo della Zisa, Palermo 27071
Museo dell'Abazzia di San Colombano, Bobbio 25433
Museo dell'Abbazia, Casamari-Veroli 25700
Museo dell'Abbazia, Castiglione a Casauria 25759
Museo dell'Abbazia, San Benedetto Po 27614
Museo dell'Abbazia di Farfa, Fara Sabina 26086
Museo dell'Abbazia di San Nilo, Grottaferrata 26374
Museo dell'Abbazia di Santa Maria in Sylvis, Sesto al Reghena 27822
Museo dell'Accademia d'Arte C. Scalabrino, Montecatini Terme 26813
Museo dell'Accademia di Belle Arti, Perugia 27149
Museo dell'Accademia Etrusca e della Città di Cortona, Cortona 25988
Museo dell'Accademia Ligustica di Belle Arti, Genova 26303
Museo dell'Aeronautica G. Caproni, Trento 28058
Museo dell'Agricoltura e del Mondo Rurale, San Martino in Rio 27663
Museo dell'Agro Falisco, Civita Castellana 25920
Museo dell'Agro Veientano, Formello 26232
Museo dell'Alta Val Venosta, Curon Venosta 26025
Museo dell'Alto Medioevo, Roma 27481
Museo dell'Apicoltura, Renon 27396
Museo dell'Apicultura, Abbiategrasso 25139
Museo dell'Arciconfraternita, Tolentino 27985
Museo dell'Arciconfraternita dei Genovesi, Cagliari 25583
Museo dell'Ardesia, Cicagna 25886
Museo dell'Aria dello Spazio, Due Carrare 26049
Museo dell'Arredo Contemporaneo, Russi 27580
Museo dell'Arte Classica, Sezione di Archeologia, Roma 27482
Museo dell'Arte Contadina, Marciana 26603
Museo dell'Arte Contemporanea e dei Pittori dell'Emigrazione, Recanati 27378
Museo dell'Arte del Cappello, Ghiffa 26332
Museo dell'Arte del Vino, Staffolo 27897
Museo dell'Arte della Seta, Catanzaro 25780
Museo dell'Arte della Tornitura del Legno, Pettenasco 27181
Museo dell'Arte e della Tecnologia Confetteria, Sulmona 27919
Museo dell'Arte Mineraria, Iglesias 26396
Museo dell'Arte Sacra, Aldino 25190
Museo dell'Arte Serica e Laterizia, Malo 26579
Museo dell'Arte Votiva, Cesena 25831
Museo dell'Artigianato Tessile e della Seta, Reggio Calabria 27381
Museo dell'Attore Napoletano, Napoli 26889
Museo dell'Automobile, San Martino in Rio 27664
Museo dell'Automobile Carlo Biscaretti di Ruffia, Torino 28008
Museo dell'Automobile e della Tecnica, Farra d'Isonzo 26089
Museo dell'Automobile L. Bonfanti, Romano d'Ezzelino 27551
Museo dell'Aviazione, Rimini 27408

Museo dell'Avifauna Appenninica, Sarnano 27741
Museo delle 3 Religioni Promoteiste, Bertinoro 25408
Museo delle Anime dei Defunti, Roma 27483
Museo delle Antiche Vie, Urbana 28118
Museo delle Armi, Terni 27963
Museo delle Armi Antiche, Martinsicuro 26614
Museo delle Armi Antiche, Offagna 26971
Museo delle Armi Antiche, San Marino 38004
Museo delle Armi Antiche e Fortezza Medievale, Acquaviva Picena 25148
Museo delle Armi, Armature e Strumenti di Tortura, Ischia Ponte 26412
Museo delle Armi e del Martello, Sarnano 27742
Museo delle Armi e delle Uniformi, Urbisaglia 28129
Museo delle Armi Moderne, San Marino 38005
Museo delle Arti e delle Tradizioni, Treviso 28072
Museo delle Arti e delle Tradizioni di Puglia, Latiano 26464
Museo delle Arti e delle Tradizioni Popolari, Canepina 25644
Museo delle Arti e delle Tradizioni Popolari, Sassoferrato 27764
Museo delle Arti e Tradizioni Popolari, Civitanova Marche 25922
Museo delle Arti e Tradizioni Popolari, Correggio 25981
Museo delle Bilance e dei Pesi, Monterchi 26846
Museo delle Cappelle Medicee, Firenze 26172
Museo delle Carrozze, Catanzaro 25781
Museo delle Carrozze, Firenze 26173
Museo delle Carrozze, Napoli 26890
Museo delle Carrozze, Trani 28048
Museo delle Carrozze dell'Ottocento, Verona 28234
Museo delle Centuriazioni, Granze 26364
Museo delle Ceramiche, Tolentino 27986
Museo delle Ceramiche Acerbo, Loreto Aprutino 26523
Museo delle Cere, Roma 27484
Museo delle Cere Anatomiche Luigi Cattaneo, Bologna 25456
Museo delle Cere - Strumenti di Tortura, San Marino 38006
Museo delle Cripta, Poggiardo 27266
Museo delle Culture, Castagnola 41283
Museo delle Culture del Mondo, Genova 26304
Museo delle Culture Extraeuropee Dinz Rialto, Rimini 27409
Museo delle Curiosità, San Marino 38007
Museo delle Due Guerre Mondiali, Loro Piceno 26525
Museo delle Erbe, Teggiano 27949
Museo delle Erbe, Veroli 28227
Museo delle Ferrovie, Monserrato 26788
Museo delle Ferrovie in Sardegna, Cagliari 25584
Museo delle Genti d'Abruzzo, Pescara 27163
Museo delle Genti delle Valli di Lanzo, Ceres 25807
Museo delle Genti di Montagna, Palazzuolo sul Senio 27062
Museo delle Grigne, Esino Lario 26065
Museo delle Guerre, Grassobbio 26365
Museo delle Icone, Peccioli 27129
Museo delle Icone e della Tradizione Bizantina, Frascineto 26248
Museo delle Marionette, Campomorone 25641
Museo delle Miniere, Valle Aurina 28144
Museo delle Mummie, Ferentillo 26096
Museo delle Mura, Roma 27485
Museo delle Musiche dei Popoli, Genova 26305
Museo delle Navi e delle Antiche Carte Geografiche, Bologna 25457
Museo delle Navi Romane, Fiumicino 26206
Museo delle Navi Romane, Nemi 26918
Museo delle Origini, Roma 27486
Museo delle Palafitte, Molina di Ledro 26767
Museo delle Pentole nella Storia, Rozzano 27576
Museo delle Porcellane, Firenze 26174
Museo delle Rimembranze, Cosenza 25991
Museo delle Scienze della Terra e del Lavoro Contadino, Piandimeleto 27191
Museo delle Sinopie, Pisa 27243
Museo delle Statue Menhir, Laconi 26438
Museo delle Statue - Stele Lunigianesi, Pontremoli 27293
Museo delle Tavolette di Biccherna, Siena 27842
Museo delle Terme, Bagni di Lucca 25346
Museo delle Terre Marchigiane, San Lorenzo in Campo 27657
Museo delle Tradizioni Agroalimentari della Sardegna, Siddi 27834
Museo delle Tradizioni ed Arti Contadine, Picciano 27198
Museo delle Tradizioni Popolari, Città di Castello 25908
Museo delle Tradizioni Popolari, Offida 26975
Museo delle Tradizioni Popolari, San Vito di Cadore 27690
Museo delle Trame Mediterranee, Gibellina 26338
Museo delle Truppe Alpine Mario Balocco, Biella 25419
Museo delle Valli, Argenta 25273
Museo delle Valli, Comacchio 25947
Museo delle Valli Valdesi, Torre Pellice 28040
Museo dell'Emigrante Casa Giannini, Favale di Malvaro 26091
Museo dell'Emigrazione G.B. Scalabrini, Francavilla Angitola 26244
Museo dell'Energia Elettrica, Brescia 25537
Museo dell'Enoteca Regionale Piemontese, Grinzane Cavour 26369
Museo dell'Homo Salvadego, Cosio Valtellino 25993
Museo dell'Illustrazione, Ferrara 26112
Museo dell'Immagine, Cesena 25832
Museo dell'Informazione, Senigallia 27802
Museo dell'Ingegno e della Tecnologia Preindustriale, Colorno 25944
Museo dell'Intreccio Mediterraneo, Castelsardo 25753
Museo dell'Istituto di Anatomia Patologica, Padova 27038
Museo dell'Istituto di Patologia del Libro, Roma 27487
Museo dello Sbarco, Anzio 25245
Museo dello Sbarco Alleato, Nettuno 26922
Museo dello Scalpellino Povese, Pove del Grappa 27319
Museo dello Scarpone e della Scarpa Sportiva, Montebelluna 26808
Museo dello Sfilato Siciliano, Chiaramonte Gulfi 25853
Museo dello Spazzacamino, Santa Maria Maggiore 27706
Museo dello Strumento Musicale, Reggio Calabria 27382
Museo dell'Occhiale, Pieve di Cadore 27209
Museo dell'Olio, Cisano 25900
Museo dell'Olio in Sabina, Castelnuovo di Farfa 25747
Museo dell'Olivo, Imperia 26405
Museo dell'Olivo e dell'Olio, Torgiano 27994
Museo dell'Ombrello e del Parasole, Gignese 26340
Museo dell'Opera del Duomo, Orvieto 27007
Museo dell'Opera del Duomo, Prato 27337
Museo dell'Opera del Santuario, Tolentino 27987
Museo dell'Opera di Santa Croce, Firenze 26175
Museo dell'Opera di Santa Maria del Fiore, Firenze 26176
Museo dell'Opera di Santa Niccolao, Buggiano 25562
Museo dell'Opera e del Territorio, Caserta 25704
Museo dell'Opera Metropolitana, Siena 27843
Museo dell'Opificio delle Pietre Dure, Firenze 26177
Museo dell'Orologio da Torre G.B. Bergallo, Tovo San Giacomo 28047
Museo dell'Orso, Villavallelonga 28282
Museo dell'Osservatorio, Monte Porzio Catone 26801
Museo dell'Osservatorio Astronomico di Capodimonte, Napoli 26891
Museo dell'Uomo, Serrapetrona 27815
Museo dell'Uomo, Susegana 27928
Museo dell'Uomo e dell'Ambiente, Terra del Sole 27966
Museo dell'Utensileria, Lanzo Torinese 26453
Museo Delso, L'Alfàs del Pi 38838
Museo Demantropologico, Cervaro 25818
Museo Demoantropologico, Pastena 27112
Museo Demologico dell'Economia, del Lavoro e della Storia Sociale Silana, San Giovanni in Fiore 27644
Museo Départemental A. Demard, Le Haut-du-Them 13593
Museo des Origenes, La Habana 09142
Museo di Anatomia Comparata Battista Grassi, Roma 27488
Museo di Anatomia e Ostetricia, Bologna 25458
Museo di Anatomia Patologica, Roma 27489
Museo di Anatomia Patologica e Teratologia Veterinaria, Ozzano dell'Emilia 27026
Museo di Anatomia Umana, Napoli 26892
Museo di Anatomia Umana, Torino 28009
Museo di Anatomia Umana Normale, Parma 27101
Museo di Anatomia Veterinaria, Napoli 26893
Museo di Antichita, Torino 28010
Museo di Antichità Etrusche e Italiche, Roma 27490
Museo di Antonio da Sangallo il Giovane, Montefiascone 26824
Museo di Antropologia, Bologna 25459
Museo di Antropologia, Napoli 26894
Museo di Antropologia Criminale, Torino 28011
Museo di Antropologia ed Etnografia, Padova 27039
Museo di Antropologia ed Etnografia, Torino 28012
Museo di Antropologia ed Etnologia, Firenze 26178
Museo di Antropologia G. Sergi, Roma 27491
Museo di Apicoltura Guido Fregonese, Magera di Oderzo 26564
Museo di Archeologia Sacra e Profana, Roma 27492
Museo di Archeologia Urbana e dei Commerc Antichi in Abruzzo, Lanciano 26447
Museo di Archeologia Urbana Fiorelli, Lucera 26538
Museo di Architettura, Ferrara 26113
Museo di Architettura Militare, Bologna 25460
Museo di Arte Cinese ed Etnografico, Parma 27102
Museo di Arte Contadina, Montegallo 26833
Museo di Arte Contemporanea, Stia 27903
Museo di Arte Contemporanea e del Novecento, Monsummano Terme 26791
Museo di Arte e Giacimenti Minerari, Latina 26466
Museo di Arte e Storia delle Miniere, Massa Marittima 26631
Museo di Arte e Tradizione Contadina, Olevano di Lomellina 26980
Museo di Arte Medioevale e Moderna, Arezzo 25267
Museo di Arte Religiosa, Oleggio 26979
Museo di Arte Sacra, Agnone 25156
Museo di Arte Sacra, Bientina 25421
Museo di Arte Sacra, Castrovillari 25769
Museo di Arte Sacra, Longiano 26517
Museo di Arte Sacra, Sabbioneta 27584
Museo di Arte Sacra, Saracena 27739
Museo di Arte Sacra, Suvereto 27932
Museo di Arte Sacra, Vibo Valentia 28247
Museo di Arte Sacra del Castello, Genga 26287
Museo di Arte Sacra San Giuseppe, Rogliano 27437
Museo di Arti Decorative della Fondazione Pietro Accorsi, Torino 28013
Museo di Auto e Moto d'Epoca, Vigolzone 28270
Museo di Biologia Marina, Fano 26085
Museo di Biologia Marina, Porto Cesareo 27302
Museo di Biologia Marina, Taranto 27942
Museo di Biologia Marina & Paleontologia, Reggio Calabria 27383
Museo di Blenio, Lottigna 41575
Museo di Botanica, Firenze 26179
Museo di Capodimonte, Napoli 26895
Museo di Casa Vasari, Arezzo 25268
Museo di Castelgrande, Bellinzona 41193
Museo di Castelvecchio, Verona 28235
Museo di Chimica, Roma 27493
Museo di Cimeli Storico Militari, Chiaramonte Gulfi 25854
Museo di Civiltà Preclassiche della Murgia Meridionale, Ostuni 27023
Museo di Criminologia Medievale, San Gimignano 27639
Museo di Criminologico, Roma 27494
Museo di Cultura Contadina, Oriolo 26994
Museo di Cultura Contadina dell'Alta Val Trebbia, Montebruno 26811
Museo di Cultura Popolare, Egna 26051
Museo di Dipinti Sacri Ponte dei Greci, Venezia 28194
Museo di Ecologia e Storia Naturale, Marano sul Panaro 26601
Museo di Etnografico e Antropologia Giovanni Podenzana, La Spezia 26435
Museo di Etnomedicina A. Scarpa, Genova 26306
Museo di Etnomedicina A. Scarpa, Piazzola sul Brenta 27197
Museo di Etnopreistoria, Napoli 26896
Museo di Fisica, Bologna 25461
Museo di Fisica, Napoli 26897
Museo di Fossili, Ronca 27553
Museo di Fotografia Contemporanea, Cinisello Balsamo 25896
Museo di Gallese e Centro Culturale Marco Scacchi, Gallese 26267
Museo di Geologia, Camerino 25619
Museo di Geologia, Napoli 26898
Museo di Geologia, Roma 27495
Museo di Geologia e Paleontologia, Firenze 26180
Museo di Geologia e Paleontologia, Padova 27040
Museo di Geologia e Paleontologia, Pavia 27121
Museo di Geologia e Paleontologia, Torino 28014
Museo di Leventina, Giornico 41433
Museo di Merceologia, Roma 27496
Museo di Milano, Milano 26704
Museo di Minerali e Fossili, Montefiore Conca 26826
Museo di Mineralogia, Massa Marittima 26632
Museo di Mineralogia, Palermo 27072
Museo di Mineralogia, Roma 27497
Museo di Mineralogia, di Petrologia e Geologia, Modena 26751
Museo di Mineralogia e Paleontologia, Saint Vincent 27589
Museo di Mineralogia e Petrografia, Parma 27103
Museo di Mineralogia e Petrografia, Pavia 27122
Museo di Mineralogia e Petrografia L. Bombicci, Bologna 25462
Museo di Mineralogia e Petrologia, Padova 27041
Museo di Mineralogia L. De Prunner, Cagliari 25585
Museo di Muggia e del Territorio, Muggia 26875
Museo di Nostra Signora di Bonaria, Cagliari 25586
Museo di Palazzo Altieri, Oriolo Romano 26995
Museo di Palazzo d'Arco, Mantova 26595
Museo di Palazzo Davanzati, Firenze 26181
Museo di Palazzo De Vio, Gaeta 26260
Museo di Palazzo della Penna, Perugia 27150
Museo di Palazzo Ducale, Mantova 26596
Museo di Palazzo Gamilli, Nocera Umbra 26934
Museo di Palazzo Mirto, Palermo 27073
Museo di Palazzo Mocenigo, Venezia 28195
Museo di Palazzo Pepoli Campogrande, Bologna 25463
Museo di Palazzo Piccolomini, Pienza 27201
Museo di Palazzo Pretorio, Certaldo 25815
Museo di Palazzo Reale, Napoli 26899
Museo di Palazzo Reale Genova, Genova 26307
Museo di Palazzo Strozzi, Firenze 26182
Museo di Palazzo Vecchio, Firenze 26183
Museo di Paleobotanica ed Etnobotanica dell'Orto Botanico, Napoli 26900
Museo di Paleontologia, Milano 26705
Museo di Paleontologia, Modena 26752
Museo di Paleontologia, Napoli 26901
Museo di Paleontologia, Roma 27498
Museo di Paleontologia dell'Accademia Federiciana, Catania 25774
Museo di Paleontologia e Mineralogia, Campomorone 25642
Museo di Paleontologia e Paletnologia, Maglie 26567
Museo di Paleontologia e Speleologia E.A. Martel, Carbonia 25676
Museo di Paleontologia G. Buriani, San Benedetto del Tronto 27610
Museo di Paletnologia, Polignano a Mare 27272
Museo di Pantelleria, Pantelleria 27088
Museo di Peppone e Don Camillo, Brescello 25532
Museo di Pittura Murale, Prato 27338
Museo di Prali e della Val Germanasca, Prali 27324
Museo di Preistoria e del Mare, Trapani 28050
Museo di Preistoria e Protostoria della Valle del Fiume Fiora, Manciano 26582
Museo di Preistorico e Paleontologico, Sant'Anna d'Alfaedo 27723
Museo di Rievocazione Storica, Mondavio 26776
Museo di Rocca Fregoso, Sant'Agata Feltria 27715
Museo di Rodoretto, Prali 27325
Museo di Roma, Roma 27499
Museo di Roma in Trastevere, Roma 27500
Museo di San Domenico, Bologna 25464
Museo di San Francesco, Aversa 25333
Museo di San Francesco, Montella 26835
Museo di San Marco, Firenze 26184
Museo di San Marco, Venezia 28196
Museo di San Martino, Olivone 41705
Museo di San Michele, Sagrado 27586
Museo di San Petronio, Bologna 25465
Museo di San Pietro, Assisi 25309
Museo di Santa Cecilia, San Lazzaro di Savena 27651
Museo di Santa Maria degli Angioli, Lugano 41584
Museo di Santa Maria delle Grazie, San Giovanni Valdarno 27648
Museo di Santa Maria di Castello, Genova 26308
Museo di Sant'Agostino, Genova 26309
Museo di Sant'Antonino, Piacenza 27188
Museo di Santo Stefano, Bologna 25466
Museo di Scienza dela Terra, San Gemini 27634
Museo di Scienza della Terra, Bari 25365
Museo di Scienze Archeologiche e d'Arte, Padova 27042
Museo di Scienze della Terra U. Baroli, Crodo 26010
Museo di Scienze Naturali, Belvi 25392
Museo di Scienze Naturali, Bolzano 25489
Museo di Scienze Naturali, Brescia 25538
Museo di Scienze Naturali, Camerino 25620
Museo di Scienze Naturali, Cesena 25833
Museo di Scienze Naturali, Città della Pieve 25902
Museo di Scienze Naturali, Malnate 26576
Museo di Scienze Naturali, Pavia 27123
Museo di Scienze Naturali, Rosignano Marittimo 27562
Museo di Scienze Naturali A. Brandolini, Ravenna 27373
Museo di Scienze Naturali del Collegio, Lodi 26509
Museo di Scienze Naturali Don Bosco, Alassio 25171
Museo di Scienze Naturali ed Umane, L'Aquila 26455
Museo di Scienze Naturali L. Paolucci, Offagna 26972
Museo di Scienze Naturali Tommaso Salvadori, Fermo 26100
Museo di Sculture Iperspaziali, Bomarzo 25491
Museo di Speleologia, Borgo Grotta Gigante 25503
Museo di Speleologia e Carsismo A. Parolini, Valstagna 28151
Museo di Speleologia Vincenzo Rivera, L'Aquila 26456
Museo di Stato, San Marino 38008
Museo di Storia Contadina, Fontanabona di Pagnacco 26220
Museo di Storia Contemporanea, Milano 26706
Museo di Storia, d'Arte e d'Antichità Don Florindo Piolo, Serravalle Sesia 27818
Museo di Storia della Fisica, Padova 27043
Museo di Storia della Medicina, Roma 27501
Museo di Storia dell'Agricoltura, Cesena 25834
Museo di Storia dell'Agricoltura, Urbania 28120
Museo di Storia dell'Agricoltura e della Pastorizia, Morano Calabro 26867
Museo di Storia delle Mezzadria, Senigallia 27803
Museo di Storia e Arte, Gorizia 26352
Museo di Storia e Civiltà, Medesano 26646

Museo di Storia e Cultura Contadina, Genova 26310
Museo di Storia e Cultura della Val del Biois, Vallada Agordina 28143
Museo di Storia Locale, Velturno 28174
Museo di Storia Naturale, Cremona 26004
Museo di Storia Naturale, Follonica 26216
Museo di Storia Naturale, Genova 26311
Museo di Storia Naturale, Macerata 26553
Museo di Storia Naturale, Merate 26661
Museo di Storia Naturale, Parma 27104
Museo di Storia Naturale, Parma 27105
Museo di Storia Naturale, Perugia 27151
Museo di Storia Naturale, Piacenza 27189
Museo di Storia Naturale, Senna Lodigiana 27806
Museo di Storia Naturale, Stroncone 27914
Museo di Storia Naturale, Sulmona 27920
Museo di Storia Naturale, Venezia 28197
Museo di Storia Naturale A. Orsini, Ascoli Piceno 25295
Museo di Storia Naturale A. Stoppani, Venegono Inferiore 28177
Museo di Storia Naturale Aquilegia, Assemini 25303
Museo di Storia Naturale del Cilento, Laureana Cilento 26469
Museo di Storia Naturale della Lunigiana, Aulla 25328
Museo di Storia Naturale della Maremma, Grosseto 26372
Museo di Storia Naturale dell'Accademia dei Fisocritici, Siena 27844
Museo di Storia Naturale Don Bosco, Torino 28015
Museo di Storia Naturale e del Territorio, Calci 25593
Museo di Storia Naturale e dell'Uomo, Carrara 25690
Museo di Storia Naturale Faraggiana Ferrandi, Novara 26950
Museo di Storia Naturale S. Ferrari, Bedonia 25387
Museo di Storia Naturale - Sezione Mineralogia, Firenze 26185
Museo di Storia Naturale - Sezione Zoologica La Specola, Firenze 26186
Museo di Storia Naturali e Archeologia, Montebelluna 26809
Museo di Storia Quarnese, Quarna Sotto 27349
Museo di Strumenti del Conservatorio Statale di Musica Giuseppe Verdi, Torino 28016
Museo di Torcello, Venezia 28198
Museo di Tossicia, Tossicia 28046
Museo di Tradizioni Popolari e del Costume d'Epoca, San Pietro Avellana 27679
Museo di Val Verzasca, Sonogno 41889
Museo di Vallemaggia, Giumaglio 41435
Museo di Valmaggia, Cevio 41285
Museo di Villa Beatice d'Este, Cinto Euganeo 25897
Museo di Villa Cagnola, Gazzada 26280
Museo di Villa Faraggiana, Albissola 25187
Museo di Vita Artigiana e Contadina, Scarperia 27788
Museo di Vita Cappuccina, Genova 26312
Museo di Vita Quotidiana e Lavoro in Villa, Orgiano 26988
Museo di Zoologia, Bari 25366
Museo di Zoologia, Bologna 25467
Museo di Zoologia, Napoli 26902
Museo di Zoologia, Roma 27502
Museo di Zoologia e Mineralogia, Siddi 27835
Museo di Zoologia Pietro Doderlein, Palermo 27074
Museo di Zoologico e di Anatomia, Torino 28017
Museo Diacesano, Enna 26057
Museo Didáctico Antonini, Nasca 34280
Museo Didáctico Artiguista, Maldonado 54380
Museo Didáctico de Prehistoria, Alba de Tormes 38801
Museo Didattico, Morimondo 26870
Museo Didattico, Niscemi 26928
Museo Didattico, Tivoli 27978
Museo Didattico Archeologico, Laterza 26463
Museo Didattico d'Arte e Vita Preistorica, Capo di Ponte 25658
Museo Didattico del Mare, Napoli 26903
Museo Didattico della Riserva Regionale Incisioni Rupestri, Ceto 25841
Museo Didattico della Scuola Medica Salernitana, Salerno 27596
Museo Didattico della Seta, Como 25952
Museo Didattico delle Maioliche, Grottaglie 26375
Museo Didattico Giuseppe Pellizza, Volpedo 28307
Museo Didattico Zoologico, Oria 26992
Museo Diefenbach, Capri 25666
Museo Diego Aragona Pignatelli Cortes, Napoli 26904
Museo Diego Rivera Anahuacalli, México 31110
Museo Diffuso, Pennabilli 27132
Museo Dillmann S. Bullock, Angol 07427
Museo d'Interesse Locale, Crespadoro 26007
Museo Diocesà, Tarragona 40100
Museo Diocesà i Comarcal, Lleida 39476
Museo Diocesano, Agrigento 25163
Museo Diocesano, Albenga 25177
Museo Diocesano, Ancona 25229
Museo Diocesano, Andria 25234
Museo Diocesano, Ariano Irpino 25277
Museo Diocesano, Ascoli Piceno 25296
Museo Diocesano, Bari 25367
Museo Diocesano, Benevento 25395
Museo Diocesano, Bisceglie 25425
Museo Diocesano, Brugnato 25553
Museo Diocesano, Capua 25669
Museo Diocesano, Castelvecchio Subequo 25755
Museo Diocesano, Catanzaro 25782
Museo Diocesano, Ciudad Real 39189
Museo Diocesano, Córdoba 39206
Museo Diocesano, Cortona 25989
Museo Diocesano, Cuenca 39235
Museo Diocesano, Faenza 26075
Museo Diocesano, Fermo 26101
Museo Diocesano, Gaeta 26261
Museo Diocesano, Gerace 26328
Museo Diocesano, Gubbio 26389
Museo Diocesano, Imola 26401
Museo Diocesano, Jesi 26426
Museo Diocesano, Lamezia Terme 26444
Museo Diocesano, Lanciano 26448
Museo Diocesano, Lanusei 26450
Museo Diocesano, Mantova 26597
Museo Diocesano, Mazara del Vallo 26644
Museo Diocesano, Milano 26707
Museo Diocesano, Monopoli 26780
Museo Diocesano, Nola 26936
Museo Diocesano, Osimo 27012
Museo Diocesano, Palencia 39771
Museo Diocesano, Piazza Armerina 27194
Museo Diocesano, Pienza 27202
Museo Diocesano, Pistoia 27252
Museo Diocesano, Pitigliano 27256
Museo Diocesano, Salerno 27597
Museo Diocesano, San Miniato 27673
Museo Diocesano, Santa Lucia del Mela 27702
Museo Diocesano, Sant'Agata de Goti 27714
Museo Diocesano, Sarzana 27751
Museo Diocesano, Sassari 27753
Museo Diocesano, Spoleto 27894
Museo Diocesano, Squillace 27896
Museo Diocesano, Sulmona 27921
Museo Diocesano, Trani 28049
Museo Diocesano, Tropea 28105
Museo Diocesano, Tudela 40174
Museo Diocesano, Urbania 28121
Museo Diocesano, Valle del Espíritu Santo 54599
Museo Diocesano, Vallo della Lucania 28148
Museo Diocesano, Velletri 28172
Museo Diocesano A. Bernareggi, Bergamo 25403
Museo Diocesano Albani, Urbino 28126
Museo Diocesano Antonio Bergamaschi, Pennabilli 27133
Museo Diocesano-Catedralicio, Barbastro 38950
Museo Diocesano-Catedralicio, Segorbe 40046
Museo Diocesano-Catedralicio, Valladolid 40238
Museo Diocesano Catedralicio de Arte Sacro, Orihuela 39744
Museo Diocesano d'Arte Sacra, Alghero 25198
Museo Diocesano d'Arte Sacra, Brescia 25539
Museo Diocesano d'Arte Sacra, Caltanissetta 25609
Museo Diocesano d'Arte Sacra, Grosseto 26373
Museo Diocesano d'Arte Sacra, Nicotera 26926
Museo Diocesano d'Arte Sacra, Orte 27000
Museo Diocesano d'Arte Sacra, Pinerolo 27222
Museo Diocesano d'Arte Sacra, Piteglio 27254
Museo Diocesano d'Arte Sacra, Pordenone 27299
Museo Diocesano d'Arte Sacra, Rossano 27565
Museo Diocesano d'Arte Sacra, Scaria di Lonzo 27786
Museo Diocesano d'Arte Sacra, Treviso 28073
Museo Diocesano d'Arte Sacra, Trivento 28101
Museo Diocesano d'Arte Sacra A. Luciani, Vittorio Veneto 28300
Museo Diocesano d'Arte Sacra Santa Apollonia, Venezia 28199
Museo Diocesano de Arte Antiguo, Sigüenza 40079
Museo Diocesano de Arte Religioso Nuestra Señora del Cisne, Loja 10418
Museo Diocesano de Arte Sacro, Bilbao 39045
Museo Diocesano de Arte Sacro, Moguer 39654
Museo Diocesano de Arte Sacro, Las Palmas de Gran Canaria 39800
Museo Diocesano de Arte Sacro, Teruel 40121
Museo Diocesano de Huesca, Huesca 39400
Museo Diocesano de Jaca, Jaca 39418
Museo Diocesano de San Sebastián, Donostia-San Sebastián 39252
Museo Diocesano de Tui, Tui 40176
Museo Diocesano della Basilica del Crocefisso, Amalfi 25216
Museo Diocesano di Arte Sacra, Chiavari 25859
Museo Diocesano di Arte Sacra, Genova 26313
Museo Diocesano di Arte Sacra, Lodi 26510
Museo Diocesano di Arte Sacra, Palermo 27075
Museo Diocesano di Arte Sacra, Santa Severina 27710
Museo Diocesano di Arte Sacra, Sarsina 27747
Museo Diocesano di Arte Sacra, Susa 27927
Museo Diocesano di Arte Sacra, Volterra 28310
Museo Diocesano di Arte Sacra del Duomo, Arezzo 25269
Museo Diocesano di Arte Sacra San Gregorio Barbarigo, Padova 27044
Museo Diocesano di Santo Stefano, Firenze 26187
Museo Diocesano e Catedralicio de Lugo, Lugo 39498
Museo Diocesano e Catedralicio Santos Sancristobal, Mondoñedo 39661
Museo Diocesano e Collezione dei Presepi, Bressanone 25544
Museo Diocesano e Gallerie del Tiepolo, Udine 28113
Museo Diocesano e Oratorio di San Bernardino, Siena 27845
Museo Diocesano e Tesoro della Cattedrale, Troia 28103
Museo Diocesano Intercomunale di Arte Sacra, Castignano 25764
Museo Diocesano Intercomunale di Arte Sacra, Comunanza 25959
Museo Diocesano Intercomunale di Arte Sacra, Grottammare 26376
Museo Diocesano Intercomunale di Arte Sacra, Monteprandone 26843
Museo Diocesano Intercomunale di Arte Sacra, Ripatransone 27417
Museo Diocesano Intercomunale di Arte Sacra, Rotella 27566
Museo Diocesano Lucas Guillermo Castillo, Falcón 54572
Museo Diocesano Regina Coeli, Santillana del Mar 40034
Museo Diocesano San Pietro, Teggiano 27950
Museo Diocesano Teatino, Chieti 25875
Museo Diocesano Tridentino, Trento 28059
Museo Diocesano y Catedralicio, Salamanca 39943
Museo Diocesano y Cedralicio, Calahorra 39094
Museo División del Norte, Canutillo 30749
Museo División del Norte, Lerdo 31004
Museo do Humor, Fene 39295
Museo do Mosteiro de San Paio de Antealtares, Santiago de Compostela 40025
Museo do Moucho, Cerceda 39162
Museo do Pobo Galego, Santiago de Compostela 40026
Museo Doctor Francia, Yaguarón 34104
Museo Documentario della Città, Gradisca d'Isonzo 26358
Museo Doganale Svizzero, Gandria 41396
Museo Dolores Olmedo, México 31111
Museo Domenicani, Taggia 27935
Museo Don Pio Palla, Leonessa 26485
Museo Doña Rosario, Santiago de Cuba 09321
Museo Donazione Luxoro, Genova 26314
Museo Donazione U. Mastroianni, Arpino 25285
Museo Donizettiano, Bergamo 25404
Museo Dr. Alfredo Gutierrez Valenzuela, Sucre 04249
Museo Dr. Atl, Pihuamo 31259
Museo Dr. José Gómez Panaco, Balancán 30734
Museo Duca di Martina, Napoli 26905
Museo Ducati, Bologna 25468
Museo Dupré, Fiesole 26128
Museo E. Bernardi, Padova 27045
Museo e Biblioteca della Resistenza, Sansepolcro 27699
Museo e Casa Morandi, Bologna 25469
Museo e Certosa di San Martino, Napoli 26906
Museo e Chiostri Monumentali di Santa Maria Novella, Firenze 26188
Museo E. Galessi, Montevarchi 26857
Museo e Galleria di Villa Borghese, Roma 27503
Museo e Galleria Mozzi Bardini, Firenze 26189
Museo E. Leandro, San Cesario di Lecce 27623
Museo e Pinacoteca Basilio Cascella, Pescara 27164
Museo e Pinacoteca Civica, Alessandria 25193
Museo e Pinacoteca Comunali, Gubbio 26390
Museo e Pinacoteca della Basilica di San Paolo, Roma 27504
Museo e Pinacoteca della Città, Todi 27981
Museo e Pinacoteca della Comunità di Fiemme, Cavalese 25788
Museo e Pinacoteca Diocesana, Camerino 25621
Museo e Pinacoteca Nazionale di Palazzo Mansi, Lucca 26535
Museo e Pinacoteca S. Gentili, San Ginesio 27642
Museo E. Renzi, Coriano 25976
Museo e Villaggio Africano, Basella di Urgnano 25377
Museo Ebraico, Asti 25316
Museo Ebraico, Bologna 25470
Museo Ebraico, Ferrara 26114
Museo Ebraico, Firenze 26190
Museo Ebraico, Livorno 26498
Museo Ebraico, Soragna 27877
Museo Ebraico di Roma, Roma 27505
Museo Ebraico - Jüdisches Museum, Merano 26659
Museo Ecclesiastico, Caorle 25651
Museo Ecuatoriano de Ciencias Naturales, Quito 10443
Museo Eduardo Barreiros, Valdemorillo 40187
Museo Eduardo Carranza, Villavicencio 08788
Museo Eerola, Tuulos 11410
Museo Efrain Martínez Zambrano, Popayán 08728
Museo Egipcio, Montevideo 54402
Museo Egizio, Torino 28018
Museo Egnazia, Monopoli 26781
Museo El Cañón, Puerto Boniato 09259
Museo El Castillo Diego Echavarría, Medellín 08678
Museo el Centenario, Garza García 30873
Museo El Chapulín, Saltillo 31316
Museo El Hombre y la Naturaleza, San Juan 00587
Museo El Hurón Azul, La Habana 09143
Museo El Jonotal, Playa Vicente, Veracruz 31261
Museo El Minero, Santa Bárbara 31375
Museo El Pariancito, San Luis Potosí 31350
Museo el Pausílipo, Las Tablas 34076
Museo El Piñero, Nueva Gerona 09236
Museo El Polvorín, Monclova 31174
Museo El Reencuentro, Darregueira 00352
Museo El Remate, Comala 30789
Museo El Sol de las Botellas, Barcelona 54522
Museo Elder, Las Palmas de Gran Canaria 39801
Museo Elías Nandino, Cocula 30779
Museo Elisarion, Minusio 41630
Museo elle Arti e Tradizioni Contadine, Roccasecca dei Volsci 27434
Museo Emidiano, Agnone 25157
Museo Emilia Ortiz, Tepic 31443
Museo Emilio Greco, Catania 25775
Museo Emilio Greco, Orvieto 27008
Museo Enologico Grasso, Milazzo 26728
Museo Enrico Poli, Lima 34257
Museo Entomológco Mariposas del Mundo, San Miguel 00594
Museo Entomológico, León 33286
Museo Entomológico, Lima 34258
Museo Entomológico Luis E. Peña Guzman, Santiago de Chile 07459
Museo Epper, Ascona 41130
Museo Erbario, Roma 27506
Museo Eritreo Bottego, Parma 27106
Museo Ermita de Santa Elena, Irún 39412
Museo Ernest Hemingway, San Francisco de Paula 09278
Museo Ernesto Treccani, Milano 26708
Museo Escolar de Pusol, Elche 39267
Museo Escolar de Sitio de la Matamba, Jamapa 30971
Museo Escolar María Goretti, Pasto 08709
Museo Escolar y Comunitario de Acamalín, Xico 31562
Museo Escultórico de Geles Cabrera, México 31112
Museo Específico de la Academia de Artillería, Segovia 40054
Museo Específico de la Academia de Caballería, Valladolid 40239
Museo Específico de la Academia de Ingenieros, Hoyo de Manzanares 39395
Museo Específico de la Academia General Militar, Zaragoza 40346
Museo Específico de la Brigada Paracaidista, Alcalá de Henares 38814
Museo Específico de la Legión, Ceuta 39171
Museo Estatal de Culturas Populares, Monterrey 31185
Museo Estatal de Patología, Puebla 31286
Museo Estudio Diego Rivera, México 31113
Museo Ethnográfico, Florencia 08641
Museo Ethnográfico de las Obras Misionales Pontificias, Buenos Aires 00229
Museo Ethnografico Don L. Pellegrini delle Civiltà Contadina, San Pellegrino in Alpe 27676
Museo Ethnohistórico, Soatá 08757
Museo Etiopico G. Massaia, Frascati 26245
Museo Etnico-Agricolo-Pastorale ed Artigiano, Alcamo 25189
Museo Etnico Arbereshe, Civita 25919
Museo Étnico de los Seris, Hermosillo 30930
Museo Étnico de los Yaquis, Obregón 31230
Museo Etnico della Civiltà Salentina Agrilandi Museum, Brindisi 25546
Museo Etno-Antropologico, Calatafimi 25592
Museo Etno-Antropologico, San Cipirello 27625
Museo Etno-Antropologico, San Mauro Castelverde 27667
Museo Etno-Antropologico, Scaletta Zanclea 27782
Museo Etno-Antropologico della Terra di Zabut, Sambuca di Sicilia 27605
Museo Etno-Antropologico della Valle del Belice, Gibellina 26339
Museo Etno-Antropologico delle Madonie, Geraci Siculo 26330
Museo Etnoarqueológico del Beni Kenneth Lee, Trinidad 04254
Museo Etnobotánico, Tuluá 08771
Museo Etnográfico, Albán 08526
Museo Etnográfico, Alcubilla del Marqués 38830
Museo Etnografico, Aprica 25252
Museo Etnografico, Aritzo 25279

M

Museo Etnográfico, Atauta 38914
Museo Etnografico, Bomba 25492
Museo Etnográfico, Cajamarca 34129
Museo Etnográfico, Castrojeriz 39155
Museo Etnográfico, Cilleros 39183
Museo Etnografico, Collinas 25941
Museo Etnografico, Cortina d'Ampezzo 25985
Museo Etnografico, Faeto 26080
Museo Etnografico, Fossalta di Portogruaro 26237
Museo Etnografico, Fratta Polesine 26250
Museo Etnográfico, Grado 39353
Museo Etnográfico, Grandas de Salime 39372
Museo Etnográfico, Inzá 08662
Museo Etnografico, Lusevera 26549
Museo Etnográfico, Madruga 09199
Museo Etnografico, Malborghetto Valbruna 26573
Museo Etnografico, Montodine 26861
Museo Etnografico, Novalesa 26948
Museo Etnografico, Novara di Sicilia 26956
Museo Etnografico, Oneta 26983
Museo Etnografico, Ortonovo 27006
Museo Etnografico, Ossimo Superiore 27014
Museo Etnografico, Parma 27107
Museo Etnografico, Ponte in Valtellina 27283
Museo Etnográfico, Quintana Redonda 39886
Museo Etnografico, Roccalbegna 27431
Museo Etnográfico, Romanillos de Medinaceli 39912
Museo Etnografico, Roseto Capo Spulico 27558
Museo Etnográfico, Santa Lucía 09302
Museo Etnográfico, Santa Marta 08753
Museo Etnografico, Schilpario 27791
Museo Etnográfico, Tuxtla Gutiérrez 31515
Museo Etnografico, Vione 28286
Museo Etnografico Africo-Mozambico, Bari 25368
Museo Etnográfico Andrés Barbero, Asunción 34097
Museo Etnografico Antico Mulino ad Acqua Licheri, Fluminimaggiore 26207
Museo Etnografico Archeologico, Palau 27057
Museo Etnográfico Benigno Eiriz, Alija del Infantado 38847
Museo Etnografico Cerginolano, Cerignola 25810
Museo Etnográfico Comarcal de Azuaga, Azuaga 38932
Museo Etnografico Comunale, Premana 27343
Museo Etnografico Coumboscuro della Civiltà Provenzale in Italia, Sancto Lucia de Coumboscuro 27692
Museo Etnográfico da Limia, Vilar de Santos 40289
Museo Etnográfico de Autilla, Autilla del Pino 38917
Museo Etnográfico de Cantabria, Muriedas 39705
Museo Etnográfico de Colombia, Bogotá 08579
Museo Etnográfico de Don Benito, Don Benito 39246
Museo Etnográfico de el Cerrp de Andèvalo, El Cerro de Andévalo 39163
Museo Etnográfico de Oyón-Oion, Oyón 39762
Museo Etnográfico de Quirós y Comarca, Quirós 39887
Museo Etnográfico de Valderredible, Valderredible 40194
Museo Etnografico del Cansiglio Servadei, Tambre 27938
Museo Etnografico del Cologio Nacional Mejia, Quito 10444
Museo Etnografico del Coumboscuro, Monterosso Grana 26849
Museo Etnografico del Ferro, Bienno 25420
Museo Etnográfico del Oriente de Asturias, Llanes 39471
Museo Etnografico del Pinerolese e Museo del Legno, Pinerolo 27223
Museo Etnografico del Ponente Ligure, Cervo 25823
Museo Etnográfico del Reino de Pamplona, Ollo 39725
Museo Etnografico della Civiltà Contadina, Marianopoli 26607
Museo Etnografico della Cultura Contadina, Morigerati 26869
Museo Etnografico della Lunigiana, Villafranca in Lunigiana 28277
Museo Etnografico della Piana del Dragone, Volturara Irpina 28313
Museo Etnografico della Provincia die Belluno, Cesiomaggiore 25840
Museo Etnografico della Torre, Albino 25181
Museo Etnografico della Trinità, Botticino 25518
Museo Etnografico della Val Varatella, Toirano 27982
Museo Etnografico della Valle, Ultimo 28117
Museo Etnografico della Valle Brembana, Zogno 28318
Museo etnografico della Valle di Muggio, Cabbio 41279
Museo Etnografico dell'Alta Valle Seriana, Ardesio 25262
Museo Etnografico dell'Etna, Linguaglossa 26492
Museo Etnografico di Servola, Trieste 28093
Museo Etnografico di Sorrentini, Patti 27117
Museo Etnografico Don Luigi Pellegrini, Castiglione di Garfagnana 25762
Museo Etnográfico e da Historia de San Paio de Narla, Friol 39310
Museo Etnografico e del Folklore Valsesiano, Borgosesia 25509
Museo Etnografico e della Cultura Materiale, Aquilonia 25255
Museo Etnografico e della Stregoneria, Triora 28100
Museo Etnografico e di Scienze Naturali, Torino 28019
Museo Etnográfico e do Viño de Cambados, Cambados 39109
Museo Etnografico Extremeño González Santana, Olivenza 39724
Museo Etnografico Francesco Bande, Sassari 27754
Museo Etnografico G. Carpani, Lizzano in Belvedere 26503
Museo Etnografico Galluras, Luras 26548
Museo Etnografico Gino di Benedetto, Torricella Sicura 28043
Museo Etnográfico Huichol Wixarica, Zapopan 31590
Museo Etnografico Il Cicle della Vita, Quartu Sant'Elena 27351
Museo Etnográfico Juán B. Ambrosetti, Buenos Aires 00230
Museo Etnografico La Steiva, Piverone 27259
Museo Etnográfico Leonés Ildefonso Fierro, León 39458
Museo Etnográfico Madre Laura, Medellín 08679
Museo Etnográfico Miguel Angel Builes, Medellín 08680
Museo Etnográfico Minero, Oruro 04232
Museo Etnográfico Municipal Dámaso Arce, Olavarría 00480
Museo Etnográfico Olimio Liste, San Cristovo de Cea 39953
Museo Etnográfico Piedad Isla, Cervera de Pisuerga 39167
Museo Etnografico Romagnolo Benedetto Pergoli, Forlì 26226
Museo Etnográfico Rural, Fuentepinilla 39318
Museo Etnográfico Rural y Comarcal, Prioro 39873
Museo Etnografico San Domu de is Ainas, Armungia 25280
Museo Etnográfico Santa Rosa de Ocopa, Concepción 34153
Museo Etnografico Siciliano Giuseppe Pitrè, Palermo 27076
Museo Etnográfico Sotelo Blanco, Santiago de Compostela 40027
Museo Etnografico sulla Lavorazione del Legno, San Vito di Leguzzano 27691
Museo Etnográfico Tanit, San Bartolomé de Lanzarote 39951
Museo Etnográfico Textil Pérez Enciso, Plasencia 39844
Museo Etnografico Tiranese, Madonna di Tirano 26563
Museo Etnografico Tiranese, Tirano 27973
Museo Etnografico U. Ferrandi, Novara 26951
Museo Etnografico Vallivo, Val Masino 28134
Museo Etnográfico y Archivo Histórico Enrique Squirru, Azul 00132
Museo Etnográfico y Colonial Juan de Garay, Santa Fé 00615
Museo Etnográfico y de Arqueología Industrial, Chillarón de Cuenca 39174
Museo Etnográfio Casa do Patrón, Lalín 39441
Museo Etnohistórico de Ayabacara, Ayabacara 34118
Museo Etnológica de Casa Mazo, Valle de Hecho 40248
Museo Etnológica del Valle Medio del Ebro, Belchite 39017
Museo Etnológico, Barca 38951
Museo Etnológico, Cehegín 39159
Museo Etnológico, San Juan de Plan 39960
Museo Etnológico Aljibes de Gasparito, Rojales 39911
Museo Etnológico de Amazonas, Puerto Ayacucho 54594
Museo Etnológico de Canet d'En Berenguer, Canet d'En Berenguer 39119
Museo Etnológico de Horcajuleo de la Sierra, Horcajuelo de la Sierra 39392
Museo Etnológico de la Diputación, Castellón de la Plana 39149
Museo Etnológico de la Huerta de Murcia, Alcantarilla 38817
Museo Etnológico de la Laguna Grande, Beteta 39034
Museo Etnológico de Morella y del Meastrazgo, Morella 39685
Museo Etnológico de Navarra Julio Caro Baroja, Ayegui 38925
Museo Etnológico de Pavías, Pavías 39818
Museo Etnológico de Ribadavia, Ribadavia 39897
Museo Etnológico de Sechura, Sechura 34307
Museo Etnologico della Apuane, Massa 26621
Museo Etnológico la Posada, Chinchón 39178
Museo Etnológico Misional C.M.F., Santa Isabel 10557
Museo Etnologico Missionario, Castelnuovo Don Bosco 25749
Museo Etnológico Popular, Fuensanta 39314
Museo Etnológico y Arqueológico Municipal, Manzanares el Real 39618
Museo Etnolólogico Anselmo Buil, Blecua y Torres 39052
Museo Etnomusicale I Gigli di Nola, Nola 26937
Museo Etrusco, Asciano 25292
Museo Etrusco Gasparri, Piombino 27231
Museo Etrusco M. Guarnacci, Volterra 28311
Museo Ettore Pomarici Santomasi, Gravina di Puglia 26367
Museo Europeo degli Studenti, Bologna 25471
Museo Euskal Herria, Gernika-Lumo 39333
Museo Evaristo Valle, Gijón 39339
Museo Evita, Buenos Aires 00231
Museo Evocativo y de Bellas Artes Osvaldo Gasparini, San Antonio de Areco 00565
Museo Ex Convento Agustino de San Pablo, Yuriria 31573
Museo Ex Convento Agustino Siglo XVI, México 31114
Museo Ex-Convento de San Juan Bautista, Tlayacapan 31482
Museo Ex Convento del Carmen, Guadalajara 30903
Museo Ex Hacienda San Gabriel de Barrera, Guanajuato 30919
Museo Ex Templo de San Agustín, Zacatecas 31578
Museo Ex Voto di Santa Maria della Portella, Cervara di Roma 25817
Museo Ex Votos Sr. de la Misericordia, Tepatitlán de Morelos 31431
Museo-Exposición de Creencias y Religiosidad Popular, Abizanda 38782
Museo-Exposición de Oficios y Artes Tradicionales, Aínsa 38798
Museo Exposición El Hombre, Bogotá 08580
Museo-Exposición Marín Bosqued, Aguarón 38788
Museo-Exposición Permanente de Geológico, Laspaúles 39445
Museo-Exposición Permanente de Juegos Tradicionales, Campo 39111
Museo Exposición Permanente de Juguetes, Albarracín 38808
Museo Exposición Permanente de la Electricidad, Murillo de Gállego 39706
Museo Exposición Permanente de Pedagogía, Huesca 39401
Museo-Exposición Permanente Diocesano, Albarracín 38809
Museo Extremeño e Iberoamericano de Arte Contemporáneo, Badajoz 38936
Museo F.A. Marcucci, Ascoli Piceno 25297
Museo F. Assetto, Frontino 26251
Museo F. Renzi, Borghi 25501
Museo Fallero, Valencia 40210
Museo Farallón Dillon, Ballenita 10373
Museo Farmacéutico, Matanzas 09215
Museo Farmacia Cinquecentesca, Roccavaldina 27435
Museo Farmacia Conventuale, Venezia 28200
Museo Farmacobotánica Juan A. Domínguez de la Universidad de Buenos Aires, Buenos Aires 00232
Museo Faro a Colón, Santo Domingo 10355
Museo Faryluk, Aristóbulo del Valle 00127
Museo Felisa Rincón de Gautier, San Juan 36072
Museo Félix de Jesús, México 31115
Museo Félix Murga, Amurrio 38872
Museo Fernán Félix de Amador, Luján 00436
Museo Fernando García, Canelones 54376
Museo Ferrocarrilero, Empalme 30853
Museo Ferrovario, Trieste 28094
Museo Ferroviario, Alcázar de San Juan 38818
Museo Ferroviario, Cuneo 26020
Museo Ferroviario, Tacna 34313
Museo Ferroviario de Gualeguaychú, Gualeguaychú 00384
Museo Ferroviario del Verbano, Luino 26547
Museo Ferroviario Piemontese, Savigliano 27767
Museo Ferroviaro, Bogotá 08581
Museo Ferrucciano, Gavinana 26277
Museo Ferrucciano, San Marcello Pistoiese 27658
Museo Filatélico, San José 08821
Museo Filatélico del Banco de la República, Medellín 08681
Museo Filemón Gutiérrez Ramírez, Ameca 30715
Museo Filipe Santiago Gutiérrez, Toluca 31492
Museo Fin de la Tierra, Francisco I. Madero, Coahuila 30869
Museo Finca El Abra, Siguanea 09332
Museo Fiorentino di Preistoria, Firenze 26191
Museo Fioroni, Legnago 26480
Museo Fittile, Ripatransone 27418
Museo Flesia, Coronel Pringles 00344
Museo Florencio de la Fuente, Huete 39403
Museo Florencio Molina Campos, Moreno 00473
Museo Folklórico, La Rioja 00420
Museo Folklórico Araucano de Cañete Juan A. Ríos M., Cañete 07431
Museo Folklórico Regional de Humahuaca, Humahuaca 00387
Museo Fomento Cultural Banamex, México 31116
Museo Fondazione C. Faina e Museo Civico Palazzo Faina, Orvieto 27009
Museo Fondazione Famiglia Piccolo di Calanovella, Capo d'Orlando 25660
Museo-Fondazione P. Conti, Fiesole 26129
Museo Forense de la Morgue Judicial, Buenos Aires 00233
Museo Forestal de la Policía Ecológica, Jesús María 34197
Museo Forestale C. Siemoni, Poppi 27295
Museo Fortaleza de San Carlos de la Cabaña, La Habana 09144
Museo Fortaleza de San Felipe, San Felipe de Puerto Plata 10336
Museo Fortuny, Venezia 28201
Museo Fotográfico, Calarcá 08605
Museo Fotográfico de Nanchital, Nanchital de Lázaro Cárdenas del Río 31209
Museo Fournier de Naipes de Álava, Vitoria-Gasteiz 40324
Museo Fragua Martiana, La Habana 09145
Museo Francescano, Roma 27507
Museo Francescano Padre Aurelio Menin, Chiampo 25849
Museo Francescano - Raccolta d'Arte, Gubbio 26391
Museo Francesco Borgogna, Vercelli 28221
Museo Francesco Cilea e Nichola Antonio Manfroce, Palmi 27086
Museo Francisco Campos, Guayaquil 10405
Museo Francisco Goitia, Zacatecas 31579
Museo Francisco Huerta Rendón en la Casona Universitaria, Guayaquil 10406
Museo Francisco Javier Sauza, Tequila 31450
Museo Francisco José de Caldas, Bogotá 08582
El Museo Francisco Oller y Diego Rivera, Buffalo 47123
Museo Francisco Sarabia, Lerdo 31005
Museo Franco Azzinari, Altomonte 25209
Museo Franco Azzinari, Altomonte 25210
Museo Francisco Lázaro, Chiapa de Corzo 30763
Museo Franz Mayer, México 31117
Museo Fray Bernardo Padilla, Acámbaro 30687
Museo Fray Luís de León, Belmonte 39021
Museo Frey Pedro Bedón, Ibarra 10414
Museo Frida Kahlo, México 31118
Museo Frissel de Arte Zapoteca, Santa Ana del Valle 31374
Museo Friulano di Storia Naturale, Udine 28114
Museo Fuentes de Loreto y Guadelupe, Palmar de Bravo 31248
Museo Fuerte Conde de Mirasol, Vieques 36082
Museo Fuerte de la Loma, Victoria de las Tunas 09345
Museo Fuerte Independencia, Tandil 00641
Museo Fundación Antonio Maura, Madrid 39559
Museo Fundación Casa de Alba, Madrid 39560
Museo Fundación Díaz Caneja de Arte Contemporaneo, Palencia 39772
Museo Fundación Estero Beach, Ensenada 30860
Museo Fundación Eugenio Granell, Santiago de Compostela 40028
Museo Fundación Jesús Otero, Santillana del Mar 40035
Museo Fundación Liste, Vigo 40275
Museo Fundación Naum Komp, Buenos Aires 00234
Museo G. Arcucci, Ariano Irpino 25278
Museo G.B. Filippa, Rimella 27405
Museo G.D. Cassini, Perinaldo 27140
Museo G. Filangieri, San Potito Sannitico 27680
Museo G. Marconi, Pontecchio Marconi 27288
Museo G. Spontini, Maiolati Spontini 26570
Museo G. Whitaker, Marsala 26613
Museo Gabriel García Márquez, Aracataca 08529
Museo Gabriel Mateus, Sabanagrande 08741
Museo Gabriela Mistral de Vicuña, Vicuña 07473
Museo Galego de Arte Contemporánea Carlos Maside, Arès 38888
Museo Galego do Xoguete, Allariz 38849
Museo Galería Manuel Lepe, Puerto Vallarta 31297
Museo Galileo, Firenze 26192
Museo Garibaldino, Caprera 25663
Museo Garibaldino, Genova 26315
Museo Garibaldino della Campagna dell'Agro Romano per la Liberazione di Roma 1867, Mentana 26654
Museo Garnelo, Montilla 39679
Museo Gauchesco Ricardo Güiraldes, San Antonio de Areco 00566
Museo Gemológico, Bogotá 08583
Museo General, Yanahuanca 34328
Museo General Augusto César Sandino, San Rafael del Norte 33305
Museo General Belgrano, Buenos Aires 00235
Museo General de Arequito, Arequito 00126
Museo General Francisco Villa, San Juan del Río 31340
Museo General Ignacio Zaragoza, Puebla 31287

M

Museo General Porfirio Rubio, Landa de Matamoros 30995
Museo Geo-Paleontologico, Ferrara 26115
Museo Geologico, Castell'Arquato 25736
Museo Geológico de la Universidad Nacional de Ingeniería del Perú, Lima 34259
Museo Geológico del Seminari, Barcelona 38969
Museo Geológico del Uruguay, Montevideo 54403
Museo Geologico e Paleontologico G. Capellini, Bologna 25472
Museo Geologico G.G. Gemmellaro, Palermo 27077
Museo Geológico Georg Petersen, Lima 34260
Museo Geológico José Royo y Gómez, Bogotá 08584
Museo Geológico Marino Arce Herrera, Bucaramanga 08601
Museo Geológico Petrolero, Neiva 08696
Museo Geominero, Madrid 39561
Museo Geopaleontologico Alto Aventino, Palena 27063
Museo Geopaleontologico Cava Bomba, Cinto Euganeo 25898
Museo Geopaleontologico dei Fossili della Lessinia, Velo Veronese 28173
Museo Geopaleontologico del Castello, Lerici 26486
Museo Geopaleontologico Naturalistico Antropico e Ornitologico Brancaleoni, Piobbico 27227
Museo Germán Guglielmetti, Benito Juárez 00147
Museo Gianmaria da Tusa, Cefalù 25795
Museo Giannettino Luxoro, Genova 26316
Museo Gianni Visentin, Carpanè di San Nazario 25683
Museo Giovanni Antonio Sanna, Sassari 27755
Museo Giovanni Boldini, Ferrara 26116
Museo Giuseppe Scarabelli, Imola 26402
Museo Glauco Lombardi, Parma 27108
Museo Gli Orsanti, Compiano 25957
Museo Gonzalo Carrasco, Otumba 31239
Museo Gráfico de la Mujer Guerrenese, Chilpancingo 30773
Museo Gráfico El Peruano, Cercado 34142
Museo Gran Casino, Bayamo 09035
Museo Gran Mariscal de Ayacucho, Cumaná 54568
Museo Gregoriano Egizio, Città del Vaticano 54503
Museo Gregoriano Etrusco, Città del Vaticano 54504
Museo Gregoriano Profano, Città del Vaticano 54505
Museo Gregorio Prieto, Valdepeñas 40191
Museo Guaraní, Santa Cruz 04241
Museo Guerrero Malagón, Urda 40182
Museo Guillermo Perez Chiriboga del Banco Central, Quito 10445
Museo Guillermo Spratling, Taxco de Alarcón 31407
Museo Gustavo de Maeztu, Estella 39286
Museo Hacienda San Pedro, General Zuazua 30877
Museo Hendrik Christian Andersen, Roma 27508
Museo Herbario de la Facultad de Ciencias, México 31119
Museo Hermanas Giralt, Cienfuegos 09066
Museo Hermanas Mirabal, Conuco 10326
Museo Hermann Hesse, Montagnola 41633
Museo Hermanos López Rayón, Tlalpujahua 31469
Museo Hermanos Nacif Weiss, Rivadavia 00539
Museo Hermenegildo Galeana, Tecpan de Galeana 31415
Museo Hermila Domínguez de Castellanos, Comitán de Domínguez 30794
Museo Heroes y Martires Revolución Sandinista en Estelí, Estelí 33277
Museo Hidalgo, Salamanca 31311
Museo Hidráulico del Sistema de Drenaje Profundo, México 31120
Museo Hidráulico los Molinos del Río Segura, Murcia 39698
Museo História de la Medicina, Valencia 40211
Museo Historia de las Ciencias Carlos J. Finlay, La Habana 09146
Museo Historia del Ferrocarril de Antioquia, Medellín 08682
Museo Historia Municipal, Antilla 09026
Museo Historiador Hernández Galiano, Tostado 00650
Museo Histórica Ex Convento de Tepoztlán, Tepoztlán 31448
Museo Histórico, El Carmen de Bolívar 08634
Museo Histórico, Pinar del Río 09249
Museo Historico, Riobamba 10457
Museo Historico, San Antonio de los Baños 09274
Museo Histórico, Zapala 00687
Museo Histórico Andrés Avelino Cáceres, Ayacucho 34121
Museo Histórico Andrés Avelino Cáceres, Huamanga 34173
Museo Histórico Anton García de Bonilla, Ocaña 08699
Museo Histórico Arqueológico, Doña Mencia 39247
Museo Histórico Arqueológico Guillermo Magrassi, Mar del Plata 00445
Museo Histórico, Arqueológico y de Arte Pedro Balduín, San Pedro de Jujuy 00605
Museo Histórico-Artístico, Cuacos de Yuste 39228
Museo Historico Bae Calderon, Guayaquil 10407
Museo Histórico BBVA, Bilbao 39046
Museo Histórico Brigadier General Cornelio de Saavedra, Buenos Aires 00236
Museo Histórico Casa de la Convención de Rionegro, Rionegro 08739
Museo Historico Casa de los Tratados, Cuenca 10393
Museo Histórico, Colonial y de Bellas Artes y Museo de Ciencias Naturales Regional Mesopotámico, Mercedes 00461
Museo Histórico Comunal, Villa Trinidad 00685
Museo Histórico Comunal de San Genaro Norte, San Genaro Norte 00578
Museo Histórico Comunal y de la Colonización Judía Rabino, Moisés Ville 00470
Museo Histórico Conventual San Carlos, San Lorenzo 00591
Museo Histórico Cultural Juan Santamaría, Alajuela 08808
Museo Histórico de 26 de Julio, Santiago de Cuba 09322
Museo Histórico de Acapulco, Acapulco 30692
Museo Histórico de Almedinilla, Almedinilla 38855
Museo Histórico de Antioquia, Medellín 08683
Museo Histórico de Arenaza, Arenaza 00125
Museo Histórico de Arrecifes, Arrecifes 00129
Museo Histórico de Ayotzinapa, Tixtla de Guerrero 31461
Museo Histórico de Bani, Bani 10323
Museo Histórico de Belmez y del Territorio Minero, Belmez 39019
Museo Histórico de Cañuelas, Cañuelas 00289
Museo Histórico de Cartagena Casa de la Inquisición, Cartagena 08619
Museo Histórico de Cera, Buenos Aires 00237
Museo Histórico de Chacamarka, Chacamarka 34143
Museo Historico de Ciego de Avila, Ciego de Avila 09062
Museo Histórico de Corrientes Manuel Cabral de Mayo y Alpoín, Corrientes 00348
Museo Histórico de Entre Ríos Martiniano Leguizamón, Paraná 00486
Museo Histórico de Guachinango, Guachinango 30881
Museo Histórico de Guatraché, Guatraché 00385
Museo Histórico de la Casa de la Cultura, Ambalema 08528
Museo Histórico de la Ciudad, Torreón 31500
Museo Histórico de la Ciudad, Valencia 40212
Museo Histórico de la Colonia San Carlos, San Carlos Centro 00569
Museo Histórico de la Educación Tecnológica en México, Chilpancingo 30774
Museo Histórico de la Honorable Cámara de Diputados de la Nación, Buenos Aires 00238
Museo Histórico de la Policía de la Provincia de Misiones, Posadas 00499
Museo Histórico de la Policía Nacional, La Paz 04219
Museo Histórico de la Policía National, Bogotá 08585
Museo Histórico de la Prefectura Naval Argentina, Tigre 00645
Museo Histórico de la Provincia de Catamarca, San Fernando del Valle de Catamarca 00573
Museo Histórico de la Provincia Dr. Orestes di Lullo, Santiago del Estero 00629
Museo Histórico de la Revolución Mexicana, Chihuahua 30771
Museo Histórico de la Sierra Gorda, Jalpan 30969
Museo Histórico de Laborde, Laborde 00421
Museo Histórico de Múzquiz, Múzquiz 31207
Museo Histórico de Navarro, Navarro 00476
Museo Histórico de Osorno, Osorno 07443
Museo Histórico de Palma Soriano, Palma Soriano 09240
Museo Histórico de Ranchos, Ranchos 00517
Museo Histórico de Reynosa, Reynosa 31304
Museo Histórico de San Francisco, Santa Fé 00616
Museo Histórico de San Miguel de Allende, San Miguel de Allende 31356
Museo Histórico de San Nicolás de los Garza, San Nicolás de los Garza 31361
Museo Histórico de San Roque, San Roque 39970
Museo Histórico de Sargadelos, Cervo 39168
Museo Histórico del Banco Provincia Dr. Arturo Jauretche, Buenos Aires 00239
Museo Histórico del Bierzo, Ponferrada 39853
Museo Historico del Norte, Salta 00558
Museo Histórico del Oriente de Morelos, Cuautla 30813
Museo Histórico del Regimiento de Granaderos a Caballo General San Martín, Buenos Aires 00240
Museo Histórico del Transporte Carlos Hillner Decoud, Quilmes 00510
Museo Histórico Dr. Rafael Angel Calderón Guardia, San José 08822
Museo Histórico el Dique, Puerto Real 39881
Museo Histórico Etnológico de la Villa de Mijas, Mijas 39648
Museo Histórico Ex Aduana de Ciudad Juárez, Juárez 30983
Museo Histórico Ferroviario Escribano Alfredo Rueda, Rueda 00551
Museo Histórico Franciscano, San Salvador de Jujuy 00608
Museo Histórico Fuerte Barragan, Ensenada 00365
Museo Histórico Fuerte de San José el Alto, Campeche 30743
Museo Histórico General Julio de Vedia, 9 de Julio 00115
Museo Histórico Hipólito Unanue, Ayacucho 34122
Museo Histórico José Hernández-Chacra Pueyrredón, Villa Ballester 00672
Museo Histórico "Juan Bautista Ambrosetti", Gualeguay 00381
Museo Histórico Judío y del Holocausto Tuvie Maizel, México 31121
Museo Histórico La Casona de Santa Rosa, Liberia 08810
Museo Histórico La Gallareta Forestal, La Gallareta 00403
Museo Histórico Local, Puente Genil 39875
Museo Histórico Local de Monturque, Monturque 39683
Museo Histórico Manuel A. Moreira, Laboulaye 00422
Museo Histórico Mexicano de Cananea, Cananea 30746
Museo Histórico Militar, Asunción 34098
Museo Histórico Militar, La Paz 04220
Museo Histórico Militar de Melilla, Melilla 39643
Museo Histórico Militar de Valencia, Valencia 40213
Museo Histórico Minero Don Felipe de Borbón y Grecia, Madrid 39562
Museo Histórico Minero Francisco Pablo Holgado, Almadén 38851
Museo Histórico Municipal, Baena 38939
Museo Historico Municipal, Bahía Blanca 00135
Museo Histórico Municipal, Cañete de las Torres 39121
Museo Histórico Municipal, Fuente Tójar 39316
Museo Histórico Municipal, Montilla 39680
Museo Histórico Municipal, Ocotlán 31234
Museo Histórico Municipal, San Fernando 39954
Museo Histórico Municipal, Taxco de Alarcón 31408
Museo Histórico Municipal Alfredo E. Múlgura, General Belgrano 00377
Museo Histórico Municipal Andrés A. Roverano, Santo Tomé 00633
Museo Histórico Municipal Brigadier General Juan Martín de Pueyrredón, San Isidro 00581
Museo Histórico Municipal de Armstrong, Armstrong 00128
Museo Histórico Municipal de Cañada de Gómez, Cañada de Gómez 00288
Museo Histórico Municipal de Écija, Écija 39257
Museo Histórico Municipal de Guanabacoa, Guanabacoa 09088
Museo Histórico Municipal de Lincoln, Lincoln 00428
Museo Histórico Municipal de Luque, Luque 00439
Museo Histórico Municipal de Pergamino, Pergamino 00490
Museo Histórico Municipal de Priego de Córdoba, Priego de Córdoba 39872
Museo Histórico Municipal de Regla, Regla 09264
Museo Histórico Municipal de Teba, Teba 40107
Museo Histórico Municipal de Villa del Río, Villa del Río 40295
Museo Histórico Municipal de Villa Gesell, Villa Gesell 00680
Museo Histórico Municipal de Villamartín, Villamartín 40304
Museo Histórico Municipal General Levalle, General Levalle 00378
Museo Histórico Municipal Guillermo Zegarra Meneses, Arequipa 34113
Museo Histórico Municipal Hércules J. Rabagliati, Ramallo 00516
Museo Histórico Municipal Juan Lavalle, Baradero 00142
Museo Histórico Municipal La Para, La Para 00404
Museo Histórico Municipal Monte Grande, Monte Grande 00472
Museo Histórico Municipal Víctor E. Míguez, Mercedes 00462
Museo Histórico Municipal Villa Clara, Villa Clara 00675
Museo Histórico Municipal Villa del Rosario, Villa del Rosario 00678
Museo Histórico Municipal y Colección de Arte Precolombino Arminio Weiss, Rafaela 00514
Museo Histórico Nacional, Buenos Aires 00241
Museo Histórico Nacional, Santiago de Chile 07460
Museo Histórico Nacional Casa del Virrey Liniers, Alta Gracia 00120
Museo Histórico Nacional - Casa Rivera, Montevideo 54404
Museo Histórico Nacional del Cabildo de Buenos Aires y de la Revolucion de Mayo, Buenos Aires 00242
Museo Histórico Naval, Cienfuegos 09067
Museo Histórico Naval Central, México 31122
Museo Historico Okinawa Bolivia, Santa Cruz 04242
Museo Histórico Policial de La Plata, La Plata 00414
Museo Historico Provincial, Cienfuegos 09068
Museo Histórico Provincial, Sancti-Spíritus 09287
Museo Histórico Provincial Agustín V. Gnecco, San Juan 00588
Museo Histórico Provincial Brig. Estanislao López, Santa Fé 00617
Museo Histórico Provincial Colonel Manuel José Olascoaga, Chos Malal 00306
Museo Histórico Provincial de Guantánamo, Guantánamo 09092
Museo Histórico Provincial de Rosario Dr. Julio Marc, Rosario 00545
Museo Histórico Provincial Marqués de Sobremonte, Córdoba 00330
Museo Histórico Regional, Alpachiri 00119
Museo Histórico Regional - Casa Garcilaso Cusco, Cusco 34165
Museo Histórico Regional de Ayacucho, Ayacucho 00131
Museo Histórico Regional de Gaiman, Gaiman 00375
Museo Histórico Regional de la Colonia San José, San José 00583
Museo Histórico Regional De la Isla, Isla del Cerrito 00390
Museo Histórico Regional de Magallanes, Punta Arenas 07446
Museo Histórico Regional de Río Cuarto, Río Cuarto 00528
Museo Histórico Regional de Tacna, Tacna 34314
Museo Histórico Regional de Urdinarrain, Urdinarrain 00659
Museo Histórico Regional de Villa La Angostura, Villa La Angostura 00681
Museo Histórico Regional Ex Cuartel de la Compañía Fija, Ensenada 30861
Museo Histórico Regional Gabriel Campomar Cervera, Salliqueló 00553
Museo Histórico Regional Ichoalay, Resistencia 00525
Museo Histórico Regional Municipal de Magdalena, Magdalena 00440
Museo Histórico Regional Pablo Argilaga, Santo Tomé 00634
Museo Histórico Regional Padre Francisco Cremasco, Villa Concepción del Tío 00676
Museo Histórico San Ignacio de Loyola, San José 00584
Museo Histórico Sanmartíniano, Mendoza 00456
Museo Histórico Sarmiento, Buenos Aires 00243
Museo Histórico y Antropológico de Nindirí, Nindirí 33300
Museo Histórico y Antropológico Mauricio Van de Maele, Valdivia 07469
Museo Histórico y Arqueológico, Belén de Umbría 08542
Museo Histórico y Arqueológico, Charalá 08624
Museo Histórico y Arqueológico Dr. Estrada Torres, Azua 10322
Museo Histórico y de la Tradiciones Populares de la Boca, Buenos Aires 00244
Museo Histórico y Natural de Lavalle, Lavalle Villa Tulumaya 00427
Museo Histórico y Numismático del Banco Central de la República Argentina "Dr. José Evaristo Uriburu", Buenos Aires 00245
Museo Histórico y Numismático del Banco de la Nación Argentina, Buenos Aires 00246
Museo Histórico y Regional Pico Truncado, Pico Truncado 00491
Museo Histórico y Tradicional de Barracas al Sud, Avellaneda 00130
Museo Histório de Aspe, Aspe 38907
Museo Hotel Marquetá, Mariquita 08673
Museo Houssay de Historia de la Ciencia y la Técnología, Buenos Aires 00247
Museo Huehuetla-tolli, Comala 30790
Museo Hunucmá de Sisal, Hunucmá 30952
Museo I Luoghi del Lavoro Contadino, Buscemi 25566
Museo I. Nievo, Fossalta di Portogruaro 26238
Museo I. Salvini, Cocquio Trevisago 25930
Museo Iberoamericano de Arte Moderno de Popayán, Popayán 08729
Museo Ibleo delle Arti e Tradizioni Popolari - Guastella, Modica 26759
Museo Iconográfico, Itaugua 34103
Museo Iconográfico del Quijote, Guanajuato 30920
Museo Icticola del Beni, Trinidad 04255
Museo Ideale Leonardo da Vinci, Vinci 28284
Museo Iglesia del Salvador, Oña 39731
Museo Iglesia San Juan de Dios, Murcia 39699
Museo Ignacio Zuloaga, Pedraza de la Sierra 39821
Museo Igneológico del Cuerpo Ofical de Bomberos, Bogotá 08586
Museo Illichiano, Castell'Arquato 25737
Museo Iloilo, Iloilo 34389
Museo Imágenes de la Frontera, Dajabón 10327
Museo in Erba, Bellinzona 41194
Museo In Situ, Sulmona 27922

Museo Inca de la Universidad Nacional San Antonio Abad, Cusco 34166
Museo Independencia, Córdoba 00331
Museo Indígena de Colombia Ethnia, Bogotá 08587
Museo Indígena de Guatavita, Guatavita 08654
Museo Indígena Regional Enin, Mitú 08687
Museo Indigenista, Esquel 00370
Museo Indigeno, San José 08823
Museo Industrial de Santa Catarina El Blanqueo, Santa Catarina 31376
Museo Inés Mercedes Gómez Álvarez, Guanare 54574
Museo Inka, Cusco 34167
Museo Instituto Cultural México Israel, México 31123
Museo Insular de la Palma, Santa Cruz de la Palma 39992
Museo Inter-étnico de la Amazonía, Florencia 08642
Museo Interactivo de Ciencia, Tecnología y Medio Ambiente Sol del Niño, Mexicali 31035
Museo Interactivo de Ciencias Naturales, Ituzaingó 00392
Museo Interactivo de Ciencias-Puerto Ciencia, Paraná 00487
Museo Interactivo de la Ciencia y Tecnología - Principia, Málaga 39601
Museo Interactivo de la Música, Málaga 39602
Museo Interactivo del Centro de Ciencias de Sinaloa, Culiacán 30828
Museo Interactivo del Medio Ambiente, México 31124
Museo Interactivo El Rehilete, Pachuca 31244
Museo Interactivo La Avispa, Chilpancingo 30775
Museo Interattivo delle Scienze, Foggia 26211
Museo Internacional de Arte Contemporáneo, Arrecife 38896
Museo Internacional de Arte Naïf, Jaén 39423
Museo Internacional de Electrografía, Cuenca 39236
Museo Internazionale del Folklore e Civiltà Contadina, Atina 25320
Museo Internazionale del Presepio, Morrovalle 26873
Museo Internazionale della Campana, Agnone 25158
Museo Internazionale della Caricatura, Tolentino 27988
Museo Internazionale della Croce Rossa, Castiglione delle Stiviere 25760
Museo Internazionale della Fisarmonica, Castelfidardo 25724
Museo Internazionale delle Ceramiche, Faenza 26076
Museo Internazionale delle Marionette A. Pasqualino, Palermo 27078
Museo Internazionale dell'Etichetta, Cupramontana 26024
Museo Internazionale dell'Immagine Postale, Belvedere Ostrense 25391
Museo Interparroquial de Alaejos, Alaejos 38800
Museo Interparroquial de Arte Sacro, Medina de Rioseco 39636
Museo Irineo Germán, Azoyú 30731
Museo Irpino, Avellino 25331
Museo Irureta, Tilcara 00648
Museo Isabel Rubio, Isabel Rubio 09185
Museo Isidro Fabela, Atlacomulco 30726
Museo It Akean, Kalibo 34396
Museo Italgas, Torino 28020
Museo Italiano della Ghisa, Longiano 26518
Museo Italoamericano, San Francisco 52785
Museo Itinerante de Arte Contemporáneo del Mercosur, Buenos Aires 00248
Museo Ittico, Pescara 27165
Museo Ittico Augusto Capriotti, San Benedetto del Tronto 27611
Museo Ixchel del Traje Indígena, Guatemala 23300
Museo J.M. Cruxent, Los Teques 54579
Museo Jacinto Guerrero, Ajofrín 38799
Museo Jacinto Jijón y Caamaño, Quito 10446
Museo Jacobo Borges, Caracas 54554
Museo Jarocho Salvador Ferrando, Tlacotalpan 31464
Museo Jesuítico de Sitio, Yapeyú 00686
Museo Jesuitico Nacional de Jesus María, Córdoba 00332
Museo Jesútico Nacional de Jesús María, Jesús María 00393
Museo Joaquín Peinado de Ronda, Ronda 39918
Museo Jorge Eliécer Gaitán, Puerto Gaitán 08732
Museo Jorge Pasquini López, San Salvador de Jujuy 00609
Museo José Antonio Páez, Acarigua 54515
Museo José de Jesús Almaza, Xicoténcatl 31563
Museo José Domingo de Obaldía, David 34072
Museo José Fernando Ramírez, Durango 30843
Museo José Guadalupe Posada, Aguascalientes 30702
Museo José Luis Cuevas, Colima 30781
Museo José Luis Cuevas, México 31125
Museo José Manuel Estrada, Suipacha 00637
Museo José María Luis Mora, Ocoyoacac 31235
Museo José María Morelos y Pavón, Cuautla 30814
Museo José María Velasco, Toluca 31493
Museo José Martí, La Habana 09147
Museo José Natividad Correa Toca Teapa, Teapa 31410
Museo José Ortega, Arroba de los Montes 38897
Museo José y Tomás Chávez Morado, Silao 31391
Museo Juan Cabré, Calaceite 39093
Museo Juan Carlos Iramain, San Miguel de Tucumán 00600
Museo Juan Lorenzo Lucero, Pasto 08710
Museo Juan Manuel Blanes, Montevideo 54405
Museo Juárez del Jardín Borda, Cuernavaca 30818
Museo Juárez e Historia de Mapimí y Ojuela, Mapimí 31014
Museo Juarista de Congregación Hidalgo, Entronque Congregación Hidalgo 30862
Museo Judío de Buenos Aires Dr. Salvador Kibrick, Buenos Aires 00249
Museo Julio Romero de Torres, Córdoba 39207
Museo Kan Pepen, Teabo 31409
Museo Kartell, Noviglio 26962
Museo Ken Damy di Fotografia Contemporanea, Brescia 25540
Museo Kusillo, La Paz 04221
Museo L. Garaventa, Chiavari 25860
Museo L. Minguzzi, Milano 26709
Museo la Antigua Casa del Agua, Hunucmá 30953
Museo La Casa del Libro, San Juan 36073
Museo La Chole, Petatlán 31257
Museo La Cinacina, San Antonio de Areco 00567
Museo La Flor de Jimulco, Torreón 31501
Museo La Fudina, Comelico Superiore 25950
Museo La Isabelica, Santiago de Cuba 09323
Museo La Pesca, Benito Juárez 30735
Museo La Vieja Estación, María Teresa 00449
Museo La Vigía, Matanzas 09216
Museo-Laboratorio, Vernio 28224
Museo Laboratorio del Tessile, Soveria Mannelli 27883
Museo Laboratorio di Archeologia, Monsampolo del Tronto 26787
Museo Laboratorio di Arte Contemporanea, Roma 27509
Museo Laboratorio Santa Barbara, Mammola 26581
Museo Ladino di Fassa, Pozza di Fassa 27321
Museo Ladino-Fodom, Livinallongo del Col di Lana 26495
Museo Laguna del Caimán, Tlahualilo de Zaragoza 31466
Museo Lapidario, Ferrara 26117
Museo Lapidario, Urbino 28127
Museo Lapidario del Duomo di Sant'Eufemia, Grado 26359
Museo Lapidario della Canonica della Cattedrale, Novara 26952
Museo Lapidario e del Tesoro del Duomo, Modena 26753
Museo Lapidario Estense, Modena 26754
Museo Lapidario Maffeiano, Verona 28236
Museo Lapidario Marsicano, Avezzano 25335
Museo Lara, Ronda 39919
El Museo Latino, Omaha 51553
Museo Lázaro Galdiano, Madrid 39563
Museo Legado Hermanos Álverez Quintero, Utrera 40184
Museo Legislativo los Sentimientos de la Nación, México 31126
Museo Leonardiano, Vinci 28285
Museo Leoncavallo, Brissago 41257
Museo Leoniano, Carpineto Romano 25688
Museo L'Iber, Valencia 40214
Museo Libisosa, Lezuza 39460
Museo Lítico de Pucara, Pucara 34296
Museo Local de Acámbaro, Dolores Hidalgo 30836
Museo Local de Antropología e Historia de Compostela, Compostela 30797
Museo Local de Valle de Santiago, Valle de Santiago 31527
Museo Local Tuxteco, Santiago Tuxtla 31387
Museo Locale, Termeno 27958
Museo Lodovico Pogliaghi, Varese 28163
Museo Lombardo di Storia dell'Agricoltura, Sant'Angelo Lodigiano 27720
Museo López Claro, Santa Fé 00618
Museo Lorenzo Coullaut-Valera, Marchena 39623
Museo Loringiano, Málaga 39603
Museo Los Amantes de Sumpa, Guayaquil 10408
Museo Luciano Rosero, Pasto 08711
Museo Lucio Balderas Márquez, Landa de Matamoros 30996
Museo Luigi Ceselli, Subiaco 27916
Museo Luigi Varoli, Cotignola 25996
Museo Luis A. Calvo, Agua de Dios 08522
Museo Luis Alberto Acuña, Villa de Leyva 08784
Museo Luis Donaldo Colosio, Francisco I. Madero, Coahuila 30870
Museo M. Antonioni, Ferrara 26118
Museo Madre Caridad Brader, Pasto 08712
Museo Madres Benedictinas, Sahagún 39933
Museo Maestro Alfonso Zambrano, Pasto 08713
Museo Malacologico, Erice 26064
Museo Malacologico delle Argille, Cutrofiano 26029
Museo Malacologico Piceno, Cupra Marittima 26023
Museo Malvinas Argentinas, Río Gallegos 00532
Museo Manoblanca, Buenos Aires 00250
Museo Manuel de Falla, Alta Gracia 00121
Museo Manuel F. Zárate, Guarraré 34073
Museo Manuel López Villaseñor, Ciudad Real 39190
Museo Manuel Ojinaga, Manuel Ojinaga 31012
Museo Manuela Sáenz, Quito 10447
Museo Manzoniano, Lesa 26487
Museo Manzoniano, Milano 26710
Museo Marcelo López del Instituto Hellen Keller, Córdoba 00333
Museo Marcha del Pueblo Combatiente, La Habana 09148
Museo Marchigiano del Risorgimento G. e D. Spadoni, Macerata 26554
Museo Marcos Redondo, Pozoblanco 39864
Museo Mariano, La Habana 09149
Museo Mariano Acosta, Ibarra 10415
Museo Mariano Matamoros, Jantetelco 30972
Museo Marina Núñez del Prado, Lima 34261
Museo Marinaro Gio Bono Gerrari, Cairo Montenotte 25591
Museo Marinaro Tommasino-Andreatta, San Colombano Certenoli 27627
Museo Marino d'Arte Sacra Contemporanea, Comacchio 25948
Museo Marino de Tecolutla, Tecolutla 31414
Museo Marino Marini, Firenze 26193
Museo Marino Marini, Milano 26711
Museo Marino Marini, Pistoia 27253
Museo Mario Abreu, Maracay 54585
Museo Mario Praz, Roma 27510
Museo Marítimo de Asturias, Gozón 39352
Museo Marítimo de Ushuaia y Presidio, Ushuaia 00661
Museo Marítimo del Cantábrico, Santander 40011
Museo Marítimo "Seno de Corcubión", Corcubión 39201
Museo Marítimo Torre del Oro, Sevilla 40074
Museo Marítimo y Naval de la Patagonia Austral, Río Gallegos 00533
Museo Marsiliano, Bologna 25473
Museo Martín Cárdenas, Cochabamba 04208
Museo Martín Gusinde, Puerto Williams 07445
Museo Martin Perez, Montevideo 54406
Museo Martini di Storia dell'Enologia, Pessione 27177
Museo Mascagnano, Livorno 26499
Museo Maschera, Folklore e Civiltà Contadina, Acerra 25140
Museo Massó, Bueu 39060
Museo Maurice Minkowski, Buenos Aires 00251
Museo Máximo Gómez, La Habana 09150
Museo Medardo Burbano, Pasto 08714
Museo Medardo Rosso, Barzio 25376
Museo Mediceo della Petraia e Giardino, Firenze 26194
Museo Medievale Borbonico, Salle 27598
Museo Melezet, Bardonecchia 25358
Museo Melia Cayo Coco y Museo Tryp Colonial, Cayo Coco 09059
Museo Memoria Coronel Leoncio Prado, Callao 34134
Museo Memorial 12 de Septiembre, La Habana 09151
Museo Memorial 26 de Julio, Victoria de las Tunas 09346
Museo Memorial Antonio Guiteras Holmes, Pinar del Río 09250
Museo Memorial El Morrillo, Matanzas 09217
Museo Memorial Ernesto Che Guevara, Santa Clara 09297
Museo Memorial José Martí, La Habana 09152
Museo Memorial La Demajagua, Manzanillo 09208
Museo Memorial Los Malagones, Moncada 09226
Museo Memorial Mártires de Barbados, Victoria de las Tunas 09347
Museo Memorial Vicente García, Victoria de las Tunas 09348
Museo Memorias de Yucundé, San Pedro y San Pablo Teposcolula 31370
Museo Memorie Pollentine, Pollenza 27274
Museo Meo-Evoli, Monopoli 26782
Museo Mercantile, Bolzano 25490
Museo Mercedes Sierra de Pérez El Chico, Bogotá 08588
Museo Meteorológico Nacional Dr. Benjamín A. Gould, Córdoba 00334
Museo Metropolitano, Buenos Aires 00252
Museo Metropolitano de Monterrey, Monterrey 31186
Museo Michelangelo, Caprese 25664
Museo Michoacano de las Artesanías, Morelia 31200
Museo Miguel Álvarez del Toro, Tuxtla Gutiérrez 31516
Museo Militar, Melilla 31684
Museo Militar de Cartagena, Cartagena 39138
Museo Militar de Colombia, Bogotá 08589
Museo Militar de Menorca, Es Castell 39143
Museo Militar Regional, A Coruña 39222
Museo Militar Regional, Palma de Mallorca 39786
Museo Militar Regional de Burgos, Burgos 39071
Museo Militar Regional de Canarias, Santa Cruz de Tenerife 39996
Museo Militar Regional de Sevilla, Sevilla 40075
Museo Mille Voci e Mille Suoni, Bologna 25474
Museo Mineralogico, Bortigiadas 25512
Museo Mineralogico, Valle Aurina 28145
Museo Mineralogico Campano, Vico Equense 28260
Museo Mineralogico e Naturalistico, Bormio 25511
Museo Mineralogico e Paleontologico delle Zolfare, Caltanissetta 25610
Museo Mineralogico Permanente, Carro 25693
Museo Mineralógico Prof. Manuel Tellechea, Mendoza 00457
Museo Mineralógico Salón Tulio Ospina, Medellín 08684
Museo Mineralogico Sardo, Iglesias 26397
Museo Mineralógico y Antropológico, Ibagué 08660
Museo Mineralogico y Geológico, Oruro 04233
Museo Minerario Alpino, Cogne 25935
Museo Minero, La Unión 40181
Museo Minero de Riotinto, Minas de Riotinto 39650
Museo Minero Ferrería de San Blas, Sabero 39928
Museo Minimo, Corigliano Calabro 25977
Museo Miniscalchi Erizzo, Verona 28237
Museo Mirador de Quistococha, Iquitos 34193
Museo Miscellaneo Galbiati, Brugherio 25552
Museo Misional de Nuestra Señora de Regla, Chipiona 39179
Museo Missionario, Padova 27046
Museo Missionario Cinese e di Storia Naturale, Lecce 26472
Museo Missionario delle Grazie, Rimini 27410
Museo Missionario Etnologico, Città del Vaticano 54506
Museo Missionario Francescano, Fiesole 26130
Museo Mitre, Buenos Aires 00253
Museo Mohicca, Mohicca 31171
Museo Molino de Papel La Villa, Buenos Aires 00254
Museo Molino de Teodoro, Cieza 39182
Museo Molino Nant Fach, Trevelin 00655
Museo Monástico Medival y de Farmacia, Santo Domingo de Silos 40041
Museo Monográfico de Cástulo, Linares 39463
Museo Monográfico de Clunia, Peñalba de Castro 39828
Museo Monográfico de El Cigarralejo, Mula 39690
Museo Monográfico de la Alcudia, Elche 39268
Museo Monográfico de la Colonia Celsa, Velilla de Ebro 40258
Museo Monográfico de la Villa Romana de la Olmeda, Saldaña 39946
Museo Monográfico de Tiermes, Montejo de Tiermes 39674
Museo Monográfico del Azafrán, Monreal del Campo 39669
Museo Monográfico del Castro de Viladonga, Castro de Rei 39153
Museo Monográfico y Necrópolis Púnica de Puig des Molins, Eivissa 39260
Museo Monseñor Agustín Blujaki, Asunción 34099
Museo Monseñor José Fagnano, Río Grande 00536
Museo Monseñor Juan Sinforiano Bogarin, Asunción 34100
Museo Monte de Piedad, Buenos Aires 00255
Museo Monumental de la Municipalidad Distrital de Huaura, Huaura 34184
Museo Monumento al Deportato, Carpi 25687
Museo Morando Bolognini, Sant'Angelo Lodigiano 27721
Museo Morelos, Carácuaro 30750
Museo Multidisciplinario del Colegio La Salle, Lima 34262
Museo Mundial del Tango, Buenos Aires 00256
Museo Mundo de Ambar, Santo Domingo 10356
Museo Mundo Yámana, Ushuaia 00662
Museo Mundocaña, Cali 08613
Museo Mundocaña, Palmira 08702
Museo Municipal, Alcázar de San Juan 38819
Museo Municipal, Algeciras 38841
Museo Municipal, Alzira 38867
Museo Municipal, Antequera 38875
Museo Municipal, Aroche 38895
Museo Municipal, Ayllón 38926
Museo Municipal, Cabra 39079
Museo Municipal, Calatayud 39098
Museo Municipal, Carmona 39132
Museo Municipal, Ciego de Avila 09063
Museo Municipal, Consuegra 39200
Museo Municipal, Holguín 09183
Museo Municipal, Iquitos 34194
Museo Municipal, Mier 31164
Museo Municipal, Nerva 39716
Museo Municipal, Ourense 39756
Museo Municipal, Palma del Río 39795
Museo Municipal, Pizarra 39843
Museo Municipal, El Puerto de Santa María 39880
Museo Municipal, Puertollano 39882
Museo Municipal, Requena 39892

M

Museo Municipal, Sancti-Spíritus 09288
Museo Municipal, Valdepeñas 40192
Museo Municipal, Zuheros 40360
Museo Municipal Alex Urquiola, El Salvador 09270
Museo Municipal Arco Ajújar, Medina de Rioseco 39637
Museo Municipal Barão de Santo Angelo, Rio Pardo 04776
Museo Municipal C.I.P.A.S., Salto 00562
Museo Municipal Cabildo de Montevideo, Montevideo 54407
Museo Municipal Carmen Funes, Plaza Huincul 00495
Museo Municipal Casa de Alfaro, Montecristi 10421
Museo Municipal de 10 de Octubre, La Habana 09153
Museo Municipal de Abreus, Abreus 09021
Museo Municipal de Aguada de Pasajeros, Aguada de Pasajeros 09022
Museo Municipal de Alto Songo y La Maya, Alto Songo 09023
Museo Municipal de Amancio, Amancio 09025
Museo Municipal de Arqueología y Etnología, Segorbe 40047
Museo Municipal de Arqueología y Etnología, Villajoyosa 40303
Museo Municipal de Arqueología y Etnológica, Bejís 39016
Museo Municipal de Arte Angel María de Rosa, Junín 00395
Museo Municipal de Arte Contemporáneo de Madrid, Madrid 39564
Museo Municipal de Arte Decorativo Firma y Odilo Estévez, Rosario 00546
Museo Municipal de Arte en Vidrio de Alcorcón, Alcorcón 38825
Museo Municipal de Arte Juan Carlos Castagnino, Mar del Plata 00446
Museo Municipal de Arte Moderno, Cuenca 10394
Museo Municipal de Arte Moderno de Mendoza, Mendoza 00458
Museo Municipal de Artes Plásticas, Olavarría 00481
Museo Municipal de Artes Plásticas, Rivera 54430
Museo Municipal de Artes Plásticas Faustino Jorge Bonadeo, Rauch 00518
Museo Municipal de Artes Plásticas Pompeo Boggio, Chivilcoy 00305
Museo Municipal de Artes Visuales, Concordia 00316
Museo Municipal de Artes Visuales, Quilmes 00511
Museo Municipal de Artes Visuales Sor Josefa Díaz y Clucellas, Santa Fé 00619
Museo Municipal de Arucas, Arucas 38905
Museo Municipal de Báguano, Báguano 09027
Museo Municipal de Bahía Honda, Bahía Honda 09028
Museo Municipal de Baraguá, Baraguá 09031
Museo Municipal de Bartolomé Masó, Bartolomé Masó 09032
Museo Municipal de Bayamo, Bayamo 09036
Museo Municipal de Baza, Baza 39010
Museo Municipal de Bellas Artes, Bahía Blanca 00136
Museo Municipal de Bellas Artes, La Calera 00397
Museo Municipal de Bellas Artes, La Paz 00405
Museo Municipal de Bellas Artes, Santa Cruz de Tenerife 39997
Museo Municipal de Bellas Artes, Tandil 00642
Museo Municipal de Bellas Artes de Valparaíso, Valparaíso 07472
Museo Municipal de Bellas Artes Dr. Genaro Perez, Córdoba 00335
Museo Municipal de Bellas Artes Dr. Genaro Pérez, Río Cuarto 00529
Museo Municipal de Bellas Artes Dr. Urbano Poggi, Rafaela 00515
Museo Municipal de Bellas Artes Juan B. Castagnino, Rosario 00547
Museo Municipal de Bellas Artes Juan Manuel Blanes, Montevideo 54408
Museo Municipal de Bellas Artes Lucas Braulio Areco, Posadas 00500
Museo Municipal de Bellas Artes Prof. Amelio Ronco Ceruti, Metán 00464
Museo Municipal de Bolivia, Bolivia 09040
Museo Municipal de Boyeros, La Habana 09154
Museo Municipal de Buey Arriba, Buey Arriba 09041
Museo Municipal de Cabaiguán, Cabaiguán 09042
Museo Municipal de Cabana, Cabana 34124
Museo Municipal de Cacomun, Cacomun 09043
Museo Municipal de Caibarién, Caibarién 09044
Museo Municipal de Caimanera, Caimanera 09045
Museo Municipal de Calahorra, Calahorra 39095
Museo Municipal de Calimete, Calimete 09047
Museo Municipal de Camajuaní Hermanos Vidal Caro, Camajuaní 09053
Museo Municipal de Campechuela, Campechuela 09054
Museo Municipal de Candelaria, Candelaria 09055
Museo Municipal de Carlos Manuel de Céspedes, Carlos Manuel de Céspedes 09057
Museo Municipal de Cauto Cristo, Cauto Cristo 09058
Museo Municipal de Cerámica, Paterna 39816
Museo Municipal de Chambas, Chambas 09060
Museo Municipal de Chancay, Chancay 34145
Museo Municipal de Chincha, Chincha Alta 34147
Museo Municipal de Ciencias, Bahía Blanca 00137
Museo Municipal de Ciencias Naturales, Valencia 40215
Museo Municipal de Ciencias Naturales Carlos Ameghino, Mercedes 00463
Museo Municipal de Ciencias Naturales Carlos Darwin, Punta Alta 00509
Museo Municipal de Ciencias Naturales Lorenzo Scaglia, Mar del Plata 00447
Museo Municipal de Cifuentes, Cifuentes 09069
Museo Municipal de Ciro Redondo, Ciro Redondo 09070
Museo Municipal de Colón José R. Zulueta, Colón 09072
Museo Municipal de Consolación del Sur, Consolación del Sur 09073
Museo Municipal de Contramaestre, Contramaestre 09074
Museo Municipal de Corralillo, Corralillo 09075
Museo Municipal de Cruces, Cruces 09076
Museo Municipal de Cueto, Cueto 09078
Museo Municipal de Cumanayagua, Cumanayagua 09079
Museo Municipal de Daimiel, Daimiel 39239
Museo Municipal de El Trébol, El Trébol 00360
Museo Municipal de Eldorado, Eldorado 00362
Museo Municipal de Esmeralda, Esmeralda 09080
Museo Municipal de Estepona, Estepona 39289
Museo Municipal de Etnografía, Villadiego 40296
Museo Municipal de Etnología, Castellón de la Plana 39150
Museo Municipal de Florencia, Florencia 09081
Museo Municipal de Florida, Florida 09082
Museo Municipal de Fomento, Fomento 09083
Museo Municipal de Gibara, Gibara 09086
Museo Municipal de Guamá, La Plata 09195
Museo Municipal de Guane, Guane 09089
Museo Municipal de Guantánamo - Antigua Cárcel, Guantánamo 09093
Museo Municipal de Guayaquil, Guayaquil 10409
Museo Municipal de Guisa, Guisa 09094
Museo Municipal de Habana del Este, La Habana 09155
Museo Municipal de Historia, Necolí 08694
Museo Municipal de Historia del Arte, Montevideo 54409
Museo Municipal de Historia Malitzín de Huiloapan, Huiloapan de Cuauhtémoc 30948
Museo Municipal de Historia Natural, General Alvear 00376
Museo Municipal de Historia Natural, San Rafael 00606
Museo Municipal de Historia y Arqueología, González 30880
Museo Municipal de Historia y Arqueología, Rivera 54431
Museo Municipal de Huaytará, Huaytará 34186
Museo Municipal de Imías, Playitas y Cajobabo, Cajobabo 09046
Museo Municipal de Jagüey Grande, Jagüey Grande 09187
Museo Municipal de Jesús Menéndez, Jesús Menéndez 09190
Museo Municipal de Jiguaní, Jiguaní 09191
Museo Municipal de Jovellanos Domingo Mujica, Jovellanos 09194
Museo Municipal de la Ciudad de Carcarañá, Carcarañá 00291
Museo Municipal de la Ciudad de Rosario, Rosario 00548
Museo Municipal de la Lisa, La Habana 09156
Museo Municipal de La Sierpe, La Sierpe 09331
Museo Municipal de La Yaya, La Yaya 09357
Museo Municipal de Lajas, Lajas 09196
Museo Municipal de Los Arabos Clotilde García, Los Arabos 09197
Museo Municipal de Los Palacios, Los Palacios 09198
Museo Municipal de Madrid, Madrid 39565
Museo Municipal de Maisí, Maisí 09200
Museo Municipal de Majagua, Majagua 09201
Museo Municipal de Majibacoa, Majibacoa 09202
Museo Municipal de Manatí Jesús Suárez Gayol, Manatí 09203
Museo Municipal de Manicaragua, Manicaragua 09204
Museo Municipal de Mantua, Mantua 09205
Museo Municipal de Manuel Tames, Manuel Tames 09206
Museo Municipal de Manzanillo, Manzanillo 09209
Museo Municipal de Marianao, La Habana 09157
Museo Municipal de Martí, Martí 09210
Museo Municipal de Mayarí, Mayarí 09220
Museo Municipal de Media Luna, Media Luna 09222
Museo Municipal de Minas, Minas 09223
Museo Municipal de Minas de Matahambre, Minas de Matahambre 09224
Museo Municipal de Moa, Moa 09225
Museo Municipal de Morón, Morón 09227
Museo Municipal de Najasa, Sibanicú 09329
Museo Municipal de Nazca y Casa Museo María Reiche, Nazca 34281
Museo Municipal de Niceto Pérez, Niceto Pérez 09228
Museo Municipal de Niquero, Niquero 09229
Museo Municipal de Nuevitas, Nuevitas 09239
Museo Municipal de Palma Soriano, Palma Soriano 09241
Museo Municipal de Pedro Betancourt, Pedro Betancourt 09243
Museo Municipal de Perico, Perico 09244
Museo Municipal de Pilón, Pilón 09245
Museo Municipal de Pintura, Villadiego 40297
Museo Municipal de Placetas, Placetas 09255
Museo Municipal de Poblado de Jimaguayú, Jimaguayú 09192
Museo Municipal de Ponteareas, Ponteareas 39856
Museo Municipal de Porteña, Porteña 00496
Museo Municipal de Primero de Enero, Primero de Enero 09257
Museo Municipal de Quemado de Güines, Quemado de Güines 09261
Museo Municipal de Rafael Freyre, Rafael Freyre 09262
Museo Municipal de Ranchuelo, Ranchuelo 09263
Museo Municipal de Río Cauto, Río Cauto 09266
Museo Municipal de Rodas, Rodas 09267
Museo Municipal de Ronda, Ronda 39920
Museo Municipal de Sagua de Tánamo, Sagua de Tánamo 09268
Museo Municipal de Sagua la Grande, Sagua la Grande 09269
Museo Municipal de San Cristobal, San Cristobal 09277
Museo Municipal de San Juan y Martínez, San Juan y Martínez 09279
Museo Municipal de San Luis 29 de Abril, San Luis 09280
Museo Municipal de San Miguel del Padrón, La Habana 09158
Museo Municipal de San Telmo, Donostia-San Sebastián 39253
Museo Municipal de Sandino, Sandino 09291
Museo Municipal de Santa Cruz del Sur, Santa Cruz del Sur 09301
Museo Municipal de Santiponce, Santiponce 40038
Museo Municipal de Santo Domingo, Santo Domingo 09328
Museo Municipal de Sibanicú, Sibanicú 09330
Museo Municipal de Sierra de Cubitas, Cubitas 09077
Museo Municipal de Taguasco, Taguasco 09333
Museo Municipal de Trinidad, Trinidad 09337
Museo Municipal de Unión de Reyes Juan G. Gómez, Unión de Reyes 09339
Museo Municipal de Urbano Noris, Urbano Noris 09340
Museo Municipal de Varadero, Varadero 09341
Museo Municipal de Vedado, La Habana 09159
Museo Municipal de Venezuela, Venezuela 09342
Museo Municipal de Vertientes, Vertientes 09343
Museo Municipal de Vigo Quiñones de León, Vigo 40276
Museo Municipal de Viñales Adela Azcuy Labrador, Pinar del Río 09251
Museo Municipal de Yaguajay, Yaguajay 09354
Museo Municipal de Yara, Yara 09355
Museo Municipal de Yateras, Yateras 09356
Museo Municipal del Cerro, La Habana 09160
Museo Municipal del Deporte, San Fernando del Valle de Catamarca 00574
Museo Municipal del III Frente, Santiago de Cuba 09324
Museo Municipal del Teatro, Lima 34263
Museo Municipal Dr. Rodolfo Doval Fermi, Sastre 00635
Museo Municipal Dreyer, Dreyer 34170
Museo Municipal Eduard Camps Cava, Guissona 39382
Museo Municipal Elisa Cendrero, Ciudad Real 39191
Museo Municipal Emilio Daudinot, San Antonio del Sur 09275
Museo Municipal Estados del Duque, Malagón 39608
Museo Municipal Fernando García Grave de Peralta, Puerto Padre 09260
Museo Municipal Forte de San Damián, Ribadeo 39898
Museo Municipal Francisco Javier Balmaseda, Santa Clara 09298
Museo Municipal Histórico Fotográfico de Quilmes, Quilmes 00512
Museo Municipal Histórico Regional Almirante Brown, Bernal 00148
Museo Municipal Histórico Regional Santiago Lischetti, Villa Constitución 00677
Museo Municipal Ignacio Balvidares, Puan 00503
Museo Municipal IV Centenario, Pinos 31260
Museo Municipal Jatibonico, Jatibonico 09189
Museo Municipal Jerónimo Molina, Jumilla 39434
Museo Municipal Jesús González Herrera, Matamoros 31018
Museo Municipal Jesús H. Salgado, Teloloapan 31421
Museo Municipal José A. Mulazzi, Tres Arroyos 00653
Museo Municipal José Manuel Maciel, Coronda 00341
Museo Municipal José Reyes Meza, Altamira 30712
Museo Municipal Juan M. Ameijeiras, El Salvador 09271
Museo Municipal Lino Enea Spilimbergo, Unquillo 00658
Museo Municipal López Torres, Tomelloso 40143
Museo Municipal Los Guachimontones y Museo de Arquitectura Prehispánica, Teuchitlán 31451
Museo Municipal Manuel Torres, Marín 39625
Museo Municipal Marceliano Santa María, Burgos 39072
Museo Municipal Mariano Benlliure, Crevillente 39227
Museo Municipal Mateo Hernández, Béjar 39014
Museo Municipal Nueva Gerona, Nueva Gerona 09237
Museo Municipal Precolombino y Colonial, Montevideo 54410
Museo Municipal Primeros Pobladores, San Martín de los Andes 00593
Museo Municipal Punta Hermengo, Miramar 00469
Museo Municipal Quetzalpapalotl, Teotihuacán 31428
Museo Municipal Ramón María Aller Ulloa, Lalín 39442
Museo Municipal Roberto Rojas Tamayo, Colombia 09071
Museo Municipal Rosendo Arteaga, Jobabo Dos 09193
Museo Municipal San Juan de Iris, San Juan de Iris 34303
Museo Municipal Ulpiano Checa, Colmenar de Oreja 39199
Museo Municipal Victorio Macho, Palencia 39773
Museo Municipal Dr. Santos F. Tosticarelli, Casilda 00297
Museo Mural Diego Rivera, México 31127
Museo Mural Homenaje a Benito Juárez, México 31128
Museo Muratoriano, Modena 26755
Museo Musicale d'Abruzzo, Ortona 27003
Museo Na-Bolom, San Cristóbal de las Casas 31329
Museo Nacional Antropológico Eduardo López Rivas, Oruro 04234
Museo Nacional Casa Guillermo Valencia, Popayán 08730
Museo Nacional Centro de Arte Reina Sofía, Madrid 39566
Museo Nacional de Aeronáutica, Morón 00475
Museo Nacional de Aeropuertos y Transporte Aéreo, Málaga 39604
Museo Nacional de Agricultura, Texcoco de Mora 31453
Museo Nacional de Antropología, Bogotá 08590
Museo Nacional de Antropología, Madrid 39567
Museo Nacional de Antropología, México 31129
Museo Nacional de Antropología, Arqueología e Historia del Perú, Lima 34264
Museo Nacional de Arqueología, La Paz 04222
Museo Nacional de Arqueología Marítima, Cartagena 39139
Museo Nacional de Arqueología y Etnología, Guatemala 23301
Museo Nacional de Arquitectura, Madrid 39568
Museo Nacional de Arquitectura, México 31130
Museo Nacional de Arte, La Paz 04223
Museo Nacional de Arte, México 31131
Museo Nacional de Arte Decorativo, Buenos Aires 00257
Museo Nacional de Arte Fantástico, Córdoba 30802
Museo Nacional de Arte Moderno, Guatemala 23302
Museo Nacional de Arte Oriental, Buenos Aires 00258
Museo Nacional de Arte Popular, Mérida 31028
Museo Nacional de Arte Romano, Mérida 39646
Museo Nacional de Artes Decorativas, La Habana 09161
Museo Nacional de Artes Decorativas, Madrid 39569
Museo Nacional de Artes Decorativas de Santa Clara, Santa Clara 09299
Museo Nacional de Artes e Industrias Populares, México 31132
Museo Nacional de Artes Visuales, Montevideo 54411
Museo Nacional de Bellas Artes, Asunción 34101
Museo Nacional de Bellas Artes, Buenos Aires 00259
Museo Nacional de Bellas Artes, La Habana 09162
Museo Nacional de Bellas Artes, Montevideo 54412

Museo Nacional de Bellas Artes de Santiago de Chile, Santiago de Chile 07461
Museo Nacional de Camilo Cienfuegos, Sancti-Spíritus 09289
Museo Nacional de Caza, Riofrío 39902
Museo Nacional de Cerámica y de las Artes Suntuarias González Martí, Valencia 40216
Museo Nacional de Ciencia y Tecnología, Madrid 39570
Museo Nacional de Ciencias Naturales, Madrid 39571
Museo Nacional de Colombia, Bogotá 08591
Museo Nacional de Costa Rica, San José 08824
Museo Nacional de Culturas Populares, México 31133
Museo Nacional de Escultura, Valladolid 40240
Museo Nacional de Etnografía y Folklore, La Paz 04224
Museo Nacional de Ferrocarriles Mexicanos, Puebla 31288
Museo Nacional de Historia, Guatemala 23303
Museo Nacional de Historia, México 31134
Museo Nacional de Historia Colonial, Omoa 23329
Museo Nacional de Historia de la Medicina Eduardo Estrella, Quito 10448
Museo Nacional de Historia Natural, La Habana 09163
Museo Nacional de Historia Natural, La Paz 04225
Museo Nacional de Historia Natural, Santiago de Chile 07462
Museo Nacional de Historia Natural, Santo Domingo 10357
Museo Nacional de Historia Natural y Antropología, Montevideo 54413
Museo Nacional de Informática, Lima 34265
Museo Nacional de la Acuarela, México 31135
Museo Nacional de la Alfabetización, La Habana 09164
Museo Nacional de la Cerámica, La Habana 09165
Museo Nacional de la Cerámica, Tonalá, Jalisco 31496
Museo Nacional de la Cultura Peruana, Lima 34266
Museo Nacional de la Dirección Regional del Banco Central del Ecuador, Quito 10449
Museo Nacional de la Estampa, México 31136
Museo Nacional de la Historia del Traje, Buenos Aires 00260
Museo Nacional de la Inmigración, Buenos Aires 00261
Museo Nacional de la Máscara, San Luis Potosí 31351
Museo Nacional de la Música, La Habana 09166
Museo Nacional de la Revolución, México 31137
Museo Nacional de las Culturas, México 31138
Museo Nacional de las Intervenciones, México 31139
Museo Nacional de las Telecomunicaciones, Bogotá 08592
Museo Nacional de los Ferrocarriles Mexicanos, México 31140
Museo Nacional de Nicaragua, Managua 33290
Museo Nacional de Reproducciones Artísticas, Madrid 39572
Museo Nacional de San Carlos, México 31141
Museo Nacional de Virreinato, Tepotzotlán 31445
Museo Nacional del Cobre, Salvador Escalante 31320
Museo Nacional del Grabado, Buenos Aires 00262
Museo Nacional del Hombre del Instituto Nacional de Antropología y Pensamiento Latinoamericano, Buenos Aires 00263
Museo Nacional del Ministerio del Interior, La Habana 09167
Museo Nacional del Petróleo, Comodoro Rivadavia 00310
Museo Nacional del Petróleo Samuel Schneider Uribe, Barrancabermeja 08538
Museo Nacional del Prado, Madrid 39573
Museo Nacional del Sombrero, Aguadas 08524
Museo Nacional del Teatro, Almagro 38852
Museo Nacional del Teatro, Buenos Aires 00264
Museo Nacional del Títere, Huamantla 30938
Museo Nacional Ferroviario, Buenos Aires 00265
Museo Nacional Justo José de Urquiza, Concepción del Uruguay 00313
Museo Nacional Mariano, Chiquinquira 08625
Museo Nacional Paleontológico y Arqueológico, Tarija 04251
Museo Nacional y Centro de Investigación de Altamira, Santillana del Mar 40036
Museo Nacional y Zoológico La Aurora, Guatemala 23304
Museo Nahim Isaias, Guayaquil 10410
Museo Nájerillense, Nájera 39710
Museo Napoleónico, La Habana 09168
Museo Napoleonico, Rivoli Veronese 27424
Museo Napoleonico, Roma 27511
Museo Napoleonico, Tolentino 27989
Museo Napoleonico G. Antonelli, Arcole 25260
Museo Natural, Pensilvania 08718
Museo Natural Dr. Carlos A. Marelli, Colón 00309
Museo Naturali e Appennino, Revi nel Lazio 27398
Museo Naturalistica-Archeologico, Vicenza 28254
Museo Naturalistico, Champdepraz 25846
Museo Naturalistico, Chieri 25867
Museo Naturalistico, Lesina 26488
Museo Naturalistico, Morino 26871
Museo Naturalistico, Picinisco 27199
Museo Naturalistico, Portoferraio 27307
Museo Naturalistico Archeologico M. Locati, Lama dei Peligni 26441
Museo Naturalistico Archeologico Paolo Barrasso, Caramanico Terme 25672
Museo Naturalistico Centro Natura, Pescasseroli 27168
Museo Naturalistico del Frignano, Pavullo nel Frignano 27128
Museo Naturalistico del Parco, San Romano di Garfagnana 27681
Museo Naturalistico della Riserva Naturale Orientata di Onferno, Gemmano 26283
Museo Naturalistico della Vallata del Bidente, Bagno di Romagna 25348
Museo Naturalistico e della Civiltà Contadina, Berceto 25398
Museo Naturalistico F. Minà Palumbo, Castelbuono 25722
Museo Naturalistico G. Zanardo, Tambre 27939
Museo Naturalistico Libero Gatti, Copanello 25966
Museo Naturalistico Missionario, Chiusa Pesio 25882
Museo Naturalistico N. De Leone, Penne 27135
Museo Naturalistico O. Perini, Borca di Cadore 25496
Museo Naucalpan de la Cultura Tlatilca, Naucalpan de Juárez 31212
Museo Naval, Catia la Mar 54561
Museo Naval, Madrid 39574
Museo Naval, Montevideo 54414
Museo Naval, Puno 34299
Museo Naval de Cartagena, Cartagena 39140
Museo Naval de Ferrol, Ferrol 39297
Museo Naval de Ilo, Ilo 34190
Museo Naval de la Nación, Tigre 00646
Museo Naval de San Fernando, San Fernando 39955
Museo Naval del Perú, Callao 34135
Museo Naval del Puerto de la Cruz, Puerto de la Cruz 39877
Museo Naval Puerto Belgrano, Puerto Belgrano 00504
Museo Navale, Napoli 26907
Museo Navale Archeologico N. Lamboglia, La Maddalena 26428
Museo Navale di Pegli, Genova 26317
Museo Navale Didattico, Milano 26712
Museo Navale Internazionale del Ponente Ligure, Imperia 26406
Museo Navale O. Zibetti, Caravaggio 25674
Museo Navale Romano, Albenga 25178
Museo Nazionale, Ravenna 27374
Museo Nazionale Alinari della Fotografia - MNAF, Firenze 26195
Museo Nazionale Archeologico, Altamura 25206
Museo Nazionale Archeologico, Civitavecchia 25924
Museo Nazionale Archeologico, Ferrara 26119
Museo Nazionale Archeologico, Sarsina 27748
Museo Nazionale Archeologico, Tarquinia 27943
Museo Nazionale Archeologico di Egnazia, Fasano-Savelletri 26090
Museo Nazionale Archeologico di Vulci, Canino 25645
Museo Nazionale Atestino, Este 26066
Museo Nazionale Concordiese, Portogruaro 27311
Museo Nazionale d'Abruzzo, L'Aquila 26457
Museo Nazionale d'Arte Medievale e Moderna, Matera 26640
Museo Nazionale d'Arte Orientale, Roma 27512
Museo Nazionale degli Strumenti Musicali, Roma 27513
Museo Nazionale del Bargello, Firenze 26196
Museo Nazionale del Cinema, Torino 28021
Museo Nazionale del Palazzo di Venezia, Roma 27514
Museo Nazionale del Risorgimento Italiano, Torino 28022
Museo Nazionale del San Gottardo, Airolo 41097
Museo Nazionale della Attività Subacquee, Marina di Ravenna 26608
Museo Nazionale della Certosa, Calci 25594
Museo Nazionale della Fotografia, Brescia 25541
Museo Nazionale della Montagna Duca degli Abruzzi, Torino 28023
Museo Nazionale della Residenza Napoleonica dei Mulini, Portoferraio 27308
Museo Nazionale della Scienza e della Tecnica Leonardo da Vinci, Milano 26713
Museo Nazionale della Siritide, Policoro 27270
Museo Nazionale dell'Agro Picentino, Pontecagnano 27287
Museo Nazionale dell'Alta Val d'Agri, Grumento Nova 26379
Museo Nazionale dell'Antartide F. Ippolito, Genova 26318
Museo Nazionale dell'Arma della Cavalleria, Pinerolo 27224
Museo Nazionale delle Arti e Tradizioni Popolari, Roma 27515
Museo Nazionale delle Arti Naïves C. Zavattini, Luzzara 26550
Museo Nazionale delle Paste Alimentari, Roma 27516
Museo Nazionale dell'Età Neoclassica in Romagna, Faenza 26077
Museo Nazionale di Canossa N. Campanini, Ciano d'Enza 25884
Museo Nazionale di Casa Giusti, Monsummano Terme 26792
Museo Nazionale di Castel Sant'Angelo, Roma 27517
Museo Nazionale di Palazzo Reale, Pisa 27244
Museo Nazionale di Reggio Calabria, Reggio Calabria 27384
Museo Nazionale di San Matteo, Pisa 27245
Museo Nazionale di Santa Maria delle Monanche e Paleolitico, Isernia 26414
Museo Nazionale di Villa Guinigi, Lucca 26536
Museo Nazionale di Villa San Martino, Portoferraio 27309
Museo Nazionale Domenico Ridola, Matera 26641
Museo Nazionale Etrusco di Villa Giulia, Roma 27518
Museo Nazionale Etrusco Pompeo Aria, Marzabotto 26615
Museo Nazionale Ferroviario, Napoli 26908
Museo Nazionale Garibaldino, La Maddalena 26429
Museo Nazionale Jatta, Ruvo di Puglia 27582
Museo Nazionale Preistorico ed Etnografico Luigi Pigorini, Roma 27519
Museo Nazionale Romano-Terme di Diocleziano, Roma 27520
Museo Néstor, Las Palmas de Gran Canaria 39802
Museo Nezahualcóyotl Acomiztli, Nezahualcóyotl 31215
Museo ng Bangko Sentral ng Pilipinas, Manila 34445
Museo ng Batangas, Lipa 34404
Museo ng Buhay Pilipino, Pasay 34473
Museo ng Kalinangang Pilipino, Pasay 34474
Museo ng Katipunan, Lipa 34405
Museo ng Makati, Makati 34415
Museo ng Maynila, Manila 34446
Museo ng Rebolusyon, San Juan 34498
Museo Ñiace, Tepelmeme de Morelos 31435
Museo Nicanor Piñole, Gijón 39340
Museo Ñico Lopez, Bayamo 09037
Museo Niño Minero, Vetagrande 31539
Museo Norace, Rosignano Marittimo 27563
Museo Norawa, Guachochi 30883
Museo Normanno-Bizantino, San Marco d'Alunzio 27659
Museo Norte de Potosí, Potosí 04237
Museo Notarial Argentino, Buenos Aires 00266
Museo Notario Roberto Mario Arata, Buenos Aires 00267
Museo Novarese d'Arte e Storia R. Fumagalli, Novara 26953
Museo Nuestra Señora de Carrodilla, Luján de Cuyo 00437
Museo Nuestro Pueblo y Museo Etnográfico de Tecpatán, Tecpatán 31416
Museo Nueva Cádiz, La Asunción 54577
Museo Numantino, Soria 40093
Museo Numismático, La Habana 09169
Museo Numismático del Banco Continental, Arequipa 34114
Museo Numismático del Banco Wiese, Lima 34267
Museo Numismatico della Zecca, Roma 27521
Museo Numismático y Filatélico del Banco Central, Santo Domingo 10358
Museo Nuovo - Museo del Conservatori, Roma 27522
Museo Obispo Fray José Antonio de San Alberto, Córdoba 00336
Museo Obispo Vellosillo, Ayllón 38927
Museo Ocarina e degli Strumenti in Terracotta, Budrio 25560
Museo O'Higginiano y de Bellas Artes de Talca, Talca 07466
Museo Omar Huerta Escalante, Jonuta 30977
Museo Omero, Ancona 25230
Museo Oncativo, Oncativo 00482
Museo Onsernonese, Loco 41574
Museo Ora Ton, Chamula 30758
Museo Oraziano, Licenza 26491
Museo Orazio Comes, Portici 27301
Museo Organologico Didattico, Cremona 26005
Museo Organológico Folklórico Musical, Bogotá 08593
Museo Oriental, Valladolid 40241
Museo Orientale, Tagliacozzo 27936
Museo Orlando Binaghi, Moreno 00474
Museo Ornitológico, Aragua 54516
Museo Ornitologico, Poppi 27296
Museo Ornitologico, San Gimignano 27640
Museo Ornitologico C. Beni, Stia 27904
Museo Ornitologico Ferrante Foschi, Forlì 26227
Museo Ornitologico Naturalistico S. Bambini, Pietralunga 27204
Museo Ornitológico Patagónico, El Bolsón 00358
Museo Orocoveño de Familia Avilés, Orocovis 36044
Museo Orve, Chaupimascos 34146
Museo Oscar Fernandez Morera, Sancti-Spíritus 09290
Museo Oscar María de Rojas, Cárdenas 09056
Museo Osservanza, Bologna 25475
Museo Ostetrico, Padova 27047
Museo Ostiense, Ostia Antica 27017
Museo Othoniano, San Luis Potosí 31352
Museo P. Calderini, Varallo Sesia 28157
Museo P. Vagliasindi, Randazzo 27364
Museo Pablo Bush Romero, Puerto Aventura 31294
Museo Pablo Casals, San Juan 36074
Museo Pablo Gargallo, Zaragoza 40347
Museo Pablo Serrano, Zaragoza 40348
Museo Padiglione delle Carozza, Città del Vaticano 54507
Museo Padre Coll, Buenos Aires 00268
Museo Paisa, Caicedonia 08604
Museo Palacio de Don Pedro I, Astudillo 38912
Museo Palacio de la Canal, San Miguel de Allende 31357
Museo-Palacio Spínola, Teguise 40111
Museo Palatino, Roma 27523
Museo Palazzina Storica, Peschiera del Garda 27169
Museo Palenteológico de la Laguna, Torreón 31502
Museo Paleocristiano, Aquileia 25254
Museo Paleontologico, Amandola 25218
Museo Paleontologico, Bova 25519
Museo Paleontologico, Fluminimaggiore 26208
Museo Paleontologico, L'Aquila 26458
Museo Paleontológico, Mariquita 08674
Museo Paleontologico, Mondaino 26774
Museo Paleontologico, Montevarchi 26858
Museo Paleontológico, Murcia 39700
Museo Paleontológico, Pinar del Río 09252
Museo Paleontologico Archeologico V. Caccia, San Colombano al Lambro 27626
Museo Paleontológico, Arqueológico e Histórico de la Ciudad de Deán Funes, Deán Funes 00353
Museo Paleontológico de Elche, Elche 39269
Museo Paleontologico della Rocca di Monfalcone, Monfalcone 26779
Museo Paleontológico Egidio Feruglio, Trelew 00651
Museo Paleontologico M. Gortani, Portogruaro 27312
Museo Paleontologico Parmense, Parma 27109
Museo Paleontologico Rinaldo Zardini, Cortina d'Ampezzo 25986
Museo Paleontologico S. Lai, Ceriale 25808
Museo Paleontológico - Sala Longinos Navás, Zaragoza 40349
Museo Paleontológico - Sala Lucas Mallada, Zaragoza 40350
Museo Paleontológico y Petrolero Astra, Comodoro Rivadavia 00311
Museo Paleoveneto, Pieve di Cadore 27210
Museo Palladio, Vicenza 28255
Museo Pambata, Manila 34447
Museo Pampeano de Chascomús, Chascomús 00300
Museo Pancho Coímbre, Ponce 36053
Museo Pantaleón Panduro, Tlaquepaque 31473
Museo Panteón San Fernando, México 31142
Museo Parigino a Roma, Roma 27524
Museo Parma Gemma, Casarza Ligure 25702
Museo Parque de Esculturas, Montevideo 54415
Museo Parque de la Cristianía, Chapala 30759
Museo Parque de la Muralla, Lima 34268
Museo Parque Los Tres Ojos, Santo Domingo 10359
Museo Parrocchiale, Bionaz 25422
Museo Parrocchiale, Brescello 25533
Museo Parrocchiale, Campo Tures 25634
Museo Parrocchiale, Challand Saint Victor 25843
Museo Parrocchiale, Chambave 25844
Museo Parrocchiale, Champorcher 25847
Museo Parrocchiale, Fié 26125
Museo Parrocchiale, Gressoney Saint Jean 26368
Museo Parrocchiale, La Salle 26431
Museo Parrocchiale, Ornavasso 26998
Museo Parrocchiale, Ponte Buggianese 27282
Museo Parrocchiale, Saint Vincent 27590
Museo Parrocchiale, San Marco d'Alunzio 27660
Museo Parrocchiale, Sarre 27746
Museo Parrocchiale, Valgrisenche 28142
Museo Parrocchiale, Valtournenche 28154
Museo Parrocchiale d'Arte Sacra, Dozza 26047
Museo Parrocchiale del Santuario del Santissimo Crocifisso, Buggiano 25563
Museo Parrocchiale della Chiesa Collegiata, Mercatello sul Metauro 26663
Museo Parrocchiale di Arte Sacra, Ponte in Valtellina 27284
Museo Parrocchiale di Arte Sacra, Valstrona 28152
Museo Parrocchiale e Museo P. Mascagni, Bagnara di Romagna 25344

Museo Parrocchiale Madonna dei Raccomandati, Gessopalena 26331
Museo Parrocchiale San Giacomo, Massignano 26637
Museo Parrocchiale San Pietro Martire, Venezia 28202
Museo Parroquial, Alquezar 38865
Museo Parroquial, El Barco de Ávila 39008
Museo Parroquial, Bocairent 39054
Museo Parroquial, Celanova 39161
Museo Parroquial, Chillón 39175
Museo Parroquial, Écija 39258
Museo Parroquial, Espejo 39281
Museo Parroquial, Ezcaray 39292
Museo Parroquial, Hondarribia 39391
Museo Parroquial, Lucena del Cid 39497
Museo Parroquial, Paradas 39811
Museo Parroquial, Pastrana 39815
Museo Parroquial, San Sebastián del Oeste 31371
Museo Parroquial, Santa María del Campo 40003
Museo Parroquial, Valdemoro 40188
Museo Parroquial Colegiata de San Miguel, Aguilar de Campoo 38792
Museo Parroquial de Arte Religioso, Turmequé 08778
Museo Parroquial de Arte Sacro y Etnología, Anso 38873
Museo Parroquial de Calatañazor, Calatañazor 39097
Museo Parroquial de Frómista, Frómista 39311
Museo Parroquial de la Asunción de Nuestra Señora de Navarrés, Navarrés 39713
Museo Parroquial de Monterroso, Monterroso 39676
Museo Parroquial de San Cosme y San Damián, Covarrubias 39224
Museo Parroquial de San Martiño de Mondoñedo, Foz 39308
Museo Parroquial de San Martiño de Mondoñedo, San Martiño-Foz 39964
Museo Parroquial de Santa Cruz, Madrid 39575
Museo Parroquial de Santa Eulalia, Paredes de Nava 39813
Museo Parroquial de Santa María, Becerril de Campos 39012
Museo Parroquial de Valderas, Valderas 40193
Museo Parroquial de Yanguas, Yanguas 40329
Museo Parroquial de Zurbarán, Marchena 39624
Museo Parroquial Etnográfico, La Ercina 39277
Museo Parroquial Nuestra Sacra de la Asunción, Palomares del Campo 39803
Museo Parroquial Santa María del Salvador, Chinchilla de Monte-Aragón 39177
Museo Participativo de Ciencias, Buenos Aires 00269
Museo Particular Los Abuelos, Franck 00373
Museo Partigiano, Carpasio 25684
Museo Partigiano Alfredo Di Dio, Ornavasso 26999
Museo Pasos de Semana Santa, Bilbao 39047
Museo Patio del Moro, Guaduas 08652
Museo Patio Herreriano, Valladolid 40242
Museo Patria Chica de Mohamed Diaz, Realicó 00520
Museo Pavese di Scienze Naturali, Pavia 27124
Museo Pecharromán, Pasarón de la Vera 39814
Museo Pedagógico Ciencias Naturales Jesús María Hernando, Valladolid 40243
Museo Pedagógico de Arte Infantil, Madrid 39576
Museo Pedagógico de Chile, Santiago de Chile 07463
Museo Pedagógico de Historia del Arte, Caracas 54555
Museo Pedagógico José Pedro Varela, Montevideo 54416
Museo Pedagógico Textil, Madrid 39577
Museo Pedagóxico de Galicia, Santiago de Compostela 40029
Museo Pedrin Troya, San Nicolás de Bari 09281
Museo Pedro Coronel, Zacatecas 31580
Museo Pedro Nel Gomez, Medellín 08685
Museo Penitenciario Argentino Antonio Ballvé, Buenos Aires 00270
Museo Peñón Blanco, Peñón Blanco 31256
Museo per la Matematica - Giardino di Archimede, Priverno 27346
Museo per la Storia dell'Università di Pavia, Pavia 27125
Museo per l'Arte e l'Archeologia del Vastese, Gissi 26344
Museo Pérez Comendador-Leroux, Hervás 39390
Museo Perfumeria Gal, Alcalá de Henares 38815
Museo Perrando, Sassello 27757
Museo Petatlán, Petatlán 31258
Museo Peten Ak, Valladolid 31524
Museo Petrográfico del Servicio Nacional de Geología y Minería, Quito 10450
Museo Picasso, Buitrago del Lozoya 39061
Museo Picasso Málaga, Málaga 39605
Museo Piersanti, Matelica 26638
Museo Pietro Canonica, Roma 27525
Museo Pietro della Vedova, Rima San Giuseppe 27404
Museo Pietro Micca e dell'Assedio, Torino 28024
Museo-Pinacoteca, Visso 28290
Museo Pinacoteca A. Salvucci, Molfetta 26765
Museo Pinacoteca Comunale G. De Nittisu, Barletta 25374
Museo-Pinacoteca de Arte Contemporáneo, Zarzuela del Monte 40356
Museo Pinacoteca della Santa Casa, Loreto 26519
Museo Pinacoteca dell'Arte dei Bambini, Smerillo 27860
Museo Pinacoteca Parrocchiale di Santa Maria, Mogliano Marche 26761
Museo - Pinacoteca San Francesco, San Marino 38009
Museo Pintor Antonio Rodríguez Luna, Montoro 39682
Museo Pio Clementino, Città del Vaticano 54508
Museo Pio Collivadino, Banfield 00140
Museo Pio Cristiano, Città del Vaticano 54509
Museo Pio IX, Senigallia 27804
Museo Pio X, Salzano 27603
Museo Planetario, Garza García 30874
Museo Plaza de Revolución General Antonio Maceo, Santiago de Cuba 09325
Museo Plebano, Agno 41092
Museo Polare Etnografico Silvio Zavatti, Fermo 26102
Museo Poldi Pezzoli, Milano 26714
Museo Policial, Ávila 38922
Museo Polifacético Regional Vottero, Laguna Larga 00423
Museo Polivalente, Chiaramonte Gulfi 25855
Museo Polivalente, Vittoria 28295
Museo Polivalente de Palmira, Palmira 09242
Museo Pomposiano, Codigoro 25931
Museo Ponchielliano, Paderno Ponchielli 27031
Museo Popol Vuh, Guatemala 23305
Museo Popoli e Culture, Milano 26715
Museo Popular de Arte Contemporáneo, Villafamés 40299
Museo Porfirio Corona Covarrubias, El Grullo, Jalisco 30847
Museo Poschiavino, Poschiavo 41731
Museo Posta de Yatasto, Metán 00465
Museo Postal Cubano, La Habana 09170
Museo Postal Eduardo Santos, Bogotá 08594
Museo Postal y Filatélico del Perú, Lima 34269
Museo Postal y Telegráfico, Madrid 39578
Museo Postal y Telegráfico Dr. Ramón J. Cárcano, Buenos Aires 00271
Museo Postale e Telegrafico della Mitteleuropa, Trieste 28095
Museo Postale, Filatelico-Numismatico, San Marino 38010
Museo Prehistórico de Tepexpan, Tepexpan 31438
Museo Preindustrial de la Caña de Azúcar, Motril 39687
Museo Preistorico, Paceco 27030
Museo Preistorico Balzi Rossi, Ventimiglia 28217
Museo Preistorico dell'Isolino, Biandronno 25415
Museo Preistorico e Archeologico A.C. Blanc, Viareggio 28244
Museo Preistorico N. Lamboglia e Val Varatello, Toirano 27983
Museo Presidente José Evaristo Uriburu, Salta 00559
Museo Presidio Modelo, Nueva Gerona 09238
Museo Pringles de Arqueología e Historia, Coronel Pringles 00345
Museo Privado de Arqueología Regional Julio Gironde, Carmen de Patagonés 00294
Museo Prof. Arturo Reyes Viramontes, Jalpa 30967
Museo Prof. Dr. Juan Augusto Olsacher, Zapala 00688
Museo Prof. Moisés Sáenz Garza, Apodaca 30721
Museo Prof. Ninfa María Báez, Acoyapa 33271
Museo Prof. Ricardo Vega Noriega, San Ignacio 31332
Museo Profano, Città del Vaticano 54510
Museo Provincial, Ciego de Avila 09064
Museo Provincial, Lugo 39499
Museo Provincial, Teruel 40122
Museo Provincial Arqueológico Eric Boman, Santa María de Catamarca 00623
Museo Provincial de Arqueología Wagner, Santiago del Estero 00630
Museo Provincial de Artes de La Pampa, Santa Rosa 00624
Museo Provincial de Bellas Artes, Salta 00560
Museo Provincial de Bellas Artes Dr. Juan Ramon Vidal, Corrientes 00349
Museo Provincial de Bellas Artes Dr. Pedro E. Martínez, Paraná 00488
Museo Provincial de Bellas Artes Emiliano Guiñazu - Casa de Fader, Luján de Cuyo 00438
Museo Provincial de Bellas Artes Emilio A. Caraffa, Córdoba 00337
Museo Provincial de Bellas Artes Ramón Gómez Cornet, Santiago del Estero 00631
Museo Provincial de Bellas Artes Rene Brusau, Resistencia 00526
Museo Provincial de Bellas Artes Rosa Galisteo de Rodríguez, Santa Fé 00620
Museo Provincial de Bellas Artes Timoteo Eduardo Navarro, San Miguel de Tucumán 00601
Museo Provincial de Ciencias Naturales Dr. Angelo Gallardo, Rosario 00549
Museo Provincial de Ciencias Naturales Florentino Ameghino, Santa Fé 00621
Museo Provincial de Ciencias Naturales y Oceanográfico, Puerto Madryn 00507
Museo Provincial de Historia Natural de La Pampa, Santa Rosa 00625
Museo Provincial de Holguín La Periquera, Holguín 09184
Museo Provincial de la Historia de Pinar del Río, Pinar del Río 09253
Museo Provincial de Matanzas, Matanzas 09218
Museo Provincial de Villa Clara y Museo Municipal, Santa Clara 09300
Museo Provincial do Mar, Cervo 39169
Museo Provincial Don Emilio Bacardí Moreau, Santiago de Cuba 09326
Museo Provincial Dora Ochoa de Masramón, San Luis 00592
Museo Provincial General Vicente García, Victoria de las Tunas 09349
Museo Provincial Ignacio Agramonte, La Vigía 09351
Museo Provincial José María Morante, Arequipa 34115
Museo Provincial Mario Brozoski, Puerto Deseado 00505
Museo Provincial y Parroquial, Piña de Campos 39841
Museo Provinciale, Catanzaro 25783
Museo Provinciale Campano, Capua 25670
Museo Provinciale d'Arte Contempranea, Crotone 26014
Museo Provinciale degli Usi e Costumi, Brunico 25556
Museo Provinciale della Caccia e della Pesca, Racines 27355
Museo Provinciale della Ceramica, Vietri 28264
Museo Provinciale della Vita Contadina, San Vito al Tagliamento 27689
Museo Provinciale delle Miniere, Racines 27356
Museo Provinciale delle Miniere, Vipiteno 28287
Museo Provinciale delle Tradizioni Popolari Abbazia Cerrate, Lecce 26473
Museo Provinciale di Storia Naturale, Foggia 26212
Museo Provinciale di Storia Naturale, Livorno 26500
Museo Provinciale Sigismondo Castromediano, Lecce 26474
Museo Pucciniano, Torre del Lago 28037
Museo Pueblo de Luis, Trelew 00652
Museo Puertorriqueño del Deportes, Salinas 36055
Museo Quattroruote, Rozzano 27577
Museo Quezaltenango, Quezaltenango 23307
Museo Quinta Amalia Simoni, Camagüey 09052
Museo Quinta Casa de Juan Montalvo, Ambato 10368
Museo Quinta de San Pedro Alejandrino, Santa Marta 08754
Museo Quinta Juan Leon Mera, Ambato 10369
Museo R. Campelli, Pievebovigliana 27218
Museo R. Cassolo, Mede 26645
Museo R. Cuneo, Savona 27779
Museo R. De Cesare, Spinazzola 27889
Museo Rabelo, La Habana 09171
Museo-Raccolta di Fisica, Bari 25369
Museo Radiofónico de la Ciudad de Boulogne, Boulogne 00149
Museo Rafael Coronel, Zacatecas 31581
Museo Rafael Escriña, Santa Clara de Saguier 00612
Museo Rafael Zabaleta, Quesada 39885
Museo Raffaello Pagliaccetti, Giulianova 26347
Museo Raíces, Jesús María 08664
Museo Raíces de Satevó, Satevó 31388
Museo Rainis e Aspazija, Castagnola 41284
Museo Ralli, Marbella 39620
Museo Ralli, Punta del Este 54429
Museo Ralli, Santiago de Chile 07464
Museo Ramón Gaya, Murcia 39701
Museo Ramón López Velarde, México 31143
Museo Ramón Pérez Fernández, Rivadavia 00540
Museo Ramos Generales, Villa María 00682
Museo Ratti dei Vini d'Alba, La Morra 26430
Museo Rauzi, Palmira 00483
Museo Rayo, Roldanillo 08740
Museo Real Colegiata de San Isidoro, León 39459
Museo Recoleto, Quezon City 34494
Museo Regional, San Juan Bautista Tuxtepec 31335
Museo Regional Amazónico, Iquitos 34195
Museo Regional Aníbal Cambas, Posadas 00501
Museo Regional Arqueológico Rossbach, Chichicastenango 23294
Museo Regional Carlos Funes Derieul, Coronel Dorrego 00342
Museo Regional Castelli, Castelli 00298
Museo Regional Cayetano Alberto Silva, Venado Tuerto 00666
Museo Regional Comunitario Altepepialcalli, México 31144
Museo Regional Cuauhnáhuac, Cuernavaca 30819
Museo Regional Daniel Hernández Morillo, Huancavelica 34175
Museo Regional de Acambay, Acambay 30688
Museo Regional de Actopán, Actopán 30696
Museo Regional de Aldama, Aldama 30710
Museo Regional de Antropología, Resistencia 00527
Museo Regional de Antropología Carlos Pellicer Cámara, Villahermosa 31554
Museo Regional de Antropología e Historia de Baja California Sur, La Paz 30989
Museo Regional de Antropología J. Jesús Figueroa Torres, Sayula 31390
Museo Regional de Arqueología, Huamanga 34174
Museo Regional de Arqueología de Junín, Chupaca 34151
Museo Regional de Arqueología de Tuxpan, Tuxpan de Rodríguez Cano 31510
Museo Regional de Arte e Historia, Jerez de García Salinas 30974
Museo Regional de Arte Popular La Huatápera, Uruapán 31521
Museo Regional de Atacama, Copiapó 07437
Museo Regional de Bolillo, Cartagena 39141
Museo Regional de Campeche, Campeche 30744
Museo Regional de Casma Max Uhle, Casma 34139
Museo Regional de Chiapas, Tuxtla Gutiérrez 31517
Museo Regional de Cinco Saltos, Cinco Saltos 00307
Museo Regional de Claromecó, Balneario Claromecó 00139
Museo Regional de Cuetzalan, Cuetzalán 30822
Museo Regional de Durango y Museo del Niño, Durango 30844
Museo Regional de Guadalajara, Guadalajara 30904
Museo Regional de Guanajuato Alhóndiga de Granaditas, Guanajuato 30921
Museo Regional de Guerrero, Chilpancingo 30776
Museo Regional de Hidalgo, Pachuca 31245
Museo Regional de Historia, Hermosillo 30931
Museo Regional de Historia de Aguascalientes, Aguascalientes 30703
Museo Regional de Historia de Colima, Colima 30782
Museo Regional de Historia de Tamaulipas, Victoria 31544
Museo Regional de Ica María Reiche Gross Newman, Ica 34189
Museo Regional de Iguala, Iguala de la Independencia 30956
Museo Regional de Iquique, Iquique 07439
Museo Regional de la Araucania, Temuco 07467
Museo Regional de la Arqueológico de Tiwanaku, Tiwanaku 04252
Museo Regional de la Cerámica, Tlaquepaque 31474
Museo Regional de la Ciudad de La Paz, La Paz 00406
Museo Regional de la Laguna, Torreón 31503
Museo Regional de las Culturas Occidente, Guzmán 30926
Museo Regional de Mascota, Mascota 31016
Museo Regional de Nayarit, Tepic 31444
Museo Regional de Nuevo León, Monterrey 31187
Museo Regional de Pigüé, Pigüé 00493
Museo Regional de Pintura José Antonio Terry, Tilcara 00649
Museo Regional de Puebla, Puebla 31289
Museo Regional de Querétaro, Querétaro 31302
Museo Regional de Río Verde, Río Verde 31306
Museo Regional de Sacanta, Sacanta 00552
Museo Regional de San Carlos, San Carlos 54432
Museo Regional de San Martín, Moyobamba 34279
Museo Regional de Sinaloa, Culiacán 30829
Museo Regional de Sonora, Hermosillo 30932
Museo Regional de Telecomunicaciones Tomás Guzmán Cantú, La Paz 30990
Museo Regional de Tierra Caliente, Coyuca de Catalán 30807
Museo Regional de Tlaxcala, Tlaxcala 31478
Museo Regional de Ucayali, Pucallpa 34294
Museo Regional de Yucatán, Mérida 31029
Museo Regional del Valle del Fuerte, Ahome 30705
Museo Regional Doctor Leonardo Oliva, Ahualulco de Mercado 30706
Museo Regional Dr. Adolfo Alsina, Carhué 00292
Museo Regional Dr. Federico Escalada, Río Mayo 00537
Museo Regional Dr. José Luro, Pedro Luro 00489
Museo Regional El Fuerte, El Fuerte 30846
Museo Regional Francisco Mazzoni, Maldonado 54381
Museo Regional Guayupe de Puerto Santander, Puerto Santander 08733
Museo Regional Huasteco, Valles 31530
Museo Regional José G. Brochero, Santa Rosa de Río Primero 00626
Museo Regional Las Varillas, Las Varillas 00426
Museo Regional Lázaro Cárdenas, Tlapehuala 31472
Museo Regional Luciano Márquez, Zacatlán 31586
Museo Regional Malargüe, Malargüe 00442
Museo Regional Maracó, General Pico 00379

Museo Regional Michoacano, Morelia 31201
Museo Regional Municipal de El Calafate, El Calafate 00359
Museo Regional Particular Epifanio Saravia, Santa Catalina 00611
Museo Regional Patagónico, Comodoro Rivadavia 00312
Museo Regional Potosino, San Luis Potosí 31353
Museo Regional Provincial Padre Manuel Jesús Molina, Río Gallegos 00534
Museo Regional Rescate de Raíces, Ahualulco de Mercado 30707
Museo Regional Rincón de Vida, Villa Mirasol 00683
Museo Regional Rufino Muñiz Torres, Ocampo 31231
Museo Regional Rumi Mayu (Río de Piedra), Sanagasta 00610
Museo Regional Salesiano de Rawson, Rawson 00519
Museo Regional Salesiano Maggiorino Borgatello, Punta Arenas 07447
Museo Regional Trevelin, Trevelin 00656
Museo Regional Valcheta, Valcheta 00664
Museo Regional Valparaíso, Valparaíso 31531
Museo Regional y de Arte Marino, Puerto San Julián 00508
Museo Regionale, Messina 26668
Museo Regionale della Ceramica, Caltagirone 25606
Museo Regionale della Ceramica, Deruta 26032
Museo Regionale delle Centovalli e del Pedemonte, Intragna 41493
Museo Regionale di Camarina, Ragusa 27360
Museo Regionale di Scienze Naturali, Saint Pierre 27588
Museo Regionale di Scienze Naturali, Torino 28025
Museo Religioso de Guatavita, Guatavita 08655
Museo Religioso Santo Domingo, Bolivar 10376
Museo Religioso y Colonial de San Francisco, Cali 08614
Museo Renacimiento Indígena, Huamuxtitlán 30940
Museo Renato Brozzi, Traversetolo 28051
Museo Ricardo J. Alfaro, Bella Vista 34060
Museo Richard Ginori della Manifattura di Doccia, Sesto Fiorentino 27825
Museo Ripley, México 31145
Museo Risorgimentale, Montichiari 26860
Museo Risorgimentale, Pieve di Cadore 27211
Museo Rivarossi dei Treni in Miniatura, Como 25953
Museo Roca, Buenos Aires 00272
Museo Rocsen, Nono 00478
Museo Rodera Robles, Segovia 40055
Museo Rodolfo Valentino, Castellaneta 25733
Museo Rognoni Salvaneschi, Novara 26954
Museo Romagnolo del Teatro, Forlì 26228
Museo Romano, Tortona 28044
Museo Romano de Astorga, Astorga 38911
Museo Romano Oiasso, Irún 39413
Museo Romántico, Madrid 39579
Museo Romántico, Trinidad 09338
Museo Romántico - Casa de Montero, Montevideo 54417
Museo Rómulo Raggio, Vicente López 00667
Museo Ron Arehucas, Arucas 38906
Museo Rosendo Mejica, Iloilo 34390
Museo Rosminiano, Stresa 27911
Museo Rubén Herrera, Saltillo 31317
Museo Rudolf Stolz, Sesto in Pusteria 27826
Museo Runa de Antropología, Tafí del Valle 00640
Museo Rural, Pozorrubio de Santiago 39865
Museo Rural de la Posta de Sinsacate, Jesús María 00394
Museo S. Fancello, Dorgali 26045
Museo Sa Domu De Donna Nassia, Lotzorai 26527
Museo Sac-be, Yaxcabá 31572
Museo Sacrario delle Bandiere, Roma 27526
Museo Sacrario Galileo Ferraris, Livorno Ferraris 26501
Museo Sacro, Città del Vaticano 54511
Museo Sacro, Pompei 27278
Museo Sacro de Caracas, Caracas 54556
Museo Sacro de la Catedral, San Miguel de Tucumán 00602
Museo Sala, San Juan Bautista Tuxtepec 31336
Museo Sala de Arte, Tijuana 31458
Museo Sala de Arte Moderno, Juchitán de Zaragoza 30984
Museo Sala de Arte Público David Alfaro Siqueiros, México 31146
Museo Sala de Exposición Banco Central de Esmeraldas, Esmeraldas 10396
Museo-Sala de los Símbolos Patrios, Hermosillo 30933
Museo Salesiano, Huancayo 34178
Museo Salvatore Ferragamo, Firenze 26197
Museo Salzillo, Murcia 39702
Museo San Antonio de Padua, Córdoba 00338
Museo San Francisco, La Paz 04226
Museo San Giuseppe da Copertino, Osimo 27013
Museo San Gregorio de Polanco, San Gregorio de Polanco 54433
Museo San José el Mogote, San José el Mogote 31334
Museo San Juan de la Cruz, Úbeda 40179
Museo San Juan de Letrán, Chucuito 34149
Museo San Juan en la Historia, San Juan del Río 31341
Museo San Miguel del Padrón, La Habana 09172
Museo San Miguel Teotongo, México 31147
Museo San Nicolò, Militello in Val di Catania 26731
Museo Santa Cabrini, Sant'Angelo Lodigiano 27722
Museo Santa Clara de Murcia, Murcia 39703
Museo Santa Clara la Real, Tunja 08777
Museo Santa Iglesia Catedral, Plasencia 39845
Museo Sant'Andrea, Clusone 25928
Museo-Santuario de San Pedro Claver, Cartagena 08620
Museo Santuario Nuestra Señora Natividad del Guayco, La Magdalena 10419
Museo Santuarios Andinos de la Universidad Católica de Santa María, Arequipa 34116
Museo Sardo di Antropologia ed Etnografia, Cagliari 25587
Museo Sardo di Geologia e Paleontologia D. Lovisato, Cagliari 25588
Museo Schaferrer, Oxapampa 34284
Museo Scientifico L. Guidi, Pesaro 27157
Museo Scout-Guía, Madrid 39580
Museo Scuola Odin-Bertot, Angrogna 25240
Museo Sefardí, Toledo 40136
Museo Selinuntino, Castelvetrano 25757
Museo Sella e Mosca, Alghero 25199
Museo Senador Domingo Faustino Sarmiento, Buenos Aires 00273
Museo Senen Mexic, Acatlán de Osorio 30694
Museo Señorita Elupina Cordero, Santa Teresa 10337
Museo Serfín de Indumentaria Indígena, México 31148
Museo Servando Cabrera Moreno, La Habana 09173
Museo Siciliano delle Tradizioni Religiose, San Salvatore di Fitalia 27682
Museo Siderúrgico de Colombia, Nobsa 08697
Museo Silvanus G. Morley y de las Estelas, El Petén 23295
Museo Silvestre Vargas, Tecalitlán 31411
Museo Simón Bolívar, Ziortza-Bolibar 40358
Museo Sinagoga de Santa María la Blanca, Toledo 40137
Museo Sistino Vescovile, Montalto Marche 26798
Museo Sitio Arqueológico de San Andrés, Valle de Zapotitán 10556
Museo Sitio Huellas de Acahualinca, Acahualinca 33270
Museo Sito Arqueológico Guayabo de Turrialba, Turrialba 08826
Museo Social Argentino de Universidad, Buenos Aires 00274
Museo Sol y Mar, Matanzas 09219
Museo Soldati della Battaglia, San Biagio di Callalta 27615
Museo Sor María de Jesús de Agreda, Agreda 38787
Museo Sorolla, Madrid 39581
Museo Soumaya, México 31149
Museo Speleopaleontologico, Genga 26288
Museo Speologico, Sgonico 27833
Museo Sperimentale d'Arte Contemporanea Muspac, L'Aquila 26459
Museo Statale, Anghiari 25238
Museo Statale, Mileto 26730
Museo Stauros di Arte Sacra Contemporanea, Isola del Gran Sasso 26416
Museo Stibbert, Firenze 26198
Museo Storia Memoria, Sossano 27882
Museo Storico, Montese 26853
Museo Storico, Voghera 28304
Museo Storico-Aeronautica Militare, Bracciano 25525
Museo Storico Alfa Romeo, Arese 25263
Museo Storico Aloisiano, Castiglione delle Stiviere 25761
Museo Storico Alpino, Antrodoco 25243
Museo Storico Archeologico, Cicagna 25887
Museo Storico Archeologico, Segni 27797
Museo Storico Archeologico del Territorio, Bagni di Lucca 25347
Museo Storico Archeologico dell'Antica Nola, Nola 26938
Museo Storico Archeologico G. Gabetti, Dogliani 26039
Museo Storico Artistico, Cassino 25712
Museo Storico Artistico, Città del Vaticano 54512
Museo Storico Badogliano, Bari 25370
Museo Storico C. Musazzi, Parabiago 27089
Museo Storico Cappuccino, Camerino 25622
Museo Storico Carlo Levi, Aliano 25201
Museo Storico Civico Cuggionese, Cuggiono 26016
Museo Storico Dante Foschi, Forlì 26229
Museo Storico degli Truppe Alpini, Trento 28060
Museo Storico dei Bersaglieri, Roma 27527
Museo Storico dei Granatieri di Sardegna, Roma 27528
Museo Storico dei Vigili del Fuoco, Mantova 26598
Museo Storico del Castello di Miramare, Trieste 28096
Museo Storico del Combattente, Palazzolo Milanese 27060
Museo Storico del Mutuo Soccorso, Pinerolo 27225
Museo Storico del Nastro Azzurro, Salò 27600
Museo Storico del Risorgimento della Società Economica, Chiavari 25861
Museo Storico del Soldatino M. Massacesi, Bologna 25476
Museo Storico del Trotto, Civitanova Marche 25923
Museo Storico del Vetro e della Bottiglia J.F. Mariani, Montalcino 26796
Museo Storico dela Caccia e del Territorio, Cerreto Guidi 25812
Museo Storico della Brigata Sassari, Sassari 27756
Museo Storico della Città, Bergamo 25405
Museo Storico della Didattica Mauro Laeng, Roma 27529
Museo Storico della Fanteria, Roma 27530
Museo Storico della Filarmonica Sestrese, Genova 26319
Museo Storico della Guardia di Finanza, Roma 27531
Museo Storico della Guerra 1915-1918, Roana 27426
Museo Storico della Liberazione, Roma 27532
Museo Storico della Linea Gotica, Auditore 25326
Museo Storico della Motorizzazione Militare, Roma 27533
Museo Storico della Resistanza S. Maneschi, Neviano degli Arduini 26923
Museo Storico della Resistenza, Stazzema 27900
Museo Storico della Tappezzeria, Bologna 25477
Museo Storico dell'Arma dei Carabinieri, Roma 27534
Museo Storico dell'Arma del Genio, Roma 27535
Museo Storico dell'Arte Sanitaria, Roma 27536
Museo Storico delle Armi, Civitella del Tronto 25926
Museo Storico delle Poste e Telecomunicazioni, Roma 27537
Museo Storico delle Saline, Margherita di Savoia 26604
Museo Storico dell'Oro Italiano, Predosa 27342
Museo Storico di Architettura Militare, Roma 27538
Museo Storico Dittico Alto Livenza, Sacile 27585
Museo Storico e Topografico, Firenze 26199
Museo Storico Etnografico, Bruscoli 25557
Museo Storico Etnografico, Sampeyre 27607
Museo Storico-Etnografico, Sanluri 27695
Museo Storico Etnografico della Bassa Valsesia, Romagnano Sesia 27549
Museo Storico Etnografico delle Alpi Liguri, Mendatica 26653
Museo Storico Etnografico Naturalistico, Chiesa in Valmalenco 25870
Museo Storico G. Garibaldi, Como 25954
Museo Storico in Trento Onlus, Trento 28061
Museo Storico Italiano della Guerra, Rovereto 27571
Museo Storico Italo-Ungherese, Vittoria 28296
Museo Storico Michele Ghislieri, Vicoforte 28261
Museo Storico Militare, Comun Nuovo 25958
Museo Storico Militare e Sacrario di Redipuglia, Fogliano Redipuglia 26213
Museo Storico Militare Umberto I, Turate 28108
Museo Storico Minerario, Perticara 27142
Museo Storico Minerario di Perticara, Novafeltria 26947
Museo Storico Multimediale B. d'Alviano, Alviano 25212
Museo Storico Navale, Venezia 28203
Museo Storico Nazionale di Artiglieria, Torino 28026
Museo Storico Novarese Aldo Rossini, Novara 26955
Museo Storico Perugina, Perugia 27152
Museo Storico Vaticano, Città del Vaticano 54513
Museo Storico Villa Ciani, Lugano 41585
Museo Stradivariano, Cremona 26006
Museo Sudtirolese della Frutticoltura, Lana 26446
Museo Sveviano, Trieste 28097
Museo T. Nuvolari, Mantova 26599
Museo Taller Arte Africano, San Juan 36075
Museo Taller de Escultura, Victoria de las Tunas 09350
Museo Taller del Moro, Toledo 40138
Museo Taller Litográfico, Cádiz 39092
Museo Taller Luis Nishizawa, Toluca 31494
Museo Tamayo Arte Contemporáneo, México 31150
Museo Taminango de Artes y Tradiciones Populares de Nariño, Pasto 08715
Museo Tarike, Chihuahua 30772
Museo Taurino, Cali 08615
Museo Taurino, Córdoba 39208
Museo Taurino, Hellín 39386
Museo Taurino, Huamantla 30939
Museo Taurino, La Línea de la Concepción 39466
Museo Taurino, Madrid 39582
Museo Taurino, Murcia 39704
Museo Taurino, Salamanca 39944
Museo Taurino, Valencia 40217
Museo Taurino de Béjar, Béjar 39015
Museo Taurino de Bogotá, Bogotá 08595
Museo Taurino de la Plaza de Acho, Lima 34270
Museo Taurino de Mijas, Mijas 39649
Museo Taurino Municipal, Antequera 38876
Museo Teatrale alla Scala, Milano 26716
Museo Teatrale del Burcardo, Roma 27539
Museo Teatro La Casa de los Títeres, Monterrey 31188
Museo Tecnico Navale, La Spezia 26436
Museo Técnico Pedagógico de la Región de Pasco, San Juan 34302
Museo Tecnologia, Cultura e Civiltà Contadina, Guardia Lombardi 26381
Museo Tecnológico Aeroespacial, Las Higueras 00425
Museo Tecnológico de la Comisión Federal de Electricidad, México 31151
Museo Tecnológico Ingeniero Eduardo Latzina, Buenos Aires 00275
Museo Telefónico Victoria, México 31152
Museo Templete de los Heros Nacionales, Quito 10451
Museo Teniente José Azuela, Zihuatanejo 31597
Museo Teodoro Ramos Blanco, La Habana 09174
Museo Tepeapulco, Tepeapulco 31433
Museo Teresiano, Alba de Tormes 38802
Museo Teresiano, Ávila 38923
Museo Teresiano Monasterio de la Encarnación, Ávila 38924
Museo Territoriale del Lago di Bolsena, Bolsena 25481
Museo Territoriale dell'Agro Foronovano, Torri in Sabina 28042
Museo Tesoro Basilica di San Francesco e Collezione F. M. Perkins, Assisi 25310
Museo-Tesoro Catedralicio, Toledo 40139
Museo-Tesoro de la Cofradía de la Expiración, Málaga 39606
Museo Tesoro del Tempio della Beata Vergine, Reggio Emilia 27391
Museo Tesoro della Collegiata di San Lorenzo, Chiavenna 25864
Museo Tesoro Parroquial, Lebrija 39447
Museo Tetetzontlilco, Tizayuca 31463
Museo Textil La Trinidad, Santa Cruz Tlaxcala 31377
Museo Thermalia, Caldes de Montbui 39099
Museo Thyssen-Bornemisza, Madrid 39583
Museo Tiflológico, Madrid 39584
Museo Timoteo L. Hernández, Villaldama 31555
Museo Tipologico del Presepe, Macerata 26555
Museo Tlallán, Tala 31397
Museo Tohue, Vicente Guerrero, Durango 31542
Museo Tolone Azzariti, Girifalco 26343
Museo Toma de Zacatecas o Cerro de la Bufa, Zacatecas 31582
Museo Tomás Morel, Santiago de Los Caballeros 10340
Museo Tomaskitla, Epazoyucan 30863
Museo Tomochic, Vicente Guerrero 31541
Museo Tonallán, Tonalá, Jalisco 31497
Museo Tonatiuh, Zempoala 31595
Museo Torlonia, Roma 27540
Museo Tornambe de la Universidad Nacional de San Juan, San Juan 00589
Museo Torre di Porta Romana, Gubbio 26392
Museo Torres García, Montevideo 54418
Museo Torricelliano, Faenza 26078
Museo Toti dal Monte, Pieve di Soligo 27215
Museo Towi, Guachochi 30884
Museo Tradicional Ik Al Ojov, Zinacantán 31599
Museo Tradiciones y Leyendas del Coronel Arechavala, León 33287
Museo Tribunal de la Santa Inquisición y del Congreso, Lima 34271
Museo Tridentino di Scienze Naturali, Trento 28062
Museo Túenl Casa Parker, Angangueo 30716
Museo Tuscolano, Frascati 26246
Museo U. Gumbu, Cervo 25824
Museo Ucraniano, Apóstoles 00124
Museo Ugo Carà, Muggia 26876
Museo Universitario, Medellín 08686
Museo Universitario, Pasto 08716
Museo Universitario Alejandro Rangel Hidalgo, Nogueras 31216
Museo Universitario Antigua Casa de los Muñecos, Puebla 31290
Museo Universitario Contemporáneo de Arte, México 31153
Museo Universitario de Arqueología de Manzanillo, Manzanillo 31013
Museo Universitario de Artes, Mendoza 00459
Museo Universitario de Artes Populares, Colima 30783
Museo Universitario de Ciencias, Zacatecas 31583
Museo Universitario de Ciencias y Artes, México 31154
Museo Universitario de la Universidad Nacional Federico Villarreal, Lima 34272
Museo Universitario de Paleontología, México 31155

Museo Universitario de San Carlos, Guatemala 23306
Museo Universitario del Chopo, México 31156
Museo Universitario Dr. Ricardo Bohorquez R., Potosí 04238
Museo Universitario Florentino y Carlos Ameghino, Rosario 00550
Museo Universitario La Memoria Colectiva, Llallagia 04228
Museo Universitario Lariab de Historia Antigua de la Huaxteca, Tamuín 31400
Museo Universitario Pio López Martínez, Cayey 36035
Museo Universitario y Museo de Anatomía General, Yanacancha 34327
Museo Univérsitas, El Cerrito 08637
Museo Urdaneta Histórico Militar, Maracaibo 54582
Museo Usina Vieja, San Antonio de Areco 00568
Museo V.G. Rossi, Santa Margherita Ligure 27703
Museo V.S. Breda, Padova 27048
Museo V. Venturi, Loro Ciuffenna 26524
Museo Valdese, Rorà 27555
Museo Valdese della Balziglia, Massello 26636
Museo Valenciano de la Ilustración y la Modernidad, Valencia 40218
Museo Valenzuela, Valenzuela 34520
Museo Vallerano Bregagliotto, Stampa 41902
Museo Vallivo, Valfurva 28140
Museo Valtellinese di Storia e Arte, Sondrio 27873
Museo Vasariano di San Croce, Bosco Marengo 25516
Museo Vasco de Historia de las Ciencias y de la Medicina, Leioa 39451
Museo Vasco de Quiroga, Madrigal de las Altas Torres 39590
Museo Vasco del Ferrocarril, Azpeitia 38931
Museo Vásquez Fuller, Otavalo 10422
Museo Verbum - Casa das Palabras, Vigo 40277
Museo Vesuviano G. B. Alfano, Pompei 27279
El Museo Viajero, Buenos Aires 00276
Museo Víctor Alejandro Jaramillo, Ibarra 10416
Museo Víctor Manuel, La Habana 09175
Museo Victorio Macho, Toledo 40140
Museo Villa Colloredo Mels, Recanati 27379
Museo Villa dei Cedri, Bellinzona 41195
Museo Villa Medicea, Poggio a Caiano 27268
Museo Villa Romana de Navatejera, Villaquilambre 40305
Museo Villa Urania-Antiche Maioliche di Castelli, Pescara 27166
Museo Villa Urpo, Siivikkala 11311
Museo Vincenzo Vela, Ligornetto 41566
Museo Virgen de Zapopan, Zapopan 31591
Museo Virgiliano, Virgilio 28289
Museo Virreinal de Zinacantepec, San Miguel Zinacantepec 31360
Museo Virtual de Artes el Pais, Montevideo 54419
Museo Vitaliano Zuccardi, Sincelejo 08756
Museo Vivo de al-Andalus, Córdoba 39209
Museo Vivo della Scienza, Napoli 26909
Museo Vostell-Malpartida de Cáceres, Malpartida de Cáceres 39609
Museo Vulcanologico dell'Osservatorio Vesuviano, Ercolano 26059
Museo Wakuatay, Playas de Rosarito 31262
Museo Weilbauer, Quito 10452
Museo Wifredo Lam, La Habana 09176
Museo Wilhelm Schmid, Brè sopra Lugano 41252
Museo Würth La Rioja, Agoncillo 38783
Museo Xochiltepec, Magdalena 31010
Museo Xólotl, San Bartolo Tenayuca 31323
Museo Xul Solar, Buenos Aires 00277
Museo y Archivo Dardo Rocha, La Plata 00415
Museo y Archivo Histórico de la Facultad de Derecho, Buenos Aires 00278
Museo y Archivo Histórico Localista Almirante Guillermo Brown, Adrogué 00116
Museo y Artesanía el Molino, Agüimes 38790
Museo y Biblioteca de Ciencias Stephen Hawking, San Salvador 10554
Museo y Centro Cultural AGADU, Montevideo 54420
Museo y Centro Cultural Victoria Ocampo, Mar del Plata 00448
Museo y Centro de Estudios Humanísticos, Gurabo 36040
Museo y Centro Didáctico del Encaje de Castilla y León, Tordesillas 40148
Museo y Centro Estudios Históricos de la Facultad de Odontología, Buenos Aires 00279
Museo y Jardín Botánico Profesor Atilio Lombardo, Montevideo 54421
Museo y Monumento Histórico Las Bóvedas, Uspallata 00663
Museo y Parque Arqueológico de Obando, Obando 08698
Museo y Parque Libres del Sur, Dolores 00356
Museo y Parque Nacional, Montenegro 08692
Museo Yanque, Yanque 34329
Museo Yoloxóchitl, Cuernavaca 30820
Museo Zacatecano, Zacatecas 31584
Museo Zambiano, Castelfidardo 25725
Museo Zoloxúchitl, Zacualpan 31588
Museo Zoologico, Catania 25776
Museo Zoologico, Padova 27049
Museo Zoologico Cambria, Messina 26669
Museo Zoologico César Dominguez Flores, Tuxtla Gutiérrez 31518
Museo Zoológico Dámaso A. Larrañaga, Montevideo 54422
Museo Zoologico degli Vertebrati, Avellino 25332
Museo Zoologico dell'Istituto Nazionale per la Fauna Selvatica, Ozzano dell'Emilia 27027
Museo Zoologico G. Scarpa, Treviso 28074
Museo Zuloaga, Segovia 40056
Museo Zumalacárregui, Ormaiztegi 39747
Museoalus Sigyn, Turku 11396
Muséobus, Naninne 03909
Muséobus, Otterburn Park 06557
Muséocabinet, Saint-Jérôme 06831
Museográfica Antonio Martínez, La Roda 39906
Museogràfica de la Parròquiad l'Assumpció, Montesa 39677
Museográfica de la Villa, Camporrobles 39113
Museográfica de los Campamentos Romanos et Petavonium, Santibáñez de Vidriales 40031
Museogràfica Permanent de Forcall, Forcall 39305
Museokeskus Vapriikki, Tampere 11354
Museokylä Kalluntalo, Laukaa 11055
Museon, Den Haag 32083
Museon Arlaten, Arles 11686
Museonder, Otterlo 32643
Muséoparc de la Préhistoire, Saint-Vrain 15542
Museos Arqueológico, Portoviejo 10423
Museos de la Atalaya, Jerez de la Frontera 39427
Museos de la Universidad Nacional de Cajamarca, Cajamarca 34130
Museos del Gaucho y de la Moneda, Montevideo 54423
Museos Integrales de Antofagasta de la Sierra y Laguna Blanca, San Fernando del Valle de Catamarca 00575
Museos Oro del Perú - Armas del Mundo, Lima 34273
Museos Universitarios de Bolivia, La Paz 04227
Muséoscope du Lac, Rousset 15127
Muséosite de l'Amphithéâtre Gallo-Romain, Lillebonne 13762
Muséosite Gallo-Romain, Pithiviers-le-Vieil 14829
Museotalo Warpunen, Anjalankoski 10711
Museu, Pallejà 39774
O Museu, Porto 35843
Museu a Casa do Povo, Ferro 35604
Museu a Cidade do Açúcar, Funchal 35610
Museu Académico de Coimbra, Coimbra 35528
Museu Aeroespacial, Rio de Janeiro 04697
Museu Afro-Brasileiro, Salvador 04788
Museu Afro Brasileiro de Sergipe, Laranjeiras 04533
Museu Agrícola da Escola Profissional Agricultura de Cister, Alcobaça 35392
Museu Agrícola de Entre Douro e Minho, Vila do Conde 35988
Museu Agrícola de Riachos, Riachos 35885
Museu Agrícola e Museu de Arte Sacra, Ribeira Chã 35887
Museu Alfredo Andersen, Curitiba 04428
Museu Alfredo Varela, Jaguarão 04507
Museu Amsterdam Sauer de Pedras Preciosas, Rio de Janeiro 04698
Museu Anchieta Ciencias Naturais, Porto Alegre 04639
Museu Angelo Agostini / Casa de Cultura Laura Alvim, Rio de Janeiro 04699
Museu Aníbal Ribeiro Filho, Paranaguá 04610
Museu Anjos Teixeira, Sintra 35945
Museu Antoniano, Aljezur 35407
Museu Antoniano, Lisboa 35669
Museu Antonino, Faro 35595
Museu Antonio Granemann de Souza, Curitibanos 04443
Museu Antonio Lago, Rio de Janeiro 04700
Museu Antonio Parreiras, Niterói 04590
Museu Antropológico, Coimbra 35529
Museu Antropológico, Goiânia 04472
Museu Antropológico Caldas Júnior, Santo Antônio da Patrulha 04818
Museu Antropológico Diretor Pestana, Ijuí 04491
Museu Antropológico do Instituto Superior de Cultura Brasileira, Rio de Janeiro 04701
Museu Antropológico do Rio Grande do Sul, Porto Alegre 04640
Museu Arquediocesano Dom Joaquim, Brusque 04374
Museu Arqueològic Comarcal, Banyoles 38947
Museu Arqueològic de la Ciutat de Dénia, Dénia 39244
Museu Arqueològic de l'Esquerda, Roda de Ter 39908
Museu Arqueològic del Vendrell, El Vendrell 40261
Museu Arqueològic d'Ontinyent i de la Vall d'Albaida, Ontinyent 39738
Museu Arqueològic i Etnogràfic Municipal Soler Blasco, Xàbia 40325
Museu Arqueològic i Paleontològic, Baldomar 38943
Museu Arqueològic i Paleontològic, Moià 39656
Museu Arqueològic Municipal, Banyeres de Mariola 38945
Museu Arqueològic Municipal, Bocairent 39055
Museu Arqueològic Municipal, Els Prats de Rei 39868
Museu Arqueològic Municipal, La Vall d'Uixo 40225
Museu Arqueológico, Lagoa Santa 04528
Museu Arqueológico, Oliva 39723
Museu Arqueológico da Citânia de Sanfins, Sanfins de Ferreira 35896
Museu Arqueológico de Barcelos, Barcelos 35455
Museu Arqueológico de Sambaqui Joinville, Joinville 04514
Museu Arqueológico de São Miguel de Odrinhas, Sintra 35946
Museu Arqueológico de Serpa, Serpa 35928
Museu Arqueológico do Carmo, Lisboa 35670
Museu Arqueológico e Etnográfico de Setúbal, Setúbal 35935
Museu Arqueológico e Etnológico António Tomás Pires, Elvas 35556
Museu Arqueológico e Lapidar do Infante D. Henrique, Faro 35596
Museu Arqueológico Municipal, Gouveia 35631
Museu Arqueológico Regional de Castro Marim, Castro Marim 35517
Museu Arquidiocesano de Mariana, Mariana 04560
Museu Arquivo Histórico da Santa Casa, Belém 04332
Museu Arquivo Histórico Florense, Flores da Cunha 04451
Museu Arxiduc Lluís Salvador d'Austria de son Marroig, Deià 39242
Museu Arxiu Comarcal, Moià 39657
Museu Arxiu de l'Arboç, L'Arboç 38881
Museu-Arxiu de Santa María de Mataró, Mataró 39632
Museu Arxiu de Vilassar de Dalt, Vilassar de Dalt 40291
Museu-Arxiu Municipal de Calella, Calella 39102
Museu-Arxiu Parroquial, Canet de Mar 39118
Museu-Arxiu Tomàs Balvey i Bas, Cardedeu 39129
Museu Banespa, São Paulo 04864
Museu Barão de Mauá, Mauá 04565
Museu Barrau, Santa Eulària del Río 39999
Museu Bernardino Machado, Vila Nova de Famalicão 36004
Museu Bi Moreira, Lavras 04534
Museu Biblioteca Anselmo Braamcamp Freire, Santarém 35908
Museu-Biblioteca da Casa de Bragança, Vila Viçosa 36016
Museu Biblioteca Fundació Mauri, La Garriga 39327
Museu Bocage da Zoológico e Antropológico, Lisboa 35671
Museu Botânico, Coimbra 35530
Museu Botânico da Escola Superior Agrária de Beja, Beja 35462
Museu Botânico Dr. João Barbosa Rodrigues, São Paulo 04865
Museu Calmon Barreto, Araxá 04318
Museu Calouste Gulbenkian, Lisboa 35672
Museu Câmara Cascudo, Natal 04582
Museu Capela Santa Luzia, Vitória 04956
Museu Capitular de la Seu Nova, Lleida 39477
Museu Carlos Costa Pinto, Salvador 04789
Museu Carmen Miranda, Rio de Janeiro 04702
Museu Carpológico do Jardim Botânico do Rio de Janeiro, Rio de Janeiro 04703
Museu Casa Alphonsus de Guimaraens, Mariana 04561
Museu Casa Cel. Joaquim Lacerda, Lapa 04530
Museu Casa da Hera, Vassouras 04948
Museu Casa d'Areny-Plandolit, Ordino 00100
Museu Casa de Benjamin Constant, Rio de Janeiro 04704
Museu Casa de Portinari, Brodowski 04373
Museu Casa de Rui Barbosa, Rio de Janeiro 04705
Museu Casa de Rui Barbosa, Salvador 04790
Museu Casa do Anhanguera, Santana de Parnaíba 04814
Museu Casa do Assento, Friande 35608
Museu Casa do Pontal, Rio de Janeiro 04706
Museu Casa do Sertão, Feira de Santana 04449
Museu Casa dos Contos, Ouro Preto 04599
Museu Casa Fritz Alt, Joinville 04515
Museu Casa Guignard, Ouro Preto 04600
Museu Casa Guimarães Rosa, Cordisburgo 04417
Museu Casa Pedro Américo, Areia 04322
Museu Casa Rull, La Massana 00097
Museu-Casa Verdaguer, Vallvidrera 40251
Museu Castro Alves, Salvador 04791
Museu Cau Ferrat, Sitges 40083
Museu Central das Forças Armadas, Luanda 00110
Museu Cerdà, Puigcerdà 39884
Museu Ceroplástico Augusto Esteves, São Paulo 04866
Museu Civico-Religioso Padre Cicero, Juazeiro do Norte 04518
Museu Clara Nunes, Caetanópolis 04379
Museu Claretià, Vic 40268
Museu Comajoan Fauna del Montseny, Santa María de Palautordera 40002
Museu Comarcal de Berga, Berga 39028
Museu Comarcal de la Conca de Barberà, Montblanc 39670
Museu Comarcal de la Garrotxa, Olot 39727
Museu Comarcal de l'Anoia-Museu de la Pell, Igualada 39408
Museu Comarcal de l'Horta Sud, Torrente 40159
Museu Comarcal de l'Urgell, Tárrega 40106
Museu Comarcal de Manresa, Manresa 39614
Museu Comendador Ananias Arruda, Baturit 04331
Museu Comendador José Dias de Andrade, Arceburgo 04321
Museu Comercial e Arqueológico do Sindicato dos Empregados, Maceió 04542
Museu Comunitário, Ipumirim 04492
Museu Comunitário, Itapiranga 04499
Museu Condes de Castro Guimarães, Cascais 35512
Museu Convento dos Lóios, Santa Maria da Feira 35902
Museu Corina Novelino, Sacramento 04784
Museu Coritiba Football Club, Curitiba 04429
Museu Cosme Bauça, Felanitx 39294
Museu CTT/TLP, Lisboa 35673
Museu D. João VI, Rio de Janeiro 04707
Museu da Abolição, Recife 04661
Museu da Agricultura de Fermentões, Guimarães 35638
Museu da Agua da Epal, Lisboa 35674
Museu da Alfaia Agrícola, Estremoz 35576
Museu da Alfaia Agrícola, Mafra 35749
Museu da Alfaia Agrícola, Montemor-o-Novo 35774
Museu da Alfaia Agrícola, Santo Amaro 35913
Museu da Arte Brasileira, Maceió 04543
Museu da Arte Pierre Chalita, Maceió 04544
Museu da Associação Nacional dos Veteranos, São João Del Rei 04834
Museu da Bacia do Paraná, Maringá 04563
Museu da Baleia, Caniçal 35505
Museu da Biblioteca Nacional, Lisboa 35675
Museu da Borracha, Rio Branco 04677
Museu da Brigada Militar, Porto Alegre 04641
Museu da Caça, Vila Viçosa 36017
Museu da Cachaça, Paty do Alferes 04616
Museu da Capela de Sã Frutuoso de Montélios, São Jerónimo de Real 35918
Museu da Carris, Lisboa 35676
Museu da Casa Brasileira, São Paulo 04867
Museu da Casa Cadaval, Évora 35584
Museu da Casa da Cultura da Orada, Orada 35794
Museu da Casa da Cultura de Teresina, Teresina 04929
Museu da Casa do Povo do Raminho, Raminho 35880
Museu da Casa Grande, Freixo de Numão 35607
Museu da Cera, Fátima 35601
Museu da Cerâmica, Caldas da Raínha 35496
Museu da Cerâmica de Sacavém, Sacavém 35893
Museu da Cidade, Almada 35414
Museu da Cidade, Campinas 04386
Museu da Cidade, Coimbra 35531
Museu da Cidade, São Paulo 04868
Museu da Cidade de Lisboa, Lisboa 35677
Museu da Cidade de Recife, Recife 04662
Museu da Ciência e Indústria, Porto 35844
Museu da Companhia das Lezírias, Salvaterra de Magos 35895
Museu da Companhia Independente do Palácio Guanabara, Rio de Janeiro 04708
Museu da Companhia Paulista, Jundiá 04525
Museu da Cortiça da Fábrica do Inglês, Silves 35939
Museu da Cultura e Arte Popular, João Pessoa 04512
Museu da Curia Metropolitana, Porto Alegre 04642
Museu da Educação e do Brinquedo, São Paulo 04869
Museu da Electricidade, Lisboa 35678
Museu da Eletricidade, Rio de Janeiro 04709
Museu da Emigração, das Comunidades e da Luso-Descendência, Fafe 35591
Museu da Energia, Itu 04502
Museu da Escola de Engenharia, Rio de Janeiro 04710
Museu da Escola de Minas Gerais, Belo Horizonte 04345
Museu da Escola Naval, Rio de Janeiro 04711
Museu da Escola Regional de Artesanato de Santo Amaro, Santo Amaro 35914
Museu da Escravatura, Luanda 00111
Museu da Estrada de Ferro Madeira-Mamoré, Porto Velho 04657
Museu da Estrada de Ferro Sorocabana, Sorocaba 04919
Museu da Fábrica de Cimentos Lis, Maceira 35746
Museu da Faculdade de Direito, São Paulo 04870
Museu da Faculdade de Odontologia, Araraquara 04316
Museu da Família Colonial, Blumenau 04353
Museu da Farmácia, Lisboa 35679

Museu da Farmáciada Santa Casa da Misericórdia do Rio de Janeiro, Rio de Janeiro 04712
Museu da Fauna, Rio de Janeiro 04713
Museu da Fazenda Federal, Rio de Janeiro 04714
Museu da Força Expedicionária Brasileira, Rio de Janeiro 04715
Museu da Freguesia de Estômbar, Estômbar 35573
Museu da Fundação Castro Alves, Vila Nova de Famalicão 36005
Museu da Fundação Cupertino de Miranda, Vila Nova de Famalicão 36006
Museu da Fundação da Casa de Mateus, Vila Real 36011
Museu da Fundação Nacional de Saúde, Brasília 04359
Museu da Graciosa, Santa Cruz da Graciosa 35897
Museu da Gravura, Curitiba 04430
Museu da Gravura Brasileira, Bagé 04325
Museu da Guarda, Guarda 35636
Museu da História da Medicina Maximiano Lemos, Porto 35845
Museu da Igreja Matriz, Gondomar 35630
Museu da Imagem, Campinas 04387
Museu da Imagem, Franca 04467
Museu da Imagem e do Som, Campo Grande 04393
Museu da Imagem e do Som, Rio de Janeiro 04716
Museu da Imagem e do Som de Alagoas, Maceió 04545
Museu da Imagem e do Som de Campinas, Campinas 04388
Museu da Imagem e do Som de Campos, Campos dos Goitacazes 04399
Museu da Imagem e do Som de Cascavel, Cascavel 04407
Museu da Imagem e do Som de Curitiba, Curitiba 04431
Museu da Imagem e do Som de e da Imprensa da Bahia, Salvador 04792
Museu da Imagem e do Som de Goiás, Goiás 04478
Museu da Imagem e do Som de Piracicaba, São Paulo 04871
Museu da Imagem e do Som de Resende, Resende 04673
Museu da Imagem e do Som de Santa Catarina, Florianópolis 04453
Museu da Imagem e do Som de São Paulo, São Paulo 04872
Museu da Imagem e do Som de Taubaté, Taubaté 04923
Museu da Imagem e do Som do Ceará, Fortaleza 04462
Museu da Imagem e do Som do Pará, Belém 04333
Museu da Imagem e do Som do Paraná, Curitiba 04432
Museu da Imagem em Movimento, Leiria 35658
Museu da Imigração, Santa Bárbara d'Oeste 04806
Museu da Imprensa, Brasília 04360
Museu da Imprensa de Fafe, Fafe 35592
Museu da Inconfidência, Ouro Preto 04601
Museu da Indústria Baleeira, São Roque do Pico 35925
Museu da Indústria da Cortiça, Santa Maria da Feira 35903
Museu da Indústria Têxtil, Calendário 35498
Museu da Latoaria, Tomar 35963
Museu da Liga dos Combatentes da Grande Guerra, Lisboa 35680
Museu da Limpeza Urbana - Casa de Banhos de Dom João VI, Rio de Janeiro 04717
Museu da Literatura, São Paulo 04873
Museu da Loucura, Barbacena 04328
Museu da Margalha, Gavião 35625
Museu da Marioneta, Lisboa 35681
Museu da Matriz, Ponta Delgada 35827
Museu da Mina de Pirites, Aljustrel 35411
Museu da Música, Lisboa 35682
Museu da Música Portuguesa, Estoril 35575
Museu da Paróquia, Redondo 35881
Museu da Paróquia da Benedita, Alcobaça 35393
Museu da Pessoa, São Paulo 04874
Museu da Polícia Civil, Rio de Janeiro 04718
Museu da Polícia Judiciária, Barro 35459
Museu da Polícia Militar de Pernambuco, Recife 04663
Museu da Porcelana Senhor Adelino dos Santos Gouveia, Pedreira 04619
Museu da Quinta das Cruzes, Funchal 35611
Museu da Quinta de Santiago, Leça da Palmeira 35657
Museu da Rádio, Lisboa 35683
Museu da Rainha Dona Leonor, Beja 35463
Museu da Regiao Flâviense, Chaves 35523
Museu da República, Rio de Janeiro 04719
Museu da República e Resistência, Lisboa 35684
Museu da Revolução, Maputo 31706
Museu da Santa Casa da Misericórdia da Vila da Ericeira, Ericeira 35560
Museu da Santa Casa da Misericórdia de Alvor, Alvor 35428
Museu da Sé de Angra do Heroísmo, Angra do Heroísmo 35438
Museu da Secção de Tecnologia do Serviço Florestal, Rio de Janeiro 04720
Museu da Sociedade Imparcial 15 de Janeiro de 1898, Alcochete 35396
Museu da Televisão, Lisboa 35685
Museu da Terra de Miranda, Miranda do Douro 35764
Museu da Terra e da Vida, Mafra 04550
Museu da UFRGS, Porto Alegre 04643
Museu da Universidade Federal do Pará, Belém 04334
Museu da Venrável Ordem Terceira de São Francisco da Penitência, Rio de Janeiro 04721
Museu da Vida Submarina e da História Submersa, Lisboa 35686
Museu da Vinha e do Vinho de Reguengos, Reguengos de Monsaraz 35884
Museu Darder, Banyoles 38948
Museu d'Arqueologia de Catalunya, Barcelona 38970
Museu d'Arqueologia de Catalunya, Olèrdola 39722
Museu d'Arqueologia de Catalunya-Empúries, L'Escala 39278
Museu d'Arqueologia de Catalunya Girona, Girona 39343
Museu d'Arqueologia de Catalunya Ullastret, Ullastret 40180
Museu d'Arqueologia Salvador Vilaseca, Reus 39894
Museu d'Arqueològia Submarina, Pals 39805
Museu d'Arquitectura de la Real Càtedra Gaudí, Barcelona 38971
Museu d'Art, Girona 39344
Museu d'Art Casa Turull, Sabadell 39924
Museu d'Art Contemporani, Pego 39823
Museu d'Art Contemporani de Barcelona, Barcelona 38972
Museu d'Art Contemporani de Mallorca, Sa Pobla 39846
Museu d'Art Contemporani d'Eivissa, Eivissa 39261
Museu d'Art Espanyol Contemporani, Palma de Mallorca 39787
Museu d'Art Frederic Marès, Montblanc 39671
Museu d'Art i Història de Reus, Reus 39895
Museu d'Art Modern, Tarragona 40101
Museu d'Art Modern Jaume Morera, Lleida 39478
Museu d'Art Municipal, Cadaqués 39086
Museu d'Art Popular, Horta de Sant Joan 39393
Museu d'Art Sacre de Santa Eulàlia, Encamp 00088
Museu das Alfaias, Cachoeira 04376
Museu das Alfândegas, Lisboa 35687
Museu das Armas, Lapa 04531
Museu das Bandeiras, Goiás 04479
Museu das Ciências da Terra, Estremoz 35577
Museu das Comunicações, Lisboa 35688
Museu das Crianças, Lisboa 35689
Museu das Flores, Santa Cruz das Flores 35898
Museu das Monções, Porto Feliz 04655
Museu das Oferendas ao Soldado Desconhecido, Batalha 35460
Museu das Reduções, Ouro Preto 04602
Museu das Rendas de Bilros, Vila do Conde 35989
Museu das Técnicas Rurais e Museu Municipal de Oliveira de Frades, Oliveira de Frades 35793
Museu das Tropas Aerotransportadas, Praia do Ribatejo 35876
Museu de Afife, Afife 35384
Museu de Alberto Sampaio, Guimarães 35639
Museu de Alhandra, Alhandra 35405
Museu de Anatomia Humana Prof. Dr. Alfonso Bovera, São Paulo 04875
Museu de Anatomia Veterinária Prof. Dr. Plínio Pinto e Silva, São Paulo 04876
Museu de Angra do Heroísmo, Angra do Heroísmo 35439
Museu de Antropologia do Vale do Paraíba, Jacareí 04506
Museu de Armas Ferreira da Cunha, Petrópolis 04628
Museu de Armas Major Lara Ribas, Florianópolis 04454
Museu de Arqueologia, Montemor-o-Novo 35775
Museu de Arqueologia, Sesimbra 35931
Museu de Arqueologia, Vila Viçosa 36018
Museu de Arqueologia de Mértola, Mértola 35760
Museu de Arqueologia e Artes Populares, Paranaguá 04611
Museu de Arqueologia e Etnologia, Salvador 04793
Museu de Arqueologia e Etnologia da Universidade de São Paulo, São Paulo 04877
Museu de Arte, Belém 04335
Museu de Arte, Goiânia 04473
Museu de Arte, Joinville 04516
Museu de Arte, Londrina 04538
Museu de Arte, Santa Maria da Feira 35904
Museu de Arte Africana da Azambuja, Azambuja 35450
Museu de Arte Assis Chateaubriand, Campina Grande 04383
Museu de Arte Brasileira, São Paulo 04878
Museu de Arte Contemporãea, Jataí 04509
Museu de Arte Contemporânea, Feira de Santana 04450
Museu de Arte Contemporânea, Funchal 35612
Museu de Arte Contemporânea da Universidade de São Paulo, São Paulo 04879
Museu de Arte Contemporânea de Niterói, Niterói 04591
Museu de Arte Contemporânea de Pernambuco, Olinda 04596
Museu de Arte Contemporônea, Curitiba 04433
Museu de Arte Contemprânea da Bienal de Cerveira, Vila Nova de Cerveira 36002
Museu de Arte da Bahia, Salvador 04794
Museu de Arte da Pampulha, Belo Horizonte 04346
Museu de Arte da Universidade Fedral do Ceará, Fortaleza 04463
Museu de Arte de Santa Catarina, Florianópolis 04455
Museu de Arte de São Paulo Assis Chateaubriand, São Paulo 04880
Museu de Arte do Paraná, Curitiba 04434
Museu de Arte do Rio Grande do Sul Aldo Malagoli, Porto Alegre 04644
Museu de Arte e Cultura Canjira, Itaobim 04498
Museu de Arte e Cultura Popular da UFMT, Cuiabá 04422
Museu de Arte e Etnografia da Ribeira Chã, Ribeira Chã 35888
Museu de Arte e História, Natal 04583
Museu de Arte e Histórico, Jaboticabal 04505
Museu de Arte Mágica e Ilusionismo João Peixoto dos Santos, São Paulo 04881
Museu de Arte Moderna, Resende 04674
Museu de Arte Moderna da Bahia, Salvador 04795
Museu de Arte Moderna de São Paulo, São Paulo 04882
Museu de Arte Moderna do Rio de Janeiro, Rio de Janeiro 04722
Museu de Arte Popular, Lisboa 35690
Museu de Arte Popular da Fortaleza dos Reis Magos, Natal 04584
Museu de Arte Primitiva Moderna, Guimarães 35640
Museu de Arte Religiosa, Vitória 04957
Museu de Arte Sacra, Aljezur 35408
Museu de Arte Sacra, Amarante 35432
Museu de Arte Sacra, Castelo Branco 35514
Museu de Arte Sacra, Esposende 35568
Museu de Arte Sacra, Évora 35585
Museu de Arte Sacra, Funchal 35613
Museu de Arte Sacra, Gouveia 35632
Museu de Arte Sacra, Mértola 35761
Museu de Arte Sacra, Pedrógão Grande 35813
Museu de Arte Sacra, Rio de Janeiro 04723
Museu de Arte Sacra, Salvador 04796
Museu de Arte Sacra, São Cristóvão 04826
Museu de Arte Sacra, São João Del Rei 04835
Museu de Arte Sacra, São Luís 04843
Museu de Arte Sacra, São Paulo 04883
Museu de Arte Sacra, São Sebastião 04913
Museu de Arte Sacra, Taubaté 04924
Museu de Arte Sacra, Vila do Conde 35990
Museu de Arte Sacra, Vila Viçosa 36019
Museu de Arte Sacra da Catedral-Basílica, Salvador 04797
Museu de Arte Sacra da Misericórdia do Fundão, Fundão 35621
Museu de Arte Sacra da Sé de Portalegre, Portalegre 35833
Museu de Arte Sacra da Universidade, Coimbra 35532
Museu de Arte Sacra da Universidade Lusíada, Vila Nova de Famalicão 36007
Museu de Arte Sacra de Alcoutim, Alcoutim 35401
Museu de Arte Sacra de Angra dos Reis, Angra dos Reis 04307
Museu de Arte Sacra de Arouca, Arouca 35445
Museu de Arte Sacra de Boa Morte, Goiás 04480
Museu de Arte Sacra de Iguape, Iguape 04490
Museu de Arte Sacra de Monsaraz, Monsaraz 35771
Museu de Arte Sacra de Santos, Santos 04820
Museu de Arte Sacra do Carmo / Paróquia do Pilar, Ouro Preto 04603
Museu de Arte Sacra do Estado de Alagoas, Marechal Deodoro 04559
Museu de Arte Sacra do Pará, Belém 04336
Museu de Arte Sacra Dom Ranulfo, Maceió 04546
Museu de Arte Sacra e Arqueologia, Porto 35846
Museu de Arte Sacra e Arqueologia de Vila Alva, Vila Alva 35985
Museu de Arte Sacra e Etnografia Religiosa, Horta 35643
Museu de Arte Sacra e Etnologia de Fátima, Fátima 35602
Museu de Arte Sacra Padre Roberto Barbalho, Recife 04664
Museu de Artes Decorativas, Évora 35586
Museu de Artes e Ofícios, Belo Horizonte 04347
Museu de Artes e Ofícios Francisco Távora, Escalhão 35564
Museu de Artes Plásticas, Taubaté 04925
Museu de Artes Plásticas Quirino da Silva, Mococa 04568
Museu de Artes Sacras Iria Josepha da Silva, Mococa 04569
Museu de Artes Visuais, São Luís 04844
Museu de Artes Visuais Ruth Schneider, Passo Fundo 04612
Museu de Astronomia e Ciências Afins, Rio de Janeiro 04724
Museu de Aveiro, Aveiro 35447
Museu de Badalona, Badalona 38937
Museu de Barroso, Montalegre 35772
Museu de Belas-Artes, Cataguases 04409
Museu de Biodiversidade do Cerrado, Uberlândia 04941
Museu de Biologia Prof. Mello Leitão, Santa Teresa 04813
Museu de Bonecos e Ventrílocos, São Paulo 04884
Museu de Cabrera, Palma de Mallorca 39788
Museu de Carrosses Fúnebres, Barcelona 38973
Museu de Cartão, São Paulo 04885
Museu de Cera, Barcelona 38974
Museu de Cera, Binissalem 39050
Museu de Cerâmica, Barcelona 38975
Museu de Cerâmica, Borba 35471
Museu de Ceràmica de l'Alcora, Alcora 38823
Museu de Ceràmica de Manises, Manises 39612
Museu de Ciência e Técnica da Escola de Minas, Ouro Preto 04604
Museu de Ciência e Tecnologia, Salvador 04798
Museu de Ciências da Terra - DNPM, Rio de Janeiro 04725
Museu de Ciências Morfológicas, Pampulha 04608
Museu de Ciências Naturais, Guarapuava 04484
Museu de Ciências Naturais, Porto Alegre 04645
Museu de Ciências Naturais, Recife 04665
Museu de Ciências Naturais, Triunfo 04935
Museu de Ciências Naturais PUC Minas, Belo Horizonte 04348
Museu de Ciències Naturals, Costitx 39223
Museu de Ciències Naturals, Onda 39734
Museu de Ciències Naturals de Barcelona, Barcelona 38976
Museu de Comunicação Social Hipólito José da Costa, Porto Alegre 04646
Museu de Corrente, Corrente 04419
Museu de Coses del Poble, Collbató 39198
Museu de Curiositats Marineres, Vilanova i la Geltrú 40284
Museu de D. Lopo de Almeida, Abrantes 35383
Museu de Design e Artes Decorativas, Lisboa 35691
Museu de Documentos Especiais, Praia 07419
Museu de Electricidade, Fornelos 35606
Museu de Escola Prática de Artilharia, Vendas Novas 35981
Museu de Escultura Comparada, Mafra 35750
Museu de Esportes Mané Garrincha, Rio de Janeiro 04726
Museu de Etnografia e Arqueologia Industrial, Torres Novas 35968
Museu de Etnografia e História de Póvoa de Varzim, Póvoa de Varzim 35873
Museu de Evora, Évora 35587
Museu de Ex-votos do Senhor do Bonfim, Salvador 04799
Museu de Farmácia, São Paulo 04886
Museu de Física, Coimbra 35533
Museu de Folclore Édison Carneiro, Rio de Janeiro 04727
Museu de Folclore Saul Martins, Vespasiano 04950
Museu de Formentera, Formentera 39306
Museu de Francisco Tavares Proença Júnior, Castelo Branco 35515
Museu de Gavà, Gavà 39329
Museu de Geociências, São Paulo 04887
Museu de Geociências da Universidade de Brasília, Brasília 04361
Museu de Geologia, Barcelona 38977
Museu de Geologia, Maringá 04564
Museu de Geologia, Porto Alegre 04647
Museu de Geologia, Sant Celoni 39976
Museu de Geologia, Vila Real 36012
Museu de Geologia de la Universitat de València, Burjassot 39074
Museu de Geologia Valenti Masachs, Manresa 39615
Museu de Granollers, Granollers 39373
Museu de Granollers-Ciències Naturals, Granollers 39374
Museu de Grão Vasco, Viseu 36023
Museu de História e Ciências Naturais, Além Paraíba 04303
Museu de História Natural, Belo Horizonte 04349
Museu de História Natural, Campinas 04389
Museu de História Natural, Coimbra 35534
Museu de História Natural, Crato 04421
Museu de História Natural, Maputo 31707
Museu de História Natural, Porto 35847
Museu de História Natural de Sines, Sines 35941
Museu de História Natural Prof. Antonio Pergola, Atibaia 04324

M

Museu de Homem do Norte, Manaus 04554
Museu de Homem Sergipano, Aracaju 04313
Museu de Imagem e do Som de Pernambuco, Recife 04666
Museu de Images do Inconsciente, Rio de Janeiro 04728
Museu de Índio, Uberlândia 04942
Museu de Instrumentos de Cálculo Numérico, São Carlos 04825
Museu de Instrumentos Musicais do Conservatório Nacional, Lisboa 35692
Museu de Janes e Malveira, Alcabideche 35387
Museu de Jazigos Minerais Portugueses, São Mamede de Infesta 35923
Museu de José Malhoa, Caldas da Raínha 35497
Museu de la Casa Codorniú, Sant Sadurní d'Anoia 39990
Museu de la Catedral, Barcelona 38978
Museu de la Ciència i de la Tècnica de Catalunya, Terrassa 40118
Museu de la Ciència Pare Eduard Vitòria, Alcoy 38828
Museu de la Ciutat, Valls 40250
Museu de la Ciutat Casa de Polo, Villarreal 40308
Museu de la Conca Dellà, Isona 39415
Museu de la Farmàcia Catalana, Barcelona 38979
Museu de la Festa, Casal de Sant Jordi, Alcoy 38829
Museu de la Fusta, Areu 38889
Museu de la Història de Fontcoberta, Fontcoberta 39304
Museu de la Jugueta de Mallorca, Sa Pobla 39847
Museu de la Mar, Peñíscola 39834
Museu de la Marina de Vilassar, Vilassar de Mar 40293
Museu de la Música, Barcelona 38980
Museu de la Noguera, Balaguer 38942
Museu de la Pesca, Palamós 39768
Museu de la Rajoleria de Paiporta, Paiporta 39766
Museu de la Taronja, Burriana 39076
Museu de la Universitat d'Alacant, Alicante 38846
Museu de la Vall de Lord, Sant Lorenç de Morunys 39988
Museu de la Valltorta, Tírig 40124
Museu de la Vila, Vilar-Rodona 40290
Museu de l'Acordió, Arsèguel 38898
Museu de l'Adet, Torelló 40149
Museu de l#Aigua, Lleida 39479
Museu de l'Aigua Col·lecció Gavarró, Vilanova del Camí 40283
Museu de Lamego, Lamego 35656
Museu de Lanifícios, Covilhã 35553
Museu de l'Aquarela, Llança 39470
Museu de l'Art de la Pell, Vic 40269
Museu de l'Artiste Faller, Valencia 40219
Museu de l'Arxiprestal, Morella 39686
Museu de l'Automòbil, Barcelona 38981
Museu De L'automoció Roda Roda, Lleida 39480
Museu de l'Ebre, Tortosa 40165
Museu de l'Empordà, Figueres 39298
Museu de les Arts Decoratives, Barcelona 38982
Museu de les Arts Escèniques, Barcelona 38983
Museu de les Ciències Príncipe Felipe, Valencia 40220
Museu de les dues rodes, Canillo 00087
Museu de l'Escola de Gràmatica Ramon Llull, Algaida 38839
Museu de l'Estable i Troballes, Guimerà 39381
Museu de l'Estampació de Premià de Mar, Premià de Mar 39869
Museu de l'Horta d'Almàssera, Almàssera 38853
Museu de l'Institut de Criminologia, Barcelona 38984
Museu de l'Institut Paleontològic Miquel Crusafont, Sabadell 39925
Museu de Llavaneres, Sant Andreu de Llavaneres 39975
Museu de Lloret de Mar, Lloret de Mar 39488
Museu de Malacologia Cau del Cargol, Vilassar de Dalt 40292
Museu de Mallorca, Palma de Mallorca 39789
Museu de Maquetes d'Art Romànic, Escaldes-Engordany 00091
Museu de Marinha, Lisboa 35693
Museu de Mataró, Mataró 39633
Museu de Medallística Enrique Giner, Nules 39719
Museu de Medicina, Curitiba 04435
Museu de Minerais e Rochas, Uberlândia 04943
Museu de Mineralogia, Congonhas 04413
Museu de Mineralogia Luiz Englert, Porto Alegre 04648
Museu de Miniaturas de Arte e Ofícios, Cartaxo 35510
Museu de Moixent, Mogente 39652
Museu de Monforte, Monforte 35770
Museu de Monte Redondo, Monte Redondo 35773
Museu de Montgrí i del Baix Ter, Torroella de Montgrí 40162
Museu de Montserrat, Monistrol de Montserrat 39666
Museu de Nampula, Nampula 31714
Museu de Nines Russes, Escaldes-Engordany 00092
Museu de Nossa Senhora da Glória do Outeiro, Rio de Janeiro 04729
Museu de Odontologia Dr. Solon de Miranda Galvão, Natal 04585
Museu de Odontología Professor Salles Cunha, Rio de Janeiro 04730
Museu de Olaria, Barcelos 35456
Museu de Ornitologia, Goiânia 04474
Museu de Ovar, Ovar 35799
Museu de Paleontologia, Monte Alto 04572
Museu de Patologia do Instituto Oswaldo Cruz, Rio de Janeiro 04731
Museu de Pedras Ramis Bucair, Cuiabá 04423
Museu de Pesca, Santos 04821
Museu de Pintura, Lourinhã 35739
Museu de Pintura e Escultura Martins Correia, Golegã 35628
Museu de Poblet, Monestir de Poblet 39663
Museu de Polícia Militar, São Paulo 04888
Museu de Pré-História Professora Márcia Angelina Alves, Perdizes 04625
Museu de Relleu, Relleu 39891
Museu de Relógio Prof. Dimas de Melo Pimenta, São Paulo 04889
Museu de Rocafort, El Pont de Vilomara i Rocafort 39855
Museu de Roque Gameiro, Minde 35763
Museu de Rua Alto Paraná Ontem e Hoje, Alto Paran 04304
Museu de Rubí, Rubí 39922
Museu de Sant Cugat, Sant Cugat del Vallès 39979
Museu de Santa Maria, Vila do Porto 35993
Museu de Santa Maria Adelaide, Arcozelo 35443
Museu de Santa Maria de Lamas, Santa Maria de Lamas 35907
Museu de Santo André, Santo André 04817
Museu de São Jorge, Calheta de São Jorge 35499
Museu de São Miguel de Entre-Ambos-os-Rios, Ponte da Barca 35830
Museu de São Roque, Lisboa 35694
Museu de Saúde Pública Emílio Ribas, São Paulo 04890
Museu de Serralves, Porto 35848
Museu de Sertão, Petrolina 04627
Museu de Setúbal - Convento de Jesus - Galeria de Arte Quinhentista, Setúbal 35936
Museu de Telecomunicações, Rio de Janeiro 04732
Museu de Terrassa, Terrassa 40119
Museu de Tradições do Nordeste, Brasília 04362
Museu de Valores, Rio de Janeiro 04733
Museu de Valores do Banco Central do Brasil, Brasília 04363
Museu de Venâncio Aires, Venâncio Aires 04949
Museu de Vila Real, Vila Real 36013
Museu de Vila Verde de Ficalho, Vila Verde de Ficalho 36015
Museu de Vilafamés Vicente Aguilera Cerni, Vilafamés 40279
Museu de Vilafranca, Vilafranca del Penedès 40280
Museu de Zoologia, São Paulo 04891
Museu d'Eines Antigues, Torregrossa 40155
Museu d'Eines del Camp Masia de Can Deu, Sabadell 39926
Museu d'Eines del Pagès, Sant Climent de Llobregat 39977
Museu del Blat, Taradell 40097
Museu del Blat i la Pagesia, Cervera 39165
Museu del Camp, Vilabella 40278
Museu del Càntir, Argentona 38890
Museu del Castell de Onda, Onda 39735
Museu del Centre d'Estudis Contestans, Cocentaina 39197
Museu del Cinema, Girona 39345
Museu del Convent de la Beata sor Fracinaina de la Mare de Déu dels Dolors, Sencelles 40057
Museu del Dolmen, Llagostera 39468
Museu del Ferrocarril, Vilanova i la Geltrú 40285
Museu del Joguet de Catalunya, Figueres 39299
Museu del Monestir de Sant Bartomeu, Inca 39409
Museu del Monestir de Santa Maria de L'Estany, L'Estany 39285
Museu del Montsec, Artesa de Segre 38902
Museu del Montsià, Amposta 38870
Museu del Pagès, Castelldans 39146
Museu del Pagès, Linyola 39467
Museu del Pagès, Torroella de Fluvià 40161
Museu del Perfum, Barcelona 38985
Museu del Perfum, Escaldes-Engordany 00093
Museu del Pin, Ordino 00101
Museu del Port, Tarragona 40102
Museu del Prat, El Prat de Llobregat 39866
Museu del Suro, Palafrugell 39767
Museu del Tabac - Antiga Fàbrica Reig, Sant Julià de Lòria 00105
Museu del Taulell Manolo Safont, Onda 39736
Museu del Temple Expiatori de la Sagrada Familia, Barcelona 38986
Museu del Vi, Vilafranca del Penedès 40281
Museu del Vidre, Algaida 38840
Museu del Volcans, Olot 39728
Museu dels Autòmats, Barcelona 38987
Museu dels Raiers, El Pont de Claverol 39854
Musèu d'Era Val d'Aran, Vielha e Mijaran 40272
Museu des Comunicações Marechal Rondon, Ji-Paran 04511
Museu d'Estris Antics de l'Agricultura, Bellvís 39018
Museu d'Etnografia d'Eivissa, Santa Eulària del Río 40000
Museu d'Etnologia, Pego 39824
Museu d'Etnologia i Arqueologia, Callosa d'En Sarrià 39105
Museu Deu, El Vendrell 40262
Museu d'Història, Sabadell 39927
Museu d'Història de Calp, Calpe 39107
Museu d'Història de Cambrils, Cambrils 39110
Museu d'Història de Catalunya, Barcelona 38988
Museu d'Història de la Ciudad, Sant Feliu de Guíxols 39982
Museu d'Història de la Ciutat, Girona 39346
Museu d'Història de la Ciutat de Barcelona, Barcelona 38989
Museu d'Història de la Ciutat de Palma, Palma de Mallorca 39790
Museu d'Història de la Joguina, Sant Feliu de Guíxols 39983
Museu d'Història de l'Hospitalet de Llobregat, L'Hospitalet de Llobregat 39394
Museu d'Història de Manacor, Manacor 39610
Museu d'Història de Nules, Nules 39720
Museu d'Historia de Tarragona, Tarragona 40103
Museu d'Història del Calçat, Barcelona 38990
Museu d'Història dels Jueui, Girona 39347
Museu di Arte Contemporânea, Americana 04305
Museu Didáctico do Instituto de Arqueologia, Coimbra 35535
Museu Diocesà de Barcelona, Barcelona 38991
Museu Diocesà de Girona, Girona 39348
Museu Diocesà de Mallorca, Palma de Mallorca 39791
Museu Diocesà de Menorca, Ciutadella de Menorca 39193
Museu Diocesà d'Urgell, La Seu d'Urgell 40059
Museu Diocesà i Comarcal, Solsona 40088
Museu Diocesà Santa Maria, Eivissa 39262
Museu Dionís Bennàssar, Pollença 39850
Museu do Ar, Alverca do Ribatejo 35426
Museu do Arroz, Comporta 35546
Museu do Automóvel, Fafe 35593
Museu do Automóvel, Vila Nova de Famalicão 36008
Museu do Automóvel Antigo, Paço d'Arcos 35801
Museu do Automóvel, Arte & História, São Francisco de Paula 04828
Museu do Banco de Crédito Real de Minas Gerais, Juiz de Fora 04520
Museu do Banco de Portugal, Lisboa 35695
Museu do Barreiro e Museu Industrial da Quimiparque, Barreiro 35457
Museu do Barro, Redondo 35882
Museu do Batalhão de Sapadores Bombeiros, Porto 35849
Museu do Bombeiro, Lisboa 35696
Museu do Bombeiro Manuel Valdés Sobral, Valença 35978
Museu do Bonde, Rio de Janeiro 04734
Museu do Brinquedo, Évora 35588
Museu do Brinquedo, São Paulo 04892
Museu do Brinquedo de Sintra, Sintra 35947
Museu do Café Cel. Francisco Schmidt, Ribeirão Preto 04676
Museu do Café de Novadelta, Campo Maior 35502
Museu do Campo da Lacam, Carapinheira 35507
Museu do Cangaço, Triunfo 04936
Museu do Caramulo, Caramulo 35506
Museu do Carnaval, Rio de Janeiro 04735
Museu do Carnaval Hilton Silva, Florianópolis 04456
Museu do Carro Eléctrico, Porto 35850
Museu do Castelo, Ourém 35796
Museu do Castelo de Castro Marim, Castro Marim 35518
Museu do Ceará, Fortaleza 04464
Museu do Centro Científico e Cultural de Macau, Lisboa 35697
Museu do Chiado, Lisboa 35698
Museu do Chocalho, Alcáçovas 35389
Museu do Cinema, Lisboa 35699
Museu do Círio de Nazaré, Belém 04337
Museu do Colégio Mauá, Santa Cruz do Sul 04807
Museu do Colono, Santa Leopoldina 04808
Museu do Convento de São Domingos, Montemor-o-Novo 35776
Museu do Crime, São Paulo 04893
Museu do Diamante, Diamantina 04444
Museu do Douro, Peso da Régua 35821
Museu do Estado da Bahia, Salvador 04800
Museu do Estado de Pernambuco, Recife 04667
Museu do Estado do Pará, Belém 04338
Museu do Eucalipto, Rio Claro 04678
Museu do Expedicionário, Curitiba 04436
Museu do Fado e da Guitarra Portuguesa, Lisboa 35700
Museu do Ferro e da Região de Moncorvo, Torre de Moncorvo 35967
Museu do Forte de Santiago, Funchal 35614
Museu do Fuzileiro, Palhais 35805
Museu do Gás, Rio de Janeiro 04736
Museu do Grupo Folclórico, Santa Eulália de Barrosas 35901
Museu do Homem, Curitiba 04437
Museu do Homem do Nordeste, Recife 04668
Museu do Índio, Florianópolis 04457
Museu do Índio, Lagoa Seca 04529
Museu do Índio, Manaus 04555
Museu do Indio, Rio de Janeiro 04737
Museu do Instituto Arqueológico, Histórico e Geográfico Pernam Bucano, Recife 04669
Museu do Instituto Butantan, São Paulo 04894
Museu do Instituto de Anatomia, Porto 35851
Museu do Instituto de Anatomia Patológica, Coimbra 35536
Museu do Instituto de Antropologia Câmara Cascudo, Natal 04586
Museu do Instituto de Botânica Gonçalo Sampaio, Porto 35852
Museu do Instituto de Medicina Legal, Lisboa 35701
Museu do Instituto Geografico e Historico da Bahia, Salvador 04801
Museu do Instituto Geográfico Português, Lisboa 35702
Museu do Instituto Histórico, Carpina 04405
Museu do Instituto Histórico e Geográfico Brasileiro, Rio de Janeiro 04738
Museu do Instituto Histórico e Geográfico de Alagoas, Maceió 04547
Museu do Instituto Histórico e Geográfico do Rio Grande do Norte, Natal 04587
Museu do Instituto Oceanográfico, São Paulo 04895
Museu do Jogo, Oeiras 35788
Museu do Lar dos Veteranos Militares, Torres Vedras 35971
Museu do Mamulengo, Olinda 04597
Museu do Mar, Sesimbra 35932
Museu do Mar, Vila do Conde 35991
Museu do Mar - Rei D. Carlos, Cascais 35513
Museu do Minstério do Trabalho, Brasília 04364
Museu do Monte da Guerreira, Tavira 35959
Museu do Monumento Nacional aos Mortos da Segunda Guerra Mundial, Rio de Janeiro 04739
Museu do Mosteiro de Alcobaça, Alcobaça 35394
Museu do Mosteiro de São Martinho de Tibães, Braga 35478
Museu do Mosteiro dos Jerónimos, Lisboa 35703
Museu do Neo-Realismo, Vila Franca de Xira 35996
Museu do Nordeste, Nordeste 35783
Museu do Ouro, Sabará 04783
Museu do Ouro de Travassos, Travassos - Póvoa de Lanhoso 35974
Museu do Papel Moeda, Porto 35853
Museu do Papel Terras de Santa Maria, Paços de Brandão 35802
Museu do Pescador, Montijo 35777
Museu do Piauí, Teresina 04930
Museu do Pio XII, Novo Hamburgo 04595
Museu do Porco, Pampilhosa 35807
Museu do Porto de Belém, Belém 04339
Museu do Porto do Rio, Rio de Janeiro 04740
Museu do Primeiro Reinado, Rio de Janeiro 04741
Museu do Rádio, Juiz de Fora 04521
Museu do Rancho Folclórico das Papoilas do Campo, Cela 35520
Museu do Rancho Folclórico de Alenquer, Alenquer 35403
Museu do Rancho Folclórico de São Miguel do Milharado, Milharado 35762
Museu do Rancho Folclórico do Calvário, Estômbar 35574
Museu do Regimento de Infantaria 14, Viseu 36024
Museu do Rio, Alcoutim 35402
Museu do Santuário de Nossa Senhora da Abadia, Bouro 35477
Museu do Scrimshaw, Horta 35644
Museu do Seridó, Caicó 04380
Museu do Seridó, Natal 04588
Museu do Sol, Penápolis 04622
Museu do STF, Brasília 04365
Museu do Superior Tribunal de Justiça, Brasília 04366
Museu do Teatro Nacional de São Carlos, Lisboa 35704
Museu do Teatro Romano, Lisboa 35705
Museu do Telephone, Rio de Janeiro 04742
Museu do Theatro Municipal, São Paulo 04896
Museu do Trabalho Michel Giacometti, Setúbal 35937
Museu do Traje de Viana do Castelo, Viana do Castelo 35983
Museu do Trajo, São Brás de Alportel 35917
Museu do Trem, Rio de Janeiro 04743
Museu do Trem, São Leopoldo 04841
Museu do Trem da Cidade de Recife, Recife 04670
Museu do Tribunal de Justiça do Estado de São Paulo, São Paulo 04897
Museu do Tropeiro, Castro 04408
Museu do Vaqueiro, Morada Nova 04577
Museu do Vidro, Marinha Grande 35756

Museu do Vinho, Funchal 35615
Museu do Vinho, Pico da Pedra 35823
Museu do Vinho, Videira 04954
Museu do Vinho dos Biscoitos, Biscoitos 35468
Museu Dom Avelar Brandão Vilela, Teresina 04931
Museu Dom Diogo de Souza, Bagé 04326
Museu Dom João VI, Rio de Janeiro 04744
Museu Dom José, Sobral 04918
Museu Dom Vital, Recife 04671
Museu dos Baleeiros, Lajes do Pico 35653
Museu dos Biscaínhos, Braga 35479
Museu dos Bombeiros, Vila do Conde 35992
Museu dos Bombeiros Voluntários de Bucelas, Bucelas 35489
Museu dos Caminhos de Ferro de Lousado, Lousado 35743
Museu dos Capelinhos, Horta 35645
Museu dos Cordofones, Tebosa 35961
Museu dos Cristos, Borba 35472
Museu dos Cristos, Sousel 35957
Museu dos Fósforos, Tomar 35964
Museu dos Inhamuns, Tauá 04921
Museu dos Relógios Antigos, Serpa 35929
Museu dos Rios e das Artes Marítimas, Constância 35550
Museu dos Teatros do Rio de Janeiro, Rio de Janeiro 04745
Museu dos Terceiros, Ponte de Lima 35831
Museu dos Transportes e Comunicações, Porto 35854
Museu dos Transportes Públicos Gaetano Ferolla, São Paulo 04898
Museu dos Transportes Urbanos, Coimbra 35537
Museu Dr. José Faibes Lubianca, Porto Alegre 04649
Museu Dr. Marcelino Moules, Cinco Ribeiras 35525
Museu Durán i Sanpere, Cervera 39166
Museu e Arquivo Histórico, Itu 04503
Museu e Arquivo Histórico do Centro Cultural Banco do Brasil, Rio de Janeiro 04746
Museu e Arquivo Histórico Municipal, Guapor 04483
Museu e Arquivo Municipal, Bom Jesus 04355
Museu e Arquivo Municipal, Lagoa 35649
Museu e Arquivo Público Municipal de Campo Belo, Campo Belo 04391
Museu e Biblioteca do Instituto Geográfico e Histórico do Amazonas, Manaus 04556
Museu e Biblioteca Pública Pelotense, Pelotas 04621
Museu Ecologia, Nanuque 04579
Museu Educativo Gama d'Eça, Santa Maria 04810
Museu Egipci de Barcelona, Barcelona 38992
Museu Els Cups, Montbrió del Camp 39672
Museu Eng. António de Almeida, Porto 35855
Museu Enric Monjo, Vilassar de Mar 40294
Museu Epigráfico Egitanense, Idanha-a-Nova 35646
Museu Episcopal de Vic, Vic 40270
Museu-Escola de Artes Decorativas Portuguesas, Lisboa 35706
Museu-Escola Esacro do Estado da Paraiba, João Pessoa 04513
Museu Escolar de Marrazes, Leiria 35659
Museu Estadual do Carvão, Arroio dos Ratos 04323
Museu Etnogáfico e Arqueológico de Ourém, Ourém 35797
Museu Etnogràfic Andino-Amazònic dels Caputxins, Barcelona 38993
Museu Etnogràfic Casa Cristo, Encamp 00089
Museu Etnogràfic de Ripoll, Ripoll 39904
Museu Etnográfico, Santa Cruz das Flores 35899
Museu Etnográfico, Santa Maria da Feira 35905
Museu Etnográfico da Arada, Arada 35441
Museu Etnográfico da Casa do Povo, Almeirim 35419
Museu Etnográfico da Freguesia de Almoster, Almoster 35422
Museu Etnográfico da Lousã, Lousã 35741
Museu Etnográfico da Madeira, Ribeira Brava 35886
Museu Etnográfico da Pedreira, Pedreira 35811
Museu Etnográfico da Povoação, Povoação 35874
Museu Etnográfico da Praia de Mira, Praia de Mira 35875
Museu Etnográfico da Região do Vouga, Trofa 35975
Museu Etnográfico da Sociedade de Geografia de Lisboa, Lisboa 35707
Museu Etnográfico de Campelos, Campelos 35501
Museu Etnográfico de Carregueiros, Carregueiros 35509
Museu Etnográfico de Espiche, Luz 35744
Museu Etnográfico de Serpa, Serpa 35930
Museu Etnográfico de Vila Franca, Lisboa 35708
Museu Etnográfico de Vila Franca de Xira, Vila Franca de Xira 35997
Museu Etnográfico de Vilarinho da Furna, Terras de Bouro 35962
Museu Etnográfico do Cabouco, Cabouco 35490
Museu Etnográfico do Espinheiro, Espinheiro 35565
Museu Etnográfico do Espírito Santo, Vila do Porto 35994
Museu Etnográfico do Grupo Típico de Ançã, Ançã 35436
Museu Etnográfico e Arqueológico do Dr. Joaquim Manso, Nazaré 35780
Museu Etnográfico e Arte Sacra de São Torcato, São Torcato 35926
Museu Etnográfico e Museu Casa Tradicional, Glória do Ribatejo 35626
Museu Etnografico Nacional, Bissau 23317
Museu Etnográfico Regional do Algarve, Faro 35597
Museu Etnográficu de la Llechería, Morcín 39684
Museu Etnològic, Benaguasil 39023
Museu Etnològic de Barcelona, Barcelona 38994
Museu Etnològic de Dénia, Dénia 39245
Museu Etnologic de Guadalest, Guadalest 39376
Museu Etnològic de l'Alcúdia, L'Alcúdia (Valencia) 38834
Museu Etnològic de Muro, Muro 39707
Museu Etnològic del Montseny La Gabella, Arbúcies 38882
Museu Etnològic i Arqueològic, Juneda 39435
Museu Etnológic Local, Betxí 39035
Museu Etnològic Molí de Dalt, Sant Lluis 39987
Museu Etnològic Municipal, Torrella 40157
Museu Eugênio Teixeira Leal, Salvador 04802
Museu Farroupilha, Triunfo 04937
Museu Fazenda, Nanuque 04580
Museu Ferreira de Almeida, Nampula 31715
Museu Ferreira de Castro, Sintra 35948
Museu Ferroviário de Arco de Baúlhe, Arco de Baúlhe 35442
Museu Ferroviário de Chaves, Chaves 35524
Museu Ferroviário de Pires do Rio, Pires do Rio 04635
Museu Florestal Octávio Vecchi, São Paulo 04899
Museu Folclorico da Divisão de Discoteca e Biblioteca de Música de São Paulo, São Paulo 04900
Museu Frederic Marès, Barcelona 38995
Museu Frei Galvão, Guaratinguetá 04486
Museu Fundação Amalia Rodrigues, Lisboa 35709
Museu Fundação Arquivo Paes Telles, Ervedal 35561
Museu Fundação Laura Santos, Gouveia 35633
Museu Fundació Emili Vilá, Llagostera 39469
Museu Futbol Club Barcelona President Núñez, Barcelona 38996
Museu Galeria Manuel Cabanas, Vila Real de Santo António 36014
Museu GEAL, Lourinhã 35740
Museu Geológico, Lisboa 35710
Museu Geológico da Bahia, Salvador 04803
Museu Geológico Valdemar Lefèvre, São Paulo 04901
Museu Getúlio Vargas, São Borja 04823
Museu Goiano Zoroastro Artiaga, Goiânia 04475
Museu Guido Straube, Curitiba 04438
Museu Gustavo Barroso, Fortaleza 04465
Museu H. Stern, Rio de Janeiro 04747
Museu Henrique e Francisco Franco, Funchal 35616
Museu Henriqueta Catharino, Salvador 04804
Museu Hernández Mora-Hernández Mora, Mahón 39592
Museu Històric de la Seu, Manresa 39616
Museu Històric Municipal, Riudoms 39905
Museu Historic Municipal de Polinyà, Polinyà 39849
Museu Histórico, Cambé 04381
Museu Histórico, Divinópolis 04445
Museu Histórico, Igarassu 04488
Museu Histórico, Lavras 04535
Museu Histórico, Londrina 04539
Museu Histórico, Nova Ponte 04593
Museu Histórico, Pará de Minas 04609
Museu Histórico, Rio Pomba 04777
Museu Histórico, São Francisco do Sul 04829
Museu Histórico, São Paulo 04902
Museu Histórico, Sertanópolis 04916
Museu Histórico, Siqueira Campos 04917
Museu Histórico Abílio Barreto, Belo Horizonte 04350
Museu Histórico Aurélio Dolabella, Santa Luzia 04809
Museu Histórico Bárbara Heliodora, São Gonçalo do Sapucaí 04832
Museu Histórico Beato José de Anchieta, Pedreira 04620
Museu Histórico Casa do Imigrante, Bento Gonçalves 04352
Museu Histórico Corália Venites Maluf, Sacramento 04785
Museu Histórico Cultural, Rio do Sul 04774
Museu Histórico da Cidade, Brasília 04367
Museu Histórico da Cidade do Rio de Janeiro, Rio de Janeiro 04748
Museu Histórico da Imigração Japonesa no Brasil, São Paulo 04903
Museu Histórico da Universidade Federal, Viçoasa 04952
Museu Histórico da Vista Alegre, Ílhavo 35647
Museu Histórico de Araxá - Don Beja, Araxá 04319
Museu Histórico de Brejo Madre de Deus, Brejo Madre de Deus 04372
Museu Histórico de Jataí Francisco Honório de Campos, Jataí 04510
Museu Histórico de Sergipe, São Cristóvão 04827
Museu Histórico de Witmarsum, Palmeira 04607
Museu Histórico Desembargador Edmundo Mercer Júnior, Tibagi 04934
Museu Histórico do Corpo de Bombeiros Militar do Estado do Rio de Janeiro, Rio de Janeiro 04749
Museu Histórico do Exército e Forte de Copacabana, Rio de Janeiro 04750
Museu Histórico do Piauí, Teresina 04932
Museu Histórico do Senado Federal, Brasília 04368
Museu Histórico Dona Mariana Joaquina da Costa, Vespasiano 04951
Museu Histórico Dr. Sellmann Nazareth, São José do Rio Preto 04838
Museu Histórico e Artístico, Quitandinha 04659
Museu Histórico e Artístico do Maranhão, São Luís 04845
Museu Histórico e de Artes, Santana do Ipanema 04815
Museu Histórico e de Ordem Geral Plínio Travassos dos Santos, Rebeirão Preto 04660
Museu Histórico e Diplomático do Itamaraty, Rio de Janeiro 04751
Museu Histórico e Geográfico de Monte Sião, Monte Sião 04575
Museu Histórico e Geográfico de Poços de Caldas, Poços de Caldas 04637
Museu Histórico e Geral da Cidade de São Vicente, São Vicente 04915
Museu Histórico e Pedagógico, Mococa 04570
Museu Histórico e Pedagógico Conselheiro Rodrigues Alves, Guaratinguetá 04487
Museu Histórico e Pedagógico Dr. Washington Luís, Batatais 04330
Museu Histórico e Pedagógico Fernão Dias Paes, Penápolis 04623
Museu Histórico e Pedagógico Marechal Cândido Rondon, Araçatuba 04315
Museu Histórico e Pedagógico Monteiro Lobato, Taubaté 04926
Museu Histórico e Pedagógico Pe. Manoel da Nóbrega, São Manuel 04846
Museu Histórico e Pedagógico Prof. Lourenço Filho, Porto Ferreira 04656
Museu Histórico e Pedagógico Zequinha de Abreu, Santa Rita do Passa Quatro 04812
Museu Histórico Farroupilha, Piratini 04634
Museu Histórico Kiyoshi Takano, Ura 04946
Museu Histórico Militar, Maputo 31708
Museu Histórico Muncipal José Chiachiri, Franca 04468
Museu Histórico Municipal, Carangola 04402
Museu Histórico Municipal, Dois Irmãos 04446
Museu Histórico Municipal, Montenegro 04576
Museu Histórico Municipal, Telêmaco Borba 04928
Museu Histórico Municipal e da Imigração Italiana Oswaldo Samuel Massei, São Caetano do Sul 04824
Museu Histórico Municipal Tuany Toledo, Pouso Alegre 04658
Museu Histórico Nacional, Rio de Janeiro 04752
Museu Histórico Pedagógico Conselheiro Dr. João da Silva Carrão, Americana 04306
Museu Histórico Pedagógico Índia Vanuíre, Tupã 04938
Museu Histórico Prof. Carlos de Silva Lacaz, São Paulo 04904
Museu Histórico Professor José Alexandre Vieira, Palmas 04606
Museu Histórico Regional, Apucarana 04310
Museu Histórico Regional, Itambacuri 04496
Museu Histórico Regional, Passo Fundo 04613
Museu Histórico Tenente Coronel PM Augusto de Almeida Garrett, Curitiba 04439
Museu Histórico Visconde de São Leopoldo, São Leopoldo 04842
Museu Homo Sapiens, Nanuque 04581
Museu i Centre d'Estudis de l'Esport Dr. Melcior Colet, Barcelona 38997
Museu i Centre d'Estudis Juniperians, Petra 39838
Museu i Fons Artistic, Porreres 39860
Museu i Necròpolis Paleocristians, Tarragona 40104
Museu Iconogràfic, Ordino 00102
Museu Imperial, Petrópolis 04629
Museu Inaldo de Lyra Neves-Manta, Rio de Janeiro 04753
Museu Integrado de Roraima, Boa Vista 04354
Museu Internacional de Arte Naïf do Brasil, Rio de Janeiro 04754
Museu Internacional de Titelles d'Albaida, Albaida 38806
Museu Jaguaribano, Aracati 04314
Museu João de Deus, Lisboa 35711
Museu João Pedro Nunes, São Gabriel 04831
Museu Joaquim José Felizardo, Porto Alegre 04650
Museu José Bonifácio, São Paulo 04905
Museu José Luciano de Castro, Anadia 35434
Museu Julio de Castilhos, Porto Alegre 04651
Museu Júlio Dinis, Ovar 35800
Museu Krekovic, Palma de Mallorca 39792
Museu la Granja, Santa Perpètua de Mogoda 40005
Museu Laboratório Mineralógico e Geológico, Porto 35856
Museu Lasar Segall, São Paulo 04906
Museu Leprológico, Curitiba 04440
Museu Local Arqueològic, Artesa del Lleida 38903
Museu Local do Pico da Pedra, Pico da Pedra 35824
Museu Luso-Hebraico Abraham Zacut, Tomar 35965
Museu Maçônico Mário Verçosa, Manaus 04557
Museu Maçonico Português, Lisboa 35712
Museu Malinverni Filho, Lages 04526
Museu Manolo Hugué, Caldes de Montbui 39100
Museu Marès de la Punta, Arenys de Mar 38886
Museu Mariano Procópio, Juiz de Fora 04522
Museu Maricel de Mar, Sitges 40084
Museu Marítim de Barcelona, Barcelona 38998
Museu Marítima Almirante Ramalho Ortigão, Faro 35598
Museu Marítimo, Funchal 35617
Museu Marítimo e Regional de Ílhavo, Ílhavo 35648
Museu Marítimo Joaquín Saludes, Valencia 40221
Museu Mazzaropi, Taubaté 04927
Museu Medina, Braga 35480
Museu Memória do Bixiga, São Paulo 04907
Museu Meridional, Porto Alegre 04652
Museu Metropolitano de Arte, Curitiba 04441
Museu Militar, Barcelona 38999
Museu Militar, Ponta Delgada 35828
Museu Militar de Bragança, Bragança 35486
Museu Militar de Coimbra, Coimbra 35538
Museu Militar de Lisboa, Lisboa 35713
Museu Militar do Baixo Alentejo, Beja 35464
Museu Militar do Buçaco, Buçaco 35488
Museu Militar do Forte do Brum, Recife 04672
Museu Militar do Porto, Porto 35857
Museu Mineiro, Belo Horizonte 04351
Museu Mineiro, São Pedro da Cova 35924
Museu Mineralógico da Escola de Minas, Ouro Preto 04605
Museu Mineralógico e Geológico, Coimbra 35539
Museu Mineralógico e Geológico, Lisboa 35714
Museu Moacir Andrade, Manaus 04558
Museu Molí Paperer de Banyeres de Mariola, Banyeres de Mariola 38946
Museu Molí Paperer de Capellades, Capellades 39124
Museu Mollfulleda de Mineralogía, Arenys de Mar 38887
Museu Monàstic, Vallbona de les Monges 40244
Museu-Monestir de Pedralbes, Barcelona 39000
Museu Monografic de Pol Lèntia, Alcúdia (Baleares) 38833
Museu Monogràfic del Castell de Llinars, Llinars del Vallés 39482
Museu Monográfico de Conimbriga, Condeixa-a-Nova 35548
Museu Monográfico do Bárrio e Campo Arqueológico, Bárrio 35458
Museu Monográfico do Fundão, Fundão 35622
Museu Monstruário de Manica, Manica 31704
Museu Muncipal de Victor Graeff, Victor Graeff 04953
Museu Municipal, Alcover 38826
Museu Municipal, Amadora 35430
Museu Municipal, Barbacena 04329
Museu Municipal, Cachoeira do Sul 04378
Museu Municipal, Caxias do Sul 04411
Museu Municipal, Conquista 04414
Museu Municipal, Ferreira do Alentejo 35603
Museu Municipal, Garibaldi 04471
Museu Municipal, Itambaracá 04497
Museu Municipal, Llívia 39486
Museu Municipal, Manlleu 39613
Museu Municipal, Missal 04567
Museu Municipal, Paulínia 04617
Museu Municipal, Pedras Grandes 04618
Museu Municipal, Rafelcofer 39888
Museu Municipal, Rolândia 04779
Museu Municipal, Rosário de Oeste 04780
Museu Municipal, São João Del Rei 04836
Museu Municipal, Sobralinho 35956
Museu Municipal, Tortosa 40166
Museu Municipal, Tossa de Mar 40167
Museu Municipal, Uberlândia 04944
Museu Municipal, Visconde do Rio Branco 04955
Museu Municipal Abade de Pedrosa, Santo Tirso 35916
Museu Municipal Adão Wolski, Contenda 04416
Museu Municipal Adolfo Eurich, Turvo 04939
Museu Municipal Alcácer do Sal, Alcácer do Sal 35388
Museu Municipal Alípio Vaz, Cataguases 04410
Museu Municipal Amadeo de Souza-Cardoso, Amarante 35433
Museu Municipal Armindo Teixeira Lopes, Mirandela 35765
Museu Municipal Atilio Rocco, São José dos Pinhais 04840
Museu Municipal Can Xifreda, Sant Feliu de Codines 39981

M

Museu Municipal Capitão Henrique José Barbosa, Canguçu 04400
Museu Municipal Carlos Reis de Torres Novas, Torres Novas 35969
Museu Municipal Carmen Miranda, Marco de Canaveses 35755
Museu Municipal da Azambuja, Azambuja 35451
Museu Municipal da Chamusca, Chamusca 35522
Museu Municipal da Fotografia João Carpinteiro, Elvas 35557
Museu Municipal da Moita, Moita 35768
Museu Municipal da Ribeira Grande, Ribeira Grande 35890
Museu Municipal d'Agramunt, Agramunt 38786
Museu Municipal d'Almassora, Almazora 38854
Museu Municipal David Canabarro, Sant'Anna do Livramento 04816
Museu Municipal de Alcochete-Núcleo de Arte Sacra, Alcochete 35397
Museu Municipal de Alcochete-Núcleo do Sal, Alcochete 35398
Museu Municipal de Alcochete - Núcleo Sede, Alcochete 35399
Museu Municipal de Alvaiázere, Alvaiázere 35425
Museu Municipal de Antônio Prado, Antônio Prado 04308
Museu Municipal de Arqueologia, Albufeira 35386
Museu Municipal de Arqueologia, Aljezur 35409
Museu Municipal de Arqueologia, Amadora 35431
Museu Municipal de Arqueologia, Monte Alto 04573
Museu Municipal de Arqueologia de Silves, Silves 35940
Museu Municipal de Arqueologia e Etnologia, Sines 35942
Museu Municipal de Arte e História, Nova Era 04592
Museu Municipal de Arte Moderna Abel Manta, Gouveia 35634
Museu Municipal de Avis, Avis 35449
Museu Municipal de Baião, Baião 35454
Museu Municipal de Bobadela, Bobadela 35469
Museu Municipal de Bombarral, Bombarral 35470
Museu Municipal de Ca l'Arrà, Cabrils 39080
Museu Municipal de Caminha, Caminha 35500
Museu Municipal de Campo Maior, Campo Maior 35503
Museu Municipal de Castro Daire, Castro Daire 35516
Museu Municipal de Ciutadella, Ciutadella de Menorca 39194
Museu Municipal de Esposende, Esposende 35569
Museu Municipal de Estremoz, Estremoz 35578
Museu Municipal de Fotografia Carlos Relvas, Golegã 35629
Museu Municipal de Grândola, Grândola 35635
Museu Municipal de la Festa d'Elx, Elche 39270
Museu Municipal de la Pagesia, Castellbisbal 39145
Museu Municipal de Loulé, Loulé 35737
Museu Municipal de Loures, Loures 35738
Museu Municipal de Mafra, Mafra 35751
Museu Municipal de Marvão, Marvão 35758
Museu Municipal de Molins de Rei, Molins de Rei 39658
Museu Municipal de Montijo, Montijo 35778
Museu Municipal de Moura, Moura 35779
Museu Municipal de Nàutica, El Masnou 39630
Museu Municipal de Óbidos, Óbidos 35784
Museu Municipal de Olhão, Olhão 35790
Museu Municipal de Paços de Ferreira, Paços de Ferreira 35803
Museu Municipal de Palmela, Palmela 35806
Museu Municipal de Paredes, Paredes 35808
Museu Municipal de Paredes de Coura, Paredes de Coura 35809
Museu Municipal de Penafiel, Penafiel 35817
Museu Municipal de Penamacor, Penamacor 35819
Museu Municipal de Peniche, Peniche 35820
Museu Municipal de Pinhel, Pinhel 35825
Museu Municipal de Pollença, Pollença 39851
Museu Municipal de Portalegre, Portalegre 35834
Museu Municipal de Portimão, Portimão 35835
Museu Municipal de Porto de Mós, Porto de Mós 35869
Museu Municipal de Sant Pol de Mar, Sant Pol de Mar 39989
Museu Municipal de Santarém, Santarém 35909
Museu Municipal de São José dos Campos, São José dos Campos 04839
Museu Municipal de Torres Vedras, Torres Vedras 35972
Museu Municipal de Vale de Cambra, Vale de Cambra 35977
Museu Municipal de Viana do Castelo, Viana do Castelo 35984
Museu Municipal de Vila Flor, Vila Flor 35995
Museu Municipal de Vila Franca de Xira, Vila Franca de Xira 35998
Museu Municipal de Vila Franca do Campo, Vila Franca do Campo 35999
Museu Municipal de Vouzela, Vouzela 36026
Museu Municipal Deolindo Mendes Pereira, Campo Mourão 04396
Museu Municipal d'Història i Arqueologia, Cullera 39238
Museu Municipal Dias de Oliveira, Valongo 35980
Museu Municipal do Cadaval, Cadaval 35492
Museu Municipal do Crato, Crato 35554
Museu Municipal do Folclore, Penápolis 04624
Museu Municipal do Funchal - História Natural, Funchal 35618
Museu Municipal do Fundão, Fundão 35623
Museu Municipal do Sabugal, Sabugal 35892
Museu Municipal do Seixal, Seixal 35927
Museu Municipal Domingos Battistel, Nova Prata 04594
Museu Municipal Dr. António Gabriel Ferreira Lourenço, Benavente 35466
Museu Municipal Dr. João Calado Rodrigues, Mação 35745
Museu Municipal Dr. José Formosinho, Lagos 35651
Museu Municipal Dr. Santos Rocha, Figueira da Foz 35605
Museu Municipal Elisabeth Aytai, Monte Mor 04574
Museu Municipal Embaixador Hélio A. Scarabôtolo, Marília 04562
Museu Municipal Francisco Manoel Franco, Itaúna 04501
Museu Municipal Francisco Veloso, Cunha 04425
Museu Municipal Guilleries, Sant Hilari Sacalm 39984
Museu Municipal Hipólito Cabaço, Alenquer 35404
Museu Municipal Joan Pla i Gras, Llinars del Vallés 39483
Museu Municipal Joaquim de Bastos Bandeira, Perdões 04626
Museu Municipal Josep Aragay, Breda 39057
Museu Municipal Karl Ramminger, Mondaí 04571
Museu Municipal Miquel Soldevila, Prats de Lluçanès 39867
Museu Municipal Montcada, Montcada i Reixac 39673
Museu Municipal Napoleão Joele, São Sebastião de Paraíso 04914
Museu Municipal Oswaldo Russomano, Bragança Paulista 04356
Museu Municipal Padre Antônio Ribeiro Pinto, Urucânia 04947
Museu Municipal Pedro Laragnoit, Miracatu 04566
Museu Municipal Polinucleado de Tomar, Tomar 35966
Museu Municipal Prof. Álvaro Viana de Lemos, Lousã 35742
Museu Municipal Prof. Hugo Daros, Gramado 04482
Museu Municipal Severo Portela, Almodôvar 35421
Museu Municipal Sobral Mendes, Penafiel 35818
Museu Municipal Vicenç Ros, Martorell 39627
Museu Municipal Visconde de Guarapuava, Guarapuava 04485
Museu Municipal Wenceslau Braz, Itajubá 04494
Museu Municipal Wülson Jehovah Lütz Farias, Frederico Westphalen 04470
Museu Nacional, Rio de Janeiro 04755
Museu Nacional, São Tomé 38013
Museu Nacional Arqueològic de Tarragona, Tarragona 40105
Museu Nacional da Ciência e da Técnica, Coimbra 35540
Museu Nacional da Imprensa, Jornais e Artes Gráficas, Porto 35858
Museu Nacional da Literatura, Porto 35859
Museu Nacional da Moeda, Maputo 31709
Museu Nacional d'Art de Catalunya (MNAC), Barcelona 39001
Museu Nacional de Agricultura, Coimbra 35541
Museu Nacional de Antropologia, Luanda 00112
Museu Nacional de Arqueologia, Benguela 00106
Museu Nacional de Arqueologia, Lisboa 35715
Museu Nacional de Arte, Maputo 31710
Museu Nacional de Arte Antiga, Lisboa 35716
Museu Nacional de Belas Artes, Rio de Janeiro 04756
Museu Nacional de Etnologia, Lisboa 35717
Museu Nacional de Geologia, Maputo 31711
Museu Nacional de História Natural, Luanda 00113
Museu Nacional de História Natural e Museu de Ciência, Lisboa 35718
Museu Nacional de Imigração e Colonização, Joinville 04517
Museu Nacional de l'Automòbil, Encamp 00090
Museu Nacional de Machado de Castro, Coimbra 35542
Museu Nacional de Soares dos Reis, Porto 35860
Museu Nacional do Azulejo, Lisboa 35719
Museu Nacional do Desporto, Lisboa 35720
Museu Nacional do Mar - Embarações Brasileiras, São Francisco do Sul 04830
Museu Nacional do Teatro, Lisboa 35721
Museu Nacional do Traje, Lisboa 35722
Museu Nacional do Vinho, Alcobaça 35395
Museu Nacional dos Coches, Lisboa 35723
Museu Nacional Ferroviário, Entroncamento 35558
Museu Naval de Amazônia, Belém 04340
Museu Naval e Oceanográfico, Rio de Janeiro 04757
Museu Navio Comandante Bauru, Rio de Janeiro 04758
Museu Nikolaï Siadristy, Ordino 00103
Museu Nogueira da Silva, Braga 35481
Museu Nossa Senhora Aparecida, Aparecida 04309
Museu Nossa Senhora d'Aires, Viana do Alentejo 35982
Museu Numismático e Filatélico, Rio de Janeiro 04759
Museu Numismático Português, Lisboa 35724
Museu O Mundo Ovo de Eli Heil, Florianópolis 04458
Museu obert de geologia 'Roques al Carrer', Andorra la Vella 00086
Museu Oceanográfico, Setúbal 35938
Museu Oceanográfico do Valo do Itajaí, Itajaí 04493
Museu Oceanográfico Prof. Eliézer de C. Rios, Rio Grande 04775
Museu Oceanográfico Prof. Luiz Saldanha, Azeitão 35453
Museu Oswaldo Aranha, Alegrete 04302
Museu Padre Joseph Cornellius Maria Demam, Coronel Fabriciano 04418
Museu Padre Júlio Maria, Russas 04782
Museu Palácio Nacional da Ajuda, Lisboa 35725
Museu Paleontològic, Valencia 40222
Museu Paleontologica Juna Cano Forner, San Mateo 39966
Museu Paraense Emílio Goeldi, Belém 04341
Museu Paranaense, Curitiba 04442
Museu Parc Cultural de la Muntanya de Sal, Cardona 39130
Museu Paroquial, Moncaparracho 35769
Museu Paroquial, Queiriga 35878
Museu Paroquial de Óbidos, Óbidos 35785
Museu Paroquial de Oliveira de Azeméis, Oliveira de Azeméis 35791
Museu Paroquial de São Sebastião, Guimarães 35641
Museu Paroquial de Vila Praia de Âncora, Vila Praia de Âncora 36010
Museu Parque Malwee, Jaraguá do Sul 04508
Museu Parroquial, Altafulla 38866
Museu Parroquial, Castellò d'Empúries 39147
Museu Parroquial, Santa Pau 40004
Museu Parroquial de Sant Julià, Campos 39114
Museu Parroquial de Sant Llorenç des Cardassar, Sant Llorenç des Cardassar 39986
Museu Parroquial de Santa Joan Baptista, Deià 39243
Museu Parroquial de Santa Maria de Sineu, Sineu 40082
Museu Parroquial de Santa Maria la Major d'Inca, Inca 39410
Museu Parroquial dels Sants Joans, Estivella 39291
Museu Particular de Armas de Arlindo Pedro Zatti, Porto Alegre 04653
Museu Particular del Pastor Joaquim Gassó, Fornells de la Muntanya 39307
Museu Patríco Corrêa da Câmara, Bagé 04327
Museu Pau Casals, El Vendrell 40263
Museu Paulista da Universidade de São Paulo, São Paulo 04908
Museu Paulo Firpo, Dom Pedrito 04447
Museu Paulo Machado de Carvalho, São Paulo 04909
Museu Pedro Cruz, Pedrógão Grande 35814
Museu Pedro Ludovico, Goiânia 04476
Museu Picasso, Barcelona 39002
Museu Pinacoteca, Igarassu 04489
Museu Pio XII, Braga 35482
Museu-Poblat Ibèric, Tornabous 40150
Museu Popular Beatriz Costa, Malveira 35754
Museu Postal, Ordino 00104
Museu Postal e Telegráfico da ECT, Brasília 04369
Museu Professor Hugo Machiado da Silveira, Patrocínio 04615
Museu Rafael Bordalo Pinheiro, Lisboa 35726
Museu Regional, Vitória da Conquista 04959
Museu Regional Abade de Baçal, Bragança 35487
Museu Regional Carlos Machado, Ponta Delgada 35829
Museu Regional D. Bosco, Campo Grande 04394
Museu Regional da Fauna e Flora do Itatiaia, Itatiaia 04500
Museu Regional da Huila, Lumbango 00114
Museu Regional d'Artá, Artá 38899
Museu Regional de Arqueologia D. Diogo de Sousa, Braga 35483
Museu Regional de Arqueologia e Etnografia de Arganil, Arganil 35444
Museu Regional de Cabinda, Cabinda 00107
Museu Regional de Etnografia do Lobito, Lobito 00109
Museu Regional de Leiria, Leiria 35660
Museu Regional de Olinda, Olinda 04598
Museu Regional de Oliveira de Azeméis, Oliveira de Azeméis 35792
Museu Regional de Paredes de Coura, Paredes de Coura 35810
Museu Regional de São João Del Rei, São João Del Rei 04837
Museu Regional de Sintra, Sintra 35949
Museu Regional do Alto Uruguai, Erechim 04448
Museu Regional do Huambo, Huambo 00108
Museu Regional do Sul de Minas, Campanha 04382
Museu Regional Dr. Alípio de Araújo Silva, Rio Preto 04778
Museu Regional Manuel Nobre, Azambujeira 35452
Museu Regional Olívio Otto, Carazinho 04404
Museu Renato Lobo Garcia, Colares 35545
Museu Republicano Convenção de Itu, Itu 04504
Museu Restauraçao, Lisboa 35727
Museu Rocaguinarda de Terrissa dels Països Catalans, Oristà 39746
Museu Rodrigues Lapa, Anadia 35435
Museu Romàntic Can Llopis, Sitges 40085
Museu Romantic Can Papiol, Vilanova i la Geltrú 40286
Museu Romàntic Delger, Caldes de Montbui 39101
Museu Romântico da Quinta da Macieirinha, Porto 35861
Museu Rondon, Cuiabá 04424
Museu Rural, Alcáçovas 35390
Museu Rural - Centro de Arte e Cultura, Reguengo Grande 35883
Museu Rural da Casa do Povo de Santa Maria, Estremoz 35579
Museu Rural de Beringel, Beringel 35467
Museu Rural de Boticas, Boticas 35476
Museu Rural de Vila Franca do Campo, Vila Franca do Campo 36000
Museu Rural de Villa-Lobos, Borba 35473
Museu Rural do Grupo dos Amigos de Estremoz, Estremoz 35580
Museu Rural e do Vinho do Concelho, Cartaxo 35511
Museu Rural e Etnográfico, São João da Ribeira 35920
Museu S. Pedro da Palhaça, Palhaça 35804
Museu Sacro da Igreja de São Sebastião, Araxá 04320
Museu Sacro São José de Ribamar, Aquiraz 04311
Museu Sala de Armas, Almeida 35417
Museu Salles Cunha, Rio de Janeiro 04760
Museu Salvador, Barcelona 39003
Museu Santa Casa da Misericórdia, Coimbra 35543
Museu Santa Maria Rosa Molas, Jesús 39431
Museu Santos Barosa da Fábrica do Vidro, Marinha Grande 35757
Museu São João Batista, Caratinga 04403
Museu São Norberto, Pirapora do Bom Jesus 04633
Museu Sargento Edésio de Carvalho, Sousa 04920
Museu Sebastião da Gama, Vila Nogueira de Azeitão 36001
Museu Sitio Arqueológico Casa dos Pilões, Rio de Janeiro 04761
Museu Sobre a Pesca do Atum, Tavira 35960
Museu Solar Monjardim, Vitória 04958
Museu Sub-Tenente António Bernardino Apolónio Piteira, Arraiolos 35446
Museu Taurí de la Monumental, Barcelona 39004
Museu Taurino, Alcochete 35400
Museu Teatro, Rio de Janeiro 04762
Museu Técnico-Científicio do Instituto Oscar Freire, São Paulo 04910
Museu Temático do Vinho, Almeirim 35420
Museu Tempostal, Salvador 04805
Museu Teresa Bernardes Adami de Carvalho, Santo Antônio do Monte 04819
Museu Territorial do Amapá, Macapá 04540
Museu Textil i d'Indumentària, Barcelona 39005
Museu Théo Brandão de Antropologia de Folclore, Maceió 04548
Museu Theodormiro Carneiro Santiago, Itajubá 04495
Museu Tinguí-Cuera, Araucária 04317
Museu Torre Balldovina, Santa Coloma de Gramenet 39991
Museu-Tresor Parroquial, Olot 39729
Museu Universitário, Campinas 04390
Museu Universitário de Arte, Uberlândia 04945
Museu Universitário Gama Filho, Rio de Janeiro 04763
Museu Universitário Professor Oswaldo Rodrigues Cabral, Florianópolis 04459
Museu Valencia del Joguet, Ibi 39405
Museu Valencià d'Etnologia, Valencia 40223
Museu Valencià d'Història Natural, Valencia 40224
Museu Víctor Balaguer, Vilanova i la Geltrú 40287
Museu Victor Bersani, Santa Maria 04811
Museu Victor Meirelles, Florianópolis 04460
Museu Viladomat d'Escultura, Escaldes-Engordany 00094
Museu Villa-Lobos, Rio de Janeiro 04764
Museu Vivo da Comutação Manual, Vilar 36020
Museu Vivo da Memória Candanga, Brasília 04370
Museu Vivo de Vilar do Pinheiro, Vilar do Pinheiro 36021
Museu von Martius, Teresópolis 04933
Museu Vulcano-Espeleológico Machado Fagundes, Angra do Heroísmo 35440
Museu Zé Didor, Campo Maior 04395
Museu Zoobotânico Augusto Ruschi, Passo Fundo 04614

Museu Zoológico, Coimbra 35544
Museu Zoológico, Pombal 35826
Museum, Bagheria 25341
Museum, Coswig 17728
The Museum, Dunwich 44068
The Museum, Greenwood 49087
Museum, Großpösna 18801
Museum, Mühlacker 20397
The Museum, Newton Stewart 45252
Museum, Zofingen 42061
Museum 1806, Jena 19384
Museum 1915-1918, Kötschach-Mauthen 02215
Museum 1940-1945, Dordrecht 32151
Museum 3. Dimension, Dinkelsbühl 17868
Museum '40-'45, Slochteren 32785
Museum Aan de Stroom, Antwerpen 03295
Museum Aargau, Lenzburg 41554
Museum Aargau, Lenzburg Castle, Lenzburg 41554
Museum Abdij Mariënlof, Kerniel 03752
Museum Abdij van Egmond, Egmond Binnen 32182
Museum Abdijkerk, Thorn 32821
Museum Abodiacum, Denklingen 17821
Museum Abri Satriamandala, Jakarta 24477
Museum Abtei Liesborn, Wadersloh 22263
Museum Abteiberg, Mönchengladbach 20357
Museum Abtsküche, Heiligenhaus 19105
Museum Adam Malik, Jakarta 24478
Museum Adlerturm, Dortmund 17923
Museum Adorf, Adorf 16433
Museum Affandi - Museum Seni, Yogyakarta 24583
Museum Africa, Johannesburg 38660
Muséum Agricole et Industriel de Stella Matutina, Piton-Saint-Leu 36090
Museum Aguntum, Dölsach 01823
Museum Albert van Dyck, Schilde 04005
Museum Ald Slot, Wergea 32936
Museum Alex Mylona, Athinai 22928
Museum Alexandrowka, Potsdam 21095
Museum Alpin Pontresina, Pontresina 41726
Museum Alte Brettersäge, Rappottenstein 02561
Museum Alte Kirche, Reken 21258
Museum Alte Lateinschule Großenhain, Großenhain 18789
Museum Alte Münze, Stolberg (Harz) Kurort 21907
Museum Alte Pfefferküchlerei, Weißenberg 22447
Museum Alte Textilfabrik, Weitra 02907
Museum alte Ziegelei Westerholt, Wardenburg 22334
Museum Altes Bad Pfäfers, Bad Ragaz 41139
Museum Altes Bürgerhaus, Stolberg (Harz) Kurort 21908
Museum Altes Gymnasium, Schweinfurt 21684
Museum Altes Land, Jork 19403
Museum Altes Rathaus, Leingarten 19886
Museum Altes Rathaus, Pirmasens 21041
Museum 'Altes Zollhaus', Baltrum 16904
Museum Altomünster, Altomünster 16524
Museum Alzey, Alzey 16530
Museum am Burghof, Lörrach 20016
Museum am Dom, Würzburg 22717
Museum am Lindenbühl, Mühlhausen/Thüringen 20403
Museum am Lindenplatz, Weil am Rhein 22376
Museum am Markt, Schiltach 21556
Museum am Maxplatz, Rehau 21239
Museum am Meer, Büsum 17640
Museum am Modersohn-Haus, Worpswede 22696
Museum am Odenwaldlimes, Elztal 18135
Museum am Ostwall, Dortmund 17924
Museum am Schölerberg, Osnabrück 20925
Museum am Thie e.V., Göttingen 18694
Museum am Widumhof, Urbach 22186
Museum Amandinhuis, Herk-de-Stad 03699
Museum Amden, Amden 41110
Museum Amelinghausen mit Zinnfigurenausstellung, Amelinghausen 16535
Museum Amtspforte Stadthagen, Stadthagen 21856
Museum an der Krukenburg, Bad Karlshafen 16758
Museum Anatomicum, Marburg 20193
Museum Anatomy FK UI, Jakarta 24479
Museum, Ancient Synagogue and Necropolis, Beit Shearim 24938
Museum and Archaeological Sites of Anamur, Anamur 42692
Museum and archive of popular music, Praha 09789
Museum and Archive of Town History Admiral William Brown, Adrogué 00116
Museum and Archives of Georgia Education, Milledgeville 50691
Museum and Art Gallery, Burdwan 24105
Museum and Art Gallery, Forfar 44242
Museum and Art Gallery, Hooghly 24209
Museum and Art Gallery, Loreto 26519
Museum and Art Gallery, Minden 30610
Museum and Art Gallery, Newport, Gwent 45240
Museum and Art Gallery, Nuneaton 45305
Museum and Art Gallery, Pulau Pinang 30613
Museum and Art Gallery, Shah Alam 30618
Museum and Art Gallery of the Northern Territory, Darwin 01030
Museum and Arts Center in the Sequim Dungeness Valley, Sequim 53037
Museum and Exhibition Center, Dalnegorsk 36872
Museum and Exhibition Center, Ivanovo 36952
Museum and Exhibition Center - Diorama, Kirov 37042
Museum and Exhibition Center - Kremlin of Volokolamsk, Volokolamsk 37966
Museum and Gallery for optical Illusions and Holography, Ittigen 41497
Museum and Gallery of Art and Archaeology, Dharwad 24164
Museum and Gallery of Northern Pilsen Region, Mariánská Tynice - Kralovice 09648
Museum and Gallery of Podgorica, Podgorica 31656
Museum and Institute of the Roerich Family, Sankt-Peterburg 37679
Museum and Institute of Zoology of the Polish Academy of Sciences, Warszawa 35248
Museum and Study Collection, London 44920
Museum Anna Nordlander, Skellefteå 40848
Museum Anna Paulowna, Breezand 32008
Museum Anthon, Ammerbuch-Breitenholz 16539
Museum-Appartement of A. Tvardovsky in Smolensk, Smolensk 37770
Museum-Appartment of Abdulla Shaig, Bakı 03162
Museum-Appartment of Niyazi, Bakı 03163
Museum Appartment of the Mironov and Menaker Actors' Family, Moskva 37314
Museum Appeleshof, Gechingen 18548
Museum Appenzell, Appenzell 41118
Muséum-Aquarium de Nancy, Nancy 14357
Museum-Aquarium of the Pacific Research Fisheries Centre, Vladivostok 37942
Museum Arbeit am Rhein, Elchesheim-Illingen 18115
Museum Arbeitswelt Steyr, Steyr 02769
The Museum, Archaeological Institute of Kashihara, Kashihara 28677
Museum-Armoury, Liw 34858
Museum Arnold Vander Haeghen, Gent 03633
Museum Arnoldstein, Arnoldstein 01727
Museum-Art Collection of the National Bank, Athinai 22929
Museum Artha Suaka, Jakarta 24480
Museum Aschenbrenner, Garmisch-Partenkirchen 18541
Museum Asia Africa, Bandung 24441
Museum Asimbugo, Bima 24454
Museum Asmat, Jakarta 24481
Museum Aspern-Essling 1809, Wien 03030
Museum Asserajah El Hasyimiah Palace, Bengkalis 24451
Museum Association of Arizona, Phoenix 55161
Museum Association of Israel, Rishon Le-Zion 54893
Museum Association of Montana, West Glacier 55162
Museum Association of New York, Troy 55163
Museum Association of Newfoundland and Labrador, Saint John's 54742
The Museum at Drexel University, Philadelphia 51855
Museum at Mountain Home, Johnson City 49634
Museum at Prophetstown, Battle Ground 46646
Museum at Southwestern Michigan College, Dowagiac 48173
Museum at the Fashion Institute of Technology, New York 51221
Museum at the Vydunas Culture Centre in Kintai, Kintai 30275
The Museum at Warm Springs, Warm Springs 53817
Museum Atelier Franz Zeh, Heidenreichstein 02061
Museum Auerbach, Auerbach, Vogtland 16608
Museum auf dem Burghof, Springe 21846
Museum auf dem Sachsenberg, Schwerin 21696
Museum auf der Burg, Raron 41747
Museum August Kestner, Hannover 19000
Museum Autovision, Altlußheim 16512
Museum Bad Abbach, Bad Abbach 16665
Museum Bad Arolsen, Bad Arolsen 16668
Museum Bad Arolsen, Bad Arolsen 16669
Museum Bad Arolsen, Bad Arolsen 16670
Museum Bad Arolsen, Bad Arolsen 16671
Museum Bad Orb, Bad Orb 16802
Museum Bärengasse, Zürich 42091
Museum Bahari, Jakarta 24482
Museum Bahnhof Bexbach, Bexbach 17317
Museum Bakkerij Mendels, Middelstum 32555
Museum Balanga, Palangkaraya 24535
Museum Bangga & Lore, Poso 24548
Museum Bargteheide, Bargteheide 16928
Museum Barockschloß Delitzsch, Delitzsch 17812
Museum Barockschloß Rammenau, Rammenau 21175
Museum Bassermannhaus für Musik und Kunst, Mannheim 20168
Museum Batara Guru, Palopo 24539
Museum Batara Guru, Pare-Pare 24541
Museum Batik, Pekalongan 24542
Museum Batik, Yogyakarta 24584
Museum Battle of the Silver Helmets, Halen 03666
Museum Bautzen, Bautzen 16940
Museum Bayerische Metallwarenfabrik (BMF), Nürnberg 20780
Museum Bayerisches Vogtland, Hof 19220
Museum Bebinghehoes, Exloo 32227
Museum Beeckestijn, Velsen Zuid 32882
Museum Beelden aan Zee, Den Haag 32084
Museum - Begegnung der Kulturen, Krenglbach 02235
Museum bei der Kaiserpfalz, Ingelheim am Rhein 19335
Museum beim Markt, Karlsruhe 19443
Museum Bellerive, Zürich 42092
Museum Bergér, Eichstätt 18083
Museum Berggruen, Berlin 17192
Museum Bernhard Brühl, Landsberg, Saalekreis 19798
Museum Bevrijdende Vleugels, Best 31971
Museum Biologi UGM, Yogyakarta 24585
Museum Bischofsheim, Bischofsheim 17362
Museum Bitola, Bitola 30500
Museum.BL, Liestal 41564
Museum Blaafarveværket, Åmot i Modum 33388
Museum Blauer Anger, Bergen, Kreis Traunstein 17011
Museum Blindenwerkstatt Otto Weidt, Berlin 17193
Museum Bockwindmühle, Falkenhain 18290
Museum Bodmann, Bodman-Ludwigshafen 17415
Museum Boerhaave, Leiden 32484
Museum BOjnice, Castle, Bojnice 38201
Museum Bragegården, Vaasa 11422
Museum Bramberg "Wilhelmgut", Bramberg 01793
Museum Brandhorst, München 20464
Museum Brawijaya, Malang 24521
Museum Bredius, Den Haag 32085
Museum Breitenfurt, Breitenfurt 01800
Museum Brigade en Garde Prinses Irene, Oirschot 32611
Museum Briner und Kern, Winterthur 42031
Museum Broeker Veiling, Broek op Langedijk 32012
Museum Bruder Klaus, Sachseln 41786
Museum Bürgstadt, Bürgstadt 17639
Museum Bundo Kandung, Bukittinggi 24462
Museum Buren en Oranje, Buren, Gelderland 32025
Museum Burg Abenberg, Abenberg 16418
Museum Burg Bederkesa, Bad Bederkesa 16676
Museum Burg Brome, Brome 17591
Museum Burg Creuzburg, Creuzburg 17748
Museum Burg Eisenhardt, Belzig 16986
Museum Burg Falkenstein, Falkenstein, Harz 18292
Museum Burg Frankenberg, Aachen 16404
Museum Burg Golling, Golling an der Salzach 01959
Museum Burg Heidenreichstein, Heidenreichstein 02062
Museum Burg Hohnstein, Hohnstein 19249
Museum Burg Hornberg, Neckarzimmern 20566
Museum Burg Kriebstein, Kriebstein 19714
Museum Burg Linn, Krefeld 19706
Museum Burg Mylau, Mylau 20540
Museum Burg Posterstein, Posterstein 21078
Museum Burg Pottenstein, Pottenstein 21114
Museum Burg Querfurt, Querfurt 21155
Museum Burg Ramsdorf, Velen 22214
Museum Burg Ranis, Ranis 21179
Museum Burg Ronneburg, Ronneburg, Hessen 21368
Museum Burg Scharfenstein, Scharfenstein 21533
Museum Burg Stein, Hartenstein 19018
Museum Burg Stickhausen, Detern 17836
Museum Burg Wendelstein, Vacha 22195
Museum Burghalde, Lenzburg 41555
Museum Buurt Spoorweg, Haaksbergen 32291
Museum by the Bay, Lewisporte 06169
Museum Caledon River, Smithfield 38732
Museum Çanakkale, Çanakkale 42748
Museum de Casteelse Poort, Wageningen 32919
Museum Castellum, Mainz-Kastel 20152
Museum Castle of Eppstein, Eppstein 18186
Museum - Castle of the Dukes of Pomerania, Darłowo 34620
Museum Castle Red Rock, Častá 38244
Museum Catharijneconvent, Utrecht 32849
Museum Center, Maysville 50539
Museum Center Heritage of Chukotka, Anadyr 36757
Museum Center of the Bydgoszcz Railway Junction, Bydgoszcz 34580
Museum Centre Vapriikki, Tampere 11354
Museum Ceuclum, Cuijk 32041
Museum Charles Delporte, Damme 03534
Museum Charlotte Zander, Bönnigheim 17426
Museum Charlottenburg-Wilmersdorf, Berlin 17194
Museum Chasa Jaura, Valchava 41972
Museum Cheann A'Loch, Laxay 44649
Museum Chornhuus, Gränichen 41447
Museum Chris Van Niekerk, Boshof 38572
Museum-Church of the Decembrists, Čita 36870
Museum Chuvash School of Simbirsk - Appartement-Museum of I. Jakovlev, Uljanovsk 37906
Museum Classisal Secondary School, Uljanovsk 37907
Museum Clenze Das Blaue Haus, Clenze 17712
Museum Collection, Ajdovščina 38396
Museum Collection, Imotski 08882
Museum Collection, Knić 38104
Museum Collection and Old Pharmacy of the Friars Minor, Dubrovnik 08866
Museum Collection of the Dominican Monastery, Bol 08847
Museum Collection of the Franciscan Monastery, Imotski 08883
Museum Complex Izmailovo, Moskva 37379
Museum Connection Die, Cape Town 38588
Museum Constant Permeke, Jabbeke 03740
Museum Çorum, Çorum 42755
Museum Council, Moskva 55001
Museum, Culture and Exhibition Complex, Dnipropetrovs'k 43036
Museum Daerah Bangkalan, Bangkalan 24446
Museum Daerah Blambangan, Banyuwangi 24450
Museum Daerah Sumenep, Sumenep 24563
Museum Dara Yuanti, Sintang 24560
Museum de 5000 Morgen, Hoogeveen 32405
Museum De Bres, Ruisbroek, Antwerpen 03988
Museum de Burghse Schoole, Burgh-Haamstede 32027
Museum De Cateraars, Groningen 32284
Museum De Cruquius, Cruquius 32039
Museum De Doornboom, Hilvarenbeek 32385
Museum De Eendracht, Anjum 31908
Museum de Géologie Provençale, La Roque-d'Anthéron 13439
Museum de Gevangenpoort, Den Haag 32086
Museum De Ghulden Roos, Roosendaal 32700
Museum De Goudse Glazen - Sint Janskerk, Gouda 32267
Museum De Grutterswinkel, Leeuwarden 32475
Museum De Helpoort, Maastricht 32525
Museum de Historische Drukkerij, Maastricht 32526
Museum De Kijkuit, Merksem 03870
Museum De Kluis, Boekel 31978
Museum De Koloniehof, Frederiksoord 32234
Museum de Koperen Knop, Hardinxveld-Giessendam 32314
Museum de Maurits 1940-1945, Doesburg 32133
Museum De Moriaan, Gouda 32268
Museum De Paviljoens, Almere 31803
Museum De Ronde Venen, Vinkeveen 32889
Museum De Roode Tooren, Doesburg 32134
Museum De Schilpen, Maasland 32521
Museum de Speeltoren, Monnickendam 32559
Museum de Stratemakerstoren, Nijmegen 32589
Museum de Striid tsjin it Wetter, Wommels 32963
Museum de Timmerwerf, De Lier 32048
Museum de Trije Gritenijen, Grou 32290
Museum de Valkhof, Hellendoorn 32356
Museum de Verzamelaar, Zuidhorn 33012
Museum De Vier Quartieren, Oirschot 32612
Museum De Weem, Westeremden 32941
Museum De Wieger, Deurne 32117
Muséum de Zoologie et d'Anthropologie de l'ULB, Bruxelles 03496
Museum de Zwarte Tulp, Lisse 32503
Museum Delfzijl, Delfzijl 32067
Museum der 50er Jahre, Bad Bevensen 16696
Museum der 50er und 60er Jahre, Vaihingen an der Enz 22198
Museum der Alltagskultur - Schloss Waldenbuch, Waldenbuch 22287
Museum der Anthropologie, Zürich 42093
Museum der Arbeit, Hamburg 18946
Museum der Arbeit, Hann. Münden 18987
Museum der Barbarossastadt Gelnhausen, Gelnhausen 18571
Museum der Belgischen Streitkräfte in Deutschland, Soest 21796
Museum der Berliner Bürgerbräu Brauerei, Berlin 17195
Museum der bildenden Künste Leipzig, Leipzig 19921
Museum der Brotkultur, Ulm 22166
Museum der Cistercienserinnen-Abtei Lichtenthal, Baden-Baden 16885
Museum der Deutschen Binnenschifffahrt, Duisburg 18022
Museum der Deutschen Spielzeugindustrie, Neustadt bei Coburg 20675
Museum der deutschen Sprachinselorte bei Brünn, Erbach, Alb-Donau-Kreis 18187
Museum der Donauschwaben, Herbrechtingen 19135
Museum der Freiwilligen Feuerwehr, Heiligenhaus 19106
Museum der Freiwilligen Feuerwehr, Laxenburg 02274
Museum der Göttinger Chemie, Göttingen 18695
Museum der Grafschaft Mark, Altena 16487
Museum der Grafschaft Rantzau, Barmstedt 16929
Museum der Granitindustrie, Haselbachtal 19022
Museum der Havelländischen Malerkolonie, Schwielowsee 21707
Museum der Heimatvertriebenen, Vöcklabruck 02866
Museum der Heimatvertriebenen, Wels 02916

M

Museum der historischen Waschtechnik, Ostbevern 20929
Museum der Katakombenstichting, Valkenburg, Limburg 32865
Museum der Klosterruine mit Lapidarium, Rüeggisberg 41780
Museum der Kommunikationstechnik und Bürogeschichte, Bamberg 16913
Museum der Kulturen Basel, Basel 41169
Museum der Landschaft, Saanen 41784
Museum der Landschaft Eiderstedt, Sankt Peter-Ording 21509
Museum der Landschaft Hasli, Meiringen 41619
Museum der Landsmannschaft Ostpreußen, Gelsenkirchen 18578
Museum der Landtechnik, Knittelfeld 02207
Museum der Marktgemeinde Sankt Johann in Tirol, Sankt Johann in Tirol 02657
Museum der Mechitharisten-Congregation, Wien 03031
Museum der Moderne Salzburg Mönchsberg, Salzburg 02620
Museum der Natur, Gotha 18719
Museum der Natur und Umwelt Cottbus, Cottbus 17733
Museum der Oberzent, Beerfelden 16978
Museum der Photographie, Fischamend 01891
Museum der Porzellan-Manufaktur Meissen, Meißen 20265
Museum der Puppengeschichte - Joyce Merlet, Köln 19618
Museum der Schildbürger, Schildau, Gneisenaustadt 21550
Museum der Schwalm, Schwalmstadt 21662
Museum der Schweizer Kapuzinerprovinz, Sursee 41925
Museum der Seefahrt, Geiselhöring 18557
Museum der Staatlichen Münze Berlin, Berlin 17196
Museum der Stadt Alfeld (Leine), Alfeld 16467
Museum der Stadt Aue, Aue 16605
Museum der Stadt Bad Berleburg, Bad Berleburg 16692
Museum der Stadt Bad Gandersheim, Bad Gandersheim 16738
Museum der Stadt Bad Ischl, Bad Ischl 01755
Museum der Stadt Bad Neuenahr-Ahrweiler, Bad Neuenahr-Ahrweiler 16795
Museum der Stadt Bad Schwartau, Bad Schwartau 16835
Museum der Stadt Bad Staffelstein, Bad Staffelstein 16843
Museum der Stadt Bensheim, Bensheim 16997
Museum der Stadt Bludenz, Bludenz 01790
Museum der Stadt Boppard, Boppard 17475
Museum der Stadt Borna, Borna 17485
Museum der Stadt Butzbach im Solms-Braunfelser Hof, Butzbach 17671
Museum der Stadt Calw, Calw 17680
Museum der Stadt Dommitzsch, Dommitzsch 17889
Museum der Stadt Eberbach, Eberbach 18033
Museum der Stadt Eschborn, Eschborn 18233
Museum der Stadt Ettlingen, Ettlingen 18280
Museum der Stadt Füssen, Füssen 18501
Museum der Stadt Gladbeck, Gladbeck 18650
Museum der Stadt Gronau (Leine), Gronau, Leine 18772
Museum der Stadt Güstrow, Güstrow 18829
Museum der Stadt Hagenow, Hagenow 18862
Museum der Stadt Kapfenberg, Kapfenberg 02151
Museum der Stadt Korneuburg, Korneuburg 02218
Museum der Stadt Kraichtal, Kraichtal 19696
Museum der Stadt Lahnstein im Hexenturm, Lahnstein 19767
Museum der Stadt Lahr, Lahr 19773
Museum der Stadt Langen, Langen, Hessen 19811
Museum der Stadt Lauffen am Neckar, Lauffen am Neckar 19856
Museum der Stadt Lennestadt, Lennestadt 19956
Museum der Stadt Lienz, Lienz 02301
Museum der Stadt Ludwigsfelde, Ludwigsfelde 20050
Museum der Stadt Lünen, Lünen 20103
Museum der Stadt Miltenberg und Porzellansammlung Kreis Dux, Miltenberg 20327
Museum der Stadt Mittweida "Alte Pfarrhäuser", Mittweida 20346
Museum der Stadt Nauen, Nauen 20551
Museum der Stadt Neu-Isenburg Haus zum Löwen, Neu-Isenburg 20582
Museum der Stadt Neustadt an der Weinstraße, Neustadt an der Weinstraße 20672
Museum der Stadt Neustrelitz, Neustrelitz 20684
Museum der Stadt Neutraubling, Neutraubling 20685
Museum der Stadt Parchim, Parchim 20983
Museum der Stadt Pasewalk und Künstlergedenkstätte Paul Holz, Pasewalk 20987
Museum der Stadt Pegau, Pegau 20999
Museum der Stadt Ratingen, Ratingen 21191
Museum der Stadt Schkeuditz, Schkeuditz 21559
Museum der Stadt Schopfheim, Schopfheim 21622
Museum der Stadt Schwaz auf Burg Freundsberg, Schwaz 02726
Museum der Stadt Steyr, Steyr 02770
Museum der Stadt Troisdorf, Troisdorf 22108
Museum der Stadt Villach, Villach 02859
Museum der Stadt Vils, Vils 02863
Museum der Stadt Waiblingen, Waiblingen 22278
Museum der Stadt Weinheim, Weinheim 22431
Museum der Stadt Worms im Andreasstift, Worms 22685
Museum der Stadt Zerbst mit Sammlung Katharina II., Zerbst 22788
Museum der Stadtpfarrkirche St. Xaver - museum sacrum, Leoben 02286
Museum der Stadtwerke, Pirmasens 21042
Museum der Trauerarbeitsplätze, Klagenfurt 02189
Museum der unerhörten Dinge, Berlin 17197
Museum der Universität Tübingen MUT, Tübingen 22121
Museum der Votivkirche, Wien 03032
Museum der Weltkulturen mit Galerie 37, Frankfurt am Main 18381
Museum der Westlausitz Kamenz, Kamenz 19421
Museum der Winterthur-Versicherungen, Winterthur 42032
Museum Dermbach, Dermbach 17825
Museum des Augustiner Chorherrenstifts, Reichersberg 02572
Museum des Blindenwesens, Wien 03033
Museum des Chorherrenstifts Klosterneuburg, Klosterneuburg 02202
Museum des ehemaligen Kreises Stuhm/Westpr., Bremervörde 17580
Museum des Geschichtsvereins Keltisches Noreia, Mühlen 02410
Museum des Haager Landes, Haag in Oberbayern 18845
Museum des Handwerks, Bad Bederkesa 16677
Museum des Heimatkundlichen Vereins, Scheßlitz 21545
Museum des Historischen Vereins Freising, Freising 18451
Museum des Kreises Plön mit norddeutscher Glassammlung, Plön 21065
Museum des Landes Glarus, Näfels 41672
Museum des Nötscher Kreises, Nötsch im Gailtal 02451
Museum des Ohrekreises, Wolmirstedt 22676
Muséum des Sciences Naturelles de Belgique, Bruxelles 03497
Museum des Stiftes Admont, Admont 01693
Museum des Teltow, Zossen 22820
Museum des Todesmarsches im Belower Wald, Wittstock/Dosse 22630
Muséum des Volcans, Aurillac 11756
Museum des Wiener Männergesang-Vereins, Wien 03034
Museum des Zisterzienserstifts mit Margret Bilger-Galerie, Schlierbach 02707
Museum Development Service, Ipswich 55075
Museum Dewantara Kirti Griya, Yogyakarta 24586
Museum Dharma Wiratama, Yogyakarta 24587
Muséum d'Histoire Naturell, Auxerre 11778
Muséum d'Histoire Naturelle, Autun 11766
Muséum d'Histoire Naturelle, Bagnères-de-Bigorre 11821
Muséum d'Histoire Naturelle, Blois 12036
Muséum d'Histoire Naturelle, Chartres 12426
Muséum d'Histoire-Naturelle, Cherbourg-Octeville 12506
Muséum d'Histoire Naturelle, Dijon 12744
Muséum d'Histoire Naturelle, Gras 13108
Muséum d'Histoire Naturelle, Grenoble 13139
Muséum d'Histoire Naturelle, Le Havre 13599
Muséum d'Histoire Naturelle, Lyon 13903
Muséum d'Histoire Naturelle, Marseille 14011
Muséum d'Histoire Naturelle, Neuchâtel 41682
Muséum d'Histoire Naturelle, Nice 14440
Muséum d'Histoire Naturelle, Perpignan 14793
Muséum d'Histoire Naturelle, Rouen 15122
Muséum d'Histoire Naturelle, Saint-Denis 36092
Muséum d'Histoire Naturelle, Troyes 16002
Muséum d'Histoire Naturelle de Bordeaux, Bordeaux 12074
Muséum d'Histoire Naturelle de la Ville de Bayonne, Bayonne 11880
Muséum d'Histoire Naturelle de la Ville de Genève, Genève 41427
Muséum d'Histoire Naturelle de Toulouse, Toulouse 15946
Muséum d'Histoire Naturelle de Tours, Tours 15971
Museum Dhondt-Dhaenens, Deurle 03546
Museum Dingolfing-Herzogsburg, Dingolfing 17865
Museum-Diorama "Orel Offensive Operation in 1943", Orël 37509
Museum Diponegoro, Magelang 24518
Museum Dirgantara Mandala, Yogyakarta 24588
Museum DKM, Duisburg 18023
Museum Documentation Centre, Zagreb 09009
Museum Donkmeer, Berlare 03362
Museum Dorestad, Wijk bij Duurstede 32948
Museum Dorpsbehoud Papendrecht, Papendrecht 32666
Museum Dr. Guislain, Gent 03634
Museum du Coquillage, Les Sables-d'Olonne 13716
Museum du Parc Tsimbazaza, Antananarivo 30538
Museum Echt, Echt 32165
Museum Eckernförde, Eckernförde 18053
Museum Education Roundtable, Washington 55164
Museum ehemalige Klöppelschule Tiefenbach, Tiefenbach, Oberpfalz 22060
Museum Ehingen im Heilig-Geist-Spital, Ehingen 18068
Museum Eisenhüttenwerk Peitz, Peitz 21005
Museum Elbinsel Wilhelmsburg, Hamburg 18947
Museum Elisabeth Weeshuis, Culemborg 32043
Museum en oud Kempisch Dorp, Heist-op-den-Berg 03692
Museum Engiadine Bassa, Scuol 41852
Museum Enschedé, Haarlem 32298
Museum Ephraim-Palais, Berlin 17198
Museum Erding, Erding 18194
Museum Erzgebirgische Volkskunst "Buntes Holz" im Postgut am Altmarkt, Hohenstein-Ernstthal 19245
Museum Eschenz, Eschenz 41360
Museum-estate of Lev Tolstoy "Yasnaya Polyana", Jasnaja Poljana (Tula) 36975
Museum-Estate of Literature and Art Priyutino, Vsevoložsk 37979
Museum-Estate of N.A. Durova, Elabuga 36906
Museum-Estate of N.K. Roerich, Izvara 36958
Museum-Estate of the Poet Dmitry Venevitinov, Novoživotinnoe 37492
Museum-Estate of the Writer S.T. Aksakov, Aksakovo 36750
Museum-Estate Ostafjevo "Russian Parnas", Ostafjevo 37518
Museum-Estate Tvardovsky in Zagorye, Zagorje 37984
Museum Esterhazy Castle, Fertőd 23557
Museum Ethnografic Park in Wdzydze, Wdzydze Kiszewskie 35294
Museum Ethnographers Group, Exeter 55076
Museum Europäischer Kulturen, Berlin 17199
Museum Fahrzeug-Technik-Luftfahrt, Bad Ischl 01756
Museum Fahrzeug - Technik - Luftfahrt, Lauffen 02267
Museum Falak-ol-Aflak, Khorram-Abaad 24623
Museum Fatahillah, Jakarta 24483
Museum Félix de Boeck, Drogenbos 03563
Museum Festung Dresden-Kasematten, Dresden 17962
Museum Fethiye, Fethiye 42784
Museum Feuerbachhaus Speyer, Speyer 21836
Museum Fischerhaus, Öhningen 20862
Museum 't Fiskershúske, Moddergat 32558
Museum Flederwisch, Furth im Wald 18516
Museum Flehite Amersfoort, Amersfoort 31814
museum FLUXUS gemeinnützige GmbH, Potsdam 21096
Museum Folkwang, Essen 18263
Museum for African Art, Long Island City 50184
Museum for animal lifestock 'Julius Kühn', Halle, Saale 18901
Museum for Beduin Culture, Kibbutz Lahav 25048
Museum for Children, Manila 34447
Museum for Communication, Frankfurt am Main 18383
Museum for Contemporary Art, Gent 03644
Museum for Costume and Lace, Bruxelles 03483
Museum for Diaconical Work, Stockholm 40891
Museum for epilepsy and the history of epilepsy, Kehl 19495
Museum for Figurative Art "De Buitenplaats", Eelde 32174
Museum for Handweaving, Bad Münstereifel 16788
Museum for Historic Military Technology, Röthenbach an der Pegnitz 21359
Museum for History of Aviation, Laatzen 19755
Museum for Industrial Culture, Nürnberg 20784
Museum for Local Culture, Beipu 42147
Museum for National History and Archaeology, Constanța 36305
Museum for Photography, Amsterdam 31849
Museum for sepulchral culture, Kassel 19470
Museum for the Victims of Political Repression, Ulaanbaatar 31625
Museum for time - Palatina Tower Clocks Museum, Rockenhausen 21340
Museum for Wood Culture, Muraoka 28946
Museum Ford and Bulgaria, Sofia 05183
Museum Fort Count Mirasol, Vieques 36082
Museum Fort Kijkduin, Den Helder 32111
Museum Fort Liezele, Puurs 03956
Museum Forum Arenacum Rindern, Kleve 19566
Museum Forum der Völker, Werl 22490
Museum Foundation, Lübeck 54841
Museum Frankomarchia, Frankenmarkt 01899
Museum Frankonian Switzerland, Pottenstein 21113
museum franz gertsch, Burgdorf 41275
Museum Franz Schubert und sein Freundeskreis, Atzenbrugg 01736
Museum Freriks, Winterswijk 32957
Museum Frerikshuus, Aalten 31785
Museum Freudenthaler Glassammlung, Weißenkirchen im Attergau 02903
Museum-Freundschaften in der Kulturkunststatt Prora, Prora 21133
Museum Frieder Burda, Baden-Baden 16886
Museum Friedhof Ohlsdorf, Hamburg 18948
Museum Friedrichshagener Dichterkreis, Berlin 17200
Museum Friesenheim, Ludwigshafen am Rhein 20053
Museum Fritz von Graevenitz, Stuttgart 21954
Museum für Abgüsse Klassischer Bildwerke, München 20465
Museum für Angewandte Kunst, Frankfurt am Main 18382
Museum für Angewandte Kunst, Köln 19619
Museum für Angewandte Kunst Gera, Gera 18594
Museum für Antike Schiffahrt des Römisch-Germanischen Zentralmuseums, Mainz 20148
Museum für Archäologie des Kantons Thurgau, Frauenfeld 41378
Museum für Archäologie und Volkskunde, Kösching 19665
Museum für Archäologie Westfälisches Außenstelle, Olpe 20904
Museum für Asiatische Kunst, Berlin 17201
Museum für bäuerliche Arbeitsgeräte, Imst 02109
Museum für bäuerliche Arbeitsgeräte des Bezirks Oberfranken, Bayreuth 16961
Museum für bäuerliche und sakrale Kunst, Ruhpolding 21448
Museum für bäuerliches Handwerk und Kultur, Wilhelmsdorf, Kreis Ravensburg 22573
Museum für bergmännische Volkskunst Schneeberg, Schneeberg 21591
Museum für Beschirrung und Besattelung, Hufbeschlag und Veterinär-Orthopädie, Wien 03035
Museum für bildende Kunst im Landkreis Neu-Ulm, Nersingen 20574
Museum für Billard- und Kaffeehauskultur, Wien 03036
Museum für Bodenschätzung, Eickendorf 18087
Museum für brandenburgische Kirchen- und Kulturgeschichte, Ziesar 22802
Museum für das Fürstentum Lüneburg, Lüneburg 20100
Museum für Design der 60er & 70er Jahre im PopDom, Köln 19620
Museum für deutsche Fernsehgeschichte, Wiesbaden 22556
Museum für die Archäologie des Eiszeitalters, Neuwied 20687
Museum für dörfliche Alltagskultur Rubenheim, Gersheim 18615
Museum für Druck, Mosbach 20391
Museum für Druckgrafik, Rankweil 02559
Museum für Druckkunst Leipzig, Leipzig 19922
Museum für ehemals Großherzoglich Badische Schulen, Offenburg 20886
Museum für Eisenkunstguß, Hirzenhain 19200
Museum für Energiegeschichte(n), Hannover 19001
Museum für Fabergé Kunst, Baden-Baden 16887
Museum für Fotografie - Sammlung Fotografie, Kunstbibliothek, Berlin 17202
Museum für Frühgeschichte des Landes Niederösterreich, Traismauer 02818
Museum für Frühislamische Kunst, Bamberg 16914
Museum für Gegenwartskunst, Basel 41170
Museum für Gegenwartskunst Siegen, Siegen 21755
Museum für Geologie und Paläontologie, Tübingen 22122
Museum für Gerberei- und Stadtgechichte, Hirschberg, Saale-Orla-Kreis 19198
Museum für Gestaltung Basel, Basel 41171
Museum für Gestaltung Zürich, Zürich 42094
Museum für Gestaltung Zürich - Sammlungen für Design, Grafik, Plakat und Kunstgewerbe, Zürich 42095
Museum für Glaskunst, Lauscha 19861
Museum für Hamburgische Geschichte, Hamburg 18949
Museum für Haustierkunde Julius Kühn, Halle, Saale 18901
Museum für Heimatgeschichte und Nordpfalzgalerie, Alsenz 16480
Museum für historische Schreib- und Rechenmaschinen, Schaan 30193
Museum für historische Wehrtechnik, Röthenbach an der Pegnitz 21359
Museum für Höhlenkunde, Laichingen 19779
Museum für Holzhandwerk, Sinzig 21788
Museum für Islamische Kunst im Pergamonmuseum, Berlin 17203
Museum für Jagdtier- und Vogelkunde des Erzgebirges, Augustusburg 16641
Museum für Kaffeetechnik, Emmerich am Rhein 18150
Museum für Kinematographie, Vollbüttel 22257
Museum für klösterliche Kultur, Weingarten, Kreis Ravensburg 22427

Museum für Kloster- und Heimatgeschichte, Harsefeld 19016
Museum für Klosterkultur, Weingarten, Kreis Ravensburg 22428
Museum für Kommunikation, Bern 41212
Museum für Kommunikation, Hamburg 18950
Museum für Kommunikation Berlin, Berlin 17204
Museum für Kommunikation Frankfurt, Frankfurt am Main 18383
Museum für Kommunikation Nürnberg (Verkehrsmuseum), Nürnberg 20781
Museum für Konkrete Kunst, Ingolstadt 19344
Museum für Krippen, Eß- und Trinkkultur und Brauereigeschichte, Burghausen 17654
Museum für Kunst in Steatit, Frankfurt am Main 18384
Museum für Kunst und Geschichte, Hüfingen 19287
Museum für Kunst und Gewerbe Hamburg, Hamburg 18951
Museum für Kunst und Kulturgeschichte Dortmund, Dortmund 17925
Museum für Kunst und Technik des 19. Jahrhunderts, Baden-Baden 16888
Museum für Kutschen, Chaisen, Karren, Heidenheim an der Brenz 19091
Museum für Lackkunst, Münster 20516
Museum für ländliches Kulturgut im Dorfstadel Leupolz, Wangen im Allgäu 22326
Museum für ländliches Kulturgut, Landwirtschaft, Forsten und Jagd, Ulrichstein 22171
Museum für Landtechnik und Landarbeit, Emmerthal 18154
Museum für Landwirtschaftliche Maschinen, Erbach, Alb-Donau-Kreis 18188
Museum für Licht und Beleuchtung, Arnsberg 16572
Museum für Literatur am Oberrhein, Karlsruhe 19444
Museum für Luftfahrt und Technik, Wernigerode 22498
Museum für Magie und Hexenverfolgung, Penzlin 21009
Museum für Medizin-Meteorologie, Zwettl 03135
Museum für Militär- und Zeitgeschichte, Kolitzheim 19675
Museum für Militär- und Zeitgeschichte. e. V, Stammheim/Gde. Kolitzheim 21862
Museum für Mission- und Völkerkunde, Sankt Wendel 21513
Museum für Modelleisenbahnen und Modellautos, Steißlingen 21896
Museum für Moderne Kunst Frankfurt am Main, Frankfurt am Main 18385
Museum für modernes Glas, Öhringen 20865
Museum für Mühlenbautechnik, Syrau 21997
Museum für Musikautomaten, Seewen 41857
Museum für Musikinstrumente, Kürnbach 19740
Museum für Nachrichten- und Hochfrequenztechnik, Aachen 16405
Museum für Natur und Umwelt, Lübeck 20079
Museum für Naturkunde, Berlin 17205
Museum für Naturkunde, Chemnitz 17700
Museum für Naturkunde, Dortmund 17926
Museum für Naturkunde, Magdeburg 20127
Museum für Naturkunde der Stadt Gera, Gera 18595
Museum für Naturkunde und Vorgeschichte, Dessau 17830
Museum für Neue Kunst, Freiburg im Breisgau 18433
Museum für Niederrheinische Sakralkunst, Kempen 19508
Museum für Nummernschilder, Verkehrs- und Zulassungsgeschichte, Großolbersdorf 18799
Museum für Österreichische Zeitgeschichte, Scharnstein 02699
Museum für Optik, Meßtechnik und mathematische Geräte, Nothweiler 20750
Museum für Ortsgeschichte, Mutterstadt 20538
Museum für Ortsgeschichte und Weinbau, Zellertal 22786
Museum für Ostasiatische Kunst, Köln 19621
Museum für Outsiderkunst, Schleswig 21567
Museum für Photographie, Braunschweig 17527
Museum für Puppen, Bären und Spielzeug, Solingen 21804
Museum für Rechtsgeschichte, Pöggstall 02510
Museum für Regionalgeschichte und Volkskunde, Gotha 18720
Museum für Sächsische Volkskunst mit Puppentheatersammlung, Dresden 17963
Museum für Sakrale Kunst und Liturgie, Heidelberg 19077
Museum für schlesische Landeskunde im Haus Schlesien, Königswinter 19659
Museum für Schnitzkunst und Kulturgeschichte, Oberammergau 20800
Museum für Seenotrettungsgeräte, Neuharlingersiel 20616
Museum für Sepulkralkultur, Kassel 19470
Museum für Stadtgeschichte, Alpirsbach 16478
Museum für Stadtgeschichte, Breisach am Rhein 17536
Museum für Stadtgeschichte, Höhr-Grenzhausen 19215
Museum für Stadtgeschichte, Neuenburg am Rhein 20605
Museum für Stadtgeschichte Dessau, Dessau 17831
Museum für Stadtgeschichte im Adam-und-Eva-Haus, Paderborn 20972
Museum für Stadtgeschichte, Naturkunde und Völkerkunde Julius Riemer, Lutherstadt Wittenberg 20117
Museum für Stadtgeschichte und Volkskunde, Heppenheim 19129
Museum für Thüringer Volkskunde Erfurt, Erfurt 18207
Museum für Uhren und mechanische Musikinstrumente, Oberhofen am Thunersee 41696
Museum für Uhren und Schmuck, Frankfurt am Main 18386
Museum für Unterhaltungskunst, Wien 03037
Museum für Unterwasserarchäologie, Sassnitz 21519
Museum für Unterwasserarchäologie, Sassnitz 21520
Museum für Ur- und Frühgeschichte, Eichstätt 18084
Museum für Ur- und Frühgeschichte, Stillfried 02780
Museum für Ur- und Frühgeschichte, Wieselburg an der Erlauf 03101
Museum für Ur- und Frühgeschichte Thüringens, Weimar 22410
Museum für Urgeschichte, Koblach 02209
Museum für Urgeschichte(n), Zug 42126
Museum für verfolgte Kunst-Israel, Ashdod 24929
Museum für visuelle Kommunikation, Köln 19622
Museum für Völkerkunde, Burgdorf 41276
Museum für Völkerkunde, Wien 03038
Museum für Völkerkunde Dresden, Dresden 17964
Museum für Völkerkunde Hamburg, Hamburg 18952
Museum für Volkskultur, Spittal an der Drau 02752
Museum für Volkskunst, Haslach 02056
Museum für Vor- und Frühgeschichte, Berlin 17206
Museum für Vor- und Frühgeschichte, Gunzenhausen 18840
Museum für Vor- und Frühgeschichte, Langenau 19820
Museum für Vor- und Frühgeschichte, Saarbrücken 21457
Museum für Vor- und Frühgeschichte sowie Stadtgeschichte, Egeln 18059
Museum für Waage und Gewicht, Balingen 16901
Museum für Waffentechnik, Ausrüstung, Auszeichnungswesen, Vilshofen 22245
Museum für Wattenfischerei, Wremen 22702
Museum für Weinbau und Stadtgeschichte, Edenkoben 18055
Museum für Weinkultur, Deidesheim 17809
Museum für Westfälische Literatur, Oelde 20869
Museum für Wohnkultur des Historismus und des Jugendstils, Hilterfingen 41482
Museum für Zeit - Pfälzisches Turmuhrenmuseum, Rockenhausen 21340
Museum für Zeitgenössische Glasmalerei, Langen, Hessen 19812
Museum Furthmühle, Pram 02523
Museum Garching an der Alz, Garching an der Alz 18536
Museum Gayo, Aceh Tengah 24432
Museum Gedong Kirtya, Singaraja 24559
Museum Gedung Arca, Gianyar 24470
Museum Gedung Joang '45, Jakarta 24484
Museum Geelvinck Hinlopen Huis, Amsterdam 31861
Museum gegenstandsfreier Kunst, Otterndorf 20951
Museum Geiserschmiede, Bühlertal 17633
Museum-Gemaal de Hoogte, Nieuwolda 32584
Museum Gensungen, Felsberg 18309
Museum Geologi, Bandung 24442
Museum Geologie/Paläontologie, Heidelberg 19078
Museum Georg Schäfer, Schweinfurt 21685
Museum Gesigt van't Dok, Hellevoetsluis 32360
Museum GGZ-Drenthe, Assen 31938
Museum Giersch, Frankfurt am Main 18387
Museum Goa Bala Lompoa, Sungguminasa 24564
Museum Goch, Goch 18666
Museum Godshuis Belle, Ieper 03729
Museum Göltzsch, Rodewisch 21346
Museum Goldener Steig, Waldkirchen 22298
Museum Goldenes Dachl, Innsbruck 02124
Museum Gottfried Silbermann, Frauenstein 18413
Museum Griesheim, Griesheim 18759
Museum Grootseminarie, Brugge 03421
Museum Grossauheim, Hanau 18980
Museum Großklein, Großklein 02003
Museum Großkrotzenburg, Großkrotzenburg 18796
Museum Gülden Creutz, Leinefelde-Worbis 19880
Museum Gugging, Maria Gugging 02350
Museum Gula, Klaten 24515
Museum Gunnar-Wester-Haus, Schweinfurt 21686
Museum Gust de Smet, Deurle 03547
Museum Gustavianum, Uppsala 41009
Museum Gut Grottenhof, Leibnitz 02280
Museum H.R. Giger, Gruyères 41461
Museum Hacibektaş, Hacibektaş 42792
Museum Haldensleben, Haldensleben 18883
Museum Halle, Halle / Westfalen 18910
Museum Hameln, Hameln 18967
Museum Haus Cajeth, Heidelberg 19079
Museum Haus Esters, Krefeld 19707
Museum Haus Kast, Gaggenau 18527
Museum Haus Kasuya, Yokosuka 29621
Museum Haus Lange, Krefeld 19708
Museum Haus Löwenberg, Gengenbach 18583
Museum Haus Ludwig für Kunstausstellungen, Saarlouis 21463
Museum Haus Martfeld, Schwelm 21690
Museum Havezate Mensinge, Roden 32695
Museum Headquarter of V.K. Bluecher, Tjumen 37850
Museum Headquarters of the Russian Army 1877-1878, Pordim 05135
Museum Hedendaagse Glaskunst, Vledder 32894
Museum Hedendaagse Grafiek, Vledder 32895
Museum Heimathaus Irmintraut, Ottersberg 20953
Museum Heimathaus Münsterland, Telgte 22023
Museum Heineanum, Halberstadt 18878
Museum Henriette Polak, Zutphen 33019
Museum Heraldiek Benelux, Temse 04054
Museum Herbarium Bogoriensis, Bogor 24456
Museum - Herberg de Ar, Westerbork 32939
Museum Herisau, Herisau 41481
Museum Herman De Cuyper, Blaasveld 03371
Museum Hermann der Lahme, Altshausen 16526
Museum Herzogsburg, Braunau am Inn 01794
Museum Hessenstube, Baunatal 16937
Museum Het Alaam, Genk 03620
Museum Het Gebied van Staden, Staden 04047
Museum Het Groot Graffel, Warnsveld 32923
Museum Het Oude Raadhuis, Megen 32547
Museum Het Petershuis, Gennep 32246
Museum Het Prinsenhof, Delft 32062
Museum Het Rembrandthuis, Amsterdam 31862
Museum Het Schip, Amsterdam 31863
Museum het Tramstation, Schipluiden 32761
Museum Het Valkhof, Nijmegen 32590
Museum Het Verscholen Dorp, Vierhouten 32888
Museum Heylshof, Worms 22686
Museum Hindeloopen, Hindeloopen 32391
Museum Historische Verzameling Nederlandse Politie, Zaandam 32977
Museum Historischer Bierkrüge, Hainfeld 02040
Museum Höxter-Corvey, Höxter 19219
Museum Hoffmann'sche Sammlung, Kohren-Sahlis 19672
Museum Hofmühle, Immenstadt im Allgäu 19328
Museum Hohenau an der March, Hohenau an der March 02090
Museum Hohenzollern in Franken, Kulmbach 19749
Museum Holstentor, Lübeck 20080
Museum House, Csongrád 23512
Museum Huelsmann, Bielefeld 17342
Museum Hüsli-Sammlung Schwarzwälder Volkskunst, Grafenhausen 18730
Museum Humanum, Waldkirchen an der Thaya 02891
Museum Huta Bolon Simanindo, Pematangsiantar 24544
Museum Huta Bolon Simanindo, Samosir 24551
Museum Huthaus Einigkeit, Brand-Erbisdorf 17504
Museum "Huthaus" Zinnwald, Altenberg 16495
Museum Huthaus Zinnwald, Altenberg 16495
Museum Ianchelevici, Goudriaan 32272
Museum Idar-Oberstein, Idar-Oberstein 19309
Museum Illingen, Illingen, Kreis Neunkirchen 19318
Museum im Adler, Benningen am Neckar 16996
Museum im alten Brauhaus, Homberg (Ohm) 19265
Museum im alten Pfarrhaus, Wiedensahl 22539
Museum im Alten Rathaus, Bad Brückenau 16705
Museum im Alten Rathaus, Mönchberg im Spessart 20355
Museum im Alten Rathaus, Neckargemünd 20562
Museum im Alten Rathaus, Waghäusel 22270
Museum im Alten Schloß, Altensteig 16503
Museum im Alten Schloß, Neckarbischofsheim 20561
Museum im alten Zeughaus, Bad Radkersburg 01764
Museum im Ammes Haus, Hatzfeld 19044
Museum im Amtshaus, Grünsfeld 18816
Museum im Amtshausschüpfla, Erlangen 18223
Museum im Astorhaus, Walldorf 22308
Museum im Ballhaus, Imst 02110
Museum im Bellpark, Kriens 41518
Museum im Bierlinghaus, Bad Bayersoien 16675
Museum im Blaahaus am Unteren Römerweg, Kiefersfelden 19526
Museum im Bock, Leutkirch im Allgäu 19973
Museum im Boyneburgischen Schloß, Sontra-Wichmannshausen 21824
Museum im Bürgerspital, Drosendorf an der Thaya 01830
Museum im Deutschhof, Heilbronn 19099
Museum Im Dorf, Reutlingen 21285
Museum im Dorf, Wittingen 22622
Museum im ehemaligen Augustiner-Chorherrenstift, Herrenchiemsee 19159
Museum im 'Fressenden Haus' - Burgkasten Weißenstein, Regen 21213
Museum im Frey-Haus, Brandenburg an der Havel 17511
Museum im Fürstenstöckl, Ebenau 01833
Museum im Goetz-Haus, Leipzig 19923
Museum im Goldschmiedehaus, Ahlen 16439
Museum im Gotischen Haus, Bad Homburg v.d.Höhe 16751
Museum im Graulturm, Gau-Algesheim 18546
Museum im Greuterhof, Islikon 41495
Museum im Grünen Haus, Reutte 02583
Museum im Heimathaus, Traunstein 22078
Museum im Herrenhaus, Hausach 19048
Museum im Hexenturm, Markdorf 20202
Museum im Hirsch, Remshalden 21271
Museum im Hollerhaus, Dietfurt an der Altmühl 17855
Museum im Jost-Sigristen-Haus, Ernen 41358
Museum im Kastenturm, Bischofshofen 01786
Museum im Kinzigtorturm, Gengenbach 18584
Museum im Kleihues-Bau, Kornwestheim 19690
Museum im Klösterle, Peiting 21004
Museum im Kloster Grafschaft, Schmallenberg 21584
Museum im Kloster - Museum des Landkreises Osnabrück, Bersenbrück 17308
Museum im Koffer, Nürnberg 20782
Museum im Kornhaus, Altheim 16505
Museum im Kornhaus, Rorschach 41776
Museum im Kräuterkasten, Albstadt 16455
Museum im Kreuzgang, Landshut 19804
Museum im Kulturspeicher, Würzburg 22718
Museum im Lagerhaus, Sankt Gallen 41807
Museum im Ledererhaus, Purgstall an der Erlauf 02540
Museum im Leuchttum, Putgarten 21143
Museum im Malhaus, Wasserburg, Bodensee 22347
Museum im Marstall, Winsen, Luhe 22606
Museum im Mendelssohn-Haus, Leipzig 19924
Museum im Mesnerhaus, Pfaffenhofen an der Ilm 21020
Museum im Mönchenkloster, Jüterbog 19407
Museum im Palais Kulturhistorische Sammlung, Graz 01979
Museum im Pflegschloss, Schrobenhausen 21640
Museum im Porsche Werk Leipzig, Leipzig 19925
Museum im Prediger, Schwäbisch Gmünd 21649
Museum im Rathaus, Gleisdorf 01934
Museum im Rathaus Möhringen, Tuttlingen 22137
Museum im Ritterhaus, Offenburg 20887
Museum im Ritterhaus, Osterode am Harz 20937
Museum im Römerbad, Heidenheim an der Brenz 19092
Museum im Schäferhaus, Wahlsburg 22273
Museum im Schafstall, Neuenstadt am Kocher 20612
Museum im Schlößle, Freiberg am Neckar 18419
Museum im Schlössle, Kressbronn am Bodensee 19712
Museum im Schloss, Bernstadt, Alb-Donau-Kreis 17306
Museum im Schloß, Frohburg 18487
Museum im Schloss, Fürstenberg, Kreis Holzminden 18491
Museum im Schloß Bad Pyrmont, Bad Pyrmont 16804
Museum im Schloß Lützen, Lützen 20106
Museum im Schloss Wolfenbüttel, Wolfenbüttel 22654
Museum im Schmidt-Haus, Nabburg 20541
Museum im Schottenstift, Wien 03039
Museum im Schulhaus, Niederrohrdorf 41686
Museum im Schweizer Hof, Bretten 17585
Museum im Seelhaus, Bopfingen 17472
Museum im Stadtturm, Groitzsch 18771
Museum im Stadtturm, Waidhofen an der Ybbs 02889
Museum im Stasi-Bunker, Machern 20119
Museum im Steinhaus, Bad Wimpfen 16867
Museum im Steintor, Anklam 16550
Museum im Steintorturm, Brandenburg an der Havel 17512
Museum im Stern, Warburg 22332
Museum im Stockalperschloß, Brig 41256
Museum im Strumpferhaus, Oberstaufen 20841
Museum im Tabor, Feldbach 01878
Museum im Trafohaus, Leipzig 19926
Museum im Troadkostn zu Giem, Feldbach 01879
Museum im Turm, Neuleiningen 20628
Museum im Turmhof, Steckborn 41908
Museum im Umweltschutz-Informationszentrum Oberfranken mit Kinder-Erlebnis-Museum, Bayreuth 16962
Museum im Vorderen Schloss, Mühlheim an der Donau 20408
Museum im Wächterhaus, Nürnberg 20783
Museum im Wasserkraftwerk, Wahlsburg 22274
Museum im Wasserschloß Kottingbrunn, Kottingbrunn 02220

Museum im Wasserwerk, Berlin 17207
Museum im Wehener Schloss, Taunusstein 22015
Museum im Westrich, Ramstein-Miesenbach 21176
Museum im Wetterbergschen Adelshof, Bad Münder am Deister 16785
Museum im Wittelsbacher Schloss Friedberg, Friedberg, Kreis Aichach-Friedberg 18469
Museum im Zendenrathaus, Ernen 41359
Museum im Zeughaus, Innsbruck 02125
Museum im Zeughaus, Schaffhausen 41829
Museum im Zeughaus, Vechta 22207
Museum im Zisterzienserkloster Lehnin, Kloster Lehnin 19577
Museum im Zwinger, Goslar 18713
Museum in Bruntal, Bruntál 09445
Museum in Commemoration of Pēteris Upītis, Dobele 30001
Museum in de Stiftskerk, Thorn 32822
Museum in den Berglen, Berglen 17023
Museum in den Halven Maen, Rockanje 32692
Museum in der Adler-Apotheke, Eberswalde 18047
Museum in der Alten Schule, Efringen-Kirchen 18058
Museum in der Bruderkirche, Altheim 16506
Museum in der Burg, Bad Driburg 16712
Museum in der Burg, Coppenbrügge 17725
Museum in der Fronfeste, Neumarkt am Wallersee 02436
Museum in der Kaiserpfalz, Paderborn 20973
Museum in der Majolika, Karlsruhe 19445
Museum in der Osterburg, Weida 22365
Museum in der Präparandenschule, Höchberg 19208
Museum in der Remise, Bad Harzburg 16745
Museum in der Scheuergasse, Zwingenberg, Kreis Bergstraße 22840
Museum in der Stadthalle Stern (Interim), Riesa 21329
Museum in Docklands, London 44921
Museum in Progress, Wien 03040
Museum in shoemakeryard, Naila 20546
Museum In 't Houten Huis, De Rijp 32049
Museum in The Bled Castle, Bled 38400
Museum in the Cistercian monastery Lehnin, Kloster Lehnin 19577
Museum in the Community, Hurricane 49452
Museum in the little Shed, Erlangen 18223
The Museum in the Park, Stroud 45794
Museum in the State Nature Reservation of Cepkeliai - Tadas Ivanauskas Zoology Museum, Marcinkonys 30306
Museum in the State Nature Reservation of Zuvintas - Tadas Ivanauskas Zoology Museum, Aleknonių 30202
Museum in the World of Fairy Tales, Smolensk 37774
Museum Indonesia, Jakarta 24485
Museum Industriekultur, Osnabrück 20926
Museum Industriekultur mit Motorradmuseum, Nürnberg 20784
Museum Innviertler Volkskundehaus und Galerie der Stadt Ried, Ried im Innkreis 02584
Museum Insel Hombroich, Neuss 20659
Museum-Institute of the Armenian Genocide, Yerevan 00747
Museum Internationales Baum-Archiv, Winterthur 42033
Museum Istana Mangkunegaran, Surakarta 24568
Museum Istana Sultan Ternate, North Malaku 24532
Museum Istant Siak Sri Indapura, Bengkalis 24452
Museum It Kokelhûs van Jan en Sjut, Earnewald 32164
Museum Jacobs van den Hof, Amersfoort 31815
Museum Jacques van Mourik, Mook 32561
Museum Jagd und Wild auf Burg Falkenstein, Falkenstein, Kreis Cham 18295
Museum Jagdschloß Kranichstein, Darmstadt 17793
Museum Jakob Smits, Mol 03876
Museum Jamu Nyonya Menir, Semarang 24555
Museum Jan Boon, De Rijp 32050
Museum Jan Cunen, Oss 32640
Museum Jan Heestershuis, Schijndel 32757
Museum Jan Karlsgården och Vita Björn, Sund 11334
Museum Jan Lont, Hippolytushoef 32393
Museum Jan Vaerten, Beerse 03355
Museum Jan van der Togt, Amstelveen 31822
Museum Jannink, Enschede 32211
Museum Jerusalem Panorama Kreuzigung Christi, Altötting 16519
Museum "Jews in Latvia", Riga 30097
Museum Joang '45, Medan 24525
Museum Joang '45, Surabaya 24565
Museum Josephine Baker, Castelnaud-la-Chapelle 12314
Museum Joure, Joure 32432
Museum Judengasse, Frankfurt am Main 18388
Museum Judenplatz Wien, Wien 03041
Museum jüdischer Geschichte und Kultur, Aschaffenburg 16587
Museum Junge Kunst, Frankfurt (Oder) 18405
Museum Junkerhaus, Lemgo 19948
Museum "Kaffeemühle", Wolgast 22673
Museum Kalimantan Tengah, Palangkaraya 24536
Museum Kalt-Heiß, Deckenpfronn 17803
Museum Kameleon Dorp en Museum Hans Brinker, Terherne 32815
Museum Kampa, Praha 09764
Museum Kampa - Jan and Meda Mládek Foundation, Praha 09764
Museum Karl-Marx-Haus, Trier 22100
Museum Kartause Astheim, Volkach 22256
Museum Kartini, Jepara 24511
Museum Kartini Rembang, Rembang 24550
Museum Kasper, Norrtälje 40775
Museum Kasteel Wijchen, Wijchen 32946
Museum Katendrecht, Rotterdam 32713
Museum Kateri Tekakwitha, Kahnawake 06034
Museum Katharinenhof mit Mühlenturm und Stadtscheune, Kranenburg 19699
Museum Kebangkitan Nasional, Jakarta 24486
Museum Kebudayaan dan Kemajuan Asmat, Irian Jaya 24473
Museum Kebun Raya Bogor, Bogor 24457
Museum Kedokteran, Jakarta 24487
Museum Kellinghusen, Kellinghusen 19502
Museum Kempenland, Eindhoven 32191
Museum Kempenland, Lommel 03829
Museum Kendil Riau, Tanjungpandan 24575
Museum Kennemerland, Beverwijk 31972
Museum Kepolisian Negara Republik Indonesia, Jakarta 24488
Museum Kereta, Ambarawa 24435
Museum Kerkelijke Kunst, Workum 32965
Museum Kesselhaus Herzberge, Berlin 17208
Museum Kierling, Kierling 02162
Museum Kinderwereld, Roden 32696
Museum Kindheit und Jugend, Berlin 17209
Museum Kirche zum Heiligen Kreuz, Zittau 22813
Museum Kiscell - Municipal Picture Gallery, Budapest 23404
Museum Kiscell of the Budapest History Museum, Budapest 23421
Museum Kitzbühel, Kitzbühel 02177
Museum Kleines Klingental, Basel 41172
Museum Kleines Schloß, Blankenburg 17373
Museum Klösterli, Ettiswil 41366
Museum de Klompenmaker, Keijenborg 32444
Museum Kloster Asbach, Asbach 16579
Museum Kloster Bentlage, Rheine 21298
Museum Kloster Hude, Hude 19284
Museum Kloster Kamp, Kamp-Lintfort 19424
Museum Kloster Stift zum Heiligengrabe, Heiligengrabe 19103
Museum Kloster Zeven, Zeven 22797
Museum Knoblauchhaus, Berlin 17210
Museum Knochenstampfe, Zwönitz 22841
Museum Kölner Karnevalsorden, Köln 19623
Museum Komodo, Jakarta 24489
Museum Komplex of the Tomsk Polytechnical University, Tomsk 37856
Museum Kraftwerk Plessa, Plessa 21059
Museum Kranenburgh, Bergen 31968
Museum Kraton Kasepuhan, Cirebon 24465
Museum Kraton Yogyakarta, Yogyakarta 24589
Museum Kronstadt Fortress, Kronštadt 37086
Museum Kruijsenhuis, Oirschot 32613
Museum Krupka, Krupka 09611
Museum Kuenburggewölbe, Werfen 02923
Museum Künstlerhaus Gasteiger, Utting am Ammersee 22194
Museum Küppersmühle für Moderne Kunst, Duisburg 18024
Museum Kulturgeschichte der Hand, Wolnzach 22678
Museum Kunst der Westküste, Alkersum 16471
Museum Kunstraum Neuss, Neuss 20660
Museum Der Kunststall, Dahlem 17769
Museum Kurá Hulanda, Willemstad 09359
Museum Kurhaus Kleve, Kleve 19567
Museum Kutschen-Wagen-Karren, Gescher 18622
Museum La Congiunta, Giornico 41434
Museum La Diligence, Heerlen 32337
Museum La Truaisch, Sedrun 41854
Museum Lambert Van Meerten, Delft 32063
Museum Lammert Boerma, Borgercompagnie 31988
Museum Langenargen/ Bodensee, Langenargen 19816
Museum Langenthal, Langenthal 41527
Museum Langes Tannen, Uetersen 22152
Museum Langmatt, Baden 41148
Museum Laufental, Laufen 41531
Museum Lauriacum, Enns 01870
Museum Laxenburg, Laxenburg 02275
Museum Le Mayeur, Sanur 24552
Museum Leben am Meer, Esens 18242
Museum Lettergieten 1983, Westzaan 32942
Museum Leuchtenburg, Seitenroda 21733
Museum-Library of National Languages, Penza 37530
Museum Lichtenberg im Stadthaus, Berlin 17211
Museum Lidice a Lidice Galerija, Lidice 09627
Museum Lindengut, Winterthur 42034
Museum Lindwurm, Stein am Rhein 41915
Museum Liner Appenzell, Appenzell 41119
Museum Lionardo da Vinci Ideale, Karlskrona 40636
Museum Löffingen, Löffingen 20015
Museum Löwenburg, Ederswiler 41340
Museum Loka Budaya, Jayapura 24509
Museum London, London 06189
Museum Ludwig, Köln 19624
Museum Maarssen, Maarssen 32517
Museum Malerwinkelhaus, Marktbreit 20215
Museum Malikusaleh, Aceh Utara 24434
Museum Mandala Bakti, Semarang 24556
Museum Mandala Wangsit Siliwangi, Bandung 24443
Museum Manggala Wanabakti, Jakarta 24490
Museum Mannersdorf, Mannersdorf 02348
Museum Mariastein, Kirchbichl 02169
Museum Mayer van den Bergh, Antwerpen 03296
Muséum Mécanique Naturelle, Giverny 13070
Museum mechanische Klangfabrik, Haslach 02057
Museum Meerhem, Gent 03635
Museum Meermanno-Westreenianum, Den Haag 32087
Museum Meiji-Mura, Inuyama 28573
Museum Meinhardt, Mainhardt 20137
Museum Memorial Sultan Ternate, Ternate 24578
Museum Mengeringhausen, Bad Arolsen 16672
Museum Menkemaborg, Uithuizen 32836
Museum Mensch und Natur, München 20466
Museum Menshikov Palace - State Hermitage Museum, Sankt-Peterburg 37675
Museum Mesdag, Den Haag 32088
Museum Meteorological Station of Simbirsk, Uljanovsk 37903
Museum Military Intelligence, Chicksands 43833
Museum Militer, Jakarta 24491
Museum Mini Korem, Sintang 24561
Museum Mistra, Mystras 23108
Museum Mitterteich, Mitterteich 20344
Museum Mobil, Bogor 24458
Museum Modern Art, Hünfeld 19289
Museum Moderner Kunst Kärnten MMKK, Klagenfurt 02190
Museum Moderner Kunst Stiftung Ludwig Wien, Wien 03042
Museum Moderner Kunst - Wörlen, Passau 20991
Museum Modest Huys, Zulte 04170
Museum Mödling, Mödling 02395
Museum Møhlman, Appingedam 31918
Museum - Monastery Santa Maria de L'Estany, L'Estany 39285
Museum-Monument Spas na Krovi, Sankt-Peterburg 37705
Museum Moorseer Mühle, Nordenham 20731
Museum Morsbroich, Leverkusen 19976
Museum MPU Tantular, Surabaya 24566
Museum Münster, Münster 41662
Museum Municipal, Beira 31703
Muséum Municipal d'Histoire Naturelle Henri-Lecoq, Clermont-Ferrand 12552
Museum Murten, Murten 41668
Museum Musica, Stadskanaal 32798
Museum Nagele, Nagele 32565
Museum Naila im Schusterhof, Naila 20546
Museum Nan Eilean, Lionacleit 44723
Museum Nan Eilean, Steornabhagh, Stornoway 45767
Museum Nanning Hendrik Bulthuis, Leeuwarden 32476
Museum Nasional, Jakarta 24492
Muséum National d'Histoire Naturel Arboretum de Chèvreloup, Rocquencourt 15076
Muséum National d'Histoire Naturelle, Paris 14741
Museum National d'Histoire Naturelle, Rabat 31692
Museum Natur und Mensch, Greding 18742
Museum Nature of the Sea and its Protection, Institute of Ocean Biology of the Far Eastern Dept. of the Russian Academy of Sciences, Vladivostok 37947
Museum Necca and New England Center for Contemporary Art, Brooklyn 47065
Museum-Necropol of the Demidov Family and Exhibition Hall Tula Metal, Tula 37869
Museum Nederlandse Cavalerie, Amersfoort 31816
Museum Negeri Bali, Denpasar 24466
Museum Negeri Jambi, Jambi 24508
Museum Negeri Java Tengah, Semarang 24557
Museum Negeri Jawa Barat, Bandung 24444
Museum Negeri Kalimantan Barat, Pontianak 24547
Museum Negeri Kalimantan Timur Mulawarman, Tenggarong 24577
Museum Negeri Lampung, Bandarlampung 24439
Museum Negeri Nusa Tenggara Barat, Mataram 24524
Museum Negeri of Aceh, Banda Aceh 24438
Museum Negeri of Bengkulu, Bengkulu 24453
Museum Negeri of Central Sulawesi, Palu 24540
Museum Negeri of Irian Jaya, Jayapura 24510
Museum Negeri of La Galigo, Ujungpanang 24581
Museum Negeri of Lambung Mangkurat, Banjarbaru 24448
Museum Negeri Propinsi Nusa Tenggara Timur, Kupang 24517
Museum Negeri Propinsi Sumatera Selatan Balaputra Dewa, Palembang 24537
Museum Negeri Sulawesi Tenggara, Kendari 24514
Museum Negeri Sulawesi Utara, Manado 24522
Museum Negeri Sumatera Barat, Padang 24533
Museum Negeri Sumatera Utara, Medan 24526
Museum Negeri Timor-Timur, Dili 10365
Museum Nekara, Selayar 24553
Museum Neue Mühle, Erfurt 18208
Museum Neuhaus, Biel 41231
Museum Neuhaus – Sammlung Liaunig, Neuhaus 02428
Museum Neukirchen-Vluyn, Neukirchen-Vluyn 20627
Museum Neukölln, Berlin 17212
Museum Neuruppin, Neuruppin 20650
Museum Neustadt an der Orla, Neustadt an der Orla 20667
Museum Nicolaihaus, Berlin 17213
Museum Nienburg, Nienburg 20718
Museum Niesky, Niesky 20722
Museum Nikolaikirche, Berlin 17214
Museum Nordenham, Nordenham 20732
Museum NordJura, Weismain 22444
Museum Nostalgie der 50er Jahre, Burgpreppach 17661
Museum Nusantara, Delft 32064
Museum Nymphenburger Porzellan - Sammlung Bäuml, München 20467
Museum Ober-Ramstadt, Ober-Ramstadt 20798
Museum Obere Saline mit Bismarck-Museum, Bad Kissingen 16759
Museum Oberes Donautal, Fridingen an der Donau 18466
Museum Oberschützen, Oberschützen 02459
Museum Oculorum, Buenos Aires 00280
Museum od Musical Instruments, Dušanbe 42495
Museum of A.M. Gorky's Childhood - Kashirin's House, Nižnij Novgorod 37454
Museum of a shoemaker's workshop, Schrobenhausen 21641
Museum of Aboriginal Affairs, Gombak 30553
Museum of Accounting, Vilnius 30400
Museum of Acre, Acre 24919
Museum of Actors and Theater Producers Brothers Amtmaņi, Valle 30138
Museum of Adomas Petrauskas - Kupiškis Ethnographic Museum, Uoginiai 30389
Museum of African American Art, Los Angeles 50251
Museum of African Arts, Beograd 38060
The Museum of African Tribal Art, Portland 52066
Museum of Afro-American History, Boston 46939
Museum of Agios Stephanos, Meteora 23093
Museum of Agricultural Implements, Estremoz 35576
Museum of Agricultural Implements, Mafra 35749
Museum of Agriculture and Melioration of the Latvian Republic, Mālpils 30047
Museum of Alaska Transportation and Industry, Wasilla 53954
Museum of Alpine Farming, Bohinjsko Jezero 38402
Museum of Amana History, Amana 46249
Museum of Ambleside, Ambleside 43299
Museum of American Architecture and Decorative Arts, Houston 49396
Museum of American Finance, New York 51222
Museum of American Frontier Culture, Staunton 53277
Museum of American Glass at WheatonArts, Millville 50696
Museum of American Historical Society of Germans from Russia, Lincoln 50100
The Museum of American Illustration at the Society of Illustrators, New York 51223
Museum of American Political Life, West Hartford 54058
Museum of American Presidents, Strasburg 53332
Museum of Anatolian Civilisations, Ankara 42695
Museum of Anatomy, Glasgow 44290
Museum of Anatomy at the Moscow Agrarian Timiryazev Academy, Moskva 37260
Museum of Anatomy of the Department of Anatomy, Athinai 22930
Museum of Ancient and Contemporary Art, Épinal 12840
Museum of Ancient and Modern Art, Penn Valley 51762
Museum of Ancient Art, Bruxelles 03462
Museum of Ancient Artifacts, Chicago 47490
Museum of Ancient Artifacts, Las Vegas 49976
Museum of Ancient Beekeeping, Stripeikiai 30375
Museum of Ancient Culture, Kilmartin 44580
Museum of Ancient Mazovian Metallurgy, Pruszków 35052
Museum of Ancient Metallurgy, Nowa Słupia 34949
Museum of Ancient Raileay Parenzena, Izola 38423
Museum of Ancient Sculpture and Architecture, Litomyšl 09633
Museum of Andrej Kmet, Martin 38296
Museum of Andrey Tarkovsky, Jurjevec 36985

M

Museum of Andrzej Strug, Warszawa 35235
Museum of Anhui Province, Hefei 07776
Museum of Animal Husbandary and Veterinary, Guwahati 24193
Museum of Anthracite Mining, Ashland 46390
Museum of Anthropology, Ann Arbor 46308
Museum of Anthropology, Chico 47528
Museum of Anthropology, Columbia 47717
Museum of Anthropology, Denver 48084
Museum of Anthropology, Detroit 48126
Museum of Anthropology, Fullerton 48808
Museum of Anthropology, Goudi 23008
Museum of Anthropology, Highland Heights 49279
Museum of Anthropology, Lawrence 49990
Museum of Anthropology, Pullman 52198
Museum of Anthropology, Taipei 42406
Museum of Anthropology, Tempe 53469
Museum of Anthropology, Vancouver 07217
Museum of Anthropology, Winston-Salem 54246
Museum of Anthropology and Archaeology, Tshwane 38750
Museum of Antique Armour Helmets and Swords, Kyoto 28852
Museum of Antiquities, Alnwick 43287
Museum of Antiquities, Jamnagar 24237
Museum of Antiquities, Newcastle-upon-Tyne 45230
Museum of Antiquities, Saskatoon 06920
Museum of Antropology, History and Art, Rio Piedras 52355
Museum of Appalachia, Norris 51377
Museum of Applied Art, Beograd 38074
Museum of Applied Art, Sankt-Peterburg 37673
Museum of Applied Art of the Lithuanian Art Museum, Vilnius 30434
Museum of Applied Art of Uzbekistan, Toshkent 54480
Museum of Applied Arts, Budapest 23416
Museum of Applied Arts, Poznań 35044
Museum of Aqaba Antiquities, Aqaba 29661
Museum of Archaeological Site Moenjodaro, Dokri 34000
Museum of Archaeology, Batumi 16335
Museum of Archaeology, Durham 44073
Museum of Archaeology, La Habana 09130
Museum of Archaeology, Legon 22861
Museum of Archaeology, Vadodara 24420
Museum of Archaeology and Ethnography of the Far East State University, Vladivostok 37943
Museum of Archaeology and Ethnography of the Syktykvar State University, Syktyvkar 37818
Museum of Archaeology and Ethnology, Burnaby 05504
Museum of Archaeology and Fortifications, Khojand 42502
Museum of Archaeology and History, Turku 11387
Museum of Archaeology Canton Thurgau, Frauenfeld 41378
Museum of Archaeology, Ethnography and Ecology of Siberia, Kemerovo State University, Kemerovo 37033
Museum of Archaeology of the Institute of Language, Literature and History of the Komi Science Center of the Ural Division of the Russian Academy of Sciences, Syktyvkar 37817
Museum of Architecture, Samara 37617
Museum of Architecture, Wrocław 35328
Museum of Architecture and Archaeology Tower of Burana, Tokmok 29967
Museum of Architecture and of the Architectural Development of Siberia, Novosibirsk 37485
Museum of Armed Struggle and Martyrology in Palmiry, Warszawa 35276
Museum of Armenian-Russian Friendship, Abovyan 00689
Museum of Arms and City Fortification, Bratislava 38208
Museum of Army Flying, Middle Wallop 45138
Museum of Art, Ann Arbor 46309
Museum of Art, Durham 48225
Museum of Art, Ein Harod 24945
Museum of Art, Fort Lauderdale 48638
Museum of Art, Orono 51603
Museum of Art, Providence 52170
Museum of Art, Pullman 52199
Museum of Art, Saint Louis 52589
Museum of Art, Sion 41876
Museum of Art, Skopje 30521
Museum of Art, Žytomyr 43211
Museum of Art and Archaeology, Columbia 47718
Museum of Art and Far Eastern Antiquities, Ulricehamn 40995
The Museum of Art and History, Santa Cruz 52868
Museum of Art and History, Weston 54102
Museum of Art, Archaeology and Folklore, Mysore 24330
Museum of Art at Brigham Young University, Provo 52188
Museum of Art, Ehime, Matsuyama 28904
Museum of Art Fakes, Wien 02986
Museum of Art in Łódź, Łódź 34884
Museum of art in Steatite, Frankfurt am Main 18384
The Museum of Art, Kochi, Kochi 28743
Museum of Art Lucerne, Luzern 41593
Museum of Art (MoA), Seoul 29904
Museum of Arthropoda, Pune 24361
Museum of Arts and Crafts, Berlin 17169
Museum of Arts and Crafts, Bratislava 38229
Museum of Arts and Crafts, Escalhão 35564
Museum of Arts and Crafts, Lucknow 24297
Museum of Arts and Crafts, Zagreb 09008
Museum of Arts and Design, New York 51224
The Museum of Arts and Sciences, Daytona Beach 47994
Museum of Arts and Sciences, Macon 50367
Museum of Asian Art, Corfu 22978
Museum of Asian Art, Kuala Lumpur 30582
Museum of Asparagus, Schrobenhausen 21638
Museum of Audio Visual Technology, Wellington 33250
Museum of Automobile Art and Design, Syracuse 53401
Museum of Automobile Transport, Kursk 37096
Museum of Automobile Transport, Sankt-Peterburg 37654
Museum of Automobiles, Morrilton 50848
Museum of Aviation, Paraparaumu 33178
Museum of Aviation at Robins Air Force Base, Warner Robins 53819
Museum of Aviation in Košice, Košice 38271
Museum of Bad Art, Dedham 48025
Museum of Balvi Region, Balvi 29987
Museum of Bamboo Art, Nantou 42280
Museum of Banknotes, Porto 35853
Museum of Banknotes of the Ionian Bank, Corfu 22979
Museum of Barbarism, Lefkoşa 33367
Museum of Barnstaple and North Devon, Barnstaple 43387
Museum of barricades of 1991, Rīga 30063
Museum of Baseball, Caracas 54551
Museum of Bath at Work, Bath 43406
Museum of Beauty, Melaka 30598
Museum of Bedřich Hrozný, Lysá nad Labem 09644
Museum of Bee-keeping in Slovakia, Kráľová pri Senci 38276
Museum of Beekeeping of the Živanović Family, Sremski Karlovci 38153
Museum of Belize, Belize City 04176
Museum of Bells, Alcáçovas 35389
Museum of Berkshire Aviation, Woodley 46065
Museum of Bērzgale Parish History, Bērzgale 29992
Museum of Biblical Art, New York 51225
Museum of Blue Dyeing, Pápa 23753
Museum of Bobbin Laces, Vila do Conde 35989
Museum of Bohemian Karst Beroun, Branch Žebrák, Žebrák 09935
Museum of Bohemian Karst in Beroun, Beroun 09404
Museum of Bolivian Writers, La Paz 04213
Museum of Books and Printing, Kyïv 43092
Museum of Bosnian Serb Republic, Banja Luka 04257
Museum of Božena Němcová, Česká Skalice 09459
Museum of Bread, Sankt-Peterburg 37724
Museum of Brewing, Burton-upon-Trent 43688
The Museum of Broadcast Communications, Chicago 47491
Museum of Bronx History, Bronx 47050
Museum of Brushs, Bankeryd 40452
Museum of Bryansk Forest, Brjansk 36828
Museum of Bucarest, București 36213
Museum of Buffalo and Horse-Breeding, Šumen 05217
Museum of Bulgarian National Revival and the National Liberation Struggles, Elena 05027
Museum of Bus Transportation, Hershey 49264
Museum of Byzantine Culture, Thessaloniki 23217
Museum of Cannock Chase, Hednesford 44436
Museum of Cape Breton Heritage, North East Margaree 06475
Museum of Carlsbad Porcelain, Březová u Karlových Var 09419
Museum of Carmelite Mission in Africa, Krceszowice 34827
Museum of Carousel Art and History, Sandusky 52837
Museum of Carpets, Kilims and Flatweaves, İstanbul 42820
Museum of Carriages, Studénka 09854
Museum of Carriages, Wien 03085
Museum of Cars, Lagan 40664
Museum of Casts and Archaeological Collection, Thessaloniki 23218
Museum of Central Australia, Alice Springs 00799
Museum of Central Bohemia, Roztoky u Prahy 09828
Museum of Central Connecticut State University, New Britain 51028
Museum of Central Finland, Jyväskylä 10894
Museum of Central Kalimantan, Palangkaraya 24536
Museum of Central Pomerania, Słupsk 35112
Museum of Central Slovakia, Banská Bystrica 38187
Museum of Ceramic Art, Hyogo, Sasayamashi 29233
Museum of Ceramics, Bolesławiec 34568
Museum of Ceramics, East Liverpool 48256
Museum of Changhua, Changhua 42157
Museum of Chia-Yi City, Chai-Yi 42154
Museum of Chicot County Arkansas, Lake Village 49893
Museum of Childhood, Edinburgh 44132
Museum of Childhood, Masterton 33142
Museum of Childhood, Sudbury, Derbyshire 45796
Museum of Childhood, Edith Cowan University, Perth 01411
Museum of Childhood Memories, Beaumaris 43427
Museum of Children, Santo Domingo 10354
Museum of Children Literature - Maironis Lithuanian Literature Museum, Kaunas 30267
Museum of Children's Art, Oakland 51468
Museum of Children's Creation, Sankt-Peterburg 37674
Museum of Chinese Chive Farming, Cingshuei 42168
Museum of Chinese in the Americas, New York 51226
Museum of Chodsko, Domažlice 09496
Museum of Chungju National University, Chungju 29791
Museum of Church History and Art, Salt Lake City 52692
Museum of Cieszyn Silesia, Cieszyn 34610
Museum of Cinematography of Uzbekistan, Toshkent 54477
Museum of Circus Art, Sankt-Peterburg 37672
Museum of City History Leipzig, Leipzig 19937
Museum of City History Leipzig-New Building, Leipzig 19939
Museum of City History Leipzig-Old City Hall, Leipzig 19938
Museum of Civil Defence in Helsinki, Helsinki 10784
The Museum of Clallam County Historical Society, Port Angeles 52026
Museum of Classical Antiquities, Lund 40706
Museum of Classical Archaeology, Adelaide 00779
Museum of Classical Archaeology, Cambridge 43730
Museum of Classical Balinese Painting, Klungkung 24516
Museum of Coast and Anti-Aircraft Artillery, Green Point 38636
Museum of Combat Fame, Kaluga 36995
Museum of Communication, Burntisland 43682
Museum of Communication, Den Haag 32091
Museum of Communism, Praha 09771
Museum of Comparativ Anatomy, Roma 27488
Museum of Comparative Zoology, Cambridge 47184
Museum of Computing Technique and Computer Science, Rīga 30090
Museum of Connecticut History, Hartford 49197
Museum of Contemporary Art, Andros 22883
The Museum of Contemporary Art, Atlanta 46445
Museum of Contemporary Art, Beograd 38076
Museum of Contemporary Art, Chicago 47492
Museum of Contemporary Art, Cleveland 47634
Museum of Contemporary Art, Fort Collins 48610
Museum of Contemporary Art, Lyon 13883
Museum of Contemporary Art, Miami 50634
Museum of Contemporary Art, Novi Sad 38123
Museum of Contemporary Art, Roskilde 10244
Museum of Contemporary Art, Tehrān 24671
Museum of Contemporary Art, Tilburg 32830
Museum of Contemporary Art, Turku 11389
Museum of Contemporary Art, Washington 53894
Museum of Contemporary Art Denver, Denver 48085
Museum of Contemporary Art Kiasma - Finnish National Gallery, Helsinki 10792
Museum of Contemporary Art Leipzig, Leipzig 19905
Museum of Contemporary Art London, London 44922
Museum of Contemporary Art Los Angeles, Los Angeles 50252
Museum of Contemporary Art of Antwerp, Antwerpen 03301
Museum of Contemporary Art of Marseille, Marseille 13995
Museum of Contemporary Art of the Jewish Autonomous Region, Birobidžan 36808
Museum of Contemporary Art San Diego - Downtown, San Diego 52740
Museum of Contemporary Art San Diego - La Jolla, La Jolla 49846
Museum of Contemporary Art Sapporo, Sapporo 29220
Museum of Contemporary Art, Skopje, Skopje 30520
Museum of Contemporary Art Sydney, Sydney 01553
Museum of Contemporary Art, Taipei, Taipei 42407
Museum of Contemporary Art - The National Museum of Art, Design and Architecture, Oslo 33733
Museum of Contemporary Art, Tokyo, Tokyo 29498
Museum of Contemporary Articrafts, Tehrān 24683
Museum of Contemporary Canadian Art, Toronto 07124
Museum of Contemporary Craft, Portland 52080
Museum of Contemporary Glass, Vledder 32894
Museum of Contemporary Japanese Poetry, Tanka and Haiku, Kitakami 28710
The Museum of Contemporary Photography, Chicago 47493
Museum of Contemporary Prints, Vledder 32895
Museum of Contemporary Religious Art, Saint Louis 52590
Museum of Contemporary Sacred Art, Saint-Pierre-de-Chartreuse 15457
Museum of Cosmetology, Chicago 47494
Museum of Cosmonautics S. Korolyov, Žytomyr 43212
Museum of Costume, Bath 43407
Museum of Costume, São Brás de Alportel 35917
Museum of Costume and Textiles, Nottingham 45294
Museum of Craft and Folk Art, San Francisco 52786
Museum of Crafts, Ånge 40426
Museum of Cres, Cres 08856
Museum of Cretan Ethnology, Vori 23240
Museum of Croatian Archaeological Monuments, Split 08961
Museum of Cuban Historical Background, Santiago de Cuba 09309
Museum of Cultural and Natural History, Mount Pleasant 50884
Museum of Cultural Heritage, Kyïv 43093
Museum of Cultural History, Lund 40703
Museum of Cultures, Castagnola 41283
Museum of Cultures of Vietnam's Ethnic Groups, Thai Nguyen 54636
Museum of Cumania Minor, Kiskunfélegyháza 23666
Museum of Cycladic and Ancient Greek Art, Athinai 22931
Museum of Czech Forest, Tachov 09859
Museum of Czech Garnet, Třebenice 09874
Museum of Czech Literature, Praha 09786
Museum of Czech Porcelain, Klášterec nad Ohří 09584
Museum of Dacian and Roman Civilisation, Deva 36344
Museum of Daejeon History, Daejeon 29801
Museum of Danish Cartoon Art, København 10140
Museum of Danish Resistance 1940-1945, København 10131
Museum of Dartmoor Life, Okehampton 45312
Museum of Death, Los Angeles 50253
Museum of Decorative and Applied Art, Sumy 43189
Museum of Decorative and Applied Art of the Ural, Čeljabinsk 36844
Museum of Decorative and Applied Folk Art, Belozersk 36799
Museum of Decorative, Applied and Industrial Art, Moskva 37270
Museum of Decorative Applied Arts, Rīga 30064
Museum of Decorative Arts, Chicago 47495
Museum of Decorative Arts, Dresden 17951
Museum of Decorative Arts and Design - The National Museum of Art, Design and Architecture, Oslo 33730
Museum of Decorative Arts in Prague, Praha 09797
Museum of Delphic Celebrations of Angelos and Eva Sikellianou, Athinai 22932
Museum of Dentistry, Liverpool 44741
Museum of Design Zurich, Zürich 42094
Museum of Design Zurich - Poster, Design, Applied Art and Graphics Collections, Zürich 42095
Museum of Development of Moscow Power Supply, Moskva 37296
Museum of Dinosaurs, Salas de los Infantes 39945
Museum of Dionysios Solomos and Eminent Zakynthians, Zakynthos 23245
Museum of Discovery, Little Rock 50140
Museum of Discovery and Science, Fort Lauderdale 48639
Museum of Divine Treasures, Genkai 28456
Museum of Dobruška, Dobruška 09493
Museum of Dolls and Costumes, Tampere 11355
Museum of Domestic Design and Architecture, Barnet 43382
Museum of Dominican Monastery, Dubrovnik 08867
Museum of Dominican St. Catherine's Monastery, Tallinn 10621
Museum of Dr. Sun Yat-Sen, Zhongshan 08506
Museum of Dr. Vincas Kudirka - National Museum of Lithuania, Kudirkos Naumiestis 30290
Museum of Drinking Water, Taipei 42408
Museum of Druskininkai, Druskininkai 30222
Museum of Early Southern Decorative Arts, Winston-Salem 54247
Museum of Early Trades and Crafts, Madison 50385
Museum of Earth Science of the M.V. Lomonosov Moscow State University, Moskva 37386
Museum of Earth Science of the Polish Academy of Sciences, Warszawa 35284
Museum of East Alabama, Opelika 51573
Museum of East Anglian Life, Stowmarket 45773
Museum of East Asian Art, Bath 43408
Museum of East Bosnia, Tuzla 04290

The Museum of East Texas, Lufkin 50318
Museum of Eastern Bohemia in Hradec Kralove, Hradec Králové 09534
Museum of Eastern Slovakia, Košice 38275
Museum of Ecclesiastical Art, Samos 23177
Museum of Ecclesiastical Art of Serrai and Nigrita, Serres 23180
Museum of Ecclesiastical Art of the Metropolis of Trikki and Stagoi, Trikala 23229
The Museum of Edinburgh, Edinburgh 44133
Museum of Education, Beograd 38084
Museum of Education, Korçë 00020
Museum of Education, Szczecin 35164
Museum of Education and Documents Library, Cairo 10502
Museum of Education and Pedagogy, Bratislava 38227
Museum of Education in Flanders, Ieper 03732
Museum of Electricity, Christchurch 43841
Museum of Electricity, Lima 34242
Museum of English Rural Life, Reading 45468
Museum of Engravings and Graphic Arts, Athinai 22933
Museum of Entertaining Art, Wien 03037
Museum of Entertainment, Whaplode 45992
Museum of Entrepreneurs, Maecenas and Benefactors, Moskva 37349
Museum of Epirot Folk Art, Metsovo 23095
Museum of Erotic Art, Paris 14669
Museum of Estonian Architecture, Tallinn 10623
Museum of Ethnographic Costumes on Dolls, Moskva 37274
Museum of Ethnography, Accra 22846
Museum of Ethnography, Budapest 23455
Museum of Ethnography, Gaziantep 42787
Museum of Ethnography, Reghin 36559
Museum of Ethnography, Stockholm 40892
Museum of Ethnography and Folklore, Tel Aviv 25117
Museum of Ethnology, Tai Po Kau 08262
Museum of Ethnology and Folk Art, Ratnapura 40381
Museum of Eton Life, Windsor 46043
Museum of European Cultures, National Museums in Berlin, Berlin 17199
Museum of Evolution, Uppsala 41005
Museum of Evolution of the Palaeobiological Institute of the Polish Academy of Science, Warszawa 35240
Museum of Evolution, Paleontology, Uppsala 41006
Museum of Faience and of the Arts of Entertaining, Samadet 15615
Museum of Fallen Soldiers, Stepanakert 03199
Museum of Family, Chasselay 12428
Museum of Fantastic Art Vienna, Wien 03056
Museum of Far Eastern Antiquities, Stockholm 40924
Museum of Farming Life, Ellon 44169
Museum of Farming Tools, Somero 11321
Museum of Farnham, Farnham 44218
Museum of Fiber Science and Technology, Tokyo University of Agriculture and Technology, Koganei 28759
Museum of Fife and Drum, Ivoryton 49553
Museum of Fights for Freedom - Lazdijai Area Museum, Lazdijai 30299
Museum of Filipino Life, Pasay 34473
The Museum of Films for Children, Weston 54103
Museum of Fine Art, Las Cruces 49959
Museum of Fine Art, Novokuzneck 37481
The Museum of Fine Art, Gifu, Gifu 28457
Museum of Fine Art of the Republic Udmurtiya, Iževsk 36956
Museum of Fine Arts, Beijing 07551
Museum of Fine Arts, Budapest 23477
Museum of Fine Arts, Chandigarh 24110
Museum of Fine Arts, Gent 03639
Museum of Fine Arts, Hobart 01155
The Museum of Fine Arts, Houston 49397
Museum of Fine Arts, Komsomolsk-na-Amure 37054
Museum of Fine Arts, Missoula 50756
Museum of Fine Arts, Rostov-na-Donu 37600
Museum of Fine Arts, Saint Petersburg 52640
Museum of Fine Arts, Santa Fe 52880
Museum of Fine Arts, Springfield 53226
Museum of Fine Arts and Cultural Center, Alexandria 10475
Museum of Fine Arts, Boston, Boston 46940
Museum of Fine Arts of the Museums Association Art Culture of the Russian North, Archangelsk 36768
Museum of Finnish Architecture, Helsinki 10821
Museum of Finnish Tourism, Imatra 10857
Museum of Fire, Edinburgh 44134
Museum of First Romanian School of Braşov, Braşov 36181
Museum of Fishery and Navigation, Liepāja 30041
Museum of Fishing, Santos 04821
Museum of Flight, Seattle 53004
Museum of Florida History, Tallahassee 53427
Museum of Flying, Santa Monica 52899
Museum of Foklore of the Mariupol Museum of Local Lore, Mariupol' 43146
Museum of Folk Architecture, Sanok 35100
Museum of Folk Architecture and Folklore of the Perejaslav National Historical and Ethnographic Preserve, Perejaslav-Chmel'nyc'kyj 43164
Museum of Folk Architecture - Ethnographic Park in Olsztynek, Olsztynek 34980
Museum of Folk Architecture of Ukraine, Kyïv 43098
Museum of Folk Art, Uljanovsk 37904
Museum of Folk Art of the Sloboshanshchyna, Charkiv 43021
Museum of Folk Arts and Crafts, Penza 37534
Museum of Folk Craft and Applied Arts, Trojan 05232
Museum of Folk Culture and Art, Poznań 35034
Museum of Folk Decorative Art, Černigiv 43006
Museum of Folk Education, Penza 37533
Museum of Folk Education in the Simbirsk Government in the 70th and 80th of the 19th Century, Uljanovsk 37905
Museum of Folk Graphics, Moskva 37339
Museum of Folk Musical Instruments, Szydłowiec 35174
Museum of Folklore, Brugge 03425
Museum of Folklore and Popular Art, Sifnos 23186
Museum of Folklore Arts and Culture, Salamina 23174
Museum of Foreign Art, Rīga 30062
Museum of Forestry, Moskva 37328
Museum of Forestry and Timber Industry, Nazarje 38488
Museum of Forestry, Hunting and Fisheries, Hluboká nad Vltavou 09516
Museum of Fragrance, Iwata, Toyoda 29526
Museum of Franciscan Monastery, Dubrovnik 08868
Museum of Freemasonry, London 44923
Museum of Fulham Palace, London 44924
Museum of Garden History, London 44925
Museum of Gardening and Environment, National Museum of Agriculture in Valtice, Valtice 09897
Museum of Gems and Minerals, Ratnapura 40382
Museum of Genocide Victims, Vilnius 30407
Museum of Geodesy and Cartography, Moskva 37277
Museum of Geologic Collection, Kielce 34746
Museum of Geology, Edmonton 05750
Museum of Geology, Rapid City 52253
Museum of Geology, Toshkent 54476
Museum of Geology and Mineralogy - Dept. of Natural Sciences, Vilnius University, Vilnius 30408
Museum of Geology of Lithuania, Vievis 30397
Museum of Geology of the Institute of Geological Sciences of the Polish Academy of Sciences, Kraków 34794
Museum of Geology, Oil and Gas, Tjumen 37845
Museum of Geometric and Madi Art, Dallas 47936
Museum of German Peasants War, Böblingen 17416
Museum of Giannoulis Halepas, Pyrgos 23162
Museum of Glass, Marcinelle 03851
Museum of Glass, Tacoma 53412
Museum of Glass and Ceramics, South Portland 53179
Museum of Glass and Jewellry, Jablonec nad Nisou 09545
Museum of Glass - Grünenplan, Grünenplan 18812
Museum of Gold, Travassos - Póvoa de Lanhoso 35974
Museum of Gorica, Nova Gorica 38489
Museum of Greek Children's Art, Athinai 22919
Museum of Greek Folk Art, Athinai 22918
Museum of Greek Folk Art, Athinai 22934
Museum of Greek Folk Art-Pottery Collection, Athinai 22935
Museum of Greek Musical Instruments, Thessaloniki 23219
Museum of Greek Popular Musical Instruments, Athinai 22924
Museum of Guides and Scouts, Ogre 30051
Museum of Gujarat Vidyasabha, Ahmedabad 24050
Museum of Hamamatsu, Hamamatsu 28484
Museum of Handicraft, Architecture and Allday Life, Kaluga 37000
Museum of Harjamäki Mental Hospital, Siilinjärvi 11310
Museum of Harlow, Harlow 44396
Museum of Hartlepool, Hartlepool 44408
Museum of Health Care, Kingston 06087
Museum of Health Services, Toshkent 54481
Museum of Hercegovina Trebinje, Trebinje 04287
Museum of Herman Blode, Nida 30320
Museum of Herzegovina, Mostar 04270
Museum of Highlands, Třešt 09878
Museum of Highlands Jihlava, Jihlava 09561
Museum of Historic Clocks, Bratislava 38207
Museum of Historic Firefighting Technique, Chrastava 09483
Museum of Historical Agricultural Engineering, Wageningen 32920
Museum of Historical Aircraft in Flying, Madrid 39536
Museum of Historical Archaeology, Kingston 28333
Museum of Historical Geology and Geological Survey of Pakistan, Quetta 34039
Museum of Historical Jewelry of Ukraine, Kyïv 43090
Museum of History, Dobrič 05019
Museum of History, Jambol 05043
Museum of History, Olten 41706
Museum of History and Art, Belozersk 36798
Museum of History and Art, Kaliningrad 36989
Museum of History and Art, Ontario 51568
Museum of History and Art, Serpuchov 37765
Museum of History and Art, Tichvin 37842
Museum of History and Culture of the Moscow Quarter, Nižnij Novgorod 37455
Museum of History and Culture of Votkinsk, Votkinsk 37978
Museum of History and Ethnography, Melaka 30603
Museum of History and Folk Lore, Matsusaka 28900
Museum of History and Folk, Uozu, Uozu 29562
Museum of History and Industry, Seattle 53005
Museum of History, Anthropology and Art, San Juan 36076
Museum of History at the Moscow State University, Moskva 37291
Museum of History of Katowice, Katowice 34725
Museum of History of Katowice - Dept. Ethnography of the Town, Katowice 34726
Museum of History of Katowice - Pawel Steller Museum, Katowice 34727
Museum of History of Money, Köln 19604
Museum of History of Orenburg, Orenburg 37513
Museum of History of Peterburg Conservatory, Sankt-Peterburg 37682
Museum of History of Riga and Navigation, Rīga 30111
Museum of History of Russian Military Schools (Cadet Corps), branch of the Military-Historical Museum of Artillery, Engineer and Signal Corps, Sankt-Peterburg 37715
Museum of History of the City of Gdańsk, Gdańsk 34638
Museum of History of the City of Łódź, Łódź 34876
Museum of History of the County Nógrád, Salgótarján 23791
Museum of History of the Prioksky Quarter, Nižnij Novgorod 37459
Museum of History of the University of Latvia, Rīga 30091
Museum of History of Varna, Varna 05243
Museum of History Saint Petersburg University, Sankt-Peterburg 37686
Museum of History Silistra, Silistra 05163
Museum of Holography, Chicago 47496
Museum of Hops Cultivation, Žalec 38539
Museum of Horse-drawn Railway in Bujanov, Kaplice 09572
Museum of Horyuji Mural Paintings Reproductions, Nagakute 28951
Museum of household appliances, Ziębice 35371
Museum of Human Sciences, Harare 54669
Museum of Hungarian Agriculture, Budapest 23439
Museum of Hungarian Aluminium Industry, Székesfehérvár 23856
Museum of Hungarian Cartography - Museum and Institution of Military History, Budapest 23444
Museum of Hungarian mopeds, Bodrogkeresztúr 23374
Museum of Hunting, Sibiu 36611
Museum of Hunting, Zagreb 09002
Museum of Hunting and Fishing, Moskva 37342
Museum of Hunting and Game, Vila Viçosa 36017
Museum of Hunting and Horsemanship, Warszawa 35257
Museum of Hunting and Nature, Paris 14649
Museum of Hunting of the Irkutsk Agricultural Academy, Irkutsk 36939
Museum of Hygiene, Cairo 10503
Museum of Icons and Byzantine Traditions, Frascineto 26248
Museum of Icons of the National Revival, Radomir 05145
Museum of Idaho, Idaho Falls 49462
Museum of Ikebana, Kyoto 28815
Museum of Illustration, Ferrara 26112
Museum of Images from the Unconscious, Rio de Janeiro 04728
Museum of Imaging Technology, Bangkok 42546
Museum of Independence, Warszawa 35262
Museum of Independent Telephony, Abilene 46122
Museum of Indian Arts and Culture, Santa Fe 52881
Museum of Indian Culture, Allentown 46229
Museum of Industrial Archaeology and Textile, Gent 03638
Museum of Industrial Railways, Zbýšov 09931
Museum of Industrial Worker's Dwellings, Imatra 10858
Museum of Industry, Stellarton 07008
Museum of Industry, Tokyo 29485
Museum of Industry and Technic, Wałbrzych 35212
Museum of Inhaca Island, Maputo 31712
Museum of Insects, San José 08817
Museum of Installation, London 44926
Museum of Instrument, Urueña 40183
Museum of Instruments, London 44927
Museum of Interial Manor, Ożarów 35001
Museum of International Ceramic Art Denmark, Middelfart 10188
Museum of International Folk Art, Santa Fe 52882
Museum of International Friendship, Beijing 07552
Museum of International Friendship, Saransk 37742
Museum of Irish Transport, Killarney 24851
Museum of Iron and Darby Furnace, Coalbrookdale 43860
Museum of Iron Ore Mining, Częstochowa 34616
Museum of Islamic Art, Cairo 10504
Museum of Islamic Art (MIA), Doha 36087
Museum of Islamic Ceramic, Cairo 10505
Museum of Island History, Newport, Isle of Wight 45244
Museum of Islay Life, Port Charlotte 45411
The Museum of Israeli Art, Ramat Gan 25085
Museum of Istrian History, Pula 08935
Museum of Ivanovo Calico, Ivanovo 36950
Museum of Jacques Lipchitz in Druskininkai, Druskininkai 30223
Museum of Japanese Steel, Yasugi 29589
Museum of Japanese Traditional Art Crafts, Tokyo 29449
Museum of Jewish Culture, Bratislava 38230
Museum of Jewish Heritage - A Living Memorial to the Holocaust, New York 51227
Museum of Jewish History and Culture, Kraków 34790
Museum of Jindřich Jindřich, Domažlice 09497
Museum of Jomon Art, Nakaniida 29018
Museum of Jurassic Technology, Culver City 47904
Museum of Kaigarayama Shell Mound, Kiyosu 28720
Museum of Kainuu, Kajaani 10903
Museum of Kalymnos, Kalymnos 23031
Museum of Karlovy Vary, Karlovy Vary 09575
Museum of Kashubian and Pomeranian Literature and Music, Wejherowo 35296
Museum of Kent Life, Sandling 45612
Museum of Kerala History, Edapally 24168
Museum of Kerman, Kerman 24618
Museum of Kežmarok and Town Castle, Kežmarok 38267
Museum of Komotini, Komotini 23060
Museum of Korean Buddhist Art, Seoul 29905
Museum of Korean Commercial History, Yongin 29944
Museum of Korean Embroidery, Seoul 29906
Museum of Korean Traditional Music, Seoul 29907
Museum of Krkonoše Region, Jilemnice 09563
Museum of Kyoto, Kyoto 28838
Museum of Kyushu Sangyo University, Fukuoka 28435
Museum of Lakeland Life, Kendal 44559
Museum of Lancashire, Preston 45445
Museum of Landscape Painting, Ples 37567
Museum of Latin American Art, Long Beach 50175
Museum of Laukminiskes Village - Kupiškis Ethnographic Museum, Laukminiškes 30296
Museum of Lead-Crystal Glass, Colle di Val d'Elsa 25938
Museum of Lead Mining, Wanlockhead 45927
Museum of Leading Persons of Ukrainian Culture, Kyïv 43099
Museum of Leather Art, Vic 40269
Museum of Lesya Ukrainka, Kyïv 43094
Museum of Lies, Kyritz 19754
Museum of Lighting, Edinburgh 44135
Museum of Ligurian Archaeology, Genova 26296
Museum of Lillydale, Lillydale 01230
Museum of Lincolnshire Life, Lincoln 44714
Museum of Literary Life in the Jaroslavl Region, Jaroslavl 36973
Museum of Literature, Melaka 30602
Museum of Literature and Music, Banská Bystrica 38183
Museum of Literature and Theatre Art and Gallery MAK, Sarajevo 04279
Museum of Lithuanian Fighting's for Independence, Merkinė 30312
Museum of Lithuanian Mathematics at the Dpt. of Mathematics and Informatics of the Vilnius University, Vilnius 30417
Museum of Lithuanian Sport of the Great War Museum, Kaunas 30257
Museum of Liverpool Life, Liverpool 44742
Museum of Local History, Burnley 43677
Museum of Local History, Hastings 44416
Museum of Local History and Astronomy, Tokyo 29409
Museum of Local History and Industry, Llandrindod Wells 44755
Museum of Local History and Nature, Ekaterinburg 36900
Museum of Local Lore Mariupol, Mariupol' 43144
Museum of London, London 44928
Museum of Long Island Natural Sciences, Stony Brook 53314

Museum of Lower Silesian Weaving in Kamienna Góra, Kamienna Góra 34712
Museum of Macedonia - Archaeological, Ethnological and Historical, Skopje 30519
Museum of Macedonia - Gallery of Icons, Skopje 30517
Museum of Majdanpek, Majdanpek 38115
Museum of Making Music, Carlsbad 47247
Museum of Malawi, Blantyre 30544
Museum of Man in the Sea, Panama City 51693
Museum of Manuscripts and Calligraphy, İstanbul 42821
Museum of Marine Life of Cyprus, Aghia Napa 09360
Museum of Maritime Science, Tokyo 29386
Museum of Maroccan Arts, Tanger 31699
Museum of Martsial Mineral Water, Petrozavodsk 37562
Museum of Martynas Jankus, Bitenai 30217
Museum of Martyrdom, Bielsk Podlaski 34555
Museum of Martyrdom, Poznań 35039
Museum of Mary Costumes, Andocs 23335
Museum of Maryland History, Baltimore 46582
Museum of Masharov's House, Tjumen 37844
Museum of Masovien Village, Sierpc 35107
Museum of Meat Industry, Budapest 23415
Museum of Mechanical Music, Varkaus 11453
Museum of Mechanical Musical Instruments, Tokyo 29456
Museum of Medical Equipment, Bangkok 42547
Museum of Medical History, Bruxelles 03468
Museum of Medical History, Seltjarnarnes 24038
Museum of Medical History, Shanghai 08175
Museum of Medical History, Vänersborg 41024
Museum of Medical History in Uppsala, Uppsala 41014
Museum of Medical Missionary in Remote Areas, Puli 42296
Museum of Medicinal Plants and Pharmacy, Bruxelles 03478
Museum of Medicine and Pharmacology of Tajikistan, Isfara 42500
Museum of Medieval Stockholm, Stockholm 40940
Museum of Medieval Torture Instruments, Praha 09765
Museum of Meteorology, Tainan 42363
Museum of Methodism, London 44929
Museum of Migrating People, Bronx 47051
Museum of Mikola Lisenko, Kyïv 43095
Museum of Military History, Budapest 23408
Museum of Military History, Gorna Studena 05037
Museum of Military History, Maracaibo 54582
Museum of Military Medicine, Lahti 11035
Museum of Mineralogy of the Faculty of Science - Charles University, Praha 09762
Museum of Mineralogy, Petrology and Geology, Zografou 23249
Museum of Minerals, Vilnius 30425
Museum of Minerals, Ores, precious stones "A surprising essence of stone", Teberda 37840
Museum of Mining and Metallurgy, Bor 38093
Museum of Mining and transportation of Limestone, Beroun 09405
Museum of Mining of Metals and Minerals, Rudabánya 23789
Museum of Missing Soldiers, Stepanakert 03200
Museum of Mobile, Mobile 50770
Museum of Modern and Contemporary Art, Aalborg 09956
Museum of Modern and Contemporary Fine Arts, Sosnovyj Bor 37791
Museum of Modern Art, Arnhem 31929
Museum of Modern Art, Cairo 10506
Museum of Modern Art, Dubrovnik 08874
Museum of Modern Art, Ljubljana 38454
Museum of Modern Art, Niepołomice 34945
Museum of Modern Art, Olten 41707
Museum of Modern Art, Radom 35074
Museum of Modern Art, Szczecin 35157
Museum of Modern Art, Troyes 15998
Museum of Modern Art, Yerevan 00748
Museum of Modern Art Gunma, Takasaki 29330
Museum of Modern Art, Ibaraki, Mito 28918
Museum of Modern Art, Kamakura and Hayama, Hayama 28500
Museum of Modern Art Ostend, Oostende 03933
Museum of Modern Art, Saitama, Saitama 29190
Museum of Modern Art, Shiga, Otsu 29166
The Museum of Modern Art, Toyama, Toyama 29522
Museum of Modern Art, Toyoshina, Toyoshina 29532
Museum of Modern Art Wakayama, Wakayama 29575
Museum of Modern Ceramic Art, Gifu, Tajimi 29311
Museum of Modern History and Culture of North Osetiya, Vladikavkaz 37931
Museum of Modern Japanese Literature, Tokyo 29448
Museum of Mohamed Mahmoud Khalil and his wife, Giza 10528
Museum of Money, Tbilisi 16373
Museum of Mongolian Ethnography, Ulaanbaatar 31626
Museum of Montaneus Banat, Reşiţa 36562
Museum of Mosaics, Devnja 05012
Museum of Mosovce Artisans, Mošovce 38309
Museum of Motoori Norinaga, Matsusaka 28898
The Museum of Museums, Waregem 04131
Museum of Music and the Theatre, Kislovodsk 37049
Museum of Music Boxes and Automata, Sainte-Croix 41796
Museum of Music History of the Hungarian Academy, Budapest 23494
Museum of Musical Instruments, Sankt-Peterburg 37700
Museum of musical instruments Markneukirchen, Markneukirchen 20207
Museum of Musical Instruments, Musashino Academia Musicae, Tokyo 29388
Museum of Naive Art, Jagodina 38100
Museum of NaJu Pears, Naju 29854
Museum of National Antiquities, Dušanbe 42496
Museum of National Antiquities, Stockholm 40899
Museum of National Antiquities, Stockholm 40937
Museum of National Applied Art, Şeki 03195
Museum of National Architecture and Urban Life of Gyumri, Gyumri 00707
Museum of National Culture, Kazan 37022
The Museum of National History at Fredriksborg Castle, Hillerød 10076
Museum of Nativity, Třebĕchovice pod Orebem 09873
Museum of Natural and Cultural History, Eugene 48435
Museum of Natural History, Amherst 46269
Museum of Natural History, College Los Baños 34380
Museum of Natural History, Detroit 48127
Museum of Natural History, Jerusalem 25000
Museum of Natural History, Marshall 50506
Museum of Natural History, Nanaimo 06411
Museum of Natural History, Neskaupstadur 24010
Museum of Natural History, Providence 52171
The Museum of Natural History, Stevens Point 53289
Museum of Natural History, Ulaanbaatar 31627
Museum of Natural History, Vestmannaeyjar 24043
Museum of Natural History and Archaeology, Trondheim 33932
Museum of Natural History and Forest, Smołdzino 35113
Museum of Natural History and Historical Photography, Polygyros 23155
Museum of Natural History and Primitive Technology, Somers 53144
Museum of Natural History and Technology, Shiraz 24647
Museum of Natural History of the City of Geneva, Genève 41427
Museum of Natural History of the Łódź University, Łódź 34882
Museum of Natural History Potsdam, Potsdam 21098
Museum of Natural History Thurgau, Frauenfeld 41379
Museum of Natural History, Tohoku University, Sendai 29247
Museum of Natural History, Wrocław University, Wrocław 35337
Museum of Natural Resources in Xiaoxinganling, Yichun, Heilongjiang 08442
Museum of Natural Science, Baton Rouge 46638
Museum of Natural Sciences, Bruxelles 03497
Museum of Natural Sciences, Saskatoon 06921
Museum of Natural Sciences, Tiranë 00033
Museum of Nature, Čerepovec 36850
Museum of Nature, Cherson 43025
Museum of nature, Chur 41300
Museum of Nature, Kazimierz Dolny 34734
Museum of Nature, Olten 41708
Museum of Nature, Smolensk 37775
Museum of Nature, Žytomyr 43213
Museum of Nature and Environment, Lübeck 20079
Museum of Nature and Human Activities, Hyogo, Sanda 29212
Museum of Nature of the Academy of Sciences of the Republic of Kazakhstan, Almaty 29690
Museum of Nature of Wielkopolski National Park, Mosina 34929
Museum of Nautical Art, Symi 23201
Museum of Nebraska Art, Kearney 49717
Museum of Nebraska History, Lincoln 50101
Museum of Neon Art, Los Angeles 50254
Museum of Net Manufacture, Bridport 43614
Museum of New Art, Pontiac 52021
Museum of New Hampshire History, Concord 47775
Museum of New Mexico, Santa Fe 52883
The Museum of Newport History, Newport 51322
Museum of Nonconformist Art, Sankt-Peterburg 37703
Museum of North Carolina Minerals, Spruce Pine 53244
Museum of North Craven Life, Settle 45640
Museum of North Idaho, Coeur d'Alene 47672
Museum of Northern Arizona, Flagstaff 48551
Museum of Northern British Columbia, Prince Rupert 06671
Museum of Northern History at the Sir Harry Oakes Château, Kirkland Lake 06098
Museum of Northwest Art, La Conner 49831
Museum of Northwest Colorado, Craig 47865
The Museum of Nursing History, Philadelphia 51856
Museum of Óbuda, Budapest 23456
Museum of Occupations, Tallinn 10640
Museum of Odontology, Helsinki 10777
Museum of Odontology, Madrid 39551
Museum of Old Photographs and Museum of Unearthed Artifacts, Nagasaki 28976
Museum of Old Silesian Trade, Świdnica 35147
Museum of Old Western Time, Sindian 42335
Museum of Olympic Glory, Toshkent 54478
Museum of One Picture, Penza 37538
Museum of optical illusions, Gent 03629
Museum of Ore-Mining and Mining Construction, Madan 05088
Museum of Orient and East Marion History, Orient 51587
The Museum of Oriental Ceramics Osaka, Osaka 29153
Museum of Oriental Manuscripts, Prizren 38135
Museum of Osaka Prints, Osaka 29129
The Museum of Outdoor Arts, Englewood 48400
Museum of Oxford, Oxford 45333
Museum of Painting and Sculpture, İstanbul 42835
Museum of Palace Interior, Choroszcz 34602
Museum of Palekh Art, Palech 37526
Museum of Paleontology, Berkeley 46758
Museum of Palestinian Popular Heritage, Al-Bireh 34046
Museum of Paper Making, Duszniki Zdrój 34627
Museum of Parapsychology, Moskva 37346
Museum of Partisan Struggle, Polichno 35021
Museum of Pathology, Sydney 01554
Museum of Patrons and Benefactors of Russia, Moskva 37332
Museum of peat and watermanagement, Vinkeveen 32889
Museum of Pedagogics, Kaunas 30264
Museum of Pediatric Surgery, Bangkok 42548
Museum of Peoples and Cultures, Provo 52189
Museum of Perfume, Escaldes-Engordany 00093
Museum of Persecuted Art-Israel, Ashdod 24929
Museum of Petroleum Industry in Bóbrka, Chorkówka 34601
Museum of Pharmacy, Bratislava 38209
Museum of Pharmacy, Rīga 30066
Museum of Pharmacy, Collegium Medicum at the Jagiellonian University, Kraków 34793
Museum of Pharmacy - Mazeikiai Museum, Viekšniai 30396
Museum of Philippine Art, Manila 34448
Museum of Philippine Humanities, Pasay 34474
Museum of Philippine Political History, Manila 34449
Museum of Philippine Traditional Cultures, Pasay 34476
Museum of Photographic Arts, San Diego 52741
Museum of Photography, Berlin 17202
Museum of Physical Culture, Beograd 38063
Museum of Physical Culture and Sport, Moskva 37276
Museum of Physical Culture in Slovak Republic, Bratislava 38228
Museum of Physical Education and Sports, Budapest 23483
Museum of Physical Education and Sports - Exhibition Ward, Budapest 23484
Museum of Physics of the Vilnius University, Vilnius 30404
Museum of Picardie, Amiens 11588
Museum of Pigs, Pampilhosa 35807
Museum of Pilgrimage to Holy Places, Moskva 37344
Museum of Pioneer Settlement, Kibbutz Yifat 25059
Museum of Piping, Glasgow 44291
Museum of Piracy, Santiago de Cuba 09315
Museum of Police Techniques, Solna 40872
Museum of Polish Arms, Dept. of History of Polish Arms, Kołobrzeg 34752
Museum Of Polish Arms, Dept. of the Town History, Kołobrzeg 34753
Museum of Polish Military Health Services, Łódź 34881
Museum of Popular Culture - Waldenbuch Castle, Waldenbuch 22287
Museum of Porn in Art, Zürich 42096
Museum of Post and Telecommunication, Škofja Loka 38515
Museum of Post and Telecomunication and Mokhaberat Museum, Tehrān 24684
Museum of Posters, Río Grande 36054
Museum of Postindependence Athens, Athinai 22936
Museum of Pottery, Kostelec nad Černými Lesy 09598
Museum of Precious Stones, Kaunas 30242
Museum of Prehistory, Jerusalem 25001
Museum of Prehistory, Shaar Hagolan 25093
Museum of Prehistory and History, Russellville 52489
Museum of Primitive Art and Culture, Peace Dale 51742
Museum of Primitive Methodism, Englesea Brook, Crewe 43922
Museum of Printing History, Houston 49398
Museum of Prints and Printmaking, Albany 46160
Museum of Private Collections, Moskva 37329
Museum of Psalms, Jerusalem 25002
Museum of Public Health, Madrid 39552
Museum of Public Health, Sankt-Peterburg 37712
Museum of Puerto Rican Music, Ponce 36051
Museum of Puppetry, Chrudim 09485
Museum of Puppets, Monterrey 31188
Museum of Puppets and Children Books Wonderland, Ekaterinburg 36895
Museum of Quebec Masters and Artisans, Montréal 06360
Museum of Radio and Technology, Huntington 49437
Museum of Rail and Truck, Ängelholm 40409
Museum of Rail Travel, Ingrow 44516
Museum of Railway Engineering, Toshkent 54482
Museum of Railway Technics, Sankt-Peterburg 37713
Museum of Rape Making, Älvängen 40405
Museum of Rare Stones, Mosėdis 30315
Museum of Reading, Reading 45469
Museum of Recent History Celje, Celje 38406
Museum of Reconstruction, Hammerfest 33531
Museum of Regional and Mediterranean Archaeology, Nir-David 25074
Museum of Regional Ethnology, Antwerpen 03319
Museum of Regional History, Türkmenbaşi 42984
Museum of Regional History of San Lorenzo and Gallery of the Liberator General San Martín, San Lorenzo 00590
Museum of Religious Art, Haarlem 32300
Museum of Religious Art, La Habana 09133
Museum of Religious Art, Lemvig 10177
Museum of Religious Art, Saint Louis 52591
Museum of Religious Art, Summerside 07034
Museum of Repressive Measures Victim Memory, Toshkent 54479
Museum of Rev. Kazimieras Jaunius, Lembo 30303
Museum of Richmond, Richmond, Surrey 45493
Museum of Roma Culture in Slovakia, Martin 38297
Museum of Romani Culture, Brno 09437
Museum of Romanian Literature, Iaşi 36421
Museum of Romanticism, Opinogóra 34982
Museum of Rovinj's Heritage, Rovinj 08943
Museum of Rural History and Culture, Genova 26310
Museum of Rural Life, Denton 48059
Museum of Rural Life in Wallonie, Saint-Hubert 03995
Museum of Russian Art, Minneapolis 50740
Museum of Russian Arts, Yerevan 00749
Museum of Russian Culture, San Francisco 52787
Museum of Russian Felt Boots, Moskva 37356
Museum of Sacral Sculpture of the 18th c. - Work of Ivan Pinzel, Lviv 43135
Museum of Sacred Art, Macau 08002
Museum of Sacred Icons, Athinai 22937
Museum of Saint Albans, Saint Albans 45547
Museum of Saint-Dizier, Saint-Dizier 15244
Museum of Salamanca, Salamanca 39940
Museum of Samoa, Apia 38000
Museum of Samogitian Village, Telšiai 30380
Museum of Sarajevo 1878-1914, Sarajevo 04280
Museum of Saxon-Bohemian ore Mountains, Marienberg 20198
Museum of School and the Private Culture Foundation, Malkara 42901
Museum of Science, Boston 46941
Museum of Science and History of Jacksonville, Jacksonville 49584
Museum of Science and Industry, Chicago 47497
Museum of Science and Industry, Tampa 53434
Museum of Science and Industry in Manchester, Manchester 45092
Museum of Science and Technology, Accra 22847
Museum of Science and Technology, Beograd 38070
Museum of Science and Technology, Tshwane 38751
Museum of Science at the Toledo Zoo, Toledo 53522
Museum of Scotland, Edinburgh 44136
Museum of Scottish Country Life, East Kilbride 44089
Museum of Scottish Lighthouses, Fraserburgh 44254
Museum of Sculpture, Sankt-Peterburg 37708
Museum of Sculpture and Figurines of Loukia Georganti, Athinai 22938
Museum of Sculptures of S.T. Konenkov, Smolensk 37771

Museum of Seminole County History, Sanford 52846
Museum of Setúbal-Convent of Jesus-16th Century Art Gallery, Setúbal 35936
Museum of Sex, New York 51228
Museum of Shadow Puppet, Gangshan 42191
Museum of Shanxi Province, Taiyuan 08273
Museum of Shells of the Greek Seas, Néa Moudiana 23124
Museum of Shipbuilding, Beograd 38075
Museum of Shipbuilding and the Fleet, Mykolaïv 43149
Museum of Silk Art, Catanzaro 25780
Museum of Silver, Firenze 26169
Museum of Sisiyat Culture, Nanjhuang 42279
Museum of Sisters of Sigurata Monastery, Dubrovnik 08869
Museum of Site, Most I, Hong Kong 07809
Museum of Site, Most II, Hong Kong 07810
Museum of Škofja Loka, Škofja Loka 38514
Museum of Slavonia Osijek, Osijek 08923
Museum of Slovak Red Cross, Martin 38298
Museum of South-Eastern Moravia, Zlín 09941
Museum of South Somerset, Yeovil 46096
Museum of South Texas History, Edinburg 48291
Museum of South Yorkshire Life, Doncaster 43994
Museum of South Zealand, Vordingborg 10320
Museum of Southern African Rock Art, Johannesburg 38661
Museum of Southern History, Jacksonville 49585
The Museum of Southern History, Sugarland 53357
Museum of Southern Railways, Šentjur 38511
Museum of Southwestern Biology, Albuquerque 46182
Museum of Soviet Repression, Tbilisi 16374
Museum of Spanish Abstract Art, Cuenca 39231
Museum of Spanish Colonial Art, Santa Fe 52884
Museum of Special Documents, Praia 07419
Museum of Special Educational System, Levoča 38283
Museum of Speed, Pendine 45355
Museum of Spiš, Spišská Nová Ves 38359
Museum of Sport and Tourism, Łódź 34883
Museum of Sport Glory of the Central Army Sport Club, Moskva 37364
Museum of Sport History at the Russian Academy of Physical Culture, Moskva 37220
Museum of Sports, Cienfuegos 09065
Museum of Sports, Malmö 40720
Museum of Sports and Tourism, Warszawa 35270
Museum of Sports History, La Habana 09134
Museum of Srem, Sremska Mitrovica 38151
Museum of St. Petersburg Railway Line, Lahti 11031
Museum of Stefan Żeromski's School Years, Kielce 34741
Museum of Stone and Wood Carving, Konya 42890
Museum of Stone Masonry, Akşehir 42684
Museum of Strife for Independence, Kraków 34788
Museum of Struggle and Martyrdom, Warszawa 35275
Museum of Study and Conquest of the European North, Apatity 36764
Museum of Surveying, Lansing 49944
Museum of Sussex Archaeology, Lewes 44700
Museum of Swedish Shoe Industry, Kumla 40658
Museum of Sydney, Sydney 01555
Museum of Taipa & Coloane History, Macau 08003
Museum of Tartu University History, Tartu 10659
Museum of Tatra National Park, Tatranská Lomnica 38370
Museum of Taxes, Jerusalem 25003
Museum of Technology, Helsinki 10829
Museum of Technology in Warsaw, Warszawa 35273
Museum of Teenage Art, Housatonic 49379
Museum of Telecommunications, Néa Kifissia 23122
Museum of Telecommunications, Rīga 30074
The Museum of Television and Radio, New York 51229
Museum of Tetovo Region, Tetovo 30532
Museum of Texas Tech University, Lubbock 50312
Museum of Textile and Clothing Industry, Budapest 23485
Museum of Textile Art of the Moscow State Textile University n.a. A.N. Kosygin, Moskva 37267
Museum of Textile Industry, Sliven 05168
Museum of Thai Herbal Medicine and Thai Pharmaceutical History, Bangkok 42549
Museum of the 10th Pavilion of the Warszaw Citadel, Warszawa 35282
Museum of the 16th Winter Olympic Games, Albertville 11534
Museum of the 19th Century Tartu Citizens, Tartu 10644
Museum of the 40's, Point Clear Bay 45395
Museum of the Afghanistan War, Moskva 37287
Museum of the Agricultural and Food Industrial Trade School, Szabadkígyós 23830
Museum of the Agricultural College and Research Institute, Coimbatore 24122
Museum of the Albanian Costume, Frascineto 26247
Museum of the Albemarle, Elizabeth City 48338
Museum of the Aleutians, Unalaska 53642
Museum of the Almanach of Rusalka Dnistrova, Lviv 43130
Museum of the Alphabet, Waxhaw 54000
Museum of the Altaj Republic named after A.V. Anochin, Gorno-Altajsk 36927
Museum of the American Cowboy, Colorado Springs 47704
Museum of the American Hungarian Foundation, New Brunswick 51035
Museum of the American Numismatic Association, Colorado Springs 47705
Museum of the American Piano, New York 51230
Museum of the American Quilter's Society, Paducah 51669
Museum of the American West, Los Angeles 50255
Museum of the Americas, Lomas de Zamora 00430
Museum of the Ancient Memorials of Lviv, Lviv 43133
Museum of the Ancient Near East, Berlin 17279
Museum of the Arabic World Institute, Paris 14673
Museum of the Archdiocese, Wrocław 35327
Museum of the Archdiocese Gniezno, Gniezno 34672
Museum of the Archeology of Prikamsk Region, Perm 37548
Museum of the Archipelago, Lappoby-Åland 11052
Museum of the Arkansas Grand Prairie, Stuttgart 53351
Museum of the Armed October Rebellion, Voronež 37973
Museum of the Armenian Genocide, Yerevan 00757
Museum of the Art and Life in the GULAG, Moskva 37376
Museum of the Art of Ancient Ukrainian Books, Lviv 43132
Museum of the Artist Morozov, Ivanovo 36951
Museum of the Association of Sports Dynamo Moscow, Moskva 37363
Museum of the Augustinian Abbey of Klosterneuburg, Klosterneuburg 02202
Museum of the Aviation in the North, Archangelsk 36771
Museum of the Aviation Pioneer Aleksandras Griskevicius - Mazeikiai Museum, Viekšniai 30395
Museum of the Banat, Timişoara 36689
Museum of the Bank of Lithuania, Vilnius 30414
Museum of the Bārda Family Rumbiņi, Katvaru 30030
Museum of the Battle of Grunwald, Stębark 35134
Museum of the Battle of Legnica, Legnickie Pole 34849
Museum of the Bedford Historical Society, Bedford 46688
Museum of the Benua Family, Sankt-Peterburg 37707
Museum of the Berkeley Springs, Berkeley Springs 46762
Museum of the Beskid Mountains, Wisła 35303
Museum of the Beskydy, Frýdek-Místek 09504
Museum of the Bible, Alūksne 29985
Museum of the Big Bend, Alpine 46238
Museum of the Black Eagle Pharmacy, Székesfehérvár 23854
Museum of the Black Sea Fleet, Moskva 37289
Museum of the Blind - Typhlology, Zagreb 09013
Museum of the Bohemian Paradise, Turnov 09881
Museum of the Bolivian Coast, La Paz 04218
Museum of the Bulgarian National Revival, Varna 05241
Museum of the Butyrska Prison, Moskva 37263
Museum of the Cape Fear Historical Complex, Fayetteville 48532
Museum of the Čapek Brothers, Malé Svatoňovice 09646
Museum of the Cariboo Chilcotin, Williams Lake 07337
Museum of the Castle, Baranów Sandomierski 34534
Museum of the Center for Meteorite Studies, Tempe 53470
Museum of the Central Bank of the Philippines, Manila 34445
Museum of the Central Marine Fisheries Research Station, Mandapam Camp 24306
Museum of the Centre of Europe, Vilnius 30403
Museum of the Chapter, Soignies 04035
Museum of the Cherokee Indian, Cherokee 47423
Museum of the Cherokee Strip, Enid 48401
Museum of the Children Social Institution of Turkey, Ankara 42714
Museum of the Church of Saint Peter and Paul, Tjumen 37843
Museum of the City Brussels, Bruxelles 03451
Museum of the City of Athens, Athinai 22939
Museum of the city of Bad Berleburg, Bad Berleburg 16692
Museum of the City of Coimbra, Coimbra 35531
Museum of The City of Gdynia, Gdynia 34657
Museum of the City of Havana, La Habana 09136
Museum of the City of Kratovo, Kratovo 30504
Museum of the City of Lake Worth, Lake Worth 49896
Museum of the city of Lüdenscheid, Lüdenscheid 20088
Museum of the City of Manila, Manila 34446
Museum of the City of Negotino, Negotino 30510
Museum of the City of New York, New York 51231
Museum of the City of Skopje, Skopje 30518
Museum of the City of Split, Split 08960
Museum of the City Sarajevo, Sarajevo 04277
Museum of the College of Engineering, Chennai 24118
The Museum of the Confederacy, Richmond 52335
Museum of the Conspirativ Flat of the Russian Social Democratic Party of 1904-1906, Uljanovsk 37902
Museum of the Convent of Agios Andreas, Agios Andreas Arkadias 22869
Museum of the Cork Industry, Santa Maria da Feira 35903
Museum of the Cramp of Beverlo, Leopoldsburg 03785
Museum of the Cultural Palace of National Minorities, Beijing 07553
Museum of the Culture of Astrakhan, Astrachan 36785
Museum of the Culture of Carpathian Germans, Bratislava 38222
Museum of the Culture of Hungarians in Slovakia, Bratislava 38223
Museum of the Cumbraes, Millport 45148
Museum of the Cumhuriyet Era, Ankara 42700
Museum of the Custom, Vilnius 30426
Museum of the Customs and Finance Guard, Budapest 23489
Museum of the Czech Countryside, Kutná Hora 09615
Museum of the Deaf, Helsinki 10796
Museum of the Decembrists, Minusinsk 37137
Museum of the Decembrists - House of the Volkonskyjs, Irkutsk 36936
Museum of the Decembrists - House of Trubetskoy, Irkutsk 36935
Museum of the Democratic Movements, Machida 28877
Museum of the Development of Metallurgy, Moskva 37351
Museum of the Diplomatic Corps, Vologda 37962
Museum of the Dobrzyń Region, Golub-Dobrzyń 34675
Museum of the Dog, Saint Louis 52592
Museum of the Dolenjsko Region, Novo Mesto 38490
Museum of The Duke of Wellington's Regiment, Halifax 44385
Museum of the Dutch Clock, Zaandam 32978
Museum of the Education of the Komi Region at the Syktyvkar State University, Syktyvkar 37819
Museum of the Egyptian People's Assembly, Cairo 10507
Museum of the Elbe Region, Poděbrady 09724
Museum of the Eliasi Family Zīlēni, Platone 30055
Museum of the Environmental Sciences - Institute for Agricultural and Forest Environment - Polish Academy of Sciences, Poznań 35046
Museum of the Estonian Swedes, Haapsalu 10565
Museum of the European Year 1997, Atpazarı 42718
Museum of the Evenks, Tura 37877
Museum of the Evgeny Vakhtangov Theatre, Moskva 37369
Museum of the Evolution of Brain at the Scientific Research Institute of Brain at the Russian Academy of Medical Sciences, Moskva 37275
Museum of the Faculty of Agriculture, Sapporo 29215
Museum of the Faculty of Arts, Alexandria 10476
Museum of the Făgăras County Valeriu Literat, Făgăraş 36363
Museum of The Fairs Foundation, Medina del Campo 39638
Museum of the FC Red Star, Beograd 38064
Museum of the Fire Brigade, Teremia Mare 36686
Museum of the Fire Department, Ponce 36052
Museum of the Fire Department Italia 5, Bellavista 34123
Museum of the Flood Disaster, Ouwerkerk 32661
Museum of the Franciscan Convent, Jerusalem 25004
Museum of the Franciscan Monastery, Rovinj 08942
Museum of the Franciscan Monastery Košljun, Punat 08936
Museum of the Fur Trade, Chadron 47327
Museum of the Galician People, Santiago de Compostela 40026
Museum of the Gas Industry, Bratislava 38231
Museum of the Geological Sciences, Blacksburg 46845
Museum of the George and Nefeli Giabra Pierides Collection, Lefkosia 09380
Museum of the Glory of the Fighters, Astrachan 36783
Museum of the Gonia Monastery, Kissamos 23056
Museum of the Gorki Filmstudios, Moskva 37305
Museum of the Grandfather Durov, Moskva 37377
Museum of the Great Plains, Lawton 50000
Museum of the Grove of the Village Shrine, Osaka 29135
Museum of the Gulf Coast, Port Arthur 52029
Museum of the Hajdúság Region, Hajdúböszörmény 23603
Museum of the Heroic Defense and Liberation of Sevastopol, Sevastopol 43177
Museum of the Highwood, High River 05986
Museum of the Historical Society of Trappe, Collegeville, Perkiomen Valley, Trappe 53552
Museum of the History and Art of the Holy City of Missolonghi, Mesolongion 23090
Museum of the History and Culture of the Middle Prikamje, Sarapul 37746
Museum of the History and Ethnography of the Asov Greeks, Mariupol' 43145
Museum of the History, Culture and Life of Kola Saamis, Lovozero 37113
Museum of the History of Adler District of Sochi, Soči 37781
Museum of the History of Aksay, Aksaj 36749
Museum of the History of Architecture and Industrial Technology of the Ural Region, Ekaterinburg 36892
Museum of the History of Azerbaijan, Bakı 03144
Museum of the History of Buzuluk, Buzuluk 36832
Museum of the History of Calculation and Computer Technology, Moskva 37301
Museum of the History of Catalonia, Barcelona 38988
Museum of the History of Children's Games and Books, Myrina 23106
Museum of the History of Communication of the Tatarian Republic, Kazan 37017
Museum of the History of Cypriot Coinage, Lefkosia 09381
Museum of the History of Dagestan Fishing Industry, Machačkala 37117
Museum of the History of Department of Internal Affairs of the Kaluga Region, Kaluga 36999
Museum of the History of Development of Norilsk Industrial Region, Norilsk 37474
Museum of the History of Ekaterinburg, Ekaterinburg 36893
Museum of the History of Folk Arts and Crafts, Nižnij Novgorod 37456
Museum of the History of Houses in the 19th and 20th Centuries, Tjumen 37847
Museum of the History of Irkutsk, Irkutsk 36937
Museum of the History of Jewish People, Girona 39347
Museum of the History of Kasan University, Kazan 37016
Museum of the History of Khmelnitski, Chmel'nyc'kyj 43029
Museum of the History of Kolmya, Kolomija 43071
Museum of the History of Kyïv, Kyïv 43089
Museum of the History of Lithuania Medicine and Pharmacy, Kaunas 30255
Museum of the History of Lublin, Lublin 34902
Museum of the History of Medicine, Astrachan 36784
Museum of the History of Medicine, Bakı 03164
Museum of the History of Medicine, Varna 05242
Museum of the History of Money and Banking, Moskva 37290
Museum of the History of Moscow, Moskva 37294
Museum of the History of Moscow District Military Regiments, Moskva 37300
Museum of the History of Political Banishment, Bratsk 36823
Museum of the History of Political Banishment in Yakutia, Jakutsk 36963
Museum of the History of Printing, Sankt-Peterburg 37681
Museum of the History of Russian Entrepreneurship, Moskva 37297
Museum of the History of Schlüsselburg, Šlisselburg 37768
Museum of the History of Science, Oxford 45334
Museum of the History of Science, Sydney 01556
Museum of the History of Sochi, Soči 37783
Museum of the History of Stavropol, Stavropol 37804
Museum of the History of Strasbourg, Strasbourg 15831
Museum of the History of Termurids at the Academy of Sciences of Uzbekistan, Toshkent 54489
Museum of the History of the Blind, Hellerup 10056
Museum of the History of the City of Poznań, Poznań 35036
Museum of the History of the District, Belozersk 36801
Museum of the History of the Herzen State Pedagogical University of Russia, Sankt-Peterburg 37685
Museum of the History of the JSC Television Plant named after V.I. Lenin, Nižnij Novgorod 37458

M

Museum of the History of the Moscow Agricultural Timiryazev Academy, Moskva 37293
Museum of the History of the N.A. Alekseev Psychiatric Hospital, Moskva 37288
Museum of the History of the Perm Region, Perm 37549
Museum of the History of the Poltical Police of Russia, Sankt-Peterburg 37683
Museum of the History of the Preobrashenskiy Giljarovskiy Psychiatric Hospital, Moskva 37295
Museum of the History of the Revolutionary-Democratic Movement of the 1880-1890 Years, Sankt-Peterburg 37684
Museum of the History of the Stae Marine University named after G.I. Nevelskyi, Vladivostok 37945
Museum of the History of the Theatres of Dagestan, Machačkala 37118
Museum of the History of Tiberias and the Lake Kinnereth, Tiberias 25129
Museum of the History of Tjumen, Tjumen 37848
Museum of the History of Tourism, Sankt-Peterburg 37687
Museum of the Holmesdale Natural History Club, Reigate 45481
Museum of the Holy House of Mercy, Macau 08004
Museum of the Holy Land, Santiago de Compostela 40023
Museum of the Holy Shroud, Torino 28007
Museum of the Holy Trinity Church of Ethiopia, Addis Ababa 10679
Museum of the Hudson Highlands, Cornwall-on-Hudson 47815
Museum of the Hungarian Culture & Danubian Region, Komárno 38269
Museum of the Hungarian Petroleum Industry, Zalaegerszeg 23972
Museum of the Hussite, Žlutice 09945
Museum of the Ice Age, Moskva 37366
Museum of the Imperial Collections, Tokyo 29425
Museum of the Imperial Palace of Manchu State, Changchun 07589
Museum of the Institute of African Studies, Ibadan 33322
Museum of the Institute of African Studies, Oshogbo 33340
Museum of the Institute of Ethnology, Taipei 42409
Museum of the Institute of History and Philology, Taipei 42410
Museum of the Institute of Marine Biology of the Far Eastern Branch of the Russian Academy of Sciences, Vladivostok 37944
Museum of the Iron Age, Andover 43303
Museum of the Isles, Isle-of-Skeye 44541
Museum of the Japan Monkey Centre, Inuyama 28578
Museum of the Jewellery Quarter, Birmingham 43508
Museum of the Jewish Community, Tovačov 09871
Museum of the Jewish Historical Institute, Warszawa 35285
Museum of the Jimmy Carter Library, Atlanta 46446
Museum of the Kalisz Region, Kalisz 34710
Museum of the Kampinoska Forest, Kampinos 34714
Museum of the Kaunas Archdiocese, Kaunas 30248
Museum of the Kaunas Doll Theatre, Kaunas 30250
Museum of the Kazan Chemical Secondary School, Kazan 37019
Museum of the Kitamae-Bune Cargo Ships, Echizen 28414
Museum of the Kłodzko Region, Kłodzko 34747
Museum of the Kórnik Library, Kórnik 34756
Museum of the Kozara Region, Prijedor 04271
Museum of the Krapina Prehistoric Man, Krapina 08900
Museum of the Krkonoše Mountains, Vrchlabí 09914
Museum of the Krotoszyn Region, Krotoszyn 34832
Museum of the Kujawy and Ziemia Dobrzyńska, Włocławek 35311
Museum of the Kysuce Village, Nová Bystrica 38319
Museum of the Latvia Agriculture University, Jelgava 30020
Museum of the Liberation of Pleven 1877, Pleven 05118
Museum of the Light Infantry, Charleroi 03517
Museum of the Literary Life of the Ural in the 19th Century, Ekaterinburg 36897
Museum of the Literature of the XVIIth - XIXth Ceneturies, Moskva 37330
Museum of the Literature of the XXth c, Moskva 37355
Museum of the Lithuanian Radio and Television, Vilnius 30420
Museum of the Llano Estacado, Plainview 51975
Museum of the Lubusz Region, Zielona Góra 35372
Museum of the Manchester Regiment, Ashton-under-Lyne 43332
Museum of the March district, canton Schwyz, Vorderthal 41993
Museum of the Marsh of the Village Fighter, La Habana 09148
Museum of the Martyrdom of Allied Prisoners of War, Żagań 35352
Museum of the Mausoleum of King Nanyue, Guangzhou 07734
Museum of the Medical College, Mysore 24331
Museum of the Meiji Restoration, Kagoshima 28632
Museum of the Merchant Marine, Piraeus 23146
Museum of the Metropolis Square, Naxos 23120
Museum of the Mexican Army and Air Force at Barrack Colorado, Guadalajara 30899
Museum of the Mid-Suffolk Light Railway, Wetheringsett 45984
Museum of the Middle Otava Area, Strakonice 09849
Museum of the Ministry of Internal Affairs, Sofia 05186
Museum of the Mitsubishi Heavy Industries, Nagasaki 28977
Museum of the Mologa Region, Rybinsk 37605
Museum of the Moravian Heights, Nové Město na Moravě 09679
Museum of the Mordovian Culture, Saransk 37743
Museum of the Moscow Academic Artist Theatre, Moskva 37336
Museum of the Moscow-Concert Agency, Moskva 37334
Museum of the Mosfilm Filmstudios, Moskva 37306
Museum of the Mossovet-Theatre, Moskva 37367
Museum of the Mountain Man, Pinedale 51925
Museum of the Mussorgsky State Academic Opera and Ballet Theatre, Sankt-Peterburg 37677
Museum of the N.E. Bauman Moscow State Technical University, Moskva 37337
Museum of the Natalya Sats Children's Music Theatre, Moskva 37271
Museum of the National Academy of Art, Sofia 05185
Museum of the National Bank of Belgium, Bruxelles 03466
Museum of the National Center of Afro-American Artists, Boston 46942
Museum of the National Old Theatre, Kraków 34810
Museum of the National Palace, Guatemala 23299
Museum of the National Park of the Losiny Island, Moskva 37282
Museum of the Nature of Buryatya, Ulan-Udė 37896
Museum of the Naval Fleat, Moskva 37384
Museum of the Nenetsk Autonomous Region, Narjan-Mar 37443
Museum of the Never Seen, Puymaurin 14962
Museum of the 'New Academy of Fine Art', Sankt-Peterburg 37704
Museum of the Nizhni Novgorod' Intelligentsia, Nižnij Novgorod 37464
Museum of the North, Fairbanks 48472
Museum of the Northamptonshire Regiment, Northampton 45269
Museum of the Northern Great Plains, Fort Benton 48593
Museum of the Norwegian Knitting Industry, Salhus 33807
Museum of the Oaş Land, Negreşti-Oaş 36489
Museum of the Occupation of Latvia (1940 - 1991), Riga 30084
Museum of the Ocean Fleat, Moskva 37333
Museum of the Old Poland Basin, Sielpia 35104
Museum of the Old Serbian Orthodox Church, Sarajevo 04281
Museum of the Olive and Greek Olive Oil, Spárti 23197
Museum of the Olomučany ceramics, Olomučany 09691
Museum of the Opera House, Budapest 23433
Museum of the Opole Countryside, Opole 34987
Museum of the Orava Village Zuberec, Zuberec 38393
Museum of the Order of Saint John, London 44930
Museum of the Origin of the Polish State, Gniezno 34673
Museum of the Peasant Revolts, Gornja Stubica 08876
Museum of the Percy Tenantry Volunteers 1798-1814, Alnwick 43288
Museum of the Plains Indian, Browning 47092
Museum of the Poet S. Orlov, Belozersk 36802
Museum of the Police of the Czech Republic, Praha 09773
Museum of the Police of the Slovak Republic, Bratislava 38226
Museum of the Policy Academy, Ljubljana 38456
Museum of the Považie Region, Žilina 38391
Museum of the Poznań Army, Poznań 35032
Museum of the Poznań-Citadel, Poznań 35045
Museum of the Presidential Center of Culture of the Republic of Kazakhstan, Astana 29692
Museum of the Printing Arts Leipzig, Leipzig 19922
Museum of the Puck Region, Puck 35063
Museum of the Puerto Rican Family of the 19th Century, San Juan 36077
Museum of the Red River, Idabel 49461
Museum of the Reformed College and Ecclesiastical Art, Debrecen 23520
Museum of the Regiments, Calgary 05526
Museum of the Region of Podlasie, Siedlce 35103
Museum of the Region of Pyzdry, Pyzdry 35069
Museum of the Region of the Upper Hron, Brezno 38238
Museum of the Region Wallachia in Vsetin, Vsetín 09918
Museum of the Revolution, La Habana 09137
Museum of the River Daugava, Salaspils 30122
Museum of the River Fleet of the Volga State Academy of Naval Transport, Nižnij Novgorod 37465
Museum of the Rockies, Bozeman 46982
Museum of the Romanaţi, Caracal 36259
Museum of the Rose, Kazanlăk 05059
Museum of the Royal College of Physicians, London 44931
Museum of the Royal College of Surgeons of England, London 44932
Museum of the Royal Houses, Santo Domingo 10351
Museum of the Royal Leicestershire Regiment, Leicester 44678
Museum of the Royal Military School of Music, Twickenham 45887
Museum of the Royal Palace, Gödöllő 23565
Museum of the Royal Pharmaceutical Society of Great Britain, London 44933
Museum of the Royal Thai Air Force, Bangkok 42550
Museum of the Royal Tombs of Sipán, Lambayeque 34204
Museum of The Royal Westminster Regiment Historical Society, New Westminster 06437
Museum of the Rural Municipality of Roland, Roland 06771
Museum of the Russian Academy of Art, Sankt-Peterburg 37706
Museum of the Sacrifice of the People of Kalavryta, Kalavryta 23028
Museum of the Sarvajanik Wachanalaya, Nasik 24336
Museum of the Sculpture Studio, Victoria de las Tunas 09350
Museum of the Sea Fleet of the Far East Sea Shippery, Vladivostok 37946
Museum of the Serbian Orthodox Church, Beograd 38078
Museum of the Serbian Orthodox Church, Beograd 38077
Museum of the Serbian Orthodox Church, Dubrovnik 08870
Museum of the Shwenawdaw Pagoda, Bago 31716
Museum of the Silesian Piasts, Brzeg 34571
Museum of the Silver Century - The Brjusov House, Moskva 37361
Museum of the Slovak National Councils, Myjava 38311
Museum of the Slovakian National Uprising, Banská Bystrica 38184
Museum of the Slovakian Village, Martin 38299
Museum of the Slovinian Village in Kluki, Smołdzino 35114
The Museum of the South Dakota State Historical Society, Pierre 51915
Museum of the Southern Electrical Power Network, Jelgava 30016
Museum of the Southern Jewish Experience, Jackson 49563
Museum of the Southwest, Midland 50673
Museum of the Stanislavsky Theatre, Moskva 37368
Museum of the State Academic Maly Theatre, Moskva 37281
Museum of the Statesman Eleftherios Venizelos and the Corresponding Historical Period, Athinai 22940
Museum of the Stock Company Vulkans, Kuldīga 30035
Museum of the Stroganov University of Fine and Industrial Art, Moskva 37335
Museum of the Struggle against Bandits, Trinidad 09336
Museum of the Students Struggle, Camagüey 09051
Museum of the Sugar Industry, Abel Santamaría 09020
Museum of the Sugar Industry, San Miguel de Tucumán 00599
Museum of the Sulmierzyce Region, Sulmierzyce 35139
Museum of the Swiss Abroad, Pregny-Chambésy 41735
Museum of the Taipei Municipal Cheng Kung Senior High School, Taipei 42411
Museum of the Textile Enterprise Danilov Manufctures, Moskva 37299
Museum of the Theatre of the Russian Army, Moskva 37372
Museum of the Theatre of Young Generation, Moskva 37370
Museum of the Thirties, Boulogne-Billancourt 12090
Museum of the Tkachev Brothers, Brjansk 36827
Museum of the Town, Penza 37531
Museum of the Town, Radovljica 38502
Museum of the Town of Bad Ischl, Bad Ischl 01755
Museum of the Traditional Provençal Frabrice, Tarascon (Bouches-du-Rhône) 15856
Museum of the Transdanubian Reformed Church District, Pápa 23751
Museum of the Turkish Revolution 1839, İstanbul 42840
Museum of the TV Technical and Programming History, Budapest 23481
Museum of the Twentieth Century, Hoorn 32409
Museum of the Underground Prisoners-Acre, Acre 24920
Museum of the University of Connecticut Health Center, Farmington 48514
Museum of the Ural State Technical University, Ekaterinburg 36899
Museum of the Vitkovice Metal Works, Ostrava 09698
Museum of the Walewski Family in Tubadzin, Wróblew 35321
Museum of the Waxhaws and Andrew Jackson Memorial, Waxhaw 54001
Museum of the Weathersfield Historical Society, Weathersfield 54016
Museum of the Western Jesuit Missions, Hazelwood 49232
Museum of the Western Prairie, Altus 46245
Museum of the Weston Historical Society, Weston 54106
Museum of the White Mountains of Samaria, Chania 22965
Museum of the Women, Echt 32166
Museum of the Worcestershire Regiment, Worcester 46071
Museum of the World Ocean, Kaliningrad 36992
Museum of the Writer A.P. Bibik, Mineralnye Vody 37133
Museum of the Writers of Orël, Orël 37510
Museum of the Yapi Kredi Culture Center, İstanbul 42848
Museum of the Young Defenders of the Homeland, Kursk 37097
Museum of the Yugoslav Cinema, Beograd 38068
Museum of the Zawkrzańska Region, Mława 34927
Museum of Theatre Art of Serbia, Beograd 38073
Museum of Theatre Dolls, Moskva 37374
Museum of Theatre Puppets by A.A. Veselov, Voronež 37974
Museum of Theatrical, Musical and Cinematographic Art of Ukraine, Kyïv 43077
Museum of Thomas Godec, Bohinjska Bistrica 38401
Museum of Tinian Artists, Pyrgos 23163
Museum of Tolerance, Los Angeles 50256
Museum of Tools, Delft 32056
Museum of Tourism and History of Teberda, Teberda 37839
Museum of Town Construction and Everyday Life of Taganrog, Taganrog 37825
Museum of Town Council, Tjumen 37846
Museum of Town Life Style, Ruse 05153
Museum of Trade, Bratislava 38225
Museum of Trade, Ilava 38264
Museum of Trade and Industry, Warszawa 35268
Museum of Trademen's Life, Ekaterinburg 36896
Museum of Traditional Costumes, Alor Gajah 30548
Museum of Traditional Nan Bel Music and Theater, Changhua 42158
Museum of Traditional Toys, San Francisco de los Romo 31331
Museum of Transport, Bratislava 38221
Museum of Transport, Glasgow 44292
Museum of Transport and Technology, Auckland 33056
Museum of Transport - Greater Manchester, Manchester 45093
Museum of Transportation, Saint Louis 52593
Museum of Transportation and Rural Industries, Boyanup 00886
Museum of Tropical Queensland, Townsville 01598
Museum of Tropinin and His Contemporaries, Moskva 37375
Museum of Tuchola Wood, Tuchola 35200
Museum of Türi Technical and Rural Economy School and Türi Municipality, Särevere 10616
Museum of Turkish and Islamic Art, Edirne 42768
Museum of Turkish and Islamic Art, İstanbul 42845
Museum of Turkish Tiles and Artisan Works, İstanbul 42847
Museum of Turkmen Values, Aşgabat 42971
Museum of Turopolja, Velika Gorica 08977
Museum of Ukrainian-Ruthenian Culture, Svidník 38367
Museum of Underground and Partizan Struggle in the Mykolaïv Region during the Great Patriotic War 1941-1945, Mykolaïv 43148
Museum of Underground Struggle, Nueva Gerona 09234
Museum of Unique Dolls, Moskva 37378
Museum of V.N. Goryaev, Moskva 37278
Museum of Velká Bíteš, Velká Bíteš 09899

Museum of Vertebrate Zoology, Berkeley 46759
Museum of Veterinary, Guwahati 24194
Museum of Veterinary Pathology and Teratology, Ozzano dell'Emilia 27026
Museum of Victorian Reed Organs and Harmoniums, Shipley 45666
Museum of Victorian Whitby, Whitby 45995
Museum of Vilnius Sport History, Vilnius 30441
Museum of Vinegar, Handa 28497
Museum of Vision, San Francisco 52788
Museum of Vision Science and Optometry, Waterloo 07288
Museum of Visual Art, Galați 36379
Museum of Visual Impairment, Jiris 10871
Museum of Vladimir Dal, Moskva 37269
Museum of Volynian Orthodox Icons, Luc'k 43112
Museum of Vyatsk Folk Arts and Crafts, Kirov 37041
Museum of Waldensian History, Valdese 53684
Museum of Warsaw Gaswork, Warszawa 35243
Museum of Watchmaker, Artisans and Precision, Warszawa 35283
Museum of Waxen Figures, Sankt-Peterburg 37711
Museum of Weapons, Gjirokastër 00017
Museum of Weaving and Silk, Reggio Calabria 27381
Museum of Wellington, City and Sea, Wellington 33251
Museum of Welsh Life, Saint Fagans 45565
The Museum of Western Art, Kerrville 49751
Museum of Western Australian Sport, Claremont, Western Australia 00977
Museum of Western Colorado, Grand Junction 48994
Museum of Western Kashubia, Bytów 34585
Museum of Wine, Prešov 38336
Museum of Wine and Viticulture, Aigle 41094
Museum of Witchcraft, Cornwall 43894
Museum of Wonders, Imbil 01172
Museum of Wood Carving and Icon Paintings, Trjavna 05230
Museum of Woodcarving, Shell Lake 53069
Museum of Woodrafting, Javnik 38424
Museum of Woodwork and Ornamental Turning, Ashburton 33038
Museum of Worcester Porcelain, Worcester 46072
Museum of Work, Norrköping 40769
Museum of Work, Setúbal 35937
Museum of Workers' Struggles, Manzanillo 09207
Museum of World Culture, Göteborg 40558
Museum of World Cultures, Frankfurt am Main 18381
Museum of World Cultures, Genova 26304
Museum of World Religions, Yungho 42484
Museum of Wschowa Region, Wschowa 35347
Museum of Yachting, Newport 51323
Museum of Yarmouth History, Yarmouth 54324
Museum of Yarmukian, Kibbutz Shaar Hagolan 25058
Museum of Yilan Government History, Yilan 42468
Museum of Ylä-Savo, Iisalmi 10848
Museum of Yokohama Urban History, Yokohama 29602
Museum of York County, Rock Hill 52403
Museum of Yorkshire Dales Lead Mining, Earby 44078
Museum of Yugoslav History, Beograd 38066
Museum of Yugoslav History - Flowers House, Beograd 38055
Museum of Zoology, Ann Arbor 46310
Museum of Zoology, Lund 40709
Museum of Zoology, Latvia University, Rīga 30092
Museum of Zoology of the A.I. Herzen State Pedagogical University, Sankt-Peterburg 37737
Museum of Zoology of the Sankt-Petersburg State University, Sankt-Peterburg 37736
Museum of Zubrnice Railway Line, Zubrnice 09950
Museum Old Village, Sirogojno 38142
Museum on Artillery, Cavalery and Infantery from 1870-1920, Seclin 15695
Museum on Main Street, Pleasanton 51986
Museum on the Boyne, Alliston 05319
Museum on the Mound, Edinburgh 44137
Museum on the Seam, Jerusalem 25005
Museum Ons' Lieve Heer op Solder, Amsterdam 31864
Museum Ons Mijnverleden, Houthalen 03719
Museum Oorlog en Vrede, Breda 32005
Museum Orth, Orth an der Donau 02473
Museum Orvelte, Orvelte 32636
Museum Oskar Reinhart am Stadtgarten, Winterthur 42035
Museum Osterzgebirgsgalerie im Schloß, Dippoldiswalde 17872
Museum Osthusschule, Bielefeld 17343
Museum Otto Ludwig, Eisfeld 18108
Museum Otto Schäfer, Schweinfurt 21687
Museum Otto Weidt's Workshop for the Blind, Berlin 17193
Museum Otzberg, Otzberg 20963
Museum Oud-Asperen, Asperen 31933
Museum Oud Overschie, Rotterdam 32714
Museum Oud-Rijnsburg, Rijnsburg 32687
Museum Oud Soest, Soest 32793
Museum Oud Vriezenveen, Vriezenveen 32911
Museum Oud Westdorpe, Westdorpe 32938
Museum Oude Boekdrukkunst, Almen 31802
Museum Our Lord in the Attic, Amsterdam 31864
Museum Pachen, Rockenhausen 21341
Museum Pachten, Dillingen, Saar 17864
Museum Palace of Peter the Great, Sankt-Peterburg 37636
Museum Palagan, Ambarawa 24436
Museum Palthehof, Nieuwleusen 32582
Museum Pancacsila Sakti, Jakarta 24493
Museum Pangeran Diponegoro, Yogyakarta 24590
Museum Papageorgiou, Limenaria 23077
Museum Papiermühle Homburg, Markt Triefenstein 20214
Museum Passeier - Andreas Hofer, San Leonardo in Passiria 27654
Museum Pater Valentinus Paquay-Heilig Paterke, Hasselt 03684
Museum Paul Delvaux, Sint-Idesbald 04016
Museum Paul Tétar Van Elven, Delft 32065
Museum Pemda Balige, Balige 24437
Museum Pemirintah Daerah Grobogan, Purwodadi Grobogan 24549
Museum Pendidikan Islam, Yogyakarta 24591
Museum Perjuangan, Bogor 24459
Museum Perjuangan, Yogyakarta 24592
Museum Perjuangan Bukit Barisan, Medan 24527
Museum Pers, Surakarta 24569
Museum Pers Antara, Jakarta 24494
Museum Petersberg, Petersberg bei Halle, Saale 21014
Museum Pfalzgalerie, Kaiserslautern 19412
Museum Pfleggerichtshaus, Anras 01723
Museum Plagiarius, Solingen 21805
Museum Plantin-Moretus / Prentenkabinet, Antwerpen 03297
Museum Pol Mara, Gordes 13085
Museum Polović, Dobova 38410
Museum Pos & Giro, Bandung 24445
Museum Prabu Geusan Ulun, Sumedang 24562
Museum Preserve Kremlin of Kazan, Kazan 37025
Museum Presidency, Bloemfontein 38565
Museum Priest Houses, Zwickau 22834
Museum Priesterhäuser, Zwickau 22834
Museum Professionals Group, Duxford 55077
Museum Prüm, Prüm 21136
Museum Pugung Ulago Sembah, Donggala 24469
Museum Pura Mangkunegara, Surakarta 24570
Museum Purbakala Mojokerto, Mojokerto 24529
Museum Purbakala Trowulan, Mojokerto 24530
Museum Puri Lukisan, Ubud 24580
Museum Quintana - Archäologie in Künzing, Künzing 19738
Museum Raabe-Haus, Eschershausen 18235
Museum Rade am Schloß Reinbek, Reinbek 21251
Museum Radio-Wereld, Diever 32129
Museum Radya Pustaka, Surakarta 24571
Museum Raesfeld, Raesfeld 21166
Museum Railway Blonay-Chamby, Blonay 41242
Museum Ratna Wartha, Gianyar 24471
Museum Ravensburger, Ravensburg 21204
Museum Region Neulengbach, Neulengbach 02435
Museum Regional, Savognin 41824
Museum Regiunal Surselva Casa Carniec, Ilanz 41492
Museum Reich der Kristalle, München 20468
Museum Reichenau, Reichenau 21244
Museum Reichenfels, Hohenleuben 19238
Museum Rekor Indonesia Jamu Jago, Semarang 24558
Museum Reksa Artha, Jakarta 24495
Museum René de Clercq, Deerlijk 03537
Museum Retz, Retz 02579
Museum Rietberg Zürich, Zürich 42097
Museum Rietberg Zurich, Zürich 42097
Museum Rijn- en Binnenvaart, Antwerpen 03298
Museum Rijswijk Het Tollenshuis, Rijswijk, Zuid-Holland 32690
Museum Ritter, Waldenbuch 22288
Museum Roemervilla, Bad Neuenahr-Ahrweiler 16796
Museum Römervilla, Grenzach-Wyhlen 18754
Museum Rolf Werner, Bansin, Seebad 16926
Museum Rosenegg, Kreuzlingen 41514
Museum Rudolf Wachter, Kißlegg 19559
Museum Rufferdinge, Landen 03776
Museum Ruman Bolon, Pematang 24543
Museum Rupelklei, Terhagen 04056
Museum Sacrum, Tiel 32824
Museum Sächsisch-Böhmisches Erzgebirge, Marienberg 20198
Museum Saet en Cruyt, Andijk 31906
Museum Saigerhütte Olbernhau- Kupferhammer, Olbernhau 20893
Museum Salz und Moor, Grassau 18740
Museum Salzbütte, Huttwil 41491
Museum Sankt Blasien, Sankt Blasien 21490
Museum Sankt Dionys, Esslingen 18276
Museum Sankt Ingbert, Sankt Ingbert 21500
Museum Sankt Leon-Rot, Sankt Leon-Rot 21504
Museum Sankt Veit, Sankt Veit an der Glan 02691
Museum Sankt Wendel, Sankt Wendel 21514
Museum Saoraja Lapawowoi, Watampone 24582
Museum Sarganserland, Sargans 41818
Museum Sasmita Loka, Yogyakarta 24593
Museum Sasmita Loka A. Yani, Jakarta 24496
Museum Satria Mandala, Jakarta 24497
Museum Scheuerle, Backnang 16660
Museum Scheveningen, Den Haag 32089
Museum Schietkamp-Harskamp, Harskamp 32320
Museum Schillerhaus, Mannheim 20169
Museum Schlösschen im Hofgarten, Wertheim 22504
Museum Schloß Adelsheim, Berchtesgaden 17002
Museum Schloß Bernburg, Bernburg 17295
Museum Schloss Bruck, Lienz 02302
Museum Schloß Burgk, Burgk 17658
Museum Schloss Colditz, Colditz 17724
Museum Schloß Ehrenstein, Ohrdruf 20890
Museum Schloss Fasanerie, Eichenzell 18079
Museum Schloss Fellenberg, Merzig 20301
Museum Schloss Friedrichsfelde, Berlin 17215
Museum Schloß Glücksburg, Römhild 21356
Museum Schloss Hardenberg, Velbert 22211
Museum Schloss Hellenstein, Heidenheim an der Brenz 19093
Museum Schloss Herzberg, Herzberg am Harz 19167
Museum Schloß Hohenlimburg, Hagen 18859
Museum Schloß Hohentübingen, Tübingen 22123
Museum Schloss Homburg, Nümbrecht 20753
Museum Schloß Klippenstein, Radeberg 21158
Museum Schloß Kuckuckstein, Liebstadt 19986
Museum Schloss Kyburg, Kyburg 41521
Museum Schloß Lackenbach, Lackenbach 02250
Museum Schloss Lembeck, Dorsten 17911
Museum Schloss Lichtenberg, Fischbachtal 18322
Museum Schloß Lichtenburg, Prettin 21122
Museum Schloß Moritzburg, Zeitz 22777
Museum Schloss Münsingen, Münsingen 41660
Museum Schloss Netzschkau, Netzschkau 20580
Museum Schloss Neuenburg, Freyburg (Unstrut) 18459
Museum Schloss Oberschwappach, Knetzgau 19582
Museum Schloss Ottersbach, Großklein 02004
Museum Schloss Peigarten, Dobersberg 01820
Museum Schloß Ratibor, Roth 21404
Museum Schloß Rochlitz, Rochlitz 21338
Museum Schloß Rochsburg, Lunzenau 20109
Museum Schloß Schwarzenberg, Schwarzenberg 21675
Museum Schloss Seefeld, Seefeld 21715
Museum Schloss Steinau an der Straße mit Brüder-Grimm-Gedenkstätte, Steinau an der Straße 21878
Museum Schloß Steinheim, Hanau 18981
Museum Schloss Tenneberg, Waltershausen 22318
Museum Schloss und Gut Liebenberg, Löwenberger Land 20019
Museum Schloss Waldegg, Feldbrunnen 41369
Museum Schloß Weesenstein, Müglitztal 20396
Museum Schloß Wilhelmsburg, Schmalkalden, Kurort 21580
Museum Schmelzra, Scuol 41853
Museum Schmiede am Burgplatz, Backnang 16661
Museum Schneggli, Reinach (Aargau) 41753
Museum Schnütgen, Köln 19625
Museum Schöne Stiege, Riedlingen 21319
Museum Schöngrabern, Hollabrunn 02096
Museum Schokland, Ens 32209
Museum Schorfheide, Schorfheide 21624
Museum Schwab, Biel 41232
Museum Schwarzes Ross, Hilpoltstein 19189
Museum Science and Technology, Jakarta 24498
Museum Seitingen-Oberflacht, Seitingen-Oberflacht 21734
Museum Sejarah Mesjid Banten, Bogor 24460
Museum Sender Dobl, Dobl 01822
Museum Seni, Denpasar 24467
Museum Seni, Jakarta 24499
Museum Seni, Medan 24528
Museum Seni, Padang 24534
Museum Seni Rupa dan Keramik, Jakarta 24500
Museum Sensenwerk, Deutschfeistritz 01813
Museum Sepakat Segenap, Aceh Tenggara 24433
Museum-Ship, Krasnojarsk 37076
Museum Ship Dar Pomorza, Gdynia 34662
Museum Ship Norgoma, Sault Sainte Marie 06931
Museum Ship Sigyn, Turku 11396
Museum Ship Sołdek, Gdańsk 34651
Museum Simalungun, Pematangsiantar 24545
Museum Sint Bernardushof, Aduard 31790
Museum Sint Sellian, Sellingen 32768
Museum Siwa Lima, Ambon 10363
Museum Slag bij Heiligerlee, Heiligerlee 32347
Museum Slag der Zilveren Helmen, Halen 03666
Museum Slager, 's-Hertogenbosch 32373
Museum Smallingerland, Drachten 32155
Museum Smidt van Gelder, Antwerpen 03299
Museum Smolensk Region in the Years of the Great Patriotic War, Smolensk 37772
Museum Soltau, Soltau 21814
Museum Sønderjylland - Museum of Fine Arts Brundlund Castle, Aabenraa 09954
Museum Sono Budoyo, Yogyakarta 24594
Museum Sonyine Malige, Halmahera Tengah 10366
Museum Sophie La Roche, Bönnigheim 17427
Museum Spaans Gouvernement, Maastricht 32527
Museum Spasskiy Church, Tjumen 37849
Museum Specken, Bad Zwischenahn 16877
Museum SPUR Cham, Cham 17690
Museum Stad Appingedam, Appingedam 31919
Museum Stadt Bad Hersfeld, Bad Hersfeld 16747
Museum Stadt Königsberg, Duisburg 18025
Museum Stammertal, Unterstammheim 41963
Museum Stapelen, Boxtel 31997
Museum Starnberger See, Starnberg 21864
Museum Stedhûs Sleat, Sloten 32787
Museum Steinarbeiterhaus, Hohburg 19232
Museum Steinau...das Museum an der Straße, Steinau an der Straße 21879
Museum Steinmetzhaus, Eggenburg 01844
Museum Stellingwerff-Waerdenhof, Hasselt 03685
Museum Stemmler, Schaffhausen 41830
Museum Stichting van het Rotterdamsche Tramweg Maatschappij, Ouddorp 32649
Museum Stift Börstel, Berge 17006
Museum Stift Melk, Melk 02375
Museum Stoffels Säge-Mühle, Hohenems 02093
Museum Stoomdepot, Rotterdam 32715
Museum Stoomgemaal Winschoten, Winschoten 32952
Museum Strauhof, Zürich 42098
Museum Strom und Leben, Recklinghausen 21209
The Museum Studio and Arts Centre, Nairobi 29748
Museum Suaka Budaya, Surakarta 24572
Museum Subak Bali, Kediri 24513
Museum Sudirman, Magelang 24519
Museum Sulawesi Tengah, Palu 24540
Museum Sumpah Pemuda, Jakarta 24501
Museum Suriname, Amsterdam 31865
Museum Sursilvan, Trun 41953
Museum Swaensteyn, Voorburg 32906
Museum Synagoge Gröbzig, Gröbzig 18768
Museum Synthese, Hallbergmoos 18887
Museum 't Coopmanshûs, Franeker 32232
Museum 't Gevang, Doetinchem 32136
Museum 't Kniphof, Essen 03589
Museum 't Nieuwhuys Hoegaarden, Hoegaarden 03705
Museum 't Schilderhuis, Driebergen-Rijsenburg 32157
Museum Taf Wallet, Koksijde 03754
Museum Takstil, Jakarta 24502
Museum Taman Laut Ancol, Jakarta 24503
Museum Taman Prasati, Jakarta 24504
Museum Taras Shevtchenko's Kobzarya, Čerkasy 43003
Museum Taruna Akbari Udarat, Magelang 24520
Museum Tauernbahn, Schwarzach 02718
Museum Tegernseer Tal, Tegernsee 22017
Museum Telefonzentrale Magden, Magden 41602
Museum The Grocery, Leeuwarden 32475
Museum The Kennedys, Berlin 17216
Museum The Other Art, Moskva 37272
Museum Theo Jans, Klarenbeek 32450
Museum Theo Kerg, Schriesheim 21637
Museum Theulegium, Tholey 22056
Museum Thonet, Frankenberg, Eder 18344
Museum Thurnfels, Völs 02869
Museum Tinguely, Basel 41173
Museum Tni A.L. Loka Jala Carana, Surabaya 24567
Museum Ton Schulten, Ootmarsum 32632
Museum Tre Kronor, Stockholm 40916
Museum - Treasures from the National Archives, Kew 44571
Museum Trustee Association, Washington 55165
Museum Tucher Mansion and Hirsvogel Hall, Nürnberg 20792
Museum Tugu Nasional, Jakarta 24505
Museum Tysmans, Houthalen 03720
Museum U.P.T. Balitung, Tanjungpandan 24576
Museum uf the History of the Russian Post, Moskva 37298
Museum und Begegnungsstätte Alter Krug, Zossen 22821
Museum und Besucherbergwerk, Goslar 18714
Museum und Besucherbergwerk der Graphit Kropfmühl AG, Hauzenberg 19053
Museum und Galerie der Stadt Schwabmünchen, Schwabmünchen 21646
Museum und Galerie im Prediger, Schwäbisch Gmünd 21650
Museum und Galerie in Uigendorf, Unlingen 22174
Museum und Kunstsammlung Schloß Hinterglauchau, Glauchau 18657
Museum und Kupferzeche Hüttau, Hüttau 02100
Museum und Stiftung Anton Geiselhart, Münsingen 20501
Museum unter der Y-Burg, Kernen im Remstal 19519
Museum Uranbergbau, Bad Schlema 16827
Museum Uslar, Uslar 22193
Museum Valkenheide, Maarsbergen 32516
Museum Valse Kunst, Vledder 32896

Museum Van Bommel Van Dam, Venlo 32884
Museum van de Abdij van Roosenberg, Waasmunster 04123
Museum van de Kanselarij der Nederlandse Orden, Apeldoorn 31913
Museum van de Koninklijke Marechaussee, Buren, Gelderland 32026
Museum van de Kruisbooggilde Sint-Joris, Brugge 03422
Museum van de Onze-Lieve-Vrouwkerk, Brugge 03423
Museum van de Sint-Jakobskerk, Antwerpen 03300
Museum van de Stichting Bisdom van Vliet, Haastrecht 32305
Museum van de Twintigste Eeuw, Hoorn 32409
Museum van de Vrouw, Echt 32166
Museum van de Willebroekse Vaart, Willebroek 04152
Museum van Deinze en de Leiestreek, Deinze 03539
Museum van Egmond, Egmond aan Zee 32180
Museum Van Gerwen-Lemmens, Valkenswaard 32869
Museum van Hedendaagse Kantwerken, Sint-Truiden 04028
Museum van Hedendaagse Kunst Antwerpen, Antwerpen 03301
Museum Van Hemessen, Woubrugge 32968
Museum van het Abtenhuis, Geraardsbergen 03652
Museum van het Ambacht, Den Haag 32090
Museum van het Heilig Bloed, Brugge 03424
Museum van het Kamp van Beverlo, Leopoldsburg 03785
Museum van het Nederlandse Uurwerk, Zaandam 32978
Museum van het Vlaams Studentenleven, Leuven 03788
Museum van Knipkunst, Westerbork 32940
Museum van Lien, Fijnaart 32230
Museum Van Loon, Amsterdam 31866
Museum van Zwerfstenen in de Rosmolen, Zeddam 32991
Museum Veluwezoom, Doorwerth 32147
Museum Verbindingsdienst, Ede 32170
Museum Viadrina, Frankfurt (Oder) 18406
Museum Victoria, Carlton South 00957
Museum Viechtach, Viechtach 22226
Museum Villa Erckens, Grevenbroich 18756
Museum Villa Haiss, Zell am Harmersbach 22779
Museum Villa Rot - Kunstsammlung Hoenes-Stiftung, Burgrieden 17662
Museum Villa Sarabodis, Gerolstein 18611
Museum Villa Stahmer, Georgsmarienhütte 18592
Museum Villa Stuck, München 20469
Museum Village, Monroe 50793
Museum Vítkovických Železáren, Ostrava 09698
Museum Vitznau-Rigi, Vitznau 41991
Museum Vlaamse Minderbroeders, Sint-Truiden 04029
Museum Vleeshuis, Antwerpen 03302
Museum Vliegbasis Deelen, Deelen 32054
Museum Volunteers of the Philippines, Dasmariñas 54986
Museum voor Anesthesie Verantare, Antwerpen 03303
Museum voor Communicatie, Den Haag 32091
Museum voor de Geschiedenis van de Wetenschappen, Gent 03636
Museum voor de Oudere Technieken, Grimbergen 03661
Museum voor Figuratieve Kunst De Buitenplaats, Eelde 32174
Museum voor Folklore, Zottegem 04168
Museum voor Geschiedenis van de Geneeskunde, Gent 03637
Museum voor Heem- en Oudheidkunde, Kontich 03757
Museum voor Heem- en Volkskunde, Wenduine 04143
Museum voor Heemkunde Almelo, Almelo 31799
Museum voor het Kruideniersbedrijf, Utrecht 32850
Museum voor het Radiozendamateurisme Jan Corver, Budel 32017
Museum voor Industriele Archeologie en Textiel, Gent 03638
Museum voor Keramiek Pablo Rueda Lara, Rotterdam 32716
Museum voor Kerkelijke Kunst Sint-Carolus, Antwerpen 03304
Museum voor Klederdracht en Oud Speelgoed, Warnsveld 32924
Museum voor Moderne Kunst, Arnhem 31929
Museum voor Naaldkunst, Wedde 32931
Museum voor Natuur- en Wildbeheer, Doorwerth 32148
Museum voor Religieuze Kunst, Uden 32833
Museum voor Schone Kunsten Gent, Gent 03639
Museum voor Sleutel en Slot, Oostduinkerke 03925
Museum voor Torhouts Aardewerk, Torhout 04079
Museum voor Volkskunde, Brugge 03425
Museum voor Volkskunde, Dendermonde 03541
Museum voor Vrede en Geweldloosheid, Amsterdam 31867
Museum Voswinckelshof, Dinslaken 17870
Museum Vrolik, Amsterdam 31868
Museum "Währungsreform 1948" und Konklave in Rothwesten, Fuldatal 18514
Museum Waemena, Irian Jaya 24474
Museum Wäschefabrik, Bielefeld 17344
Museum Wald und Umwelt, Ebersberg 18045
Museum Waldhof, Bielefeld 17345
Museum Walzengravieranstalt Guntramsdorf, Guntramsdorf 02019
Museum Wanua Paksinanta, Manado 24523
Museum - Warship Blyskawica, Gdynia 34659
Museum Wasseralfingen, Aalen 16413
Museum Wasseramt, Halten 41470
Museum Wasserburg, Wasserburg am Inn 22345
Museum Wasserburg Anholt, Isselburg 19368
Museum Wateringhuis, Lommel 03830
Museum Wattens - Industrie & Vorgeschichte, Wattens 02896
Museum Wayang, Jakarta 24506
Museum Weiler Textilgeschichte, Weil am Rhein 22377
Museum Weißenfels Schloß Neu-Augustusburg, Weißenfels 22457
Museum Weltjugendherberge, Altena 16488
Museum Weltkulturen, Mannheim 20170
Museum Wendener Hütte, Wenden 22476
Museum "Werkstatt Plaumer", Öhringen 20866
Museum Western Macedonia during the National Liberation War, Kičevo 30503
Museum Wierdenland, Ezinge 32229
Museum Wiesbaden, Wiesbaden 22557
Museum Wildberg, Wildberg 22567
Museum Willem van Haren, Heerenveen 32333
Museum Willet-Holthuysen, Amsterdam 31869
Museum Wilmersdorf - Forum für Stadtkultur und Geschichte, Berlin 17217
Museum Wilnsdorf, Wilnsdorf 22589
Museum Windmill 'The New Palmtree', Schiedam 32754
Museum Wirtles Haus, Rottenacker 21418
Museum Wolfhalden, Wolfhalden 42045
Museum Wolfram von Eschenbach, Wolframs-Eschenbach 22656
Museum Workshop and Stores, Grangemouth 44343
Museum-workshop of I. Kavaleridze, Kyïv 43096
Museum World of Water of Saint-Petersburg, Sankt-Peterburg 37699
Museum Würth, Künzelsau 19736
Museum Wurzen mit Ringelnatzsammlung, Wurzen 22751
Museum Wustrow, Wustrow, Wendland 22757
Museum Wuyts-Van Campen en Baron Caroly, Lier 03819
Museum Yadnya, Tangawi 24574
The Museum Yamato Bunkakan, Nara 29032
Museum Yerseke, Yerseke 32973
Museum Yogya Kembali, Yogyakarta 24595
Museum Zahal, Hakirya 24970
Museum Zeitreise Mensch, Cortaccia 25983
Museum Zeit(T)räume, Walldürn 22310
Museum Zella-Mehlis, Zella-Mehlis 22785
Museum Zeughaus, Mannheim 20171
Museum Zinkenbacher Malerkolonie, Sankt Gilgen 02649
Museum Zitadelle Jülich, Jülich 19405
Museum zu Allerheiligen, Schaffhausen 41831
Museum Zum Arabischen Coffe Baum, Leipzig 19927
Museum zum Schiff, Laufenburg 41533
Museum Zunftturm, Bacharach 16654
Museum zur brandenburg-preußischen Geschichte - Zinnfigurenmuseum, Gusow-Platkow 18842
Museum zur Geschichte der Herrnhuter Brüdergemeinde, Königsfeld im Schwarzwald 19650
Museum zur Geschichte der Juden im Ostalbkreis, Bopfingen 17473
Museum zur Geschichte der Juden in Kreis und Stadt Heilbronn, Obersulm 20843
Museum zur Geschichte Hohenheims, Stuttgart 21955
Museum zur Geschichte von Christen und Juden, Laupheim 19859
Museum zur Kloster- und Stadtgeschichte, Steinheim an der Murr 21891
Museum zur Ronmühle, Schötz 41843
Museum zur Stadtgeschichte, Königslutter am Elm 19654
Museum Zwaantje Hans Stockman's Hof, Schoonebeek 32763
Museum Zwijgershoek, Sint-Niklaas 04022
museumbickel, Walenstadt 41997
Museumboerderij, Leens 32469
Museumboerderij, Paasloo 32665
Museumboerderij, Staphorst 32801
Museumboerderij de Karstenhoeve, Ruinerwold 32739
Museumboerderij de Tip, Herveld 32381
Museumboerderij de Zwemkolk, Markelo 32538
Museumboerderij Erve Hofman, Hellendoorn 32357
Museumboerderij Eungs Schöppe, Markelo 32539
Museumboerderij Gilde Koat, Zeddam 32992
Museumboerderij New Greenwich Village, Zuidwolde, Drenthe 33015
Museumboerderij Scholten, Den Ham 32108
Museumboerderij 't Olde Maat Uus, Giethoorn 32254
Museumboerderij Tante Jaantje, Callantsoog 32034
Museumboerderij Vreeburg, Schagen 32747
Museumboerderij Wendezoele, Ambt-Delden 31805
Museumboerderij Westfrisia, Hoogwoud 32407
Museumbrouwerij de Hemel, Nijmegen 32591
Museumbrouwerij De Roos, Hilvarenbeek 32386
Museumcenter Leoben, Leoben 02287
Museumgemaal Caners, 's-Hertogenbosch 32374
MuseumgoudA, Gouda 32269
Museumhaven Zeeland, Zierikzee 33004
'T Museumke, Handel 32309
Museumkelder Derlon, Maastricht 32528
Museumkerk Sint Pieter, Rekem 03964
Museummolen, Schermerhorn 32751
Museummolen de Nieuwe Palmboom, Schiedam 32754
Museummolen Jan Pol, Dalen 32045
Museums Alaska, Anchorage 55166
Museums and Galleries Disability Association, Hove 55078
Museums Aotearoa, Wellington 54964
Museums Association, London 55079
Museums Association of India, Delhi 54885
Museums Association of Namibia, Windhoek 54951
Museums Association of Nigeria, Lagos 54967
The Museums Association of Pakistan, Karachi 54979
Museums Association of Saskatchewan, Regina 54743
Museums Association of Slovenia, Ljubljana 55013
Museums Association of the Caribbean, Bridgetown 54705
Museums Australia, Civic Square 54691
Museums-Bergwerk Schauinsland, Freiburg im Breisgau 18434
Museums Collection of the Nationbal Ossoliński Institute, Wrocław 35344
Museums Gold of Peru - Arms of the World, Lima 34273
Museums-Gutshof Sonnekalb, Bad Kösen 16764
Museums in Essex Committee, Chelmsford 55080
Museums, Libraries and Archives Council, London 55081
Museums, Libraries and Archives North East, Newcastle-upon-Tyne 55082
Museums North, Hartlepool 55083
Museums of Filipinana and Rizaliana, Sampaloc 34496
Museums of Furesø, Farum 10019
The Museums of Oglebay Institute, Wheeling 54124
Museums of Religions, Marburg 20194
Museums of South Lanarkshire, Ferniegair 44224
Museums of the Perm State University, Perm 37547
Museums ship Cap San Diego, Hamburg 18955
Museums Weapons Group, Leeds 55084
Museumsanlage Kulturstiftung Landkreis Osterholz, Osterholz-Scharmbeck 20935
Museumsbäckerei Imsweiler, Imsweiler 19331
Museumsbahn Bremerhaven-Bederkesa, Bad Bederkesa 16678
Museumsbauernhof Wennerstorf, Wennerstorf 22479
Museumsberg Flensburg, Flensburg 18328
Museumsboerderij Oud Noordwijk, Noordwijk, Zuid-Holland 32603
Museumsbund Österreich, Leonding 54696
Museumscenter Hanstholm, Hanstholm 10054
Museumschiff Elbe 1, Cuxhaven 17756
Museumschip Mercuur, Den Haag 32092
Museumsdampfzug Rebenbummler, Freiburg im Breisgau 18435
Museumsdorf Baruther Glashütte, Baruth 16932
Museumsdorf Bayerischer Wald, Tittling 22065
Museumsdorf Cloppenburg, Cloppenburg 17713
Museumsdorf Düppel, Berlin 17218
Museumsdorf Krumbach, Krumbach 02241
Museumsdorf Trattenbach, Ternberg 02803
Museumsdorf Volksdorf mit Spiekerhus, Hamburg 18953
Museumseisenbahn, Hamm, Westfalen 18971
Museumseisenbahn Payerbach-Hirschwang, Hirschwang 02081
Museumseisenbahn Schwalm-Knüll, Fritzlar 18484
Museumsfarm Tomita, Nakafurano 29015
Museumsfeuerschiff Amrumbank/Deutsche Bucht, Emden 18139
Museumsfriedhof Tirol, Kramsach 02222
Museumsgalerie, Altomünster 16525
Museumsgalerie Allerheiligenkirche, Mühlhausen/Thüringen 20404
Museumsgården, Læsø 10172
Museumshafen Oevelgönne, Hamburg 18954
Museumship Pommern, Mariehamn 11100
Museumshof, Bad Oeynhausen 16800
Museumshof, Neuruppin 20651
Museumshof, Rahden 21167
Museumshof, Roßtal 21384
Museumshof am Sonnenluch, Erkner 18216
Museumshof auf dem Braem, Gescher 18623
Museumshof Emmerstedt, Helmstedt 19124
Museumshof Ernst Koch, Wernigerode 22499
Museumshof Großkoschen, Großkoschen 18795
Museumshof Historischer Moorhof Augustendorf, Gnarrenburg 18665
Museumshof-Senne, Bielefeld 17346
Museumshof und Heimatmuseum, Falkenstein, Harz 18293
Museumshof Zirkow, Zirkow 22808
Museumskirche Sankt-Katharinen, Lübeck 20081
Museumsland Donauland Strudengau, Waldhausen 54697
Museumslandschaft Deilbachtal, Essen 18264
Museumslandschaft Hessen Kassel, Kassel 54842
Museumslogger AE7 Stadt Emden, Emden 18140
Museumsmühle, Sankt Julian 21503
Museumsmühle, Woldegk 22644
Museumsmühle Hasbergen, Delmenhorst 17815
Museumsmühle mit heimatkundlicher Sammlung, Varel 22205
Museumspädagogische Beratung, Kassel 19471
Museumspädagogische Gesellschaft e.V., Köln 54843
Museumspark, Rüdersdorf bei Berlin 21433
Museumspoorlijn Star, Stadskanaal 32799
Museumsscheune, Brest 17582
Museumsscheune, Helmenzen 19120
Museumsscheune Fränkische Schweiz, Hollfeld 19254
Museumsscheune Kremmen, Kremmen 19710
Museumsschiff Cap San Diego, Hamburg 18955
Museumsschiff FMS Gera, Bremerhaven 17572
Museumsschiff Mannheim, Mannheim 20172
Museumsschmiede, Winnweiler-Langmeil 22604
Museumsspinnerei Neuthal, Bäretswil 41150
Museumsstadl, Bernried, Kreis Deggendorf 17303
Museumstjenesten, Viborg 54782
Museumstoomtram Hoorn-Medemblik, Hoorn 32410
Museumstramway Mariazell-Erlaufsee, Mariazell 02358
Museumsverband Baden-Württemberg e.V., Villingen-Schwenningen 54844
Museumsverband des Landes Brandenburg e.V., Potsdam 54845
Museumsverband für Niedersachsen und Bremen e.V., Hannover 54846
Museumsverband Mecklenburg-Vorpommern e.V., Güstrow 54847
Museumsverband Sachsen-Anhalt e.V., Bernburg 54848
Museumsverband Schleswig-Holstein e.V., Rendsburg 54849
Museumsverband Thüringen e.V., Erfurt 54850
Museumsverbund Pankow - Heynstraße, Berlin 17219
Museumsverbund Pankow - Prenzlauer Allee, Berlin 54851
Museumsverbund Südniedersachsen e.V., Göttingen 54852
Museumsverein Bahnhofmuseum e.V, Benneckenstein 16994
Museumsverein Dokumentation 2. Weltkrieg Hürtgenwald e.V, Hürtgenwald 19294
Museumsverein Feuerwehr Obererbach, Obererbach 20805
Museumsverein Mönchengladbach e.V, Mönchengladbach 54853
Museumswohnung WBS 70, Berlin 17220
Museumszentrum Lorsch, Lorsch 20035
Museumszug Hespertalbahn, Essen 18265
Museumwerf 't Kromhout, Amsterdam 31870
Museumwinkel Albert Heijn, Zaandam 32979
Museus de História Natural e Etnologia Indígena, Juiz de Fora 04523
Museus Municipal de Almada, Almada 35415
Musexpo, Clairfayts 12536
Muséyé Chay-e Lahijan, Lahijan 24627
Muséyé Honarha-ye Melli, Tehrān 24685
Muséyé Kerman, Kerman 24618
Muséyé Mardom Shenassi, Tehrān 24686
Mushakoji Saneatsu Memorial Hall, Chofu 28406
Music and Time - Private Museum of John Mostoslavsky, Jaroslavl 36974
Music Box Museum of Izu, Ito 28594
Music Box of Zami Museum, Metulla 25065
The Music House Museum, Acme 46132
Music Information Museum, Kaohsiung 42240
Music-Memorial Museum of Igor Stravinsky, Ustyloog 43195
Music-Memorial Museum of Solomia Krushelnytska, Lviv 43136
Music Museum, La Habana 09166
Music Museum, Rīga 30098
Music Museum of the Slovak National Museum, Bratislava 38216
Musical Box Society International Museum, Houston 49399

M

Musical Boxes Museum, Kobe 28737
Musical Instruments Museum, Bruxelles 03477
Musical Instruments Museum, Jerusalem 25006
Musical Museum, London 44934
Musical Wonder House, Wiscasset 54263
Musik- und Wintersportmuseum, Klingenthal 19572
Musikhistorische Sammlung Jehle, Albstadt 16456
Musikhistorisk Museum og Carl Claudius' Samling, København 10142
Musikinstrumente-Museum der Völker, Sankt Gilgen 02650
Musikinstrumenten-Ausstellung des Händel-Hauses, Halle, Saale 18902
Musikinstrumenten-Museum, Berlin 17221
Musikinstrumenten-Museum, Markneukirchen 20207
Musikinstrumenten- und Puppenmuseum, Goslar 18715
Musikinstrumentenmuseum Lißberg, Ortenberg-Lißberg 20915
Musikinstrumentenmuseum Schloss Kremsegg, Kremsmünster 02234
Musikinstrumentensammlung, Würzburg 22719
Musikinstrumentensammlung der Universität, Göttingen 18696
Musikinstrumentensammlung Hans und Hede Grumbt, Hattingen 19039
Musikmuseet, Stockholm 40917
Musikmuseum, Basel 41174
MUSIS, Graz 54698
Muskegon County Museum, Muskegon 50925
Muskegon Museum of Art, Muskegon 50926
Muskogee War Memorial Park and Military Museum, Muskogee 50929
Muskoka Heritage Place, Huntsville 06003
Muskoka Lakes Museum, Port Carling 06631
Muskoka's Steamship Museum, Gravenhurst 05908
Musquodoboit Railway Museum, Musquodoboit 06405
Musselshell Valley Historical Museum, Roundup 52473
Mustafa Kamil Museum, Cairo 10508
Mustansiriya School Collections, Baghdad 24716
Mustard Shop Museum, Norwich 45281
Musteum of the History of Altai Literature, Art and Culture, Barnaul 36791
Mustiala Agricultural Museum, Mustiala 11127
Mustialan Maataloushistoriallinen Museo, Mustiala 11127
Muswellbrook Regional Arts Centre, Muswellbrook 01341
Mutare Museum, Mutare 54676
Mutinės Muziejus, Vilnius 30426
Mutsen en Poffermuseum Sint-Paulusgasthuis, Sint Oedenrode 32777
Muttart Art Gallery, Calgary 05527
Mutter Museum, Philadelphia 51857
Mutter. Museum für Kunst. Sammlung Berger, Amorbach 16542
Mutua Museum, Imphal 24226
Muurla Arts and Crafts Museum, Muurla 11128
Muurlan Kotiseutumuseo, Muurla 11128
Muuruveden Kotiseutumuseo, Muuruvesi 11129
Muuseum, Tartu 10653
Művészetek Háza - Vass Gyűjtemény, Veszprém 23961
Művészetek Kincsesháza - Vaszary Emlékház, Kaposvár 23631
Művészetek Kincsesháza - Vaszary Képtár, Kaposvár 23632
MUWA - Museum der Wahrnehmung, Graz 01980
Muzee Scheveningen, Den Haag 32093
Muzei Geroïčnoï Oborony i Vyzvolennja Sevastopolja, Sevastopol 43177
Muzei Permskogo Gosudarstvennogo Universiteta, Perm 37547
Muzej 700 let - Landskrona, Nevskoe Ustjě, Nienšanc, Sankt-Peterburg 37668
Muzej A.A. Kiseleva, Tuapse 37863
Muzej 'A Muzy ne Molčali', Sankt-Peterburg 37669
Muzej A.V. Suvorova, Velikij Novgorod 37920
Muzej Abaja, Abaj 29669
Muzej Afričke Umetnosti, Beograd 38060
Muzej Aleksandăr Stambolijski, Slavovica 05164
Muzej Aleksandăr Stambolijski, Sofia 05182
Muzej Al'manachu Rusalki Dnistrovoï, Lviv 43130
Muzej Alpinizma, Turizma i Istorii Kurorta Teberda, Teberda 37839
Muzej Anatomii, Moskva 37260
Muzej Antropologii i Ėtnografii im. Petra Velikogo (Kunstkamera), Sankt-Peterburg 37670
Muzej Antropologii im. D.N. Anučina Moskovskogo Gosudarstvennogo Universiteta im M.V. Lomonosova, Moskva 37261
Muzej Archeologii, Almaty 29685
Muzej Archeologii, Syktyvkar 37817
Muzej Archeologii i Ėtnografii, Syktyvkar 37818
Muzej Archeologii i Ėtnografii, Ufa 37887
Muzej Archeologii i Ėtnografii, Vladivostok 37943
Muzej Archeologii Moskvy, Moskva 37262
Muzej Archeologii Prikamja, Perm 37548
Muzej Archeologija, Ėtnografija i Ėkologija Sibiri Kemerovskogo Gosudarstvennogo Universiteta, Kemerovo 37033
Muzej Architektury i Archeologii Vez Burana, Tokmok 29967
Muzej Architektury i Byta Narodov Nižegorodskogo Povolžja, Nižnij Novgorod 37453
Muzej Architektury i Stroitelstva, Samara 37617
Muzej-Archiv D.I. Mendeleeva, Sankt-Peterburg 37671
Muzej Arsenal, Lviv 43131
Muzej Arsenal - Muzej Velikoj Otečestvennoj Vojny, Voronež 37971
Muzej Automobila, Beograd 38061
Muzej Aviacii Severa, Archangelsk 36771
Muzej Avtogravov imeni Imperatora Nikolaja Vtorogo, Novosibirsk 37484
Muzej Avtomobilnogo Transporta, Kursk 37096
Muzej Baki Urmanče, Kazan 37013
Muzej Banjičkog Logora, Beograd 38062
Muzej Belišće, Belišće 08843
Muzej Bitola, Bitola 30500
Muzej Boevogo i Trudovogo Podviga 1941-1945, Saransk 37741
Muzej Boevoj Slavy, Astrachan 36783
Muzej E.A. Boratynskogo, Kazan 37014
Muzej Bratjev Tkačevych, Brjansk 36827
Muzej Brdovec, Prigorje Brdověko 08931
Muzej Brjanskij Les, Brjansk 36828
Muzej Brodskog Posavlja, Slavonski Brod 08953
Muzej I.A. Bunina, Orël 37508
Muzej Butyrskoj Tjurmy, Moskva 37263
Muzej V.I. Čapaeva, Čeboksary 36840
Muzej Carskoselskaja Kollekcija, Puškin 37583
Muzej Cerkov Dekabristov, Čita 36870
Muzej Cerkov Petra i Pavla, Tjumen 37843
Muzej Cetinjske Krajine, Sinj 08950
Muzej Cetinjskog Manastira, Cetinje 31642
Muzej Chlebnoe Delo v Moskve, Moskva 37264
Muzej-chram Svjatitelja Nikolaja v Tolmačach, Moskva 37265
Muzej Chrama Christa Spasitelja, Moskva 37266
Muzej Chudožestvennych Tkanej, Moskva 37267
Muzej Cirkovogo Iskusstva, Sankt-Peterburg 37672
Muzej-čitalnja N.V. Fëdorova, Moskva 37268
Muzej-čitalnja na Nacionalnych Jazykach, Penza 37530
Muzej Čovek i negovoto Zdrave, Varna 05239
Muzej Crne Metalurgije Sartid A.D., Smederevo 38143
Muzej Đakovština, Đakovo 08857
Muzej V.I. Dalja, Moskva 37269
Muzej Dekabristov, Minusinsk 37137
Muzej Dekabristov - Dom Trubeckogo, Irkutsk 36935
Muzej Dekabristov - Dom Volkonskich, Irkutsk 36936
Muzej Dekorativno-Prikladnogo i Promyšlennogo Iskusstva, Moskva 37270
Muzej Dekorativno-prikladnogo Iskusstva, Sankt-Peterburg 37673
Muzej Dekorativno-prikladnogo Iskusstva Urala, Čeljabinsk 36844
Muzej Dekoratyvno-prykladnogo Mystectva, Sumy 43189
Muzej Derevjannogo Zodčestva Malye Karely, Archangelsk 36772
Muzej Detskogo Muzykalnogo Teatra im. N. Sats, Moskva 37271
Muzej Detskogo Tvorčestva, Sankt-Peterburg 37674
Muzej Detstva A.M. Gorkogo - Domik Kaširina, Nižnij Novgorod 37454
Muzej-Diorama Orlovskaja Nastupatelnaja Operacija, Orël 37509
Muzej Diplomatičeskogo Korpusa, Vologda 37962
Muzej Dom Mašarova, Tjumen 37844
Muzej Dominikanskog Samostana, Bol 08847
Muzej Dominikanskog Samostana, Dubrovnik 08867
Muzej Drugoe Iskusstvo, Moskva 37272
Muzej A.A. Durova, Taganrog 37824
Muzej Durylina S.N., Korolev 37058
Muzej Dvorec A.D. Menšikova, Sankt-Peterburg 37675
Muzej Ėkslibrisa, Moskva 37273
Muzej Emfiedžieva Kăšta, Kjustendil 05066
Muzej Ėtnografičeskogo Kostjuma na Kuklach, Moskva 37274
Muzej Ėtnografii, Istorii i Kultury Narodov Baškortostana, Ufa 37888
Muzej Ėvoljucii Mozga, Moskva 37275
Muzej Evolucije i Nalazište Pračovjeka Hušnjakovo, Klanjec 38429
Muzej-Filial Nadi Ruševoj, Kyzyl 37098
Muzej Filiki Eterija, Odesa 43157
Muzej Fizičeskoj Kultury i Sporta, Moskva 37276
Muzej Fizičke Kulture, Beograd 38063
Muzej FK Crvena Zvezda, Beograd 38064
Muzej Ford i Bălgarija, Sofia 05183
Muzej Franjevačkog Samostana, Dubrovnik 08868
Muzej Franjevačkog Samostana, Rovinj 08942
Muzej Franjevačkog Samostana Košljun, Punat 08936
Muzej Gabdulla Tukaja, Kazan 37015
Muzej Gacka, Otočac 08924
Muzej-Galerija Kavadarci, Kavadarci 30502
Muzej Gasilstva Dravinjske doline, Slovenske Konjice 38520
Muzej Geodezii i Kartografii, Moskva 37277
Muzej Geologii, Almaty 29686
Muzej Geologii, Čeljabinsk 36845
Muzej Geologii, Toshkent 54476
Muzej Geologii Centralnoj Sibirii, Krasnojarsk 37075
Muzej Geologii, Nefti i Gaza, Tjumen 37845
Muzej Georgi Velčev, Varna 05240
Muzej Get'manstva, Kyïv 43088
Muzej Glavna Kvartira na Ruskata Armija 1877-1878, Pordim 05135
Muzej V.N. Gorjaeva, Moskva 37278
Muzej A.M. Gorkogo, Moskva 37279
Muzej Goroda Jurjevca, Jurjevec 36984
Muzej Goroda Penzy, Penza 37531
Muzej Gorodskaja Duma, Tjumen 37846
Muzej Gosudarstvennogo Akademičeskogo Bolšogo Teatra Rossii, Moskva 37280
Muzej Gosudarstvennogo Akademičeskogo Malogo Teatra, Moskva 37281
Muzej Gosudarstvennogo Akademičeskogo Marijnskogo Teatra Opery i Baleta, Sankt-Peterburg 37676
Muzej Gosudarstvennogo Akademičeskogo Teatra Opery i Baleta imeni Mussorgskogo, Sankt-Peterburg 37677
Muzej Gosudarstvennogo Prirodnogo Nacionalnogo Parka Losinogo Ostrova, Moskva 37282
Muzej Grada Beograda, Beograd 38065
Muzej Grada Koprivnice, Koprivnica 08892
Muzej Grada Novog Sada, Novi Sad 38121
Muzej Grada Pazina, Pazin 08927
Muzej Grada Perasta, Perast 31651
Muzej Grada Rijeke, Rijeka 08937
Muzej Grada Sarajeva, Sarajevo 04277
Muzej Grada Splita, Split 08960
Muzej Grada Trogira, Trogir 08970
Muzej Grada Zagreba, Zagreb 09004
Muzej Grada Zenice, Zenica 04294
Muzej Gradostroitelstva i Byta Taganroga, Taganrog 37825
Muzej Grebnych Vidov Sporta, Moskva 37283
Muzej Hercegovine, Mostar 04270
Muzej Herzegovine-Trebinje, Trebinje 04287
Muzej Hmeljarstva, Žalec 38539
Muzej Hrvatske Književnosti i Kazališne Umjetnosti, Zagreb 09005
Muzej Hrvatskih Arheoloških Spomenika, Split 08961
Muzej Hvarske Baštine, Hvar 08880
Muzej I. Karpenka-Karogo, Kirovograd 43069
Muzej i Obščestvennyj Centr Mir, Progress, Prava Čeloveka im. Andreja Sacharova, Moskva 37284
Muzej Igora Stravinskogo, Ustyloog 43195
Muzej Igorja Talkova, Moskva 37285
Muzej Ikona, Korčula 08895
Muzej im. Auėzova Kzachskogo Dramatičeskogo Teatra, Almaty 29687
Muzej im. S.M. Kirova, Sankt-Peterburg 37678
Muzej im. B.M. Kustodieva, Ostrovskoe 37521
Muzej-institut Semji Rerichov, Sankt-Peterburg 37679
Muzej Instituta Biologii Morja, Vladivostok 37944
Muzej Internacionalnoj Družby, Saransk 37742
Muzej Ipavčeva Hiša, Šentjur 38510
Muzej Iskusstva Narodov Vostoka, Moskva 37286
Muzej Istočne Bosně, Tuzla 04290
Muzej Istorii Afganskoj Vojny, Moskva 37287
Muzej Istorii Aksaja, Aksaj 36749
Muzej Istorii Architektury i Architekturnogo Obrazovanija Sibiri, Novosibirsk 37485
Muzej Istorii Architektury i Promyšlennoj Techniki Urala, Ekaterinburg 36892
Muzej Istorii Bolnicy im. N.A. Alekseeva, Moskva 37288
Muzej Istorii Burjatii imeni Changalova, Ulan-Udė 37894
Muzej Istorii Černomorskogo Flota, Moskva 37289
Muzej Istorii Deneg i Bankovskogo Dela, Moskva 37290
Muzej Istorii Doma XIX-XX vekov, Tjumen 37847
Muzej Istorii Ekaterinburga, Ekaterinburg 36893
Muzej Istorii Goroda Irkutska, Irkutsk 36937
Muzej Istorii Goroda Joškar-Oly, Joškar-Ola 36978
Muzej Istorii Goroda-Kurorta Soči, Soči 37783
Muzej Istorii Goroda Naberežnye Čelny, Naberežnye Čelny 37437
Muzej Istorii Goroda Šlisselburga, Šlisselburg 37768
Muzej Istorii Goroda Stavropolja, Stavropol 37804
Muzej Istorii Goroda Tjumeni, Tjumen 37848
Muzej Istorii i Kultury Goroda Votkinska, Votkinsk 37978
Muzej Istorii i Kultury Moskovskogo Rajona, Nižnij Novgorod 37455
Muzej Istorii i Kultury Srednego Prikamja, Sarapul 37746
Muzej Istorii Izučenija i Osvoenija Evropejskogo Severa, Apatity 36764
Muzej Istorii Jaroslavlja, Jaroslavl 36972
Muzej Istorii Kamnereznogo i Juvelirnogo Iskusstva, Ekaterinburg 36894
Muzej Istorii Kazanskogo Universiteta, Kazan 37016
Muzej Istorii Kolskich Saamov, Lovozero 37113
Muzej Istorii Kraja, Belozersk 36801
Muzej Istorii Milicii, Sankt-Peterburg 37680
Muzej Istorii Mista Chmel'nyc'kogo, Chmel'nyc'kyj 43029
Muzej Istorii Mista Dniprodzeržins'ka, Dniprodzeržins'k 43030
Muzej Istorii Mista Kolomyï, Kolomija 43071
Muzej Istorii Mista Kyjeva, Kyïv 43089
Muzej Istorii Morskogo Universiteta im. G.I. Nevelskogo, Vladivostok 37945
Muzej Istorii Moskovskogo Gosudarstvennogo Universiteta im. M.V. Lomonosova, Moskva 37291
Muzej Istorii Moskovskogo Metropolitena, Moskva 37292
Muzej Istorii Moskovskoj Selskochosjajstvennoj Akademii im K.A. Timirjazeva, Moskva 37293
Muzej Istorii Moskvy, Moskva 37294
Muzej Istorii Narodnych Chudožestvennych Promyslov, Nižnij Novgorod 37456
Muzej Istorii OAO Gaz, Nižnij Novgorod 37457
Muzej Istorii OAO Nižegorodskij Televizionnyj Zavod im. V.I. Lenina, Nižnij Novgorod 37458
Muzej Istorii Orenburga, Orenburg 37513
Muzej Istorii Osvoenija i Razvitija Norilskogo Promyšlennogo Rajona, Norilsk 37474
Muzej Istorii Pečati Sankt-Peterburga, Sankt-Peterburg 37681
Muzej Istorii Permskogo Rajona, Perm 37549
Muzej Istorii Peterburgskoj Konservatorii, Sankt-Peterburg 37682
Muzej Istorii Političeskoj Policii Rossii - Gorochovaja 2, Sankt-Peterburg 37683
Muzej Istorii Političeskoj Ssylki, Bratsk 36823
Muzej Istorii Politssylki v Jakutii, Jakutsk 36963
Muzej Istorii Poljarnych Olimpiad, Murmansk 37435
Muzej Istorii Preobraženskoj Bolnicy im. V.A. Giljarovskogo, Moskva 37295
Muzej Istorii Priokskogo Rajona, Nižnij Novgorod 37459
Muzej Istorii Prosveščenija Komi Kraja, Syktyvkar 37819
Muzej Istorii Razvitija Mediciny, Astrachan 36784
Muzej Istorii Razvitija Mosenergo, Moskva 37296
Muzej Istorii Revoljucionno-demokratičeskogo Dviženija 1880-1890 gg, Sankt-Peterburg 37684
Muzej Istorii Rossijskogo Gosudarstvennogo Pedagogičeskogo Universiteta im A.I. Gercena, Sankt-Peterburg 37685
Muzej Istorii Rossijskogo Predprinimatelstva, Moskva 37297
Muzej Istorii Russkoj Počty, Moskva 37298
Muzej Istorii Rybnoj Promyšlennosti Dagestana, Machačkala 37117
Muzej Istorii Sankt-Peterburgskogo Universiteta, Sankt-Peterburg 37686
Muzej Istorii Svjazi Respubliki Tatarstan, Kazan 37017
Muzej Istorii ta Etnografiï Grekiv Priazov'ja, Mariupol' 43145
Muzej Istorii ta Kul'tury Mista Lugans'ka, Lugans'k 43117
Muzej Istorii Teatrov Dagestana, Machačkala 37118
Muzej Istorii Tekstilnoj Fabriki Danilovskaja Manufaktura, Moskva 37299
Muzej Istorii Turizma, Sankt-Peterburg 37687
Muzej Istorii UVD Kalužskoj Oblasti, Kaluga 36999
Muzej Istorii Vojsk Moskovskogo Voennogo Okruga, Moskva 37300
Muzej Istorii Vyčislitelnoj Techniki, Moskva 37301
Muzej Istorije Jugoslavije, Beograd 38066
Muzej Istoryčnych Koštovnostej Ukraïny, Kyïv 43090
Muzej Ivana Gončara, Kyïv 43091
Muzej Ivanovskogo Sitca, Ivanovo 36950
Muzej Izobrazitel'nych Iskusstv, Ašgabat 42972
Muzej Izobrazitelnych Iskusstv, Komsomolsk-na-Amure 37054
Muzej Izobrazitelnych Iskusstv, Rostov-na-Donu 37600
Muzej Izobrazitelnych Iskusstv im. A.S. Puškina, Moskva 37302
Muzej Izobrazitelnych Iskusstv Karelii, Petrozavodsk 37561
Muzej Izobrazitelnyh Iskusstv Respubliki Marij Ėl, Joškar-Ola 36979
Muzej Jantarja, Kaliningrad 36991
Muzej Jevreja Bosne i Hercegovine, Sarajevo 04278
Muzej Josipa Jurčića, Muljava 38485
Muzej Jovana Cvijića, Beograd 38067
Muzej Jovana Jovanovića Zmaja, Sremska Kamenica 38149
Muzej Jugoslavenska Kinoteka, Beograd 38068
Muzej Jugoslovenskih Vazduhoplovstva, Beograd 38069

Muzej Junye Zaščitniki Rodiny, Kursk 37097
Muzej Južne Železnice, Šentjur 38511
Muzej K. Ė. Ciolkovskogo, Aviacii i Kosmonavtiki, Kirov 37040
Muzej Kajuma Nasyri, Kazan 37018
Muzej Käkrinsko Chanče, Käkrina 05044
Muzej Karamzinskaja Obščestvennaja Biblioteka, Uljanovsk 37901
Muzej Kazakstan Respublikasynyn Prezidenttik Mėdeniet Ortalyfy, Astana 29692
Muzej Kazanskoj Chimičeskoj Školy, Kazan 37019
Muzej Keramiki i Usadba Kuskovo XVIII Veka, Moskva 37303
Muzej Kino, Moskva 37304
Muzej Kinoiskusstva Uzbekistana, Toshkent 54477
Muzej Kinostudii im. A.M. Gorkogo, Moskva 37305
Muzej Kinostudii Mosfilm, Moskva 37306
Muzej Klassičeskogo i Sovremennogo Iskusstva - Dom Burganova, Moskva 37307
Muzej V.O. Ključevskogo, Penza 37532
Muzej Knigi, Almaty 29688
Muzej Knigi, Moskva 37308
Muzej Knigi i Drukarstva Ukraïny, Kyïv 43092
Muzej Književnosti i Pozorišne Umjetnosti Bosne i Hercegovine i Galerija MAK, Sarajevo 04279
Muzej "Kobzarja" T.G. Ševčenka, Čerkasy 43003
Muzej A. Kolcova, Voronež 37972
Muzej Kolju Fičeto, Drjanovo 05025
Muzej Konevodstva, Moskva 37309
Muzej Konspiritavnaja Kvartira Simbirskoj Gruppy RSDRP 1904-1906, Uljanovsk 37902
Muzej Koški, Moskva 37310
Muzej Kosmonavtiki im. S. Koroljova, Žytomyr 43212
Muzej Kraevedenija g. Elabuga, Elabuga 36905
Muzej Krajine Negotin, Negotin 38116
Muzej Krapinskog Pračovjeka, Krapina 08900
Muzej Krejsera Avrora, Sankt-Peterburg 37688
Muzej-krepost Baba Vida, Vidin 05262
Muzej Kriminalistiki, Barnaul 36792
Muzej-kripta Aleksandăr Nevski, Sofia 05184
Muzej Kronštadtskaja Krepost, Kronštadt 37086
Muzej Kronštadtskogo Morskogo Zavoda, Kronštadt 37087
Muzej Krutickoe Podvore, Moskva 37311
Muzej Krylova, Tula 37868
Muzej Kuca Bore Stankovica, Vranje 38161
Muzej Kukol i Detskoj Knigi "Strana Čudes", Ekaterinburg 36895
Muzej Kukol 'Kukolnyj Dom', Moskva 37312
Muzej Kul'turnoï Spadščyny, Kyïv 43093
Muzej Kultury Astrachani, Astrachan 36785
Muzej Kupečeskogo Byta, Ekaterinburg 36896
Muzej Kutuzovskaja Izba, Moskva 37313
Muzej-kvartira Aktërskoj Semi M.V. i A.A. Mironovych i A.S. Menakera, Moskva 37314
Muzej-kvartira Alliluevych, Sankt-Peterburg 37689
Muzej-kvartira I.I. Brodskogo, Sankt-Peterburg 37690
Muzej-kvartira A.S. Čaplygina, Sankt-Peterburg 37691
Muzej-kvartira M.I. Cvetaevoj v Bolševo, Korolev 37059
Muzej-kvartira Dirižëra N.S. Golovanov, Moskva 37315
Muzej-kvartira F.M. Dostoevskogo, Moskva 37316
Muzej-kvartira Elizarovych, Sankt-Peterburg 37692
Muzej-kvartira A.B. Goldenvejzera, Moskva 37317
Muzej-kvartira A.M. Gorkogo, Moskva 37318
Muzej-kvartira A.M. Gorkogo, Nižnij Novgorod 37460
Muzej-kvartira G.D. Krasilnikova, Iževsk 36954
Muzej-kvartira A.I. Kuindži, Sankt-Peterburg 37693
Muzej-kvartira L. Lapcuja, Salechard 37609
Muzej-kvartira V.S. Mejercholda, Moskva 37319
Muzej-kvartira Musa Džalilja, Kazan 37020
Muzej-kvartira Naziba Žiganova, Kazan 37021
Muzej-kvartira V.I. Nemiroviča-Dančenko, Moskva 37320
Muzej-kvartira A.V. Neždanovoj, Moskva 37321
Muzej-kvartira A. Puškina, Sankt-Peterburg 37694
Muzej-kvartira Svjatoslava Richtera, Moskva 37322
Muzej-kvartira A.D. Sacharova, Nižnij Novgorod 37461
Muzej-kvartira F.I. Šaljapina, Sankt-Peterburg 37695
Muzej-kvartira T.G. Ševčenko, Sankt-Peterburg 37696
Muzej-Kvartira Ju.P. Spegalskogo, Pskov 37574
Muzej-kvartira I.D. Sytina, Moskva 37323
Muzej-kvartira K.A. Timirjazeva, Moskva 37324
Muzej-kvartira A.N. Tolstogo, Moskva 37325
Muzej-kvartira A.T. Tvardovskija v Smolenske, Smolensk 37770
Muzej-kvartira V. Tytova, Lugans'k 43118
Muzej Lavka Čechovych, Taganrog 37826
Muzej-ledokol Angara, Irkutsk 36938
Muzej-ledokol Bajkal, Angarsk 36762
Muzej Lefortovo, Moskva 37326
Muzej V.I. Lenina, Moskva 37327
Muzej Lesi Ukraïnki, Kyïv 43094
Muzej Lesovodstva, Moskva 37328
Muzej Ličnych Kollekcij, Moskva 37329
Muzej Literaturnaja Žizn' Urala XIX Veka, Ekaterinburg 36897
Muzej Literaturnoj Žizni Jaroslavskogo Kraja, Jaroslavl 36973
Muzej Literatury Burjatii im. Choca Namsaraeva, Ulan-Udė 37895
Muzej Literatury XVII - XIX vekov, Moskva 37330
Muzej Ljudevit Gaj, Krapina 08901
Muzej Ljudske Revolucije Trbovlje, Trbovlje 38528
Muzej M.V. Lomonosova, Sankt-Peterburg 37697
Muzej M.V. Lisenka, Kyïv 43095
Muzej-majsternja I.P. Kavaleridze, Kyïv 43096
Muzej Maksimilijana Vanka, Korčula 08896
Muzej Marcialnye Vody, Petrozavodsk 37562
Muzej-masterskaja Gapara Ajtieva, Biškek 29958
Muzej Matija Skurjeni, Zaprešić 09017
Muzej Mebeli v Moskve, Moskva 37331
Muzej Mecenatov i Blagotvoritelej Rossii, Moskva 37332
Muzej Medjimurja, Čakovec 08852
Muzej Mengeš, Mengeš 38480
Muzej Meteorologii Glavnoj Geofizičeskoj Observatorii im. A.I. Voejkova, Sankt-Peterburg 37698
Muzej Meterologičeskaja Stancija Simbirska, Uljanovsk 37903
Muzej Mimara, Zagreb 09006
Muzej Mineralov, Rud, Samocvetov - Udivitelnoe v Kamne, Teberda 37840
Muzej Miniaturnych Kukol Malenkij Mir, Semipalatinsk 37760
Muzej Mir Vody Sankt-Peterburga, Sankt-Peterburg 37699
Muzej Mirovogo Okeana, Kaliningrad 36992
Muzej Moderne i Suvremene Umjetnosti, Rijeka 08938
Muzej Mologskogo Kraja, Rybinsk 37605
Muzej Mordovskoj Kultury, Saransk 37743
Muzej Morskogo Flota, Moskva 37333
Muzej Morskogo Flota Dalnevostočnogo Morskogo Parochodstva, Vladivostok 37946
Muzej Moskoncerta, Moskva 37334
Muzej Moskovskogo Chudožestvenno-promyšlennogo Universiteta im. S.G. Stroganova, Moskva 37335
Muzej Moskovskogo Chudožestvennogo Akademičeskogo Teatra, Moskva 37336
Muzej Moskovskogo Gosudarstvennogo Techičeskogo Universiteta im. N.Ė. Baumana, Moskva 37337
Muzej Moslavine, Kutina 08904
Muzej Muzykalnoj i Teatralnoj Kultury na Kavkazskich Mineralnych Vodach, Kislovodsk 37049
Muzej Muzykalnych Instrumentov, Sankt-Peterburg 37700
Muzej Mystectv imeni Bogdana ta Varvari Chanenkiv, Kyïv 43097
Muzej Mystectva Davnoï Ukraïns'koï Knigi, Lviv 43132
Muzej na Aviacijata, Krumovo 05081
Muzej na Bivolarstvoto i Konevădstvoto, Šumen 05217
Muzej na Blejskem Gradu, Bled 38400
Muzej na Chudožestvenata Akademija, Sofia 05185
Muzej na Chudožestvenite Zanajati i Priložnite Izkustva, Trojan 05232
Muzej na Grad Kratovo, Kratovo 30504
Muzej na Grad Negotino, Negotino 30510
Muzej na Grad Skopje, Skopje 30518
Muzej na Gradskij Bit Oslekova Kăšta, Koprivštica 05074
Muzej na Gradskija Bit, Ruse 05153
Muzej na Kotlenskite Văzroždenci i Panteon na Georgi S. Rakovski, Kotel 05077
Muzej na Makedonija - Archeološki, Etnološki i Istoriski, Skopje 30519
Muzej na Ministerstvoto na Vnatrešnite raboti, Sofia 05186
Muzej na Mozaikite, Devnja 05012
Muzej na Nacionalnata Opera i Balet, Sofia 05187
Muzej na Naroden Teatar Ivan Vazov, Sofia 05188
Muzej na Prostem, Rogatec 38508
Muzej na Prostem Jurjeva Domačija, Dobrnič 38412
Muzej na Prostem pri Kartuziji Pleterje, Drča 38417
Muzej na Rezbarskoto i Zografsko Izkustvo, Trjavna 05230
Muzej na Rodopskija Karst, Čepelare 05001
Muzej na Rudodobiva i Minnoto Delo, Madan 05088
Muzej na Sovremenata Umetnost Skopje, Skopje 30520
Muzej na Tekstilnata Industrija, Sliven 05168
Muzej na Tetovskiot Kraj, Tetovo 30532
Muzej na Umetnosta na Makedonija, Skopje 30521
Muzej na Văzraždaneto, Varna 05241
Muzej na Văzraždeneto i Nacionalno-osvoboditelnite Borbi, Elena 05027
Muzej V.V. Nabokova, Sankt-Peterburg 37701
Muzej Nacionalnoj Kultury, Kazan 37022
Muzej Naivne Umetnosti, Jagodina 38100
Muzej Naivne Umetnosti Ilianum, Šid 38141
Muzej Najdavnišych Pamjatok Lvova, Lviv 43133
Muzej Narodne Osvoboditve Maribor, Maribor 38476
Muzej Narodnogo Chudožnika Rossii A.I. Morozova, Ivanovo 36951
Muzej Narodnogo Dekorativnogo Mystectva, Černigiv 43006
Muzej Narodnogo Derevjannogo Zodčestva Vitoslavlicy, Velikij Novgorod 37921
Muzej Narodnogo Iskusstva, Moskva 37338
Muzej Narodnogo Mystectva Slobožanščyny, Charkiv 43021
Muzej Narodnogo Obrazovanija Oblasti, Penza 37533
Muzej Narodnogo Pobutu, Mariupol' 43146
Muzej Narodnogo Tvorčestva, Penza 37534
Muzej Narodnogo Tvorčestva, Uljanovsk 37904
Muzej Narodnoho Mystectva Gucul'ščini ta Pokuttja im. J. Kobryns'koho, Kolomija 43072
Muzej Narodnoï Architektury i Pobutu u Lvovi, Lviv 43134
Muzej Narodnoï Architektury ta Pobutu, Perejaslav-Chmel'nyc'kyj 43164
Muzej Narodnoï Architektury ta Pobutu Ukraïny, Kyïv 43098
Muzej Narodnoj Grafiki, Moskva 37339
Muzej Narodnoje Obrazovanie Simbirskoj Gubernii v 70-80 gody 19 veka, Uljanovsk 37905
Muzej Narodnych Muzykalnych Instrumentov Kazachstana im. Ychlasa, Almaty 29689
Muzej Nauke i Tehnike, Beograd 38070
Muzej Nekropol Demidovych i Vystavočnyj Zal Tulskij Metall, Tula 37869
Muzej-nekropol Literatorskie Mostki, Sankt-Peterburg 37702
Muzej Nikole Tesle, Beograd 38071
Muzej Niżegorodskaja Radiolaboratorija, Nižnij Novgorod 37462
Muzej Nižegorodskij Ostrog, Nižnij Novgorod 37463
Muzej Nižegorodskoj Intelligencii, Nižnij Novgorod 37464
Muzej NOB - Pustova Hiša, Slovenske Konjice 38521
Muzej Nonkonformistskogo Iskusstva, Sankt-Peterburg 37703
Muzej Novejše Zgodovine Celje, Celje 38406
Muzej Novejše Zgodovine Slovenije, Ljubljana 38455
Muzej 'Novoj Akademii Izjaščnych Iskusstv', Sankt-Peterburg 37704
Muzej Obretaja Svobodu, Moskva 37340
Muzej Obščestvennogo Pitanija, Moskva 37341
Muzej Ochotovedenija, Irkutsk 36939
Muzej Ochoty i Rybolovstva, Moskva 37342
Muzej Oktjabrskogo Vooružennogo Vosstanija, Voronež 37973
Muzej Oleksi Dovbuša, Ivano-Frankivs'k 43043
Muzej Olimpijskoj Slavy, Toshkent 54478
Muzej Orijentalnih Rukopisa, Prizren 38135
Muzej Osvoboždenieto na Pleven 1877, Pleven 05118
Muzej Otečestvennaja Kultura, Moskva 37343
Muzej Ozkotirne Železnice Parenzana, Izola 38423
Muzej Paje Jovanovića, Beograd 38072
Muzej Palomničestva po Svjatym Mestam, Moskva 37344
Muzej Pamjati Žertv Repressij, Toshkent 54479
Muzej-Pamjatnik Chram Voskresenija Christova (Spas na Krovi), Sankt-Peterburg 37705
Muzej-panorama Borodinskaja Bitva, Moskva 37345
Muzej Parapsichologii, Moskva 37346
Muzej Parka Iskusstv na Krymskoj Nabereżnoj, Moskva 37347
Muzej-Parochod Svjatitel Nikolaj, Krasnojarsk 37076
Muzej Pčelarstva Porodice Živanović, Sremski Karlovci 38153
Muzej Pčelovodstva, Moskva 37348
Muzej Pejzaža, Ples 37567
Muzej Pidpil'nyj ta Partizans'kyj Ruch na Mykolaïvščyni u Roki Velikoï Vitčyznjanoï Vijny 1941-1945 rr, Mykolaïv 43148
Muzej Pisanka, Kolomija 43073
Muzej Pisatelej-orlovcev, Orël 37510
Muzej Pisatelja A.P. Bibika, Mineralnye Vody 37133
Muzej Podviga Ivana Susanina, Susanino 37780
Muzej Poėta S. Orlova, Belozersk 36802
Muzej Policijske Akademije, Ljubljana 38456
Muzej Političnih Zapornikov, Internirancev in Izgnancev Grad Rajhenburg, Brestanica 38403
Muzej Polovič, Dobova 38410
Muzej Ponišavlja Pirot, Pirot 38130
Muzej Pošte in Telekomunikacij, Škofja Loka 38515
Muzej Pozorišne Umetnosti Srbije, Beograd 38073
Muzej Predprinimatelej, Mecenatov i Blagotvoritelej, Moskva 37349
Muzej Premogovništva Slovenije, Velenje 38533
Muzej Prigorja, Sesvete 08947
Muzej Prikladnogo Iskusstva Uzbekistana, Toshkent 54480
Muzej Primenjene Umetnosti, Beograd 38074
Muzej Prioratskij Dvorec, Gatčina 36921
Muzej Priroda Morja i eë Ochrana, Vladivostok 37947
Muzej Prirody, Almaty 29690
Muzej Prirody, Čerepovec 36850
Muzej Prirody Burjatii, Ulan-Udė 37896
Muzej S.S. Prokofjeva, Krasne 43074
Muzej S.S. Prokofjeva, Moskva 37350
Muzej Pryrody, Žytomyr 43213
Muzej A.S. Puškina, Toržok 37861
Muzej A.S. Puškina v Bernove, Bernov 36804
Muzej Radio im. A.S. Popova, Ekaterinburg 36898
Muzej Radničkog Pokreta i Narodne Revolucije, Novi Sad 38122
Muzej Vladimira Raevskogo, Bogoslovka 36813
Muzej Razvitija Metallurgii, Moskva 37351
Muzej Rečnog Brodarstva, Beograd 38075
Muzej Rečnogo Flota, Nižnij Novgorod 37465
Muzej Remesla, Architektury i Byta, Kaluga 37000
Muzej Republike Srpske, Banja Luka 04257
Muzej Revoljucionnogo Dviženija, Altanbulag 31615
Muzej Rossijskogo Krasnogo Kresta, Moskva 37352
Muzej Rossijskoj Akademii Chudožestv, Sankt-Peterburg 37706
Muzej Nikolaja Rubcova, Moskva 37353
Muzej N.G. Rubinštejna - Muzej Moskovskoj Konservatorii, Moskva 37354
Muzej Rudarstva i Metalurgije, Bor 38093
Muzej Rumănski Voin, Pordim 05136
Muzej Russkaja Literatura XX Veka, Moskva 37355
Muzej Russkie Valenki, Moskva 37356
Muzej Russko-Armjanskoj Družby, Rostov-na-Donu 37601
Muzej Russkoj Garmoniki Alfred Mireka, Moskva 37357
Muzej Russkoj Pesni im. A. Averkina, Sasovo 37755
Muzej Russkoj Usadebnoj Kultury Kuzminki, Moskva 37358
Muzej Šachmata, Moskva 37359
Muzej Sakena Sejfullina, Astana 29693
Muzej Sakral'noï Skul'ptury XVIII st. - Tvorčist' Ivana Pinzelja, Lviv 43135
Muzej Salicha Sajdaševa, Kazan 37023
Muzej M.E. Saltykova-Ščedrina, Tver 37878
Muzej Samocvety, Moskva 37360
Muzej Samostana Sestara od Sigurate, Dubrovnik 08869
Muzej Sarajeva 1878-1914, Sarajevo 04280
Muzej Šarifa Kamala, Kazan 37024
Muzej Savremene Likovne Umetnosti, Novi Sad 38123
Muzej Savremene Umetnosti, Beograd 38076
Muzej Sceničeskogo Iskusstva im. V.Ė. Mejercholda, Penza 37535
Muzej Seljačkih Buna, Gornja Stubica 08876
Muzej Semberija, Bijeljina 04260
Muzej Semji Benua, Sankt-Peterburg 37707
Muzej Serebrjanogo Veka - Dom V.Ja. Brjusova, Moskva 37361
Muzej Simbirskaja Čuvašskaja Škola - Kvartira I.Ja. Jakovleva, Uljanovsk 37906
Muzej Simbirskaja Klassičeskaja Gimnazija, Uljanovsk 37907
Muzej Skulptury, Sankt-Peterburg 37708
Muzej Skulptury S.T. Konenkova, Smolensk 37771
Muzej Skulptury pod Otkrytym Nebom, Biškek 29959
Muzej Slavonije Osijek, Osijek 08923
Muzej Smolenščina v Gody Velikoj Otečestvennoj Vojny 1941 -1945, Smolensk 37772
Muzej Smolenskij Len, Smolensk 37773
Muzej Snovidenij Zigmunda Frejda, Sankt-Peterburg 37709
Muzej Solane Tuzla, Tuzla 04291
Muzej Sovremennogo Iskusstva Evrejskoj Avtonomnoj Oblasti, Birobidžan 36808
Muzej Sovremennogo Izobrazitelnogo Iskusstva im. A.A. Plastova, Uljanovsk 37908
Muzej Sovremennoj Istorii i Kultury Severnoj Osetii, Vladikavkaz 37931
Muzej Spasskaja Cerkov, Tjumen 37849
Muzej Sporta v Lužnikach, Moskva 37362
Muzej Sportivnogo Obščestva Dinamo, Moskva 37363
Muzej Sportivnoj Slavy CSKA, Moskva 37364
Muzej Srema, Sremska Mitrovica 38151
Muzej Srpske Pravoslavne Crkve, Beograd 38077
Muzej Srpske Pravoslavne Crkve, Dubrovnik 08870
Muzej Srpske Prvoslavne Crkve, Beograd 38078
Muzej Štab-kvartira V.K. Bljuchera, Tjumen 37850
Muzej Stara Livnica, Kragujevac 38107
Muzej Staražytna Belaruskaj Kultury, Minsk 03254
Muzej Stare Srpske-Pravoslavne Crkve, Sarajevo 04281
Muzej Staro Selo, Sirogojno 38142
Muzej-studija Radioteatra, Moskva 37365
Muzej Sudnobuduvannja ta Flotu, Mykolaïv 43149
Muzej Suvremene Povijesti, Dubrovnik 08871
Muzej Suvremene Umjetnosti, Zagreb 09007
Muzej Svjazi Amurskoj Oblasti, Blagoveščensk 36810
Muzej Talcev, Begunje na Gorenjskem 38398
Muzej S.I. Taneeva v Djutkovo, Zvenigorod 37990
Muzej Andreja Tarkovskogo, Jurjevec 36985
Muzej-teatr Lednikovyj period, Moskva 37366
Muzej Teatra, Sankt-Peterburg 37710

Muzej Teatra im. Mossoveta, Moskva 37367
Muzej Teatra im. K.S. Stanislavskogo, Moskva 37368
Muzej Teatra im. Evg. Vachtangova, Moskva 37369
Muzej Teatra Junogo Zritelja, Moskva 37370
Muzej Teatra Operetty, Moskva 37371
Muzej Teatra Rossijskoj Armii, Moskva 37372
Muzej Teatra Satiry, Moskva 37373
Muzej Teatralnoj Kukly A.A Veselova, Voronež 37974
Muzej Teatralnych Kukol, Moskva 37374
Muzej A.K. Tolstogo, Krasnyj Rog 37083
Muzej L.N. Tolstogo na stancii Lev Tolstoj, Stancija Lev Tolstoj 37794
Muzej Tomaža Godca, Bohinjska Bistrica 38401
Muzej Tome Rosandića, Beograd 38079
Muzej Tropinina i Moskovskich Chudožnikov ego Vremeni, Moskva 37375
Muzej Tulskie Samovary, Tula 37870
Muzej Tulskij Kreml, Tula 37871
Muzej Turopolja, Velika Gorica 08977
Muzej Tverskogo Byta, Tver 37879
Muzej Tvorčestva i Byta v GULAGe, Moskva 37376
Muzej u Arandjelovcu, Arandjelovac 38030
Muzej u Majdanpeku, Majdanpek 38115
Muzej u Smederevu, Smederevo 38144
Muzej Uezdnogo Goroda, Čistopol 36867
Muzej Ugolok Deduški Durova, Moskva 37377
Muzej Unikalnych Kukol, Moskva 37378
Muzej Unsko-Sanskog Kantona, Bihać 04259
Muzej Uralskogo Gosudarstvennogo Techničeskogo Universiteta, Ekaterinburg 36899
Muzej-usadba A.T. Bolotova, Dvorjaninovo 36882
Muzej-usadba V.E. Borsova-Musatova, Saratov 37750
Muzej-usadba N.G. Černyševskogo, Saratov 37751
Muzej- Usadba F.M. Dostoevskogo Darovoe, Darovoe 36873
Muzej-usadba N.A. Durovoj, Elabuga 36906
Muzej-usadba M.I. Glinki, Novospasskoe 37490
Muzej-usadba im. N.K.702Rericha, Izvara 36958
Muzej-usadba Izmajlovo, Moskva 37379
Muzej-usadba "Muranovo" im. F.I. Tjutčeva, Lugovskoe 37114
Muzej-usadba Ostafjevo "Russkij Parnas", Ostafjevo 37518
Muzej-usadba poėta D. Venevitinova, Novoživotinnoe 37492
Muzej-usadba I.E. Repina, Repino 37590
Muzej-usadba Roždestveno, Roždestveno 37603
Muzej-usadba V.I. Surikova, Krasnojarsk 37077
Muzej-usadba A.V. Suvorova, Končanskoe-Suvorovskoe 37055
Muzej-usadba A.N. Tolstogo, Samara 37618
Muzej-usadba L.N. Tolstogo Nikolskoe Vjazemskoe, Nikolskoe-Vjazemskoe 37449
Muzej-Usadba L.N. Tolstogo v Chamovnikach, Moskva 37380
Muzej-usadba 'A.T. Tvardovskij na Chutore Zagorje', Zagorje 37984
Muzej Usadba Vučetiča, Moskva 37381
Muzej "V Mire Skazok", Smolensk 37774
Muzej Vasil Levski, Karlovo 05050
Muzej Vasil Levski, Lovеč 05087
Muzej Konstantina Vasiljeva, Moskva 37382
Muzej Văzraždane i Učreditelno Săbranie, Veliko Tărnovo 05255
Muzej Văzraždenski Ikoni, Radomir 05145
Muzej Velenje na Velenjskem Gradu, Velenje 38534
Muzej Vereščaginych, Čerepovec 36851
Muzej Visoko, Visoko 04293
Muzej Vjatskie Narodnye Chudožestvennye Promysly, Kirov 37041
Muzej Vody, Moskva 37383
Muzej Voenno-morskogo Flota, Moskva 37384
Muzej Vojniško Văstanie 1918, Radomir 05146
Muzej Vojvodine, Novi Sad 38124
Muzej Volyns'koj Ikony, Luc'k 43112
Muzej Voskovych Figur, Sankt-Peterburg 37711
Muzej Vrbovec - Muzej Gozdarstva in Lesarstva, Nazarje 38488
Muzej Vuka i Dositeja, Beograd 38080
Muzej Vydatnych Dijačiv Ukraïns'koï Kul'tury, Kyïv 43099
Muzej Vyzvol'nych Zmagan' Prikarpats'koho Kraju, Ivano-Frankivs'k 43044
Muzej za Istorija na Medicinata, Varna 05242
Muzej za Istorija na Varna, Varna 05243
Muzej za Umjetnost i Obrt, Zagreb 09008
Muzej Zapadna Makedonija vo NOV, Kičevo 30503
Muzej-zapovednik Abramcevo, Abramcevo 36747
Muzej-zapovednik Dmitrovskij Kreml, Dmitrov 36876
Muzej-zapovednik Kazanskij Kreml, Kazan 37025
Muzej-zapovednik Pamjatniki Materialnoj Kultury Mangystau i Ustjurta, Aktau 29671
Muzej-zapovednik Pisatelja S.T. Aksakova, Aksakovo 36750
Muzej Zavoda Crvena Zastava, Kragujevac 38108
Muzej Zavoda Moskvič, Moskva 37385
Muzej Zdravoochranenija, Sankt-Peterburg 37712
Muzej Zdravoochranenija, Toshkent 54481
Muzej Železniki, Železniki 38541
Muzej Železnodorožnoj Techniki, Sankt-Peterburg 37713
Muzej Železnodorožnoj Techniki, Toshkent 54482
Muzej Zemlevedenija Moskovskogo Gosudarstvennogo Universiteta im. M.V. Lomonosova, Moskva 37386
Muzej Žrtava Genocida, Beograd 38081
Muzeji i Galerije Podgorica, Podgorica 31656
Muzejno-vystavočnyj Centr, Dalnegorsk 36872
Muzejno-vystavočnyj Centr, Ivanovo 36952
Muzejno-vystavočnyj Centr - Diorama, Kirov 37042
Muzejno-vystavočnyj Centr Ego, Sankt-Peterburg 37714
Muzejno-vystavočnyj Centr Tulskie Drevnosti, Tula 37872
Muzejno-vystavočnyj Kompleks Volokolamskij Kreml, Volokolamsk 37966
Muzejnyj Centr Nasledie Čukotki, Anadyr 36757
Muzejnyj Kompleks Tomskogo Politechničeskogo Universiteta, Tomsk 37856
Muzejnyj Kulturno-vystavočnyj Kompleks, Dnipropetrovs'k 43036
Muzejnyj - Vystavočnyj Kompleks Školy Akvareli Sergeja Andrijaki, Moskva 37387
Muzejs "Ebreji Latvijā", Rīga 30097
Muzejska Jedinica Odžaci, Odžaci 38127
Muzejska Zbirka, Ajdovščina 38396
Muzejska Zbirka, Imotski 08882
Muzejska Zbirka, Knić 38104
Muzejska Zbirka Franjevačkog Samostana, Imotski 08883
Muzejska Zbirka Laško, Laško 38442
Muzejska Zbirka NOB, Zagorje ob Savi 38537
Muzejske Zbirke Fojničkog Samostana, Fojnica 04263
Muzejski Dokumentacijski Centar, Zagreb 09009
Muzejsko Društvo na Makedonija, Skopje 54934
Muzejsko Društvo SR Crne Gore - Cetinje, Cetinje 54948
Muzejsko Društvo Srbije, Beograd 55006
Muzeu Arkeologjik, Durrës 00013
Muzeu Arkeologjik, Tiranë 00031
Muzeu Etnografik, Berat 00010
Muzeu Etnografik, Elbasan 00015
Muzeu Etnografik, Gjirokastër 00018
Muzeu Etnografik, Kavaja 00019
Muzeu Etnografik, Krujë 00025
Muzeu Etnografik, Vlorë 00034
Muzeu Gjergj Kastrioti Skënderbeu, Krujë 00026
Muzeu Historik Fier, Fier 00016
Muzeu Historik i Lushnjës, Lushnja 00027
Muzeu Historik i Vlorës, Vlorë 00035
Muzeu Historik Kombëtar, Tiranë 00032
Muzeu Historik Shkodër, Shkodër 00028
Muzeu i Arsimit, Korçë 00020
Muzeu i Artit Mesjetar, Korçë 00021
Muzeu i Filmit Shkodër, Shkodër 00029
Muzeu i Kosoves / Muzej Kosova i Metohije, Priština 38133
Muzeu i Kulturës Popullore, Durrës 00014
Muzeu i Pavarësise, Vlorë 00036
Muzeu i Shkencave Natyrore, Tiranë 00033
Muzeu Ikonografik Onufri, Berat 00011
Muzeu Kombëtar Arkeologjik, Korçë 00022
Muzeu Prehistorik, Korçë 00023
Muzeul Agriculturii din România, Slobozia 36629
Muzeul Apelor Mihai Băcescu, Fălticeni 36366
Muzeul Arheologic Adamclisi, Adamclisi 36096
Muzeul Arheologic Istria, Istria 36431
Muzeul Arheologic Sarmizegetusa, Sarmizegetusa 36588
Muzeul Arhitecturii Populare din Gorj, Bumbești-Jiu 36236
Muzeul Aristide Caradja, Grumăzești 36391
Muzeul Arta lemnului, Câmpulung Moldovenesc 36255
Muzeul Aurului, Brad 36165
Muzeul Aviației, București 36195
Muzeul Badea Cârțan, Cârțișoara 36267
Muzeul Banatului, Timișoara 36689
Muzeul Banatului Montan, Reșița 36562
Muzeul Banatului. Secția de Artă, Timișoara 36690
Muzeul Banatului. Secția de Științele Naturii, Timișoara 36691
Muzeul Băncii Naționale a României, București 36196
Muzeul Brăilei, Brăila 36170
Muzeul Brăilei-Secția de Etnografie și Artă Populară, Brăila 36171
Muzeul Brăilei-Secția Științele Naturii, Brăila 36172
Muzeul Bran, Bran 36173
Muzeul Breslelor, Tîrgu Secuiesc 36694
Muzeul Căilor Ferate Române, București 36197
Muzeul Câmpiei Băileștitor, Băilești 36131
Muzeul Câmpiei Boianului, Drăgănești-Olt 36353
Muzeul Carpaților Răsăriteni, Sfântu Gheorghe 36600
Muzeul Casa Mureșenilor, Brașov 36177
Muzeul Casa Săsească, Livezile 36446
Muzeul Castelul Corvinestilor, Hunedoara 36404
Muzeul Ceasului Nicolae Simache, Ploiești 36529
Muzeul Civilizației Dacice și Române, Deva 36344
Muzeul Civilizatiei Gumelnița Oltenița, Oltenița 36499
Muzeul Colecțiilor de Artă, București 36198
Muzeul Comunei Buteni, Buteni 36239
Muzeul Conacul Bellu, Urlați 36715
Muzeul Cornel Medrea, București 36199
Muzeul Curtea Veche, București 36200
Muzeul de Arheologie, Săveni 36593
Muzeul de Arheologie, Topalu 36697
Muzeul de Arheologie Callatis Mangalia, Mangalia 36465
Muzeul de Arheologie și Artă Religioasă, Băile Govora 36129
Muzeul de Arheologie și Etnografie, Corabia 36311
Muzeul de Arheologie și Istorie, Baia Mare 36124
Muzeul de Artă, Arad 36113
Muzeul de Artă, Botoșani 36162
Muzeul de Artă, Brașov 36178
Muzeul de Artă, Drobeta-Turnu Severin 36359
Muzeul de Artă, Iași 36417
Muzeul de Artă, Medgidia 36468
Muzeul de Artă, Râmnicu Vâlcea 36552
Muzeul de Artă, Roman 36567
Muzeul de Artă, Satu Mare 36590
Muzeul de Artă, Sfântu Gheorghe 36601
Muzeul de Artă, Târgoviște 36657
Muzeul de Artă, Târgu Jiu 36664
Muzeul de Artă, Tulcea 36703
Muzeul de Artă Centrul Artistic Baia Mare, Baia Mare 36125
Muzeul de Artă Comparată, Sângeorz-Băi 36584
Muzeul de Artă Constanța, Constanța 36302
Muzeul de Artă Craiova, Craiova 36324
Muzeul de Artă Decorativă, Dorohoi 36349
Muzeul de Artă Dinu și Sevasta Vintilă, Topalu 36698
Muzeul de Artă Ion Irimescu, Fălticeni 36367
Muzeul de Artă Piatra-Neamț, Piatra-Neamț 36518
Muzeul de Artă Populară, Constanța 36303
Muzeul de Artă Populară, Ploiești 36530
Muzeul de Artă Populară Prof. Dr. Nicolae Minovici, București 36201
Muzeul de Artă Populară și Etnografie, Tulcea 36704
Muzeul de Artă Religioasă al Mănăstirii Sfinții Petru și Pavel, Huși 36407
Muzeul de Artă Religioasă Nicolae Iorga, Vălenii de Munte 36720
Muzeul de Artă și Artă Populară, Calafat 36247
Muzeul de Artă Veche Apuseană Dumitru Minovici, București 36202
Muzeul de Artă Vizuală, Galați 36379
Muzeul de Cultură și Artă Relgioasă, Borzești 36158
Muzeul de Etnografie, Botoșani 36163
Muzeul de Etnografie, Comarnic 36299
Muzeul de Etnografie, Toplița 36700
Muzeul de Etnografie al Văii Teleajenului, Vălenii de Munte 36721
Muzeul de Etnografie Brașov, Brașov 36179
Muzeul de Etnografie Casa Băniei, Craiova 36325
Muzeul de Etnografie Gheorghe Cernea, Rupea 36576
Muzeul de Etnografie Piatra-Neamț, Piatra-Neamț 36519
Muzeul de Etnografie Săsească Emil Sigerus, Sibiu 36606
Muzeul de Etnografie și a Regimentului de Graniță, Caransebeș 36263
Muzeul de Etnografie și Artă, Drăgășani 36354
Muzeul de Etnografie și Artă Bisericească, Fântânele 36369
Muzeul de Etnografie și Artă Contemporană, Comănești 36298
Muzeul de Etnografie și Artă Populară, Baia Mare 36126
Muzeul de Etnografie și Artă Populară, Florești-Stoenești 36371
Muzeul de Etnografie și Artă Populară, Reghin 36559
Muzeul de Etnografie și Artă Populară Orăștie, Orăștie 36506
Muzeul de Etnografie Universală Franz Binder, Sibiu 36607
Muzeul de Icoane și Obiecte Bisericesti, Sibiu 36608
Muzeul de Istorie, Bicaz 36150
Muzeul de Istorie, Onești 36500
Muzeul de Istorie, Scornicești 36595
Muzeul de Istorie, Sibiu 36609
Muzeul de Istorie, Siret 36622
Muzeul de Istorie, Târgoviște 36658
Muzeul de Istorie, Târgu Jiu 36665
Muzeul de Istorie, Teremia Mare 36685
Muzeul de Istorie a Clisurii Dunării, Moldova Nouă 36481
Muzeul de Istorie a Evreilor din România, București 36203
Muzeul de Istorie a Orașului Hârlău, Hârlău 36397
Muzeul de Istorie a Teatrului și a Orașului Oravița, Oravița 36508
Muzeul de Istorie al Moldovei, Iași 36418
Muzeul de Istorie Augustin Bunea, Blaj 36155
Muzeul de Istorie, Etnografie și Artă Plastică, Lugoj 36450
Muzeul de Istorie Gherla, Gherla 36383
Muzeul de Istorie Locală și Etnografie, Brad 36166
Muzeul de Istorie Militară, Constanța 36304
Muzeul de Istorie Național și Arheologie, Constanța 36305
Muzeul de Istorie Naturală din Sibiu, Sibiu 36610
Muzeul de Istorie Naturală Iași, Iași 36419
Muzeul de Istorie și Arheologie, Tulcea 36705
Muzeul de Istorie și Artă Roman, Roman 36568
Muzeul de Istorie și Etnografie, Târgu Neamț 36676
Muzeul de Istorie Sighișoara, Sighișoara 36617
Muzeul de Istorie Turda, Turda 36709
Muzeul de Istorie Valea Hârtibaciului, Agnita 36101
Muzeul de Mineralogie Baia Mare, Baia Mare 36127
Muzeul de Sculptură Ion Jalea, Constanța 36306
Muzeul de Științe Naturale, Piatra-Neamț 36520
Muzeul de Științele Naturii, Craiova 36326
Muzeul de Științele Naturii, Dorohoi 36350
Muzeul de Științele Naturii, Focșani 36372
Muzeul de Stiințele Naturii, Roman 36569
Muzeul de Științele Naturii, Suceava 36644
Muzeul de Științele Naturii Aiud, Aiud 36102
Muzeul de Științele Naturii și Cinegetică, Vatra Dornei 36727
Muzeul de Șttiinte Naturale, Tulcea 36706
Muzeul de Vânătoare August von Spiess, Sibiu 36611
Muzeul Deltei Dunării, Tulcea 36707
Muzeul Doftana, Doftana 36347
Muzeul Dunării de Jos, Călărași 36248
Muzeul Episcopiei Romano-Catolice, Oradea 36502
Muzeul Etnografic, Săcele 36577
Muzeul Etnografic, Vatra Dornei 36728
Muzeul Etnografic al Bucovinei, Suceava 36645
Muzeul Etnografic al Moldovei, Iași 36420
Muzeul Etnografic al Transilvaniei, Cluj-Napoca 36287
Muzeul Etnografic Ioan Tugui, Câmpulung Moldovenesc 36256
Muzeul Etnografic Lupșa, Lupșa 36454
Muzeul Etnografic Tehnici Populare Bucovinene, Rădăuți 36547
Muzeul Fierului, Hunedoara 36405
Muzeul Grăniceresc Năsăudean, Năsăud 36486
Muzeul Haáz Rezső, Odorheiu Secuiesc 36497
Muzeul Haszmann Pál, Cernatu de Sus 36274
Muzeul Hrandt Avakian, București 36204
Muzeul în Aer Liber, Negrești-Oaș 36488
Muzeul în Aer Liber din Dumbrava Sibiului, Sibiu 36612
Muzeul Ion Luca Caragiale, Ploiești 36531
Muzeul Județean Alexandru Ștefulescu, Târgu Jiu 36666
Muzeul Județean Argeș, Pitești 36526
Muzeul Județean Bistrița-Năsăud, Bistrița, Năsăud 36152
Muzeul Județean Botoșani, Botoșani 36164
Muzeul Județean Buzău, Buzău 36244
Muzeul Județean de Artă, Ploiești 36532
Muzeul Județean de Artă și Etnografie, Bacău 36121
Muzeul Județean de Etnografie, Slatina 36627
Muzeul Județean de Istorie, Galați 36380
Muzeul Județean de Istorie, Piatra-Neamț 36521
Muzeul Județean de Istorie Brașov, Brașov 36180
Muzeul Județean de Istorie și Arheologie Iulian Antonescu Bacău, Bacău 36122
Muzeul Judetean de Istorie și Arheologie Prahova, Ploiești 36533
Muzeul Județean de Istorie și Artă Zalău, Zalău 36742
Muzeul Județean de Istorie Teleorman, Alexandria 36107
Muzeul Județean de Științele Naturii Prahova, Ploiești 36534
Muzeul Județean Ialomița, Slobozia 36630
Muzeul Județean Maramureș, Baia Mare 36128
Muzeul Județean Mureș, Târgu Mureș 36669
Muzeul Județean Mureș-Secția de Artă, Târgu Mureș 36670
Muzeul Județean Mureș-Secția de Etnografie și Artă Populară, Târgu Mureș 36671
Muzeul Județean Mureș-Secția de Științele Naturii, Târgu Mureș 36672
Muzeul Județean Olt, Slatina 36628
Muzeul Județean Satu Mare, Satu Mare 36591
Muzeul Județean Ștefan cel Mare, Vaslui 36726
Muzeul Județean Teohari Antonescu, Giurgiu 36386
Muzeul Județean Vâlcea, Râmnicu Vâlcea 36553
Muzeul Krikor H. Zambaccian, București 36205
Muzeul Literaturii Române, Iași 36421
Muzeul Locomotivelor cu Abur, Reșița 36563
Muzeul Maghiar, Bogdand 36157
Muzeul Mănăstirii Bistrița, Bistrița, Neamț 36153
Muzeul Maramureșului, Sighetu Marmației 36615
Muzeul Mării, Constanța 36307
Muzeul Marinei Române, Constanța 36308
Muzeul Memorial Ady Endre, Căuaș 36270
Muzeul Memorial Ady Endre, Oradea 36503

Muzeul Memorial Alexandru Ioan Cuza, Ruginoasa 36575
Muzeul Memorial Alexandru Vlahuţă, Dumbrăveni 36361
Muzeul Memorial Aurel Vlaicu, Geoagiu 36381
Muzeul Memorial Avram Iancu, Avram Iancu 36116
Muzeul Memorial Bojdeuca Ion Creangă, Iaşi 36422
Muzeul Memorial C.I. & C.C. Nottara, Bucureşti 36206
Muzeul Memorial Calistrat Hogaş, Piatra-Neamţ 36522
Muzeul Memorial Costache Negri, Costache Negri 36315
Muzeul Memorial Dimitrie & Aurelia Ghiaţă, Bucureşti 36207
Muzeul Memorial Dr. Victor Babeş, Bucureşti 36208
Muzeul Memorial Emil Isac, Cluj-Napoca 36288
Muzeul Memorial George Călinescu, Bucureşti 36209
Muzeul Memorial George Coşbuc, Coşbuc 36314
Muzeul Memorial George Enescu, Dorohoi 36351
Muzeul Memorial Gheorghe M. Tattarascu, Bucureşti 36210
Muzeul Memorial Henri Catargi, Scăeşti 36594
Muzeul Memorial Ioan Slavici şi Emil Monţia, Şiria 36623
Muzeul Memorial Ion Pop Reteganul, Reteag 36564
Muzeul Memorial Iosif Vulcan, Oradea 36504
Muzeul Memorial János Arany, Salonta 36580
Muzeul Memorial Liviu Rebreanu, Năsăud 36487
Muzeul Memorial Mihail Kogălniceanu, Iaşi 36423
Muzeul Memorial Nicolae Bălcescu, Nicolae Bălcescu 36491
Muzeul Memorial Nicolae Grigorescu, Câmpina 36250
Muzeul Memorial Octavian Goga, Ciucea 36283
Muzeul Memorial Paul Constantinescu, Ploieşti 36535
Muzeul Memorial Poni-Cernătescu, Iaşi 36424
Muzeul Memorial Sándor Petőfi, Coltău 36295
Muzeul Mihai Eminescu, Iaşi 36425
Muzeul Militar Dolj, Craiova 36327
Muzeul Militar Naţional, Bucureşti 36211
Muzeul Mineritului, Petroşani 36515
Muzeul Mineritului Avrifer Roşia Montană, Roşia Montană 36572
Muzeul Mitropolit Antim Ivireanul, Bucureşti 36212
Muzeul Municipal, Huşi 36408
Muzeul Municipal, Râmnicu Sărat 36549
Muzeul Municipal Câmpulung, Câmpulung 36252
Muzeul Municipal Câmpulung-Secţia de Etnografie şi, Câmpulung 36253
Muzeul Municipal Curtea de Argeş, Curtea de Argeş 36336
Muzeul Municipal de Artă, Turnu Măgurele 36710
Muzeul Municipal de Istorie, Roşiorii de Vede 36573
Muzeul Municipal Dej şi Galeria de Artă, Dej 36341
Muzeul Municipal Ioan Raica Sebeş, Sebeş 36596
Muzeul Municipal Medias, Mediaş 36469
Muzeul Municipiului Bucureşti, Bucureşti 36213
Muzeul Naţional al Bucovinei, Suceava 36646
Muzeul Naţional al Hărţilor şi Cărţii Vechi, Bucureşti 36214
Muzeul National al Literaturii Române, Bucureşti 36215
Muzeul Naţional al Petrolului, Ploieşti 36536
Muzeul Naţional al Pompierilor, Bucureşti 36216
Muzeul Naţional al Satului Dimitrie Gusti, Bucureşti 36217
Muzeul Naţional al Unirii, Alba Iulia 36106
Muzeul Naţional Brukenthal, Sibiu 36613
Muzeul Naţional Cotroceni, Bucureşti 36218
Muzeul Naţional de Artă al României, Bucureşti 36219
Muzeul Naţional de Artă Cluj-Napoca, Cluj-Napoca 36289
Muzeul Naţional de Artă Contemporană al României, Bucureşti 36220
Muzeul National de Arte Plastice, Chişinău 31600
Muzeul National de Etnografie si Istorie Naturala, Chişinău 31601
Muzeul National de Istorie a Moldovei, Chişinău 31602
Muzeul Naţional de Istorie a României, Bucureşti 36221
Muzeul Naţional de Istorie a Transilvaniei, Cluj-Napoca 36290
Muzeul National de Istorie Naturala Grigore Antipa, Bucureşti 36222
Muzeul Naţional Geologic, Bucureşti 36223
Muzeul Naţional George Enescu, Bucureşti 36224
Muzeul Naţional Peleş, Sinaia 36620
Muzeul National Secuisec, Sfântu Gheorghe 36602
Muzeul Nicolae Iorga, Vălenii de Munte 36722
Muzeul Obiceiurilor Populare din Bucovina, Gura Humorului 36394
Muzeul Olteniei, Craiova 36328
Muzeul Operei Române, Bucureşti 36225
Muzeul Orăşanesc Fălticeni, Fălticeni 36368
Muzeul Orăşenesc, Adjud 36097
Muzeul Orăşenesc, Baraolt 36137
Muzeul Orăşenesc, Carei 36264
Muzeul Orăşenesc, Lipova 36443
Muzeul Orăşenesc, Târnăveni 36678
Muzeul Orăşenesc, Tăşnad 36679
Muzeul Orăşenesc, Tecuci 36680
Muzeul Orăşenesc de Istorie, Aiud 36103
Muzeul Orăşenesc Molnar Istvan, Cristuru Secuiesc 36332
Muzeul Oraşului Avrig, Avrig 36117
Muzeul Parohial de Icoane pe Sticlă, Sibiel 36603
Muzeul Pedagogic Republican, Chişinău 31603
Muzeul Podgoriei Drăgăşani, Drăgăşani 36355
Muzeul Poliţiei Române, Târgovişte 36659
Muzeul Pompierilor, Teremia Mare 36686
Muzeul Portului Constanţa, Constanţa 36309
Muzeul Primei Şcoli Româneşti din Scheii Braşovului, Braşov 36181
Muzeul Regiunii Porţilor de Fier, Drobeta-Turnu Severin 36360
Muzeul Rezervaţiei Bucegi, Sinaia 36621
Muzeul Romanaţiului, Caracal 36259
Muzeul Romanaţiului Secţia de Artă Plastică, Caracal 36260
Muzeul Sării, Slănic Prahova 36626
Muzeul Sătesc, Arcani 36114
Muzeul Sătesc, Bănia 36136
Muzeul Sătesc, Bărbăteşti 36139
Muzeul Sătesc, Bârca 36140
Muzeul Sătesc, Bechet 36145
Muzeul Sătesc, Botoşana 36159
Muzeul Sătesc, Broşteni 36184
Muzeul Sătesc, Castranova 36268
Muzeul Sătesc, Celaru 36271
Muzeul Sătesc, Cernăteşti 36273
Muzeul Sătesc, Cornu 36312
Muzeul Sătesc, Dănciuleşti 36337
Muzeul Sătesc, Drăcşenei 36352
Muzeul Sătesc, Dranic 36357
Muzeul Sătesc, Frăteşti 36375
Muzeul Sătesc, Ghighera 36385
Muzeul Sătesc, Goicea 36387
Muzeul Sătesc, Gornea 36388
Muzeul Sătesc, Gostavăţu 36389
Muzeul Sătesc, Grumezoaia 36392
Muzeul Sătesc, Leleşti 36437
Muzeul Sătesc, Leu 36440
Muzeul Sătesc, Mehadica 36470
Muzeul Sătesc, Sălişte 36579
Muzeul Sătesc, Tăcuta 36650
Muzeul Sătesc, Teiu 36681
Muzeul Sătesc, Valea Stanciului 36716
Muzeul Sătesc, Vultureşti 36738
Muzeul Sătesc, Vutcani 36739
Muzeul Sătesc Constantin T. Cumpănaşu, Orlea 36509
Muzeul Sătesc Prăjeşti, Traian 36701
Muzeul Satului Bănăţean-Pădurea Verde, Timişoara 36692
Muzeul Satului din Bucovina, Suceava 36647
Muzeul Satului Vâlcean, Bujoreni 36234
Muzeul Scriitorilor Târgovişteni, Târgovişte 36660
Muzeul Secuiesc al Ciucului, Miercurea-Ciuc 36472
Muzeul şi Rezervaţia Arheologică de la Castrul Roman Jidava, Câmpulung 36254
Muzeul Sportului din România, Bucureşti 36226
Muzeul Staţiunii Băile Herculane, Băile Herculane 36130
Muzeul Ştiinţei şi Tehnicii Ştefan Procopiu, Iaşi 36426
Muzeul Şvăbesc, Petreşti 36514
Muzeul Ţăranului Român, Bucureşti 36227
Muzeul Ţăranului Român, Hotarele 36403
Muzeul Tarii Beiusului, Beiuş 36146
Muzeul Ţării Codrului, Valea Vinului 36718
Muzeul Ţării Crişurilor, Oradea 36505
Muzeul Ţării Făgăraşului Valeriu Literat, Făgăraş 36363
Muzeul Ţării Oaşului, Negreşti-Oaş 36489
Muzeul Tarisznyás Márton, Gheorghieni 36382
Muzeul Teatrului, Iaşi 36427
Muzeul Teatrului Naţional, Bucureşti 36228
Muzeul Teatrului Naţional, Craiova 36329
Muzeul Tehnic Prof. Ing. Dimitrie Leonida, Bucureşti 36229
Muzeul Ţesutului şi Postăvăritului, Buhuşi 36233
Muzeul Theodor Aman, Bucureşti 36230
Muzeul Theodor Pallady, Bucureşti 36231
Muzeul Tiparului şi al Cartii Vechi Româneşti, Târgovişte 36661
Muzeul Unirii, Iaşi 36428
Muzeul Universitar de Istoria Medicinei şi Farmaciei Victor Gomoiu, Târgu Jiu 36667
Muzeul Universitatii Babeş-Bolyai, Cluj-Napoca 36291
Muzeul Universităţii Bucureşti, Bucureşti 36232
Muzeul Văii Superioare a Teleajenului, Măneciu 36463
Muzeul Vasile Pârvan, Bârlad 36142
Muzeul Viei şi Vinului, Miniş 36475
Muzeul Viticol, Huşi 36409
Muzeul Viticol, Odobeşti 36496
Muzeul Vivariul Bacău, Bacău 36123
Muzeul Vrancei, Focşani 36373
Muzeul Zoologic al Universităţii Babeş-Bolyai, Cluj-Napoca 36292
Muzeum, Koszalin 34761
Muzeum 24. Pułku Ułanów, Kraśnik 34823
Muzeum 7 Pułku Ułanów Lubelskich, Mińsk Mazowiecki 34923
Muzeum a Galerie Litomyšl, Litomyšl 09632
Muzeum a Galerie Orlických Hor, Rychnov nad Kněžnou 09832
Muzeum a Galerie Severního Plzeňska, Mariánská Tynice - Kralovice 09648
Muzeum a Památník Jana Ámoše Komenského, Horní Branná 09525
Muzeum a Pojizerská Galerie, Semily 09836
Muzeum Adwokatury Polskiej, Warszawa 35233
Muzeum Afrykanistyczne im. Dr. Bogdana Szczygła, Olkusz 34973
Muzeum Akademii Medycznej, Poznań 35029
Muzeum Akademii Sztuk Pięknych, Warszawa 35234
Muzeum Akademii Sztuk Pięknych, Wrocław 35325
Muzeum Aloise Jiráska, Hronov 09538
Muzeum Aloise Jiráska a Mikoláše Alše, Praha 09766
Múzeum Andreja Kmet'a, Martin 38296
Múzeum Andreja Sládkoviča Krupina, Krupina 38280
Muzeum Andrzeja Struga, Warszawa 35235
Muzeum Antického Sochařství a Architektury, Litomyšl 09633
Muzeum Antonína Dvořáka, Praha 09767
Muzeum Archeologiczne, Gdańsk 34636
Muzeum Archeologiczne, Poznań 35030
Muzeum Archeologiczne, Wrocław 35326
Muzeum Archeologiczne i Etnograficzne, Łódź 34873
Muzeum Archeologiczne Środkowego Nadodrza, Świdnica, k Zielonej Góry 35148
Muzeum Archeologiczne w Biskupinie, Gąsawa 34630
Muzeum Archeologiczne w Krakowie, Kraków 34784
Muzeum Archeologiczne w Krakowie, Oddział w Nowej Hucie, Kraków 34785
Muzeum Archeologiczno-Historyczne, Głogów 34668
Muzeum Archidiecezjalne, Katowice 34724
Muzeum Archidiecezjalne, Poznań 35031
Muzeum Archidiecezjalne, Przemyśl 35054
Muzeum Archidiecezjalne, Szczecin 35158
Muzeum Archidiecezjalne, Wrocław 35327
Muzeum Archidiecezjalne Lubelskie, Lublin 34901
Muzeum Archidiecezjalne w Krakowie, Kraków 34786
Muzeum Archidiecezjalne w Oliwie, Gdańsk 34637
Muzeum Archidiecezji Gnieźnieńskiej, Gniezno 34672
Muzeum Archidiecezji Łódzkiej, Łódź 34874
Muzeum Archidiecezji Warszawskiej, Warszawa 35236
Muzeum Architektury, Wrocław 35328
Muzeum Armii Krajowej im. Generala Emila Fieldorfa "Nila", Kraków 34787
Muzeum Armii Poznań, Poznań 35032
Muzeum Arthura Fleischmanna, Bratislava 38220
Muzeum Artystów, Łódź 34875
Muzeum Azji i Pacyfiku, Warszawa 35237
Múzeum Bábkarských Kultúr a Hračiek Hrad Modry Kameň, Modrý Kameň 38307
Muzeum Barwy i Oręża Arsenał, Zamość 35361
Múzeum Beckov, Beckov 38198
Muzeum Bedřicha Hrozného, Lysá nad Labem 09644
Muzeum Bedřicha Smetany, Praha 09768
Muzeum Benátky nad Jizerou, Benátky nad Jizerou 09400
Muzeum Beskidzkie im. Andrzeja Podżorskiego, Wisła 35303
Muzeum Beskyd Frýdek-Místek, Frýdek-Místek 09504
Muzeum Betlémů, Karlštejn 09578
Muzeum Biblioteki Kórnickiej, Kórnik 34756
Muzeum Biograficzne Władysława Orkana, Niedźwiedź 34944
Muzeum Bitwy Grunwaldzkiej, Stębark 35134
Muzeum Bitwy Legnickiej, Legnickie Pole 34849
Muzeum Bitwy nad Bzurą, Kutno 34838
Muzeum Blansko, Blansko 09409
Muzeum Bolesława Prusa, Nałęczów 34938
Muzeum Borów Tucholskich, Tuchola 35200
Muzeum Boskovicka, Boskovice 09411
Muzeum Boženy Němcové, Česká Skalice 09459
Muzeum Bratří Čapků, Malé Svatoňovice 09646
Muzeum Brněnska, Šlapanice 09839
Muzeum Broni Pancernej, Poznań 35033
Muzeum Broumovska, Broumov 09443
Muzeum Budownictwa Ludowego, Sanok 35100
Muzeum Budownictwa i Techniki Wiejskiej, Bogdaniec 34567
Muzeum Budownictwa Ludowego - Park Etnograficzny w Olsztynku, Olsztynek 34980
Muzeum Byłego Obozu Zagłady w Chełmnie nad Narem, Chełmno nad Nerem 34592
Múzeum Čachtice, Čachtice 38241
Muzeum Cechu Rzemiosł Skórzanych im. Jana Kilińskiego, Warszawa 35238
Muzeum Ceramiki, Bolesławiec 34568
Muzeum Ceramiki Kaszubskiej Neclów, Chmielno 34594
Múzeum Červený Kameň, Častá 38244
Múzeum Červený Kláštor, Červený Kláštor 38245
Muzeum Českého granátu, Třebenice 09874
Muzeum Českého Krasu, Hořovice 09528
Muzeum Českého Krasu v Berouně, Beroun 09404
Muzeum Českého Krasu v Berouně - pobačka muzeum v Žebráku, Žebrák 09935
Muzeum Českého Lesa, Tachov 09859
Muzeum Českého Porcelánu, Klášterec nad Ohří 09584
Muzeum Českého Ráje, Turnov 09881
Muzeum Českého Venkova, Kutná Hora 09615
Muzeum Chełmskie, Chełm 34587
Muzeum Chełmskie Dział Sztuki Dawnej, Chełm 34588
Muzeum Chodska, Domažlice 09496
Muzeum Cystersów, Wąchock 35206
Muzeum Częstochowskie, Częstochowa 34615
Muzeum Czynu Niepodległościowego, Kraków 34788
Muzeum Czynu Partyzanckiego, Polichno 35021
Muzeum Czynu Powstańczego, Góra św. Anny 34678
Muzeum Czynu Zbrojnego Pracowników Huty im. Tadeusza Sendzimira, Kraków 34789
Muzeum Dawnego Kupiectwa, Świdnica 35147
Muzeum Diecezjalne, Płock 35013
Muzeum Diecezjalne, Siedlce 35102
Muzeum Diecezjalne, Tarnów 35176
Muzeum Diecezjalne, Włocławek 35308
Muzeum Diecezjalne Sztuki Kościelnej, Sandomierz 35096
Muzeum Diecezjalne w Opolu, Opole 34985
Muzeum Diecezji Pelplińskiej, Pelplin 35005
Muzeum Dom Urbańczyka, Chrzanów 34605
Muzeum Dom Wincentego Witosa, Wierzchosławice 35302
Múzeum Dopravy, Bratislava 38221
Muzeum Dr. Aleše Hrdličky, Humpolec 09541
Muzeum Dr. Bohuslava Horáka, Rokycany 09818
Muzeum Drukarstwa Warszawskiego, Warszawa 35239
Muzeum Dworek Zabytkowy, Stryszów 35136
Muzeum Dworu Polskiego, Płochocin 35012
Muzeum Dyplomacji i Uchodźstwa Polskiego, Bydgoszcz 34576
Muzeum Dzieje i Kultura Żydów, Kraków 34790
Muzeum Emeryka Hutten-Czapskiego w Kamienicy Łozińskich, Kraków 34791
Muzeum Emila Zegadłowicza, Wadowice 35208
Muzeum Energetyki Jeleniogórskiej, Szklarska Poręba 35169
Muzeum Etnograficzne, Poznań 35034
Muzeum Etnograficzne, Włocławek 35309
Muzeum Etnograficzne, Wrocław 35329
Muzeum Etnograficzne im. Franciszka Kotuli, Rzeszów 35092
Muzeum Etnograficzne im. M. Znamierowskiej-Prüfferowej, Toruń 35193
Muzeum Etnograficzne im. Seweryna Udzieli, Kraków 34792
Muzeum Etnograficzne w Zielonej Górze z Siedzibą w Ochli, Ochla 34968
Muzeum Ewolucji Instytutu Paleobiologii PAN, Warszawa 35240
Muzeum Farmacji, Kraków 34793
Muzeum Farmacji, Poznań 35035
Muzeum Farmacji im. Antoniny Leśniewskiej, Warszawa 35241
Muzeum Farmacji przy Aptece pod Łabędziem, Bydgoszcz 34577
Muzeum Feliksa Nowowiejskiego, Barczewo 34535
Múzeum Ferdiša Kostku, Stupava 38363
Muzeum Filumenistyczne, Bystrzyca Kłodzka 34583
Muzeum Fojtství, Kopřivnice 09595
Muzeum Fryderyka Chopina, Warszawa 35242
Muzeum Gazownictwa, Paczków 35003
Muzeum Gazownictwa, Sobótka 35115
Muzeum Gazownictwa przy Gazowni Warszawskiej, Warszawa 35243
Muzeum Geologiczne, Kraków 34794
Muzeum Geologiczne, Wrocław 35330
Muzeum Geologiczne Państwowego Instytutu Geologicznego, Warszawa 35244
Muzeum Geologiczne Uniwersytetu Szczecińskiego, Szczecin 35159
Muzeum Geologii Złóż im. Czesława Poborskiego, Gliwice 34664
Muzeum Górnictwa i Hutnictwa Kopalina Złota, Złoty Stok 35375
Muzeum Górnictwa Podziemnego, Nowa Ruda 34947
Muzeum Górnictwa Węglowego, Zabrze 35349
Muzeum Górnicze, Bogatynia 34566
Muzeum Górnośląskie, Bytom 34584
Muzeum Gross-Rosen, Wałbrzych 35211

M

Muzeum Gross-Rosen w Rogoźnicy, Rogoźnica 35084
Múzeum-Ház, Csongrád 23512
Muzeum Henryka Sienkiewicza, Oblęgorek 34967
Muzeum Henryka Sienkiewicza, Wola Okrzejska 35315
Muzeum Historické Hasičské Techniky, Chrastava 09483
Muzeum Historických Motocyklů, Kašperské Hory 09580
Muzeum historie, Kolín 09590
Muzeum Historii Akademii Górniczo-Hutniczej im. S. Staszica, Kraków 34795
Muzeum Historii Fotografii, Kraków 34796
Muzeum Historii Fotografii im. Prof. Władisława Bogackiego, Kraków 34797
Muzeum Historii i Tradycji Żołnierzy Suwalszczyzny, Suwałki 35142
Muzeum Historii Kanału Augustowskiego, Augustów 34530
Muzeum Historii Katowic, Katowice 34725
Muzeum Historii Katowic - Dział Etnologii Miasta, Katowice 34726
Muzeum Historii Medycyny i Farmacji, Szczecin 35160
Muzeum Historii Miasta, Szczecin 35161
Muzeum Historii Miasta, Zduńska Wola 35366
Muzeum Historii Miasta Łodzi, Łódź 34876
Muzeum Historii Miasta Lublina, Lublin 34902
Muzeum Historii Miasta - Podziemna Trasa Turystyczna, Rzeszów 35093
Muzeum Historii Miasta Poznania, Poznań 35036
Muzeum Historii Młynarstwa i Wodnych Urządzeń Przemysłu Wiejskiego w Jaraczu, Jaracz 34694
Muzeum Historii Polskiego Rucha Ludowego, Piaseczno koło Gniewa 35006
Muzeum Historii Polskiego Ruchu Ludowego, Warszawa 35245
Muzeum Historii Polskiego Ruchu Ludowego w Warszawie, Oddział w Sandomierzu, Sandomierz 35097
Muzeum Historii Przemysłu, Opatówek 34981
Muzeum Historii Ratusza i Trybunału Koronnego, Lublin 34903
Muzeum Historii Spółdzielczości w Polsce, Nałęczów 34939
Muzeum Historii Włocławka, Włocławek 35310
Muzeum Historii Żydów Polskich, Warszawa 35246
Muzeum Historyczne, Białystok 34549
Muzeum Historyczne, Przasnysz 35053
Muzeum Historyczne Miasta Gdańska, Gdańsk 34638
Muzeum Historyczne Miasta Krakowa, Kraków 34798
Muzeum Historyczne Miasta Krakowa - Oddział Pomorska, Kraków 34799
Muzeum Historyczne Miasta Stołecznego Warszawy, Warszawa 35247
Muzeum Historyczne Miasta Tarnobrzega, Tarnobrzeg 35175
Muzeum Historyczne - Pałac w Dukli, Dukla 34626
Muzeum Historyczne w Sanoku, Sanok 35101
Muzeum Historyczne w Wrocławiu, Wrocław 35331
Muzeum Historyczno-Archeologiczne, Ostrowiec Świętokrzyski 34995
Muzeum Historyczno-Etnograficzne, Chojnice 34599
Muzeum Hlavního Města Prahy, Praha 09769
Muzeum Hraček, Praha 09770
Muzeum Hrnčířství, Kostelec nad Černými Lesy 09598
Muzeum Hutnictwa i Górnictwa Rud Żelaza, Częstochowa 34616
Muzeum Hymnu Narodowego, Nowy Karczma 34955
Muzeum i Instytut Zoologii PAN, Warszawa 35248
Muzeum Ignacego Jana Paderewskiego i Wychodźstwa Polskiego w Ameryce, Warszawa 35249
Muzeum im. Adama Mickiewicza w Śmiełowie, Żerków 35369
Muzeum im. Aleksandra Kłosińskiego, Kęty 34739
Muzeum im. Anny i Jarosława Iwaszkiewiczów w Stawisku, Podkowa Leśna 35018
Muzeum im. Edmunda Bojanowskiego w Grabonogu, Grabonóg 34682
Muzeum im. J.G. Herdera w Morągu, Morąg 34928
Muzeum im. Jacka Malczewskiego w Radomiu, Radom 35073
Muzeum im. Jana Dzierżona, Kluczbork 34748
Muzeum im. Jana Kasprowicza, Inowrocław 34691
Muzeum im. Jana Nikodema Jaronia, Olesno 34972
Muzeum im. Jerzego Dunin-Borkowskiego, Krośniewice 34828
Muzeum im. Juliana Fałata, Bystra Śląska 34582
Muzeum im. Kazimierza Pułaskiego, Warka-Winiary 35215
Muzeum im. Michała Kajki, Ogródek 34969
Muzeum im. Orła Białego, Skarżysko-Kamienna 35108
Muzeum im. Oskara Kolberga, Przysucha 35058
Muzeum im. prof. Władysława Szafera, Ojców 34970
Muzeum im. Stanisława Noakowskiego, Nieszawa 34946
Muzeum im. Stanisława Staszica, Hrubieszów 34688
Muzeum im. Władysława Orkana, Rabka 35070
Muzeum im. Władysława St. Reymonta, Lipce Reymontowskie 34857
Muzeum im. Wojciecha Kętrzyńskiego, Kętrzyn 34738
Muzeum Inkluzji w Bursztynie, Gdynia 34655
Muzeum Instrumentów Muzycznych, Poznań 35037
Muzeum Inżynierii Miejskiej, Kraków 34800
Múzeum J.M. Petzvala, Spišská Béla 38355
Muzeum Jana Ámoše Komenského, Uherský Brod 09888
Muzeum Jana Kasprowicza, Zakopane 35356
Múzeum Jána Kollára, Mošovce 38308
Múzeum Janka Kráľa, Liptovský Mikuláš 38287
Muzeum Jihovýchodní Moravy, Zlín 09941
Muzeum Jindřicha Jindřicha, Domažlice 09497
Muzeum Jindřicha Šimona Baara, Klenčí pod Čerchovem 09586
Muzeum Jindřichohradecka, Jindřichův Hradec 09565
Muzeum Jižního Plzeňska v Blovicích, Blovice 09410
Muzeum Josefa Dobrovského, Chudenice 09487
Muzeum Josefov, Jaroměř 09551
Muzeum Józefa Ignacego Kraszewskiego, Romanów 35086
Muzeum Juszkiewiczów, Mława 34926
Muzeum Kamieni Szlachetnych, Polanica Zdrój 35019
Muzeum Kamienica Łozińskich, Kraków 34801
Muzeum Kamienica Orsettich w Jarosławiu, Jarosław 34696
Muzeum Karkonoskie, Jelenia Góra 34704
Muzeum Karkonoskie w Jeleniej Górze, Jelenia Góra 34705
Muzeum Karola Szymanowskiego w Willi Atma w Zakopanem, Zakopane 35357
Muzeum Karykatury, Warszawa 35250
Muzeum-Kaszubski Park Etnograficzny w Wdzydzach Kiszewskich, Wdzydze Kiszewskie 35294
Muzeum Kaszubskie, Kartuzy 34718
Muzeum Kaszubskie, Kościerzyna 34758
Muzeum Katedralne Jana Pawła II, Kraków 34802
Muzeum Katyńskie, Warszawa 35251
Múzeum Keramiky Mestské Múzeum, Holíč 38259
Muzeum Kinematografii w Łodzi, Łódź 34877
Muzeum Klasztorne na Jasnej Górze, Częstochowa 34617
Muzeum Klasztorne OO Cystersów, Szczyrzyc 35167
Muzeum Klimkovice, Klimkovice 09587
Muzeum Knihy, Žďár nad Sázavou 09933
Muzeum Kolei Wąskotorowej, Sochaczew 35118
Muzeum Kolei Wąskotorowej w Wenecji, Wenecja 35297
Muzeum Kolejnictwa, Warszawa 35252
Muzeum Kolekcji im. Jana Pawła II, Warszawa 35253
Muzeum Komenského v Přerově, Přerov 09804
Muzeum Komunismu, Praha 09771
Muzeum Koněspřežné Cráhy v Bujanově, Kaplice 09572
Muzeum Kornela Makuszyńskiego, Zakopane 35358
Muzeum Kościuszkowskie Dworek Zacisze, Miechów 34915
Muzeum Kouřimska, Kouřim 09599
Muzeum Kowalstwa w Warszawie, Warszawa 35254
Muzeum Krajky, Prachatice 09732
Muzeum Krajky, Vamberk 09898
Muzeum Kresów w Lubaczowie, Lubaczów 34893
Muzeum Kroměřížska, Kroměříž 09610
Muzeum Krupka, Krupka 09611
Muzeum Książąt Czartoryskich, Kraków 34803
Muzeum Książki Artystycznej, Łódź 34878
Múzeum Kultúry Karpatských Nemcov, Bratislava 38222
Muzeum Kultury Kurpiowskiej, Ostrołęka 34992
Muzeum Kultury Łemkowskiej w Zyndranowej, Tylawa 35203
Muzeum Kultury Ludowej, Kolbuszowa 34749
Muzeum Kultury Ludowej, Osiek nad Notecią 34990
Muzeum Kultury Ludowej, Węgorzewo 35295
Múzeum Kultúry Maďarov na Slovensku, Bratislava 38223
Múzeum Kultúry Rómov na Slovensku, Martin 38297
Muzeum Kultury Szlacheckiej, Łopuszna 34891
Múzeum Kysuckej Dediny, Nová Bystrica 38319
Muzeum Lachów Sądeckich im. Zofii i Stanisława Chrząstowskich, Podegrodzie 35017
Muzeum Lasu i Drewna przy Lesnym Zakładzie Doświadczalnym SGGW, Rogów 35083
Muzeum Lat Szkolnych Stefana Żeromskiego, Kielce 34741
Muzeum Latarnia Morska, Władysławowo 35306
Muzeum Łazienki Królewskie, Warszawa 35255
Múzeum Letectva Košice, Košice 38271
Muzeum Lidových Staveb, Kouřim 09600
Muzeum Literackie Henryka Sienkiewicza, Poznań 35038
Muzeum Literackie im. Józefa Czechowicza, Lublin 34904
Muzeum Literatury im. Adama Mickiewicza, Warszawa 35256
Muzeum Literatury im. Jarosława Iwaszkiewicza, Sandomierz 35098
Muzeum Litovel, Litovel 09635
Muzeum Lotnictwa Polskiego, Kraków 34804
Muzeum Loutek Milana Knížáka, Děčín 09490
Muzeum Loutkářských Kultur, Chrudim 09485
Muzeum Łowiectwa i Jeździectwa, Warszawa 35257
Muzeum Lubelskie, Lublin 34905
Múzeum Ludovíta Štúra, Modra 38306
Muzeum Ludowych Instrumentów Muzycznych, Szydłowiec 35174
Muzeum Luhačovického Zálesí, Luhačovice 09643
Múzeum M.R. Štefánika, Košariská 38270
Múzeum Mad'arskej Kultúry a Podunajska, Komárno 38269
Muzeum Małego Miasta, Bieżuń 34563
Muzeum Marii Dąbrowskiej, Warszawa 35258
Muzeum Marii Konopnickiej, Suwałki 35143
Muzeum Marii Konopnickiej w Żarnowcu, Jedlicze 34701
Muzeum Marii Skłodowskiej Curie, Warszawa 35259
Muzeum Martyrologiczne w Żabikowie, Luboń 34912
Muzeum Martyrologiczno-Historyczny, Biała Podlaska 34542
Muzeum Martyrologii, Bielsk Podlaski 34555
Muzeum Martyrologii Alianckich Jeńców Wojennych, Żagań 35352
Muzeum Martyrologii i Walki Radogoszcz, Łódź 34879
Muzeum Martyrologii Pod Zegarem, Lublin 34906
Muzeum Martyrologii (Sonnenburg), Słońsk 35110
Muzeum Martyrologii Wielkopolan Fort VII, Poznań 35039
Muzeum Marynarki Wojennej, Gdynia 34656
Muzeum Mazowieckie w Płocku, Płock 35014
Muzeum Mazurskie, Ostrzeszów 34996
Muzeum Mazurskie, Szczytno 35168
Muzeum Města Brna, Brno 09436
Muzeum města Duchcova, Duchcov 09498
Muzeum Města Ústí nad Labem, Ústí nad Labem 09892
Muzeum Městec Králové, Městec Králové 09651
Muzeum Městské Ústí nad Orlicí, Ústí nad Orlicí 09893
Muzeum MHD - vozovna Střešovice, Praha 09772
Muzeum Miar Głównego Urzędu Miar, Warszawa 35260
Muzeum Miasta Gdyni, Gdynia 34657
Muzeum Miasta i Rzeki Warty, Warta 35292
Muzeum Miasta Jaworzna, Jaworzno 34700
Muzeum Miasta Ostrowa Wielkopolskiego, Ostrów Wielkopolski 34994
Muzeum Miasta Pabianice, Pabianice 35002
Muzeum Miasta Zgierza, Zgierz 35370
Múzeum Michala Greisigera, Spišská Béla 38356
Muzeum Miedzi, Legnica 34848
Muzeum Miejskie, Nowa Sól 34950
Muzeum Miejskie, Świętochłowice 35150
Muzeum Miejskie, Zabrze 35350
Muzeum Miejskie im. Maksymiliana Chroboka, Ruda Śląska 35087
Muzeum Miejskie Sztygarka im Stanisława Staszica, Dąbrowa Górnicza 34619
Muzeum Miejskie w Wadowicach, Wadowice 35209
Muzeum Miejskie Wrocławia, Wrocław 35332
Muzeum Mikołaja Kopernika, Frombork 34629
Muzeum Mikołaja Kopernika, Toruń 35194
Múzeum Milana Dobeša, Bratislava 38224
Muzeum Militariów Arsenał Miejski, Wrocław 35333
Muzeum Mineralogiczne, Szklarska Poręba 35170
Muzeum Mineralogiczne, Wrocław 35334
Muzeum Misji Karmelitańskich w Afryce, Kreczowice 34827
Muzeum Misyjne, Krosno 34829
Muzeum Misyjne Misjonarzy Oblatów MN, Święty Krzyż 35151
Muzeum Misyjne Ojców Sercanów, Polanica Zdrój 35020
Muzeum Misyjno-Etnograficzne, Pieniężno 35007
Muzeum Mladaboleslavska, Mladá Boleslav 09657
Muzeum Młynarstwa Powietrznego, Bęsia 34540
Múzeum moderného umenia Andy Warhola, Medzilaborce 38304
Muzeum Mohelnice, Mohelnice 09661
Muzeum Morskie, Szczecin 35162
Múzeum Mošovských Remeselníkov, Mošovce 38309
Muzeum Motocyklů, Lesná u Znojma 09622
Muzeum Motoryzacji, Otrębusy 34999
Muzeum Motoryzacji Automobilklubu Wielkopolskiego, Poznań 35040
Muzeum Nadwiślański Park Etnograficzny, Babice 34532
Muzeum Nadwiślańskie, Kazimierz Dolny 34733
Muzeum Narodowe, Szczecin 35163
Muzeum Narodowe Rolnictwa i Przemysłu Rolno-Spożywczego w Szreniawie, Komorniki 34754
Muzeum Narodowe w Gdańsku, Gdańsk 34639
Muzeum Narodowe w Gdańsku, Gdańsk 34640
Muzeum Narodowe w Gdańsku, Gdańsk 34641
Muzeum Narodowe w Gdańsku, Gdańsk 34642
Muzeum Narodowe w Gdańsku, Gdańsk 34643
Muzeum Narodowe w Gdańsku, Nowy Barkoczyn 34954
Muzeum Narodowe w Gdańsku, Stary Targ 35133
Muzeum Narodowe w Kielcach, Kielce 34742
Muzeum Narodowe w Krakowie, Kraków 34805
Muzeum Narodowe w Poznaniu, Poznań 35041
Muzeum Narodowe w Poznaniu, Oddział w Rogalinie, Rogalin 35081
Muzeum Narodowe w Warszawie, Warszawa 35261
Muzeum Narodowe we Wrocławiu, Wrocław 35335
Muzeum Narodowe Ziemi Przemyskiej, Przemyśl 35055
Muzeum Neolitycznego Górnictwa i Rezerwat Archeologiczny, Krzemionki koło Ostrowca 34834
Muzeum Niepodległości, Warszawa 35262
Muzeum Nikifora, Krynica 34833
Muzeum Novojičínska, Nový Jičín 09685
Múzeum Obchodu, Ilava 38264
Múzeum Obchodu Bratislava, Bratislava 38225
Muzeum Oceanograficzne Mir, Gdynia 34658
Muzeum Odlewqnictwa Artystycznego, Gliwice 34665
Muzeum Odona Bujwida, Kraków 34806
Muzeum of A. Dvořák, Zlonice 09943
Museum of Sport Education and Sport, Praha 09785
Muzeum Okręgowe, Konin 34755
Muzeum Okręgowe, Krosno 34830
Muzeum Okręgowe, Nowy Sącz 34959
Muzeum Okręgowe, Ostrołęka 34993
Muzeum Okręgowe, Piotrków Trybunalski 35010
Muzeum Okręgowe, Rzeszów 35094
Muzeum Okręgowe, Sandomierz 35099
Muzeum Okręgowe, Suwałki 35144
Muzeum Okręgowe, Tarnów 35177
Muzeum Okręgowe, Zamość 35362
Muzeum Okręgowe im. Leona Wyczółkowskiego, Bydgoszcz 34578
Muzeum Okregowe Mazowsza Zachodniego, Żyradów 35381
Muzeum Okręgowe - Muzeum Etnograficzny, Tarnów 35178
Muzeum Okręgowe w Lesznie, Leszno 34850
Muzeum Okręgowe w Lesznie Dział Judaistyczny, Leszno 34851
Muzeum Okręgowe w Sieradzu, Sieradz 35105
Muzeum Okręgowe w Tarnowie, Tarnów 35179
Muzeum Okręgowe w Toruniu, Toruń 35195
Muzeum Okręgowe Ziemi Kaliskiej, Kalisz 34710
Muzeum Okresu Benešov, Benešov 09401
Muzeum Okręt Błyskawica, Gdynia 34659
Muzeum Olomouckých Tvarůžků, Loštice 09638
Muzeum Olomučanské Keramiky, Olomučany 09691
Múzeum Oravskej Dediny, Zuberec 38393
Muzeum Orderu Uśmiechu, Rabka 35071
Muzeum Oręża Polskiego, Kołobrzeg 34751
Muzeum Oręża Polskiego w Kołobrzegu, Kołobrzeg 34752
Muzeum Oręża Polskiego w Kołobrzegu, Kołobrzeg 34753
Muzeum Oświatowe, Puławy 35064
Muzeum Oświaty, Bydgoszcz 34579
Muzeum Oświaty Ziemi Łódzkiej, Łódź 34880
Muzeum Otokara Březiny, Jaroměř nad Rokytnou 09554
Muzeum Pałac w Wilanowie, Warszawa 35263
Muzeum Pamiątek po Janie Matejce "Koryznówka", Nowy Wiśnicz 34965
Muzeum Pamięci Narodowej, Kraków 34807
Muzeum Pamięci Narodowej lata 1939 -1956, Kielce 34743
Muzeum Pamięci św. Maksymiliana Kolbego, Zduńska Wola 35367
Muzeum Papiernictwa w Dusznikach Zdroju, Duszniki Zdrój 34627
Muzeum Parafialne, Dobra 34623
Muzeum Parafialne, Grybów 34686
Muzeum Parafialne, Tropie 35198
Muzeum Parafialne, Widawa 35298
Muzeum Parafialne, Złota 35373
Muzeum Parafialne im. Jana Piechoły, Iwkowa 34693
Muzeum Parafialne Sztuki Ludowej im. Ks. Edwarda Nitki, Paszyn 35004
Muzeum Pawła Stellera, Katowice 34727
Muzeum Pedagogiczne im. Prof. K. Sośnickiego, Gdańsk 34644
Muzeum Piastów Śląskich, Brzeg 34571
Muzeum Pienińskie im. Józefa Szalaya, Szczawnica 35155
Muzeum Pierwszych Piastów na Lednicy, Lednogóra 34846

M

Muzeum Piśmiennictwa i Muzyki Kaszubsko-Pomorskiej, Wejherowo 35296
Muzeum Plakatu w Wilanowie, Warszawa 35264
Muzeum Początków Państwa Polskiego, Gniezno 34673
Muzeum Poczty i Telekomunikacji, Wrocław 35336
Muzeum Poczty i Telekomunikacji Oddział w Gdańsku, Gdańsk 34645
Muzeum Podhalańskie PTTK, Nowy Targ 34963
Muzeum Podkrkonoší, Trutnov 09880
Muzeum Podlaskie w Białymstoku, Białystok 34550
Muzeum Pódrożników, Toruń 35196
Muzeum Pojezierza Łęczyńsko-Włodawskiego, Włodawa 35313
Muzeum Policie České Republiky, Praha 09773
Múzeum Polície Slovenskej Republiky, Bratislava 38226
Muzeum Politechniki Warszawskiej, Warszawa 35265
Muzeum Polná, Polná 09729
Muzeum Północno-Mazowieckie, Łomża 34890
Muzeum Polskiej Wojskowej Służby Zdrowia, Łódź 34881
Muzeum Południowego Podlasia, Biała Podlaska 34543
Muzeum Pomorza Środkowego, Słupsk 35112
Muzeum Porcelánu Pirkenhammer, Březová u Karlových Var 09419
Muzeum Powstania Chochołowskiego, Chochołów 34595
Muzeum Powstania Poznańskiego - Czerwiec 1956, Poznań 35042
Muzeum Powstania Warszawskiego, Warszawa 35266
Muzeum Powstania Wielkopolskiego 1918-1919, Poznań 35043
Muzeum Pożarnictwa, Alwernia 34528
Muzeum Pożarnictwa, Lidzbark 34854
Muzeum Pożarnictwa Ziemi Olkuskiej, Olkusz 34974
Muzeum Pozytywizmu im. Aleksandra Świętochowskiego, Gołotczyzna 34674
Muzeum-Prącownia Literacka Arkadego Fiedlera, Puszczykowo 35067
Muzeum Prasy Śląskiej, Pszczyna 35060
Muzeum Prof. Josepha Wittiga, Nowa Ruda 34948
Muzeum Prostějovska, Prostějov 09810
Muzeum Prowincji Ojców Bernardynów, Leżajsk 34853
Muzeum Průmyslových Železnic Brno, Zbýšov 09931
Muzeum Prywatne Diabła Polskiego Przedpiekle, Warszawa 35267
Muzeum Przemysłu, Warszawa 35268
Muzeum Przemysłu i Techniki, Wałbrzych 35212
Muzeum Przemysłu Naftowego w Bóbrce, Chorkówka 34601
Muzeum Przyrodnicze, Jelenia Góra 34706
Muzeum Przyrodnicze, Kazimierz Dolny 34734
Muzeum przyrodnicze, Mosina 34929
Muzeum Przyrodnicze, Władysławowo 35307
Muzeum Przyrodnicze Babiogórskiego Parku Narodowego, Zawoja 35364
Muzeum Przyrodnicze Bieszczadzkiego Parku Narodowego, Ustrzyki Dolne 35204
Muzeum Przyrodnicze im. Prof. Andrzeja Myrchy, Białystok 34551
Muzeum Przyrodnicze Instytutu Systematyki i Ewolucji Zwierząt PAN, Kraków 34808
Muzeum Przyrodnicze Uniwersytetu Łódzkiego, Łódź 34882
Muzeum Przyrodnicze Uniwersytetu Wrocławskiego, Wrocław 35337
Muzeum Przyrodnicze Wielkopolskiego Parku Narodowego, Puszczykowo 35068
Muzeum Przyrodnicze Wolińskiego Parku Narodowego im. Prof. Adama Wodziczki, Międzyzdroje 34918
Muzeum Przyrodniczo-Leśne, Białowieża 34544
Muzeum Przyrodniczo-Leśne Słowińskiego Parku Narodowego, Smołdzino 35113
Muzeum Przyrodniczo-Leśne Świętokrzyskiego Parku Narodowego, Święty Krzyż 35152
Muzeum Puszczy Kampinoskiej, Kampinos 34714
Muzeum Reformacji Polskiej, Mikołajki 34921
Muzeum Regionalne, Barlinek 34537
Muzeum Regionalne, Bełchatów 34539
Muzeum Regionalne, Brodnica 34570
Muzeum Regionalne, Brzeziny 34572
Muzeum Regionalne, Cedynia 34586
Muzeum Regionalne, Chojnów 34600
Muzeum Regionalne, Człuchów 34618
Muzeum Regionalne, Głogówek 34669
Muzeum Regionalne, Główno 34670
Muzeum Regionalne, Gostyń 34681
Muzeum Regionalne, Iłża 34689
Muzeum Regionalne, Jarocin 34695
Muzeum Regionalne, Jasło 34697
Muzeum Regionalne, Jawor 34699
Muzeum Regionalne, Kościan 34757
Muzeum Regionalne, Kozienice 34763
Muzeum Regionalne, Kraśnik 34824
Muzeum Regionalne, Krasnystaw 34826
Muzeum Regionalne, Kutno 34839
Muzeum Regionalne, Łęczna 34844
Muzeum Regionalne, Lubań 34894
Muzeum Regionalne, Lubartów 34895
Muzeum Regionalne, Łuków 34913
Muzeum Regionalne, Międzychód 34916
Muzeum Regionalne, Mielec 34920
Muzeum Regionalne, Myślibórz 34935
Muzeum Regionalne, Nowe Miasto nad Pilicą 34952
Muzeum Regionalne, Nysa 34966
Muzeum Regionalne, Ojców 34971
Muzeum Regionalne, Opoczno 34983
Muzeum Regionalne, Pińczów 35009
Muzeum Regionalne, Przeworsk 35056
Muzeum Regionalne, Pułtusk 35066
Muzeum Regionalne, Pyzdry 35069
Muzeum Regionalne, Radomsko 35076
Muzeum Regionalne, Siedlce 35103
Muzeum Regionalne, Sławków 35109
Muzeum Regionalne, Słupca 35111
Muzeum Regionalne, Środa Śląska 35126
Muzeum Regionalne, Szczecinek 35166
Muzeum Regionalne, Szydłów 35173
Muzeum Regionalne, Wągrowiec 35210
Muzeum Regionalne, Wiślica 35304
Muzeum Regionalne, Wodzisław Śląski 35314
Muzeum Regionalne, Wronki 35345
Muzeum Regionalne Dom Grecki, Myślenice 34934
Muzeum Regionalne Dom Szewca, Pszczew 35059
Muzeum Regionalne im. Adama Fastnachta, Brzozów 34573
Muzeum Regionalne im. Albina Nowickiego, Ryn 35090
Muzeum Regionalne im. Andrzeja Kaubego, Wolin 35316
Muzeum Regionalne im. Antoniego Minkiewicza, Olkusz 34975
Muzeum Regionalne im. Dzieci Wrzesińskich, Września 35346
Muzeum Regionalne im. Hieronima Łwainczaka, Krotoszyn 34832
Muzeum Regionalne im. Janusza Petera, Tomaszów Lubelski 35183
Muzeum Regionalne im. Marcina Rożka, Wolsztyn 35318
Muzeum Regionalne im. Seweryna Udzieli, Stary Sącz 35132
Muzeum Regionalne im. Władysława Kowalskiego, Dobczyce 34622
Muzeum Regionalne im Wojciechy Dutkiewicz, Rogoźno 35085
Muzeum Regionalne Kromerówka - Baszta Kowalska, Biecz 34554
Muzeum Regionalne PTTK, Golub-Dobrzyń 34675
Muzeum Regionalne PTTK, Muszyna 34932
Muzeum Regionalne PTTK, Puławy 35065
Muzeum Regionalne PTTK, Starachowice 35128
Muzeum Regionalne PTTK, Zagórze Śląskie 35353
Muzeum Regionalne PTTK i Urzędu Gminy, Iwanowice 34692
Muzeum Regionalne PTTK im. Ignacego Łukasiewicza, Gorlice 34679
Muzeum Regionalne w Sosnowcu, Sosnowiec 35123
Muzeum Regionalne w Stęszewie, Stęszew 35135
Muzeum Regionalne w Świebodzinie, Świebodzin 35149
Muzeum Regionalne Ziemi Limanowskiej, Limanowa 34856
Muzeum Regionalne Ziemi Sadowieńskiej, Sadowne 35095
Muzeum Regionalne Ziemi Sulmierzyckiej, Sulmierzyce 35139
Muzeum Regionalne Ziemi Zbąszyńskiej PTTK, Zbąszyń 35365
Muzeum Rekordů a Kuriozit, Pelhřimov 09709
Muzeum Roberta Kocha, Wolsztyn 35319
Muzeum Rolnictwa im. Ks. Krzysztofa Kluka, Ciechanowiec 34608
Muzeum Romantyzmu, Opinogóra 34982
Muzeum Romské Kultury, Brno 09437
Muzeum Rumburk, Rumburk 09831
Muzeum Rybackie, Jastarnia 34698
Muzeum Rybołówstwa, Hel 34687
Muzeum Rybołówstwa Morskiego, Świnoujście 35153
Muzeum Rzemiosł Ludowych - Zagroda Sitarska, Biłgoraj 34564
Muzeum Rzemiosła, Krosno 34831
Muzeum Rzemiosła Tkackiego, Turek 35201
Muzeum Rzeźby Alfonsa Karnego, Białystok 34552
Muzeum Rzeźby im. Xawerego Dunikowskiego Królikarnia, Warszawa 35269
Muzeum Sakralne Kolegiaty Zamojskiej, Zamość 35363
Múzeum Samuela Jurkoviča Sobotište-Nyaryiovský kaštiel', Sobotište 38354
Muzeum Skansen Kolejnictwa, Kościerzyna 34759
Muzeum Skansen-Zagroda Pszczyńska, Pszczyna 35061
Muzeum Skla, Harrachov 09507
Muzeum Skla a Bižuterie, Jablonec nad Nisou 09545
Múzeum Školstva a Pedagogiky Bratislava, Bratislava 38227
Muzeum J.V. Sládka, Zbiroh 09929
Muzeum Śląska Cieszyńskiego w Cieszynie, Cieszyn 34610
Muzeum Śląska Opolskiego, Opole 34986
Muzeum Śląskie, Katowice 34728
Muzeum Ślężańskie im. Stanisława Dunajewskiego, Sobótka 35116
Múzeum Slovenského Červeného Kríža, Martin 38298
Muzeum Slovenského Narodného Povstania, Banská Bystrica 38184
Múzeum Slovenskej Dediny, Martin 38299
Múzeum Slovenských Národných Rád, Myjava 38311
Muzeum Sokolov, Krásno 09605
Muzeum Sopotu, Sopot 35120
Múzeum Špecialneho Školstva, Levoča 38283
Múzeum Spiša, Spišská Nová Ves 38359
Muzeum Społeczne, Strzyżów 35137
Muzeum Sportu i Turystyki, Łódź 34883
Muzeum Sportu i Turystyki, Warszawa 35270
Muzeum Sportu i Turystyki Regionu Karkonoszy, Karpacz 34715
Muzeum Sportu i Turystyki Ziemi Gdańskiej, Gdańsk 34646
Muzeum Sprzętu Gospodarstwa Domowego, Ziębice 35371
Muzeum Śremskie, Śrem 35125
Muzeum Śrowiska Przyrodniczegi i Łowiectwa, Uzarzewo 35205
Muzeum Stanisława Staszica, Piła 35008
Muzeum Stanisława Wyspiańskiego w Kamienicy Szołayskich, Kraków 34809
Muzeum Starego Teatru, Kraków 34810
Muzeum Starożytnego Hutnictwa Mazowieckiego, Pruszków 35052
Muzeum Starożytnego Hutnictwa Świętokrzyskiego im. prof. M. Radwana, Nowa Słupia 34949
Muzeum Stefana Żeromskiego, Nałęczów 34940
Muzeum Středního Pootaví, Strakonice 09849
Muzeum Strojíren Poldi, Kladno 09582
Muzeum Stylu Zakopiańskiego im. Stanisława Witkiewicza - Willa Koliba, Zakopane 35359
Muzeum Šumavy, Kašperské Hory 09581
Muzeum Šumavy, Sušice 09856
Muzeum Svatopluka Čecha, Liteň 09628
Muzeum Szlachty Mazowieckiej, Ciechanów 34607
Muzeum Sztuk Użytkowych, Poznań 35044
Muzeum Sztuki Cmentarnej, Wrocław 35338
Muzeum Sztuki Dziecka, Warszawa 35271
Muzeum Sztuki Książki, Wrocław 35339
Muzeum Sztuki Medalierskiej, Wrocław 35340
Muzeum Sztuki Mieszczańskiej, Wrocław 35341
Muzeum Sztuki Nowoczesnej w Niepołomicach, Niepołomice 34945
Muzeum Sztuki Sakralnej, Bardo 34536
Muzeum Sztuki Sakralnej, Żnin 35377
Muzeum Sztuki w Łodzi, Łódź 34884
Muzeum Sztuki Współczesnej, Radom 35074
Muzeum Sztuki Złotniczej, Kazimierz Dolny 34735
Muzeum T.G. Masaryka, Lány 09618
Muzeum T.G. Masaryka, Rakovník 09814
Múzeum Tatranského Národného Parku, Tatranská Lomnica 38370
Muzeum Tatrzańskie im. Tytusa Chałubińskiego, Zakopane 35360
Muzeum Teatralne, Warszawa 35272
Muzeum Teatralne im. Stanisława Wyspiańskiego, Kraków 34811
Muzeum Technik Ceramicznych, Koło 34750
Muzeum Techniki w Warzawie, Warszawa 35273
Muzeum Techniki Włókienniczej, Bielsko-Biała 34561
Muzeum Telč, Telč 09861
Múzeum Telesnej Kultúry v SR, Bratislava 38228
Muzeum Těšínska, Český Těšín 09470
Muzeum těžby a dopravy vápence v Českém krasu, Beroun 09405
Muzeum Tkactwa Dolnośląskiego w Kamiennej Górze, Kamienna Góra 34712
Muzeum Tradycji Niepodległościowych, Łódź 34885
Muzeum Třešť, Třešť 09878
Muzeum Třineckých Železáren a Města Třince, Třinec 09879
Muzeum Týn nad Vltavou, Týn nad Vltavou 09885
Muzeum Tynecké Keramiky, Týnec nad Sázavou 09886
Muzeum Ubezpieczeń, Kraków 34812
Múzeum Ukrajinsko-Rusínskej Kultúry, Svidník 38367
Múzeum Umeleckých Remesiel, Bratislava 38229
Muzeum umění a designu, Benešov 09402
Muzeum Umění Olomouc, Olomouc 09689
Muzeum Uniwersytetu im. Ks. Jana Władzińskiego, Lublin 34907
Muzeum Uniwersytetu Jagiellońskiego - Collegium Maius, Kraków 34813
Muzeum Uniwersytetu Marii Curie-Sklodowskiej, Lublin 34908
Muzeum Uniwersytetu Warszawskiego, Warszawa 35274
Muzeum Uniwersytetu Wrocławskiego, Wrocław 35342
Muzeum Uzbrojenia - Cytadela Poznańska, Poznań 35045
Muzeum v Bruntále, Bruntál 09445
Muzeum v Ivančicích, Ivančice 09543
Múzeum v Kežmarku a Mestský Hrad, Kežmarok 38267
Múzeum Včelárstva na Slovensku, Král'ová pri Senci 38276
Muzeum ve Štramberku, Štramberk 09850
Muzeum ve Žluticích, Žlutice 09945
Muzeum Velké Meziříčí, Velké Meziříčí 09902
Múzeum Vín, Prešov 38336
Múzeum vo Svätý Antone, Svätý Anton 38364
Múzeum Vojtecha Löfflera, Košice 38272
Muzeum Voskových Figurín České Historie, Karlštejn 09579
Muzeum Voskových Figurín České Historie, Praha 09774
Muzeum Východních Čech v Hradci Králové, Hradec Králové 09534
Muzeum Vyškovska, Vyškov 09919
Muzeum Vysočiny, Třebíč 09875
Muzeum Vysočiny Jihlava, Jihlava 09561
Muzeum Vysočiny Jihlava, Telč 09861
Muzeum Vysočiny Pelhřimov, Pelhřimov 09710
Muzeum w Bielsku-Białej, Bielsko-Biała 34562
Muzeum w Bielsku Podlaskim, Bielsk Podlaski 34556
Muzeum w Bochni im. Prof. Stanisława Fischera, Bochnia 34565
Muzeum w Chorzowie, Chorzów 34604
Muzeum w Chrzanowie im. Ireny i Mieczysława Mazarakich, Chrzanów 34606
Muzeum w Dobrzycy, Dobrzyca 34625
Muzeum w Elblągu, Elbląg 34628
Muzeum w Gliwicach, Gliwice 34666
Muzeum w Gorzowie Wielkopolskim, Gorzów Wielkopolski 34680
Muzeum w Grudziądzu, Grudziądz 34685
Muzeum w Kwidzynie, Kwidzyń 34840
Muzeum w Lęborku, Lębork 34843
Muzeum w Łęczycy, Łęczyca 34845
Muzeum w Łowiczu, Łowicz 34892
Muzeum w Międzyrzeczu, Międzyrzecz 34917
Muzeum w Przeworsku, Przeworsk 35057
Muzeum w Raciborzu, Racibórz 35072
Muzeum w Rybniku, Rybnik 35089
Muzeum w Stargardzie, Stargard 35129
Muzeum w Szklarskiej Porębie Hoffmanna, Szklarska Poręba 35171
Muzeum w Tarnowskich Górach, Tarnowskie Góry 35180
Muzeum w Tomaszowie Mazowieckim, Tomaszów Mazowiecki 35184
Muzeum w Tykocinie, Tykocin 35202
Muzeum w Wałbrzychu, Wałbrzych 35213
Muzeum w Żywcu, Żywiec 35382
Muzeum Waldemara Babinicza, Nałęczów 34941
Muzeum Walewskich w Tubądzinie, Wróblew 35321
Muzeum Walk o Wał Pomorski przy Miejsko-Gminnym Ośrodku Kultury, Mirosławiec 34925
Muzeum Walki i Męczeństwa 1939-1945, Warszawa 35275
Muzeum Walki i Męczeństwa w Palmirach, Warszawa 35276
Muzeum Walki i Męczeństwa w Treblince, Kosów Lacki 34760
Muzeum Warmii i Mazur, Olsztyn 34978
Muzeum Warmińskie, Lidzbark Warmiński 34855
Muzeum Wartownia nr 1 na Westerplatte Sucharskiego, Gdańsk 34647
Muzeum Weterynarii, Ciechanowiec 34609
Muzeum Wiedzy o Środowisku, Poznań 35046
Muzeum Więzienia Pawiak, Warszawa 35277
Muzeum Wikliniarstwa i Chmielarstwa, Nowy Tomyśl 34964
Muzeum Wikliny, Olkusz 34976
Muzeum Wisły, Tczew 35182
Muzeum Władysława Broniewskiego, Warszawa 35278
Muzeum Władysława Sikorskiego, Warszawa 35279
Muzeum Wnętrz Dworskich, Ożarów 35001
Muzeum Wnętrz Pałacowych w Choroszczy, Choroszcz 34602
Muzeum Wnętrz Stylowych, Oporów 34988
Muzeum Wnętrz Zabytkowych, Dębno 34621
Muzeum Wojska, Białystok 34553
Muzeum Wojska Polskiego, Warszawa 35280
Muzeum Woli, Warszawa 35281
Muzeum Wsi Kieleckiej, Kielce 34744
Muzeum Wsi Lubelskiej, Lublin 34909
Muzeum Wsi Mazowieckiej, Sierpc 35107
Muzeum Wsi Opolskiej, Opole 34987
Muzeum Wsi Radomskiej, Radom 35075
Muzeum Wsi Słowińskiej w Klukach, Smołdzino 35114

Muzeum X Pawilonu Cytadeli Warszawskiej, Warszawa 35282
Muzeum Zabawek ze Zbiorow Henryka Tomaszewskiego, Karpacz 34716
Muzeum Zabawkarstwa, Kielce 34745
Muzeum Zábřeh, Zábřeh 09924
Muzeum Zachodnio-Kaszubskie, Bytów 34585
Muzeum Zagłębia, Będzin 34538
Muzeum Zagłębia Staropolskiego w Sielpi, Sielpia 35104
Muzeum zahradnictví a životního prostředí, Valtice 09897
Muzeum Zakładu Historii Farmacji, Lublin 34910
Muzeum Zamek, Baranów Sandomierski 34534
Muzeum-Zamek Górków, Szamotuły 35154
Muzeum - Zamek Książąt Pomorskich, Darłowo 34620
Muzeum-Zamek w Bolkowie, Bolków 34569
Muzeum-Zamek w Gołuchowie, Gołuchów 34676
Muzeum Zamek w Łańcucie, Łańcut 34842
Muzeum Zamkowe, Pszczyna 35062
Muzeum Zamkowe w Malborku, Malbork 34914
Muzeum Zamoyskich w Kozłówce, Kamionka 34713
Muzeum Zbiorów Geologicznych, Kielce 34746
Muzeum Zbrojownia, Liw 34858
Muzeum Zdeňka Buriana, Štramberk 09851
Múzeum Ždiarsky Dom, Ždiar 38388
Muzeum Zegarów, Rzemiosł Artystycznych i Precyzyjnych, Warszawa 35283
Muzeum Zegarów Wieżowych, Gdańsk 34648
Muzeum Zespoł Zamkowy w Niedzicy, Niedzica 34943
Muzeum Židovské Komunity, Tovačov 09871
Múzeum Židovskej Kultúry, Bratislava 38230
Muzeum Ziemi Augustowskiej, Augustów 34531
Muzeum Ziemi Chełminskiej, Chełmno 34591
Muzeum Ziemi Dobrzyńskiej, Rypin 35091
Muzeum Ziemi Kępińskiej, Kępno 34737
Muzeum Ziemi Kłodzkiej, Kłodzko 34747
Muzeum Ziemi Kociewskiej, Starogard Gdański 35130
Muzeum Ziemi Krajeńskiej, Nakło nad Notecią 34937
Muzeum Ziemi Krzeszowickiej, Krzeszowice 34836
Muzeum Ziemi Kujawskiej i Dobrzyńskiej, Włocławek 35311
Muzeum Ziemi Lubawskiej, Nowe Miasto Lubawskie 34951
Muzeum Ziemi Lubuskiej, Zielona Góra 35372
Muzeum Ziemi Mińskiej, Mińsk Mazowiecki 34924
Muzeum Ziemi Mrągowskiej, Mrągowo 34930
Muzeum Ziemi Nadnoteckiej im. Wiktora Stachowiaka, Trzcianka 35199
Muzeum Ziemi Otwockiej, Otwock 35000
Muzeum Ziemi Pałuckiej, Żnin 35378
Muzeum Ziemi Piskiej, Pisz 35011
Muzeum Ziemi Polskiej Akademii Nauk, Warszawa 35284
Muzeum Ziemi Prudnickiej, Prudnik 35051
Muzeum Ziemi Puckiej, Puck 35063
Muzeum Ziemi Rawickiej, Rawicz 35079
Muzeum Ziemi Rawskiej, Rawa Mazowiecka 35078
Muzeum Ziemi Sochaczewskiej i Pola Bitwy nad Bzurą, Sochaczew 35119
Muzeum Ziemi Średzkiej Dwór w Koszutach, Koszuty 34762
Muzeum Ziemi Suskiej, Sucha Beskidzka 35138
Muzeum Ziemi Wałeckiej, Wałcz 35214
Muzeum Ziemi Wieluńskiej, Wieluń 35301
Muzeum Ziemi Wschowskiej, Wschowa 35347
Muzeum Ziemi Zaborskiej i Galeria Sztuki Ludowej, Wiele 35299
Muzeum Ziemi Zawkrzańskiej, Mława 34927
Muzeum Ziemi Złotowskiej, Złotów 35374
Muzeum Zlatý Klíč, Karlovy Vary 09576
Muzeum Zofii i Wacława Nałkowskich, Wołomin 35317
Muzeum Zoologiczne Instytutu Zoologii Uniwersytetu Jagiellońskiego, Kraków 34814
Muzeum Żup Krakowskich Wieliczka, Wieliczka 35300
Muzeum Żuraw, Gdańsk 34649
Muzeum Żydowskiego Instytutu Historycznego w Polsce, Warszawa 35285
Muziejus, Kintai 30275
Muziejus Čepkelių Valstybiniame Rezervate, Marcinkonys 30306
Muziejus Žuvinto Valstybiniame Rezervate, Aleknonių 30202
Muziekinformatie- en documentatiecentrum Ton Stolk, Vlaardingen 32890
Muziekinstrumenten-Museum, Peer 03948
Muziekinstrumentenmakersmuseum, Tilburg 32827
Mūzikas Muzejs, Rīga 30098
Muzikų Brolių Motiejaus, Broniaus ir Antano Budriūnų Gimtinė, Pabiržė 30324
Muzium Alor Gajah, Alor Gajah 30549
Muzium Angkatan Tentera, Kuala Lumpur 30577
Muzium Arkeologi, Merbok 30609
Muzium Brunei, Bandar Seri Begawan 04965
Muzium dan Galeri Seni, Minden 30610
Muzium Darul Ridzwan, Ipoh 30555
Muzium Diraja Abu Bakar, Johor Bahru 30558
Muzium Diraja Perak, Kuala Kangsar 30570
Muzium Jasin, Jasin 30557
Muzium Kecantikan, Melaka 30598
Muzium Layang-Layang, Melaka 30599
Muzium Negara, Kuala Lumpur 30578
Muzium Negeri Kedah, Alor Setar 30552
Muzium Negeri Kelantan, Kota Bharu 30564
Muzium Negeri Perlis, Kangar 30559
Muzium Negeri Sarawak, Kuching 30591
Muzium Negeri Selangor, Shah Alam 30619
Muzium Negeri Sembilan, Seremban 30616
Muzium Negeri Trengganu, Kuala Terengganu 30589
Muzium Numismatik Maybank, Kuala Lumpur 30579
Muzium Perak, Taiping 30622
Muzium Polis Diraja Malaysia, Kuala Lumpur 30580
Muzium Rakyat, Melaka 30600
Muzium Samudera, Melaka 30601
Muzium Sastera, Melaka 30602
Muzium Sejarah & Etnografi, Melaka 30603
Muzium Sejarah National, Kuala Lumpur 30581
Muzium Seni Asia, Kuala Lumpur 30582
Muzium Teknologi Melayu, Bandar Seri Begawan 04966
Muzium Tekstil, Kuala Lumpur 30583
Muzium Tentera Laut Diraja Malaysia, Melaka 30604
Muzium Tuan Yang Terutama Melaka, Melaka 30605
Muzj Istorii Rossijskich Voennych Učilišč, Sankt-Peterburg 37715
Muzyčno-memorialnyj Muzej Solomiï Krušel'nic'koï, Lviv 43136
Muzyka i Vremja - Častnyj Muzej Džona Mostoslavskogo, Jaroslavl 36974
MVG-Museum, München 20470
My Old Kentucky Home, Bardstown 46607
Myanan Sankyaw Shwe Nandaw, Mandalay 31718
Myers Museum of Egyptian and Classical Art, Eton 44190
Mykolaïv Local Museum, Mykolaïv 43151
Mykolaïv Vereshchagin Museum of Fine Arts, Mykolaïv 43150
Mykolaïvs'kyj Oblasnyj Chudožnij Muzej im. Vereščagina, Mykolaïv 43150
Mykolaïvs'kyj Oblasnyj Krajeznavčyj Muzej, Mykolaïv 43151
Mykolas Zilinskas Art Gallery of the M.K. Čiurlionis National Art Museum, Kaunas 30261
Mykolo Žilinsko Dailės Galerija, Kaunas 30261
Mykonos Folklore Museum, Mykonos 23104
Myllarheimen, Rauland 33777
Myllykolu, Birthplace of F.E. Sillanpää, Hämeenkyrö 10740
Myllykolu, F.E. Sillanpään Syntymäkoti, Hämeenkyrö 10740
Myllymuseo, Juuka 10882
Myllysaaren Museo, Valkeakoski 11437
Myntkabinett, Oslo 33734
Myntsafn Sedlabanka og Thjódminjasafns, Reykjavík 24025
Myohyang-san Museum, Hyangsan 29755
Myong Ji University Museum, Seoul 29908
Myra Powell Art Gallery and Gallery at the Station, Ogden 51502
Myrbergsgården Local Museum, Vörå 11475
Myreton Motor Museum, Aberlady 43253
Myrskylän Kotiseutumuseo, Myrskylä 11132
Myrtleville House Museum, Brantford 05470
Mysore Silk Museum, Bangalore 24074
Mystic Art Center, Mystic 50934
Mystic Seaport, Mystic 50935
MZM - Museumszentrum Mistelbach, Mistelbach an der Zaya 02386
Mzuzu Regional Museum, Mzuzu 30546

N

N.A. Nekrasov Apartment Museum, Sankt-Peterburg 37661
N.A. Nekrasov House-Museum, Čudovo 36871
N.A. Nekrasov Memorial Museum Preserve of Literature, Karabicha 37002
N.A. Ostrovskiy Museum and Humanitarian Center Preodolenie, Moskva 37198
N.A. Yaroshenko Memorial Estate Museum, Kislovodsk 37048
N.D. Kusnetsov Centre for the History of Aeromotors, Samara 37611
N.E. Zhukovsky Memorial House-Museum, Orechovo 37502
N.E. Zhukovsky Memorial Museum, Moskva 37388
N.G. Chernyshevski Museum-Estate, Saratov 37751
N.G. Poletaev Regional Museum of Tuapse, Tuapse 37864
N. Hadjikyriakos-Ghikas Gallery, Athinai 22941
N.M. Djakonov Literary and Theater Museum, Syktyvkar 37816
N.M. Przevalsky Regional Museum, Prževalsk 29964
N.N. Shukov House-Museum, Elec 36908
N. Ostrovskyj Memorial Museum of Literature, Soči 37782
N.S. Leskov House Museum, Orël 37505
N.S. Muchin Appartment Museum, Joškar-Ola 36976
N.V. Fedorov Library-Museum, Moskva 37268
N.V. Gogol House Museum - City Library - Culture Centre, Moskva 37164
N.V. Gogol Literary Memorial Museum, Velyki Soročynci 43201
N.V. Gogol Preserve-Museum, Gogoleve 43040
N.Y. Vaptsarov Memorial House, Bansko 04980
Naaimachinemuseum Dordrecht, Dordrecht 32152
Naantali Museum, Naantali 11133
Naantalin Museo, Naantali 11133
NAB Gallery, Chicago 47498
Nábytkové Múzeum, Markušovce 38292
Nace's House, Škofja Loka 38516
Nacetova Hiša, Škofja Loka 38516
Nacionalen Archeologičeski Rezervat Kabile, Jambol 05042
Nacionalen Cărkoven Istoriko-archeologičeski Muzej, Sofia 05189
Nacionalen Etnografski Muzej, Sofia 05190
Nacionalen Istoričeski Muzej, Sofia 05191
Nacionalen Istoriko-archeologičeski Rezervat i Muzej Veliki Preslav, Veliki Preslav 05248
Nacionalen Istoriko-archeologičeski Rezervat Pliska, Pliska 05123
Nacionalen Literaturen Muzej, Sofia 05192
Nacionalen Muzej Bojanska Crkva, Sofia 05193
Nacionalen Muzej Georgi Dimitrov, Sofia 05194
Nacionalen Muzej na Bălgarskata Architektura, Veliko Tărnovo 05256
Nacionalen Muzej na Fizičeskata Kultura i Sport, Sofia 05195
Nacionalen Muzej na Obrazovanieto, Gabrovo 05033
Nacionalen Muzej na Transporta i Săobščenijata, Ruse 05154
Nacionalen Muzej Parachod Radecki, Kozloduj 05080
Nacionalen Muzej Rilski Manastir, Rilski Manastir 05150
Nacionalen Muzej Roženski Manastir, Melnik 05092
Nacionalen Muzej Zemjata i Chorata, Sofia 05196
Nacionalen Park-muzej Samuilova krepost, Petrič 05114
Nacionalen Park-muzej Šipka-Buzludža, Kazanlăk 05060
Nacionalen Politechničeski Muzej, Sofia 05197
Nacionalen Prirodonaučen Muzej, Sofia 05198
Nacionalen Selskostopanski Muzej, Sofia 05199
Nacionalen Voenno-morski Muzej, Varna 05244
Nacionalinis Mikalojaus Konstantino Čiurlionio Dailės Muziejus, Kaunas 30262
Nacionalna Chudožestvena Galerija, Sofia 05200
Nacionalna Galerija za Čuždestranno Izkustvo, Sofia 05201
Nacionalna Galerija za Dekorativni Izkustva, Sofia 05202
Nacionalna Izložba na Narodnite Chudožestveni Zanajati i Priložni Izkustva, Orešak 05100
Nacionalnaja Galereja Respubliki Komi, Syktyvkar 37820
Nacionalnij Muzej u Lvovi, Lviv 43137
Nacionalnij Muzej u Lvovi - Viddil Narodnogo Mystectva, Lviv 43138
Nacionalnij Muzej u Lvovi - Viddil Starodrukiv ta Rukopysiv, Lviv 43139
Nacionalnij Muzej u Lvovi - Viddil Sučasnogo Mystectva, Lviv 43140
Nacionalnyj Chudožestvennyj Muzej Respubliki Belarus, Minsk 03255
Nacionalnyj Chudožestvennyj Muzej Respubliki Sacha (Jakutija), Jakutsk 36964
Nacional'nyj Chudožnij Muzej Ukraïny, Kyïv 43100
Nacional'nyj Istoriko-etnografičnyj Zapovidnyk Perejaslav, Perejaslav-Chmel'nyc'kyj 43165
Nacional'nyj Kyjvo-Pečerskyj Istoryko-kul'turnyj Zapovidnyk, Kyïv 43101
Nacional'nyj Muzej Černobyl', Kyïv 43102
Nacional'nyj Muzej Istoriï Ukraïny, Kyïv 43103
Nacional'nyj Muzej Istoriï Velykoï Vitčyznjanoï Vijny 1941-1945 Rokiv, Kyïv 43104
Nacionalnyj Muzej Kabardino-Balkarskoj Respubliki, Nalčik 37442
Nacional'nyj Muzej Literatury Ukraïny, Kyïv 43105
Nacional'nyj Muzej Medycyny Ukraïny, Kyïv 43106
Nacionalnyj Muzej Respubliki Baškortostan, Ufa 37889
Nacionalnyj Muzej Respubliki Komi, Syktyvkar 37821
Nacionalnyj Muzej Respubliki Marij Ėl im. T. Evseeva, Joškar-Ola 36980
Nacionalnyj Muzej Respubliki Tatarstan, Kazan 37026
Nacional'nyj Muzej Tarasa Ševčenka, Kyïv 43107
Nacionalnyj Muzej Udmurtskoj Respubliki im. K. Gerda, Iževsk 36955
Nacional'nyj Muzej Zbrojnych Syl Ukraïny, Kyïv 43108
Nacional'nyj Zapovidnyk Chersones Tavrijs'kyj, Sevastopol 43178
Nacional'nyj Zapovidnyk Chorticja, Zaporižžja 43208
Nádasdi Ferenc Múzeum, Sárvár 23796
Nádasdy Castle, Nádasdladány 23722
Nádasdy-Kastély, Nádasdladány 23722
Nadbaltyckie Centrum Kultury w Gdansku, Gdańsk 34650
Nadelburgmuseum, Lichtenwörth 02298
Nader Mausoleum, Mashad 24633
Nadlerhaus, Stützengrün 21923
Nääs Slott & Slöjdseminarium, Floda 40524
Nähmaschinen-Museum, Steckborn 41909
Nähmaschinen-Museum Gebrüder Mey, Albstadt 16457
Näitustesaal Rotermanni Soolalaos, Tallinn 10637
Näkövammaismuseo, Jiris 10871
Närpes Hembygdsmuseum, Närpes 11134
Næs Iron works Museum, Tvedestrand 33938
Næs Jernverksmuseum, Tvedestrand 33938
Naeshirogawa Folk Pottery Museum, Higashi-Ichiki 28503
Naeshirogawa Minkan-yaki Bijutsukan, Higashi-Ichiki 28503
Næstved Museum, Næstved 10190
Nafplion Folklore Museum, Nafplion 23117
Nagano-ken Shinano Bijutsukan, Nagano 28957
Nagano-kenritsu Rekishi Hakubutsukan, Chikuma 28398
Nagano Prefectual Museum of History, Chikuma 28398
Nagano Prefectural Shinano Art Museum, Nagano 28957
Nagaoka City Industrial Exhibition Hall, Nagaoka 28962
Nagaoka Gear Museum, Nagaoka 28959
Nagaoka Local History Museum, Nagaoka 28961
Nagaoka Municipal Science Museum, Nagaoka 28960
Nagaoka-shiritsu Hakabutsukan, Nagaoka 28961
Nagaoka-shiritsu Kogyo Hakurankai, Nagaoka 28962
Nagasaki Atomic Bomb Museum, Nagasaki 28978
Nagasaki-ken Bijutsukan, Nagasaki 28979
Nagasaki Kite Museum, Nagasaki 28980
Nagasaki Municipal Museum, Nagasaki 28983
Nagasaki Municipal Nyokodo Nagai Memorial Museum, Nagasaki 28984
Nagasaki Museum of History and Culture, Nagasaki 28981
Nagasaki Prefectural Art Museum, Nagasaki 28979
Nagasaki Rekishi no Bunka Hakubutsukan, Nagasaki 28981
Nagasaki Science Museum, Nagasaki 28982
Nagasaki-shiritsu Hakabutsukan, Nagasaki 28983
Nagasaki-shiritsu Nyokodo Nagai Kinenkan, Nagasaki 28984
Nagashima Museum, Kagoshima 28633
Nagatoro Kyukokan, Nagatoro 28988
Nagatoto Archaeological Museum, Nagatoro 28988
Nagelmuseum, Löchgau 20014
Nagelschmiede Sulz, Sulz 41921
Nagelschmiedemuseum, Losenstein 02338
Nagi-cho Gendai Bijutsukan, Nagi 28990
Nagi Museum, Cairo 10509
Nagi Museum of Contemporary Art, Nagi 28990
Nagoya-Boston Museum of Fine Arts, Nagoya 29003
Nagoya City Art Museum, Nagoya 29005
Nagoya City Hideyoshi and Kiyomasa Memorial Museum, Nagoya 29004
Nagoya City Museum, Nagoya 29007
Nagoya City Science Museum, Nagoya 29008
Nagoya-shi Bijutsukan, Nagoya 29005
Nagoya-shi Daigaku Hakubutsukan, Nagoya 29006
Nagoya-shi Hakubutsukan, Nagoya 29007
Nagoya-shi Kagakukan, Nagoya 29008
Nagoya University Museum, Nagoya 29006
Nagu Hembyggdsmuseum/Nauvon Kotiseutumuseo, Nauvo 11138
Nagy Ferenc Galéria, Tab 23908
Nagy István Képtár, Baja 23346
Nagy Lajos Emlékkiállítás, Apostag 23336
Nagy László Emlékház, Iszkáz 23621
Nagykunság Region House, Karcag 23640
Nagykunsági Tájház, Karcag 23640
Nagytétényi Kastélymúzeum, Budapest 23454
Nahkl Fort Museum, Nahkl 33993
Nahsholim Museum, Hof Dor 24973
Nahum Goldmann Museum of the Jewish Diaspora, Tel Aviv 25099
Nahum Gutman's Museum, Tel Aviv 25118
Naicam Museum, Naicam 06406
Nail Museum in Jokioinen, Jokioinen 10875
Nain Museum, Nain 06407
Nairn Fishertown Museum, Nairn 45189
Naismith Memorial Basketball Hall of Fame, Springfield 53227
Naito Kinen Kusuri Hakubutsukan, Kawashima 28697
Naito Museum of Pharmaceutical Science and Industry, Kawashima 28697

Naive Art Gallery - Ivan Generalić Collection, Hlebine 08877
Naive art museum of Latvia, Riga 30099
Nakagawa Chaini-zu Kindai Bijutsukan, Fukuyama 28448
Nakagawa Museum of Chinese Modern Art, Fukuyama 28448
Nakagawa Photo Gallery, Kyoto 28853
Nakama City Folklore Museum, Nakama 29017
Nakama-shiritsu Minzoku Hakubutsukan, Nakama 29017
Nakambale Museum, Ondangwa 31751
Nakamura Kinen Bijutsukan, Kanazawa 28661
Nakamura Memorial Art Museum, Kanazawa 28661
Nakano City History and Folk Custom Museum, Nakano 29021
Nakano Historical Museum, Tokyo 29441
Nakano-shi Rekishi Minzoku Shiryokan, Nakano 29021
Nakata Bijutsukan, Onomichi 29119
Nakata Museum of Art, Onomichi 29119
Nakhchivan Autonomous Republic Djalil Mammadkulizade Literature Museum, Naxçivan 03186
Nakhchivan Carpet Museum, Naxçivan 03187
Nakhchivan State History Museum, Naxçivan 03188
Nakhon Si Thammarat National Museum, Nakhon Si Thammarat 42585
Nakkilan Kotiseutumuseo ja Juustomeijerimuseo, Nakkila 11136
Nakusp & District Museum, Nakusp 06408
Nalsia Museum, Švenčionys 30376
Nalšios Muziejus, Švenčionys 30376
Nam June Paik Art Center, Yongin 29945
Nam Nao, Petchabun 42590
Namban Bunkakan, Osaka 29136
Namban Culture Museum, Osaka 29136
Nambu Toji Denshokan, Ishidoriya 28585
Namgyal Institute of Tibetology, Gangtok 24175
Namık Kemal Zindanı ve Müzesi, Gazimağusa 33352
Namikawa Banri Bukkyo Bijutsu Shashin Gallery, Hiroshima 28541
Namikawa Cloisonne Museum of Kyoto, Kyoto 28854
Namjin Art Museum, Imhoe 29831
Namoc-National Art Museum of China, Beijing 07554
Namsaraev Museum of the Literature of Buryatya, Ulan-Udė 37895
Namsdalsmuseet, Namsos 33691
Namsskogan Bygdatun, Namsskogan 33693
Nan Yue Wang Palace Museum, Guangzhou 07735
Nanaimo Art Gallery, Nanaimo 06412
Nanaimo District Museum, Nanaimo 06413
Nanao Art Museum, Nanao 29024
Nan'ao Coast Defence City Museum, Nan'ao 08027
Nanao-shiritsu Bijutsukan, Nanao 29024
NANB Nursing History Centre, Fredericton 05833
Nance Museum, Lone Jack 50171
Nanchang Museum, Nanchang 08029
Nanchang Uprising Museum, Nanchang 08030
Nancheng Museum, Nancheng 08034
Nancy Fine Arts Museum, Nancy 14354
Nancy Island Historic Site, Wasaga Beach 07280
Nandan Museum-Vichittra and Art Gallery, Santiniketan 24378
Nandewar Historical Museum, Barraba 00834
Nanhai Museum, Nanhai 08035
Naniwa-no Umino Jikukan, Osaka 29137
Nanji Guan, Qingdao 08106
Nanjiang Museum, Nanjiang 08036
Nanjing Folklorish Museum, Nanjing 08043
Nanjing International Culture and Arts Exchange Center, Nanjing 08044
Nanjing Museum, Nanjing 08045
Nanjing Museum, Nanjing, Zhangzhou 08052
Nanjing Museum of Palaeontology, Nanjing 08046
Nanjingdi Zhen Science Museum, Nanjing 08047
Nanjingdi Zhi Museum, Nanjing 08048
Nankang County Museum, Nankang 08053
Nanning Museum, Nanning 08056
Nanortalik Local Museum, Nanortalik 23258
Nanortallip Katersugaasivia, Nanortalik 23258
Nanping Museum, Nanping 08058
Nansong Guanstove Museum, Hangzhou 07765
Nansong Qian Currency Museum, Hangzhou 07766
Nantgarw China Works Museum, Nantgarw 45190
Nanton Lancaster Air Museum, Nanton 06415
Nantong Bowu House, Nantong 08059
Nantong Museum, Nantong 08060
Nantong Textile Museum, Nantong 08061
Nantou Ceramic Museum, Nantou 42281
Nantou County Natural History Museum, Ren-Ai 42303
Nantou History Museum, Nantou 42282
Nantucket Historical Association Museum, Nantucket 50946
Nantucket Life-Saving Museum, Nantucket 50947
Nantucket Lightship Basket Museum, Nantucket 50948
Nantwich Museum, Nantwich 45191
Nanxiong Museum, Nanxiong 08062
Nanyang Guan, Nanyang 08063
Nanyang Museum, Nanyang 08063
Nanzhang Museum, Nanzhang 08066
Nanzhao Prehistoric Pithecanthrope Museum, Nanzhao 08067
Nanzhou Nationality Museum, Kaili 07910
Naomi Wood Collection at Woodford Mansion, Philadelphia 51858
Napa Valley Museum, Yountville 54355
Naper Settlement Museum, Naperville 50950
Naples Museum of Art, Naples 50952
Napoleon- und Heimatmuseum, Deutsch Wagram 01812
Napoleonic Museum G. Antonelli, Arcole 25260
Napoleonic Society of America, Clearwater 47610
Napoleonmuseum Thurgau, Salenstein 41798
Napotnikova Galerija, Šoštanj 38524
Náprstek Museum of Asian, African and American Culture, Praha 09775
Náprstkovo Muzeum Asijských, Afrických a Amerických Kultur, Praha 09775
Nara City Museum of Calligraphy in Honor of Kason Sugioka, Nara 29033
Nara City Museum of Photography, Nara 29034
Nara-kenritsu Bijutsukan, Nara 29035
Nara-kenritsu Kashihara Kokogaku Kenkyujo Fuzoku Hakubutsukan, Kashihara 28677
Nara-kenritsu Man'yo Bunkakan, Asuka 28380
Nara-kenritsu Mingeikan, Yamato-koriyama 29586
Nara-kokuritsu Hakubutsukan, Nara 29036
Nara National Museum, Nara 29036
Nara Prefectural Museum of Art, Nara 29035
Nara Prefectural Museum of Folklore, Yamato-koriyama 29586
Nara Prefecture Complex of Man'yo Culture, Asuka 28380
Naracoorte Art Gallery, Naracoorte 01347
Naramata Heritage Museum Society, Naramata 06418
Narberth Museum, Narberth 45193
Narcissa Prentiss House, Prattsburgh 52133
Narembeen Historical Museum, Narembeen 01348
Nariman Narimanov Memorial Museum, Baki 03165
Naritasan Calligraphy Museum, Narita 29040
Naritasan Reikokan and Reikoden Museum, Narita 29041
Narkose Schmerztherapie Museum, Potsdam 21097
Narlıkuyu Mosaic Museum, Silifke 42940
Narlıkuyu Mozaik Müzesi, Silifke 42940
Naroden Muzej, Gevgelija 30501
Naroden Muzej, Ohrid 30512
Naroden Muzej, Sveti Nikole 30531
Naroden Muzej, Veles 30533
Naroden Muzej D-r Nikola Nezlobinski, Struga 30527
Naroden Muzej Kumanovo, Kumanovo 30507
Narodna Galerija, Ljubljana 38457
Narodna Galerija Čifte Amam, Skopje 30522
Národní Galerie v Praze, Praha 09776
Národní Galerie v Praze - Palác Kinských, Praha 09777
Národní Kulturní Památka Vyšehrad, Praha 09778
Národní Kulturní Památník - Valy, Mikulčice 09652
Narodni Muzej, Bački Petrovac 38031
Narodni Muzej, Bela Crkva 38033
Narodni Muzej, Kragujevac 38109
Narodni Muzej, Leskovac 38114
Narodni Muzej, Pančevo 38128
Narodni Muzej, Požarevac 38132
Narodni Muzej, Smederevska Palanka 38145
Narodni Muzej, Štip 30526
Narodni Muzej, Užice 38158
Narodni Muzej, Vranje 38162
Narodni Muzej, Zadar 08983
Narodni Muzej, Zaječar 38164
Narodni Muzej, Zrenjanin 38166
Narodni Muzej Čačak, Čačak 38094
Narodni Muzej Crne Gore, Cetinje 31643
Narodni Muzej Kikinda, Kikinda 38102
Narodni Muzej Kraljevo, Kraljevo 38111
Narodni Muzej Labin, Labin 08906
Narodni Muzej Niš, Niš 38118
Narodni Muzej Novi Vinodolski, Novi Vinodolski 08917
Narodni Muzej Ras, Novi Pazar 38119
Narodni Muzej Sabac, Sabac 38138
Narodni Muzej Slovenije, Ljubljana 38458
Narodni Muzej Toplice, Prokuplje 38136
Narodni Muzej u Beogradu, Beograd 38082
Narodni Muzej Valjevo, Valjevo 38160
Národní Muzeum, Praha 09779
Národní Technické Muzeum, Praha 09780
Národní Zemědělské Muzeum, Praha 09781
Narodnyj Geologičeskij Muzej Volgageologija, Nižnij Novgorod 37466
Narodnyj Istoriko-kraevedčeskij Muzej Tatarska, Tatarsk 37838
Národopisná Expozícia SARM Bardejovské Kúpele, Bardejovské Kúpele 38196
Národopisná Pobočka, Železný Brod 09937
Národopisné Muzeum, Nejdek 09671
Národopisné Múzeum, Smižany 38353
Národopisné Múzeum Liptova, Liptovský Hrádok 38285
Národopisné Muzeum Plzeňska, Plzeň 09717
Národopisné Muzeum Třebíz, Třebíz 09876
Národopisné Oddělení, Praha 09782
Narok Museum, Narok 29750
Narooma Lighthouse Museum, Narooma 01349
Narovčatskij Rajonnyj Kraevedčeskij Muzej, Narovčat 37446
Narrabri Old Gaol Museum, Narrabri 01350
Narragansett Historical Society Museum, Templeton 53475
Narrenmuseum, Gengenbach 18585
Narrenmuseum, Rheinfelden 21300
Narrenschopf, Bad Dürrheim 16720
Narrogin Gallery, Narrogin 01353
Narrogin Old Courthouse Museum, Narrogin 01354
Narrow Gauge Railway Museum, Sochaczew 35118
Narrow Gauge Railway Museum, Tywyn 45891
Narryna Heritage Museum, Battery Point 00841
Narsaq Museum, Narsaq 23259
Narsarsuaq Museum, Narsarsuaq 23260
Narukawa Art Museum, Hakone 28479
Narukawa Bijutsukan, Hakone 28479
Narva Muuseum, Narva 10598
Narwhal Inuit Art Gallery, London 44935
NASA Glenn Research Center's Visitor Center, Cleveland 47635
Nasher Museum of Art at Duke University, Durham 48221
Nash's House and New Place, Stratford-upon-Avon 45783
Nashua Center for Arts, Nashua 50958
Nasionale Afrikaanse Letterkundige Museum en Navorsingsentrum, Bloemfontein 38566
Nasiriya Museum, Nasiriya 24722
Nasjonalgalleriet, Oslo 33735
Nasjonalmuseet for Kunst, Arkitektur og Design, Oslo 33736
Nassau County Museum of Art, Roslyn 52463
Nassau Public Library and Museum, Nassau 03209
Nastolan Kotiseutumuseo, Nastola 11137
Nasu Classic Car Museum, Nasu 29048
Nasu Orgel Museum, Nasu 29049
Nasu Orugoru Hakubutsukan, Nasu 29049
Nasu Queen's Museum, Nasu 29050
Natadera Fumon-kaku Homotsukan, Komatsu 28766
Natadera Fumon-kaku Treasure Museum, Komatsu 28766
Natal Museum, Pietermaritzburg 38706
Natale Labia Museum, Muizenberg 38692
Natalie and James Thompson Gallery, San Jose 52806
Natchez Trace Parkway, Tupelo 53614
Nathan Denison House, Forty Fort 48721
Nathan Manilow Sculpture Park, University Park 53656
Nathaniel W. Faison Home and Museum, La Grange 49844
Nation Museum, Beijing 07555
Nationaal Ambulance- en Eerste Hulpmuseum, Kampen 32439
Nationaal Archief, Den Haag 32094
Het Nationaal Autominiaturen Museum, Asperen 31934
Nationaal Baggermuseum, Sliedrecht 32781
Nationaal Beiaardmuseum, Asten 31941
Nationaal Bevrijdingsmuseum 1944-1945, Groesbeek 32277
Nationaal Borstelmuseum, Izegem 03737
Nationaal Brandweermuseum, Hellevoetsluis 32361
Nationaal Brilmuseum Amsterdam, Amsterdam 31871
Nationaal Coöperatie Museum, Schiedam 32755
Nationaal Fietsmuseum Velorama, Nijmegen 32592
Het Nationaal Gevangenismuseum, Veenhuizen 32876
Nationaal Glasmuseum, Leerdam 32471
Nationaal Historisch Orgelmuseum Elburg, Elburg 32199
Nationaal Houtambachtenmuseum De Wimpe, Herenthout 03698
Nationaal Jenevermuseum, Hasselt 03686
Nationaal Landschapskundig Museum, Dordrecht 32153
Nationaal Likeur- en Frisdrankenmuseum Isidorus Jonkers, Hilvarenbeek 32387
Nationaal Luchtvaart-Themapark Aviodome, Schiphol 32760
Nationaal Modelspoor Museum, Sneek 32790
Nationaal Monument Kamp Vught, Vught 32915
Nationaal Museum en Archief van Douane en Accijnzen, Antwerpen 03305
Nationaal Museum Historisch Landbouwtechniek, Wageningen 32920
Nationaal Museum van de Speelkaart, Turnhout 04099
Nationaal Museum van Speelklok tot Pierement, Utrecht 32851
Nationaal Museum Verpleging en Verzorging, Zetten 32997
Nationaal Oorlogs- en Verzetsmuseum, Overloon 32663
Nationaal Reddingmuseum Dorus Rijkers, Den Helder 32112
Nationaal Rijtuigmuseum, Leek 32467
Nationaal Ruimtevaart Museum, Lelystad 32494
Nationaal Schaakmuseum, Den Haag 32095
Nationaal Scheepvaartmuseum, Antwerpen 03306
Nationaal Schoeiselmuseum, Izegem 03738
Nationaal Schoolmuseum, Rotterdam 32717
Nationaal Scoutsmuseum, Leuven 03789
Nationaal Sleepvaart Museum, Maassluis 32523
Nationaal Smalspoormuseum - Stoomtrein Valkenburgse Meer, Valkenburg, Zuid-Holland 32867
Nationaal Tabaksmuseum, Wervik 04147
Nationaal Tinnen Figuren Museum, Ommen 32619
Nationaal Vakbondsmuseum De Burcht, Amsterdam 31872
Nationaal Veeteelt Museum, Beers bij Cuyk 31955
Nationaal Visserijmuseum van Oostduinkerke, Oostduinkerke 03926
Nationaal Vlas-, Kant- en Linnenmuseum, Kortrijk 03761
Nationaal Vlechtmuseum, Noordwolde 32605
Nationaal Volkssportmuseum voor de Vinkensport, Harelbeke 03678
Nationaal Wielermuseum, Roeselare 03976
National 1798 Visitor Centre, Enniscorthy 24831
National Academy Museum, New York 51232
National Afro-American Museum and Cultural Center, Wilberforce 54158
The National Agricultural Center and Hall of Fame, Bonner Springs 46901
National Agricultural Museum, Sofia 05199
National Air and Space Museum, Washington 53895
National Aivazovsky Picture Gallery, Feodosija 43039
National Anthem Museum in Będomin, Nowy Barkoczyn 34954
National Apple Museum, Biglerville 46803
National Archaeological Museum, Athinai 22920
National Archaeological Museum, Lisboa 35715
National Archaeological Museum, Napoli 26883
National Archaeological Reserve Cabyle, Jambol 05042
National Archives of Zambia, Lusaka 54659
National Army Museum, Camberley 43718
National Army Museum, London 44936
National Army Museum Sandhurst Departments, Sandhurst, Berkshire 45611
National Art Gallery, Bhaktapur 31770
National Art Gallery, Colombo 40373
National Art Gallery, Dhaka 03227
National Art Gallery, Islamabad 34008
National Art Gallery, Kabul 00004
National Art Gallery, Minsk 03256
National Art Gallery, Sarajevo 04283
National Art Gallery, Sofia 05200
National Art Gallery, Tórshavn 10699
National Art Gallery Malaysia, Kuala Lumpur 30573
National Art Gallery of Armenia - Sisian Branch, Sisian 00716
National Art Gallery of Namibia, Windhoek 31764
National Art Museum of Sport, Indianapolis 49505
National Art Museum of the Republic of Belarus, Minsk 03255
National Art Museum of Ukraine, Kyïv 43108
National Art Museum of Ukraine, Kyïv 43100
National Arts Club, New York 51233
National Arts Education Archive, Wakefield 45909
National Arts Museum, Tehrān 24685
National Association of Miniature Enthusiasts, Carmel 47252
National Association of Women Artists Gallery, New York 51234
National Astra Museum, Sibiu 36604
National Atomic Museum, Albuquerque 46183
National Automobile Museum, Reno 52287
National Automobile Museum of Tasmania, Launceston 01222
National Automotive and Truck Museum of the United States, Auburn 46467
National Aviation Park Avidome, Schiphol 32760
National Bank of Slovakia - Museums of Coins and Medals, Kremnica 38279
National Baseball Hall of Fame and Museum, Cooperstown 47791
National Bicycle Museum Velorama, Nijmegen 32592
National Birendra Art Gallery, Kathmandu 31775
National Border Patrol Museum and Memorial Library, El Paso 48327
National Broadcasting Museum, Minsiong 42277
National Bronze Art Museum, Lalitpur 31781
National Building Museum, Washington 53896
National Cach Museum, Lisboa 35723
National Canal Museum, Easton 48270
National Capital Trolley Museum, Silver Spring 53107
National Carriage Museum, Leek 32467
National Center for the American Revolution, Valley Forge 53694

N

National Central University Art Center, Jhongli 42221
National Ceramics Museum, Safi 31693
National Cheng Kung University Museum, Tainan 42364
National Chengchi University Museum of Ethnology, Taipei 42412
National Chiang Kai-shek Memorial Hall, Taipei 42413
National Chiao Tung University Art Center, Hsinchu 42212
National Children's Museum, Delhi 24144
National Children's Museum, Washington 53897
National Children's Theatres Museum, Moskva 37197
National Chung Hsing University Art Center, Taichung 42351
National Civil Rights Museum, Memphis 50581
National Coal Mining Museum for England, Wakefield 45910
National Commission for Museums and Monuments, Abuja 54968
National Copper Museum, Salvador Escalante 31320
National Coracle Centre, Cenarth 43784
National Corvette Museum, Bowling Green 46974
National Costume Center of Finland, Jyväskylä 10893
National Council of Science Museums, Kolkata 54886
National Cowboy and Western Heritage Museum, Oklahoma City 51519
National Cowgirl Museum and Hall of Fame, Fort Worth 48718
National Cryptologic Museum, Fort Meade 48654
National Cultural History Museum, Tshwane 38752
National Cycle Collection, Llandrindod Wells 44756
National Czech and Slovak Museum, Cedar Rapids 47306
National D-Day Museum, New Orleans 51088
National Dairy Museum, Ashurst, Southampton 43334
National Dinosaur Museum, Nicholls 01368
National Dr. Sun Yat-sen Memorial Hall, Taipei 42414
National Dragonfly Biomuseum, Ashton, Oundle 43330
National Educational Museum and Library, Budapest 23462
National English Literary Museum, Grahamstown 38631
National Ethnographic Museum, Krujë 00025
National Ethnographical Museum of the Bulgarian Academy of Sciences, Sofia 05190
National Exhibition of Folk Artistic Crafts and Applied Arts, Orešak 05100
National Exhibits by Blind Artists, Philadelphia 51859
National Factory's Museum, Lisboa 35729
National Fairground Archive, Sheffield 45654
National Farm Toy Museum, Dyersville 48229
National Film Archive of Iceland Collection, Hafnarfjörður 23999
National Fine Arts Museum, Rio de Janeiro 04756
National Fire Fighters' Museum, București 36216
National Firearms Museum, Fairfax 48478
National Fishing Heritage Centre, Great Grimsby 44355
National Folk Art Museum of Armenia, Yerevan 00750
National Folk Museum, Seoul 29900
National Folk Museum, Seoul 29909
National Folk Museum of Lesbos, Mytilini 23111
National Football Museum, Canton 47228
National Football Museum, Preston 45446
National Fresh Water Fishing Hall of Fame, Hayward 49231
National Frontier Trails Center, Independence 49479
National Gallery, Bangkok 42551
National Gallery, London 44937
National Gallery Bulawayo, Bulawayo 54665
National Gallery Chifte Amam, Skopje 30522
National Gallery for Decorative Arts, Sofia 05202
National Gallery for Foreign Art, Sofia 05201
National Gallery in Prague, Praha 09795
National Gallery of Armenia, Yerevan 00751
National Gallery of Armenia - Alaverdi Branch, Alaverdi 00690
National Gallery of Art, Washington 53898
National Gallery of Australia, Canberra 00946
National Gallery of Canada, Ottawa 06551
National Gallery of Corfu, Skripero 23191
National Gallery of Iceland, Reykjavík 24022
National Gallery of Ireland, Dublin 24807
National Gallery of Jamaica, Kingston 28334
National Gallery of Modern Art, Delhi 24145
National Gallery of Modern Art, Mumbai 24319
National Gallery of Prague, Permanent Collection of Asian Art, Praha 09776
National Gallery of Scotland, Edinburgh 44138
National Gallery of Slovenia, Ljubljana 38457
National Gallery of the Komi Republic, Syktyvkar 37820
National Gallery of Victoria, Melbourne 01288
National Gallery of Zimbabwe, Harare 54670
National Gallery - The National Museum of Art, Design and Architecture, Oslo 33735
National Gandhi Museum and Library, Delhi 24146
The National Gas Museum, Leicester 44679
National Geological Museum, Maputo 31711
National Glass Centre, Sunderland 45800
National Glass Museum, Leerdam 32471
National Great Rivers Museum, Alton 46241
National Hall of Fame for Famous American Indians, Anadarko 46281
National Handicraft Museum, Rawang 30615
National Hearing Aid Museum, London 44938
National Heisey Glass Museum, Newark 51292
National Heritage, Hawkhurst 55085
National Heritage Museum, Lexington 50061
National Hispanic Art Museum, Albuquerque 46184
National Historic Oregon Trail Interpretive Center, Baker City 46534
National Historical and Archeological Reserve Madara, Madara 05089
National Historical and Archeological Reserve Pliska, Pliska 05123
National Historical-Archaeological Reserve and Museum Veliki Preslav, Veliki Preslav 05248
National Historical Museum, Athinai 22942
National Historical Museum, Montevideo 54404
National Historical Museum, Tiranë 00032
National Historical Museum of Ukraine, Kyiv 43103
National History Museum, Campinas 04389
National History Museum, Kuala Lumpur 30581
National Horseracing Museum, Newmarket 45239
National Infantry Museum, Fort Benning 48591
National Institute of Industrial Safety Museum, Tokyo 29422
National Inventors Hall of Fame, Akron 46147
National Jesuit Museum Jesus Mary, Jesús María 00393
National Jewelry Treasury of Iran, Tehrān 24675
National Kyiv-Pechersk Historical and Cultural Museum- Preserve, Kyiv 43101
National Liberty Museum, Philadelphia 51860
National Library, Valletta 30642
National Library, Warszawa 35217
National Library Museum, Thimphu 04203
National lifeboat Museum, Den Helder 32112
National Literary and Memorial Museum-Complex of Sabit Mukanov and Gabit Musrepov, Almaty 29681
National McKinley Birthplace Memorial, Niles 51352
National Maritime Museum, Alexandria 10477
The National Maritime Museum, Amsterdam 31885
National Maritime Museum, Antwerpen 03306
National Maritime Museum, Galle 40378
National Maritime Museum, Haifa 24965
National Maritime Museum, London 44939
National Maritime Museum, Mokpo 29853
National Maritime Museum, Stockholm 40933
National Maritime Museum Cornwell, Falmouth 44209
National Maritime Museum of Ireland, Dún Laoghaire 24826
National Medal of Honor Museum of Military History, Chattanooga 47412
National Media Museum, Bradford 43581
National memorial, Kielce 34743
National Memorial of the Fortress of Breendonk, Willebroek 04153
National Military Museum, București 36211
National Mining Hall of Fame and Museum, Leadville 50006
National Mining Museum, Kwekwe 54673
National Mining Museum, Rumelange 30491
National Model Aviation Museum, Muncie 50905
National Monument, Jakarta 24476
National Motor Museum, Beaulieu 43424
National Motor Museum, Birdwood 00871
National Motor Racing Museum, Bathurst 00840
National Motorcycle Museum, Bickenhill 43475
National Motorcycle Museum, Nabiac 01344
National Museum, Aden 54643
National Museum, Alofi 33348
National Museum, Bački Petrovac 38031
National Museum, Bangkok 42552
National Museum, Høyvik 10687
National Museum, Kragujevac 38109
National Museum, Kuala Lumpur 30578
National Museum, Leskovac 38114
National Museum, Malé 30623
National Museum, Monrovia 30182
National Museum, Nagara Pathama 42581
National Museum, Oshogbo 33341
National Museum, Pančevo 38128
National Museum, Port Moresby 34085
National Museum, Požarevac 38132
National Museum, Poznań 35041
National Museum, Praha 09779
National Museum, Ras-Al-Khaimah 43222
National Museum, Riyadh 38017
National Museum, Sanaa 54650
National Museum, Smederevska Palanka 38145
National Museum, Štip 30526
National Museum, Szczecin 35163
National Museum, Tehrān 24679
National Museum, Tehrān 24687
National Museum, Užice 38158
National Museum, Valjevo 38160
National Museum, Veles 30533
National Museum, Victoria 38168
National Museum, Vranje 38162
National Museum, Zadar 08983
National Museum, Zrenjanin 38166
National Museum, Abuja, Abuja 33313
National Museum and Art Gallery of Trinidad and Tobago, Port-of-Spain 42615
National Museum at Ruwi, Ruwi 33996
National Museum Bauchi, Bauchi 33317
National Museum Benin, Benin City 33318
National Museum Bloemfontein, Bloemfontein 38567
National Museum Chernobyl, Kyïv 43102
National Museum & Cultural Centre, Honiara 38547
National Museum, Department of Entomology, Praha 09744
National Museum Directors' Conference, London 55086
National Museum Esie, Esie 33321
National Museum for Egyptian Civilization, Cairo 10510
National Museum for Footwear, Izegem 03738
National Museum for Trading, Riyadh 38018
National Museum George Enescu, București 36224
National Museum Gidan Makama, Kano, Kano 33332
National Museum Ibadan, Ibadan 33323
National Museum Ile-Ife, Ife 33325
National Museum, Imo State, Owerri 33342
National Museum in Belgrade, Beograd 38082
National Museum in Cracow, Kraków 34805
National Museum in Gdansk - Department of Ethnography - Abbots' Granary, Gdańsk 34640
National Museum in Gdansk - Department of Modern Art - Abbots' Palace, Gdańsk 34642
National Museum in Gdansk - Department of the Green Gate, Gdańsk 34643
National Museum in Gdańsk, Dept. of Ancient Art, Gdańsk 34641
National Museum in Gdansk - Gdansk Gallery of Photography, Gdańsk 34639
National Museum in Gdansk - Museum of the Tradition of Polish Nobility, Stary Targ 35133
National Museum in Kielce, Kielce 34742
National Museum in Lviv, Lviv 43137
National Museum in Lviv, Dept of Folk Art, Lviv 43138
National Museum in Lviv, Dept of Old Printings and Handwritings, Lviv 43139
National Museum in Lviv - Memorial and Art Museum of Ivan Trush, Lviv 43119
National Museum in Lviv - Memorial and Art Museum of Leopold Levytskyj, Lviv 43120
National Museum in Lviv - Memorial and Art Museum of Oleksa Novakivskyj, Lviv 43121
National Museum in Lviv - Memorial and Art Museum of Olena Kulchitska, Lviv 43122
National Museum in Warsaw, Warszawa 35261
National Museum in Wroclaw, Wrocław 35335
National Museum-Institute of Architecture, Yerevan 00752
National Museum Kabul, Kabul 00005
National Museum Kaduna, Kaduna 33331
National Museum, Lagos, Onikan 33337
National Museum Makurdi, Makurdi 33335
National Museum Minna, Minna 33336
National Museum Nairobi, Nairobi 29749
National Museum Niš, Niš 38118
National Museum of African Art, Washington 53899
National Museum of Agriculture, Praha 09781
National Museum of Agriculture and Food Industry, Komorniki 34754
National Museum of airports and Air Transportation, Málaga 39604
National Museum of American Art, Washington 53900
National Museum of American History, Washington 53901
National Museum of American Illustration, Newport 51324
National Museum of American Jewish History, Philadelphia 51861
National Museum of American Jewish Military History, Washington 53902
National Museum of Antiquities, Leiden 32487
National Museum of Archaeology, Valletta 30643
National Museum of Archeology and Ethnology, Guatemala 23301
National Museum of Art and History, Yangon 31736
National Museum of Art, Architecture and Design, Oslo 33736
National Museum of Art of Romania, București 36219
National Museum of Art, Osaka, Osaka 29130
National Museum of Art, Osaka, Osaka 29144
National Museum of Arts and Popular Traditions, Roma 27515
National Museum of Australia, Canberra 00947
National Museum of Australian Pottery, Wodonga 01660
National Museum of Bashkortostan, Ufa 37889
National Museum of Bhutan, Paro 04201
National Museum of Bosnia and Herzegovina, Sarajevo 04284
National Museum of Brushware, Izegem 03737
National Museum of Bulgarian Architecture, Veliko Tărnovo 05256
National Museum of Cacak, Čačak 38094
National Museum of Calabar, Calabar 33319
National Museum of Cambodia, Phnom Penh 05287
National Museum of Catholic Art and History, New York 51235
National Museum of Ceramic Art and Glass, Baltimore 46583
National Museum of Ceramics, Leeuwarden 32478
National Museum of Cham Sculpture, Da Nang 54604
National Museum of Chinese History, Beijing 07556
National Museum of Civil War Medicine, Frederick 48758
National Museum of Colonial History, Aba 33310
The National Museum of Communications, Irving 49534
National Museum of Contemporary Art, Athinai 22943
National Museum of Contemporary Art, Gwacheon 29820
National Museum of Contemporary Art, Lisboa 35698
National Museum of Contemporary History, Ljubljana 38455
National Museum of Damascus, Damascus 42132
National Museum of Dance and Hall of Fame, Saratoga Springs 52921
National Museum of Decorative Arts, Trondheim 33925
National Museum of Denmark, København 10144
National Museum of Ecclesiastical History and Archaeology, Sofia 05189
National Museum of Education, Gabrovo 05033
National Museum of Emerging Science and Innovation, Tokyo 29418
National Museum of Enugu, Enugu 33320
National Museum of Ethiopia, Addis Ababa 10680
National Museum of Ethnography and Natural History, Chişinău 31601
National Museum of Ethnology, Islamabad 34009
National Museum of Ethnology, Leiden 32488
National Museum of Ethnology, Minpaku, Suita 29295
National Museum of Fine Arts, La Habana 09162
National Museum of Fine Arts, Valletta 30644
National Museum of Fine Arts of the Saha Republic (Yakutia), Jakutsk 36964
National Museum of Finland, Helsinki 10820
National Museum of Flight, North Berwick 45264
National Museum of Funeral History, Houston 49400
National Museum of Gardening, Helston 44445
National Museum of Ghana, Accra 22848
National Museum of Health and Medicine, Washington 53903
National Museum of History, Sofia 05191
National Museum of History, Taipei 42415
National Museum of Iceland, Reykjavík 24030
National Museum of Ilorin, Ilorin 33327
National Museum of India, Delhi 24147
National Museum of Industrial History, Bethlehem 46782
National Museum of Ireland - Archaeology, Dublin 24808
National Museum of Ireland - Country Life, Castlebar 24756
National Museum of Ireland - Decorative Arts & History, Dublin 24809
National Museum of Ireland - Natural History, Dublin 24810
National Museum of Japanese History, Sakura 29210
National Museum of Kashan, Kashan 24616
National Museum of Korea, Seoul 29910
National Museum of Literature, Sofia 05192
National Museum of Lithuania, Vilnius 30419
National Museum of Man, Bhopal 24093
National Museum of Mandalay, Mandalay 31719
National Museum of Marine Biology, Checheng 42159
National Museum of Marine Science and Technology, Keelung 42249
National Museum of Maritime History, Varna 05244
National Museum of Medicine of Ukraine, Kyïv 43106
National Museum of Medieval Art, Korçë 00021
National Museum of Modern Art, Guatemala 23302
National Museum of Modern Art Kyoto, Kyoto 28843

N

The National Museum of Modern Art of Tokyo, National Film Center, Tokyo 29493
The National Museum of Modern Art, Tokyo, Tokyo 29491
National Museum of Modern Chinese Literature, Beijing 07557
National Museum of Mongolian History, Ulaanbaatar 31628
National Museum of Music, Beijing 07558
National Museum of Namibia, Windhoek 31765
National Museum of Natural History, Colombo 40374
National Museum of Natural History, Delhi 24148
National Museum of Natural History, La Habana 09163
National Museum of Natural History, Luanda 00113
National Museum of Natural History, Mdina 30631
National Museum of Natural History, Sofia 05198
National Museum of Natural History, Washington 53904
National Museum of Natural Science, Taichung 42352
National Museum of Naval Aviation, Pensacola 51767
National Museum of Nepal, Kathmandu 31776
National Museum of Pakistan, Karachi 34014
National Museum of Pasta Food, Roma 27516
National Museum of Polo and Hall of Fame, Lake Worth 49897
National Museum of Prehistory, Taitung 42444
National Museum of Racing and Hall of Fame, Saratoga Springs 52922
National Museum of Roller Skating, Lincoln 50102
National Museum of Science and Technology, Lahore 34025
National Museum of Science and Technology, Stockholm 40945
National Museum of Science, Planning and Technology, Haifa 24966
National Museum of Singapore, Singapore 38177
National Museum of Slovenia, Ljubljana 38458
National Museum of Sport, Sofia 05195
National Museum of Taiwanese History, Tainan 42365
National Museum of Taiwanese Literature, Tainan 42366
National Museum of the American Indian, New York 51236
National Museum of the American Indian, Washington 53905
National Museum of the Army, Luanda 00110
National Museum of the Health Insurance, Lormont 13824
National Museum of the History and Culture of Belarus, Minsk 03250
National Museum of the History of the Great Patriotic War 1941-1945, Kyïv 43104
National Museum of the Kabardino-Balkarian Republic, Nalčik 37442
National Museum of the Komi Republic, Syktyvkar 37821
National Museum of the Middle Ages, Paris 14729
National Museum of the Morgan Horse, Shelburne 53064
National Museum of the Ogun State, Abeokuta 33311
National Museum of the Ondo State, Akure 33314
National Museum of the Philippines, Manila 34450
National Museum of the Republic Udmurtiya, Iževsk 36955
National Museum of the Romanian Literature, Bucureşti 36215
National Museum of the Tatarian Republic, Kazan 37026
National Museum of the Ust-Ordynsk Buryat Autonomous Region, Ust-Ordynsk 37913
National Museum of Tiles, Lisboa 35719
National Museum of Transport and Communication, Ruse 05154
National Museum of Turkmenistan, Aşgabat 42973
National Museum of Ukrainian Folk Decorative Art, Kyïv 43078
National Museum of Ukrainian Literature, Kyïv 43105
National Museum of Vietnamese History, Ha Noi 54616
National Museum of Watercolours, México 31135
The National Museum of Western Art, Tokyo 29419
National Museum of Wildlife Art, Jackson 49573
National Museum of Women in the Arts, Washington 53906
National Museum of Woodcarving, Custer 47916
National Museum of Workers' Halls, Riihimäki 11273
National Museum Oron, Oron 33338
National Museum Uyo, Uyo 33347
National Museum Zurich, Zürich 42105
National Museums and Galleries of Northern Ireland, Belfast 55087
National Museums and Monuments of Zimbabwe, Harare 55201
National Museums in Berlin - Prussian Cultural Heritage Foundation - Directorate-general, Berlin 54858
National Museums in Berlin - Prussian Cultural Heritage Foundation - New National Gallery, Berlin 17225
National Museums of Scotland, Edinburgh 44139
National Music Museum, Bloemfontein 38568
National New York Central Railroad Museum, Elkhart 48347
National Oil Museum, Ploieşti 36536
National Opera and Ballet Museum, Sofia 05187
National Ornamental Metal Museum, Memphis 50582
National Packard Museum, Warren 53825
National Palace Museum, Taipei 42416
National Paper Museum, Malmedy 03844
National Park-Museum Samuil-Castle, Petrič 05114
National Park Museum Shipka-Budloudzha, Kazanlăk 05060
National Philatelic Museum, Delhi 24149
National Picture Gallery and Alexander Soutzos Museum, Athinai 22909
National Pioneer Women's Hall of Fame, Alice Springs 00800
National Pipe Museum, Amsterdam 31880
National Police Museum and Ancient Police Museum, Cairo 10511
National Polytechnical Museum, Sofia 05197
National Portrait Gallery, Canberra 00948
National Portrait Gallery, London 44940
National Portrait Gallery, Montacute 45169
National Portrait Gallery, Washington 53907
National Postal Museum, Washington 53908
National Preserve of Tauric Chersonesos, Sevastopol 43178
National Presidential Wax Museum, Keystone 49772
National Print Museum, Dublin 24811
National Prison Museum, Veenhuizen 32876
National Pushkin Museum, Sankt-Peterburg 37732
National Rail Museum, Delhi 24150
National Railroad Museum, Green Bay 49040
National Railroad Museum and Hall of Fame, Hamlet 49140
National Railway Museum, Port Adelaide 01422
National Railway Museum, York 46105
National Railways of Zimbabwe Museum, Raylton 54677
National Resistance Museum, Karpenissi 23038
National Revival and Constituent Assembly, Veliko Tărnovo 05255
National Road/Zane Grey Museum, Norwich 51441
National Roman Legionary Museum, Caerleon 43707
National Route 66 Museum and Old Town Museum, Elk City 48341
National Science and Technology Museum, Kaohsiung 42241
National Science Centre, Delhi 24151
National Science Museum, Tokyo 29416
National Science Museum of Korea, Daejeon 29802
National Science Museum, Shinju-ku Branch, Tokyo 29417
National Scout Museum of New Zealand, Kaiapoi 33132
National Scouting Museum, Irving 49535
National Sculpture Society, New York 51237
National Slate Museum, Llanberis 44754
National Soaring Museum, Elmira 48380
The National Soccer Hall of Fame, Oneonta 51564
National Society of the Children of the American Revolution Museum, Washington 53909
National Softball Hall of Fame and Museum Complex, Oklahoma City 51520
National Space Centre, Leicester 44680
National Space Museum, Lelystad 32494
National Sport Museum, Lisboa 35720
National Sportsmuseum, Patiala 24344
National Sprint Car Hall of Fame and Museum, Knoxville 49809
National Steinbeck Center, Salinas 52677
National Széchényi Library Museum, Budapest 23463
National Taiwan Museum, Taipei 42417
National Taiwan Museum of Fine Arts, Taichung 42353
National Taiwan Science Education Center, Taipei 42418
National Taras Shevchenko Museum, Kyïv 43107
National Technical Museum, Praha 09780
National Theater Museum, Ljubljana 38464
The National Time Museum, Chicago 47499
National Tin Figurines Museum, Ommen 32619
National Tobacco Museum, Wervik 04147
The National Toy Train Museum, Strasburg 53328
National Tramway Museum, Matlock 45115
National Transport Museum, Dublin 24812
National Trust for Historic Preservation, Washington 53910
National Trust Museum, Streaky Bay 01533
National Tsing Hua University Arts Center, Hsinchu 42213
National Viceroyalty Museum, Tepotzotlán 31445
National Vietnam Veterans Art Museum, Newhaven 01367
National Vintage Wireless and Television Museum, Harwich 44412
National Wallace Monument, Stirling 45750
National War- and Resistance Museum, Overloon 32663
National War Museum of Scotland, Edinburgh 44140
National War Museum of Umauhia, Umauhia 33346
National Warplane Museum, Horseheads 49369
National Watch and Clock Museum, Columbia 47720
National Waterfront Museum, Swansea 45815
National Waterways Museum, Gloucester 44317
National Waterways Museum - Ellesmere Port, Ellesmere Port 44168
National Wine Museum, Alcobaça 35395
National Winemuseum, Eger 23542
National Women's Hall of Fame, Seneca Falls 53035
National Wood Working Museum, Bhaktapur 31771
National Wool Museum, Geelong 01104
National Wool Museum, Llandysul 44761
National Wrestling Hall of Fame, Stillwater 53297
National Yiddish Book Center, Amherst 46270
Det Nationale Fotomuseum, København 10143
La Nationale Galerie d'Art Alyssa, Le Bardo 42617
Nationalgalerie Friedrichswerdersche Kirche - Schinkelmuseum, Berlin 17222
Det Nationalhistoriske Museum på Frederiksborg, Hillerød 10076
Nationalmuseet, København 10144
Nationalmuseum Stockholm, Stockholm 40918
Nationalpark-Besucherzentrum TorfHaus, Altenau 16491
Nationalpark-Haus, Butjadingen 17669
Nationalpark-Haus Berchtesgaden, Berchtesgaden 17003
Nationalparkhaus Sankt Andreasberg, Sankt Andreasberg 21488
Nationalparkzentrum, Zernez 42054
Native American Exhibit, Fonda 48584
Native American Heritage Museum at Highland Mission, Highland 49277
Native American Museum, Terre Haute 53481
Native American Resource Center, Pembroke 51755
Native Cultural Arts Museum, Grouard 05915
Native Museum, San José 08823
Native Museum of Mashteuiatsh, Mashteuiatsh 06245
Natsagdorj Museum, Ulaanbaatar 31629
Náttúrufrædaistofnun Islands, Akureyrarsetur, Akureyri 23984
Natturufrædistofa Kopavogs, Kópavogur 24009
Náttúrufræðlistofnun Íslands, Reykjavík 24026
Náttúrugripasafnid i Neskaupstad, Neskaupstadur 24010
Natur-Heimat-Museum, Langenburg 19826
Natur-Museum Luzern, Luzern 41595
Natur-Museum mit Bauerngarten, Goldberg 18703
Natur- und Heimatmuseum Vellberg, Vellberg 22215
Natural History and Ethnological Native Museums, Juiz de Fora 04523
Natural History Collection of Komló, Komló 23681
Natural History Museum, Arcata 46338
Natural History Museum, Århus 09965
Natural History Museum, Bangkok 42553
Natural History Museum, Bangor, Gwynedd 43377
Natural History Museum, Belfast 43442
Natural History Museum, Belogradčik 04984
Natural History Museum, Beograd 38085
Natural History Museum, Bratislava 38232
Natural History Museum, Colchester 43875
Natural History Museum, Gorgan 24606
Natural History Museum, Grenoble 13139
Natural History Museum, Hamedan 24608
Natural History Museum, Helsinki 10801
Natural History Museum, Ife 33326
Natural History Museum, Ilam 24609
Natural History Museum, Isfahan 24613
Natural History Museum, Jelenia Góra 34706
Natural History Museum, Karachi 34015
Natural History Museum, Kathmandu 31777
Natural History Museum, Khorram-Abaad 24624
Natural History Museum, Kingston 28335
Natural History Museum, Kotel 05078
Natural History Museum, Lahore 34026
Natural History Museum, Las Cuevas 04179
Natural History Museum, Lawrence 49991
Natural History Museum, London 44941
Natural History Museum, Luzhou 07988
Natural History Museum, Medani 40394
Natural History Museum, Neuchâtel 41682
Natural History Museum, Nottingham 45295
Natural History Museum, Nukus 54450
Natural History Museum, Port Louis 30684
Natural History Museum, Port Sudan 40397
Natural History Museum, Portales 52056
Natural History Museum, Portsmouth 45435
Natural History Museum, Porvoo 11228
Natural History Museum, Rijeka 08940
Natural History Museum, Santa Maria 52894
Natural History Museum, Sanyuan 08153
Natural History Museum, Solothurn 41886
Natural History Museum, Tehrān 24688
Natural History Museum, Tring 45876
Natural History Museum, Tripoli 30190
Natural History Museum, Urmieh 24701
Natural History Museum, Yangon 31737
Natural History Museum, Yazd 24704
Natural History Museum and Institute, Chiba, Chiba 28392
Natural History Museum Basel, Basel 41175
Natural History Museum Complex, Constanţa 36300
Natural History Museum of Amarousio, Amarousio 22878
Natural History Museum of Babia Góra National Park, Zawoja 35364
Natural History Museum of Bakony Mountains, Zirc 23979
Natural History Museum of Barcelona, Barcelona 38976
Natural History Museum of Bombay Natural History Society, Mumbai 24320
Natural History Museum of Cephalonia and Ithaca, Argostolion 22891
Natural History Museum of Elassona, Elassona 22991
Natural History Museum of Kópavogur, Kópavogur 24009
Natural History Museum of Latvia, Rīga 30077
Natural History Museum of Los Angeles County, Los Angeles 50257
Natural History Museum of Montenegro, Podgorica 31657
Natural History Museum of Oeta, Ipati 23017
Natural History Museum of Roman, Roman 36569
Natural History Museum of the Aegean, Samos 23178
Natural History Museum of the National Museum, Praha 09791
Natural History Museum of the University of Basrah, Basrah 24718
Natural History Museum of Tilos, Tilos 23224
Natural History Museum of Toulon and Var, Toulon (Var) 15925
Natural History Museum of Zimbabwe, Bulawayo 54666
Natural History Museum Rotterdam, Rotterdam 32718
Natural History Museum, Tokai University, Shimizu 29262
Natural History Society and Folk Museum, Bacup 43349
The Natural Science Center of Greensboro, Greensboro 49067
Natural Science Collections Alliance, Washington 55167
Natural Science Museum, Marawi 34460
Natural Science Museum, Plovdiv 05132
Natural Science Museum, Split 08962
Natural Science Museum, Tulcea 36706
Natural Science Museum, Victoria 30653
Natural Science Museum of the Institute of Systematics and Evolution at the Polish Academy of Sciences, Kraków 34808
Natural Sciences Museum, Burgas 05000
Natural Sciences Museum, Černi Osăm 05002
Natural Sciences Museum, Kibbutz Hulata 25045
Natural Sciences Museum, Varna 05246
Natural Sciences Museum Complex, Galaţi 36378
Natural-Scientific Museum of Macedonia, Skopje 30523
Naturama, Svendborg 10291
Naturama Aargau, Aarau 41083
Nature and Forest Museum of the Białowieża National Park, Białowieża 34544
Nature and Orne Estuary Center, Sallenelles 15603
Nature Discovery Center, Waren (Müritz) 22335
Nature Hall The Ark, Pori 11219
Nature in Art, Gloucester 44318
Nature Museum at Grafton, Grafton 48983
Nature Park and Galleries, Jerusalem 25007
Nature Science and Hunting Museum, Cegléd 23501
Nature-Study and Folk Art Museum of Loutra Almopias, Loutraki 23082
Naturens Hus, Stockholm 40919
Natureum am Schloß Ludwigslust, Ludwigslust 20061
Natureum Niederelbe, Neuhaus/Balje 20617
Naturhiscope, Oropesa del Mar 39748
Naturhistorisches Museum, Admont 01694
Naturhistorisches Museum, Heiden 41477
Naturhistorisches Museum, Mainz 20149
Naturhistorisches Museum, Nürnberg 20785
Naturhistorisches Museum, Wien 03043
Naturhistorisches Museum Basel, Basel 41175
Naturhistorisches Museum der Burgergemeinde Bern, Bern 41213
Naturhistorisches Museum Schloss Bertholdsburg, Schleusingen 21572

Naturhistorisk Museum, Århus 09965
Naturhistorisk museum, Oslo 33737
Naturhistoriska Riksmuseet, Stockholm 40920
Naturhistoriske Samlinger, Bergen 33431
Naturkunde-Museum, Bielefeld 17347
Naturkunde-Museum, Coburg 17718
Naturkunde-Museum Bamberg, Bamberg 16915
Naturkundemuseum, Altdorf 41106
Naturkundemuseum, Bayreuth 16963
Naturkundemuseum, Dobersberg 01821
Naturkundemuseum, Glees 18658
Naturkundemuseum, Kassel 19472
Naturkundemuseum, Potsdam 21098
Naturkundemuseum, Reutlingen 21286
Naturkundemuseum Erfurt, Erfurt 18209
Naturkundemuseum Freiberg, Freiberg, Sachsen 18423
Naturkundemuseum im Marstall, Paderborn 20974
Naturkundemuseum im Tierpark Hamm, Hamm, Westfalen 18972
Naturkundemuseum Leipzig, Leipzig 19928
Naturkundemuseum Niebüll, Niebüll 20698
Naturkundemuseum Ostbayern, Regensburg 21230
Naturkundliche Bildungsstätte Nordeifel, Simmerath 21770
Naturkundliche Sammlung, Altusried 16528
Naturkundliche Sammlung, Königsbrunn 19646
Naturkundliche Sammlung Oberhasli, Meiringen 41620
Naturkundliches Bildungszentrum Stadt Ulm, Ulm 22167
Naturkundliches Museum, Wiesenfelden 22560
Naturkundliches Museum in der Harmonie, Schweinfurt 21688
Naturkundliches Museum und Schulungsstätte "Alte Schmiede", Handeloh 18984
Naturmuseum, Augsburg 16631
Naturmuseum, Freiburg im Breisgau 18436
Naturmuseum, Neuberg an der Mürz 02425
Naturmuseum, Sankt Gallen 41808
Naturmuseum Lüneburg, Lüneburg 20101
Naturmuseum Olten, Olten 41708
Naturmuseum Solothurn, Solothurn 41886
Naturmuseum Thurgau, Frauenfeld 41379
Naturmuseum Winterthur, Winterthur 42036
Naturparkhaus Stechlin, Stechlin 21873
Naturschau in der Zehntscheune, Thallichtenberg 22045
Naturschutzausstellung der Naturschutzstation, Neschwitz 20575
Naturschutzzentrum, Nettersheim 20578
Naturum Stendörren, Tystberga 40989
Naturwissenschaftliche Sammlung, Stans 41904
Naturwissenschaftliche Sammlungen, Berlin 17223
Naturwissenschaftliche Sammlungen, Innsbruck 02126
Naturwissenschaftliche Sammlungen am Landesmuseum Joanneum, Graz 01981
Naturwissenschaftliches Museum, Aschaffenburg 16588
Naturwissenschaftliches Museum der Stadt Flensburg, Flensburg 18329
Naturwissenschaftliches Museum Duisburg, Duisburg 18026
Naturzentrum Ternell, Eupen 03591
Natuur Historisch Museum en Heemkunde Centrum, Meerssen 32546
Natuurdiorama Holterberg, Holten 32401
Natuurhistorisch en Volkenkundig Museum, Oudenbosch 32653
Natuurhistorisch Museum Boekenbergpark, Deurne 03548
Natuurhistorisch Museum de Peel, Asten 31942
Natuurhistorisch Museum het Diorama, Nunspeet 32608
Natuurhistorisch Museum Maastricht, Maastricht 32529
Natuurhistorisch Museum Rotterdam, Rotterdam 32718
Natuurmuseum Ameland, Nes 32569
Natuurmuseum Brabant, Tilburg 32828
Natuurmuseum de Wielewaal, Lopik 32507
Natuurmuseum Dokkum, Dokkum 32139
Natuurmuseum E. Heimans, Zaandam 32980
Natuurmuseum Enschede, Enschede 32212
Natuurmuseum Groningen, Groningen 32285
Natuurmuseum het Drents-Friese Woud, Wateren 32929
Natuurmuseum Mar en Klif, Oudemirdum 32652
Natuurmuseum Nijmegen, Nijmegen 32593
Natuurpunt Museum, Turnhout 04100
Natuurwetenschappelijk Museum, Antwerpen 03307
NAU Art Museum, Flagstaff 48552
Naučno-memorialnyj Muzej N.E. Žukovskogo, Moskva 37388
Naučnyj Morskoj Muzej, Kaliningrad 36993
Naujasis Arsenalas, Vilnius 30427
Naujenes Novadpētniecības Muzejs, Naujene 30049
Naumann-Museum Köthen, Köthen 19667
Naumburger Domschatzgewölbe, Naumburg, Saale 20554
Naumkeag House, Stockbridge 53301
Nausicaa, Boulogne-sur-Mer 12095
Nautelankosken Museo, Lieto 11068
Nautical Museum, Castletown 43776
Nautical Museum of Galaxidi, Galaxidi 23005
Nauticus, Norfolk 51367
Nautiko Mouseio Chiou, Chios 22970
Nautikon Mouseiontis Ellados, Piraeus 23147
Nauvoo Historical Society Museum, Nauvoo 50992
Navajo Nation Museum, Window Rock 54225
Naval Academy Museum, Jinhae 29843
Naval and Maritime Museum, Ballina 00830
Naval Museum, Cienfuegos 09067
Naval Museum, İstanbul 42814
Naval Museum, Karlskrona 40635
Naval Museum, Madrid 39574
Naval Museum, Rio de Janeiro 04757
Naval Museum, Vancouver 07218
Naval Museum of Alberta, Calgary 05528
Naval Museum of Québec, Québec 06695
Naval Museum of the North Fleet, Murmansk 37436
Naval Shipyard Museum, Portsmouth 52110
Naval Undersea Museum, Keyport 49768
Naval War College Museum, Newport 51325
Navan Centre, Armagh 43319
Navarro County Historical Society Museum, Corsicana 47825
Nave Museum, Victoria 53741
Naviscope Alsace, Strasbourg 15835
Navy Museum, Gdynia 34656
The Navy Museum, Washington 53911
Nayong Pilipino Museum, Pasay 34475
Nayuma Museum, Limulunga 54654
Nazib Shiganov Museum-Appartement, Kazan 37021
NBS-Múzeum Mincí a Medailí, Kremnica 38279
NCCL Galleries of Justice, Nottingham 45296
Néa Moni Collection, Chios 22971
Neanderthal Museum, Mettmann 20314
Neath Museum, Neath 45196
Nebenaigoching Heritage, Batchawana Bay 05394
Nebraska Conference United Methodist Historical Center, Lincoln 50103
Nebraska Museums Association, Aurora 55168
Nebraska Prairie Museum, Holdrege 49312
Nécropole de Louis Cauvin, Gareoult 13034
Ned Shed Museum, Meldrum Bay 06260
Nederlands Architectuurinstituut, Rotterdam 32719
Nederlands Artillerie Museum, 't Harde 32310
Nederlands Bakkerijmuseum Het Warme Land, Hattem 32323
Nederlands Baksteen en Dakpanmuseum de Panoven, Zevenaar 33000
Nederlands Biermuseum de Boom, Alkmaar 31794
Nederlands Centrum voor Biodiversiteit Naturalis, Leiden 32485
Nederlands Centrum voor Handwerken, Breda 32006
Nederlands Drogisterij Museum, Maarssen 32518
Nederlands Economisch Penningkabinet, Rotterdam 32720
Nederlands Fotomuseum, Rotterdam 32721
Nederlands Goud-, Zilver- en Klokkenmuseum, Schoonhoven 32764
Nederlands Graanmuseum/Olie- en Korenmolen Woldzigt, Roderwolde 32697
Nederlands Instituut voor Beeld en Geluid-Colleties, Hilversum 32389
Nederlands Instituut voor Mediakunst - NIMk, Amsterdam 31873
Nederlands Kachelmuseum, Alkmaar 31795
Nederlands Kansspelmuseum, Den Haag 32096
Nederlands Leder- en Schoenenmuseum, Waalwijk 32918
Nederlands Militair Kustverdedigingsmuseum, Hoek van Holland 32395
Nederlands Museum van Knipkunst, Schoonhoven 32765
Nederlands Openluchtmuseum, Arnhem 31930
Nederlands Parfumflessenmuseum, Winkel 32951
Nederlands Politie Museum, Apeldoorn 31914
Nederlands Spoorwegmuseum, Utrecht 32852
Nederlands Steendrukmuseum, Valkenswaard 32870
Nederlands Stoommachinemuseum, Medemblik 32543
Nederlands Strijkijzer-Museum, Noordbroek 32601
Nederlands Tegelmuseum, Otterlo 32644
Nederlands Textielmuseum, Tilburg 32829
Nederlands Uitvaartmuseum, Amsterdam 31874
Het Nederlands Vestingmuseum, Naarden 32564
Nederlands Waterleidingmuseum, Utrecht 32853
Nederlands Wijnmuseum, Arnhem 31931
Nederlands Zouavenmuseum, Oudenbosch 32654
De Nederlandse Museumvereniging, Amsterdam 54961
Nedervetil Hembygdsgård och -museum, Nedervetil 11139
Neeses Farm Museum, Neeses 50999
Nef des Jouets, Soultz-Haut-Rhin 15802
Negev Museum, Be'ersheva 24936
Negotin National Museum, Negotin 38116
Negri Sembilan State Museum, Seremban 30616
Negri Sembilan State Museum, Seremban 30617
The Negros Museum, Bacolod 34337
Nehru Children's Museum, Kolkata 24278
Nehru Memorial Museum and Library, Delhi 24152
Nehru Science Centre, Mumbai 24321
Nei Mongol Zhelimu Meng Guan, Tongliao 08300
Neidhart Fresken, Wien 03044
Neil Armstrong Air and Space Museum, Wapakoneta 53815
Neill-Cochran Museum House, Austin 46516
Neill Museum, Fort Davis 48612
Neiraku Art Museum, Nara 29037
Neiraku Bijutsukan, Nara 29037
Neisser Haus - Heimathaus und Archiv, Hildesheim 19184
Neixiang Museum, Neixiang 08068
Nejarine-Musée, Fès 31672
Neka Museum and Gallery, Gianyar 24472
Neligh Mills, Neligh 51001
Nelimarkka Museo . Etelä-Pohjanmaan Aluetaidemuseo ja Nelimarkka-resedenssi, Alajärvi 10708
Nelimarkka Museum, Regional Art Museum of Southern Ostrobothnia, Alajärvi 10708
Nels Berggran Museum, Imperial 06010
The Nelson, Davis 47976
The Nelson-Atkins Museum of Art, Kansas City 49695
Nelson Collection at Lloyd's, London 44942
Nelson & District Museum, Nelson 06422
Nelson Mandela Metropolitan Art Museum, Port Elizabeth 38711
Nelson Mandela Museum, Umtata 38770
Nelson Monument, Edinburgh 44141
Nelson Museum and Local History Centre, Monmouth 45168
Nelson Pioneer Farm and Museum, Oskaloosa 51617
Nelson Provincial Museum, Nelson 33153
Nelson Tower, Forres 44246
Nemes-Nagel Emlékszoba, Pécsvárad 23782
Nemes-Nagel Memorial Room, Pécsvárad 23782
Német Nemzetiségi Gyűjtemény, Gyula 23599
Németh László Emlékkiállítás, Hódmezővásárhely 23613
NEMO, Amsterdam 31875
Nemours Mansion and Gardens, Wilmington 54196
Nemzeti Bormúzeum, Eger 23542
Nenagh Heritage Centre, Nenagh 24883
Nene Valley Railway, Stibbington 45746
Neneckij Okružnoj Kraevedčeskij Muzej, Narjan-Mar 37443
Nenthead Mines Heritage Centre, Nenthead 45200
Neofit Rilski House Museum, Bansko 04979
Neot Kedumim, Lod 25061
Nepean Historical Museum, Sorrento 01509
Nepean Museum, Nepean 06425
Népi Iparművészeti Gyűjtemény, Kecskemét 23652
Népi Műemlékház, Zalalövo 23973
Népművészetek Háza, Szentendre 23878
Népmuvészeti Kiállítás, Heves 23607
Néprajzi Ház, Tápiószele 23912
Néprajzi Kiállítás, Pécs 23769
Néprajzi Kiállítás, Sopron 23814
Néprajzi Múzeum, Budapest 23455
Neptune Township Historical Museum, Neptune 51005
Népviseletes Babamúzeum - Történelmi Panoptikum, Keszthely 23662
Ner Hole Museum, Åndalsnes 33389
Nergal Gate Museum, Mosul 24721
Nerima Art Museum, Tokyo 29442
Nerima Bijutsukan, Tokyo 29442
Neriman Nerimanovun Xatire Muzeyi, Bakı 03165
Neringa History Museum, Nida 30321
Neringa History Museum - Fisherman's Ethnographic Homestead, Nida 30323
Neringos Istorijos Muziejus, Nida 30321
Nes Lensemuseum, Skiptvet 33837
Nesch-Museet, Ål 33378
Nesna Bygdemuseum, Nesna 33699
Ness Historical Society Museum, Port-of-Ness 45412
Nesstofusafn, Seltjarnarnes 24038
Nest Ume Tzimbari Gabizza, Giazza-Selva di Progno 26336
Netaji Museum, Kolkata 24279
Netherland Inn House Museum, Kingsport 49786
Netherlands Centre for Biodiversity Naturalis, Leiden 32485
Netherlands Institute of Sound and Vision-Collections, Hilversum 32389
Netherlands Media Art Institute - NIMk, Amsterdam 31873
Netherlands Museums Association, Amsterdam 54961
Network of European Museum Organisations - NEMO, Berlin 54854
Neuberger Museum of Art, Purchase 52201
Neuberin-Museum, Reichenbach im Vogtland 21246
Neubistritzer Heimatstube, Reingers 02576
Neudorfer Suppenmuseum, Sehmatal 21727
Neue Galerie, Dachau 17766
Neue Galerie am Landesmuseum Joanneum, Graz 01982
Neue Galerie der Karl-Hofer-Gesellschaft, Berlin 17224
Neue Galerie im Höhmannhaus, Augsburg 16632
Neue Galerie - Malerei und Skulptur ab 1750, Kassel 19473
Neue Galerie New York, New York 51238
Neue Kammern, Potsdam 21099
Neue Nationalgalerie, Berlin 17225
Neue Pinakothek, München 20471
Neue Residenz Bamberg, Bamberg 16916
Neue Sächsische Galerie, Chemnitz 17701
Neue Sammlung, Passau 20992
Neuer Berliner Kunstverein, Berlin 17226
Neuer Pavillon im Schlossgarten Charlottenburg, Berlin 17227
Neuer Sächsischer Kunstverein, Dresden 17965
Neuer Shed im Eisenwerk, Frauenfeld 41380
Neues Kunsthaus Ahrenshoop, Ahrenshoop, Ostseebad 16443
Neues Museum, Berlin 17228
Neues Museum, Nürnberg 20786
Neues Museum Schwechat, Schwechat 02729
Neues Museum Weimar, Weimar 22411
Neues Palais, Potsdam 21100
Neues Schloss Bayreuth, Bayreuth 16964
Neues Schloss Herrenchiemsee, Herrenchiemsee 19160
Neues Schloss Meersburg, Meersburg 20250
Neues Schloss Schleißheim, Oberschleißheim 20834
Neues Schloß, Schloßmuseum, Tettnang 22041
Neues Schloss Wallerstein, Wallerstein 22314
Neues Stadtmuseum Landsberg am Lech, Landsberg am Lech 19796
Neugablonzer Industrie- und Schmuckmuseum im Isergebirgs-Museum, Kaufbeuren 19490
Das Neumann Haus Museum, Laidley 01212
Neumarkter Brauereimuseum im Glossner-Bräu, Neumarkt in der Oberpfalz 20633
Neustadt Museum of Tiffany Art, New York 51239
Neustädter Rathaus, Hanau 18982
Neutral Ground, Regina 06721
Neuwerk Kunsthalle, Konstanz 19681
Nevada Historical Society Museum, Reno 52288
Nevada Museum of Art, Reno 52289
Nevada State Fire Museum and Comstock Firemen's Museum, Virginia City 53761
Nevada State Museum, Carson City 47260
Nevada State Museum, Las Vegas 49977
Nevada State Railroad Museum, Carson City 47261
Neversink Valley Area Museum, Cuddebackville 47898
Neville Blakey Museum of Locks, Keys and Safes, Brierfield 43615
Neville Public Museum of Brown County, Green Bay 49041
Nevşehir Devlet Güzel Sanatlar Galerisi, Nevşehir 42916
Nevşehir Müzesi, Nevşehir 42917
Nevşehir Museum, Nevşehir 42917
Nevşehir State Fine Art Gallery, Nevşehir 42916
New Archaeological Museum, Mytilini 23112
New Arsenal, Vilnius 30427
New Art Centre, East Winterslow 44096
The New Art Gallery Walsall, Walsall 45920
New Bedford Art Museum, New Bedford 51013
New Bedford Whaling Museum, New Bedford 51014
The New Berlin Historical Society Museum, New Berlin 51017
New Braunfels Conservation Society Museum, New Braunfels 51023
New Braunfels Museum of Art & Music, New Braunfels 51024
New Britain Museum of American Art, New Britain 51029
New Britain Youth Museum, New Britain 51030
New Britain Youth Museum at Hungerford Park, Kensington 49735
New Brunswick Art Bank, Fredericton 05834
New Brunswick College of Craft and Design Gallery, Fredericton 05835
New Brunswick Healthcare Museum, Fredericton 05836
New Brunswick Mining and Mineral Interpretation Centre, Petit-Rocher 06597
New Brunswick Museum, Saint John 06838
New Brunswick Sports Hall of Fame, Fredericton 05837
New Canaan Historical Society Museum, New Canaan 51037
New Castle Historical Society Museum, New Castle 51039
New Castle of Banská Štiavnica, Banská Štiavnica 38192
The New Children's Museum, San Diego 52742
New Denmark Memorial Museum, New Denmark 06426
New Detroit Science Center, Detroit 48128
New Echota, Calhoun 47167

New England Air Museum of the Connecticut Aeronautical Historical Association, Windsor Locks 54230
New England Brass and Iron Lace Foundry Museum, Uralla 01615
New England Carousel Museum, Bristol 47028
New England College Gallery, Henniker 49255
New England Fire and History Museum, Brewster 47009
New England Maple Museum, Pittsford 51970
New England Maple Museum, Rutland 52494
New England Museum Association, Arlington 55169
New England Quilt Museum, Lowell 50306
New England Regional Art Museum, Armidale 00814
New England Ski Museum, Franconia 48731
New England Wireless and Steam Museum, East Greenwich 48241
New Forest Museum and Visitor Centre, Lyndhurst 45056
New Gallery of Dachau, Dachau 17766
New Hall, Dymchurch 44077
New Hall Art Collection, Cambridge 43731
New Hampshire Antiquarian Society Museum, Hopkinton 49365
New Hampshire Farm Museum, Milton 50701
New Hampshire Institute of Art, Manchester 50424
New Harmony Gallery of Contemporary Art, New Harmony 51049
New Harmony State Historic Site, New Harmony 51050
New Haven Colony Historical Society Museum, New Haven 51053
New Hungarian Gallery, Székesfehérvár 23863
New Iceland Heritage Museum, Gimli 05866
New Image Art, West Hollywood 54065
New International Cultural Center, Antwerpen 03308
New Jersey Association of Museums, Newark 55170
New Jersey Center for Visual Arts, Summit 53364
New Jersey Children's Museum, Paramus 51698
New Jersey Historical Society Museum, Newark 51286
New Jersey Museum of Agriculture, North Brunswick 51385
New Jersey Naval Museum, Hackensack 49111
New Jersey State House, Trenton 53560
New Jersey State Museum, Trenton 53561
New Juaben Palace Museum, Koforidua 22854
New Lanark Visitor Centre, New Lanark 45203
New Langton Arts, San Francisco 52789
New London County Historical Society Museum, New London 51067
New London Historical Society Museum, New London 51069
New London Public Museum, New London 51070
New Marine Centre, Sidney 06971
New Market Battlefield State Historical Park, New Market 51072
New Mexico Association of Museums, Albuquerque 55171
New Mexico Bureau of Geology Mineral Museum, Socorro 53140
New Mexico Farm and Ranch Heritage Museum, Las Cruces 49960
New Mexico Museum of Natural History and Science, Albuquerque 46185
New Mexico Museum of Space History, Alamogordo 46151
New Mexico State University Art Gallery, Las Cruces 49961
New Mexico State University Museum, Las Cruces 49962
New Mexico Wing-Commemorative Air Force, Hobbs 49305
New Milford Historical Society Museum, New Milford 51073
New Mills Heritage and Information Centre, New Mills 45204
New Museum of Contemporary Art, New York 51240
New Museum of European and Eastern Art, Athinai 22944
New Norcia Museum and Art Gallery, New Norcia 01360
New Orleans Artworks Gallery, New Orleans 51089
New Orleans Fire Department Museum, New Orleans 51090
New Orleans Museum of Art, New Orleans 51091
New Orleans Pharmacy Museum, New Orleans 51092
New Otani Art Museum, Tokyo 29443
New Realms, London 44943
New Realms at Adams Street, London 44944
New South Wales Lancers Memorial Museum, Parramatta 01407
New Sweden Farmstead Museum, Bridgeton 47020
New Sweden Historical Museum, New Sweden 51104
New Tretyakov Gallery on Krymsky Val, Moskva 37185
New Virgin Nunnery, Moskva 37389
New Visions Gallery, Marshfield 50513
New Walk Museum and Art Gallery, Leicester 44681
New Windsor Cantonment State Historic Site and National Purple Heart Hall of Honor, Vails Gate 53683
New York City Fire Museum, New York 51241
New York City Police Museum, New York 51242
New York Hall of Science, Flushing 48578
The New York Historical Society Museum, New York 51243
New York Museum of Transportation, West Henrietta 54062
New York Public Library for the Performing Arts, New York 51244
New York State Military Museum, Saratoga Springs 52923
New York State Museum, Albany 46161
New York Studio School of Drawing, Painting and Sculpture - Gallery, New York 51245
New York Transit Museum, Brooklyn 47080
New Zealand Academy of Fine Arts Gallery, Wellington 33252
New Zealand Centre for Photography, Wellington 33253
New Zealand Cricket Museum, Wellington 33254
New Zealand Fighter Pilots Museum, Wanaka 33232
New Zealand Marine Studies Centre, Dunedin 33093
New Zealand National Maritime Museum, Auckland 33057
New Zealand Police Museum, Porirua 33185
New Zealand Portrait Gallery, Wellington 33255
New Zealands Rugby Museum, Palmerston North 33174
New Zone Virtual Gallery, Eugene 48436
Newark Air Museum, Newark-on-Trent 45211
The Newark Museum, Newark 51287
Newark Museum, Newark-on-Trent 45212
Newark Town Treasures and Art Gallery, Newark-on-Trent 45213
Newark Valley Depot Museum, Newark Valley 51297
Newarke Houses Museum, Leicester 44682
Newart's Millgate Museum, Newark-on-Trent 45214
Newaygo County Museum, Newaygo 51298
Newbridge House, Donabate 24774
Newburn Hall Motor Museum, Newburn 45219
Newburyport Maritime Society Museum, Amesbury 46262
Newby Hall, Ripon 45499
Newcastle Gaol Museum, Toodyay 01590
Newcastle Region Art Gallery, Newcastle 01364
Newcastle Regional Museum, Newcastle 01365
Newcomb Art Gallery, New Orleans 51093
Newcomen Engine House, Dartmouth 43954
Newell Museum, Newell 51311
Newfoundland Science Centre, Saint John's 06850
Newham Grange Leisure Farm Museum, Middlesbrough 45142
Newham Heritage Centre, London 44945
Newhaven Fort, Newhaven 45236
Newhaven Heritage Museum, Edinburgh 44142
Newlyn Art Gallery, Penzance 45360
Newman House, Dublin 24813
Newport Art Museum, Newport 51326
Newport Mansions, Newport 51327
Newport Pagnell Historical Society Museum, Newport Pagnell 45247
Newport Roman Villa, Newport, Isle of Wight 45245
Newry and Mourne Arts Centre and Museum, Newry 45250
Newseum, Arlington 46358
Newsome House Museum, Newport News 51335
Newspaper Kkeskisuomalainen's Museum, Jyväskylä 10896
Newspark - The Japan Newspaper Museum, Yokohama 29603
Newsstand Museum, Osaka 29138
Newstead Abbey Historic House and Gardens, Nottingham 45297
Newton Abbot Town and Great Western Railway Museum, Newton Abbot 45251
Newton History Museum at Jackson Homestead, Newton 51341
Newton Museum, Newton 51339
Newtonmore Museum, Newtonmore 45255
Newtown Historic Museum, Newtown 51346
Newtown Textile Museum, Newtown, Powys 45256
Nexø Museum, Nexø 10191
Nexus Foundation for Today's Art, Philadelphia 51862
Nez Perce County Museum, Lewiston 50035
Nez Perce National Historical Park, Spalding 53191
Nezu Bijutsukan, Tokyo 29444
Nezu Museum of Fine Arts, Tokyo 29444
Nghe-Tinh Museum, Vinh 54640
Ngurdoto Gate Museum, Arusha 42508
Nhabe Museum, Nhabe 04300
NHK Broadcasting Museum, Tokyo 29445
NHK Hoso Hakubutsukan, Tokyo 29445
Niagara Apothecary, Niagara-on-the-Lake 06452
Niagara County Historical Center, Lockport 50158
Niagara Falls Art Gallery - William Kurelek Art Collection, Niagara Falls 06446
Niagara Fire Museum, Niagara-on-the-Lake 06453
Niagara Gorge Discovery Center, Niagara Falls 51347
Niagara Historical Museum, Niagara-on-the-Lake 06454
Niagara Power Project Visitors' Center, Lewiston 50039
Nianhua Museum, Mianzhu 08017
Niavaran Palace, Tehrān 24689
Nibelungenmuseum, Worms 22687
Nichi-Mongoru Mingeikan, Tantou 29351
Nichido Contemporary Art, Tokyo 29446
Nicholas Roerich Museum, New York 51246
Nichols House Museum, Boston 46943
Nicholson Museum, Sydney 01557
Nici Self Museum, Centennial 47312
Nickels-Sortwell House, Wiscasset 54264
The Nickle Arts Museum, Calgary 05529
Nicola Valley Museum, Merritt 06269
Nicolas Denys Museum, Saint Peters 06874
Nicolaus August Otto Museum, Holzhausen an der Haide 19258
Nicolaus Copernicus Museum, Frombork 34629
Nicolaus Copernicus Museum of the District Museum In Toruń, Toruń 35194
Nicolaus-Schmidt-Künzel-Gedenkstätte, Tanna 22006
Nicolaysen Art Museum, Casper 47272
Nicosia Municipal Arts Centre, Lefkosia 09382
Nidderdale Museum, Pateley Bridge 45346
Nidwaldner Museum, Stans 41905
Niederbayerisches Feuerwehrmuseum, Eggenfelden 18062
Niederbayerisches Landwirtschaftsmuseum, Regen 21214
Niederbayerisches Vorgeschichtsmuseum, Landau an der Isar 19787
Niederbergisches Museum Wülfrath, Wülfrath 22703
Niederebersdorfer Heimat- und Archivstuben, Tutzing 22140
Niederlausitz-Museum Luckau, Luckau, Niederlausitz 20040
Niederlausitzer Heidemuseum, Spremberg 21843
Niederlausitzer Sorbisches Dorfmuseum Bloischdorf, Felixsee 18303
Niederösterreichisches Dokumentationszentrum für Moderne Kunst, Sankt Pölten 02684
Niederösterreichisches Feuerwehrmuseum, Tulln 02831
Niederösterreichisches Freilichtmuseum, Haag, Niederösterreich 02030
Niederösterreichisches Landesmuseum, Sankt Pölten 02685
Niederösterreichisches Museum für Volkskultur, Groß Schweinbarth 01997
Niederrheinisches Freilichtmuseum, Grefrath 18744
Niederrheinisches Motorradmuseum, Moers 20363
Niederrheinisches Museum für Volkskunde und Kulturgeschichte, Kevelaer 19524
Niedersächsisches Deichmuseum, Dorum 17931
Niedersächsisches Kleinbahn-Museum Bruchhausen-Vilsen, Bruchhausen-Vilsen 17592
Niedersächsisches Kutschenmuseum, Lilienthal 19990
Niedersächsisches Landesmuseum Hannover, Hannover 19002
Niedersächsisches Münzkabinett der Deutschen Bank, Hannover 19003
Niedersächsisches Museum für Kali- und Salzbergbau, Ronnenberg 21371
Niedersächsisches Spargelmuseum, Nienburg 20719
Nielstrup Museum, Hadsten 10051
Niemeyer Tabaksmuseum, Groningen 32286
Niemeyer Tobacco Museum, Groningen 32286
Nietzsche-Archiv, Weimar 22412
Nietzsche Archive, Weimar 22412
Nietzsche-Haus, Naumburg, Saale 20555
Nietzsche-Haus, Sils Maria 41870
Nietzsche House, Sils Maria 41870
Nieuw Land Poldermuseum, Lelystad 32495
Nieuwe Republiek Museums - Vryheid, Vryheid 38775
Nieuweschans Naaimachinemuseum, Nieuweschans 32579
NIFCA - Nordic Institute for Contemporary Art, Helsinki 10805
Niğde Müzesi, Niğde 42918
Nigde Museum, Niğde 42918
Niguliste Muuseum, Tallinn 10638
Nihon Biyoin Hakubutsukan, Kyoto 28855
Nihon Camera Hakubutsukan, Tokyo 29447
Nihon Gangu Hakubutsukan, Kodera 28753
Nihon Hakubutsukan Kyokai, Tokyo 54910
Nihon Jidosha Hakubutsukan, Futatsunashi 28455
Nihon Kindai Bungakukan, Tokyo 29448
Nihon Kogeikan, Tokyo 29449
Nihon Kyodogangu-Kan, Kurashiki 28788
Nihon Mingeikan, Osaka 29139
Nihon Mingeikan, Tokyo 29450
Nihon Minka Shuraku Hakubutsukan, Toyonaka 29530
Nihon Monki Senta Hakubutsukan, Inuyama 28578
Nihon Oni Hakubutsukan, Oe 29089
Nihon Shodo Bijutsukan, Tokyo 29451
Nihon Ukiyo-e Hakubutsukan, Matsumoto 28896
Niigata Bandaijima Art Museum, Niigata 29055
Niigata City Aizu Yaichi Memorial Museum, Niigata 29058
Niigata City Art Museum, Niigata 29059
Niigata City History Museum, Niigata 29057
Niigata-kenritsu Bandaijima Bijutsukan, Niigata 29055
Niigata-kenritsu Kindai Bijutsukan, Nagaoka 28963
Niigata-kenritsu Rekishi Hakubutsukan, Nagaoka 28964
Niigata Prefectural Government Memorial Hall, Niigata 29054
Niigata Prefectural Museum of History, Nagaoka 28964
Niigata Prefectural Museum of Modern Art, Nagaoka 28963
Niigata Science Museum, Niigata 29056
Niigata-shi Hakubutsukan, Niigata 29057
Niigata-shiritsu Aizu Yaichi Kinenkan, Niigata 29058
Niigata-shiritsu Bijutsukan, Niigata 29059
Niimi Bijutsukan, Niimi 29066
Niimi Museum of Art, Niimi 29066
Niitsu House, Niigata 29060
Nijmeegs Volkenkundig Museum, Nijmegen 32594
Niki Bijutsukan, Nasu 29051
Niki Museum, Nasu 29051
Nikko Futarasan-jinja Hakubutsukan, Nikko 29068
Nikko Futarasan Shrine Museum, Nikko 29068
Nikko Toshogu Treasure Museum, Nikko 29069
Niko Pirosmanashvili State Museum, Tbilisi 16375
Nikodem Jaroń Regional Museum, Olesno 34972
Nikol Duman Memorial Museum, Ghshlagh 03177
Nikola Tesla Museum, Beograd 38071
Nikola Vaptsarov Memorial House, Sofia 05179
Nikolaj Rubcov Museum, Moskva 37353
Nikolaj Rubinstein Museum of the Moscow Conservatoire, Moskva 37354
Nikolaos Perantinos Museum of Sculpture, Athinai 22945
Nikolaus Matz Bibliothek (Kirchenbibliothek), Michelstadt 20320
Nikoloz Baratashvili Memorial Museum, Tbilisi 16376
Nikos Nikolaides Theatrical Archives, Lemesos 09389
Nils Aas Kunstverksted, Inderøy 33575
Nilsiän Kotiseutumuseo, Nilsiä 11143
Nimbin Museum, Nimbin 01369
Ninart Centro de Cultura, México 31157
Ninety-Six National Historic Site, Ninety-Six 51354
Ningbo Baoguo Si Display Center, Ningbo 08070
Ningbo Museum, Ningbo 08071
Ningde Museum, Ningde 08073
Ningde Zone Shezu Museum, Ningde 08074
Ningdu Museum, Ningdu 08075
Ningxia Hui Autonomous Region Museum, Yinchuan 08447
Ningyang Museum, Ningyang 08076
Ninja Museum of Iga-ryu, Ueno 29556
Ninohe Historical Folk Art Museum, Ninohe 29071
Ninohe-shiritsu Rekishi Mingeikan, Ninohe 29071
Ninomaru Museum of Art, Kakegawa, Kakegawa 28637
Nipawin Living Forestry Museum, Nipawin 06459
Nipigon Museum, Nipigon 06460
Nipissing Township Museum, Nipissing 06461
Nippon Doreikan, Shirotori 29282
Nippon Kokeshikan, Naruko 29042
Nisa Historical Site, Nusaý 42980
Nishijin Textile Center, Kyoto 28856
Nishinomiya-shiritsu Otani Bijutsu Kinenkan, Nishinomiya 29074
Nishna Heritage Museum, Oakland 51473
Nitobe Kinenkan, Hanamaki 28493
Nitobe Memorial Museum, Hanamaki 28493
Nitrianska Galéria, Nitra 38314
Nitten - Japan Fine Arts Exhibition, Tokyo 29452
Nitz Gallery, Nuuk 23261
NIU Art Museum, DeKalb 48039
Nivågårds Malerisamling, Nivå 10192
Nivala Museum Katvala, Nivala 11145
Nivalan Museo Katvala, Nivala 11145
Nizami ad Dövlet Azerbaycan Edebiyyati Muzeyi, Baki 03166
Nizami Ganjavi State Museum of Azerbaijan Literature of the Azerbaijan National Academy of Sciences, Baki 03166
The Nizam's Museum, Hyderabad 24215
Nižegorodskij Chudožestvennyj Muzej, Nižnij Novgorod 37467
Nižegorodskij Gosudarstvennyj Istoriko-architekturnyj Muzej-zapovednik, Nižnij Novgorod 37468

N

Nižegorodskij Kreml s Archangelskim Soborom, Nižnij Novgorod 37469
Nizhni Novgorod Art Museum, Nižnij Novgorod 37467
Nizhni Novgorod Kremlin with Archangel Cathedral, Nižnij Novgorod 37469
Nizhni Novgorod State Reserve Museum of History and Architecture, Nižnij Novgorod 37468
Nizhny Tagil Museum-Reserve of Mining and Metallurgy of the Middle Urals, Nižnij Tagil 37473
Nizhny Tagil Town Museum of Fine Arts, Nižnij Tagil 37472
Nižnetagilskij Municipalnyj Muzej Izobrazitelnych Iskusstv, Nižnij Tagil 37472
Nižnetagilskij Muzej-zapovednik Gornozavodskoj Ural, Nižnij Tagil 37473
Njegošev Muzej, Cetinje 38098
Njegoš's Museum, Cetinje 38098
Njepila-Hof Rohne, Rohne 21363
Njudungs Hembygdsmuseum, Vetlanda 41056
Nk'Mip Desert & Heritage Centre, Osoyoos 06526
No Man's Land Historical Museum, Goodwell 48972
Noah Webster House - Museum of West Hartford History, West Hartford 54059
Noank Museum, Noank 51355
Nobel Technical Museum, Batumi 16336
Nobelmuseet och Bofors Industrimuseum, Karlskoga 40632
NobelMuseum, Stockholm 40921
The Noble Maritime Collection, Staten Island 53265
Nobleboro Historical Society Museum, Nobleboro 51356
Nobles County Historical Society, Worthington 54306
Nobynäs Säteri, Aneby 40423
Noctalis-Welt der Fledermäuse, Bad Segeberg 16837
Noctalis - World of Bats, Bad Segeberg 16837
Nogata Municipal Coal Memorial Museum, Nogata 29081
Nogata-shi Sekitan Kinenkan, Nogata 29081
Nógrádi Történeti Múzeum, Salgótarján 23791
Noguchi Hideyo Kinenkan, Inawashiro 28571
Nogyo Hakubutsukan, Miyazaki 28929
Noh Theater Museum of Sado, Ryotsu 29174
Nohli Croft, Laitila 11040
Nohwike' Bagowa, Fort Apache 48589
Noirmont Command Bunker, Saint Brelade 45561
Nokia Local History Museum, Nokia 11147
Nokian Kotiseutumuseo Hinttala, Nokia 11147
Nokomis and District Museum, Nokomis 06462
Nolan Gallery, Tharwa 01581
Nolde Stiftung Seebüll, Berlin 17229
Nolde Stiftung Seebüll, Neukirchen, Kreis Nordfriesland 20625
Nolhin Torppa, Laitila 11040
Nomura Art Museum, Kyoto 28857
Nomura Bijutsukan, Kyoto 28857
Nonnahús, Akureyri 23985
Nonseum, Herrnbaumgarten 02074
Nonthaburi Natural History Museum, Nonthaburi 42588
Noord-Hollands Motoren Museum, Nieuwe Niedorp 32575
Noordbrabants Museum, 's-Hertogenbosch 32375
Noordelijk Busmuseum, Winschoten 32953
Noordelijk Scheepvaartmuseum, Groningen 32287
Het Noorderhuis, Zaandam 32981
De Noordwester, Vlieland 32898
Noosa Regional Art Gallery, Tewantin 01579
Noosa Shire Museum, Pomona 01421
Nor West Bend Museum, Morgan 01320
Nor' Wester and Loyalist Museum, Williamstown 07339
Nora Barnacle House Museum, Galway 24837
Nora Eccles Harrison Museum of Art, Logan 50161
Nora Museum, Gyttorp 40570
Nord-Jarlsbergmuseene, Holmestrand 33554
Nord-Troms Museum, Sørkjosen 33852
Nord-Trøndelag Fylkesgalleri, Namsos 33692
Nordamerika Native Museum-NONAM, Zürich 42099
Norddal Museum, Valldal 33957
Norddeutsches Vogelmuseum, Osterholz-Scharmbeck 20936
Nordenfjeldske Kunstindustrimuseum, Trondheim 33925
Nordens Ark, Hunnebostrand 40606
Norderneyer Fischerhausmuseum, Norderney 20733
Nordfjord Folkemuseum, Sandane 33810
Nordfriesisches Museum Ludwig-Nissen-Haus, Husum 19299
Nordfyns Museum, Bogense 09980
Nordhallands Hembygdsförening, Kungsbacka 40659
Nordhannoversches Bauernhaus-Museum, Isernhagen 19363
Nordic Heritage Museum, Seattle 53006
Nordic House, Reykjavík 24027
Nordica Homestead Museum, Farmington 48516
Nordico - Museum der Stadt Linz, Linz 02320
Nordische Botschaften Gemeinschaftshaus/ Felleshus - Galerie, Berlin 17230
Nordiska Akvarellmuseet, Skärhamn 40843
Nordiska Museet, Stockholm 40922
Nordiska Travmuseet i Årjäng, Årjäng 40430
Nordjyllands Kunstmuseum, Aalborg 09956
Nordkappmuseet, Honningsvåg 33560
Norðoya Fornminnissavn, Klaksvík 10690
The Nordland Museum, Bodø 33445
Nordland Red Cross War Museum, Narvik 33695
Nordland Røde Kors Krigsminnemuseum, Narvik 33695
Nordlandsmuseet, Bodø 33445
Nordli Bygdemuseum, Nordli 33702
Nordmøre Museum, Kristiansund 33615
Nordnorsk Fartøyvernsenter og Båtmuseum, Gratangen 33517
Nordnorsk Kunstmuseum, Tromsø 33918
Nordpfälzer Heimatmuseum, Rockenhausen 21342
Nordre Husan, Alvdal 33386
Nordseemuseum, Bremerhaven 17573
Nordsjaellandsk Folkemuseum, Hillerød 10077
Nordsjøfartmuseet, Tælavåg 33899
Nordsømuseet, Hirtshals 10079
Nordstrand Visual Arts Gallery, Wayne 54006
Nordvärmlands Jakt och Fiskemuseum, Sysslebäck 40967
Nordwestdeutsches Schulmuseum Friesland, Zetel 22791
Nordwolle Delmenhorst, Delmenhorst 17816
Nore og Uvdal Bygdetun, Uvdal i Numedal 33951
Noreview Folk and Heritage Museum, Bennettsbridge 24744
Norfolk and Suffolk Aviation Museum, Flixton 44234
Norfolk Arts Centre, Simcoe 06980
Norfolk Historical Museum, Norfolk 51360
Norfolk Historical Society, Norfolk 51368
Norfolk Island Museum, Kingston 33349
Norges Birøkterlags Museet, Billingstad 33436
Norges Fiskerimuseet, Bergen 33432
Norges Hjemmefrontmuseum, Oslo 33738
Norges Museumsforbund, Oslo 54972
Norges Olympiske Museum, Lillehammer 33638
Norilsk Art Gallery, Norilsk 37475
Norilskaja Chudožestvennaja Galereja, Norilsk 37475
Norman Cleveland County Historical Museum, Norman 51375
Norman No. 1 Oil Well Museum, Neodesha 51003
Norman R. Eppink Art Gallery, Emporia 48393
Norman Rockwell Museum at Stockbridge, Stockbridge 53302
Norman Wells Historical Centre, Norman Wells 06463
Normanby Hall, Scunthorpe 45626
Normanby Park Farming Museum, Scunthorpe 45627
Normannischer Turm auf dem Ruinenberg, Potsdam 21101
Normanton Church Museum, Oakham 45307
Normisist Folk Inventions Museum, Butuan 34355
Norræna Húsið, Reykjavík 24027
Norrahammars Industrimuseum, Norrahammar 40768
Norrbottens Järnvägsmuseum, Luleå 40698
Norrbottens Museum, Luleå 40699
Norrbyskärs Museum, Hörnefors 40599
Norris Arm Heritage Museum, Norris Arm 06464
Norris Museum, Saint Ives, Cambridgeshire 45574
Norrköping Museum of Art, Norrköping 40773
Norrköpings Konstmuseum, Norrköping 40773
Norrköpings Stadsmuseum, Norrköping 40774
Norrtälje Konsthall, Norrtälje 40776
Norseman Historical and Geological Collection, Norseman 01371
Norsewood Pioneer Cottage Museum, Norsewood 33160
Norsk Barnemuseum, Stavanger 33868
Norsk Bergverksmuseum, Kongsberg 33603
Norsk Bremuseum, Fjærland 33493
Norsk Emballasje Museum, Oslo 33739
Norsk Farmasihistorisk Museum, Oslo 33740
Norsk Fiskeindustrimuseum, Melbu 33663
Norsk Fiskerværsmuseet, Sørvågen 33855
Norsk Fjellmuseum, Lom 33646
Norsk Folkemuseum, Oslo 33741
Norsk Grafisk Museum, Stavanger 33869
Norsk Hagebruks Museum, Grimstad 33520
Norsk Hermetikkmuseum, Stavanger 33870
Norsk Husflidsmuseum, Moen 33672
Norsk ICOM, Oslo 54973
Norsk Industriarbeidermuseet, Rjukan 33785
Norsk Jernbanemuseum, Hamar 33530
Norsk Kartmuseum, Hønefoss 33558
Norsk Kjøretøyhistorisk Museum, Lillehammer 33639
Norsk Klippfiskmuseum, Kristiansund 33616
Norsk Landbruksmuseum, Ås 33402
Norsk Luftfartsmuseum, Bodø 33446
Norsk Oljemuseum, Stavanger 33871
Norsk Reiselivsmuseum, Balestrand 33414
Norsk Seminmuseum, Stange 33862
Norsk Sjøfartsmuseum, Oslo 33742
Norsk Skogmuseum, Elverum 33481
Norsk Speidermuseum, Oslo 33743
Norsk Teknisk Museum, Oslo 33744
Norsk Telemuseum - Kristiansand, Kristiansand 33611
Norsk Telemuseum - Lødingen, Lødingen 33642
Norsk Telemuseum - Sørvågen, Sørvågen 33856
Norsk Telemuseum - Stavanger, Stavanger 33872
Norsk Telemuseum - Trondheim, Trondheim 33926
Norsk Tindemuseum, Åndalsnes 33390
Norsk Tollmuseum, Oslo 33745
Norsk Trikotasjemuseum, Salhus 33807
Norsk Utvandrermuseum, Ottestad 33762
Norsk Vasskraft- og Industristadmuseum, Tyssedal 33944
Norsk Vegmuseum, Fåberg 33486
Norsk Veterinærmedisinsk Museum, Oslo 33746
Norske Grenselosers Museum, Bjørkelangen 33441
Norske Kunst- og Kulturhistoriske Museer, Oslo 54974
Norske Museumspedagogiske Forening, Bodø 54975
North American Black Historical Museum, Amherstburg 05335
North Andover Historical Museum, North Andover 51379
North Anna Nuclear Information Center, Mineral 50725
North Atlantic Aviation Museum, Gander 05848
North Ayrshire Museum, Saltcoats 45608
North Bengal Science Centre, Siliguri 24395
North Berwick Museum, North Berwick 45265
North Bohemian Museum Liberec, Liberec 09625
North Bohemian Museum of Art Litoměřice, Litoměřice 09631
North Carolina Central University Art Museum, Durham 48222
North Carolina Maritime Museum, Beaufort 46666
North Carolina Museum of Art, Raleigh 52232
North Carolina Museum of Forestry, Whiteville 54133
North Carolina Museum of History, Raleigh 52233
North Carolina Museum of Life and Science, Durham 48223
North Carolina Museum of Natural Sciences, Raleigh 52234
North Carolina Museums Council, Raleigh 55172
North Carolina Pottery Center, Seagrove 52990
North Carolina State University Gallery of Art and Design, Raleigh 52235
North Carolina Transportation Museum, Spencer 53201
North Carr Lightship, Dundee 44049
North Castle Historical Society Museum, Armonk 46361
North Cornwall Museum and Gallery, Camelford 43738
North Country Museum of Arts, Park Rapids 51709
North Dakota Museum of Art, Grand Forks 48989
North Devon Maritime Museum, Appledore, Devon 43308
North Down Heritage Centre, Bangor, County Down 43375
North East Aircraft Museum, Sunderland 45801
North Florida Community College Art Gallery, Madison 50376
North german museum of minor railways, Bruchhausen-Vilsen 17592
North Haven Historical Society Museum, North Haven 51394
North Hertfordshire Museums, Letchworth 44696
North Hertfordshire Museums Resources Centre, Hitchin 44475
North Highland Community Museum, Cape North 05551
North Hills Museum, Granville Ferry 05904
North House Museum, Lewisburg 50034
North Huron District Museum, Wingham 07349
North Karelian Museum, Joensuu 10873
North Kazakh Regional Fine Arts Museum, Petropavl 29721
North Kerry Museum, Ballyduff 24737
North Lanark Regional Museum, Almonte 05325
North Lee County Historic Center and Santa Fe Depot Museum Complex, Fort Madison 48652
North Leverton Windmill, North Leverton 45266
North Lincolnshire Museum, Scunthorpe 45628
The North Museum of Natural History and Science, Lancaster 49929
North Naval Museum, Archangelsk 36770
North Norfolk Railway Museum, Sheringham 45662
North Norwegian Vessel Preservation Center and Boatmuseum, Gratangen 33517
North of England Museums Libraries and Archives Council, Newcastle-upon-Tyne 55088
North-Osetian Museum of History, Architecture and Literature, Vladikavkaz 37932
North Otago Museum, Oamaru 33162
North Pacific Cannery Village Museum, Port Edward 06637
North Plainfield Exempt Firemen's Museum, North Plainfield 51400
North Platte Valley Museum, Gering 48896
North Sea Museum, Hirtshals 10079
The North Shore Arts Association, Gloucester 48947
North Somerset Museum, Weston-super-Mare 45981
North Star Mining Museum, Grass Valley 49025
North Star Museum of Boy Scouting and Girl Scouting, North Saint Paul 51403
North Stradbroke Island Historical Museum, Dunwich 01050
North Sumatera Government Museum, Medan 24526
North Sydney Museum, North Sydney 06476
North Thompson Museum, Barriere 05388
North Tonawanda History Museum, North Tonawanda 51407
North Vancouver Museum, North Vancouver 06480
North View Gallery, Portland 52081
North Weald Airfield Museum and Memorial, Epping 44182
North West Federation of Museums and Art Galleries, Ashton-under-Lyne 55089
North West Film Archive, Manchester 45094
North West Museum of Road Transport, Saint Helens 45567
North West Sound Archive, Clitheroe 43855
North Western Agricultural Museum, Warracknabeal 01630
North Wind Undersea Institute, Bronx 47052
North Woolwich Old Station Museum, London 44946
North Yorkshire Moors Railway, Pickering 45382
Northampton and Lamport Railway, Chapel Brampton 43788
Northampton County Historical and Genealogical Society Museum, Easton 48271
Northampton Museum and Art Gallery, Northampton 45270
Northart Community Arts Centre, Auckland 33058
Northborough Historical Museum, Northborough 51416
Northcliffe Pioneer Museum, Northcliffe 01378
Northcote Pottery Gallery, Thornbury 01585
Northcutt Steele Gallery, Billings 46805
Northeast Classic Car Museum, Norwich 51440
Northeast Louisiana Delta African American Heritage Museum, Monroe 50790
Northeast Mississippi Museum, Corinth 47808
Northeast Mississippi Museums Association, Corinth 55173
Northeast Oakland Historical Museum, Oxford 51647
Northeast Texas Rural Heritage Museum, Pittsburg 51937
Northeastern Illinois University Art Gallery, Chicago 47500
Northeastern Montana Threshers and Antique Association, Culbertson 47901
Northeastern Nevada Museum, Elko 48354
Northern Clay Center Galleries, Minneapolis 50741
Northern Culture Museum, Niigata 29061
Northern Galleries, Aberdeen 46117
Northern Gallery for Contemporary Art, Sunderland 45802
Northern Gateway Museum, Denare Beach 05694
Northern Illinois University Art Gallery in Chicago, Chicago 47501
Northern Indiana Arts Association, Munster 50909
Northern Indiana Center for History, South Bend 53158
Northern Ireland Museums Council, Belfast 55090
Northern Kentucky University Art Galleries, Highland Heights 49280
Northern Life Museum and National Exhibition Centre, Fort Smith 05823
Northern Lights Art Gallery, Mayville 50541
Northern Lights Centre, Watson Lake 07291
Northern Lights Military Museum, Happy Valley 05970
Northern Maritime Museum, Groningen 32287
Northern Ostrobothnia Museum, Oulu 11177
Northern Ostrobothnia Museum - SailorS Home, Oulu 11170
Northern Pacific Railway Museum, Toppenish 53539
Northern Paiwan Art and Culture Museum, Sandimen 42309
Northern Taiwan Visitors Center, Wanli 42458
Northfield Farm Museum, New Pitsligo 45206
Northfield Historical Society Museum, Northfield 51421
Northland Historical Society Museum, Lake Tomahawk 49892
Northport Historical Society Museum, Northport 51425
The Northumberland County Historical Society Museum, Sunbury 53369
Northumberland Fisheries Museum, Pictou 06603
Northwest Arctic Heritage Center, Kotzebue 49827
Northwest Art Center, Minot 50746
Northwest Gallery and Sinclair Gallery, Powell 52126

Northwest Museum of Arts and Culture, Spokane 53207
Northwest Railway Museum, Snoqualmie 53137
Northwestern Mutual Art Gallery, Milwaukee 50717
Northwestern Oklahoma State University Museum, Alva 46247
Northwestern Ontario Sports Hall of Fame, Thunder Bay 07077
Northwind Arts Center, Port Townsend 52044
Northwood Art Space, Columbus 47756
Norton Conyers, Ripon 45500
Norton Museum of Art, West Palm Beach 54073
Norton Priory Museum, Runcorn 45530
Norton Simon Museum of Art, Pasadena 51720
Norway's Resistance Museum, Oslo 33738
Norwegian Archive, Library and Museum Authority, Oslo 54969
Norwegian Aviation Museum, Bodø 33446
Norwegian Beekeepers Association Museum, Billingstad 33436
The Norwegian Canning Museum, Stavanger 33870
Norwegian Center for Constitutional History, Eidsvoll Verk 33476
Norwegian Children's Museum, Stavanger 33868
Norwegian Federation of Friends of Museums, Oslo 54976
Norwegian Fishing Village Museum, Sørvågen 33855
Norwegian Forest Museum, Elverum 33481
Norwegian Industrial Workers' Museum, Rjukan 33785
Norwegian Laft Hus, Red Deer 06708
Norwegian Map Museum, Hønefoss 33558
Norwegian Maritime Museum, Oslo 33742
Norwegian Mining Museum, Kongsberg 33603
Norwegian Mountain Museum, Lom 33646
Norwegian Museum of Agriculture, Ås 33402
Norwegian Museum of Architecture - The National Museum of Art, Design and Architecture, Oslo 33716
Norwegian Museum of Hydropower and Industry, Tyssedal 33944
Norwegian Museum of Science and Technology, Oslo 33744
Norwegian Railway Museum, Hamar 33530
Norwegian Road Museum, Fåberg 33486
Norwich and District Museum, Norwich 06489
Norwich Arts Council Gallery, Norwich 51437
Norwich Gallery, Norwich 45282
Norwich Historical Society, Norwich 51443
Norwich University Museum, Northfield 51423
Norwood Historical Association Museum, Norwood 51444
Norwood Museum, Holm 44477
Nosaka Automata Museum, Ito 28597
Nosaka O-tomatto Hakubutsukan, Ito 28597
Nose Creek Valley Museum, Airdrie 05311
Nostalgie auf Rädern - Fahrzeugmuseum, Großklein 02005
Nostalgie Museum, Beetzseeheide 16980
Nostalgie-Museum, Wörth an der Donau 22643
Nostalgisches Musikparadies, Oberhasli 41694
Nostell Priory, Wakefield 45911
Nøstetangen Museum, Hokksund 33550
Noszlopy Kúria, Újvárfalva 23941
Noszlopy Mansion, Újvárfalva 23941
Notburga-Museum, Eben am Achensee 01831
De Notelaer, Hingene 03703
Nothe Fort Museum of Coastal Defence, Weymouth 45988
Notojima Glass Art Museum, Notojima 29082
Notranjska Museum, Postojna 38496
Notranjski Muzej Postojna, Postojna 38496
Notre Dame of Jolo College Museum, Jolo 34392
Nottingham Castle, Nottingham 45298
Nottingham Transport Heritage Centre, Ruddington 45523
Nottingham University Museum of Archaeology, Nottingham 45299
Notukeu Heritage Museum, Ponteix 06622
Nousiainen Museum and Topoinen Open Air Museum, Nousiainen 11149
Nousiaisten Kotiseutumuseo ja Topoisten Ulkomuseoalue, Nousiainen 11149
Nova Scotia Museum, Halifax 05943
Nova Scotia Museum of Natural History, Halifax 05944
Novalis-Literaturgedenkstätte, Weißenfels 22458
Novalis-Museum, Wiederstedt 22541
Novato History Museum, Novato 51447
Novgorod-Siverskyj Istoriko-kulturnyj Muzej-zapovidnyk Slovo o Polku Igorevim, Novgorod-Sivers'kyj 43153
Novgorod State Museum Reservation, Velikij Novgorod 37922
Novgorodskij Gosudarstvennyj Muzej-zapovednik, Velikij Novgorod 37922
Novi Sad City Museum, Novi Sad 38121
Novočerkasskij Muzej Istorii Donskich Kazakov, Novočerkassk 37478
Novocherkassk Museum of the History of the Don Cossacks, Novočerkassk 37478
Novodevičij Monastyr, Moskva 37389
Novohradská Galéria, Lučenec 38289
Novohradské Múzeum, Lučenec 38290
Novokuzneckij Chudožestvennyj Muzej, Novokuzneck 37481
Novokuzneckij Kraevedeskij Muzej, Novokuzneck 37482
Novorossijskij Gosudarstvennyj Istoričeskij Muzej Zapovednik, Novorossijsk 37483
Novorossiysk State Historical Preserve, Novorossijsk 37483
Novosibirsk Museum of the Nature, Novosibirsk 37487
Novosibirsk Picture Gallery, Novosibirsk 37486
Novosibirsk Regional Museum, Novosibirsk 37488
Novosibirskaja Kartinnaja Galereja, Novosibirsk 37486
Novosibirskij Muzej Prirody, Novosibirsk 37487
Novosibirskij Oblastnoj Kraevedčeskij Muzej, Novosibirsk 37488
Novotroick History Museum, Novotroick 37491
Novotroickij Istoriko-kraevedčeskij Muzej, Novotroick 37491
Novum Forum - Schule der Sinne, Neumarkt in Steiermark 02439
Nový Zamok Banská Štiavnica, Banská Štiavnica 38192
Nowata County Historical Society Museum, Nowata 51448
Nowosądecka Mała Galeria, Nowy Sącz 34960
Nowra Museum, Nowra 01384
Noyes and Read Gallery and Herndon Gallery, Yellow Springs 54330
Noyes Art Gallery, Lincoln 50104
The Noyes Museum of Art, Oceanville 51493
NRW-Forum Düsseldorf, Düsseldorf 18002
NRW-Forum Kultur und Wirtschaft, Düsseldorf 18003
NS-Dokumentationszentrum der Stadt Köln, Köln 19626
NSCC Art Gallery, Seattle 53007
NTT InterCommunication Center, Tokyo 29453
Nubbekullens Hembygdsmuseum, Motala 40757
Nubia Museum, Aswan 10481
Nubien Museum, Aswan 10482
Núcleo Arqueológico da Rua dos Correeiros, Lisboa 35728
Núcleo de Arte Sacra, Évora 35589
Núcleo do Moinho de Maré de Corroios, Corroios 35551
Núcleo Etnográfico de Cedros, Cedros 35519
Núcleo Museológico da Bemposta, Bemposta 35465
Núcleo Museológico da Capela do Espírito Santo dos Mareantes, Sesimbra 35933
Núcleo Museológico da Casa do Infante, Porto 35862
Núcleo Museológico da Cultura do Pastel dos Açores, Ribeira Chã 35889
Núcleo Museológico da Faculdade de Engenharia, Porto 35863
Núcleo Museológico da Faculdade de Farmácia, Porto 35864
Núcleo Museológico da Junta de Freguesia da Orada, Borba 35474
Núcleo Museológico da Misericórdia de Aveiro, Aveiro 35448
Núcleo Museológico da Nacional, Lisboa 35729
Núcleo Museológico da Paróquia da Beirã, Beirã 35461
Núcleo Museológico da Quinta do Prado, Celorico de Basto 35521
Núcleo Museológico da Roda dos Expostos, Almeida 35418
Núcleo Museológico das Ruínas de Milreu, Faro 35599
Núcleo Museológico das Salinas Naturais de Rio Maior, Rio Maior 35891
Núcleo Museológico de Alverca, Alverca do Ribatejo 35427
Núcleo Museológico de Amieira do Tejo, Nisa 35781
Núcleo Museológico de Arte Sacra, Nisa 35782
Núcleo Museológico de Cachopo, Cachopo 35491
Núcleo Museológico de Espinho, Espinho 35566
Núcleo Museológico de Metrologia, Évora 35590
Núcleo Museológico de S. Cucufate, Vila de Frades 35986
Núcleo Museológico do Castelo do Lindoso, Braga 35484
Núcleo Museológico do Centro Cultural de Borba, Borba 35475
Núcleo Museológico do Hospital de Santo António dos Capuchos, Lisboa 35730
Núcleo Museológico do IBTAM, Funchal 35619
Núcleo Museológico do Lagar de Azeite, Mafra 35752
Núcleo Museológico do Lagido, São João 35919
Núcleo Museológico do Santuário do Senhor Jesus da Pedra, Óbidos 35786
Núcleo Museológico do Tempo, Santarém 35910
Núcleo Museológico dos Escuteiros, Torres Vedras 35973
Núcleo Museológico dos Moinhos de Carreço, Carreço 35508
Núcleo Museológico Etnográfico José Ramires Saraiva, Ervidel 35562
Núcleo Museológico Medieval / Moderno, Almada 35416
Núcleo Rural de Ervidel, Ervidel 35563
Núcleos Museológicos da Lagoa, Lagoa 35650
Nuijamaan Museo, Lappeenranta 11047
Nukarin Koulumuseo, Nukari 11150
Nukke- ja Pukumuseo - Hatanpää Mansion, Tampere 11355
Nukkemuseo Suruton, Savonlinna 11302
Nukketeatteri- ja Lelumuseo Unelma, Pieksämäki 11203
Nukumuuseum, Tallinn 10639
Number 1 Royal Crescent Museum, Bath 43409
Numismatic Museum, Beijing 07577
Numismatic Museum, Kathmandu 31778
Numismatic Museum, Reykjavík 24025
Numismatic Museum, Athens, Athinai 22946
Nunatta Katersugaasivia Allagaateqarfialu, Nuuk 23262
Nunatta Sunakkutaangit Museum, Iqaluit 06020
Nundah and Districts Historical Museum, Nundah 01385
Nuova-Icona, Venezia 28204
Nuovo Museo Provinciale Sannitico, Campobasso 25636
Nurmeksen Museo, Nurmes 11152
Nurmon Museo, Nurmo 11153
Nurom Hat Museum, Suame 22863
Nurse Bennett Heritage House, Daniel's Harbour 05672
Nursery School Museum, Martonvásár 23692
NUS Museum, Singapore 38178
Nussknackermuseum, Neuhausen / Erzgebirge 20620
Nutcote Museum, Neutral Bay 01358
Nuutajärvi Glass Museum, Nuutajärvi 11154
NVA Museum Prora in der Kulturkunststatt Prora, Prora 21134
Ny Ålesund By- og Gruvemuseum, Ny-Ålesund 33706
Ny Carlsberg Glyptotek, København 10145
Nyahokwe Ruins Site Museum, Inyanga 54672
Nyborg og Omegns Museer, Nyborg 10199
Nyírségi Erdészeti Zrt. Debreceni Erdészet - Bánki Tájház, Debrecen 23519
NYK Kaiun Hakubutsukan, Yokohama 29604
NYK Maritime Museum, Yokohama 29604
Nykarleby Museum och Konstsamlingen Kalevala, Nykarleby 11156
Nylander Museum, Caribou 47242
Nýlistasafnið Nýlo, Reykjavík 24028
Nyman House and Pharmacy Museum, Tohmajärvi 11379
Nymindegab Museum, Nørre Nebel 10194
Nymindegab Redningsbådsmuseum, Nørre Nebel 10195
Nynäs Slott, Tystberga 40990
Nynäshamns Järnvägsmuseum, Nynäshamn 40783
Nyoirin-ji Homotsukan, Yoshino 29628
Nyoirin-ji Treasure House, Yoshino 29628
Nyslotts Landskapsmuseum, Nyslott 11157
Nyströmska Gården, Köping 40648
Nyúzó Gáspár Fazekas Tájház, Tiszafüred 23931
Nyúzó Gáspár Pottery Region House, Tiszafüred 23931
NZH Vervoers Museum, Haarlem 32299

O

O Art Museum, Tokyo 29454
O Bijutsukan, Tokyo 29454
O. Dovbush Museum, Ivano-Frankivs'k 43043
O. Henry Museum, Austin 46517
O. Kelaptrishvili Museum, Tbilisi 16377
O. Kobyljanska Literary Memorial Museum, Černivci 43012
O Museu do Marajó, Cachoeira do Arari 04377
O. Shovkunenko Museum of Fine Art, Cherson 43022
Oak Creek Pioneer Village, Oak Creek 51452
Oak Hall, Niagara Falls 06447
Oak House Museum, West Bromwich 45967
Oak Ridge Art Center, Oak Ridge 51458
Oakdale Museum and History Center, Oakdale 51459
Oakham Castle, Oakham 45308
Oakland Aviation Museum, Oakland 51469
Oakland Museum of California, Oakland 51470
Oaklands Historic House Museum, Murfreesboro 50914
Oakleigh and District Museum, Oakleigh 01389
Oakleigh House Museum, Mobile 50771
Oakley Pioneer Museum, Oakley 51475
Oaks House Museum, Jackson 49564
Oakville Galleries, Centennial, Oakville 06495
Oakville Galleries, Gairloch, Oakville 06496
Oakville Museum at Erchless Estate, Oakville 06497
Oakwell Hall Country Park, Birstall 43519
Het Oale Meestershuus, Slagharen 32780
Oale Smederie, Hellendoorn 32358
Oasis Gallery, Marquette 50500
Oatlands, Leesburg 50020
Ob Luang, Chiang Mai 42569
Obalne Galerije - Galerija Loža, Koper 38433
Obalne Galerije - Mestna Galerija, Piran 38491
Oban War and Peace Museum, Oban 45310
Obara-mura Kami no Bijutsukan, Obara 29084
Obara Paper Art Museum, Obara 29084
Obecné múzeum a Galéria Jana Hálu, Važec 38385
Obecní Dům, Praha 09783
Obelisk Museum, Al-Matariyya 10466
Oberer Torturm-Ausstellung, Marbach am Neckar 20180
Oberfränkisches Bauernhofmuseum, Zell, Oberfranken 22783
Oberfränkisches Textilmuseum, Helmbrechts 19119
Oberharzer Bergwerksmuseum, Clausthal-Zellerfeld 17709
OberhausMuseum, Passau 20993
Oberhofen Castle, Oberhofen am Thunersee 41697
Oberkärntner Brauchtums- und Maskenmuseum, Oberdrauburg 02452
Oberkrain Museum, Jesenice 38425
Oberkrain Museum - Birthplace of Dr. France Prešeren, Žirovnica 38544
Oberkrain Museum - The Birthplace of Finžgar, Žirovnica 38545
Oberkrain Museum - The Birthplace of Matije Čop, Žirovnica 38543
Oberkrain Museum - The Liznjek House, Kranjska Gora 38439
Oberlausitzer Sechsstädtebund- und Handwerksmuseum Löbau, Löbau 20013
Oberlin College Gallery, Cleveland 47636
Oberlin Heritage Center, Oberlin 51479
Oberösterreichische Fotogalerie, Linz 02321
Oberösterreichische Landesmuseen, Linz 02322
Oberösterreichischer Musealverein, Linz 54699
Oberösterreichischer Steingarten, Vorchdorf 02876
Oberösterreichisches Feuerwehrmuseum St. Florian, Sankt Florian 02638
Oberösterreichisches Jagdmuseum, Sankt Florian 02639
Oberösterreichisches Schulmuseum, Bad Leonfelden 01759
Oberpfälzer Fischereimuseum, Tirschenreuth 22062
Oberpfälzer Flußspat-Besucherbergwerk Reichhart-Schacht, Stulln 21925
Oberpfälzer Freilandmuseum Neusath-Perschen, Nabburg 20542
Oberpfälzer Handwerksmuseum, Rötz 21361
Oberpfälzer Künstlerhaus, Schwandorf 21664
Oberpfälzer Volkskundemuseum, Burglengenfeld 17660
Oberrheinische Narrenschau, Kenzingen 19518
Oberrheinisches Bäder- und Heimatmuseum, Bad Bellingen 16679
Oberrheinisches Tabakmuseum, Mahlberg 20133
Oberschlesisches Landesmuseum, Ratingen 21192
Oberschwäbische Galerie Kloster Heiligkreuztal, Altheim 16507
Oberschwäbisches Museumsdorf Kreisfreilichtmuseum Kürnbach, Bad Schussenried 16831
Obersimmentaler Heimatmuseum, Zweisimmen 42127
Oberweißbacher Bergbahn, Mellenbach-Glasbach 20273
Obinitsa Põhikooli Muuseum, Obinitsa 10599
Obir-Tropfsteinhöhlen, Bad Eisenkappel 01743
Objedinenie Fotocentr Sojuza Žurnalistov, Moskva 37390
Oblasnyj Istoryčnyj Muzej im. V. Tarnovs'kogo, Černigiv 43007
Oblasnyj Krajeznavčyj Muzej, Žytomyr 43214
Oblasnyj Krajeznavčyj Muzej, Cherson 43024
Oblasten Istoričeski Muzej, Jambol 05043
Oblastnaja Kartinnaja Galereja Obraz, Kaluga 37001
Oblastní Galerie, Liberec 09624
Oblastní Galerie Vysočiny v Jihlavě, Jihlava 09562
Oblastní Muzeum, Most 09667
Oblastní Muzeum v Chomutově, Chomutov 09478
Oblastní Muzeum v Lounech, Louny 09641
Oblastnoj Chudožestvennyj Muzej, Petropavlovsk-Kamčatskij 37555
Oblastnoj Istoriko-Kraevedčeskij Muzej, Aktöbe 29673
Oblastnoj Kraevedčeskij Muzej, Ekaterinburg 36900
Oblastnoj Muzej im. N.M. Prževalskogo, Prževalsk 29964
Oblastnoj Muzej Izobrazitelnych Iskusstv, Semey 29730
Oblastnoj Muzej Literatury i Iskusstva im. Buchar Žyrau, Pavlodar 29717
Oblastnoj Muzejnyj Centr, Sankt-Peterburg 37716
Oboronna Bašta Malachova Kurgana, Sevastopol 43179
Obrazárna Pražského Hradu, Praha 09784
Obščinski Istoričeski Muzej, Etropole 05028
Observatoriemuseet, Stockholm 40923

Observatory Museum, Grahamstown 38632
Observatory Museum of the Royal Swedish Academy of Sciences, Stockholm 40923
Obstbaumuseum, Werder 22487
Obstmuseum, Sörup 21794
Obu Municipal Folk Art Museum, Obu 29086
Obu-shiritsu Mingeikan, Obu 29086
Óbuda Picture Gallery, Budapest 23457
Óbudai Múzeum, Budapest 23456
Óbudai Pincegaléria, Budapest 23457
Óbudai Társaskör Galéria, Budapest 23458
Obuvnické Muzeum, Zlín 09942
OCAD Student Gallery, Toronto 07125
Ocak Köyü Özel Müzesi, Kemaliye 42877
Occupation Tapestry Gallery and Maritime Museum, Saint Helier 45573
Ocean City Art Center, Ocean City 51486
Ocean City Historical Museum, Ocean City 51487
Ocean County Historical Museum, Toms River 53528
Ocean Star Offshore Drilling Rig & Museum, Galveston 48850
Oceania Design, Wellington 33256
Oceanographic Museum, Nha Trang 54635
Oceanographic Museum and Aquarium, Gdynia 34658
Océanopolis, Brest 12165
Oceanside Museum of Art, Oceanside 51492
Océarium du Croisic, Le Croisic 13578
Ocmulgee National Monument, Macon 50368
Oconto County Historical Society Museum, Oconto 51494
OCT-LOFT Art Terminal, Shenzhen 08223
The Octagon Center for the Arts, Ames 46259
The Octagon House, Hudson 49413
Octagon House, Watertown 53976
The Octagon Museum, Washington 53912
Octava del Corpus, Peñalsordo 39829
Octave Chanute Aerospace Museum, Rantoul 52249
October Railway Central Museum, Sankt-Peterburg 37628
Odalstunet Gårdsmuseum, Skarnes 33830
Odapark, Venray 32885
Odawara Castle History Museum and Dungeon, Odawara 29088
Odawara-jo Rekishi Kenbunkan no Tenshukaku, Odawara 29088
Oddělení dějin tělesné výchovy a sportu, Praha 09785
Oddentunet, Os i Østerdalen 33712
Odder Museum, Odder 10208
Oddział Literacki im. Marii Dąbrowskiej w Russowie, Russów 35088
Oddział Zbiorów Graficznych i Kartograficznych, Kraków 34815
O'Dell House Museum, Annapolis Royal 05343
Odense City Museum, Odense 10209
Odenwälder Freilandmuseum, Walldürn 22311
Odenwald- & Spielzeugmuseum, Michelstadt 20321
Oderlandmuseum, Bad Freienwalde 16733
Oderljunga Mölla, Perstorp 40807
Odes'kyj Chudožnij Muzej, Odesa 43158
Odes'kyj Istoriko-krajeznavčyj Muzej, Odesa 43159
Odes'kyj Literaturno-memorial'nyj Muzej K. Paustovskogo, Odesa 43160
Odes'kyj Literaturnyj Muzej, Odesa 43161
Odes'kyj Muzej Zachidnogo ta Schidnogo Mystectva, Odesa 43162
Odessa Archaeological Museum, Odesa 43163
Odessa Fine Arts Museum, Odesa 43158
Odessa Local History Museum, Odesa 43159
Odessa Local History Museum - Memorial of the Heroic Defense of Odessa, Odesa 43155
Odessa Museum of Literature, Odesa 43161
Odessa State Museum of European and Oriental Art, Odesa 43162
Odigia Ikonen-Museum, Den Haag 32097
Odinani Museum, Awka 33316
Odoevsky Gallery, Ekaterinburg 36886
Odontological Museum, London 44947
Odontology Museum Prof. Salles Cunha, Rio de Janeiro 04730
Odsherreds Kulturhistoriske Museum, Nykøbing Sjælland 10206
Odsherreds Kunstmuseum, Asnæs 09969
Odsherreds Museum, Asnæs 09970
ODTÜ Müzesi, Ankara 42707
Odyssée des Bâtisseurs, Alma 05322
Odyssey Maritima Discovery Museum, Seattle 53008
Ödemiş Müzesi, Ödemiş 42920
Ödenburger Heimatmuseum, Bad Wimpfen 16868
Öjskog Park Pharmacy Museum, Närpes 11135
Öjskogsparkens Museiapoteket, Närpes 11135
Oekumenische Sammlung Gertrud Weinhold 'Das Gottesjahr und seine Feste', Oberschleißheim 20835
Öle-Museum, Münsingen 41661
Ölmühle Dörzbach, Dörzbach 17885
Ölmühle Michelau, Rudersberg 21430
Ölmühle Neckargröningen, Remseck am Neckar 21270
Ölmühlenmuseum, Ottweiler 20960
Der Ölschlag, Kirchberg in Tirol 02168
Önningeby Museum, Mariehamn 11101
Önningebymuseet, Mariehamn 11101
Öntödei Múzeum, Budapest 23459
Örebro County Museum, Örebro 40789
Örebro Konsthallen, Örebro 40788
Örebro Läns Museum, Örebro 40789
Oerka Irene Verbeek Museum, Raard 32678
Örnsköldsviks Museum & Konsthall, Örnsköldsvik 40794
Örökség Galéria, Budapest 23460
Oertijdmuseum De Groene Poort, Boxtel 31998
Östasiatiska Museet, Stockholm 40924
Österreichisches Brückenbaumuseum, Edelsbach bei Feldbach 01840
Östergötland County Museums, Linköping 40684
Östergötlands Länsmuseum, Linköping 40684
Österlens Museum, Simrishamn 40841
Österreichische Filmgalerie, Krems 02231
Österreichische Nationalbibliothek, Wien 03045
Oesterreichischer Alpenverein-Museum, Innsbruck 02127
Österreichischer Skulpturenpark, Unterpremstätten 02842
Österreichisches Donau- und Fischereimuseum, Orth an der Donau 02474
Österreichisches Felsbildermuseum, Spital am Pyhrn 02750
Österreichisches Filmmuseum Wien, Wien 03046
Österreichisches Forstmuseum Silvanum, Großreifling 02010
Österreichisches Freilichtmuseum Stübing, Stübing bei Graz 02790
Österreichisches Freimaurermuseum, Schloß Rosenau 02708
Österreichisches Gartenbaumuseum, Wien 03047
Österreichisches Gebäckmuseum, Wels 02917
Österreichisches Gesellschafts- und Wirtschaftsmuseum, Wien 03048
Österreichisches Jüdisches Museum, Eisenstadt 01862
Österreichisches Kriminalmuseum, Scharnstein 02700
Österreichisches Luftfahrtmuseum, Feldkirchen bei Graz 01883
Österreichisches Museum für Volkskunde, Wien 03049
Österreichisches Papiermacher- und Druckereimuseum, Steyrermühl 02779
Österreichisches Pfahlbaummuseum und Museum Mondseeland, Mondsee 02406
Österreichisches Spiele Museum, Leopoldsdorf im Marchfelde 02295
Österreichisches Sprachinselmuseum, Wien 03050
Österreichisches Theatermuseum, Wien 03051
Österreichisches Wandermuseum und Roseggers Waldschule, Alpl 01706
Österreichisches Zuckermuseum, Tulln 02832
Östersund City Museum, Östersund 40798
Östra Södermanlands Museijärnväg, Mariefred 40733
Oeteldonks Gemintemuzejum, 's-Hertogenbosch 32376
Ötztaler Heimat- und Freilichtmuseum, Längenfeld 02251
Övermark Hembygdsmuseum, Övermark 11159
Övertorneå Hembygdsmuseum, Övertorneå 40800
Övralids Lantbruksmuseum, Motala 40758
O'Fallon Historical Museum, Baker 46533
Ofen- und Keramikmuseum Velten, Velten 22217
Ofenmuseum und Baumgalerie, Allmersbach im Tal 16474
Ofenplattensammlung der Schwäbische Hüttenwerke, Aalen 16414
Oficina Museu da Universidade Estácio de Sá, Rio de Janeiro 04765
Ofoten Museum - Museum Nord, Narvik 33696
Ogden Historical Society Museum, Spencerport 53202
Ogden House, Fairfield 48481
Ogden Union Station Museums, Ogden 51503
Oginskių Kultūros Istorijos Muziejus, Rietavas 30345
Oglethorpe University Museum of Art, Atlanta 46447
Ogni Moskvy - Muzej Istorii Gorodskogo Osveščenija, Moskva 37391
Ogniwo Polish Museum, Winnipeg 07372
Ogre History and Art Museum, Ogre 30052
Ogres Vēstures un Mākslas Muzejs, Ogre 30052
Ogunquit Museum of American Art, Ogunquit 51507
Ohara Art Reference Center, Kobe 28738
Ohara Bijutsukan, Kurashiki 28789
Ohara Museum of Art, Kurashiki 28789
Ohef Sholom Temple Archives, Norfolk 51369
Ohi Museum & 10th Ohi Chozaemon Ware Gallery, Kanazawa 28662
Ohio Ceramic Center, Roseville 52462
Ohio Historical Society Museum, Columbus 47757
Ohio Museums Association, Columbus 55174
Ohio Railway Museum, Worthington 54307
Ohio River Museum, Marietta 50476
Ohio University Art Gallery, Saint Clairsville 52536
The Ohr-O'Keefe Museum of Art, Biloxi 46812
Ohsabanan Museijärnväg, Växjö 41039
The Ohsha'joh Museum of Art, Hiroshima 28542
Oikawa Museum of Sound Instruments, Higashi-kurume 28504
Oil and Gas Exhibition Centre, Mina Al-Fahal 33986
Oil Museum of Canada, Oil Springs 06503
Oil Patch Museum, Russell 52487
Oil Sands Discovery Centre, Fort McMurray 05817
Oinoussian Maritime Museum, Oinoussai 23127
Oise Living Museum, Beauvais 11908
Oisu Municipality Museum, Oisu 10601
Õisu Tehnikumi Muuseum, Halliste 10567
Oisu Valla Muuseum, Oisu 10601
Oita-kenritsu Geijutsu Kaikan, Oita 29096
Oita-kenritsu Rekishi Hakubutsukan, Usa 29564
Oita Prefectural Art Hall, Oita 29096
Oita Prefectural Museum of History, Usa 29564
Ojai Center for the Arts, Ojai 51510
Ojai Valley Museum, Ojai 51511
Ojala Museum, Pattijoki 11195
Ojāra Vācieša Memoriālais Muzejs, Rīga 30100
Ojars Vacietis Memorial Museum, Riga 30100
Ojeblikket, København 10146
OK Offenes Kulturhaus Oberösterreich, Linz 02323
Okains Bay Maori and Colonial Museum, Okains Bay 33163
Okanagan Military Museum, Kelowna 06055
Okanoyama Bijutsukan, Nishiwaki 29077
Okanoyama Museum of Art, Nishiwaki 29077
Okaukuejo Museum, Okaukuejo 31748
Okawa Bijutsukan, Kiryu 28703
Okawa Museum of Art, Kiryu 28703
Okawa-suji Samurai Residence, Kochi 28745
Okaya Museum, Okaya 29097
Okayama-kenritsu Bijutsukan, Okayama 29100
Okayama-kenritsu Bizen-yaki Hakubutsukan, Bizen 28390
Okayama-kenritsu Hakubutsukan, Okayama 29101
Okayama Orient Museum, Okayama 29102
Okayama Prefectural Bizen Ceramic Museum, Bizen 28390
Okayama Prefectural Museum, Okayama 29101
Okayama Prefectural Museum of Art, Okayama 29100
Okayama-shiritsu Oriento Bijutsukan, Okayama 29102
Okazaki Municipal Art Museum, Okazaki 29105
Okazaki-shiritsu Bijutsukan, Okazaki 29105
Okefenokee Heritage Center, Waycross 54002
Okeresztény Mauzóleum, Pécs 23770
Okinawa-ken Heiwa Kinen Shiryokan, Itoman 28603
Okinawa-kenritsu Hakubutsukan & Gendai Bijutsukan, Naha 29014
Okinawa Prefectural Museum & Art Museum, Naha 29014
Okinawa Prefectural Peace Memorial Museum, Itoman 28603
Okkupasjonsmuseet, Eidsvoll Verk 33478
Okladnikov Museum of Archaeology of Khabarovsk, Chabarovsk 36857
Oklahoma City Museum of Art, Oklahoma City 51521
Oklahoma Firefighters Museum, Oklahoma City 51522
Oklahoma Forest Heritage Center Forestry Museum, Broken Bow 47038
Oklahoma Museum of African American Art, Oklahoma City 51523
Oklahoma Museum of History, Oklahoma City 51524
Oklahoma Museums Association, Oklahoma City 55175
Oklahoma Route 66 Museum, Clinton 47661
Oklahoma Territorial Museum, Guthrie 49108
Økomuseum Grenseland, Sarpsborg 33819
Økomuseum Samsø, Tranebjerg 10300
Okražen Istoričeski Muzej, Chaskovo 05005
Okražen Istoričeski Muzej, Montana 05095
Okražen Istoričeski Muzej, Pernik 05111
Okražen Istoričeski Muzej, Plovdiv 05130
Okražen Istoričeski Muzej, Sliven 05169
Okražen Istoričeski Muzej, Smoljan 05171
Okražen Istoričeski Muzej, Veliko Tărnovo 05257
Okražen Istoričeski Muzej, Vidin 05263
Okražen Istoričeski Muzej, Vraca 05265
Okražen Muzej na Văzraždaneto i Nacionalno-osvoboditelnite Borbi, Plovdiv 05131
Okražna Chudožestvena Galerija, Kardžali 05046
Okražna Chudožestvena Galerija, Stara Zagora 05210
Okražna Chudožestvena Galerija, Vraca 05266
Okražna Chudožestvena Galerija Dimităr Dobrovič, Sliven 05170
Okražna Chudožestvena Galerija Vladimir Dimitrov-Majstora, Kjustendil 05067
Okresní Muzeum, Děčín 09491
Okresní Muzeum, Litoměřice 09630
Okresní Muzeum, Vlašim 09910
Okresní Muzeum Praha-východ, Brandýs nad Labem 09413
Øksnes Bygdemuseum, Alsvåg 33383
Oktibbeha County Heritage Museum, Starkville 53258
Oku Kyodokan, Oku 29107
Oku Municipal Museum, Oku 29107
Okubank Group Museum, Helsinki 10806
Okukiyotsu Electric Power Museum, Yuzawa 29635
Okupatsioonimuuseum, Tallinn 10640
Okura Shukokan Museum of Fine Arts, Tokyo 29455
Okutama Watershed Museum, Okutama 29108
Okuuchi Bijutsukan, Toyonaka 29531
Okuuchi Museum, Toyonaka 29531
Okyo and Rosetsu Art Museum, Kushimoto 28797
Okyo no Rosetsu Bijutsukan, Kushimoto 28797
Olaf-Gulbransson-Museum für Graphik und Karikatur, Tegernsee 22018
Olana State Historic Site, Hudson 49411
Olav Bjaaland Museum, Morgedal 33681
Olav Holmegaards Samlinger, Mandal 33659
Olavinlinna Castle, Olavinlinna 11161
Old Aircraft Museum, Arnemuiden 31920
Old Arsenal, Vilnius 30431
Old Arsenal Museum, Solothurn 41883
Old Arts Gallery, Tshwane 38753
Old Automobiles Collection, Bordány 23377
Old Bank of New Brunswick Museum, Riverside-Albert 06759
Old Barracks Heritage Centre, Cahersiveen 24750
Old Barracks Museum, Trenton 53562
Old Bell Museum, Montgomery 45170
Old Bergen Museum, Bergen 33426
Old Bethpage Village Restoration, Old Bethpage 51532
Old Blythewood, Pinjarra 01415
Old Bohemia Historical Museum, Warwick 53833
Old Borroloola Police Station Museum, Borroloola 00879
Old Bridge House Museum, Dumfries 44038
Old Brown's Mill School, Chambersburg 47332
Old Brutus Historical Society Museum, Weedsport 54020
The Old Byre Heritage Centre, Dervaig 43977
Old Cable Station Museum, Apollo Bay 00806
Old Capitol, Iowa City 49518
Old Capitol Museum of Mississippi History, Jackson 49565
Old Carleton County Court House, Woodstock 07404
Old Carriages Museum, Roma 27465
Old Castle Jevišovice - Moravian Museum Brno, Jevišovice 09558
Old Castle Museum, Baldwin City 46540
Old Castle of Banská Štiavnica, Banská Štiavnica 38194
Old Cathedral Museum, Saint Louis 52594
Old City Hall and Prison of Bejucal, Bejucal 09038
Old City Park - The Historical Village of Dallas, Dallas 47937
Old Clock Museum, Pharr 51808
Old Colony Historical Museum, Taunton 53453
Old Constitution House, Windsor 54229
Old Conway Homestead Museum and Mary Meeker Cramer Museum, Camden 47194
Old Council Chambers Museum, Cleve 00983
Old Council Chambers Museum, Norton Summit 01380
Old Court House Museum, Drysdale 01044
Old Court House Museum, Durban 38609
Old Court House Museum, Guysborough 05922
Old Court House Museum-Eva Whitaker Davis Memorial, Vicksburg 53734
Old Courthouse Heritage Museum, Inverness 49514
Old Courthouse Museum, Santa Ana 52853
Old Courtyard Museum, Kibbutz Ein Shemer 25039
Old Cowtown Museum, Wichita 54147
Old Crofton School Museum, Crofton 05660
Old Crown Court and Cells, Dorchester 44002
Old Depot Museum, Ottawa 51629
Old Derbent - Museum-Preserve of History, Architecture and Art, Derbent 36874
Old Dominion Railway Museum, Richmond 52336
Old Dominion University Gallery, Norfolk 51370
Old Dorchester State Historic Site, Summerville 53363
Old Dubbo Gaol, Dubbo 01047
Old Dutch Parsonage, Somerville 53150
Old Economy Village Museum, Ambridge 46254
Old English Court Museum, Moskva 37394
Old Exchange and Provost Dungeon, Charleston 47374
Old Falls Village, Menomonee Falls 50590
Old Fire House and Police Museum, Superior 53379
Old Firehouse Museum, Greenville 49073
Old Forge, Hellendoorn 32358
Old Fort Bissell, Phillipsburg 51894
Old Fort Erie, Niagara Falls 06448
Old Fort Garland, Fort Garland 48621
Old Fort Harrod Museum, Harrodsburg 49189
Old Fort House Museum, Fort Edward 48619
Old Fort Jackson, Savannah 52943
Old Fort Johnson, Fort Johnson 48631
Old Fort Lauderdale Museum of History, Fort Lauderdale 48640
Old Fort Meade Museum, Fort Meade 48655

Old Fort Niagara, Youngstown 54351
Old Fort Number 4 Associates, Charlestown 47382
Old Fort Western, Augusta 46486
Old Fort William Historical Park, Thunder Bay 07078
Old Foundry Museum, Kragujevac 38107
Old Gala House, Galashiels 44261
Old Gaol Museum, Albany 00787
Old Gaol Museum, Buckingham 43666
Old Gaol Museum, York 54336
Old Government House, Fredericton 05838
Old Government House, Paramatta 01403
The Old Governor's Mansion, Frankfort 48741
Old Grammar School, Castletown 43777
Old Greer County Museum and Hall of Fame, Mangum 50429
The Old Guard Museum, Fort Myer 48660
Old Guildhall Museum, East Looe 44091
Old Haa, Yell 46093
Old Harbour Museum, Hermanus 38641
Old Highercombe Hotel Museum, Tea Tree Gully 01575
Old House, Hereford 44457
Old House Museum, Bakewell 43355
Old House Museum, Durban 38610
Old House of Keys, Castletown 43778
Old House of the Marquises of Campo Florido, San Antonio de los Baños 09272
Old Hoxie House, Sandwich 52840
Old Independence Regional Museum, Batesville 46627
Old Iron County Courthouse Museum, Hurley 49448
Old Ironworks in Josefov Valley, Adamov 09394
The Old Jail Art Center, Albany 46167
Old Jail House & Old Court House, Croydon 01014
Old Jail Museum, Albion 46171
The Old Jail Museum, Crawfordsville 47871
The Old Jail Museum, Warrenton 53831
The Old Jameson Distillery, Dublin 24814
Old Jefferson Town, Oskaloosa 51618
Old Kuopio Museum, Kuopio 11007
Old Lifeboat Museum, Poole 45406
Old Lighthouse Museum, Michigan City 50647
Old Lighthouse Museum, Stonington 53312
Old Lincoln County Courthouse Museum, Lincoln 50110
Old Log Church Museum, Whitehorse 07326
Old Log Jail and Chapel Museums, Greenfield 49055
Old Mackinac Point Lighthouse & Colonial Michilimackinac, Mackinaw City 50355
Old Manchester Museum, Manchester 50418
The Old Manse, Concord 47769
Old Market House, Galena 48828
Old Massett Village Council's Heritage, Old Massett 06507
Old Masters Picture Gallery, Dresden 17946
Old Meeting House, Barrington 05391
Old Melbourne Gaol and Penal Museum, Melbourne 01289
Old Merchant's House and Row 111 Houses, Great Yarmouth 44359
Old Mill Museum, Cimarron 47550
Old Mill Museum, San Marino 52820
Old Mine, Løkken Verk 33643
Old Mission Santa Ines Collection, Solvang 53143
Old Monterey Jail, Monterey 50805
Old Mornington Post Office Museum, Mornington 01323
Old Museum Arts Centre, Belfast 43443
Old Operating Theatre, Museum and Herb Garret, London 44948
Old Ordinary House Museum, Hingham 49298
Old Parliament House, Canberra 00949
Old Police Station Museum, Mount Barker 01326
Old Police Station Museum Brookton, Brookton 00908
Old Post Office Museum, Turriff 45883
The Old Prison, Victoria 30654
Old Prison Museum, Deer Lodge 48030
Old Prison of Jaruco, Jaruco 09188
Old Racing Collection, Dvůr Králové nad Labem 09501
Old Railway Station Museum, Northam 01375
Old Rectory Museum, Loughborough 45034
Old Religious Art Collection, Reşiţa 36561
Old Royal Palace, Praha 09794
Old Saint Edward's Anglican Church, Clementsport 05622
Old Saint Ferdinand's Shrine, Florissant 48572
Old Salem, Winston-Salem 54248
Old School, Kolín 09592
Old School House, York 54337
Old School Museum, Merimbula 01295
Old Schwamb Mill, Arlington 46349
Old Shawnee Town, Shawnee 53058
Old Slavonic Ringwall - National Cultural Monument, Mikulčice 09652
Old South Meeting House, Boston 46944
Old Spanish Fort Museum, Pascagoula 51725
Old Speech Room Gallery, London 44949
Old Springsure Hospital Museum, Springsure 01524
Old State Bank, Vincennes 53750
Old State Capitol, Springfield 53219
Old State House, Boston 46945
Old State House, Hartford 49198
Old State House Museum, Little Rock 50141
Old Stone Fort Museum Complex, Schoharie 52963
The Old Stone House, Washington 53913
The Old Stone House Museum, Brownington 47093
Old Stone House Museum, Windsor 54227
The Old Stone Jail Museum, Palmyra 51687
Old Stone Mill Museum, Delta 05693
Old Strathcona Model and Toy Museum, Edmonton 05751
Old Sturbridge Village, Sturbridge 53344
Old Tamba Pottery Museum, Sasayama 29230
Old Tavern Museum, Tuscaloosa 53619
Old Timers Traeger Museum, Alice Springs 00801
Old Town Hall Arts Centre, Hemel Hempstead 44447
Old Town Hall History House, Portland 01439
Old Town Hall Museum, Fifield 48540
Old Town Hall Museum, Thamesville 07071
Old Town Museum, Old Town 51538
The Old Town - National Open Air Museum of Urban History and Culture, Århus 09963
Old Trail Museum, Choteau 47546
Old Village Hall Museum, Lindenhurst 50118
Old Ware House Gallery, Maryborough, Queensland 01268
Old West Museum, Sunset 53375
Old West Wax Museum, Thermopolis 53493
Old Westbury Gardens, Old Westbury 51539
Old Wool and Grain Store Museum, Beachport 00843
Old World Wisconsin, Eagle 48230
Old York Historical Museum, York 54338
Old York Road Historical Society, Jenkintown 49620
Old Zhi Display Center, Nanchang 08031
Olde Colonial Courthouse, Barnstable 46613
'T Olde Ras, Doesburg 32135
Oldenburg-Museum, Unna 22176
Oldenburger Fahrradmuseum, Oldenburg 20899
Oldest Building of Panevezys, Panevėžys 30333
The Oldest House Museum, Key West 49765
Oldest House Museum Complex, Saint Augustine 52523
Oldest Stone House Museum, Lakewood 49908
Oldham County History Center, La Grange 49842
Oldman River Cultural Centre, Brocket 05485
Oldtimer Museum, Sternberg 21902
Oldtimercentrum Select, Joure 32433
Oldtimermuseum, Blindenmarkt 01789
Oldtimermuseum, Kleinwetzdorf 02195
Oldtimermuseum De Rijke, Oostvoorne 32631
Oldtimermuseum De Ronkel, Kloosterburen 32451
Oldtimermuseum Kröpfl, Hartberg 02052
Oldtimermuseum Meßkirch, Meßkirch 20306
Oldtimermuseum Stift Ardagger, Ardagger 01726
Dat ole Hus, Aukrug 16642
Ole Rømer Museet, Taastrup 10295
Olens Radiomuseum, Olen 03921
Ølgod Museum, Ølgod 10222
Oliemolen de Zoeker, Zaandam 32982
Oliemolen Het Pink, Koog aan de Zaan 32456
Oliewenhuis Art Museum, Bloemfontein 38569
Olin Art Gallery, Gambier 48852
Olin Fine Arts Center, Washington 53942
Olin Hall Galleries, Salem 52671
Olive DeLuce Art Gallery, Maryville 50524
Olive Grove and Oil Museum, Baena 38938
Olive Hyde Art Gallery, Fremont 48785
Olive Schreiner House, Cradock 38600
Oliver Cromwell's House, Ely 44173
Oliver H. Kelley Farm, Elk River 48344
Oliver Heritage Society Museum, Oliver 06511
Oliver House Museum, Penn Yan 51764
Oliver House Museum, Saint Martinville 52605
Oliver Tucker Historic Museum, Beverly 46787
Olkijoen Rauhanpirtti, Pattijoki 11194
Olkijoki Treaty Museum, Pattijoki 11194
Olle Olsson-Huset, Solna 40871
Olmstead Place State Park, Ellensburg 48360
Olmsted County Museum, Rochester 52389
Olofsfors Bruksmuseum, Nordmaling 40767
Olomouc Museum of Art, Olomouc 09689
Oloneckaja Kartinnaja Galereja, Olonec 37495
Oloneckij Nacionalnyj Muzej Karelov-Livvikov im. N.T.Prilukin, Olonec 37496
Olorgesailie Prehistoric Site Museum, Olorgesailie 29751
Olustee Battlefield, Olustee 51541
Olustvere Muuseum, Olustvere 10602
Olympic Hall of Fame and Museum, Calgary 05530
The Olympic Museum, Lausanne 41550
Olympic Museum, Wellington 33257
Olympic National Park Visitor Center, Port Angeles 52027
Oma-Freese-Huus, Dornum 17909
Omachi Alpine Museum, Omachi 29109
Omaha Center for Contemporary Art, Omaha 51554
Omaha Children's Museum, Omaha 51555
Oman Museum, Muscat 33990
Oman Natural History Museum, Muscat 33991
Omani-French Museum Bait Fransa, Muscat 33992
Omaruru Museum, Omaruru 31749
Ombalantu Baobab Tree, Ombalantu 31750
Ome City Museum, Ome 29113
Ome Kimono Museum, Ome 29111
Ome Municipal Museum of Art, Ome 29112
Ome-shiritsu Bijutsukan, Ome 29112
Ome-shiritsu Hakubutsukan, Ome 29113
Omega Gallery, Jefferson City 49615
Omega Museum, Biel 41230
Omi Natural History Museum, Omi 29115
Omnisphere and Science Center, Wichita 54148
Omoide Museum, Kyoto 28858
Omsk Fine Art Museum Mikhail Vrubel, Omsk 37501
Omsk Literature Museum F.M. Dostoevsky, Omsk 37500
Omskij Gosudarstvennyj Istoriko-kraevedčeskij Muzej, Omsk 37499
Omskij Gosudarstvennyj Literaturnyj Muzej im. F.M. Dostoevskogo, Omsk 37500
Omskij Oblastnoj Muzej Izobrazitelnych Iskusstv im. M.A. Vrubelja, Omsk 37501
Omuta Coal Industry and Science Museum, Omuta 29117
Omuta-shi Tanko oyo Kagakukan, Omuta 29117
On the Hill Cultural Arts Center, Yorktown 54345
On Yedinci Yüzyıl Osmalı Evi, Bursa 42745
Ondergronds Museum, Kanne 03749
Onderwijsmuseum Educatorium, Ootmarsum 32633
Oneida Community Mansion House, Oneida 51560
Oneida County Historical Society Museum, Utica 53677
Oneida Nation Museum, De Pere 47999
Ongerup and Needilup District Museum, Ongerup 01392
Onggi Folk Museum and Institute, Seoul 29911
Onion Museum, Checheng 42160
Onishi Seiwemon Museum, Kyoto 28859
Onomichi City Museum of Art, Onomichi 29120
Onomichi-shiritsu Bijutsukan, Onomichi 29120
Onondaga Historical Association Museum, Syracuse 53402
Ons Museum, Giessenburg 32249
Ons Museum, Wintelre 32955
Onslow County Museum, Richlands 52309
Onslow Goods Shed Museum, Onslow 01393
Ontario County Historical Society Museum, Canandaigua 47214
Ontario Electric Railway Museum, Milton 06291
Ontario Museum Association, Toronto 54744
Ontario Police College Museum, Aylmer 05368
Ontario Provincial Police Museum, Orillia 06513
Ontario Science Centre, Toronto 07126
De Ontdekhoek Kindermuseum, Rotterdam 32722
Ontmoetingscentrum de Veldkei, Expositie, Havelte 32325
Ontonagon County Historical Society Museum, Ontonagon 51571
Onyang Folk Museum, Asan 29770
Onyang Minsok Pakmulgwan, Asan 29770
Oorlogs- en Verzetsmuseum Johannes Post, Ridderkerk 32685
Oorlogsmuseum, Bloemfontein 38570
Oorlogsverzetsmuseum Rotterdam, Rotterdam 32723
Oost-Indisch Huis, Amsterdam 31876
Oostends Historisch Museum De Plate, Oostende 03935
Opatska Riznica Sv. Marka, Korčula 08897
Opelousas Museum, Opelousas 51574
Opelousas Museum of Art, Opelousas 51575
Open-Air Dept. of the Jēkabpils History Museum, Jēkabpils 30013
Open-Air Ethnographic Collection, Szenna 23870
Open-Air Ethnographic Museum, Nagyvázsony 23730
Open-Air Ethnographic Museum, Tihany 23927
Open-air Ethnographic Museum in Palotavaros, Székesfehérvár 23857
Open Air Mining Museum, Banská Štiavnica 38188
Open-air Mining Museum, Tatabánya 23921
Open-Air-Museum, Karlskrona 40637
Open Air Museum, Kihniö 10955
Open Air Museum, Stará Lubovňa 38362
Open Air Museum, Villa Carlos Paz 00673
Open Air Museum, Wolsztyn 35320
Open-Air Museum at the Pleterje Carthusian Monastery, Drča 38417
Open Air Museum for Archeology, Mrówki 34931
Open Air Museum Jurjev Farmstead, Dobrnič 38412
Open-Air Museum of Ethnography, Gabrovo 05029
Open Air Museum of Ethnography, Tbilisi 16378
Open Air Museum of Fishing, Jūrmala 30027
Open-Air Museum of Folk Architecture, Bardejovské Kúpele 38197
Open Air Museum of Folk Architecture and Folklore, Humenné 38261
Open-Air Museum of Lithuania, Rumšiškės 30347
Open Air Museum of Oceanic Fishery, Gdynia 34661
Open Air Museum of Old Japanese Farmhouses, Toyonaka 29530
Open Air Museum of Sóstó, Sóstógyógyfürdő 23824
Open Air Museum of the Oaş Land, Negreşti-Oaş 36488
Open Air Museum of the Sibiu Grove, Sibiu 36612
Open Air Museum Old Dobrich, Dobrič 05017
Open-Air Museum Rogatec, Rogatec 38508
Open Air Railway Museum, Zduńska Wola 35368
Open-Air Water-Power Museum, Dimitsana 22984
Open Eye Gallery, Liverpool 44743
Open Haven Museum, Amsterdam 31877
Open Museum, Migdal-Tefen 25067
Open Museum, Omer 25075
Open Space Gallery, Victoria 07256
Open Studio, Toronto 07127
Openbaar Vervoer Museum, Doetinchem 32137
Openbaar Vervoer Museum, Rotterdam 32724
Openlucht Laagveenderij Museum Damshûs, Nij Beets 32585
Openluchtmuseum Bachten de Kupe, Alveringem 03269
Openluchtmuseum De Duinhuisjes, Rockanje 32693
Openluchtmuseum Ellert en Brammert, Schoonoord 32767
Openluchtmuseum Erve Kots, Lievelde 32502
Openluchtmuseum Front Line Hooghe, Ieper 03730
Openluchtmuseum Het Hoogeland, Warffum 32922
Openluchtmuseum Nieuw Amsterdam, Nieuw Amsterdam 40399
Openluchtmuseum Ootmarsum, Ootmarsum 32634
Opera Library & Museum, Paris 14574
Operetta Theatre Museum, Moskva 37371
Oplenova Hiša, Srednja Vas v Bohinju 38526
Oplen's House, Srednja Vas v Bohinju 38526
Opolian Silesia Museum, Opole 34986
Opotiki Museum, Opotiki 33164
Oppdal Bygdemuseum, Oppdal 33707
Optik Industrie Museum, Rathenow 21190
Optisches Museum der Ernst-Abbe-Stiftung Jena, Jena 19385
Opus 40 and the Quarryman's Museum, Saugerties 52927
Ópusztaszer National Historical Memory Park, Opusztaszer 23736
Ópusztaszeri Nemzeti Történeti Emlékpark, Opusztaszer 23736
Orang Asli Museum, Melaka 30606
Orange County Historical Museum, Hillsborough 49291
Orange County Museum of Art, Newport Beach 51332
Orange County Museum of Art - Orange Lounge, Costa Mesa 47840
Orange County Regional History Center, Orlando 51592
Orange Empire Railway Museum, Perris 51778
Orange Regional Gallery, Orange 01395
Orangedale Railway Museum, Orangedale 06512
Orangerie, Gera 18596
Orangerie, 's-Hertogenbosch 32377
Orangerie Belvedere, Weimar 22413
Orangerieschloss und Turm, Potsdam 21102
Orangetown Historical Museum, Pearl River 51744
Oranien-Nassau-Museum, Diez 17858
Oranien-Nassauisches Museum, Dillenburg 17861
Orava Gallery, Dolný Kubín 38250
Orava Gallery, Námestovo 38312
Oravská Galéria, Dolný Kubín 38250
Oravská Galéria, Námestovo 38312
Oravské Múzeum Pavla Orszägha Hviezdoslava, Dolný Kubín 38251
Orawski Park Etnograficzny, Zubrzyca Górna 35379
Orbán Ház, Szilvásvárad 23895
Orbán House, Szilvásvárad 23895
Orcas Island Historical Museum, Eastsound 48272
Orchard Gallery, Londonderry 45023
Orchard House, Concord 47770
Orchard Park Historical Society Museum, Orchard Park 51584
Order of Saint John Museum, Christchurch 33075
Ordrupgaard, Charlottenlund 09986
Ordsall Hall Museum, Salford 45600
Ordu Devlet Güzel Sanatlar Galerisi, Ordu 42921
Ordu Etnografya Müzesi, Ordu 42922
Ordu State Fine Art Gallery, Ordu 42921
Ordubad Regional History Studies, Ordubad 03191
Ore-Petrographic Museum of the Institute of Geology of Ore Deposits, Petrography, Mineralogy and Geochemistry of RAS, Moskva 37405
Øregaard-Museum, Hellerup 10058
Oregon Air and Space Museum, Eugene 48437
Oregon Coast History Center, Newport 51316
Oregon Electric Railway Historical Society Museum, Lake Oswego 49889
Oregon Electric Railway Museum, Salem 52670
Oregon Historical Society, Portland 52082
Oregon-Jerusalem Historical Museum, Oregon 51585
Oregon Military Museum, Clackamas 47570
Oregon Museum of Science and Industry, Portland 52083
Oregon Museums Association, Portland 55176
Oregon Sports Hall of Fame and Museum, Portland 52084

Oregon Trail Museum, Gering 48897
Oregon Trail Regional Museum, Baker City 46535
Orel Regional Museum of Fine Art, Orël 37512
Orenburg Museum of Fine Arts, Orenburg 37515
Orenburg Regional Museum, Orenburg 37514
Orenburgskij Oblastnoj Kraevedčeskij Muzej, Orenburg 37514
Orenburgskij Oblastnoj Muzej Izobrazitelnych Iskusstv, Orenburg 37515
Organ Pipe Cactus Museum, Ajo 46144
Organization of Military Museums of Canada, Gloucester 54745
Organization of Museums, Monuments and Sites of Africa OMMSA, Accra 54863
Orgel-Art-Museum Rhein-Nahe, Windesheim 22596
Orgel Museum, Moji-ko, Kitakyushu 28714
Orgelbaumuseum, Ostheim vor der Rhön 20940
Orgelbaumuseum, Weißenfels 22459
Orgelmuseet i Fläckebo, Salbohed 40829
Orgelmuseum Borgentreich, Borgentreich 17477
Orgelmuseum Kelheim, Kelheim 19501
Orhan Kemal Müzesi, İstanbul 42833
Oriamu Museum, Izumiotsu 28615
Oriel Davies Gallery, Newtown, Powys 45257
Oriel Gallery, Mold 45164
Oriel Mostyn Gallery, Llandudno 44760
Oriel Ynys Môn, Llangefni 44766
Oriental Institute Museum, Chicago 47502
Oriental Kazakh Regional History Museum, Petropavl 29720
Oriental Museum Durham, Durham 44074
Orientalisches Münzkabinett Jena (OMJ), Jena 19386
Orientteppich-Museum, Hannover 19004
Original- und Abgußsammlung der Universität Trier, Trier 22101
Orillia Museum of Art and History, Orillia 06514
Orimattila Local History Museum, Orimattila 11164
Orimattilan Kotiseutumuseo, Orimattila 11164
Orimattilan Taidemuseo, Orimattila 11165
Orimattilas Art Museum, Orimattila 11165
Orinasukan, Kyoto 28860
Oripään Kotiseutumuseo, Oripää 11167
Oriskany Battlefield, Oriskany 51588
Orissa State Museum, Bhubaneshwar 24097
Oriveden Paltanmäen Museo, Orivesi 11168
Orivesi Paltanmäki Museum, Orivesi 11168
Orkdal Bygdemuseum, Svorkmo 33897
Orkney Farm and Folk Museum, Birsay 43518
Orkney Farm and Folk Museum, Harray 44397
Orkney Museum, Kirkwall 44615
Orkney Wireless Museum, Kirkwall 44616
Orland Historical Museum, Orland 51589
Orlando Brown House, Frankfort 48742
Orlando Museum of Art, Orlando 51593
Orlando Science Center, Orlando 51594
Orleans House Gallery, Twickenham 45888
Orlické Muzeum, Choceň 09476
Orlogsmuseet, København 10147
Orlovskij Oblastnoj Kraevedčeskij Muzej, Orël 37511
Orlovskij Oblastnoj Muzej Izobrazitelnych Iskusstv, Orël 37512
Ormeau Baths Gallery, Belfast 43444
Ormond Memorial Art Museum, Ormond Beach 51600
Ornithology and Entomology Museum, Cairo 10512
Ornitological Collection, Reghin 36558
Ornunga Museum, Vårgårda 41051
Orono Museum of Art, Bangor 46598
Oroszlány Miners' Museum Foundation, Oroszlány 23743
Oroszlányi Bányász Múzeum Alapítvány, Oroszlány 23743
Oroville Chinese Temple, Oroville 51606
Őrség Popular Monuments - Szalafo-Pityerszer, Szalafő 23833
Őrségi Népi Műemlékegyüttes - Szalafo-Pityerszer, Szalafő 23833
Orskij Istoriko-kraevedčeskij Muzej, Orsk 37516
Országos Földtani Múzeum, Budapest 23461
Országos Pedagógiai Könyvtár és Múzeum, Budapest 23462
Országos Széchényi Könyvtár, Budapest 23463
Országos Színháztörténeti Múzeum és Intézet, Budapest 23464
Orta Dogu Teknik Universitesi Sanat Galleri, Ankara 42708
Orthodox Church Museum of Finland, Kuopio 11010
Orthodox Ecclesiastic Art Collection, Kecskemét 23653
Orthopedics Museum, Bangkok 42554
Ortodoksinen Kirkkomuseo, Kuopio 11010
Ortodox Egyházművészeti Gyűjtemény, Kecskemét 23653
Orton Geological Museum, Columbus 47758
Orts- und Weinbaumuseum, Neftenbach 41673
Orts- und Wohnmuseum, Marthalen 41606
Ortsgeschichtliche Sammlung, Bovenden 17490
Ortsgeschichtliche Sammlung, Oberrieden 41698
Ortsgeschichtliche Sammlung Rebstein, Rebstein 41748
Ortsgeschichtliches Museum, Siegenburg 21758
Ortskundliches Museum Jois, Jois 02145
Ortsmusem Wittenbach im Schloss Dottenwil, Wittenbach 42043
Ortsmuseum, Amriswil 41112
Ortsmuseum, Andwil 41116
Ortsmuseum, Bergün 41199
Ortsmuseum, Binningen 41235
Ortsmuseum, Brachwitz-Saalkreis 17493
Ortsmuseum, Bülach 41268
Ortsmuseum, Bütschwil 41271
Ortsmuseum, Häggenschwil 41467
Ortsmuseum, Jona 41499
Ortsmuseum, Kaltbrunn 41500
Ortsmuseum, Laax 41522
Ortsmuseum, Liesberg 41560
Ortsmuseum, Meilen 41618
Ortsmuseum, Merenschwand 41628
Ortsmuseum, Nackenheim 20544
Ortsmuseum, Neuburg am Rhein 20596
Ortsmuseum, Nürensdorf 41688
Ortsmuseum, Oftringen 41704
Ortsmuseum, Opfikon 41712
Ortsmuseum, Rafz 41740
Ortsmuseum, Roggwil 41771
Ortsmuseum, Rüti 41782
Ortsmuseum, Sankt Stephan 41815
Ortsmuseum, Schänis 41827
Ortsmuseum, Thalwil 41933
Ortsmuseum, Unterengstringen 41958
Ortsmuseum, Untersiggenthal 41962
Ortsmuseum, Vaz 41979
Ortsmuseum, Vnà 41992
Ortsmuseum, Wängi 41995
Ortsmuseum, Walenstadt 41998
Ortsmuseum, Weikertschlag 02900
Ortsmuseum Albisrieden, Zürich 42100
Ortsmuseum Altes Rathaus, Balgach 41151
Ortsmuseum Altstetten, Altstetten 41109
Ortsmuseum Beringen, Beringen 41200
Ortsmuseum Brittnau, Brittnau 41258
Ortsmuseum Diessenhofen, Diessenhofen 41328
Ortsmuseum Dietikon, Dietikon 41330
Ortsmuseum Eglisau, Eglisau 41341
Ortsmuseum Erlenbach, Erlenbach 41355
Ortsmuseum Eschlikon, Eschlikon 41361
Ortsmuseum Flawil, Flawil 41372
Ortsmuseum Frenkendorf, Frenkendorf 41382
Ortsmuseum für Heimatkunde, Schlieren 41838
Ortsmuseum Gaiserwald, Sankt Josefen 41810
Ortsmuseum Grächen, Grächen 41446
Ortsmuseum Hinwil, Hinwil 41483
Ortsmuseum Höngg, Zürich 42101
Ortsmuseum Kefiturm, Belp 41197
Ortsmuseum Kilchberg, Kilchberg (Zürich) 41503
Ortsmuseum Küsnacht, Küsnacht 41519
Ortsmuseum mit Karl-Jauslin-Sammlung, Muttenz 41671
Ortsmuseum Mollis, Mollis 41632
Ortsmuseum Mühle, Maur 41617
Ortsmuseum Neunkirch, Neunkirch 41683
Ortsmuseum Oberes Bad, Marbach 41605
Ortsmuseum Oberuzwil, Oberuzwil 41700
Ortsmuseum Oetwil, Oetwil am See 41702
Ortsmuseum Pratteln, Pratteln 41733
Ortsmuseum Rüschlikon, Rüschlikon 41781
Ortsmuseum Schleitheim, Schleitheim 41836
Ortsmuseum Schmitten, Schmitten, Albula 41840
Ortsmuseum Spreitenbach, Spreitenbach 41896
Ortsmuseum Steinmaur, Sünikon 41919
Ortsmuseum Sust, Horgen 41488
Ortsmuseum Trotte, Arlesheim 41125
Ortsmuseum Urdorf, Urdorf 41965
Ortsmuseum Ursulastift, Gerstetten 18617
Ortsmuseum Vaz, Lenzerheide 41556
Ortsmuseum Vechigen, Boll 41244
Ortsmuseum Wallisellen, Wallisellen 41999
Ortsmuseum Weiach, Weiach 42004
Ortsmuseum Wetzikon, Wetzikon 42012
Ortsmuseum Wiesendangen, Wiesendangen 42015
Ortsmuseum Wila, Wila 42018
Ortsmuseum Wilchingen, Wilchingen 42019
Ortsmuseum Wilderswil, Wilderswil 42021
Ortsmuseum Wollishofen, Zürich 42102
Ortsmuseum Zollikon, Zollikon 42062
Ortsmuseum zur Farb, Stäfa 41900
Ortsmuseum zur Hohlen Eich, Wädenswil 41994
Ortssammlung Wettelsheim, Treuchtlingen 22090
Ortsstube, Bolligen 41245
Orugoru no Chiisana Hakubutukan, Tokyo 29456
Oružejnaja Palata, Moskva 37392
Orvos- és Gyógyszerészettörténeti Kiállítás, Kecskemét 23654
Orvostörténeti és Patikakiállítás, Karcag 23641
Orwell Corner Historic Village, Orwell 06519
Osage County Historical Society, Lyndon 50332
Osage County Historical Society Museum, Pawhuska 51733
Osage Historic Village, Lamar 49911
Osaka Castle Museum, Osaka 29140
Osaka City Museum of Modern Art, Osaka 29152
Osaka-furistu Sayamaike Hakubutsukan, Osakasayama 29159
Osaka-furitsu Gendai Bijutsu Senta, Osaka 29141
Osaka-furitsu Senboku Koko Shiryokan, Sakai 29197
Osaka Human Rights Museum, Osaka 29142
Osaka International Peace Center, Osaka 29143
Osaka-kenritsu Chikatsu-Asuka Hakubutsukan, Kanan 28653
Osaka-kenritsu Chu Hakabutsukan, Mino 28910
Osaka-kenritsu Kokuritsu Bijutsukan, Osaka 29144
Osaka-kenritsu Waha Kamigata Bungeikan, Osaka 29145
Osaka Lottery Dream Museum, Osaka 29146
Osaka Maritime Museum, Osaka 29137
Osaka Municipal Museum of Art, Osaka 29151
Osaka Museum of History, Osaka 29147
Osaka Museum of Housing and Living, Osaka 29148
Osaka Museum of Natural History, Osaka 29149
Osaka Nippon Mingeikan, Suita 29296
Osaka Prefectural Chikatsu-Asuka Museum, Kanan 28653
Osaka Prefectural Insect Museum, Mino 28910
Osaka Prefectural Museum of Kamigata Comedy and Performing Arts, Osaka 29145
Osaka Prefectural Museum of Yayoi Culture, Izumi, Osaka 28614
Osaka Prefectural Sayamaike Museum, Osakasayama 29159
Osaka Regional Contemporary Art Center, Osaka 29141
Osaka Science Museum, Osaka 29150
Osaka Senboku Area Archaelogical Museum, Sakai 29197
Osaka-shi Kagaku Hakubutsukan, Osaka 29150
Osaka-shiritsu Bijutsukan, Osaka 29151
Osaka-shiritsu Kindai Bijutsukan Kensetsu Jumbishitsu, Osaka 29152
Osaka-shiritsu Toyo Toji Bijutsukan, Osaka 29153
Osakarovka Historical and Regional Museum, Osakarovka 29715
Osakarovskij Rajonnyj Istoriko-Kraevedčeskij Muzej, Osakarovka 29715
Osbane Museet, Fana 33489
Osborn-Jackson House, East Hampton 48248
Osborne House, East Cowes 44084
Oscar Anderson House Museum, Anchorage 46290
Oscar Getz Museum of Whiskey History, Bardstown 46608
Oscar Howe Art Center, Mitchell 50764
Oscarsborg Festningsmuseum, Oscarsborg 33713
Oscarshall Slott, Oslo 33747
Osen Bygdemuseum, Osen 33714
Osen Bygdetun, Steinsdalen 33880
Osgoode Township Museum, Vernon, Ontario 07239
Osh United Historical Cultural Museum-Preserve, Oš 29963
Oshawa Community Museum and Archives, Oshawa 06522
Oshawa Sydenham Museum, Oshawa 06523
Oshkosh Public Museum, Oshkosh 51615
Osinskij Kraevedčeskij Muzej, Osa 37517
Osipov-Woolf House-Museum, Puškinskie Gory 37586
Oskar Luts Majamuuseum, Tartu 10654
Oskar Schlemmer Archiv, Stuttgart 21956
Oskar Schlemmer Theatre Estate and Collection, Oggebbio 26977
Oskara Kalpaka Muzejs Airīšu Piemiņas Vietā, Zirņi 30152
Oskarshamns Sjöfartsmuseum, Oskarshamn 40803
Osler Historical Museum, Osler 06525
Osman Hamdi Bey Müzesi, Gebze 42788
Osmani Museum, Sylhet 03237
Osorkhonai Akademiiai Ulumi Jumhurii Tojikiston, Dušanbe 42497
Osorkhonai Bostonshinosii Khujand, Chudžand 42492
Osorkhonai Davlatii Ta"rikhi - Kishvarshinosi va Sa"nati Tasvirii Jumhurii Tojikiston ba nomi Kamoliddin Behzod, Dušanbe 42498
Osorkhonai Jumkhuriiavii Ta"rikhi - Kishvarshinosii ba nomi Abuabdullo Rudaki, Pandžakent 42505
Osorkhonai Ta"rikhi - Kishvarshinosii Isfara, Isfara 42499
Osorkhonai Ta"rikhi - Kishvarshinosii Shahri Uro-Teppa, Uro-Teppa 42506
Osorkhonai Ta"rikhi - Kishvarshinosii Viloiati Khujand, Chudžand 42493
Osorkhonai Ta"rikhi - Kishvarshinosii Viloiati Kŭlob, Kŭlob 42503
Osorkhonai Tibb va Giëhhoi Shifobakhshi Tojikiston, Isfara 42500
Osorkhonai Viloiatii Ta"rikhi - Kishvarshinosii, Chorog 42491
Osoyoos Art Gallery, Osoyoos 06527
Osoyoos Museum, Osoyoos 06528
Ośrodek Biograficzny Komisji Turystyki Górskiej, Kraków 34816
Osrodek Muzealny Bydgoskiego Wezla Kolejowego, Bydgoszcz 34580
Ośrodek Propagandy Sztuki, Łódź 34886
Ossining Historical Society Museum, Ossining 51621
Oŝskij Objedinennyj Istoriko-Kulturnyj Muzej-Zapovednik, Oš 29963
Ost-, Südostdeutsche Heimatstuben, Ulm 22168
Ostarrichi-Kulturhof, Neuhofen an der Ybbs 02429
Ostasiatika-Sammlung Ehrich, Hattingen 19040
Ostdeutsche Heimatsammlung, Ravensburg 21205
Ostdeutsche Heimatstube, Bad Zwischenahn 16878
Ostdeutsche Heimatstube, Fellbach 18306
Ostdeutsche Heimatstube, Weinstadt 22439
Ostdeutsche Heimatstube, Zweibrücken 22829
Ostdeutsche Kultur- und Heimatstuben mit Schönbacher Stube, Heppenheim 19130
Ostdeutsche Museumsräume in der Heimatstube, Kaiserslautern 19413
Osterburgsammlung, Bischofsheim an der Rhön 17363
Ostereimuseum, Sonnenbühl 21820
Osterley Park House, Isleworth 44542
Osterøy Museum, Lonevåg 33648
Osterville Historical Museum, Osterville 51622
Osterzgebirgsmuseum, Lauenstein 19850
Ostfriesen Bräu, Großefehn 18787
Das Ostfriesische Teemuseum und Norder Heimatmuseum, Norden 20730
Ostfriesisches Landesmuseum, Emden 18141
Ostfriesisches Landwirtschaftsmuseum, Krummhörn 19732
Ostfriesisches Schulmuseum Folmhusen, Westoverledingen 22518
Osthaus Museum Hagen, Hagen 18860
Osthofentor-Museum, Soest 21797
Ostholstein-Museum, Eutin 18285
Ostindiefararen Götheborg, Göteborg 40553
Ostpreußisches Landesmuseum, Lüneburg 20102
Ostravské Muzeum, Ostrava 09699
Ostrobothnia Australis, Vaasa 11425
Ostrobothnian Museum, Vaasa 11426
Ostrogožskij Rajonnyj Istoriko-chudožestvennyj Muzej im. I.N. Kramskogo, Ostrogožsk 37520
Ostrovsky House Museum, Moskva 37166
Østsamisk Museum, Neiden 33697
Østsjællands Museum, Stevns Museum, Store Heddinge 10285
Osuuspankkimuseo, Helsinki 10806
Oswego Historical Museum, Oswego 51623
Oswestry Transport Museum, Oswestry 45323
Oświatowe Muzeum, Szczecin 35164
Oszla Region House, Cserépfalu 23508
Oszlai-Tájház, Cserépfalu 23508
Ōta Kinen Bijutsukan, Tokyo 29457
Ota-kuritsu Mingeikan, Tokyo 29458
Ōta Memorial Museum of Art, Tokyo 29457
Otago Military Museum, Dunedin 33094
Otago Museum, Dunedin 33095
Otago Settlers Museum, Dunedin 33096
Otago Vintage Machinery Club Museum, Dunedin 33097
Otakou Marae Museum, Dunedin 33098
Otani Memorial Art Museum, Nishinomiya, Nishinomiya 29074
Otar Lordkipanidze Vani Archaeology Miseum-Reserve, Vani 16389
Otaru Municipal Art Museum, Otaru 29163
Otaru Museum, Otaru 29161
Otaru Music Boxes Museum & Antique Museum, Otaru 29160
Otaru Science Center, Otaru 29162
Otaru-shi Hakubutsukan, Otaru 29161
Otaru-shi Seishonen Kagakugijutsukan, Otaru 29162
Otaru-shiritsu Bijutsukan, Otaru 29163
Otaru Venice Museum, Otaru 29164
Otautau and District Local History Museum, Otautau 33165
Otello Kalesi, Gazimağusa 33353
Othello's Tower, Gazimağusa 33353
Otley Museum, Otley 45324
Otoe County Museum of Memories, Syracuse 53396
OTP Bank - Monetary Historical Collection, Budapest 23465
Otrar Memlekettik Archaeologicalik Korik Muzej, Čimkent 29699
Otrar State Archaeological Museum-Preserve, Šaulder 29726
Otrar State Archaeological Park Museum, Čimkent 29699
Otrarskij Gosudarstvennyj Archeologičeskij Muzej-Zapovednik, Šaulder 29726
Otraženie - Muzej Voskovych Figur, Moskva 37393
Otsu-e Bijutsukan, Otsu 29165
Otsu Picture Art Museum, Otsu 29165
Otsuka Bijutsukan, Naruto 29044
Otsuka Museum of Art, Naruto 29044
Ottawa Art Gallery, Ottawa 06552
Ottawa County Historical Museum, Port Clinton 52031

O

Ottawa House By-The-Sea Museum, Parrsboro 06571
Ottawa Scouting Museum, Ottawa 51628
Ottenby Naturum, Degerhamn 40486
Otter Tail County Museum, Fergus Falls 48534
Otterton Mill Centre and Working Museum, Budleigh Salterton 43669
Otterup Museum, Otterup 10224
Ottmar-Mergenthaler-Museum, Bad Mergentheim 16784
Otto Dill-Museum, Neustadt an der Weinstraße 20673
Otto-Ditscher-Galerie, Neuhofen 20622
Otto-Dix-Haus, Gera 18597
Otto-Dix-Haus Hemmenhofen, Gaienhofen 18532
Otto-Flath-Kunsthalle, Bad Segeberg 16838
Dat Otto Huus, Emden 18142
Otto-Klages-Sammlung, Königslutter am Elm 19655
Otto-König von Griechenland-Museum, Ottobrunn 20957
Otto Lilienthal Gedenkstätte, Gollenberg / OT Stölln 18705
Otto-Lilienthal-Museum, Anklam 16551
Otto Modersohn Museum, Ottersberg 20954
Otto-Pankok-Museum, Bad Bentheim 16683
Otto Pankok Museum, Hünxe 19293
Otto-Schwabe-Museum, Hochheim am Main 19205
Otto-Ubbelohde-Haus, Lahntal 19768
Otto-von-Guericke-Museum in der Lukasklause, Magdeburg 20128
Otto Wagner Hofpavillion Hietzing, Wien 03052
Otto Wagner Pavillion Karlsplatz, Wien 03053
Ottoman House, Bursa 42745
Ottoman Imperator Museum, Gebze 42788
Ottoman Poetry Museum, İstanbul 42815
Otwock Agricultural Museum, Otwock 35000
Oud Amelisweerd Museum, Bunnik 32018
Ouddorps Raad- en Polderhuis, Ouddorp 32650
De Oude Aarde, Giethoorn 32255
Oude Ambachten en Speelgoedmuseum, Terschuur 32819
Oude Bakkerij, Medemblik 32544
De Oude Drostdy, Tulbagh 38765
Oude Kerk Volksmuseum, Tulbagh 38766
't Oude Raadhuis, Beek en Donk 31953
Het Oude Raadhuis, Leerdam 32472
Het Oude Raadhuis, Oud-Beijerland 32645
Het Oude Raadhuis, Urk 32840
Het Oude Slot Heemstede, Heemstede 32328
Het Oude Stadhuis, Hasselt 32321
Oudheidkamer, Bruinisse 32015
Oudheidkamer, Den Burg 32068
Oudheidkamer, Koudekerk 32458
Oudheidkamer, Steenwijk 32803
Oudheidkamer, Wijngaarden 32949
Oudheidkamer Beilen, Beilen 31957
Oudheidkamer Bleiswijk, Bleiswijk 31973
Oudheidkamer Boerderij Strunk, Raalte 32676
Oudheidkamer Brederwiede, Vollenhove 32904
Oudheidkamer Buisjan Enter, Enter 32216
Oudheidkamer Dantumadeel, Zwaagwesteinde 33021
Oudheidkamer de Oude Pastory, Heiloo 32348
Oudheidkamer Doorn, Doorn 32144
Oudheidkamer Eemnes, Eemnes 32175
Oudheidkamer Geervliet, Geervliet 32237
Oudheidkamer Heerde, Heerde 32329
Oudheidkamer Hellevoetsluis, Hellevoetsluis 32362
Oudheidkamer Hoolt'n, Holten 32402
Oudheidkamer Horst, Horst 32416
Oudheidkamer IJsselstreek, Wijhe 32947
Oudheidkamer in het Stadhuis, Bolsward 31980
Oudheidkamer Leiderdorp, Leiderdorp 32491
Oudheidkamer Lemster Fiifgea, Lemmer 32498
Oudheidkamer Leshuis, Dedemsvaart 32053
Oudheidkamer Lunteren, Lunteren 32512
Oudheidkamer/ Museum Mr. Andreae, Kollum 32454
Oudheidkamer Nederlandse Kaap-Hoornvaarders, Hoorn 32411
Oudheidkamer Nunspeet, Nunspeet 32609
Oudheidkamer Renswoude, Renswoude 32682
Oudheidkamer Ridderkerk, Ridderkerk 32686
Oudheidkamer Rozenburg, Rozenburg 32736
Die Oudheidkamer tot Medenblick, Medemblik 32545
Oudheidkamer Wassenaar, Wassenaar 32928
Oudheidkamer Weststellingwerf, Wolvega 32962
Oudheidkamer Wieringerwaard het Polderhuis, Wieringerwaard 32945
Oudheidkamer Willem Van Strijen, Zevenbergen 33001
Oudheidkundeige Verzameling, Schinveld 32758
Oudheidkundig Museum, Sint Michielsgestel 32773
Oudheidkundig Museum Arnemuiden, Arnemuiden 31921
Oudheidkundig Museum Sint-Janshospitaal, Damme 03535
Oudheidkundig Museum van het Begijnhof, Turnhout 04101
Oudheidkundige Verzameling, Sluis 32788
Oudheidskamer Vreeswijk, Nieuwegein 32578
Oulaisten Kotiseutumuseo, Oulainen 11169
Oulu Car Museum, Oulu 11171
Oulu City Art Museum, Oulu 11172
Oulun Automuseo, Oulu 11171
Oulun Taidemuseo, Oulu 11172
Oulun Yliopiston Eläinmuseo, Oulu 11173
Oulun Yliopiston Geologinen Museo, Oulu 11174
Oulun Yliopiston Kasvimuseo, Oulu 11175
Oulunsalon Kotiseutumuseo, Oulunsalo 11179
Oundle Museum, Oundle 45325
Our Heritage Museum, Adelaide 38550
Our House State Memorial, Gallipolis 48840
Our India Project Museum, Narendrapur 24335
Our Lady of Mercy Museum, Port-au-Port 06629
Oushuu-shokoku Mingeikan, Teshikaga 29360
Out of the Blue Gallery, Leek 44671
Outagamie Museum, Appleton 46331
Outback Regional Gallery, Winton 01657
Outeniqua Railway Museum, George 38625
Outer Space Museum, Garni 00699
Outjo Museum, Outjo 31753
Outlook and District Heritage Museum and Gallery, Outlook 06558
Outokummun Kaivosmuseo, Outokumpu 11180
Ouvrage de la Ligne Maginot, Leutenheim 13727
Ouyen Local History Resource Centre, Ouyen 01397
Ovens Natural Park and Museum, Riverport 06758
Overbeck-Museum, Bremen 17559
Overbeck's Museum, Salcombe 45596
Overfield Tavern Museum, Troy 53576
Overholser Mansion, Oklahoma City 51525
Overland Trail Museum, Sterling 53284
Overstreet-Kerr Historical Farm, Keota 49747
Óvodamúzeum, Martonvásár 23692
Owatonna Arts Center, Owatonna 51635
Owela Display Centre, Windhoek 31766
Owen Sound Marine and Rail Museum, Owen Sound 06562
Owens Art Gallery, Sackville 06788
Owens-Thomas House, Savannah 52944
Owensboro Area Museum of Science and History, Owensboro 51639
Owensboro Museum of Fine Art, Owensboro 51640
Owls Head Transportation Museum, Owls Head 51641
Owo Museum, Owo 33343
Owyhee County Historical Museum, Murphy 50916
Oxford Bus Museum Trust, Long Hanborough 45026
Oxford Library Museum, Oxford 51645
Oxford Museum, Oxford 33168
Oxford Museum, Oxford 51646
The Oxford Story, Oxford 45335
Oxford University Museum of Natural History, Oxford 45336
Oxford University Press Museum, Oxford 45337
The Oxfordshire Museum, Woodstock 46067
Oxon Cove Park Museum, Oxon Hill 51659
Oyamazumi-jinja Kokuhokan, Omishima 29116
Oyamazumi Jinja Treasure House, Omishima 29116
Øyfjell Bygdemuseum, Vinje 33982
The Oyster and Maritime Museum of Chincoteague, Chincoteague 47539
Oyuncak Müzesi, Ankara 42709
Ozark Folk Center, Mountain View 50895
Ozaukee Art Center, Cedarburg 47308
Ozaukee County Historical Society Pioneer Village, Saukville 52930
Özbekiston Respublikasi San'at Akademiyasi, Toshkent 54483
Özbekiston San'at Davlat Muzeyi, Toshkent 54484
Özbekiston Tarikhi Davlat Muzeyi, Toshkent 54485
Ózd Town Museum, Ózd 23744
Ózdi Városi Múzeum, Ózd 23744
Ozu Yasujiro Seihunkan Museum, Matsusaka 28899

P

P.A. Krasikov House-Museum, Krasnojarsk 37068
P. and Al. Canellopoulos Museum, Athinai 22947
P.D. Korin House-Museum, Palech 37525
P.H. Sullivan Museum and Genealogy Library, Zionsville 54372
P.I. Tchaikovsky Museum-Estate, Votkinsk 37977
P.I. Tchaikovsky State Memorial House, Klin 37052
P.K. Koslov Memorial Apartment Museum, Sankt-Peterburg 37660
P.P. Chistyakov House-Museum, Puškin 37578
P. Ramlee Memorial, Kuala Lumpur 30584
P.S. 1 Contemporary Art Center, Long Island City 50185
P. Stulgos Lietuvių Tautinės Muzikos Muziejus, Kaunas 30263
P. Stulgos Lithuanian Folk Music Museum, Kaunas 30263
P.T. Boat Museum, Germantown 48899
P.V. Alabin History and Regional Museum, Samara 37622
P.W. Vorster Museum, Middelburg 38689
Paamiune Katersugaasivik, Paamiut 23263
Paarl Museum, Paarl 38700
Paattinen District Museum, Paattinen 11181
Paattisten Kotiseutumuseo, Paattinen 11181
Paavolan Kotiseutumuseo, Paavola 11182
PAC - Padiglione d'Arte Contemporanea, Milano 26717
Pacem in Terris, Warwick 53834
Pacific Arts Center, Seattle 53009
Pacific Asia Museum, Pasadena 51721
Pacific Coast Air Museum, Santa Rosa 52907
Pacific County Historical Museum, South Bend 53162
Pacific Design Center, West Hollywood 54066
Pacific Grove Art Center, Pacific Grove 51665
Pacific Grove Museum of Natural History, Pacific Grove 51666
Pacific Islands Museums Association, Suva 54792
Pacific Museum, Yongin 29946
Pacific Northwest Truck Museum, Brooks 47088
Pacific Science Center, Seattle 53010
Pacific University Museum, Forest Grove 48585
Packwood House, Lapworth 44641
Packwood House Museum, Lewisburg 50030
Paço dos Duques de Bragança, Guimarães 35642
Paço Imperial, Rio de Janeiro 04766
Padasjoen Kotiseutumuseo, Padasjoki 11183
Padasjoen Museo Nyystölä, Nyystölä 11158
Padasjoen Museo Torittu, Torittu 11382
Paddle Steamer Industry Museum, Renmark 01457
Padre Burgos National Museum, Vigan 34523
Padstow Museum, Padstow 45341
Päämajamuseo, Mikkeli 11116
Pädagogisches Museum der Universität Bielefeld, Bielefeld 17348
Paekakariki Rail and Heritage Museum, Paekakariki 33170
Pängelanton-Eisenbahnmuseum, Münster 20517
Pärnu Muuseum, Pärnu 10604
Paeroa and District Historical Society Museum, Paeroa 33171
Pätärin Talomuseo, Ylijärvi 11480
PAF Museum, Karachi 34016
Paffenroter Dorfmuseum, Marxzell 20232
Pagode de Chanteloup, Amboise 11579
Pahiatua and District Museum, Pahiatua 33172
Pahl Museum, Mainhardt 20138
Paignton and Dartmouth Steam Railway, Paignton 45344
Paikkari Cottage, Sammatti 11298
Paikkarin Torppa, Sammatti 11298
Paimion Kotiseutumuseo, Paimio 11184
Paine Art Center, Oshkosh 51616
Paine-Gillam-Scott Museum, Saint Johns 52553
Paint Creek Center for the Arts Exhibitions, Rochester 52387
Painted Bride Art Center Gallery, Philadelphia 51863
Painting and Sculpture Museum, İzmir 42857
The Painting Center, New York 51247
Paintings and Sculpture Museum, Ankara 42710
Paipoonge Historical Museum, Thunder Bay 07079
Paisley Museum and Art Galleries, Paisley 45345
Paiwan Carving Gallery, Pingtung 42292
Paja Jovanovic Memorial Museum, Beograd 38072
Pajaro Valley Historical Museum, Watsonville 53986
Pakistan Air Force Museum Peshawar, Peshawar 34035
Pakistan Army Museum, Rawalpindi 34041
Pakistan Museum of Natural History, Islamabad 34010
Pákozdi Csata Emlékmüve, Pákozd 23746
Pakruojis Manor House of the Šiauliai Aušra Museum, Pakruojis 30325
Pakruojo Dvaro Rūmai, Pakruojis 30325
Paks Gallery, Paks 23747
Paksi Képtár, Paks 23747
Pal Kinizsi Castle Museum, Nagyvázsony 23728
Pál Molnár C. Studio Museum, Budapest 23452
Pál Péter Urban House, Szentes 23885
Pałac Radziwiłłów w Nieborowie, Nieborów 34942
Pałac Sztuki w Krakowie, Kraków 34817
Pałac w Rogalinie, Świątniki 35146
Palace and Museum of Antiquities, Padmanabhapuram 24339
The Palace Armoury, Valletta 30645
Palace Collection, Roskilde 10245
The Palace Museum, Beijing 07559
Palace Museum, Ulaanbaatar 31630
Palace Museum, Zanzibar 42523
Palace-Museum Betnava, Maribor 38475
Palace Museum of Peter the III at the Museum-Preserve, Lomonosov 37107
Palace Museum of Wilanów, The Royal Residence, Warszawa 35263
Palace of Holyroodhouse, Edinburgh 44143
Palace of Peter the Great, Strelna 37809
Palace of Physics, Kazanlăk 05061
Palace of Royal Jewelleries, Zizinia 10544
Palace of the Governors, Santa Fe 52885
Palace of the Grand Masters, Rhodos 23173
Palace of Vouni, Lefke 33366
Palace of Westminster, London 44950
Palacete del Embarcadero, Santander 40012
Palacio Arzobispal, Cusco 34168
Palácio Conde dos Arcos, Goiás 04481
Palacio Condesa de Lebrija, Sevilla 40076
Palácio da Bolsa, Porto 35865
Palácio das Laranjeiras, Rio de Janeiro 04767
Palácio de Estói, Estói 35572
Palacio de la Granja, Orihuela 39745
Palácio de Monserrate, Sintra 35950
Palacio de Montemuzo, Zaragoza 40351
Palacio de Viana, Córdoba 39210
Palacio del Mar, Donostia-San Sebastián 39254
Palacio del Mate, Posadas 00502
Palácio dos Coruchéus, Lisboa 35731
Palácio dos Marqueses de Fronteira, Lisboa 35732
Palácio Guanabara, Rio de Janeiro 04768
Palácio Gustavo Capanema, Rio de Janeiro 04769
Palacio La Quinta, Madrid 39585
Palácio Nacional da Pena, Sintra 35951
Palacio Nacional de Queluz, Queluz 35879
Palácio Nacional de Sintra, Sintra 35952
Palacio Real de Aranjuez, Aranjuez 38879
Palacio Real de El Pardo, El Pardo 39812
Palacio Real de Madrid, Madrid 39586
Palacio Real de Riofrío, Riofrío 39903
Palacio Real y Museos de Tapices, San Ildefonso 39958
Palacio Taranco, Montevideo 54424
Palácio Valenças, Sintra 35953
Palaeobotanisch Museum, Utrecht 32854
Palaeontological Museum University of Zurich, Zürich 42103
Paläontologisches Museum der Universität Zürich, Zürich 42103
Paläontologisches Museum Dr. Herbert Schaffer, Offenhausen 02470
Paläontologisches Museum München, München 20472
Paläontologisches Museum Nierstein, Nierstein 20720
Palaepaphos Museum, Kouklia 09364
Palæsamlingerne, Roskilde 10245
Palais Bénédictine, Fécamp 12915
Palais de Glace, Buenos Aires 00281
Palais de la Découverte, Paris 14742
Palais de la Miniature, Saint-Pourçain-sur-Sioule 15469
Palais de l'Ile, Annecy 11626
Palais de Tokyo, Paris 14743
Palais des Beaux-Arts, Bruxelles 03498
Le Palais des Beaux-Arts, Bruxelles 03499
Palais des Expositions Chinagora, Alfortville 11553
Palais des Papes, Avignon 11796
Le Palais du Chocolat, La Côte-Saint-André 13352
Palais du Roure, Avignon 11797
Palais du Tau, Reims 15006
Palais Ducal, Nevers 14419
Palais für aktuelle Kunst, Glückstadt 18664
Palais Grand Ducal, Luxembourg 30485
Palais Jacques Coeur, Bourges 12121
Palais Liechtenstein, Feldkirch 01881
Palais Minéral, Granville 13106
Palais Montcalm, Québec 06696
Palais Princier, Monaco 31614
Palais Synodal Musée, Sens 15720
Palamas Museum, Athinai 22948
Palamuse O. Lutsu Kihelkonnakoolimuuseum, Palamuse 10606
Palamuse Parish School Museum, Palamuse 10606
Palanga Amber Museum of the Lithuanian Art Museum, Palanga 30329
Palangos Gintaro Muziejus, Palanga 30329
Palata na Fizikata, Kazanlăk 05061
Palatinate Museum of the City of Heidelberg, Heidelberg 19076
Palatul Scroviștea, Periș 36513
Palaty Starogo Anglijskogo Dvora, Moskva 37394
Palau Ducal dels Borja, Gandía 39325
Palau Reial de l'Almudaina, Palma de Mallorca 39793
Palawan Museum, Puerto Princesa 34482
Palawan National Museum, Quezon 34484
Palawan State University Museum, Puerto Princesa 34483
Palazzina Marfisa, Ferrara 26120
Palazzo Apostolico ed Archivio Storico della Santa Casa, Loreto 26520
Palazzo Armeria, Zurrieq 30663
Palazzo Attems-Petzenstein, Gorizia 26353
Palazzo Barolo, Torino 28027
Palazzo Belgiojoso, Lecco 26479
Palazzo Belmonte Riso, Palermo 27079
Palazzo Bianco, Genova 26320
Palazzo Borromeo, Stresa 27912
Palazzo Bricherasio, Torino 28028
Palazzo Castelmur, Stampa 41903
Palazzo Castiglioni, Cingoli 25893
Palazzo del Principe, Genova 26321
Palazzo del Senato, Pinerolo 27226
Palazzo della Ragione, Padova 27050
Palazzo della Triennale, Milano 26718
Palazzo delle Esposizioni, Roma 27541
Palazzo Ducale, Genova 26322
Palazzo Ducale, Venezia 28205
Palazzo Falson Historic House Museum, Mdina 30632

Palazzo Foro Boario, Modena 26756
Palazzo Grassi, Venezia 28206
Palazzo Medici Riccardi, Firenze 26200
Palazzo Orlandi, Busseto 25570
Palazzo Parisio, Naxxar 30634
Palazzo Pfanner, Lucca 26537
Palazzo Pubblico, San Marino 38011
Palazzo Reale, Torino 28029
Palazzo Reale-Reggia di Caserta, Caserta 25705
Palazzo Rosso, Genova 26323
Palazzo Sandretta Re Rebaudengo, Guarene 26387
Palazzo Silva, Domodossola 26043
Palazzo Taffini d'Acceglio, Savigliano 27768
Palazzo Tursi, Genova 26324
Palazzo Vertemate Franchi, Piuro 27258
Palazzo Vescovile, Reggio Emilia 27392
Palazzo Zabarella, Padova 27051
Paleis aan de Blijmarkt, Zwolle 33027
Paleis Het Loo - Nationaal Museum, Apeldoorn 31915
Paleontological and Geological Museum of the University of Athens, Athinai 22949
Paleontological Museum, Asenovgrad 04972
Paleontological Museum, Tuxtla Gutiérrez 31514
Paleontologičen Muzej, Asenovgrad 04972
Paleontologičeskij Muzej im. Ju.A. Orlova, Moskva 37395
Paleontologisches Museum, Burg-Reuland 03507
Paleontologo-stratigrafičeskij Muzej pri Kafedre Istorii Geologii, Sankt-Peterburg 37717
Paleontology Museum, Moskva 37395
Paleorama - Museo Itinerante, Don Torcuato 00357
Paleozoological Museum of China, Beijing 07560
Palėpės Galerija, Kintai 30276
Palestinian Archaeological Museum, Birzeit 34052
The Palette and Chisel Academy, Chicago 47503
Palette & Chisel Academy of Fine Arts Gallery Exhibition, Chicago 47504
The Paley Center for Media, Beverly Hills 46792
Palio Mikro Chorio Museum, Mikro Chorio 23096
Pallant House Gallery, Chichester 43830
Palm Beach Institute of Contemporary Art, Lake Worth 49898
Palm Beach Photographic Center, Delray Beach 48047
Palm Springs Art Museum, Palm Springs 51681
Palma Rosa Museum, Hamilton, Brisbane 01142
Palmer/Gullickson Octagon Home, West Salem 54080
Palmer House, Northfield 51422
Palmer Museum, Jewell 49629
Palmer Museum of Art, University Park 53659
Palmse Manor, Viitna 10673
Palmse Mõis, Viitna 10673
Palo Alto Art Center, Palo Alto 51688
Palo Alto Junior Museum, Palo Alto 51689
Palóc Ethnographical Private Collection and Doll Exhibition, Mátrafüred 23695
Palóc Museum of Popular Art, Eger 23543
Palóc Múzeum - Palóc Ház, Balassagyarmat 23350
Palóc Népművészet, Eger 23543
Palóc Néprajzi Magángyűjtemény és Palóc Baba Kiállítás, Mátrafüred 23695
Palomuseo, Helsinki 10807
Palomuseo, Sairaalamäki 11294
Palos Verdes Art Center, Rancho Palos Verdes 52242
Palotavárosi Skanzen, Székesfehérvár 23857
Pålsjö Kvarn, Helsingborg 40591
Pålsjö Mill, Helsingborg 40591
Paltamon Kotiseutumuseo, Paltamo 11186
Paltrockwindmühle Brehna, Brehna 17534
Pałucka Izba Muzealna, Kcynia 34736
Pamätná Izba Franza Schuberta, Želiezovce 38371
Pamätná Izba Frica Kafendu, Mošovce 38310
Pamätná Izba Jozefa Murgaša, Tajov 38369
Pamätná Izba Sama Chalupku, Horná Lehota 38260
Pamätník Ludwiga van Beethovena, Dolná Krupá 38248
Pamätný Dom Kálmána Mikszátha, Sklabiná 38351
Pamätný Dom Pavla Jozefa Šafárika, Kobeliarovo 38268
Památník Adolfa Heyduka, Písek 09713
Památník Adolfa Kašpara, Loštice 09639
Památník Antonína Dvořáka, Nelahozeves 09672
Památník Antonína Dvořáka, Rožmitál pod Třemšínem 09826
Památník Antonína Dvořáka, Zlonice 09943
Památník Bedřicha Smetany, Jabkenice 09544
Památník Bedřicha Smetany, Obříství 09687
Památník Bible Kralické, Kralice nad Oslavou 09602
Památník Bitvy 1866 na Chlumu, Všestary 09917
Památník Bratří Mrštíků, Diváky 09492
Památník Dr. Emila Holuba, Holice v Čechách 09522
Památník Dr. Františka Křižíka, Plánice 09715
Památník Františka Palackého, Hodslavice 09520
Památník II. Světové Války, Hrabyně 09531
Památník Jana Ámoše Komenského, Bílá Třemešná 09406
Památník Jana Ámoše Komenského, Fulnek 09506
Památník Jiřího Melantricha z Aventýna, Rožďalovice 09823
Památník Josefa Suka, Křečovice 09607
Památník Karla Čapka, Stará Hut u Dobříše 09845
Památník Karla Havlíčka Borovského, Havlíčkův Brod 09512
Památník Karla Hynka Máchy, Doksy 09494
Památník Karla Jaromíra Erbena, Miletín 09654
Památník Karla Václava Raise, Lázně Bělohrad 09620
Památník Krále Jiřího z Poděbrad a Lapidarium, Poděbrady 09723
Památník Leoše Janáčka, Brno 09438
Památník Leoše Janáčka, Hukvaldy 09540
Památník Lidice, Lidice 09627
Památník Města, Police nad Metují 09725
Památník Mikoláše Alše a Matěje Kopeckého, Mirotice 09656
Památník Mistra Jana Husa, Husinec 09542
Památník Mohyla Míru, Prace u Brna 09731
Památník Národního Písemnictví, Praha 09786
Památník Otakara Štáfla, Havlíčkův Brod 09513
Památník Petra Bezruče, Kostelec na Hané 09597
Památník Prokopa Diviše, Znojmo 09948
Památník Svatopluka Čecha, Obříství 09688
Památník Svatopluka Čecha, Ostředek 09702
Památník Terezín, Terezín 09864
Památník Terezín - Bývalé Krematorium na Židovském Hřbitově, Terezín 09865
Památník Terezín - Kolumbárium, Obřadní Místnosti a Ústřední Márnice, Terezín 09866
Památník Terezín - Magdeburská Kasárna, Terezín 09867
Památník Terezín - Malá Pevnost, Terezín 09868
Památník Terezín - Muzeum Ghetta, Terezín 09869
Památník Terezy Novákové, Proseč u Skutče 09809
Památník Václava Beneše Třebízského, Třebíz 09877
Památník W.A. Mozarta a Manželů Duškových, Praha 09787
Památník Zapadlých Vlastenců, Paseky nad Jizerou 09708
Památník Životické Tragedie, Havířov 09509
Pambasang Museo ng Pilipinas, Manila 34450
Pamětní Síň Jana Ámoše Komenského, Brandýs nad Orlicí 09415
Pametnik na Kulturata Zemenski Manastir, Zemen 05267
Pamplin Historical Park and the National Museum of the Civil War Soldier, Petersburg 51799
PAN kunstforum niederrhein, Emmerich am Rhein 18151
Panachrantos Monastery, Andros 22884
Panaddaman-Cagayan State University Museum, Tuguegarao 34519
Panamin Museum, Pasay 34476
Panarcadic Archaeological Museum of Tripolis, Tripoli 23231
Panas Myrnyj Literary and Memorial Museum, Poltava 43170
Panayot Hitov Memorial House, Sliven 05167
Panerių Memorialinis Muziejus, Vilnius 30428
Panevėžio Kraštotyros Muziejus, Panevėžys 30332
Panevezys Local Lore Museum, Panevėžys 30332
Panhandle-Plains Historical Museum, Canyon 47230
Panke Museum, Berlin 17231
Pankhurst Centre, Manchester 45095
Pannonhalmi Főapátság Gyűjteménye, Pannonhalma 23750
Pannonisches Heimatmuseum, Neusiedl am See 02443
Panorama de la Bataille de Waterloo, Braine-l'Alleud 03396
Panorama Defence of Sevastopol 1854-1855, Sevastopol 43180
Panorama Kreuzigung Christi, Einsiedeln 41344
Panorama Mesdag, Den Haag 32098
Panorama Museum, Bad Frankenhausen 16730
Panorama Oborona Sevastopolja 1854-1855 rr., Sevastopol 43180
Panorama of the Aussult of the Sapun Gora Mtn. on 7th of May 1944, Sevastopol 43181
Panorama Plevenska Epopeja 1877, Pleven 05119
Panorama Racławicka, Wrocław 35343
Panorama Šturm Sapun-gori 7 Travnja 1944 r., Sevastopol 43181
Panoramamuseum, Salzburg 02621
Panos Aravantinos Theatrical Museum of Painting and Stage Design, Piraeus 23148
Pansarmuséet, Axvall 40450
Panssarimuseo, Parola 11193
Państwowa Galeria Sztuki w Sopocie, Sopot 35121
Państwowe Muzeum Archeologiczne, Warszawa 35286
Państwowe Muzeum Etnograficzne, Warszawa 35287
Państwowe Muzeum im. Przypkowskich, Jędrzejów 34702
Państwowe Muzeum na Majdanku, Lublin 34911
Państwowe Muzeum Stutthof, Sztutowo 35172
Państwowe Muzeum Stutthof w Sztutowie, Sopot 35122
Państwowe Zbiory Sztuki, Sułoszowa 35140
Panteão da Pátria Tancredo Neves, Brasília 04371
Panthéon, Paris 14744
Pao Galleries at Hong Kong Art Centre, Hong Kong 07811
Papadopoulos Picture Gallery, Tinos 23228
Papakura and District Historical Society Museum, Papakura 33177
Paper Industry Museum of the Diósgyőr Papermill, Miskolc 23701
Paper Museum, Espoo 10731
Paper Museum, Paços de Brandão 35802
Paper Museum, Tokyo 29407
Paperimuseo, Espoo 10731
Paphos District Archaeological Museum, Paphos 09392
Papierfabrik Hohenofen, Sieversdorf-Hohenofen 21762
Papiergeschichtliche Sammlung, Bergisch Gladbach 17018
Papiermacher- und Heimatmuseum, Frankeneck 18350
Papiermolen de Schoolmeester, Westzaan 32943
Papiermühle Weddersleben, Weddersleben 22351
Papiermuseum, Düren 17983
Papiliorama-Nocturama Foundation, Kerzers 41501
Papplewick Pumping Station, Ravenshead 45465
Papua New Guinea Display Centre, Crafers 01007
Papua New Guinea National Museum and Art Gallery, Boroko 34080
Papyrusmuseum und Papyrussammlung, Wien 03054
Paqgoeta Park Information Center, Aia 38796
Para/Site Art Space, Hong Kong 07812
Paradise Mill Silk Industry Museum, Macclesfield 45066
Parafia Ewangelicko - Augsburska Wang, Karpacz 34717
Parafialne Muzeum Regionalne, Krasnobród 34825
Parainen District Museum, Parainen 11187
Parainen Industrial Museum, Parainen 11188
Paramita Museum, Komono 28767
Parasite Museum, Bangkok 42555
Parasol Unit, London 44951
Parc Archéologique de la Pointe-du-Buisson, Melocheville 06263
Parc Archéologique de Larina, Hières-sur-Amby 13216
Parc Archéologique Européen de Bliesbruck-Reinheim, Bliesbruck 12027
Parc Ardèche Miniature, Soyons 15806
Parc Botanique de la Fosse, Montoire-sur-le-Loir 14234
Parc & Château de Valencay, Valençay 16037
Parc de la Préhistoire, Tarascon-sur-Ariège 15858
Parc de l'Art Préhistoire, Tarascon-sur-Ariège 15859
Parc de l'Artillerie, Québec 06697
Parc de Préhistoire de Bretagne, Malansac 13929
Parc de Reconstitution Archéologique, Villeneuve-d'Ascq 16207
Parc Historique Pointe-du-Moulin, Notre-Dame-de-l'Île-Perrot 06490
Parc Howard Museum and Art Gallery, Llanelli 44762
Parc Miniature Alsace-Lorraine, Saint-Ame 15164
Parc National de la Culture des Ethnies, Vientiane 29976
Parc Préhistorique des Grottes de Fontirou, Castella 12306
Parc Préhistorique Imaginaire, Chamoux 12382
Parc Zoologique Arche de Noé et Amazonia, Saint-Clément-des-Baleines 15217
Parco Archeologico di Castel Seprio, Castelseprio 25754
Parco Monumentale di Pinocchio, Collodi 25942
Parco Museo Minerario, Abbadia San Salvatore 25138
Parco Nazionale delle Incisioni Rupestri, Capo di Ponte 25659
Parco Scherrer, Morcote 41642
Parcours-découverte de l'écurie n1, Le Pin-au-Haras 13632
Parcul Etnografic Romulus Vuia, Cluj-Napoca 36293
Pardee Home Museum, Oakland 51471
Parham Airfield Museum and Museum of the Resistance Organisation Museum, Framlingham 44252
Parham House & Gardens, Storrington 45768
Parikkalan Kotiseutumuseo, Parikkala 11190
Paris Gibson Square Museum of Art, Great Falls 49031
Parish Museum, Dobra 34623
Parish Museum, Grybów 34686
Parish Museum, Paszyn 35004
Parish Museum, Zábřeh 09923
Parish Museum, Złota 35373
Parishville Museum, Parishville 51705
Park City Historical Society & Museum, Park City 51707
Park Farm Museum, Milton Abbas 45152
Park Gallery, Falkirk 44206
Park House Museum, Amherstburg 05336
The Park-McCullough House, North Bennington 51383
Park Museum Monuments of the Culture of Mangystau and Ustjurt, Aktau 29671
Park-Museum Skobelev, Pleven 05121
Park-muzej General V.Z. Lavrov, Gorni Dabnik 05038
Park-muzej im. A.K. Tolstogo, Brjansk 36829
Park-Muzej Vladislav Varnenčik, Varna 05245
Parkanon Kotiseutumuseo, Parkano 11191
Parkanon Metsämuseo, Parkano 11192
Parkdale-Maplewood Community Museum, Barss Corner 05393
Parker-O'Malley Air Museum, Ghent 48906
Parker Tavern Museum, Reading 52260
Parkhöhle Weimar, Weimar 22414
Parkland College Art Gallery, Champaign 47336
Parks Canada Visitor Reception Centre, Churchill 05615
Parksepa Keskkooli Muuseum, Parksepa 10607
Parkside Cottage Museum, Narrandera 01351
Parkzicht, Ermelo 32223
Parlamentsausstellung des Deutschen Bundestages "Wege, Irrwege, Umwege", Berlin 17232
Parliament House Art Collection, Canberra 00950
Parliament In Exile Museum, Kauhajoki 10926
Parnell Museum, Rathdrum 24890
Parochialkirche, Berlin 17233
Parque Arqueológico Cueva Pintada, Gáldar 39321
Parque Cultural, Zerain 40357
Parque de las Ciencias, Granada 39370
Parque Etnográfico do Río Arnoia, Allariz 38850
Parque-Museo La Venta, Villahermosa 31555
Parris Island Museum, Parris Island 51714
The Parrish Art Museum, Southampton 53185
Parroquial Arxiprestal, San Mateo 39967
Parry Mansion Museum, New Hope 51062
Pars Museum, Shiraz 24648
Parson Fisher House, Blue Hill 46874
Parson Historical Commission Museum, Parsons 51717
Parsonsfield-Porter History House, Porter 52059
The Parthenon, Nashville 50972
Partisan Hospital Franja of the Idrija Municipal Museum, Dolenji Novaki 38415
Partizanska Bolnica Franja, Dolenji Novaki 38415
Paruyr Sevak House-Museum, Zangakatun 00763
Parys Museum, Parys 38701
Pasadena City College Art Gallery, Pasadena 51722
Pasadena Museum of California Art, Pasadena 51723
Pasadena Museum of History, Pasadena 51724
Pascack Historical Society Museum, Park Ridge 51711
Pasinger Fabrik - Ausstellungen, München 20473
Pasmans Huus, Ruinen 32738
Pass of Killiecrankie Visitor Centre, Pitlochry 45385
Passages Centre d'Art Contemporain, Troyes 16003
Passaic County Community College Galleries, Paterson 51728
Passaic County Historical Society Museum, Paterson 51729
Past Students Association Museum, Gatton 01099
Pasvalio Krašto Muziejus, Pasvalys 30335
Pasvalys Area Museum, Pasvalys 30335
Pásztói Múzeum, Pásztó 23755
Pásztormúzeum, Bugac 23498
Pataka, Porirua 33186
Patan Museum, Lalitpur 31782
Pate Museum of Transportation, Cresson 47876
Patee House Museum, Saint Joseph 52563
Patek Philippe Museum, Genève 41428
Pateniemen Sahamuseo, Oulu 11176
Pateniemi Sawmill Museum, Oulu 11176
Patenschaftsmuseum Goldap in Ostpreußen, Stade 21851
Patent Model Museum, Fort Smith 48686
Pater Haw-Museum, Schweich an der römischen Weinstraße 21680
Pater Jon Sveinsson Memorial Museum, Akureyri 23985
Paterson Court House, Adamstown 00769
Paterson Museum, Paterson 51730
Pathein Museum, Pathein 31727
Pathological Museum, London 44952
Pathologisch-anatomisches Bundesmuseum, Wien 03055
Pathology and Bacteriology Museum, Lucknow 24298
Pathology Museum, Herston 01152
Pathology Museum, Mumbai 24322
Pathology Museum, Tshwane 38754
Patika-ház, Sopron 23815
Patio Centrum Sztukii, Łódź 34887
Patkó Imre Gyűjtemény, Győr 23585
Patna Museum, Patna 24346
Patona Potter's House, Sümeg 23827
Patonai Fazekasház, Sümeg 23827
Patriarch's Palace at the Moscow Kremlin, Moskva 37396
The Patrick and Beatrice Haggerty Museum of Art, Milwaukee 50718
Patrick Collection Motor Museum, Birmingham 43509

Patrick Taylor Cottage Folk Museum, Albany 00788
Patrijaršie Palaty, Moskva 37396
Patrimoine Historique et Études du Repassage, Sebourg 15694
Patriots Point Naval and Maritime Museum, Mount Pleasant 50888
Patterson Homestead, Dayton 47987
Patterson's Spade Mill, Templepatrick 45834
Pattijoen Kotiseutumuseo Ojala, Pattijoki 11195
Patton Museum of Cavalry and Armor, Fort Knox 48633
Paul Anton Keller-Museum, Lockenhaus 02331
Paul Bunyan Logging Camp, Eau Claire 48278
Paul Dresser Memorial Birthplace, Terre Haute 53482
Paul Ernst-Gedenkstätte, Sankt Georgen an der Stiefing 02642
Paul Gugelmann-Museum, Schönenwerd 41842
Paul H. Karshner Memorial Museum, Puyallup 52207
Paul Jönska Gården, Viken 41061
Paul Laurence Dunbar State Memorial, Dayton 47988
Paul Revere House, Boston 46946
Paul-Röder-Museum, Marktoberdorf 20220
Paul Stradin Museum of the History of Medicine, Riga 30101
Paul W. Bryant Museum, Tuscaloosa 53620
Paul Whitney Larson Art Gallery, Saint Paul 52627
Paula Modersohn-Becker-Stiftung, Bremen 17560
Paula Stradiņa Medicīnas Vēstures Muzejs, Riga 30101
Pauline E. Glidden Toy Museum, Ashland 46385
Pauliner Church, Goettingen State and University Library, Göttingen 18697
Paulinerkirche, Göttingen 18697
Paulise de Bush Costume Collection, Broadclyst 43643
Pauselijk Zouavenmuseum, Roeselare 03977
Paved Arts, Saskatoon 06922
Pavel Dobsinsky Memorial Site, Slavošovce 38352
Pavel Horov Literary Museum, Bánovce nad Ondavou 38182
Pavel Kuznetsov House-Museum, Saratov 37748
Pavilhão do Conhecimento, Lisboa 35733
Pavilion of Knowledge, Lisboa 35733
Paviljon Cvijeta Zuzorića, Beograd 38083
Paviljon Katalnoj Gorki, Lomonosov 37111
Pavillion Horta-Lambeaux, Bruxelles 03500
Pavillon Chinois, Tour Japonaise et Musée d'Art Japonais, Bruxelles 03501
Pavillon de l'Arsenal, Paris 14745
Pavillon de Vendôme, Aix-en-Provence 11520
Pavillon des Arts, Paris 14746
Pavillon Flaubert, Canteleu 12271
Pavillon of Art, Podgorica 31658
Pavillon-Press Museum for Print Graphic Weimar, Weimar 22415
Pavillon-Presse Druckgrafisches Museum Weimar, Weimar 22415
Pavillon Saint-Arnaud, Trois-Rivières 07179
Pavilon Anthropos, Brno 09439
Pāvilosta Museum of Local Studies, Pāvilosta 30053
Pāvilostas Novadpētniecības Muzejs, Pāvilosta 30053
Pavle Beljanski Memorial Collection, Novi Sad 38125
Pavlodar Historical and Regional Potanin Museum, Pavlodar 29719
Pavlodar Museum of Fine Arts, Pavlodar 29718
Pavlodarskij Oblastnoj Chudožestvennyj Muzej, Pavlodar 29718
Pavlodarskij Oblastnoj Istoriko-Kraevedčeskij Muzej im. G.N. Potanina, Pavlodar 29719
Pavlos Vrellis Museum of Greek History, Ioannina 23016
Pavlovskij Istoričskij Muzej, Pavlovo 37527
Pavol Jozef Shafarik Memorial Site, Kobeliarovo 38268
Pawiak Prison Museum, Warszawa 35277
Pawilon Wystawowy - Centrum Gemini, Gdynia 34660
Pawnee Bill Ranch, Pawnee 51736
Pawnee City Historical Society Museum, Pawnee City 51737
Pawnee Indian Village, Republic 52296
Pawnshop Museum, Macau 08005
Paxton House Grannery, Berwick-upon-Tweed 43463
Payne Gallery, Bethlehem 46783
Paynes Creek Historic State Park, Bowling Green 46972
Paznauner Bauernmuseum, Mathon 02362
Peabody Essex Museum, Salem 52657
Peabody Museum of Archaeology and Ethnology, Cambridge 47185
Peabody Museum of Natural History, New Haven 51054
Peabody Place Museum, Memphis 50583
Peace Memorial Museum, Zanzibar 42524
Peace Museum, Bradford 43582
The Peace Museum, Chicago 47505
Peace Museum, Heping 07782
Peace Museum Project, Amsterdam 31867
Peace River Centennial Museum, Peace River 06575
Peacock Gallery, Cochran 47667
Peacock Heritage Centre, Chesterfield 43822
Peacock Visual Arts, Aberdeen 43245
Peak District Mining Museum, Matlock 45116
Peak Rail Museum, Matlock 45117
Pearce Collections Museum, Corsicana 47826
Pearce Institute, Glasgow 44293
Pearl Button Museum, Muscatine 50922
Pearl S. Buck Birthplace, Hillsboro 49290
Pearl S. Buck House, Perkasie 51776
Pearn's Steam World, Westbury 01645
Pearse Museum, Dublin 24815
Pearson Air Museum, Vancouver 53707
The Pearson Museum, Springfield 53220
The Peary-MacMillan Arctic Museum, Brunswick 47103
Peasant House, Święty Krzyż 35152
Peasant House Museum, Laukaa 11055
Peasant Museum, Kälviä 10901
Peat Museum, Kihniö 10954
Pebble Hill Plantation, Thomasville 53500
Pechermuseum, Hernstein 02072
Peckover House, Wisbech 46048
Pecos Park, Pecos 51745
Pedagogic Museum of Galicia, Santiago de Compostela 40029
Pedagogical and Region Historical Collection of the 18. District, Budapest 23492
Pedagogical Museum of A.S. Makarenko, Moskva 37397
Pedagogičeskij Muzej A.S. Makarenko, Moskva 37397
Pedagogické Muzeum Jana Ámoše Komenského, Praha 09788
Pedagoginis Muziejus, Kaunas 30264
Pedagoški Muzej, Beograd 38084
Peekskill Museum, Peekskill 51748
Peel Castle, Peel 45351
Peel Heritage Complex, Brampton 05456
Peel Mansion Museum, Bentonville 46744
Peelmuseum, Ospel 32639
Peeter I Maja, Tallinn 10641
Peggy Guggenheim Collection, Venezia 28207
Peggy Notebaert Nature Museum, Chicago 47506
Pei County Museum, Peixian 08077
Pei Kuan Crab Museum, Toucheng 42456
Pejepscot Historical Society Museum, Brunswick 47104
Pékmúzeum, Sopron 23816
Pelastusarmeijan Museo, Helsinki 10808
Pelee Island Heritage Centre, Pelee Island 06577
Pelgrom Museum Poorterswoning, Antwerpen 03309
Pelham Art Center, Pelham 51749
Pelham Historical Resource Centre, Fenwick 05793
Pelkosenniemen Kotiseutumuseo, Pelkosenniemi 11196
Pella Historical Village, Pella 51750
Pellenz-Museum für Archäologie, Nickenich 20691
Pellisier House Museum, Bethulie 38560
Pellon Kotiseutumuseo, Pello 11197
Pellonpään Talonpoikaistalo, Inkere 10860
Peltola Cotters Museum, Tammijärvi 11347
Peltolan Mäkitupalaismuseo, Tammijärvi 11347
Peltotyökalumuseo, Somero 11321
Pemba Museum, Chake Chake 42511
Pember Museum of Natural History, Granville 49019
Pemberton & District Museum, Pemberton 06579
Pembina Lobstick Historical Museum, Evansburg 05789
Pembina State Museum, Pembina 51753
Pembina Threshermen's Museum, Winkler 07350
Pembroke Historical Society Museum, Pembroke 51754
Pembroke Lodge Patterson Heritage Museum and Family History Centre, Birchington 43488
Pen and Brush Museum, New York 51248
Pena-Peck House, Saint Augustine 52524
Penang Forestry Museum, Pulau Pinang 30614
Pencarrow House, Bodmin 43551
Pence Gallery, Davis 47977
Pendarvis Historic Site, Mineral Point 50726
Pendeen Lighthouse, Pendeen 45354
Pendennis Castle, Falmouth 44210
Pendjikent Archaeological Museum-Preserve, Pandžakent 42504
Pendle Heritage Centre, Barrowford 43390
Pendleton District Agricultural Museum, Pendleton 51759
Pendon Museum, Long Wittenham 45028
Penetanguishene Centennial Museum, Penetanguishene 06582
Pengsjö Nybyggarmuseum, Umeå 40998
Peninsula and Saint Edmunds Township Museum, Tobermory 07086
Peninsula Fine Arts Center, Newport News 51336
Penkill Castle, Girvan 44271
Penland Gallery, Penland 51761
Penlee House Gallery and Museum, Penzance 45361
Penneshaw Maritime and Folk Museum, Penneshaw 01409
Pennsbury Manor, Morrisville 50860
Pennsylvania Academy of the Fine Arts Gallery, Philadelphia 51864
Pennsylvania Anthracite Heritage Museum, Scranton 52983
Pennsylvania College of Art and Design Galleries, Lancaster 49930
Pennsylvania Dutch Folk Culture Society Museum, Lenhartsville 50023
Pennsylvania Federation of Museums and Historical Organizations, Harrisburg 55177
Pennsylvania German Cultural Heritage Center, Kutztown 49829
Pennsylvania Lumber Museum, Galeton 48836
Pennsylvania Military Museum and 28th Division Shrine, Boalsburg 46881
Pennsylvania Trolley Museum, Washington 53943
Pennypacker Mills, Schwenksville 52965
Pennyroyal Area Museum, Hopkinsville 49364
Penobscot Marine Museum, Searsport 52992
Penrhyn Castle, Bangor, Gwynedd 43378
Penrith Museum, Penrith 45357
Penrith Regional Gallery and The Lewers Bequest, Emu Plains 01073
Pensacola Historical Museum, Pensacola 51768
Pensacola Museum of Art, Pensacola 51769
Penselmuseet i Bankeryd, Bankeryd 40452
Penshurst Place and Toy Museum, Penshurst 45358
Penticton Art Gallery, Penticton 06583
Penwith Galleries, Saint Ives, Cornwall 45576
Penyo Penev House-Museum, Dimitrovgrad 05013
Penzenskaja Kartinnaja Galerija im. K.A. Savickogo, Penza 37536
Penzenskij Gosudarstvennyj Obedinennyj Kraevedčeskij Muzej, Penza 37537
Penzenskij Muzej Odnoj Kartiny, Penza 37538
Pénzügytörténeti Gyűjtemény, Budapest 23465
People's Gallery, El Paso 48328
The People's Gallery, Stalybridge 45734
People's History Museum, Manchester 45096
The People's Museum, Belfast 43445
People's Museum, Gevgelija 30501
People's Museum, Melaka 30600
People's Museum, Sveti Nikole 30531
People's Museum Dr. Nikola Nezlobinski, Struga 30527
People's Museum Kumanovo, Kumanovo 30507
People's Museum Kumanovo - Art Gallery, Kumanovo 30508
People's Museum Kumanovo - House Museum of People's Hero Cristian T. Karposh, Kumanovo 30506
People's Palace Museum, Glasgow 44294
People's Story Museum, Edinburgh 44144
Peoria Historical Society, Peoria 51774
Peppers Art Gallery, Redlands 52275
Perak Royal Museum, Kuala Kangsar 30570
Perak State Museum, Taiping 30622
Perbadanan Muzium Melaka, Melaka 30607
Perc Tucker Regional Gallery, Townsville 01599
Perch Proshyan House-Museum, Ashtarak 00692
Percy Pilcher Museum, Lutterworth 45049
Perejaslav National Historical and Ethnographic Preserve, Perejaslav-Chmel'nyc'kyj 43165
Pères Blancs - Saint Anne, Jerusalem 25008
Pereslavl-Zalesskij Gosudarstvennyj Istoriko-architekturnyj i Chudožestvennyj Muzej-zapovednik, Pereslavl-Zalesskij 37542
Pereslavl-Zalessky Museum-Preserve of History, Art and Architecture, Pereslavl-Zalesskij 37542
Perfect Little Girl's and History Museum of Verfeil, Verfeil 16114
Performance Space, Eveleigh 01078
Performing Arts Collection, Melbourne 01290
Performing Arts Collection of South Australia, Adelaide 00780
Perfume Museum, Niagara-on-the-Lake 06455
Pergamonmuseum, Berlin 17234
Perge Müzesi, Aksu 42685
Perhon Kotiseutumuseo, Perho 11199
Perkins Bull Collection, Peterborough 06594
Perkins Center for the Arts, Moorestown 50833
Perkins County Historical Society Museum, Grant 49015
Perkins History Museum, Watertown 53971
Perkinson Gallery, Decatur 48017
Perlis State Museum, Kangar 30559
Perm State Art Gallery, Perm 37550
Permanent Exhibition of the Painter József Horváth, Sopron 23810
Permanenten Vestlandske Kunstindustrimuseum, Bergen 33433
Permskaja Gosudarstvennaja Chudožestvennaja Galereja, Perm 37550
Permskij Oblastnoj Kraevedčeskij Muzej, Perm 37551
Perniön Museo, Perniö 11200
Perpignan Museum, Perpignan 14793
Perranzabuloe Folk Museum, Perranporth 45363
Perrault's Museum, Val Marie 07195
Perry County Lutheran Historical Society, Altenburg 46239
Perry-Downer House and Costume Gallery, Monterey 50806
Perry's Cider Mills, Ilminster 44513
Perry's Victory and International Peace Memorial, Put in Bay 52202
Perryville Battlefield Museum, Perryville 51783
Persepolis Museum, Marvdashd 24630
Persepolis Museum, Shiraz 24649
Persian Carpet Museum, Raas 29950
Persmuseum, Amsterdam 31878
Perspective Gallery, Blacksburg 46846
Perth Institute of Contemporary Arts, Northbridge, Western Australia 01377
Perth Museum, Perth 06587
Perth Museum and Art Gallery, Perth 45366
Perttelin Kotiseutumuseo, Inkere 10861
Peshawar Museum, Peshawar 34036
Peshtigo Fire Museum, Peshtigo 51790
Peštera Rabiša Muzej, Belogradčik 04983
Pesterzsébeti Múzeum, Budapest 23466
Petalax Hembygdsmuseum, Petalax 11202
Petaluma Wildlife and Natural Science Museum, Petaluma 51791
Pete and Susan Barrett Art Gallery, Santa Monica 52900
Peter and Paul Fortress, Sankt-Peterburg 37719
Peter Anson Gallery, Buckie 43664
Péter Benedek Memorial House, Uszód 23942
Peter Benoitmuseum, Harelbeke 03679
Peter Conser House, Heavener 49236
Peter I. House, Tallinn 10641
Peter J. McGovern Little League Baseball Museum, South Williamsport 53184
Péter Pál Polgárház, Szentes 23885
Peter Paul Luce Gallery, Mount Vernon 50891
Peter Rice Homestead, Marlboro 50493
Peter Rosegger-Geburtshaus, Alpl 01707
Peter Rosegger-Museum, Krieglach 02237
Peter Scott Gallery, Lancaster 44635
Peter the Great House Museum, Sankt-Peterburg 37635
Peter the Great Museum of Anthropology and Ethnography - Kunstkamera, Sankt-Peterburg 37670
Peter Van den Braken Centrum, Sterksel 32807
Péter Veres Memorial Exhibition, Balmazujváros 23360
Peter Wentz Farmstead, Worcester 54304
Peter Whitmer Farm, Waterloo 53967
Peter-Wiepert-Museum, Fehmarn 18299
Peter Yegen Jr. Yellowstone County Museum, Billings 46806
Peter-Zirbes-Haus Niederkail, Landscheid 19799
Pētera Upīša Piemiņas Muzejs, Dobele 30001
Peterborough Centennial Museum and Archives, Peterborough 06595
Peterborough Historical Society Museum, Peterborough 51792
Peterborough Museum and Art Gallery, Peterborough 45368
Petergof - Gosudarstvennyj Muzej Zapovednik, Sankt-Peterburg 37718
Peterhof State Museum Reserve, Sankt-Peterburg 37718
Petersburg Area Art League, Petersburg 51800
The Petersburg Museums, Petersburg 51801
Petersburg National Battlefield, Petersburg 51802
Petersen Automotive Museum, Los Angeles 50258
Petersfield Museum, Petersfield 45374
Peterstormuseum, Zwiefalten 22836
Petes Museigård, Havdhem 40584
Petit Musée, Clair 05617
Petit Musée, Conakry 23312
Petit Musée Ambulant de la Musique Mécanique, Villeneuve-la-Garenne 16212
Petit Musée d'Art Camerounais, Yaoundé 05303
Petit Musée d'Automat et du Vin de Pauillac, Pauillac 14764
Petit Musée de France, Albaret-Sainte-Marie 11530
Le Petit Musée de la Bicyclette, La Batie-Montgascon 13314
Petit Musée de la Chartreuse Saint-Hugon, Arvillard 11710
Petit Musee de la Table, Paris 14747
Petit Musée de l'Argenterie, Paris 14748
Petit-Musée de Vin au Cave de la Dive Bouteille, Bourgueil 12126
Petit Musée des Abeilles et du Miel-De la Fleur à la Table, Alle-sur-Semois 03264
Petit Musée des Chailloux, Chevagny-les-Chevrieres 12507
Petit Musée des Silos de Jouques, Jouques 13284
Petit Musée du Bizarre, Villeneuve-de-Berg 16210
Petit Musée Fantastique de Guignol, Lyon 13904
Le Petit Musée Maison de Mariette, Saint-Laurent-du-Pont 15362
Petit Musée Minéraux et Faune de l'Alpe, Bourg-Saint-Maurice 12111
Petit Musée Rural, Varaire 16073
Petit Museu Parroquial, Calonge de Mar 39106
Petit Palais, Genève 41429

P

Petit-Palais, Paris 14749
La Petite Anglicane, Forestville 05805
Petite Maison de Le Corbusier, Corseaux 41315
La Petite Provence du Paradou, Le Paradou 13629
Les Petites Franciscaines de Marie, Baie-Saint-Paul 05373
Petko and Pencho Slaveykov Memorial House, Sofia 05181
Petőfi Museum of Hungarian Literature, Budapest 23467
Petőfi Emlékmúzeum, Kiskőrös 23664
Petőfi House and Memorial Museum, Kiskőrös 23664
Petőfi Irodalmi Múzeum, Budapest 23467
Petőfi Múzeum, Aszód 23338
Petőfi Sándor Emlékház, Dömsöd 23527
Petőfi Sándor Emlékmúzeum, Szalkszentmárton 23834
Petőfi Sándor Emlékszoba, Kiskunfélegyháza 23668
Petone Settlers Museum, Hutt City 33127
Petone Settlers Museum, Wellington 33258
Petra Museum, Petra 29665
Petran Museum, Haltdalen 33527
Petrefaktensammlung, Bad Staffelstein 16844
Petrie Museum of Egyptian Archaeology, London 44953
Petrified Creatures Museum of Natural History, Richfield Springs 52305
Petrified Forest of the Black Hills, Piedmont 51914
Petro Cvirkos Memorialinis Muziejus-Sodyba, Veliuona 30393
The Petroleum Museum, Midland 50674
Petrolia Discovery, Petrolia 06598
Petropavlovskaja Krepost - Istoriko-kulturnyj Zapovednik, Sankt-Peterburg 37719
Petrov-Vodkin Memorial Art Museum, Chvalynsk 36864
Pettaquamscutt Museum, Kingston 49794
Petter Dass-Museet, Sandnessjøen 33816
Petterson Museum of Intercultural Art, Claremont 47573
Pettis County Historical Society, Sedalia 53024
The Petworth Cottage Museum, Petworth 45377
Petworth House and Park, Petworth 45378
Pevnost Josefov - Podsemní Chodby, Jaroměř 09552
Pewabic Pottery Museum, Detroit 48129
Pewsey Heritage Centre, Pewsey 45380
Peyo Yavorov Memorial House, Sofia 05180
Pfälzisches Bergbaumuseum, Imsbach 19330
Pfahlbaumuseum Unteruhldingen, Uhldingen-Mühlhofen 22158
Pfahlbausammlung Dr. h.c. Carl Irlet, Twann 41954
Pfalzmuseum, Forchheim, Oberfranken 18337
Pfalzmuseum für Naturkunde, Bad Dürkheim 16716
Pfarr- und Wallfahrtsmuseum, Mariapfarr 02356
Pfarrmuseum, Dießen am Ammersee 17852
Pfarrmuseum, Flintsbach am Inn 18330
Pfarrmuseum, Sankt Georgen im Attergau 02646
Pfefferminzmuseum, Eichenau 18076
Pfeifenmuseum, Sankt Aegyd am Neuwalde 02633
Pfeifenmuseum, Scheifling 02705
Pfeifenmuseum, Windigstein 03115
Pfeifermuseum im alten Rathaus, Werbach 22482
Pfeilburg Museum, Fürstenfeld 01910
Pferde-Eisenbahn Museum, Rainbach im Mühlkreis 02554
Pfingstritt-Museum Kötzting, Kötzting 19668
Pfinzgaumuseum, Karlsruhe 19446
Pforzheim Jewellery Museum, Pforzheim 21029
Pha That Luang, Vientiane 29977
phæno gGmbH, Wolfsburg 22666
Phänomenta Lüdenscheid, Lüdenscheid 20089
Phaliashvili Museum, Kutaisi 16345
Phantasten Museum Wien, Wien 03056
The Pharaonic Village, Giza 10529
Pharma- und Drogistenmuseum, Wien 03057
Pharmaceutical and Medical Museum of Bohuslav Lavička, Ljubljana 38448
Pharmaceutical History Exposition, Sibiu 36605
Pharmacy House, Póvoa de Lanhoso 35871
Pharmacy House, Sopron 23815
Pharmacy Museum, Kouvola 10989
Pharmacy Museum at the Pharmacists Under the Swan, Bydgoszcz 34577
Pharmaka Gallery, Los Angeles 50259
Pharmazie-Historisches Museum der Universität Basel, Basel 41176
Phathai Cave, Lampang 42576
Phelps House, Burlington 47134
Phelps Store Museum, Palmyra 51686
Philadelphia Art Alliance, Philadelphia 51865
Philadelphia Mummers Museum, Philadelphia 51866
Philadelphia Museum of Art, Philadelphia 51867
Philadelphia Museum of Judaica, Philadelphia 51868
Philadelphia Society for the Preservation of Landmarks, Philadelphia 51869
Philatelic Centre of French Polynesia, Papeete 16323
Philatelic Museum, Athinai 22952
Philatelic Museum, Tokyo 29412
Philbrook Museum of Art, Tulsa 53610
Philip and Muriel Berman Museum of Art, Collegeville 47691
Philip L. Wright Zoological Museum, Missoula 50757
Philipp-Matthäus-Hahn-Museum, Albstadt 16458
Philipp-Reis-Haus und Hugenottenmuseum, Friedrichsdorf 18476
Philipp Schäfer II Museum, Riedstadt 21326
Philippine Air Force Aerospace Museum, Pasay 34477
Philippine Army Museum, Makati 34416
Philippine House Museum, Shepparton 01503
Philippine Navy Museum, Cavite City 34363
Philippine Presidential Museum, Malacañang 34419
Philippine Radiation Science Museum, Manila 34451
Philippine Science Centrum, Manila 34452
Philippine Science Centrum, Marikina 34461
Het Philips Gloeilampenfabriekje anno 1891, Eindhoven 32192
Philips Museum Eindhoven, Eindhoven 32193
Philipse Manor Hall, Yonkers 54332
Phillip Island Community Art and Craft Gallery, Cowes 01005
The Phillips Collection, Washington 53914
Phillips Countryside Museum, Brokerswood 43652
Phillips County Museum, Helena 49238
Phillips County Museum, Holyoke 49326
Phillips Historical Society Museum, Phillips 51893
The Phillips Museum of Art, Lancaster 49931
The Philmont Museum, Seton Memorial Library and Kit Carson Museum, Cimarron 47551
Phiphitthaphanthasathan Haeng Chat Ram Khamhaeng Sukhothai, Sukhothai 42603
Phipitapan Haeng Chart Phra Pathom Chedi, Nakhon Pathom 42582
Phipitapan Laeng Chart, Suphan Buri 42605
Phipitapantasatan Haeng Chart Nakhon Si Thammarat, Nakhon Si Thammarat 42585
Phipittapan Tasatan Haengchart Khon Kaen, Khon Kaen 42574
Phippen Museum of Western Art, Prescott 52134
Phoebe Apperson Hearst Museum of Anthropology, Berkeley 46760
Phoenix Art Museum, Phoenix 51906
Phoenix Cultural Artifacts Center, Tainan 42367
Phonothèque Québécoise - Musée du Son, Montréal 06377
Photo Antiquities, Pittsburgh 51955
Photoarchive of the Union of Photographers of Russia, Moskva 37171
Photogallery Brothers Lumiere, Moskva 37172
The Photographers' Gallery, London 44954
Photographia-Museu Vicentes, Funchal 35620
Photographic Archives, Lexington 50055
Photographic Investments Gallery, Atlanta 46448
Photographic Resource Center, Boston 46947
Die Photographische Sammlung/SK Stiftung Kultur, Köln 19627
Photography Museum, Charleroi 03515
Photography Museum of the Šiauliai Aušra Museum, Šiauliai 30356
Photohistorische Sammlung, Penk 02484
Photomuseum, Zarautz 40353
Photomuseum des Landes Oberösterreich, Bad Ischl 01757
Photoraum VISIBLE, Villach 02860
Phra Borommathat National Museum, Nakhon Si Thammarat 42586
Phra Chetuponwimonmangkhalaram National Museum, Bangkok 42556
Phra Pathom Chedi National Museum, Nakhon Pathom 42582
Phra Phuttachinnarat National Museum, Phitsanulok 42595
Phukradung, Loei 42579
Phulsanda Religious Arts Museum, Phulsanda 24348
Phuthadikobo Museum, Mochudi 04299
Phyletisches Museum, Jena 19387
Physics Museum, Saint Lucia 01485
Pi Yandong Xue Yi Zhi Display Center, Guilin 07744
Pianguan Museum, Pianguan 08079
Piano Collection, Hereford 44458
Pianoforte-Museum, Meißen 20266
Pianola Museum, Amsterdam 31879
Piaohao Museum, Pingyao 08090
Piatra-Neamț Art Museum, Piatra-Neamț 36518
Piatra-Neamț Ethnography Museum, Piatra-Neamț 36519
Piatt Castles, West Liberty 54068
Piatt County Museum, Monticello 50820
Pibe Mølle, Helsinge 10062
Pic du Midi, La Mongie 13393
Picasso Museum, Antibes 11638
Picasso-Museum, Luzern 41596
Piccolo Antiquarium, Fossato di Vico 26240
Piccolo Museo dei Popoli, Lodi 26511
Piccolo Museo della Moneta, Vicenza 28256
Piccolo Museo di San Paolo, Reggio Calabria 27385
Piccolo Museo Navale, Laives 26440
Piccolo Museo Parrocchiale, Torgnon 27995
Piccolo Museo Parrocchiale, Valsavarenche 28150
Pick Szalámi and Szeged Paprika Museum, Szeged 23848
Pick Szalámi és Szegedi Paprika Múzeum, Szeged 23848
Pickens County Museum of Art History, Pickens 51912
Picker Art Gallery, Hamilton 49137
Pickering Museum Village, Pickering 06599
Pickford's House Museum, Derby 43967
Picton Community Museum, Picton 33182
Picture Exposition of A. Tycina of the Siauliai Region Literature Museum, Gruzdžiai 30229
Picture Gallery, Bhaktapur 31772
Picture Gallery, Kam'janec'-Podil's'kyj 43052
Picture Gallery, Sarnano 27743
Picture Gallery of Dachau, Dachau 17764
Picture Gallery of Minusinsk, Minusinsk 37135
Picture Gallery of Olonetsk, Olonec 37495
Picture Gallery of P.M. Gretchishkin, Stavropol 37803
Picture Gallery of Tchajkovsky, Čajkovskij 36833
Picture Gallery of the Adygean Republic Ministry of Culture, Majkop 37124
Picture Gallery of the Painter Veronika Sleivyte - Kupiškis Ethnographic Museum, Viktariskes 30398
Picture Gallery of the Serbian Cultural Association, Novi Sad 38120
Piddig Museum, Piddig 34479
Piece Hall Art Gallery, Halifax 44386
Piedmont Arts Museum, Martinsville 50520
Piedra Lumbre Visitors Center, Abiquiu 46129
Pielaveden Kotiseutumuseo, Pielavesi 11205
Pielavesi Local History Museum, Pielavesi 11205
Pielisen Museo, Lieksa 11067
Pienmäen Talomuseo, Niemisjärvi 11140
Pienmäki Farm Buildings Museum, Niemisjärvi 11140
The Pier-2, Kaohsiung 42242
Pier 21, Halifax 05945
Pier Arts Centre, Stromness 45792
Pier Pander Museum, Leeuwarden 32477
Pierce Manse, Concord 47776
Pierides Museum, Larnaka 09369
Pierides Museum of Ancient Cypriot Art, Athinai 22950
Pierides Sculpture Garden, Larnaka 09370
Pierre Gianadda Foundation, Martigny 41608
Pierre Gildesgame Maccabi Sports Museum, Ramat Gan 25086
Pierre Menard Home, Ellis Grove 48369
Pierrepont Museum, Canton 47223
Pietarin Radan Museo, Lahti 11031
Pieter Vermeulen Museum, IJmuiden 32424
Piętro Wyżej, Katowice 34729
Pigeon Island, Castries 37995
Pigeon Valley Steam Museum, Wakefield 33231
Pighouse Folk Museum, Cornafean 24771
Pihlajaveden Kotiseutumuseo, Keuruu 10953
Pihtiputaan Kotiseutumuseo, Pihtipudas 11206
Pijpenkabinet & Smokiana, Amsterdam 31880
Pikakannu Põhikooli Muuseum, Pikakannu 10608
Pikakannu School Museum, Pikakannu 10608
Pike County Museum, Milford 50689
Pikšas - Memorial Museum of Kārlis Ulmanis, Bērze 29991
Pila Museum, Pila 34480
Pilgrim Hall Museum, Plymouth 51995
Pilgrim Monument and Provincetown Museum, Provincetown 52182
Pilgrim's Rest Museum, Pilgrim's Rest 38709
Pilies Muzieus, Klaipėda 30284
Pilon Gallery, Ajdovščina 38397
Pilonova Galerija, Ajdovščina 38397
Pilot Station, Uusikaupunki 11419
Pilots Cottage Museum, Kiama 01196
Pilz-Insekten-Vogel-Museum, Ribnitz-Damgarten 21311
Pilzmuseum Taura, Schildau, Gneisenaustadt 21551
Pima Air and Space Museum, Tucson 53595
Pimeria Alta Historical Society Museum, Nogales 51358
Pin Art Gallery, Klaaswaal 32447
Pinacoteca, Castelfiorentino 25726
Pinacoteca, Chiomonte 25878
Pinacoteca, Córdoba 00339
Pinacoteca, Teramo 27956
Pinacoteca, Varallo Sesia 28158
Pinacoteca Ambrosiana, Milano 26719
Pinacoteca Andrea Alfano, Castrovillari 25770
Pinacoteca Barão de Santo Ângelo, Porto Alegre 04654
Pinacoteca Benedicto Calixto, Santos 04822
Pinacoteca Cantonale Giovanni Züst, Rancate 41741
Pinacoteca Carmelita del Convento del Carmen, Trujillo 34323
Pinacoteca Cascella, Ortona 27004
Pinacoteca Chiesa di San Giovanni, Pieve Torina 27217
Pinacoteca Chiesa di San Tomaso Becket, Padova 27052
Pinacoteca Civica, Abano Terme 25136
Pinacoteca Civica, Ascoli Piceno 25298
Pinacoteca Civica, Baiardo 25351
Pinacoteca Civica, Cento 25802
Pinacoteca Civica, Cepagatti 25804
Pinacoteca Civica, Crotone 26015
Pinacoteca Civica, Fermo 26103
Pinacoteca Civica, Iesi 26395
Pinacoteca Civica, Imola 26403
Pinacoteca Civica, Imperia 26407
Pinacoteca Civica, Montelupone 26841
Pinacoteca Civica, Monza 26865
Pinacoteca Civica, Pieve di Cento 27214
Pinacoteca Civica, San Gimignano 27641
Pinacoteca Civica, Santarcangelo di Romagna 27727
Pinacoteca Civica, Savona 27780
Pinacoteca Civica, Spello 27886
Pinacoteca Civica, Vado Ligure 28132
Pinacoteca Civica A. Ricci, Monte San Martino 26804
Pinacoteca Civica Amedeo Modigliani, Follonica 26217
Pinacoteca Civica Bruno Molajoli, Fabriano 26071
Pinacoteca Civica d'Arte Moderna, Latina 26467
Pinacoteca Civica e Gipsoteca U. Gera, Ripatransone 27419
Pinacoteca Civica F. Duranti, Montefortino 26830
Pinacoteca Civica Francesco Podesti, Ancona 25231
Pinacoteca Civica G. Cattabriga, Bondeno 25494
Pinacoteca Civica Melozzo degli Ambrogi, Forlì 26230
Pinacoteca Civica Tosio Martinengo, Brescia 25542
Pinacoteca Comunale, Assisi 25311
Pinacoteca Comunale, Bevagna 25413
Pinacoteca Comunale, Bosa 25513
Pinacoteca Comunale, Castiglion Fiorentino 25758
Pinacoteca Comunale, Cesena 25835
Pinacoteca Comunale, Città di Castello 25909
Pinacoteca Comunale, Deruta 26033
Pinacoteca Comunale, Faenza 26079
Pinacoteca Comunale, Locarno 41570
Pinacoteca Comunale, Manciano 26583
Pinacoteca Comunale, Massa Fermana 26622
Pinacoteca Comunale, Massa Marittima 26633
Pinacoteca Comunale, Matelica 26639
Pinacoteca Comunale, Narni 26915
Pinacoteca Comunale, Ostra 27021
Pinacoteca Comunale, Quistello 27352
Pinacoteca Comunale, Ripe San Ginesio 27420
Pinacoteca Comunale, San Benedetto del Tronto 27612
Pinacoteca Comunale, San Severino Marche 27686
Pinacoteca Comunale, Sarnano 27743
Pinacoteca Comunale, Spoleto 27895
Pinacoteca Comunale, Verucchio 28239
Pinacoteca Comunale A. Moroni, Porto Recanati 27303
Pinacoteca Comunale Alberto Martini, Oderzo 26970
Pinacoteca Comunale D. Stefanucci, Cingoli 25894
Pinacoteca Comunale E. Giannelli, Parabita 27090
Pinacoteca Comunale e Museo A. De Felice, Terni 27964
Pinacoteca Comunale F. Galante Civera, Margherita di Savoia 26605
Pinacoteca Comunale Foresiana, Portoferraio 27310
Pinacoteca Comunale Francesco Cozza, Stilo 27906
Pinacoteca Comunale Vincenzo Bindi, Giulianova 26348
Pinacoteca D. Inzaghi, Budrio 25561
Pinacoteca Dantesca Fortunato Bellonzi, Torre de Passeri 28035
Pinacoteca d'Arte Antica, Gemona del Friuli 26284
Pinacoteca d'Arte Contemporanea, Smerillo 27861
Pinacoteca d'Arte Francescana, Lecce 26475
Pinacoteca d'Arte Moderna, Avezzano 25336
Pinacoteca d'Arte Moderna e Contemporanea "AM International", Bivongi 25431
Pinacoteca Davide Bergh, Calice al Cornovoglio 25600
Pinacoteca de la Escuela Provincial de Bellas Artes Dr. Figueroa Alcorta, Córdoba 00340
Pinacoteca de la Facultad de Bellas Artes, Manizales 08671
Pinacoteca de la Universidad de Concepción, Concepción 07436
Pinacoteca de Nuevo León, Monterrey 31189
Pinacoteca dei Cappuccini, Voltaggio 28309
Pinacoteca del Ateneo, Saltillo 31318
Pinacoteca del Banco Central de Nicaragua, Managua 33291
Pinacoteca del Centro de la Profesa, México 31158
Pinacoteca del Duomo, Cittadella 25912
Pinacoteca del Estado Juan Gamboa Guzmán, Mérida 31030
Pinacoteca del Pio Monte della Misericordia, Napoli 26910
Pinacoteca del Templo de la Compañía, Guanajuato 30922

Pinacoteca dell' Accademia Albertina di Belle Arti, Torino 28030
Pinacoteca della Cassa di Risparmio, Cesena 25836
Pinacoteca della Certosa, Firenze 26201
Pinacoteca della Rocca Ubaldinesca, Sassocorvaro 27759
Pinacoteca dell'Accademia Carrara, Bergamo 25406
Pinacoteca dell'Accademia dei Concordi e del Seminario, Rovigo 27575
Pinacoteca dell'Accademia di Belle Arti, Carrara 25691
Pinacoteca di Brera, Milano 26720
Pinacoteca di Corinaldo, Corinaldo 25978
Pinacoteca di San Silvestro, Montecompatri 26819
Pinacoteca Diego Rivera, Xalapa 31560
Pinacoteca Diocesana di Arte Sacra, Senigallia 27805
Pinacoteca do Estado de São Paulo, São Paulo 04911
Pinacoteca e Biblioteca Rambaldi, Sanremo 27697
Pinacoteca e Musei Civici, Jesi 26427
Pinacoteca e Musei Comunali, Macerata 26556
Pinacoteca e Museo Civici, Camerino 25623
Pinacoteca e Museo Civico, Bettona 25411
Pinacoteca e Museo Civico, Volterra 28312
Pinacoteca e Museo de Napoli, Terlizzi 27957
Pinacoteca e Museo delle Arti di Locri Epizephiri, Locri 26506
Pinacoteca E. Notte, Ceglie Messapica 25796
Pinacoteca Eduardo Ramírez Castro, Aranzazu 08531
Pinacoteca G. A. Levis, Racconigi 27354
Pinacoteca G. Spano, Ploaghe 27263
Pinacoteca Giovanni e Marella Agnelli, Torino 28031
Pinacoteca Giovanni Morscio, Dolceacqua 26040
Pinacoteca Giuseppe Stuard, Parma 27110
Pinacoteca in Palazzo Volpi, Como 25955
Pinacoteca Internazionale dell'Età Evolutiva A. Cibaldi, Rezzato 27399
Pinacoteca L. Répaci, Palmi 27087
Pinacoteca Malaspina, Pavia 27126
Pinacoteca Manfrediana, Venezia 28208
Pinacoteca Municipal de Guadalupe, Guadalupe 30909
Pinacoteca Municipal de la Provincia de Corongo, Corongo 34154
Pinacoteca Municipal Ignacio Merino, Lima 34274
Pinacoteca Municipal Leoncio Lugo, Paucartambo 34286
Pinacoteca Municipal Miguel Ângelo Pucci, Franca 04469
Pinacoteca-Museo Beato Sante, Mombaroccio 26770
Pinacoteca Nazionale, Bologna 25478
Pinacoteca Nazionale, Cagliari 25589
Pinacoteca Nazionale, Ferrara 26121
Pinacoteca Nazionale, Siena 27846
Pinacoteca Parrocchiale, Castroreale 25767
Pinacoteca Parrocchiale, Corridonia 25982
Pinacoteca Provinciale Corrado Giaquinto, Bari 25371
Pinacoteca Repossi, Chiari 25856
Pinacoteca Rossetti Valentini, Santa Maria Maggiore 27707
Pinacoteca Universitaria, Colima 30784
Pinacoteca Vaticana, Città del Vaticano 54514
Pinacoteca Virreinal de San Diego, México 31159
Pinacoteca Zelantea, Acireale 25143
Pinakothek der Moderne, München 20474
Pinakothiki Kouvoutsaki, Kifissia 23052
Pinal County Historical Museum, Florence 48567
Pinchbeck Marsh Engine and Land Drainage Museum, Spalding 45722
Pincher Creek & District Museum, Pincher Creek 06606
Pine County Historical Society Museum, Askov 46396
Pine Creek Museum, Pine Creek 01414
Pine Grove Historic Museum, Pontiac 52022
Pine Islet Lighthouse, Mackay 01247
Pinewood Museum, Wasagaming 07281
Ping Jin Zhan Yi, Old Zhi Display Center, Yangliuqing 08423
Ping Yang Wood Sculpture Museum, Hualien 42217
Pingdu Museum, Pingdu 08081
Pinggu Shangzhai Cultural Display Center, Beijing 07561
Pinghe Museum, Pinghe 08082
Pinghu Museum, Pinghu 08084
Pingliang Museum, Pingliang 08085
Pingliang Zone Museum, Pingliang 08086
Pinglin Tea Museum, Pinglin 42287
Pinglu Museum, Pinglu 08087
Pingshun Museum, Pingshun 08088
Píngtán Bówūguǎn, Shanghai 08176
Pingtung Fo Guang Yuan Art Gallery, Pingtung 42293
Pingtung Hakka Museum, Jhutian 42225
Pingtung Museum of Natural History, Hengchun 42200
Pingxiang Museum, Pingxiang 08089
Pingyuan Museum, Pingyuan 08092
Pinhey's Point, Ottawa 06553
Pinnacles Gallery, Thuringowa 01586
Pinnaroo Heritage Museum, Pinnaroo 01417
Pinnaroo Printing Museum, Pinnaroo 01418
Pinson Mounds State Archaeological Area, Pinson 51929
Pintér Muvek Hadtörténeti Múzeum, Hadtörténeti Park, Kecel 23642
Pinzgauer Heimatmuseum, Saalfelden am Steinernen Meer 02607
Pioneer Air Museum, Fairbanks 48473
Pioneer Arizona Living History Village, Phoenix 51907
Pioneer Auto Museum, Murdo 50910
Pioneer Corner Museum, New Holstein 51060
Pioneer Farm Museum and Ohop Indian Village, Eatonville 48275
Pioneer Florida Museum, Dade City 47919
Pioneer Heritage Center, Cavalier 47285
Pioneer Heritage Center, Shreveport 53086
Pioneer Historical Connors Museum, Connors 05639
Pioneer-Krier Museum, Ashland 46379
Pioneer Log Cabin, Manhattan 50436
Pioneer Memorial Association of Fenton and Mounty Townships, Fenton 48533
Pioneer Museum, Fabius 48468
Pioneer Museum, Fairbanks 48474
Pioneer Museum, Sayward 06935
Pioneer Museum, Silverton 38729
Pioneer Museum, Watford City 53980
Pioneer Museum, Wild Rose 54159
Pioneer Museum and Vereins Kirche, Fredericksburg 48762
Pioneer Museum of Alabama, Troy 53571
Pioneer Museum of Motorcycles, Tacoma 53413
Pioneer Museum Society of Grande Prairie and District, Grande Prairie 05901
Pioneer Rum Museum, Beenleigh 00853
Pioneer Settlement Museum, Swan Hill 01540
Pioneer Town, Wimberley 54217
Pioneer Village, Farmington 48520
Pioneer Village, Silverdale 33201
Pioneer Village Museum, Burnie 00924
Pioneer Woman Statue and Museum, Ponca City 52018
Pioneerimuseo, Ala-Pihlaja 10707
Pioneers of Aviation Museum, Kimberley 38672
Pioneers, Trail and Texas Rangers Memorial Museum, San Antonio 52719
Pioniermuseum, Klosterneuburg 02203
Piopio and District Museum, Piopio 33183
Pipaongo Etnografia Museoa, Pipaón 39842
Piper Aviation Museum, Lock Haven 50155
Pipestone County Historical Museum, Pipestone 51930
Piqua Historical Area State Memorial, Piqua 51931
Pirate Contemporary Art Oasis, Denver 48086
Piratenamüseum, Wilhelmshaven 22579
La Pirogue Museum, Cheticamp 05609
Piroska Szántó Memorial Museum, Budapest 23476
Pirot Ethnographical Museum, Pirot 38130
Piskarevskoe Memorialnoe Kladbišče - Muzej, Sankt-Peterburg 37720
Pitäjämuseo ja Nymann Talo ja Apteekkimuseo, Tohmajärvi 11379
Pitäjäntupa-Museo, Joutseno 10879
Piteå Museum, Piteå 40809
Piteå Sparkmuseum, Piteå 40810
Pitkänpellon Talomuseo, Kangasniemi 10913
Pitkäpelto House Museum, Kangasniemi 10913
Pitlochry Festival Theatre Art Gallery, Pitlochry 45386
Pitot House Museum, New Orleans 51094
Pitstone Green Farm Museum, Pitstone 45387
Pitt Meadows Museum, Pitt Meadows 06607
Pitt Rivers Museum, Oxford 45338
Pittencrieff House Museum, Dunfermline 44056
Pittock Mansion, Portland 52085
Pittsburg State University Natural History Museum, Pittsburg 51935
Pittsburgh Center for the Arts, Pittsburgh 51956
Pittsburgh Children's Museum, Pittsburgh 51957
Pittsford Historical Society Museum, Pittsford 51971
Pittsworth Pioneer Historical Museum, Pittsworth 01419
Pitzhanger Manor-House and Gallery, London 44955
Pivovarské Muzeum, Plzeň 09718
Pivovarski Muzej, Ljubljana 38459
Pixelprojekt_Ruhrgebiet, Gelsenkirchen 18579
Pixian Museum, Pixian, Sichuan 08094
Pizhou Museum, Pixian, Jiangsu 08093
Pjatigorsk Regional Museum, Pjatigorsk 37565
Pjatigorskij Kraevedčeskij Muzej, Pjatigorsk 37565
Pjóðmenningarhúsið, Reykjavík 24029
Pjóðminjasafn Islands, Reykjavík 24030
PLA Naval Museum, Qingdao 08107
Place of Rememberance of Forced Labour in the Volkswagen Factory, Wolfsburg 22661
Placentia Area Museum, Placentia 06608
Plaine des Dinosaures, Mèze 14113
Plains Art Museum, Fargo 48511
Plains Indians and Pioneers Museum, Woodward 54293
Plains Vintage Railway and Historical Museum, Ashburton 33039
Plainsman Museum, Aurora 46497
Plamondon and District Museum, Plamondon 06611
Plan Incliné de Ronquières, Ronquieres 03984
Plan-Weseritzer Heimatstuben mit Archiv, Tirschenreuth 22063
Planet Bordeaux, Beychac-et-Caillau 11970
Planet Earth Museum, Newhaven 45237
Planète Bordeaux, Beychac-et-Caillau 11970
Planète de la Découverte, Beirut 30162
Planetron - Aards Paradijs, Dwingeloo 32162
Planšarski Muzej, Bohinjsko Jezero 38402
Plantation Agriculture Museum, Scott 52973
Plantation Historical Museum, Plantation 51978
Plantation of Ulster Visitor Centre, Draperstown 44022
Plas Mawr, Conwy 43887
Plas Newydd, Llanfairpwll 44765
Plas Newydd, Llangollen 44772
Plassenburg, Kulmbach 19750
Plastic Club, Philadelphia 51870
Plastov Museum of Modern Art, Uljanovsk 37908
Plastov Picture Gallery, Uljanovsk 37899
Platenhäuschen, Erlangen 18224
Platform 1 Heritage Farm Railway, Littlehampton 01235
Platform - Centre for Photographic and Digital Arts, Winnipeg 07373
Platt and Borstein Galleries, Los Angeles 50260
Platt R. Spencer Memorial Archives and Special Collections Area, Geneva 48882
Plattner Bienenhof Museum, Soprabolzano 27875
Plattsburgh Art Museum, Plattsburgh 51984
Plauener Spitzenmuseum, Plauen 21054
Play Museum Soltau and Foundation Play, Soltau 21815
Playhouse Gallery, Orleans 51598
Plaza de la Raza - Boathouse Gallery, Los Angeles 50261
Pleasant Point Railway and Historical Society Museum, Pleasant Point 33184
Please Touch Museum, Philadelphia 51871
Plečnik Collection of the Architecture Museum Ljubljana, Ljubljana 38460
Plečnikova Zbirka, Ljubljana 38460
Plenty and District Museum, Plenty 06613
Plesskij Gosudarstvennyj Istoriko-Architekturnyj i Chudožestvennyj Muzej-Zapovednik, Ples 37568
Pleven Art Gallery - Donation Svetlin Russev Collection, Pleven 05115
Plimsoll Gallery, Hobart 01156
Płocka Galeria Sztuki, Płock 35015
plug in-Kunst und Neue Medien, Basel 41177
Pluimvee Museum, Barneveld 31950
Plum Coulee and District Museum, Plum Coulee 06614
Plum Grove Historic Home, Iowa City 49519
Plumas County Museum, Quincy 52211
Plumb Farm Museum, Greeley 49035
Plumber Museum, Tomar 35963
Plumershuuske, Diepenheim 32126
Plunket Museum of Irish Education, Dublin 24816
Plymouth Plantation Museum, Plymouth 51996
Plymouth Antiquarian Society Museum, Plymouth 51997
Plymouth Arts Centre, Plymouth 45392
Plymouth Historical Museum, Plymouth 51999
Plynárenské Múzeum, Bratislava 38231
Poble Espanyol, Palma de Mallorca 39794
Pocahontas County Iowa Historical Society Museum, Laurens 49986
Pocahontas County Museum, Marlinton 50497
Pocarjeva Domačija v dolini Radovne, Kranjska Gora 38440
Pocar's Farmhouse, Kranjska Gora 38440
Počvenno-agronomičeskij Muzej im. Akademika V.R. Viljamsa, Moskva 37398
Podgórze Dom Kultury Galeria, Kraków 34818
Podhorácké Muzeum - Klášter Porta Coeli, Předklášteří 09802
Podium Kunst, Schramberg 21632
Podjavorinské Múzeum, Nové Mesto nad Váhom 38320
Podještědské Muzeum, Český Dub 09466
Podlipanské muzeum v Českém Brodě, Český Brod 09465
Podpolianske Múzeum Detva, Detva 38247
Podpolnaja Tipografija CK RSDRP 1905-1906, Moskva 37399
Podřipské Muzeum, Roudnice nad Labem 09821
Podsreda Castle, Podsreda 38494
Podtatranské Múzeum, Poprad 38332
Poeh Museum, Santa Fe 52886
Pöljä Local History Museum, Pöljä 11210
Pöljän Kotiseutumuseo, Pöljä 11210
Pörtom Hembygdsmuseum, Pörtom 11211
Poet Jovaras House of the Šiauliai Aušra Museum, Šiauliai 30359
Poet Maironis Birthplace, Bernotai 30211
Poeto Jono Mačiulio-Maironio Gimtinė, Bernotai 30211
Poeto Jovaro Namas, Šiauliai 30359
Pöytyän Kotiseutumuseo, Pöytyä 11212
Pohjanmaan Museo, Vaasa 11426
Pohjois-Karjalan Museo, Joensuu 10873
Pohjois-Pohjanmaan Museo, Oulu 11177
Pohronské Múzeum v Novej Bani, Nová Baňa 38318
Point Amour Lighthouse, L'Anse Amour 06138
Point Ellice House, Victoria 07257
Point Leamington Heritage Interpretation Centre, Point Leamington 06615
Point of Honor, Lynchburg 50329
Point of Remembrance, Hódmezővásárhely 23611
Point Pelee Natural Park Visitor Centre, Leamington 06155
Pointe-à-Callière - Musée d'Archéologie et d'Histoire de Montréal, Montréal 06378
Pointe Coupee Museum, New Roads 51101
Pointe-Noire - Parc Marin du Saguenay, Baie-Sainte-Catherine 05374
Pokrajinski Muzej, Maribor 38477
Pokrajinski Muzej, Murska Sobota 38487
Pokrajinski Muzej, Strumica 30528
Pokrajinski Muzej Celje, Celje 38407
Pokrajinski Muzej Kočevje, Kočevje 38432
Pokrajinski Muzej Koper, Koper 38434
Pokrajinski Muzej Ptuj, Ptuj 38499
Pokrov Church Museum, Moskva 37154
Pokrovskij Sobor-chram Vasilija Blažennogo, Moskva 37400
Pola Fuzoku Hakubutsukan, Tokyo 29459
Pola Museum Annex, Tokyo 29459
Pola Museum of Art, Hakone 28480
Polabské Muzeum, Poděbrady 09724
Polarmuseet, Andenes 33393
Polarmuseet i Tromsø, Tromsø 33919
Polarmuseum in Tromsø, Tromsø 33919
Poldark Mine and Heritage Complex, Helston 44446
Poldermuseum, Antwerpen 03310
Poldermuseum, Puttershoek 32674
Poldermuseum de Hooge Boezem Achter Haastrecht, Haastrecht 32306
Poldermuseum Den Huijgen Dijck, Heerhugowaard 32335
Poldermuseum Het Grootslag - Nationaal Saet en Cruytmuseum, Andijk 31907
Poldowrian Museum of Prehistory, Coverack 43905
Polemiko Mouseio, Athinai 22951
Polenmuseum Rapperswil, Rapperswil 41746
Polenz-Museum, Cunewalde 17753
Polesden Lacey House, Dorking 44007
Poli Museo della Grappa, Bassano del Grappa 25382
Polibinskij Memoralnyj Muzej-usadba S.V. Kovalevskoj, Polibino 37570
Police Headgear Museum, Slochteren 32786
Police History Museum, Budapest 23472
The Police Museum, Belfast 43446
Police Museum, Hong Kong 07813
Police Museum, Imphal 24227
Police Station and Courthouse Collections, Auburn 00818
Polish American Museum, Port Washington 52048
Polish Army Museum, Kołobrzeg 34751
Polish Association of Members of the Museum Professions, Kraków 54988
Polish Aviation Museum, Kraków 34804
Polish Bar Museum, Warszawa 35233
Polish Cultural Institute, London 44956
Polish Estate Museum, Plochocin 35012
Polish Maritime Museum, Gdańsk 34631
Polish Military Museum, Warszawa 35280
Polish Museum of America, Chicago 47507
Polish Museums Association, Kraków 54990
Polish Scenography Center - Andrzej Majewski's Theatrical Gallery, Katowice 34719
Polishistoriska Museet, Stockholm 40925
Polistekniska Museet, Solna 40872
Politechničskij Muzej, Moskva 37401
Political Life Museum, Amman 29659
Political Museum, Shanghai 08177
Politie-Petten Museum, Slochteren 32786
Politiemuseum, Wommelgem 04157
Politiemuseum Oudaan, Antwerpen 03311
Politimuseet i Oslo, Oslo 33748
Polizeihistorische Sammlung Berlin, Berlin 17235
Polizeimuseum, Aarau 41084
Polizeimuseum, Basel 41178
Polk County Heritage Gallery, Des Moines 48098
Polk County Historical Museum, Bartow 46622
Polk County Historical Museum, Cedartown 47309
Polk County Historical Museum, Osceola 51610
Polk County Historical Society, Des Moines 48099
Polk County Memorial Museum, Livingston 50151
Polk County Museum, Balsam Lake 46543
Polk County Museum, Crookston 47886

P

Polk Museum of Art, Lakeland 49902
Pollismolen, Bos-en Pijpenmuseum, Bree 03404
Pollock Gallery, Dallas 47938
Pollock's Toy Museum, London 44957
Pollok House, Glasgow 44295
Polokwane Art Museum, Polokwane 38710
Polson Park and Museum, Hoquiam 49366
Poltava Art Museum, Poltava 43167
Poltava Museum of Aviation and Cosmonautics, Poltava 43172
Poltava Regional Museum, Poltava 43168
Poltavs'kyj Chudožnij Muzej, Poltava 43167
Poltavs'kyj Krajeznavčyj Muzej, Poltava 43168
Poltavs'kyj Literaturno-memorial'nyj Muzej I.P. Kotljarevskogo, Poltava 43169
Poltavs'kyj Literaturno-memorial'nyj Muzej Panasa Myrnogo, Poltava 43170
Poltavs'kyj Literaturno-memorial'nyj Muzej V.G. Korolenka, Poltava 43171
Poltavs'kyj Muzej Aviacii ta Kosmonavtiki, Poltava 43172
Põltsamaa Muuseum, Põltsamaa 10609
Põlva Talurahvamuuseum, Karilatsi 10580
Polvo Art Studio, Chicago 47508
Poly Art Museum, Beijing 07562
Polygnotos Vagis Museum, Thassos 23204
Polynesian Cultural Center, Laie 49873
Polytechnical Museum, Moskva 37401
Polyteekkarimuseo, Espoo 10732
Pomarkun Kotiseutumuseo, Pomarkku 11218
Pomázer Heimatstube, Sinsheim 21785
Pomeroy Living History Farm, Yacolt 54318
Pommeranian Military Museum, Bydgoszcz 34581
Pommersches Bettenmuseum, Peenemünde 20997
Pommersches Landesmuseum, Greifswald 18750
Pomona College Museum of Art, Claremont 47574
Pomorski i Povijesni Muzej Hrvatskog Primorja, Rijeka 08939
Pomorski Muzej, Dubrovnik 08872
Pomorski Muzej, Zadar 08984
Pomorski muzej Orebić, Orebić 08920
Pomorski Muzej Sergej Mašera, Piran 38492
Pomorski Muzej u Kotoru, Kotor 31647
Pomorskie Muzeum Wojskowe, Bydgoszcz 34581
Pompallier House, Russell 33199
Pompejanum, Aschaffenburg 16589
Pompey Museum of Slavery and Emancipation, Nassau 03210
Pomskizillious Museum of Toys, Xaghra 30659
Ponce de León's Fort Museum, Higüey 10328
Ponce DeLeon Inlet Lighthouse, Ponce Inlet 52019
Ponce History Museum, Ponce 36050
Ponce Museum of Art, Ponce 36049
Pondicherry Museum, Pondicherry 24351
Pondmill in the franconian forrest, Steinwiesen 21895
Pongauer Heimatmuseum im Schloß Goldegg, Goldegg 01957
Poni-Cernătescu Memorial Museum, Iaşi 36424
Ponitrianske Múzeum, Nitra 38315
Ponshu-kan-Echigo Sake Museum, Yuzawa 29636
De Pont, Tilburg 32830
Pont-à-Mousson City Museum 'Au fil du papier', Pont-à-Mousson 14877
Pontefract Castle Museum, Pontefract 45397
Pontefract Museum, Pontefract 45398
Pontifical Biblical Institute Museum, Jerusalem 25009
Pontypool and Blaenavon Railway, Blaenavon 43535
Pontypridd Museum, Pontypridd 45403
Pony Express Museum, Saint Joseph 52564
Pope County Historical Museum, Glenwood 48942
Pope's Tavern Museum, Florence 48564
Poplar Grove Historic Plantation, Wilmington 54204
Popmuseum, Praha 09789
Popol Vuh Archaeological Museum, Guatemala 23305
Poppelsdorfer Heimatsammlung Stöcker, Bonn 17464
Poppen en Poppenhuis Museum 't Duvelke, Breda 32007
Poppen- en Speelgoedmuseum, Tilburg 32831
Poppenhuis Carmen, Utrecht 32855
Poppenhuismuseum, Heesch 32341
Poppenmuseum, Ter Apel 31910
Poppenmuseum, Maastricht 32530
Poppenmuseum Christus Koning, Wetteren 04150
Poppenspe(e)lmuseum, Vorchten 32908
Poppentheatermuseum, Tienen 04069
Popppenhuismuseum Alida's Kleine Wereldje, Veendam 32871
Popta Castle, Marssum 32541
Poptaslot Heringa-State, Marssum 32541
Popular Art Centre, Al-Bireh 34047
Popular Arts Museum, San Juan 36078
Popular Culture Museum, Heredia 08809
Popular Monument House, Zalalövo 23973
Popular Traditions Museum Qasrelazem, Damascus 42133
Porcelain Museum, Herend 23605
Porcelain Museum, Santo Domingo 10348
Porcelánmúzeum, Hollóháza 23615
Porcupine Plain and District Museum, Porcupine Plain 06623
Porfyrmuseet, Älvdalen 40407
Pori Art Museum, Pori 11220
Porilaismuseo, Säkylä 11290
Porin Taidemuseo, Pori 11220
Porirua Hospital Museum, Porirua 33187
Porkeri Museum, Porkeri 10695
Porkeris Bygdasavn, Porkeri 10695
Porlammi Local History Museum, Porlammi 11224
Porlammin Kotiseutumuseo, Porlammi 11224
Porsanger Museum, Lakselv 33623
Porsche Automuseum Helmut Pfeifhofer, Gmünd, Kärnten 01942
Porsche Museum, Stuttgart 21957
Porsche Traktorenmuseum, Hitzendorf 02086
Porselein Dierenpark, Wildervank 32950
Porsgrunn Bymuseum, Porsgrunn 33770
Port Albert Maritime Museum, Port Albert 01426
Port Angeles Fine Arts Center, Port Angeles 52028
Port Arthur Historic Site, Port Arthur 01427
Port-au-Choix, Port-au-Choix 06628
Port Chalmers Museum, Port Chalmers 33188
Port Clements Museum, Port Clements 06632
Port Colborne Historical and Marine Museum, Port Colborne 06633
Port Columbus National Civil War Naval Center, Columbus 47740
Port Community Art Centre, Port Adelaide 01423
Port Coquitlam Heritage and Cultural Museum, Port Coquitlam 06634
Port de Québec au XIXe Siècle, Saint-André 06792
Port Discovery Children's Museum, Baltimore 46584
Port Discovery - Children's Museum in Baltimore, Baltimore 46585
Port Dover Harbour Museum, Port Dover 06636
Port Elizabeth Museum, Port Elizabeth 38712
Port Fairy Historic Lifeboat Station, Port Fairy 01429
Port Fairy History Centre, Port Fairy 01430
Port Gamble Historic Museum, Port Gamble 52032
Port Hardy Museum, Port Hardy 06638
Port Huron Museum, Port Huron 52035
Port-la-Joye-Fort Amherst, Rocky Point 06770
Port MacDonnell and District Maritime Museum, Port MacDonnell 01433
Port Macquarie Historic Museum, Port Macquarie 01434
Port McNeill Museum, Port McNeill 06644
Port Moody Station Museum, Port Moody 06646
Le Port-Musée, Douarnenez 12776
Port Natal Maritime Museum, Durban 38611
Port of Morgan Historic Museum, Morgan 01321
Port Phillip City Collection, Saint Kilda, Victoria 01481
Port Pirie National Trust Museum, Port Pirie 01435
Port Pirie Regional Art Gallery, Port Pirie 01436
Port Royal, Annapolis Royal 05344
Port-Said Museum, Port Said 10538
Port Stephens Shell Museum, Corlette 01001
Port Sunlight Heritage Centre, Port Sunlight 45416
Port Townsend Marine Science Center, Port Townsend 52045
Port Union Museum, Port Union South 06649
Port Victor Gallery, Victor Harbor 01617
Port Victoria Maritime Museum, Port Victoria 01437
Portage County Historical Society Museum, Ravenna 52256
Portage County Museum, Stevens Point 53290
Portal of the Folded Wings, North Hollywood 51395
Portal zur Geschichte, Bad Gandersheim 16739
Portanje's Vespa Scooter en Nostalgie Collectie, Bunnik 32019
Portchester Castle, Portchester 45419
La Porte County Historical Society Museum, La Porte 49853
Porter County Museum, Valparaiso 53699
Porter-Phelps-Huntington House Museum, Hadley 49118
Porter Thermometer Museum, Onset 51567
Porterville Historical Museum, Porterville 52060
Porthcawl Museum, Porthcawl 45420
Porthcurno Telegraph Museum, Porthcurno 45421
Porthmadog Maritime Museum, Porthmadog 45422
Portholes Into the Past, Medina 50558
Portico Gallery, Helensburgh 44439
Portico Library and Gallery, Manchester 45097
Portikus, Frankfurt am Main 18389
Portland Art Museum, Portland 52086
Portland Basin Museum, Ashton-under-Lyne 43333
Portland Castle, Portland 45424
Portland Children's Museum, Portland 52087
Portland Harbor Museum, South Portland 53180
Portland Maritime Discovery Centre, Portland 01440
Portland Museum, Louisville 50295
Portland Museum, Portland 45425
Portland Museum of Art, Portland 52067
Portland State University Galleries, Portland 52088
Portmoneum - Muzeum Josefa Váchala, Litomyšl 09634
Portobello Gallery, Saint James 03240
Portsmouth Athenaeum, Portsmouth 52098
Portsmouth Museum, Portsmouth 52104
Portsmouth Museum of Military History, Portsmouth 52111
Portuguese Artillery School Museum, Vendas Novas 35981
Portuguese Freemanson Museum, Lisboa 35712
Portuguese Post and Telephone Company Museum, Lisboa 35673
Porvoon Museo, Porvoo 11230
Porzellan-Museum Höchst, Frankfurt am Main 18390
Porzellanmuseum Reichmannsdorf, Reichmannsdorf 21249
Porzellansammlung, Dresden 17966
Posavje Museum at Castle Brežice, Brežice 38404
Posavski Muzej Brežice, Brežice 38404
Poshalostin Memorial House-Museum, Solotča 37787
Posion Kotiseutumuseo, Posio 11234
POSK Gallery, London 44958
Post and Telecommunication Museum, Wrocław 35336
Post-Archaeological Institute, Jackson 49567
Post-horn Gallery, Miskolc 23713
Post Museum, Fort Leavenworth 48643
Post Museum, Helsinki 10809
Post Museum, Pécs 23771
Post Museum, Vyšší Brod 09922
Post Museum Banská Bystrica, Banská Bystrica 38185
Post Office Museum, Hollókö 23617
Post Office Museum, Nagyvázsony 23729
Post Office Museum, Opusztaszer 23737
Post Rock Museum, La Crosse 49834
Post & Tele Museum, København 10148
Postakürt Galéria, Miskolc 23713
Postal and Philatelic Museum, Tel Aviv 25119
Postal Museum, Amman 29660
Postal Museum, Balatonszemes 23355
Postal Museum, Budapest 23468
Postal Museum, Cairo 10513
Postal Museum, Cheonan 29785
Postal Museum, Praha 09790
Postal Museum, Shanghai 08178
Postal Museum, Storby 11330
Postal Museum, Taipei 42419
Postal Museum and Philatelic Library, Manila 34453
Postal Museum Botswana, Gaborone 04297
Postamúzeum, Balatonszemes 23355
Postamúzeum, Budapest 23468
Postamúzeum, Hollókö 23617
Postamúzeum, Nagyvázsony 23729
Postamúzeum, Opusztaszer 23737
Postapalota, Pécs 23771
Postcard Museum, Kobe 28724
Poste de Traite Chauvin, Tadoussac 07058
Poster Museum, Hoorn 32408
Poster Museum at Wilanów, Warszawa 35264
Postimuseo, Helsinki 10809
Postmaster Museum, Vyra 37983
Póstminjasafn Íslands, Hafnarfjörður 24000
Postmuseum, Stockholm 40926
Postmuseum des Fürstentums Liechtenstein, Vaduz 30200
Postmuseum Mettingen, Mettingen 20310
Postmuseum Rheinhausen, Oberhausen-Rheinhausen 20811
Postojannaja Požarno-techničeskaja Vystavka, Sankt-Peterburg 37721
Poštové Múzeum Banská Bystrica, Banská Bystrica 38185
Poštovní Muzeum, Praha 09790
Poštovní Muzeum, Vyšší Brod 09922
Postrotemuseet, Storby 11330
Postville Courthouse Museum, Lincoln 50087
Potchefstroom Museum, Potchefstroom 38714
Poterie Dubois-Musée la Grange aux Potiers, Bouffioulx 03390
Potland Museum, Basildon 43393
Potsdam-Museum, Potsdam 21103
Potsdam Public Museum, Potsdam 52115
Pottenbakkerij Het Ovenhuis, Ruinerwold 32740
Pottenbakkerij Museum, Kattendijke 32442
Potter County Historical Society Museum, Coudersport 47848
Potter House, Tihany 23926
The Potteries Museum and Art Gallery, Stoke-on-Trent 45762
Potters Bar Museum, Potters Bar 45441
Potter's House at Csúcs, Hódmezövásárhely 23610
Pottery Museum, Argentona 38890
Pottery Museum, Barcelos 35456
Pottery Museum, Montalegre 35772
Pottery Museum, Raeren 03960
Pottery Museum, Redondo 35882
Pottery Museum, Somero 11323
Potts Inn Museum, Pottsville 52118
Pottsgrove Manor, Pottstown 52117
Pouce Coupe and District Museum, Pouce Coupe 06652
Pouch Cove Museum, Pouch Cove 06653
Pound Ridge Museum, Pound Ridge 52125
Pouryun Museum, Guanyun 07738
Považie Art Gallery, Žilina 38390
Považská Galéria Umenia, Žilina 38390
Považské Múzeum, Žilina 38391
Povijesni Muzej Istre, Pula 08935
Powder Magazine, Beechworth 00849
Powder Magazine, Charleston 47375
Powderham Castle, Kenton 44560
Powell County Museum, Deer Lodge 48031
Powell River Historical Museum, Powell River 06654
Power Plant Contemporary Art Gallery, Toronto 07128
Power Station- and Engineering Museum, Mynämäki 11131
Power Station Museum of Helsinki City Museum, Helsinki 10838
Powerhouse House and Car Museum, Portland 01441
Powerhouse Museum, Ultimo 01609
Powerhouse Museum Yanco, Yanco 01676
Powers Museum, Carthage 47268
Powis Castle, Welshpool 45962
Pownalborough Court House, Dresden 48182
Powysland Museum and Montgomery Canal Centre, Welshpool 45963
Pra Site Museum, Bogor 24461
Prabhas Patan Museum, Prabhas Patan 24352
Prachatické Muzeum, Prachatice 09733
Prácheňské Muzeum, Písek 09714
Pracownia - Muzeum Józefa Ignacego Kraszewskiego, Poznań 35047
Praedicantenbibliothek der Nikolaikirche, Isny im Allgäu 19367
Prähistorische Privatsammlung Walzer, Mühlen 02411
Prähistorische Siedlung Pestenacker, Weil, Kreis Landsberg am Lech 22382
Prähistorisches Heimatmuseum, Veringenstadt 22221
Prämonstratenser-Chorherrenstift mit Stiftssammlungen, Geras 01929
Prague Castle Gallery, Praha 09784
Prahova District Museum of History and Archaeology, Ploieşti 36533
Prairie Acres Museum, Picture Butte 06604
Prairie Homestead, Philip 51891
Prairie Memories Museum, Irvine 06024
Prairie Mountain Regional Museums Collection, Shoal Lake 06966
The Prairie Museum of Art and History, Colby 47680
Prairie Panorama Museum, Czar 05670
Prairie Pioneer Museum, Craik 05656
Prairie River Museum, Prairie River 06655
Prairie Trails Museum of Wayne County, Corydon 47837
Prairie Village, Madison 50386
Prairie West Historical Centre, Eston 05785
Pranas Domšaitis Gallery and Cultural Centre, Klaipėda 30277
Sri Pratap Singh Museum, Srinagar 24397
Pratermuseum, Wien 03058
Praters Mill, Varnell 53710
Pratt Building, Cohasset 47677
Pratt County Historical Society Museum, Pratt 52132
Pratt Manhattan Gallery, New York 51249
Pratt Museum, Homer 49330
Praxis Art Actuel, Sainte-Thérèse 06902
Precious Stones Museum, Polanica Zdrój 35019
Predjama Castle, Postojna 38497
Predjamski Grad, Postojna 38497
Prefectural Museum of Kissidougou, Kissidougou 23313
Prefectural Sakitama Archaeological Hall, Gyoda 28460
Prefectural Science Museum Joetsu, Joetsu 28621
Prefectural Torii Memorial Museum, Naruto 29045
Préhisto-Parc et Grottes, Réclère 41749
Préhistoire et Protohistoire, Jujurieux 13291
Le Préhistorama, Rousson 15130
Prehistoric Museum, Saint-Porchaire 15467
Prehistorical Museum, Korçë 00023
Prehistory Museum, Tripoli 30191
Prehistory Museum of Valencia, Valencia 40206
Preiļi History and Applied Art Museum, Preiļi 30056
Preiļu Vēstures un Lietišķās Mākslas Muzejs, Preiļi 30056
Prempeh Jubilee Museum, Kumasi 22859
Prentenkabinet Universiteit Leiden, Leiden 32486
Presbyterian Historical Society Museum, Montreat 50828
Presbyterian Historical Society Museum, Philadelphia 51872
Prescot Museum, Prescot 45442
Prescott House Museum, Port Williams 06650
Presentatieruimte de Overslag, Eindhoven 32194
Presentation House Gallery, Vancouver 07219
Prešernov Spominski Muzej, Kranj 38438
Presidency Museum, Tehrān 24690
President Andrew Johnson Museum, Greeneville 49052

P

President Benjamin Harrison Home, Indianapolis 49506
President Chester A. Arthur Historic Site, Fairfield 48487
President Pretorius Museum, Potchefstroom 38715
The Presidential Museum, Odessa 51497
Presidential Pet Museum, Williamsburg 54172
President's Cottage Museum, White Sulphur Springs 54131
President's Museum, Astana 29694
President's of Lithuania Kazys Grinius Memorial Museum, Marijampolė 30307
Presidio La Bahia, Goliad 48968
Presnja Istoriko-memorialnyj Muzej, Moskva 37402
Presque Isle County Historical Museum, Rogers City 52442
Presqu'île Provincial Park Museum, Brighton 05477
Press Museum, Amsterdam 31878
Press Museum, İstanbul 42807
Pressoirs des Ducs de Bourgogne, Chenôve 12501
Preston Hall Museum, Stockton-on-Tees 45757
Preston Manor, Brighton 43623
Preston Mill and Phantassie Doocot, East Linton 44090
Prestongrange Museum, Prestongrange 45448
Prestwould Foundation, Clarksville 47595
Pretoria Art Museum, Tshwane 38755
Prêtres de Saint-Sulpice de Montréal, Montréal 06379
Preus Museum, Horten 33564
Preußen-Museum Nordrhein-Westfalen, Minden 20337
Preußen-Museum Nordrhein-Westfalen, Wesel 22507
Preveli Monastery, Myrthios 23107
Price Tower Arts Center, Bartlesville 46619
Prichard Art Gallery, Moscow 50865
Pricketts Fort, Fairmont 48490
Prickwillow Drainage Engine Museum, Prickwillow 45449
Prienai Area Museum, Prienai 30340
Priental-Museum, Aschau im Chiemgau 16594
Prienų Krašto Muziejus, Prienai 30340
Priešgaisrinės Apsaugos Muziejus, Šiauliai 30360
Priest House, West Hoathly 45969
Priester Poppe Museum, Temse 04055
Priest's House, Easton-on-the-Hill 44107
Priest's House Museum, Wimborne Minster 46020
Prieuré du Vieux Logis, Nice 14441
Prignitz-Museum am Dom Havelberg, Havelberg 19054
Prignitz-Museum at Cathedral Havelberg, Havelberg 19054
Prilukin National Museum of the Karelian Livviki people, Olonec 37496
Primer Museo de las Abejas en México, Colonia Condesa 30786
Primorskaja Gosudarstvennaja Kartinnaja Galereja, Vladivostok 37948
Primorskaja State Picture Gallery, Vladivostok 37948
Primorskij Gosudarstvennyj Muzej im. V.K. Arseneva, Vladivostok 37949
Prince Albert Arts Centre, Prince Albert 06663
Prince Albert Historical Museum, Prince Albert 06664
Prince Chichibu Memorial Sports Museum, Tokyo 29376
Prince Edward Island Museum, Charlottetown 05595
Prince Edward Island Potato Museum, O'Leary 06510
The Prince Edward Island Regiment Museum, Charlottetown 05596
Prince Edward Island Shellfish Museum, Bidford 05424
Prince Georg Railway and Forestry Museum, Prince George 06668
Prince Milosh's Residence, Beograd 38054
Prince of Thurn and Taxis Museums, Regensburg 21222
Prince of Wales Fort, Churchill 05616
Prince of Wales Martello Tower, Halifax 05946
Prince of Wales Museum of Western India, Mumbai 24323
Prince of Wales Northern Heritage Centre, Yellowknife 07411
Prince Rupert Fire Museum, Prince Rupert 06672
Prince Township Museum, Sault Sainte Marie 06932
Princess Czartoryski Museum - National Museum in Cracow, Kraków 34803
Princess Ljubica's Residence, Beograd 38053
Princess of Wales' Own Regiment Military Museum, Kingston 06088
Princess of Wales's Royal Regiment and Queen's Regiment Museum, Dover 44016
Princess Patricia's Canadian Light Infantry Museum, Calgary 05531
Princessehof Leeuwarden, Leeuwarden 32478
Princeton and District Museum, Princeton 06674
Princeton Museum, Princeton 52150
Princeton University Art Museum, Princeton 52151
Princeton University Museum of Natural History, Princeton 52152
Pringle Cottage Museum, Warwick 01635
Prins Eugens Waldemarsudde, Stockholm 40927
Prins Hendrik de Zeevaarder Museum, Egmond aan Zee 32181
Print and Picture Collection, Philadelphia 51873
The Print Center, Philadelphia 51874
Print Collection of the National and University Library, Zagreb 08992
Print Consortium, Kansas City 49696
Print Room of the Polish Academy of Arts and Sciences, Kraków 34773
Print Room of the Warsaw University Library, Warszawa 35225
Printing House Museum, Cockermouth 43868
Printing Museum, Esbjerg 09999
Printing Museum, Kungsgården 40660
Printing Museum, Nantes 14363
Printing Museum, Tokyo 29408
Printmaking Council of New Jersey, Somerville 53151
Prioratsky Palace of the State Museum-Reserve Gatchina, Gatčina 36921
Prirodnjački Muzej, Beograd 38085
Prirodnjački Muzej Crne Gore, Podgorica 31657
Prirodnyčyj Muzej, Cherson 43025
Prirodonaučen Muzej, Belogradčik 04984
Prirodonaučen Muzej, Burgas 05000
Prirodonaučen Muzej, Černi Osăm 05002
Prirodonaučen Muzej, Kotel 05078
Prirodonaučen Muzej, Plovdiv 05132
Prirodonaučen Muzej, Varna 05246
Prirodonaučen Muzej na Makedonija, Skopje 30523
Prirodoslovni Muzej, Rijeka 08940
Prirodoslovni Muzej, Split 08962
Prirodoslovni Muzej Slovenije, Ljubljana 38461
Prirodovedčeskij Muzej, Smolensk 37775
Přírodovědecké Muzeum, Praha 09791
Prírodovedné Múzeum, Bratislava 38232
Prison Gate Museum, Den Haag 32086
Prison History Museum, Jūrmala 30022
Prison Model Museum, Nueva Gerona 09238
Prison Museum, Hämeenlinna 10750
Prittlewell Priory Museum, Southend-on-Sea 45712
Private Ethnographic Museum, Góra Kalwaria 34677
Private Museum of Grammophones an Phonographs, Sankt-Peterburg 37626
Private Museum of Sculptural Culture of the Azat Akimbek Fond, Almaty 29676
Private Sammlung Mechanischer Musikinstrumente, Markneukirchen 20208
Private School Museum, Ankara 42701
Privatizējamās Valsts Vukā Muzejs, Kuldīga 30035
Privatmuseum "Feuerwaffen einst und jetzt" im Reckturm, Wiener Neustadt 03098
Privatmuseum Hans Klein, Prichsenstadt 21125
Privatmuseum Im Blauen Haus, Appenzell 41120
Privatsammlung Bergkristallmuseum, Saalbach 02606
Privatsammlung Burg Runkel, Runkel 21450
Privatsammlung Hermine Brandstetter, Ostermiething 02475
Privatsammlung Leo Gesell, Triefenstein 22094
Pro Arts at Oakland Art Gallery, Oakland 51472
Pro Rodeo Hall of Fame and Museum of the American Cowboy, Colorado Springs 47706
Pro Urba, Orbe 41714
Probstei-Museum, Schönberg, Holstein 21599
Prochorovo Field - Museum-preserve of Military History, Prochorovka 37571
Proclamation of Independence Memorial, Melaka 30597
Proctor House Museum, Brighton 05478
Produzentengalerie im Wasserturm Eutin, Eutin 18286
Prof. Adomas Hrebnickis Memorial Museum, Dūkšto 30227
Prof. Aleksandra Bieziņa Muzejs, Madona 30046
Prof. Bertlein Schulmuseum, Worms 22688
Prof. E.V. Jershov Geological Museum of the Moscow State Mining Institute, Moskva 37180
Prof. Eidenberger-Museum, Niederwaldkirchen 02449
Prof. G.R. Sharma Memorial Archaeology Museum, Allahabad 24062
Prof. Hans Gerstmayr - Museum, Mauthausen 02374
Profanierte Marienkapelle, Ludwigsstadt 20064
Profesoriaus Povilo Matulionio Memorialinis Muziejus, Kupiškis 30293
Profesoriaus Stasio Ušinsko Memorialinis Kambarys, Pakruojis 30326
Professor Li Mei-Shu Memorial Museum, Sansia 42313
Professoriaus Adomo Hrebnickio Memorialinis Muziejus, Dūkšto 30227
Profil Auto- und Technikmuseum, Mittelbiberach 20341
Profiles Public Art Gallery, Saint-Albert 06791
Programme for Museum Development in Africa, Mombasa 54916
Project Art-University of Iowa and Clinics, Iowa City 49520
Project Arts Center, Dublin 24817
Prokofyev-Museum, Krasne 43074
Prokofyev-Museum, Moskva 37350
Prokop Diviš Memorial, Znojmo 09948
Promont House Museum, Milford 50688
Pronkkamer Uden-Museum Hedendaagse Kunst, Uden 32834
Propst House and Marple Grove, Hickory 49268
Prora Tierreich, Prora 21135
Prorokov House-Museum, Ivanovo 36944
Proserpine Historical Museum, Proserpine 01443
Prospect Hill Historical Museum, Meadows 01272
Protestants Historisch Museum Abraham Hans, Horebeke 03712
Proussos Monastery Museum, Prousos 23159
Prouty-Chew Museum, Geneva 48880
Provan Hall House, Glasgow 44296
Provand's Lordship, Glasgow 44297
Proviant-Eisen Museum, Gresten 01990
Providence Art Club, Providence 52172
Providence Athenaeum, Providence 52173
Providence Children's Museum, Providence 52174
Providence Preservation Society, Providence 52175
Providence University Art Center, Shalu 42321
Province House, Charlottetown 05597
Provincetown Art Association and Museum, Provincetown 52183
Provincetown Heritage Museum, Provincetown 52184
Provinciaal Archeologisch Museum, Velzeke-Ruddershove 04104
Provinciaal Archeologisch Museum-Ename, Oudenaarde 03941
Provinciaal Archeologisch Museum van Zuid-Oost-Vlaanderen, Zottegem 04169
Provinciaal Domein Rivierenhof Galerij, Deurne 03549
Provinciaal Museum Bulskampveld, Beernem 03353
Provinciaal Museum Emile Verhaeren, Sint-Amands 04014
Provinciaal Museum Stijn Streuvels, Ingooigem 03733
Provinciaal Openluchtmuseum Bokrijk, Genk 03621
Provinciaal Suske en Wiske Kindermuseum, Kalmthout 03748
Provinciaal Veiligheidsinstituut, Antwerpen 03312
Provincial Museum, Strumica 30528
Provincial Museum Kočevje, Kočevje 38432
Provincial Museum of History of Pinal del Río, Pinar del Río 09253
Provincial Museum of Kymenlaakso, Kotka 10986
Provincial Museum of Lapland, Rovaniemi 11276
Provincial Museum of Newfoundland and Labrador at The Rooms, Saint John's 06851
Provincial Museum of Tornion River Valley, Tornio 11385
Provincial Museums of Gorizia, Gorizia 26351
Provincial Natuurhistorisch Museum Natura Docet, Denekamp 32116
Provincial Open-Air Museum, Genk 03621
Provincial Regional-Historical Museum, Külob 42503
Provinzmuseum der Franziskaner, Wien 03059
Provost Skene's House, Aberdeen 43246
Prøysenhuset, Rudshøgda 33801
Prudence Crandall Museum, Canterbury 47220
Prywatne Muzeum Etnograficzne, Góra Kalwaria 34677
Prywatne Muzeum Etnograficzno-Historyczne, Bielsk Podlaski 34557
Psienièčių Muziejus, Vilnius 30429
Pskov Memorial Museum of V.I. Lenin, Pskov 37576
Pskov Museum-Preserve of History, Architecture and Art, Pskov 37575
Pskovskij Gosudarstvennyj Istoričesko-architekturnyj i Chudožestnennyj Muzej-zapovednik, Pskov 37575
Pskovskij Memorialnyj Muzej V.I. Lenina, Pskov 37576
Psychiatrie-Museum, Bern 41214
Psychiatrie-Museum, Emmendingen 18147
Psychiatrie Museum im Philippshospital, Riedstadt 21327
Psychiatriemuseum Haina, Haina (Kloster) 18870
Psykiatrihistoriska Museet, Uppsala 41010
Psykiatrihistoriskt Muséum, Växjö 41040
Psykiatrimuseet, Hisings Backa 40593
Ptolemais Museum, Tolmeitha 30184
PTT Museum of Denmark, København 10148
PTT Muzej Zajednice Jugoslovenskih Pošta, Telegrafa i Telefona, Beograd 38086
Pu Le Temple, Chengde 07613
Pubei Museum, Pubei 08095
Public Art Gallery of Chios, Chios 22972
Public Library Gallery, Llanelli 44763
Public Security Museum, Gongan 07717
Public Transport Museum - Tram Depot Střešovice, Praha 09772
Public Transportmuseum Brussel, Bruxelles 03487
PUC Minas Museum of Natural Sciences, Belo Horizonte 04348
Pucheng Museum, Pucheng 08096
Pucuk Rebung-Gallery Museum, Kuala Lumpur 30585
Pudasjärven Kotiseutumuseo, Pudasjärvi 11235
Pueblo County Historical Society Museum, Pueblo 52192
El Pueblo de Los Angeles, Los Angeles 50262
Pueblo Grande Museum, Phoenix 51908
Puerta de Isabel II Gallery, Manila 34454
Puerto Galera National Museum, Puerto Galera 34481
Puerto Rican Museum of Sports, Salinas 36055
Püspöki Kastély, Fertőrákos 23561
Püspöki Palota, Sümeg 23828
Puffing Billy Railway Museum, Belgrave 00855
Pugachev Museum, Uralsk 29735
Puget Sound Coast Artillery Museum at Fort Worden, Port Townsend 52046
Puhoi Historical Society Museum, Puhoi 33189
PuK- Museum für PuppentheaterKultur, Bad Kreuznach 16766
Puke Ariki, New Plymouth 33158
Pukkila Manor, Piikkiö 11207
Pukkilan Kartanomuseo, Piikkiö 11207
Pukkilan Kotiseutumuseo, Pukkila 11236
Pulandian Museum, Dalian 07644
Puli Wine Museum, Puli 42297
Pulkauer Gewerbemuseum, Pulkau 02538
Pulkautaler Weinbaumuseum, Hadres 02035
Pulkkilan Kotiseutumuseo, Pulkkila 11237
Pullman Museum, Chicago 47509
La Pulperie de Chicoutimi/Musée du Saguenay-Lac-Saint-Jean, Chicoutimi 05612
Pulszky Society - Hungarian Museums Association, Budapest 54877
Pump House Center for the Arts, Chillicothe 47536
Pump House Gallery, London 44959
Pump House Regional Arts Center, La Crosse 49836
Pump House Steam Museum, Kingston 06089
Pumphouse Educational Museum, London 44960
Punct Muzeal, Balş 36135
Punct Muzeal, Bezdead 36149
Punct Muzeal, Buneşti 36237
Punct Muzeal, Caţa 36269
Punct Muzeal, Comana de Jos 36297
Punct Muzeal, Cuciulata 36333
Punct Muzeal, Fundata 36376
Punct Muzeal, Gostavăţu 36390
Punct Muzeal, Maieru 36456
Punct Muzeal, Malu cu Flori 36459
Punct Muzeal, Mateiaş 36466
Punct Muzeal, Moroeni 36482
Punct Muzeal, Moţăieni 36483
Punct Muzeal, Pucheni 36542
Punct Muzeal, Râu Alb 36557
Punct Muzeal, Romoş 36571
Punct Muzeal, Rotbav 36574
Punct Muzeal, Sângeorz-Băi 36585
Punct Muzeal, Tărlungeni 36677
Punct Muzeal, Ulmi 36714
Punct Muzeal, Văleni 36719
Punct Muzeal, Viştea 36731
Punct Muzeal, Vlădeni 36732
Punct Muzeal Dimitrie Anghel, Miroslava 36477
Punct Muzeal Gospodăria Tărănească, Enisala 36362
Punct Muzeal Turnu Măgurele, Turnu Măgurele 36711
Puning Museum, Puning 08097
Punjab Archives Museum, Lahore 34027
Punkaharjun Kotiseutumuseo, Punkaharju 11239
Punkt Etnograficzny w Rogierówku, Rogierówko 35082
Punkt Muzealny w Pałacu Opatów, Supraśl 35141
Puolakan Talomuseo, Valkeala 11440
Puppen-Stuben-Museum, Zimmerschied 22803
Puppen- und Spielzeugmuseum, Lichtenstein/Sachsen 19983
Puppen- und Spielzeugmuseum, Mödling 02396
Puppen- und Spielzeugmuseum, Rottweil 21427
Puppen- und Spielzeugmuseum, Solothurn 41887
Puppen- und Spielzeugmuseum mit Zinfigurenkabinett, Dornstetten 17907
Puppen- und Spielzeugmuseum Sammlung Katharina Engels, Rothenburg ob der Tauber 21409
Puppenhausmuseum, Basel 41179
Puppenmuseum, Bad Breisig 16704
Puppenmuseum, Bad Schönborn 16829
Puppenmuseum, Künzelsau 19737
Puppenmuseum, Lienz 02303
Puppenmuseum, Nordrach 20745
Puppenmuseum, Villach 02861
Puppenmuseum Falkenstein, Hamburg 18956
Puppenmuseum Hanisch, Waldkirchen an der Thaya 02892
Puppenmuseum Jeannine, Kreuzlingen 41515
Puppenmuseum Kärntner Eisenwurzen, Hüttenberg 02104
Puppenmuseum Laufeld, Laufeld 19853

Puppenstubenmuseum der 30er und 50er Jahre, Jakobsweiler 19374
Puppentheater Museum Berlin, Berlin 17236
Puppentheatermuseum, Kaufbeuren 19491
Puppentheatermuseum, Mistelbach an der Zaya 02387
Puppet Centre, North York 06488
Puppet Exhibition, Bosta 23378
Puppet Museum, Tallinn 10639
Puppet Museum House of Dolls, Moskva 37312
Puppetry Art Center of Taipei, Taipei 42420
Puratatva Sangrahalaya, Gorakhpur 24180
Purdue University Calumet Library Gallery, Hammond 49142
Purdue University Galleries, West Lafayette 54067
Puri Gamelan Suar Agung, Denpasar 24468
La Purisima Mission, Lompoc 50167
Purmerends Museum, Purmerend 32672
Purnananda Memorial Museum, Hazarikapra 24207
Purrmann-Haus, Speyer 21837
Purton Museum, Purton 45452
Purvattatva Sangrahalaya Madhya Pradesh Shasan, Damoh 24124
Pusat Latihan Kesenian & Pertukangan Tangan Brunei, Bandar Seri Begawan 04967
Pushkin Apartment Museum, Sankt-Peterburg 37694
Pushkin Country House Museum, Puškin 37581
Pushkin Museum, Toržok 37861
Pushkin Museum in Bernov, Bernov 36804
Pushkin Museum of Fine Arts, Moskva 37302
Pusta, Katowice 34730
Pustozerskij Kompleksnyj Istoriko-prirodnyj Muzej, Narjan-Mar 37444
Pusulan Museo, Pusula 11241
Putian Museum, Putian 08098
Putkiradiomuseo, Kouvola 10991
Putnam Cottage, Greenwich 49083
Putnam County Historical Society and Foundry School Museum, Cold Spring 47681
Putnam County Historical Society Museum, Kalida 49669
Putnam Museum of History and Natural Science, Davenport 47970
Putney Historical Society, Putney 52203
Putney School Gallery, Putney 52204
Puuhkalan Talomuseo, Ylivieska 11483
Puukohola Heiau Ruins, Kawaihae 49712
Puyang Museum, Puyang 08099
Pyhäjärven Kotiseutumuseo, Pyhäjärvi 11243
Pyhäjärvi Local History Museum, Pyhäjärvi 11243
Pyhäjoki Local History Museum, Pyhäjoki 11244
Pyhämaan Kotiseutumuseo, Pyhämaa 11245
Pyhäntän Kotiseutumuseo, Pyhäntä 11246
Pyhtään Kotiseutumuseo, Pyhtää 11247
Pykäri Nuutajärven Lasimuseo, Nuutajärvi 11154
Pyramid Hill and District Historical Museum, Pyramid Hill 01444
Pyramid Hill Sculpture Park and Museum, Hamilton 49139
Pythagoras Museum, Norrtälje 40777
Pyynikinlinna, Tampere 11356

Q

Qalipakkanik Katersugaasivik, Ilulissat 23256
QANTAS Founders Outback Museum, Longreach 01240
Qaqortup Katersugaasivia, Qaqortoq 23265
Qashqadaryo Viloyat Ülkashunoslik Muzeyi, Qarshi 54454
Qasigiannguit Katersugaasiviat, Qasigiannguit 23267
Qasr Al-Gawharah, Cairo 10514
Qasr Al-Ibrahim, Cairo 10515
Qasr Ali Ibrahim, Cairo 10516
Qatar National Museum, Doha 36088
QCA Gallery, Brisbane 00898
QCC Art Gallery, Bayside 46657
Qedem Museum, Kedumim 25027
Qeqertarsuaq Museum, Qeqertarsuaq 23268
Qi County Folkways Museum, Qixian 08127
Qi Gucheng Yi Zhi Guan, Linzi 07960
Qianjiang Museum, Qianjiang 08100
Qiankou Minzai Museum, Huangshan 07835
Qianling Museum, Qianxian 08103
Qiannan Zhousandu Nationality Museum, Sandu 08147
Qiannanzhou Museum, Douyun 07674
Qianshan Museum, Qianshan, Anhui 08101
Qijiang Stone Engraving Museum, Gunan 07747
Qijinshang Culture Museum, Qixian 08128
Qimen Museum, Qimen 08104
Qin Dai Zhan Guan, Lintong 07953
Qin Shi Huan Bingmayong Museum, Xian 08358
Qin Shi Huang Terracotta Warriors and Horses Museum, Lingtong 07945
Qinan Museum, Qinan 08105
Qing Tycoon Yashu Bowu Yuanji, Hohhot 07791
Qingdao Art Gallery, Qingdao 08108
Qingdao City Folklorish Museum, Qingdao 08109
Qingdao Municipal Museum, Qingdao 08110
Qingdao Painting Gallery, Qingdao 08111
Qinghai Logroll Museum, Huzhu 07852
Qinghai Provincial Museum, Xining 08397
Qingliu Museum, Qingliu 08112
Qingpu Museum, Shanghai 08179
Qingyang Museum, Qingyang 08114
Qingyang Zone Museum, Xifeng 08384
Qingyuan Museum, Qingyuan 08118
Qingyunpu Museum, Nanchang 08032
Qingzhou Museum, Qingzhou 08120
Qinshihuang Military Forces Yong Museum, Lintong 07954
Qinyang Museum, Qinyang 08121
Qinzhou Museum, Qinzhou 08123
Qiongshan Museum, Qiongshan 08124
Qishan Museum, Qishan 08125
Qiz Qalasy Gallery, Bakı 03167
Qom Museum, Qom 24639
Qo'qon Viloyat Ülkashunoslik Muzeyi, Qo'qon 54458
Qoraqalpoğiston Respublikasi Ülkashunoslik Muzeyi, Nukus 54451
Qoraqalpoğiston Tarikhi Muzeyi, Nukus 54452
Qozhaya Museum, Zgharta 30177
Quaco Museum and Archives, Saint Martins 06867
Quadreria Cesarini, Fossombrone 26242
Quadreria Comunale, Offida 26976
Quadreria Comunale, Prato 27339
Quadreria Comunale, San Constanzo 27628
Quadreria dei Girolamini, Napoli 26911
Quadreria della Cassa Depositi e Prestiti, Roma 27542
Quadreria della Società Economica, Chiavari 25862
Quadreria e Pinacoteca Civica, San Giovanni in Persiceto 27647
Quai Branly Museum, Paris 14700
Quaid-i-Azam Birthplace and Museum, Karachi 34017
Quaker Meeting House, Mount Pleasant 50887
Quaker Museum, Ballytore 24741
Quaker William Ray House, Dartmouth 05677
Quàng Ninh Museum, Ha Long 54605
Quanzhou Maritime Museum, Quanzhou 08130
Quanzhou Museum, Quanzhou 08131
Quanzhou Overseas-Relations History Museum, Quanzhou 08132
Quanzhou Painting Gallery, Quanzhou 08133
Quarry Bank Mill & Styal Estate, Styal 45795
Quartel de Santa Cruz, Rio de Janeiro 04770
Quarzsandbergwerk-Museum, Buchs (Zürich) 41267
Quaternary Glacial Vestige of China Exhibition Hall, Beijing 07563
Queanbeyan and District Historical Museum, Queanbeyan 01446
Quebec House, Westerham 45978
Queen Elizabeth II Army Memorial Museum, Waiouru 33222
Queen Elizabeth National Park Museum, Lake Katwe 42994
Queen Elizabeth's Hunting Lodge, London 44961
Queen Emma Gallery, Honolulu 49351
Queen Emma Summer Palace, Honolulu 49352
Queen Mary Museum, Long Beach 50176
Queen Street Mill Textile Museum, Burnley 43678
Queen Victoria Museum, Harare 54671
Queen Victoria Museum and Art Gallery at Inveresk, Launceston 01223
Queen Victoria Museum and Art Gallery at Royal Park, Launceston 01224
Queen Victoria Rifles Museum - Hill 60, Zillebeke 04162
Queens County Farm Museum, Floral Park 48562
Queens County Museum, Liverpool 06178
Queens County Museum, Tilley House and Court House, Gagetown 05846
The Queen's Gallery, Edinburgh 44145
The Queen's Gallery, London 44962
Queens Historical Society, Flushing 48579
Queen's Lancashire Regiment Museum, Preston 45447
Queens Library Gallery, Jamaica 49592
Queens Museum of Art, Flushing 48580
Queen's Own Cameron Highlanders of Canada Regimental Museum, Winnipeg 07374
The Queen's Own Hussars Museum, Warwick 45937
Queen's Own Rifles of Canada Regimental Museum, Toronto 07129
Queen's Own Royal West Kent Regimental Museum, Maidstone 45074
Queen's Royal Irish Hussars Museum, Eastbourne 44100
The Queen's Royal Lancers Regimental Museum, Grantham 44346
Queen's Royal Surrey Regiment Museum, Guildford 44374
Queen's Royal Surrey Regiment Museum, West Clandon 45968
Queen's York Rangers Regimental Museum, Toronto 07130
Queenscliffe Historical Museum, Queenscliff 01448
Queenscliffe Maritime Museum, Queenscliff 01449
Queensferry Museum, South Queensferry 45696
Queensland Air Museum, Caloundra 00936
Queensland Art Gallery, South Brisbane 01512
Queensland Fire Museum, Kurwongbah 01209
Queensland Maritime Museum, South Brisbane 01513
Queensland Military Memorial Museum, Fortitude Valley 01088
Queensland Museum, South Bank 01510
Queensland Police Museum, Brisbane 00899
Queensland University of Technology Art Museum, Brisbane 00900
Queenstown and Frontier Museum, Queenstown 38720
The Queenstown Story, Cobh 24765
Queichtalmuseum, Offenbach an der Queich 20884
Quellen Museum, Bad Wildungen 16863
Quest - Center for Earth Science And Discovery, Makati 34417
Questacon-The National Science and Technology Centre, Canberra 00951
Quetico Provincial Park Heritage Pavilion, Atikokan 05361
Quevedo Zornoza House Museum, Zipaquirá 08790
Quex Museum, House and Gardens, Birchington 43489
Quezonia Museum, Manila 34455
Quidi Vidi Battery, Saint John's 06852
Quiet Valley Living Historical Farm, Stroudsburg 53340
Quijano Museum, Mandaue 34428
Quilpie Museum, Quilpie 01451
Quilters Hall of Fame, Marion 50484
Quilters' Ruild Resource, Halifax 44387
Quincy and Adams County Museum, Quincy 52215
Quincy Art Center, Quincy 52216
Quincy Historical Museum, Quincy 52220
The Quincy Museum, Quincy 52217
Quinta Municipal de Nossa Senhora da Piedade, Póvoa de Santa Iria 35872
Quinte Educational Museum, Ameliasburgh 05330
Quirindi and District Historical Cottage and Museum, Quirindi 01452
Qujiang Museum, Qujiang 08135
Qurm Museum, Qurm 33995
Qutb Shahi Museum, Hyderabad 24216
Quwo Museum, Quwo 08136
Quzhou Museum, Quzhou 08137
Qwensel House and Pharmacy Museum, Turku 11388

R

R.E. Olds Transportation Museum, Lansing 49945
R.E. Stevenson Museum, Colenso 38595
R.M. Bohart Museum of Entomology, Davis 47978
R.N. Atkinson Museum, Penticton 06584
R.S. Barnwell Memorial Garden and Art Center, Shreveport 53087
The R.W. Norton Art Gallery, Shreveport 53088
Råå Museum för Fiske och Sjöfart, Råå 40811
Raabe-Haus, Braunschweig 17528
Raahe Museum, Raahe 11248
Raahen Museo, Raahe 11248
Raasay Heritage Museum, Kyle 44626
Rábaközi Muzeális Gyűjtemény, Kapuvár 23637
Rabasova galerie, Rakovník 09815
Rabaul Museum, Rabaul 34086
Rabekk Museum, Moss 33684
Rabindra Bharati Museum, Kolkata 24280
Rabindra Bhavan Art Gallery, Delhi 24153
Rabindra Bhavana, Santiniketan 24379
Rabiša Cave Museum, Belogradčik 04983
Raby Castle, Darlington 43950
Raccolata Comunale G.C. Corsi, Cantiano 25648
Raccolta Agorà dell'Arte, Sersale 27820
Raccolta Archeologica, Narni 26916
Raccolta Archeologica, Terni 27965
Raccolta Archeologica, Tolentino 27990
Raccolta Archeologica Alberto Pisani Dossi, Corbetta 25969
Raccolta Archeologica Comunale, Celenza Valfortore 25799
Raccolta Archeologica e Paleontologica, Corciano 25972
Raccolta Civica, Città di Castello 25910
Raccolta Civica d'Arte Contemporanea, Molfetta 26766
Raccolta Civiche di Storia, Albino 25182
Raccolta Comunale, Acquasparta 25146
Raccolta Comunale, Sigillo 27849
Raccolta Comunale d'Arte, Frontone 26252
Raccolta d'Arte, Città della Pieve 25903
Raccolta d'Arte, Roccalbegna 27432
Raccolta d'Arte C. Lamberti, Codogno 25933
Raccolta d'Arte Contemporanea R. Pastori, Calice Ligure 25601
Raccolta d'Arte della Provincia, Modena 26757
Raccolta d'Arte e Archeologia, Piazza Armerina 27195
Raccolta d'Arte Pagliara, Napoli 26912
Raccolta d'Arte Sacra, Figline Valdarno 26132
Raccolta d'Arte Sacra, Scarperia 27789
Raccolta d'Arte Ubaldiana, Piandimeleto 27192
Raccolta dei Padri Passionisti, Paliano 27082
Raccolta della Chiesa di Santa Maria Maggiore, Alatri 25173
Raccolta dell'Avifauna delle Marche, Montefortino 26831
Raccolta dell'Avifauna Lombarda, Arosio 25282
Raccolta delle Piastrelle di Ceramica, Sassuolo 27765
Raccolta dell'Opera del Duomo, Oristano 26997
Raccolta di Cose Montesine, Montese 26854
Raccolta di Fisarmoniche d'Epoca, Camerano 25614
Raccolta di Fossili Francesco Angellotti, Ostra 27022
Raccolta di Ingegneria Navale, Genova 26325
Raccolta di Opere d'Arte, Solarolo 27862
Raccolta di Sant'Urbano, Apiro 25251
Raccolta di Scienze Naturali, Vicenza 28257
Raccolta E. Guatelli, Collecchio 25939
Raccolta Etnografica del Centro Studi Pugliesi, Manfredonia 26588
Raccolta Etnografica della Civiltà Contadina, San Giorgio Piacentino 27643
Raccolta G. Manzù, Ardea 25261
Raccolta Guatelli, Ozzano Taro 27028
Raccolta Kalefati, Oria 26993
Raccolta Lapidaria Capitolina, Roma 27543
Raccolta Materiale Archeologico, Bevagna 25414
Raccolta Memoria e Tradizioni Religiose, Serramanna 27814
Raccolta Naturalistica, Tagliolo Monferrato 27937
Raccolta O. Recchione, Palena 27064
Raccolta Osservatorio di Apicoltura Giacomo Angeleri, Pragelato 27322
Raccolta Paleontologica e Preistorica, Narni 26917
Raccolta Parocchiale, Valpelline 28149
Raccolta Privata Toraldo di Francia, Tropea 28106
Raccolta Russo-Ortodossa, Merano 26660
Raccolta Storica della Vita Materiale dell'Antico, Altopascio 25211
Raccolta Temporanea, Cosenza 25992
Raccolte d'Arte dell'Ospedale Maggiore di Milano, Milano 26721
Raccolte del Dipartimento di Scienze Radiologiche, Roma 27544
Raccolte dell'Istituto di Clinica Otorinolaringoiatrica, Roma 27545
Raccolte di Geologia, Mineralogia e Vulcanologia, Catania 25777
Raccolte Frugone, Genova 26326
Racemotormuseum Lexmond, Lexmond 32499
Rachana Open Air Museum, Rachana 30171
Rachel Carson Homestead, Springdale 53213
Racine Art Museum, Racine 52223
Racine Heritage Museum, Racine 52224
Rackstadmuseet, Arvika 40439
Rada Galerii České Republiky, Praha 54775
Ráday Castle, Pécel 23756
Ráday-Kastély, Pécel 23756
Ráday Museum of the Danubian District of the Hungarian Reformed Church, Kecskemét 23645
Radboud Castle, Medemblik 32542
Radbrook Culinary Museum, Shrewsbury 45674
Råde Bygdetun, Råde 33774
Rademacher Forges, Eskilstuna 40503
Rademachersmedjorna, Eskilstuna 40503
Radetski-Ship National Museum, Kozloduj 05080
Radetzky-Gedenkstätte Heldenberg, Kleinwetzdorf 02196
Radford University Art Museum, Radford 52225
Radijo ir Televizijos Muziejus, Šiauliai 30361
Radio Amateur Museum, Reusel 32683
Radio and Television Museum of the Šiauliai Aušra Museum, Šiauliai 30361
Radio and TV Museum, Lahti 11032
Radio- en Speelgoed Museum, Onstwedde 32622
Rádió és Televízió Múzeum, Diósd 23525
Radio- ja TV-Museo, Lahti 11032
Radio Matterhorn Museum, Zermatt 42053
Radio Museum Köln, Köln 19628
Radio- og Motorcykel Museet, Stubbekøbing 10287
Radio-Television Museum, Bowie 46971
Radio Transmission Museum, Siófok 23804
Radiomuseet i Göteborg, Göteborg 40554
Radiomuseet i Jönköping, Jönköping 40615
Radiomuseum, Bad Tatzmannsdorf 01768
Radiomuseum, Borculo 31985
Radiomuseum, Grödig 01993
Radiomuseum, Linsengericht 20011
Radiomuseum, Rottenburg an der Laaber 21423
Radiomuseum, Waldbronn 22284
Radiomuseum, Zürich 42104
Rádiós Hírközlési Múzeum, Siófok 23804
Radishchev Art Museum, Saratov 37747
Radmuseum anno dazumal, Altmünster 01717
Radnorshire Museum, Llandrindod Wells 44757
Radstädter Museumsverein, Radstadt 02552
Radstock Museum, Radstock 45455

Radvila Palace of the Lithuanian Art Museum, Vilnius 30430
Radvilų Rūmai, Vilnius 30430
Radziwill Palace in Nieborów, Nieborów 34942
Rääkkylän Kotiseutumuseo, Rääkkylä 11250
Rælingen Bygdetun, Fjerdingby 33494
Ränkimäen Ulkomuseo, Lapua 11053
Rätisches Museum, Chur 41304
Rättviks Konstmuseum och Naturmuseum, Rättvik 40812
Raevangla Fotomuuseum, Tallinn 10642
Ragailong Museum, Imphal 24228
Ragged School Museum, London 44963
Raglan and District Museum, Raglan 33192
Ragley Hall, Alcester 43265
Raha- ja Mitalikokoelma, Helsinki 10810
Rahmi M. Koç Müzesi, İstanbul 42834
Rahr West Art Museum, Manitowoc 50443
Raiffeisen-Museum, Flammersfeld 18326
Rail Museum, Cairo 10517
Rail Transport Museum, Bassendean 00836
Railroad and Heritage Museum, Temple 53474
Railroad House Historical Museum, Sanford 52849
Railroad Museum of Long Island, Riverhead 52367
Railroad Museum of Pennsylvania, Strasburg 53329
Railroader's Memorial Museum, Altoona 46242
Rail's End Gallery - Haliburton Highlands Guild of Fine Arts, Haliburton 05928
Railway Age, Crewe 43923
Railway and Industry Museum, Hagfors 40575
Railway Coastal Museum, Saint John's 06853
The Railway Exposition, Covington 47859
Railway Historical Collection, Mátészalka 23694
Railway Museum, Athinai 22926
Railway Museum, Beograd 38092
Railway Museum, Kalamata 23025
Railway Museum, Livingstone 54656
Railway Museum, Ljubljana 38469
Railway Museum, Nynäshamn 40783
Railway Museum, Paks 23749
Railway Museum, Ravenglass 45462
Railway Museum, Saitama 29188
Railway Museum, Šiauliai 30357
Railway Museum, Vilnius 30406
Railway Museum, Wylam 46088
Railway Museum Zlonice, Zlonice 09944
Railway Preservation Society of Ireland, Whitehead 46002
Railway Station Museum, Castlegar 05576
Railway Station Museum, Uitenhage 38769
Railway Station Museum, Wonthaggi 01666
Railway Worker's Home Museum, Kouvola 10992
Railways to Yesterday, Rockhill Furnace 52419
Railworld, Peterborough 45369
Raimund-Gedenkstätte, Gutenstein 02025
Raiņa Literatūras un Mākslas Vēstures Muzejs, Rīga 30102
Raiņa Memoriālāis Muzejs Jasmuiža, Preiļi 30057
Raiņa Muzejs Tadenava, Dunavas 30004
Raiņa un Aspazijas Māja, Rīga 30103
Raiņa un Aspazijas Memoriālā Vasarnīca, Jūrmala 30026
Rainbow - Culture Centre and Exhibition Hall, Uljanovsk 37900
Rainermuseum, Salzburg 02622
Rainham Hall, Rainham 45456
Rainis and Aspazia Memorial Museum, Jūrmala 30026
Rainis and Aspazija House, Rīga 30103
Rainis Museum, Preiļi 30057
Rainis Museum of the History of Literature and Art, Rīga 30102
Rainis Museum Tadenava, Dunavas 30004
Rainy Hills Pioneer Exhibits, Iddesleigh 06005
Rainy River District Women's Institute Museum, Emo 05775
Rais Ali Delvary Museum, Bushehr 24601
Raitioliikennemuseo, Helsinki 10811
Raja Dinkar Kelkar Museum, Pune 24362
Rajalahden Talomuseo, Virrat 11471
Rajalahti House Museum, Virrat 11471
Rajamuseo, Imatra 10856
Rajkiya Sangrahalaya, Udaipur 24415
Rajoloteca Salvador Miquel, Vallromanes 40249
Rajputana Museum, Ajmer 24055
Rakennuskulttuuritalo Toivo ja Korsmanin talo, Pori 11221
Rakhine State Museum, Sittwe 31728
Rakhtshuy-Khaneh Museum, Zanjan 24707
Rakiura Museum, Stewart Island 33202
Rakkestad Bygdetun, Rakkestad 33775
Rákóczi Múzeum, Sárospatak 23792
Rákospalotai Múzeum, Budapest 23469
Rakstniecības un Mūzikas Muzejs, Rīga 30104
Rakstnieka Herberta Dorbes Māja-Muzejs Senču Putekļi, Ventspils 30143
Raku Museum, Kyoto 28861
Rakushu Rakuchu Gallery, Kyoto 28862
Rakvere Castle, Rakvere 10611
Rakvere Citizen's Museum, Rakvere 10610
Rakvere Exhibition House, Rakvere 10612
Rakvere Linnakodaniku Muuseum, Rakvere 10610
Rakvere Linnus-Muuseum, Rakvere 10611
Rakvere Näitustemaja-Muuseum, Rakvere 10612
Raleigh City Museum, Raleigh 52236
Sri Rallabandi Subbarao Government Museum, Rajahmundry 24369
Rallarmuseet, Moskosel 40749
Ralli Museum, Caesarea 24939
Ralls Historical Museum, Ralls 52237
Ralph Allen Memorial Museum, Oxbow 06564
Ralph Foster Museum, Point Lookout 52008
Ralph Milliken Museum, Los Banos 50271
Ralph Waldo Emerson House, Concord 47771
Ram Khamhaeng National Museum, Sukhothai 42603
Ram Mala Museum, Comilla 03216
Rama IX Art Museum, Bangkok 42557
Raman Science Centre, Nagpur 24333
Ramla Museum, Ramla 25089
Ramlingappa Lamture Museum, Osmanabad 24338
Rammerscales, Lockerbie 44777
Ramón Magsaysay Memorabilia Museum, Manila 34456
Ramsay House, Alexandria 46212
Ramsay Museum, Honolulu 49353
Ramsey Center for Arts, Shoreview 53082
Ramsey House Museum Plantation, Knoxville 49822
Ramsey Rural Museum, Huntingdon 44502
Ramsgate Maritime Museum, Ramsgate 45457
Ramsgate Museum, Ramsgate 45458
Ran-In-Ting Museum, Taipei 42421
Rana Museum, Mo i Rana 33667
Rancho Los Alamitos, Long Beach 50177
Rancho Los Cerritos, Long Beach 50178
Randaberg Bygdemuseum, Randaberg 33776
Randall House Museum, Wolfville 07399
Randall Museum, San Francisco 52790
Randeck-Museum, Mannweiler-Cölln 20176
Randers Kunstmuseum, Randers 10230
Randolph County Historical Museum, Winchester 54219
Randolph County Museum, Beverly 46789
Randolph Historical Society, Randolph 52246
Rangau-Handwerkermuseum, Markt Erlbach 20212
Rangiora Museum, Rangiora 33193
Rangpur Archaeological Museum, Rangpur 03235
Rani Laxmi Bai Palace Sculpture Collection, Jhansi 24244
Rankin House Museum, Miramichi 06302
Rankin House State Memorial, Ripley 52356
Rankin Museum, Ellerbe 48363
Rankin Museum, Rankin 52247
Rannarootsi Muuseum, Haapsalu 10565
Rannikkotykistömuseo, Helsinki 10812
Ransom County Historical Society Museum, Fort Ransom 48673
Rantasalmen Museo, Rantasalmi 11253
Rantsilan Kotiseutumuseo, Rantsila 11254
Ranua Parsonage Museum, Ranua 11255
Ranuan Pappila- ja Pitäjämuseo, Ranua 11255
Ranzan Historical Museum, Ranzan 29170
Raoping Museum, Raoping 08138
Rapid City Museum, Rapid City 06703
Rappresentazione Scenografica Gioacchino Murat, Pizzo 27262
Raritätensammlung Bruno Gebhardt, Zwönitz 22842
Rariteiten- en Ambachtenmuseum 't Krekelhof, Koksijde 03755
Ras-al-Tin Palace Museum, Ras-al-Tin 10539
Raseiniai Area History Museum, Raseiniai 30342
Raseinių Krašto Istorijos Muziejus, Raseiniai 30342
Rashid Al-Oraifi Museum, Muharraq 03213
Rasht Museum, Rasht 24640
Raşidbey Efandiyev Historical-Regional Ethnography Museum, Şeki 03196
Rasmussen Art Gallery, Angwin 46301
Rassan Arabzadeh Carpet Museum, Tehrān 24691
Rassegna Attrezzi e Oggetti del Passato, Belluno 25390
Rastrakavi Govind Pai Museum, Udupi 24416
Rašytojo Tomo Mano Memorialinis Muziejus, Nida 30322
Rašytojos Šatrijos Raganos Memorialinis Muziejus, Židikai 30452
Rašytojos Žemaitės Memorialinis Muziejus, Bukante 30218
Ráth György Museum, Budapest 23470
Rathaus-Galerie, Berlin 17237
Rathaus Galerie Balingen, Balingen 16902
Rathausgalerie, Castrop-Rauxel 17682
Rathausgalerie München, München 20475
Rathausgalerie Munster, Munster 20532
Rathausgalerie Vellmar, Vellmar 22216
Rathausmuseum, Sempach 41860
Rathausmuseum Nieder-Ohmen, Mücke 20394
Rathaussammlung, Stein am Rhein 41916
Rathgen-Forschungslabor, Berlin 17238
Rátkay Endre - Átlók Galéria, Budapest 23471
Ratnapura National Museum, Ratnapura 40383
Raton Museum, Raton 52255
Ratsuväkimuseo, Lappeenranta 11048
Rattlesnake Museum, Albuquerque 46186
Ratvolden, Røros 33795
Rauchfangkehrermuseum, Wien 03060
Rauhaniemen Kotiseutumuseo, Sulkava 11331
Rauma Art Museum, Rauma 11261
Rauman Museo, Rauma 11260
Rauman Taidemuseo, Rauma 11261
Rauriser Talmuseum, Rauris 02566
Rautalammin Museo, Rautalampi 11263
Rautalampi Museum, Rautalampi 11263
Rautatieläiskotimuseo, Kouvola 10992
Rautenstrauch-Joest-Museum, Köln 19629
Raution Kotiseutumuseo, Rautio 11264
Ravalli County Museum, Hamilton 49132
Raverty's Motor Museum, Echuca 01058
Ravintolamuseo, Lahti 11033
Ravmuseet, Oksbøl 10220
Rawhide Old West Museum, Scottsdale 52976
Rawls Museum Arts, Courtland 47856
The Ray Drew Gallery, Las Vegas 49964
Ray E. Powell Museum, Grandview 49013
Raymond M. Alf Museum of Paleontology, Claremont 47575
Raymore Pioneer Museum, Raymore 06704
Raynham Hall Museum, Oyster Bay 51662
Razstavni Salon Rotovž, Maribor 38478
RCAF Memorial Museum, Trenton 07161
RCMP Centennial Celebration Museum, Fairfield 05790
Reaction and Lesson Palace and National Palace, Tehrān 24676
Reader's Digest Art Gallery, Pleasantville 51990
Reading Abbay Gateway, Reading 45470
Reading Historical Society, Reading 52267
Reading Historical Society Museum, Reading 52261
Reading Public Museum and Art Gallery, Reading 52265
Reading Railroad Heritage Museum, Reading 52266
Readsboro Historical Society, Readsboro 52269
Real Cartuja, Valldemosa 40245
Real Cartuja de Miraflores, Burgos 39073
Real Casa del Labrador, Aranjuez 38880
Real Fábrica de Tapices, Madrid 39587
Real Monasterio de San Jerónimo, Granada 39371
Real Monasterio de Santa María de El Puig, El Puig de Santa María 39883
Real Museo Mineralogico, Napoli 26913
Real TART Community Gallery, New Plymouth 33159
Real World Computer Museum, Boothwyn 46912
Rebbaumuseum am Bielersee-Musée de la Vigne au Lac de Bienne, Ligerz 41565
Rebecca Nurse Homestead, Danvers 47957
Reblandmuseum, Baden-Baden 16889
Rechnermuseum, Furtwangen im Schwarzwald 18521
Rechnermuseum der Gesellschaft für wissenschaftliche Datenverarbeitung, Göttingen 18698
Rechtshistorisches Museum, Karlsruhe 19447
Recinto Casa Benito Juárez, Saltillo 31319
Record Office for Leicestershire, Leicester and Rutland, Leicester 44683
Red Barn Museum, Morristown 50857
Red Bay National Historic Site, Red Bay 06705
Red Brick Arts Centre and Museum, Edson 05766
Red Bull Hangar-7, Salzburg 02623
Red Clay State Historical Park, Cleveland 47645
Red Deer and District Museum, Red Deer 06709
red dot design museum Deutschland, Essen 18266
Red Granary, Archaeological and Ethnographic Dept., Bydgoszcz 34574
Red Hill-Patrick Henry National Memorial, Brookneal 47087
Red House Glass Cone Museum, Stourbridge 45769
Red House Museum, Gomersal 44326
Red House Museum and Art Gallery, Christchurch 43842
Red Lake Regional Heritage Centre, Red Lake 06711
Red Location Museum, New Brighton 38694
Red Lodge, Bristol 43638
Red Mill Museum Village, Clinton 47658
Red Monastery, Červený Kláštor 38245
Red River Historical Society Museum, Clay City 47597
Red River Valley Museum, Vernon 53723
Red Rock Museum, Church Rock 47549
Red Sea Maritime Museum, Eilat 24944
Redbridge Museum, Ilford 44509
Redcliff Museum, Redcliff 06712
Redcliffe Museum, Redcliffe 01454
Reddingsmuseum Abraham Fock, Hollum 32400
Redhill Museum, Redhill 01455
Redington Museum, Waterville 53978
Redland Museum, Cleveland 00984
Redningsselskapets Museum, Horten 33565
Redoña Residence Museum, Tacloban 34510
Redoubt Fortress - Military Museum of Sussex, Eastbourne 44101
Redpath Museum, Montréal 06380
Redpath Sugar Museum, Toronto 07131
Redwater and District Museum, Redwater 06713
Redwood County Museum, Redwood Falls 52279
Redwood Library and Athenaeum, Newport 51328
Redwood Museum, Crescent City 47875
Reed Gallery, Presque Isle 52138
Reed Gold Mine, Midland 50668
Reed Gold Mine State Historic Site, Stanfield 53252
Reed Whipple Cultural Center, Las Vegas 49978
Reedville Fishermen's Museum, Reedville 52281
Reese Bullen Gallery, Arcata 46339
The Reeves Center, Lexington 50072
Refectory of the Hosios Loukas Monastery, Boeotia 22958
Refettorio dell'ex-Convento di Sant'Onofrio detto di Fuligno, Firenze 26202
Reflex Miniatuur Museum voor Hedendaagse Kunst, Amsterdam 31881
Református Kollégiumi és Egyházművészeti Múzeum, Debrecen 23520
Le Refuge de Grasla, Les Brouzils 13695
Refuge Fortifié Muséalisé, Dossenheim-sur-Zinsel 12773
Refugio County Museum, Refugio 52282
Regementsmuseet med 91:an, Halmstad 40580
Régészeti Osztály és Állandó Régészeti Kiállítás, Pécs 23772
Reghthuys, Giessenburg 32250
Regimental Museum of 1st The Queen's Dragoon Guards, Cardiff 43752
Regimental Museum of the 13th/18th Royal Hussars and Kent Dragoons, Cawthorne 43782
Regimental Museum of the 9th/12th Royal Lancers, Derby 43968
Regimental Museum of the Oxfordshire and Buckinghamshire Light Infantry, Oxford 45339
Regimental Museum of the Royal Dragoon Guards and the Prince of Wales's Own Regiment of Yorkshire, York 46106
Regimental Museum of the Royal Scots Dragoon Guards, Edinburgh 44146
Regimental Museum of The Royal Welsh (Brecon), Brecon 43599
Regimental Museum The Highlanders, Inverness 44526
The Regina A. Quick Center for the Arts, Saint Bonaventure 52530
Regina Firefighters Museum, Regina 06722
Regina General Hospital Museum, Regina 06723
Regina Gouger Miller Gallery, Pittsburgh 51958
Regina Plains Museum, Regina 06724
Reginald F. Lewis Museum of Maryland African American History and Culture, Baltimore 46586
Reginalds Tower, Waterford 24908
Region Historical Collection, Várpalota 23951
Region Historical Collection at the Derkovits Cultural Center, Tiszaújváros 23932
Region Historical Exhibition Space in the District 11, Budapest 23400
Region Historical Museum, Komló 23682
Regionaal Museum 1940-1945 Schagen en Omstreken, Schagen 32748
Regionaal Natuurmuseum Westflinge, Sint Pancras 32778
Regional Art Gallery, Kaluga 37001
Regional Art Gallery, Kardžali 05046
Regional Art Gallery, Liberec 09624
Regional Art Gallery, Razgrad 05147
Regional Art Gallery, Sliven 05170
Regional Art Gallery, Stara Zagora 05210
Regional Art Gallery, Vraca 05266
Regional Art Museum, Uljanovsk 37909
Regional Children's Picture Gallery, Vladivostok 37938
Regional Gallery, Stary Sącz 35131
Regional Gallery of Fine Arts in Zlín, Zlín 09940
Regional Gallery of Modern Art, Hradec Králové 09533
Regional Historical Museum, Chaskovo 05005
Regional Historical Museum, Pazardžik 05109
Regional Historical Museum, Pernik 05111
Regional Historical Museum, Ruse 05155
Regional Historical Museum, Sliven 05169
Regional Historical Museum, Smoljan 05171
Regional Historical Museum, Vidin 05263
Regional Historical Museum, Vraca 05265
Regional Historical Museum of Varna, Varna 05247
Regional History and Ethnography Museum, Aktöbe 29673
Regional History Collection, Isaszeg 23620
Regional History Museum, Pleven 05120
Regional History Museum, Targovište 05225
Regional History Museum, Veliko Tărnovo 05257
Regional House, Balatonszentgyörgy 23357
Regional Literary Museum, Chmel'nyc'kyj 43028
Regional Museum, Balachna 36787
Regional Museum, Berane 31637
Regional Museum, Biecz 34554
Regional Museum, Borisoglebsk 36816
Regional Museum, Cedynia 34586
Regional Museum, Čeljabinsk 36843
Regional Museum, Chomutov 09478
Regional Museum, Djakovica 38099
Regional Museum, Dobczyce 34622
Regional Museum, Elec 36910
Regional Museum, Enisejsk 36915
Regional Museum, Glazov 36924
Regional Museum, Głogówek 34669

Regional Museum, Główno 34670
Regional Museum, Herceg-Novi 31646
Regional Museum, Irkutsk 36933
Regional Museum, Iskitim 36941
Regional Museum, Ivanovo 36947
Regional Museum, Jurjevec 36984
Regional Museum, Kaliningrad 36990
Regional Museum, Karacol 29962
Regional Museum, Kościan 34757
Regional Museum, Kozienice 34763
Regional Museum, Kraśnik 34824
Regional Museum, Krasnokamsk 37079
Regional Museum, Krasnystaw 34826
Regional Museum, Kutno 34839
Regional Museum, Lubań 34894
Regional Museum, Łuków 34913
Regional Museum, Magadan 37119
Regional Museum, Mielec 34920
Regional Museum, Murska Sobota 38487
Regional Museum, Narovčat 37446
Regional Museum, Novokuzneck 37482
Regional Museum, Orël 37511
Regional Museum, Penza 37537
Regional Museum, Perm 37551
Regional Museum, Puškin 37580
Regional Museum, Puškino 37584
Regional Museum, Sergač 37761
Regional Museum, Środa Śląska 35126
Regional Museum, Staraja Russa 37797
Regional Museum, Stary Sącz 35132
Regional Museum, Staryj Oskol 37801
Regional Museum, Szczecinek 35166
Regional Museum, Tambov 37833
Regional Museum, Taraz 29733
Regional Museum, Tarnów 35177
Regional Museum, Tomsk 37859
Regional Museum, Valdaj 37914
Regional Museum, Velikie Luki 37918
Regional Museum, Vetluga 37927
Regional Museum, Volgograd 37959
Regional Museum, Voronež 37976
Regional Museum, Wągrowiec 35210
Regional Museum, Wronki 35345
Regional Museum, Zaječar 38164
Regional Museum and Art Gallery, Jičín 09559
Regional Museum and Gallery, Česká Lípa 09457
Regional Museum at Maribor's castle, Maribor 38477
Regional Museum at Ptuj castle, Ptuj 38499
Regional Museum Celje, Celje 38407
Regional Museum Center, Sankt-Peterburg 37716
Regional Museum in Chrudim, Chrudim 09486
Regional Museum in Teplice, Teplice 09863
Regional Museum Iron Gates, Drobeta-Turnu Severin 36360
Regional Museum Jagodina, Jagodina 38101
Regional Museum Kristianstad, Kristianstad 40655
Regional Museum - Museum of the Battle at Bzurą, Kutno 34838
Regional Museum of Andrzej Kaube, Wolin 35316
Regional Museum of Boskovice, Boskovice 09411
Regional Museum of Dachau, Dachau 17763
Regional Museum of Eternal Frost - Permafrostmuseum, Igarka 36930
Regional Museum of Fine Arts, Semey 29730
Regional Museum of Hadsund, Hadsund 10053
Regional Museum of History, Kardžali 05047
Regional Museum of History, Kjustendil 05068
Regional Museum of History, Loveč 05086
Regional Museum of History, Montana 05095
Regional Museum of History, Plovdiv 05130
Regional Museum of History, Stara Zagora 05211
Regional Museum of Kamtchatka Region, Petropavlovsk-Kamčatskij 37553
Regional Museum of Lgov, Lgov 37101
Regional Museum of Local Lore, Donec'k 43038
Regional Museum of Local Lore, Žytomyr 43214
Regional Museum of Lomonosov, Lomonosov 37110
Regional Museum of Miass, Miass 37130
Regional Museum of Mineralnye Vody, Mineralnye Vody 37132
Regional Museum of Natural History, Bhopal 24094
Regional Museum of Pinczov, Pińczów 35009
Regional Museum of South-Eastern Moravia, Napajedla 09670
Regional Museum of Spartanburg County, Spartanburg 53194
Regional Museum of Świebodzin, Świebodzin 35149
Regional Museum of Tarnowskie Góry, Tarnowskie Góry 35180
Regional Museum of the Bulgarian National Revival period and Liberation Movement, Plovdiv 05131
Regional Museum of the City Wrocław, Wrocław 35332
Regional Museum of the Jewish Autonomous Region, Birobidžan 36807
Regional Museum of the Kemerovo region, Kemerovo 37031
Regional Museum of the North Ladoga Lake Region, Sortavala 37789
Regional Museum of the Republic Khakassiya, Abakan 36745
Regional Museum of the Republic Mordovia, Saransk 37740
Regional Museum of the Zbąszyń Region, Zbąszyń 35365
Regional Museum of Völgység, Bonyhád 23376
Regional Museum of Yeghegnadzor, Yeghegnadzor 00722
Regional Museum Sokolov, Sokolov 09844
Regional Museum Specialized in Gold, Jílové u Prahy 09564
Regional Muzeum of the Požeška Valley, Požega 08930
Regional Parish Museum, Krasnobród 34825
Regional Science and Discovery Center, Horseheads 49370
Regional Science Centre, Bhopal 24095
Regional Science Centre, Guwahati 24195
Regional Science Centre, Lucknow 24299
Regional Science Centre, Tirupati 24408
Regional Science Centre and Planetarium, Kozhikode 24286
Regional Science Centre Bhubaneswar, Bhubaneshwar 24098
Regional- und Telegrafenmuseum, Stegersbach 02762
Regionale Heemmusea Bachten de Kupe, Izenberge 03739
Regionalen Istoričeski Muzej, Kardžali 05047
Regionalen Istoričeski Muzej, Kjustendil 05068
Regionalen Istoričeski Muzej, Pazardžik 05109
Regionalen Istoričeski Muzej, Pleven 05120
Regionalen Istoričeski Muzej, Ruse 05155
Regionalen Istoričeski Muzej, Stara Zagora 05211
Regionalen Istoričeski Muzej, Šumen 05218
Regionalen Istoričeski Muzej Varna, Varna 05247
Regionales Heimatmuseum für das Renchtal, Oppenau 20905
Regionalgeschichtliches Museum, Lutherstadt Eisleben 20112
RegionalHistorical Museum, Šumen 05218
Regionalmuseum, Bad Lobenstein 16781
Regionalmuseum Alsfeld, Alsfeld 16481
Regionalmuseum Alte Schule, Kaufungen 19493
Regionalmuseum Bad Frankenhausen, Bad Frankenhausen 16731
Regionalmuseum Binn, Binn 41233
Regionalmuseum Burg Beeskow, Beeskow 16979
Regionalmuseum des Amtes Malliß, Neu Kaliß 20584
Regionalmuseum Fritzlar, Fritzlar 18485
Regionalmuseum Harkerode, Harkerode 19014
Regionalmuseum Langnau im Chüechlihus, Langnau im Emmental 41529
Regionalmuseum Leben und Arbeiten im Blauen Ländchen, Nastätten 20549
Regionalmuseum Neubrandenburg, Neubrandenburg 20592
Regionalmuseum Reichelsheim Odenwald, Reichelsheim 21243
Regionalmuseum Schlangenhaus, Werdenberg 42006
Regionalmuseum Schwarzwasser, Schwarzenburg 41848
Regionalmuseum Stockstadt, Stockstadt am Rhein 21906
Regionalmuseum Wolfhager Land, Wolfhagen 22655
Regionalmuseum Xanten, Xanten 22761
Regionalna Izba Muzealna, Poddębice 35016
Regionalne Muzeum Młodej Polski - Rydlówka, Kraków 34819
Regionalni Muzej, Djakovica 38099
Regionalni Muzej Doboj, Doboj 04262
Regionální Muzeum, Český Krumlov 09468
Regionální Muzeum, Mělník 09650
Regionální Muzeum, Mikulov na Moravě 09653
Regionální Muzeum, Náchod 09669
Regionální Muzeum, Žďár nad Sázavou 09934
Regionální Muzeum a Galerie, Jičín 09559
Regionální Muzeum K.A. Polánka, Žatec 09926
Regionální muzeum v Chrudimi, Chrudim 09486
Regionální Muzeum v Jílovém u Prahy, Jílové u Prahy 09564
Regionální Muzeum v Kolíně, Kolín 09591
Regionální Muzeum v Teplicích, Teplice 09863
Regionální Muzeum ve Vysokém Mýtě, Vysoké Mýto 09920
Regionalnyj Muzej Severnogo Priladožja, Sortavala 37789
Regionmuseet Kristianstad, Kristianstad 40655
Regionmuseum Västra Götaland, Vänersborg 41025
Het Regthuijs, Abbekerk 31789
Regthuis Oudkarspel, Oudkarspel 32659
Regthuis t'Schou, Schipluiden 32762
Rehoboth Art League, Rehoboth Beach 52284
Rehoboth Museum, Rehoboth 31754
Reial Acadèmia Catalana de Belles Arts de Sant Jordi, Barcelona 39006
Reiat-Museum, Thayngen 41934
Reichsdorf-Museum Gochsheim, Gochsheim 18667
Reichskammergerichtsmuseum, Wetzlar 22530
Reichspräsident-Friedrich-Ebert-Gedenkstätte, Heidelberg 19080
Reichsstadtmuseum, Rothenburg ob der Tauber 21410
Reichsstadtmuseum im Ochsenhof, Bad Windsheim 16873
Reichsstadtmuseum Weißenburg, Weißenburg in Bayern 22450
Reichsstädtisches Archiv, Rathaushalle und Ratsstube, Mühlhausen/Thüringen 20405
Reichsstädtisches Museum, Bad Wimpfen 16869
Reichstagsmuseum, Regensburg 21231
Reid Concert Hall Museum of Instruments, Edinburgh 44147
Reigate Priory Museum, Reigate 45482
Reinberger Galleries, Cleveland 47637
Reineke-Fuchs-Museum, Linden-Leihgestern 20004
Reinhard-Blauth-Museum, Weilerbach 22390
Reipin Museo, Pirkkala 11209
Reisjärven Kotiseutumuseo, Reisjärvi 11265
Reiss-Engelhorn-Museen (rem), Mannheim 20173
Reitz Home Museum, Evansville 48457
Reitzin Säätiön Kokoelmat, Helsinki 10813
Rekishi Minzoku Shiryokan, Shimosuwa 29271
Rekishi no Bunka Hakubutsukan, Matsusaka 28900
Rekishi Shiryokan, Ranzan, Ranzan 29170
Religieus Museum Kijk-je Kerk-Kunst, Gennep 32247
Religionskundliche Sammlung der Philipps-Universität, Marburg 20194
Religious Art, Gold and Silver of Nin, Nin 08915
Religious Art Museum of the Bishoprie of Husi, Huşi 36407
Religious Museum, Ano Mera 22886
Rembrandt House Museum, Amsterdam 31862
REME Museum of Technology, Arborfield 43310
Remember Museum 39-45, Thimister-Clermont 04062
Remenyik Zsigmond Emlékszoba, Dormánd 23528
Remick Country Doctor Museum and Farm, Tamworth 53439
Remington Carriage Museum, Cardston 05562
Remington Firearms Museum, Ilion 49467
Renaissance Museum, Catanzaro 25779
Renaissance-Schloss Demerthin, Gumtow 18834
Renaissance Schloss Greillenstein, Röhrenbach 02591
The Renaissance Society at the University of Chicago, Chicago 47510
Renaissanceschloss Rosenburg, Rosenburg am Kamp 02599
Renava Manor - Mazeikiai Museum, Renavas 30344
Renavo Rūmai, Renavas 30344
Rendorség-történeti Múzeum, Budapest 23472
Rene Descartes House and Museum, Descartes 12716
Reneszánsz Kőtár, Pécs 23773
Renfrew Museum, Waynesboro 54013
Rengebu-ji Temple, Ogi 29092
Renhua Museum, Renhua 08139
Renishaw Hall Museum and Art Gallery, Renishaw 45483
Renmin Kang Ri Zhan Zhen Memorial Hall, Beijing 07564
Rennebu Bygdemuseum, Rennebu 33779
Renners Mechanisches Musikinstrumenten Museum, March 20197
Rennesøy Bygdemuseum, Rennesøy 33780
Reno County Museum, Hutchinson 49455
Renoir Museum, Cagnes-sur-Mer 12239
Rensselaer County Historical Society Museum, Troy 53575
Rensselaer Russell House Museum, Waterloo 53962
Renton Museum, Renton 52295
Renville County Historical Society Museum, Mohall 50777
Renwick Gallery, Washington 53915
Renwick Museum, Renwick 33194
Replikate der Welt-Kunst im Schloss, Miltach 20325
Replot Hembygdsmuseum, Replot 11267
Reppuniemen Ulkomuseo, Pöytyä 11213
Repslagarmuseet, Älvängen 40405
Republic County Historical Society Museum, Belleville 46700
Republic Museum of Education, İstanbul 42813
Republic Museum of Education, Karataş 42865
Republican Pedagogical Museum, Almaty 29691
Research Museum of the Institute of Zoology, Taipei 42422
Resenhof, Bernau im Schwarzwald 17292
Réserve Géologique de Haute-Provence, Digne-les-Bains 12728
Réserve Naturelle Géologique, Sisteron 15766
Reserve Officer School Museum, Hamina 10754
Réserves Nationales des Arts et Traditions Populaires Picardie, Saint-Riquier 15498
Reserviupseerikoulun Museo, Hamina 10754
Reshaw Exhibit, Evansville 48458
Résidence d'Artistes, Dompierre-sur-Besbre 12764
Residency Museum, York 01684
Residenz Ansbach, Ansbach 16561
Residenz Ellingen, Ellingen 18117
Residenz Kempten, Kempten 19515
Residenz und Hofgarten Würzburg, Würzburg 22720
Residenzgalerie Salzburg, Salzburg 02624
Residenzmuseum, München 20476
Residenzmuseum im Celler Schloss, Celle 17689
Residenzschloß Arolsen, Bad Arolsen 16673
Residenzschloß Oettingen, Oettingen in Bayern 20879
Resim ve Heykel Müzesi, Ankara 42710
Resim ve Heykel Müzesi, İstanbul 42835
Resim ve Heykel Müzesi, İzmir 42857
Resistance Museum, Akko 24924
Resistance Museum, Bondues 12051
Resistance Museum, Nueva Gerona 09233
Resistance Museum, Santiago de Cuba 09314
Resistance Museum Friesland, Leeuwarden 32480
Resource Centre for the Arts, Saint John's 06854
Rėspublikanskaja Mastackaja Galerėja, Minsk 03256
Respublikanskij Pedagogičeskij Muzej, Almaty 29691
Respublikanskij Vystavočnyj Zal, Joškar-Ola 36981
Respublikinis Unikalių Amenų Muziejus, Mosėdis 30315
Restaurant Museum, Lahti 11033
Restigouche Gallery, Campbellton 05543
Restigouche Regional Museum, Dalhousie 05671
Reston and District Historical Museum, Reston 06734
Restoration House, Mantorville 50455
Rétközi Múzeum, Kisvárda 23673
Rétro-Mobile Club Drouais, Dreux 12793
Returned Services League War Museum, Stawell 01528
Retzer Erlebniskeller, Retz 02580
Reubacher Heimatmuseum, Rot am See 21398
Reuben and Edith Hecht Museum, Haifa 24967
Reuben H. Fleet Science Center, San Diego 52743
Reuel B. Pritchett Museum, Bridgewater 47022
Réunion des Musées Nationaux, Paris 54824
Réunion des Musées Nationaux - Nice, Nice 54825
Reuter-Villa mit Wagner-Sammlung, Eisenach 18096
Reutov Picture Gallery, Reutov 37591
Reutovskaja Gorodskaja Kartinnaja Galereja, Reutov 37591
Rev. Krzysztof Kluk Museum of Agriculture, Ciechanowiec 34608
Revelstoke Museum, Revelstoke 06735
Revelstoke Railway Museum, Revelstoke 06736
Reventlow-Museet Pederstrup, Horslunde 10102
Revermont departmental Museum, Treffort 15979
Revolution House, Chesterfield 43823
Revolution Museum, Harbin 07772
Revolutionary Museum, Altanbulag 31615
Revolutionary Museum Ho Chi Minh City, Ho Chi Minh City 54625
Revolving Museum, Lowell 50307
Rewa's Village, Kerikeri 33137
Reykjanes Art Museum, Reykjanesbær 24013
Reykjanes Museum, Reykjanesbaer 24012
Reykjavik Art Museum - Harbour House, Reykjavík 24018
Reykjavik Art Museum - Kjarvalsstaðir, Reykjavík 24019
Reykjavík Electricity Museum, Reykjavík 24033
Reykjavík Museum of Photography, Reykjavík 24024
Reynard the Fox, Linden-Leihgestern 20004
Reynold Rapp Museum, Spalding 06999
Reynolda House, Winston-Salem 54249
Reynolds-Alberta Museum, Wetaskiwin 07312
Reynolds Gallery, Stockton 53307
Reynolds Homestead, Critz 47883
Reyum Institute of Arts and Culture, Phnom Penh 05288
Reza Abbasi Museum, Tehrān 24692
Reza Shah Museum, Johannesburg 38662
Rezervația Arheologică, Cucuteni 36334
Rezervația Arheologică Cotnari, Cotnari 36319
Rezervația Arheologică Târgșoru Vechi, Târgșoru Vechi 36662
Rezervația de Arhitectură Vrânceană, Focșani 36374
Rezerwat Archeologiczny na Zawodziu, Kalisz 34711
Rezydencja Księży Młyn, Łódź 34888
Rhein-Museum Koblenz, Koblenz 19588
Rheingauer Weinmuseum Brömserburg, Rüdesheim am Rhein 21436
Rheinisches Bildarchiv, Köln 19630
Rheinisches Eisenkunstguss Museum, Bendorf 16989
Rheinisches Fastnachtsmuseum Koblenz, Koblenz 19589
Rheinisches Feuerwehrmuseum, Erkelenz 18213
Rheinisches Freilichtmuseum und Landesmuseum für Volkskunde Kommern, Mechernich 20243
Rheinisches Industriemuseum, Euskirchen 18283
Rheinisches Industriemuseum, Oberhausen 20810

R

Rheinisches Industriemuseum, Solingen 21806
Rheinisches Landesmuseum für Industrie- und Sozialgeschichte, Ratingen 21193
Rheinisches Landesmuseum Trier, Trier 22102
Rheinisches Malermuseum, Bonn 17465
Rheinisches Schützenmuseum, Neuss 20661
Rheinland-Pfälzisches Freilichtmuseum Bad Sobernheim, Bad Sobernheim 16839
Rheinmuseum, Emmerich am Rhein 18152
Rhinebeck Aerodrome Museum, Rhinebeck 52301
Rhineland Industrial Museum, Engelskirchen 18171
Rhinelander Logging Museum, Rhinelander 52302
Rhinopolis, Gannat 13029
Rhode Island Black Heritage Society Museum, Providence 52176
Rhode Island Foundation Art Gallery, Providence 52177
Rhode Island Historical Society Exhibition, Providence 52178
Rhode Island Museum Network, Pawtucket 55178
Rhodes College, Memphis 50584
Rhodes Cottage Museum, Muizenberg 38693
Rhodes Memorial Museum and Commonwealth Centre, Bishop's Stortford 43524
Rhodes Museum of Classical Antiquities, Grahamstown 38633
Rhön-Museum, Fladungen 18325
Rhöner Naturmuseum, Tann 22005
Rhondda Heritage Park, Trehafod 45873
Rhondda Museum, Llwynypia 44776
Rhuddlan Castle, Rhuddlan 45485
Rhyl Library, Museum and Arts Centre, Rhyl 45486
Rialto Museum, Rialto 52303
RIBA Drawings and Archives Collections, London 44964
Ribarsko-Biološki Muzej, Nov Dojran 30511
Ribchester Roman Museum, Ribchester 45487
Ribe Kunstmuseum, Ribe 10232
Ribe Raadhussamling, Ribe 10233
Ribes Vikinger Museum, Ribe 10234
Rice County Museum of History, Faribault 48512
Rice Museum, Comporta 35546
The Rice Museum, Crowley 47894
Rice Museum, Dajia 42174
The Rice Museum, Georgetown 48895
Rice University Art Gallery, Houston 49401
Ricewolrd Museum, Los Baños 34407
Richard Billinger-Gedenkraum, Sankt Marienkirchen bei Schärding 02667
Richard-Brandt-Heimatmuseum Bissendorf, Wedemark 22355
Richard E. Peeler Art Center, Greencastle 49047
Richard F. Brush Art Gallery and Permanent Collection, Canton 47224
Richard Gallery and Almond Tea Gallery, Cuyahoga Falls 47917
Richard H. Schmidt Museum of Natural History, Emporia 48394
Richard-Haizmann-Museum, Niebüll 20699
Richard Jefferies Museum, Coate 43865
Richard Salter Storrs House, Longmeadow 50192
Richard-Simoncic-Museum, Rabensburg 02547
Richard Sparrow House, Plymouth 51998
Richard-Wagner-Museum, Bayreuth 16965
Richard Wagner-Museum, Luzern 41597
Richard-Wagner-Museum Graupa, Pirna 21045
Richardson-Bates House Museum, Oswego 51625
Richborough Castle - Roman Fort, Sandwich 45616
Richey Historical Society, Richey 52304
Richibucto River Museum, Richibucto 06739
Richland County Historical Museum, Wahpeton 53785
Richland County Museum, Lexington 50066
Richmond Art Center, Richmond 52311
Richmond Art Gallery, Richmond 06744
Richmond Art Museum, Richmond 52314
Richmond County Historical Society Museum, Melbourne 06259
Richmond Gaol Museum, Richmond, Tasmania 01458
Richmond Main Mining Museum, Pelaw Main 01408
Richmond Museum, Richmond 06745
Richmond Museum, Richmond 38722
Richmond Museum of History, Richmond 52312
Richmond National Battlefield Park, Richmond 52337
Richmond River Historical Museum, Lismore 01233
Richmond Vale Railway Museum, Richmond Vale 01459
Richmondshire Museum, Richmond, North Yorkshire 45490
Richter Múzeum, Budapest 23473
Ric's Art Boat, Bruxelles 03502
Riddarholmskyrkan, Stockholm 40928
Riddarhuset, Stockholm 40929
Ridderhof Martin Gallery, Fredericksburg 48771
Ridderhofstad Gunterstein, Breukelen 32010
Riddoch Art Gallery, Mount Gambier 01327
Rideau Canal Museum, Smiths Falls 06987
Rideau District Museum, Westport 07309
Rideau Hall, Ottawa 06554
Rider University Art Gallery, Lawrenceville 49997
Ridge House Museum, Ridgetown 06749
Riding Mountain National Park Interpretive Centre, Wasagaming 07281
Riding School Exhibition Hall, Praha 09753
Rieck-Haus, Hamburg 18957
Riedenburger Bauernhofmuseum, Riedenburg 21318
Riedmuseum, Rastatt 21184
Riel House, Saint-Vital 06885
Riemland Museum, Heilbron 38640
Riesengebirgsmuseum, Marktoberdorf 20221
Riesengebirgsstube, Würzburg 22721
Riesenrundgemälde Schlacht am Bergisel, Innsbruck 02128
Rieser Bauernmuseum, Maihingen 20134
Rieskrater-Museum, Nördlingen 20725
Rietgaverstede, Nevele 03912
De Rietgors, Papendrecht 32667
Rietveld Schröderhuis, Utrecht 32856
't Rieuw, Nuis 32607
Rievaulx Abbay, Rievaulx 45496
Rıfat Ilgaz Culture Center, Kastamonu 42871
Rıfat Ilgaz Kültür Merkezi, Kastamonu 42871
Riffe Gallery, Columbus 47759
Riffmuseum im Bahnhof, Gerstetten 18618
Rifle Creek Museum, Rifle 52351
Rifles (Berkshire and Wiltshire) Museum, Salisbury 45605
Riga Art Nouveau museum, Rīga 30106
Riga Dome Church, Rīga 30105
Riga Film Museum, Rīga 30107
Riga Motor Museum, Rīga 30108
Riga Porcelain Museum, Rīga 30109
Riga Technical University Museum, Rīga 30110
Rīgas Doma Baznīca, Rīga 30105
Rīgas Jūgendstila centrs, Rīga 30106
Rīgas Kino Muzejs, Rīga 30107
Rīgas Motormuzejs, Rīga 30108
Rīgas Porcelāna Muzejs, Rīga 30109
Rīgas Tehniskās Universitātes Muzejs, Rīga 30110
Rīgas Ūdensapgādes Muzejs, Ādaži 29980
Rīgas Vēstures un Kuģniecības muzejs, Rīga 30111
Riihimäen Kaupunginmuseo, Riihimäki 11268
Riihimäen Taidemuseo, Riihimäki 11269
Riihimäki Art Museum, Riihimäki 11269
Riihimäki City Museum, Riihimäki 11268
Riihimuseo, Peräseinäjoki 11198
Riihipiha Museo, Vuolijoki 11476
Rijdend Electrisch Tram Museum, Amsterdam 31882
Rijksarchief, Brugge 03426
Rijksmuseum, Amsterdam 31883
Rijksmuseum Hendrik Willem Mesdag, Den Haag 32099
Rijksmuseum Twenthe, Enschede 32213
Rijksmuseum van Oudheden, Leiden 32487
Rijksmuseum voor Volkenkunde, Leiden 32488
Rijksprentenkabinet, Amsterdam 31884
Rijssens Museum, Rijssen 32689
Rijtuigmuseum, Bree 03405
Rijwiel- en Bromfietsmuseum, Zoutkamp 33010
Rikosmuseo, Vantaa 11447
Riksidrottsmuseet, Stockholm 40930
Rikuzentakata City Museum, Rikuzentakata 29171
Rikuzentakata-shi Hakubutsukan, Rikuzentakata 29171
Rila Monastery National Museum, Rilski Manastir 05150
Riley County Historical Museum, Manhattan 50437
Riley House Museum of African American History and Culture, Tallahassee 53428
Rim Country Museum, Payson 51741
Rimrocker Historical Museum of West Montrose County, Naturita 50990
Rimsky-Korsakov Memorial House, Tichvin 37841
Rimsky-Korsakov Memorial Museum-Reserve Loubensk-Vechasha, Pljussa 37569
Rincón de Historia, Salto 00563
Rindal Bygdemuseum, Rindal 33781
Ringe Museum, Ringe 10235
Ringerikes Museum, Hønefoss 33559
Ringkøbing Museum, Ringkøbing 10236
Ringsaker Vekter- og Brannhistoriske Museum, Moelv 33670
Ringsted Museum, Ringsted 10237
Ringve Museum, Trondheim 33927
Ringwood Miners Cottage, Ringwood 01460
Rinno-ji Jokodo Homotsukan, Nikko 29070
Rinno-ji Jokodo Treasure House, Nikko 29070
Rinsing Museum, Qingshui 08113
Rio Grande Valley Museum, Harlingen 49170
Rio Hondo College Art Gallery, Whittier 54137
Riordan Mansion, Flagstaff 48553
Ripley Castle, Ripley, North Yorkshire 45498
Ripley County Historical Society Museum, Versailles 53729
Ripley's Believe it or not!, Key West 49766
Ripley's Believe It or Not Museum, Blackpool 43530
Ripley's Believe It or Not Museum, Guadalajara 30905
Ripley's Believe It or Not Museum, Hong Kong 07814
Ripley's Believe It or Not Museum, København 10149
Ripley's Believe It or Not Museum, México 31160
Ripley's Believe It or Not Museum, Niagara Falls 06449
Ripley's Believe It or Not Museum, Pattaya 42589
Ripley's Believe It or Not Museum, Shah Alam 30620
Ripley's Believe It or Not! Museum, Surfers Paradise 01538
Ripon College Art Gallery, Ripon 52358
Ripon Historical Society Museum, Ripon 52359
Ripon Prison and Police Museum, Ripon 45501
Ripon Workhouse Museum, Ripon 45502
Rippl-Rónai Emlékház, Kaposvár 23633
Rippon Lea Estate, Elsternwick 01068
Riserva Naturale Speciale del Sacro Monte di Varalio, Varallo Sesia 28159
Rishon Le-Zion Museum, Rishon Le-Zion 25092
Rising Sun Tavern, Fredericksburg 48772
Risør Museum, Risør 33782
Rissa Bygdemuseum, Rissa 33784
Risten-Lakviks Järnväg, Åtvidaberg 40448
RiSuPia Panasonic Digital Network Museum, Tokyo 29460
Rittergut Haus Laer, Bochum 17405
Ritterhaus Bubikon, Bubikon 41264
Rittman Historical Society Museum, Rittman 52360
Riuttala Peasant Museum, Karttula 10922
Riuttalan Talonpoikaismuseo, Karttula 10922
River and Rowing Museum, Henley-on-Thames 44449
River Arts Center, Clinton 47651
River Brink, Queenston 06701
River Heritage Museum, Paducah 51670
River Hills Park Museum, Wildwood 54160
River Legacy Living Science Center, Arlington 46353
River Museum, Wellsville 54033
River Museum of the Southern Sea, Salango 10458
River Valley School Museum, Virden 07269
Riverdale Mansion, Riverdale Park 52365
Riverhurst F.T. Hill Museum, Riverhurst 06757
Riverside Art Museum, Riverside 52373
Riverside Municipal Museum, Riverside 52374
Riverside Museum, La Crosse 49837
Riverside Museum at Blake's Lock, Reading 45471
Riverton Museum, Riverton 52377
Riverton Transportation & Heritage Centre, Riverton 06760
Riverview at Hobson Grove, Bowling Green 46975
Riverview Ethnographic Museum, Bear River 05404
Rivnenski Museum of Regional Studies, Rivne 43173
Rivnens'kyj Krajeznavčyj Muzej, Rivne 43173
Riza, Sergiev Posad 37763
Rizal Technological and Polytechnic Institute Museum, Morong 34462
Rize Devlet Güzel Sanatlar Galerisi, Rize 42926
Rize State Fine Art Gallery, Rize 42926
Rizhao Museum, Rizhao 08140
Riznica Crkvene Umjetnosti - Zlato i Srebro Nina, Nin 08915
Riznica Franjevačkog Samostana Split, Split 08963
Riznica i Galerija Svetista Gospe Trsatske, Rijeka 08941
Riznica Katedrale Svetog Tripuna, Kotor 31648
Riznica Samostana Svetog Frane, Zadar 08985
Riznica Stolne Crkve u Dubrovniku, Dubrovnik 08873
Riznica Zagrebačke Katedrala, Zagreb 09010
Rjazanskij Gosudarstvennyj Istoriko-architekturnyj Muzej-zapovednik, Rjazan 37595
Rjazanskij Oblastnoj Chudožestvennyj Muzej im. I.P. Požalostina, Rjazan 37596
RLB Kunstbrücke, Innsbruck 02129
RMIT Gallery & RMIT University's Art Collection, Melbourne 01291
RNA Signals Museum, Ede 32170
Road Museum, Helsinki 10830
Road & Traffic Museum of Central Finland, Kintaus 10961
Roald Amundsens Hjem Uranienborg, Svartskog 33894
Roald Amundsens Minne, Torp 33912
Roan Bygdetun, Roan 33787
Roanoke Island Festival Park, Manteo 50453
Roaring Twenties Antique Car Museum, Hood 49358
Rob Roy and Trossachs Victor Centre, Callander 43714
Robben Island Museum, Robben Island 38724
Robbins Hunter Museum, Granville 49023
Robbins Museum of Archaeology, Middleborough 50649
Roberson Museum and Science Center, Binghamton 46816
Robert A. Bogan Fire Museum, Baton Rouge 46639
Robert A. Peck Gallery, Riverton 52378
Robert A. Vines Environmental Science Center, Houston 49402
Robert and Mary Montgomery Armory Art Center, West Palm Beach 54074
Robert B. Ferguson Museum of Mineralogy, Winnipeg 07375
Robert Brady Museum, Cuernavaca 30821
Robert Burns Centre, Dumfries 44039
Robert Burns House, Dumfries 44040
Robert C. Williams American Museum of Papermaking, Atlanta 46449
Robert Crown Center for Health Education, Hinsdale 49300
Robert E. Lee Memorial Association, Stratford 53336
Robert E. Wilson Art Gallery, Huntington 49424
Robert Edward Hart Memorial Museum, Souillac 30685
Robert Frost Farm, Derry 48094
Robert Fulton Birthplace, Quarryville 52209
Robert Gordon University Collection, Aberdeen 43247
Robert Hamerling - Museum, Kirchberg am Walde 02166
Robert Hull Fleming Museum, Burlington 47144
Robert Hytha-Museum, Oberwölbling 02465
Robert-Koch-Museum, Berlin 17239
Robert Koch Museum, Wolsztyn 35319
Robert Langen Art Gallery, Waterloo 07289
Robert Louis Stevenson House, Monterey 50807
Robert Louis Stevenson Memorial Cottage, Saranac Lake 52911
Robert Louis Stevenson Museum, Vailima 38001
Robert Louis Stevenson Silverado Museum, Saint Helena 52548
Robert McLaughlin Gallery, Oshawa 06524
Robert Mills Historic House and Park, Columbia 47730
Robert Musil-Literatur-Museum, Klagenfurt 02191
Robert Newell House, Saint Paul 52631
Robert O'Hara Burke Memorial Museum, Beechworth 00850
Robert Owen Memorial Museum, Newtown, Powys 45258
Robert R. McCormick Museum at Cantigny, Wheaton 54121
Robert S. Kerr Museum, Poteau 52114
Robert S. Peabody Museum of Archaeology, Andover 46298
Robert-Schumann-Haus Zwickau, Zwickau 22835
Robert-Sterl-Haus, Struppen 21922
Robert-Struppmann-Museum, Lorch, Rheingau 20033
Robert Toombs House, Washington 53936
Robert V. Fullerton Art Museum, San Bernardino 52734
Robert W. Ryerss Museum, Philadelphia 51875
Robertinum, Halle, Saale 18903
Roberto Mata Taller de Fotografía, Caracas 54557
Roberts County Museum, Miami 50639
Robertson Museum, Millport 45149
Robeson Center, Newark 51288
Robin Hood's Bay and Fylingdale Museum, Robin Hood's Bay 45504
Robinson Visitors Center, Hartsville 49205
Robson Gallery, Selkirk 45637
Le Roc des Harmonies: Palais de Mineralogique, Granville 13107
Rocanville and District Museum, Rocanville 06765
Rocca Borromeo, Angera 25236
Rocca Sanvitale, Fontanellato 26221
Roccolta d'Arte Sacra, San Piero a Sieve 27678
Rochdale Art and Heritage Centre, Rochdale 45505
Rochdale Pioneers Museum, Rochdale 45506
Rocher Saint-Léon, Dabo 12696
Rochester Art Center, Rochester 52390
Rochester Hills Museum at Van Hoosen Farm, Rochester 52388
Rochester Historical and Pioneer Museum, Rochester 01463
Rochester Museum, Rochester 52398
Rochow-Museum Reckahn, Kloster Lehnin 19578
Rock Art Centre, Tanumshede 40969
Rock County Historical Society Museum, Janesville 49603
Rock Creek Station State Historic Park, Fairbury 48476
Rock House Museum, Wytheville 54315
Rock Inscriptions Museum, Çifteler 42753
Rock Island Arsenal Museum, Rock Island 52409
Rock Island County Historical Museum, Moline 50779
Rock 'n Roll & Blues Heritage Museum, Clarksdale 47588
Rock 'n' Roll Hall of Fame and Museum, Cleveland 47638
Rock Springs Historical Museum, Rock Springs 52411
Rockbourne Roman Villa, Fordingbridge 44239
Rockbridge Historical Society, Lexington 50073
Rockcavern, Beechworth 00851
Rockefeller Archaeological Museum, Jerusalem 25010
Rockefeller Museum, Jerusalem 25011
Rockford Art Museum, Rockford 52416

R

Rockford College Art Gallery/Clark Arts Center, Rockford 52417
Rockglen Rolling Hills Museum, Rockglen 06767
Rockhampton and District Historical Museum, Rockhampton 01465
Rockhampton Art Gallery, Rockhampton 01466
Rockhill Trolley Museum, Rockhill Furnace 52420
Rockingham, Kingston 49788
Rockingham, Princeton 52153
Rockingham Castle, Market Harborough 45110
Rockingham Free Public Library and Museum, Bellows Falls 46716
Rockland Center for the Arts, West Nyack 54070
Rockland County Museum, New City 51043
rock'n'pop museum, Gronau, Westfalen 18774
Rockoxhuis, Antwerpen 03313
Rockport Art Association Galleries, Rockport 52423
Rockwell Museum of Western Art, Corning 47811
Rockwood Museum, Wilmington 54197
Rocky Ford Historical Museum, Rocky Ford 52433
Rocky Lane School Museum, Fort Vermilion 05825
Rocky Mount Arts Center, Rocky Mount 52436
Rocky Mount Children's Museum, Rocky Mount 52437
Rocky Mount Museum, Piney Flats 51927
Rocky Mountain College of Art and Design Galleries, Denver 48087
Rocky Mountain Conservation Center, Denver 48088
Rocky Mountain House Museum, Rocky Mountain House 06769
Rocky Mountain National Park Museum, Estes Park 48427
Rocky Mountain Quilt Museum, Golden 48961
Rocky Mountain Rangers Museum, Kamloops 06037
Rocky Reach Dam Museum, Wenatchee 54036
Rød Bygdetunet, Uskedalen 33947
Rodgers Tavern, Perryville 51784
Rodin Museum, Philadelphia 51876
Rodinné muzeum pohledů – Dům U Červené židle, Praha 09792
Rodman Hall Arts Centre, Saint Catharines 06807
Rodney Fox Shark Experience-Museum, Glenelg 01115
Rodný Dom Jána Hollého, Borský Mikuláš 38202
Rodný Dom L'udovita Štúra a Alexandra Dubčeka, Uhrovec 38384
Rodny Dom Pavla E. Dobšinského, Slavošovce 38352
Rodný Domek Aloise Jiráska, Hronov 09539
Rodný Domek Otokara Březiny, Počátky 09722
Rodný dům Adalberta Stiftera, Horní Planá 09526
Rodný Dům Josefa Hoffmanna, Brtnice 09444
Roebourne Old Goal Museum, Roebourne 01467
Roedde House Museum, Vancouver 07220
Röhsska Museet, Göteborg 40555
Roelantsmuseum Oude Abdij Heemmuseum, Hemiksem 03694
Römer und Bajuwaren Museum Burg Kipfenberg, Kipfenberg 19541
Roemer- und Pelizaeus-Museum, Hildesheim 19185
Römerbad, Schwangau 21667
Römerbadmuseum, Hüfingen 19288
Römerhalle, Bad Kreuznach 16767
Römerhaus Walheim, Walheim 22306
Römerkastell Saalburg, Bad Homburg v.d.Höhe 16752
Römerkeller Ottmarsheimer Höhe, Mundelsheim 20527
Römermuseum, Heitersheim 19115
Römermuseum, Mainhardt 20139
Römermuseum, Mengen 20283
Römermuseum, Obernburg am Main 20817
Römermuseum, Tulln 02833
Römermuseum, Wien 03061
Römermuseum Bedaium, Seebruck 21714
Römermuseum Favianis-Sankt Severin, Mautern 02369
Römermuseum Flavia Solva, Wagna 02880
Römermuseum im Haus Isenburg, Eisenberg, Pfalz 18102
Römermuseum in der Schule, Ehingen 18069
Römermuseum Kastell Boiotro, Passau 20994
Römermuseum Osterburken, Osterburken 20932
Römermuseum Stettfeld, Ubstadt-Weiher 22141
Römermuseum Teurnia, Lendorf 02283
Römermuseum Wallsee-Sindelburg, Wallsee 02893
Römermuseum Weißenburg, Weißenburg in Bayern 22451
Römermuseum Xanten, Xanten 22762
Römerpark Köngen, Köngen 19636
Römersiedlung, Friesenheim 18482
Römerthermen Zülpich - Museum der Badekultur, Zülpich 22826
Römervilla e.V., Möckenlohe 20348
Römerzeitliches Museum, Ratschendorf 02562
Römisch-Germanisches Museum, Köln 19631
Römisch-Germanisches Zentralmuseum, Mainz 20150
Römische Badruine, Badenweiler 16893
Römische Bäder, Potsdam 21104
Römische Baureste Am Hof, Wien 03062
Römische Kalkbrennerei Iversheim, Bad Münstereifel 16790
Das Römische Kastell in Waldmüssingen, Schramberg 21633
Römische Thermen, Weißenburg in Bayern 22452
Römische Villa Nennig, Perl 21012
Römische Villa Urbana Longuich, Longuich 20032
Römischer Keller in Sulz am Neckar, Sulz am Neckar 21982
Römischer Weinkeller Oberriexingen, Oberriexingen 20829
Römisches Freilichtmuseum, Hechingen 19063
Römisches Haus, Weimar 22416
Römisches Heiligtum für Isis und Mater Magna, Mainz 20151
Römisches Museum, Augsburg 16633
Römisches Museum Remagen, Remagen 21261
Römisches Museum und Naturkunde-Museum, Kempten 19516
Röntgen-Gedächtnisstätte, Würzburg 22722
Roentgen-Museum, Neuwied 20688
Rörbäcksnäs Hembygdsgård, Lima 40678
Rörstrand Museum, Linköping 40685
Roerstreekmuseum, Sint Odiliënberg 32774
Rößler Museum, Untermünkheim 22182
Rogaland Bilmuseum, Vigrestad 33977
Rogaland Fiskerimuseum, Åkrehamn 33376
Rogaland Krigshistorisk Museum, Sola 33849
Rogaland Kunstmuseum, Stavanger 33873
Rogaland Museum of Fine Arts, Stavanger 33873
Rogendorfer-Ausstellung, Pöggstall 02511
Roger Guffey Gallery, Kansas City 49697
Roger Raveelmuseum, Machelen 03839
Roger Tory Peterson Institute of Natural History Museum, Jamestown 49598
Roger Williams National Memorial, Providence 52179
Rogers Historical Museum, Rogers 52440
Rogers House Museum and Gallery, Ellsworth 48371
Rogue Gallery and Art Center, Medford 50551
Rohnstädter Heimatstube, Weilmünster 22393
Roja Museum of Maritime Fishing, Roja 30117
Rojas Jūras un Zvejniecības Muzejs, Roja 30117
Rojstna Hiša Blaženega Antona Martina Slomška, Šentjur 38512
Rojstna Hiša Matije Čopa, Žirovnica 38543
Rojstna Hiša Pesnika Dr. Franceta Prešerna, Žirovnica 38544
Rojstna Hiša Pisatelja Frana Saleškega Finžgarja, Žirovnica 38545
Rojstna Hiša Rista Savina, Žalec 38540
Rojstna Hiša Simona Gregorčiča, Vrsno 38536
Rokeby Museum, Ferrisburgh 48537
Rokiškio Krašto Muziejus, Rokiškis 30346
Rokiskis Area Museum, Rokiškis 30346
Rokuzan Art Museum, Hotaka 28549
Rokuzan Bijutsukan, Hotaka 28549
Roland Gibson Gallery, Potsdam 52116
Røldal Bygdemuseum, Røldal 33790
Rolf Bergendorffs Radio Museum, Lessebo 40672
Rolla Minerals Museum, Rolla 52447
Rollag Bygdetun, Rollag 33791
Rollettmuseum, Baden bei Wien 01776
Rollin Arts Centre, Port Alberni 06627
Rollo Jamison Museum, Platteville 51981
Rolls-Royce Museum, Dornbirn 01828
Rol'nícky Dom Vlkolínec, Ružomberok 38345
Római Katolikus Egyházi Gyűjtemény, Sárospatak 23793
Római Kötár, Szentendre 23879
Római Kori Kotár, Sopron 23817
Romain De Saeghermuseum, Sint-Amands 04015
Roman Army Museum, Greenhead 44364
Roman Bath, Schwangau 21667
Roman Bath House, Lancaster 44636
Roman Baths Museum, Jena 43410
Roman-Catholic Ecclesiastical Collection in Sárospatak, Sárospatak 23793
Roman Edificie with Mosaic at the Museum for National History and Archaeology, Constanța 36301
Román Kori Kőtár, Pécs 23774
Roman Museum, Canterbury 43743
Roman Painted House, Dover 44017
Roman Soldier Museum, Pordim 05136
Roman Stone Store, Sopron 23817
Roman Theatre of Verulamium, Saint Albans 45548
Roman Villa Museum at Bignor, Pulborough 45451
Romanian Agricultural Museum, Slobozia 36629
Romanian Ethnic Art Museum, Cleveland 47639
Romanian History National Museum, București 36221
Romanian National Museum of Contemporary Art, București 36220
Romanian Naval Museum, Constanța 36308
Romanian Peasants Museum, București 36227
Romanov's Chambers in Zaryadye, Moskva 37403
Romanovskie Palaty v Zarjade, Moskva 37403
Romantikerhaus, Jena 19388
Romany Folklore Museum, Selborne 45633
Rome Art and Community Center, Rome 52452
Rome Historical Society Museum, Rome 52453
Rommel Museum, Blaustein 17383
Romney Toy and Model Museum, New Romney 45207
Romsdalsmuseet, Molde 33678
Ron Morel Memorial Museum, Kapuskasing 06043
Ronald Reagan Boyhood Home, Dixon 48146
Ronald Reagan Presidential Library and Museum, Simi Valley 53110
Ronathahon:ni Cultural Centre, Cornwall 05644
Rondeau Provincial Park Visitor Centre, Morpeth 06393
Rong County Museum, Rongxian 08142
Rongshui Museum, Rongshui 08141
Rongyu Guan, Jiangyin 07865
Roniger Memorial Museum, Cottonwood Falls 47845
Roodepoort Museum, Roodepoort 38726
Rooseum, Malmö 40726
Roosevelt Campobello, Lubec 50314
Roosevelt County Museum, Portales 52057
Root House Museum, Marietta 50471
Roots of Norfolk at Gressenhall, Gressenhall 44366
Rooty Bread Exhibition, Komáron 23679
Roparshaugsamlinga, Isdalstø 33578
Rørosmuseet, Røros 33796
Rosalie House Museum, Natchez 50985
Rosalie Shire Historical Museum, Goombungee 01119
Rosalie Whyel Museum of Doll Art, Bellevue 46709
Roscoe Village Foundation, Coshocton 47839
Roscommon County Museum, Roscommon 24892
Roscrea Heritage Centre, Roscrea 24893
Rose Art Museum, Waltham 53810
Rose Center Museum, Morristown 50859
Rose Hill Manor Park & Children's Museum, Frederick 48759
Rose Hill Mansion, Geneva 48881
Rose Hill Museum, Bay Village 46655
Rose Hill Plantation State Historic Site, Union 53651
Rose Lawn Museum, Cartersville 47265
Rose Museum at Carnegie Hall, New York 51250
Rose Seidler House, Wahroonga 01622
Rose Valley and District Heritage Museum, Rose Valley 06772
Roseau County Historical Museum, Roseau 52456
Rosebud Centennial Museum, Rosebud 06773
Rosebud County Pioneer Museum, Forsyth 48588
Rosemont Art Gallery, Regina 06725
Rosemont Plantation, Woodville 54289
Rosemount Museum, Pueblo 52193
Rosenbach Museum, Philadelphia 51877
Rosenberg Gallery, Baltimore 46587
Rosenborg Castle, København 10150
Rosenborg Slot, København 10150
Rosenbruch Wildlife Museum, Saint George 52545
Den Rosendahlske Bibelsamling, Esbjerg 10003
Rosendal Palace, Stockholm 40931
Rosendals Slott, Stockholm 40931
Rosenholm Slot, Hornslet 10094
Rosenlew-museo, Pori 11222
Rosenlöfs Tryckerimuseum, Kungsgården 40660
Rosenmuseum Steinfurth, Bad Nauheim 16791
Rosenthal Gallery of Art, Caldwell 47161
Rosenwald-Wolf Gallery, Philadelphia 51878
Rosersberg Castle, Rosersberg 40816
Rosersbergs Slott, Rosersberg 40816
Rosetown & District Museum, Rosetown 06775
Rosewall Memorial Shell Museum, Port Lincoln 01432
Rosewood Scrub Historical Museum, Marburg 01261
Roseworthy Agricultural Museum, Roseworthy 01469
Rosgartenmuseum, Konstanz 19682
Roshdestveno Museum-Estate, Roždestveno 37603
Rosicrucian Egyptian Museum, San Jose 52807
Rosily Arts Museum, Taipei 42423
Roskilde Domkirkemuseum, Roskilde 10246
Roskilde Museum, Roskilde 10247
Roskilde Museums Købmandsgård, Roskilde 10248
Roslagens Sjöfartsmuseum, Väddö 41023
Roslagsmuseet, Norrtälje 40778
Roslavl Historical and Art Museum, Roslavl 37597
Roslavlskij Istoriko-chudožstvennyj Muzej, Roslavl 37597
The Ross C. Purdy Museum of Ceramics, Westerville 54091
Ross Castle, Killarney 24852
Ross County Historical Society Museum, Chillicothe 47537
Ross Farm Museum, New Ross 06434
Ross House Museum, Winnipeg 07376
Ross Memorial Museum, Saint Andrews 06799
Ross Museum, Foresters Falls 05804
Ross-Thomson House Museum, Shelburne 06954
Rossburn Museum, Rossburn 06776
Rossendale Footware Heritage Museum, Rawtenstall 45466
Rossendale Museum, Rossendale 45515
Rossfelder Heimatstuben, Crailsheim 17737
Rossijskij Ėtnografičeskij muzej, Sankt-Peterburg 37722
Rossijskij Komitet Meždunarodnogo Soveta Muzeev (IKOM Rossii), Moskva 55002
Rossijskij Muzej Lesa, Moskva 37404
Rossland Historical Museum, Rossland 06777
Rosslyn Chapel, Roslin 45514
Rosson House Museum, Phoenix 51909
Rostejn Castle, Telč 09860
Rostov Kremlin - State Museum Reserve, Rostov 37599
Rostovskij Kreml - Gosudarstvennyj Muzej-zapovednik, Rostov 37599
Rostovskij Oblastnoj Muzej Kraevedenija, Rostov-na-Donu 37602
Rostropovich Museum, Bakı 03168
Roswell Museum and Art Center, Roswell 52470
Rot-Kreuz-Museum, Geislingen an der Steige 18562
Rot-Kreuz-Museum, Regenstauf 21235
Rotary Museum of Police and Corrections, Prince Albert 06665
Rotch-Jones-Duff House and Garden Museum, New Bedford 51015
Roter Haubarg, Witzwort 22636
Róth Miksa Emlékház, Budapest 23474
Rothe House Museum, Kilkenny 24849
Rotherham Art Gallery, Rotherham 45518
Rothko Chapel, Houston 49403
Rothschild House, Port Townsend 52047
Rothschild Petersen Patent Model Museum, Cazenovia 47288
Rotkreuz-Museum, Nürnberg 20787
Rotkreuz-Museum Berlin, Berlin 17240
Rotkreuzmuseum, Hofheim in Unterfranken 19228
Rotonda di Via Besana, Milano 26722
Rotorua Museum of Art and History, Rotorua 33196
Rotovž Art Gallery, Maribor 38478
Rottauer Museum für Fahrzeuge, Wehrtechnik und Zeitgeschichte, Pocking 21070
Rotterdams Radio Museum, Rotterdam 32725
Rottingdean Grange Art Gallery and Museum, Brighton 43624
Rottneros Park, Rottneros 40817
Rottnest Island Museum, Rottnest Island 01471
The Rotunda, Charlottesville 47399
Rotunda Gallery, Brooklyn 47081
Rotunda Museum of Archaeology and Local History, Scarborough 45623
Rough and Tumble Engineers Museum, Kinzers 49798
Rough Point, Newport 51329
Rouleau and District Museum, Rouleau 06779
Roulston Museum, Carstairs 05570
Round Top Center for the Arts - Arts Gallery, Damariscotta 47950
Rouse Hill Estate, Rouse Hill 01472
Roussi Chorbadzhi House-Museum, Žeravna 05270
Rovaniemen Kotiseutumuseo, Rovaniemi 11278
Rovaniemen Taidemuseo, Rovaniemi 11279
Rovaniemi Art Museum, Rovaniemi 11279
Rovaniemi Local Museum, Rovaniemi 11278
Rowan Museum, Salisbury 52681
Rowan Oak, William Faulkner's Home, Oxford 51648
Rowley Historical Museum, Rowley 52475
Rowsell House Interpretation Center, Harrington Harbour 05972
Roy Boyd Gallery, Chicago 47511
Roy Rogers-Dale Evans Museum, Branson 46997
Royal Academy of Arts, London 44965
Royal Academy of Arts, London 44966
Royal Academy of Fine Arts, Stockholm 40905
Royal Air Force Air Defence Radar Museum, Norwich 45283
Royal Air Force Museum, London 44967
Royal Air Force Museum, Shifnal 45664
Royal Air Force Museum 201 Squadron, Saint Peter Port 45593
Royal Air Force Museum Reserve Collection, Stafford 45729
Royal Albert Memorial Museum and Art Gallery, Exeter 44195
Royal Alberta Museum, Edmonton 05752
Royal Anglian Regiment Museum, Duxford 44076
Royal Armouries, London 44968
Royal Armouries at Fort Nelson, Fareham 44211
Royal Armouries Museum, Leeds 44665
Royal Armoury, Stockholm 40909
Royal Army Dental Corps Historical Museum, Aldershot 43273
Royal Armymuseum, Delft 32059
Royal Art Museum, Berlin 17241
Royal Arts Foundation, Newport 51330
Royal Australian Air Force Museum, Point Cook 01420
Royal Australian Infantry Corps Museum, Singleton 01506
Royal Barges National Museum, Bangkok 42558
Royal Berkshire Yeomanry Cavalry Museum, Windsor 46044
Royal Borough Museum Collection, Windsor 46045
Royal Brierley Crystal Museum, Dudley 44031
Royal British Columbia Museum, Victoria 07258
Royal British Society of Sculptors, London 44969

Royal Brunei Armed Forces Museum, Bandar Seri Begawan 04968
Royal Cambrian Academy of Art, Conwy 43888
Royal Canadian Artillery Museum, Shilo 06965
Royal Canadian Geographical Society, Ottawa 54747
Royal Canadian Military Institute Museum, Toronto 07132
Royal Canadian Mounted Police Museum, Regina 06726
Royal Canadian Ordnance Corps Museum, Montréal 06367
The Royal Canadian Regiment Museum, London 06190
Royal Carriage Museum, Bulaq-el-Dakrur 10483
Royal Castle in Warsaw, Warszawa 35291
Royal Castle of Săvârşin, Săvârşin 36592
Royal Ceremonial Dress Collection, London 44970
The Royal Coin Cabinet, Stockholm 40904
Royal Collection of Coins and Medals, København 10136
Royal College of Art, London 44971
Royal College of Obstetricians and Gynaecologists Collection, London 44972
Royal College of Surgeons in Ireland Museum, Dublin 24818
Royal Cornwall Museum, Truro 45879
Royal Crown Derby Museum, Derby 43969
Royal Danish Arsenal Museum, København 10156
Royal Danish Naval Museum, København 10147
Royal Engineers Museum, Chatham 43798
Royal Engineers Museum, Gillingham, Kent 44269
Royal Flying Doctor Service Visitors Centre, Edge Hill 01061
Royal Fusiliers Museum, London 44973
Royal Glasgow Institute Kelly Gallery, Glasgow 44298
The Royal Governor's Mansion, Perth Amboy 51786
Royal Green Jackets Museum, Winchester 46032
Royal Gunpowder Mills, Waltham Abbey 45925
Royal Hamilton Light Infantry Heritage Museum, Hamilton 05962
Royal Hampshire Regiment Museum, Winchester 46033
Royal Highland Fusiliers Regimental Museum, Glasgow 44299
Royal Holloway College Picture Gallery, Egham 44164
Royal Home for Retired Military Personnel and Bronbeek Museum, Arnhem 31928
Royal Hospital Chelsea Museum, London 44974
Royal Inniskilling Fusiliers Regimental Museum, Enniskillen 44180
Royal Irish Fusiliers Museum, Armagh 43320
Royal Irish Regiment Museum, Ballymena 43365
Royal Lazienki Museum, Warszawa 35255
Royal Lincolnshire Regiment Museum, Lincoln 44715
Royal Logistic Corps Museum, Deepcut 43962
Royal London Hospital Archives and Museum, London 44975
Royal London Wax Museum, Victoria 07259
Royal Malaysian Air Force Museum, Kuala Lumpur 30586
Royal Malaysian Navy Museum, Lumut 30592
Royal Malaysian Navy Museum, Melaka 30604
Royal Manufacturers of Tapestry De Wit, Mechelen 03859
Royal Marines Museum, Portsmouth 45436
The Royal Mews, London 44976
Royal Military Academy Sandhurst Collection, Camberley 43719
Royal Military College of Canada Museum, Kingston 06090
Royal Military Police Museum, Chichester 43831
Royal Mint Museum, Kongsberg 33599
Royal Mint Museum, Llantrisant 44775
Royal Monastery of Brou, Bourg-en-Bresse 12107
Royal Museum, Edinburgh 44148
Royal Museum, Kota Bharu 30563
Royal Museum, Seoul 29912
Royal Museum and Art Gallery, Canterbury 43744
Royal Museum for Central Africa, Tervuren 04058
Royal Museum Sri Menanti, Seremban 30617
Royal Museums of Art and History, Bruxelles 03495
Royal National Lifeboat Institution Headquarters Museum, Poole 45407
Royal Naval Museum Portsmouth, Portsmouth 45437
Royal Naval Patrol Service Association Museum, Lowestoft 45043
Royal Navy Submarine Museum, Gosport 44336
Royal New Zealand Army Medical Corps Museum, Christchurch 33076
Royal New Zealand Navy Museum, Devonport 33086
Royal Newfoundland Constabulary Museum, Saint John's 06855
Royal Norfolk Regimental Museum, Norwich 45284
Royal Norwegian Navy Museum, Horten 33563
Royal Nova Scotia Historical Society, Halifax 05947
Royal Observatory Greenwich, London 44977
Royal Observer Corps Museum, Eastleigh 44105
Royal Ontario Museum, Toronto 07133
Royal Palace Amsterdam, Amsterdam 31855
Royal Pavilion, Brighton 43625
Royal Pharmaceutical Society Late Victorian Pharmacy, Edinburgh 44149
The Royal Photographic Society, Fenton House, Bath 43411
Royal Picture Gallery Mauritshuis, Den Haag 32082
Royal Place Museum, Luang Prabang 29968
Royal Prince Alfred Hospital Museum, Camperdown 00943
Royal Pump Room Museum, Harrogate 44402
Royal Regalia Gallery, Bandar Seri Begawan 04961
Royal Regiment of Fusiliers Museum, Warwick 45938
Royal Research Ship Discovery, Dundee 44050
Royal Residence, Trondheim 33928
Royal Saint John's Regatta Museum, Saint John's 06856
Royal Saskatchewan Museum, Regina 06727
The Royal Scots Regimental Museum, Edinburgh 44150
Royal Scottish Academy, Edinburgh 44151
Royal Scottish Academy Collections, Edinburgh 44152
Royal Shakespeare Company Collection, Stratford-upon-Avon 45784
Royal Signals Museum, Blandford Forum 43541
Royal Summer Palace, Praha 09757
Royal Tapestry Factory, Madrid 39587
Royal Thai Navy Museum, Samut Prakan 42598
Royal Tyrrell Museum of Palaeontology, Drumheller 05713
Royal Ulster Rifles Regimental Museum, Belfast 43447
Royal Welch Fusiliers Regimental Museum, Caernarfon 43710
Royal West of England Academy New Gallery, Clifton 43853
Royal Western Australian Historical Society Museum, Nedlands 01357
Royal Wiltshire Yeomanry Museum, Swindon 45818
Royal Winnipeg Rifles Museum, Winnipeg 07377
Royall House, Medford 50547
Royals Museum, Toronto 07134
Royalton Historical Society, Royalton 52477
Roykstovan, Kirkjubøur 10689
Røyrvik Bygdatun, Limingen 33641
Royston and District Museum, Royston 45521
Róza Szegedy House - Literary Memorial Museum, Badacsony 23344
Rozelle House Galleries, Ayr 43348
Rozen Monastery National Museum, Melnik 05092
Rubell Family Collection, Miami Beach 50643
Rubelle and Norman Schafler Gallery, Brooklyn 47082
Rubenianum, Antwerpen 03314
Rubens House, Antwerpen 03315
Rubens Maskinhistoriska Samlingar, Götene 40559
Rubenshuis, Antwerpen 03315
Rubin Museum, Tel Aviv 25120
Ruby Green Contemporary Arts Foundations, Nashville 50973
Ruchill Centre for Contemporary Arts, Glasgow 44300
Ruční Papírna - Muzeum Papíru, Velké Losiny 09901
Rudarska in Etnologoška Zbirka v Črni na Koroškem, Črna na Koroškem 38409
Rudd House Museum, Kimberley 38673
Ruddell Gallery, Spearfish 53200
Ruddick's Folly, Suffolk 53354
Ruddington Framework Knitters' Museum, Ruddington 45524
Ruddington Village Museum, Ruddington 45525
Rudno-petrografičeskij Muzej, Moskva 37405
Rudolf-Scharpf-Galerie, Ludwigshafen am Rhein 20054
Rudolf Steiner Archiv, Dornach 41334
Rudolf Weinwurm-Museum, Göpfritz 01954
Rūdolfa Blaumaņa Memoriālais Muzejs Braki, Ērgļi 30008
Rudolfs Blaumanis Memorial Museum - Braki Estate, Ērgļi 30008
Rudolph E. Lee Gallery, Clemson 47616
Rudolph Tegners Museum, Dronningmølle 09993
Rübeländer Tropfsteinhöhlenmuseum, Elbingerode, Harz 18113
Rübesams Da Capo Oldtimermuseum, Leipzig 19929
Rügen-Museum in der Kulturkunststatt Prora, Binz, Ostseebad 17358
Rügener Puppen- und Spielzeugmuseum, Putbus 21141
Rüstkammer, Dresden 17967
Rueylong Museum, Cyonglin 42171
Rufford Old Hall, Rufford 45526
Rugao Museum, Rugao 08143
Rugby Art Gallery and Museum, Rugby 45527
Rugby School Museum, Rugby 45528
Ruggles House, Columbia Falls 47737
Ruhlaer Uhrenmuseum, Ruhla 21441
Ruhmeshalle und Bavaria, München 20477
Ruhnu Muuseum, Ruhnu 10614
Ruhr Museum, Essen 18267
Ruhrtalmuseum, Schwerte 21700
Ruichang Museum, Ruichang 08144
Ruicheng Museum, Ruicheng 08145
Ruinas del Tazumal, Chalchuapa 10545
Ruïne van Brederode, Santpoort 32743
Ruins of Gour and Pandua, Malda 24305
Ruiskumestarin Talo, Helsinki 10814
Rūjienas Exhibition Hall, Rūjiena 30119
Rūjienas Izstāžu Zāle, Rūjiena 30119
Rukmini Devi Museum, Tiruvanmiyur 24410
Rumänisches Institut, Freiburg im Breisgau 18437
Rumeli Fortress Museum, İstanbul 42836
Rumelihisarı Müzesi, İstanbul 42836
Rumford Historical Museum, North Woburn 51411
Rumšiškių Jono Aisčio (Jono Aleksandravičiaus) Muziejus, Rumšiškės 30348
Rundāle Palace Museum, Pilsrundāle 30054
Rundāles Pils Muzejs, Pilsrundāle 30054
Rundetaarn, København 10151
Rundfunkmuseum der Stadt Fürth, Fürth 18499
Rundfunkmuseum "Manfred von Ardenne", Backnang 16662
Rundfunkmuseum Schloß Brunn, Emskirchen 18163
Rundgang Urchigs Terbil, Törbel 41944
Rundlet-May House, Portsmouth 52099
Rundlingsmuseum Wendlandhof-Lübeln, Küsten 19741
Runeberg's Cottage, Jakobstad 10867
Runebergs stugan, Jakobstad 10867
Runestone Museum, Alexandria 46197
Rungehaus, Wolgast 22674
Runnels Gallery, Portales 52058
Ruokolahden Kotiseutumuseo, Ruokolahti 11280
Ruokolahti Local History Museum, Ruokolahti 11280
Ruoveden Kotiseutumuseo, Ruovesi 11282
Ruovesi Local History Museum, Ruovesi 11282
Rupe Ethnographic Museum, Dubrovnik 08862
Rupert Museum, Stellenbosch 38737
Rural Life Centre, Farnham 44219
Rural Life Museum, Steenwerck 15807
Rural Life Museum and Windrush Gardens, Baton Rouge 46640
Rural Museum on Folk Art and Archaeology, Bagnan 24069
Rush County Historical Society Museum, Rushville 52482
Rushen Abbey, Ballasalla 43358
Rushworth Museum, Rushworth 01474
Rusk County Historical Society Museum, Ladysmith 49862
Ruskin Library, Lancaster 44637
Ruskin Museum, Coniston 43885
Ruskon Kotiseutumuseo ja Rindellin Mäkitupamuseo, Rusko 11283
Russell Cave Museum, Bridgeport 47014
Russell-Cotes Art Gallery and Museum, Bournemouth 43567
Russell Gallery, Peterborough 06596
Russell Hill Rogers Gallery, San Antonio 52720
Russell Museum, Russell 33200
Russian Exhibition Hall, Moskva 37413
Russian Historical and Ethnographic Musuem, Toržok 37862
Russian Museum of Estonia, Tallinn 10631
Russian Museum of Ethnography, Sankt-Peterburg 37722
Russian Museum of Photography, Nižnij Novgorod 37470
Russian Museum of the Forest, Moskva 37404
Russian Museums Association, Tula 54998
Russian Red Cross Museum, Moskva 37352
Russian State Museum of the Arctic and Antarctic, Sankt-Peterburg 37723
Russkij Gosudarstvennyj Muzej Arktiki i Antarktiki, Sankt-Peterburg 37723
Russkij Muzej Fotografii, Nižnij Novgorod 37470
Rust en Vreugd Museum, Cape Town 38589
Rustam Mustafayev ad. Azerbaycan Dövlet İncesenet Muzeyi, Bakı 03169
Rustam Mustafayev Azerbaijan State Museum of Fine Arts, Bakı 03169
Rustington Heritage Exhibition Centre, Rustington 45531
Rusty Relics Museum, Carlyle 05565
Ruth and Charles Gilb Arcadia Historical Museum, Arcadia 46336
Ruth Chandler Williamson Gallery, Claremont 47576
Ruth Eckerd Hall, Clearwater 47611
Ruth Hall Museum of Paleontology, Abiquiu 46130
Ruth Youth Wing, Jerusalem 25012
Rutherford B. Hayes Presidential Center, Fremont 48788
Rutherford House, Edmonton 05753
Ruthin Goal, Ruthin 45532
Ruthmere House Museum, Elkhart 48348
Ruthven Park, Cayuga 05582
Rutland County Museum, Oakham 45309
Rutland Railway Museum, Cottesmore 43896
Ruununmylly, Hämeenlinna 10747
Ruurd Wiersma Hûs, Burdaard 32022
Ruusmäe Koduloomuuseum, Ruusmäe 10615
Ruusmäe Museum of Local Lore, Ruusmäe 10615
RV/MH Heritage Foundation, Elkhart 48349
Ryan Fine Arts Center, Abilene 46125
Ryazan Historical-Architectural Museum Reservation, Rjazan 37595
Ryazan Regional Art Museum, Rjazan 37596
Rybinsk Historical, Architectural and Art Museum-Preserve, Rybinsk 37606
Rybinskij Gosudarstvennyj Istoriko-architekturnyj i Chudožestvennyj Muzej-zapovednik, Rybinsk 37606
Rydal Mount, Ambleside 43300
Rydals Museum, Rydal 40819
Rye Art Gallery, Rye 45535
Rye Castle Museum, Rye 45536
Rye Heritage Centre Town Model, Rye 45537
The Rye Historical Society and Square House Museum, Rye 52495
Ryedale Folk Museum, Hutton-le-Hole 44505
Ryfylkemuseet, Sand 33809
Rygge Museum, Larkollen 33626
Rygnestadtunet, Rysstad 33803
Ryhope Engines Museum, Sunderland 45803
Rymättylä Local History Museum, Rymättylä 11284
Rymättylän Kotiseutumuseo, Rymättylä 11284
Ryoma Sakamoto Memorial Museum, Kochi 28747
Ryoma's Hometown Museum, Kochi 28746
Ryosan-cho Rekishi Hakubutsukan, Kyoto 28863
Ryosan Museum of History, Kyoto 28863
Ryotsu Folk Museum, Ryotsu 29173
Ryšių Istorijos Muziejus, Kaunas 30265
Ryuga-Do Cave Museun, Tosayamada 29514
Ryushi Kinenkan, Tokyo 29461
Ryushi Memorial Museum, Tokyo 29461

S

S.C.R.A.P. Gallery, Indio 49509
S.H. Ervin Gallery, Sydney 01558
S.I. Taneev Museum in Dyutkovo, Zvenigorod 37990
S.I. Vasylkivskyj City Art Gallery, Charkiv 43020
S.M. Kirov Memorial Appartment Museum, Vladikavkaz 37930
S.M. Kirov Museum, Sankt-Peterburg 37678
S. Merkurov House-Museum, Gyumri 00708
S.P. Korolev House Museum, Moskva 37229
S. Ray Miller Auto Museum, Elkhart 48350
S.S. Meteor Maritime Museum, Barkers Island 46609
S.S. Moyie National Historic Site, Kaslo 06047
S. Stanevicius Memorial Farmstead-Museum, Kanopenai 30237
S.V. Rachmaninov House-Museum, Ivanovka 36943
SA Breweries World of Beer, Johannesburg 38663
Saab Museum, Trollhättan 40985
Saad Zaghlul Museum, Munira 10537
SAAF Museum Swartkop, Tshwane 38756
SAAF Museum Ysterplaat, Ysterplaat 38780
Saalfelder Feengrotten, Saalfeld/Saale 21451
Saamelaismuseo Siida, Inari 10859
Saanich Historical Artifacts Society, Saanichton 06785
Saaremaa Muuseum, Kuressaare 10587
Saarijärven Museo, Saarijärvi 11286
Saarländischer Museumsverband e.V., Ottweiler 54855
Saarländisches Bergbaumuseum, Bexbach 17318
Saarländisches Künstlerhaus e.V., Saarbrücken 21458
Saarländisches Schulmuseum, Ottweiler 20961
Saarländisches Uhrenmuseum, Püttlingen 21137
Saarländisches Zweiradmuseum, Wadgassen 22265
Saarlandmuseum, Saarbrücken 21459
Saaser Museum, Saas Fee 41785
Saatchi Gallery, London 44978
Sabac National Museum, Sabac 38138
Sabah Art Gallery, Kota Kinabalu 30566
Sabah State Museum, Kota Kinabalu 30565
Sabancı Culture House, İmit 42797
Sabancı Kültür Sitesi, İmit 42797
Sabā's House, Tehrān 24693
Saçaklı Ev - Kültür ve Sanat Merkezi, Lefkoşa 33372
Sacavém Municpal Pottery Museum, Sacavém 35893
Sacha-Art Gallery at the Constant Agency of the Republic Sacha, Moskva 37177
Sachem Historical Society Museum, Holbrook 49310
Sachs Harbour Museum, Sachs Harbour 06787
Sackets Harbor Battlefield, Sackets Harbor 52497
Saco Museum, Saco 52498
Sacral Art Collection and Saints' Corpses at the Church of Saint Blaise, Vodnjan 08980
Sacramento Mountains Historical Museum, Cloudcroft 47664
Sacred Circle Gallery of American Indian Art, Seattle 53011
Sacrewell Farm and Country Centre, Thornhaugh 45852

Sacro Convento Castillo de Calatrava la Nueva, Aldea del Rey ... 38836
Sa'dabad Museums, Tehrān ... 24694
Sadberk Hanım Müzesi, İstanbul ... 42837
Saddleworth Museum and Art Gallery, Uppermill ... 45900
Sądecki Park Etnograficzny, Nowy Sącz ... 34961
Sadelmakare Öbergs Hus, Solna ... 40873
Sadnikar Collection, Kamnik ... 38428
Sadnikarjeva Muzejska Zbirka v Kamniku, Kamnik ... 38428
Sado Archaeology Museum, Ogi ... 29093
Sado Hakabutsukan, Sawata ... 29237
Sado Kokokan, Ogi ... 29093
Sado-koku Ogi Folk Museum, Ogi ... 29094
Sado Museum-Art and Natural History of Sado, Sawata ... 29237
Sado Nohgaku-no-sato, Ryotsu ... 29174
Sadriddin Aini House-Museum, Samarqand ... 54462
Sadriddin Aini Yodgorlik Muzeyi, Samarqand ... 54462
Sääksmäen Kotiseutumuseo, Sääksmäki ... 11287
Sääksmäki Museum, Sääksmäki ... 11287
Säätyläiskotiseutumuseo, Kolkanlahti ... 10976
Sæby Museum and Sæbygaard Manor Museum, Sæby ... 10255
Sächsischer Museumsbund e.V., Dresden ... 54856
Sächsisches Apothekenmuseum Leipzig, Leipzig ... 19930
Sächsisches Bademuseum Bad Elster, Bad Elster ... 16722
Sächsisches Eisenbahnmuseum, Chemnitz ... 17702
Sächsisches Feuerwehrmuseum Zeithain, Zeithain ... 22774
Sächsisches Industriemuseum Chemnitz, Chemnitz ... 17703
Sächsisches Lehr- & Besucherbergwerk-Himmelfahrt-Fundgrube, Freiberg, Sachsen ... 18424
Sächsisches Nutzfahrzeugmuseum, Hartmannsdorf ... 19020
Sächsisches Psychiatriemuseum, Leipzig ... 19931
Sächsisches Schmalspurbahn-Museum Rittersgrün, Breitenbrunn ... 17541
Sächsisches Strafvollzugsmuseum, Waldheim ... 22296
Säffle Marinmotor Museum, Säffle ... 40821
Sägehammer Hofwies, Windhaag bei Freistadt ... 03113
Sägemühle, Thallwitz ... 22046
Sägereimuseum-Alte Sagi, Richterswil ... 41759
Sägmühlmuseum Marhördt, Oberrot ... 20830
Säimenen Mylly, Rökönvaara ... 11274
Säkylän Kotiseutumuseo, Säkylä ... 11291
Sängermuseum, Feuchtwangen ... 18315
Säräisniemen Kotiseutumuseo, Säräisniemi ... 11292
Särestad Lantbruksmuseum, Grästorp ... 40564
Särestöniemi-Museo, Kaukonen ... 10931
Säters Film & Biografmuseum, Säter ... 40822
Säters Hembygdsmuseum, Säter ... 40824
Sætersgårds Samlinger Dølmotunet, Tolga ... 33909
Säyneisten Kotiseutumuseo, Säyneinen ... 11293
Safe Haven Museum & Education Center, Oswego ... 51626
Safety Harbor Museum of Regional History, Safety Harbor ... 52512
Saffron Walden Museum, Saffron Walden ... 45540
Safier Heimatmuseum, Safien ... 41787
Safn, Reykjavík ... 24031
Safn Ásgrims Jónssonar, Reykjavík ... 24032
Safn Rafmagnsveitu Reykjavíkur, Reykjavík ... 24033
Safnahusid Húsavík, Húsavík ... 24004
Safnasafnið, Akureyri ... 23986
Sag Harbor Whaling and Historical Museum, Sag Harbor ... 52514
Saga-kenritsu Bijutsukan, Saga ... 29179
Saga-kenritsu Hakubutsukan, Saga ... 29180
Saga-kenritsu Kyushu Toji Bunkakan, Arita ... 28365
Saga-kenritsu Nagoya Jou Hakubutsukan, Chinzei ... 28402
Saga Prefectural Art Museum, Saga ... 29179
Saga Prefectural Museum, Saga ... 29180
Saga Prefectural Nagoya Castle Museum, Chinzei ... 28402
Sagalund Museum, Kimito ... 10959
Sagalunds Museum, Kimito ... 10959
Sagamihara Citizen's Gallery, Sagamihara ... 29183
Sagamihara City Museum, Sagamihara ... 29184
Sagamihara-shi Hakubutsukan, Sagamihara ... 29184
Sagamore Hill National Historic Site, Oyster Bay ... 51663
Sagasco Historical Group Museum, Brompton ... 00905
Sagawa Art Museum, Moriyama ... 28940
Sagawa Bijutsukan, Moriyama ... 28940
Saginaw Art Museum, Saginaw ... 52516
Sagstua Skolemuseum, Sagstua ... 33805
Saguache County Museum, Saguache ... 52517
Såguddens Museum, Arvika ... 40440
Sahan Museo, Puumala ... 11242
Şahbuz District Regional History Museum, Şahbuz ... 03193
Sahitya Parishad Museum, Midnapore ... 24311
Saigo Kinenkan, Kagoshima ... 28634
Saigon Open City Studio Space, Ho Chi Minh City ... 54626
Saigou Memorial Museum, Kagoshima ... 28634
Saijo City Folklore Museum, Saijo ... 29186
Saijo-shi Minzokugaku Hakubutsukan, Saijo ... 29186
Sailing History Museum, Zebegény ... 23976
Sailor's House, Uusikaupunki ... 11420
Sailor's Memorial Museum, Islesboro ... 49543
Saimaa Canal Museum, Lappeenranta ... 11049
Saimaan Kanavan Museo, Lappeenranta ... 11049
Sainsbury Centre for Visual Arts, Norwich ... 45285
Saint Albans Historical Museum, Saint Albans ... 52519
Saint Albans Organ Museum, Saint Albans ... 45549
Saint Andrews Cathedral Museum, Saint Andrews ... 45554
Saint Andrews College Museum, Christchurch ... 33077
Saint Andrews Museum, Saint Andrews ... 45555
Saint Andrews Preservation Museum, Saint Andrews ... 45556
Saint Andrews University Museum Collections, Saint Andrews ... 45557
Saint Asaph Cathedral Treasury, Saint Asaph ... 45559
Saint Augustine Historical Society Museum, Saint Augustine ... 52525
Saint Augustine Lighthouse and Museum, Saint Augustine ... 52526
Saint Augustine's Abbey, Canterbury ... 43745
Saint Barbe Museum and Art Gallery, Lymington ... 45055
Saint Barnabas Icon and Archaeological Museum, Gazimağusa ... 33354
Saint Bartholomew's Hospital Museum, London ... 44979
Saint Basil Cathedral, Moskva ... 37400
Saint Basile Chapel Museum, Saint-Basile ... 06803
Saint-Boniface Museum, Winnipeg ... 07378
Saint Bride's Crypt Museum, London ... 44980
Saint Catharines Museum, Saint Catharines ... 06808
Saint Catherine's National School Museum, Aughrim ... 24731
Saint Cecilia's Hall Museum of Instruments, Edinburgh ... 44153
Saint Charles County Historical Society, Saint Charles ... 52535
Saint Charles Heritage Center, Saint Charles ... 52531
Saint Clair County Historical Museum, Belleville ... 46699
Saint Claude Museum, Claude ... 05621
Saint Clements Island-Potomac River Museum, Colton Point ... 47711
Saint David's Hall, Cardiff ... 43753
Saint Dominic Church Museum, Macau ... 08006
Saint Francis County Museum, Forrest City ... 48587
Saint Gaudens National Historic Site, Cornish ... 47812
Saint George Art Museum, Saint George ... 52546
Saint George's Heritage Museum, Brigus ... 05481
Saint George's Historical Society Museum, Saint George's ... 04198
Saint Georges Regional Museum, Hurstville ... 01169
Saint George's United Methodist Church Museum, Philadelphia ... 51879
Saint Helens History Room, Saint Helens ... 01476
Saint Irene Church Museum, İstanbul ... 42804
Saint Ives Museum, Saint Ives, Cornwall ... 45577
Saint Ives Society of Artists Members Gallery (Norway Gallery) and Mariners Gallery, Saint Ives, Cornwall ... 45578
Saint James Cavalier, Valletta ... 30646
Saint James' House, Fredericksburg ... 48773
Saint James Museum and Library, Jerusalem ... 25013
Saint Janhuis Mill & Koelewei Mill, Brugge ... 03428
Saint John and Coningsby Medieval Museum, Hereford ... 44459
Saint John Jewish Historical Museum, Saint John ... 06839
Saint John Sports Hall of Fame, Saint John ... 06840
Saint John's Church and Parish Museum, Hampton ... 49153
Saint John's Co-Cathedral and Museum, Valletta ... 30647
Saint John's College Museum, Winnipeg ... 07379
Saint John's Northwestern Military Academy Museum, Delafield ... 48042
Saint John's Schoolhouse Museum, Reid ... 01456
Saint John's Theatre and Arts Centre, Listowel ... 24868
Saint Johnsbury Athenaeum, Saint Johnsbury ... 52555
Saint Joseph College Art Gallery, West Hartford ... 54060
Saint Joseph Island Museum Village, Richard's Landing ... 06738
Saint-Joseph Museum, Saint-Joseph ... 06858
Saint Joseph Museum, Saint Joseph ... 52565
Saint Joseph's College Museum, Tiruchirappalli ... 24406
Saint Joseph's University Gallery, Philadelphia ... 51880
Saint Lawrence College Art Gallery, Kingston ... 06091
Saint Lewis Heritage Museum, Saint Lewis ... 06863
Saint Louis Art Museum, Saint Louis ... 52595
Saint Louis Artists' Guild Museum, Saint Louis ... 52596
Saint Louis Car Museum, Saint Louis ... 52597
Saint Louis County Historical Museum, Duluth ... 48199
Saint Louis Science Center, Saint Louis ... 52598
Saint Louis University Museum of Arts and Culture, Baguio ... 34342
Saint Lucie County Historical Museum, Fort Pierce ... 48668
Saint Luis University, Louisiana ... 50279
Saint Luke's Gallery, Paxton ... 51740
Saint Mamas Manastırı İkon Müzesi, Güzelyurt ... 33362
Saint Mamas Monastery Icon Museum, Güzelyurt ... 33362
Saint Margaret's Museum, Saint Margaret's Bay ... 45582
Saint Mary Magdalene Treasury, Newark-on-Trent ... 45215
Saint Marys and Benzinger Township Museum, Saint Marys ... 52607
Saint Marys Community Arts Centre, Saint Marys ... 01487
Saint Marys Museum, Saint Marys ... 06869
Saint Mary's Museum of Maxstone, Maxstone ... 06252
Saint Mary's University Art Center, San Antonio ... 52721
Saint Mary's University Art Gallery, Halifax ... 05948
Saint Mawes Castle, Truro ... 45880
Saint Mel's Diocesan Museum, Longford ... 24870
Saint Michael's Loft Museum, Christchurch ... 43843
Saint Michael's Museum, Miramichi ... 06303
Saint Mullins Heritage Centre, Saint Mullins ... 24896
Saint Mungo Museum of Religious Life and Art, Glasgow ... 44301
Saint Neots Museum, Saint Neots ... 45585
Saint Nicolas' Church Museum, Tallinn ... 10638
Saint Norbert Arts Centre, Winnipeg ... 07380
Saint Norbert Provincial Heritage Park, Winnipeg ... 07381
Saint Patrick's Church Museum, Sydney ... 07050
Saint Patrick's Trian, Armagh ... 43321
Saint Paul Historical Museum, Saint Paul ... 06870
Saint Paul Museum, Randleman ... 52243
Saint Peter Hungate Church Museum, Norwich ... 45286
Saint Peter's Church, Sunderland ... 45804
Saint Peter's College Art Gallery, Jersey City ... 49628
Saint Petersburg Museum of History, Saint Petersburg ... 52641
Saint Petersburg State Museum of Theatre and Music, Sankt-Peterburg ... 37725
Saint Philip's Fortress Museum, San Felipe de Puerto Plata ... 10336
Saint Photios Greek Orthodox National Shrine, Saint Augustine ... 52527
Saint Robert's Cave, Knaresborough ... 44620
Saint Sharbel Museum, Annaya ... 30153
Saint Simons Island Lighthouse Museum, Saint Simons Island ... 52648
Saint Sophia of Kiev, Kyïv ... 43109
Saint Tammany Art Gallery, Covington ... 47860
Saint Theresa's College Folklife Museum, Cebu ... 34372
Saint Therese's House, Lisieux ... 13776
Saint Thomas-Elgin Public Art Centre, Saint Thomas ... 06880
Saint Thomas' Hospital Mini Museum, London ... 44981
Saint Thomas More Art Gallery, Saskatoon ... 06923
Saint Vigeans Museum, Arbroath ... 43314
Saint Vincent Church Museum, Isandlwana ... 38644
Saint Vital Historial Society Museum, Winnipeg ... 07382
Saint Volodymyr Ukrainian Catholic Centre Museum, Winnipeg ... 07383
Saint Walburg and District Historical Museum, Saint Walburg ... 06887
Sainte Anne Museum, Lachine ... 06122
Sainte Genevieve Museum, Sainte Genevieve ... 52651
Sainte Maison de la Flocellière, La Flocellière ... 13369
Sainte-Marie Among the Hurons, Midland ... 06279
Sainte Marie among the Iroquois, Liverpool ... 50146
Sairaalamuseo, Lahti ... 11034
De Saisset Museum, Santa Clara ... 52863
Sait Faik Müzesi, Burgaz Adası ... 42739
Sait Faik Museum, Burgaz Adası ... 42739
Saitama-kenritsu Hakubutsukan, Saitama ... 29189
Saitama-kenritsu Kindai Bijutsukan, Saitama ... 29190
Saitama-kenritsu Sakitama Shiryokan, Gyoda ... 28460
Saitama-kenritsu Shizenshi Hakubutsukan, Nagatoro ... 28989
Saitama Museum of Natural History, Nagatoro ... 28989
Saitama Prefectural Museum of History and Folklore, Saitama ... 29189
Saitama-shiritsu Manga Bijutsukan, Saitama ... 29191
Saito Ho-on Kai Museum of Natural History, Sendai ... 29244
Saito Ho-on Kai Shizenshi Hakubutsukan, Sendai ... 29244
Sajanskaja Kartinnaja Galereja, Sajansk ... 37607
Sakai City Museum, Sakai ... 29199
Sakai Cutting Tools Museum, Sakai ... 29200
Sakai Greenery Museum Harvest Hill, Sakai ... 29198
Sakai House, Osaka ... 29154
Sakai-shi Hakubutsukan, Sakai ... 29199
Sakai-shiritsu Hamono Hakubutsukan, Sakai ... 29200
Sakaide Civic Art Museum, Sakaide ... 29202
Sakaide-shi Bijutsukan, Sakaide ... 29202
Sakalhe's Cultural Art Museum, Majia ... 42269
Sakamoto Ryoma Kinenkan, Kochi ... 28747
Sakarya Müzesi, Sakarya ... 42928
Sakata Museum of Art, Sakata ... 29206
Sakata-shiritsu Bijutsukan, Sakata ... 29206
Sake Museum, Ishidoriya ... 28585
Sake Museum, Sakata ... 29207
Saken Sejfullin Museum, Astana ... 29693
Sakıp Sabancı Müzesi, İstanbul ... 42838
Sakkmúzeum, Heves ... 23608
Sakrale Kunst in der Sankt-Anna-Kapelle, Cham ... 17691
Sakralna Baština Muzej, Senj ... 08946
Saksische en Museumboerderij Erve Brooks Niehof, Gelselaar ... 32241
Saku Museum of Modern Art, Saku ... 29208
Saku-shiritsu Bijutsukan, Saku ... 29208
Sakura City Museum-Memorial of Kampo Arai, Ujiie ... 29561
Sakura City Museum of Art, Sakura ... 29211
Sakura-shiritsu Bijutsukan, Sakura ... 29211
Sakura-shiritsu Kampo Arai Kinenkan, Ujiie ... 29561
Sala A. De Carolis, Montefiore dell'Aso ... 26828
Sala Cadafe, Caracas ... 54558
Sala d'Arqueologia i Gabinet Numismàtic, Lleida ... 39481
Sala de Arqueología del Municipio de Mújica, Múgica ... 31204
Sala de Arqueologia e Pré-Historia, Porto ... 35866
Sala de Arte Carlos F. Sáez, Montevideo ... 54425
Sala de Arte Prehispánico, Santo Domingo ... 10360
Sala de Exhibitiones Temporales del Banco Central, San José ... 08825
Sala de Exposiciones, Barakaldo ... 38949
Sala de Exposiciones de la Universidad del Pacífico, Lima ... 34275
Sala de Exposiciones de las Alhajas, Madrid ... 39588
Sala de Exposiciones Edificio Historico, Oviedo ... 39761
Sala de Exposiciones Rekalde, Bilbao ... 39048
Sala de los Bomberos, La Habana ... 09177
Sala de Memorias Culturales, Mogotes ... 08688
Sala de Memórias de Chapada dos Guimarães, Chapada dos Guimarães ... 04412
Sala de Recuerdos de la Escuela Militar de Montaña, Jaca ... 39419
Sala de Troféus do S.L. Benfica, Lisboa ... 35734
Sala de Zoologia, Porto ... 35867
Sala del Centro Cultural Provincial, Málaga ... 39607
Sala del Costume e delle Tradizioni Popolari, Corinaldo ... 25979
Sala d'Exposicions, Ponts ... 39858
Sala Díaz, San Antonio ... 52722
Sala Exposição Museu José Brito da Luz, Estremoz ... 35581
Sala Exposición Permanente Municipal, Cáceres ... 39084
Sala Ferrani Martí, Sant Cugat del Vallès ... 39980
Sala Fotográfica El Pasaje, Alcázar de San Juan ... 38820
Sala Girona de la Fundació la Caixa, Girona ... 39349
Sala Histórica de la Facultad de Medicina, Mitras ... 31170
Sala Histórica del Regimiento de Infantería Mecanizado 24, Río Gallegos ... 00535
Sala Historica General Savio, Buenos Aires ... 00282
Sala Homenaje a Juárez, Guelatao de Juárez ... 30924
Sala Ipostel, Caracas ... 54559
Sala Josefa Rodriguez del Fresno, Santa Fé ... 00622
Sala Melchor Ocampo, Morelia ... 31202
Sala Mendoza, Caracas ... 54560
Sala Montcada de la Fundació la Caixa, Barcelona ... 39007
Sala Muncunill, Terrassa ... 40120
Sala Museo de la Casa de la Cultura Martín Gamarra, Gamarra ... 08644
Sala Museo Literario, Bogotá ... 08596
Sala Museo Pablo Sarasate, Pamplona ... 39809
Sala Museu de Arqueologia, Mogadouro ... 35767
Sala Museu Fialho de Almeida, Cuba ... 35555
Sala Museu Jenny Mendes de Brito, Entroncamento ... 35559
Sala Nacional de Cultura Mariano Fuentes Lira, Cusco ... 34169
Sala Nacional de Exposiciones, San Salvador ... 10555
Sala Numismática del Banco Central de Nicaragua, Managua ... 33292
Sala Numismática del Banco de Crédito del Peru y Sala de Acuarelas Pancho Fierro, Lima ... 34276
Sala Piano Museum, Cebu ... 34373
Sala Pietro Canonica, Stresa ... 27913

S

Sala Sergio Larrain, Santiago de Chile 07465
Sala Traktormuseum, Sala 40827
Salacgrīvas Muzejs, Salacgrīva 30121
Salamanca Arts Centre, Hobart 01157
Salamanca Rail Museum, Salamanca 52653
Salangen Bygdetun, Sjøvegan 33828
Salar Jung Museum, Hyderabad 24217
Salas Municipales de Arte Sanz-Enea, Zarautz 40354
Salcombe Maritime Museum, Salcombe 45597
Sale Historical Museum, Sale 01489
Sale Monumentali della Biblioteca Marciana, Venezia 28209
Salem County Historical Society Museum, Salem 52662
Salem Historical Society Museum, Salem 52663
Salem Maritime Houses, Salem 52658
Salem Museum, Salem 52672
Salem Pioneer Village, Salem 52659
Salem Witch Museum, Salem 52660
Salford Museum and Art Gallery, Salford 45601
Salgo Trust for Education, Port Washington 52049
Salich Sajdashev Museum, Kazan 37023
Salida Museum, Salida 52673
Salihli Culture Center, Salihli 42929
Salihli Kültür Merkezi, Salihli 42929
Salina Art Center, Salina 52674
Salinas Pueblo Missions National Monument, Mountainair 50896
Saline County Historical Society Museum, Dorchester 48153
Saline- und Heimatmuseum, Bad Sulza 16849
Salinenmuseum im Gradierbau, Bad Dürkheim 16717
Salinenmuseum Unteres Bohrhaus, Rottweil 21428
Salisbury and South Wiltshire Museum, Salisbury 45606
Salisbury Folk Museum, Salisbury 01490
Salisbury Historical Society Museum, Salisbury 52683
Salisbury House, Des Moines 48100
Salisbury Mansion, Worcester 54301
Salisbury State University Galleries, Salisbury 52678
Sallan Kotiseutumuseo, Salla 11295
Sallands Landbouwmuseum de Laarman, Luttenberg 32514
Salle Alfred Pellan, Laval 06149
Salle Augustin-Chénier, Ville-Marie 07267
Salle de Traditions de la Garde Républicaine, Paris 14750
La Salle des Étiquettes, Épernay 12837
Salle des Souvenirs, Lisieux 13779
Salle Mémorial Franco Américaine, Sommepy-Tahure 15780
Salle Panoramique de Notre-Dame-de-Monts, Notre-Dame-de-Monts 14475
Salle Souvenir du 24ème BCA, Villefranche-sur-Mer 16195
Salles du Tresor, Prades 14937
Salling-Fjends Museum, Fur 10034
Sally Lunn's Refreshment House and Kitchen Museum, Bath 43412
Salmagundi Museum of American Art, New York 51251
Salmo Museum, Salmo 06904
Salmon Arm Museum, Salmon Arm 06905
Salmon Brook Historical Society Museum, Granby 48984
Saloinen Local History Museum, Arkkukari 10712
Saloisten Kotiseutumuseo, Arkkukari 10712
Salomeja Neris Museum of the Maironis Lithuanian Literature Museum, Kaunas 30266
Salomėjos Nėries Memorialinis Muziejus, Kaunas 30266
Salomons Museum, Southborough 45709
Salón de la Fama del Beisbol Profesional de México, Monterrey 31190
Salón de la Fama Nicaragüense, Rivas 33302
Salon de l'Automobile Miniature, Ally 11559
Salón Electo Brizuela, Concepción del Uruguay 00314
Salon Muzeja Savremene Umetnosti, Beograd 38087
Salon Muzyczny im. Fryderyka Chopina, Antonin 34529
Salone Villa Romana, Firenze 26203
Salonik Chopinów, Warszawa 35288
Salonika Front Museum, Troissereux 15991
Salons voor Schone Kunsten, Sint-Niklaas 04023
Salpalinja-Museo, Miehikkälä 11110
Salt and Moor, Grassau 18740
Salt and Potash Mining Museum, Uslar 22192
Salt and Sugar Museum, Aguadulce 34059
Salt Archaeological Museum, Salt 29666
Salt Folklore Museum, Salt 29667
Salt Lake Art Center, Salt Lake City 52693
Salt Museum, Alcochete 35398
Salt Museum, Delden 32055
Salt Museum, Liverpool 50147
Salt Museum, Northwich 45273
The Salt Queen Foundation, New York 51252
Salt Road Museum, Itoigawa 28600
Salt Spring Island Museum, Ganges 05849
Saltash Heritage Centre, Saltash 45607
Saltdal Museum, Rognan 33789
Salter Museum, Argonia 46346
Saltram House, Plymouth 45393
Saltstraumen Museum, Saltstraumen 33808
Saltworks Museum Tuzla, Tuzla 04291
Saltykv-Shchedrin House-Museum, Kirov 37037
Saltykv-Shchedrin Museum, Tver 37878
Saltyning Zone Museum, Xianning 08373
Salvador Dalí Museum, Saint Petersburg 52642
Salvage Fisherman's Museum, Salvage 06906
Salvation Army George Scott Railton Heritage Centre, Toronto 07135
Salvation Army Heritage Centre, Bern 41206
Salvation Army International Heritage Centre, London 44982
Salvation Army Museum, Helsinki 10808
The Salvation Army Museum, Toronto 07136
Salvation Army Southern Historical Museum, Atlanta 46450
The Salvation Army Territorial Archives and Museum, Wellington 33259
Salvatorianerkloster, Gurk 02020
Salveymühle, Gartz 18544
Salzbergwerk Berchtesgaden, Berchtesgaden 17004
Salzburg Cathedral Museum, Salzburg 02609
Salzburg Museum, Salzburg 02625
Salzburg Open-air Museum, Großgmain 02001
Salzburger Barockmuseum, Salzburg 02626
Salzburger Freilichtmuseum, Großgmain 02001
Salzburger Landes-Skimuseum, Werfenweng 02924
Salzkammer, Pratteln 41734
Salzkammergut-Lokalbahn-Museum, Mondsee 02407
Salzmagazin, Stans 41906
Salzmuseum, Bad Sooden-Allendorf 16842
Salzmuseum, Bad Sülze 16847
Salzwelten Altaussee, Altaussee 01709
Salzwelten Hallstatt, Hallstatt 02048
Salzwelten Salzburg-Bad Dürrnberg, Bad Dürnberg 01742
Sam Bell Maxey House, Paris 51704
Sam Brook Gallery at the Vysotskiy Cultural Centre and Museum, Moskva 37178
Sam Brown Log House, Browns Valley 47094
Sam Houston Historical Schoolhouse, Maryville 50526
Sam Houston Memorial Museum, Huntsville 49446
Sam Noble Oklahoma Museum of Natural History, Norman 51376
Sam Rayburn House Museum, Bonham 46899
The Sam Rayburn Museum, Bonham 46900
Sam Roi Yod, Phachuab Khiri Khan 42592
SAM - Schweizerisches Architekturmuseum, Basel 41180
Sam Tung Uk Museum, Hong Kong 07815
Sam Waller Museum, The Pas 07072
Samara Art Gallery Maria, Samara 37619
Samara Art Museum, Samara 37621
Samara Eparchial Church History Museum, Samara 37620
Samarkand Museum of Regional Studies, Samarqand 54463
Samarqand Ülkashunoslik Muzeyi, Samarqand 54463
Samarra Museum, Samarra 24725
Samarskaja Chudožestvennaja Galereja Marija, Samara 37619
Samarskij Eparchialnyj Cerkovno-istoričeskij Muzej, Samara 37620
Samarskij Oblastnoj Chudožestvennyj Muzej, Samara 37621
Samarskij Oblastnoj Istoriko-kraevedčeskij Muzej im. P.V. Alabina, Samara 37622
Samford and District Historical Museum, Samford 01491
Sámi Museum Siida, Inari 10859
De Samiske Samlinger, Karasjok 33588
Sammenslutningen af Grønlandske Lokalmuseer, Sisimiut 54870
Sammler- und Glücksschweinmuseum, Bad Wimpfen 16870
Sammlung alter Musikinstrumente, Wien 03063
Sammlung Angewandte Kunst, Kassel 19474
Sammlung Antiker Kleinkunst, Jena 19389
Sammlung Bodingbauer, Steyr 02771
Sammlung Crous im Alten Kurhaus, Aachen 16406
Sammlung des Ägyptologischen Instituts, Heidelberg 19081
Sammlung Domnick, Nürtingen 20794
Sammlung Dr. Irmgard von Lemmers-Danforth, Wetzlar 22531
Sammlung Ernst Schneider, Düsseldorf 18004
Sammlung Fossilien des Jura, Lichtenfels 19980
Sammlung Friedhof Hörnli, Riehen 41765
Sammlung Friedrichshof, Zurndorf 03133
Sammlung für Papier- und Druckgeschichte, Frankenberg, Sachsen 18348
Sammlung Gauselmann, Espelkamp 18245
Sammlung Geyer zu Lauf, Emmendingen 18148
Sammlung Goetz, München 20478
Sammlung handgeschriebener Choralbücher, Kiedrich 19525
Sammlung Hauser & Wirth, Henau 41479
Sammlung historischer Biscuitdosen, Basel 41181
Sammlung historischer Maschinen und Geräte, Calau 17675
Sammlung historischer Mikroskope von Ernst Leitz, Wetzlar 22532
Sammlung historischer Prägestempel, Wien 03064
Sammlung historischer Tasteninstrumente, Wassenach 22342
Sammlung historischer Tasteninstrumente Neumeyer-Junghanns-Tracey, Bad Krozingen 16770
Sammlung Hoffmann, Berlin 17242
Sammlung Holzinger, München 20479
Sammlung industrielle Gestaltung, Berlin 17243
Sammlung Irmgard Friedl, Bad Griesbach 16742
Sammlung landwirtschaftlicher Geräte, Unterpleichfeld 22183
Sammlung Ludwig in Bamberg, Bamberg 16917
Sammlung Meinrad Burch-Korrodi, Sarnen 41821
Sammlung Mittelalterliche Kunst in Thüringen, Eisenach 18097
Sammlung Moderne Kunst, München 20480
Sammlung Nuss, Weinstadt 22440
Sammlung Oskar Reinhart 'Am Römerholz', Winterthur 42037
Sammlung Patt, Willisau 42023
Sammlung Picasso Donation Rosengart, Luzern 41598
Sammlung Prinzhorn, Heidelberg 19082
Sammlung Scharf-Gerstenberg, Berlin 17244
Sammlung Schloß Aichberg, Rohrbach an der Lafnitz 02595
Sammlung und Kunstraum Grässlin, Sankt Georgen im Schwarzwald 21495
Sammlung Volkskunde, Kassel 19475
Sammlung Waagen und Gewichte, Albstadt 16459
Sammlung zur Geschichte der Benediktinerabtei Tholey, Tholey 22057
Sammlung zur Geschichte der Geburtshilfe, Göttingen 18699
Sammlung zur Landeskunde Ost- und Westpreußens "Es war ein Land...", Oberschleißheim 20836
Sammlungen Burg Bernstein, Bernstein 01783
Sammlungen der Benediktiner-Abtei Braunau, Rohr in Niederbayern 21364
Sammlungen der Stadt Amstetten auf Schloß Ulmerfeld, Ulmerfeld-Hausmening 02836
Sammlungen der Universität für angewandte Kunst Wien mit Oskar-Kokoschka-Zentrum, Wien 03065
Sammlungen des Archäologischen Instituts der Universität, Göttingen 18700
Sammlungen des Augustiner Chorherrenstiftes, Sankt Florian 02640
Sammlungen des Instituts für Hochschulkunde, Würzburg 22723
Sammlungen des Kunstvereins, Ibbenbüren 19304
Sammlungen des Marktamtes, Wien 03066
Sammlungen des Stiftes Sankt Lambrecht, Sankt Lambrecht 02659
Sammlungen des Winckelmann-Instituts - Lehrsammlung Klassische Archäologie, Berlin 17245
Sammlungen Schloß Moosham, Unternberg 02841
Sammlungen Zisterzienserstift Stams, Stams 02761
Sammy Marks Museum, Kungwini 38680
Sammy Miller Motorcycle Museum, New Milton 45205
Samo Chalupka Memorial Site, Horná Lehota 38260
Samobor Museum, Samobor 08944
Samoborski Muzej, Samobor 08944
Samogitian Art Museum, Plungė 30338
Samogitian Diocese Museum, Varniai 30391
Samogitian-Japanese Homestead Museum of Alfredas Jonusas, Telšiai 30378
Samoilov Memorial Apartment Museum, Sankt-Peterburg 37659
Samoy Art Museum, Bacoor 34338
Samson-Haus, Leer 19871
Samson V Maritime Museum, New Westminster 06438
Samsun Arkeoloji ve Etnografya Müzesi, Samsun 42931
Samsun Atatürk Müzesi, Samsun 42932
Samsun Devlet Güzel Sanatlar Galerisi, Samsun 42933
Samsun State Fine Art Gallery, Samsun 42933
Samsung Children's Museum, Seoul 29913
Samsung Foundation of Culture, Seoul 29914
Samu Géza Múzeum, Pécsvárad 23783
Samuel Cupples House, Saint Louis 52599
Samuel Dorsky Museum of Art, New Paltz 51097
Samuel Johnson Birthplace Museum, Lichfield 44706
Samuel K. Fox Museum, Dillingham 48141
Samuel P. Harn Museum of Art, Gainesville 48816
Samuel S. Fleisher Art Memorial, Philadelphia 51881
San Agustin Museum Intramuros, Manila 34457
San Angelo Museum of Fine Arts, San Angelo 52704
San Antonio Children's Museum, San Antonio 52723
San Antonio Conservation Society Museum, San Antonio 52724
San Antonio Missions Museum, Jolon 49645
San Antonio Missions Visitor Center, San Antonio 52725
San Antonio Museum of Art, San Antonio 52726
San Benito County Museum, Hollister 49320
San Bernardino County Museum, Redlands 52276
San Buenaventura Mission Museum, Ventura 53714
San Cozy Ceramic Museum, Puli 42298
San Diego Air & Space Museum, San Diego 52744
San Diego Art Institute, San Diego 52745
San Diego Automotive Museum, San Diego 52746
San Diego Command Museum, San Diego 52747
San Diego Hall of Champions Sports Museum, San Diego 52748
San Diego Historical Society Museum, San Diego 52749
San Diego Maritime Museum, San Diego 52750
San Diego Model Railroad Museum, San Diego 52751
San Diego Museum of Art, San Diego 52752
San Diego Museum of Man, San Diego 52753
San Diego Natural History Museum, San Diego 52754
San Diego Viceregal Art Gallery, México 31159
San Dieguito Heritage Museum, Encinitas 48396
San el-Haggar Museum, Sharqiya 10540
San Francisco Airport Museums, San Francisco 52791
San Francisco Camerawork, San Francisco 52792
San Francisco Fire Department Museum, San Francisco 52793
San Francisco Maritime National Historical Park, San Francisco 52794
San Francisco Museum of Modern Art, San Francisco 52795
San Gabriel Mission Museum, San Gabriel 52802
San Jacinto Museum, San Jacinto 52803
San Jacinto Museum of History, La Porte 49855
San Joaquin County Historical Museum, Lodi 50159
San Jose Institute of Contemporary Art, San Jose 52808
San Jose Museum of Art, San Jose 52809
San Jose Museum of Quilts & Textiles, San Jose 52810
San Juan Bautista Mission Museum, San Juan Bautista 52812
San Juan County Historical Museum, Silverton 53109
San Juan County Museum, Bloomfield 46852
San Juan Historical Society Museum, Friday Harbor 48794
San Juan Island National Historical Park, Friday Harbor 48795
San Luis Obispo Art Center, San Luis Obispo 52815
San Luis Obispo County Historical Museum, San Luis Obispo 52816
San Marco Gallery, San Rafael 52829
San Marcos de Apalache Historic State Park, Saint Marks 52603
San Mateo County History Museum, Redwood City 52278
San Miguel Mission Church, Santa Fe 52887
San Pedro Museo de Arte, Puebla 31291
San Telmo Museum, Donostia-San Sebastián 39253
Sanandaj Museum, Sanandaj 24642
Sanati Museum, Kerman 24619
Sanchez Adobe, Pacifica 51668
Sancta Birgitta Klostermuseum, Vadstena 41020
Sanctuary Wood Museum - Hill 62, Zillebeke 04163
Sand Bakkens Medisinske Museum, Flekkefjord 33498
Sand Springs Cultural and Historical Museum, Sand Springs 52831
Sandal Castle, Sandal 45609
Sandauer Heimatstube, Arzberg 16577
Sande Historielags Bygdesamlinger, Sande 33811
Sandefjord Bymuseum, Sandefjord 33813
Sandefjord Maritime Museum, Sandefjord 33814
Sandefjord Sjøfartsmuseum, Sandefjord 33814
Sandefjord Town Museum, Sandefjord 33813
Sandellmuseet, Ramkvilla 40813
Sandelsches Museum, Kirchberg an der Jagst 19543
Sandgate and District Museum, Sandgate 01493
Sandilands Forest Centre and Museum, Hadashville 05923
Sandnes Museum, Sandnes 33815
Sandon Museum, New Denver 06427
Sándor Kántor Pottery House, Karcag 23639
Sándor Kisfaludy Memorial-House, Sümeg 23825
Sándor Petőfi Memorial Hall, Kiskunfélegyháza 23668
Sándor Petőfi Memorial House, Dömsöd 23527
Sándor Petőfi Memorial Museum, Szalkszentmárton 23834
The Sandor Teszler Gallery, Spartanburg 53195
Sandown Historical Museum, Sandown 52834

Sandpoint Lighthouse, Escanaba 48415
Sandringham House Museum, Sandringham 45614
Sands Bygdasavn, Sandur 10696
Sands-Willets House, Port Washington 52050
Sandsteinmuseum, Bad Bentheim 16684
Sandsteinmuseum Kloster Cornberg, Cornberg 17726
Sandtorg Bygdetun, Sørvik 33857
Sanduotang Museum, Taigu 08265
Sandwich Glass Museum, Sandwich 52841
Sandwich Historical Society Museum, Center Sandwich 47314
Sandwich Historical Society Museum, Sandwich 52838
Sandy Bay Museums, Rockport 52424
Sandy Spring Museum, Sandy Spring 52845
Sanford Museum, Cherokee 47421
Sanford Museum, Sanford 52847
Sangerberger Heimatstube, Regenstauf 21236
Sangji Vocational College Museum, Andong 29767
Sangli State Museum, Sangli 24377
Sangre de Cristo Arts Center and Buell Children's Museum, Pueblo 52194
Sanilac County Historical Museum, Port Sanilac 52041
Sanitärhistorisches Museum, Wien 03067
Sanjay Sharma Museum, Jaipur, Rajasthan 24235
Sankei-en Garden, Yokohama 29605
Sankt Marien, Müntzergedenkstätte, Mühlhausen/ Thüringen 20406
Sankt Matthäus-Kirche im Kulturforum, Berlin 17246
Sankt Nikolaus-Hospital Cusanusstift, Bernkastel-Kues 17299
Sankt Pauli-Museum, Hamburg 18958
Sankt-Peterburgskij Muzej Chleba, Sankt-Peterburg 37724
Sankt-Peterburgskij Muzej Teatralnogo i Musykalnogo Iskusstva, Sankt-Peterburg 37725
Sankt Veiter Geschichtenhaus, Sankt Veit im Mühlkreis 02692
Sankta Karins Kyrkoruin, Visby 41071
Sankturbanhof Sursee, Sursee 41926
Şanlıurfa Devlet Güzel Sanatlar Galerisi, Şanlıurfa 42934
Şanlıurfa Müzesi, Şanlıurfa 42935
Şanlıurfa State Fine Art Gallery, Şanlıurfa 42934
Sanmen Museum, Sanmen 08148
Sanmenxia Museum, Sanmenxia 08149
Sanming Museum, Sanming 08150
Sannidal Bygdetun, Sannidal 33817
Sano Art Museum, Mishima 28914
Sano Bijutsukan, Mishima 28914
Sano Gallery, Kokubunji 28762
Sanquhar Tolbooth Museum, Sanquhar 45618
Sanshui Museum, Sanshui 08151
Sansu Museum, Meishan 08012
Santa Ana College Art Galleries, Santa Ana 52854
Santa Babara Maritime Museum, Santa Barbara 52856
Santa Barbara Contemporary Arts Forum, Santa Barbara 52857
Santa Barbara Historical Museum, Santa Barbara 52858
Santa Barbara Museum of Art, Santa Barbara 52859
Santa Barbara Museum of Natural History, Santa Barbara 52860
Santa Cruz Art League, Santa Cruz 52869
Santa Cruz Museum of Natural History, Santa Cruz 52870
Santa Fe Children's Museum, Santa Fe 52888
Santa Fe Community Museum, Santa Fe 34501
Santa Fe Gallery, Gainesville 48817
Santa Fe Trail Museum, Larned 49955
Santa Fe Trail Museum, Trinidad 53569
Santa Fe Trail Museums of Gray County, Ingalls 49510
Santa Giulia - Museo della Città, Brescia 25543
Santa Maria Museum of Flight, Santa Maria 52895
Santa Maria Valley Historical Society Museum, Santa Maria 52896
Santa Monica Museum of Art, Santa Monica 52901
Santa Rosa Junior College Art Gallery, Santa Rosa 52908
Santarella Museum, Tyringham 53637
Santiago do Cacém Museu Municipal, Santiago do Cacém 35912
Santiaoshi History Museum, Tianjin 08282
Santo Niño Shrine and Heritage Museum, Tacloban 34511
Santuario Casa Santa Caterina, Siena 27847
Santuario de Nuestra Senora de Guadalupe, Santa Fe 52889
Santuario della Beata Vergine della Salute, Solarolo 27863
Santuario Eucaristico, Bolsena 25482
Sanuki Folk Art Museum, Takamatsu 29319
Sanuki Museum of Soy Sauce Art, Sakaide 29203
Sanxingdui Museum, Guanghan 07720
Sanya Museum, Sanya 08152
Sanying Pagoda and Museum of Dinosaur, Shaoguan 08208
Sanyuan Bowu Guan, Sanyuan 08153
Sapporo Art Park Museum, Sapporo 29222
Sapporo Clock Tower Exhibition, Sapporo 29221
Sapporo Geijutsu no Mori Bijutsukan, Sapporo 29222
Sapporo Museum Activity Center, Sapporo 29225
Sapporo Museum of Sculpture, Sapporo 29223
Sapporo Science Center, Sapporo 29224
Sapporo-shiritsu Hakubutsukan, Sapporo 29225
Sapporo Waterworks Memorial Museum, Sapporo 29226
Sapporo Winter Sports Museum, Sapporo 29227
Sapulpa Historical Museum, Sapulpa 52910
Sapumal Foundation Gallery, Colombo 40375
Sara Hildén Art Museum, Tampere 11357
Sara Hildénin Taidemuseo, Tampere 11357
Sarafkina Kašta Gradski Bit, Veliko Tǎrnovo 05258
Sarah Campbell Blaffer Foundation, Houston 49404
Sarah Campbell Blaffer Gallery, Houston 49405
Sarah Moody Gallery of Art, Tuscaloosa 53621
Sarah Orne Jewett House, South Berwick 53164
Sarah Spurgeon Gallery, Ellensburg 48361
Sarakatsani Folklore Museum, Serres 23181
Sarasota Classic Car Museum, Sarasota 52915
Sarat Smriti Granthagar, Panitras 24342
Saratoga Museum, Saratoga 52919
Saratoga National Historical Park, Stillwater 53295
The Saratoga Springs History Museum, Saratoga Springs 52924
Saratov Museum of Ethnography, Saratov 37752
Saratov Museum of Local History, Nature and Culture, Saratov 37754
Saratov Museum of the Glory of the Fighters, Saratov 37753
Saratovskij Ėtnografičeskij Muzej, Saratov 37752
Saratovskij Gosudarstvennyj Muzej Boevoj Slavy, Saratov 37753
Saratovskij Kraevedčeskij Muzej, Saratov 37754
Sarawak Police Museum, Kuching 30590
Sarawak State Museum, Kuching 30591
Sardar Patel University Museum, Vallabhvidyanagar 24422
Sardar Vallabhbhai Patel Museum, Surat 24399
Sardarapat Museum, Sardarapat 00715
Sarehole Mill, Birmingham 43510
Sargent House Museum, Gloucester 48948
Sarguinetti House Museum, Yuma 54364
Saris Museum, Bardejov 38195
Šarišská Galéria, Prešov 38337
Šarišské Múzeum, Bardejov 38195
Sarjeant Gallery - Te Whare o Rehua, Wanganui 33234
Sárospatak Gallery of the Herman Ottó Museum, Sárospatak 23794
Sárospataki Képtár, Sárospatak 23794
Sárospataki Református Kollégium Tudományos Gyűjteményei Múzeum, Sárospatak 23795
Sarpy County Historical Museum, Bellevue 46703
Sarrat Museum, Sarrat 34504
Şarur District Regional History Museum, Şarur 03194
SASA Gallery, Adelaide 00781
Sasayama Museum of History and Art, Sasayama 29231
Sasayama-shiritsu Bijutsu oyo Rekishi Hakubutsukan, Sasayama 29231
Saskatchewan Arts Board Collection, Regina 06728
Saskatchewan Baseball Hall of Fame and Museum, Battleford 05401
Saskatchewan Craft Council Gallery, Saskatoon 06924
Saskatchewan Military Museum, Regina 06729
Saskatchewan Pharmacy Museum, Regina 06730
Saskatchewan River Valley Museum, Hague 05924
Saskatchewan Science Centre, Regina 06731
Saskatchewan Sports Hall of Fame and Museum, Regina 06732
Saskatchewan Western Development Museum, Yorkton 07415
Sassnitzer Fischerei- und Hafenmuseum, Sassnitz 21521
Satakunnan Museo, Pori 11223
Satirical Art Cabinet, Mechelen 03865
Sato Art Museum Toyama, Toyama 29520
Satoe Kindai 21-seiki Bijutsukan, Kazo 28699
Satoe Memorial Art Museum of 21Century, Kazo 28699
Satrosphere, Aberdeen 43248
Sattar Bahlulzadeh Exhibition Center, Baki 03170
Satterlee Clark House, Horicon 49367
Satu Mare District Museum, Satu Mare 36591
Sauda Museum, Sauda 33821
Sauerland-Museum, Arnsberg 16573
Saugus Iron Works, Saugus 52928
Sauhutupamuseo, Somero 11322
Sauk County Historical Museum, Baraboo 46606
Sauk Prairie Area Historical Society Museum, Prairie du Sac 52131
Saukkojärven Kotiseutu- ja Koulumuseo, Ranua 11256
Saukkojärvi Local and School Museum, Ranua 11256
Saules muzejs, Rīga 30112
Sault Sainte Marie Canal, Sault Sainte Marie 06933
Sault Sainte Marie Museum, Sault Sainte Marie 06934
Saunders County Historical Museum, Wahoo 53784
Saunders Memorial Museum, Berryville 46769
Saurer Museum, Arbon 41123
Saurier-Museum, Frick 41393
Sauriermuseum Aathal, Aathal-Seegräben 41087
Saurierpark Jurassic Island, Traismauer 02819
Sautee-Nacoochee Museum, Sautee 52937
Sauvon Kotiseutumuseo, Sauvo 11301
Sauwald-Heimathaus, Sankt Roman 02687
Sava Filaretov House-Museum, Žeravna 05271
Savannakhet provincial Museum, Savannakhet 29971
Savaria Múzeum, Szombathely 23902
Savenvalajamuseo, Somero 11323
Savez Muzejskih Društva Hrvatske, Zagreb 54768
Savignéen Museum, Savigné-sur-Lathan 15677
Savigny-le-Temple Ecomuseum, Savigny-le-Temple 15679
Savings Banks Museum, Ruthwell 45533
Savio-museet, Kirkenes 33592
Savitsky nomidagi Qoraqalpoğiston Respublikasi Davlat San"at Muzeyi, Nukus 54453
Savitsky Picture Gallery of Penza, Penza 37536
Savitsky State Art Museum of Karakalpakstan, Nukus 54453
Savonlinna Provincial Museum, Savonlinna 11303
Savonlinnan Maakuntamuseo, Savonlinna 11303
Savremena Galerija, Pančevo 38129
Savremena Galerija Umetničke Kolonije Ečka, Zrenjanin 38167
Savupirttimuseo, Kuopio 11011
SAW Gallery, Ottawa 06555
Sawanoi Museum of Traditional Japanese Ornaments, Ome 29114
Sawanworanayok National Museum, Sukhothai 42604
Sawhill Gallery, Duke Hall, Harrisonburg 49187
Sawin Museum, Dover 48165
Sawmill Workers' Home-Museum, Puumala 11242
Sawyer's Sandhills Museum, Valentine 53690
Saxonia Spa Museum Bad Elster, Bad Elster 16722
Saxtead Green Post Mill, Saxtead 45620
Sayansk Picture Gallery, Sajansk 37607
Sayner Gießhalle, Bendorf 16990
Sayville Historical Society Museum, Sayville 52952
Sayward-Wheeler House, York Harbor 54343
SBW Terra Mystica & Montana, Bad Bleiberg-Kreuth 01740
SCA Gallery, Rozelle 01473
SCAD Museum of Art, Savannah 52945
Scalamandre, Long Island City 50186
Scalloway Museum, Scalloway 45621
Scandinavia House, New York 51253
Scandinavian Museum Association, Icelandic Section, Reykjavík 54881
Scantic Academy Museum, East Windsor 48265
Scapa Flow Museum, Hoy 44495
Scaplen's Court Museum, Poole 45408
Scarabantia Forum, Archeological Exhibition Room, Sopron 23818
Scarabantia Fóruma, Régészeti Kiállítóhely, Sopron 23818
Scarborough Art Gallery, Scarborough 45624
Scarborough Historical Museum, Scarborough 06938
Scarborough Historical Museum, Scarborough 52953
Scarfone and Hartley Galleries, Tampa 53435
Scarsdale Historical Society Museum, Scarsdale 52954
Scavi Archeologici di Velia, Ascea 25289
Scavi di Ercolano, Ercolano 26060
Scavi di Oplontis, Torre Annunziata 28034
Scavi di Ostia, Ostia Antica 27018
Scavi di Pompei, Pompei 27280
Scavi di Stabiae, Castellammare di Stabia 25731
Schaár Erzsébet Gyűjtemény, Székesfehérvár 23858
Schaár Erzsébet - Utca, Pécs 23775
Schachmuseum Ströbeck, Halberstadt 18879
Schack-Galerie, München 20481
Schäfertanz-Kabinett, Rothenburg ob der Tauber 21411
Schaeffer House Museum, Grafton 01128
Schaffhauser Weinbaumuseum, Hallau 41469
schafhof - europäisches künstlerhaus oberbayern, Freising 18452
Schamser Talmuseum, Zillis 42057
Schatkamer Sint Lambertuskerk, Horst 32417
Schatkamer Sint Walburgis Basiliek, Arnhem 31932
Schatkamer van de Basiliek van Onze Lieve Vrouwe, Maastricht 32531
Schatkamer van de Kathedrale Basiliek Sint-Bavo, Haarlem 32300
De Schatkamer van de Onze-Lieve-Vrouwekerk, Sint-Truiden 04030
Schatkamer van de Sint-Salvatorskathedraal, Brugge 03427
Schatkamer van de Sint-Servaasbasiliek, Maastricht 32532
Schatkammer Sint-Catharina, Maaseik 03838
Schattleitenmühle, Garsten 01925
Schatzhaus in der Lausitz-Museum für Morgenlandfahrer, Göda 18668
Schatzkammer, Wien 03068
Schatzkammer der Basilika St. Ludgerus, Essen 18268
Schatzkammer der Evangelisch-Lutherischen Sankt Andreaskirche, Weißenburg in Bayern 22453
Schatzkammer der Katholischen Pfarrkirche Sankt-Servatius, Siegburg 21751
Schatzkammer der Residenz München, München 20482
Schatzkammer der Wallfahrtskirche, Hohenpeißenberg 19240
Schatzkammer der Wallfahrtskirche Maria Schutz, Maria Schutz am Semmering 02354
Schatzkammer des Münsters, Reichenau 21245
Schatzkammer Grafenrheinfeld, Grafenrheinfeld 18732
Schatzkammer Kloster Sankt Marienstern, Panschwitz-Kuckau 20978
Schatzkammer - Turmmuseum der Pfarrkirche, Sankt Wolfgang im Salzkammergut 02694
Schatzkammer und Museum des Deutschen Ordens, Wien 03069
Schau- und Erlebnissägewerk Mühlauersäge, Fusch 01914
Schaubergwerk, Leogang 02291
Schaubergwerk Arzberg, Arzberg 01730
Schaubergwerk Büchenberg, Elbingerode, Harz 18114
Schaubergwerk Grillenberg, Payerbach 02481
Schaubergwerk - Historische Silbergruben, Oberzeiring 02467
Schaubergwerk Hochfeld, Neukirchen am Großvenediger 02430
Schaubergwerk Kupferplatte, Jochberg 02144
Schaubergwerk Seegrotte, Hinterbrühl 02078
Schaubergwerk Teufelsgrund, Münstertal 20523
Schaubergwerk Zum Tiefen Molchner Stolln, Pobershau 21068
Schaudenkmal Gaszentrale, Unterwellenborn 22185
Schauglashütte, Uetendorf 41955
Schaulager, Münchenstein 41659
Schaumanufaktur Ostsee-Schmuck, Leopoldshagen 19964
Schaumuseum Alte Schmiede, Gmünd, Niederösterreich 01946
Schauraum im ehemaligen Stift Göß, Leoben 02288
Schauwerk Sindelfingen, Sindelfingen 21777
Schauwerkstatt Historische Handweberei, Oederan 20861
Scheepswerf De Delft, Rotterdam 32726
Het Scheepvaartmuseum, Amsterdam 31885
Scheepvaartmuseum, Bornem 03387
Scheepvaartmuseum, Gasselternijveen 32235
Scheepvaartmuseum Baasrode, Baasrode 03338
Schefflenztal-Sammlungen, Schefflenz 21538
The Schein-Joseph International Museum of Ceramic Art, Alfred 46216
Scheldemuseum de Notelaer, Bornem 03388
Schelpen Museum, Zaamslag 32974
Schenectady County Historical Society Museum, Schenectady 52961
Schenectady Museum, Schenectady 52962
Schenkbarsches Haus Museum, Biedenkopf 17336
Schepenzaal Stadhuis, Zwolle 33028
Schering-Museum, Berlin 17247
Scheringa Museum of Realist Art, Spanbroek 32797
Scheringa Museum voor Realisme, Spanbroek 32797
Scherjon's Klompenmakerij en Museum, Noardburgum 32597
Schick Art Gallery, Saratoga Springs 52925
Schiefer- und Ziegelmuseum Dörfles-Esbach, Dörfles-Esbach 17884
Schiefermuseum, Ludwigsstadt 20065
Schiele Museum of Natural History, Gastonia 48870
Schiffahrt- und Marinemuseum, Tessenow 22032
Schiffahrts- und Schiffbaumuseum, Wörth am Main 22640
Schiffahrtsmuseum der Oldenburgischen Weserhäfen, Brake 17498
Schiffahrtsmuseum mit Nordseeaquarium Nordseeheilbad Langeoog, Langeoog 19836
Schiffahrtsmuseum Nordfriesland, Husum 19300
Schiffbau- und Schifffahrtsausstellung Roßlau, Roßlau 21383
Schiffbau- und Schifffahrtsmuseum, Rostock 21395
Schiffermuseum, Zehdenick 22771
Schifferstadt Architectural Museum, Frederick 48760
Schifffahrtsmuseum, Haren (Ems) 19013
Schifffahrtsmuseum, Nierstein 20721
Schifffahrtsmuseum der Marktgemeinde Spitz an der Donau, Spitz 02753
Schifffahrtsmuseum Haßmersheim, Haßmersheim 19033
Schifffahrtsmuseum Heimethus, Männedorf 41601

Schifffahrtsmuseum im Schloss Greinburg, Grein an der Donau 01988
SchifffahrtsMuseum im Schlossturm, Düsseldorf 18005
Schiffleutmuseum, Stadl-Paura 02754
Schiffmühle Höfgen, Grimma 18762
Schiffmühle Minden, Minden 20338
Schiffsmuseum Seitenradschleppdampfer "Württemberg", Magdeburg 20129
Schilder- en Bakkerijmuseum 't Steenhuis, Niebert 32570
Schilderijenzaal Prins Willem V, Den Haag 32100
Schiller-Museum Bauerbach, Bauerbach 16935
Schiller-Nationalmuseum, Marbach am Neckar 20181
Schillerhäuschen, Dresden 17968
Schillerhaus, Leipzig 19932
Schillerhaus, Ludwigshafen am Rhein 20055
Schillers Gartenhaus der Friedrich-Schiller-Universität, Jena 19390
Schillers Geburtshaus, Marbach am Neckar 20182
Schiller's Home, Weimar 22417
Schillers Wohnhaus, Weimar 22417
Schingoethe Center for Native American Cultures, Aurora 46493
Schirn Kunsthalle Frankfurt, Frankfurt am Main 18391
Schleifermuseum, Solingen 21807
Schleifmühle Schwerin, Schwerin 21697
Schleimuseum, Kappeln 19430
Schleipehammer, Kierspe 19538
Schlesisch-Oberlausitzer Dorfmuseum Markersdorf, Markersdorf 20203
Schlesisches Museum zu Görlitz, Görlitz 18680
Schleswig-Holstein Agricultural Museum, Meldorf 20270
Schleswig-Holsteinisches Freilichtmuseum, Molfsee 20369
Schleswig-Holsteinisches Landwirtschaftsmuseum und Dithmarscher Bauernhaus, Meldorf 20270
Schleuse 16 - Forum für experimentelle Arbeiten, Böblingen 17420
Schliekau-Museum, Bad Bevensen 16697
Schlösschen Borghees, Emmerich am Rhein 18153
Schlößchen Vorder Bleichenberg Biberist, Biberist 41227
Schloss Achberg, Achberg 16422
Schloss Agathenburg, Agathenburg 16435
Schloß Ambras, Innsbruck 02130
Schloß Augustusburg, Brühl 17607
Schloss Babelsberg, Potsdam 21105
Schloß Baldern, Bopfingen-Baldern 17474
Schloss Belvedere, Weimar 22418
Schloß Berleburg, Bad Berleburg 16693
Schloß Brake - Das Weserrenaissance-Museum, Lemgo 19949
Schloß Bürgeln, Schliengen 21573
Schloss Bürresheim, Sankt Johann bei Mayen 21501
Schloss Burg, Solingen 21808
Schloß Burgau, Burgau 01808
Schloss Callenberg, Coburg 17719
Schloss Cappenberg, Selm 21741
Schloss Caputh, Schwielowsee 21708
Schloss Cecilienhof, Potsdam 21106
Schloss Charlottenburg, Berlin 17248
Schloss Charlottenhof, Potsdam 21107
Schloss Deutschkreutz, Deutschkreutz 01815
Schloss Drachenburg, Königswinter 19660
Schloß Egg, Bernried, Kreis Deggendorf 17304
Schloss Ehrenburg, Coburg 17720
Schloss Elisabethenburg, Meiningen 20260
Schloß Faber-Castell, Stein bei Nürnberg 21874
Schloß Falkenlust, Brühl 17608
Schloß Farrach, Zeltweg 03130
Schloß Favorite, Rastatt 21185
Schloss Freienwalde mit Walther-Rathenau-Gedenkstätte, Bad Freienwalde 16734
Schloss Freudenberg, Wiesbaden 22558
Schloss Glienicke, Berlin 17249
Schloss Glücksburg, Glücksburg 54857
Schloss Greinburg, Grein an der Donau 01989
Schloss Güstrow, Güstrow 18830
Schloß Hämelschenburg, Emmerthal 18155
Schloss Halbturn, Halbturn 02041
Schloß Hallwyl, Seengen 41856
Schloss Harburg, Harburg 19009
Schloss Hegi, Winterthur 42038
Schloß Heidelberg, Heidelberg 19083
Schloß Herberstein, Sankt Johann bei Herberstein 02656
Schloß Heubach, Heubach 19175
Schloss Höchstädt, Höchstädt an der Donau 19211
Schloss Hof, Schlosshof 02709
Schloss Hohenschwangau, Schwangau 21668
Schloss Hundisburg-Sammlung Apel, Haldensleben 18884
Schloß Jegenstorf, Jegenstorf 41498
Schloss Johannisburg mit Schlossgarten und Pompejanum, Aschaffenburg 16590
Schloss Kirchheim, Kirchheim unter Teck 19550
Schloss Kochberg mit Park und Liebhabertheater, Großkochberg 18794
Schloss Königs Wusterhausen, Königs Wusterhausen 19638
Schloß Leitzkau, Leitzkau 19944
Schloß Lichtenstein, Lichtenstein 19982
Schloss Linderhof, Ettal 18279
Schloß Loosdorf mit Prunkräumen und Zinnfigurensammlung, Loosdorf, Bez. Mistelbach 02336
Schloss Ludwigsburg, Ludwigsburg 20047
Schloss Ludwigslust, Ludwigslust 20062
Schloss Luisium, Dessau 17832
Schloss Lustheim, Oberschleißheim 20837
Schloß Marienburg, Pattensen 20995
Schloss Miltach, Miltach 20326
Schloß Mörsburg, Stadel, Winterthur 41899
Schloss Moritzburg und Fasanenschlösschen, Moritzburg 20386
Schloss Mosigkau, Dessau 17833
Schloß Murau / Eisenmuseum, Murau 02417
Schloß Museum, Pram 02524
Schloß-Museum, Reidling 02574
Schloss-Museum, Wolfshagen 22670
Schloss Neuburg, Neuburg an der Donau 20599
Schloß Neunhof, Nürnberg 20788
Schloss Neuschwanstein, Schwangau 21669
Schloss Ökonomie Gern, Eggenfelden 18063
Schloss Oranienbaum, Oranienbaum 20907
Schloss Paretz, Ketzin 19523
Schloß Parz, Grieskirchen 01991
Schloss Pfaueninsel, Berlin 17250
Schloß Reinbek, Reinbek 21252
Schloss Rheda-Wiedenbrück, Rheda-Wiedenbrück 21293
Schloss Rheinsberg, Rheinsberg 21306
Schloss Rosenau, Rödental 21352
Schloss Sachsenburg, Frankenberg, Sachsen 18349
Schloss Sacrow, Potsdam 21108
Schloss Salem, Salem, Baden 21472
Schloß Sanssouci, Potsdam 21109
Schloss Sargans - Schlossmuseum und Schlossgalerie, Sargans 41819
Schloß Schallaburg, Loosdorf, Bez. Melk 02335
Schloss Schallaburg, Schallaburg 02697
Schloss Schlettau, Schlettau 21571
Schloß Schönbrunn, Wien 03070
Schloss Schönhausen, Berlin 17251
Schloß Schwetzingen, Schwetzingen 21703
Schloss Seehof, Memmelsdorf 20278
Schloß Solitude, Stuttgart 21958
Schloß Spiez, Spiez 41893
Schloss Stolzenfels, Koblenz 19590
Schloß Strünkede, Herne 19156
Schloß Tarasp, Tarasp 41928
Schloß Thürnthal, Fels am Wagram 01886
Schloß Thunstetten, Thunstetten 41942
Schloss Tiefurt, Weimar 22419
Schloss Tratzberg, Jenbach 02142
Schloß und Festung Senftenberg, Senftenberg 21744
Schloß- und Goldbergbaumuseum, Großkirchheim 02002
Schloß- und Heimatmuseum, Sigharting 02741
Schloss und Hofgarten Dachau, Dachau 17767
Schloss- und Kirche Sankt Bartholomä, Berchtesgaden 17005
Schloss und Park Schönbusch, Aschaffenburg 16591
Schloss und Schlossgarten Weilburg, Weilburg 22386
Schloß und Schloßpark Bad Homburg, Bad Homburg v.d.Höhe 16753
Schloss und Schlosspark Charlottenburg mit Schinkelpavillon, Belvedere und Mausoleum, Berlin 17252
Schloß und Schloßpark Wilhelmsthal, Calden 17677
Schloß- und Spielkartenmuseum, Altenburg 16499
Schloss Unteraufsess, Aufseß 16610
Schloß Urbach, Urbach 22187
Schloss Veitshöchheim, Veitshöchheim 22209
Schloß Vianden, Vianden 30496
Schloß Villa Ludwigshöhe, Edenkoben 18056
Schloß vor Husum, Husum 19301
Schloß Weesenstein, Weesenstein 22358
Schloß Weikersheim, Weikersheim 22372
Schloss Weissenstein, Pommersfelden 21075
Schloss Werdenberg, Werdenberg 42007
Schloß Wernigerode, Wernigerode 22500
Schloss Wiederau, Pegau 21000
Schloss Wildegg, Wildegg 42020
Schloß Wilflingen, Langenenslingen 19829
Schloss Wilhelmshöhe-Weißensteinflügel, Kassel 19476
Schloß Wiligrad, Lübstorf 20084
Schloss Wörlitz, Wörlitz 22639
Schloss Wolkenburg, Limbach-Oberfrohna 19994
Schlossbergmuseum Chemnitz, Chemnitz 17704
Schloßgalerie Kastenscheuer, Eberdingen 18036
Schloßmuseum, Arnstadt 16574
Schloßmuseum, Bad Bentheim 16685
Schlossmuseum, Bad Iburg 16755
Schlossmuseum, Burgdorf 41277
Schlossmuseum, Ellwangen 18124
Schloßmuseum, Erbach, Alb-Donau-Kreis 18189
Schlossmuseum, Höchstädt im Fichtelgebirge 19212
Schloßmuseum, Kirchberg an der Jagst 19544
Schlossmuseum, Kirchentellinsfurt 19546
Schloßmuseum, Kronburg 19725
Schlossmuseum, Linz 02324
Schlossmuseum, Nauders 02423
Schlossmuseum, Oberhofen am Thunersee 41697
Schloßmuseum, Ortenburg 20916
Schlossmuseum, Peuerbach 02497
Schloßmuseum, Quedlinburg 21152
Schloßmuseum, Schrozberg 21642
Schloßmuseum, Texing 02807
Schlossmuseum, Weimar 22420
Schloßmuseum Amerang, Amerang 16538
Schlossmuseum Arnstadt, Arnstadt 16575
Schlossmuseum Aulendorf-Kunst des Klassizismus-Altes Spielzeug, Aulendorf 16643
Schlossmuseum Beromünster, Beromünster 41225
Schlossmuseum Darmstadt, Darmstadt 17794
Schloßmuseum der Stadt Aschaffenburg, Aschaffenburg 16592
Schloßmuseum Heidegg, Gelfingen 41397
Schloßmuseum Hohenlohe-Schillingsfürst, Schillingsfürst 21554
Schloßmuseum Hubertusburg, Wermsdorf 22492
Schloßmuseum Ismaning, Ismaning 19365
Schloßmuseum Jagsthausen, Jagsthausen 19373
Schlossmuseum Jever, Jever 19397
Schloßmuseum Landeck, Landeck 02254
Schlossmuseum Langenburg, Langenburg 19827
Schloßmuseum Mespelbrunn, Mespelbrunn 20303
Schloßmuseum Murnau, Murnau am Staffelsee 20535
Schloßmuseum Niederleis, Niederleis 02446
Schlossmuseum Oranienburg, Oranienburg 20910
Schloßmuseum Rimpar, Rimpar 21335
Schlossmuseum Schwerin, Schwerin 21698
Schlossmuseum Sondershausen, Sondershausen 21817
Schloßmuseum Thun, Thun 41938
Schloßmuseum und Park Molsdorf, Erfurt 18210
Schlossmuseum Weitra, Weitra 02908
Schloßparkmuseum, Bad Kreuznach 16768
Schlüsselburg Fortress Oreshek, Šlisselburg 37769
Schlumberger Wein- und Sektkellerei, Wien 03071
Schmalspurbahn-Museum, Frojach 01908
Schmelenhaus Museum, Windhoek 31767
Schmidt House Museum, Grants Pass 49017
Schmiede im Hammergraben, Helfenberg 02068
Schmiede- und Schlossermuseum Schwalbach, Schwalbach, Saar 21660
Schmiedegesellenhaus, Göstling an der Ybbs 01956
Schmiedemuseum, Fulpmes 01912
Schmiedemuseum, Hatten 19035
Schmiedemuseum, Hohen Demzin 19233
Schmiedemuseum, Lölling 02333
Schmiedemuseum, Sersheim 21747
Schmiedemuseum Arfeld, Bad Berleburg 16694
Schmiedemuseum Bremecker Hammer, Lüdenscheid 20090
Schmiedemuseum Kirchlauter, Kirchlauter 19554
Schminck Memorial Museum, Lakeview 49904
Schmuckmuseum Pforzheim im Reuchlinhaus, Pforzheim 21029
Schnapsglas-Museum, Echsenbach 01851
Schnapsmuseum, Kötzting 19669
Schnarch-Museum, Alfeld 16468
Schnauferlstall, Ruhpolding 21449
Schneebergmuseum, Puchberg 02536
Schneider Museum of Art, Ashland 46388
Schnopfhagen-Museum, Oberneukirchen 02456
Schnupftabakmuseum, Grafenau, Kreis Freyung-Grafenau 18728
Schocken Institute of Jewish Research, Jerusalem 25014
Schöffer Museum, Kalocsa 23627
Schoemansdal Museum, Louis Trichardt 38683
Schönberg-Haus in Mödling, Mödling 02397
Schoenbrunn Village State Memorial, New Philadelphia 51098
Schoharie Colonial Heritage Association Museum, Schoharie 52964
Schoharie Crossing, Fort Hunter 48627
Schokoladenmuseum Köln, Köln 19632
Schokoladenmuseum-Musée du Chocolat, Eupen 03592
Schokoland Alprose, Caslano 41282
SchokoMuseum in SchokoLand, Peine 21002
Scholte House Museum, Pella 51751
Scholz Park Museum, Riverton 01461
Schomburg Center for Research in Black Culture, New York 51254
School Days' Museum, Fredericton 05839
School Historical and Clerical Art Museum, Cegléd 23502
School History Collection of the József Katona Museum, Kecskemét 23646
School House Museum, Deep River 05688
School Museum, Békés 23363
School Museum, Littoinen 11076
School Museum, Nukari 11150
School Museum, Somero 11320
School Museum, Trjavna 05231
School Museum, Villeneuve-d'Ascq 16203
School Museum, Yttilä 11485
School Museum of Helsinki City Museum, Helsinki 10794
School Museum of the Visually Handicapped, Jyväskylä 10889
School of Art and Art History Galleries, Denver 48089
School of Art Gallery, Baton Rouge 46641
School of Art Gallery, Bozeman 46983
School of Art Gallery and Museum, Aberystwyth 43259
School of Earth Sciences Museum, Townsville 01600
School of Fine Arts Gallery, Bloomington 46866
School of Nations Museum, Elsah 48381
School of Political and Social Inquiry Museum, Clayton 00981
School of the Art Institute of Chicago Gallery, Chicago 47512
Schoolarchief en Museum, Sint-Katelijne-Waver 04017
Schoolhouse Museum, Ridgewood 52350
Schoolmuseum Michel Thiery, Gent 03640
Schooltijd Schoolmuseum, Terneuzen 32816
Schoonewelle, Zwartsluis 33023
Schoonhovens Edelambachtshuys, Schoonhoven 32766
Schopper- und Fischermuseum, Aschach an der Donau 01731
De Schotse Huizen, Veere 32878
Schott GlasMuseum, Jena 19391
Schraube Museum, Halberstadt 18880
Schreibmaschinenmuseum, Wattens 02897
Schrift-, Heimat- und Druckereimuseum Bartlhaus, Pettenbach 02496
Schriftmuseum J.A. Dortmond, Amsterdam 31886
Schubartstube, Blaubeuren 17380
Schubert Club Museum of Musical Instruments, Saint Paul 52628
Schubert Geburtshaus, Wien 03072
Schubert Sterbewohnung, Wien 03073
Schüttesägemuseum, Schiltach 21557
Schützenmuseum der Königlich privilegierten Feuerschützengesellschaft Weilheim, Weilheim in Oberbayern 22391
Schützenmuseum im Kanzlerturm, Eggenburg 01845
Schützenscheibenmuseum, Scheibbs 02703
Schützenscheibensammlung, Mainbernheim 20135
Schuhmachermuseum, Ladbergen 19758
Schul- und Heimatmuseum, Mark Landin 20201
Schule für Gestaltung Basel, Basel 41182
Schulgeschichtliche Sammlung, Magdeburg 20130
Schulgeschichtliche Sammlung im Main-Taunus-Kreis, Kriftel 19715
Schulhausmuseum, Hollingstedt 19255
Schulheimatmuseum, Asbach-Bäumenheim 16580
Schulhistorische Sammlung Cruismannschule, Bochum 17406
Schullandheim Sassen, Sassen 21517
Schulmuseum, Haldensleben 18885
Schulmuseum, Immerath 19329
Schulmuseum, Königsfeld im Schwarzwald 19651
Schulmuseum, Lockenhaus 02332
Schulmuseum, Neumark 20632
Schulmuseum, Obersulm 20844
Schulmuseum, Schneverdingen 21594
Schulmuseum Bergisch Gladbach, Bergisch Gladbach 17019
Schulmuseum Bremen e.V., Bremen 17561
Schulmuseum des Bezirks Unterfranken, Bad Bocklet 16700
Schulmuseum Friedrichshafen am Bodensee, Friedrichshafen 18477
Schulmuseum Fronau, Roding 21348
Schulmuseum Lilienthal, Lilienthal 19991
Schulmuseum Mozartschule Rheingönheim, Ludwigshafen am Rhein 20056
Schulmuseum Mühlebach, Amriswil 41113
Schulmuseum Nordwürttemberg in Kornwestheim, Kornwestheim 19691
Schulmuseum Nürnberg, Nürnberg 20789
Schulmuseum Schulstubn Glocknhäusl, Sankt Peter am Wimberg 02679
Schulmuseum Steinhorst, Steinhorst 21893
Schulmuseum und Hallenhaus, Middelhagen 20322
Schulmuseum - Werkstatt für Schulgeschichte Leipzig, Leipzig 19933
Schultehuis, Diever 32130
Schulze-Delitzsch-Haus, Delitzsch 17813
Schumacher Gallery, Columbus 47760
Schumann-Haus, Leipzig 19934
Schumannhaus Bonn, Bonn 17466
Schunkelhaus, Obercunnersdorf 20802
Schussenrieder Bierkrugmuseum, Bad Schussenried 16832

Schutterskamer, Sneek 32791
Schuyler County Historical Society Museum, Montour Falls 50822
Schuyler-Hamilton House, Morristown 50856
Schuyler Jail Museum, Rushville 52481
Schuyler Mansion, Albany 46162
Schwäbisches Bauern- und Technikmuseum, Eschach 18232
Schwäbisches Bauernhofmuseum Illerbeuren, Kronburg 19726
Schwäbisches Handwerkermuseum, Augsburg 16634
Schwäbisches Krippenmuseum im Jesuitenkolleg, Mindelheim 20329
Schwäbisches Schnapsmuseum Bönnigheim, Bönnigheim 17428
Schwäbisches Schützenmuseum, Kronburg 19727
Schwäbisches Turmuhrenmuseum, Mindelheim 20330
Schwäbisches Volkskundemuseum Oberschönenfeld, Gessertshausen 18627
Schwartz Collection of Skiing Heritage, Tiburon 53505
Schwartz Heritage House, Altona 05327
Schwartzsche Villa, Berlin 17253
Schwarz-Afrika-Museum, Vilshofen 22246
Schwarzachtaler Heimatmuseum, Neunburg vorm Wald 20637
Schwarzenberger Skulpturenpark, Schwarzenberg am Böhmerwald 02721
Schwarzwälder Freilichtmuseum Vogtsbauernhof, Gutach 18843
Schwarzwälder Mühlenmuseum, Grafenhausen 18731
Schwarzwälder Skimuseum, Hinterzarten 19195
Schwarzwälder Trachtenmuseum, Haslach im Kinzigtal 19026
Schwarzwald-Museum, Triberg im Schwarzwald 22092
Schwazer Silberbergwerk, Schwaz 02727
Schwedenspeicher-Museum Stade, Stade 21852
Schweinfurth Memorial Art Center, Auburn 46470
Schweizer Jazzmuseum, Uster 41967
Schweizer Kamm-Museum, Mümliswil 41655
Schweizer Kindermuseum, Baden 41149
Schweizer Museum für Wild und Jagd, Utzenstorf 41968
Schweizer Zither-Kultur-Zentrum, Trachselwald 41946
Schweizerische Nationalbibliothek, Bern 41215
Schweizerische Theatersammlung, Bern 41216
Schweizerisches Agrarmuseum Burgrain, Alberswil 41099
Schweizerisches Alpines Museum, Bern 41217
Schweizerisches Dampfmaschinenmuseum Vaporama, Thun 41939
Schweizerisches Feuerwehrmuseum Basel, Basel 41183
Schweizerisches Freilichtmuseum für ländliche Kultur, Hofstetten 41485
Schweizerisches Gastronomie-Museum, Thun 41940
Schweizerisches Militärmuseum Full, Full 41394
Schweizerisches Nationalmuseum, Zürich 42105
Schweizerisches Schützenmuseum, Bern 41218
Schweizerisches Tanksäulenmuseum, Gänsbrunnen 41395
Schwemmgut-Museum, Finsing 18319
Schwenkfelder Library and Heritage Center, Pennsburg 51765
Schwerkolt Cottage Museum, Mitcham 01308
Schwules Museum, Berlin 17254
Sci-Port Discovery Center, Shreveport 53089
Sci-Tech Center of Northern New York, Watertown 53973
Science Activity Centre, Sirsa 24396
Science Activity Corner, Gwalior 24199
Science and Technology Museum of Atlanta, Atlanta 46451
Science and Technology Museum, University of Science, Tokyo 29389
Science Center, Napoli 26909
Science Center, Trollhättan 40982
Science Center of Connecticut, West Hartford 54061
Science Center of Iowa, Des Moines 48101
Science Center of Pinellas County, Saint Petersburg 52643
Science Center of Thessaloniki and Technology Museum, Thessaloniki 23220
Science Center of West Virginia, Bluefield 46877
Science Center Spectrum, Berlin 17255
Science Centre for Education, Bangkok 42559
Science City Museum, Kolkata 24281
Science Discovery Center of Oneonta, Oneonta 51565
The Science Factory Children's Museum, Eugene 48438
The Science Fiction Museum and Hall of Fame, Seattle 53012
Science Hall of Penghu County, Magong 42268
Science Imaginarium, Waterloo 53963
Science Learning Center, Upton 53668
Science Museum, Alexandria 10478
The Science Museum, Cebu 34374
Science Museum, Dhaka 03228
Science Museum, Giza 10530
Science Museum, London 44983
Science Museum, Japan Science Foundation, Tokyo 29404
Science Museum of Long Island, Manhasset 50430
Science Museum of Minnesota, Saint Paul 52629
Science Museum of Virginia, Richmond 52338
Science Museum of Western Virginia, Roanoke 52382
Science Museum Wroughton, Wroughton 46086
Science North, Sudbury 07029
Science Park, Granada 39370
Science Spectrum, Lubbock 50313
Science Station Museum, Cedar Rapids 47307
Science & Technology Museum, Beijing 07565
Sciencenter, Ithaca 49550
Scienceworks Museum, Spotswood 01521
Scientific Collections, Forssa 10738
Scientific Collections of the Sárospatak Calvinist School, Sárospatak 23795
Scientific Oceanographic Museum, Kaliningrad 36993
Scientific Researches Museum, Cairo 10518
Scitech, Aurora 46494
Scituate Historic Sites, Scituate 52966
Scituate Maritime and Irish Mossing Museum, Scituate 52967
Sciworks, Winston-Salem 54250
Scollay Square Gallery, Boston 46948
Scolton Manor Museum, Haverfordwest 44422
Scone and Upper Hunter Museum, Scone 01498
Scone Palace, Perth 45367
Scotch Whisky Heritage Centre, Edinburgh 44154
Scotchtown, Beaverdam 46683
Scotia-Glenville Children's Museum, Scotia 52971
Scotia Museum, Scotia 52969
Scotland Heritage Chapel and Museum, Scotland 52972
Scotland Street School Museum, Glasgow 44302
Scotland's Secret Bunker, Saint Andrews 45558
The Scott Gallery, Hawick 44427
Scott Monument, Edinburgh 44155
Scott Polar Research Institute Museum, Cambridge 43732
Scottish Fisheries Museum, Anstruther 43306
Scottish Football Museum, Glasgow 44303
Scottish Industrial Railway Centre, Dalmellington 43947
Scottish Jewish Museum, Glasgow 44304
Scottish Maritime Museum, Irvine 44535
Scottish Mining Museum, Newtongrange 45253
Scottish Museums Council, Edinburgh 55091
Scottish Museums Federation, Edinburgh 55092
Scottish National Gallery of Modern Art Gallery, Edinburgh 44156
Scottish National Portrait Gallery, Edinburgh 44157
Scottish Railway Museum, Bo'ness 43560
Scottish Rugby Union Museum, Edinburgh 44158
Scotts Bluff National Monument, Gering 48898
Scottsdale Museum of Contemporary Art, Scottsdale 52977
Scout Heritage Centre of Western Australia, West Perth 01643
Scout Museum of Finland, Turku 11400
Scouting Museum de Ducdalf, Rotterdam 32727
Scouting Museum Haagse Randstad, Den Haag 32101
Scouts Museums Centre, Torres Vedras 35973
Scrimshaw Museum, Horta 35644
Scryption, Tilburg 32832
Ščučinskij Oblastnoj Muzej Literatury i Iskusstva, Kokšetau 29704
Scuderie del Quirinale, Roma 27546
Scugog Shores Historical Museum Village, Port Perry 06647
Sculpture at Schoenthal, Langenbruck 41524
Sculpture Center, Long Island City 50187
Sculpture Collection, Dresden 17970
Sculpture Galena, Galena 48829
Sculpture Open Air Museum, Biškek 29959
Sculpturparc, Dietikon 41329
Scuola Dalmata dei Santi Giorgio e Trifone, Venezia 28210
Scuola Grande Arciconfraternità di San Rocco, Venezia 28211
Scuola Grande di San Giovanni Evangelista, Venezia 28212
Scurry County Museum, Snyder 53139
Ščusev Museum, Chişinău 31604
Scythe forging museum/Museum of Sound, Micheldorf 02379
Sea Cliff Village Museum, Sea Cliff 52985
Sea Folk Museum, Toba 29361
Sea Museum, Petrozavodsk 37560
Sea Museum, Sesimbra 35932
Sea Shell Museum, Scarness 01496
Sea World, Durban 38612
Seabee Museum, Port Hueneme 52034
Seabrook Village, Midway 50678
Seafood Museum, Qingdao 08106
Seaford Historical Museum, Seaford 52988
Seaford Museum of Local History, Seaford 45629
Seaforth Highlanders of Canada Regimental Museum, Vancouver 07221
Seal Island Light Museum, Barrington 05392
Sealife Scheveningen, Den Haag 32102
Seaport Art Center, Cingshui 42169
SEARCH-Centre, Gosport 44337
Searchlight Historic Museum and Mining Park, Searchlight 52991
Seashore Trolley Museum, Kennebunkport 49729
Seaton Delaval Hall, Seaton 45630
Seattle Art Museum, Seattle 53013
Seattle Asian Art Museum, Seattle 53014
Seba Beach Heritage Museum, Seba Beach 06940
Sebastian-Kneipp-Museum, Bad Wörishofen 16874
Sebastopol Open Air Museum, Tokat 42953
Sebastopolis Açıkhava Müzesi, Tokat 42953
Sebežskij Kraevedčeskij Muzej, Sebež 37757
Sebnitzer Kunstblumen- und Heimatmuseum Prof. Alfred Meiche, Sebnitz 21710
Secção Museológica da CP de Estremoz, Estremoz 35582
Secção Museológica da CP de Lagos, Lagos 35652
Secção Museológica da CP de Macinhata do Vouga, Macinhata do Vouga 35747
Secção Museológica da CP de Santarém, Santarém 35911
Secção Museológica da CP de Valença, Valença 35979
Sección Arqueología, Ceuta 39172
Sección de Bellas Artes, Ceuta 39173
Secession, Wien 03074
Second Street Gallery, Charlottesville 47402
Second Turkish Grand National Assembly Museum, Ankara 42694
Second World War Experience Centre, Horsforth 44487
Second Yesemek Sculpture Atelier and Open Air Museum, İslahiye 42798
Secret Christians Museum, Nagoya 29009
Secrets de Soie Musée, Campel 12258
Secţia de Etnografie-Bastion, Timişoara 36693
Secwepemc Museum and Heritage Park, Kamloops 06038
Sed Gallery, Athens 46414
Sederholm House of Helsinki City Museum, Helsinki 10815
Sederholmin Talo, Helsinki 10815
Sedgewick Archives Museum and Gallery, Sedgewick 06942
Sedgwick-Brooklin Historical Museum, Sedgwick 53026
Sedgwick Museum of Earth Sciences, Cambridge 43733
Sedona Arts Center, Sedona 53027
Šeduvos Kraštotyros Ekspozicija, Seduva 30351
See Science Museum, Manchester 50425
See-Vogel-Museum Herbert Vargyas, Rust 02601
Seedamm Kulturzentrum, Pfäffikon (Schwyz) 41721
Seelackenmuseum, Sankt Veit im Pongau 02693
Seemuseum in der Kornschütte, Kreuzlingen 41516
Seenot-Rettungskreuzer Georg Breusing, Emden 18143
Şefik Bursali Müze Evi, Ankara 42711
Segantini-Museum, Sankt Moritz 41813
Segedunum Roman Fort, Baths and Museum, Wallsend 45917
Segontium Roman Museum, Caernarfon 43711
Seguine-Burke Plantation, Staten Island 53266
Şehir Müzesi, İstanbul 42839
Seigfred Gallery, Athens 46418
Seiji Togo Memorial Sompo Japan Museum of Art, Tokyo 29462
Seikado Bunko Bijutsukan, Tokyo 29463
Seikado Bunko no Bijutsukan, Tokyo 29464
Seikado Library and Art Museum, Tokyo 29463
Seikado Library and Art Museum, Tokyo 29464
Seinäjoen Luonto-Museo, Seinäjoki 11305
Seinäjoki Natural History Museum, Seinäjoki 11305
Seinan Gakuin Daigaku Hakubutsukan, Fukuoka 28436
Seiryuden Art Museum, Matsushima 28903
Seiryuden Bijutsukan, Matsushima 28903
Seison-kaku, Kanazawa 28663
Seiyun in Wadi Hadhramaut Museum, Seiyun 54651
Sejarah Mesjid Banten Museum, Banten 24449
Sejong College Museum, Seoul 29915
Sekai Gangu Hakubutsukan, Karuizawa 28670
Sekai Oota Kokaikan, Yokaichi 29590
Sekido Bijutsukan, Tokyo 29465
Sekido Museum of Art, Tokyo 29465
Sekiguchi Koh Museum Yuzawa, Yuzawa 29637
Sekisho Art Museum, Misumi 28915
Sekisho Bijutsukan, Misumi 28915
Sekisui Bijutsukan, Tsu 29538
Sekisui Museum of Art, Tsu 29538
Sekitan Kogyo Hakubutsukan, Ube 29549
Sekkeh Museum, Kerman 24620
Sekkeh Museum, Tehrān 24695
Sekken Museum, Sekken 33822
Sektor I, Katowice 34731
Selbu Bygdemuseum, Selbu 33823
Selbu Strikkemuseum, Selbu 33824
Selby Gallery, Sarasota 52916
Selçuk Müzesi, Konya 42890
Self Help Graphics and Art Gallery, Los Angeles 50263
Selkirk College Mansbridge Kootenay Collection, Castlegar 05577
Sellner Glashütte, Lohberg 20021
Selly Manor Museum, Birmingham 43511
Selsey Lifeboat Museum, Selsey 45639
Selsor Gallery of Art, Chanute 47343
Šeltozerskij Vepsskij Ėtnografičeskij Muzej, Šeltozero 37758
Şemaki Evi Müzesi, Yenisehir 42966
Şemaki House Museum, Yenisehir 42966
Semed Vurğunun Ev-Muzeyi, Bakı 03171
Semenovskij Gosudarstvennyj Kraevedčeskij Istoriko-chudožestvennyj Muzej, Semenov 37759
Seminole Canyon State Historical Park, Comstock 47765
Seminole Nation Museum, Wewoka 54114
Semmelweis Museum for the History of Medicine, Budapest 23475
Semmelweis Museum, Library and Archives of Medical History - Golden Eagle Pharmacy, Budapest 23383
Semmelweis Orvostörténeti Múzeum, Budapest 23475
Sen. MacArthur Memorabilia, Tacloban 34512
Sen-oku Hakukokan, Kyoto 28864
Senasis Arsenalas, Vilnius 30431
Senate House, Kingston 49791
Senator George Baldwin Hotel, Klamath Falls 49808
Senator George Norris State Historic Site, McCook 50344
Senator John Heinz Pittsburgh Regional History Center, Pittsburgh 51959
Senćanski Gradski Muzej, Senta 38139
Senckenberg-Forschungsinstitut und Naturmuseum, Frankfurt am Main 18392
Senckenberg Museum für Naturkunde Görlitz, Görlitz 18681
Senckenberg Museum of Natural History Görlitz, Görlitz 18681
Senckenberg Naturhistorische Sammlungen Dresden, Dresden 17969
Sendai City Museum, Sendai 29246
Sendai Music Box Museum, Sendai 29245
Sendai-shi Hakubutsukan, Sendai 29246
Sender- und Funktechnikmuseum, Königs Wusterhausen 19639
Seneca County Museum, Tiffin 53507
Seneca Falls Historical Society Museum, Seneca Falls 53036
Seneca-Iroquois National Museum, Salamanca 52654
Senhouse Roman Museum, Maryport 45113
Seniausias Panevėžio Pastatas, Panevėžys 30333
Senj Municipal Museum, Senj 08945
Senjehesten Kystforsvarsmuseum, Moen 33673
Sennerei-Museum und Heimatmuseum, Unterwasser 41964
Sensen-Schmiede-Museum/ Klangwelten, Micheldorf 02379
Sensen- und Heimatmuseum, Achern 16423
Sensenmuseum Geyerhammer, Scharnstein 02701
Senshu Bunko Museum, Tokyo 29466
Sensler Museum, Tafers 41927
Senta Museum, Senta 38139
Seoul Art Museum, Seoul 29916
Seoul Design Museum, Seoul 29917
Seoul Education Museum, Seoul 29918
Seoul Metropolitan Museum of Art, Seoul 29919
Seoul National Science Museum, Seoul 29920
Seoul National University Museum, Seoul 29921
Seoul Olympic Museum, Seoul 29922
Seoul Station Hall, Seoul 29923
Sepänmäen Käsityömuseo, Mäntsälä 11092
Sepänmäki Handicraft Museum, Mäntsälä 11092
Septem Maria Museum, Adria 25154
Sequoyah Cabin, Sallisaw 52684
Serbian Church Museum, Szentendre 23882
Serbski muzej - Sorbisches Museum, Bautzen 16941
Serbski muzej / Wendisches Museum, Cottbus 17734
Šeremetevskij Dvorec - Fontannyj Dom, Sankt-Peterburg 37726
Serengeti Museum, Arusha 42509
Sergačskij Kraevedčeskij Muzej, Sergač 37761
Sergei Parajanov House-Museum, Yerevan 00753
Sergej Borodin Muzeyi, Toshkent 54486
Sergej Esenin Museum-Preserve, Konstantinovo 37056
Sergej Esenin Muzeyi, Toshkent 54487
Sergey Borodin House-Museum, Toshkent 54486
Sergiev-Posad State History and Art Museum-Reserve, Sergiev Posad 37764

Sergievo-Posadskij Gosudarstvennyj Istoriko-chudožestvennyj Muzej Zapovednik, Sergiev Posad 37764
Seriemuseet Comicland, Ullared 40994
Serpent Mound Museum, Peebles 51747
Serpentine Gallery, London 44984
Serpentine Vintage Tractors and Machinery Museum, Serpentine 01500
Serpentinsteinmuseum Zöblitz, Zöblitz 22815
Serpuchovskij Istoriko-chudožestvennyj Muzej, Serpuchov 37765
Serra Museum, San Diego 52755
Serrallés Castle Museum, Ponce 36048
Sertão House Museum, Feira de Santana 04449
Service Historique de la Police, Bruxelles 03503
Serviceton Historic Railway Station, Serviceton 01501
Servoninė Bitininkystės Muziejus, Stripeikiai 30375
Session Cottage Museum, Turriff 45884
Setagaya Art Museum, Tokyo 29468
Setagaya-kuritsi Riterashi Hakubutsukan, Tokyo 29467
Setagaya-kuritsu Bijutsukan, Tokyo 29468
Setagaya Literary Museum, Tokyo 29467
Setesdalsbanen Stiftelsen, Vennesla 33971
Setesdalsmuseet, Rysstad 33804
Seto Inland Sea Folk History Museum, Takamatsu 29320
Seto Muuseumitaro, Obinitsa 10600
Seto-Ohashi Memorial Bridge Museum, Kurashiki 28790
Seto Rekishi Mingeikan, Takamatsu 29320
Seto-shi Rekishi Minzoku Shiryokan, Seto 29251
Seton Hall University Museum, South Orange 53177
Setor de Malacologia, Juiz de Fora 04524
Setsuro Takahashi Bijutsukan, Azumino, Azumino 28387
Setsuryosha Art Museum, Niigata 29062
Setsuryosha Bijitsukan, Niigata 29062
Settlers Museum, Brewarrina 00891
Settlers Museum, Himeville 38642
Seurasaaren Ulkomuseo, Helsinki 10816
Seurasaari Open Air Museum, Helsinki 10816
Ševčenkivskyj Nacional'nyj Zapovidnyk, Kaniv 43060
Seven Oaks House Museum, Winnipeg 07384
Sevenoaks Museum and Gallery, Sevenoaks 45642
Severinhaus Basilika Sankt Laurenz, Enns 01871
Severn Valley Railway, Bewdley 43470
Severo-Kazachstanskij Oblastnoj Istoriko-Kraevedčeskij Muzej, Petropavl 29720
Severo-Kazachstanskij Oblastnoj Muzej Izobrazitelnogo Iskusstva, Petropavl 29721
Severo-osetinskij Gosudarstvennyj Obedinennyj Muzej Istorii, Architektury i Literatury, Vladikavkaz 37932
Severo-osetinskij Literaturnyj Muzej im. K.L. Četagurova, Vladikavkaz 37933
Severočeská Galerie Výtvarného Umění v Litoměřicích, Litoměřice 09631
Severočeské Muzeum v Liberci, Liberec 09625
Seville Great House and Heritage Park, Ocho Rios 28339
Sevnica Castle, Sevnica 38513
Sevniški Grad, Sevnica 38513
Sewall-Belmont House, Washington 53916
Seward County Historical Society Museum, Goehner 48952
Seward House, Auburn 46471
Seward Museum, Seward 53042
Sewerby Hall Art Gallery and Museum, Bridlington 43611
Sex Machines Museum, Praha 09793
Sexmuseum Amsterdam Venustempel, Amsterdam 31887
Sexton's Cottage Museum, Crows Nest, New South Wales 01012
Seyitgazi Müzesi, Seyitgazi 42938
Seymour and District Historical Museum, Seymour 01502
Seymour Art Gallery, North Vancouver 06481
Seymour Community Museum, Seymour 53045
Seyun Iron Museum, Eumseong 29803
Sezon Museum of Modern Art, Karuizawa 28671
SFA Galleries, Nacogdoches 50936
De Sfeer van Weleer, Rekken 32680
Sha Museum, Shaxian 08213
Shaanxi Chinese Painting Gallery, Xian 08359
Shaanxi History Museum, Xian 08360
Shaanxilong Books Museum, Longxian 07973
Shackerstone Railway Museum, Shackerstone 45643
Shades of the Past, Carbonear 05559
Shadi Abdel Salam Museum, Alexandria 10479
The Shadows-on-the-Teche, New Iberia 51064
Shaftesbury Abbey Museum and Garden, Shaftesbury 45644
Shaftesbury Town Museum, Shaftesbury 45645
Shaftsbury Historical Society, Shaftsbury 53048
Shah Alam Art Gallery, Shah Alam 30621
Shaheed-e-Azam Bhagat Singh Museum, Khatkar Kalan 24259
Shahrood Museum, Semnan 24644
Shahryar Museum, Tabriz 24653
Shaker Heritage Society Museum, Albany 46163
Shaker Historical Society Museum, Shaker Heights 53049
Shaker Museum, New Gloucester 51046
The Shaker Museum, Old Chatham 51534
Shaker Museum at South Union, South Union 53183
Shaker Village of Pleasant Hill, Harrodsburg 49190
Shakespeare Birthplace Trust, Stratford-upon-Avon 45785
Shakespeare's Globe Exhibition and Tour, London 44985
Shakir Ali Museum, Lahore 34028
Shakowi Cultural Center, Oneida 51561
Shalimar Palace, Dubai 43219
Shalva Amiranasvili Museum of Art, Tbilisi 16379
Shalyapin Hall in the Bogoyavlenskij Clock Tower, Kazan 37008
Shalyapin House Museum, Moskva 37167
Shalyapin House Museum, Sankt-Peterburg 37695
Shambellie House Museum of Costume, New Abbey 45202
Shambles Museum, Newent 45234
Shambyu Museum, Shambyu 31755
Shamrock and Cotehele Quay Museum, Saint Dominick, Cornwall 45563
Shamrock Museum, Shamrock 06947
Shan Rong Cultural Display Center, Beijing 07566
Shan State Museum, Taunggyi 31730
Shand House Museum, Ferry Hill 05796
Shandong Painting Gallery, Jinan 07883
Shandong Provincial Museum, Jinan 07884
Shandong Stone Inscription Art Museum, Jinan 07885
Shandy Hall, Coxwold 43910
Shandy Hall, Geneva 48883
Shanggao Museum, Shanggao 08154
Shanghai Art Museum, Shanghai 08180
Shanghai Arts and Crafts Museum, Shanghai 08181
Shanghai Duolun Museum of Moder Art, Shanghai 08182
Shanghai History Museum, Shanghai 08183
Shanghai Museum, Shanghai 08184
Shanghai Museum of Contemporary Art, Shanghai 08185
Shanghai Museum of Folk Collectibles, Shanghai 08186
Shanghai Museum of Natural History, Shanghai 08187
Shanghai Museum of Public Security, Shanghai 08188
Shanghai Science & Technology Museum, Shanghai 08189
Shanghai Urban Planning Exhibition Hall, Shanghai 08190
Shanghai Zendai Museum of Modern Art, Shanghai 08191
Shanghang Museum, Shanghang 08200
Shangjing Ruins Museum, Ning'an 08069
Shangqiu Museum, Shangqiu 08201
Shangqiu Zone Museum, Shangqiu 08202
Shangrao County Museum, Shangrao 08203
Shangrao Zone Museum, Shangrao 08204
Shangrila Hakka Museum, Zaociao 42487
Shangzhou Museum, Shangzhou 08205
Shankar's International Dolls Museum, Delhi 24154
Shantou Archaeology Museum, Shantou 08206
Shantou Museum, Shantou 08207
Shantytown, Greymouth 33115
Shanxi Painting Gallery, Taiyuan 08274
Shaoguan Museum, Shaoguan 08209
Shaolin Temple Religion Museum, Yuanli 42478
Shaowu Museum, Shaowu 08210
Shaoxing Museum, Shaoxing 08212
Sharadin Art Gallery, Kutztown 49830
Sharia Museum, Cairo 10519
Sharif Kamala Museum, Kazan 37024
Sharjah Archaeology Museum, Sharjah 43225
Sharjah Art Museum, Sharjah 43226
Sharjah Heritage Museum, Sharjah 43227
Sharjah Islamic Museum, Sharjah 43228
Sharjah Natural History Museum, Sharjah 43229
Sharjah Science Museum, Sharjah 43230
Sharlot Hall Museum, Prescott 52135
Sharon Arts Center, Sharon 53051
Sharon Museum, Emek Hefer 24947
Sharon Museum Emek Hefer, Midreshet Ruppin 25066
Sharon Temple Museum Historic Site, Sharon 06948
Sharpsteen Museum, Calistoga 47168
Shaw Island Historical Society Museum, Shaw Island 53055
Shaw Ocean Discovery Centre, Sidney 06972
Shawano County Historical Society Museum, Shawano 53056
Shawneetown Historic Site, Old Shawneetown 51537
Shawqi Museum, Cairo 10520
Shaw's Corner, Ayot-Saint-Lawrence 43343
Shchelykovo - Memorial Museum-Estate of Aleksander Ostrovsky, Ščelykovo 37756
Shchuchin Regional Museum of Literature and Art, Kokšetau 29704
Shchusev State Museum of Architecture, Moskva 37207
Shearwater Aviation Museum, Shearwater 06951
Sheboygan County Historical Museum, Sheboygan 53062
Sheehan Gallery at Whitman College, Walla Walla 53798
Sheelin Antiques and Irish Lace Museum, Enniskillen 44181
Shefton Museum of Greek Art and Archaeology, Newcastle-upon-Tyne 45231
Sheikan Museum, El-Obeid 40385
Sheikh Saeed's House, Dubai 43220
Shekspear Ghost Town Museum, Lordsburg 50200
Shelburne County Museum, Shelburne 06955
Shelburne Museum, Shelburne 53065
The Sheldon Art Galleries, Saint Louis 52600
Sheldon Jackson Museum, Sitka 53125
Sheldon Memorial Art Gallery and Sculpture Garden, Lincoln 50105
Sheldon Museum, Haines 49125
Sheldon Peck Museum, Lombard 50165
Shell Exhibit at Chi Chin Seashore Park, Kaohsiung 42243
Shell Lake Museum, Shell Lake 06956
Shelter House, Emmaus 48389
Shelter Island Historical Society Museum, Shelter Island 53071
Shemer Art Center and Museum, Phoenix 51910
Shenyang Palace Museum, Shenyang 08217
Shenzhen Art Gallery, Shenzhen 08224
Shenzhen Museum, Shenzhen 08225
Shenzhen Painting Gallery, Shenzhen 08226
Shephela Museum, Kibbutz Kfar Menahem 25047
Shepherd's Museum, Bugac 23498
Shepherd's Museum, Hortobágy 23619
Sheppard Fine Arts Gallery, Reno 52290
Shepparton Art Gallery, Shepparton 01504
Sheppy's Farm and Cider Museum, Bradford-on-Tone, Taunton 43584
Shepton Mallet Museum, Shepton Mallet 45658
Sheqi Museum, Sheqi 08228
Sherborne Castle, Sherborne 45659
Sherborne House, Sherborne 45660
Sherborne Museum, Sherborne 45661
Sherbrooke Village, Sherbrooke, Nova Scotia 06957
Sherburne County Historical Museum, Becker 46684
Sheremetyev Palace - Fountain House, Sankt-Peterburg 37726
Sheriff Andrews House, Saint Andrews 06800
Sheringham Museum, Sheringham 45663
Sherlock Holmes Museum, London 44986
Sherlock Holmes-Museum, Meiringen 41621
The Sherman House, Lancaster 49921
Sherrier Resources Centre, Lutterworth 45050
Sherwin Miller Museum of Jewish Art, Tulsa 53611
Sherwood-Davidson House, Newark 51293
Sherwood Forest Plantation, Charles City 47351
Sheszkasgan Historical and Regional Museum, Žezkazgan 29737
Shetland Croft House Museum, Dunrossnes 44062
Shetland Museum, Lerwick 44693
Sheung Yiu Folk Museum, Hong Kong 07816
Shevchenko Apartment-Museum at Museum of the Russian Academy of Art, Sankt-Peterburg 37696
Shevchenko Museum, Fort Shevchenko 29701
Shi Liao Display Center, Nanjing 08049
Shi-San-Hang Museum of Archaeology, Bali 42142
Shibayama Haniwa Hakubutsukan, Shibayama 29255
Shibayama Tomb and Clay Figures Museum, Shibayama 29255
Shibden Hall Museum and West Yorkshire Folk Museum, Halifax 44388
Shibunkaku Bijutsukan, Kyoto 28865
Shibunkaku Museum of Art, Kyoto 28865
Shibuya Bijutsukan, Fukuyama 28449
Shibuya Museum of Art, Fukuyama 28449
Shicheng Museum, Shicheng 08229
Shido-ji Homotsukan, Shido 29257
Shido-ji Treasure House, Shido 29257
Shiga-kenritsu Kindai Bijutsukan, Otsu 29166
Shiga-kenritsu Shiryokan, Azuchi 28386
Shiga Prefectural Archaeological Museum, Azuchi Castle, Azuchi 28386
Shigaraki-cho Dentouteki Bijutsu Senta, Shigaraki 29259
Shigaraki Pottery Museum, Shigaraki 29260
Shigaraki Traditional Art Center, Shigaraki 29259
Shigaraki-yaki Bijutsukan, Shigaraki 29260
Shihezi Junken Museum, Shihezi 08230
Shijiazhuang Museum, Shijiazhuang 08233
Shiko Munakata Memorial Museum of Art, Aomori 28361
Shikoku Folk House Museum and Gallery, Takamatsu 29321
Shikoku-mura Minka Hakubutsukan no Bijutsukan, Takamatsu 29321
Shikumen Open House Museum, Shanghai 08192
Shillong Tribal Research Institute, Shillong 24387
Shiloh Museum of Ozark History, Springdale 53212
Shiloh National Military Park and Cemetery, Shiloh 53078
Shima Marineland, Ago 28348
Shimane-kenritsu Hakubutsukan, Matsue 28889
Shimane Museum of Ancient Izumo, Matsue 28889
Shimane Prefectural Yakumodatsu Fudoki-no-oka Museum, Matsue 28890
Shimazu Foundation Memorial Hall, Kyoto 28866
Shimen Museum, Shimen 08234
Shimogamo Hanki-cho Kyoto, Kyoto 28867
Shimonoseki City Art Museum, Shimonoseki 29269
Shimonoseki Marine Science Museum, Shimonoseki 29270
Shimonoseki-shi Bijutsukan, Shimonoseki 29269
Shimonoseki-shiritsu Kaikyokan, Shimonoseki 29270
Shin-Yokohama Ramen Hakubutsukan, Yokohama 29606
Shinagawa Historical Museum, Tokyo 29469
Shinano Drawing Museum, Ueda 29552
Shinchon Museum, Shinchon 29761
Shingenkou Treasure Hall Historical Museum, Enzan 28417
Shinshushin-machi Hakubutsukan, Shinshushin 29273
Shinshushin Museums, Shinshushin 29273
Shinuiju Historical Museum, Shinuiju 29762
Shinyanga Mazingira Museum, Shinyanga 42519
Shinyoung Cinema Museum, Namwon 29856
Shio Aragvispireli House Museum, Tbilisi 16380
Shio-no-michi Hakabutsukan, Itoigawa 28600
Shiogama-jingu Homotsukan, Shiogama 29274
Shiogama Shrine Museum, Shiogama 29274
Shiojiri Tanka Verse Museum, Shiojiri 29276
Shiojiri Tankakan, Shiojiri 29276
Shiozawa Tsumugi Commemorative Museum, Shiozawa 29278
Ship Museum, Eskifjörður 23994
Shipley Art Gallery, Gateshead 44265
Shippensburg Historical Society Museum, Shippensburg 53081
Ships of the Sea Maritime Museum, Savannah 52946
Shipwreck Heritage Centre, Hastings 44417
Shipwreck Museum, Bredasdorp 38573
Shipwreck Museum, Girne 33356
Shirahata Shiro World Mountains Photograph Collection, Yuzawa 29638
Shirakaba Bijutsukan, Onomichi 29121
Shirakaba Museum of Art, Onomichi 29121
Shiraz House Museum, Gyumri 00709
Shire Hall Gallery, Stafford 45730
Shirehall Museum, Little Walsingham 44728
Shiretoko Museum, Shari 29254
Shirley-Eustis House, Boston 46949
Shirley Plantation, Charles City 47352
Shiseido Art House, Kakegawa 28638
Shishi Kaikan, Takayama 29343
Shishou Museum, Shishou 08235
Shitamachi Museum, Tokyo 29470
Shizen Kyoiku En, Tokyo 29471
Shizenshi Hakubutsukan, Sendai 29247
Shizenshi Hakubutsukan, Shimizu 29262
Shizukuishi Historical Folk Art Museum, Shizukuishi 29283
Shizukuishi-machi Rekishi Mingeikan, Shizukuishi 29283
Shizuoka Art Gallery, Shizuoka 29287
Shizuoka City Toro Museum, Shizuoka 29288
Shizuoka-kenritsu Bijutsukan, Shizuoka 29289
Shizuoka Prefectural Museum of Art, Shizuoka 29289
Shkodër Historical Museum, Shkodër 00028
Shloss Nöthnitz, Bannewitz 16924
Shoah Memorial, Paris 14625
Shoah Museum, Jerusalem 25015
Shoal Creek Living History Museum, Kansas City 49698
Shoal Lake Police and Pioneer Museum, Shoal Lake 06967
Shodo Hakubutsukan, Tokyo 29472
Shoe Museum, Street 45790
Shoe Museum, Zlín 09942
Shogetsu Art Museum, Wakakusa 29571
Shogetsu Bijutsukan, Wakakusa 29571
Shohada Museum, Ghazvin 24603
Shoji Ueda Museum of Photography, Kishimoto 28705
Shoko Shuseikan Historical Museum, Kagoshima 28635
Sholem Asch House, Bat Yam 24934
Sholokhov Museum-Reserve, Vešenskaja 37925
Sholom Aleichem Museum, Tel Aviv 25121
Shore Line Trolley Museum, East Haven 48250
Shore Road Pumping Station, Birkenhead 43491
Shore Village Museum, Rockland 52421
Shoreham Aircraft Museum, Shoreham 45667
Shoreham Airport Visitor Centre, Shoreham-by-Sea 45669
Shoreline Historical Museum, Seattle 53015

S

Shorter Mansion Museum, Eufaula 48429
Shoshone Tribal Cultural Center, Fort Washakie 48697
Shosoin Treasure Repository, Nara 29038
The Shoto Museum of Art, Tokyo 29473
Shotts Heritage Centre, Shotts 45670
Shou Museum, Shouxian 08240
Shouguang Museum, Shouguang 08239
Shouhaku Art Museum, Nara 29039
Shouhaku Bijutsukan, Nara 29039
Shoundo Bijutsukan, Ube 29550
Shoundo Museum of Art, Ube 29550
Show Gallery, Toronto 07137
Showa Bijutsukan, Nagoya 29010
Showa Museum of Art, Nagoya 29010
Shpakovsky Regional Museum of Essentuki, Essentuki 36917
Shreemanthi Bai Memorial Government Museum, Mangalore 24308
Shrewsbury Castle, Shrewsbury 45675
Shrewsbury Museum and Art Gallery, Shrewsbury 45676
Shrewsbury Windle House, Madison 50381
Shri Chhatrapati Shivaji Maharaj Museum, Satara 24381
Shri Girdharbhai Sangrahalaya, Amreli 24065
Shrikrishna Science Centre, Patna 24347
Shrine of the Book, Jerusalem 25016
Shropshire Regimental Museum, Shrewsbury 45677
Shtekelis Prehistory Museum, Haifa 24968
Shtepia muze e piktorit Vangjush Mio, Korçë 00024
Shuanglinsi Painted Sculpture Art Museum, Pingyao 08091
Shugborough Estate Museum, Milford, Staffordshire 45146
Shukshin Memorial Preserve, Srostki 37793
Shunan City Museum of Art and History, Shunan 29292
Shunan-shi Bijutsukan nara Hakubutsukan, Shunan 29292
Shunchang Museum, Shunchang 08242
Shung Ye Museum of Formosan Aborigines, Taipei 42424
Shushenskoye State Museum-Preserve of History and Etnography, Šušenskoe 37811
Shuswap Lake Provincial Park Nature House, Kamloops 06039
Shuttleworth Collection, Old Warden 45313
Siatista Botanical Museum, Siatista 23182
Siatista Paleontological Museum, Siatista 23183
Šiauliai Aušra Museum, Šiauliai 30362
Šiauliai Region Literature Museum, Naisiai 30316
Šiaulių Aušros Muziejus, Šiauliai 30362
Šiaulių Miesto Vandentiekio ir Kanalizacijos Muziejus, Šiauliai 30363
Šiaulių Rajono Literatūros Muziejus, Naisiai 30316
Šiaulių Universiteto Istorijos Muziejus, Šiauliai 30364
Sibbesgården Museo, Sipoo 11313
Sibbo Agricultural Museum, Nikkila 11142
Sibbo Jordbruksmuseum, Nikkila 11142
Sibeliuksen Syntymäkoti, Hämeenlinna 10748
Sibelius Museum, Turku 11397
Sibiu Natural History Museum, Sibiu 36610
Sibley County Historical Museum, Henderson 49248
Sibley House, Mendota 50588
Siboney Farm Museum, Santiago de Cuba 09311
Sibsagar College Museum, Sibsagar 24392
Sicamous and District Museum, Sicamous 06969
Sichuan Museum, Chengdu 07619
Sichuan University Museum, Chengdu 07620
Sickingen-Museum, Landstuhl 19806
Sid Richardson Collection of Western Art, Fort Worth 48719
Side Müzesi, Side 42939
Side Museum, Side 42939
Sidenväveri Museum, Stockholm 40932
Sidmouth Museum, Sidmouth 45679
Sidney and Gertrude Zack Gallery, Vancouver 07222
Sidney Historical Association Museum, Sidney 53098
Sidney Historical Museum, Sidney 06973
Sidney Lanier Cottage, Macon 50369
Sidney Mishkin Gallery of Baruch College, New York 51255
Sieben-Keltern-Museum, Metzingen 20315
Sieben-Schwaben-Museum, Türkheim 22129
Siebenberg House, Jerusalem 25017
Siebenbürgisches Museum Gundelsheim, Gundelsheim, Württemberg 18836
Siebengebirgsmuseum, Königswinter 19661
Sieblos-Museum Poppenhausen, Poppenhausen 21076
Siebold Kinenkan, Nagasaki 28985
Siebold Memorial Museum, Nagasaki 28985
Siebold-Museum, Würzburg 22724
Siege Museum, Petersburg 51803
Siegerlandmuseum mit Ausstellungsforum Haus Oranienstraße, Siegen 21756
Siegfried Charoux Museum, Langenzersdorf 02262
Siegfried H. Horn Archaeological Museum, Berrien Springs 46767
Siegfrieds Mechanisches Musikkabinett, Rüdesheim am Rhein 21437
SiemensForum, München 20483
Sierra Arts Foundation, Reno 52291
Sierra Leone National Museum, Freetown 38169
Sievin Kotiseutumuseo, Sievi 11308
Sigaar Musea, Geraardsbergen 03653
Sigdal og Eggedal Museum, Prestfoss 33773
Siggebohyttans Bergmansgård, Nora 40765
Siglhaus, Sankt Georgen bei Salzburg 02644
Sigmund Freud Museum, Sankt-Peterburg 37709
Sigmund Freud Museum, Wien 03075
Signal Hill Arts Centre, Weyburn 07315
Signal Hill National Historic Site, Saint John's 06857
Signal Museum of Finland, Riihimäki 11272
Signatarų Namai, Vilnius 30432
Sigtuna Museer, Sigtuna 40835
Sigtuna Rådhus, Sigtuna 40836
Sigurhaedir - Hús Skáldsins Museum, Akureyri 23987
Sigurjón Ólafsson Museum, Reykjavík 24023
Sihui Museum, Sihui 08243
Siikaisten Kotiseutumuseo, Siikainen 11309
Sikar Museum, Sikar 24393
Sikinos Archaeological Collection, Sikinos 23187
Sikkens Schildersmuseum, Sassenheim 32745
Siksika Nation Museum, Siksika 06975
Silale Area Museum, Šilalė 30367
Šilalės Krašto Muziejus, Šilalė 30367
Silas Wright House, Canton 47225
Silberbergwerk, Ramingstein 02556
Silbereisenbergwerk Gleißinger Fels, Fichtelberg 18316
Silberstollen Geising, Geising 18560
Silberwaren- und Bijouteriemuseum, Schwäbisch Gmünd 21651
Silcher-Museum des Schwäbischen Chorverbandes, Weinstadt 22441
Síldarminjasafnið á Siglufirði, Siglufjörður 24039
Silesian Museum, Opava 09692
Silesian Museum - Arboretum Nový Dvůr - Botanical Garden, Stěbořice-Nový Dvůr 09846
Silesian Museum - Czechoslovak Fortification Complex Hlucin-Darkovicky, Hlučín-Darkovičky 09517
Silesian Museum - Historical Exhibition Building, Opava 09693
Silesian Museum of Görlitz, Görlitz 18680
Silesian Museum - Petr Bezruc Log Cabin, Ostravice 09701
Silesian Museum - Petr Bezruc Memorial, Opava 09694
Silesian Museum - Second World War Memorial, Hrabyně 09532
Silifke Atatürk Evi, Silifke 42941
Silifke Müzesi, Silifke 42942
Silifke Museum, Silifke 42942
Siliman University Anthropology Museum, Dumaguete 34386
Siljansfors Skogsmuseum, Mora 40746
Siljustøl Museum, Paradis 33768
Silk Mill - Derby's Museum of Industry and History, Derby 43970
Silk Museum, Gangnae 29805
Silk Museum, Komoro 28768
Silk Museum, Soufli 23193
Silk Museum, Yokohama 29600
Silkeborg Kulturhistoriske Museum, Silkeborg 10256
Silkeborg Kunstmuseum, Silkeborg 10257
Silkeborg Museum of Art, Silkeborg 10257
Silkmuseum, Meliskerke 32548
Silla University Museum, Busan 29781
Sillaotsa Farm Museum, Rapla 10613
Sillaotsa Talumuuseum, Rapla 10613
Siloam Springs Museum, Siloam Springs 53099
Silomuseum, Waidhofen an der Thaya 02885
Silpakorn University Art Centre, Bangkok 42560
Silute Museum, Šilutė 30369
Šilutės Melioracijos Statybos Valdybos Istorijos Muziejus, Šilutė 30368
Šilutės Muziejus, Šilutė 30369
Silver City Museum, Silver City 53103
Silver Cliff Museum, Silver Cliff 53105
Silver Eye-Center for Photography, Pittsburgh 51960
The Silver Ledge Hotel, Ainsworth Hot Springs 05310
Silver Mines Museum, L'Argentière-la-Bessée 13526
Silver Museum, Al-Khobar 38014
Silver River Museum, Ocala 51484
Silver Stream Railway Museum, Hutt City 33126
Silvercreek Museum, Freeport 48782
Silvermine Guild Arts Center, New Canaan 51038
Silvermuseet, Arjeplog 40431
Silverton Gaol Museum, Silverton 01505
Silverton Outdoor Mining Exhibit, Silverton 06977
Silverware Museums Collection, Brașov 36176
Silvery Slocan Museum, New Denver 06428
Símaminjasafnið, Reykjavík 24034
Simcoe County Museum, Minesing 06296
Simeon Perkins House, Liverpool 06179
Simferopol Art Museum, Simferopol' 43186
Simferopolskij Chudožestvennyj Muzej, Simferopol' 43186
Simmons Visual Center, Gainesville 48822
Simón Bolívar's Birthplace, Caracas 54526
Simon Fraser Gallery, Burnaby 05505
Simon-Gfeller-Gedenkstube, Heimisbach 41478
Simon Janashia Museum of Georgia, Tbilisi 16381
Simon Kotiseutumuseo, Simo 11312
Simon Paneak Memorial Museum, Anaktuvuk Pass 46284
Simon van Gijn-Museum aan Huis, Dordrecht 32154
Simono Stanevičiaus Memorialinė Sodyba-muziejus, Kanopenai 30237
Simon's Town Museum, Simon's Town 38730
Simplicissimus-Haus, Renchen 21272
Simsbury Historical Society Museum, Simsbury 53112
Le Simserhof, Siersthal 15755
Sincheonji Art Museum, Aeweol 29764
Sinclair Inn, Annapolis Royal 05345
Sinclair Lewis Museum, Sauk Centre 52929
Sind Provincial Museum, Hyderabad 34005
Sindringer Heimatmuseum - Stadtmühle, Forchtenberg 18340
Sinebrychoff Art Museum - Finnish National Gallery, Helsinki 10817
Sinebrychoffin Taidemuseo, Helsinki 10817
Singapore Art Museum, Singapore 38179
Singapore Science Centre, Singapore 38180
Singer Laren, Laren, Noord-Holland 32465
Singhofener Heimatstuben, Singhofen 21782
Singida Museum, Singida 42520
Singleton Historical Museum, Singleton 01507
Singsås Museum, Singsås 33826
Sining Makiling Gallery, Los Baños 34408
Sinka István Emlékház - Helytörténeti kiállítás, Vészto 23964
Sinop Devlet Güzel Sanatlar Galerisi, Sinop 42943
Sinop Müzesi, Sinop 42944
Sinop Museum, Sinop 42944
Sinop State Fine Art Gallery, Sinop 42943
Sint-Baafskathedraal Gent, Gent 03641
Sint-Godelievemuseum, Gistel 03657
Sint-Janshuismolen & Koeleweimolen, Brugge 03428
Sint Jansmuseum De Bouwloods, 's-Hertogenbosch 32378
Sint-Pieters Museum op de Lichtenberg, Maastricht 32533
Sint-Plechelmusbasiliek met Schatkamer, Oldenzaal 32617
Sintra Museu de Arte Moderna, Sintra 35954
Sintra Museum of Modern Art-Berado Collection, Sintra 35954
Sintra Toys Museum, Sintra 35947
Sioux City Art Center, Sioux City 53113
Sioux City Public Museum, Sioux City 53114
Sioux Empire Medical Museum, Sioux Falls 53121
Sioux Indian Museum, Rapid City 52254
Sioux Lookout Museum, Sioux Lookout 06981
Siouxland Heritage Museums, Sioux Falls 53122
Sipahi Ay. Trias Bazilikası ve Kantara Kalesi, Kantara 33364
Sipalaseequtt Museum Society, Pangnirtung 06567
Siping Museum, Siping 08244
Sippola Museum, Sippola 11314
Sippolan Kotiseutumuseo, Sippola 11314
Sir Alexander Galt Museum, Lethbridge 06162
Sir Alfred Munnings Art Museum, Dedham 43961
Sir Andrew MacPhail Foundation, Orwell 06520
Sir Choturam Memorial Museum, Sangaria 24376
Sir Edgeworth David Memorial Museum, Kurri Kurri 01207
Sir Francis Cook Gallery, Augrès 43336
Sir Frank Hutson Sugar Museum, Saint James 03241
Sir Henry Jones Museum, Llangernyw 44767
Sir Henry Royce Memorial Foundation, Paulerspury 45347
Sir Hermann Black Gallery, Sydney 01559
Sir Isaac and Lady Edith Wolfson Museum, Jerusalem 25018
Sir John Soane's Museum, London 44987
Sir Joseph Banks Conservatory, Lincoln 44716
Sir Max Aitken Museum, Cowes 43908
Sir Seewoosagar Ramgoolam Memorial Centre, Plaine Verte 30681
Sir Walter Scott's Courtroom, Selkirk 45638
Sir Wilfred Grenfell College Art Gallery, Corner Brook 05642
Sir William F. Coaker Heritage Foundation, Port Union 06648
Sırçalı Medrese Müzesi, Konya 42891
Sırçalı School Museum, Konya 42891
Sisal Industry Historical Exhibition, Hengchun 42201
Sisian Museum of History N. Adonts, Sisian 00717
Sisimiut Katersugaasiviat, Sisimiut 23269
Sisimiut Museum, Sisimiut 23269
Siskin Museum of Religious and Ceremonial Art, Chattanooga 47413
Siskiyou County Museum, Yreka 54359
Site Archéologique La Cave Peinte, Brain-sur-Allonnes 12142
Site-Chantier de Fouilles Archéologiques, Charavines 12406
Site du Bois du Cazier, Marcinelle 03852
Site du Moulin des Pionniers de la Doré, La Doré 06110
Site et Musée du Regourdou, Montignac 14216
Site Fortifié des Bois de Cattenom, Cattenom 12321
Site Gallo-Romain de Grand, Grand 13096
Site Gallo-Romain du Fâ, Barzan 11855
Site Gallo-Romaine de Sanxay, Sanxay 15627
Site Historique de l'île-des-Moulins, Terrebonne 07068
Site-Historique du Banc-de-Paspébiac, Paspébiac 06573
Site Historique Matamajaw, Causapscal 05580
Site Mary-Travers dite - La Bolduc, Newport 06440
Site of the 1st National Congress of the CCP, Shanghai 08193
Site of the First National Congress of the Communist Party of China, Shanghai 08194
Site Paléochrétien de Saint-Seurin, Bordeaux 12075
Site Santa Fe, Santa Fe 52890
Sitio Arqueológico Huaca Malena, Capilla de Asia 34137
Sitio Arqueológico de Cumbemayo, Cumbemayo 34155
Sitio Arqueológico Joya de Cerén, San Juan 10547
Sítio Roberto Burle Marx, Rio de Janeiro 04771
Sitka Historical Society Museum, Sitka 53126
Sitorai Mochi-Khosa, Buxoro 54445
Sittingbourne and Kemsley Light Railway, Sittingbourne 45683
Sittingbourne Heritage Museum, Sittingbourne 45684
Situation Kunst, Bochum 17407
Siuntion Kotiseutumuseo, Siuntio 11315
Sivas Devlet Güzel Sanatlar Galerisi, Sivas 42946
Sivas State Fine Art Gallery, Sivas 42946
Six Nations Indian Museum, Onchiota 51558
Sixth Floor Museum at Dealey Plaza, Dallas 47939
Sizdah Aban Museum, Tehrān 24696
Sjöfartmuseet, Oskarshamn 40804
Sjöfartsmuseet, Kristiinankaupunki 10997
Sjöfartsmuseet Akvariet, Göteborg 40556
Sjöhistoriska Institutet vid Åbo Akademi, Turku 11398
Sjöhistoriska Museet, Stockholm 40933
Sjøfartsmuseet i Porsgrunn, Porsgrunn 33771
Sjølingstad Uldvarefabrik, Mandal 33660
Sjóminja- og Vélsmiðjusafn, Reykjavík 24035
Sjóminjasafn Austurlands, Eskifjörður 23994
Sjóminjasafn Íslands, Reykjavík 24036
Sjóminjasafnid á Eyrarbakka, Eyrarbakki 23996
Sjukehusmuseet i Molde, Molde 33679
Sjukehusmuseet i Skien, Skien 33835
Ska-Nah-Doht Iroquoian Village and Museum, Mount Brydges 06398
Skaalurensamlinga - Skipsbyggingsmuseet i Rosendal, Rosendal 33800
Skärgårdens Sågverksmuseum, Holmsund 40601
Skärgårdsmuseet, Lappoby-Åland 11052
Skärgårdsmuseet, Stavsnäs 40880
Skärhamns Sjöfartsmuseum, Skärhamn 40844
Skagafjördur Folk Museum, Varmahlid 24041
Skagen By- og Egnsmuseum, Skagen 10262
Skagens Museum, Skagen 10263
Skagit County Historical Museum, La Conner 49832
Skagway Museum, Skagway 53128
Skanderborg Museum, Skanderborg 10264
Skandinavisk Museumförbund - Skandinaavinen Museoliitto, Finnish Section, Helsinki 54794
Skandinavisk Museumsforbund, Danske Afdeling, København 54783
Skandinavisk Museumsforbund, Norwegian Section, Tønsberg 54977
Skandinavisk Museumsforbund, Swedish Section, Stockholm 55034
Skansen Archeologiczny w Mrówkach, Mrówki 34931
Skansen Budownictwa Ludowego Zachodniej Wielkopolski, Wolsztyn 35320
Skansen Górniczy Królowa Luiza, Zabrze 35351
Skansen i Muzeum Pszczelarstwa im prof. Ryszarda Kosteckiego, Swarzędz 35145
Skansen Kultury Ludowej Pogórza Sudeckiego, Kudowa Zdrój 34837
Skansen Kurpiowski im. Adama Chętnika, Nowogród 34953
Skansen Lokomotyw, Zduńska Wola 35368
Skansen Open Air Museum, Stockholm 40895
Skansen Rybołowstwa Morskiego, Gdynia 34661
Skånska Lasses Hus, Mjölby 40741
Skanzen Březno u Loun, Louny 09642
Skanzen - Múzeum L'udovej Architektúry, Bardejovské Kúpele 38197
Skara Järnvägsmuseum, Skara 40845
Skara Railway Museum, Skara 40845
Skate Canada Hall of Fame & Museum, Ottawa 06556
Skattkammaren, Stockholm 40934

S

Skattkammaren - Uppsala Domkyrkas Museum, Uppsala 41011
Skaun Bygdamuseum, Børsa 33449
Skellefteå Museum, Skellefteå 40849
Skënderbeg Museum, Krujë 00026
Skeppsvarvet Sjöfartsutställning, Västanfjärd 11434
Skerryvore Museum, Hynish, Isle of Tiree 44506
Ski Museum, Lahti 11026
Ski Museum and Museum of Finnish Baseball, Vimpeli 11468
Ski- und Heimatmuseum, Sankt Anton am Arlberg 02635
Ski- und Heimatmuseum Upfingen, Sankt Johann, Württemberg 21502
Skidby Windmill and Museum of East Riding Rural Life, Skidby 45686
Skimuseet, Oslo 33749
Skimuseum, Oberreute 20827
Skimuseum, Vaduz 30201
Skinner Museum, South Hadley 53171
Skiptvet Bygdemuseum, Skiptvet 33838
Skirball Cultural Center, Los Angeles 50264
Skirball Museum, Cincinnati 47563
Skirball Museum of Biblical Archaeology, Jerusalem 25019
Skironio Centrum Kifissa, Néa Kifissia 23123
Skironio Museum Polychronopoulos, Megara 23089
Skissernas Museum, Lund 40708
Skive Kunstmuseum, Skive 10265
Skive Museum, Skive 10266
Skive Museum of Art, Skive 10265
Skjern-Egvad Museum, Skjern 10270
Skjern Vindmølle, Skjern 10271
Skjoldborgs Hus, Vesløs 10312
Sklářské Muzeum, Kamenický Šenov 09571
Sklářské Muzeum Moser, Karlovy Vary 09577
Sklářské Muzeum Nový Bor, Nový Bor 09683
Sklenka Muzeum v Železném Brodě, Železný Brod 09938
Skobelev Park-muzej, Pleven 05121
Škoda Auto Museum, Mladá Boleslav 09658
Skövde City Museum, Skövde 40852
Skövde Konsthall och Konstmuseum, Skövde 40851
Skövde Stadsmuseum, Skövde 40852
Skogar Museum, Hvolsvöllur 24006
Skogsmuseet i Lycksele, Lycksele 40710
Skogsmuseum, Åsele 40442
Skogsmuseum, Sunne 40961
Skoindustrimuseet, Kumla 40658
Skokloster Castle, Skokloster 40854
Skokloster Motormuseum, Skokloster 40853
Skoklosters Slott, Skokloster 40854
Skolmuseet, Fjälkinge 40522
Skolmuseum Bunge, Fårösund 40518
Školoto Muzej, Trjavna 05231
Skottvångs Gruva, Mariefred 40734
Skovbrugsmagasinet på Corselitze, Nykøbing Falster 10201
Skovgaard Museet i Viborg, Viborg 10316
Skovgaard Museum, Viborg 10316
Skowhegan History House, Skowhegan 53129
Skriaudžių Buities Muziejus, Skriaudžiai 30370
Skrolsvik Fiskebruksmuseum, Moen 33674
Skulptoriaus Broniaus Pundziaus Memorialinis Muziejus, Pieveniai 30337
Skulpturenmuseum Glaskasten Marl, Marl, Westfalen 20224
Skulpturenmuseum im Hofberg, Landshut 19805
Skulpturenpark, Bad Nauheim 16792
Skulpturens Hus, Stockholm 40935
Skulpturensammlung, Dresden 17970
Skulpturenweg, Emmendingen 18149
Skulpturhalle Basel, Basel 41184
Skuodas' Museum, Skuodas 30372
Skuodo Muziejus, Skuodas 30372
Skupnost Muzejev Slovenije, Ljubljana 55013
Skyscraper Museum, New York 51256
Skyttalan Museo, Parainen 11189
Slaait'n Hoes, Onstwedde 32623
Slabsides, West Park 54076
Sládečkovo Vlastivědné Muzeum v Kladně, Kladno 09583
Slag van de Somme Museum, Schagen 32749
Slagelse Museum for Handel, Håndværk og Industri, Slagelse 10276
Slagterbutikken anno 1920, Roskilde 10249
Slate Run Living Historical Farm, Ashville 46395
Slate Valley Museum, Granville 49020
Slater Memorial Museum, Norwich 51438
Slater Mill Historic Site, Pawtucket 51738
Slaughter Ranch Museum, Douglas 48156
Slave Lodge, Cape Town 38590
Slaveikov House Museum, Trjavna 05229
Slawenburg Raddusch, Vetschau 22223
Sledmere House, Driffield 44023
Sleire Skulemuseum, Hosteland 33566
Slesvigske Vognsamling, Haderslev 10050
Slevarenske Muzeume Komarov, Komárov 09593
Slevogthof Neukastel, Leinsweiler 19887
Slezské zemské muzeum, Hlučín-Darkovičky 09517
Slezské zemské muzeum, Hrabyně 09532
Slezské zemské muzeum, Opava 09692
Slezské zemské muzeum, Opava 09693
Slezské zemské muzeum, Opava 09694
Slezské zemské muzeum, Ostravice 09701
Slezské zemské muzeum, Stěbořice-Nový Dvůr 09846
Slidell Art Center, Slidell 53130
Slidell Museum, Slidell 53131
Sliedrechts Museum, Sliedrecht 32782
Slifer House, Lewisburg 50031
Šlisselburgskaja Krepost Orešek, Šlisselburg 37769
Sloan Museum & Longway Planetarium, Flint 48560
Sloane-Stanley Museum and Kent Furnace, Kent 49737
Slocumb Galleries, Johnson City 49635
Slöjdmuseet, Ånge 40426
Sloss Furnaces, Birmingham 46824
Slot Loevestein, Poederoijen 32670
Slot Zeist, Zeist 32994
Slot Zuylen, Oud-Zuilen 32648
Slough Museum, Slough 45690
Slovácké Muzeum, Uherské Hradiště 09887
Slovácko Museum, Uherské Hradiště 09887
Slovak Agricultural Museum, Nitra 38316
Slovak Glass Museum, Lednické Rovne 38281
Slovak House, Tótkomlós 23936
Slovak Mining Museum, Banská Štiavnica 38193
Slovak Museum of Nature Protection and Speleology, Liptovský Mikuláš 38288
Slovak National Gallery, Bratislava 38233
Slovak National Gallery - Schaubmar Mill - Gallery of Native Art, Pezinok 38326
Slovak National Literary Museum of the Slovak National Library, Martin 38300
Slovak National Museum, Bratislava 38234
Slovak Region House, Szarvas 23835
Slovak Region House, Tótkomlós 23937
Slovak Technical Museum, Košice 38273
Slovakian Museum of Local History, Békéscsaba 23372
Slovakian Region House of the Petőfi House and Memorial Museum, Kiskőrös 23665
Slovene Ethnographic Museum, Ljubljana 38463
Slovenian Cinematheque, Ljubljana 38462
Slovenian Museum of Firefighting, Metlika 38482
Slovenian Museum of Natural History, Ljubljana 38461
Slovenian Museum of Religion, Ivančna Gorica 38422
Slovenian School Museum, Ljubljana 38465
Slovenska Kinoteka, Ljubljana 38462
Slovenská Národná Galéria, Bratislava 38233
Slovenská Národná Galéria - Galéria Insitného Umenia - Schaubmar Mill, Pezinok 38326
Slovenská Národná Galéria - Kaštieľ Strážky, Spišská Béla 38357
Slovenská Národná Galéria - Zvolenský Zámok, Zvolen 38395
Slovenské Banské Múzeum, Banská Štiavnica 38193
Slovenské Múzeum A.S. Puškina, Brodzany 38239
Slovenské Múzeum Ochrany Prírody a Jaskyniarstva, Liptovský Mikuláš 38288
Slovenské Národné Literárne Múzeum, Martin 38300
Slovenské Národné Múzeum, Bratislava 38234
Slovenské Národné Múzeum - Múzeum Betliar, Betliar 38199
Slovenské Národné Múzeum - Múzeum Bojnice, Bojnice 38201
Slovenské Polnohospodárske Múzeum, Nitra 38316
Slovenské Sklárske Múzeum, Lednické Rovne 38281
Slovenské Technické Múzeum, Košice 38273
Slovenski Etnografski Muzej, Ljubljana 38463
Slovenski Gasilski Muzej Dr. Branka Božiča, Metlika 38482
Slovenski Gledališki Muzej, Ljubljana 38464
Slovenski Šolski Muzej, Ljubljana 38465
Slovenski Verski Muzej, Ivančna Gorica 38422
Slovensko Konservatorsko Društvo, Ljubljana 55014
Slovensko Muzejsko Društvo, Ljubljana 55015
SM City Art Center, Cebu 34375
Smålands Bil, Musik och Leksaksmuseum, Rydaholm 40818
Smalands Cars, Toys and Musicmuseum, Rydaholm 40818
Smålands Konstarkiv, Värnamo 41029
Smålands Militärhistoriska Museum, Eksjö 40496
Smålands Museum, Växjö 41041
Small Cars Museum, Fontenay-sur-Mer 12971
Small Galery, Ljubljana 38450
Small Museums Association, Clinton 55179
Smallwood's Retreat, Marbury 50464
Smart Museum of Art, Chicago 47513
Smedemuseet, Øster-Assels 10223
Smederij Museum Wijlen dhr Verkley, Nieuwkoop 32581
Smidt Múzeum, Szombathely 23903
Smilgiai Ethnographic Farmstead, Smilgiai 30374
Smilgių Etnografinė Sodyba, Smilgiai 30374
Smiterlöwsche Sammlung, Franzburg 18408
Smith Art Gallery, Brighouse 43616
Smith College Museum of Art, Northampton 51414
Smith County Historical Society, Tyler 53634
Smith-McDowell House Museum, Asheville 46375
Smith Robertson Museum, Jackson 49566
Smith-Zimmermann Museum, Madison 50387
Smithers Public Art Gallery, Smithers 06984
Smithery Museum of the Lithuania Minor History Museum, Klaipėda 30278
Smithills Hall Museum, Bolton 43556
Smith's Cove Historical Society Museum, Smith's Cove 06985
Smiths Falls Railway Museum of Eastern Ontario, Smiths Falls 06988
Smith's Fort Plantation, Surry 53383
Smithson International Truck Museum, Rimbey 06751
Smithsonian American Art Museum, Washington 53917
Smithsonian Institution, Washington 55180
Smithtown Historical Society Museum, Smithtown 53134
The Smithy-Pioneer Gallery, Cooperstown 47792
Smoke Cabin Museum, Kuopio 11011
Smoke House Museum, Somero 11322
Smoki Museum, Prescott 52136
Smoky Hill Museum, Salina 52675
Smoky Mountain Trains, Bryson City 47108
Smøla Museum, Smøla 33844
Smolensk Art Gallery, Smolensk 37776
Smolensk Cathedral, Smolensk 37779
Smolensk Historical Museum, Smolensk 37777
Smolensk State Museum-Preserve, Smolensk 37778
Smolenskaja Chudožestvennaja Galereja, Smolensk 37776
Smolenskij Istoričeskij Muzej, Smolensk 37777
Smolenskij Muzej-Zapovednik, Smolensk 37778
Smolenskij Sobor, Smolensk 37779
Smrcka's House - Hussite Museum, Soběslav 09843
Smrčkův Dům, Soběslav 09843
sms contemporanea, Siena 27848
SM's - Stedelijk Museum's-Hertogenbosch, 's-Hertogenbosch 32379
Smyth County Museum, Marion 50490
Snake River Heritage Center, Weiser 54022
Snappertuna Folk Museum, Snappertuna 11316
Snappertunan Forngårdsmuseum-Talomuseo, Snappertuna 11316
Snåsa Bygdamuseum, Snåsa 33845
Snibston Discovery Park, Coalville 43863
The Snite Museum of Art, Notre Dame 51445
Snoqualmie Valley Historical Museum, North Bend 51382
Snowden House, Waterloo 53964
Snowshill Manor, Broadway 43648
Snug Harbor Cultural Center, Staten Island 53267
The Snyder County Historical Society Museum, Middleburg 50650
Snyder Museum and Creative Arts Center, Bastrop 46624
Soap Museum, Saïda 30173
Søby Brunkulslejer og Brunkulsmuseet, Herning 10073
Søbygård, Søby Ærø 10278
Social History Collections, Cape Town 38591
Social History Curators Group, Croydon 55093
Socialmuseet Det Gamle Syge-Plejehjem, Hinnerup 10078
Società Geografica Italiana, Roma 27547
Società per le Belle Arti ed Esposizione Permanente, Milano 26723
Societas Museologia, Český Těšín 54776
Société Canadienne d'Éducation par l'Art, Toronto 07138
Société des Musées Québécois, Montréal 54746
Société Géographique Royale du Canada, Ottawa 54747
Society for Arts and Technology, Montréal 06381
Society for Contemporary Craft Museum, Pittsburgh 51961
Society for Contemporary Photography, Kansas City 49699
Society of Arts and Crafts-Exhibition Gallery, Boston 46950
Society of County Museum Officers, Taunton 55094
Society of Decorative Art Curators, Liverpool 55095
Society of Museum Archaeologists, London 55096
The Society of the Cincinnati Museum, Washington 53918
The Society of the Four Arts Gallery, Palm Beach 51678
Sočinskij Chudožestvennyj Muzej, Soči 37784
Socrates Sculpture Park, Long Island City 50188
Sod House Museum, Aline 46220
Sod House Museum, Arctic Bay 05351
Sod House Museum, Gothenburg 48979
Sodankylän Kotiseutumuseo, Sodankylä 11317
Sodbuster Archives Museum, Strome 07023
Sodus Bay Historical Society Museum, Sodus Point 53141
Söderbärke Hembygdsgård, Söderbärke 40860
Söderhamns Stadsmuseum, Söderhamn 40861
Söderlångvik Museum, Dragsfjärd 10719
Söderlångvikin Museo, Dragsfjärd 10719
Södertälje Art Gallery, Södertälje 40863
Södertälje Konsthall, Södertälje 40863
Söğüt Müzesi, Söğüt 42947
Sölje Bygdegård, Glava 40536
Sölvesborgs Museum, Sölvesborg 40867
Sörmlands Museum, Nyköping 40782
Sörmlandsgården, Eskilstuna 40504
Søfarts og Fiskerimuseet, Læsø 10173
Søfartssamlingerne i Troense, Svendborg 10292
Sofia City Art Gallery, Sofia 05203
Sofiero Slott, Helsingborg 40592
Sofijska Gradska Chudožestvena Galerija, Sofia 05203
Sofijski Istoričeski Muzej, Sofia 05204
Sogetsu Art Museum, Tokyo 29474
Sogetsu Bijutsukan, Tokyo 29474
Søgne Bygdemuseum, Søgne 33846
Sogo-Bunka Art Gallery, Amagasaki 28357
Sogo Museum of Art, Yokohama 29607
Soho House Museum, Birmingham 43512
Soho Photo Gallery, New York 51257
Soieries C.J. Bonnet - Musée de la Soie, Jujurieux 13292
Soil and Agronomy Museum W.R. Williams, Moskva 37398
Soinin Museo, Soini 11319
Sojeon Art Museum, Siheung 29936
Sokang University Museum, Seoul 29924
Sokličev Muzej, Slovenj Gradec 38519
Soknedal Bygdemuseum, Soknedal 33847
Sol Koffler Graduate Gallery, Providence 52180
Solar Boats Museum, Giza 10531
Solar Condes de Resende, Canelas 35504
Solar de Dom João VI, Rio de Janeiro 04772
Solar Grandjean de Montigny, Rio de Janeiro 04773
Solar Museu Quinta dos Cónegos, Maia 35753
Le Soldat Artiste, Nouvron-Vingré 14476
Soldiers and Sailors Memorial Building, Huntington 49432
Soldiers' Memorial Military Museum, Saint Louis 52601
Soldiers National Museum, Gettysburg 48904
Soldiers of Gloucestershire Museum, Gloucester 44319
Soldiers' Uprising in 1918, Radomir 05146
Solent Sky, Southampton 45705
Soli Archaeological Site, Lefke 33365
Solikamsk Municipal Museum, Solikamsk 37785
Solikamskij Gorodskoj Kraevedčeskij Muzej, Solikamsk 37785
Sollefteå Museum, Sollefteå 40868
Søllerød Museum, Holte 10093
Solna Hembygdsmuseum, Solna 40874
Solomon R. Guggenheim Museum, New York 51258
Solomos Museum, Corfu 22980
Soloveckij Gosudarstvennyj Istoriko-Architekturnyj i Prirodnyj Muzej-Zapovednik, Soloveckij 37788
Solovetsky State Historical, Architectural and Natural Museum-Reserve, Soloveckij 37788
Solway Aviation Museum, Crosby-on-Eden 43933
Soma Gyofu Kinenkan, Itoigawa 28601
Soma Gyofu Memorial Museum, Itoigawa 28601
Somali National Museum, Mogadishu 38549
Sombor Municipal Museum, Sombor 38148
Sombra Township Museum, Sombra 06990
Somdet Phra Narai National Museum, Lopburi 42580
Somers Historical Society Museum and Museum of the Early American Circus, Somers 53145
Somers Mansion, Somers Point 53147
Somerset and Dorset Railway Trust Museum, Washford 45945
Somerset Brick and Tile Museum, Bridgwater 43607
Somerset County Museum, Taunton 45829
Somerset Cricket Museum, Taunton 45830
Somerset East Museum, Somerset East 38733
Somerset Historical Center, Somerset 53148
Somerset Military Museum, Taunton 45831
Somerset Place, Creswell 47881
Somerset Rural Life Museum, Glastonbury 44308
Somervell County Museum, Glen Rose 48924
Somerville Museum, Somerville 53149
Somesville Museum, Mount Desert 50874
Somme Heritage Centre, Newtownards 45262
Sommerhuber-Kachelofenmuseum, Steyr 02772
Sømna Bygdetun, Sømna 33850
Somogy Kereskedelme Anno..., Kaposvár 23634
Somogyi Megyei Múzeumok Igazgatósága, Kaposvár 23635
Sonargaon Folk Art and Craft Museum, Sonargaon 03236
Søndhordland Motormuseum, Valevåg 33956
Sondre Nordheimstova, Morgedal 33682
Song Qingling Former Residence, Shanghai 08195
Song Qingling Tongzhi Former Residence, Beijing 07567
Songam Art Museum, Incheon 29834
Songjiang Museum, Shanghai 08196

S

Songtangzhai Museum, Beijing 07568
Songyang Museum, Songyang 08245
Songzi Fuwu Guan, Songzi 08246
Songzi Service Museum, Songzi 08246
Sonkajärven Kotiseutumuseo, Sonkajärvi 11326
Sonnenberg Mansion, Canandaigua 47215
Sonoma County Museum, Santa Rosa 52909
Sons of the American Revolution, Roselle 52460
Sons of the American Revolution Museum, Louisville 50296
Sontg Hippolytus, Veulden 41982
Sony Wonder Technology Lab, New York 51259
Sonya Art, Brooklyn 47083
Soo Line Historical Museum, Weyburn 07316
Soo Visual Arts Center, Minneapolis 50742
Sood Sangvichien Prehistoric Museum, Bangkok 42561
Sooke Region Museum, Art Gallery and Historic Moss Cottage, Sooke 06991
Sookmyung Women's University Museum, Seoul 29925
Sooner Ga Museum, Dhaka 03229
Sophienburg Museum, New Braunfels 51025
Sopron Lutheran Museum, Sopron 23819
Sopron Museum - Mithrász-sanctuary, Fertőrákos 23560
Sopron Roman Catholic Clerical Collection, Sopron 23821
Soproni Evangélikus Egyházközség Múzeuma, Sopron 23819
Soproni Múzeum, Sopron 23820
Soproni Római Katolikus Egyházművészeti Gyűjtemény, Sopron 23821
Sør-Senja Museum, Stonglandseidet 33884
Sør-Troms Museum, Evenskjer 33484
Sorbische Bauernstube, Heinersbrück 19111
Sordoni Art Gallery, Wilkes-Barre 54163
Sørlandets Kunstmuseum, Kristiansand 33612
Sørli Museum, Sørli 33853
Sørlistøa Fløtermuseet, Osen 33715
Sorø Amts Museum, Sorø 10281
Soros Center for Contemporary Art, Almaty 29677
Sørvágs Bygdasavn, Sørvágur 10697
Sosa-Carillo-Fremont House Museum, Tucson 53596
Sosnovoborskij Chudožestvennyj Muzei Sovremennogo Iskusstva, Sosnovyj Bor 37791
Sosnowieckie Centrum Sztuki - Zamek Sielecki, Sosnowiec 35124
Sóstói Múzeumfalu, Sóstógyógyfürdő 23824
Sotamuseo, Helsinki 10818
Sotavallan Tuulimylly, Lempäälä 11060
Sotavalta Windmill, Lempäälä 11060
Sotilaslääketieteen Museo, Lahti 11035
Sotkamon Makasiinimuseo, Sotkamo 11327
Sotterley Plantation Museum, Hollywood 49323
Sotterranei dell'Arte, Monte Carasso 41634
Soun Museum, Ashikaga 28374
Sound of the Century Museum, Macau 08007
Sound School House Museum, Mount Desert 50875
Sound Visitor Centre, Port Saint Mary 45414
Source of Life Monastery, Poros 23157
El Sourdog Hex, Berlin 17256
Souter Johnnie's Cottage, Kirkoswald 44614
Souterroscope des Ardoisières, Caumont-l'Eventé 12326
South African Jewish Museum, Gardens 38622
South African Mint Museum-Coin World, Tshwane 38757
South African Museum, Cape Town 38592
South African Museums Association (SAMA), Centrahil 55018
South African National Gallery, Cape Town 38593
South African National Museum of Military History, Johannesburg 38664
South African National Railway & Steam Museum, Auckland Park 38553
South African Police Service Museum, Tshwane 38758
South Arkansas Arts Center, El Dorado 48310
South Australian Aviation Museum, Port Adelaide 01424
South Australian Maritime Museum, Port Adelaide 01425
South Australian Museum, Adelaide 00782
South Australian Police Museum, Thebarton 01583
South Bannock County Historical Center, Lava Hot Springs 49989
South Bend Regional Museum of Art, South Bend 53159
South-Bohemian Motorbike Museum, České Budějovice 09463
South-Bohemian Museum in České Budějovice, České Budějovice 09464
South Boston-Halifax County Museum of Fine Arts and History, South Boston 53165
South Canterbury Museum, Timaru 33212
South Cariboo Historical Museum, Clinton 05624
South Carolina Artisans Center, Walterboro 53807
South Carolina Confederate Relic Room and Museum, Columbia 47731
South Carolina Cotton Museum, Bishopville 46834
South Carolina Federation of Museums, Columbia 55181
South Carolina Historical Society Museum, Charleston 47376
South Carolina Law Enforcement Officers Hall of Fame, Columbia 47732
South Carolina Military Museum, Columbia 47733
South Carolina National Guard Museum, Sumter 53365
South Carolina State Museum, Columbia 47734
South Carolina Tennis Hall of Fame, Belton 46726
South Central Museum Association, Wood Mountain 54748
South Charleston Museum, South Charleston 53166
South Coast Railroad Museum, Goleta 48965
South County Museum, Narragansett 50955
South Dakota Amateur Baseball Hall of Fame, Lake Norden 49888
South Dakota Art Museum, Brookings 47059
South Dakota Discovery Center, Pierre 51916
South Dakota Hall of Fame, Chamberlain 47330
South Devon Railway, Buckfastleigh 43660
South-East Armed Force Museum, Ho Chi Minh City 54627
South Eastern Federation of Museums and Art Galleries, Bromley 55097
South Florida Museum, Bradenton 46987
South Florida Museum of Natural History, Dania Beach 47955
South Florida Science Museum, West Palm Beach 54075
South Grey Museum and Historical Library, Flesherton 05799
South Hero Bicentennial Museum, South Hero 53173
South Hill Park Arts Centre, Bracknell 43574
South Holland Historical Museum, South Holland 53174
South Karelia Artmuseum, Lappeenranta 11046
South Karelia Museum, Lappeenranta 11045
South Kazakh Historical and Regional Museum, Čimkent 29698
South Lake Tahoe Museum, South Lake Tahoe 53175
South London Gallery, London 44988
South Midlands Museums Federation, King's Lynn 55098
South Milwaukee Historical Society Museum, South Milwaukee 53176
South Molton and District Museum, South Molton 45693
South Moravian Museum, Znojmo 09947
South Notts Hussars Museum, Bulwell 43670
South Otago Historical Museum, Balclutha 33062
South Park City Museum, Fairplay 48491
South Pass City State Historic Site, South Pass City 53178
South Peace Centennial Museum, Beaverlodge 05408
South Peace Historical Museum, Dawson Creek 05685
South Ribble Museum and Exhibition Centre, Leyland 44702
South River Meeting House, Lynchburg 50330
South Sask Photo Museum, Arcola 05350
South Shields Museum and Art Gallery, South Shields 45698
South Shore Art Center, Cohasset 47678
South Similkameen Museum, Keremeos 06064
The South Street Seaport Museum, New York 51260
South Sutton Old Store Museum, South Sutton 53182
South Taranaki District Museum, Patea 33180
South Texas Museum, Alice 46218
South Tipperary County Museum, Clonmel 24763
South Tynedale Railway Preservation Society, Alston 43290
South Wales Miner's Museum, Port Talbot 45418
South Wales Police Museum, Bridgend 43602
South West Museums, Libraries and Archives Council, Taunton 55099
South Western Federation of Museums and Art Galleries, Penzance 55100
South Wood County Historical Museum, Wisconsin Rapids 54265
South Yorkhire Transport Museum, Aldwarke Rotherham 43274
Southampton Art Gallery, Southampton 06998
Southampton City Art Gallery, Southampton 45706
Southampton Historical Museum, Southampton 53186
Southampton Maritime Museum, Southampton 45707
Southchurch Hall Museum, Southend-on-Sea 45713
Southeast Alaska Indian Cultural Center, Sitka 53127
Southeast Arts Center, Atlanta 46452
Southeast Missouri State University Museum, Cape Girardeau 47234
Southeast Museum, Brewster 47010
Southeast Museum of Photography, Daytona Beach 47995
Southeastern Center for Contemporary Art, Winston-Salem 54251
Southeastern Museums Conference, Baton Rouge 55182
Southeastern Railway Museum, Duluth 48192
Southend Central Museum, Southend-on-Sea 45714
Southend Pier Museum, Southend-on-Sea 45715
Southern African Development Community Association of Museums and Monuments, Harare 55202
Southern Alberta Art Gallery, Lethbridge 06163
Southern Alleghenies Museum of Art, Altoona 46243
Southern Alleghenies Museum of Art, Loretto 50201
Southern Alleghenies Museum of Art at Johnstown, Johnstown 49642
Southern Arts Federation, Atlanta 46453
Southern Exposure Gallery, San Francisco 52796
Southern Forest World Museum, Waycross 54003
Southern Highland Craft Guild at the Folk Art Center, Asheville 46376
Southern Museum of Flight, Birmingham 46825
Southern Newfoundland Seamen's Museum, Grand Bank 05887
Southern Ohio Museum and Cultural Center, Portsmouth 52103
Southern Oregon Historical Society Museum, Medford 50552
Southern Oregon University Museum of Vertebrate Natural History, Ashland 46389
Southern Ostrobothnia Museum, Seinäjoki 11304
Southern Plains Indian Museum, Anadarko 46282
Southern Troms Museum - The Department of Skånland Museum, Evenskjer 33484
Southern Ute Museum and Cultural Center, Ignacio 49466
Southern Vermont Art Center, Manchester 50428
Southern Vermont College Art Gallery, Bennington 46740
Southern Vermont Natural History Museum, Marlboro 50496
The Southland Art Collection, Dallas 47940
Southland Fire Service Museum, Invercargill 33129
Southland Museum and Art Gallery, Invercargill 33130
Southold Historical Society Museum, Southold 53188
Southold Indian Museum, Southold 53189
Southsea Castle, Portsmouth 45438
Southside House, London 44989
Southward Museum, Paraparaumu 33179
Southwest Florida Museum of History, Fort Myers 48664
Southwest Graduate University Museum, Fukuoka 28436
Southwest Museum of Science and Technology, Dallas 47941
Southwest Museum of the American Indian, Los Angeles 50265
Southwest Virginia Museum, Big Stone Gap 46800
Southwestern Häme Museum, Forssa 10737
Southwestern Saskatchewan Oldtimer's Museum, Maple Creek 06234
Southwestern University Museum, Cebu 34376
Southwestern Utah Art Gallery, Saint George 52547
Southwick Hall, Peterborough 45370
Southwold Museum, Southwold 45720
Sovereign Hill, Ballarat 00829
Sovietinių Skulptūrų Grūto Parkas, Druskininkai 30225
Sovremennaja Galereja-muzej Nikor, Moskva 37406
Spaarnestad Fotoarchief, Haarlem 32301
Space 101 Gallery, Pittsburgh 51962
Space Center Houston, Houston 49406
Space Expo, Noordwijk, Zuid-Holland 32604
Space Farms Zoological Park and Museum, Sussex 53386
A Space Gallery, Toronto 07139
Space One Eleven, Birmingham 46826
Space Science Museum, Ikoma 28565
Spaces, Cleveland 47640
Spadina Museum, Toronto 07140
Spakenburgs Museum 't Vurhuus, Bunschoten Spakenburg 32021
Spalding Gentlemen's Society Museum, Spalding 45723
SPAM Museum, Austin 46499
Spanish Citrus Historical Museum, Burriana 39076
Spanish Government Museum, Maastricht 32527
Spanish Governor's Palace, San Antonio 52727
The Spanish Institute, New York 51261
The Spanish Quarter Museum, Saint Augustine 52528
Spanish Village, Palma de Mallorca 39794
Sparebankmuseet, Oslo 33750
Sparekassemuseet, Korsør 10169
Spark Gallery, Denver 48090
Sparkassen-Museum, Greding 18743
Sparkassen-Museum der Erste Bank AG, Wien 03076
Sparkassenmuseum Leipzig, Leipzig 19935
Sparresholm Gods og Vognsamling, Holme-Olstrup 10088
Spartanburg County Museum of Art, Spartanburg 53196
Spartanburg Science Center, Spartanburg 53197
Spårvägsmuseet, Stockholm 40936
Spathareion Museum of the Shadow Theatre, Maroussi 23083
Spazio Oberdan, Milano 26724
Spectacular Folk Cultural Collection Museum, Longtan 42254
Speed Art Museum, Louisville 50297
Speelgoedmuseum De Brug, Eerbeek 32177
Speelgoedmuseum De Kijkdoos, Hoorn 32412
Speelgoedmuseum Deventer, Deventer 32121
Speelgoedmuseum Mechelen, Mechelen 03860
Speelgoedmuseum Op Stelten, Oosterhout 32629
Speicherstadtmuseum, Hamburg 18959
Speke Hall, Liverpool 44744
Speldjesmuseum, Klaaswaal 32447
Spellman Museum of Stamps and Postal History, Weston 54107
Spelman College Museum of Fine Art, Atlanta 46454
Spelthorne Museum, Staines 45732
Spencer Entomological Museum, Vancouver 07223
Spencer House, London 44990
Spencer Museum of Art, Lawrence 49992
Spencer-Peirce-Little Farm, Newbury 51303
Spengler-Museum, Sangerhausen 21485
Sperm Whale Museum, Tainan 42368
Sperrgebiet Museum, Oranjemund 31752
Sperrin Heritage Centre, Plumridge 45389
Spertus Museum, Chicago 47514
Spessartmuseum, Lohr am Main 20027
Spetses Museum, Spetses 23198
Spichlerz, Olsztyn 34979
Spicy's Gewürzmuseum, Hamburg 18960
Spicy's Spice Museum, Hamburg 18960
spiegel, München 20484
Spielkartensammlung Piatnik, Wien 03077
Spielmuseum Soltau und Stiftung Spiel, Soltau 21815
Spielzeug anno dazumal - Museum Beilngries, Beilngries 16984
Spielzeug-Museum, Aeugstertal 41091
Spielzeugmuseum, Alsfeld 16482
Spielzeugmuseum, Bad Lauterberg 16776
Spielzeugmuseum, Davos Platz 41322
Spielzeugmuseum, Görlitz 18682
Spielzeugmuseum, Hemau 19126
Spielzeugmuseum, Hinterzarten 19196
Spielzeugmuseum, Schkeuditz 21560
Spielzeugmuseum, Wittmund 22627
Spielzeugmuseum, Wolfurt 03121
Spielzeugmuseum, Dorf- und Rebbaumuseum, Riehen 41766
Spielzeugmuseum im Alten Rathausturm, München 20485
Spielzeugmuseum im Alten Schloß, Sugenheim 21975
Spielzeugmuseum im Stadtmuseum, Ingolstadt 19345
Spielzeugmuseum - Museum Lydia Bayer, Nürnberg 20790
Spielzeugmuseum Salzburg, Salzburg 02627
Spielzeugmuseum - Sammlung Depuoz, Zürich 42106
Spielzeugmuseum Solz, Bebra 16969
Spielzeugmuseum Trier, Trier 22103
Spiers Gallery, Brevard 47007
Spillemands-Jagt og Skovbrugsmuseet i Rebild, Skørping 10272
Spindletop and Gladys City Boomtown Museum, Beaumont 46679
Het Spinozahuis, Rijnsburg 32688
Spirit of '76 Museum, Elyria 48387
Spirit of '76 Museum, Wellington 54028
Spirit River & District Museum, Spirit River 07000
Spiro Mounds Archaeological Center, Spiro 53204
Spišské Múzeum, Levoča 38284
Spišský Hrad, Spišské Podhradie 38360
Spitäle an der Alten Mainkirche, Würzburg 22725
Spital Hengersberg, Markt Hengersberg 20213
Spitfire and Hurricane Memorial Museum, Ramsgate 45459
Splavarski Muzej, Javnik 38424
SPLIA Gallery, Cold Spring Harbor 47683
Split Archaeological Museum, Split 08954
Split Ethnographical Museum, Split 08955
Split Rock Lighthouse, Two Harbors 53631
Spode Museum, Stoke-on-Trent 45763
Spokane House Interpretive Center, Nine Mile Falls 51353
Spoke Wheel Car Museum, Charlottetown 05598
Społeczne Muzeum Konstantego Laszczki, Dobre 34624
Spomen Kuća Bitke na Sutjesci, Tjentište 04285
Spomen Kuća Laze K. Lazarevića, Beograd 38088
Spomen Muzej Biskupa Josipa Jurja Strossmayera, Đakovo 08858
Spomen Muzej Ive Andrića, Beograd 38089

S

Spomen Park Danica, Koprivnica 08893
Spomen-Park Kragujevački Oktobar, Kragujevac 38110
Spomen-Zbirka Pavla Beljanskog, Novi Sad 38125
Spominska Soba Dragotina Ketteja na Premu, Prem 38498
Spominska Soba Giuseppe Tartini, Piran 38493
Spominska Soba Ivana Cankarja, Ljubljana 38466
Spominska Soba Jurija Vege, Dol pri Ljubljani 38414
Spominska Zbirka Pisatelja Ivana Cankarja, Vrhnika 38535
Spominski Muzej Prežihovega Voranca, Ravne na Koroškem 38504
Spontanart Artforum, Courlevon 41316
Spoor- en Tramweg Verzameling, Loppersum 32508
Spoorwegmuseum van Miniaturmodellen, Heist-op-den-Berg 03693
Sport Museum at Luzhniki, Moskva 37362
Sportmuseum Berlin, Berlin 17257
Sportmuseum der Stadt Frankfurt/Oder, Frankfurt (Oder) 18407
Sportmuseum Leipzig, Leipzig 19936
Sportmuseum Schweiz, Basel 41185
Sportmuseum Vlaanderen, Hofstade 03708
Sportmúzeum, Kaposvár 23636
Sports Car Museum of Japan, Gotemba 28459
Sports Museum Borås, Borås 40467
Sports Museum of Finland, Helsinki 10822
Sports Museum of Latvia, Riga 30085
The Sports Museum of New England, Boston 46951
Sports Museum Pierre de Coubertin, La Falda 00402
Sportsmuseum, Stockholm 40930
Sportsmuseum Leipzig, Leipzig 19936
Sporveismuseet Vognhall 5, Oslo 33751
Sporvejsmuseet Skjoldenæsholm, Jystrup Midtsj 10112
Spotsylvania Historical Association and Museum, Spotsylvania 53208
Spreewald-Museum Lübbenau, Lübbenau 20071
Sprengel Museum Hannover, Hannover 19005
Spring Mill State Park Pioneer Village and Grissom Memorial, Mitchell 50760
Spring Museum, Yangchun 08421
Spring Street Historical Museum, Shreveport 53090
Spring Valley Community Museums, Spring Valley 53210
Springbank Visual Arts Centre, Mississauga 06313
Springfield Armory Museum, Springfield 53228
Springfield Art Gallery, Springfield 53221
Springfield Art Museum, Springfield 53233
Springfield Historical Society Museum, Springfield 53234
Springfield Museum, Springfield 53238
Springfield Museum of Art, Springfield 53236
Springfield Museum of Old Domestic Life, High Point 49276
Springfield Science Museum, Springfield 53229
Springhill Costume Museum, Moneymore 45165
Springhill House, Magherafelt 45071
Springhill Miner's Museum, Springhill 07002
Springs Museum, Springs 53241
Springvale and District Historical Museum, Springvale 01525
Springville Museum of Art, Springville 53243
Spruce Lane Farm, Oakville 06498
Spruce Row Museum, Waterford 07283
Spruce Woods Provincial Heritage Park Museum, Carberry 05557
Spruill Gallery, Atlanta 46455
Spurlock Museum, Urbana 53669
Spurn Lightship, Kingston-upon-Hull 44599
Spy Museum, Tampere 11368
Spydeberg Bygdetun, Spydeberg 33859
Squam Lakes Natural Science Center, Holderness 49311
Squamish Valley Museum, Garibaldi Heights 05851
Squerryes Court, Westerham 45979
Srinakharinwirot Art Exhibition Hall, Bangkok 42562
SS Great Britain Museum, Bristol 43639
St. Johann Baptist, Aachen 16407
St John's House, Warwick 45939
St. Stefan University Agricultural Tool and Machine Development Museum, Gödöllő 23568
Staatliche Antikensammlungen und Glyptothek, München 20486
Staatliche Bücher- und Kupferstichsammlung Greiz, Greiz 18753
Staatliche Graphische Sammlung München, München 20487
Staatliche Kunsthalle Baden-Baden, Baden-Baden 16890
Staatliche Kunsthalle Karlsruhe, Karlsruhe 19448
Staatliche Kunstsammlungen Dresden, Dresden 17971
Staatliche Münzsammlung, München 20488
Staatliche Museen zu Berlin - Stiftung Preußischer Kulturbesitz, Berlin 54858
Staatliche Museumsberatung für Nordhessen, Kassel 19477
Staatlicher Schloßbetrieb Schloß Nossen / Kloster Altzella, Nossen 20748
Staatliches Museum Ägyptischer Kunst, München 20489
Staatliches Museum für Naturkunde Karlsruhe, Karlsruhe 19449
Staatliches Museum für Naturkunde Stuttgart, Stuttgart 21959
Staatliches Museum für Völkerkunde München, München 20490
Staatliches Naturhistorisches Museum, Braunschweig 17529
Staats- und Stadtbibliothek, Augsburg 16635
Staatsarchiv Bamberg, Bamberg 16918
Staatsbibliothek Bamberg, Bamberg 16919
Staatsbibliothek zu Berlin, Berlin 17258
Staatsburgh State Historic Site, Staatsburg 53245
Staatsgalerie im Hohen Schloß, Füssen 18502
Staatsgalerie im Neuen Schloß, Oberschleißheim 20838
Staatsgalerie in der Benediktiner-Abtei, Ottobeuren 20956
Staatsgalerie in der Burg, Burghausen 17655
Staatsgalerie in der Katharinenkirche, Augsburg 16636
Staatsgalerie in der Neuen Residenz, Bamberg 16920
Staatsgalerie in der Residenz, Ansbach 16562
Staatsgalerie in der Residenz Würzburg, Würzburg 22726
Staatsgalerie Moderne Kunst, Augsburg 16637
Staatsgalerie Neuburg - Flämische Barockmalerei, Neuburg an der Donau 20600
Staatsgalerie Stuttgart, Stuttgart 21960
Staatsopernmuseum, Wien 03078
Staatspark Fürstenlager, Bensheim 16998
Staatspark Wilhelmsbad, Hanau 18983
Stabbursnes Naturhus og Museum, Indre Billefjord 33576
Stabler-Leadbeater Apothecary Museum, Alexandria 46213
Stables Art Gallery of Taos Art Association, Taos 53443
Stacy's Tavern Museum, Glen Ellyn 48921
Stadhuis, Gent 03642
Stadhuis van Antwerpen, Antwerpen 03316
Stadhuismuseum, Zierikzee 33005
Stadsboerderij Het Wevershuisje, Almelo 31800
Stadsgalerie Engels Tuin, Ootmarsum 32635
Stadsgalerie Gouda, Gouda 32270
Stadsgalerij Heerlen - Museum voor Moderne en Hedendaagse Kunst, Heerlen 32338
Stadsmuseet och Ahlbergshallen, Östersund 40798
Stadsmuseum Doetinchem, Doetinchem 32138
Stadsmuseum Groenlo, Groenlo 32275
Stadsmuseum i Gråbrödraklostret, Ystad 41077
Stadsmuseum IJsselstein, IJsselstein 32428
Stadsmuseum STAM, Gent 03643
Stadsmuseum 't Schippershof, Menen 03868
Stadsmuseum Woerden, Woerden 32959
Stadsmuseum Zoetermeer, Zoetermeer 33007
Stadt-Museum, Weil der Stadt 22381
Stadt- und Bädermuseum, Bad Salzuflen 16822
Stadt- und Bädermuseum Bad Doberan, Bad Doberan 16710
Stadt-und Bergbaumuseum der Stadt Staßfurt, Staßfurt 21866
Stadt- und Bergbaumuseum Freiberg, Freiberg, Sachsen 18425
Stadt- und Brauereimuseum Pritzwalk, Pritzwalk 21128
Stadt- und Burgmuseum, Eppstein 18186
Stadt- und Dampfmaschinenmuseum, Werdau 22485
Stadt- und Fachwerkmuseum Alte Universität, Eppingen 18183
Stadt- und Festungsmuseum im Ludwigstor, Germersheim 18605
Stadt- und Freilichtmuseum Sinsheim, Sinsheim 21786
Stadt- und Heimatmuseum, Heidenreichstein 02063
Stadt- und Heimatmuseum, Malchow 20156
Stadt- und Heimatmuseum der Stadt Marl, Marl, Westfalen 20225
Stadt- und Heimatmuseum Kusel, Kusel 19752
Stadt- und Heimatmuseum mit Galerie, Waischenfeld 22279
Stadt- und Hochstiftmuseum, Dillingen an der Donau 17863
Stadt- und Industriemuseum, Wetzlar 22533
Stadt- und Industriemuseum Rüsselsheim, Rüsselsheim 21438
Stadt- und Kreisgeschichtliches Museum, Hünfeld 19290
Stadt- und Kulturgeschichtliches Museum Torgau, Torgau 22073
Stadt- und Landesbibliothek Dortmund, Artothek, Dortmund 17927
Stadt- und Manfred-Kyber-Museum, Löwenstein 20020
Stadt- und Multschermuseum, Vipiteno 28288
Stadt- und Regionalmuseum Lübben, Lübben 20069
Stadt- und Schulmuseum, Ronneburg, Thüringen 21369
Stadt- und Turmmuseum, Bad Camberg 16708
Stadt-und Wagenmuseum Oschatz, Oschatz 20918
Stadt- und Wallfahrtsmuseum, Walldürn 22312
Stadtarchiv mit städtischen Sammlungen, Lauf an der Pegnitz 19852
Stadtarchiv und Ehemals Reichsstädtische Bibliothek Lindau, Lindau 20000
Stadtbibliothek Reutlingen, Reutlingen 21287
Stadtbibliothek Trier - Schatzkammer und Ausstellung, Trier 22104
Stadtgalerie, Osnabrück 20927
Stadtgalerie Alte Post, Sylt 21994
Stadtgalerie Altena, Altena 16489
Stadtgalerie Altötting, Altötting 16520
Stadtgalerie am Minoritenplatz, Wolfsberg 03119
Stadtgalerie Bamberg - Villa Dessauer, Bamberg 16921
Stadtgalerie Bern, Bern 41219
Stadtgalerie im alten Herrenhaus, Ternitz 02804
Stadtgalerie im Elbeforum Brunsbüttel, Brunsbüttel 17612
Stadtgalerie Kiel, Kiel 19534
Stadtgalerie Klagenfurt, Klagenfurt 02192
Stadtgalerie Lünen, Lünen 20104
Stadtgalerie Saarbrücken, Saarbrücken 21460
Stadtgalerie Schwaz, Schwaz 02728
Stadtgeschichtliche Sammlung, Gaggenau 18528
Stadtgeschichtliche Sammlung, Velbert 22212
Stadtgeschichtliches Museum, Kehl 19498
Stadtgeschichtliches Museum, Ottweiler 20962
Stadtgeschichtliches Museum, Pfullingen 21034
Stadtgeschichtliches Museum, Tangermünde 22003
Stadtgeschichtliches Museum, Waren (Müritz) 22336
Stadtgeschichtliches Museum im Landrichterhaus, Karlstadt 19455
Stadtgeschichtliches Museum Leipzig, Leipzig 19937
Stadtgeschichtliches Museum Leipzig - Altes Rathaus, Leipzig 19938
Stadtgeschichtliches Museum Leipzig - Neubau, Leipzig 19939
Stadtgeschichtliches Museum Schabbellhaus, Wismar 22613
Stadtgeschichtliches Museum Spandau, Berlin 17259
Stadtgeschichts- und Schradenmuseum, Ortrand 20917
Stadthaus am Dom, Wetzlar 22534
Stadthaus-Galerie, Münster 20518
Stadthaus Olten, Olten 41709
Stadthaus Ulm, Ulm 22169
Stadthistorisches Museum Flohburg, Nordhausen 20739
Stadthistorisches Sole, Salz und Kali-Bergbaumuseum, Bad Salzdetfurth 16820
Stadtmauermuseum Nördlingen, Nördlingen 20726
Stadtmuseum, Bad Bergzabern 16687
Stadtmuseum, Gmünd, Niederösterreich 01947
Stadtmuseum, Hornberg 19275
Stadtmuseum, Jena 19392
Stadtmuseum, Mattersburg 02365
Stadtmuseum, Murau 02418
Stadtmuseum, Neuburg an der Donau 20601
Stadtmuseum, Riesa 21330
Stadtmuseum, Stadtschlaining 02757
Stadtmuseum, Traiskirchen 02817
Stadtmuseum, Waidhofen an der Thaya 02886
Stadtmuseum Aarau, Aarau 41085
Stadtmuseum Aichach, Aichach 16445
Stadtmuseum "Alte Burg" Wittenberge, Wittenberge 22621
Stadtmuseum Alte Hofmühle, Hollabrunn 02097
Stadtmuseum Alte Post, Ebersbach an der Fils 18042
Stadtmuseum Amberg, Amberg 16533
Stadtmuseum Amtsturm, Lübz 20085
Stadtmuseum Andernach, Andernach 16545
Stadtmuseum Apolda, Apolda 16567
Stadtmuseum Arelape-Bechelaren-Pöchlarn, Pöchlarn 02506
Stadtmuseum Bad Cannstatt, Stuttgart 21961
Stadtmuseum Bad Soden, Bad Soden am Taunus 16841
Stadtmuseum Bad Wildungen, Bad Wildungen 16864
Stadtmuseum Baden-Baden, Baden-Baden 16891
Stadtmuseum Battenberg, Battenberg 16934
Stadtmuseum Baunatal, Baunatal 16938
Stadtmuseum Beckum, Beckum 16974
Stadtmuseum Bergen, Bergen auf Rügen 17007
Stadtmuseum Bergkamen, Bergkamen 17021
Stadtmuseum Bocholt, Bocholt 17396
StadtMuseum Bonn, Bonn 17467
Stadtmuseum Borken, Westfalen, Borken, Westfalen 17479
Stadtmuseum Brakel, Brakel 17499
Stadtmuseum Breuberg-Neustadt, Breuberg 17588
Stadtmuseum Brilon, Brilon 17590
Stadtmuseum Bruck an der Leitha, Bruck an der Leitha 01803
Stadtmuseum Bruck an der Mur, Bruck an der Mur 01804
Stadtmuseum Bruck an der Mur, Bruck an der Mur 01805
Stadtmuseum Burgdorf, Burgdorf 17649
Stadtmuseum Burghausen, Burghausen 17656
Stadtmuseum Burgstädt, Burgstädt 17664
Stadtmuseum Coesfeld, Coesfeld 17723
Stadtmuseum Cottbus, Cottbus 17735
Stadtmuseum Cuxhaven, Cuxhaven 17757
Stadtmuseum Damme, Damme 17778
Stadtmuseum Deggendorf, Deggendorf 17806
Stadtmuseum Ditzingen, Ditzingen 17878
Stadtmuseum Döbeln/Kleine Galerie, Döbeln 17882
Stadtmuseum Dornbirn, Dornbirn 01829
Stadtmuseum Dresden, Dresden 17972
Stadtmuseum Eilenburg, Eilenburg 18089
Stadtmuseum Einbeck mit Radhaus im Stadtmuseum, Einbeck 18092
Stadtmuseum Eisenberg, Eisenberg, Saale-Holzland-Kreis 18103
Stadtmuseum Erlangen, Erlangen 18225
Stadtmuseum Eupen, Eupen 03593
Stadtmuseum Euskirchen, Euskirchen 18284
Stadtmuseum Fellbach, Fellbach 18307
Stadtmuseum Fembohaus, Nürnberg 20791
Stadtmuseum Freilassing, Freilassing 18445
Stadtmuseum Friesach, Friesach 01907
Stadtmuseum Fürstenfeldbruck, Fürstenfeldbruck 18493
Stadtmuseum Gardelegen, Gardelegen 18538
Stadtmuseum Gehrden, Gehrden bei Hannover 18552
Stadtmuseum Geithain, Geithain 18564
Stadtmuseum Gera, Gera 18598
Stadtmuseum Gerlingen - Gerlinger Heimatmuseum - Museum der Deutschen aus Ungarn, Gerlingen 18603
Stadtmuseum Gerolzhofen, Gerolzhofen 18612
Stadtmuseum Giengen, Giengen an der Brenz 18633
Stadtmuseum Grafenau, Grafenau, Kreis Freyung-Grafenau 18729
Stadtmuseum Graz, Graz 01983
Stadtmuseum Groß-Gerau, Groß-Gerau 18776
Stadtmuseum Gütersloh, Gütersloh 18833
Stadtmuseum Gunzenhausen, Gunzenhausen 18841
Stadtmuseum Hadamar, Hadamar 18850
Stadtmuseum Hall in Tirol, Hall in Tirol 02044
Stadtmuseum Hartberg, Hartberg 02053
Stadtmuseum Hattingen, Hattingen 19041
Stadtmuseum Haus Kupferhammer, Warstein 22341
Stadtmuseum Herrenmühle, Hammelburg 18973
Stadtmuseum Herzogenaurach, Herzogenaurach 19168
Stadtmuseum Hildburghausen, Hildburghausen 19179
Stadtmuseum Hofgeismar, Hofgeismar 19224
Stadtmuseum Hofheim am Taunus, Hofheim am Taunus 19226
Stadtmuseum Hohe Lilie, Naumburg, Saale 20556
Stadtmuseum Holzminden, Holzminden 19261
Stadtmuseum Horb, Horb am Neckar 19272
Stadtmuseum Hornmoldhaus, Bietigheim-Bissingen 17349
Stadtmuseum im Alten Forstamt, Stockach 21904
Stadtmuseum im Augustinerkloster Bad Langensalza, Bad Langensalza 16773
Stadtmuseum im Gelben Haus, Esslingen 18277
Stadtmuseum im Historischen Rathaus, Staufen im Breisgau 21868
Stadtmuseum im Kammerhof, Eisenerz 01855
Stadtmuseum im Knochenhauer-Amtshaus, Hildesheim 19186
Stadtmuseum im Schlössle, Weingarten, Kreis Ravensburg 22429
Stadtmuseum im Schloss, Münsingen 20502
Stadtmuseum im Spital, Crailsheim 17738
Stadtmuseum Ingolstadt im Kavalier Hepp, Ingolstadt 19346
Stadtmuseum Innsbruck, Innsbruck 02131
Stadtmuseum Iserlohn, Iserlohn 19360
Stadtmuseum Judenburg, Judenburg 02146
Stadtmuseum Karlsruhe, Karlsruhe 19450
Stadtmuseum Kassel, Kassel 19478
Stadtmuseum Kaufbeuren, Kaufbeuren 19492
Stadtmuseum Kelsterbach, Kelsterbach 19505
Stadtmuseum Killingerhaus, Idstein 19311
Stadtmuseum Klostermühle, Bad Urach 16854
Stadtmuseum Klosterneuburg, Klosterneuburg 02204
Stadtmuseum Kremayr, Rust 02602
Stadtmuseum Kröpelin, Kröpelin 19717
Stadtmuseum Landeshauptstadt Düsseldorf, Düsseldorf 18006
Stadtmuseum Leimen - Sankt Ilgen, Leimen 19878
Stadtmuseum Leonberg, Leonberg 19963
Stadtmuseum Leun, Leun 19967
Stadtmuseum Lichtenfels, Lichtenfels 19981
Stadtmuseum Lindau, Lindau 20001
Stadtmuseum Ludwig Erhard, Fürth 18500

S

Stadtmuseum Ludwigshafen, Ludwigshafen am Rhein 20057
Stadtmuseum Meersburg, Meersburg 20251
Stadtmuseum Meißen, Meißen 20267
Stadtmuseum Melk, Melk 02376
Stadtmuseum Meppen, Meppen 20287
Stadtmuseum mit Weinmuseum, Bad Vöslau 01769
Stadtmuseum Mosbach, Mosbach 20392
Stadtmuseum Mühlberg, Elbe, Mühlberg 20398
Stadtmuseum Mühlheim, Mühlheim am Main 20407
Stadtmuseum Mülheim-Kärlich, Mülheim-Kärlich 20416
Stadtmuseum Münster, Münster 20519
Stadtmuseum Mutzschen, Mutzschen 20539
Stadtmuseum Neumarkt, Neumarkt in der Oberpfalz 20634
Stadtmuseum Neuötting, Neuötting 20643
Stadtmuseum Neustadt an der Waldnaab, Neustadt an der Waldnaab 20668
Stadtmuseum Nittenau, Nittenau 20723
Stadtmuseum Nördlingen, Nördlingen 20727
Stadtmuseum Norderstedt, Norderstedt 20735
Stadtmuseum Nürtingen, Nürtingen 20795
Stadtmuseum Obermühle, Braunfels 17515
Stadtmuseum Oldenburg, Oldenburg 20900
Stadtmuseum Penzberg, Penzberg 21008
Stadtmuseum Pforzheim, Pforzheim 21030
Stadtmuseum Pinneberg, Pinneberg 21039
Stadtmuseum Pirna, Pirna 21046
Stadtmuseum Pößneck, Pößneck 21071
Stadtmuseum Povelturm, Nordhorn 20743
Stadtmuseum Pulsnitz, Pulsnitz 21139
Stadtmuseum Purkersdorf, Purkersdorf 02542
Stadtmuseum Quakenbrück, Quakenbrück 21144
Stadtmuseum Rastatt, Rastatt 21186
Stadtmuseum Rheinfelden, Rheinfelden 21301
Stadtmuseum Rothenburg, Rothenburg/Oberlausitz 21415
Stadtmuseum Rottweil, Rottweil 21429
Stadtmuseum Saalfeld im Franziskanerkloster, Saalfeld/Saale 21452
Stadtmuseum Sachsenheim, Sachsenheim 21466
Stadtmuseum Sankt Pölten, Sankt Pölten 02686
Stadtmuseum Schelklingen, Schelklingen 21541
Stadtmuseum Schladming, Schladming 02706
Stadtmuseum Schleswig, Schleswig 21568
Stadtmuseum Schloß Hoyerswerda, Hoyerswerda 19282
Stadtmuseum Schloss Wolfsburg, Wolfsburg 22667
Stadtmuseum Schlüsselfeld, Schlüsselfeld 21578
Stadtmuseum Schongau, Schongau 21620
Stadtmuseum Schorndorf, Schorndorf 21627
Stadtmuseum Schramberg, Schramberg 21634
Stadtmuseum Schwabach, Schwabach 21645
Stadtmuseum Schwandorf mit Falkenauer Heimatstube, Schwandorf 21665
Stadtmuseum Schwedt, Schwedt / Oder 21678
Stadtmuseum Siegburg, Siegburg 21752
Stadtmuseum Simeonstift Trier, Trier 22105
Stadtmuseum Sindelfingen, Sindelfingen 21778
Stadtmuseum Stadtoldendorf, Stadtoldendorf 21859
Stadtmuseum Steinfurt, Steinfurt 21887
Stadtmuseum Stolpen, Stolpen 21912
Stadtmuseum Sulzbach-Rosenberg, Sulzbach-Rosenberg 21985
Stadtmuseum Teterow, Teterow 22037
Stadtmuseum Traismauer, Traismauer 02820
Stadtmuseum Trostberg, Trostberg 22111
Stadtmuseum Tübingen, Tübingen 22124
Stadtmuseum und Fürstlich Starhembergisches Familienmuseum, Eferding 01841
Stadtmuseum und Tuchmachermuseum Pinkafeld, Pinkafeld 02499
Stadtmuseum Vogtturm, Zell am See 03127
Stadtmuseum Waldkraiburg, Waldkraiburg 22300
Stadtmuseum Wasungen, Wasungen 22350
Stadtmuseum Wehr, Wehr 22361
Stadtmuseum Weilheim, Weilheim in Oberbayern 22392
Stadtmuseum Weimar, Weimar 22421
Stadtmuseum Wendlingen, Wendlingen am Neckar 22478
Stadtmuseum Werne, Werne 22494
Stadtmuseum Wiener Neustadt, Wiener Neustadt 03099
Stadtmuseum Wienertor Hainburg, Hainburg an der Donau 02039
Stadtmuseum Wil, Wil 42017
Stadtmuseum Zehentstadel, Nabburg 20543
Stadtmuseum Zistersdorf, Zistersdorf 03132
Stadtmuseum Zweibrücken, Zweibrücken 22830
Stadtmuseum Zwettl, Zwettl 03136
Stadtraum Vilshofen, Vilshofen 22247
Stadtteilmuseum Nowaweser Weberstube, Potsdam 21110
Stadttormuseum, Wehrheim 22363
Stadtturmgalerie, Gmünd, Kärnten 01943
Stadtturmgalerie, Innsbruck 02132
Städel Museum, Frankfurt am Main 18393
Städtische Galerie, Bietigheim-Bissingen 17350
Städtische Galerie, Dreieich 17937
Städtische Galerie, Ettlingen 18281
Städtische Galerie, Paderborn 20975
Städtische Galerie, Schieder-Schwalenberg 21546
Städtische Galerie, Traunstein 22079
Städtische Galerie, Überlingen 22145
Städtische galerie ada Meiningen, Meiningen 20261
Städtische Galerie am Markt Wurzen, Wurzen 22752
Städtische Galerie Böblingen, Böblingen 17421
Städtische Galerie Bremen, Bremen 17562
Städtische Galerie Delmenhorst - Haus Coburg - Sammlung Stuckenberg, Delmenhorst 17817
Städtische Galerie Die Fähre, Bad Saulgau 16824
Städtische Galerie Dresden - Kunstsammlung, Dresden 17973
Städtische Galerie Eichenmüllerhaus, Lemgo 19950
Städtische Galerie Filderhalle, Leinfelden-Echterdingen 19885
Städtische Galerie Filderstadt, Filderstadt 18318
Städtische Galerie Fruchthalle Rastatt, Rastatt 21187
Städtische Galerie Grimma, Grimma 18763
Städtische Galerie Haus Seel, Siegen 21757
Städtische Galerie im Alten Rathaus, Fürstenwalde 18494
Städtische Galerie im Cordonhaus, Cham 17692
Städtische Galerie im Königin-Christinen-Haus, Zeven 22798
Städtische Galerie im Kornhaus, Kirchheim unter Teck 19551
Städtische Galerie im Kulturforum, Offenburg 20888
Städtische Galerie im Lenbachhaus und Kunstbau München, München 20491
Städtische Galerie im Park, Viersen 22232
Städtische Galerie im Rathauspark, Gladbeck 18651
Städtische Galerie im Schloßpark Strünkede, Herne 19157
Städtische Galerie im Spital zum Heiligen Geist, Riedlingen 21320
Städtische Galerie im Theater Ingolstadt, Ingolstadt 19347
Städtische Galerie in der Alten Schule, Sigmaringen 21765
Städtische Galerie In der Badstube, Wangen im Allgäu 22327
Städtische Galerie in der Reithalle, Paderborn 20976
Städtische Galerie Iserlohn, Iserlohn 19361
Städtische Galerie Kaarst, Kaarst 19409
Städtische Galerie Karlsruhe, Karlsruhe 19451
Städtische Galerie Leerer Beutel, Regensburg 21232
Städtische Galerie Lehrte, Lehrte 19875
Städtische Galerie Liebieghaus, Frankfurt am Main 18394
Städtische Galerie "Lovis-Kabinett", Villingen-Schwenningen 22240
Städtische Galerie Lüdenscheid, Lüdenscheid 20091
Städtische Galerie Mennonitenkirche, Neuwied 20689
Städtische Galerie Neunkirchen, Neunkirchen, Saarland 20642
Städtische Galerie Nordhorn - Kunstwegen, Nordhorn 20744
Städtische Galerie Ostfildern, Ostfildern 20939
Städtische Galerie Rastatt, Rastatt 21188
Städtische Galerie Reutlingen, Reutlingen 21288
Städtische Galerie Rosenheim, Rosenheim 21378
Städtische Galerie Sohle 1, Bergkamen 17022
Städtische Galerie Sonneberg, Sonneberg 21819
Städtische Galerie Stapflehus, Weil am Rhein 22378
Städtische Galerie Villa Streccius, Landau in der Pfalz 19791
Städtische Galerie Villa Zanders, Bergisch Gladbach 17020
Städtische Galerie Waldkraiburg, Waldkraiburg 22301
Städtische Galerie Wesseling, Wesseling 22513
Städtische Galerie Wolfsburg, Wolfsburg 22668
Städtische Gemäldegalerie, Füssen 18503
Städtische Kunstgalerie im Deutschordenshaus, Donauwörth 17899
Städtische Kunstgalerie Torhaus Rombergpark, Dortmund 17928
Städtische Kunsthalle, Recklinghausen 21210
Städtische Kunstsammlung, Murrhardt 20537
Städtische Kunstsammlung Schloß Salder, Salzgitter 21475
Städtische Museen Heilbronn, Heilbronn 19100
Städtische Museen Heilbronn/Archäologie, Heilbronn 19101
Städtische Museen Wangen im Allgäu, Wangen im Allgäu 22328
Städtische Sammlungen, Adelsheim 16429
Städtische Sammlungen Freital, Freital 18453
Städtische Wessenberg-Galerie Konstanz, Konstanz 19683
Städtisches Heimatmuseum, Meßkirch 20307
Städtisches Heimatmuseum Ballenstedt, Ballenstedt 16903
Städtisches Heimatmuseum Höchstadt, Höchstadt an der Aisch 19209
Städtisches Heimatmuseum im Waldemarturm, Dannenberg 17780
Städtisches Heimatmuseum Landau, Landau in der Pfalz 19792
Städtisches Heimatmuseum Lippstadt, Lippstadt 20012
Städtisches Heimatmuseum Mengen, Mengen 20284
Städtisches Heimatmuseum Reinfeld, Reinfeld 21253
Städtisches Hellweg-Museum, Geseke 18626
Städtisches Hutmuseum, Lindenberg im Allgäu 20005
Städtisches Kramer-Museum, Kempen 19509
Städtisches Kunstmuseum Singen, Singen 21781
Städtisches Kunstmuseum Spendhaus Reutlingen, Reutlingen 21289
Städtisches Lapidarium, Stuttgart 21962
Städtisches Museum, Aschersleben 16596
Städtisches Museum, Bad Reichenhall 16808
Städtisches Museum, Bruchsal 17598
Städtisches Museum, Eisenhüttenstadt 18107
Städtisches Museum, Herborn 19133
Städtisches Museum, Kalkar 19417
Städtisches Museum, Kitzingen 19561
Städtisches Museum, Menden 20281
Städtisches Museum, Pfungstadt 21036
Städtisches Museum, Salzgitter 21476
Städtisches Museum, Schärding 02696
Städtisches Museum, Überlingen 22146
Städtisches Museum, Welzheim 22472
Städtisches Museum, Werl 22491
Städtisches Museum, Wiesloch 22563
Städtisches Museum, Zeulenroda-Triebes 22794
Städtisches Museum Braunschweig, Braunschweig 17530
Städtisches Museum Engen und Galerie, Engen 18172
Städtisches Museum Fürstenwalde, Fürstenwalde 18495
Städtisches Museum Göppingen im Storchen, Göppingen 18677
Städtisches Museum Göttingen, Göttingen 18701
Städtisches Museum Grevesmühlen, Grevesmühlen 18758
Städtisches Museum Hann. Münden, Hann. Münden 18988
Städtisches Museum Haus Letmathe, Iserlohn 19362
Städtisches Museum im ehemaligen Heiliggeistspital, Munderkingen 20529
Städtisches Museum im Hospital, Nidderau 20694
Städtisches Museum im Kornhaus, Kirchheim unter Teck 19552
Städtisches Museum im Kornhaus Bad Waldsee, Bad Waldsee 16857
Städtisches Museum im Spiegelschen Palais, Halberstadt 18881
Städtisches Museum Kamen, Kamen 19419
Städtisches Museum Ludwigsburg, Ludwigsburg 20048
Städtisches Museum Peterskirche, Vaihingen an der Enz 22199
Städtisches Museum Rosenheim, Rosenheim 21379
Städtisches Museum Saarlouis, Saarlouis 21464
Städtisches Museum Schloss Rheydt, Mönchengladbach 20358
Städtisches Museum Seesen, Seesen 21724
Städtisches Museum Wesel, Wesel 22508
Städtisches Museum Wesel, Wesel 22509
Städtisches Museum Wesel, Wesel 22510
Städtisches Museum Zirndorf, Zirndorf 22810
Städtisches Naturkundliches Museum, Göppingen 18678
Städtisches Ortsmuseum Schwamendingen, Zürich 42107
Städtisches Schulmuseum, Lohr am Main 20028
Ständerhaus, Buus 41278
Staffanstorps Konsthall, Staffanstorp 40878
Staffin Museum, Staffin 45727
Stafford Castle and Visitor Centre, Stafford 45731
Staffordshire Regiment Museum, Lichfield 44707
Stafsjö Bruksmuseum, Falkenberg 40510
Stagecoach Inn Museum, Newbury Park 51304
Stahlpark Riedersbach, Sankt Pantaleon 02676
Stained Glass Museum, Ely 44174
Stájer-házi Erdészeti Múzeum, Kőszeg 23684
Stała Ekspozycja Muzealna Ziemi Sierakowickiej, Sierakowice 35106
Stalheim Folkemuseum, Stalheim 33861
Stalin Museum, Batumi 16337
Stamford Brewery Museum, Stamford 45736
Stamford Historical Society Museum, Stamford 53248
Stamford Museum, Stamford 45737
Stamford Museum and Nature Center, Stamford 53249
Stamp Museum, Budapest 23388
Stan Hywet Hall and Gardens, Akron 46148
Standard Bank Gallery, Johannesburg 38665
Standing Conference on Archives and Museums, Ipswich 55101
Stanislav Dospevski Memorial House, Pazardžik 05108
Stanislav Dospevsky Art Gallery, Pazardžik 05105
Stanisław Noakowski Museum, Nieszawa 34946
Stanisław Wyspiański Museum in the Szolayski House - The National Museum in Cracow, Kraków 34809
Stanisław Wyspiański Theatrical Museum, Kraków 34811
The Stanley and Gerald Rubin Center for the Visual Arts, El Paso 48329
Stanley Museum, Kingfield 49780
The Stanley Picker Gallery, Kingston-upon-Thames 44603
Stanley Spencer Gallery, Cookham 43889
Stanley-Whitman House, Farmington 48515
Stanly County Historic Museum, Albemarle 46169
Stanmer Rural Museum, Brighton 43626
Stansbury Museum, Stansbury 01526
Stansted Mountfitchet Windmill, Stansted Mountfitchet 45739
Stanthorpe Regional Art Gallery, Stanthorpe 01527
Stanton County Museum, Pilger 51917
Star City Heritage Museum, Star City 07006
Star Mound School Museum, Snowflake 06989
Star of the Republic Museum, Washington 53947
Star-Spangled Banner Flag House and 1812 Museum, Baltimore 46588
Stara Galeria ZPAF, Warszawa 35289
Stará Hut' v Josefovském Údolí, Adamov 09394
Stará Polenská Škola, Polná 09730
Stará škola, Kolín 09592
Stará Škola, Police nad Metují 09726
Stark Museum of Art, Orange 51578
Staro Selo Kumrovec, Kumrovec 08903
Staročerkasskij Istoriko-architekturnyj Muzej-zapovednik, Staročerkassk 37798
Staroladožskij Istoriko-Architekturnyj i Archeologičeskij Muzej-Zapovednik, Staraja Ladoga 37795
Starooskolskij Kraevedčeskij Muzej, Staryj Oskol 37801
Starorusskij Kraevedčeskij Muzej, Staraja Russa 37797
Startqing Museum, Zhaoqing 08494
Starved Rock State Park, Utica 53674
Starý Královský Palác, Praha 09794
Starý Zámek, Jevišovice 09558
Starý Zamok Banská Štiavnica, Banská Štiavnica 38194
Stasi Die Ausstellung zur DDR-Staatssicherheit, Berlin 17260
Stasimuseum, Berlin 17261
Stasio Girėno Gimtinė, Vytogala 30449
Stasio Vainiūno Namai-Muziejus, Vilnius 30433
Stasys Vainiunas House-Museum, Vilnius 30433
Statarmuseet, Bara 40453
Statarmuseum Mejeriet, Örebro 40790
State A.S. Pushkin Museum, Moskva 37202
State Academic Bolshoi Theatre Museum, Moskva 37280
State Academic Mariinsky Theatre Museum, Sankt-Peterburg 37676
State Agricultural Heritage Museum, Brookings 47060
State Archaeological Museum, Kolkata 24282
State Archaeological Museum, Warszawa 35286
State Art Collection, Sułoszowa 35140
State Art Gallery, Ahmedabad 24051
State Art Museum, Bakı 03172
State Art Museum of the Altai Region, Barnaul 36790
State Art Museum of the Kalmyk Republik, Ėlista 36911
State Authority of Museums, Rīga 54925
State Ayurvedic College Museum, Hardwar 24204
State Capital Publishing Museum, Guthrie 49109
State Central Museum, Ulaanbaatar 31632
State Central Museum of Contemporary History of Russia, Moskva 37189
State Central Theatre Museum A.A. Bakhrushin, Moskva 37190
State Ceramics Museum and 18th Century Kuskovo Estate, Moskva 37303
State Coal Mine Museum, Wonthaggi 01667
State Collection of Unique Musical Instruments, Moskva 37183
State Cultural Center Palace, Balčik 04975
State Ethnographic Museum, Warszawa 35287
State Exhibition Hall at Peresvetok Pereulok, Moskva 37214
State Exhibition Hall Belyaevo, Moskva 37208
State Exhibition Hall Chodynka, Moskva 37209
State Exhibition Hall Kovcheg, Moskva 37210
State Exhibition Hall - Museum Heritage, Moskva 37211
State Exhibition Hall New Manezh, Moskva 37256
State Exhibition Hall On the Kashirka, Moskva 37212
State Exhibition Hall On the Solyanka, Moskva 37213
State Exhibition Hall Solntsevo, Moskva 37215

State Exhibition Hall Zamoskvorechye, Moskva 37216
State Gallery, Banská Bystrica 38186
State Gallery of Contemporary Cypriot Art, Lefkosia 09383
State Gallery of Fine Arts, Cheb 09474
State Health Institute Museum, Lucknow 24300
State Hermitage Museum, Sankt-Peterburg 37640
State Historical and Architectural Museum-reserve Shirvan Shahs' Palace, Bakı 03173
State Historical and Cultural Museum-Preserve Azret Sultan, Turkestan 29734
State Historical and Memorial Museum Smolny, Sankt-Peterburg 37641
State Historical-Ethnographic and Architectural Museum-Preserve The old Sarepta, Volgograd 37954
State Historical Museum of Kyrgyzstan, Biškek 29954
State Historical Museum-Preserve, Kam'janec'-Podil's'kyj 43051
State Historical Society of Iowa Museum, Des Moines 48102
State Historical Society of Missouri, Columbia 47719
State Historical Society of North Dakota Museum, Bismarck 46839
State History and Art Museum, Bolšoe Muraškino 36815
State History, Art and Architecture Museum-Preserve, Lomonosov 37108
State Kala Akademi Museum, Imphal 24229
State Library and Museum, Kyaukpyu 31717
State Library and Museum, Mawlamyine 31723
State Literature and Memorial Anna Akhmatova Museum at Fountain House, Sankt-Peterburg 37642
State Literature Museum, Moskva 37195
State Memorial A.V. Suvorov Museum, Sankt-Peterburg 37643
State Museum, Bhopal 24096
State Museum, Hyderabad 24218
State Museum, Kohima 24264
State Museum, Mandalay 31720
State Museum, Trichur 24412
State Museum Association Artistic Culture of the Russian North, Archangelsk 36769
State Museum at Majdanek, Lublin 34911
State Museum Gatchina Palace and Park, Gatčina 36920
State Museum of A.M. Gorky, Nižnij Novgorod 37452
State Museum of Abkhasia, Suchumi 16350
State Museum of Armenian History, Yerevan 00754
State Museum of Art, Rīga 30114
State Museum of Azerbaijan Carpets and Applied Art Letif Kerimov, Bakı 03159
State Museum of Azerbaijan Musical Culture, Bakı 03174
State Museum of Ethnography, Arts and Crafts, Lviv 43141
State Museum of Fine Arts, Aşgabat 42972
State Museum of Folk Art of Karelia, Archangelsk 36765
State Museum of Georgian National Music Instruments, Tbilisi 16382
State Museum of History, Architecture, Art and Nature Tsaritsyno, Moskva 37143
State Museum of History of Uzbekistan, Toshkent 54485
State Museum of Natural History of NAS of Ukraine, Lviv 43142
State Museum of Nature, Yerevan 00755
State Museum of Oriental Art, Moskva 37286
State Museum of Pennsylvania, Harrisburg 49181
State Museum of Political History of Russia, Sankt-Peterburg 37648
State Museum of Prehistory, Halle, Saale 18900
State Museum of South Sumatra Province Balaputra Dewa, Palembang 24537
State Museum of the Defense and Siege of Leningrad, Sankt-Peterburg 37644
State Museum of the East Nusa Tenggara Province, Kupang 24517
State Museum of the History of Aviation, Sankt-Peterburg 37645
State Museum of the History of Religion, Bakı 03175
State Museum of the History of Religion, Sankt-Peterburg 37646
State Museum of the History of Saint Petersburg, Sankt-Peterburg 37647
State Museum of the Writers of the Ural, Ekaterinburg 36891
State Museum of Theatre, Music and Cinema, Tbilisi 16383
State Museum of Wood Carving, Yerevan 00756
State Museum-Panorama Battle of Stalingrad, Volgograd 37955
State Museum Pavlovsk Palace and Park, Pavlovsk 37528
State Museum-Preserve of History, Gorki Leninskie 36926
State Museum-Preserve of History and Literature Alexander Pushkin with Estates Vjasemy and Zakharovo, Bolšie Vjazemy 36814
State Museum-Reserve Kolomenskoye, Moskva 37191
State Museum Reserve of History, Architecture, Nature and landscape - Izborsk, Izborsk 36953
State Museum Tsarsky village - Alexander Palace, Puškin 37577
State Museum Tsarsky Village - Palace of Cathrine II. the Great, Puškin 37579
State Museum Vyborg Castle, Vyborg 37982
State Museums of Upper Austria, Ansfelden 01724
State of the Art Gallery, Ithaca 49551
State Open Air Museum Kizhi of Architecture, History and Ethnography, Kiži 37050
State Opera Museum, Wien 03078
State Park Museum on the Monuments of Ancient Taraz, Taraz 29732
State Russian Museum - Benois Wing, Sankt-Peterburg 37650
State Russian Museum - Marble Palace, Sankt-Peterburg 37666
State Russian Museum - Mikhailovsky Palace, Sankt-Peterburg 37649
State Russian Museum - Saint Michael's or Engineer's Palace, Sankt-Peterburg 37651
State Russian Museum - Stroganov Palace, Sankt-Peterburg 37727
State Studio Museum of the Sculptor A.S. Golubkina, Moskva 37200
State Stutthof Museum, Sopot 35122
State Stutthof Museum, Sztutowo 35172
State Tretyakov Gallery, Moskva 37184
State Tribal Museum, Chhindwada 24119
Statek-Muzeum Dar Pomorza, Gdynia 34662
Statek-Muzeum Sołdek, Gdańsk 34651
Staten Island Children's Museum, Staten Island 53268
Staten Island Historical Society Museum, Staten Island 53269
Staten Island Institute of Arts and Sciences, Staten Island 53270
Štatenberg Castle, Makole 38474
Statens Historiska Museum, Stockholm 40937
Statens Museum for Kunst, København 10152
Statens Naturhistoriske Museum (SNM), København 10153
Station 15, Norton 51429
The Station Gallery, Whitby 07319
Station House Studio and Gallery Society, Williams Lake 07338
Station Museum, North Vancouver 06482
Štátna Galéria, Banská Bystrica 38186
Státní Galerie Výtvarného Umění, Cheb 09474
Státní Hrad a Zámek, Frýdlant v Čechách 09505
Státní Hrad a Zámek, Horšovský Týn 09529
Státní Hrad Bouzov, Bouzov 09412
Státní Zámek Vizovice, Vizovice 09909
STATT-Museum, Herbstein 19137
Statthaus Böcklerpark, Berlin 17262
Statue Museum Balai Arca, Nganjuk 24531
Statue of Liberty National Monument and Ellis Island Immigration Museum, New York 51262
Staufergedächtnisstätte und Museum Wäscherschloß, Wäschenbeuren 22267
Stauffenberg-Erinnerungsstätte, Stuttgart 21963
Stauffenberg-Gedenkstätte, Albstadt 16460
Staufferhaus - Sammlung Alt Unterentfelden, Unterentfelden 41959
Staunton Augusta Art Center, Staunton 53278
Staurothek, Domschatz und Diözesanmuseum Limburg, Limburg an der Lahn 19998
Stauth Memorial Museum, Montezuma 50812
Stavanger Brannmuseum, Stavanger 33874
Stavanger Maritime Museum, Stavanger 33876
Stavanger Museum, Stavanger 33875
Stavanger Sjøfartsmuseum, Stavanger 33876
Stavropol Museum of Fine Arts, Stavropol 37806
Stavropolskij Kraevedčeskij Muzej im. G.N. Prozriteleva i G.K. Prave, Stavropol 37805
Stavropolskij Muzej Izobrazitelnych Iskusstv, Stavropol 37806
Stawell Historical Society Museum, Stawell 01529
Steam Engine Museum, Mabel 50338
Steam Mill, Halfweg 32308
The Steam Museum, Straffan 24900
Steam Museum of the Great Western Railway, Swindon 45819
Steam Museum Strabally, Port Laoise 24886
Steam Railroading Institute, Owosso 51644
Steamboat Arabia Museum, Kansas City 49700
Steamboat Bertrand Museum, Missouri Valley 50758
Steamboat Dock Museum, Keyport 49767
Steamrailway Goes-Borsele, Goes 32259
Steamship Collection, Baltimore 46589
Steamship William G. Mather Museum, Cleveland 47641
Steamtown National Historic Site, Scranton 52984
Steamtown Railway Museum, Carnforth 43764
Stearns Collection of Musical Instruments, Ann Arbor 46311
Stearns History Museum, Saint Cloud 52540
Stedelijk Archeologisch Museum, Oudenburg 03943
Stedelijk Museum, Alkmaar 31796
Stedelijk Museum, Oudenaarde 03942
Stedelijk Museum Aalst, Aalst 03259
Stedelijk Museum Aarschot, Aarschot 03260
Stedelijk Museum Alfons Blomme, Roeselare 03978
Stedelijk Museum Amsterdam, Amsterdam 31888
Stedelijk Museum Brusselpoort, Mechelen 03861
Stedelijk Museum Bureau Amsterdam, Amsterdam 31889
Stedelijk Museum De Lakenhal, Leiden 32489
Stedelijk Museum Diest, Diest 03552
Stedelijk Museum Diksmuide, Diksmuide 03554
Stedelijk Museum Drenthe's Veste, Coevorden 32038
Stedelijk Museum Het Domein, Sittard 32779
Stedelijk Museum Het Toreke, Tienen 04070
Stedelijk Museum-Hoevemuseum, Sint-Truiden 04031
Stedelijk Museum Hof van Busleyden, Mechelen 03862
Stedelijk Museum Hoogstraten, Hoogstraten 03711
Stedelijk Museum Huize Ernest Claes, Scherpenheuvel 04004
Stedelijk Museum Ieper, Ieper 03731
Stedelijk Museum Kampen, Kampen 32440
Stedelijk Museum Lokeren, Lokeren 03827
Stedelijk Museum Schatkamer van Sint-Pieter, Leuven 03790
Stedelijk Museum Schepenhuis, Mechelen 03863
Stedelijk Museum Schiedam, Schiedam 32756
Stedelijk Museum Vianen, Vianen 32887
Stedelijk Museum voor Actuele Kunst, Gent 03644
Stedelijk Museum voor Folklore en Regionale Geschiedenis, Ronse 03986
Stedelijk Museum voor Pijp en Tabak, Harelbeke 03680
Stedelijk Museum Zutphen, Zutphen 33020
Stedelijk Museum Zwolle, Zwolle 33029
Stedelijk Onderwijsmuseum Ieper, Ieper 03732
Stedelijk Textielmuseum, Ronse 03987
Stedelijke Academie voor Schone Kunsten, Roeselare 03979
Stedelijke Musea, Torhout 04080
Stedelijke Oudheidkamer, Genemuiden 32243
Stedelijke Oudheidkamer De Roos, Geertruidenberg 32236
Stedman Art Gallery, Camden 47196
Steele County Museum, Hope 49360
Steenbakkerijmuseum 't Gelleg te Rumst, Rumst 03990
Steenkolenmijn Valkenburg, Valkenburg aan de Geul 32861
Steenmuseum, Gent 03645
Steep and Brew Gallery, Madison 50394
Stefan cel Mare County Museum, Vaslui 36726
Stefan Fadinger Museum, Sankt Agatha 02634
Stefan-George-Museum im Stefan-George-Haus, Bingen am Rhein 17354
Stefan Procopiu Science and Technics Museum, Iași 36426
Stefan Żeromski Museum, Nałęczów 34940
Steffen Thomas Museum of Art, Buckhead 47109
Steglitz Museum, Berlin 17263
Steiglitz Court House, Steiglitz 01530
Steilacoom Historical Museum, Steilacoom 53282
Stein- und Bauernmuseum Großdöllnerhof, Rechberg 02567
Stein- und Beinmuseum, Weiterstadt 22467
Stein- und Fossiliensammlung Albert, Sulzdorf an der Lederhecke 21989
Stein und Wein, Aspach 16599
Steinhauermuseum, Randersacker 21177
Steinhauermuseum Mühlbach, Eppingen 18184
Steinhausen-Museum, Frankfurt am Main 18395
Steinhuder Spielzeugmuseum, Wunstorf 22735
Steinkohlen-Besucherbergwerk Rabensteiner Stollen, Ilfeld 19314
Steinmetzmuseum, Burgschleinitz 01809
Steinmuseum, Solothurn 41888
Steinsburg-Museum, Römhild 21357
Steinscher Hof Kirberg, Hünfelden 19292
Steinsland Klyngetun, Harstad 33534
Steinzeitdorf Kussow, Moor-Rolofshagen 20378
Steinzeitmuseum, Korb 19685
Steinzeitmuseum, Thunstetten 41943
Steirisches Feuerwehrmuseum, Gross Sankt Florian 01996
Steirisches Glaskunstzentrum und Glasmuseum Bärnbach, Bärnbach 01778
Steirisches Holzmuseum, Sankt Ruprecht 02688
Stellenbosch Museum, Stellenbosch 38738
Stelzhamer-Gedenkstätte, Pramet 02528
Stenenexpositie de Molen - Museum Batjuchin, Borculo 31986
Stenersenmuseet, Oslo 33752
Stengel-True Museum, Marion 50489
Steninge Slott Kulturcenter, Märsta 40712
Stenneset Bygdetun, Mo i Rana 33668
Stenneset Open Air Museum, Mo i Rana 33668
Steno Museet, Århus 09966
Steno Museum, Århus 09966
Stepan Shahumyan House-Museum, Stepanavan 00718
Stepan Zoryan House-Museum, Vanadzor 00720
Stephanie Ann Roger Gallery, Frostburg 48802
Stephansson House, Edmonton 05754
Stephen A. Douglas Tomb, Chicago 47515
Stephen C. Foster State Park, Fargo 48507
Stephen Foster Folk Culture Center, White Springs 54129
Stephen G. Beaumont Museum, Wakefield 45912
The Stephen Girard Collection, Philadelphia 51882
Stephen Lawrence Gallery, London 44991
Stephen Leacock Museum, Orillia 06515
Stephen Phillips Memorial House, Salem 52661
Stephens African-American Museum, Shreveport 53091
Stephens Collection, London 44992
Stephens County Historical Society Museum, Duncan 48205
Stephens Museum, Fayette 48526
Stephenson County Historical Society Museum, Freeport 48783
Stephenson Railway Museum, North Shields 45267
Stephenville Museum, Stephenville 53283
Stepono Dariaus Gimtinė, Dariaus 30220
Stepping Stones Museum for Children, Norwalk 51433
Steppingstone Museum, Havre de Grace 49220
Sterbehaus Ferdinand Raimunds, Pottenstein 02515
Sterckshof Silver Museum, Antwerpen 03321
Sterkstroom Museum, Sterkstroom 38743
Sterling and Francine Clark Art Institute, Williamstown 54175
Sterling Hill Mining Museum, Ogdensburg 51505
Sterling Historical Society and Little Red Schoolhouse Museum, Sterling 53286
Sterling-Rock Falls Historical Society Museum, Sterling 53285
Sternberg Museum of Natural History, Hays 49225
Šternberský Palác, Praha 09795
Sterne-Hoya Museum, Nacogdoches 50937
Stettler Town and Country Museum, Stettler 07009
Steuben House Museum, River Edge 52362
Steuben Memorial State Historic Site, Remsen 52286
Stevenage Museum, Stevenage 45742
Stevens-Coolidge Place, North Andover 51380
Stevens County Gas and Historical Museum, Hugoton 49416
Stevens County Historical Society Museum, Morris 50849
Stevens Museum, Salem 52655
Stevenson-Hamilton Memorial Information Centre, Skukuza 38731
Steves Homestead Museum, San Antonio 52728
Steveston Museum, Richmond 06746
Stevington Windmill, Stevington 45743
Stewart Collection, Pocklington 45394
Stewart Hay Memorial Museum, Winnipeg 07385
Stewart Historical Museum, Stewart 07010
Stewart Indian Cultural Center, Carson City 47262
Stewart M. Lord Memorial Museum, Burlington 47137
Stewart Museum at the Fort Ile Sainte-Hélène, Montréal 06376
Stewarton and District Museum, Stewarton 45744
The Stewartry Museum, Kirkcudbright 44611
Steyning Museum, Steyning 45745
Steyrtal heritage railway, Steyr 02773
Steyrtal-Museumsbahn, Steyr 02773
Stichting Atlas Van Stolk, Rotterdam 32728
Stichting De Appel, Amsterdam 31890
Stichting De Brakke Grond, Amsterdam 31891
Stichting George Grard, Gijverinkhove 03656
Stichting Huis Bergh, 's-Heerenberg 32331
Stichting Joden Savanna, Paramaribo 40400
Stichting Oosterkerk, Amsterdam 31892
Stichting Sjenters en Kantjes, Maastricht 32534
Stichting tot Beheer Museum Boijmans Van Beuningen, Rotterdam 32729
Stichting Veranneman, Kruishoutem 03764
Stichting Zeeuwse Schaapkudde, Heinkenszand 32351
Stickereimuseum Eibenstock, Eibenstock 18074
Stickereimuseum Lustenau, Lustenau 02342
Stickmaskinsmuseum, Glemminebro 40537
Stiefelmachermuseum, Rechnitz 02568
Stierhübelteichhaus, Karlstift 02155
't Stift, Susteren 32812
Stift Fischbeck, Hessisch Oldendorf 19171
Stift Herzogenburg, Herzogenburg 02076
Stift Obernkirchen, Obernkirchen 20823
Stiftelsen Militärfordons Museum, Malmköping 40717
Stiftelsen Musikkulturens Främjande, Stockholm 40938
Stiftlandmuseum, Waldsassen 22303
Stifts- und Heimatmuseum Schloß Seeburg, Seekirchen am Wallersee 02736
Stiftsbibliothek, Einsiedeln 41345
Stiftsgalerie Seitenstetten, Seitenstetten 02738

Stiftsgården, Trondheim 33928
Stiftskirchenmuseum, Himmelkron 19192
Stiftsmuseum, Aschaffenburg 16593
Stiftsmuseum, Bad Buchau 16707
Stiftsmuseum, Garsten 01926
Stiftsmuseum, Millstatt 02383
Stiftsmuseum, Vomp 02872
Stiftsmuseum, Xanten 22763
Stiftsmuseum, Zwettl 03137
Stiftsmuseum Mattsee, Mattsee 02367
Stiftsmuseum Praemonstratenser-Chorherren Stift Schlägl, Aigen im Mühlkreis 01700
Stiftsmuseum Schatzhaus Kärntens, Sankt Paul 02678
Stiftsmuseum Steyr-Gleink, Steyr-Gleink 02775
Stiftssammlungen Lambach, Lambach 02252
Stiftssammlungen Wilten, Innsbruck 02133
Stiftung B.C. Koekkoek-Haus, Kleve 19568
Stiftung Bauhaus Dessau, Dessau 17834
Stiftung Bohlenständerhaus Schrofen, Amriswil 41114
Stiftung Brandenburgische Gedenkstätten, Oranienburg 20911
Stiftung Bundeskanzler-Adenauer-Haus, Bad Honnef 16754
Stiftung Domäne Dahlem, Berlin 17264
Stiftung für Eisenplastik, Zollikon 42063
Stiftung für Konkrete Kunst, Reutlingen 21290
Stiftung für Konkrete Kunst Roland Phleps, Freiburg im Breisgau 18438
Stiftung Fürst-Pückler-Museum, Cottbus 17736
Stiftung Gerhart-Hauptmann-Haus, Deutsch-osteuropäisches Forum, Düsseldorf 18007
Stiftung Historische Museen Hamburg, Hamburg 18961
Stiftung Hubert Aratym, Gutenstein 02026
Stiftung Käthe Kollwitz Haus, Moritzburg 20387
Stiftung Keramion, Frechen 18416
Stiftung Kloster Michaelstein-Museum, Blankenburg 17374
Stiftung Kohl'sche Einhorn-Apotheke, Weißenburg in Bayern 22454
Stiftung Mecklenburg, Ratzeburg 21198
Stiftung Moritzburg, Halle, Saale 18904
Stiftung museum kunst palast mit Sammlung Kunstakademie und Glasmuseum Hentrich, Düsseldorf 18008
Stiftung Museum Schloss Moyland, Bedburg-Hau 16975
Stiftung Naturschutzgeschichte, Königswinter 19662
Stiftung Neue Synagoge Berlin - Centrum Judaicum, Berlin 17265
Stiftung Opelvillen, Rüsselsheim 21439
Stiftung Preußische Schlösser und Gärten Berlin-Brandenburg, Potsdam 54859
Stiftung Römermuseum Homburg-Schwarzenacker, Homburg 19268
Stiftung S BC - pro arte, Biberach an der Riß 17321
Stiftung Sammlung E.G. Bührle, Zürich 42108
Stiftung Scheibler-Museum Rotes Haus, Monschau 20376
Stiftung Schleswig-Holsteinische Landesmuseen Schloß Gottorf, Schleswig 21569
Stiftung Schloss Ahrensburg, Ahrensburg 16441
Stiftung Schloss Eutin, Eutin 18287
Stiftung Schloss Neuhardenberg, Neuhardenberg 20614
Stiftung Schloss und Park Benrath, Düsseldorf 18009
Stiftung Simonshof, Gleisweiler 18660
Stiftung Stadtmuseum Berlin, Landesmuseum für Kultur und Geschichte Berlins, Berlin 17266
Stiftung Starke, Berlin 17267
Stiftung Stift Neuzelle, Neuzelle 20690
Stiftung Wilhelm Lehmann, Niederhelfenschwil 41685
Stiftung Wilhelm Lehmbruck Museum, Duisburg 18027
Stiftung Ziegelei-Museum, Cham 41288
Stiftung Zollverein, Essen 18269
Stiklestad Nasjonal Kultursenter, Verdal 33973
Stiklestad National Cultural Center, Verdal 33973
Still National Osteopathic Museum, Kirksville 49801
Das Stille Museum, Berlin 17268
Stille Nacht-Museum, Hallein 02047
Stille-Nacht- und Heimatmuseum Bruckmannhaus, Oberndorf bei Salzburg 02455
Stills Gallery, Edinburgh 44159
Stinzenmuseum, Beers 31954
Stirk Cottage, Kalamunda 01184
Stirling Old Town Jail, Stirling 45751
Stirling Smith Art Gallery and Museum, Stirling 45752
Stjepan Gruber Museum, Županja 09019
Stjernsunds Slott, Askersund 40445
Stjørdal Museum KF, Stjørdal 33881
Stoani Haus der Musik, Gasen 01928
Stobi Archaeological Open Air Museum, Negotino 30509
Stockbridge Library Historical Collection, Stockbridge 53303
Stockholm City Museum, Stockholm 40942
Stockholm Historical Society Museum, Stockholm 53304
Stockholm School Museum, Stockholm 40941
Stockholms Läns Museum, Stockholm 40939
Stockholms Medeltidsmuseum, Stockholm 40940
Stockholms Skolmuseum, Stockholm 40941
Stockholms Stadsmuseum, Stockholm 40942
Stockholms Universitet Scheffierska huset, Stockholm 40943
Stockport Art Gallery, Stockport 45754
Stockport Museum, Stockport 45755
Stockwood Park Museum, Luton 45047
Störtebeker Park, Wilhelmshaven 22580
Stofnun Arna Magnússonar, Reykjavík 24037
Stokke Bygdetun, Stokke 33882
Stoltze-Museum der Frankfurter Sparkasse, Frankfurt am Main 18396
Stondon Museum, Lower Stondon 45040
Stone Chamber Ethnography Museum, Burdur 42738
Stone Fort Museum, Nacogdoches 50938
Stone grinding mill, Schwerin 21697
Stone House - Czech Museum of Silver, Kutná Hora 09614
Stone House Gallery, Fredonia 48774
Stone House Museum, Belchertown 46693
Stone Inscription Art Museum, Beijing 07569
Stone Museum, Monroe Township 50794
Stone Sacristy, Mikkeli 11114
Stone-Tolan House, Rochester 52399
Stonefield Historic Site, Cassville 47275
Stonehouse Museum, Vale 53688
Stones River National Battlefield, Murfreesboro 50915
Stone's Tavern Museum, Ligonier 50081
Stones'n Bones Museum, Sarnia 06908
Stonewall Jackson House, Lexington 50074
Stonewall Jackson Museum at Hupp's Hill, Strasburg 53333
Stonnington Local History Collection - Malvern, Malvern, Victoria 01254
Stonnington Local History Collection - Prahran, Prahran 01442
Stonor Park, Henley-on-Thames 44450
Stony Plain and District Pioneer Museum, Stony Plain 07015
Stony Point Battlefield State Historic Site, Stony Point 53316
Stoomcentrum, Maldegem 03841
Stoomgemaal Halfweg, Halfweg 32308
Stoomgemaal Hertog Reijnout en Bezoekerscentrum Arkemheen, Nijkerk 32586
Stoomgemaal Mastenbroek, Genemuiden 32244
Stoomhoutzagerij, Groenlo 32276
Stoomtrein Goes-Borsele, Goes 32259
Stoottroepen Museum, Assen 31939
Stor-Elvdal Museum, Koppang 33605
Stora Nyckelvikens Gård, Nacka 40762
Storchenturm-Museum, Zell am Harmersbach 22780
Storm King Art Center, Mountainville 50898
Storm P.-Museet, Frederiksberg 10023
Stormarnsches Dorfmuseum, Altes Bauernhaus am Thie, Hoisdorf 19250
Stormont, Dundas and Glengarry Highlanders Regimental Museum, Cornwall 05645
Storno Collection, Sopron 23822
Storno-ház, Sopron 23822
Storoževoj Korabl 'Krasnyj Vympel' 1911 g., Vladivostok 37950
Storrowton Village Museum, West Springfield 54083
Storstrøms Kunstmuseum, Maribo 10186
Story of Castle Cornet, Saint Peter Port 45594
Story of People, Yorkton 07416
Stott Park Bobbin Mill, Newby Bridge 45222
Stoughton and District Museum, Stoughton 07017
Stoughton Historical Society Museum, Stoughton 53322
Stourhead House, Stourton 45770
Stow House, Goleta 48966
Stow West 1825 School Museum, Stow 53324
Stowarzyszenie Muzeów na Wolnym Powietrzu, Toruń 54989
Stowarzyszenie Zwiazek Muzeów Polskich, Kraków 54990
Stowitts Museum, Pacific Grove 51667
The Stoy Museum of the Lebanon County Historical Society, Lebanon 50013
Strachur Smiddy Museum, Strachur 45775
Strafvollzugsmuseum, Ludwigsburg 20049
Strahov Gallery, Praha 09796
Strahovská Obrazárna, Praha 09796
Stranahan House, Fort Lauderdale 48641
Stranda Hembygdsförening, Mönsterås 40744
Strandgaarden Museum, Ulfborg 10301
Strandhalle Ahrenshoop, Ahrenshoop, Ostseebad 16444
Strandingsmuseum St. George, Ulfborg 10302
Strang Print Room, London 44993
Strangers Hall Museum of Domestic Life, Norwich 45287
Stranraer Museum, Stranraer 45777
Strasbourg and District Museum, Strasbourg 07018
Strasburg Museum, Strasburg 53334
Straßenbahn-Museum Thielenbruch, Köln 19633
Strassenbaumuseum, Sayda 21527
Strategic Air & Space Museum, Ashland 46383
Strater Hotel, Durango 48213
Stratfield Saye House and Wellington Exhibition, Stratfield Saye 45778
Stratford and District Museum, Stratford 01531
The Stratford Historical Society and Catherine B. Mitchell Museum, Stratford 53335
Stratford-Perth Museum, Stratford 07020
Strathalbyn National Trust Museum, Strathalbyn 01532
Strathclair Museum, Strathclair 07021
Strathcona County Heritage Museum, Sherwood Park 06964
Strathearn Museum, Simi Valley 53111
Strathnaver Museum, Bettyhill 43465
Strathpeffer Spa Pumping Room Exhibition, Strathpeffer 45789
Strathroy Middlesex Museum, Strathroy 07022
Strathspey Railway, Aviemore 43339
Strathwork Galerie, Ladbergen 19759
Straumsnes Bygdemuseum, Kanestraum 33586
Straw and Grass Handicraft Museum, Seoul 29926
Strawberry Culture Museum, Dahu 42172
Strawbery Banke Museum, Portsmouth 52100
Mr. Straw's House, Worksop 46077
Stražky Chateau of the Slovak National Gallery, Spišská Béla 38357
Strecker Museum's Collection, Waco 53779
Středočeské Muzeum, Roztoky u Prahy 09828
Stredoslovenské Múzeum, Banská Bystrica 38187
Streekheemmuseum den Aanwas, Ossendrecht 32641
Streekhistorisch Centrum, Stadskanaal 32800
Streeklandbowmuseum Agrimunda, Sint Anna ter Muiden 32770
Streekmuseum Alphen, Alphen, Noord-Brabant 31804
Streekmuseum De Groote Sociëteit, Tiel 32825
Streekmuseum de Meesthof, Sint Annaland 32771
Streekmuseum de Moennik, Helden 32353
Streekmuseum De Oude Wolden, Bellingwolde 31960
Streekmuseum de Tolbrug, Bakel 31947
Streekmuseum De Vier Ambachten, Hulst 32423
Streekmuseum Goeree en Overflakkee, Sommelsdijk 32796
Streekmuseum Het Admiraliteitshuis, Dokkum 32140
Streekmuseum Het Dorp van Bartje, Rolde 32699
Streekmuseum Het Land van Axel, Axel 31943
Streekmuseum Het Rondeel, Rhenen 32684
Streekmuseum Hoeksche Waard, Heinenoord 32349
Streekmuseum Jan Anderson, Vlaardingen 32891
Streekmuseum Land van Valkenburg, Valkenburg, Limburg 32866
Streekmuseum Leudal, Haelen 32307
Streekmuseum Ommen, Ommen 32620
Streekmuseum Opsterland, Gorredijk 32264
Streekmuseum Oudheidkamer Reeuwijk, Reeuwijk 32679
Streekmuseum Schippersbeurs, Elsloo 32202
Streekmuseum Stevensweert/ Ohé en Laak, Stevensweert 32808
Streekmuseum van Klein-Brabant De Zilverreiger, Weert 04142
Streekmuseum/Volkssterrenwacht Burgum, Burgum 32029
Streekmuseum voor de Krimpenerwaard Crimpenerhof, Krimpen aan den IJssel 32460
Streekmuseum West Zeeuws-Vlaanderen, IJzendijke 32429
Street Museum, Henri-Chapelle 03695
Street Museum of Heslinki City Museum, Helsinki 10791
Streetlife Museum, Kingston-upon-Hull 44600
Streichinstrumentensammlung, Einsiedeln 41346
Stretham Old Engine, Stretham 45791
Stretton Water Mill, Farndon 44215
Strich-Chapell-Zimmer, Sersheim 21748
Strieffler-Haus, Landau in der Pfalz 19793
Strigel-Museum und Antoniter-Museum, Memmingen 20280
Strindberg-Museum Saxen, Saxen 02695
Strindbergsmuseet, Stockholm 40944
Striped House Museum of Art, Tokyo 29475
Strömforsin Iron Works Museum, Ruotsinpyhtää 11281
Strömforsin Teollisuusmuseo, Ruotsinpyhtää 11281
Strömsholm Castle, Kolbäck 40651
Strömsholms Slott, Kolbäck 40651
Strötzbacher Mühle, Mömbris 20353
Stroganov Church, Nižnij Novgorod 37471
Stroganovskaja Roždestvenskaja Cerkov, Nižnij Novgorod 37471
Stroganovskij Dvorec, Sankt-Peterburg 37727
Strohhaus, Muhen 41665
Strohhaus und Dorfmuseum, Kölliken 41508
Stromness Museum, Stromness 45793
Strømstads Museum, Strømstad 40955
Strong National Museum of Play, Rochester 52400
Strood Library Gallery, Rochester 45509
Stroom Den Haag, Den Haag 32103
Strossmayer Gallery of Old Masters, Zagreb 09011
Strossmayerova Galerija Starih Majstora, Zagreb 09011
Stroud Mansion, Stroudsburg 53341
Strübhaus-Haus der Malkunst, Veringenstadt 22222
Struer Museum og Johs. Buchholtz Hus, Struer 10286
Struts Gallery and Faucet Media Arts Centre, Sackville 06789
Struwwelpeter-Museum, Frankfurt am Main 18397
Stuart Collection, La Jolla 49847
Stuart House City Museum, Mackinac Island 50352
Stuart Town Gaol Museum, Alice Springs 00802
Stubnitz Gallery, Adrian 46139
Stuck-Museum, Freiburg im Breisgau 18439
Studebaker National Museum, South Bend 53160
Student Center Gallery, Wilmore 54207
Studentenkarzer, Heidelberg 19084
Studiensammlung historischer Ziegel, Irdning 02138
The Studio Annex, New York 51263
Studio Bildende Kunst, Berlin 17269
Studio Channel Islands Art Center, Camarillo 47171
Studio Eugène Ysaye, Liège 03817
Studio Gallery, Washington 53919
Studio im Hochhaus, Berlin 17270
Studio-Museo di Giuseppe Pellizza, Volpedo 28308
The Studio Museum in Harlem, New York 51264
Studio San Giuseppe Art Gallery, Cincinnati 47564
Studiogalerie Kaditzsch, Grimma 18764
The Study Gallery, Poole 45409
Stuhlbau- und Heimatmuseum, Rabenau 21157
Stuhlmuseum Burg Beverungen, Beverungen 17316
Stuhr Museum of the Prairie Pioneer, Grand Island 48992
Sturdivant Hall, Selma 53032
Sturgeon River House, Sturgeon Falls 07024
Sturgis Station House Museum, Sturgis 07025
Sturman Institute Museum of Regional Science, Afula 24921
Stutsman County Memorial Museum, Jamestown 49594
Stuttgarter Gesellschaft für Kunst und Denkmalpflege, Stuttgart 21964
Stuttgarter Kunstverein e.V., Stuttgart 21965
Stutzhäuser Brauereimuseum, Luisenthal 20108
Su Ho Paper Museum, Taipei 42425
Su-no-sato Hakubutsukan, Handa 28497
Subak Museum, Banjar Senggulan 24447
Subalyuk-Múzeum, Cserépfalu 23509
Subcarpathian Museum of Art J. Bokshay, Užgorod 43200
Subiaco Museum, Subiaco 01535
Submarine Adventure World, Sassnitz 21518
Submarine Force Museum and Historic Ship Nautilus, Groton 49095
Submarine Vesikko, Helsinki 10819
Subway Museum, Tokyo 29378
Sucharduv Dům, Nová Paka 09677
Sudan National Museum, Khartoum 40392
Sudan Natural History Museum, Khartoum 40393
Sudbury Region Police Museum, Sudbury 07030
Sudeley Castle, Winchcombe 46022
Sudhaus Sorgendorf, Bleiburg 01787
Sudley House, Liverpool 44745
Südbahn, Mürzzuschlag 02414
Süddeutsches Eisenbahnmuseum, Heilbronn 19102
Südharzer Eisenhüttenmuseum, Bad Lauterberg 16777
Südhessisches Handwerksmuseum, Roßdorf 21382
Südmährer Heimatmuseum, Laa an der Thaya 02249
Südmährisches Landschaftsmuseum, Geislingen an der Steige 18563
Südostbayerisches Naturkunde- und Mammut-Museum, Siegsdorf 21759
Südsauerlandmuseum, Attendorn 16603
Südschwäbisches Archäologiemuseum im Jesuitenkolleg, Mindelheim 20331
Südtiroler Landesmuseum für Kultur- und Landesgeschichte Schloss Tyrol, Tirolo di Merano 27976
Sülchgau-Museum, Rottenburg am Neckar 21421
Sümegi Vármúzeum, Sümeg 23829
Suermondt-Ludwig-Museum, Aachen 16408
Suffolk County Historical Society Museum, Riverhead 52368
Suffolk County Vanderbilt Museum, Centerport 47315
Suffolk Museum, Suffolk 53355
Suffolk Punch Heavy Horse Museum, Woodbridge 46061
Suffolk Regiment Museum, Bury Saint Edmunds 43696
Suffolk Resolves House, Milton 50700
Sugar Industry Museum, Ciaotou 42165
Sugar Museum, Klaten 24515
Sugarmuseum, Szerencs 23886
Sugimoto Art Museum, Mihama 28907

Sugimoto Bijutsukan, Mihama 28907
Sugino Costume Museum, Tokyo 29476
Sugino Gakuen Isho Hakubutsukan, Tokyo 29476
Sui Tangyao Zhi Display Center, Chengdu 07621
Suichuan Museum, Suichuan 08247
Suide Museum, Suide 08248
Suifu-Meitoku Shokoh-kan Tokugawa Museum, Mito 28922
Suining Museum, Suining, Jiangsu 08249
Suining Museum, Suining, Sichuan 08250
Suishouen Ceramic Museum, Tokuro Memorial Museum, Nagoya 29011
Suishouen Toji Shiryokan, Nagoya 29011
Suixi Museum, Suixi 08251
Suk Joo-sun Memorial Museum of Korean Folk Arts, Seoul 29927
Suka Art Space, Busan 29782
Suka Gogoda Art Museum, Nagasaki 28986
Suka Gogoda Bijutsukan, Nagasaki 28986
Sukagawa Municipal Museum, Sukagawa 29298
Sukagawa-shiritsu Hakubutsukan, Sukagawa 29298
Sukanen Ship Pioneer Village and Museum of Saskatchewan, Moose Jaw 06387
Sukellusvene Vesikko, Helsinki 10819
Sukkulenten-Sammlung Zürich, Zürich 42109
Suksimuseo ja Pesäpallomuseo, Vimpeli 11468
Sukuma Museum, Mwanza 42518
Sulabh International Museum of Toilets, Delhi 24155
Suleiman Stalsky Literature and Memorial Apartment-Museum, Ašaga-Stalsk 36776
Sulitjelma Gruvemuseum, Sulitjelma 33890
Sullivan County Historical Society Museum, Hurleyville 49449
Sulphide Street Station Railway and Historical Museum, Broken Hill 00904
Sulphur Springs Station, Ancaster 05339
Sultan Abdul Aziz Royal Gallery, Kelang 30560
Sultan Abu Bakar Museum Pahang, Pekan 30611
Sultan Alam Shah Museum, Shah Alam 30619
Sultan Ali Dianr Museum, El-Fasher 40384
Sultan Mahmud Badaruddin II Museum, Palembang 24538
Sultan's Armed Forces Museum, Ruwi 33997
Sulu National Museum, Jolo 34393
Sumelocenna - Römisches Stadtmuseum, Rottenburg am Neckar 21422
Sumitomo Collection, Kyoto 28864
Sumiya Motenashi Museum of Art and Culture, Kyoto 28868
The Summer Palace, Beijing 07570
Summerland Museum, Summerland 07031
Summerlee Heritage Park, Coatbridge 43864
Summit County Historical Society Museum, Akron 46149
Sumner/Redcliffs Historical Museum, Christchurch 33078
Sumo Hakubutsukan, Tokyo 29477
Sumo Museum of the Japan Sumo Association, Tokyo 29477
Sumove Museum, Suqian 08252
Sumpu Museum, Shizuoka 29290
Sums'kyj Chudožnij Muzej im. N. Onac'kogo, Sumy 43190
Sums'kyj Oblasnyj Krajeznavčyj Muzej, Sumy 43191
The Sumter County Museum, Sumter 53366
Sumter Gallery of Art, Sumter 53367
Sumy Regional Museum of Local Lore, Sumy 43191
Sun Gallery, Hayward 49230
Sun-Kong Factory Museum, Su-Ao 42345
Sun Museum, Riga 30112
Sun Prairie Historical Museum, Sun Prairie 53368
Sun Valley Center for the Arts, Ketchum 49754
Sun Yat-Sen University Museums, Guangzhou 07736
Sunbury Planatation House, Saint Phillip 03244
Sunbury Shores Arts and Nature Centre, Saint Andrews 06801
Sund Bygdemuseum, Skogsvåg 33839
Sund Fisheries Museum, Sund i Lofoten 33891
Sund Fiskerimuseum, Sund i Lofoten 33891
Sundby Samling Bryggergården, Nørresundby 10197
Sundby Samling Raschgården, Nørresundby 10198
Sundbybergs Museum, Sundbyberg 40957
Sunderland Museum, Sunderland 45805
Sundomin Kotiseutumuseo, Sundom 11335
Sundsbergs Gård Museum och Konsthall, Sunne 40962
Sundsvalls Museum, Sundsvall 40960
Sung-Am Archives of Classical Literature, Seoul 29928
Sung Kyun Kwan University Museum, Seoul 29929
Sungkok Art Museum, Seoul 29930
Sungshin Women's University Museum, Seoul 29931
Sunken History Exhibition, Yarmouth 46092
Sunndal Bygdemuseum, Sunndalsøra 33892
Sunnfjord Museum, Førde 33503
Sunnhordland Museum, Stord 33885
Sunnmøre Museum, Ålesund 33381
Sunnybrook Farm Museum and Interpretive Centre, Red Deer 06710
Sunnyhurst Wood Visitor Centre, Darwen 43955
Sunnyside Historical Museum, Sunnyside 53371
Sunnyside Museum, Hillside 44469
Sunnyvale Heritage Park Museum, Sunnyvale 53373
Sunrise Museum, Charleston 47378
Sunrise Trail Museum, Tatamagouche 07061
Sunritz Hattori Bijutsukan, Suwa 29302
Sunritz Hattori Museum of Art, Suwa 29302
Sunshine Coast Museum, Gibsons 05864
Suntory Bijutsukan, Tokyo 29478
Suntory Hakubutsukan, Osaka 29155
Suntory Museum, Osaka 29155
Suntory Museum of Art, Tokyo 29478
Sunwatch Indian Village - Archaeological Park, Dayton 47989
Suodenniemi Kotiseutumuseo, Suodenniemi 11336
Suojeluskuntamuseo, Seinäjoki 11306
Suolahden Museo, Suolahti 11338
Suomen Finnish Honeybee Museum, Juva 10885
Suomen Ilmailumuseo, Vantaa 11448
Suomen Jääkiekkomuseo, Tampere 11358
Suomen Järvikalastusmuseo ja Kerimäen Ulkomuseoalue, Kerimäki 10948
Suomen Jalkapallomuseo, Valkeakoski 11438
Suomen Joutsen, Turku 11399
Suomen Käsityön Museo, Jyväskylä 10897
Suomen Kansallismuseo, Helsinki 10820
Suomen Kansansoitinmuseo, Kaustinen 10933
Suomen Kellomuseo, Espoo 10733
Suomen Koulumuseo, Tampere 11359
Suomen Lasimuseo, Riihimäki 11270
Suomen Matkailumuseo, Imatra 10857
Suomen Merimuseo, Kotka 10988
Suomen Metsästysmuseo, Riihimäki 11271
Suomen Museoliitto, Helsinki 54795
Suomen Nyrkkeilymuseo, Tampere 11360
Suomen Partiomuseo, Turku 11400
Suomen Rakennustaiteen Museo, Helsinki 10821
Suomen Rautatiemuseo, Hyvinkää 10846
Suomen Tietojenkäsittelymuseo, Jyväskylä 10898
Suomen Urheilumuseo, Helsinki 10822
Suomen Värimuseo, Kemi 10942
Suomen Valimomuseo ja Högforsin Masuuni, Karkkila 10919
Suomen Valokuvataiteen Museo, Helsinki 10823
Suomen Viestimuseo, Riihimäki 11272
Suomenlinna-Museo, Helsinki 10824
Suomenlinna Museum, Helsinki 10824
Suomenlinna - Sveaborg, Helsinki 10825
Suomenlinna Toy Museum, Helsinki 10826
Suomenlinnan Lelumuseo, Helsinki 10826
Suomusjärven Kotiseutumuseo, Suomusjärvi 11339
Suonenjoen Kotiseutumuseo, Suonenjoki 11340
Supa-Ngwao Museum Centre, Francistown 04295
The Supreme Court of the United States Museum, Washington 53920
Suquamish Museum, Poulsbo 52124
Suquamish Museum, Suquamish 53380
Surahammars Bruksmuseum, Surahammar 40964
Surface Gallery, Nottingham 45300
Surgeons' Hall Museum, Edinburgh 44160
Surinaams Museum, Paramaribo 40401
Surnadal Bygdemuseum, Surnadal 33893
Surratt House Museum, Clinton 47653
Surrey Art Gallery, Surrey 07037
Surrey Heath Museum, Camberley 43720
Surrey Museum, Surrey 07038
Survey Museum, Roorkee 24374
Surxondaryo Vilojati Arxeologiya Muzeyi, Termiz 54466
Susan B. Anthony House, Rochester 52401
Susannah Place Museum, Sydney 01560
Sushi Performance and Visual Art, San Diego 52756
Susquehanna Art Museum, Harrisburg 49182
Susquehanna County Historical Society, Montrose 50831
Susquehanna Museum of Havre de Grace, Havre de Grace 49221
Sussex Combined Services Museum, Eastbourne 44102
Sussex County Historical Society Museum, Newton 51343
Sussex Farm Museum, Heathfield 44433
Sutama-cho Bijutsukan, Sutama 29299
Sutama-cho Rekishi Hakubutsukan, Sutama 29300
Sutama Museum of Art, Sutama 29299
Sutama Town Museum of History, Sutama 29300
Sutcliffe Gallery, Whitby 45996
The Suter, Te Aratoi o Whakatu, Nelson 33154
Sutherland Steam Mill Museum, Denmark 05696
Sutton-Ditz House Museum, Clarion 47582
Sutton Park, Sutton-on-the-Forest 45807
Sutton Windmill and Broads Museum, Sutton, Norwich 45806
Suur-Miehikkälän Kylämuseo, Suur-Miehikkälä 11341
Suur-Savon Museo, Mikkeli 11117
Suwa City Art Museum, Suwa 29303
Suwa-ko Orgel Museum, Shimosuwa 29272
Suwa-ko Somei-kan (Orugoru Hakubutsukan), Shimosuwa 29272
Suwa-shi Bitjutsukan, Suwa 29303
Suwałki Province Museum, Suwałki 35144
Suworow-Museum, Schwanden 41847
Suzhou Art Gallery, Suzhou 08254
Suzhou Beike Guan, Suzhou 08255
Suzhou Engraved Stone Museum, Suzhou 08255
Suzhou Folkways Museum, Suzhou 08256
Suzhou Gardens Museum, Suzhou 08257
Suzhou Museum, Shanghai 08197
Suzhou Museum, Suzhou 08258
Suzhou National Treasure Numismatics Museum, Suzhou 08259
Suzhou Silk Museum, Suzhou 08260
Suzhou Sportqu Museum, Suzhou 08261
Suzu-yaki Kyodo Bijutsukan, Suzu 29306
Suzuki Bokushi Kinenkan, Shiozawa 29279
Suzuki Bokushi Memorial Museum, Shiozawa 29279
Svalbard-Museum, Longyearbyen 33650
Svaneholms Slotts Museum, Skurup 40856
Svaneti History-Ethnography Museum, Mestia 16347
Svedino's Automobile and Aviation Museum, Slöinge 40858
Svedinos Bil- och Flygmuseum, Slöinge 40858
Sveindal Museum, Kollungtveit 33598
Svelvik Museum, Svelvik 33895
Svendborg og Omegns Museum, Svendborg 10293
Svenska Statens Porträttsamling, Mariefred 40735
Sverev Center of Modern Art, Moskva 37432
Sveriges Järnvägsmuseum, Gävle 40533
Sveriges Jörnvögsmuseum, Ängelholm 40409
Sveriges Museimannaförbund, Nacka 55035
Sveriges Rundradiomuseum, Motala 40759
Sveriges Sjömanshusmuseum, Uddevalla 40993
Sveriges Teatermuseum, Nacka 40763
Sveriges Tennis Museum, Båstad 40454
Sveriges VVS-Museum, Katrineholm 40641
Svetska Galerija na Karikaturi, Skopje 30524
Svijashsk Architectural and Artistic Museum-Preserve, Svijažsk 37812
Svijažskij Architekturno-chudožestvennyj Muzej, Svijažsk 37812
Svjato-Troickij Sobor, Sankt-Peterburg 37728
Svjatogorskij Monastyr, Puškinskie Gory 37588
Svrzina Kuća, Sarajevo 04282
Svyatogorsk Monastery, Puškinskie Gory 37588
Svyatoslav Richter Museum-Appartment at the Pushkin Museum of Fine Arts, Moskva 37322
Swaffham Museum, Swaffham, Norfolk 45808
Swakopmund Museum, Swakopmund 31757
Swalcliffe Barn, Swalcliffe 45809
Swaledale Museum, Reeth 45479
Swan Coach House Gallery, Atlanta 46456
Swan Hill Regional Art Gallery, Swan Hill 01541
Swan Valley Historical Museum, Swan River 07043
Swanage Museum, Swanage 45810
Swanage Railway, Swanage 45811
Swansea Museum, Swansea 45816
Swanton Mill, Ashford 43328
Swarovski Kristallwelten, Wattens 02898
Swarthout Memorial Museum, La Crosse 49838
Swayambunath Museum and Art Gallery, Kathmandu 31779
Swaziland National Museum, Lobamba 40403
Swedish Amber Museum, Höllviken 40595
Swedish American Museum Association of Chicago Museum, Chicago 47516
Swedish Association of Museum Curators, Nacka 55035
Swedish Federation of Friends of Museums, Enhorna 55036
Swedish House of Photography, Sundsvall 40958
Swedish Museum of Natural History, Stockholm 40920
Swedish Museum of Textile History, Borås 40471
Swedish National Portrait Gallery, Mariefred 40735
Swedish Railway Museum, Gävle 40533
Swedish Seamen's Shipping Office Museum, Uddevalla 40993
Swedish Tennis Hall of Fame, Båstad 40454
Swedish Toy Theatre Museum, Nyköping 40781
Sweeney Art Gallery, Riverside 52375
Sweet Briar College Art Gallery, Sweet Briar 53391
Sweet Briar Museum, Sweet Briar 53392
Sweet Shop Museum, Donghe 42179
Sweetwater County Historical Museum, Green River 49043
Swensbylijda, Svensbyn 40966
Swenson Memorial Museum of Stephens County, Breckenridge 47003
Swift County Historical Museum, Benson 46741
Swift Current Museum, Swift Current 07045
Swigart Museum, Huntingdon 49423
Swindler's Ridge Museum, Benton 46742
Swindon Museum and Art Gallery, Swindon 45820
Swinford Museum, Filkins 44228
Swisher County Archives and Museum Association, Tulia 53606
Swiss Alpine Museum, Bern 41217
Swiss Customs Museum, Gandria 41396
Swiss Fire Museum Basle, Basel 41183
Swiss Gas Pump Museum, Gänsbrunnen 41395
Swiss Historical Village, New Glarus 51045
Swiss Institute - Contemporary Art, New York 51265
Swiss Museum for Ceramics and Glass, Genève 41416
Swiss Museum of Stained Glass and Glass Art, Romont 41773
Swiss Museum of Transport, Luzern 41600
Swiss National Library, Bern 41215
Swiss Science Center Technorama, Winterthur 42039
Swiss Sports Museum, Basel 41185
Swiss-Type Automatic Lathe and History Museum of Moutier, Moutier 41652
Swissminiatur, Melide 41622
Switzerland County Historical Society Museum, Vevay 53731
Swope Art Museum, Terre Haute 53483
Swords and Ploughshares Museum, Kars 06044
Swords Into Plowshares Peace Center and Gallery, Detroit 48130
Sydhimmerlands Museum, Hobro 10083
Sydney and Louisburg Railway Museum, Louisbourg 06199
Sydney Children's Museum, Merrylands 01299
Sydney Heritage Fleet, Pyrmont 01445
Sydney Jewish Museum, Darlinghurst 01026
Sydney Observatory, Sydney 01561
Sydney Tramway Museum, Loftus 01236
Sydsjællands Museum, Vordingborg 10320
Sygun Museum of Wales, Beddgelert 43431
Sykepleiemuseum, Oslo 33753
Syker Vorwerk, Syke 21992
Sykkylven Naturhistorisk Museum, Sykkylven 33898
Sylter Heimatmuseum, Sylt 21995
Sylvia Plotkin Judaica Museum, Scottsdale 52978
Synagóga - Centrum Súčasného Umenia, Trnava 38380
Synagoge, Bourtange 31993
Synagoge, Worms 22689
Synagoge Saffig, Saffig 21467
Synagoge Stommeln, Pulheim-Stommeln 21138
Synagoge und Jüdisches Museum Ermreuth, Neunkirchen am Brand 20639
Synagoge Wittlich, Wittlich 22624
Synagogen-Museum, Urspringen 22189
Synagogue for the Arts, New York 51266
Syndicat Mixte de Brouage, Hiers-Brouage 13217
Synskadades Museum, Enskede 40499
Syon House, London 44994
Syracuse University Art Collection, Syracuse 53403
Sysmäen Kotiseutumuseo, Sysmä 11342
Syzranskij Gorodskoj Kraevedčeskij Muzej, Syzran 37823
Szabadtéri Bányászati Múzeuma, Tatabánya 23921
Szabadtéri Néprajzi Gyűjtemény, Szenna 23870
Szabadtéri Néprajzi Múzeum, Nagyvázsony 23730
Szabadtéri Néprajzi Múzeum, Szentendre 23880
Szabadtéri Néprajzi Múzeum, Tihany 23927
Szabadtéri Néprajzi Múzeum - Tájház, Bakonybél 23349
Szalay Collection of the Miskolc Gallery of Fine Arts - Petró House, Miskolc 23714
Szalay-Gyűjtemény - Petróház, Miskolc 23714
Szálkahalma Guard House of the Hortobágy National Park, Balmazujváros 23359
Szálkahalmi Orháza, Balmazujváros 23359
Szántó Kovács János Területi Múzeum, Orosháza 23740
Szántó Piroska Emlékmúzeum, Budapest 23476
Szatmárcseke Region House, Szatmárcseke 23838
Szatmárcsekei Tájház, Szatmárcseke 23838
Szatmári Múzeum, Mátészalka 23693
Széchenyi István Emlékmúzeum, Nagycenk 23723
Széchenyi Patika, Győr 23586
Széchenyi Pharmacy Museum, Győr 23586
Szegedy Róza Ház Irodalmi Emlékmúzeum, Badacsony 23344
Székesegyház Treasury and Diocesan Collection, Vác 23945
Székesegyházi Kincstár és Egyházmegyei Gyűjtemény, Vác 23945
Székesfehérvári Egyházmegyei Múzeum, Székesfehérvár 23859
Szekler National Museum, Sfântu Gheorghe 36602
Szent István Király Múzeum, Székesfehérvár 23860
Szent István Király Múzeum - Rendház, Székesfehérvár 23861
Szentendre Gallery, Szentendre 23881
Szentendrei Képtár, Szentendre 23881
Szépművészeti Múzeum, Budapest 23477
Szerb Egyházy Múzeum, Szentendre 23882
Szerecsen Patikamúzeum, Pécs 23776
Szerecsen Pharmacy Museum, Pécs 23776
Színháztörténeti és Színészmúzeum, Miskolc 23715
Szlovák Emlékház, Tótkomlós 23936
Szlovák Tájház, Békéscsaba 23372
Szlovák Tájház, Kiskőrös 23665

S

Szlovák Tájház, Szarvas 23835
Szlovák Tájház, Tótkomlós 23937
Szolnay-Museum, Potzneusiedl 02517
Szolnok Gallery, Szolnok 23898
Szolnoki Galéria, Szolnok 23898
Szombathely Roman Catholic Collection and Treasury, Szombathely 23905
Szombathelyi Képtár, Szombathely 23904
Szombathelyi Római Katolikus Egyházmegyei Gyűjtemény és Kincstár, Szombathely 23905
Szőnyi István Emlékmúzeum, Zebegény 23977
Szórakaténusz Játékmúzeum, Kecskemét 23655
Szórakaténusz Toy Museum, Kecskemét 23655
Sztuka Dalekiego Wschodu, Toruń 35197
Szuz Mária Kegy- és Emléktárgyak Gyűjteménye, Gyula 23600

T

T.B. Ferguson Home, Watonga 53983
T.C. Agricultural Bank Museum, Ankara 42712
T.C. Steele State Historic Site, Nashville 50961
T.C. Ziraat Bankası Müzesi, Ankara 42712
T.Ch. Malsagov Regional Museum of Ingushetiya, Nazran 37448
T. Evseev National Museum of the Republic Marij El, Joškar-Ola 36980
T.F. Chen Cultural Center, New York 51267
T.G. Masaryk's House, Čejkovice 09453
T.N. Chrenikov House-Museum, Elec 36907
T.N. Granovsky House-Museum, Orël 37504
T. Tumanishvili Children Toy Museum, Tbilisi 16384
T.W. Wood Gallery and Arts Center, Montpelier 50824
Ta' Kola Windmill, Xaghra 30660
Taalintehtaan Ruukinmuseo, Dalsbruk 10718
Tabak Musea, Geraardsbergen 03654
Tabak-Museum der Stadt Hockenheim, Hockenheim 19207
Tabak- und Zigarrenmuseum, Menziken 41626
Tabakhistorische Sammlung Reemtsma im Museum der Arbeit, Hamburg 18962
Tabakmuseum, Lorsch 20036
Tabakmuseum, Oberzeiring 02468
Tabakmuseum, Schwedt / Oder 21679
Tabako to Shio no Hakubutsukan, Tokyo 29479
Tabakpfeifenmuseum und Museum für Stadtgeschichte, Ruhla 21442
Tabakspeicher, Nordhausen 20740
Tabán Múzeum, Budapest 23478
Tabán Region House, Szolnok 23899
Tabáni Tájház, Szolnok 23899
Taber and District Museum, Taber 07053
Table Rock Historical Society Museum, Table Rock 53404
Tabley House, Knutsford 44624
Tabor Opera House Museum, Leadville 50007
Táborvárosi Múzeum, Budapest 23479
Tabusintac Centennial Museum, Tabusintac 07054
Tachihara Michizo Kinenkan, Tokyo 29480
Taco Bell Discovery Science Center, Santa Ana 52855
Tacoma Art Museum, Tacoma 53414
Tactual Museum, Kallithea 23029
Tadas Ivanauskas Zoology Museum, Kaunas 30252
Tadeusz Kulisiewicz Centre of Graphics and Drawings, Kalisz 34707
Tadre Mølle, Hvalsø 10106
Tändsticksmuseet, Jönköping 40616
The Taft Museum of Art, Cincinnati 47565
Taganrog Municipal Museum, Taganrog 37829
Taganrog Picture Gallery, Taganrog 37827
Taganrogskaja Kartinnaja Galereja, Taganrog 37827
Taganrogskij Gosudarstvennyj Literaturnyj i Istoriko-architekturnyj Muzej-zapovednik, Taganrog 37828
Taganrogskij Kraevedčeskij Muzej, Taganrog 37829
Taganrogskij Muzej im. A.P. Čechova, Taganrog 37830
Tagawa Bijutsukan, Tagawa 29309
Tagawa Museum of Art, Tagawa 29309
Tagelöhner- und Brennereimuseum Faißt, Bad Peterstal-Griesbach 16803
Tagore Memorial Museum, Santiniketan 24379
Taha Hussein Museum, Cairo 10521
Tahir Pasa Konagi, Mudanya 42911
Tahsis Museum, Tahsis 07059
Tahydromiko Moyseio, Athinai 22952
Tai Hao Ling Museum, Huaiyang 07829
Tai'an Museum, Tai'an 08263
Taicang Museum, Taicang 08264
Taichung Distillery Museum, Taichung 42354
Taidekeskus Mältinranta, Tampere 11361
Taidekoti Kirpilä, Helsinki 10827
Taidemuseo, Kokkola 10975
Taieri Historical Museum, Outram 33166
Taihape and District Museum, Taihape 33203
Taihe Museum, Taihe 08266
Taikomosios Dailės Muziejus, Vilnius 30434
Tailai Museum, Tailai 08267
Tain and District Museum, Tain 45822
Tainan County History Museum, Syuejia 42346
Tainan Municipal Cultural Center, Visual Art Section, Tainan 42369
Tainan National University of the Arts Museum, Kuantien 42252
Tainan Natural History Museum, Zuojhen 42489
Tainan Salt Museum, Cigu Township 42166
Taining Museum, Taining 08268
Taipa House Museum, Macau 08008
Taipei Astronomical Museum, Taipei 42426
Taipei Children's Art Museum, Taipei 42427
Taipei Fine Arts Museum, Taipei 42428
Taipei Story House, Taipei 42429
Taiping Kingdom History Museum, Nanjing 08050
Tairawhiti Museum, Gisborne 33109
Taishan Doll Museum, Taishan 42443
Taishan Museum, Taishan 08269
Taishun Museum, Taishun 08270
Taitung Aboriginal Exhibition Room, Taitung 42445
Taitung Museum of Natural History, Chengkung 42162
Taivalkosken Kotiseutumuseo, Taivalkoski 11344
Taivassalon Museo, Taivassalo 11345
Taiwan Aboriginal Culture Park, Majia 42270
Taiwan Coal Mine Museum, Pingsi 42289
Taiwan Craft Center, Taipei 42430
Taiwan Democratic Movement Museum, Wujie 42461
Taiwan Folk Arts Museum, Taipei 42431
Taiwan Folk Culture Museum, Bali 42143
Taiwan Folklore Museum, Taichung 42355
Taiwan Forestry Exhibition Hall, Taipei 42432
Taiwan High Speed Rail Exhibition Center, Jhubei 42224
Taiwan Historica, Nantou 42283
Taiwan International Visual Arts Center, Taipei 42433
Taiwan Medical Museum, Kaohsiung 42244
Taiwan Museum of Art Sanri, Sanyi 42318
Taiwan Mushroom Museum, Wufeng 42460
Taiwan Nature Lacquer Culture Museum, Taichung 42356
Taiwan Silk Culture Museum, Gongguan 42193
Taiwan Sugar Industry Museum, Tainan 42370
Taiwan Television Exposition, Taipei 42434
Taiwan Temple and Shrine Art Museum, Douliou 42184
Taiwan Theater Museum, Yilan 42469
Taiyuan Painting Gallery, Taiyuan 08275
Taizhou Museum, Taizhou 08276
Taizi Harada Art Museum, Suwa 29304
Taizi Harada Bijutsukan, Suwa 29304
Taizz National Museum, Taizz 54652
Tájház, Balatonszentgyörgy 23357
Tájház, Buzsák 23499
Tajmyr Regional Museum, Dudinka 36880
Tajmyrskij Okružnoj Kraevedčeskij Muzej, Dudinka 36880
Takachiho-cho Rekishi oyo Minzoku Hakubutsukan, Takachiho 29313
Takachiho History and Folklore Museum, Takachiho 29313
Takahama-shiritsu Kawara Bijutsukan, Takahama 29314
Takahashi Setsuro Art Museum of Azumino, Azumino 28387
Takamatsu City Museum of Art, Takamatsu 29322
Takamatsuzuka-Hekigakan, Asuka 28381
Takamatsuzuka Wall Painting Museum, Asuka 28381
Takamura's Cottage Memorial Museum, Hanamaki 28494
Takaoka Art Museum, Takaoka 29326
Takaoka Municipal Museum, Takaoka 29327
Takaoka-shi Bijutsukan, Takaoka 29326
Takaoka Shiritsu Hakubutsukan, Takaoka 29327
Takarazuka Daigaku Hakabutsukan, Takaraduka 29328
Takarazuka University Museum, Takaraduka 29328
Takasaki City Gallery, Takasaki 29332
Takasaki Museum of Art, Takasaki 29333
Takasaki-shi Bijutsukan, Takasaki 29333
Takasawa Folk Craft Museum, Niigata 29063
Takayama Jinya, Takayama 29344
Takayama Museum of Local History, Takayama 29345
Takayama-shi Kyoudokan, Takayama 29345
Takehisa Yumeji Ikaho Kinenkan, Ikaho 28561
Takenaka Carpentry Tools Museum, Kobe 28739
Takenouchi Kaido Rekishi Hakubutsukan, Taishi 29310
Takenouchi Road Historical Museum, Taishi 29310
Takht-e-Jamshid, Marvdashd 24630
Takht-e-Jamshid, Shiraz 24649
Tako no Hakubutsukan, Tokyo 29481
Tal Museum Engelberg, Engelberg 41352
Talana Museum and Heritage Park, Dundee 38601
Talbot Arts and Historical Museum, Talbot 01567
Talbot House, Poperinge 03954
Talbot Rice Gallery, Edinburgh 44161
Taldykorgan Historical and Regional Museum, Taldykorgan 29731
Taldykorganskij Istoriko-Kraevedčeskij Muzej, Taldykorgan 29731
Tale of Genji Museum, Uji, Kyoto 29559
Tales of the Old Gaol House, King's Lynn 44588
Talkeetna Historical Society Museum, Talkeetna 53421
Tallahassee Museum of History and Natural Science, Tallahassee 53429
Tallbo, Järbo 40611
Taller Puertoriqueno, Philadelphia 51883
Taller Tanamá y Casa Jesús T. Piñero, Canóvanas 36034
Tallinn City Museum, Tallinn 10633
Tallinn City Museum, Tallinn 10643
Tallinna Linnamuuseum, Tallinn 10643
Talmuseum Hüttschlag, Hüttschlag 02105
Talmuseum Kaunertal, Feichten 01876
Talmuseum Lachitzhof, Klein Sankt Paul 02194
Talmuseum Samnaun, Samnaun Compatsch 41801
Talmuseum Ursern, Andermatt 41115
Talomuseo, Kuortane 11013
Talomuseo, Polvijärvi 11217
Talonpoikaismuseo Yli-Kirra, Punkalaidum 11240
Talonpojanmuseo, Kälviä 10901
Talsi Regional Museum, Talsi 30128
Talsu Novada Muzejs, Talsi 30128
Talvinen Cottage, Humppila 10844
Talvisen Työläismökki, Humppila 10844
Talvisotanäyttely, Kuhmo 10999
Tam O'Shanter Experience, Alloway 43284
Tama County Historical Museum, Toledo 53519
Tamagawa Kindai Bijutsukan, Tamagawa 29349
Tamagawa Modern Art Museum, Tamagawa 29349
Taman Warisan Tasek Merimbun, Mukim Rambai 04969
Tamara Hanum Yodgorlik Muzeyi, Toshkent 54488
Tamara Khanum House-Museum, Toshkent 54488
Tamastslikt Cultural Institute, Pendleton 51756
Tamba Tachikui Traditional Ceramic Center, Sasayama 29232
Tamborine Mountain Heritage Centre, Eagle Heights 01051
Tambov Picture Gallery, Tambov 37832
Tambovskaja Oblastnaja Kartinnaja Galereja, Tambov 37832
Tambovskij Oblastnoj Kraevedčeskij Muzej, Tambov 37833
Tamkang University Maritime Museum, Tamsui 42449
Tammiharjun Sairaalamuseo, Ekenäs 10723
Tampa Bay History Center, Tampa 53436
Tampa Museum of Art, Tampa 53437
Tampere Art Museum Moominvalley, Tampere 11365
Tampere Art Museum - Regional Art Museum of Pirkanmaa, Tampere 11364
Tampere Mineral Museum, Tampere 11351
Tampere Museum of Contemporary Art, Tampere 11363
Tampere Natural History Museum, Tampere 11362
Tampereen Luonnontieteelinen Museo, Tampere 11362
Tampereen Nykytaiteen Museo, Tampere 11363
Tampereen Taidemuseo - Pirkanmaan Aluetaidemuseo, Tampere 11364
Tampereen Taidemuseon Muumilaakso, Tampere 11365
Tamralipta Museum and Research Center, Tamluk 24400
Tamworth Castle Museum, Tamworth 45823
Tamworth Regional Gallery, Tamworth 01569
Tamyang Bamboo Museum, Tamyang 29937
Tan Oe Pang Art Studio, Singapore 38181
Tana Museum, Tana 33900
Tanabe Bijutsukan, Matsue 28891
Tanabe Museum of Art, Matsue 28891
Tanah Art Museum, San Antonio 04180
Tanaka Bijutsukan, Suzaka 29305
Tanaka Museum of Art, Suzaka 29305
Tanakamaru Collection, Fukuoka 28437
Tanakami Mineral Museum, Otsu 29167
Tandanya, Adelaide 00783
Tanegashima Teppokan, Nishinomote 29076
Tang Teaching Museum and Art Gallery, Saratoga Springs 52926
Tangdai Art Museum, Xian 08361
Tangjia Glaswegiansu Guan, Xunyi 08415
Tangjia Village Folkways Museum, Xunyi 08415
Tango Chirimen Rekishi Hakubutsukan, Nodagawa 29080
Tangshan Museum, Tangshan 08277
Tanimura Art Museum, Itoigawa 28602
Tanimura Bijutsukan, Itoigawa 28602
Tank Museum, Axvall 40450
Tank Museum, Bovington 43570
Tank Museum, Parola 11193
Tank Museum, Poznań 35033
Tanks Museum, Avlona 22956
The Tanks of Aden, Aden 54644
Tanner House - House of Artisans, Debrecen 23521
Tanta Museum, Tanta 10542
Tantaquidgeon Indian Museum, Uncasville 53643
Tante Blanche Museum, Madawaska 50373
Tanum Museum of Rock Carvings, Tanumshede 40968
Tanums Hällristningsmuseum - Underslös, Tanumshede 40968
Tanyamúzeum, Szabadkígyós 23831
Tanyamúzeum, Tótkomlós 23938
Tanzimat Müzesi, İstanbul 42840
Taos Center for the Arts, Taos 53444
Taos Historic Museums, Taos 53445
Taoyuan County Museum of Natural History, Taoyuan 42453
Het Tapijtmuseum, Genemuiden 32245
Tapisseries Jean Lurçat, Rocamadour 15063
Tappantown Historical Society Museum, Tappan 53446
Tapulimuseo, Lempäälä 11061
Tar Tunnel, Coalport 43862
Tara and District Pioneer Memorial Museum, Tara 01571
Taralga Historical Museum, Taralga 01572
Taras H. Shevchenko Museum & Memorial Park Foundation, Toronto 07141
Tarble Arts Center, Charleston 47354
The Tareq Rajab Museum, Hawelli 29947
Tari László Múzeum, Csongrád 23513
Taro Okamoto Bijutsukan, Kawasaki 28695
Taro Okamoto Kinenkan, Tokyo 29482
Taro Okamoto Memorial Museum, Tokyo 29482
Taro Okamoto Museum of Art, Kawasaki 28695
Tarpon Springs Cultural Center, Tarpon Springs 53449
Tarrant County College Southeast The Art Corridor, Arlington 46354
Tarrytowns Museum, Tarrytown 53452
Tarskij Istoriko-kraevedčeskij Muzej, Tara 37834
Tarsus Müzesi, Tarsus 42949
Tarsus Museum, Tarsus 42949
Tartous Museum, Tartous 42139
Tartu Art House, Tartu 10655
Tartu Art Museum, Tartu 10656
Tartu County Museum, Elva 10562
Tartu Kunstimaja, Tartu 10655
Tartu Kunstimuuseum, Tartu 10656
Tartu Linnamuuseum, Tartu 10657
Tartu Mänguasjamuuseum, Tartu 10658
Tartu Toy Museum, Tartu 10658
Tartu Ülikooli Ajaloomuuseum, Tartu 10659
Tartu Ülikooli Geoloogiamuuseum, Tartu 10660
Tartu Ülikooli Kunstimuuseum, Tartu 10661
Tartu Ülikooli Zooloogiamuuseum, Tartu 10662
Tartu University Museum of Geology, Tartu 10660
Tartu University Museum of Zoology, Tartu 10662
Tartumaa Muuseum, Elva 10562
Tarusa Museum of the Tsvetaeva Family, Tarusa 37837
Tarusa Picture Gallery, Tarusa 37835
Tarusskaja Kartinnaja Galereja, Tarusa 37835
Tarusskij Kraevedčeskij Muzej, Tarusa 37836
Tarusskij Muzej Semji Cvetaevych, Tarusa 37837
Tarvasjoen Kotiseutumuseo, Tarvasjoki 11371
Tarvastu Kihelkonnamuuseum, Tarvastu 10663
Taryhy we Etnografiki Muzeý, Mary 42978
Taryhy we Etnografiki Muzeý, Türkmenabat 42982
Taryhy Ýer Abiwerd, Kaka 42976
Taryhy Ýer Gadymy Merw, Mary 42979
Taryhy Ýer Gökdepe Galasy, Gökdepe 42975
Taryhy Ýer Könürgenç, Könürgenç 42977
Taryhy Ýer Nusaý, Nusaý 42980
Taryhy Ýer Saragt, Saragt 42981
Taş Eserler Müzesi - Lapidari, Lefkoşa 33373
Tas Oda Etnografya Müzesi, Burdur 42738
Tasaki Bijutsukan, Karuizawa 28672
Tasaki Museum of Art, Karuizawa 28672
Tasek Merimbun Heritage Park, Mukim Rambai 04969
Tasiilap Katersugaasivia, Tasiilaq 23270
Tåsinge Skipperhjem og Folkemindesamling, Svendborg 10294
Tasmania Distillery Museum, Hobart 01158
Tasmanian Medical History Museum, South Hobart 01514
Tasmanian Museum and Art Gallery, Hobart 01159
Tasmanian Wool Centre, Ross 01470
Tasneem Arts Gallery, Peshawar 34037
Tassenmuseum Hendrikje, Amstelveen 31823
Tasviriy San'at Muzeyi - Andijon Qal'asi, Andijon 54442
Tata Castle, Tata 23919
Tatabánya Múzeum, Tatabánya 23922
Tatai Vár, Tata 23919
Tatar State Museum of Fine Arts, Kazan 37027
Tatarian Town - Archaeological Museum-Preserve, Stavropol 37807
Tatarna Monastery Museum, Granitsa 23010
Tatarskij Gosudarstvennyj Muzej Izobrazitelnych Iskusstv, Kazan 37027
Tatarskoe Gorodišče - Archeologičeskij i prirodnyj Muzej-zapovednik, Stavropol 37807
Tate Britain, London 44995
Tate Geological Museum, Casper 47273
Tate House, Portland 52068

S
T

Tate Liverpool, Liverpool 44746
Tate Modern, London 44996
Tate Saint Ives, Saint Ives, Cornwall 45579
Tateshina World Arts & Crafts Museum, Tateshina 29352
Tateyama Hakubutsukan anna Toyama-ken, Tateyama, Toyoma 29354
Tateyama Museum of Toyama, Tateyama, Toyoma 29354
Tatham Art Gallery, Pietermaritzburg 38707
Tatra Technical Museum, Kopřivnice 09596
Tatranská Galéria, Poprad 38333
Tatsuno Bijutsukan, Tatsuno 29355
Tatsuno Museum of Art, Tatsuno 29355
Tatsuuma Archaeological Museum, Nishinomiya 29075
Tatsuuma Kokokan, Nishinomiya 29075
Tatton Park, Knutsford 44625
Tattoo Museum, Amsterdam 31893
Tatura Irrigation and Wartime Camps Museum, Tatura 01574
Tauberfränkisches Landschaftsmuseum, Tauberbischofsheim 22010
Tauberländer Dorfmuseum, Weikersheim 22373
Tauernstraßen-Museum, Eben im Pongau 01832
Tauragnai Area Museum, Tauragnai 30377
Tauragnų Krašto Muziejus, Tauragnai 30377
Tauranga Museum, Tauranga 33206
Tauriska-Galerie, Neukirchen am Großvenediger 02431
Tauro Apygardos Partizanu ir Tremties Muziejus, Marijampolė 30309
Tavastilan Kotiseutumuseo, Mietoinen 11111
Taweelah Museum of Mummies, Mahweet 54648
Tawhiti Museum, Hawera 33121
Tawi-Tawi Ethnological Museum, Bongao 34348
Taxameter-Museum, Hamburg 18963
Taxandriamuseum, Turnhout 04102
Taylor Brown and Sarah Dorsey House, Midland 50675
Taylor-Grady House, Athens 46415
Taymori Hall, Isfahan 24613
TBA Exhibition Space, Chicago 47517
Tbilisi State Museum of Anthropology and Ethnography, Tbilisi 16385
Tchechov Memorial Museum, Jalta 43049
Tcherkasy Museum of Fine Art, Čerkasy 43001
Tcherkasy Regional Museum of Local Lore, Čerkasy 43002
Tcherkech Museum of Political Banishment in Yakutia, Čerkech 36853
Tchita Regional Art Museum, Čita 36868
Te Amorangi Trust Museum, Rotorua 33197
Te Awamutu Museum, Te Awamutu 33207
Te Kuiti and District Historical Society Museum, Te Kuiti 33208
Te Manawa Art, Palmerston North 33175
Te Manawa Life Mind, Palmerston North 33176
Te Papa, Wellington 33260
Te Tuhi - The Mark, Auckland 33059
Te Whare Taonga O Akaroa, Akaroa 33033
Te Whare Whakaaro o Pito-one, Hutt City 33127
Tea Ceremony School Museum, Kyoto 28869
Tea Museum, Bois-Cheri 30676
Tea Museum of Iran, Lahijan 24627
Teackle Mansion, Princess Anne 52142
Teatermuseet i Hofteatret, København 10154
Teatermuseet i Oslo, Oslo 33754
Teatermuseet og Interkulturelt Museum, Oslo 33755
Teatralnaja Galereja na Maloj Ordynke, Moskva 37407
Teatralnyj Muzej, Kazan 37028
Teatralnyj Salon na Tverskom Bulvare, Moskva 37408
Teatre-Museu Dalí, Figueres 39300
Teatterimuseo, Helsinki 10828
Tecck an Phiarsaigh, Rosmuck 24894
The Tech Museum of Innovation, San Jose 52811
Techni-Macedonian Artistic Association Kilkis, Kilkis 23054
Technical Museum, Zagreb 09012
Technical Museum of Slovenia, Bistra pri Vrhniki 38399
Technical Museum of Slovenia, Ljubljana 38467
Technical Museum Paper Mill Zwoenitz, Zwönitz 22843
Technichus, Härnösand 40573
Technické Muzeum, Petřvald u Karviné 09711
Technické Muzeum Tatra, Kopřivnice 09596
Technické Muzeum v Brně, Brno 09440
Techniek Museum Delft, Delft 32066
Techniekmuseum Heim, Hengelo, Overijssel 32368
Technik anno dazumal - Museum Kratzmühle, Kinding 19539
Technik-Modell-Museum Rügen e.V., Samtens 21481
Technik-Museum, Soest 21798
Technik Museum Speyer, Speyer 21838
Technik u. Zweiradmuseum Dargen/Usedom, Dargen 17783
Technik- und Verkehrsmuseum, Stade 21853
Technikmuseum, Magdeburg 20131
Technikmuseum Hugo Junkers, Dessau 17835
Technikmuseum U-Boot "Wilhelm Bauer", Bremerhaven 17574
Technikon Natal Gallery, Durban 38613
Techniksammlung Backnang, Backnang 16663
Technikschau, Wismar 22614
Techningsmuseet, Laholm 40665
Techniquest, Cardiff 43754
Technische Sammlungen der Stadt Dresden, Dresden 17974
Technisches Denkmal Brikettfabrik Louise, Domsdorf bei Bad Liebenwerda 17890
Technisches Denkmal "Gaswerk", Neustadt, Dosse 20678
Technisches Denkmal Kupferhammer, Thießen 22053
Technisches Denkmal Museum Kalkwerk Lengefeld, Lengefeld 19951
Technisches Denkmal Neue Hütte, Schmalkalden, Kurort 21581
Technisches Denkmal Tobiashammer, Ohrdruf 20891
Technisches Denkmal Ziegelei Hundisburg, Haldensleben 18886
Technisches Eisenbahnmuseum, Lieboch 02299
Technisches Halloren- und Salinenmuseum, Halle, Saale 18905
Technisches Kulturdenkmal Ölmühle, Marbach am Neckar 20183
Technisches Landesmuseum Mecklenburg-Vorpommern, Wismar 22615
Technisches Museum Alte Wasserkunst, Bautzen 16942
Technisches Museum Alte Ziegelei und Turmwindmühle, Falkenstein, Harz 18294
Technisches Museum der Bandweberei, Großröhrsdorf 18803
Technisches Museum der Pforzheimer Schmuck- und Uhrenindustrie, Pforzheim 21031
Technisches Museum Frohnauer Hammer, Annaberg-Buchholz 16556
Technisches Museum Holzschleiferei Weigel, Breitenbrunn 17542
Technisches Museum Ölmühle, Pockau 21069
Technisches Museum Papiermühle, Zwönitz 22843
Technisches Museum Silberwäsche Antonsthal, Breitenbrunn 17543
Technisches Museum Wien, Wien 03079
Technisches Museum Zechau, Rositz 21380
Technisches Schaudenkmal Gießerei Heinrichshütte, Wurzbach 22750
Technology Museum, Petřvald u Karviné 09711
Technopolis, Mechelen 03864
Technoseum, Mannheim 20174
Teddy Bear Museum, Hof 19221
Teddy Bear Museum, Seogwipo 29862
The Teddy Bear Museum, Stratford-upon-Avon 45786
Teddy Bear Museum of Naples, Naples 50953
Teddy Bear Shop and Museum, Ironbridge 44533
Teddy Museum Hof, Hof 19221
Teddybär Museum, Luzern 41599
Teddybärmuseum, Sempach 41861
Teddymuseum, Klingenberg am Main 19569
Tees Archaeology, Hartlepool 44409
Teeterville Pioneer Museum, Teeterville 07062
Téglamúzeum, Veszprém 23962
Tehkummah Township Little Schoolhause and Museum, Tehkummah 07063
Tehni Art Gallery, Thessaloniki 23221
Tehnički Muzej, Zagreb 09012
Tehnični Muzej Rudarstva, Zagorje ob Savi 38538
Tehniški Muzej Slovenije, Bistra pri Vrhniki 38399
Tehniški Muzej Slovenije, Ljubljana 38467
Teifi Valley Railway, Henllan 44451
Teignmouth and Shaldon Museum, Teignmouth 45832
Teine Memorial Museum, Sapporo 29228
Teipei Fo Guang Yuan Art Gallery, Taipei 42435
Teishin Sogo Hakubutsukan, Tokyo 29483
Teisko Museum, Tampere 11366
Teiskon Kotiseutumuseo, Tampere 11366
Tejipari Emléktár, Székesfehérvár 23862
Tekakwitha Fine Arts Center, Sisseton 53124
Tekfur Sarayi, İstanbul 42841
Tekiho Memorial Museum of Dolls, Ogose 29095
Tekirdağ Devlet Güzel Sanatlar Galerisi, Tekirdağ 42950
Tekirdağ Müzesi, Tekirdağ 42951
Tekirdağ Museum, Tekirdağ 42951
Tekirdağ State Fine Art Gallery, Tekirdağ 42950
Tekisui Bijutsukan, Ashiya-shi 28377
Tekisui of Artseum, Ashiya-shi 28377
Tekniikan Museo, Helsinki 10829
Teknikens Hus, Luleå 40700
Teknikens och Sjöfartens Hus, Malmö 40727
Tekniska Kvarnen, Örebro 40791
Tekniska Museet, Stockholm 40945
Tekovské Muzeum v Leviciach, Levice 38282
Tel Aviv Museum of Art, Tel Aviv 25122
Tel Hai Museum, Tel Hai 25126
Tele-Gallery, Pécs 23777
Tele-Museum, Qaqortoq 23266
Teleborgs Slott, Växjö 41042
Telecommunication Museum, Reykjavík 24034
Telecommunications Museum, LaSalle 06142
Telecommunications Museum, Madrid 39549
Telefónia Múzeum, Budapest 23480
Telefonmuseet, Hellerup 10059
Telefonmuseum Hittfelder Bahnhof, Seevetal 21726
Telefonmuseum im Greuterhof, Islikon 41496
Telefoonmuseum, Drouwen 32160
Telegaléria, Pécs 23777
Telekessy Patikamúzeum, Eger 23544
Telekessy Pharmacy Museum, Eger 23544
Telemark Kunstnersenter, Porsgrunn 33772
Telemark og Grenland Museum, Skien 33836
Teleorman District Museum of History, Alexandria 36107
Telephone Museum, Budapest 23480
Telephone Museum, Ellsworth 48372
Telephone Pioneer Museum of New Mexico, Albuquerque 46187
Televízió Technikatörténeti és Műsor Múzeum, Budapest 23481
Telfair Museum of Art, Savannah 52947
Telkwa Museum, Telkwa 07064
Tell-Museum, Bürglen 41270
Tell Qasile Archaeological Site, Tel Aviv 25123
Telling the Menonite Story - Visitor Centre, Saint Jacobs 06824
Telluride Historical Museum, Telluride 53461
Tēlnieka Voldemāra Jākobsona Memoriālāis Muzejs, Bebri 29990
Teloglion Foundation of Art, Thessaloniki 23222
Telstra Museum Brisbane, Clayfield 00979
TELUS World of Science, Edmonton 05755
Temecula Valley Museum, Temecula 53462
Temel Sanat Kolleksiyon, İstanbul 42842
Temiskaming Art Gallery, Haileybury 05926
Temlett House, Grahamstown 38634
Temmeksen Kotiseutumuseo, Temmes 11372
Tempe Center for the Arts, Tempe 53471
Tempe Historical Museum, Tempe 53472
Tempelhof Museum, Berlin 17271
Tempelmuseum Frauenberg, Seggauberg 02737
Temperance Tavern, Newcomerstown 51309
Tempietto Rossiniano della Fondazione Rossini, Pesaro 27158
Tempio Voltiano, Como 25956
Templar and Medieval Museum, Cogolin 12568
Temple des Mille Bouddhas, La Boulaye 13319
Temple Gallery, Philadelphia 51884
Temple Judea Museum, Elkins Park 48353
Temple Museum of Religious Art, Beachwood 46658
The Temple Museum of Religious Art, Cleveland 47642
Temple Newsam House, Leeds 44666
Temple of Aphrodite, Kouklia 09365
Temple Réformé Mulhouse, Mulhouse 14330
Templerhaus, Amorbach 16543
Templin Historical Village, Boonah 00876
Temporäre Galerie Schloß Neersen, Willich 22583
Téms Swiya Museum, Sechelt 06941
Temuka Courthouse Museum, Temuka 33209
Ten Broeck Mansion, Albany 46164
Ten Duinen 1138, Koksijde 03756
Ten Sing Ceramic Art Village, Jiji 42231
Tenala Hembygdsmuseum, Tenala 11373
Tenbury and District Museum, Tenbury Wells 45835
Tenby Museum and Art Gallery, Tenby 45836
Tenement House, Glasgow 44305
Teng County Museum, Tengxian 08279
Tengzhou Museum, Tengzhou 08280
Tenino Depot Museum, Tenino 53477
Tenis museum, Rīga 30113
Tenisa muzejs, Rīga 30113
Tenmon Bijutsukan, Hirakata 28525
Tenmon Museum of Art, Hirakata 28525
Tennent Art Foundation Gallery, Honolulu 49354
Tennessee Association of Museums, Piney Flats 55183
Tennessee Central Railway Museum, Nashville 50974
Tennessee Historical Society Museum, Nashville 50975
Tennessee Museum of Aviation, Sevierville 53040
Tennessee State Museum, Nashville 50976
Tennessee Valley Art Center, Tuscumbia 53625
Tennessee Valley Railroad Museum, Chattanooga 47414
Tenri Daigaku Sankokan, Tenri 29358
Tenri University Annex Museum, Tenri 29358
Tenshi Bijutsukan, Nasu 29052
Tenshi Museum, Nasu 29052
Tenshin Memorial Museum of Art, Ibaraki, Kitaibaraki 28708
Tensta Konsthall, Spånga 40876
Tenterden and District Museum, Tenterden 45840
Tenterfield Centenary Cottage, Tenterfield 01576
Tentoonstellingsruimte 't Oute Hus, Burgh-Haamstede 32028
Tentoonstellingzaal Hessenhuis, Antwerpen 03317
Teodoro P. Resurreccion Memorial Museum, Luna 34409
Teollisuustyöväen Asuntomuseo, Imatra 10858
Tepco Denryokukan, Tokyo 29484
Tepco Denryokukan, Yokohama 29608
Tepco Electric Energy Museum, Tokyo 29484
Tepia, Kikai Sangyo Kinenkan, Tokyo 29485
Teplitz-Schönauer Heimatmuseum, Frankfurt am Main 18398
Terada Torahiko Memorial Museum, Kochi 28748
Terengganu State Museum, Kuala Terengganu 30589
Teresia and Rafael Lönnström Home Museum, Rauma 11262
Teresia ja Rafael Lönnströmin kotimuseo, Rauma 11262
Terezín Memorial, Terezín 09864
Terezín Memorial - Columbarium, Ceremonial Halls and Central Morgue, Terezín 09866
Terezín Memorial - The Crematorium at the Jewish Cemetery, Terezín 09865
Terezín Memorial - The Ghetto Museum, Terezín 09869
Terezín Memorial - The Lesser Fortress, Terezín 09868
Terezín Memorial - The Magdeburg Barracks, Terezín 09867
Teriade Museum, Mytilini 23113
Term City Museum, Xiangcheng 08367
Természet Háza, Kecskemét 23656
Természettudományi Osztály, Pécs 23778
Termuridi Tarikhi Muzeyi, Toshkent 54489
Ternitzer Stahlstadtmuseum, Ternitz 02805
Ternopil Regional Museum of Art, Ternopil' 43193
Ternopil Regional Museum of Local Lore, Ternopil' 43194
Ternopil's'kyj Oblasnyj Chudožnij Muzej, Ternopil' 43193
Ternopil's'kyj Oblasnyj Krajeznavčyj Muzej, Ternopil' 43194
Terra Foundation for American Art, Giverny 13068
Terra Museum of American Art, Chicago 47518
Terra Nova National Park Museum, Glovertown 05877
Terra Sancta Museum, Nazareth 25072
Terra Sigillata Museum, Rheinzabern 21308
Terra-Vinea, Portel-des-Corbières 14922
Terrace Gallery, Orlando 51595
Terrace Hill Historic Site and Governor's Mansion, Des Moines 48103
Terracota Museu de Terrissa i Ceràmica Industrial, La Bisbal d'Empordà 39051
Terracotta Warriors Museum, Dorchester 44003
Terrain Gallery, New York 51268
Terrebonne Museum, Houma 49378
Territoires du Jazz, Marciac 13956
Territorial Capital-Lane Museum, Lecompton 50015
Territorial Statehouse State Museum, Fillmore 48542
Territory Town USA, Old West Museum, Okemah 51512
Terror Háza Múzeum, Budapest 23482
Tersløsegaard, Dianalund 09987
Tervajärven Talomuseo, Sotkamo 11328
Terwilliger Museum, Waterloo 53968
Tesoro del Duomo, Messina 26670
Tesoro del Duomo, Milano 26725
Tesoro del Duomo, Monreale 26784
Tesoro del Duomo, Vigevano 28267
Tesoro della Cattedrale, Palermo 27080
Tesoro della Cattedrale, Rimini 27411
Tesoro della Santa Maria della Scala, Chieri 25868
Tesoro di Santa Maria La Stella, Militello in Val di Catania 26732
Tesouro da Sé Patriarcal de Lisboa, Lisboa 35735
Tesouro-Museu da Catedral, Viseu 36025
Tesouro-Museu da Sé de Braga, Braga 35485
Tesouros de Santa Maria e São Pedro de Óbidos, Óbidos 35787
Tessai Bijutsukan, Takaraduka 29329
Tessai Museum, Takaraduka 29329
Tessedik Sámuel Múzeum, Szarvas 23836
Tessiner Mühlenmuseum, Tessin 22033
Testnevelési és Sportmúzeum, Budapest 23483
Testnevelési és Sportmúzeum Kiállítóterme, Budapest 23484
Tetbury Police Museum, Tetbury 45841
Tetley's Brewery Wharf, Leeds 44667
Teulon and District Museum, Teulon 07070
Teutopolis Monastery Museum, Teutopolis 53487
Teuvan Museo, Teuva 11374
Teviot District Museum, Roxburgh 33198
Tewkesbury Museum, Tewkesbury 45844
Texana Museum, Edna 48298
Texarkana Museums System, Texarkana 53489
Texas Air Museum, Rio Hondo 52354
Texas Association of Museums, Austin 55184
Texas Baptist Historical Center Museum, Brenham 47006
Texas City Museum, Texas City 53490
Texas Energy Museum, Beaumont 46680
Texas Forestry Museum, Lufkin 50319
Texas Governor's Mansion, Austin 46518

Texas Heritage Museum, Hillsboro 49288
Texas Historical Museum and Tobacco Museum, Texas 01580
Texas Maritime Museum, Rockport 52427
Texas Memorial Museum, Austin 46519
Texas Military Forces Museum, Austin 46520
Texas Music Museum, Austin 46521
Texas Pharmacy Museum, Amarillo 46253
Texas Ranger Hall of Fame and Museum, Waco 53780
Texas Seaport Museum, Galveston 48851
Texas Sports Hall of Fame, Waco 53781
Texas Woman's University Art Galleries, Denton 48062
Textil- és Textilruházati Ipartörténeti Múzeum, Budapest 23485
Textil- und Heimatmuseum, Sennwald 41862
Textil- und Rennsportmuseum, Hohenstein-Ernstthal 19246
Textilarkivet Västernorrland, Sollefteå 40869
Textildruckmuseum Mittelweiherburg, Hard 02050
Textile Conservation Centre, Winchester 46034
Textile Museum, Blönduós 23988
Textile Museum, Kuala Lumpur 30583
Textile Museum, Prato 27335
Textile Museum, Thimphu 04204
The Textile Museum, Washington 53921
Textile Museum of Canada, Toronto 07142
Textile Museum of TIBA, Česká Skalice 09460
Textile Printing Museum of Premià de Mar, Premià de Mar 39869
Textilforum, Herning 10074
Textilmuseet, Borås 40471
Textilmuseum, Hauptwil 41472
Textilmuseum, Sankt Gallen 41809
Textilmuseum Brennet, Wehr 22362
Textilmuseum Die Scheune, Nettetal-Hombergen 20579
Textilmuseum Sandtnerstiftung, Mindelheim 20332
Textilní Muzeum, Česká Skalice 09460
Textilsammlung Max Berk, Heidelberg 19085
Teylers Museum, Haarlem 32302
Teys Art Gallery, Miles 01303
Teysen's Woodland Indian Museum, Mackinaw City 50356
Thackray Museum, Leeds 44668
Thai Royal Carriages and Paraphenalia, Bangkok 42563
Thalassa - Aghia Napa Municipal Museum, Aghia Napa 09361
Thames Art Gallery, Chatham, Ontario 05603
Thames Museum, Newport 51331
Thames Police Museum, London 44997
Thames School of Mines and Mineralogical Museum, Thames 33210
Thames Valley Museum School, Burgessville 05495
Thameside Aviation Museum, East Tilbury 44095
Thamshavnbanen, Løkken Verk 33644
Thanassis Frissiras Museum, Athinai 22953
Thanjavur Art Gallery, Thanjavur 24401
The Archaeological Museum of Innsbruck - Collection of Plaster Casts and Originals University of Innsbruck, Innsbruck 02113
The Art Collections Stockholm University, Stockholm 40943
The Bernadotte Library, Stockholm 40886
The Danish Museum of Photography, Herning 10070
The Gertrud Weinhold Collection "The Religious Year and its Festivals", Oberschleißheim 20835
The Great Copper Mountain Minemuseum, Falun 40516
The Historic Green Vault and The New Green Vault, Dresden 17948
The House of Cuypers, Roermond 32698
The International Design Museum Munich, Weiden in der Oberpfalz 22366
The Mission Museum, Stavanger 33867
The New Masters Gallery, Dresden 17944
The Print Room, Amsterdam 31884
The Renovation Centre Toivo and the Korsman House, Pori 11221
The Riddarholmen Church, Stockholm 40928
The Royal Palace in Stockholm, Stockholm 40906
The South Tyrol Museum of Folk Traditions, Brunico 25556
The State Historical Museum, Moskva 37193
The State Museum of Natural History Stuttgart, Stuttgart 21959
The Tolbooth Museum, Aberdeen 43249
The Tre Kronor Museum (Three Crowns), Stockholm 40916
The Treasury, Stockholm 40934
Thea G. Korver Visual Art Center, Orange City 51581
Theater-Atelier, Zürich 42110
Theater Museum, Amsterdam 31894
Theaterfigurenmuseum, Lübeck 20082
Theatergeschichtliche Sammlung und Hebbel-Sammlung, Kiel 19535
Theatermuseum der Landeshauptstadt Düsseldorf, Düsseldorf 18010
Theatermuseum Hannover, Hannover 19006
Theatermuseum Zauberwelt der Kulisse, Meiningen 20262
Theaterwissenschaftliche Sammlung, Köln 19634
Théâtre d'Automates-Le Village Enchanté, Bellefontaine (Manche) 11927
Theatre Historical and Actor Museum, Miskolc 23715
Theatre Museum, Helsinki 10828
Theatre Museum, Warszawa 35272
Theatre Museum Canada, Toronto 07143
Theatre Museum in the Court Theatre, København 10154
Theatre Museum of the Kasan Academic Bolshoy Theatre of Drama V.I. Kachalov, Kazan 37028
De Theefabriek, Houwerzijl 32418
Thelma Miles Museum, Matheson 06249
Themapark Openluchtmuseum De Spitkeet, Harkema 32315
Theo-Steinbrenner-Turm-Museum, Sommerach 21816
Theo Swagemakers Museum, Haarlem 32303
Theo Thijssen Museum, Amsterdam 31895
Theodor-Fontane-Archiv, Potsdam 21111
Theodor-Heuss-Haus, Stuttgart 21966
Theodor-Heuss-Museum, Brackenheim 17495
Theodor-Mommsen-Gedächtnisstätte, Garding 18539
Theodor-Storm-Haus, Husum 19302
Theodor-Zink-Museum und Wadgasserhof, Kaiserslautern 19414
Theodore Roosevelt Birthplace, New York 51269
Theodore Roosevelt Inaugural National Historic Site, Buffalo 47124
Theofilos Museum, Mytilini 23114
Theomin Gallery, Dunedin 33099
Theophilos Kairis Museum, Andros 22885
Theophilos Museum, Varia Lesbos 23232
Theresienthaler Glasmuseum, Zwiesel 22838
Theresienthaler Heimatstube, Gaildorf 18534
Therisso Museum of Eleftherios Venizelos, Chania 22966
Thermae Maiores - Bath Museum of the Budapest History Museum, Budapest 23405
Thermen-Museum Juliomagus, Schleitheim 41837
Thermenmuseum, Heerlen 32339
Thermometermuseum, Geraberg 18599
Thessaloniki Museum for the Macedonian Struggle, Thessaloniki 23223
Theta-Museet, Bergen 33434
Thetford Collection, Thetford 45848
Thetford Historical Society Library and Museum, Thetford 53496
Thiel Gallery, Stockholm 40946
Thielska Galleriet, Stockholm 40946
Thieves Tower, Riquewihr 15054
Thingbaek Kalkminer Bundgaards Museum, Skørping 10273
Thinktank, Birmingham 43513
Third Cavalry Museum, Fort Carson 48604
Thirlestane Castle, Lauder 44645
Thirsk Museum, Thirsk 45849
This Century Art Gallery, Williamsburg 54171
This is the Place Heritage Park, Salt Lake City 52694
Thomas Bewick's Birthplace, Mickley 45136
Thomas Carlyle's Birthplace, Ecclefechan 44110
Thomas Center Galleries, Gainesville 48818
Thomas Clarke House-Princeton Battlefield, Princeton 52154
Thomas E. McMillan Museum, Brewton 47011
Thomas Edison Birthplace Museum, Milan 50681
Thomas Edison House, Louisville 50298
Thomas Fisher Rare Book Library, Toronto 07144
Thomas-Foreman Home, Muskogee 50930
Thomas Foster Memorial, Uxbridge 07190
Thomas Griswold House and Museum, Guilford 49104
Thomas Handforth Gallery, Tacoma 53415
Thomas Hart Benton Home and Studio, Kansas City 49701
Thomas J. Boyd Museum, Wytheville 54316
Thomas J. Walsh Art Gallery and Regina A. Quick Center for the Arts, Fairfield 48482
Thomas K. Lang Gallery, Wien 03080
Thomas L. Kane Memorial Chapel, Kane 49674
Thomas-Legler-Haus, Diesbach 41326
Thomas-Mann-Archiv, Zürich 42111
Thomas Mann Museum, Nida 30322
Thomas McCulloch Museum, Halifax 05949
Thomas-Müntzer-Gedenkstätte, Heldrungen 19116
Thomas Muir Museum, Bishopbriggs 43521
Thomas Newcomen Museum, Exton 48467
Thomas P. Kennard House, Lincoln 50106
Thomas Price House, Woodruff 54279
Thomas Sappington House Museum, Crestwood 47880
Thomas T. Taber Museum, Williamsport 54173
Thomas Warne Historical Museum, Matawan 50533
Thomas Warne Historical Museum, Old Bridge Township 51533
Thomas Whaley House Museum, San Diego 52757
Thomas Wolfe Memorial, Asheville 46377
Thomaston Historical Museum, Thomaston 53498
Thomasville Cultural Center, Thomasville 53501
Thonet-Museum, Friedberg 01906
Thong Nhat Palace, Ho Chi Minh City 54628
The Thoreau Institute at Walden Woods, Lincoln 50093
Thorma János Múzeum, Kiskunhalas 23671
Thornbury and District Museum, Thornbury, South Gloucestershire 45850
Thorne-Sagendorph Art Gallery, Keene 49720
Thorney Heritage Museum, Thorney 45851
Thornhill Gallery, Kansas City 49702
Thornhill Historic Site and 19th Century Village, Chesterfield 47437
Thornton W. Burgess Museum, Sandwich 52842
Thorvaldsens Museum, København 10155
Thorvaldsens Samlingen på Nysø, Vordingborg 10321
Thousand Islands Museum of Clayton, Clayton 47602
Thracian Tomb, Kazanlăk 05062
Thrai Kao, Battani 42566
Thrasher Carriage Museum, Frostburg 48803
Three Rivers Museum, Muskogee 50931
Three Rivers Museum of Local History, Rickmansworth 45495
Three Village Historical Society Museum, Setauket 53039
Threshing Museum, Mellilä 11105
Threshing Museum, Mullinahone 24879
Threshing Museum, Peräseinäjoki 11198
Thronateeska Heritage Foundation, Albany 46156
Thua Thien Hue Museum, Thua Thien Hue 54639
Thünen-Museum-Tellow, Warnkenhagen 22340
Thüringer Freilichtmuseum Hohenfelden, Hohenfelden 19235
Thüringer Korbmachermuseum, Bad Berka 16690
Thüringer Landesmuseum Heidecksburg, Rudolstadt 21431
Thüringer Museum Eisenach, Eisenach 18098
Thüringer Universitäts- und Landesbibliothek Jena, Jena 19393
Thüringer Wald - Kreativ Museum mit 1. Deutschem Kloßpressenmuseum, Großbreitenbach 18784
Thunder Bay Art Gallery, Thunder Bay 07080
Thunder Bay Military Museum, Thunder Bay 07081
Thunder Bay Museum, Thunder Bay 07082
Thunderbird Museum, Merrillan 50603
Thurles Famine Museum, Thurles 24903
Thurn und Taxis Museum, Regensburg 21233
Thurrock Museum, Grays 44351
Thursford Collection, Fakenham 44204
Thurso Heritage Museum, Thurso 45857
Thury György Múzeum, Nagykanizsa 23725
Thwaite Mills Watermill, Leeds 44669
Thyssen-Bornemisza Art Contemporary, Wien 03081
Tiagarra Aboriginal Museum, Devonport 01037
Tianjin Art Museum, Tianjin 08283
Tianjin Folklorish Museum, Tianjin 08284
Tianjin Glaswegiansu Guan, Tianjin 08284
Tianjin History Museum, Tianjin 08285
Tianjin Museum of Natural History, Tianjin 08286
Tianjin Painting Gallery, Tianjin 08287
Tianjin Science Museum, Tianjin 08288
Tianjin Theatre Museum, Tianjin 08289
Tianshui Museum, Tianshui 08292
Tiantai Museum, Tiantai 08293
Tianyi Ge Museum, Ningbo 08072
Tibet House Museum, Delhi 24156
Tibet House Museum, New York 51270
Tibet Museum, Gruyères 41462
Tibet Museum, Lhasa 07929
Tibet Songtsen House, Zürich 42112
Tibooburra Local Aboriginal Land Council Keeping Place, Tibooburra 01588
Tibro Museum, Tibro 40970
Ticho House, Jerusalem 25020
Tidaholms Konsthall, Tidaholm 40971
Tidaholms Museum, Tidaholm 40972
Tie-dying Museum, Nagoya 28993
Tieling Museum, Tieling 08294
Tiemuseo, Helsinki 10830
Tiermuseum, Lenggries 19955
Tierpark Sababurg Museum, Hofgeismar 19225
Tierwelt-Panorama, Ebikon 41337
Tieteelliset Kokoelmat, Forssa 10738
Tiflološki Muzej, Zagreb 09013
Tikanoja Art Museum, Vaasa 11427
Tikanojan Taidekoti, Vaasa 11427
Tikotin Museum of Japanese Art, Haifa 24969
Tilbury Fort, Tilbury 45858
Tile Heritage Foundation, Healdsburg 49234
Tiled Kiosk Museum, İstanbul 42812
Till-Eulenspiegel-Museum, Schöppenstedt 21618
Tillamook County Pioneer Museum, Tillamook 53509
Tillamook Naval Air Station Museum, Tillamook 53510
tim - Bayerisches Textil- und Industriemuseum, Augsburg 16638
Tímárház - Kézmuvesek Háza, Debrecen 23521
Timber Creek Police Station Museum, Timber Creek 01589
Timber Museum of New Zealand, Putaruru 33191
Timber-rafting Museum, El Pont de Claverol 39854
Timber Village Museum, Blind River 05431
Timbertown Museum, Wauchope 01637
Time Bygdemuseum, Undheim 33946
Time Machine, Weston-super-Mare 45982
The Time Museum, Newark-on-Trent 45216
Time Museum, Tehrān 24697
Time & Tide - Museum of Great Yarmouth Life, Great Yarmouth 44360
Timeball Tower, Deal 43960
Times Past - Bygones of Yester Year, Blindley Heath 43544
Times Square Gallery, New York 51271
Timespan Heritage Centre and Art Gallery, Helmsdale 44440
Timexpo Museum, Waterbury 53957
Timken Museum of Art, San Diego 52758
Timmermans-Opsomerhuis, Lier 03820
Timmins Museum, Timmins 07084
Tin Museum - Gedung Raja Abdullah, Kelang 30561
Tin Sheds Gallery, Sydney 01562
Tin Toy Museum & Motion Display Museum, Yokohama 29609
Tinelmuseum, Sinaai-Waas 04012
Tineriba Tribal Gallery and Museum, Hahndorf 01140
Tingvoll Museum, Tingvoll 33906
Tingwall Agricultural Museum, Gott 44338
Tinker Swiss Cottage Museum, Rockford 52418
Tinley Park Historical Society Museum, Tinley Park 53511
Tinn Museum, Rjukan 33786
Tintern Abbey, Chepstow 43813
Tintic Mining Museum, Eureka 48443
Tioga County Historical Society Museum, Owego 51636
Tioga Point Museum, Athens 46419
Tippecanoe Battlefield, Battle Ground 46647
Tippecanoe County Historical Museum, Lafayette 49865
Tipton County Museum, Covington 47861
Tipton-Haynes Museum, Johnson City 49636
Tiptree Museum, Tiptree 45859
Tipu Sahib Museum, Srirangapatna 24398
Tire Müzesi, Tire 42952
Tire Museum, Tire 42952
Das Tirol Panorama, Innsbruck 02134
Tiroler Bauernhausmuseum, Kitzbühel 02178
Tiroler Bergbau- und Hüttenmuseum, Brixlegg 01801
Tiroler Kaiserjägermuseum Obergricht, Serfaus 02739
Tiroler Künstlerschaft - Kunstpavillon, Innsbruck 02135
Tiroler Landesmuseum Ferdinandeum, Innsbruck 02136
Tiroler Volkskunstmuseum, Innsbruck 02137
Tirpitz-Stillingen, Blåvand 09979
Tischbein-Haus-Museum, Haina (Kloster) 18871
Tischlerei Museum, Bremen 17563
Titan Missile Museum, Sahuarita 52518
Tittot Glass Art Museum, Taipei 42436
Titus Brandsma Museum, Bolsward 31981
Tiverton Museum of Mid Devon Life, Tiverton 45860
Tjolöholms Slott, Fjärås 40523
Tjumen Picture Gallery, Tjumen 37851
Tjumen Region Museum named after I.Ya. Slovtsov, Tjumen 37852
Tjumen Regional Museum of Fine Art, Tjumen 37853
Tjumenskaja Kartinnaja Galereja, Tjumen 37851
Tjumenskij Oblastnoj Kraevedčeskij Muzej im. I.Ja. Slovcova, Tjumen 37852
Tjumenskij Oblastnoj Muzej Izobrazitelnych Iskusstv, Tjumen 37853
TM Gallery, Crestview 47879
To-ji Homotsukan, Kyoto 28870
To-ji Temple Treasure House, Kyoto 28870
Toad Hole Cottage Museum, Ludham 45044
Tobacco and Match Museum, Stockholm 40947
Tobacco and Salt Museum, Tokyo 29479
Tobacco Farm Life Museum, Kenly 49726
Tobacco Museum, Ljubljana 38468
Tobacco Museum - Old Reig Factory, Sant Julià de Lòria 00105
Tobačni Muzej, Ljubljana 38468
Tobaks- och Tändstickmuseum, Stockholm 40947
Tobias Community Historical Society Museum, Tobias 53517
Tobias-Mayer-Museum, Marbach am Neckar 20184
Tobias Smollett Museum, Alexandria 43275
Tobinodai Historical Park Museum, Funabashi 28453
Tobinodai Shiseki Koen Hakubutsukan, Funabashi 28453
Tobolsk Picture Gallery, Tobolsk 37854
Tobolsk State Historical and Architecture Museum-Estate, Tobolsk 37855
Tobolskaja Kartinnaja Galereja, Tobolsk 37854

T

Tobolskij Gosudarstvennyj Istoriko-architekturnyj Muzej-zapovednik, Tobolsk 37855
Tochigi-kenritsu Bijutsukan, Utsunomiya 29566
Tochigi-kenritsu Hakubutsukan, Utsunomiya 29567
Tochigi Prefectural Museum, Utsunomiya 29567
Tochigi Prefectural Museum of Fine Arts, Utsunomiya 29566
Todd Madigan Gallery, Bakersfield 46538
Todmorden Mills Heritage Museum and Arts Centre, Toronto 07145
Todor Kableshkov Memorial House, Koprivštica 05073
Toei Movie Land Toei Uzumasa Eigamura Movie Museum, Kyoto 28871
Töllinmäen Museo, Hämeenkyrö 10741
Töllinmäki Museum, Hämeenkyrö 10741
Töllstorps Industrimuseum, Gnosjö 40541
Töpferei und Museum im Kannenofen, Höhr-Grenzhausen 19216
Töpfereimuseum, Speicher 21832
Töpfereimuseum Langerwehe, Langerwehe 19837
Töpfereimuseum Raeren, Raeren 03960
Töpfermuseum, Duingen 18011
Töpfermuseum, Kohren-Sahlis 19673
Töpfermuseum, Rödermark 21353
Töpfermuseum, Stoob 02784
Töpfermuseum Im alten Kannenofen, Ransbach-Baumbach 21180
Töpfermuseum Thurnau, Thurnau 22058
Törngrens Krukmakeri, Falkenberg 40511
Török Ház, Szigetvár 23891
Töysän Museo, Töysä 11377
Tofield Historical Museum, Tofield 07087
Togari-Ishi Jomon Kokokan, Chino 28401
Togari-Ishi Jomon Ruins Archaeological Museum, Chino 28401
Toggenburger Museum, Lichtensteig 41559
Toggenburger Schmiede- und Werkzeug-Museum, Bazenheid 41188
Toguri Bijutsukan, Tokyo 29486
Toguri Museum of Art, Tokyo 29486
Tohoku Daigaku Hakubutsukan, Sendai 29248
Tohoku History Museum, Tagajou 29308
Tohoku Kindai Toji oyo Yaki Bijutsukan, Nakaniida 29019
Tohoku Modern Pottery and Porcelain Museum, Nakaniida 29019
Tohoku Rekishi Shiryokan, Tagajou 29308
Tohoku University Museum, Sendai 29248
Tøjhusmuseet, København 10156
Tokai Daigaku Kaiyo Kagaku Hakubutsukan, Shizuoka 29291
Tokaj-Hegyalja Geological Collection, Tállya 23910
Tokaji Múzeum, Tokaji 23934
Tokat Müzesi ve Örenyerleri, Tokat 42954
Tokat Museum and Ruins, Tokat 42954
Token Hakubutsukan, Tokyo 29487
Toki City Mino Ceramic History Museum, Toki 29363
Toki-shiritsu Mino-yaki Rekishi Hakubutsukan, Toki 29363
Tokmok Historical and Regional Museum, Tokmok 29966
Tokomairiro Historical Society Museum, Milton 33146
Tokomaru Steam Engine Museum, Tokomaru 33215
Tokoname City Folk Museum and Ceramic History, Tokoname 29365
Tokoname-shiritsu Mingeikan oyo Toji Rekishi, Tokoname 29365
Tokugawa Art Museum, Nagoya 29012
Tokugawa Bijutsukan, Nagoya 29012
Tokushima-kenritsu Hakubutsukan, Tokushima 29366
Tokushima-kenritsu Kindai Bijutsukan, Tokushima 29367
Tokushima-kenritsu Torii Kinen Hakubutsukan, Naruto 29045
Tokushima Modern Art Museum, Tokushima 29367
Tokushima Prefectural Local Culture Hall & Awa Deko-puppet Museum, Tokushima 29368
Tokushima Prefecture Museum, Tokushima 29366
Tokyo Central Art Museum, Tokyo 29488
Tokyo Chuo-ku Bijutsukan, Tokyo 29488
Tokyo Fuji Art Museum, Hachioji 28467
Tokyo Kasei Daigaku Hakubutsukan, Tokyo 29489
Tokyo Kasei University Museum, Tokyo 29489
Tokyo Kokuritsu Hakubutsukan, Tokyo 29490
Tokyo Kokuritsu Kindai Bijutsukan, Tokyo 29491
Tokyo Kokuritsu Kindai Bijutsukan Kogeikan, Crafts Gallery, Tokyo 29492
Tokyo Kokuritsu Kindai Bijutsukan, National Film Center, Tokyo 29493
Tokyo Kokusai Bijutsukyoukai, Tokyo 29494
Tokyo Metropolitan Art Museum, Tokyo 29497
Tokyo Metropolitan Museum of Photography, Tokyo 29499
Tokyo Metropolitan Teien Art Museum, Tokyo 29500
Tokyo National Museum, Tokyo 29490
Tokyo Noko Daigaku Kogakubu Fuzoku Sen'i Hakubutsukan, Koganei 28759
Tokyo Opera City Art Gallery, Tokyo 29495
Tokyo Station Gallery, Tokyo 29496
Tokyo-to Bijutsukan, Tokyo 29497
Tokyo-to Gendai Bijutsukan, Tokyo 29498
Tokyo-to Shashin Bijutsukan, Tokyo 29499
Tokyo-to Teien Bijutsukan, Tokyo 29500
Tolbooth Art Centre, Kirkcudbright 44612
Tolbooth Museum, Stonehaven 45765
Toledo Museum of Art, Toledo 53523
Tolerance Centre, Vilnius 30435
Tolerancijos Centras, Vilnius 30435
Tolgus Tin Mill and Streamworks, Redruth 45478
The Tolhouse, Great Yarmouth 44361
Tolland County Jail and Warden's Home Museum, Tolland 53525
Tolmin Museum, Tolmin 38527
Tolminski Muzej, Tolmin 38527
Tolpuddle Martyrs Museum, Tolpuddle 45863
Tolsey Museum, Burford 43673
Tolson Memorial Museum, Huddersfield 44497
Toltec Mounds Archeological State Park, Scott 52974
Tolvmansgården, Hopsala 10841
Tom Brown's School Museum, Uffington 45895
Tom Collins House, Swanbourne 01542
Tom Mix Museum, Dewey 48134
Tom Thomson Art Gallery, Owen Sound 06563
Tom Tits Experiment, Södertälje 40864
Tomakomai Museum, Tomakomai 29510
Tomakomai Science Museum, Tomakomai 29511
Tomakomai-shi Hakubutsukan, Tomakomai 29510
Tomakomai-shi Kagagukan, Tomakomai 29511
Tomarps Kungsgård, Kvidinge 40662
Tomasulo Gallery, Cranford 47867
Tomball Community Museum Center, Tomball 53526
Tombe Reali di Casa Savoia, Torino 28032
Tome Parish Museum, Tome 53527
Tomelilla Konsthall, Tomelilla 40974
Tomimoto Kenkichi Kinenkan, Ando 28359
Tomintoul Museum, Tomintoul 45864
Tomo-no-Ura Museum of History and Folklore, Fukuyama 28450
Tomo-no-Ura Rekishi oyo Minzoku Hakubutsukan, Fukuyama 28450
Tomsk Museum of Wooden Art, Tomsk 37857
Tomsk Regional Museum of Art, Tomsk 37858
Tomskij Muzej Derevjannogo Zodčestva, Tomsk 37857
Tomskij Oblastnoj Chudožestvennyj Muzej, Tomsk 37858
Tomskij Oblastnoj Kraevedčeskij Muzej, Tomsk 37859
Ton Duc Thang Museum, Ho Chi Minh City 54629
Ton Sak Yai, Uttaradit 42609
Ton Smits Huis, Eindhoven 32195
Tonawandas Museum, Tonawanda 53529
Tonbandmuseum, Korneuburg 02219
Tonbergbaumuseum Westerwald, Siershahn 21761
Tong-Jin Irrigation Folk Museum, Kimje 29846
Tongan Museum, Tongan 08296
Tongarra Museum, Albion Park 00791
Tongass Historical Museum, Ketchikan 49752
Tongcheng Museum, Tongcheng 08297
Tongdosa Museum, Yangsan 29940
Tonglushan Ancient Metallurgy Museum, Daye 07655
Tongzhou Museum, Beijing 07571
Toni-Merz-Museum, Sasbach 21515
Tonina Hiša v Sveti Petru, Portorož 38495
Tonina House in St. Peter, Portorož 38495
Tono Kyodo Hakubutsukan, Tono 29513
Tono Municipal Museum, Tono 29513
Toowoomba Regional Art Gallery, Toowoomba 01594
Top of Oklahoma Historical Museum, Blackwell 46847
Le Tophet de Salammbô à Carthage, Salammbô 42648
Topkapı Palace Museum, İstanbul 42843
Topkapı Sarayı Müzesi, İstanbul 42843
Topographie des Terrors, Berlin 17272
Topography of terror, Berlin 17272
Toppenish Museum, Toppenish 53540
Toppler-Schlößchen, Rothenburg ob der Tauber 21412
Topsfield Historical Museum, Topsfield 53541
Topsham Museum, Topsham 45865
Torahiko Terada Kinenkan, Kochi 28748
Torajiro Kojima Memorial Hall, Kurashiki 28783
Torbay Museum, Torbay 07088
Torbogenmuseum, Königsbronn 19643
Torekällbergets Friluftsmuseum, Södertälje 40865
Torekovs Sjöfartsmuseum, Torekov 40975
Torenmuseum, Goedereede 32257
Torenmuseum, Mol 03877
Torenverplaatsingsmuseum, Bocholt 03377
Torfaen Museum, Pontypool 45402
Torfmuseum, Neustadt am Rübenberge 20665
Torfschiffwerftmuseum, Worpswede 22697
Torhalle Frauenchiemsee und Vikarhaus, Frauenchiemsee 18412
Torhaus Dölitz, Leipzig 19940
Torhaus Markkleeberg, Markkleeberg 20205
Torhaus Otterndorf, Otterndorf 20952
Tornaritis Pierides Municipal Museum of Paleontology, Larnaka 09371
Tornio Area Parish Museum, Tornio 11384
Tornionlaakson Maakuntamuseo, Tornio 11385
Tornyai János Múzeum, Hódmezővásárhely 23614
Toronto Aeropspace Museum, Toronto 07146
Toronto Board of Education, Toronto 07147
Toronto Dominion Gallery of Inuit Art, Toronto 07148
Toronto Police Museum and Discovery Centre, Toronto 07149
Toronto Sculpture Garden, Toronto 07150
Toronto's First Post Office, Toronto 07151
Torpedo Factory Art Center, Alexandria 46214
Torppamuseo, Somero 11324
Torquay Museum, Torquay 45866
Torre Abbey Historic House and Gallery, Torquay 45867
Torre del Conde, San Sebastián de La Gomera 39972
Torre dell'Orologio, Venezia 28213
I Torre Guaita, San Marino 38012
Torre Monumental, Buenos Aires 00283
Torreón Fortea, Zaragoza 40352
Torres Strait Museum, Thursday Island 01587
Torrington Historical Society Museum, Torrington 53544
Torrington Museum, Torrington 45869
Torsby Finnkulturcentrum, Torsby 40976
Torsby Fordonsmuseum, Torsby 40977
Torum Maa Ethnographic Open Air Museum, Chanty-Mansijsk 36860
Tørvikbygd Bygdemuseum, Tørvikbygd 33913
Toshiba Kagakukan, Kawasaki 28696
Toshiba Science Institute, Kawasaki 28696
Totem Heritage Center, Ketchikan 49753
Toten Ecomuseum, Bøverbru 33451
Toten Ecomuseum, Kapp 33587
Toten Økomuseum, Bøverbru 33451
Toten Økomuseum, Kapp 33587
Totius House Museum, Potchefstroom 38716
Totnes Costume Museum, Totnes 45871
Totnes Elizabethan House Museum, Totnes 45872
Tottori Folk Art Museum, Tottori 29516
Tottori-kenritsu Gigei Bijutsukan, Nakayama 29023
Tottori-kenritsu Hakubutsukan, Tottori 29515
Tottori Mingeikan, Tottori 29516
Tottori Museum of Modern Arts & Crafts, Nakayama 29023
Tottori Prefectural Museum, Tottori 29515
Touchstone Center for Crafts, Farmington 48519
Tougaloo College Art Collection, Tougaloo 53545
Tour de Constance, Aigues-Mortes 11504
Tour de la Motte-Forte, Arnay-le-Duc 11691
Tour de l'Horloge, Dinan 12749
Tour de Moricq, Angles 11615
Tour des Voleurs, Riquewihr 15054
Tour du Luxembourg, Ligny-en-Barrois 13748
Tour-Musée de Velaux, Velaux 16093
Touring Exhibitions Group, London 55102
Touristik-Museum, Unterseen 41960
Towada Kyodo Hakubutsukan, Towada 29519
Towada Municipal Folk Museum, Towada 29519
Tower Bridge Exhibition, London 44998
Tower Fine Arts Gallery, Brockport 47035
Tower Museum, Derry 43974
Tower Museum of Prosphorion, Ouranoupoli 23128
Tower of David, Jerusalem 25021
Tower of senses, Nürnberg 20793
Tower of the Livonian Order Castle, Tukums 30132
Town and Crown Exhibition, Windsor 46046
Town and Regional Museum of Skagen, Skagen 10262
Town Creek Indian Mound Historic Site, Mount Gilead 50877
Town Exhibition Hall, Petrozavodsk 37557
Town Exposition Hall Baroness Kežmarok, Kežmarok 38266
Town Gallery, Gyula 23602
Town Gallery - Deák Collection, Székesfehérvár 23864
Town Hall, Aachen 16400
Town Hall Art Gallery, Kranj 38436
Town House, Culross 43939
Town House, East Hampton 48249
Town House Museum of Lynn Life, King's Lynn 44589
Town Museum, East Grinstead 44085
Town Museum, Gödöllő 23566
Town Museum, Ledeč nad Sázavou 09621
Town Museum, Moravská Třebová 09662
Town Museum, Nurmes 11152
Town Museum, Paks 23748
Town Museum, Šibenik 08949
Town Museum, Skuteč 09837
Town Museum, Stříbro 09853
Town Museum, Tapolca 23914
Town Museum, Valašské Klobouky 09895
Town Museum, Zbiroh 09928
Town Museum of Příbor, Příbor 09805
Town Museum of the Art of Omsk, Omsk 37497
Town Museum of Tokaji, Tokaji 23934
Town Museum of Ústí nad Labem, Ústí nad Labem 09892
Town Museum of Vrsac, Vršac 38163
Town Museum Rijeka, Rijeka 08937
Town of Clarence Museum, Clarence 47580
Town of Manlius Museum, Manlius 50448
Town of North Hempstead Museum, Westbury 54086
Town of Okotoks Museum and Archives, Okotoks 06505
Town of Warwick Museum, Warwick 53835
Town of Yorktown Museum, Yorktown Heights 54349
Town Tower Gallery, Gmünd, Kärnten 01943
Town War Museum for Peace, Trieste 28081
Towneley Hall Art Gallery and Museums, Burnley 43679
Towner Art Gallery and Local Museum, Eastbourne 44103
Townsend Harris Kinenkan, Shimoda 29264
Townsend Harris Memorial Museum, Gyokusenji Temple, Shimoda 29264
Township Museum, Lwandle 38684
Townsville Museum, Currajong 01018
Toy and Miniature Museum, Stellenbosch 38739
Toy and Miniature Museum of Kansas City, Kansas City 49703
Toy Horse Museum, Grandate 26362
Toy Museum, Ankara 42709
Toy Museum, Bad Lauterberg 16776
Toy Museum, Bruxelles 03484
Toy Museum, Davos Platz 41322
Toy Museum, Moirans-en-Montagne 14138
Toy Museum, Nürnberg 20790
The Toy Museum, Soultz-Haut-Rhin 15802
Toy Museum, Stockholm 40900
The Toy Museum, Stow-on-the-Wold 45771
Toy Museum, Tottori 29517
Toy Museum, Valletta 30648
Toy Museum Prag, Praha 09770
Toy-Toy Museum, Rotterdam 32730
Toyama Kenminkaikan Bijutsukan, Toyama 29521
Toyama-kenritsu Kindai Bijutsukan, Toyama 29522
Toyama Kinenkan, Kawajima 28689
Toyama Memorial Museum, Kawajima 28689
Toyama Municipal Folkcraft Village, Toyama 29523
Toyama Museum, Toyama 29524
Toyama Pottery Museum, Toyama 29525
Toyama-shiritsu Yakimono Bijutsukan, Toyama 29525
Toymuseum Deventer, Deventer 32121
Toyohashi City Art Museum, Toyohashi 29529
Toyohashi Museum of Natural History, Toyohashi 29528
Toyohashi-shi Shizenshi Hakubutsukan, Toyohashi 29528
Toyohashi-shiritsu Bijutsukan, Toyohashi 29529
Toyoshina-machi Kindai Bijutsukan, Toyoshina 29532
Toyota Automobile Museum, Nagakute 28952
Toyota Commemorative Museum of Industry and Technology, Nagoya 29013
Toyota Folk Crafts and Ceramic Museum, Toyota 29534
Toyota Kuragaike Commemorative Hall and Art Salon, Toyota 29533
Toyota Municipal Museum of Art, Toyota 29535
Toyota-shi Mingei oyo Toji Shiryokan, Toyota 29534
Toyota-shiritsu Bijutsukan, Toyota 29535
Toyozo Commemorative Museum, Kani 28664
Toyozo Shiryoukan, Kani 28664
Toys Museum, Évora 35588
Toys Museum, Hoorn 32412
Trabzon Devlet Güzel Sanatlar Galerisi, Trabzon 42956
Trabzon Müzesi, Trabzon 42957
Trabzon Museum, Trabzon 42957
Trabzon State Fine Art Gallery, Trabzon 42956
Trachtemyriv - Deržavnyj Istoriko-kul'turnyj Zapovidnyk, Kaniv 43061
Trachtemyriv - Historical and Cultural Preserve, Kaniv 43061
Trachten- und Heimatmuseum, Weiltingen 22394
Trachten- und Mühlenmuseum, Pfullingen 21035
Trachten- und Volkskunstmuseum, Seebach 21713
Trachtenhaus Jatzwauk, Hoyerswerda 19283
Trachtenmuseum, Ochsenfurt 20854
Tracy Maund Historical Collection, Carlton South 00958
Trade of Somogy Anno..., Kaposvár 23634
Trader Windmill, Sibsey 45678
Traditiekamer Regiment Stoottroepen, Assen 31940
Traditiekamer Typhoon, Volkel 32903
Tradition Room of the Air Force Signals School, Tikkakoski 11375
Traditional Chinese Culture Society of Montréal, Montréal 06382
Traditional Heritage Museum, Sheffield 45655
Traditional House of the Butchery, Limoges 13765
Traditional Village, Virrat 11472
Træe, Bryne 33458
Tragor Ignác Múzeum, Vác 23946
Trail End State Historic Museum, Sheridan 53074

T

Trail Museum and Sports Hall of Memories, Trail 07159
trail of peace, Purgstall an der Erlauf 02541
Traill County Historical Society Museum, Hillsboro 49285
Trailside Nature and Science Center, Mountainside 50897
Trailside Nature Center and Museum, Cincinnati 47566
Trailside Nature Museum, Cross River 47891
Train Museum, Ambarawa 24435
Train Station of Bejucal, Bejucal 09039
Trainland U.S.A., Colfax 47684
Les Trains de Saint-Eutrope, Evry 12890
Trakai Historical Museum, Trakai 30382
Trakijska Grobnica, Kazanlăk 05062
Traktoren-Museum Kempen, Horn-Bad Meinberg 19273
Traktorenmuseum, Vordernberg 02879
Traktorenmuseum, Westerkappeln 22515
Traktorveteranensammlung Dorf an der Pram, Wendling 02919
Trakų Istorijos Muziejus, Trakai 30382
Tram Museum, Rio de Janeiro 04734
Tram Museum of Helsinki City Musuem, Helsinki 10811
Tram Museum Rotterdam, Rotterdam 32731
Tram Museum Zürich, Zürich 42113
Trammell and Margaret Crow Collection of Asian Art, Dallas 47942
Trammuseum, Schepdaal 04003
Tramway Museum, Bendigo 00861
Tramway Museum, Coimbra 35537
Tramway-Museum Graz, Graz 01984
Tramway Museum Tramshed 5, Oslo 33751
Tranby House, Maylands, Western Australia 01271
Tranquilino Sandalio de Noda Museo de Ciencias Naturales, Pinar del Río 09254
Tranquilino Sandalio de Noda Museum of Natural Science, Pinar del Río 09254
Transcona Historical Museum, Winnipeg 07386
Transgariep Museum, Philippolis 38703
TransNamib Museum, Windhoek 31768
Transnet Heritage Foundation Library, Johannesburg 38666
Transport Museum, Budapest 23426
Transport Museum, Stockholm 40936
Transportation Museum - Kiskőrös Museum, Kiskőrös 23663
Transports and Communications Museum, Porto 35854
Transvaal Museum, Tshwane 38759
Transylvania Museum, Lexington 50056
Trapezium House, Petersburg 51804
Trapholt, Kolding 10164
Trappers Museum, Marten River 06242
Traquair House, Innerleithen 44517
Trastad Samlinger, Borkenes 33448
Tratzberg castle, Jenbach 02142
Travel Town Transportation Museum, Los Angeles 50266
Traveler's Rest State Historic Site, Toccoa 53518
Travellers Rest Plantation and Museum, Nashville 50977
Traverse County Historical Society Museum, Wheaton 54123
Traversothèque, Mulhouse 14331
Tread of Pioneers Museum, Steamboat Springs 53280
Treadgolds Museum, Portsmouth 45439
Treasure House of the Eiheiji Temple, Eiheiji 28416
Treasurer's House, York 46107
Treasures of the Earth, Fort William 44248
Treasures of the Sea Exhibit, Georgetown 48890
Treasury Grafenrheinfeld, Grafenrheinfeld 18732
Treasury in the Munich Residenz, München 20482
Treasury of Arts - Vaszary Gallery, Kaposvár 23632
Treasury of Arts - Vaszary Memorial House, Kaposvár 23631
Treasury of Sainte Foy, Conques 12615
Treasury of the Basilica of Sanct Servatius, Maastricht 32532
Treasury of the Cathedral, Kotor 31648
Treasury of the Teutonic Order, Wien 03069
Treasury of Zagreb Cathedral, Zagreb 09010
Treasury - Uppsala Cathedral Museum, Uppsala 41011
Treaty House, Waitangi, Auckland 33228
Treaty Site History Center, Saint Peter 52633
Třebechovické Muzeum Betlémů, Třeběchovice pod Orebem 09873
Tredegar House and Park, Newport, Gwent 45241
Treherne Museum, Treherne 07160
Het Trekkermuseum, Nisse 32596
Trelleborgs Museum, Trelleborg 40981
Trembowla Cross of Freedom Museum, Dauphin 05680
Tremont Gallery, Boston 46952
Trenčianske Múzeum, Trenčín 38376
Trentham Agricultural and Railway Museum, Trentham 01602
Trenton City Museum, Trenton 53563
Trenton Historical Museum, Trenton 53555
Trepassey Area Museum, Trepassey 07162
Trerice, Newquay 45249
Tresfjord Museum, Tresfjord 33915
Tresoar, Frisian Historic and Literature Centre, Leeuwarden 32479
Tresoar, Frysk Histoarysk en Letterkundich Sintrum, Leeuwarden 32479
Trésor de la Basilique Saint-Materne, Walcourt 04127
Tresor de la Catedral, Girona 39350
Trésor de la Cathédrale de Liège, Liège 03818
Trésor de la Cathédrale Notre-Dame, Moulins (Allier) 14308
Trésor de la Cathédrale Notre-Dame, Tournai 04089
Trésor de la Cathédrale Notre-Dame-du-Glarier, Sion 41878
Trésor de la Cathédrale Saint-Nicolas, Fribourg 41392
Trésor de la Collégiale, Huy 03725
Trésor de la Collégiale de Visé, Visé 04119
Trésor de la Collégiale Sainte-Waudru, Mons 03889
Trésor de l'Eglise Primaire Saint-Sebastien, Stavelot 04051
Trésor de Pleaux, Pleaux 14836
Trésor de Sainte Foy, Conques 12615
Trésor d'Église, Marseille 14012
Trésor d'Église, Narbonne 14381
Trésor du Frère Hugo d'Oignies, Namur 03907
Trésor et Fouilles Archéologiques de l'Abbaye, Saint-Maurice 41792
Trésor et Musée de la Collegiale Sainte-Begge, Andenne 03275
Trésor Saint-Gervais, Avranches 11803
Tresorama, La Ferté-Macé 13361
Tresure of Lisbon's Patriarchal Cathedral, Lisboa 35735
Tresure Room Gallery of the Interchurch Center, New York 51272
Trev Deeley Motorcycle Museum, Richmond 06747
Trevi Flash Art Museum, Trevi 28065
Trevithick Cottage, Camborne 43722
Tri-Cities Museum, Grand Haven 48991
Tri-County Historical Society and Museum, Herington 49258
Tri-State Mineral Museum, Joplin 49651
Triangle Rush Exhibition Hall, Yuanli 42479
Triangle Tattoo and Museum, Fort Bragg 48597
Trianon Museum, Várpalota 23953
Trianon Museum and Art Gallery, Denver 48091
Trianon Múzeum, Várpalota 23953
Trias-Museum, Ochsenfurt 20855
Tribal Museum, Ahmedabad 24052
Tribal Museum, Grand Ronde 49011
Tribečské Múzeum Topoľčany, Topoľčany 38372
Tribhuvan Memorial Museum, Kathmandu 31780
Trick Art Museum, Nasu 29053
Trick Bijutsukan, Nasu 29053
Tridaya Eka Dharma Museum, Bukittinggi 24463
Triestingtaler Heimatmuseum, Weissenbach an der Triesting 02902
Triglavska Muzejska Zbirka, Mojstrana 38483
Trignolles - Espace Arthur Masson, Treignes 04094
Trillarium, Cleebronn 17710
Trinity Church in Nikitniki, Moskva 37156
Trinity Heights - Saint Joseph Center-Museum, Sioux City 53115
Trinity House, Newcastle-upon-Tyne 45232
Trinity House National Lighthouse Centre, Penzance 45362
Trinity Interpretation Centre, Trinity 07166
Trinity Museum, Trinity 07167
Trinity Museum of the Parish of Trinity Church, New York 51273
Triton Museum of Art, Santa Clara 52864
Troadkasten, Freinberg 01902
Troadkasten, Sauerlach 21524
Troadkasten, Schardenberg 02698
Troadkasten Fornach, Fornach 01895
Trochu and District Museum, Trochu 07168
Trøgstad Bygdemuseum, Trøgstad 33916
Troickij Sobor, Serpuchov 37766
Trollenäs Slott, Eslöv 40507
Trolley Museum of New York, Kingston 49792
Trolleybus Museum at Sandtoft, Doncaster 43995
Trompetenmuseum, Bad Säckingen 16818
Het Tromp's Huys, Vlieland 32899
Troms Forsvarsmuseum, Bardu 33416
Tromsø Folkemuseum, Folkeparken, Tromsø 33920
Tromsø Universitets Museum, Tromsø 33921
Trondarnes Distriktmuseum, Harstad 33535
Trøndelag Folkemuseum, Trondheim 33929
Trondheim Kunstmuseum, Trondheim 33930
Trondheim Politimuseum, Trondheim 33931
Tropenmuseum Junior, Amsterdam 31896
Trophäensaal, Altenfelden 01711
Trosterud Skolemuseum, Rømskog 33793
Trout Gallery, Carlisle 47244
Trova Foundation, Saint Louis 52602
Trowbridge Museum, Trowbridge 45878
Trowulan Archeological Museum, Mojokerto 24530
Troxell-Steckel House and Farm Museum, Egypt 48305
Troy-Hayner Cultural Center, Troy 53577
Troy Museum and Historic Village, Troy 53572
Troy Open Air Museum, Çanakkale 42749
Troyzky Cathedral, Serpuchov 37766
TRT Müzesi ve Ulusal Kurtulus Sergisi, Ankara 42713
TRT Museum and Ulus Deliverance Exhibition, Ankara 42713
Truck Centre, Calgary 05532
True's Yard Fishing Museum, King's Lynn 44590
Trützschler's Milch- und Reklamemuseum, Hildburghausen 19180
Trummlehus, Langenthal 41528
Truppenmuseum, Straß in Steiermark 02786
Truva Müzesi Örenyeri, Çanakkale 42749
Try Museum, Dronninglund 09991
Tryart Gallery, Louisville 50299
Tryon Palace, New Bern 51020
Trysil Bygdemuseum, Trysil 33934
Tržiška Galerija Ferdo Majer, Tržič 38531
Tržiški Muzej z Galerijo, Tržič 38532
TS-Auto- & Viestintämuseo, Turku 11401
TS Car & Communication Museum, Turku 11401
Tsa Mo Ga Memorial Museum, Plains 51974
Ts'ai-Liao Fossil Museum, Zuojhen 42490
Tsing Hua Stone Teapot Arts Center, Yungho 42485
Tsiolkovsky House-Museum, Kaluga 36994
Tsiolkovsky State Museum of the History of Cosmonautics, Kaluga 36996
Tsitsernakaberd, Yerevan 00757
Tskhinvali Museum, Tskhinvali 16387
Tsou-Ma-Li Country Gallery, Danei 42176
Tsubouchi Memorial Theatre Museum, Waseda University, Tokyo 29503
Tsuchida Bakusen Art Museum, Sawata 29238
Tsuchida Bakusen Bijutsukan, Sawata 29238
Tsui Museum of Art, Hong Kong 07817
Tsukuba Bijutsukan, Tsukuba 29540
Tsukuba Museum of Art, Tsukuba 29540
Tsumeb Cultural Village, Tsumeb 31759
Tsumeb Museum, Tsumeb 31760
Tsurui Bijutsukan, Niigata 29064
Tsurui Museum of Art, Niigata 29064
Tsuwano Gendai Photo Gallery, Tsuwano 29547
Tsuyama-furitsu Rekishi Hakubutsukan, Tsuyama 29548
Tsvetaeva House Museum, Moskva 37162
Tsvetaeva Museum-Appartement in Bolshevo, Korolev 37059
TTD Museum, Tirupati 24409
Tu-Endie-Wei State Park, Point Pleasant 52011
Tuam Mill Museum, Tuam 24905
Tuapsinskij Kraevedčeskij Muzej im. N.G. Poletaeva, Tuapse 37864
Tuatapere Bushmans Museum, Tuatapere 33216
Tubac Center of the Arts, Tubac 53580
Tube Receiver Museum, Kouvola 10991
Tuberculosemuseum Beatrixoord, Appelscha 31917
Tubman African-American Museum, Macon 50370
Tuch und Technik Textilmuseum Neumünster, Neumünster 20636
Tucherschloss mit Hirsvogelsaal, Nürnberg 20792
Tuchmacher Museum Bramsche, Bramsche 17500
Tuchmuseum Lennep der Anna-Hardt-Stiftung, Remscheid 21266
Tuck Museum, Hampton 49147
Tucker House Museum, Saint George's 04199
Tucker Tower Nature Center Park Museum, Ardmore 46345
Tucson Children's Museum, Tucson 53597
Tucson Museum of Art and Historic Block, Tucson 53598
Tucumcari Historical Research Institute Museum, Tucumcari 53604
Tuddal Bygdetun, Tuddal 33935
Tudor House, Margate 45107
Tudor House, Weymouth 45989
Tudor House Museum, Southampton 45708
Tudor Place Museum, Washington 53922
Tüftler-Werkstatt-Museum, Frickingen 18465
Tüötten-Museum, Mettingen 20311
Türbeler Müzesi, İstanbul 42844
Türi Muuseum, Türi 10666
Türi Valla ja Türi Tehnika- ja Maamajanduskooli Muuseum, Särevere 10616
Türk İslam Eserleri ve Etnografya Müzesi, Erzurum 42778
Türk ve Islâm Eserleri Müzesi, Edirne 42768
Türk ve Islam Eserleri Müzesi, İstanbul 42845
Türkiye Çocuk Esirgeme Kurumu Müzesi, Ankara 42714
Türkmenistanyn Milli Muzeýi, Aşgabat 42973
Türkmenistanyt Gahrymany Gurbansoltan eje Adyndaky Türkmen Halsynyt Muzeýi, Aşgabat 42974
Türr István Múzeum, Baja 23347
Tüshaus-Mühle, Dorsten 17912
Tuffstein-Museum "Steinmetzbahnhof", Weibern 22364
Tufts University Art Gallery, Medford 50548
Tugu Nasional Museum, Jakarta 24507
Tukaj House-Museum, Košlauč 37060
Tukaj Memorial Museum Complex, Novyj Kyrlaj 37493
Tukuma Audēju Darbnīcas, Tukums 30133
Tukuma Muzejs, Tukums 30134
Tukums Museum, Tukums 30134
Tukums Weavers' Workshops, Tukums 30133
Tula Cremlin, Tula 37871
Tula Museum of Fine Arts, Tula 37874
Tula Museum of Regional Studies, Tula 37875
Tula Museum of Weapons, Tula 37873
Tula Samovar Museum, Tula 37870
Tulane University Art Collection, New Orleans 51095
Tulare County Museum, Visalia 53763
Tularosa Basin Historical Society Museum, Alamogordo 46152
Tullgarn Castle, Vagnhärad 41046
Tullgarns Slott, Vagnhärad 41046
Tullie House - City Museum and Art Gallery, Carlisle 43760
Tullimuseo, Helsinki 10831
Tullis-Toledano Manor, Biloxi 46813
Tullmuseum, Stockholm 40948
Tullner Museen im Minoritenkloster, Tulln 02834
Tullow Museum, Tullow 24906
Tulsi Sangrahalaya Ramvan, Satna 24382
Tulskij Gosudarstvennyj Muzej Oružija, Tula 37873
Tulskij Muzej Izobrazitelnych Iskusstv, Tula 37874
Tulskij Oblastnoj Kraevedčeskij Muzej, Tula 37875
Tumbergs Skol- och Hembygdsmuseum, Vårgårda 41052
Tumen Museum, Tumen 08305
Tumut and District Historical Museum, Tumut 01606
Tun Abdul Razak Memorial, Kuala Lumpur 30587
Tunabygdens Gammelgård, Borlänge 40477
Tunbridge Wells Museum and Art Gallery, Tunbridge Wells 45881
Tune Bygdemuseum, Sarpsborg 33820
Tung Salaeng Luang, Phitsanulok 42596
Tunghai University Art Center, Taichung 42357
Tungxiang Museum, Tongxiang 08303
Tunica-Biloxi Native American Museum, Marksville 50492
Tunku Abdul Rahman Putra-Memorial, Kuala Lumpur 30588
Tunxi Museum, Huangshan 07836
Tuoketuo Museum, Tuoketuo 08306
Tuol Sleng Museum, Phnom Penh 05289
Tuomarinkylä Museum - Helsinki City Museum, Helsinki 10832
Tuomarinkylän Museo, Helsinki 10832
Tupou College Museum, Nuku'alofa 42614
Tupperville School Museum, Tupperville 07182
Turabo University Museum, Gurabo 36040
Turaida Museum Reserve, Sigulda 30124
Turaidas Muzejrezervāts, Sigulda 30124
Turčianska Galéria, Martin 38301
Turenne-Museum Sasbach, Sasbach 21516
De Turfschuur, Kolhorn 32453
Turiec Gallery, Martin 38301
Turkansaaren Ulkomuseo, Oulu 11178
Turkansaari Open Air Museum, Oulu 11178
Turkeyenhof, Bredene 03401
Turkish House, Szigetvár 23891
Turkish Islamic Art and Ethnography Museum, Erzurum 42778
Turkish Sports Museum, İstanbul 42846
Turkmen Carpet Museum Gurbansoltan Eje, Aşgabat 42974
Turku Art Museum, Turku 11405
Turku Biological Museum, Turku 11402
Turku Cathedral Museum, Turku 11407
Turku Health Care Museum, Turku 11406
Turku Historical Museum, Turku Castle, Turku 11403
Turku Provincial Museum, Turku 11404
Turkuvaz Art Gallery, Ankara 42715
Turkuvaz Sanat Galerisi, Ankara 42715
Turm 9 - Stadtmuseum Leonding, Leonding 02293
Turm der Sinne, Nürnberg 20793
Turmmuseum, Breitenbrunn, Neusiedlersee 01799
Turmmuseum, Öhringen 20867
Turmmuseum Oetz, Oetz 02469
Turmmuseum Sankt Magnus, Esens 18243
Turmmuseum Schloß Mengerskirchen, Mengerskirchen 20285
Turmmuseum Stadt Blankenberg, Hennef 19128
Turmuhren- und Heimatmuseum Bockenem, Bockenem 17409
Turmuhrenmuseum, Naunhof 20557
Turmuhrenmuseum, Neulußheim 20630
Turmuhrenmuseum Rammensee, Gräfenberg 18725
Turner House Gallery, Penarth 45352
Turner House Museum, Hattiesburg 49213
The Turner Museum, Sarasota 52917
Turner Valley Gas Plant, Turner Valley 07183
Turnpike Gallery, Leigh, Lancashire 44685
Turquoise Museum, Albuquerque 46188
Turtle Mountain Chippewa Heritage Center, Belcourt 46694

T

Turtola Village Museum, Turtola 11409
Turtolan Kylämuseo, Turtola 11409
Turton Tower, Turton Bottoms 45885
Turun Biologinen Museo, Turku 11402
Turun Kaupungin Historiallinen Museo, Turun Linna, Turku 11403
Turun Maakuntamuseo, Turku 11404
Turun Taidemuseo, Turku 11405
Turun Terveydenhuoltomuseo, Turku 11406
Turun Tuomiokirkkomuseo, Turku 11407
Tutankhamun Exhibition, Dorchester 44004
Tute-Natura, Bennekom 31963
Tuttlinger Haus, Tuttlingen 22138
Tuupalan Museo, Kuhmo 11000
Tuusniemen Kotiseutumuseo, Tuusniemi 11412
Tuvinskij Respublikanskij Kraevedčeskij Muzej im. Aldan Maadyr, Kyzyl 37099
Tűzoltástörténeti kiállítás, Barcs 23362
Tűzoltó- és Községi Emlékház, Vál 23949
Tűzoltó Múzeum, Bonyhád 23375
Tűzoltó Múzeum, Budapest 23486
Tűzoltó Múzeum, Vörs 23967
TV Toys Dieren, Dieren 32127
TVDART, Berlin 17273
Tvedestrand Museum, Tvedestrand 33939
Tveit Bygdemuseum, Tveit 33940
Tveitens Samlinger, Eggedal 33473
Tver Regional Art Gallery, Tver 37880
Tver State Museum, Tver 37881
Tverskaja Oblastnaja Kartinnaja Galereja, Tver 37880
Tverskoj Gosudarstvennyj Obedinennyj Muzej, Tver 37881
Tvorčeskij Centr i Vystavočnyj Zal Fëdor, Sankt-Peterburg 37729
Tvøroyrar Bygda- og Sjósavn, Tvøroyri 10700
Tvøroyri Heritage and Ship Museum, Tvøroyri 10700
Tvrz Žumberk, Žumberk 09951
Tweed and Area Heritage Centre, Tweed 07184
Tweed Museum of Art, Duluth 48200
Tweed River Art Gallery, Murwillumbah 01340
Tweeddale Museum, Peebles 45349
Tweewielermuseum Tankstop, Workum 32966
Twickenham Museum, Twickenham 45889
Twierdza Wisłoujście, Gdańsk 34652
Twillingate Museum, Twillingate 07185
Twin Falls County Museum, Twin Falls 53628
Twin Falls Museum, Mullens 50900
Twinsburg Historical Society Museum, Twinsburg 53629
Two Hills and District Museum, Two Hills 07186
Two Rivers Gallery of Prince George and Region, Prince George 06669
Ty Mawr Wybrnant, Dolwyddelan 43990
Tybee Museum and Tybee Island Light Station, Tybee Island 53632
Tydal Museum, Tydal 33941
Tyler Art Gallery, Oswego 51627
Tyler Galleries, Elkins 48352
Tyler Museum of Art, Tyler 53635
Tylldalen Bygdetun, Tylldalen 33942
Tymperleys Clock Museum, Colchester 43876
Tyndale Galleria, Upland 53661
Tynset Bygdemuseum, Tynset 33943
Työväenasuntomuseo, Helsinki 10833
Työväenasuntomuseo ja Kotiseututalo, Kuusankoski 11018
Työväenmuseo Werstas, Tampere 11367
Typaldos-Iakobatos Museum, Lixouri 23080
Typewritermuseum, Parcines 27091
Typorama, Bischofszell 41241
Tyresö Slott, Tyresö 40988
Tyrnävän Kotiseutumuseo, Tyrnävä 11416
Tyrvää Regional Museum, Vammala 11442
Tyrvään Seudun Museo, Vammala 11442
Tyrwhitt-Drake Museum of Carriages, Maidstone 45075
Tyseley Loco Works Visitor Centre, Birmingham 43514
Tytus Chałubiński Tatra Museum, Zakopane 35360
Tytuvėnų Miško Muziejus, Tytuvėnai 30385
Tytuvėnų Žmės Ūkio Mokyklos Kraštotyros Muziejus, Tytuvėnai 30386
Tytyri Limestone Mining Museum, Lohja 11078
Tytyrin Kalkkikaivosmuseo, Lohja 11078
Tyutchev Museum-Estate Muranovo, Lugovskoe 37114

U

The U. Nahon Museum of Italian Jewish Art, Jerusalem 25022
U-Thong National Museum, Suphan Buri 42605
U Zlaté Koruny, Brno 09441
UBC Fish Museum, Vancouver 07224
Učebno-chudožestvennyj Muzej Slepkov im. I.V. Cvetaeva, Moskva 37409
Uchiko Hakubutsukan, Uchiko 29551
Uchiko Museum, Uchiko 29551
Uchu Kagakukan, Ikoma 28565
Uckermärkische Heimatstuben Fürstenwerder, Fürstenwerder 18496
Uckermärkisches Volkskundemuseum Templin, Templin 22030
UCL Art Collections, London 44999
UCL Museums and Collections, London 45000
Ucross Foundation Art Gallery, Clearmont 47609
Udmurtskij Respublikanskij Muzej Izobrazitelnych Iskusstv, Iževsk 36956
UDT-SEAL Museum, Fort Pierce 48669
Übersee Museum Bremen, Bremen 17564
Überwälder Heimatmuseum, Wald-Michelbach 22280
Uehara Kindai Bijutsukan, Shimoda 29265
Uehara Museum of Modern Art, Shimoda 29265
Ueno Royal Museum, Tokyo 29501
Ürgüp Müzesi, Ürgüp 42958
Ürgüp Museum, Ürgüp 42958
Uetiker Museum, Uetikon am See 41956
Ufa Picture Gallery, Ufa 37890
Uferhallen, Berlin 17274
Uffenheimer Gollachgaumuseum, Uffenheim 22155
Ufimskaja Kartinnaja Galereja, Ufa 37890
Uganda Museum, Kampala 42993
Ugbrooke House, Chudleigh 43844
Uhren- und Musikgerätemuseum-Museum zu Putbus, Putbus 21142
Uhrenindustriemuseum, Villingen-Schwenningen 22241
Uhrenmuseum, Bad Grund 16744
Uhrenmuseum, Bad Iburg 16756
Uhrenmuseum, Regensburg 21234
Uhrenmuseum, Wien 03082
Uhrenmuseum Beyer Zürich, Zürich 42114
Uhrenmuseum Lorenz-Berlin, Berlin 17275
Uhrensammlung Kellenberger, Winterthur 42040
Uhrenstube Aschau, Oberschützen 02460
Uhrentechnische Lehrschau Hennig, Tharandt 22050
Uilenspiegel Museum, Damme 03536
Uiryeong Museum, Uiryeong 29938
Uitamonkosken Myllymuseo, Multia 11123
Új Magyar Képtár, Székesfehérvár 23863
Uji Kanbayashi Kinenkan, Uji, Kyoto 29560
Uji-Kanbayashi Memorial House, Uji, Kyoto 29560
Újpest Region Historical Collection, Budapest 23487
Újpesti Helytörténeti Gyűjtemény, Budapest 23487
Újpesti Lepkemúzeum Alapítvány, Budapest 23488
Ukai Museum, Kawaguchi-ko 28686
Ukiyo-e and Pottery Museum, Osaka 29156
Ukmerge Local Lore Museum, Ukmergė 30388
Ukmergės Kraštotyros Muziejus, Ukmergė 30388
Ukraina Museum, Saskatoon 06919
Ukraine National Gallery, Kyïv 43110
Ukrainian-American Museum, Hamtramck 49155
Ukrainian-American Museum, Warren 53821
Ukrainian Canadian Archives and Museum of Alberta, Edmonton 05756
Ukrainian-Canadian Art Foundation, Toronto 07152
Ukrainian Catholic Women's League Museum, Edmonton 05757
Ukrainian Cultural & Education Centre Oseredok, Winnipeg 07387
Ukrainian Cultural Heritage Village, Edmonton 05758
Ukrainian Homestead Museum, Winnipeg Beach 07395
Ukrainian Institute of Modern Art, Chicago 47519
Ukrainian Museum, Cleveland 47643
Ukrainian Museum, Gardenton 05850
The Ukrainian Museum, New York 51274
Ukrainian Museum, Torrens Park 01596
Ukrainian Museum of Canada, Saskatoon 06925
Ukrainian Museum of Canada, Vancouver 07225
Ukrainian Museum of Canada - Alberta Branch, Edmonton 05759
Ukrainian Museum of Canada - Manitoba Branch, Winnipeg 07388
Ukrainian National Museum, Chicago 47520
Ukrainian Peoples Home of Ivan Franco, Angusville 05341
Ulaanbaatar City Museum, Ulaanbaatar 31631
Uleybury School Museum, One Tree Hill 01391
Uljanovskij Oblastnoj Chudožestvennyj Muzej, Uljanovsk 37909
Uljanovskij Oblastnoj Kraevedčeskij Muzej im. I.A. Gončarova, Uljanovsk 37910
Ullapool Museum, Ullapool 45896
Ullensaker Bygdemuseum, Jessheim 33580
Ullevål Sykehus Museum, Oslo 33756
Ulmer Museum, Ulm 22170
Ulricehamns Konst- och Östasiatiska Museum, Ulricehamn 40995
Ulrika museum, Ulrika 40996
Ulriksdal Palace and Orangery Museum, Solna 40875
Ulriksdals Slott samt Orangerimuseet, Solna 40875
Ulster-American Folk Park, Omagh 45317
Ulster County Historical Society Museum, Marbletown 50463
Ulster Folk and Transport Museum, Holywood 44482
Ulster History Park, Omagh 45318
Ulster Museum, Belfast 43448
Ulsyn Töw Muzei, Ulaanbaatar 31632
Ultner Talmuseum, Ulten 28116
Ulug-Beg Memorial Museum, Samarqand 54464
Ulug-Beg Yodgorlik Muzeyi, Samarqand 54464
Ulverstone Local History Museum, Ulverstone 01610
Ulverton Wool Mill, Ulverton 07187
Ulysses S. Grant Cottage State Historic Site, Wilton 54216
Umatilla County Historical Society Museum, Pendleton 51757
Umeå Energicentrum med Kraftverksmuséet, Umeå 40999
Umekoji Steam Locomotive Museum, Kyoto 28872
Uměleckoprůmyslové Muzeum v Praze, Praha 09797
Umetnička Galerija, Kruševac 38113
Umetnička Galerija, Kumanovo 30508
Umetnička Galerija Nadežda Petrović, Čačak 38095
Umetnička Galerija Strumica, Strumica 30529
Umetnička Galerija Velimir A. Leković, Bar 31635
Umetnička Zbirka Flögel, Beograd 38090
Umetnostna Galerija Maribor, Maribor 38479
Umeyama Ancient Pottery Museum, Tobe 29362
Umeyama Furui Yakimono Bijutsukan, Tobe 29362
Umi-no Hakubutsukan, Toba 29361
U'mista Cultural Centre, Alert Bay 05315
Umjetnička Galerija, Dubrovnik 08874
Umjetnička Galerija, Skopje 30525
Umjetnička Galerija Bosne i Hercegovine, Sarajevo 04283
Umjetnička Galerija Josip-Bepo Benkovic, Herceg-Novi 31645
Umjetnička Galerija Nikola I i Galerija Ilija Sobajić, Nikšić 31649
Umjetnički Paviljon, Podgorica 31658
Umjetnički Paviljon u Zagrebu, Zagreb 09014
UMKC Gallery of Art, Kansas City 49704
Umlauf Sculpture Garden and Museum, Austin 46522
Umm Qais Archaeological Museum, Umm Qais 29668
Umoona Opal Mine and Museum, Coober Pedy 00996
Umpqua Discovery Center, Reedsport 52280
Umstädter Museum Gruberhof, Groß-Umstadt 18778
Umuzi-Afrika-Haus, Freiberg am Neckar 18420
Un Village se Raconte, Lourdios-Ichère 13853
Uncle Remus Museum, Eatonton 48273
Uncle Tom's Cabin Historic Site, Dresden 05711
Underground Gold Miners Museum, Alleghany 46223
Underground Passages, Exeter 44196
Underground Prisoners Museum, Jerusalem 25023
Underhill Museum, Idaho Springs 49465
Ungarie Museum, Ungarie 01611
Ungarndeutsche Heimatstuben, Langenau 19821
Ungarndeutsches Heimatmuseum, Backnang 16664
Unguri Manor, Raiskums 30058
Ungurmuiža, Raiskums 30058
Uniacke Estate Museum Park, Mount Uniacke 06401
Unicorn Trust Equestrian Centre, Stow-on-the-Wold 45772
Unidad Museológica Municipal de La Banda, La Banda 00396
Union Art Gallery, Baton Rouge 46642
Union-Art Gallery, Milwaukee 50719
Union Brewery Museum, Ljubljana 38459
Union County Heritage Museum, New Albany 51011
Union County Historical Complex, Creston 47878
Union County Historical Foundation Museum, Union 53652
Union County Historical Society Museum, Blairsville 46848
Union County Historical Society Museum, Marysville 50523
Union County Museum, Lewisburg 50032
Union County Museum, Union 53650
Union County Public Library Union Room, Monroe 50792
Union Mills Homestead and Grist Mill, Westminster 54099
Union of American Hebrew Congregation, New York 51275
Union of Museums in Slovakia, Banská Bystrica 55011
Union Pacific Railroad Museum, Council Bluffs 47851
Union Station Kansas City, Kansas City 49705
Union Street Gallery, Chicago 47521
UNISA Art Gallery, Tshwane 38760
Uniseum, Freiburg im Breisgau 18440
Unité Musée Numismatique, Port-au-Prince 23324
United Counties Museum in the Wood House, Cornwall 05646
United Kingdom National Committee of the International Council of Museums, London 55103
United States Air Force Academy Museum, El Paso 48314
United States Air Force Museum, Wright-Patterson Air Force Base 54311
United States Army Air Defense Artillery Museum, Fort Bliss 48595
United States Army Aviation Museum, Fort Rucker 48677
United States Army Center of Military History, Museum Division, Fort McNair 48651
United States Army Chaplain Museum, Fort Jackson 48629
United States Army Communications-Electronics Museum, Fort Monmouth 48657
United States Army Engineer Museum, Fort Leonard Wood 48647
United States Army Finance Corps Museum, Fort Jackson 48630
United States Army Heritage and Education Center, Carlisle 47245
United States Army Medical Department Museum, Fort Sam Houston 48679
United States Army Military Police Corps Museum, Fort Leonard Wood 48648
United States Army Museum of Hawaii, Fort DeRussy 48613
United States Army Ordnance Museum, Aberdeen Proving Ground 46118
The United States Army Quartermaster Museum, Fort Lee 48645
United States Army Signal Corps and Fort Gordon Museum, Fort Gordon 48623
United States Army Transportation Museum, Fort Eustis 48620
United States Army Women's Museum, Fort Lee 48646
United States Brig Niagara, Homeport Erie Maritime Museum, Erie 48412
United States Capitol Visitor Center, Washington 53923
United States Cavalry Museum, Fort Riley 48676
United States Coast Guard Museum, New London 51068
United States Department of the Interior Museum, Washington 53924
United States Federation of Friends of Museums, Baltimore 55185
United States Grant Ancestral Homestead, Dungannon 44058
United States Grant's Home, Galena 48830
United States Hockey Hall of Fame, Eveleth 48459
United States Holocaust Memorial Museum, Washington 53925
United States-International Council on Monuments and Sites, Washington 55186
United States Marine Corps Air-Ground Museum, Quantico 52208
United States Marine Corps Museum, Washington 53926
United States Mint-Philadelphia, Philadelphia 51885
United States National Committee of the ICOM, Washington 55187
United States National Ski Hall of Fame and Museum, Ishpeming 49538
United States Naval Academy Museum, Annapolis 46321
United States Navy Art Gallery, Washington 53927
United States Navy Memorial Foundation and Naval Heritage, Washington 53928
United States Navy Supply Corps Museum, Athens 46416
United States Senate Commission on Art Collection, Washington 53929
United States Space and Rocket Center, Huntsville 49445
Unites States Air Force Academy Visitor Center, USAF Academy 53672
Unity and District Heritage Museum, Unity 07189
Unity House, Detroit 48107
Universalmuseum Joanneum, Graz 01985
Universeum, Göteborg 40557
Universitäts- und Landesbibliothek Bonn, Bonn 17468
Universitäts- und Landesbibliothek Darmstadt, Darmstadt 17795
Universitätsbibliothek, München 20492
Universitätsbibliothek der Humboldt-Universität zu Berlin, Berlin 17276
Universitätsbibliothek Johann Christian Senckenberg, Frankfurt am Main 18399
Universitätsbibliothek Salzburg, Salzburg 02628
Universitätsbibliothek Wien, Wien 03083
Universitätsmuseum, Altdorf bei Nürnberg 16484
Universitätsmuseum Heidelberg, Heidelberg 19086
Universitätssammlung antiker und nachantiker Münzen und Medaillen, Tübingen 22125
Universitätssammlungen - KunstTechnik der TU Dresden, Dresden 17975
Universiteitsbibliotheek Gent, Gent 03646
Universiteitsmuseum, Groningen 32288
Universiteitsmuseum De Agnietenkapel, Amsterdam 31897
Universiteitsmuseum Utrecht, Utrecht 32857
Universitetets Oldsaksamling, Oslo 33757
University Archaeology Museum, Birmingham 43515

T U

University Art Galleries, Akron 46150
University Art Galleries, Murray 50918
University Art Galleries, Vermillion 53720
University Art Gallery, Bridgeport 47019
University Art Gallery, Carson 47258
University Art Gallery, Chico 47529
University Art Gallery, Florence 48565
University Art Gallery, Fort Worth 48720
University Art Gallery, Jackson 49569
University Art Gallery, Jacksonville 49586
University Art Gallery, La Jolla 49848
University Art Gallery, Mount Pleasant 50885
University Art Gallery, Pittsburgh 51963
University Art Gallery, Potchefstroom 38717
University Art Gallery, Rohnert Park 52444
University Art Gallery, San Diego 52759
University Art Gallery, Stellenbosch 38740
University Art Gallery, Stony Brook 53315
University Art Gallery, Terre Haute 53484
University Art Gallery, Turlock 53615
University Art Gallery & Cabinet of Natural History, New Bedford 51016
University Art Gallery of the University of the South, Sewanee 53041
University Art Museum, Albany 46165
University Art Museum, Albuquerque 46189
University Art Museum, Binghamton 46817
University Art Museum, Lafayette 49868
University Art Museum, Long Beach 50179
University Art Museum, Santa Barbara 52861
University Art Museum, Kyoto City University of Arts, Kyoto 28804
The University Art Museum, Tokyo National University of Fine Arts and Music, Tokyo 29392
University at Buffalo Anderson Art Gallery, Buffalo 47125
University at Buffalo Art Galleries, Buffalo 47126
University College Dublin Classical Museum, Dublin 24819
University College Gallery, College Park 47687
University College of Cape Breton Art Gallery, Sydney 07051
University College Zoological Museum, Cork 24770
University Fine Art Collection, Hobart 01160
University Galleries, Clarion 47583
University Galleries, Edinburg 48292
University Galleries, Normal 51372
University Gallery, Ada 46135
University Gallery, Alexandria 46196
University Gallery, Amherst 46271
University Gallery, Boca Raton 46887
University Gallery, Commerce 47764
University Gallery, Gainesville 48819
University Gallery, Hays 49226
University Gallery, Newark 51282
The University Gallery, Newcastle-upon-Tyne 45233
University Gallery, Oxford 51649
University Gallery Leeds, Leeds 44670
University Historical Museum, Musuan 34464
University History Museum, Guangzhou 07737
University Medical Museum, Charleston 47377
University Museum, Anseong 29768
University Museum, Carbondale 47241
The University Museum, Edwardsville 48301
The University Museum, Indiana 49485
University Museum, Martin 50516
University Museum, Port Harcourt 33345
The University Museum, Tokyo 29502
University Museum, Vallabhvidyanagar 24423
University Museum and Art Gallery, Hong Kong 07818
University Museum for the History of Medicine and Pharmacy Victor Gomoiu, Târgu Jiu 36667
University Museum of Anthropology, Quezon City 34495
University Museum of Archaeology and Anthropology, Cambridge 43734
University Museum of Science and Culture, Aligarh 24057
University Museum of Zoology, Cambridge 43735
University Museum - Sasol Art Museum, Stellenbosch 38741
University Museum Utrecht, Utrecht 32857
University Museums, Newark 51283
University Museums, Oxford 51650
University Museums Group, Birmingham 55104
University of Alberta Dental Museum, Edmonton 05760
University of Alberta Museums, Edmonton 05761
University of Arizona Mineral Museum, Tucson 53599
University of Arizona Museum of Art, Tucson 53600
University of Arizona Student Union Galleries, Tucson 53601
University of Arkansas at Little Rock Art Galleries, Little Rock 50142
University of Arkansas Collections, Fayetteville 48528
University of Art and Design Linz, Linz 02313
University of Brighton Gallery, Brighton 43627
University of Bristol Theatre Collection, Bristol 43640
University of Calgary Museum of Zoology, Calgary 05533
University of Colorado Museum, Boulder 46961
University of Dundee Exhibitions, Dundee 44051
University of Dundee Museum Collections, Dundee 44052
University of Essex Exhibition Gallery, Colchester 43877
University of Fort Hare Museum, Alcie 38551
University of Hawaii Art Gallery, Honolulu 49355
University of Helsinki Agricultural Museum, Helsinki 10786
University of Helsinki, The K.E. Kivirikko Collection of Birds and Mammals, Helsinki 10789
University of Iowa Museum of Art, Iowa City 49521
University of Iowa Museum of Natural History, Iowa City 49522
University of Johannesburg Art Gallery, Auckland Park 38554
University of Kansas Ryther Printing Museum, Lawrence 49993
University of Kentucky Art Museum, Lexington 50057
University of Lethbridge Art Gallery, Lethbridge 06164
University of Liverpool Art Gallery, Liverpool 44747
University of Manchester Medical School Museum, Manchester 45098
University of Mandalay Collections, Mandalay 31721
University of Nebraska Art Gallery at Omaha, Omaha 51556
University of Nebraska State Museum, Lincoln 50107
University of New Brunswick Art Centre, Fredericton 05840
University of North Dakota Zoology Museum, Grand Forks 48990
University of North Texas Art Gallery and Cora Stafford Gallery, Denton 48063
University of Northern Iowa Museum Museums & Collections, Cedar Falls 47294
University of Northern Philippines Museum, Vigan 34524
University of Nueva Caceres Museum, Naga 34467
University of Oulu, Zoological Museum, Oulu 11173
University of Pennsylvania Museum of Archaeology and Anthropology, Philadelphia 51886
University of Pretoria Art Collection, Tshwane 38761
University of Queensland Art Museum, Brisbane 00901
University of Rhode Island Fine Arts Center Galleries, Kingston 49795
University of San Carlos Anthropology Museum, Cebu 34377
University of San Carlos Biological Museum, Cebu 34378
University of Santo Tomas Museum of Arts and Sciences, Manila 34458
University of South Carolina at Spartanburg Art Gallery, Spartanburg 53198
University of South Carolina Beaufort Art Gallery, Beaufort 46668
University of Southern Philippines Museum, Cebu 34379
University of Southern Queensland Art Gallery & Art Collection, Toowoomba 01595
University of Swaziland Library Museums Collection, Kwaluseni 40402
University of Sydney Art Collection, Sydney 01563
University of Toronto Art Centre, Toronto 07153
University of Virginia Art Museum, Charlottesville 47400
University of Waterloo Art Gallery, Waterloo 07290
University of West Florida Art Gallery, Pensacola 51770
University of Wisconsin Zoological Museum, Madison 50395
University of Wyoming Anthropology Museum, Laramie 49949
University of Wyoming Art Museum, Laramie 49950
University of Zambia Library Museums Collection, Lusaka 54660
University Place Art Center, Lincoln 50108
University Saint Thomas Art Collection and Exhibitions, Saint Paul 52630
Universum Bremen, Bremen 17565
Universum, Museo de las Ciencias de la UNAM, México 31161
UNLV Barrick Museum, Las Vegas 49979
Uns lütt Museum, Dargun 17784
Uns lütt Schiffsmuseum, Cuxhaven 17758
UNSW Art Collection, Sydney 01564
Untamalan Kotiseutumuseo, Laitila 11041
Unterfränkisches Grenzmuseum, Bad Königshofen 16761
Untersbergmuseum Fürstenbrunn, Grödig 01994
Unterwarter Heimathaus, Unterwart 02843
UnterWasserReich, Schrems 02715
Untzi Museoa - Museo Naval, Donostia-San Sebastián 39255
Uozu-shi Rekishi no Mingei Hakubutsukan, Uozo 29562
Upernavik Museum, Upernavik 23271
Upham Mansion, Marshfield 50514
UPIDIV Galerija, Novi Sad 38126
Upīte Cultural History Museum, Šķilbēnu 30125
Upītes Kultūrvēstures Muzejs, Šķilbēnu 30125
Upminster Mill, Upminster 45898
Upminster Tithe Barn Museum of Nostalgia, Upminster 45899
Uppark Exhibition, Petersfield 45375
Upper Canada Village, Morrisburg 06395
Upper East Region Museum, Bolgatanga 22849
Upper Musselshell Museum, Harlowtown 49171
Upper Paxton Township Museum, Millersburg 50693
Upper Room Chapel Museum, Nashville 50978
Upper Silesian Museum, Bytom 34584
Upper Snake River Valley Historical Museum, Rexburg 52300
Upper Waitaki Pioneer Museum and Art Gallery, Kurow 33138
Upper Wharfedale Folk Museum, Grassington 44349
Upper Yarra Valley Museum, Yarra Junction 01679
Upperclass Home Museum, Kolkanlahti 10976
Upperlands Eco-Museum, Derry 43975
Uppertown Firefighters Museum, Astoria 46404
Upplandsmuseet, Uppsala 41012
Uppsala Konstmuseum, Uppsala 41013
Uppsala Medicinhistoriska Museum, Uppsala 41014
Uppsala Universitet Myntkabinett, Uppsala 41015
Uppsala Universitets Konstsamlingar, Uppsala 41016
Uppsala University Art Collection, Uppsala 41016
Uppsala University Museum, Uppsala 41009
Upstairs Art Gallery, Los Alamos 50206
Upton House and Gardens, Banbury 43371
Ur-Donautal-Museum, Wellheim 22469
Ur- und Frühgeschichtliche Sammlung, Erlangen 18226
Urahoro-cho Kyodo Hakubutsukan, Urahoro 29563
Urahoro Municipal Museum, Urahoro 29563
Urajärven Kartanomuseo, Urajärvi 11417
Urajärvi Manor, Urajärvi 11417
Ural Geological Museum V.V. Vakhrushev, Ekaterinburg 36901
Ural Tansykbaev House-Museum, Toshkent 54490
Ural Tansykbaev Yodgorlik Muzeyi, Toshkent 54490
Uralskij Geologičeskij Muzej V.V. Vachrušev, Ekaterinburg 36901
Urawa Art Museum, Saitama 29192
Urawa-kuritsu Bijutsukan, Saitama 29192
Urban Institute for Contemporary Arts, Grand Rapids 49006
Urban Shaman, Winnipeg 07389
UrbanArts Institute, Boston 46953
Ure Museum of Greek Archaeology, Reading 45472
Urgeschichtemuseum Niederösterreich, Asparn an der Zaya 01734
Urgeschichtliche Sammlung, Engerwitzdorf 01868
Urgeschichtliches Museum, Mauer 20238
Urgeschichtliches Museum, Peggau 02483
Urgeschichtliches Museum und Galerie Vierzigtausend Jahre Kunst, Blaubeuren 17381
Urgeschichtssammlung und Heimatmuseum, Jedenspeigen 02140
Urho Kekkonen Museo Tamminiemi, Helsinki 10834
Uri and Rami Nechushtan, Kibbutz Ashdot Yaakov 25032
Urios College Museum, Butuan 34356
Urjala Museum, Urjala 11418
Urjalan Museo, Urjala 11418
Urmensch-Museum, Steinheim an der Murr 21892
Uroomieh Museum, Uroomieh 24702
Urquhart Castle, Drumnadrochit 44027
Urskog-Hølandsbanen, Sørumsand 33854
Ursula Blickle Stiftung, Kraichtal 19697
Ursuline Hallway Gallery, San Antonio 52729
Urumchi City Museum, Urumchi 08307
Urushi Art Museum, Wajima, Wajima 29570
Urwelt-Museum Hauff, Holzmaden 19260
Urwelt-Museum Oberfranken, Bayreuth 16966
Urweltmuseum Aalen, Aalen 16415
Urzeitmuseum Nußdorf ob der Traisen, Traismauer 02821
US Patent and Trademark Office - Museum, Arlington 46359
Uşak Archeological Museum, Uşak 42960
Uşak Arkeoloji Müzesi, Uşak 42960
Usher Gallery, Lincoln 44717
Usk Rural Life Museum, Usk 45901
USM Art Gallery, Gorham 48976
Usman Jusupov Memorial Museum, Yangiyo'l 54493
Usman Jusupov Yodgorlik Muzeyi, Yangiyo'l 54493
Uspenskij Sobor, Moskva 37410
USS Alabama Battleship Memorial Park, Mobile 50772
USS Arizona Memorial, Honolulu 49356
USS Bowfin Submarine Museum, Honolulu 49357
USS Constellation Museum, Baltimore 46590
USS Constitution Museum, Boston 46954
USS Lexington Museum on the Bay, Corpus Christi 47822
USS Radford National Naval Museum, Newcomerstown 51310
Ussishkin House, Kibbutz Dan 25035
Ust-Ilimsk Picture Gallery, Ust-Ilimsk 37912
Ust-Ilimskaja Kartinnaja Galereja, Ust-Ilimsk 37912
Utah Field House of Natural History State Park, Vernal 53722
Utah Museum of Fine Arts, Salt Lake City 52695
Utah Museum of Natural History, Salt Lake City 52696
Utah Museums Association, Salt Lake City 55188
Utah State Historical Society, Salt Lake City 52697
Ute Indian Museum, Montrose 50830
Utena Local Lore Museum, Utena 30390
Utenio Kraštotyros Muziejus, Utena 30390
Utica Museum, Hobson 49306
Útisavnið í Hoyvík, Hoyvík 10688
Utstein Kloster, Mosterøy 33685
Utsukushigahara Kogen Bijutsukan, Takeshi 29346
Utsukushigahara Open-Air Museum, Takeshi 29346
Utsunomiya Bijutsukan, Utsunomiya 29568
Utsunomiya Museum of Art, Utsunomiya 29568
Uttar Pradesh State Museum, Lucknow 24301
Uttoxeter Heritage Centre, Uttoxeter 45902
Utvandrarnas Hus, Växjö 41043
Uudenkaupungin Kultuurihistoriallinen Museo, Uusikaupunki 11421
Uummannap Katersugaasivia, Uummannaq 23272
Uummannaq Museum, Uummannaq 23272
Uusikaupunki Cultural History Museum, Uusikaupunki 11421
UVAN Historical Museum, Winnipeg 07390
Uwajima Date Museum, Uwajima 29569
Uwajima-shiritsu Date Hakubutsukan, Uwajima 29569
Uxbridge Historical Centre Museum, Uxbridge 07191
Uxbridge - Howick's Creative Centre, Auckland 33060
Uzbekistan Art Academy, Toshkent 54483

V

V.A. Rusanov House-Museum of the Orël Regional Museum, Orël 37506
V.G. Belinski State Memorial Museum-Estate, Belinskij 36797
V.G. Korolenko House-Museum, Gelendžik 36922
V.G. Korolenko Literary and Memorial Museum, Poltava 43171
V.G. Korolenko Literary and Memorial Museum, Žytomyr 43216
V. Gross Doll Museum, Warren 07279
V.I. Chapaev Museum, Čeboksary 36840
V.I. Lenin House-Museum, Samara 37614
V.I. Nemirovich-Danchenko Apartment Museum, Moskva 37320
V.I. Nemirovich-Danchenko Memorial Museum of the Donetsk Regional Museum of Local Lore, Neskučne 43152
V.I. Surikov Art Museum, Krasnojarsk 37069
V.I. Surikov Museum-Estate, Krasnojarsk 37077
V. Ivasyuk Memorial Museum, Černivci 43016
V.M. Vasnetsov House Museum, Moskva 37169
V.P. Sukachev Regional Art Museum, Irkutsk 36932
V.S. Meyerhold Museum, Moskva 37319
V.S. Vysotskiy Cultural Centre and Museum, Moskva 37224
V. Samadova Exhibition Center, Bakı 03176
V.Tarnovsky Historical Museum, Černigiv 43007
V. Tytov Museum-Apartment of the Local Museum of History and Culture, Lugans'k 43118
V.V. Bianki Local Museum, Bijsk 36805
V.V. Mayakovsky State Museum, Moskva 37199
VA Museum of Childhood, Bethnal Green, London 45001A
Vaajakosken Työläiskotimuseo, Jyväskylä 10899
Vaal Teknorama Cultural Museum, Vereeniging 38772
Vaasa Car- and Motormuseum, Vaasa 11428
Vaasan Auto- ja Moottorimuseo, Vaasa 11428
Vaasan Merimuseo, Vaasa 11429
Vacaville Museum, Vacaville 53679
Vachel Lindsay Home, Springfield 53222
Vachott Sándor Town Library, Gyöngyös 23575
Vachott Sándor Városi Könyvtár, Gyöngyös 23575
Vaclovo ir Elenos Venskų Etnografinė Sodyba, Tytuvėnai 30387
Váczy Péter Gyűjtemény, Győr 23587
Vadim Kozin Memorial Museum, Magadan 37120
Vadim Sidur State Museum, Moskva 37203
Vadsbo Museum, Mariestad 40738
Vadsø Museum - Ruija Kvenmuseum, Vadsø 33953
Vadstena Slott, Vadstena 41021
Vadstena Stadsmuseum, Vadstena 41022
Vääksyn Vesimylly- ja Piensähkölaitosmuseo, Vääksy 11431
Vähänkyrön Museo, Vähäkyrö 11432
Väike-Männiku Talumuuseum, Kõpu 10585

Väinontalo - Järviseudun Museo, Vasikka-Aho 11457
Vämöparken, Karlskrona 40637
Vänermuseet, Lidköping 40676
Vänersborgs Museum, Vänersborg 41026
Vännäs Motormuseum, Vännäs 41027
Världskulturmuseet, Göteborg 40558
Værløse Museum, Værløse 10304
Värmlands Museum, Karlstad 40639
Värtsilän Myllymuseo, Värtsilä 11433
Väsby Kungsgård, Sala 40828
Västerås Flyg Museum, Västerås 41031
Västerås Konstmuseum, Västerås 41032
Västerås Skolmuseum, Västerås 41033
Västerbottens Museum, Umeå 41000
Västergötlands Museum, Skara 40846
Västernorrland Textile Archive, Sollefteå 40869
Västerviks Museum, Västervik 41037
Västmanlands Läns Museum, Västerås 41034
Vättehult - Pälle Nävers Diktarstuga, Vetlanda 41057
Växjö Konsthall, Växjö 41044
Vagnshistoriska Museet, Fristad 40527
Vagonářské Muzeum, Studénka 09854
Vágs Bygdasavn, Vágur 10701
Vágur Museum, Vágur 10701
Vajnorský L'udový Dom - Etnografická Expozícia, Bratislava 38235
Vaikų Literatūros Muziejus, Kaunas 30267
Vaile Mansion - Dewitt Museum, Independence 49480
Vajda János Emlékház, Vál 23950
Vakıf İnşaat ve Sanat Eserleri Müzesi, İstanbul 42847
Vakoilumuseo, Tampere 11368
Val Gardena Heritage Museum, Ortisei 27001
Valaam Island Scientific Museum Preserve of Religion, Archeology and Nature, Sortavala 37790
Valaamskij Naučno-issledovatelskij Cerkovno-archeologičeskij i Prirodnyj Muzej-zapovednik, Sortavala 37790
Valašské Muzeum v Přírodě, Rožnov pod Radhoštěm 09827
Vâlcea District Museum, Râmnicu Vâlcea 36553
Valdajskij Kraevedčeskij Muzej, Valdaj 37914
Valdemārpils Forest Museum, Valdemārpils 30136
Valdemārpils Mežu Muzejs, Valdemārpils 30136
Valdez Museum, Valdez 53685
Valdivia Museo, Valdivia 10463
Valdosta State University Fine Arts Gallery, Valdosta 53687
Valdres Folkemuseum, Fagernes 33487
Vale and Downland Museum, Wantage 45929
Valence House Museum and Art Gallery, Dagenham 43943
Valencian Museum of Ethnology, Valencia 40223
Valencian Toys Museum, Ibi 39405
Valentia Island Heritage Centre, Knightstown 24859
Valentin-Karlstadt-Musäum, München 20493
Valentin Serov Memorial Museum of Art in Domotkanovo, Krasnaja Nov 37064
Valentine Richmond History Center, Richmond 52339
Valentown Museum, Victor 53739
Våler Torvdriftsmuseum, Våler i Solør 33955
Valga Muuseum, Valga 10667
Valhalla Museum, Tresco 45875
The Valiant Soldier, Buckfastleigh 43661
Valka Museum of Local Studies, Valka 30137
Valkas Novadpētniecības Muzejs, Valka 30137
Vallachian Open-Air Museum, Rožnov pod Radhoštěm 09827
Vallby Friluftsmuseum, Västerås 41035
Valle de los Caídos, Valle de Cuelgamuros 40246
Vallejo Naval and Historical Museum, Vallejo 53691
Valley Art Center, Clarkston 47591
Valley Community Historical Society Museum, Valley 53692
Valley County Pioneer Museum, Glasgow 48915
Valley Forge National Historical Park, Valley Forge 53695
Valley Museum, McBride 06208
Valmiera Gallery, Valmiera 30139
Valmiera Museum of Local Studies, Valmiera 30140
Valmieras Novadpētniecības Muzejs, Valmiera 30140
Valsts Mākslas Muzejs, Rīga 30114
Valsts Mākslas Muzejs Izstazu Zale Arsenāls, Rīga 30115
Valstybinis Vilniaus Gaono Žydų Muziejus, Vilnius 30436
Valtakunnallinen Työväentalomuseo, Riihimäki 11273
Valtion Taidemuseo, Helsinki 10835
Vám- és Pénzügyőr Múzeum, Budapest 23489
Vampulan Museo, Vampula 11443
Het Van Abbemuseum, Bonheiden 03380
Van Abbemuseum, Eindhoven 32196
Van Andel Museum Center, Grand Rapids 49007
Van Buren County Historical Society Museum, Keosauqua 49746
Van Cortlandt House Museum, Bronx 47053
Van de Poll-Stichting, Zeist 32995
Van-der-Werf's Wedgwoodmuseum, Zuidwolde, Groningen 33016
Van Devlet Güzel Sanatlar Galerisi, Van 42961
The Van Egmond Foundation, Egmondville 05768
Van Gogh Documentatiecentrum, Nuenen 32606
Van Gogh Museum, Amsterdam 31898
Van Gybland-Oosterhoff Collection, Tshwane 38762
Van Meter State Park, Miami 50638
Van Müzesi, Van 42962
Van Museum, Van 42962
Van Nostrand-Starkins House, Roslyn 52464
Van Riper-Hopper House, Wayne 54009
Van State Fine Art Gallery, Van 42961
Van 't Lindenhoutmuseum, Nijmegen 32595
Van Tilburg Collection, Tshwane 38763
Van Wyck Homestead Museum, Fishkill 48547
Vanadzor Museum of Fine Art, Vanadzor 00721
Vanaja Local History Museum, Hämeenlinna 10749
Vanajan Kotiseutumuseo, Hämeenlinna 10749
Vanatehnikaklubi "Retrom" Tuletorjemuuseum, Kärevere 10579
Vancouver Art Gallery, Vancouver 07226
Vancouver Arts Centre, Albany 00789
Vancouver Holocaust Education Centre, Vancouver 07227
Vancouver Island Military Museum, Nanaimo 06414
Vancouver Maritime Museum, Vancouver 07228
Vancouver Museum, Vancouver 07229
Vancouver Naval Museum, Qualicum Beach 06675
Vancouver Police Centennial Museum, Vancouver 07230
Vandalia State House, Vandalia 53709
Vanderbilt Kennedy Center Arts Gallery, Nashville 50979
Vanderbilt Mansion National Museum, Hyde Park 49460
Vanderbilt University Fine Arts Gallery, Nashville 50980
Vanderbilt University Sarratt Gallery, Nashville 50981
Vang Church, Karpacz 34717
Vangi - The Sculpture Garden Museum, Nagaizumi 28950
Vangjush Mio Museum, Korçë 00024
Vanguard Centennial Museum, Vanguard 07232
Vanha-Rantalan Talomuseo, Pertunmaa 11201
Vanhalinna Museum, Vanhalinna 11444
Vankilamuseo, Hämeenlinna 10750
Vann House, Chatsworth 47405
Vantaa City Museum, Vantaa 11449
Vantaa City Museum - Visual Arts, Vantaa 11450
Vantaan Kaupunginmuseo, Vantaa 11449
Vantaan Kaupunginmuseo - Kuvataideasiat, Vantaa 11450
Vanuatu Cultural Centre and National Museum, Port Vila 54495
Vapentekniska Samlingarna, Eskilstuna 40505
Varala Museum of Gymnastics, Tampere 11369
Varalan Liikuntamuseo, Tampere 11369
Varanger Museum Sør, Kirkenes 33593
Varaždin City Museum, Varaždin 08975
Varde Artillerimuseum, Varde 10308
Varde Museum, Varde 10307
Varden Redningsmuseum, Narbø 33694
Vardømuseene, Vardø 33961
Varendra Research Museum, Rajshahi 03234
Varga Mátyás Színháztörténeti Kiállítása, Szeged 23849
Vargas Museum and Art Gallery, Mandaluyong 34427
Varistaipale Canal Museum, Varistaipale 11451
Varistaipaleen Kanavamuseo, Varistaipale 11451
Varjjat Sámi Musea, Varangerbotn 33960
Varkauden Museot, Varkaus 11454
Varkauden Taidemuseo, Varkaus 11455
Varkaus Art Museum, Varkaus 11455
Varkaus Museum, Varkaus 11454
Varley Art Gallery of Markham, Unionville 07188
Várműzeum, Dunaföldvár 23529
Várműzeum, Esztergom 23555
Várműzeum, Gyula 23601
Várműzeum, Pécsvárad 23784
Várműzeum, Siklós 23800
Várműzeum, Simontornya 23801
Várműzeum, Veszprém 23963
Várműzeum - Kotár, Szeged 23850
Varner-Hogg Plantation, West Columbia 54052
Varnum Memorial Armory and Military Museum, East Greenwich 48242
Városi Galéria, Mezőkövesd 23700
Városi Helytörténeti Gyűjtemény, Szigetszentmiklós 23890
Városi Helytörténeti Múzeum, Komló 23682
Városi Képtár, Gyula 23602
Városi Képtár, Orosháza 23741
Városi Képtár - Deák Gyűjtemény, Székesfehérvár 23864
Városi Könyvtár és Múzeum, Tapolca 23914
Városi Muvészeti Múzeum - Napóleon-ház, Győr 23588
Városi Művészeti Múzeum - Városi Képtár, Győr 23589
Városi Múzeum, Kőszeg 23685
Városi Múzeum, Paks 23748
Várostörténeti Múzeum, Pécs 23779
Varpaisjärven Kotiseutumuseo, Varpaisjärvi 11456
Várpalota Mining History Collection, Várpalota 23954
Várpalotai Bányászattörténeti Gyűjtemény, Várpalota 23954
Varpo - Viurinės Mokyklos Istorijos Muziejus, Klaipėda 30285
Varpunen House Museum, Anjalankoski 10711
Varteig Bygdemuseum, Varteig 33964
Varusschlacht im Osnabrücker Land, Bramsche 17501
Vasa Lutheran Church Museum, Vasa 53711
Vasa Museum, Stockholm 40949
Vasalopps Museet Mora, Mora 40747
Vasalopps Museet Sälen, Transtrand 40979
Vasamuseet, Stockholm 40949
Vasarely Múzeum, Budapest 23490
Vashon Maury Island Heritage Museum, Vashon 53712
Vasi Múzeumfalu, Szombathely 23906
Vasil Levski Memorial House, Loveč 05087
Vasil Levski Museum, Karlovo 05050
Vasily Eroshenko Memorial Museum of Literature, Staryj Oskol 37800
Vassdragsmuseet Labro, Skollenborg 33840
Vastseliina Gümnaasiumi Muuseum, Vastseliina 10669
Vastseliina Koduloomuuseum, Vastseliina 10670
Vastseliina Local Lore Museum, Vastseliina 10670
Vasúti Helytörténeti Gyűjtemény, Mátészalka 23694
Vasúti Múzeum, Paks 23749
Vat Xiengkhouane, Vientiane 29978
Vaucluse House, Vaucluse 01616
Vaud Military Museum, Morges 41644
Vaxholm Fortress Museum, Vaxholm 41054
Vaxholms Fästnings Museum, Vaxholm 41054
Vay Ádám Múzeum, Vaja 23947
Vecdaugava Regional Studies and Fishing Museum, Rīga 30116
Vecdaugavas Novadpētniecības un Zvejniecības Muzejs, Rīga 30116
Veen Park, Barger Compascuum 31949
Het Veenkoloniaal Museum, Veendam 32872
Veenmuseum het Vriezenveense Veld, Vriezenveen 32912
Vefsn Museum, Mosjøen 33683
Vega Bygdemuseum, Vega 33968
Vegårshei Bygdetun, Vegårshei 33969
Végh-kúria, Kiskunhalas 23672
Vegmuseet - Rogaland, Stavanger 33877
Vegreville Regional Museum, Vegreville 07235
Vehikelsammlung Eppelheim, Eppelheim 18180
Vehmaan Kotiseutumuseo ja Kivityömuseo, Vehmaa 11458
Vejen Kunstmuseum, Vejen 10309
Vejle Art Museum, Vejle 10310
Vejle Kunstmuseum, Vejle 10310
Vejle Museum, Vejle 10311
Vekkilän Kotiseutumuseo, Tammela 11346
Velenje Museum at Velenje Castle, Velenje 38534
Veletržní Palác, Praha 09798
Velfjord Bygdemuseum, Hommelstø 33556
Velikolukskij Kraevedčeskij Muzej, Velikie Luki 37918
Velikoustjugskij Gosudarstvennyj Istoriko-Architekturnyj i Chudožestvennyj Muzej-Zapovednik, Velikij Ustjug 37923
Velimir Chlebnikov House-Museum, Astrachan 36781
Veliš Lekovič Art Gallery, Bar 31635
Veliuonos Krašto Istorijos Muziejus, Veliuona 30394
Velnių Muziejus, Kaunas 30268
Veluws Museum Nairac, Barneveld 31951
Veluws Museum van Oudheden - Stadsmuseum, Harderwijk 32313
Veluws Streekmuseum Hagedoorns Plaatse, Epe 32218
Velykosoročynskyj Literaturno-memorialnyj Muzej Gogolja, Velyki Soročynci 43201
Venango Museum of Art, Science and Industry, Oil City 51509
Venclauskiai House of the Šiauliai Aušra Museum, Šiauliai 30365
Venclauskių Namai, Šiauliai 30365
Venclovų Namai-Muziejus;, Vilnius 30437
Vendsyssel Historiske Museum, Hjørring 10080
Vendsyssel Kunstmuseum, Hjørring 10081
Venetsanos Museum of Modern Art, Piraeus 23149
Vengerskij Kulturnyj Centr, Moskva 37411
Vennel Gallery, Irvine 44536
Vennesla Bygdemuseum, Vennesla 33972
Venrays Museum, Venray 32886
Vent Haven Museum, Fort Mitchell 48656
Ventnor Heritage Museum, Ventnor 45904
Ventspils Jūras Zvejniecības Brīvdabas Muzejs, Ventspils 30144
Ventspils Museum, Ventspils 30145
Ventspils Muzejs, Ventspils 30145
Ventspils Open-Air Museum of Maritime Fishing, Ventspils 30144
Ventura College Art Galleries, Ventura 53715
Ventura County Maritime Museum, Oxnard 51658
Ventura County Museum of History and Art, Ventura 53716
Vepsskiy Ethnographical Museum, Šeltozero 37758
Verantare Museum of Anesthesia, Antwerpen 03303
Verband der Museen der Schweiz, Zürich 55038
De Verbeelding Kunst Landschap Natuur, Zeewolde 32993
Das Verborgene Museum, Berlin 17277
Verbund Oberösterreichischer Museen, Leonding 54700
Verdant Works, Dundee 44053
The Verdier House, Beaufort 46669
Verdmont Historic House Museum, Smith's Parish 04200
Verein "Chemnitzer Gewölbegänge" e. V., Chemnitz 17705
Verein für Heimatpflege, Viersen 22233
Verein Kunstsammlung Unterseen, Unterseen 41961
Verein Shedhalle, Zürich 42115
Vereinigung der Mund- und Fussmalenden Künstler in aller Welt e.V., Schaan 30194
Vereinigung Westfälischer Museen e.V., Münster 54860
Vereniging van Rijksgesubsidieerde Musea, Amsterdam 54962
Veres Péter Emlékkiállítás, Balmazujváros 23360
Vereschagin Museum, Čerepovec 36851
Verfmolen de Kat, Zaandam 32983
Vērgale Parish Museum, Vērgale 30146
Vērgales Pagasta Muzejs, Vērgale 30146
Verger's Museum, Loppi 11081
Veria Byzantine Museum, Veria 23235
Verkehrsdrehscheibe Schweiz und unser Weg zum Meer, Basel 41186
Verkehrshaus der Schweiz, Luzern 41600
Verkehrsmuseum, Frankfurt am Main 18400
Verkehrsmuseum Dresden GmbH, Dresden 17976
Verkehrsmuseum Karlsruhe, Karlsruhe 19452
Verla Mill Museum, Verla 11459
Verlan Tehdasmuseo, Verla 11459
Vermesova Vila, Dunajská Streda 38252
Vermessen in Bayern, München 20494
Vermilion County Museum, Danville 47960
Vermont Folklife Center, Middlebury 50654
Vermont Historical Society Museum, Montpelier 50825
Vermont Marble Museum, Proctor 52156
Vermont Museum and Gallery Alliance, Vergennes 55189
The Vermont State House, Montpelier 50826
Vernacular Clothed Doll's Museum - Historical Waxworks, Keszthely 23662
Vernadsky State Geological Museum, Moskva 37192
Verner von Heidenstams Hem, Motala 40760
Vernon County Historical Museum, Viroqua 53762
Vernon Historians, Vernon 53724
Vernon Public Art Gallery, Vernon, British Columbia 07238
Vernonia Historical Museum, Vernonia 53726
Vero Beach Museum of Art, Vero Beach 53728
Verpleeghuis Cornelia, Zierikzee 33006
Verpleegkundig Historisch Bezit, Amersfoort 31817
Verran Museum, Malm 33656
Verrerie La Mailloche, Québec 06698
Vértes Múzeum, Csákvár 23504
Vértesszőllős Open-air Exhibition, Vértesszőllős 23957
Vértesszőllősi Bemutatóhely, Vértesszőllős 23957
Verulamium Museum, Saint Albans 45550
Verwood Community Museum, Verwood 07240
Very Special Arts Gallery, Albuquerque 46190
Verzameling Steenbakker Ten Bokkel Huinink, Neede 32567
Verzets Museum Zuid-Holland, Gouda 32271
Verzetsmuseum Amsterdam, Amsterdam 31899
Verzetsmuseum Friesland, Leeuwarden 32480
Vest-Agder Fylkesmuseum, Kristiansand 33613
Vest-Lindsey House, Frankfort 48743
Vest-Telemark Museum, Dalen i Telemark 33463
Vestal Museum, Vestal 53730
Vesterålsmuseet, Melbu 33664
Vesterheim Norwegian-American Museum, Decorah 48022
Vestfold Fylkesmuseum, Tønsberg 33911
Vestfyns Hjemstavnsgård, Glamsbjerg 10039
Vesthimmerlands Museum, Års 09967
Vesting Bourtange, Bourtange 31994
Vestingmuseum, Nieuweschans 32580
Vestingmuseum Oudeschans, Oudeschans 32656
Vestisches Museum, Recklinghausen 21211
Vestlandske Skolemuseum, Stavanger 33878
Vestry House Museum, London 45002
Vestsjællands Kunstmuseum, Sorø 10282
Vesztő-Mágor Historical Memorial Location, Vészto 23965
Vesztő-Mágor Történelmi Emlékhely, Vészto 23965

Vetelin Museo, Veteli 11460
Veterans Federation of the Philippines Museum, Taguig 34514
Veterans' Memorial Military Museum, Kensington 06060
Veterans Museum, Samsun 42930
Veterinärhistoriska Museet, Skara 40847
Veterinärmedizinhistorisches Museum der Tierärztlichen Hochschule Hannover, Hannover 19007
Veterinary Medicine Museum, Ciechanowiec 34609
Veterinary Museum, London 45003
Veterinary Parasitology Museum, Bangkok 42564
Vetlanda Museum, Vetlanda 41058
Vetlanda Skolmuseum, Vetlanda 41059
Vetlužskij Kraevedčeskij Muzej, Vetluga 37927
Větrný Mlýn, Rudice 09830
Veturimuseo, Toijala 11381
Vevelstad Bygdetun, Vevelstad 33976
Vi Gurnis-Vieux Greniers, Bertrix 03366
Viborg Stiftsmuseum, Viborg 10317
Vicarage Museum of Icons on Glass, Sibiel 36603
Vicente Manansala Museum, Binangonan 34346
Vicksburg National Military Park, Vicksburg 53735
Victor and Apollinaris Vasnetsov Regional Art Museum, Kirov 37038
Victor and Appolinari Vasnetsov Memorial House-Museum, Rjabovo 37592
Victor Heritage Museum, Victor 53737
Victor-Schultze-Institut für christliche Archäologie und Geschichte der kirchlichen Kunst, Greifswald 18751
Victor Valley Museum and Art Gallery, Apple Valley 46328
Victor Vasarely Múzeum, Pécs 23780
Victoria and Albert Museum, London 45004
Victoria and Albert Museum, Mumbai 24324
Victoria Art Gallery, Bath 43413
Victoria County Archiv and Museum, Baddeck 05371
Victoria County Historical Society Museum, Lindsay 06173
Victoria Falls Information Centre, Livingstone 54657
Victoria Jubilee Museum, Cawthorne 43783
Victoria Jubilee Museum, Vijayawada 24429
Victoria Mansion, Portland 52069
Victoria Memorial Hall, Kolkata 24283
Victoria Police Historical Museum, Victoria 07260
Victoria School Museum, Carleton Place 05564
Victoria School Museum and Archives, Edmonton 05762
Victoria Settlement Provincial Historic Site, Saint Paul 06871
Victoria Tower, Huddersfield 44498
Victoria West Regional Museum, Victoria West 38774
Victorian College of the Arts Gallery, Southbank 01519
Victoria's Tramway Museum, Bylands 00930
VIDA Museum, Borgholm 40474
Video History Museum, Grass Valley 49026
Vidovszky Béla Helytörténeti Gyűjtemény, Gyomaendrőd 23578
Viechtauer Heimathaus Neukirchen, Neukirchen bei Altmünster 02433
Viehstrichmuseum, Wörth am Rhein 22642
La Vieille Maison, Belliveau Cove 05417
Vieille Maison Traditionelle, Saint-Véran 15534
Viekšnių Vaistinės Muziejus, Viekšniai 30396
Vielstedter Bauernhaus, Hude 19285
Vien Bao Tang Lich Sa Viet Nam, Ha Noi 54616
Vienna University Library, Wien 03083
Vieques Historical Conservation Trust, Vieques 36081
Vieremän Kotiseutumuseo, Vieremä 11461
Vierschaar Museum, Veere 32879
Viestikeskus Lokki, Mikkeli 11118
Viet Nam History Museum Ho Chi Minh City, Ho Chi Minh City 54630
Viet Nam National Fine Arts Museum, Ha Noi 54617
Viet Nam Revolution Museum, Ha Noi 54618
Viet Nam Women's Museum, Ha Noi 54619
Vietnam Era Educational Center, Holmdel 49325
Vietnam Museum of Ethnology, Ha Noi 54608
Vietnam Veterans Museum, San Remo 01492
Vieux Grand Port Archaeological Site, Old Grand Port 30680
Le Vieux Moulin, Sainte-Jeanne-d'Arc 06895
Vieux Presbytère de Batiscan, Batiscan 05398
Vigado Galeria, Budapest 23491
Vigan House Museum, Vigan 34525
Vigatunet, Hjelmeland 33549
Vigeland-Museet, Oslo 33758
The Vigeland Museum, Oslo 33758
Vigo County Historical Museum, Terre Haute 53485
Vihannin Kotiseutumuseo, Vihanti 11462
Vihanti Local History Museum, Vihanti 11462
Vihavuoden Myllymuseo, Sappee 11299
Vihdin Museo, Vihti 11463
Vihorlatské Múzeum, Humenné 38262
Viirilän Kotiseutumuseo, Kuninkaankylä 11005
Viitainahon Kotiseutumuseo, Piippola 11208
Viitasaaren Metsätyömuseo, Viitasaari 11466
Vijs'kovo-istoryčnyj Muzej, Černigiv 43008
Vikedal Bygdemuseum, Vikedal 33980
Vikens Sjöfartsmuseum, Viken 41062
Viking Historical Museum, Viking 07262
The Viking Museum, Højbjerg 10085
Viking Ship Museum, Roskilde 10250
Viking Ship Museum, University Museum of Cultural History, Oslo 33759
Viking Union Gallery, Bellingham 46712
Vikingemuseet Moesgård, Højbjerg 10085
Vikingeskibsmuseet, Roskilde 10250
Vikingmuseet på Borg, Bøstad 33450
Vikingsholm Museum, Tahoma 53420
Vikingskipshuset, Oslo 33759
Vikram Kirti Mandir Museum, Ujjain 24417
Viktor Rydbergmuseet, Jönköping 40617
Vija Plūdoņa Memoriālā Māja Muzejs, Ceraukste 29995
Vila Tugendhat, Brno 09442
Vilis Pludons Memorial House Museum, Ceraukste 29995
Viljamakasiinimuseo, Vihti 11464
Viljandi Muuseum, Viljandi 10674
Viljandimaa Gümnaasiumi Muuseum, Viljandi 10675
Villa Arson, Nice 14442
Villa Bergzicht/Brouwerijmuseum Raaf, Heumen 32382
Villa Carlotta, Tremezzo 28053
Villa Cicogna Mozzoni, Bisuschio 25428
Villa Collezione Panza, Varese 28164
Villa delle Rose, Bologna 25479
Villa des Arts, Casablanca 31661
Villa d'Este, Tivoli 27979
Villa du Parc Centre d'Art Contemporain, Annemasse 11627
Villa Eschebach, Dresden 17977
Villa et Jardins Ephrussi de Rothschild, Saint-Jean-Cap-Ferrat 15317
Villa Farnesina, Roma 27548
Villa Flora Winterthur, Winterthur 42041
Villa Foscari La Malcontenta, Mira 26736
Villa Gallo-Romaine de Saint-Ulrich, Haut-Clocher 13188
Villa Gallo-Romaine de Séviac, Montréal-du-Gers 14252
Villa Grècque Kérylos, Beaulieu-sur-Mer 11896
Villa Gyllenberg, Helsinki 10836
Villa Gyllenberg Art Collection, Helsinki 10836
Villa Hügel, Essen 18270
Villa Imperiale, Pesaro 27159
Villa Julie College Gallery, Stevenson 53291
Villa Kruger, Clarens 41306
Villa Lattes, Istrana 26420
Villa Louis Historic Site, Prairie du Chien 52129
Villa Majorelle, Nancy 14358
Villa Manin, Codroipo 25934
Villa Mansi, Capannori 25654
Villa Merkel und Bahnwärterhaus, Esslingen 18278
Villa Mirabello, Varese 28165
Villa Montezuma, San Diego 52760
Villa Museo Puccini, Viareggio 28245
Villa Oppenheim, Berlin 17278
Villa Pelsser, Henri-Chapelle 03696
Villa Pisani, Strà 27909
Villa Publica Romkert, Zalalövo 23974
Villa Publica Ruin Garden, Zalalövo 23974
Villa Romana, Desenzano del Garda 26036
Villa Romana del Casale, Piazza Armerina 27196
Villa San Michele, Anacapri 25224
Villa Sinnenreich, Rohrbach/OÖ 02597
Villa Strassburger, Deauville 12703
Villa Terrace Decorative Arts Museum, Milwaukee 50720
Villa Torrigiani, Camigliano 25625
Villa Trips, Kerpen 19520
Villa Vauban, Luxembourg 30486
Villa Vauban, Luxembourg 30487
Villa Vauban - Luxembourg City Art Museum, Luxembourg 30487
Villa Verdi, Villanova sull'Arda 28278
Villa Vermes - Slovakian National Gallery, Dunajská Streda 38252
Villa Zebra Museum for Children, Rotterdam 32732
Villa Zebra Museum voor Kinderen, Rotterdam 32732
Villacher Fahrzeugmuseum, Villach 02862
Village Club Private Museum, Kemaliye 42877
Village d'Artistes, Rablay-sur-Layon 14978
Le Village des Pêcheurs, Canet-en-Roussillon 12266
Village du Miniature, Pailhares 14552
Village Historical Society Museum, Harrison 49185
Village Historique Acadien, Caraquet 05555
Village Lacustre Gletterens, Gletterens 41440
Village Miniature, Marchesieux 13952
Village Minier de Bourlamaque, Val-d'Or 07194
Village-Musée, Fos-sur-Mer 12982
Village Museum, Arcani 36114
Village Museum, Bănia 36136
Village Museum, Bărbăteşti 36139
Village Museum, Bârca 36140
Village Museum, Bechet 36145
Village Museum, Castranova 36268
Village Museum, Celaru 36271
Village Museum, Cernăteşti 36273
Village Museum, Dănciuleşti 36337
Village Museum, Dar-es-Salaam 42513
Village Museum, Drăcşenei 36352
Village Museum, Dranic 36357
Village Museum, Frăteşti 36375
Village Museum, Gighera 36385
Village Museum, Goicea 36387
Village Museum, Gornea 36388
Village Museum, Gostavăţu 36389
Village Museum, Grumezoaia 36392
Village Museum, Hollókő 23616
Village Museum, Kfar Tavor 25030
Village Museum, Kouřim 09600
Village Museum, Leleşti 36437
Village Museum, Leu 36440
Village Museum, Mehadica 36470
Village Museum, Sălişte 36579
Village Museum, Skjern 10267
Village Museum, Stellenbosch 38742
Village Museum, Tăcuta 36650
Village Museum, Teiu 36681
Village Museum, Valea Stanciului 36716
Village Museum, Vultureşti 36738
Village Museum, Vutcani 36739
Village Museum of Göcsej, Zalaegerszeg 23970
Village Museum of Local History, Chihaya-Akasaka 28397
Village of Elsah Museum, Elsah 48382
Village of Living History, Turku 11392
Village Québécois d'Antan, Drummondville 05715
Village Site Museum, Eddystone 05736
Villanova University Art Gallery, Villanova 53745
Vilna Gaon Lithuanian State Jewish Museum, Vilnius 30436
Vilniaus Apskrities Centrinio Pašto Muziejus, Vilnius 30438
Vilniaus Dailės Akademijos Muziejus, Vilnius 30439
Vilniaus Krašto Etnografijos Muziejus, Nemenčinės 30318
Vilniaus Paveikslų Galerija, Vilnius 30440
Vilniaus Sporto Istorijos Muziejus, Vilnius 30441
Vilniaus Universiteto Adomo Mickevičiaus Muziejus, Vilnius 30442
Vilniaus Universiteto Chemijos Fakulteto Muziejų, Vilnius 30443
Vilniaus Universiteto Medicinos Fakulteto Medicinos Istorijos Muziejus, Vilnius 30444
Vilniaus Universiteto Mokslo Muziejus, Vilnius 30445
Vilniaus Universiteto Zoologijos Muziejus, Vilnius 30446
Vilnius Area Ethnography Museum, Nemenčinės 30318
Vilnius Art Academy Museum, Vilnius 30439
Vilnius District Central Post-Office Museum, Vilnius 30438
Vilnius Picture Gallery of the Lithuanian Art Museum, Vilnius 30440
Vilnius University Zoological Museum, Vilnius 30446
Vilppulan Museo, Vilppula 11467
Vilu Open-Air Museum, Vilu 38548
Vin- och Sprithistoriska Museet, Stockholm 40950
Vina Cooke Museum of Dolls and Bygone Childhood, Newark-on-Trent 45217
The Vinalhaven Historical Society Museum, Vinalhaven 53746
Vinarium, Cleebronn 17711
Vincas Kreve-Mickevicius Memorial Museum, Vilnius 30447
Vincas Mykolaitis-Putinas Memorial Flat-Museum, Vilnius 30448
Vincas Mykolaitis-Putinas Memorial Museum, Skriaudžiai 30371
Vincent Art Gallery, Townsville 01601
Vincent Price Art Museum, Monterey Park 50810
Vinco Grybo Memorialinis Muziejus, Jurbarka 30234
Vinco Krėvės-Mickevičiaus Memorialinis Muziejus, Vilnius 30447
Vinco Mykolaičio-Putino Memorialinis Butas-muziejus, Vilnius 30448
Vinco Mykolaičio-Putino Memorialinis Muziejus, Marijampolė 30310
Vinco Mykolaičio-Putino Memorialinis Muziejus, Skriaudžiai 30371
Vindolanda Fort and Museum Chesterholm, Bardon Mill 43379
Vindonissa-Museum, Brugg 41261
Vineta-Museum, Barth 16931
Vineyard Museum, Drăgăşani 36355
Vingelen Kirke- og Skolemuseum, Vingelen 33981
Vingerhoedmuseum, Veenendaal 32874
Viniani Museum, Pala Viniani 23131
Vinius University Adam Mickiewicz Museum, Vilnius 30442
Vinius University Chemistry Department Museum, Vilnius 30443
Vinius University Museum of Medicine History, Vilnius 30444
Vinius University Science Museum, Vilnius 30445
Vinjestoga, Vinje 33983
Vinnitsa Art Museum, Vinnycja 43205
Vinnitsa Museum of Local Lore, Vinnycja 43206
Vinnitsa National Pirogov's Museum-Estate, Vinnycja 43204
Vinnyc'kyj Literaturno-memorial'nyj Muzej M.M. Kocjubyns'koho, Vinnycja 43203
Vinnyc'kyj Nacional'nyj Muzej-sadiba im. Pirogova, Vinnycja 43204
Vinnyc'kyj Oblasnyj Chudožnij Muzej, Vinnycja 43205
Vinnyc'kyj Oblasnyj Krajeznavčyj Muzej, Vinnycja 43206
Vinohrad Exposition at Apponyiho Palac, Bratislava 38236
Vinohradnícka Expozícia Apponyiho Palác, Bratislava 38236
Vinorama - Musée du Vin, Bordeaux 12076
Vintage Aircraft Museum, Masterton 33143
Vintage Car and Carriage Museum, Inishowen 24843
Vintage Wireless Museum, London 45005
Vintschger Museum, Sluderno 27858
Viranşehir Culture Center, Viranşehir 42963
Viranşehir Kültür Merkezi, Viranşehir 42963
Virgil I. Grissom State Memorial, Mitchell 50761
Virgilkapelle, Wien 03084
Virgin Mary Relic Collection, Gyula 23600
Virginia Air and Space Center, Hampton 49154
Virginia Association of Museums, Richmond 55190
Virginia Aviation Museum, Richmond 52340
Virginia Baptist Historical Society, Richmond 52341
Virginia Beach Maritime Museum, Virginia Beach 53758
Virginia City Madison County Historical Museum, Virginia City 53760
The Virginia Discovery Museum, Charlottesville 47401
Virginia Groot A. Foundation, Chicago 47522
Virginia Historical Museum, Richmond 52342
Virginia Institute of Marine Science, Fish Collection, Gloucester Point 48949
Virginia Living Museum, Newport News 51337
Virginia Marine Science Museum, Virginia Beach 53759
Virginia Military Institute Museum, Lexington 50075
Virginia Museum of Fine Arts, Richmond 52343
Virginia Museum of Natural History, Martinsville 50521
Virginia Museum of Transportation, Roanoke 52383
Virginia Sports Hall of Fame and Museum, Portsmouth 52112
The Virginia War Museum, Newport News 51338
Virginia Wright Fund, Seattle 53016
Virgo Labour Museum, Oravais 11163
Virkki Käsityömuseo, Helsinki 10837
Virkki Museum of Handicrafts, Helsinki 10837
Viroconium Museum, Wroxeter 46087
Virolahden Kotiseutumuseo, Virolahti 11469
Virovitica Municipal Museum, Virovitica 08978
Virserums Konsthall, Virserum 41065
Virserums Möbelindustrimuseum, Virserum 41066
Virtain Perinnekylä, Virrat 11472
Virtasalmen Kotiseutumuseo, Virtasalmi 11473
Visafslag Elburg, Elburg 32200
Visceglia Art Gallery, Caldwell 47163
Vischpoort en Kazematten, Elburg 32201
Viseum Wetzlar, Wetzlar 22535
Visionarium, Centro de Ciência do Europarque, Santa Maria da Feira 35906
Visitor Center of the Central Bank of Hungary - Banknote and Coin Collection, Budapest 23440
Viski Károly Múzeum, Kalocsa 23628
Visserij- en Cultuurhistorisch Museum, Woudrichem 32969
Visserijmuseum, Breskens 32009
Visserijmuseum, Bruinisse 32016
Visserijmuseum, Vlaardingen 32892
Visserijmuseum Zoutkamp, Zoutkamp 33011
Vista Alegre Historic Museum, Ílhavo 35647
Vista Historical Museum, Vista 53765
Vistula River Museum, Tczew 35182
Visual Aid, San Francisco 52797
Visual Art Gallery, Santa Fe 52891
Visual Arts and Galleries Association, Ely 55105
Visual Arts Annex Gallery, San Antonio 52730
Visual Arts Center, Portsmouth 52113
Visual Arts Center of Northwest Florida, Panama City 51694
Visual Arts Centre, Westmount 07308
Visual Arts Centre of Clarington, Bowmanville 05450
Visual Arts Gallery, Birmingham 46827
Visual Arts Gallery, New York 51276
Visual Arts Gallery, Pensacola 51771
Visual Arts Museum, New York 51277
Visvesvaraya Industrial and Technological Museum, Bangalore 24075
Vitenskapsmuseet, Trondheim 33932
Viticultural Exhibition, Teremia Mare 36683

Viticultural Museum, Huşi 36409
Viticultural Museum, Odobeşti 36496
Vitlycke Museum, Tanumshede 40969
Vitoslavlitsy - Museum of Folk Wood Carving, Velikij Novgorod 37921
Vitra Design Museum, Weil am Rhein 22379
Vitromusée Romont, Romont 41773
Vivarium, Quito 10453
Vivarium du Moulin, Lautenbach-Zell 13537
Vizcaya Museum, Miami 50635
Vízellátás-történeti Muzeális Kiállítóhely, Orosháza 23742
Vjatskaja Kunstkamera, Kirov 37043
Vjekoslav Karas Gallery of the Karlovac Municipal Museum, Karlovac 08885
Vlaams Filmmuseum en -archief, Kortrijk 03762
Vlaams Mijnmuseum, Beringen 03361
Vlaams Tram- en Autobusmuseum, Berchem 03360
Vlaamse Museumvereniging, Antwerpen 54716
Vladimir Nabokov Museum, Sankt-Peterburg 37701
Vladimir Raevsky Museum, Bogoslovka 36813
Vladimir Serov Memorial House-Museum, Émmaus 36914
Vladimir-Suzdal Museum of Art, History and Architecture, Vladimir 37934
Vladimir-Volynski Historical Museum, Volodymyr-Volyns'kij 43207
Vladivostok Fortress - Military Historical Museum of Fortification, Vladivostok 37951
Vladivostokskaja Krepost - Voenno istoričeskij Fortifikacionnyj Muzej, Vladivostok 37951
Vladymyr Dal House-Museum of the Local Museum of History and Culture, Lugans'k 43116
Het Vlaemsche Erfgoed, Groede 32274
Vlasbewerkingsmuseum It Braakhok, Ee 32171
Vlastivědné Muzeum, Bělá pod Bezdězem 09399
Vlastivedné Múzeum, Hanušovce nad Topl'ou 38257
Vlastivědné Muzeum, Kyjov 09616
Vlastivedné Múzeum, Nové Zámky 38322
Vlastivědné Muzeum, Nymburk 09686
Vlastivedné Múzeum, Považská Bystrica 38334
Vlastivědné Muzeum, Slaný 09838
Vlastivědné Muzeum, Šumperk 09855
Vlastivedné Múzeum, Trebišov 38374
Vlastivědné Muzeum, Žirovnice 09939
Vlastivědné Muzeum a Galerie, Česká Lípa 09457
Vlastivědné Muzeum Dr. Hostaše v Klatovech, Klatovy 09585
Vlastivědné Muzeum Jesenicka, Jeseník 09557
Vlastivědné Muzeum pro Vysoké nad Jizerou a Okolí, Vysoké nad Jizerou 09921
Vlastivedné Múzeum v Galante, Galanta 38255
Vlastivedné Múzeum v Hlohovci, Hlohovec 38258
Vlastivědné Muzeum v Olomouci, Olomouc 09690
De Vleeshal, Middelburg 32553
Vleeshuis, Antwerpen 03318
Vleeshuismuseum, Dendermonde 03542
Vliegend Museum Lelystad, Lelystad 32496
Vliegend Museum Seppe, Bosschenhoofd 31991
Voas Museum, Minburn 50722
The Vocational Museum, Seika 29239
Vodné Kasárne, Bratislava 38237
Völgységi Múzeum, Bonyhád 23376
Völkerkundemuseum der Archiv- und Museumsstiftung der VEM, Wuppertal 22747
Völkerkundemuseum der Universität Zürich, Zürich 42116
Völkerkundemuseum der von Portheim-Stiftung, Heidelberg 19087
Völkerkundemuseum Herrnhut, Herrnhut 19163
Völkerkundesammlung, Lübeck 20083
Völkerkundliche Sammlung, Winterthur 42042
Völkerkundliche Sammlung der Philipps-Universität, Marburg 20195
Völkerkundliche Sammlung der Steyler Missionare, Aulendorf 16644
Völkerkundliches Museum, Witzenhausen 22634
Völkerschlachtdenkmal / Forum 1813, Leipzig 19941
Voenno-istoričeski Muzej, Gorna Studena 05037
Voenno-istoričeski Muzej, Pleven 05122
Voenno-istoričeski Muzej Osvoboditelna Vojna 1877-1878, Bjala 04988
Voenno-istoričeskij Muzej Artillerii, Inženernych Vojsk i Vojsk Svjazi, Sankt-Peterburg 37730
Voenno-istoričeskij Muzej Tichookeanskogo Flota, Vladivostok 37952
Voenno-medicinskij Muzej, Sankt-Peterburg 37731
Voenno-morskoj Muzej Severnogo Flota, Murmansk 37436
Voergård Slot, Dronninglund 09992
Voerman Museum Hattem, Hattem 32324
Vörösmarty Mihály Emlékmúzeum, Kápolnásnyék 23629
Vörsi Talpasház, Vörs 23968
Vötters Oldtimermuseum, Kaprun 02154
Vogelmuseum, Aigen im Mühlkreis 01701
Vogelmuseum, Waging am See 22272
Vogelsberger Heimatmuseum, Schotten 21628
Vogelschutzmuseum, Biberach an der Riß 17322
Vogteimuseum mit Blumenauer Heimatstube, Aurach 16646
Vogtländisches Dorfmuseum, Erlbach 18227
Vogtländisches Freilichtmuseum, Erlbach 18228
Vogtländisches Freilichtmuseum Landwüst, Markneukirchen 20209
Vogtlandmuseum Plauen, Plauen 21055
Voimalamuseo, Helsinki 10838
Voipaalas Taidekeskus, Sääksmäki 11288
Voithenberghammer, Furth im Wald 18517
Vojaški Muzej Tabor, Lokev 38473
Vojenské Múzeum, Piešt'any 38328
Vojenské Múzeum Trenčín, Trenčín 38377
Vojenské technické museum Lešany, Krhanice 09608
Vojni Muzej, Logatec 38472
Vojni Muzej Beograd, Beograd 38091
Vojtech Löffler Museum, Košice 38272
Vojvodina Museum, Novi Sad 38124
Volcan de Lemptégy, Saint-Ours-les-Roches 15437
Volcano Art Center, Volcano 53766
Volda Bygdetun og Garverimuseet, Volda 33984
Volendam Windmill Museum, Milford 50687
Volendams Museum, Volendam 32902
Volgograd Art Museum, Volgograd 37958
Volgograd History Museum of the State Museum-Panorama Battle of Stalingrad, Volgograd 37957
Volgogradskij Memorialno-istoričeskij Muzej, Volgograd 37957
Volgogradskij Muzej Izobrazitelnych Iskusstv, Volgograd 37958
Volgogradskij Oblastnoj Kraevedčeskij Muzej, Volgograd 37959
Volkening Heritage Farm at Spring Valley, Schaumburg 52959
Volksbuurtmuseum, Den Haag 32104
Volksbuurtmuseum Wijk, Utrecht 32858
Volkskultur Niederösterreich, Atzenbrugg 54701
Volkskulturmuseum Raußmühle, Eppingen 18185
Volkskunde-Museum, Rudolstadt 21432
Volkskunde Museum Schleswig, Schleswig 21570
Volkskunde- und Freilichtmuseum, Konz 19684
Volkskunde- und Mühlenmuseum, Großschönau 18807
Volkskundemuseum, Antwerpen 03319
Volkskundemuseum, Drognitz 17980
Volkskundemuseum, Mödling 02398
Volkskundemuseum, Ostrach 20944
Volkskundemuseum, Schönberg, Mecklenburg 21601
Volkskundemuseum am Landesmuseum Joanneum, Graz 01986
Volkskundemuseum des Bezirks Unterfranken, Bad Bocklet 16701
Volkskundemuseum Deurne, Deurne 03550
Volkskundemuseum Salzburg, Salzburg 02629
Volkskundemuseum Spiralschmiede, Lasberg 02266
Volkskundemuseum Treuchtlingen, Treuchtlingen 22091
Volkskundemuseum Wyhra, Borna 17486
Volkskundliche Sammlung, Mömbris 20354
Volkskundliche Sammlung alter bäuerlicher Geräte, Ludesch 02339
Volkskundliche Sammlung des Fichtelgebirgsvereins, Weidenberg 22369
Volkskundliche Sammlungen, Straßburg 02787
Volkskundliche Sammlungen, Witzenhausen 22635
Volkskundliches Freilichtmuseum im Stadtpark Speckenbüttel, Bremerhaven 17575
Volkskundliches Gerätemuseum, Arzberg 16578
Volkskundliches Museum, Deutschlandsberg 01817
Volkskundliches Museum Alois Alphons, Hirschwang 02082
Volksmuseum Deurne, Antwerpen 03320
Volo Antique Auto Museum, Volo 53768
Volodymyr-Volyns'kij Istoryčnyj Muzej, Volodymyr-Volyns'kij 43207
Vologda Historical, Architectural and Artistic Museum Reserve, Vologda 37964
Vologda Regional Art Gallery, Vologda 37963
Vologodskaja Oblastnaja Kartinnaja Galereja, Vologda 37963
Vologodskij Gosudarstvennyj Istoriko-Architekturnyj i Chudožestvennyj Muzej-Zapovednik, Vologda 37964
Volta Regional Museum, Ho 22853
Volunteer Committees of Art Museum, New Orleans 55191
Volunteer Firemen's Mall and Museum of Kingston, Kingston 49793
Volynian Regional Museum, Luc'k 43113
Volyns'kyj Krajeznavčyj Muzej, Luc'k 43113
Vom Kloster zum Dorf, Creglingen 17744
Von der Heydt-Museum, Wuppertal 22748
The von Liebig Art Center, Naples 50954
Vonderau Museum, Fulda 18512
Vonderau Museum Fulda, Fulda 18513
Voorhistorisch Museum, Zonhoven 04165
Voormalig Stoombierbrouwerij De Keijzer N.A. Bosch, Maastricht 32535
Voortrekker Monument Heritage Site, Tshwane 38764
Voortrekker Museum, Pietermaritzburg 38708
Vor- und Frühgeschichtliche Sammlung, Kassel 19479
Vor- und Frühgeschichtliches Museum, Thalmässing 22047
Vorarlberger Landesmuseum, Bregenz 01796
Vorarlberger Militärmuseum, Bregenz 01797
Vorarlberger Museumswelt, Frastanz 01900
Vorbasse Museum, Vorbasse 10319
Vorderasiatisches Museum im Pergamonmuseum, Berlin 17279
Vorderösterreichisches Museum, Endingen am Kaiserstuhl 18168
Vorgeschichtsmuseum der Oberpfalz Amberg, Amberg 16534
Vorgeschichtsmuseum im Grabfeldgau, Bad Königshofen 16762
Vorkuta Regional Museum, Vorkuta 37968
Vorkutinskij Mežrajonnyj Kraevedčeskij Muzej, Vorkuta 37968
Voronezh Art Museum I.N Kramskoy, Voronež 37975
Voronežskij Chudožestvennyj Muzej im. I.N. Kramskogo, Voronež 37975
Voronežskij Oblastnoj Kraevedčeskij Muzej, Voronež 37976
Vorres Museum, Paiania 23129
Võrumaa Muuseum, Võru 10677
Voss Folkemuseum, Skulestadmo 33842
Vostočno-Kazachstanskij Etnografičeskij Muzej, Özkemen 29711
Vostočno-Kazachstanskij Istoriko-Kraevedčeskij Muzej, Özkemen 29712
Vostočno-Kazachstanskij Muzej Iskusstv, Özkemen 29713
Votivsammlung der Wallfahrtskirche Maria Kirchental, Sankt Martin bei Lofer 02669
Votivschatz der Wallfahrtskirche St. Erhard, Sankt Jakob-Breitenau 02652
Vouros Eftaxias Museum, Athinai 22954
Vox Populi Gallery, Philadelphia 51887
Voyageur Heritage Centre, Mattawa 06251
Vredesmuseum IJzertoren, Diksmuide 03555
Vredespaleis, Den Haag 32105
Vrindavan Research Institute Museum, Vrindavan 24430
Het Vrouwenhuis, Zwolle 33030
VSE Elektro-Museum, Illingen, Kreis Neunkirchen 19319
Vserossijskij Istoriko-ėtnografičeskij Muzej, Toržok 37862
Vserossijskij Memorialnyj Muzej-zapovednik V.M. Šukšina, Srostki 37793
Vserossijskij Muzej Dekorativno-Prikladnogo i Narodnogo Iskusstva, Moskva 37412
Vserossijskij Muzej Puškina, Sankt-Peterburg 37732
Vserossijskij Vystavočnyj Centr, Moskva 37413
Vsevoložskij Gosudarstvennyj Istoriko-kraevedčeskij Muzej, Vsevoložsk 37980
VU Centre de Diffusion et de Production de la Photographie, Québec 06699
Vuchetich Estate Museum, Moskva 37381
Vughts Historisch Museum, Vught 32916
Vuk and Dositej Museum of the National Museum, Beograd 38080
Vukani Zulu Cultural Museum, Eshowe 38617
Vuni Sarayi, Lefke 33366
Vuollerim 6000 år - Arkeologiskt Museum, Vuollerim 41074
Vuorela's Author Home, Keuruu 10950
VW-Käfer Museum, Gaal 01916
Východočeská Galerie Pardubicích, Pardubice 09705
Východočeská Galerie v Pardubicích, Pardubice 09706
Východočeské Muzeum v Pardubicích, Pardubice 09707
Východoslovenská Galéria, Košice 38274
Východoslovenské Múzeum, Košice 38275
Vylym-Hütte, Userin 22191
The Vyne, Basingstoke 43396
Vyskupo Motiejaus Valančiaus Gimtinės Muziejus, Nasrėnų 30317
Vysočna County Gallery, Jihlava 09562
Výstavní Síň Musaion, Havířov 09510
Výstavní Síň Muzea, Ústí nad Orlicí 09894
Výstavní Síň Muzea Těšínska, Český Těšín 09471
Výstavní Síň Muzea Těšínska, Orlová 09695
Výstavní síň Muzeum Těšínska, Jablunkov 09548
Výstavní Síň - Synagoga, Hranice 09537
Vystavočnye Zaly na Starom Arbate, Moskva 37414
Vystavočnye Zaly Rossijskoj Akademii Chudožestv, Moskva 37415
Vystavočnye Zaly v Dome Aksakovych, Moskva 37416
Vystavočnyj Centr -Galereja, Iževsk 36957
Vystavočnyj Centr Rosstrojekspo, Moskva 37417
Vystavočnyj Centr Sankt-Peterburgskogo Otdelenija Sojuza Chudožnikov Rossii, Sankt-Peterburg 37733
Vystavočnyj Zal, Čerepovec 36852
Vystavočnyj Zal, Ekaterinburg 36902
Vystavočnyj Zal, Kirov 37044
Vystavočnyj Zal, Kirov 37045
Vystavočnyj Zal, Krasnoturinsk 37082
Vystavočnyj Zal, Tula 37876
Vystavočnyj Zal, Volgograd 37960
Vystavočnyj Zal, Vologda 37965
Vystavočnyj Zal Belye Palaty na Prečistenke, Moskva 37418
Vystavočnyj Zal Biblioteki im. A.A. Bloka, Sankt-Peterburg 37734
Vystavočnyj Zal Biblioteki Iskusstv im. A.P. Bogoljubova, Moskva 37419
Vystavočnyj Zal Bogorodskoe, Moskva 37420
Vystavočnyj Zal Doma Družby s Narodami Zurebežnych Stran, Moskva 37421
Vystavočnyj Zal Doma Učenych, Novosibirsk 37489
Vystavočnyj Zal g. Smolenska, Smolensk 37780
Vystavočnyj Zal im. L.I. Kuindži, Mariupol' 43147
Vystavočnyj Zal Izmajlovo, Moskva 37422
Vystavočnyj Zal Karelskoj Gosudarstvennoj Filarmonii, Petrozavodsk 37563
Vystavočnyj Zal-masterskaja D.I. Nalbandjana, Moskva 37423
Vystavočnyj Zal Moskovskogo Fonda Kultury, Moskva 37424
Vystavočnyj Zal Moskovskogo Sojuza Chudožnikov, Moskva 37425
Vystavočnyj Zal Narodnaja Galereja, Moskva 37426
Vystavočnyj Zal Permskogo Oblastnogo Kraevedčeskogo Muzeja, Perm 37552
Vystavočnyj Zal Raduga, Čeboksary 36841
Vystavočnyj Zal Rossijskoj Gosudarstvennoj Biblioteki po Iskusstvu, Moskva 37427
Vystavočnyj Zal - Russkie Chudožniki v Samarskoj Gubernii, Žigulevsk 37989
Vystavočnyj Zal Sojuza Chudožnikov Rossii, Sankt-Peterburg 37735
Vystavočnyj Zal Spasskaja Cerkov, Irkutsk 36940
Vystavočnyj Zal Tverskoj Oblastnoj Kartinnoj Galerei, Tver 37882
Vystavočnyj Zal Tvorčestvo, Moskva 37428
Vystavočnyje Zaly Vserossijskogo Fonda Kultury, Moskva 37429
Vytautas The Great War Museum, Kaunas 30269
Vytautas Valiusis Ceramics Museum, Leliūnai 30302
Vytauto Didžiojo Karo Muziejus, Kaunas 30269
Vytauto Kazimiero Jonyno Galerija, Druskininkai 30226
Vytauto Valiušio Keramikos Muziejus, Leliūnai 30302
Výtvarné Centrum Chagall, Ostrava 09700

W

W. and J. Kulczyckich Art Gallery, Zakopane 35354
W.C. Handy Home Museum and Library, Florence 48566
W.D. Trotter Anatomy Museum, Dunedin 33100
W.H. Over Museum, Vermillion 53721
W.H. Smith Museum, Newtown, Powys 45259
The W.H. Stark House, Orange 51579
W.K.P. Kennedy Gallery, North Bay 06471
W.M. Keck Museum, Reno 52292
Het Waaierkabinet, Amsterdam 31900
Het Waalres Museum, Waalre 32917
Waba Cottage Museum, White Lake 07322
Wabash Frisco and Pacific Association, Glencoe 48925
Wabasha County Museum, Reads Landing 52268
Wabasso Historical Society Museum, Wabasso 53770
Wabaunsee County Historical Museum, Alma 46235
Wabowden Historical Museum, Wabowden 07270
Wachaumuseum, Weißenkirchen in der Wachau 02905
Wachiraprasat Museum, Phetchaburi 42593
Wachsstöcklkabinett - Wachszieher- und Lebzelter-Museum, Viechtach 22227
Waddesdon Manor, Waddesdon 45905
Wade House Stagecoach Inn and Wesley Jung Carriage Museum, Greenbush 49045
Wadena and District Museum and Nature Centre, Wadena 07271
Wadowice Cita Museum, Wadowice 35209
Wadsworth Atheneum, Hartford 49199
Wadsworth-Longfellow House, Portland 52070
Waechtersbach Ceramic Museum, Brachttal 17492
Wäinö Aaltonen Museo, Turku 11408
Wäinö Aaltonen Museum of Art, Turku 11408
Wäschepflegemuseum, Rainbach im Mühlkreis 02555
Wäscherei-Museum Omas Waschküche, Eberswalde 18048
Wafangdian Museum, Wafangdian 08310
Waffenkammer im Munot-Turm, Schaffhausen 41832
Waffenmuseum, Oberndorf am Neckar 20820
Waffenmuseum Suhl, Suhl 21977
Waffensammlung Schloss Ebelsberg, Linz 02325
Waffensammlung Willibald Folger, Bruck an der Mur 01806

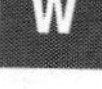

Wagenburg im Schloss Schönbrunn, Wien 03085
Wagenmakersmuseum, Leiden 32490
Wagga Wagga Art Gallery, Wagga Wagga 01620
Wagin Historical Village Museum, Wagin 01621
Wagnalls Memorial, Lithopolis 50131
Wagner Free Institute of Science, Philadelphia 51888
Wahkiakum County Historical Museum, Cathlamet 47282
Waiheke Community Art Gallery, Auckland 33061
Waiheke Island Historical Society Museum, Waiheke Island 33217
Waihi Arts Centre and Museum Association, Waihi 33218
Waikato Museum of Art and History, Hamilton 33117
Waikawa District Museum, Tokanui 33214
Waikouaiti Museum, Waikouaiti 33220
Waimate Historical Museum, Waimate 33221
Wainwright Museum, Wainwright 07272
Waipara County Historical Society Museum, Hawarden 33120
Waipu Heritage Centre, Waipu 33224
Waipukurau Museum, Waipukurau 33225
Wairoa Museum Kopututanga, Wairoa 33226
Waitomo Museum of Caves, Waitomo Caves 33230
Wajawithayawat Museum, Bangkok 42565
Wajima Urushi Bijutsukan, Wajima 29570
Wakasa History Museum, Obama 29083
Wakasa Rekishi Minzoku Shiryokan, Obama 29083
Wakaw Heritage Museum, Wakaw 07274
Wakayama-kenritsu Hakubutsukan, Wakayama 29574
Wakayama-kenritsu Kindai Bijutsukan, Wakayama 29575
Wakayama Prefectural Museum, Wakayama 29574
Wake Forest College Bithplace Society Museum, Wake Forest 53789
Wake Forest University Fine Arts Gallery, Winston-Salem 54252
Wakefield Art Gallery, Wakefield 45913
Wakefield Historical Museum, Wakefield 53791
Wakefield House Museum, Blue Earth 46873
Wakefield Museum, Wakefield 45914
Wakefield Museum Association, Wakefield 53790
Wakoh Hakubutsukan, Yasugi 29589
Wald- und Forstmuseum, Herschberg 19165
Wald- und Lönsmuseum, Heinade 19110
Wald- und Moormuseum Berumerfehn, Großheide 18793
Waldarbeits-Museum und Historische Wappenschmiede, Elmstein 18127
Waldbauernmuseum, Gutenstein 02027
Waldgeschichtliches Museum Sankt Oswald, Sankt Oswald-Riedlhütte 21507
Waldhufen-Heimatmuseum Salmbach, Engelsbrand 18170
Waldmuseum, Burg, Dithmarschen 17644
Waldmuseum, Furth im Wald 18518
Waldmuseum, Mehlmeisel 20255
Waldmuseum, Münstertal 20524
Waldmuseum, Surwold 21990
Waldmuseum, Zwiesel 22839
Waldmuseum Göhrde, Göhrde 18669
Waldmuseum im Wildpark Neuhaus, Holzminden 19262
Waldmuseum Reichental, Gernsbach 18608
Waldmuseum Stendenitz, Neuruppin 20652
Waldmuseum Wassermühle, Wingst 22598
Waldmuseum Watterbacher Haus, Kirchzell 19556
Waldnaabtal-Museum in der Burg Neuhaus, Windischeschenbach 22597
Waldschmidt-Ausstellung, Eschlkam 18236
Waldviertler Bauernhaus-Museum, Weitra 02909
Waldviertler Bauernhof-Museum, Gföhl 01933
Waldviertler Eisenbahnmuseum, Sigmundsherberg 02742
Walhalla, Donaustauf 17893
Walkaway Station Museum, Walkaway 01623
The Walker African American Museum, Las Vegas 49980
Walker Art Center, Minneapolis 50743
Walker Art Collection, Garnett 48866
The Walker Art Gallery, Liverpool 44748
The Walker Hill Art Center, Seoul 29932
Walker Street Gallery, Dandenong 01024
Walker's Point Center for the Arts, Milwaukee 50721
Wall Roman Site and Museum - Letocetum, Lichfield 44708
Wallace and Area Museum, Wallace 07275
The Wallace Collection, London 45006
Wallace District Mining Museum, Wallace 53799
Wallace Early Settlers Museum, Riverton 33195
Wallace House, Somerville 53152
Wallaceburg and District Museum, Wallaceburg 07276
Wallaroo Heritage and Nautical Museum, Wallaroo 01624
Wallenfels'sches Haus, Abteilung Vor- und Frühgeschichte, Archäologie und Völkerkunde, Gießen 18641
Wallfahrts- und Heimatmuseum, Altötting 16521
Wallfahrtsbasilika Mariazell, Mariazell 02359
Wallfahrtsmuseum, Bruckmühl 17601
Wallfahrtsmuseum, Maria Taferl 02355
Wallfahrtsmuseum, Neukirchen beim Heiligen Blut 20624
Wallfahrtsmuseum, Steingaden 21888
Wallfahrtsmuseum Inchenhofen, Inchenhofen 19332
Wallfahrtsmuseum Maria Steinbach, Legau 19872
Wallingford Historical Society Museum, Wallingford 53800
Wallingford Museum, Wallingford 45915
Wallington, Cambo 43721
Wallis Museum at Connors State College, Warner 53818
Walliser Reb- und Weinmuseum, Salgesch 41799
Wallmuseum, Oldenburg in Holstein 20902
Wallqu Museum, Yuanqu 08466
Wallraf-Richartz-Museum & Fondation Corboud, Köln 19635
Wally Findlay Galleries, New York 51278
Walmer Castle, Walmer 45918
Walmstedtska Gårdens Museivåning, Uppsala 41017
Walnut Grove Plantation, Roebuck 52438
Walpurgishalle Thale, Thale (Harz) 22043
Walsall Leather Museum, Walsall 45921
Walsall Local History Centre, Walsall 45922
Walsall Museum, Walsall 45923
The Walsenburg Mining Museum and Fort Francisco Museum of La Veta, Walsenburg 53805
Walser Heimatmuseum, Triesenberg 30196
Walserhaus Gurin, Bosco 41246
Walsermuseum, Alagna Valsesia 25170
Walsermuseum, Riezlern 02590
Walsermuseum Lech-Tannberg, Lech 02278
The Walt Disney Family Museum, San Francisco 52798
Walt Whitman Birthplace, Huntington Station 49440
Walt Whitman House, Camden 47197
Walter and McBean Galleries, San Francisco 52799
Walter Anderson Museum of Art, Ocean Springs 51489
Walter Elwood Museum, Amsterdam 46276
Walter F. George Law Office Museum, Vienna 53744
Walter P. Chrysler Boyhood Home and Museum, Ellis 48368
Walter P. Chrysler Museum, Auburn Hills 46474
Walter Phillips Gallery, Banff 05382
Walter Rothschild Zoological Museum, Tring 45877
Walter Runeberg Sculpture Collection, Porvoo 11231
Walter Runebergin Veistoskokoelma, Porvoo 11231
Walters Art Museum, Baltimore 46591
Waltham Historical Society Museum, Waltham 53811
The Waltham Museum, Waltham 53812
Walton Hall Heritage Centre, Warrington 45935
Walton Hall Museum, Linford 44718
Walton House Museum, Centerville 47317
Walton Maritime Museum, Walton-on-the-Naze 45926
Waltscher Heimatstube, Neckargemünd 20563
Waltzing Matilda Centre and Qantilda Museum, Winton 01658
Walvis Bay Museum, Walvis Bay 31761
Wan Fang Arts Museum, Taipei 42437
Wan Li Brick Museum, Yuanli 42480
Wan Zhou District Museum, Chongqing 07632
Wanapum Dam Heritage Center, Beverly 46788
Wandle Industrial Museum, Mitcham 45160
Wandsworth Museum, London 45007
Wang Hsiu Chi Sculpture Park, Taipei 42438
Wangaratta Exhibitions Gallery, Wangaratta 01627
Wangaratta Museum, Wangaratta 01628
Wangfu Museum, Naiman Qi 08026
Wangjian Tomb Museum, Chengdu 07622
Wangjiang Museum, Wangjiang 08311
Wangmiao Display Center, Dujiangyan 07675
Wang's Fine Art Museum, Kaohsiung 42245
Wanho Shrine Museum, Taichung 42358
Wannenmacher-Museum, Emsdetten 18161
Wansanxia Museum, Wanxian 08313
Wantagh Historical & Preservation Society, Wantagh 53814
Wanuskewin Heritage Park, Saskatoon 06926
Wanxi Museum, Lu'an 07976
Wapello County Historical Museum, Ottumwa 51631
War and Hospital Museum, Ridderkerk 32685
War in the Pacific National Historical Park, Asan 23286
War in the Pacific National Historical Park, Piti 23288
War Memorial, Dublin 24820
War Memorial Carillon Tower and Military Museum, Loughborough 45035
War Memorial Gallery of Fine Arts, Sydney 01565
War Museum, Addis Ababa 10681
War Museum, Athinai 22951
War Museum, Siem Reap 05292
War Museum, Valletta 30649
War Museum, Yangon 31738
War Museum of the Boer Republics, Bloemfontein 38570
War Remnants Museum, Ho Chi Minh City 54631
War Room and Motor House Collection, Harrogate 44403
Warabekan, Tottori 29517
Ward County Historical Society Museum, Minot 50747
Ward Hall, Georgetown 48892
The Ward Museum of Wildfowl Art, Salisbury 52679
Ward O'Hara Agricultural Museum of Cayuga County, Auburn 46472
Warden's House Museum, Stillwater 53294
Wardown Park Museum, Luton 45048
Ware Museum, Ware 45930
Wareham Town Museum, Wareham 45931
Warehouse Gallery, Lee 50016
Warkums Erfskip, Workum 32967
Warkworth and District Museum, Warkworth 33239
Warmia and Mazuries Museum, Olsztyn 34978
Warmuseum The Consciousness, Aalten 31784
Warner Archive, Milton Keynes 45155
Warner Museum, Springville 53242
Warooka and District Museum, Warooka 01629
Warracknabeal Historical Centre, Warracknabeal 01631
Warragul and District Historical Museum, Warragul 01632
Warrandyte Old Post Office Museum, Warrandyte 01633
Warren County Historical Society Museum, Lebanon 50012
Warren County Museum, Warren 53827
Warren Historical Museum, Warren 53820
Warren ICBM and Heritage Museum, Cheyenne 47446
Warren Rifles Confederate Museum, Front Royal 48801
Warrick County Museum, Boonville 46909
Warrington Museum and Art Gallery, Warrington 45936
Warrnambool Art Gallery, Warrnambool 01634
Warsaw Agriculture University Experimental Forest Station - Museum of Forestry and Wood, Rogów 35083
Warsaw Historical Museum, Warsaw 53832
Warsaw Uprising Museum, Warszawa 35266
Warta and River Warta Museum, Warta 35292
Wartburg, Eisenach 18099
Warther Museum, Dover 48168
Warwick Castle, Warwick 45940
Warwick Doll Museum, Warwick 45941
Warwick Museum of Art, Warwick 53837
Warwick Regional Art Gallery, Warwick 01636
Warwickshire Museum, Warwick 45942
Warwickshire Museum of Rural Life, Moreton Morrell 45176
Warwickshire Yeomanry Museum, Warwick 45943
Wasch-en Strijkmuseum, Boxtel 31999
Waschhaus Potsdam, Potsdam 21112
De Wascht en Strekt, Gilze 32256
Wasco County Historical Museum, The Dalles 47946
Waseca County Museum, Waseca 53839
Waseda Daigaku Tsubouchi Hakushi Kinen Engeki Hakubutsukan, Tokyo 29503
Washakie Museum, Worland 54305
Washburn County Historical Society Museum, Shell Lake 53070
Washburn Historical Museum, Washburn 53841
Washington Center for Photography, Washington 53930
Washington County Historical Association Museum, Fort Calhoun 48602
Washington County Historical Society, Washington 53938
Washington County Historical Society Museum, Portland 52089
Washington County Historical Society Museum, Washington 53944
Washington County Historical Society Museum, West Bend 54044
Washington County Museum, Akron 46145
Washington County Museum, Sandersville 52833
Washington County Museum of Fine Arts, Hagerstown 49122
Washington Crossing Historic Park, Washington Crossing 53949
Washington Dolls House and Toy Museum, Washington 53931
Washington Historical Museum, Washington 53937
Washington Museum, Washington 53939
Washington Museum Association, Bellevue 55192
Washington National Cathedral, Washington 53932
Washington Old Hall, Washington 45947
Washington Pavillion of Arts and Science, Sioux Falls 53123
Washington State Capital Museum, Olympia 51545
Washington State History Museum, Tacoma 53416
Washington's Headquarters, Newburgh 51301
Washington's Lands Museum and Sayre Log House, Ravenswood 52257
Waskada Museum, Waskada 07282
Wassenberg Art Center, Van Wert 53703
Wasser- und Fischkundemuseum, Kuhlen-Wendorf 19743
Wasserburg Haus Kemnade, Hattingen 19042
Wasserkraftmuseum, Ziegenrück 22799
Wasserkunst von 1535, Bad Arolsen 16674
Wasserleitungsmuseum, Reichenau an der Rax 02570
Wasserleitungsmuseum, Wildalpen 03103
Wassermühle Höfgen, Grimma 18765
Wassermühle Kuchelmiß, Kuchelmiß 19733
Wasserschloß Groß Leuthen, Märkische Heide 20121
Wasserschloß Mitwitz, Mitwitz 20347
Waste Products Museum, Zwolle 33025
Wat Ko Museum, Phetchaburi 42594
Wat Phra Thart Lampang Luang Museum, Lampang 42577
Wat Si Saket, Vientiane 29979
The Watari Museum of Contemporary Art, Tokyo 29504
Watari-Um, Tokyo 29504
Watashi no Shigotokan, Seika 29239
Watch and Clock Museum, Mechelen 03857
Watches and Pocket Watches Museum, Berlin 17275
Watchet Market House Museum, Watchet 45948
Water Barracks of the Slovak National Gallery, Bratislava 38237
Water Distribution and Plumbing History Museum, Šiauliai 30363
Water Mill, Mikkeli 11112
Water Mill Exposition, Ginučiai 30228
Water Mill Museum, Water Mill 53955
Water Museum, Moskva 37383
Water Ski Hall of Fame, Polk City 52014
Water Supply Museum, Weymouth 45990
Water Supply Museum of Riga, Ādaži 29980
Water Supply Museum Piece Exhibition Ward, Orosháza 23742
Water Tower Museum, Gunnedah 01136
Waterford Historical Museum, Waterford 53959
Waterford Museum of Treasures, Waterford 24909
Waterfront Gallery, Westport 24910
Waterfront Museum, Poole 45410
Waterfront Museum and Showboat Barge, Brooklyn 47084
Waterland Neeltje Jans, Vrouwenpolder 32913
Waterloo Area Farm Museum, Grass Lake 49024
Waterloo Center for the Arts, Waterloo 53965
Watermen's Museum, Yorktown 54346
Watermill Museum, Vääksy 11431
Watersnoodmuseum, Ouwerkerk 32661
Watertown Historical Society Museum, Watertown 53969
Watervliet Arsenal's Museum of the Big Guns, Watervliet 53979
Waterways Visitor Centre, Dublin 24821
Waterworks Museum - Hereford, Hereford 44460
Waterworks Visual Arts Center, Salisbury 52682
Watford Museum, Watford 45950
Watkins Community Museum of History, Lawrence 49994
Watkins Gallery, Washington 53933
Watkins Institute Art Gallery, Nashville 50982
Watkins Museum, Taylorsville 53456
Watkins Woolen Mill, Lawson 49998
Watson Crossley Community Museum, Grandview 05902
Watson-Curtze Mansion, Erie 48413
Watson Farm, Jamestown 49600
Watson Museum, Rajkot 24370
Watson's Mill, Manotick 06230
Das Wattenmeerhaus, Wilhelmshaven 22581
Watters Smith Memorial, Lost Creek 50275
Watts Gallery, Compton, Surrey 43882
Watts Towers Arts Center, Los Angeles 50267
Wauchope District Museum, Wauchope 01638
Wauconda Township Museum, Wauconda 53988
Waukesha County Museum, Waukesha 53990
Waveland Museum, Lexington 50058
Wawel Royal Castle - State Art Collections, Kraków 34821
Wawota and District Museum, Wawota 07292
Wax-Honey-Museum, Viechtach 22227
Wax Museum, Byblos 30168
Wax Museum, Dublin 24822
Wax Museum, Fátima 35601
Wax Museum, Santa Claus 52865
Wax Museum Český Krumlov, Český Krumlov 09469
Wax Museum Karlštejn, Karlštejn 09579
Wax Museum Prague, Praha 09774
Waxworks Museum, Moskva 37393
Wayland Historical Museum, Wayland 54004
Wayne Art Center, Wayne 54012
Wayne Center for the Arts, Wooster 54295
Wayne County Historical Museum, Richmond 52315
Wayne County Historical Society Museum, Lyons 50336
Wayne County Historical Society Museum, Wooster 54296
Wayne County Museum, Honesdale 49334
Waysi Museum, Fufeng 07696

W

Wayside Folk Museum, Zennor 46114
Wayville Latvian Museum, Brooklyn Park 00906
Weald and Downland Open Air Museum, Chichester 43832
Weapons collection, Eskilstuna 40505
Weardale Museum, Weardale 45952
Weather Discovery Center, North Conway 51390
Weatherspoon Art Museum, Greensboro 49068
Weaver's Cottage, Kilbarchan 44575
Weavers' Triangle Visitor Centre, Burnley 43680
Weaving and Silk Museum, Bussières 12210
Weaving Craft Museum, Fengyuan 42190
Weaving Museum, Heidenreichstein 02064
The Webb-Deane-Stevens Museum, Wethersfield 54112
Webb House Museum, Newark 51294
Weber State University Art Gallery, Ogden 51504
Weberei- und Heimatmuseum Laichingen, Laichingen 19780
Weberei- und Heimatmuseum Ruedertal, Schmiedrued 41839
Webereimuseum, Breitenberg 17539
Webereimuseum, Haslach 02058
Webereimuseum, Sindelfingen 21779
Weberhaus Marlesreuth, Naila 20547
Weberhausmuseum, Schauenstein 21535
Webermuseum, Biberach an der Riß 17323
Webermuseum, Heidenreichstein 02064
Weberstube Jonsdorf, Jonsdorf, Kurort 19401
Webster County Historical Museum, Red Cloud 52271
Webster House Museum, Elkhorn 48351
Wedding and Custom Museum, Kuah 30569
Wedgwood Museum, Stoke-on-Trent 45764
Wednesbury Museum and Art Gallery, Wednesbury 45953
Weegmuseum de Oude Waag, Someren 32795
Weeks Air Museum, Miami 50636
Weems-Botts Museum, Dumfries 48203
Weeping Water Valley Historical Society Museum, Weeping Water 54021
Weg des Friedens, Purgstall an der Erlauf 02541
Wegmachermuseum, Wasserburg am Inn 22346
Wehrgeschichtliches Museum, Salzburg 02630
Wehrgeschichtliches Museum Rastatt, Rastatt 21189
Wehrtechnische Studiensammlung des Bundesamtes für Wehrtechnik und undBeschaffung (BWB), Koblenz 19591
Wehrturm, Niederstetten 20714
Wei Chin-Yuan Musical Instruments Museum, Taipei 42439
Wei County Museum, Weixian 08321
Weibertreu-Museum, Weinsberg 22434
Weicheltmühle, Hartmannsdorf-Reichenau 19021
Weidmann Cottage Heritage Centre, Muswellbrook 01342
Weifang Kite Museum, Weifang 08314
Weifang Museum, Weifang 08315
Weihai Museum, Weihai 08317
Weihui Museum, Weihui 08318
Weil Art Gallery, Corpus Christi 47823
Weimar-Haus, Weimar 22422
Wein- und Heimatmuseum, Durbach 18030
Wein-und Heimatmuseum, Leiwen 19945
Wein- und Heimatmuseum, Mehring 20256
Wein- und Heimatmuseum, Winningen 22602
Weinan Center Museum, Weinan 08320
Weinbau in Mundelsheim, Mundelsheim 20528
Weinbau-Museum, Chur 41305
Weinbau Museum Peter, Wachenheim an der Weinstraße 22259
Weinbau- und Fahnenschwingermuseum, Neckenmarkt 02424
Weinbau- und Heimatmuseum Klingenberg, Klingenberg am Main 19570
Weinbau- und Landwirtschaftsmuseum, Hadersdorf 02032
Weinbau- und Landwirtschaftsmuseum, Oetwil am See 41703
Weinbaumuseum, Asperg 16601
Weinbaumuseum, Flörsheim-Dalsheim 18331
Weinbaumuseum, Ortenberg, Baden 20913
Weinbaumuseum, Sitzenberg-Reidling 02746
Weinbaumuseum Alte Kelter, Erlenbach, Kreis Heilbronn 18230
Weinbaumuseum im Herrenhof, Neustadt an der Weinstraße 20674
Weinbaumuseum Meersburg, Meersburg 20252
Weinbaumuseum Stuttgart-Uhlbach, Stuttgart 21967
Weinberg Nature Center, Scarsdale 52955
Weindorfmuseum, Horrweiler 19278
Weingart Gallery, Los Angeles 50268
Weingreen Museum of Biblical Antiquities, Dublin 24823
Weingutmuseum Hoflößnitz, Radebeul 21161
Weinlandmuseum, Asparn an der Zaya 01735
Weinman Mineral Museum, White 54127
Weinmuseum, Güssing 02016
Weinmuseum, Vaihingen an der Enz 22200
Weinmuseum Burgenland - Eisenstadt, Eisenstadt 01863
Weinmuseum im Kloster Machern, Bernkastel-Kues 17300
Weinmuseum Kitzeck, Kitzeck im Sausal 02179
Weinmuseum Prellenkirchen, Prellenkirchen 02531
Weinmuseum Schlagkamp-Desoye Senheim, Senheim 21745
Weinmuseum Schloß Gamlitz, Gamlitz 01922
Weinstadtmuseum Krems, Krems 02232
Weinstadtmuseum Poysdorf, Poysdorf 02519
Weinviertler Museumsdorf, Niedersulz 02448
Weinviertler Oldtimermuseum Poysdorf, Poysdorf 02520
Weir Farm, Wilton 54213
Weissenhofmuseum im Haus Le Corbusier, Stuttgart 21968
Weißenhorner Heimatmuseum, Weißenhorn 22461
Weissenstein-Museum, Weissenstein bei Solothurn 42005
Weißgerbermuseum, Doberlug-Kirchhain 17880
Weißgerberwalk, Biberach an der Riß 17324
Weisův Dům, Veselí nad Lužnicí 09904
Weitwandermuseum, Eibiswald 01850
Weizmann Archives and House, Rehovot 25090
The Welch Regiment Museum (41st/ 69th Foot) of the Royal Regiment of Wales, Cardiff 43755
Welholme Galleries, Grimsby 44367
Welkom Museum, Welkom 38776
Wella Museum, Darmstadt 17796
Welland Historical Museum, Welland 07294
Wellbrook Beetling Mill, Cookstown 43890
Wellcome Library, London 45008
Wellcome Museum of Anatomy and Pathology, London 45009
Wellesbourne Wartime Museum, Wellesbourne 45955
Wellesley Historical Society, Wellesley Hills 54024
Wellfleet Historical Society Museum, Wellfleet 54026
Wellingborough Heritage Centre, Wellingborough 45957
Wellings Landsbymuseum, Lintrup 10178
Wellington Arch, London 45010
Wellington Arts Centre, Wellington 33261
Wellington Aviation Museum, Moreton-in-Marsh 45175
Wellington B. Gray Gallery, Greenville 49075
Wellington County Museum, Fergus 05794
Wellington Courthouse Museum, Wellington 01640
Wells Auto Museum, Wells 54030
Wells County Historical Museum, Bluffton 46878
Wells County Museum, Fessenden 48539
Wells Fargo History Exhibits, Portland 52090
Wells Fargo History Museum, Los Angeles 50269
Wells Fargo History Museum, Minneapolis 50744
Wells Fargo History Museum, Phoenix 51911
Wells Fargo History Museum, Sacramento 52510
Wells Fargo History Museum, San Diego 52761
Wells Fargo History Museum, San Francisco 52800
Wells Fargo History Museum - Old Sacramento, Sacramento 52511
Wells Museum, Wells 07296
Wells Museum, Wells 45958
Wells Walsingham Light Railway, Wells-next-the-Sea 45960
Welser Puppenweltmuseum, Wels 02918
Welsh Highland Railway Museum, Porthmadog 45423
Welshpool and Llanfair Light Railway, Llanfair Caereinion 44764
Welterbe Kloster Lorsch, Lorsch 20037
Welterbemuseum Hallstatt, Hallstatt 02049
Weltkulturerbe Völklinger Hütte, Völklingen 22249
Welwyn Roman Baths Museum, Welwyn 45964
Wen Tianxiang Temple, Beijing 07572
Wenatchee Valley Museum, Wenatchee 54037
Wenceslao Vinzons Historical Landmark Museum, Vinzons 34526
Wendell Gilley Museum, Southwest Harbor 53190
Wendeng Museum, Wendeng 08322
Wengyuan Museum, Wengyuan 08323
Wenham Museum, Wenham 54038
Wenmiao Museum, Tianjin 08290
Wenshanzhou Museum, Wenshan 08324
Wenshui Museum, Wenshui 08325
Wentworth-Coolidge Mansion, Portsmouth 52101
Wentworth Gardner and Tobias Lear Houses, Portsmouth 52102
Wentzingerhaus, Freiburg im Breisgau 18441
Wenxi Museum, Wenxi 08326
Wenzel-Hablik-Museum, Itzehoe 19370
Wenzhou Museum, Wenzhou 08327
Weoley Castle, Birmingham 43516
Werberger Stuben, Motten 20393
Werdenfelser Museum, Garmisch-Partenkirchen 18542
Wereld Bodem Museum, Wageningen 32921
Wereldmuseum Rotterdam, Rotterdam 32733
Werkbundarchiv - Museum der Dinge, Berlin 17280
Werksmuseum der Firma Linke-Hofmann-Busch, Salzgitter 21477
Werksmuseum der MTU, München 20495
Werksmuseum der Waechtersbacher Keramik, Brachttal 17492
Werksmuseum Koenig & Bauer AG, Würzburg 22727
Werkspoor Museum, Amsterdam 31901
Werner Berg Museum, Bleiburg 01788
Werner-Egk-Begegnungsstätte, Donauwörth 17900
Werner Weckwerth Museum, Friedrich-Wilhelm-Lübke-Koog 18473
Wernher Collection at Ranger's House, London 45011
Werra-Kalibergbau-Museum, Heringen (Werra) 19143
Werribee and District Historical Museum, Werribee 01642
Wertheim Glassmuseum, Wertheim 22502
Wertpapierwelt, Olten 41710
Weserburg Museum für Moderne Kunst, Bremen 17566
Weslaco Museum, Weslaco 54040
Wesselstuerne, København 10157
West Allis Historical Society Museum, West Allis 54042
West Baton Rouge Museum, Port Allen 52025
West Bend Art Museum, West Bend 54045
West Berkshire Museum, Newbury 45221
West Blatchington Windmill, Hove 44494
West Bohemian Gallery in Plzeň, Plzeň 09719
West Bohemian Museum, Plzeň 09720
West Chicago City Museum, West Chicago 54051
West Coast Historical Museum, Hokitika 33123
West Coast Museum of Flying, Sidney 06974
West Coast Pioneers Museum, Zeehan 01687
West Cork Model Railway Village, Clonakilty 24761
West Cork Regional Museum, Clonakilty 24762
West Gate Towers, Canterbury 43746
West Hants Historical Society Museum, Windsor, Nova Scotia 07343
West Highland Museum, Fort William 44249
West Java Museum, Bandung 24444
West Kalimantan Museum, Pontianak 24547
West Kazakh Regional and Historical Museum, Oral 29714
West Kilbride Museum, West Kilbride 45970
West Midlands Police Museum, Birmingham 43517
West-Moravian Museum in Třebíč - Museum Jemnice, Jemnice 09555
West Norway Museum of Decorative Art, Bergen 33433
West of the Pecos Museum, Pecos 51746
West Overton Museums, Scottdale 52975
West Park Museum, Macclesfield 45067
West Parry Sound District Museum, Parry Sound 06572
West Pasco Historical Society Museum, New Port Richey 51099
West Point Lighthouse Museum, West Point 07302
West Point Museum, West Point 54078
West River Museum, Philip 51892
West Somerset Railway, Minehead 45156
West Somerset Rural Life Museum, Allerford 43280
West Stow Anglo-Saxon Village, Bury Saint Edmunds 43697
West Sumatrian Museum, Padang 24533
West Torrens Railway, Signal Telegraph and Aviation Museum, Brooklyn Park 00907
West Valley Art Museum, Surprise 53381
West Vancouver Museum, West Vancouver 07304
West Virginia Independence Hall, Wheeling 54125
West Virginia Northern Community College Alumni Association Museum, Wheeling 54126
West Virginia State Farm Museum, Point Pleasant 52012
West Virginia State Museum, Charleston 47380
West Virginia University Mesaros Galleries, Morgantown 50846
West Wales Arts Centre, Fishguard 44231
West Wycombe Motor Museum, West Wycombe 45974
West Wycombe Park House, West Wycombe 45975
Westallgäuer Heimatmuseum, Weiler-Simmerberg 22389
Westbury Manor Museum, Fareham 44212
Westchester Art Workshop Gallery, White Plains 54128
Western Approaches, Liverpool 44749
Western Archaeological and Conservation Center, Tucson 53602
Western Art Gallery and Museum, Dillon 48143
Western Australian Maritime Museum, Fremantle 01095
Western Australian Medical Museum, Subiaco 01536
Western Australian Museum Geraldton, Geraldton 01107
Western Australian Museum Perth, Perth 01412
Western Canada Aviation Museum, Winnipeg 07391
Western Colorado Center for the Arts, Grand Junction 48995
Western Development Museum, Saskatoon 06927
Western Front, Vancouver 07231
Western Gallery, Bellingham 46713
Western Hennepin County Pioneers Museum, Long Lake 50190
Western Heritage Center, Billings 46807
Western Hotel Museum, Lancaster 49916
Western Illinois University Art Gallery, Macomb 50364
Western Kentucky University Gallery, Bowling Green 46976
Western Manitoulin Island Historical Society Museum, Gore Bay 05883
Western Museum, Abilene 46123
Western Museum of Mining and Industry, Colorado Springs 47707
Western Museums Association, Berkeley 55193
Western New Mexico University Museum, Silver City 53104
Western Pacific Railroad Museum, Portola 52092
Western Power World of Energy, Fremantle 01096
The Western Railway Museum, Suisun City 53358
Western Reserve Historical Society Museum, Cleveland 47644
Western Rhode Island Civic Museum, Coventry 47857
Western Ski Sport Museum, Donner Pass 48151
Western Springs Museum, Western Springs 54088
Western Wyoming College Art Gallery, Rock Springs 52412
Westerwald Museum Motorrad & Technik, Steinebach 21885
Westerwolds Crashmuseum, Oude Pekela 32651
Westfälisch-Niederländisches Imkereimuseum, Gescher 18624
Westfälischer Kunstverein, Münster 20520
Westfälisches Feuerwehrmuseum, Hattingen 19043
Westfälisches Glockenmuseum, Gescher 18625
Westfälisches Schieferbergbau- und Heimatmuseum Holthausen, Schmallenberg 21585
Westfälisches Schulmuseum, Dortmund 17929
Westfalia-Auto-Museum, Rheda-Wiedenbrück 21294
Westfield Heritage Village, Rockton 06768
Westfries Museum, Hoorn 32413
Westgate Museum, Winchester 46035
Westlands Museum voor Streek- en Tuinbouwhistorie, Honselersdijk 32403
Westman's Cottage, Jakobstad 10868
Westmansmors stugan, Jakobstad 10868
Westminster Abbey Museum, London 45012
Westminster College Art Gallery, New Wilmington 51106
Westminster Dragoons Collection, London 45013
Westminster Historical Society, Westminster 54100
Westmoreland Museum of American Art, Greensburg 49070
Weston Park, Weston-under-Lizard 45983
Weston Park Museum, Sheffield 45656
Westonzoyland Pumping Station Museum, Bridgwater 43608
Westover House, Charles City 47353
Westpac Museum, Sydney 01566
Westpfälzer Musikantenmuseum, Mackenbach 20120
Westphalian State Museum of Industrial Heritage Gernheim Glassworks, Petershagen 21016
Westphalsches Haus, Markkleeberg 20206
Westport Historical Museum, Westport 54110
Westport House, Westport 24911
Westport Maritime Museum, Westport 54111
Westpreussisches Landesmuseum, Münster 20521
Westsächsisches Textilmuseum Crimmitschau, Crimmitschau 17750
Westslovakisches Museum in Trnava, Trnava 38381
Westville Historic Handicrafts Museum, Lumpkin 50322
Westwall-Bunker, Beckingen 16973
Westwallmuseum, Irrel 19353
Westwallmuseum Gerstfeldhöhe, Pirmasens 21043
Westzaanse Bodemvondsten Kok-Voogt, Westzaan 32944
Wetaskiwin and District Museum, Wetaskiwin 07313
Wethersfield Museum, Wethersfield 54113
Wetterau-Museum, Friedberg, Hessen 18468
Wettermuseum Lindenberg, Tauche 22012
Weverijmuseum, Geldrop 32240
Wexford Arts Centre, Wexford 24914
Wexford County Historical Museum, Cadillac 47157
Wexford County Museum, Enniscorthy 24832
Wexner Center for the Arts, Columbus 47761
Weyers-Sampson Art Gallery, Greenville 49077
Weygang-Museum, Öhringen 20868
Weymouth Museum, Weymouth 45991
Weymouth Woods-Sandhills Nature Preserve Museum, Southern Pines 53187
Whakatane District Museum and Gallery, Whakatane 33263
Whale Museum, Caniçal 35505
Whale Museum, Friday Harbor 48796
Whale World, Albany 00790
Whanganui Regional Museum, Wanganui 33235
Whanganui Riverboat Centre Museum, Wanganui 33236
Whangarei Art Museum, Whangarei 33265

W

Whangarei Museum, Whangarei 33266
Whangaroa County Museum, Kaeo 33131
Whanki Museum, Seoul 29933
Wharton County Historical Museum, Wharton 54116
Wharton Esherick Museum, Malvern 50411
Whatcom Museum of History and Art, Bellingham 46714
Wheat Ridge Sod House and Museum, Log Cabin, First Post Office and Tool Shed, Wheat Ridge 54117
Wheaton History Center, Wheaton 54122
Wheeler Gallery, Providence 52181
Wheeler Historic Farm, Salt Lake City 52698
Wheels O' Time Museum, Peoria 51775
Wheelwright Museum of the American Indian, Santa Fe 52892
Whetung Craft Centre and Art Gallery, Curve Lake 05668
Whipple House Museum, Ashland 46386
Whipple Museum of the History of Science, Cambridge 43736
Whistler House Museum of Art, Lowell 50308
Whistler Museum and Archives Society, Whistler 07317
Whitaker Center for Science and the Arts, Harrisburg 49183
Whitbourne Museum, Whitbourne 07318
Whitburn Community Museum, Whitburn 45993
Whitby Abbey Museum, Whitby 45997
Whitby Museum, Whitby 07320
Whitby Museum, Whitby 45998
Whitchurch Silk Mill, Whitchurch 45999
Whitchurch-Stouffville Museum, Gormley 05884
White County Historical Museums, Carmi 47253
White County Historical Society Museum, Cleveland 47619
White County Historical Society Museum, Monticello 50821
White Deer Land Museum, Pampa 51690
White Elephant Museum, Makkovik 06221
White Fox Museum, White Fox 07321
White Gallery, Portland 52091
White Granary, Crafts Exhibition, Bydgoszcz 34575
White Hall Historic Site, Richmond 52317
The White House, Washington 53934
White House, Westbury 01646
White Memorial Conservation Center, Litchfield 50129
White Mill Rural Heritage Centre, Sandwich 45617
White Pillars Museum, De Pere 48000
White Pine Public Museum, Ely 48385
White River Museum, Meeker 50560
White River Valley Museum, Auburn 46473
White Rock Museum, White Rock 07323
White Water Gallery, North Bay 06472
whiteBOX e.V., München 20496
Whitechapel Art Gallery, London 45014
Whitefield House Museum, Nazareth 50994
Whitehall, Cheam 43800
Whitehead Memorial Museum, Del Rio 48040
Whitehern Historic House and Garden, Hamilton 05963
Whiteley Museum, Saint Paul's River 06873
Whitemouth Municipal Museum, Whitemouth 07330
Whitewater Canal Historic Site, Metamora 50614
Whitewater Historical Museum, Whitewater 54135
Whitewood Historical Museum, Whitewood 07331
Whithorn - Cradle of Christianity, Whithorn 46004
Whithorn Priory and Museum, Whithorn 46005
Whitingham Historical Museum, Whitingham 54136
Whitley County Historical Museum, Columbia City 47736
Whitman House Museum, Canso 05548
Whitney Museum of American Art, New York 51279
Whitney Museum of American Art at Altria, New York 51280
Whitney Museum of American Art at Champion, Stamford 53250
Whitney Pier Historical Museum, Sydney 07052
Whitstable Museum and Gallery, Whitstable 46007
Whitstable Oyser and Fishery Exhibition, Whitstable 46008
Whittier Fine Arts Gallery, Wichita 54149
Whittlesey Museum, Whittlesey 46009
The Whitworth Art Gallery, Manchester 45099
Whitworth Historical Society Museum, Whitworth 46010
Whowhatwherewhenwhy W5, Belfast 43449
Whyalla Maritime Museum, Whyalla 01649
Whyte Museum of the Canadian Rockies, Banff 05383
Wichita Art Museum, Wichita 54150
Wichita Center for the Arts, Wichita 54151
Wichita Falls Museum and Art Center, Wichita Falls 54154
Wichita-Sedgwick County Historical Museum Association, Wichita 54152
Wick Heritage Centre, Wick 46011
Wickersham House Museum, Fairbanks 48475
Wickliffe Mounds, Wickliffe 54156
Wickman Maritime Collection, Zabbar 30661
Wide Bay and Burnett Historical Museum, Maryborough, Queensland 01269
Wide Bay Hospital Museum, Maryborough, Queensland 01270
Widener Gallery, Hartford 49200
Widener University Art Collection and Gallery, Chester 47432
Widukind-Museum, Enger 18173
Wiechers Woon Oase, Dwingeloo 32163
The Wiegand Gallery, Belmont 46718
Wieland-Archiv, Biberach an der Riß 17325
Wieland Estate in Ossmannstedt, Oßmannstedt 20928
Wieland-Gedenkzimmer, Achstetten 16424
Wieland-Museum Biberach, Wieland-Gartenhaus, Biberach an der Riß 17326
Wielandgut Oßmannstedt mit Wieland-Museum, Oßmannstedt 20928
Wielkopolski Park Etnograficzny, Lednogóra 34847
Wielkopolskie Military Museum, Poznań 35049
Wielkopolskie Muzeum Pożarnictwa PTTK, Rakoniewice 35077
Wielkopolskie Muzeum Walk Niepodległościowych w Poznaniu, Poznań 35048
Wielkopolskie Muzeum Wojskowe, Poznań 35049
Wien Museum Karlsplatz, Wien 03086
Wienbibliothek im Rathaus, Wien 03087
Wiener Feuerwehrmuseum, Wien 03088
Wiener Glasmuseum, Wien 03089
Wiener Kriminalmuseum, Wien 03090
Wiener Tramwaymuseum, Wien 03091
Wiener Ziegelmuseum, Wien 03092
Wienerwaldmuseum, Eichgraben 01851
't Wienkeltje van Wullempje, Hoedekenskerke 32394
Wieringer Boerderij, Den Oever 32113
Wiesenmeisterei, Neustadt-Glewe 20680
Wiesenthäler Textilmuseum, Zell im Wiesental 22781
Wieża Ratuszowa, Kraków 34820
Wigan Pier, Wigan 46014
Wiggertaler Museum, Schötz 41844
Wightwick Manor, Wolverhampton 46059
Wignacourt Museum, Rabat 30637
Wignall Museum of Contemporary Art, Rancho Cucamonga 52239
Wijnklder Soniën, Overijse 03946
Het Wijnkopersgildehuys, Amsterdam 31902
Wijnmuseum Maastricht, Cadier en Keer 32033
Wikinger Museum Haithabu, Busdorf 17668
Wilber Czech Museum, Wilber 54157
Wilberforce House, Kingston-upon-Hull 44601
Wilbur D. May Museum, Reno 52293
Wilbur Wright Birthplace and Interpretive Center, Hagerstown 49119
Wild Geese Heritage Museum, Portumna 24887
Wilder Memorial Museum, Strawberry Point 53337
Wilderer-Museum Sankt Pankraz, Sankt Pankraz 02674
Wilderness Road Regional Museum, Newbern 51299
Wildlife Education Centre, Entebbe 42988
Wildlife Museum, Eksjö 40495
Wildlife Museum, Rio de Janeiro 04713
Wildlife Museum, Ulaanbaatar 31633
Wildlife Wonderlands-Giant Earthworm Museum, Bass 00835
Wildling Art Museum, Los Olivos 50274
Wildnispark Zürich, Sihlwald 41868
Wildwood Center, Nebraska City 50996
Wile Carding Mill Museum, Bridgewater 05475
Wilfrid Israel Museum of Oriental Art and Studies, Kibbutz Hazorea 25044
Wilhelm-Busch-Geburtshaus, Wiedensahl 22540
Wilhelm-Busch-Gedenkstätte, Seesen 21725
Wilhelm-Busch-Mühle, Ebergötzen 18038
Wilhelm-Busch-Museum Hannover, Hannover 19008
Wilhelm Busch Zimmer, Dassel 17799
Wilhelm-Fabry-Museum, Hilden 19181
Wilhelm-Hack-Museum Ludwigshafen am Rhein, Ludwigshafen am Rhein 20058
Wilhelm-Kienzl-Museum, Paudorf 02479
Wilhelm Kienzl-Stüberl, Waizenkirchen 02890
Wilhelm-Morgner-Haus, Soest 21799
Wilhelm-Ostwald-Gedenkstätte, Großbothen 18783
Wilhelm-Panetzky-Museum, Rammelsbach 21174
Wilhelm Wagenfeld Haus - Design im Zentrum, Bremen 17567
Wilhelmietenmuseum, Huijbergen 32419
Wilkes Art Gallery, North Wilkesboro 51410
Wilkie and District Museum, Wilkie 07336
Wilkin County Historical Museum, Breckenridge 47002
Wilkinson County Museum, Woodville 54290
Will Rogers Memorial Museums, Claremore 47579
Willa Cather State Historic Site, Red Cloud 52272
Willard House and Clock Museum, Grafton 48981
Willem Prinsloo Museum, Rayton 38721
Willemoesgårdens Mindestuer, Assens 09972
Willenhall Museum, Willenhall 46016
Willi Dickhut Museum, Gelsenkirchen 18580
Willi Sitte Galerie, Merseburg 20295
William A. Farnsworth Art Museum and Wyeth Center, Rockland 52422
William A. Quayle Bible Collection, Baldwin City 46541
The William Benton Museum of Art, Storrs 53320
William Blizard Gallery, Springfield 53230
William Bonifas Fine Arts Center, Escanaba 48416
William Booth Birthplace Museum, Nottingham 45301
William Breman Jewish Heritage Museum, Atlanta 46457
William Carey Museum, Leicester 44684
William Clark Market House Museum, Paducah 51671
William Cullen Bryant Homestead, Cummington 47909
William Fehr Collection, Cape Town 38594
William H. Harrison Museum/ Grouseland, Vincennes 53751
William H. Van Every jr. and Edward M. Smith Galleries, Davidson 47973
William Hammond Mathers Museum, Bloomington 46867
William Heath Davis House Museum, San Diego 52762
William Henry Steeves Museum, Hillsborough 05988
William Holmes McGuffey Museum, Oxford 51656
William Howard Taft National Historic Site, Cincinnati 47567
William Humphreys Art Gallery, Kimberley 38674
William King Regional Arts Center, Abingdon 46126
William Lamb Memorial Studio, Montrose 45174
William M. Colmer Visitor Center, Ocean Springs 51490
William McKinley Presidential Library and Museum, Canton 47229
William Morris Gallery and Brangwyn Gift, London 45015
The William Morris Society, London 45016
William Pryor Letchworth Museum, Castile 47277
William Rea Museum, Murfreesboro 50912
William S. Hart Museum, Newhall 51314
William S. Webb Museum of Anthropology, Lexington 50059
William Scarbrough House, Savannah 52948
William Trent House, Trenton 53564
William Tyndale Museum, Vilvoorde 04115
William Weston Clarke Emison Museum of Art, Greencastle 49048
William Whipple Art Gallery, Marshall 50507
William Whitley House, Stanford 53254
William Woods University Art Gallery, Fulton 48809
Williams College Museum of Art, Williamstown 54176
Williams County Historical Museum, Montpelier 50823
Williamson Art Gallery and Museum, Birkenhead 43492
Williamson County Historical Society Museum, Marion 50482
Williamstown Museum, Williamstown 01652
Willibaldsburg, Eichstätt 18085
Willinger Milchmuhseum, Willingen 22586
Willis Museum, Basingstoke 43397
Willistead Manor, Windsor, Ontario 07345
Willoughby-Baylor House, Norfolk 51371
Willoughby Historical Museum, Niagara Falls 06450
Willow Bunch Museum, Willow Bunch 07340
Wilmette Historical Museum, Wilmette 54186
Wilmington Railroad Museum, Wilmington 54205
Wilno Heritage Society Museum, Wilno 07341
Wilson Castle, Proctor 52157
Wilson County Historical Society Museum, Fredonia 48775
Wilson Historical Museum, Wilson 54212
Wilson MacDonald Memorial School Museum, Selkirk 06944
Wilson's Creek National Battlefield, Republic 52298
Wilton Historical Museums, Wilton 54214
Wilton House, Wilton, Salisbury 46019
Wilton House Museum, Richmond 52344
Wilton Windmill, Wilton, Marlborough 46018
Wiltondale Pioneer Village, Bonne Bay 05440
Wiltshire Fire Defence and Brigades Museum, Potterne 45440
Wiltshire Heritage Museum and Gallery, Devizes 43979
Wimbledon Lawn Tennis Museum, London 45017
Wimbledon Society Museum of Local History, London 45018
Wimbledon Windmill Museum, London 45019
Wimmera Mallee Pioneers Museum, Jeparit 01178
Wimpole Hall and Home Farm, Royston 45522
Wincanton Museum, Wincanton 46021
Winchcombe Folk and Police Museum, Winchcombe 46023
Winchcombe Railway Museum, Winchcombe 46024
Winchelsea and District Historical Records Centre, Winchelsea 01655
Winchelsea Museum, Winchelsea 46025
Winchendon Historical Museum, Winchendon 54218
Winchester Cathedral Triforium Gallery, Winchester 46036
Winchester City Museum, Winchester 46037
Winchester College Treasury, Winchester 46038
Winchester-Frederick County Historical Society, Winchester 54222
Winckelmann-Museum, Stendal 21899
Wind Museum, L'Aldea 38835
Wind Museum, Hijikawa 28508
Windermere Steamboat Museum, Bowness-on-Windermere 43572
Windermere Steamboat Museum, Windermere 46039
Windermere Valley Museum, Invermere 06016
Windham Textile and History Museum, Willimantic 54179
Windmill and Milling Museum, Polegate 45396
Windmill Island Municipal Park Museum, Holland 49318
Windmill Museum, Harrström 10759
Windmühle, Podersdorf 02502
Windmühle, Retz 02581
Windmühle Waldfeucht, Waldfeucht 22293
Windsor Castle, Windsor 46047
Windsor Wood Carving Museum, Windsor, Ontario 07346
Windsor's Community Museum, Windsor, Ontario 07347
Wine and Spirits Historical Museum, Stockholm 40950
Wine Museum, Ehnen 30471
Wine Museum, Tolcsva 23935
Wine Museum, Vilafranca del Penedès 40281
Wine Museum in Osaka, Osaka 29127
Wine Museum of Greyton H. Taylor, Hammondsport 49144
Wine Museum Paris, Paris 14703
Winedale Historical Center, Round Top 52472
Wing Luke Asian Museum, Seattle 53017
Wings of Freedom Museum, Huntington 49427
Wings of History Air Museum, San Martin 52821
Wings of Love, Clinton 47654
Winkelriedhaus, Stans 41907
Winnebago Area Museum, Winnebago 54235
Winnetka Historical Museum, Winnetka 54236
Winnie Davis Museum of History, Gaffney 48812
Winnipeg Art Gallery, Winnipeg 07392
Winnipeg Police Museum, Winnipeg 07393
Winnipeg Railway Museum, Winnipeg 07394
Winnipegosis Museum, Winnipegosis 07396
Winns Historic Bakehouse Museum, Coromandel Valley 01002
Winona County Historical Museum, Winona 54239
Winser Museumshof, Winsen, Aller 22605
Winslow Crocker House, Yarmouth Port 54326
Winston Churchill Memorial and Library in the United States, Fulton 48810
Winter War Museum, Kuhmo 10999
Winterset Art Center, Winterset 54257
Wintersport-Museum, Davos Platz 41323
Winter!Sport!Museum!, Mürzzuschlag 02415
Winterthur Museum, Winterthur 54258
Winterton Boat Building Museum, Winterton 07397
Winthrop University Galleries, Rock Hill 52404
Winzermuseum Bachem, Bad Neuenahr-Ahrweiler 16797
Winzermuseum Rauenberg, Rauenberg 21199
Wipperkotten, Solingen 21809
Wiregrass Museum of Art, Dothan 48155
Wireless Hill Telecommunications Museum, Applecross 00808
Wirrabara Forestry Museum, Wirrabara 01659
Wirral Museum, Birkenhead 43493
Wirral Transport Museum and Bierkenhead Tramways, Birkenhead 43494
Wirtschaftsgeschichtliches Museum Villa Grün, Dillenburg 17862
Wirtz Gallery, Miami 50637
Wisbech and Fenland Museum, Wisbech 46049
Wisconsin Automotive Museum, Hartford 49201
Wisconsin Federation of Museums, Wausau 55194
Wisconsin Historical Museum, Madison 50396
Wisconsin Maritime Museum, Manitowoc 50444
Wisconsin National Guard Memorial Library and Museum, Camp Douglas 47205
Wisconsin Union Gallery, Madison 50397
Wisconsin Veterans Museum King, King 49779
Wisconsin Veterans Museum Madison, Madison 50398
The Wise Art Gallery, Tyler 53636
Wise County Heritage Museum, Decatur 48020
Wiseman and Firehouse Galleries, Grants Pass 49018
Wistariahurst Museum, Holyoke 49328
Withernsea Lighthouse Museum, Withernsea 46050
Witney and District Museum and Art Gallery, Witney 46053
Witte de With, Rotterdam 32734
Witte Museum, San Antonio 52731
Wittelsbachermuseum Aichach, Aichach 16446
Wittenberger Freilichtmuseum, Edewecht 18057
Witter Gallery, Storm Lake 53317
Witte's Museum, Axel 31944
Wittumspalais, Weimar 22423
Wixhäuser Dorfmuseum, Darmstadt 17797

Wladyslaw Orkan's House, Niedźwiedź 34944
Wladyslaw Orkan's Museum, Rabka 35070
Woburn Abbey, Woburn 46054
Wocher-Panorama der Stadt Thun um 1810, Thun 41941
Wocoal Art Center, Tokyo 29505
Wohl Archaeology Museum Herodian Mansions, Jerusalem 25024
Wohn- und Porzellanmuseum Agentenhaus, Horgen 41489
Wohnmuseum Egga, Visperterminen 41989
Woksens Samlinger, Rørvik 33798
Wolcott House Museum Complex, Maumee 50536
Wolf House Museum, Manhattan 50438
Wolf Kahlen Museum, Bernau bei Berlin 17290
Wolf Point Area Historical Society, Wolf Point 54267
Wolfgang-Bonhage-Museum, Korbach 19686
Wolfson Museum, Jerusalem 25025
The Wolfsonian, Miami Beach 50644
Wolfsoniana, Genova 26327
Wolfsteiner Heimatmuseum im Schramlhaus, Freyung 18462
Woljeon Art Museum, Seoul 29934
Wolkoff House Museum, Lappeenranta 11050
Wolkoffin talomuseo, Lappeenranta 11050
Wollaston Museum, Wollaston 46056
Wollondilly Heritage Centre, The Oaks 01390
Wollongong City Gallery, Wollongong 01663
Wolseley and District Museum, Wolseley 07400
Wolverhampton Art Gallery, Wolverhampton 46060
Wolverine Hobby and Historical Society Museum, Spy Hill 07003
Woman Made Gallery, Chicago 47523
Women and Their Work, Austin 46523
Women's Art Library - Make, London 45020
Women's History Museum, West Liberty 54069
The Women's Museum, Dallas 47943
Women's museum, Hittisau 02084
Women's Museum in Denmark, Århus 09964
Women's Museum of South Vietnam, Ho Chi Minh City 54632
Women's Technical Education Museum, Ankara 42704
Won Kwang College Museum, Iri 29835
Wondai Heritage Museum, Wondai 01664
Wonders of the Sea, Jdeidet El-Metn 30170
Wongan Ballidu Museum, Wongan Hills 01665
Wonnerup House and Old School, Busselton 00928
Wonsan Historical Museum, Wonsan 29763
Woo-Bong Art Museum, Daegu 29796
Woocoo Museum, Brooweena 00911
Wood Carving Museum, Yerevan 00758
Wood Carving Museum and Taiwan Sericultural Museum, Zaociao 42488
Wood County Historical Center, Bowling Green 46978
Wood End Museum, Scarborough 45625
Wood Library-Museum of Anesthesiology, Park Ridge 51710
Wood Mountain Post, Wood Mountain 07401
Wood Mountain Ranch and Rodeo Museum, Wood Mountain 07402
Wood Street Galleries, Pittsburgh 51964
Woodbine International Fire Museum, Woodbine 54272
Woodbridge Museum, Woodbridge 46062
Woodburn Plantation, Pendleton 51760
Woodchester Villa, Bracebridge 05453
Woodchurch Village Life Museum, Woodchurch 46063
Wooden House Museum, De Rijp 32049
Wooden Nickel Historical Museum, San Antonio 52732
Wooden Shoe Museum Gebr. Wietzes, Eelde 32173
Wooden Shoes Museum, Porcheresse 03955
Wooden Shoes Museum, Soucht 15792
Woodend History Resource Centre, Woodend 01668
Woodenshoe Museum "De Plattyn", Best 31970
Woodhall Spa Cottage Museum, Woodhall Spa 46064
Woodhorn, Ashington 43329
Woodland Cultural Centre, Brantford 05471
Woodland Heritage Museum, Westbury 45976
Woodlands Pioneer Museum, Woodlands 07403
Woodlawn Museum, Ellsworth 48373
Woodlawn Plantation, Alexandria 46215
Woodloes Homestead Folk Museum, Cannington 00952
Woodmere Art Museum, Philadelphia 51889
Woodone Bijutsukan, Yoshiwa 29629
Woodone Museum of Art, Yoshiwa 29629
Woodrow Wilson House, Washington 53935
Woodrow Wilson Presidential Library at his Birthplace, Staunton 53279
Woodruff Museum of Indian Artifacts, Bridgeton 47021
Woods Hole Oceanographic Institution Ocean Science Exhibit Center, Woods Hole 54280
Woodside House, Rochester 52402
Woodside National Historic Site, Kitchener 06104
Woodson County Historical Society, Yates Center 54327
Woodstock Art Gallery, Woodstock 07405
Woodstock Artists Gallery, Woodstock 54284
Woodstock Historical Society, Woodstock 54288
Woodstock Historical Society Museum, Bryant Pond 47106
Woodstock Historical Society Museum, Woodstock 54282
Woodstock Museum, Woodstock 07406
Woodstock Museum of Shenandoah County, Woodstock 54286
Woodstock School of Art Gallery, Woodstock 54285
Woodville Museum, Rogue River 52443
Woodville Pioneer Museum, Woodville 33268
Woody Keihoku, Keihoku 28701
Wookey Hole Cave Diving and Archaeological Museum, Wookey Hole 46068
Woolaroc Museum, Bartlesville 46620
Woolpit and District Museum, Woolpit 46069
Woomera Heritage Centre, Woomera 01671
Woonbootmuseum, Amsterdam 31903
Worcester Art Museum, Worcester 54302
Worcester City Museum and Art Gallery, Worcester 46073
Worcester Historical Museum, Worcester 54303
Worcester Museum, Worcester 38779
The Worcestershire and Sherwood Foresters Regimental Museum, Nottingham 45302
Worcestershire County Museum, Kidderminster 44573
Worcestershire Yeomanry Cavalry Museum, Worcester 46074
Words and Pictures Museum, Northampton 51415
Wordspring Discovery Center, Orlando 51596
Wordsworth House, Cockermouth 43869
Work Gallery, Ann Arbor 46312
Worker Housing Museum of Helsinki City Museum, Helsinki 10833
Workers Arts and Heritage Centre, Hamilton 05964
Worker's Home Museum, South Bend 53161
Workers' Museum, Kuusankoski 11018
Workers' Museum and Hairdresser Museum, Kemi 10941
Workhouse Museum, Derry 43976
Working-Class Home Museum in Vaajakoski, Jyväskylä 10899
Workman and Temple Family Homestead Museum, City of Industry 47569
The Works, Newark 51295
Works of Art Collection, Rosebank 38728
Worksop Museum, Worksop 46078
World Cultural Heritage Site Völklinger Hütte, Völklingen 22249
World Federation of Friends of Museums, Paris 54812
World Figure Skating Museum and Hall of Fame, Colorado Springs 47708
World Gallery of Caricatures - Museum of the City of Skopje, Skopje 30524
World Golf Hall of Fame, Saint Augustine 52529
World in Wax Museum, Echuca 01059
World Kite Museum, Yokaichi 29590
World Kite Museum and Hall of Fame, Long Beach 50180
World Methodist Museum, Lake Junaluska 49886
World Museum Liverpool, Liverpool 44750
World Museum of Mining, Butte 47153
World Museum of Natural History, Riverside 52376
World O' Tools Museum, Waverly 53997
World of Country Life, Exmouth 44198
World of Energy at Keowee-Toxaway, Seneca 53034
World of Glass, Saint Helens 45568
World of the Bible Museum, Budapest 23396
World of Wings Pigeon Center Museum, Oklahoma City 51526
The World Organization of China Painters' Museum, Oklahoma City 51527
World Police Museum, Gueishan 42197
World Rugby Museum, Twickenham 45890
World Soil Museum, Wageningen 32921
World Toy Museum, Karuizawa 28670
World War II Museum, Betikama 38546
Wormsloe State Historic Site, Savannah 52949
Worpsweder fotografische Werkstätten, Worpswede 22698
Worpsweder Kunsthalle, Worpswede 22699
Worpsweder Mühle, Worpswede 22700
Worsbrough Mill Museum, Barnsley, South Yorkshire 43386
Worthing Museum and Art Gallery, Worthing 46079
Worthington Historical Society, Worthington 54308
Wortley Top Forge, Thurgoland 45856
Wosinsky Mór Megyei Múseum, Szekszárd 23869
Wotton Heritage Centre, Wotton-under-Edge 46080
Wouldle Museum, Jiangle 07858
Wowan and District Museum, Wowan 01672
Wrackmuseum, Cuxhaven 17759
Wrackmuseum Cuxhaven, Cuxhaven 17760
Wrangell Museum, Wrangell 54309
Wrather West Kentucky Museum, Murray 50919
Wray Museum, Wray 54310
Wren's Nest House Museum, Atlanta 46458
Wrest Park, Silsoe 45680
Wrexham Arts Centre, Wrexham 46084
Wrexham County Borough Museum, Wrexham 46085
Wright Brothers National Memorial, Kill Devil Hills 49775
Wright County Historical Museum, Buffalo 47113
Wright Historical Museum, Swift Current 07046
Wright Museum, Wolfeboro 54270
Wright Museum of Art, Beloit 46725
Wright State University Art Galleries, Dayton 47990
Wright's Ferry Mansion, Columbia 47721
Wriston Art Center Galleries, Appleton 46332
Writers Museum, Dublin 24824
The Writers' Museum, Edinburgh 44162
Wroclaw University Museum, Wrocław 35342
Wuchuan Museum, Wuchuan 08329
Würfelspielhaus - Mensch.Macht.Leben, Frankenburg 01897
Würth Art Room, Böheimkirchen 01792
Württemberg State Museum, Ludwigsburg 20046
Württemberg State Museum, Ludwigsburg 20045
Württemberg State Museum, Stuttgart 21950
Württembergische Landesbibliothek, Stuttgart 21969
Württembergischer Kunstverein Stuttgart, Stuttgart 21970
Württembergisches Psychiatriemuseum, Zwiefalten 22837
Wüstenroter Glas- und Heimatmuseum, Wüstenrot 22729
Wuhan Museum, Wuhan 08331
Wuhole Museum, Xiaogan 08378
Wuhou Museum, Chengdu 07623
Wuhua Museum, Wuhua 08334
Wujiang Museum, Wujiang 08335
Wujin Museum, Wujin 08336
Wulai Atayal Museum, Wulai 42462
Wulastook Museums, Fredericton 05841
Wuping Museum, Wuping 08338
Wuppertaler Uhrenmuseum, Wuppertal 22749
Wuwei Museum, Wuwei 08339
Wuwei Zone Museum, Wuwei 08340
Wuxi Beike Guan, Wuxi 08341
Wuxi City Museum, Wuxi 08342
Wuxi Engraved Stone Display Center, Wuxi 08341
Wuxi Folk Arts Museum, Wuxi 08343
Wuxi Glaswegianjian Guan, Wuxi 08343
Wuxi Revolutionary Display Center, Wuxi 08344
Wuyi Museum, Wuyi 08346
Wuyishan Museum, Wuyishan 08347
Wuyishan Nature Museum, Wuyishan 08348
Wuyuan Museum, Wuyuan 08349
Wuzhangyuan-Zhu Geliangmiao Museum, Qishan 08126
Wuzhi Museum, Wuzhi 08350
Wuzhou Museum, Wuzhou 08351
Wyalkatchem C.B.H. Agricultural Museum, Wyalkatchem 01673
Wyalong Park Private Museum, Maddington 01248
Wyandot County Historical Society Museum, Upper Sandusky 53667
Wyandotte County Museum, Bonner Springs 46902
Wyandotte Museum, Wyandotte 54312
Wyannie Malone Historical Museum, Hope Town 03201
Wyatt Earp Birthplace, Monmouth 50783
Wyck Museum, Philadelphia 51890
The Wyckoff Farmhouse Museum, Brooklyn 47085
Wycombe Museum, High Wycombe 44466
Wye Marsh Wildlife Centre, Midland 06280
Wylie House Museum, Bloomington 46868
Wymondham Heritage Museum, Wymondham 46089
Wyndham Lewis Society, London 45021
Wyndham Museum, Wyndham 33269
Wynn Collection, Las Vegas 49981
Wyoming Arts Council Gallery, Cheyenne 47447
Wyoming Dinosaur Center, Thermopolis 53494
Wyoming Pioneer Home, Thermopolis 53495
Wyoming Pioneer Memorial Museum, Douglas 48159
Wyoming State Museum, Cheyenne 47448
Wyoming Territorial Prison, Laramie 49951
Wysing Arts Centre, Bourn 43564
Wyspa - Institue of Art, Gdańsk 34635
Wystan Hugh Auden-Dokumentationsräume, Kirchstetten 02174
Wythenshawe Hall, Manchester 45100

X

Xá:ytem Museum, Mission 06307
Xantus János Múzeum, Győr 23590
Xavier University Museum (Museo de Oro), Cagayan de Oro City 34358
Xawery Dunikowski Museum of Sculpture-Królikarnia Palace, Warszawa 35269
Xi Bao Celadon Museum, Shenzhen 08227
Xi County Museum, Huangshan 07837
Xiajiang Museum, Xiajiang 08353
Xiamen Museum, Xiamen 08356
Xian Banpo Museum, Xian 08362
Xian Beilin Museum, Xian 08363
Xian Forest of Stone Tablets Museum, Xian 08364
Xian Highway Display Center, Xian 08365
Xiangfan Museum, Xiangfan 08368
Xiangfen Museum, Xiangfen 08370
Xiangtan Museum, Xiangtan 08371
Xiangwall Museum, Xianyuan 08376
Xiangxi Zone Museum, Jishou 07901
Xiangxiang Museum, Xiangxiang 08372
Xianyang Museum, Xianyang 08374
Xianyou Museum, Xianyou 08375
Xiaogan Museum, Xiaogan 08379
Xiaomen Geology Hall, Siyu 42342
Xiaoshan Museum, Xiaoshan 08380
Xiaoyanta Bag Guan Museum, Xian 08366
Xiaoyi Museum, Xiaoyi 08381
Xifeng Museum, Xifeng 08385
Xijia Mountain Folkways Museum, Jiang'an 07855
Xijiashan Glaswegiansu Guan, Jiang'an 07855
Xilin Museum, Xilin 08386
Xin Hai Shouyi Mausoleum of Fallen Heroes, Wuhan 08332
Xin Metropolis Revolution Museum, Weihui 08319
Xin'an Qiantang Zhizhai Museum, Xin'an 08387
Xincheng District Museum, Nanning 08057
Xincity Tusi Museum, Xincheng 08388
Xinfeng Museum, Xinfeng, Guangdong 08389
Xinfeng Museum, Xinfeng, Jiangxi 08390
Xingan-meng Museum, Wulanhoate 08337
Xingan Museum, Xingan 08391
Xinghua Museum, Xinghua 08392
Xingjing Yandao Gucheng Ruins Museum, Yingjing 08449
Xingning Museum, Xingning 08393
Xinhui Museum, Xinhui 08396
Xinjian Museum, Xinjian 08398
Xinjiang Painting Gallery, Urumchi 08308
Xinjiang Uighur Autonomous Region Museum, Urumchi 08309
Xinjiang Xian Museum, Xinjiang 08399
Xinle Yi Zhi Ruins Museum, Shenyang 08218
Xinmi Museum, Xinmi 08400
Xinping Bowuguan, Xinping 08402
Xinping Natural History Museum, Xinping 08402
Xintai Museum, Xintai 08403
Xinxiang Museum, Xinxiang 08404
Xinyi Museum, Xinyi, Guangdong 08405
Xinyi Museum, Xinyi, Jiangsu 08406
Xinyu Museum, Xinyu 08407
Xinzhou Zone Museum, Xinzhou 08408
Xishui Museum, Xishui 08409
Xiu Rock Museum, Xiuyan 08411
Xiuning Museum, Xiuning 08410
Xizhou Yandu Ruins Museum, Beijing 07573
Xperiment Huset, Växjö 41045
Xu Bei-Hong Museum, Beijing 07574
Xuanzhou Museum, Xuanzhou 08412
Xuchang Museum, Xuchang 08413
Xunyang Museum, Xunyang 08414
Xunyi Museum, Xunyi 08416
Xuyi Museum, Huai'an 07823
Xuzhou Museum, Xuzhou 08418
XVIII. Kerületi Pedagógiai és Helytörténeti Gyűjtemény, Budapest 23492
Xylon Museum Werkstätten, Schwetzingen 21704

Y

Ya Pian Zhan Zhen Museum, Dongguan 07669
Yacimiento-Museo Arqueologico, Ambrona 38868
Yacimiento Museo Arqueológico de Ambrona, Miño de Medinaceli 39651
Yacimiento-Museo Arqueológico de Numancia, Garray 39326
Yad Labanim Museum and Memorial Center, Petah Tikva 25078
Yad Lashiryon Museum, Latrun 25060
Yad-Lebanim Memorial Center, Rehovot 25091
Yad Vashem Art Museum, Jerusalem 25026
Yad Yaari Museum of Jewish Youth Movement, Givat Haviva 24949
Yaeyama Museum, Ishigaki 28586
Yagai Hakubutsukan, Gashozukuri Minkaen, Shirakawa 29281
The Yager Museum, Oneonta 51566
Yakage Museum, Yakage 29577
Yakima Valley Museum, Yakima 54320
Yakyu Taiiku Hakubutsukan, Tokyo 29506
Yaldhurst Museum of Transport and Science, Christchurch 33079
Yale Center for British Art, New Haven 51055
Yale University Art Gallery, New Haven 51056
Yale University Collection of Musical Instruments, New Haven 51057
Yalvaç Müzesi ve Örenyerleri, Yalvaç 42965
Yalvaç Museum and Ruins, Yalvaç 42965
Yamagata Bijutsukan, Yamagata 29578
Yamagata-kenritsu Hakubutsukan, Yamagata 29579
Yamagata Museum of Art, Yamagata 29578
Yamagata Prefectural Museum, Yamagata 29579
Yamaguchi Hakubutsukan, Yamaguchi 29580

W

Yamaguchi Hoshun Memorial Hall, Hayama 28501
Yamaguchi-kenritsu Bijutsukan, Yamaguchi 29581
Yamaguchi Museum, Yamaguchi 29580
Yamaguchi Prefectural Museum of Art, Yamaguchi 29581
Yamamoto Isoroku Kinenkan, Nagaoka 28965
Yamamoto Isoroku Memorial Museum, Nagaoka 28965
Yamamoto Kanae Memorial Museum, Ueda 29553
Yamanashi-kenritsu Bijutsukan, Kofu 28754
Yamanashi-kenritsu Hakubutsukan, Fuefuki 28419
Yamanashi-kenritsu Kagaku Senta, Kofu 28755
Yamanashi-kenritsu Koko Hakubutsukan, Kofu 28756
Yamanashi Prefectural Museum, Fuefuki 28419
Yamanashi Prefectural Museum of Archaeology, Kofu 28756
Yamanashi Prefectural Museum of Art, Kofu 28754
Yamanashi Prefectural Science Center, Kofu 28755
Yamatane Bijutsukan, Tokyo 29507
Yamatane Museum of Art, Tokyo 29507
Yamato Hakubutsukan, Kure 28794
Yamato Museum, Kure 28794
Yamauchi Collection of Art Museum, Kochi 28749
Yamauchi Library, Yokohama 29610
Yamba Museum, Yamba 01675
Yanaga Hokkaido Museums, Sapporo 29229
Yanaginogosho Kokokan, Hiraizumi 28523
Yanaginogosho Museum, Hiraizumi 28523
Yanbian Museum, Yanji 08430
Yancheng Museum, Yancheng 08419
Yanchi Museum, Yanchi 08420
Yang County Antiquity Museum, Yangxian 08426
Yang Ming Oceanic Culture and Art Museum, Keelung 42250
Yang Ming Studio, Taipei 42440
Yangjiang Museum, Yangjiang 08422
Yangliuqing Museum, Tianjin 08291
Yangshan Museum, Yangshan 08424
Yangxin Museum, Yangxin 08427
Yangzhou Hantomb Museum, Yangzhou 08428
Yangzhou Museum, Yangzhou 08429
Yanhuang Art Museum, Beijing 07575
Yanka Kupala Literary Museum, Minsk 03251
Yankalilla District Historical Museum, Yankalilla 01677
Yankee Air Museum, Belleville 46702
Yankee Candlemaking Museum, South Deerfield 53168
Yannis Tsarouchis Museum, Maroussi 23084
Yanshi Marketplace Museum, Yanshi 08431
Yantai Arts Museum, Yantai 08432
Yantai Museum, Yantai 08433
Yanzhou Museum, Yanzhou 08434
Yao County Museum, Yaoxian 08436
Yao'an Museum, Yao'an 08435
Yaoban Yaostove Museum, Tongchuan 08298
Yaoyao Wangshan Museum, Yaoxian 08437
Yaozu Self-Rule Museum, Jinxiu 07898
Yapı Kredi Vedat Nedim Tör Müzesi, İstanbul 42848
Yard Gallery, Nottingham 45303
Yarmouth County Museum, Yarmouth 07410
Yaroslavski Yakut State Museum of History and Culture of Northern Peoples, Jakutsk 36962
Yarraman Heritage Centre, Yarraman 01680
Yass Railway Museum, Yass 01683
Yataro Noguchi Bijutsukan, Nagasaki 28987
Yataro Noguchi Memorial Museum, Nagasaki 28987
Yatsugatake Bijutsukan, Hara 28498
Yatsugatake Museum of Art, Hara 28498
Yatsuhashi-An & Shishu-Yakata Gallery, Kyoto 28873
Yatsushiro Kyodo Hakubutsukan, Yashiro 29587
Yatsushiro Municipal Museum, Yashiro 29587
Yayoi Museum, Tokyo 29508
Yazılıkaya Müzesi, Çifteler 42753
Yazoo Historical Museum, Yazoo City 54328
Ybor City State Museum, Tampa 53438
Ydessa Hendeles Art Foundation, Toronto 07154
Yeats Tower - Thoor Ballylee, Gort 24840
Yedekule Fortress Museum, İstanbul 42849
Yedikulehisarı Müzesi, İstanbul 42849
Yeghegnadzor Art Gallery, Yeghegnadzor 00723
Yegishe Charents Literary House-Museum, Yerevan 00759
Yegishe Charents Literature and Arts Museum, Yerevan 00760
Yeiser Art Center, Paducah 51672
Yeleswasam Pavilion, Hyderabad 24219
Yellow Medicine County Historical Museum, Granite Falls 49014
Yellowstone Art Museum, Billings 46808
Yellowstone Gateway Museum of Park County, Livingston 50149
Yelverton Paperweight Centre, Yelverton 46095
Yeongil Folk Museum, Pohang 29859
Yerba Buena Center for the Arts, San Francisco 52801
Yerebatan Cistern Museum, İstanbul 42850
Yerebatan Sarnıcı, İstanbul 42850
Yerevan History Museum, Yerevan 00761
Yervand Kochar Museum, Yerevan 00762
Yeshiva University Museum, New York 51281
Yester-Year Artifacts Museum, Rowley 06782
Yester-Years Community Museum, Dinsmore 05704
Yesterday Museum, Bay Roberts 05402
Yesterday's Children Antique Doll & Toy Museum, Vicksburg 53736
Yesteryear Costume Gallery, Orroroo 01396
Yesteryear House-Central Mine, Clarkston 47590
Yesteryear Museum, Captains Flat 00953
Yesteryear Museum, Pamplin 51691
Yesteryears Doll and Toy Museum, Sandwich 52843
Yeungnam University Museum, Gyeongsan 29828
Yi City Museum, Yicheng, Hubei 08440
Yi Jun Peace Museum, Den Haag 32106
Yi Zhi Shi Liao Guan, Shanghai 08198
Yiannoulis Halepas Museum and Museum of Tinian Artists, Panormos Tinou 23133
Yibin Museum, Yibin 08438
Yichang Museum, Yichang 08439
Yicheng Museum, Yicheng, Shandong 08441
Yichun Museum, Yichun, Jiangxi 08443
Yifeng Museum, Yifeng 08445
Yigal Allon Center, Ginnosar 24948
Yilan County Historical Museum, Yilan 42470
Yilan County Natural History Museum, Luodong 42266
Yilan Fo Guang Yuan Art Gallery, Yilan 42471
Yilan Museums Association, Yilan 55040
Yıldız Sarayı Müzesi ve Tiyatrosu ve Sahne Kostümkeri Müzesi, İstanbul 42851
Yili Zone Museum, Yili 08446
Ying City Museum, Yingcheng 08448
Ying-Min JHS Folk Art Museum, Kaohsiung 42246
Yingde Cultureal Museum, Qingyuan 08119
Yingge Ceramics Museum, Yingge 42473
Yingkou Museum, Yingkou 08450
Yingtan Museum, Yingtan 08451
Yinque Shanhan Tomb Zhujian Museum, Linyi, Shandong 07956
Yiwu Museum, Yiwu 08452
Yixing Ceramic Museum, Yixing 08453
Yiyang Museum, Yiyang, Hunan 08454
Yiyang Museum, Yiyang, Jiangxi 08455
Yizheng Museum, Yizheng 08456
Ykhlas Kazakh Museum of Popular Musical Instruments, Almaty 29689
Ylä-Savon Kotiseutumuseo, Iisalmi 10848
Ylämaan Jalokivimuseo, Ylämaa 11477
Yläne Local Histoy Museum, Yläne 11478
Yläneen Kotiseutumuseo, Yläne 11478
Yli-Kirra Museum, Punkalaidum 11240
Yli-Laurosela Farmhouse Museum, Ilmajoki 10851
Yli-Lauroselan Talomuseo, Ilmajoki 10851
Ylihärmän Kotiseutumuseo, Ylihärmä 11479
Ylikiimingin Kotiseutumuseo, Ylikiiminki 11481
Ylistaron Kotiseutumuseo, Ylistaro 11482
Ymir Museum, Ymir 07412
Yo Sin-ci Stone Sculpture Museum, Sioulin 42341
Yogi Berra Museum, Little Falls 50135
Yokohama Archives of History, Yokohama 29613
Yokohama Bijutsukan, Yokohama 29611
Yokohama Curry Museum, Yokohama 29618
Yokohama Doll Museum, Yokohama 29612
Yokohama History Museum, Yokohama 29617
Yokohama Kaiko Shiryokan, Yokohama 29613
Yokohama Maritime Museum and Nippon-Maru Memorial Park, Yokohama 29614
Yokohama Museum of Art, Yokohama 29611
Yokohama Museum of EurAsian Cultures, Yokohama 29615
Yokohama Portside Gallery, Yokohama 29616
Yokohama Rekishi Hakabutsukan, Yokohama 29617
Yokohama-shi Kare-Hakubutsukan, Yokohama 29618
Yokohama-shi Sutori-Toka-Hakubutsukan, Yokohama 29619
Yokohama Streetcar Museum, Yokohama 29619
Yokosuka City Museum, Yokosuka 29622
Yokosuka-shi Hakubutsukan, Yokosuka 29622
Yokota Hakubutsukan, Tsukude 29541
Yokota Museum, Tsukude 29541
Yokoyama Memorial Manga Museum, Kochi 28750
Yokoyama Taikan Kinenkan, Tokyo 29509
Yokoyama Taikan Memorial Hall, Tokyo 29509
Yoldosh Oxunboboev Yodgorlik Muzeyi, Marg'ilon 54449
Yolo County Historical Museum, Woodland 54278
Yomitanson Bijutsukan, Yomitanson 29624
Yomitanson History and Folklore Museum, Yomitanson 29625
Yomitanson Museum, Yomitanson 29624
Yomitanson Rekishi oyo Minzoku Hakubutsukan, Yomitanson 29625
Yonago City Museum of Art, Yonago 29626
Yonago-shi Bijutsukan, Yonago 29626
Yonezawa Municipal Uesugi Museum & Uesugi Shrine Keisyo Palace Museum, Yonezawa 29627
Yongan Museum, Yongan 08457
Yongchun Museum, Yongchun 08458
Yongding Museum, Yongding 08459
Yongfeng County Museum, Enjiang 07679
Yongji Museum, Yongji 08460
Yongtai Museum, Yongtai 08461
Yonsei University Museum, Seoul 29935
Yorba and Slaughter Families Adobe Museum, Chino 47540
Yordan Yovkov Memorial House, Žeravna 05269
York and Lancaster Regimental Museum, Rotherham 45519
York Art Gallery, York 46108
York County Museum, York 54342
York Courthouse Complex, York 01685
York Minster Undercroft Treasury and Crypt, York 46109
York Museum, Toronto 07155
York Quay Gallery at Harbourfront Centre, Toronto 07156
York Racing Museum, York 46110
York Redoubt, Halifax 05950
York Region District School Board Heritage Schoolhouse, Markham 06240
York Story, York 46111
York Sunbury Museum, Fredericton 05842
Yorkshire Air Museum, Elvington 44171
Yorkshire and Humberside Federation of Museums and Art Galleries, Doncaster 55106
Yorkshire Museum, York 46112
Yorkshire Museum of Carriages and Horse Drawn Vehicles, Wensleydale 45965
Yorkshire Museum of Farming, York 46113
Yorkshire Museums Council, Leeds 55107
Yorkshire Sculpture Park, West Bretton 45966
Yorkshire Waterways Museum, Goole 44330
Yorkton Arts Council, Yorkton 07417
Yorktown Battlefield Visitor Center Museum, Yorktown 54347
Yorktown Colonial Park, Yorktown 54348
Yorktown Historical Museum, Yorktown 54344
Yorozu Tetsugoro Kinen Bijutsukan, Towa 29518
Yorozu Tetsugoro Memorial Museum of Art, Towa 29518
The Yosemite Museum, Yosemite National Park 54350
Yoshimoto Bungeikan, Osaka 29157
Yoshinogawa Sake Museum Hisago-tei, Nagaoka 28966
Yoshizawa Kinenkan, Joetsu 28623
Yoshizawa Memorial Museum, Joetsu 28623
Yoshizawa Memorial Museum of Art, Sano (Kuzuu), Kuzuu 28801
Youjiang Nationality Museum, Baise 07496
Youli Zhouyi Museum, Tangyin 08278
Young at Art Children's Museum, Davie 47974
Young's Brewery Museum, London 45022
Youth Club Museum, Muhniemi 11120
Youth Museum of Southern West Virginia & Beckley Exhibition Coal Mine, Beckley 46685
YouthLink Calgary, Calgary 05534
Youxi Museum, Youxi 08462
Youzhou Youci Museum, Pingdingshan 08080
Yozgat Devlet Güzel Sanatlar Galerisi, Yozgat 42967
Yozgat State Fine Art Gallery, Yozgat 42967
Ypäjän Kotiseutumuseo, Ypäjä 11484
Ypsilanti Historical Museum, Ypsilanti 54357
Yrjar Heimbygdslag, Opphaug 33708
Yrjö A. Jäntin Taidekokoelma, Porvoo 11232
Yrjö A. Jäntti Art Collection, Porvoo 11232
Yrjö Liipola Art Collection, Koski 10985
Yrjö Liipolan Taidekokoelma, Koski 10985
Ystads Konstmuseum, Ystad 41078
Ystads Militärmuseum, Ystad 41079
Yttilän Museokoulu, Yttilä 11485
Yturri-Edmunds Historic Site, San Antonio 52733
Yu.A. Fedkovich Literary Memorial Museum, Černivci 43011
Yu.P. Spegalsky Museum-Apartement, Pskov 37574
Yuan Shuo Shi Han Lin Cong Huei Stone Sculpture Museum, Sincheng 42332
Yuanling Museum, Yuanling 08464
Yuanping Museum, Yuanping 08465
Yuanyehao Animal Specimen Museum, Puli 42299
Yucaipa Adobe, Yucaipa 54362
Yudu Museum, Yudu 08467
Yufuin-cho Hakubutsukan, Yufuin 29631
Yufuin-cho Kindai Bijutsukan, Yufuin 29632
Yufuin Modern Art Museum, Yufuin 29632
Yufuin Museum, Yufuin 29631
Yugawara Art Museum, Yugawara 29633
Yugawara-machi Bijutsukan, Yugawara 29633
Yukara-Ori Folkcraft Museum, Asahikawa 28371
Yukara-Ori Kogeikan, Asahikawa 28371
Yuki Bijutsukan, Osaka 29158
Yuki Museum of Art, Osaka 29158
Yukon Arts Centre Gallery, Whitehorse 07327
Yukon Beringia Interpretive Centre, Whitehorse 07328
Yukon Historical and Museums Association, Whitehorse 54749
Yukon Transportation Museum, Whitehorse 07329
Yuldosh Ahunbabaev Yodgorlik Muzeyi, Toshkent 54491
Yuldosh Akhunbabaev House-Museum, Toshkent 54491
Yulin Museum, Yulin 08468
Yuma Art Center and Galleries, Yuma 54365
Yuma Territorial Prison Museum, Yuma 54366
Yumeji Art Museum, Okayama 29103
Yumeji Bijutsukan, Okayama 29103
Yumeji Takehisa Memorial Museum, Ikaho, Ikaho 28561
Yumen Museum, Yumen 08469
Yumiko Igarashi Bijutsukan, Kurashiki 28791
Yumiko Igarashi Museum, Kurashiki 28791
Yun Brocade Museum, Nanjing 08051
Yun County Museum, Shiyanyun 08238
Yunan Museum, Yunan 08470
Yuncheng Museum, Yuncheng 08471
Yuncheng Zone Hedong Museum, Yuncheng 08472
Yunfu Museum, Yunfu 08473
Yunmeng Museum, Yunmeng 08474
Yunnan Nationality Museum, Kunming 07916
Yunnan Provincial Museum, Kunming 07917
Yunus Emre Müzesi, Eskişehir 42782
Yunus Emre Museum, Eskişehir 42782
Yupiit Piciryarait Museum, Bethel 46773
Yurii Gagarin Memorial Museum, Gagarin 36918
Yurin Museum, Kyoto 28808
Yushan Museum, Yushan 08475
Yushe Reliquary Museum, Yushe 08476
Yusif Mammadaliyev Home Museum, Ordubad 03192
Yusupov's Palace, Sankt-Peterburg 37653
Yuyu Yang Museum, Taipei 42441
Yuzawa History and Folk Museum, Yuzawa 29639
Yuzhang Folklorish Museum, Nanchang 08033
Yuzhang Glaswegiansu Bowu Yuan, Nanchang 08033
Yuzhong Museum, Yuzhong 08479
Yuzo Kondo Memorial Museum, Kyoto 28828
YYZ Artists' Outlet, Toronto 07157

Z

Z. Paliashvili House Museum, Tbilisi 16386
Z33, Hasselt 03687
Zaanlandse Oudheidkamer Honig Breet Huis, Zaandijk 32986
Zaans Museum, Zaandam 32984
De Zaanse Schans, Zaandam 32985
Zabbar Sanctuary Museum, Zabbar 30662
Zabol Museum, Zabol 24705
Zabytkowa Huta Żelaza, Chlewiska 34593
Zabytkowa Kopalnia, Tarnowskie Góry 35181
Zabytkowa Kuźnia Wodna, Gdańsk 34653
Zabytkowa Kuźnia Wodna w Starej Kuźnicy, Stara Kuźnica 35127
Zachęta - Narodowa Galeria Sztuki, Warszawa 35290
Zacheta - National Gallery of Art, Warszawa 35290
Zadužbina Kralja Petra I Karadjordjevića, Topola 38157
Zahari Stoyanov House-Museum, Ruse 05152
Zahedan Museum, Zahedan 24706
Zahnärztliches Museum, Tübingen 22126
Zahnmuseum Linz, Linz 02326
Záhorská Galéria Senica, Senica 38349
Záhorské Múzeum, Skalica 38350
Zajsan Historical and Regional Museum, Zajsan 29736
Zajsanskij Gorodskoj Istoriko-Kraevedčeskij Muzej, Zajsan 29736
Zakarpatsk Museum of Folk Architecture and Folklore, Užgorod 43199
Zakarpatsk Regional Museum, Užgorod 43198
Zakarpats'kyj Kraeznavčyj Muzej, Užgorod 43198
Zakarpats'kyj Muzej Narodnoï Architektury ta Pobutu, Užgorod 43199
Zakarpats'kyj Oblasnyj Chudožnij Muzej im. J. Bokšaja, Užgorod 43200
Zakhari Stpyanov House-Museum, Medven 05090
Het Zakkendragershuisje, Rotterdam 32735
Zakład Narodowy im. Ossolińskich, Wrocław 35344
Zakynthos Museum, Zakynthos 23244
Zakynthos Museum of the Natural Heritage, Zakynthos 23246
Zaldivar Museum, Albuera 34332
Žaliūkiai Windmill of the Šiauliai Aušra Museum, Šiauliai 30366
Žaliūkių Vėjo Malūnas, Šiauliai 30366
Zaltbommels Stoom- en Energiemuseum, Zaltbommel 32988
Zalud House, Porterville 52061
Zambala Art Museum, Shanhua 42322
Zamboanga National Museum, Zamboanga 34527
Zámecké Muzeum, Kravaře 09606
Zámecké Muzeum, Tovačov 09872
Zámek, Hradec nad Moravicí 09535
Zámek, Jirkov 09566
Zámek Brandýs, Brandýs nad Labem 09414
Zámek Bučovice, Bučovice 09446
Zámek Budišov, Budišov u Třebíče 09447
Zámek Chropyně, Chropyně 09484
Zámek Hrubý Rohozec, Turnov 09882
Zámek Kinski, Chlumec nad Cidlinou 09475
Zámek Kinských, Valašské Meziříčí 09896

Zamek Królewski na Wawelu - Państwowe Zbiory Sztuki, Kraków 34821
Zamek Królewski w Warszawie, Warszawa 35291
Zamek Książąt Pomorskich, Szczecin 35165
Zámek Libochovice, Libochovice 09626
Zámek Líčkov - Galerie O. Brázda, Žatec 09927
Zámek Lysice, Lysice 09645
Zámek Manětín, Manětín 09647
Zámek Moravský Krumlov - Galerie Slovanská Epopej Alfonse Muchy, Moravský Krumlov 09665
Zámek - Pamětní Síň Josefa Mánesa, Čechy pod Košířem 09452
Zamek Piastowski, Gliwice 34667
Zámek Slavkov - Austerlitz, Slavkov u Brna 09841
Zámek Troja, Praha 09799
Zámek Vsetín, Vsetín 09918
Zamek w Czersku, Czersk 34612
Zamek w Gniewie - Muzeum Regionalne, Gniew 34671
Zamek w Lipowcu, Babice 34533
Zamogitian Art Museum - Memorial Museum of the Writer Zemaite, Bukante 30218
Zanabazar Museum of Fine Arts, Ulaanbaatar 31634
Zanavykai Area Museum, Sakiai 30350
Zanavykų Krašto Muziejus, Sakiai 30350
Zandenbos, Nunspeet 32610
Zandlopermuseum Glanerbrug, Enschede 32214
Zandvoorts Museum, Zandvoort 32989
Zane Grey Museum, Lackawaxen 49858
Zanesville Art Center, Zanesville 54368
Zante Museum, Zante 30150
Zanzibar National Archives and Museums Department, Zanzibar 42525
Zaoyang Museum, Zaoyang 08480
Zaozhuang Museum, Zaozhuang 08481
Zapadno-Kazachstanskij Oblastnoj Istoriko-Kraevedčeskij Muzej, Oral 29714
Západočeská Galeria v Plzni, Plzeň 09719
Západočeské Muzeum, Plzeň 09720
Západomoravské Muzeum v Třebíči - Muzeum Jemnice, Jemnice 09555
Západoslovenské Múzeum v Trnave, Trnava 38381
Zaporiz'kyj Chudožnij Muzej, Zaporižžja 43209
Zaporiz'kyj Krajeznavčyj Muzej, Zaporižžja 43210
Zaporozhe Local Museum, Zaporižžja 43210
Zaporozhe Museum of Art, Zaporižžja 43209
Zapovidnyk-muzej Gogolja, Gogoleve 43040
Zara Clark Folk Museum, Charters Towers 00964
Zarajsk Kremlin - History, Architecture, Archeological and Art Museum, Zarajsk 37986
Zarasai Area Museum, Zarasai 30450
Zarasai Area Museum - Kazimieras Buga Museum, Paziegė 30336
Zarasų Krašto Muziejus, Zarasai 30450
Zarauzko Arte eta Historia Museoa, Zarautz 40355
Zászlómúzeum, Budapest 23493
Zavičajna Muzejska Zbirka, Čajetina 38096
Zavičajna Zbirka Đelekovec, Đelekovec 08859
Zavičajni Muzej, Aleksinac 38029
Zavičajni Muzej, Bar 31636
Zavičajni Muzej, Biograd na Moru 08844
Zavičajni Muzej, Buzet 08851
Zavičajni Muzej, Danilovgrad 31644
Zavičajni Muzej, Herceg-Novi 31646
Zavičajni Muzej, Nikšić 31650
Zavičajni Muzej, Ogulin 08919
Zavičajni Muzej, Pljevlja 31652
Zavičajni Muzej, Ruma 38137
Zavičajni Muzej, Travnik 04286
Zavičajni Muzej, Ulcinj 31659
Zavičajni Muzej Baška, Baška 08842
Zavičajni Muzej Grada Rovinja, Rovinj 08943
Zavičajni Muzej Kozara, Prijedor 04271
Zavičajni Muzej Našice, Našice 08912
Zavičajni Muzej Ozlja, Ozalj 08925
Zavičajni Muzej Poreštine, Poreč 08929
Zavičajni Muzej u Jagodini, Jagodina 38101
Zavicajni Muzej Zemuna, Zemun 38165
Zavizajni Muzej, Varaždinske Toplice 08976
Zavod za Zaštita na Spomenicite na Kulturata i Muzej - Strumica, Strumica 30530
Zbiory Historyczno-Etnograficzne, Oświęcim 34998
Zbiory Sprzętu Ratownictwa Kolejowego, Poznań 35050
Zbiory Sztuki, Włocławek 35312
Zbirci Umjetnina u Crkvi Sv. Ivana Krstitelja, Trogir 08971
Zbirka Anke Gvozdanović, Zagreb 09015
Zbirka Baltazara Bogišića, Cavtat 08854
Zbirka Crkvene Umjetnosti, Šibenik 08948
Zbirka Drave, Gola 08875
Zbirka Franjevacki Samostan, Orebić 08921
Zbirka Franjevačkog Samostana, Sinj 08951
Zbirka Franjevačkog Samostana Visovac, Visovac 08979
Zbirka Klašter Sv. Nikolaus, Korčula 08898
Zbirka Sakralne Umjetnini Župe Sveti Blaža, Vodnjan 08980
Zbirka Umjetnina Andra Vida i Katarine Mihičić i Zbirka Starij Majstora Piperata, Mali Lošinj 08909
Zbirka Umjetnina Crkve Sv. Domenika, Trogir 08972
Zbirka Umjetnina Franjevačkog Samostana, Hvar 08881
Zbirka Umjetnina Kairos, Trogir 08973
Zbirka v Jamarskem Domu, Domžale 38416
Zdravotnické Muzeum, Praha 09800
Zebulon B. Vance Birthplace, Weaverville 54018
Zeche Zollern, Dortmund 17930
Zee- en Havenmuseum de Visserijschool, IJmuiden 32425
Zeeuws Biologisch Museum, Oostkapelle 32630
Zeeuws Museum, Middelburg 32554
Zeeuws Poppen- en Klederdrachten Museum, Zoutelande 33009
Zeews Maritiem MuZEEum, Vlissingen 32900
Zehnthaus, Jockgrim 19398
Zeidel-Museum, Feucht 18311
Zeiler Foto- und Filmmuseum, Zeil am Main 22773
Zeilopleidingsschip Mercator, Oostende 03936
Žeimelio Žiemgalos Muziejus, Žeimelis 30451
Zeimelis Ziemgala Museum of the Šiauliai Aušra Museum, Žeimelis 30451
Zeiselmairhaus, Schrobenhausen 21641
Zeiss-Großplanetarium Berlin, Berlin 17281
Zeitbrücke Museum, Gars am Kamp 01924
Zeitgeschichtliches Forum Leipzig, Leipzig 19942
Zeitreise - Renner Villa, Gloggnitz 01939
Zeitspurenmuseum Ochzethaus, Altheim 01714
zeittor - Museum der Stadt Neustadt in Holstein, Neustadt in Holstein 20681
Zeittunnel Wülfrath, Wülfrath 22704
Zekelman Family Holocaust Memorial Center, Farmington Hills 48521
Železnički Muzej, Beograd 38092
Železniční Muzeum Výtopna Jaroměř, Jaroměř 09553
Železniční Muzeum Zlonice, Zlonice 09944
Železniški Muzej, Ljubljana 38469
Zelienople Historical Museum, Zelienople 54369
Zemaiciai Museum Alka, Telšiai 30379
Žemaičių Dailės Muziejus, Plungė 30338
Žemaičių Muziejus Alka, Telšiai 30379
Žemaičių Vyskupystės Muziejus, Varniai 30391
Žemaitijos Kaimo Muziejus, Telšiai 30380
Zemaljski Muzej Bosne i Hercegovine, Sarajevo 04284
Zemgale Farmstead and Agricultural Machinery Museum, Uzvara 30135
Zemgales Sētas un Lauksaimniecības Mašīnu Muzejs, Uzvara 30135
Zemljepišni Muzej Slovenije, Ljubljana 38470
Zempléni Múzeum, Szerencs 23887
Zemplínske Múzeum, Michalovce 38305
Zenetörténeti Múzeum, Budapest 23494
Zengcheng Museum, Zengcheng 08482
Zenica City Museum, Zenica 04294
Zentral-und Landesbibliothek Berlin, Berlin 17282
Zentralarchiv, Berlin 17283
Zentrale Kustodie und Universitätsmuseum, Halle, Saale 18906
Zentralfilmarchiv Laxenburg, Laxenburg 02276
Zentrum für Internationale Lichtkunst, Unna 22177
Zentrum Paul Klee, Bern 41220
ZEPHYR - Raum für Fotografie!, Mannheim 20175
Zeppelin-Museum, Neu-Isenburg 20583
Zeppelin Museum Friedrichshafen, Friedrichshafen 18478
Zeppelin-Museum Meersburg, Meersburg 20253
Zerain Cultural Park, Zerain 40357
Zetland Lifeboat Museum, Redcar 45474
Zetlin Museum, Ramat Gan 25087
Zettl Langer Collection, Sopron 23823
Zettl Langer Gyűjtemény, Sopron 23823
Zeze-Ware Art Museum, Otsu 29168
Zeze-yaki Bijutsukan, Otsu 29168
Žezkazganskij Istoriko-Kraevedčeskij Muzej, Žezkazgan 29737
Zgodovinski Arhiv i Muzej Univerze v Ljubljani, Ljubljana 38471
Zhabei Revolutionary Shi Liao Display Center, Shanghai 08199
Zhan Tianyou Former Residence, Wuhan 08333
Zhan Tianyou Memorial Hall, Beijing 07576
Zhangheng Museum, Nanyang 08064
Zhangjiachuan Museum, Zhangjiachuan 08483
Zhangjiagang City Gallery, Zhangjiagang 08484
Zhangjiakou Museum, Zhangjiakou 08485
Zhangping Museum, Zhangping 08486
Zhangpu Museum, Zhangpu 08487
Zhangqiu Arts and Crafts Guan, Jinan 07886
Zhangqiu Museum, Zhangqiu 08488
Zhangshu Museum, Zhangshu 08489
Zhangye Museum, Zhangye 08490
Zhangzhongjing Museum, Nanyang 08065
Zhangzhou Museum, Zhangzhou 08491
Zhanjiang Museum, Zhanjiang 08492
Zhaoling Museum, Liquan 07961
Zhaotong Museum, Zhaotong 08495
Zhejiang Natural Museum, Hangzhou 07767
Zhejiang Provincial Museum, Hangzhou 07768
Zhengding County Display Center, Zhengding 08496
Zhengjiatun Museum, Shuangliao 08241
Zhengzhou Museum, Zhengzhou 08500
Zhenjiang Museum, Zhenjiang 08502
Zhenjiang Revolution History Museum, Zhenjiang 08503
Zhi City Museum, Zhicheng 08504
Zhi Li Victory Museum, Baoding 07499
Zhijiang Museum, Zhijiang 08505
Zhongguo Qianbi Bowuguan, Beijing 07577
Zhongshan Museum, Zhongshan 08507
Zhou Hejing County Museum, Hejing 07779
Zhoukou Zone Museum, Zhoukou 08508
Zhoushan Museum, Zhoushan 08509
Zhouyuan Museum, Fufeng 07697
Zhuanglang Museum, Zhuanglang 08510
Zhucheng Museum, Zhucheng 08511
Zhuhai Museum, Changning 07594
Zhuhai Museum, Zhuhai 08513
Zhuzhou Museum, Zhuzhou 08515
Zhytomyr District Museum of Literature, Žytomyr 43215
Zibo Museum, Zibo 08516
Zichy Mihály Emlékmúzeum, Zala 23969
Židovské Múzeum, Prešov 38338
Židovské Muzeum v Praze, Praha 09801
Ziegel- und Kalkmuseum, Winzer 22609
Ziegelei Benzin, Kritzow 19716
Ziegeleimuseum, Jockgrim 19399
Ziegeleipark Mildenberg, Zehdenick 22772
Ziekenhuis- en Verpleegkundig Museum, Deventer 32122
The Zigler Museum, Jennings 49621
Zigong Salt Industry History Museum, Zigong 08518
Zigong Saurian Museum, Zigong 08519
Zijdemuseum, Meliskerke 32548
Zijin Museum, Zijin 08520
Zijper Museum, Schagerbrug 32750
Zilla Samgrahasala, Purulia 24365
Zille Museum, Berlin 17136
Zillertaler Goldschaubergwerk, Zell am Ziller 03128
Zilvermuseum Sterckshof Provincie Antwerpen, Antwerpen 03321
Zimbabwe Military Museum, Gweru 54667
Zimmer Children's Museum, Los Angeles 50270
Zimmerei-Museum, Feldkirch 01882
Zimmermannshaus Lackinger, Windhaag bei Freistadt 03114
Zimmertoren, Lier 03821
Zinkhütter Hof, Stolberg, Rheinland 21910
Zinnfiguren-Klause im Schwabentor, Freiburg im Breisgau 18442
Zinnfiguren-Museum, Bad Bergzabern 16688
Zinnfiguren-Museum Goslar, Goslar 18716
Zinnfigurenkabinett im Haspelturm, Eberbach 18034
Zinnfigurenmuseum, Grüningen 41458
Zinnfigurenmuseum, Zürich 42117
Zinnfigurenmuseum Aach, Aach 16390
Zinnfigurensammlung, Rheinsberg 21307
Zinnfigurenwelt, Katzelsdorf, Leitha 02157
Zinngießerhaus, Mattighofen 02366
Zinngrube Ehrenfriedersdorf, Ehrenfriedersdorf 18071
Zinnkeller, Weidenberg 22370
Zinovyev House-Museum, Djagilevo 36875
Zion Historical Museum, Zion 54370
Zion Human History Museum, Springdale 53214
Zippo and Case Visitors Center Company Museum, Bradford 46989
Zirkelmuseum, Wilhelmsdorf 22571
Zirkus und Varieté Archivsammlung Reinhard Tetzlaff, Hamburg 18964
Zisterzienserabtei Mehrerau, Bregenz 01798
Zisterzienserkloster Bebenhausen, Tübingen 22127
ZisterzienserMuseum Kloster Walkenried, Walkenried 22307
Zisterziensermuseum Riddagshausen, Braunschweig 17531
Zisterzienserstift, Lilienfeld 02306
Zisterzienserstift Rein, Rein 02575
Žitnoostrovné Múzeum, Dunajská Streda 38253
Ziya Gökalp Müzesi, Diyarbakır 42765
Ziya Gökalp Museum, Diyarbakır 42765
ZKM | Center for Art and Media Karlsruhe, Karlsruhe 19453
ZKM | Zentrum für Kunst- und Medientechnologie Karlsruhe, Karlsruhe 19453
Znojemský Hrad, Znojmo 09949
Zoar State Memorial, Zoar 54373
Zoll- und Finanzgeschichtliche Sammlung, Linz 02327
Zoll- und Heimatmuseum, Perwang 02493
Zollmuseum, Mandelbachtal 20161
Zollmuseum, Wegscheid 22360
Zollmuseum Friedrichs, Aachen 16409
Zoltán Kodály Memorial Museum, Budapest 23424
Zomerkostal uut 't Wold, Ruinerwold 32741
Zompenmuseum, Enter 32217
Zona arkeologjike e Butrintit, Butrinti 00012
Zonal Anthropological Museum, Shillong 24388
Zonal Museum, Dehradun 24132
Zonengrenz-Museum Helmstedt, Helmstedt 19125
Zonguldak Devlet Güzel Sanatlar Galerisi, Zonguldak 42968
Zonguldak Ereğli Müzesi, Zonguldak Ereğli 42969
Zonguldak State Fine Art Gallery, Zonguldak 42968
De Zonnehof, Centrum voor Moderne Kunst, Amersfoort 31818
Zoölogisch Museum Amsterdam, Amsterdam 31904
Zoolgical Museum N.M. Kulagin, Moskva 37430
Zoological Museum, Addis Ababa 10682
Zoological Museum, Biškek 29960
Zoological Museum, Bukittinggi 24464
Zoological Museum, Dublin 24825
Zoological Museum, Faizabad 24174
Zoological Museum, Gorakhpur 24181
Zoological Museum, Hamburg 18965
Zoological Museum, Hardwar 24205
Zoological Museum, Helsinki 10774
Zoological Museum, Jaunpur 24240
Zoological Museum, Lahore 34029
Zoological Museum, Meerut 24310
Zoological Museum, Pematangsiantar 24546
Zoological Museum, Tehrān 24698
Zoological Museum, Tel Aviv 25124
Zoological Museum, Zografou 23250
Zoological Museum at Agricultural College, Karaj 24615
Zoological Museum of the Babeş-Bolyai University, Cluj-Napoca 36292
Zoological Museum of the Far East State University, Vladivostok 37953
Zoological Museum of the Jagiellonian University, Kraków 34814
Zoological Museum of the Kasan University, Kazan 37029
Zoological Museum of the M.V. Lomonosov Moscow State University, Moskva 37431
Zoological Museum of the State Pedagogical University of Samara, Samara 37623
Zoological Museum of the Syktyvkar State University, Syktyvkar 37822
Zoological Museum of the Tomsk State University, Tomsk 37860
Zoological Research Museum Alexander Koenig, Bonn 17469
Zoologičeskij Muzej, Biškek 29960
Zoologičeskij Muzej, Kazan 37029
Zoologičeskij Muzej, Samara 37623
Zoologičeskij Muzej, Syktyvkar 37822
Zoologičeskij Muzej, Vladivostok 37953
Zoologičeskij Muzej im. N.M.Kulagina, Moskva 37430
Zoologičeskij Muzej Kafedry Zoologii Pozvonočnych Sankt-Peterburgskogo Gosudarstvennogo Universiteta, Sankt-Peterburg 37736
Zoologičeskij Muzej Moskovskogo Gosudarstvennogo Universiteta im. M.V. Lomonosova, Moskva 37431
Zoologičeskij Muzej Rossijskogo Gosudarstvennogo Pedagogičeskogo Universiteta im. A.I. Gercena, Sankt-Peterburg 37737
Zoologičeskij Muzej Rossijskoj Akademii Nauk, Sankt-Peterburg 37738
Zoologičeskij Muzej Tomskogo Gosudarstvennogo Universiteta, Tomsk 37860
Zoologische Sammlung der Universität Rostock, Rostock 21396
Zoologische Sammlung des Zentralmagazins Naturwissenschaftlicher Sammlungen, Halle, Saale 18907
Zoologische Schausammlung, Tübingen 22128
Zoologisches Forschungsinstitut und Museum Alexander Koenig, Bonn 17469
Zoologisches Museum der Universität Göttingen, Göttingen 18702
Zoologisches Museum der Universität Hamburg, Hamburg 18965
Zoologisches Museum der Universität Heidelberg, Heidelberg 19088
Zoologisches Museum der Universität Zürich, Zürich 42118
Zoologisches Museum zu Kiel, Kiel 19536
Zoologisches und Tiermedizinisches Museum der Universität Hohenheim, Stuttgart 21971
Zoologisk Museum, Grindsted 10045
Zoologisk Museum, København 10158
Zoologiska Museet, Lund 40709
Zoologiska Utställingar, Uppsala 41018
Zoology and Botany Museum, Ernakulam 24169
Zoology Museum, Annamalai Nagar 24068
Zoology Museum, Ibadan 33324
Zoology Museum, Kanpur 24254
Zoology Museum, Kingston 28336
Zoology Museum, Legon 22862
Zoology Museum, Lucknow 24302
Zoology Museum, Muzaffarnagar 24327
Zoology Museum, Saint Lucia 01486
Zoology Museum of National Taiwan University, Taipei 42442
Zoology Section, Museum of Evolution, Uppsala 41018

Zoolohičnyj Muzej imeni Benedikta Dybows'koho, Lviv 43143
Zoom Children's Museum, Wien 03093
Zoom Kindermuseum, Wien 03093
The Zora Neale Hurston Museum, Eatonville 48274
Zornmuseet, Mora 40748
Het Zotte Kunstkabinet, Mechelen 03865
Zoutmuseum Delden, Delden 32055
Zrinyi Miklos Castle Museum, Szigetvár 23892
Zrinyi Miklós Vármúzeum, Szigetvár 23892
Zsigmond Kun Folk Art Collection, Budapest 23430
Zsigmond Móricz Memorial House, Prügy 23785
Zsigmond Móricz Memorial House, Tiszacsécse 23929
Zsigmond Remenyik Memorial Room, Dormánd 23528
Zsolnay Kerámia Kiállítás, Pécs 23781
Zubrnická Museální Železnice, Zubrnice 09950
Zucchi Collection, Milano 26726
Zucker-Museum, Berlin 17284
Zuckmantler Heimatstube, Bietigheim-Bissingen 17351
Zündapp-Museum der Brauerei Zoller-Hof, Sigmaringen 21766
Zürcher Abwassermuseum Kläranlage Werdhölzli, Zürich 42119
Zürcher Spielzeugmuseum, Zürich 42120
Zuid-Limburgse Stoomtreinmaatschappij, Simpelveld 32769
Zuiderkerk, Amsterdam 31905
Zuiderzeemuseum, Enkhuizen 32208
Zuidwestbrabants Museum, Halle 03670
Zuivelmuseum, Blankenberge 03372
Zululand Historical Museum, Eshowe 38618
Zululand Mission Museum Chapel, Eshowe 38619
Zum Puppenhaus, Immenstaad am Bodensee 19327
Zunfthaus zur Meisen, Zürich 42121
Zunftstube, Gernsbach 18609
Zunftstube im Torli, Endingen am Kaiserstuhl 18169
Županijski Muzej Šibenik, Šibenik 08949
Zurab Cereteli Art Gallery at the Russian Academy of Arts, Moskva 37174
Zurich Succulent plant collection, Zürich 42109
Zväz múzeí na Slovensku, Banská Bystrica 55011
Zvejniecības Brīvdabas Muzejs, Jūrmala 30027
Zvejniecības un Kugniecības Vestures Muzejs, Liepāja 30041
Žvejo Etnografinė Sodyba, Nida 30323
Zvenigorodskij Istoriko-architekturnyj i Chudožestvennyj Muzej, Zvenigorod 37991
Zverevskij Centr Sovremennogo Iskusstva, Moskva 37432
Zwack Gyümölcspálinka -Fozde és Kiállítás, Kecskemét 23657
Zwack Pálinka Distillery and Exhibition, Kecskemét 23657
Zwack Unicum Múzeum, Budapest 23495
Zwanenbroedershuis, 's-Hertogenbosch 32380
Zweibrücken City Museum, Zweibrücken 22830
Zweigmuseum des Staatlichen Museum für Völkerkunde München, Oettingen in Bayern 20880
Zweirad- und Technikmuseum, Werder 22488
Zweiradmuseum, Berlin 17285
Zweiradmuseum, Pleidelsheim 21057
Zwierzyniec Art Salon, Kraków 34822
Zwierzyniecki Salon Artystyczny, Kraków 34822
Zwijvekemuseum, Dendermonde 03543
Zwingligeburtshaus, Wildhaus 42022
Žytomyr Oblasnyi Literaturnyj Muzej, Žytomyr 43215
Žytomyrs'kyj Literaturno-memorial'nyj Muzej V.G. Korolenka, Žytomyr 43216

Z

Museum Staff

Directors, curators, presidents and academic staff of museums

A

Aalund, Dorothy 50777
Aamodt, David 49479
Aaraas, Olav 33740, 33741
Aarhus, Aasne 33436
Aarsen, John W. 48598
Aarsrud, Christian 41025
Aarstrand, Kjetil 33646
Aarts, H.J. 33029
Aartsen, Drs. Judith 32083
Aas, Scientist Christian K. 33737
Aasberg, Dagny 33511
Aaserud, Anne 33918
Aashagen, Kjersti Haugen 33783
Aasland, Odd 05657
Aasted Schanz, Elsebeth 09963
Aav, Marianne 10771
Abad, Dr. Andres 10392
Abad, Dr. Vicent 39076
Abadesa, R.M. 39066
Abadese, M. 39634
Abadie, Myriam 14335
Abadom, Monica 33329
Abadzic, Sabaheta 04284
Abadzieva, Sonia 30520
Abakumov, Vladimir 37217
Abal de Russo, Prof. Clara 00453
Abaño, Isidro C. 34458
Abaraonye, Dr. Felicia Ihuoma 33318
Abate, Richard d' 52065, 52070
Abazid, RNDr Daniel 09842
Abba, Sylvie 12318
Abbas, Muhammad 34027
Abbassi, Q. 24605
Abbate, Dr. Vincenzo 27065
Abbott, Christine 33123
Abbott, Diane 46950
Abbott, Philip 44665
Abbott, Phyliss 46649
Abbott, Sonia 00894
Abbott, Will 49311
Abcouwer, Dr. N.F. 32495
Abderhalden, Michele 41582
Abdo Mahgoub, Prof. Dr. Mohamed 10476
Abdolalipour, Mohammad 24655, 24658, 24662, 24664, 24666, 24669, 24673, 24678, 24681, 24694
Abdoolrahaman, S. 30677, 30680, 30681, 30684, 30685, 30686
Abdulla, K. 42522
Abdullah, Mohamed Yusuf 30589
Abdullin, Rif Mudarisovič 37883
Abdullina, Svetlana Aleksandrovna 37982
Abdunov, Georgi 05102
Abe, Hiroshi 28587
Abegg, Dr. Angelika 21565
Abejide, Ola 33343
Abel, Marianne 17747
Abel, Dr. Susanne 20399
Abel, Tiina 10624
Abelanet, Jean 15868
Abele, Susan 51341
Abeler, Jürgen 22749
Abell, Carl T. 50556
Abell, Sherry S. 50556
Abella, Carmencita T. 34456
Abella, Jordi 39290
Abels, Peggy 49716
Abenov, Kalkabaj Altaevič 29708
Abensour, Dominique 14966
Åberg, Anders 40766
Abgottspon, Elisabeth 41519
Abgrall 11840
Abildgaard, Hanne 10123
Abita, Salvatore 25990
Abitz, Gerald 49755
Ablinger, Andreas 02043
Abner, Steve 47597
Abraham, Debbie 00877
Abraham, Julian 43412
Ábrahám, Levente 23635
Abraham-Silver, Linda 47631
Abrahams, Katy 38624
Abrahams, Ruth K. 50494
Abrahams-Willis, Gabeba 38591
Abrahamson, Dana 48888
Abram, Ruth J. 51206
Abramek, Bogusław 35301
Abramović-Miletić, Ljiljana 38074
Abrams, Linda C. 50192
Abramsson, Görel 40634
Abrão, Flávio Jorge 04389
Abravanel, Claude 25006
Abreu, Jorge M. de 35969
Abreu, Juana Ines 31088
Abreu Nunes, Dr. Maria Luisa 35690
Abric, Christian 12477
Abril, Joana Mari 40291
Abrojan, Ėlvira Vartanovna 36908
Abruzzi, Sarah K. 51662
Absi, Nuha 29658
Absmeier, Christine 21192
Abston, Fleet 50579
Abu Hadba, Abdul Aziz H. 34046
Abubakar, Ali Mohamed 29746
Abueg, Dr. Nelle F. 34482
Abylkasymov, Mengiš Maulenovič 29723
Acalija, Sanja 08887
Acampora, Nick 52036
Accardi, Dr. Giuseppina 25757
Acevedo Basurto, Sara 34266
Aceves Romero, David 31057
Achaah Ndumbi, Patrick 05295
Achatz, Erika 01946
Achenbach, Dr. Nora von 18951
Acher, Jean-Pierre 15119
Achi, Y. 25023
Achleitner, Gunda 03029
Achuri, Julio César 08661, 08662
Acidini, Dr. Cristina 26177
Ackerman, Andrew S. 51148
Ackermann, Cornelia 41532
Ackermann, Felix 42001
Ackermann, Dr. Marion 17997
Ackermann, Paul R. 54078
Ackerson, Rex D. 53530
Ackley, Clifford S. 46940
Ackroyd, Paul 44937
Ackroyd, Sally 44466
Acorn, Eleanor 06662
Acosta Guerrero, Elena 39799
Acosta Mohalem, José de Jesús 08704
Acquavina, Dr. Rosario 25566
Acquisti, Jon 52785
Actman, Phyllis 51837
Acton, David 54302
Adachi, Hogara 28688
Adachi, Takanori 29588
Ådahl, Dr. Karin 40913
Adair, Mary J. 49990
Adam, Patricia 03514
Adam, Thomas 17598
Adamczyk-Pogorzelska, Bożena 34894
Adami, Valerio 26649
Adamo, Giuseppe 27278
Adams, Anne 49394
Adams, Dr. Beverly 51906
Adams, Celeste M. 49003
Adams, Cheryl 47982
Adams, Chuck 53584
Adams, Cindy 49874
Adams, Connie 49535
Adams, David 44273
Adams, Diana 01316
Adams, Eloise 50936
Adams, F. 05584
Adams, G. Rollie 52400
Adams, Jay 46486
Adams, Jeremy 51157
Adams, Kathy 52017
Adams, Kenneth 06569
Adams, Dr. Kenneth 48615
Adams, Larry 46906
Adams, Lorraine Margaret 54671
Adams, Mazie M. 49908
Adams, Mel 52015
Adams, Nancy 46938
Adams, Patricia 52557
Adams, Patsy 51074
Adams, Roxana 50414
Adams, Stephen 43294
Adams, Virgil 50370
Adams, Will 44757
Adams Muranaka, Dr. Therese 52755
Adamsom, Noel 01429
Adamson, Alberta 54122
Adamson, J. 44237
Adamson, John 53334
Adamson, Tony 43286
Adamyan, M.S. 00755
Adán García, Angel Manuel 40213
Adar, Yossi 24923
Addenbrooke, Rosie 43902
Adderley, D.E. 07242
Addison, David 44166
Addison, K.A. 22847
Addison, Laura 52880
Addoms, Andrew 49297
Adelantar, Jonas O. 34506
Adelsberger, Paul 18194
Aderman, Ella 52965
Adesina, Bode 33325
Adey, Elizabeth 45046
Adey, Dr. Elizabeth 45048
Adiletta, Dawn C. 49194
Adisumarto, Dr. Soenartono 24489
Adkin, Tammy 06187
Adlbauer, Dr. Karl 01981
Adleman, Debra 50831
Adler, Frances Beatty 51167
Adler, Peter 01339
Adler, Richard 41212
Adler, Tracy L. 51193
Adlı, Nadir 42767
Adolfsson, Stig 40560
Adolphs, Dr. Volker 17460
Adorno, Prof. Francesco 26191
Adriaens Pannier, Anne 03463
Adriaensens, Annemie 03321
Adriaenssens, Werner 03495
Adrian, Dolab Jorge 00178
Adriani, Prof. Dr. Götz 22118
Adriansen, Inge 10279
Adriázola, Teresa 04223
Adrimi-Sismani, Vassiliki 23237
Adsett, Virginia 43810
Adunka, Roland 01715
Adžemlerski, Svetoslav 05140
Aeberli, Matthias 41563
Aebersold, Marianne 42069
Aebi, Richard E. 41862
Aerts, Dirk 03282
Aeschlimann, Trudi 41277
Aescht, Dr. Erna 02310
Affolter, Ruth 42088
Afonso, Elsa 35399
Afuhs, Eva 42081, 42092, 42094, 42095
Afzal, Dr. Muhammad 34010
Agar, Will 50728
Agarwal, R.C. 24233
Ageeva, Marina Vladimirovna 37472
Ageeva, Zinaida Michajlovna 37288
Agejkina, Marija Vladimirovna 29713
Agenjo, Xavier 40008
Agliati Ruggia, Dr. Mariangela 41741
Agnelli, Marella 28031
Agnelli Caracciolo, Marella 28206
Agner, Raul Dumas 34509
Agnesi, Prof. V. 27077
Agostini, Giuseppe 27133
Agostini, Dr. Grazia 26121
Agostini, Dr. Silvano 27063
Ágota, Kasper 23979
Ågotnes, Dr. Anne 33425, 33427
Agrech, F. 15654
Agrenius, Helen 40614
Agresti, Giannicola 25234
Agten, Beate 20736
Agudo Cadarso, Christina 40229
Agudo García, P. Javier 39515
Ague, Jay 51054
Aguero Moras, Alfonso 34135
Aguiar Branco, Dr. Fernando 35855
Aguilar, Dr. Hugo 08817
Aguilar, Dr. Pedro 35954
Aguilar Piedra, Raúl 08808
Aguilera, Kristin 51222
Aguiló, Anna 38837
Aguiló Ribas, Catalina 39778
Aguiló Victory, Magdalena 39783
Aguinaldo, Rosalinda 34398
Aguirre, Gustavo 08603
Aguirre, Joel O. 30754
Agúndez García, José Antonio 39609
Agusti, Jorge 39925
Agustín, Sandy 50734
Ahern, Maureen 49720
Ahlberg, Leif 40659
Ahlbrecht, Helmut 21728
Ahlefeldt Laurvig Bille, Count Michael 10170
Ahlers, Dr. Jens 19532
Ahlers, Silvia 04565
Ahlfort, Barbro 40547
Ahlqvist, Ingemar 40818
Ahlsten, Eva 41014
Ahlstrand, J.T. 40708
Ahmad, Mushtaq 34020
Ahmad, Dr. Razi 24345
Ahmady, B. 24692
Ahmaogak, George N. 46284
Ahmas, Kristina 10974
Ahmedzade, T.Z. 03147
Ahola, Tuija 11457
Ahonon, Leonard 04182
Ahrens, Carsten 17566
Ahrens, Claus 20900
Ahrens, Cornelia 17600
Ahrens, Regi 49688
Ahrndt, Prof. Dr. Wiebke 17564
Ahronson, Meir 25085
Åhsberger, Douglas 40513
Ahtela, Eero 11077
Ahtola-Moorhouse, Leena 10769
Ahumada Valencia, Jacob 07472
Aida, Susumu 29097
Aidara, Abdoul Hadir 38027
Aide, Inese 30143
Aigneaux, Jean d' 15574
Aigner, Carl 02685
Aigner, Josef 02039
Aigner, Sonja 02779
Aigner, Dr. Thomas 03087
Aigner, Uli 20457
Aigremont, Robert d' 13556
Aihara, Chris 50225
Aiken, Dr. Edward A. 53399
Aikens, C. Melvin 48435
Aikey, Michael 52923
Aillagon, Jean-Jacques 28206
Ailonen, Riitta 10770, 10816, 10824, 10834, 10963, 11298
Aimé About, Assamoa 08832
Ainhirn, Karola 01739
Airey, Laura 46192
Airlie, Louise 45411
Aistars, John 47286
Aitchison, Julie 01075
Aitken, Catherine 41137
Aitken, Terry 54311
Aizpurua, Inaki 38796
Ajmetov, Ramis Kijamovič 37024
Ajzenberg, Prof. Dr. Elza 04879
Akam, Prof. Michael 43735
Akamati-Lilimbaki, Dr. Maria 23141
Akanbiemu, Martin Ogunayo 33330
Akasaka, Norio 28350
Akasaki, Yoko 29493
Akasheh, Anahid 51221
Akazijeva, Tetjana Abaštajevna 43023
Akbarnia, Ladan 47069
Akbulut, Nazım 42902
Akers, Charlene 53139
Akers, Lloyd 52012
Akersten, Dr. William A. 52006
Åkesson, Bernt-A. 40840
Akgül, Nebi 42802
Akgün, Ünal 42871
Akimbek, Azat 29676
Akın, Gülşen 42780
Akin, Joseph 52539
Akinade, Dr. Olalekan 33332
Akinpelu, David Ano 33330
Akins, Ken 47967
Akins, Rebecca 50604
Akira, Nagoya 29394
Akira, Tatehata 29144
Akishige, Kunikazu 29297
Akiyama, Yoshi 50426
Akkaya, Mustafa 42805, 42810
Akker, C. 32767
Akmamedova, Maja 42982
Akomolafe, Tunde 33339
Akpan, Aniefiok Udo 33338
Akramov, N. 54464
Akre, Lorne 06942
Aksajskij, Ivan Grigorjevič 37842
Aksënova, Alisa Ivanovna 37934
Aksoy, Sule 42845
Aksu, Selahattin Eyüp 42717
Al-Arifi, Dr. Nassir S. 38015
Al-Bubu, Sana' 29666
Al-Dahes, Mohammed 29948
Al-Diban, Martina 20551
Al-Esawi, Talaat 10530
Al-Farsi, Said Ali Said 33991
Al-Hameed, Hazim A. 24720
Al-Maamari, Moussa 30163
Al-Makhawi, Shatha R. 43230
Al-Qaisi, Rabie 24715
Al-Robaae, Dr. Khalaf 24718
Al-Saeed, Susan 33988
Al-Saud, Abdullah S. 38017
Al-Shibli, Alae 24713
Al-Sid, Siddig M. Gasm 40392
Al-Wohaibi, Dr. Fahed 29949
Alam, Shamim 34011
Alam, Dr. Shamsul 03225, 03231, 03233, 03235
Alam, Syamsir 24537
Alaminos López, Eduardo 39564
Alarcó, Paloma 39583
Alarcón, Enrique 38974
Alarcón Lotero, José Félix 08717
Alarcón Osorio, Gonzalo 31055
Alasalmi, Jikka 11476
Alaurin, Alice A. 34383
Alausy, Abdalla 29752
Alba, Dominique 14745
Alba Alvarez, Dionisio 40334
Albacete, M.J. 47227
Alban, Marc 14088
Albani, Julia 41180
Albani, Lino 25945
Albany, James 46229
Albarian, Kimberly 49845
Albendea Solis, Juan Manuel 40060
Albendea Solis, Juan Maurel 40129
Alber, Merike 10627
Alberdi, Juan B. 00671
Alberigi, Merry 52828
Alberrán, Fernando 31589
Albers, Birgit 01795

A

Albers, Dr. M. 32332, 32333, 32787
Albers, Mary 06701
Albers, Thomas 18774
Albert, Karen T. 49246
Albert, Peggy 15342
Albert, Shea 38622
Albertelli, Prof. Giancarlo 26311
Alberti, Lluís 39975
Alberti, Dr. Walter 26697
Albertin, Anne-Marie 12952
Albertini, L. 41127
Albertis, Serge de 14445
Alberto, Dr. Giordano 26261
Alberts, Rod 48005
Albertsson, Kristinn J. 23984
Albertus, Dana 19918, 19939
Alberty, Beth 47067
Albinson, Cassandra 51055
Albiol Sampietro, Vicente F. 39686
Albisetti, Guiseppe 41092
Alborino, Dr. V. 18058
Albrecht, Donald 51156
Albrecht, Hansjörg 20650
Albrecht, Heike 16491
Albrecht, Marvin 54116
Albrecht, Roland 17197
Albrecht, Silke 21084
Albrecht-Weinberger, Dr. Karl 03008, 03041
Albrechtsen, Bo 23271
Albright, Liz 51569
Albright, Shirley 53561
Albright, Terrence 51387
Albufera, E. d' 14212
Alcolea Blanch, Santiago 38965
Aldeguer Gordiola, Daniel 38840
Aldema, Yehuda Levy 25018
Alder, Barbara 41564
Alder, M. 44379
Alderdice, Dr. David 01556
Alderfer, Joel D. 49168
Alderton, Michéle 06711
Aldhizer, H.T. 48669
Aldin, Shiraz 00005
Aldini, Tobia 26231
Aldrich, William M. 51443
Aldridge, Carolyn 45796
Aldridge, G.E. 43670
Aldridge, Martha 47305
Aldunate del Solar, Carlos 07451
Alebo, Lena 40841
Alegria, Dr. Ricardo E. 36068
Aleksandrova, Marija 05270
Aleksandrova, Nataša 05144
Aleksandrovič, Natalja Jurjevna 37068
Aleksandrovna, Marija 05269, 05271
Alekseev, Nikolaj Makarovič 37605
Alekseev, Vladimir Petrovič 36824
Alekseeva, Aida Nikolaevna 36963
Alekseeva, Galina Ivanovna 36798
Aleksiuk, Galina A. 37949
Aleksoska-Baceva, Zaharinka 30520
Alemany, Uiso 40299
Alemany, Véronique 13915
Alencar, Ana Maria M. de 04744
Alencar, Vera de 04680, 04688
Alessi, Vincent 00922
Alessio, Alberto 28028
Alexander, Beatrix 19612
Alexander, Darvan 46888
Alexander, G. 46025
Alexander, Helene 44845
Alexander, Irene O. 47715, 47716
Alexander, Jake F. 52682
Alexander, Jane 01217, 45671
Alexander, Jean 53526
Alexander, Kimberly 52657
Alexander, Kristi 47668
Alexander, Michelle 45939
Alexander, Nancy 52284
Alexander, Prof. Richard D. 46310
Alexander, Robert 01020
Alexander, Terry Pink 46755
Alexandre, Claude 16122
Alexandru, I. 36557
Alexis, Gerald 23321
Alexis Grace, Trudie 47681
Alfageme Sánchez, Vitaliano 40332
Alfaro, Ana 40204
Alfaro Castillo, Andrés José 08571
Alfen, Peter van 51116
Alfoldy, Andy 05781
Alfoldy, Elaine 05781
Alfred, Darrin 48072
Alfter, Dr. Dieter 16804
Algeri, Dr. Guiliana 26294
Alguacil Ruiz, Pedro 38799
Alhaainen, Kari M. 10927
Alho, Anna-Liisa 10821
Ali, H.-A. 24714
Ali, Mazhar 34045
Ali, Mohamed 42511
Ali, Samirah Mohd 30587
Ali, Sanaa 10467
Alibaeva, Sara Kuandykovna 29698
Alibhai, Dr. Abdulssatar Mussá 31705
Alicante, Apolinario 34460
Aliev, Kazi Magomet Ibragimovich 36854
Alig, Joyce L. 47310
Alimonti, Keith 06051
Alin, Margareta 40703
Alisch, Joachim 17645
Aliyasak, Munasor 30560
Alkema, J. 31929
Allaire, Kay 51156
Allan, Laurie 46429
Allan, Randy 46789
Allan, Shona 45345
Allanfranchini, Patrice 41248
Allansson, Jón 23980
Allard, Jeanine 52551
Allard, Sarah 43965
Allart, Dr. B. 32857
Allas, Anu 10624
Allason-Jones, Lindsay 45230, 45231
Alleaume 12260
Allegrini, Prof. Piero 26638
Allemand, Eolo 25357
Allemand, Evelyne-Dorothée 15953
Allemand, Jacqueline 15609
Allemann, Remo 41895
Allen, Anne 46350
Allen, Babara 53303
Allen, Barbara 53585
Allen, Brian 54175
Allen, Brian T. 46296
Allen, Bruce 53085
Allen, Catherine 47686
Allen, Christopher 44857
Allen, Dale 48034
Allen, Dick 53692
Allen, Donald T. 50375
Allen, Kay 50222
Allen, Kim 06306
Allen, Louise 44315
Allen, Lucy 49565
Allen, Dr. Marti 49647
Allen, Dr. Marti L. 52189
Allen, Nancy L. 51408
Allen, Richard 47940, 49406
Allen, Susan 52911
Allen, Timothy S. 53574
Allen, Trudy J. 48947
Allen, Virginia S. 46909
Allen, Zack 50859
Allen Fleming, Kathy 52526
Allerman, Cheryl 49581
Allevi, Dr. Gabriele 25403
Allgood, Louise 01586
Allie, Stephen J. 48642
Allikvee, Anu 10624
Allington, Peter 45563
Allione, Patricia 13323
Allison, Brian 01282
Allison, C. 46012
Allison, David 33048
Allison, Glenn 07285
Allison, N. 33146
Allison Rogers, Traude 54676
Allman, William G. 53934
Alloisi, Dr. Sivigliano 27453
Allsop, Julie 45601
Allwood, Rosamond 44695
Alm, Göran 40886
Almarri, Ali 43228
Almeda, Ermelo 34466
Almeev, Robert V. 54444
Almeida, Joel 35688
Almeida, José Maria 07419
Almeida Campos, Diogenes de 04725
Almeida-Matos, Lúcia 35843
Almeras 13721
Almirante, Julián 40204
Almosara, Emelita V. 34449
Almquist, Katarina 40825
Almquist, Kathleen 53650
Alms, Barbara 17817
Alonso, Alejandro G. 09165
Alonso, Alicia 39511
Alonso, Jésus 40323
Alonso Martín, Juan José 39586
Alòs Trepat, Carme 38942
Alowan, Zoe 51762
Alper, Mara 50944
Alperin-Hochman, Marlene 47071
Alpers, Rojann R. 53463
Alphons, Alois 02082
Alpözen, T. O. 42731
Alram, Dr. Michael 03019
Alsam, Ali 44937
Alsobrook, Dr. David E. 47688
Alston, Joseph 48985
Alsvik, Elling 33929
Alswang, Hope 53065
Altamont, Earl of 24911
Alteen, Glenn 07212
Alteneder, Prof. Susanne 03033
Altenkirch, Gunter 18615
Alterio, Italo 08786
Altermatt, Franz 41689
Althaus, Marceline 41684
Altherr, Hans Werner 17602
Althoff, Bruno 22594
Althoff, Ralf 18020
Altmann, Ferdinand 02384
Altmann, Katja 21158
Altorfer, Dr. Andreas 41214
Altringer, Lothar 17462
Altripp, Prof. Dr. Dr. Michael 18751
Alvarado, Janet 17186
Alvarado-Dupuy, Dr. André 03019
Alvarado Gonzalo, Manuel de 39082
Álvares Sánchez, Roberto 38922
Alvarez, Aliciz Lidia 00357
Alvarez, Helena 40912
Alvarez, José Carlos 35721
Alvarez, Josefina 00560
Alvarez, Juan S. 48254
Alvarez, Sergio Antonio 00146
Alvárez Arévalo, Miguel 23303
Alvárez Masso, Montserrat 38983
Alvárez Polo, Giesela 09217
Alvarez Rojas, Antonio 39089
Alves, C. Douglass 53142
Alves, Hélio Osvaldo 35637
Alves, Manuel 35802
Alves, Dr. Sónia 35834
Alves Costa, Catarina 35717
Alves de Sousa, Dr. Pio G. 35485
Alves Gomes, José António 04552
Alves Pinto, José F. 35637
Alvina, Corazon S. 34444
Alvina, Corazón S. 34450
Alybekova, Aina 29957
Alymer, John 43585
Alzamarova, Raušan Ergazievna 29704
Amack, Rex 50995
Amadei, Daniel 06551
Amakasu, Ken 29057
Amalla, Joesilo C. 34353
Amamra 00041
Aman, Karen 54314
Amanberdiýev, Annamyrat 42981
Amann, Bernd 02093
Amano, Yukichi 28906
Amaral, Maria Conceicão Alves 35940
Amarelli, Pina 27564
Amaro das Neves, António 35637
Amato, Ann 52088
Amato, Prof. Domenico 26765
Amato, Pietro 54501, 54507, 54513
Amberg, Annette 41165
Ambroise, Guillaume 14757
Ambrose, David 53824
Ambrose, Pamela E. 52599
Ambrosetti, Dr. Giancarlo 27386
Ambrosini, Lynne 47565
Ambrosino, Guy 52872
Ambrosio, Prof. Augusto C. 27293
Amdre, Leopold 02504
Amdreoli, Prof. Annararie 26272
Amelin, Olov 40921
Amell, Alexander 48224
Amemiya, Tadashi 29376
Amend, Brad 48311
Amend, Debbie 48311
Amendolagine, Dr. Beatrice 26641
Amer, Prof. Mordechai 25081
Ames, Gerhard 21456
Ametrano, Dr. Silvia 00413
Amias, Elizabeth 43780
Amic, S. 14248
Amick, Alison 51521
Amick, Henry 48113
Amidon, Catherine 52000
Amidon, Eleanor 49157
Aminova, Svetlana Gašimovna 37117
Amir, H. 30682
Amir, Mohammad 24457
Amiranišvili, S.J. 16379
Amiri, M. 24646
Amlow, Lothar 18048
Ammann, Christopher 41499
Ammann, Prof. Dr. Gert 02177
Ammann, Hugo 41140, 41142
Ammann, Sandra 41137
Amnehäll, Jan 40558
Amo, Dr. Toshino 29366
Amon, Prof. Dr. Karl 01739
Amonson, John 53799
Amor, Kebbour 00073
Amoretti, Guido 28024
Amoriñ, Prof. Dr. José L. 00232
Amoroso, Dr. Domenico 25606
Amourette, Philippe 11773, 11776, 11777
Amouyel, Prof. 13755
Amprimoz, M.F.X. 15923
Ampuero, Gonzalo 07440
Amrein, Dr. Heidi 42105
Amsallem, Christiane 15797
Amselgruber, Prof. Dr. 21971
Amsellem, Patrick 47069
Amsler, Cory 48180
Amstad, Hans 42100
Amstutz, Damian 41600
An, Tran Huy 54622
An, Yingnan 08430
An, Zemin 08466
Anand, Dr. Y.P. 24146
Anati, Prof. Emmanuel 25657
Anaversa, Inginera 35668
Anawalt, Patricia 50223
Ancheotti, Dr. Angelo 26109
Ančiporov, A.I. 37571
Ancuța, Maria 36355
Anda, Raquel de 52774
Anda, Tone 33381
Andami, Parisa 24682
Anday, Anton 53290
Ander, Dr. Angelika 03045
Anderegg, Fred 41618
Andergassen, Dr. Leo 25544
Anderjack, George M. 52858
Anderl, Dick 54304
Anderle, Charlotte 02237
Anderlik, Heidemarie 17072
Anderlik, M.A. Heidemarie 17530
Anderlinger, Mag. Inge 02055
Anders, Ines 18679
Anders, Win 07377
Andersen, Britta 09974
Andersen, Gerd 19217
Andersen, H.C. 10049
Andersen, Håkon 33925
Andersen, Jeffrey W. 51536
Andersen, Jens 10054
Andersen, Jesper Munk 10102
Andersen, John 10031
Andersen, Kjell 33507
Andersen, Lise 10052, 10053
Andersen, Michael 10136
Andersen, Nikki 54298
Andersén, Peter 11134
Andersen, Randi 33647, 33648
Andersen, Thyge 10201
Andersen, Verginia 04611
Anderson, Annette 47649
Anderson, Betsy 48093, 50169
Anderson, Bill 53590
Anderson, Camille 52346
Anderson, Carol 46313
Anderson, Christine 53118
Anderson, Claudia 53471
Anderson, Craig 47498
Anderson, Dale W. 49201
Anderson, David 51528
Anderson, Donald B. 52467
Anderson, Douglas 47033
Anderson, Duane 52881
Anderson, Dr. Duane 52883
Anderson, Elizabeth 05575
Anderson, Ellen 07103
Anderson, Eve 53498
Anderson, Fiona 45202
Anderson, Garry 05658
Anderson, Harold 01129
Anderson, Hilary 50061
Anderson, J. 32891
Anderson, Jaime A. 46419
Anderson, Joe 50000
Anderson, Dr. John P. 54249
Anderson, Judy 50460, 50462
Anderson, Kathi 50093
Anderson, Kris 49905
Anderson, Linda 52250
Anderson, Lynn 53416
Anderson, Michael 49515, 49516
Anderson, Neil 43329
Anderson, Owen 01584
Anderson, Pam 54159
Anderson, Peter 53020
Anderson, Philip J. 47516
Anderson, Raymond L. 53518
Anderson, Richard R. 53137
Anderson, Rolf 53958
Anderson, Scott 52135
Anderson, Steve 52295
Anderson, Steven R. 46363
Anderson, Susan K. 46609
Anderson, Susan M. 46869
Anderson, W.M. 53071
Anderson-Densmore, Mari C. 49832
Andersson, Beritne 10735
Andersson, Karin 40797
Andersson, Lars 40626, 40658

Andersson, Solveig 40564
Andersson Flygare, Irene 41012
Andersson Møller, Vibeke 10165
Andes, Brad 54241
Andjal, Gabriela Vasylivna 43199
Andjelković Dmitrijević, Branislava 38076
Ando, György 23368
Andorfer, Gregory Paul 46578
Andrade, Rosemary 40403
Andral, Jean-Louis 11638, 11641
Andrási, Gábor 23457, 23458
André, Dr. Georges 03532
André, Mervyn 01074
Andre, Simone 43636
André-Salvini, Béatrice 14698
André-Schellner, Prof. Brigitte 03033
Andreadaki-Vlasaki, Maria 22961, 23165
Andreas, Düspohl 16399
Andreasen, Vagn 10183
Andreassen, Dag 33744
Andreassi, Dr. Giuseppe 25362, 27941
Andreasson, Sven-Glov 40455
Andreescu, Laurentiu 36478
Andreeva, Antonina Borisovna 37923
Andrefsky, William 52198
Andrejev, Dr. Aleksandr 37660
Andrell, Pauline 33182
Andren, Martin 40703
Andres, Louis M. 50315
Andresen, Thomas Bjorkan 33844
Andreu, Dr. Jaime Sancho 40203
Andreu Chazarra, Antonio 39697
Andreu-Lanoe, Guillemette 14698
Andrew, Marcia 53762
Andrews, Bernice 05546
Andrews, Gail 46822
Andrews, Janine 05761
Andrews, Dr. Peter 43539
Andrews, Rebecca 00773
Andrews, Richard 53000
Andrews, Thomas F. 50249
Andrews, Toni 53395
Andriamanantena, Ranaivoarivelo 30537
Andrich, Dr. Uwe 18762, 18764
Andrien, Joseph 03894
Andries, Pool 03287
Andrieu, Esther 00493
Andrishak, Steve 06667
Andrist, Jean-François 41242
Andronic, Tasia 36392
Androutsou, Elina 22946
Andrušaite, Dzintra 30083
Andújar Persinal, Carlos 10352
Anelli, Dr. Giuseppe 27831
Anene, Francis 33314
Anfang, Werner 02043
Angama, Henri 08827
Angarita Jiménez, Edgar 08697
Angel Aranaz, Miguel 39332
Angel de Marco, Dr. Miguel 00220
Angelett, Prof. Luciana Rita 27501
Angelis, Gudrun 17454
Angell, Callie 51279
Angelo, Therese 33067
Angeloni, Araldo 27656, 27657
Angelov, Anastas 05012
Angelov, Emil 05239
Angelucci, Sara 07107
Anger, Maksymilian 34854
Anger, Virginia 06633
Angerbauer, Dr. W. 20843
Angerer, Dr. Birgit 20542
Angerer, Ferdl 02726
Angerer, Dr. Martin 21224, 21226, 21231, 21232
Angert, Rina 25056
Anghel, Elena 36434
Anglin Burgard, Timothy 52772, 52781
Anguiano Lozano, Juan Manuel 31120
Angus, Barbara J. 48322
Angus, E. 45857
Anh, Tran Hong 54632
Anheuser, Kilian 41418
Anibas, Martin 03134
Anić, Hanja 08954
Anılır, Burhanettin 42789
Aniol, Roland 21565
Anisi, Ali Reza 24686
Aniz Montes, Mercè 39283
Anken, Thomas 41364
Ankert, Kerstin 40515
Ankudinova, Elena Andreevna 36968
Anlagan, Çetin 42837
Anna, Dr. Susanne 18006
Annear, Judy 01544
Annemans, Jan 03338
Annibaldi, Dr. Cesare 27423, 28206
Annicchiarico, Silvana 26683
Annison, F. 33137
Anosovskaja, Alla Vasiljevna 36970
Anraku, Yoshio 29234
Ansel, Dr. Valerie 15827
Ansell, Andy 43530
Ansell, Erynne 49408
Anselm, Gudrun 40782
Anselm, Paul S. 51487
Ansenk, E.L. 32710
Ansenk, Emily 32797
Ansett, Tony 01103
Ansorge, Gabriele 20089
Anspach, Beate 18942
Anspach, Isabelle 03464
Anthenaise, Claude d' 12379, 14649
Anthoine, Robert 51138
Anthony, Dr. David 51566
Anthony, Ginger 53641
Anthony, James 44456
Anthony, Vincent 46430
Antier, Caroline 15318
Antihi, Adriana 36616, 36617
Antillano Armas, Sergio 54545
Antipas, Archimandrit 23137
Antl, Prof. Gerhard 02780
Antognini, Marco 41582
Antoime, Janeen 52765
Antoine, J.-L. 03899
Antoli Candela, Fernando 40299
Antonakos, Mary 47334
Antonelli, Dr. Angelo 27367
Antonen, Rusty 49888
Antoniadou, Sofia 22950
Antonić, Dr. Dragomir 38035
Antonietti, Thomas 41504, 41874
Antonín, RNDr. Vladimír 09421
Antonino, Dr. Biancastella 25443, 25473
Antoniuk, Aleksander 34549
Antonoglou, Manolis 23172
Antonov, Anton 05022
Antonov, Teodosi 05094
Antonova, Irina A. 37302
Antonova, Ljubov Vladimirovna 37918
Antonsen, Lasse B. 51016
Antonsson, Markus Örn 24029
Antonuccio, Domenico 27435
Antti-Poika, Jaakko 10825
Anttila, Ritva 10880
Antton, Goicoechea 12528
Antunes, José R. 35493, 35494
Antweiler, Dr. Wolfgang 19181
Anwander, Dr. Theresia 01796
Anyanwu, Dr. Starling E.N. 33317
Anzai, Ikuro 28844
Anzola, Sonia 54545
Anzuoni, Robert 48623
Ao, Bingquan 08026
Aoki, Yoshiharu 28923
Aoyagi, Masanori 29419
Apalkov, Jurij Aleksandrovič 36816
Apap Bologna, Daniela 30647
Aparin, Dr. Boris F. 37629
Apel, Gefion 17840
Apel, Dr. Michael 20466
Apelbaum, Laura C. 53887
Apetrei, Maria 36520
Apfel, Prof. Kurt 02956
Aplan, Peg 48655
Aponte, Dr. Nilda E. 36043
Aponte Galvis, Alonso 08772
Apostolou, Eva 22946
Apostolov, Šanko 05245
Apostolova, Anna 04987
Apostolova, Lili 05037
Apovnik, Dorothea 18391
Appel, G. 32540
Appel, H.H. 10156
Appel, Dr. Michaela 20490
Appelboom, Prof. Thiérry 03468
Appelhof, Ruth 48245
Appenzeller, Hans-Ingo 21784
Appignanesi, Paolo 25891
Appleby, Roger 44824
Appleton, Maggie 45048
April, Jeniffer P. 48701
Aprill, Gregg 47237
Aprison, Dr. Barry 47497
Apschnikat, Kenneth 49307
Apsega, Alejo 00511
Apsemetov, Azhibay 29732
Aquilera Cerni, Vicente 40299
Aquiloni, Laura 26858
Aquilué, Dr. Xavier 39278
Aquilué i Abadies, Xavier 39722
Aquin, Stéphane 06357
Aquino, Amelia M. 34422
Arab, Mona 00059
Arai, Toshizo 28609
Arakčeev, Boris Serafimovič 37647, 37719
Aralica, Miodrag 38163
Aramă, Ivona Elena 36417
Aramaki, Teiichi 28838
Arambaşa, Dumitru 36145
Aramova, Lidija Grigorjevna 37987, 37988, 37989
Aranburu, Xabier 39029
Aranburu Lasa, Juan Jose 40355
Aranda-Alvarado, Rocio 49627
Aranega, Prof. Dr. César I. 00319
Arango, Jaime 08692
Arango, Placido 39573
Arango Trujillo, Ignacio 08614
Aranne, Heikki 10904
Arapoyanni, Xeni 22892
Araújo, Emmanoel 04911
Araújo, Dr. José António de 35962
Araujo, Marcelo 04906
Araújo, Maria Helena 35620
Araújo, Ricardo 35618
Araujo, Samuel M. 04687
Aravantinos, Dr. Vassilis 23205
Arbart, Kazimierz 34628
Arbelo de Mazzaro, Prof. Aurora Catalina 00347
Arbogast, Pierre 11590
Årbu, Grete 33717
Arbuckle, John B. 50034
Arbucó, Susana B. 00268
Arbulla, Dr. Deborah 28089
Arbury, Steve 52225
Arcand, Jessica 51938
Arcangel, Norberto S. 34425, 34426
Arcangeli, Lino 25296
Arce Torres, Susana 34189
Arces, Prof. Antonio 26375
Arch, Nigel 44970
Archabal, Nina M. 52623
Archambault, Alan H. 48649
Archambault, B.Sc. Guylaine 06145
Archangelskij, Venjamin 37468
Archer, James Ronald 33629
Archer, Kent 06915
Archibald, Dr. Joanna 43811
Archibald, Dr. Robert R. 52587
Archimanaritis, Alexios 23092
Archino, Sarah 51193
Archontidou-Argyri, Aglaia 22967, 23105, 23109
Arciniegas de Giraldo, Julia 08768
Arco-Zinneberg, Ulrich Graf 02593
Arcuri, Arthur 04522
Arcuşi, Marin 36337
Ardila, Marsha T. 47732
Ardizzon, Dr. Paolo 25876
Arduini, Dr. Franca 26141
Arefyeva, Galyna V. 43190
Arell, Berndt 10792
Arena, Albert A. 53812
Arena, Dr. Maria Stella 27481
Arenas Abello, Rosaema 08578
Arenas i Sampera, Josep A. 38978
Arenas Morales, Lía Esther 08660
Arends, Dr. T. 32473
Arens, A. 18626
Arenz, Lic. Susana 00264
Ares, Felix 39250
Arévalo Pabón, Dr. Emildo 08644
Arevshatyan, Ruben 00737
Arfelli, Guido Arch 27966
Arfmann, Peter 21977
Argun, A.A. 16350
Argungu, Emir of 33315
Arhaldo, Liboni 26921
Arhip, Cristian 36414
Ariano, Terry 53145
Arickx-George, M. 03900
Arih, Prof. Aleš 38499
Aringer, Dr. Klement 20494
Aristizábal Giraldo, Tulio 08620
Arizzoli-Clémentel, Pierre 16126, 16129
Arkadjevna, A.A. 37906
Arkio-Laine, Leena 10776, 10782, 10791, 10794, 10798, 10811, 10814, 10815, 10832, 10833, 10838
Arkley, J. 45537
Arko-Pijevac, Milvana 08940
Arlauskienė, Laimutė 30218
Arman, Kıvanç 42815
Armand-Delille, Lionel 13918
Armater, Raymond J. 52122
Armella Maza, Virginia 31100
Armenta Camacho, Juan 31289
Armin, Karlen 41944
Arminio, Roberta Y. 51621
Armirail, Sandrine 13728
Armitage, Dr. Philip L. 43641
Armour, Erica 48504
Armstrong, Alison C. 44550
Armstrong, Charlotte 51218
Armstrong, Chris 00959
Armstrong, Elizabeth 51332
Armstrong, Fredrick 47380
Armstrong, Joan T. 50490
Armstrong, Richard 51942
Armstrong, Steve 52437
Armstrong-Gillis, Kathy 52720, 52729
Armstrong Touchine, Maxine 47549
Arna, Dr. Angela 22733
Árnadóttir, Jóhanna Guðkr()rún 24024
Arnaldo, Javier 39583
Arnaud, Jean-Yves 13022
Arnauld, Claude 06109
Arnauld, Marie-Paul 14728
Arnault, Jean-Charles 12504
Arnăutu, Nicoleta 36142
Arndt, Elke 22323
Arnemann, Heinrich 16758
Arneodo, Sergio 27692
Arnesson, Ingemar 40903
Arnesson, Kerstin 40512
Arnheim, Avigail 25122
Arnhold, Dr. Hermann 20512
Arning, Bill 47179
Arnold, Beate C. 22692
Arnold, Dr. Charles D. 07411
Arnold, Dorothea 51212
Arnold, Francis 49741
Arnold, Gotthard 21783
Arnold, J. 45203
Arnold, Karl Heinz 18384
Arnold, Lukas 41871
Arnold, Mary Bea 48969
Arnold, Paul 41270
Arnold, Peter S. 48230
Arnold, Dr. Sven 17128
Arnold, Ute U. 22794
Arnold-Becker, Dr. Alice 18469
Arnoldi, Clelia 28004
Arnolli, G. 32473
Arnon, Carmella 25036
Arnot, Sally 48440
Arnott, Alastair 45707, 45708
Arnould, Marie-Agnès 12476
Arnoult, Hubert 12299
Arnoux, François 12398
Arnst, Dr. Caris-Beatrice 17028
Arobba, Daniele 26138
Arocha Rovira, José A. 09326
Arodaky, Badr-Eddine 14673
Aron, Kathy 49699
Arons, Dr. Kārlis-Ēriks 30101
Aronson, Julie 47555
Aronsson, Annika 40586
Arpin, Odette 12185
Arpin, Pierre 07241
Arpin, Roland 06687
Arraras, Maria Teresa 36073
Arrasmith, Anne 46826
Arrate-Hernandez, Danilo M. 09278
Arrecis, April 48038
Arredondo, Joe R. 50311
Arriaga, Ana 40358
Arriaga, Brenda 47403
Arrieta, Katrin 21394
Arriola, Juan Carlos Juárez 39465
Arriola, Julia 53596
Arroyo, Marta 40201
Arroyo Martín, Dr. José Víctor 39046
Arsak, Mychajlo Mychajlovyč 43071
Arsenault, Linda 06640
Arseneau, Steve 48173
Arseneault, Gérald 03695, 03709
Arslan, Dr. Ermanno A. 26677, 26703
Art, Sue Backhouse 01159
Arteaga, Agustín 00164, 31106
Arteaga, Agustin 36049
Artëmov, Evgenij 37648
Arterburn, Lesa 46663
Artero-Pamplona, Rafael 39734
Arthur, John E. 49452
Arthur, Lloyd 05911
Articus, Dr. Rüdiger 18936
Artman jr, Paul C. 49073
Artmann, Dr. 02235
Artykov, B.R. 54454
Aru, Krista 10645
Arunga Sumba, Owen 29742
Aruz, Joan 51212
Arvi, Barbara 50265
Arvidsson, Elisabeth 40669
Arvola, Silja 33593
Arx, Christine von 41555
Ary, Max L. 51517
Arzadeh, Kamal 24691
Arzur, Anna-Vari 14862
Asakura, Tetsuo 28759
Ašanin, Snezana 38094
Asbury, Michael J. 45380
Ascencio, Michaelle 54543
Aščeulova, Tatjana Stanislavovna 37480
Asch, Georges 14869
Aschauer, Manfred 02292
Aschengreen Piacenti, Dr. Kirsten 26198
Aschinger 17232
Aschkenes, Anna M. 51932
Asensi Catalán, Maria Dolores 39587
Asercion, Allisen 52310
Ash, Jeanelle 52008

Ash, Nancy 51867
Ash Ezersky, Lorinda 51169
Ashdjian, Peter H. 09369
Ashe, James S. 49991
Ashe, Janet 47624
Asher, Dave 33195
Asher, Dorothy J. 48376
Asher, Toni 49836
Ashfaq, Prof. Dr. Muhammad 34001
Ashgar, Amanollah Mohammad 00008
Ashley, Brian 45298
Ashley, Catherine 48183
Ashley, Raymond E. 52750
Ashooh Lazos, Marylou 50423
Ashraf, Dr. Mohammad 34043
Ashton, Alan 45060, 45062
Ashton, Joanne G. 51018
Ashton, Kate 01278
Ashton, Lynn 50287
Ashton, Lynne 45591
Ashton, Rand 54058
Ashton, Richard 44075
Ashton, Dr. S. 43727
Ashton, Vicki 52031
Ashworth, Jim 01161
Ashworth, Meg 45601
Ašimbaeva, N. 37656
Askali, Nurhusih 34392
Aslamatzidou, Z. 23062, 23115
Aslani, Q. 24627
Asmaramah, Rasanubari 30572
Asmus, Cathy 53284
Asmuss, Dr. Burkhard 17072
Asngari, Ir. Moch. 24490
Aso, Keiko 29522
Asolati, Dr. Michele 27036
Asomaddin, Abdugayyumi 42499
Asparuchov, Milko 05119
Asparužov, Dr. Milko 05122
Aspe, Virginia Armella de 31159
Aspenwall, Jim 50886
Aspes, Dr. Alessandra 28232
Asplund, Lars 40647
Asprov, Ognian 05079, 05111
Ass, Karl-J. 21126
Assaàd, Khaled 42137
Assadulina, Maŷa 42978
Assaf, Amnon 25050
Assamoa Aime, About 08835
Assamoi, Aimé About 08833
Assatouroff, Corinne ter 03483
Assel, Dr. Marina von 16958
Assémat, Jean 15602
Assen, Simone 41212
Aßkamp, Dr. Rudolf 18912
Assmann, Dr. Peter 02310, 02317
Assmann, Mag. Dr. Peter 02322
Assmann, Dr. Peter 02324
Assoghba, Romain-Philippe 04189
Assouline, Hadassah 24983
Assunção, Dr. Ana Paula 35738, 35893
Assunto, Richard 49751
Ast, Hiltraud 02027
Astaa, Mikko 10778
Astachova, Irina Valerjevna 37946
Aste, Richard 47069
Aster, Robert 51019
Astika, Tjokorda Bagus 24580
Astini, Piero 26546
Astley Cooper, Patricia 45007
Astore, Silvio 27265, 27266
Astorino, Jean 46464
Astramskas, Arunas 30332
Astro, C. 14438
Åström, Kjell 40868
Åströmad, Kenneth 40504
Astrua, Dr. Paola 28002
Astruc, Franck 16283
Astruc, Jean-Yves 14913
Astrup, Yngve 33715
Astrup Bull, Knut 33925
Astui, Aingeru 39030
Asunción Mateo, María 39878
Asyrov, Ýagsymyrad 42977
Asztalos, István 23338
Atac, Robert 53356
Atanasova Tsoneva, Neli 05104
Atanga, Peter S. 05293
Atasever, Yaşar 42895
Ateş, Sevim 42747
Athanasiadis, Tasos 22948
Athanasiadis, Yanni 00996
Athani, Shankar S. 24120
Athayde, Sylvia 04794
Athens, Joe 47549
Atherton, Lois 51998
Atik, Dr. Şeniz 42823
Atiti, Abel 29743
Atiyeh, Walid 44897
Atkins, David 46003
Atkinson, Elmer F. 51840
Atkinson, J. 01557
Atkinson, Lesley 45404
Atlan, Eva 18379
Atsuo, Shioiri 29407
Attack, Paul H. 06289
Attale, Véronique 16093
Attali, Zohara 24928
Atter, Kerri 48082
Attewell, Alex 44849
Attila, Dimény 36694
Attum, Elaine 48281
Attuquayefio, Dr. Dan K. 22862
Attwood, Maryon 54300
Atvur, Orhan 42939
Atwater, William F. 46118
Atwell, Richard 43715
Atwood, Leland 47684
Atwood, Valdine C. 50349
Atze, Dr. Marcel 03087
Aubagnac, Gilles 12788
Aubert, Ben 48308
Aubert, Fabrice 14547
Aubert, Glenda 05440
Aubert, Laurent 41421
Aubin, Maride 14057
Aubrecht, Dr. Gerhard 02310, 02310, 02322
Aubrun, Max 12488
Aubry, Françoise 03488
Aucella, Frank J. 53935
Auchterlonie, Fiona 44110
Auclair, Mathias 14574
Auderset, Patrick 41423
Audinet, Gérard 14646
Audisio, Aldo 28023
Audley, Paul 47017
Auel, Lisa B. 48814
Auer, Dr. Alfred 02130, 03019
Auer, Dr. Anita 22235, 22236
Auer, Helmut 01695
Auer, Rolf 22547
Auer, Wilfried 21578
Auerbach, Dr. Konrad 21729, 21730
Auffant, Luisa 10342
Auffret, M.J. 13578
Aufgebauer, Peter 17490
Aufrecht, Evelyn 21657
Aufseß, Eckhart Baron von 16610
Augé, Jean-Louis 12318, 12320
Auger, Michele 50308
Augier, Jean-Pierre 13729
Augustat, Dr. Claudia 03038
Augusti, Dr. Adriana 28182
Augustin, Josef 01761
Augustin, Stephan 19163
Aujard-Catot, Eliane 11794
Aulagner, Renaud 14895
Aulan, Comte d' 11743
Aulich, C. 22498
Ault, Lee 43646
Aumasson, Pascal 12776
Aumüller, Prof. Dr. Gerhard 20189, 20193
Aunela, Jorma A. 11229
Aupetitallot, Yves 13130
Auping, Michael 48717
Auquier, Jean 03442
Aurandt, David 06807
Aurél, Szakál 23671
Aurell, Pirkko 11463
Auserve, Philippe 14336, 14337
Ausfeld, Margaret Lynne 50815
Ausiàs, Vera 12896
Ausich, Prof. Bill 47758
Auska, Adolf 01954
Ausman, Mary Beth 47393
Austad, Dan 48910, 53345
Austermayer, H. Josef 19159, 19160
Austigard, Bjørn 33678
Austin, Fran 48879
Austin, Pat 49820
Auth, Dr. Susan 51287
Autio, Tuula 41004, 41012
Autry, Brick 51918
Autzen, Henning 21960
Auvergnat, Dr. Arielle 15936
Auwärter, Konrad 19785
Auwärter-Brodbeck, Susanne 21935
Auwera, Joost Vande 03462
Auxerre, Alain 12785
Avalos Valenzuela, Ana 07471
Avant, Joanne 52712
Avdeeva, M.V. 37481
Avdeeva, Taisija Dmitrievna 37803
Avendaño, Clarissa L. 34458
Averjanova, Ėlvira Valentinovna 37050, 37559
Averoff-Ioannou, Tatiana 22902, 23094
Avery, Dr. D.M. 38592
Avery, Dr. G.A. 38592
Avery, Joseph T. 51183
Avery, Kim 05922
Avery, Laura 52916
Avery, Louise 06105
Avery, Susan 54280
Avesani, Ricciarda 06913, 26049
Avest, H.P. ter 32316
Avić, Mirsad 04284
Avila Bardaji, José R. 39531
Avila de Monroy, María 08672
Avila Martínez, Jairo Hernando 08674
Avramenko, Mychajlo Pavlovyč 43181
Avrandt, David 06524
Avril, Ellen 49548
Awald, John C. 47060
Awwal, Dr. Iftikhar ul 03217, 03220
Ax, Goeran 40828
Ax, Michael 10262
Axel, Hungerbeuhler 53603
Axsom, Richard H. 49003
Ayalon, Dr. E. 25114, 25115
Aydın, M. Sezai 42733
Ayer, Jacques 41729
Ayerbe, Santiago 08726
Ayers, Betty 05570
Ayers, Marilyn L. 46729
Ayerve, José M. 10379
Ayesha, Abdulla 43218
Aykaş, Satı 42742
Ayling, Emma 46020
Aymon, Marie-F. 41607
Ayob, Dato Muslim 30586
Ayrault, Jacqueline 11583, 11584
Ayres, William 53313
Ayscue, Gene 54199
Aytar, Sefer 42963
Aytiyev, Satar Gaparovich 29958
Ayvaz, Melahat 42950
Ayvilera, Fernando Gómez 40108
Azarcon, Romualdo A. 34469
Azarmehr, G. 24612
Azcue, Leticia 39573
Azeredo, Paula 35860
Azevedo, José Henrique 35644
Azevedo Abreu, Arienei 04651
Azevedo de Silva, Dr. Nearch 04720
Azimi, Esmail 24684
Ažimova, Irina 37948
Aziz, Kapt Nazri 30586
Azizbekova, P.A. 03144
Aznar-Alfonso, Dr. Lydia 34376
Azón i Masoliver, Marisa 38994
Azuar, R. 38843
Azuma, Hisao 29593

B

Baad Pedersen, Margit 10055
Baagøe, Jette 10100
Baalbergen, W. 32603
Baare, Dr. Heike 18023
Baartman, H. 38626
Baartvedt, Randi 33944
Baatz, Henning 16483
Bab, Bettina 17454
Babajceva, Nadežda Borisovna 37833
Babanazarova, Marinika 54453
Babans'ka, Natalija 43083
Babătă, Ioan 36723
Babboni, Dr. Marina 25689
Babcock, Catherine 49942
Babcock, Elizabeth 52346
Babcock, Sidney H. 51216
Babel, Jerzy 34834
Babelon, Jean-Pierre 12955
Babenko, Valerij Volodymyrovyč 43056
Baber, Charlotte 45706
Babias, Marius 17226
Babič, Prof. Franka 08869
Babić, Mirko 04260
Babin, Anikca 08887
Babini, Tino 27579
Babiński, Leszek 34630
Bacallado Aránega, Juan José 39995
Bacalles, Lorraine 51132
Baccetti, Dr. Nicola 27027
Baccheschi, Dr. Edi 26303
Bacchetta, Véronique 41407
Baccini, Dr. Stefano 26793
Baccrabere, G. 15933
Bačeva, Irina 05178
Bach, Beverly 51656
Bach, Dana 20918
Bach, Kathy 53525
Bach, Dr. Thomas 19377
Bach-Nielsen, Hanne 10273
Bach Østergaard, Bjarne 10152
Bachčevanov, Ivan 05207
Bachelder, Laura 52418
Bacher, Fabian 16507
Bacher, Wilhelmine 01884
Bachhiesl, Mag. Dr. Christian 01972
Bachl-Hofmann, Mag. Christina 02952
Bachman, Dona R. 47241
Bachmann, Birgit 02751
Bachmann, Prof. Lutz 33737
Bachmann-Tonelli, Karen 07084
Bachmisov, Ėduard Konstantinovič 36837, 36838
Bachner, Franziska 01988, 01989, 17719
Bachorz, Hanna 35346
Bacigalupi, Don 53523
Bacin, Mark S. 51658
Bâciu, Dr. Florea 36310, 36311
Backer, Julie P. 47612
Backer, Kathleen L. 53838
Backes, Joan 51013
Backes, Peter 22249
Backhaus, Fritz 18379, 18388
Backhouse, Andrew 45795
Backhurst, Timothy D. 50873
Backlund, Connie 48555
Backström, Ragnar 10987
Bacon, Weldon 05706
Bacourou, Emilia 23108
Bacquet, Emmanuel 14323
Bačvarov, Ivan 05036, 05163
Bączyk, Jarosław 35032
Badaró Nadal, Enrique 54390
Badas, Ubaldo 28279
Badault, Albert 12380
Badder, Susan 53855
Bade, Mary 49779, 50398
Bader, Joerg 41406
Bader, Judith 22079
Badermann, Robert 19419
Badham, Mark 06084
Badoúin, Gerhard 21201
Badovinac, Zdenka 38450, 38454
Badre, Dr. Leila 30157
Badura-Triska, Dr. Eva 03042
Bächtinger, M.A. Anke 21283
Baeckler, Sabine 21393
Bähr, Gustav-Adolf 20674
Baehrend, John 47440
Baeke, Jan 31841
Bækgaard, Bjarne 09962
Bækkelund, Björn 33481
Baele, Annemie 04168
Baena Alcántara, María Dolores 39204
Baensch, Antje 18716
Baensch, Norbert 18692
Bär, Dr. Frank P. 20764
Baer, Jan Hodges 49255
Baer, Dr. Rudolf 41123
Bärnighausen, Hendrik 20396
Baerwaldt, Wayne 07128
Baeten, Herman 03948
Baets, August de 03866
Baetz, Tracy L. 49727
Báez Plaza, Anastasio 39014
Baganz, Bruce P. 53921
Bagdanavičienė, Asta 30299
Bagdonienė, Rūta 30373
Baggaley, Sean 45314
Bagge, Christian 18900
Baggenstoss, Josef 41431
Bagheri-Zenouz, Dr. E. 24615
Bagle, Eyvind 33744
Bagnall, Dinah Ryle 45548
Bagnatori, Paola 52785
Bagnell, Warren 06199
Bagnoli, Dr. Alessandro 27846
Bahadar Khan, Mohammad 34002
Bahadur Shah, Prof. Karan 31777
Baharin Bin Buyong, Dr. Kamarul 30581, 30609, 30622
Bahe, Michelle 46488
Bahena García, Jaime Arturo 31087
Bahm, Linda 46189
Bahn, Erich 03100
Bahn, Dr. Peter 17585
Bahnmaier, Paul M. 50015
Bahrami, M. 24689
Bahtsetzis, Sotirios 41563
Baijot, Joseph 03674
Baik, Jo Gang 29763
Bail, Anne le 16310
Bailer, Dr. Brigitte 02978
Bailey, Angela 00902
Bailey, Ann S. 51299
Bailey, Billie K. 53961, 53962, 53963, 53964
Bailey, Chris H. 47027
Bailey, Colin B. 51178
Bailey, Courtney 52227
Bailey, David 48994
Bailey, Elizabeth 47235
Bailey, Erin 46439
Bailey, Jack C. 53254
Bailey, Jann L.M. 06035
Bailey, Jeniffer A. 53728
Bailey, Margaret 51384
Bailey, Michael 43347
Bailey, Patricia 05464
Bailey, Patricia E. 49019
Bailey, Peter 45235
Bailey, Prof. Reeve M. 46310

Bailey, Sven 53599
Bailey, Tom 46376
Bailleux-Delbecq, Martine 12885
Bailly-Maitre, Marie-Christine 13489
Bain, Heather 50298
Bainbridge, Carol 51351
Bainbridge, David S. 53157, 53158, 53161
Bainbridge, H.M. 45479
Bainbridge, R.A. 45479
Baiocco, Simone 28004
Baioni, Marco 26276
Baird, Vaughan L. 07355
Bairrad Ruiro, Marina 35666
Bairsto, Rachel 44801
Baisier, Claire 03296
Baitchiba-Afalna, Kloumtouin 07426
Baj, Janina 34748
Bajada, John 30654
Bajanik, Dr. Stanislav 38217
Bajcurová, Dr. Katarína 38233
Bajic, Svetlana 04284
Bajko, Daria 51274
Bajpayee, Aparna 24090
Bajramović, Muhamed 04294
Bajuk, Laura 50273
Bak, Aase 09956
Bak, Jolán 23557
Bak, Dr. Saskia 32473
Bakajutova, Ljudmila M. 37630
Bakal, Jeniffer 51187
Bakalarz, Krystyna 34950
Bakalov, Michail 05206
Bakanauskas, Alvydas 30338
Bakanova, Irina Viktorovna 37272
Bakărdžieva, Cenka 05067
Bakărdžijev, Dimităr 05048
Bakarić, Lidija 08986
Bakarich, Carolyn C. 53736
Bakarich, Michael N. 53736
Bakay, Prof. Dr. Kornél 23685
Bakels, Dr. P. 31989
Baker, Alasdair 43985
Baker, Alexander 51864
Baker, Allan 07133
Baker, Alyson 50188
Baker, Barbara J. 46202
Baker, Billy P. 48923
Baker, Bob 49809
Baker, Charlotte 46236
Baker, Chris 43951
Baker, Christopher 44138
Baker, Chuck 47741
Baker, David 47227
Baker, Diane 01594
Baker, Gary E. 51361
Baker, James 51809, 52650, 52651
Baker, James W. 48226
Baker, Janet 46075
Baker, Dr. Janet 51906
Baker, Janet 52852
Baker, Jeanne 54020
Baker, John 45220, 45753
Baker, Kendall 47163
Baker, Larry L. 46852
Baker, Linda 05429
Baker, Paul 45291
Baker, Peggy M. 51995
Baker, Rachel 47487
Baker, Raymond 54020
Baker, Dr. Robert 50312
Baker, Rusty 49182
Baker, Scott 53879
Baker, Sue 47876
Baker, Susan K. 53990
Baker, Warren L. 46202
Baki, Péter 23650
Bakić, Ljiljana 38163
Bakke, Prof. Tor A. 33737
Bakker, C. 32413
Bakker, G. 31833
Bakker, Robert J. Roozemond den 32097
Bakker, W. 31957
Bakkevig, Sverre 33863
Bakouti, Moncef 42652
Bakr, Nemat 10465
Bakreeva, Ljubov Anatoljevna 37134
Baksa-Soós, Dr. Veruschka 17060
Bákula Budge, Cecilia 34227
Bal, Danielle 13134
Balabanov, Dimitar 05156
Balagtas, Alex L. 34421, 34423
Balakina, Natalja Michajlovna 37690
Bălan, Filofteia 36356
Balan, Sandrine 14181
Bălan, Zamfir 36167, 36168
Balanza, Ferran 39626, 39627
Balao Gonzáles, Angel 39586
Balard, Prof. Michel 15836
Balas, Mary 51509
Balas, Shawna 07387
Balaštíková, Eva 09828
Balatti, Ambrogio 25864
Balázs, Dr. György 23455
Balbaugh, Joe 49669
Balčaitienė, Rasa 30205
Balciunas, Vytautas 30440
Bald, H. 20027
Baldaia, Peter J. 49444
Baldarrago Umpire, Prof. Miguel 34109
Baldauf, Manfred 22249
Baldellou, Vicente 39398
Baldemair, Vinzenz 02367
Balden, Friedesine 18772
Baldeón, Dr. Amelia 40321
Balderi, Felice 26533
Balderstone, Michael 01369
Baldeschi, Nino 28122
Baldewijns, Jeannine 03643
Baldini, Marino 08929
Baldino, Renzo 41313
Baldit, Alain 14983
Baldoni, Lorenzo 26760
Baldry, Anita 45620
Baldwin, Deirdre 48888
Baldwin, Graham 45839
Baldwin, Jane 46559
Baldwin, Nicholas 44862
Baldwin Rehbein, Valerie B. 52304
Balen, Dubravka 08986
Balen, Jacqueline 08986
Balestra, Nino 27551
Balevičius, Mindaugas 30404
Baley, Susan G. 51374
Balher, Gisela Carina 00484
Balic, Haris 03051
Balie, Dr. Isaac 38623
Baligand, Françoise 12775
Balint, Lăcrămioara 36411
Bálint, Magdolna 23485
Bálint, Tamás 23403
Balint, Valerie 49411
Balistrieri, Bruce 51730
Balkanska, Veneta 05093
Ball, Cobie 53172
Ball, Dr. Daniela 41856
Ball, Daniela 42123
Ball, David 52966
Ball, Elizabeth 05577
Ball, Felicity 45819
Ball, Glenys 01621
Ball, JoAnn 51973
Ball, Rex 52716
Ball, Sabine 19054
Ballantine, Dr. David 36043
Ballantine, Thomas 05927
Ballantyne, Dorothy 49303
Ballantyne, Weyman 33191
Ballard, Mark 49574
Ballbé, Xavier 39100
Balle, M.A. Gereon 19092
Ballerini, Rosangela 26533
Ballerstedt, Mareike 17549
Ballesta, J. 40276
Ballesta Gómiz, Pedro 39696
Ballesteros, Prof. Miguel 00547
Ballestra 14078
Balletto, Prof. Emilio 28017
Ballinger, J.C. 50489
Ballinger, James K. 51906
Ballot, Prof. Dr. G.M. 38717
Balmer, John 54317
Balog, Jelena 08997
Balogh, Daniela 02200
Balogh, Mihály 23462
Balogh, Sándor 23496
Balsamo, Prof. Renato 25984
Balshaw, Dr. Maria 45099
Baltasar Coll Tomas, M.I.S.D. 39785
Baltensperger, Marianne 41906, 41907
Baltes, Peter 17759
Băltuță, Steliana 36163
Baltz, Gerald W. 52580
Baltzegar, Mack 49087
Baltzer, Doralies 16776
Balza, Raimundas 30325, 30352, 30353, 30362, 30365, 30451
Balzardi, Elena 41215
Balzekas, Robert 47456
Balzekas, Stanley 47456
Bamala, Louis 08802
Bambara, Parfait Z. 05278
Bamberger, Gerald 17335
Bambery, Anneke 43965
Bamesa, Prof. Tshungu 08799
Bamforth, Janice 46721
Ban, Gheorghe 36605, 36610, 36611
Bana, Giovanni 25282
Banach, Wiesław 35101
Banach, Witold 34994
Banai, Merav 24965
Bańbura, Prof. Dr. Jerzy 34882
Banchieri, Daria 28161
Bancroft, Evelyn 01063
Bandaranaike, Sunethra 40369
Bandeira, Dr. Dione da Rocha 04514
Bandes, Dr. Susan J. 48253
Bandrowski, Margaret 53234
Bandt, J. 01315
Bandurraga, Peter L. 52288
Bandy, Mary Lea 51215
Bandyopadhyay, Dr. Banya 24069, 24424
Banerji, Dr. Jayasri 24291
Banfi, Enrico 26691
Banfield, Audray M. 00792
Bang, Mogens 10056
Bang Andersen, Sveinung 33863
Bang Larsen, Bent 10255
Bangcas, Salamon L. 34386
Bangcaya, Regino Z. 34494
Bange, Maja 33729
Bangert, Dr. Susanne 10102
Banghard, Karl 20874
Banghart, Bob 49656, 53125
Banker, Amy 48066
Banker, Maureen 52229
Bankert, John 47228
Banks, Peter 05734
Banks, R. 01153
Banks, Robert 53532
Bannicke, Elke 17051
Bannister, R. 43544
Bannon, Anthony 52395
Banova, Vera 05262, 05263
Banović, Miroslav 38109
Bantel, Prof. Dietrich 20814
Bantle, Tom 49007
Banu Syed, Rehana 31776
Banz, Richard 54341, 54342
Banzato, Dr. Davide 25286, 27032, 27034, 27050
Bao, Yingjian 07753
Baptista, Dr. Teresa 35544
Baptista Pereira, Fernando António 35936
Baquedano, Enrique 38812
Baquero, Dr. Angeles Ramos 34071
Baquero Velásquez, Ramón 08595
Bar-Gera, Dov 24929
Bar-Gera, Kenda 24929
Bar-Or, Galia 24945
Bar Semech, Vered 24940
Barabaş, Dr. Neculai 36120, 36123, 36546
Barahona, Rolando 08816
Baral, Jody 50239
Barale, Joseph 13117
Baran, Iwona 35173
Barandun, Plasch 41982
Barangulova, Farida F. 37886
Baranova, M.N. 37739
Bárány, Anders 40921
Baraton, Christelle 13370
Barbagallo, Dr. Alfredo I. 00241
Barbalić, Željko 08939
Barbanera, Prof. Marcello 27482
Barbara, Sibille 14482, 14483
Barbare, Inese 30074
Barbaret, Hervé 14698
Barbe, Sylbie 14834
Barbeau, Bruce 07106
Barber, Charles D. 51834
Barber, Dan 50901
Barber, Diane 49385
Barber, John 01680
Barber, R.G. 45617
Barber, T. 44781
Barberet-Girardin, E. 16037
Barberini, Maria Giulia 27514
Barbier, Jean 36093, 36095
Barbier, Marjolijn 03290
Barbier, Michel 03916
Barbier-Ludwig, Georges 12650
Barbier-Mueller, Jean-Gabriel 41417
Barbieri, Frances T. 53036
Barbieux, José 15951
Barbina, Prof. Alfredo 27443
Barbizet, Laure 14690
Barblan, Otto 41992
Barbosa, Artemio 34403
Barbosa Godefroid, Leonardo 04604
Barbour, Mark 47257
Barbour Strickland, C. 49074
Barbry jr, Earl J. 50492
Barbu, Cristian 36227
Barca, Ana 36217
Barcal, Alexandra 42073
Barcel, Ellen 53189
Barcellos, Lauro 04775
Barclay, Craig 44073
Barclay, Don 48144
Barclay, Mitzi 51790
Barcley, Craig 44074
Barcroft, Arnold 43349
Barcus, Leisha 48097
Bardelli, Alvaro 25269
Bardgette, S. 44881
Bardiaux, David 15791
Bardiés-Frarty, Isabelle 14729
Bardies-Fronty, Isabelle 14102
Bardiová, Dr. Mariana 38183
Bardo, Antoine 13515
Bardwell, Sarah 44871
Bareinz, Reinhard 17675
Barendsen, J.E. 32323
Barens, H. 51504
Barents, Els 31849
Barger, Judith M. 49902
Bargero, Vicente 34178
Barghini, Sandra 43398
Bargholz, Christina 18946
Bargiel, Réjane 14662
Barilleaux, Rene Paul 49560
Barina, Miloslav 09677
Barisch, G. 17338
Barisch, M. 17338
Barisione, Silvia 26327
Barison, Mara 25339
Baritault du Carpia, M. de 14051, 14052
Barjou, François 13723
Barkefeld, Anna 21820
Barkemeyer, Dr. Werner 18329
Barker, Alex 47932
Barker, Deborah 51629
Barker, G. 01552
Barker, Garry 50841
Barker, Jane 46610
Barker, Dr. Nicolas 43910
Barker, Ray 46098
Barker, Rod 48213
Barker, Sally L. 52166
Barker, Stuart 43661
Barker, Wendy 48419
Barketov, Vladimir Aleksandrovič 37753
Barkhofen, Dr. Eva-Maria 17049
Barkley, Terry 47022
Barkoun, Gennady 03247
Barksdale, Daryl 50471
Barkwill, Ken 46904
Bârlea, Gheorghe Mihai 36614
Barlek, Josip 08989
Barlier, Gabriel 13007
Barlow, Erin 51169
Barlow, Geraldine 00980
Barlow, Howard 54054
Barlow, Keith 45266
Barlow, Martin 44760
Barlow, Patricia S. 47710
Barlow, Victoria 44799
Barmore, Jim 47260
Barndollar, Lue 47673, 47674
Barnea, Dr. Jackob 02389
Barnes, Althemese 53428
Barnes, James H. 48669
Barnes, Judy 50551
Barnes, Lolly 46810
Barnes, Lucinda 46752
Barnes, Marion 43424
Barnes, Richard 49416
Barnet, Peter 51152, 51212
Barnett, Eric B. 48301
Barnett, George 44680
Barnett, James F. 50984
Barnett, June 53068
Barnett, Lawrence 53767
Barnett, Patricia 51178
Barnett, Steve 49486
Barney, Edres 51919
Barnfather, Margaret 43329
Barnham, George 44529
Barnhart, Audrey 54181
Barnhill, Michael 46675
Barnhill-West, Barbara 51573
Barnisch, Prof. Manfred 17448
Barnosky, Antony 46758
Barns, Laurie 48114
Barnum, Lorie 52401
Barnwell, Andrea D. 46454
Barocelli, Prof. Francesco 27110
Baroff, Deborah 50000
Barollo, G. 26363
Baron, David 06727
Baron, Donna 54112
Baron, Evelyne 15231
Baron, Dr. Jacques 14329
Barón, Javier 39573
Baron, Richard 48316
Baroncini, Dr. Claudia 26403
Barone, Constance B. 52497
Baroni, Fabio 25706
Baroni, Laura 26046
Baronne, Gerlool 51089
Barons, Richard I. 53186
Barovier, Jacopo 28186
Barr, Caroline 44280, 44302
Barr, Cheryl B. 46753

B

Barr, Edward 53537
Barr, Mike 43263
Barr, Norma 05694
Barr, Dr. Peter J. 46138
Barr, Tom 48652
Barracane, Gaetano 25367
Barrachina, Jaime 39836
Barrack, Pamela 46661
Barral Iglesias, Alejandro 40022
Barrand, Janine 01280, 01290
Barreiros, Mariluz 40187
Barrera, Alberto 34079
Barrera Bassols, Marco 31076
Barrere, Eliza 14432
Barrere, Sylvie 12062
Barreto, Ricardo D. 46953
Barrett, Candace 50217
Barrett, Joanna 44590
Barrett, Maurine 53207
Barrett, Michael 00797, 49707
Barrett, Wendy 52683
Barrett-Lennard, Amy 01480
Barreyre, J.Y. 14578
Barrie, Brooke 49133
Barrientos, J. Oscar 23302
Barriga, Prof. Fernando J.A.S. 35714
Barringer, Jane S. 52849
Barrington, Robin 06259
Barrois, Alain 13981
Barron, G.D. 33099
Barron, Stephanie 50244
Barroso Panatieri, Angélica 04634
Barrowclough, Dr. George F. 51115
Barry, Alexis 53380
Barry Byrne, Brenda 05784
Barryte, Bernard 53253
Barsalou, Victoria 06575
Barsch, Dr. Barbara 17141, 17142
Barsch, Wolfgang 16494
Barskaja, Marina Iljinična 37408
Barson, Jerry 52872
Barsotti, Gianfranco 26500
Barstow, Norman 49117
Bârsu, Mariana 36282
Barsuglia, Dr. Bruno 26533
Barszcz, Janusz 35084, 35211
Barszczewska, Krystyna 35349
Bart, Barbara M. 49440
Barta, Andrew 54181
Barta, Dr. Ilsebill 03005
Bárta, Karel 09910
Barta-Fliedl, Dr. Ilsebill 03009
Bartels, Dr. Christoph 17398
Bartenev, Aleksandr Igorevič 37673
Barth, Dr. Angelika 22040
Barth, Frank 18932
Bárth, Dr. János 23646, 23648, 23651, 23654
Barth, Katharina 02620
Barth, Linda J. 50266
Barth, Dr. Robert 41869
Barth, Dr. Robert L. 52096
Barth, Dr. Stephanie 20902
Bartha, Dr. István 23862
Barthel, J. 20539
Barthelson, Karen 50934
Bartholomew, Ed 46105
Bartík, Dr. Juraj 38204
Bartków-Domagala, Jadwiga 35336
Bartkowski, Gregor 17601
Bartkus, Stephen J. 53969
Bartl, Anna 41161
Bartl, Johanna 17828
Bartlett, Prof. Christopher 53551
Bartlett, Diane 50196
Bartlett, Margaret 00979
Bartley, James A. 54234
Bartlit, Nancy 50205
Bartolf, M. 06564
Bartolini Bussi, Prof. Maria Giuseppina 26739
Bartolomé Arraiza, Alberto 39569
Barton, Anita S. 50467
Barton, C. Michael 53469
Barton, Elizabeth 49327
Barton, William R. 51692
Barton Billman, Hilary 49724
Barton Thurber, T. 49162
Bartoś, Magdalena 34928
Bartos, Paul I. 48958
Bartoš, Václav 09811
Bartosz, Anna 35178
Bartosz, Dr. Antoni 34792
Bartram, Rita 48797
Bartsch, Joel A. 49390
Bartsch, Monika 20462
Bartuś, Marta 34554
Bartz, Rolf-Peter 22340
Barwell, Victoria 43409
Barwick, Kent 51219
Bary, Bertrand de 15129
Barzakova, Maria 05001
Barzen, Jeb 46605
Bas, Michael le 43539
Basamakov, Christo 04971
Basamakov, Ilko 05132
Basanieri, Giovanni 25989
Basante, Enar 08709
Basart 04881
Basbous, Alfred 30171
Bascha, Nadya 16392
Baschnagel, H. 18731
Bascom, Mansfield 50411
Bascou, Marc 14698
Başer, Ǝlsmail 42916
Basile, Corrado 27855
Basile, Kenneth A. 52678, 52679
Basir, Abdul 24590
Baškirova, Marija Michajlovna 36904
Bašnjak, Jurij Michajlovič 36804
Basque, Maurice 06320
Bass, Anneta Jakovlevna 37621
Bass, Carol 54256
Bass, Jeffery 52735
Bassegoda Nonell, Juan 38971
Bassett, Donald E. 47692
Bassett, Fannie 47442
Bassett, Dr. M.G. 43748
Bassett, Nancy W. 47692
Bassett, Ralph 51598
Bassi, George 49985
Bassi, Dr. Roberto 28190
Bassi Neri, Raffaella 26518
Bassin, Aleksander 38445, 38452
Bassinger, Marilyn 51496
Bassler, Roseli Fischer 04428
Basso, Dr. Laura 26694
Bassús, María M. 00179
Bast, Douglas G. 46908
Bastard, Laurent 15965, 15966
Bastek, Dr. Alexander 20072
Basterretxea-Moreno, Dr. Amaia 39042
Bastia, France 03456
Bastidas, Hugo Xavier 49626
Bastin, Donald 52312
Basu Ray Chaudhuri, S. 24337
Bata, Philippe 12840, 14868
Bata, Sonja 07093
Bataillard, Marianne 41339
Batchelor, Elisabeth 49695
Batchelor, Trevor 06568
Bateman, Prof. Richard 44941
Bateman, Suellen 48987
Baten, F.O. Abdul 03215
Batenin, Nikita Sergeevič 36921
Bater, John 48736
Bates, Brenda 47891
Bates, Charles 50084
Bates, Craig D. 54350
Bates, F. 45674
Bates, Geoffrey 50157
Bates, Michael L. 51116
Bates, Tom 50694
Bateva, Mimi 05084
Bath, Marquess of 45934
Bathe, I. 06521
Batista, Alvaro 35383
Batkin, Jonathan 52892
Batkin, Norton 46314
Batori, Dr. Armicla 27487
Batova, Anna Nikolaevna 37537
Batrd, Nicola 45619
Battaglini, Dr. Giuseppe 27310
Battali, Kathleen 48552
Battcher, Ray 47031
Battel, Angelo 27688
Batterson, Lorna 50121
Battinou, Zanet 22910
Battis, Nicholas 51249
Battle, Dr. Thomas C. 53880
Batty, Theresa 04176
Batty, Tim 43998, 44003
Baty, Catherine 54098
Bauch, Herbert 19811
Bauch, Hermann 02239
Bauche, Dean 06465
Bauchiller, Dr. Rafael 39529
Baucum-Fields, Penny 51671
Baud, Arlette 41135
Baud, Christiane 41187
Baud-Berthier, Gilles 12089
Baudement, Marie-Laure 11778
Bauder, Charles C. W. 48881
Baudet, Jean-Charles 16046
Baudet, Jean-Michel 14832
Baudin-Reneau, Katia 15815, 19624
Baudisson, C. 13442
Bauduin, Jean-Loup 13432, 13433, 13436
Bauer, M.A. Anja 20414
Bauer, Arthur 02982
Bauer, M.A. Christoph 21781
Bauer, Douglas F. 48575
Bauer, Fritz 22839
Bauer, Gudrun 19126
Bauer, Hans 17293
Bauer, Hans-Rudolf 41712
Bauer, Heidrun 21996, 21997
Bauer, Dr. Helmut 20462
Bauer, Herbert 16520
Bauer, Dr. Markus 18680
Bauer, Peter 20006
Bauer, Richard 53472
Bauer, Sigrud 17293
Bauer, Stéphane 17170
Bauer, Susan 06273
Bauer, Dr. Thomas 18371
Bauer-Feigel, Gabriele 18067
Bauer-Friedrich, Thomas 17698
Bauermeister, Janneke 20519
Bauernfeind, M.A. Günther 20624
Baughan, Clark 52891
Baulard, Hervé 12407
Baule, John A. 54320
Baum, Prof. Dr. Eckhard 22634
Baum, Kristina 18210, 22320
Baum, Wolfgang 21751
Bauman, Victor-Henrich 36705
Bauman Taylor, Paula 47446
Baumane, Inta 30023
Baumann, Anton 02490
Baumann, Evelies 17727
Baumann, Klaus 18745
Baumann, Dr. Maria 21216, 21219
Baumann, Rudolf 41527, 41528
Baumbach, Susan M. 45393
Baume, Nicholas 46933
Baumeister, Dr. Ralf 16706
Baumgärtel, Dr. Bettina 18008
Baumgärtner, Iris 21184, 21186, 21188
Baumgardner, George 47260
Baumgardner, Stan 47380
Baumgarten, Dr. K. 16801
Baumgarten, Ute 19281
Baumgartner, Guido 41997
Baumgartner, Mag. Lothar 01829
Baumstark, Dr. Brigitte 19451
Baumstark, Prof. Dr. Reinhold 16920, 20421, 20425, 20471, 20480, 20481
Baur, Andreas 18278
Bausart, Lucie 03292
Bause, George S. 51710
Bausier, Karine 03328
Bauske, Clay R. 49475
Baussan, Magali 16063
Baußmann, Dr. Edda 20519
Bautrėnas, Justinas 30420
Bauziene, Morta 30401
Bavnshøj, Peter 09973
Bavoni, Dr. Umberto 28310
Bawden, Dr. Garth L. 46181
Baxpöhler, Monika 16438
Baxter, Alan 43697
Baxter, April 49845
Baxter, Clare 43287, 43288
Baxter, David 06551
Baxter, Ellen 51942
Baxter, Marilynn 52052
Bayard, Jean 07206
Bayard, Pierre 05297
Bayarri, Carlos 39950
Baybut, John 52102
Bayden, Ian 53798
Bayer, Dagmar 22287
Bayer-Niemeier, Dr. Eva 19738
Bayet 04087
Baylis Ashby, Lisa 48559
Bayliss, Joan 43378
Bayliss, Dr. Paul 38645
Baymuradov, N. 42505
Bayona, Wilhelm A. 34520
Bayou, Charlie 14991
Bayram, Metin 42796
Baysal, H. Hüseyin 42759
Bayuelo Castellar, Soraya 08634
Baz, Samir 30169
Bazančuk, Galina 37337
Bazanov, Valerij Anatoljevič 36877
Bazely, Sue 06080
Bazennerye, Dr. Jean 16173
Bazin, J. 14133
Bazin de Jessey, Hervé 12438
Bažkenova, Nesipbiken Orazchanovna 29706, 29722
Bazley, Shirley 01295
Bazzini, Marco 27332
Bazzoni, Maurizio 27371
Bazzotti, Prof. Ugo 26592
Beach, James 49991
Beadle, Kristine 48011
Beady, Charles H. 51928
Beairsto, Murray 52635
Beal, Graham W.J. 48114
Beale, Carolyn 47883
Beall, William 49815
Beam, Michael J. 50892
Beam, Michael N. 47427
Beamis, Frances D. 47110
Beams, Daniel J. 52963
Bean, Jeanet 51476
Bean, Dr. Susan 52657
Beanchard, Laurent 13157
Bearcroft, Don 43255
Beard, David 51843
Beard, Dr. James S. 50521
Beard-Simpkins, Doreen 53407
Beardsley, Robert 48348
Beardsley, Theodore S. 51192
Beardsmore, R.J. 33166
Beasley, James M. 53233
Beasley, Pam 53132
Beaton, Stuart 05441
Beattie, Keith 43366
Beattie, Natalie 01410
Beatty, Brett 49453
Beatty, Thomas P. 53753
Beatyy, John T. 52431
Beaud, Marie-Claude 30478
Beaudin Saeedpour, Dr. Vera 47077
Beaufort, Dr. Christian 03004
Beaufort-Spontin, Dr. Christian 03019
Beaumont, Robert N. 53800
Beausoleil, Raymond R. 07027
Beauvau, Princesse de 13183
Beavers, Randell A. 49402
Beavin, H.A. 44471
Beavis, Dr. Ian 45881
Bebout, Terry L. 50974
Beccaglia, Valdemaro 27332
Beccari, Giuseppe 28304
Beccaria, Marcella 27423
Beccart, Christine 13098
Becchi, Casimiro 27167
Bećević, Azra 04284
Bech, Marianne 10244
Bechelet, Florence 45588
Becher, Inge 18592
Bechers, H.H 32831
Bechet, Georges 30484
Bechhoefer, Melissa 49849
Bechly, Dr. Günter 21959
Bechtelheimer, Daryl 52496
Bechtold, Dr. Ekkehard 01827
Bečić, Berina 04284
Beck, Dr. Andreas 20430
Beck, Bettina 17326
Beck, Elisha 49222
Beck, Gary 52757
Beck, Jane C. 50654
Beck, Jerry 50307
Beck, Mike 49435
Beck, Noldi 30201
Beck, Rolf 22532
Beck, Stefan 41722
Beck, Tom 46544
Beck, Walter 30193
Beck-Lipsi, Gabrielle 15106
Beckemeyer, Klemens 20310
Becken, Ellen 09982
Becker, Dr. Annette 18361
Becker, Christine 19939
Becker, Dr. Christoph 42083
Becker, Daniel 20478
Becker, Edwin 31898
Becker, Gail R. 50290
Becker, Dr. H. 18537
Becker, Heinrich Georg 22400
Becker, Dr. Herbert 18538
Becker, Horst R. 18373
Becker, Dr. Ingeborg 17056
Becker, Dr. Jack 50963
Becker, Jürgen 18246
Becker, Julia 49029
Becker, Lawrence 54302
Becker, Lisa 50309, 50337
Becker, Dr. Michael 02001
Becker, Dr. Peter-René 17564
Becker, Dr. Rayda 38655
Becker, Sara 46213
Becker, Scott R. 53943
Becker, Dr. Ulrich 01961
Becker-Jákli, Dr. Barbara 19626
Beckers, Harald 16402
Beckers, Marion 17277
Beckerson, Dr. John 44419, 45964
Beckert, Harriet 51346
Beckett, D. 44167
Beckett, Michael D. 49998
Beckford, Chip 54044
Beckley, Frank J. 45511
Beckmann, Isrun 17106
Beckmann, Michael 01176
Beckmann, Dr. Ralf 18307
Beckmann, Dr. Uwe 18858
Beckmann, Werner J. 18015
Becks, Jürgen 22508, 22509, 22510
Becks, Dr. Leonie 19599

Beckstrøm, Alf 33869
Becucci, Dr. Sandra 25746, 27884
Bedal, Albrecht 21655
Bedal, Prof. Dr. Konrad 16871, 16872
Bedaux, Dr. R. 32488
Bedbur, Dr. Eckart 19528
Bedell, Lisa 51218
Bedenham, M.D. 43803
Bedenig, Dr. K. 42111
Bedford, Emma 38593
Bedia García, Juana 39396
Bednarek, Monika 34799
Bedouin, Regan 48217
Bedoya, José 04223
Bedoya de Rendón, Martha Elena 08719
Bedula, Jane 50853
Bee, Dr. Andreas 18385
Beebe, John 48412
Beebe, Mary L. 49847
Beech, Mike 44250
Beeck, Eva in der 17008
Beeck, Stefan in der 17008
Beedenbender, Britt 47316
Beeks, Bonny 48962
Beer, Dr. Manuela 19625
Beer, Petra 18209
Beerli, Klaus 41656
Beernaert, Veerle 03711
Beerts, Harry 04004
Beesch, Ruth 51200
Beesley, Bruce A. 53748, 53750
Beevers, David 43625
Begay, Clarenda 54225
Beggs, Jo 45099
Begonja, Zlatko 08984
Begouin, Bernard 12238
Begović, Akad. Miroslav 08994
Begue, Christelle 15902
Béguin, Béatrice 41535
Béguin, Gilles 14638
Béguin, Pierre Henri 41249
Beha, Wolfram 19560
Behler, Mark 54037
Behlmann, Cletus 52721
Behm, Stefan 02016
Behne, Dr. Axel 20950
Behnke, Arleen 51790
Behnke, Martin 18793
Behr, Lothar 22199
Behrends, Carey F. 51497
Behrendt, Martina 17257
Behrens, Friedrich 21536
Behrens, Sarah 51706
Behrens, Todd 49902
Behrman, Sharon 49222
Behrndt, Helle 10137
Behroozy, N. 25117
Beier, Aasmund 33770, 33771
Beier, Dr. H.-J. 22485
Beier, Stefan 19577
Beier-de Haan, Prof. Dr. Rosmarie 17072
Beiersdörfer, Dr. Kurt 20969
Beiersdorf, Maciej 34796
Beighton, James 45141
Beil, Judith 53451
Beil, Dr. Ralf 17790
Beilmann-Schöner, Dr. Mechthild 21298
Beimel, Alice 52607
Bein, Peter von 22506
Beiranvand, Asadollah 24621
Beirer, Ingrid 17062
Beirne, T. 44410
Beisel, P. 20561
Beisken, Angela 19037
Béistegui de Robles, Dolores 31049
Beitchman, Marsha 51112
Beitin, Andreas F. 19453
Beitl, Matthias 03049
Beitz, Dr. Uwe 18053
Bejbosynov, Kuzembaj Bajbosynovič 29733
Bejenaru, Ionel 36351, 36445, 36593
Bekmuradov, I. 54458
Bekouan, Hortense Zagbayou 08830
Bektureeva, Roza Abutalipovna 29685
Bela, Dr. Zbigniew 34793
Belai, George 23384
Belaja, Z.A. 37806
Belart, Curdin 41116
Belbin, J. 01356
Belčeski, Rubinčo 30513
Belčeva, Tonja 05023, 05024, 05025
Belcher, Helen 49446
Belder, Marie de 04036
Belenis 23064
Beleznay, Andor 23380
Belfield, John 01352
Belfry, Craig 06647
Belgado Camblor, Victoria 39122
Belger, Kathleen 19892
Belgin, Dr. Tayfun 18857, 18860
Belie, Liesbeth de 03462
Belier, Corinne 14728
Belinskis, Andrejs 30108
Belinskyj, Valeriy 43013
Belisario, Alessandro 26200
Beljaev, A.A. 37814
Beljaeva, Nadežda V. 37550
Beljaeva, Svetlana A. 37820
Belk, Brad 49649, 49651
Bell, Allan 00896
Bell, Brendan 38707
Bell, Chee Chee 49464, 49465
Bell, Christine 43369
Bell, Elizabeth 47264
Bell, Eunice 51486
Bell, Dr. Graham A.C. 06380
Bell, Greg 53411
Bell, Gregory 49790
Bell, Lynn 46506
Bell, Margaret 33157
Bell, Michael 06541
Bell, Michael S. 52516
Bell, Nicholas 47590
Bell, P.J. 10168
Bell, Rae 46223
Bell, Robert 00946
Bell, Theresa H. 50527
Bell, Valerie 48468, 50146
Bella, Miguel Angel di 00544
Bellah, Suzanne 51657
Bellaigue, Dominique de 12049
Bellais, Leslie 50396
Bellamore, Stan 33179
Bellamy, Gene 49032
Bellamy, Dr. Martin 45608
Bellando, Francesco 25357
Bellanger, Françoise 14565
Bellani, Giancarlo 26388
Bellanová, S. 38391
Bellar, Gail 49172
Bellas, Gerald M. 47564
Bellaton, Jean-Louis 15280
Bellavia, Kim 49807
Bellazzi, Dr. Pietro 28267
Bellec, Henri 05621
Bellehumeur, Alain 05422
Bellenger, Sylvain 47624
Beller, Janet 50302
Bellerby, Greg 07207
Bellettini, Dr. Pierangelo 25438, 25446, 25448
Belli, Dr. Gabriella 27567, 27570, 28056
Belli, Larry 49775
Bellier, C. 04092
Bellifemine, Graziano 25429
Bellinger, Donald F. 49672
Bellingeri, Dott. Luca 26742
Bellini, Lucette 12268
Bello Dièguez, José M. 39217
Bello Morales, Marinelly 54533
Bellod, Anne 14641
Belloni, Luigi Mario 27015
Bellucci, Alberto Guillermo 00259
Bellucci, Dr. Bruno 27672
Belluscio, Lynne J. 50002
Bellwald, Tony 41504
Belmondo, Dr. Rosalba 27766
Belmonte, Patty 51544
Bělochová, Olga 09570
Belonovič, Éleonora Nikolaevna 37750
Belonovich, Galina Ivanovna 37052
Belotti, Enrico 25181
Belotti, Dr. Ione 25856
Beloubek-Hammer, Dr. Anita 17172
Belousova, Natalj Aleksandrovna 37033
Belov, Vladimir Ivanovič 37956
Belova, N.H. 37708
Belser, Eduard J. 41168
Belsham, J. W. 45723
Belsley, Kathy 51774
Belt, Virginia 46482
Belting, Dr. Isabella 20462
Beltrán, Carlos 30904
Beltran, Hélène-Sybille 13885
Beltran, Josel 34469
Beltrán Figueredo, Martin 08687
Beltrán Lloris, Miguel 40343, 40344
Beltrani, Francesco 26623, 26624
Beltz, Dr. Johannes 42097
Belušić, Andreja 08997
Belver, Manuel H. 39576
Belvin, Robert 51032
Bem, Marek 35313
Beman, Lynn S. 46272
Bemis, Todd 53063
Bemis, William 46269
Bemporad, Dr. Dora Liscia 26190
Ben, Dr. Arie 25111
Ben-Barak, I. 25043
Ben-Chaôbane, Abderrazzak 31678
Ben-Gal, Dr. Ely 25033
Ben-Nuftali, Aya 25127
Benachetti, Sandra 16050
Benalioua, Selim 13837
Benavides Becquis, Maria del Carmen 08562, 08737
Benayahu, Dr. Y. 25124
Benbow, Colin 04191
Benda, Susanne 53536
Bender, Joseph 53241
Bender, Patricia L. 48370
Bender, Phil 48086
Bender-Bardenhagen, Dipl.-Päd. Claudia 18006
Bendersky, Lydia 53848
Bendig, William C. 49552
Bendiksen, Katriina 33737
Bendixen, Maren 19269
Bendixen, Uwe 19269
Bendyukova, Natalia Oleksandrivna 43176
Benedetti, Prof. Laudomia 26632
Benedetti, Sergio 24807
Benedetti, Susan 52936
Benedict, Andreas 18027
Benedict, Hope 52685
Benedik, Dr. A. Christian 02929
Benedik, Yvonne 03045
Benediktová, Jana 09673
Beneke, Dr. Sabine 17072
Beneš, Dr. Luděk 09657
Benesch, Dr. Evelyn 02942
Benetti, Attilio 28173
Benezra, Neal 52795
Bengston, Rod 46150
Bengtson, Keith 49286
Bengtson, Prof. S.A. 40709
Bengtsson, Bengt 40878
Bengtsson, Eva-Lena 40905
Benington, Jon 43413
Benítez, Antonio Maria 08758
Benitez, Moya 34432
Benitez-Johannot, Purissima 34413, 34414
Benito, Fernando 40204
Benito Goerlich, Dr. Daniel 40208
Benjadol, Piyaluk 42532
Benjamin, Bart 52364
Benjamin, Brent 52595
Benjamin, Jenny E. 52788
Benjamin, Jeremy 51160
Benjamin, Dr. Tritobia H. 53879
Benke, Dr. Harald 21913
Benke, Uwe 17070
Benn, Carl 07121
Benner, Sue 00897
Bennerhag, Carina 40699
Bennet, Alex 52995
Benneth, Dr. Solbritt 40940
Bennett, Alan 44099
Bennett, Betsy 48983
Bennett, Dr. Betsy 52234
Bennett, Carolyn J. 54365
Bennett, Charles 52883, 52885
Bennett, Christobel 01535
Bennett, Dorthy E. 47189
Bennett, Fernanda 52463
Bennett, Graham 45904
Bennett, Harold D. 47189
Bennett, Isabel 24738
Bennett, James 48916
Bennett, James A. 45334
Bennett, Kate 52398
Bennett, Linda 45064
Bennett, Michael 47624
Bennett, Michael W. 50159
Bennett, Dr. Nicholas 44710
Bennett, Sally B. 50858
Bennett, Sandra 46455
Bennett, Sharon 47359
Bennett, Shelley 52819
Bennett, Swannee 50139
Bennett, Tom 46287
Bennett, Tony 03201
Bennett, Valerie 46517
Bennett, William P. 49270
Benninghoff, Wolfgang 41601
Bennington, Jeanette 49338
Bennison, B. 45147
Bennyman, James C. 53625
Benoist, Guillaume 13397
Benoit, Jean-Louis 16285
Benova, Katarina 38233
Beňovský, Jozef 38300
Benscheidt, Dr. Anja 17570, 17572
Bensen, Pat 52919
Benson, Barbara E. 54191
Benson, Benjamin Foley 52905
Benson, Claire 46413
Benson, Dave 05602, 06433
Benson, Joanne 45056
Benson, Lisa 33742
Benson, Mike 44505
Benson, Patricia 48333
Benson, Tammy 48634
Bentabet, Mohamed 00041
Bente, Prof. Dr. Klaus 19919
Bentheim, Ferdinand Erbprinz zu 16685
Bentini, Dr. Jadranka 26076
Bentlage, Deb 52134
Bentley, Alan 44430
Bentley, Rachel 45340
Bentz, Abigail L. 48074
Bentz, Sylviane 15077
Benvenuti, Prof. Carlo 27240
Benvie, Rachel 45173, 45174
Benyahia, Boualem 00069
Benyovsky, Lucija 08997
Benz, Barbara 22502
Benz, Edward 46914
Benz, Paul 21492
Benz, Robert G. 54287
Benzel, S. 19544
Benzner, Sonja 16402
Beran, Antonia 18586
Beran, Helmut 16962
Beránek, Petr 09696
Beránková, Dr. Helena 09427
Beránková, Dr. Jana 09621
Beranová Vaicová, Romana 09844
Bérard-Azzouz, Odile 11492
Béraud, Isabelle 13001
Berbakov-Wimmer, Marijana 02403
Berbee, Dr. Mary 07199
Berbion, Phillipe 12062
Berchoux, Pierre 12210
Berchtold, Adriana 41933
Berchtold, Marianne 41207
Berdjajeva, Tatjana Jurjevna 37789
Bereczky, Dr. Loránd 23441
Berendsen, Th. J. 32151
Berens, Ingo 16944, 16964
Berents, Dr. Catharina 18663
Beresford, P. 46030
Bereska, Norbert 20550
Bereza, Vasyl Panasovyč 43057
Berežanskij, Aleksandr Samuilovič 37105
Berg, B. W. van den 32905
Berg, Bettina 17551
Berg, Dan 46818
Berg, Hans-Jürgen 21847
Berg, Kaare 33725
Berg, Karin 33749
Berg, Kristian 40899, 40937
Berg, Manfred 16997
Berg, Peter 16815
Berg, Ralph B. 52802
Berg, Dr. Richard B. 47152
Berg, Roger 49703
Berg, Prof. Dr. Stephan 17460
Berg Lofnes, Solveig 33831, 33833
Bergamaschi, Cirillo 26393
Bergamini, Dr. Giuseppe 28110
Bergbauer, Bertrand 12809
Bergdahl-Nörby, Kerstin 40863
Bergé, Pierre 14057, 31678
Bergen, Terry 46331, 47307, 49711
Bergenthal, Dr. Theo 19359
Bergeon, Annick 12066
Berger, Aaron 46155
Berger, Andrea 01783
Berger, Dr. Andreas 19318
Berger, Claudio 41630
Berger, Erwin 16744
Berger, Dr. Eva 20923
Berger, Dr. Frank 18377
Berger, Georg 18083
Berger, Ingrid 33810
Berger, Jerry A. 53233
Berger, Karin 16744
Berger, Mag. Karl 02137
Berger, Michael 16542
Berger, Moshe Tzvi 25002
Berger, Paul 41234
Berger, Shelly 54146
Berger, Teresa J. 50650
Berger, Theodor 20056
Berger, Tobias 33041
Berger, Torge 16744
Berger, Dr. Ursel 17117
Berger, Vicki L. 53464
Berger-Klingler, Martina 02620
Bergeret, Anne-Marie 13220, 13221, 13222, 13223, 13224
Bergeret, Jean 13778, 15286
Bergero, Susana 00340
Bergeron, André 06356, 06492
Bergès, Catherine 11492
Berges, Sylvie 11999
Berggård, Ingeborg 33941
Berggren, Jan 40948
Bergh, Susan 47624
Bergheger, Brian F. 48375
Bergholm, Synnöve 11230
Berghout, Dr. P.-J. 32182
Bergiers, Marcel 04129
Bergil, Catharina 40558

B

B

Bergin, Christer 40515
Bergman, Ingela 40431
Bergman, Kaj 40534
Bergmann, Eva-Maria 20722
Bergmann, Jóhanna 23993
Bergmann, Prof. Marianne 18700
Bergmann, Sabine 17301
Bergmann, Sven 17998
Bergroth, Tom 11404
Bergs, Hans 22596
Bergs, Lilita 52430
Bergstedt, Dr. Clemens 22802
Bergsten, Wayne 50724
Bergström, Anna-Lena 40782
Bergström, Anne 11187, 11188, 11189
Bergström, Dr. Carin 40854, 40870, 40875, 40897, 40906, 40916, 40928, 40931, 40934
Bergström, Eva-Lena 40918
Bergström, Lars 40526
Bergström, Matti 10846
Bergstrøm, Ture 10142
Bergundthal, Hermann 41475
Beringer, Emile 12928
Berke, Debra 53924
Berkel, R.R. 32832
Berkes, Peter 41170
Berkey, Vernon 53148
Berkhahn, Bernhart 19880
Berkmen, Çetin 42739
Berkovitz, Sam 51960
Berlepsch, Heide von 20537
Berlichingen, Hans Reinhard Freiherr von 19373
Berliet, Paul 13881
Berlin, K.W. 01199
Berliner, Nancy 52657
Berlmont, Karleen 53705
Berman, Jerôme 46791
Berman-Hardman, Richard 01289
Berman-Miller, Cynthia 49322
Bern, José Manuel 34067
Berná Garcia, María T. 38907
Bernabò Brea, Dr. Luigi 27940
Bernabò Brea, Dr. Maria 27099
Bernard, A. 03474
Bernard, Charles 12942
Bernard, Christian 41415
Bernard, Goetz 14184
Bernard, Jean 12715
Bernard, Loic 06168
Bernard, Marie-Thérèse 15564
Bernard, Michel 03607, 12892
Bernard, Raymond 14344
Bernardi, Dr. Vito 27957
Bernardini, Dr. Carla 25469
Bernardini, Dr. Maria Grazia 26741
Bernardini, Paolo 27724
Bernardinis, Domenico de 41448
Bernasconi, John G. 44598
Bernat, Clark 06454
Bernaudeau, Chloé 13775
Bernauer-Keller, Christine 17776
Berndes, Claske 53834
Berndt, Harald 21106
Berndt, Randall 50391
Berndt, Siegfried 22467
Berner, Dr. Hermann 20365
Bernesconi, César 54434
Bernhardt, Harold O. 47274
Bernhardt, Marcia A. 47274
Bernhausen, Sara 17067
Bernier, Ronald R. 54163
Bernier, Suzie 05639, 06456
Berning, Dr. Björn 02310
Berninger, Gudrun 19570
Berninger, M.A. Ulrike 20792
Bernini, C. 42111
Bernklau, Maria 03045
Bernnett, James 00773
Berns, Christiane 30483
Berns, Marla C. 50223
Bernsmeier, Dr. Uta 17549
Bernstein, Bruce 51236
Berntsen, Gunnar 33816
Beron, Prof. Petar 05198
Beroza, Barbara L. 54350
Berraute, Luis Bernardo 00582
Berres, Marge 51452
Berreth, David S. 48763
Berrettoni, Giulio 27013
Berridge, Peter 43872
Berrigan, Douglas 06129
Berrin, Kathleen 52772, 52781
Berry, Barbara 53995
Berry, Bridget 01374
Berry, D.L. 05457
Berry, Diane 48280, 48282
Berry, Ian 43921, 52926
Berry, James M. 49598
Berry, Janet 48789
Berry, Paul 06544
Berry, Susan 53103
Berry, Warren 43978
Berryman, Jill 52291
Berryman, Val Roy 48254
Bersirov, A.M. 37124
Berswordt-Wallrabe, Prof. Dr. Kornelia von 20062, 21698
Bert, James 53725
Bert, Putter 46708
Bertalan, Karin 20868
Bertani, Licia 26172
Bertaud, Henri 11571
Bertazzolo, Prof. Luigi 27553
Berthaud, Gérard 14055
Bertheaud, Michael 53949
Berthelier, Jean 13957
Berthelier, Nadine 12425
Berthelsen, David 01101
Berthod, Dr. Bernard 13884
Berthold-Hilpert, Monika 18498, 21590
Bertholi, Jean 15511
Berthramieux, André 14202
Berti, Dr. Fausto 26837
Bertie, Dr. David M. 44262, 45159, 45371
Bertini, Dr. Pierluigi 26570
Bertling, Dr. Markus 20509
Bertola, Carinne 41690
Bertoletti, Marina 27467
Bertolini, Madalen 53261
Bertolucci, Serena 28053
Bertout, Ernest 13958
Bertram, Dr. Marion 17206
Bertram-Neunzig, Dr. Evelyn 19630
Bertrand, Gérard 14375
Bertrand-Sabiani, Julie 14534
Bertrant, Albert 04071
Bertsch, Patricia 50897
Bertschi, Victor 41432
Bertucio, Barbara 50109
Bertuleit, Dr. Sigrid 21685
Bertus, Jean-Paul 13963
Beruete Calleja, Francisco 39286
Berup, Christina 40634
Berwick, Hildegarde 45150
Beryt, Andrzej 34912
Besch, Prof. Eckart 19065
Bescós, Alberto 39940
Beširov, Ilija 05266
Beskrovnaja, Nina Sergeevna 37048
Besom, Bob 53212
Bessborough, Madeleine 44096
Besse, Nadine 15251
Besse, Pamela 49103
Besseiche, Daniel 16148
Bessette, Yvonne 06056
Bessir, Mark H. C. 50037
Besson, Christine 11609
Bessone, Dr. Silvana 35723
Bessor, Joyce M. 54369
Best, Sherry L. 53532
Beštáková, Dr. Kamila 09686
Besteher, Axel 17128
Besten, G.J. den 32601
Besten, Julia 22747
Beštok, Chabas Korneevič 37439
Bestrom, Beth 52207
Betancourt, Julián 08567
Betancourt Nuñez, Prof. Dr. Claudia Gabriele 18951
Betancur, Víctor Raúl 08532
Betanio, Nemesio V. 34464
Bétemps, Alexis 27587
Bethell, Chris 44918
Bethell, Philip 46091
Bethge, Dr. Ulrich 19930
Betrim 31741
Betscher, Hermann 19969, 19970
Betsky, Aaron 32719
Betta, Dr. Carlo 25788
Bettendorf, Josef 20339
Bettens, Doug 01421
Bettison, Dr. Cynthia Ann 53104
Betton, Richard A. 49067
Betton, Thomas 28323
Betts, Jonathan 44220
Betts, Shirley 06544
Betts, Wendy 05832
Betz, Dr. Astrid 20758
Betz, Eden 51504
Betz, Julia P. 51673
Betz, Martin 50404
Beu, Liliana Marinela 36553
Beuchat, Dr. Paul 41666
Beuf, Jean-Claude 15458, 15459
Beugeling, Dr. Niels 31878
Beuke, Arnold 16822
Beukers, Prof. H. 32482
Beukes, Engela 38743
Beumer, A.C. 33001
Beumers, C.E.L. 32733
Beurskens, N. 32275
Beuster, Stephanie 21475
Beutelspacher, Prof. Dr. Albrecht 18640
Beutelspacher, Martin 18275, 18276, 18277, 20336
Beutl, Richard 02294
Beuze, Lyne-Rose 30669
Bevan, Dr. Alex 01412
Bevan, Alison 45361, 55100
Bevan, Jennifer 01009
Beverly, Roberta 51503
Bevers, Dr. Holm 17172
Bevilacqua, Jean-Paul 12346
Bevilecqua, Emanuele 27541
Bex, Florent 03301
Beyeler, Christophe 12966
Beyeler, Ernst 41763
Beyer, Patrick 06500
Beyer, René 42114
Beyer, Dr. Roland 17913
Beyer, Teresa 49951
Beyer-Rotthoff, Dr. Brigitte 17462
Beynon, Jon 45836
Bezbach, Illja Vasil'ovyč 43205
Bezemek, Dr. Ernst 02097
Bezerra Rolim, Valéria Laena 04464
Bezin, C. 16031, 16032
Bezlaj Krevel, Ljudmila 38467
Bezzola, Dr. Tobia 42083
Bhan, Kuldeep K. 24420
Bhandari, Dr. J.S. 24135
Bhargava, P.C. 24101
Bhat, Prof. H. Krishna 24416
Bhatt, Vikas 24092
Bhattacharya-Stettler, Dr. Therese 41210
Bhaumik, P.K. 24151
Bhave, Dr. S.M. 24355
Bhering Cardoso, Marcio 04697
Bhikul, Tassanee 42568
Bhreathnach-Lynch, Dr. Sighle 24807
Bhullar, Akbar Ali 34027
Biagini, Lorena 27663
Biah, C. Calixte 04182
Bian, Xisuo 07959
Bianchi, Arianna 25896
Bianchi, Loredana 41194
Bianchini, Dr. Gregorio 28067
Bianchino, Dena 50640
Bianco, Dr. Anna Lo 27452
Bianco, Prof. Vito Donato 27606
Biancolini, Daniela 25155, 28029
Bianconi, Teresa 27509
Biasini, Cathérine 21834
Biass-Fabiani, Sophie 11790, 11796
Biba, Prof. Dr. Otto 02994
Bibas, Dr. David 50215
Bibby, Miriam 44639
Biberin, Vladimir Nikolaevič 37880
Bibler, Helen Ann 49132
Bibra, G. von 01243
Bicher, Kristina 52495
Bichot, Denise 12435
Bickel, John 43328
Bickel, Susanne 41383
Bicker, Roxane 20489
Bickford, Christpher P. 49794
Bickhoff, Dr. N. 21937
Bicklmeier, Lori 05679
Bicknell, Terry 33250
Bidar, A. 14440
Biddlecombe, Julie 44511
Biderman, Ellen 52888
Bidner, Stefan 02122
Bidstrup, Wendy 50485
Bidwell, John 51216
Bidwell, Paul 45697
Biedenharn, Murray 50787
Biefel, H. 18187
Biefel, I. 18187
Biegel, Dr. Gerd 22651
Biegel, Susanne 16643
Bieger, Dr. Alfons 41113
Biehl, Martin 33730
Bielamowicz, Zbigniew 35054
Bielański, Janusz 34802
Bielby, Jane 45686
Bielejec, Stephanie 46470
Biemans, Johan 31966
Biemans, Prof. Dr. Jos A.A.M. 31886
Bienedell, Arturo Alberto 00576
Biener, Herbert 01953
Bieńkowski, Joachim 35218
Bier, F. 12820
Bierens de Haan, J.C. 32463, 32860
Bieri, Susanne 41215
Bieri Thomson, Helen 41732
Biesboer, Dr. P. 32295
Biesen, Alden 03367
Bieske, Dr. Dorothee 18328
Bieswisch, Heidi 18202
Biezaitis, M. 00906
Biffen, Lyndia 33190
Biffiger, Steffan 41910
Biganti, Dr. Tiziana 27145
Bigay, Victorino 34335
Bigda, Daisy R. 47376
Bigeck, Heather 51915
Biggart, Norman 47136
Bigger, Mary Beth 46510
Bigham, Julia 46060
Bignasca, Dr. Andrea 41154
Bihari, József 23423
Bihler, A. 18088
Bijelić, Borislav 08857
Bijelić, Dubravka 38074
Bijleveld, W. 31885
Bijleveld van Lexmond, Caspar 41501
Biktaševa, Natalija Nikolaevna 37604
Bilashini Devi, Dr. Haobam 24107
Bilek, Alena 02935, 03029
Bilenduke, Florence 06579
Biles, LeAnna S. 48689
Bilger, Judith 17080
Bilgin, Mehmet 42929
Biličenko, Valentina Andreevna 37667
Bilich, Ivanka 08887
Bilici, Mehmet 42772
Bílková, Libuše 09625
Billa, Lawrenece 38645
Billard, Marie Claude 11787
Billerbeck, Barb 51298
Billian, Egon 19772
Billier, Philippe 15666
Billig, Pamela 47042
Billing, Jerome 52594
Billings, Doug 54142
Billings, Kathy 46144
Billings, Loren 47496
Billings, Vernon N. 49938
Billingsley, Pam 52466
Billops, Camille 51191
Bills, Danny 54154
Billups, Di 45517, 45519
Biloparlović, Ivo 08940
Biloslav, Toni 38433, 38491
Bilyk, Zorjana 43127
Bin Ismail, Halal 30580
Binaud, Daniel 12021
Binder, Christian 18461
Binder, Doug 44383
Binder, Fred 54323
Binder, Gerhard 22773
Binder, Hanspeter 41717
Binder-Krieglstein, Dr. Monika 01982
Bindija, Mosar 38099
Bindley, Kate 43632
Bines, Dr. Joan P. 54105
Binette, René 06338
Bing, Jens 10023
Bingham, David S. 48671
Binikowski 22352
Binkert, Almut 22454
Binkley, David 53899
Binon, Erik 04005
Bins, Coleen 48302
Binti Sirin, Dayang Hajah Zasia 04964
Bintz, Carol 53523
Biran, Prof. Avraham 25019
Biraud, Guy 11603
Birchall, Heather 45099
Bircher, Prof. Dr. Martin 41308
Birchwood-Harper, Barbara 44091
Bird, David M. 48347
Bird, Gill 43888
Bird, Richard 47324
Bird, William 48386
Birdick, Birdy 51561
Birdseye, William S. 53864
Bireley, Ellen K. 49500
Birk, Sherry C. 53912
Birkebæk, Frank A. 10245, 10247, 10248
Birkenstock, Eva 20097
Birker, Susanne 18969
Birklid, Richard 48673
Birkmaier, Leslie E. 47017
Birkner, T. 17124
Birley, Robin 43379
Birman de Bessudo, Raquel 31094
Birnbaum, Dr. Daniel 18389
Birnie, Ian 50244
Birnudóttir, SigrÎðkr[]ur Kristín 24024
Bíró, György 23443
Birol, Çiğdem 42819
Birthälmer, Dr. Antje 22748
Birtwhistle, Marjorie 54219
Bis-Worch, Christiane 30483
Bisanz-Prakken, Dr. Marian 02929
Bisbee, Fay 49571
Bischler, Dipl. sc. pol. Univ. Gaston 18887
Bischof, Amy 49263
Bischof, Klaus 21957
Bischof, Petra 20453
Bischofberger, Emil 41110

B

Bischoff, Dieter 21822
Bischoff, Dr. Franz 17165
Bischoff, Dr. Harald 18199
Bischoff, Dr. Michael 19949
Bischoff, Prof. Dr. Ulrich 17944
Bischoff, Walter 22779
Bisconti, Chino 27299
Bisetto, Christine 49532
Bishof, John 53026
Bishop, Adrian 43768
Bishop, Dr. C. 43377
Bishop, Eliza H. 47884
Bishop, Gale A. 52253
Bishop, Henry 05674
Bishop, Janet C. 52795
Bishop, Kenny 48735
Bishop, Knox 47933
Bishop, Larry 52500
Bishop, Dr. Michael 45244, 45245
Bishop, Norman 52856
Bishop, Ron 48662
Bishop, Russell J. 53076
Bishop, Tommy 53652
Bisig, Xaver 42003
Biskupić, Margareta 08977
Bismarck, Prof. Dr. Beatrice von 19904
Bisol, David 52858
Bissell, James 47625
Bissett, George Ann 48208
Bissière, Caroline 14109
Bissinger, M.A. Eva 21283
Bisson, Gäétan 06109
Bissonette, France 07196
Bissonnette, Anne 49740
Bissonnette, Daniel 07234
Biström, Prof. Olof 10774
Biswas, Dr. T.K. 24425
Bitler, Dorothea 52233
Bitomsky, Michael 17090
Bitsch, Dr. Helmut 21214
Bitter, Adriana 50186
Bitter, Edwin 50186
Bitterli, Konrad 41805
Bittner, Jan 01347
Bivens, Peg 49810
Bivingou-Nzeingui 08806
Bizjak, Vjekoslav 08922
Bizzarri, Laura 25357
Bjarnason, Páll V. 24014
Bjarnesen, Lynne 47963
Bjelić, Dobrila 04262
Bjerg, Ida 10014
Bjerkestrand, Steinar 33528, 33529
Bjerregaard, Jørgen O. 09975
Bjerregaard, Kirsten 09993
Bjerring-Hansen, Jacob 10123
Bjerring Jensen, Knud 10256
Bjerrkjar, Anne 10098
Björk, Barbro 40960
Björklund, Anders 40892, 40933
Björkman, Anna 40515
Björkman, Sten 10727
Björkroth, Maria 40515
Björn, Eva 41007
Björnemalm, Barbro 41012
Bjørgen, Asle B. 33643, 33644
Bjork, Alan 52474
Bjork, Tim 51916
Bjørklund, Prof. Kjell R. 33737
Blaas, Karl 42007
Blaauw, J.P. 32475
Blacharska, Wiktoria 34640
Black, Barbara 53148
Black, Bettye 49937
Black, Bill 01426
Black, Prof. C.M. 44931
Black, George B. 50953
Black, Jack 48524
Black, Jane A. 47985
Black, Jay 53781
Black, Joe 48992
Black, Kenneth N. 52421
Black, L. 01224
Black, Peter 44284
Black, Robin 46122
Black, William N. 48683
Black Petersen, S. 10059
Blackaby, Anita D. 49181
Blackburn, Dr. Bob 46245, 46618, 48134, 48214, 48401, 48622, 49108, 49109, 49235, 49236, 49782, 51524, 51736, 52684
Blackburn, M.A.S. 43727
Blackburn, Rachael 49691
Blackett, Diane 44228
Blackman, Barbara W. 52736
Blackman, Joni 49597
Blackman, Leona 48568
Blackwelder, Martha 52726
Blackwood, R. Duke 53110
Bladen, Martha 53731
Blades, John M. 51677
Blades, Julian 46066
Blades, Kent 05553
Bladt, Inger K. 09955
Blæsild, Benno 09996
Blättler, Alexandra 42024
Blaettler, Roland 41416
Blagg, Margaret 46167
Blagg, Tim 49056
Blagov, Jurij Alekseevič 37028
Blahuta, Renee 48475
Blair, Allan 06435
Blair, David 50223
Blair, Kay 54139
Blair, Robert 01464
Blair, William 43364
Blaise, Nicolet 41609
Blajan, Ion 36227
Blakar, Antony 15978
Blake, Leo D. 47197
Blake, Paul 46092
Blake, Ruby 52486
Blake Roberts, Gaye 45764
Blakeley, Erik 44707
Blakey, Ellen Sue 53493
Blanc, Jean 13450
Blanc, Micheline 15887
Blanc, Monique 14682
Blanchard, Cathérine 14550
Blanchard, Edith 47306
Blanchard, Marian 33184
Blanchard, Raoul 41460
Blanchard, Vincent 11910
Blanchebarbe, Prof. Dr. Ursula 21756
Blanchegorge, Eric 12597, 12598, 12673
Blanchette, Normand 05714
Blanchette, Odile 45592
Blanchy, Béatrice 31611
Blanck, Denis 16295
Blanco, Bobby 34516
Blanco, Cristina 39544
Blanco, Hipólito S. del 00356
Blanco Sansa, Helena 38997
Bland, Bart 53270
Bland, Bruce F. 48234
Bland, Gaye K. 52440
Bland, Julia W. 51086
Bland, Sharon 50007
Blank, Dr. Claudia 20435
Blank, Maxine 49797
Blanke, Dr. H. 18884
Blankenfeld, Ruth 21565
Blankenship jr, James H. 51802
Blankenship, Julie 52797
Blankenship, Ken 47423
Blanquez, Juan 39970
Blanshard, Catherine 44659
Blanshard, Emily 45770
Bláquez, Jimena 40255
Blas, Javier 39502
Blas Pascual, Javier 39958
Blaschke, Inge 41972
Blaschke, Ursula Ruth 18910
Blaschke, Wolfgang 16548
Blasco, Santiago Mateu 39291
Blass, Hillary 51279
Błaszczyk, Bożena 35076
Blatt, Martin 46922
Blattner, Dr. Evamaria 22124
Blauert, Elke 17165
Blaugrund, Dr. Annette 51232
Blavia, Francisco 54524
Błażewicz, Sławomir 35251
Blazewiczea, Stanisław 34626
Blazwick, Iwona 45014
Blazy, Guy 13887, 13893, 15432
Blazy, Simone 13895
Bleathman, William 01159
Bled, Max 15789
Bleecker Blades, Margaret 48701
Bleiberg, Edward 47069
Bleiberg, Elizabeth W. 50359
Bleibler, Jürgen 18478
Bleich, Gerhard 17649
Bleile, Dr. Ralf 21565
Bleistein, Joachim 20777
Blendea, Ion 36399
Bles, H. de 32110
Blesa de la Parra, Miguel 39588
Blesser, Carol 46717
Blessing, Jennifer 51258
Blethen, H. Tyler 47903
Bley, Karl 20723
Bleymehl-Eiler, Dr. Martina 16833
Blicher-Moritz, Tine 10145
Blicke, Linda 47111
Blickle, K.H. 19060
Blickle, Ursula 19697
Bliemeister, Jeffrey 54013
Bliermeister, Simone 18881
Blinderman, Barry 51372
Blinov, Viktor Filippovič 36793
BŁrlinš, Guntis 30025
Blitz, Amy 51181
Bljudova, Ljudmila Grigorjevna 36880
Bloch, Dorete 10698
Bloch Ravn, Thomas 09963
Blocher, Miriam 41181
Block, Christian 12066
Block, Diana 48063
Block, Isabel 50640
Block, Merle 48979
Block, Phillip 51194
Blöchlinger, Peter 41969
Bloem, H. 32209
Bloem, M. 31888
Bloemsma van Nes, Lauren 53461
Blohm Pultz, Janet 48022
Blok, Rhea Sylvia 14598
Blok, Dr. W. 32918
Blom, J. 31990
Blom, P. 32879
Blomberg, Dr. Katja 17127
Blomberg, Maj-Gun 40517, 40518
Blomberg, Nancy 48072
Blome, Dr. Astrid 20143
Blome, Jens 19154
Blome, Prof. Peter 41154
Blommaert, André O. 03712
Blommaert, Georges 03704
Blommel, Henry H. 47780
Blomquist, Fredrik 40782
Blomstedt, Severi 10821
Blomstrom, Bruce 51721
Blond, Anne 10299
Blondel, Madeleine 12737, 12738
Blondin, Linda 07185
Bloodgood, Craig 48227
Bloomfield, Sara J. 53925
Bloor, Paul M. 54033
Blosser, Jack 51586
Blount, Robert 53435
Blue, John 07020
Blue Stanfield, Linda 50870
Blübaum, Dr. Dirk 18478, 21694
Blühm, Dr. Andreas 19635
Blümel, Karin 42076
Bluhm, Claudia 16925
Bluhm, Hans-Georg 19502
Blum, Dilys 51867
Blum, Jürgen 19289
Blume, Dr. Eugen 17123
Blume, Peter F. 50902
Blume, Sharon 53249
Blumenstein, Christian 21098
Blumenthal, Arthur R. 54254
Blumenthal, Prof. Dr. W. Michael 17147
Blumentritt, R. 21541
Blundstone, Judith 01082
Blustain, Malinda S. 46298
Bluthardt, Robert 52701
Blyth, Mary 45663
Blyth-Hill, Victoria 50244
Boa, Valerie N.S. 44365
Boal, Robert 51474
Boal Lee, Mathilde 46880
Boano, Giovanni 25679
Boardman, Penelope 47478
Boarman, Don 49247
Boarowy, Innocenta 35207
Boas, Heinz 21412
Boas, Natasha 52909
Boase, Colin 01624
Boase, Mike 45951
Boast, Mair 45418
Boatright, Neva 48095
Bobar, Maja 04283
Bober, Anna 34568
Bober, Jonathan 46511
Boberg, Scott 49950
Bobinger, Wanda L. 50151
Bobkina, Tatjana Gennadjevna 37917
Bobkov, Konstantin Vasilevič 37129
Bobovec, Ana 08904
Bobzieri, Geraldine 01362
Bočarov, Lev Nikolaevič 37942
Bocca, Massimo 25846
Boccanera, Prof. Giacomo 25621
Boccardo, Piero 26302, 26309, 26320, 26323, 26324
Bocchieri, Prof. Franco 28088
Boccolari, Prof. Giorgio 26755
Bochmann, Deb 46232
Bochnakov, Nikola 05135
Bock, Jessica 19933
Bock, Michael 21952
Bock, Monika 41811
Bock, Dr. Robert 18355
Bockelman, James 53043
Bockius, Dr. Ronald 20148, 20150
Bockrath, Dr. Cornelia 02186, 17926
Bockstefl, Mag. Gregor-Anatol 02260, 02262
Bockwoldt, Michael 48004
Bocquier 12105
Bocvarova-Plavevska, Marika 30520
Bod, Roy 47511
Bode, Monika 18900
Bode-Hempton, Sharon 49957, 49959
Boden, Karen 50445
Bodenmüller, Carola 21946
Bodenstein, Anton 02904
Bodenstein, John Gilbert 20728
Bodilsen, Inge 10029
Bodine jr., William B. 51941, 51949
Bodingbauer, Prof. Adolf Karl 02771
Bodini, Prof. Arturo 26687
Bodini, Prof. Floriano 25691
Bodini, Mauro 27870
Bodish, Glenn 47150
Bodley, Alison 44184
Bodner, Connie 50901
Bodner, Helen 06438
Bodó, Dr. Sándor 23392, 23425
Bodoky, Dr. Tamás 23397
Bodorová, Olga 38340
Bodrova, Zoja Evgenjevna 37752
Bodsch, Dr. Ingrid 17453, 17467
Bøe, Kaare 33412
Boebel, Ute 16774
Böcher, Friederike 16765
Böcher, Hans-Georg 19072
Böck, Elfriede 18973
Boeck, Lorraine 53255
Böckle, Karlheinz 21783
Bödecker, Ehrhardt 22755
Bödewadt, Uwe 40903
Böge, Rainer 16475
Boeggild Johannsen, Dr. Kristine 10155
Boeglin, Yves 15874
Böhle, Hans-Theo 17837
Böhle, Rainer 16812
Böhler, Dr. Bernhard 02979
Böhm 21318
Böhm, Dr. András 38389
Boehm, Barbara 51152
Böhm, Helmut 02983
Böhm, Klaus-Dieter 16848
Böhm, Dr. Max 19338
Böhme, Dr. Ernst 18701
Boehme, Prof. Dr. Madelaine 22122
Boehme, Dr. Sarah E. 47671
Böhme, Prof. Dr. Wolfgang 17469
Böhme-Netzel, Susanne 22699
Böhmer, Sylvia 16408
Böhnert, Miriam M. 18360
Böhnke, Bärbel 18126
Böhnlein, E. 20724
Boelen, Jan 03687
Boelens, Christ 32775
Böller, Susanne 20491
Bömerová, Jana 09758
Böndel, Prof. Dr. Dirk 17074
Bönig, Dr. Jürgen 18946, 18962
Böning, Heinrich 21144
Bönnen, Dr. Gerold 22684
Bönninghausen, D.E.J. von 31866
Boer, A. den 32170
Boer, Dr. Anne-Marie 32083
Boer, F.G. de 32626
Boer, P. den 32142
Börjesson, Klas 40623
Boero, Prof. Ferdinando 27302
Boers, T. 31864
Börschlein, Eva-Maria 20428
Börste, Dr. Norbert 20966, 20970
Böse, Hans-Joachim 22620
Böser, Wolfgang 18644, 21893
Boesmans, Annick 03621
Böth, Dr. Gitta 18858
Böthig, Dr. Peter 21302, 21304
Boëthius, Lena 40547
Böttcher, Andreas 17528
Boettcher, Chris 52790
Boettcher, Claudia 04643
Boettcher, Graham C. 46822
Böttcher, Dr. Ronald 21959
Böttner, Waltraud 19224
Böttrich, Prof. Dr. Christfried 18749
Boffa, Prof. Giovanni 41092
Bofill, Teresa Blanch 39921
Bogan, Alan T. 48660
Bogardus, Lorraine B. 50857
Bogatyrev, E. 37202, 37244
Bogdanov, Pavel 37157
Bogdanova, Lidija Aleksandrovna 37878
Bogdanova, Marija 04988
Bogdanovna, Vera Ivanovna 37796
Bogdanowicz, Prof. Wiesław 35248
Bøge Henriksen, Kirsten 10237
Boggio, Stefano 28156
Boggs, Glen 54279

Boggs, Mac 53193
Boggs, Rex 48603
Bogman, Suzanne 31898
Bognanni, Filippo 27194
Bogner, Mag. Sonja 01772
Bogoljubskaja, A.V. 37977
Bogurat, W. 35184
Bogus, Vasile 36619
Bogusky, Alfred M. 06103
Boguszewicz, Ireneusz 34581
Bogutz, Ruth 49113
Boguwolski, Ryszard 34683, 34684, 34685
Boheme, Prof. Dr. Enrico 27536
Bohlmann, Dieter-Theodor 19403
Bohlmann, Dr. Heinz 21673
Bohnenkamp-Renken, Prof. Dr. Anne 18370
Bohnert, G. 19771, 19773
Bohórquez, Jorge Alfonso 08665
Bohr, G. Ted 51549
Boie, Charles 50715
Boileau, Nanette E. 50279, 52589
Boillot, M. 14095
Boire, Janine 46891
Bois, M. de 31937
Boissard, Jerome de 13042
Boisselle, Lynn 06586
Boissevain, C.F.C.G. 32120, 32121
Boissière, Yves 13011
Boitani, Dr. F. 27518
Boits, Paul 03309
Bojani, Prof. Gian Carlo 27153
Bojarczuk, Helena 34688
Bojarskij, Dr. Viktor I. 37723
Boje, Corinna E. 21960
Bojko, Elena Michajlovna 37799
Bojko, Nina Vladimirovna 37079
Bojkova, Irina Petrovna 37969
Bojkova, Margarita Alekseevna 36889
Bojović, Nevenka 38094, 38094
Bojović, Radivoje 38094
Bok, Franciszek 35151
Bok, Klaus 19271
Bokbot, Dr. Youssef 31685
Bokchun, Youngja 29865
Bokum, Anne-Marie ten 03321
Bol, Marsha C. 52880
Boland, Raymond 47205
Bolandrini, Santina 41731
Bolas, Gerald D. 47344
Bolderman, J.C.L. 31928
Boldt, Erwin 16836, 16838
Bolduan, Anka 17564
Boley, Mark 50692
Bolge, George S. 46883
Bolger, Doreen 46551
Bolger, Dr. Francis 05594
Bolin, Jerry 50357
Boling, Ron 52706
Bolipata, Alfonso Corpus 34497
Bolívar, Julián Gil 08542
Bolla, Dr. Margherita 28230, 28235, 28236
Bolla, Stefano 41705
Bolland, Lionel 45416
Bolle, Sandrine 03744
Bollendorf, Jeanne M. 49274
Bolliger Schreyer, Sabine 41207
Bolling, Richard G. 49067
Bollinger, Dr. Thomas 41087
Bologa, Adela 36307
Bologna, Alberto 26480
Bolognani, Betty 52269
Bolohiş, Ioan 36435
Bolongaita, Prof. Johny P. 34355
Bolotov, Valentin A. 37292
Bolotova, G.P. 37635
Bolster, Jane 51111
Bolt, Ed L. 51912
Boltar, Barbara 38501
Bolten Rempt, Dr. H. 32489
Boltesilverman, Diane 48484
Bolton, Bruce D. 06329, 06376
Bolton, Chris 53628
Bolton, David 44195
Bolufer Marqués, Joaquim 40325
Boly-Lecocq, Anny 03958
Bolz, Dr. Peter 17081
Bombard, LuAnn 54051
Bomberger, Bruce 49928
Bommerlyn, Loren 47874
Bon, Dr. Mauro 28197
Bona Dea, Artemis 46286
Bonaccorsi, R. 13442
Bonačić Mandinić, Maja 08954
Bonafede, Irja 50815
Bonagura, Pasquale 28206
Bonami, Francesco 25934, 28000
Bonamini, Eros 30194
Bonanno, Dr. Carmela 26056
Bonansinga, Kate 48323, 48329
Bond, Anthony 01544
Bond, Debra 51770
Bond, Hallie 46876
Bond, Karen 48871
Bond, Kaye 51781
Bond, Wanda L. 49089
Bond-Fahlberg, Margareta 40921
Bondar, Dr. Vitalij Vjačeslavovič 37066
Bondareva, Valentina Nikolaevna 36915
Bondaz, Gilles 15895
Bondil, Nathalie 06357
Bondjers, Rune 40515
Bone, Laurie 48560
Bone, Oliver 45846
Bonelius, Elina 11364, 11365
Bonemaison, Michel 13882
Bonet Armengol, Jordi 39006
Bonet Rosado, Helena 40206
Bonev, Ivan 05219
Boney, Roger 45311
Bongaarts, J. 32363
Bongaerts, Ursula 27442
Bongard, Rolf B. 17318
Bonham, Rebecca 53160
Bonias, Zisis 23046
Boniface, Didier 15695
Bonifay, Suzanne 15230
Bonilla, Irene 40201
Bonilla, Mary Yamilet 54584
Bonin, Flavio 38492
Bonisławski, Wojciech 34641
Bonn, Greg 49889, 52670
Bonnard, Jacques 14604
Bonnard Yersin, Pascale 41987
Bonnefille, J. Lovis 15204
Bonnefoi, Geneviève 13063
Bonnema, Hollie 49210
Bonner, Deanna 46485
Bonner, Ian 44503
Bonnet, P. 32047
Bonnet Borel, Francoise 41971
Bonnett, Dr. Richard B. 49434
Bonnevialle, Gilles 15219
Bonnier, Bernadette 03906
Bonnin, Mirta 00321
Bono, Sam 46183
Bonofiglio, Marta Maria 00538
Bonoli, Prof. Fabrizio 25455
Bonomi, Prof. Alfredo 27141
Bonomi, Dr. S. 25153
Bonstecht, Andreas 17179
Bonvalet, Louis-Georges 14049
Bonzel, Helen 18511
Booknight, Alice 47728
Boomstra Van Raak, M.A.C. 32472
Boonstra, Dr. J.R. 31834
Boos, Dr. Andreas 21224
Boos, Kurt 17517
Boot, C. 32829
Boot, M. 31888
Booth, Amanda 45248
Booth, Beverley 33075
Booth, C.M. 45510
Booth, Helen 06033
Booth, William H. 53521
Booty, Peter 47314
Bora, Dr. D.K. 24231
Borák, PhDr. Mečislav 09692
Borbolla, Dr. Sol Rubin de la 31133
Borchardt, Dr. Albert 19071
Borchardt, Ernstine 03591
Borchardt, Dr. Stefan 19050
Borchardt, Susan A. 50530
Borchers, Karl-Hans 22205
Borchert, Hans 19840
Borchert, Till Holger 03406, 03414, 03415
Borčić, Goran 08960
Bordallo Huidrobo, José Antonio 39586
Bordaz, Odile 16248
Borden, Tamika 51100
Bordier, Alain 41462
Bordinskich, Gennadij Aleksandrovič 37785
Bordon, Bob 54053
Bore, Ove Magnus 33865, 33866, 33875
Boreham, Peter 45507
Borel, Dr. Gilles 41542
Borel-Menut 13378
Borelli, Dr. Ing. Ferdinando 28026
Borellini, Alessia 41283
Borén, Ingjerd 40555
Borg, Prof. Dr. Vince 30630
Borgegård, Eva 41032
Borgen, Prof. Liv 33737
Borges, Grethe 33518
Borges, Richard 46739
Borges, Wolfgang 19808
Borges Rosales, Prof. J.C. 54582
Borgeson, Jaquelyn L. 47342
Borggrefe, Dr. Heiner 19949
Borghi, Gian Paolo 26111
Borgmann, Jan 18809
Borick, Carl P. 47359
Borisenko, Irina Gavrilovna 37389
Borisenko, Viktor Ivanovič 37869
Borisov, Boris 05142
Borisova, Diana 05017, 05018, 05019
Borisova, Tanja 05137
Borisova, Valentina Vladimirovna 36763
Borja-Santiago, Monina Katherina 34446
Borja-Villel, Manuel J. 38972
Börjesson, Sven 40975
Bork, Prof. Dr. Hans Rudolf 21248
Borlinghaus, M.A. Natalie 20414
Borman, R.Th.A. 31929
Born, J.G.M. 32339
Bornacelli Guerrero, Humberto 08752
Bornecrantz, Mona 40960
Bornefeld, Christian 22121
Bornfriend, Carl 48798
Bornhorst, Dr. Theodore J. 49377
Bornman, A.R. 38555
Bornscheuer, Dr. Marion 18027
Borobia, Mar 39583
Borodavkin, Andrej Viktorovič 37497
Borodina, Valentina Ivanovna 36864
Boroghină, Emil 36329
Borošak Marjanović, Jelena 08997
Borovich, Yael 25122
Borovinskich, Nadežda Petrovna 37858
Borowski, Batya 24980
Borowski, Wiesław 35227
Borowsky, Gwen 51860
Borpuzari, D. 24194
Borrachero, Germán M. 39619
Borremans, Dr. R. 03670
Borriello, Dr. Mariarosaria 26883
Borromeo, Dr. Georgina 52170
Borrowdale-Cox, Deborah 50057
Borsanyi, Jackie 50564
Borsboom, Leo 32958
Borščenko, Lidija Mychajlivna 43114
Borsch, Prof. Thomas 17053
Borsdorf, Prof. Dr. Ulrich 18264, 18267
Borselli, Lucilla 27787
Borsig, Josef 09395
Bortolan, Gino 28199
Borut, Križ 38490
Borys, Stephen 51478
Borysova, Galyna Stepanivna 43209
Borzan, Maria 36559
Bos, Agnès 12809
Bos, W. 32782
Bosano Ansaldo, Dr. Carlos de 00116
Bosch i Cornellà, Salvador 39984
Boscheinen, Simone 18391
Boschi, Prof. Enzo 26059
Boschiero, Dr. Gemma 25312, 25314, 25315
Bose, Krishna 24279
Bose, Prof. Sugata 24279
Boser, Dr. Elisabeth 17764, 17766
Boserup, Dr. Ivan 10126
Boshoff, Jaco 38591
Boshouwers, F. 32277
Bosi, Erio 26533
Bosioković, Sava 38093
Boškova, Tatjana 30523
Bošković, Dora 08997
Bosma, M. 32844
Bosman, C. 31885
Bosmans, Wim 03477, 03495
Bosnar Salihagić, Željka 09013
Bošniak, Jurij Michailovič 37881
Boss, George 44076
Boss, Dr. M. 18219
Bossa Herazzo, Donaldo 08619
Bossaert 03825
Bossan, Marie-Josèphe 15088
Bossart, Hans Peter 42083
Bossé, Laurence 14646
Bosshard, Daniel 41720
Bosshard, Reymond 41650
Bosshard, Valerie 41165
Bosshard-Frischknecht, Hulda 41964
Bossine, Lesley 44569
Boßler, Sascha 22264
Bostanci, Gülten 33368
Bostander, Nellie 38774
Bostic, Connie 46370
Bostick, Christine 52085
Bostin, Franky 04166
Boston, David 45978
Bostrem, Lidija Andreevna 36772
Boswell, Mary Rose 49861
Boswell, Peter 50631
Boswell Minnerly, Vet 48259
Bot, Nadia 41410
Bot Riera, Issa 40292
Botea, Alin 36509
Botelho, Joce M. 52169
Botelle Ortega, Daniel 39496
Botero, Dr. Clara Isabel 08577
Botha, R. 38560
Bothe, Günter 16745
Bothe, Heinrich 17409
Bothmer, Dietrich von 51212
Botineau, Pierre 12064
Botirov, I.T. 54466
Botsaris, Markos S. 22899
Botsen, Jean-Pierre 14809
Botte, Prof. Virgilio 26902
Bottengård, Else 33921
Bottini, Dr. Angelo 26167
Bottlaender, Michael 18969
Bottrel Tostes, Vera Lúcia 04752
Botwinick, Michael 54331
Bouchard, Guy 06761
Bouchard, Marie 14635
Bouchard, Yann 15665
Bouchard-Acodas, Louise 05697
Boucher, Celine 05563
Boucher, George 48250
Boucher, Michel 12141, 12142
Boucher, Roger 05555
Boucher, Sid 01422
Bouchet, Denis 13700
Bouchet, J.-C. 11568
Bouchon, Juan Santiago 00478
Bouck, Jill 48285
Bouckaert, Odile-Marie 03420
Bouda, Jiří 09410
Boudet, Sylvie 14174
Boudier, Pierre 12426
Boudou, Dominique 14292
Boudreau, Luce Marie 05607
Boudrot, Donna 05352
Bouffault, Colette 14514
Boughton, Jim 52493
Boughton, Peter 43818
Bouglé, Frédéric 15880
Bouillon, Philippe 03769
Boujamid, Azzouz 31668
Boukli, Aïcha 00044
Boulanger, Patrick 13997
Boulanger, Sylvie 12475
Boulbes, M. 12861
Bould, Gillian 45646
Boulet, Jean 16052
Boulliard, Jean-Claude 14583
Bounous, Jewell P. 53684
Bourassa, M.Sc. Annie 06145
Bourassa, Marie 06611
Bourbon, Prof. Giulio 25696
Bourbon Busset, Jean-Louis de 11831
Bourbonnais, Alain 05398
Bourbonnais, Caroline 12719
Bourcier, Paul 50396
Bourdais, Jany 12674
Bourdon, Don 05383
Bourdon, Kate 06745
Bourdon, Michel 12507
Boureima, Konfé 05276
Bourel, Yves 15431
Bouret, Brigitte 15059
Bourgade, Véronique 11756
Bourgat, Robert 14793
Bourgeau, Jef 52021
Bourgeois, Jean 04133
Bourgeois, Valérie 07171
Bourget, Steve 41421
Bourgoin, Colette 05543
Bourgon, Jacques 14172
Bourgougnon, Eric 14222
Bourgoz, Michel 41134
Bourke, Marie 24807
Bourlet, Michel 12718
Bournazel, Comtesse de 14943
Bourne, Susan 43679
Bouše, Petr 09649
Bouska, Marianne 47544
Bousmanne, Bernard 03438
Bousquet, Nicolas 11555
Boussard 14514
Bousska, Prof. Dr. Vladimira 02969
Boustany, Bernadette 13455
Bouteiller, Hubert 12692
Bouthat, Chantal 06339
Boutry, Pierre 37993
Bouttemy, Marie-Françoise 13756
Bouvelot, Q. 32082
Bouvette, Henry d'Otreppe de 04112
Bouwman, J.W.M. 31785
Bouxin, Marc 15005
Bouyack, Kay 52103
Bouzón Gallego, Avelino 40175
Bouzy, Olivier 14533
Bovewizer, Austin 24891
Bovey, Patricia E. 07392
Bowden, Cindy 46449
Bowdery, J. 43872
Bowdion Gil, Carol 47573
Bowditch, John 46303
Bowdrey, Jeremy 43875
Bowen, Allyson 51168

Bowen, Duncan 44344
Bowen, Graham 43487
Bowen, Jane 43301
Bowen, Linnell R. 46315
Bower, Alan 45824
Bower, Barbara A. 48620
Bower, Corinna 45949
Bower, Finn 06955
Bower, Mary 48457
Bower, Thomas 48452
Bowers, David 45157
Bowes, Murray 01634
Bowes-Crick, C.P. 44973
Bowie, Dr. Gavin 43683
Bowle, Jean 44200
Bowles, Evelyn 48300
Bowles, G. 43803, 43804
Bowman 44647
Bowman, Jack 48195
Bowman, Leslie Green 54258
Bowman, Midge 52999
Bowman, Travis 47618
Bownan, JoAnn 50856
Bowron, Julian 01300
Box, David 01177
Boxall, Michael 43476
Boxberg, Hendrik von 18263
Boxer, David 28328
Boxer, Graham 44742
Boxhall, Patricia 00924
Boxley, Robert F. 54222
Boyce, Connie 47008
Boyce, David Y. 50020
Boychev, Petr 05233, 05234
Boyd, Beatrice 53729
Boyd, Sir John 44803
Boyd, L. 06013
Boyd, Peter 45676
Boyd, Robert 46733
Boyd, Willard L. 49522
Boydell, Owen 44097
Boyden, Dr. P. 44936
Boyer, Anne Marie 12960
Boyer, Cindy 52393, 52399
Boyer, Jean-Marc 15742
Boyer, John Vincent 49195
Boyer, Sylvain 11789
Boyette, Todd 46374
Boylan, James 53312
Boylan, Olive S. 50908
Boyle, Catherine 47246
Boyle, Dorothy 52675
Boyle, Gail 43632
Boyle, William J.S. 07156
Boynton, Hugh 44421
Boysen, Jesper 10093
Boysen, Renate 19831
Boysson, Bernadette de 12068
Boz Balmer, Katia 41988
Bozan, Maria 36607
Božič, Bojan 38435
Božko, Boris Stepanovyč 43171
Bozkurt, Batuhan 42897
Bozok, Erdoğan 42824
Braack, Craig R. 46719
Braam, Dr. P. 32183
Braarud, Reni 33709
Braat-Voorhuis, T.M.A. 33002
Braaten, Alicia L. 47580
Braaten, Ivar Gunnar 33381
Braaten, Trond 33413
Brabant, Diane 06742
Brabec, Jiří 09812
Bracegirdle, Hilary 45879
Bracegirdle, Robert 45033
Brachaczek, Zofia 35150
Bracht, Dr. Christian 20187
Brack, Dr. P. 42071
Bracke, W. 03438
Bracke-Logeman, P. 31968
Brackenreg, Philip 00815
Bracker de Hugo, M. 15440
Brădățan, Aura 36640, 36713
Bradbury, Christopher 46028
Bradbury, David 44866
Bradbury, J. Robert 06432
Bradbury, Jade 50273
Bradbury, Karen 44735
Braddick, Lyndon 44149
Braddock, Pam 44333, 45419
Brade, Dr. Johanna 18680
Brademann, Margret 17620
Bradfield, Gerald 49883
Bradford, Anna 05953, 05956
Bradford, Dr. Bruce C. 47998
Bradford, Jan 51081, 51082
Bradford, Jolene 49258
Bradford, W.J. 05726
Bradley, Allen 51298
Bradley, Anne 45389
Bradley, Betsy 49560
Bradley, Bob 45106
Bradley, Dean 06730
Bradley, Fiona 44124
Bradley, Graeme 33149
Bradley, J. 45317
Bradley, J.M. 43977
Bradley, Larry 53721
Bradley, Laurel 51418
Bradley, Richard 50928
Bradley, Dr. Robert 50312
Bradley, Ron 07255
Bradman, Susan 43913
Bradner, Anne 52201
Bradshaw, Graham 45226
Bradt, Rachelle 51281
Brady, Carolyn 52285
Brady, Christine 44594
Brady, Edith 49275
Brady, Jim 52062
Brady, Tess 53255
Braekeleer, Catherine de 03766
Brändle, Christian 42081, 42092, 42094, 42095
Bräuer, Jörg 16553, 16555, 16556
Brafford, C.J. 50830
Bragagnolo, Dr. Pietro 26130
Bragg, Cheryl H. 46322
Bragg, Gary O. 51775
Braham, George O. 07168
Brahim, Natalia Majluf 34232
Brahm-Lindberg, Lisa 54305
Brahma Chary, P. 24255
Braica, Silvio 08955
Braid, Dennis 51644
Braillard, Marie-Christine 15610
Braipson, Roger 04073
Brakke, Jaap 19002
Bralić, Zdenko 08959
Bramah, Edward 44798
Bramberger, Alfred 01925
Brancaccio, Jim 50052
Brancaleoni, V. 04045
Brancati, Prof. Antonio 27154
Brancato, Prof. Francesco 27070
Brancel, Esther 54095
Brancelj, Andreja 38481
Brancucci, Dr. Michel 41175
Brand, M.A. Elke 17673
Brand, Dr. Joachim 17165
Brand, Margrith 41784
Brand, Dr. Michael 52343
Brand, Dr. Ruth 41261
Brand-Claussen, Dr. Bettina 19082
Brânda, Nicolae 36146
Brandão, Daniela 35802
Brandão, Fátima 35802
Brandaó, Maria Luiza 04479
Brandão Aguirre, Maria Beatriz 04864
Brandel, Katrin 17200
Brandeller, Paul 03963
Brandén, Peter 40923
Brandenburg, Dr. Hajo 20734
Brandenburg, Dr. Winfried 21500
Brandenburg-Eisele, Dr. Gerlinde 20888
Brandenstein, Ameli von 21577
Brander, Prof. A. 10736, 10738
Brander, Alli 10876
Brandes, Paula 52746
Brandis, Wolfgang 22542
Brandl, Andrea 21682
Brandl, M.A. Andrea 21683
Brandl, Franz 01787
Brandmeyer, Dennis 53637
Brandoli, Susan 07238
Brandon, Cath 00973
Brandon, G. 05629
Brandson, Lorraine 05614
Brandt, Dr. Jochen 18936
Brandt, K. 31885
Brandt, Prof. Dr. Michael 19182
Brandt, Monique 05721
Brandt, Dr. Reinhard 22453
Brandt, Thomas 20657
Brandt, Tova 48022
Brandt, Wayne 48476
Branscome, Curtis 53310
Bransom, Jennifer 47937
Brantl, Dr. Markus 20424
Brantley, R. Brit 47407
Braquilanges, Guy de 12336
Brascamp, M.H. 32448
Brasel, Ronald 47339
Brasie, Jeff 47554
Brasil do Amaral, Paulo Cesar 04644
Braslavsky, Emma 17088
Braslavsky, Noam 17088
Brassel, Dirk 17499
Brassine, Edouard 04136
Bratčenko, Tetjana Georgiïvna 43024
Bråthen, Magne 33600
Bratscher, Renzo 41434
Bratschi-Tschirren, Max 41356
Brattig, Dr. Patricia 19619
Brattmyhr, John 40895
Bratuž, Ciril 38492
Brauer, Gerald J. 48038
Braun, Dr. Andreas 02898, 20428
Braun, Annette 13301
Braun, Dr. Claudia 20173
Braun, Dr. Emanuel 18081
Braun, Hannes 02631
Braun, Prof. Dr. Harald 18458
Braun, Dr. J.K. 51376
Braun, Joseph 13300
Braun, Karen 52224
Braun, Märta 41061
Braun, Mary E. 52613
Braun, Matthew 49546
Braun, Monika 19449
Braun, Peter 02617
Braun, Thomas 16854
Braun, Volker 17029
Braun, Wilhelm 20380
Braunegger, Reinhold 02496
Braunlein, John 48859, 51096
Braunová, Helena 09571
Brauns, Dmitri G. 52787
Braunstein, Susan L. 51200
Braunsteiner, Dr. Michael 01693
Braut, Else 33637
Bravo, John 52981
Braxton, Gail G. 48767, 48770, 48772, 48773
Bray, Alida 55138
Bray, Bill 48572
Bray, Rick 47951
Bray, Tamara 48126
Bray, Xavier 44937
Brayer, Hermione 13694
Brayner, Vânia 04668
Braz Teixeira, Dr. Madalena 35722
Brazauskas, Pijus 30370
Brázdová, Marie 09927
Brearley, Anna 44345
Brebeck, Wulff E. 17637
Brecciaroli, Dr. Luisa 28010
Brecht, Fatima 51220
Bredenkamp, Dalene 38575
Brederek, Wolfgang 00999
Bredow, Florentine C. 19221
Bredsdorff, Eva B. 44773, 45256, 45963
Breede, Claus 05879
Breen, Evelyn 24749
Breen, John D. 52940
Breese, Dr. Kathryn 49564
Bregar, Marjeta 38490
Bregeaut, Pierre 12245
Bregovac Pisk, Marina 08997
Brehm, Georgia 50317
Brehm, Dr. Thomas 20764
Brehme, Sylvia 17036
Breidbach, Prof. Dr. Dr. O. 19377
Breiler, Ronald 21411
Breinegaard, Jens 10063
Breiter, Fritz 41525
Breithaupt, Brent H. 49947
Breitkreuz, Petra 18396
Breitmeier, Uwe 17787
Brekelmans-Bijveld 32955
Breker, Uwe H. 19622
Brekke, Borghild 33711
Brembeck, Josef 20343
Bremer, D. Neil 48374
Bremer, Stanley 32733
Bremner, R.W. 44026
Brendel-Pandich, Susanne 53451
Brendstrup, Dagmar 09997
Brener, Lidija Davydivna 43131
Brengle, Anne B. 51014
Brenke, Johannes 16818
Brennan, Anita 50856
Brennan, Anne G. 54202
Brennan, Edward 51251
Brennan, Nancy 51996
Brennan, Robert 24826
Brennan, Terry 52744
Brenneman, Jan 48719
Brenner, Itzhak 25092
Brenner, Dr. Klaus-Peter 18696
Brenner, Stefan 01707
Brenner-Wilczek, Dr. Sabine 17991
Brenner Wilczek, Dr. Sabine 18500
Brent, Marty 06486
Bréon, Emmanuel 12090
Bresc, Geneviéve 14698
Breščak, Mateja 38457
Bresching, Christian 20593
Brescia, Michael M. 53584
Breslauer, David 50853
Bresler, Zinaida Ivanovna 36760
Bresnahan, Bob 50600
Bresse, Robert 14667
Bresser, G.J. de 32380
Bressi, Dr. Nicola 28089
Bressor, Carol 48984
Breth, Renata 52908
Breton, Gérard 13599
Bretón, Prof. Raúl 31054
Bretscher, Peter 41377
Bretschneider, Jürgen 19021
Brett, Ruth 05775
Brettner, Friedrich 01939
Breuel, Margitta 17375
Breuer, Dieter 22551
Breuer, Heiko 18900
Breuker, P. 32231
Breunig, Robert G. 48551
Breuze, Marc 13980
Brevik, Bernt 33860
Breville, Tristan 30683
Brewer, Billy 46251
Brewer, Dr. Douglas J. 53669
Brewer, Esther Vivian 48123
Brewer, R. 43748
Brewster, Carrie 50838
Brewster, Christine 43486
Brewster, Pete 05381
Brewster, Rae 01533
Brewster, Suzanne G. 46126
Brewster, William F. 50398
Brey, Karen 52278
Breyer, Emilie 02620
Brez, Wendy 48578
Breza, Bogusław 35296
Breza, Michael 51615
Briand, Henri 15394
Brice, Alison 43468
Brichet, Michel 03467
Brichtová, Dobromila 09653
Brick, Elizabeth M. 51406
Bricka, Olena Mykolaïvna 43027
Brickman, Penny 52954
Brickwood, Kevin 05510
Bridenstine, James A. 49666
Bridger, Tozzy 44216
Bridges, Dr. E.M. 44203
Bridges, Edwin C. 50813
Bridges, Mary 46238
Bridges, Tim 46074
Bridgford, Todd C. 49154
Bridgman, Norma 38628
Bridwell, Shawn 50119
Briedenhann, M.M. 38564, 38565
Briedis, Agris 30066
Briend, J. 14648
Briere, Florence 13572
Briese, Jodie G. 54272
Briest, Francis 16135
Brigger, Lilli 41446
Briggs, Kate Halle 53190
Briggs, Dr. Peter 50312
Briggs-Jude, Wendy 07309
Briggs Murray, James 51254
Brigham, Ann 48201
Brigham, David 54302
Brigham, David R. 46225
Bright, Bill 48738
Bright, Paul 54243
Brighton, Terence 44346
Brillonet, Jean-Marc 15501
Brimm, Steve 53199
Brincard, Marie Therese 52201
Brind, Rosemary 43434
Brindeau, Irène 13581
Brindley, Kate 45141
Brindley, Lynne 44802
Brindley, Ronan 43691
Briner, Dr. Thomas 41886
Bringager, Frithjof 33735
Brink, Bob 54318
Brink, Siegfried 22517
Brinklow, R. 44047
Brinkmaier, Ulrich 49199
Brinkman, L.D. 49750
Brinkmann, Hermann 22661
Brinkmann, Prof. Dr. Vinzenz 18394
Brinkmann, PD Dr. W. 42103
Brintzinger, Dr. Klaus-Rainer 20492
Brissard, Michel 16154
Brissette, Serge 12935
Brisson 14765
Brisson, Steven C. 50351
Brito Joao, Benedito 31714
Brito Moreira, Alvaro de 35916
Britschgi, Dr. Jorrit 42097
Britton, Marion 50163
Britz, Günther 16971
Brivio, Dr. Ernesto 26698
Brix, Thorsten 14279, 14281
Břízová, Zuzana 09504
Brizzi, Prof. Gian Caolo 25471
Brkić, Dragana 04283
Brkić, Staniša 38110
Broach, Barbara K. 48563, 48566
Broad, Sandy 06990

B

Broadey, R. 43565
Broady, Duncan 45083
Brobbel, Joni 45507
Brobst, William A. 52117
Brocas, Jon 33138
Brock, Mary 51920
Brocke, Dr. Edna 18247
Brockhaus, Prof. Dr. Christoph 18027
Brockhoff, Joan 07023
Brockman, Prof. Christian 33737
Brocksch, Dr. Dieter 20813
Brodanec, Vlado 08989
Brodbeck, Andrea 21960
Brodeur, Bill 06581
Brodie, Ian J. 33232
Brodl, Michaela 03045
Brodneva, Adelja Vladimirovna 37074
Brody, Aaron 46750
Broege, Wallace W. 52368
Broeker, Dr. Holger 22665
Broekgaarden, A. 31852
Brogiolo, Prof. G. Pietro 27599
Broglie, Charles Edouard Prince de 12376
Brohel, Edward R. 51984
Broidl, Erich 02785
Brok, Dr. Barbara den 41564
Broker, David 01087
Brokke, Harris 52346
Brolsma, Allen 47886
Brom, Jodi 54238, 54239
Bromberg, Dr. Anne 47931
Brombierstäudl, Andreas 19352
Bromley, Fran 49176
Bromovsky, Fabia 45905
Brondeau, Armelle de 11538
Brook, Griff W.T. 01145
Brooke, Janet M. 06073
Brooke, Patrick 45860
Brookes-Sullivan, Rebecca 43549
Brookman, Philip 53855
Brooks, Anthony T.P. 54162
Brooks, April 45578
Brooks, Carol 54364
Brooks, Carolyn 50284
Brooks, Daniel F. 46820
Brooks, Eric 50045
Brooks, Dr. Jim 50982
Brooks, Leonard L. 51046
Brooks, Lindsay 45599
Brooks, Nick 45526
Brooks, Peggy 48887
Brooks, Philip R.B. 46088
Brooks, Ruth 54084
Brooks jr., William F. 50651
Brooks Joyner, J. 51552
Brooks-Myers, Inez 51470
Brooslin, Michael 53227
Brose, Barbara H. 47923
Brose, Dr. David S. 46857
Brose, Lawrence F. 47120
Brosens, Annemarie 33017
Brosi, Dr. Sibylle 19440, 19448
Brosnahan, Seán 33096
Brosnan, Susan 51052
Brost, Jeff 52304
Brost, Leif 40595
Brosthaus, Dr. Karl-Heinz 20224
Brothers, Ann 01287
Brothers, Leslie A. 54354
Brotherton, Barbara 53013
Broughton, Jamie 43325
Brouillet, Johanne 06959
Broun, Dr. Elizabeth 53917
Brousseau, Francine 05861, 05862
Brouwer, M. 32494
Brouwer, Dr. M. 32590
Brouwer, Norman 51260
Brouwer, T. de 32606
Brovčenkova, Valentina Ivanovna 36840
Brovinsky, Boris 38467
Brower, Ken 54072
Brower, Paul 46713
Browes, Judi 43902
Brown, A. 01159
Brown, Alastair 33464
Brown, Alex 45854
Brown, Ann 48989
Brown, Anne 52299
Brown, Ashley 46411
Brown, Bern Will 05633
Brown, Bruce 52425
Brown, Charles B. 49732
Brown, Charles T. 50687
Brown, Dr. Charlotte V. 52235
Brown, Dr. Christopher 45327
Brown, Dr. Claudia 51906
Brown, Curtis E. 51852
Brown, Dan 52679
Brown, Daniel 52345
Brown, Dr. Dann 52056
Brown, David 28328, 47273
Brown, David C. 51323
Brown, David J. 54251
Brown, Don 49016
Brown, Dorothy 07202
Brown, Doug 49038
Brown, Elizabeth A. 53000
Brown, Eric 43519
Brown, Forrest 51419
Brown, Geneva 48123
Brown, Geoffrey I. 54225
Brown, Georgina 05707
Brown, Gordon 45463, 45464
Brown, Hannah 53619
Brown, Harold 06651, 49305
Brown, Harold L. 49177
Brown, Helen 43810
Brown, Ian 45637
Brown, James A. 49209
Brown, Jane 47371
Brown, Jean 44648
Brown, Jocelyn 33177
Brown, Johanna 54247
Brown, Karon 53480
Brown, Kate 45949
Brown, Kevin 44779, 49148
Brown, Laura K. 46415
Brown, Leanne 53782
Brown, Lee N. 14454
Brown, Leslie 46947
Brown, M. 05368
Brown, Margaret 43481, 43482, 43484
Brown, Marlene 53675
Brown, Marlin 05818
Brown, Michael K. 49397
Brown, Nancy 48776
Brown, Neil 07043
Brown, Nicolas 52166
Brown, Patricia 47620
Brown, Peter B. 46102
Brown, R.J. 52549
Brown, Rich 51940
Brown, Rick 49835, 49838
Brown, Robert 43427
Brown, Robert M. 50753
Brown, Ron 51736
Brown, Rondalyn 06768
Brown, Rowan 55092
Brown, Dr. S. 43504
Brown, Scott 46110
Brown, Shirley 49819
Brown, Tamara 49061
Brown, Tammi 53385
Brown, Terrence 51223
Brown, Tim 47298
Brown, Vinita 48582
Brown, Vivian A. 34484
Brown, W. 45270
Brown, William 45269, 49131
Brown, Dr. William 49340
Brown Hirst, Kaye 52681
Brownawell, Christopher J. 48267
Browne, Charles C. 52554
Browne, Mary-Louise 33050
Browne, Patrick 48228
Brownell, Carlton C. 50132
Browning, Dawn C. 50539
Browning, Mark 50682
Brownlee, Reuben 53785
Brownlow, Mary 44272, 44617
Brownson, Ron 33042
Broy, Erich 19889
Brubaker, Bob 54009
Brubaker, Sandra 51451
Bruce, Donald 51100
Bruce, Elspeth 45367
Bruce, Michael 49108, 49109
Bruch, Claudia 18283
Bruchhäuser, A. 17316
Bruck, Anton 03059
Brueck, Siegfried 21812
Brückner, Sigrid 22003
Brückner, Uwe 22455
Brüderlin, Prof. Dr. Markus 22665
Brüdgam, Iris 18829
Brüggerhoff, Dr. Stefan 17398
Brühlmann, Fritz 41144
Brülisauer, Annemarie 41726
Bruer, Dr. Stephanie-Gerrit 21897, 21899
Bruesehoff, Alison 48930
Brugge, Dr. J.P. ter 32892
Bruggeman, Jean 16205
Brugger, Dr. Ingried 02942
Brugnoli, Renzo 27571
Brugnoli Bailoni, Francisco 07453
Brugo, Carlo 27549
Bruguière, Philippe 14656
Bruhin, Felix 41381
Bruhn, Thomas P. 53320
Bruijn, J.A. 32118
Bruin, Dr. R.E. de 32844
Bruinjes, Wilhelm 18037
Bruintjes, Jaap C.N. 32155
Bruist, Gwen 45841
Brukwicki, Jerzy 35228
Brumbauch, Mary 50326
Brumberg, Esther 51227
Brumder, Robert B. 53996
Brumelis, Andrew 07119
Brumgardt, Dr. John R. 47359
Brummer, Hans Henrik 40927
Brun-Frandji, Hélène 14011
Brunacci, Prof. Dr. Aldo 25306
Brundage, Barbara 53697
Brune, Thomas 19091, 21950, 22287
Brunecker, Frank 17319
Brunel, Georges 14639
Brunet, P. 14280
Brunette, T. 07220
Brunetti, Armando 25866
Brungard, Charlene K. 46698
Brunhart, Arthur 30195, 30199
Bruni, Giovanni 25783
Bruni, Luca 41705
Brunin, Ph. 04088
Brunius, Staffan 40892
Brunn-Lundgren, Maria 40543
Brunnemann, Eric 23288
Brunner, Alois 20988
Brunner, Dieter 19097, 19100
Brunner, Dorla 46807
Brunner, H. 21644
Brunner, Dr. Max 20993
Brunner, Dr. Michael 22145, 22146
Brunner, Peter 41500
Bruno, Agnès 13292
Bruno, Prof. Gianfranco 26303
Brunovský, Ferdinand 38376
Bruns, James 46426
Brunscheen, Scott 48100
Brunschwiler, Willi 41372
Brunschwyler, Greta 49977
Brunson, Jeana 53427
Brunson, Neal E. 49624
Brunson, Theodore 49624
Brunson Frisch, Marianne 51990
Brunt, Kevin 05823
Brupbacher, Adolf 41618
Brusca, Richard C. 53583
Bruschetti, Dr. Paolo 25988
Bruschi, Luisa 26435
Bruscia, Maria Giuseppina 27467
Bruse, Ralf 20592
Brusila, Johannes 11397
Brussa, Carlos A. 54421
Brussee, G.J. 32644
Brust, Alexander 41169
Bruton, Prof. David L. 33737
Bruun Beyer, Mette 10135
Brüvere, Solvita 30057
Bruwier, Marie-Cécile 03893
Bruyère, Edgard 03887
Bruyn, Dr. Jean-Pierre de 03631
Bruyn, M.A. de 32752
Bruyn, Vera de 33927
Bruyn, Dr. Wolfgang de 18404
Bryan, Anne 52230
Bryan, Dr. Betsy 46546, 46572
Bryan, Dr. Charles F. 52342
Bryan, Tracey 52342
Bryan Hood, Mary 51640
Bryant, Arthur H. 51580
Bryant, James 52374
Bryant, Jim 52054, 52058
Bryck, Jack 46539
Bryde, John A. 33796
Bryden, Jim 06019
Bryer, Stephen 45999
Bryman, Donna 05692
Bryson, Mary 53570
Bryzensky, Nicole 01527
Brzeziński, Prof. Tadeusz 35160
Brzeziński, Dr. Wojciech 35286
Brzon, Narveen 52296
Brzozowska, Anna 17989
Brzozowski, Jerzy 35144
Brzozowski, Leszek 34679
Bu, He 08337
Bu, Yangwu 07789
Buberl, Dr. Brigitte 17925
Buccellato, Prof. Laura 00183
Buccheri, Prof. G. 27077
Bucci, Jonathan 53933
Bucciferro, Rose 50156
Buccino Grimaldi, Marchese Luigi 27881
Buccleuch, Duke of 45634, 45636
Buch, Carl E. 33538, 33539
Buchan, Andrew 44701
Buchanan, Dennis 43671
Buchanan, John 01136
Buchanan jr., John E. 52086
Buchanan, Karen 50329
Buchel, M. 31875
Bucher, Friedrich 01730
Bucher, Prof. Dr. H. 42103
Bucher, Hans 18466
Bucher, Regina 41633
Bucher, Stefan 41682
Bucher-Häberli, Elisabeth 41464
Buchholtz, Annegret 19624
Buchholz, Sabine 19471
Buchholz-Todoroska, Małgorzata 35120
Buchi, Danielle 12134
Buchmann, Anna 41746
Buchmann, Vagn 10123
Buchmayr, Dr. Friedrich 02695
Buchner, Dr. Gabriele 21055
Buchner, U. Ernest 05787
Buchroithner, Dr. Sonia 02136
Buchs, Denis 41272
Buchtmann, Hans-Georg 22204
Bučinskaja, Valentina Stanislavovna 29683
Buck, Angela 46788
Buck, Dennis 46489
Buck, Prof. Dr. Elmar 19634
Buck, Kurt 20979
Buck, Lisa 47558
Buckaloo, Terence 53285
Buckellew, Shari Spaniol 46858
Buckie, Valerie 05336
Buckingham, Bill 53589
Buckley, Annette 43419
Buckley, Kerry W. 51413
Buckley, Laurence 51348
Buckley, Laurene 48580, 51029
Buckley, Tracy 47570
Bucknill, Greg 45370
Buckridge, Martin 45402
Buckson, Deborah 51495
Bucur, Corneliu-Ioan 36604, 36606, 36607, 36612
Buczkowski, Hans 02515
Buczynski, Bodo 17052
Budarov, Viktor Viktorovič 37970, 37972
Buday, Kathleen 53553
Buddle, Allan 49512
Buderer, Dr. Hans-Jürgen 20173
Budiastra, Putu 24466
Budiati, Dr. Tinia 24483
Budintis 30436
Budka, Mikołaj 34555
Budko, Prof. Dr. Anatolij A. 37731
Budney, Jen 06545
Budriunas, Algimantas 30324
Budriūnas, Algimantas 30413
Budrys, Milda 47486
Budrys, Romualdas 30415
Budzynski, Jim 47793
Büche, Wolfgang 18904
Büchler, Dr. Hans 41559
Büchler-Mattmann, Dr. Helene 41223
Bücken, Véronique 03462
Buecker, Thomas R. 47868
Bückling, Dr. Maraike 18394
Bühl, Gudrun 53865
Bühl, Herbert 41083
Buehler, J. Marshall 54265
Bühler, Dr. Kathleen 41210, 41299
Buehler, Ken 48198
Bühlmann, Isolde 41348
Bühner, Dr. Jan-A. 21929
Buelens, Paul 03548
Buell, Pamela 06656
Bueltjer, Dr. Uwe 17807
Buendía García, Mercedes 40243
Buenrostro Alba, Manuel 31102
Büren, Daniela von 41453
Büren, Guido von 19405
Bürge, M. 42064
Bürger, Werner 16560
Bürgin, Christoph 41602
Bürgin, Dr. Toni 41808
Buerlein, Robert A. 52321
Buermeyer, Chris 49111
Buesing, Kay 50180
Bütikofer, Prof. Alfred 42038
Büttner, Dr. Philippe 41763
Bützer, Michael 18003
Büyükçapar, Ahmet 42861
Bufano, Ralph A. 53004
Buff, Urs 41639
Buffa, Gérard 14930
Buford, Elizabeth F. 52233
Bugaeva, Galina Alekseevna 37811
Bugajov, Oleksandr Sergijovyč 43152
Bugallo, Bernardino 39861
Bugalová, Dr. Edita 38216
Buganu, Melitina 36356
Bugár-Mészáros, Károly 23435
Bugarski, Astrida 04284
Bugg, Mathias 41818
Buggeland, Tord 33637
Buggle, Claudia 22431
Buhl, Henry 51944
Buhl Andersen, Christine 10160
Buhler, Leslie L. 53922

Buhlmann, Dr. Britta E. 19412
Buhr, Johannes 22479
Buhrmann, Connie 50514
Buhrs, Michael 20469
Buijaen, E. 32082
Buijs, H.W.J. 32876
Buijs, Hans 14598
Buisson, Fernand 14885
Buisson, René 06373
Bujakowska, Beata 34689
Bujić, Marija 38074
Bujok, Dr. Elke 20490
Bujold, Michel 06616
Bukač, Václav M. 09524
Bukačová, Dr. Irena 09647, 09648
Bukowska, Halina Lubomiła 34938
Bulaty, Dr. Milan 17276
Bulau, Cheryl 50130
Bulei, Gheorghe 36317, 36384, 36543, 36653, 36654, 36656, 36657, 36658, 36660, 36661, 36736
Bulfon, Christiane 01884
Bulgarelli, Grazia Maria 27519
Bulger, Ralph 07302
Bulgin, Lloyd 05728
Buljević, Zrinka 08954
Bull, Milan 48480
Bullard, E. John 51091
Buller, Osmo 11344
Buller-Focht, Brenda 52374
Bullock, Margaret 52086
Bulova, Gretchen M. 46206
Bultereys, Veerle 03624
Bulthuis, P. 32987
Bultmann, Horst 17547
Bumgardner, Tim 49549
Bumiller, Manfred 16914
Bunch, Lonnie G. 47466
Bunch, Nancy B. 47725
Bunch, Rosemary 54291
Bund, Dr. Konrad 18746
Bundesen, Laura 48052
Bung, Maria-Hildegard 16885
Bunger, James 53374
Bungert, Petra 03441
Bunke, Jim 53019
Bunker-Lohrenz, Jackie 52492
Bunkichi, Ito 29061
Bunyan, Vicki 01263
Bunzel, PD Dr. Wolfgang 18370
Buol-Wischenau, Mag. Karin 01982
Buonanno, Prof. Roberto 27462
Buonocore, Sergio 07078
Burak, Marek 35333
Buran, Dr. Dusan 38233
Buranelli, Dr. Francesco 54497, 54501
Burbano, Ignacio 08714
Burch, Danielle 49238
Burch, Ivan 06367
Burch, Wanda 49638
Burchard, Helen 46687
Burchenal, Margaret 46934
Burchfield, Stanley G. 52656, 52659
Burckhardt, Dr. Daniel 41175
Burckhardt, Nicole 41125
Burdak, Elisabeth 42122
Burdick, Funi 48463
Burdick, Richard L. 52455
Burdina, Lidija Nikolaevna 37047
Burdychová, Milena 09486
Burel, Bernard 14902
Burford, Beverley 44863
Burg, Dr. Tobias 18263
Burganov, Aleksandr 37307
Burge, Roisin 44458
Burge, Sarah 43578
Burger, Hannes 41533
Burger, Harald 21629, 21634
Burger, Richard L. 51054
Burger, Serge 16300
Burger, W. 19322
Burgère, André 12641
Burges-Smith, Karen 48465
Burgess, Brian 01183
Burgess, Cynthia 53771
Burgess, Jerry G. 48646
Burgess, Jo 46868
Burgess, Robert 51105
Burgess, Susan M. 44169
Burghardt, Glenn 51459
Burgin, William 49599
Burgiss, Ken 33103
Burgos Cantor, Roberto 08557
Burgraff, Joseph 03647
Buri, Dr. phil. Andreas 41590
Burik, Tatjana Borisovna 37801
Burkard, Adolf 41264
Burkarth, Dr. Axel 21950
Burke, Bonnie 05486
Burke, Daniel 51846
Burke, Darrell 06602
Burke, Gregory 33156
Burke, Jo 48039
Burke, Joan T. 47416
Burke, Kathleen M. 49300
Burke, Marcus B. 51192
Burke, Susan M. 06102
Burkel, Don 46626
Burkert, Richard A. 49640, 49641
Burkhalter, Françoise 41638
Burkhardt, Elisabeth 41879
Burkhardt, Guido 17001
Burkhardt, Pia 22323
Burkhardt-Aebi, Rosmarie 41491
Burkhart, Linda 47421
Burki, Erika 41276
Burleson, D. 05944
Burlet, Dr. Jürg 41396
Burlingham, Cynthia 50227
Burluaux-Blanchard, Odile 14646
Burman, Rickie 44889
Burmann, Peter 40840
Burmeister, Artur 19815
Burmeister, Helmut 19224
Burmeister, Dr. Ralf 17049
Burn, Dr. L. 43727
Burn-Murdoch, R.I. 45574
Burnand, Frédérique 41651
Burnet, Marie-Jeanne 41790
Burnett, Ed 45743
Burnett, Kay Taylor 50983
Burnett, La Jean 49304
Burnett, Richard 44339
Burnett, Roger K. 52469
Burnett, Wayne 52550
Burnette, David 50888
Burnham, Laura 48094, 49619
Burnichioiu, Sevastița 36257
Burnie, John 43560
Burnod, Véronique 12257
Burns, Courtney 52923
Burns, Diane 00997
Burns, Emma-Kate 43603
Burns, James 53472
Burns, Jane 48346
Burns, Kelly S. 49452
Burns, Laura 46587
Burns, McAndrew 46400, 46403, 46404
Burns, Nancy 51762
Burns, Norman O. 50962
Burns, Pamela 06393
Burns, Richard 43691
Burnside, Madeleine H. 49764
Burny, Jeannine 03491
Buron, Gildas 11862
Buron, René 15445, 15446
Burpee, Jane 05720
Burrell, Jane 45221
Burresi, Dr. Mariagiulia 27245
Burri, Katri 42024
Burri, Mario 42058
Burroughs, Stuart 43406
Bursey, M.L. 38615
Burt, Ann 51600
Burt, Steve 44665
Burtch, Michael 06928
Burton, John 48552, 49900
Burton, Mary 54055
Burton, Val 00909
Burwick, Georg 52369
Busack, Dr. Stephen 52234
Busceme, Greg 46671
Busch, Anna 40921
Busch, Fritz B. 22648
Busch, Dr. Ina 17789
Busch, Ingo 18999
Busch, Jean 53541
Busch, Wolf-Rüdiger 18550
Buschhoff, Dr. Anne 17556
Busching, Dr. Dr. rer. nat. habil. Wolf-Dieter 19667
Buschmann, Harald 18191
Bush, Dianne 06249
Bush, Kate 44791
Bush, Michelle 06847
Bush, Walter 51918
Bush Tomio, Kimberley 53635
Bushman, David 51229
Busine, Laurent 03713, 03714
Busjan, Béatrice 22613
Busk Laursen, Bodil 10129
Buskirk, Dr. William H. 52313
Buskov, Poul 10318
Busquet, Andoni 39750
Buss, Gisela 19869
Buss, Leo W. 51054
Bussac, Monique 11620
Busse, Prof. Rido 21805
Busse, Valerie 50511
Bussel, Gerard W. van 03038
Bussers, Dr. Helena 03460, 03462
Bussinger, Julia 46734
Busslinger, Ernst 41623
Bussy, Diana de 00916
Bustard, Wendy 50939
Bušueva, Margarita Viktorovna 37512
Busuladžić, Adnan 04284
But, Aurel 36283
Buta, Sanda Maria 36177
Butala, Daniel 08886
Butave, Denis 15133
Butchko, Tom 48338
Butkevičius, Gintaras 30216
Butko, Trina 00789
Butler, Amena 51518
Butler, Bill 54365
Butler, Charles Thomas 47739
Butler, David 45172
Butler, Gay 47403
Butler, John 43477, 45893
Butler, Joyce 52070
Butler, Maria V. 51769
Butler, Mary 46644
Butler, R. 38698
Butler, Roger 00946
Butler, S.R. 44699
Butler, Samuel 51244
Butler, T. 44530
Butler, Tony 45773
Butnaru, Ioan 36500
Butt, Raymond 45417
Buttacavoli, Eva 46503
Buttar, Lyn 33259
Buttarazzi, Silvestro 25700
Buttel, Larry 48782
Butter, Jelena 17207
Butterfield, Tom 46846
Butterweck, Dagmar 03049
Buttini, Eric 30484
Buttlar-Elberberg, Dr. Gertrud 03098
Button, Kerri 07311
Butts, H. Daniel 50451
Butz, Dr. Herbert 17201
Buurma, Hilbrand 32786
Buurman, H. 32073
Buus, Henriette 10042
Buxton, Judith 05799
Buyandelger, U. 31633
Buyko, Malgorzata 34769
Buys, C. 32488
Buyssens, Danielle 41425
Buzenschi, Florentina 36519
Buziałkowski, Jerzy 34620
Buzzi, Alessandro 27294
Buzzoni, Dr. Andrea 26107, 26116, 26117, 26120
Byambasüren, P. 31620
Byck, Dr. David B. 52938
Byčkova, Marina Nikolaevna 37900
Byčkova, Olga Nikolaevna 37293
Bye, Joellen 51010
Byers, Bill 47007
Byh, Per-Olov 40638
Bykov, Dr. Aleksandr 37962
Bykova, Jekaterina Vladimirovna 37458
Bykovceva, L.P. 37279, 37318
Bylicki, Tomasz 35224
Byman, Bertyl 11100
Bynum Robertson, Marci 50140
Byong, Ha Min 29929
Byrczek, Jakub 34730
Byrd, Eugene 53257
Byrd, Gloria 50001
Byrn, Brian 48346
Byrne, Joseph 49200
Byrne, Lisa 00888, 01259
Byrne, Mary Jo 50503
Byrne, Sue 44315
Byrtusová, Jana 09471
Byström, Lars 40915
Bytkovskij, Oleg 37516
Byvanck, Dr. F.V. 32554
Byzewski, Leah 48987

C

Caballero Barnard, Prof. José M. 31485
Cabanes, Francisco 40217
Cabanis, Prof. E.A. 14642
Cabano-Dempsey, Annali 38554
Cabaret, Michel 15013
Cabel, Oswalda A. 34393
Cabello Carro, Dr. Paz 39532
Cabello Carro, Paz 39586
Cabezas, Hervé 13358, 15474
Cable, Christopher B. 46289
Cabot, Ian 45587
Cabot, Paul 07146
Cabral, Dr. Elisabeth 35670, 35690
Cabrera, E. 48497
Cabrera, Geles 31112
Cabrera Cachón, Teresa 31511
Caccia, Antonietta 27785
Cachola Schmal, Peter 18361
Cad, Gabriel 34382
Cadavid López, Susana 08761
Cadbury, Tom 44195
Cadby-Sorensen, Robin 53503
Cadenet, Anne 12062
Cadenne 12416
Cadenne, Ivan 12417, 12562, 15779
Cadier, Gérard 14940
Cady, Don 46770, 50697
Caes, Ward 03284
Caffarelli, Prof. Roberto Vergara 27241
Caffrey, James 46903, 46905
Cagigós i Soró, Antonio 40059
Cagli, Prof. Bruno 27158
Çağman, Dr. Filiz 42843
Cahen, J. 31848
Cahen, Joël J. 31850
Cahen-Delhaye, Anne 03469, 03482, 03495, 03500, 03501
Cahill, Ann 45304
Cahill, Kevin M. 51114
Cai, Changkui 08284
Cai, Hetian 08269
Cai, Mingkang 07875
Cai, Xiaofang 07583
Cai, Zhengquan 07767
Caiado, José Pedro 35717
Caicedo, Iván 08631
Caicedo de Gómez, Hilda María 08607
Caicedo Lourido, María Isabel 08616
Caille, R.T. 11845, 11847
Cain, Glena 51908
Cain, Prof. Dr. Hans-Ulrich 19893
Caines, M. 44068
Caisley, Sherry 50863
Caivano, Felice 49200
Caji, Jaroslav 38031
Čajka, Vladimir Michailovič 37939, 37950, 37952
Čajkovskij, Bogdan M. 43126
Çakmak, Hacı 42720
Caladio, Prof. Cosimo 26464
Calahan, Deborah 46828
Calame, Caroline 41571
Calandra, Eliana 27076
Calatayud Cases, José Vte 39896
Calbet, Jean-Marie 13664
Calbiac, Christoph de 13799
Calboutin, A. 38777
Caldeira, Ernesto 54290
Caldeira, Prof. Dr. Helena 35533
Caldeira, Vasco 04880
Calder, Jacqueline 50825
Calderan, Gianni 27530
Calderer Serra, Joaquim 40088
Caldwell, Brenda B. 53626
Caldwell, Desiree 47767
Caldwell, Frank 45967
Caldwell, Dr. J.P. 51376
Caldwell, James 47190
Caldwell, Janet 46689
Caldwell, Joan G. 51095
Caldwell, Judy 50568
Caldwell, K. 43942
Caldwell, Karen 44207
Caldwell, Roy 46758
Calhoun, Angela 48750
Calhoun, Ralph 50787
Caliaa, Prof. Roberto 25189
Caligiuri, Olimpia 41681
Calisi, Dr. Marinella 27462
Calkins, Jack 05762
Call, Linda 54327
Callaghan, Dr. Amanda 45467
Callaghan, Madelein 06938
Callaghan, Richard 44100, 44102
Callahan, Ann 50960
Callahan, Christine 51326
Callahan, Colin J. 47773
Callahan, Colleen 52339
Callahan, David 51346
Callahan, Elizabeth 48257
Callahan, Martin L. 48678
Callan, Josi Irene 53412
Callanan, Mary 24840
Callataÿ, François de 03440
Callatÿ, François de 03438
Callebert, Fernand 03407
Calleen, Dr. Justinus Maria 22623
Callegher, Dr. Bruno 27036
Calleja Ansótequi, Zoilo 40319
Callejo, Marcilyn 51568
Callen, Bonnie 05794
Callewaert, Ferdy 03976
Callies, Prof. Dr. H. 21846
Callis, Charles 33257
Callis, Eleanor 51474
Callis, Marion M. 54178

C

Callison, Chris 53670
Calloway, Jim 51551
Calloway, William 47734
Calogheros, Tracy 06666
Calomiris, Ellen 50178
Calonne, Sylvain 16206
Caltik, Heidi 03026
Calvarin, Margaret 12203, 12648
Calvesi, Prof. Maurizio 25905
Calvi di Bergolo, Ippolito 27730
Calvin, Prudence K. 53975
Calvo, B. 39562
Calvo, Ernesto 08816
Calvo, Luis María 00615
Calvocoressi, Richard 44119, 44156
Calzada, S. 38969
Camacho, Dr. João Carlos 35513
Camacho, Dr. Paula Cristina 35702
Camacho Cadaval, Fernando 40074
Camara, Dr. Abdoulaye 38026
Camara, Calvin 46328
Camara, Prof. Edeltrudes 04348
Čambálová, Dr. Daniela 38378
Camber, Diane W. 50641
Cameron, Dan 51240
Cameron, E. 33043
Cameron, Elizabeth 44043
Cameron, Gary L. 50512
Cameron, Jane 45748
Cameron, John 43536
Cameron, Joyce M. 53207
Cameron, Rosanna 01160
Cameron, Ross 00854
Cameron of Lochiel, Sir Donald 45725
Camfield, Paul 48762
Camhi, Prof. Jeffrey 25007
Camilleri, Carla 03079
Camilo, Eva 10347
Camin, Philippe 14158, 15615
Camm, Simon 45477
Cammarata, Paolo 27593
Cammarota, Dr. Giampiero 25463
Cammuso, Philomena M. 53036
Camnitzer, Luis 51167
Camoys, Lord 44450
Camp, Kimberly 50601
Campagna, Damon A. 53837
Campagna, Dr. Francesca 26668
Campagnolle, Delphine 11948
Campagnolo, Matteo 41418
Campanelli, Dr. Adele 25871, 25872
Campanelli, Elide 26070
Campanero, Amy 47010
Campbell, Bruce 45240
Campbell, C.A. 45747
Campbell, Dr. Caroline 44830
Campbell, Chris 50973
Campbell, Colin 54167
Campbell, Connie 06452
Campbell, D. 44432
Campbell, Daniel T. 53147
Campbell, Deborah 46302
Campbell, Donald 01372
Campbell, Francis D. 51116
Campbell, George 06059
Campbell, Ina 46851
Campbell, Kathleen 52282
Campbell, Kay 01517
Campbell, Loraine 53572
Campbell, Lorne 44937
Campbell, Mabel 06578
Campbell, Mei 50312
Campbell, Mognon 01574
Campbell, Mungo 44284
Campbell, Nancy 54012
Campbell, Nathan 46678
Campbell, Neil 06206
Campbell, Richard 50739
Campbell, Roy G. 52234
Campbell, Sara 51720
Campbell, Shearon 47547
Campbell, Tacie N. 48189
Campbell, Wallace 51742
Campbell Coyle, Heather 54189
Campbell-Ferguson, Gail 47005
Campbell McMurray, H. 45437
Campbell-Shoaf, Heidi 48756
Camphausen, Ute 19911
Campins, Lluís 39978, 39979
Campione, Prof. Dr. F. Paolo 41283
Campo, Marta 38960
Campodonico, Pierangelo 26293
Campofiorito, Italo 04591
Campos, Prof. Antonio Celso de Arruda 04572
Campos, Maria Lucia Rocha 04887
Campos, Nelson 35967
Campredon, Elisabeth 12701
Camps Extremera, Antonio 39194
Campuzano Ruiz, Enrique 40034
Camus, Marie-France 13505
Canales, Rocio Gonzalez de 31180
Cañamás, Maita 40201
Canaves, Margalida 39851
Canbey, Maşuk 42961
Canca Lara, Juan Enrique 40271
Cândea, Ionel 36167, 36168, 36169, 36170, 36171, 36172
Candeloro, Ilaria 27573
Cândido, Maria João 35936
Candon, Anthony 45318
Candon, Tony 24756
Candrowicz, Krzysztof 34875
Candy, Ron 07236
Caneer, Ann 52019
Cangialosi, Patricia 00173
Canini, Alberto 27484
Canizales Montoya, Luis Fernando 08720
Cankova, Tatjana 05031
Cannan, Gwynedd 51273
Cannata, Dr. Roberto 27455
Cannatella, David C. 46519
Cannegieter, Dr. D.A.S. 32213
Canning-Lacroix, Johane 06830
Cannistraro, Dr. Philip V. 51198
Cannon, Annetta 51436
Cannon, Janet 05912
Cannon, Sara 50245
Cannon-Brookes, Peter 44624
Cano, Delphine 13795, 14374, 15228, 15979
Cano, Luis Martín 31158
Cano, Rose Marie 50261
Cano de Martín, Marta 00324
Cano Rivero, Ignacio 40069
Cano Salas, Gilda 31095
Canogar, Rafael 40299
Canonne, Xavier 03515
Canovai, Dr. Roberto 26533
Canright, Steve 52794
Canseliet, Ovide 03722
Cant, Dr. John G.H. 36059
Cantinho, Dr. Manuela 35707
Cantor, Helen 48747
Canty, Scott 50245, 52242
Canulli, Richard G. 53745
Canuto, Marcello A. 51054
Cao, Chuansong 07967
Cao, Sanzhuang 08515
Cao, Zhongkui 08352
Caouette, Muriel 48862
Capa, José 39787
Capa Eiriz, Dr. José 39231
Capaloi, Franco 14897
Capanna, Prof. Ernesto 27488
Capano, Dr. Antonio 28214
Capazza, Gérard 14347
Capdevila, Dr. Ricardo 00202
Capella Molas, Anna 39298
Capenberghs, Joris 03416, 03417
Capers, Pierre 04139
Capin, Djordje 31645, 31646
Capin, Jean 15921
Capitani, Silvano 26106
Capitanio, Fabrizio 25404
Čapková, Dr. Jana 09528
Čaplinska, Olga Mychajlivna 43144
Capomolla, Soccorso 26848
Capon, Barry 43652
Capon, Edmund 01544, 01545
Capon, S.H. 45976
Capone, Dr. Patricia 47185
Capot, Stéphane 13766
Capovilla, Prof. Sergio Carlos 00570
Cappelletti, Prof. Vincenzo 27237
Cappola, Dr. Gianpietro 25208
Capps, Gene 53789
Capps, Jennifer 49506
Capps, Larry 49445
Capralos, Souli 22866
Caproni di Taliedo Guasti, Firmina 27459
Carabo, Lucille 46737, 46738
Caradec, Marie-Anne 12693
Caramazza, O.G. 26634
Caramelle, Dr. Franz 02115
Caramelli, Fernando 27216
Caramore, B. 28272
Carazzetti, Prof. Riccardo 41569, 41570, 41574
Carbajal, Londi 52888
Carbajal, Sergio 31108
Carballo, José Rodriguez 40023
Carbert, R.W. 06288
Carbery, David 24832
Carbone, Teresa 47069
Card, Nan 48788
Cardarelli, Andrea 26745, 26746
Cardelli, Alessandra 27519
Cardelli, Patrice 13904
Carden, Willie F. 47566
Cardenas, Dr. Lourdes B. 34380
Cardenas, Mary R. 51744
Cárdenas Marcaida, María Caridad 09067
Cárdenas Tovar, Jairo Simón 08782
Cardenas Troncoso, Gloria 07431
Carder, Julia 44733
Cardilli, Dr. Luisa 27475
Cardim Ribeiro, Dr. José 35946
Cardinaels, Marie-Jeanne 03402
Cardini, Giuliano 27912
Cardon, Véronique 03463
Cardona, Hermana Martha 08756
Cardone, Veronique 11963, 13561
Cardoner, F. 38953
Cardoni, Edmund 47121
Cardoso, Prof. Dr. António 35433
Cardoso Botelho, Americo G. 35451
Cardot, Annie 41975
Carduner, Michel 15372
Cardwell, E. 45317
Carengna, Dr. Daniela 27583
Carevic Rivera, Alvaro 07438
Carew, John E. 07136
Carey, Betsy 53443
Carfano, M. Susan 54353
Caridad, Jorge 10353, 10356
Cariello, Dr. Rubina 27881
Carignani, Dr. Mario 26887
Carion, P. 03466
Cariou, André 14971
Carl, Christine S. 48845
Carlano, Annie 52883
Carles, Gilbert 15960
Carlestam, Håkan 40570
Carleton, Dr. Don 46506, 53678
Carli, Dr. Laura 27925
Carlier, Sylvie 16196
Carlier, Yves 12966
Carlile, Janet 05355
Carlosama Mora, Rosa Cecilia 08663
Carlsen, Dr. Charles 51632
Carlsen, Margaret 49788, 52153
Carlson, Anne 46899
Carlson, Erling 21135
Carlson, Jim 53103
Carlson, Joseph 48345
Carlson, Linda 48608
Carlson, Norman P. 49597
Carlson, Robert 46901
Carlson, Dr. Shawn 53947
Carlson, Shirley 47274
Carlson-Reynolds, Alexis 48448
Carlsson, Ann-Britt 40992
Carlsson, Eva 40515
Carlström, Dr. Ann-Kristin 40769
Carlstrom, Terry 49174, 53053, 53871, 53872, 53891, 53913
Carlton, Jamon 54375
Carlton, Rosemary 53125
Carlton, Stewart 52974
Carlweitz, Pontus 40990
Carman, Andrea 50815
Carmen Lema Pensado, María del 54416
Carmer, Craig 50202
Carmona Lancheros, Carlos Alberto 08604
Carnap-Bornheim, Prof. Dr. Claus von 17668, 21565, 21569
Carnazzi, Umberto 25419
Carneiro, Dr. Deolinda Maria Veloso 35873
Carneiro, Paula Dias 35860
Carnerini, Carla 27110
Carnes, Griffiths 49749
Carnes, John 50084
Carney, Barbara 53485
Carney, Dr. Margaret 46216
Carney, Terry J. 44351
Carolan, Janet 43989
Carollo, Albert H. 53769
Carollo, Eugene R. 53769
Caron, Euclid 51614
Caron, Jacqueline 06031
Carowick, Sandra L. 49421
Carpenter, Dean 48275
Carpenter, Elizabeth 50743
Carpenter, Frank 43994
Carpenter, Joan 54357
Carpenter, Dr. M.A. 43733
Carpenter, Matthew 46331
Carpenter, T.F. 43799
Carpenter Correa, Laura 47053
Carpenter Troccoli, Joan 48072
Carpentier, Régine 03437
Carpio, Agustín 00487
Carr, Carolyn 53907
Carr, Christian 53392
Carr, Dr. Dawson 44937
Carr, Josh 51181
Carr, Lucia C. 50369
Carr, Shelvie 47991
Carr, Ted 44059
Carr-Griffin, Paul 44765
Carr-Whitworth, Caroline 43991
Carradice, Prof. Ian A. 45557
Carrara, Dr. Giancarlo 25870
Carraro, Dr. Francine 49573
Carré, Dr. Enrique González 34264
Carreau, Alain 15686
Carreau, Raphaële 12483, 12484
Carreck, Marjorie W. 44448
Carregha, Luz 31353
Carreira, Dr. Isabel 35544
Carrel, Michel 15479
Carrel-Lautelme, Suzanne 13795
Carreño, Irving 54530
Carreño Díaz, Prof. Esboardo 30837
Carrera Hontana, Enrique de 39550
Carretero, Andrés 39558
Carretero Rebés, Salvador 40009
Carrier, Alan 51605
Carrier, Nicole M. 49549
Carrlee, Ellen 49657
Carrlee, Scott 49656
Carrocera Fernández, Elias 39711
Carrol, Brian 47095
Carroll, Daine L. 53559
Carroll, M. Lyda 49087
Carroll, Rowan 33188
Carroll, Sue 52181
Carroll, Vicky 45002, 45015
Carroll, William B. 44481
Carron, Christian G. 49007
Carruthers, Karen 43542
Carruthers, Ray 07061
Carruzzo, S. 41984
Carson, Heather 44929
Carson, John 52256
Carson, Michael 47515
Carson, Sully 33093
Carstensen, Dr. Jan 17840
Carswell, A.L. 44140
Cartacci, Marcello 27537
Cartagena, María Fernanda 10437, 10440
Carter, Anne 50140
Carter, Beth L. 53273
Carter, Dr. Curtis L. 50718
Carter, Earl N. 49096
Carter, Hazel 50484
Carter, John S. 51843
Carter, Jon 43336, 45569, 45573, 45580, 45583
Carter, Jonathan 45570, 45571
Carter, Kathy 49843
Carter, Martha C. 53208
Carter, Michael 51152
Carter, Penelope 44138
Carter, R. 00904
Carter, Susan 53433
Carter, Warrick L. 47493
Cartwright, Colin M. 43467
Cartwright, Derrick R. 49162
Cartwright, Thomas 48751
Carty, Paul 24794
Caruana, Sebastian 30662
Carucci, Arturo 27597
Carus, Benno 17135
Carusi, Corina 47478
Caruso, Dr. Ida 25924
Caruso, Romy 54019
Caruso, Sarah 52622
Carustrom, Terry 46357
Carvalho, Adelaide 35860
Carvalho, Luís Manuel Mendonça 35462
Carvalho, Maria Judite de 35822
Carvallo, Henri 16163
Carver, Kate 44485
Carvey-Stewart, Elizabeth 52406
Cary, PeeWee 48205
Carzou, Jean-Marie 13941
Čas, Alenka 38406
Casabán, Consuelo Císcar 40201
Casadei, Dr. Sauro 26079
Casado Soto, José Luis 40011
Casagrande, Louis B. 46920
Casagrande, Rodrigo Fernando 04937
Casagrande, Tiziana 26093, 26094
Casal, Gabriel 34430
Casal Tatlock, Alvaro 54398
Casaldi, Silvano 26922
Casale, Tommaso 27936
Casalini, Dr. Antonio 27098
Casalino, Vittorio 26312
Casanelles, Eusebi 40118
Casanovas, María Antonia 38975
Casaro, Massimo 26715
Casarola, Albino 26859
Casartelli, Marisa Helena Cestari 04821
Cascales Ayala, Manuel 38875
Casciu, Dr. Stefano 26161, 26173
Caselli Wijsbek, H.E.J. 32861
Casetti, Christoph 41302
Casey, Laura 53058
Casey, Lynn R. 48856
Casey, Michael 24885
Casey, Prof. Michael T. 24875
Cash, Laurisa 49526
Cash, Sarah 53855

Cashbaugh, Todd M. 46306
Cashman, William 46682
Casini, Dr. Stefania 25401
Caskey, Phyllis 50231
Caskey, Tom 48173
Câşlaru, Constantin 36131
Casler, Jennifer R. 48715
Caslin, Jean 49388
Caspary, Dr. Gundula 21752
Caspers, Martha 18377
Caspersen, Kim 44118
Cass, Naomi 01080
Cass, Wilfred 44328
Cass-Hassol, Nicholas 51239
Cassan, Dagmar de 02295
Cassan, Ferdinand de 02295
Cassanelli, Dr. Roberto 26865
Cassata, Giovanni 27079
Cassell, Valerie 49383
Cassels, Patrick 43400
Casseminho Meira, Mariana 04651
Casser, Anja 19436
Cassiano, Dr. Antonio 26474
Cassidy, Christina 52901
Cassin, Michael 54175
Cassin, Virginia 51454
Cassway, Nick 51862
Castagnoli, Prof. Pier Giovanni 28001
Căstăian, Mihai 36506
Castañeda, Lochin 34168
Castaño Cruz, Hermana Cecilia 08609
Castañón Albo, Luis 39531
Castanyer Bachs, Narcis 40263
Castel, Marie-Chantal 14816
Castel-Branco Pereira, Dr. João 35672
Castele, Theodore J. 47632
Castell, Edmon 38835
Castellà Reial, Josep 39868
Castellana, Dr. Gruiseppe 25161
Castellano i Tresserra, Anna 39000
Castellano Zapater, Eustaquio 38808
Castelletti, Dr. Lanfredo 25951, 25954, 25955, 25956
Castello, Rosa María 38845
Castells, Ruben 40342, 40345
Castelnuovo, Sheri 50393
Casteret, Norbert 12958
Castex, Jessica 14646
Castex, Nicole 12858
Castier, Annie 13756
Castilla, Francisca 40204
Castillo, Anna 52868
Castillo Reyes, Samuel 34124
Castillo Tovar, Carlos Arturo 08646
Castillos, Prof. Juan José 54402
Castle, Charles E. 49846
Castle, Lynn P. 46670
Castle, Myra 49086
Caston, Dr. Ellie 53778
Castor, A. 14792
Castor González, Jesús 31083, 31104
Castrillón Urrera, Carlos Alberto 08541
Castrischer, Claudia 41564
Castro, Prof. Dr. Aníbal Pinto de 36003
Castro, Dr. Inés de 19185
Castro, Prof. Dr. Inés de 21951
Castro, Maria da Graça Salema de 35900
Castro, Milagros 36079
Castro Benítez, Daniel 08544, 08546
Castro Mejía, María Eugenia 08540
Castro Meza, Raúl 31377
Castro Otero, Gumersindo 40030
Castronovo, Simona 28004
Catala, Claude 14573
Català Gorgues, Angel 40197
Cataldi, Dr. Maria 27943
Catchpole, Julie 33175, 33176
Cates, David L. 03203
Catford, B.L. 01396
Cathcart, Anne 53298
Cathcart, Rosemary 44181
Cathrin, Marcel 15416
Cathy, Salter-Jansen 38730
Catleugh, Jon 44835
Catlin, Erica 47078
Catlin-Legutko, Chinnamon 47869
Cato, John S. 53365
Cato, Dr. Paisley 52754
Catolin, Eufemia B. 34527
Catron, Joanna D. 48763
Catroppa, Frank 46443
Catrysse, Bernard 29125
Catta, Javier 10419
Cattaneo, Claudia 42027
Cattaneo, Pascal 41629
Cattelain, P. 04091, 04092
Catton, Jonathan 44351
Cau, Ettore 28307, 28308
Cauchie, S. 03951
Caudoux, Michel 16081
Cauk, Françoise 11819
Caulkins, Bethel I. 52308
Caurie, Philippa 46077
Čauševa, Dora 05050
Caussinus, Dany 11598
Cauwe, Nicolas 03495
Cavagnis Sotgiu, Dr. Maria Carla 27471
Cavaillès, Maria 14753
Cavalcante, Raquel 04370
Cavalchini, Pieranna 46934
Cavalho-Roos, Dr. T. Rosa de 31915
Cavalier, Odile 11789
Cavalier, Sebastien 13901
Cavalieri, Corlis 51288
Cavalli, Attilio 26701
Cavalli, Germano 28277
Cavalli-Björkman, Åsa 40889
Cavanaugh, Joe 48761
Cavatrunci, Claudio 27519
Cavelti, Dr. Urs J. 41443
Cavendish, Kim L. 48639
Cavens, Travis 49723
Cavish, Jackie 51658
Cavnor, Julie Ann 46575
Cavour Aramayo, Ernesto 04214
Cavruc, Valeriu 36600
Cay Henry, Mary 47843
Cayer, Richard 46230
Çaylak, £lsmail 42750
Cayphas, Jean-Paul 03735
Cazacu, Silvia 36227
Cazakoff, Ingrid 06949
Cazan, Mag. Ildikó 03038
Cazé, Sophie 13251
Cazenare, Sabine 11588
Cazenave, Sabine 11587
Cazes, Daniel 15945
Ceballos, Orlando Toro 08742
Ceballos Carrero, Rosa Maria 30929
Cebucoprea, Dr. Alexandru 36198
Ceccarelli, Giampiero 27894
Cecchetto, Dr. Giacinto 25729
Ceccini de Dallo, Ana María 00622
Cech, Christine 03079
Cecil, William A.V. 46369
Čecov, Prof. Dr. V.P. 37261
Cederholm, Katherine 50351
Cederlund, Dr. Johan 40748, 41016
Cedersand, Kerstin 40892
Cederstrøm, Elisabeth 10152
Cedro, Dr. Bernard 35159
Ceesay, B.A. 16331
Ceesay, Hassoum 16331
Čega, Fani Celio 08970
Čekerinac, Bora 38151
Čeko, Franko 08948
Celant, Germano 51258
Celayır, £lsmail Hakkı 42816, 42851
Celej, Leszek 34923, 34924
Céleyran, Tapié de 14384
Celi, Maria 51160
Celio, Franco 41433
Celio (-Scheurer), Marie Eve 41647
Celis, Augustin de 40299
Celli, Jeanann 53264
Celma, Cécile 30665
Celuzza, Dr. Mariagrazia 26371, 26373
Cembellín, Juan Manuel Gonzalez 39045
Cembrano, Rita 34351, 34354
Cembrola, Bob 51325
Cencic, Jasmine 01741
Cendo, Nicolas 13993
Ceni, Giuseppe Riccardo 28234
Ceniceros Herreros, Javier 39710
Cenni, Mario 26533
Centellas Bodas, Aurelio 40328
Cepec, Dr. Martin 17565
Čepič, Taja 38468
Cepon, Miloska 38467
Čeran, Branko 09004
Cerar, Estera 38467
Čerba, Larisa Valerjanovna 37876
Cercega, Petru 36277
Cerepak, Mag. Susanne 02951
Cerezo Aranda, Dr. José Antonio 39679
Ceriana, Dr. Matteo 28184
Cérino, Christophe 13821
Ceriotti, Guido 25572
Cerişer, Nicolae 36404
Čerkalin, Sergej Dmitrievič 37606
Čerkasova, Olga Anatoljevna 36958
Cerméle, Dominic M. 51882
Cermesoni, Barbara 26058
Černá, Margit 09658
Černavin, Lev Davidovyč 37688
Černin, Dr. Svatomír 09854
Černjak, Éduard Isaakevič 37859
Černobai, Prof. Dr. Juri 43142
Černova, Vita Vasylivna 43063
Cerny, Zbynek 09473
Ceron, Clodualdo N. 34467
Cerón, Pablo 31081
Cerón Mireles, Prof. José Luis 31079
Ceroni, Nadia 27372
Cerovac, Branko 08938
Cerović, Dr. Ljubivoje 38124
Cerretelli, Claudio 27337, 27338
Ceruti, Mary 50187
Cerutti, Dr. Antonio 27357
Cerutti, Lilia P. de 00685
Cervellino, Miguel 07437
Červenka, Eduard 09849
Cervetti, Dr. Valerio 27092, 27097
Cervini, Dr. Fulvio 27997
Cesaretto, Mauro 27574
Cesarino, José 54545
Cesbron, Pierre 15354
Češková, Stanislava 09809
Česnova, Ljudmila Alekseevna 36897
Cession Louppe, Jacqueline 03893
Cetinić, Ljiljana 38066
Ceulemans, Anne-Emmanuelle 03477
Ceulemans, Louisa 03697
Ceunynck, Christoph de 12676, 12677
Ceut 13060
Cevrim, Dr. Mustafa 17347
Ceynowa, Dr. Klaus 20424
Ceynowa, Dr. Tatjana 16441
Chabard, Dominique 11766
Chabert, Noëlle 14740
Chabot, Jacques 13942
Chačikjan, Svetlana Melkonovna 37601
Chadbourne, Janice 46923
Chadli, Mohammed 31670
Chadwick, Adam 44915
Chadwick, Joseph 46995
Chadwick, P.J. 05637
Chadwick, Phyllis 51611
Chadwick-Case, Wendy 49597
Chadžiev, N. 05189
Chafe, Dr. K.S. 33331
Chaguaceda Toledano, Ana 39934
Chai, Paris K. 49346
Chaik, Pak In 29761
Chaimerez, M. G. 25060
Chainé, Françoise 07177
Chaisson, Rodney 06019
Chait, Karen 51181
Chakarova, Veselina 05081
Chakhmouridian, Dr. Anton 07375
Chakraborti, R.M. 24321
Chakravarti, Rasmohan 03216
Chakravarti, Dr. S.K. 24275
Chald, Elisabeth 45560
Chalfant, Rhonda 53024
Chalker, Mary 01572
Challen, John 43858, 43860
Challender, Nancy 49152
Challener, Elisbeth 52918
Challis, Liz 01032
Challis, Paul 01032
Chalmers, Kim 46976
Chalmers, Peggy 44938
Chalupa, Josef 09678
Chalupar, Erwin 02638
Chamay, Jaques 41152
Chambareau, René 11739
Chambaud, Serge 14683
Chamberlain, Amelia E. 51312
Chamberlain, Lori 06880
Chamberlin, Edward M. 48853
Chamberlin, Ginny 52136
Chambers, Carol 45832
Chambers, Coleen B. 49482
Chambers, Cullen 53632
Chambers, Elizabeth 52086
Chambers, Emma 06450
Chambers, Keith 00991
Chambers, Kevin 51488
Chambers, Natalie 43370
Chambon, Cati 13575
Chambon, Fabrice 14264
Chambon, Marion S. 51075
Chamil, Iskhakov 09419
Champanil, Vira 42535
Chan, Elise 53972
Chan, Peng Sam 07998
Chance, Delores 47303
Chancel-Bardelot, Béatrice de 12115, 12118, 12120
Chandès, Hervé 14597
Chandžieva, Evelina 05016
Chandžieva, Radka 04999
Chanel, Sylvie 14075
Chang, In-Kyung 29803
Chang, Zhenming 08014
Chanler, James 51305
Channing, Susan R 47640
Channon, Ann 43799
Chanson, Robert 41823
Chantrain, Jean-Pascal 03520
Chantrenne, Claire 03477
Chanzit, Gwen 48072
Chapalain, Jean-Philippe 13798
Chapin, Angela 06758
Chapin, Jean 53842
Chapin, Mona 47555
Chaplin, Dr. Simon 45008
Chapman, A.T. 05461
Chapman, Alan 43323
Chapman, Allison 50403
Chapman, Ashley 53363
Chapman, Dennis 54309
Chapman, Gordon 01626
Chapman, J.R. 45489
Chapman, Dr. Jefferson 49818
Chapman, Jim 45234
Chapman, Peter 49040
Chapman, Prof. Stanley 45524
Chapman, Dr. Sydney T. 45357
Chapman Banks, Barbara 50098
Chapnick, Harold 48578
Chapoose, Betsy 48618
Chapp, Belena S. 51282
Chappaz, Jean-Luc 41418
Chappey, Frédéric 13781
Chappuis Waeber, Dominique 41389
Chapron, Philippe 11875
Chapuis, Sophie 41423
Chaput, Leo W. 49887
Charauel, P. 13891
Charbonneau, Diane 06357
Charbonneau, Reine 06321
Charest, France 06493
Charest, Marcie 50417
Charette, Joanne 06551
Charette, Luc A. 06315
Chariot, C. 04116
Chariot, Constantin 03796
Charitatos, Katerina 23102
Charitonov, Christo 05255, 05257
Charland, Stuart 49602, 49603
Charles, Dorothée 14682
Charles, Mary 06595
Charles, Steve 53011
Charlet, Jean-Michel 41974
Charlier, Paul 03911
Charlot, Claude 14661
Charlot, Colette 14239
Charlton, Dr. Christopher 43931, 43932
Charlton, K. 33222
Charmayne Young, J. 52542
Charmelo, Julie 47501
Charpentier, Anne 06341
Charpentier, Corinne 12705
Charpentier, Dominique 11857
Charrière, Edmond 41293
Chartier, Catherine 15364
Chartrain, Michèle 13105
Charzynska, Jadwiga 34632
Chasansky, Matthew 46488
Chase, Ed 46382
Chase, Harriet 52246
Chase, Jim 47312
Chaserant, Françoise 13605, 13607, 13608
Chassaing, Jean-François 13269
Chassé, Sonia 06175
Chasseval, Geneviève de 13322
Château, Georges 15369
Chateau, Micheline 13128
Chateigne, Yann 12062
Chatelain, A. 41984
Chatelain, Prof. E. 13961
Chatelain, Paul 12349
Chatenet, Georges 15940
Chatfield, Linda 48162
Chatland, Gordon 48542
Chatman, Patrina 48109
Chatterjee, Dr. H.J. 44861
Chatterjee, Dr. Sankar 50312
Chatterjee, Sunil Kumar 24383
Chaturvedi, Naresh 24320
Chaucer, S. 45242
Chaudhry, Jamil Ahmed 34007
Chauffeteau, Claude 12067
Chaussee, Gérard 14484
Chauvel, Marie Claire 13327
Chauvin, Dr. Robert S. 47998
Chauviniére, Comte de la 15852
Chavanne, Blandine 14365
Chavannes, Virgile Alexandre 00584
Chavarria, Antonio 52881
Chaves, Dr. Julio César 34104
Chaves, Manuel Wenceslao 34098
Chavez, Anja 54023
Chavez, Michael 48959
Chavez, Thomas 52885
Chavez, Thomas E. 46184
Chávez Chávez, Dr. José Antonio Rodolfo 34116
Chavez Posada, Zoraida 08790
Chazal, Gilles 14749
Cheatham, Jim 48132
Chebanova, N. 54486
Chebitwey, Jane 29741
Checefsky, Bruce 47637
Chefdebien, Anne de 14723
Cheikh, Aboubakar 08792

C

Chek, Dr. Deborah 53719
Chelbi, Fethi 42620
Chelius, Margot 20583
Chellah, Mwimanji Ndota 54653
Chemera, Lyn 48236
Chen, Bilin 08229
Chen, Changgen 08462
Chen, Changjin 07578
Chen, Chu-Ming 42248
Chen, Chunfa 07603
Chen, Chung-Tsang 42180
Chen, Chuyuan 07688
Chen, Dayuan 07980
Chen, Demao 08345
Chen, Deshi 08240
Chen, Dongquan 07652
Chen, Jian Ming 07597
Chen, Jianbiao 08242
Chen, Jiang 08045
Chen, Jianhong 08129
Chen, Jieshi 08118
Chen, Jinsheng 08509
Chen, Jui-Wen 42242
Chen, Julie 51267
Chen, Jun-Ling 42270
Chen, Junqing 07939
Chen, K.C. 42429
Chen, Kevin B. 52778
Chen, Lucia 51267
Chen, Pao-Chung 42373
Chen, Po 07749
Chen, Quanxi 07676
Chen, Rongqi 07975
Chen, S.S.H. 32078
Chen, Shaoming 08463
Chen, Shenghua 07831
Chen, Shih-Bey 42418
Chen, Shilong 07742
Chen, Stone 42151
Chen, Wei 08453
Chen, Xianchun 07684
Chen, Xinming 07671
Chen, Xixin 08286
Chen, Xue'ai 08251
Chen, Zhengyu 08083
Chen, Zhiliang 07693
Chen, Zhimin 08473
Chen, Zhixin 08138
Chen, Zhuxuan 08454
Chen, Zu-E 08259
Chen-Lee, Margaret 42394
Chenault, Amanda 47665
Cheney, Dotty 48835
Cheney, Rebecca 43798
Cheng, Jian 07835
Cheng, Sibao 08412
Chenoufi, Brahim 00083
Chenoweth, Bob 53191
Chenoweth, Diana 52728
Chenoweth, Richard 52896
Chepp, Mark 53236
Chera, Constantin 36096, 36301, 36305
Cherchi Usai, Paolo 52395
Cherek, Dr. Janina 34597, 34599
Cherif, Housine Med 00079
Cherix, Christophe 41403
Cherix, Dr. Daniel 41543
Chermak, Glenda L. 49863
Cherrier, Anne-Marie 14777
Cherry, Stella 24767
Chesebrough, David 47118
Chesi, Prof. Gert 02725
Chesneau-Dupin, Laurence 12565, 12944
Chester, Timothy J. 49007
Chesters, Dr. Robin 44751
Cheval, François 12359, 12360
Chevalier, Alain 15877, 16264
Chevalies, Ségolène le 12751
Chevallier, Bernard 13236, 15139
Chevallier, Denis 14003
Chevillot, Catherine 14692
Chew, Nigel 45612
Chew, Ron 53017
Chew Matsuda, Fay 51226
Chez Checor, José 10345
Chia, Khoo Boo 30613
Chiacchierini, Prof. Ernesto 27496
Chiapperino, Dr. Paola 25565, 28180, 28195
Chiappini, Dr. Marco 25283
Chiara Bettini, Dr. Maria 25680
Chiaramonte, Dr. Giuseppe 27076
Chiarchiaro, Charles 51641
Chiarelli, Dr. Caterina 26158
Chiarucci, Dr. Giuseppe 25175, 26451
Chiba, Mizuo 29518
Chicharro Chamorro, José Luis 39422
Chidavaenzi, R.P. 54666
Chiego, William J. 52717
Chiesi, Dr. Giuseppe 41193
Chikhaoui-Trachene, Prof. Henia 31671
Chikkamare Gowda, B.M. 24073
Chilcote, Doug 52562, 52563
Chilcote, Gary 52562, 52563
Childress, Ann 47913
Childrey, Brooke 49541
Childs, Victoria W. 52566
Chilton, Meredith 07110
Chimay, Prince de 03523
Chimello, Sylvain 15884
Chimier, Brigitte 16025
Chiminelli, Baldassare 25730
Chimisliu, Cornelia 36326
Chin, Song-gi 29838
Chin, Wanda 48472
Chin Foo, Brenda 16325
Chinea Brito, Carmen Dolores 39439
Chinen, Satoru 28537
Chinery, Derek 45480
Chintăuan, Ioan 36446, 36584
Chinwe Anyaegbuna, Felicia 33320
Chiocchetti, Dr. Fabio 27321
Chiong, Dr. Chew Tuan 38180
Chipperfield, Donna 51314
Chiriac, Aurel 36503, 36504, 36505, 36580
Chiriac, Mihai 36196
Chirilă, Dorina 36674
Chiriță, Lenuta 36426
Chisholm, Rachel 45255
Chişiu, Ambrozia 36548
Chitajlenko, Nikolaj Nikolaevič 37380
Chitaley, Dr. Shya 47625
Chitoshi, Matsuo 28635
Chiu, Lei-Mei Julia 28998
Chiu, Melissa 51128
Chivers, Jim 33036
Chižnjak, Olga Romanivna 43045
Chlebarova, Marija 04993
Chlebecek, Friedrich 02162
Chlebinskas, Antanas 30447
Chlobystin, Andrej 37703
Chłodnicki, Dr. Marek 35030
Chlopinska, Ljudmyla Dmytrivna 43151
Chmara, Gitta von 19270
Chmelinova, Dr. Katarina 38233
Chmelnickaja, Ljudmila 03257
Chmelyk, Roman 43141
Chmielewska, Wiesława 35090
Chmielewski-Hagius, Dr. Anita 19369
Chmurzyński, Wojciech 35250
Cho, Byoung Soon 29928
Choate, J.R. 49225, 49225
Chocholáč, Ivo 09676
Chochorowska, Elżbieta 34784
Chodasiewicz, Jerzy 35152
Chodžaev, Muchtar Bachadirovič 29726
Choe, Sung Eun 29874
Choetf, Betsy 49481
Choi, Prof. Duck-Keun 29921
Choi, Eric 07992
Choi, Il Sung 29791
Choi, Kim Sang 29756
Choi, Man-Li 29820
Choinière, France 06337
Choiselat, Roland 06460
Choisne, Franck 15663
Cholette, Yvon 07056
Cholewa, Anita 50735
Cholewa, Kurt 22539
Cholodinskienė, Adele 30311
Chomenko, Tamara Jevgenivna 43089
Chomič, Svetlana Viktorovna 37054
Chong, Alan 46934
Chongkol, Chira 42552
Choremi, Dr. Alkestis 22917
Choremis, Alketis 22947
Chorunženko, V.A. 36917
Chou, Ju-Hsi 47624
Choudhury, B. 24186
Choudhury, Dr. R.D. 24188
Choulet, Jean-Marie 13024
Chow, Christina 51247
Chow, Kim-Fung 07805
Chowdhuri, Dr. M.K. 24272
Chowdhury, Prof. Dr. Saifuddin 03234
Chrest, Gwen 47651
Chrétien, Isabelle 15961
Chris, Dercon 44996
Chrisanov, Petr Aleksandrovič 37784
Chrisman, D. Kent 52381
Christ, Alexa-Beatrice 17788
Christ, Hans 19207
Christ, Hans D. 21970
Christ, Magdalene 17018
Christ, Nelda M. 46863
Christen, Benoît 11966
Christen, François 41684
Christen, Gian Franco 41239
Christensen, Barbara 50921
Christensen, Christa 47383
Christensen, Henry 33141
Christensen, J.O. 10156
Christensen, Lars K. 10165
Christensen, Lisbeth 10049
Christensen, Margaret 06016
Christensen, Peter M. M. 10093
Christensen, Ruth 30194
Christensen, Svend 10093
Christenson, Elroy 53007
Christenson, Goran 40723
Christiaens, Daniel 03291
Christian, Jay 48514
Christian Pedersen, Sune 10148
Christiansen, Calvin 53967
Christiansen, Jette 10145
Christiansen, Sherie 53055
Christie, David 51762
Christie, Gordon A. 44235
Christie, Nancy 51762
Christman, David C. 49246
Christmann, John 07137
Christmann, Werner 20120
Christmeier, Martina 20758
Christodoulakos, Yannis 22961
Christoffel, Udo 17217
Christoffer, Ed 05819
Christoffersen, Erna 10075
Christoforou, Dionysia 45099
Christophersen, Axel 33932
Christou, Dr. Demos 09393
Christov, Christo 05026
Christov-Bakargiev, Carolyn 27423
Christova, Emilia 05053
Christy, Strum 52886
Chrzanowski, Jerzy 35366
Chu, Dr. Christina 07794
Chu, Doris 46952
Chu, Guiqing 07701
Chu, Shibin 07921
Chu, Xinjia 08218
Chubbuck Weinstein, Elizabeth 46633
Chubrich, Michael 52098
Chudjakova, Lidija Aleksandrovna 36891
Chudy, Tomasz 35223
Chughtai, Arif Rahman 34018
Chuma-Ibe, Gloria Chianu 33333
Chumchal, Milan 09804
Chunchu, Ai 07518
Chung, Estella 50255
Chung, Prof. Hsiung Yih 42326
Chung, Prof. Jing-Won 29775
Chung, Yang-Ho 29910
Chuoc, Cu Van 54615
Chupin, Olivier 11616, 12244
Church, Dorian 45653
Church, Rena J. 46490
Church, Susan 49471
Churcher, Andrea May 01322
Chushupal, E.N. 29966
Chusnutdinova, Rušanija Salichovna 37018
Chutiwongs, Dr. N. 32488
Chvanov, Michail Andreevič 37884
Chvátal, Marek 09512
Chwałek, Roman 35096
Chwaliński, Andrzej 34964
Chwalisz, Zofia 35318
Chwalzcyck, Diana 17688
Chylak, Ana 43315
Ciaffoni, Mirko 26768, 26769
Ciamos, Christian 11560
Ciancio, Dr. Angela 26341
Cianciulli, John 48748
Ciannella, Joeen 49302
Ciapponi Landi, Bruno 26563
Ciaran, Dr. Fiona 33211
Ciardi, Dr. Roberto Paolo 27673
Ciarla, Dr. Roberto 27512
Ciavolino, Prof. Giuseppe 28036
Cibulová, Dr. Vlasta 09787
Cicala, Francesco 27085, 27086
Cicala Campagna, Dr. Francesca 27853
Cicali, Francesco 27087
Çiçeksız, £lbrahim 42803
Cicelyn, Eduardo 26881
Cichecka, Barbara 35292
Ciciero, Carla 46759
Cicierski, Ernest 07387
Čičimov, Aleksandar 30530
Cicimurri, David 47613
Cicognani, Giorgio 25548
Cidenova, Rita Badmaevna 37892
Ciechanowska, Joanna 44958
Ciecior, Gabriele 18995
Ciepiela, Elżbieta 35352
Cierna, Katarína 38233
Cieślak, Bogdan 35321
Cieslicki, Dorothy 48271
Cieslik, Gérard 11602
Cieślikowska-Ryży, Julita 34703
Cifelli, Dr. R.L. 51376
Cifuentes, Joaquín 31119
Ciganok, Nina 43030
Cigarroa, Melissa 49952
Cigliano, Flavia 46943
Cihan Özsayıner, Zübeyde 42821
Čikin, Vladimir Iljič 43174
Čilašvili, L.A. 16381
Cilona, Filippo Ferdinando 26092
Cimdiņa, Ruta 30070
Činč-Juhant, Dr. Breda 38461
Cinc-Mars, François 07073
Cindi, F. 38675
Cini, Amato 28126
Cini, Stephen 30650
Činovec, Igor 09836
Cinquepalmi, Angela 26090
Cintulova, Erika 38265, 38267
Ciobanu, Adrian 36567
Cioce, Dr. Pasquale 25430
Cioffari, Gerardo 25363
Cioppi, Dr. Elisabeth 26180
Ciorăscu, Nicolae 36393
Cipoletti, Christopher 47705
Cipollone, Dr. Rosa 26995
Ciprian, Dr. Pavel 09431, 09436, 09442
Cirani, Simona 27613
Cirbian, Susanna 40260
Cîrciu, Gratiela Ionelia 36312
Cirill, T. 23750
Cirillo, Lena 01379
Cirjakovic, Ivana 38094
Čirkova, Larisa I. 37663
Ciro Galotta, Brás 04905
Cirule, Astrīda 30026
Ciseri, Dr. Ilaria 26792
Cisewski, Gene 49448
Cisilino, Dr. Daniela 25325
Cisneros, Jose 49618
Cisneros, Mario Toledo 31328
Cisneros, Rosa María 33285
Cist, O. 02306
Ciubotă, Dr. Viorel 36157, 36270, 36514, 36589, 36590, 36591, 36717, 36718
Ciubotaru, Neculai 36738
Ciucă, Răzvan 36629
Ciugudean, Horia Ion 36106
Civai, Mauro 27836
Číž, Marian 38364
Claassen, Garth 47161
Clabby, Paul 51051
Claes, Gabrielle 03481
Claessens Peré, Nanette 03321
Claesson, Anna Maria 40614, 40846
Clair, Cynthia 51038
Claliūna, Alfons 41800
Clam-Martinic, Dr. M.C.D. 02193
Clancey, Erin 50264
Clancy, John L. 53738
Clap, Sylvestre 11791
Clapperton, Raymond 33102
Clarck, Trinkett 46268
Clare, Roy 44939, 44977
Clarés, María Jesús 40204
Claret, Sawador 40081
Claret de Gouvêia, Antônio M. 04605
Clarin jr., Oligario L. 34406
Clark, Alexia 44457
Clark, Andrew 01512
Clark, Becky 51475
Clark, Bill 48151
Clark, Carroll 51342
Clark, Charles T. 47606
Clark, Elaine R. 49050
Clark, Eleanor 44963
Clark, Gayle 52349
Clark, H. Nicholas B. 46266
Clark, Jackie 51003
Clark, Jane E. 48745
Clark, Jeanne 49909
Clark, Joann 50677
Clark, Kathy M. 51585
Clark, L.R. 01224
Clark, Martin 45575, 45579
Clark, Michael J. 53682, 53683
Clark, Nancy 46339
Clark, Nora 47216
Clark, Patrick S. 52950
Clark, Peter 48776
Clark, Peter P. 47791
Clark, Phillip 50192
Clark, Richard H. 54078
Clark, Robert 43949
Clark, Robert A. 43915
Clark, Robin 49846
Clark, Roger 43583
Clark, Scott 52561
Clark, Dr. Thomas D. 50051
Clark, Tom 53707
Clark, Vicky A. 51956
Clark-Bedard, Sharon 06469
Clark-Langager, Sarah 46713
Clarke, A.H. 00914
Clarke, Alicia 52847

Clarke, B. 43727
Clarke, Cliff 05772
Clarke, David 33035
Clarke, Deb 45762
Clarke, Ignatius 24871
Clarke, Jeremy 45507
Clarke, Jo 45601
Clarke, Judith 45357
Clarke, Lance 49022
Clarke, Michael 44138
Clarke, Robert 43261
Clarke, Ron 43902, 44398
Clarke, Sandi 00935
Clarke, T. Henry 54002
Clarke Di Vico, Carol 46613
Clary, Owen 46140
Clason, Anders 40549
Clatworthy, J.C. 43506
Claude, Yves 30463
Claus, Elisabeth 16586
Clausen, Claus 10080
Clausen, H. Erhardt 09972
Clausen, Jochen 19807
Clausen, Kim 10067, 10267, 10269, 10270, 10271
Clausen, Kirstin 05482
Clausen, Williy 41359
Clausnitzer, Regina 18780
Claussen, Hans-Joachim 22177
Clavedetscher, Silvia 41743
Clavel, Sylvie 14650
Claveria, Josef 03015
Clavier, Michel 03874
Clavijo, Martha Isabel 08765
Clavijo Colom, Moraima 09162
Clawson, Rita 46663
Clay Chace, Helen 51178
Clayden, Stephen 06838
Clayton, George C. 48473
Clayton, Rachelle 05319
Cleal, Rosamund 43338
Clear, Celia 44995, 44996
Clearwater, Bonnie 50634
Cleary, Chris 01459
Cleasby, I.W.J. 43310
Clegg, Dr. W.H. 45861
Cleland, Yvonne 43418
Clemens, Darlene 52028
Clemens, M. 13972
Clemens, William 46758
Clement, Constance 51055
Clément, Lise 05611
Clement, M. 14226
Clement, Morag 44558
Clement, T.A. 46059
Clemente, Milagros 34345
Clements, Diane 44923
Clements, Elaine 46297
Clements, Gayle 53608
Clements, Louis 52300
Clements, Lynn B. 53759
Clendaniel, Dianne 46603
Clephas, Egon 20517
Clergue, Yolande 11682
Clerke, Rachel 45556
Clette, Michele 03476
Cleve, Thomas 19825
Cleveland, James C. 52299
Cleveland, Richard L. 52999
Cleven, E. 32000
Clevenger, Martha R. 50193
Clews, Stephen 43410
Cleyet-Merle, J.J. 13699
Cliffe, L. 45675
Clifford, Bridget 44968
Clifford, Lord Thomas 43844
Clifton, Dan 48238
Clifton, James 49397
Clifton, Dr. James 49404
Clifton, Julia 52881
Clifton-Harvey, Melanie 53293
Clin, Marie-Véronique 14533, 14691
Cline, Garry 46278
Cline, Raymond 51448
Clinefelter, Pamela 49913
Clinger, Melinda 52384
Clinkingbeard, Brion 50288
Clitheroe, Gordon 45381
Clizbe, William C. 49100
Clodfelter, Lee 47052
Cloître, M.C. 14854
Clos Llombart, Jordi 38992
Cloşcă, Gabriela 36171
Close, Elisabeth 01029
Close, Freddy 03568
Close, Helen 45048
Close, Oliver 00913
Close, Timothy 46890
Cloudman, Ruth 50297
Clouter, Carl 05848
Cluff, Raydene 53491
Clute, Richard 46237
Cluver, Dr. M.A. 38597
Clynckemaillie, Olivier 03525
Clyve, Lynda 46597
Cnops, Pieter 03447
Coakley, Stephanie 53598
Coaña deValicourt, Yago Pico de 39586
Coates, G. 05908
Coates, Ross 52199
Cobarg, Dr. Merete 20591
Cobb, Elvan 49291
Cobb, Michael 49151
Cobb, Myrna 05576
Cobban, Janet 05975
Cobbe, Toy L. 50520
Cobble, Kelly 52218
Cobbold, A.G.B. 43696
Cobert, William 51114
Cobo, Maria 47078
Cobos Colmenares, Carlos Eduardo 08732
Cobos Wilkins, Juan 39653
Cobridges, Robert 50846
Coburn, Dr. Carol 49702
Coburn, Oakley H. 53195
Cobzaru, Dumitru 36370
Cocartă, Adrian 36364, 36365, 36366, 36367, 36368
Cocea, Sergio Eduardo 00688
Cochat, L. 15893
Cochener, Margaret M. 52380
Cochet, Vincent 12966
Cochran, Dorothy 51272
Cochran, Glenn 53516
Cochrane, G. 44021
Cochrane, Dr- Timothy 48999
Cockrill, Pauline 00796
Cockrum, Terry 51526
Cocordas, Eleni 51199
Cocuz, Prof. Ioan 36458, 36622, 36640, 36642, 36644, 36713
Codagnone, Brian 46951
Coday, Jean C. 50450
Codding, Mitchell A. 51192
Codega, Samuele 27343
Codicé Pinelli, Dr. Ferruccio 25559
Codomi, Shelley 48294
Codreanu, Elena 36482
Coe, Erin M. 48935
Coe, John G. 47447
Coekelbergs, Prof. R. 04153
Coelewij, L. 31888
Coelho, Conceição 35879
Coen, Dr. Vittoria 25440
Coenen, Luc 03819, 03820
Coënen, Michel 16299
Coenen-Staß, Prof. Dr. Dieter 20101
Coers, Frits 32845
Coetzee, F.P. 38750
Coetzee, Marc 50643
Cofax, Debbie 52350
Coffey, Jeanne 49296
Coffey, John W. 52232
Coffin, Anne 51195
Coffrini, Ermes 25530
Cogan, Stanley 48577, 48579
Cogar, William 50935
Cogdell, Sabine 51461
Cogeval, Guy 06357
Cohalan, Mary Lou 48251
Cohen, A. 25087
Cohen, Christopher 00835
Cohen, David 51245
Cohen, Françoise 14446, 14450
Cohen, Jack 51866
Cohen, Janie 47144
Cohen, Judith 25093
Cohen, Marvin H. 50661
Cohen, Nancy 53364
Cohen, Patricia 00262
Cohen, Ronald A. 50679
Cohen, Sandor 46514
Cohen-Straytner, Barbara 51244
Cohn, Arthur B. 53717
Cohn, Marjorie B. 47177
Cohn, Terri 46752
Cohon, Dr. Robert 49695
Cohrs, Karin 19300
Cohrs, Peter 19300
Coibanu, Doina 36244
Cojocaru, Adriana 36648
Cokendolpher, James 50312
Coker, Gylbert 53501
Čolakov, Dimităr 05224
Colantuono, Patrizio 25245
Colardelle, Michel 14727
Colardelle, Renée 13132
Colato, Juan 50269
Colazo, Susana 00527
Colbert, Brigitte de 14244
Colburn, Bolton 49871
Colby, James 49596
Coldren, James 48365
Cole, Felicity 45251
Cole, Garrett 46596
Cole, Geoffrey 43352
Cole, Glory 49308
Cole, Jane 06412
Cole, K. 38615
Cole, Mark 47752
Cole, Natalie 46058
Cole, Polly 43628
Cole, Richard 44934
Cole, Vicki 52817
Cole-Will, Rebecca 46600
Coleman, C. Allen 51912
Coleman, Don 48573
Coleman, Evelyn 24856
Coleman, Mary Ruth 46201
Coleman, Rhonda 52713
Coleman, Sally 45761
Coleman, Winifred E. 54060
Coles, Alec 45227, 45229, 45697, 45805
Coles, Kerrie 01362
Coley, Dwaine C. 49272
Colin, N. 03608
Colina, Gay 53955
Colitti, Raffaele 25635
Coll, Tom 54345
Coll Conesa, Dr. Jaume 40216
Coll Gormley, Jean 49233
Colla, Francis 03599
Collado, Elisa 39758
Collas, Josiane 12762
Colle, Jean-Pierre 13063
Colle, Marcella 03701
Collens, David R. 50898
Collett, Keith L. 00907
Collett, Morine 49526
Collette, Dr. Alfred T. 53399, 53403
Collette, Alice 49386
Collette, Louis Jeffery 50623
Collier, H. 45481
Collier, Ric 52693
Collin, Fernand 03602
Collin, Dr. Gérard 11977
Collin, Ludo 03641
Colling, Guy 30484
Collings, Donna 06323
Collins, Anne 45961
Collins, Craig 46912
Collins, D. Cheryl 50431, 50433, 50436, 50437, 50438
Collins, David 51042
Collins, George 49052
Collins, J. Thomas 49264
Collins, James H. 48145, 50778
Collins, Lin 45218
Collins, Louanne 45067
Collins, Lynn 01616
Collins, Mary 52198
Collins, Michael 50318
Collins, Paul W. 52860
Collins, Philip 45132
Collins, Phillip 47924
Collins, Toni I. 48404
Collins, Vickie 51859
Collins Blackmon, Mary 46507
Collins Coleman, Clara 52584
Collinson, Howard 49521
Collison, Nika 06982
Collu, Cristiana 26965
Colman, Devin A. 52620
Colom, Connie 50333
Colombara, Franco 25898
Colombi Ferretti, Dr. Anna 26077
Colombié, Ervé 15427
Colombino, Carlos 34089, 34096
Colombo, Dr. Silvano 25763
Colombo Dougoud, Roberta 41421
Colomer, Laia 38826
Colon, Anibal 52541
Colon, Constance B. 53311
Colon, Evangelina 36074
Colon, Helen 45594
Colonna, Cécile 15114
Colonna, Vittoria 25625
Colonna Ceccaldi, Michel 16125
Colossi, David 51187
Colpaert, Hilde 03737, 03738
Colton, Chris 07161
Colton, Fiona 44427, 44548, 45638
Columbo, Colleen 05880
Colvia, Dr. Anna 27503
Colvin, Richard D. 50402
Colwell, Bruce 51421
Comanducci, Dr. Francesco 27698
Comas Estatella, Dr. Antonio 40281
Comba, Steve 47574
Combe, Anette 41713, 41714
Combe, Jean 41976
Combe, M. 15031
Combs, Jim 49747
Combs, Robert K. 48647
Comeault, Maurice 06866
Comedown, Alberta 53817
Comer, Karen 46432
Comer, Michael 51071
Cometti, Jorge 00182
Cometti, Marta 41283
Comingore, B.J. 53029
Comlan Godonou, Alain 04188
Commandeur, Beatrix 17017
Commock, Tracy 28328, 28335
Communeau, Matthieu 12704
Compagno, Dr. L.J.V. 38592
Compagno-Roeleveld, M.A. 38592
Compaore, Benilde 05278
Comparin, Guy 13125
Compston, Nici 00773
Compton, Richard 45499
Comte, Philippe 14755
Comte, Samantha 01081
Comtois, George S. 50060
Cona, Dr. Cherubino 27723
Conable, Mary 53832
Conaway, Mary Ellen 48863, 48885
Conceição Correia, Maria da 35575
Conceição Duarte Rosa, Maria da 04599
Conconi, Maria Pia 41770
Conde, Victor 39552
Condeescu, Dan Alexandru 36186, 36187, 36188, 36189, 36190, 36215, 36723
Conder, Marjorie D. 52692
Conder, Tony 44805
Condos, Lena 01022, 01023
Condurache, Val 36424, 36428, 36575
Cone Busch, Ellen 51660
Cone Skelton, Annette 46445
Cones-Browne, Rodney 01364
Coneva, Tanja 05173
Confer, Deborah 46961
Conforti, Michael 54175
Cong, Pham Van 54625
Conger, Ivan A. 51642
Conger, Ken 53062
Conihout, Isabelle de 14571
Conijn, G.H. 32167
Conijn, N.W. 31915
Conkle Davis, Gloria 46408
Conklin, David W. 48107
Conklin, Jo-Ann 52164
Conkright, Lance 54050
Conlan, Lynne 43851
Conley, Alston 47441
Conley, Maria 46133
Conlon, Beth 47601
Conlon, Helen 45591
Connaughton, Vincent 24870
Conneau, Jean-Claude 12608
Connell, Andrew 44160
Connell, Dr. D.P. 44593
Connell, John 06303
Connemann, Elfi 17686
Connemann, Hans-Joachim 17686
Conner, Michael 47335
Conner, Roberta 51756
Connolly, Angela 48098
Connolly, Thomas 48435
Connor, Brian 33094
Connor, James T. H. 53903
Connor, Kathy 52395
Connor, Lorraine 50382
Connors, Dennis 53402
Connors, Dr. Thomas 47291
Conocido, Domicilio 30837
Conover, Margaret 53039
Conrad, Dr. Christofer 21960
Conrad, Geoffrey 46867
Conrad, Dr. J. 18642
Conrads, Margaret 49695
Conradt, Dirk-Michael 22658
Conrath-Scholl, Gabriele 19627
Conrot, Valérie 30478
Conroy, Christopher J. 46759
Consagre, Franesca 52595
Consey, Kevin E. 46752
Considine, Brian 50237
Console, Dr. Ester Maria 54501
Constable, Linda 43567
Constable, S. 45270
Constant, Carol 49328
Constant Noanti, Mahoun 04184
Constantin, Alupei 36158
Constantinescu, Corina 36299
Constantino De La Penna, Carol 40087
Constantopoulos, Dr. Jim 52055
Consuelo Dominguez Cuanalo, Delia del 31291
Contarino, Dr. Giuseppe 25143

Contassot, Cécile 13126
Conte, Cynthia 47397
Conte, Dr. Robert S. 54131
Conte, Roxana Diaz 00520
Contessa, Gregorio 26584
Conti, Prof. Dr. Luigi Francesco 26280
Conti, Richard C. 51367
Conti, Roberto 26864
Conticelli, Dr. Valentina 26157
Conticello, Prof. Dr. Baldassare 25517, 27887
Conticello, Dr. Baldo 26735
Contiguglia, Georgianna 48069, 53566, 53567, 53569
Contini, Dr. Roberto 17116
Contreras, Mamerto 34481
Contreras, Pilar 31085
Contreras Martínez, José Eduardo 31505
Contro, Antonia 47488
Converse, Diane 49330
Conway, Allen C. 51009
Conway, Cari 51306
Conway, James 48113
Conway, Rosemary 53765
Conway, Steve 04199
Conway, Steven 04196
Conwell, Idella M. 50454
Conyers, Wayne 50371
Conyngham, Frank E.P. 54162
Conzelmann, Peter 17419
Conzen, Friedrich G. 17987
Conzen, Dr. Ina 21960
Cook, Darrell J. 47892
Cook, Deborah 45692
Cook, Emma 45953, 45967
Cook, Janel 50335
Cook, Jennifer 48959
Cook, Dr. Joseph A. 46182
Cook, Joyce A. 48579
Cook, Mary Jo 50906
Cook, Prof. Nigel 33737
Cook, Patrick 46017
Cook, Robert 51704, 54052
Cook, Silas B. 52075
Cook, Steven 05711
Cook, Thomas 47721
Cook, Wendy 46072
Cooke, Bill 50050
Cooke, Carolyn 49508
Cooke, Glenn 01512
Cooke, Lynne 46660, 51166
Cooke, Michael 28325, 28329, 28331, 28341
Cooke, Pat 24805, 24815
Cooke, Patrick 43997
Cooke, Prof. Robin 01151
Cooke, Dr. Steven 01518
Cooke, Susan 45031
Cooke, Vina 45217
Cookendorfer, Betty 49185
Cool, F. 03975
Coolidge, Hope 48729
Coolidge, Jennifer 47996
Coolidge, John 49929
Coombe, JoAnne 48199
Coombs, Gary B. 48965
Coombs, Margery C. 46269
Coombs, Robert 54136
Coomer, Robert 48369, 53709
Cooney Frelinghuysen, Alice 51212
Cooper, Andrea 47691
Cooper, Avis 33101
Cooper, Bernard F. 53115
Cooper, Charles E. 44001
Cooper, Grant 33079
Cooper, Harry 47177
Cooper, Helen 51056
Cooper, Dr. James H. 51273
Cooper, Jo-Ann 01281
Cooper, John 44741
Cooper, Julie 47318
Cooper, Katie 49033
Cooper, Linda G. 54349
Cooper, Norman 06212
Cooper, Rhonda 53315
Cooper, Robert L.D. 44126
Cooper, Ronna 51814
Cooper, Rosemary 45243
Cooper, S. 44497
Cooper, Timothy O. 47108
Cooper, Valerie 05514
Cooper, Virginia 07100, 50921
Cooper Case, Carolyn 52318
Coopmans, H. 31924
Coote, Sylvia 45786
Cope, Derek 45989
Cope, Michael W. 44516
Copeland-Burns, Carla 50064
Coppens, Marguerite 03495
Coppernoll, Lee 50718
Coppieters, Pierre 03887
Coppock, C. 45352
Coppola, Gianfranco 27622
Coppola, Regina 46271
Coppus, G.M. 32804
Copson, R. 44167
Coquery, Emmanuel 15998, 15999, 16000, 16001
Coral, Laurent de 15325
Corazza, Dr. Carla 26108
Corbat, Marc 41725
Corbett, Agnes 47199
Corbier, René 12267
Corbin, Dr. James E. 50938
Corbin, Kendall W. 50735
Corbishley, Grant 33181
Corblet de Fallerans, J. 15947
Corbu, Emilia 36630
Corcoran, M. 24812
Corcos, Sidney 25000
Corcuera, Catalina 31036
Cordaro, Dr. Michele 27447
Cordell, Prof. Linda S. 46961
Corder, Russ 51988
Cordero, Dr. Mario 26019
Cordes, David 48103
Cordewener, Raphaël 03959
Córdoba, Elsie D. 34059
Cordonnier, Aude 12798
Cordova, Rudy 53805
Cordova-Gonzalez, Julia 07430
Coreth, Peter 02891
Corey, Mark 49050
Coria, Rodolfo 00495
Corio, Amedeo 28220
Corkern, Wilton C. 46131
Corkum, Kate 51015
Cormack, Malcolm 52343
Cormier, Angèle 06317
Cormier, Bernard J. 06835
Cormier, George Ann 52038
Cornamusaz, François 41451
Cornejo, Stephanie 52508
Cornejo Gutiérrez, Miguel Angel 34165
Cornelis, J.P. 03938
Cornelissen, Leo M. 31891
Cornelius, Betty L. 51712
Cornelius, Robert E. 49537
Cornelius, Dr. St. 38695
Cornelius, Steffi 17312
Cornet, Josep 38973
Cornetto, Elisabeth 11859
Cornez, G. 03980
Corniani, Maurizio 26593
Cornick, N. 03954
Cornini, Dr. Guido 54501
Cornu, Bernard 13932
Cornwell, John 01196
Cornwell, M.J. 45605
Coroama, Radu 36221
Coroiu, Dr. Ioan 36292
Corominas de Basso, Raquel 00420
Coronel, Maria Delia 34372
Coronilla Castro, Manuel 39201
Corrain, Prof. C. 26363, 27898, 28118, 28272
Corre, Caroline 16110
Correa Sutil, Dr. Sofía 07460
Correia, Dr. Alberto 36022
Correia, Dr. Margarida Rebelo 35860
Correia, Dr. Maria João Pinto 36023
Correia, Dr. Virgílio H. 35548
Corrigan, David J. 49197
Corrin, Lisa 53013, 54176
Corriveau, Linda 51762
Corry, Corrine 06744
Corry, L. 45317
Corser, Elizabeth 05630
Corsini, Principessa 26155
Corsini, Deborah 52810
Corsolini, Prof. Rodolfo P. 00145
Cort, Cynthia 49022
Cort, Louise 53849, 53873
Cortenova, Prof. Giorgio 28228
Cortes, Prof. Erlinda S. 34519
Cortés Gomez, Carlos 31039
Corteville, Julie 14217
Cortez, Miguel 47508
Cortina, Leonor 31058
Cory, Jeff 47460
Corzo, Miguel-Angel 51878
Cosandier, Juliane 41539
Coşereanu, Valentin 36429
Cosgrove, Patricia 46473
Coski, Dr. John M. 52335
Cosma, Aurelia 36217
Cosmetatos, Helen 22890
Cossa, Egidio 27519
Cossart, Graham 00919
Cossignani, Dr. Tiziano 26023
Cosson, Maurice 12434
Costa, Diane 52843
Costa, Elisabeth 35690
Costa, Dr. Gian Paolo 26074
Costa, Prof. Giovanni 25776
Costa, John 52301
Costa jr., Luís da 31711
Costa, Dr. Susana Faro da 35498
Costamagna, Dr. Alba 27503
Costamagna, Liliana 27892
Costantini, Dr. Lorenzo 27512
Costar, John 33206
Costard, Matthew P. 45561
Costas Loyola, José E. 36048
Coste, Pierre 11848
Costello, Lisa 47336
Costello, Noel 01186
Costelloe, Jean 24889, 24890
Costely, Nigel 45863
Costersian, John 48921, 54236
Costescu, Cornelia 36573
Costeur, Dr. Loïc 41175
Costoulas, Costas 23013
Cosulich Canarutto, Sarah 25934
Coté, Michel 13903
Cote, Richard 53859, 53860
Cotie, Donna 06874
Cott, Peter 44353
Cotter, Patrick 24769
Cotter, Suzanne 45332
Cotterill, F.D.P. 54666
Cottet, Jean-Pierre 14288
Cottino, Prof. Alberto 28013
Cottino, Peter 25224
Cottle, Gerry W. 46068
Cotton, Areca 54374
Cotton, Carol 48915
Cotton, Gordon A. 53734
Cottrell, Robert 53439
Cotty, Céline 13477
Couaille, C. le 14241
Couch, Janet 50279
Couderc, Sylvie 11587, 11588
Coudray, Robert 13792
Coudrot, Jean-Louis 12474
Coudurier, Yvonne 14348
Coughlan, M. 01356
Coughlin, Richard J. 49800
Coughran-Rayden, Sharon 53715
Coulibaly, Seydou 08828
Coulombe, Marie-Christine 07193
Coulombier, Jean-Luc 15334
Coulon, François 15016
Coulon, Gérard 13411, 15145
Counot, Bernard 13744
Count, Mary Ellen 50793
Cour Dragsbo, Peter la 10279
Courbe, Daniel 03893
Courson, Bruce 52841
Courtelarre, Henry 12254
Courtial, M.C. 13860
Courtiour, Jon 45868
Courtney, Al 53037
Courtney, Vernon 54158
Courtois, Jean-Pierre 03648
Couser, Carol 50140
Cousin, M. 11922
Cousino, Beate 11880
Cousinou, Olivier 13993
Coutagne, Denis 11518
Coutheillas, Didier 12874
Couto, Dr. Matilde Tomáz do 35497
Coutts, Herbert 44162
Coutts, Howard 43381
Couturas, Francis 14778
Couture, André 06677
Couture, Llyr E. 48166
Covaci, Heruvima 36473
Covato, Prof. Carmela 27529
Coventry, Susan 44894
Covey, Dustin 05749
Cowan, Connie 49508
Cowan, Joan 05776
Cowan, Lina 49597
Cowart, Jack 53855
Cowburn, Ian 13526
Cowden, Chris 46523
Cowell, Lorraine 53790
Cowell, R. 44750
Cowell, Tom 07330
Cowie, Kate 00949
Cowie, Robert G. 51699
Cowley, Paul 44040
Cowlin, Wendy 45980
Cowling, A. 45270
Cowling, Dr. Mary 44164
Cowling, Sue 51577
Cox, Arrington 47726, 47729
Cox, Dan 50469
Cox, Dr. Douglas C. 52187
Cox, Janet 47619
Cox, Janson L. 46834
Cox, Katherine 49435
Cox, Marilyn 50829
Cox, Nigel R. 44316
Cox, Dr. Patrick 46900, 53678
Cox, Richard 43750, 47075
Coxe, Trudy 51327
Coxson, Michael 46094
Coy, John 50576
Cozzolino, Giorgio 27374
Crabtree, Andrew 43562
Crăciun, Georgeta 36423
Cracket, Alan 45756
Craft, John J. 51823
Crago, Jody A. 54119
Craig, Currell 53603
Craig, Joanna 47201
Craig, Len 01450
Craig, Martin 54067
Craig, Neil 45751
Craighead, Linda 51688
Crailsheim, Freiherr Ortholf von 16538
Crain, Sandra 49503
Cramer, Carole 52160
Cramer, Jeffrey S. 50093
Cramer, Wendy 51060
Cramer-Petersen, Lars 09984
Crampton, Julian 43627
Crampton, Lewis S.W. 52413
Crampton, Sharon 38567, 38569
Crandall, Daniel T. 52085
Crane, Louise 53498
Crane, Prof. Peter 44570
Crane, Susan 52977
Cranson, Diane 00791
Cranston, Peter S. 47978
Craughwell-Varda, Kathy 52954
Craven, Jane 49580
Craven, K. 45601
Craven, Tim 45706
Crawford, Bob 52750
Crawford, C. 06738
Crawford, Henry B. 50312
Crawford, James 47211
Crawford, Joan 47010
Crawford, Louise J. 50774
Crawford, Madelyn 50207
Crawford, Nina 05678
Crawford, Dr. Richard 48990
Crawford, Shirley 46193
Crawford, Vivian 28328
Crawford-Gore, Betsy 49955
Crawley, Tracey 45737
Crawshaw, Clare 44560
Creal, Carolyn 53194
Creamer, Winifred 48037
Crean, Kristen 52068
Creary, Marilou 51434
Creasey, Michael 52179
Creddes, Jean 05999
Credland, Arthur G. 44499, 44592, 44599
Credle, Jamie 52939
Creech, C.L. 49812
Creecy, Kerry 01132
Creemers, Guido 04075
Crégut, Dr. Evelyne 11795
Creighton, Martha 53806
Creighton, Dr. Philip D. 48585
Crelot, Stefan 03872
Cremer, Kurt 16409
Cremers, Filip 04099
Cremoncini, Roberta 44844
Crenzien, Helle 10103
Crépin-Leblond, Thierry 12030, 12032, 12035, 12809
Crépy, Edvard de 16141
Crespo, Michael 46641
Crespo-Burgos, Dr. Antonio 10448
Crespo Gutiérrez, Victoria 39578
Cressey, Dr. Pamela J. 46198
Cresswell, J. 43565
Creutzburg, Dipl. Biol. Frank 18595
Creutzburg, Gerlinde 16443
Crevecœur, Claude F. 04144
Crew, Keith 51891
Crews, Gerald 05887
Cribb, J. 44803
Crick, Micheal 44567
Crick, Monique 41411
Crippa, Flavio 26274
Crippen, Donna 48313
Crişan, Aurelia 36678
Crişan, Ciprian Petru 36607
Crişan, Eva 36285
Crişan, Veronica 36110
Crispini, Prof. Giuseppe 27642
Crissman, Erin 46158
Cristache, Florin 36455
Criste, Colleen 51194
Cristofoli, Raoul 28006
Cristofoli, Vincent 13383, 14466, 14468
Critchlow, William H. 06262
Crites, Dr. Gary 49818
Crittin, Jean-François 41289
Crivelli, Paolo 41279
Crnković, Prof. Vladimir 08995

Crociata, Mariano 26612
Crocker, Margaret S. 46815
Crocker, Piers 33870
Croft, Brenda L. 00946
Croft, Evelyn 52517
Crofts, Nicholas 41418
Croke, Fionnuala 24807
Croker, R.C. 33057
Crokett, James 46563
Croll, Helen 00943
Cromie, Cheryl-Anne 38646
Cronin, Michael 51242
Cronin, Thomas E. 53798
Cronk, Dr. Quentin 07199
Cronlund, Carl-Johan 40925
Crook, Rosemary 44382
Crook, Rosie 44386
Cropera, Dr. Carla 26123
Cropp, Aubrey 33066
Cros, Caroline 14646
Crosby, Philippa 44363
Crosman, Christopher B. 52422
Cross, Christine 46330
Cross, Dave 07407
Cross, Louise 48931
Cross, Sharon 43720
Cross, Tony 43292, 43293, 45373
Crossland, Leonard B. 22861
Crossland, Maynard 50025
Crosson, David 55138
Crosson, Diana 47687
Crothers, George M. 50059
Croton, G. 33116
Crotti, Pierre 41541
Crouch 01007
Crouch, Ned 47592
Crouch, Philip 44564, 46075
Crous, Dr. D. 38586
Crouse, George 49684
Crouter, Philip 43593
Crouy-Chanel, Bernard de 13382
Crow, Pamela B. 53775
Crowdy, Hannat 43818
Crowe, K.L. 45714
Crowe, Lilah J. 49009
Crowell, Claudia 48995
Crowell, Darryl 05523
Crowell, Stephen 52207
Crowell, Stewart A. 51096
Crowell, Todd 47546
Crowley, Noel 24830
Crowley, Peter D. 46263
Crowley, Terry 01117
Croy, David E. 47845
Crozat, Oliver 14703
Crozemarie, Gilles 12521
Crozzolin, Umberto 28073
Crué, Emmanuel 12541
Cruikshanks, Randal 52816
Cruise, Sandra 45515
Crul, Yves 03365
Crum, Katherine 53185
Crum, Rebecca 53944
Crum McCabe, Elsie 50184
Crump, Caroline 38659
Crusan, Ronald L. 51066
Crusoe, Lewis D. 50352
Crutchfield, Margo 47634
Crutchfield, Sharon 46458
Cruvinel, Márcio Borges 04940
Cruz, Amparo Magdaleno de la 40227
Cruz, Fe M. de la 34445
Cruz, Martin 40251
Cruz, Prudenciana C. 34434, 34455
Cruz, Sumra I. de la 34396
Csaplár, Dr. Ferenc 23418
Csar, Mary 46882
Cséfalvay, Pál 23553
Cserei, Dr. Ladislav 38343
Csergö, Tibor-András 36382
Cseri, Dr. Miklós 23880
Cserményi, Vajk 23860
Cservenyák, László 23693
Csillag, Péter 23452
Csobádi, Jozef 38342
Csorba, Dr. László 23442
Csütörtöky, Jozef 38269
Čtvrtník, Dr. Pavel 09922
Cuadrado, Emeterio 39690
Čubareva, Roza Petrovna 37513
Cubberley, Peddyr 43920
Cubells, Raymond 13665
Cubillo, Ángeles Vera 38932
Čubrik, Liljana 38080
Cucchi, Franco 27833
Cuccuini, Dr. Piero 26179
Čučković, Lazo 08886
Cuda, Dr. Maria Teresa 25842
Cudahy, Michael J. 50707
Cudbill, Carolyn 43606
Čudnova, Lidija Aleksandrovna 37793
Cudworth, Keith R. 50129
Cuello, Juan Pablo 54422
Cuendet, Loïse 41535
Cuervo de Jaramillo, Elvira 08591
Cuevas, Elena 34453
Cuevas y Lara, Mónica 31139
Čuhelová, Blanka 09857
Cui, Baiting 08210
Cui, Junfeng 08483
Cui, Youliang 07635
Cui, Zhenkun 07922
Cukić, Milica 38074
Ćuković, Petar 31643
Cukrov, Tončika 09009
Culbertoon, Thomas J. 48788
Čulig, Igor 08886
Cullen, Deborah 51220
Cullen, Eilish 52780
Cullen, M.L. 45899
Cullen, Maggie 43965
Culligan, Jenine 49435
Cullinan, Deborah 52778
Cullingford, Neville A. 44105
Cullman, Lewis B. 53185
Cullmann, Dorothy 53185
Culot, Maurice 03446, 03461
Culp, David A. 46240
Culpepper, Marilyn 50771
Culver, Ph.D. Michael 50952
Culver, Michael 51507
Cumbey, Susan G. 46203
Cumming, Glen 07344
Cummings, Bonnie 47625
Cummings, Dorothy A. 38169
Cummings, Ernst E. 48498
Cummings, Fern P. 47137
Cummings, Mary Lou 51052
Cummins, Alissandra 03242
Cummins, Heather 45685
Cummins, Joan 47069
Cummins McNitt, R. 46242
Cunard, Gail C. 48978
Cunia, Claudia 03095
Cunliffe-Lister, E.S. 43686
Cunningham, Deirdre 52395
Cunningham, Denyse M. 48636
Cunningham, Eleanor W. 49059
Cunningham, Jewel 06844, 06857
Cunningham, Jo 45805
Cunningham, Kim 49588
Cunningham, Noreen 45250
Cunningham, Patrick J. 47117
Cunningham, Scott 49588
Cunningham Lavdar, Colleen 48271
Cunz, Dr. Reiner 19003
Cupp, Franzas 53393
Curbastro, Riccardo Ricci 25668
Curbelo, Dr. Jorge 54427, 54428
Curd, Rolin 47325
Curda, Dr. Andreas 02987
Curdy, Philippe 41873
Curiger, Bice 42083
Curnow, Jack 01071
Curran, Donald 49270
Curran, Emily 46944
Curran, John 51748
Currat, Joseph 41976
Currell, M. 00834
Currey, David 50977
Currie, Gloria 06904
Currin, Robert 48836
Curry, Larrie Spier 49190
Curry, Michael 50167
Curry, Michael P. 49150
Curtet, Jean-Christophe 41402
Curtin, Donna D. 51997
Curtis, Brian 50605
Curtis, Dr. J. 44803
Curtis, John 46221
Curtis, Linda 05313
Curtis, Nan 47690, 49041, 52076
Curtis, Neil 43244
Curtis, Nellie K. 51485
Curtis, Penelope 44661
Curtis, Ron 00886
Curtis, Sue 43943
Curtis, Vaughn 45488
Curtiss, Courtney P. 46883
Curtius, Dr. Andreas 20763
Curto, Argeo 08943
Curto, Dr. Carles 38976
Curto, Domenico 25678
Curto i Homedes, Albert 40166
Curtz, Arley 46964
Ćus Rukonić, Jasminka 08856, 08913
Cusack-McVeigh, Holly 49330
Čušenkov, Valerij Gennadjevič 37436
Cush, Dr. Jane 01125
Cushing, Dr. Paula E. 48075
Cushing, Raymond L. 47630
Cushing, Stanley E. 46919
Cushman, Brad 50142
Cushman, Paula N. 50070
Cusiac, Dragoş 36547
Cusick, Bonnie A. 48194
Cusming, J.R. 44204
Čusova, Elena 37186
Cussner, Alice 54321
Cutler, Scott 48317
Cutler, Walter L. 53892
Čvančara, Alois 09625
Cvetkova, Natalija 05028
Cvetković, Branislav 38101
Cvetković, Marija 38045
Cvjetićanin, Tatjana 38040, 38082
Čvorović, Božena 04284
Cybulski, Antoni 35315
Cygański, Janusz 34978
Cyncar, Nadia 05757
Cypher, Paul 46552
Cypress, Billy L. 46794
Cyr, Jean-Claude 05436
Cytacki, Joe 48639
Cywiński, Dr. Piotr M.A. 34997
Czajka, Jacek 35126
Czajkowski, Józef 35245
Czankus, Ruth 46720
Czaplewski, Dr. N.J. 51376
Czapska, Zofia 35103
Czarny, Eugenusz 34695
Czchodzińska-Przybysławska, Anna 35086
Czerannowski, Dr. Barbara 17831
Czére, Dr. Andrea 23477
Czerniak, Robert Michał 35069
Czernin, Dr. Martin 03039
Czerw, Ewa 35304
Czerwinski, Bogdan 35135
Czichos, Raymond L. 54217
Czoma, Dr. László 23660
Czopek, Sylwester 35094
Czoßek, Jens 19421
Czub, Zofia 35099
Czubaczyński, Ryszard 34876
Czuma, Stanislaw 47624
Czymmek, Dr. Götz 19635

D

D' Agnencia, Jill 50236
Da Pra Galanti, Dr. Lorenzo 27424
Daalder, Dr. R. 31885
Dabell, Simon 43528
Dabić, Ljubica 38091
Dabney, Mary K. 51839
Dabović, Mirjana 31643
Dąbrowska, Anna 34602
Dąbrowski, Wojciech 34625
Daby, Rebecca 01681
Daccó, Dr. Gian Luigi 26477, 26478, 26479
Dacey, Timothy J. 46589
Dachnenko, Lidija Oleksandrivna 43211
Dae, Sung Kwon 29905
Daele, Durandy van den 13518
Daelemans, Danny 03585
Daelemans, Frank 03438
Daems, Gérard 15755
Daenens, Lieven 03625
Daenzer, Denise 42099
D'Afflitto, Dr. Chiara 27249
Dagan, Zvi 24993
Dagdar, L. 24663
Dage, Carol J. 49473
Dager, Max 24027
Daggett, John E. 48666
Dagnicourt, Eric 14068
Dagon, Yoav 25118
Dagorne, Richard 14171, 14173
Daguerre de Hureaux, Alain 15939
Dahan, Arie 25067, 25075, 25125
Dahesh, Dr. A.S.M. 30156, 30159
Dahl, Erling 33766
Dahl, Jonas 40558
Dahl, Torveig 33587
Dahlbring, Magnus 40558
Dahle, Sigrio 07352
Dahlgren, Dorothy 47672
Dahlitz, Ray 00831
Dahlström, Björn 30478
Dahlström, Per 40555
Dahm, Klaus 20424
Dahmali, Ech-cherki 31689
Dahmani, Dr. Saïd 00050
Daho, Keltoum Kitouni 00057
Dai, Linyan 08335
Dai, Qingming 07777
Dai, Zengqun 08444
Daijoji, Yoshifumi 29627
Dailey, John R. 53895
Dailledouze 15512
Daily, Charles 52877
Daily, Steven L. 50712
Daim, Prof. Dr. Falko 20150
Dajani, Virginia 51111
Daker, Emma 46058
Dakin Hastings, Lynne 53550
Dalborg, Björn 40936
Dalby, Åge 33638
Dale, Alex P. 43253
Dale, Charles 52528
Dale, Michal 53232
Dale Zemlansky, Denise 49136
Dalen, Mary G. 06106
Daley, Jean 48337
Daley, Teddie 49123
Dalhousie, Earl of 44046
Dalichow, Dr. Bärbel 21087
Dalla Vecchia, Fabio M. 26779
Dallago, Antonio 28254
Dallaj, Dr. Arnalda 26677
Dallas, John 43353
Dallas, Linda 52230
Dalleo, Bruce 51039
Dallesandro, Robert 47245
Dallett, Timothy 06922
Dalli, Alberto 27719, 27721
Dallinger, Dr. Petra-Maria 02307
Dallman, Dr. John E. 50395
Dalous, Pierre 15946
Daloze, M. 03394
Dalsgaard Larsen, Keld 10256
Dalton, Bryan 47138
Dalton, Carolyn 43992
Dalton, Jake 48392
Dalton, Ron 50649
Dalton, Shirley B. 52051
Daly, Beth 53556
Daly, James 51217
Daly, Tess 33209
Daly, Thomas M. 53694, 54010
Dalyell, Kathleen 44719
Dalzell, John 52569
Dam, H. ten 32124
Dama, Prof. Salvatore Cirillo 07447
Damamme, Jeanne 14866
Damaschin, Violeta 36160
Damaskinos 23059
D'Amato, Dr. Alfonso 25464
D'Amato, Aurelia 27581
Dambaev, Munko Žartam 36748
Dambergs, Guntis 30145
D'Ambrosio, Anna T. 53676
D'Ambrosio, Dr. Antonio 27280
D'Ambrosio, Paul 47788
D'Ambrosio, Paul 47789
Damdinsuren, Tsedmaa 31624
Damdoumi, Vasiliki 23232
Dame, Susan 48147
Damen, L. 32729
Damgaard, Ellen 10175, 10176
Damgaard Sørensen, Tinna 10250
Damian, Dr. Paul 36221
Damiani, Dr. Giovanna 27096
Damianov, Nikola 05171
D'Amico, Alessandro 27443
D'Amico, Prof. Claudio 25462
Damito, Dr. Ronaldo 04753
Damjanova, Katerina 04994
Dammel, Werner 20552
D'Amore, Dr. Paola 27512
Dampman, Allan 49449
Damsgaard, Nina 10310
Damy, Ken 25540
Dana, Rae 54035
Dana, Richard 47169
Danailov, Boris 05177, 05200
Dance, Pat 07386
Dancey, Jenny 45126
Danchev, Dancho 05060
Dăncuş, Mihai 36615
Dandaura, M. 33330
Dander, Dr. Marilena 26429
Dander, Patrizia 20445
Dandwate, P.P. 24354
Danelzik-Brüggemann, Dr. Christoph 18006, 18008
Daneo, Angelica 48072
Daners, Peter 21960
Danesi, Claudio 26031
Dangel, Gerhard 18428
D'Angelo, Starlyn 46163
D'Angelo, Starlyn 51534
Danh, Prof. Dr. Trinh 54613
Daniel, Alain 11740
Daniel, Françoise 12161
Daniel, James 54102
Daniel, Mary 48041
Daniel, Rochette 03459
D'Aniello, Dr. Antonia 27671
Daniels, Christopher 49074
Daniels, John P. 51766
Daniels, R. 44409
Danielson, Jeanette 41020
Danielsson, Ing-Mari 40960

Danielsson, Ralf 40469, 40471
Danielsson, Dr. Roy 40709
Danilow, Susan 51186
Danini, Gianluca 26408
Daniylopoulou, Olga 22990
Danjo, Prof. Tatsuo 28388
Dankbar, Johannes 22091
Danker-Carstensen, Dr. Peter 21395
Dankl, Dr. Günther 02136
Dann, Alan O. 50495
Dann, Gaston 13742
Danna, Gina 49614
Danna, Tracy 46671
Danne, Rainer 19361
Danneel, M. 03466
Danner, Elisabeth 41595
Dannhorn, Beate 18362
Dannhus, H. 19946
Dano, Dr. Ján 38282
Dansako, Tetsuya 29577
Danson, Stuart 01644
Danšova, Olga Nikolajevna 37496
Dante Coronel, Hugo 00123
Dantinne, Claude 13662
Danzer, Dr. Gudrun 01982
Danzl, Barbara 02728
Dapkyunas, Jeanne 03251
Dapsens, J.-C. 03849
Darakčieva, Uljana 05095
Darbé, Louis 03580
Darby, Daniel 07409
Darde, Dominique 14449
Dardenne, Geneviève 03659
Darding, Nicole 03796
Darga, Dr. Robert 21759
Darling, Denise L. 47987
Darling, L.A. 07160
Darling, Sharon 52958
Darling-Finan, N. 45661
Darr, Alan P. 48114
Darre, Pierre 15097
Darrow, Deborah 52193
Darst, Elisabeth 48846
Dart, John 43755
Dartevelle, Patrice 03909
Darwish, Ebrahim 10469
Dary, Anne 11659, 12754, 13817, 15598
Das, Chandranath 24127
Dasbach, Theo P. 47588
Dăscălescu, Gheorghe 36367
Date, Tatumi 29434
Dattilio, Dan 49778
Daubendiek, James 49606
Daubert, Debra 51615
Dauda, J.Y. 33335
Daudelin, Robert 06336
Daudry, Damiano 25246
Dauge, Alexandra 13356
Daugelis, Osvaldas 30262
Daughdrill, Kay 47965
Daughterman-Maguire, Dr. Eunice 46572
Daum, Patrick 15016
Daum, Sigrid 19746
Daumann, Gerda 19366
Dauphin, Jean 03779
Daur, Dr. Jörg 22557
Daure, Regine 12303
Dauschek, Dr. Anja 21939, 21941, 21961, 21962
Dautermann, Dr. Christoph 19706
Dautermann Ricciardi, Dana 48730
Dauth, Louise 00950
Dautović, Andrea 04284
Dauwe, Marc 03783
Dauzenroth, Willi 20186
Dávalos de Camacho, Maya 31130
Davaras, Dr. C. 22872
Davenport, Kimberly 49401
Davenport, Leonard 47012
Davey, Arnold 45441
Davey, Ranson 24807
David, Clifford C. 48178
David, Johan 03661
David, Josiane 14106
David, Karen 48342
David, Mihai 36166
David, Norma Jean 49071
David, Romeo G. 34357
David, Wendy 14398
David, Dr. Wolfgang 20159
David-Jones, Carmina 39605
David-Weill, Hélène 14654, 14662, 14682, 14734
Davidov, Dr. Dinko 38045
Davidović, Jasna 38151
Davidović, Jelena 38110
Davidow, Joan 47928
Davidse, I. 32637
Davidson, Anne 51690
Davidson, Dan 47865
Davidson, Kathryn 46523
Davidson, Paul 33063
Davidson, Scott 05907
Davies, Andrew 44628
Davies, Brian 45403
Davies, Gerald 46029
Davies, Hugh M. 49846
Davies, Dr. Hugh M. 52740
Davies, Jane 05807, 06750, 45447
Davies, Jenny 45053
Davies, Kate 48175
Davies, Lucy 44778
Davies, Marjorie 44044
Davies, Nada 51735
Davies, Pascal 11741
Davies, Richard 45721
Davies, Rick 50662
Davies, Rita 07102
Davies, S. 01552
Davies, Simon 45308, 45309
Davies, Steve 45053
Davies, Su 44566, 44568
Davies, Suzanne 01291
Davies, Dr. W.V. 44803
Davignon, Étienne 03498
Dávila, Alejandro 00335
Davila, Julio V. 34259
Davin-Guerisse, Anne 03617
Davis, Dr. Ann 05529
Davis, Ann 46420
Davis, Audrey P. 46199
Davis, Beryl 07051
Davis, Beth 47708
Davis, Betty J. 51550
Davis, Carol A. 48491
Davis, Cheryy 46195
Davis, Darwin R. 52558
Davis, Dustin P. 48802
Davis, Elizabeth 52995
Davis, Elliot 46940
Davis, Gordon 46488
Davis, Israel 49006
Davis, Jacqueline B. 48678
Davis, Janet 07300
Davis, Jaqueline Z. 51244
Davis, Joanne 54004
Davis, Kate 06720
Davis, Katherine 51249
Davis, Kay 47669
Davis, Kieth 33126
Davis, Dr. Leroy 51582
Davis, Lesia 05539
Davis, Dr. Leslie B. 46982
Davis, M. 24968
Davis, Marsh 48842, 48851
Davis, Michael 53311
Davis, Nancy 46582
Davis, P.A.C. 33194
Davis, Phillip V. 53220
Davis, Priscilla 53131
Davis, Robert O. 52692
Davis, Roger 51682
Davis, Ron 06440
Davis, Ruth 46192
Davis, Sarah 51143
Davis, Susan 51682
Davis, Susan S. 49780
Davis, Suzanne 46307
Davis, Todd A. 07256
Davis, Walter R. 47230
Davis, William J. 51362
Davis, Zina 54057
Davis Anderson, Brooke 51113, 51171
Davis Gardner, Katie 47700
Davis MacNaughton, Mary 47576
Davison, Liane 07037
Davison, Dr. Patricia 38591
Davison, Walter 49799
Davletov, B. 54492
Davson, Victor 51284
Davy, Geukens 03763
Davy, Mark 51665
Davydova, Natalja Alekseevna 37580
Dawans, Francine 03805
Dawey, Dr. Chris 50751
Daws, Russell S. 53429
Dawson, Amy 52790
Dawson, Barbara 24797
Dawson, Beverly 48937
Dawson, Chris 47628
Dawson, David 43607
Dawson, E.A. 43787
Dawson, Frank C. 48255
Dawson, J. 43727, 43885
Dawson, Dr. Mary R. 51943
Dawson, P. 24850
Dawson, Valerie 44623
Dawson Penniman, H. 51847
Daxl, Heiko 17088
Day, Brenda 46540
Day, Dr. David 53047
Day, Edward 52365
Day, Ginette 13218
Day, Jacqueline 53313
Day, Dr. James 50056
Day, John A. 53720
Day, Karen 48368
Day, Dr. Lucille L. 46754
Day, Michael 44093, 44873
Day, Patricia 49067
Day, Richard 43808, 53749
Day, Sally 43271
Day, Teresa 54145
Day Vincent, Kris 48794
Dayan, Prof. T. 25124
Dayrit, Marina 34437
Dayson, Diane H. 51262
De Angeli, Cristina Frua 27800
De Angelis, Prof. Giuseppe 27416
De Baere, Bart 03301
De Block, Ann 03259
De Boer, A. Spil. 32544
de Bruyn, Dr. Jean-Pierre 03289
De Coster, Annie 04155
De Cuyper, Chris 03371
De Cuyper, Meve Jo 03260
De Filippis, Dr. Elena 28159
De Franceschi, Antonio Fernando 04696
De Gelder, Dr. F.A. 32851
De Goede, J. 32710
De Herdt, René 03638
De Jong, H. 32725
De Jong, Leen 03290
De Jonge, Jurgen 03576
De Jonghe, Marcel 03372
De Keerle, Monique 03561
De Kelver, Jan 03563
De Kleijn, G. 31809
de Linos Escofet, Mónica 39530
De Luca, Dr. Bianca 25873
De Luca, Dr. Francesca 26157
De Luca, Maurizio 54501
De Luca, Prof. Paolo 26900
De Luca, Renzo 28968
De Lucia Brolli, Maria Anna 25920
De Marchi, Dr. Andrea G. 27451
De Marco, Dr. Marco 26127
De Martin, Prof. Giancandido 27207, 27208, 27210, 27211
De Mier Riaño, Elisa 08582
De Paepe, Pantyika 12585
De Palma, Prof. Luigi Michele 26765
De Palma, Maria Camilla 26304, 26305
De Paolis, Rosario 26472
De Pascale, Andrea 26138
De Pasquale, Vicenzo 26644
De Polo Saibanti, Claudio 26195
De Pompeis, Claudio 27163
De Poorter, Alexandra 03495
De Reymaeker, Michel 03879, 03886
De Rochambeau, M. 15897
De Roeck, Harald 03423
De Rosa, Giorgio 25934
De Ruyter, Rob 03998
de Rycke, Jean-Pierre 04086
De Santi, Claudio 26775, 26776
De Siena, Antonio 25407
De Simone, Dr. Emilio 25215
De Smet, Urbain 04142
De Strobel, Dr. Anna Maria 54501
De Strycker, Pierre 03316
De Vincentiis, Prof. Italo 27545
De Vito, Natalie 07123
De Waal, Dr. Lydia M. 38740, 38741
De Wilde, Karel 03684
De Wilde, Karin 04064
De Wit, Jean-Louis 03391
De Witte, Hubert 03408, 03409, 03410, 03411, 03413
De Wolf, Walter 04149
Deacy-Quinn, Christa 53669
Deakin, G.T. 01239
Dealy, Anne F. 48880
Dean, Deborah 45288
Dean, Jim 50844
Dean, Kevin 52916
Dean, M. 43981
Dean Krute, Carol 49199
Dean Stock, Michele 52654
Deaner, Beth J. 51063
DeAngelo, Darlene 49438
Dearing, Vicki 47101
Dearinger, David 46919
Dearinger, Dr. David 51232
Deaton, Linda 53429
Debain, Yvette 05857
Debarge, René 13182
Deblanc, Sylva 03922
Deblanc Magnée, Marie-Paule 04107, 04108
Debo, Kaat 03294
Debo, Thomas A. 52008
Debočički, Valentin 05063, 05064, 05065, 05066, 05068
DeBow, Arthur 52078
Dębowska, Alina 34556
Debrabandère, Béatrice 14433
Debrah, Dr. Issac N. 22848
DeBrari, Frank 54313
Debray, Cécile 14646
DeBree, Gretchen C. 47646
Debrincat, Saviour 30651
DeBruyn, David L. 49007
Debruyn, Raphaël 03786
Debry, Jaques 12461
Debryn, Johan 04025
Debus-Steinberg, Astrid 21964
DeBuse Potter, Gail 47327
Deca, Eugen 36129, 36138, 36234, 36316, 36401, 36550, 36552, 36553
Decames, Jean-Claude 14031
DeCamp, Alta J. 51000
Decán Gambús, Ivanova 54563
Decavele, Dr. Johan 03633
Dechambre, Jeanne 12343
Dechaux, Carine 03972
Deck, Linda T. 52006
Decker, Carla 51300
Decker, E. 15644, 15645
Decker, Friedrich 02275
Decker, Raymond 16124
Decker, Sandy 50272
Decker, Scott E. 49482
Deckers, Ann 03287
Deckers, Yolande 03290
Deckert, Bernd 17986
Deckert, Inge 17986
Deckert, Michael 22323
Declercq, Joost 03546
DeCocker, Dean 53615
Decosterd, Denis 41796
Decoud, Dr. Marcelo 00510
Decroix, Philippe 15168
Decron, Benoît 13715
Decrouez, Danielle 41427
Decter, Avi Y. 46571
Dectot, Xavier 14729
Děd, Stanislav 09478
Dedenroth-Schou, Poul 10163
Dedíková, Soňa 09450, 09451
Deegan, Dr. Ann 47540, 54362
Deer, Naomi 49527
Deerpalsingh, Salni 30678
Deeva, Valentina Kirillovna 36950
DeFalla, Josie E. 53580
Defauwes, Georges 03533
Defeo, Prof. Rúben D.F. 34489
Defferrard, Jacky 13681
DeFloren, Rude 46134
Deford, T. M. 52992
Degel, Hermann 21960
Degel, Kirsten 10103
Degen, Kurt 21515
Degen, Dr. Thomas 18896
Deggelsegger, M. 02674
Degraeve, Willem 03442
DeGrazia, Diane 49501
Degreif, Dr. Uwe 17319
DeGroft, Aaron 52914
Degueurce, C. 13924
Deguglielmo, M. 11834
Deguillaume, Marie-Pierre 15841
Deharde, Ewald 17669
DeHart, Rob 50977
Dehlinger, Martina 18380
Dehollander, Stef 03751
Dehrkoop, Prestene 48326
Deiber-Kumm, Michele 47421
Deiker, Dr. Tom 47420
Deimel, Dr. Claus 17964, 19163, 19913
Deiser, Leopoldine 03118
Deitrich, Darcy 06234
Deitsch, Elka 51133
Deitz, Philippe 03811
Dejardin, Dr. Fiona M. 51566
Dejardin, Ian 44840
Dekiert, Dr. Markus 20425
Dekiss, Jean-Paul 11586
Dekker, Anthony P. 32440
Dekker, H. 32321
Dekker, R. 32485
Deknop, A. 03439, 03451
Del Bagno, Rober 51243
Del Falco, Vittorio 27533
Del Frate, Dr. Gabriella 27083
Del-Prete, Sandro 41497
Del Testa, Luise 22797, 22798
Dela, Feliks 34936
Delacoste, Raymond 41635
Delacote, Goery 52773

Delaender, Georges 04124
Delage, Dominique 15785
Delahant, John M. 52911
Delahaye, Marie-Claude 11771
Delaine, Joël 14325, 14328
Delaire 11920
Delaney, Chris 43762
Delangle, J.F. 11992
Delannoy, Agnès 15290
Delany, Max 00980
Delaousie, Renée 14509
Delapierre, Emmanuelle 16043
Delaquis, Christian 07356
Delarge, Alexandre 13009
Delattre, Jean-Luc 03917
Delaval, Eric 11640
Delay, Vincent 41577
Delbaere, Stefaan 31927
Delcourt, Dr. Marianne 04083
Delecouls, Sonia 13538
Delee, Kathryn 46155
Delehanty, Suzanne 50631
Delépine, Francine 03457
Del'Furia, Lucienne 14020
Delgado, Avella 39421, 39423
Delgado, Eric 11738
Delgado, James P. 07228
Delgado, Jane 51181
Delgado, Lino Torres 30697
Delgado, Sally 46417
Delgado Restrepo, Gloria 08608
Delhaise, Christophe 03396
Delić, Nermin 04258
Delić Gozze, Vesna 08865
Delile Benoit, Justin 41401
Delisle, André J. 06366
DeLisle, Godfray 05628
Deliss, Dr. Clémentine 18381
Delivorrias, Prof. Dr. Angelos 22925
Dell, Roger 48548
Dell, Sharon 33235
Della Casa, Bettina 41581
Della Fina, Giuseppe M. 27009
Della Toffola, Leomberto 28193
Dellacroce, Raymond 48491
Dell'Antonia, Dario 16217
Dellantonio, Elio 27341
Dellbeck, Johan 41012
Dellebuur, Amy 07070
Dellinger, Mary H. 48765
Dell'Oca Fiordi, Dr. Angela 27873
Delmas, Jean 15078
Delmas, Luc 16178
Delmont, Patrick 16269
Delobel, Jean 11829
Deloffre, Véronique 11871
Deloing, Evelyne 11690
Deloncle, Jacques-Gaspard 14788
DeLong, Dr. Marilyn 52616
Delormas, Jerôme 13977
DeLorme, Harry 52947
Delorme, Yves 15058
Delperdange, Roland 03344
Delporte, Luc 04097
Delroy, Ann 01412
Deltedesco, Franco 26495
Deltour-Levie, Claudine 03495
Delucchi, Gina 53305
DeLucia, Francis 52244
Delye, Emmanuel 03803
DeMaio, Bonnie 52473
Demard, Jean-Christophe 12393
Demaret, Nicole 04082
Demarez, Leonce 03335, 03374
DeMars, Louise L. 47028
DeMarsche, Michael 47699
Demarteau, Jo 32032
Dematté, Mag. Rosanna 02044
Demazure, Michel 14742
Dembiniok, Marian 34610
Dembski, Prof. Dr. Günther 03064
Dembski, Dr. Ulrike 03051
Demčenko, Ljubov Grigorjevna 36866
Demeester, Ann 31890
Demelenne, Marie 03893
Demere, Dr. Thomas 52754
Demerling, Rod 06980
DeMersman, Jim 49229
Demetis, Ioannis 23245
Demetriou, Maro 09371
Demetz, Dr. Stefan 25487
Demeurie, Dirk 03555
Demidenko, Olga Moiseeva 37434
Deminey, E. 38625, 38666
Deming, Diane 52289
Demirci, Süheyla 42945
Demke, Elisabeth 46838
Demo, Dr. Željko 08986
Dempsey, Rey 24814
Dempsey, Terrénce E. 52590, 52591
Demski, Tadeusz 35110
Demuth, Renate 21296
DeMuzio, David 51867
Denaro, Dolores 41228
Denelli, Gjergj 00029
Denford, Dr. G.T. 46035
Denford, Geoffrey 46028
Deng, Binquan 07724
Deng, Chaoyuan 07775
Deng, Liming 08018
Deng, Murong 07602
Deng, Qingyun 08351
Deng, Yongfa 08146
Dengler, Thomas 18295
Denhaen, G. 03438
Denizli, Hikmet 42695
Denman, Ron 05613
Dennerly, Peter 33086
Dennett, Anthea 44631
Denning, Kylee 52936
Dennis, Linda 52416
Dennis, Sue M. 51272
Dennison, Foster K. 06675
Dennison, Lisa 51258
Dennler, William V.A. 53522
Denny, Colin 01154
Dennys, Vincent 11610
DeNoon, Betty 47165
Densapa, Tashi 24175
Denscher, Dr. Bernhard 03029
Densen, Paul M. 52834
Dent, Dr. D.L. 32921
Dent, G.J.H. 45379
Dentressangle, Franck 12895
Denys, Lama 11710
Denysenko, Jevgen Ivanovyč 43038
Denzel, H. 01717
Deon, Luca 41588
DeOrsay, Paul B. 47682
Deparpe, Patrice 13655
Depasquale, Suzannah 30643
Depierre, Marie-Colette 15071, 15073
Depoid, Henry 12602
Deppe, Dipl.Ing. Frauke 18239
Deppeler-Lang, Walter 41929
Depreter, R. 03466
Depuox, Christian 42106
Dequidt, Luk 03953
Derbic, Rupert 01812
Derbier, Josselin 15066
Derby, Newton 53328
Derbyshire, Janine 43965, 43967
D'Ercole, Dr. Vincenzo 25872
Dercon, Chris 20445
Dercourt, J. 13200
Derda, Dr. Hans-Jürgen 17518, 22474
Derefaka, Dr. Abi Alabo 33345
Derenthal, Dr. Ludger 17165, 17202
Deriks, M. 32805
Dering, Dr. Florian 20462
Derjabin, Vladimir Ignatevič 37626
Derjužina, Alla Vasylivna 43025
Derksen, J. 32145
Derksen, J.J.V.M. 32843
Dermitzakis, Prof. Dr. M.D. 22949
Derr, Rex 51727
Derriks, Claire 03893
Dertazzo, Ludovico M. 27035
Des Rochers, Jacques 06357
Desai, Vishakha N. 51128
Desaint, Jeanne 14050
DeSalvo, Donna 51279
Desaulniers, Guy 05365
Desbenoit, Martine 12108
Desborough, Jeff 54154
Desbrosse, René 13291
Descamps, Patrick 13197
Descatoire, Christine 14729
Deschamps, Stephanie 11698
Descheemaeker, Dominique 03554
Desel, Jochen 16757
Deseyne, Alex 03928, 03929, 03932
DeSeyter, Craig 49315
Deseyve, Dr. Yvette 17551
Desfossés Le Page, Carmen 07173
DeShaw, Larry 53337
DeShon, Richard N. 52564
Deshpande, Dr. P.N. 24166
Désirat, Georges 15608
Desjardins-Ménégalli, Marie-Hélène 12913, 12914
Deslarzes, Bertrand 41286
Desmarais, Charles 47069
Desmarais, Joan 47775
Desmet, Viviane 03496
Desmett, Don 49666
Desmond, Tim 45296
Desmules, Olivier 13962
Desnier, Jean-Luc 14655
Desormeau, Monique 48559
Despotidis, Emmanouil 22879
DesRoches, C. 06250
Desse, Daniel 16140
Dessemme, Christine 12586
Dessens, H.J.A. 31885
Desserud, Thore 33476
Dessornes, Maria E. 53502
Dessy, Marco 33273
Destefanis, Joseph 31609
DeStefano, Giacomo R. 51730
Desvigne, M. 15634
Desy, Dr. Elizabeth A. 50506
Detch, Rosalie S. 50033
Deterts, Dr. Dorothea 17564
Dethlefs, Dr. Gerd 20512
Dethmann, Donald 47650
Detrani, Geoffrey 51277
Detrée, Jean-François 15527
Detrich, Kalman 51230
Dettlaff, Christian 44891
Dettmann, Julia 18859
Dettmer, Dr. Frauke 21274
Dettmer, Hans-Georg 18714
Deuchar, Stephen 44995
Deuchert, Dr. Norbert 17662
Deudon, G. 03600
Deuker, Horst 14089
Deutmann, Karl Heinrich 17923, 17925
Deutsch, Franz 03120
Deutschbauer, Franz 02721
Deutschle, Knut 21340
Dev, Anand 24153
Devanik, Dave 06388
Devé, Jean 13062
Deveau, Bertha 52834
Deveau, Carole 06314
Devendorf, Meredith 50678
Devereux, Dr. B.J. 43725
Devereux, David 44611, 44612
Deveselleer, Jacques 04035
DeVille, Sue 51574
Devin, Pierre 12778
Devine, Jackie 48103
Devine-Reed, Pat 47458
Devoghel, Philippe 13148
DeVonyar-Zansky, Jill 51990
Devory, Marian 51790
Devos, Danny 03308
Devos, E. 03985, 03986, 03987
Devries, Fred 50647
DeVries, Karl 52946, 52948
Devroye-Stilz, Anne 14429
Devynck, Danièle 11540
Dewald, Ann L. 51675
Dewald, Gabi 21735
Dewald, Rolf 42104
DeWalt, Billie R. 51943
Deweerdt, Jean 12799
Dewey, David 51604
Dewey, Judy 45915
Dewey, Stuart 45915
Dewhurst, Dr. C. Kurt 48254
Dewilde, Annick 03761
Dewilde, Jan 03727, 03728, 03729, 03731, 03732
Dewing, David 44854
Dewitz, Dr. Bodo von 19624
Dewsbery, Sara 45332
Dewsbury, R. 45293
Dexter, Susan 48547
Deyle, Betty 46736
Deysenroth, Paul 54101
Dezellus, Eric 15131
Dezort, Steve 06530
Dhaene, Sylvie 03628
Dhaliwal, Theresa 44903
Dhaulakhandi, V.N. 24349
Dholakia, P.V. 24247
Di Belgiojoso, Dr. Giuseppe B. 27720
Di Benedetto, Gino 28043
Di Carluccio, Dr. Antonietta 25197
Di Castro, Dr. Daniela 27505
Di Felice, Paola 27955
Di Felice, Dr. Paola 27956
Di Flumeri, Dr. Gabriella 27512
Di Girolami, Dr. Paola 25764, 25959, 26376, 26797, 26843, 27417, 27566, 27608
Di Grazia Costa, Ivani 04880
Di Ianni, Angelo 27176
Di Lorenzo, Maria Teresa 27679
Di Malfetta, Dr. Felice 26765
Di Marco, Antoinette 13868
Di Mario, Dr. Mario 27967
Di Martino, Maria 27224
Di Nepi, Simona 44937
Di Niro, Dr. Angela 25636
Di Paolo, Luciana 27216
Di Pascale, Paolo 25357
Di Pietrantonio, Dr. Giacinto 25400
Di Pietro, Gaetano 27165
Di Stefano, Dr. Carmela Angela 27066
Di Tommaso, Francis 51277
Diaconescu, Nicolae 36229
Diaconescu, Petru 36654
Diaconu, Ion 36739
Diakite, Sinkoun 23313
Diamantakis, Stavros 02484
Diamantopoulos, Aella 51864
Diamond, Dr. Judy 50107
Dianto, Herkus 24496
Dianzinga, Dias-Théodore 08806
Diaper, Dr. Hilary 44670
Diaper, Robin 44595
Dias, Dr. Maduro 35439
Dias, Robério 04771
Dias Duarte, Luiz Fernando 04755
Dias Pereira, Prof. Carlos 35541
Diaw, Amadou Tahirou 38022
Díaz, Dr. Jimena Perera Diaz 54418
Díaz, Joaquín 40183
Díaz, Luis H. 08628
Díaz, Miguel 36078
Díaz Arriola, Luisa 34247
Díaz de Mazzoldi, Olga 08708
Díaz Díaz, José Antonio 39458
Díaz García, Luís 39714
Díaz-Guerra Álvarez, Luis 39556
Díaz Pardo, Isaac 38888
Díaz Rodríguez, Anselmo 34254
Díaz-Tusham, Rafael 51111
Díaz Ungría, Dr. Adelaida de 54584
Dibbets, W.A.G. 31942
Dibert, James 52956
Dičeva, Vera 05194
DiChristina Hager, Joan 47545
Dick, Alvin 50894
Dick, James 52471
Dick, Jutta 18874
Dicke, James F. 51026
Dickens, Denise 52228
Dickenson, Joanne 52053
Dickenson, Victoria 06371
Dickermann, Dr. Fred 02185
Dickerson, Patricia 51527
Dickerson Lind, Anne 52935
Dickey, Bobby 48695
Dickey, Michael 46364, 46907
Dickhaut, Monique F.A. 32527
Dickie, Kerry 47207, 47208
Dickinson, Cindy 46265
Dickinson, Gary 51104
Dickinson, Pat 07094
Dickinson, Robert 45820
Dickinson, Roger 51041
Dickson, Carl A. 49179
Dickson III, Ephriam D. 48617
DiCosimo, Joanne 05860, 06536
Didier, Alain 14965
Didier, Prof. Arturo 27950
Didier, Hebert 13761
Didrichsen, Maria 10772
Didrichsen, Peter 10772
Diebel, C. 33043
Diederen, Dr. Roger 20454
Diederen, Roger 51160
Diederichs, Dr. Urs 21264
Diederichsen, Detlef 17128
Diedrich, Dr. Stephan 19624
Diegel, Julie 50108
Diehl, Jörg 41991
Diehl, Reinhard 20036
Diehl, Dr. Richard 53616, 53618
Diekmann, Rolf 21973
Dieleman, Peter A. 32596
Diemer, Dr. Kurt 16831
Diene, Baidy 38023
Dienst, Bgm.Ing. Robert 02272
Dierking, Gary R. 48675
Dießner, Dr. Petra 19934
Dietel, Volker 20610
Dieter Jr., Raymond A. 47482
Diethorn, Karie 51842
Dietl, Dr. Gerhard 21959
Dietl, Hans 21892
Dietrich, Bev 05917, 05918
Dietrich, Charlotte 21192
Dietrich, Frank 21960
Dietrich, Dr. Gerhard 19619
Dietrich, Liselotte 41614
Dietrich, Nikola 18389
Dietrich, Ute 22634
Dietz, Franca 17159
Dietz, Karen 52606
Dietz, Karl-Heinz 18026
Dietz, M.A. Stefanie 22366
Dietz, Steffen 16433
Dietz, Ulysses G. 51287
Dietze, Falk 20448
Dietzfelbinger, Dr. Eckardt 20758
Diewald, Mag. Sigrid 02451
Díez, José Luis 39573
Díez, Sofía 39605
Diez Canseco, Victoria Mujica 34273
Diez Valdez, Oscar 34133

D

DiFerdinando, Charles A. 48212
Diffily, James P. 48713
Difuccia, Mike A. 50782
Digby, Linda 05730
Digel, M.A. Brigitte 22668
Digger, Jo 45920
Dignef, Leo 03354
Dijkstra, J.G. 32585
Dijkstra, Kol.J.A. 31855
Dijkstra, Rob 32281
Dijols, Pierre 13479
Điković, Mario 38049
Dikun, Z.M. 37056
Dilas, Milica 08938
Dill, Chris 49034, 49035
Dill, Gregor 41185
Dillard Rech, Lori 51819
Diller, Douglas 52858
Diller, Dr. Stephan 21118
Dilley, Cherry 38730
Dilley, Mary Lou 50497
Dillmann, Claudia 18362
Dillon, Alice 53364
Dillon, Douglas 10373
Dillon, Idelle 51764
Dillon, John 52790
Dilly, Georges 11941
Dimaandal, Carmelita 34439
Dimaki, Sofia 22897
DiMaria, Tom 51464
DiMeo Carlozzi, Annette 46511
Dimitrijević, Bora N. 38164
Dimitrov, Prof. Dr. Bojidar 05191
Dimitrov, Ivan 05143
Dimitrova, Stanka 05241
Dimmick, Walter 49991
Dimova, Ljubov Ivanovna 37966
Dimt, Dr. Heidelinde 01870
Dimuchametova, Svetlana Aleksandrovna 37551
Dimundo, Elisabetta 27332
Dinard, N. Jean 13298
Dinçer, Hüseyin 42794
Diner, Mónica 31123
Ding, Ke 07686
Ding, Rong 07719
Ding, Yucai 08367
Dingemans-Bakels, Drs. F.N. 32529
Dingertz, Stig 40725
Dinges, Christiane 19595
Dingwall, Fiona 44244
Dinh, Le Quang 54626
Dinhobl, Dr. Günter 02414
Dinkelspiel, Florence 50181
Dinkins, Dr. Leon 51935
Dinsmoie, Elizabeth 05774
Dintaman, Jeniffer 49121
Dinu, Cristina 36707
Dinwerth, Shaun 52314
Diogo, Joanne 47195
Diogo Ribeiro, José 35548
Dion, Jean-Noël 06822
DiPaolo, Michael 50028
Dippold, Günter 16961
Dippold Vilar, Dalmo 04899
Dirmeikis, Edvardas 30287
Dirnbacher, Karl 01938
Dirnbacher, Sonja 01732
Dirnberger, Rotraut 02395, 02398
Dirnböck, Mag. Robert 02003
Dirosa, Hervé 15734
Dirscherl, Angelika 19076
Dirschka, Steffen 20500
Dirun, Moh Lativ 30583
Dishman, Jay 50614
Dismukes, Sandra B. 47427
Disney, Betty 47918
Disney, Jenny 52675
Diss, Gabriel 13985, 16143
Disston, Deborah 47766
Distel, Matt 47558
Dite, Dr. Tibor 38381
Dittberner, Stephen 52131
Dittemore, Diane 53584
Dittertová, Dr. Eva 09472, 09473
Dittmann, Britta 20073
Dittmar, Monika 22217
Dittmar, Volker 20222
Dittmer, Joerg 19921
Dittrich, Dr. Erika 18475, 18476
Dittrich, Rainer 18969
Dittrich, Ulrike 20671
Divelbiss, Maggie 52194
Divincenzo, Dr. Concetta 25374
Divola, John 52375
Dix, Andrea 18532
Dix-Williams, Traci 45833
Dixon, Barbara 46792
Dixon, Bonnie 48857
Dixon, Carol 50440
Dixon, Charlie 49562
Dixon, Christine 00946
Dixon, Claudine 50227
Dixon, David J. 46814
Dixon, Debbie 52913
Dixon, Elaine 52203
Dixon, Gerald 48972
Dixon, Hugh 45136
Dixon, Jenny 50182
Dixon, Kate 50574, 50575
Dixon, Margaret 43878
Dixon, Dr. Michael 44941
Dixon, Wendy M.C. 01216
Diz Ardid, Emilio 39741
Djabbar, Abd. 24577
Djačenko, N.D. 43186
Djačkov, Ivan Ivanovič 29725
Djadaibaev, Amir Jalinovich 29683
Djalal, Dr. Zaidir 24488
Djamba K. Shungu, Joseph 08799
Djanguenane, Nayondjoua 42612
Djokić, Danica 38132
Djordjević, Nenad 38110
Djuraboyev, Abdullajon 54440
Djurdjekanović, Sladjana 38093
Đjurić, Gordana 38100
Djuve, Kjetil 33825
Dlin, Elliott 47930
Dlugač, Vladimir Vladimirovič 36773
Długoszewska-Nadratowska, Hanna 34607
Dłużwska-Sobczak, Anna 34845
Dmitriev, Prof. I.S. 37671
Dmitrovic, Katarina 38094
Dmytrykiw, Danylo 49155
Doak, Elaine M. 49800
Doane, Erin 52924
Dobbelstein, H. 32202
Dobbie, Dave 06395
Dobbratz, Alferna 49367
Dobbs, Jerry 52684
Dobigny, Sylvaine 11587
Dobigny, Sylviane 11588
Dobke, Dr. Dirk 18923
Dobkowski, Mieczysław 34584
Doblin, Bruce 47505
Dobosi, Dr. Viola 23442, 23957
Dobras, Dr. Wolfgang 20147
Dobrea, George 47639
Dobrev, Marin 05210
Dobrianowa-Bauer, Dr. Snejanka D. 18378
Dobrik, Dr. István 23703, 23712, 23714, 23715
Dobrileit, Margitta 18009
Dobrogosz, Glenn 49067
Docavo Alberti, Ignacio 40224
Docenko, Alina Vasylivna 43022
Dočev, Spas 05085
Dočeva, Penka 05069
Dočeva-Peeva, Elka 05191
Docksey, Graham 00832
Docquier, Gilles 03893
Dodd, Dr. Donald B. 46825
Dodd, Hanut 01104
Dodd, Julie 45490
Dodd, Nick 45656
Dodd, Roger 43879
Dodds, E. 44530
Dodel, Benjamin 41275
Dodge, Laura 48392
Dodier, Virginia 47249
Dodina, Tatjana Alekseevna 37958, 37960
Dodson, Gin 53992
Dodson, Howard 51254
Dodsworth, Roger C. 44030
Dodwell, A. 44904
Dodwell, P. 44904
Döllinger, Theodor 20984
Dönmez, Asuman 42823
Dörfer, Dr. Anja 20741
Dörfer, Sylva M. 19921
Dörhöfer, Friedrich 17622
Döring, Dr. Jürgen 18951
Döring, Martel 20798
Döring, Dr. Thomas 17522
Dörnenburg, Manuela 17360
Dörr, Cornelia 18097
Dörr, Dr. Cornelia 18098, 19478
Doerr, Marjorie A. 48382
Doetsch, Dr. Rainer 19588
Doetzer-Berweger, Dr. Oliver 19153, 19156, 19157
Doezema, Marianne 53170
Doffey, Marie-Christine 41215
Dogan, Salih 42850
Dogariu, Octavian 36689, 36690, 36691, 36693
Doherty, Peggy M. 48039
Doherty, Roisin 24878, 43973, 43974
Doherty, Trafford 49143
Dohertyld, Erin 33064
Dohm, Katharina 18391
Dohnal, Karel 09540
Doiron, Sue 06817
Doix, Jean Michel 15985
Dojčev, Nejčo 05105, 05108
Dokken Urdal, Haldis 33506
Dolan, Douglas C. 48177, 48180
Dolan, Joseph J. 48239
Dolan, Susan 53598
Dolata, Jan 35199
Dolby, Joe 38593
Dolby, Malcolm J. 46078
Dolce, Dr. Sergio 28078, 28089
Dolci, Alberto 25864
Dolczewski, Zygmunt 35044
Dolenz, Dr. Heimo 02186
Dolenz, Univ.-Doz. Dr. Heimo 02500
Doletsky-Rackham, Nina 48136
Doleva, John 53227
Dolgopolova, Zinaida Stepanovna 37791
Doll, Nancy 49068
Dollins-Colton, Sonya 48270
Dolor, Danny 34404
Doltenkov, V.V. 36797
Dolženko, Fëdor Nikolaevič 36879
Domaas, Ruwe 41031
Domaine, Elmo 26095
Domanig, Andrea 02992
Domański, Edward 35265
Dombrowski, Dieter 22229
Domènech, Glòria 38956
Domenech de Celles, Yann 13138
Dominder, Christer 40774
Domingo Monteverde, Coia 39893
Domingos, Tadeu 00107
Dominguez, Antonio Franco 38936
Domínguez, German 07458
Domínguez Cabrera, Gonzalo 09218
Domínguez Cadeño, Nelson 09105
Dominjak, Aleksandr Vladimirovič 37546
Domitruk, Daryl 06391
Dommisse, E. 38696
Dompierre, Louise 05951
Dompnier, Pierre 15329
Domröse, Ulrich 17049
Don Angelo, Maria 26378
Don Che, Kyu 29781
Donabauer, Michael 20348
Donadio, Emmie 50653
Donahue, Rachel L. 48712
Donaldson, E. 00843
Donaldson, John 43730
Donaldson, Susan 53209
Donath, David A. 54287
Donath, Wolfram 17239
Donati, Prof. Bruno 41435
Donati, Dario 42116
Donches, Stephen G. 46782
Donchez Mowers, Charlene 46777, 46778, 46779, 46781
Dondero, Veronica 49230
Donders van den Aardweg, M.G. 31931
Dondo-Tardiff, Michelle 06532
Donegá, Dr. Giuseppe 28206
Donelli, Ivo 08954
Donelly, P. 44633
Donelson, Patty J. 47844
Donevski, Peti 05220, 05221, 05222, 05223
Dong, Daochang 08434
Dong, Fuping 08080
Dong, Pengyun 08318
Dong, Sen 08112
Dong, Xiaowen 07478
Dong, Ziru 07841
Dong, Zso 42185
Donkersloot, B.J. 32203
Donlon, Dr. D. 01549
Donneau, Olivien 03347
Donnellon, Maureen 00869
Donnelly, George J. 49855
Donnelly, John 53707
Donnelly, Peter 44632
Donner, Drs. E.M. 31854
Donner, Josef 02570, 03103
Donnermeyer, Reinhold 20311
Donoghue, Gail 53668
Donoghue, Michael 51054
Donohue, Deidre 51194
Donohue, Lorena 50144
Donoso, Ricardo 41001
Dony, Thierry 03994
Donzé, Frédéric 41791
Donzelli, Rodolfo 27548
Dooley, Bill 53621
Dooley, Cassandra 50409
Dooley, Max 54195
Doosry, Dr. Yasmin 20764
Dopffer, Anne 12025
Doppelfeld, Dr. Basilius 21670
Doppelfeld, Peter 42110
Doppelstein, Jürgen 22353
Doppler, Margarete 01794
Dor, Ehud 25044
Doračić, Damir 08986
Dorado Ortega, Oscar Javier 08612
Doran, John 45691
Doran, Patricia A. 49832
Dorbolo, Cheryl 53237
Dorchin, Dr. Netta 17469
Dordević, Radisav 38137
Dore, Dr. Anna 25447
Dore, Giovanni 27934
Dorée, Yves 15617
Doren, Jo-an' Van 47658
Dorgerloh, Prof. Dr. Hartmut 17043, 17145, 17185, 17227, 17248, 17249, 17250, 17252, 19523, 19638, 20910, 21080, 21081, 21083, 21085, 21086, 21088, 21090, 21091, 21094, 21099, 21100, 21101, 21102, 21104, 21105, 21106, 21107, 21108, 21109, 21306, 21708
Dorhs, Michael 19224
Doriot Galofaro, Sylvie 41553
Dorling, M. 43733
Dornan, Duncan 44089
Dornan, Vorin 51606
Dorner, Ernst 02241
Dorner, Friederike-Andrea 20919
Dorner, Markus 16766
Dornert, Nadine 12001
Dornieden, Horst 22020
Dornisch, Richard 52607
Dorociak, Heather 49030
Dorosiev, J. 05074
Dorra, Srarfi 42634
Dorsch, Dr. Klaus J. 21195, 21197
Dorsett, A.G. 43762
Dorsett, Ann 44762
Dorsey, Robert W. 47553
Dorsky, David A. 51097
Dorst, A. 32751
Dortch, Steve 48401
Dose, Dr. Hanna 16798, 16800
Doseva, Iva 05191
Dosik, Jeffrey 51262
Dossi, Dr. Barbara 02929
Dossi, Dr. Cristina 28185
Doth, Diane 07367
Dottori, Loretta 41282
Doublard du Vigneau, Simone 13159
Doubravský, PaedDr. Zdeněk 09855
Doucet, C. 16037
Doucet, Donna 05897
Doucet, Jean 12008
Doucet, M. 16036
Doudiet, Norman 47278
Dougherty, Erin 50854
Dougherty, J. Terry 48651
Dougherty, Molly 51848
Dougherty, Sandra 49480
Dougherty, Thomas A. 51200
Doughty, P.S. 43448
Doughty, Richard 44367
Douglas, Dr. B. 01321
Douglas, Bill 51292
Douglas, Deborah 47183
Douglass, Amy A. 53472
Douglass, Ardith 50560
Douglass, Larry 47024
Douglass, Michael 52129
Doulcet, Philippe 12502
Doum, Freddy 03959
Douma, Marian 32899
Douroux, Xavier 12731, 12733
Dourthe, Pacal 14309
Doval Sànchez, Gonzalo 39530
Dovgyj, Mykola Mychajlovyč 43061
Dovre, Terry 47088
Dow, George F. 51356
Doway, Stella 38168
Dowdeswell, Prof. J.A. 43732
Dowie, Peggy 45715
Dowlan, Neil 44662
Dowler, H. 44633
Downend, Allan 45536
Downer, Dr. 24745
Downes, Elizabeth 01230
Downes, Dr. Jennifer 43244
Downes, Marcia W. 51295
Downey, Martha J. 46832
Downey, Dr. Robert H. 46992
Downham, John 45447
Downie Banks, Dr. Margaret 53719
Downs, Alex 46128, 46130
Downs, Linda 47969
Downs, Dr. Robert T. 53599
Downs, Stuart 49187
Doyle, Louise 00934

Doyle, Maura 48245
Doyle, Dr. Richard 50749
Doyon, Jean-Pierre 06328
Dräger, Ulf 18904
Drăgan, Tereza 36097
Draganić, Danica 08978
Draganov, Dr. Dimitar 05042, 05043
Dragićević, Mate 04268
Drăgoi, Dr. Livia 36289
Dragomir, Petre 36431
Dragoni, Prof. G. 25461
Dragotă, Gheorghe 36363
Dragsbo, Peter 09983
Dragt, G.I.W. 32140, 32558
Draguet, Michel 03462, 03463, 03477
Drahoňovský, Jan 09637
Drahošova, Viera 38350
Drake, Ellen Mary 06150
Drake, Pia 19000
Drake, Robert M. 49008
Drake, Tommi 49018
Dran, Jean-Luc 13077
Draper, Amanda 45444
Draper, Jill 44867
Draper, Jo 45054
Draper, Steven C. 48625
Drapkin, Adrienn 49520
Drapkin, Adrienne 49517
Draps, M. Willem 03487
Drašković, Dragan 38111
Dratwa, Daniel 03490
Draughn, Raymond 52436
Draževa, Conja G. 04996
Dražić, Jovanka 38151
Dražin Trbuljak, Lada 09009
Drda, Dr Miloš 09842
Drda, Dr. Miloš 09843, 09858, 09904
Drechsel, Kerstin 17697
Drechsel, Manfred 21571
Drechsler, Dr. Wolfgang 03042
Dreckow, T. 00983
Drees, Dr. Jan 21566
Dreezen, J. 03699
Dreher, Derick 51877
Dreher, Nancy 50861
Dreier, Bruno 19482
Dreishpoon, Douglas 47114
Drejholt, Nils 40909
Drempetić-Hrcic, Nikola 08876
Drenker-Nagels, Dr. Klara 17447
Drennan, Emily 47293
Drenzek, Dennis 48631
Dresch, Dr. Jutta 19437
Dresco, Jean-Pierre 41715
Drese, Friedrich 20155
Dressel, Barry 46474, 49497
Dressel, Ina-Maria 16575
Dressler, Iris 21970
Dreuzy, Jehan de 15267
Drew, Alejandrina 48328
Drew, Jane 47265
Drew, Stephen E. 52503
Drewes, C. 38714
Drewitz, Britta 18774
Drewnowski, Benedykt 41884
Drews, Gayle 46702
Drews, Ute 21565
Drexel, Hermann 02564
Drexler, Ray 49725
Drexler, Toni 21609
Dreycus, Dominique 13812
Dreydoppel, Susan M. 50993, 50994
Dreyer, Dr. Andreas 33530
Dreyer, Chris 51450
Dreyer, Prof. L. 38754
Dreyer, Dr. Wolfgang 19536
Dreyfus, Renee 52772, 52781
Drezgić, Olivera 38151
Drias, Lakhdar 00045
Driesbach, Janice 50105
Driesel, Hans 19560
Drieselmann, Jörg 17261
Driesens, F.J.J. 32683
Driessen, Hendrik 32830
Driessens, Mevrouw Colette 03589
Driewer, Ginny L. 54339
Drignath, Nurdan 17417
Drinjaković, Milica 38094
Drinkard, Joel F. 50286
Drinnenberg, Julia 19224
Drischel, Dr. Georg F. 18432
Driscoll, David 50396
Driskell, Julia 46871
Driver, Rod 01038
Drobney, Dr. Jeffrey A. 49730
Drögemüller, Inka 18391
Dröscher, Elke 18956
Drößler, Stefan 20440, 20462
Droguet, Vincent 12966
Dromgoole, Dominic 44985
Dromgoole, P.S.B.F. 44271
Dronkers, Ben 31844
Dronkers, S.L. 31799
Dror, Mira 25003
Drosinou, Paraskevi 22961
Droste-Vischering-Galen, Hubertus 01957
Drozd, Henryk 35102
Dru Burns, Mary 46218
Druček, A.A. 37324
Drucha-Graber, Barbara 02739
Drucker, R. David 51439
Druesedow, Jean 49740
Drumm, Mel 46303
Drummond, H. Maury 46634
Drutt, Matthew 49394
Družak, Antonija 08885, 08886
Dryden, Jackie 45605
Dryfhout, John 47812
Dsuve, Kjetil 33884
Du, Gencheng 08309
Du, Jiuming 07489
Du Bau, Giorgio 28091
du Fayet de la Tour, Alain 12950
Duan, Bilin 07672
Duan, Shuangyin 07979
Duan, Xinshu 08244
Duan, Xinya 08099
Duan, Xuelin 08391
Duarte, Berta 35531
Duarte, Carlos F. 54540
Duarte Burró, Ramón M. 34099, 34100
Duarte Ferreira, Selma 04930
Duarte Rodríguez, Rafael Enrique 08584
Dubach, Margaretha 41761
Dubal Bon, Vicente 39652
Dubbi, Franz-Josef 22332
Dubé, Lisa 46225
Dubedat, Paul 15513
Dubick, David 46390, 48283
Dubois, Arlette 15116
Dubois, Bernard 03390
Dubois, Danielle 06374
Dubois, Danuta 14634
Dubois, Oliver 14988
Dubreuil, Nicole 14014, 14015
Dubrisay, Pascal 13794
Dubrović, Ervin 08937
Dubrovin, Sergej Michajlovič 37486
Dubrovin, V.M. 36761
Dubuc, André 13735
Dubuisson, Laurente 03329
Ducaroy, Agnès 15979
Ducasse, Sonia 14174
Ducastelle, Jean-Pierre 03328, 03330, 03840
Ducey, Dr. Peter K. 47829
Duchamp, Luc 13983
Duchemin, Michael 50255
Duchemin-Philippe, Catherine 12902
Duchesne, Prof. Jean-Patrick 03795, 03813
Duckers, Peter 45677
Duckworth, Elisabeth 06036
Duclos, Jean-Claude 13133, 13135
Duco, D.H. 31880
Ducor, Jérôme 41421
Ducourau, Vincent 11877, 11879
Ducreux, Anne-Claire 13944
Duda, Eugeniusz 34790
Duda, Igor Mykytovyč 43193
Dudant, Didier 03277, 03278, 03279
Dudar, Dr. Wiesław 35007
Dudeck, Dr. Volker 22813
Dudenhöffer, Franz 21835
Dūdinš, Andis 30144
Dudley, Andrew 01512
Dudley, Clarissa 49358
Dudnik, I.E. 43103
Dudź, Jerzy 35166
Dudzik, Przemysław 34700
Dueck, Lois 05408
Dückershoff, Michael 17925
Dühr, Dr. Elisabeth 22105
Duell, Charles H.P. 47373
Duell, Marshall 52853
Duenkel, Bob 48637
Dünser, Mag. Hans 01827
Duensing, Darrell 52130
Dürr, Frank 22121
Duerr, Mark 50394
Dürüce, Servet 42967
Duesberg, Baron François 03888
Düspohl, Martin 17160
Düsterwald, Dr. Brigitte 19901
Dütschler, H.R. 41939
Dufek, Mike R. 48875
Duff, Ana 47053
Duff, Doris 54021
Duff, James H. 47322
Duffin, Marion 45950
Duffy, Prof. A. 38763
Duffy, Henry J. 47812
Dufilho, Diane 53085
Dufour, Albertine 53304
Dufour, Christophe 41682
Dufour, Gary 01410
Dufour, Pierre 07194
Dufrasne, F. 04045
Dugajeva, Tetjana Igorivna 43009
Dugall, Berndt 18399
Duggan, Ervin S. 51678
Duggen Below, Ina 21039
Dugmore, I.M. 38620
Dugot, Joël 14656
Dugué, Prof. J. 14721
Dugulin, Dr. Adriano 28076, 28077, 28079, 28080, 28081, 28082, 28083, 28085, 28086
Duits, Th. te 32729
Duivesteijn, John 32104
Dujardin, Paul 03498
Dujmović-Kosovac, Ljubica 08938
Dukat, Zdenka 08986
Duke, Jacqueline 52882
Duke, Judy 47785
Dulaban, Alta 47655
Dulaney, W. Marvin 47357
Dulau, Robert 14728
Dulda, Mustafa 42672
Dulgheru, Lia Maria 36183, 36238, 36528, 36529, 36530, 36531, 36533, 36535, 36662, 36715, 36720, 36721, 36722
Dulibić, Ljerka 09011
Dulieu, René 03957
Dulin, Étienne 16182
Dumalag, Filemon R. 34336
Dumanoir, Thierry 13396, 15006
Dumaret, Isabelle 41280
Dumas, Cyril 13690, 13691, 13693, 13694
Dumas, Sylvia 20565
Dumbravă, Ion 36410
Dumbravă, Lucia 36512
Dumitrescu, Corina 36693
Dumitrescu, Horia 36257, 36361, 36372, 36373, 36374
Dumitrescu, Ilinca 36224
Dumitroaia, Gheorghe 36150, 36516, 36517, 36521, 36522, 36567, 36673, 36674, 36675, 36676
Dumitru, Țeicu 36562
Dummer, S. 44530
Dumon, Noël 03418
DuMont, Bruce 47491
Dumont, Elizabeth R. 46269
Dumont, F. 03808
Dumont, Fabienne 03443
Dumont, Marcel 41948
Dumont Saint-Priest, F. 11898
Dumortier, Christian 15306
Dumortier, Claire 03495
Dumortier, Philippe 15306
Dumouchelle, Kevin D. 47069
Dumoulin, Géry 03477
Dumoulin, J. 04089
Dumville, John P. 48487, 49409, 52477, 53239, 53326, 54229
Dunand, Michele 13427
Duncan, Jan 52704
Duncan, Jenepher 01410
Duncan, Larry 53363
Duncan, Oliver 54325
Dundass, Kay 33151
Dundon, Margo 49584
Dundore, Mary Margaret 52073
Dungan, Erica Howe 46981
Dunham, Elizabeth L. 53436
Dunina, Elena Michajlovna 37088
Dunkelman, Arthur 50645
Dunkerton, Jill 44937
Dunkley, Diane L. 53858
Dunlap, Joanne 52167
Dunlap, Margo 51472
Dunlap, Melissa L. 50158
Dunlap, Paula 49317
Dunlap, Susan 52438
Dunlop, Harry 44297, 44301
Dunlop, J.M. 44180
Dunn, Cathey 50193
Dunn, David W. 53329
Dunn, Deborah F. 49344
Dunn, Don 47958
Dunn, Forrest 53084
Dunn, Hedy 50205
Dunn, Jayne 44855
Dunn, Lawren 48518, 48841, 52009
Dunn, Lindsley A. 47582
Dunn, Madeline 48163
Dunn, Nancy 46365
Dunn, Richard 43798, 44269
Dunn, Sharon 48793
Dunne, John 24873
Dunne, Maxine 05797
Dunning, Louise 43965
Dunning, Lynn 43384
Dunstall, David 45669
Dunton, Anna Mary 47210
Duparc, F.J. 32100
Dupe, Nicole 50008
Duperray, Eve 12058, 12957, 12959
Dupeux, Cécile 15824
Duplančić, Arsen 08954
DuPont, Diana 52859
DuPont, George 49897
Dupont, Jean-Pierre 16108
Dupont, Joel 15507
Dupont, Paule 03717
Dupont, Roland 16247
Dupont-Bauverie, Caroline 11996
Duppré, Hans Jörg 17777
Duprat, Andrés 00133, 00136
Duprat, Jean-Marie 15785
Dupraz, Cathérine 12450
Dupre la Tour, Yves 13881
Dupree, Sherry 48121
Dupret, Sandra 48111, 48117
Dupriez, Françoise 04138
Dupuis-Sabron, Geneviève 12066
Duquenne, Raymond 04066
Duquesnay, Jean Claude 15517
Duquesnoye, Jeannot 03676
Dura, Ulrike 19937, 19939
Duraj, Paweł 35016
Durajková, Dr. Vladimíra 09946, 09947, 09948, 09949
Duran Alcalá, Francisco 39871
Durán Gómez, Eduardo 08598
Durán Nájera, Juan José 31025
Durán Solís, Leonel 31138
Durand, Agnes 13994, 14004
Durand, Alain 12863
Durand, André 12192
Durand, Régis 14614, 14615
Durandy van den Daele, Baroness C. 13518
Durant, Graham 00951
Durant, John 47183
Durante, Aldo 26808
Durante, Dr. Antonio 25602
Duranthon, Francis 15946
Đuraš, Prof. Mišo 08871
Durben, Silvan 51635
Durbin, Brian H. 47101
Durden, Chris J. 46519
Durden, Robert 51672
Durdyéeva, Tuvakbibi Kurbanovna 42974
Duret, Evelyne 15490
Duret, Michel 12908
Durham, A. Scott 50900
Durhône, Marise 11894
Durić, Vladimir 38034
Durkin, Brigid 47952
Durkin, D. 44882
Durko, Prof. Janusz 35247
Durlić, Paun Es. 38115
Durman, Prof. Mick 43499
Durmaz, Ş. Nihal 42921
Durnford, Louise 07068
Durnoff, M. 12987
Durocher, Kristina 48548
Duron, Maurice 11851, 12301, 12302
Durracq, Roger-Pierre 13071
Durrani, Prof. Farzaud 34032
Durrleman, Sophie 14654, 14662, 14682, 14734
Durst, Duane R. 49160
Dursum, Brian A. 47802
Durville 14560
Durye, Joseph 16219
Dus, Michael 09902
Dusar, Mark 04031
Duscher, Michael 02204
Dušejko, Stanislav Dmitrievič 37295
Dusheshe, Tanny 23288
Dutarte, Albert 13152
Duțescu, Mihai 36321, 36322, 36325, 36326, 36328
Duthion, Jacques 13669
Dutillieux, Léon 03618
Dutoit, Albert 41644
Dutra, James 53986
Dutton, Sandra 46335
Duval, Dennis 47578

D

Duval Reese, Becky 48321
Duvall, John S. 48529
Duvernois, Bruno 13180
Duvivier, Christophe 14908, 14909
Duvosquel, Jean-Marie 04133
Duwel, Emely 53469
Duyckaerts, E. 03834
Duysters, K. 31929
Dvir, Orah 25047
Dvořáčková, PhDr. Sylva 09685
Dvořáková, Dr. Hana 09427
Dwyer, Christine M. 46169
Dybbro, Inge 10114
Dybing, Leif 33470, 33471
Dyck, Angela 07294
Dye, Dr. Joseph M. 52343
Dye, Norman 50544
Dyer, David 50757
Dyer, Deborah 46602
Dyer, John T. 53926
Dyer, Michael P. 53050
Dyer, Mike 01431
Dykhuis, Peter 05929
Dymarski, Lech 35048
Dyment, Dave 07123
Dynak, Sharon 47608
Džanaeva, Alla Akimovna 37928
Dzeko, Lebiba 04284
Dzeranov, Timur Efimovič 37932
Dzhalilov, I. 42503
Dziechciaruk-Maj, Bogna 34783
Dzięciołowski, Edwin 34663
Dziedzic, Bogusław 34539
Dziembowski, Bettina von 20609
Dziewior, Dr. Yilmaz 01795
Dziurzyński, Tadeusz 34779, 34780
Dzjadik, Jevgenija Mychajlivna 43138
Džuma, Galina Ivanivna 43120

E

Eade, Coila 49156
Eadie, Bridget 44895
Eager, Barry W. 46763
Eakin, Eric 46655
Eames, Fred 01672
Earenfight, Dr. Phillip J. 47244
Earl, L.M. 01253
Earle, Edward 51194
Earle, Susan 49992
Early, J. W. 33043
Earnest, E. Peter 53883
Earnest, Ola May 51987
Earnst, Arlene E. 48159
Eastgate, S. 44610
Eastleigh, Lord 31606
Easton, Lois 49743
Easton-Moore, Barbara 47856
Eastwood, Stuart 43757
Eaton, Alex 45896
Eaton, Kenneth M. 49332
Eaton, Mary Candace 51807
Eaton, Nellie 06565
Eatts, M. 01198
Eayrs, Walter F. 47061
Ebbinghaus, R. 17124
Ēbedžans, Svetlana Georgievna 37474
Eberhard, Prof. Dr. Klaus 19906
Eberhardt, Dr. Hans-Joachim 20487
Eberhardt, Herman 49458
Eberhardt, Ursula 16890
Eberle, Josef 30196
Eberle, Lidia 35238
Eberle, M. 49225
Eberle, Richard H. 51145
Eberle, Urs 41342, 41343
Eberli, Ulrich 42126
Eberold, Dr. Bernd 42078
Ebersberger, Eva 03081
Ebert, Glenn 07305
Eberwein, Dr. Roland 02186
Ėbinger, Margarita Astemirovna 36872
Ebner, Dr. Lois 02300
Ebner, Mark 53991
Ebner, Silvia 02302
Ebneter, Curdin 41866
Ebrahimi, Nader 24688
Ebster, Dr. Silke 01769
Ebtehaj, V. 24654
Eccles, Tony 44195
Echeverria Llanos, Clemente 08744
Ecker, Dr. Berthold 03029
Ecker, Reingard 02918
Eckerbom, Jonas 40634
Eckerson, Rosemary 52207
Eckert, Bob 50417
Eckert, Horst 16616
Eckert, Jochen 18809
Eckert, Prof.Dr. Rainer 19942
Eckhardt, Mária 23431
Eckhardt, Susan 49666
Eckhel, Nerina 08989
Eckhoff, Audun 33418, 33419, 33420
Eckhout, Debbra 46745
Eckloff, Dr. Wolfram 20079
Eckmann, Sabine 52586
Eckstrom, Carolyn 48392
Edblom, Mats 40905
Eddie, I. 33172
Ėdeleva, Inessa Aleksandrovna 37008
Edelmann, Dr. Martina 22208
Eden, Dave 50792
Edenfield, W. Vernon 48766
Eder, Dr. Johanna 21959
Ederer, Walter 20690
Edgar, Angela 44127
Edgar jr., Arthur C. 50676
Edge, Yvonne 45095
Edgert, Elizabeth 50949
Edgren, Bengt 40572
Edhofer, Elisabeth M. 03045
Edidin, Stephen R. 51160
Edina, Mèri 23753
Edinger, Dorothy 33267
Edinger, Reuben 48356
Edison, Carol 52688
Edison, Robert D. 53583
Edland, Svein 50193
Edler, Dr. Doris 18577
Edler, Hans 02688
Edlinger, Matth. 02439
Edlund, D. 43843
Edmonds, Pam 07139
Edmonds, Tom 48635
Edmondson, J. 44750
Edmonson, Dr. Jim 47629
Edmunds, Allan L. 51821
Edo, Austin 34084
Edson, Gary 50312
Edson Way, Dr. J. 52875
Eduard, Ebner 01833
Edus, M.J. 32784
Edvardsen, Erik Henning 33727
Edwards, David 53416
Edwards, Denise 43821
Edwards, Geoffrey 01102
Edwards jr., Harford 48677
Edwards, Jeff 47752
Edwards, John 43235
Edwards, Karen 06487
Edwards, N. 43284
Edwards, Nancy E. 48715
Edwards, Nat 43283
Edwards, Pamela 51073
Edwards, Paula 49257
Edwards, Rita 48825
Edwards, Susan 50970
Edwards, William 44859
Edzes, M.J. 32229
Eecher, Danilo 27470
Eekhof, H.R. 32845
Eekhout, L.L.M. 32467
Effenberger, Prof. Dr. Arne 17052
Effner, Bettina 17080
Efimova, Alla 46755
Efimovskij, A.F. 37069
Efrati, Y. 25134
Efsic, Dianne 47756
Eftekhari, M. 24685
Egan, Caroline 43590
Egan, Dominic 24740
Egawa, Tokusuke 29072
Egbert jr, Francis 49623
Egelseder, Berthold 02378
Eger, Peter 21203
Egermann, Gerhard 02516, 02517
Egg, Prof. Dr. Markus 20150
Egge, Dr. Reimer 22150
Eggen, S. 03276
Egger, Dr. Franz 41160, 41161
Egger, Norbert 02184
Eggers, Bernardo 07442
Eggers, Dr. José Carlos 04842
Egging, Dr. Björn 21151
Eggl, Christine 02750
Egglezos, Evangeline T. 48575
Eggli, Dr. Urs 42109
Egholm, Aage 10095
Egilsson, Kristjan 24043
Egipciaco, Wanda 51181
Egli, Christina 41798
Egli, Markus 41615
Egli, Monika 42099
Egloffstein, Dr. Albrecht Graf von und zu 20981
Egorova, Ljudmila Semenovna 37913
Egoryčev, Viktor Vasiljevič 37143
Egounlety, Micheline 04185
Eguía, Dr. José 40320, 40324
Ehiro, Prof. Masayuki 29248
Ehis, Martin 43500
Ehling, Dr. Kay 20488
Ehlitzer, Bernd 18667
Ehls, Heiner 19224
Ehm, Rainer 21220
Ehrenberger, Mag. Anton 01924
Ehrenfellner, Dr. Karl 02625
Ehrenheim, Jacob von 40498
Ehrenhuber, Franz 01751
Ehrentraud, Adolf 02012
Ehrenwerth, Dr. Manfred 18847
Ehres, Maria 41032
Ehrhart, Bernhard 19857
Ehrmann-Schindlbeck, Anna-Maria 19378
Ehry, Carl A. 48349
Eibl, Simon 19854
Eichelberger, Allison 49929
Eichenauer, Dr. Jürgen 20882
Eichhorn jun., Georg 02928
Eichhorn, Herbert 21289
Eichhorn, Virginia 06563
Eichler, Dr. Anja 22527, 22529, 22530, 22535
Eichler, Heidrun 20207
Eichmann, Erland 42020
Eick, Dr. Simone 17568
Eickhorst, Dr. William S. 49696
Eickmann, Margaret 52087
Eickmann, Theodore 48250
Eidam, Hardy 18204, 18208
Eidemüller, Gisela 17354
Eidlin, Nathan 25095
Eidsaune, Thor Helge 33667
Eifel-Porter, Rhoda 51216
Eifrig, Jennifer S. 54112
Eigler, Gerd 19018
Eigner, Walter 01771
Eikelmann, Dr. Renate 19305, 19512, 19513, 20426, 20824, 20833, 20835, 20836
Eiken, Dr. Douglas 49860
Eikenoord, R.G. 32727
Eikermann-Moseberg M.A., Silke 17612
Eiland, William U. 46411
Eilat, Galit 24974
Eilenberger, Prof. Dr. Guido 20437
Eilertson, Orie 52131
Eillott, John 01526
Eilschou Holm, Niels 10150
Eimbeck 31739
Eimer, Prof. Gerhard 19292
Eimert, Dr. Dorothea 17982, 17983
Einarsdóttir, Björg 24044
Einberger, E. 20443
Einreinhofer, Nancy 54007
Eipeldauer, Heike 02942
Eipper, Paul-Bernhard 41140
Eiras, Prof. Jorge 35847
Eirola, Martti 31751
Eiselt, Gerhard 20094
Eisen, M.A. Markus 20423
Eisenburger, Dr. C.M.S. 31964
Eisenhofer, Dr. Stefan 20490
Eisenmann, Mark 51044
Eisenreich, Jessica 52818
Eisenstein, Daniela 21590
Eisenstein, Dr. Daniela F. 18498
Eiskonen, Satu 11045
Eisl, Sigmund 02789
Eisler, Dr. János 23477
Eisley, Patricia 53793
Eissenhauer, Dr. Michael 17202, 19467, 19469, 19477, 19479
Eißfeller, Sigrid 17522
Eißfellner, Sigrid 22474
Eitam, David 24962
Eiynck, Dr. A. 20008
Eizenberg, Arie 24937
Ek-Nilsson, Katarina 40944
Ekberg, Anna 40782
Ekblom, Bengt 41010
Ekeberg, Jonas 33564
Ekelhart-Reinwetter, Dr. Christine 02929
Ekinci, H. Ali 42737
Eklund, Linda 40782
Eklund, Dr. Sigrid 40922
Ekpo, Violeta I. 33347
Ekström, Börje 40699
Ekström, Kjell 11098, 11101
Ekström Söderlund, Birgitta 10757
Eksturm, Nancy 51892
El-Amrani, Dr. Ahmed 31701
El-Aziz, Fatma Abd 10505
El-Hajj, Amal Abu 24984
El-Hajraoui, Mohamed Abdelwahed 31688
El-Husseini, Dr. Mehrez 10477
El-Kasri, Houceine 31690
El-Moussli, Majed 42136
El-Nur, Prof. Osama A. 40388, 40391
El-Ouafi, Prof. Boubker 31698
El-Tawab, Garethe 45816
El-Younsi, Dr. Amin 31700
Eland, Steve 01404
Elas-Necel, Karol 34594
Elayyan, Khaled 34047
Elba La Gioiosa, Rosa 00379
Elbert, Dirk 21797, 21799
Elbracht-Iglhaut, M.A. Gisela 21802
Eldamaty, Dr. Mamdouh 10492
Elder, Bob 52159
Elder, E.J. 07336
Elder, Sarah M. 52565
Elderfield, John 51215
Eldredge, Bruce B. 50109, 53207
Eldridge, Ruth 43834
Eleki, Dr. Krisztina 50215
Elen, A. 32729
Elerd, Udo 20900
Eletson, Helen 45016
Elgen, Fran 46801
Eliades, Chryso G. 09391
Elias, Prof. Giacomo 28053
Elias, Margery M. 49607
Elias, Ramon Jan 49607
Elias Ortiz, Alejandrina 31399
Eliasen, Kirstin 10199
Eliason, Sara 41070
Eliassen, Sven G. 33818
Eliëns, Prof. Dr. T.M. 32073
Eliot, D. 45831
Elis, Dr. Karlpeter 01967
Elk, Sara Jane 51828
Elkin, G.U. 37156
Elkington, Marian 43300
Elkington, Peter 43300
Elkins, Scott 51279
Elkjær, Trine 09963
Elkmann, Alain 28018
Elkovitch, Pat 46469
Elleberg Petersen, Mads 09961
Elleboudt, André 03557
Ellebrecht, Pat 49799
Ellen, Marc ter 32898
Ellenbroek, Dr. F.J.M. 32828
Ellender, Dan 49876
Ellensint Kremer, M. 22188
Eller, Olivia 15831
Eller Risi, Nicole 41352
Ellerman, Deborah L. 46367
Ellermeier, Dr. Friedrich 19010
Ellingsen, Harry 33444, 33445
Ellingson, Eugene 52441
Ellington, Howard W. 54151
Elliott, Dana R. 48526
Elliott, J.H. 46428
Elliott, Keith 45121
Elliott, Lucy 48191
Elliott, Mary Gene 46428
Elliott, Patrick 44156
Elliott, Rachael 01139
Elliott, Richard 43660
Elliott, Simon 00946
Ellis, Anita 47555
Ellis, Caroline 45881
Ellis, Evelyn M. 50118
Ellis, Gillian 43511
Ellis, J.C. 49972
Ellis, Jane 44847
Ellis, Joe 47272
Ellis, Lori 05383
Ellis, Pam 47287
Ellis, Robert E. 49665
Ellis, Ruth 01467
Ellis, Vern 00801
Ellis-Schön, Jane 45606
Ellison, Dr. Curtis W. 51656
Elliston Weiner, Sarah 51214
Ellsworth, Frank L. 51199
Ellsworth, Lynn 50882
Ellwanger, Natalie 01796
Ellwanger, Ricard 54114
Ellwood, Tony 01288, 01512
Elmén Berg, Anna 40809
Elmer, Hans R. 41838
Elmes, Robert 47074
Elmore, Fletcher L. 50296
Elmore, George 49954
Elmqvist Söderlund, Inga 40923
Elms, Anthony 47477
Elorduy, Julieta Gil 30692
Elorza Guinea, Juan Carlos 39068
Eloy, Paul 03835
Elsbree, Carl J. 46854
Elsbury, Kathy 47146
Elsdon, Judith 45860
Elsen, Dr. Thomas 16619, 16620, 16632
Elsen-Schwedler, Dr. Beate 19736, 21657
Elsener, Walter G. 42032
Elser, Oliver 18361
Elsig, Patrick 41789
Elšin, Sergej Veniamovič 37039
Elsner, Dr. Tobias von 20123
Elsposch, Friedrich 18019
Elston, Tim 51762

Elsworth, Jo 43640
Elustondo, André 15326
Eluyemi, Dr. Omotoso 33310, 33313
Elven, Prof. Reidar 33737
Elvers-Švamberk, Dr. Kathrin 21459
Elvira Barba, Miguel Angel 39526
Ely, Bruno 11515, 11520
Elyasiv, Isaac 25091
Elzinga, Dr. C. 32473
Emack Cambra, Jane 53453
Emanuel, Ludovít 09788
Emanuel, Nina 19887
Emanuel D., Hough 50200
Emanuelli, Dr. Giuseppe 27188
Embach, Prof. Dr. Michael 22104
Ember, Dr. Ildikó 23477
Emberling, Geoff 47502
Emenegger, Ashley 54063
Emerick, Emily W. 50781
Emerson, Francis 54315, 54316
Emerson, Julie 53013
Emerson, Lucius J. 24895
Emerson, Myra 49918
Emerson, Pam 53547
Emerson, Philip G. 54169
Emerson, Robert L. 54351
Emery, Dominique 13008
Emig, Ulli 17790
Emigholz, Dr. Björn 22220
Emiliani, Prof. Andrea 25478
Eminger, Dr. Jürgen 22078
Emmanuel, Victoria 43508
Emmendörffer, Dr. Christoph 16629
Emmert, Dr. Claudia 18222
Emmett, Don 05578
Emonson, Janet 06948
Emont Scott, Deborah 49695
Emptoz, Dr. F. 13886
Emsden, Katharine 49856
Ena, Jan 00908
Enache, Ştefan 36325
Encinas, Isabel 39544
Enciso Núñez, José Antonio 30783
Enckell Julliard, Julie 41986
End, Reinhard 18583
Ende, Arlyn 53041
Endenburg, Constanze 32723
Ender, Benjamin 41605
Enders, Don 52544
Enders, Donald L. 52692
Enders, John 50552
Endersbe, Susan U. 52975
Endert, D. van 16532
Endicott, John 43796
Endo, Koichi 29326
Endraß, Max 16564
Endslow, Ellen 54049
Endthoff, Gertrude 53176
Endzweig, Pamela 48435
Ene, Ioana 36552
Enei, Flavio 27708
Enei, Vicente 00489
Enevoldsen, V. 10156
Engberg, Siri 50743
Engdahl, Sarah 47603, 47604
Engel, Beate 41219
Engel, Edmée 30484
Engel, Erik 40995
Engel, Michael 49991
Engel, Nicolas 14638
Engel, Robert 47039
Engel, Silke 20753
Engel, Volker 17451
Engel, Dr. Walter 18007
Engelbosch, Christa 04029
Engelbrecht, Dr. C.M. 38561
Engelbrecht, Dean Mark 46258
Engelhardt, Jeannine 22160
Engelhardt, Thomas 18225
Engell, Oluf 10261
Engelmann, Barbara 22421
Engelmann, Mag. Josef 01734
Engels, D. 09358
Engels, Kim 22548
Engels, Maria 16401
Engelsing, Dr. Tobias 19682
Engelskirchen, Lutz 18858
Engelsman, Dr. S.E. 32488
Engen, Jorunn 33475
Engen, Luc 03723, 03796
Engeseter, Aage 33589
Engfors, Christina 40884
Engh-Grady, Linda 49671
Enghoff, Dr. Henrik 10158
Engi, Dr. Andrea 41967
England, Ann 46438
England, Peter 47464
Engle, Duane E. 52266
Englebright, Steven E. 53314
Englick, Gerd 16825
English, Catherine 06735
English, Charles L. 54227
English, Donise 52119
English, Louella F. 54227
English, Marie 51647
Engström, Urban 40730
Engstrom, Mark 07133
Enguita, Nuria 38956
Enkhtsetseg, Dashdavaa 31623
Enne, Eva 02381
Enneberg, Lee 53726
Ennen, Dr. Jörg 21969
Enomoto, Toru 29311
Enriqueta del Castillo y Mejía, María 31115
Enriquez, Felonilla D. 34379
Enseki, Carol 47067
Ensign, Garald 49414
Ensslen, D. 05759
Entin, Daniel 51246
Entremont, Paul d' 07303
Entrudo, Ana Flores 35879
Entwistle, R. 44530
Enzler, Jerome A. 48188, 48189
Epp, Bear 05727
Eppensteiner, Heinz 02305
Eppich, Linda 52168
Eppinger, Virginia 47026
Epple, Sabine 19911
Eppler-Jehle, Ursula 16456
Epprecht, Dr. Katharina 42097
Epps, Renee 51206
Erasmus, Prof. Theuns 38746
Erašova, Galina 37477
Erb, Christian 18372
Erb, Dr. Martin 41604
Erbelding, Susanne 19437
Erbentraut, Dr. Regina 18830
Erbguth, Dr. Horst 21215, 21220
Ercolani, Fausto 28172
Erdem, Emin 42878
Erdenebat, P. 31627
Erdmann, Walter Johannes 18715
Erdös, Iudita 36589
Eremina, Natalija Aleksandrovna 37514
Ereoli, Mario 26800
Erf, Lisa K. 47474
Erffa, Dr. Axel von 19455
Erftemeijer, A.F.W. 32295
Ergin, Muhlis 42814
Ergino, Nathalie 16236
Erhart, Heinrich 02710
Erhart, Ingeborg 02132, 02135
Erhatic Širnik, Romana 38467
Erias Martinez, Alfredo 39033
Ericani, Dr. Giuliana 25378, 25381
Erice Lacabe, Romana 40342, 40345
Erich, Frasl 02576
Erichsen, Dr. Johannes 19744, 19749, 19750
Erichson, Ulf 21309
Ericini, Carlo 25510
Erickson, Bruce R. 52629
Erickson, Florence 06246
Erickson, Marsha 49710
Erickson, Ruth 51790
Erickson, Willis D. 46543
Eriksson, Håkan 41012
Eriksson, Karl 40454
Eriksson, Maud 40690
Eriksson, Nadja 40627
Eriksson, Susan C. 46845
Eringen, Peri 52238
Eritsyan, Elena Sergeevna 37303
Erk, Werner 18655
Erkelens, Dr. A.M.L.E. 31915
Erkens, Michel 03707
Erkinger, Helmut 02529
Erkinova, Rimma Michajlovna 36927
Erlandsen, Roger 33887
Erler-Striebel, Petra 19440
Erlindo, Vittorio 26737
Erlinger, Fritz 02634
Ermacora, Dr. Beate 02117
Ermeling, Andreas 20518
Ermert, Axel 17143
Ermolaeva, Ljudmila Nikolaevna 37136
Ernst, Dr. Alexander B. 22739
Ernst, Alfred 20390
Ernst, Dr. Antje 21815
Ernst, Else 20390
Ernst, Gene 53255
Ernst, Jürgen 19924
Ernst, Mathias 41915
Ernst, Dr. Rainer 18321
Ernst-Adolf, Kloth 19121
Ernstell, Eva-Sofi 40885
Erochin, Viktor Ivanovič 37891
Erol, Dr. Erdogan 42883, 42885, 42886, 42889, 42890, 42891
Erramah, Mourad 42631
Errotabéréa, Claude 12101
Erskine, David 43342
Eršova, Valentina Petrovna 37572
Ertel, Dr. Christine 02369
Ërtjukova, Nadežda Semenovna 36960
Ertler, Karl 02670
Erven, Wilma 48049
Erwin, Sarah 53608
Eržen, Miro 38483
Escalas, Román 38980
Escat, Monique 15363, 15795
Escauriza, Edgardo 00644
Eschapasse, Anne 14699
Eschbach, Barbara 53466
Esche, Charles 32196, 40726
Eschebach, Dr. Erika 17517, 17972
Eschebach, Dr. Insa 18490
Eschenburg, Dr. Barbara 20491
Eschenburg, Wieland 21081
Eschment, Renate 18223
Escobar, Gloria 51563
Escobar, Ticio 34089
Escobar, Washington 54435
Escoffier, André 15657
Escolà, Dr. Olegrer 38976
Escriche Jaime, Carmen 40066, 40122
Escrivà, J. Ramón 40201
Escudero, Juan Miguel 34499
Esera, T. 38000
Esguerra, Liwliwa 34352
Eshenroder, Glenda 49373
Eshleman, Janet 50842
Eshun, Elton 44883
Eskelinen, Liisa 11067
Eskildsen, Prof. Ute 18263
Eskritt, Jodi Ann 07161
Esman, J. 32829
Espejo-Saavedra Romero, Luis 39996
Espeland, Else 33718
Espinagosa Marsà, Jaume 40106
Espinola, Vera 52638
Espinós, Adela 40204
Espinosa, Deborah 52191
Espinosa, Gina Agnes 34389
Espinosa Jiménez, Eleazar 09267
Espinosa Rodriguez, Antonio 30644, 30657
Espinosa Ruiz, Antonio 40303
Espinosa Yglesias, Angeles 31265
Espinoza, Carlos A. 07456
Espinoza, Felipe 31063
Espinoza, Dr. Isabel Flores 34250
Espinoza Arrubarrena, Luis 31074
Espinoza Mella, Adelaida 08572
Espinoza Pérez, Edgar 33290
Espírito Santo Bustorff Silva, Dr. Maria João 35706
Espiritu Gaston, Lynell 34505
Espintour, Elizabeth 46396
Esposito, Prof. Vincenzo 26892
Eßbach, Elke 19572, 22844
Esseiva, Renato 41899
Essen, Dr. Manfred von 20735
Esser, Dr. Hans-Joachim 20430
Esser, Dr. Werner 20794
Essl, Prof. Karlheinz 02200
Estall i Poles, Vicent 39733, 39735
Esteban Darder, Vicente 40167
Estep, Connie 52307
Esterhuizen, Valerie 38762
Estes, Judith 49428, 49429, 49431, 49432
Esteve, Y. 13214
Estham, Inger 41007
Estrada, Cesar P. 34363
Estrada, Gerard Martí 39110
Estrada Sánchez, Ramiro 30721
Eszter, Hajdu 23412
Etcheverry, Charles 03271
Eteläinen, Suvi-Mari 11093
Etgar, Raphie 25005
Ethelberg, Per 10049
Etter, Dr. Walter 41175
Etter, Willi 41332
Ettl, Maria 02963
Eubanks, Charles 51244
Eubanks, Ray 53196
Eugster, Katharina 41966
Eugster, Willi 41816
Euler, Dr. Andrea 02324
Euler-Schmidt, Dr. Michael 19612
Eure, Linda Jordan 48284
Euskirchen, Dr. Marion 19631
Eutsler, Therese 50117
Evangelinou, Rallou 23116
Evanich, Joan 54236
Evans, Bob 06415
Evans, Catherine 47752
Evans, Cathy 47368
Evans, Ceryl 44401, 44402, 44620
Evans, David 45427, 45428, 45429, 45438
Evans, Ed 05444
Evans, Eddie L. 46686
Evans, Edward 45329, 50507
Evans, Elaine A. 49818
Evans, G.H. 43762
Evans, Graham 44655
Evans, Greg 07254
Evans, Helen H. 48743
Evans, Jaime 51375
Evans, Jonathan 44975
Evans, Leroy 49456
Evans, Linda 53370
Evans, Lynn 50351
Evans, Margaret 49141
Evans, Mark 44681
Evans, Megan E. 49130
Evans, Mervyn 00655
Evans, Michael 46644
Evans, N. 33197
Evans, Neil 45048
Evans, Paul 44665, 44968
Evans, Steven 46660, 51166
Evans-Cracknell, Victoria 04193
Evdokimov, Vladimir Aleksandrovič 37148
Evdokimov, Vladimir Nikolaevič 36909, 36910
Eveleigh, David J. 43631, 43632
Eveleigh, Valerie 45168
Evelyn, Douglas E. 51236
Evensen, Anne Bjørg 33641
Everard, R.H.A.J. 32300
Everbrand, Lars-Åke 40449
Everett, Julienne 05409
Everett, Martin 43599
Everly, Nancy 47877
Evers, Cécile 03495
Evers, Richard 49270
Evers, Dr. Ulrika 21001
Eversberg, Dr. Gerd 19302
Eversole, Diane 49921
Evgenidou, Despina 22946
Evgeniev, Todor 04997
Evstigneeva, Irina V. 37725, 37726
Evtimov, Dejan 05213
Evtukova, Olga Aleksandrovna 37433
Ewald, Jill 51420
Ewbank, Nick 44238
Ewell, Matson 48492
Ewers, William 47503
Ewert, George H. 50770
Ewertowski, Zbigniew 34951
Ewigleben, Prof. Dr. Cornelia 20045
Ewigleben, Dr. Cornelia 20046
Ewigleben, Prof. Dr. Cornelia 21950
Ewin, Dr. Tim 43632
Ewing, Phyllis 49173
Ewing, Steve 50888
Ewing, William A. 41547
Ewington, Julie 01512
Ex, K.M.T. Sjarel 32729
Exley, Stuart 52352
Exnarová, Alena 09485
Eyb, Alexander Freiherr von 22643
Eyb, Silvia Freifrau von 22643
Eychner, George 49511
Eyland, Cliff 07361
Eyler, Carolyn 48976
Eymold, Ursula 20462
Eynatten, K.C. 49403
Eyün Aksu, S. 42898
Ezquerra, Beatriz 40122
Ezsöl, Adolf 02729

F

Faass, Dr. Martin 17177
Fabbri, Fabio 27142
Fabbri, Dr. Franco 26223, 26224, 26225, 26226, 26228, 26230
Fabbri, Jennifer S. 38711
Fabeck, Sylvie 03910
Fabela, Pamela K. 53538
Fabényi, Dr. Júlia 23453
Faber, Alain 30484
Faber, Lucien 03715
Faber, Dr. Marion 20790
Faber, Dr. Michael 20243
Faber, Dr. Monika 02929
Faber, O. 31954
Faber, Ole 09977, 09978, 09979, 10193, 10219, 10220, 10306, 10307, 10308
Faber, P. 31852
Faberman, Hilarie 53253
Fabian, Dr. Claudia 20424
Fabiani, Jérôme 12840
Fabiani, Rossella 28096
Fabiankowitsch, Gabriele 03026
Fabich, Uwe 19855
Fabing, Suzannah 51414
Fabiny, Dr. Tibor 23401
Fabisiak, Wojciech 35116
Fabra Carreras, Carlos 40279

Fabre, Alexia 16259
Fabre, Eric 13111, 14657, 14658
Fabre, Jean-Louis 14917
Fabre, Regis 12790
Fabrega i Gallaguet, Marta 39657
Fabregues, M. de 14036
Fabrezi, Dr. Marissa 00556
Fabricius, Karen M. 09957, 09958, 09959, 10278
Fabris, Dr. Corrado 25302
Fabritius, Mag. Art. Elisabeth 10261
Fabritius, Dr. Ruth 21297
Fabrizi, Elisabetta 44796
Fabrizio, Fernando di 27135
Fabyan, Dr. E. Joseph 53751
Facchinetti, Claudio 25667
Facchini, Prof. Fiorenzo 25459
Faccinto, Victor 54243, 54252
Facciolongo, Sabino 25647
Fachrudin, H. 24493
Facklam, Gudrun 17673
Fadani, Dr. Andrea 22166
Fadden, John 51558
Fadden, Karen 48196
Fadden, Lee R. 48196
Fadeev, Igor 37157
Fae Kamm, Mary 47934
Fährig, Ute 18890
Faelmar-Abad, Mary 34365
Faenzi, Giovanna 26160
Färber, Steffen 19002
Færøy, Frode 33738
Faesler Bremer, Dr. Cristina 31082
Fässler, Urs 41120
Fagbohunmi, Johnson 33324
Fage, André 14602
Fagerli, Karl A. 33702
Faherty, William B. 49232
Fahey, Genevieve 01521
Fahey, John M. 53867
Fahey, Peggy 05971
Fahl, Dr. Andreas 18993
Fahl-Dreger, Axel 22207
Fahmy, Prof. Dr. Mamdouh Amin 10541
Fahrenbruch, Melanie 47834
Fahrer, Uwe 17536
Fahy, Everett 51212
Faietti, Dr. Marzia 25442, 26153
Faill, Carol E. 49931
Failla, Donatella 26299
Faille d'Huysse, B. Baron della 04167
Faine, Susan 01479
Fair, Donald 46149
Fair, Eileen 52843
Fair Sleeper, Sara 46191
Fairbank, Lynn 48024
Fairbanks, Theresa 51055
Fairclough, Oliver 43748
Fairfax, Anne 51211
Fairfax, James A. 53926
Fairfield, Douglas 46175
Fairfull, Thomas M. 48613
Fairman, Elisabeth 51055
Faison, Elizabeth 52848
Faißner, Waltraud 02317
Faivre, Bernard 13390
Fajzullina, Galija Šauketovna 29679
Falabrino, Ugo 28099
Falaschi, Prof. Pierluigi 25617, 25623
Falcão, Prof. José Antonio 35423
Falco, João 04394
Falco, Joseph 16168
Falcon, Orlando Arturo 00238
Falconbridge, Prof. Brian 44969
Falconbridge, Liz 44586
Falgayrettes, Christiane 14643
Falk, Dr. Birgitta 18251
Falk, Dr. Brigitta 18268
Falke, Emily 46536
Falke, Ulla 13287
Falkenberg, Dr. Regine 17072
Falkenberg, Dr. Uwe 17947
Falkenstien-Doyle, Cheri 52892
Falkinger, Johann 02543
Falkner, Dr. Gerd 21048
Faller, Susanne 21951
Fallet, Estelle 41420
Falletti, Dr. Franca 26159
Fallon, Steve 53781
Falls, John 47919
Falomir, Miguel 39573
Faltaits, Manos 23192
Faltings, Dr. Dina 19081
Falvey, Tom 47734
Falvy, Dr. Zoltán 23494
Falz, A. 21703
Falz, Andreas 19083, 20162
Falz, Bernd 19147
Famuyiwa, M.B. 33323
Fan, Shenglin 08388
Fan, Wanglin 08460
Fan, Xiyi 07942
Fan, Yousheng 08431
Fan, Zhiqian 07892
Fanciotto, Cristiana 27177
Fandrei, Martina 18150
Fanelli, Doris 51830
Fang, Guangyao 08380
Fang, Guoquan 07873
Fang, Madeleine 46760
Fang, Yingsan 08045
Fang, Zaifu 08373
Fangarezzi, Dr. Riccardo 26940
Fanger, Dr. Henrik 22615
Fankhauser, Christina 41655
Fanni, Dr. Ignazio 28275
Fanning, Dale A. 52184
Fanony Fulgence 30542
Fansa, Prof. Dr. Mamoun 20898
Fanti, Mario 25465
Fanucci, Dennis 49785
Fanzun, Jon 41928
Farah, Ahmed 38549
Farahmanesh, Nazi 24621
Farar, Bill 47895
Farasin, Kurt 02697, 02709
Farbaky, Dr. Péter 23421
Farhad, Dr. Massumeh 53873
Faria, Dr. João Carlos Lázaro 35388
Faria, Sofia 35802
Farid Bin Abdul Jalal, Enoik Ahmad 30611
Farina, Dr. Ermes 28211
Farina, Prof. Tina 19600
Fariña Busto, Francisco 39754, 39897
Farinelli, Dr. Leonardo 27100
Faringdon, Lord 44213
Faris, Douglas D. 53054
Faris, M.P. 24771
Farkas, Jesse 48416
Farley, Annie 53679
Farley, G. 49225
Farley, Henry G. 53546
Farley, Dr. Linda E. 48068
Farley-Harger, Sara 48740
Farmer, Dennis P. 47438
Farnden, Bob 01202
Farnell, C. 44410
Farnell, Cynthia 53792
Farnham, Priscilla 48496
Farnós y Bel, Alex 38870
Farooqi, Dr. Anish 24145
Farr, Amanda 45257
Farr, Fonda 51477
Farrajota Ataide Garcia, Dr. Maria Madalena 35690
Farrand, Dr. William R. 46308
Farrar, H.D. 51508
Farrar, Robert A. 49283
Farrell, Anne 49846
Farrell, Kathleen 52397
Farrell, Laurie 50184
Farrell, Lesley 45711
Farrell, Dr. Peter C. 52740
Farrell, Thomas L. 52128
Farrell, Walter 52475
Farrenkopf, Dr. Michael 17398
Farrington, Charles W. 51391
Farrington, Dr. Gregory C. 52768
Farrow, Victoria 44925
Farver, Bonnie 46871
Farver, Jane 47179
Farwell-Gavin, Robin 52883
Farynyk, Diane 51178
Fasching, Josef 01820
Fasciati, Luciano 41374
Fasel, André 41388
Fash, Barbara 23328
Fash, Dr. William L. 47185
Fasold, Dr. Peter 18354
Fasold, Dr. Regina 19108
Fassati, Tomáš 09402
Fassbinder, Dr. S. 18750
Fasse, Dr. Norbert 17479
Fastlicht, Dr. Samuel 31092
Fatah, Sayed Abdul 00004
Fateh, Suzanne 53842
Fatehi, F. 24611
Fatimah, T. 24450
Fatka, Jaroslav 09735, 09741, 09742, 09746, 09799
Faton, Louis 14580
Faubert, Claude 06534
Fauchille, Bernard 14185
Fauerso, Joey 52710
Faulds, W. Rod 46887
Faulkner, D. Lindsay 33216
Faulkner, John 47597
Faulkner, Peter 01585
Faull, Caroll 49359
Faull, Dr. Margaret 45910
Faure, Philippe 13637
Faure-Comte, Karen 16166
Fauria i Roma, Dr. Carme 38994
Faust, David T. 53588
Faust, Pierre-Paul 15123
Faust, Dr. Sabine 22102
Fautin, Daphne 49991
Fauver, Lynn N. 50884
Fava, Lourdes 31579
Favarato, Dr. Paolo 26496
Favaro, Gian Pietro 27316
Favata, Alexa 53431
Favell, Georg 49400
Favre, Lee M. 50160
Fawaz, Salma 30168
Fawdry, Eddy 44957
Fawzi, Ibrahim 25011
Faxon, Susan 46296
Fay, David 48506
Fay, Martin 49207
Fayard, Armand 13139
Faydali, Erol 42918
Fayeq, Abdul Zahir 00002, 00003
Fazekas, Patricia L. 51035
Fazia, Gloria 26209
Fazzi, Kim 52351
Fazzone, Mary Ann 51042
Fearing, Chris 48342
Fearn, Peter 01221
Feather, Philip H. 50013
Fechter, E. 18868
Fechter, Dr. Sabine 18324
Fedele, Prof. F.G. 26894
Federico, Jean 46205
Federle Orr, Dr. Lynn 52772, 52781
Fedič, Vasil 38261, 38262
Fediw, Mary 51859
Fedorenko, Tamara Grigorjevna 37684
Fëdorova, Ljubov Nikolaevna 37492
Fëdorova, Ljudmila Ivanovna 37000
Fëdorova, Natalja Viktorovna 36945
Fedorowicz, Władysław 34826
Fedyna, Holger 20571
Fedynsky, Andrew 47643
Feeley, Betsey 51111
Feely, Dr. Martin 24836
Feeney, Lawrence 47745
Feeney, Warren 33070
Feest, Prof. Dr. Christian 03038
Feffee, Claudette 52116
Fehér, Béla 23705
Fehér, Dr. György 23439, 23688
Fehle, Dr. Isabella 20146, 20462
Fehlemann, Dr. Sabine 22746
Fehr, Horst 20022
Fehrmann, Silvia 17128
Feichtlbauer, Brunhilde 01720
Feickert, Rudolf 22387
Feigel, Martine 14324
Feil, Fr. Carl 53527
Feinberg, Norman M. 47069
Fejfer, Mariusz 34630
Fejös, Dr. Zoltán 23455
Fekeza, Lidija 04284
Felbauer, Josef 02269
Felbauer, Maria 02269
Felbermayer, Dietrich 02532
Feldberg, Dr. Michael 51143
Feldes, Angelika 21826
Feldkamp, Dr. Jörg 17703, 17750
Feldkamp, Ursula 17569
Feldman, Arthur M. 53611
Feldman, Kaywin 50577
Feldman, M.M. 24992
Feldman Bennet, Anita 45183
Feldmane, Milda 30051
Feldmann, Dorothee 21075
Feldmer, Peter 19029
Felgenhauer-Schmiedt, Prof. Dr. Sabine 02809
Felice, Silvio 30628
Felisati, Vittorio 26110
Fellay, André 41287
Felley, Jean-Paul 41399
Fellmeth, Prof. Dr. Ulrich 21955
Fellner, Fritz 01904
Fellner, Marianne 08670
Fellner, Reed 50740
Fellner, Susan 52867
Fellows, Leslie 51221
Felsher, Lynn 50658
Felshin, Nina 01929
Felsinger, Benedikt 30481
Feltz, Georges 30646
Fenech, Michael 07784
Feng, Baoxi 07951
Feng, Feng 08142
Feng, Shaozhu 07946
Feng, Wenxue 08494
Feng, Yonghao 07673
Feng, Zhaoxun 10103
Fenger, Lars 20579
Fenkes, Rolf 32051
Fennema, A.T. 53602
Fenner, Gloria J. 00514
Fenoglio, Norma 00515
Fenoglio, Norma C.
Fenton, Dr. Frank M. 54177
Fenton, John 00932
Fenton, Nicole 53262
Fényi, Tibor 23474
Fenz, Achim 18357
Fenz, Dr. Werner 01975
Ferarresi, Roberto 26137
Ferasin, Luigi 13738
Ferber, Oskar 16788
Ferdi, Sabah 00081
Ferebee, Peggy V. 49067
Ferez Kuri, Maria del Carmen 31143
Fergola, Dr. Lorenzo 28034
Fergurson, Boyd 49587
Ferguson, Francesca 41180
Ferguson, Ian 33239
Ferguson, Mary-Catherine 48418
Ferguson, Moira 44583
Ferguson, Randy 52147
Ferguson, Ray 52428
Ferguson, Robert S. 44295
Fergusson, Ian L.C. 44972
Ferigo, Dr. Giorgio 27992
Ferleger Brades, Susan 44872
Fernan, Marissa N. 34374
Fernandes, Isabel Maria 35639
Fernandes, João 35848
Fernandes Gomes, João José 35404
Fernández, Dr. Antonio 07432
Fernandez, Joanne 51272
Fernández, Jorge H. 39259, 39260
Fernandez, Jorge Osvaldo 00212
Fernández, Kathleen M. 46895
Fernandez, Luisito F. 34485
Fernández, Maria Angélica 00517
Fernandez, Nohema 49529
Fernández, Paloma Acuña 40140
Fernández Bal, María Luisa 39897
Fernández-baldur, Fidel Torcida 39945
Fernández Castañón Carrasco, José Antonio 39759
Fernández Cid, Miguel 40014
Fernández de Calderón, Cándida 31116
Fernández de Ulibarri, Rocío 08815
Fernández García, Juan Alfonso 39336
Fernández Gómez, Fernando 40064
Fernández Lóp, Olga 40242
Fernández Martinez, Dr. Miguel Ángel 40144
Fernandez Segorra, Natalia 40028
Fernández Ugalde, Antonio 39257
Fernández Vega, Pedro Ángel 40010
Fernández Villaverde, Duque Alvaro 39961
Fernier, Jean-Jaques 14540
Fernoile, Kate 47068
Ferolla Frizzi, Dr. Vanda 25705
Ferran, Domènec 40115, 40119
Ferran Sans, Marc 39895
Ferrand, Jean 14206
Ferrandi, Dr. Giuseppe 28061
Ferrara, Giovanni 26911
Ferraresi, Roberto 26136
Ferrari, Alessio 26837
Ferrari, Amedeo 26002
Ferrari, Ciro 25761, 26597
Ferraris, Francesco 28221
Ferraris, Pietro 25170
Ferrazzini-Siegel, Petra 41282
Ferré, Dr. Rosario 36049
Ferreira, Claudia Marcia 04727
Ferreira, Ivette Senise 04870
Ferreira, João Carlos A. 04733
Ferreira, José 00114
Ferreira Brandão, Dr. Carlos Roberto 04891
Ferreira da Costa, Paulo 35717
Ferreira da Cunha, Sergio Henrique 04628
Ferrell, John 47524
Ferren, Dick 53820
Ferrence, Cheryl 46963
Ferrer jr., Dominador 34454
Ferrer Ciari, Agustín 38867
Ferrer Dalgà, María Rosa 39348
Ferrer i Godoy, Joan 39985
Ferrer-Joly, Fabien 11735
Ferrer Soria, José Luis 39325
Ferrero, Marcelo 00152
Ferreyra, Prof. Carlos Alfredo 00404
Ferrin, Russ 50190
Ferrini, Giovanni 41889
Ferris, Alison 47102
Ferriso, Brian 53610
Ferro, Dr. Inês 35952
Ferryman, Charles 50582
Fersen, Vera von 10804
Fertig-Möller, Heidelore 22493
Fesik, Kateryna Borysivna 43168
Fest, Dr. Bärbel 17235
Festanti, Dr. Maurizio 27388
Festini, Prof. Zvonko 08990
Fett, T. 33426
Fetter, Rosemary 46561
Fetzer, Fanni 41526
Fewkes, Nancy 47468
Fewster, Dr. Kevin 01561, 01609

Fey, Cass 53585
Feyler, Gabrielle 15674, 15675, 15676
Feys, Nicole 03268
Fiallo Hernández, Eduardo 10344
Fiammingo, Raphaël 41401
Fiaschetti, Maria Egizia 27509
Fibicher, Bernard 41544
Fibiger, Susanne 10079
Fichner-Rathus, Prof. Dr. Lois 53557
Fichtenbaum, Judy 47418
Fichtner, Dr. Lutz 17789
Ficker, Peter 17752
Fickert, Hildegard 17653
Fideucci, Elaine 49206
Fiedler, Dr. Elisabeth 02842
Fiedler, Marek 35067
Fiedler, Thomas 18682
Fiedler, Uwe 17704
Fiedler Streng, Evelyn 53028
Fiege, Kirsten 19736, 21657
Fiehler, Tim 46896
Field, Hannah 43334
Field, Monica 05661, 05662, 07183
Field, Peter 45268, 45270
Field, Dr. Richard 49484
Field, William 52884
Fielden, Dr. Kate 43716
Fielding, Andrew 45272
Fielding, James 07384
Fields, Edward D. 46892
Fields, Hoyt 52830
Fields, Laura 46660
Fields, Vilma S. 47406
Fierro Gossman, Julieta N. 31161
Fierz, Gaby 41169
Fietsam, Robert 46699
Fiette, Alexandre 41418
Fietz, Dr. Rudolf 20896
Fifield, Donald 53496
Fifiţă, Mariana 36371, 36386
Figari, G.B. Roberto 25626
Figdor, Anna Maria 02574
Figiel, Flavia 19437
Figueiredo, Alexandrine 12645
Figueras i Casademont, Lluís 38948
Figueres, Roger 14037
Figueroa, Elisa 07470
Figueroa, Paul 52868
Figuerola i Pujol, Prof. Iris 38979
Figuti, Midory Kimura 04863
Fike, Nancy J. 53645
Filchner, Gerhard 20831
Filieri, Dr. Maria Teresa 26535, 26536
Filimonos, Dr. Melina 23170
Filina, Larissa 37229
Filip, Aurel 36688
Filipčić-Maligec, Vlatka 08876
Filipetti, Dr. Giuseppe 26465
Filipová, Maria 09722
Filipová, Marie 09721
Filipová, PhDr. Milena 09855
Filipović, Marica 04284
Filipow, Dr. Krzystof 34553
Filipowicz, Zygmunt 35143
Filippi, Dr. Giorgio 54501
Filippo, Giuseppe 25322
Filippova, Éleonora Anatoljevna 36920
Filipunas, Winifred 49905
Fillman, Joseph B. 46226
Filtvedt, Per Kr. 33775
Fimpeler-Philippen, Annette 18005
Finaldi, Dr. Gabriele 39573
Finamore, Dr. Daniel 52657
Finamore, Prof. Nino 27748
Finardi, Luigi 25377
Finat, Joelle 13619
Finch, Jon 44714, 44715
Finch, Michelle 47038
Finch, Rick 53671
Finch, Sarah 43713
Finch-Crisp, Sarah 45817
Finch-Morales, Nikii 52822
Finckh, Dr. Gerhard 22748
Finding, Gerhard 02181
Findley, Alvin 06989
Findley, Bill Brown 52232
Fingernagel, Dr. Andreas 03045
Fings, Dr. Karola 19626
Fink, Courtney 52796
Fink, David G. 46147
Fink, Roland 03029
Fink, Prof. William L. 46310
Finke, Karl 18161
Finkele, Diana 20362
Finkelstein, Jacqueline A. 04754
Finkelstein, Lori 51218
Finks, Harvey 48156
Finkster, Paul 49240
Finlay, James 46104
Finlayson, Prof. J.C. 22864
Finley, Janet 48961
Finley, Valerie 06205
Finnegan, Paul M. 51666
Finneiser, Klaus 17028
Finnemore, Donnice 53978
Finnemore, Harry 53978
Finnerty, Sinéad 54063
Finney, Libby 44682
Finnigan, Penny 01364
Finnigan, Stephen 49095
Finotti, Franco 27569
Finweg, Sabine 22810
Finžgar, Maja 38454
Finzi, Dr. Ranny 25099
Finzi, Dr. Sonia Guetta 28188
Fionik, Doroteusz 34557
Fiorentino, Alessandro 27880
Fiorenza, Joseph 48843
Fiori, Dennis 46549, 46937
Fiori, Dennis A. 46582
Fiori, Dr. Pietro 26712
Fiorillo, Anthony 47932
Fiorio, Dr. Maria Teresa 26677, 26679, 26681, 26694
Fiouri, E. 09375
Firl, Wolfgang 22486
Firmenich, Andrea 16749
Firmin, Sandra 47126
Firnges, Jörg 19200
Firnhaber, Waltraud 20665
Firstenberg, Jean 50208
Firszt, Stanisław 34704
Firth, Neil 45792
Fischbacher, Dr. Marianne 41492
Fischer, Andrea 21500
Fischer, Angelika 18873
Fischer, Barbara 06310
Fischer, Dr. Bernhard 22402
Fischer, Christian 10256
Fischer, D. 19447
Fischer, Dr. Eckhard 22662
Fischer, Erika 17755
Fischer, Felice 51867
Fischer, Dr. Fritz 16643, 21424, 21950
Fischer, Germann 19332
Fischer, Gottlieb 20221
Fischer, M.A. Hannelore 19610
Fischer, Dr. Hartwig 18263
Fischer, Helmut 17710, 17711
Fischer, Dr. h.c. Hermann von 41498
Fischer, Ilse 18934
Fischer, Johann 21498
Fischer, Dr. Jutta 21386
Fischer, Karl-Günther 17516
Fischer, Klaus 17821
Fischer, Manfred W.K. 02455
Fischer, Dr. Manuela 17081
Fischer, Marco 18209
Fischer, Mark 49232
Fischer, Prof. Dr. Martin S. 19387
Fischer, Martina 20267
Fischer, Peter 02657, 41593
Fischer, Reinhold 20798
Fischer, Ron 47358
Fischer, Steffen 18666
Fischer, M.A. Susanne 18337
Fischer, Tatjana 17697
Fischer, Thomas 16992
Fischer, Uwe 18093
Fischer, Prof. Dr. Volker 18382
Fischer, Dr. Werner 20307
Fischer-Elfert, Prof. Dr. Hans-W. 19891
Fischer Jonge, Ingrid 10140, 10143, 10218
Fischetti, Amy 48562
Fischhaber, Martin 20323
Fischlin, Jean-Mario 41136
Fisecker, August 02094
Fish, Paul 53584
Fish, Suzanne 53584
Fishback, Pat 52338
Fisher, Alice 47861
Fisher, Claire 45855
Fisher, Clive 45410
Fisher, D.L. 33137
Fisher, Debra 47035
Fisher, Debra S. 50680
Fisher, F.S. 47353
Fisher, Frederick J. 53876
Fisher, Jay 46551
Fisher, John R. 53459
Fisher, Kathleen 50832
Fisher, Kathryn 06934
Fisher, Kay 54206
Fisher, Kevin M. 50077
Fisher, Lawrence A. 46604
Fisher, Lilian 53739
Fisher, M. 48234
Fisher, Pat 44161
Fisher, Roddy 44189
Fisher Sterling, Dr. Susan 53906
Fishman, Bernard P. 46227
Fishmann, Bernard B. 52178
Fiske, David A. 53281
Fiskesjö, Dr. Magnus 40924
Fissiaux, Jean-Claude 04063
Fisson, Bruno 12929
Fitch, Arwen 45575, 45579
Fitch, Blake 54220
Fitch, James A. 48895
Fitts, Catherine 53920
Fitz, Prof. Erwin 01797
Fitz, Péter 23421
Fitzgerald, Annamary 51542
Fitzgerald, Dr. Edward 52220
Fitzgerald, Elizabeth 52585
Fitzgerald, Emma 32372
Fitzgerald, Erika 53221
Fitzgerald, Jane 50702
Fitzgerald, Lawrence 44292
Fitzgerald, Mary Ann 50390
Fitzgerald, Maryann 52924
Fitzgerald, Paul 51389
Fitzgerald, Shannon 52573
Fitzhugh, Robin 48488
Fitzpatrick, Gareth 44567
Fitzpatrick, M. 54666
Fitzpatrick, Mike 24866
Fitzpatrick, Pat 07402
Fitzpatrick, Robert 47492
Fitzpatrick, Sheila E. 53811
Fitzpatrick, Tom 25009
Fitzsimmons, Rita 49670
Fitzsimons, Dr. J. Michael 46638
Fiumi, Prof. Enrico 28311
Fivez, Dr. Liesbeth 03312
Fjellhøi 33378
Fjermedal, Torkell 33966
Flacke, Prof. Dr. Monika 17072
Flacke, Robert 49883
Flaherty, Mary A. 46290
Flaksman, Sandy 50287
Flanagan, Anne 01544
Flanagan, Brendan 44625
Flanagan, I.D. 44080
Flanagan, Michael 54134
Flanagan, Shelia M. 50770
Flanety, Lynda 49837
Flanigan, Brian 51910
Flannery, Kent V. 46308
Flannery, Kerry 49550
Flannery, Prof. Tim 00782
Flask, David 50372
Flasza, Jan 34565
Flaud, Jackie 16013
Flavin, Samantha 44658
Flechas Corredor, Jorge Arturo 08780
Fleck, Robert 14360
Fleck, Dr. Robert 17459, 18920
Fleckenstein, Jutta 20448
Fleet, Prof. A. 44941
Fleet, Judy 48874
Flegel, Andreas 18089
Flegel-Wantier, Klaus 21728
Fleischanderl, Rudolf 01870
Fleischer, Dr. Martina 02992
Fleischhauer, Dr. Carsten 21570
Fleischman, Stephen 50393
Fleischmann-Heck, Dr. Isa 19703
Fleissner, Alfred 02358
Fleming, Dr. David 44750
Fleming, Elvis 52469
Fleming, Geoffrey 53188
Fleming, Jeff 48096
Fleming, Leslie 51466
Fleming, Marty 53936
Fleming, Sam 43363
Fleming-Pawelek, Brenda 46692
Flensburg, Birgitta 40547
Flensburg, Brigitta 40773
Flentje, Rachel Daly 46869
Flesch, Edward 49369
Flesche, Susanne 19489
Flesher, Karen 53669
Fletcher, Ann 33110
Fletcher, Carrol 50509
Fletcher, Charles 50929
Fletcher, Corinne 54208
Fletling, Walter 18417
Flett, Dr. Alistair M. 43248
Flett, Harry 44397
Fleurke, Else 33014
Fleury, Nathalie 41324
Fleury, Dr. Robert 11662
Flick, André 14493
Fliegel, Stephen N. 47624
Flieger, Erwin 20160
Flinsch, Stefanie 02625
Flinspach, Joan L. 48707
Flint, Jody 47349
Flintoff, Paul 43526, 43527
Flodin, Göran 40883
Flögel, Dr. Evelyn 22297
Flögel-Kostić, Lela 38090
Flößer, Dr. Reinhard 16716
Flon, André 15352
Flon, Dominique 14352, 14356
Flood, Ellen 50603
Flood, Leon 51099
Flood, Richard 50743
Flood, Rick 54329
Flood, Robert 50603
Flood, Timothy 47510
Floquet, Prof. Jean 16064
Florance, Robyn 01384
Florea, Ileana 36623
Florentino, M. Raquel 35677
Florenzano, Clorinda 26869
Florer, Michael R. 48901
Flores, Frank 48422
Flores, Juan C. 54583
Florescu, Aurelia 36322
Flórez, Helmer de Jesús 08762
Flórez de Grajales, Aura 08770
Florian, Carmen 36218
Floridi, Dr. Massimiliano 26321
Florineth, Rico 41852
Floris, Prof. Giovanni 25587
Floris, Lene 10087
Florizoone, Jan 03924
Flory, Maurice 13855
Floryan, Dr. Margrethe 10155
Flotow, Julia von 06488
Flour, Joke 04100
Flourentzos, Dr. Pavlos 09375
Flowers, Betty Sue 46514
Flowers, Randy 47821
Floyd, David 46640
Floyd, Roger 53025
Flucker, Turry 49566
Flude, Kevin 44948
Flückiger, Dr. Peter F. 41708
Flüe, Barbara von 41885
Flügge, Matthias 17136
Fluke, Pam 45666
Fluke, Phil 45666
Flukinger, Roy 46510
Fluksi, Josip 08986
Flumm, S. 21031
Flunger, Mag. (FH) Julia 02951
Fluri, Annette 41182
Flury, Dorothea 41182
Flury-Steiner, Barbara 41227
Flutsch, Laurent 41551
Flynn Johnson, Robert 52772, 52781
Flynt, Suzanne 48033
Foa, Piers 00851
Foa, Robyn 00851
Foanaota, Lawrence 38547
Fobbs, Archie 53903
Fochessati, Matteo 26327
Focht, John 46662
Focke, Sue 52630
Fodor, Mihály 23497
Foe, Peter 53431
Foelix, Dr. Rainer F. 41083
Foelsche, Gudrun 02395, 02398
Först, Dr. Elke 18936
Förster, Monique 18206
Förster, Roswitha 21175
Förster, Steffen 20267
Föttinger, Franz 02765
Fogarasi, Zsuzsa 23645
Fogarty, Lori 52935
Fogle, Douglas 50743
Fogliazza, Virginio 26510
Foister, Dr. Susan 44937
Fojtik, Miroslav 09697, 09908
Folch, María Jesús 40201
Folch, Stella 38958
Foldøy, Oddveig 33863
Foley, Cindy M. 52064
Foley, J.A. 33221
Foley, Jim 47437
Foley, Suzanne 47400
Folga-Januszewska, Dorota 35261
Folger, Willibald 01806
Folgić-Korjak, Angelina 38074
Folie, Dr. Sabine 02993
Folini, Mara 41129
Folk, Luther 46993
Folkins, Alice 05708
Follbert, Léon 11696
Folley, Stephanie 05884
Folsach, Dr. Kjeld von 10130
Foltin, Viktorija Tyborivna 43197
Folwell Krueger, Hilary 48702
Fonclare, Guillaume de 14784
Fong, Lawrence 48431
Fonio, Prof. Ezio 28015
Fonsmark, Anne-Birgitte 09986
Font, Walter 48703
Font Sentias, Josep 39374
Fontaine, Axel 12498
Fontaine, Christian 15926

F

Fontaine, M. 13703
Fontaine, Patrick 14846
Fontaine, Susanne 17249, 17250
Fontana, Agostino 28163
Fontana, Armon 41374
Fontana, Prof. Dr. Eszter 19912
Fontana, Lilia 50628
Fontanarossa, Dr. Raffaella 25857
Fontaneda, Cristina 38871
Fontenioue, Henri du 12968
Fontenote-Jamerson, Belinda 50251
Fontenoy, Paul 46666
Foote, W.T. 01513
Foote-Smith, Christy 50091
Fopp, Dr. Michael A. 44967
Foppa, Christian 41301
Forand, Edward J. 54061
Forand, Peggy 53182
Foray, J.-M. 11998
Foray, Jean-Michel 14436, 16051
Forbes, Duncan 44157
Forbes, Jamie 07159
Forbes, Dr. John M. 46541
Forcada Serrano, Miguel 39870
Ford, Chris 48028
Ford, David 05885
Ford, Peggy A. 49034
Ford, Richard I. 46308
Ford, Dr. S.D. 51935
Ford, Sandra 48566
Ford jr., Thomas 34387
Førde, Anne-Marie 33979
Fordham, Gerald J. 52653
Fordham, Michael 44381
Fordyce, Prof. Dr. R. Ewan 33092
Forellad, Dolors 39924
Foreman, Martin 44597
Forero, Dr. Enrique 08566
Foresman, Dr. Kerry 50757
Forest, Barbara 12248
Forest, Dominique 14682
Forest, Gilles 13207
Forest, Marie-Cécile 14711
Forgan, Dr. Sophie 45994
Forgang, David M. 54350
Forgeng, Dr. Jeffrey 54298
Forhaug, Helén 40675
Forkl, Dr. Hermann 21951
Forman, Weldon 47787
Formanová, Markéta 09718
Formery, Valerie 03768
Formicone, Luigi 26944
Fornals Villalonga, Francisco 39143
Fornara, Livio 41408, 41414
Fornasari, Giuseppe 26859
Forner Caballero, Ester 40315
Fornés i Garcia, Prof. Josep 38994
Forney, Jack D. 48077
Forni, Prof. Gaetano 27720
Forno, Dr. Carla 25313
Forray-Carlier, Anne 14682
Forrer, Dr. M. 32488
Forrest, Richard 06207
Forrester, Gillian 51055
Forrester, Hugh 43446
Forrester, Jim 45086
Forsberg, Anders 33863
Forsberg, Diane 49195
Forsberg, Helena 41000
Forsberg, Virpi 10751
Forschler, Dr. Anne 46822
Forssberg, Anna Maria 40885
Forssblad, Marianne 53006
Forster, Helen 05357
Forster, Judith 42024
Forster, Michael 16514
Forster, Dr. Peter 22557
Forstner, Christine 02620
Forsythe, Dee 49989
Fort, Thomas A. 48291
Forte, Richard V. 46809
Fortenberry, Kay 50986
Forti, Fabio 25503
Forti, Dr. Micol 54501
Fortier, Jana 50504
Fortin, Jacques 05612
Fortin, James 06176
Fortin, Verna 50373
Fortino, Sally 16770
Fortson, Judith 48219
Fortun, Pascal 04061
Fortuna, Mary 52387
Fortunat Černilogar, Damjana 38527
Fortune, C.F. 38667, 38670, 38671, 38672
Fortune, Collin 38771
Foschi, Dr. Pier Luigi 27407, 27409
Foss, Birgit 33779
Foss, Marybeth 53425
Fossati, Alessandro 41582
Fosse, Martine 12248
Fossey, John M. 06357
Fossheim, Eivind 33965
Fossum, Finn 33416
Fostekew, Ken 46065
Foster, Alasdair 01399
Foster, Clarence 05598
Foster, Cynthia 52531
Foster, Di 01254
Foster, Jeanne K. 51646
Foster, Kathleen 51867
Foster, Keith 50605
Foster, Kenneth 52801
Foster, Laura 51506
Foster, Marusia 07388
Foster, Preston 47787
Foster, Robert 50942
Foster, Stephen 45704
Foster, Velma 05472
Foster, W.N. 43461
Fotchi Chian, Mansoor 24649
Fotoghi, M. 24639
Foucart, Bruno 12087
Foucher, Jean-Pascal 15697
Foulds, Kris 05305
Foulon, Dr. Anne-Cécile 17606
Foulon, Françoise 03713
Fountouli, Mary 22946
Fouray, Monique 15122
Fourie, Dr. H. 38759
Fouriscot, Mick 13644
Fournet, Claude 14424
Fournier, Johanne 06248
Fournier, Dr. M. 32484
Fournier, Monique 13888
Fournier, Vincent 12454
Fourny-Dargere, Sophie 13631, 16225
Foust, Dorothy 50374
Fowkes, Dudley 45497
Fowler, Martin 43784
Fowler, Sophie 44168
Fowler, Stephen 00933
Fowler, Victoria 54096
Fowler Shearer, Christine 50531
Fox, Andrew 01115, 43306
Fox, Catherine 43850
Fox, Charles 53148
Fox, Christopher D. 53506
Fox, Emily 53231
Fox, Eugene 49083
Fox, Dr. Georgia 47528
Fox, Georgia 49555
Fox, Howard 50244
Fox, Jill 45728
Fox, Julie 45871
Fox, Lisa 47139
Fox, Maureen 33128
Fox, Patric 47702
Fox, Peter 45900
Fox, Rodney 01115
Fox, Dr. Ronald 50619
Fox, Susan 43410
Foyer, Marle 13573
Fraaye, Dr. R.H.B. 31998
Fraboni, Anne 05627
Fraccaro, Prof. Marco 27125
Fradregada, Bruno 28053
Frady, Jackie L. 52287
Fraga Vila, Dr. Mª Isabel 40021
Fraipont, J. de 03900
Fraisset, M.M. 11513
Frakes, Rita 50543
Frame, Ron 47330
Framke, Dr. Gisela 17918, 17925
Framnes, Svein 33663
Franc, Karl 02806
França, Prof. Alexandre 04945
Francell, Larry 46238
Francés, Fernando 39594
Francès, Henry 12573
Frances, Mary 48359
Franceschi, Branko 08938
Franceschini, Luca 26620
Franceschini, Maria Anna 26164
Francesconi, Gino 51250
Francia, Ennio 54512
Franciolli, Marco 41581
Franciosi, Dr. Ario 25835
Francis, Dr. Carl 47182
Francis, Dermot 43976
Francis, Elaine 43689, 44187
Francis, James H. 48441
Francis, Jesse 47437
Francis, Kathy 46934
Francis, Dr. R. 44243
Francis, R.W. 45960
Francis, Rachelle 48236
Francis, Ross 06537
Francis, Tim 53244
Francis, Wayne 50498
Francisco, Lisa 50566
Francisco, Paul S. 04176
Franck, Dr. Frederick 53834
Franck, Marketta 11355
Franclieu, Comte de 15278
Franco, Barbara 53875
Franco, Leoncio Verdera 39222
Franco, Dr. Maria Nobre 35954
Francois, Gary 49891
François, Lionel 12875
Franconville 16104
Francou, Dr. Carlo 25736
Francová, Zuzana 38215
Frâncu, Simona 36155
Frandsen, Lene B. 09978, 09979, 10193, 10219, 10220, 10306, 10307, 10308
Frangos, Ioannis 23242
Frank, Claude 12583
Frank, Dr. Günter 17584
Frank, Heinrich 04648
Frank, M. 22010
Frank, Ing. Miroslav 09692
Frank, Peggy 50950
Frank, Richard J. 46883
Frank, Richard W. 48642
Frank, Stuart M. 53050
Frank, Thomas 19337
Frank, Thomas H. 19739
Frank, Mag. Veronika 02010
Franke, Bernd 18803
Franke, Daniela 03051
Franke, Dr. Ute 17203
Frankel Nathanson, Marjorie 47657
Frankenberger, Anette 20452
Frankignoul, Daniel J. 03458
Franklin, Ailesia 48210
Franklin, Dr. David 06551
Franklin, Joyce 46283
Franklin, Karen S. 47047
Franklin, Kelly 50514
Franklinsdóttir, Svanfrīðkr{}ur 24022
Franko, Blanche 06999
Franková, Božena 09838, 09876
Franković, Dr. Bernard 08849
Franks, Carolyn J. 46344
Franks, Clara 50508
Franks, Curtis J. 47357
Franks, R.J. 06399
Fransen, Linda 54224
Fransen, Martin 06736
Fransson, Elisabeth 40688
Frantz, Lewis 49798
Frantzen, Ole L. 10156
Frantzen-Heger, Gaby 30496
Franulic, Markita 09009
Franz, Dr. Corinna 16754
Franz, Dr. Erich 20512
Franze, Dr. Manfred 18039
Franzel, Frank E. 50540
Franzén, Ann 40919
Franzen, Dr. Brigitte 16402, 20512
Franzen, Darlyne S. 54287
Franzén, Eva-Lotta 40616
Franzen, Helga 19140
Franzen, Merry 48929
Franzke, Dr. Jürgen 20756
Franzolin Trevisan, Amélia 04905
Franzone, Gianni 26327
Frapporti, Giovanni 24985
Fraser, Leslee 52710
Fraser, Dr. Nicholas C. 50521
Fraser, Nonie 05589, 07032
Fratelli, Dr. Maria 26674, 26711
Frauendorfer, Helmuth 17111
Fraueneder, Dr. Hildegard 02611
Fraux, Sébastien 12188
Frayling, Prof. Christopher 44971
Frazer, John 05407
Frazer, Kay J. 49825
Frazier, Ronald F. 48023
Freaney, Linda 54284
Frechen, Anne 20704
Frechette, Myles R.R. 51117
Frechou, M. 12018
Freddi, Dr. Loris 25791
Frederici, Joan 47009
Frederick, Frances 51940
Frederick, Karen 52954
Frederick, Margaretta S. 54189
Fredericks, Marie 51216
Fredj, Jacques 14625
Freece, David W. 49723
Freed, Douglass 53023
Freed, Peter 48520
Freed, R. Mitchell 48389
Freed, Rita E. 46940
Freedman, Myron 49228
Freedman, Richard 38580
Freedman, Robert A. 47611
Freegman, David 45359
Freeland, Clara 47193
Freeland, George W. 49872
Freeman, A. 45507
Freeman, B.J. 46412
Freeman, Craig C. 49991
Freeman, D. Carl 48127
Freeman, Dottie 54096
Freeman, Dr. George 49886
Freeman, Lenette 50904
Freeman, Marina 50179
Freeman, Michael 43258
Freeman, Rusty 50963
Freemantle, Andrew 45407
Freemantle, Barbara 38665
Freer, Elene J. 05452, 05453
Freeze, Nancy 50586
Frehner, Dr. Matthias 41210, 41211
Frei, Prof. Dr. Alfred Georg 22705
Frei, Thomas 41856
Freidberg, Dr. A. 25124
Freidel, Karl 18522
Freimark, Dan 46764
Freises, M. 15735
Freitag, Dr. Anne 41543
Freitag, Matthias 21226
Freitag, Michael 18904
Freitag, Winfried 18045
Freitag-Mair, Claudia 21638, 21639, 21640, 21641
Freitas, Dr. Henrique de 04475
Freixinet, Mercè 39472
Freland, Colin 33052
Frémaux, Thierry 13902
Frembgen, Dr. Jürgen W. 20490
Fremd, T. 49776
French, Ann 45099
French, B.G. 46123
French, Brenda 43957
French, David 44828
French, Mary 47717
Freng, Prof. Martin J. Earwicker 44983
Frenkley, Helen 25061
Frennesson Westblad, Karin 40848
Frennet, Raymond 03275
Frenssen, Dr. B. 18750
Frenz, Achim 18377
Frenz, Bonnie 47605
Frenz, Dr. Hans G. 20150
Frerichs, M.A. Ilse 18141
Frerichs, Dr. Klaus 17673
Freschi Conti, Dr. Graziella 25509
Frese, Dr. Annette 19076
Fresk, Klas 40864
Fresneau, Estelle 13541, 13542
Freudenberg, Dr. Mechtild 21565
Freund, Franzika 02905
Freund, Leslie 46760
Frevert, Astrid 20636
Frew, Dr. David 48413
Frey, Barbara 47764
Frey, Dietrich 22472
Frey, Dominique 41085
Frey, Dr. Eberhard 19449
Frey, Dr. Paul 03019
Frey, Pierre 41534
Freyberg, Wolfgang 18116
Freyburger, J.-Claude 14359
Freytag, R. 17851
Friary, Donald R. 48032
Frias, Jennifer 52375
Frías Acosta, Alvaro Francisco 08791
Friberg, Gösta 40486
Friborg, Flemming 10145
Frick, William R. 50712
Fricke, Evelyn 18941
Fricke, Inge 19797, 19798
Fricke, Dr. Ronald 21959
Frickman, Linda 48609
Fridfinnsson, Jóhann 24042
Frieb, M.T. 16911
Friebel, V. 19856
Fried, S. 40894
Fried, Wendy 05664
Friedberg, Erin H. 48818
Friedberg, Rachel 49110
Friedel, Prof. Dr. Helmut 20491
Friedhofen, Barbara 16989
Friedl, Gerlinde 16742
Friedli, Michael 41112
Friedman, Dr. Alan J. 48578
Friedman, Ann 51543
Friedman, Maxine 53263, 53269
Friedman, Robert 47715, 47716
Friedrich, Hanns 16761
Friedrich, Heinz-Jürgen 22788
Friedrich, Holger 21786
Friedrich, Julia 19624
Friedrich, Dr. Sven 16951, 16954, 16965
Friedrich-Sander, Harald 19223
Friel, Dr. I. 43825, 43828
Frielinghaus, Volker 17405
Friemerding, Wolfgang 17778
Friend, Amy 51430
Frierson, Amy H. 47410
Fries, Dee 46736

Friese, Dr. Elisabeth 19508, 19509
Friesen, Steve 48956
Friesenbichler, Gerhard 02563
Friesenecker, Josef 02673
Friesinger, Prof. Dr. Herwig 02813
Frieß, Jörg 17072
Friess, Peter 52811
Friggeri, Dr. Rosanna 27520
Friis, Birte 10082, 10083
Friis-Hansen, Dana 46503
Friis Møller, Søren 10805
Frinchillucci, Dr. Gianluca 26100, 26102
Frinsko, Linda M. 48975
Frisch, Patricia L. 53140
Fritsch, Martin 17151
Fritts, Teresa 48205
Fritz, Beatrice 19077
Fritz, Dr. Ingomar 01981
Fritz, Dr. Norbert 19077
Fritz, Prof. Dr. Uwe 17969
Fritzell, Charles 53851
Fritzsche, Elmar 17574
Frizzi, Dr. Vanda 25704
Frlan, Damodar 08989, 08989
Froberville, de 12618
Froehlich, Conrad G. 47340, 47341, 47342, 47343
Frohl, Matthias 17509
Frohmann, Dr. Sieglinde 02584
Frohriep, Monika 21485
Froitzheim, Dr. Eva-Marina 17418, 17421
Frolcova, Dr. Milada 09887
Frolec, Dr. Ivo 09887
Frolova, Aleksandra Anatoljevna 36876
Frolova, T.V. 37976
From-Brown, Kristy 49441
Frommelt, Volker 30195, 30199
Frommhage-Davaz, Gudrun-Sophie 22540
Fromont, Chantal 11893
Fromont, Dr. Jane 01412
Froning, Prof. Dr. H. 20185
Frontalini, Costantino 25892
Fronty, Colette 45582
Frorud, Harald 33837, 33838
Frost, Cheryl A. 49466
Frost, Nada 53555
Frost, Natalie 45229
Frost, Sarah 17523
Frost-Pennington, Peter 45460
Frotscher, Matthias 19845
Frottier, Dr. Elisabeth 03065
Froud, Alan 00946
Frumuşelu, Teodor 36139
Fry, Judith 01308
Fry, Stephen 05949
Fryberger, Betsy 53253
Frýda, Dr. František 09717, 09720
Frye, J. Paul 51927
Frye, Todd 52185
Fryer, John 43600
Fryer, Jonathan 45647
Fryer, Molly 48485
Fu, Songliang 08356
Fu, Suanbao 08131
Fu, Dr. Tse Tak 07801
Fu, Yinan 07839
Fu, Zhongxian 08064
Fuchs, Anni 02981
Fuchs, E. 34871
Fuchs, Friedrich von 20004
Fuchs, Dr. Friedrich 21216, 21219
Fuchs, Hofrat DI Georg 02505
Fuchs, Dr. Gerald 02245
Fuchs, Dr Ingrid 02994
Fuchs, Matthias 02687
Fuchs, Max 41344
Fuchs, Monique 15831
Fuchs, Dr. Rainer 03042
Fuchs, Ronald 02412
Fuchs, Tanja 02092
Fuchs, Thomas 41481
Fuchssteiner, Ingrid 02088
Füköh, Dr. Levente 23574
Füldner, Dr. Kai 19472
Fülep, Dr. Katalin 23463
Fülepp, Ingeborg 17088
Füllner, Dr. Karin 17991
Fülöp, Dr. Gyula 23746, 23851, 23852, 23853, 23854, 23857, 23858, 23860, 23861, 23909
Fünfschilling, Sylvia 41137
Fuente Andrés, Félix de la 40269
Fuentes, Isabel 38822
Füreszné-Molnár, Anikó 23922
Fürle, Gisela 22219
Fürst, Sabine 19448
Füvessy, Anikó 23930
Fugami, Tracey 51167
Fugaro, Anthony F. 53718
Fugate, Charlotte 51570
Fugate, Dody 52881
Fugate, Gary 53688
Fugazza, Prof. Stefano 27183
Fugazzola Delpino, Maria Antonietta 27519
Fuge, Rex 00970
Fugger-Babenhausen, Markus Graf 16651
Fuglie, Gordon L. 50242
Fuglsang, Harald 10031
Fuhlbrügge, Dr. Heike 17242
Fuhr, Anette 19308
Fuhr, Dr. Michael 03022, 18328
Fuhrman, Loretta 52010
Fuhry, Eva 19533
Fujii, Kazuo 28516
Fujii, Miyako 28808
Fujii, Yoshitugu 28808
Fujii, Zenzaburou 28808
Fujisaki, Aya 28537
Fujita, Chikako 29126
Fujita, Hirohiko 29055
Fujita, Yukio 28426
Fukagawa, Masafumi 28693
Fukuda, Hiroko 28537
Fukuda, Mamoru 28610
Fukuhara, Yoshiharu 29499
Fukuji, Eiji 28501
Fulconis, Marie-Madeleine 15255, 15256, 15257
Fulfer, Glenn 50896
Fulga, Ligia 36179, 36576, 36577
Fulghum, Joyce 48774
Fuller, Benjamin A. G. 52992
Fuller, C.D. 32219
Fuller, Candy 48633
Fuller, Elaine L. 47902
Fuller, Elizabeth E. 51877
Fuller, Janae 50062
Fullerton, Jane 06838
Fulop, Lajos 36332
Fulton, Dr. R. 49323
Fulton, William 45801
Fumero, Patricia 33301
Funck, Ludwig 17587
Funk, Dr. Frank E. 54205
Funk, Linda 06971
Funk, Michael 21016
Funk, Tamara 50726
Funk, Uli 17093
Funke, Gerhard 19491
Funsho Adedayo, Oluremi 33341
Furber, Rich 52190
Furey, Linda 51425
Furger, Dr. Alex 41137
Furgol, Dr. Edward 53911
Furiesi, Dr. Alessandro 28312
Furingsten, Agne 40653
Furlong, James 51205
Furlow, Elizabeth 53005
Furman, Robert 53235
Furness, Jane 44156
Furnival, Gloria 51394
Furr, Kathlyn 53064
Furrer, Gerti 01829
Furrer, Dr. H. 42103
Furrer, Heinz 41687
Furry, Nancy 53397
Fursdon, E.D. 43706
Furshong, Peg 48510
Fursova, Irina Alekseevna 37491
Furtner, Alois 16521
Furtwengler, U. 16882
Fuscarino, Elise 53151
Fusco, Maria Antonella 27441
Fusi, Pietro 25758
Fuso, Dr. Silvio 26877, 28201
Fussey, Dr. George D. 44189
Fustig, Manuela 18632
Futivić, Vlasta 09018
Futoransky, Donna J. 48500
Futter, Christina 49695
Futter, Ellen V. 51115
Fux, Franz 01933
Fyfe, R. 33069
Fylypčuk, Dr. Olga Ivanivna 43011

G

Ga, Dr. Paolo 27299
Gaaff, G. 32047
Gaál, Dr. Attila 23865, 23869
Gaálné, Dr. Merva Mária 23566
Gaand, Jacques 03847
Gaasch, Cynnie 48778
Gabarra, Edward 51491
Gabarra, Josep Feliu 38889
Gabay, Oshra 25068
Gabbard, Dennis 48713
Gabbitas, Victoria 43956
Gaber, Vizeleutnant Josef 02786
Gabet, Francis 41550
Gabet, Oliver 14692
Gabici, Dr. Franco 27373
Gable, Edward 46853
Gabler, Dr. Josephine 20991
Gábor, Komáromi 23965
Gabra, Dr. Gwadat 10490
Gabriel, Carol A. 49138
Gabriel, Prof. Reinhold 02199
Gabriel, Siegfried 18398
Gabrielson, Bengt 40628
Gabrielsson, Carl-Eric 40882
Gabršek Prosenc, Meta 38478, 38479
Gabyševa, Asja Lvovna 36964
Gachet, Louis Jean 12372
Gachet, Louis-Jean 12375
Gačková, Dr. Lýdia 38305
Gadd, Richard W. 50803
Gaddis, Eugene R. 49199
Gaddy, Kenneth 53620
Gadebusch, Raffael D. 17201
Gądor, Józef 35065
Gäbler, Kristin 17973
Gaede, Carola 20447
Gädeke, Dr. Thomas 21566
Gähler, Walter 41847
Gaensheimer, Dr. Susanne 18385, 20491
Gärtner, Monika 02127
Gäßler, Prof. Dr. Ewald 20895
Gaete Harzenetter, Rodrigo 39415
Gafert, Alfred 19130
Gaffino Rossi, Rodolfo 28008
Gaffney, T.J. 52035
Gafner, Alain 42060
Gagarina, Dr. Elena Jurjevna 37140, 37141, 37142, 37155, 37194, 37392, 37396, 37410
Gage, Mary Margaret 46276
Gage, Mike 49037
Gagea, Pavelina 36570
Gagel, Renate 20779
Gagge, Carl-Magnus 41034
Gagnaison, Jean-Claude 15677
Gagneur, Didier 11658
Gagneux, Dominique 14646
Gagneux, Yves 14618
Gagnon, Kathryn 05723
Gagnon, Louis 06015
Gagnon, Paulette 06352
Gagro, Ivan 08986
Gaiber, Maxine 54190
Gaida, Dr. Maria 17081
Gaier, Dr. Claude 03804
Gaigalat, Michael 20809
Gaillard, Dr. Karin 32478
Gain, Prof. D.B. 38633
Gaines, Robert 48380
Gair, Michael 05380
Gais Muller, Deborah del 52170
Gaisbauer, Ernst 01739
Gaisberger, Karl 01739
Gaissert Jackson, Mary 46200
Gaitán de Santamaria, Magdalena 08633
Gaither, Barry 46942
Gajárszky, Vera 23632
Gajda, Odette 13029
Gajdošík, Martin 09692
Gajdukova, Nadežda Michajlovna 37012
Gajewski, Dr. Pawel 26636
Gajski, Zora 08997
Galan, Brenda de 03652
Galanova, Raisa Aleksandrovna 37856
Galanti Papola, Mary 51850
Galassi, Peter 51215
Galassi, Susan Grace 51178
Galasso, Prof. Elio 25394
Galasso, Giampiero 25207
Galasso, Dr. Giovanna 25389
Galasso, Prof. Maria 00361
Galati, Dr. Roberto 25405
Galbert, Antoine de 14624
Galbraith, Marie 53956
Galčenko, Sergij A. 43107
Gale, Marilyn 54090
Galea, Victoria 30660
Galeazzi, Dr. Claudio 27292
Gałecki, Zdzisław 35254
Galendo, Dr. Pedro G. 34457
Gales, Melinda 52335
Gałęza, Jarosław 34987
Galicer, Jorgelina 00446
Galicki, Piotr 34544
Galiègue, Josette 11910
Gálig, Dr. Zoltán 23904
Galili, Avner 25059
Galimberti, Dr. Paolo 26721
Galindo, Luis 54545
Galindo Rosales, Hortensia 31052
Galindo Steffens, Margarita 08539
Galizia, Anna Lisa 41195
Galkina, Inna Konstantinovna 36790
Galkina, Tatjana Ivanovna 37581
Gall, Bert 47493
Gall, Dr. Wolfgang M. 20887
Gallagher, Charles R. 51024
Gallagher, Marlys M. 46741
Gallagher, Marsha V. 51552
Gallagher, Mary 24837
Gallagher, Sheila 24837
Gallant, Aprile 51414
Gallant, Cécile 06305
Gallant, Greg 05596
Gallardo, Evonne 50263
Gallardo, Luis 31154
Gallasch, Harold E. 01140
Gallatin, Morgan Dean 53828
Gallavam, Teresa 47425
Galleano, Dr. Franco 26273
Gallego Londoño, María Elena 08735
Gallegos Franco, Francisco 31430, 31431
Gallemí, Dr. Jaume 38976
Galler, Peter 41765
Gallert, Andrea 22704
Galli, Dr. Cinzia 26004
Galli, Fiorenzo 26713
Galli, Mario 41616, 41617
Gallico, Prof. Claudio 26590
Gallie, Jim 43420
Gallifant, David W. 45972
Galliffet, P. de 13692
Gallimore, Janet 53987
Galliot, Gérard 11958
Gallivan, Dr. P.A. 01494
Galljamova, Dolaris G. 36903, 36905
Gallo, Prof. Pietro 27489
Gallois, J.L. 11577
Galloway, Dr. Ian 01510
Galloway, Mary Lou 54138
Galluzzi, Prof. Paolo 26192
Galočkina, Svetlana Jurjevna 37924
Galonska, Ann 53319
Galopim de Carvalho, Prof. A.M. 35718
Galpin, Ami 52752
Galscott, Joseph 01120
Galtier, Jacques 14448
Galusha, Emily 50741
Galuška, PhDr. Jan 09790
Galvani, Bill 49768
Galvin, John E. 50378
Galvin, Maureen 47113
Galvis, Hermana Martha Lucía 08679
Galway, Mary 51702
Galyckyj, Mikola Oleksandrovyč 43064
Gamache 12601
Gamazo Barrveco, Mercedes 39550
Gambetta Uría, Fredy 34311
Gambin, Kenneth 30655
Gambiroza, Roland 09013
Gamboa, Lyn 34505
Gamboa, Marilyn B. 34337
Gameiro, Dr. Ana Cristina 35502
Gamet, Prosper 16009
Gammersvik, Ågot 33502, 33637, 33949
Gamolin, Vladimir Danilovič 37522
Gamperer, Franz 02366
Gamsjäger, Rudolf 02049
Gamsjäger, Siegfried 02462
Gamwell, Lynn 46817
Gamzatova, Patimat Saidovna 37115, 37116
Gan, Jianbo 08507
Gancarski, Jan 34830
Gandee, Cynthia 53433
Gandhi, K.M. 24384
Gandhi, M. 24314, 24428
Gandioso, Carmen Roman 04927
Gandon, Gérard 11548
Gandy, Doris 49204
Gane, Constance 46767
Ganebnaja, Nina Evgenjevna 36884
Gangi, Sabine 16133
Gangkofner, M. 22838
Gangloff, Roland 48472
Gangopadhyay, Paula 49007
Gangwere, Stanley K. 48127
Ganiev, A. 54462
Ganijon, Alijonov 54449
Ganino, Luigi 25248
Ganje, Melissa 48374
Gannett, Michael R. 47813
Gannon, Thomas 48198
Ganong, Dr. Overton G. 52523
Ganozova, Radoslava 05183
Ganpat, Ganesh 46252
Gansz, Kevin 53122
Gant, Bob 47444
Gantenberg, Anette 20809
Gantner, Bernard 13337
Gantt, Judy M. 46440
Gantt, Nancy M. 53032
Ganz, James A. 54175
Ganzelewski, Dr. Michael 17398
Ganzer, Dr. Gilberto 27297, 27298
Ganzkow, Martin 19251

Gao, Prof. Chong-Wen 07505
Gao, Hua 07858
Gao, Jun 08212
Gao, Liren 07853
Gao, Maojie 08465
Gao, Qiang 08319
Gao, Xia 08363
Gao, Xianzhong 07634
Gao, Zenglin 08289
Gao, Zongyu 07916
Gappmayer, Helga 02371
Garanne Modise, Rogoff 04301
Garaudy, Roger 39209
Garavaglia, Gianluigi 26016
Garavaglia, Giuseppe 26672
Garay, Nora Cecilia 08533
Gârbacea, Adriana Rodica 36484
Garbarini Islas, Dr. Guillermo 00274
Garbe, Dr. Detlef 18943
Garbes, Pam 33134
Garbrecht, Dr. Jörg 17229
Garbutt, Kirstie 43487
Garces, Donna M.P. 34401
Gárces Julve, Vicente 39225
Garcia, Anna 46830
Garcia, Damian 52195
Garcia, Denis 16065
Garcia, Elizabeth F. 50365
García, Dr. Eulàlia 38976
García, Joaquín Emilio 08743
Garcia, José 50631
Garcia, Julian 39079
García, Lola 40204
Garcia, Magdalena A. 51553
Garcia, Miki 52857
Garcia, Noah 52715
Garcia, Rene R. 34514
García, Susana 40342
Garcia, Susana 40345
Garcia-Anton, Katya 41404
García Arévalo, Manuel Antonio 10360
García Chacón, Daniela 04209
García de Del Valle, Prof. Delia 54584
García de la Rosa, Ignacio 39440
Garcia Forner, Anna 39074
García-Frias Checa, Carmen 40246
García Fuertes, Pilar 40285
García Limón, Fernando 31284
García López, Dr. J.R. 39352
Garcia-Lorca de los Ríos, Laura 39362
García Martinez, Carlos 40026
García Menárguez, Antonio 39379
García Moreno, Beatriz 08556
García Mota, Francisco 39597
García Muriano, Felipe O. 39025
Garcia-Nakata, Lorraine 52783
García Quiros, Paz 39337
García Rozas, Rosario 40335
García Ruíz, María Esther 30762
Garcia Serrano, Rafael 40095
García Serrano, Rafael 40125, 40131, 40133, 40134
García Soto, Rubén 34287
García-Tuñon, Guillermo 39339
García Vanegas, Rubén Darío 08683
García Velasco, Francisco 39281
Garcías y Truyols, Joan 38899
Gard, Jean-Noël 12164, 14437, 14724, 14919, 15069, 15072, 15929
Gardent, Bruno 13381
Gardi Dupré, Maria Grazia 26128
Gardiner, Mary N. 46388
Gardner, David 01420
Gardner, John 49565
Gardner jr., William F. 50376
Gardner Broske, Janet 51282, 51283
Gardner Gates, Mimi 53013, 53014
Garduño Pulido, Blanca 31060, 31113
Garduño Ruíz, Dr. Everardo 31031
Garey, Rose 53396
Garfield, Leonard 53005
Gari Lacruz, Dr. Angel 38782
Garibaldi, Patrizia 26296, 26310, 26317
Gariboldi, Leopoldo 26864
Garifdžanova, Roza G. 36867
Garimorth, Julia 14646
Garland, Jeff 49092
Garland, Joan 50396
Garland, Kathleen 49695
Garland, Steve 43553
Garlinski, Majan 41421
Garlits, Donald G. 51483
Garncarzová, Hana 09445
Garner, I.I.D. 44645
Garner, Jennifer 48076
Garner, Les 50891
Garner, Sheila 01606
Garnham, Peter 06973
Garniche, Marie José 13339
Garnier 15350
Garnier, J.F. 16221
Garnier, Laurent 12525
Garnier, Maurice 14633
Garnier, Nicole 12399
Garnier, Renée 13018
Garnier, Sylvie 12716
Garniewicz, Rex 49497
Garrels, Gary 51215
Garret, Patty 47165
Garrett, Anne 06994
Garrett, Jim 50330
Garrett, Jural J. 50257
Garrett, Rose M. 46769
Garrido, Emilio M. 34442
Garriga, Francesc 39981
Garrigan, Curt 50969
Garrigue, Philippe 15595
Garrison, Elizabeth 52086
Garrity, Lois 07079
Garrott, Martha 52349
Garrut, Dr. J.M. 38953
Garside, Sioux 01563, 01565
Garside, Vivienne 38617
Garstenauer, Karl 02573
Garton, Diane 06575
Garua, Marina Prieto 38813
Garuti, Alfonso 25685
Garweg, Udo 22748
Garwood, Susan 48512
Gary, Grace 54196
Garza, Michael A. 48650
Garza Camino, Dr. Mercedes de la 31129
Gasc, Michel 14126
Gasco, Yves 11887
Gascon, France 06030
Gash, Lyn 55075
Gąsior, Stanisław 35010
Gąska, Wiesław 35095
Gaskell, Ivan 47177
Gaspar, Burghard 01844
Gaspar, Filorena 35383
Gaspari, Prof. Gianmarco 26710
Gaspari, Kristin H. 54072
Gasparini, Gauchesco Luis 00565
Gašparović, Miroslav 09008, 09015
Gasperment, Dr. Philippe 15565
Gass, Monika 19214
Gasser, Dr. Christoph 25880
Gasser, Ferdinand 25881
Gasser, Dr. Martin 42026
Gasset y Salafranca, Josep M. 39473
Gaßner, Prof. Dr. Hubertus 18932
Gassner, Therese 03078
Gast, Dr. Otto 20764
Gastal, Ney 04645
Gastaut, Amélie 14662
Gaston, Elizabeth 50716
Gastou, François-Régis 15935
Gatchell, Melanie 49014
Gateau, Fabienne 13737
Gatellier, Michel 12971
Gately, Barbara 47702
Gatenbröcker, Dr. Silke 17522
Gates, Charles 46492
Gates, Daphnea 33131
Gates, David 43724
Gates, Jay 53914
Gates, Jim 47791
Gates, Sue 46116
Gates, William 47757
Gatewood, Bill 50141
Gattemair-Farthofer, Ingrid 18444
Gatti, Dr. Sandra 27081
Gau, Sönke 42115
Gaubert, Prof. Jean 31613
Gaubert, Philippe 13418
Gaudet-Blavignac, Richard 41309
Gaudiau, Edmond 12838
Gaudichon, Bruno 15111
Gaudio, Jennifer 50333
Gaudlitz, Winfried 21560
Gaudnek, Prof. Dr. Walter 16523
Gaudzinski-Windheuser, Dr. Sabine 20150
Gaudzinski-Windheuser, Prof. Dr. Sabine 20687
Gauffin, Sten 40796
Gaugele, Roland 18676
Gauhar, Hasan 34039
Gaul, Roland 30465
Gauld, Gail 00911
Gauldine, K.C. 52997
Gaulke, Dr. Karsten 19458
Gault, Sam 43766
Gaulupeau, Yves 15121
Gaumé, Jean-Charles 14303
Gausling, Helga 16436
Gaustad, Randi 33730
Gauthier, Benoît 07178
Gauthier, Catherine 15679
Gauthier, Jacques A. 51054
Gauthier, Virgile 04049, 04050
Gautier, Jean 04034
Gautschi, Karl 41665
Gava, Hsueh 42353
Gavagan, Mary 44635
Gavazzoci, Dr. Maria Laura Tomea 26954
Gavin, Glenda 53512
Gavin, John 47127
Gavín Arnal, Alfredo 39444
Gaviria, Alba 08605
Gavish, Galia 24994
Gavran, Dr. Ignacije 04292
Gavril, Victoria 36208
Gavrilaki, Eirini 22961
Gavrilova, Marija Jurjevna 37461
Gavrilović, Miladin 38157
Gawlick, Henry 18862
Gay, Dr. Hebe Dina 00326
Gay, Millicent 52431
Gay, Pierre-Raymond 12568
Gayler, Billie Sue 53606
Gayot, Lise 14680
Gayraud, Gilbert 14950
Gazengel, Alcide 14121
Gazlay, Cheryl 51428
Gazzaway, Ofiera 54292
Ge, Daigui 08438
Gea Calatayud, Manuel de 39910
Gealt, Adelheid M. 46864
Gear, Dr. Gilian 44792
Gearinger, Robert E. 48758
Geary, Adam 44579
Gebert, Walter 22155
Gebhardt, Hans 16950
Gebhardt, Matt 52361, 52362
Gebhardt, Dr. Ute 19449
Gebühr, Dr. Michael 21565
Gečeva, Magdalena 05225
Geddes, Jim 33111
Geddes, Victoria J. 44511
Geddry Sr., Mike 52895
Geden, Dennis 06471
Gedeon, Lucinda H. 53728
Gedgaudienė, Eleonora 30448
Gedvilas, Algirdas 30395
Gee, Lisa 46076
Geens, Catherine 39519
Geerdinck, J.J.M. 32501
Geerken, B. 32485
Geeroms, Luc 03259
Geerts, Dr. T. 32213
Geeslin, Phyllis 49506
Gefahun, Dr. Abebe 10682
Geffers, Dr. Andrea 20019
Gehör, Dr. Seppo 11174
Gehri, Emil 41456, 41458
Gehrig, Bruno 41235
Gehrmann, Rudi 17394
Gehrt, Gertrude 46235
Geibig, Dr. Alfred 17717
Geidner, Oskar 22656
Geier, Bob 51502
Geier, Katrin 18929
Geiger 20829
Geiger, Gisela 21008
Geiger, H. 19719, 19720, 19721
Geiger, Malies 19366
Geiger, Maurice 12138
Geiger, Uwe 18042
Geiges, Michael 42089
Geil, Bernd 19767
Geiselman, Anette 48606
Geisenhofer, Paul 21376
Geiser, Anne 41549
Geisler, David A. 50910
Geisler, John 50910
Geisler, Dr. Jonathan 53272
Geisler, Jürgen 17812
Geißel, Helmut 22587
Geisser, Dr. Hannes 41379
Geisser, Hans 41121
Geissl, Gerhard 03096
Geißler, G. 21199
Geissler, Josef 02448
Geißler, Ramona 21330
Geist, Dr. Joseph E. 48525
Geitner, Mary Elizabeth 49267
Gelao, Clara 25371
Gelbrich, Gabriele 19850
Gelburd, Dr. Gail 50016
Geldenhuys, M. 38566
Geldin, Sherri 47761
Gelfi, Yves 14098
Gelin, Pär 40424
Gélinas, Cécile 06962
Gelis, Joseph 03560
Gelius, Dr. William 10155
Geller-Grimm, Fritz 22557
Gelo, Josip 04267
Gelsthorpe, D. 45090
Gelszinnis, Hans Joachim 21356
Gelzer, Bev 51359
Gembroll, Robert 49562
Gemmell, Michael 07012
Gemmill, Moira 45004
Gemmingen, Freiherr von 19032
Gencer, Necdet 42880
Gencsi, Zoltán 23619
Gendre, Cathérine 16128
Gendron, Christian 14455, 14457, 14458
Generalli, Genevieve 47648
Genev, Hristo 05056
Genger, Angela 18001
Gengler, Guy 30483
Gennari, Massimo 25814
Geno, Jaiy 54001
Genorio, Rado 38470
Genoud, Jean-Claude 41548
Gens, María L. Alonso 39688
Gentele, Glen P. 52584
Gentet, Alfred 41793
Gentil, M. Bruno 14617
Gentile, Dr. Gerardo 27162
Gentili, Dr. Elio 28177
Gentili, Lamberto 27891, 27895
Gentis, Thierry 47033
Gentle, Bill 47445, 48158, 48600
Gentry, Nola 46646
Gentzel, Heiko 18094
Genys, Dr. Jonas 30283
Geoffroy, Pierre 15899, 15900
Geominy, Dr. Wilfred 17442
Geoppa, Judith 47966
Georg, Rolf 21811
George, Dave 52711
George, Dr. Donny 24715
George, Elizabeth 50283, 50292
George, Hardy 51521
George, Kerrianne 01566
George, M.R. 01224
George, Prof. Ron 45909
George, Sarah B. 52696
George, Tamsen E. 46949
George, Terri 48202
Georgel, Chantal 14692
Georgescu, Valeriu 36465
Georgeson, Melinda 53302
Georget, Luc 14002
Georghiades, A. 09375
Georghiou, G. 09375
Georgieva, Elene 05015
Georgieva, Tanička 05154
Georgsson, Agust 24036
Geppert, K. 21421, 21422
Geraci, Ester 25952
Gérard, Peter 22430
Gerasimčuk, Olga 37707
Gerasimidi, Elena Ivanovna 36785
Gerbauckas, Maryanne 54071
Gerber, Jan 47865
Gerber, Olive 53063
Gerber, Dr. Peter R. 42116
Gerber, Theodore 46921
Gerchow, Dr. Jan 18377
Gercke, Irmgard 16402
Gerdes, Edward 48438
Gerdes, Peggy 53015
Gerdur-Gunnardóttir, Gudny 24014
Gereau, Rebecca 48934
Gergel, Belinda 47727
Gergely, Dr. Katalin 23409
Gerhards, Dr. Eva 18436
Gerhardt, Barbara 41485
Gerhardt, Dr. Gunther 22316
Gerhardy, Wilhelm 18037
Gerharz, Michael 21180
Gerichhausen, Heinz 22359
Gerini, Fiorenzo 25177
Gerke, Lucien 04137
Gerken, Dr. Thomas 17469
Germain, Christiane 15732
Germain, P. 12786
Germann, Edwin 41467
Germann-Bauer, Dr. Peter 21224, 21226, 21231
Germano, Anna 28171
Germanova, Galina 05191
Germiny, Comte Gabriel de 14904
Gerner, Dr. Cornelia 17133
Gerrard, Olga 50252
Gerrietts, Frank 46673
Gerritsen, J. C. 31970
Gershon, Stacey 51146
Gershuny, Justin 50229
Gerstein, Dr. Alexandra 44830
Gerstheimer, Christian 48321
Gertjejansen, Doyle 51079
Gertsch, Fritz 41552
Gertsch, Liliane 41796
Gervais, Alex 05316
Gervais, Louise 47912
Geschke, Marcus 17198
Gesemann, Dr. Björn 20150

Gęsicki, Waldemar 34570
Gesing, Dr. Martin 16974
Gesquière, Dominique 13613, 14069
Gess, Dr. F. W. 38629
Gessener, Andrew 51162
Gesser, Susanne 18377, 18380
Gessie, Dwight 07076
Gestin, Jean-Pierre 15509
Gestrin, Tryggve 10727
Gestring, Ron 52271
Gestsson, Magnús 23991
Getchell, Katherine 46940
Gether, Christian 10108
Getman, Z.D. 37695
Getson, Ralph 06202
Gette, Timothy J. 50521
Gettinger, Maria 02224
Gettler, Peter 20893
Geurts, Joe 05862, 06539
Geus, Prof. Dr. Armin 20598
Geuze, M.A. 32771
Gevas, Sophia 48479
Gex, Catherine 41983
Geyer, Prof. Dr. A. 19375
Geyer, Prof. Dr. Angelika 19389
Geyer, Jeannette 38639
Geyer, Nancy 46955
Geyer, Dr. Roderich 02834
Gfeller, Irma E. 50116
Gfrereis, Dr. Heike 20179, 20181
Ghaeni, Farzaneh 24654
Ghambaryan, Armida 00692
Ghandgar, J. 24650
Gharrity, Mary Lou 50686
Ghedini, Prof. E. Francesca 27042
Gherardi, Dr. Pierluigi 27205
Ghesquières, L.-E. 06601
Ghez, Prof. Claude 41429
Ghez, Susanne 47510
Ghiara, Prof. Maria Rosaria 26913
Ghidini, Elena 26959, 26960
Ghini, Dr. Giuseppina 26918
Ghini, John 50406
Ghiorsi-Hart, Carol 53134
Ghiotto, Dr. Roberto 26816
Ghirlanda, Silvia 41279
Ghiron, Yves 13932
Ghislain, Bourdon 04065
Ghiță, Ioana 36247
Ghoniem, Atef 10499
Ghorbani, M. 24702
Ghosh, Indrani 24280
Giachino, Dr. Roberto Sandri 28022
Giacometti, Emmanuelle 41537
Giacomini, Dr. Giovanni 25740
Giacomini, Dr. Pietro 25888
Giaimo, Catherine L. 51851
Giaj, Ezio 27223
Giamarchi, Paulette 13440
Giambuti, Lia 28277
Giampaolo, Francesca 26497
Gianadda, Léonard 41608, 41611, 41612
Gianaroli, Onofrio 25475
Gianelli, Ida 27423
Gianfranco, Armando 25521
Gianfreda, Dr. Sandra 42029
Gianmoena, Willi 20916
Giannetti, Prof. Antonio 27414, 27415, 27419
Giannetti, Elio 27004
Gianni, Dr. Ginsberti 25459
Giannoulia, Eirini 23019
Gianora, Giulietta 41859
Giaquinto, Michael 48057
Giarmoleo, Giuseppe 27516
Gibb Roff, Samantha 48193
Gibbard, Dr. Russell 33169
Gibbon, Imogen 44157
Gibbon, Wilda V. 07009
Gibbons, Cindy 06705
Gibbons, Michael L. 46548
Gibbs, Ivor 45520
Gibbs, Margaret 48340
Gibbs, Marie 24894
Gibbs, Maurice E. 50947
Gibbs, Roger 49861
Gibernau, Dr. Jesús Gil 40144
Gibert-Balboni, Françoise 16003
Gibson, Charlotte 53571
Gibson, Hugh 43969
Gibson, Keith E. 50075, 51072
Gibson, Ken 50397
Gibson, Paula 43234
Gibson, Ralph 54153
Gibson Garvey, Susan 05933
Gică, Secție-Bâeștean 36588
Gichere, Hjuguna 29745
Giebelhaus, L. 07235
Gieben, Servus 27507
Gier, Dr. Helmut 16613, 16635
Gier, Dr. Klaus 18915
Giere, Prof. em. Dr. Wolfgang 18366
Gierisch, Nita 47847
Gierkinger, Gerda 01696
Gieschen, Margarete 18665
Giese, Dr. Dale F. 53104
Giesen, Alison 49953
Giesenschlag, Bill 47164
Giesert, Ted 49784
Giffault, Michèle 11734
Gigandet, Eric 41749
Giglierano, Geoff 53957
Giglio, Alessia 41252, 41583
Gigou, Laure 13706, 14808
Giguet, Aurore 49979
Gil, Dr. Euzebiusz 34837
Gil de los Reyes, Soledad 39131
Gil de Prado, Eva 39446
Gil Imaz, Maria Cristina 40347
Gil Rodriggez, Juan Juis 39084
Gilardi Polar, Hernando 08550
Gilberger, Ruth 18027
Gilbert, André 06699
Gilbert, Barbara 50264
Gilbert, Courtney 49754
Gilbert, David 45518
Gilbert, Estelle 45958
Gilbert, Frank 53887
Gilbert, Jennifer 50306
Gilbert, Sylvie 06343
Gilbertson, Elsa 46136
Gilbertson, Laurann 48022
Gilch, Eva 17656
Gilchrist, Rod 52769
Giles, Mel 06344
Giles, Sue 43632
Gilfillan, Richard M. 54027
Gilg, Karen 46484
Gilgen Hamisultane, Ruth 41210
Gili, Antonio 41284, 41585
Gill, Angela 43654
Gill, B. 33043
Gill, Cecil 50639
Gill, J. D. 43383
Gill, Manfred 22650
Gill, Peter 01341
Gill, Sandra 06784
Gill, Stephen J. 30180
Gill, Vince 50964
Gill, Wendy 44274
Gille, Martha 31754
Gillerist, Christopher H. 53718
Gilles, Dr. Karl-Josef 22102
Gillespie, Gordon 52155
Gillespie, Margaret 47562
Gillespie, Maria 43559
Gillespie, Marilyn 49973
Gillespie, Rosemarie G. 46753
Gillespie, Sarah 24763
Gillespie, Shane 24839
Gillett, Marnie 52792
Gillette, David D. 48551
Gilliam, Jan 54164
Gilliam-Beckett, Elizabeth 49713, 50079
Gilliand, Michel 41719
Gillieaux, L. 03475
Gillies, A. 46013
Gillies, Scott 06979
Gillis, Jean 52839
Gillis, Nancy S. 46592
Gillis, Stuart 44366
Gillman, Derek 51864
Gillon, Patrick 16291
Gilman, Robert G. 49243
Gilman, Sondra 51279
Gilmore, Jerry 46957
Gilmore-House, Gloria 50202
Gilmour, Rosemary 43825, 43828
Gilpin, Melinda 50487
Giltaij, J. 32729
Gimenez, Carmen 51258
Giménez Carrazana, Dr. Manuel 04243, 04245
Gimeno Salueña, Joaquín 39313
Gimmi, Karin 42094
Ginalski, Jerzy 35100
Ginette, Eric 47156
Ginger, Greg 46706
Ginlund, Hans 10718
Ginter, Donald W. 51393
Ginton, Ellen 25122
Gioacchini, Delfo 27000
Giobbi, Maria Paola 25297
Giombini, Dr. L. 25909
Gionio, Maria Teresa 26720
Giordano, Prof. Christian 41385
Giordano, Claudio 25665
Giorgetti Franscini, Pia 41582
Giorgi, Dr. Angelo 25528
Giorgi, Benedetta 41581
Giorgi, Dr. Marisa 27512
Giorolano, Fabris 28074
Giovannoni, Günther 41283
Giovetti, Dr. Paola 25447
Giovine, Dr. Francesco Paolo 28169
Gipson Adams, Shelley 52138
Giradeau, Ted 05531
Girard, Alain 11825, 11826, 14893, 14894, 16214
Girardin, Daniel 41547
Girardon, Eric 12730
Giraud, Milou 14774
Girgenti, Cecilia 00144
Gironde, Herberto 00294
Girouard, Terry 47707
Giroud, Jean-Charles 41401
Giroud, Magali 15611
Girousse, Laurent 11908
Giroux, Patrice 06111
Girsberger, Jürg 41341
Girst, Roswitha 18376
Girvan-Brown, Chris 01092
Girveau, Bruno 14590
Gisbert Santonja, Josep Antoni 39244
Gísladóttir, Ilmur Dögg 24027
Gisler, H. 41877
Gisler-Jauch, Dr. Rolf 41104
Gittelman, Steve 47315
Giubilei, Maria Flora 26289
Giubilei, Dr. Maria Flora 26316, 26326
Giuliano, Prof. Carlo 28030
Giuliano, Dr. Laura 27512
Giulierini, Dr. Paolo 25988
Giuntini, Trey 50580
Giurescu, Dinu C. 36227, 36403
Giuricco, Judith 51262
Giusti, Dr. Annamaria 26177
Giusti, Carla 26909
Giusti, Dr. Giovanna 26157
Givens, Stefanie 53418
Givnish, Gerry 51863
Gizatulin, Marat R. 37541
Gizzi, Dr. Federico 28035
Gjaurov, Toško 04975
Gjedsted, Kirsten 10171, 10172, 10173
Gjerdi, Trond 33720
Gjipali, Ilir 00031
Gjmacan 40113
Gjurinovic Canevaro, Pedro 34231
Gladstone, Prof. M. 24117
Gläßel, Adolf 22303
Glaister, Jane 44550
Glaizot, Dr. Olivier 41543
Glanzer, Harvey W. 52308
Glanzer, Michelle 47060
Glaser, Prof. Dr. Franz 01935, 02186, 02283
Glaser, Dr. Michael 21457
Glaser-Schnabel, Dr. Silvia 20764
Glasgow, Istiharoh 50211
Glaskova, Vera Evgenjevna 37935
Glass, Brent D. 48424, 53516, 53901
Glass, Catherine 49935
Glass, Christian 22160
Glass, Frank 06775
Glass, Jennifer 48702
Glass, Patrice 47407
Glassel, Jim 48583
Glasser, Dale 51275
Glasser, Susan 52328, 52608
Glasser, W. 01580
Glassman, Elizabeth 13068, 47518
Glassman, Susan 51888
Glassmoyer, Jeniffer 50191
Glassner, Prof. Dr. Gottfried 02375
Glasson, Michael 45921
Glasswell, Samantha 45484
Glaubinger, Jane 47624
Glauser, Daniel 41797
Glauser, Julien 41681
Glavočić, Daina 08938
Glaze, George 46847
Gleason, Mat 55126
Gleaton, Sonja 50999
Gleeson, Elizabeth 01034
Gleeson, Val 01628
Gleffe, Waltraud 16550
Glehr, Dr. Reinhold 02053
Gleirscher, Dr. Paul 02186
Gleissner, Stephen 54150
Gleixner, Johann 22261
Gleizer, Raquel 04908
Glenholme, Ian 44667
Glenn, Rosalyn 53456
Glenn, Sherrain 49904
Glevenko, Elena Jurjevna 37661
Glick, Dr. Shmuel 25014
Glidden, Joy 47072
Gligor, Ierusalima 36696
Gliozheni, Lorenc 00021
Glistak, John 01461
Gliszczyńska, Elżbieta 34535
Gliwitzky, Dr. Christian 20486
Glock, Walter 21326
Glocker, Dr. Jürgen 17470, 18730, 21490
Glöckner, Prof. Harry 02961
Gloor, Dr. Lukas 42108
Gloria, Dr. Heidi K. 34385
Gloria, Rogelia 34415
Gloßner, F.X. 20633
Glósz, Dr. József 23869
Glot, Claudine 12605
Glotzhober, Robert 47757
Gloudemans, M. 32612
Glover, Jack 53375
Glover, Mike 43693
Glover, Sarah Elizabeth 05404
Glover, Susan 46923
Glover, Thomas 49135
Glover-Wilson, Mary Beth 53073
Gloza-Rausch, Florian 16837
Gluchova, Tatjana Vladimirovna 36956
Gluck, Daniel 51228
Glüber, Dr. Wolfgang 17789
Glückert, Ewald 19852
Glueckert, Stephen 50754
Glušacov, Viktor Ivanovyč 43160
Glušenkova, Ljudmila Nikolaevna 36825
Gluskova, Marina 37040
Głyda, Krzysztof 35042
Glynne-Percy, Colin 45514
Gmelch, Christa 21287
Gnecco, Prof. María Julia 00588
Gnidovec, Dale 47758
Gnilšak, Ida 38500
Go, Sugil 29788
Gobbi, Elmar 26655, 26657
Gobel, Mary 49663
Goberna, María Victoria 40201
Goble, Carole 47145
Goble, Rebecca 48018
Gochenour, Anne 50138
Gockel-Böhner, Christoph 20972
Gockov, Ilija 04990
Gócsáné-Móró, Csilla 23911
Gocz, Teodor 35203
Godart, Freddy 03530
Godbey, Martin 52915
Godbille, Lara 52034
Godbout, Dominique 06115
Goddard, Stephen 49992
Godefroy, Hubert 15373
Godelet, Andre 03724
Godfrey, Honor 45017
Godfrey, Judith 46124
Godfrey, Laurie 46269
Godgenger, Jean-Marc 12924
Godoy, Arnaldo Augusto 04350
Godoy, Eliana 51123
Godoy, José 41418
Godoy Muñoz, Maria Antonieta 23291
Godtsenhoven, Karen van 03294
Godwin, Carmen 48535
Godwin, Mary 45421
Godwin, Sharon 07080
Goebel, Susanne 16452, 16454, 16455, 16456, 16458, 16459, 16460
Göbel, Prof. Dr. Dr. Ulf 17239
Goebel, Wolf 18779
Göçmen, Ali Bülent 42968
Goedbloed, J. 31787
Goedicke, Dr. Christian 17238
Goehring, Alicia L. 49852
Gökten, Serdar 42700
Göldner, Andrea 22444
Göpfrich, Jutta 20881
Görcke, Bärbel 20664
Gördüren, Petra 19531
Görgner, Ernst 17830
Goerke, Udo 18951
Görlich, Manfred 03088
Görner, Dr. Irina 19479
Görner, Klaus 18385
Görner, Dr. Rolf 22590
Görner, Dr. Veit 18994
Goertzen, Peter 05900
Goes, B.R. 31880
Goes, Dr. Georg 16932
Goessen, J. 32534
Gösseringer, Irene 02103
Gößwald, Udo 17212
Göthberg, Elisabeth 40738
Goethert, Dr. Klaus-Peter 22101
Götte, Dr. Gisela 17560
Göttner, Dr. Jens 18898
Götz, Angela 20786
Götz, Matthias 41171
Götz, Reinhard 02817
Götz, Rudolf 41995
Goetz, Scott 53178
Götz, Dr. Ulrike 18451
Götze, Barbara 17283

G

G

Götze, Dr. Bettina 21190
Goetze, Lisel 48426
Götze, Robby Joachim 18657
Goetzman, Richard 49538
Goff, Nick 45981
Goffe, Gwendolyn H. 49397
Goffinet, Bernard 03847
Goffman Cutler, Judy 51324
Goguen, Georges 06316
Goheen, Marion 52306
Gohlke, Gerrit 21084
Goj, Margarete 21228
Golay, Laurent 41548
Golay, Muriel 41863
Golbin, Pamela 14654
Gold, Dr. Helmut 18383
Gold, Jack A. 52776
Gold, Jack Anthony 52175
Gold, Ruth C. 53146
Gold-Levi, Vicki 46460
Goldbacher, Alexander 02940
Goldberg, Beth 52827
Goldberg, Beverly 53656
Goldberg, Elizabeth 53038
Goldberg, Marsea 54065
Goldberg, Paul 53267
Goldberg, Shalom 46886
Goldblum, Bonnie 51085
Golden, John 54203
Golden, Rorey 51141
Golden, Thelma 51264
Goldenkova, Vera Anatoljevna 37745
Goldenvejzer, Elena Ivanovna 37317
Golder, Sheri 47809
Goldfarb, Hilliard 06357
Goldfinger, I. 24951
Goldfinger, Yaffa 25122
Goldfuhs, Ingeborg 01841
Golding, Norman 33189
Goldman, Steven 51594
Goldnagl, Friedrich 01819
Goldner, George R. 51212
Goldowsky, Alexander 54297
Goldschmidt, Lawrence P. 53909
Goldsmith, Jeremy 45738
Goldsmith, John 44500
Goldsmith, Robert B. 51178
Goldstein, Ann 31888
Goldstein, Gabriel M. 51281
Goldstein, Jon 53309
Goldstein, Lisa 48392
Goldstein, Sid 52595
Goldwasser, Edwin 53902
Goleas, Janet 48251
Goley, Mary Anne 53868
Golf, Anamari 49602, 49603
Golias, Allison 46676
Golikova, Ljudmyla Vasylivna 43179
Golinelli, Jean 16094
Golinski, Dr. Hans Günter 17401, 19036, 19039, 19040, 19042
Golisek, Lisa 49656
Goll, Dr. Jürg 41288
Goll, Dr. Roland 01669
Golloway, Zoe C. 52212
Golofast, Larisa 43178
Gologranc Zakonjšek, Bronislava 38406
Golonka, Jan 34617
Golosovska, A. 22712
Golovčenko, Ljudmila 43178
Golshan, J. 24672
Goltz, R. von der 17174
Gołubiew, Zofia 34805
Gomá Lanzón, Javier 39520
Gombert, Florence 13756
Gombotz, Christian 01693
Gomel, Luc 14452
Gomendio Kindelán, Montserrat 39571
Gomes, José Antonio Alves 04553
Gomes de Pinho, António 35848
Gomes Leite, Gisélia Antonia 04800, 04801
Gómez, Arley 08668
Gomez, Carla 52873
Gómez, Carlo Humberto 08570
Gómez, José Luis 00357
Gomez, Nadine 12727
Gómez, Roque Manuel 00465, 00558, 00559
Gomez, Victor 48315
Gómez-Alba, Dr. Julio 38976
Gómez Arnanz, Miguel 38926
Gómez Gómez, Alicia 39530
Gómez Hildawa, Isidro 34471, 34474
Gómez Macías, Juan Antonio 39970
Gomig, Dr. Leo 01823
Gomis, Joan Carles 39778
Gomozkova, Marina 37195
Gompertz, P. 45735
Gomringer, Prof. Eugen 21238
Gomringer, Dr. Nortrud 21238
Gonçalves, Cristina Martins 35466
Goncalves, Fabian 53848
Gonçalves, Maria Manuel 35802
Gonçalves, Prof. Dr. Maria Teresa Serra Silva 35530
Gonçalves, Vereador Renato 35778
Gončar, Petro Ivanovyč 43091
Gončarenko, Larisa Fedorivna 43048
Gončarova, Elena Alekseevna 36944
Goncerut Estèbe, Véronique 41400
Gong, Dazhong 07983
Gong, Ping 08256
Gong, Stephen 46752
Gong, Xiaoan 08371
Gong, Yunxue 08241
Gonos, Arthur 48694
Gonseth, Marc-Olivier 41681
Gonzaga Bastidas Ortiz, Rodrigo 08641
Gonzales, Ben 34468
Gonzales, Michael E. 51513
Gonzales Baldoui, Mariano 40327
Gonzales-Belleza, Erlinda 34402
Gonzalez, Anita 52624
Gonzalez, Arturo Gonzalez 31315
González, Aurelio 39449
Gonzalez, Christine 12982
González, Doña Ligia 33298
Gonzalez, Gilbert 48788
González, Jaime 08764
González, Jorge 08543
Gonzalez, Dr. Juan G. 36043
González, Julieta 54543
Gonzalez, Manuel 51146
González, Milagros 54543
Gonzalez, Sylvie 15236
Gonzalez, Victor 36076
González Alonso, Alberto 39550
González Amezúa, Mercedes 39547
González Antón, Dr. Rafael 39994
González Aranda, Lucilia 08599
Gonzalez Azcoaga, Miguel Fernando 00348
González Cevallos, María A. 00179
Gonzalez-Cid, Malena 52714
González de Aspuru Hidalgo, Sara 40322
Gonzalez de Durana, Javier 40317
González de León, Antonio 31059
Gonzalez de Santiago, Ignacio María 39069
González Entidad, María Inocencia 08741
Gonzaléz Gonzàlez, Ignacio 31454
Gonzalez Gonzalez, Mildgros 39618
González Gozalo, Dr. Elvira 39279
González-Hontoria y Allendesalazar, Guadalupe 39534
González Lorente, Margarita 09098
González Padrón, Antonio María 40112
González Parra, Samuel 08536
González Penagos, Claudio 08525
Gonzalez Pons, Esteban 39405
González Rodríguez, Rosalía 39426
González Romero, Santiago 39963
González-Sancho, Eva 12734
Gonzalo Andrade, Luis 08775
Good, Les 43632
Good, Liz 51092
Good, Patricia 50085
Goodale, Thomas 48452
Goodband, Lara 46108
Goodchild, Anne 45651
Goode, Dr. David 07022
Goodell, Greg 48902
Goodger, Margaret 43695
Goodin, Kitty 49655
Gooding, Brian 44724
Goodlett, Chris 50287
Goodman, A. 05455
Goodman, Barbara 49578
Goodman, Carl 46399
Goodman, Gwen 51861
Goodman, Norman 46008
Goodman, Susan 51200
Goodman, Wallace 53500
Goodnight, Gwenyth 51636
Goodrich, Dr. James W. 47719
Goodrich, Judith 50587
Goodridge, Janet 44463
Goodridge, Wendy 45812
Goodsell, Leo J. 46766
Goodwin, John 48397, 48399
Goodwin, Murray 33108
Goodwin, Sue 53139
Goodwinl, Patricia 50933
Goody, Dick 52385
Goodyear jr., Frank H. 51904
Goos, D. 17610
Goossens, Dr. E.J.H.P. 31855
Goossens, Gerda 04152
Gor-Arie, D. 25078
Gorbenko, Anatolij Aleksandrovič 36786
Gorbey, Ken 33260
Gorbunova, Ėleonora Grigorjevna 36947
Gorczyca, Krzyštof 34931
Gorczynski, Bohdan 47507
Gordenker, E.S. 32082
Gordenker, Emilie 44138
Gordon, Alec 44028
Gordon, Alexander 43373
Gordon, B. 38596
Gordon, David 46536, 50711
Gordon, Dr. Dillian 44937
Gordon, Don 06943
Gordon, Donnarae 52294
Gordon, Iain 43863
Gordon, Lindsay 43245
Gordon, Sharon 47720
Gordon, Stanley 53804
Gordon, Sue 43815
Gordon-Smith, Peter 43991
Gorduza, Victor 36127
Gordy Edwards, Esther 48125
Gore, Michele 49894
Goren, Dr. M. 25124
Goren, Nili 25122
Gores, Mary 49571
Goreta, Aleksandra 08886
Goretti, Ndayarinze M. 05280
Gorewitz, Shalom 50400
Gorgerat, Laurent 41154
Gorham, Laura T. 04192
Gorham, Linda 46542
Gorham, Marilyn R. 50099
Gori, Beatrice 25792, 25793
Gorini, Prof. Giovanni 27036
Gorini Esmeraldo, Eugênia 04880
Gorjão, Sérgio 35784, 35785, 35786
Gorman, Jill 46709
Gorman, Dr. Martyn 43236
Gormley, John 53581
Gormley, Michelle 46951
Gormley, Nina Z. 53190
Gormsen, Gudrun 10266
Gorny, Dagmar 20972
Goroch, Tansija Vladimirovna 37861
Gorochov, Vladimir Aleksandrovič 37086
Gorodnij, Aleksandr Ivanovič 37936
Gorove, Margaret 51649
Gorr, Claudia 20793
Gorroño de Crespo, Maria del Carmen 00406
Górski, Mirosław 34749
Gorski, Stefan 34942
Gorsline, Marie 12212
Gortzak, Roos 41165
Gosden, Dr. Chris 45338
Gosell, Julie 49470
Goshen, R. 25044
Goshow, Nancy 51217
Gosling, Susan 45642
Goss, Fred 49062
Gosse, Rita 53045
Gosselin, C.A. Claude 06334
Gossy, Reinhard 02664
Gostick, Misha 07187
Gostisha, Don 46234
Gostomski, Michelle L. 47155
Gostomski, Sonja von 19304
Gostwick, Martin 43928
Goswami, Raj Ratna 24352
Gotgelf, Lev K. 36753
Gothe, Nina 19449
Goto, Hidehiko 29563
Goto, Ikuo 28572
Gottfried, Claudia 21193
Gottfried, Heinz 02408
Gottfried, Dr. Michael D. 48254
Gottlieb, Jaromir 09559
Gottlieb, Paul 51138
Gottschalk, Werner 21098
Gottskálksdóttir, Júlíana 24021
Goudy, David 51442
Goueiva Santos, Dr. Antonio Alberto 35854
Gough, David 48121
Gough, Joan 00937
Gouinguenet, Hélène 11657
Goujat, Bérangère 11660
Goula, Gemma 38994
Goulaine, Marquis de 13189
Goulandris, Niki Angelos 23051
Gould, Alac 54347
Gould, Alec 49601
Gould, Claudia 51844
Gould, Gail 07118
Gould, Jonathan H. 54030
Gould, Marilyn 54214
Gould, Richard 52296
Gould, William 48957
Goulden, F. 44433
Goumand, Bruno 14037
Gounari, Dr. Emmanuela 23218
Gourdin, Michel 03393
Gouriotis, Lena 23073
Gourlin, Thierry 11533
Gout, Jacques 12457
Goutayer, Virginie 12562
Goutorche, Christian 12697
Gouveia, Alexandre Patrício 35868
Gouveia, Tiago Patrício 35506
Gouy-Gilbert, Cécile 13496
Govan, Michael 46660, 51166
Gove, Domingos Zefanias 31712
Govindaraj, C. 24353
Goward, Trevor 07199
Gowing, Laura 48463
Goycea, Ana Maria 00415
Gozak, Roman 43123
Gozani, Tal 50264
Gozdowski, Krzysztof 35145
Gozzi, Dr. Fausto 25800, 25802
Graaf, F.M. de 32959
Graas, Lucette 03324
Grabar, Olga 43178
Grąbczewski, Wiktoryn 35267
Grabenstatter 14986
Graber, Cleon 48780
Graber, Marcel 41237
Grabert, Walter 18825
Grabham, Chris 45048
Grabher, Gerhard 01796
Grabka, Dr. Marion 17795
Grabner, FOL Adolf 02010
Grabner, Dr. Sabine 02951
Grabovec'kyj, Volodymyr Vasil'ovyč 43043
Grabow, Astrid 17838
Grabowska, Genowefa 35089
Grabowska-Chałka, Janina 35172
Grabowski, Dr. Jörn 17283
Graça, Filipe 35927
Grachos, Louis 47114
Gracyalny, Devan 54042
Gradin, Eva 40699
Gradova, Dina Aleksandrovna 37633
Gradstein, Prof. Felix 33737
Gradwohl, Hans 01765
Graells, Eulàlia 38953
Graen, Dr. Dennis 19389
Graen, Dr. Monika 20966, 20972
Graepler, Dr. Daniel 18700
Gräser, Dr. 22473
Grässer, Irina 20154
Graesslé, Isabelle 41424
Grässlin, Thomas 21495
Graeve-Ingelmann, Dr. Inka 20425
Graf, Dr. Angela 18951
Graf, Prof. Dr. Bernhard 17143
Graf, Bernhard 41766
Graf, Brigitte 17270
Graf, Diane 54331
Graf, Dr. Rudolf 41101
Grafe, Steve 51519
Grafenauer, Michael 01740
Graff, Allison 47831
Graff, Nicholas 52527
Graff, Terry 06918
Graffagnino, J. Kevin 50825
Graffam, Olive 53858
Grafton, Duke of 45847
Graham, Lady 45500
Graham, Angela 45760
Graham, Barry 44842
Graham, Bruce 52621
Graham, Douglass J.M. 52917
Graham, Fiona 44682
Graham, J. 44167
Graham, Sir James 45500
Graham, John 01474
Graham, John R. 50364
Graham, Kenneth 24886
Graham, Michael S. 51046
Graham, Dr. Sean 07199
Graham, Shelby 52867
Graham, Tom 47355
Graham LeBlanc 05554
Graham Wade, Karen 47569
Grahan, Dr. John 47118
Grahn, Göran 40938
Grahn, Martin 43902
Gram, Dr. Magdelena 40922
Gramatikov, Georgi 05005
Gramatikov, Jančo 05133
Grambye, Lars 40722
Gramer, Kevin 48065, 48079
Gramm, Jean-Jacques 41769
Grammel, Sören 01970
Grammenos, Dr. D. 23209
Grammer, Mag. Manfred 02596
Gramolina, N.N. 37808
Grams, Greg 53768
Gramsz, Bożena 34706
Grančarov, Michail 05118, 05120, 05121
Grančarova, Kamelia 04991
Grancea, Maria 36117
Grand, Stanley I. 47234
Grande, Prof.Dr. Nuno 35922
Grander, Johann 02144
Grandidier, Daniel 15241
Grandjean, Gilles 13643, 14591
Grandmont, Jean-François 06120
Grando, Constance 46539

Grandrud, Dr. Reba 48156
Grandy, Patrick 46989
Grandy, Peter 41881
Grange, Bill 43965
Grangé, Claire 11534
Grange, Sylvie 12329, 12331
Granger, Brenda 48294
Granier, Jean-Marie 14720
Granitto, Andrew 54320
Granitz, Gerhard 02402
Granlund, Mark E. 52089
Grannan, Jill 48113
Grannis, Sue Ellen 50539
Granomont, Josée 07175
Granqvist, Eirik 11228, 15130
Grant, A.E. 43308
Grant, Anne 01439
Grant, Bronwyn 33185
Grant, Carolyn 47247
Grant, Christin 52401
Grant, Elinah 04299
Grant, Frank F. 06171
Grant, Gavin 45555
Grant, Gillian 33034
Grant, Hugh 48080
Grant, J. 43359
Grant, Jaqueline 47815
Grant, Laurence 05917, 43339
Grant, Rachel 44310
Grant, Ronald 44823
Grant Morrisey, Marena 51593
Grantčarov, Michail 05038
Granville, Lee Z. 53129
Grapin, Claude 11554
Gras, Mag. Claudia 02979
Grasberger-Lorenz, H. 22346
Graskoski, Debra 51435
Grasmann, Lambert 22243
Grasmug, Dr. Rudolf 01878
Grass, Hans Rudolf 41603
Grass, Yasmin 00777
Grasse, Carola 18145
Grasse, Marie-Christine 13109, 13114, 13115
Grasselt, Dr. Th. 22410
Grassi, Alfredo 00264
Grassi, Dr. Renzo 27841
Grassmayr, Johannes 02118
Grasso, Giacomo 26308
Graswinkel, J. 32058
Gratacós i Teixidor, Georgina 38948
Grathwohl-Scheffel, Christiane 18433
Gratl, Dr. Franz 02136
Grattan, Patricia 06842
Gratte, Stan 01623
Gratz, Dr. Reinhard 02609
Gratzer, Heinz 03038
Grau, Salvador 40291
Grau Lobo, Luís A. 39455, 39456
Graudina, Anna 52430
Grauer, Michael R. 47230
Gravatt, Kristin 46804
Grave, Marion 45897
Gravel, Mabel 05385
Graves, Tolley 50044
Graves-Brown, Carolyn 45812
Gray, A. 45090
Gray, Alan 33187
Gray, Dr. Anna 00946
Gray, Bonnie 06230
Gray, Brian 50252
Gray, Campbell B. 52188
Gray, Cherry 43260, 46067
Gray, Gabrielle 51638
Gray, Ian 05526
Gray, Jo 44201, 45616, 45918
Gray, John 33236
Gray, John L. 50255
Gray, K. 46032
Gray, Les 01201
Gray, Mary H. 47759
Gray, Muriel 01201
Gray, Peter 01490, 45448
Gray, Sam 51535
Gray, Sara 52290
Gray, Sarah 43342
Gray, Dr. Sharon R. 53243
Gray, Susan Inskeep 48477
Gray, Susie 44959
Gray, Xanthi 52093
Gray, Y. 44562
Grayburn, Patricia 44372
Grayson, Neil 51159
Graze, Sue 46500
Grąziewicz-Chludzińska, Barbara 35295
Graziola, Claudio 26564
Grazzini, Patricia J. 50739
Grdanjski, Dipl.-Ing. Mirjana 20423
Greaves, Debbie 45840
Greaves, Sarah 43731
Grebe, Erhard 18779
Grébert, Eliane 15554
Grečenko, Ljudmila Pavlovna 37077
Grech, Frances 30653
Greco-Kaufmann, Dr. Heidy 41216
Greeeson, Laura 47405
Greeff, I.B. 38581
Green, Adrian 45606
Green, Alison 43691
Green jr., Byron E. 50768
Green, Cheryl 51233
Green, Chris 43459, 43460, 45543, 45547, 45550
Green, Dr. D. 45090
Green, Erica 00772
Green, James 45444
Green, Jeremy 01412
Green, John F. 53197
Green, Prof. Jonathan 52369
Green, Louise 53071
Green, M. J. 45302
Green, Mike 04181
Green, Nancy 49548
Green, Pat 48158
Green, R.M. 01077
Green, Richard 46108
Green, Tina 47153
Green, Dr. William 46724
Green, William D. 46652
Green, William Davis 52490
Green, Dr. William E. 47230
Greenan, Althea 45020
Greenbaum, Elizabeth R. 46869
Greenberg, Judy A. 53885
Greenberg, Suzy 50742
Greene, A. Wilson 51799
Greene, Gerard 44509
Greene, Marilyn 44870
Greene, Dr. Patrick 00956
Greene, Virginia 51886
Greener, Peter 43335
Greenfield, Mark Steven 50245
Greenhill, John 45835
Greenhow, Dr. Desna 43669
Greenlaw, William 05943
Greenlee, Hal 47837
Greenlee, Nina 50486
Greenlee, Sandy 47741
Greeno, Judy 07278
Greenway, Lord 45235
Greenwood, Gail 00776, 00846
Greenwood, Tom 49756
Greer, Alberta 52394
Greer, Judy 52247
Greer, Kim E. 52677
Greer, Rina 07150
Greer, William 07121
Greffrath, Dr. Bettina 22660, 22663, 22667
Grefstad, Ola 33929, 33929
Gregersen, Bo 10254
Gregersen, R. 18921
Gregersen, Thomas 48046
Gregg, David 47033
Gregg, David W. 54107
Gregg, Rosalie 48020
Gregl, Dr. Zoran 08986
Gregoire, Jean 11653
Gregoire, Noël 03781
Gregoor, Yves 31891
Gregori i Berenguer, Joan 40223
Gregori Sebatià, Núria 39530
Gregors, Tommy 46156
Gregory, Joseph 46758, 51826
Gregory, Linda 36079
Gregory, Susan 46757
Gregory, Tabitha 53685
Gregson, Brian 01136
Greguski, Eva 53313
Greif, B. 17740
Greif, Steve 51381
Greig, John 43935
Greimel, Susanne 02620, 02620
Grein, Gerd J. 20963
Greinwald, Dr. Elisabeth 01746
Greisenegger, Dr. Vana 03051
Grelier, Joseph 15357
Grella, Consalvo 25331
Grella, Prof. P. 27047
Grellier, Gérard 13585
Gremillet, E. 15850
Grémont, Pascale 15494
Grenda, Edward 06078
Grendi Ilharreborde, Paola 07445
Grenon, Prof. Ernesto A. 00633
Grenter, Steve 46085
Grentmann, Christian 42008
Grenville, John H. 06074
Grépier, Gabriel 12710
Greppi, Graciela B. 00541
Grepstad, Ottar 33568
Greschny, Nicolaï 13984
Greshner, Barb 06621
Gressenbauer, Roland 01906
Greve, Jürgen 18969
Grever, T.F. 31866
Grevisse, Edgard 03920
Grew, Dr. Priscilla C. 50107
Grewenig, Dr. Meinrad Maria 22249
Grexa, Dr. Susanne 17165
Grezet, Cédric 41137
Grezinger, August 19997
Gribble, Michael 33147
Gribbon, Deborah 50237
Grice, Jean 43589
Gricenko, Vladimir Petrovič 37866
Griebel, Dr. Rolf 20424
Griebler, Reinhold 02580
Grieder, Susanne 41672
Grieger, Dr. Manfred 22661
Griepink, Dr. B. 32658
Grießmair, Dr. Hans 25595
Griff, Hanna 51169
Griffeuille, Marie F. 15302
Griffeuille, Marie-Françoise 15300
Griffin, Dave 48071
Griffin, David J. 24798
Griffin, Elizabeth P. 49398
Griffin, Ian 45092
Griffin, James 47251
Griffin, Jennifer 48162
Griffin, Jonathan 44209
Griffin, Kjerstin 33863
Griffin, O.T. 44712
Griffin, Priscilla 06873
Griffin Gregory, Diane 49388
Griffioen, B. K. P. 32060
Griffis, Larry L. 50120
Griffith, Amanda 46366
Griffith, Ken 46533
Griffith, Lee Ellen 48779
Griffiths, A. 44803
Griffiths, Bill 45154
Griffiths, Caitlin 46020
Griffiths, Coralie 01214
Griffiths, D. 05652
Griffiths, D.F. 44763
Griffiths, Dr. G. 43634
Griffiths, Jayne 43604
Griffiths, Jean 43881
Griffiths, Mary 45099
Griffiths, Pauline 45193
Griffiths, Peter 45274
Griffiths, R. John 45442
Griffo, Dr. Pietro 26735
Griggs, David W. 47254
Griggs, K. 01272
Grigo, Dr. M. 19605
Grigore, Nina 36183
Grigorian, Anelka 00738
Grigorian, Prof. Dr. Gurgen G. 37401
Grigorjeva, Tamara Andriïvna 43068
Grigoryan, Avedik 00763
Grigsby, Roger 47809
Gril du Guern, Cajetan 02595
Grimaldi, Dr. David 51115
Grimaldi, Floriano 26519, 26520
Grimaldi, Gloria 23332
Grimaldi, Louis A. 48264
Grimaldi, Stéphane 12229
Grimaud, Patricia 14433
Grimes, John 52657
Grimes, Richard 01335
Grimm, Albert 41999
Grimm, Dr. Alfred 20489
Grimm, Prof. Dr. Günter 22101
Grimm, Herbert 18209
Grimm, Jakob 41703
Grimm, Tess C. 48457
Grimmer, Dietgard 02613
Grimod, Dr. Ivana 27588
Grimshaw, Sir Nicholas 44966
Grimsrud, Halvard 33785
Grinage, Jeanine 46161
Grinbaum, Blanche 12505
Grincevičienė, Zita 30260
Grindell, Beth 53584
Grinenko, Lucy 43178
Grinëv, Nikolaj N. 37922
Grinnell, Debbie 50950
Grinnell, Nancy 51326
Grinsted, Stephen 44982
Gripari 23084
Grišina, Ekaterina Vasiljevna 37578
Griss, Victor 00927
Grissom, Don 50346
Griswold, Sarah 54275
Gritt, Stephen 06551
Grivegnee, M.R. 04010
Grjaznova, Ann 37581, 37694
Grob, Dr. B.W.J. 32484
Grobler, I. 38731
Groce, Prof. Aldo A. 00619
Groce, Dr. W. Todd 52941
Groden, Anke 21678
Grodskova, Tamara 37747
Grodwohl, Marc 16017
Grodzicki, Prof. Andrzej 35330
Gröger, Walter 02248
Groen, Ir.B. 31915
Groen, M.F. 32711
Groenendijk, Drs. G.H. 32059
Groeneveld, A. 32733
Groeneveld, A.C. 32759
Groenewald, A. 38566
Groenewald, M. 38678
Groeneweg, H. 32473, 32480
Groenewold, Rod 53764
Groenewoud, G.J.M.M. 32089
Grönholm, Kirsti 10788
Grötsch, Dr. Kurt 40073
Groff, John M. 51890
Grohnert, René 18250
Grohs, Kurt 18038
Groiß, Dr. Albert 01710
Groleau, Michel 06820
Gromova, Ljudmila Ivanovna 37662
Grondman, Dr. A.C.M. 32119
Grondman, Drs. A.C.M. 32352, 33027
Gronert, Dr. Stefan 17460
Grøngaard Jeppesen, Torben 09968, 10209, 10210, 10213, 10214, 10215
Groninga, Janet 46122
Grønn Iversen, Trine 33710
Gronne, Claus 10145
Grønnegaard Jeppesen, Torben 10216
Grønvold, Ulf 33716
Gronychová, PhDr. Marie 09855
Groom, Simon 44746
Groos, Dr. Ulrike 21948
Grooss, Dr. K.S. 32484
Groot, E. de 32729
Groot, J.P. de 32289
Groot, Rolf de 20822
Groot, Dr. Susan 32478
Grootveld, Cor 31892
Grootveld-Parrée, C. 32703
Gros, Christophe 41421
Gros, Dr. Patrick 02615
Grosboll, Sue 47294
Groschner, Dr. Gabriele 02624
Grosdidier de Matons, Benoît 13087
Grosfeld, J.F. 32002
Grosfeld, Dr. J.F. 32003
Groß, Helmut 18179
Gross, Jennifer 51056
Gross, Martial 41613, 41951
Gross, Miriam 51144
Gross, Dr. Raphael 18379, 18388
Gross, Renee 51144
Gross, Silvia 01796
Groß, Simone 22288
Groß, Thea 20246
Groß-Morgen, Markus 22096
Grossbard, Elayne 46755
Grosshans, Dr. Rainald 17116
Grossi, George 51791
Großjohann, Heidrun 17344
Großkinsky, Dr. Manfred 18387
Grossman, Cissy 51155
Grossman, David J. 47693
Grossman, Grace Cohen 50264
Grossmann, Anna 34630
Großmann, Prof. Dr. G. Ulrich 20764, 20765, 20788
Grossmann, Robert 15810
Grosspietsch, J. 21988
Grøtan, Geir A. 33691
Grote, Dr. Dietmar 20925
Grote, Dr. Hans-Henning 22654
Grote, Marina 19660
Grote, Dr. Udo 20506, 20507, 22763
Grotkamp-Schepers, Dr. Barbara 21801
Grotz, Helmut 22226
Grouwet, Carl 18008
Grove, David 52583
Grove, Jeffrey 47624
Grover, Ruth 47409
Grover, Ryan 48161
Groves, Bonnie Sue 48152
Groves, Elke 50272
Grower, Steve 00939
Grozin, Vladimir 37196
Gruat, Philippe 13486, 14263, 15741
Grubb, Tom 48531
Grube, Prof. Dr. NiKolai 17443
Grube, Susanne 17458
Grube, Walter 20530
Gruben, Dr. Eva-Maria 18009
Gruber, F. 02246
Gruber, Franz 02009, 02247
Gruber, Dr. Gabriele 17789
Gruber, Herbert 01791

Gruber, Dr. Herbert 18030
Gruber, Petra 20993
Gruber, Rosina 02008
Gruber, Sonja 02943, 03029
Gruber, Wolfgang 01803
Grubessi, Prof. Dr. Odino 27497
Grubler, Mitchell 48577
Grubola, Prof. James 50280
Grubola, James 50293
Grüger, Dr. Andreas 21914
Grüner, Almut 44668
Grünewald, Dr. Mathilde 22685
Grünwald, Karin 20123, 20130
Grünwald, Manfred 02534
Grünwald, Prof. Michael 01913
Grünzweil, K. 45270
Grütter, Daniel 41831
Grütter, Theo 18267
Grütters, Prof. Monika 17186
Gruetzmacher, Mark 53436
Gruetzner, Sara Jane 05518
Gruffydd, Alun 43426, 44766
Gruhl, Ulrike 21088, 21105
Gruhs, Ernie 05997
Grujić, Dragana 38093
Grunberg, Evelina 04661
Grund, M. Maria 17108
Grund, Dr. Rainer 17961
Grundler, Raymond 13508
Grundmann, Dr. Kornelia 20193
Grundmann, Thomas 17450
Grunenberg, Christoph 44746
Grunz, Karin 17096
Grunzweil, Karen 45048
Grusman, V.M. 37722
Gruzdeva, Tatjana Aleksandrovna 37810
Gruzska, Larysa Petrivna 43214
Grygiel, Prof. Ryszard 34873
Grygorenko, Iryna Vasylivna 43019
Gryka, Jan 34898
Grynszpan, Anna 34699
Grynsztejn, Madeleine 52795
Grys, Dr. Iwona 35270
Gryseels, Guido 04058
Gryspeert, Hans 03657
Grzechca-Mohr, Dr. Ursula 16751
Grzegorczyk, Dr. Detlef 20513
Grzesiak, Marion 49627
Grześkowiak, Andrzej 34681
Grzymkowski, Andrzej 34927
Grzymski, Krzysztof 07133
Gschwantler, Dr. Kurt 02980, 03019
Gschwendtner, Stefan 02324
Gstrein, Josef 02109
Gu, Feng 08429
Gu, Jianxiang 08045
Gu, Wenxia 08253
Guadarrama Guevara, Jorge 31077
Gual Dolmau, Oriol 38964
Gualdoni, Flaminio 28161
Guardia, Gabriel de la 34067
Guardiola, Juan 38972
Guarino, Dr. Ariberto 27521
Guarnaccia, Monica 41564
Guati Rojo, Prof. Alfredo 31135
Guay, Serge 06754
Guayasamín, Pablo 10442
Gubb, A.A. 38616
Gubbels, J.H.M. 32446
Gubčevskaja, Ljudmila Aleksandrovna 37795
Gube, Gundula 19401
Gubel, Eric 03495
Gubin, V.V. 37190
Gubler, Bernhard 41723
Gubser, Paul 41998
Gudelis, V. 30441
Guderzo, Dr. Mario 27316
Gudina, Jurij Volodymyrovyč 43133
Gudmundsdottir, Anna Lísa 24014
Gudnadòttir, Kristin 24020
Gudnason, Thorarinn 23999
Gudonytė, Alma 30424
Guedj, Gabrielle 13790
Gügel, Dominik 41798
Güiraldes, Consuelo 00473
Gül, Sebahat 42845
Guelbert de Rosenthal, Eva 00470
Güler, M. Güven 42853
Gülicher, Dr. Nina 20058
Güllüce, Hamza 42785
Güllüoğlu, F. Nevin 42934
Günaydın, Celal 42758
Gündogdu, Veysel 42753
Gündüz, Yaşar 42899
Guenette, Irene 07024
Guennec, Aude Le 12524
Günter, Werner 20851
Güntert, Prof. Dr. Marcel 41213
Günther, Albert J. 18084
Guenther, Bruce 52086
Günther, Clas 40892
Günther, Kurt 21413
Günther, Michael 19680
Günther, Rolf 18453
Günther, Dr. Sigrid 01855
Guenther, Todd 49933
Günther, Wolfgang 21803
Günzler, Lilo 18375
Guerbabi, Ali 00080
Guerber, Steve 46893
Guerci, Prof. Antonio 26306
Guerdy Lissade, Joseph 23324
Guerin, Charles A. 53600
Guerin, Kerry 01394
Guermont, Mauricette 15496
Guerra, Jairo 08784
Guerra, Yolanda 08784
Guerra Pizòn, Olga 10409
Guerrero, Doris 08613
Guerrero Cestero, Felipe 39354
Guerreschi, Prof. Antonio 27798
Guerri, Dr. Roberto 26700, 26704, 26706
Guersen, Pascale 14332
Gürtler, Dr. Eleonore 02136
Gürtler, Sonja 21818
Güse, Dr. Ernst-Gerhard 22405, 22413
Guest, Fiona 44618
Guest, Raechel 53956
Guevara Hernández, Jorge 31462
Guex, Stéphanie 41572
Güzel, Selahattin 42783
Guezi, Dr. Angela 27960
Guffee, Eddie 51975
Guffee, Patti 51975
Gugelberger, Rachel 51277
Guggenberger, Hannes 02188
Guggenberger, Hans 02222
Gugier, Pierre 11963
Guheen, Elizabeth 47609, 50885
Guhl, Klaus 41330
Guibal, Jean 11649
Guibert, Etienne 11844
Guichard, Christiane 15406
Guichard, Marie-Claire 13248
Guichard, Vincent 14190
Guichaumon, Jean-Louis 13956
Guichon, Françoise 11522
Guidon, Suzanne 12633
Guidotti, Prof. Francesco 26654
Guidotti, Nives 25569
Guidry, Keith 51575
Guiducci, Dr. Anna Maria 27846
Guiffre, Florence 48584
Guigo, Elsa 14646
Guigon, Emmanuel 15821
Guihaumé, Antoine 15409
Guiheneuf, Bernrd 15380
Guilardian, D. 03470
Guilbaud, Laurence 16100
Guilbert, Manon 05862
Guilbot 13276
Guilbride, Kurt 06126
Guild, John 53269
Guild, John W. 53263
Guilhot, Jean-Olivier 12852
Guillard, Alain 13791
Guillard, Maire-Joe 13330
Guillaume 04141
Guillaut, Laurent 12242
Guillemeteaud, François 12062
Guillemin, Daniel 11679, 13168
Guillemin, Jean-Claude 11947
Guilleminot, Jacques 12400
Guillemot, Hélène 13558
Guillen, Dr. Carlos B. 10554
Guillén Jiménez, Marilú 31518
Guillerme, Lucienne 14317
Guillermo, Joseph 34333
Guilliem Arroyo, Salvador 31096
Guilliod, Raymond 23283
Guillot 13629
Guimarães, Prof. Dr. Edi Mendes 04361
Guinand, Jean 41473
Guinid, Rosario 34399
Guinness, Alison 48421
Guinness, Robert 24900
Guiotto, Prof. Dr. Mario 27909
guiraud, Louis 16172
Guisolan, Dr. M. 41916
Gulbina, Anna Alekseevna 37947
Gulbrandson, Alvhild 33640
Guldemond, J. 32729
Guljaev, Vladimir Andreevič 37412
Gullan, Penny J. 47978
Gullickson, Nancy 48191
Gullickson, Rachel 53070
Gullsvåg, Per Morten 33968
Gulyásné Gömöri, Anikó 23407
Gulyes, Andreas 03048
Gumă, Nicoleta 36262, 36263
Gumnit, Daniel 50734
Gumpert, Lynn 51187
Gunarsa, Dr. Nyoman 24516
Gundacker, Reinhard 03136
Gundel, Dr. Marc 19097, 19099, 19100, 19101
Gunderson, Catherine F. 06920
Gunderson, Dan 47997
Gundry, Beverly F. 49153
Gunér, Tullan 40633, 40634, 40637
Gunjić, Ranko 08870
Gunn, Lorna 06050, 06051
Gunnarsson, Ann-Marie 40515
Gunnarsson, Annika 40915
Gunnarsson, Torsten 40918
Gunning, Judy 01512
Gunnlaugsdóttir, Elin 23984
Guntermann, Erika 19933
Gunther, Uwe 38761
Guo, Erfu 07863
Guo, Kang 08271
Guo, Qinghua 08015
Guo, Xiankun 07590
Guo, Xingjian 07628
Guo, Yu'an 08315
Guo, Zhenghong 08248
Guohua, Gu 08048
Gupta, Chitta Ranjandas 24102
Gupta, Preeti 48578
Gupta, V.N. 24258
Gupta, Dr. Vibha 24431
Gupta Wiggers, Namita 52080
Guralnik, Nehama 25122
Guratzsch, Prof. Dr. Herwig 17625, 18769, 21566, 21569
Gurba, Norma 49915, 49916
Gurevič, Dr. E. 37354
Gurkina, Tatjana Aleksandrovna 37125
Gurley, Diana 05378
Gurney, George 53917
Gurney, Helen 44462
Gurney, Marilyn 05939
Gurney-Pringle, Chantal 00812
Gursky, André 18894
Gurtner, Max 02253
Gusarov, Ass. Prof. Vladimir 33737
Guse, Ann 50542
Gusenleitner, Fritz 02310
Gusev, Vladimir 37649, 37650
Gushulak, Ruth 05801
Gusjeva, Svitlana Ivanivna 43147
Gust, Donald F. 46705
Gustafson, Donna 47657
Gustafson, Kelli 53130
Gustafsson, Malin 40547
Gustafsson, Pierre 41037
Gustafsson Ekberg, Kent 40436
Gustavson, Carrie 46830
Gustavson, Todd 52395
Gustin, Marcelo 00521
Gut, Andreas 18120
Gut, Heinrich 42011
Gut, Sibylle 41926
Gut, Ursula 41841
Gut, Werner 41949
Gutbrod, Dr. Helga 20587
Gutek, Dr. František 38195
Guth-Dreyfus, Dr. Katia 41162
Guthardt, Dr. Wolfgang 22666
Guthier, Mark 50397
Guthrie, Alyce N. 48899
Guthrie, Giles 45073
Guthrie, Jeanne 47165
Guțică-Florescu, Laurențiu 36134, 36276, 36523, 36627, 36628
Gutiérrez, Ana 39573
Gutiérrez, Guillermo Giraldo 08673
Gutiérrez, José A. 00206
Gutiérrez, Mario 52508
Gutiérrez, Raquel 40201
Gutiérrez, Reynaldo 04223
Gutiérrez Alba, D. Julio 40074
Gutiérrez Campos, Yolanda Milagros 09351
Gutiérrez Carcia, Inés 09169
Gutiérrez Coto, Fernando 08810
Gutiérrez García, María Angeles 39692
Gutiérrez Salamanca, Carlos Julio 08688
Gutierrez-Solana, Carlos 47080
Gutiérrez Velásquez, Amado 08602
Gutker de Geus, Sven E. 32648
Gutknecht, Eugene 46841
Gutman, Kevel Jane 48176
Gutmann, Gabriela 02944
Guttormsen, Sissel 33927
Gutzmann, Dr. Ulrike 22661
Guy, Dr. A.J. 44936
Guy, Frances 43830
Guy, Grant 07352
Guy, Troger 14090
Guyader, Raymond 14961
Guye, Francis 41794
Guyonneau, Olivier 13393
Guyot, Philippe 14961
Guyot-Corteville, Julie 11991
Guzmán, Antonio 14350
Guzman, Antonio 16042
Guzmán, Manuel Gil 31075
Guzman, Rene de 52801
Guzzi, B. Anthony 49262
Guzzo, Enrico Maria 28231
Guzzo, Prof. Pietro Giovanni 25731, 26060, 27280, 28034
Gvišiani, G.M. 16362
Gwan, Pak Yong 29762
Gwara, Czesław 35118
Gweki Booi, Vuyani 38551
Gwiazda, Bronisław 35013
Gwilim, Ian 43231
Gwyn, Brenda 50490
Gwynva, Rumball 01643
Gyaltso, Lenka 09776
Gye, Peggy 45111
Gyergyádesz, László 23644
Gygax, Christian 41132
Gygax, Raphael 42087
Gyger, Patrick 42048
Gyger, Robert 41977
Gyllenberg Pernevi, Bodil 40555
Gyllenhaal, C. Edward 47107
Gyllkrans, Monika 40972
Gyöngyössy, Márton 23440
Györe, Zsófia 23334
Györgyné, Dr. Varga László 23516
Gyorody, Ninette 06066
Gysin, Jacques 41669, 41671
Gyucháné Veres, Krisztina 23369
Gyula, Dr. Ernyey 23437
Gyulay, Endre 23842

H

Ha, Hong-Chien 42350
Ha, Paul 52573
Haack, Jan 18946
Haag, Angela 53531
Haag, Olinde S. 51101
Haag, Dr. Sabine 03019
Haag, Tilla 19736
Haak, Dr. Bill 48943
Haakanson, Sven 49823
Haake, Susan M. 53217
Haakestad, Jorunn 33433
Haaland, Anders 33432
Haamer, Kaia 10638
Haan, David de 44532
Haan, Maartje de 32088
Haan, Drs. Peter de 32857
Haapala, Anu 11469
Haar, Dr. Frauke von der 17549, 17558
Haas, Prof. Dr. Alexander 18965
Haas, Gordon 48472
Haas, Prof. Dr. Helmuth 02975
Haas, James R. 48029, 48030, 48031
Haas, Josef 09497
Haas, Margarete 01988, 01989
Haas, Dr. Mechthild 17789
Haas, Dr. Richard 17081
Haas, Ruth 54201
Haas, Sabine 19791
Haas, Dr. U. 19684
Haase, Dr. Evelin 17530
Haase, Werner 22721
Haavaldsen, Per 33863
Habarta, Gerhard 03056
Habartova, Dr. Romana 09887
Habditch, Max 45054
Habegger, Dr. Ueli 41597
Habel-Schablitzky, Andrea 19612
Habenicht, Wilfried 16678
Haberhauer, Günther 16866, 16867, 16869
Haberman, Patty 50607, 53471
Habermann, Dr. Sylvia 16952
Habermann, Yvonne 21451
Habersatter, Dr. Thomas 02624
Haberstroh, Sarah 50706, 50720
Habes, Scott 47685
Habeshaw, Julia 45368
Habichler 02767
Habrich, Prof. Dr. Dr. Christa 19340
Habsburg, Francesca von 03081
Hacala, Mark T. 53928
Hachet, Michel 15922
Hachmer, Hendrik 32872
Hack, Dr. C. 32391
Hack, Janice 49882
Hack, Shery N. 47221
Hackenschmidt, Sebastian 03026
Hacker, Heinrich 18324
Hackett, Brian 47986
Hackford, Clive 43978
Hackl, Prof. Dr. A. 02907
Hackler, Cornelia 21690

Hackman, John 11395
Hacksley, M.M. 38631
Hacksley, Malcolm 38600
Hackstein, Hermann 18854
Haddad, Regina Clélia 04943
Hadders, Rieks 31841
Haddington, Earl of 44331
Haddock, Brent 52141
Hadfield, Jessica 43916
Hadjicosti, Dr. M. 09375
Hadjisavvas, Dr. Sophocles 09368, 09378, 09385, 09387
Hadjusarnas, S. 09367
Hadland, Brian 06672
Hadland, Tony 45929
Hadley, Diana 53584
Hadley, V.J. 43905
Hadsbjerg, Lene 10237
Haduch, Henryka 34532
Haeck, Jules 03602
Häcki, Iren 42072
Häder, Dr. Ulf 17638
Haedo, Dr. José Antonio 00598
Häffner, Martin 22110
Häfliger, Toni 41588
Häfner, Dr. Yvonne 17325, 17326
Haegen, Erik van der 04149
Häggman, Gunvor 11162
Häggström, Ralf 10756
Häggström, Stig 10917
Haenel, Johanna 01752
Hänggi, Dr. Ambros 41175
Hänggi, Dr. Christoph E. 41857
Hänggi, René 41261
Hänisch, Sabine 20387
Haenni, Jean-Paul 41682
Hänsel, Dr. Alix 17206
Hänssler, Erica 42110
Häring, Dr. Friedhelm 18636, 18638, 18641
Häring, Simona 22652
Häse, Dr. Clivia 20617
Haese, Josette d' 12275
Haese, Jürgen 18659
Haesler, Andreas 22822
Häuser, Dr. Christoph 21959
Häusle, Magdalena 01796
Häusler, Anita 41395
Häusser, Peter 19360
Häußler, Theodor 16652
Hafen, M.A. Thomas 18843
Hafenecker, Ingo 18040
Hafer, Mary K. 53737
Haffenden, Peter 01264
Haffner, Dr. Marianne 42101
Haffner, Zsoltné 23586
Hafner, Dr. Mark S. 46638
Hafok, Barbara 03079
Hagan, Annemarie 06309
Hagan, Marylee 53482, 53485
Hagberg, Magnus 40428, 40898, 40909
Hagdorn, Dr. Hans 19333
Hagedorn Olsen, Claus 10097
Hagedorn-Saupe, Prof. Monika 17143
Hagel, Frank von 17143
Hagemeier, Anna Mae 53284
Hagen, Dr. Harold H. 52984
Hagen, John 49124
Hagen, Philipp 42034
Hagen, Rien 31841
Hagendorfer, Dr. Johann 02087
Hagenmaier, Peter 19189
Hager, H. 38638
Hager, Helmuth 21172
Hager, Joyce 50905
Hagerup, Asbjørn 33692
Haggart, M.W. 06254
Haggenmiller, Sharon 49248
Haggerty, John 44879
Haglund, Elisabet 40468
Hagmark-Cooper, Dr. Hanna 11099
Hagnau, Dr. Carola 19625
Hagström, Lars 40679
Hague, Alan 44234
Hahling, Albert 41095
Hahn, Dr. Andreas 17444
Hahn, Cheryl H. 54319
Hahn, Prof. Dr. Dietbert 22722
Hahn, Donna 52002
Hahn, M.A. Manfred 16633
Hahn, Dr. Sylvia 18450
Hahn Pedersen, Morten 10002
Hahner, Cathi 46901
Hahnl, Prof. Dr. Adolf 02367
Haider, Adolf 03138, 03139
Haider, Prof. Gertrude 02823
Haider, Johann 03113
Haigh, Brian 43415
Haight, Pete 54235
Haiko, Pertti 11058
Haila, Leena 10988
Hain, Dr. Beatrix 03079
Haine, Heiko 18108
Haine, Marie-Louise 03495
Haines, Alan P. 46987
Haines, Jeniffer 46471
Hainfellner, Gerhard 02804
Hair, Denny G. 49391
Hair, Julia 49179
Haja, Simion 36727, 36728
Hajdamach, Charles 44470
Hajdamach, Charles R. 44030
Hajduk, Stjepan 08976
Hájiček, Dr. Jaroslav 38244
Hájková, Olga 09914
Hakamata, Akihiro 28485
Håkansson, Roger 40558
Hakasalo, Kalevi 11122, 11123
Hake, Helga 21036
Hakes, Jay E. 46446
Hakins, Judy 46278
Hakobyan, Lilit 00759
Hakoda, Chiho 28904
Halaas, Kristin 33683
Halama, Jakub 09855
Halamish, Uzi 25048
Halaš, Jakub 09817
Halberstadt, Benjamin 17937
Halbmayerl, Prof. Dr. Ernst 20195
Halbreich, Kathy 50743
Haldemann, Dr. Anita 41166
Haldemann, Dr. Matthias 42125
Haldimann, Marc-André 41418
Haldner, Bruno 41171
Haldo, Soeren 33645
Haldy, Lanny 46249
Hale, Donald G. 51903
Hale, James K. 44442
Hale, Pat 48205
Halén, Widar 33730
Halevy, Maya 24981
Haley, Melissa 51273
Halford-Forbes, Emma 45364
Halfpenny, Geoff 54192
Hall, Alma 54267
Hall, Anne M. 52800
Hall, Brian 43680
Hall, Charles W. 49404
Hall, Eli 33926
Hall, Fay 01643
Hall, Gail 50065
Hall, Gary 43900
Hall, Herbert 52104
Hall, Jenny 43416
Hall, Jim 48644
Hall, Dr. Joe B. 50124
Hall, Mark 45366, 50525
Hall, Marlene 47194
Hall, Melissa 45214
Hall, Michael D. 50624
Hall, Michael S. 49369
Hall, Nancy 54221
Hall, Peter J. 38657
Hall, Philip 45347
Hall, Robert 44496
Hall, Robin 45116
Hall Miller, Thelma 52341
Hall-Patton, Mark P. 49250
Halla, Dr. Jaana 10775
Hallam, Doug 06796
Hallan, Knut Christian 33739
Hallas-Murula, Karin 10623
Hallatt, Kim 52396
Hallberg, Sara 51478
Halldorsson, Gudni 24002, 24004
Hallé, Antoinette 15743
Halle, Dr. Uta 17549
Haller, Prof. Dr. Heinrich 42054
Haller, Dr. Kerstin 17565
Haller, P. 16446
Hallgrímsdóttir, Margrét 24030
Hallin, Jens 10093
Halling, Finn 33854
Hallman, Charles V. 53659
Hallman, Christina 40920
Hallmark, Dr. Donald P. 53215
Hallock, Robert 47863
Hallonová, K. 38391
Hallová, Dr. Markéta 09607, 09672
Halls, E.J. 43453
Halma, Sidney 51342
Halmágyi, Dr. Pál 23689, 23690
Halmová, Dr. Mária 38294, 38299
Halmyradow, Ahmed 42976
Halonen, Maija-Liisa 11343
Halota, Marianna 34812
Halouzková, Marcela 09920
Halsey Gardiner, Margaret 51211
Halstead, Graham 43505
Halter, Marc 13228
Halusa, Gerhard 03048
Halvorsen, Prof. Rune 33737
Ham, Mrs. Robert 48414
Hama, Atsushi 28401
Hamada, Naotsugu 29246
Hamaguchi, Dan 52778
Hamanishi, Koji 29505
Hamann, Dr. Klaus 18984
Hamano, Tokuei 29255
Hamberger, M.A. Heike 19583
Hambolu, Dr. M.O. 33337
Hamby, Elaine 54183
Hamed, Dr. Dawi Musa 40393
Hamed, Mohammed 40389
Hamel, Bettina 41658
Hamel, W.A. 32105
Hamelin, Joël 14837
Hamelmann, M.A. Cordula 17870
Hamer, Diana 00534
Hamer, Fritz 47734
Hamil, Lisa 49116
Hamilton, Duke of 44376
Hamilton, Barbara 45366
Hamilton, Catherine 44834
Hamilton, Dan 47933
Hamilton, David 47363
Hamilton, Elaine 33202
Hamilton, James 43498
Hamilton, Dr. James 43504
Hamilton, James 43515
Hamilton, John 00912
Hamilton, Kelly 50396
Hamilton, Laura 44278
Hamilton, Mary 47238
Hamilton, Mary Ellin 46608
Hamilton, Mort 46343
Hamilton, Olivia 45405
Hamilton, R.V. 01224
Hamilton, Roy 50223
Hamilton, W. 33113
Hamilton-Smith, Katherine 53987
Hammell, Peter 52922
Hammer, Anders 40720
Hammer, Bettina 17795
Hammer, Dr. Karen Elisabeth 20935, 20936, 22690
Hammer, Dr. Karen-Elisabeth 22691
Hammer, Victoria 02119
Hammer, William R. 52408
Hammerl, M. A. Tobias 16419
Hammerle-McGuire, Peggy 51235
Hammermann, Dr. Gabriele 17765
Hammerschmidt, Elmar 19358
Hammerschmied, Gerhard 02832
Hammersla, Keith E. 50519
Hammerstrom, Kirsten 52178
Hammond, Barbara M. 47556
Hammond, Eleanor A. 46966
Hammond, Gary R. 46539
Hammond, Jennifer 44474, 49263
Hammond, Karen L. 48824
Hammond, Dr. Michael 51680
Hammond, Niall 43520
Hammond, Stacie 51661
Hammond, Theresa N. 49063
Hammonds, Kit 44988
Hamoniaux, Jacky 12748
Hamooya, C. 54659
Hampe, Henrike 22160
Hampshire, Camilla 44195
Hampson, Louise 46109
Hampton, Jan 33077
Hampton, Joel L. 47839
Hampton, John 46062
Hampton-Porter, Joan 53232
Hamraeva, G. 54474
Hamran, Ulf 33395, 33397
Hamre, Harald 33876
Hamrin, Örjan 40515
Hamshere, John 45648, 45652
Hamzabegović, Amra 04284
Hamzaev, A. 54447
Han, Guangyun 07970
Han, Jinke 07696
Han, Liji 08321
Han, Yong 07517, 07569
Han, Zhenfei 07711
Han, Zongxiang 07482
Han Keuk 29793
Hanau, E. 38751
Hancock, Robert 52335
Hand Evans, Dana 52344
Handel, Elke 03065
Handler, Jeremy 47642
Handley, Jerry 47742
Handley, Neil 44804
Handley, Pamela 06470
Handlos, Helga 20002
Hando, Justin 42509
Handy, Ilke 21580
Hanejko, Eugeniusz 35183
Hanemann, Dr. Regina 16910, 16917, 16921
Hanemza, Jarett 52488
Hanes, Alice C. 52107, 52110
Hanfstengel, E. 18085
Hangl 22344
Hanhardt, John 51258
Hanhart, Karin 07010
Hanisak, Dr. Dennis 48667
Hanisch, Heinz 16402
Hankel, Lillian 00976
Hanken, James 47184
Hankewych, Jaroslaw J. 47520
Hankins, Evelyn 47144
Hankins, Roger 54299
Hanley, Joanne 48518
Hanley, Robin 44587, 44589
Hanley, Theresa 51568
Hanna, Annette Adrian 48092
Hanna, Dan 48381
Hanna, Emily 46822
Hanna, George 46647
Hanna, John 54346
Hanna, Martha 06535, 06551
Hanna, Michael 49792
Hanna, Richard 51239
Hannah, Dr. Michael 33242
Hannan, Sarah 47678
Hannay, Alf 43309
Hannay, Rosemary 45349
Hannecart, Jean 12536
Hanner, Frank 48591
Hannesen, Dr. Hans Gerhard 17029
Hannibal, Dr. Joseph 47625
Hannick, P. 03323, 03324, 03325
Hanniffy, Constance 24734
Hanniffy, Kieran 24734
Hannula, Betsy 48549
Hannus, L. Adrien 53119
Hanor, Stephanie 49846
Hanot, Nicole 03700
Hánová, Markéta 09776
Hanow, Orazmyrat 42975
Hanrahan, Kim 51322
Hanreich, Georg 02522, 02524
Hanreich, Liselotte 02522
Hans, Annette 18942
Hansch, Dr. Wolfgang 19099
Hansel, Anna-Elisabeth 18783
Hanselman, David S. 48620
Hansen, Arlette 48002
Hansen, Arne 05871
Hansen, Barbara 52144
Hansen, David K. 53706
Hansen, Don 51790
Hansen, Dr. Dorothee 17556
Hansen, E.P.H. 04029
Hansen, Emma 47671
Hansen, Eva Maria 51894
Hansen, Frank 05424
Hansen, Dr. Harriet M. 10189
Hansen, Jesper Vang 10151
Hansen, Jörg 18093
Hansen, Klaus Georg 23269
Hansen, Larry 48952
Hansen, Laura R. 48252
Hansen, Maiken 10236, 10301
Hansen, Mogens 09967, 10043, 10044, 10044, 10319
Hansen, Moya 48069
Hansen, Olaf 21273
Hansen, Palle Birk 10190
Hansen, Prof. Pearl 54006
Hansen, Shannon 52029
Hansen, Sidsel 33812, 33813, 33814
Hansen, Dr. Steven D. 53114
Hansen, Thor Gunnar 33660
Hansen, Prof. Dr. Dr. h.c. Volkmar 17990
Hansen, Wayne 47444
Hansford, Michele 47268
Hansing, Anke 20895
Hansley, Kevin 51177
Hanslok, Andreas 17880
Hanson, Brent 52547
Hanson, Chris 49752
Hanson, Collen W. 53038
Hanson, Georgia 46398
Hanson, Gloria 48303
Hanson, Dr. John 49314
Hanson, Luther D. 48645
Hanson, Stephen 47367
Hanssen, L.H. 32733
Hanssen-Bauer, Françoise 33735
Hanssens, Marjolaine 04008
Hansson, Lars 41043
Hanta, Dr. Rajoharison 30540
Hantschmann, Dr. Katharina 20426, 20833
Hanus-Möller, Christl 21535
Hanzel-Wachter, Dr. Lieselotte 02709
Hao, Liangzhen 07758
Hao, Liguo 07876
Hapkemeyer, Andreas 25485
Happe, Michael 19235

H

Haque, Dr. Enamul 03237
Hara, Fumihiko 29012
Hara, Toshio 29256, 29395
Harada, Heisaku 28904
Harada, Minoru 28527
Harbige, Sarah 43340
Harbin, Melanie M. 48949
Hardacre, John 46036
Hardardottir, Asa S. 23989
Hardberger, Linda 52717
Harden, Bob 53522
Harden, Mike 46379
Harden, R. 43407
Harden, Stan R. 49284
Harder, H. 32435
Harder, Katie 05702
Harder, Dr. Matthias 17137, 18664
Harder, Dr. Ruth E. 41837
Hardesty, Sarah 53587
Hardiati, Dr. Endang 24492
Hardie, Andrea 44425
Hardiman, Thomás 24835
Hardiman, Tom 52098
Hardin, Jennifer 52640
Harding, Keith 45271
Harding, Maureen 54047
Harding, Shelley 06624
Harding-Davis, Elise 05335
Hardingham, Dennis 44071
Hardjonagoro, K.R.T. 24571
Hardman, Genee 52468
Hardmeier, Daniela 41745
Hardt, Nis 17782
Hardung, Klaus 21752
Hardwick, Tammy 05659
Hardy, Alwyn 45932
Hardy, Anne 46058
Hardy, Channing C. 48339
Hardy, Dr. Florent 46635
Hardy-Marais, Laurence 14032
Hare, Sandra J. 49748
Harent, Sophie 14354
Harger, Sara 48742
Hargittai, Emil 23419
Harinck, Prof. Dr. G. 31846
Harinen, Dot 06646
Harju, Jouni 10997
Harker, Robert 53062
Harkins, Michael J. 46616
Harkiw, Michael 43941
Harleman, Kathleen 46706
Harlen, Willie 52655
Harley, Winifred 43785
Harling, Nick 43526, 43527
Harlow, Ray 45615
Harltow, Harvey 06995
Harman, Fred C. 51674
Harman, Malcolm 43489
Harman, Robert L.B. 48504
Harmon, Darla 48848
Harmon, Horace E. 50067
Harmon, Dr. J. Scott 46321
Harmon, Mark S. 48174
Harms, Annette 22743
Harms, Dr. Eduard 19449
Harmsworth, Lord 44839
Harn, Alan D. 50040
Harnesk, Helena 41012
Harni, Heli 11408
Harnly, Marc 50237
Harnoncourt, Anne d' 51867, 51876
Haro, Jambo 29750
Harold, Steve 50439
Harpel, Dr. Judith 07199
Harper, Avis 46621
Harper, Mary Kay 52431
Harper, Robert W. 52520, 52522
Harper Lee, Laura 46973
Harpley, Melissa 01410
Harr, Jemima 48440
Harrand, Peggy 51790
Harre Hérin, A. 03869
Harreither, Dr. Reinhardt 01870
Harreld Love, Josephine 48120
Harrell, Deloris 50453
Harrer, Anton 02376
Harrer, Nicole 03048
Harrewijn, H. 32250
Harriett, Rebecca 49167
Harrington, Bill 01309
Harrington, Caren Anne 46693
Harrington, Dennis 52367
Harrington, Gary 49528
Harrington, Katie 49237
Harrington, Sian 44542
Harris, Adam Duncan 49573
Harris, Amy S. 46305
Harris, Dr. C. 45338
Harris, Carter 47720
Harris, Dee 52675
Harris, Donna 52237
Harris, Dr. E.C. 04194
Harris, Ed 46695
Harris, Florence E. 53255
Harris, Gene E. 47322
Harris, George 06669
Harris, Ivan 07101
Harris, James T. 46605
Harris, Jane 51118
Harris, Dr. Jennifer 45099
Harris, Joe 49283
Harris, Julie 51670
Harris, Kara 52516
Harris, Kaye 01141
Harris, Kevin 52043
Harris, L. 45090
Harris, Mac R. 49960
Harris, Miriam B. 48605
Harris, Ottawa 48064
Harris, René 48322
Harris, Rhian 44851
Harris, Richard 43832
Harris, Sandra 46760
Harris, Scott H. 51072
Harris, Tina B. 49605
Harris, Tom 48730
Harris-Fernandez, Al 53113
Harrison, Camilla 46105
Harrison, Christina 45182
Harrison, Don 51617
Harrison, Elizabeth 50901
Harrison, Gordon 46231
Harrison, Harry 51810
Harrison, Dr. J. 38759
Harrison, Jefferson 51361
Harrison, John 45873
Harrison, Keith 44684
Harrison, Loretta 51316
Harrison, Lynette 01574
Harrison, Marie 33220
Harrison, Michael 43728
Harrison, Monica 50396
Harrison, Myra 47062, 47063
Harrison, Stephen 24916, 24917, 43358, 43774, 43776, 43777, 43778, 44650, 45350, 45351, 45413, 45414
Harrison, Susan 44480, 45496
Harriss, Joan 07049
Hărşan, Aurel 36369
Harsch, René 41130
Hart, Brooks 06474
Hart, John 46161
Hart, Karen 49378
Hart, Katherine 49162
Hart, Macy B. 49563
Hart, Nancy 48733
Hart, Peter R. 44078
Hart, Robert L. 48433
Hart, Susan E. 54200
Hart, William 52570
Hart Hennessey, Maureen 53302
Harte, Jeremy 44192
Harte, Miriam 43422
Hartelman, Bart 32459
Hartemink, A.E. 32921
Hartfield, Libby 49561
Harth, Arno 22249
Harthorn, Sandy 46890
Hartke, Becki 52587
Hartl, Christa 20992
Hartl, Prof. F. 02273
Hartl, Georg L. 17304
Hartl, Dr. Nina 17062
Hartleb, Silvia Waltraud 03130
Hartleib, Dr. Rudolf 18386
Hartley, C. 43727
Hartley, David B. 51915
Hartley, Keith 44156
Hartman, Bruce 51632
Hartman, C. 38683
Hartman, Diane 53600
Hartman, Janice 46242
Hartmann, Andy 41817
Hartmann, Bonnie 53349
Hartmann, Emil 41834
Hartmann, Felix 41817
Hartmann, Joseph 18316
Hartmann, K. 21372
Hartmann, Matthias 18209
Hartmann, Dr. Ralf F. 17104
Hartmann, M.A. Susanne 16406
Hartnigk-Kümmel, Dr. Christa 19439
Harto, Edward S. 50695
Hartog, Dr. Arie 17551
Hartshorn, Gary 52077
Hartshorn, Willis 51194
Hartung, Lee 48938
Hartweger, Hans 03048
Hartwell, Carroll T. 50739
Hartwell, Jane 44919
Hartwich, Bernd 21014
Hartwick, Sue 48590
Hartwig, Laurie 51925
Hartwig, Paul B. 36062
Hartz, Jill 47400
Hartz, Dr. Sönke 21565
Hartz, Susanne 10103
Hartzell, Matthew D. 48055
Hartzold, Susan 46861
Harun, Rahimie 30573
Harvey, A.M. 38693
Harvey, Liz 50179
Harvey, Dr. Mark 01412
Harvey, Matt 45591
Harvey, Paul 47787
Harvey, Phyllis 51223
Harvey, Ragnheidur 24027
Harville, Vickie 50321
Harwell, Marcelle 51234
Harwood, Barry 47069
Haryu, Tsutomu 28799
Haschke, Heiner 21732
Haşdeu, N. Titus 36178
Hasegawa, Koichi 28649
Hasegawa, Mariko 29396
Hasegawa, Saburo 29243
Hasegawa, Sakae 29454
Hasegawa, Tei 29190
Hasegawa, Tokushichi 28675
Hasert, Eva-Maria 22500
Hashagen, Joanna 43381
Hashimoto, Akinobu 28354
Hashimoto, Kentarou 29452
Hashimoto, Masao 28799
Hashimoto, Ryutaro 29487
Hashimov, Bahodir Djuraevič 54446
Hashitomi, Hiroki 28795
Hashtroodi, A. 24650
Haskell, Barbara 51279
Haskell, Dr. Eric T. 47571
Haskell, Heather 53225, 53226
Håskoll Haugen, Liv 33599, 33601, 33603
Haslauer, Dr. Wilfried 02616
Haslehner, Hubert 02497
Hasler, Margrit 41442
Hasprunar, Jill 48746
Hasque, Jürgen de 20758, 20778
Haß, M.A. Heike 19071
Hassan, Syed Jamal 24136, 24142
Hassbecker, Egon 19079
Hassdenteufel, Dr. Erwin 20142
Hasse, Dr. Claus-Peter 20123
Hasselbalch, Kurt 47183
Hassinen, Esa 11028
Hassler, Donna 53298, 53575
Hassler, Hermann 30200
Hassmann, Dr. Elisabeth 03085
Hassol, Milton D. 51239
Hasson, Rachel 24999
Hassrick, Peter H. 48072
Hastaba, Dr. Ellen 02136
Hastedt, Catherine A. 47689
Hastenteufel, Michel 13528
Hastings, Lord 45630
Hastings, Bill 50144
Hasvold, Carol 48022
Hata, Nobuyuki 28514
Hatchell, Steven J. 47704, 47706
Hatcher, Alexandra 06790
Hatcher, Alison 46860
Hatcher, Jennifer 47190
Hatcher, Stephen 43922
Hatebur, Beatrice 41165
Hatfield, P. 46043
Hathaway, Merle 01163
Hatibu, Hadji 42519
Hatlen, Jean 05765
Hatoum, Hassan 42138
Hatschek, Christoph 03001
Hattam, Edward 01228
Hatterer, Sylviane 15831
Hattingh, J. 38570
Hatton Davidson, J. 45258
Hatzitaki-Kapsomenou, Prof. Ch. 23213
Hau, Hal 49796
Hauck, Richard 51505
Hauck, Dr. Wolfgang 17615
Haudiquet, Annette 13598
Haudot, Charles 13402, 13403
Hauer, Armin 18405
Hauer, Karl 02964
Hauer, Ulrich 18883
Hauerberg, Louiswe 10103
Hauerová, Dr. Stanislava 09879
Hauff, Rolf Bernhard 19260
Haufschild, Kurt 21733
Haug, Prof. Dr. Andreas 22719
Haug, M.A. Elisabeth 21867
Haug, Dr. Ute 18932
Hauge, Tore 33671, 33673, 33674, 33806, 33852
Haugen, Charleen 52456
Hauger, Therese 33488
Haugg, Hans 17672
Hauglid, Anders Ole 33502, 33637
Hauk, Roman 34981
Hauksdóttir, Kristín 24024
Haumesser, Giselher 19426
Haunerland, Prof. Dr. W. 20455
Haupenthal, Dr. Uwe 19301, 20699
Haupt, Eva 19488
Haupt, Dr. Herbert 03019
Hauptmann, Prof. Dr. Andreas 17398
Hauregard, Léon J. 03274
Haurie, Michel 14542
Haus, Anny-Claire 15809
Hausammann, Hans Peter 41908
Hauschild, Dr. Jan-Christoph 17991
Hauschke, Dr. Norbert 18896
Hauschke, Dr. Sven 17717, 21351
Hausding, Michaela 17950
Hauser, Claire 41919
Hauser, Emil 41490
Hauser, Erin 53784
Hauser, Hansgeorg 19704
Hauser, Janina 41377
Hauser, Larry 49253
Hauser, Monika 21378
Hauser, Peter 42079
Hauser, Robert 51014
Hauser, Rudolf 41687
Hauser, Urban 41255
Hauser, Dr. Walter 20376
Hauser, Dr. Walther 20809
Hauser, Dr. Zoltán 23538
Hauser-Schäublin, Prof. Dr. Brigitta 18690
Hauss, M. 13261
Haussonville, Comte d' 41313
Haustein-Bartsch, Dr. Eva 21208
Hauswald, Dr. Kenneth 46380
Hautekeete, Stefaan 03462
Hautmann, Dr. M. 42103
Havard, Carine 03939
Havas, Sandra H. 51500
Havasi, Bálint 23659
Havassy, Dr. Péter 23591, 23592
Havassy, Péter 23593
Havassy, Dr. Péter 23596, 23597, 23598, 23599, 23601, 23602
Havel, Joseph 49397
Havelková, Miloslava 09750, 09778
Havelund, Jørgen 10235
Havens, Patricia 53111
Havens, Ron 49053
Haveri-Nieminen, Marja-Liisa 11304
Haverkamp, Frode 33735
Havlíček, Jaroslav 09840
Havlíčková, Dr. Miloslava 09650
Havlik, Betty 49289
Havlik, Philipe 22122
Havlík, Dr. Vlastimil 09460
Havron, Rina 24935
Haw, Jane 51728
Hawadi, T. 54668
Hawelka, Dr. Walter 18602
Hawell, Dr. Stevan 52994
Hawes, Irene 53647
Hawk, Alan 53903
Hawke, Jenny 38618, 38619
Hawke-Smith, Cameron 43723
Hawkes, Dr. Michael 07199
Hawkey, Diane 53469
Hawkins, Andrea 45099
Hawkins, Lynda 47857
Hawkins, Marilyn 50343
Hawkins, Renee 50195
Hawkins, Dr. Seth C. 52619
Hawksworth, Susan 52675
Hawley, Anne 46934
Hawley, Greg 49700
Hawley, Liza 51829
Hawley, Pam 05809
Haworth, Brenda 49276
Hawvermale, Hettie G. 46762
Haxhiu, Prof. Dr. Idriz 00033
Haxton, Jason 49801
Hay, Irma 44790
Hay, Malcolm 44950
Hay, Mary 44539
Hay, Patricia 46666
Hay, Susan 52170
Haya, Dan 53438
Hayashi, Hirotumi 29422
Hayashi, Kiichiro 28593
Hayashi, Masayoshi 29622
Hayashi, Misa 29390
Hayashi, Toshihiko 29522
Hayashibara, Ken 29098
Haycock, Dr. Lorna 43979
Hayel, Janice 45936
Hayes, C. Daniel 48467
Hayes, David 43765

Hayes, Emily 48445
Hayes, Fiona 44294
Hayes, Neila C. 53098
Hayes, Randall 51548
Hayes, Sally 43468
Hayes, Virginia B. 53255
Hayhow, Simon 44232
Haynes, Bruce 05970
Haynes, Jackie 46299
Haynes, John H. 45724
Haynes, Peter 00944, 01581
Haynes, Tom 44197
Haynes, Valerie 52018
Haynie, Ron 53933
Hayoz, Roger 41652
Hays, Bonnie J. 51683, 51685, 51686
Hays, George W.S. 52663
Hays-Gilpin, Kelley 48551
Haysom, David 45810
Hayter, Stephen 05462
Hayward, Joanna 45568
Hayward, M.R. 44530
Hayward, Ron 33039
Hayward, Susan P. 45794
Haywood, Dr. C. 24819
Hayworth, Ross 43247
Hazelrigg, Scott 46383
Hažír, Dušan 38336
He, Fanf 07513
He, Hongyuan 07881
He, Kezhou 08020
He, Nonglin 07653
He, Yinzhong 08250
He, Yu 07665
He, Zhongyong 08137
Headley, Dr. David 52033
Heady, Karen 51618
Healey, Mary 53230
Heaman, Ken 05953, 05956, 05963
Hearn, Brian 51521
Hearn, Redell 50214
Hearn, Robert 48890
Heath, Betty 53283
Heath, Hether 06998
Heath, Sophie 43861
Heathcote, Richard 01068, 01522
Heather, Robert 01246
Heaton, Alice 46607
Heaton, Colin 45115
Heaume, R.L. 44241
Heaver, Stephen G. 50323
Heazlewood, Brenton 01645
Hebblethwaite, Christine 43980
Hebden, M.J. 44356
Hebestreit, Prof. Dr.-Ing. Andreas 19897
Hebestreit, Dr. Franz 20483
Hebler, Ingo 17607
Hebron, Victoria 46038
Hecher, Markus 22085
Hećimović, Branko 09005
Heck, M.A. Brigitte 17594
Heck, Brigitte 19437
Heck, Petra 31873
Heckelmann, Maya 20219
Heckenbenner, Dominique 15643
Heckenberg, Dr. Norman 01485
Hecker, Ilka 16879
Heckert, Dr. Uwe 22278
Heckes, Jürgen 17398
Hecking, Detlef 41383
Heckl, Prof. Dr. Wolfgang M. 17452, 20433, 20434, 20831
Heckmann, Dr. Stefanie 16975
Heckscher, Morrison H. 51212
Hectors, Jan 03586, 03587
Heddon, Gordon 44771
Heddy, Suzanne 47993
Hedegaard, Esben 10288, 10290, 10292, 10293
Hedel-Samson, Brigitte 11998
Hedelin, Jean-Louis 11744
Heder, Terence 50600
Hedgecock, Deborah 44807
Hedger, Michael 00940
Hedges, Ken 52753
Hedin, Marika 40921
Hedin, Svante 40814
Hedinger, Prof. Dr. Bärbel 18938
Hedlund, John 47257
Hedrick, Nancy 50042
Heege, Dr. Elke 18092
Heel, Lena 18674
Heeley, H.F. 45211
Heer, Ed de 31862
Heerebout, Dr. G.R. 32258
Heerebout, G.R. 32441
Heerink, B. 32291
Heermann, Dr. Ingrid 21951
Heese, Dr. Thorsten 20923
Hefermehl, Norbert 17334
Hefferhan, Tom 01302
Heffern, Marion 06906
Heffley, Scott 49695
Heffner, Jinger 50212
Heffner, Sarah 49168
Heflin, Kirk 48660
Hefti, Dr. Ernst 42068
Hegan, M. 43365
Hegel, Andrea von 17072
Hegeman, Carol A. 48901
Heger, Gabriele 02769
Heger, Wolfgang 18904
Heggade, Dr. D. Veerendra 24162
Heggen, Bert 32865
Hegidio, Margaret 53743
Hegnar von Ubisch, Frederikke 33572, 33573
Hegseth Garberg, Ann Siri 33929
Heguerte de Soria, María del Carmen 54405
Hegyi, Lorand 15252
Hehr, Dr. John 48528
Heidbrink, Dr. Ingo 17569
Heide, Dr. Birgit 20146
Heide, Diana 08989
Heide, Prof. Dr. Gerhard 18421
Heideman, William N. 50602
Heidenreich, Barbara 34603
Heiðkr{}dal, Dagný 24022
Heidrich, Bill 49320
Heij, Dr. J.J. 31937
Heijmans, Dr. H.G. 32057, 32066
Heil, Gerd 21390
Heilbrun, Françoise 14692
Heilbrunn, Margaret 51243
Heiligenstein, Louis 15691
Heiliger, Don 49779
Heiligmann, Dr. Jörg 19676
Heilman, Anna Bea 50345
Heilman, Patrick 46660
Heilmann, Dr. Angela 20001
Heiman, Mary-Ellen 05469
Heimann, Dr. Curry 40852
Heimann, Sybille 22324, 22325
Heimann-Jelinek, Felicitas 03008
Heimerdinger, Luc 41418
Hein, Carolyn 06246
Hein, Ivar-Kristjan 10624
Hein, Jørgen 10122
Hein, Lisbet 10101
Hein, Louise 48871
Heinämies, Ms. Kati 10785
Heinänen, Seija 10893
Heine, Sabine 17469
Heinecke, Friedhelm 21480
Heinemann, Dr. Christiane 20127
Heinemann, Ingo F. 19284
Heinerud, Jans 41000
Heinicke, Gabriele 16499
Heinien, Jouko 11035
Heinimaa, Mia 11353
Heinisch, Prof. Dr. H. 18896
Heinlein, Roland 16705
Heinlein, Dr. Stefan 21459
Heinmiller, Lee D. 49124
Heinonen, Jouko 11026, 11027, 11028, 11029, 11030, 11032, 11034
Heinrich, Christoph 48072
Heinrich, Paul 30474
Heins, Dr. 33072
Heintzelman, Scott A. 53369
Heintzman, John 51772
Heinz, Dr. Marianne 19473
Heinz, Stefan 17145, 20070, 20071
Heinz, Ulrich von 17140
Heinzelmann, Dr. Markus 19976
Heinzelmann, Prof. Michael 41202
Heinzl, Richard 03076
Heise, Bernd 17957
Heise, Dr. Karin 20292
Heiser, Jerry 46666
Heising, Cornelia 17998
Heisner, Ron 51907
Heiss, Alanna 50185
Heiss, Herbert 41357
Heitele, Thomas 19729
Heitlinger, Maria Theresia 19736
Heitmann, Dr. Claudia 19587
Heitz, Henri 15675
Heitzenröder, Dr. Wolfram 21438
Heizmann, Dr. Elmar 21959
Hejdová, Dagmar 09498
Helavuori, Hanna-Leena 10828
Helbig, Gabriele 18291
Helbig, Heike 21442
Helbling Gerster, Regine 41906, 41907
Held, Elwir 20120
Held, Peter 53465
Heldt, Helma 21306
Helfenstein, Marie-Therese 41860
Helfrich, Kurt 52861
Helfrich, Miguel 17118
Helg, Lukas 41346
Helgason, Gísli 24024
Helgesson, Michael 40634
Helke, Dr. Gabriele 03019
Hellandsjö, Karin 33569
Helleland, Allis 10152
Hellemans, Marijke 03297
Heller, Barbara 48114
Heller, Dr. Erna 51225
Heller, Laura 52203
Heller, Margot 44988
Heller-Karneth, Dr. Eva 16530
Hellerstein, Bruce S. 47311
Hellfaier, Detlev 17838
Hellmann, Birgitt 19392
Hellmund, Dr. M. 18895
Hellrung, Reinhard 18576, 18577
Hellstern, Dr. John R. 46194
Hellwig, Frauke 17561
Helm, Charles 47761
Helm, Cherie 53372
Helm, Erika 03124
Helm, Lorna 47580
Helm, Dr. Reinhardt 22392
Helme, Mehis 10589
Helmecke, Gisela 17203
Helmerding, Willy 16799
Helmerson, Klas 40933, 40949
Helmig, Loeda 50170
Helminen, Hilkka 10863
Helminen, Liisa 10797
Helminen, Ritva 11322
Helms, Cheryl V. 51328
Helms, Hans 20385
Helms, Dr. Leo 09359
Helmus, L.M. 32844
Helstrom, Linnea 51103
Helton, Debra M. 53090
Helzer, Richard 46983
Héman, S. 31888
Hemdal, Jay F. 53522
Hemdorff, Olle 33863
Hemm, Marsha 52156
Hemm, Michael 01791
Hemmendorff, Ove 40796
Hemmer, J. 32001
Hemmerijckx, Dr. Rik 04014
Hemmerle, Andreas 20428
Hemmerlein, Tamara 47870
Hemmes, H.H. 32845
Hemmi-Ragaz, Christian 41950
Hemsley, Jacqueline 00903
Hemsworth, Sandra 48785
Henar, A.F. 32525
Henbery, Edward 43917
Henchoz, Jean-Frédéric 41292
Hendel, Dr. Margarete 20603
Hendeles, Ydessa 07154
Henderby, D. 44037
Henderson, Cherel 49817
Henderson, David 45758
Henderson, Diana 43324
Henderson, E. Ray 47567
Henderson, Graeme 01095
Henderson, Jai 50214
Henderson, Justin 53680
Henderson, Kevin 33178
Henderson, Margaret 01233
Henderson, Marie 06733
Henderson, Robbin 46751
Henderson, Stan 06970
Henderson Fahnestock, Andrea 51231
Hendren, Eileen 47977
Hendricks, Albert J. 53543
Hendricks, Dr. Alfred 20513
Hendricksh, M. 38690
Hendrickson, Dean A. 46519
Hendricx, Hilde 04026
Hendrikman, Dr. Lars 32524
Hendrikx, R. 33020
Hendrix, Lee 50237
Hendrix, Mittie B. 52645
Hendry, R. 33233
Hendryx, Michael 54359
Heneberry, June L. 51096
Hengstenberg, Thomas 21741
Hengster, Sabine 18979, 18981
Hengstman, G. 32107
Henke, James 47638
Henke, Jerry D. 48724
Henkel, Jens 21419
Henkel, Katharina 18138
Henkel, Dr. Matthias 20779, 20784, 20790, 20791, 20792
Henkens-Lewis, Helen 43800
Henmi, Yoshiharu 29066
Henn, Catherine E. C. 46944
Henn, Char 52273
Hennaut, Eric 03461
Hennebert, Diane 03767
Hennedy, Jane M. 53453
Henneman, Inge 03287
Hennessey, Dr. William 51361, 51366, 51371, 53752
Hennessy, Thomas F. 53486
Hennig, Alexandra 03081
Hennig, Lutz 20396
Hennig, Dr. Michael 20512
Hennig, Ulrich 21263
Henning, Darrell D. 48022
Henning, Dirk 21452
Henrich, Sarah 50918
Henricksen, Paul 50505
Henriksen, Kenneth Roy 10064
Henrikson, Steve 49656
Henrikssen, Tracey 07080
Henrion, Pierre 03813
Henriot, Elisabeth 12637
Henriques, Paulo 35719
Henry, Barbara 51260, 51470
Henry, Cassandra L. 50313
Henry, James Pepper 46288
Henry, John B. 48559
Henry, Karen A. 05501
Henry, Patricia 47531
Henry, Patrick 44743
Henry, Dr. Sarah 51231
Henry-Barnaudière, Elise 16307
Henrykowska, Anna 35066
Henrysson, Harald 40476
Hens, Pierre 03573
Henschel, Sylvia 22036
Hensel, C. 18113
Hensel, Diethardt 20663
Henseler, Petra 20634
Henseler, Rainer 21447
Henshall, Barbara E. 47342
Henshall, W.A. 46106
Henshaw Jones, Susan 51231
Hensinger, Irene D. 52515
Hensley, Dr. Dannie 36043
Henson, Leonora 47124
Henson, Trevis 46566
Henstra, Klaas R. 32029
Hentschel, Dr. Martin 19705
Hentz, Dr. Károly 23568
Henwood, Peter C. 33132
Heon, Laura 51378
Hepburn, Lorna 44305, 44438
Hepp, Prof. Dr. Frieder 19076
Hepp, Dr. Frieder 19085
Heralus, Dr. Stefanie 17345
Héran, Emmanuelle 14692
Heraud, François 15199
Herb, Reinhard 22397
Herbaugh, Karen 50303
Herber, Bob 53654
Herberg, Mayde 52854
Herbert, Jacky 15710
Herbert, Lynn M. 49383
Herbert, Sharon C. 46307
Herbst, Andreas 17113
Herbst, Detlev 22192
Herbst, John 49492
Herbuveaux, Arlette 15865
Herce, Ana I. 40122
Herda, Isabel 18433
Herder, Dr. Fabian 17469
Herdits, H. 01861
Heredia Cortés, Gabriel 31566
Herger-Kaufmann, Josef 41894
Herholdt, Nancy 47380
Herje, Torunn 33797, 33798
Herkenhoff, Paulo 04756
Herkt, Silvia 03065
Herlihy, Berry H. 50229
Hermalyn, Dr. Gary D. 47044, 47050
Herman, Bernard J. 46536
Herman, Judy 49215
Herman, Monica 48719
Herman, Richard 48667
Hermand, Michel 14523
Hermand, Nathalie 03780
Hermann, Adolf 22092
Hermann, Dr. Claudia 41600
Hermann, Daniel 44156
Hermann, Reyh 21553
Hermann, Dr. Siegfried W. 02512
Hermann-Szabó, Sara 10343
Hermansdorfer, Mariusz 35335, 35339
Hermansson, Jan 40993
Hermellin, G. 15713
Hermsdorf, Franziska 16498
Hernandez, Annie 15553
Hernández, Carlos Arturo 08554
Hernandez, John 50000
Hernández, Rodolfo 31147
Hernandez, Sofia 51117
Hernández Carralón, Gema 39542
Hernández Carrión, Emiliano 39434
Hernandez-Cruz, Luis 52355
Hernández Gutiérrez, Jorge 08600

H

Hernández Hervás, Emilia 39932
Hernández Nieves, Román 38935
Hernàndez Pérez, Mauro 38846
Herold, G. 20787
Herold, Dr. Inge 20166
Herold, Joyce L. 48075
Herold, Prof. Roland 02955
Herold, Prof. Roland-Peter 02158
Herr, Daniela 17106
Herrán, Alfonso Carcero 08659
Herranen, Merja 11230
Herrbach-Schmidt, Dr. Brigitte 19437
Herren, Hans 41244
Herren, Walter 41478
Herrera Acosta, Elvira 31148
Herrera de La Muela, Ignacio 39916
Herrera Salazar, Germán 08640
Herrero, Prof. Eutimio 39383
Herreshoff, Halsey C. 47034
Herrick, Phyllis M. 49525
Herrick O'Hara, Virginia 47322
Herrin, Kari 52945
Herring, Madeline 46624
Herring, Sarah 44937
Herrity, Emma 44108
Herrmann, Dr. Elisabeth 03019
Herrmann, Gerd-Ulrich 21720
Herrmann, Josef 20033
Herrmann, Jürgen 22308
Herrmann, Dr. Klaus 21930
Herrmann, Uwe 19429
Herrmann Fiore, Dr. Kristina 27503
Herron, Christopher 48080
Herron, James 51715
Herron, Mary 49461
Herselman, D.E. 38737
Hersey, Irwin 51144
Hershfield, Shauna 50806
Herskowitz, Sylvia A. 51281
Herta, Ana 36102
Hertel, Friderike 18900
Hertel, Kathleen 46288
Hertford, Marquess of 43265
Hertig, HP. 41696
Hertiš, Iris 38406
Hertling, Nele 17029, 17062
Hertner, Max 41842
Hertwig, Dr. Stefan 41213
Hertzlieb, Gregg 53698
Herz, Carrie 46865
Herzer, Wolfgang 22367
Herzmanatus, Klaus 18575
Herzog, M.A. Claudia 21239
Herzog, Dr. Georges 41498
Herzog, Hannes 41740
Herzog, Dr. Hans-Michael 04682, 42069, 42070
Herzog, John E. 51222
Herzog, Melinda 50414, 53779
Herzog, Monika 18490
Herzog, Rainer 18054
Herzogenrath, Prof. Dr. Wulf 17556
Hesche, Eva-Maria 02092
Hesketh-Campbell, Lorraine 45134
Hesl, Wolfgang 16518
Heslewood, Emma 45444
Hesling, Mary 44404
Hess, Carsten 10110
Hess, Dr. Daniel 20764
Hess, Roman 42098
Heß-Emmert, Claudia 21552
Hessaby, Mohandes Iraj 24670
Hesse, Gisela 19364
Hesse, Dr. Sabine 20045, 21950
Hesse, Verena 21388
Hessel, Rita 41996
Hesselbom, Ted 40555
Hessellund, Birgit 09956
Heth, Charlotte 51236
Hetterich, Jochen 21948
Hettinga, Mary Jane 53994
Heubaum, Dr. Regine 20738
Heuberger, Vera 41697
Heuck-Neelsen, Dietrich 21506
Heuckeroth, Erwin 20258
Heuer, Bernd 17783
Heuer, Christoph 21728
Heufelder, Jochen 19602
Heumann, Dr. Konrad 18370
Heunemann 21857
Heuschen, Imelola 03842
Heusden, G. van 32307
Heuser, Prof. Dr. August 18364, 19998
Heuser, Douglas 50425
Heuser, Frederick J. 51872
Heuser, Dr. Mechthild 41215
Heußner, Karen 22208
Heuwinkel, Christiane 17341
Heuwinkel, Dr. Hubert 18009
Heuwinkel, Martin 17742
Hevers, Dr. J. 17529
Hevia, Agustin 39760
Hew Wing, Micheal 51226
Hewelt, James 45423
Hewitt, Karen 47335
Hewitt, Marion 45094
Hewitt, Susan 01618
Hewitt, Timothy M. 51497
Hewitt-Dower, Bill 00989
Hewitt-Soar, Hugh David 45315
Hewson, Alan 43256
Hewson, Karen E. 06494
Hewson, Rose 33167
Hewson, Shirley O. 01665
Heyburn, Tim 44530
Heyd, Paul 20863, 20865
Heyden, Dr. Thomas 20786
Heyen, Lora 46533
Heyes, John 44132
Heyl, Cornelius A. von 22686
Heyl, Dr. Matthias 18490
Heym, Dr. Sabine 20482
Heymans, Dr. Hubert 03837
Heymer, Kay 18008
Heyneman, C. 03422
Heynen, Arnold 42013
Heynen, Dr. Julian 17998
Heyning, Dr. John 50257
Heystek, A.C.W. 32387
Heyward, Carroll 47723
Heywood, Dawn 45692
Hibbits, Paul E. 53819
Hibel, Edna 49660
Hibler, Dolores 02852
Hickey, Janet 50440
Hickey, Leo J. 51054
Hickey, Michael 05532
Hicklin, Karen 47531
Hickman, Carole 46758
Hickman, Warren 53824
Hickock, Paul 49374
Hicks, Christine 14936
Hicks, Hugh Francis 46581
Hicks, Louis 46199
Hicks, Oberstabsfeldwebel Paul 18297
Hicks-Connors, Robin 53342
Hidasi, Prof. José 04474
Hidber, Alfred 41141
Hideo, Kurita 28373
Hidook, Tatjana 43039
Hiebel, Kaye 50499
Hiebler, Prof. Dr. Herbert 02878
Hieblinger, Faith 06101
Hiekisch-Picard, Sepp 17401, 19042
Hiesinger, Kathryn B. 51867
Hiesmayr, Herbert 02689
Higashi, Shunro 29536
Higginbotham, Evangeline J. 52506
Higgins, James 53508
Higgins, Kate 47104
Higgins, Sarah 46651
Higgs, Jane 44111
High, Dennis 47250
High, Laura 46346
High, Steven S. 52289
Hightower, Anne Marie 49866
Hightower, John B. 51334
Higman, David 44756
Hignette, Michel 14726
Higuchi, Dr. Takayasu 28864
Higuchi, Yoshikazu 28622
Hiis Søftestad, Hanne 33387
Hijiwara, Takshi 29116
Hilaire, Michel 14242, 14248
Hilarides, John 47390
Hilberg, Thomas D. 50557
Hilberg, Dr. Volker 21565
Hilbun, Dorinda 46637
Hildebradtová, Barbora 09831
Hildebran, Julia 47665
Hildebrand, Dr. Josephine 17168, 17169
Hildebrand, Walter 01921
Hildebrandt, Alexandra 17184
Hildebrandt, Dr. Frank 18951
Hildebrandt, Sandra 51304
Hildebrant, Paweł 34630
Hildenbrand, Manfred 19025
Hiles, Wayne 53985
Hilgers, T. 32562
Hilkenbach, Rainer 19525
Hilkhuijsen, J. 32059
Hill, Andrew 51054
Hill, Andrew F. 43374
Hill, Becky 48788
Hill, Brian O. 53563
Hill, Carla 46625
Hill, Carolyn 51521
Hill, Charles 06551
Hill, Christine C. 49186
Hill, Claire 45742
Hill, Heidi L. 47618
Hill, John 51524
Hill, Dr. John 54226
Hill, Joyce 48458
Hill, June 44382
Hill, Karen 48154
Hill, Lindsay 06631
Hill, N.B. 45079
Hill, Peter 44494
Hill, Ray 50151
Hill, Robert D. 49180
Hill, Robin 44573
Hill, Sergej Ivanovič 36750
Hill, Shannen 48089
Hill, Steve 46377
Hill Carter, Charles 47352
Hill-Festa, Lisa 53006
Hill Perrell, Franklin 52463
Hillard, Elbert R. 49558
Hillebrand, Dr. Melanie 38711
Hiller, Fritz 21757
Hiller, Dr. Hubertus 21616
Hiller, N. 33069
Hiller, Robert J. 47698
Hiller von Gaertringen, Dr. Julia 17838
Hiller von Gaertringen, Dr. Rudolf 19890, 19896
Hillerman, Barbara 03080
Hillesheim, Dr. Jürgen 16613
Hillhouse, Susan 52864
Hilliard, Elbert R. 49559
Hillier, Kenneth 43326
Hillis, Florence E. 49814
Hillsman, Matthew 52053
Hillstrom, Richard L. 50737
Hilpert, Dr. Wilhelm 20424
Hilpert Stuber, Susanne 41638
Hiltermann, Elisabeth 16806, 17004
Hilti, Caroline 41564
Hilti, Joseph 41444
Hiltula, Leena 11449
Himanen, Jaakko 11361
Himber, René 12321
Himmelbauer, Robert 02080
Himmelsbach, Sabine 20894
Himmelstrand, Erik 41054
Himmler, Radim 09804
Hincapié, Alberto 08651
Hincapié Santa Maria, Jaime 08707
Hinck, Petra 19462
Hind, Charles 44964
Hindelang, Prof. Eduard 19816
Hindman, James 50208
Hinds, Roger 46776
Hine, Charles Henri 52642
Hine, Marjorie 07180
Hine, N.A. 43816
Hinerman, Lenore 48915
Hines, Christine 06951
Hines, Tommy 53183
Hinghaus, Gabriele 20260
Hingst, Eckhard 19213
Hinkley, Marilyn J. 54324
Hinman, Alison 46294
Hinman, Marjory 49176
Hinnant, Vandorn 49066
Hinrichsen, Prof. Dr. Torkild 18915, 18935
Hinrichsen, Dr. Torkild 18957
Hinsch, Werner 19849
Hinsching, Susanne 20737
Hinson, Tom E. 47624
Hinton, Dr. Brian 44255
Hinton, William 50278
Hinz, Dr. Hans-Martin 17072
Hinz, Sally 01303
Hiott, William D. 47614, 47615
Hippe, Margarethe 02590
Hipps, Will 46577
Hipwell, John 07268
Hir, Dr. János 23755
Hirabayashi, Kunio 29109
Hirai, Takuya 28882
Hirakawa, Minami 29210
Hirako, Akihiko 28457
Hiramatsu, Hideki 28405
Hirano, Irene 50238
Hirano, Itaru 29190
Hirata, Shozo 29341
Hirata, Yutaka 29607
Hirayama, Miyako 29190
Hirci, Andrej 38457
Hird, Maureen 45040
Hirner, Dr. René 19090
Hiroshi, Doi 28528
Hiroshi, Sugaya 28604
Hiroya, Yukiatu 29161
Hirsch, Hans Joachim 20167
Hirsch, Helen 41936
Hirsch, Ilana 52211
Hirsch, Jan 53087
Hirsch, John 46512
Hirsch, Marjorie 53704
Hirsch, Dr. Martin 20488
Hirsch, Dr. Vanessa 18915
Hirschbiegel, Dieter 19702
Hirschel, Anthony G. 47513
Hirschfeld, Prof. DR. Gerhard 21969
Hirschler, Christa 21817
Hirschlieb, Brandy 52386
Hirst, M. 38675
Hirst, Marlys 46648
Hirth, Dr. Antoine 11987
Hirvonen, Anna-Riikka 10951, 10952
Hisa, Asuka 52901
Hiscock, Donna 06841
Hiscock, E. 43525
Hiscock, Rosella 06649
Hiscox, Margaret 01430
Hiss, Elfi 41657
Hitchmough, Dr. Wendy 44229
Hite, Andy 51931
Hite, Jessie Otto 46511
Hitov, Peicho 05167
Hittaller, Hans Martin 01778
Hitz-Walser, Cordula 41505
Hitzing, Jörg 21279
Hiukka, Toini 11168
Hjartarson, Siguður 24003
Hjermind, Jesper 10317
Hjorth, Carl-Gustav 40964
Hladky, Michael 51213
Hladky, Sylvia 20434
Hlaváčková, Dr. Miroslava 09820
Hlaváčková, Silvie 09873
Hlôška, L. 38391
Hlusko, Leslea 46758
Hnateyko, Olha 51274
Ho, Keimei 28792
Ho, Lynn 46431
Hoag, Everett H. 51015
Hoagland, Glenn D. 48860
Hoard, Constance G. 47099
Hoare, Adrian D. 46089
Hoare, Nell 46034
Hobaugh, J.H. 52931
Hobbs, M.D. 44183
Hobbs, Patricia A. 50071
Hobein, Beate 18856
Hoberg, Dr. Annegret 20491
Hobin, Ellen 05955
Hobl, PhDr Dalibor 09600
Hoboltth, Nina 09956
Hobson, Marlena 53276
Hoch, Manfred 21538
Hochberg, Dr. F.G. 52860
Hochberger, Ernst 19441
Hochdörfer, Achim 03042
Hoche, Frank 19054, 20931
Hochleitner, Dr. Martin 02317, 02322
Hochreiter, Otto 01969, 01983
Hochscheid, Christian 18009
Hochstrasser, Markus 41888
Hochuli, Prof. Dr. P. 42103
Hochuli-Gysel, Dr. Anne 41138
Hockett, Esther 48155
Hockman, Winnie 06757
Hodel, Daniel E. 41586
Hodel, R. 32473
Hodge, Anne 24807
Hodge, Dr. G. Jeannette 51076
Hodge, Katherine 49306
Hodges, Greg A. 52336
Hodges, Rhoda 48312
Hodgins-Wollan, Sonya 06023
Hodgson, Beverly 49131
Hodgson, Clarence 05802
Hodgson, Kenneth O. 51355
Hodgson, Polly 44172
Hodgson, Tom 43872, 43874
Hodkinson, Wendy 45090
Hodslavská, Lenka 09595, 09596
Hodson, Janice 47180
Höbler, Dr. Hans-Joachim 19919
Höck, Anton 02136
Höck, Dr. Christoph 19435
Höfchen, Dr. Heinz 19412
Hoefer, Denise 53637
Hoefer, Dieter 17977
Höfer, Dr. Hubert 19449
Höfert, Dr. Dorothee 20166
Höffer, Dr. Volker 21388
Höffner, Richard 20505
Höfler, Werner Oskar 19114
Höfling, Eckart Hermann 04721
Hoefnagels, J. 03400
Høeg, Asger 10057
Hoegen, Friedrich 20287
Högl, Dr. Günther 17922
Höglander, Agneta 40782
Höglmeier, M.A. Barbara 20601
Höhl, Dr. Claudia 19182
Höhne, Annett 18760
Höibo, Roy 33809
Hoekstra-van Vuurde, C.J. 32079
Hoel, Arild 33707
Hoel, Randall 53051
Höller, Anette 41180
Höller, Eva 02779
Höller, Silvia 02129

Höllinger, Mag. Nina 01836
Hoelscher, Dr. Monika 16481
Höltken, Dr. Thomas 19631
Höltl, Georg 20989, 22065
Hoeltzel, Susan 47048
Hölz, Marjatta 22135
Hölzel, Josef 01766
Hoelzeman, Buddy 50848
Hölzer, Dr. Adam 19449
Hölzinger, Nora 17795
Hölzl, Dr. Christian 03019
Hölzl, Martin 17113
Hölzl, Dr. Regina 03019
Höneisen, Markus 41831, 41837
Hönggi, Ambras 41689
Hoenig, Thomas M. 49694, 49697
Hönigmann-Tempelmayr, Mag. Alexandra 02945, 02947, 02948, 03000, 03007, 03072, 03073
Höper, Dr. Corinna 21960
Höpfner, Martin 16582, 16587, 16592, 16593
Hoerig, Dr. Karl 48589
Hörker, Alois 02096
Hörmann, Barbara 19164
Hörmann, Fritz 02923
Hörner García, Gabriel 31300
Hoerschelmann, Dr. Antonia 02929
Hösch, Jürgen 18386
Höser, Dr. Norbert 16498
Hoesgen, Jennifer 06001
Hoesli, Hans Rudolf 41254
Höylä, Risto 10961
Hofbauer, Erwin 01896
Hofbauer, John D. 46791
Hofbauer, Dipl.-Ing. Mag. Raimund 02238
Hofer, Elke 02057
Hofer, Ernst 41833
Hofer, Hans 41821
Hofer, Manfred 02138
Hofer, Rudolf 02255
Hofer, Verena 41578
Hoff, Dr. Marlise 17161
Hoff, Prof. Dr. Ralf von den 18426
Hoff, Svein Olav 33636
Hoff Jørgensen, Jan 33480
Hoffberger, Rebecca 46545
Hoffer, Andreas 02200
Hofferer, Robert 02343
Hoffert, Tonia 53507
Hoffman, Angela 52242
Hoffman, Dan 06425
Hoffman, Jill 53772
Hoffman, Kate 49438
Hoffman, Dr. Kerstin 18022
Hoffman, Marilyn 53055
Hoffman, Michael E. 51138
Hoffman, Dr. Richard S. 50521
Hoffman, Steeven A. 53123
Hoffman, Thomas 52844
Hoffmann, Almut 17206
Hoffmann, Dr. Andreas 18918
Hoffmann, Anja 17921
Hoffmann, Audrey 01636
Hoffmann, Christa 22703
Hoffmann, Christian 11405
Hoffmann, Dr. David Marc 41870
Hoffmann, Dipl.-Rest. Frauke 18009
Hoffmann, Dr. H. 17878
Hoffmann, Helmut W. 20672
Hoffmann, Dr. Ingrid-Sibylle 22276
Hoffmann, Jens 52779
Hoffmann, Joel M. 50635
Hoffmann, John 47496
Hoffmann, Paul 02134
Hoffmann, Dr. Per 17569
Hoffmann, Dr. Silke 17990
Hoffmann, Susanne 20985
Hoffmann, Dr. Tobias 19344
Hoffmann, Walter 21371
Hoffmann, Will 30491
Hoffpauer, Diane 47894
Hofgartner, Mag. Heimo 01965
Hofmacher, Anna 02540
Hofman, Jack 49990
Hofman, O. 25078
Hofmann 21078
Hofmann, Dr. Alexander 17201
Hofmann, Andreas 18567
Hofmann, Dr. B. 41213
Hofmann, Beate Alice 18979, 18980
Hofmann, Bernard 46979
Hofmann, C. 31760
Hofmann, Christa 03045
Hofmann, Fabian 18391
Hofmann, Gertrud 19490
Hofmann, Helmut 02355
Hofmann, Irene 51332
Hofmann, Jürgen 19317
Hofmann, Lea 41181
Hofmann, Lois 47966
Hofmann, Martina 20398
Hofmann, M.A. Rainer 21113
Hofmann, Ramona 22825
Hofmann, Susanne 41868
Hofmann-Kastner, Dr. Iris 22826
Hofmeister, Walter 19240
Hofschen, Dr. Heinz-Gerd 17549
Hofstad, Turid 33973
Hofstetter, Adolf 20993
Hofstetter-Sprunger, Heidi 41514
Hofwolt, Gerald 49357
Hogan, Jacqueline 46817
Hogben, Barbara 01112
Hogben, Ian 01112
Hoge, Robert 51116
Hogen, Tim 46961
Hogenhaug, Gunnar 33888
Høgestøl, Mari 33863
Hogue, Kate 49796
Hoguet, Anne 14670
Hohbaum, R. 18600
Hoheisel, Tim 53974
Hoheisel-Huxmann, Reinhard 17569
Hohenberg, Anita Fürstin von 01728
Hohenlohe, Prinz Constantin von 21554
Hohenlohe-Bartenstein, Ferdinand Fürst zu 21642
Hohenlohe-Langenburg, Fürst zu 19823, 19827
Hohenlohe-Oehringen, Katharina Fürstin zu 20613
Hohensee, Falko 17901
Hohl, Erhard 01778
Hohl, J. 22428
Hohmann, H.-J. 19143
Hohmann, Markus 20029
Hohrath, Daniel 17072
Hohri, Sasha 51129
Hohut, Khrystyna 05756
Høibo, Roy 33549
Høie, Bjørn K. 33687
Hoisington, Rena 49199
Højer Petersen, Anne 10186
Højgaard, Johnny 10018
Hokanson, Randee 50849
Hol Haugen, Bjørn Sverre 33477
Holahan, Mary F. 54189
Holahan, Paula M. 50395
Holben Ellis, Margaret 51216
Holbrook, Christopher 47790
Holbrook, Maureen 01301
Holcomb, Grant 52396
Holcomb, Julie 47826
Holcomb, Michael 48437
Holcombe, David C. 47626
Holcombe, Holly 47641
Holdcraft, Rose 47185
Holden, Charles 44354
Holden, Florence 47776
Holden, Rebecca 53036
Holden, Sally 43738
Holden, Steve 48832
Holden, Thomas 48197
Holdenbusch, Heinrich 21435
Holderbaum, Linda 46643
Holding, John 07237
Holdren, Greg 46787
Holdstock, Barry C. 44490
Holdsworth, Jane A. 51331
Hole, Frank 51054
Holečková, Dana 09499
Holeman, Claude A. 47981
Holen, Dr. Stephen 48075
Holeva, Daniel 50673
Holian, Sarah 49003
Holland, J. 01552
Holland, Kim 51419
Holland, M. 38643
Holland, Neil 43259
Holland, Rudolf 22827
Hollander, Barbara J. 46243
Hollander, Stacy 51113, 51171
Hollander, Volker 17810
Hollberg, Dr. Cecilie 17530
Hollein, Max 18391, 18393, 18394
Hollender, G. 42111
Holler, Prof. Dr. Wolfgang 16935, 18794, 20928, 21924, 22396, 22398, 22401, 22403, 22409, 22411, 22412, 22416, 22417, 22418, 22419, 22420, 22423
Holler-Schuster, Mag. Günther 01982
Hollerbach, Dr. R. 19605
Hollerová, Petra 09403
Holleschorsky, Catherine und Thomas 20220
Holliday, Anne 51667
Holliger, C. 41260
Hollinetz, Alfred 02874
Hollingworth, Dr. Brad 52754
Hollingworth, John 44808
Hollingworth, Mar 50214
Hollister, Robert 50252
Hollitzer, Tobias 19909
Holló, Dr. Szilvia Andrea 23916, 23917, 23918
Hollomon, James W. 49151
Hollosy, Gyuri 50597
Holloway, Donald 49002
Holloway, James 44157
Hollows, Victoria 44281
Hollwedel, Dr. Alke 20048
Holm, Axel 51358
Hólm, Geir 23994
Holm, M.A. Nina 20478
Holm, Torben 10286
Holm Sørensen, Bodil 10117
Holman, Graham 07299
Holman, Kurt 51783
Holman, Robin 50036
Holman, William M. 48154
Holmberg, Lori 50763
Holmes, Carolyn C. 47395
Holmes, Dugald 01358
Holmes, Meryle 52804
Holmes, Saira 43505
Holmes, Selean 47472
Holmes, Willard 49199
Holmes Helgren, Heidi 47218
Holmesland, Liv 33843
Holmgrain, Ardith 51111
Holmgren, Bengt 40806
Holmlund, Bertil 11162
Holmsted Olesen, Chritian 10129
Holmstrand-Krueger, Lena 40851
Holo, Dr. Selma 50222
Holoch, Jürgen 40959
Holodňáková, Dr. Radmila 09926
Holowacz, Eric V. 46668
Holozet, Andrée 16323
Holpainen, Anja 10938
Holsbeke, Mireille 03295
Holst, Dr. Daniela 20687
Holst, Gerd 22702
Holst, Lars 10066
Holstein, Lars 41000
Holsten, Christina 40576
Holsten, Dr. Siegmar 19448
Holt, Daniel D. 46120
Holt, Frank 51591, 51595
Holt, Loucinda 47641
Holt, Sharon 51260
Holtby, J.Mick 44346
Holte, Bettye 47594
Holten, Arnfinn J. 33405
Holten, Johan 16890, 19073
Holten, Lars 10174
Holthaus, Dr. Katrin 21016
Holthof, Marc 03286
Holthuis, Dr. Gabriele 21649, 21650, 21651
Holtman, Marge 53165
Holton, Linda 48971
Holtry, Cara 47243
Holtz, Jerzy 40934
Holtz, Wolfgang 17263
Holtzapple, John C. 47735
Holverson, John H. 53018
Holvět, Brigitte 13349
Holz, Francisc 36432
Holz, Uwe 17370, 22650
Holzach, C. 21031
Holzach, Cornelie 21029
Holzer, Dr. Elmar 22388
Holzer, Mag. Isabella 02183
Holzer-Weber, Ruth 42120
Holzhey, Dr. Magdalena 17995
Holzinger, Hans 20479
Holzleitner, Mag. Laurin 01870
Holzman, Rae 51468
Holzner, Dr. Anna 02046, 02047
Homa-Rożek, Ewa 34618
Homann, Dr. Klaus 22180
Homberger, Lorenz 42097
Homburg, Cornelia 52595
Homen, Prof. Zoran 08902
Homer, Russell 51684
Homering, Liselotte 20173
Hommell, Michel 13802
Homolova, Dr. Alexandra 38233
Homs-Chabbert, Arlette 14136
Honadel, Elroy 51452
Honaker, Corney A. 53457
Honcoopova, Dr. Helena 09776
Honda, Masamitsu 28655
Honda, Shizuo 28907
Honerød, Anne Grete 33799
Hong, Trinh Van 54622
Honkanen, Pekka 10822
Honour, Peggy 45739
Honzíková, Barbara 09682
Hoo, Gloria Chun 47207, 47208
Hoober, David 51901
Hood, Betty 47259, 53952
Hood, Michael 49485
Hood, Sadie Ann 54205
Hoofstat, Jan van 04002
Hoog, Phil 52753
Hoogenboom, W. 31962
Hoogendonk, Mabel 32296
Hook, I.D. 43806
Hook, Lu 48895
Hooker, Saralinda 47213
Hooks, Rosie Lee 50267
Hooks, Terri 47784
Hoon, Wilhelm 16683
Hooper, Dr. John 01510
Hoopes, John 49990
Hoops, Erik 17569
Hoorn Bakker, A. 32739
Hoover, Bessie 52268
Hoover, Rhonda J. 53827
Hooymayers, Dr. J.M.M. 32484
Hoozee, Dr. Robert 03639
Hopfinger, Rebecca 47600
Hopfner, Dr. Rudolf 03019
Hopfner, HR Dr. Rudolf 03063
Hopgood, Dr. James F. 49279
Hopkins, Dave 06785
Hopkins, S.K. 45611
Hopkins, Terri M. 50522
Hopkins Coughlin, Joan 54026
Hopkinson, Susan 45626, 45627
Hopmann, Gerd 18158
Hopp, Dr. Andrea 21610
Hoppe, Barbara G. 49828
Hoppe, Michael 17928
Hoptman, Laura 51942
Horacio Aranguren, Juan José 00581
Horat, H. 41587
Horat, Dr. Heinz 41592
Hordyński, Piotr 34815
Horeftosz, Konstantin 23653
Horgan, Judy 51957
Horgen, Randi 33572
Horký, Milan 09459
Horman, Trevor 00784
Horn, Damian 45587
Horn, Elizabeth 52152
Horn, Elizabeth G. 06674
Horn, Gabriele 17173
Horn, Dr. Karsten 18196
Horn, Ken 50234
Horn, Murray 51964
Horn, Dr. Reinhard 20424
Horn, Robert 47498
Horn, Tom 21265
Hornback, Kristina 48791
Hornby, D. Mary 43854
Horndasch, Mickey 46360
Horne, Daphne 01138
Horne, Gene 33267
Horne, Graham 01096
Horne, Jerry G. 51921
Horner, Gerald 48080
Horner, John R. 46982
Hornig, Dieter 22697
Hornik, Dr. Heidi 53776
Hornstrup, Nils 10057
Hornung, Herb 52732
Hornung, Prof. Dr. Maria 03050
Horoiu, Otilia 36118
Horpeniak, Dr. Vladimír 09581, 09856
Horray, Henrietta 48977
Horrigan, William 47761
Horsfield, Louise 43869
Horsley, Juliet 45799, 45805
Horsnæs, Helle W. 10136
Horst, Dr. Katarina 19437
Horst, Randy 48143
Horstik, Deborah 49945
Horsting, Archana 46756
Horstrup, Ellen 21834
Horton, Constance 49736
Horton, Lon 47308
Horton, Melissa 50925
Hortz, Laura 53122
Horvath, Barbara 03081
Horváth, Dr. Béla 23555
Horvath, Eugen 02331
Horváth, György 23441
Horváth, Dr. István 23549, 23550
Horváth, Dr. László 23725
Horvath, Ursula 02413
Horwath, Alexander 03046
Horwood, Michelle 33235
Horz, Ulrich Chr. 21264
Horžen, Vesna 38505
Hosack, George 54160
Hosaka, Kenjiro 29491
Hosar, Kåre 33637
Hoschek, Mag. Eleonore 02655
Hose, Henry 53558
Hoskin, Rebecca 00871

Hoskins, Rebecca 50851
Hoskins, Susan 51930
Hoskins, William J. 53122
Hoskinson, Susan 51979
Hoskisson, Don 50784
Hošková, Jitka 09729, 09861, 09878
Hošková, PhDr. Miloslava 09690
Hosley, William 49192
Hosoe, Eikoh 29325
Hosokawa, Morisada 29385
Hosokawa, Toshiro 29428
Hossain, Shamsul 03214
Hossaka, Luiz 04880
Høst-Madsen, Lene 10135
Hoštálek, Petr 09463
Hostetter, David 49867
Hostyn, Norbert 03930
Hotchner, Holly 51112, 51224
Hoti, Prof. Afrim 00013
Hotz, Benedict 41595
Hotz, Claus-Dieter 18739
Hotz, Dr. Gerhard 41175
Hou, Fuxin 07605
Hou, Jiming 08372
Hou, Liancheng 07901
Hou, Wenyu 07582
Hou, Yancheng 08314
Hou, Yumei 07949
Houby-Nielsen, Dr. Sanne 40913
Houdamer, Chloé 12114
Hough, Douglas R. 51693
Hough, Katherine 51681
Hough, Melissa E. 51822
Houghtling, Kim 07044
Houghton, J. 50926
Houghton-Bowen, Linda Kay 49892
Hougue, Veronique de la 14682
Houkjaer, Ulla 10129
Houlden, Penny 06851
Houle, Joy 46542
Houlihan, E.T. 50051
Houliston, Laura 44894
Hoult, R. 33076
House, Emma 43381
House, George M. 46151
House, Kirk 49143
House, Patricia 48807
Houseago, Avril 45281
Housego, Bill 01036
Housen, Jean 03813
Houssinot, Jean-Claude 15177
Houston, David 47616
Houston, Joe 46855
Houston, John 52354
Houston, L.S. 50622
Houston, Dr. Margaret 49988
Houyez, Marcel 03809
Houzet, Bernard 12685
Hovagimian, Sarkis Vahan 03031
Hovell, Bob 44046
Hoveng, Dot 46362
Hovens, P. 32488
Hovers, Dr. Erella 25001
Hovi, Jouko 10713
Høvsgaard, Thomas 10283
Howard, Betty L. 54088
Howard, Helen 44937
Howard, J. 38552
Howard, James 50916
Howard, Jan 52170
Howard, Jane 44877
Howard, Jean 45037
Howard, Jerome 06508
Howard, Jerry 50605
Howard, Julia 43800
Howard, Louise 50950
Howard, Nora 46526
Howard, Paul 46112
Howard, Stephen R. 47440
Howarth, Mjr. Alan 43993
Howarth, Crispin 00946
Howarth, Frank 01546
Howdle, Joanne 44518
Howe, Angela 45552
Howe, Eileen 52252
Howe, Henry D. 48630
Howe, Katherine S. 49397
Howe, Kathleen 46015, 47574
Howe, Philip 33212
Howell, Alan C. 45591
Howell, Beverly 49955
Howell, Busser 53576
Howell, Karen 44948
Howell, Melva 52293
Howells, David 45784
Howes, Stephen 47908
Howland, Robert 53167
Howoldt, Dr. Jenns E. 18932
Hoyd, Ed 47815
Hoyer, Dr. Eva Maria 19911
Hoyer, Prof. Jan 18044
Hoyer, Klaus 16583
Hoyer, Lindy 51555
Høyersten, Erlend G. 33612
Hoyos, Cristina 40073
Hoyos, Hans 02599
Hoyos Galarza, Melvin 10409
Hoyt, Christine 47546
Hoyt, Marilyn 48578
Hoyum, Wanda 46730
Hozumi, Toshiaki 29219
Hrabánková, Dr. Světlana 09612, 09614
Hrebíčková, Marta 38337
Hrisenková, Nataša 38304
Hristov, Dr. Martin 05191
Hristova, Mariana 05193
Hristova-Radoeva, Vessela 05203
Hroch, David 09835
Hron, Vincent 46872
Hrtánková, Valéria 38233
Hrushetska, Maryna 50213
Hruška, Václav 09597, 09810
Hsiao, Tsung-huang 42417
Hsieh, Chia-Liang 42283
Hsieh, Kun-Shan 30194
Hsu, Claire 07793
Hu, Daxin 08459
Hu, Jiqiao 07481
Hu, Qingyun 08246
Hu, Shiliang 08511
Hu, Xiaojiang 07932
Hu, Yaoqing 07765
Hu, Yuanhong 08407
Hu, Zhisheng 07500
Huan, André 15036
Huang, Chongyue 08225
Huang, Chun-Min 42183
Huang, Dong 07786
Huang, Fengquan 07717
Huang, Gaoqian 07532
Huang, Huaixin 08152
Huang, Jianbang 07658
Huang, Juming 08035
Huang, Kuang-Nan 42415
Huang, Liang 07897
Huang, Ning 08200
Huang, Peiqi 07745
Huang, Qiang 08023
Huang, Qishan 08054
Huang, Tsai-Lang 42428
Huang, Weixiong 07624
Huang, Xizheng 07596
Huang, Yaohua 08110
Huang, Yongbang 07496
Huang, Yunzhong 08056
Huard, Micheline 06945
Huart, Nicole d' 03479
Hubach, Lynn 46498
Hubbard, Dr. Charles M. 49191
Hubbard, Dr. Jane 46049
Hubbard, Jeremy 33056
Hubbard Mecham, Denny 52990
Hubbs, Heather 47517
Hubbs, Jenice 05329
Huber, Arthur 41256
Huber, Dr. Bernhard 17469
Huber, Christoph Anselm 03449
Huber, Franz 02864
Huber, Markus 41830
Huber, Mary 47293
Huber, Dr. Waldemar 01926
Huber, Walter 41350
Huber, Mag. Wolfgang 02202
Huberman, Anthony 14743
Hubert, François 03419, 15012
Hubert-Jeannin 12371
Hubinger, Horst 01733
Huck, Thomas 18717, 18720
Hudák, Julianna 23577
Huddleston, David 06953, 06954
Hudeczek, Dr. E. 02737
Hudgins, Don 52996
Hudson, Briony 44933
Hudson, Dr. Julian D. 47595
Hudson, Leslie 46156
Hudson, Louise 52243
Hudson, Mark S. 48756
Hudson, Dr. Neville 33049
Hue, Michel 12609, 12804, 13672, 13782, 14128, 14130, 16041
Hueber, Régis 12582
Hübl, Prof. Dr. Richard 02834
Hübner, Heinz 01691
Hübner, Dr. John 01987
Hübner, Dr. Ute 18531
Hübsch, Norbert 16945
Hückstädt, Antje 21119
Hügel, Dr. Peter 36111
Hüls, Johannes 17393
Hülsewig-Johnen, Dr. Jutta 17341
Hülsmann, Gaby 21298
Huemer, Dr. Peter 02126, 02136
Hümme, Dr. Julia 18285
Hünnekens, Prof. DR. Ludger 16886
Hünninghaus-Böckmann, Ute 02775
Hürkey, Dr. Edgar J. 18351
Huerta, Benito 46351
Huerta, Enrike 39611
Hürzeler, Adelheid 41896
Hüsam, Gernot 17725
Huescar, Duque de 39522
Hüsken, Wim 03861, 03862, 03863
Hüßner, Reinhard 19350
Huet, Dr. A. 13886
Huete, Alberto 34078
Hueter, Diana 47941
Hueter, Ken 46185
Hütt, Dr. Michael 22235, 22236
Hüttel, Dr. Richard 19921
Hütter, Prof. Dr. Hans Walter 17455
Hüttermann, Prof. Dr. Armin 20184
Hüttner, Rudolf 02876
Huff, Richard E. 49533
Huffaker, Shirley 50608
Huffman, Joan 49189
Huffman, John 53323
Huffman, Lorilee C. 47241
Huffman, Vincent 06255
Hufnagl, Prof. Dr. Florian 20436
Hufnagle, Sue 47317
Hug, Guy 13730
Hug, Hannes 41155
Hug, Hans 41507
Hug, Shamsul 03236
Hug, Dr. Thomas 41394, 41756
Hugel, André 13302, 15052, 15054
Hugentobler, Hanspeter 41496
Huggert, Anders 41000
Hughes, Dr. David 54146
Hughes, Debra 54193, 54194
Hughes, Gareth 45538, 45680
Hughes, Gary 06838
Hughes, Helga 44326
Hughes, James T. 53966, 53968
Hughes, John 44767
Hughes, K.C. 46042
Hughes, Marianne 51040
Hughes, Mark 45250
Hughes, Michael 44600
Hughes, P. 01159
Hughes, Richard Ll. 44759
Hughes, Sheila 43341
Hughes, Susan 07145, 43818
Hughes, Thomas A. 51301
Hugo 38577
Hugomiot, Jean-Yves 11976
Hugonich, J. Yves 15582, 15583, 15584, 15585
Huguet, Jean-Pierre 41677
Huhtamäki, Tuula 11479
Huhtamäki, Ulla 10817
Hui, Hwang Sun 29759
Huijs Janssen, Dr. Paul 32375
Huijts, Stijn 32779
Huizi, María Elena 54543
Huizing, Pat 50876
Hujala, Susanna 11408
Hull, Prof. Andrea 01519
Hull, Dale 51806
Hull, Howard 43884
Hull, Karen 45469, 45471
Hull, Priscilla 43668
Hulló, István 38156
Hulse, E. 43594
Hulse, Sir Edward 43593
Hulse, Robert 44809
Hulser, Kathleen 51243
Hulshof, G. 31936
Hulston, Nancy 49680
Hultsch, Ute 18805
Humbert, Jean-Baptiste 34057
Humbert, Vincent 13474
Hume, J. Wilbert 05565
Hume, Ted 05864
Humer, Franz 01741, 02494, 02495
Humes, Alex B. 51839
Hummel, Alfred 20235
Hummer, Hermann 03049
Humphrey, Chevy 51900
Humphrey, L. 43722
Humphrey, M. 01552
Humphrey, Sylvia 45227
Humphreys, Mary 43600
Humphreys, Richard 44995
Humphries, Alan 44668
Humphries, Martin 44823
Humpl, Christine 02200
Hundertmark, Cordula 18490
Hundt, W.-D. Graf von 16617
Hunger, Mattli 41787
Hunt, Alice 52279
Hunt, C. 45714
Hunt, Clare 45710
Hunt, Dr. D.A. 44702
Hunt, David 51578
Hunt, Dr. Donald W. 46419
Hunt, Eugene A. 49715
Hunt, Harrison 52464
Hunt, Jacqueline 45333
Hunt, Michael 48265
Hunt, Perdita 43882
Hunt, Phillip 46922
Hunt, Rhonda 01341
Hunt, Susan 50591
Hunt, Tony 33107
Hunt, Wilma 47839
Hunt Chamberlain, Carola 52944
Hunt Miller, Lee 52781
Huntebrinker, Dr. Jan Willem 18993
Hunter, Charles 54368
Hunter, Elizabeth 05329, 06600, 43316
Hunter, Elizabeth Ives 48057
Hunter, Fredericka 50466
Hunter, Jamie 06277
Hunter, Jim 43842
Hunter, Judith 46046
Hunter, Ken 45729
Hunter, Lance 51637
Hunter, Michael 44084
Hunter, Paul 51170
Hunter, Richard 51578
Hunter, Terry 49211
Hunter, Thomas 53402
Hunter, Valerie 44138
Hunter Carsch, C. Morag 45254
Hunter Larsen, Jessica 49031
Hunter-Stiebel, Penelope 52086
Huntoon, Katherine 51370
Huntzinger, John 50685
Hunziker, Armando T. 00320
Hunziker, John 52389
Huo, Qiming 07955
Huo, Prof. Wei 07620
Hupe, Dr. Joachim 22102
Hupponen, Elina 11280
Huret, J. 15879
Hurezan Pascu, Georghe 36113, 36442, 36443, 36475, 36623, 36740
Hurkmans, Riek 32242
Hurler, R. 16446
Hurley, Arthur L. 47381
Hurry, S.P. 43984
Hursey, Dick 50081
Hurst, Henry 43730
Hurst, Rachel 44085
Hurt, V. 03792
Hurtado Alfaro, Miguel Ángel 39808
Hurtado Londoño, Gloria 08563
Hurum, Prof. Jørn H. 33737
Hurvitz, Uri 25126
Hušáková, Ivana 09838
Husband, John 33128
Husband, Michael B. 49421
Husband, Timothy 51152
Huseboe, Arthur R. 53116
Husedžinović, Meliha 04283
Huskisson, Christine 48891
Husmann, Karin C.E. 32604
Husmeier-Schirlitz, Dr. Uta 20654, 20655, 20656
Huss, John 50419
Hussaini, Syed Mohammad 24631
Hussein, Datò Ibrahim 30612
Hussein, Elena 36736
Husseini, Frederic 30160
Husseini, Frédéric 30167
Hussenot, Eric 12165
Hussey, Noelene 33136
Hussey-Arntson, Kathy 54186
Husslein-Arco, Dr. Agnes 02949, 02950, 02951, 02952
Husted, Charlene 48871
Husty, Mag. Peter 02625
Huszár, Zoltán 23757, 23758, 23759, 23762, 23763, 23767, 23768, 23770, 23772, 23773, 23778, 23779, 23780, 23781
Hutchcrot, Sally 49810
Hutchings, Lamar 50021
Hutchins, Patricia 47278
Hutchins Bobyrk, Dawn 53112
Hutchinson, J. Dwight 50660
Hutchinson, John 24783
Hutchinson, Susan 01508
Huterberger, Renate 02422
Huth, Nancy 50902
Huthwohl, Joël 14572

Hutin, Albert 47346
Hutt, Marilyn 53284
Hutt, Richard B. 01568
Hutt, Sarah 46948
Hutter, Dr. Ernestine 02625, 02629
Hutterer, Dr. Karl 52860
Hutterer, Dr. Rainer 17469
Huttlinger, John B. 49891
Hutto, Dr. Richard 50757
Hutton, Susan 53105
Huttunen, Marianna 11009
Hutz, Dr. Ferdinand 01802
Huuskola, Arto 11347
Huuskonen, Kari 10871
Huvenne, Patrick 03706
Huvenne, Dr. Paul 03290
Huwer, Dr. Elisabeth 19071
Huwyler, Dr. Edwin 41485
Hux, P. Meinrado 00432
Huxley, Geralyn H. 51938
Huybrechts, R. 04068
Huyge, Dirk 03495
Huygens, Ch.E.C. 32733
Huynh, Michel 14729
Huysza, Hans 03107
Huzel, Karl 19882
Hwan Kim, Myung 29781
Hwang, Pae Gang 29869
Hyams, Prof. Dr. Helge-Ulrike 20190
Hyde, Hilary 45259
Hyde, J.R. 50581
Hyde, Lex 24849
Hyer, Warren W. 54307
Hyland, Alice 48227
Hyland, Douglas 51029
Hylmarsová, Eva 09646
Hyman, Dr. Paul 45048
Hyman, Ryan 50853
Hynes, Patricia 51112
Hynes, William Q. 53080
Hyross, Dr. Peter 38203, 38207, 38208, 38209, 38213, 38214, 38215, 38218, 38219, 38229, 38235
Hysell, Barbara 46793
Hyšková, Věra 09525
Hyson, Dr. John M. 46561
Hyss, Dr. Lothar 20521
Hytha, Prof. Dr. Robert K. 02465
Hyvönen, Marja-Liisa 10892
Hyytiäinen, Maija 11323, 11324

I

Iaconis, Héctor José 00115
Iacopi, Dr. Irene 27440, 27523
Iancu, Emilia 36463, 36534, 36626
Iancu, Pavel 36397
Iannelli, Dr. Maria Teresa 28246
Iannucci, Heather 54332
Ibanez, Jean-Louis 12311
Ibarra, Jorge A. 23298
Ibbertson, Prof. H.K. 33046
Ibbotson, Michael P.S. 33135
Ibing-Harth, Krista 16967
Ibolya, Dr. Szathmári 23516
Ibragimov, Token Smailovič 29729
Ibragimov, Ulykbek Šarachynovič 29692
Ibragimova, Gulnar 29708
Ibrahim 29744
Ibrahim, Abbas 30623
Ibrahim, Dr. Asma 34014
Ibsen, Tim 21565
Icart, Françoise 11729
Ice, Dr. Joyce 52882, 52883
Ichijyo, Akiko 29491
Ichim, Dorinel 36121
Ickowicz, Pierre 12721
Idänheimo, Raija 11093
Ide, Evan 47064
Ide, Richard 21620
Idemitsu, Shosuke 29400
Idleman, Paul 47333
Ido, Makoto 29269
Idris, Hassan Hussein 40392
Idshinnorov, S. 31628
Idzikowski, Waldemar 35024
Iemura, Katsuyuki 29563
Iercoşan, Neţa 36264
Ieronimidis, M. 23248
Ieronymidou, M. 09375
Ievleva, Nina Vasiljevna 37612
Iff, Ursula 41169
Ifkovits, Dr. Kurt 03051
Ifland, Dorothee 17050
Igel, Dr. Walter 18436
Igitian, Henry 00748
Iglesias Grande, Manuel 39161
Igna, Ion 36295
Ignatev, Viktor Jakovlevič 37062
Ignatjeva, Tatjana Ivanovna 37267
Igrakova, Natalja Ivanovna 37563
Ihász, István 23442
Ihrenberger, Elisabeth 02620
Iida, Kishiro 28573
Ikeda, Hosai 28525
Ikeda, Sakae 29348
Ikefuji, Kiyoshi 29461
Ikejima, Kazuo 29567
Ikibaeva, Roza Eržanovna 29717
Ikitian, H. 00730
Iles, Chissie 51279
Ilg, Werner 41512
Ilg, Dr. Wolfgang 01881
Ilić, Dragica 38074
Ilich-Klančnik, Breda 38454
Iliescu, Prof. Constantin 36597
Iliev, Christo 05002
Iliev, Petăr 05101
Ilieva, Marija 05010
Ilisch, Dr. Peter 20512
Iljasova, Ryšžan Iljasovna 29673
Iljin, Vadim Jurjevič 36829
Iljina, Ljudmila Ivanovna 36777
Ilkjær, Marianne 10081
Ilkosz, Jerzy 35328
Ille-Kopp, Regina 17349
Illich, Dr. Inge 02615
Illner, Dr. Eberhard 22742
Illuzzi, Dr. Michael C. 52980
Ilming, Thomas 03001
Ilse-Neuman, Ursula 51112
Ilshöfer, B. 22200
Imai, Yoko 29492
Imanse, G. 31888
Imbeault, Nisk 06317
Imber, Sofia 54541
Imbernor, Laurence 15016
Imbert, Christiane 12295, 12296, 12297, 12298
Imhof, Dirk 03297
Imhoff, Bert C. 06886
Imiela, Edward 35128
Immel, Dr. Isabell 21801
Immel, Dr. Wolfgang 17807
Immenkamp, Dr. Andreas 17921
Immerzeel, A.A.G. 32403
Immisch, T.O. 18904
Immler, Verena 20448
Immonen, Kari 11405
Immonen, Olli 11046, 11047
Imponente, Anna 26457
Impraim, Mary 22848
Imura, Yoshihiro 28816
Imychelova, E.C. 37896
Inaba, Hisao 29069
Inaba, Yoshinori 29190
Inai, Keijirô 29399
Inauen, Roland 41118
Inazuka, Hiroko 29522
Inbar, Janet 25122
Inbar, Yehudit 24990, 25026
Incardona, Dr. Angela 25165
İnce, R. Sefa 42946
Inch, John 33084
Inckmann, Lisa 20511
Indahl, Trond M. 33433
Indāns, Juris 30079
Indeck, Dr. Jeff 47230
Indra, Dr. W.I. Pandji 24482
Ineba, Pilar 40204
Infante, Victor Manuel 00323
Ingber, Abie 47561
Ingber, Barbara 51126
Ingebretsen, Gry 33576
Ingebrigtsen, Dr. Kristian 33746
Ingelheim, gen. Echter von Mespelbrunn, Graf zu 20303
Inghels, Cyriel 03305
Ingilby, Sir Thomas C.W. 45498
Ingle, R. 43797
Inglis, Elspeth 49667
Inglis, Marion L. 07182
Inglis, Robin 06480, 06482
Inglis, Stephen 05862
Ingmarsson, Niklas 40703
Ingo, Emi 10734
Ingold, Ed 48544
Ingraham, Joannie 54095
Ingram, Cathy 53872
Ingram, Daniel 49640, 49641
Ingram, Doris 47434
Ingram, J. 43396
Ingram, Jack 33205
Ingram, Jack E. 48654
Ingvordsen, Jens 09963
Iniesta González, Montserrat 40280
Innes, J.F. 06521
Innes, Kathryn V. 52277
Innes, Moira 44663
Innocenti, Pierluigi 25235
Ino, Koushin 29194
Inojatov, K.Ch. 54472
Inostroza Saavedra, Jorge E. 07469
Inoue, Kouki 28433
Inoue, Tadashi 28559
Insam, Peter 21223
Inselmann, Andrea 49548
Inserra, Margaret 53478
Insley, Debra 49088
Intas, Vaclovas 30315
Internoscia, David 50478
Intihar Ferjan, Jana 38454
Inui, Kazuo 28590
Inversini, Eva 41453
Io, Lau Si 08008
Iolen, Cadi 44754
Ion, Prof. Dr. I. 36419
Ion, Toma 36714
Ionescu, Ruxandra 36250, 36532
Ionesov, Anatoly 54461
Ioniţă, Ionel 36191, 36192, 36199, 36200, 36201, 36206, 36208, 36210, 36213
Iotta, Ivana 26003
Iovănuţ, Camelia 36682, 36683, 36684
Iovinelli, Maria Teresa 27539
Iprajian, Avedis 25013
Ipsen, Dr. Anne 16837
Irazú, José María 00568
Ireland, Lynne 50344
Ireland, Shasta 52023
Ireland, Terese 48129
Irén, Kállai 23373
Iriarte, Prof. Elba Esther 00566
Iribarren Avilés, Rodrigo 07473
Irie, Yoshiro 29493
Irie, Yoshiyuki 28638
Irinina, Elena 04978, 04979, 04980
Irish Jr., Mahlon 49331
Irle, Lisa 53830
Irlich, Franz 21710
Irrgang, Judith 16886
Irša, Dr. Rudolf 38291
Irvin, Mary A. 46787
Irvin, Stan 46508
Irving, Dorien 43462
Irving, Gillian 43902
Irving, James 44980
Irving, Mary L. 53474
Irving-Dusting, L. 01449
Irwin, Charles W. 46903, 51499
Irwin, Dr. Darren 07199
Irwin, Ruth E. 49936
Irwin Meyer, Elaine 51377
Irwinnson, Charles W. 46905
Isaac, Gwyneira 53469
Isaac, Preston 43838
Isabelle, M.Sc. Martine 06145
Isakova, Ksenija Viktorivna 43036
Isaksen, Knud Fink 10050
Isawumi, Prof. Dr. M.A. 33326
Isbrandt, Lester 51062
Isbruch, Ursula 21377
Isdale, John 33210
Iseki, Masaaki 29500
Isensee, Dr. Sabine 20900
Iser, Wunibald 20548
Isernhagen, Karin 42099
Isert, Ingo Rüdiger 21940
Işfănoni, Doina 36217
Ishag, Ali 34038
Ishibashi, Kentaro 28537
Ishiga, Harue 28524
Ishihama, Beniko 29137
Ishii, Hisataka 28622, 29483
Ishii, I. 28470
Ishikawa, Chiyo 53013
Ishikawa, Koichi 28764
Ishikawa, M. 28474
Ishikawa, Tetsuko 28537
Ishimaru, Shoun 28509
Ishizaki, Katsumoto 29536
Ishizuka, Kaname 28481
Ishtiaq Khan, Muhammad 34012
Isifidis, Kaplanis 23047
Iske, Gary 46703
Isken, Suzanne 50252
Iskender, Prof. Kemal 42835
Islamova, Tatjana Michajlovna 37852
Isler, Andreas 42116
Isler, Ernst 21327
Isler, Jacques 42044
Ismai, Reza 24693
Isman, Eileen 07400
Isogai, Hideo 28443
Ison, Judy C. 52196
Ison, Susan P. 50301
Isoz, Nicolas 41094
Israel, Sue 50483
Israfilovglu, I.R. 03166
Isshi, Kaihei 28549
Istas, Dr. Yvonne 21904
Istha, Marieke 31873
István, Dr. Berta 23426
Itai, Josef 25040
Itakura, Fumiaki 29493
Iten, Brigitte 41599, 41861
Ito, Masakazu 29097
Ito, Tomoyuki 29112
Itoh, Hiroji 29056
Itter, Wolfgang 19654
Ittmann, John 51867
Iuga, Georgeta Maria 36126
Iurman, Emilio 27102
Ivančević, Nataša 08938
Ivanenko, Pjotr Ivanovič 43062
Ivannikova, Ljudmila 37274
Ivanov, Aleksandr Aleksandrovič 37122
Ivanov, Ivan 05141, 05148
Ivanov, Stefan 05076
Ivanov, Vladimir Alexandrovič 37596
Ivanova, Elena 46855
Ivanova, Galina Olegovna 37034
Ivanova, Katja 05195
Ivanova, Maja I. 37597
Ivanova, Natalja Vasiljevna 36844
Ivanova, Silvia 46817
Ivanova, Tatjana Nikolaevna 37984
Ivanovskaja, Dr. Irina Nikolaevna 37235
Ivanuš, Rhea 08997
Ivask, Krista 10624
Ivaškina, Ljudmila 37901
Ivčević, Sanja 08954
Ivelic, Milan 07461
Ivens, R. 32511
Iversen, Mette 10317
Iverson, Dr. John B. 52313
Iverson, Stephen C. 50178
Ives, Alicia 46193
Ives, Dr. William 48404
Iveson, Jon 44013
Ivić, Prof. Vinko 09004
Ivica, Križ 38490
Ivičić, Ratomir 08848
Ivinskis, Antanas 30391
Ivkanec, Ivanka 08989
Ivory, Ken 33231
Ivory, Margaret 33168
Iwaki, Ken-ichi 28843
Iwakiri, Masanori 28928
Iwamoto, Mitsuo 28578
Iwasaki, Yohachiro 28552
Iwata, Tadao 28575
Izaqirre, Manu 38797
Izdebska, Barbara 35275, 35277
Izquerdo Pascual, José María 39538
Izsatovska, Dora 04985
Izui, Hidekazu 29190

J

Jaacks, Prof. Dr. Gisela 18949
Jaakkola, Markku 10753
Jaatinen, Aija 11009
Jaatinen, Carina 10727
Jaatinen, Toimi 11348, 11349, 11354, 11362
Jaber Al-Jaber, Ibrahim 36088
Jablonská, Dr. Beáta 38233
Jabłońska, Teresa 34595, 35354, 35355, 35358, 35360
Jablovska, Gaida 30103
Jacanović, Dragan 38132
Jachimowicz, Krzysztof 34634
Jachson, Haidee 45297
Jacklitch, Prof. Paul 46748
Jackowska, Anna 35191
Jackson, Angela 49574
Jackson, Barry 43783
Jackson, Bee 05864, 06941
Jackson, Bev 46170
Jackson, Bryan 05812
Jackson, Henry 44219
Jackson, Jerry 52436
Jackson, Kenneth T. 51243
Jackson, Margaret 47802
Jackson, Paul 37995
Jackson, Sailor 46635
Jackson, Steven B. 46982
Jackson, Toby 44996
Jackson, Tony 46039
Jackson-Forsberg, Eric 51348
Jacob, Dr. Christina 19100, 19101
Jacob, Eric 13232
Jacob, François 41413
Jacob, Dr. João Manuel Neto 35487
Jacob, Milner 33213

Jacob, Pascal 03332
Jacob, Walter 51936
Jacob-Friesen, Dr. Holger 19448
Jacobacci, Emilia 27509
Jacobi, Dr. Fritz 17225
Jacobs, Carin 46755
Jacobs, Charles 50536
Jacobs, Edwin 32640
Jacobs, F.J. 38570
Jacobs, Flora Gill 53931
Jacobs, John 49041
Jacobs, Joseph 51287
Jacobs, Karen 52846
Jacobs, Lisa D. 49944
Jacobs, Marty 51231
Jacobs, Mike 53584
Jacobs, Dr. Stephanie 19900
Jacobs, Warwick 44335
Jacobsen, Anne Marie 10049
Jacobsen, Bent Sonne 10191
Jacobsen, Gaute 33637
Jacobsen, Harald 33863
Jacobsen, Robert 50739
Jacobsen, Signe 10310
Jacobsen, Wolfgang 17069
Jacobson, Harold 49726
Jacobson, L. 38670
Jacobson, LouAnn 48150
Jacobson, Sally 48402
Jacobson, Thomas N. 51778
Jacobson, Thora E. 51881
Jacobsson, Ingela 40981
Jacobsson, Yvonne 40902
Jacoby, Christina 30197
Jacoupy, Charles 15760
Jacqmard, Jules 03771
Jacqmard, Philippe 03770
Jacqué, B. 15057
Jacquemin, Hervé 15249
Jacquemyn, Erik 03864
Jacques, Claude 13005
Jacques, R.A. 15837
Jacquesson, Alain 41425
Jacquet, Alain 15963
jacquinot, Marc 14985
Jacquot, Dominique 15829
Jadid-al-Eslam, Rasoul 24653
Jae, Jun Young 50241
Jaeger, Anders 09980
Jäger, Dr. Hagen 18095
Jäger, Hans 02469
Jäger, Dr. Joachim 17123
Jäger, Dr. M. 17933
Jaeger, Steven D. 50719
Jäger, Susanne 18986
Jäger, Veronika 19224
Jaegere, Isabelle De 03759
Jaeggi, Dr. Annemarie 17041
Jaehner, Inge 20921
Jämbäck, Juha 11230
Jaensch, Daniela 02124
Jäppinen, Jere 10782
Järliden, Barbro 40639
Järvinen, Marja-Liisa 11288
Järvinen, Vesa 10760
Jaffe, David 44937
Jaffe, Linda 46595
Jaffe, Steven 51243
Jagenteufel, Hermann 03129
Jager, Chuck 46743
Jager, Drs. G.W.M. 32764
Jäger, Viktor 23705
Jaggi, Fred 41619
Jaggi, Martin 41395
Jagodin, Karen 10623
Jahan-Giry, Dr. Ali 24626
Jahn, Thomas 19486
Jahn, Dr. Wolfgang 18141
Jahnigen, Rick 47556
Jahnke, Dr. Elke 21834
Jahns, Achim 18714
Jaimes Camargo, Jairo Antonio 08703
Jain, Dr. Jyotindra 24139
Jaindl, Franz 02848
Jajčević, Zdenko 09000
Jakeuchi, Yoichi 28742
Jakeway, Robert 52418
Jaki, Dr. Barbara 38457
Jakibčuk, Mykola Mychajlovyč 43041
Jakimińska, Grazyna 34902
Jakob, Andrea 20260
Jakober, Yannick 38831
Jakobi, Paul 20333
Jakobs, Dr. Ingrid 20301
Jakobsen, Björn M. 40596
Jakobsen, Egil Olav 33489
Jakobsen, S.E. 10073
Jakobsons, Imants 29980
Jakomin, Dušan 28093
Jakoubĕová, Dr. Vladimíra 09881
Jakovleva, G.A. 37338
Jakovleva, Marina Dmitrievna 37724
Jakovleva, Nadežda Petrovna 37914
Jakovljevič, Goran 08845
Jakubowski, Dr. Krzysztof 35284
Jakunskienė, Audronė 30203
Jakuševa, Valentina Maksimovna 37983
Jalal, Masoody 24704
Jalandoni, Carlos L. 34389
Jalava, Mari 10907, 11419, 11420, 11421
Jalilian, S.J. 24675
Jalon, Rita 03306
Jalšić Ernečić, Draženka 08892
Jambon, Marcel 13313, 13314
James Jr., Alton 51233
James, Beth 48310
James, Betty 05785
James, Debbie 43404
James, Frank 44917
James, Jon G. 54266
James, Louise 54293
James, Marion E. 48224
James, Rodney 01322
James, Sigrid 05803
James-Sarazin, Ariane 14671
Jamieson, Louise 01637
Jamieson, Malcom E. 47350
Jamieson, P.M. 06389
Jaminon, Martine 03800
Jamison, Wayne 48056
Jamkhedkar, Dr. A.P. 24377
Jammot, Dr. Dominique 14536
Jampen, Ulrich 42046
Jamro, Ron D. 50951
Jamščikova, Natalja Borisovna 37926
Jan, Darja 38406
Janák, PhDr. Dušan 09692
Janaloo H., Hough 50200
Janas, Stanisław 35004
Jančár, Dr. Ivan 38211
Jančeková, Viera 38379, 38380
Jancsy, Peter 38189
Janda, Don 50442
Janda, Martin 02728
Jandl, Ferdinand 01726
Jandorová, Monika 09932
Jané, Joseph 38881
Janecek, James 52161
Janes, Todd 05748
Janev, Georgi 05209
Jangoux, Prof. M. 03496
Janics, Dr. Hunert 02015
Janik, Michael 02186
Janisch, Mag. P. August 02575
Janjić, Gordan 38116
Janke, Volker 21695
Janker, B. 20633
Janki, Ruth 42000
Jankóné Pajor, Ildikó 23478
Jankov, Dr. Peter 22167
Jankova, Julia 05039
Jankova Kazakova, Rumjana 05011
Jankovskyj, Anatolij Andrijovyč 43018
Jankowski, Andrzej 34743
Janků, Miloslav 09479
Jankulovski, Plamen 05116
Janneau, O. 12213
Jannek, Norbert 19407, 19580
Jannet, Monique 14434
Jannke, William 53976
Janović, Ognjana 08917
Janovy, Karen 50105
Janovyč, Oleksandr Vasyljovyč 43166
Janscó, Istvan 04853
Janse, K.P.J. 31879
Jansen, F.W. 32398
Jansen, G. 32729
Jansen, Dr. Gregor 17995
Jansen jr, John D. 47722
Jansen, Kristian 33499
Jansen, Leo 31898
Jansén, Maija 11393
Jansen, Siri 33421, 33429
Janser, Andres 42094
Janßen, B. 16708
Janssen, Dr. Elsje 03282
Janssen, Dr. J.A.M.M. 32794
Janssen, Johannes 16749
Janssen, Laura 54084
Janssen, Peter G. 19731
Janssen-Reinen, M.A. M. 32561
Janssens, Dominique-Charles 11769
Janssens, Jan 03912
Janssens, Sandra 03414
Jansson, Björn 40872
Jansson, Tarja 11056
Jantsch, Karl 02027
Jantzen, Dr. Detlef 21900
Janus, Jerzy 34600
Janzen, Thomas 19705
Janzweert, Jan 18867
Jaouen, Jean-Pierre 11951
Jappy, Lorna 44440
Jara, Dr. Holguer 10427
Jarábik, Gabriella 38223
Jaramillo Agudelo, Darío 08565
Jaramillo de Carrión, María Mercedes 10441
Jaramillo Salazar, Juan Andrés 08725, 08729
Jarchow, Jens 22336
Jareon, Cécile 15244
Jarh, Dr. Orest 38399, 38467
Jaritz, Josef 02691
Jarmoljuk, Ol'ga Mikolaivna 43124
Jarocka-Bieniek, Jolanta 35155
Jaros, John R. 46489
Jaros, Urszula Maria 35002
Jaroševskaja, Valentina Michajlovna 37072
Jarošová, Lenka 09692
Jaroszewicz, Mieczysław 35112
Jarova, Galina Ivanivna 43088
Jarquín Pacheco, Ana María 31013
Jarratt, Ardell 52145
Jarratt, Dervla 45478
Jarrige, Jean-François 14609, 14725
Jarron, Matthew 44052
Jarvie, Robert 45366
Jarvis, Anthea 45088
Jarvis, Antony 44711
Jarzembowski, Prof. Dr. E. 45073
Jaschke, Mag. Beatrice 02202
Jašić, Ivana 08861
Jasim, Dr. Sabah 43225
Jasiuk, Jerzy 35104, 35273
Jasiūnas, Dr. Henrikas 30417
Jaskanis, Paweł Olaf 35263
Jaskowiak, Jennifer 50222
Jaskulski, Janusz 35037
Jastram-Porsche, Sabine 16442
Jastrebova, Raisa 37479
Jastrzębski, Jerzy 34890
Jatti Bredekamp, Prof. Henry C. 38578, 38579, 38585
Jauker, Dr. Birgit 17789
Jaukkuri, Maaretta 10792
Javed, Raja Muhammad 34029
Jávor, Dr. Anna 23441
Jawlensky-Bianconi, Angelica 41568
Jaworowicz, Hanna 34707
Jaworowska-Mazur, Ewa 34943
Jay, Hilary 51827
Jaysane, A. Patricia 49995
Jean, Louis 43269
Jeanbrau, Hélène 13693
Jeandron, Angela 49767
Jeanes, K.I. 38625
Jeanmart, Jacques 03903
Jeannot, Bernard 14666, 14744
Jeanselme, Brigitte 15968
Jeansson Kock, Brigitta 40934
Jebreil Zadeh, A. 24638
Jebsen, Mary 52880
Jecha, James 49834
Jecht, Heidrun 19437
Jeckel, J. 32845
Jedda McNab, Barbara 52592
Jedlička, Jiří 09512
Jędrzejewski, Antoni 35167
Jędrzejewski, Grzegorz 34729
Jefferies, Dr. Chris 48204
Jefferies, Susan 07110
Jeffery, Gordon 07181
Jeffries, Bill 05505, 07219
Jeffries, Bob 44997
Jeffries, Donna 00890
Jegorov, Boris Michajlovič 36766
Jehin, Paul 04042
Jehl, Dr. Rainer 19354
Jeitler, Johannes 02885
Jeker, Pierre-Alain 41169
Jekov, Vasko 05192
Jelavić, Zeljka 08989
Jeletich, Dr. F. 32197
Jelfs, Ann 33193
Jeličić, Janko 08997
Jelínek, Petr 09468
Jelinková, Jaroslava 09937, 09938
Jeliseeva, Tatjana 43112
Jelvakova, Irena A. 37231
Jemison, Gerald Peter 53738
Jenal, Christian 41801
Jenied, Carol 52869
Jenisch, R. Dietlinde 18675
Jenkins, Adrian 43381
Jenkins, Carol 51794
Jenkins, Dr. Deborah 44892
Jenkins, Donald 52086
Jenkins, Frances Susong 48892
Jenkins, Susan 44784
Jenkner, Ingrid 05942
Jenko, Helena 38459
Jenko, Lilija Dmytrivna 43040
Jenko, Mojca 38457
Jenks, Tracey 46329
Jenne, Hans-Jörg 18146
Jenneman, Eugene A. 53553
Jenni, Franziska 41169
Jenni, Rudi 00975
Jennings, Anne B. 49245
Jennings, Bill 49245
Jennings, David 43552
Jennings, Linda 07199
Jennings, Lucille 51476
Jensch, Dr. Rainer 22328
Jensen, B. 33203
Jensen, Ellen Elisabeth 10304
Jensen, Erik 10148
Jensen, Herbert 19310
Jensen, James 49336, 49337
Jensen, Jennifer R. 51763
Jensen, Johan M. 10135
Jensen, Kathleen A. 53238
Jensen, Ken 50501
Jensen, Leslie D. 54078
Jensen, M. Wayne 53509
Jensen, Mona 09962
Jensen, Niels Erik 10166
Jensen, Ole G. 23270
Jensen, Otto 10303
Jensen, Pam 48035
Jensen, Sterling 06426
Jensen, Tom 33719
Jensen, Vivi 10163
Jensen, William 51352
Jensner, Magnus 40585
Jenssen, Ragnar 33454
Jenvold, Birgit 10163
Jenzen, Dr. Igor A. 17963
Jepbarov, Rejep 42979
Jeppesen, Hans 10064
Jeppsson, Kristina 40473
Jepson, David 45301
Jerin, Robert M. 48266
Jerković, Dragan 38091
Jerlich, Walter 01743
Jerome, Jeff 46562
Jerris, Rand 48506
Jerrome, Peter 45377
Jervois, W. 38630
Jésman, Prof. Czesław 34881
Jesretić, Miroslav 38151
Jessat, Mike 16498
Jessewitsch, Dr. Rolf 21802
Jessop, L. 45227
Jeszenszky, Dr. Sándor 23434
Jett, Cheryl 47694
Jette, Carol 46277
Jetzer, Gianni 41804, 51265
Jeufroy, Pierre 13054
Jevdokimova, Jemilija Mykolaïvna 43046
Jevsejeva, Olga Jevgenivna 43021
Jevtović, Mila 38074
Jewell, Anne 50291
Jewell, Brian 44403
Jewell, Robert 50920
Jezierski, Richard 14515
Jeziorowski, Tadeusz 35049
Jezler, Peter 41361
Jezler, Prof. Dr. h.c. Peter 41831
Jezmann, Meike 22035, 22037
Ježowitzová, Jiřina 09548
Jha, Dr. Uma Raman 24430
Ji, Gon Gil 29823
Ji, Liancheng 07497
Ji, Qiang 07538
Ji, Zhongliang 07869
Jian, Li 47984
Jiang, Chuanhao 07709
Jiang, Guangyou 07739
Jiang, Hua 08455
Jiang, Huimin 07971
Jiang, Shidong 07667
Jiang, Tingyu 08055
Jiang, Wei 07608
Jiang, Weiming 08339
Jiang, Xukui 08017
Jiang, Yaonan 07647
Jiang, Yibing 08519
Jiang, Yuanjin 07965
Jiang, Zhili 07880
Jianu, Magda 36416
Jiao, Desen 07885
Jibah, Haji Matassim Bin Haji 04961, 04963, 04965
Jicărean, Monica 36442
Jie, Youjun 07990
Jilhjalmsson, Larus 24028
Jiliá, Pablo 38856
Jillings, Sarah 44890
Jillis, Nellie 38639
Jiménez, Angel 39982
Jimeñez, Mar Villafranca 39358, 39367
Jimenez, Mercedes 31175
Jiménez, Pedro Lópe z 39216
Jiménez Cadena, Gustavo 08710
Jiménez Chávez, Alma Rosa 31156
Jiménez Espejo, Francisco J. 39680
Jiménez Losantos, Encarna 40201

Jiménez Martínez, Dr. Rafael 39401
Jimeno Santoyo, Myriam 08590
Jimison, Tom 50913
Jimmie, Charles 49124
Jin, Baidong 08327
Jin, Dechang 08305
Jin, Hongde 08311
Jin, Jon Moon 29757
Jin, Yongfeng 07926
Jin, Yongtian 07495
Jing, Guo 07917
Jing, Zhanyuan 08081
Jing Weng, Liao 07574
Jirásek, Prof. PhDr. Zdeněk 09692
Jirglová, Lucie 09727
Jiřík, Jaroslav 09714
Jirmusová, Hana 09467
Jirsová, Blanka 09849
Joans, Dr. Barbara 51466
Jobke, Ryszard 34935
Jobst, Franz 17106
Jobst-Rieder, Marianne 03045
Joch, Dr. Peter 17792
Jochem, Marlene 19414
Jochims, Jennifer 53317
Jochims, Richard 18554, 19112
Jochum, Franz 02452
Jockel, Nils 18951
Jodłowski, Prof. Antoni 35300
Joeckel, Reecanne 46350
Joehnk, Dr. Carsten 18141
Joenniemi, Sirpa 11363, 11364
Jönsson, Bertil 40506
Jönsson, Tomas 40639, 40770
Jöris, Dr. Olaf 20687
Joffre, Roland 13642
Joger, Prof. Dr. Ulrich 17529
Johannes, Janet 49222
Johannessen, Arild 33497
Johannessen, Kåre 10275
Johannessen, Preben L. 33559
Johannesson, Birgitte 10133
Johannsen, Prof. Dr. Carl I. 20369
Johannsen, Frerk 19830
Johannsen, Ilona 22606
Johannsen, Karin 18028
Johannsen, Ruby 49450
Johannson, Sigfus 06201
Johansen, Annette 10263
Johansen, Jens 10093
Johansen, Michael 10080
Johansen, Rikke 10317
Johanson, Bruce 51571
Johansson, Börje 40405
Johansson, Conny 40703
Johansson, Eric J. 47049
Johansson, Hjördis 40960
Johansson, Jan 40691
Johansson, Kennet 40718, 40727
Johansson, Ketty 10197, 10198
Johansson, Kjell 40653, 40911
Johansson, Lars G. 40676
Johansson, Marie 40974, 40992
Johansson, Sonny 40983
Johansson, Sven-Eve 40687
Johansson, Tina 40960
Johansson, Ulf 11163
Johansson, Ulrica 41000
John, Holverson 53179
John, J. 09708
John-Willeke, Bettina 17794
Johncock, M. 01188
Johnes, John V. 46427
Johns, Daniel 49425
Johns, Kevin L. 50220
Johns, Sharleen 48205
Johns, Valerie 48993
Johnsen, Aksel 33890
Johnsen, Berit Åse 33588
Johnsen, Ron 52250
Johnson, Alfred E. 49990
Johnson, Amy Jo 52007
Johnson, Andrew 44115
Johnson, Anne 53847
Johnson, Arden 49214
Johnson, Bert 48263
Johnson, Brooks 51361
Johnson, Byron 50916
Johnson, Byron A. 53780
Johnson, C. Clayton 52103
Johnson, Carl 06755
Johnson, Christina 50220
Johnson, Christine 33170
Johnson, Craig 52612, 52618
Johnson, D. Thomas 46645
Johnson, Dale 43370
Johnson, David 51948
Johnson, David A. 53623
Johnson, Diane 47907
Johnson, Douglas 46860
Johnson, Drew 51470
Johnson, Edward 53270
Johnson, Dr. Eileen 50312
Johnson, Frank 38663
Johnson, Gerardo 23329
Johnson, Gina 53295
Johnson, Grace 52753
Johnson, Harold 52950
Johnson, Dr. Heide-Marie 50605
Johnson, James 47426
Johnson, Jean 47390
Johnson, Jeanne 52842
Johnson, Jennifer L. 47751
Johnson, Jeremy 00827, 00829
Johnson, Jerome Martin 49869
Johnson, Joan 45882
Johnson, Dr. John R. 52860
Johnson, Judi 47134
Johnson, Judy 47131, 47133
Johnson, Karen 52904
Johnson, Kathleen E. 53450
Johnson, Kaytie 49046, 49047, 49048
Johnson, Kenneth R. 48834
Johnson, Dr. Kirk R. 48075
Johnson, Laurel 00826
Johnson, LaVern M. 50334
Johnson, Libba 47746
Johnson, Lisa 48515
Johnson, Lydia 47289
Johnson, Mark M. 50815
Johnson, Marlo L. 49714
Johnson, Mary A. 52691
Johnson, Maurianna 50000
Johnson, Melinda 50925
Johnson, Mirma A. 46154
Johnson, Nancy T. 50790
Johnson, Nichola 45285
Johnson, Patricia 01142, 52504
Johnson, Pearl 49254
Johnson, Penny 44860
Johnson, Polly 49804
Johnson, Dr. Richard E. 52197
Johnson, Robert S. 48653
Johnson, Robin 43902
Johnson, Dr. Ron 07287
Johnson, Sandra L. 51768
Johnson, Sharon 47920
Johnson, Sheila A. 07406
Johnson, Steven 48022
Johnson, Stuart 50915
Johnson, Suni 46574
Johnson, Terry 50616
Johnson, Thomas B. 54338
Johnson, Tom 54336
Johnson, Willard 51518
Johnson, William C. 48874
Johnson Bowles, Kay 48523
Johnston, Curtis 50701
Johnston, Eileen 53879
Johnston, Phillip M. 52859
Johnston, Renée 05964
Johnston, Sally 46588
Johnston, Sona 46551
Johnston, Stephen 45334
Johnston, Virginia 48140
Join-Diéterle, Catherine 14708
Joiner, Janet 53743, 53744
Joiner, S. 05726
Jojo, Eva 50748
Jokio, Mikko 10710, 10711
Jokulo Cooper, Jimmie 46569
Jolliff, Dawna 49249
Jolly, Dr. Anna 41767
Jolly, Dianne 01307
Jolly, Linda Lee 53577
Jolly, Renee 52883
Joly, M. 15076
Jomppanen, Tarmo 10859
Jonaitis, Aldona 48472
Jonaitis, J. 30368
Jónatansdóttir, Guðkr{}rún Dís 24017
Jones, Alan 45705
Jones, Amanda C. 50463
Jones, Amy 43731
Jones, Andy B. 54139
Jones, Ann 44127
Jones, Anne 44218
Jones, Barbara L. 49070
Jones, Betty Lou 50599
Jones, C.R. 47788, 47789
Jones, Dr. Cheri A. 48075
Jones, Clive 43829
Jones, Cynthia L. 49125
Jones, D. 43711
Jones, D.A. 43357
Jones, D. A. 46090
Jones, David 44530, 47451
Jones, Diana 01412
Jones, Donna P. 53141
Jones, Dr. Douglas S. 48813
Jones, Duffie 49633
Jones, E.B. 33225
Jones, E. Michael 04198
Jones, Edward 48804
Jones, Eric 43841
Jones, Erin 49078
Jones, Glenn 49025
Jones, Harvey 51470
Jones, Hugh R. 50324
Jones, Huw 43902
Jones, Jeanne 45030
Jones, Jeff 51639
Jones, Jill 53003
Jones, Joanna 45716
Jones, Josephine 44070
Jones, Joy 06209
Jones, Julie 51212
Jones, Kathy 05545, 48279
Jones, Keith 01041, 44325
Jones, Kevin 01425, 46022
Jones, Larry A. 50275
Jones, Lester 43829
Jones, Lial A. 52505
Jones, Linda 47320
Jones, Lisa 00899
Jones, Lowell N. 47538
Jones, Marilyn 49558, 52124, 53380
Jones, Mark 45004
Jones, Nia 45163
Jones, P. 44081
Jones, R. 06387
Jones, Rhian 46179
Jones, Richard 43751
Jones, Robert 00835
Jones, Dr. Robert 49561
Jones, Sharain 07072
Jones, Sian 45708
Jones, Stephen 46375, 52011
Jones, Sue 49844
Jones, Suzi 46288
Jones, Thomas 50673
Jones, Prof. Tom 43502
Jones, Tom 44215
Jones, Trevor 44015, 49041
Jones, W.J. 33222
Jones Cox, Beverly 53907
Jones Harper, Suzanne 50366
Jonet, Sylviane 13048
Jong, A. de 32857
Jong, Andy de 05731
Jong, C.C. de 32413
Jong, Dr. H. de 31904
Jong, Dr. M.A. de 32098
Jong-Wook, Prof. Bae 29779
Jongh, Marja de 32684
Jonghe, Mieke de 03535
Jongs, Carel 05811
Jongstra, Jaap 32478
Jonin, Marie-Thérèse 12464
Jonker, Drs. A. 32112
Jonker, Drs. M. 31854
Jonsson, Prof. Goran 40847
Jonsson, Ingrid 40867
Jónsson, Ólafur Ingi 24022
Jónsson, Petur 23990
Jonušienė, Eleonora 30321
Jonušienė, Vitalija Teresė 30322
Joo-Sik, Kim 29843
Joohee, Kim 29889
Joos, Cristian 42057
Joos, Erwin 03285
Joos, Marleen 03267
Jooss, Dr. Birgit 20764
Joppien, Dr. Rüdiger 18951
Joralmon, Mary 52762
Joraslafsky, Helen 00800
Jordaens, Lode 03586, 03587
Jordan, Dr. Alexander 21189
Jordan, Brenda 49835, 49837, 49838
Jordán, Dr. František 09919
Jordan, Jai 53447
Jordan, Mary Alice 52832, 52833
Jordan, Peter 41888
Jordan, Robert 05673, 18756
Jordan, Rosemarie 22488
Jordan, Sandra 02869
Jordan, Thomas E. 48908
Jordan Wass, Catherine H. 51361
Jordanov, Ivan 05134
Jordanović, Branislava 38084
Jordon, G. 00930
Jordon, J. Greg 48392
Jørgensen, Anja 10166
Jørgensen, Erik 10049
Jørgensen, Mogens 10145
Jorgensen, Roy 48547
Jorikson, Håkan 40562
Joris, Philippe 03796
Joris, Yvònne G.J.M. 32379
Jorquera Caro, Sonia Alexandra 31228
Jorrand, C. 13521
Jorrín García, J.A. 39705
Josanu, Vitalie 36676
Joselu, Sergio 25501
Joseph, Carolyn 49217
Joseph, Jean-Louis 11653
Joseph, Lorraine 47790
Joseph, Sarah 49005
Joseph-Debaque, Sandrine 12512
Josephson, Eva 40781
Josephson, Magnus 40782
Joshi, Prof. B.D. 24205
Joshi, Mallickarjun 24426
Joshi, Vishwanath N. 24340
Jošić, Zoran 38030
Joslin, Stewart 53525
Jossen, Dr. Erwin 41878
Josset, Dr. Patrice 14704
Jossifova, Maria 05191
Jost, Gerda 18571
Jost, Loren 52377
Jost, Stephan 51467
Jost-Prześlakowska, Maria 35174
Jósvai, Dr. Katalin Dankó 23792
Joubert, Alain 13384, 13385, 14017, 14474
Jouen, Jean-Pierre 13344
Joüon des Longrais, F. 15011
Jouffroy 11937
Joulin, Guy 14260
Jourdan, Aleth 11549, 11550
Jourdan, Jean-Pierre 15295
Jourdan, Patrick 14275, 14276
Journiac, Virginie 12236, 12239
Jous, Léon 03570
Jouve, Gerard 12455
Jouzeau, Marie-Hélène 14368
Jovanov, Dr. Jasna 38125
Jovanović, Igor 38093
Jovel Castro, Walter E. 08823
Jovelić, Marijana 38091
Jover Maestre, Francisco Javier 39839
Jović, Vukašin 38093
Joyce, Dan 49734
Joye, Jean 03455
Joyeux, Y. 03394
Joyner, Elizabeth 53732, 53735
Joyner, Roger D. 49067
Ju, Calvin 42163
Ju, Jin-Soon 29824
Ju, Wancang 08125
Juárez Nieto, M. Magdalena Sofía 31283
Juaye, Comte Pierre du Manoir de 12383
Juchli, Cornelia M. 41893
Juchniewich, Daniel 50443
Juck, Marcel 38225
Jucuti, Ioan 36245
Judd, Michael 46133
Judge, Joseph M. 51363
Judson, E. 38752
Judson, Etta 38721
Judson, Jack 52718
Judson, William D. 51942
Jueg, Uwe 20061
Juelich, Clarence 54123
Jülich, Dr. Theo 17789
Jüllig, Carola 17072
Jürries, Wolfgang 19741
Jufa, Marija 37557
Juffinger, Dr. Roswitha 02624
Jugie, Sophie 12741
Juhász, Gyula 23701
Juhl, Kirsten 33863
Jul, Ole 10299
Jules Kouloufoua, Jean Gilbert 08805
Julia, I. 14731
Julia y Macias, Manuel 39658
Julien-Laferriére, Alain 15961
Julin, Richard 40910
Julio, Roberto 25640
Julius, Dieter 20615
Juliusburger, Thomas 47019
Julkunen, Pirjo 11115
Jullian, Nathalie 14585
Jumalon, Julian N. 34348
Jumalon, Prof. Julian N. 34371
Jumară, Dan 36413
Jumelet, J. 32015
Jun, Yong-Bok 28937
Juncosa, Enrique 24801
Juneau, Madeleine 06372
Junek, PhDr. David 09727
Junek, David 09728
Jung, H. 32608
Jung, Ilse 03019
Jung, Ludwig 21324
Jung, M. 32339
Jung, Dr. Michael 27512
Jung, Dr. Norbert 16906
Jung, Dr. Sabine 22751, 22752
Jung, Seung Sun 29824
Jung, Dr. Werner 19626
Jungbauer-Radax, Andrea 03026
Jungblut, Maria-Paule 30482
Junge, Brigitte 16647

J

Junge, Cornelia 19896
Junge, Dr. Michael 19247
Junge, Dr. Peter 17081
Junge-Stevnsborg, Kirse 40585
Junginger, Ernst 19818, 19820
Junglas, Uta 22421
Jungmann, Judith 03026
Junhem, Inga 40947
Junier, Caroline 41680
Junior, Peter 19150
Junker, Patricia 48709
Junno, Johanna 11180, 11470, 11471, 11472
Junod, Barbara 42094, 42095
Juntikka, Ilse 11177
Juola, Emmy 11242
Juranek, Dr. Christian 22500
Jurca, Emanuela 36100
Jurčenko, Oleg Grygorovyč 43067
Jurdana, Ela 08997
Jurecka, Małgorzata 35075
Jurevičius, Juozapas 30269
Jurgemeyer, Marne 48659
Jurić, Radomir 08914, 08981
Jurić, Spomenka 08952
Jurica, Varina 08876
Jurkāne, Anna 30124
Jurkiewicz, Jan Gustaw 35351
Jurriaans-Helle, Dr. G. 31825
Jussila, Neil 46805
Just, Arthur 03048
Just, Berthold 17118
Just, Helmut 02957
Just, Rüdiger 18905
Justice, Ed jr. 48184
Justice, Gloria 52035
Justus, Dirk 18924
Justus, Mary Beth 46410
Jusupova, Dr. Majram Ažibekovna 29956
Juton, Alain 14127
Jutras, Michel 07170
Juul Holm, Michael 10103
Juva, Antti 11212, 11213
Juvigny, Sophie de 15220
Juvonen-Eskola, Annamari 10727
Juzwenko, Dr. Adolf 35344

K

Kaak, Dr. Joachim 20425
Kaarto, Sari 11260
Kaasen, Svanhild Alver 33509
Kaba, Sory 23308
Kabić, Dušan 38093
Kábrtova, Jiřina 09699
Kabrud, Jack A. 50732
Kačalova, Alisa D. 36819
Kačírek, Luboš 38218
Kacso, Carol 36124
Kaczarová, Iveta 38290
Kaczmarczyk, Witold 35136
Kaczmarek, Drs. J. 32563
Kader, Dr. Ingeborg 20465
Kadlec, Dr. František 09738, 09753, 09754, 09757, 09761, 09784, 09794
Kadowaki, Akira 29514
Kadurina, Elena 05149
Kadyrov, A.S. 42493
Käär, Pekka 11402
Käch, Robert 41751
Käck, Fredrik 40581
Kähler, Kai 17571
Kaelcke, Wolfgang 20982, 20983
Kaelin, Léon 41483
Kämpf, Michael 19756
Kämpfner, Gabriele 17485
Kaenel, Gilbert 41541
Känsälä, Risto 10933
Käppeli, Bruno 41628
Kärgling, Dr. Karlheinz 20123, 20126
Kärki, Sirpa 40767
Kärnefelt, Prof. Ingvar 40701
Käs, Rudolf 20769
Käs, M.A. Rudolf 20791
Kaeser, Prof. Dr. Marc-Antoine 41474
Kaeser, Olivier 41399
Kaesler, Roger L. 49991
Kästner, Ursula 17036
Kästner, Dr. Volker 17036
Käyhkö, Unto 10940
Kafetsi, Anna 22943
Kagioka, Masanori 29100
Kahan, Mitchell 46146
Kahl, Elizabeth 51626
Kahle, Patricia 51064
Kahle, Shawn M. 48128
Kahlen, Prof. Wolf 17290
Kahler, Caroline 50122
Kahn, David M. 49193
Kahn, Leo 52889
Kahn, Tammie 49382
Kaijt, W.P.A.M. 31902
Kailashnathan 24047
Kaim, M.-L. 20443
Kaindl, Heimo 01966
Kainz, Mag. Irmengard 01805
Kaiqu, Liu 07527
Kaire, Inese 30061
Kaiser, F.W. 32073
Kaiser, Friederike 20420
Kaiser, H. 18886
Kaiser, Dr. Hermann 17713
Kaiser, Katharina 17164
Kaiser, Michael 20762
Kaiser, Miriam 31146
Kaiser, Peter 41706
Kaiser, Philipp 41166
Kaiser, Dr. Stephan 21192
Kaiser-Reka, Berol 17506
Kaißer, Paul 22267
Kaizuka, Tsuyoshi 29373
Kaján, Imre 23551
Kajanus, Pirkko 10847
Kajikawa, Yoshitomo 28818
Kakaras, Gunaras 30291
Kakko, Leena 11353
Kakourou-Chroni, Dr. Georgia 23195
Kakuda, Arata 28537
Kakusei, Miki 28680
Kal, W.H. 31852
Kalaber, Ladislau 36558
Kalaitzis, Myron 23166
Kalajainen, Jason 49316
Kalantzes, Isabelle 52695
Kalašnik, Tatjana Nikolajevna 36953
Kalašnikov, Nikolaj Ivanovič 37798
Kalčev, Kamen 05033
Kalčev, Petar 05211
Kalchthaler, Peter 18428
Kaldewei, Prof. Dr. Gerhard 17816
Kaldor, Cynthia 50541
Kaldy, Robert 03037
Kalela-Brundin, Maarit 40710
Kalenberg, Ágel 54411
Kalenberg, Angel 54412
Kalenitchenko, Liliane 15138
Kalicki, Jerzy 34761
Kaligin, Pavel N. 37740
Kalina, Dr. Walter 03001
Kalina-Vidić, Ivana 38074
Kalinová, Dr. Alena 09427
Kalinsky, Nicola 44157
Kaljanac, Muhiba 04276
Kaljnyj, Dmytro Pavlovyč 43172
Kalkhoff, Gabriele 20899
Kalkmann, Hans-Werner 16819
Kallas, Nick 53644
Kállay, György 23498, 23656
Kallberg, Ulla 11390
Kallicas, Antonia 09360, 09361
Kalligas, P. 22898
Kallio, Kalle 11367
Kallio, Maija-Riitta 11030
Kalmár, Ágnes 23655
Kalmikova, Elena 37332
Kalmite, Lelde Alida 50646
Kalmring, Sven 21565
Kalnins, Indulis 52510
Kalogeros-Chattan, Charlie 50206
Kalokerinos, Alexis 23019
Kaloutsakis, Agisilaos 23019
Kalpakova, Akgul 42983
Kalpickaja, Elena Jakovlevna 37651
Kalšan, Prof. Vladimir 08852
Kalt, Ewald 41921
Kaltenberger, Johann 02666
Kaltenbrunner, Regina 02626
Kaltofen, Dr. Andrea 20286
Kaltsas, Dr. Nikolaos 22920
Kaltved, David 48428
Kalus, Dr. Jaromír 09694
Kamal, Hassan 42133
Kamarun, Prof. Ham Rabeah 30618
Kamber, Ljubomir 38110
Kamber, Pia 41161
Kamber, Robert 41531
Kamber, Urs 41090
Kamburg, Petra 19041
Kamel, Linda A. 53881
Kameneva, Marina Rimovna 37945
Kamer, S.J. 32571
Kamerling, Leonard 48472
Kames, Holger 19902
Kamii, Monsho 29558
Kamijo, Kazuo 29275
Kamino, Yoshiharu 28752
Kaminski, Jane 17044
Kamiya, Yukio 28495
Kamm, David 48021
Kamm, James J. 50571
Kammel, Dr. Frank Matthias 20764
Kammerhoff, Alexander 20478
Kamp 21441
Kamp, Dr. Diane D. 53454
Kamp, Michael 20007
Kamp, Michal 09512
Kampe, Allen 51103
Kampe, Katrine 09954
Kamper, Gerard de 38745
Kamras, Barbro 40474
Kamras, Borge 40474
Kamsky, Virginia A. 51150
Kamuhangire, Dr. E. 42993
Kamyšova, Antonina Abramovna 36833
Kan, G. 32966
Kan, Marina 30097
Kana Anderson, Gail 51374
Kanabe, Albert 05773
Kanai, Takao 28619
Kanapkienė, Vida 30288
Kanaseki, Hiroshi 29075
Kanavrov, Darin 04976
Kandeler-Fritsch, Martina 03026
Kandl, Veronica 49007
Kane, Allen 53908
Kane, Debra 33207
Kane, Gordon 45135
Kane, Katherine 49194
Kane, Linda 05667
Kane, Mary Ann 47830
Kane, Patricia 51056
Kaneichi, Ninomiya 28567
Kaneko, Kenji 29492
Kaneko, Yuriko 29190
Kanetaka, Kaoru 29612
Kanetuki, Takeshi 28889
Kanev, Ivajlo 05191
Kang, Prof. Joeng-Won 29921
Kang, Utchang 29824
Kangas, Faye 49413
Kania, Rolf 20809
Kania-Schütz, Monika 18809
Kankaala, Katri 11121
Kann, Lisa von 52555
Kann, Yanhua 07774
Kannan, Dr. R. 24115
Kannu, T. 38176
Kano, Kenji 28726
Kano, Takumi 29508
Kant Monteith, Joanita 50764
Kanter, Dr. Karin 20123
Kanter, Laurence B. 51212
Kantvilas, G. 01159
Kao, Deborah M. 47177
Kao, Li-Chen 42142
Kao, Shu-Kwei 42372
Kapfer, Otto 02449
Kapfer, Rudolf 02838
Kapinan, Catherine 53857
Kapinčeva, Dr. Iskra 05242
Kaplan, Dave 50135
Kaplan, Ilee 50179
Kaplan, Luděk 09522
Kaplan, Dr. Susan A. 47103
Kaplan, Wendy 50244
Kapler, Joseph 50396
Kappelmark, Ragnhild 40965
Kapronczay, Dr. Károly 23383, 23475
Kaptain, H.J. 09992
Kapteijn, Christine 44217
Kaptein, Helen 01541
Kapusta, Jan 09668
Kapustina, Nadežda Ivanovna 43031, 43033, 43034, 43035
Karačevcev, Vladimir F. 36922
Karadi, Gabriella 07121
Karadontis, Fleurette 22883
Karajotov, Dr. Ivan 04995
Karakas, Steve 52071
Karamaliki, Nota 22961
Karamanou, Dr. Aglaia 22978
Karamitrou-Mentessidi, Dr. Georgia 22875
Karamut, Dr. Ŀlsmail 42812, 42817, 42823
Karamzin, Alex 52787
Karanitsch, Peter 02395
Karantzali, Efthymia 23070
Kararadeva, Svetozara 05168
Karas, Lourdes 51956
Karasawa, Masahiro 29492
Karaseva, Alevtina Pavlovna 36779
Karaska, Dušan 38251
Karastamati, Dr. Helen 22918
Karbacher, Ursula 41809
Karbaum, Roswitha 20704
Kardinal, Royce 54155
Kardon, Nina 48699
Kardong, Don 53205
Kardosh, Judy 07215
Karegar, M. 24679
Karel, Johannes 03029
Karetsou, Al 23018
Karey, Carol 53948
Karg, Dr. Heike 22365
Kargar 24674, 24687
Karge, Annelen 21393
Kargere, Lucretia 51152
Karhoff, Brent 48289
Karhunen, Lars 10773
Karibo, Lou 48739, 48743
Karinen, Helge 40877
Karingi, Lina 29748
Karinja, Snježana 38492
Karis, Sirje 10657
Kārkliņa, Rūta 30073
Karl, Barbara 03026
Karlinger, Friedrich 02844
Karlsdottir, Anna Stella 33803, 33804
Karlsson, Anders 41000
Karlsson, Christer 40746
Karlsson, Eva-Lena 40735
Karlsson, Lars Olov 40409
Karlsson, Stig 40446
Karlstedt, Harry 11162
Karman Ohlberger, Christina 40778
Karmazin, N.I. 37107, 37109, 37111
Karnath, Sonja 21823
Karnauchova, Galina Georgievna 36780
Karnes, Andrea 48717
Karneth, Dr. Rainer 16530
Karojeva, Ljudmyla Rostislavivna 43206
Karon, Stephen A. 53400
Karosas, Algis 30428
Karosas, Gintaras 30403
Karov, Zdravko 05129
Karp, M. 51107, 51137
Karpf, Kurt 02600
Karpińska, Zofia 35189
Karpova, M.I. 37759
Karppinen, Eira 10986
Karr, Ed F. 51739
Karsch, Sylvia 20109
Karshens, Xander 32296
Karskens, Xander 32295
Karsmizki, Ken 47946
Karstadt, Bruce N. 50727
Karsten, Joan 52357
Karstens, E. 38723
Kartio, Kai 10767
Karttunen, Matti 10946, 11114, 11116, 11117, 11118
Karu, Mihkel 10623
Karwatzki, Dr. Barbara 21286
Karwowski, Lech 35157, 35161, 35162, 35163
Kasal, Václav 09516
Kasbi, Yasmine 03590
Kaschkat, Dr. Hannes 20832
Kase, Karin 17044
Kashima, Teishin 29628
Kashiwabara, Shinji 28608
Kasilag, Emma Y. 34485
Kaširin, Oleg Semënovič 37865, 37871, 37875
Kasparavičius, Stasys 30379
Kasperovič, K.K. 42998
Kasperska, Helena 34832
Kaspruk, Stepan Dmytrovyč 43044
Kass, Amalie 46937
Kass, Emily S. 53437
Kast, Dr. Raimund 22169
Kaštelionovaitė, Zofija 30421
Kastler, Daniel 11836
Kastler, Dr. Raimund 02625
Kastriot, Dervishi 00010, 00011
Kastrup, Dr. U. 42071
Kasumov, Nazym Kasumovič 36874
Kasym, Sofija Vasiljevna 37668
Kaszak, Carole 45521
Kąsziołka, Stanisław 35373
Kaszubkiewicz, Andrzej 34846
Katagishi, Shoji 29522
Kátai, Dr. László 23675
Katakami, Syusaku 29186
Katalin, Dr. Keserü 23399
Katanekwa, K.V. 54655
Katanekwa, N.M. 54656, 54657
Katayama, Takashi 29320
Katerinkino, Lidija P. 37104
Katevenidou, Tassoula 22961
Kathane, M. Y. 24332
Katili, Dr. Ekki Husein 24476
Katins, Anne-Marie 18008
Kato, Bruce 49656
Kato, Jatsuo 29321
Kato, Sadao 28918, 29430
Kató, Zoltán 36144
Katoh, Yuji 28693
Katona, Dr. András 23426, 23451
Katona, Dr. Stephen 46603
Katsalaki, Georgia 23019
Katsiff, Bruce 48179
Katsimpalis, Tom 50301

Katsura, Nobuo 29222
Katz, Alain 12522, 12523, 12524
Katz, Paul 46253
Katzenbach, Dr. Annette 18141
Katzensteiner, Peter 02927
Katzinger, Dr. Willibald 02319, 02320
Kauder, Hans-Peter 02201
Kauder-Steiniger, Dr. Rita 20519
Kauders, Audrey S. 49717
Kauer, Dr. Andrea 41304
Kauffmann, André 11654, 11655
Kauffmann, Bernd 20614
Kaufhold, Dr. Tobias 20411
Kaufman, Donald G. 51653
Kaufmann, Burkhard 17368
Kaufmann, Carol 38593
Kaufmann, Christoph 19939
Kaufmann, Don 47777
Kaufmann, Hans 41686
Kaufmann, Dr. José Luis 00408
Kaufmann, Dipl.-Ing. Mechthild 20423
Kaufmann, Patrick 41441
Kaufmann, Dr. Sylke 19420
Kaukiainen, Hannu 10754
Kaukiainen, Heli 10989, 10990, 10992
Kaukoreit, Doz. Dr. Volker 03045
Kaulbach, PD Dr. Hans-Martin 21960
Kaumanns, Kurt 16987
Kaupang, Brit 33730
Kaupinis, Juozas 30312, 30313
Kaus, Dr. K. 02740
Kauschat, Gerda 18166
Kauth, K.O. 49450
Kava, Ritva 10758
Kavanaugh, Martha 05547
Kavasch, E. Barrie 53950
Kawabe, Masami 29074
Kawaguchi, Kei 28993
Kawaguchi, Kimio 29419
Kawaguchi, Masako 29419
Kawai, Masaki 29529
Kawai, Suyako 28821
Kawakami, Kunihiko 28677
Kawamithu, Sigeo 28603
Kawanabe, Hiroya 28796
Kawanabe, Kusumi 29576
Kawano, Katsuhiko 28886
Kawasaki, Matsuju 28543
Kawasaki, Shoichiro 29379
Kawashima, Akira 29603
Kawashima, Etsuko 29139
Kawiorski, Wacław 34833, 34956, 34959, 34961, 35017
Kay, Hannah 43524
Kay, Joann 51241
Kaya, Settar 42866
Kayaselcuk, Mary 51335
Kaye, Jennifer 07142
Kayem, Abdul 00009
Kayombo, Dr. Norbert Andrew 42512
Kaysel, Dr. Roger 41149
Kayser, Christine 13857
Kayser, David 52658
Kayser, Gisela 17098
Kayser, Horst 20669
Kayumov, Bekmurad Djumayevič 42972
Kazbekov, F.P. 37929
Kazik, Mariusz 34630
Kazlienė, Viktorija 30314
Kaźmierczyk, Adam 34686
Kazus, Igor A. 37207
Kazutoshi, Kato 28409
Kazychanova, Marisa Grigorjevna 37953
Ke, Fengmei 08098
Ke, Shengwu 08421
Ke, Shilun 08027
Kean, John 53649
Kean, Roger 52143
Kearney, Jarod 52366
Kearney, John 47469
Kearney, Lynn 47469
Keasal, Debra L. 45148
Keates, Eve 05317
Keating, Neal B. 49408
Keaveney, Raymond 24807
Keck, Gabriele 41207
Keck, John 49101, 53321
Keckeisen, Robert J. 53536
Keddie, Robert 05790
Kędzierski, Sławomir 35278
Kędziora, Prof. Dr. Andrzej 35046
Keeble, Phyllis 43897
Keefer, Dr. Erwin 21950
Keegan, Daniel T. 52809
Keehner, Jill 53536
Keel, Othmar 41383
Keen, Lewis 48700
Keena, Katherine 52942
Keene, John 38664
Keene, Rochelle 38646
Keene Fields, Catherine 50128
Keene Muhlert, Jan 53659
Keery, John 45790
Keesecker, Martin 50519
Keeve, Valencia 49932
Kefer, Kons.OStR. Mag. Wilfrid 01747
Kehrer, Helma 16527
Kehrli, lic. phil. Manuel 41498
Keil, Franka 19692
Keil, Katherine 49613
Keil, Toralf 16499
Keim, Dr. Helmut 16536
Keinänen, Timo 10821
Keinan, Ofra 24921
Keinath, Rudi 16459
Keiper, Elke 22300, 22301
Keiper, Dr. Joe 47625
Keisuke, Iwamoto 28379
Keitel, Anja 17171
Keith, Barbara W. 48641
Keith, Dorothy 50410
Keith, Jack 46461
Keith, Larry 44937
Keith, Uwe 22260
Kejariwal, Dr. O. P. 24152
Kejzar, Josef 02691
Kekäle, Leena 11258
Kekeke, Daniel O. 33321
Kekic, Dr. M. 01554
Kéler, Yves 06476
Kell, Dr. Klaus 19268
Kellams, Marty 48335
Kellar, Jane 49791
Kellein, Dr. Thomas 17341
Kellen, Madonna M. 49899
Keller, Dana 50618
Keller, Esther 41161
Keller, Jaroslav 09470
Keller, Linn 49169, 49170
Keller, Prof. Lukas 42118
Keller, Pascale 41209
Keller, Dr. Peter 02609
Keller, Pierre 41755
Keller, Sylvia M. 51943
Keller, Tim 53285
Kellerer, Anna 02025
Kellerhoff, Reinhard 22490
Kellett, John 46550
Kelley, Joyce J. 52258
Kelley, Robert P. 47955
Kelley, Samantha 46822
Kelley, Timothy 51645
Kellgren, Ola 41000
Kelling, Paisley 51749
Kellner, Bernd 18148
Kellner, Heike 02417
Kellokumpu, Mariina 11045
Kelly, Chris 47122
Kelly, Dr. Christopher 43729
Kelly, D. 05689
Kelly, Eamonn 24808
Kelly, Elaine 46815
Kelly, Gemey 06788
Kelly, Gillian 44060
Kelly, Jim 53001
Kelly, John 53600
Kelly, Karen 51166
Kelly, Linda 06952
Kelly, Michelle 43299
Kelly, Nancy 51556
Kelly, Patricia 52641
Kelly, R.J. 43272
Kelly, Samantha 50815
Kelly, Sheila 06732
Kelly, Vonda 46870
Kelly, William 46161
Kelly Elder, Catherine 49323
Kelm, Dr. Bonnie G. 52861
Kelm, Dr. Rüdiger 16451
Kelnberger, Raimund 21354
Kelsey, Brian 48910, 53345
Kelsey Foley, Kathy 53993
Kelso, Karen 51259
Kelso, Kirsty 45332
Kelso, Marcie 47391
Keltti, Pia 10716
Keltz, Ursula 20491
Kemdal, Anne Louise 40945
Kemkes, Martin 16412
Kemkes, Dr. Martin 20829, 21424
Kemm, B. 41226
Kemmerer, Allison 46296
Kemp, Alan 45523
Kemp, Blanch 50347
Kemp, Dawn 44160
Kemp, Jane 48021
Kemp, Jo 45292
Kemp, N.R. 01159
Kemp, Richard 45146
Kemp, Dr. T.S. 45336
Kemp, W. Bradley M. 49991
Kemper, Heinrich 18022
Kemper, Juli 49613
Kemper, Julie Maio 52576
Kemper Fields, Laura 49686
Kemperdick, Dr. Stephan 41166
Kempf, Jürgen 11453
Kempf, Klaus 20424
Kempf, Peter 21763
Kempler, Cheryl A. 53853
Kempling, Steven 05712
Kenagy, Dr. James 52994
Kendall, Andrew 46393
Kendall, Andy 47909
Kendall, Clifford P. 48522
Kendall, LeeAnn 49556
Kendall, Shirley J. 47015, 53419, 53420
Kendell, Shirley J. 53578
Kendrigan, Ernie 01213
Kenen, Susan 51031
Kenichi, Yuyama 29036
Kenichiro, Makino 28991
Kenig, Piotr 34561
Kennedy, Arlene 06188
Kennedy, Caroline 52025
Kennedy, Cheryl 50399
Kennedy, Clay W. 47098
Kennedy, Dr. Craig 44128
Kennedy, Elaine 44036
Kennedy, Elizabeth 47518
Kennedy, Greg 54048
Kennedy, Ian 49695
Kennedy, Mary 46363
Kennedy, Nancy 49825
Kennedy, R.H. 07130
Kennedy, Dr. Robert 50944
Kennedy, S.B. 43448
Kennedy, Sally E. 52305
Kennedy, Prof. W.J. 45336
Kennedy, William 47982
Kennedy Harty, Leanne 53318
Kenner, Dr. Rex 07199
Kenney, Kimberley 47229
Kenney, Stephen 46927
Kennis 31965
Kenny, Jessica 43877
Kenny, Kevin 48584
Kenny, Michael 24809
Kensler, Terry 36059
Kent, David 45607
Kent, Jerry 52585
Kent, Joan 52050
Kent, Liz 49466
Kent, Rachel 01553
Kenworthy, M. 43653
Kenwright, Carole 45977
Kenwright, Peter 45178
Kenyon, Colleen 54283
Kenyon, Sir George 43932
Kenyon, Kathleen 54283
Kenyon, Sarah 45540
Kenyon, Therese 01258
Keo, Pich 05286
Keogh, Anthony J. 38711
Keough, Beverly 49728
Keough, Donald R. 51114
Keough, Tracey 52171
Kepford, Kris S. 51825
Kephart, Betsy 51703
Keppert, Jozef 38255
Kerbourc'h, Emmanuel 11570
Kerefov, Betal Muradinovič 37442
Kerekesh, Michael 52109
Keren, S. 24970
Kerényi, Péter 23495
Kerimova, L.L. 54443
Kerkhoff, Dr. Jacqueline A.M. 32342
Kerkmeijer, W. 32287
Kermode, Deborah 43505
Kern, Alfred 41932
Kern, Emil S. 41489
Kern, James E. 53691
Kern, Robert 16808
Kern, Ursula 18377
Kern, Volker 20262
Kern Paster, Gail 53869
Kernan, Thomas J. 50136
Kerndler, Werner 01739
Kerner, Johann 19821
Kernhan, John 06444
Kerr, Andrea J. 45345
Kerr, Fred 01053
Kerr, John 44106
Kerr, Paul K. 46723
Kerr, Peter 01387
Kerr-Wilson, Ian 05957
Kerschbaum, Rupert 03082
Kerschbaumer, Erich 02545
Kersing-Blok, Drs. I. 32055
Kerssenbrock-Krosigk, Dedo von 47810
Kersten, Hendrik 22249
Kersting, Dr. Hannelore 20357
Kert, Christian 11514
Kervern, Jeannine 11995
Kervran, Sophie 15400
Kervyn, Benoît 03427
Kes, Vicki 50340
Kesaeva, E.A. 37933
Kessel, Jane 44210
Kesselman, Steve 52928
Kessler, Adolf 41694
Kessler, Agnes 41484
Kessler, Earl 24155
Kessler, M. A. Horst 16625
Kessler, Dr. Michael 41176
Kessler, Monika 17793
Kessler, Nava 25022
Kessler, Dr. Ulrich 17052
Kessler, Werner 41468
Kestel, Michael 21895
Kesteman, Prof. Emile 03456
Kestenbaum, Stuart J. 48027
Kester, Marilyn 47188
Ketelsen, Cheryl 47581
Ketelsen, Dr. Thomas 17954
Ketner, Joseph D. 53810
Ketteler, Georg 17396
Kettemann, Dr. Otto 19726, 19727
Kettula, Suvi 10727
Ketzer, Dr. Hans-Jürgen 17486
Keulen, W.J. 31820
Keunecke, Dr. Hans-Otto 18221
Keurs, P. ter 32488
Keuter, J.H.G. 31949
Keutgens, Eric 03360
Keweloh, Hans-Walter 17569
Key, Eric 54144
Keyaerts, Geneniève 15420
Keyaerts, Magali 15420
Keyes, George S. 48114
Keyse, Val 01434
Keyser, Ralph 47282
Keyworth, Gretchen 47036
Keyworth, John 44788
Kezich, Giovanni 27669
Khabibullaev, Nozim 54489
Khaelss-Khaelssberg, Alexander 02942
Khalid, Syed Ali 30578
Khan, Arif 52875
Khan, Aurangzeb 34036
Khan, Gulzar Muhammad 34044
Khan, Muhammad Afzal 34022
Khan, Nazir Ahmad 34042
Khan Jatoi, Saeed 34031
Khanhoushi, Mustapha 42630
Kharissov, Ildar 17101
Khasanov, S.R. 54467
Khasawneh, Alia 29662
Khatib Eghdami, Roshanak 24667
Khazam, Sandra 01479
Khevenhüller-Metsch, Ing. Maximilian 02268
Khewhok, Sanit 49341
Khom, Helmut 01907
Khotko, P.S. 03250
Khreis, Dr. Khalid 29655
Khulani-Mkhize, Leo 38642
Khuldashev, Tulkin 54438, 54440, 54441, 54442
Kibritova, Vanja 05106
Kicherer, Michael 18532
Kida, Junichiro 29598
Kida, Takuya 29492
Kidd, Alan 52748
Kidd, Jane 45345
Kidrick, Jim 52744
Kiebert, Susan 52835
Kieblinger, Hans 02123
Kiedel, Klaus-Peter 17569
Kiefer, Theresia 20058
Kieffer-Olsen, Jakob 10231, 10233
Kiehl, David 51279
Kiel, Dennis 47555
Kielak, Bernard 34993
Kielholz, Walter B. 42083
Kielland, Morten 33974
Kielmayer, Oliver 42028
Kiely, Ann 45965
Kiely, David 45965
Kiemele, E. 18232
Kiendl, Anthony 05382
Kiener-Hofer, Trudi 41940
Kienholz, Kathleen 51111
Kieninger, Ernst 02937
Kienitz, Gerlinde 20684
Kientz, J.J. 12197
Kienzle, Karen 52863
Kiereś, Małgorzata 35303
Kiermeier-Debre, Prof. Dr. Joseph 20280
Kiers, J. 31864
Kiest, Rhonda 51433
Kietlińska-Michalik, Barbara 34794
Kievit, P.R. de 32093
Kieweg 02767
Kift, Dr. Dagmar 17921

K

Kiger, Robert W. 51950
Kihiyo, Jackson M. 42513
Kiiski-Finel, Päivi 11408
Kijima, Prof. Shunsuke 29330
Kikić, Prof. Vesna 08916
Kilaghbian, Elia 28191
Kilani, Huda 29653, 29654
Kilby, Virginia 49637
Kildee, Magee 47171
Kilger, Dr. Gerhard 17917
Kilham, D.D. 49175
Kilian, Dr. Ruth 20134
Kiliti, Zsuzsanna 23670
Killguss, Hans-Peter 19626
Killias, Augustin 41522
Killoran, Elizabeth 51740
Kilner, Josefin 40555
Kilroy, Frank 45061
Kilsdonk, Betty 48426
Kim, Chi-Eun 29920
Kim, Christine Y. 51264
Kim, Hongnam 29876
Kim, Jin-Hyuck 29796
Kim, Jong Chull 29794
Kim, Judy 47069
Kim, Jung Soo 29862
Kim, Dr. Ke Chung 53657
Kim, Kumja Paik 52766
Kim, Sun-Jung 29825, 29864
Kim, Sung-Rae 29848
Kim, Sunggu 29844
Kim, Wook-Han 29825
Kim, Yong-Ha 29824
Kim, Yongsan 29936
Kim, Yun Soon 29942
Kimball, Cathy 52808
Kimball, Cindy 51589
Kimball, Kathryn A. 50978
Kimfoko-Madoungou, Joseph 08807
Kimmel, Dominik 20150
Kimsey, Dr. Lynn S. 47978
Kimura, Hidenori 29358
Kin, Wook-Han 29864
Kincaid, Jean E. 05366
Kinderen, Ronny de 03870
Kindgren, Hans 40992
Kindlen, Helen 36438
Kindler, Dr. Gabriele 19437
Kindrachuk, William 07273
Kindred, Amy 49495, 54371
Kindschy, Errol 54079, 54080
Kindt, Chantal 03658
Kindt, Diane 50773
King, Alan 51441
King, Andy 43632, 43633
King, Beverly 50602
King, Brad 44875
King, Carolyn 24750, 47449
King, Charles 47203
King, Douglas R. 52598
King, Dr. Duane H. 50265
King, Edward 43571, 44557, 44559
King, Elspeth 45752
King, Evelyn 01363
King, G.J. 01224
King, George G. 52874
King, Graham 43894
King, Harry 49821
King, Hugh R. 48246
King, John 33097
King, Jonathan 44803
King, Karma S. 49821
King, Lyndel 50731
King, Margaret H. 44242
King, Margery 51938
King, Martha 48688
King, Michael D. 44020
King, Muriel 44275
King, Tamara 50753
King, William C. 50589
Kingan, Beryl A. 46080
Kingdon, Charles 06298
Kingery, James 48932, 51291
Kingsbury, Bernard 33083
Kingsford Hale, James 44443
Kingsley, April 48253
Kingston, Jean T. 54209
Kinii, Shoko 29522
Kinnear, J. 45365
Kinney, Emilie 50880
Kinno, Seiichi 28935
Kinoshita, Kyuichiro 29394
Kinoshita, Masafumi 28895
Kinsell, Tracey 50585
Kinsey, Danna 47709
Kinsman, Jane 00946
Kinsolving, Lucie 51232
Kinson, Alan 00819
Kintisch, Christine 02944
Kintzel, Jürgen 21740
Kinyon, Catherine 53513
Kinzinger, Nancy 53355
Kipiani, M.A. 16360
Kipkorir, Daniel 29738
Kipnis, Jeffrey 47761
Kippes, Dr. Wolfgang 03009, 03070
Kipping, Jens 16498
Kiprop, Nancy 29739
Király, Elvira 23454
Kirby, C. 44389
Kirby, Jo 44937
Kirby, John 05904
Kirby, Kris 51382
Kirby, Michael 54171
Kircher, Ashley 53126
Kircher, Gunter 19553
Kirchhauser, Hannes 19449
Kirchmayer, Christian 02230
Kirchmeier Andersen, J.C. 10156
Kirchner, Dr. Ferdinand 02622
Kirchner, Gertrud 02088
Kiričenko, E.I. 37044
Kiričenko, Galina Vladimirovna 37070
Kirigin, Branko 08954
Kirilova, Lena 05017, 05018
Kirk, Natalie 53817
Kirk, Ruth 47736
Kirk, Wendy 44855
Kirkbride, Cinoy 05778
Kirkby, Patrick 46031
Kirkeby Andersen, Dorte 10257
Kirkham, Thomas G. 43445
Kirklin, Deborah 52908
Kirkman, John 01073
Kirkpatrick, Andrea 06838
Kirkpatrick, David 43933
Kirkpatrick, Eleanor 51517
Kirkpatrick, Dr. Gail B. 20504
Kirkpatrick, John 51517
Kirkpatrick, May 43363
Kirkpatrick, Roger 53948
Kirkpatrick, Stephanie 45493
Kirkwood, Michael 48325
Kirmeß, Max 20834, 20837
Kirnbauer, Dr. Martin 41161, 41174
Kirr, Katharina 19682
Kirrane, Siobhan 45048
Kirsch, Eva 52734
Kirsch, Dr. Jeffrey W. 52743
Kirsch, Dr. John A.W. 50395
Kirsch, Olaf 18951
Kirsch, Dr. Steven 49929
Kirschbaum, Barbara 19626
Kirschberg-Grossman, Reva G. 51155
Kirschstein, Jörg 20910
Kirstein, Michaela 19986
Kirsten, Alex F. 06278
Kisapáti, István 23377
Kisëv, Ërdan 05161
Kishioka, Yukio 28367
Kisić, Anica 08872
Kisielewicz, Marcin 35260
Kisjova, Vera 05128
Kisler, Mary 33042
Kisler, Shuli 25122
Kisling, Ing. Andreas 02788
Kisman, A.K. 32136, 32138
Kiss, Andrei 36691
Kiss, K. 44833
Kissack, Terence 52775
Kisser, Gerhard 01930
Kissi, Flavio 27012
Kissinger, Scot 06478
Kissiov, Kostadin 05124
Kißro, Reinhard 20917
Kistner, Dr. Wolfgang 18779
Kitagawa, Kazuo 28858
Kitagawa, Seika 29068
Kitamura, Hitomi 29492
Kitamura, Sadao 28795
Kitamura, Yoshiro 29136
Kitamura, Yutaka 29505
Kitauo, Tsuguto 28955
Kitazawa, Toshio 29301
Kitchen, Dr. M. 44069
Kitchin, Cameron 53755
Kitchin, Simon 43825, 43828
Kitching, Matthew 38652
Kite Powell, Rodney 53436
Kitschen, Dr. Friederike 22169
Kittas, Prof. Dr.. Chr. 23008
Kittel, Gerti 02956
Kittelmann, Udo 17032, 17192, 17222, 17244
Kittleson, Rosalie 46543
Kittner, Ed 49369
Kitto, Suzanne 44665
Kittredge, Cindy 49030
Kiturka, Jury 03246
Kityo, Robert Martin 42989
Kitz, Hilary 53608
Kitzler, Gottfried 02445
Kitzlinger, Dr. Christine 18951
Kitzmann, Andrew 53397
Kiuchi, Kazuhiro 28371
Kivimäe, Juta 10624
Kivimäki, Kati 11259, 11262
Kiyoaki, Morita 28920
Kiyohiko, Sakurai 29413
Kızıltan, Zeynep 42823
Kjeldbæk, Esben 10131
Kjeldgaard, Inge Merete 10000
Kjeldsberg, K. 33562
Kjeldsberg, Peter Andreas 33927
Kjeldsen, Kjeld 10103
Kjellberg, Anne 33730
Kjems, Folke 10090, 10092
Kjerström Sjölin, Eva 40703
Kjupčupov, Georgi 05169
Kładna, Dr. Aleksandra 35160
Klages, Dr. Claudia 17462
Klang Eriksson, Eva 40703
Klanicová, Dr. Evženie 09416, 09417
Klapczyński, Stanisław 34592
Klar, Alexander 19758
Klar, Annika 40633, 40634, 40637
Klar, Herma 20545
Klarić, Josip 09010
Klassnik, Robin 44914
Klasson, Gunnar 40992
Klatt, David 06385, 06468, 06909, 06927, 07415, 07416
Klatt, Dr. Ingaburgh 20076
Klatte, Matthias 22432
Klattenberg, Hinrich 20616
Klaue, Karin 20579
Klaus, Dietmar 16498
Klaus, Franz Josef 02305
Klaus, Judy 33148
Klaus, Werner 41487
Klauser, K.D. 03999
Kleckerová, Marcela 09883
Klee, Sonja 19736, 21657
Kleeblatt, Norman 51200
Kleefisch, Stephen J. 51980, 51981
Kleeman, Drs. E. 31813
Kleemann, Kurt 21261
Kleespies, Gavin 50889
Klei, Mary 50012
Kleiger, Christian 51470
Kleijn, G. de 31818
Klein, A. 32466
Klein, Dr. Angela 17518, 22474
Klein, Anton 03025
Klein, Beatrixe 22548
Klein, Carolyn 48366
Klein, Christophe 14154
Klein, Dr. Emanuel 50959
Klein, Mag. phil. Eveline 03099
Klein, Holger A. 47624
Klein, Janice B. 48451, 48454
Klein, Jean-Claude 15197
Klein, Jean-Raymond 11590
Klein, Matthias 16574
Klein, Dr. Michael 20146
Klein, Peter 53368
Klein, Thomas 17871, 17872
Klein, Valérie 14135
Klein-Langner, Wolfgang 22724
Kleinbongartz, M.A. Sigrid 18006
Kleindorfer-Marx, Dr. Bärbel 22291
Kleine, Dr. Joachim 22795
Kleiner, Gerald 50794
Kleiner, Michael 22214
Kleinhans, M. 38767, 38768, 38769
Kleinová, Olga 38329
Kleinschmidt, Torsten 19375
Kleinschmidt-Altpeter, Dr. Irene 17460
Kleinschrot, Dr. Dorothée 22715
Kleinsteuber, Wolfgang Friedrich R. 17241
Kleipaß, Dr. Herbert 18152
Kleisner, Ted J. 54131
Kleist, Monika von 16695
Klejman, N.I. 37304
Klemm, Dr. Christian 42083
Klemm, Federico 00160
Klemm, Hélène 41920
Klemm, Randy 49038
Klemm, Susanne 18313
Klencke, Lippold von 18155
Klenner, Ingrid 21946
Klenow, Rosemary E. 48261
Klepač, Koraljka 08940
Klepatsch, Alfred 03109, 03113
Klerk, Prof. Dr. L.F.W. de 32373
Klesman, Kees 31856
Klettner, Emmi 01957
Kley, Dr. Stefan 20781
Kliege, Dr. Melitta 20786
Klim, T.B. 37190
Klimaszewski, Cathy 49548
Klimek, Larry 50758
Klimenova, Tamara Aleksandrovna 36860
Klimesch, Dr. Gertraud 01856
Klimm, Erich 21116
Klimov, Vladislav Petrovič 36845
Klimova, Anna Danilovna 36873, 37985, 37986
Klimova, Elena Michajlovna 37837
Klímová, Jarmila 09804
Klimtová, PhDr. Zdenka 09776
Kline, Dr. John 48106
Kline, Katy 47102
Kline, Margaret 48754
Kline, Phil 47392, 47393
Kling, Burkhard 21877, 21879
Klingelhöfer, Claudia 21750
Klinger, Dr. H.C. 38592
Klinghardt, G. 38592
Klinghardt, Gerald 38591
Klingman, Catherine 48113
Klingsöhr-Leroy, Dr. Cathrin 19592
Klink, Alfonso Caballero 39187
Klink, Peter 51924
Klink, R. 22112
Klinka, Isabelle 14535, 14537
Klinkhardt, Bernd 22655
Klinkow, Michaela 20896
Kljajić, Damijan 38534
Kljueva, Valentina Ivanovna 36882
Klobe, Tom 49355
Klobučníková, Magdaléna 38321
Kloch, Ove 10315
Klochko, Deborah 52741
Klocker, Hubert 03133
Kloiber, Augustin 02648
Kloosterboer, A.E. 32690
Klootwijk, L.P. 32693
Klooz, Donan 50769
Kloppenborg, David N. 48147
Klos, Günther 19258
Klose, Dr. Dietrich 20488
Klose, Simon 00857
Klosek, Melesa 47749
Klostermeier, David 52532
Klotzer, Dr. Georg 26656
Kloub, Ola 29667
Klucas, Kevin 47954
Klüver, M.A. Dennis 21601
Klüver, Heinrich 19990
Klug, PD Dr. C. 42103
Kluge, Dr. Arnd 19220
Kluge, Prof. Arnold G. 46310
Kluge, Prof. Dr. Bernd 17051
Kluge, Joachim 22336
Kluge, Martin 41158
Kluge, W. 20210
Kluger, Reinhard 01807
Klukina, Anna I. 37157
Klumpp, Sibylle 14330
Klupp, Krzysztof 35047
Kluska, Adam 35137
Kluss, Dr. Maciej 35062
Kluver, William C. 53358
Kluza, Radosław 34915
Klydie, Thomas 52333
Klykova, Galina Ivanovna 37517
Kmieck, Anne T. 47626
Kmínek, Petr 09935
Knabe, Dr. Hubertus 17111
Knackert, Bruce J. 50717
Knape, Gunilla 40544
Knapen, I.A.M. 32667
Knapp, Clark 48417
Knapp, D.R. 44768
Knapp, Marylin 46288
Knapp, S. 44768
Knapp, Trevor 45925
Knappertsbusch, Dr. Michael 41175
Knauer, Wilfried 22652
Knauss, Carol 46303
Knauss, Dr. Florian 20486
Knauss, Dipl. Geogr. Jürgen 17749
Knebelsberger, Caroline 46712
Knecht, Dr. Richard 53642
Kneubühl, Dr. Urs 41217
Kneuss, Martine 41890
Knezevich, Alan 53899
Knibiely, Philippe 15375, 15376
Knick, Dr. Stanley 51755
Kniefeld, Lothar 22031
Knieriem, Dr. Michael 22741
Knierriem, Peter 19943, 21338
Kniesz, Jürgen 22336
Knigge, Prof. Dr. Volkhard 22399
Knight, Brian L. 50427
Knight, Cynthia 54274
Knight, David J. 49280
Knight, Gregory 47465
Knight, Irene 06497
Knight, Jeremy 44489
Knight, Jim 47734
Knight, Katie 49241
Knight, Michael 52766

K

Knight, Robert 46808
Knight, V.M. 44866
Knights, Mary 00781
Knill, Phyllis 48367
Knipe, Peter 51697
Knipp, Otto 19295
Knipprath, Ute 17462
Kniprath Smith, JaNeen 47669
Knisch, Griseldis 17531
Knittel, Pierre 16298
Knittel, W. 20621
Knížák, Dr. Milan 09777
Knízák, Milan 09798
Knížek, Col. Aleš 09608, 09734, 09759
Knoch, Francis 30475
Knoche, Norma 48153
Knocke, Erich 18261
Knode, Marilu 50710, 52977
Knodel, Bernard 41681
Knöchlein, Dr. Bernhard 21768
Knödel, Dr. Susanne 18952
Knödler-Kagoshima, Brigitte 19435
Knöpfel, Sylvia 18545
Knofler, Dr. Monika 03020
Knogler, Helmut 02555
Knol, Egge 32281
Knol, M. 32844
Knoll, Andrew H. 47173
Knoll, Annegret 19688
Knolle, Dr. P. 32213
Knop, Matthias 17988
Knop, Monika 20908
Knop Kallenberger, Christine 53610
Knoppers, K. 32000
Knorowski, Mariusz 34989
Knott de Arechaga, Deborah 52320
Knowles, A.M. 43820
Knowles, Elizabeth 45360
Knowles, Elizabeth P. 50274
Knowles, Dr. P. 24754
Knowlson, Paul 43586
Knox, J. 43447
Knox, Judy 46594, 46994
Knox, Keith 06724
Knox, Lu 51412
Knox, R. 44803
Knox, Tim 44987
Knudsen, Ann Vibeke 10047, 10241, 10242
Knudsen, Henning 10125
Knudsen, Herluf 10224
Knudsen, Karsten 10268
Knudsen, Marie Hedegaard 43331
Knudsen, Svend Aage 09985
Knüppel, Stefan 18301
Knuth, Dr. Detlef 21098
Knuth, Dr. Michael 17052
Knutsen, Jim 53200
Knutson-Hedetun, Anna 40856
Ko, David C. 42457
Ko, Seong-Jun 29055
Kob, Karin 41137
Kobayashi, Akira 29284
Kobayashi, Fumio 29506
Kobayashi, Kazuhiko 28464
Kobayashi, Kiroku 29215
Kobayashi, Koichi 29355
Kobayashi, Saburo 29432
Kobayashi, Shigeko 29466
Kobayashi, Tadashi 28393
Kobayashi, Dr. Tatsuo 28964
Kobayashi, Toshinobu 28666
Kobayashi, Yasuyuki 29055
Kobayashi, Yotaro 29450
Kobe, Michael 18898
Koblmiller, Gertraud 02057
Kobusiewicz, Zofia 34925
Koç, Sema 42702
Koca, Osman 42908
Koççoban, Cahit 42857
Kocebenkova, Vera Ivanovna 36988
Kočerženko, Jevgenija Ivanivna 43189
Kočevar, Sanda 08886
Koch, Dr. Alexander 21834
Koch, Claudia 02992
Koch, Cynthia M. 49458
Koch, Eva 17398
Koch, Fred 50320
Koch, Friederike 19421
Koch, Dr. Lars-Christian 17081
Koch, MA Matthias 18009
Koch, Dr. Michael 20426
Koch, Werner 17414, 19001
Koch, Wilhelm 20290
Koch, Will 52865
Kochančuk, Vitalij Mykolajovyč 43059
Kochanova, Ljudmila Petrovna 37915
Kochanowicz, Roman 34982
Kochendörfer, Reinhard 19858
Kocher, Prof. Dr. Dr. h.c. Gernot 01972
Koches, Susan 46573
Kočík, Zdeněk 09469, 09579, 09774
Kock, G.M. de 32823, 32825
Kock, Prof. Hans 19529
Kock, R.D. de 38635
Kocsis, Dr. Gyula 23503
Koczur, Anton 01998
Koda, Harold 51212
Kodani, Susan 49345
Köb, Edelbert 03042
Koebbel, Wolfgang 17530
Köberl, Prof. Mag. Rudolf 02499
Köberl, Sieglinde 01739
Koechlin, Bernard 41423
Koefoed, Lizette 10169
Koegel, Alice 21960
Köger, Dr. M.A. Annette 19881
Kögl, Ing. Alois 01857
Kögl, Reinhard 21588
Kögler, Johannes 18468
Kögler, Dipl.-Geol. Reinhard 22730
Köhler, Birte 18951
Köhler, Dr. Georg 02910
Köhler, Dr. Hubert 18858
Köhler, Dr. Jörn 17789
Köhler, Dr. Juliana 22574, 22581
Köhler, Lothar 18131
Köhler, Neeltje 32295
Köhler, Otto 19237
Köhler, Thomas 22335
Köhn, Regine 17818
Köhne, Dr. Eckart 22102
Koehnline, Lyn 47344
Kölbl, S. 17381
Kölbl-Ebert, Dr. Martina 18082
Köll, Helmut 02694
Köllhofer, Dr. Thomas 20166
Koelliger, Marcus 41333
Költő, Dr. László 23635
Könenkamp, Dr. Wolf 20269, 20270
König, Cornelia 22069, 22070, 22071, 22072, 22073
Koenig, Davison 53584
König, G. 17366
König, Dr. Gabriele 18511
König, H. 16844
König, Hans 42018
König, Heinz 19119
König, Helmut 19224
König, Josef 02965
König, Prof. Kasper 19624
König, Norbert 17136
König, Reginald 21948
Koenig, Robert J. 53476
König, Prof. Dr. Viola 17081, 17150
König-Rainer, Mag. Julia 03087
Koenigsmarková, Dr. Helena 09797, 09882
Köntös, László 23751
Köpke, Ingeborg 18479
Köpke, Prof. Dr. Wulf 18952
Köppl, Günther 01907
Koerber, Martin 17069
Koerber, Susannah 52379
Koerhuis, E. 33015
Környei, Dr. Attila 23723, 23809, 23822
Koesoemowarni, Endang 24549
Köster, Hein 17243
Koevend, Mary Helen de 50990
Koffi, Bilé Yapo 08836
Kofler, Claudia 03026
Kofoeb, Jakob 10080
Kofuku, Akira 29419
Koga, Kim 50254
Kogan, Lee 51113, 51171
Kogler, Leopold 02906
Kogure, Susumu 28561
Koh, Dr. Eugen 01406
Kohan, Christopher 46248
Kohl, Barbara 21415
Kohlberg, Ilona 18458
Kohler, Ruth 53061
Kohlschmidt, Ulrich 20618
Kohlstruck, Gerd 21359
Kohn Loncarica, Prof. Dr. Alfredo G. 00194, 00247
Kohnhorst, Bridgette 50981
Kohnke, Dr. Hans-Georg 17511, 17512
Kohút, Dr. Teodor 38263
Koike, Tomio 29012
Koivisto, Mikko 10969
Koja, Dr. Stephan 02951
Kojaev, M. 29699
Koji, Yoneda 28979
Kojić, Liljana 38100
Kok, Harry de 04101, 04102
Kokisch, André 17111
Kokki, Kari-Paavo 10765, 10766
Kokkotidis, Dr. Klaus Georg 21950
Kokošinek, Nataša 38425
Kokoska, Monika 18245
Kokott, Dr. Jeanette 18952
Kokroko, Grace Juliet 22853
Kokuryo, Masao 29505
Kolarić, Dr. Juraj 08988
Kolb, Jennifer 50396
Kolb, Nancy D. 51871
Kolb, Rainer 16583
Kolbach, B. 32349
Kolbasina, Larisa Aleksandrovna 37483
Kolbe, Joachim 19812
Kold, Anders 10103
Kolde, H. 19408
Kolenberg, Hendrik 01544
Koles, L.A. 01407
Kolesnikova, Ljudmila Arsenjevna 36978
Kolesnikova, Ljudmila Petrovna 37191
Kolesov, S.N. 37812
Koletsky, Sue 47642
Koletsky, Susan 46658
Kolev, Cvetan 05227, 05230
Koll, Kersti 10619, 10624
Kolland, Dr. Dorothea 17162
Kollar, Mary 50118
Kollár, Dr. Štefan 38185
Kollath, Jeffrey 50398
Kollbaum, Marc 52579, 52582
Kollbaum-Weber, Jutta 22759
Koller, Karl 02972, 03092
Koller, Margaret H. 48042
Koller-Jones, Kimberly B. 51042
Kollintzas Drikakis, Elaine 47479
Kollmann, Dr. Karl 18237, 19961
Kollmann-Caillet, Virginie 11676
Kolm, Herbert 02509
Kolmos, Dr. Alfred S. 54135
Kolner, Brian A. 46992
Kolodzei, Natalia 51696
Kolodziej, Dr. Axel 21808
Kolodziejak, Adeline 06265
Kołodziejski, Dr. Adam 35148
Kolomceva, Raisa Michajlovna 37269
Kolonas, L. 22873, 22889, 23138
Kolonia, Rozina 22982
Kolonoas, Lazaros 23091
Koloskova, T.G. 37327
Kolosov, Viktor 36869
Kolosova, Ekaterina Matveevna 37685
Kolsteren, Steven 32281
Kolt, Ingrid 07037
Kolt, Sheryl 07386
Kolter, Horst 19034
Koltz, Jean-Luc 30483
Koltzer, Ulrich 21081
Koludrović, Ivana 08977
Kolveši, Željka 09004
Kolyesnik, Ljudmila Michajlovna 36933
Komanecky, Michael 47984, 51906
Komarek, Beverly 49027
Komarov, Ilja Iljič 37257
Komatu, Nario 29204
Kombu, Anatolij Stepanovič 37098, 37099
Komenan, Robert Yao 08838
Komick, J. 07235
Komissarov, Eha 10624
Komnenović, Ilija 38121
Komor, Valerie 51243
Komzak, Wolfgang 02460
Komzyuk, Vera 43070
Konan, Louise Constante N'Guesssan 08840
Končakovs'kyj, Volodymyr Petrovyč 43086
Koncevaja, Tatjana Leonidovna 37781, 37783
Kondakova, Nina Viktorovna 37868
Kondo, Tetsuo 28547
Kondratenko, Tatjana Michailovna 37065
Kondziela, Mariusz 35125
Konečná, Katarína 38318
Konecny, Dr. Andreas 01741
Konevskaja, Tatjana Ivanovna 37602
Konieczny, Marian 34974
Konieczny, Władysław 35348
Koning, Elizabeth 53373
Konishi, Masafumi 29031
Konishi, Tamao 29055
Konkle, Scott 05338
Konnikov, Boris Aleksandrovič 37501
Konno, Hirofumi 28516
Kono, Osamu 28774
Kononov, Alexander 38679
Kononova, Natalia 38679
Konopka, Lawrence 48619
Konova, Ljubava 05191
Konrad, Dr. Claudia 19219
Konst, Thomas 47256
Konstam, Nigel 25708
Konstantinos, Manioudakis 22964
Konstantinus, Lahurakis 23229
Konstantios, Dr. Dimitris 22904
Konukiewitz, Insa 17592
Konuma, Hirokazu 28758
Konvalinka, Danilo 54263
Konz, Klaus 19963
Kooiman, R. 32866
Kooiskra, L. 32264
Kooistra-Manning, Kevin 46807
Koolkin, Kenneth 24937
Koontz, Darlene 48028
Koop, Kathy 51106
Koopmans, J.J.G. 32731
Koopmans, Y. 31929
Koopmans-Witt, C.E. 31975
Koos, Greg 46861
Kopálová, Dr. Danka 38324
Kopbosinova, R.T. 29683
Kopco, Mary A. 48002, 48003
Kopczak, Dr. Chuck 50215
Kopec, Diane 46600
Kopecky, Carle J. 52963
Kopeinig, Josef 02792
Koperkiewicz, Adam 34638, 34652
Koperski, Kate 51348
Kopf, Sibylle 16969
Kopf, Vicki C. 54251
Kopiasz, Jarosław 34630
Kopil, Olga Andrïivna 43212
Kopis, Mildred 51789
Kopówka, Edward 34760
Kopp, Dr. Marianne 16793
Kopp, Siegfried 02566
Kopp, Werner 17237
Koppel, Greta 10624
Koppel, John 09971
Koppel, Merike 10624, 10638
Koppel de Ramírez, Elsa 08588
Koppelman, Dorothy 51268
Koppensteiner, Dr. Erhard 02625
Kopplin, Dr. Monika 20516
Koppy, Ann 51381
Kopydłowski, Janusz 34585
Korandová, Martina 09818
Korb, Uwe 20126
Korbacher, Dr. Dagmar 17172
Korbelářová, PhDr. Irena 09692
Korčagin, E.N. 37632
Korčevnikova, I.L. 37336
Kordiš, Ivan 38432
Kordos, Dr. László 23461
Kordowska, Barbara 35314
Koref, Barbara 16665
Koren, Elisabeth 33402
Koren, Elisabeth S. 33742
Koreny, János 23481
Korhonen, Paavo 11351
Korhoven, Teppo 10786
Korina, Olga Aleksandrovna 37232
Korir, Daniel 29740
Korkmasova, Maimusat Arslan-Alievna 37118
Kormano, Riitta 11408
Korn, Brigitte 16417, 16418
Korn, Prof. Dr. Lorenz 16914
Korn, Susanne 18344
Kornacki, Krzysztof 34713
Kornatek, Anna 35215
Kornblum, Aaron T. 46755
Kornejeva, Galina Viktorivna 43130
Kornetchuk, Elena 53044
Kornhofer, Elisabeth 02882
Kornhoff, Dr. Oliver 21259
Korobar, Klime 30518
Kőrös, Dr. Endre 23605
Kőrösi, Krisztina 23728
Koroly, József 23615
Korotaeva, Ljudmila Dmitrievna 37964, 37965
Korotkij, Aleksandr Nikolaevič 37112
Korovina, Vera Vasiljevna 37568
Korppi, Svea 11224
Korpusik-Jelonkowa, Zofia 35374
Korres, Manolis 22914
Koršikov, Anatolij Semënovič 37518
Kort, Betty 52272
Kort, Drs. Toon 31810
Korteweg, Drs. A. J. 32078
Kortländer, Prof. Dr. Bernd 17991
Kortlander, Christopher 48869
Kortschak, Dr. Werner 02926
Kortunova, M. 37677
Korz, Johannes-Rudi 18372
Korzec, Patricia 52371
Korzeniewska, Iwona 34791
Kos, Dr. Peter 38458
Kos, Stanisław 35097
Kos, Dr. Wolfgang 03003, 03086
Košacký, Petr 09557
Kosanji, Kozo 29253
Kosasih, Dikdik 24442
Koschatko, Alois 02873
Koschnik, Dr. Leonore 17072
Kościńska, Halina 34573
Kościołko, Eugeniusz 34901
Kosica, Marilyn K. 46988
Kosicka, Anna 35186, 35192, 35194, 35196, 35197
Kosinski, Dr. Dorothy 47931
Kosinski, Joan 47507

K

Koski, Ann 50396
Koski, Elizabeth 46394
Koski, Janne 11261
Koskijoki, Maria 10730
Koskimies-Envall, Marianne 11426
Koskinen, Pekka 11442
Kośko, Maria 35034
Kosler, Barbara 20449
Kosmadaki, Polina 22883
Kosok, Prof. Dr. Lisa 18946
Kosonen, Elina 11239
Kosonen, Juhani 11075
Kosova, Parim 38134
Kossatz, Dr. Tilman 22714
Kossmann, M.A. Joachim 18917
Kost', Ljubov Vasylivna 43122
Kosta, Bernard 49196
Kostadinova, Snežana 05146
Kostanian, Rachel 30410
Kostede, Ulrich 19418
Kostelny, Elizabeth 52331, 53133, 53384, 53754, 53757
Koster, Dr. A. 32590
Koster, F. 32227
Koster, Dr. J. 32430
Koster, John 53719
Koster, Koos 32518
Koster, Richard de 54077
Kostet, Juhani 11181, 11388, 11392, 11402, 11403, 11404
Kostina, N.V. 37816
Kostková, Magda 38363
Kostołowski, Andrzej 35369
Kostov, Georgi 05252
Kostova, Penka 04974
Kostrowicki, Andrzej 35020
Kostyál, Dr. László 23971
Kostynowicz, Roman 35158
Kosunen, Mirja 11157, 11303
Kosvincev, Boris Jurjevič 36808
Kosvinceva, Tatjana Dmitrievna 36807
Koszkul, Jan 35132
Kosznik, Danuta 34736
Koszutski, Oskar 35018
Kotai-Ewers, Trisha 01542
Kotalík, Jan 09714
Kotasiak, Wojciech 34995
Kotei, Ebenezer 54194
Kothe, O. 19878
Kothe, Robert J. 48909
Kotilainen, Eija-Maija 10795
Kotilainen, Kari 11286
Kotilainen, Simo 10897
Kotitonenko, Nadija Oleksiïvna 43188
Kotivuori, Hannu 11276
Kotkin, Lauren 46133
Kotljakova, N.V. 37127
Kotoň, Friedrike 21688
Kotooka, Mitsushige 28772
Kotouč, Friederike 21682, 21683
Kotrošan, Dražen 04284
Kotsch, Natalie 49439
Kotschwar, Sandra 02317
Kotsiubynsky, Igor 43005
Kottcamp, Edward 48467
Kottenbrink, Magdalene 21167
Kotterer, Michael 21228
Kottmann, Ingeborg 22241
Kotula, Krystyna 35312
Kotzan, Peter 19985
Kotzur, Dr. H.-J. 20141
Kouakou, Thierry Konan 08837
Kouba, Jess 52283
Kouchi, Osami 29523
Koukal, PhDr. Petr 09692
Koukal, Dr. Vitezlaw 09827
Koumantakis, Prof. I. 23249
Kourinou, Eleni 22920
Kourouma, Moussa 23311
Koutsomallis, Kyriakos 22883
Kouvaras, Fotini 23199
Kouvoutsakis, Panagiotis 23052
Kouzin, E. 36996
Kouzmin, Alexander 37643
Kovačev, Dimitar 04972
Kovačević, Dr. Bojan3a 38065
Kovačić, Goranka 08876, 08900
Kovačić, Marcelo 08940
Kovačić, Vladimir 08929
Kovács, Dr. Anna 23510, 23790, 23791
Kovács, Irén 23468
Kovács, Józsefné 23499
Kovács, Dr. Tibor 23552
Kovácsné Bircher, Dr. Erzsébet 23812
Kovácsné Ládi, Katalin 23702
Kovacsovics, Dr. Wilfried 02608, 02625
Kovalenko, Donna 52999
Kovalskij, Sergej Viktorovič 37624
Kovanen, Marja-Leena 11001
Kovarik, Jiri 09582
Kovrakis, Stavros 23124
Kovtun, Vladimir Vasilјevič 37991
Kowalczyk, Longin 34913
Kowalczyk, Wiktor 34734
Kowald, Dr. Margaret 01510
Kowalewska, Dr. Agnieszka 35311
Kowalski, Beth 50396
Kowalski, The Very Reverend James A. 51140
Kowalski, M.A. Michael 19340
Kowalski, Zygfryd 34929
Kowarik, Prof. Wilfried 02375
Kowark, Dr. Hannsjörg 21969
Kowerko, Oleh 47519
Kowski, Robert 49065
Kox, Jo 30476
Koyama, Atsushi 28888
Koyama, Miturn 28423
Kôyama, Noboru 29096
Kožachmet, Chasen 29689
Kozáková, K. 38391
Koželuh, Josef 09410
Kozen, Hiroshi 28842
Kozerski, Paweł 34571
Kozij, Galyna Ivanivna 43164
Kožišková, Lenka 09654
Kozlova, T.D. 37636
Kozlova, Vera Danilovna 36821
Kozlovskaja, Nadežda Borisovna 36826
Kozlowski, Kazimiera 47220
Kozubek, Gwalbert 34622
Kožucharov, Dimitar 05096
Kozyra, Peter H. 05406
Kozyreff, Chantal 03495
Kraber, Rosa 53352
Kraciak, Aniela 35061
Kräftner, Dr. Johann 03024
Krämer, M.A. Anja 21968
Krämer, Dr. Annette 21951
Krämer, Dirk 18023
Krämer, Frank 22249
Krämer, Gerd 17016, 21198
Kraemer, Jörg 17477
Krafczyk, Alois 19026
Kraft, Joe C. 51006
Kraft, Karl Ludwig 16760
Kraft, Dr. Perdita von 17732
Kragelund, Patrick 10127
Kragh, Mogens 10305
Kraglund, Ivar 33738
Kragness, Kurt K. 46684
Krahn, Dr. Volker 17052
Kraiński, Maciej 35133
Krajasisch, Dr. P. 02762
Krajewski, Sara 50393
Krajneva, Anna Anatoljevna 37131
Krajniak, Paul J. 50707
Krakauer, Dr. Thomas H. 48223
Kral, Dr. Martina 41598
Kral, Dr. Silke 16553
Kral, Silke 16555
Kral, Dr. Silke 16556
Kralevska, Julijana 30520
Kraljević, Marija 08894
Kraljni, Ariana 08993
Kramer, Prof. Cheryl 49547
Kramer, Dr. Dieter 02501
Kramer, Dr. E. 32473
Kramer, Finn Erik 10077
Kramer, Hermann 02830
Kramer, Lovetta 50176
Kramer, M. 21028
Kramer, Dr. Mario 18385
Kramer, Sonja 17163
Kramer, Trudy C. 53185
Kramme, Dr. Mike 50883
Kramp, Dr. Mario 19587
Kranjc, Igor 38454
Kranjc, Katja 38454
Kranner, Romana 02063
Kranz, Betsy 49487
Kranz, M.A. Diethelm 18141
Kranz, Prof. Dr. P. 18219
Kranz, Tomasz 34911
Kranzelbinder, Norbert 01834
Krapež, Mateja 38457
Krapf, Eva 02946
Krapf, Dr. Michael 02951
Krasilnikov, Grigorij Vladimirovič 36746
Kraske, Bernd M. 21251, 21252
Kraskin, Sandra 51255
Krasnokutskaja, Lidiya Ivanovna 37565
Krasovskaja, E.S. 37162
Krasovskaja, Nonna 43178
Krasowska, Dorothy 51162
Krasteva, Dimitijka 05191
Krăsteva, Natasha 05052
Krátká, Drahomíra 09834
Kratochvil, Jiří 09691
Kratochvilová, Alena 09813
Kratochvilová, Irena 09523
Kratochwil, Franz 02965
Kraus, E. 26785
Kraus, Gerhard 16467
Kraus, Jeannie W. 46666
Kraus, Dr. Jens 21113
Kraus, Dr. Jürgen 19339
Kraus, Dr. Michael 02708
Kraus, Peter 22171
Kraus, Dr. Stefan 19613
Krause, Ines 20050
Krause, Joanne 48998
Krause, Dr. Jürgen 20512
Krause, Manfred 21806
Krause, Melissa 51022
Krause, Dr. Siegfried 20764
Krause, Walt 52302
Krause-Willenberg, Horst 18844
Krauß, M. 19400
Krauß M.A., Marianne 18223
Krautgartner, Toni 02527
Krautschneider, Astrid Britt 01282
Krautwurst, Dr. Miriam 18196, 18197
Kravcov, Igor Nikolaevič 37804
Kravcova, Ljudmila Petrovna 37031
Kraxner, Ing. Hubert 02691
Krazmien, Mindy 49592
Kreamer, Todd A. 50770
Krebernik, Prof. Dr. Manfred 19381
Křeček, Ing. Vilém 09897
Krech, Shepard 47033
Kreem, Tiina-Mall 10624
Kregar, Tone 38406
Krehe, Lynette 47524
Kreibich, Matthias 18071
Kreijn, Drs. L. 32642
Kreil, Lisa 02942
Kreim, Dr. Isabella 19342
Kreindl, Richard 02266
Kreis, Johanna 01796
Kreiser-Francis, Deborah 47769
Krejči, Irena 09898
Krejci, Jim 54339
Kremer, Dr. Roberta 07227
Krempel, León 20445
Krempel, Prof. Dr. Ulrich 19005
Kremsreiter, Hans 21024
Kren, Thomas 50237
Krenkel, Ljudmila Ėrnestovna 37322
Krenn, Mag. Katharina 02827
Krens, Thomas 51188, 51258
Krenzlin, Kathleen 17106
Kreplin, Charles 54020
Kreps, Dr. Christina 48084
Kress, Brady 47192, 47983
Kreß, Hannelore 18796
Kress, Dr. Theo 18838
Krestin, Steffen 17734, 17735
Kret, Robert A. 47411, 51654
Kreth, PhD Rasmus 10096
Krethlow-Benziger, Dr. Donata 41589
Kretschmer, Karl 20391
Kretser, David de 45720
Kretzmer-Diepold, Monika 22147
Kretzschmar, Ulrike 17072
Kreuger, S.C.P. 32564
Kreul, Dr. Andreas 17556
Kreulen, Ivast 32792
Kreunen, Jean 32366
Kreuter, M.A. Bernd 18006
Kreutzer, Teresa 49211
Kreuzberger, Robert 52793
Kreuzer, Dr. Barbara K. 53172
Kreuzer, Hermine 03021
Kreuzer, Dr. Stefanie 19976
Kreyenbühl, Barbara 41212
Krick, Robert K. 48764
Kriebel, Svenjy 19412
Kriebner, Prof. Dr. 21159
Krieg, M. 11854
Krieger, Edino 04716
Kriegerbecková, RNDr. Karin 09404
Kriegs, Dr. Jan Ole 20513
Kriegseisen, Dr. Jacek 34641
Kriek, Johan 38576
Krier, Jean 30483
Krier, Patricia 48435
Krijnen, Marloes 31842
Kriller, Dr. Beatrix 03019
Krings, Dr. Michael 20472
Krippner, Regina 21591
Krischel, Dr. Roland 19635
Krischer, René 18787
Krischka, Karl 02279
Krishnappa, Dr. M.V. 24073
Krishtalka, Dr. Leonard 49991
Kristament, Werner 02369
Kristan, Dr. Markus 02929
Kristensen, Hans H. 10279
Kristensen, Tenna 10049
Kristfinnsson, Örlygur 24039
Kristiansen, Peter 10122
Kristic, Dr. Danilo 23882
Kristinsdottir, Gudrun 23983
Kristjansdottir, Gudbjorg 24008
Kristoffersen, Åge 33501
Kristy, Ben 48708
Kritzas, Charalambos 22908
Kritzman, George 50265
Krivoshejev, Vladimir 38160
Krivova, N.N. 37467
Kříž, Ivo 09899
Krizanac, Milica 38074
Krizner, L.J. 51243
Krjakunova, Nadežda PProkofjevna 36887
Krjukov, Anatolij Charlampievič 36923
Kroča, Květoslav 09575, 09576, 09945
Kroczak, Czesław 34586
Krody, Sumru 53921
Kröger, Dr. Michael 19140
Kroeger, Ted 46909
Kröhnert, Gesine 21695
Krönke, Rudolf 19657
Krönsjö, Marianne 40817
Kroeper, Dr. Karla 17028
Krösche, Klaus 18009
Kroeze, Hans 31909
Krog, Paul 42117
Krogedal, David 47901
Krogh, Leila 10028
Krogseng, Gari 52932
Krogstad, Bjørn R. 33666
Krohn-David, Robin 46811
Król, Adam 35003
Krolevec', Sergyj Pavlovyč 43101
Kroll, Stefanie 02942
Kromann, Erik 10187
Kronberger, Dr. Michaela 02939, 03044, 03061, 03062, 03084
Kronbichler, Dr. Johann 02681
Kronenfeld, Daniel 51181
Kronenwett, M.A. Heike 16891
Kronig, Karl 41212
Kronkright, Dale 52883
Kronquist, Bo 11155
Kroodsma, Donald E. 46269
Kroon, Rita 10624
Kropf, Dr. Christian 41213
Kropff, Heike 18263
Kropivka, Georgij Andrijovyč 43208
Kroqh, Finn E. 33871
Krotky, Gene 50125
Krouthen, Mats 33927
Kroy, Eadwierd 51554
Krpan, Sonja 08986
Krstić, Nina 38100
Krucký, Marian 09889
Kruczyński, Dr. Andrzej 35272
Kruegel, Rosalie 53210
Krüger, Frank 19116
Krüger, Dr. Gundolf 18690
Krüger, Hans 19483
Krueger, Hilary 53949
Krüger, Prof. Dr. Kersten 21389
Krüger, Dr. M. 38759
Krüger, Martin 20086
Krueger, Marylin 51726
Krueger, Phyllis 53795
Krueger, Thomas 18491
Krüger, U. 20047
Krueger, Ute 20854
Krüper, Jochen 18269
Krug, Prof. Dr. Dr. Klaus 20291
Krug, Martina 18988
Krug, Wolfgang 02685
Kruger, Gert 31761
Kruitwagen, W.H. 33000
Kruk, Karol 35316
Krukowski, Wojciech 35220
Krul, James 50152
Krull, Regine 18173
Krummenacker, Carolyne 15624
Krumova, Božura 05185
Krupa, Alain-Gérard 03807
Krupa, Dr. Vladimír 38327
Krupp, Dr. phil. M.A. Ingrid 22424
Kruppa, Gail 53544
Kruschen, Franziska 07398
Kruse, Anette 10246
Kruse, Hans 10119
Kruseman, Pauline W. 31826, 31869
Krusenstiern, Konstantin von 46998
Kruska, Dr. Peter 19534
Kruszynski, Dr. Anette 17997
Kruter, Stefan 18889
Krutisch, Dr. Petra 20764
Krutova, Marina Anatoljevna 36754
Kruus, Ülle 10619, 10624
Krużel, Krzysztof 34773
Krylov, V.M. 37715, 37730
Krylova, Raisa Nikolaevna 37794
Krystof, Doris 17998
Krzemiński, Jerzy 35098
Krzemiński, Prof. Wiesław 34808

Krzesinska, Caroline 45653
Krzeszowiak, Aleksander 35138
Krzmarzick, Pam 51105
Krznarić, Stjepan 08843
Krzykowska, Zofia 35180
Kseniak, Dr. Mieczysław 34909
Kuang, Maosheng 08119
Kuasnicka, Mellanee 52272
Kubáček, Jiří 38221
Kubach-Reutter, Dr. Ursula 20763
Kubadinow, Irina 03019
Kubassek, Dr. János 23548
Kube, Dr. Alfred 17570, 17572
Kube, Jan K. 16873, 21975
Kube, Manuela 21975
Kubešová, RNDr. Svatava 09421
Kubiak, Anna 34541
Kubičková, Marcela 09541
Kubitz, Peter Paul 17069
Kubizek, Mathilde 01751
Kublanovskaja, Natalija Alekseevna 37476
Kubli, Dr. Renate 19851
Kubo, Hiroji 29059
Kuboschek, Andreas 18713
Kuboshima, Seiichiro 29552
Kubota, Gaylord 52205
Kubota, Dr. Tadashi 29291
Kubrak, Zygmunt 34893
Kučerová, Marie 09489
Kucharski, Karen 48378
Kuchenbecker, Antje 16731
Kucherawy, Jack 06665
Kuchorenko, Konstantin Gavrilovič 36993
Kuchyňka, Zdeněk 09583
Kučinskij, Stanislav A. 37646
Kuckertz, Josefine 17028
Kučuk, Julija Valerjevna 37110
Kuczek, Irena 34934
Kuczyńska, Urszula 34953
Kudelová, Mária 09855
Kudielka, Prof. Dr. Robert 17029
Kudo, Masahide 28529
Kudrjavcev, Aleksandr Abakarovič 37807
Kudrjavceva, Olga 37634
Kübler, Dr. Gabriele 21290
Kübler, Sabine 16791
Küchler, Max 41383
Kuecker, Susan 47299
Kuefstein, Elisabeth Gräfin 02591
Kügler, Dr. Martin 18680
Kühebacher, Dr. Egon 27619
Kühhas, Margareta 02882
Kuehhas, Thomas A. 51661
Kuehl, Tracey K. 46784
Kühling, Karin 19939
Kühn, Christine 17165, 17202
Kühnel, Dr. Anita 17165
Kühnert, Alexandra 19905
Kuehnert, Leonard A. 46239
Kuelpman, Darrell 50363
Kümin, Dr. Beatrice 42116
Kümmel, Dr. Birgit 16668, 16669, 16670, 16671
Künnapuu, Kalver 10664
Künnecke, Otto 17408
Künnemann, Thorsten-D. 42039
Kueny, Michel 15162
Kuenzel, Kurt 48864
Künzl, Uta 16499
Künzle, Fredy 41558
Kürten, Hans Peter 21260
Küßner, Dr. M. 22410
Küster, Babette 19911
Küster, Prof. Dr. Bernd 16862, 17677, 19457, 19458, 19461, 19463, 19464, 19471, 19473, 19474, 19475, 19476, 20897, 54842
Küster, Dr. Ulf 41763
Küster-Heise, Dr. Katharina 20045, 21950
Küttner, Sybille 18936
Kufel, Tadeusz 34980
Kufeld, Klaus 20051
Kugel, Natalija Arkadjevna 37535
Kugina, Marija Borisovna 36795
Kugler, Andrea 20727
Kugler, Hans 18076
Kugler, Dr. Lieselotte 17204
Kugler, Dr. Walter 41334
Kuhfeld, Ellen 50729
Kuhl, Hartmut 18850
Kuhl, Dr. Uta 21566
Kuhler, Astrid 21565
Kuhlmann-Hodick, Dr. Petra 17954
Kuhn, Prof. Dr. Axel 19962
Kuhn, Dieter 42044
Kuhn, Hans Jörg 41270
Kuhn, Marko 19939
Kuhn, Peter 42099
Kuhn, Reinhilde 19435
Kuhn, Richard 41457
Kuhn, Thomas G. 47537
Kuhn, Werner 20572
Kuhnlein, Erhard 21894
Kuhnt-Saptodewo, Dr. Jani 03038
Kuijers, Arie 38563
Kuiper, A. 32412
Kuiper, J. 32068
Kuipers, J. H. 32265
Kuipers, T. 32846
Kuitunen, Antero 10878
Kujukina, Tatjana Savvateevna 37880
Kujundžić, Zilka 04284
Kukk, Inge 10661
Kukko, Timo 11238
Kukla, Gabriele 17103
Kukla, Dr. Jon 47087
Kuklik, Mirosław 35063
Kuklová, Zdenka 09943
Kukovalska, Nelya M. 43109
Kukuruzova, Nina V. 37653
Kula, Krysztof 34722
Kulakov, Viktor Evgenevič 36863
Kuldna, Kersti 10624
Kulęgowski, Mieczyslaw 35011
Kulemeyer, Jorge Alberto 00609
Kulesa, Chester 46390, 48283, 52983
Kulešov, Viktor Ivanovič 03256
Kulichikchin, Vladimir 43032
Kulik, Gary 54258
Kulik, Zbigniew 34715
Kulinčenko, Galina Aleksandrovna 37113
Kulinič, V'jačeslav Jakovyč 43079
Kuljachtina, Ljubov Fëdorovna 36919
Kulkarni, B.S. 24163
Kulke, Willi 19764
Kullenberg, Ulrika 40819
Kulling, Catherine 41548
Kullnigg, Johann 02877
Kulmala, Matti 11414
Kulnytė, Birutė 30419
Kulsarina, Alma 29957
Kulygina, Svetlana Sergeevna 37498
Kumada, Tsukasa 29152
Kumar, Jitendra 24301
Kumar, Niraj 24149
Kumar, Samarendra 24287
Kumar, Dr. Sarvesh 24130
Kumasaka, Yoshihiro 28716
Kumazawa, Hiroshi 29392
Kume, Atsuko 29424
Kumela, Marjut 10768
Kumirai, Albert 54663, 54666
Kumler, Deloris M. 47596
Kumler, Phyllis 46710
Kummer, Edelbert 41762
Kummer, Prof. Dr. Stefan 22714, 22723
Kunac, Prof. Ana 08907
Kunaev, Diar Askarovič 29680
Kunar, Dr. Arun 24129
Kundar, Cynthia A. 51027
Kung-Haga, Vivian 50571
Kung Haga, Vivian 52741
Kuniholm, Thor H. 31694
Kunst, Beate 17044
Kunštek, Vesna 08899, 08901
Kuntz, Edward 47787, 51631
Kunz, Hartmut 02582
Kunz, John 53973
Kunz, Jürgen 20424
Kunz, Mary Ellyn 50860
Kunz, Richard 41169
Kunz, Veronika 02582
Kunze, Kahtrin 19320
Kunze, Kathrin 19321
Kunze, Prof. Dr. Max 21899
Kunze, Dr. Ronald 18898
Kunze, Prof. Dr. Walter 02406
Kunzelmann, Steffen 19249
Kunzemann, Dr. Thomas 21124
Kuononoka, Américo António 00112
Kuosmanen, Hannes 10848
Kupec, Dr. Tomas 09780
Kupiec, Janina 35302
Kuprijanov, Aleksandr Alekseevič 37333
Kuragina, Ljudmila Anatoljevna 37043
Kuraishi, Dr. Mahmud 03219
Kuraku, Yoshiyuki 29159
Kurant, Dr. Bogdan 35253
Kuras, Peter 17835
Kuraya, Mika 29491
Kurban, Michael J. 47490, 49976
Kurčakova, Olga Mykolaïvna 43167
Kurdiovsky, Horst W. 01901
Kurella, Dr. Doris 21951
Kuri, Prof. Vilson 00032
Kuribara, Masami 29192
Kuring, Alfons 20977
Kuring, Gertrud 20977
Kuroda, Hideo 29331
Kuroda, Raiji 28431
Kuronen, Darcy 46940
Kurosawa, Hiroshi 29432
Kuroyanagi, Tetsuko 29377
Kurpik, Maria 35264
Kurre 19171
Kurt, Halim 42696
Kurtz, Jeffrey J. 50922
Kurtz, Dr. John K. 47111
Kurtz Schaefer, Guillermo S. 38933
Kurućev, Dragana 38163
Kurz, A. 21430
Kurz, Birgit 21948
Kurz, Elmar 02362
Kurz, Friedhelm 20400
Kurz, Harry 16845
Kurz, Jürgen 18177
Kurz, Manfred 22563
Kurz, Toni 02099
Kurze, Gisela 19940
Kurzel-Runtscheiner, Dr. Monica 03019, 03085
Kurzenberger, Jim 53150, 53152
Kusch, Heike 17532
Kushwaha, M.R.S. 24298
Kusnecova, Anna Grigorevna 36747
Kuso, Mag. Martina 03086
Kussé, Kristel 03574
Kusserow, Karl 47866
Kussi, Kurt 22780
Kussmann-Hochhalter, Andreas 20819, 20820
Kustelny, Elizabeth 53383
Kuster, Emma Maria 16736
Kusuma, Agung 24583
Kusumajaya, Dr. Made 24470
Kutcher, Ellen M. 54245
Kutluay, Sevgi 42845
Kutsioğlu, F. Suzan 42723
Kutter, Fritz 01746
Kutuev, Anas Abdrahmanovič 36806
Kutz, Artur 16913
Kuuioka, Yasuo 29515
Kuusk, Sven 01194
Kuuskemaa, Jüri 10624
Kuutsa, Tommi 10918
Kuvšinskaja, Ljudmila Anatoljevna 36979
Kuyama, Masaru 28785
Kuzeev, Rail Gumerovič 37887
Kuzina, Marina Nikolaevna 37874
Kužma, Pavel 38243
Kuźma, Teresa 34739
Kuzmanić Pierotić, Ivo 07429
Kuzmenkova, Natalija Vasylivna 43213
Kuzmić, Zdenko 09004
Kuźmicki, Mieczysław 34877
Kuzmin, Aleksej 37729
Kuzmin, Anatolij Ivanovyč 43002
Kuzmin, Kirill Stanislavovič 37657
Kuzmits, Dr. Wolfgang 01859, 01860
Kuznecov, Nikolaj 37479
Kuznecova, Ljudmila Valentinovna 37622
Kuznecova, Natalija 37672
Kvalø, Frode 33742
Kvaran, Gunnar B. 33717
Kvaran, Dr. Olafur 24032
Kvisle, Colleen 07334
Kvist, Kari 10751
Kwack, Dong Suk 29817
Kwadwo, Osei 22858
Kwakernaak, A.D. 32690
Kwami, Dr. Ataa 22857
Kwang-ho, Dr. Hahn 29885
Kwashigroh, David 49864
Kwaśniewicz, Dr. Włodzimierz 34852
Kwasnik, Elizabeth I. 43347, 43348
Kwiadas-Wierzbicka, Ludwika 34714
Kwiatkowski, Prof. Marek 35255
Kwiatkowski, Phillip C. 49943
Kwiatkowski, Tanja 22577
Kwiatkowski, Zygmunt 34675
Kwiecinski, William 48315
Kwok, Dr. Kenson 38172
Kwok, Kian Chow 38179
Kyas, Dr. Vojtěch 09438
Kyek, Mag. Martin 02615
Kyle, Peter 44985
Kyler, Duane 47578
Kyllönen, Eeva 10727
Kyø Hermansen, Cathrine 10019
Kyparissi-Apostolika, Dr. Aikaterini 22897, 23070
Kyriazi, Nelli 22927
Kyriazopoulos, Prof. Basil 23104
Kyriazopoulou, Katerina 22961
Kyroudi, Prof. Dr. A. 23008
Kytr, Hobe 49468

L

La Bianca, Domenico 26055
La Fayette, Aubert 16268
La Gamba, Prof. Filippo 25157
La Malfa, Francesca 26643
La Morgia, Dr. Jasmine 26533
La Porte, John Thomas 49919
La Roche, A.J. 33173
La Rue, Eva 53531
La Tour, Bruno de 11559
La Vove, Eleanor 52897
Laabs, Dr. Annegret 20124
Laajo, Eero 10854
Laakso, Kathy 53377
Laaksonen, Tapio 11006, 11007
Laas, Anne 10623
Labaig Gallardo, Esther 39748
Labanič, Eugen 38271, 38273
Labarre, Angélique de 12314
Labaste, Jacqueline 14571
Labat, Béatrice 16158
Labaugh, Sally 50292
Labaune, Patrick 16038
L'Abbate, Vito 25965
Labbé 13540
Labbe, Armand 52852
Labbe, Bruno 15186
Labbe, Daniel 15968
Laber, Philip 50524
LaBier, Brenda 53755
Labonne, Michel 11897
Laborderie, Gérard 15849
LaBorwit, Melanie 49963
LaBounty, Robert 53377
Labourdette, Anne 16123
Labovvić Marinković, Liljana 04256
Labrador, Ana Maria Theresa P. 34492
Labrails, Marie-Dominique 15946
Labrosse, Gilles 12730
Labuda, Dr. Jozef 38188, 38189, 38190, 38191, 38192, 38193, 38194
Labuschagne, L. 38691
Lacabanne, Marie 05646
Lacabe, Jean-Marc 15931
Lacasse, Johanne 06958
Lacaussade, James de 11869
Lacchi, G. 26380
Lāce, Mara 30114
Lace, Simon 45073, 45075
Lacena, Daniel L. 34381
Lacerna, Daniel L. 34394
Lacey, Eileen A. 46759
Lacey, Michael 01061
Lacey, Susan J. 46618
Lachatenere, Diana 51254
Lacher, Gerhard 03071
Lacher, Dr. Reimar 18877
Lachish, Z. 25106
Lachmann, Dr. Stefan 22126
Lachmuth, Uwe 18059
Łaciuk, Adam 34569
Lackner, Dr. Helmut 03079
LaComb, Gerald 51444
Lacoureux, Marie-Aude 03893
Lacovara, Dr. Peter 46444
Lăcrămioara, Marin 36122
Lacy, Kyleen 49510
Lacy, Tandy 53669
Laczko, Valerie 48493
Ladd, Audrey 51305
Ladenburger, Dr. Michael 17448
Ladier, E. 14175
Ladleif, Dr. Christiane 16582, 16585
Ladleif, Dr- Christiane 16587
Ladleif, Dr. Christiane 16592, 16593
Ladlow, Nick 44672, 44673, 44675, 44682
Ladouceur, H. 06828
LaDouceur, Philip Alan 54368
Laenen, Marc 04027
Länsivaara, Ilkka 10707
Läufer, Egbert 18953
Laevskaja, Galina Leonidovna 37489
Lafargue, Guy 12061
Lafer, Celso 04857
Laffon, Gèraud de 13074
Laffon, Juliette 14636, 14646, 14646
Lafon, Alain 13823
Lafont-Couturier, François 12066
Lafont-Couturier, Hélène 15502
Lafontant Vallotton, Chantal 41679, 41680
Laforet, Andrea 05862
Laforge, Rosanne 05697
LaFountain, W. Weston 48225
Lafragette, Florence 15544
LaFrance, Liselle 46158
Lagabrielle, Sophie 14729
Lagasse, Norm 46285
Lagercrantz, Jacob 40816
Lagercrantz, Kristian 40892
Lagerkrans, Jacob 40488, 40489, 41046
Lagerkvist, Cajsa 40558
Lagerwaard, Dr. Cornelieke 21514
Lagés, Hélène 16220

K L

Laghouati, Sofiane 03893
Lagier, Valérie 16253, 16255, 16256, 16257
Łagiewski, Dr. Maciej 35331
Lagiglia, Dr. Humberto A. 00606
Lagler, Dr. Annette 16402
Lagnel, Cosette 41549
Lagoria, Georgiana 49336, 49337
Lagos, Marta E. 48107
Lagos, Patricio Acevedo 07441
Lagoutte, Claudine 11809
LaGrave, Michelle 50417
Lagree, Dr. Kevin 49507
Laguiche, Comte R. de 11681
Lagumdžija, Ermana 04284
Lagvik, Elisabeth 40502, 40503, 40505
Lahcéne, Zghidi Mohamed 00061
Laher, Herman 01711
Lahikainen, Dean 52657
Lahn, Tore 33480
Lahs-Gonzales, Olivia 52600
Lahtinen, Maija 11311
Lai, Charles 51226
Laï, Christine 14651
Lai, Guoyun 07968
Lai, Hsiang-Ling 42407
Lai, Piero 26214
Lai, Weng-Hsian 42308
Lai, Yingxin 07490
Laidlaw, William K. 47757
Laing, D.W. 43702
Laing, Henry 04197
Laing, Lloyd 45299
Lainhoff, Thomas A. 50530
Lair, George 53623
Laird, Ruth 45834
Lais, Bruno 21597
Laitinen, Suvi 10772
Laitman, Nanette 51112
Lajevs'kij, Sergij Lazarevič 43007
Lajos, Asperjan 23627
Lajos, Dargay 23627
Lajoux, Jacques 13061
Lakatos, Adél 23624, 23626
Lake, Linda 49903
Lake, Paul 45757
Lake, Simon 44681
Lake, Susan 53877
Laken, Lorraine 51405
Lakhani, Prof. S.R. 01152
Lakin, Ian 43242
Lakin, W.G. 45353
Lakson, Stephen 46961
Lakstīgala, Aivars 30127
Lal Gaur, Banwari 24138
Lalande, Bernard 14950
Lalaounis, Ioanna 22915
Lalev, Ivan 05044, 05086, 05087
Lalevic, Olga 04284
Lallemant, Marie-Françoise 12317
Lally, Dr. Janice 00814
Lally, Michael H. 50308
Lalošević, Aleksander 31651
Laloue, Christine 14656
Laluc 14516
Lam, Andrew 07809, 07810
Lam, Anita 50401
Lam, Joseph 46311
Lam, Prof. Peter Y.K. 07792
Lam, Tiffany 07802
Lama, Luis 34211
Lamaestre, Laurence de 13055
Lamb, Donna 46470
Lamb, Dorothy 44752
Lamb, Nancy 52971
Lamb, Rebecca B. 46375
Lamb, Susan 45575, 45579
Lamb, Vaughn 52821
Lambacher, Lothar 17168, 17169
Lambaša, Gojko 08949
Lambel, T. de 12946
Lambers, Dr. P. 32857
Lambert, Al 50422
Lambert, Anne 50388
Lambert, Dan 46540
Lambert, Elodie 12386
Lambert, Françoise 41985
Lambert, Dr. Frederick A. 54124
Lambert, Jean Marie 16075
Lambert, John 44067
Lambert, Kirby 49242
Lambert, Marianne 46456
Lambert, Marie-Jeanne 12387, 13815
Lambert, Dr. Miles 45082
Lambert, Phyllis 06330
Lambert, Ruth 45069
Lambert, Thierry 03843, 03844
Lambert, Véronique 03760
Lambertson, John Mark 49479
Lambertye, Jean de 12616
Lambeth, Linda 05635
Lambić, Sunčica 38167
Lambilly, Comtesse Louis de 15575
Lambon, Corinne 15613
Lambraki-Plaka, Prof. Dr. Marina 22909
Lambrecht, Miriam 03495
Lambrigger, Josef 41662
Lambropoulos, Iraklis 23153
Lamesch, Chantal 12548
Laminou, Souley 33309
Lammassaari, Taina 10746
Lammertse, F. 32729
Lamonaca, Marianne 50644
Lamontagne, Nicole 05858
Lamort, Jean-Pierre 03655
Lamouret, Prof. Alfredo Daniel 00157
Lampard, D. 44530
Lampe, Dr. Karl-Heinz 17469
Lampl, Johann 02878
Lampl, Dr. Sixtus 22202
Lamprecht, Anett 19911
Lamprecht, M.S. 38713, 38714
Lamprecht, Wolfgang 02942
Lamprechter, Mag. Cornelia 02225, 02226
Lamprechter, Cornelia 02227
Lamprechter, Mag. Cornelia 02228
Lamprechter, Cornelia 02697
Lamschus, Dr. Christian 20096
Lamy, Frank 16259
Lamy, Laurence 16007
Lan, Kaigeng 08092
Lân, Nguyên Khã 54631
Lan, Zhongzhao 08338
Lancelot, Antoine 15167
Lancmanis, Imants 30018, 30054
Landauer, Susan 52809
Lande Moe, Tore 33885
Landen, A. Wayne 06561
Landert, M. 41201
Landert, Markus 42001, 42002
Landes, Carolyn 50609
Landis, Prof. Dr. Anna Maria 26397
Landis, Mary Ann 46254
Landman, Dr. Neil H. 51115
Landmann, Jörg 16609
Landmann, Jürgen 20615
Landois, Marie-Ange 12022
Landová, Blanka 38311
Landress, Helen T. 46624
Landrum-Bittle, Jenita 46418
Landry, Frédéric 06008
Landry, Mario 05626
Landsbergen, Michael 01609
Landsmann, Hannah 03041
Lane, Dr. Arne H. 07259
Lane, H.E. 43732
Lane, Joshua W. 48032
Lane, Margaret Mary 49435
Lane, Martin 06546
Lane, Richard B. 52538
Lane, Steve 43902
Lanfermann, Dr. des. Petra 17350
Lang, Alexandra 20793
Lang, Anton 02958
Lang, Bernhardt 21895
Lang, Brian J. 51954
Lang, Christoph 16445
Lang, Elisabeth 15834
Lang, Ewald 18800
Lang, Frank 21950, 22287
Lang, Huguette 15831
Lang, Irma 38146
Lang, Merike 10630
Lang, Dr. Odo 41345
Lang, Paul 41418
Lang, Rolf 17200
Lang, Sabine 20501
Lang, Sherryl 52942
Lang, Sherryl P. 52213
Lang, Sonia 41778
Lang, Tom 52851
Lang, Ulrich 18385
Lang Pierce, Beverly 53877
Lang Walker, Barbra 05512
Langauer, Dipl.-Päd. Fritz 02887
Langberg, Dr. Harald 10157
Langberg, Signe 10080
Lange, Dr. Albert de 20877
Lange, Allynne 49790
Lange, Amanda E. 48032
Lange, Andrzej 35108
Lange, Carmen 22630
Lange, Christian 20284
Lange, Dr. Christine 20454
Lange, Cortina 40895
Lange, Dietmar 22341
Lange, Jane 52827
Lange, Dr. Justus 19463
Lange, Karl Wilhelm 18485
Lange, Martina 17546
Lange, Monika 20173
Lange, Nicolas 03348
Lange, Rich 53584
Lange Malinverni, Jonas R. 04526
Langelaan, M.M.C. 32427, 32428
Langelier, Jacques 15668
Langellier, John 52135
Langen-Crasemann, Sabine 20658
Langenegger, Christoph 41114
Langenthal, Antje 16837
Langer, Agnes von 23823
Langer, Axel 42097
Langer, Brian 01569
Langer, Dr. Brigitte 17651, 20599
Langer, Dr. Heiderose 21426, 22211
Langer, Dr. Kai 20199
Langer, Margit 18843
Langford, Jan 48921
Langhammer, Peter 02218
Langhammerová, Dr. Jiřina 09782
Langhof, Jörgen 40519
Langhoff, Helmut 22507
Langhoff, Thomas 17029
Langhorne, Richard 45767
Langini, Alex 30469
Langkafel, Sonja 19139
Langland, Leo 50605
Langley, Lloyd 43721
Langley-Ware, Lyntia 47783
Langlois, Daniel 06872
Langman, David 33253
Langner, Kerstin 19914
Langos, Barbara 16489
Langrell, Opal 07403
Langsbury, Hoani 33098
Långvik-Huomo, Marianne 10727
Lanigan Wood, Helen 44178
Lankford, Éamon 24897
Lanko, Valeri 47867
Lannerbro Norell, Maria 40515
Lanning, Bruce 52980
Lannon, John 46919
Lannoo, Luc 03477, 03495
Lannoye, Henri 03384
Lanphier, Barbara 53644
Lanser, Dr. Klaus Peter 20513
Lansky, Aaron 46270
Lantos, Titus 02245, 02501
Lantz, Jerry A. 48837
Lantz, Lynn 48036
Lantz, Marcel 15374
Lanyon, Scott M. 50735
Lanz, Dr. Hanspeter 42121
Lanz, Vreni 41620
Lanzarini, Viviana 25439, 25457, 25458, 25460
Lanzinger, Dr. Michele 26767, 28062
Lapa, Dr. Pedro 35698
Lapalus, Marie 13907, 13908
Lape, Dr. Peter 52994
Lapėnienė, Vitalija 30406
Lapetra, Damaso 38873
Lapeyre, Jean-François 15946
Lapham, Jennifer 46862
Lapi Ballerini, Dr. Isabella 26194
Łapiak, Dorota 34608, 34609
Lapious, Sydney 51143
Laplana, Josep de C. 39666
Lappalainen, Ari 11271
Lappel, Mag. Edith 01775
LaPrelle, Robert 47925
Laqueste, Marc 12694
Laquindanum, Felicidad 34480
Lara, Maria 53678
Lara, Rita 47999
Laranjeira de Schroeder, Prof. Viviana Nuñes 00385
Larcher, Pedro Bruno 04523
Larco de Alvarez Calderon, Isabel 34225
Lardi, Guido 41730
Larey, Frank 44022
Large, Carol E. 51539
Large, Christine 43543
Large, Joan T. 51856
Large, Stephen 01427
Larguem, Salima 00075
Lark, Thomas C. 46175
Larken Royds, Meriel Victoria 34299
Larkin, Annie 46830
Larkin, Graham 06551
Larkins, Catherine 54002
Larl, Rupert 02116
Larocque, Peter 06838
LaRose, Helen 06144
LaRowe, Annette 49071
Larraburu, Margo 00666
Larrañaga, Juan Ramón 39449
Larrañaga, Maria Isabel de 00185
Larry, Billings 47106
Lars, Roger 13498
Larsch, Georg 18323
Larsen, Carsten U. 10124, 10144
Larsen, Karen 54100
Larsen, Lynne 50140
Larson, Dr. Andre P. 53719
Larson, Barbara 53820
Larson, Bradley 51615
Larson, Eric 50018, 53287
Larson, John 49366
Larson, Dr. Judy L. 52379
Larson, Julie 52665
Larson, Larry 50811
Larson, Michael E. 47338
Larson, Nathan 50142
Larson, Neal L. 49283
Larson, Sidney 47719
Larson, Silvia 07313
Larsson, Åke 40430
Larsson, Göran 40724
Larsson, Hans-Ove 40895
Larsson, Hugo 40782
Larsson, Lars-Erik 40887
Larsson, Thorbjörn 40891
Larsson, Ulf 40921
Larsson Modin, Anita 40969
LaRue, Dr. Hélèna 45328
LaRue, Dr. Hélène 45338
Lárusdóttir, Unnur 24014
Lary, Christian de 14041
Lasansky, Jeannette 50032
Lasansky, Leonardo 52617
Lasfargues, Jacques 13896
Lash, Michael 47499
Lasher, Leigh 52437
Lasheras Corruchaga, José Antonio 40036
Lasić, Dr. Stanko 08873
Lasić, Višnja 08886
Lasius, Dr. Angelika 20263
Laska, Dr. Vera 54106
Laskey, Jane 45113
Laskowski, Rainer 19549, 19552
Lasmane, Daina 30122
Lasowa, Teresa 35294
Lassen, Thomas W. 10184, 10185
Lassere, Odile 14403
Lasson, Dr. E. 20879
Lassotta, Dr. Arnold 17921
Lastic, Henri de 15172
Laštovková, Dr. Věra 09624
Lastra, Cristina 00195
Lastra Belgrano, Prof. M. Cristina 00264
Lastu, Didier 15971
Laszlo, Želimir 09009
Latady, Bill 46962
Latanza, Dr. Antonio 27513
Latchford, Becky 43805
Latchford, Norma 01185
Lategan, L. 38566
Latham, Charles 53496
Latham, Stanley A. 53774
Latimer, Dr. Bruce 47625
Latour, Patrick 14571
Latourelle, Philippe 15492
Latsay, Brenda 06282
Latschar, Dr. John A. 48901, 48902
Lattanzi, Dr. Elena 26011, 26012, 27384, 28246
Lattanzi, Gregory 53561
Lattanzi, Vito 27519
Lattey, Gordon 46164
Latuconsina, Dr. Achmad 24501
Latus, Urs 20790
Latvala, Anja 11282
Lau, Nina 21565
Lauand, Débora 04880
Laub, Dr. Peter 02625
Laub, Dr. Richard S. 47118
Laubach, Simone 19370
Laubard, Charlotte 12062
Laube, Kurt 20491
Laube, Robert 19038
Laubie, M. 16074
Laubier, Lucien 14579
Laudat, František 09783
Lauder, Dawn 05814
Lauder, William T. 46274
Lauderdale, David 49576
Lauenborg, Michael 09999
Lauer, Dr. Bernhard 19459
Lauer, Keith G. 46425
Lauermann, Dr. Ernst 01734, 02818
Lauermann, Frank 50479
Laughlin, Patricia J. 53282
Laughlin, Phyllis 48309
Laughton, Rodney 52953
Laugier, Pierre-Henri 11635
Lauk, Petra 22262
Laukkanen, Rauni 11436, 11437
Laurence, Sue 44571
Laurent, Dominique 05859
Laurent, Elizabeth M. 51882
Laurenza, Joseph 48980
Laurette, Sandra 46670
Lauridsen, Henning Ringgaard 10314
Lauridsen, Inger 10086, 10298
Laurière, Claudine 14296, 14299
Laurinavičius, Vidmantas 30414

Lauring, Kåre 10064
Lauritzen, Ruth 49043
Lauro, Dr. Maria Giuseppina 27480
Laus, Harry 04455
Lausson, Mirja 40950
Lautanen-Raleigh, Marcia 48375
Lauter, Dr. Marlene 22718
Lautz, Thomas 19604
Lauwers, J. 03307
Laux, Regula 41532
Lauxerois, Roger 16149, 16150, 16152
Lauzière, Yves 06961
Lavalle Argudin, Mario 31122
Lavanchy, John 48156
Lavandera López, José 39800
Lavandeyra-Schöffer, Eléonore de 14567
Lavarnway, Carrie 47060
Laveck, Jocelyn 07329
LaVelle, Nancy 48432
Lavelle, Phyllis 46990
Lavender, Prof. Tony 45709
LaVere, Jeanette 51657
LaVerne, Thomas 53097
Laverty, Bruce 51816
Lavery, Peter 00900
Lavery, Sally Ann 46261
Lavie, Pierre 13734
Lavigue, Emma 14656
Laville, Yann 41681
Lavin, Lucienne 53950
Lavino, Jane 49573
Laviolette, Suzanna 52604, 52605
Lavis, Tom 05864
Lavoie, Suzanne 06829
Lavoy, Carl L. 05603
Lavoyer, Martin 12194
Lavrenjuk, Benedykt Antonovyč 43194
Lawitschka, Valérie 22115
Lawler, Jo 44332
Lawless, Ann 54228
Lawlor, Seamus 24744
Lawrance, Sarah 43891
Lawrason, Helena 05670
Lawrence, Brad 06718
Lawrence, D.R. 01447
Lawrence, John 49841, 50358
Lawrence, John H. 51083
Lawrence, Kirk 06108
Lawrence, Sarah 49093
Lawrence, Wade 47362
Lawrence-Adams, Mechelle 52813
Lawrimore, Katrina P. 48894
Lawson, H. William 54352
Lawson, Jane 45004
Lawson, Julie 44157
Lawson, Karol 50328
Lawson, Pam 01168
Lawson, Scott J. 52211
Lawton, Gregory 49451
Lawton, Peter 45439
Lawton, Rebecca 48709
Lawton, Sheryl 48919
Lawton, Simon R. 45842, 45843
Lay, Chan Kim 30593
Layher, Hermann 21783, 21838
Layne, Janice 48840
Layne, Valmont Edward 38582
Layton, Dwight 46795
Laz, Lauren 41986
Lazareva, Lilija Georgievna 36861
Lázaro López, Agustin 39067, 39070
Lazárová, Eva 38199
Lazarus, Fred 46577
Lazer, Laurie 52782
Lažetić, Predrag 38091
Lazier, Isabelle 13136
Lázničková, Dr. Ilona 09437
Lazov, Gavrail 05191
Lăzureanu, Emil 36147
Lazzarini, Dr. Antonio 27243
Le, Ma Luong 54609
Le, Trung 54630
Le Bail, Jean-Marc 14841
Le Balc'h, Jean 12163
Le Blanc, Lisette 32948
Le Blanc, Tony 43268
Le Boulanger, Yann 15541
Le Brun, Foni 30483
Le Cieux, L. 12889
Le Cieux, Laurence 36091
Le Dinh, Diana 41548
Le Foll, Typhaine 14272
Le Gall, Régine 14362
Le Gendre, Bernard 13543
Le Grelle, Daniël Graaf 03310
Le Harivel, Adrian 24807
Le Leyzour, Philippe 11808, 15964
Le Loup, Willy 03425, 03428
Le Maguet, Jean Paul 15015
Le Maître, Claude 12088
Le Nouëne, Patrick 11607, 11608, 11609, 11612, 11613
Le Pelley, Paul 43779
Le Restif, Claire 13256
Le Rouzic-Giovanneli, Gisèle 13796
Le Saux, Marie-Françoise 16068, 16069
Le Stum, Philippe 14970
Le Tissier, Paul 45591
Le Van Ho, Mireille 14571
Le Vay, Jonathan 45164
Lea, Sandra 44792
Lea, Vanessa 45333
Leach, Mark 47392, 47393
Leach, Patricia 46714, 49262
Leach, Sally 46510
Leader, Prof. Mary 24818
Leafdale, Norris 49177
League, Alice 48138
Leah, Chris 43985
Leah, Mervyn 44688
Leahu, Victor 36412
Leahy, K.A. 45628
Leahy, Mary 50826
Leal, José H. 52850
Leal Spengler, Eusebio 09136
Leamon, David L. 53532
Leander, Dr. Brian 07199
Lear, Paul 51624
Leard Coolidge, Lindsay 47174
Leatherman, Marvin 53995
Leavesley, Dr. James 01536
Leavey, Jane 46457
Leavitt, Tracey 06286
Lebailly, Florence 03609
Lebas, Frédéric 18951
Lebeau, Monique 14334
Lebedeva, Alla Markovna 37460
Lebegern, Robert 22068
LeBihan, Olivier 12069
LeBlanc, Denise 46133
LeBlanc, Guy R. 53353
Leblanc, Jacqueline 15495
Leblanc, Pascal 16001, 16002
LeBlanc, Suzanne 49975
Lebouteiller, Claire 16270
Lebrasseur, Marie-Josée 06573
Lebrero Stals, José 40061
Lebrun, Jacques 16120
Lebrun, Jean-Marc 03579
Lebsack, Ron 06706
Lecasseur, Marie 15619
Lečev, Dečko 05089, 05123, 05212, 05214, 05215, 05216, 05218
Lechner, Georg 18787
Lechner, Dr. Gregor Martin 01913
Lechowski, Andrzej 34550
Lechuga, Dr. Ruth 31061
Leclair, Alain 13492
Leclair, Tom 06519
Leclef, Louis 03300
LeCleire Taylor, Gay 50696
Leclerc, Claudette 07370
Leclerc, Marie-Andrée 06322
Leclerc, Patrick 12145
Leclerc, Paul 51244
Leclerc, Paul-André 06113
Leclercq, Jean-Paul 14654
Leclert, Hélène 15092
LeCompte, Sarah 46534
Leconte, Philippe 03823
L'Ecotais, Emmanuelle de 14646
Lecouturier, Yves 12230
Lecuillier, Michel 11872
Lecuyer, Françoise 14462
Ledbetter, William C. 50914
Leder, Alan 48449
Leder, Carolyn R. 44949
Lederer, Carrie 53802
Lederer, Irene 19168
Lederer, Nan 46961
Ledermair, Martin 02727
Ledet, Claudette 06685
Ledford, Thomas G. 50327
Ledoux, François 06226
LeDrew-Day, Karen 05451
Leduc, Dawn 05325
Lee, Bok Hyung 29847
Lee, Chris 45345
Lee, Christopher 46880
Lee, Colin 44704
Lee, Cynthia 51226
Lee, Dae-Sup 29824
Lee, David L. 52695
Lee, Eric M. 51374
Lee, Hee-Jae 29925
Lee, Jackie 53831
Lee, Jane 50948
Lee, Jean 51048
Lee, Jim 48781
Lee, Prof. Jong-Sang 29921
Lee, Karen 06462
Lee, Linda 44704
Lee, Michel 43408
Lee, Molly 48472
Lee, Olivia 07199
Lee, Ra Hee Hong 29943
Lee, Robert 51129
Lee, Seung-Koo 29802
Lee, Steven 46556, 46567
Lee, William D. 47060
Lee, Wuh-Kuen 42353
Lee Reid, Katharine 47624
Lee Song, C.J. 32106
Leebrick, Gilbert 49075
Leech, James R. 48170
Leech, William J. 46881
Leeder, P.G. 45845
Leedy, Harriet 48872
Leedy, Sherry 49692
Leen, Frederik 03463
Leenes, Tony 32497
Leerintveld, A. 32905
Lees, D. 44881
Lees, Diane 45001, 45086
Lees, Traci 51525
Leeth, Sara 53535
Leeuw, G. de 32162
Leeuw, Prof. Drs. R. de 31883
Lefarth, Ute 22665
Lefebvre, Eric 14638
Lefebvre, Pascal 03799
Lefebvre-Carter, Michelle 47579
Leferman, Mariana 36321
Lefever, Joel 49317
Lefevre, Gérard 14255
LeFevre, John 47708
Lefèvre, Patrick 03438
Lefèvre, Prof. Patrick 03494
LeFevre Stratton, Leslie 48859, 51096
Leff, Cathy 50644
Leff, Deborah 46935
Lefftz, M. 03834
Lefki, Michaelidou 09380, 09381
Lefour, Pierre 15914
Lefrançois, Michèle 12090, 12091
Lefrançois, Thierry 13426, 13428, 13430
Lefrancq, Janette 03495
Legac, Mirjana 08940
Legarda, Eduardo 34417
Legarda, Rosalita R. 34417
Legendre, Françoise 12271, 15120
Legêne, Dr. S. 31852
Legg, Dr. G. 43619
Legget, J.A. 44678
Legleiter, Kim 50812
Legout, Gérard 11701
Łęgowska-Radosz, Dr. Anna 35117, 35242, 35288
Legrand, Sybille 03393
Legrand, Véronique 15485
Legrand-Rossi, Sylvie 14734
Legro, Patrice 53889
Legros, Charles 04111
Lehberger, Prof. Dr. Reiner 18933
Lehman, Arnold L. 47069
Lehmann, Gérard 13727
Lehmann, Dr. Hans-Ulrich 17954
Lehmann, Hilmar 22734
Lehmann, Kai 21580
Lehmann, Dr. Karin 19144
Lehmann, Klaus 19201
Lehmann, Markus 41685
Lehmann, Randy 48879
Lehmann, Dr. Stefan 18903
Lehmann, Ursula 22492
Lehnemann, Dr. Wingolf 20103
Lehner, Adam 51167
Lehner, Astrid 02992
Lehner, Claudia 02315
Lehner, Jan Nepomuk 09818
Lehner, Dr. Manfred 01963
Lehnert, Marie Josée 30488
Lehnherr, Dr. Yvonne 41384
Lehodey, Colette 15007
Lehr, Mag. Johann 01837
Lehr, William 49182
Lehtinen, Dr. Martti 10775
Lei, Anjun 07477
Lei, Weixin 08104
Lei, Xianzhong 08436
Leiber, Christian 18813, 18814
Leibnitz, Dr. Thomas 03045
Leich, Jeffrey R. 48731
Leicht, Walter 21379
Leichty, Erle 51886
Leidl, Mag. Bettina 03018
Leidner, Udo 20675
Leidy, Susan 50421
Leighton, John 32088, 32099, 44138, 44156
Leijen, A.J.M. 31967
Leijerzapf, Dr. I.Th. 32486
Leijten, C.J.M. 32225
Leilund, Helle 10123
Leim, Nico 22056
Leimu, Tuula 11444
Lein, Sharon 48053
Leiner, Dr. Srećko 08998
Leinfelder, Prof. Dr. Reinhold 17205
Leinicke, Kris G. 52409
Leinonen, Paula 11348
Leinweber, Dr. Ulf 19475
Leinz, Dr. Gottlieb 18027
Leiseth, Sylvia 53980
Leismann, Burkhard 16438
Leistenschneider, Dr. Eva 22170
Leister, Claudia F. 48163
Leister, Michael D. 48170
Leistner, Dr. Gerhard 21228
Leite, Ana Cristina 35669, 35677, 35726
Leite, Prof. Michael 47326
Leitgeb, Dr. Franz 01878
Leitgeb, Otto 02194
Leitinger, Prof. Josef 02839, 02840
Leitmeyer, Wolfgang 21834
Leitnaker, Connie 49921
Leitner, Andreas 02829
Leitner, Cynthia 48400
Leitner, Dr. Friedrich W. 02186
Leitner, Mag Irene 01703
Leitner, Norbert 02696
Leitner-Böchzelt, Susanne 02285, 02287
LeJambre, L.A. 46913
Lejeune, Louis 03323
Lejnieks, Dr. Janis 30076
Lekavičienė, Angelė 30265
Lekens, Luc 03544
Lekholm, Kerstin 40555
Lelaurain 13591
Leloup, Michel 12903
Lemaire, Jean-Michel 12617
Lemaire, Michèle 12117
Lemaire, Philippe 15740
Lemal Mengeot, Chantal 03516
Lemańska, Stefania 34657
LeMarie, M.J. 43290
Lemay, Brian 46945
Lembke, Dr. Katja 19185, 19186
Lemcke, J. Peter 17693
Lemelle, François 14570
Lemercier, Christian 12080
Lemeunier, A. 03725
Lemeunier, Albert 03796, 03806
Lemežienė, Laima 30339
LeMieux, Linda D. 47707
Leming, C.F. 45710
Lemle, Robert 48857
Lemmens, Claude 03435
Lemmon, Dr. Alfred 51083
LeMoine, Dr. Geneviève 47103
Lemoine, Serge 14692
Lemonedes, Heather 47624
Lemons, C.R. 48633
Lemos, Nikolas S. 23127
Lemothe, Peter Thomas 51053
Lenaers, J.H. 32178
Lenaghan, Patrick 51192
Lenarczyk, Leszek 34668
Lenarduzzi, Nelson 00393, 00394
Lenart-Juszczewska, Krystyna 34986
Lenarz, Michael 18379
Lencova, Valentina Bronislavovna 37165
Lencquesaing, May-Eliane de 14760
Lenglet, Marie-Cornélie 13752
Lenglin, Catherine 13107
Lengyel, Dr. Beatrix 23442
Lengyel-Kiss, Katalin 23459
Lenhart, Ralf 02217
Lenk, Elfriede 02518
Lenkiewicz, Jolanta 34764, 34803
Lennard, John Barret 01010
Lennard, Mark 01106
Lennig, Dr. Petra 17044
Lennoy, Kim 48955
Lensen, Jean-Pierre 04118, 04119
Lenski, Jean-Pierre 15694
Lenssen, Dr. Jürgen 22256, 22709, 22717
Lent, Eric 47205
Lenten, Joke 31867
Lentino, Miguel 54530
Lentz, Ingo 17999
Lentz, Lamar 52471
Lentz, Thomas W. 47177
Lenz, M.A. Gerhard 17478
Lenz, Helmut 18368
Lenz, Iris 21945, 21946
Lenz, Mary Jane 51236
Lenz, Prof. Dr. Norbert 19449
Lenz, Vera 18694
Lenz-Weber, Dr. Diana 18969

L

Lenze, Susanne 17256
León, Imelda de 30783
Leon, Lamgen 51226
León Rivera, Jorge de 31053
León Rivera, Melba Rocío 08580
Leonard, Amy 46133
Leonard, Ben 52632, 52633
Leonard, Claude 14225
Leonard, Douglas 05383
Léonard, Georges 03606
Leonard, Glen M. 52692
Leonard, Jessie 50704
Leonard, Mark 50237
Leonard, Pascal 03273
Leonard, Robert 50411
Leonardo-Finger, Jascin 50945
Leoncini, Dr. Luca 26307
Leone, Carol 53817
Leonenko, Gajane 30223, 30435
Leonhardt, Günter 19755
Leonhardt, M.A. Monika 42114
Leonhardt, Rainer 20342
Leonhardt, Dr. Ute 19391
Leonori, Dr. Maria Chiara 26098, 26103
Leontiev, Andrej 37599
Leontjeva, Sofija Jakovlevna 37494
Leopardi, Anna 27375
Léotoing, Comte de 15954
Lepădatu, Ioana 36528
Lepage, Jean 14377, 14379, 14380
Lepard, David 48884
Lepdor, Catherine 41544
Lepe, Marcella 31297
Lepičić, Janko 38098
Lépine, André 03512
Lépine, Brigitte 11754
Lépine, Luc A. 06795
Leplat, R. 03379
Lepley, John G. 48592, 48593
Leppala, Vesa 10761, 11299
Lepper, Beate 22395
Lepper, Dr. Katharina B. 18027
Lepper, Dr. Verena 17028
Leppiniemi-Järvinen, Anita 10924
Leproust, Evelyne 14972
Lepší, Miroslav 09573
Leray, Victor 13854
Lerberghe, Anja Van 03477
Lerch, Michaela 14632
Lerebours, Lionel 23322
Leresche, Anne 41548
Léri, Jean-Marc 14593, 14637
Leritt, Sarah 44673
Lernaut, Françoise 11587
Lernbaß, H. 19740
Lerner, Peter 47051
Lernout, Françoise 11588
Lerolle, Aggy 14698
Leron-Lesur, Pierre 15486
Lerouge-Benard, Jean-Pierre 13113, 14430
Leroux, Doug 06582
Leroux, Jean-François 16176
Lerov, Trifon 05041
LeRoy, Angelina 47270
Leroy, Gilbert 13811
Leroy, Jacques 11770
Leroy, Vincent 01949
Lerrick, Dr. Yacob 24517
Lervy, Annick 14123
Lesaffre, Jean-Marie 13757
Leschhorn, Prof. Dr. Wolfgang 17522
Lescombes, Germain 12243
Lescroat-Cazenave, Elizabeth 14520
Lesdorf, Svetlana Pavlivna 43180
Lesher, Ronald E. 52611
Leshikar, Howard B. 53473
Lesko, Diane B. 52944, 52947
Leskova, Marina 30520
Leskovar, Jutta 02324
Leskovar, Peter 38519
Leskovec, Ivana 38415, 38421
Leslie, James 43367
Leslie, Mike 51519
Leśniewska, Józefa Ewa 34844
Lesouef, Jean Yves 12158
Lesser, Ann 53826
Lesseur, Catherine 11609
Leßmann, Dr. Sabina 17460
Lester, Alan K. 46936
Lester, Ellen 49824
Letenayova, Dr. Zdenka 38228
Letendre, Julie 48024
Letertre, J.B. 15845
Leth, Gwen I. 52307
Lethen, Paulien 47076
Letourneur, Céline 12901
Letschert, Edmund 22385
Lettany, Lieve 03861
Letts, A. 38726
Letz, Andrea 19580
Leu, Maja 41314
Leu-Barthel, Angelika 18920
Leuba, Marion 11903, 11904, 11905
Leumann, Alexander 41377
Leung, Corinne 49417
Leuschner, Dr. Jörg 21475
Leutenegger-Kröni, Tanja 41452
Leuzinger, Dr. Urs 41378
Lev-Ari, Y. 25035
Levá, Kimmo 10910
Levallet, Didier 14182
Levander, Karin 41032
LeVangie, Patrick 48749
Levčenko, Vitalij Vasiljevič 37094
Levee, Catherine 49639
Levens, Dwight R. 48110
Leverton, Raymond 49538
Lévesque, André 06893
Levey, Georges 15675
Levi, A. 25038
Levièvre, Francine 06378
Levin, Annika 14612
Levin, Dalia 24972
Levin, Mai 10624
Levin, Nikolaj Iljič 37792
Levin, Pamela 52978
Levine, Beverly 48699
Levine, Brett 46827
Levine, Janet 51262
Levine, Kenneth 51256
Levine, Dr. Louis 51227
Levine, Phyllis 51194
Levinho, José Carlos 04737
Levitt, S. 44676, 44678
Levitt, Sarah 44681
Levkin, Jaroslav Borisovič 37562
Levy, David C. 53855, 53856
Levy, Harry A. 50626
Levy, Sue M. 51822
Lew, Dr. Irena 34968
Lewallen, Constance 46752
Lewandowski, Leszek 34731
Lewandowski, Marian 35031
Lewandowski, Ryszard 34946
Lewenhaupt, Dr. Inga 40763
Lewicki, Ireneusz 34918
Lewin, S.J. 43792
Lewis, Amy E. 47942
Lewis, Bethan 43707
Lewis, Carolyn Chinn 49992
Lewis, Deborah 52825
Lewis, Frank C. 46332
Lewis, Greg 53663
Lewis, Karen 46672
Lewis, Larry 54283
Lewis, Lesley 07126
Lewis, Louise 51426
Lewis, Maisie 00845
Lewis, Mario 34067
Lewis, Merri 46183
Lewis, Oli 49970
Lewis, Sally-Anne 01598
Lewis, Stephanie 53211
Lewis, Vel A. 42615
Lewis, William R. 51895
Lewitzki, Wolfgang 19597
Lewry, Joan 46056
Lewyllie, Ingrid 03649
Ley, Dr. Astrid 20908
Ley, Sabrina van der 18932
Leyge, François 14120
Leyoudec, Maud 15693
Leypold, Denis 15830
Leys, K. 03436
Leys, Kathleen 03450
Leysen, Veerle 03829
Leyten, C.J.M. 32839
Leyten, Vik 03290
Leyva, Maria 53848
Lezarrdiere, A. de 13336
Lhagvasüren, I. 31626
L'Hostis, Denez 12776
Li, Bingzhong 08449
Li, Chaozhen 07633
Li, Chia-Wei 42352
Li, Chun 08091
Li, Chunfang 08163
Li, Chunrong 08115
Li, Denggao 08076
Li, Deyang 08467
Li, Fuxin 08440
Li, Gengshan 08252
Li, Guicheng 08147
Li, Hancai 07840
Li, He 07843
Li, Hongri 07826
Li, Jiahua 07636
Li, Jianeng 08324
Li, Jianxin 08121
Li, Jichun 08075
Li, Jinghan 08285
Li, Jinglin 08513
Li, Jubao 07700
Li, Junping 08517
Li, Juzheng 08468
Li, Kexiang 08470
Li, Kuanhong 08235
Li, Kunsheng 07917
Li, Lifu 07589
Li, Linna 07734
Li, Peiyuan 08143
Li, Qifu 08149
Li, Qiliang 07479
Li, Qingrong 07493
Li, Rongji 07878
Li, Shangshu 07614
Li, Shecan 08396
Li, Shunjian 07783
Li, Shutao 07625
Li, Tai 08410
Li, Te'er 07659
Li, Tianbao 08376
Li, Weidong 08400
Li, Weigui 08479
Li, Weiguo 08497
Li, Wensheng 07870
Li, Xi 07937
Li, Xia 08077
Li, Xianheng 07726
Li, Xiliang 07902
Li, Xinling 07849
Li, Xiongkun 08334
Li, Xuewen 08370
Li, Xumei 08505
Li, Yan 08059, 08060
Li, Yangke 08510
Li, Yanming 08279
Li, Yecheng 07484
Li, Yinde 07740
Li, Yinsheng 08088
Li, Youcheng 08408
Li, Yousheng 08450
Li, Yuanqing 08448
Li, Zegao 08504
Li, Zhongshan 08330
Li, Zuyao 07978
Liabeuf, Brigitte 15881
Liagouras, Angelos 23091
Liang, Binggui 07898
Liang, Hengtang 08325
Liang, Huimin 07924
Liang, Jimin 07729
Liang, Juxiu 08369
Liang, Nongxi 08009
Liang, Sichong 08493
Liang, Tingbao 08036
Liang, Xiaomei 08090
Liang, Yaoji 07660
Liang, Zhaobi 07852
Liang, Zuozhen 08126
Liaño, Marta 54543
Liao, Jinxiong 08236
Liao, Kui-Yin 42378
Liapis, Kostas 22870
Liardon, Corinne 41423
Liavas, L. 22924
Libbus Boon, Edna 05716
Libby, Gary R. 47994
Liberatore, Laurie 48378
Libert, Laura L. 51851
Liby, Håkan 41004, 41012, 41012, 41017
Lică, Paul 36259, 36260
Lichnerova, Rut 38190
Licht, Ruth 41126
Lichte, Dr. Claudia 22713
Lichtenberg, James J. 52261
Lichter, Dr. Clemens 19437
Lichtin, Christoph 41593
Lichtlé, Francis 13302
Lichtscheidl, Mag. Olivia 03009
Lichty, Robert C. 47226
Lickleder, Dr. Hermann 19501
Lickteig, Dorothy L. 48865
Licky, Maurice 07421
Liczkov, Sergej Nikolajevič 36747
Liddell, Dave 54076
Liddle, Peter 43996
Liddle, Dr. Peter 44487
Lideń, Magnus 41008
Lidequist, Dr. Christina 40859
Lidman, Roger 51908
Lidner, Andreas 19902
Lidtke, Thomas D. 54045
Līduma, Ilze 30050
Lieb, Hartmut 18849
Liebelt, Glenn T. 01235
Liebenberg, O.J. 38566
Lieber, Vincent 41691
Lieberman, Christina 53601
Lieberman, Richard K. 50183
Lieberman, William S. 51212
Liebgott, Dorion 07095
Liebgott, Niels-Knud 10122
Liebhart, Prof. Dr. Wilhelm 16524, 16525
Liebich, Haimo 20450
Liebling, Carolina 41546
Liebscher, Sabine 20123
Liechtenstein, Friedrich 02587
Liechti, Werner G. 41900
Liechty, Ronald 51992
Liedekerke Beaufort, Comte Hadelin de 03511
Lieder, Cecilia 46731
Liedtke, Peter 18579
Liekonen, Marita 10943
Lien, Mary 52005
Lienau, Jennifer 52870
Lienen, Birgit 20579
Lienhard, Kurt 41183
Lienhard, Ulrich 42061
Liesbrock, Dr. Heinz 17488
Liesch, F. 20324
Liesens, Liliane 03461
Lieser, Wolf 17075
Lieth, Elke-Luise von der 17158, 17217
Lif, Malka 25100
Lifjeld, Prof. Jan T. 33737
Lifshitz, S. 25053
Light, Brad 54232
Light, Brian 45558
Light, Dr. Helen 01479
Light, Natalie de 20754
Lightbourn, Brian 04190
Lightfoot, Ricky 47827
Lighthall, Penny L. 07180
Ligne, Michel Prince de 03358
Lignori, Prof. Laura 28007
Ligon, Doris H. 47712
Ligúde, Ginsa 30020
Liikkanen, Hilkka 11279
Liken, Paula 50605
Liknes, Fern 05789
Lilienthal, Dr. Georg 18848
Lillehammer, Grete 33863
Lilley, Clare 45966
Lillington, Sophie 44961
Lilly, Bruce 52624
Lilly, Stephanie 47380
Lilly, Tina 50377
Lillystone, Jane 43810
Lim, Hyo-Taek 29777
Lim, Sarah 50595
Lim, Yujin 50240
Lim-Yuson, Dr. Nina 34447
Lima, Dr. Francisco 35439
Lima Greene, Alison de 49397
Limat, Marc 41564
Limbacher, Dr. Gabor 23840
Limbeck-Lilienau, Elisabeth 03079
Limbree, Christian 03973
Limburg-Stirum, Comtesse F. de 15901
Limme, Luc 03495
Limón Delgado, Antonio 40068
Limpo Pérez, Luis Alfonso 39724
Lin, Chien-Ling 42381
Lin, Chunxin 07601
Lin, Huanzhen 08024
Lin, Jiafu 08052
Lin, Lee Chor 38177
Lin, Maoyuan 07877
Lin, Shenggui 08073
Lin, Weiqi 08058
Lin, Wen-Ren 42404
Lin, Xiaoming 08196
Lin, Xing'er 07851
Lin, Zhaoren 07848
Lin, Zhifang 08336
Lin, Zuliang 08098
Linaeque, Dr. P.M. 31748
Liñán, Prof. Eladio 40350
Linciano, Dr. Antonio 26262
Lincoln, Lee 48040
Lincoln, Louise 47470
Lind, Axel 10260
Lind, Dr. Christoph 20173
Lind, Ed 52188
Lind, Evgenij Alekseevič 37669
Lind, Gabriele 02880
Lind-Sinanian, Gary 53970
Lind-Sinanian, Susan 53970
Lindahl, Per 40589, 40591
Lindauer, H. 33200
Lindauist, Christopher 54092
Lindberg, Annika 41000
Lindberg, David R. 46758
Lindberg, Janelle 51548
Lindberg, Kathy M. 48534
Lindblom, Michelle 46836
Lindbom, tina 40696
Lindeman, Jean 54108
Lindemann, Prof. Dr. Bernd Wolfgang 17116
Lindemann, Marilyn 48446
Linden, Grace 53114
Linden, Jaap ter 32717
Lindén, Lena M. 40606
Linden, Peppy G. 47401

Lindenhofer, Harald 03048
Linder, Karin 40569
Linder, Maria 40639
Linder, Marja-Liisa 11364
Linderer, Steve 49954
Linderman, Mike 48455
Lindert, Dietmar 02823
Lindh, Anders 40769
Lindh, Marianne 41051
Lindhartseni, Ole 33961
Lindholm, Dan 10721
Lindhorst, André 20924
Lindley, Ralph 45501, 45502
Lindmier, Tom 49951
Lindner, Gerd 16730
Lindner, Dr. Helmut 17974
Lindner, Mathias 17701
Lindner, Dr. Robert 02615
Lindner, Walter 18580
Lindquist, Beth 47851
Lindquist, Everett 53711
Lindquist, Sven-Olof 41068
Lindqvist, Catharina 40921
Lindqvist, Per-Inge 40635
Lindqvist, Svante 40921
Lindsay, Frances 01288
Lindsay, Judy 43999
Lindsay, Dr. Laura F. 46636
Lindsay, Mike 46405
Lindsay Baker, Dr. T. 49288
Lindsberger, Josef F. 02002
Lindseth, Knuth 33658
Lindsey, David 44543
Lindsey, Jack 51853, 51867
Lindsey, Jane 49657, 49658
Lindsey, Dr. Rob 49864
Lindström, Anneli 10827
Lindström, Inger 40684
Lindström, Leif 40784
Lindstrom, Janet 51037
Lindstrom, Dr. Sandra 07199
Lindvall, Karin 40782
Lineros, Ricardo 39132
Lines, Sue 43962
Ling, Bo 08045
Ling, Rencai 07759
Lingier, Rafael 03545
Lingstadt, Kirsty 44392
Linhares, Philip 51470
Linhart, Dr. Eva 18382
Lininger, Kris 53667
Link, Dr. Alexander 19478
Link, Wayne 05378
Linke, Irmgard 02507
Linkov, Krasimir 05127
Linkowski, Zdzislaw 34680
Linley, Jim 44520
Linn, Joe Dale 52146
Linn, Dr. M. 51376
Linna, Markku 11376
Linnehan, Sharon 48139
Linnel, Eric 46874
Linnen, Robert 07286
Linnertz, Rudolf 17515
Linničenko, Antonina 43111
Linnie, Martyn J. 24825
Linnikova, Nina Jakovlevna 37702
Lino, Elizalde 40359
Linou, Marie-Josée 15045, 15046
Lins, P. Andrew 51867
Linsbauer, Mag. Michael 02594
Linsenmaier, Dr. Walter 41337
Linsky, Carol 52423
Linsmann, Dr. Maria 22108
Linsmayer, Thomas 20473
Linthorne, Marguerite 05437
Linton, Dr. M. 01089
Linton, Meg 50212
Lintott, B. 33073
Linville, Tony 53297
Liot, David 15001
Liotta, Jorge 00603
Lipchinsky, Zelek L. 46746
Lipfert, Nathan 46628
Lipkens, Myriam 03685
Lipo, Frank 51455
Lipońska-Sajdak, Jadwiga 34725
Lipovanski, Georgi 05201
Lippe-Weißenfeld, Dr. Hagen W. 17997
Lippens, Kathleen 03476
Lippert, M.A. Anja 16582, 16587, 16592, 16593
Lippi, Angelo 26531
Lippi, Dr. Emilio 28068, 28069
Lippiett, John 45434
Lippincott, Louise 51942
Lippman, Judith 46579
Lippmann, Renate 17724
Lippold, Björn 40850
Lipps, Jere 46758
Lips-Liechti, Ursula 41606
Lipschutz, Jeff 51612
Lipscombe, Bruce 53775
Liptuga, Tatjana Ivanivna 43161
Lira, Dr. Raúl 00326
Lisai, Virginia 54100
Lisberg Jensen, Ole 10147
Liščáková, Daniela 09674
Lisenko, Mychajlo Lukyč 43169, 43170
Lisicyn, Viktor Borisovič 37961
Lisk, Susan J. 49118
Lisk-Hurst, Corrie 49061
Liška, Peteras 38370
L'Isle, Viscount de 45358
Lisnjak, Katerina Jevgenivna 43065
Lisowski, Silvia 22495
Liss, David 07124
Lissague, Bernard 12140
Lissens, Paul 03465
List, Manfred 01971
Listen, Simon 44082
Lister, Alison 44885
Listoe, Philip 06917
Litak, Anna 34810
Lithén, Hans-Olof 11425
Litochleb, Dr. Jiří 09791
Litscher, Mag. Manfred 03001
Litteral, Kate 50605
Little, Ann 50003
Little, Faye 43810
Little, Stephen 49343
Little, Susan 48535
Little, Tricia 49266
Litvak, Silvia 00342
Litwin, Dr. Jerzy 34631, 34649, 34651, 34662, 34687, 35182
Litwin, Józefa 35357
Litzenburg, Thomas V. 50072
Liu, Bailing 07984
Liu, Chuan 07935
Liu, Dejun 08481
Liu, Fo'an 07868
Liu, Guang Rui 07629
Liu, Guang Tang 07641
Liu, Guangyu 07936
Liu, Guangzhong 07941
Liu, Guojin 08123
Liu, Haichao 07703
Liu, Hong 08383
Liu, Hongbo 07785
Liu, Hongshi 07933
Liu, Hui 08263
Liu, Ji 08122
Liu, Jianli 08205
Liu, Jianzhong 08485
Liu, Jionghai 07773
Liu, Liben 08128
Liu, Lingyun 07682
Liu, Ning 08230
Liu, Qi'an 08441
Liu, Qingyan 07718
Liu, Shun 07488
Liu, Suiqun 07895
Liu, Wenquan 08201
Liu, Wie 07580, 08503
Liu, Wujun 08516
Liu, Yan 08078
Liu, Yi 07888
Liu, Ying 07486
Liu, Yirong 08100
Liu, Yiwen 07884
Liu, Yuchuan 07630
Liu, Yulin 08086
Liu, Zhiquan 07699
Liu, Zhiyi 07627
Liu, Zhongxin 08079
Liu, Zubei 07891
Lium, Randi N. 33930
Lively, Carter C. 46319
Liverani, Dr. Paolo 54501, 54502, 54505, 54508
Livernette 13716
Livezey, Dr. Bradley C. 51943
Livingston, Glenda A. 53783
Livingston, Jan 52315
Livingston, Dr. Valerie 53030
Livingstone, Judy A. 05691, 06350
Livland, Haakon 33602, 33911
Livneh, M. 25035
Livsey, Karen E. 49597
Lizarazo, Plata 08537
Lizon, Sylvian 14900
Lizzadro, John S. 48376
Ljiljak, Aleksandar 04279
Ljøkjell, J. 33586
Ljubaščenko, Anna Ivanovna 37834
Ljubimov, Vladimir Alekseevič 36767
Ljubović, Prof. Blaženka 08945
Ljungberg, Louise 40832
Ljunggren, Lieutenant colonel Ingmar 41079
Ljungström, Lars 40934
Llabrés, Pere-Joan 39791
Llabrés Martorell, Pere-Joan 39567
Llagostera Martínez, Dr. Agustín 07449
Llanas, Joan 40291
Llanes, Manuel 40127
Llanes-Myers, Susan 49387
Llewellyn, Mike 44428
Llobet, Guy 12573
Llongarriu, Marissa 39728
Llorente, Tina 49397
Lloveras, Fernando 36045
Lloyd, Alan 01544
Lloyd, Deanna S. 50458
Lloyd, Kristin B. 46211
Lloyd, Stephen 44157
Lloyd Harvey, Peter 45399
Lluch, Montserrat 40259
Lmas, Jean de 12855
Lobanova, Galina Pavlovna 36885, 36893
Lobaõ, Dr. Paulo 35439
Lobas, Aleksandr 29737
Lobb, Muriel 50113
Lobb, Sylvia 49597
Lobjakas, Kai 10627
Lobkareva, Antonina Vasilevna 37898
Lobner, Walter 02773
Lobo, Dr. Wibke 17081
Lobos Lápera, Victor 07435
Loburev, Vladimir Vasiljevič 37273
Locher, Prof. Dr. Hubert 20187
Lochman, Dr. Tomas 41154, 41184
Lochmann, Reinhart 16428
Lock, Victoria A. 49852
Lock-Weir, Tracey 00773
Locke, Gertraud 20792
Locke, Dr. H. 06430
Locke, Michelle 47818
Lockett, Daniel 45045
Lockwood, D.C. 33218
Lockwood, Dave 48171
Lockwood, David 44036, 44039, 44040
Lockwood, Dr. Yvonne R. 48254
Locmelis, Prof.em. Dr. Jazeps 30074
Loderer, Klaus J. 16664
Lodewijk, W. 32313
Lodge, Penny 51957
Lodge, Stuart 43274
Lodowski, Dr. Jerzy 35326
Löb, Heidrun 42099
Löbel, Andrea 03016
Loeben, Dr. Christian E. 19000
Löber, Eva 20113
Löbert, Dr. Horst W. 21972
Löbke, Dr. Matthia 19098
Löckemann, M.A. Karsten 20478
Löcken, Monika 16603
Löffler, Dr. Erzsébet 23537, 23539
Löffler, Dr. Michael 22834
Loeffler, Robert J. 50835
Löffler, Ruth 18020
Löfgren, Pia 40493
Löfgren Ek, Anita 40684
Löfström, Lena. A 40957
Löher, Jochen 17925
Löhle, Martina 17688
Löhr, Dr. Alfred 17549
Löhr, H. 18903
Löhr, Dr. Hartwig 22102
Loehr, Scott W. 46475
Löliger-Henggeler, Elisabeth 41733
Lönnebo, Lovisa 40915
Lönnström, Yvonne 11434
Loera Chávez y Peniche, Dr. Margarita 31134
Lösch, Bernhard 27619
Lösche, Wolfgang 20442
Lötsch, Prof. Dr. Bernhard 03043
Lövgren, Amanda 40860
Loew, Benedikt 21464
Löwenkopf, A.Z. 24941
Loewy, Dr. Hanno 02092
Lofgren, Donald 47575
Loft, Elaine P. 49365
Lofthus, Brian 48989
Loftin, Jack 46341
Loftin, Jim 48811
Loftis, Lynnea 49556
Loftus, John 21700
Loftus, Sydney L. 51559
Logan, Brad 49990
Logan, Burt 46954
Logan, Joyce 06623
Logan, Judy 49076
Loge, Øystein 33467
Logger, R. 31882
Logghe, Marc 04078, 04079, 04080
Logunov, Dr. D. 45090
Lohberg, Dr. Gabriele 22099
Lohiniva, Leena 11279
Lohman, Prof. Jack 44921
Lohman, Jack 44928
Lohmann, J.H.A. 32330
Lohmann, Uwe 22425
Lohner, Beate 18809
Lohrenz, Mary 49559
Loichemol, Hervé 12922
Loichinger, Hannes 20097
Loidl, Josef 01756
Loikkanen, Kirsi 11161
Loisy, Antoine de 11663
Loizou Hadjigavriel, Loukia 09379
Lojdová, Ivona 09853, 09859
Løken, Trond 33863
Lokensgard, Dr. Lynne 46674
Loker, Dr. Eric S. 46182
Lokin, Dr. D.H.A.C. 32062, 32063, 32064
Lokki, Juhani 10801
Lomaheftewa-Slock, Tatiana 52877
Lomakov, Dmitrij Aleksandrovič 37225
Loman, D.E. 32280
Lomatter, Alice 41785
Lomax, James 44666
Lomax, Jenni 44812
Lombardi, Gloria R. 51337
Lombardi, Michele 16224
Lombardo, Candice 53280
Lombardo, Dr. Giovanna 27512
Lombardo, Marta Alicia 00377
Lombardo, Romualda 34064
Lombino, Mary-Kay 50179
Łonak, Wojciech 34972
Lønborg Andersen, Vivi 10123
London, Elizabeth 51148
Londoño, Jaime Manzur 08569
Londos, Eva 40614
Londt, Dr. Jason 38706
Lonergan, Gillian 45506
Long, Charles 52719
Long, Greg 54296
Long, J. Grahame 47359
Long, Jena 53467
Long, John D. 52672
Long, Linda 47737
Long, Philip 44156
Long, Phillip C. 47565
Long, Steve 51206
Long, Xibin 08234
Longaven, Dorris 48503
Longbourne, Dr. D. 45128
Longcore, Robert R. 51343
Longenecker, Martha W. 48420
Longenecker, Vickie 48646
Longevialle, Ghislain de 13075
Longeway, Tricia 06012
Longhauser, Elsa 52901
Longhi, Alberto 27578
Longhi, Janice 43647
Longhi, L.A. 43644
Longhi, Rolf 42124
Longley, Dione 50659
Longuein, Paul 15531
Longwell, Alicia 53185
Lonnberg, Thomas R. 48456
Loodts, Isabelle 04006
Loomeijer, F.R. 32712
Loomis, John 52806
Looper, Gail S. 53135
Loos, Margit 18093
Loose, Annegret 18877
Loose, Helmut 16666
Loosli, Jean-Pierre 41639
Lootens, Michel 03657
Lopata, Robert 46780
Łopatkiewicz, Zdzisław 34701
Lopatkin, Michail Vasiljevič 37788
Lopes, Dr. Joaquim José de Castro 35735
Lopes, Manuel Teixeira 36009
Lopes, Manuel José Ferreira 35873
López, Al Ignatius G. 34488
López, Ana María 40136
Lopez, Donald 53895
López, Edgar 08622
López, Javier Fernandez 39337
López, Jorge 31128
Lopez, Julie Anna 52873
López, Magdalena 39544
López, María Jose Muñoz 39206
López, Mónica Alejandra 00146
López, Sebastián 04682
López Álvárez, Joaquín 39338
López de Arriba, Mercedes 39544
López Fernández, Francisco Javier 39191
López Franco, Amparo 08666
López Hervás, Maria Victoria 39550
López Pimentel, María del Carmen 09099
López Portillo Romano, Carmen 31093
Lopez Rodriguez, Jose Ramón 40065
López-Vidriero, Luisa 39586
López Wehrli, Silvia 40316
Lopiene, Regina 30422
Lopreato, Paola 25916
Loquenz, Norbert 02578
Loránd, Klára 23643
Lorbeer, Marie 17180

L

Lorblanchet, Hélène 14238
Lorck, Bjorn 31757
Lord, Anne 01601
Lord, Gary T. 51423
Loren, Dr. Diane 47185
Lorenčak, Marja 38457
Lorenz, Dr. Angelika 20512
Lorenz, Barbara 01773
Lorenz, Dr. Harald 18706
Lorenz, Jens D. 17275
Lorenz, Margaret 44591
Lorenz, Dr. Ulrike 18596, 18597, 20166
Lorenzi, Franco 26682
Lorenzi, Georges 15765
Lorenzo Rosales, Elaine 09202
Lorenzo Villamor, Felicitas 39333
Lorenzoni, Dr. Mario 27843
Loret, John 50430
Lorian, F. 16627
Lorimer, Wayne 33114
Loring Brace, C. 46308
Lorquin, Olivier 14719
Lorys, Jan 47507
Lorz, Julienne 20445
Łos, Ewa 34904
Los, Dr. W. 31904
Losada, Santiago 39544
Losavio, Sam 46633
Losch, Michael 54097
LoSchiavo, Joseph A. 52530
Loshi, Ygal 25076
Losonczy, Attila 23417
Lotfullin, Galim F. 37493
Lotorev, N.D. 37351
Lott-Büttiker, Dr. Elisabeth 41503
Lottermann, W. 16708
Lotz, Rouven 18855
Lotze, Manfred 18559
Lou, Xia 08273
Louati, Ali 42664
Loud, Patricia C. 48715
Louédec, Patricia 13499
Lõugas, Anne 10624
Loughridge, Sally 54262
Loughton, Gregory 07132
Louis, Hubert 12098, 12099
Louma, Mark 50907
Lounibous, Leroy 52903
Lounsbury, Kathryn 46121
Lourdes Barreto, Maria de 04450
Louring, Aage 10038
Loustau, Alain 13722
Louvencourt, Amaury de 12338
Louvi, Dr. Aspasia 22984, 23193, 23197, 23219
Louw, Prof. D.J. 38736
Louw, J.P. 38699
Louw, P. 38677
Louwagie-Neyens, Linda 06879
Louwman, E.V.N. 32677
Lovász, Stephanie 41169
Lovdžieva, Marija 05181
Love, Jackie 51386
Love, Mindi 48286
Lovell, Albert 33144
Lovell, Carol 50083, 53335
Lovell, Charles M. 53441
Lovell jr., Russell A. 52842
Lovén, Lucie 33754
Lovenz, Marianne 49971
Lovett, Margaret 50083
Lovick, Emily 48683
Loving, Charles R. 51445
Lovis, Dr. William A. 48254
Low, Dr. James A. 06087
Low, William 50037
Lowagie, Luc 03594
Lowe, Carole 52014
Lowe, Dr. D. 44952
Lowe, Janet 05624
Lowe jr., John 48183
Lowe, Joseph 47755
Lowe, Susan 30659
Lowe, Vivian J. 50086
Lowell, Molly W. 48180
Lowenfeld, David 47635
Lowkunsombat, Chavalit 42576
Lowly, Tim 47459
Lowrey, Annie 54149
Lowrey, Carol 51233
Lowrey, Dr. Timothy K. 46182
Lowry, Dale 05560
Lowry, Glenn D. 51215
Loxa, Juan de 39317
Loyau, Anne 11885, 11886
Loynes, G. 46063
Loyrette, Henri 14698
Lozada Arias, José 08527
Lozado de Solla, Noemi 00120
Lozano, Jacob 50127
Lozano, Prof. Luis-Martín 31070
Lozano Peña, Carlos 08788
Lozuk, Josip 08953
Lsayeva, Olga Nikolaevna 43058
Lu, Benshan 07838
Lu, Cangxin 07944
Lu, Chenghan 08095
Lu, Chimu 08427
Lu, Dongzhi 07541
Lu, Wensheng 07884
Lu, Xiaodang 08320
Lu, Xincai 08445
Lu, Yinghong 07546
Lubac, Marcel 13010
Lubbe, T.A. 38568, 38572, 38703, 38732, 38776, 38778
Lubic, James E. 49184
Lubimova, L.M. 37204, 37205
Lubina, Michail 35087
Lubitz, Dr. Telse 19066
Lubowsky Talbott, Susan 48096
Luca, Claudia 36709
Lucas, Annemarie 20508
Lucas, Cynthia L. 49051, 49052
Lucas, Daniel 11851, 12301, 12302, 14102
Lucas, Enriqueta T. 34517
Lucas, Judith S. 47563
Lucas, Martyn 44685
Lucas, Mike 43423
Lucas, Palma B. 51866
Lucas, Ritchie 47800
Lucchese, Dr. Angiolo del 28217
Lucchesi Ragni, Dr. Elena 25534, 25542, 25543
Lucchini, Mario 41433
Luce, Debbie 47817
Luce, Donald T. 50735
Lucewicz, Cher 51514
Lucey, Denis 24851
Luchetti, Dr. Glauco 26134
Luchsinger, Cornelia 42078
Luchter-Krupińska, Lidia 34621
Luciani, Luciano 27531
Lucio, Cilia 28208
Lucke, Dr. Arne 19202
Luckeneder, Walter 02701
Luckey, George 44396
Luckhardt, Prof. Dr. Jochen 17522
Luckhardt, Dr. Ulrich 18932
Luckhurst, Ailene 47586
Luckow, Dr. Dirk 19531
Łuczkowski, Jan 34983
Luderowski, Barbara 51953
Ludescher, Inge 18253, 18255
Ludewig, Dr. Thomas 20654, 20655, 20656
Ludington, Link 50380
Ludvigsen, Lucia 23272
Ludvigsen, Peter 10123
Ludwig, Andreas 18105
Ludwig, Dr. Annette 20143
Ludwig, David 53172
Ludwig, Dr. Jochen 18433
Ludwig, Jürgen 17866
Ludwig, Dr. Renate 19076
Ludwig, Rudolf 02169
Ludwig, Sheila 18209
Ludwig, Walter 51292
Ludwig-Weber, Birgit 20752, 20753
Ludyga, Johann 16820
Luebbers, Dr. Leslie 50566
Lübbers-Ukena, A. 17836
Lübcke, Peter 18552
Lübeck, Elisabet 40630
Lübeck, Karla-Kristine 21695
Lübke, Steffen 20244
Lüchinger, Paul 41965
Lüdemann, Luise 22058
Lüdi, Heidi 41565
Lüdtke, A. 18431
Lüdtke, Prof. Dr. Hartwig 20174
Lueg-Werner, Dr. Gabriele 19619
Lührs, Günter 18454
Lüken, Dr. Sven 17072
Lüll, Martina 21969
Lüpkes, Dr. Vera 19949
Lüps, Dr. Peter 41968
Lüscher, Hans 42119
Lüscher, Regula 41809
Luesma, Teresa 39397
Lütgens, Dr. Annelie 22665
Lüth, Dr. Friedrich 21520
Lueth, Virgil W. 53140
Lüthi, Werner 41274, 41277
Lüthy, Jürg 41711
Lütkemeyer, Ursula 21298
Lütken, Ilona 21956
Lüttichau, Dr. Mario-Andreas von 18263
Luft Dávalos, Rafaela 31255
Lugger, Leopold 02351
Luginbill, Troy 50331
Luginbühl Wirz, Regula 41207
Lugo, Flavia Marichal 36076
Lugosi, Dr. József 23642
Ługowski, Ryszard 35231
Lugtenburg, Anneke 38748
Luh, Willi 17628
Luhta, Caroline 47633
Luijten, G. 31884
Luijten, Ger 14598
Luisi, Vincent 48209
Luisi-Potts, Billy 53035
Lukas, Dr. Clemens 16947
Lukasch, Dr. Bernd 16551
Lukasiewicz, Nancy 46413
Łukaszewski, Andrzej 35111
Lukaszuk-Ritter, Joanna 02930
Lukatis, Dr. Christiane 19464
Luke, Gregorio 50175
Luke, Suzanne 07289
Lukeš, František 09849
Lukeš, Dr. Michal 09779
Lukesch, I. 03002
Lukić, Ivan 38095
Lukičeva, Alla Stepanovna 37201
Lukina, Inna Nikolaevna 37576
Lukowczyk, Evelyn 18904
LukšicĘ, Hilar 08942
Luksic, Tugomir 09006
Lulav, S. 24941
Lum, Shawn 53679
Luma, Ramon E.S. 34487
Lumb, Dorothy 47823
Lumley, SimonWhite 44790
Lummel, Dr. Peter 17086, 17264
Luna, Juan J. 39573
Lunardi, Prof. Ernesto 26295
Lund, Arnstein 33929
Lund, Elaine H. 50502
Lund, Mikael 10112
Lund, Tora Jervell 33480
Lundberg, Åsa 41000
Lundberg, Carl-Anders 10996
Lundbergh, Hans 41022
Lundblad, Lena 40558
Lunde, Oeivind 33923
Lundgreen, James 50446
Lundgren, Britta 41000
Lundgren Nielsen, Hanne 10069
Lundin, Kennet 40548
Lundkvist, Anders 40517, 40518
Lundrigan, Dr. Barbara 48254
Lundskow, Pete 49330
Lundström, Sven 40577, 40578
Lundtofte, Anne Sofie 10001
Lungarotti, Dr. Maria Grazia 27993
Lunger, Brian 06020
Lunger-Valentini, Eva 02254
Lunghi, Enrico 30476
Lungu, Alexandru 36605, 36609, 36610, 36611, 36613
Lungu, Cornel 36609
Lungu, Hariton 36719
Lunn, Will 51564
Lunning, Elizabeth 49394
Lunsford, P. Celina 18367
Lunsingh Scheurleer, Dr. R.A. 31825
Lunt, Sara 44359, 44359, 44708, 46087
Luo, Anhe 07966
Luo, Biaoyuan 07741
Luo, Jianzhi 07648
Luo, Tingru 07674
Luo, Xizhang 07694
Luo, Xuewang 07925
Luo, Xuzhong 08237
Luo, Yisheng 08486
Luo, Zhangui 07550
Luojus, Susanna 11405
Luong, Tran 54626
Luoto, Jukka 11045, 11048
Lupal, Kimberly A. 05500
Luperini, Prof. Ilario 27235
Lupiy, Dr. Tetiana 43137
Luppino, Tony 05743
Luptáková, Eva 38250, 38312
Lupton, Ellen 51156
Lupton, Jo 00908
Lupton, John M. 54110
Lupu, Sabina 36663
Lupu, Valentin 36123
Luraghi, Prof. Raimondo 28026
Luria, Michael 53597
Lurie, Doron J. 25122
Lusa Monforte, Guillermo 38967
Lusardy, Martine 14611
Lusby, D.A. 33198
Luse, Vel 49746
Lushaba, Yvonne K. 38645
Lushear, Jen 51092
Lussi, Kurt 41366, 41592
Lussky, Richard 47994
Luštinec, Jan 09563
Luterbach-Renggli, Hansjörg 41844
Lutes, Betty 06318
Luth, Jan 49953
Luther, Charles 47805
Luti, Claudio 26962
Lutnes, Kjell 33446
Lutnesky, Dr. Marvin 52056
Lutterodt, Isabelle 49848
Luttropp, Pernilla 40558
Lutyens-Humfrey, A. 43803
Lutz, Dr. Albert 42097
Lutz, Dr. Gerhard 19182
Lutz, Gottlob 41472
Lutz, Marian 49397
Lutz, Peter 22561
Lutz, Tina 53613
Lutz, Dr. Volker 02770
Lutz, William 48702
Lux, D. Ruth 51610
Lux, Simonetta 27509
Lux, Ulla 17996
Luxford, Bob 33174
Luyken, Dr. Gunda 19531
Luz, David W. 51765
Luz, Henrique 04880
Luz Paula Marques, Maria da 35841
Luz Ramirez Estrada, Maria de la 09126
Luzin, Sergej Stepanovič 37046
Luzina, Lubomir 38188
Lyall, Clare 45878
Lyall, Dr. Dorothee-Camilla 21969
Lydakis, Prof. Dr. Stelios 22939
Lydamore, Chris 44396
Lydon, Andrea 24807
Lydon, Kate 51961
Lyke, Linda 50268
Lyle, Charles T. 48868
Lyles, Pelham 54237
Lynas, Donna 43564
Lynch, Cindy 47743
Lynch, Geraldine 24874
Lynch, Harry P. 46148
Lynch, Kevin 44708
Lynch, Sheila 54137
Lynch, Thomas F. 47105
Lynch, Thomas J. 46980
Lyndaker, Vernon 47885
Lynesmith, P. 45167
Lynn, Bob 33038
Lynn, Devell 48638
Lynn, Judy G. 47854
Lynn, Kitty 43971
Lynn, Michael Anne 50074
Lynn, Ralph 06277
Lynx, David 54320
Lyon, David Peter 01390
Lyon, Katharine 50137
Lyon, Richard 48939
Lyons, Don 48172
Lyons, Jana 53919
Lyons, Joan 48172
Lyons, Dr. John D. 50395
Lyons, Marilyn 51489
Lyons, Michael 46005
Lyons, Michele 46775
Lyons, Rex 48734
Lyra, Fernando Soares 04668
Lysdahl, Per 10080
Lysenko, Nikolaj Stepanovič 29720
Lysnjak, Kateryna Jevgenivna 43069
Lysogórski, Krzysztof 34664
Lyssaris, Agelos 23189
Lytjen, Lokey 49571
Lytle Lipton, Susan 51200
Lytton Cobbold, Martha 44621
Lytwyn, Stephen 07396
Lyu, Ma-ri 29853

M

Ma, Fazhu 07535
Ma, Hong 07914
Ma, Honglin 08420
Ma, Jiayu 07619
Ma, Kuijun 07788
Ma, Liangyun 07622
Ma, Long 08295
Ma, Peilin 08403
Ma, Wei-du 07539
Ma, Weidou 07540
Ma, Weiyi 07772
Ma, Wenwei 07664
Ma, Zhengyuan 07871
Maanit, G. 25044
Maarek, Lucien 15773
Maas, Dr. A.J.P. 32484
Maas, Klaus 18023
Maas, Wilhelm 20627
Maaskant, M. 32710
Maass, R. Andrew 52637

Maassen, Dr. Walter 16406
Maat, Erik 32251
Maaz, Prof. Dr. Bernhard 17946, 17954
Mabee, Kim 52961
Mabungule Chissano, Alberto 31713
Mac Phee, S.D. 05779
MacAdam, Barbara J. 49162
McAfee, Michael J. 54078
McAffee, Dionne 07157
McAinsh, Deirdre S. 45883, 45884
Macalister, Helen 52407
McAlister, Judith 46966, 48505
McAllister, Lowell 51506
McAlpine, Dr. Donald 06838
Macalpine Foshay, Susan 05941
McAnally, Fiona 49822
McAneney, Helen 45165
McArdle, Jean 43501
McArthur, Randy 50611
MCarty, Kenneth W. 51851
Macat, Andreas 20409
MacAulay, Donald 05371
MacAulay, Margot 05371
McAuley, Joyce 46532
McAuley, Kathleen A. 47044, 47050
McAuliffe, Dr. Chris 01284
McBay, Ann 45384
McBean, Jude 01127
MacBeath, Doug 44137
McBratney-Stapleton, Deborah 46291
MacBride, Frank 00895
McBride, Heather 46126
McBride, Joe 46281
McBride, Michael A. 49102
McBride, Peggy 52221
McCabe, Jennifer 52786
McCabe, Tomas 52767
McCafferty, Estelle 53238
McCall, Janet L. 51961
McCall, Louise 43840
McCalla, Mary Bea 48839
McCallum, Cheryl 49382
MacCallum, Malcom 45446
McCallum, Sue 51793
McCalmot, Faye 53100
McCammon, Alice 50273
McCamphill, Debbie 45261
McCandless, Barbara 48709
McCann, Mac 47160
McCarroll, Stacey 46925
McCarter, John W. 47473
McCarthy, Brian 06303
McCarthy, Christine 52183
McCarthy, Darlene 48954
McCarthy, Frances 47374
McCarthy, Indi 49529
McCarthy, J. 01662
McCarthy, John J. 52883
McCarthy, Judith 24862
McCarthy, K. 33069
McCarthy, Michael J. 45912
McCarthy, Dr. Stephen 52616
McCarthy, Wayne 53811
McCartney, Dr. Nancy G. 48528
McCarty, Cara 52595
McCarty, Gayle 52259
McCarty, Kimberly 50860
McCarty, Laura 48729
McCarty, R. Paul 48619
McCaskey, Jane 50823
McCauley, John F. 46916
McCay, Bill 49609
Macchi, M. 15772
Macciotta, Prof. Giampaolo 25585, 25588
McCleave, John 06739
McClelland, Jesse 47488
McClenahan, Jane 51511
McClenny, Bart 46245
McCloskey, Anne M. 50154
McClung, Elizabeth 50663
McClung-Oakes, Katherine 43691
McClure, Donna 50819
McClure, Kent 50819
McClure, Nancy 53074
McClure, Rod 52092
McCluskey, Holly H. 54124
McClusky, Pamela 53013
McCollum, Jay 47697
McComb, Joseph 50847
MacConnel, Cal 48588
McConnell, Bridget 44288
McConnell, Glenn 51529
MacConnell, Raymond 46749
McConnell, Ruth 06203
McCooey, Patrick 05889
McCord, Dr. Robert 50605
McCord-Rogers, Suzette 49277
McCormack, Anne 44602
McCormack, Peter 44613
McCormick, Donald 51244
McCormick, Edgar L. 49738
McCormick, Todd 46697
McCorquodale, Ross 01527
McCosh, John 33105
McCourt, Frank A. 51620
McCourt, Steve 00842
McCowan, Kathy 53763
McCowan, Kurt 52455
McCown, Donna 51989
McCoy, Bill 54017
McCoy, James P. 51812
McCoy, Jeff 52485, 52487
McCoy, Kathy 50795
McCoy, Nancy 50198
McCracken, Katherine 05918
McCreary, Kathy 49327
McCree, Timothy 46168
McCue, Debbie 46793
McCue, Donald 52274
McCue, Kaz 53484
McCuistion, Mary Ann 53723
McCullick, Nancy 50537
MacCulloch, Alexandra 43342
McCulloch, Bruce 33162
McCulloch, Judith 48945
McCullough, Catherine 43317
McCully, Kathleen J. 47187
McCurtis, Barbara S. 52747
McCutcheon, Caroline 46045
McDaniel, George W. 47362
McDaniel, Robert 48212
McDaniels, Garvin 50941
McDaniels, Warren E. 51090
McDermott, Eugene 47931
McDianmid, Gordon 05458
MacDonald, Andrew 43919
McDonald, Bonnie 46323
MacDonald, Catherine 44656
MacDonald, Colin 45397, 45914
MacDonald, Colinn 43771
McDonald, Craig 06957
McDonald, Donald M. 01324
McDonald, Doris 06528
McDonald, Douglass W. 47557
MacDonald, Elizabeth H. 51870
McDonald, Garry 05711
MacDonald, Dr. George 52994
MacDonald, Gerald J. 51192
Macdonald, I.H. 45744
McDonald, J.P. 50318
MacDonald, Jessie 06874
MacDonald, Judith A. 51994
MacDonald, Kate 45825
McDonald, Dr. L. 01465
McDonald, Marty 47135
MacDonald, Mary V. 48006, 48007
McDonald, Michael 48183
MacDonald, Ned 06017
McDonald, Paul 46471
MacDonald, Peter 44616
McDonald, Ruth 43609
MacDonald, S. 44953
MacDonald, Stanley 06510
McDonald, Sussanne 05806
McDonald Crowley, Amanda 51172
McDonald Pang, Toni 49530
McDonnell, Dr. Patricia 53414
McDonough, Stephanie 46127
McDonough, Stephen 46261
MacDougal, Bruce 52709, 52724, 52733
McDougall, Jamie 05443
McDougall, Kim 05862
McDowell, Anna 45644
McDowell, Ina 53489
MacDowell, Dr. Marsha 48254
McDowell, Susanne 17688
McDowell, Suzanne 47903
McEachern, Mark 53544
Macedo, Maria de Fátima 35860
Macedonio, Prof. Giovanni 26059
Macek, Stanislav 09502
McElhone, William K. 46829
McElroy, L.J. 50955
McEneaney, Eamonn 24909
Maceroni, Prof. Giovanni 27403
McEwen, Gloria 49540
McEwin, Florence 52412
McFadden, David 51112, 51224
McFadden, Lila 05634
McFadden, P. 38601
McFalls, W.J. 05967
McFarland, Seth 47551
MacFarlane, Aynsley 05369
MacFarlane, Nathalie 06982
Macfarlane, Sheena 45314
McFeely, Marie 24807
McFerrin, Jean 52702
McGahen, DeAnn 46220
McGarrie, Mary 44166
McGarry, John H. 50924, 50925
McGarry, Susan 51622
McGary, Nancy 51200
McGaugh, Houston 53947
McGavin, Dr. G. 45336
McGee, Prof. Mike 48806
McGee, Patrick 24792
McGhie, H. 45090
McGill, Forrest 52766
McGill, Shirley 48165
McGillivray, Allan 07191
McGillivray, Don 06419
McGillivray, Dr. W. Bruce 05752
McGinley, Dr. Jack 24800
McGinnis, Howard 47864
McGinnis, Melissa S. 53407
McGinnity, Betsy L. 53971
McGirr, Margo 53278
McGivern, Andrew J. 53993
McGlumphy, D. 53609
McGlynn, Robert 51244
McGovern, Robert M. 47280
McGowan, Jeanie L. 47242
McGowan, Rebecca 07139
McGrady, Dr. Patrick J. 53659
McGrath, Deborah 01234
McGrath, Joe 01275
McGrath, Steph 54119
McGreevy, Candace 53338, 53339, 53341, 54203
Macgregor, Elizabeth Ann 01553
MacGregor, Neil 44803
McGrew, Rebecca 47574
McGuffey, Patrick 48847
McGuire, Donna 06503
McGuire, Janice 48725
McGuire, Dr. Jim A. 46638
McGuire, Jimmy A. 46759
McGuire, Raymond J. 51264
MacGunnigle, Bruce C. 48240, 48242
Mach, Jiří 09493
Machacek, R.J. 47927
Machan, Dallas 44608
Macheret, Elisabeth 41308
Machhammer, Helmut 01853
Machinek, Hubert 21754
Machneva, Margarita A. 37036
Machnicki, Monika 22165
Machon, Abdurremane 31704
McHugh, Eileen 46469
McHugh, Joseph 07226
McHughs, David 48727
MChuny, V.B. 38625, 38666
Macías Ramos, María Teresa 40253
Maciejewski, Andrzej 35249
Maciejunes, Nannette V. 47752
MacIlreith, Rob 05777
McInerney, Gay 46650
McInerney, Michaela 48732
MacInnis, Karen 50462
McIntosh, Jean 06624
McIntosh, Mildred 50538
McIntosh, Muriel 44311
McIntyre, Art 06764
McIntyre, Barbara 06867
McIntyre, Chris 44418, 45546
McIntyre, Stella 43541
Mačionis, Z. 30443
MacIsaac, Kim 51743
McIver, J.H. 46999
Mack, Angela 47364
Mack, Gary 47939
Mack, Jeanne P. 48924
Mack, Vanessa 01552
McKague, Margot 06046
McKain, Martha 49913
McKale, William 48676
McKamey, Sheldon 46982
Mackay, Andrew 45688
McKay, B. 44833
McKay, Barrie 49478
McKay, Dr. David 50703
Mackay, Jim 46211
Mackay, John L. 47387, 47388
McKay, Kelly 05842
MacKay, Kevin 05590
McKay, Maria 53069
McKay, Mary Alice 51239
McKay, Patrick J. 52388
MacKay, Robert B. 47683
MacKay, Sarah 43242
Mackay, W.A.F. 06726
Mackay-Collins, Dorothy 52548
McKean, Virginia 50766
Mackechnie, John 44282
McKee, Barry 05962
McKee, Dorothy 50360
McKee, Francis 44276
McKee, Kevin 46409
McKee, Marshall 44482
McKeil, David 06603
MacKellar, Dr. John 01317
McKellar, Stacey 05467
McKenna, George L. 49695
McKenna, James 51532
McKenna, Sharon L. 51743
McKenrick, Fremont 48279
McKenzie, Cameron 44027
McKenzie, Colin 44229
McKenzie, Molly 47158
Mackenzie, Philippa 44385
MacKenzie, Stuart C. 47541
McKenzie, Wendy 06037
McKenzie Smith, Dr. Ian 44151
McKeown, Jack J. 48951
McKernan, Robert L. 47540, 52276, 54361, 54362
Mackey, Brian 44725
Mackey, David M. 49362
Mackey, Marjorie R. 46405
Mackie, Louise W. 47624
McKie, S. Zurie 49590
Mackie, Thomas 47839
McKillop, Joan 45316
McKinlay, Dawn 07264
McKinley, Nancy R. 50672
McKinney, Evans 49291
McKinney, Henry C. 53928
McKinney, Jack N. 53773
McKinney, Janice 54148
McKinney, Rick 48527
McKinney-Tandy, Ruth 48408
MacKinnon, Ann. M 44541
McKinnon, Karl L. 54211
McKitterick, Michael 10377, 53027
Mackow, Leigh 49927
Maćkowiak, Dr. Janryk 34754
Mackowiak, Reinhild 17352, 20092
Maclagan, Dr. Helen 45941, 45942
McLain, Brenda 46622
McLain, Charles 49769
McLain, Guy 53224
McLallen, Helen M. 49777
McLaren, Hazel 46611
McLaren, M. 01219
McLauchlin, Mary Nell 47109
McLaughlin, David 50735
McLaughlin, Heather Jo 49265
McLaughlin, Penny 49636
McLaughlin, Robert 53240
McLaughlin, Stacey B. 52457
McLaughlin, William F. 48124
McLaughlir, Dr. John 05829
McLaurin, Genette 51254
McLea 52687
McLean, Claire 54172
McLean, Dr. Janice V. 46237
MacLean, John 47600
McLean, Kathleen 06485
McLean, Libby 44058
McLean, Linda E. 49411
Maclean, Marge 05909
MacLean, Pauline 06019
McLean, Rita 43500
MacLean, Roderick 05930
MacLean, Rusty 45528
McLean, Steve 45227
McLean Ward, Dr. Barbara 52097
MacLeish, Bruce 51329
McLellan, Darlene 48295
MacLellan, Heather 06609
MacLellan, Iain 50420
McLellan, Richard A. 47323
Maclennan, F. 44626
Macleod, Chris 44575
McLeod, Cynthia 52333
MacLeod, Cynthia 52337
McLeod, Deborah 50362
Macleod, Dr. Ian 01412
MacLeod, Karen 50018
MacLeod, Kathleen 05551
MacLeod, N. 44941
MacLeod, Patricia 52660
McLeod, Roddy 44291
McLeod, Susan M. 48276
McLerran, Jennifer 46417
MacLoy, David 00875
McMahan, Jackie 52897
McMahon, Cynthia 47509
McMahon, James 49263
McMahon, James E. 48906
MacMahon, Rose 44772
McMahon, Stephen 53300, 53301
McMains, John 52821
McManus, Greg W. 33196
McManus, Marcia 48629
McManus Zurko, Kathleen 54294
McMasters, Thomas O. 48679
McMeekin, G. 46026
McMeekin, Jim 47842
McMeekin, Rebecca B. 52245
McMenamin, Brendan 45023
McMichael, Sindy 48050
MacMillan, Arthur 44638
McMillan, Edward 47656
McMillan, Erin 48131
McMillan, Patsy H. 49806
McMillan, Dr. R. Bruce 53216

McMullen, Darlene 06509
McMurtrie, I. 44134
McNabb, Debra 07008
McNabb, Liz 43556
McNair, Cynthia 50648
McNally, Ginger 54351
McNamara, C. J. 48561
McNamara, Carole 46309
McNamara, Mary E. 51272
McNaught, Lewis 44911
McNealy, Terry 51062
McNeely, Don 53252
MacNeil, Malcolm 44537
McNeill, Anne E. 43580
McNeill, C. 01607
McNeilly, Kath 00984
McNelly, Dennis 51880
McNett, Sarah 49398
McNichol, Susan 06587, 06986
McNichols, Patricia 51461
McNicol, Jean 07301
McNicol, Morag 45148
McNulty, E.C. 48443
McNulty, F.M. 48741
McNulty, J.L. 48443
Macon, Robert 51767
McOuat, Gordon 06291
McOwan, Daniel 01144
MacPhee, Dana 44723
MacPhee, Dr. Ross D.E. 51115
McPherson, Beth Ann 51716
McPherson, Dan 47915
MacPherson, Kathy 50212
McPherson, Loretta 01076
McPherson, Tommy 51716
McQuaid, Matilda 51156
Macquart, Bruno 14722
Macquart-Moulin, X. 14416
McQueen, Jack 52818, 52820
McQuillis Iscrupe, Shirley 50082
McRae, Jennifer 49062
McReynolds, Anne 43443
McShane, Michael 43937
McVay, Darice 53701
McVey, Dorothy 53458
McWayne, Barry 48472
McWilliams, Mary 47177
Macyszyn, Dr. Jacek 35280
Mczakez, Franz 02956
Maczulis, Norbert 34718
Madanian, M. 24636
Madawick, Paula 48052
Madden, Jennifer 52839
Madden, Pamela 47240
Maddison, Dr. Wayne 07199, 07199
Maddox, Iva Jean 48668
Maddox, Margaret 49816
Madec 14850
Madeline, Laurence 14692
Mader, Dr. Gerald 02756
Maderek, Dorota 35085
Madert, Johannes 20300
Madesta, Dr. Andrea 02190
Madin, John 44195
Madisia, Joeseph 31764
Madjarov, Mitko 05006
Madkour, Christopher 50428
Madonia, Ann C. 54170
Madonsela, Dudu 38775
Madrid, Candy L. 52437
Madrid Rojas, Juan José 08740
Madsen, Inge 10048
Madsen, Jan Skamby 10084
Madsen, Lennart S. 10049
Madsen, Orla 10049
Madsen, Per Kristian 10311
Madshus, Eva 33716
Madurini, Dr. Giulio 26687
Madžarac, Nada 38454
Maeda, Kenichiro 29034
Maeda, Koichiro 28536
Maeda, Shoso 28455
Maeder, Andreas 41687
Maeder, Arnaud 41295
Maeder, Edward F. 48032
Maeght, Adrien 14297
Mähl, Andrea 22650
Mäkelä, Asko 10823
Mäkelä, Sirkku 11091, 11092
Mäkeläinen, Reijo 11113
Mäki, Riitta 11044
Mäkinen, Antti 10903
Mäkinen, Marketta 10890
Mael, Stephen 44689
Mämmetnurow, Öwezmuhammet 42973
Mänd, Anu 10638
Maeng, Injae 29845
Männchen, Prof. Dr. Julia 18749
Mäntele, Dr. Martin 22170
Mänttäri, Roy 10821
Mänz, Peter 17069
Maergerman, Roger 04093
Maerten, Michel 15186
Maerz, Jürgen 18403
Maes, Jean-Michel 33286
Maes, Leopold 04017
Maes de Wit, Yvan 03859
Maesen, Mathieu 03964
Maessen, R. 32822
Maestro Guerrero, Dr. Fernando 39111
Mäuser, Dr. Matthias 16915
Maeyama, Mikio 29520
Maeyama, Yuji 29190
Maffei, Prof. Lamberto 27548
Maffia, Prof. Cesare 26820
Maffioli, Dr. Monica 26195
Mafre, Christian 11541
Magacz, Joanna 17949
Magajne, Miljojka 38421
Magalhães, Carlos 04906
Magalhães, Dr. Ivone 35569
Magaña Cota, Gloria E. 30914
Magdich, Bob 53522
Magee, Eileen M. 51816
Magee, Eugene 06199
Magee, Henry J. 51834
Magendans, J.R. 32487
Mager, Gerald P. 21428
Mager, Gerald Paul 21425, 21429
Mager, Dr. Karl 02112
Maggi, Ersilio 27349
Maggi, Gianfranco 25174
Maggi, Roberto 25858
Maggini, Roberto 41981
Magglo-Serra, Dr. Rosanna 06916
Maggs, Richard 45455
Magidson, Phyllis 51231
Magill, Tina M. 52024
Magilton, Jack 06083
Magini, Gianluca 26856
Magiru, Maria 36303
Maglica, Nadja 08989
Magnan, Oscar 49628
Magnetti, Daniela 28028
Magnifico, Marco 28164
Magnin 11820
Magnin, Bénédicte 11821
Magnin-Gonze, J. 41545
Magnini, Prof. Marsilio 26033
Magnussen Svare, M. 10142
Magnusson Staaf, Björn 40703
Mago, Norma A. 34526
Magolfi, Dr. Miria 26215
Magomedova, Seir Seitovna 37839
Magrane, Nigle 46101
Magras, M. 23279
Magrath, Philip 44211
Magré, Piet 32424
Magrini, Boris 41634
Magro, Antonio 25263
Magro, Pasquale 25310
Magsaysay, Dr. Eulogio Amang 34439
Maguindanao, Macmod 34459
Maguire, Brian 43995
Maguire, Cathy 52922
Maguire, Dr. Eunice 46546
Maguire, Dr. Hugh 24864
Maguire, Nancy 47196
Magun', Marija Karpivna 43129
Magurany, J.D. 52579, 52582
Mahamadou, Kelessé 33308
Mahé, Lucien 13156
Mahecha Useche, Ernesto 08654, 08655
Mahéo, Noël 11588
Maher, Kathleen 47016
Maher, Kristie 51916
Maher, Tina 43279
Mahida, Bhamini A. 24399
Mahilom, Mila M. 34378
Mahlknecht, Dr. Eduard 02182
Mahlman, Jennifer 51234
Mahlstedt, Thomas 53791
Mahmoud, Assad 42134
Mahmudov, N. 54460
Mahmutović, Džafer 04259
Mahon, Rose 45486
Mahran, Younes 10496
Mahy, Ivan 03791
Mai, Do Thi Tuyet 54626
Mai, Hai 08151
Mai Hung, Prof. Dr. Pham 54618
Maia e Castro, Catarina 35860
Maicki, Richard J. 47622
Maier, Daniela 18280, 18281
Maier, Erhard 02667
Maier, Jakob 16537
Maier, Marco 41911
Maier, Stefan 21918
Maier, Thomas 25008
Maier-Schöler, Barbara 22273
Maierhofer, Fritz 02209
Maikweki, James 29749
Mailer, Allison 07205
Mailhot, Philippe R. 06804
Maillard, Anne 14056
Maillé, Jacqueline 15234
Maillot, Suzanne 36089
Mailloux, Eleanor F. 49244
Main, Sally 51093
Mainds, Paul E. 44449
Mainella, Fran P. 46327
Mainieri Piedade, Leticia 04714
Mainland, Katrina 43518
MIntosh, George 52398
Mair, Walter 01945
Maire, Marcel 14522
Mairinger, Jutta 02925
Mairot, Philippe 15599
Maisak, Dr. Petra 18370
Maisano, Jack 51150
Maischberger, Dr. Martin 17036
Maisel, Jeff 16959
Maisel, Mag. Thomas 02933
Maistrello, Roberto 26578
Maitilasso, Mario 28103
Maitland, Richard 53387
Maitre-Allain, Olivier 14463
Maiwald, Dr. Christine 18951
Maiwald, Gabriele 20209
Maiwald, Simone 21424
Maixner, Dr. Birgit 21565
Majača, Antonia 08891
Majahad, Carol 51379
Majboroda, Ljudmyla Oleksiïvna 43008
Majburova, Irina Borisovna 36895
Majcen, Jože 38442
Majewska, Grażyna 34988
Majka, Mary 06759
Majnik, Gerlinde 02614
Major, Jürgen 21911
Major, Richard 52125
Majorov, Aleksandr Gennadjevič 36929
Majstorovic, Prof. Božo 08956
Majstorović, Ljuba 38091
Majstorski, Georgi 05248
Majtanov, Bachytžan Kauaskanovič 29680
Majumdar, Prof. Swapan 24379
Makabe, Tadahiko 28787
Makarewicz, Anna 35205
Makariou, Sophie 14698
Makarov, Nikolaj Michajlovič 36832
Makarova, N.B. 37190
Makeeva, I.F. 37907
Maketz-Anderle, Sabine 02237
Makholm, Kristin 50738
Maki, Linda 47592
Makin, Jean 53465
Makkonen, Esko 11274
Makoev, Feliks H. 37764
Makoni, P. 54666
Makovičková, Marie 09665
Makowiecki, Wojciech 35027
Maksić, Nemanja 38151
Maksić, Slobodan 38151
Maksimova, L.D. 37595
Makwana, R.K. 24418
Malahoff, Dr. Alex 33125
Malak, Gregory 47579
Malakar, Diana 48878
Malao, N. 38752
Malaspina, Bartolomeo 28019
Malatare, Marilyn 53538
Malatesta, Giuseppe 25293
Malausssena, Paul-Louis 14427
Malceva, Ljudmila Dmitrievna 37600
Malcolm, John 33986
Malcolm, Megan 33196
Malcolm, N. 33160
Malden, John 43463
Maldent, Dominique 12084
Maldon Fitch, M. Derek 45076
Maldonado, Eduard M. 47467
Maldonado, Laural R. 52442
Maleev, Dr. Michail 05196
Malek, Peter 02781
Malenfant, Robert 06753
Malenke, Patti 47838
Malet Ybern, Rosa María 38957
Malgeri, Dina G. 50405
Malgin, Andrej 43184, 43185
Malgorzewicz, Anna 01030, 01283
Malherbe, C. 38759
Malicky, Michael 02310
Maliet, Vincent 11967
Malik, Sajid Anwar 34025
Malikova, D. 38391
Malinar, Branka 48266
Malinauskas, V. 30225
Malinová, Miluše 09711
Malinverni, Pierluigi 26545
Mališeva, Valentina Pavlovna 37055
Maliszewski, Juliusz Michał 19907
Maljković, Rada 38045
Malkočev, Angel 04982, 04983
Małkowski, Dr. Krzysztof 35240
Mallalieu, William P. 03239
Mallamo, J. Lance 47315
Mallard, Michael 49569
Malley, Diane 44024
Malliet, Muriel 48489
Mallin, Risa 51898
Mallinson, Justin 43687
Mallo Area, Alberto 39625
Malloy, Dean 07105
Malmberg, Brita 41062
Malme, Arnfinn 33628
Malme, Heikki 10769
Malmlöf, Maria 40624
Malmquist, Hilmar 24009
Malnati, Dr. Luigi 26615
Malniča, Andrej 38489
Malone, Gail 06856
Malone, Mary 24741
Malone, Peter 47066
Malonek, Maryla 20556
Malonly, Rachelle 53630
Malov, Nikolaj Michajlovič 37754
Malsch, Dr. Friedemann 30197
Maltais, Marie E. 05840
Maltby, Janet 07047, 07048, 07049, 07050
Maltezou, Prof. Chryssa 28194
Malura, Andrew 22178
Malushi, Petrit 00016
Małuszyńska, Krystyna 34616
Maluțan, Cornelia 36341
Malý, Dr. František 09438
Malý, Karel 09561, 09860
Maly, Reinhold 02058
Malyševa, Tatjana Vasiljevna 37035
Mamaev, Aleksandr A. 36781
Mamalakis, Kostis 23019
Mamaserikova, Šyryn Bukeevna 29710
Mambetova, D. 29962
Mambrini, Prof. Caterina 26263
Mamer, Helen 07314
Mammadov, Emin 03167
Mamuro, Yoshitake 29497
Man, Grigore 36126, 36128
Man, Xueyuan 07476
Manabe, Masahiko 28880
Manakidou, Dr. Eleni 23218
Manakou, Paraskevi 23216
MNaught, William 51587
Mancaș, Ioan 36338, 36346, 36633, 36726
Manceau, Sophie 13628
Manchak, Barry 05758
Mancinelli, Dr. Fabrizio 54497
Mancini, Prof. Franco 27981
Manconi, Dr. Dorica 27147
Manconi, Francesca 27304
Mancusi-Ungaro, Carol 51279
Mancuso, Francesco 27067
Mandanas Esquito, Joven 34443
Mandeville, Jeffrey 50066
Mandeville, Joyce 50824
Mandic, Dr. Davor 08935
Mandis, Alexander 22888, 22894
Mândran, Ion 36586
Mândru, Doina 36479
Manera, Brigida 14787
Manes, John 48660
Maness, David C. 51337
Manetsch, Lic. phil. Letizia 41217
Manfield, Nigel 45452
Manfredini, Prof. Alessandra 27486
Manfrina, Mario 41493
Mangado, Denis 12927
Mangan, Dianne 01627
Mangan, Richard 44912
Mangani, Elisabetta 27519
Mangels, W. Gilbert 52015
Mangia, Dr. Paola 27503
Mangiagalli, B. 38744, 38779
Mangiagalli, Tizzie 38744
Mangion, Eric 13990, 14442
Mangnus, G. 32593
Mangold, Dr. Meret 41203
Mangucci, Celso 35583
Manguso, John M. 48678
Manhart, Patty 48787
Manhès, Bruno 13462
Mania, Dr. Thomas 18774
Manibusan, Rose 23288
Manicastri, Dr. Claudio 27468
Manieri Elia, Dr. Giulio 28184
Maniez, Erica S. 49544
Maniguet, Thierry 14656
Maninger, Peter 18541
Manini, Paula 53566, 53567, 53569
Maniurka, Dr. Piotr 34985
Manka, Riitta 11185
Manke, Elke 20965
Mankin Kornhauser, Elizabeth 49199
Mańkowska, Ewa 34831
Mankus, Edward 47456

M

Mann, David 54109
Mann, Esther 44906
Mann, George E. 51572
Mann, Harald J. 21168
Mann, Dr. James 49971
Mann, Jim 49161
Mann, Judith 52595
Mann, LaDonna 54000
Mann, Dr. Stephan 18666
Mann, Theresa 53196
Mann, Vivian 51200
Mannberg Wretin, Karin 41012
Manne, Robert 52969
Manneby, Dr. Hans 40992
Mannerheim, Lowissa 40530
Manni, Dr. Riccardo 27498
Manning, Anne 46511
Manning, Dr. P. 45090
Mannini, Maria Pia 27334, 27339
Mannion, Dr. Susan 24733
Manns, John 01666
Mano, Livio 26019
Manojlović, Miroljub 38132
Manoli, Britt 13409
Manring, Yvonne 54091
Mansell, Jefferson 46669
Manser, Jurg 41587
Mansfield, Dr. Bernard M. 48838
Mansfield, Lisa 51160
Mansfield, Nicholas 45096
Manske, Beate 17567
Manson, John 43283
Manson, Terry 33227
Manson, Wayne 49805
Mansour Naguib, Sylvia 46446
Mansperger, Linda W. 51619
Manssen, W.J. 31951
Mantel, Ken 44935
Mantovani, Prof. Giuseppe 28039
Mantzke, Irmgard 19616
Manuel, R.S. 06584
Manuelas, Paul 50729
Manuhutu, W. 32848
Manus, Dieter 19336
Manyon, Ellen 54121
Manzanares, Rick 48621
Manzanilla, Dr. Linda 31103
Manzato, Prof. Eugenio 28071
Manzi, Prof. Giorgio 27491
Manzini Gazzani, Giorgio Mario 08716
Mao, Liqiang 07862
Mao, Peiqi 07552
Mao, Yanwen 07943
Mao, Yuehong 08393
Mapfarakora, Jacques 05281
Maphet, L. Ernestine 48871
Maple, Dr. William 50944, 50945
Maples, Philip G. 52392
Maquelin, Chloé 41681
Maquet, Nadine 03807
Maquiso, Dr. Teresita R. 34463
Mara, Kanfela 23315
Mara, Kevin 47953
Mara, Maria 13085
Maračić, Antun 08874
Marafioti, Dr. Sergio 27085, 27086, 27087
Marais, Caroline 44960
Marais, Caroline de 48257
Maráky, Dr. Peter 38234
Marandel, J. Patrice 50244
Marandu, Gallus 42510
Marangon, Maria 23168
Maraniello, Prof. Gianfranco 25450
Marapana, P.B. 40381, 40382
Maraszek, Dr. Regine 18900
Maraval, P. 11917
Maravelaki, Noni 22961
Marazzi, Laura 28162
Marbach, Wilhelm 17709
Marble, Lori L. 51004
Marc, Denis 16029
Marc, Dorel 36700
Marcaccini, Ignazio 26070
Marcantel, Gregory 49621
Marcato, Prof. Paolo Stefano 27026
Marcel, Louis 13193, 13522
Marcet, Roger 38998
March and Kinrara, Earl of 43827
Marchado, Philippe 13472
Marchand, Charlotte 16140
Marchand, Lionel 41795
Marchand, Scott 53595
Marchant, Dr. R. 41542
Marchant Smith, Dick 46018
Marchante, Carlos M. 09145
Marchbank, Frances A. 49765
Marchbank, James 07029
Marchenko, Leonid V. 43178
Marchesi, M. 32729
Marchesi, Dr. Marinella 25447
Marchesseau, Daniel 14665
Marcheville, de 13533
Marchini, Gian Paolo 28237
Marchione, Kenneth 53249
Marchois, Bernard 14705
Marciari, John 52752
Marciari Alexander, Julia 52752
Marciniak, Leszek 35347
Marció, Pietro 26856
Marco, Sixio 39265
Marcon, Gaby 14210
Marcos, Dr. Dieter 19587
Marcos, Ludovic 14653
Marcos, Mariano 15861
Marcos V., Emilio 39759
Marcoux, Charles 41724
Marcovici, Vladimir 36687
Marcu, Aurora 36378
Marcu, Maria 36261
Marcum, Donald 48242
Marcum-Estes, Leah 51458
Marcus, Evelyn 49162
Marcus, Jed 51256
Marcus, Joyce 46308
Marcus, Mette 10103
Marcus, Susan 47514
Maréchal, Dominique 03463
Marek, Danuta 34676
Marek, Stanisław 34963
Marenco, Jean-Jacques 15110
Marengo, Simeone 28032
Mareos Ruiz, P. 38919
Marepo Eoe, Soroi 34080
Mares, Dr. Michael A. 51376
Marešić, Jagoda 08954
Maret, J.F. 32543
Maretzky, Dieter 19626
Marfà i Riera, C. 39631, 39633
Marghieri, Roland 14439
Margita, Laura 06922
Margkudilaga, Sufwandi 24502
Margles, Judy 52083
Margos, Rina 03851
Margrey, Peter 47238
Margulies, Stephen N. 47400
Mariani, Prof. Franco 26069
Mariani, Giuseppe Antonio 25875
Máriássy, Eva-Maria von 18753
Mariátegui Ezeta, José Carlos 34207
Marić, Dušan 31636
Marichal, Rémy 07163
Maričić, Dušanka 38091
Maricich, Andrea 49267
Maricki-Ostojić, Vesna 38151
Maries, John 43903
Marijan, Jassmina 38457
Marin, Prof. Dr. Emilio 08910, 08954
Marin, Jean-Yves 12231
Marin, Dr. Jean-Yves 41400
Marin, Jean-Yves 41426
Marin, Jeanne 15140
Marin, Lăcrămioara 36233
Marin Torres, Dr. Maria Teresa 39702
Marín Vázquez, Gilberto 31051
Marinac, Bogdana 38492
Marinazzo, Dr. Angela 25545
Marincig, Harald 03091
Marincola, Michele 51152
Marinelli, Dr. Giampiero Gabrio 26286
Marínez García, Luisa 38825
Marini, Gabriele 26455
Marini, Dr. Giorgio 28235
Marini, Mariagrazia 27772
Marini, Marinella 27772
Marini, Dr. Paola 28233, 28235
Marinis, Dr. Giuliano de 25227
Marino, Christy 46679
Marino, Mari 04883
Marino, Ruta 52530
Marinoiu, Vasile 36236, 36399, 36663, 36664, 36665, 36666, 36733, 36734
Marinov, Evstati 05261
Marinska, Dr. Rouja 05184
Mariotti, Prof. G. 27495
Marits, Ülle 10573
Mark, Guido 02871, 02896
Mark, John 48343
Mark R., Thompson 53180
Markel, Stephen A. 50244
Markelova, Nina Konstantinovna 37435
Marker, Peggy 50748
Markert, Leo 18808
Markey, Teresa 48308
Markham, Philip 33252
Markiewicz, Katarzyna 35058
Markina, Natalja Petrovna 37835
Markishtum, Merlee 53011
Markko, Helvi 10862
Markland, Eleanor 45048
Markley, Penny 49129
Marko, Dr. Eva 01979
Markoe, Glenn 47555
Markoulaki, Stavroula 22961
Marková, Radka 09727
Markovčič, Albin 38497
Marković, Mirjana 38145
Marković, Prof. Slavica 09001
Markovič, Srdjan 38114
Markowitz, John 51770
Markowitz, Zev 47070
Marks, Dede 46110
Marks, Edward 53289
Marks, Jacqueline 22524
Marks, John 47479
Marks, John C. 48880
Marks, Katherine Duer 50123
Marks, Michael 22524
Marks, Michael J. 49294
Marks, William 48635
Marley, Rita 28324
Marlow, Martha 48190
Marlowe, Dean 53176
Marloyannis, Georgi 23144
Marmolejo, Martha Lucía 08637
Maroević, Igor 08977, 08987
Marović, Metodija 08986
Marøy Hansen, Arild 33423
Marquardt, Claudia 20786
Marquardt, Jean F. 53444
Marquardt, Dr. Markus 16582
Marquart, Dr. Markus 16587, 16592, 16593
Marques, António Fernando 35585
Marques, Luiz 04880
Marquet, Helmut 03272
Marquet, Jean-Claude 13587, 14021
Marquet, Sybil 15248
Marquete, Ronaldo 04703
Márquez Martínez, Teresa 31140
Márquez Rojas, Margoth 08671
Marquis, Jean M. 41280
Marr, Louise 45624
Marr, Matthias 19523
Marre-Muls, Anne 03510
Marrin, Jim 53696
Marriott, Janine 43400
Marriott, Wayne P. 33153
Marsac, Michel 15299
Marsan, Geneviève 13852
Marsch, Peter 33262
Marschall, Jolanta 35080
Marsden, Dale 50348
Marsden, Melinda 49121
Marsden, Peter 44417
Marsden, Susan 06670, 06671
Marsden-Atlass, Lynn 51864
Marseille, Jeremias 25128
Marsh, Joanna 49199
Marsh, John S. 51788
Marsh, Natalie 47750
Marsh, Tracy 05632
Marsh, Warwick 00796
Marsh-Matthews, Dr. F. 51376
Marshall, Alan 13885
Marshall, Ann E. 51904
Marshall, Betty 47430
Marshall, Dr. Cora 51028
Marshall, Danny 52552
Marshall, Francis 45444
Marshall, Jerry 50930
Marshall, Karen 49832
Marshall, Dr. Larry 50605
Marshall, Marjorie 05578
Marshall, Robert A. 49114
Marshall, Ron 46251
Marshall, Steven 45055
Mársico de Larribité, Viviana 00137
Marson, Luigi 28299
Marston, Elizabeth 48334
Marsure-Rosso, Danila 25376
Marszalec, Roman 34825
Marszałek, Zygmunt 35243
Marszewski, Tadeusz 35082
Martel, Philippe 12378
Marteljanova, Elisaveta Vladimirovna 37527
Martens, Caspar 32281
Martens, H.E.D. 31929
Martens, Jean 03375
Martens, Melissa 46571
Martens, Ralph 41395
Martens, Warburg 16443
Martensen, Robert 46775
Martensen, Robert L. 49680
Marth, Dr. Regine 17522
Marthalie, Furber 46217
Martí, Maria 00105
Martí Aixalà, Joseph 40100
Martí Armengol, Teresa 38957
Martí Bonet, José María 38991
Martí i Llambrich, Miquel 39768
Martig, Prof. Dr. Peter 41498
Martin, Albert 06823
Martin, Alvia D. 50533, 51533
Martin, Amanda 45584
Martín, Andrés Peláez 38852
Martin, Andy 01648
Martin, Anna 51227
Martin, Anthea 38602
Martin, Antoinette 49675
Martin, Aurora 39343, 40180
Martin, Carol 07096
Martin, Danny 45207
Martin, Darryl 44147
Martin, Erin L. 51043
Martin, Êtienne 15828
Martin, Eva 20786
Martin, Fabian 36686
Martin, François 12432, 12433, 13391, 15159, 16077
Martin, Frank 51582
Martin, Glynn B. 50247
Martin, Jacques 05763
Martin, James E. 52253
Martin, Jean-Claude 11547
Martin, Jenni 52805
Martin, Jennifer 50854
Martin, Jerry 54276
Martin, JoAnne 05349
Martin, Dr. Joanne M. 46566
Martin, Jörg 21868
Martin, Juliet 51904
Martin, Larry 49991
Martin, Loretta 53568
Martin, Dr. Lutz 17279
Martin, M. 44679
Martín, Macarena Alés 39464
Martin, Marilyn 38593
Martin, Mary 48395
Martin, Maurice 13234
Martin, Dr. Michael 19792
Martin, Michael 51559
Martin, Otto-Walter 17317
Martin, Pascal 14267
Martin, Paul 48604
Martin, Pierre de 15808
Martin, Rebecca J. 50985, 53536
Martin, Richard 48051
Martin, Rick 48395
Martin, Robert 45618
Martin, Roger 05428
Martin, Ros 33264
Martin, Sandra 45088
Martin, Dr. Shéonagh 44125
Martin, Stéphane 14700
Martin, Dr. Sylvia 19705
Martin, Terry 48809
Martin, Theo 41382
Martin, William 54106
Martin, William J. 52339
Martin, Wilson 52697
Martín Flores, Alfonso 39550
Martín Merás, Luisa 39574
Martín Pascual, Crescencio 39638
Martin-Vignes, Nicole 11517
Martina, Dr. Alessandra 26356
Martindale, Bruce 07066
Martindale, Wendy 06709
Martínek, Jaroslav 09555, 09663, 09875
Martinelli, Massimo 26430
Martinelli, Dr. Roberta 27308, 27309
Martinet, Jean-Louis 13475
Martínez, Dr. Armando Luis 31097
Martinez, Connie 52805
Martinez, Daniel 49356
Martínez, Gonzalo 08647
Martinez, Jean-Luc 14698
Martínez, José Luis 31107
Martínez, José Roberto 36063
Martínez, Margarita 54530
Martinez, Pita 48083
Martínez, Ramiro 31150
Martinez, Ray 46129
Martínez, Ricardo Ariel 08722
Martínez Alvárez, Dr. José María 39646
Martínez Arvizu, Jesus 31225
Martínez Cañabate, Juan Carlos 39272
Martinez Carril, Manuel 54391
Martínez Cortés, Ariel 08748
Martínez de Aguilar, Dr. Ana 39566
Martínez Fuentes, Dr. Antonio J.F. 09110
Martínez Mestre, Miguel 31114
Martínez Rodríguez, Andrés 39492
Martínez Sarria, Victoria 08728
Martínez Vargas, Enrique 31013
Martini, Prof. Dr. Erlend 21076
Martiniani-Reber, Marielle 41418
Martínková, Jana 09662
Martino, Giuseppe 26039
Martino, Mike 52434
Martino, Dr. Ronald L. 49434
Martinović, Jovan 31647
Martins, Helga 21707
Martins, Josephine 49717

M

Martins, Michael 48499
Martins Carneiro, José Manuel 35951
Martins Catharino, Henriqueta 04804
Martins e Silva, José 35693
Martins Fernandes, Francisco R. 35637
Martinson, Gordon 47002
Martinson, Helle-Reet 10629
Martis, Ela-Heigi 10659
Martis, Val 47456
Martone, Dr. Patrick 07199
Martonová, T. 09519
Martorell, Jaume 39776
Marty, Hanspeter 42083
Marty, John F. 51045
Marty, Theodor 41702
Marty-Francart, Marie-Jo 15296
Martykánová, Marie 09887
Martynjuk, Oleg Ivanovič 29721
Martynova, Nina Petrovna 37038
Martynova, Raisa Fëdorovna 36955
Maruschke, Thomas 17504
Marusic, Irena 38467
Maruszeczka, Dave 05992
Maruyama, Mariko 28447
Maruyama, Masatake 29488
Maruyama, Tamiko 29522
Marvin, Anne 53057
Marwell, Dr. David G. 51227
Marwick, Fiona C. 44249
Marx, Dr. Erich 02608, 02625, 02627, 02629
Marx, Günther 22249
Marx, Dr. Petra 20512
Marx, Prof. R. 38677
Mary Bess K., Paluzzi 46219
Marzahn, Dr. Joachim 17279
Marzatico, Dr. Franco 25410, 27901, 28054
Marzio, Frances 49397
Marzio, Peter C. 49381, 49397
Marzolf, John G. 46487
Mas, Monique 14291
Más Vilaplana, Antonio 39467
Masa, Karin 41850
Masaaki, Ueda 28829
Masafumi, Shimooka 28750
Masahide, Maekawa 29211
Masako, Sakashita 28765
Masanek, Wolfgang 21165
Masanori, Komiya 28696
Masaracı, Yasemin 42839
Masato, Ueno 28876
Mašátová, Ivana 09408
Masau Dan, Dr. Maria 28084
Masayuki, Ogata 28413
Mascarenhas, Dr. Aubrey A. 24317
Mascheroni, Silvia 25896
Mascia, Sara 53452
Maselli Scotti, Dr. Franca 25253, 25254, 28075
Masemore, Dr. Ira 50155
Masen, Larz 50738
Mashhadi, Syed Gulzar 34023
Masiero, Elena 27574
Masing, Maia 10638
Masino, H. 22284
Masjoan, Prof. Estela 00273
Maske, Andrew 52657
Maslić, Fatima 04286
Maslov, V.I. 37388
Mason, B. 01152
Mason, Bonnie N. 51654
Mason, Carl 46328
Mason, Darielle 51867
Mason, Dorothy 48501
Mason, Erik 50193
Mason, G. 45227
Mason, Gina 52557
Mason, Julia 43602
Mason, Dr. Kevin 43546
Mason, Laurie 06383
Mason, Michael E. 48301
Mason, Ngahirika 33042
Mason, R.P. 44150
Mason, Richard 45973
Mason, Wally 46598, 51603
Mason, William J. 49115
Mason-Chaney, Lisa 48364
Masoodi, Omara Khan 00005
Masriera, Dr. Alicia 38976
Masriera, Alicia 38977
Massada, Satê 29661
Massagra, Dr. Maria Grazia 27478
Massari, Dr. Stefania 27515
Massaro, Marilyn 52162, 52171
Massart, Claire 03495
Massart, Roger 03369
Massas, Natacha 03495
Massé, Eric 15135
Massenbach, Elisabeth von 20461
Massey, Kristine 53592
Massey, Lewis 46451
Massey, Rachel 45622
Massey, Timothy 47035
Massicotte, Jean 06223
Massie Lane, Rebecca 53391
Massier, John 47121
Massimbe, Julieta 31710
Masso, Kaisa 10623
Massó Carballido, Jaume 39894
Massobrio, Dr. Giulio 25193
Måsson, Anette 10237
Masson, Colette 06122
Masson Rusinowski, Fabienne 03701
Massotti, Jeffery R. 52174
Massy, Jean-Baptiste 41990
Mastemaker, Lorie 49146
Master-Karnik, Paul 50090
Masterson, Dorothy 47936
Masterson, Judy 48461
Mastorgio, Carlo 25287
Mastorito, Susana 00362
Mastrogiuseppe, Joy 52198
Mastrovic, Mikica 08992
Masuda, Hideaki 28502
Masuda, Rei 29491
Masuda, Yoko S. 29051
Mašulovic, Branimir 38148
Masuo, Tano 28741
Masuyer, François 12702
Mata, Dr. Aida Maria Reis da 35766
Mata, Rene Luis S. 34454
Mata González, Juan C. de la 39586
Matajis, Andrijana 38074
Matakieva, Prof. Dr. Teofana 05191
Matamoros, Joséphine 12340, 12575
Matarazzo, Eva 00264
Matasci, Mario 41930
Matat, Leonid Nykonorovyč 43077
Mataušić, Nataša 08997
Mateescu, Ilie 36631
Matei, Rodica 36219
Matejková, Adriana 38193
Matenga, Edward 54675
Mater, B. 31937
Maternati-Baldouy, Danielle 14009
Materón Salcedo, Edgar 08615
Maters, Drs. Petra 32872
Matešic, Suzanne 21565
Mateu y Miró, Josep 39145
Mateva, Borjana 05040
Matharan, Paul 12066
Matheis, Adam 20138
Matheny, Paul 47734
Mathes, Wayne 51463
Matheson, Susan 51056
Matheson, Trueman 06191
Mathews, George K. 51337
Mathews, Greg 51058
Mathews, Nancy 49734
Mathews, Susan 44174
Mathewson, Jon 50652
Mathez, Dr. Edmond A. 51115
Mathias, Gérard 12915
Mathiassen, Tove E. 09963
Mathie, William 48290
Mathieu, Caroline 14692
Mathieu, Frederic 13995
Mathieu, Dr. Kai R. 18570, 21737
Mathis, Bernhard 41351
Mathis, Brandi 46956
Mathis, Walti 41592
Mathison, Judy K. 53762
Mathusa, Parker D. 53031
Matijaca, Ivo 08897
Matijašić, Prof. Sonja 08927
Matijevič, Jožef 38490
Matile, Michael 42073
Matilla, José Manuel 39573
Matilsky, Barbara 47344
Matino, Dr. Lucia 26717
Matiskainen, Dr. Heikki 11268, 11270
Matits, Ferenc 23350
Matland, Susan 33635
Mato, Nancy 51678
Matos Moctezuma, María Fernanda 30889
Matos Reis, Dr. António 35984
Matoušková, Martina 09803
Matrica, Vasilica 36104
Matschiner, Dieter 02690
Matskási, Dr. István 23445
Matson, Jaff 49360
Matson, Dr. Timothy 47625
Matsubayashi, Hosai 29557
Matsuda, Hiroshi 28537
Matsuda, Motoi 29103
Matsuda, Yoshiho 28458
Matsumoto, Takeshi 28953
Matsumoto, Tohru 29491
Matsuo, Nobuhiro 29140
Matsuoka, Masao 28983
Matsuoka, Mieko 29427
Matsura, Akira 28519
Matsushita, Kensuke 28349
Matsuura, Kunio 29284
Matsuzaki, Yukiko 29392
Matsuzawa, Shigefumi 29600
Matsuzono, Makio 29295
Matt, Edvin 49999
Matt, Dr. Gerald 03017, 03018
Mattausch-Schirmbeck, Dr. Roswitha 16751
Matteotti, Mario 27421
Matter, K. 41696
Matter, Kuno 41839
Mattes, Glenda 48165
Matteson, Judith M. 48628
Mattet, Laurence 41417
Matteucci, Prof. Ruggero 27498
Matthäi, Bernhard 19613
Matthaes, Gottfried 26693
Matthes, Lilian 10049
Matthes, Olaf 18961
Matthew, A.T. 38620
Matthew, Malcom 52727
Matthewman, Su 46105
Matthews, Dr. B. 51376
Matthews, Dr. B.E. 44123
Matthews, Bert 50966
Matthews, Christy 48109
Matthews, Claire 50934
Matthews, Keith 46099
Matthews, Marsha 52082
Matthews, Nancy 53892
Matthews, Ronald 49793
Matthews Hoffmann, Catherine 50063
Matthies, Wilhelm 19974
Matthiesen, Kjell Marius 33637
Matthiessen, Joy A. 48104
Matthieu, Bernard 15658
Matti, Arnold 42127
Matticks, Rebecca 48311
Mattila, Karen 06398
Mattioni, Fernando 27216
Mattl-Wurm, Dr. Sylvia 03087
Mattle, Anton 01920
Mattlin, Håkan 10722
Mattner, Daphne 17688
Matton, A. 03679, 03680
Matton, Ann 03677
Matton, Olivier 06493
Mattox, Boyd 53459
Mattsson, Christina 40922
Matukova, Marija Bajramovna 36980
Matuschka, Dagmar Gräfin von 21524
Matuska, Mike von 40900
Matuss, Dr. Márta 23533
Matuszczyk, Andrzej 34816
Matuszewski, Nina 19626
Matveeva, Tatjana Sergeevna 36770
Matvenov, Evgenij Rostitovič 36913
Matyniak, Margaret 53833
Matz, Darrell 53415
Matzke, Klaus 21559
Matzke, Dr. Michael 41161
Matzon, Ea 10281
Maubano, M. 13595
Mauch, Michael 41163
Maucq, R. 03572
Maucuer, Michel 14638
Maudet, Serge 30194
Mauer, Dr. Alexander 25633
Mauerhan, Joëlle 11959
Mauersberger, Dr. Klaus 17955, 17975
Mauersich, Angelika 02799, 02800
Maugain, Gregory 14289
Maugham, Joan 00882
Mauk, Dannielle 48805
Maul, Dr. Bärbel 21438
Mauldin, Dr. Barbara 52882
Mauldin, Barbara 52883
Mauler, Norbert 01990
Maulsby, Richard 46359
Maumont, A. 15946
Maung, U Min Khin 31722
Maunoury, Catherine 13564
Mauoloux, Herman 04048
Maurer, Eleonore 02078
Maurer, Prof. Dr. Friedemann 19050
Maurer, Gudrun 39573
Maurer, Lena 02951
Maurer, Dr. O. 31814
Maurer, Dr. Philipp 03011
Maurer, Dr. Rudolf 01776
Maurer, Sherry C. 52405
Maurer, Simon 42077
Mauri, Dr. Gemma 26217
Maurice-Chabard, Brigitte 11764
Maurin, Ange 14956
Maurin, Bernard 15623
Maurin, Ray 49528
Maurirere, Naomi 33109
Maurizio, Dr. Remo 41902
Mauro, Dr. Marta 26220
Mauron, Christophe 41272
Mautner-Markhof, Dr. Marietta 02929
Mauzi Martí, Alfred 39330
Mavric, Kaye 00993
Mavrodiev, Zdravko 05202
Mavrotas, Takis 23007
Mavuso, Makana R. 40402
Maw, Dr. Martin 45337
Mawdsley, Simon 33467
Mawe, Theodore 34081
Mawleke, M. 38666
Maxfield, Dorothy 54212
Maxfield, Mike 45091
Maxham, R.S. 48677
Maxheimer, Greta 48035
Maxholmes, Thomas 01686
Maxim, Dr. Elena 36203
Maxwell, Jennifer 45788
Maxwell, Robert N. 49463
Maxwell, Robyn 00946
Maxwell Stuart, Catherine 44517
May, Denis 01582
May, Dr. Elisabeth 18860
May, Guy 30493
May, Harry 06664
May, Jane 44681, 44682
May, Jörg 17525
May, John M. 47703
May, Sally 01412
May, Thérèse 30493
May Dobbins, Melissa 48490
Mayberry, Matthew 47700
Mayer, August 02711
Mayer, Dr. Gregory C. 50395
Mayer, Dr. Günter 02915
Mayer, Hedwig 01830
Mayer, Herbert 02984, 03045
Mayer, Jannie 14694
Mayer, Klaus 22443
Mayer, M. 20279
Mayer, Olafa 20233
Mayer, Ronald 53167
Mayer, Walter 47116
Mayerhofer, M.A. Susanne 17673
Mayes, Catherine 48227
Mayes, James 52427
Mayes-Wright, Virginia 44254
Mayfield, Signe 51688
Mayhew, Kim 51297
Mayland-Mason, Paul 44677
Mayle, A.H. 44853
Maynard, Helen B. 48893
Maynard, Marie-Noëlle 12285
Maynard, Rex 46726
Mayo, Dr. Margaret Ellen 52343
Mayo, Martha 50305
Mayo, Marti 49383
Mayor, Grégoire 41681
Mayor, N. 12112
Mayorga Sanchón, D. Juan 40135
Mayou, Roger 41423
Mayr, Karin 18619
Mayr, Manfred 17852
Mayr, Peter 02923
Mayr, Thomas 21781
Mayr, Walter 02554
Mayr-Oehring, Dr. Erika 02624
Mayrhofer, OSB Martin 02738
Mayrhofer, Meinrad 02523
Mayrhofer, Monika E. 03126
Mayrhuber, Petra Gerti 02890
Mayrock, Roger 19511
Mays, N. 00817
Maysonnave, J. 16309
Maytk, Paul 54015
Maywald, Dr. Claus 20143
Maza, Roberto Ramos 31517
Mazeika, Dr. Rasa 06312
Mazepov, Jurij Ivanovyč 43177
Maziarka, Włodzimierz 34693
Mazibuko, Elizabeth 38618, 38619
Mazitova, Svetlana Vladimirovna 37064
Mazlish, Anne 53153
Mazlumoğlu, Hasan 42776
Maznev, Kostadin 05061
Mazonowicz, Douglas 51182
Mazor, Robin 49706
Mazourek, Paul 01649
Mazow, Dr. Leo 53659
Mazur, Leonid Viktorovič 37979
Mazur, Tracey 52862
Mazza, Germana 25695
Mazza, Thomas C. 51153
Mazzanti, Laura 48587
Mazzari, Roberto 25433
Mazzei, Dr. Lapo 26175
Mazzei, Marina 26587
Mazzeo, Dr. Donatella 27512
Mazzerioli, Paolo 27149
Mazzini, Dr. Laura 26403
Mazziotti Gillan, Maria 51728
Mazzocchi, G. 27577
Mazzolti, Prof. Mario 27370
Mazzoni, Anne-Hélène 15094
Mazzoni de Bravo, Helga 00286

M

Mazzotta, Gabriele 26684
Mazzotti, Dr. Stefano 26108
Mazzotti, Tullio 25183
Mbe, John C. Jacques 45643
Mbodji, Abdoulaye 38028
Mbokashanga, Nyimi Kuete 08803
McAlister, R. 44881
McAlpine, Sue 44868
McCusker, Carol 52741
McDonnell, Dr. Patricia 54140
McFadden, J.J. Kegan 07373
McGarry, James R. 49766
McIntosh, Jane 48560
McIntyre, Dr. Nancy 50312
McKillop, Beth 45004
McLean, Al 45664
McNeill, Richard 07099
McPhail, Dr. J. Donald 07199
McPherson, Tommy A. 50769
Meabon, Linda 46989
Meachum, Gil 52061
Mead, Tina 53932
Meade, C. Wade 52491
Meade, Diane 48236
Meade, John 50004
Meador-Woodruff, Robin 46307
Meadors, William 50168
Meadows, Larry G. 47597
Meagher, George E. 47356
Meagher, Kathy 05837
Meagher, Tom 06197
Mealing, David 33127, 33254
Mealings, Bob 44336
Mears, Peter 46510
Mearwah, Reena 24198
Mecham, Denny 46812
Mechan, Dallas 43662, 43681, 44609
Mechell, Barbara 53777
Mechler, Sabine 19082
Mechler, Dr. Wolf Dieter 18993
Mecking, Dr. O. 22410
Mecklin, Kai 11376
Mecredy, Stephen D. 06077
Medaković, Anica 38163
Medard-Blondel, Anne 14011
Medeiros, Francisco 35653
Medek, Prof. Robert 02971
Meder, Katrin 17523
Mederski, Mark 51913
Medhi, Prof. Birinchi K. 24186
Medhurst, Alex 45664
Medica, Dr. Massimo 25449
Medici, Felix Nazar 00134
Medicus, Mag. Christine 02615
Medina, Dennis H.J. 46120
Medina i Montes, Susana 40120
Medinger, Bob 53126
Medley, Mark 50964
Medlock, Leon 52020
Medlycott, Sir Mervyn 45610
Medri, Prof. Gianluca 26078
Medri, Dr. Litta 26163, 27268
Medvedev, Nikolaj Jurjevič 37624
Medvedow, Jill 46933
Medzihradská, Dr. Elena 38287
Mee, Dr. Jane 45435
Meehan, Brian 06183, 06189
Meehan, Joseph M. 48550
Meek, Kenneth 46620, 46620
Meek, Noel 33104
Meeke, Dr. Noel 44460
Meekers, Christa 03687
Meena, K.L. 24242
Mees, Dr. Allard 20148
Meester de Heyndonck, B. de 03334
Meeusen, M. 04076
Meeuwsen, Jeffrey 49006
Meewis, Menno 03292
Meglio, Luciano di 26411
Mehaska, Dr. Walter 49181
Mehler, Sarah 42115
Mehmann, Moritz 41084
Mehra, Sandeep 24150
Mehren Gorev, Leslie 52614
Mehring, Albert G. 49181
Mehring, Karen 48581
Mehring, Melanie 21192
Mehta, Jawahar G. 24064
Mei, Runxin 07606
Meiborg, E. 32555
Meidell, Bjarne A. 33431
Meidenbauer, Roy 51017
Meidinger, Ronald J. 46392
Meier, Andreas 41721
Meier, Anton C. 42047
Meier, Dr. Anton M. 41454
Meier, Bernhard 41258
Meier, Ernest 41560
Meier, Heinz 41349
Meier, Herbert 18315
Meier, Jakob 42009
Meier, Jürgen 21198
Meier, Michelle 51277
Meier, Rupert 41188
Meier, Stefan 41158
Meier, Wolfgang 18336
Meighörner, Dr. Wolfgang 02136
Meijer, M. 31929
Meiler, Olaf 18733
Meillat, Eric 11563
Meillat, Joël 14211
Meillon, Anne 13149
Meilutienė, Gražina 30348
Meinderts, Drs. A.J.M. 32074
Meindre, Colette 13827, 15142
Meinelschmidt, C. 19228
Meiners, Dr. Jochen 17683
Meiners, Prof. Dr. Uwe 17713
Meinig, U. 16640, 16641
Meir, Cecilia 25112
Meire, Martine 03935
Meireles, Alexa Bastos Gambogi 04391
Meiring, Barbara 48674
Meiring, Jacob 41306
Meirose, Jan-Dirk 18057
Meirovics, Prof. Imauts 30082
Meisch, Jim 30484
Meisser, Dr. N. 41542
Meissimilly-Rosin, Martine 16134
Meissner, Dipl.-Ing. Irene 20423
Meissner, Roman 35029
Meißner, Uwe 18905
Meissonnier, J. 12824
Meister, Arthur 41265
Meister, Mark 47982, 47989
Meister, Nicolette 46724
Meister, Pamela 47386
Meisterman, Ingrid 40454
Meiszies, Dr. Winrich 18010
Meixner, Brigitte 21150
Mejdahl, Ulla 10001
Mejer, Robert Lee 52214
Mejia, Javier 08747, 08754
Mejía Maya, Rafael 08783
Mekas, Tauras 30255
Méla, Prof. Dr. Charles 41308
Melamed, Nuril 25113
Meland, Marte 33493
Melandri, Leonardo 25408
Melandri, Lisa 52901
Melcher, Dr. Ralph 21459
Melchert, Dr. Monika 17033
Melchionne, Kevin 51884
Melchior, Dr. Christoph 22739
Meldgaard, Bjarne 10070
Mele, Dr. Marko 01985
Meleško, Dr. V.P. 37698
Melgarejo Cabrera, Elizabet 09286
Meli, Pietro 05592
Melichar, Dr. Peter 01796
Meline, Anne 52540
Melisi, Prof. Francesco 26888
Melissa, Talachy 52886
Melke, Gunta 30000
Melki, Dr. Loutfalla 30161
Melkonyan, Hasmik 00728
Mella, Joseph S. 50980
Mellander, Barbo 40410
Mellander, Barbro 40655
Mellby, Julie 51033
Mellbye, Maren-Otte 33916
Mellemsether, Hanna 33929
Meller, Dr. Harald 18900
Mellinger, A. Clair 49186
Mello, Luiz Carlos 04728
Mellon, Lori A. 48088
Mellor, Jackie 45754
Mellott, Tina 47029
Mellow, Ralph 05634
Melly, Anne-Lyse 41990
Melly, R.G. 44858
Melman, Herman 53770
Melničuk, Andrej Fëdorovič 37548
Melnikova, Iraida Fëdorovna 29728
Melnikova, Natalija Ivanovna 36939
Melnikova, Tamara Michajlovna 37100
Mel'nyk, Anatolij Ivanovič 43100
Melo, Jorge Orlando 08549, 08551
Melo, Dr. Manuela 35844
Melo Miranda, Selma 04783
Melody, Paul 33139
Melomina, Olga Alekseevna 36782
Meloni, Alberto C. 52565, 53950
Melrose, A. 01159
Melton, Babs B. 53330, 53331, 53332, 53333
Melton, R. Neal 53040
Meltzer, J.L. 38594
Meltzer, Lalou 38591
Melzer, Ellen 18752
Melzer, Gustav 02506
Melzer, Dr. Miloš 09661, 09891, 09924
Melzer, Dr. Walter 21795
Memoire, Natalie 12074
Mena, Manuela 39573
Mena Segarra, Prof. Enrique 54404
Menaker Rothschild, Deborah 54176
Menalo, Romana 08860
Ménard, Jean 15401
Menard, Jean-Daniel 15908, 15909
Menardi, Dr. Herlinde 02137
Menasse-Wiesbauer, Dr. Elisabeth 03093
Menato, Prof. Giuliano 28136
Menconeri, Kate 54283
Mencucci, Angelo 27804
Mendel, Shlomoh 25088
Mendelssohn, Dr. Gabriele 19335
Mendelssohn, Prof. H. 25124
Mendenhall, Winnifred 47669
Mendes, Albano 23317
Mendes Bürgi, Dr. Bernhard 41166, 41170
Mendes Gonçalves, Maria Arlete 04732
Mendes Marques, Dr. Maria da Concerção 35642
Mendez, Ivan 36076
Méndez, Dr. Marion 18768
Méndez Lecanda, Dr. Elia Josefina 31151
Méndez López, Nohra Isabel 08632
Mendizabal, Tomás 34065
Mendoza Ramírez, Patricia 31062
Menefee, Antonio 46928
Menegazzi, Dr. Alessandra 27042
Meneguzzo, Giobatta 26577
Meneses, Adriana 54554
Menezes Gomes, Geraldo de 04888
Meng, Fancai 07911
Meng, Jianyao 08072
Meng, Qingjin 07639
Meng, Ung Vai 07997
Menge, Heide 17564
Mengede, Dr. Bernd 18253, 18255
Mengele, Mark 53362
Mengis, Tom 50338
Menis, Gian Carlo 28113
Menk, Jeanette 21005
Menke, Hubertus 19404
Menke, Wilhelm 19012
Menna, Edith 52691
Menne, Gilbert 03616
Mennel, Suzanne 41207
Mennicke-Schwarz, Christiane 17953
Mennicken, Ralph 03960
Mennucci, Dr. Antonello 27637, 27641
Mens, Drs. I.J.A. 32999
Mentel, Anette 21145
Menthoonex, Jean 41317
Mentzelopoulos, Constantine 23224
Menu, Daniel 16203
Menu, Louis 03871
Menz, Dr. Cäsar 41403, 41408, 41414, 41416, 41418, 41420
Menz, Christopher 00773
Menz, Martin 21728
Menz-Bächle, Anke 21960
Menzel, Joachim 21845
Menzel, Steffen 20203
Menzi, Renate 42094, 42095
Menzie, Mary 51969
Menzies, Donald 54273
Menzies, G.M. 43250
Menzies, Jackie 01544
Menzies, Sheila A. 49234
Merabet, Née 00044
Meral, A. Seda 42697
Merc, Viktor Karlovič 29719
Mercadal i Fernández, Oriol 39884
Mercado, Stephanie 55126
Mercado Lopez, Eugenio 31201
Mercandino Jucker, Gabriella 06797
Mercede, Nevin 54330
Mercer, Anna 45585
Mercer, Bill 52086
Mercer, Régis 16241
Mercer, Valerie 48114
Mercier, Poppi 46884
Mercuri, Joan B. 47475, 51453
Mercuri, Mariano 27180
Merdjanov, Plamen 05077
Merdžanov, Plamen 05090
Meredith, Jill 46268, 46268
Meredith, Mary Ellen 53417
Merhib, David 47059
Merhiou, Olivier de 14077
Mérilleau 11840
Merino Acosta, Dr. Luz 09162
Mèriot, Christian 12071
Mériot, Christian 12691
Merk, Jan 20417
Merk, Zsuzsa 23347
Merkel, Dr. Michael 18936
Merker, Matthias 22415
Merkert, Prof. Jörn 17049
Merland, Bénédicte 03798
Merlin, Georges 11807
Merlin-Anglade, Véronique 14778
Mermet, Christian 14317
Merode-Westerloo, Princess de 15284
Meronen, Mikko 11390
Merrell, Ed 48338
Merrell Domingo, Merle 34405
Merriam, Robert W. 48241
Merrill, Kimberly 46151
Merrill, Michael 53106
Merrill-Mirsky, Dr. Carol 50230
Merriman, Larry 49202
Merriman, Dr. Nicholas 45090
Merrin, Keith 44545
Merrit, Carrie 48941
Merritt, Jody A. 46459
Merritt, Roxanne M. 48599
Mersseman, Denise 04109
Merten, Jürgen 22102
Mertens, Carlos Frederico 00445
Mertens, Dr. Eberhard 19014
Mertens, Heike Catherina 17247
Mertens, Dr. Rainer 20756
Mertens, Dr. Sabine 20146
Mertens, Dr. Veronika 16453
Mertens, Wim 03294
Mertes, Lorie 51838
Mertin, Dr. Michael 19378
Merveldt, Catherine Gräfin von 17911
Merwe, Somine van der 38676
Mery, Estelle 16281
Merz, Werner 41701
Merzweiler, Heinz H. 41186
Mesa, José de 04210, 04212, 04216, 04218
Mesa, Laura 50775
Mesbahi, Sh. 24663
Meščerina, Nadežda Alekseevna 37521
Meschede, Dr. Friedrich 17062
Meschede, Prof. Dr. Martin 18748
Mesejo Maestre, Eduardo 09101
Mešerjakov, Vladimir 37675
Meshke, R. 05656
Meskin, Maria 05115
Meslin, Chantal 13771
Mesnil du Buisson, Laurent du 16179
Mesquida García, Dr. Mercedes 39816
Mesquita, Ivo 04882
Messel, Nils 33735
Messenger, Jane 00773
Messenger, Janet M. 51701
Messenger, Margaret 05553
Messer, Jennifer 52410
Messer, Michael 16634
Messer, Renate 20847
Messer, Richard 50258
Messerli, Dr. Jakob 41207
Messerli, Niggi 41563
Messerli, Ursula 41848
Messiaen, Marcel 03269, 03739
Messika, Natalie 25089
Messmer, Dorothee 42002
Messner, Alexander 28288
Messner, Kathrin 03040
Messner, Reinhold 25721, 25885
Meštan, Prof. Pavol 38230
Mészáros, Dr. Ferenc 23445
Mészáros, Júlia 23580, 23584, 23587, 23589
Metallinou, Garyfalia 22974
Metcalf, Charles D. 54311
Metcalf, Craig 06515
Metcalf, Elizabeth 51167
Metcalf, Mike 53101
Metcalfe, Edward C. 50496
Metcalfe, John 44005
Metcalfe, Robin 05948
Meteku Zekeke, Endalew 10684
Metherall, John 45052
Methuen-Campbell, James 43895
Methven, John 01492
Metropolis, William C. 47182
Mets, Esta 10667
Mettala, Teri 51510
Metthews, Ray 50043
Mettler-Frercks, M.A. Beatrix 20606
Metz, Donna 52558
Metz, Katharina 17696, 17697
Metz, Lance 48270
Metz, Thomas 19585
Metz, Wolfgang 19936
Metzger, Amy 49122
Metzger, Frederick W. 07200
Metzler, Jeannot 30483
Metzler, Dr. Sally 47487
Metzner, Paula 49667
Meulemeester, Jean Luc 03943
Meulenbeld, B.C. 31852
Meuter, Günter 20660
Meûter, Ingrid de 03495
Mewborn, Frankie 53816
Mewes, Claus 18940
Mewton-Hynds, Sally 45918

M

Mey, Thomas De 03477
Meyer 17358, 21130, 21134
Meyer, Alain 14282
Meÿer, Bob 31851
Meyer, Dr. Christian 02397, 02934
Meyer, Dr. Christian A. 41175
Meyer, Ellen L. 46423
Meyer, Fifi 38595
Meyer, Franz 16822
Meyer, G. 21668
Meyer, Hans Rudolf 41626
Meyer, Helen 50003
Meyer, Hélène 41418
Meyer, Hubert 15704
Meyer, Irin von 18951
Meyer, J.B. 32331
Meyer, Jackie-Ruth 11536
Meyer, Jochanan 24933
Meyer, Karl 20576
Meyer, Katharina 41276
Meyer, Kimberli 50250, 54064
Meyer, Kurt 21131, 21133
Meyer, Lioba 20900
Meyer, M. 31923
Meyer, Marc 30484
Meyer, Marie-Emmanuelle 16112
Meyer, Maurice 11590
Meyer, Niels 09963
Meyer, Rachel 51689
Meyer, Dr. Reiner 21232, 21791
Meyer, Robert 51226
Meyer, Simone 22375
Meyer, Stefan 21337
Meyer, Steve 49407
Meyer, Thomas 42067
Meyer, Werner 18674
Meyer Bühler, Monica 41229
Meyer-Friese, Dr. Boye 18915
Meyer-Stoll, Christiane 30197
Meyer Utne, Janneke 33636
Meyer zu Bargholz, Wolf 41600
Meyer zu Schwabedissen, Doris 20562
Meyerdierks, Judith 47275
Meyere, J.A.L de 32853
Meyers, Dr. Albert 17443
Meyers, Amy 51055, 52819
Meyerson, Barbara 50604
Meyerson, Randy 53522
Meyrick, Robert 43259
Meysner 22005
Mezan Barrera, Jesús Manuel 31035
Mezger, Prof Klaus 20514
Mezhlumyan, G.B. 00735
Mezil, Eric 11786
Mežinski, Jelena 38045
Mezvinski-Smithgall, Elsa 53914
Mezzadri, Prof. Maria Grazia 27104, 27105, 27106
Mhire, Herman 49868
Mhute, Josiah Rungano 54673
Mialkowsky, Patricia 06919
Miano, Ederick C. 34370
Miao, Duo'an 08127
Miaoulis, Ioannis 46941
Michael, Dean 01210
Michael, Dr. Eckhard 20100
Michael, Dr. Elisabeth 20424
Michael, Helene 42057
Michael, Janet 43714
Michael A., Flannery 46819
Michaelidon, Maria 23171
Michaelis, Dr. Rainer 17116
Michaels, Bonni-Dara 51281
Michaels, Dr. Sonja 18406
Michaelsen, Miah 46465, 51424, 53617
Michailov, Christo 04986
Michailov, Igor 37680
Michajlov, A.B. 37862
Michajlov, Delčoe 30524
Michajlova, Ljudmila Vasiljevna 37683
Michajlova, Zdravka 05083, 05217
Michajlovna, Jelena 37195
Michalaki-Kollia, Maria 22896
Michálek, RNDr. Jaroslav 09844
Michalevskij, Aleksandsandr Vlaimirovič 29712
Michalka, Matthias 03042
Michard, Dr. S.G. 14608
Michaud, Florine 41868
Michaud, François 14646
Michaud, Jean-Marc 13582, 15074
Michaud, Jocelyne 06825
Michaud, Raynald 06370
Michavila, Joaquín 40299
Michel, Dr. Ch. 03793
Michel, D. 14247
Michel, Ernest W. 51170
Michel, Gundega 30084
Michel, Helmut 22363
Michel, James 53077
Michel, Jean-François 11523, 12471, 12472, 12473
Michel, Kaspar 41849
Michel, Samuel 41455
Michelangelo, Dr. 25960
Michelberger, M.A. Larissa 20478
Michelet, Annick 14951
Michelina, Prof. Dr. 25332
Michelmann, Nina 18979
Michelon, Olivier 15067
Michelotti, Dr. Francesca 38002, 38004, 38008, 38012
Michels, Dr. Anette 22114
Michels, Dr. Hubertus 17840
Michels, Dr. Norbert 17826
Michelsen, Pia 40707
Michie, Thomas 52170
Michieli, Catalina Teresa 00118
Michnová, Věra 09692
Michoutouchkine, N. 54494
Mickenberg, David 54023
Mickle-Smaczny, Lynne 48104
Micklich, Judy 52673
Micklich, Dr. Norbert 17789
Micko, Ingeborg 02912
Miclescu, Constanța 36722
Micol, P.M. 12956
Mičurová, M. 38391
Micynska, Tamara Petrivna 43028
Middag, B.M. 32465
Middelbasch, Max 32675
Middlemost, Wayne 01619
Middleton, J. 01159
Middleton, J.W. 43737
Middleton, S.I. 43737
Miedema, B. 31979
Mielatz, Yvonne 17789
Mieles, André 18362
Mielke, Dr. Heinz-Peter 18744
Mierkiewicz, Edward 35268
Mierzwiński, Mariusz 34914
Miesbeck, Peter 21374
Miesenberger, Martin 02437
Miessgang, Dr. Thomas 03018
Mießl, Peter 02185
Mieth, Dr. Matias 19383
Mieth, Matias 19392
Mieth, Thomas 17478
Miethe, Susann 16443
Mietke, Dr. Gabriele 17052
Miettinen, Sinikka 11455
Miggitsch, Jörg 02392
Miglbauer, Dr. Renate 02911, 02912, 02916, 02917
Miglinas, Skirmante 47486
Migliori, Gabriele 26837
Mignot, D. 31888
Migritone Gondwe, Michael 30546
Miguet, Daniéle 12411, 12412
Mihai, Raul 36173
Mihail, Benoît 03503
Mihăilă, Sanda 36338
Mihailović, Milica 38051
Mihalić, Jana 08886
Mihelić, Sanjin 08986
Mihevc, Bibijana 38470
Miho, Terao 28635
Mihovilić, Kristina 08932
Mihut, Elena 36230
Mijić, Suzanalav 38093
Mijnlieff, E.M. 32268, 32269
Mikaelian, Kajik 00702
Mikaelsen, Laila 23267
Mikelson, Mary 06996
Mikes, Johannes H. 01929
Mikesková, Magdalena 09618
Miki, Akiko 14743
Mikkalo, June 47778
Mikkelsen, Dr. Egil 33722, 33726, 33734, 33757, 33759
Mikkelsen, Falk 10037
Mikkola, Ari 11127
Mikkola Huk, Eira 11077
Mikołajska, Ewa 35140
Mikulaštík, Dr. Tomáš 09918
Mikus, Joan 48640
Mikuž, Marjeta 38455
Milani, Arch. 51422
Milani, Dr. Domenico 27107
Milano, Barbara R. 52920
Milano, Dean 48377
Milbank, Bill 33234
Milburn, Alan 44263
Milchram, Gerhard 03041
Milde, Brigitta 17697
Mildner, Dr. Paul 02186
Miles, Blanche 52692
Miles, Carolyn P. 52567
Miles, Carolyn W. 46140
Miles, Christine M. 46157
Miles, David A. 54287
Miles, Ed 46813
Miles, Ellen 53907
Miletić, Jelena 38093
Mileusnić, Slobodan 38077
Mileusnić, Snježana Radovanlija 09009
Milev, Svetoslav 05170
Milhau, Muriel 15296
Milhazes, Dr. Maria Claúdia C. 35456
Miliadzidou-Ioannou, Dr. E. 23212
Milian, Mihai 36272, 36704
Milivojević, Radmila 38107
Mill, Dennis J. 51584
Millan, William I. 43414
Millar, C. 44047
Millar, Jim 06646
Millar, Joyce 06617
Millard, Ellen 07061
Millard, John 44750
Millard, Rhoda 05695
Millberg, Per-Olof 40613, 40614
Mille, Dr. Diana Dimodica 48482
Millen, A.J. 05356
Millenbruch, Gary L. 46225
Millender, Lynn R. 46275
Miller, Alan 00905
Miller, Dr. Albrecht 18117, 20773
Miller, Alison 43387, 49994
Miller, Andree 46227, 47280, 48305
Miller, Angus 44347
Miller, Ann 50533
Miller, Archie W. 06436
Miller, C.J. 49106
Miller, Carl R. 49837
Miller, Charles 51604, 51606
Miller, Dan 53472
Miller, Dana 51279
Miller, David 52925
Miller, David M. 50596
Miller, Denise 47468
Miller, Douglas 43274, 50860
Miller, E.M.J. 43390, 45059
Miller, Erika 48939
Miller, Prof. George 49227
Miller, George 51985
Miller, Gordon 47494
Miller, Harriet 51296
Miller, Heidi 46201
Miller, Howard L. 49197
Miller, John 45837
Miller, John F. 49094
Miller, Dr. Kathy Ann 53409
Miller, Dr. Kelly B. 46182
Miller, Kevin R. 47995
Miller, Kristin 46210
Miller, Lenore D. 53861
Miller, Leueen 49058
Miller, Linda L. 48350
Miller, Lisa 46225
Miller, Lois 49396
Miller, Margaret A. 53431
Miller, Marguerite 46222
Miller, Dr. Markus 18079
Miller, Mary 06912, 50288
Miller, Dr. Mattthias 17072
Miller, Michael 52073, 53184
Miller, Mike 46502
Miller, Nancy Lee 46958
Miller, Pamela 52140
Miller, Paul F. 51327
Miller, R. 45453
Miller, Randy 06838
Miller, Dr. Rick 46152
Miller, Rosemary 01157
Miller, Sammy 45205
Miller, Sandy 53327
Miller, Sarah 01377
Miller, Stephen 49884
Miller, Steve 50854
Miller, Steven 45530
Miller, Terry J. 48830
Miller, Tom 05868
Miller, Tracy 46749
Miller, Walter 53401
Miller-Betts, Christine 46479
Millereau, Marie-Pierre 12045
Millet, Florence 13534
Millet, Teresa 40201
Milli, Konrad 19399
Milligan, Dr. Frank D. 50946
Milligan, Robert B. 51318
Millin, Laura J. 50754
Miller, Marcus 06555
Millman, David 49977
Millroth, Thomas 41078
Mills, Brendon 49890
Mills, Dan 50029
Mills, Dinah 53741
Mills, Ellsworth 49281
Mills, John 44163
Mills, John K. 52154
Mills, Josephine 06164
Mills, R.G. 45938
Mills, Susan 43282
Milne, Jane 44554
Milne, Julie 45229
Milne, Lee 01143
Milne, V. 00785
Milner, Dr. M.J. 45551
Milner, Renate 53686
Milnor, Nancy 48846
Milojković, Jelica 38131
Milonjić, Miroslav 08944
Milosavljević, Branka 38091
Milošević, Prof. Ante 08961
Milošević-Đerek, Svjetlana 08862
Milosevich, Joe B. 49644
Milotd, Jocelyne 06692
Miloux, Yannick 13763
Milovanović, Dušan 38074
Milović, Olivera 38132
Milstreu, Dr. Gerhard 40968
Miltner, Viola 20453
Milton, Lester R. 46610
Milton, Peter 45289
Milton-Tee, Ann 33109
Milyov, Simeon 05191
Min, Choi Hyong 29755
Min, Susette 51167
Mina, Dr. Gianna A. 41566
Minar, Rena 53426
Minard, Barbara 49468
Minard jr., Richard A. 47774
Minarini, Dr. Laura 25447
Minassian, Levon 24610
Minchinton, J. 45270
Mindell, Dr. David 52768
Mindell, Prof. David P. 46310, 46310
Minder, Nicole 41732
Mindhoff, Otto 21704
Mine, William K. 43374
Mineau, Marc 06993
Mineeva, Valentina Alekseevna 36839
Minelli, Donato 25277
Miner, Curt 49181
Minerva, Michael 51227
Minetti, Dr. Alessandra 27749
Minger, Karin 41209
Minighetti, Clément 30478
Minish, Marveen 50588
Minke, Prof. Dr. Alfred 03593
Minkenberg, Dr. Georg 16396
Minkkinen, Aimo 11353
Minmeister, J.C. 12777
Minney, Michael J. 49925
Minns, Dr. Jonathan E. 44492
Mino, Dr. Yutaka 29151
Minoja, Dr. Marco 25301
Minse, C. 32304
Minshew, Cheryl 48307
Minto, Don 49600
Minto, Heather 49600
Minton, Dwayne 23288
Minton-Elmore, Kara 51410
Mintz, Ann 54049
Mintz, Deborah Martin 48530
Minx, Jim 51522
Mion-Jones, Esta 45706
Mir, Dr. Ronen 46494
Mir-khalaf, A. 24611
Mir Sancho, Joaquín 39628
Mira, Vegar 08844
Mira Bonomi, Angelo Vittorio 26266, 27823
Mirabdolbaghi, Zia 16096
Mirabella Roberti, Prof. Mario 28165
Miranda, Gaudenzio 26978
Mirelez, Cathy 48912
Mires, Cherry 46535
Mirmoghtadaie, S. 24612
Mirnik, Dr. Ivan 08986
Miró i Cuberes, Enric 38994
Mirolo, Diana 41130
Miron, Dr. Andrei 21457
Miron, Dr. O. 25044
Mironneau, Paul 14759
Mironowicz-Panek, Maria 34940
Miroslavljević, Dragisa 38158
Mirror, Louise 51243
Mirzejewska, Maria 35235
Mirzoyan, Pharaon 00751
Misago, Dr. Kanimba 37992
Misailova, Valentina Ivanovna 37628
Miscault, René de 13523
Mischke, Petra 19104
Mischkewitz, Franz 20592
Mischkulnig, John 41282
Mišečkina, Marija V. 36930
Mishima, Tadao 29082
Mishima, Yashuhisa 29116
Mishima, Yoshinori 29116
Mishne, Merle H. 50558
Mishra, H. P 24095
Mišina, Ljudmila Aleksandrovna 37590
Misjuga, Bogdan Vasil'ovič 43140

Miskowski, Anthony 53561
Mislej, Irene 38397
Misner, Mary 05536
Misočnik, Svetlana Michajlovna 37786
Misof, Dr. Bernhard 17469
Miss, Dr. Stig 10155
Misterek, René 21046
Mistrot, Vincent 12066
Mitanov, Petăr 05204
Mitar, Mladen 08904
Mitchell, Archie 53647
Mitchell, Charles 51764
Mitchell, D.I. 45714
Mitchell, Freda 47003
Mitchell, Harold 00957
Mitchell, James 47380
Mitchell, Jena 52457
Mitchell, Joan 01057
Mitchell, Joanne 46212
Mitchell, John 50022
Mitchell, Leah 49594, 51289
Mitchell, Marjorie 52374
Mitchell, Paul 05654
Mitchell, Dr. Peter 45338
Mitchell, Renee 49781, 49782
Mitchell, Scott 51484
Mitchell, Starr 50139
Mitchell, Sue 01573
Mitchinson, David 45183
Mitev, P. 05150
Miteva, Kremena 05020
Mitkevič, Maja 36768, 36769
Mitkovs'ka, Tetjana Sergiïvna 43149
Mitrev, Dimităr 05107, 05109, 05164
Mitsides, Dr. Andreas 09372
Mitten, David G. 47177
Mittendorfer, Gabriele 02671
Mitterwenger, Dr. Christine 02486, 02487, 02488, 02489
Mittmann, Dr. Hans-Walter 19449
Mittwoch, Wolfgang 18311
Mitu, Aneta 36498
Mitzel, Horst 22468
Mitzel, June 05788
Mitzscherlich, Dr. Birgit 16939
Miwa, Kenjin 29491
Miyagawa, Ellen 51687
Miyaji, Akira 29289
Miyake, Akiko 28711
Miyamoto, Akio 28624
Miyamoto, Makiko 28537
Miyaoka, Kiyomi 29584
Miyatani, Masayuki 29316
Miyazaki, Takashi 29035
Miyazawa, Akimichi 29005
Miyazima, Hisao 29130
Mizerová, Hana 09732
Mizerski, Dr. Włodzimierz 35244
Mizina, Ljudmila V. 37817
Mizinga, Flexon Moono 54658, 54661
Mizukami, Iwao 28387
Mizukami, Takeshi 29364
Mizuno, Keizaburo 28963
Mizuno, Takashi 29190
Mizuta, Dr. Akira 29032
Mizutani, Takeshi 29491
Mizzan, Dr. Luca 28197
Mjärdner, Stellan 40553
Mjasiščeva, Olena Georgiïvna 43196
Mjøs, Lise 33758
Mlecz, Richard S. 52305
Mlecz, Stella C. 52305
Mleczko, Radosław 35269
Mlynarčík, Dr. Jozef 38296
Mlynarski, Ryszard 35342
Mnasaganyan, Susanna 00705
Mndaweni, Aurelian 38618, 38619
Mo, Charles 47392
Mo, Guofu 07604
Mo, Huamin 07909
Moacir Maia, Prof. Pedro 04796
Moberg, Jack 50475
Mobley, Ree 52882
Moc, Joan S. 50543
Mochi Onori, Dr. Lorenza 28123
Mochola, Tatiana 34705
Mochoruk, Mike 06459
Mochova, Jelena Fillipovna 36747
Mock, Janelle 48081
Mock, Rayonna 46235
Modegi, Masaaki 29481
Modersohn, Antje 20954
Modest, Wayne 28328
Modi, Amrutbhai 24048
Modl, David 01985
Modonesi, Dr. Denise 28236
Modotti, Luigi 25917
Modrzejewska, Dr. Irena 34540
Moe, Barbara 49768
Moe, Helene 33703, 33704
Möbius, Ingrid 20386
Moed, G.J. 32592
Moeferdt, Dr. Horst 17874
Möhlenkamp, Dr. Klaus 20358
Moehlmann, Nathan 51342
Möhlmeyer, Erich 20933
Möhring, Dr. Helmut 21410
Moehring, Markus 20016
Möhrle, Reinhard 41236
Möller, Hilke Gesine 20091
Möller, Dr. Kathrin 22614
Möller, Kristina 40863
Moeller, Prof. Dr. Magdalena M. 17058
Möller, Marie von 33418, 33419, 33420
Möller, Peter 40992
Möller-Arnsberg, Birgit 20453
Möllers, Dr. Sebastian 21848, 21849, 21850, 21852
Möllmann, Dr.Dr. Elke 17926
Moen, Eivind 33909
Mönig, Dr. Roland 19567
Mönkkönen, Mikko 11173
Mönnig, Dr. Eckhard 17718
Moens, Dominique 03931, 04024, 04158
Moens, Karel 03318
Möntl, Karin 03051
Mörgeli, Prof. Dr. Christoph 42085
Mörgeli, Markus 41654
Mörike, Dr. Doris 21959
Moers, Rainer 18889
Mörter, Hermann 42086
Mörtinger, Franz 01789
Moes, K. 31908
Moesgaard, Jens Christian 10136
Moeske, Ulrich 17927
Moesser, Suzanne 50536
Mössinger, Ingrid 17697
Moffat, Jill 48792
Moffat, John 43298, 44429
Mofford, Juliet 46297
Moga, Jack 49038
Mogensen, Ove 10299
Moget, Layla 15756
Mohamadi, Naghme 24642
Mohammadrezaie, P. 24620
Mohan, S. 40399
Mohl, Dr. Max 16923
Møhlmann, Rob 31918
Mohr, Jeanette 22747
Mohr, Julie 45098
Mohr, Margit 20761
Mohr, Dr. Peter 45098
Mohr, Vibeke 33836
Mohrenberger, Ingeborg 01971
Mohring, Marilyn 52428
Moigne, Roland de 12122
Moilanen, Arja 11361
Moinet, Eric 14352, 14356
Moinet, Josette 12066
Moise, Leonida 36304
Moisen Lechuga, Patricia 31126
Moises, Loly Keith 34400
Moisio, Roberto 28025
Moissaides, Christina 48245
Mojica, Brenda 34397
Mojseeva, T.M. 37697
Mojzer, Miklós 23477
Mojzes, Ildikó 23469
Možíšová, Dr. Olga 09544, 09768
Mokarram Hossain, Mahmud 03230
Mokre, Jan 02996, 03045
Mol 04085
Mol, I. 32047
Mol, W. 32005
Moland, Arnfinn 33738
Molas, Gustau 38994
Moldenhauer, Susan 49950
Moldon, John 06257
Molen, Dr. J.R. ter 31915
Molenaar, Corine 32757
Molesworth, Kris 49831
Molesworth-Saint Aubyn, Lady Iona 43551
Molina, Prof. María Fernanda 00530
Molina Cabal, Alvaro José 08636
Molina Pico, Horacio 00646
Molinari, Danielle 14621, 45592
Molinari, Giovanni 26336
Molinaro, Ettore 25524
Molinaro, Marcene 47314
Molist, Miguel 38970
Mollano, Prof. Olga 00486
Mollay, Catherine 06239
Møller, Dan E. 10317
Møller, Lars Kærulf 10046
Møller, Stig B. 09955
Møller, Thorkild 10013
Møller-Hansen, Keld 10320
Møller Jensen, Johnny 09963
Møller Pedersen, Kurt 09966
Moller-Racke, Heidi 20145
Mollet, Fran 53561
Mollfulleda i Borrell, Joaquin 38887
Mollo, Boris 44113
Molloy, Dr. Peter M. 51903
Molloy, Tristan 44554
Molnar, August J. 51035
Molnar, Cindy Lou 53907
Molnár, Dr. Erzsébet 23666, 23667
Molnár, Ildikó 23616
Molnar, Michael 44852
Molnár, Sándor 23947
Molnar, Veronika 20975
Molnár-Sipos, Judit 23440
Molnos, Alison 43986
Mols, P. 32880
Molsbergen, Bert 32083
Molteni, A. 13697
Molteni, Dr. Gionfranco 25564
Moltesen, Mette 10145
Moltke-Hansen, David 51841
Molvær, Edvard 33381
Molzberger, Ansgar 19597
Momar Toboso, Manuel 39922
Momas, Jacques 12149
Mombrun, C. 23274
Momčilov, Dimčo 05051
Momin, Shamim 51279
Momoitio Astoria, Iratxe 39332
Momonkulov, Džumaaly Momonkulovič 29954
Monachon, Danuta 14599
Monaco, Dr. Giuseppe 27462
Monaghan, A. 45601
Monaghan, Bryan 05746
Monaghan, Katherine 53594
Monaghan, Nigel 24810
Monahan, Gloria 06214
Monbron, Comte de 13387
Monch, Manfred 21435
Moncrief, Dr. Nancy D. 50521
Moncur, Shannon 05435
Mond, Ulrike 17454
Mondini, Massimo 25525
Mondragón, Lourdes 31042
Mondragón, Martín Antonio 31422
Monds, Kathy 52026
Monedoñedo Espinoza, Juan 34269
Monego-Venzin, Barbara 41854
Monery, Jean-Paul 15525
Mones, Prof. Alvaro 54413
Money, Sarah 45535
Monfort, Catherine 03962
Mongan, Barbara 47771
Monge, Janet 51886
Monge, Maria de Jesus 36016, 36017, 36018
Mongellaz, Jacqueline 15661
Mongon, Alain 11694
Moni, Daniela 25905
Moniak, Eugeniusz 35379
Monigan, Michael F. 53926
Monk, Suny 53390
Monkhouse, Christopher 50739
Monmousseau, Patrice 15660
Monnani, Walter 01746
Monnet, C. 13754
Monnier, Georges 41863
Monnier, Jean-Laurent 14773
Monod, Frédéric 14142
Monok, István 23463
Monot, Michel 16243
Monrad, Kasper 10152
Monroe, Dan L. 52657
Monroe, Dr. Sam 52029
Monschaw, Elline von 24807
Monsef, Gity 44846
Monsieur, Dr. Patrick 03622
Montagne, Drs. Cor 32083
Montagu of Beaulieu, Lord 43649
Montaldo, Dr. Anna Maria 25578, 25582
Montaner, Jaume 38981
Montañés García, Carmen 40340
Montanti, Deborah 49648
Montaut 15405
Montebello, Philippe de 51212
Monteil, Esclarmonde 12166
Montel, Philippe 15437
Montenegro Carneiro, Dr. Maria Margarida V. 35750
Montérémal, R. 13652
Montero, Alberto E. 07427
Montero, Prof. Paula 04877
Montero Agüerz, Ildefonso 39208
Montero de Povarché, Elena 00277
Montesano, Frank 47822
Montesdeoca García-Sáenz, Antonio Daniel 39802
Montet-White, Anta 49990
Montezoma, Anna de 05683
Montfort, Helen G. 49112
Montgomery, A. 44519
Montgomery, Alexander 07110
Montgomery, Brian 45927
Montgomery, Cathie 45775
Montgomery, Dr. John 52053
Montgomery, Küllike 41003
Montgomery, N. 38566
Montgomery, S.E. 45940
Montgomery, Dr. Sally 43449
Montgomery, Prof. W.I. 43442
Monticelli, Dr. Paolo 26952
Montijo, Mark 52750
Montileaux, Paulette 52254
Montin, Kim 11398
Montjalin, Olaf de 15672
Montmany, Marta 38982
Montoya Soto, Sonia 08626
Monzó, Josep Vicent 40201
Moodie, Kathleen 52868
Moodie, Sandra 28332
Moody, Dr. Barry 05343
Moody, Eric N. 52288
Moody, Tammy 48147
Moogan, Kerri 44276
Mooij, Dr. Charles de 32375
Moombolah-Goagoses, Esther U. 31762, 31765, 31766
Moon, Diane 44910
Moons, L.J.M. 32303
Moor, Rebecca 53815
Moore, Anne 45177
Moore, Barbara 51643
Moore, Bennett 49147
Moore, Bill 05862, 44268
Moore, Carleton B. 53470
Moore jr., Dan 46421
Moore, Dan 46421
Moore, David 43597, 46666
Moore, Dona 06053
Moore, Dorothy 49360
Moore, Duane 53604
Moore, Edra 49914
Moore, Elke aus dem 21946
Moore, Ellen K. 52470
Moore, Fran 05686
Moore, Gary F. 53608
Moore, Geoff 01050
Moore, George 48642
Moore, Harold D. 53124
Moore, J. Kenneth 51212
Moore, J. Roderick 48538
Moore, James C. 46175
Moore, Jennifer 49060
Moore, John 47384
Moore, Juanita 49682
Moore, Kevin 45446
Moore, Linda 50931
Moore, Michael R. 52319
Moore, Richard 01222
Moore, Richard H. 48922
Moore, Rick 46291
Moore, Rod 43867
Moore, Rowan 44785
Moore, S. 05734
Moore, Susan J. 49368
Moore, Thomas 47299
Moore, Prof. Thomas E. 46310
Moore, Dr. Viola 51695
Moore, Wayne 46898
Moore, William S. 48127
Moore Booker, Margaret 50943
Moore Ede, Minna 44937
Moore Stucker, Kathryn 51923
Moorefield, Amy G. 52322
Moores-Franklin, Lolisa 51976
Moos, Norbert 19601
Moosbauer, Christine 16736
Mootoosamy, Morghan 41573
Moppett, Ronald 05521
Moqimi-zadeh, Hamid 24618
Mor, Judith 25132
Mora, Alejandra 31223
Mora, Dr. Rafael A. 00203
Mora García, Carlos 31544
Moraes Branco, Pércio de 04647
Moraes Sophia, Eunice 04880
Moragas, Cora Romy 07439
Moraitou-Apostolopoulou, Dr. Maria 23250
Morales, Dr. Michael 48391
Morales-Coll, Eduardo 36058
Morales Gómez, Antonio 31210
Morales Gómez, David 31211
Morales Marenco, Eliázar 33279
Morales Pérez, Eduardo 31514
Morales Ramón, José María 39619
Moran, Maryann 52979
Moran, Paula G. 46149
Moran, Vivienne 01597
Moran-Zerda, Leticia 34452, 34461
Moraña, Prof. Oscar 54536
Morand, J. 05729
Morand, Marie Claude 41873, 41874, 41875
Morandini, Dr. Carlo 28114
Morando, Paul 47075
Morante López, Dr. Rubén Bernardo 31557
Morash, Terrence 46947

Morath-Vogel, Dr. Wolfram 18196, 18197, 18198
Moravčík, J. 38391
Moravec, Tomaš 09698
Morawietz, Doris 19509
Morawińska, Agnieszka 35290
Mórawski, Karol 35281
Morawski, Mike 49759
Morbach, Philippe 14718
Morcom, G. 38615
Morcos Gayyed, Mervat 10526
Mordal, Christen A. 33481
Mordi, Pietro 26437
More-Molyneux, Michael 44373
Moreau, Michel 13205
Morehouse, Dorothy V. 50115
Moreira, Ana Arbués 35947
Moreira, João Arbués 35947
Morel, Michel 06809
Morel, Nicolas 13609
Morel, Pascal 15240
Morel, Sylvie 05862
Morel-Deledalle, Myriame 14005
Morelli, Gianni 27372
Morello, Dr. Giovanni 54510, 54511
Morelock, Dr. Jerry 48810
Morena López, José Antonio 38939
Moreno, Anne Marie 34507
Moreno, Barry 51262
Moreno, Enrique 39544
Moreno, Fred 51108
Moreno, María Cristina 08667
Moreno, Mercedes 39940
Moreno, William 52783
Moreño de Medellín, Patricia 31529
Moreno Garrido, Maria Jesús 39204
Moreno Pérez, D. Jorge 39788
Moreno Ulloa, Graciela 08814
Moret, J.-L. 41545
Moret, Jean-Charles 41290
Moretti, Dr. A. Maria 27518
Moretti, Dr. Giovanni 26071
Moretti, Dr. Mario 25819
Moretti, Prof. Mario 27684, 27686
Moretti, Dr. Mario 28292
Morgan, B.V. 06437
Morgan, Chris 43704, 45208, 48622
Morgan, Dahlia 50621
Morgan, Dave 50770
Morgan, David 43983
Morgan, Dina 50229
Morgan, Dr. Gary 01093
Morgan, Dr. J.D. 52014
Morgan, Jeremy 06728
Morgan, K. 33042
Morgan, Martin 46339
Morgan, Dr. Michele 47185
Morgan, Peter J. 45682
Morgan, Peter M. 50967
Morgan, Ron 43899
Morgan-Welch, Beverly 46939
Morgany, Jacquie 06481
Morgenbrod, Klaus 20260
Morgenegg, Hans 41323
Morgenroth, Heiner 21310
Morgenroth, Dr. Ulrich 22210
Morgenstern, Undine 16498
Morgner, Bernd 19181
Morhun, Sue 05813
Mori, Ichiro 29536
Mori, Kei 29247
Mori, Masatoshi 28551
Moriarty, M.J. 43824
Morigi Govi, Dr. Cristiana 25447
Morillo-Cajiao, Pablo 08715
Morillo León, Anabel 40062
Morimoto, Prof. Mitsuo 28916
Morimura, Hideyuki 29614
Morin, Robert G. 50923
Morin, Yvan 13636
Morinelli, Dom Leone 25786
Morishigue, Akifumi 29120
Morisod, Dominique 41769
Morita, Hajime 29192
Morita, Takashi 28978
Moritz, Craig 46759
Moritz, Franz 19645
Moritz, Heidi 18052
Moritz, Dr. Marina 18207
Moritz, Per 40699
Moriyama, Mayumi 29447
Moriyama, Ryoichi 28409
Mork, Eileen 49214
Morl, Dr. Claudia 26215
Morley, Stephen H. 47107
Morlok, Dr. Wolfgang 21627
Morman, C. 32017
Morna, Dr. Teresa 35694
Morneburg, Otto 16943
Moro, Alberto E. 00402
Moro, Luigi 25149
Moroder, Robert 27001
Morofushi, Masami 28694
Moroyama, Masanori 29492
Moroz, Vira Dmytrivna 43216
Morozov, Ivan Ivanovič 37078
Morozov, Vladimir Vasilevič 37397
Morozova, Ludmila Ivanovna 37564
Morozova, T.V. 36997
Morral, Eulàlia 40116
Morrall, Dr. Andrew 51132
Morran, Gerald 05902
Morré, Jörg 17066
Morretta, Rosemary 52348
Morrill, Hope 54025
Morrin, Peter 50297
Morris, C.I. 44316
Morris, Catherine J. 47069
Morris, Charles 47380
Morris, Christopher 44315
Morris, Clive 43752
Morris, David 45099, 52300
Morris, Don 51304
Morris, Henry 44888
Morris, Jack 06047
Morris, Jeffory 52182
Morris, Joan 44643
Morris, Johnathan 01067
Morris, Lynda 45282
Morris, Nell 49880
Morris, Pamela 52086
Morris, Patricia A. 52648
Morris, Robbie 46361
Morris, Valerie 45869
Morris Hambourg, Maria 51212
Morrison, Alan 45108
Morrison, Andrew 43458, 45827, 45997
Morrison, Charlotte K. 48050
Morrison, Cindi 49927
Morrison, D. 01505
Morrison, David 44038
Morrison, David A. 05862
Morrison, Garth 44375
Morrison, James F. 49639
Morrison, John 52013
Morrison, Kim 49299
Morrison, Lynn 45540
Morrison, Lynne 49922
Morrison, Sandra 51707
Morrissey, Dr. Kristine 48254
Morrissey, Tom 50112
Morrow, David 46319
Morrow, Frances 46195
Morsch, Prof. Dr. Günter 20908, 20911
Morsch, Nathalie 05431
Morsch, Sigrid 20846
Morscheck, Charles 51855
Morscher, Dr. Lukas 02124, 02131
Morse, Carla 05959
Morse, F.E. 47550
Morse, Paula 49417
Morseth, Kari 33089
Morsiani, Paola 49383
Mort, Mark 49991
Mortel, Dr. Hans van de 32698
Mortensen, Ingrid 10313
Mortensen, Lene 17230
Mortenson, Marianne 50205
Morteveille, Gérard 15581
Morteyrol, Amélie 15889
Morton, Craddock 00947
Morton, E. 45227
Morton, Evan 07184
Morton, Fiona 45048
Morton, Geneva 51700
Morton, Lisa 48218
Morton, Sandra 48850
Morucci, Alberto 27600
Moś, Ignacy 35038
Mosbrugger, Prof. Dr. Volker 18392
Moscatelli, Daniel 00521
Moschig, Günther 02177
Mosco, Dr. Marilena 26169, 26174
Mosconi, Prof. Andrea 26001, 26006
Moscoso Möller, M.A. Fernando 23301
Mosdal, Brita 10041
Moseley, Bruce M. 51560
Mosena, David R. 47497
Mosenfelder, Mrs. Robert 48415
Mosenstine, Ken 47204
Moser, Audrey 48183
Moser, Dr. Bernd 01981
Moser, Hans-Rudolf 41665
Moser, Dr. Juliane 18287
Moser, Kenneth 47069
Moser-Gautrand, Claire 12189
Moser-Menna, Maria 41396
Moses, Stephen A. 51081, 51082
Moses, Vincent 52374
Moses Nesmith, Cynthia 47730
Moshenson, Edna 25122
Mosica, Phillippe 15445
Mosier, Milton 53340
Moskovaja, Ljudmila Ivanovna 37519, 37520
Moslavac, Slavica 08904
Mosler, Dr. Bettina 19612
Mosneagu, Marian 36308
Mosquera, Lic. Maria 00264
Moss, Karen 51332
Moss, Kenneth 51217
Moss, Michael E. 54078
Moss, Rick 51462
Moss, Dr. Roger W. 51816
Mossman, Kate 05842
Mostegl, Sabine 02943
Moster-Hoos, Dr. Jutta 20895
Mostert, Antoinette 31742
Mothe, Florence 14923, 14924
Mothe, Hubert 13546
Motlotle, Ntikwe Pene 04299
Motornaja, Larisa Ivanovna 37454
Motoyama, Yoshiao 29412
Motre, Doina 36239
Mott, Diane 52772, 52781
Mott, Joyce 50034
Mottart, Eugène 03939
Motte, Vic 04057
Mottram, Graham 46097
Motts, Warren E. 49099
Motyka, Jerzy 34606
Motyka, Maria 34829
Mouche, Alexandre 41430
Moudry, Mary Lou 46991
Mouely-Koumba, Marie-Agnès 16329
Mougel, René 13305
Moughdad, S. 42129
Mougin, Denis 13593
Mould, Laurel 05979
Moulias, Christos 23140
Moulin, Hélène 16040
Moulin, Pierre 11750
Moult, Lynne 44691
Mounier, Bernard 15847
Mountfort, Dick 46803
Moura, Fernando 35688
Moura Delphim, Angelo Alberto de 04534
Mourão, R. 04601
Mourlet, Gérard 15572
Mousley, Patricia K. 49371
Moussa Nene, Djamil 07424
Mousseigne, Alain 15934
Moussellard, Renaud Pascal 12294
Mousset, Jean-Luc 30483
Moutashar, Michèle 11685
Moutsopoulos, Thanassis 22953
Mouw, Christine 54047
Movillon, Mario M. 34407
Movsisyan, Vergine 00739
Mowat, Dr. P. 45317
Mowll Mathews, Nancy 54176
Mowrey, Debbie 46371
Mowry, Chrystal 06103
Mowry, Robert 47177
Moxeme, Oscar O. 52066
Moxley, Richard W. 52320
Moy, Henry 49461
Moye, L.M. 50322
Moye, Matthew M. 50322
Moyes, Nick 43965
Moyle, N. 45391
Moyle, Nicola 45390, 46094
Moynes, Debbie 05330
Moyroud, Raymond 12154
Mpouna Coline, Samuel 23314
Mrabet, Ridha 42647
Mrachacz, Maria 19632
Mravlincić, Ivan 08967
Mraz, Dr. Gottfried 02998
Mroczkowski, Dennis P. 48658
Mrozinski, Mary Anne 49591
Mrugalska-Banaszak, Magdalena 35036
Mrva, Dr. Marian 38246
Mrva, Dr. Marián 38391
Mrvová, J. 38391
Msemwa, Dr. Paul J. 42512
Msimanga, A. 54666
Mtshiza, Zola 38593
Mu-Liepmann, Véronique 16327
Muceus, Cheryl 46130
Muceus, Cheryl L. 46128
Muchametčina, Lejla Munirovna 37060
Muchanov, Gennadij Stepanovič 37026
Muchitsch, Hr. Dr. Wolfgang 01977
Mucnjak, Prof. Alois 02962
Mudd, Douglas 47705
Mudd, Margaret 50755, 50756
Muddiman, Susi 01340, 01620
Mühlbacher, Anton 01958
Mühlbacher, Peter 21234
Mühlbauer, Heinrich 19872
Mühldorfer-Vogt, Ch. 20996
Mühlegger-Henhapel, Dr. Christiane 03051
Mühlemann, Lorenz 41946
Mühlemann, Yves 41304
Mühlenbrock, Dr. Josef 19155
Mühlhäußer, Werner 18840, 18841
Mühlhausen, Priv.-Doz. Dr. Walter 19080
Muehlig, Linda 51414
Mühne, Dr. Christian 21045
Mührenberg, Dr. Anke 19446
Mülhaupt, Freya 17049
Müllejans-Dickmann, Dr. Rita 18554, 19112
Müller, Alois 41470
Müller, Andreas 18442
Müller, Dr. Anette 22835
Müller, Arndt 16563
Müller, Astrid B. 17143
Müller, Benno 41513
Müller, Berit 22693
Müller, Brigitte 41631
Müller, Bruno 01832
Müller, Dr. Christian 41166, 41167
Müller, Dr. Claudius 20490
Müller, Prior Conrad 18483
Müller, Erwin 20320
Müller, Eugen 02306
Müller, Prof. Dr. Felix 41207
Müller, Mag. Florian Martin 02113
Müller, Gabriele 17564
Müller, Gerda 18839
Müller, Giesela 18413
Müller, Dr. Gino 41545
Müller, Gisela 21183
Mueller, Dr. Gordon H. 51088
Müller, H. 17480
Müller, Hans 02924
Müller, Dr. Hans-Peter 19893
Müller, Heinz-Willi 16674
Müller, Hermann 19747, 19748, 41240
Müller, Prof. Dr. Irmgard 17404
Mueller, Jane 47112
Mueller, Jo 49650
Mueller, John 50339
Mueller, Judith 51189
Müller, Dr. Jürgen 19626
Müller, M. A. Julia 21954
Müller, K. 41696
Müller, Prof. Dr. Dr. Karlheinz 22712
Müller, Karsten 18926
Müller, Klaus 17498
Müller, Dr. Klaus-Peter 20896
Müller, Prof. Dr. Markus 20510
Mueller, Prof. Marlene 54006
Müller, Dr. Martin 18593, 18598, 22760
Müller, Matthias 19232
Müller, Dr. Norbert 02575
Müller, Peter 21993, 21995
Müller, Dr. Peter 23464
Müller, Roland 41353, 41615
Müller, Rudolf 42010
Müller, Sabine 18415, 19001
Mueller, Shirley 54344
Müller, Dr. Siegfried 17398, 20897
Müller, Stefan 20388, 20392
Müller, M.A. Stefanie 20166
Müller, Prof. Telmo Lauro 04842
Müller, Dr. Thomas 16400, 19339
Müller, Thomas 19931, 42062
Müller, M. A. Thomas T. 20401, 20402, 20403, 20404, 20406
Müller, Udo 22488
Müller, Dr. Volker 22231
Müller, Walter 16424
Müller, Prof. Dr. Werner 21389
Müller, Wolfgang 17912
Müller, Zsuzsanna 40892
Müller-Bahlke, Dr. Thomas 18891
Müller-Gaebele, Erich H. 18477
Müller-Gerbl, Prof. Dr. med. Magdalena 41153
Müller-Späth, Esther 19224
Müller-Spindler, Ines 19421
Müller-Tamm, Dr. Pia 19448
Müller-Westermann, Dr. Iris 40915
Müllner, Dr. Jenő 23410
Münch, Dipl.Lehrerin Annegret 16552
Münch, Eugen 22130
Münch, Dr. Roger 22264
Muench Navarro, Pablo 31511, 31514, 31518
Münch-Severin, Kornelia 17089
Muenkel, Harold 50348
Münzner, Harald 19417
Muertegui, Charles 34366
Müsch, Dr. Imgard 21950
Muffet, A. 45043
Mufti, Uxi 34009
Mugavero, C.J. 46224
Muggeridge, Rosalind 51218
Muheeb Zada, Mohammad Yahia 00005
Muhl, Dr. Arnold 18900
Muhlbauer, Mic 52058
Muir, Eleanor 45750
Muir, J. Duncan 47553

Muirhead, Frank H. 43930
Muirhead, George 45665
Mujanović, Prof. Fehim 04290
Mujta, Dr. Józef Stanisław 34750
Mukai, Tetsuya 29339
Mukai, Tomoko 29392
Mukanova, Najlja Arystanovna 29681
Mukasa, Josephine 42992
Mukaševa, Šajzada Tlegenovna 29673
Mukela, Manyando 54654
Mukulu 31750
Mulcahy, Michael 24857
Mulder, Drs. D. 32406
Mulder, Dr. E.W.A. 32116
Mulder, Jenny 32049
Mulhauser, Blaise 41682
Mulholland, Hugh 43444
Mulinas, Cristina 40201
Mulk, Dr. Inga-Maria 40618
Mulkey, Terry 50866, 50867
Mull, Astrid 15051
Mullally, Matt 52478
Mullen, Beverly J. 47536
Mullen, Karen 52293
Mullen, Kathleen 52050
Muller, Donald 47027
Muller, Jean-Paul 14327
Muller, Joe 49375
Muller, John 48644
Muller, Judith 51183
Mullers Bast, G.J. 32312
Mulligan, Dermot 24751
Mullin, Amanda 51827
Mullin, D. 43454
Mullin, Diane 50731
Mullin, Marsha A. 49262
Mullink, Annette 31873
Mullins, James 24845
Mullins, Lashe 52317
Mullins, Margaret 24833
Mullins, Rose 43557
Mullins, Sam 44909
Mullis, Connie 50818
Mulqueen, Grace 24860
Mulsow, Dörte 21393
Mulvaney, Rebecca 46117
Mulvany, Dr. John 47468
Mulvay, Roger 33150
Mummery, Brian 44514
Mun, Kyong Hyun 29795
Mun-lee, Lin 42416
Munakata, Pariji 28645
Munakata, Shinsaku 28715
Munalov, Temir 29671
Munch, Jytte 10294
Munch, Paul Bernard 13227
Munck, Jacqueline 14646
Munday, Graham K. 43588
Munday, Val 44176
Munder, Heike 42087
Mundorff, Angelika 18493
Mundus, Doris 19937, 19939
Mundwiller, Mollye C. 49261
Mundy, James 52121
Munk, Dr. Jan 09864
Munk, Jan 09869
Munk Pedersen, Hans 10024, 10025, 10027
Munksgaard, Jan Henrik 33613
Munktell, Ing-Marie 41009
Munns, Judith 53128
Muno, Ed 51519
Muno Jordan, Iris 48294
Muñoz, Adriana 40558
Muñoz, Concha Carmona 39623
Muñoz-Alonso, Laura 17062
Munro, Dr. David M. 44606
Munro, Gale 53927
Munro, J.A. 43727
Munro, Lewis 44495
Munroe, Alexandra 51199
Munshi, Prof. Supriya 24080
Munsick, Lee R. 51691
Munson, Larry 53066
Munson, Robin 48845
Munstermann, Leonard 51054
Munt, Martin 45613
Muntean, Ioan 36603
Muntean, Vasile 36449
Munteanu, Simona 36287, 36293
Munteanu, Vasile 36420
Munthe, Adam 44989
Muntildeoz Rivera, Lorena 08581
Munyer Branyan, Cheryl 53940
Munyikwa, D. 54666
Munzenrieder, Claire 52883
Muolo, Vicenzo 26780
Mur, Pilar 39048
Mura-Smith, Collin 51277
Mura Sommella, Dr. Anna 27439, 27458, 27522, 27543
Murai, Yasuhiko 28845
Murakami, Isamu 28537
Muramatsu, Masafumi 29442
Murariu, Dr. Dumitru 36222
Murase, Masao 28428, 29473
Murat, Francine 23320
Murat de Chasseloup-Laubat, Prince 13965
Murata, Tetsuro 28875
Muratore, P. Umberto 27911
Murauchi, Michimasa 28466
Murch, Cliff 06134
Murchie, John 06789
Murchison, Morag 45566
Murdin, Alexander 43569
Murdoch, John 52819
Murdock, Gordon R. 50735
Murdock, Helen 01629
Murenina, Galina Platonovna 37751
Murgueítio de Montoya, Lucy 08621
Murillo Toro, Edificio 08594
Murken, Prof. Dr. Jan 20957
Murko, Matthias 20784
Murlasits, Mag. Elke 01965
Murphey, F. Warren 49617
Murphy, Alistair 43929
Murphy, Anne 52045, 52596
Murphy, Arlene 50839
Murphy, Beth 50729
Murphy, D. 44150
Murphy, Emily 52658
Murphy, Dr. Jack A. 48075
Murphy, Joe 24868
Murphy, John 44001
Murphy, Kenny 46563
Murphy, Patricia 51479
Murphy, Patrick 46728, 47772
Murphy, Patrick T. 24777
Murphy, Paula 46244
Murphy, Randy 54355
Murphy, Russell 01538
Murphy, Shawna 47530
Murphy, Sue 46510
Murphy, Tony 44314
Murphy, Tricha J. 24770
Murphy, Valary 43432, 44902
Murra-Regner, Georg 17908
Murray, Alisdair 45988
Murray, Ann H. 51430
Murray, Graeme 43238
Murray, Janice 46105
Murray, Jeffrey 49045
Murray, Johanne 05347
Murray, Karin G. 53687
Murray, Kay 24822
Murray, Linda 49129
Murray, M. 05940
Murray, Mandy 43240
Murray, Mary 50803, 53676
Murray, Dr. Peter 00799
Murray, Peter 24768, 45966
Murray, Roddy 45766
Musa, George 05490
Musa, Peter 05298
Musameh, Abdulrahman S. 03212
Musango, Simon Peter 42993
Muscă, Luciana 36658
Muscă, Mihai 36125
Muschwitz, Tanja 17022, 22232
Muscolino, Celly 27374
Mușețeanu, Dr. Crișan 36221
Musil-Jantjes, Angela 33730
Muširovski, Volodymyr Makarovyč 43173
Muškalov, Sergej Michajlovič 37089
Musketa, Dr. Konstanze 18902
Mussari, Antonio 26903
Musser, Bernice 48048
Musser, Hugo 00316
Musset, Daniele 13936
Mustaine, Beverly 49896
Musto, David F. 51054
Musty, Tony 43952, 44214, 44378, 45343, 45424
Muswasi, P. 40402
Musy-Ramseyer, Sylviane 41294
Muszała-Ciałowicz, Joanna 35072
Muth, Dr. Wolfgang 20074
Muxeneder, Therese 02934
Muya Wa Bitanko, Dr. Kamuanga 08801
Muzii, Dr. Rossana 26906
Mužík, Jiří 09553
Mveng, Engelbert 05301
Mvinjelwa, A. 38757
Mya, U San 31730
Mychajlunow, Lorraine 05739
Mycock, Chris 45144
Myers, Dr. Arnold 44147
Myers, Arnold 44153
Myers, Bob 46186
Myers, Dr. Charles W. 51115
Myers, Jane 48709
Myers, Kenneth 53873
Myers, Mark 44405
Myers, Maryann 48234
Myers, Mickey 53325
Myers, Pam 49910, 49911
Myers, Pamela L. 46368
Myers, Prof. Philip 46310
Myers Jr., Barry W. 52522
Myette, Ellen M. 53915
Mygh, Liselotte 10200
Myhrvold, Sissel 33730
Myllyharju, Taina 11172, 11364, 11365
Myllylä, Pälvi 11435
Myllymaa, Marju 10724
Mylona, Alex 22928
Mylona, Zoe A. 23244
Myltygaševa, Ljudmila Petrovna 36745
Mylyadi, Dr. Wulan Rujiati 24475
Myon, Kim Ik 29754
Mytton-Famié, Elisabeth 12783
Myzgina, Valentina Vasylivna 43017
Myzina, Larisa Ivanovna 37032

N

Naab, Michael 49752, 49753
Naar, Harry I. 49997
Nabačikov, Dr. V.A. 37286
Nabaura, Trudy 24986
Nabot, Anna 25305
Nachama, Prof. Dr. Andreas 17272
Nachtigäller, Roland 19140, 20744
Nacimento Brito, M.F. do 04722
Nacke, Josef 18635
Nacking, Åsa 40705
Nad-Jerković, Erika 08852
Nadaillac, Jean de 14927
Nadan, José 13583
Nadel, D. 24968
Nader, Georges S. 23323
Nadkarni, Dnyaneshwar 24318
Nadler Thomas, Genny 46639
Nadyrova, Galja O. 29675
Nadzeika, Bonnie-Lynn 50850
Naef, Weston 50237
Naegel, Marie-Luise 41810
Nägele, Dr. Reiner 20424
Näse, Li 10959
Näslund, Erik 40890
Naess, Jenny-Rita 33863
Naeßl-Doms, Vinzenz 20247
Näther, M.A. Ulrike 17594
Nätscher, Dr. Wolfgang 20985
Nafziger, Helen 46346
Nagai, Akio 28537
Nagai, Taizan 28476
Nagai, Yoshitaka 28512
Nagasaki, Motohiro 29097
Nagashima, Kosuke 28633
Nager, Hans-Werner 41115
Naghawy, Aida 29652
Nagiełło, Zbigniew 34857
Nagl, Michaela 02935, 03029
Nagorskij, Nikolaj Viktorovič 37652, 37705
Nagueira, Dr. Lenita W. Mendes 04384
Nagy, Dr. András 23385
Nagy, Dr. Aniko S. 23438
Nagy, Dr. Árpád 23477
Nagy, Endre 23500
Nagy, Dr. Imre 23609, 23611, 23614
Nagy, Dr. István 23477
Nagy, Michael 46450
Nagy, Rebecca 48816
Nagy, Dr. Zoltán 23674
Nagy-Molnár, Dr. Miklós 23638
Nagyné-Martyin, Emília 23594
Naheb, Abél 11587
Nahon, Pierre 16097
Nahser, Dr. Siegmar 17081
Nairne, Andrew 45332
Nairne, Sandy 44940
Naish, Carolyn 44898
Naito, Makio 29303
Najafi, M. 24648
Naji Sari, Ahmed 54650
Nakabayashi, Kazuo 29491
Nakabe, Yoshitaka 29032
Nakabo, Tetsuji 28837
Nakagawa, Hajime 29254
Nakagawa, Misao 29522
Nakahara, Yusuke 28728
Nakajima, Prof. Jimi Naoki 04941
Nakamura, Hiroshi 29196
Nakamura, Hitoshi 28872, 29131
Nakamura, Junsuke 29037
Nakamura, Kenjiro 29476
Nakamura, Makoto 29190
Nakamura, Minoru 29448
Nakamura, Nobuo 28711
Nakamura, Reiko 29491
Nakamura, Saihachi 28390
Nakamura, Satoru 28392
Nakano, Masaki 28442
Nakano, Seijiro 29357
Nakano, Yuzo 46756
Nakashima, Kunio 28360
Nakashima, Michiko 29004
Nakazawa, Eigorou 28727
Nakrem, Prof. Hans Arne 33737
Naksuetrong, Somporn 42589
Namiki, Masashi 29012
Nan, Baosheng 08113
Nance, Krista 50980
Nance, Paul J. 50171
Nanfeldt, Philippa 40547
Nanni, Dr. Romano 28283
Nanni, Romano 28285
Nannup, Brett 01008
Nanu, Dan Basarab 36379
Nanzer, Angel Omar 00590
Naoki, Togawa 29394
Napier, Ellen B. 51833
Napier, John H. 50814
Náplavová, Dr. Marie 09887
Nápoli, Prof. Dr. José 00216
Naqshbandi, Ghulam 00001
Naqvi, Prof. Asif A. 24057
Naramore, Bruce E. 47618
Narbonne, Peter J. 52095
Narbutaite, Gražina 30359
Nardelli, Lucia 25490
Nardi, Christopher J. 48498
Nardin, A. 14668
Narinskaja, Natalija Efimovna 37608
Narkiewicz, Czesław 34706
Narkiewicz-Laine, Christian K. 52957
Narr, Claudia 21580
Narro, Alicia 34127
Narzt, Karl 02672
Nasalski, Zygmunt 34905
Nasarov, N. 42506
Nasby, Judith M. 05919
Nascimento, Leo 04500
Nash, Steven 51681
Nasi, Stefano 26818
Naski-Multanen, Marke 11413
Naski-Multanen, Marki 11415
Nason, Dr. James 52994
Nason Davis, Robert 52421
Naspleda i Arxer, Joan 39350
Nasset, Yvette 48508
Nassi, Talma 24949
Nasu, Takayoshi 29149
Nasybullin, Émir Salimovič 37023
Natali, Dr. Antonio 26157
Nater, Dr. Carol 41883
Nater-Schönenberger, Sandra 41912
Nathan, Gail 47041
Nathan, Jaqueline S 46977
Nathans, David 47659
Nathanson, Morris 52163
Nathiri, N. Y. 48274
Natier, Michel 13858
Natter, Dr. Tobias G. 01796
Naturel, Michèle 12459
Natuschke, Peter 16732
Natzer, Dr. E.-M. 20443
Nauck, Dr. Corinna 22589
Naudascher, Josef 20133
Nauderer, Ursula Katharina 17763
Nauhaus, Dr. Julia M. 17530
Naukkarinen, Tiina 10796
Naumann, Gerd 21055
Naumann, Heike 17130
Naumann, Johannes 22056
Naumann, Peter 20754
Naumann, Rolf 19394
Naumann, M.A. Thomas 22311
Naumann, Thomas 22312
Naumann-Steckner, Dr. Friederike 19631
Naumer, Sabine 54842
Naumova, Nadija Grygorivna 43087
Naupp, Thomas 02872
Nause, Jean Pierre 12255
Nauta, Dr. J.F. 32460
Nauwerck, Prof. Dr. Arnold 02643
Naval Mas, Antonio 39400
Navarro, Anna Dolores 38861
Navarro, Daisy A. 34445
Navarro, Edward 50262
Navarro, Elías Terés 38920
Navarro, Dr. Fausta 26146, 26173
Navarro, Dr. Fausto 26161
Navarro, Mariana Varela 08561
Navarro García, Ramón 39552
Navarro Guitart, Jesús 39478
Navarro Linares, Antonio 39359
Navas Gutiérrez, Mariano 08733
Navascués, Pedro 39568
Navascués Benlloch, Pilar 39061
Navascués Benlloch, Pilar de 39527
Nave, Dr. Francine de 03297
Naveau, Jacques 13288

M

N

Naveaux, Ralph 50791
Navia, Olga Lucía 08610
Nawafleh, Sami 29665
Nawka, Tomasz 16941
Nawrocki, Czesław 34528
Nawrot, Anna 34898
Nawroth, Dr. Manfred 17206
Naylor, Dr. Bruce G. 05713
Naylor, Eleanor 53288
Nazarov, Roman Grygorovyč 43029
Nazarov, Temirchan 29734
Naziebło, Dr. Jerzy 35381
Nazionale, Nina 51243
Ndacyayisenga, Albert 37994
Ndege, Conchita F. 49066
Neacşu, Epraxia 36141
Neagu, Gheorghe 36320
Neagu, Marian 36248
Neal, George 49188
Neal, Dr. John R. 47532
Neale, Susan 06595
Nealis, Sharon 47905
Néar, Marie-José 15980
Neault, Caroline 47744
Neave, R. 44047
Nebel, Dr. Martin 21959
Nebes, Prof. Dr. Norbert 19386
Nebl, Rudolf 02706
Nebuloni, Dr. Gabriela 26482
Nečajeva, Praskovija Mychajlivna 43016
Nečasová, Eva 09409
Necker, Hans 16771
Necker, Louis 15894
Nedelčeva, Cvetanka 05205
Nédélec-Bellevenue, Diane 14091
Nedeljković, Dragana 38151
Nedelkovska-Dimitrovska, Liljana 30520
Nédellec, François 14179
Nedeva, Galja 05268
Nedoma, Petr 09749
Neeb, Vanessa 53522
Needham, Dr. Claude 51762
Needham, Karen 07199
Needham, Melanie 07325
Neel, David 49023
Neely, Alastair 06185
Neergaard, Mogens de 10161
Nees Holcomb, Sara 50047
Neff, Emily 49397
Neff, Jean W. 46272
Neff, Jerry W. 50131
Nefzi, Tarak 42618
Negrisin, Dr. Barbara 26875
Negro, Dr. Angela 27452
Negro Cobo, Marta 39068
Neich, R. 33043
Neidert, David 46292
Neidhart, Herbert 02511
Neidhöfer, Andrea 02943
Neidigh, Ronald 48297
Neighbors, Dale 50912
Neighbours, Dr. K.F. 51307
Neil, Erik 51093
Neil, Ph.D. Erik H. 49430
Neill, Teda W. 48612
Neilsen, Henry L. 53835
Neinhuis, Prof. Dr. Ch. 21044
NeJame, Art 48047
Nejdl, Josef 09496
Neka, Suteja 24472
Nekrasov, Aleksandr Michailovič 37583
Nekrasov, Sergej Michailovič 37732
Nel, M. 38566
Nelgares Raye, José 39420
Nelhams, Mike 45875
Nelis, Laurence 03616
Nelišerová, Dr. Eva 38232
Nelje, Kerstin 40699
Nelley, Betty 33144
Nellis, Mike 46501
Nelly, Catherin 14883
Nelsen, Rick 53460
Nelson, Amanda 49524
Nelson, Bob 52411
Nelson, Bryan 05910
Nelson, Cindy J. 49733
Nelson, Cortland 46962
Nelson, Daniel A. 49948
Nelson, Darrel 48002
Nelson, Diane Bassette 49512
Nelson, Doug 48299
Nelson, Ellie 05447
Nelson, Eric 54355
Nelson, Esther 47788, 47789
Nelson, Geoff 46133
Nelson, Gwen 48147
Nelson, Henry 45198
Nelson, James R. 47824
Nelson, John 44561, 49120
Nelson, Julie D. 52216
Nelson, Dr. Larry 51782
Nelson, Lee Ann 49053
Nelson, Lori 06057
Nelson, Mike 52902
Nelson, Mindy Honey 51294
Nelson, Natalie 47977
Nelson, Philip 44561
Nelson, Ray 47887
Nelson, Rebecca 53421
Nelson, Robert 50220
Nelson, Ron 50174
Nelson, Sarah E. 46227
Nelson, Stephen R. 52824
Nelson, Susan 50009
Nelson, Taylor 47372
Nelson-Balcer, Mona 54184
Nelsson, Sigvard 40744
Němeček, Jaroslav 09492
Nemeckova, Klara 17159
Nemes, András 23807, 23811, 23813, 23814, 23815, 23816, 23820
Nemeth, Alice 07345
Németh, Dr. Péter 23733
Nemeth Ehrlich, Dorica 08986
Nemetz, Norbert 22678
Nenciu, Paul 36149
Nenno, Dr. Rosita 20881
Nenonen, Yrjo 11400
Nenov, Dr. Nikolai 05152, 05153, 05155
Nentwig, Dr. Franziska 17086, 17181, 17198, 17210, 17214, 17215, 17218, 17223, 17257, 17266
Nenz, Cornelia 21870
Nenzioni, Gabriele 27650
Nepi, Dr. Chiara 26179
Nepočatych, Viktor Alekseevič 36865
Nepomuceno, Susan 46683
Nepputh, Fritz 18091
Néray, Katalin 23432
Nerdinger, Prof. Dr. Winfried 16611, 20423
Neretniece, Anastasija 30036
Nerheim, Gunnar 33744
Neri, Joanne 46085
Nero, Dr. Donnie 53818
Nesbitt, Dr. Elizabeth 52994
Nesbitt, Mark 45491
Nesbitt, Melissa 53488
Neshushtan, M. 25032
Nesin, Jeffrey D. 50578
Nesje Bjørlo, Aslaug 33810
Nesme, H. 13453
Nesmith, Michael 52878
Nespoli, T. 31904
Nešporová, Markéta 09564
Ness, Dr. Arlin E. 46173
Ness, Gary C. 50472, 50476
Ness, Dr. Gary C. 51098
Nessell Colglazier, Gail 50423
Nesselrath, Dr. Arnold 54498, 54501, 54514
Nessi, Daniela 41633
Nesta, Prof. Paolo 25878
Nestler-Zapp, Dr. Angela 16767, 16768
Nestorova, Ani 05003
Neszmelyi, Adrienne 48179
Neto, Luis 35936
Netta, Irene 20491
Netter, Esther 50270
Netterlund, Karin 40869
Netuschil, C.K. 17791
Netzer, Dr. Nancy 47441
Netzer, Dr. Susanne 17168, 17169
Neu, Noreen 06716
Neubauer, Alexandra 18378
Neubert, André 19244
Neubert, Christhard-Georg 17246
Neubert, George W. 52726, 52726
Neubert, Hermann 20327
Neubert, Dr. Michaela 22723
Neuburger, Dr. Susanne 03042
Neuendorf, Petra 20689
Neuer, D. 19074
Neufeld, Alice 07315
Neufeld, Carol E. 53408
Neufeld, Helmut 05364
Neugebauer, Rud. 16559
Neugebauer-Maresch, Dr. J.-W. 02821
Neugent, Susan E. 46433, 46434
Neuhaus, A.F. 32455
Neuhaus, Daniel 41310, 41311
Neuhaus, Hanspeter 41146
Neuhoff, Stephan 19607
Neuhuber, Prof. Dr. W.L. 18218
Neuland-Kitzerow, Dr. Dagmar 17199
Neulinger, Adolf 01902
Neulinger, Maximilian 02252
Neuman, David 40910
Neuman, Thom 46681
Neumann, B. 01251
Neumann, Christoph 20379
Neumann, Dr. Dieter 02859
Neumann, Hanka 21779
Neumann, Herbert 01901
Neumann, Manfred 17580
Neumann, Timothy C. 48033
Neumann, Dr. U. 22120
Neumann, Ursula 51224
Neumann-Eisele, Petra 19499
Neumayer, Dr. Heino 17206
Neumayer, W. 01717
Neuner, Ernst 03131
Neuner, Meinhard 02125, 02136
Neuner, Mag. Wolfgang 02126
Neuner, Wolfgang 02136
Neunzert, Hartfrid 19794, 19796
Neusius, Gabriele 17299
Neuwirth, Karin 03051
Neuwirth, Prälat Wilhelm 01724
Neve, André 03522
Nevermann, Ulrike 19001
Neves, Julio 04880
Neves, Maria da Soledade S. 35637
Neveux, Yannick 14266
Neville Blakey, W.H. 43615
Nevin, Michael E. 53761
Nevraumont, Marcel 03651
Nevskaja, Galina N. 36819
Nevskij, Aleksej 37195
Newbery, Chris 45436
Newbery, David 45433
Newbery, Maria 45632
Newbury, Barbara 43563, 43684
Newcomb, Nancy 51243
Newhard-Farrar, Nancy 47649
Newhouse, Larry 54356
Newkirk, Leslie 53214
Newkirk, Sally 51008
Newland, Judy 46725
Newman, Bruce 48752
Newman, David 48122
Newman, Ed 05962
Newman, Lilly 52316
Newman, Megan M. 50074
Newman, R. D. 51107, 51137
Newnham, Kate 43632
Newsome, Mary 48210
Newsome, Steven C. 53845
Newton, Bette 47921, 49418
Newton, Christina 52323
Newton, Gael 00946
Newton, Paula H. 48355
Newton Gibson, Yasabel 51584
Nex, Jenny 44927
Neyer, Dr. Hans Joachim 19008
Neylon, Nicholas 51732
Neyses-Eiden, Mechthild 22102
Neysters, Dr. Silvia 18008
Nezu, Kohichi 29444
Nga, Dang Viet 54602
N'Gele, Edward O. 30181
Nghien, Nguyen Thi 54619
Ngoubeli, Joseph 08805
Ngozi, Niobe 49382
Nguyen, Phi 51193
Nguyên, T.P. 41398
Nhlane, Dr. Med 30544, 30547
Ni, Ming 08045
Nias, Peter 43582
Nica, Cristina Mihaela 36659
Nica, Viorica 36296
Nicandri, David L. 53416
Nicasius, Sandu 42518
Nichitoiu, Tinca 36510
Nicholas, Darcy 33186
Nicholas, Jeremy 45919
Nicholis, Alexandra 50531
Nicholls, David 46288
Nichols, Charlotte 49033, 53177
Nichols, Elaine 47734
Nichols, John 52902
Nichols, Lawrence W. 53523
Nichols, Madeleine 51244
Nichols, Marian 45048
Nichols, Sarah 51942
Nichols, William O. 53735
Nicholson, Claudia J. 51403
Nicholson, Joseph 51889
Nicholson, Marilyn L. 53766
Nicholson, Robert 24804, 24824
Nicht, Dr. Christoph 16619
Nichterlein, Peter 22496
Nick, Lloyd 46447
Nickel, Douglas 53585
Nickel, Manfred 21363
Nickell, Jeff 46537
Nickens, Harry 52382
Nickerson, Cindy 47846
Nickl, Dr. B.E. 17284
Nickles, Karen 48293
Nicod, Annelise 41418
Nicodemus, Dr. Victoria 31746
Nicola, A. 32893
Nicolae, Nina 36531
Nicolai, Dr. Bernd 18878
Nicolai, David 51471
Nicolaisen, C. 10274
Nicolaisen, Dr. Jan 19921
Nicolas, Alain 14001
Nicolas, Brigitte 14918
Nicolas, Jean 11628
Nicolas, Jean-Marc 41636, 41640
Nicolau, Antoni 38989
Nicolau, Irina 36227
Nicolau, Sanda Gabriela 36715
Nicolaÿ, Louis-Jean de 13604
Nicolaysen, Katja 33579
Nicolescu, Costion 36227
Nicoletta, Catherine 12585
Nicolini, Michèle 16093
Nicolini, Renato 27541
Nicolis, Irene 27933
Nicoll, Jessica 52067
Nicosia, Angelo 25256
Nicosia, Prof. Dr. Francesco 27755
Nicu, Mircea 36680
Nie, Fuhe 08209
Niederer, Robert 41480
Niederhöfer, Hans-Jörg 21959
Niedermaier, Robert 21070
Niedzielenko, Andrzej 34848, 34849
Niehaus, Dr. Andrea 17452
Niehoff, Dr. Franz 19801, 19803, 19804
Niehoff, Matthias 17278
Nield, Michael 45222
Nield, Thenetta 53121
Niell-LLabrés, Francisca 39779
Nielo Alcaide, D. Victor 39547
Nielsen, Anna Kathrine 09991
Nielsen, Anne-Marie 10145
Nielsen, Christin 40668
Nielsen, David R. 46209
Nielsen, Debbie 06801
Nielsen, Erik 10287
Nielsen, Ervin 10212
Nielsen, Hans 10259
Nielsen, Jan 51498
Nielsen, Johannes 10238
Nielsen, Jytte 10105, 10296
Nielsen, Keld 09976
Nielsen, Kim 53911
Nielsen, Kori L. 47850
Nielsen, Michele 52276
Nielsen, Niels Jul 10135
Nielsen, Ole 10049
Nielsen, Poul Bryde 10276
Nielsen, Teresa 10309
Nielsen, Wendy 06747
Niemann, Dr. Carsten 19006
Niemann, Willi 20515
Niemeier, Dr. Jörg-Peter 20994
Niemetz, Dr. Michael 19859
Niemeyer, Kirk 51515
Niemi, Anneli 11406
Niemi, Marja 11260
Nier, Eberhard 16903
Nieraad, Karla 22169
Nierengarten-Smith, Dr. Beej 46176
Nierhaus, Dr. Andreas 03052, 03053
Niermann, Katrin 17489
Niesel, Bernd 18514
Niesner, Dr. Tomáš 09445, 09609
Nieswandt, Dr. Helge 20503
Nieszczerzewska, Elżbieta 34969
Nieuwstraten, Frits 32294
Nievera, Dr. Eduardo 34451
Nievergelt, Frank 42016
Nieves, Marysol 51117
Nieweg, J. 32410
Niewęgłowski, Stanisław 35127
Niewerth, Herbert 22319
Niezabitowski, Michał 34798
Nigbur, MR Werner 17604
Niggemann, Ralf 22535
Niggemann, Dr. Stefan 19358
Niggli, Stefan 41459
Nigratschka, Kurt 19291
Nigro, Dr. Lorenzo 54501
Niinistö-Alanne, Arja 11200
Niizeki, Kimiko 29392
Nijazbekova, Sajrapan Jakubbekovna 29959
Nijkamp, J.H. 32911
Nijsten, J. 32202
Nikbakhsh, M. 24638
Nikiforov, Valerij Ivanovič 37390
Nikiforov, Viktor S. 43162
Nikita, Eleni S. 09383
Nikitović, Lidija 38094
Niklasch, Wilfried 17800
Nikočević, Lidija 08926
Nikolaev, V.A. 37625
Nikolaeva, Irina Alekseevna 37345
Nikolaeva, Lidija Dmitrievna 37447

N

Nikolaidis, J. 31659
Nikolajenko, Galina 43178
Nikolaou, Yorka 22946
Nikolasch, Dr. Franz 02383
Nikoleišvili, M.V. 16344
Nikolić, Dejan 38074
Nikolić, Dragan 38091
Nikolić, Dušica 38093
Nikolić, Jadranka 08863
Nikolić, Tatjana 38049
Nikolov, Prof. Dr. Vasil 05172
Nikolova, Asja 05045
Nikolova, Prof. Dr. Diana 05191
Nikolova, Erika 05110
Nikolova, Petja 05191
Nikoltsios, Vasilios 23223
Nikonov, Aleksandr K. 37151
Nill, Annegreth T. 47752
Nilsen, Rolf Erik 33519
Nilsen, Tore L. 33423
Nilson, Göran 40548
Nilsson, Bo 40908
Nilsson, Christer 33481
Nilsson, Christin 40634
Nilsson, Håkan 41075, 41077
Nilsson, Harald 41015, 41015
Nilsson, Isabella 40673
Nilsson, Karin 40634, 40637
Nilsson, Marianne 40956
Nilsson, Rune 40594
Nilsson, Torsten 40769
Nimac, Dr. Dragan 08979
Nimri, Mohamed 42667
Nimz, Tim 50144
Ning, Yinchang 08488
Ning, Yongqing 08019
Nini, Dr. Roberto 26915, 26916, 26917
Niniou-Kindeli, Vanna 22961
Nirša, Alla Mychajlivna 43154
Nisbet, A. 46093
Nisbet, Peter 47177
Nishi, Dr. Genjirou 29285
Nishi, Ken-ichiro 28434
Nishi, Yukitaka 28799
Nishida, Aquio 28909
Nishida, Dr. Hiroko 29444
Nishidera, Masaya 29312
Nishikawa, Kyotaro 29597
Nishimura, Atsushi 29585
Nishitakaesuji, Nobuyoshi 28407
Nishitani, Noboru 29455
Nishizawa, Jun-ichi 29244
Nishizawa, Keiichi 29095
Niskz, Siiri 11264
Nisman, Dr. V. 31904
Nissen, Carl 20451
Nissen, Dan 10128
Nissen, M.A.J.M. 32194
Nit, A. 32322
Niță, Gabriela 36661
Nitsch, Hubert 02312
Nitta, Hideo 28905
Nitter, Marianne 33863
Nittmann, Karl 01874
Nittve, Lars 40915
Nitzsche, Dr. Gunther 22024
Nitzschke, Katrin 17939
Niu, Lisuo 08145
Niven, Catharine 44525
Nivière, Marie-Dominique 11494
Nix, Dr. J.C. 32728
Nix, M. 44405
Nix-Hauck, Nicole 20642
Nixon, Chris 48167
Nixon, Isabelle 06137
Nixon, James D. 48167
Nixon, Prof. Jim 01151
Nixon, Nigel 43263
Nixon, W.A. 01209
Nizamiev, Muzip Tazievič 37015
Nižegorodova, Elena Vladimirovna 36803
Nko'o, Hulio 05302
Noack, W. 17851
Noack, Wita 17187
Noah, Randy 52973
Nobilisová, Jana 09818
Nobis, Dr. Norbert 19005
Noble, Jo 47168
Noble, Phil 48600
Nóbrega, Enrique 54543
Noda, Renate 03038
Noda, Dr. Renate 17564
Noda Gómez, Onaysi 09107
Nodges, David 44474
Nodine, Jane 53198
Noe, Adrianne 53903
Noe, Per 10202, 10203, 10204, 10223
Noeding, Tom 53440
Nöhammer, Josef 01926
Nöhmayer, Andreas 02293
Noel, Albert 03527
Noël, Ètienne 15010
Noël, Jacques 14363
Nörenberg, Bernd 16466
Nöth, Dr. Stephan 16918
Noetinger, Robert-Mare 14790
Noever, Prof. Peter 02995
Noever, Peter 03026
Noguchi, Mitsunari 28467
Noguchi, Reiichi 29392
Nogueira, Abrelino 33045
Nogueira, Sandy 33045
Nogueira da Costa, Maria Inés 31709
Nohsoh, Toshio 28441
Noirot, André 12102
Noko, Oarabilke J. 04298
Nolan, Edward V. 53416
Nolan, John 49078
Nolan, Julia 43893
Nolan, Patrick B. 49446
Nolan, Peg 50465
Nolan Jones, Nancy 47621
Noland, Dr. Martin R. 52572
Nold, Carl 46932, 48464, 48944, 51065, 52094, 52099, 52933, 53163, 53164
Noli, Gino 31743
Noll, Linda M. 49220
Noll, Roger 46681
Nollar, Philip 53679
Nolle, Karl 17958
Nollée, Armand 03266
Nollert, Dr. Angelika 20786
Nolley, Lance 50140
Nolte, Joseph 47307
Nolte-Fischer, Dr. Hans-Georg 17795
Nomayo, Stephanie 19561
Nombluez, Pieter 03693
Nome, Alexander 46229
Nommensen, Dr. Sven 17522
Nomura, Harukata 28857
Nonesteid, Mark 48237
Nong, Wentian 07894
Nonnenmacher, Eugen 17423, 17424
Noon, Patrick 50739
Noon, Rosemary 54104
Nooter Roberts, Mary 50223
Noppe, Catherine 03893
Nørager Pedersen, Erik 09962
Nørbach, Lars Chr. 10196
Nørbach, Lars Christian 09955
Nørballe, Inge 10022
Norberg, Deborah 52809
Norby, Lory 50387
Nordal, Bera 40843
Nordberg, Henri 11385
Nordberg, Olle 40700
Nordbrock, Carol 46491
Norden, Linda 47177
Nordgren, Sune 33733, 33735, 33736
Nordheim, Lutz von 22785
Nordiguian, Levon 30158
Nordin, Maxine 05985
Nordine, Dr. Malki 00071
Nordlinder, Gunnar 40926
Nordlund, Ivar 11316
Nordmo, Per E. 33531
Nordmo, Sverre 33517
Nordon, Deborah 51221
Nordqvist, Jørgen 09981
Nordstrand, Polly 48072
Nordström, Lena 40794
Nordstrom, Alison 52395
Nordstrom, Ralf 11424
Nordtorp-Madson, Shelly 52630
Norell, Dr. Mark A. 51115
Norgaard Larsen, Peter 10152
Norheim, Kari 33525
Norheim, Svein 33525
Noriega, Natividad 34495
Noring, Ann-Sofi 40915
Norkute, J. 30237
Norland, Lissie 33513
Norley, Mark 49671
Norlund, Yvonne 40871
Norman, Dr. David B. 43733
Norman, Heather L. 48833
Norman, Jan 49097
Norman, Margaret Jane 54114
Normann, Mark 45327
Normann, Øystein 33535
Normark, Ben 46269
Norovtseren, Oyuntegsh 31634
Norrington, S. 01552
Norris, B. 33069
Norris, Baden N. 33074
Norris, Johnny 46993
Norris, Dr. Patrick 49667
Norseng, Per 33607
Norseng, Per G. 33742
Northam, J. 06703
Northcote, P.S. 33219
Northcutl, Ginia 50194
Northrop Adamo, Amy 51177
Northwood, John 01305
Norting, Jack 49013
Nortmann, Dr. Hans 22102
Norton, Elaine 47198
Norton, Fran 53360
Norton, George 54107
Norton, I.S. 01224
Norton, Jessica A. 46272
Norton, Kelly 48926
Norton, Lyle 49915
Norton, Maria 50921
Norton, Mary Lynn 52403
Norton jr., Richard W. 53088
Norton, Scott L. 48648
Norton, Susan E.S. 53867
Norton Moffatt, Laurie 53302
Norvaišas, Ignas 30397
Norvell, Edward 52681
Norwood, Nancy 52396
Nory, Jean-Marc 15152
Noryńska, Barbara 35109
Nose, Dr. Martin 20472
Nosek, Elizabeth 49905
Nosič, Dr. Stipe 08868
Noskova, Alla Anatoljevna 37038
Nostrala, Justin 49507
Nota, Dr. Maria Teresa 27463
Notari, Bernard 12966
Notet, Jean-Claude 14191
Nothnagl, Hannes 02415
Nothnagl, Reinhold 02753
Notin, Véronique 13765, 13770
Notley, Ben 45385
Nottage, James H. 49490
Notteghem, Patrice 13576
Notter, Eveline 41407
Nougarède, Martine 14453
Nougué, Cendrine 14201
Nouvel, Odile 14682
Nouyrit, H. 12219
Novak, Diane 52280
Novak, Gabriela 02439
Novák, Dr. László 23727
Novák, Dr. Pavel 09615
Novak, Stanislav 08864
Nováková, Hana 09776
Nováková, Dr. Lenka 09427
Nováková, Zuzana 09842
Novakovska, Nada 30512
Novellis, Mark de 45888
Novershtern, Prof. Avraam 25121
Novgorodova, Prof. M.I. 37250
Novick, Leslie 48319
Novikov, Timur 37704
Novo, Marla 52868
Novoselova, Rimma Vjačeslavovna 37545
Novotná, Dr. Mária 38284
Novotná, Mariana 38279
Novrie, Robert 49277
Nowack, Meg 53935
Nowacki, Janusz 35024
Nowacki, Marek 35149
Nowaczyk, Szymon 34630
Nowak, Cornelia 18196
Nowak, Holger 19382, 19384
Nowak, Joanna 35081, 35146
Nowak, Ralf 18209
Nowak, Romuald 35343
Nowakyniak, Dr. Edmund 34984
Nowobilski, Dr. Józef Andrej 34786
Nowotny, Erich 02240
Noyes, Julia 50104
Nozaka, Motoyoshi 28499
Nozet, Hervé 16015
Ntisana, N. 38697
Ntsema Khitsane, Julia 30179
Nu, Li 08045
Nuber, Dr. W. 20529
Nuculov, Georgi 05139
Nudelman, Robert 50233
Nuding, Dr. Matthias 20764
Nüesch, Ernst 41151
Nüesch, Rosmarie 41932
Nützel, Dr. Alexander 20472
Nugent, M. 05825
Nuki, Tatuto 28641
Nummelin, Esko 11220
Nummelin, Rauni 11345
Numonova, Gulchehra Tojitinovna 42498
Nunes, Dave 52032
Nunes, Dr. Graçia Soares 35998
Nunes, Isabella 04682
Nunes de Vasconcelos junior, Walter 04746
Nunes de Oliveira, Adilson 04447
Núñez, César Marku 40280
Núñez Centella, Ramón 39213, 39215
Nunge 12442
Nunley, John 52595
Nunn, Chris 45806
Nunn, Dr. Tey Marianna 52882, 52883
Nunziati, Anna 25681
Nupen, Nikki 31752
Nurabullaeva, Svetlana 54451
Nurfeisova, Nurshamal Bekvossunovna 29683
Nurger, Heli 10583
Nuri, Kedir 10686
Nurmi, Virpi 11260
Nurmi-Nielsen, Anna 11260
Nursan, Alimbaj 29678
Nurtaev, Kaken Bejsenbekovič 29731
Nurullayev, Mansur 54437
Nusbaum, Pat 48388
Nussbaum, Patricia 41653, 41707
Nussbaum, Prof. R.A. 46310
Nussbaumer, Christiana 02808
Nußbaummüller, Winfried 01795
Nußer-Wagner, Maria 16749
Nusupova, D.N. 29952
Nutkins, Geoff 45667
Nutt, Ruth J. 53013
Nuttall, R. J. 38567
Nuttall, Rick 38569
Nutting, Ryan 52498
Nutzi, Hans 41591
Nuyttens, M. 03426
Nwaneri, Justin C. 33336
Nyakas, Dr. Miklós 23603
Nyamabondo, Constantinus M. 42507
Nyamenya, Peter 29742
Nyblod, Ruth Ann 46359
Nybom, Jeny 11405
Nyce, Doug 54185
Nye, Steve 45507
Nye, Tom 47043
Nyegaard, Georg 23265
Nyein, U Kyaw 31721, 31725
Nyerges, Alexander Lee 47984
Nygaard, Jon Erik 33706
Nygaard, Knut M. 33491
Nygårds, Eva 40547
Nygren, Kris 05822
Nykänen, Veijo 11233
Nykolyshyn, Z. 07364
Nylén, Lars 41023
Nyman, Ola 40907
Nyman, Sven-Eric 40998
Nys, Greta 05365
Nys, Wim 03321
Nystad, Marianne 33906
Nyström, Hazze 10122
Nysveen, Aagot 49285
Nyunt, Kyaw 31737

O

Oaklander, Christine L. 46225
Oakley, Lisa 49817
Oakley, Steve 46848
Oaks, Martha 48948
Oalka-Konieczko, Małgorzata 34811
Oandasan, Crescencion 34395
Oanh, Phan 54627
Oates, Sarah 46398
Oathout, Kathleen C. 52638
Oatley, Val 45871
Oatman, Michelle 51025
Obal, Franc 38486
O'Bannion, Tami 50669
O'Baoill, Eilis 45260
O'Barr, James B. 47875
Obedat, Nassim 29640
Obenaus, Maria 17975
Obenhaus, Maria 17955
Oberchristl, Monika 02317
Oberer, Brigitte 02412
Oberfrank, Willi 02341
Oberg, Paul Chancy 33374
Oberhänsli, Silvia 41361
Oberhänsli, Dr. This 41600
Oberhaus, Betty 47880
Oberholenzer, Gunther 02200
Oberlin Steel, Virginia 47196
Obermeier, Christoph 20643
Obermeyer Simmons, Helen 48549
Obernosterer, Beatrix 02180
Obernosterer, Mag. Beatrix 02192
Obhof, Dr. Ute 19435
Oblin, Annie 13756
Obojes-Rubatscher, Johanna 02457
O'Bomsawin, Nicole 06499
Obradović, Slobodan 31637
Obrascova, Maja Aleksandrovna 37614
Obrecht, Elsa 41886
O'Brian, John 48603
O'Brien, Barry 01495
O'Brien, David 24879
O'Brien, Eileen 50423
O'Brien, Freda 33268

O'Brien, James 54094
O'Brien, Kevin 49865
O'Brien, Maureen 52170
O'Brien, Michael J. 47717
O'Brien, Patrick J. 53714
O'Brien, Ruairí 17938, 17942
Obrist, Dr. Gabrielle 42016
Obrist, Hans-Ulrich 14646
O'Bryen, Antoinette 24772
Ocaña, Maria Teresa 39002
Ocaña i Goma, M. Teresa 39001
O'Carroll, Helen 24904
Ocello, Claudia 51286
Ochi, Catherine Megumi 29383
Ochi, Yujiro 28728
Ochonko, Nikolaj Anatoljevič 37805
Ochsenmeier, Peter 18515
Ochsner, Henrik 09962
Očirova, Tsiren-Handa V. 37894
O'Connell, Daniel M. 51966
O'Connell, Michael 24762
O'Connell, Shirley 51892
O'Connor, Prof. Barry M. 46310
O'Connor, Emma 44700
O'Connor, Karen 48590
O'Connor, M. 33182
O'Connor, Michael J. 48398
O'Connor, Thomas P. 24786
O'Connor, Tony 45924
O'Conor Nash, Marguerite 24757
Odavić, Djordjo 04287
Odawara, Kaname 28798
Odden, Tor 33786
Odding, A. 32471
Ode, Robert 12305
Odegaard, Nancy 53584
Odehnal, Petr 09895
O'Dell, Kathryn 54041
Odendaal, E.A. 38738
Odenhoven, Dr. S.L. 32814
Odenthal, Joachim 19660
Oder, Karla 38503, 38504
Odermatt, Eduard 41089
Odess, Daniel 48472
Odin, Samy 14660
Odoj, Romuald 35134
O'Donnell, Carolyn 52870
O'Donnell, Janice 52174
O'Donnell, John 48030
O'Donnell, Molly 49614
O'Donnell, P. 45317
O'Donnell, Robert M. 50444
Odrowąż-Pieniążek, Janusz 35256
Odum, R. Andrew 53522
Oechslin, Dr. Ludwig 41296
Oegema, D. 38755
Öhlknecht, Rupert 02742
Öhman, Nina 40946
Öjemyr, Hans 40941
Ökten, Haydar 42913
Oellers, Dr. Adam C. 16404, 16408
Oelschlägel, Dr. Petra 17020
Oelschlaeger-Garvey, Barbara 50399
Oerichbauer, Edgar S. 49513
Oerlemans, Magi 03990
Örsi, Dr. Julianna 23939
Oertel, Michael 19920
Oertlin, Jean-Paul 14498
Oesch, Joseph 41383
Oeschger, Dr. Bernhard 19437, 21867
Östberg, Maths 41063
Östberg, Wilhelm 40892
Oester, Hanspeter 41088
Österholm-Granqvist, Sonja 11425
Österlund, Petter 40958
Oesterwind, Dr. Bernd C. 20241
Oetterli, Christoph 42005
Oettingen-Wallerstein, Fürst zu 17474, 19009, 22314
Oettinger, Dr. Marion 52726
Oexle, Dr. Judith 17956
Özbey, Şahin 42777
Özcan, Hilmi 42875
Özçelik, Serpil 42820
Özdemir, Abdulhalik 42727
Özerdem, Hürrem 42689
Özeren, Öcal 42910
Özgen, Emine 42730
Özgerey, Mehmet 42840
Özkütük, Zafer 42736
Özsağır, Ahmet 42858
Öztürk, Erdal 42847
Offers, Donald 41137
Offringa, D.C. 38755
Offroy, Martine 15932
Offutt, Cathy 49480
O'Flaherty, Rosemary 05726
O'Foighil, Prof. Diarmaid 46310
Ogata, Hisanobu 29543
Ogawa, Toshiro 29367
Ogden, Dale 49497
Oggenfuss, Daniel 20491
Ogilive, Peter 45601
Ogle, Brian 43641
Ogle, Nigel 33121
Ogle, Susan 54188
Ogliari, Prof. Francesco 27362
Ogny, Jean d' 13332
O'Gorman, Dr. Jodie A. 48254
O'Gorman, Tim 48645
O'Grady, Brian 50808
Ogris, Mag. Kerstin 02414
Ogrodniczak, Tomasz 35033
O'Grody, Jeannine 46822
O'Halloran, Jim 33208
O'Hanlon, Dr. Michael 45338
O'Hara, Bruce 53545
O'Hara, David 07114
O'Hara, Michael 45931
Ohashi, Koji 28365
Ohashi, Yasushi 28780
Ohe, Dr. Werner von der 17687
O'Hear, John 50750
O'Hern, John D. 48378
Ohgane, Takuya 29162
Ohl, Thomas 22507
Ohlmann, Dan 13897
Ohlsén, Marie 40684
Ohlsen, Dr. Nils 18138
Ohlsson, Birger 40458
Ohm, Dr. Matthias 21950
Ohmann, Agathe 10321
Ohmura, Etumasa 28938
Ohneberg, Dr. Nicole 02050
Ohngemach, Dr. Ludwig 18068
Ohno, Akira 29599
Ohno, Masaji 29041
Ohrt, Karsten 10217
Ohrt, Nils 10192
Ohta, Yoshinori 28456
Ohtsubo, Kenji 29522
Øien, Elisabeth Adeleide 33484
Oilliamson, Comtesse d' 12962
Oinas, Tapani 11277
Ojalvo Prieto, Roberto 08686
Ojeda Sato, Alonso 08690
Ojha, M.B. 24198
Okabe, Nobuyuki 29578
Okada, Akiko 28562
Okada, Hidenori 29493
Okada, Hirashi 29008
Okada, Itsuki 28477
Okajima, Hisashi 29493
Okániková, M. 38391
Okano, Kiheita 29286
Okano, Shun-ichiro 29402
Okawa, Eiji 28703
Okaya, Michiko 48269
Okazaki, Kyoichi 29578
Okazaki, Dr. Masaki 29185
Okazawa, Kunro 29077
O'Keefe, Barbara K. 47803
O'Keefe, John F. 51805
O'Keefe, Raymond 51206
O'Keeffe, Tony 43903
O'Kelly, Barry 24778
Oki, Shinichi 29104
Okimoto, Hiroshi 28534
Okkenhaug, Eli 33418, 33419, 33420
Økland, Olaug N. 33470
Okonogi, Miyoko 28703
Okoshi, Hisako 29190
Okubo, Takishi 28708
Okubo, Tokiko 29588
Okumura, Yoshitsugu 28932
Okuoka, Shigeo 29220
Okura, Haruhiko 28811
Olachea, Elena 31045
Olaczola, Juanjo 38931
Ólafsdóttir, Sigríðkr{}ur Melrós 24022
Olalla Mercade, Leandro 39603
Olano Trujillo, Manuel José 08727
Ólason, Vésteinn 24037
Olausson, Magnus 40735, 40918
Olbrantz, John 52666
Ołdakowski, Jan 35266
Olde Dubbelink, R.A. 32615
Olden, Sam 54328
Oldenburg, Rie 23259
Oldham, Terry L. 52560
Oldknow, Tina 47810
O'Leary, Daniel E. 52067
Olejniczak, Józef 35008
Oleščins'ka, Alla Pavlivna 43000
Olfra, Michelle 50786
Olijnyk, Michael 51953
Oliva, Francisco Miguel González de la 39357
Olivarelli, Sebastiano 27085, 27086, 27087
Olivares, Isabel 34108
Olivarez, Fátima 04223
Olivas, Art 52883
Olivas, Arthur 52885
Olive Garrard, Jacqueline 00652
Olivé Serret, Prof. Enric 40103
Oliveira, Jorge de 35758
Oliveira, Dr. José Carlos 35463
Oliveira, Natércia 35802
Oliveira, Nicolas de 44926
Oliveira, Pedro 35802
Oliveira Caetano, Dr. Joaquim 35587
Oliveira Mesquita Spränger, Paulo Roberto 04757
Oliveira Rodrigues, Adélia E. de 04341
Oliver, Dr. Graham 43748
Oliver, Katherine C. 50369
Oliver, Katie 49824
Oliver, Marvin 52994
Oliver, Miguel 39777
Oliver, Robyn 01170
Oliver, Virginia 48531
Oliveri, Meg 48856
Oliveri del Castillo, Patrizia 28289
Olivero García, Lucía 23300
Olivié, Jean-Luc 14682
Olivier, Louis 15358
Olivo, Sandra 50653
Ollenburger, Jeff 49454
Ollikainen, Anja 11454
Olmedo, Freddy 10400
Olmedo Patiño, Dolores 31110
Olmer, Dr. Beate 22175
Olmo Fernández, Jesús del 39072
Olmo Ilarta, Francisco Javier del 40012
Olmstead, Dr. Richard 52994
Olmsted, Betty 51717
Olofsson, Jan 40485
O'Looghaire, Dr. Niamh 07153
Olschowski, Petra von 21949
Olsen, Dr. Arnold 49242
Olsen, Betsy 48570
Olsen, Ed 49553
Olsen, Johanne 23268
Olsen, John 33633
Olsen, Phyllis J. 48965
Olsen, Sandra H. 47125, 47126
Olsen, Tammery 47982
Olsnes, Aanund 33570
Olson, Alice 47285
Olson, David A. 51341
Olson, Don 53405
Olson, Dona A. 51970, 52494
Olson, Dorothy 46437
Olson, Janis 46714
Olson, Joan 53097
Olson, Kathleen 50069
Olson, Kathryne 51634
Olson, Kathy 51639
Olson, Kristina 50845, 50846
Olson, Link 48472
Olson, Marilyn 51790
Olson, Nicky L. 47922
Olson, Oliver 53255
Olson, Roberta 52929
Olson, Ruth Ann 49989
Olson, Suzanne 52062
Olson, Thomas H. 52494
Olson, William 06612
Olson Peters, Ruth 49955
Olsson, Daniels Sven 40516
Olsson, Murray 00771
Olsson, Susanne 40639
Olsson, Tommy 40466
Olszak, Eliane 13012
Olszanowski, Zdzisław 35340
Oltean, Dr. Vasile 36181
Olthoff, Gabriele 22491
Oltmanns, Kay 46959
Oltmans, Kay 46958
Olucha Montins, F. 39148
O'Malley, Dennis M. 52165
O'Malley, Jeanette 51724
O'Malley, Tom 54300
Oman, Dr. Hiltrud 02643, 02644
Oman, Richard G. 52544
Omar, Hadmar 42525
Omar, Hamad H. 42511, 42522, 42523, 42524
Omar, Jana 09841
Omar, R.H. 38607, 38609, 38610, 38611
O'Mara, John F. 06845
O'Meara, John 46395
O'Meara, Nora 24883
Omedes, Dr. Anna 38976
Omeñaca, Jesus 39806
Omer, Gil 24977
Omer, Prof. Mordechai 25122
Ometto, Giulio 28013
Omi, Hideo 29055
Omieczyński, Witold 34850
Ommen, Dr. Eilert 20718, 20719
Omtvedt, Petter M. 33742
Omurbekova, Ainagul 29955
Onaga, Stephen 52205
Ondevilla, Alex 34346
Ondish, Andrea 53655
Ondrejčeková, Zora 38233
Ondrušeková, Dr. Anna 38333
Onea, Nicolae 36172
O'Neill, Anthony 44412
O'Neill, Colleen 05642
O'Neill, Heidi 51167
O'Neill, John 51192
O'Neill, John J. 46554
O'Neill, Mark 44287
O'Neill, Peter 01663
O'Neill, Vicki 01392
Ono, Masaharu 29166
Onodera, Hiroshi 29085
Onodera, Reiko 29392
Onorevoli, Dr. Giuseppe 27966
Onuki, Dr. Yoshi 34304
Onuma, Sunao 29374
Onyejekwe, Dr. Umebe N. 33346
Onyejiako, Joy 44808
O'Connor, David 00773
Oortmerssen, Frank F. van 32870
Oost, T. 04021, 04022, 04023
Oost, Tony 04020
Oostdijk, Dr. Alexandra E. 32906
Oosterbaan, A. 32047
Oosterhof, H. 32829
Oostra, Dr. R.J. 31868
Ootukasa, Mitukuni 29116
Ooyama, Masaaki 28912
Op de Beeck, Bart 03438
Op de Beeck, Jan 03865
Op de Beeck, Jozef 03857
Op de Beek, R. 32014
Opaleye, A.O. 33327
Opaleye, Yemi 33327
Opdenberg, Laurence 03445
Operta, Mevlida 04284
Ophir, Daniel 25102
Opinel, Annick 15348
Opolă, Valentina 30062
Opp, Edmund 48442
Oppenhimer, Ann 55134
Opperman, G. 38764
Oppizzi, Rosita V. 41983
Oprea, Emanuela 36345
Opstad, Jan-L. 33925
Oquendo, Dante 34498
Oquiche, Antonio 34278
Orač-Stipperger, Dr. Roswitha 01986
Oram, Dr. Richard W. 46510
Oratowska, Barbara 34906
Oražem, Vito 18266
Orbons, Dr. A. 32398
Orchard, Dr. Karin 19005
Ordine, Mauriziano 27915
Ordoñez Delgado, Napoleón 08574
Ordoš, Dr. Ján 38335
Ore, Phyllis 06765
Oredsson, Prof. Lars-Göran 40704
Orefici, Giuseppe 34280
Oreglia, Luciano 25393
Orehovec, Martina 38467
O'Reilly, Margaret 53561
O'Reilly-Lawrence, Priscilla 51083
Oremus, Frederick L. 51002
Oren, Dan 24932
Orf, Susanne 18324
Orfali, Dalila 00046
Orgeix, Charles-Louis d' 13828
Orgill, Peter 43688
Orgonikidze, I. 16358
Oriol, Calvo 38890
O'Riordan, E. 24872
O'Riordan, Ian 44116
Orkin, Frederic I. 49281
Orlandi, Giuseppe 28303
Orlando, Assunta 26567
Orlando, Fran 51344
Orlando González, Jorge 08552
Orlić, Alfons 08985
Orloski, Julie 52980
Orlov, Evgenij Michajlovič 37703
Orlov, Viktor 36805
Orłowski, Jarosław 34696
Orlson Agbo, Raymond 22850
Ormenyi, Steven 50229
Ormerod, Robin 06098
Ormhaug, Knut 33418, 33419, 33420
Ormond, Dr. Rupert 45149
Ormonde, Dr. Helena 35439
Ornsby, Philip 44414
Oroianu, Silvia 36672
Oropeza Segura, Manuel 31302
Oroveanu, Mihai 36220
Orozco, Prof. Margarita V. de 30885
Orozco, Sylvia 46515
Orozco Móron, Edna Maria 31137
Orphal, Christina 20069
Orr, Brenda P. 06319

Orr, Charlene 50612
Orr, Dr. Patrick J. 24836
Orr-Cahall, Christina 54073
Orrick, Mary Norton 51752
Orro, Leena 10884
Orsi, Dr. Oriana 26403
Orsini, Ph. 15925
Orski, Marek 35122
Ørskou, Gitte 09962
Orsler, Christine 45501
Ortal, Dr. R. 24991
Ortega, Enric 40291
Ortega, Luis Angel 39970
Ortega DeSantis, Diana 52885
Ortega Rivera, Enric 40291
Ortiz, Deborah V. 52508
Ortiz, Gustavo A. 08558
Ortiz, Rosa 38987
Ortiz Crespo, Alfonso 10434
Ortiz de Zevallos, Felipe 34275
Ortiz Lan, José E. 30744
Ortiz Muñoz, Idania 08643
Ortiz Volio, Melania 08824
Ortlepp, Thomas 18474
Ortmeier, Dr. Martin 20236, 20240
Ortner, Josef 03040
Ortner, Dr. M. Christian 03001
Orton, Anna 45962
Orton, Donald 53242
Oruch, Sam 51522
Orzschig, Cathrin 19927, 19939
Orzsi, Zoltán 23361
Osawa, Tomomichi 28994
Osborne, Bernard 54227
Osborne, Duncan 43280
Osborne, Lois I. 47645
Osborne, Lori 48450
Osborne, Peter 52037
Osborne, Richard 48796
Osborne, S. 44744
Osborne, Victoria 46108
Oscarsson, Ulla 40796
Oselies, David 05312
Oset, Hans Petter 33563
O'Shaughnessy, Margaret 24834
O'Shea, Deborah H. 52452
O'Shea, John 46308
O'Shea, Patrick B. 24859
O'Shea, William A. 06197
Oshel, Laurie 48854
Oshima, Seiji 29468
Osimitz, Stefanie 41288, 41664
Oslen, John C. 05738
Oslund, Christopher 05925
Osman, Prof. Ali Mohamed 40386
Osmann, Gudrun 18096
Osmond, Lynn 47461, 47462
Ospina Cubillos, Carlos Enrique 08589
Ospina de Muñoz, Esperanza 08702
Ospina Navia, Francisco 08751
Ospino de la Rosa, José 08638
Osses, Dietmar 17403
Ostapova, Svitlana Ivanivna 43156
Østargård, Niels Jøan 10272
Ostašenko, E.J. 37410
Ostaszewska, Stanisława 35168
Osten, Dr. Till 21959
Ostendorf, Eleanor 46257
Ostendorf, Dr. Thomas 22022, 22023
Ostenrieder, Dr. Petra 20878
Østerås, Bodil 33879
Osterhus, Cynthia B. 52680
Ostermiller, Jerry 46401
Osterrith, Martine 03917
Østervig, Torsten 10178
Osticresi, Dr. Patrizio 26176
Ostlund, Ralph 51103
Ostojić, Milan 38163
Ostrenko, Wit 53434
Ostritz, Dr. S. 21357, 22410
Ostrosky, Amelia 54074
Ostrouchova, Marina Georgievna 37101
Ostrovskaja, Raissa Iljinična 37163
Ostrow, Mindy 51627
Ostrow, Victor 52976
Ostrowski, Prof. Jan 34821
Ostrum, Meg 50654
Ostwald, Thomas 17520
O'Suilleabhain, Sean 24735
O'Sullivan, Dr. A.M. 24912
O'Sullivan, Keith M.C. 43740
Osyra, Tadeusz 34762
Ota, Diane 46923
Ōta, Motoji 29457
Otáhal, Ivo 09587
Otani, Kazuhiko 29443
Otani, Shogo 29491
Otavová, Marie 09925
Otero, Álvaro 08623
Otomo, Tsutomu 28989
O'Toole, Joseph 50336
O'Toole, Judith H. 49070
Otsuka, Masahito 29044
Otsuka, Ronald 48072
Otsuka, Yoshio 29540
Ott, Hanspeter 41783
Ott, Irene 18419
Ott, John H. 50061
Ott, Lili R. 51534
Ott, Tanja 21081
Ott, Wolfgang 22461
Otte, Prof. 03803
Otte, Els 04125
Otte, Stacey A. 46524
Otte, Wulf 17518, 22474
Otterspeer, Prof. Dr. W. 32481
Ottillinger, Dr. Eva 03005
Ottinger, Bénédicte 15715, 15716, 15717
Ottini, Alicia 00571
Ottmann, Helmut 20758, 20778
Ottnad, Clemens 21284
Otto, Henry S. 51069
Otto, Dr. Julia 17688
Otto, Kristin 22776, 22777
Otto, Martha 47757
Otto, Veronika 21485
Otto-Hörbrand, Martin 21951
Otto-Mller, Jackie 51590
Ottomeyer, Prof. Dr. Hans 17072
Otton, Dr. William G. 47818
Ottosson, Jon Gunnar 24026
Ottou Onana, Irène 05296, 05299
Ottowitz, Arthur 01788
Ottrubay, Dr. Stefan 01894
Otttaviani, Jean-Claude 11546
Ou, Prof. Shan-Hui 42360
Ouarzazi, Abdul Wahed 39089
Oudet, Benoît 14419
Ouellet, Nicole 06680, 06697
Ouellet, Raynald 06325
Ourecky, Irma 54157
Ourmieres, Jean-Guy 15523
Outerbridge, Amanda 04200
Outridge Field, Stephanie 01086
Ovalle Neira, Pedro Ubert 08745
Ovaskainen, Heikki 11441
Ovečkinaova, Olga 37676
Overdick, Dr. Thomas 18327
Overdiep, Dr. G. 32130
Overgaard, Iben 10316
Overland, Christian 48009
Overlie, Jill 50193
Overmyer, Lorraine 48211
Overstreet, Laura 46212
Overton, Jonathan 43412
Oviedo de Coria, Beatriz 00585
Ovsjankina, Ljudmila Vladimirovna 37853
Øvstebø, Solveig 33417
Owczarski, Marian 51583
Oweis, Iman 29664
Owen, Ann 44770
Owen, Brian 45891
Owen, Gwilym 44770
Owen, Paula 52720, 52729
Owen, Thomas S. 53547
Owens, David 45560
Owens, Heather 50892
Owens, Marshall 49356
Owens, Sue Ellen 48013
Owens, Suzanne 51516
Owens-Celli, Morgyn 50172
Owers, Lorna 44054
Owingayak, Winnie 05375
Owyoung, Steven D. 52595
Oxley, Nicola 44926
Oya, Mina 29419
Oygur, Dr. Sule 51287
Oyuela, Raul M. 47801, 50629
Ozaki, Masaaki 29491, 29492, 29493
Ozanam, M. Yves 14693
Ozay, Alev 42845
Ožegova, Ljudmila Ivanovnā 37045
Ozola, Agrita 30134
Ozola, Ārija 30011
Ozolina, Vita 30062
Ozolins, Diana 49551

P

Paap, Hans-Uwe 19836
Paarlberg, Bill 53174
Paarlberg, S. 32149
Paas, Dr. Sigrun 18056, 20146
Paaverud, Merl 46835, 50559, 54180
Paaverud, Merlan E. 46837, 46839, 51753
Pabst, Fieke 31898
Pabst, Stewart J. 47784
Paccoud, Stéphane 11486
Pace, Dr. Lorenzo 53665
Pacere, Tifinga Frédéric 05275
Pačev, Nikolaj 05080
Pache, Pierre 42049
Pacheco, Anelise 04719
Pacheco, Gustavo 04665
Pacholik, Kazmimierz 35298
Pachomova, Nadežda Petrovna 36894
Pachter, Marc 53907
Pacini, Marina 50569, 50577
Pacjukov, Vladimir V. 37114
Packard, Andrea 53389
Packard, Elinor 48960
Packard, Robert W. 51667
Packard, Vance 47787
Packauskas, R. 49225
Packer, Dr. Claude 28330
Packer, Nancy 51827
Packheiser, Michael 20351, 20352
Paczkowski, Dr. Jörg 22503, 22504
Padalino, Nick 31838
Padalkina, Olga Viktorovna 36789
Paddock, Eric 48069, 48072
Paddock, Dr. J.M. 43847
Paderni, Loretta 27519
Padfield, Joe 44937
Padgett, Michael 52151, 52363
Padgett, Sharon 47896
Padhy, Dr. K.P. 24251, 24284, 24363
Padian, Kevin 46758
Padovan, Dr. Stefanie 25877
Padovani, Dr. Serena 26161, 26173
Padurariu, Nicolae 36227
Paemelaere, Jean-Pierre 03949
Pärdi, Heiki 10656
Paesano, Dr. Ciro 26893
Paetzold, Angelika 20937
Pagac, Gerald J. 50761
Pagano, Dr. Denise Maria 26904
Pagano, Enrico 28155
Pagano, Ezio 25341
Pagano, Sergio 54496
Page, Eric 49681
Page, John 43411
Page, Jutta Annette 53523
Page, Loa 50088
Pagé, Suzanne 14646
Page Harrington, A. 53916
Pagella, Dr. Enrica 28004
Pagès, Solange 13625
Paget-Woods, David 45390
Pagnotta, Ph. 12363, 12365
Pagot, Serge 14376
Pagsinohin, May M. 34461
Pagulayan, Teresita L. 34435, 34441
Pahl, James R. 51492
Pahlke, Dr. Rosemarie E. 17924
Pahlow, Martina 22081
Þahýn, Seracettin 42845
Paiardini, Lino 27227
Paidasch, Christoph 03019
Paige Linn, Nancy 52146
Paillard, Ph. 13890
Pailler, Henri 12869, 14188
Paillet, Antoine 15804
Paillet, Patrick 15393
Paillot, Noël 15109
Paine, Fiona 44165
Paine, Tony 33068
Paine, Wesley M. 50972
Païni, Dominique 15444
Paino, Prof. Giuseppe 26893
Painter, Bill 49665
Païpetú, Andreas 23065
Pais, Teresa 35611
Pais de Brito, Joaquim 35717
Pais Pinto, Joaquim 00106
Paisie, Cinar 36668
Paiva, Ricardo Luís 35453
Paiva Cavalanti, Lauro Augusto de 04766
Paizis-Paradelis, Constantin 23147
Pajović, Maja 38105
Pajskrová, Dr. Jana 09513
Pakesch, Peter 01976
Pakkala, Taru 10706
Pakoma, Katariina 10886
Pakuningrat, Maulana 24465
Pakzad, Dr. Amir 19004
Pal, Raj 45953
Pal, S.N. 24273
Pala, Valter 27638
Palacios i Manuel, Josep 39099, 39101
Palacpac, Remedios 34411
Palacz, Tomasz 34989
Palágyi, Dr. Sylvia 23959
Palas, Jan 09743
Palata, Oldřich 09625
Palavonov, D. 42491
Palella, Dr. Graciela Elizabeth 00337
Paletta, Céline 11948
Pálffy, András 03074
Palgi, Adva 25057
Palisse, A. 15905
Páll, Dr. István 23824
Pall, Martina 01971
Palladino-Craig, Allys 53422
Pallan, Norbert 02533
Pallanca, Dr. Claude 31612
Pallavicini, Maria Camilla 27445, 27448
Pallestrang, Kathrin 03049
Pallottino, Paola 26112
Pallu, Jocelyn 22249
Pálmason, Ólafur 24025
Palmberg Eriksson, Stina 40782
Palme, Prof. Dr. Bernhard 03054
Palmeiri, John 47034
Palmenaer, Els de 03295
Palmer, Lord 44063
Palmer, Cheryl 47392
Palmer, Lindsay 51910
Palmer, Mike 43342
Palmer, Oonagh 44212, 44334
Palmer, Dr. Oonagh 44420
Palmer, Peregrine 46040
Palmer, Ros 45866
Palmer, Sharon S. 49777
Palmer, Stephen 44206
Palmer, Sue 46000
Palmgren, Christine 40558
Palmgren, Kaj 10723
Palmgren, Ola 40634, 40637
Palmiero, Franesco 26048
Palmquist, Janice 52667
Palmquist, Dr. Lena 40922
Palmquist, Linda 47777
Palmy-Brandstätter, Lotte 42014
Palo-oja, Ritva 11359
Paloheimo, George B. 52873
Palokangas, Markku 10803, 10812, 10818, 10819
Palombo, Aldo 51409
Palombo, Ionah 24984
Palomero, Santiago 40136
Palomo, Antonio 23284
Palop, José Alcántara 38896
Palou Sampol, Joana M. 39789
Palov, Dr. József 23836
Palsa, Raija 11072, 11073, 11074
Pálsson, Lýdur 23995, 23996
Paltrinieri, Roberto 25437
Palzerová, Hana 09508, 09509
Pamart, Jean-Luc 14476
Pambrun, Jean-Marc 16327
Pamperin, David L. 46876
Pamuk, Ömer 42933
Pan, Giuseppe 26996
Pan, Junxiang 08183
Pan, Yue 07707
Pan, Zhaolin 07923
Pan, Zheng 08189
Pan, Zhenzhou 07556
Panabière, Louis 31044
Panafieu, Bruno de 15339
Panagiotopoulos, Prof. Dr. Diamantis 19069
Pañares, Tonette 34373
Panaskin, Vladimir Vasiljevič 36828
Panawatta, Senarath 40380
Panayotova, Dr. S. 43727
Panazza, Dr. Gaetano 26513
Pančeva, Petrana 04973
Pančíková, Eva 09410
Pančuk, Fedir Vasil'ovyč 43203
Panczenko, Dr. Russell 50388
Panda, Prof. C. 24283
Pandey, A.K. 24334
Pandey, Dr. S.N. 24099, 24161
Pandilovski, Melentie 00775
Pandolfi, Sylvia 31153
Pandžić, Ankica 08997
Pane, Josh 52508
Panea, Traian 36710, 36711
Panek, Kornelia 18171
Paner, Henryk 34636
Panese, Prof. Dr. Francesco 41546
Pangburn, D.A. 52350
Pange, Roland de 14559
Panić, Djordje 38150
Paniccia, Antonio 28226
Panizza, Marco 27933
Panjuškin, A.D. 37188
Pankok, Eva 19293
Pannabecker, Rachel K. 51397
Panneton, Jean 07177
Pannewitz, Otto 21774
Panni, Frédéric 13169
Pansi, Dr. Heinz 02071
Pant, Kurt 03067
Pant, Dr. Madhu 24144
Panteghini, Ivo 25539
Pantellini, Giovanni 41257
Panter, Dr. Armin 21652
Panter, John 52749
Pantermalis, Prof. Demetrios 23222

Pantoja, Dianne 47200
Pantoja, Hernán Gil 08681
Panţuru, Prof. N. Ioan 36216
Panvini, Dr. Rosalba 25607
Pany, Franz Norbert 20590
Panzanaro, Mauro 26456
Panzig, Dr. Christel 20114
Paoletta, Dr. Donald 50094
Paoli, Arturo 25613
Paoli, Feliciano 28119
Paolillo, Dr. Antonio 26009
Paolillo, Giuseppe 25373
Paolozzi Strozzi, Dr. Beatrice 26196
Papadakis, N. 22872
Papadakis, Dr. Nikos 23188
Papademetriou, Dr. Eleni 09373
Papadopoulos, A. 09375
Papadopoulos, Prof. N. 22930
Papadopoulou, Eleni 22961
Papadopoulou-Pangou, K. 23054
Papageorgiou, Ioannis 23077
Papagno, Dr. Francesco Saverio 27151
Papait, Jean 12405
Papakonstantinou, Maria-Photini 23070
Papangeli, Kalliopi 22992
Papantoniou, Ioanna 23117
Paparella Treccia, Raffaelle 27166
Paparone, Guiseppe 27935
Papazian, Aline 51728
Papazoglou, Lena 22920
Papco, Ján 38201
Papendieck, Dr. Christine 22321
Papendorf, Dorothee C. 21618
Papet, Edouard 14692
Papetti, Dr. Massimo 27608
Papetti, Prof. Stefano 25298, 27728
Papike, Dr. James J. 46179
Papillard, Marie Laure 16078
Papin, Gregory J. 47186
Papin-Drastik, Y. 12661
Papone, Elisabetta 26292
Papounaud, Benoît-Henry 15270
Papp, Dr. Claudia 21466
Pappi, Evangelia 23115
Papst, Helga 02151
Papuc, Gheorghe 36431, 36697
Papuš, Ivan Agafonovyč 43145
Paquelot, Françoise 15367
Paquette, Adrian 51738
Paradis, Michèle 06457
Paradisopoulos, Sofronis 43157
Páramo, Amanda Lucía 08528
Paramo Sureda, Ernesto 39370
Paramonov, Vladimir Aleksandrovič 37873
Paraschos, Georgios 23151
Paraskevas-Gizis, Evaggelos 23225, 23227, 23228
Parberg, Dee 52150
Parchman, Bryn 46585
Pardilla, Helena 08773
Pardini, Prof. Edoardo 26178
Pardo, Alfredo 08555
Pardo, Isabel 39544
Pardo, Rengifo 08547
Pardue, Diana 51904
Pardue, Diana R. 51262
Pare, Glenn S. 52839
Paredis, Albert 03402
Parent, Béatrice 14646
Parent, Geraldine 06583
Parent, Dr. Thomas 17921, 17930
Parenti, Dr. Daniela 26159
Paret, John J. 46487
Pargner, Dr. Birgit 20435
Parham, Herbert S. 49159
Pariente, Myriam 13769
Pariessien, Steven 45944
Parikka, Anne 10743
Parillo, James D. 52924
París Pérez, Dr. Joaquín 39017
Parisch, Manfred 02520
Parise, Dr. Roberta 27036
Parisella, Prof. Antonio 27532
Pariset, Jean-Daniel 14694
Parish, David W. 48877
Parisi, Chiara 11899
Park, Janette A. 44615
Park, Mee-Jung 29933
Park, You Chul 29784
Park Curry, Dr. David 52343
Park Leggett, Barbara 48235
Parke, Julie 50289
Parker, Angela 44789
Parker, David 48616
Parker, Donna 46973
Parker, Elizabeth 52571
Parker, Eris 33065
Parker, Harry S. 52772, 52781
Parker, Herschell 50965
Parker, John 00952
Parker, Mackenzie 07199
Parker, Margaret 52016
Parker, Niles 50946
Parker, P. 45041
Parker, Ray 46704
Parker, Robert 48867
Parker, Sandi 46685
Parker, Tim 06442, 06449
Parker, Dr. W.A. 45025
Parkes, Susan 24816
Parkhurst, Jo 48564
Parkin, Peg 00856
Parkinson, C.D. 43273
Parkinson, Glenn 48731
Parkinson, Gordon 45856
Parks, Dennis 53004
Parks, Gary W. 50031
Parks, Heather 51278
Parks, Patty 47412
Parks, Robert E. 51216
Parks, Ron 47852
Parks, Shirley 54218
Parlic-Barisic, Slobodanka 30520
Parma, Raphaël 14831
Parmal, Pamela A. 46940
Parmalee, Dr. Paul 49818
Parman, Susan 48808
Parmentier, J.-P. 13753
Parmentier, Stéphane 03605
Parnell, Dr. Geoffrey 44968
Parnes, Stuart 48421
Parod, Dr. Barbara 01985
Parola, Publio 00513
Parpagiola, Dr. Giuseppe 26968
Parr, Jesse 50538
Parr, Melissa 51545
Parra, Raúl 31152
Parra Casamun, Gerardo 31124
Parra Santos, Alirio 08630
Parreiras Horta, Maria de Lourdes 04629
Parreño Rivera, Prof. Manuel 30194
Parrick, Allen 00908
Parrilla de Alvarez, Estela 30858
Parrinello, Prof. Nicolo 27074
Parris, Benjamin J. 48858
Parrish, M. 44556
Parrish, Ron 48155
Parrott, Robert W. 50523
Parrott, Shirley 52473
Parrow, Enid 53726
Parry, Wendy 45527
Parsch, Dipl.-Rest. Marianne 20478
Parsekiau, Robert 53970
Paršina, Lidija Semënovna 36817
Paršina, Ljudmila Vladimirovna 37921
Parsons, Brian 01003
Parsons, Ernie 06198
Parsons, Jeffrey R. 46308
Parsons, K. 01550
Parsons, Louise 51512
Parsons, Ralph 45732
Parsy, Paul-Hervé 14503
Partanen, Päivi M. 11045, 11048, 11050
Partin, Marc 54312
Parton, Neville 01158
Partridge, Victoria 43435
Partyka, Józef 34970
Parusel, Janina 35364
Parušev, Todor 05244
Pârvan, Lucia 36160, 36161, 36162, 36163, 36164, 36349, 36350, 36351, 36445, 36474, 36593, 36635
Parviainen, Helena 11026
Pârvulescu, Dorina 36690
Paša, Natko 08989
Pascal, Huston 52232
Pascall, Chris 45918
Paschen, Dr. Joachim 18928
Paschke, Ilona 20085
Pascual, Jordi 33307
Pascual, José 39405
Pascucci, Marisa 50815
Păsculescu, Gică 36509
Pasetti, Elena 27399
Pasiciel, Stanisław 34673
Paskova, Spaska 05035
Pasquali, Prof. Flavia 27831
Pasquet 13326
Pasquier, Thierry du 14663
Pasquini-Barisi, Isabella 27979
Pass, Robert 46466
Passaglia, Elio 26751
Passarin, Dr. Mauro 28253
Passe, Eugene 52268
Passi, Leena 10708
Paßlick, Tonio 22375, 22378
Passoja, Merja 10973
Pasté de Rochefort, Aymeri 12607
Pastellas, Dominiue 14714
Paster, Carol 52020
Pasternak, Natalija Anisimovna 37540
Pasti, Sara 51097
Pastor, Paloma 39957
Pastor Vázquez, Prof. J.F. 40230
Pastore, Heather 48374
Pastore, Prof. Michele 27942
Pastorek, Ivan 38258
Pastori, Jean-Pierre 41637
Pastoriza, Juan A. 36040
Pasture, B. 03885
Paszylka-Glaza, Małgorzata 34642
Patà, Franco 41889
Pataki, Dezsö 23529
Patané, Dr. Rosario 25803
Patay, Dr. Franz 03014
Patchen, Dr. Jeffrey 49488
Pátek, Jiří 09807
Patek, Sheila 46758
Paten, Richard 45369
Patenaude, Val 06237
Pater, Dr. Józef 35327
Pater, Philippe 03831
Paterson, Anne 45166
Paterson, John 31747
Paterson, Neil 44916
Pathak, Dr. Bindeshwar 24155
Pathak, G.M. 24418
Patin, Sylvie 14692
Patnode, J. Scott 53206
Paton, W.J. 33080, 33081
Patorska, Joanna 34917
Patrick, Bill 01376
Patrick, Brian 33095
Patrick, J.A. 43509
Patrick, Pat 53565
Patrick, Ruth 51809
Patrick, Stephen E. 46967, 46968, 46969, 46970, 46971
Patry, J. Michel 50038
Patry, Sylvie 14692
Patterschal, Donald 47418
Patterson jr., Aubrey B. 53612
Patterson, David 45138
Patterson, Ken 05433
Patterson, Lillian 46199
Patterson, Dr. Mark A. 47128
Patterson, Richard 53562
Patterson, Robert B. 52639
Patterson, V.A. 49557
Patterson, Vivian 54176
Patterson-O'Regan, J.J. 43488
Patteson, Rita 53771
Patti, Lynn 53966
Pattie, M. 33194
Pattinier, Philippe 11908
Pattinson, G. 45614
Pattison, Alison 44322
Patton, Becky 51907
Patton, James L. 46759
Patton, LuAnn W. 46197
Patton, Robert 53576
Patton, S.M. 46372, 46373
Patwa, Elizabeth 11956
Patzak, Dr. Beatrix 03055
Patzer, George 05967
Patzig, Eberhard 19911
Paugh, James J. 54303
Pauka, T.R. 31914
Paukert, Karel 47624
Paukšte, Regina 29982
Paul, Dr. April 51144
Paul, C.D. 45295
Paul, Eli 49693
Paul, Frédéric 11993
Paul, Gary A. 48815
Paul, Gayle 52106
Paul, Renee 33196
Paul, Ryan 47290
Paul, Shimrath 33095
Paul, Yvonne 45597
Păuleanu, Dr. Doina 36302, 36306, 36468, 36698
Pauletić, Mirjana 08851
Pauley, Ed 51713
Pauley, Edward E. 48511
Pauli-Gabi, Thomas 41554
Paulicka-Nowak, Dr. Lucja 34755
Paulo, Dālia 35596
Paulocik, Linda 07319
Paulová, Dr. Eva 09736
Paulsen, John Q. 48511
Paulsen, Nils 33493
Paulson, Christopher 52224
Paulson, Dr. Dennis R. 53409
Paulson, Warren 05360
Paulsrud, Geir 33486
Paulter, Charles D. 50133
Paulus, Gottfried 02620
Pauly, Carola 21707
Pauly, Dorothy 47328
Pausch, Gottfried 01838
Pausch, Hans-Jürgen 17540
Pause, Dr. Carl 20654
Paust, Dr. Bettina 16975
Pautet, C. 14046
Pautrizel, Françoise 11982
Pauwels, Alexandra 03862, 03863
Pavaloi, Dr. Margareta 19087
Pavel, RNDr. Václav 09681
Pavelčík, Dr. Jiří 09887
Pavelcikova, Monika 38245, 38361, 38362
Pavelec, Karen Marie 48434
Pavelka, Dr. Karel 09896, 09918
Pavia, Prof. Giulio 28014
Pavic, Stjepan 04292
Pavić, V. 08924
Pavić, Vladimira 09009
Pavičić, Snježana 08997
Pavlát, Dr. Leo 09801
Pavletić, Mira 08848
Pavlíčková, Pavla 09727
Pavlides, Paul 12374
Pavliňák, Dr. Petr 09700
Pavljuk, Vasil Mykolajovyč 43014
Pavlov, Aleksandr Anatoljevič 36822
Pavlov, Slavi 05081
Pavlova, Irina Vasiljevna 37904
Pavlu, Viktor 03013
Pavlunina, Ljudmila Fëdorovna 37978
Pavúková, Olga 38306
Pavuljak, Jaroslav Ivanovyč 43192
Pawłowska, Krystyna Iwona 35309
Pawłowski, Dr. Antoni J. 34840
Pawlowski, Pawel 34752, 34753
Pawlowski, Sebastian 09763
Paxton, Janet 44342
Payet, Hugues 36090
Payling, Catherine 27456
Payne, Prof. Anthony 44290
Payne, Elizabeth 05653
Payne, James 53797
Payne, Mary 50012
Payne, R.G. 45714
Payne, Prof. Robert B. 46310
Payne, Susan 45366
Payrich, Wolfgang 02076
Payton, Charles 45457
Payton, Sidney 48085
Paz, Dr. Boaz 17238
Paz, Diego Alonso 08724
Paz, Olivier 14094
Paz, Valeria 04223
Paz Morales Osorio, Maria de la 31073
Pazmandy Horvath, Margaret 52039
Pazoutova, Katerina 09562
Pčelinceva, Nelli Sergeevna 37057
Peabody, Ann F. 49735
Peabody Turnbaugh, Sarah 51742
Peach Lynch, Ann 52452
Peacock, Pat 53351
Peacop, James 45181
Peak, Marianne 52218
Peake, Jane 54112
Pearce, Andrew 45505
Pearce, Barry 01544
Pearce, John N. 48768
Pearce, Pelham G. 47319
Pearce, Robert J. 06186
Pearsall, Barbara 51394
Pearson, A. Faye 05966
Pearson, Dave 46401
Pearson, Fiona 44156
Pearson, Katherine 45800
Pearson, Marion 53618
Pearson, Rudy 06654
Pearsons, Don 07353
Peashock, Kristine 53586
Peaucelle, Denis 11621
Peavy, John M. 49579
Peball, Dr. Helga 02238
Pečara, Dr. Zdravka 38061
Peces Rata, Felipe Gil 40078, 40079
Pech, Edwin 34717
Pech, Dr. Jürgen 17606
Pecháček, Ivo 09603
Pechacek, Wilson 51473
Pechal, Dorothy 53473
Pecharromán y Morales, Ricardo 39814
Pechová, Dr. Jarmila 09427
Pecht, M.A. Ulrike 19076
Peci, Innocenzo 27418
Pečiukonienė, Mariona 30221
Peck, James F. 53610
Peck, Katie 53280
Peck, Scott 47927
Peck Bracy, Tralice 50780
Peckham, Cynthia A. 52424
Pecoraro, Joseph P. 47450
Peddell, Lesley 01587
Pedder, John 45057
Peden, Harold R. 53662
Pedersen, Birgit 09962

Pedersen, Jørgen 09963
Pedersen, K.S. 10156
Pedersen, Kenno 10005, 10065, 10297
Pedersen, Knud 10134
Pedersen, Lisbeth 10113
Pedersen, Lykke 10165
Pedersen, Morten 09955
Pedersen, T. 10156
Pederson, Curt 50727
Pederson, Jennifer 47534
Pedevillano, Diane 52123
Pedlar, Arthur 54041
Pedley, Maggie 43577
Pedneault, Richard 07261
Pedretti, Cornelia 41813
Pedrini, Dr. Claudia 26398, 26399, 26400, 26402, 26403
Pedrocco, Dr. Filippo 28179, 28188
Peduzzi, Francis 12246
Peebles, Cheryl 07295
Peek, Andre 53089
Peel, Prof. John S. 41005, 41006, 41018
Peele, Anna M. 48696
Peeperkorn-van Donselaar, Dr. L.A. 32702
Peer, Dr. Peter 01982
Peers, Elida 06991
Peers, Dr. Laura 45338
Peers-Gloyer, Axel 21115
Peet, Catherine 43455
Peeters, Herman 03548
Peeters, L. 03320
Peeters, Ludo 03550
Peeters Saenen, E. 03552
Peffer, Susanne 17173
Peggie, David 44937
Peić Čaldarović, Dubravka 08997
Peiffer, Jacques 13208
Peik, Karl-Heinz 18946
Peikštenis, Eugenius 30407
Peill, James 43827
Pein, Max Gerrit von 21934
Peine, Arnold 16912
Peiry, Lucienne 41535
Peisch, Dr. Sándor 17100
Peischl, Erich 03056
Peitler, Karl 01978
Peitler, Mag. Karl 01985
Pejčinova, Ekaterina 04981
Pejčochová, Michaela 09776
Pejković, Božidar 08888
Peković, Mirko 38091
Peláez Arbeláez, Leopoldo 08587
Peláez Tremols, Dr. Lucía 39335, 39340
Pélatan, Martial 11502
Pelch, Andrea 52346
Pełczyk, Antoni 34847
Pelegrini, Dr. Béatrice 41422
Péligry, Christian 14571
Pelikánová, Jaroslava 09679
Pélissié, Gérald 12491
Pelkey, Ann 51971
Pella, Nicholas W. 51546
Pellard, Sandra-Diana 12444
Pellegrinetti, Jerri 07088
Pellegrino, Fabrizio 26048
Pelleman, Lode 03747
Pellen, Robin 43939
Pellengahr, Dr. Astrid 19492
Pellew, Robin 44504
Pellmann, Dr. Hans 20127
Pelon-Ribeiro, Laurence 12689
Pelong, Dany 11984
Pelosio, Prof. Giuseppe 27109
Pelser, R. 38772
Pelsers, Dr. L. 32213
Peltier, Cynthia 50029
Pelz, H. 17597
Pemberton, E. 01159
Pena, Dr. Elizabeth 47118
Peña Castillo, Agustín 31029
Peña Javier, Estrellita 34491
Peña Reigosa, Maria Cristina 09088
Peñalver Gómez, Henriqueta 54584
Penard, Michel 16246
Penati, Fabio 26868
Pence, Debra 49219
Pence, Debra L. 47711
Pence, Noel 49415
Penčev, Prof. Dr. Vladimir 05191
Penčeva, Radka 05254
Penda, Octavian Ion 36214
Pender, Branko 08954
Pendleton, Chris 48661
Penent, Jean 15943, 15944
Peng, Fang 08520
Peng, Guirong 08034
Peng, Qi 07710
Peng, Shifan 08028
Peng, Yunqiu 08089
Pengzeng, Yuguan 07738
Pénicaud, Pierre 12552
Penick, George 49177
Penick, Steven 52540
Penin, David R. 53430
Pénisson, Elisabeth 14779
Penka, Bradley 49833
Penketh, Sandra 45415
Penkina, Elena Ivanovna 36989
Penkov, Petio 05249, 05250, 05251, 05258
Penn, Joyce 47811
Penn, Michael 45724
Penna, Liisa 46403
Penna, Dr. Vasso 22905
Pennant 04085
Penndorf, Jutta 16497
Penner, E. Paul 07284
Penner, Jim 07007
Penner, Liisa 46400, 46404
Penney, David 48114
Penney, S.F.J. 06416
Penning, Dr. Mark 38612
Pennington, Claudia L. 49758, 49762, 49763
Pennino, Gaetano 27059
Penno-Burmeister, Karin 19760
Pennock, David 49462
Penny, Nicholas 44937
Pensabene, Dr. Giovanni 26533
Pensom, Carole 06082
Pentangelo, John 46590
Pentrella, Dr. Ruggero 27517
Penvose, Jane 53529
Penz, Marta 28137
Penzin, Viktor 37339
Pèon, Marie-Thérèse 13905
Peoples, Jo 00780
Peora, Prof. Oscar Carlos 00262
Pepall, Rosalind 06357
Pepić, Eva 30192
Pepich, Bruce W. 52223
Pepin, Didier 13464
Pepion, Loretta 47092
Pepper, Myles 44231
Pepper, Victoria 45596
Peppler, Tina 05491
Perac, Jelena 38074
Perales Piqueres, Rosa 39082
Peralta, Juan 34211
Peralta, Prof. Tole 07436
Peralta Juárez, Juan 38804
Peralta Vidal, Gabriel 07443
Perchinske, Marlene 47718
Percival, Anthony 06808
Percival, Arthur 44221
Perco, Daniela 25840
Percy, Ann B. 51867
Percy, David O. 47376
Perego, Graziella 26362
Pereira, Maria Isabel de Sousa 35447
Pereira, Prof. Sonia Gomes 04744
Pereira Cabral, Augusto J. 31707
Pereira de Araujo, J. Hermes 04751
Pereira Herrera, David M. 04206
Pereira Neto, Prof. João 35707
Pereireira Reis, Brasiliano 04498
Perelló, Antonia M. 38972
Perepelkina, Aleksandra Fëdorovna 37787
Peretti, Pierre de 14240, 14246
Pereyra Salva de Impini, Prof. Nydia 00620
Pérez, Fernando Alberto 08721
Pérez, Juan Fernando 10449
Pérez, Júlia 38994
Perez, Patricia 34368
Pérez, Rosa María 23304
Pérez Avilés, José Javier 40189, 40192
Pérez Basurto, Alejandro 31078
Pérez Camps, Josep 39612
Perez Castro, Pedro 39936
Pérez Espinosa, Juan Fidel 31271
Perez Gollan, Dr. José Antonio 00230
Pérez González, Melba 09280
Pérez Gutiérrez, Arturo 31098
Pérez Maldonado, Raúl 23294
Pérez Mejía, Dr. Angela 08548
Pérez Mesuro, Dolores 40152
Pérez Moreno, Rafael 39525
Pérez Navarro, Amalia 39550
Pérez Ornes, Nora 10341
Pérez Outeiriño, Bieito 40018
Pérez Quintana, José Andrés 09106, 09137
Pérez Recio, Manuel 39454
Pérez Rodríguez, Francisco Javier 39770
Pérez Ruán, Liliana 08785
Pérez Villarreal, Pedro 08759
Perg, Franz 01803
Pergament, Allen 53146
Pergola, Prof. Philippe 28216
Pería, Alejandra 31106
Périer-d'Ieteren, C. 03444
Perifanakis, J. 42064
Perillat, Shirley 05717
Perin, Giovanni 27822
Périn, Dr. Patrick 15291
Périnet, Francine 06495, 06496
Perini, Mino 27394
Perisho, Sally L. 48066
Perišin, Stane 08854
Perissinotto, Antonio 26564
Perkins, Bruce C. 54258
Perkins, Deborah 53033
Perkins, Diane 45797
Perkins, Karen 45048
Perkins, Martin 48230
Perkins, Sophie 44868
Perkins, Stephen 49039
Perkko, Mariliina 10727, 10729
Perl, Alexander 17621
Perlee, A. 31618, 31625, 31632
Perlein, Gilbert 14428
Perline, Carol 52845
Perlis, Natasha 52771
Perlman, Patsy 46559
Perlov, Dr. Diane 50215
Pernas Ojeda, Maria Emma 31072
Pernicka, Prof. Dr. Ernst 20173
Pernot, M.M. 53564
Pero, Linda 53302
Péron, Christian 14236
Peron, Claude 14828
Perot, Jacques 12025, 12599
Perra, Dr. Carla 25675
Perram, Richard 00838
Perraudin, Marguerite 41576
Perrault, Lise 07195
Perreault, Michel 06018
Perrefort, Dr. Maria 18969
Perren, Stephan 42053
Perrenoud, Raymond 41307
Perret, Gilles 41680
Perret, Jean-Claude 13705
Perret, Pascale 41546
Perret-Sgualdo, Janine 41678
Perrier, Dr. Danièle 16724
Perrigo, Thomas E. 01685
Perrine-Dubost 12345
Perron, Jean L. 06858
Perron, Michelle M. 48108
Perrot, Annick 14735
Perrot, C. 14305
Perrot, Mollie 51628
Perroud, Paul 13348
Perry, Andy 50282
Perry, Dr. Barbara 47392
Perry, Bob 46593
Perry, Carole 51449
Perry, Ed 53727
Perry, Gregory P. 51033
Perry, James 53685
Perry, John 54028
Perry, Karen E. 51364
Perry, Michael L. 48994
Perry, Nancy 49683, 50362, 52106
Perry, Peter 00960
Perry, Rianne 47417
Perry, Sheila 50529
Persan, Marquis de 13078
Perse, Marcell 19405
Persengieva, Marina 05098
Peršin, Vladimir 08886
Persson, Ann-Charlotte 40769
Persson, Bertil 40698
Persson, Curt 40699
Persson, Dr. Per-Edvin 11446
Persson, Sven 40554
Persson, Thomas 40634
Pertin, B. 24231
Pertola, Esko 10969
Péru, Dr. Laurent 14357
Peruga, Iris 54543
Pesarini, Dr. Fausto 26108
Pescaru-Rusu, Adriana 36330, 36344, 36381
Peschel, Dr. Patricia 19550, 22372
Peschel, Tina 17199
Peschel-Wacha, Claudia 03049
Peschler, Malgorzata 42098
Pescia, Sergio 41898
Pešek, Dr. Ladislav 09626
Pesenecker, Marita 18761
Pesi, Dr. Rossana 26533
Pešková, K. 38391
Peso, Jean-Michel del 15379
Pesolillo, Carlo 41624
Pesqueira de Esesarte, Prof. Alicia 31222
Pesquer, Dr. Ismael Angulo 38798
Pesqueur, Jean-Yves 12251, 12252
Pessa, Joanna 51217
Pessa, Loredana 26316
Pessemier 's Gravendries, Paul de 03772
Pessey-Lux, Aude 11545
Pessoa, Miguel 35548
Pessy, Emanuelle 15688
Pestel, Michael 51946
Pestiaux, Pierre 03846
Petchey, Tim 46024
Peter, Carolyn 50227
Peter, Christopher 38727, 38728
Peter, Elmar 02166
Peter, Dr. Frank-Manuel 19598
Peter, Jack 52529
Péter, Dr. László 23752
Peter, Dr. Markus 41137
Peter, Dr. Michael 41767
Petercsák, Dr. Tivadar 23536, 23541, 23543, 23544
Petermann, Jörg 18103
Petermann, Ralf 16605
Peterposten, Carlo 41097
Peters, Dr. Dirk J. 17569
Peters, Georg 18991
Peters, Dr. Gustav 17469
Peters, Heinrich 17524
Peters, Irmtrude 18009
Peters, Karen 44828
Peters, M. 32149
Peters, Michael 22618
Peters, Ted H. 51976
Peters, Dr. Ursula 20764
Peters, W. 05516
Peters Bowron, Edgar 49397
Peters-Schildgen, Dr. Susanne 21192
Petersen, Christopher 45870
Petersen, Flemming 09976
Petersen, Jim 46764
Petersen, John 52758
Petersen, Michele 06772
Petersens, Magnus Af 40915
Peterson, Alan 53565
Peterson, Allan 51771
Peterson, Allan E. 52761
Peterson, Brent 53294
Peterson, Brian 48179
Peterson, Dr. Cheryl D. 49610
Peterson, Dennis 53204
Peterson, Glenn 51221
Peterson, James L. 52629
Peterson, John 52698
Peterson, Karin 49737
Peterson, Leslie 49282
Peterson, Mark F. 54238, 54239
Peterson, Merlin 48942
Peterson, Mildred 46722
Peterson, Pam 50460, 50462
Peterson, Paula J. 51392
Peterson, Roger 52838
Peterson, Thomas 46853
Peterson, Tom 52544, 53582
Peterson, Townsend 49991
Peterson, William 50935
Peterson-Moussaid, Kathleen 46516
Petey-Hache, Philippe 15896
Petit, Jean-Paul 12027
Petit, Nathalie 03919
Petit, René 12326
Petit, Prof. Th. 15826
Petke, Debra 49195
Petkov, Dimităr 05186
Petkov, Pavel 05046, 05047
Petkov, Vladimir 05157
Petout, Philippe 15388, 15389, 15390
Petr-Marčec, Prof. Smiljana 08852
Petranov, Borislav 02999
Petre, John L. 33183
Petrelli, Dr. Flavia 26905
Petrenko, Aleksej Stanislavovič 37757
Petrenko, Vladimir Grigorjevič 37121
Petri, Gerda 10122
Petri, Susanne 19921
Petrica, Vasile 36561
Petridis, Constantine 47624
Petrie, Brian 45394
Petrie, Hugh 44821
Petrins, Aivars 30092
Petrinski, Ivan 05191
Petritsch, Dr. Ernst 02985
Petritsch, Mag. Tanja 02625
Petrone, P. Paolo 26894
Petroskey, Dale A. 47791
Petrov, Angel 04989
Petrov, Evgenij 42999
Petrov, Najden 05159
Petrova, Lidija 05237
Petrova, Nadežda 36967
Petrova, Velička 05054
Petrović, Dejan 38094
Petrović, Dragutin 38163
Petrović, Marijana 38074
Petrovič, Vladimir 38259
Petrović-Raić, Marijana 38074
Petrovnina, Galina Michajlovna 37542
Petrovski, Zoran 30520

P

Petrovszky, Dr. Richard 21834
Petrozzello, Carol 53600
Petruccelli, Lic. M. Rosa 00264
Petrucci, Stefano 27172
Petruchno, Alla Sergeevna 36751
Petrus, Jerzy T. 34821
Petrus, Magdolna 23969
Petrusevičienė, Vilmantė 30204
Petrussen, Thomas 23257
Petruța, Bliderişanu 36569
Petry, Michael 44926
Petryshyn, M. 06004
Petschar, Dr. Hans 03045
Petschek-Sommer, Birgitta 17805, 17806
Petsini, Penelope 22953
Pett, Joachim 16598
Pettengill, Shirley 51708
Pettersson, Jan 40621
Pettersson, Kerstin 40952
Pettersson, Lennart 40547
Pettibone, Jody 53846
Pettibone, John W. 48946
Pettit, Marilyn 47068
Pettit, Steve 53442
Petty, Dr. Edgar 51526
Petty, Ian 46053
Petty, Laura 53634
Petty-Johnson, Gayle 54342
Petuchov, Jurij Michajlovič 37210
Pétursdóttir, Brynhildur 23985
Pétursdóttir, Rakel 24022
Pétursson, Björn 23997
Pétursson, Halldór 23984
Petz, Reinhold 01701
Petzäll, Leif 40995
Petzal, Monica 44850
Petzold, A. 17882
Petzold, Lorraine 06173
Peuckert, Valérie 03325, 03326
Peukert, Jörg 18459
Peuzin, Jean 14982
Pevereit, S. 45270
Pevernagie 03502
Pewitt, Dr. J. Dudley 46825
Peyer, Ricard de 45066
Peyer, Richard de 45065
Peylhard, Anne-Marie 11788
Peyre, P. 14071
Peyton-Jones, Julia 44984
Pezarro, D. 31807
Pezzino, Prof. Antonio 25777
Pfäffli, Barbara 41137
Pfaff, Bettina 22323
Pfaffe, Andreas 21521
Pfaffenberger, Dr. Gary 52056
Pfaffenhuemer, Dr. Johannes 02379
Pfanner, Dario 26537
Pfanner, Ute 01796
Pfanz, Donald C. 48764
Pfau, F.D. 21573
Pfau, Gisela 04541
Pfeifer, Anton 01842
Pfeiffenberger, Deborah 51030
Pfeiffer, Brian 46949
Pfeiffer, Günter 04003
Pfeiffer, Dr. Ingrid 18391
Pfeiffer, Rainer 21726
Pfeiffer-Rupp, Prof. Dr. Rüdiger 18746
Pfeifhofer, Elisabeth 01942
Pfeifhofer, Helmut 01942
Pfeil, Elke 17033, 17055
Pfennings, Edgar 07465
Pferdehirt, Dr. Barbara 20150
Pfister, Emil 41305
Pfister, Rose 17562
Pfister, Sara 41660
Pflanz, Paul H. 53402
Pfleger, Prof. Dr. Susanne 22668
Pflüger, Helmut 22600
Pflug, Dr. Hermann 19069
Pflumm, Manfred 22239
Pfnorr, Reinhard 20692
Pfnuer, Linda 17000
Pfoser, Dr. Alfred 03087
Pfosser, Dr. Martin 02310
Pfotenhauer, Louise 49041
Pfrunder, Dr. Peter 42026
Phagan, Patria 52121
Phares, Jane 50674
Phayana, Leah 38558
Phelan, Wynne 49397
Phéline, Jean-Michel 15614
Phelps, Kathleen D. 54530
Phelps, Martin C. 50555
Phelps, Patricia 07083
Pheney, Kathleen 46631
Pheysey, Dawn 52188
Phil, D. 24283
Philbin, Ann 50228
Philbrick, Harry 52347
Philbrick, Nathaniel 50943
Philbrook, Wyman 53746
Philipon, Peter 07028
Philipp, Werner 17120
Philippart, Jean-Paul 03796, 03802, 03812
Philippe, Pierre 41324
Philippon, Annie 15080
Philippot 14177
Philippot, Alain 13865
Philipson, Jean 01345
Philipson, Peter 44910
Phillip, D.M. 33082
Phillips, Amanda 44660
Phillips, Christopher 51194
Phillips, Colin 01515
Phillips, Dale 53747
Phillips, Elizabeth 03015
Phillips, G.R. 38685
Phillips, J. Michael 43429
Phillips, James 47234
Phillips, John 43545
Phillips, Joseph R. 46485
Phillips, Dr. Kenneth E. 50215
Phillips, Lisa 51240
Phillips, Prof. Ruth B. 07217
Phillips, Dr. Sandra S. 52795
Phillips, Steven 48748
Phillips, Tim 48626
Phillips Olmedo, Carlos 31111, 31118
Phillipson, Anthony 42834
Phillipson, Prof. David W. 43734
Philllips, Jane 45814
Phillp, John 43554
Philo, P. 45473
Philo, Phil 45139
Philotheou, G. 09375
Philp, B.J. 44017
Philp, Brian 45320
Philpot, Chris 43485
Philpott, Colin 43581
Philpott, Pat 06527
Philpott, Vandean 52764
Philson, William C. 50596
Phinney, Gail 52242
Phinney Kelley, Sarah 50030
Phipps, Bbetty K. 49062
Phipps, Rebecca J. 49994
Phleps, Dr. Roland 18438
Phuong, Pham Xuân 54639
Phuong, Tran Thi Thuy 54629
Piacentini, Dr. Paola 27512
Pianca, Dr. Vittorino 28298
Pianea, Dr. Elena 27602
Piasecka, Grażyna 35201
Piastrella, Dr. Carlo 25999
Piatek, Bruce J. 51679
Piątek, Jacek 35050
Piatt, Margaret 54068
Piatti, Verena 02336
Piaty, Karl 02888
Picard, Dennis 54083
Picard, Michel 14826
Picard, Thomas 51251
Piccand, Marcel 15673
Piccardi, Dr. Marco 26178
Picci, Prof. Dr. Piero 25441
Picciano, Eric 47537
Piccioli, Prof. Rossana 26435
Piccirilli, Guido 27922
Piccirillo, M. 25004
Piccoli, Prof. Adalberto 25791
Picelj, Zdenko 38490
Picher, Claude 05585
Pichler, Dr. Isfried H. 01700
Pichler, Mag. Jutta M. 02226
Pichler, Lucia 01436
Pichorner, Dr. Franz 03019
Pickard, Kelli L. 48716
Pickering, Mona 50456
Pickford, Stephanie 44839
Pickin, John 45776, 45777
Pickles, Owen 33229
Pickup, A. 44251
Picollet, Auguste 11534
Picon, Carlos 51212
Picot 11906
Picozza, Prof. Paolo 27444
Picpican, Isikias T. 34342
Pidault, Monique 14814
Pido, Asuncion Q. 34429
Pidpala, Nila Andriïvna 43080
Piek, W. 32962
Pieke, Dr. Gabriele 17441
Pieke, Hannelore 16623
Pienimaa, Vilho 10869
Pieplow, Jane 48502
Piepmeyer, Prof. Dr. Lothar 18521
Pierce, Dr. Alexandria 05960
Pierce, Dr. Charles E. 51216
Pierce, Donald L. 52662
Pierce, Donna 48072
Pierce, Dr. Donna 52873
Pierce, Sally 46919
Piergollini, Alberto 26973
Pierides, Demetrios Z. 09362, 09370
Pierlot, Claire 15986
Pierottavio, Daniele 26929
Pierrangeli, Prof. Carlo 27450
Pierré, Fernand 13459
Pierret, Philippe 03490
Pierrot-Bults, Dr. A.C. 31904
Piers, John C. 52082
Piersantini, Dr. Michele 28125
Pieschacón González, Gabriel 08777
Pieschel, Janet 05534
Pieters, Dr. Georges 03635
Pieterson, A. 32368
Pietilä-Juntura, Katriina 11383
Pietri de Caldera, Alicia 54550
Pietropaolo, Dr. Lisa 26538
Pietrzak, Stanisław 35198
Pietrzak, Ted 47119
Pietrzyk, Prof. Zdisław 34815
Pietsch, Dr. Martina 18680
Pietsch, Dr. Ted 52994
Pietsch, Dr. Ulrich 17966
Pietsch, Uwe 19225
Pietscher, Charles 49745
Pietschmann, Wolfgang 19253
Piette, Jacques 14464
Piezzi, Aron 41285
Pigeat, Jean-Paul 12486
Pigg, Gary 48789
Pigott, Tayloe 49572
Pihema, Joseph 33109
Pihlak, Jaak 10674
Pihlgren, Kjell 40568
Piirainen, Sari 10727
Piiroinen, Päivikki 10995
Pijeaux, Dr. Lawrence J. 46821
Pijl, B.H. 32463
Pijl, J. 31819
Pika-Biolzi, M. 42071
Pikal, Dr. Jaroslav 09565
Pike, Alan 38556
Pike, J.P. 06090
Pike, Julia 05681
Pike, Stephen J. 50579
Pikett, Vivian 48236
Pilar, Jacquelin 48790
Pilar Muñoz Jiménez, María del 08575
Pilati von Thassul-Filo della Torre, Francesca Gräfin 02051, 02586
Pilato, Dominique 15562
Pilchowski, Władysław 34921
Pilcz, Prof. Dr. Karlheinz 02395
Pileckas, Atanas 30308
Pili, Leleala E. 00084
Pilides, Dr. D. 09375
Pilioko, A. 54494
Pilkilton, Sue 53624
Pilkinton, Hunter M. 53997
Pillaert, E. Elizabeth 50395
Piller, Ferdinand 41218
Piller, Dr. Gudrun 41161
Piller, Josienne N. 51963
Pillichshammer, Johann 02432
Pillsbury, Dr. Edmund P. 47935
Pilotti, Bianca Maria 25921, 25982
Pils, Walter 02327
Pilven, Denis 12121
Pilz, Anette 22676
Pilz, Winfried 16403
Pimentel, Cristina 35850
Pincemin, Cecilia 06071
Pindinelli, Prof. Elio 26268
Pineau, Jacky 13708
Piñeiro, Alberto Gabriel 00236
Pinelli, Prof. Antonio 27239
Pinette, Matthieu 15708
Pingel, Carol 49057
Pingeot, Anne 14692
Pingree, Shelda 48799
Pinheiro, Maria Joao 35688
Pini, Lucia 26690
Pinilla, José 08650
Pinner, Keith 45549
Pinnix, Cleve 54132, 54233
Pino Morán, Mónica 07475
Pinon, Mary 48939
Pinsker, Dr. Bernhard 17789
Pinsky, Dave 52907
Pinson, Dr. Patricia 51489
Pintarić, Adam 08967
Pintarić, Snježana 09007
Pintarić, Tanja 08977
Pintea, June 48411
Pintea, Rodica 36383
Pinto, Dr. Bianca Alessandra 27454
Pinto, Dr. Ettore 25772
Pinto, Matt 51256
Pinto de Matos, Maria Antonia 35662
Pinto Simon, Elena 51132
Pinzl, Christoph 22677
Pinzón, Carlos 08789
Pinzón, Rito 08766
Pionk, Richard 51251
Piot, Marc-Etienne 41339
Piotrovski, Michail 37640
Piotrowski, Jerzy 34563
Piotrowski, Wojciech 34630, 34630
Pipek, K. 09507
Piper, Catharina 41030
Piper, William 52576
Pipino, Dr. Giuseppe 27342
Pipkin, Roy 53365
Piquera, F. 13606
Piquereddu, Dr. Paolo 26966, 26967
Piquero López, Blanca 39547
Pirenne, R. 04096
Pires Martins, Prof. Calebe 04958
Piriyakul, Yukol 42559
Pirker, Christina 19366
Pirker, Rudolf 02862
Pirker-Aurenhammer, Dr. Veronika 02951
Pirkmajer, Darja 38407
Pirkonen, Pekka 11017
Pirlot, Anne-Marie 03461
Pirnat, Miha 38457
Pirnie, M. 45388
Pirotte, Daniel 03965
Pirotte, Philippe 41209
Pisa, L.J. 32807
Pisa, Regina M. 53149
Pisani, Camille 03497
Pisani, Ètienne du 38627
Pisano, Dr. Jane G. 50257
Pisarev, S.V. 37016
Pişiren, Tuncay 42912
Piskal, Irina Leonidovna 37800
Piskur, Bojana 38454
Piso, Prof. Dr. Ioan 36290
Pistono Grand, Mario 27731
Pištora, Jiří 09461
Pitcher, Stephen 54297
Piterskaja, Larisa Karlovna 37637
Piteša, Ante 08954
Pithie, Helen 00793
Pitkänen, Erkki 10754
Pitman, Bonni 47931
Pitoń, Helena 35354
Pitot 04084
Pitre, Marc 06123
Pitrey, Thierry 15603
Pitruk, Dmitrij L. 37944
Pitscheider, Erich 27858
Pitsios, Prof. Dr. Th. K. 23008
Pitt, James 23273
Pitt Jones, Priscilla 48164
Pitterle SVD, Mag. Elmar 02393
Pittner, K. Heinz 21853
Pittock, Tina 43270
Pitts, Priscilla 33090, 33096
Pitts, Terence 47301
Pittz, Franeis 47698
Pitxot, Antoni 39300
Pitzen, Jutta 22232
Pitzen, Marianne 17454
Pitzer, Robert W. 54111
Piuma, Dr. Fabio 28216
Pixley, Patricia 51550
Pixner, Albin 27654
Pizitz, Jackie 51304
Pizzigoni, Prof. Gianni 28219
Pizzinini, Dr. Meinrad 02125, 02136
Pizziolo, Marina 26708
Pizzo, Tony 52948
Pizzollo, Sissy 51854
Pjankova, Svetlana Anatoljevna 36846
Pla, Josefina 34092
Plaat, Jaanus 10647
Plada, Mabel 54381
Plaen, Guy de 04052
Plaismone, Ehza Cilio 26282
Planas, Marta 00104
Planas, Ramón 38985
Planas Marcé, Silvia 39347
Planchard, Ch. 15161
Planchon, Jacques 12720
Planck, Prof. Dr. Dieter 19676
Planisek, Dr. Sandra 50353
Planjer, W.E.J.M. 31811
Plank, Benedikt 02659
Plánka, Dr. Ivan 09941
Plankensteiner, Dr. Barbara 03038
Plantalamor, Luis 39591
Plante, Stella 06193
Plantin, Colonel Kjell 41079
Planzer-Steiner, Rita Lisa 41850
Plášek, RNDr. Vítězslav 09692
Plasser, Dr. Gerhard 02625
Plaßmann, Dr. Otmar 21584
Plaßmeyer, Dr. Peter 17951, 17959
Plata Vega, Edith 08779
Platen, Dr. Michael 18890, 19380, 19390
Platero, Ramon 39760

P

Plath, Dr. Carina 20520
Platonova, T.V. 36914
Platow, Raphaela 53810
Platt jr., Geoffrey 52334
Platt, Ron 46822
Platt, Ronald 49068
Platteter, Janet 49862
Plattner, Fritz 02763
Platz, Daniela 22693
Platzgummer, Hanno 01829
Plavan, Tamara 43178
Plavec-Butković, Lidija 08852
Plazotta, Carol 44937
Pledienė, Daina 30300, 30301
Plenković, Ivo 08847
Plenter, Dr. H.A. 32894, 32895, 32896
Pleská, Leona 09833
Plessis, H. 38755
Plessis, M. du 38720
Pletnjov, Valentin 05247
Ploch, Dr. Gregor 21192
Plöckinger, Dr. Veronika 02175
Plößl, Dr. Elisabeth 18627, 20602
Ploetz, Franziska 19574
Plomp, Dr. M. 32302
Płonka, Krzysztof 35039
Płotkowiak, Wiesław 34694
Plotoagă, Gabriela 36300, 36307
Ploug, Mariann 09978, 09979, 10193, 10219, 10220, 10306, 10307, 10308
Plouin, Suzanne 11988
Ploum, Y. 31809
Plourde, Nelie 46522
Plowright, Georgina 43839, 43891, 44491
Pluciński, Dr. Józef 35153
Pluhařová, Jana 09909
Plukavec, Lina 09010
Plume, Thierry 12854
Plummer, Dr. Ellen 50138
Plummer, Marguerite R. 53086
Plummer, Mary Ann 48746
Plummer, Sue J. 46483
Plummer, Val 46058
Plump, Darlene 51790
Plunkett, Stephanie 53302
Pluntke, Margitta 16498
Plutat-Zeiner 18921
Pluymaekers, Anne 03851
Pochman, Kathryn K. 50944
Pochwała, Stanisław 34800
Pocius, D. 01382
Počivalova, Elena Ivanovna 36952
Počivavšek, Prof. Marija 38406
Pocreau, Dominique 13865
Pocsainé Eperjesi, Eszter 23795
Podany, Jerry 50237
Podedworny, Carol 07290
Podestá, Horacio 00275
Podlesnik, Mateja 38452
Podmaniczky, Christine B. 47322
Podnar, Franc 38514
Podos, Lisa Beth 51132
Pögelt, Erika 18890
Poel, Dr. Peter te 32524
Pölcher, Bertold 18101
Pöll, Ing. Helmut 01920
Pöllmann, Rotraud 21321
Poelman, Shannon 07329
Poelmans, Marc 04117
Poenaru, Constantin 36629
Pöppelmann, Dr. Heike 22474
Pöpping, Johannes 20089
Pörnbacher, Prof. Dr. Hans 21888
Pörnbacher, Dr. Karl 19484
Pöschel, Helmut 22705
Pöss, Ondrej 38222
Poester, Teresa 04654
Pöstinger, Eberhard 02741
Poette, Simonne 16082
Pöttler, E. 02790
Pöttler, Prof. Dr. V.H. 02790
Pötzsch, Herbert 20182
Poggi, Patrizia 27372
Poggi, Dr. Roberto 26297
Poggiani-Keller, Raffaella 25659
Pogliani, Dr. Marisa 28131, 28132
Pogonec', Olena Dmytrivna 43092
Pogorzelska, Renata 34611
Pogt, Dr. Herbert 22748
Pohjakallio, Lauri 10737
Pohl, Burkhart 53494
Pohl, G. 17529
Pohl, Gerhard 20279
Pohl, Dr. Klaus-Dieter 17789
Pohl, Dr. Norman 18422
Pohlman, Lynette 46256, 46257
Pohowsky, Carolyn 50809
Pointal, Jean-Louis 12036
Poirel, Evelyne 15114
Poirier, Daphne 06930
Poirier, Diane 05607
Poissant, Margaret 49380
Poisson, Philippe 13653
Poitevin, B. 13804
Poivet, Clémence 15957, 15958
Pokorná, Blanka 09616
Pokorná, Libuše 09667
Pokorný, Marek 09432, 09433, 09434
Pokorný, Zdeněk 09511
Pokorny-Nagel, Kathrin 03026
Pokorski, Janice 50307
Poláčková, Dr. Dagmar 38233
Polacsek, John 48113
Polacsek, John F. 48116
Polak, Dr. J.J.H. 32882
Polak, Dr. Jerzy 34562
Polakovic, Dr. Vincent 38205
Polanco de Laverde, Cecilia 08657
Polanía Ramírez, Tiberio 08625
Polanich, Judith K. 47015, 53419, 53420, 53578
Polanská, Zuzana 09733
Polapink, Tom 52301
Polat, Cemal 42773
Poldermans, Dr. M. 32293
Poldervaart, Dr. A.A. 32892
Pole, Merle 01397
Polenov, Juri A. 36901
Poleske, Lee E. 53042
Poletti, Catherine 12381
Polfer, Michel 30483
Poli, Cristina Tessari 25382
Poliakoff, Miranda 44924
Polich, Debra 46304
Polichetti, Maria Luisa 27803
Polichetti, Dr. Massimilano A. 27512
Polikarpova, Marina Alekseevna 36982
Polioudovarda, Anna M. 22957
Poliščuk, Natalja 43158
Poliščuk, Svetlana Alekseevna 36812
Politopoulos, Catherine 50272
Polivka, Marcus 49335
Poljakova, Valentina Alekseevna 37957
Poljanceva, I.B. 37277
Poll, Lothar C. 17171
Pollák, Dr. Róbert 38275
Polláková, Petra 09776
Polland, Annie 51169
Pollard, Ian 45659
Pollard, S. 33069
Pollard, Shari E. 49982
Pollard, Ted 54011
Pollei, Dane F. 50564
Polley, Amy E. 46352
Polli, Kadi 10624, 10632
Polli, Dr. Vittorio 28318
Pollini Paltrinieri, Lucia 41582
Pollock, Mallory 47584
Polnick, Bill 06931
Polo, Laura W. 50545
Polonovski, Max 14688
Polott, E. Leslie 49412
Poloznev, Dmitrij F. 37002
Polozova, Natalja Ivanovna 37951
Pols, Dr. Ivor 38708
Polverari, Dr. Michele 25231
Polvi, Arvo 10847
Polyarnaya, Janna 37639
Polychronopoulos, Mari Elleneca 23123
Polymerou-Kamilakis, Dr. Aik. 22913
Polzine, Roxann L. 54306
Pomar, Maria Teresa 30783
Pomarede, Vincent 14698
Pomeroy, Dan E. 50976
Pomeroy, Marilyn 54282
Pomeroy Draper, Stacy 53575
Pommeranz, Dr. Johannes 20764
Pommies, Marie-Anne 12065
Pompa, Vince 46553
Pomponi, Luigi 27886
Ponce de León, Carolina 52774
Poncho, Roland A. 50150
Ponciş, Gheorghe 36515
Pond, Jennifer 52398
Ponder, Gary 52008
Pongracz, Dr. Patricia 51225
Ponomareva, Galina B. 37316
Ponomareva, Tatjana 37701
Pons, Marc 40291
Pons-Pons, Fermí 39926
Ponsonnard, Christian 13612
Ponsoy, Teresita 34523
Pont, Jeanne 41418
Ponte, Dr. António 35992
Ponte, Emily 49389
Ponte, Raffaella 26301
Ponte, Dr. Susanne de 20435
Ponton d'Amecourt, Jean de 13296
Pontual, Sylvia 04667
Pontvianne, Jean-Paul 13472
Poodt, J.G.M. 32379
Poolaw, Linda 46280
Poole, Helen 43916, 43917, 44380
Poole, Helen E. 45668
Poole, J.E. 43727
Poole, Malcolm 45077
Pooran Yosef, Simin 24667
Poore, Henry 53274
Poorter, Nora de 03314
Poot, J.A. 31888
Pop, Valer 36669, 36670
Pop, Dr. Valer 36671
Pop, Valer 36672
Popa, Gabriela 36620
Popa, Nicolae 36673
Popa, Onusia 36730
Pope, Clare 46052
Pope, Linda 52866, 53991
Popejoy, Tammy L. 50432
Popelier, Betty 06598
Popelka, Dr. Pavel 09888
Popescu, Dr. A. 43727
Popescu, Camelia 36635
Popescu, Doina 36720
Popescu, Ilie 36459
Popescu, Ioana 36227
Popescu, Viorel 36692
Popivoda, Mileva 38110
Poplack, Robert 46718
Popoiu, Dr. Paulina 36217
Popolla, Don Gianluca 27927
Popov, Aleksandr Nikolaevič 37943
Popov, Gennadij Viktorovič 37146
Popova, L.B. 37515
Popova, Nina Ivanovna 37642
Popovčak, Borivoj 09011
Popović, Bojana 38074
Popovic, Miroslav 30520
Popović, Pavle 38151
Popovic, Tina 08997
Popovici, Elena 36294
Popovici, Georgeta 36118, 36119, 36121, 36545
Poppa, Dr. Rudolf 17863
Poppi, Dr. Cesare 27321
Popski, Dragoljub 38086
Porada, Jerzy 34952
Þórðkr{}ardóttir, Gunnhildur 23998
Porenta, Tita 38532
Porezag, Karsten 21811
Porignaux, Jonathan 03688
Porinchak, Mary Ann 53825
Pornin, Claudie 16001
Poroszlai, Dr. Ildikó 23839
Porr, Dr. Martin 21146, 21148, 21152
Porra-Kuteni, Valérie 11921
Porras, César Augusto 08693
Porrelli, D.M. 45219
Porsay, François 12062
Porsmose, Erland 10116
Porstmann, Dr. Gisbert 17973
Port, Jane L. 51995
Portal, Jane 46940
Portanje, B.K. 32019
Porte, Patrick 14730
Portell, Rosa 53249
Porteous, Ken 07381
Porter, Ann 48986
Porter, Barry 53507
Porter, Frances J. 50499
Porter, Herman 06635
Porter, John R. 06694
Porter, Dr. Joseph 52233
Porter, Julian 43471
Porter, Marty 49679
Porter, Paul 52665
Porter, Paul R. 52570
Porter, Phil 50351, 50354
Porter, Phillip 05699
Porter, Richard T. 51567
Porter, Russ 01585
Porter, Suzanne 33158, 48220
Portet, Angel 39854
Portiabohn, Ivece 51248
Portigliatti Barbos, Prof. Mario 28011
Portillo, Monica 30478
Porto, Prof. Dr. Nuno 35529
Porttier, W. 03302
Portugal-Humeau, Marie-Joëlle 13143
Portús, Javier 39573
Portz, Willi 21510
Porzenheim, Maria 17846
Posada, Teresa 39573
Posani, Dr. Miguel 54545
Posch, Eva 18412
Poser, Adriana 51128
Poser, Steffen 19937, 19939, 19941
Pošienė, Irena 30450
Posinjak, Jurij Michajlovič 37861
Pospíšilova, Dr. Alena 09598
Poßegger, Dr. Iris 17407
Possehl, Gregory L. 51886
Posselt, Dr. Martin 19486
Possémé, Evelyne 14682
Possenti, Dr. Elisa 26969
Post, Dr. K. 32953
Post, Melissa G. 47393
Post, Dr. William 47359
Postek, Roman 34858
Postel, Michel 11981
Postel, Mitchell P. 51668, 52278
Postles, Suella 45289
Poston, Jonathan 47367
Poston, Roger D. 47730
Potapov, Roald Leonidovič 37738
Poteau, Gérard 12988
Poteete, Timothy 50000
Potet, Jean-Pierre 16234
Potgieter, M.J. 38766
Pothan, Scott 33265
Potocka, Maria Anna 34945
Potorczyn, Pawel 44956
Potratz, Rainer 20199
Potsch 01782
Pott, Karin 17126
Pott, Dr. Ute 18877
Potter, Byrn 50265
Potter, Edmund D. 53279
Potter, John 45892, 50822
Potter, Ted 52322
Pottier, Marie Hélène 12932, 12933
Potts, Albert 52309
Potts, Prof. Daniel T. 01557
Potts, Jennifer 46992
Potts, Dr. Timothy 43727
Potts, Timothy 48715
Potucek, Delvin 53822
Potužáková, Dr. Jana 09719
Potvin-Barakatt, M.Sc. Julie 06145
Poueymirou, Frank 51187
Pouget, Daniel 12931, 15353
Poulain, Ivan 13151
Poulet, Anne L. 51178
Poulis, Helen 06072
Poullain, Christine 13134
Poulliot, Elizabeth A. 51363
Poulos, Ann 53472
Poulos, Paul 53639
Poulsen, Hanne 10064
Poulter, Gill 44050, 44053
Pounis Guntavid, Joseph 30565
Pourshirazi, Dr. Katja 17559
Pourtalès, Laure de 11564
Pourtalet, P. 05945
Pourtaud, François 13299
Povilionienė, Vita 30335
Poviliūnas, Virgilijus 30382
Povolný, František 09814
Povšič, Sabina 38454
Powar, N.V. 24381
Powders, Dr. Vernon 46247
Powell, Bobbie 52373
Powell, Derek 45955
Powell, Earl A. 53898
Powell, JoAnne 46666
Powell, Maurice F. 19468
Powell, Michael 45081
Powell, Robert 44605, 45255
Powell, Shana 48823
Powell, Shona 45290
Powell, William V. 49811
Powell Harold, Fran 52942
Power, Cathy 44820, 45886
Power, Christopher 01360
Power, Dennis 51470
Power, Nicola 45601
Powers, Carolyn T. 48505
Powers, Catherine 46381
Powers, Claire 06472
Powers, Dale 52519
Powers, James L. 49442
Powers, Leland 49224
Powers, Linda 49850
Powers, Richard 49902
Poyner, Noelle 01206
Poziljonov, Avazbek 54442
Pozsonyi, József 23358
Pozzi, Dr. Raffaella 26084
Pozzuoli, Prof. María Cristina 00412
Prackwieser, Walter 26124
Prada Pérez, Isaac 08645
Prader, Angela 41322
Prader, Jürg 41322
Prado, César O. 34173
Prätzel, Maritta 21330
Prag, Dr. A.J.N.W. 45090
Prahase, Mircea 36152, 36314, 36446, 36487, 36564, 36584
Prakapcov, Vladimir I. 03255

P

Prall, Steve 46383
Pralong-Salamin, Rachel 41865
Pralormo, Marcella 28031
Prammer, Dr. Johannes 21918
Prang, Werner 20385
Pranza, Denise M. 52310, 52511
Prasch, Ing. Baldur 02691
Prasch, Dr. Hartmut 02752
Prášek, Jiří 09714
Prasolova, Elena Ivanovna 37693
Prasser, Gabriele 20785
Praßl, Johann 01878, 01879
Praszelik-Kocjan, Grażyna 34973
Pratesi, Dr. Giovanni 26185
Prati, Dr. Luciana 26223, 26224, 26225, 26226, 26228, 26230
Pratt, John 45267
Pratt Brown, Alice 51279
Prause, W. 17773
Prayitno, Dr. R. Joko 24524
Praz, Jean-Claude 41875
Praženilová, Dr. Viera 38334
Préaux, Alain 15117
Prebushewsky Danyliuk, Janet 06925
Preciado, Kimberly 52749
Preece, Geoff 43992
Preininger, Kristina 38457
Preising, Dr. Dagmar 16395, 16408
Preiss, Prof. Heinz 02234
Preiss, Slawomir 35129
Preiße, Erika 19638
Preissler, Leonilda Maria 04491
Premack, Ellen 48081
Premerl, Nada 09004
Premoli Silva, Adolfo 27193
Premoli Silva, Dr. Isabella 26705
Premović-Aleksić, Dragica 38119
Prendergast, Mary B. 24755
Přenosilová, Dr. Vera 09751
Prentiss, Margo 51985
Presgrave, A.D. 01118
Presl, Igor 38492
Presnell-Weidner, Denise 52001
Pressl, Hannes 01739
Pressley, Sheila 52772
Presson, Virginia 54005
Prest, Véronique 14646
Prestaux, J.M. 11697
Preston, Bill 49031
Preston, Bryan 47305
Preston, Carol 52790
Preston, Dr. Charles R. 47671
Preston, P. Anthony 33071
Preston, Dr. P.M. 44123
Preston, Tantum 46822
Presuhn, Dr. Sabine 22169
Preszler, Robert E. 52252
Preter, Marie-Dominique de 41423
Pretola, John P. 53229
Pretorius, A.F. 38674
Pretorius, Mimi 38562
Pretsch, Dr. Peter 19450
Pretty, Carol Anne 05535
Pretzel, Elke 20591
Pretzl, Manuel 20432
Preuß, Hartmut 18107
Preuß, Helmut 19676
Preuß, M.A. Patricia 21984
Prévost-Bault, Marie-Pascale 14784
Prévost-Bouré, Pascal 14443
Prevosti, Marta 39869
Prevratil, Joseph F. 50176
Prew, R. 43793
Prewer, Steven 43392
Prewett, Vickie 47548
Preyer, Josef 02519
Pría, Melba 31132
Přibáňová, Dr. Svatava 09438
Pribbernow, Marcus 20127
Přibil, Vladimír 09703
Přibyla, Radomír 09535
Price, Alan 45314
Price, Anita 05676, 05677
Price, John 50426
Price, Kari 46865
Price, Margaret 52116
Price, Dr. Marla J. 48717
Price, Marti 47458
Price, Paula T. 52357
Price, Renée 51238
Price, Sally Irwin 46653
Price, Terri 50770
Price, Thomas E. 47735
Price Shanis, Carole 51865
Prickett, Marcus 31740
Prickett, N.J. 33043
Priddle, Th. 46114
Priddy, Dr. Bennie 22263
Prie, Jean-Yves 14839
Priego Fernández del Campo, Carmen 39518, 39565
Priem, Dr. R. 32590
Priem, Ruud 32849
Prière, Marc 11804
Prieto, Gabriel 08592
Prieur, Abel 16238
Prieur, P. 41250, 41449
Prikolota, Olga Vasylivna 43117
Přikrylová, Ludmila 09665
Priljeva, Srdjan 38166
Prillinger, Judith 02409
Primozic, Melanija 38530, 38532
Prince, David 53399, 53403
Prince, J.R. 45078
Prince, Nigel 43505
Prince, Robert 06296
Prince, Shannon 06473
Prince, Todd 47290
Pringle, Roger 45779, 45780, 45781, 45782, 45783, 45785
Prinz, Gerhard 21950
Prinz, Dr. Ina 17445
Prinz, Peter 46826
Prinz, Dr. Ursula 17049
Prinz-Büchl, Mag. Margarethe 01905
Prinzky, Manfred 21946
Prinzler, Hans Helmut 17029
Prior, Annette 40555
Prior, Claudine 05496
Prior, Terrence M. 51625
Pripačkin, I.A. 37093
Prister, Boris 08997
Prister, Lada 08997
Pritchard, Frances 45099
Pritchard, Philip 07116
Pritchard, Susan 45191
Pritikin, Renny 47976
Prittwitz, Dr. Hans-Hoyer von 17462
Pritzker, Elisa 49278
Priuli, Dr. Ausilio 25658
Privette, Carol B. 53807
Privette, David 50125
Prix, Jörg 01888
Prjamikova, Vera Pavlovna 37063
Probst, Heinz 20570
Probst, Peter 41880
Probst, Dr. Rudolf 41215
Probst, Dr. Volker 18827
Probst, Volker 18828
Procacci, Ugo 26171
Procelli, Dr. Enrico 27361
Prochaska, James J. 47320
Prochazka, Denise 48309
Prochina, Irina Evgenjevna 37350
Procopé Ilmonen, Susanne 11097
Procopiou, E. 09375
Proctor, Jacob 46309
Proctor, Rebecca 43723, 43727
Proctor, Vicki 48330
Pródér, István 23952
Prod'Hom, Chantal 41540
Proebst, Walter 16432
Prösch, Dr. Christoph 22767
Prøhl, Henning 40968
Prokisch, Dr. Bernhard 01870, 02322, 02324, 02327
Prokop, J. 09684
Prol, Elbertus 52353
Proll, Jochen 16929
Pronti, Dr. Stefano 27185
Prophet, R. 46071
Prøsch Danielsen, Lisbeth 33863
Proschwitz, Ted von 40548
Proskynitopoulou, Rosa 22920
Proß, Christian 20594
Prosser, Jo 45004
Prost, Darin 52217
Prota, Prof. Romolo 05572
Protektore, Evilena 30097
Prothero-Brooks, C.C. 06965
Proto Pisani, Dr. Rosanna Caterina 26147, 26149, 26181, 26202
Protz, Edward L. 48849
Proud, Hayden 38593
Proud, Hayden Russell 38587
Proust, Georges 14652
Prouty, Elisha 48982
Provencio, Bobbie 49959
Provick, A.M. 05782
Providi, Dr. Ioanna 22941
Provo, Dan 51524
Provorov, Viktor Petrovič 37271
Prpa Stojanac, Ika 08954
Prudhomme, James 50987
Prudhomme, Mary Louise 46635
Prudhomme, Suzanne M. 50789
Pruett, Herbert E. 54183
Prugh, Burton 47132
Pruitt, John E. 53208
Prum, Richard O. 49991
Prunet, Pierre 15335
Prusinovsky, Rupert 20955
Pruska, Mariola 34843
Pryor, Eric 53364
Prys-Jones, Dr. R. 45876
Przeczewski, P. Marek 26245
Przekazinski, Andrzej 35236
Przigoda, Dr. Stefan 17398
Przybył, Grażyna 34666
Przybylek, Leslie 48685
Przybylek, Stephanie 52962
Przybylska, Grażyna 35088
Przybysz, Jane 52810
Przychowski, Alexandra von 42097
Przygoda, Udo 18369
Przypkowski, Piotr Maciej 34702
Psaropoulou, Betty 22905
Psenner, Dr. Barbara 02116
Psota, Dr. Thomas 41207
Pstrag, Renata 12073
Pucciarelli, Bepi 26355
Puccini, Simonetta 28037
Puchner, Manfred 02321
Puckett, Jakobus 19307
Puckett, James 16649
Puckett, Lori 53554
Puckett, Robert A. 54152
Puckett, Tom 48385
Puckitt, Sarah 55138
Puddy, Audrey 43600
Pudelska, Aleksandra 35031
Pudil, Martin 09625
Puech, Laurent 13666
Pück, Thorsten 18952
Puentes de Valderrama, Ana Rosa 08757
Pürro, Gertrud 41263
Puerta, Prof. J. 39509
Puertas Tricas, Rafael 39599, 39600
Puerto, Dr. F. Javier 39545
Pütz, Ursula 20968
Püüa, Endel 10584, 10586, 10587
Puga, Rosa D. 07452
Pugačevskaja, Alla Aleksandrovna 37362
Puget, Catherine 14880
Puggaard, Ole 10099
Pugi, Guido 27335
Puharich, Nick 33085
Puhle, Prof. Dr. Matthias 20123, 20126
Puhlmann, Dr. Helga 20426
Puig, Felipe 54403
Pujol, Albert 38996
Pujol, Gloria 00094
Puklowski, Tracy 33140
Pule, D.E. 04297
Pule, Tickey 04296
Pulfer, Eleanor 45690
Pulinka, Steven M. 51495
Pulle, Mag. Thomas 02686
Pullen, Bernice 51601
Pullen, Reg 46594
Pullen, Sue 44510
Pullin, Robert 44112
Pultnones, Lucio 34379
Pultz, John 49992
Pulvenis de Séligny, Marie-Thérèse 14435
Pumbaya, Ismael 34459
Pumphrey, Jim 47253
Puncher, Sebastian 43962
Pungerčar, Ivana 38490
Puntman, Paul 32957
Pupăză, Grigore 36114
Pupo de Mogallón, Yolanda 08618
Purba, Suruhen 24526
Purcell, Beth 47697
Purcell, John K. 47252
Purdy, Dorothea 47146
Purdy, John 48633
Purdy, Margaret 49485
Pure Via, Luis 40281
Purin, Bernhard 20448
Purini, Sergio 03495
Purmal, Kathie 49803
Purpus, Dr. Elke 19615, 19630
Pursley, David W. 52546
Purves, W.M. 44135
Puscher 19672
Puschnig, Dr. André 41175
Pushman, Tracie 01008
Pusineri, Prof. Adelina 34097
Pusineri Scala, Prof. Carlos Alberto 34095
Pusterla-Cambin, Patrizia 41575
Putin, Olga 36687
Putnam, Christy 51279
Putnam, Michael 46605
Putnam, Polly 44666
Putney, Carolyn 53523
Puts, Claude 03775
Putsch, Dr. Jochem 21806
Puttock, Dr. Sonia 01484
Putyńska, Elżbieta 34572
Putz, Karl 21983
Putze, Alfred 19319
Putzert, MA Nadja 18009
Puzanova, N.A. 29683
Puzik, Olga Vitalijovna 43074
Puzo, Joseph 47828
Puzrina, Oksana Jurjevna 37138
Py, Cyille 15128
Py, Jacques 15851
Py, Jean-Samuel 41797
Pyatt, John 33122
Pycke, J. 04089
Pyhrr, Stuart 51212
Pyka, Henryk 34724
Pykäläinen, Mari 11379
Pylkkänen, Heikki 10948
Pym-Hember, Peter 06531
Pyrrou-Drivalou, Ourania 22907
Pysher, Ray 50549
Pytlik, Wincenty 35005

Q

Qaavigaq, David 23264
Qëndro, Gëzim 00030
Qi, Junjie 08316
Qian, Xiaoping 08260
Qiao, Delong 07752
Qiao, Yuguang 07791
Qiao, Zheng'an 07957
Qin, Xinghan 07531
Qiu, Aidong 07821
Qiu, Damin 07751
Qiu, Fang 07512
Qiu, Jinkun 08405
Qiu, Yongsheng 08417
Qu, Liping 08451
Quaas, Achim 18946
Quackenbush, Bill 07296
Quackenbush, Laura J. 50022
Quackenbush, Molly 47124
Quackenbush, W.G. 05384
Quaeitzsch, Christian 20439
Quaeitzsch, Dr. Christian 20476, 20477
Quaghebeur, Marc 03433
Quail, Sarah 45430
Qualizza, Dr. Marino 28112
Qualmann, Kenneth 46037
Quam, Jeffery 50512
Quan, Benning 08094
Quan, Zhongde 07931
Quann, Byron 49183
Quanten, Maarten 03477
Quarcoopome, Nii 51287
Quarles Van Ufford, Dr. C.C.G. 32860
Quartero, Dr. Luigi 28268
Quarterone, Dr. Anna 25773
Quatember, Dr. Wolfgang 01836
Quattrocchi, A. 06587
Quattrociocchi, Maria Pia 27174
Quayle, Frederick M. 53382
Quéau, Jean-François 12253
Queckenstedt, Dr. Hermann 20919
Quellenberg, Björn 42083
Queree, Jennifer 33069
Querejazu Calvo, Dr. Jorge 04244
Quermann, Dr. Andreas 21922
Quermann, Carolin 17973
Quero Castro, Salvador 39550
Quetard, Patrice 11762
Quevedo, Emilio 08564
Quevedo, Jacinto 39801
Quick, Betsy D. 50223
Quien, Prof. Guido 08957
Quijano, Marcela 22162
Quijano, Ritchie 34428
Quinata, Tom 23287
Quinlan, Karen 00859
Quinlan, M. 43451
Quinn, Alexandra 52909
Quinn, Betty 24776
Quinn, Thomas 24844
Quinn, Vivienne 43701
Quinquis, Pierre 16170
Quinsonas, M. de 13656
Quint, Eduard 32419
Quintana Reyes, Remedios de 39951
Quintanilla, Faustino 46657
Quintanilla, Mimi 52731
Quintanilla, Sonia 52752
Quintas de Sousa, Dr. Cipriano 00109
Quintelier, Johan 03652
Quintens, Patricia 03432
Quintens, Roger 03789
Quintero Durán, Miguel 08699
Quirce, Sylvie 13238
Quirke, Stephen 44953
Quist, Bob M. 48344
Quist, Kris N. 50807
Quoc Binh, Prof. Dr. Truong 54617
Quoc Quan, Phạm 54616

R

Raab, Loren 53600
Raab, DI Walter 02691
Raabe, Dr. Christiane 20447
Raabe, Dr. Eva Ch. 18381
Raahauge, Kristine 23258
Raassen Kruimel, E.J.C. 32465
Raastad, Ass. Prof. Jan Emil 33737
Raat, Drs. R. 32857
Raath, A. 38747
Rabal Merola, Victoria 39124
Rabano, Isabel 39561
Rabben, Anne Lise 33735
Rabbidge, Ted 33269
Rabeder, Dr. G. 01739
Rabeling, C.M. 32134
Raber, Christie 46978
Rabie, L.N. 38621
Rabin Siegal, Robert L. 36082
Rabino Massa, Prof. Emma 28012
Rabinovitch, Victor 05862
Rabinowitz, Rivka 25086
Rabner, Mario 31145
Rabner, Mauricio 30905, 31160
Rabner Cymerman, Mauricio 30892
Rabold, Dr. Joachim Martin 16966
Raboud-Schüle, Isabelle 41272
Raby, Dr. Julian 53849, 53873
Rachbauer-Lehenauer, Mag. Karin 02610
Rachewiltz, Dr. Siegfried de 27975, 27976
Rachimova, Akajša Rachimovna 29669
Rachinger, Dr. Johanna 03045
Rachman Zeth, H.R.A. 24538
Rachmankulov, Ilgiz Šamiljevič 37017
Racine, Bruno 14576, 14722
Racine, Ellen 51416
Rack, Theo 18613
Rackow, Marcia 51268
Racky, Kurt 20915
Raczka, Laurel 51863
Raczka, Robert 50546
Rada, Judith K. 53517
Radam, Brian 45718
Radawetz, Natalia 07383
Radčenko, Olga Ivanovna 37620
Raddatz, Birgit 17694
Radday, Helmut 17709
Radecki, Martin J. 49501
Radel, Chuck 52215
Rademacher, Dipl.Museol. Gudrun 17048
Rademacher, Henning 18959
Radford, Ron 00946
Radić, Danka 08970
Radić, Mladen 08923
Radicke, Rüdiger 21602
Radigois, Bertrand 13758, 13759
Radigonda, Ron 51520
Radin, Saul 47054
Radins, A. 30093
Radivojec, Milenko 04271
Radivojević, Milan 38092
Radley, Seán 24876
Radman-Livaja, Ivan 08986
Radojičić, Milica 04257
Radovanović, Dragan 31654
Radović, Andelija 38091
Radovič, Roman Bogdanovyč 43134
Radrizzani, Dominique 41986
Radschinski, Margitta 22812
Radtke, Edward 34855
Radulescu, Speranta 36227
Radulović, Ksenija 38073
Radut, Ion 36323
Radwan, Krzysztof 34804
Radycki, Dr. Diane 46783
Radzak, Lee 53631
Radzina, K. 30111
Raeder, Dr. Joachim 19527
Raemdonck, André 03484
Raetzer, Maren 18978
Rafal, Jasko 34634
Rafanelli, Dr. Simona 28241
Rafey, Souhad 51111
Raffaelli, Prof. Mauro 26179
Raffel, Suhanya 01512
Rafferty, Pauline 07250, 07258
Rafuse, Linda 06178, 06179
Ragan, Charlotte 47223
Ragbir, Lise 50833
Ragionieri, Dr. Pina 26143
Ragni, Dr. Elena 25534
Ragon, Jacques 15368
Ragsdale, Tammie 50357
Raguindin, Ethelinda A. 34339
Rahaman, F. 24394
Rahe, Diane 49326
Rahe, Gebhard 21489
Rahe, Dr. Thomas 20023
Rahim, Abdul Gaum Abdul 00006
Rahlff, Reiner 18298, 18299
Rahman Chuchu, Abdul 04970
Rahman-Steinert, Uta 17201
Rahmani, Dr. A.R. 24320
Rahmatulaev, I. 54455
Rahoult, Lena 40884
Raič, Kristijan Dragan 08867
Raiger, Johann 20428
Raihle, Jan 40515
Raimond, Jean-Bernard 11925
Raimondi, Dr. Ludovico 26348
Rainbird, Sean 21960
Rainbow, Prof. P. 44941
Rainer, Mag. Gerhard Florian 03038
Raineri, Maguerite 49839
Raines, James L. 52692
Raines, K. 06238
Rainger, Don 45089
Rainsbury, Anne 43812
Rais, Dagmar 06402
Raisor, Jerry 52316
Raiss, Gerhard 18233
Rait, Eleanor 49246
Raith, Robert 17893
Raithel, Andreas 20540
Rajaram, K. 24351
Rajčeva, Vasilka 05145
Rajchevska, Tsvetana 05191
Rajić, Delfina 38094
Rajkinski, Dr. Ivan 05264, 05265
Rajkowic-Kamara, Dr. Julia 18381
Rajkumar, Kimcha 05905
Rak, Natalija Egorovna 37445
Rakonjac, Srdjan 38074
Rákossy, Anna 23393
Rakotoarisoa, Jean-Aimé 30534
Rakowitz, Dietlinde 01736
Raku XV, Kichizaemon 28861
Ral'čenko, Iryna Grygorivna 43004
Ralčev, Prof. Georgi 05199
Ralli, Efterpi 22946
Ralls, Nick 43495
Ralph, Honeycutt 47166
Ralston, B. 33043
Ramachandran, V.S. 24286
Ramade, Patrick 12233
Ramaekers, M.T. 04043, 04044
Ramalho, Prof. Dr. Miguel Magalhães 35710
Raman, Anna 42819
Rambelli, Faustolo 26608
Rambourg, Bernard 14996
Rambow, Charles 48655
Ramesh, J. 24049
Ramesmayer, Cattrin 02620
Ramet, Guy 04128
Ramet, Nathan 03858
Ramey, Marilyn 47818
Ramharter, Dr. Johannes 02834
Ramhorn, Amber 49567
Ramírez, Angel Antonio 08653
Ramirez, Jan 51243
Ramirez, Joe 47272
Ramírez, José Juan 40108
Ramírez, Marcela E. 31058
Ramirez, Mari Carmen 49397
Ramirez, Dr. Rey 54039
Ramírez, Suny 30895
Ramirez Casali, María Eliana 07462
Ramírez de Lara, Ligia 08629
Ramírez Hernández, Jose Luis 31419
Rammer, Gottfried 02212
Ramon, René 44032
Ramón Vilarasau, Sandra 54536
Ramond, Serge 16121
Ramond, Sylvie 13889
Ramoneda, Josep 38955
Ramoneda, Prof. Luis Alberto 00386
Ramos, Dr. Manuel Francisco Castelo 35939
Ramos, Margarida 35724
Ramos, Salome D. 34513
Ramos Fernández, Dr. Rafael 39264
Ramos Horta, Dr. José 10364
Ramos Pérez, Froilán 31013
Ramos Ruíz, Martha 08691
Rampazzi, Filippo 41582
Ramsauer, Dr. Gabriele 02618, 02619
Ramsay, Gael 00825
Ramsay, Peter 01328
Ramsay, Susan 07011
Ramsden, Alan R. 06422
Ramsden, William 53338, 53341
Ramseier, Dr. Markus 41561
Ramsey, Ron 01364
Ramsey, Steve 51384
Ramseyer, Alfred 41718
Ramseyer, Anne 41682
Ramseyer, Dr. Denis 41474
Ramskjaer, Liv 33744
Ramskov, Conni 10163
Ramsour, Bruce 46497
Ramus, Lourdes I. 36070
Ramušćak, Prof. Ljubica 08852
Ramuschi, Maria Cristina 28051
Ramut, Eliza 34531
Rana, A.K. 24131
Ranade, Dr. Hari Govind 24362
Ranaldi, Prof. Franco 27317
Rand, Michael 53981
Rand, Patrick 47664
Rand, Richard 54175
Randall, M. 38760
Randall, Partricia J. 53282
Randall, R. Dennis 49410
Randall, Ross 46204, 46215
Randall, Tori 52753
Randazzo, Giuseppe 27075
Randić Barlek, Mirjana 08989
Randow, Renate von 22542
Raney, Floyd 50145
Rangaraju, Dr. N.S. 24330
Rangström, Lena 40909
Raninen-Siiskonen, Tarja 10872, 10873
Ranjbaran, Sa'eid 24657
Rank, Karin 18421
Rank-Beauchamp, Elizabeth 53051
Rankadi, Daniel 38574
Rankin, Gary 43769
Rankin, John L. 05965
Rankin, Marty 46738
Rankin jr, P.R. 48363
Rankin, Ray 44825
Rankin, Shan 48291
Ransom, Karen 51468
Ransome Wallis, Rosemary 44858
Ranšová, Eva 09683
Rantanen, Reijo 11467
Ranweiller, Margaret 52389
Rao, D.V. 24368
Rapetti, Rodolphe 14733
Rapisarda, Alfredo Pablo 00674
Rapkin, Grace 51200
Rapmund, J. 32729
Raposo, Luís 35715
Rappe-Weber, Dr. Susanne 22631
Raptou, E. 09375
Raptou, Dr. Eustathios 09392
Rardin Johnson, Sara 52103
Ras-Dürschner, Dr. Kerstin 17570
Rasajski, Javor 38163
Rasario, Dr. Giovanna 26184
Rasbury, Patricia 50976
Rasch, Elisabeth 41000
Rasch, Simone 21116
Rasch, Tone 33744
Rasche, Dr. Adelheid 17165
Rash, Roswitha M. 51801, 51804
Rasjid, A. 24521
Rask, Sven 10039
Rasmussen, Alan H. 10111
Rasmussen, Else 10136
Rasmussen, Frank Allan 10139
Rasmussen, Jan 51899
Rasmussen, Keith 49840
Rasmussen, Dr. Pamela C. 48254
Rasmussen, Robert 51767
Rasmussen, William 52342
Rasmusson, Allan 41076
Raspail, Thierry 13883
Rasson, Camille 03331
Rassweiler, Janet 51286
Rašticová, Dr. Blanka 09887
Rastorguyeva, Olga 36757
Rataiczyk, Matthias 18899
Ratajczak, Rafał 35048
Ratchford, Siobhan 44038
Ratcliffe, Shirley 43871
Rath, Andreas 03089
Rath, H. 21555
Rath, I. 21555
Rath, L. 19228
Rathbone, Eliza 53914
Rathburn, Robert R. 46824
Rathe, Daniela 22124
Rathfelder, Johannes 02899
Rathgeber, Paul 17680
Rathke, Dr. Christian 20881
Rathore, Dr. Madhulika 24090
Ratier, Hugo 00480
Ratner, Phillip 25133
Ratnikova, Marina Semënovna 37980
Rattemeyer, Dr. Volker 22557
Rattenbury, Richard 51519
Ratti, Dr. Enrico 28197
Ratti, Dr. Marzia 26433
Ratti, Pietro 26430
Rattigan, John 24829
Ratzeburg, Wiebke 22169
Ratzenböck, Hans-Jörg 02311
Ratzky, Dr. Rita 23467
Rau, Daniel 41508
Rau, Gudrun 19148
Rau, Lorraine 48539
Rau, Dr. Patrick 19676
Rauber, Irmi 18391
Rauch, Elvira 22465
Rauch, L. 41852
Rauch, Margot 02130
Rauch, Peter 41853
Rauch, Werner 14541
Rauchbauer, Judith von 16533
Rauchenwald, Alexandra 01741
Raud, Anu 10569
Raudsepa, Ingrida 30062
Raudsepp, Renita 10624
Rauer, Helmut 18159
Rauf, Barb 47879
Rauhut, Dr. Oliver 20472
Raulff, Prof. Dr. Ulrich 20179, 20181
Raunio, Erkki 10714
Rauø, Kåre William 33492
Raus, Edmund 50415
Rausch, Amalia 03133
Rausch, Felicitas 42070
Rausch, Guy 30480
Rausch, Jürgen 21002
Rauschenberg, Chris 52072
Rauscher, August 02527
Rauscher, Hannelore 22441
Rauschmann, PD Dr. Michael 18363
Rausis, Jean-Pierre 41789
Raussmüller, Urs 41828
Raussmüller-Sauer, Christel 41828
Rautjoki, Ville-Matti 11030
Rauzi, Prof. Estebar 00483
Ravanelli Guidotti, Dr. Carmen 26076
Ravaonatoandro, Aldine 30536
Ravasi, Prof. Gianfranco 26719
Ravasio, Dr. Tiziana 25385
Rave, Dr. August B. 21960
Raved, Noga 24936
Raveendran, K. 24412
Raveh, Kurt 25070
Ravel, Christian 13947
Ravelo Virgen, Margarita 31518
Ravenal, John 52343
Raverty, Norm 01058
Ravid, Zami 25065
Raviv, Efraim 25083
Ravnik Toman, Barbara 38401, 38402, 38436, 38437, 38438, 38526
Ravoire, Olivier 14532
Ravonatoandro, Aldine 30535
Raw, Jo 43329
Rawdon, David D. 38687
Rawe, Dr. Kai 20410
Rawle, Caroline 45964
Rawles III, S. Waite 52335
Rawlinson, Philippa 45375
Rawluk-Raus, Bożena 35079
Rawson, Geoff 01583
Rawson, Helen 45557
Rawstern, Sherri 46116
Rawsthorn, Alice 44837
Ray, Dulce Ivette 48716
Ray, Dr. Ella Maria 48075
Ray, Jeffrey 51817
Ray, Justice A.N. 24274
Ray, Randy W. 49392
Ray, Romita 46411
Ray, Sandra 46386
Rayburn, Ella S. 52984
Raymond, André 03961
Raymond, Dr. C. 48410
Raymond, Richard 54347
Raymond, Yves 06812
Raymond, Yvette 06041
Rayne, Angela 48460
Rayworth, Shelagh F. 07002
Razafindratsima, Clarisse 30539
Razzaq Moaz, Dr. Abdal 42132
Read, Albert J. 51565
Read, Bob 49964
Read, Judith 01005
Read, Peter 33096
Read, Robert F. 47914
Real, Nelson A. 00372
Reale, Giuseppe 26879
Reale, Dr. Isabella 28111
Reale, Sara W. 49597
Reap, Mary 52982
Reason, Robert 00773
Reaves, Dr. Stacy 52831
Rebbin, Karin 19581
Rebehn, Lars 17963
Reber, Dr. Alfred 41578
Reber, Eugene 47671
Reber, Paul 54248
Reber, Paul C. 54247
Rebhan, Dr. Helga 20424
Rebiere, Georges 11936
Rebman, Dr. Jon 52754
Rebmann, Margarete 16421
Réboli, Eliana Moro 04442
Rebolledo Godoy, Dr. Alvaro 18927
Rębosz, Ireneusz 34967
Rebuschat, Karlheinz 18574
Recalt, Gustavo 00289

Recanati, Harry 07464, 39620, 54429
Recanatini, Cesare 25229
Reccia, Ludovic 03893
Reček, Antonín 09731, 09802
Rečev, Vasil 05166
Rechberger, Erwin 02056
Rechlik, Dr. Karel 09422
Rechlin, Birgit 21702
Rechter, Jerril 01083
Recio Crespo, Miguel Angel 39961, 40146
Reck, Pat 46178
Reckitt, Helena 46424
Recouvrot, M. 15491
Rector, P. 38928
Recupero, Jacopo 25241
Reczek, Krzysztof 34763
Reddeker, Lioba 02943
Reddell, James R. 46519
Redding, Mary Anne 49961
Redding Jr., Dr. Forest 53818
Reden, Mauritz von 20995
Redenbach, Marj 06264
Redlberger, Melanie 18860
Redlich, Daniela 18378
Redlin, Jane 17199
Redman, Guy 38605
Redmond, Dr. Brian 47625
Redol, Pedro 35542
Redpath, Donna 05821
Redvale, Jolene 52276
Reece Hardy, Saralyn 52674
Reed, C. 45714
Reed, Dorothy 50554
Reed, Gerard 00770
Reed, Gretchen 49274
Reed, Ian 44171
Reed, Karla 51429
Reed, Leon B. 47848
Reed, Mark 52077
Reed, Mary 50864, 50992
Reed, Michael 49383
Reed, Philip 44822
Reed, Scott 52626
Reed, Ursula 51849
Reedie, K.G.H. 43742, 43743
Reedie, Kenneth 46007
Reekie, Clara 50928
Reel, David M. 54078
Reel, Paula 46294
Reenberg, Holger 10071
Rees, Diane 49219
Rees, Elfan ap 45980
Rees, James C. 50893
Rees, Jim 46105
Rees, Monica 43469
Reese, Dr. Beate 20414, 22718
Reese, David L. 51186
Reese, Gary 53415
Reese, Julie 51757
Reeskamp, E. 32844
Reeter, Rose 48782
Reeve, Christopher 43672
Reeve, Tori 45011
Reeve, William F. 06638
Reeves, Don 51519
Reeves, Kate A. 50919
Reeves, Lee 46279
Reeves, Dr. Nicholas 44190
Refice, Dr. Paola 25268
Refsum, Siri 33758
Regan, Brian 49498, 51216
Regan, Mary 48854
Regêncio Macedo, Carlos António 35539
Regenvanu, Ralph 54495
Reggiori, Albino 26471
Regibus, Raymond de 41826
Regier, Charles 51397
Reginster, Françoise 14417
Regli, Kurt 41469
Regnauer, Dr. A. 21714
Regnier, Gérard 14736
Regoli, Dr. Edina 27561
Regull, Enrique 40281
Reh, Tazalika te 18023
Rehberg, Dr. Rolf 21128
Rehberg, Vivian 14646
Rehberger, Prof. Dr. Dr. Karl 02640
Rehde, Anders 10010, 10012
Reher, Charles A. 49949
Reher, Dr. Uwe 19477
Rehfeld, Max 17915
Rehler, Martha 51023
Rehm, Dr. Heinrich 20912
Rehor, Ophelia 16940
Rehr-Unrath, Gabriele 20369
Rehsteiner, Dr. Ueli 41300
Řehulka, Ing. Jiří 09692
rEI?, Dr. aNSGAR 19339
Rei, M. 38705
Reibl, Přemysl 09781
Reich, Dr. Annette 19412
Reich, Christopher J. 47970
Reich, Dindy 48292
Reich, Dr. Ines 20908
Reich, Jacqueline 15532
Reich, Dr. Mike 18691
Reichard, Tim 53522
Reichardt, Marina 22421
Reichel, Antje 19054
Reichel, Brigitte 21393
Reichel, Maik 20105, 20106
Reichel, Reinhard 18653
Reichel, Robert 18009
Reichelt, Norman 21433
Reichelt, Petra 21708
Reichen, Quirinus 41207
Reichenbach, Dr. María Cecilia von 00411
Reichenbach, Svenja Gräfin von 17067
Reichert, Friedrich 17972
Reichert, Klara 20231
Reichl, Leo 02156
Reichl, Dr. Thomas 03001
Reichling, Dr. Norbert 17910
Reichmann, Dr. Christoph 19706
Reichmann, Hans-Peter 18362
Reichmann, Kurt 20915
Reickart, Harry 47607
Reid, Alan 43945
Reid, David 44035
Reid, Dennis 07091
Reid, Henry 44011
Reid, Jackie 54085
Reid, Jamie 48192
Reid, Dr. John S. 43237
Reid, Kim 06595
Reid, Marie-Claude 06490
Reid, Megan 54364
Reid, Richard 05894
Reid, Scott 45130, 45131
Reid, Prof. Seona 44283
Reid, Vivien 43381
Reid-Bavis, Glenda 05653
Reid-Cunningham, James 46919
Reid Hardman, Mimi 49895
Reid-Wallace, Dr. Carolyn 50968
Reid Williamson, J. 47192, 49504
Reidel, Dr. Hermann 21216, 21219
Reidy, Julian 41209
Reidy, Susan 53664
Reifenscheid-Ronnisch, Dr. Beate 19586
Reijmers, M. 31852
Reijnders, Dr. W. 31969
Reilly, Gerry 54125
Reilly, John W. 48230
Reilly, Margaret T. 44286
Reilly, Robin 05361
Reilly, Valerie A. 45345
Reim, Heidrun 17972
Reimann, Urs 41170
Reime, Albrecht 18989
Reimer, Andreas 20519
Reimer, Caroline 18952
Reimer, Wayne 06614
Reimets, Ülle 10628
Reimold, Gesine 18714
Reimonenq, L. 23282
Rein, Elfriede 16783
Reinbold, Friederike 17002
Reinbold, Dr. Michael 20897
Reinckens, Sharon 53845
Reinders, Drs. Carin E.M. 31911
Reinecke, Klaus 19424
Reineke, Dr. Brigitte 17072
Reinert, Frank 30483
Reinert, Hans-Walter 20811
Reingold, Daniel 47047
Reinhard, Adelheid 21325
Reinhard, Erika 19685
Reinhard, Rolf 19685
Reinhard-Felice, Dr. Mariantonia 42037
Reinhardt, Beatrix R. 52671
Reinhardt, Dr. Brigitte 22162
Reinhardt, David 49616
Reinhardt, Dr. Georg 19976
Reinhardt, Hannelore 22149
Reinhardt, Tanja 38606
Reinhart, Susi 41368
Reinhold, Mag. Bernadette 03065
Reinhold, Judith 22249
Reinhold, Katrin 17466
Reinhold, Renat 16499
Reinholdt, Bruce 53842
Reinholdt, Helle 10264
Reinig, Margot 18939
Reinikka, Satu 11408
Reininghaus, Daniela 18436
Reininghaus, Ruth 51251
Reinisch, Christine 02642
Reinke, Doris M. 48351
Reinsch, Prof. Dr. Dietmar 17526
Reinthaler, August 02210
Reis, Armindo 35393
Reis, Heitor 04795
Reis, Jennifer 50840
Reis, Dr. José 35544
Reis, Paulo 35802
Reisacher, Anton 41485
Reisetbauer, Hans 02639
Reisinger, Anton 02405
Reisinger, Iva Karolina 51158
Reisinger, Dr. Nikolaus 02414
Reisner, Peter 21366
Reiss, Mag. Johannes 01862
Reißer, M.A. Gudrun 17894, 17895
Reißer, Gudrun 17896
Reißer, M.A. Gudrun 17897
Reißer, Gudrun 17898
Reißer, M.A. Gudrun 17900
Reißland, Ingrid 20260
Reiter, Alois 02693
Reiter, Carmen 01790
Reiter, Dr. Fabian 17028
Reiter, Günther 02272
Reiter, Heribert 21020
Reiter-Büllingen, Friederike 01950
Reitere, Elina 30107
Reiterer, Bruno 02625
Reithinger, Gilles 14587
Reitner, Prof. Dr. J. 18691
Reitter, Judith 54842
Reitz, Thomas A. 06100
Reitzammer, Margit 21148
Reitzler, Anne 48229
Reiyu, Mori 28416
Rejn, J. Surya Narayara 24185
Rejune Adams, Gloria 48045
Rekawek, Theresa 47590
Reker, Les 52516
Rekker, L. 32877
Rell, Zofia 35021
Rem, Dr. P.H. 31915
Remaud 13713
Remen, Geir 33664
Remig, Nicola 19659
Remington, Charlotte 50170
Remm, Christiane 17058
Rémon, Régine 03794
Remsen, Dr. J.V. 46638
Remy 04088
Ren, Shude 07722
Ren, Xiangxian 07695
Renaux, Jean-Pierre 15807
Renda, Dr. Gerhard 17340
Render, Lorne E. 50435
Render, Shirley 07391
Renders, Bram 03321
Rendić-Miočević, Prof. Ante 08986
Renfrew, Katherine A. 51362
Renfro, Francis 48160
Renfro, Gladys 49416
Renftle, Dr. Barbara 17321
Renger, F. 44356
Rengnath, Rudi 17854
Renie, Pierre-Lin 12066
Renken, Dr. Lisa 50166
Renn, Wendelin 22240
Renner, Prof. Dr. Susanne S. 20430
Renner, Zsuzsanna 23413, 23416, 23470
Rennie, J. 44813
Rennie, Jennifer A. 43262
Rennie, Keely 47851
Rennie, Mairi C. 53771
Rennox, Munuma 31744
Renold, Karin 42090
Renouf, Nicholas 51057
Renouf, Phil 01445
Renslow, Chuck 47485
Renson, Bernoît 04057
Rentfrow, Donna 53394
Renting, Dr. A.D. 31915
Renton, Alan 45362
Renton, Helen 45287
Renton, John 45275
Rentsch, Ursula 41355
Rentzhog, Sten 40796, 40797
Renvall, Dr. Pertti 11008
Renz, Dorothea 17940
Renz, Ruth 52930
Renzl, Ludwig 02493
Renzo, Luigi 27565
Repede, Titus 36744
Repetti, Gregory 49643
Repetto, Arnaldo 05524
Rephann, Richard 51057
Reppe, C. 21139
Requejo, Angel 40301
Rérat-Oeuvray, Géraldine 41727
Reres, Sara 52985
Reris, L.S.D. 40375
Resina, Laura 39605
Resnick, Alyssa 48928
Resperger, Franz 02238
Releguier, Bernadette de 14314
Ressiter, Jeremy 45949
Ressmeyer, Richard 47380
Restall, Robin 54530
Restle, Prof. Dr. Conny 17221
Resurreccion jr., Teodoro R. 34409
Retallick, Leslie 45867
Réthly, Ákos 23449
Retière, Ch. 12750
Retière, Marie-Noëlle 14485
Retka, Sharon 50872
Retno, Dr. Sulistianingsih S. 24486
Rettenbeck, Georg 17865
Rettenmund, Jürg 41491
Rettig, Wolfram 18947
Retzlaff, Detlef 20592
Reumer, Prof. Dr. J.W.F. 32718
Reuß, Carsten 20337
Reuße, Dr. Felix 18428
Reusser, Prof. Dr. Ch. 42064
Reustle, M. Charles 47822
Reustle, Dr. S. 22601
Reuter, Dr. Astrid 19448
Reuter, Cecelia 48632
Reuter, Laurel J. 48989
Reuter, Pat 48330
Reuter, Sally 52052
Reuter, Simone 16841
Reuther, Prof. Dr. Manfred 20625
Reutimann, Daniel 41963
Revelard, Michel 03368
Révész, Dr. T. Mihály 23565
Rew, Christine 43233, 43246
Rex Svensson, Karin 41024, 41025, 41026
Rey, Jean-Pierre 13824
Rey, Kurt 41962
Rey-Delqué, Monique 15941
Rey-Vodoz, Véronique 41692
Reyer, Manfred 16968
Reyes, Manuel 23292
Reyes, Patricia 51216
Reyes Flórez, Graciela 08624
Reyes Mota, Dr. Alicia 31056
Reymann, Patrick H. 47644
Reynaert, Bieke 04146
Reynies, Béatrice de 14710
Reynolds, Catherine 51965
Reynolds, Craig 49433
Reynolds, Earl 50883
Reynolds, Edwin 50689
Reynolds, Elsie 06749
Reynolds, Fiona 44025
Reynolds, Frank 06952
Reynolds, James T. 48010
Reynolds, Janelle 06885
Reynolds, Jock 51056
Reynolds, Kimberly 46923
Reynolds, Nancy 07140
Reynolds, R.G. 43763
Reynolds, Susan 01674
Reynolds, Dr. Terry 49962
Reynolds, Valrae 51287
Reynolds Brown, C. 54202
Reys, M. 32542
Rezai, N. 24702
Rezeanu, Prof. Paul 36324
Rezler, Aleš 09613
Rezník, Branislav 38186
Rheault, Martine 07111
Rhees, David J. 50729
Rhein, Dr. Karin 21685
Rhein, Dr. Stefan 20110, 20111, 20115, 20116
Rhi, Chong-Sun 29941
Rhi, Prof. Ju-Hyung 29921
Rhie, Dr. Jong-chul 29900
Rhoades, John S. 50204
Rhoback, Kristine 46272
Rhode, Michael 53903
Rhodes, Dr. Michael 45867
Rhodes, Ulysses S. 47089
Rhodes, William 51117
Rhomberg, Anja 01796
Rhys-Michael, Cellan 43629
Riaño Bernal, Wolfram Armando 08586
Riaño Lozano, Fernando 40316
Ribas San Emeterio, Neus 38886
Ribbert, Dr. Margret 41160, 41161
Ribbrock, Dr. Gerhard 20414
Ribé Monge, Genis 39927
Ribeiro, Agostinho 35656
Ribeiro, Francisca 35936
Ribeiro de Faria, Manuel José Marques 35713
Ribeiro de Magalhães, Thales 04730
Ribeiro Martins, António 35637
Ribemont, Francis 15016
Riben, Hans 40917
Ribera, Agustí 39738
Ribero, Michel 16049
Ribes, Sonia 36092
Ribeton, Olivier 11878
Ribey, Barbara 06997
Riboreau, Brigitte 12123

R

Ricafrente, Bernardino 34364
Ricart, Dolors 38957
Riccardi, Dr. Ada 27582
Ricci, Manuela 25838
Ricci, Massimo 27696
Ricci, Pat 51077
Ricci, Dr. Stefania 26197
Ricciardi, Prof. Mario 28021
Riccòmini, Prof. Eugenio 25449, 25451, 25469
Rice, Dr. Albert R. 47572
Rice, Annette 48392
Rice, Calvin 47191
Rice, Danielle 54189
Rice, David 44315
Rice, Donald 46789
Rice, Jack 52360
Rice, Larry 53539
Rice, Nancy N. 52588
Rice, Sean 51054
Rice Irwin, John 51377
Rich, Andrea L. 50244
Rich, Beverly 53108, 53109
Rich, Hans 18439
Rich, Margaret 00825
Rich, Merle 46365
Rich, Nancy 51414
Rich, Tamara S. 50934
Richard, Christopher 51470
Richard, Jack 47917
Richard, M.L. 06517
Richard, Michel 13873, 14601
Richards, A. 44881
Richards, Beth 51831
Richards, Chuck 52084
Richards, Dolores 46894
Richards, Jane 46540
Richards, Janet 46307
Richards, L. Jane 50753
Richards, Nancy 49357
Richards, Nancy E. 51020
Richards, Pat 52927
Richards, Paul 45656
Richards, Ron 49497
Richards, W.R. 43475
Richardson, Barbara 47545
Richardson, Catherine 44530
Richardson, Don H. 53286
Richardson, E. 44530
Richardson, Edgar 52968
Richardson, Gary 44026
Richardson, I. 01152
Richardson, Dr. James B. 51943
Richardson, Jeanette 33228
Richardson, Katherine 53366
Richardson, Larry D. 46654
Richardson, Michael 48684
Richardson, Paul 52968
Richardson, Roy 33112
Richardson, Sarah 45229
Richardson, Steve 53685
Richardson, Thom 44665
Richardson, Tim 05838
Richardson, Vivian 53829
Richardson Banks, Melissa 50261
Richart, Christian 46371
Richaud, Maurice 16200
Riche-Descamps, Jeannine 03962
Richelson, Paul W. 50769
Richerson, James J. 51773
Richey, Susan 47736
Richmond, Alan 46269
Richmond, Lauren 51754
Richmond, Robert 44049
Richmond-Rex, Phyllis E. 33348
Richner, Trudi 41959
Richon, Marco 41230
Richoux, Sylvie 13998
Richter, Bettina 42094, 42095
Richter, Bob 06151
Richter, Donald 47960
Richter, Dr. Erhard 18754
Richter, Gert 18902
Richter, Gunnar 18844
Richter, Helmut 22450, 22451, 22452
Richter, Jenett 07369
Richter, Jörg 18875
Richter, Paul 01846
Richter, Prof. Dr. Stefan 21396
Richter, Dr. Susanne 19922
Richter, Susie 49853
Richter, Dr. Thomas 16582, 16587, 16588, 16592, 16593
Richter, Ursula 21381, 21382
Richthofen, Ferdinand von 17960
Richthofen, Dr. Jasper von 18679
Richtsfeld, Dr. Bruno 20490
Richwagen, Suzette 47101
Rick, Jane 44990
Rickards, Dr. R.B. 43733
Ricker, Maria 07405
Ricketts, Paul W. 53783
Ričko, Oka 08877
Rico Perrat, Gabriela 30748
Ridder, M. A. Thomas 17910
Riddington, Kate 43818
Riddle, Janet 33267
Ridet, Christophe 13325
Ridgway, Ginger 51680
Ridler, Gerda 22288
Ridley, David J. 43848
Ridley, Jackie D. 43998, 44000
Ridley, Jacqeline 44003
Ridley, Dr. Michael 44004
Ridpath, Marjorie 47188
Rieche, Christiane 18902
Rieck, Silvana 16709, 16710
Ried, Martha 19723
Riedel, Dr. Alexander 19449
Riedel, Andrea 20870
Riedel, Christiane 19453
Riedel, David 16890
Riedel, Erik 18379
Riedel, Frank 22755
Riedel, Uwe 20357
Riedel, Walter G. 51578, 51579
Rieder, Alfred 41978
Rieder, Markus 22677
Riedesel, Rikarde 16692
Riedmiller, Thomas 18501, 18503
Rief, Julia 16402
Rieger, Sigrun 20426
Rieger-Jähner, Prof. Dr. Brigitte 18405, 18406
Riegler, Frank 02733
Riegler, Ingrid 03099
Riego González, Angel 39116
Riehl, Ken 05495
Riek, Peter 18183, 18184
Riek, Rolf 19780
Rieken, Duane 49523
Riel, M. van 15178
Riemann, A. 17904
Riemersma, Tiny 32109
Riepe, Ulrike 19016
Riepl, Heiner 21664
Riepniece, Ineta 30009
Riepula, Anne 11002
Ries, Christian 30484
Riese, Dr. Ute 18637
Riesen, Dr. Jeorge M. 41316
Riesgo, Juan M. 39531
Riestra de Cuevas, Bertha 31125
Riether, Dr. Achim 20487
Riethmüller, Dr. Marianne 18509
Riethmuller, Bruce 01571
Rietmann, Urs 41208
Rietvelt, Carol 49565
Rietzke, I. 16996
Rife, Sandra B. 54173
Riffel, Carola 18133
Riffle, Tim 51290, 51293
Rifkin, Edwin 53846
Riganti, Riccardo 28066
Rigau, Adlin Rios 36065
Rigby, S. 43526, 43527
Rigdon, Lois 47132
Rigdon, Marge 48205
Rigeade, Pascal 11919
Rigert, Markus 42027
Rigg, Frank 46935
Riggall, R.A. 01470
Riggins-Ezzell, Lois S. 50976
Riggio, Leonard 51166
Riggs, Carol 50319
Riggs, Gerry 47701
Riggs, Mary Fran 48396
Riggs, Richard 50035
Riggs, Timothy A. 47344
Righetti, Christiane 16886
Righetti, Olimpia L. 00630
Righolt, Niels 40585
Rigoli, Adriano 28133
Rigoli, Roxanne 01503
Rigon, Lidia 27572
Riibak, Renate 10634
Riker, Janet 47081
Rikkoert, A. 32766
Rikkonen, Mikko 11438
Riley, Angelika 18951
Riley, Grace 44986
Riley, Ian 44738
Riley jr., Joseph P. 47361
Riley, Karen 49509
Riley, Norman 46472
Riley, Pat 48718
Riley, Peter 45602, 45603
Riley, Terence 51215
Rill, Martin 22160
Rilonek, Susanne 02620
Rimer, Greame 44665
Rimmö, Christina 41032
Rimoldi, Luciano 27060
Rimon, Ofra 24967
Rinaldi, Paolo 27965
Rinaldini, Dr. Francesco 25784
Rinberger, Sabine 20493
Rinck, Aksel 20726
Rindfleisch, Jan 47911
Rindlisbacher, Rolf 41657
Rindom Madsen, Finn 10179
Rindsfüsser, Annemarie 16786
Rinetti, Dr. Luigi 28242
Ringborg, Eva 40684
Ringgaard Lauridsen, Henning 10317
Ringler, Sarah 54043
Ringold, Judith 51864
Ringquist, Alden 48226
Rink, Christine 19904
Rinker, Dr. Dagmar 22162
Rintala, Liisa 11350
Rintoul, Dr. Gordon 44136, 44139, 44148, 45202, 45264, 45649, 45651, 45653
Rinuy, Anne 41418
Rionda, Patricia Elena 00637
Rionnet, Dr. Florence 12746
Rionnet, Florence 12747
Riopelle, Christopher 44937
Riordan, Mary Ann 49772
Riordon, Bernard 05828
Ríos de Molina, Martha 31084
Ríos Montero, Sylvia 07455
Ríos Reyes, Dr. Carlos Alberto 08601
Rios-Samaniego, Milita 52506
Riottot El-Habib, Béatrice 14746
Rioux, Alfred 06148
Ripatti, Marja-Liisa 10748
Ripley, John W. 53890
Ripoll, Rafael Romá 38829
Ripoll Isern, Luis Díaz 40048
Ripperová, Dr. Marie 09902
Rippy, Francine 52893
Rippy, Linda 51992
Risani Dossi, Dr. Carola 25969
Risch, Laurie 47858
Riseman, Henry 47065
Rish, Sarah 50084
Rishel, Joseph 51845
Rishel, Joseph J. 51867, 51876
Risk, Shannon M. 47681
Rišlink, Vladimír 09591
Risser, Mary 47025
Risskov Sørensen, Kurt 10115
Rister, Faye 05991
Risthein, Helena 10624
Ritch, Irene 06878
Ritchie, Christina 07209
Ritchie, Timothy 46823
Ritschard, Claude-Janine 41418
Ritschard, V. 41696
Ritson, Liz 45228
Rittenour, Maralyn 48243, 48244, 48247, 48248
Ritter, Beate 17697
Ritter, Martin 20843
Ritter, Nicole 17605
Ritter, Ruedi 41278
Ritter, Sandra Sophie 17159
Ritter, Shelley 47587
Ritter-Lutz, Dr. Susanne 41668
Rittmann, Josef 02592
Rittwage, Roy 05478
Ritzmann, Dietrich-Wilhelm 21592
Ritzmann, Hans 42019
Riva, Davide 25650
Riva, Maurizio 25650
Rivard, Céline 06780
Rivas, Francisco 39797
Rivas, Helen 52063
Rivas Octavio, José A. 39555
Rivatton, Bernard 15254
Rive, Aila le la 42088
Rivé, Isabelle 13880
Rivera, Dianna 47000
Rivera, George 52864
Rivera, Gigi 34500
Rivera, Mario 49461
Rivero, Dr. María Eloisa 54427, 54428
Rivero Borrell, Héctor 31117
Rivers, Gabrielle 45443
Rivier, Delphine 41738, 41739
Rivière, Danielle 13901
Rivière, Dr. Dominique 14812, 16113
Rivoira, Prof. Giuseppe 27752
Rivollet, Karin 41280
Rivolta, Barbara 53528
Rix, Peggy 47296
Rixhon, Maryse 03599
Rizzo, Antonio A. 00518
Rjabcev, Igor Mykolajovyč 43066
Rjabov, Gennadij Evgenjevič 37823
Rjazanov, Aleksandr Michajlovič 36814
Rjazanova, Lilija Aleksandrovna 36881
Rjuken, Kjetil 33481
Roach, Ed 06154
Roaldset, Prof. Dr. Elen 33737
Robb, Doug 45283
Robb, Lisa 51749
Robb-Dietzen, Jenny 52769
Robbins, Anna 00850
Robbins, Anne 44937
Robbins, Carol 47931
Robbins, Daniel 44899
Robbins, Neil 44476
Robe, Sofia 36549
Robel, Sigrid 17733
Robelen, William G. 48044
Roberson, Jeanette W. 49886
Roberson, W. Allen 47731
Robert, Henri 12668
Roberti, Dou Giuseppe 27722
Roberts, Becky 52518
Roberts, Bette 50535
Roberts, Brady 51906
Roberts, C. J. 52023
Roberts, Chad 53181
Roberts, Coles 51385
Roberts, Dr. Dafydd 44754
Roberts, Day 05684
Roberts, Esther 43376
Roberts, Gary 47433
Roberts, Greg 50294
Roberts, Hugh 44976
Roberts, James D. 53228
Roberts, Jan 47543
Roberts, Jane V. 46137
Roberts, Jill 50901
Roberts, John 43666
Roberts, Martin 43902
Roberts, Peg 46742
Roberts, Perri L. 47802
Roberts, Sir Samuel 43870
Roberts, Shona 33060
Roberts, Teresa 47828
Roberts, V. 44433
Roberts-Douglass, Kathy 52427
Róbertsdóttir, Gerdur 24014
Robertson, A. 44047
Robertson, Beatrice L. 51766
Robertson, Beverly 50581
Robertson, Bruce 50244
Robertson, Byron 48883
Robertson, Dongwol Kim 47819
Robertson, Dr. H.G. 38592
Robertson, I.G. 45611
Robertson, Lynn 47728
Robertson, Pamela B. 44284
Robertson, Pat 49442
Robertson, Preston 06403
Robertson, Rosalyn 52415
Robertson, Susan 53809
Robertson Hughes, Neville 01110
Robidoux, R.R. 51885
Robins, R. 00786, 00787, 00788
Robinson, Alistair 45802
Robinson, Beulah 47526
Robinson, Bonnell 46917
Robinson, Bruce 01130
Robinson, Christine 53295
Robinson, Cliff 00936
Robinson, Cynthia 50061
Robinson, D.N. 45037
Robinson, Dan 43818
Robinson, David 53717
Robinson, Deborah 45920
Robinson, Franklin W. 49548
Robinson, Dr. George W. 49377
Robinson, Graham 33201
Robinson, Herb 06027
Robinson, John A. 00855
Robinson, Dr. Joyce 53659
Robinson, Julie 00773
Robinson, Katharine S. 47367, 47375
Robinson, Louise 48721
Robinson, Malcolm J. 53581
Robinson, Michael 05519
Robinson, Michelle 47969
Robinson, Dr. Paul 43979
Robinson, Ruth 54182
Robinson, Sue Ann 50174
Robinson, Susan 50140
Robinson, Tara 48114
Robinson, Thomas A. 48332
Robinson, Travis 50048
Robinson, Wendy 45162
Robinson, William H. 47624
Robinson, William W. 47177
Robles, Constantino 39459
Robles, José Angel 31346
Robles, Sylvia de 34067
Roblin, Laurent 12612
Robson, Diana 00877
Robson, Judy 46394
Robson, Keith 45880

Robson, Pamela 43906
Robson, Ruth 43381
Roby, Teresa 43312
Roca, Teresa 39482
Rocchi, Dr. Silvano 27670
Roch, Dr. Heidi 18708
Roch, Neal 46187
Roch-Stübler, Dr. Heidi 18710
Rocha, Dr. Olivio 35439
Rocha Moreno, Leonor 08593
Rocha Virmond, Eduardo da 04434
Rochard, Dr. Patricia 19334
Roche, Fr.S. 24406
Roche, James 47222
Rochefort, Hervé de 12669
Rochette, Christelle 13419
Rochette, Didier 14432
Rochford, John 50962
Rochlitz, Heike 18820
Rochmad, Edi 24435
Rock, James T. 54358
Rockefeller, David 51117
Rockenbach, Susanne 19466
Rockhill, King 50863
Rockhold, Jean 52469
Rockstein, Margitta 16698
Roda, Dr. Burkard von 41157, 41160, 41161, 41168, 41174
Roda, Dr. Hortensia von 41831
Rodák, Dr. Jozef 38366
Rodarie, M. 13879
Rodd, H.C. 46021
Rodde, Manfred 21676
Roddenberry, Heather 51769
Roddick, J.H. 07262
Roddie, Ken 53629
Roddis, William H. 50296
Rodekamp, Dr. Volker 19892, 19918, 19927, 19932, 19936, 19937, 19938, 19939, 19941
Rodepeter, Dr. Ralf 20428
Roder, Dr. Hartmut 17564
Roderick, Myron 53297
Rodeschini Galati, Dr. Maria Cristina 25400
Rodger, David 45826
Rodger, Robin 45366
Rodgers, Forrest B. 46733
Rodgers, James 49853
Rodgers, Jeff 47232
Rodgers, Kenneth G. 48222
Rodgers, Maureen 05459
Rodgers, Michael W. 48623
Rodgers, Michele 47263
Rodgers, Rachel A. 47236
Rodgers, Susan 01362
Rodgers, Tommie 49985
Rodgers, Z. 24823
Rodica, Radu 36319, 36334
Rodiek, Dr. Thorsten 20077, 20078, 20080
Rodimceva, I.A. 37155
Rodin, Matis 10623
Rodina, Lucia-Letiţia 36101
Rodionov, V.A. 37184
Rodler, Willibald 01966
Rodman, Vickie 52698
Rodney, Clayton 23318
Rodolfi, Rosanna 26221
Rodon, Francesc 40294
Rodrigues, Dr. Dalila 35716
Rodrigues, Henrique A.N. 35426
Rodrigues, Dr. Paulo 04725
Rodrigues Meyer, Fernando 04639
Rodriguez, Carlos Armando 08606
Rodriguez, Carmen 47800
Rodriguez, Dionisio 33289
Rodriguez, Geno 51110
Rodriguez, José Alfonso 08611
Rodríguez, Dr. Marco Antonio 10437, 10440
Rodriguez, Roberto M. 47382
Rodriguez, Rosa 40204
Rodriguez, Teliza V. 50097
Rodriguez, Teresa 53665
Rodríguez, Tomás 54543
Rodríguez Angarita, María Esther 08545
Rodríguez-Boette, Taryn 52525
Rodríguez Culebras, Dr. Ramón 40046
Rodríguez de Montes, María Luisa 08526
Rodríguez del Camino, Marco 54543
Rodriguez Delucchi, Jorge 54407
Rodríguez García, Serafín 39964
Rodríguez Martín, Aurora 39513
Rodriguez Molina, Cap. D. Antonio 39643
Rodríguez Ruza, Concepciòn 39232
Rodriquez, Jennie 52784
Rodríquez Culebras, Ramón 40299
Roduit, Abt Joseph 41792
Rodway, Em 47742
Roe, Diane 43961
Röber, Dr. Ralph 19676
Röck, Erwin 02538
Rödel, Klaus 10026
Roeder, Corinna 20896
Röder, Heinz 20746
Roeder, Paul 47878
Röder, Dr. Sabine 19705
Rödger, Ralf 16729
Röding, Lutz 17384
Roegholt, Alice 31863
Roegiers, Prof. Dr. Jan 03788
Röhl, Johannes 16693
Röhling, Silke 17522, 22474
Roehmer, Dr. Marion 20730
Röhrig, Prof. Dr. Dr. Floridus 02202
Röhring, Micha 19224
Roelink, Géke 31841
Roeloffs, Abigail 46885
Roels, André 04122
Römer, Prof. Dr. Cornelia 03045
Römer, Dr. Jürgen 22655
Römer, Kurt 01757
Römer, Neria 41582
Römhild, Michael 19179
Rönn-Kollmann, Dr. Claudia 18438
Röper, Dr. Martin 21813
Rörig, Irene 19995
Rösch, Dr. Libuse 19679
Rösch, Mathias 20789
Roesink, Macha 31803
Röske, Dr. Thomas 19082
Roesle, Marie-Alix 42088
Roesler jr., Henry E. 50796
Rössl, Dr. Joachim 02905
Rößler, Dr. Hans 20607
Rössler, Heiner 20271
Rössler, Kaba 41085
Rössler, Prof. Leopold 02997
Rössler, Dr. R. 17700
Rössler-Köhler, Prof.Dr. Ursula 17441
Rößner, Dr. Alf 22407, 22421
Rößner, Dr. Gertrud 20472
Roest-den Hartogh, G.C. 32384
Roettig, Dr. Petra 18932
Röver-Kann, Dr. Anne 17556
Roffia, Dr. Elisabetta 26036, 27856
Rogalski, Dr. Andrzej 41584
Roganti, Gabriella 26740
Rogantini, Franco 41580
Rogdeberg, Guttorm A. 33688
Rogé, Monique 15981
Rogers, Alexander K. 52346
Rogers, Brett 44954
Rogers, Dianne 54268
Rogers, Frank 51731
Rogers, Gerry 46355
Rogers, Ginny 49817
Rogers, James 48340
Rogers, Malcolm 46940
Rogers, Peggy 48342
Rogers, Rachael 43252
Rogers, Richard 51111
Rogers, Robert 48371
Rogers jr, Roy Dusty 46997
Rogers, Ted 53202
Rogers, Victoria 55093
Rogerson, J.-C. 44016
Rogge, Marc 04104, 04169
Rogger, Prof. Iginio 28059
Roggero, Prof. Mario Federico 25807
Roggero, Sue 44371
Roggmann, Bettina 16435
Roggow, Barbara 22673, 22674
Rogić, Milena 08946
Rogina, Bojana 38454
Roglán-Kelly, Dr. Mark 47935
Rogne, Odd 33894
Rogozińska, Ewa 34537
Roháček, RNDr. Jindřich 09692
Rohan, Antoinette Duchesse de 13279
Rohde-Enslin, Dr. Stefan 17143
Rohde-Hehr, Patricia 18951
Rohde-Reith, M.A. Monika 18186
Rohdenburg, Dr. Ernest A. 47404
Rohlfs, Stefan 18215
Rohmer, Hans Jürgen 18743
Rohner, John R. 46961
Rohovit, Dr. Ron 50215
Rohr, Dr. Gerlinde 19936, 19939
Rohrbach, John 48709
Rohrer, Denis 41983
Rohrer, Susan 51545
Rohwer, Dr. Sievert 52994
Roiberts, David Alan 48453
Roig Toqués, Francisco 40284
Roige, Annabella 51883
Roiß, Dr. Hubert 03110, 03114
Roitman, Dr. Adolfo D. 25016
Rojanova, Galina Iljinična 36998
Rojas de Moreno, María Eugenia 08776
Rojas Iragorri, Ximena 08568
Rojas Toro, Elizabeth 04245
Rojas y Cruzat, Oscar M. de 09056
Rojo Dominguez, Ana Cristina 39082
Rola, Laura 35802
Rola, Stefan 34990
Roland, Dr. Berthold 16686
Roland, Debbie U. 52610
Roland, Greg 49055
Roland, Heather 47746
Roles, John 43251, 44660, 44662, 44664
Roll, Dr. Carmen 18450
Roll, Kempton 46371
Rolla, Maureen 51942
Rolland, C. 12322
Rolland, Raymond 14821
Rollenitz, Maria 02174
Roller, Scott 50396
Roller, Dr. Stefan 18394
Rolley, Jean-Pierre 11551
Rollig, Stella 02318
Rollin, Thierry 11929
Rollings, James R. 54075
Rollins, Avon William 49811
Rollins, Fred H. 53972
Rollko, Daniel 16720
Rolo Fajardo jr., Gaudioso 34501
Roloff, Prof. Dr. A. 22049
Romagnoli, Martha Beatriz 00678
Romagnoli, Prof. Myriam Ruth 00526
Romagnolo, Dr. Antonio 27575
Romagnosi, Angelo 28045
Román, Enrique 00342
Román Berrelleza, Juan Alberto 31109
Roman de Zurek, Teresita 08617
Romanelli, Prof. Giandomenico 25565, 26877, 28178, 28179, 28180, 28188, 28195, 28201, 28205, 28213
Romanenko, Maja Petrovna 37440
Romanenko, Vera Vladimirovna 36794
Romaniuc, Elvira 36394
Romano, Simona 26962
Romano, Socorro 39255
Romanova, Raisa Michajlovna 36834
Romanowska, Marta 34809
Romard, Judith 06197
Romatet, J.C. 15973
Rombauer, Tamás 23340, 23341
Romdhane, Khaled Ben 42616
Romeo, Paolo 26670
Rømer Sandberg, Ole 33525
Romero, Dr. Edgardo J. 00168
Romero, Mark 52057
Romero, Virginia de 34069
Romero, Yolanda 39360
Romero Ardila, Dídimo Ernesto 08664
Roming, Lorenz 21631
Romińska, Elina 34966
Romito, Dr. Matilde 26932, 27054, 27594, 27595, 28264
Rommé, Dr. Barbara 20519
Rommel, Dr. Gabriele 22541
Rommel, Dr. Ludwig 19173
Rommers, Sabine 03875
Romney, George G. 49476
Romo, Tere 52783
Romo R., Hermana Aura Elisa 08712
Romportlová, Simona 09438
Romsics, Imre 23628
Romualdi, Dr. Antonella 26157
Rona, Zeynep 42791
Ronaghan, Allen 06025
Ronald, William 01401
Ronchi, Dr. Benedetto 28049
Roncière, Florence de la 13925
Rončkevičová, Antónia 38355
Róncoli, Mónica 00116
Rondeau, James 47452
Ronfort, J.N. 14580
Rong, Enqi 07677
Ronig, Prof. Dr. Dr. Franz 22097
Ronin, Christophe 14698
Ronning, Peggy 50491
Ronowicz, Małgorzata 34891
Ronte, Dr. Dieter 02225
Roode, I. de 31888
Roodenburg, Teun 32979
Roodt, Frans 38710
Roohfar, Z. 24679
Rooizpekar, Mandana 41868
Rooney, Dr. Brendan 24807
Rooney, Jan 51110
Rooney, Mary 33123
Rooney, Steve 51194
Roos, Pieter N. 51329
Roosa, Jan 53730
Roose, B. 03426
Roosen, Jean-Paul 12208
Roosevelt, Christopher 50314
Roosing, Jan Jaap 22797, 22798
Root, Deane L. 51945
Root, Margaret 46307
Roots, Gerrard 44821
Roper, Donna K. 51759
Roper, Mark 43791
Ropers, Emmanuel 13256
Ropp, Ann 49632, 49635
Ropp, Harry 48183
Roppelt, Dr. Tanja 17670
Ropponen, Liisa 10727
Roque Gameiro, Maria Alzira 35763
Roquebert, Anne 14692
Roques, Thierry 06221
Rorem, Ned 51111
Rorschach, Dr. Kimerly 48221
Rosa, Jocelyn P. de la 34361
Rosa, Joseph 51942, 52795
Rosa, Mercedes 04789
Rosales, Felipe 34344
Rosales Huatuco, Dr. Odón 34181
Rosales Vergas, Manuel 31090
Rosanbo, Richard de 13519
Røsand, Toril 33615, 33616
Rosandich, Dr. Thomas P. 47965
Rosas, Moisés 31038
Rosas, Dr. Silvia 34260
Roschanzamir, Dr. Béatrice 17454
Roschmann, Elke 19282
Rose, Ben 53958
Rose, Coral 49126
Rose, Kate 46079
Rose, Randall A. 53457
Rose, Sally 44108, 44109
Rose, Wilhelm 16983
Rosefelds, Andrew 01159
Roselló, Núria 40291
Rosemann, Günther 20020
Rosemann, Thomas 42083
Rosen, Ferial Horesh 25082
Rosen, Rhoda 47512, 47514
Rosen Poley, Rita 48353
Rosenbaum, Joan 51200
Rosenbeck, Dr. Hans 16999
Rosenberg, Dr. H.I. 05533
Rosenberg, John 52139
Rosenberg, Dr. Stefan 20764
Rosenberger, Pat 47591
Rosenblattl, Christina 01820
Rosenbleeth, Herb 53902
Rosenblum, Robert 51258
Rosenblum Martin, Amy 47040
Rosenbluth, Betsy 47142
Rosenbruch-Hammer, Angie 52545
Rosendahl, Paul 10003
Rosenfeld, Daniel 53977
Rosenfield Lafo, Rachel 50090
Rosengart, Dr.h.c. Angela 41598
Rosengren, Anna 40556
Rosengren, Hans 40625
Rosengren, Jim 51480
Rosengren, Kerstin 40537
Rosenkranc, Milan 09491
Rosenkrantz, Baron Erik Christian 10094
Rosenkranz, Roger 48016
Rosenmeier, Edith Marie 10284
Rosenquist, Perry E. 47888
Rosenstock, F. 16410
Rosensweig, Larry 48046
Rosenthal, Beryl 47181
Rosenthal, Deborah 53349
Rosenthal, Dina R. 53268
Rosenthal, Ellen 48546
Rosenthal, Turpin 17492
Rosenzveig, Charles 48521
Rosenzweig, Dr. Rainer 20793
Rosero Solarte, Luciano 08711
Rosewitz, Mindy 48657
Rosher, Fiona 44424
Rosiak, Andrzej 35297, 35378
Rosich i Salvó, Mireia 40287
Rosiek, Barbara 35382
Rosing, Emil 23262
Rosing, M. 10153
Rosini, Prof. Dr. Corrado 25267
Rosinski, Dr. Rosa 17337
Rosioara, Iustin 36592
Rosito, Massimiliano G. 26165
Rosjat, Dr. Erika 16942
Rośkiewicz, Wiesław 35147
Rosljakov, Sergij Mykolajovič 43150
Rosoff, Nancy 47069
Rosoff, Susan 51593
Rosołowski, Szymon 34630
Ross, Alan 51934
Ross, Andrea 44156
Ross, Dr. Cathy 44928
Ross, David A. 48595
Ross, Donald A. 01570
Ross, Dugald Alexander 45727
Ross, Gwendolin 17790
Ross, Ian D. 05498
Ross, Jerry 48436

Ross, Mari B. 46553
Ross, Marian 53540
Ross, Pam 50932
Ross, Phill 49305
Ross, Richard 51711
Ross, Sandria B. 46316
Ross, Stewart 01004
Ross, Wendy 49708
Rossellini, Laurence 14010
Rosser, Merlin 07031
Rosset-Magnat, Chantal 15087
Rossetti de Scander, Dr. Antonio 28092
Roßfeldt, Heike 17106
Rossi, Cristina 27908
Rossi, Filli 25918
Rossi, Dr. Francesco 25406
Rossi, John S. 48522
Rossi, Rochelle 47212
Rossi, Dr. Sandra 28184
Rossi Brunori, Ignazio 26831
Rossiter, Shannon 49783
Rossiter, Sybil 47269
Rossner-Dietz, Petra 22343
Rossolimo, Dr. Olga L. 37431
Rost, Margret 19921
Rost, Sven 18799
Rostami 24680
Rostek, L. 01596
Rostholm, Hans 10072
Rostworowska, Dr. Magdalena 35329
Rosu, Georgeta 36227
Roşu, Steliana 36298
Roszko, Grzegorz 34674
Rota, Anna Francesca 25863
Rota, Dr. Laura 25652
Rotaru, Costel şi Elena 36650
Rotaru, Georgeta Monica 36454
Rotenberg, Nir 24948
Roth, Dr. Anja-Maria 19076
Roth, Dan 46416
Roth, Eric 51096
Roth, Flemming 10040
Roth, Gary G. 52231
Roth, Dr. Ir, Bram 32083
Roth, Isabelle 41868
Roth, John 47065
Roth, Josef 02863
Roth, Jules 14396
Roth, Linda H. 49199
Roth, Marc-André 41826
Roth, Prof. Dr. Martin 17971
Roth, Max 41086
Roth, Dr. Michael 17172
Roth, Neal 52025
Roth, Dr. Peter 38332
Roth, Ronald C. 52265
Roth, Dr. Thomas 19626
Roth, Dr. Ursula 17991
Roth-Oberth, Dr. Erna 18310
Rothe, Udo 21098
Rothe-Wörner, Heidrun 18603
Rothen, Heidi 41596
Rothenberg, Peter 50385
Rothenberger, Manfred 20771
Rotheneder, Martin 02375
Rother, Dietmar 20295
Rother, Dr. Rainer 17069
Rothermel, Barbara 50325
Rothfuss, Joan 50743
Rothfuss, Dr. Otto 16421
Rothman, Mark 50246
Rothove, Billi R.S. 48873
Roths, Jaylene 50874, 50875, 53153
Rothschild, Eric de 14625
Rothschild, Jan 51279
Rothschild, Baroness Philippine de 14763
Rotkale, Rita 30065
Rotovnik, Mitja 38447
Rotsart, John A. 52751
Rott, Dr. Herbert W. 20425
Rottermund, Prof. Andrzej 35291
Røtting, Dorrit 10182
Rottmann, Carmen 21392
Rottmann, Dr. Joseph 16755
Rottmann, Lee 52593
Rouchaud, Alain 15003
Roughley, Dr. R.E. 07365
Rouijel, Souâd 31677
Rouijel, Souad 31693
Rouleau, Reynald 05614
Roulston, Peggy 50408
Roumell, Lisa 51240
Rountree, Robin K. 53354
Rouquereol, N. 11753
Rouquet, Chantal 15994, 15999, 16000, 16001
Rourke, Ralph 47046
Rous, Pavel 09512
Roush, Margaret 49424
Rousseau-Arlardi, Daniele 14678
Roussel, Dominique 15774, 16229
Roussel, Louis 13927, 15738
Rousset-Rouard, Yves 14072
Roussia, Pierre 13338
Roussier, François 16267
Routhier, Nathalie 05445
Routledge, M.A. 45805
Roux, Bertrand 12592, 15499
Roux, Jean-Pierre 12897, 13112
Roux, Nathalie 12549
Rouyer, Morgan 16140
Rouyer, William 16140
Rovati, Dr. Clementina 27123
Rowan, Edward 50422
Rowe, Alan L. 33164
Rowe, Ann P. 53921
Rowe, Barbara 54201
Rowe, C. 38709
Rowe, Donald 51540
Rowe, M. 45890
Rowe, Matthew 44103
Rowe, Victoria 50161
Rowe-Shields, Michele 52918
Rowinski, Dr. Stanislaw 18951
Rowland, David B. 49620
Rowland Jones, Mark 43487
Rowley, Jennifer 43648
Rowley, Roger H.D. 52199
Rowold, Kathleen L. 46863
Rowsom, Christopher 46590
Rox-Schulz, Heinz 21453
Roxas Lim, Prof. Aurora 34486
Roxburghe, Duke of 44555
Roy, Dr. Archana 24270
Roy, Dr. Ashok 44937
Roy, Carolyn 52100
Roy, David T. 43346
Roy, Jan-Bernard 14391
Roy, Jean-Marc 05611
Roy, Kenneth J. 50986
Roy, Nathalie 15114
Roy, Noëlle 14739
Royce, Ralph 48848
Royer, Claude 13362
Royère, JC de la 03442
Royez, Robert 03955
Royle, Ted 43802
Royo, Dr. Ana Carmen 38788
Rozanov, Aleksej Jurjevič 37395
Rozenberg, Dr. Silvia 25010
Rozendaal, Prof. A. 38735
Rozenfeld, Boris Matveevič 37049
Rozewicz, Beverly L. 49243
Rozłowski, Tadeusz 34733, 34735
Roznoy, Cynthia 53250
Rózsa, Zoltán 23740
Rozsošyns'ka, Nina Leonidivna 43078
Rozzo, Dr. Ugo 28044
Rúa, Rosa Edith 08731
Ruan, Yingqi 08492
Ruas dos Santos, Dr. Margarida 35674
Rub, Jürg 41205
Rub, Markus 41208
Rub, Timothy 47555
Ruban, Nikolaj Ivanovič 36856
Ruben Cuñeo, Dr. Nestor 00651
Rubenstein, Earl 50210
Rubidge, Prof. Bruce 38650
Rubin, Jane 51200
Rubinstein, Adam 52241
Rubinstein, Ernest 51272
Rubinstein, Helena 51279
Rubio, J. David 08652
Rubio Pediata, Luis 40360
Rubio Sánchez, Pedro 30847
Rubnikowicz, Marek 35195
Rucellai, Oliva 27825
Ruch, Pam 52149
Ruchhöft, Bernd 21051
Rucka, Barbara 34932
Ruckel, Jay G. 51184
Rucker, Denise 49575
Rucker, Henrike 22456
Rucker, Ken 53107
Ruckstuhl, Dieter 41397
Ruczynski, F.C.K. 32004
Rudaz, Patrick 41291
Rudd, Karen 45604
Rudd, Robert 48270
Rudd, Shirley J. 51751
Rudder, Jennifer 07019
Ruddle, John 24748
Rudi, Dr. Thomas 19911
Rudigier, Dr. Andreas 01927, 02716
Rudin, Vladimir Vasiljevič 37790
Rudinger, Torsten 21090
Rudisill, Dr. Richard 52883, 52885
Rudkin, David J. 43826
Rudman, Karen 31753
Rudnickaite, Eugenija 30408
Rudnik, Galina Anatolevna 37766
Rudolp, Dr. Andrea 21009
Rudolph, Beth 46190
Rudolph, Jeffrey N. 50215
Rudolph, Johanna 21155
Rudolph, Ursula 21150
Rudzińska, Wanda M. 35225
Rübesam, Manfred 19929
Rückel, Robert 17064
Rücker, Dr. Thomas 02615
Rückert, Alexandra M. 41377
Rückert, Dr. Birgit 21472
Rückert, Hans 16429
Rueda, Horacio 10450
Rüdel, Dr. Holger 21567, 21568
Ruedin, Dr. Pascal 41876, 41876
Ruedrich, Mark 48596
Rüegg, Corinne 42001, 42002
Rüger, Axel 31898
Ruegg, Dr. François 41385
Rühl, Karin 18410
Rühlig, Cornelia 20360, 20361
Rühmer, Erich 19242
Rührl, Helmut 17539
Ruell, David 46386
Rümbeli, Monica 41393
Rümmelein, Andreas 20429
Rünger, Dr. Gabriele 18284
Ruesch, Carol 54370
Ruess, Andreas 16823, 16824
Rueß, Dr. Karl-Heinz 18672, 18673, 18677, 18678
Ruet, Karen 05835
Rütgen, Georg 03001
Rüther, Dr. Martin 19626
Rütsche, Dr. Claudia 42080
Ruette, Monique de 03495
Rütti, Dr. Beat 41137
Rüttimann, Franziska 42090
Ruf, Dietmar 20426
Ruf, Peter 20052, 20057
Rufe, Laurie J. 53598
Ruff, Eric J. 07410
Ruff, Joshua 53313
Ruff, T.J. 45995
Ruffo, Joseph M. 50095
Rugama, Miriam 33274
Ruggeri, Prof. Alessandro 25456
Ruggiero, Dr. Laurence J. 54253
Rugunanau, S. 38637
Rugus, Dot 52334
Ruhe, P. Wilhelm 16680, 16681
Ruhe, Rikke 10166
Ruhl, Erwin 19136
Ruhl, Hans-Eberhard 17797
Ruhl, Horst 22368
Ruhlig-Lühnen, Dagmar 19325
Ruhmann, Dr. Christiane 20968
Ruhmer, H.-H. 18765
Ruhrberg, Dr. Bettina 18712
Ruhstaller, Tod 52936, 53306
Ruisinger, PD Dr. Marion Maria 19340
Ruitenbeek, Klaas 17201
Ruiter, Dr. Marjan 32184
Ruivo, José 35548
Ruíz, Dimas Fernández-Galiano 39375
Ruiz, Jauier Luna 31328
Ruíz, Leticia 39573
Ruiz de Azua Murua, Asun 40331
Ruiz de Fischler, Dr. Carmen T. 36067
Ruiz-Martinez, José D. 23251
Ruiz Sastre, Elena 39261
Ruiz-Torres, Maricarmen 51568
Rukavina, Vjeročka 09004
Rukmini Devi, Srimatha 24410
Rules, John 44666
Ruley, Dennis 52457
Rulišek, Dr. Hynek 09515
Rullmann, Anne 22323
Rum, Piera 27365
Rumbaugh, Catherine 52553
Rumbaugh, Darlene 51894
Rumi, Satoh 29394
Rumiche, Prof. Antonio 34288
Rumjanceva, Valentina Pavlovna 37570
Rummel, Dr. Michael 16631
Rump, Elisabeth 47321
Rump, Elizabeth 52158
Rumrill, Alan F. 49718
Rumsby, J.H. 44497
Rumscheid, Prof. Dr. Frank 17442
Rund, Ing. Michael 09844
Runde, Dr. Sabine 18382
Rundhovde, Eric 33424
Rundle, Stella 04295
Rundqvist, Dmitry V. 37192
Runey, Dennis I. 54270
Runia, E. 32082
Runólfsson, Dr. Halldór Björn 24022
Runte, Markus 20972
Runton, Bourke A. 53591
Ruohonen, Risto 10835
Rupert, Dr. Manfred 02177
Rupert Enriquez, Carola 46537
Rupp, Hansruedi 41864
Rupp, Jane E. 47533, 50488
Rupprecht, Dr. Klaus 16918
Rupprechter, Hanns-Michael 21965
Ruprechtsberger, Prof. Dr. Erwin M. 02320
Rus, Valer 36177
Ruseckaitė, Aldona 30259
Rusev, Rusi 04977, 05175
Rush, Don 48676
Rush, Dr. Michael 49898
Rushton, Ray 44908
Rusibaev, D.S. 54484
Rusina, Prof. Ivan 38233
Rusinová, Dr. Zora 38233
Ruske, Wolfgang 16581
Ruskule, Skaidrite 30077
Rusnell, Wesley A. 52470
Russel, Jack 00860
Russel-Smith, Dr. Lilla 17201
Russell, Barry 05396
Russell, Brenda 05800
Russell, Prof. Bruce 44603
Russell, Douglas 47833
Russell, Fiona 46264
Russell, J. Fiona 51415
Russell, John I. 47055
Russell, Laurence J. 50681
Russell, Marilyn M. 51942
Russell, Pamela 48548
Russell, Willie 53002
Russell Howe, Dr. Stanley 46774
Russi, Graciela 00367
Russmann, Edna 47069
Russo, Prof. Antonio 26743, 26752
Russo, Pam 54281
Rust, Wilfried W. H. 48647
Rusterholz, Sabine 41437
Rustia, Augusto P. 34436
Rustichelli, Arlette 41134
Rustige, Rona 05415
Rusu, Olga 36415
Ruszczyc, Ferdynand B. 35261
Ruszczyc, Grazyna 35230
Ruszel, Dr. Krzysztof 35092
Ruszkowska-Macur, Małgorzata 34643
Ruta Serafini, Dr. Angela 26066
Rutberg, Carl 53260
Rutherford, John 01297
Rutherford, Dr. Richard 45330
Rutherford Conrad, Martha 05449, 06096
Rutishauser, Werner 41831
Rutkauskas, Dr. Petras 30438
Rutkowski, Czesław 35105
Rutledge, Bethany 52842
Rutsch, Franz 21018
Rutschmann, Hans 42004
Rutte, Helgard 21580
Rutten, B. 32379
Rutter, A. 44411
Rutter, Michael 06048
Rutter, Mike 43819
Ruttinger, Jacquelyn 49664
Ruttner, Nancy 51951
Rutz, Hans-Georg 41288
Ruuska, Aulis 11418
Rux, Sandra L. 49104
Ruy Van Beest Holle, C.J. du 32735
Ruys, Christoph 03287
Ruyssinck, Micheline 03495
Ružić, Ivan 09004
Ruzo, Victor 41641
Ryan, Amy 46923
Ryan, Dan 52899
Ryan, Daniel J. 53595
Ryan, Debora 53398
Ryan, Dr. E. 48616
Ryan, Graham 01381
Ryan, Jack Dingo 50971
Ryan, Lynne 46688
Ryan, Meryl 00877
Ryan, Michael 24781
Ryan, Dr. Michael 47625
Ryan, Paul 53276
Ryan, Robert 46910
Ryan, Thomas R. 49926
Ryba, Kurt 27091
Rybakov, Ivan Alekseevič 36747
Rybakova, Tamara Viktorovna 37247
Rybka, Walter P. 48412
Rybnikova, Larisa 37211, 37216
Rychener, Dr. Jürg 41137
Rychlíková, Marie 09885
Rychtarik, John 47059
Ryckman, Heather 06489
Rycquart, Walter 03406, 03408, 03409, 03410, 03411, 03412, 03413, 03414, 03415, 03416, 03417, 03425, 03428
Rydel, Maria 34819
Ryder-Davis, P. 46061

R

Rydergård, Evy 40782
Rydzewski, Dr. Jacek 34784, 34785
Rykels, Sam 51087
Ryland, James 50791
Rylands, Philip 28207
Ryman, Susan 00877
Rynd, Chase 53896
Rynierse, H. 32204
Ryoichi, Minegishi 28876
Ryol, Thae Pyong 29760
Rypka, Tom 46499
Ryšánková, Dr. Milada 09880
Ryščanova, Urkija Muchamedjarovna 29705
Ryser, Madeleine 41529
Rysin, Oleksandr Mykolajovyč 43155
Ryska, Norbert 20969
Rystedt, Prof. Eva 40706
Ryusawa, Aya 29012
Ryzdynski, Mark 49249
Ryžov, Stanislav 43178
Ryžova, Tamara Aleksandrovna 37452, 37460
Rzepecka, Mirosława 34996
Rzepiela, Urzula 35051
Rzeszotarski, Jan 35107

S

Sá Marques, Ana Rita 35717
Saabye, Marianne 10133
Saadé, William 11625, 11626
Saage, Wallace 49386
Saager-Bourret, Stephanie 51980, 51981
Saal, Dr. Wolfgang 20146
Saalmann, Ute 18025
Saar, A. 10620
Saaret, Tarmo 10624, 10638
Saarimaa, Hanna-Maija 11128
Saarinen, Jarmo 11390
Saarist, Tiiu 10605
Saarnio, Robert 46564, 46568
Saathoff, Joachim 20997
Saatsi, Niko 11152
Saavedra, Vicente Peña 40029
Šabalin, Rostislav Viktorovič 37544
Sabaté, Antonio 40281
Sabaté i Piqué, Gemma 40261
Saber, Mohammad Davood Mohammad 00007
Sabin, Carleen 50664
Sabin, Owen 50664
Sabina, Choclán 39463
Sabloff, Dr. Jeremy 51886
Sabogal, Bernardo 08639
Sabolic, Prof. Dubravka 08978
Sabran-Ponteves, E. Comte de 14942
Sabroe, Charlotte 10282
Šačalova, Natalja Vladimirovna 37361
Sacarés Mas, Antoni 40087
Sacco, Louise R. 48025
Saccocci, Prof. Andrea 27036
Sach, Martin 44905
Šachalova, Natalja Vladimirovna 37195
Sachanbiński, Prof. Michał 35334
Sachs, Angeli 42094
Sachs, Charles L. 47080
Sachs, Sid 51878
Sachse, Jutta 19951
Sachse, Mike 22457
Sachße, Dr. Claudia 19071
Sack, Dr. Birgit 17945
Sackett, Margot Magee 06799
Sackstetter, Susanna 17738
Saczecki, Claudia 17028
Sada, Ricardo J. 39424
Sadion, Martine 12839
Sadjukov, Nikolaj I. 36836
Sadler, Dr. Donna 48012
Sadler, Patsy 51750
Sádlo, Dr. Václav 09669
Sadok, Nedjah 00062
Sadowski, Władysław 34789
Sadowsky, Dr. Thorsten 16471
Saeed, Tahir 34017
Saegrov, Magnhild 33493
Säkkinen, Senja 11186
Sæland, Frode 33601
Sälejoki-Hiekkanen, Leena 10830
Sänger, Dr. Reinhard 19437
Sæther, Per 33381
Sæthre, Ellen 33873
Sættem, Nils 10109
Saeverud, Dag H. 33842
Safarnia, P. 24627
Saffiotti Dale, Maria 50388
Safford, Pete 48000
Safin, Françoise 03805
Safina, Zulfira Midechatovna 37437
Safoulis, Maxine 46798
Šafr, Pavel 09464
Safronova, V.V. 37507
Safsafi, Nourddine 31693
Saft, Marcia 51200
Saganic, Livio 50383
Sagar, Scott 54173
Sagdalen, Betty Jo 49770
Sage, J. 44047
Sagebiel, J. Chris 52276
Sager, Judy 48710
Saghaar-Bessière, Francoise 15435
Saghi, Mohamed 31692
Sági, Dr. Zoltán 23472
Sagin, Mike 44445
Sagmeister, Dr. Rudolf 01795
Sagov, Magomet Zakrievič 37448
Sagrera, Enric 39349
Sagristà, Ferran 40291
Sahakyan, Kamo 00722
Saheki, Ryoken 28564
Sahlstrand, James 48361
Said, Sanib 30591
Saies-Jones, Frank 05528
Sailard, Olivier 14654
Sain, Annette J. 52008
Saines, Chris 33042
Sainio, Tuulu 11388
Saint-Affrique, Olga de 13435
Saint-Bris, François 11575
Saint-Clair, Anne-Laure 12201
Saint-Cyr-Gherardi, Florence 14374
Saint Hippolyte, Mary 53138
Saint-Pierre, Adrienne 48481
Saint-Pierre, François 13671
Saint-Pierre, Marie Odil de 12261
Saint-Pol, Pierre P. 12788
Saint Romain, P.W. 51974
Sainte-Fare-Garnot, Nicolas 14714
Sáinz González, Elena 31055
Saisi, Enrique 00481
Saito, Satoshi 28687
Saja, Dr. David 47625
Sajenko, Dr. Milan 38352
Saji, Nobutada 29478
Sakaeva, Guzel Rinatovna 37020
Sakai, Makoto 29104
Sakai, Mark 06743
Sakai, Nobuo 28896
Sakai, Tadayasu 28500
Sakai, Tetsuo 28440
Sakai, Yasuo 29219
Sakamoto, Bishop 29329
Sakari, Marja 10792
Sakazume, Hideichi 29469
Sakli, Dr. Mourad 42654
Šakmirda, Marta 21598
Šakova, Marjana Kušbievna 37441
Sala, Susan M. 34373
Salachova, P.Ch. 37903
Săläjean, Ioan 36235
Salajić, Prof. Silvija 08978
Salameh, Khader 24995
Salamon, Sabina 08905
Salapi, Constantin 36434
Salardon, Catherine 11593
Salas, Jaime de 40171
Salas Burguera, Magdalena 39610
Salas Vázquez, Eduardo 39550
Salata, Jim 06514
Salatino, Kevin 50244
Salaverria, Ana 39249
Salazar, Alvaro 06034
Salazar, Ernesto 10446
Salazar, Lucila A. 34470
Salazar Soto, José Aníbal 08530, 08531
Salcedo Zambrano, Alonso 08713
Salcher, Hubert 03105
Salcher, Johann 02351
Saldo, Josip Ante 08951
Šale, Majda 08842
Salé, Marie-Pierre 14692
Saleh, Ali Ibrahim 10517
Salehi, M. 24689
Salerno, Osvaldo 34089
Sales Marques, José Luís de 07992
Salesch, Dr. Martin 22564
Saletti, Prof. Cesare 06275
Salgo, Nicolos 52049
Salib, Dr. Mahar 10490
Saliklis, Ruta 46225
Salin, Anne-Maj 11423, 11427
Salisbury, Joyce 50884
Salisbury, Ruth 45599
Salje, Prof. Dr. Beate 17279
Salkin, Asen 05055
Salkoranta, Marja 11404
Sall, Joan C. 51868
Salleh, Dr. Badriyah 30598, 30599, 30600, 30601, 30602, 30603, 30604, 30605, 30607
Sallen, Sheila 48652
Salm, Karl P. 22100
Salmen, Dr. Brigitte 20535
Salmen, Nora 47738
Salmi, Lyle 48015
Salmon, Béatrice 14654, 14662, 14682, 14734
Salmon, Irit 25020
Salmon, Patricia 53270
Salmona, Riccardo 51171
Salmons-Perez, Carl 52928
Šalna, Rimantas 30442
Salo, Prof. J. 11391
Salome, Elizabeth 49483
Salomé, Laurent 15113, 15115, 15118
Salomeia, Paul 36408
Salomon, Joel 52419
Salomón Zugbi, Alex Faruk 34274
Salomone, Mario R. 00158
Salonen, Jouni 11209
Saloniemi, Marjo-Riita 11349
Salsi, Dr. Claudio 26675, 26676, 26680
Saltamanikas, P. 23133
Salter, Ann C. 52402
Salter, Vanessa 44601
Saltiel, Solita 23214
Salton, Jenny 44497
Saltonstall, Patrick 49823
Saltveit, Ass. Prof. Svein. J. 33737
Salu, Luc 03287
Salus, Ernst R. 25034
Saluz, Prof. Eduard 18520
Salva, Gary 53095
Salvador, Fred 05586
Salvador, Josep 40201
Salvador, Mari Lyn 46181
Salvador Zazurca, Alejandro 40337, 40339, 40351, 40352
Salvatierra, Julio 40299
Salvator, Markus 01753
Salvatore, Ferraro 25151
Salvatori, Prof. Franco 27547
Salvesen, Britt 53585
Salvesen, Ole Bjørn 33778
Salveson, Doug 48543
Salvig, Jacob 10291
Salvioni, Mario Casanova 41192
Salzgeber, Christoph 41747
Salzillo, Bill 47659
Salzman, Alexej 09659
Salzmann, Prof. Dr. Dieter 20503
Samaniego, Lydia 34473, 34475
Samarati, Dr. Luigi 26507
Samardžić, Draženko 08889
Samardžić, Nadežda 38091
Samejima, Yasutoyo 29076
Samek, Tomasz 20519
Samet, Jennifer 51142
Samford, Patricia 46629
Sami-Azar, Dr. A.R. 24671
Samietz, Rainer 18719
Saminger, Herbert 02903
Sammallahti, Leena 11223
Sammons, Mark J. 52101
Samochvalova, Ljudmila Nikolaevna 37037
Samodurov, Jurij Vadimovič 37284
Samon, Josep 40291
Samoškina, Lilija Vjačeslavovnas 37473
Samoy, Antenor R. 34338
Sampaio, Heli de Almeida 04803
Sampaio, Madalena 35938
Sampaolo, Prof. Valeria 26883
Samper, Cristian 53904
Sampil, Mamadou 23310
Sampson, Bill 50500
Sampson, Michael 48378
Sams, Maria 01755
Samson, Dr. Bill 44048
Samson, Fr. Antonio S. 34358
Samson, James 33226
Samson, Rémy 12466
Samsonova, Élvira Aleksandrovna 37092
Samuels, Clifford 52602
Samuels, Philip 52602
Samuelson, Laura 51359
San Agustin, Leonora P. 34340
San Augustin-Filaretos, Pilar 12066
San Diego, Rodolfo 34466
San Quirce, Sylvie 13237
San Román, Ángel García 40170
Sana, Costantino 25204
Sanborn, Andrea 05315, 05397, 06839, 06871
Sanborn, Colin J. 47577
Sanborn, Michael 54187
Sanchez, Artero 04885
Sanchez, Basilio 04884
Sanchez, Marc 14743
Sanchez, Maria José 40006
Sánchez, Osvaldo 31068
Sanchez, Roger 52255
Sánchez Aguilera, Marco 07434
Sánchez de Velandia, Myriam 08522
Sánchez del Barrio, Antonio 39638
Sánchez Fuertes, Caayetano 38885
Sanchez Garcia, Javier 31187
Sánchez Hernández, Américo 31127
Sánchez Mastranzo, Nazario A. 31377
Sánchez Murguía, Socorro 30783
Sánchez Peña, José Miguel 39089
Sánchez Pérez, Luz María 31098
Sánchez Ramírez, Angeles 40356
Sánchez Recio, Miguel Angel 39523, 39524
Sánchez Ruiz, José Fernando 38819
Sánchez Sancho, Dr. Luis María 40341
Sánchez Trujillano, Teresa 39491
Sánchez-Villagra, Prof. Dr. M. 42103
Sanchis, Frank Emile 51219
Sancho, Emanuel A. 35917
Sanclemente Girón, Américo Alfonso 08648, 08649
Sand, Christiane 13340
Sand, Gabriele 19005
Sand, Viki 51817
Sandachi, Paul 36195
Sandahl, Jette 40558
Sandberg, Fredrik 40515
Sandberg, Jane 10058, 10164
Sandbichler, Veronika 02130
Sandell, Joanna 40986
Sandell, Margaret 45926
Sandell, Stephen 50733
Sanden, Jarle 33676, 33678
Sander, Dr. Antje 19396, 19397
Sander, Dr. Birgit 18387
Sander, Dr. Dietulf 19921
Sander, E. 18611
Sanders, Dr. Albert E. 47359
Sanders, Boyd 53633
Sanders, Bronwen 50679
Sanders, Debra 46367
Sanders, Emma 41409
Sanders, G. 31915
Sanders, James A. 49473
Sanders, Jennifer 01609
Sanders, Mildred 53983
Sanderson, Cecil 48600
Sanderson, Dr. Dewey D. 49434
Sanderson, Natalie 52861
Sanderson, R. 06965
Sandford, G. 45567
Sandin, Per 40909
Sandland, Bronwen 00766
Sandler, Ilan 05932
Sandlofer, Michael 47052
Sandmann, Oskar 19161
Sandmeyer, Marilyn 50864
Sandmo, Sigurd S. 33430
Šandová, Dr. Miroslava 09818
Sandoval, Josefina de 54526
Sandoval, Raquel 08538
Sandoval, Sonia 10443
Sandra, Galley 48372
Sandrini, Dr. Francesca 27108
Sandrock, Dr. Oliver 17789
Sandström, Paul 11162
Sandstrom, Ake 40600
Sandvoß, Hans-Rainer 17113
Saner, Gerhard 41918
Sanfelice di Bagnoli, Fabio 27449
Sanford, Dr. Beverly S. 54250
Sanford, Lamonica J. 50691
Sanftner, Phyllis 50856
Sang, Guozhen 07487
Sângeorzen, Monica 36653
Sangl, Dr. Sigrid 20426
Sangmanee, Kitti Cha 14702
Sanguinetti, María Isabel 00505
Sangvichien, Dr. Sanjai 42536
Sangvichien, Prof. Sood M.D. 42561
Sangwill, Richard 40501
Sani, Michele 26858
Sankowski, Janusz 35252
Sanna, Maria Antonia 27952
Sannibale, Dr. Maurizio 54501, 54504
Sano, Dr. Emily J. 52766
Sans, Bartolomé 38884
Sansano Costa, Lina 40000
Sansbury, Joan 53878
Sansom, Andrew 52137
Sansoy, Jaqueline 14576
Sant, Victoria P. 53898
Santa Ana Alvárez Ossorio, Florencio de 39581
Santa Elena, Barrio 08658
Santaca, Dr. Susanna 10783
Santamaria, Jose 54127
Santamaria, Prof. Ulderico 54501
Santamarina, Guillermo 31048
Santana Quintana, Luis Carlos 31050
Santander, Diego 10425
Santandreu, Edmundo 00223
Santaniello, Giovanni 26935
Santay, B. Mundade 24166
Sante Serpa, Prof. Balbín 34185
Santelli, Bob 52998
Santer, Rolf 21203
Santesso, Flavia 00607

Santiago, Isabel 35688
Santiago, Mark 46151
Santiago, Rubio 40191
Santinacci-Boitelle, Françoise 13444
Santini, Roberto Mario 04822
Santiot, Jacques 14372
Santis, Jorge 48638
Santomauro, Angelica M. 49127
Santoni, Mirta Elsa 00554
Santonja, Josep A. Gisbert 39245
Santoro, Geraldine 51262
Santoro, Dr. Paola 26566
Santos, Ana Paula 35860
Santos, Prof. Eliane Aparecida 04942
Santos, Maria José Ferreira dos 35802
Santos, Maria Soares dos 15801, 15802
Santos, Osmar 54430, 54431
Santos, Paula Mesquita 35860
Santos, Sara 35802
Santos, Turibio 04764
Santos Correia Victor, Ana Paula dos 00113
Santos Ramírez, Marco Antonio 31565
Santos Sátz Gomes, Ilmo 39230
Santos Simões, Joaquim A. 35637
Santrot, Jacques 14367
Santzevlariz, David 39727
Santzevlàriz, David 39728
Sanz, Eduardo 40299
Sanz, Javier 39551
Sanz, Joaquim 39615
Sanz Gamo, Rubí 38803
Sanz Sánchez, Erundino 38947
Sanzi, Dr. Maria Rita 25874
Šapošnikova, Ljudmila Vasiljevna 37249
Sapouna-Sakellaraki, Dr. Efi 23081
Šapovalov, Georgij Ivanovyč 43210
Sarabhai, Gira 24046
Sarajlić, Čazim 04289
Šaran, P.K. 24196
Šarapa, Olga Mychajlivna 43003
Sarashiya, Kazuko 29327
Sarauw, Torben 09955
Saraver, Ida 06000
Saravia Garrido, Pablo 39683
Saray, Michael 44479
Sarda, Marie-Anne 12107
Sardella, Filomena 26899
Sardo, Delfin 35691
Sarelius, Seppo 11390
Sargent, William 52657
Sargsyan, Zaven 00753
Sargus, David 46252
Sarholz, Dr. Hans-Jürgen 16725
Sari, Dr. Alireza 24698
Sarie, Pierre 13842
Sariego, Manuel Lucio Julio 08694
Sarioglu, M. Adnan 42728
Sariola, Helmiriitta 10769
Sarkisjan, David A. 37207
Sarkissian, Karen 44981
Sarl, P.M. 45595
Sarl, Peter M. 45589, 45591, 45593, 45594
Sarner Williams, Amy 51824
Sarosácz, György 23716
Saroyan, Ione 51243
Sarpong, Kwame 22852
Sarrail, Cécile 12291
Sarre, Jane 44868
Sarret, Françoise 13773
Sarsfield, Stanley 43946
Sarteanesi, Prof. Nemo 25906
Sartes, Minna 11387, 11389
Sarti, Fabiola 27821
Sarto, Pietro 41643
Sartori, Dean 48080
Sartori, Dr. Michel 41543
Sartorio, Matteo 26716
Sartorio Pisani, Dr. Giuseppina 27479
Sartorius, K. 17428
Sartorius, Tara 50815
Saruchanian, Saro 00750
Sary, Monique 14102
Saryan, Rouzan 00732
Saržickienė, Birutė 30423
Sas, Adrian 51122
Sasagawa, Takami 28881
Sasaki, Hideya 28934
Sasaki, Masamine 29416, 29417
Sasaki, Yoneyuki 28814
Sasayama, Kazutoshi 28735
Şaşmaz, Abdülkadir 42676
Sasse, Julie 53598, 53601
Sassen, J.H. 31888
Sasso, Rosalie 48183
Sassorossi, Larry 52248
Sastri, C.L.N. 24184
Sastry, Dr. V.V. Krishna 24210, 24369
Sasymanov, Y.A. 29964
Sather, Dean 50836
Sather, Stanley 07277
Šatkauskas, Gediminas 30380
Sato, Keichi 28776
Sato, Mitsuo 28663
Sato, Naoki 29419
Sato, Takeshi 28368
Sato, Teiichi 29490
Sato, Toyozo 29012
Sato, Yoshio 28793
Sato, Yumika 29219
Satoko, Maemura 28635
Satsuma, Masato 29392
Sattelmair, Hartwig 18193
Satterfield, Barbara 47782
Satterfield, Foad 52829
Satterthwaitt, Dr. Leann 01483
Sattlecker, Dr. Franz 03009, 03070
Satut, Jorge 54405
Satvé Oliván, Enrique 39929
Satyan, M.S. 24182
Satzer, Wolfgang 02248
Saubel, Katherine 46599
Sauciuc, Benedict 36724
Sauder David, D. 46340
Sauer, Dr. Albrecht 17569
Sauer, Daniel 04698
Sauer, Marc 11492
Sauer, Dr. Marina 16453
Sauerborn, Ulrich 16412, 16415
Sauermann, Bernhard 19746
Sauers, Dr. Richard A. 53378
Sauerwein, Herbert 02278
Sauge, Brigitte 33716
Saul, Dainiel 06038
Saul, Lydia 44596
Saulieu, Pascal de 12699
Saulnier, Lydwine 15719
Saulsbery, Jeff 46859
Saumade, Jeanine 13230
Saunders, Andrew 43914
Saunders, Angela 06202
Saunders, Dr. David 44803
Saunders, Dr. Gail 03203
Saunders, Jean 43865
Saunders, Kristin 45872
Saunders, Richard 50516
Saunders, Richard H. 50653
Saunders, Stuart 00867
Saunders, Dr. Timothy 20732
Saunier, Bernard 16177
Saurage, Hank 46639
Sautai, Daniel 12211
Sauter, Pierre 41094
Sauter, Wolfgang 19492
Sautov, Ivan Petrovič 37577, 37579
Sautter, Dr. Heribert 18305
Sauvebeuf, B. de 12810
Savage, Harland 54215
Savage, Pamela 54215
Savage, Paula 33244
Savage, Ronald 46192
Savard, Denise 07348
Savary, Lee 46189
Savčenko, Andrej Grigorjevič 36843
Savčenko, Nikolaj Jakovlevič 37838
Savel, Irena 38487
Savelja, Oleg 43178
Savellon, Romula 34369
Savery, Suzanne T. 51796, 51797, 51798, 51801, 51803, 51804
Savi, Dr. Dario 28137
Savickaja, Ljudmila Aleksandrovna 37091
Savidge, Amanda 46072
Savidis, Prof. G. 23104
Savill, Rosalind J. 45006
Saville, Ansley 52448
Saville, Reg 44640
Savini, Thomas M. 51145
Savini Nicci, Marco 25760
Savinič, Irina Borisovna 37736
Savko, Iraida Lvovna 37867
Savolainen, Kari 11027
Savouet, Silvia 39007
Savu, Elisabeta 36529
Sawa, Ryuken 28805
Sawada, Keizo 29055
Sawadogo, Alimata 05279
Sawatzki, Hubert 20612
Sawauchi, Takashi 29433
Sawchuk, Barry 07266
Sawchuk, Sherry 50886
Sawchyn, Linda 06053
Sawdey, Michael P. 46493
Sawelo, Victor 07376
Sawicka-Oleksy, Daniela 34604
Sawkins, Annemarie 50718
Sawyer, Carol 52343
Sawyer, Ed 52513
Sawyer, Kenneth C. 01015
Sawyer, Sean 47085
Sawyer, Tom 48516
Saxinger, Karl 22298
Sayer, Christopher 45013
Sayers, Andrew 00948
Sayler, John G. 49993
Sayles, Mary 48298
Sayner, Mary 52930
Sayunain, Selamat 30580
Saywell, Edward 46940
Sazdova, Katerina 30533
Sazonov, Valerij Petrovič 37536, 37538
Sazonova, T.V. 37132
Scafetta, Stefano 53917
Scaggs, Tom 53945
Scaglia, Orlando 00447
Scalamandre Bitter, Marc J. 50186
Scalamandre Bitter, Robert F. 50186
Scalera Liaci, Prof. Lidia 25365
Scalia, Dr. Fiorenza 26162
Scalini, Mario 25812
Scalora, Salvatore 53320
Scandola, Dr. Pietro 41231
Scanlon, Maureen 50005
Scannapieco, Carlos A. 00433, 00434, 00435
Scapellati, Claire 51598
Scapigliati, Juan 00591
Scărlătescu, Fl. 36336
Scarpetti, Ivo 26623
Scarratt, David 44131
Ščeboleva, Galina Fedorovna 37161
Scelsa, Dr. Joseph V. 51198
Ščerbakova, I.T. 37006
Ščerban, Aleksandr Nikolaevič 36810
Schaaf, Ania 03990
Schaaff, Dr. Holger 20150
Schaal, Gerald 46873
Schaar, Dana 47271
Schaar, Elke 19939
Schaatsenberg, W.J.M. 32157
Schaber, Manfred 16458
Schachl-Raber, Mag. Dr. Ursula 02628
Schachtner, Dr. Sabine 17017
Schachtsiek, Richard 50087
Schade, Karin 22051
Schade, Dr. Markus 22051
Schade-Didschies, Peter 22257
Schächter, Willi 19045
Schädle, Bernarda 16874
Schädler, Ulrich 41945
Schäfer, Dr. Alfred 19631
Schäfer, Andreas 41886
Schäfer, Bernd 18718
Schaefer, Dr. Claudia 18014
Schaefer, D.E. 38669
Schäfer, Denis 20448
Schäfer, Dr. Dorit 19448
Schaefer, Elizabeth Meg 47721
Schäfer, M.A. Gerd 19357
Schäfer, Gerd 19360, 19362
Schäfer, Günter 02735
Schäfer, Dr. Hans-Michael 18919
Schäfer, Dr. Irmhild 20424
Schaefer, Johnann 47530
Schäfer, Julia 19905
Schäfer, Dr. Klaus 16545
Schäfer, Klaus 20407
Schäfer, Leane 18577
Schäfer, Otto G. 21687
Schaefer, Patricia M. 51067
Schaefer, Rainer 19454
Schaefer, Scott 50237
Schaefer, Dr. Stacy 47528
Schäfer, Dr. Wolfgang 22193
Schäffel, Brigitte 02745
Schäffel, Karl 02745
Schaeffer, Alvina 46704
Schaeffer, Astrida 48225
Schäffer, Gisela 16395
Schäffer, Prof. Dr. Dr. Johann 19007
Schael, Alfredo 54553
Schaeps, J. 32486
Schärer, Dr. Martin R. 41983
Schaetzel, Désiré 12040
Schafer, Barbara L. 46779, 46781
Schafer, Carl 46225
Schafer, Colin 44580
Schaffer, Dale 47916
Schaffer, Dr. Herbert 02470
Schaffer, Dr. Nikolaus 02625
Schaffer-Hartmann, Richard 18979, 18980, 18982
Schaffers Bodenhausen, Dr. K. 32076
Schaffgotsch, Marie-Elise Gräfin 02446
Schaffner, Ingrid 51844
Schafhausen, Nicolaus 32734
Schafroth, Colleen 48963
Schagen, C. 32401
Schaible, Larry 53510
Schaible, Silvia 20408
Schaiek, Zvi 25129
Schakel, F. 31797
Schalhorn, Dr. Andreas 17172
Schalke, Prof. Dr. Thomas 17044
Schall, Jan 49695
Schall Agnew, Ellen 50328
Schallenkamp, Kay 48393
Schaller, Anne K. 50989
Schaller, Hydee 46318
Schallert, Dr. Elmar 01790
Schalles, Dr. Hans-Joachim 22762
Schallmayer, Prof. Dr. Egon 16752
Schaltenbrand, Therese 41564
Schamberger, Rolf 18505
Schamfuß, Axel 21359
Schampers, K.J.J. 32295, 32296
Schanche, Kjersti 33960
Schandor, Mag. Gaby 01910
Schantl, Alexandra 02685
Schantz, Michael W. 51889
Schanz, K.H. 20798
Schaper, Henning 22665
Schapp, Rebecca M. 52863
Schardin, Anne 01340
Scharff, Prof. Wolfram 21249
Scharloo, Dr. M. 32302
Scharmann, Detlev 22528
Scharmann, Rudolf 17252
Scharmann, Rudolf G. 17043, 17185, 17227, 17248
Scharnowski, Joachim 17135
Scharnweber, Karl 17883
Schaschl, Sabine 41380
Schaschl-Cooper, Sabine 41670
Schatorje, Dr. Jos 32883
Schatteman, Alice 50799
Schatten, Gereon 16865
Schatz, Duane 48336
Schatz, Günther 02388
Schaub, M.A. Cajetan 21981
Schaub, Didier 05294
Schaub, James 53593
Schauber, Dr. Gregor 02572
Schauer, Karl 02359
Schaufelberger-Breguet, Simone 41807
Schaukal, Dr. Barbara 01965
Schaumann, Prof. Walther 02214, 02215
Schaumburg-Lippe, Wolfgang Prinz zu 02309
Schaunaman, Lora 46116
Schaus, Dick 49665
Schawaller, Dr. W. 21959
Schawe, Dr. Martin 16636, 16920, 20425
Scheau, Prof. Dr. Mihail 36667
Schebesta, Dr. Kurt 02941
Scheck, Klaus Peter 21669
Scheckle, Marian 31743
Schedin, Pernilla 40639
Schedler, Christian 20328, 20329, 20330, 20331, 20332
Scheef, Vera 17839
Scheel, Karin 17102
Scheele, Edward 46121
Scheele, Dr. Friedrich 20900
Scheele-Schmidt, Gisela 31745
Scheelen, J. 03690
Scheen, Rolf 33724
Schefcik, Jerry A. 49966, 49968
Scheffler, Ina 17658
Scheffler, Jürgen 19947, 19948
Scheflen, Mark 51202
Schefzik, Dr. Michael 18900
Schegg, Jakob 41221
Scheibe, Joachim 21976
Scheiblechner, Margit 41137
Scheiblechner, Mag. Petra 01965
Scheicher, Dr. Harald 01788
Scheicher, Harald 02868
Scheid, Dr. Eva 19226
Scheidecker, Dipl. Birgit 20166
Scheidt, Diana 49165
Scheidt, Ulrich 18209
Scheifele Giger, Carmen 41461
Scheifinger, Otto 02141
Schein, Valentin 38496
Scheinost, Vojtěch 09785
Scheit-Koppitz, Stefanie 21755
Schelde-Jensen, Bodil 10021
Schelfhaut, Andrée 03671
Schellack, Dr. Fritz 21771
Schelle, Elise 32862
Scheller, Jens 20581
Schels, Bertin 19500
Schelshorn, E. 22573
Schelter, Dr. Alfred 20278
Schenck, Jean-Luc 15185
Schenck, Marvin 53638
Schenini, Elio 41581
Schenk, L.J.B. 31852
Schenk, Matthias 22558
Schenk, Ueli 41212
Schenk Freiherr von Stauffenberg, Franz 19829
Schenk-Weininger, Dr. Isabell 17350
Schenkenberger, Katja 20498
Schenker, Prof. Christoph 42084
Schepers, Heiner 20009

S

Schepers, Dr. Wolfgang 19000
Scheppink, Baukje 32964
Scher, Anne 51200
Scherbakov, Dr. Lev A. 37182
Scherbel, Dr. G. 20775
Scherer, Dr. Bernd M. 17128
Scherer, Ernst 41845
Scherer, H. 41696
Scherer, Jim 52003
Scherer, Dr. Kristine 19076, 19085
Scherer, Kurt 41143
Scherer, Manfred 41845
Scherer, Mark A. 49472
Scherer, Rodney 01325
Scherf, Angeline 14646
Scherf, Avner 25030
Schermuck-Ziesché, Margit 17826
Scherner, Dr. Antje 16862, 19474
Scherney, Dr. Ingrid 01924
Scherp, Dr. Astrid 20426
Scherrer, Prof. Dr. Peter 01963
Scherzer, Beatrice 19288
Schettino, Edvige 26897
Scheu, David R. 54198
Scheuerbrandt, Dr. J. 20932
Scheuerecker, Franz X. 20990
Scheuermann, Gernot 21068
Scheuren, Elmar 19661
Scheurer, Bruno 41328
Scheurer, Heidi 41206
Schiborra, Ute 18877
Schick, Dr. Karin 41321
Schick, Martin 16656
Schick, Dr. Ulrike 20951
Schicklgruber, Dr. Christian 03038
Schidlowsky, V. 12364
Schieber, Dr. Michael 20725
Schieck, Dr. Martin 18406
Schiedeck, Josef 22197
Schiek, Friedolf 21946
Schiek, Hauke 19421
Schiemann, Larry 46387
Schiepers, Sue 04098
Schierle, Dr. Sonja 21951
Schiers, Jörge 16811
Schierscher, Renè 30197
Schierz, Prof.Dr. Kai Uwe 18205
Schießl, Werner 18061
Schiestl, Dr. Michael 02146
Schietere de Lophem, A. de 03424
Schietinger, James 48015
Schietz, Ottar 33856
Schiff, Michael 47930
Schiffer, Tim 53716
Schiffino, Prospero 25627
Schiffmann, Dr. René 41377
Schiffner, Wolfgang 22655
Schijndel, Marieke van 32849
Schilbach, Michael 20872
Schild, Anne 41728
Schild, Margit 17988
Schilder, A.M.Th. 31897
Schildkrout, Dr. Enid 51115
Schill, Jörg 21717
Schillemans, R. 31864
Schiller, Joyce K. 54189
Schilli, K. 18582
Schilling, Wolfgang 21735
Schiltz, Kelly 49037
Schimek, Dr. Michael 16839, 18670
Schimmel, Dan 51832
Schimmel, Paul 50252
Schimmelpfennig, Michael 22141
Schimper, Gabriele 18532
Schimpf, Dr. Simone 21948
Schimpf-Reinhardt, Dr. Hans 16899, 16900, 16901
Schindel, Morton 54103
Schindlbeck, Dr. Ingrid 17081
Schindlbeck, Dr. Markus 17081
Schindler, Annette 41177
Schindler, Harald 22842
Schindler, Prof. Dr. Margot 02175
Schindler, Dr. Margot 03049
Schindler-Kaudelka, Dr. Elen 02500
Schink, Hendrik 21303
Schinkel, Dr. Eckhard 17921
Schinnerl, Heimo 02187
Schinnerling, Mechthild 22793
Schiøtz, Ottar 33855
Schipper, P. 32987
Schipper, Drs. Peter W. 32043, 32823, 32825
Schirmböck, Thomas 20173, 20175
Schisler jr, Lee C. 47030
Schitinger, Jim 48017
Schiwek, Thomas 22685
Schizas, Dr. Nikolaos 36043
Schladebach, Tilman 16979
Schläfli, Bruno 41704
Schlaepfer, Ralph 42006
Schlag, Dr. Gerald 02281, 02458, 02740, 02762
Schlagkamp, Dieter 21745
Schlass, Gerhard 03039
Schlatter, Alexander 41172
Schlatter, Dr. Rudolf 19928
Schlatter, Stephan 41832
Schlautmann, Rainer 18579
Schlechter, Gunther 17757
Schlegel, Adolf 22369
Schlegel, Amy Ingrid 50548
Schleger, Debra 01122
Schleher, Linda 47602
Schleich, Jean 03493
Schleider, Holger 17554
Schleiermacher, Karl-Heinz 20376
Schleif, Tom 49733
Schlemmer, C. Raman 26977
Schlenk, Christoph 02128
Schleper, Dr. Thomas 20809
Schlereth, Bernhard 19560
Schleyerbach-Breu, Christiane 20326
Schlicher, Fred J. 50547
Schlicht, Dr. Udo 17340
Schlichting, Art 52041
Schlichting, Heike 16565, 16567
Schlicting, Eunice 47970
Schlieben, Katharina 42115
Schliebs, Des 01244
Schliefsteiner, Herbert 02425
Schliekau, Jürgen 16697
Schliemann, Ulrich 17576
Schlieper, Hans 21814
Schlimmgen-Ehmke, Katharina 17840
Schlinkert, Dr. Dirk 22661
Schlinkmann, Jean 52959
Schlitter, Balz 41486
Schlombs, Dr. Adele 19621
Schloßbauer, Kurt 21182
Schlosser, Charles 13678, 13679
Schlosser, Christa 20541
Schlosser, Christoph 41375
Schloßer, Rosemarie 17800
Schloter, John E. 52640
Schluchter, Dr. André 41369
Schlüter, Dr. Andreas 21959
Schlüter, Dr. des. Arnulf 20489
Schlüter, D. 32116
Schlüter, Günter 19861
Schlüter, Prof. J. 18944
Schlüter, Maik 17527
Schlueter, Sue 50703
Schlumpf, Fernand 41967
Schlup, Daniel 41600
Schlup, Michel 41676
Schmälzle, Reinhard 21711
Schmager, Martin 22455, 22457
Schmahl, Prof. Dr. Wolfgang W. 20468
Schmalbrock, R.F. 19385
Schmalfuß, Dr. Helmut 21959
Schmalwaßer, Karl-Friedrich 17097
Schmalz, Björn 16596
Schmalz, Lydia H. 51085
Schmandt, Dr. Matthias 17353
Schmauder, Dr. Michael 17462
Schmeh, Thomas J. 49809
Schmeja, Dr. Lia 01777
Schmeling, Sören 41165
Schmelz, Prof. Dr. Bernd 18952
Schmelzer, Christian 22705
Schmetz, Marcel 04062
Schmid, A. 16830
Schmid, Dr. Anna 41169
Schmid, Debora 41137
Schmid, Edy 42052
Schmid, Elisabeth 42102
Schmid, Dr. Elmar D. 20458
Schmid, G. 21404
Schmid, Gerhard 41938
Schmid, Hanspeter 41809
Schmid, Jakob 41978
Schmid, John 41524
Schmid, Dr. Manfred 21939, 21941, 21961, 21962
Schmid, Dr. Margit 01825
Schmid, Dipl.Zool. Marion 41300
Schmid, Dr. Peter 42093
Schmid, Reinhard 22224
Schmid, St. 14243
Schmid, Dr. Ulla K. 21622
Schmid, Ute 51225
Schmid-Beurmann, Dr. Peter 20509, 20514
Schmid-Birri, Ingrid 41358
Schmid-Nafz, Gerlinde 20248
Schmidl, Dr. Hermann 01870
Schmidl, Matthias 02065
Schmidt, M.A. Aiko 18141
Schmidt, Dr. Alexander 20758
Schmidt, Andreas 16722, 22799
Schmidt, B. 22191
Schmidt, C. J. 49225
Schmidt, Christiane 19924
Schmidt, Dagmar 18898
Schmidt, Dwight 48953
Schmidt, Erwin 19765
Schmidt, Dr. Eva 21755
Schmidt, Franz Xaver 21959
Schmidt, Franziska 18998
Schmidt, Georg 18651
Schmidt, Dr. Hans-Günter 22706
Schmidt, Dr. Hans-Werner 19921
Schmidt, Harro 18996
Schmidt, Jochen 21692
Schmidt, Johannes 17973
Schmidt, Karl 17728
Schmidt, Kirsten 10093
Schmidt, Prof. Dr. Klaus 18226
Schmidt, Klaus 22727
Schmidt, Lawrence R. 51493
Schmidt, Dr. Mária 23482
Schmidt, Martin 18889
Schmidt, Muriel H. 48444
Schmidt, René 16774
Schmidt, Rolf 17291, 17292
Schmidt, Roman 20041
Schmidt, Dr. Rüdiger 19435
Schmidt, Dr. Sabine Maria 18263
Schmidt, Sebastian 19209
Schmidt, Teresa 53699
Schmidt, Dr. Theun-M. 17052
Schmidt, Thomas 17861, 17862
Schmidt, Volker 17553
Schmidt, Willi 21575
Schmidt-Bergmann, Prof. Dr. Hansgeorg 19444
Schmidt-Czaia, Dr. Bettina 19608
Schmidt-Esters, Gudrun 18416
Schmidt-Glintzer, Prof. Dr. Helwig 22653
Schmidt-Herwig, Dr. Angelika 18354
Schmidt-Lawrenz, Dr. Stefan 19062
Schmidt-Möbus, Dr. Friederike 17755
Schmidt-Rutsch, Dr. Olaf 17921
Schmidt-Surdez, Maryse 41676
Schmidts, Dr. Thomas 20148
Schmied, Dr. Hartmut 21387
Schmiegel, Karol A. 48161
Schmieglitz-Otten, Juliane 17689
Schmigalle, Dr. Günther 19435
Schmitt, Erika 22249
Schmitt, Horst 22098
Schmitt, Robert 15706
Schmitt, Dr. Ulrich 18695
Schmitz, Dr. Bettina 19185
Schmitz, Brigitte 17032
Schmitz, Dipl. phil. Brigitte 17222
Schmitz, Dr. Britta 17123
Schmitz, Engelbert 21807
Schmitz, Hans-Hermann 17009
Schmitz, Dr. Jörg 20411
Schmitz, Margret 21807
Schmitz, Dr. Michael 20149
Schmitz, Dr. Ralf 17462
Schmitz, Silvia 19001
Schmocker, Susan 01544
Schmöhl, Thomas 16986
Schmoldt, Robin 50397
Schmoller, Mag. Andreas 01836
Schmollgruber, Friedrich 02768
Schmook, Dr. R. 16733
Schmook, Dr. Reinhard 16734
Schmuck, Carolin 20723
Schmuhl, Boje 17374, 19966
Schmuttermeier, Elisabeth 03026
Schmutz, Christine 41319
Schmutz, Daniel 41207
Schmutz, Prof. Dr. Hans-Konrad 42036
Schmutz, Dr. Hans Konrad 42042
Schnabel, Helga 20557
Schnakenburg, Dr. Sabine 18920
Schneble, Dr. H. 19495
Schneeweis, Dr. Felix 02175
Schneider, Dr. Alfred 16540
Schneider, Prof. Dr. Angela 17123, 17225
Schneider, Anja 19736, 21657
Schneider, Dr. Carola 18257
Schneider, Claire 47114
Schneider, Dr. Cornelia 20143
Schneider, Dr. Ditmar 20128
Schneider, Don 51643
Schneider, Elke 18932, 18950
Schneider, Dr. Erich 21682, 21683, 21684, 21686, 21688
Schneider, Ester 20312
Schneider, Dr. G.C. 31763
Schneider, H. 19452
Schneider, Hansjörg 19676
Schneider, Hélène 14877
Schneider, Dr. K. 18907
Schneider, Karl-Heinz 21408
Schneider, Dr. Katja 18904
Schneider, Kerry 51948
Schneider, Klaus 18041
Schneider, Prof. Dr. Klaus 19629
Schneider, Malou 15819
Schneider, Dr. Matthias 17359
Schneider, Minna 51218
Schneider, Nadine 41081
Schneider, Paul 50219
Schneider, Philippe 15520
Schneider, Reinhild 21818
Schneider, Reisa Smiley 07222
Schneider, René 12209
Schneider, Roland 16857
Schneider, Dr. Thomas 20920
Schneider, Prof. Dr. Ulrich 16409, 18382
Schneider, Dr. Ulrich 21566
Schneider, Prof. Dr. Ulrich Johannes 19899
Schneider, Virgil 46337
Schneider, Dr. Wolfgang 18650
Schneider-Kempf, Barbara 17258
Schneidmiller, Kenneth A. 49067
Schnell, Christian 12005, 12006
Schnell, Dorothy 05578
Schnell, Dr. G. 51376
Schnell, Gertrude H. 48753
Schnellecke, Dr. Ilona 22665
Schnellhammer-Ocist, Norbert 02761
Schnepp, Patrick 13434
Schneyer, Rosemary 53303
Schnieder, Andreas 16436
Schnitzer, Dr. Claudia 17954
Schnitzler, Bernadette 15820
Schnormeier, Kasey 52500
Schnurbein, Dr. Rüdiger Freiherr von 17507
Schnurr, Marie 48236
Schober, Otto 01948
Schoberwalter, Anton 02197, 02198
Schobinger, Dr. Juan 00452
Schoch, Dr. R. 21959
Schöb, Kathrin 41757
Schöbel, Dr. G. 22158
Schoeller, Danielle 16212
Schöllmann, Dr. Lothar 20513
Schoemaker-Weeszenberg, M.P. 32062
Schoen, Kenneth 53784
Schön, Matthias D. 16676, 18861
Schönau, Dieter 20293
Schönberger, Dr. Angela 17168
Schöne, Dr. Anja 22022, 22023
Schöne, Jasmin 18666
Schoene, Lynn 21637
Schöne, Dr. Sally 17992, 18004
Schoeneck, Edith 20825
Schönenberg, Walter 18971
Schoenenberger, Nicola 41582
Schönewald, Dr. Beatrix 19338, 19345, 19346
Schönfelder, Ramona 18047
Schönfellner, Dr. Franz 02229, 02232
Schönhagen, Dr. Benigna 16624
Schönig, Dagmar 41710
Schoenke, Marilyn 49350
Schönweger, Astrid 26658
Schofield, Arnold 48681
Schofield, Dorothy 52841
Schokkenbroek, J. 31885
Scholbier, Noël 04033
Scholl, Prof. Dr. Andreas 17036
Scholl, Prof. Dr. Lars U. 17569
Schollenberger, Werner 20798
Scholler, Dr. Markus 19449
Schollian, Gerd 19063
Schollmeier, Dr. Axel 20519
Schollmeyer, Ullrich 16940
Scholtemeijer, G. 38755
Scholten, Lisa 47059
Scholten, Uta 17999
Scholz, Angelika 21109
Scholz, Dr. Antony 01984
Scholz, Christine 21969
Scholz, Dr. Dieter 17225
Scholz, Dr. Heike 21950
Scholz, Lothar 17438
Scholz, Margitta 17510
Scholz-Bauer, U. 21317
Scholz-Strasser, Inge 03075
Schomers, Dr. Florian 02133
Schommers, Dr. Annette 20426, 20833
Schonewille, Dr. P. 31937
Schonlau, Ree 51547
Schooler, Jim 54363
Schoon, P.J. 32149, 32154
Schoonhoven Okken, W. 32040
Schoonover, Larry 53207
Schoonover, Mary Ellen 46488
Schopen, Klaus H. 19597
Schopp, N. 00822, 00823
Schopper, Dr. Franz 17505
Schopping, H. 32844
Schorta, Dr. Regula 41767
Schoske, Dr. Sylvia 20489, 21715
Schøtt, Vibehe 10016
Schott-Smith, Christine 23285
Schotten, Christoph 22212
Schoutens, P. 31864
Schovel, Erin 48155

Schrade, Dorothea 22174
Schrade, Ewald 18066
Schrader, Kathleen 52343
Schrader, Wolfgang 21055
Schräer, Dr. Heinz 18249
Schramek, Maria-Antonia 02480
Schramer-Lorang, Fernande 30473
Schramm, Dr. Ágnes 38233
Schramm, Jeanne 54069
Schramm, Robert W. 54069
Schratter, Rudolf 02102, 02104
Schrecengost, Tammy 46799
Schreiber, Daniel J. 22118
Schreier, Dr. Christoph 17460
Schreilechner, Waltraud 02119
Schrein, Catherine 46623
Schreiner, Manuela 19676
Schreiter, Bernd 16556
Schrem, Wilfried 20589
Schrettl, Reinhold 02863
Schreuders-Derks, Annie 32166
Schrick, Annette 17017
Schriver, Stephanie 52869
Schroder, Dr. Anne 48221
Schrøder, Lise Lotte 10078
Schrøder Vesterkjær, Siri 33881
Schroeck, Petra 17057
Schroeder, Allen 47275, 50726
Schröder, M.A. Amelie 21867
Schroeder, Asher 50457
Schröder, Christoph 16493, 16495
Schroeder, Chuck 51519
Schroeder, Darlene 48953
Schroeder, Dorothy 47452
Schröder, Frank 17475
Schröder, Dr. Heike 16853
Schröder, Dr. Klaus Albrecht 02929
Schröder, Dr. Martina 21283
Schroeder, Patricia 48540
Schröder, Ralf-Peter 18053
Schroeder, Stephan 39573
Schröder, Till 18946
Schröder, Dr. U. 18750
Schröter, Jürgen 22061
Schröter, Wolfgang 21735
Schroll, Ludwig 18163
Schroth, Dr. Adella 52276
Schroth, Dr. Sarah W. 48221
Schroth, Thomas 51734
Schrub, Antoine 14811
Schube, Inka 19005
Schubert, Dr. Alexander 20173
Schubert, Beate 20915
Schubert, Jason 47551
Schubert, Werner 02958
Schubert, Wolfgang 19917
Schubiger, Letizia 41564
Schubiger, Pia 41850
Schuch, Eny 04654
Schuch, Herbert 19431
Schuch, Dipl.-Ing. Josef 02811
Schuchardt, Günter 18099
Schuchmann, Dr. Karl-Ludwig 17469
Schudel, Veronika 41837
Schüler, Dr. T. 22410
Schuelke, Chris 48534
Schülke, Ilsabe 17994
Schüllenbach, Konrad 31768
Schüly, Dr. Maria 18428
Schünemann, Dr. Gustav 16930
Schünemann, Karl 16464
Schünemann, Ulla 21706
Schüpbach, Ulrich 41234
Schürch, Theo 41952
Schürer 18074
Schürer, Dr. Ralf 20764
Schuerholz, Peter 53446
Schürmann, Dr. Kay U. 20192
Schuette, Bobbie 49036
Schütte, Gerd 18260
Schütte, Dr. Margret 21566
Schütte, Rainer 31941
Schütz, Dr. Christian 22246
Schütz, Ilse 38784
Schütz, Jan 09480
Schütz, Jean-Luc 03796
Schütz, Dr. Karl 03019, 03019
Schütz, Volker 17362
Schütz, Wolfgang 22381
Schuh, Walter 02441
Schuhwerk, Dr. Franz 20430
Schuldt, Doren 20901
Schulenburg, Dr. Rosa von der 17029
Schulenburg, Dr. Stephan von der 18382
Schulenburg, Susanne 19437
Schuler, Ewald 41091
Schuler, Romana 03022
Schuler, Shirley 51978
Schulkin, Jerrold L. 50080
Schuller, Kurt 01878
Schuller, Tom 49755
Schullerus, M. 16503
Schulson, Henry H. 47408
Schulte, Dr. Bärbel 22105
Schulte, Dr. Birgit 18857, 18860
Schulte, Elisabeth 16626
Schulte, Matthew 51383
Schulte, Wolfgang 03696
Schulte-Hobein, Dr. Jürgen 16573
Schulte-Kemper, Hubert 20225
Schulte-Wien, M.A. Walburga 19660
Schulten, Ton 32632
Schulters, Johannes 16908
Schultes, Alfred 02140
Schultes, Dr. Lothar 02317, 02324
Schultes, Peter 16524, 16525
Schultka, Prof. Dr. R. 18888
Schultz, Brian 49206
Schultz, Dr. Franz Joachim 16957
Schultz, Horst 16512
Schultz, J. Bernard 50846
Schultz, Jeffrey 49045
Schultz, Marie-Luise 18380
Schultz, Peter 40915
Schultze, Dr. Joachim 21565
Schulz, Andreas W. 20451
Schulz, Dr. Ch. 41765
Schulz, Emily L. 53918
Schulz, Hagen 16940
Schulz, Dr. Isabell 19005
Schulz, Lotte 34101
Schulz, Marion 21246
Schulz, Max 50222
Schulz, Ralph J. 50966
Schulz, Rebekah 01169
Schulz, Volker 22511
Schulz, Dr. Walter 18137
Schulz, Warren 46833
Schulz-Hoffmann, Prof. Dr. Carla 20425
Schulz-Weddigen, Dr. Ingo 19677
Schulze, Armin 18881
Schulze, Dietmar 19932
Schulze, Falk 19670
Schulze, Prof. Dr. Sabine 18951
Schulze, T. 16639
Schulze Altcappenberg, Prof. Dr. Hein-Th. 17172
Schumacher, Bett 53307
Schumacher, Dr. Franz-Josef 21457
Schumacher, Prof. Dr. Peter 41133
Schumacher, Dr. Renate 17463
Schumacker, Mare 04105
Schumann, Kerstin 17500
Schumann, Maurice 13246
Schumann, Peter 48950
Schumann, Robert 01884
Schumann, Romain 30477
Schumard, Ann 53907
Schummel, Helle 10285
Schuppli, Madeleine 41080, 41941
Schur, Chantel 06805
Schure, Edeltraut 22637
Schurig, Michael 18362
Schurig, Dr. Roland 16413
Schurkamp, Trish 46902
Schuster, Anna 01885
Schuster, Dr. Erika 01943
Schuster, Erwin 02770
Schuster, Eva 22548
Schuster, Kenneth L. 46795
Schuster, Marina 18008
Schuster, Prof. Dr. Peter-Klaus 17123, 17225
Schuster, Robin 50345
Schuster, Vicki 47058
Schuster-Cordone, Dr. Caroline 41387
Schutte, Maria Christina 00121
Schutz, Carine 14810, 16300
Schutz, David 50826
Schutz, Jean-Luc 03802, 03812
Schutzbier, Heribert 02348
Schwab, Christoph 22747
Schwab, K. 16889
Schwab, Sara 43253
Schwab, Yvan 41643
Schwab-Dorfman, Debbie 51200
Schwabach, Dr. Thomas 22435, 22436, 22437, 22438
Schwabe, Uwe 19895
Schwager, Michael 52444
Schwaiger, Wilhelm 01928
Schwalenberg, Gregory 46548
Schwaller, M. 14460
Schwaller, Marie 41922
Schwalm, Dr. Hans-Jürgen 21210, 21211
Schwalm, Helmut 21662
Schwan, Sandy L. 52515
Schwank, Prof. Dr. Benedikt 17313
Schwanzar, Dr. Christine 02324
Schwappach, Lisa 52807
Schwark, Dr. Thomas 18993
Schwarm, Larry W. 48393
Schwarts, Eleanore 53372
Schwartz, Constance 52463
Schwartz, David 46399
Schwartz, Deborah 47068
Schwartz, Elizabeth 49679
Schwartz, Gary 53505
Schwartz, Janet 07120
Schwartz, Jay 53561
Schwartz, Judith 07113, 07117
Schwartz, Michael 47627
Schwartzbaum, Paul 51258
Schwarz 21496
Schwarz, Bärbel 41508
Schwarz, Chris 34781
Schwarz, Dr. Dieter 42029
Schwarz, Ernst 02091
Schwarz, Dr. Godehard 02956
Schwarz, Heike 18055
Schwarz, Dr. Helmut 20790
Schwarz, Dr. Isabelle 19005
Schwarz, Prof. Dr. Karl 02985
Schwarz, Klaus 19388
Schwarz, Prof. Dr. Michael 17527
Schwarz, Peter 17355, 17356
Schwarz, Dr. Wolfgang 18900
Schwarzjirg, Dr. Helmuth 02260, 02262
Schwarzmaier-Wormit, Dr. Agnes 17036
Schwegler Peyer, Barbara 41099
Schweiger, Andrea 41182
Schweigert, Dr. Günter 21959
Schweiggl, Wolfgang 25983
Schwein, Florence 48317
Schweizer, Nicole 41544
Schweizer, Dr. Paul D. 53676
Schweizer, Dr. Rolf 20536
Schwelle, Dr. Franz 02385
Schwender, Judy 51669
Schwendner, Franz 16582, 16587, 16592, 16593
Schwenk, Dr. Bernhart 20425
Schwer, Dr. 22735
Schwering, Dr. Burkhard 19524
Schwerkolt, Charles L. 00816
Schwertner, Dr. Johann 02186, 02352
Schwiderowski, Dr. Peter 18247
Schwind, Stefan 17952
Schwind, Wilmont M. 52069
Schwinden, Lothar 22102
Schwinn, Wolfgang 21243
Schwintek, Monika 21581
Schwinzer, Dr. Ellen 18969
Schwitter, Josef 41436
Schwoeffermann, Catherine 46816
Schwolger, David 49940
Schwyzer, Carol 42088
Schymalla, Joachim 18292
Sci, LaVerne 47988
Sciallano, Martine 13254
Scichilone, Dr. Giovanni 25872
Scivier, J. 45432
Scoates, Christopher 50179
Scoccimarro, Fabio 28087
Scocos, John 50398
Scofield, P. 33069
Sconci, Maria Selene 27514
Scopelites, Celeste 07026
Scott, Adrienne 47528
Scott, Amy 50255
Scott, Andrew 46105
Scott, Angela 52856
Scott, B. 01532
Scott, Barbara 51309
Scott, Donna A. 51236
Scott, Elva 48231
Scott, Eric 52276
Scott, Dr. Gerry 52726
Scott, Jane 01647
Scott, John 50068
Scott, Julie 52807
Scott, Kitty 06551
Scott, Marilyn 46055
Scott, Michael 45811, 49827
Scott, Preston 05431
Scott, Rodney 44490
Scott, Ron 07072
Scott, Rose 49017
Scott, Susan 27635
Scott, Terri 48312
Scott, Timothy 51914
Scott, Vane S. 51310
Scott, Vanya 54147
Scott, Wendy 43948
Scott, William W. 53455
Scott-Childress, Katie 49044
Scott-Garrett, Pauline 43621, 44493
Scott Garrity, Noreen 47196
Scottez-de Wambrechies, Annie 13756
Scotti, Dr. Roland 41117, 41119
Scotton, Dr. Flavia 28178
Scoville, Pat 50954
Scrase, D.E. 43727
Scribner, John C.L. 46520
Scriven, Brian 47277
Scrobotă, Paul 36103
Scruggs, Bill 52060
Scudder, Dr. G.G.E. 07199
Scudder, G.G.E. 07223
Scudero, Domenico 27509
Scudieri, Dr. Magnolia 26184
Ščukina, Irina Anatolivna 43094
Sculley, John 45601
Scully, Cammie V. 53965
Scully, Robert J. 54161
Sczech, Dr. K. 22410
Seabold, Thomas 49744
Seage Person, Robin 49085
Seager, Pamela 50177
Seald-Chudzinski, Romi 52116
Seale, Sir David 03243
Searcy, Karen B. 46269
Searcy, Manera S. 53032
Searl, Majorie 52396
Searle, Ross 00901
Šeba, Vasyl Stanislavovyč 43198
Sebastián, Dr. Amparo 39570
Sebastian, Padmini 01285
Sebayang, Nas 24525
Šebek, Dr. F. 09707
Šebesta, Dr. Pavel 09473
Šebestian, RNDr. Jiří 09714
Sebire, Heather 45591
Secher Jensen, Dr. Thomas 09965
Seck, Dr. Amadou Abdoulaye 38025
Seco, Fernando 39544
Sedano, Pilar 39573
Sedberry, Rosemary 46632
Seddon, Jane 43825, 43828
Seddon, Joan 06153
Sedikova, Larisa 43178
Sedilek, F. 38391
Sedinko, Svetlana Alekseevna 37478
Sedioli, Giovanni 25452
Sedláček, Dr. Richard 09901
Sedlazek, Dr. Dietmar 20385
Sedler, Dr. Irmgard 19690, 19691
Sedman, Ken 45140
Sedoi, Gabriela 05187
Sedova, Galina Michajlovna 37694
Sedova, Irina V. 37207
Seear, Lynne 01512
Seebacher, Mag. Margit 03069
Seebeck, Eibe 17931
Seeber, Dr. Ekkehard 22203
Seeberg, Peter 02027
Seel, Johannes 16749
Seele, Ralf-Michael 20261
Seeley, Daniel 50352
Seelig, Thomas 42025
Seeliger, Dr. Matthias 19261
Seemann, Anja 20162
Seemann, Hellmut Th. 22406
Seemann, R. 22335
Seewald, Zahava 03490
Seewaldt, Dr. Peter 22102
Şefănescu, Radu 36174
Šefčík, Dr. Erich 09606
Sefcik, James F. 51087
Seffinga, M.J. 32789
Segal, Dror 25074
Segal, Merav 25090
Segalstad, Prof. Tom V. 33737
Segarra, Guido Barletta 36057
Segarra, Ninfa 51242
Segda, Jerzy 34822
Seger, Joe 50750
Segers, Prof. Dr. Dann 03636
Segessenmann, Vreni 41713, 41714
Segger, Martin 07253
Seghatoleslami, Parvine Sadre 24668
Seghers, George D. 46208
Segieth, Dr. Clelia 17301
Segl, Maria-Luise 18295
Seglie, Prof. Dario 27219, 27221, 27226
Segni Pulvirenti, Francesca 25589
Segovia, Emilia de 39462
Segovia Barrientos, Francisco 38999
Séguin, Louise 06861
Séguin, Mélanie 06815
Segura, Angela 40267
Segura, Avelino 47932
Segura Martí, J.M. 38827
Seibert, Dr. Elke 22376
Seibert, Peter S. 49923
Seidel, Elizabeth 47847
Seidel, Winfried A. 19761
Seidenberg, Ulrich 20318
Seider, Diana L. 50148
Seidl, Alfred 02035
Seidl, Andreas 02045
Seidl, Ernst 22121
Seidl, PD Dr. Ernst 22123
Seidl, R. Bryce 53010
Seidler, Bernadette 18391
Seif el Din, Dr. Mervat 10471
Seifermann, Ellen 20776
Seifert, Dr. Andreas 20259
Seifert, M.A. Holger 21192

S

Seifert, Jörg 16554
Seifert, Sabine 18215
Seiffert, Prof. E. 45336
Seifriedsberger, Anna 02528
Seifriedsberger, Georg 02528
Seiji, Kamei 29250
Seik, Lena 19905
Seilacher, Adolf 51054
Seiler, Anne 47526
Seiler, Oskar 41243
Seiler, R. 41502
Seillier, Bruno 13920
Seim, M.A. Andreas 17594
Sein, Joni L. 48296
Seip, Debra A. 53340
Seip, Ivar 33556
Seisbøll, Lise 10188
Seissl, Maria 03083
Seitter, John R. 47195
Seitz, Erich 22347
Seitz, Gisela 17062
Seitz, Oliver 16612
Seitz, Walter 19265
Seitz-Weinzierl, Beate 22560
Sejček, Zdeněk 09819
Sejdner Knudsen, Jesper 09969
Šejleva, Gergana 05034
Sekera, Dr. Jan 09737
Sekerák, Jiří 09430
Sekete, A. 38570
Sekgwele, David 38646
Sela-Zur, Bina 25062
Selbach, Josee 32281
Selbe, Nance 52483
Selbekk, Ass. Prof. Rune 33737
Selborne, Dr. Ioanna 44830
Seleli, Prof. Y. 38603
Self, Dana 49691, 49819
Selfridge, Anna B. 50084
Selheim, Dr. Claudia 20764
Selig, Gary D. 05474, 05475
Seligman, Norman 01026
Seligman, Rachel 52960
Seligman, Thomas K. 53253
Seligmann, Susan M. 53862
Seligmann, Tine 10244
Selin, Laura 11220
Selirand, Urmas 10578
Sell, Manfred 22625
Sella, Lodovico 25417
Sellers, Mary B. 48101
Sellers, P.E. 31816
Selles, Didier 14698
Sellés, José Pascual 38829
Selley, Andrea 44007
Sellick, Justine 50517
Sellick, Roger 43688
Selliere-Charon 03349
Sellinger, Dr. Günter 02782
Sellink, Manfred 03406, 03408, 03409, 03410, 03411, 03412, 03413, 03414, 03415, 03416, 03417, 03425, 03428
Selmer, Jørgen 10135
Selmer Olsen, Knud 33760
Selmić, Leposava 38120
Selph, Judy 48061
Selsing, Lotte 33863
Selvorious, Clement 26287
Selway, Robert 52853
Selwyn, Sue 44205
Selz, Nan 50140
Selzer, Erika 01739
Semakova, Tatjana Gennadjevna 37039
Semenova, Galina 37342
Semënova, Tatjana Andreevna 36949
Semff, Dr. Michael 20487
Semočkin, Aleksandr Aleksandrovič 37603
Semowich, Charles 46160
Semple, Eve 33349
Sen Gupta, Pradip Kumar 03223
Senadheera, S. 40377
Senanster, Hvon de 14849
Senchylo, Viktor Feodosiyvych 43215
Sencova, Ljubov Vladimirovna 36898
Sendra, Gemma 38972
Senften, Penny 50434
Seng, Dr. Joachim 18370
Sengele, J.-M. 14322
Sengewald, Angela 22457
Senior, Caroline 24907
Seniuk, Jake 52028
Sen'ko, Rostislav Oleksandrovyč 43096
Senn, Carol J. 47962
Sennequier, Geneviève 15114
Sens, Dr. Angelie 31878
Sensen, Stephan 16486, 16487
Seo, Chiaki 29326
Seoighe, Mainchín 24747
Seokgi, Kim 29801
Sepioł, Afred 34697
Seppänen, Anne 40515
Sequeira, Dora 08818, 08825
Sequeira, Dora Maria 08819
Serafin, Mary Beth 50477
Serafini, Flavio 26406
Serafini, Dr. Isabella 27475
Serafino, Gladys 05530
Serafińska-Domańska, Maria 34965
Serageldin, Dr. Ismail 10479
Seraja, Marina Leonidovna 37990
Serani Elliott, Daniela 07444
Šeravić, Ivana 08973
Şerban, Violeta 36518
Şerbănescu, Done 36460
Şerbănescu, Dr. Done 36499
Šerbec, Jože 38431
Šerbelj, Dr. Ferdinand 38457
Serboli, Laura 25616
Serdar, Elizabeta 08999
Serdarević, Mevlida 04277, 04278
Serena, Dominique 11686
Serexhe, Dr. Bernhard 19453
Serge, Pascal 15996
Sergeenko, Valentin Ivanovič 37555
Sergeeva, Elena Nikolaevna 37908
Sergeeva, I.N. 36907
Sergent Nemeth, Dana 46399
Seright, Susan 45513
Séris, Anne-Jean 13429
Serkownek, Edith 50592
Sermeno, Rhodora J. 34493
Serna Muñoz, Martha Judith 08635
Serota, Sir Nicholas 44995
Serpe, Louis 15638, 15639
Serra, Florenci 40291
Serra, Sophie 13509
Serra de Manresa, Valentí 38993
Serra Ferreira, Dr. Ana Margarida 35515
Serradj-Remili, Nedjma 00042
Serré-Van Wijk, Brenda 31856
Sersha, Tom 48459
Serson, Christine 03393
Sertznig, R. 30471
Sérullaz, Arlette 14732
Servais Lebrun, Françoise 03872
Servan, René 15032
Service, Dr. Alexandra 53492
Service, Pam 48439
Servitje de Lerdo de Tejada, Marinela 31105
Serwer, Jacquelyn D. 53856
Sesić, Maria 38071
Šestakova, Tamara Nikolaevna 37832
Sestan, Ivan 08989
Šestopalova, Galina Pavlovna 29718
Sethuraman, K. 24148
Setinský, Jiří 09446
Setouchi, Betsuji 28836
Šetrakova, Svetlana N. 37255
Sett, Dr. Saradesh K. 24208
Setter, Michele 53803
Settle Cooney, Mary 53625
Setzer, Heinz 16892
Seuffert, Dr. Ottmar 17899
Seugwoo, Nam 29889
Sevastjanova, Lidija Ivanovna 37003
Ševčenko, Ljudmyla Mychajlivna 43182
Ševčenko, Ol'ga Ivanivna 43081
Ševčík, RNDr. Jan 09692
Sevda, Dadasheva 03173
Ševeček, Dr. Ludvík 09940
Sevellec, Jacques 12159
Sever, Nancy 00767, 00768
Severance, Christopher C. 05595
Severens, Martha R. 49079
Ševereva, Tamara Anatoljevna 37475
Severino, Tenente Gerardo 27531
Severstad, Jamie H. 50997
Sévery, Sébastien 15008, 15009
Sevi, Uriel J. 00249
Sevier, James 48043
Sevigney, Nichola 53836
Sevilla, Guillermo A. 00282
Sevilla, Miguel J. 39535
Sevınç, Nurten 42748
Ševkomud, Igor Dmitrievič 36857
Ševyrin, Sergej Andreevič 37549
Seward Hannam, Kim 06104
Sewell, Dennita 51906
Sewell, Jack 01011
Sewell, Jane S. 54099
Sexton, Nick 43917
Seybold, Russell G. 51388
Seybold, Silke 17564
Seyfarth, Vera 17796
Seyfried, Dr. Friederike 17028
Seymour, C. 01477
Seymour, Dr. Claire 43297
Seyrl, Harald 02699, 02700, 03090
Sezen, Nuri 42716
Sezgin, Mustafa 42956
Seznec, A. 14969
Sforzi, Dr. Andrea 26372
Sfrappini, Dr. Alessandra 26552, 26554, 26556
Sgritta, Richard A. 51729
Sgromo, V. Joseph 52501
Sgubin, Dr. Raffaella 26351, 26352, 26353
Sgubini Moretti, Dr. Anna Maria 25645
Shaath, Shawqi M. 42128
Shacham, Tzvi 25096
Shackelford, George T.M. 46940
Shacter, Joseph 47506
Shady Solís, Dr. Ruth 34230
Shafer, Geri 48187
Shaffer, Leslie 46558
Shaffer, Marcy 46162
Shaffer, Randy 46326
Shaffer, Vivienne 53213
Shaffer Nahmias, Dr. Susan 50605
Shafts, Karen 46923
Shagonaby, Joyce 49164
Shah, Hiralal P. 24064, 24065
Shahi, M. 24605
Shahzad, Tasneem 34037
Shaiboub, Dr. Abdullah 30183, 30184, 30186, 30187, 30188, 30190, 30191
Shaiman, Jason 47728
Shaker, H. 24601
Shakespeare, Valerie 51109
Shalev, Avner 24990, 25026
Shalgosky, Sarah 43904
Shamis, Bob 51231
Shanahan, Carl 48876
Shanahan, J.P. 01088
Shane, Dr. Orrin 52629
Shank, Nicholas B. 50736
Shankle, Kent 53965
Shanley, Ellen 51221
Shanley, Grace 51431
Shanmugam, Dr. P. 24112
Shannon, George W. 49210
Shannon, Dr. Helen 53561
Shannon, Steve 44071
Shao, Guotian 07491
Shao, Jiuhua 08478
Shao, Qinglong 07790
Shao, Zitong 08148
Shapiro, Anna 46924
Shapiro, Denise 49967
Shapiro, Lynda 47794
Shapiro, Michael 46441
Shapiro, Dr. Michael E. 46436
Shapiro, Selma 51457
Sharan, Michael 24959
Sharer, Robert 51886
Sharma, Dr. O. P. 24077
Sharma, R.D. 24055
Sharon, Douglas 46760
Sharp, Corwin 54288
Sharp, James 50276
Sharp, Lewis I. 48072
Sharp, Lindsay 46086
Sharp, Prof. M. 05750
Sharp, Megan 46497
Sharp, R.W.D. 45898
Sharp, Randall 48357
Sharps, David 47084
Sharrock, D.J. 44562
Sharygin, G.V. 37400
Shastri, N.H. 24423
Shastri, Nandan 24422
Shatanawi, M. 31852
Shaudys, Vincent 53162
Shaughnessy, Beth 06093
Shaul, Nissim 25079
Shave, Sue 43786
Shaw, Alvin 47154
Shaw, D. 01462
Shaw, Dan 50688
Shaw, Dean 50880
Shaw, Deirdre 54056
Shaw, Elizabeth 43555, 45965
Shaw, Harry 44653
Shaw, J. 43553
Shaw, Karen 48251
Shaw, Lytle 51167
Shaw, R. Douglas 54081
Shaw, R.W. 44727
Shaw, Tad 48462
Shawhan, Jeff 50593
Shawn, Josephine 51808
Shawn Barber, Barbara 51808
Shawver, David 53410
Shea, Carol 47949
Shea, John C. 49722
Shea, Josephine 49094
Sheads, Scott 46565
Sheaffer, Ed 53081
Shean-Hammond, Edie 48383
Shearer, Linda 47558
Shearer, Thelma 01066
Shedd, Linda 46661
Sheedy, Gilles 05405
Sheehan, Liz 50037
Sheehan, Ron 49630
Sheesley, Tim 51562
Sheets, Arian 53719
Sheets, H.L. 50497
Sheets, Jeff 46119
Sheets, John W. 53829
Sheffield, Sir Reginald 45807
Shefsiek, Kenneth 48880, 51096
Sheku, Nail 00035
Sheley, David 49213
Shelhouse, Marvin 50684
Shelkrot, Elliot L. 51873
Shell, Lynn 52061
Shelley, Elise 49733
Shelley, Roger 43970
Shelly, Diane 46984
Shelly, Roger 43965
Shelswell White, E.R.G. 24743
Shelton, George 46462
Shelton, L.T. 49614
Shen, Bohong 08302
Shen, Changming 08025
Shen, Guoying 08424
Shen, Hsue-Man 53014
Shen, Jiangbo 08323
Shen, Junshan 07977
Shen, Dr. Lindsay 52616
Shen, Lumin 08264
Shendar, Yehudit 25026
Sheng, Yi Jia 08039
Sheng, Zhenggang 07763
Shengold, Ann 50797
Shepard, Betty J. 50235
Shepard, Charles A. 48705
Shepard, Joel 52801
Shepard, Rebecca 48790
Shepherd, Dr. Brian 01411
Shepherd, Rowena 43958
Shepley, Emma 44931
Shepp, James G. 50402
Sheppard, Audrey 53916
Sheppard, Beverly K. 53344
Sheppard, Jude 47541
Sheppy, Louisa 43584
Sherer, Aaron 51616
Sheridan, Joan 53811
Sherin, Pamela V. 52148
Sherman, Alan 25073
Sherman, Cathy R. 47806
Sherman, Mimi 51211
Sherman, Sean 44945
Shermann, Robert L. 47032
Sherrill, Sarah B. 51132
Sherring, Brenda 07414
Sherwood, Frederick 48250
Sherwood, Penelope 53741
Sherwood, Susan I. 46816
Shestack, Alan 53898
Shetler, John 53552
Shetty, Prof. N.R. 24143
Shexnayder, Charlotte 48201
Shi, Jiliang 07787
Shi, Jinhai 07721
Shi, Jungui 08306
Shi, Lei 07828
Shi, Meilong 08458
Shi, Mingrong 07856
Shi, Qinyang 08296
Shi, Tingyuan 07972
Shi, Tuankuo 08374
Shibuya, Taku 28703
Shickles, Tim 48560
Shieh, Suewhei 53549
Shields, Ann 53658
Shields, R. Michael 07221
Shields, Scott 52505
Shiels, Margaret M. 48182, 54260
Shiffman, Carol 52042
Shiftins, Liz 45475
Shiga, Taro 29012
Shigaki, Betty 52390
Shih, Chia-Chun 48715
Shikada, Norimitsu 29081
Shikata, Miho 29557
Shilo-Cohen, Nurit 25012
Shimada, Kazutaka 29432
Shimada, Kunio 29290
Shimada, Dr. Norio 28754
Shimasaki, Susumu 28656
Shimazu, Misato 29392
Shimizu, Minoru 29284
Shimo, Eiko 29536
Shimoda, Norio 29373
Shimp, Dr. Robert E. 47671
Shin, Chul-Won 29796
Shin, Prof. Kwang Cho 29840

Shin, Tak Keun 29770
Shin, Young Koog 29923, 29939
Shin-tsu Tai, Susan 52859
Shinkle, Bradford 50740
Shinn, Deborah 51156
Shino, Masahiro 28743
Shinoda, Aishin 28697
Shiozawa, Yukio 29397
Shipley, Jim 44431
Shipman, Douglas R. 54113
Shipp, Kim 06184
Shippee Lambert, Lisa 47230
Shirahara, Yukiko 53013
Shirahra, Yukiko 53014
Shiraishi, Masami 29536
Shirbacheh, Tahereh 24658
Shireman, Joyce A. 50991
Shirley, Donna 53012
Shirley, Pippa 45905
Shirley, Rosie 45706
Shives, Jenny 50302
Shivvers, Mel 48098
Shlomo, Eli Ben 25066
Shlomo, Eliyahu-Ben 24947
Sho, Masaki 29133
Shoaff, Jeanne 48610
Shock, Helen 49426
Shockley, Richard 06161
Shockro, Megan 52852
Shoemaker, Innis H. 51867
Shoemaker, John 49879
Shoemaker, Rene D. 46414
Shokoofi, M.R. 24697
Sholes, John 46732
Shôno, Prof. Susumu 29307
Shook, Ray 47987
Shore, Virginia 53847
Shoreman, Mark 07280
Shorey, Prof. C.D. 01550
Short, John D. 49505
Short, Juliet 44174
Short, William 52823
Shoshani, Jesheskel 48127
Shotelier, Jean-François 14361
Showalter, Natasha 46366
Shozo, Tamura 28635
Shrago, Esther 47974
Shrago, Mindy 47974
Shrake, Peter 46606
Shrigley, Ruth 45084, 45088
Shriver, James M. 54099
Shriver, Karen K. 49363
Shriver, Thomas L. 46324
Shrum, Kenneth 52458
Shu, Zhimei 08328
Shubert, Gabrielle 47080
Shubinski, Raymond 52139
Shuck, Debbie 47816
Shugart, Dr. Gary 53409
Shulman, Steven E. 53844
Shuman, Greg 52480
Shuman, Lynne 50473
Shunsuke, Ogata 29313
Shupeck, Fern 50705
Shury, Jane 05401
Shust, Maria 51274
Shuttleworth, John 49427
Shydlovskyy, Ihor 43143
Si, Luoping 08350
Siachoono, Stanford Mudenda 54662
Siampanopoulos, Konstantinos 23067
Siano, Mary Ann 47048
Sibanda, D. 54670
Sibbalds, Mary H. 52424
Siber, Dr. Hans Jakob 41087
Sibeth, Achim 18381
Sibille, Marie-Hélène 11683
Sibiril, Marceline 14064
Sibler, Urs 41786
Sibley, S.F. 46042
Sibley, Shelia M. 51341
Sibson, Tracy 33207
Sibutha, Addelis 54665
Sicard, D. 15423
Sichelstiel, Ludwig 20763
Sicilia, Fiorella 27395
Sidebottom, D.J. 44341
Sidén, Karin 40918
Siders, Phyllis 48121
Sidibé, Samuel 30624
Sidiropoulos, Kleanthis 23019
Sidner, Rob 52739
Sidorov, Konstantin Nikolaevič 36842
Sidorova, Elena Iljinična 36959
Siebarth, Angela 06580
Siebenberg, Theo 25017
Siebeneicker, Dr. Arnulf 19186
Siebenmorgen, Prof. Dr. Harald 17594, 17679, 19437, 19443, 19445, 21867
Sieber, Christian 41748
Sieber, Ellen 46867
Sieberer, Dr. Wido 02177
Siebert, Andreas 16773
Siebert, Dr. Anne Viola 19000
Siebert, Tim 53290
Sieblist, Kerstin 19937, 19939
Siebol, Michael 54320
Siebrand, Michel 04140
Siebrecht, Silke 19578
Sieburgh, J.W. 31883
Siedell, Daniel 50105
Siefert, Chris 51957
Siefert, Dr. Helge 20425, 20956
Siegal, Matthew 46940
Siegel, Bernard 53079
Siegel, Bob 50594
Siegel, Eric 48578
Siegel, Gail 50865
Siegel, Gerlinde 18804
Siegel, Nancy 49422
Siegenthaler, Gustavo Bernardo 00625
Siegfried, Clifford A. 46161
Siegman, Joe 25073
Siegrist, Urs 42061
Sielaff, M.A. Ingo 17478
Sielmann, Torsten 17295
Siembor, Natalie 51625
Siemen, Palle 10001
Siemen, Wilhelm 19234, 21735
Siemers, Dr. Sven-Hinrich 19298, 19299, 19301
Siena, Pier Luigi 25485
Siener, William 47116
Sienkiewicz, Joanna 35078
Siepmann, Werner 21632
Sierpowski, Prof. Dr. Stanisław 34756
Sierra Bello, Prof. Luz 31069
Sierra de la Calle, Blas 40241
Sierra Jones, Alvaro 08682
Sietz, Phillip 51825
Sievernich, Gereon 17183
Sievers-Flägel, Dr. Gudrun 20753
Sievert, Annette 18920
Sifkovits, Dr. Adam 01754, 01755
Šiftar, Davor 09013
Sigal, Bronia 31121
Sigal, Laurence 14644
Sigaud, Anne 13859
Sigerist, Rudolf 22327
Sigmon, Nancy 49894
Sigmond, Dr. J.P. 31883
Signorello, Fortunato Orazio 25774
Signorini, Prof. Rodolfo 26595
Sigo, Charles 53380
Sigurdardottir, Aldis 24040
Sigurdardóttir, Elin S. 23988
Sigurdardóttir, Maria Karen 24024
Sigurdardòttir, Sigridur 24041
Sigurdarson, Erlingur 23987
Sigurðkr{}ardóttir, Ólöf K. 23998
Sigurðkr{}sson, Hannes 23982
Sigurdson, Annchristine 40646
Sigurdsson, Helgi 24014
Sigurjonsdottir, Iris Olöf 23992
Sigurjónsson Jonsdottir, Sigrun Asta 24012
Sigwing, Marty 54141
Sihm, Marcel Agang-Ang 34350
Sijarić, Mirsad 04284
Šikić, Vojko 08866
Sikorska, Janina 34691
Sikorski, Marek 34583
Sikorskyj, Mychajlo Ivanovyč 43165
Šikšnienė, Roza 30369
Sila, Dr. Roland 02136
Silber, Evelyn 44285
Silber, H. 22116
Silek, Suzanne 48801
Šilhánová, Jana 09849
Šilina, Ljudmila Semënovna 37776
Sill, Robert 53216
Silla, Dr. Chiara 26152
Silla, Dr. Chiaretta 26151, 26183, 26188, 26189, 26199
Sillanpää, Pirkko 10727
Silliman, Thomas 50810
Sillner, Helmut 01767
Šilnikova, Inna Evgenjevna 37882
Šilov, Lev Alekseevič 37539
Šilov, Viktor Valentinovič 37765
Šilova, Tatjana Èduardovna 36756
Silva, Adelino Baptista e 35804
Silva, António Marques da 35802
Silva, Cyro Advincula da 04718
Silva, Dan 48940
Silva, Eduardo Pereira da 35585
Silva, Dr. Heliodoro 35439
Silva, Henrique 36002
Silva, Isabel 35483
Silva, Ivan Mariano 04575
Silva, John 34362
Silva, Dr. Leonardo D. 04669
Silva, Dr. Maria Rosa 35415, 35416
Silva, Dr. P.H.D.H. de 40379, 40383
Silva, Pilar 39573
Silva, Urbain Karim Elisio da 04187
Silva de Oliveira, António Manuel 35829
Silva Montauntildea, Margarita 08760
Silva Silva, Párroco Isaac 08689
Silveira, Prof. Maria Auxiliadora 04590
Silveira Godinho, Isabel 35725
Silverio, Prof. Franco di 27198
Silverio, Dr. Loca di 27198
Silverman, David 51886
Silverman, Lanny 47465
Silvester, Suzanne 51518
Silvestre, Ines K. 34475
Silze, Peter 18924
Sim, Angela 38178
Sim, Lee 29897
Simak, Ellen 47411
Simanjuntak, Pareli 24437
Simard, Claire 06688
Simard, Guylaine 07263
Simard, Julie 06888
Simavi, Çigdem 42819
Simcock, Dr. J.P. 33046
Simerly, Enixx 50526
Simić, Prof. Jasna 38151
Simina, Nicolae-Marcel 36596
Simion, Luca 36608
Simister, Kirsten 44594
Simkins, Barbara 06130
Simková, Dr. Anežka 09848
Šimková, PhDr. Gabriela 09776
Simma, Barbara 03081
Simmendinger, Oliver 17665
Simmonds, Paul 45581
Simmons, Arlene 50015
Simmons, Ian 45279
Simmons, Jamie 53488
Simmons, Joyce 48533
Simmons, Julie M. 07033
Simmons, Mark 45366
Simmons, Ruth 52169
Simmons, Susan 48251
Simms, Kate 44328
Simms, M.H. 46023
Simon, Dr. Barbara 18190
Simon, C.M. 51924
Simon, Christiane 04041
Simon, Gerlinde 16628
Simon, Hal 47937
Simon, Dr. László 23871, 23872, 23873, 23874, 23875, 23876, 23877, 23879, 23881
Simon, Martina 20736
Simon, Nancy 54204
Simon, Patrick 31607
Simon, Robert 17688, 48099
Simon, Rolf 16702
Simon, Ronald 51229
Simon, Prof. Dr. Stefan 17238
Simon, Susanne 18038
Simon-Hamel, Jeanne-Yvonne 15191
Simonato Rabim, Dr. Chiaza 28192
Simončič, Alenka 38457
Simončič, Vlastja 38418
Simoneau, Christiane 07172
Simonelli, Lynn 47982
Simonett, Dr. Jürg 41304
Simonetti, Carole 25896
Simonetti, Dr. Farida 26294
Simoni, Kate 08986
Simonnet, Philippe 14999
Simons, Andrew 50735
Simons, Henri 03434
Simons, Ulrich 20938
Simonson, George 50041
Simota, Vlastimil 09704
Simpel, Francis de 04133
Simpson, Ann 44156
Simpson, Colin 43493
Simpson, Helen J. 44079
Simpson, J.L. 45821
Simpson, John W. 06194
Simpson, Ken 43349
Simpson, M.A. 07305
Simpson, Paul 45534
Simpson, Shannon 53999
Simpson, Tracy 46084
Simpson, Wendell 53614
Simpson, William G. 45087
Simpson Niven, Shelly 05544
Sims, Judith 46503, 46504
Sims, Patterson 50798
Sims-Burch, Wylene 47713
Simson, Susanna von 19140
Simui, Muyoyeta 54660
Simundson, Peter 07129
Sin, Jang Jong 29758
Sinaga, T. Moesa 24545
Sinanovitch, Laura 41179
Sinchak, Cindy 52556
Sinčić, Branko 09019
Sinclair, F. 44047
Sinclair, H.M. 45805
Sinclair, Mark E. 49266
Šindelář, Vladimír 09655
Sindermann, Sabine 21399
Sinding Steinsvik, Tone 33388
Singer, Debra 51279
Singer, Ilana 24969
Singer, June 46115
Singer, Karl 19952
Singer, Dr. Marijane 51576
Singer, Robert 50244
Singer, Ruth 41831
Singh, Dr. Bhanu Pratap 24256
Singh, Himmat 24234
Singh, Prof. I.B. 24295
Singh, Madan M. 51899
Singh, S. 24103
Singh, S. Amolak 24066
Singh, Vidya Nand 24109
Singhvi, Dr. L.M. 24143
Singleton, G. David 53610
Singleton, Robert W. 52840
Sinick, Chris 52723
Sinisalo, Henna 10785
Šinjaev, Nikolaj Ulanovič 36911
Sinn, Margaret 53839
Sinn, Prof. Dr. Ulrich 22714
Sinnig-Haas, Christiane 12440
Sinnott, George 06606
Sinopoli, Carla 46308
Sinopoli, P.A. 00546
Sinsheimer, Karen 52859
Sintes, Claude 11684
Sinyard, Elizabeth 06439
Siokalo, Zoriana 48179
Šípková, Želmíra 38394
Sipkovsky, Frank 50199
Siqueira Celidonio, Helio Fernando 04363
Sirech, Alexander 09141
Siri, Marta 27110
Širin, Jurij 37479
Širkanova, Maja Konstantinovna 37502
Širmat, Mirna 09009
Širová, Marie 09933
Sirugue, Christophe 14538
Sisdal, Odelbjørn 33892
Sisi, Dr. Carlo 26156, 26193
Šišigin, Egor Spiridonovič 36962
Šiškin, Anatolij Aleksandrovič 37644
Šiškova, Živka 05013
Sisley, Alan 01395
Sitar, Mag. Gerfried 02678
Sitarski, Steve 51830
Sitch, B. 45090
Sithole, R. 54666
Sitjar, Corazon 34420
Sitkowska, Maryla 35234
Sittner-Hinz, Kerstin 17189
Sitwell, Sir Reresby 45483
Siudziński, Kazimierz 34658
Siváček, Dr. Vladimír 38307
Šivačev, Stefan 05130, 05131
Sivan, Pnina 24955
Sivcova, Natalja Viktorovna 37446
Sivec, Nina 09013
Sivertson, Al 05311
Sivkov, Dr. Janaki 05246
Sivkova, Svetlana G. 36992
Sivonen, Esa 11053
Sivrikaya, Elsmet 42926
Sivula, Leena 11141
Sivusaari, Tiina 11131
Sivyer, M.J. 01250
Siwek, Alberyk 35206
Siwek, Jan 34529
Siximons, Les 01097
Sizemore, Bernice 46691
Sjöberg-Pietarinen, Solveig 11394
Sjöblom, Seth 41050
Sjöö, Robert 40533
Sjöskog, Per 11071
Sjöström, Anne 11133
Sjöström, Mats 11404
Sjösvärd, Lars 40532
Skaarup, J.C. 10243
Skaarup, Jørgen 10252
Skae, Sabine 43388
Skala, William 50006
Skamby Madsen, Jan 10085
Skanis, M. 29989
Skar, Anne Kari 33863
Škarić, Mila 08986
Skarpia-Heupel, Prof. Dr. Xanthippi 23215
Skarsbø, Einar 33767
Skarstein, Sigrid 33953
Skarstøl, Gunnar 33468
Skartsounis Schwab, Diane 41297
Skatov, Nikolaj Nikolaevič 37658
Skavhaug, Kjersti 33560

S

Skeide, Cecilie 33636
Skelly, David 51054
Skenbäck, Urban 40922
Skerswetat, F. 18228
Skibbe, Uwe 17608
Škiljan, Maja 08997
Skinner, J.F. 45714
Skinner, John 45712, 45713
Skip, R. 48070
Skippings, Lin 44814
Skirtz, Stefan A. 47552
Skiver, Sharon 53495
Skjoldal, Bjørg 33905
Sklar, E. 05870
Sklenářová, Vilma 09903
Skobeleva, T.P. 37759
Škoberne, Želimir 09004
Skog, Fredik 40921
Skoghäll, Lars 40789
Skoglund, Eric Y.W. 40607
Skoglund, Ivar 33415
Skoglund, Urban 40849
Skonecki, Leonard 48723
Skopec, Dr. Manfred 03027
Skorikova, Svetlana Nikolaevna 36859
Skorobogatova, Nina Vasiljevna 37181
Skoruls'ka, Oksana Mykytivna 43095
Skory, Gary F. 50667
Skotheim, Hege 33596
Skott, Inga 11422
Skougaard, Mette 10076
Skov, Torben 10089, 10091
Skov Kristensen, Henrik 10227
Skove, Margaret A. 48615
Skozen, Connie 50909
Skriebeleit, Jörg 18332
Škrinjarić, Srećko 08986
Skrobanek, Dr. Kerstin 20058
Skrobov, Sergej 36899
Skudal, Tina M. 33448
Skuggen, Guri 33758
Skuja, Andrejs 30090
Škulj, Božica 09012
Skura, Christopher 51187
Škurko, Dr. A.I. 37193, 37403
Skutecki, Andreas 22460
Škvařilová, Božena 09752
Skvarla, Diane K. 53929
Škvarnová, M. 38391
Skye, Stephen 47898
Slabá, Jaroslava 09545
Slabakova, Ganka 05008, 05009
Slade, Norman 49991
Slader, John 48601
Slamenová, Dr. Zuzanah 38226
Slaney, Deb 46175
Slanina, Súsanne 21856
Slapinski, Tim 47159
Slastunin, A.A. 37027
Slattery, Fiona 45366
Slavik, Jan 40558
Slavin, Ruth 46309
Sławski, Marek 35083
Slear, Gary 50032
Śledź, Edward 34634
Slehofer, Vera 41154
Slej, Dr. Karen 40913
Slettjord, Lars 33696
Sleurs, Jean-Pierre 03673
Slichter, Jennifer 50301
Sliggers, B.C. 32302
Slim, Soumaya 31149
Slimon, Gary 06540
Slinger, Frederico 54423
Śliwiński, Eugeniusz 34757
Sloan, Alan P. 52912
Sloan, Catherine 45038
Sloan, Mark 47365
Sloan, Nancy 49965, 49978
Sloan, Sarah 30821
Slobodzian, John 05680
Slocum, Kay 47255
Sloss, Tencha 47096
Slotta, Prof. Dr. Rainer 17398
Slottved, Lis 09954
Slover, Louise 49252
Slovin, Rochelle 46399
Slowe, V.A.J. 43885
Složenikina, Valentina Ivanovna 37569
Sluis, R. ter 32288
Sluiter, Drs. G. 31874
Slyngborg, Mette 10001
Slythe, Margaret 45322
Smadar, Harpazi 25123
Smailes, Helen 44138
Smal, M. 38758
Small, Colleen 53733
Small, Jim 49050
Small, Lisa 51160
Small, Nicole G. 47932
Smallman, Anne 47686
Smart, Diana 46079
Smart, Dragana 43853
Šmatov, Victor 03254
Smedt, Raphaël de 03485
Smeets, H. 32848
Smerdel, Inja 38463
Smerling, Dr. h.c. Walter 18024
Smets, E. 32485
Smidt, D.A.M. 32488
Smidt, Erzsébet 23903
Smidt-Jensen, Jörgen 10229
Smietana, Evelyn 49952
Smiley, Kristine 50617
Smiley, Mary-Ellen 46785
Smilova, Dimitrina 05000
Smirnov, Valentin Vladimirovič 37331
Smirnova, N.V. 36871
Smirnova, Natalja Petrovna 37762
Smit, K. 32176
Smit, Kim 24807
Smit, Sandra 38557
Smit, Toril 33758
Smith, Adam 45628
Smith, Adam 51613
Smith, Alan 01056
Smith, Albert 48268
Smith, Allan 33042
Smith, Allison A. 50395
Smith, Amanda 53481
Smith, Dr. Amy C. 45472
Smith, Ann 43849, 45659, 49956, 53956
Smith, Antony 45969
Smith, Art 06734
Smith, Arthur 44310
Smith, Dr. B.J. 01224
Smith, Bea 48760
Smith, Bertyne 46615
Smith, Betsy 48180
Smith, Beverly 46286
Smith, Bill 46405
Smith, Bonnie H. 49524
Smith, Bradley 53329
Smith, Brian 50000
Smith, Brooke 46757
Smith, Bruce 47740
Smith, Calvin B. 53779
Smith, Carol 48230
Smith, Chris 33100
Smith, Christopher 47436
Smith, Chuck 51947
Smith, Cindy 47336
Smith, Clive 33133
Smith, D. Ryan 46680
Smith, Daniel J. 54044
Smith, Darryl 52782
Smith, David 06225, 50416
Smith, David K. 50418
Smith, Deborah A. 47104
Smith, Diane 53154
Smith, Dr. Donald 44129
Smith, Donna 05780
Smith, Donna M. 05393
Smith, Douglas G. 46269
Smith, Duane 49420
Smith, Dwight 48118
Smith, Earl 54003
Smith, Elizabeth 47492, 48470, 53201
Smith, Elizabeth A. 50831
Smith, Francis 05432
Smith, Fred T. 49739
Smith, Frederick G. 47416
Smith, Dr. G.T. 46772
Smith, Garrett 52921
Smith, Gary 49097
Smith, Gary N. 47937
Smith, Gary T. 52676
Smith, Graham 43271
Smith, Dr. Gregory 48394
Smith, Dr. H. Duane 52187
Smith, Heather 06386
Smith, Helen 50692
Smith, Ian 45465
Smith, J. 33069
Smith, J.B. 51733
Smith, Dr. James N.M. 07210
Smith, Jan 50443
Smith, Janet 06172
Smith, Janice 05952
Smith, Jaqueline 43969
Smith, Jay 49455
Smith, Jean 06418
Smith, Jerry 53930
Smith, Jill 51011
Smith, Joanne 43859
Smith, Joel 52121
Smith, John 45945
Smith, Julie 43371
Smith, Kelly 53576
Smith, Ken 50111
Smith, Kent 47481
Smith, Kevin P. 47033
Smith, Kym 00861
Smith, Lisa Deanne 07157
Smith, M. 06869
Smith, Prof.Dr. M.P. 43506
Smith, Mariann 47114
Smith, Marilyn 06163, 06570
Smith, Mark 44848, 45156
Smith, Mary 06560, 06562
Smith, Maureen M. 52150
Smith, Maurice 06089
Smith, Melvin R. 51523
Smith, Michael 50905
Smith, Michael J. 50303
Smith, Michael O. 48102
Smith, Michele L. 53703
Smith, Michelle 01454
Smith, Dr. Moya 01412
Smith, Nancy 24901
Smith, Ned A. 52753
Smith, Nelson 48119
Smith, Norman 50698
Smith, Owen 46487
Smith, Pamela 52885
Smith, Pat A. 47661
Smith, Patricia F. 50084, 54269
Smith, Patsie 50429
Smith, Pauline M. 46425
Smith, Peggy 49920
Smith, Peter W. 48380
Smith, R. 05520
Smith, R.E.G. 06665
Smith, Dr. R.M.H. 38592
Smith, Raechell 49689
Smith, Randi 53199
Smith, Rebecca A. 50625
Smith, Robert 44088, 45787
Smith, Robert J. 49990
Smith, Rose J. 51523
Smith, Sam 47871
Smith, Samantha 45600
Smith, Scott 52943
Smith, Shana 52032
Smith, Sharon 45895
Smith, Shirley 48800
Smith, Stephanie 47513
Smith, Steve 52295, 53168
Smith, Sue 01466
Smith, Susan 53072
Smith, Terri 50963
Smith, Thomas 51411, 51967
Smith, Tim 49418
Smith, Tobi 52898
Smith, Valerie 17128
Smith, Verna 51446
Smith, Vivian DuRant 51289
Smith, W.D. 43764
Smith, Warwick 33161
Smith, Wayne C. 49912
Smith-Christmas, Kenneth L. 52208
Smith de Tarnowsky, Andrea 50961
Smith-Ferri, Sherrie 53638
Smith-Speelman, Dr. Khristina 47654
Smith Turner, Brier 50572
Smith Walley, Robyn 01008
Smith Woods, Karen 46617
Smitherman, Joy 52986
Smithline, Howard 25055
Smits, Eduard 04156
Smitten, Al 05908
Smogór, Anna 34538
Smohay, András 23859
Smolders, F.J. 32624
Smolders, R. 32117
Smole Gavrilovic, Margita 04284
Smolek, Michael A. 52566
Smolenicka, Maria 41914
Smolik, Luboš 09585
Smoljaninova, Iryna Mykolaïvna 43084
Smolkin, Robert 46790
Smorenburg, D. 32866
Smothers, Ann E. 49518
Smrcka, Mariann 50550
Smrekar, Dr. Andrej 38457
Smucker, Linda 54367
Smulian, Rob 46424
Smurthwaite, D. 44936
Smylski, Brian 06446
Smyth, Anthony 06533
Smyth, Deirdre 44523
Smyth, R. 06190
Smyth, Victor 51254
Snape, Dr. Steven R. 44734
Snars, Frank 01261
Sneddon, Russell C. 48304
Šneiter, G.G. 37168
Snel, C. 38699
Snell, Dr. Gesa 18967
Snell, Dr. Howard L. 46182
Snell, Prof. Ted 00862
Snellen, Wayne 51203
Snelling, Graham 45239
Snichelotto, Paolo 27691
Snider, Katherine 51206
Snider, Phyllis 46353
Śniegocki, Michał 35283
Šniokienė, Nijolė 30346
Snoddy, Stephen 45920
Snoddy, Suzie 53357
Snodgrass, Dianna H. 49114
Snoeys, Johan 03309
Snørteland, Dr. Målfrid 33689
Snørteland, Målfrid 33690
Snørtzland, Dr. Målfrid 33694
Snotun, Trygve 33493
Snow, David H. 52885
Snowden, Frances 43572
Snowden, Gilda 48115
Snowden, Karen 45623, 45625
Snydal, Dave 05465
Snyder, Amy 50870
Snyder, Barbara C. 50549
Snyder, Craig 46207
Snyder, Hills 52722
Snyder, Howard 05562
Snyder, James 24997, 52249
Snyder, James S. 25010
Snyder, Jill 47634
Snyder, Jim 54350
Snyder, Marion 53991
Snyder, Rob 52199
Snyder, Scott 52416
So, Nguyen 54624
Soans, Hanno 10624
Soares, Elisa 35860
Soares de Souza, Prof. Jonas 04504
Sobčenko, Viktor Kindratovič 43001
Sobčuk, Galina 43204
Sobczyk, Pawel 35310
Sobezyk, Stéphanie 12792
Sobieraj, Edward 35012
Sobieszek, Robert 50244
Sobiesznak-Marcimak, Margaret 35259
Sobita Devi, Dr. K. 24225
Soboleva, L.L. 37061
Søborg, Hans Christian 33384
Sobotka, Joseph 49986
Sobrepeña, Enrique A. 34512
Sobreques, Dr. Jaume 40013
Sobrequés i Callió, Jaume 38988
Socarrás, Carlos 08746
Soccoro R. Samson, Marlene 34377
Socorro Mallorca, Maria 34390
Sodagar, L. 24692
Soden, Jack 50573
Soden, Joanna 44152
Sodermanns, Christina 16402
Söderlind, Lars 40793
Söderlind, Prof. Solfrid 40918
Söderlund, Göran 40927
Soeharto, Ki 24586
Söhnel, Ulrike 19421
Soejanto, B.U.W. 24446
Soekiman, Dr. Djoko 24594
Sölder, Wolfgang 02136
Söll-Tauchert, Dr. Sabine 41160, 41161
Söllner, Jürgen 21645
Sömer, Hermann 16711
Sönnichsen, Hans F. 20698
Söptei, István 23796
Sörensen, Steinar 33480
Sörries, Christine 17352, 20092
Sörries, Prof. Dr. Reiner 19470
Soesilo Hadi, Dr. Hidayat 24585
Soetjipto, Dr. 24566
Softić, Dr. Aiša 04284
Softic, Ajsa 04284
Soganci, Dr. Marion 20546
Soggin, Gian Piero 26987
Sogoba, Djaié 30626
Sohal, Neena 45700
Sohinki, Arnie 52429
Sohl, Kathrin 19937
Sohl, Katrin 19939
Sohler, Birgit 02092
Sohlern, Gilbert von 18686
Søholt, Petter 33665, 33929, 33929
Soileau, Melba 53939
Soja, Henryk 35114
Sojka, Nancy 48114
Šoka, Milan 38187
Sokač-Štimac, Prof. Dubravka 08930
Soklevski, Valentin 30500
Soklove, Deborah 53857
Sokol, Vladimir 08947
Sokól-Augustyńska, Maria 35070
Sokol Jr., William 47049
Sokolov, Dr. Aleksej 37627
Sokolov, Dr. Nikolaj N. 37167
Sokołowska, Renata 35170
Sokolowski, Thomas 51938
Sol i Bachs, Josep M. 38997
Solà Turell, Ramón 39849
Solana, Guillermo 39583

S

Solander, Terry J. 48866
Solano Quirós, Mayela 08809
Solańska-Alexandroff, Krystyna 34811
Solberg, Per 33425
Šolc, Dr. Josef 09652
Soldaini, Julio E. 00179
Soldatova, Nina Ivanovna 37761
Soldatova, Tamara Mikolaïvna 43085
Soldini, Dr. Simone 41625
Sole, Allen 31749
Sole, Laura M. 44545
Solé, Maties 39670, 39671
Solender, Katherine 51478
Soler, Notary Enrique Oscar 00266
Soler Garcia, José María 40311
Soles, Mary Ellen 52232
Solheim Pedersen, Einar 33863
Solias i Aris, Josep 39394
Šolić, Dr. Marija Edita 08908
Soliman, R. 28118
Solimano, Sandra 26298
Solimon, Ron 46178
Solis, Claudia 49380
Solis, Isabel 07474
Solms, Fürst zu 17514
Šolochov, Aleksandr Michajlovič 37925
Solomatenko, Marg 06657
Solomko, Juri 37242
Solomon, Dara 46752
Solomon, Elizabeth 51890
Solomon, Valerica 36267
Solomon Kiefer, Carol 46268
Solórzano Pérez, Diana 30897
Solovev, Aleksandr Borisovič 37645
Solovjeva, Ljubov Pavlovna 36951
Soloyeni, Joanna 23111
Soltek, Dr. Stefan 20883
Solymosi, Rosalia 23388
Soma, Akio 29218
Sombathyova, Magdalena 38191
Some, Porosper 05273
Somers, David 05456
Somers, Dominique 03287
Somers, Stephen G. 47814
Somerville, Ann 33196
Somma, Dr. Thomas P. 48769, 48771
Sommani, Dr. Donatella 28040
Sommer, Dr. Achim 17606
Sommer, Gertrud 16955
Sommer, Jeffrey 52515
Sommer, Lawrence J. 50101
Sommer, Mogens 10004
Sommer, Dr. Susanne 18020
Sommer Ribeiro, José 35666
Sommerfeld, Renate 21599
Sommerfeldt Gikas, Carol 46633
Sommerhuber, Christian 02772
Sommers, Joyce A. 49499
Sommers, Larry 50106
Somoza, Maria E. 36066
Somsuda, Khun 42531
Sonderegger, Christina 41583
Sonderegger, Cristina 41252
Sondergaard, Marten 10244
Søndergaard, Ph. d. Morten Karnøe 10262
Søndergaard, Per 10226
Søndergaard, Sidsel 10145
Song, Aizhen 07956
Song, Bing 07754
Song, Dehui 07494
Song, Jianping 08361
Song, Jiashan 07644
Song, Prof. Ki-Ho 29921
Song, Misook 29931
Sonke, Denton 46831
Sonnberger, Betty 41327
Sonnberger, Gerwald 51238
Sonnen, David G. 54271
Sonnenberger, Dr. Franz 20755, 20769
Sonnenburg, Hubert von 51212
Sonnenschein, Isabell 21565
Sonnleitner, Dr. E.T. 51614
Sonnleitner, Mag. Elena 03045
Sonntag, Dr. Dina 21950
Sonntag, Gabrielle 13871
Sonntag, Johannes 18140
Sonntag, Dr. Johannes-Hendrik 18622, 18623, 18624, 18625
Sonntag, Marilyn 51266
Sonoyama, Hiroshi 29404
Soós, Nándor 23339
Soosloff, Philip 52417
Sopóliga, Dr. Miróslav 38367
Sorber, Frieda 03294
Soregaroli, Dr. Art 05482
Soren, Joshua 52988
Sørensen, Anne Birgitte 10049
Sørensen, Anny 10015
Sørensen, Einar 33467
Sørensen, Jens 10222
Sørensen, Jens Erik 09962
Sørensen, Marianne 10310
Sørensen, Ove 10208
Sørheim, Helge 33863
Soriano, Alain 11597
Soriano, Dr. Corazon 34462
Soriano, Dr. Dolors 38994
Sorin, Gretchen 48259
Soroka, Halyna 43105
Sorokin, Igor Vladimirovič 37748
Sorokina, N.I. 37281
Sorsa, Jaakko 11410
Sosič, Sarival 38452
Sošin, Aleksandr Pavlovič 36976
Šošmin, Aleksandr Aleksandrovič 37720
Sosnovs'ka, Tatjana Viktorivna 43082
Sosnovskaja, Marina Jurjevna 36888
Sotheran, Cheryll 33260
Soto, Jules M.R. 04493
Soto, Susana 39253
Soto Bastida, Nicolás 31101
Součková, Dr. Jana 09775
Souichi, Takahata 29158
Souindoula, Simão 00111
Souka Séko, Cisse Nèe 04186
Soule, Robert M. 47414
Souplet, Michel 13491
Sourakli, Judy 53000
Sousa, Maria José Rego 35512
Sousa, Suzanne 50350
Souter, Teri 06003
South, Allison 52693
South, Will 49068
Southard, Edna Carter 51654
Southerland, Cindy 48885
Southern, John 43988
Southern, Kathy Dwyer 53897
Southwell, Carey 48073
Southworth, Edmund 43678, 44630, 44631, 44632, 44633, 44634, 44636, 44750, 45445
Sovaudjieva, Natalia 05021
Sovluk, Viktor Ivanovič 37075
Sowada, Dr. Karin 01557
Sowder, Marvin 47948
Sowers, Russ 49353
Sowinski, Larry 51052
Soyhan, Cihat 42845
Soylu, Nilüfer 42887
Sozio, Janet L. 50878
Space, Parker 53386
Špaček, Jaroslav 09455
Spackman, A.J. 44913
Spadavecchia, Dr. Fiorella 28189
Spadea, Dr. Roberto 26011, 26012
Späth, Annette 16699, 16700, 16701
Spagnolo, Francesco 46755
Spahn, Merrill 49924
Spahr, Elisabeth 41373
Spahr, P. Andrew 50421
Spalding, Jeffrey 05931
Spalek, Dr. Ladislav 38292, 38359
Spang, Veronika 21960
Spang-Babou, François 15147
Spangenberg, Kristin 47555
Spangenberg, L. 38677
Spanke, Dr. Daniel 21948
Spankeren-Gandhi, Martina van 21428
Spânu, Mihail 36645
Spargo, J. 44444
Sparks, Brian 50149
Sparks, Jim 05737
Sparks, Kevin S. 54207
Sparks, Laurel 54024
Sparre, Comte de 15948
Sparrow, John 45372
Sparrow, Judy 45372
Spas'ka, Larisa Ivanivna 43132
Spatafora, Dr. Francesca 27700, 27959
Spathari, Elizabeth 22880, 22887, 23022
Spatz, Rosemarie 21098
Spaulding, Daniel 46322
Spazier, Dr. I. 22410
Spear, June 05808
Spear, Robert N. 49433
Spearman, Grant 46284
Spears, Carolyn 50938
Spears, Dolores 49621
Spears, Kimberly 46293
Spears, Susan 50510
Specht Corsi, Martha 27902
Speciale, Giovanni 25609
Speciale, Michael P. 54230
Špecián, PhDr. Miroslav 09849
Speckels, Gabriele 20964
Specter, David 51239
Spector, Nancy 51258
Spedding, Janaki 44565
Spedding, Michael 33109
Speed, Bonnie 46444
Speicher, Merry 52453
Speierl, Dr. Charles 47658
Spelak, Lucyna 34949
Speler, Dr. Ralf-Torsten 18906
Speliers, Thierry 04132
Spellazza, William Salvador 48455
Speller, Margaret 44018
Spence, Cathryn 43403
Spence, David 44921
Spence, Millie 06628
Spencer, Elizabeth 43965, 43967
Spencer, Georgia 49561
Spencer, Jeff 51762
Spencer, Jim 47597
Spencer, Leonard H. 47156
Spencer, Ted 47791
Spencer, Wendy 52603, 53424
Spencer-Davies, Jenni 45813
Spencer Forsythe, Laurel 51070
Spencer-Longhurst, Dr. P. 43497
Spendel, Dr. Cölestin 02677
Spender, Michael 44092
Sperath, Albert 51650
Sperber, Dr. G. 05760
Sperber, Dr. Lothar 21834
Sperebet, Yilmaz 33371
Sperling, Jörg 17731
Speth, Cornelia 19999
Speth, John D. 46308
Spevec, Branko 08975
Spezia, Michael J. 49269
Spicer, Joaneath 46591
Spicer, Margaret E. 49162
Spichtig, Klara 41820
Špička, Dr. Dušan 09611, 09863
Spiegel, Dr. Beate 18627, 20602
Spiegel, Elisabeth 19631
Spiegel, Marty 53009
Spiegelberg, Dr. U. 19199
Spiegelman, Deborah 50632
Spieler, Dr. Reinhard 20058
Spielmann, Andreas 41850
Spies, Dr. Bernhard 17459
Spies, Dr. Britta 20661
Spies, Dr. Heike 17990
Spiessens, Frans 03385
Spigo, Dr. Umberto 26493
Spilka, Abby R. 51227
Spilker, Rolf 20926
Spillemaecker, Chantal 13351
Spiller, Christine 17714
Spiller, Harriet 47115
Spiller, Mortimer 47115
Spillman, Jane S. 47810
Spillmann, Dr. Andreas 42105
Spindle, Carol 52114
Spindler, Gabriele 02317
Spindler, Jean-Charles 12041
Spinelli, Claudia 41147
Spinelli-Flesch, Marie 11955
Spinner, Helmut 19048
Spinola, Dr. Giandomenico 54501, 54509
Spiridon, Maica 36108
Spiridonov, Jurij Ivanovič 37085
Spiro, Stephen B. 51445
Spitler, Bing G. 53052
Spits, E. 31885
Spitzbart, Ingrid 01951
Spivak, Lisa 48087
Spivey, Towana 48682
Splaine, Shirley 46385
Šplíchalová, Vladimíra 09826
Spliethoff, Dr. M. 31915
Splitt, Jerzy Aleksander 34710
Splitter, Dr. Rüdiger 19457
Splonskowski, Joann 48131
Spock, Dan 52623
Spohn, Clarence 48403
Spooner, Peter 48200
Spoorman, R.L.F. 31852
Sporer-Heis, Dr. Claudia 02125, 02136
Sporleder, Janet 46336
Sporn, PD Dr. Katja 18426
Sportiello, Anne 14008
Spottorno Díaz-Caro, Rafael 39588
Spotts, George 49042
Spowart, Cliff 01444
Spoz, Andrzej 35216
Sprague, John 52368
Sprain, D. 54095
Sprankling, Miriam 51304
Sprecher, Dr. Eva 41175
Sprecher, Dr. Dr. Thomas 42111
Spree, Tommy 17035
Spreer, Elmar 18241
Sprem Lovrić, Branka 08903
Spriggs, Dr. Lynne 49031
Spring, Marika 44937
Springer, Bill 47014
Springer, Dr. Elisabeth 02275
Springer, Ernst 02090
Springer, Jonathan 52267
Springer, Kathleen 52276
Springer, Dr. Tobias 20764
Springhorn, Prof. Dr. Rainer 17839
Sprink, Claus-Dieter 17131
Sproat, Dr. John G. 47726, 47729
Sproat, Michael 49446
Sprouls, Ellen 49941
Spruill, Sarah C. 47419
Spruill Redford, Dorothy 47881
Spruit, R.J. 32413
Sprung, Y. 24941
Spüasić-Djurić, Dragana 38132
Spur, Birgitta 24023
Spurgeon, Lucy 54295
Spurney, Dr. Robert 53049
Spychała, Jerzy 34874
Squier, P.H.M. 43831
Squiers, Carol 51194
Squire, S. 01437
Squires, Ann 49622
Šrajerová, PhDr. Ol'ga 09692
Sretenovic, Vesela 52164
Srimal, Sudip 24278
Srinivasan, Shanta 24154
Srisuchat, Amara 42603
Sroka, Steven D. 53722
Sroka, Teresa 34682
St. Onge, Dave 06075
Staarman, A. 32059
Staat, D. 32059
Staber, Deborah 49295
Stabrovskij, Aleksandr S. 37547
Stachel, G. 19824
Stachel, M. 19824
Stack, Joan 47718
Stack, Lotus 50739
Stack, R. 01552
Stackpole, Matthew 48285
Stackpole, Renny A. 53497
Stacy, Kaye 01507
Stade, Anette 41340
Stadelmann, Kurt 41212
Stadler, Mag. Eva Maria 02951
Stadler, Hilar 41518
Stadler, Dr. Peter 02395
Stadler, Wilhelm 02604
Stadlin, Bernhard 41627
Stadnicki, Marek 35213
Stadtler, Helmut P. 03034
Stadtman, Kenneth L. 52186
Staebell, Sandra 46973
Staebler, Claire 14743
Stäbler, Hans 41370
Staeck, Prof. Klaus 17029
Stäheli, Charles E. 41495
Stähli, Alois 41897
Stählin, Eduard 41782
Stämpfli, Markus 41245
Stäuble, Dr. Nicole 41983
Staffen, Dr. Zdeněk 09476
Stafford, Michael D. 50713
Stage, Dr. Elisabeth K. 46757
Stager, Prof. Lawrence 47178
Stagg, Barbara 52479
Stagleder, Linda 48513
Stagno, Dr. Laura 26321
Stahel, Urs 42025
Stahl, E. 21653
Stahl, Judith R. 46642
Stahl, Kathleen P. 47202
Stahl, Konrad 20383
Stahl, Marguerite 13741
Stahl, Patricia 18377, 18390
Stahl Gretsch, Laurence-Isaline 41422
Ståhlberg, Nina 11281
Stahlecker, Robert C.G.A. 32090
Stahnke, Peter 18009
Staikova, Evdokia 05191
Stainback, Charles 52926
Staker, Mark L. 52692
Staker, Ron 48494
Stakulienė, Silvija 30444
Stalcup, Wanda 50917
Stalder, Dieter 41562
Staley, Dr. Thomas F. 46510
Stalf, Oscar 22309
Stallings, Diane G. 49601
Stallwitz, Kurt 48202
Stalskaja, Lidija Mirzejusofovna 36776
Stalvey, Dorrance 50244
Stalvey, Rip 53432
Stamarski, Sheri 05775
Stamas, Christine 48112
Stambach, Dr. Rudi 41776
Stambrau, Dr. Doris 18381
Štambuk, Slobodan 08879
Stamey, Emily 54140
Stamhuis, I.M. 32836
Stamm, Jörg 41934
Stamm, Wolfgang 21722
Stammers, Michael 44740
Stammers, Mike 44742

Stan, Dr. Ana-Maria 36291
Stancati, Antonella 27503
Stanciu, Ştefan 36315, 36380
Stancu, Radu 36254, 36439, 36525
Stancu, Dr. Radu 36526
Stancu, Radu 36634
Standen, Roger 44383
Standenat, Peter 02394
Stander, Tom 49138
Standley, Janet 46050
Standring, Timothy 48072
Stanek, Barbara 34975
Stanek, Dr. Christian 03035
Staněk, PhDr. Tomáš 09692
Stanevičienė, M. 30376
Stângă, Ion 36358, 36359, 36360, 36625
Stange, Lorraine 52649
Stangl, Denise 46780
Stangl, Reinhard 20042
Stangler, Prof. Dr. Ferdinand 02395
Staniczek, Dr. Arnold 21959
Stanislaus, Richard 52983
Stanislav, Dr. Ján 38184
Stanislavska, Ljudmyla Pavlivna 43051, 43052
Štankina, Irina Valeriïvna 43006
Stankov, Vesna 38163
Stanley, B. 53863
Stanley, Diane 48387
Stanley, Eric 52909
Stanley, Michael 43505, 45153
Stanley, Tom 52404
Stansberry, Lucy 48341
Stanton, Carole 46899
Stanton, Dr. John E. 01008
Stanton, Meridith 53882
Stapel, Drs. H.K. 32112
Stapf, Arnulf 20720
Stappaerts, Greet 03292
Starcky, Emmanuel 12600, 12743
Starham, J.T. 45117
Stark, Dr. Barbara 19683
Stark, Bonnie 49165
Stark, Douglas 51321
Stark, Dr. Frank 47532
Stark, Julie 54360
Stark, Linda 06923
Stark, Ole 22353
Stark, Peter 51494
Stark, Steve 54054
Stark, William 49225
Starke, Dr. Holger 17972
Starke, Jörg 17267
Starker, Deborah 50385
Starkey, Marsha B. 49174
Starlin, Mildred 46145
Starling, Kate 44928
Starling, Peter 45187
Starodubcev, Gennadij Jurjevič 37095
Starr, Lori 50264
Starr Hart, Leslie 49637
Stasch, Dr. Gregor 18508, 18512, 18513
Stasinopoulou, Elisavet 22920
Stasiuk, Savelia 07225
Staski, Edward 49962
Stastna, Dr. Eva 38233
Statham, Cheryl 53340
Staton, James D. 53359
Stattler, Rick 52178
Staub, Bettina 41926
Staub, Edith 41139
Štauber, Dr. Bedřich 09641, 09642
Stauber, Richard 41654
Staubli, Gallus 41212
Staubli, Thomas 41383
Stauder, Heiner 20000
Staudigl, Karin 01772
Staudinger, Katrin 01971
Staudinger, Wilhelm 18358
Staudt, Benno 18448
Stauffacher, Ueli 41521
Stauffer, Nancy 52481
Stauß, Christian 20424
Stave, Pari 51131
Stavitsky, Gail 50798
Stavropoulos, Savvas 23122
Stavrou, Dr. Patroclos 23020
Stawarczyk, Prof. Tadeusz 35337
Stawarz, Dr. Andrzej 35262
Stayton, Kevin 47069
Stažić, Andjelko 09002
Steacy, Archie M. 07204
Stearns, Emily 54038
Steatham, Helen 46057
Stebbins, Theodore E. 47177
Steblecki, Edith 46946
Stecher, Anton 02110
Stecher, Dr. Hans Peter 02314
Steck, Dr. Volker 19438
Stecker, Heidi 19905
Stedje, Ass. Prof. Brita 33737
Steehler, Dr. Kirk W. 48409
Steel, David H. 52232
Steel, Gregory 46312
Steel, Martin 51196
Steele, Brian 46139
Steele, Curtis 49646
Steele, Jeffrey 49259
Steele, Joe 06461
Steele, Steven 48087
Steele, Dr. Valerie 51221
Steeley, Audrey 43438
Steen, Dr. Jürgen 18377
Steen Jacobsen, Jansten 10206
Steenbeek, Dr. T.G.D. 32907
Steenbruggen, Han 32281
Steenhaut, Dr. Woutez 03630
Steer, Christine 17132
Ştefan-Nicolin, Mihail 36211
Štefančíková, Dr. Alica 09640
Stefanelli, Claudio 27171, 27173
Stefanescu, Dan 36227
Ştefănescu, Radu 36175
Ştefănescu, Radu 36180
Stefani, Claude 15073
Stefani, L. 23234
Stefani, Gen. C.A. Luigi 28026
Stefanidou-Tiveriou, Prof. Dr. Th. 23218
Stefanko, V.M. 43073
Štefánková, Nikola 09556
Stefano, Prof. 25332
Stefanov, Georgi 05027
Stefanova, Morena 05191
Stefánsdóttir, Ragnheidur 23981
Stefánsdóttir, Sóley Björk 23982
Stefansson, Stig 40692
Steffan, Ferdinand 22345
Steffen, Carlos 04642
Steffen, Geneviève 48234
Steffen, Paul 41839
Steffian, Amy 49823
Ştefură, Elena Mihaela 36349
Stegall, Ron 51602
Stege, Frank 22772
Stegeman, E. 32000
Stegemann, Marco 16498
Steger, Adrian 42023
Stegmann, Dr. Markus 41831
Stegner, Dr. Cornelia 17719
Steguweit, Dr. Leif 18226
Steguweit, Dr. Wolfgang 17051
Stehli, Andres 41476, 41477
Stehlin, Vera 41172
Steiding, Dr. Ina 17955, 17975
Steidl, Erwin 02178
Steidl, Dr. Thomas 02366
Steiger, Ivan 09770, 20485
Steigerwald, Mark 53981
Steigerwald, Steve 49610
Steigman, Kenneth 50359
Stein, Arno 41275
Stein, Sebastian 20457
Stein, Simcha 25062
Stein, Susan R. 47398
Stein, Dr. Wolfgang 20490
Steinacher, Bettina 18579
Steinau, Marlies 21675
Steinberg, Dr. David 47091
Steinbrenner, Theophil 21816
Steinbring, Wally 05385
Steinegger, Franz 41600
Steiner, Barbara 19905
Steiner, Charles K. 54150
Steiner, Chris 46175
Steiner, Dietmar 02932
Steiner, Dr. Erich 02685, 02685
Steiner, Franz 51469
Steiner, Dr. Juri 41220
Steiner, Ralph 45542
Steiner, Riccardo 41673
Steinert, Michael 22577
Steinhart, PD Dr. Matthias 20486
Steinhauer, Dr. George 22865, 23143
Steinhaus, Federico 26659
Steinhausen, Kathrin 17080
Steinhauser, H. J. 21716
Steinhauser, Norbert 18477
Steinhof, Dr. Monika 17564
Steinhoff, Torsten 18009
Steininger, Prof. Erich 02684
Steininger, Florian 02942
Steinkeller, Prof. Piotr 47178
Steinklauber, Dr. Ulla 01962
Steinkrüger, Uwe 20809
Steinle, Dr. Christa 01982
Steinle, E. 21658
Steinle, Karl-Heinz 17254
Steinle, Dr. Robert Fin 17301
Steinmann, Dr. Axel 03038
Steinmann, Dirk 22619
Steinmann, Dr. Marc 19613
Steinmetz, Robert 11968
Steinmetz, Dr. Walter 02575
Steinmetz, Dr. Wiebke 17817
Steinmetz, Wolf-Dieter 17518, 22474
Steinmetz, Wolf Dieter 22651
Steinwendtner, Hans 02029, 02030
Stejskal, Václav 09886
Štekar-Vidic, Verena 38398, 38441, 38500, 38501, 38502
Steldinger, M. 17124
Stella, Dr. Clara 25543
Steller, Tom 51470
Štellerová, Kamila 09512
Stellmacher, Bernhard 21760
Stellwes, Carla 52708
Steltner, Andrea 21393
Stelzhamer, Ernst 02919
Štem, Galina Michajlovna 36820
Stemmer, Dr. Klaus 17027
Stemmer-Kilian, M.A. Daniela 18018
Stemmler, Dr. Sabine 17128
Stemshorn, Dr. Max 22169
Stenberg, Gunilla 40604
Stender, D. 18283
Steneberg, Suzanne 41000
Steneker, J. 31988
Stengel, Karin 19462
Stengel, Tom 50765
Stenger, Jürgen 18654
Stenger, Leni 18654
Stengert, Monika 54278
Steninger, Kim 48354
Stenkamp, Dr. Hermann Josef 17395
Stenkula, Anna 40921
Stenlake, Frances 43936
Stenman, Paul 10972, 10975
Stensholt, Morten Chr. 33895
Stenström, Anders 40857
Stenstrop, Georg 10032
Stenz, Michael 53144
Stepančič, Lilijana 38451
Stepanova, Anna Alekseevna 37841
Stepanova, Ljudmila Vasiljevna 36862
Stepanova, Maria 36918
Stepens, Nicolas 45184
Stephan, Erik 19383
Stephan, H. 21159
Stephan, Peter 22223
Stephan, Ralph 21780
Stephen, Charla 46885
Stephen Weppner, Mary Dessoie 46802
Stephens, Dawn 47060
Stephens, Gordon 00971
Stephens, Dr. Howard 43985
Stephens, John R. 48981
Stephens, R. 45032
Stephens, Retha 49731
Stephens, Ruby A. 52651
Stephens, Spencer 53091
Stephens Kruize, Pricilla G. 50622
Stephenson, Holly 49664
Stephenson, Marcus 46299
Stephenson, Roy 44928
Stephenson, Sally 52504
Stephenson, Tim 43918
Stepic, Barbara 47086
Šter, Janez 38532
Stergar, Branka 08925
Šterk, Slavko 09004
Sterly, Marita 19123, 19125
Stermer, Jenifer 50050
Stermitz, Martin 02186
Stern, Barry 47091
Stern, Gail F. 52150
Stern, Günter 03060
Stern, Hans 04747
Stern, Jean 49531
Stern, V. 32360
Sternath-Schuppanz, Dr. Marie-Luise 02929
Sternback, Mike 47849
Sternecker, Dieter 16705
Sternschulte, Agnes 17840
Sternweiler, Dr. A. 17254
Sterrett, Jill 52795
Stertz, Dr. Stephen 47050
Stets, Mary Anne 50935
Stetson, Daniel E. 49902
Stetzenbach, Dr. Klaus 49979
Steuer, Harald 22050
Steuer, Matthias 18124
Steuer, Yvette 16713
Steurbaut, W.W.G. 31913
Stevan Bathoorn, H. 32931
Steven, Jackie 47975
Stevens, Andrew 50388
Stevens, B. 45577
Stevens, Dr. Crosby 43991
Stevens, F. Scott 54338
Stevens, H. 32059
Stevens, Jane 47481
Stevens, Jenny 43302, 43303, 44239
Stevens, Larry 48551
Stevens, Lawrence 45396
Stevens, N. Lee 49181
Stevens, Norman S. 49676
Stevens, Scott 54333, 54334, 54335, 54336, 54337
Stevens, Timothy 44856
Stevens Hardeman, Dr. H.J.M. 31901
Stevenson, Candace 06401
Stevenson, Colin 33215
Stevenson, Esma 33215
Stevenson, Fenners W. 46904
Stevenson, James 07239
Stevenson, Moira 45088
Stevenson, Sara 44157
Stevenson, Victor 07197
Števko, Anton 38315
Steward, James 46309
Steward, Maureen 05926
Steward, Travis 51243
Stewardson, Joanne 44668
Stewart, Ann 24807
Stewart, Barbara 44533
Stewart, Beth A. 51999
Stewart, Brian 44208
Stewart, Dan 50392
Stewart, David 52705
Stewart, Doug 06549
Stewart, Fiona 00892
Stewart, Dr. Ian 43239
Stewart, J. Jason 46165
Stewart, James 44360
Stewart, Lorelei 47477
Stewart, Pamela 53314
Stewart, Richard 46702
Stewart, Rick 48709
Stewart, Sarah 45853
Stewart, Steven 47592
Stewart, W. Lyn 07307
Stewart-Abernathy, Judith C. 52489
Stewart-Leach, Sheila 50367
Stewart-Young, J. 44047
Steymans, Hans Ulrich 41383
Steyskal, Josef 02964
Stgiverson, Greg 46320
Sthanapati, Dr. Jayanta 24271
Stiassny, Dr. Melanie L.J. 51115
Stiawa, Wolfgang 02902
Štíbr, Dr. Jan 09629, 09631
Štíbrová, Eva 09630
Stichel, Dr. Rudolf 17795
Stidolph, Diana 33142
Stieb, Ernst 06452
Stief, Gerhard 17868
Stiefelmayer, Ulrike 18362
Stiefmiller, Helen 49108, 49109
Stiegel, Dr. Achim 17168, 17169
Stiegemann, Prof. Dr. Christoph 20968
Stiegler, Hans 02691
Stieglitz, Leo von 21950, 22287
Stiepani, Ute 17113
Stieper, Jürgen 17644
Stier, David 53229
Stier, Matthias 17976
Stieve, Mary Farrell 46606
Stiewe, Dr. Heinrich 17840
Stiffler, David L. 46132
Stijns, H.J.M.R. 32670
Štika, Dr. Jaroslav 09827
Stiles, Victoria 53015
Stilje, Emma Kristina 41074
Still, Todd 48207
Stille Neto, Bernardo 04734
Stiller, Dr. Bernd 22012
Stilling, Niels Peter 10093
Stillström, Gunilla 40960
Stilo, Cathy 07108
Stimmel, Amy 52311
Stine, Jack 51786
Stinson, Chris 07199
Stipling, Trevor 43550
Stipperger, Prof. Walter 02059
Stippich, Wolfram 19518
Stirratt, Betsy 46866
Stitely, David 50553
Stiwich, Undine 17780
Stochr, Bernard 12584
Stock, S. 01314
Stock, Wolfram 19379
Stockar, Rudolf 41582
Stockdale, David 45907
Stockdale, Glynn 44622
Stockebrand, Marianne 50466
Stocker, Gerfried 02308
Stocker, Marlis 41329
Stocker, Sara 41203
Stockhammer, Prof. Dr. Helmut 02189
Stockhaus, Heike 22353
Stockhausen, Dr. Tilmann von 18428
Stockhoff, Val 53650
Stockman, Edgard 03378

Stockman, Dr. René 03634
Stockmann, Robert 19303
Stockwin, Louise 45758
Stoddard, Leah 47402
Stoddard, Sheena 43632
Stoddart, Simeon 47671
Stöcker, Dr. Jens 19414
Stöckler, Elisabeth 02085
Stöckli, Edi 41175
Stöcklin, Konstantin 41365
Stöckmann, Birgit 17123
Stöhr, Michael 17820
Stöhr, Dr. Oliver 02615
Stöllner, Dr. Thomas 17398
Stölting, Dr. Siegfried 17569
Stölzel, Eckhard 22843
Stoerring, Hanne 10152
Stößl, Dr. Marianne 20835, 20836
Stöttinger, Christoph 02252
Stövesandt, Birka 22621
Stoff, Joshua 48855
Stofferan, Jan 53094
Stoga, Dr. Andrzej 35233
Stoica, Maria 36169, 36170
Stojanov, Georgi 05091
Stojanov, Stojan 05017, 05018, 05100
Stojanović, Nada 38033
Stojič, Jelena 38163
Stojkovih, Dragana 38058
Stojneva, Nadežda 05014
Stokes, David 05824
Stokes, Mike 45676
Stokes Sims, Lowery 51264
Stokesbury, Bria 06062
Stolberg-Roßla, Alexander Prinz zu 20914
Stolk, Ton 32890
Stoll, Alexander 17701
Stoller, Blanche 50820
Stollfuß, Mandy 22806
Stollman, S. 33069
Stolte, Roland 17233
Stolwerk, Rick 33224
Stolwijk, Chris 31898
Stolyarov, Gwen 51187
Stolzenburg, Dr. Andreas 18932
Stone, Donna K. 49364
Stone, Gaynell 47763
Stone, Karen 46389
Stone, Karen E. 53142
Stone, Linda 46620
Stone, Margaret 43617
Stone, Mike 43835
Stone, Nathan 52226
Stone, Pam 49750
Stone, Paul 06494
Stone, Robert E. 51404
Stone, Thelma 48714
Stone, Walker S. 48216
Stone, Walter 52176
Stoner, Sheila 53351
Stoops, Susan 54302
Stooss, Dr. Toni 02620
Stoppioni, Maria Luisa 25785
Stopps, N. 43626
Storch, Dr. Ursula 03058, 03086
Storch, Prof. Dr. Dr. Volker 19088
Storck, Dr. Ekkehard 22018
Storer, Rick 47485
Storer, Prof. Robert W. 46310
Storer, Russell 01553
Stork, Dr. Simone 18035
Storm, Alex 06198
Stormark, Hege 33868
Storto, Fred 17858
Storvik, Jehans 33954
Storz, Dr. Dieter 19339
Stoscheck, Dr. Jeannette 19921
Stošić, Jovica 38100
Stoß, Andreas 21003
Stoter, Dr. M. 32473
Stothoff, Richard H. 48557
Stottrop, Ulrike 18262, 18267
Stoullig, Claire 11957, 14354
Stoumann, Ingrid 10001
Stovall, Linda Joy 54116
Stover, James 46278
Støvern, Håvard 33773
Stovold, Fran 44101
Stow, Dr. Stephen 51456
Stowe, Darryl 47910
Stoyanov, Stoyan 05191
Stoyas, Yannis 22946
Stoye, M.A. Martina 17201
Stoye, Wilfried 22833
Strachova, Alevtina Gennadjevna 37525, 37526
Strackerjan, Wilfried 19285
Straczek, Dr. Tomasz 35274
Stradeski, Leigh-Anne 44384
Stradiotti, Dr. Renata 25534, 25536, 25542, 25543
Straehly, Christiane 11562
Strässler, Walter 41979
Straihammer, Dr. Roswitha 01772
Strandberg Olofsson, Prof. Margareta 40543
Strander, Esa 11260
Strandgaard, Kirsten 23255
Strang, Alice 44156
Strang, T.H. 54062
Strang, Tim 53093
Strange, Shirley 47652
Stránsky, Jiří 09484
Straßburger, Jens 21100
Strassen, Hermann zur 19724
Strasser, Olivia 41212
Straßer, Wilfried 20603
Strathy, Mark 51028
Stratis, Katie 51263
Stratmann, Dr. Wilhelm 17340
Stratton, Ceilia M. 48624
Stratton, David C. 52663
Stratton, Frances R. 49539
Straukamp, Werner 20743
Strauss, Bryson 50255
Strauss, Dorothea 42074
Strauss, Dr. Frank B. 52971
Strauss, Jacqueline 41212
Strauss, Dr. Karl W. 17708
Strauss, Linda 50255
Strauß, Michael 17634, 17635, 17636
Strayer, Jenny 51958
Strazar, Dr. Marie D. 49293
Streatfield, George C. 44319
Street, Dr. Martin 20687
Streetman, John W. 48456
Streets, Kim 45656
Strege, Gayle 47754
Strehl, Dieter 03077
Strehlen, Martina 18247
Streicher, Dr. Gebhard 16519
Streidt, Gert 17736
Streiff, Elsie 41742
Streiff, Dr. Hans Jakob 41326
Streihammer, Dr. Rudolf 03132
Streisselberger, Josef 02429
Streitberger, Christine 02339
Streitt, Ute 02324
Strekalova, Valentina Michajlovna 37821
Strelnitski, Vladimir 50945
Stremel, Ted 05772
Strémoouhoff, Catherine 15864
Strempfl, Dr. Heimo 02191
Stremsterfer, Joanne 52596
Strenge, Bernhard von 20626
Stretton, Jenny 38604
Strick, Jeremy 50252
Strickland, Alice 48964
Strickland, D.H. 47006
Strickland, Dan 07332
Strickland, Leigh V. 48786
Strickland jr., William 51952
Strickler, Susan E. 50421
Striefler, Christian 19670
Striefler, Dr. Christian 20386
Striehl, Dr. Georg 19862
Strigl, Alfred 02707
Strinati, Dr. Claudio 27514
Strine, Linda T. 52264
Stritzler-Levine, Nina 51132
Strižnikova, Svetlana S. 37199
Strnadová, Z. 09769
Strobbe, Johan 03977
Strobel, Alexandra 41592
Strobel, Barbara 17021
Strobel, Margaret 47483
Strobel, Sigruth 20940
Strobl, Dr. Andreas 20487
Strobl, M.A. Hilde 20423
Strode, Molly 49612
Strode, Wendell 46974
Stroebel, G. 38566
Ströbele, Dr. Werner 21283, 21285
Ströle-Jegge, Dr. Ingeborg 21187
Strömbäck, Stig 40622
Strömberg, Ulla 10154
Strömfors, Göran 11096
Strömsdörfer, Thomas 21674
Stroh, Dietmar 18366
Strohfeldt, Guido 18495
Strohhammer, Andreas 02320
Strohl, Katrin 17064
Strohmaier, Franz 02335
Strom, George 48386
Strom, Peter 48414, 48415
Stromberg, Ann 40548
Strombo, Cathryn J. 53376
Strommer, Gabriele 03029
Stromskaja, Tatjana Aleksandrovna 29730
Strong, Charles 46718
Strong, Dan 49090
Strong, Michael 00931
Stroobants, Dr. A. 03540, 03541, 03542, 03543
Stroobants, Bart 03861, 03862, 03863
Stroppel, Alice 53021
Strote, Joel 49974
Stroud, Michael 30645
Strsembski, Stephan 17446
Struchtrup, Mechthild 20519
Struck, Juliusz 34698
Struhs, David B. 47297
Štrumej, Lara 38454
Strumia, Prof. Franco 25593
Struth, Dr. Steffen 21393
Strutwolf, Prof. Dr. Holger 20505
Strydom, Mel 49330
Stryker, Richard R. 47820
Stryker Lewis, Evelyn 51005
Stuart, Ann 48060
Stuart, Jacqueline 05363
Stuart, Jan 53873
Stuart Elliott, David 29439
Stubbe, Els 31891
Stubbe Østergaard, Jan 10145
Stubbendieck, James 50096
Štubova, Elena Valentinovna 36892
Stuchlík, Ladislav 09619, 09620
Stucke, Dr. Frank 20199
Stucker, Noah Lon 51923
Stucker, Richard 51442
Stuckert, Carolin M. 47720
Stuckey, Robert 53701
Stucky, Anton 41819
Studencka, Jolanta 34746
Studencki, Zbigniew 35123
Studenroth, Zachary 51432
Studenroth, Zachary N. 52514
Studer, Dr. Daniel 41803
Studer, Winfried 20605
Studley, David 52699
Stueber, Nancy 52083
Stuempfle, Dr. Steve 50625
Stüning, Dr. Dieter 17469
Stütz, Michael 19642, 19643
Stüwe, Hartmut 19498
Stuffer, Ute 18997
Stuhec, Mojca 38454
Stuhlman, Theresa 51875
Stulić, Anselmo 08936
Stull, Ellen D. 47231
Stults, Jeannine 51340
Stumpe, L. 44750
Stumpf, Jean 12911
Stuppner, Dr. Alois 03006
Šturc, Libor 09796
Sturdy, Tom C. 06246
Sturgell, Ron 49026
Sturgeon, Ava 05896
Sturgeon, James 46878
Sturgeon, Keni 50785
Sturgis, Alexander 43405
Sturgis jr, Ewell G. 47733
Sturgis, Randy 53037
Sturm, Martin 02323
Sturm, Ulli 02185
Sturtz, Rodney 52975
Stussi, Edmond 16281
Stutzer, Dr. Beat 41299, 41813
Stutzinger, Dr. Dagmar 18354
Stuvøy, Ingunn 33833
Štýbrová, Miroslava 09942
Stylianou, Petros 09374
Styra, Peter 21221
Štyrkało, Jaroslav Jevgenovyč 43042
Styron, Thomas W. 49079
Styskal, Brenda 49015
Su, Guijie 07748
Su, Nan 07587
Su, Runshu 08213
Su, Ziming 07845
Suanda, Wayan 24552
Suárez, José 08736
Suárez, Margarita 09131
Suárez Zuloaga, María Rosa 40362
Subler, Craig A. 49704
Šubová, Dana 38288
Suchan, Jarosław 34884, 34888
Suchan, Laura 06523
Suchánková, PhDr. Miroslava 09692
Suchanov, Michail Konstantinovič 37405
Suchinina, Svetlana Jurjevna 37744, 37746
Suchocki, Wojciech 35041
Suchy, Ladislao 26807
Suchy, Theodore D. 50126
Šuckaja, Galina Konstantinovna 37403
Sudarmadji, Dr. 24500
Sudduth, William M. 46435
Sudendorf, Werner 17069
Sudwidja, Ketut 24559
Südekum, Dr. Karl 22706
Süel, Mustafa 42693
Sünderhauf, Dr. Esther 17072
Sünkel, Werner 21359
Sünwoldt, Sabine 21646
Sürth, Astrid 19626
Süsserova, Jarmila 09893
Sütterlin, Hans 41137
Sützl, Werner 01690
Sueur, Catherine 14698
Sueur, Catherine le 38588
Suezawa, Shoichi 29505
Suffolk, Randall 48935
Sugawara, Eikai 29070
Sugden, K. 45090
Suggitt, Mark 43575, 43576
Suggs, Patricia 46665
Sugihara, Dr. Shigeo 29431
Sugimoto, Lisa Ann 49702
Sugino, Hideki 29522
Sugonjako, Mykola Mykolajovič 43153
Sugrue, Christine 51626
Suhr, Dr. Norbert 20146
Suhrbier, Dr. Birgit 18381
Suhre, Terry 52577
Sui, Dr. Claude 20173
Suk, Joo-sun 29927
Sukpramool, Patcharin 42572
Sukumaran, Dr. C. 24098
Sulaiman, Dr. Nasruddin 24438
Sulayman, Dr. Tawfiq 42139
Šulce, Baiba 29988
Suldina, Elena Nikolaevna 37899
Şulea, Ioan 36670
Sulebust, Jarle 33381
Šuleř, Dr. Petr 09423, 09435
Suliga, Dr. Jan Witold 35287
Sullivan, B. 01224
Sullivan, Carole 46355
Sullivan, Christine M. 50699
Sullivan, Cyndi 52411
Sullivan, Irene 47052
Sullivan, Joanne 49216
Sullivan, John 50148
Sullivan, Kathryn D. 47753
Sullivan, Larry 51570
Sullivan, Dr. Lynne 49818
Sullivan, Martin E. 52609
Sullivan, Mary Louise 51812
Sullivan, Michael 50767
Sullivan, Dr. Robert M. 49181
Sullivan Sturgeon, Catherine 47527
Sully, Georges 13513
Suls, Dieter 03294
Šulskytė, Aldona 30298
Sultan, Samir M. 10485
Sultan, Terrie 49405, 53855
Sum, Dr. Chow Weng 30554
Šumacher, Larisa Nikiforovna 29724
Sumalvico, Ulrike 22477
Sumawilaga, H. Djamhir 24562
Sumberg, Dr. Barbara 52882
Sumberg, Jack 47093
Sumers, Anne R. 51698
Sumi, Naoko 28758
Sumikawa, Kiichi 28886
Summerfield, Jan 43745, 45858
Summers, Blake 52791
Summers, Carl 50153
Summers, Jacquie 43614
Summers, John 07121, 47600
Summers, Marcia J. 50940, 52878
Summers, Reece 50096
Summerwill, Kristin 47804
Šumnaja, Dr. T.G. 37189
Sumner, Kent 47836
Sumoza, Rosa María 08627
Sumrall, Robert F. 46321
Šumskaja, Ljubov Vladimirovna 37135
Sun, Changqing 07769
Sun, Chengfu 08140
Sun, Dr. Dennis 07811
Sun, Desheng 07480
Sun, Huijun 07969
Sun, Jixin 07612
Sun, Ming 08202
Sun, Mingxiao 07591
Sun, Minxin 08150
Sun, Qingfei 08456
Sun, Yigang 07662
Sun, Yinglong 07670
Sun, Yun 08061
Sun, Zhongguo 07705
Sunday Oyegbile, Olayinka 33311
Sundblad, Stefan 40862, 40865
Sundborg, Peter 40960
Sundelin, Åsa 40921
Sundell, Eva 40666
Sundell, Karin 40960
Sundher, Chanchil 50232
Sundin, Sven-Olov 40801
Sunding, Prof. Per 33737
Sundström, Susanne 41000
Suné, Prof. J.M. 38979
Sunier, Sandra 41423
Suominen, Tapio 11364
Supamena, M.S. 10363

S

Superior, P. 38686
Suppes, Paul D. 48049
Supriyatno, Y. 24588
Supry, Fred 49467
Surace, Dr. Angela 25754
Surberg-Röhr, Silke 20286
Šurdić, Borislav 38070
Sureau, Jean-Louis 12517, 15737
Sureda Vallespir, Catalina 39049
Surmann, Dr. Ulrike 19613
Surrett, Janet 53046
Surroca i Sens, Joan 39344
Surup, Axel 22632, 22633, 22635
Suryana, Otong 24443
Surzyńska-Błaszak, A. 35038
Susanka, Susanne 02620
Susanna, Dr. Alfonso 39003
Suščenko, Evgenija Michajlovna 37482
Sušić, Željka 09013
Susoj, Elena Grigorjevna 37609
Sussman, Elisabeth 51279
Šuštar, Branko 38465
Šutalo, Kata 08996
Suter, Daniel 41137, 41161, 41600
Suter, Dawn 01432
Suter, Dick 00881
Suter, Hans und Marlis 41695
Suter, Josef Ignaz 41224
Suter, Mary 48528
Suter, Paul 41371
Sutherland, Cara A. 52281
Sutherland, Maire-Claire 03989
Sutherland, Maureen 06154
Sutherland, Rene 01046
Sutherland, Dr. Stuart 07199, 07214
Suthers, Terence 44394
Sutorý, RNDr. Karel 09421
Sutoyo, R. 24567
Sutschek, Felix 17472
Sutter, Christoph 41772
Sutter, M. A. Julia Johanna 21065
Sutton, Cynthia 49673
Sutton, Darren 00847
Sutton, Peter C. 49081
Sutton-Vane, Angela 44193
Suvitie, Osmo 11240
Suvorova, Tatjana 36991
Suwanasai, Avudh 42605
Suwentra, I. Ketut 24468
Suzuki, Katsuo 29491
Suzuki, Keizo 28650
Suzuki, Koji 28381
Suzuki, Makoto 29209
Suzuki, Norio 29426
Suzuki, T. 29637
Suzuki, Takatoshi 29501
Suzuki, Yusai 29110
Švachová, Jarmila 09588
Svae, Karin 40782
Svanascini, Prof. Osvaldo 00258
Svedberg, Berit 40942
Svedfelt, Bjorn 40858
Svedlow, Andrew Jay 50424
Švedov, Valerij Gennadevič 37623
Švedova, Inga 43178
Svens, John-Erik 11474
Svenshon, Dr. Helga 17795
Svensson, Else Marie 40440
Sverrir Árnason, Gísli 24001
Svetina, Ivo 38464
Sviblova, Olga 37254
Svirčević, Rosa 38128
Sviridova, Tetjana Mykolaïvna 43116
Svitkova, Ani 05176
Swaan, Trudy 06668
Swadley, Ben H. 52973
Swagemakers, Walter 31841
Swager, Karen 47728
Swain, Adrian 50841
Swain, David 53712
Swain, Kristin 47811
Swallow, Dr. Deborah 44874
Swalve, Prof. Dr. 18901
Swan, Dr. Daniel 50567
Swan, Daniel C. 53608
Swanagan, Don Jo 47866
Swanee, Lois 49321
Swanepoel, S. 38671
Swank, Scott T. 47221
Swann, Dan 51608
Swann, John 07199
Swann, Paula 05522
Swanson, Alan 54306
Swanson, Ann 52453
Swanson, Brook 53990
Swanson, Erik 47882
Swanson, Robert 05493, 50538
Swanson, Tom 48075
Swanson, Dr. Vern G. 53243
Swantek, John E. 53979
Swartout, Dennis W. 48991
Swartz, Leslie 46920
Swartzlander, Martha 52454
Swatosch, Michael 03037
Swatschina, Wilhelm 02090
Sweasy, Lisa 48656
Sweeney, Alan 52979, 53963
Sweeney, Colin 45983
Sweeney, Dennis F. 51178
Sweeney, John 47064
Sweeney, Karen 52297
Sweeney Price, Mary Sue 51287
Sweets, Henry 49159
Sweha, Dr. Fawzi 10503
Sweitzer, Dr. Rick 48990
Świda, Piotr 35257
Swiderski, Christine 07125
Swift, M. 44530
Swift, R.G. 45887
Swift Jones, Kelley 06296
Swigart, Arthur L. Stewart 53695
Swigart, Stacey A. 53694
Swim, Ann 47219
Swingle, James 52462
Swingle, Mark 53759
Swinkels, Dr. L.J.F. 32590
Śwircz, Kazimierz 34828
Świst, Lucyna 34747
Swoboda-Körner, Christine 19224
Swody, Russell 49928
Swords, Carol 44955
Sy, Joaquin 34431
Syahali, Teuku Sahir 24503
Šychlinskaja, Sabina 03146
Sydow, Barbara 18774
Sykes, Diana A. 45553
Sykes, Russell 49319
Syljuk, Anatolij Michajlovyč 43113
Sylvester, Morten 10302
Symes, Dr. Robert F. 45679
Symmes, Edwin C. 46448
Symmes, Marilyn 51156
Symons, Allan 05687
Symons, Prof. T.H.B. 06594
Synder, Judith 47195
Syndram, Prof. Dr. Dirk 17948, 17967
Synofzik, Dr. Thomas 22835
Syre, Dr. Cornelia 20425
Syré, Dr. Ludger 19435
Syrjänen, Anneli 11178
Syrjänen, Inkeri 11370
Šyškina, Raïsa Petrivna 43047
Syson, Luke 44937
Systad, Gunnhild 33503
Sytek, Wioletta 35028
Šytikov, Viktor Mychajlovyč 43037
Syverson, Duane 48406
Syvertsen, Per Ole 33667
Szabo, Dr. Agneza 09004
Szabo, Andras 36452, 36472
Szabó, Dr. Botond Gáborjáni 23520
Szabo, Ernst 02843
Szabó, Imre 23335
Szabó, Dr. J. József 23884
Szabó, Lajos 23483, 23484
Szabó, Lilla 23490
Szabóné, Dr. Juhász Katalin 23381
Szabóné, Lázár Ibolya 23661
Szadziewski, Prof. Dr. Ryszard 34655
Szakacs, Dennis 51332
Szakáll, Sándor 23705
Szakmary, Joanna 50956
Szála, Dr. Erzsébet 23819
Szameit, Prof. Dr. Erik 01924, 02813
Szaniawska, Wanda 35239
Szaraniec, Dr. Lech 34728
Szathmáry, László 23605
Szatmári, Dr. Imre 23367, 23371
Szczebak, Dr. Władysław 35176
Szczepański, Marek 35154
Szczesiak, Rainer 20592
Szekely, Linda 53302
Szekeres, Viv 00778
Szente, Béla 23370
Szentkuti, Károly 23720
Szentpéteri, Dr. Erzsebet 23414
Szép, Dr. Zoltán 23395
Szerdahelyi, Dr. Károly 23415
Szewczuk, Mieczysław 35074
Szewczyk, Kazimierz 35212
Szewczyk, Monika 34545
Szinai, Stephan 03039
Szinvavölgyi, Oszkár 23708
Szirányi, János 23387
Szkop, Henryk 34629
Szmolicza, József 23620
Szocs, Francisc 36565
Szőke, Dr. Mátyás 23966
Szőnyi, Dr. Eszter 23590
Szpakiewicz, Alina 34637
Szpakowska, Ewa 35064
Szram, Dr. Antoni 34883
Szücs, Dr. Judit 23513
Szücs-Irén, B. 23364
Szura, Andrzej 34744
Szuromi, Béla 23789
Szygenda, Lynn 44092
Szymańska, Dobrochna 35214
Szymańska, Ewa 34633
Szymczyk, Adam 41165
Szymczyk, Maciej 34627
Szymusiak, Dominique 13569
Szyndzielorz, Albert 34669

T

Taaffe, Renee 50754
Taarup, Frank 10149
Taawo, Kjell 40782
Tabacu, Jenica 36249
Tabah, Béatrice 16001
Tabaka, M. 06443
Tabakoff, Sheil 51239
Tabakov, Michail Ivanovič 37954
Tabbernor, Christine 07372
Tabbert, Mark 50061
Tabeaud, Marcel 12894
Taber, David 50347
Tabkoff, Sheila K. 51437
Taboada Vázquez, Rafael 39219
Taborga Rivas, Matilde 04252
Taburet-Delahaye, Elisabeth 14729
Tac An, Dr. Nguyen 54635
Tackett, Natalie 49611
Tadahisa, Sakai 29542
Tadiello, Dr. Roberto 28200
Taeckens, Michel 13757
Taegert, Prof. Dr. Werner 16919
Täube, Dr. Dagmar 19625
Täubrich, Hans-Christian 20758, 20778
Tafilica, Zamir 00028
Taft Andrews, Nathalie 50295
Taft Terry, David 46586
Taga, Yoshiya 28676
Tagaev, Muchtar 29963
Tagage, L.M. 32532
Taggart, Travis 49225
Taghiyeva, Roya 03159
Taghon, David 48390
Tagiri, Takashi 29406
Tagliasacchi, Dr. Paolo 25973
Taguchi, Yasuo 28606
Taha, Dr. Hajji Adi bin Haji 30609
Taher, Salah E. 10506
Taimagambetov, Prof.Dr. Zhaken 29682
Tainturier, Cécile 14598
Tajedin-Noureldaim, Yousif 10558
Tajima, Susumu 29566
Takács, Dr. Imre 23477
Takács, Mária 23801
Takahara, Mayako 28400
Takahashi, Akiya 29419
Takahashi, Prof. Susumu 29502
Takakusa, Shigeru 28967
Takanashi, Mitsumasa 29419
Takano, Masahiro 28400
Takano Yoshizawa, Hirohisa 28400
Takase, Tamotsu 29524
Takashi, Tuda 28765
Takashina, Shuji 28783, 28789
Takata, Yoshikazu 28709
Takeda, Sharon 50244
Takehana, Rintaro 28355
Takemoto, Hiroshi 29626
Takemoto, Mikio 29503
Takenaka, Yasuhiko 29574
Takeuchi, Jun-ichi 29392
Takeuchi, Makoto 28757, 29384
Takeuchi, Tadashi 29007
Takita, Minoru 29260
Taksami, Prof. Čuner 37670
Taktak, Fehmi 42799
Tal, Felix 31900
Tal, Ilan 25031
Tal, Dr. Nissim 24952, 24961, 24965
Talaga, Sally 49334
Talalay, Lauren 46307
Talamo, Emilia 27467
Talamo, Prof. Giuseppe 27466
Talamo-Kemiläinen, Maarit 11404
Tálas, Dr. László 23896
Talazov, Veselin 05078
Talbot, Dennis 45093
Talbot-Stanaway, Susan 50926
Talbott, Tim 51530
Talerman, Margaretha 51813
Tales, Sofia 27043
Tall, J.J. 44336
Talla, Primitiva C. 34518
Talland, Valentine 46934
Talley, Dan R. 49830
Talsky de Ronchi, Prof. Alicia 00617
Talvio, Tuukka 10810
Talvitie, Tarja 11009
Tamari, Dr. Dov 25104
Tămaş, Mihaela 36153
Tamassia, Dr. Marilena 26154
Tamba, Dumitru Gheorghe 36741, 36742
Tambini, Michael 43568
Tamburini, Pietro 25481
Tamindarova, Miljauša Amirovna 37011
Tamir, Peleg 25110
Tamis, Catherine M. 52464
Tamla, Toomas 10622
Tammen, Mike 50386
Tammiksaar, Erki 10650
Tamminen, Jukka A. 11076
Tamminen, Marketta 11226, 11227, 11230, 11231
Tamminen, Osmo 10712
Tamplin, Illi-Maria 06589
Tamura, Haruhiko 29322
Tanabaeva, Sara Eskairovna 29714
Tanabe, Emi 29082
Tanaka, Akira 29345
Tanaka, Dr. Hideo 29608
Tanaka, Masayuki 29419
Tanaka, Morio 29505
Tanaka, Takeyoshi 28364
Tanaka, Toshiaki 29390
Tanaka, Toshiya 28472
Tanaka, Yoshiaki 29536
Tănăsescu, Gabriela 36536
Tanata, Charles 49593
Tanatarova, Sara Šaripbaevna 29697
Tanchyk, Alla 17146
Tandefelt, Petra 10826
Tandy, Virginia 45084, 45088, 45100
Taneva, Valentina 05158
Tang, Hanzhang 07864
Tang, Shulan 07879
Tang, Weizhe 08135
Tangco, Manny 34418
Tangen, Eigil 33774
Tangen, Kirsten 33455
Tani, Gohei 28712
Taniguchi, Harumichi 29309
Taniichi, Takashi 29102
Tanino, Dr. Akira 29593
Tanko, Majda 38490
Tankoano, Ounteni Ernest 05273
Tannehill, Suzette 46637
Tannen, Jason 47529
Tannenbaum, Barbara 46146
Tannenbaum, Judith 52170
Tanner, Julia 44413
Tanner, Matthew 43639
Tanner, Paul 42073
Tannert, Anika 16402
Tanninen-Mattila, Maija 10769, 10783
Tannous, Youssef 30177
Tantaquidgeon, Gladys 53643
Ţânţăreanu, Ecaterina 36107
Tantoco, José P. 34422
Tao, Hsin 42484
Tao, Tao 08278
Tapias Moreno, Rito Antonio 08778
Tapié, Alain 13756
Tàpies, Miquel 38956
Tapliss, Sue 43397, 44104
Tappe, Dr. Heinrich 17916, 17925
Tappe-Pollmann, Dr. Imke 17839
Tapsell, P. 33043
Tapsirul Dahum, Langca 34393
Taragin, davira S. 52223
Taran, Galyna Mychajlivna 43054
Taran, Jurij Jurijovyč 43055
Tarasek, Władysław 34623
Tarasenko, Tatjana Ivanovna 37911
Tarasov, Aleksandr Vladimirovič 37655
Tarasova, Nadežda Petrovna 36890
Taraszkiewicz-Zwolicka, Małgorzata 34639
Tarawneh, Khaled 29641
Tarbell, Mike 49408
Tarcalova, Dr. Ludmila 09887
Tardiem, Robert 13877
Taricco, Dr. Bruno 25848
Tarmann, Dr. Gerhard 02126, 02136
Tarnawesky, Prof. M. 07390
Tarnovskaja, Tatjana Nikolaevna 37485
Tarnow Ingvardson, Gitte 10136
Tarnowski, A. 43968
Taron, Dr. Douglas 47506
Tarr, Blair D. 53536
Tarrant, Josh 00988
Tarrats Bou, Francesc 40104, 40105
Tarrús Galter, Josep 38947
Tarshish, Noa 24964
Tartaglio, Michael 47020
Tartuferi, Dr. Angelo 26157
Tarus, Käpy 11302
Taruvinska, Pascall 54669
Tasanen, Terni 11381

S

T

Taschenmacher, Peter 16756
Tashi, Khenpo Phuntsok 04201
Tashmuchamedova, Ainur 54473
Tasick, Julian 50448
Task, Deborah 06216
Tasker, Diana 48357
Tašlaj, Ljudmyla Borysivna 43148
Taşlialan, Dr. Mehmet 42855, 42965
Tasma, G.L. 32233
Tassell, C.B. 01224
Tassi, Prof. Franco 25427, 25925, 28282
Tassianari, Duilio 25302
Tassignon, Jean-Marie 03509
Tassini, Francesco 25903
Tassotti, Dr. Ilaria 27608
Tatangelo, Mireille 16102
Tataru, Marius J. 18836
Tate, Barbara L. 51208
Tate, Belinda 54244
Tatewaki, M. 28343
Tatlıcan, Zerefşan 42943
Tátrai, Dr. Vilmos 23477
Tatsch, Madeleine 17726
Tatter, Paul 46177
Tatton Sykes, Baronet 44023
Taub, Liba 43736
Taube, Dr. Angelika 19656
Taubhorn, Ingo 18920
Tauerová, Dr. Jarmila 09767
Taufi 31759
Taulbee, Ann 51655
Taulé i Delor, Mariàngela 38992
Taupe, Christina 39406
Taupe, Georg Ingrid 02861
Taut, Dr. Konrad 19909, 20119
Tavares Garcia, Sonia Maria 04621
Tavassoli 24696
Tavčar, Dr. Lidija 38457
Taveneaux, Benoît 15428
Tavernier, Bertrand 13902
Tavernier, Marc 15090
Tavlakis, Ioannis 22868
Tawalba, Dia 29662
Tawfiq Gharib, Essam El-Din 10502
Tayler, Nicola 44260
Taylor, Allison 43902
Taylor, Anne 43596
Taylor, Barbara 43694
Taylor, Barbara E. 49275
Taylor, Catherine A. 52503
Taylor, Darrell 47292
Taylor, David 43274
Taylor, David G. 06907
Taylor, Deborah 49444
Taylor, Donald B. 48726
Taylor, Edith 49991
Taylor, Dr. Eric 07199
Taylor, Eric B. 07224
Taylor, Hollis 48922
Taylor, Howard 52700, 52704
Taylor, J. Lawrence 49568
Taylor, J. William 47419
Taylor, James C. 49426
Taylor, Jeanette 05538
Taylor, Jeff 48560
Taylor, Jennifer 44039
Taylor, Jill E. 45894
Taylor, Joseph A. 49234
Taylor, June 48331
Taylor, Karen 47267
Taylor, Kathryn T. 54087
Taylor, Kathy 44946
Taylor, Lillian 49144
Taylor, Linda 46279
Taylor, Lloyd 44551
Taylor, M.A. 45366
Taylor, Martin 45216
Taylor, Mary Rose 46442
Taylor, Michael 45365
Taylor, Mike 33182
Taylor, Richard 45277, 46583
Taylor, Robert L. 51630
Taylor, Robertine 51244
Taylor, Ron 07087
Taylor, Ruth 51320, 51322
Taylor, S.K. 01269
Taylor, Saddler 47728
Taylor, Sue Ellen 47680
Taylor, Susan H. 51061
Taylor, Susan M. 52151
Taylor, Thomas C. 53336
Taylor, Tom 49991
Taylor Bennett, Cheryl 53158
Taylor Dixon, Tricia 48718
Tazlari, Marianne 22502
Tazzoli, Vittorio 27121, 27122
Tckcli, Gonul 42845
Teach, Joce 48236
Teagle, Rachel 52742
Teague, John 01148
Teahan, John 49199
Teatero, Barbara E. 05500
Tebani, Shahzor Ali 34005
Tebbe, Dr. Karin 19076
Tebbutt, Michael L. 45120
Tecce, Dr. Angela 26890
Tederick, Lydia 53934
Tedeschi, Dr. Carlo 27930
Tedesco, Joseph 30649
Tedesco, Paul H. 48164, 48165
Tedesko, Frank W. 48568
Tedford, Catherine L. 47224
Teerimäki, Matti 10706
Teeson, Douglas H. 50935
Teeter, Wendy 50223
Teets, James W. 53479
Teets, Robert G. 53479
Teets, Ruth E. 53479
Teeuwisse, Dr. J.B.J. 32084
Tegawa, Tetsuro 29153
Tegelberg, Jaana 10785
Tegin, Taylan 42715
Tegou, Eva 22961
Țeicu, Dumitru 36481, 36507, 36508
Teinilä, Soile 11478
Teinonen, Markku 10842
Teisen, Michael 10159
Teitelbaum, Matthew 07091
Teixeira, Joaquina B. 52176
Tejada, Susan 50226
Tejada, Susana 47114
Tejada Jiménez, Natalia 08676
Tekampe, Dr. Ludger 21834
Tekatlian, Dominique 12210
Teleanu, Eugenia 36238
Telek, Ulrike 16940
Teles Dos Santos, Prof. Dr. Jocelio 04788
Telford, D.B. 33062
Telford, Helen 33154
Telford, Pamela 46001
Tellenbach, Dr. Michael 20173
Telleria de Gallardo, Prof. María E. 00395
Telleria Greiger, Dr. José Luís 04227
Tellgren, Anna 40915
Tellier, Arnaud 41428
Tellier, Dr. Cassandra 47760
Tellini, Dr. Alberto 28160
Tello Figueroa, Jimmer Jesús 08597
Telsemeyer, Ingrid 17921
Tělupilová, Marie 09627
Temkin, Ann 51867
Temmer, James D. 50706, 50720
Temming, Dr. S.N. 32663
Tempal, Denison 49757
Tempel, Benno 31898
Tempelaars, P.A. 32225
Temperli, R. 42113
Tempesti, Giorgio 25335
Tempestini, Luciano 27250, 27252
Temple, Ann 24775
Temple, Patty 45213
Temple, Steve 48811
Templeman, Jack 07393
Templer, Peggy 50587
Templin, M. A. Brigitte 20083
Tempone, Mario C. 54408
Tena, Asunción 40204
Tenabe, Gabriel S. 46570
Tenbergen, Dr. Bernd 20513
Tenenbaum, Ann 51166
Teneul, Thierry 13953
Teneva, Dr. Nadežda 05190
Teng, Mingrui 08011
Tengblad, Stefan 33266
Tengku, Abdul Hamid 30585
Tennant, Robert 43494
Tennent, Elaine 49354
Tennent, Mike 52646
Tennent, William 52043
Tenorio Vera, Ricardo 39365
Tentoni, Dr. Damiana 27260
Teodoru, Violeta Veturia 36409
Teofilov, Teofil 05256
Těplyj, Nikolaj Ivanovič 37638
Tera, Mitsuhiko 29535
Teran, Reba Jo 48697
Terasima, Yoko 29419
Terblanche, E. 38566
Tercan, Şenel 42786
Terčon, Nadja 38492
Terechov, Anatolij Vladimirovič 37554
Terechov, Vladimir Michajlovič 37451
Terentev, I.S. 37851
Terentjev, Vladyslav Sergijovyč 43191
Terežčuk, Volodymyr 43195
Tergun, Dr. Alimuddin 24487
Terhune, Joyce 52271
Terlouw, Dr. Willem 32478
Terlutter, Dr. Heiner 20513
Terman Frederiksen, Finn 10230
Țermure, Taisia 36473
Ternova, Natalja 37352
Teropšić, Tomaž 38404
Terpening, Virginia L. 49496
Terra, Prof. Pedro 04560
Terrail, Claude 14747
Terrasse, Jean-Marc 14698
Terreehorst, P.J. 32844
Terrel, Emmanuelle 14433
Terrell, Walter D. 52955
Terrier, Claude-Cathérine 12300
Terrier, France 42049
Terry, Blanche S. 53734
Terry, Cathy 45287
Terry, Ray 52455
Tersew, Sylvie 23277
Teruzzi, Dr. Giorgio 25409
Tervahauta-Kauppala, Liisa 10931
Terwey, Michael 55093
Terziev, Penjo 05057
Terzo, Dr. Fatima 28251
Tesch, Ilse 21882
Tesch, Sten 40834, 40835
Teschner, Adelheid 20397
Tesfatsion, Costentinos 10680
Teshigahara, Akane 29474
Teske, Christe 17533
Teske, Robert T. 50712, 53996
Tessenow, Inge 18827, 18828
Tesseraux, Hartmut 19206
Testoni, Marcus 00961
Testoni, Theresa 00961
Tétaz, Marie Noëlle 12546
Tête, Michel 12723
Tetlow, David 45083, 45489
Tetterode Ravestein, Elke 32218
Tetzlaff, Claudia 17159
Teuber, Dr. Dirk 16890
Teubler, M. 20443
Teufel, Tina 02620
Teunissen, P. 32729
Tew, Julie 44266
Texel, Mark 50851
Teysen, Kenneth 50356
Thacker, Murray 33163
Thacker, Shirley 47535
Thakeray, Dr. Francis 38759
Thakkar, V.M. 24418
Thaler, Dr. Herfried 02320
Thaler, Janice M. 48839
Thaler, Klaudia 01829
Thalheimer, Patricia 47994
Thaller, Toni 42056
Thalmann, Jacqueline 45330
Thalmann, Rolf 41182
Thalmann, Simone 41163
Thames, Todd 52501
Thangwa, Josephine 29747
Tharp, Dr. Brent W. 53272
Thaw, Michael 43251
Thaxton, Kate 45284
Thaxton Soule, Debora 52410
Théberge, Pierre 06551
Theile, Michael 19671
Theiler, Theo 41331
Thein, Andreas 17988
Thein, U Soe 31720
Theis, Heinz-Joachim 17152
Theis, Wolfgang 17254
Theisen, Lee Scott 52962
Theiss, Nancy 49842
Theissen, A. 17259
Theißen, Dr. Peter 17870
Thelen, Klaus 21191
Thelle, Mikkel 10165
Theml, Dipl. KuWi Christine 19390
Thenoux, Jean-Marc 14830
Theodorescu, Roxana 36198, 36204, 36205, 36207, 36219, 36231
Theoret, Yves 06551
Thériault, Michèle 06342
Thériault, Serge 06551
Theriot, Edward C. 46519
Therre, Marie-Line 16145
Therrien, Richard 05764
Thestrup, Poul 10211
Thestrup Truelsen, Susanne 10008
Theune-Großkopf, Dr. Barbara 19676
Theunis, Karin 03495
Thévenin, M. Claude 15217
Thevenon, Luc F. 14427, 14441
Thewes, Guy 30482
Theys, Guido 03757
Thi, Nguyen Toan 54623
Thi Nu, Ha 54636
Thiagarajan, Dr. Deborah 24325
Thibault, Barbara 46931
Thibault, Pascale 14952
Thibodeau, Linda 53125
Thibodeau, Tony 52881
Thiébaut, Philippe 14692
Thiel, Dr. Ulrich 18423, 18425
Thiele, Dr. Klaus 18740
Thielen, Daniel P. 47261
Thiels-Mertens, Lydia 03551
Thieme, Helga 18827, 18828
Thieme, Teresa 19392
Thieme, Thomas 41048
Thieme, Dr. Uwe-Ekkehart 18716
Thiemler, Dagmar 21806
Thier, Dr. Bernd 20519
Thier, Wolfgang 02081
Thierbach, Melanie 16615
Thierolf, Dr. Corinna 16620, 16637, 20425
Thierry, Olivier 12304
Thiery, Robert 14229
Thiesen, Barbara A. 51398
Thiesen, Bill 50444
Thiesen, John D. 51398
Thieven, Mike 52968
Thijsse, L. 32065
Thill, Edmond 30483
Thingvold, Terje 33676
Thir, Prof. Karl 01697
Thir, Dr. Mike 21959
Thiroy, Richard 04019
Thoben, Peter 32191
Thobois, Elisabeth 02929
Thöny, Christof 01810
Thörn, Eva 40970
Thörnberg, Magnus 40689
Thörne, Michael 40873
Thörner, Ilka 17500
Thøgersen, Mogens 10080
Thogersen, T. 33505
Thole, Dr. Bernward 20757
Thomanek, Kurt 01742
Thomann, Marcel 13973
Thomas, Angela 43553
Thomas, Anne 06551
Thomas, B. 38697
Thomas, Barry 01359
Thomas, Becci 46230
Thomas, Dr. Beth 45565
Thomas, Brad 47973
Thomas, Celestine 28328
Thomas, Charles 46854
Thomas, Charles C. 49750
Thomas, David H. 48232
Thomas, David T. 53309
Thomas, Dean V. 43883
Thomas, Don 54257
Thomas, Eleanor 45050
Thomas, Dr. Floyd R. 54158
Thomas, Gillian 50633
Thomas, Harold Pratt 47375
Thomas, Heidi 53796
Thomas, Joe 47583
Thomas, Kathryn 52583
Thomas, Kurtis 50769
Thomas, Larry 52799
Thomas, Linda 47109
Thomas, Linda L. 46263
Thomas, M. Simon 32729
Thomas, Maureen 52668, 52669
Thomas, Nancy K. 50244
Thomas, Nicholas 45731
Thomas, Raymond 11708
Thomas, Prof. Dr. Renate 19631
Thomas, Robert G. 45049
Thomas, Rosalyn 45820
Thomas, Staf 04070
Thomas, Steve 52369
Thomas, Susan 49031
Thomas, Thelma 46307
Thomas, Tim 50802
Thomas, Valerie 14353
Thomas, Valérie 14358
Thomas, William T. 47695
Thomas, Y. 13342
Thomas-Ziegler, Sabine 20243
Thomason, Barbara 52275
Thomasson, J.R. 49225
Thome, Teresa L. 49004
Thomenius, Kristina 11010
Thomes, Russell 53188
Thomina-Berrada, Alice 14692
Thommen, Lynn 51200
Thommesen, Nils H. 33394
Thompson, Ann 48173
Thompson, Anna 40921
Thompson, Anna M. 52559
Thompson, Barb 05331
Thompson, Barbara 49162, 53253
Thompson, Becke 48026
Thompson, Bernard J. 07413
Thompson, Bernell 52989
Thompson, Brian 45740
Thompson, Dale 49256
Thompson, David 52414
Thompson, Deborah 41013, 52934
Thompson, Dixie 47329
Thompson, Franklin 50052
Thompson, Gwen 50053

T

Thompson, Ida 49395
Thompson, Isobel 45820
Thompson, J. 47238
Thompson, James 48869
Thompson, Jennifer 43862
Thompson, Jo 54195
Thompson, Joe H. 52949
Thompson, John 47151, 52934
Thompson, Joseph 51378
Thompson, Kelly 49826
Thompson, Livia 49536
Thompson, Marc 48320
Thompson, Margaret 44453
Thompson, Dr. P. 44469
Thompson, Paul 33251, 43902
Thompson, Pauline 07229
Thompson, Peter 06669, 45852
Thompson, Ronda 54034
Thompson, Roy 06384
Thompson, Scott 48185
Thompson, Steve 45628
Thompson, Susan 51057
Thompson, Trish 47995
Thompson, Wilma 47905
Thompson-Klein, Mark 50198
Thomsen, Douglas 48238
Thomsen, Sancie 48238
Thomson, Beth 45458
Thomson, Bruce 05846
Thomson, Ceri 43534
Thomson, David 44033, 44277, 44535
Thomson, Frances 01599
Thomson, Frank E. 46368
Thomson, Dr. Philip 01514
Thømt, Torill 33487, 33698
Thongthew, Suvaporn 42597
Thorarinsson, Einar 24010
Thorburn, Ken 43651
Thorkildsen, Åsmund 33466, 33467
Thorlacius, Ethel 53822, 53823
Thorláksson, Eiríkur 24015
Thorláksson, Eiríkur 24018, 24019
Thormann, Dr. Olaf 19911
Thorn, Brett 43342
Thorndahl, Jytte 09976, 54777
Thorne, Lynne 04194
Thornton, Pam 51651, 51652
Thoroddsen, Asdís 24011
Thorp, John P. 46797
Thorpe, Christine 43773
Thorpe, Dana 48791
Thorpe, J. Derle 50078
Thorpe, Linda 43886
Thorpe, Timothy 44587
Thorsen, Sue 53127
Thorsteinsson, Arne 10687
Thouret, Nicole 11721, 11722
Thowsen, Atle 33423
Thrailkill, Helen 53257
Thrane, Helle 10080
Thresher, Leon 49166
Throckmorton, C. 43264
Throl, Dr. Hans-Joachim 22664
Thron, Rudolf 02592
Thümmler, Prof. Dr. Sabine 17169
Thümmler, Dr. Sabine 17677, 19461, 19467, 19469, 19474, 19476
Thünenkötter, Sabine 18774
Thüring, Silvan 41886
Thuesen, Søren T. 23263
Thum, Agnes 52972
Thum, Marvin 52972
Thuma, Kathryn 51167
Thumm, Cynthia 17426
Thuring, Basil 41175
Thurmann, Dr. Peter 19531
Thurmer, Robert 47627
Thurn, Ulrike 20428
Thurner, Ludwig 02421
Thurnton, Maggie 45844
Thúroczy, Dr. Csaba 23902
Thurow, Chuck 47480
Thuy, X. de 14253
Thwaites, Dr. Peter 43719
Thygesen, Anker 10049
Thykier, Claus 10295
Thyrring, Ulla 10068, 10072, 10074
Thys, Joseph 04148
Tiacci, Giovanni Battista 27148
Tian, Likun 07610
Tian, Mingxin 07663
Tian, Tao 07842
Tian, Xiao 07904
Tian, Xiuping 08385
Tibbs Thurlow jr., E. 53866
Tiberi, Prof. Romolo 27008
Tiberio, Dr. Artioli 27214
Tichenor, Daryl L. 53760
Tichonov, Dr. Igor L. 37686
Tichonov, Vladimir Viktorovič 36931
Tiderman, Conny 40529
Tidy, Josh 44694
Tiedemann, Klaus 21196
Tiedemann, Dr. Nicole 18915
Tiefenbach, Dr. J. 01861
Tiefenbach, Oskar 01878
Tielman, Prof. R.A.P. 32887
Tiendrebeogo, Prosper 05272, 05277
Tierney, Janet 44329, 45686
Tiesing, F. 32000
Tietjen, Hinrich 19989
Tietken, Antje 20895
Tietmeyer, Dr. Elisabeth 17199
Tietz, Annette 17105
Tietz, Jürgen 21797, 21799
Tietz, R.M. 38616
Tietzel, Prof. Dr. Brigitte 19703
Tiffany, Harold 49231
Tiffany, Sharon 53292
Tifft, Douglas C. 48379
Tigert, Jennifer 05581
Tiitsar, Ketli 10627
Tijdink, W. 31983
Til, Barbara 18008
Tilarinos, George 51800
Tilbrook, Val 00974
Tiley, Sian L. 38719
Till, Christopher 38584, 38648
Tillequin, Prof. François 14676
Tiller, Helmut 03030
Tilley, Nicola 45578
Tilley, Peter 44780
Tilliet, Marie-Françoise 03893
Tillmann, Dr. Doris 19530
Tillmann, Ulrich 19624
Tillmann, Walter 20579
Tillmanns, Eric 03583
Tilmant, France 03848
Timár, Gabriella 23396
Timbrook, Mark 50747
Timdal, Ass. Prof. Einar 33737
Timlia, Robert 47610
Timlin, Aaron 48112
Timm, Kristal 48569
Timm, Robert 49991
Timmelmayer, Andreas 03048
Timmerman, B.W. 32850
Timmermann, Dale 53708
Timmermann, Dietrich 10141
Timmermans, L. 32523
Timmins, John 33096
Timmons, P.J. 45937
Timms, Peter 48548
Timofeeva, Nina Vasiljevna 37797
Timotijevic, Milos 38094
Timpano, Anne 47559
Tinčev, Valentin 05029
Tindale, Bruce 01319
Tindall, Anne 44577
Tindel, Raymond D. 47502
Tindell, Anne 43295
Ting, Dr. Joseph 07808
Ting, Joseph S.P. 07800
Ting, Dr. Joseph S. P. 07807
Tinguely, Rofer 13712
Tinguy de la Giroulière, Dominique de 13506
Tinius, Joanna 52937
Tinjum, Larry 52127
Tinker, Dr. Keith 03210
Tinney, Donal 24899
Tinney, Harle H. 51319, 51330
Tinte, Francisco 00649
Tinter, Rudolf 22197
Tintori, Dr. Andrea 26408
Tió-Fernandez, Dr. Teresa 36027, 36028, 36039, 36064, 36065, 36069, 36077
Tip, J.C. 32634
Tippitt, Dr. V. Ann 48870
Tira, Masa Morioka 49351
Tirado, Vincent 50636
Tirelli, Dr. Margherita 27350
Tirkalova, Nina Oleksandrivna 43146
Tischer, Achim 17555
Tisdale, Brenda 48327
Tisdale, Shelby J. 53442
Tisken, Jörg 22109
Tisserand, Gérard 12550
Tissier, Gève 12588
Tissier de Mallerais, Martine 13950
Tissot, Karine 41563
Tissot, Susan 53705
Titova, Larisa Michajlovna 37053
Titova, Valentina Vasiljevna 37511
Tittoni, Dr. Maria Elisa 27446, 27499, 27500
Tittoni Monti, Dr. Elisa 07382, 27511
Titz, Susanne 20357
Tivadar, Petercsák 23540
Tixier, J.M. 14878
Tjan, R. 32018
Tjan, Ranti 32268, 32269
Tjandra, Gunawan 24458
Tjomlid, Steinar 33907
Tjossem, Susan 53680
Tjupina, Lidija Afanasjevna 36813
Tjutčeva, Valentina Sergeevna 37641
Tkačenko, Katerina Grygorivna 43183
Tkačeva, Irina Vladimirovna 37749
Tkach, Steven 53941
Tkačuk, Jaroslava Juriïvna 43072
Tkešelašvili, A.V. 16385
T'Kint de Roodenbeke, Juan Graaf 03538
Tleuberdina, Piruza Ablaevna 29690
Toase, Charles 45018
Tobe, Carol B. 50281
Tobe, Eiko 28406
Tober, Barbara 51112
Tobin, Shelley 43643
Tobler, Beatrice 41212
Tobler, Mathilde 42123
Tobón Arbeláez, Elvira 08739
Tocev, Petăr 05032
Tochigi, Akira 29493
Toczewski, Dr. Andrzej 35372
Tod, Chris 45745
Toda, Masayasu 29116
Todd de Iramain, Margarita Tula 00600
Todde, Nicola 25675
Todiraş, Silviu 36159
Todorov, Todor 05147
Todorova, Ani 05007
Todorova, Pepa 05049
Todorova-Stojanova, Kresa 05112
Toedtemeier, Terry 52086
Toejner, Poul Erik 10103
Tölle, Ch. 21294
Toelle, Wolfgang 18741
Toepfer, Lutz 16763
Törmälä, Satu 10727
Törnblom, Görel 40633
Törnig-Struck, Jutta 20281
Török, Tibor 23662
Toerper, Delaine A. 54340
Toftegaard, Kirsten 10129
Toftgaard Poulsen, Søren 10072
Togay, János Can 17060
Tognali, Dino Marino 28286
Tohjoh, Kohkoh 28797
Tohno, Haruyuki 28377
Toit, Elizabeth du 38641
Toivanen, Pekka 10865
Toivanen, Yrjö 11448
Toka, Arnita 30085
Tokarczyk, William 48882
Tokarski, Longin 34587
Tokiwa, Yutaka 28537
Tokombaeva, Tamara Alievna 29957
Tokovics, Tamás 23736
Tokuda, Toshiaki 28658
Tokugawa, Narimasa 28922
Tokugawa, Yoshitaka 29012
Toledo, Eva 34419
Tolentino, Onofre R. 34477
Tolhurst, Robert 43645
Tolksdorf, Reinhold 19772
Tollefson, Roy 06397
Tollet, André 12390
Tolmačeva, Valentina Michajlovna 36976, 36977
Tolman, Keith 48693
Tolmoff, Mati 10648
Tolnick, Judith E. 49795
Tolonen, Tatiana 11408
Tolson, Peter 53522
Tolstoj, Vladimir I. 36975
Tolu, Giuseppe 25392
Toluse, Joe 46893
Toma, Shiichi 29014
Tomamichel, Leonhard 41246
Tomar, Y.P.S. 24357
Tomaria Cortes, Manuel 40220
Tomaschett, Dr. Pius 41953
Tomašik, Gennadij Aleksandrovič 36809
Tomasina, Marie-Cécile 11768
Tomasović, Nives 08878, 08880
Tomassetti, Larissa 01941
Tomassini, Gabriella 27964
Tómasson, Dr. Þórðkr{}ur 24006
Tomaszewski, Robert 34937
Tomczak, Bogdan 35345
Tomea Gavazzoli, Dr. Maria Laura 25550, 26949, 26950, 26953
Tomey, J.K. 43856
Tomić, Violeta 38132
Tomíček, Ing. Rudolf 09844
Tomiyasu, Reiko 28400
Tomkins, John 51982
Tomkins, Ken 44456
Tomlinson, Janis A. 51283
Tomlinson, Dr. Janis A. 53852
Tomlinson, Lori 52200
Tomlison, Lyn 51094
Tomljenović, Zvjezdana 08986
Tommassen, W. 32555
Tomor, Michael 49642
Tomor, Dr. Michael 50201
Tomov, Lazar 05191
Tomova, Krasimira 05243
Tompkins, Jack D. 52274
Tomplak, Darinka 38529
Toms, Donald D. 50004
Toner, Dominga N. 46771
Toner, Eden R. 52277
Tong, Guorui 08490
Tong, Zerong 08249
Tonini, Dr. Camillo 28188
Toniolo, Elia 28229
Tonkin, Steven 00934
Tonković, Snježana 08882
Tonneau-Ryckelynck, Dominique 13122
Tønneberg, Ingrid 33861
Tonon, Dr. Marco 25538
Tony, Mollica 51349
Tooby, Michael 43748
Toogood, Peter 00880
Toohey, Jeanette 49577
Tooker, David P. 53381
Toom, Maire 10624
Toomet, Tiia 10658
Toone, Keith 05513
Toperzer, Tom R. 46250
Toplovich, Ann 50975
Topoleanu, Florian 36706
Topoleanu, Florin 36272, 36362, 36703, 36704, 36705, 36707
Topolinsky, Verna 05340
Toraason, Mary C. 53673
Torataro, Yoneyama 29463
Torbágyi, Dr. Melinda 23442
Torchia, Richard 48936
Torchia, Robert W. 50571
Torchio, Maurizio 27996
Torfs, Freya 03296
Torgersen, Ulf E. 33695
Torjusen, Bente 50010
Torki, Nasser 24607
Torm, Helge 10167
Tormo, Fernando 38805
Torno, Janet E. 46828
Torp, Lisbet 10142
Torra, Eduardo 40336
Torra, Emilio 26422
Torralba, Casiana C. 34356
Torrance, Joshua C. 48373
Torre, Antonio 27528
Torre, Prof. Danilo 26180
Torre, Fuensanta García de la 39205
Torre, Graciela de la 31131
Torre, Dr. Paola 27512
Torre Bonachera, José Luis la 40178
Torre González, Norberto de la 31352
Torremocha Silva, Antonio 38841
Torrence, Gaylord 49695
Torrent i Bagudà, Enric 40162
Torrents, Carme 39301
Torrenzieri, Claudio 27966
Torres, Elizabeth 04224
Torres, Prof. Emmanuel 34487
Torres, Eva 48790
Torres, Francesc X. 39262
Torres, Jesse 54366
Torres, John 51395, 52881, 52883
Torres, Julio 39544
Torres, Pedro Pablo Alberto 08718
Torres Correa de Migliorini, Beatriz 00601
Torres Gonzàlez, Begoña 39579
Torres Valderrama, Hernando 08583
Torsteinsrud 33378
Tortell, Victor E. 30638
Tortolero, Carlos 47489
Torunsky, Dr. Vera 17462
Tosatto, Guy 13134
Tošev, Miroslav 05099
Toševa, Svetla 05174
Tosi, Dr. Maria Teresa 27253
Tosini Pizzetti, Simona 26580
Tosswill, Richard 05423
Tostevin, Patrick 45487
Toteva, Teodora 04992
Totevski, Totjo 05232
Tóth, Dr. Endre 23442
Tóth, Hajnalka 23934
Tóth, János 23623, 23972
Toti, Dr. Enrico 27838
Toti, Dr. Odoardo 25203
Tottis, Jim 48114
Touche, Hugues de la 14079, 14080, 14081, 14082, 14083
Touhey, Paula 49734
Toulgoat, Morgane 12636
Toulmond, Dr. André 15104

T

Toumit, François 12234
Touret, Lydie 14677
Tourneux, Alain 12410
Tourta, A. 23217
Tourtillotte, William 53159
Tous, Rafael 38966
Toussaint, Christophe 12760
Toussaint, Jacques 03901
Toussaint, Yolaine 06327
Touw, M.A. Susanne 20478
Towe, Andrew C. 48029, 48031
Towell, Debbie 07244
Towell, Gerry 45190
Towl, Betsy 47346
Town, Geri 54327
Townley, Kim 33245
Townsend, Clifford 48287
Townsend, Mary Ann 47349
Townsend, Richard P. 46619
Townsend, Simon 43370
Townsend, Wayne 06774
Townsend, Wilma T. 47214
Toyama, Kazuo 28756
Toyama, Tohru 29433
Trabitsch, Dr. Thomas 03051
Trabold, Dr. Rudolf 17072
Trace, Howard C. 49491
Tracey, D. 43456
Trachsler, Ernst 41995
Trachtenberg, Daniel S. 47682
Tracy, Conrad 47696
Tracy, Duane 50957
Tracy, Susan 07096
Träger, Hans-Heinrich 20419
Träris, Susanne 20265
Tragent, Henry 50656
Tragner, Hermann 03079
Trahan, Eric 47211
Traikasem, Saengchan 42574, 42578
Trainor, Eugene N. 52967
Trakospoulou, Helen 23053
Trâmbaciu, Şefan 36253
Trămbaciu, Stefan 36252
Tramer, Nancy 48149
Tramonte, Pietro 27348
Tramp, Lana 53284
Tramposch, Dr. Bill 33260
Tranié, Pierre-Tarcisius 15991
Transier, Dr. Werner 21834
Trapnell, Dr. Simon H. 44318
Trapness, Thor 33571
Trapp, Graf Johannes 27857
Trappeniers, Drs. Maureen 32375
Trask, David 54255
Trask, Richard B. 47957
Trauffer, Walter 41485
Traugott, Joseph 52880, 52883
Traunecker, Claude 15812
Trautmann, Charles H. 49550
Travella, Eugenio A. 00545
Travers, John 45148
Travis, Betsy L. 52115
Travis, P. 33223
Travis Boley, F. 52564
Trawick, Rebecca 52239
Trazegnies, Marquis de 03526
Trebbi, Marco 33428
Trebilcock, Evelyn 49411
Trebilcok, Samantha 43632
Treble, B. 05663
Tréboit, Marcel 14034
Trecht, Daniela 41264
Trefný, Martin 09821
Trefor Jones, Iwan 43925
Trefzger, U. 18431
Tregaskis, Kate 44159
Tréguier, Frank 12726
Trekell, Kirk 46246
Tremain, Robert 05888
Tremblay, Diane 06781
Tremblay, Ermenegildo 27300
Treml, Robert 22303
Tremp, Prof. Dr. Ernst 41806
Tremper Hanover, Lisa 47691
Trentin-Meyer, Maike 16783
Trepesch, Dr. Christof 16614, 16619, 16625, 16630
Treppner, Jürgen 19745
Tresch, Karl 41855
Treskow, Dr. Sieglinde von 17508
Trešlová, Romana 09578
Tresner, Cliff 50788
Tresseras, Miquel 40270
Tretheway, Linda 48272
Trethewey, Graham 01409
Tretjakov, Nicolaj S. 37528
Tretjakova, Vera Dmitrievna 36878
Treude, Dr. Elke 17839
Treuherz, Julian 44745, 44748
Trevisani, Dr. Enrico 26108
Trevisani, Dr. Filippo 26596, 26754
Trevitt, Peter 43754
Trevor, Tom 43628
Trezzi, Daniela 28006
Triandaphyllos, Diamandis 23058
Triantafyllidis, Jutta 18223
Triantis, Alice 23091
Triapitsyn, Dr. Serguei V. 52370
Tribouillois, Brigitte 15478
Trice, Robbert 51762
Trickett, Jessica D. 54352
Trickett, Peter 43601
Tridente, Tommaso 26763
Triebel, Dr. Dagmar 20430
Triem, Katleen Heike 48905
Trier, Dr. Marcus 19631
Triesch, Dr. Carl 18952
Triffaux, J.M. 03323, 03324
Trifiletti, Dee 50199
Trifonova, Mariana 05267
Trifut, Viorica 36116
Trimbacher, Hans-Peter 02807
Trimmer, Jason 52530
Trimpe, Pam 49521
Tringali, Mario 28150
Trinidad, Miguel Angel 31191
Tripanier, Romain 06803
Tripathi, Dr. S.K.M. 24291
Trippi, Peter 51160
Trivelli, Marifrances 52826
Trividic, Claude 14882
Trnek, Dr. Helmut 03019
Trnek, Dr. Renate 02992
Troché, Bruno 13341
Tröster, Dr. G. 18702
Trofimova, Nadežda Nikolaevna 37768
Troiani, Dr. Stefano 27760, 27761, 27762, 27764
Trojak, Therese 48540
Troll, Tim 48141
Trollerud, Elin 33878
Troman, Louise 46016
Tromantine, Margaret 46630
Tromme, François 04120
Trommer, Markus 20470
Tromp, C.P. 32745
Tromp, Dr. G.H.M. 32678
Tron, Prof. Federico 26834
Tronchetti, Dr. Carlo 25579
Tronečková, Marie 09902
Tronem, Donatella 26723
Troniou, Marie-France 13914
Tronrud, Thorold 07082
Trop, Sandra 53398
Tropeano, Placido 26664
Trosper, Robert 46334
Trost, Maxine 46927
Trost, Dr. Vera 21969
Trotnow, Dr. Helmut 17031
Trott, Eileen 05463
Trott, Martin 17903
Trotter, Mark A. 48486
Trotter, Ruth 45229
Trotti-Bertoni, Prof. Anna 26078
Troubat, Catherine 12943
Trouplin, Vladimir 14675
Trout, Amy L. 51053
Trout, William 50584
Trouw, Bernward 19184
Trowill, John 50024
Trowles, Peter 44283
Trox, Dr. Eckhard 20088
Troy, Gloria 51219
Trubicyna, Tatjana 37463
Truc, Sylvie 13131
Truchina, Galina Vasiljevna 37968
Trucksis, Jane 54308
Trudel, Marianne 05337
Trudgeon, Roger 00827
True, Marion 50237
Trueb, Linda 49991
Trücher, Karl 01817
Trueman, Debbie 06409, 06413
Trümper, H.-J. 17240
Trümpler, Rico 41150
Trümpler, Dr. Stefan 41773
Trufanov, E.I. 37981
Trufini, Antonio 26806
Trulock, Sue 51922
Trulock, Trent 47225
Trulssen, Torbjorn 33919
Truman, Nell 46677
Trumble, Angus 51055
Trumble, Edward P. 46960
Trumble, Thomas E. 46960
Trummer, Manfred 03026
Trumpie, Dr. Ank 32478
Trumpler, Julie 51744
Trumps, Patricia 47761
Trumpy, Sigrid 46321
Trupp, Adam 53533
Truque, Angeles Penas 39220
Trusch, Dr. Robert 19449
Truškin, Michail Danilovič 37083
Truss, Bernard 45858
Tsafrir, Prof. Yoram 24979
Tsai, Eugenie 47069
Tsaneva, Svetla 05191
Tsang, Gerard C.C. 07798
Tsang, Kathleen 07323
Tsaoussis, Vassilis 23181
Tsatsos, Irene 50243
Tschabold, M. 41696
Tschamber, Théo 14512
Tscherter, Erwin 16055
Tschirner, Manfred 17225
Tschischwitz, Dr. Beate von 20505
Tschoeke, Dr. Jutta 20755, 20766
Tschopp, Walter 41680
Tschorsnig, Dr. Hans-Peter 21959
Tschudi, Gabriele 19082
Tse Bartholomew, Terese 52766
Tselekas, Panayotis 22946
Tselos, George 51262
Tsereteli, Zurab 37258
Tsipopoulou, Dr. M. 22872
Tsokas, Andreas 23000, 23002
Tsomondo, Tarisai 54667
Tsoukalas, P. 23220
Tsoutas, Nicholas 01670
Tsuchiya, Shuzo 29162
Tsuji, Shigebumi 29074
Tsuji, Dr. Tadashi 28348
Tsujimoto, Isamu 28359
Tsujimoto, Karen 51470
Tsujimura, Tetsuo 29491, 29492, 29493
Tsunoyama, Sakae 29199
Tsurumi, Kaori 29491
Tsutsumi, Seiji 28671
Tsuzuki, Chieko 29491
Tu, Zhigang 08346
Tua, Jean 41419
Tubaja, Roman 35193
Tucci, Franco 26367
Tuček, Helga 20040
Tuch, Gabriele 17128
Tuchel, Dr. Johannes 17113, 17115
Tucher, Bernhard Freiherr von 19416
Tucker, Anne W. 49397
Tucker, Brian 51722, 51723
Tucker, Gary 46930
Tucker, Mark S. 51867
Tucker, Michael S. 52502
Tucker, Mike 52507
Tucker, Norman P. 49787
Tucker, Prof. Priscilla K. 46310
Tucker, Renee 48713
Tucker, Sara 51166
Tucker, Sue 48544
Tucker, Susan A. 50816, 50817
Tuckett, Noreen 00887
Tucoo-Chala, Jean 14313
Tudeer, E. 11330
Tudor, Kim 45730
Tudose, Mihail 36130
Tuegel, Michele 52636
Türk, Roland 22032
Tugarina, Polina Innokentjevna 36938
Țuglui, Traian 36401, 36402, 36457
Tuhumwire, Joshua 42987
Tulaeva, Marina Aleksandrovna 37870
Tulekova, Margarita Vasiljevna 36823
Tuleškov, Vencislav 05082
Tulin, Vasil Ivanovyč 43060
Tulkku, Jorma 11243
Tulku, Doboom 24156
Tull, Earl B. 52987
Tulloch, Stuart 43529
Tully, Tim 06296
Tumanova, Anna Andreevna 37742
Tumasyan, Tatyana 43020
Tumidei, Stefano 25451
Tummon, David 43274
Tungarova, Saran-Tèrèl Sultimovna 37895
Tunheim, Arne 33514
Tunis, Dr. Angelika 17081
Tunnell, Bill 50772
Tunnicliff, Iris 45992
Tunsch, Dr. Thomas 17203
Tuomi, Marja-Liisa 10800
Tuomi, Timo 10821
Tuomi, Tuulia 10733, 10744, 10750
Tuominen, Laura 10727
Tuomola, Satu-Miia 11260
Tupan, Dr. H.R. 31937
Tupper, Christine 06290
Tupper, Elton 51705
Tupper, J. 45451
Tupper, Jon 05591
Tura, Jordi 38882
Turán, Róbert B. 23447
Turbanisch, Gérard 14310
Turcanu, Dr. Senica 36418
Turci, Dr. Mario 27725
Turdo, Mark A. 50994
Turek, Stanisław 34593
Turekanov, E.T. 29967
Turekian, Karl K. 51054
Turgeon, Christine 06691
Turicyna, T.K. 37905
Turja, Teppo 11368
Turkington, M. 06951
Turkov, Shura 24934
Turková, Alena 09855
Turkowski, Guntram 21570
Turley, Chuck 52182
Turnbow, Chris 52881
Turnbull, A. 00982
Turner, A.K. 33129
Turner, Annette 01237
Turner, Barbara L. 54277
Turner, Brenda 51757
Turner, David 48431
Turner, E. 01159
Turner, Dr. Elaine 20687
Turner, Elizabeth Hutton 53914
Turner, Grady T. 51228
Turner, J. Rigbie 51216
Turner, Kathy 49514
Turner, Dr. Kenneth R. 48972
Turner, Laura 44594
Turner, Linda 53490
Turner, Nathan 49108, 49109
Turner, Neil 45295
Turner, Peter 45459
Turner, Sam 47947
Turner, Sheridan 48939
Turner, Terry 44764
Turner, Thomas F. 46182
Turner, William H. 50798
Turnham, Stephanie 46727
Turnquist, Jan 47770
Turovski, Jevgenij 43178
Turpin, Susan 52438, 53194, 54279
Turton, Alan 43394
Turunen, Eija 11248
Turza, Witold 34782
Tuschell, Peter 03066
Tut, Dr. Ye 31736
Tutorov, Jasmina 38167
Tuttle, Barbara 48626
Tutty, Lauren 06177
Tuyttens, Deborah 07304
Tuzar, Dr. Johannes M. 01843
Tvedten, Lenny 48489
Tvers'ka, Ljudmila Volodymyrivna 43075
Twalba, Dia'a 29668
Tween, Trevor 43700, 45048
Twellman, Ron 51613
Tweneboa-Kodua, Maxwell Ohene 22856
Twiehaus, Christiane 18498, 21590
Twist, William 06131
Twitchell, Linda 06289
Tworek, Dr. Elisabeth 20459
Twyford, Mark 48256, 50887
Txabarri, Myriam Gonzalez de 40354
Tyack, Gerry V. 45175
Tyagi, Prof. K.N. 24310
Tyamzarne, G.W.T. 54677
Tybring, Tina 40756
Tych, Felix 35285
Tychenko, Volodymyr Oleksandrovyč 43093
Tychobayeva, Halina 43136
Tydesley, Carole 46014
Tykkyläinen, Mia 11405
Tyler, Dr. Alan 45321
Tyler, Fielding L. 53758
Tyler, Frances P.B. 47351
Tyler, Grant 07363, 07374
Tyler, Gregory 54226
Tyler, Ian 44563
Tyler, Jean 48980
Tyler, Karen 49996
Tyler, Kay Montgomery 47351
Tyler, Kim 00836
Tyllack, Thomas 17039
Tylleskär, Annika 40632
Tymonová, PhDr. Markéta 09692
Tyre, William 53989
Tyrefelt, Ronny 40523
Tysdal, Bobbie Jo 51308
Tyson, Rose 52753
Tzekova, Katia 05191
Tziafalias, Athanasios 22997, 23072
Tzikalov, Borislav 02136

U

Úbeda, Andrés 39573
Ubierna, César 39319
Ubl, Prof. Dr. Hansjörg 02893
Uboh, Chioba Francisca 33342
Uburğe, Baiba 30062

T

U

Uccello, Vincenza 54060
Uchida, Akihiko 29429
Uchida, Hiroyasu 29605
Uda, Makoto 28532
Udø, Ovin G. 33609
Udovičić, Ivana 04283
Uduwara, J.S.A. 40366
Udvary, Ildikó 23406, 23471
Üçbaylar, Enis 42936
Ueda, Ichiro 29213
Ueda, Koji 19609
Ueda, Shinya 29346
Ueki, Hiroshi 28480
Ueland, Hanne Beate 33717
Uelsberg, Dr. Gabriele 17462, 20413
Ueltschi, Hans-Peter 41909
Ueno, Katami 29479
Ueno, Kazuhiko 28476
Ueno, Takaaki 29581
Ueno, Tetsukazu 28376
Ürgová, Dana 08774
Uerscheln, Gabriele 18009, 18009
Uetz, Kitty 47560
Uffmann, Rüdiger 17340, 17344
Ufimcev, Valeri K. 36900
Ugarte-Peña, Alfredo 07459
Ugat, Pedro R. 34336
Ugawa, Masahiro 29202
Ugo, Prof. Gianni 00438
Ugren, Dragomir 38123
Uhen, Mark 46857
Uhlemann, Dr. Silvia 17795
Uhlenbrauck, Craig 46329
Uhlig, Rainer 17810
Uhlmann, Klaus 17091
Uhlmann, Reiner Jens 21080, 21099, 21102
Uhrig, Dr. Sandra 20535
Ujházi, György 23501
Ujiie, Seiichiro 29498
Újj, Dr. Írisz 23456
Ujváry, Dr. Péter 23477
Ukdah, Radi 42135
Ukena, H. 17836
Ul-Haq, Saleem 34022
Ulak, Dr. James 53873
Ulbert-Wild, Corinna 17656
Ulbrich, Marorie 46174
Ulbrich, Thomas 29885
Ulbricht, Dr. Ingrid 21565
Ulčar, Mirko 38400
Ulenberg, Dr. S.A. 31904
Uliančić-Vekić, Prof. Dr. Elena 08929
Ulieru, Nina 36524
Ulin, Dr. Jonna 40743
Ulinskyte, Vilija 30356
Ulitin, Valerij G. 37731
Ulkenbajuly, Nariman 29695
Ullmann, Silke 20198
Ulloa, Gina 31184
Ulloa Rojos, Sergio 07466
Ullock, Jack 06302
Ullrich, Prof. Dr. Ferdinand 21208, 21211
Ullrich, H. R. 20906
Ullrich, Karin 18529
Ullrich, Tyrena 21446
Ulm, Dagmar 02324
Ulm zu Erbach, Franz Freiherr von 18189
Ulmann, Dr. Arnulf von 20764
Ulmis, Janis 30042
Ulrich, Jean 14962
Ulrich, Matthias 18391
Ulrich-Hicks, Barbara 49455
Ulrichs, Dr. 21314
Ulrichs, Dr. Friederike 20467
Ulrichs, Dr. Ulrike 19211
Ultan, Prof. Lloyd 47050
Umberger, Eugene 49041
Umberger, Leslie 53061
Umemoto, Michio 28542
Umezu, Gen 29190
Uminowicz, Glenn 48268
Umorbekov, Baytursun Esjanovič 29683
Unbehaun, Dr. Lutz 21431, 21432, 21672
Underwood, David 49615
Underwood, Katie 50639
Undesser, Gottfried 03010
Undeutsch, Dr. Dieter 21509
Ungar, Ann 06331
Unge-Sörling, Suzanne 40913
Unger, Dr. Achim 17238
Unger, Gerhard 20131
Unger, P. Karl 02669
Ungerfeld, M. 25101
Unitt, Phil 52754
Uno, Tomohisa 29462
Unrau, Peter 06394
Unruh, Dr. Frank 22102
Unsain, José María 39255
Untch, Katharine 52343
Unterberger, Dr. Gerald 01694
Unterdörfer, Dr. Michaela 41479
Unternährer, Nathalie 41905
Unterschultz, Judy 07014
Untersteiner, Eva 03135
Untersulzner, Dr. Alexandra 25556
Unverfehrt, Dr. Gerd 18693
Unzueta Reyes, Prof. Ana Geraldina 31065
Uoti, Maunu 11193
Uotila, Dr. Pertti 10790
Updegraff, Kurt 47147
Upeniece, Daiga 30062
Uphoff, Johanna 19366
Upp, Janeanne A. 53414
Upshaw, Lisa 52675
Upton, Graham 44098
Upton, Jan 44098
Upward, Geoff 52386
Uracz, Andrzej 34971
Urakova, Ljudmila Evgenjevna 36954
Uranić, Igor 08986
Urban, Dr. Andreas 18993
Urban, Mag. Bettina 02952
Urban, Elke 19933
Urban, Erin 53265
Urban, Heinz 20253
Urban, Robin 46806
Urban, Wolfgang 21420
Urbancic, Anna María 24022
Urbanek, Erich 01959
Urbanek, Dr. Wilhelm 02954
Urbanelli, Lora 52170
Urbani, Vittorio 28204
Urbanová, V. 38391
Urbański, Andrzej 35362
Urbanski, Prof. Dr. Krysztof 34742
Urbas, Herbert 01887
Urbascheck, Dr. Stephan 20478
Urey-Weeks, Burdie 30182
Urgesi, Prof. Domenico 26666
Uribe, Gloria Zea de 08559, 08560
Uribe Ramón, Jorge 08769
Urio, Leonarda 41280
Urioste Avana, Jaime 04245
Urmas, Olli 10985
Urquhart, Margaret 45822
Urquina Llanos, Henry 08642
Urquiola, Beatriz S.A. 34440
Urrea, Ana Isabel 08787
Urrea Fernández, Jesús 40240
Urrutibehety, Clément 15439
Urs, Ilarion 36581
Ursachi, Vasile 36568
Ursch, Jacqueline 12725
Urschel, Joe 46358
Ursel, Dipl.Biol. Morgenstern 21488
Ursel, Robert d' 03571
Ursić, Desanka 07446
Ursina, Marija Viktorovna 37280
Ursoi, Ana 36165
Ursu, Mihai 31601
Urteaga, PhD Mertxe 39413
Urumov, Vanko 05236
Urumova, Rumjana 04998
Urushadze, Prof. Levan 16374
Ušakov, Prof. Sergej Aleksandrovič 37386
Ušakova, Tamara Ivanovna 37071
Usanov, Evgenij Nikolaevič 36752
Usaquén Ramos, Martha Lucía 08534
Uschalt, Susanne 20793
Uschmann, Uwe 21481
Uschner, Ralf 16778
Usik, Boris Grigorjevič 37955
Usler, Dr. Alexander 18633
Ussing Olsen, Peter 10181
Ustinov, Vladimir Jurjevič 37975
Ustvedt, Østein 33752
Ustvedt, Øytein 33735
Usundag, Nell 01008
Utech, Dr. Frederick H. 51943
Utermöhlen, Bernd 17673
Uteulina, Ajsha 29672
Uthardt, Lorenz 11095
Uthe, Monika 21818
Uthemann, Ernest W. 21460
Utili, Dr. Mariella 26895
Utriainen, Onerva 11356
Utry, Dr. Attila 23709
Utter, Dr. Brad L. 53959
Utz, Karen 46824
Uvarova, Elena Anatolevna 37880
Uzunov, Nikolai 05030

V

Vaassen, Ina 18009
Vaccaro, Dr. Luis Fernandez 00199
Vaccaro, Prof. Mario 25605
Vache, Bernd 18813, 18814
Vačiullin, Gali Fajzrachmanovič 37889
Väisänen, Raija 10889
Väkeväinen, Lea 10829
Välicki, Antti 10874
Väljas, Mait 10623
Vaelske, Urd 02625
Vaerten, Frank 03355
Värtinen, Lea 10709, 10715, 10850, 10851, 11019, 11086, 11151, 11207, 11417
Vaessen, J.A.M.F. 31930
Väterlein, Dr. Christian 16643
Vagrienė, Birutė 30437
Vahey, John 46921
Vahling, Ray 53487
Vai, Prof. G.B. 25472
Vaičeskauskienė, Aldona 30256
Vaida, Evghenia 36539
Vaida, Lucian 36486
Vaillancourt, Sandra 52871
Vaillant, L. le 33043
Vaimbois, Laurent 14355
Vainorius, Julius 30366
Vaireaux, Bernard 15075
Vairo, Carlos Pedro 00661
Vaitkute, Karile 47456
Vajda, László 23442
Vajnerman, Viktor Solomonovič 37500
Vajšnorajte, I.A. 37877
Valadés Sierra, Juan M. 39082
Valaison, Marie-Claude 14791
Valat, Georges 12623
Vălčev, Aleksandăr 05197
Valcke, Ludo 03353
Valda Martínez, Edgar Armando 04235
Valdeny da Rocha, Manoel 04918
Valderas, Elena Ruiz 39136
Valderrama de Díaz, Maritza 08695
Valdes, Nieves 34349
Valdés Figueroa, Eugenio 04682
Valdes Gonzalez Salas, Ana Luisa 31037
Valdez, Ana Luisa 31080
Valdez, E.A. 49859
Valdez, Juan Carlos 31241
Vale, Jessica 43606
Vale, Prof. Dr. Maria da Graça Pratas do 35534, 35544
Valencia, Alvaro Pío 08730
Valencia, María L. 36027
Valencia, Dr. Niels 34239
Valencia de Redondo, María Eugenia 08723
Valencia Ospina, José Aníbal 08523, 08524
Valente, Maurizio 41281
Valentien, Prof. Donata 17029
Valentin, J.C.P. 32224
Valentin, Jocelyne 15800
Valentin, Laetitia 11593
Valentin, Noel 51220
Valentine, Eunice E. 53803
Valentine, James 46758
Valentine, P. 38573
Valentini, Paolo 27988
Valentová, Dr. Jarmila 09589, 09590, 09599
Valentová, Martina 09818
Valenzuela Ríos, José Ignacio 39074
Valeria, Dr. Poletto 28184
Valerio, Michele 25368
Valette, Jean 13016
Valeza, Luisa S. 34359
Vališková, Vladimíra 09915
Valk, Marika 10619, 10624, 10632, 10634, 10636, 10638
Vall Petitpierre, Carmina 38974
Valla, T. 15910
Vallan, Dr. Denis 41595
Vallarino, Adriana L. de 34067
Vallat, Pierre 12150
Valle, Carlos 39536
Valle, Dr. Marco 25402
Valle Pérez, Dr. José Carlos 39857
Vallee, Dianne 05986
Vallejo Laso, Manuel 39885
Vallejo Triano, Antonio 39202
Vallely, Dana S. 51385
Vallery, Tullio 28210
Valles, Dave Anthony 34378
Valles-Bled, M. 13800
Vallette, Philippe 12095
Valley, Derek R. 51545
Vallianos, Dr. Christopher 23240
Valliant, John R. 52611
Vallo, Brian D. 52195
Vallvieille, Michel de 15568
Valor, E. 00642
Valorinta, Riitta 11357
Valowitz, Rhoda 53505
Valtere, Māra 30004
Valtorta, Roberta 25896
Valuska, Dr. David 49829
Valverde Canoil, Mercedes 39207
Valz, Jean-Paul 11521
Vametova, Farida H. 36906
Van Adrichem, J. 31888
Van Allen, Victor 05623
Van Alphen, Jan 03295
Van Amerongen, R.A.M. 32057
Van Arsdel, Sara 51592
Van Beek, Dr. R. 31825
Van Beekhofff, D. 31808
Van Belle, Eddy 03664
Van Belle, Jean-Louis 03397
Van Berge Henegouwen, Drs. Arno 32083
Van Boghout, Erik 03689
Van Boheemen, P. 32120, 32121
Van Boorne, Geert 03645
Van Bossche, André 03672
Van Boven, Dr. M.W. 32094
Van Boxtel, E. 32729
Van Brakel, J.H. 31852
Van Brakel, J.W. 32920
Van Broekhoven, L. 32488
Van Caloen, Veronique 03832
Van Campenhout, Nico 03827
Van Cauteren, J. 32932, 32933
Van Cleave, Richard H. 49752
Van Cleempoel, Dr. Roland 03298
Van Cleve, David 46915
Van Daele, Micke 03927
Van Daele, Dr. Rik 03827
Van Damme 03755
Van Damme, I. 03466
Van de Beuque, Guy 04706
Van de Guchte, Dr. Maarten 49577
Van de Maele, Maurice 07445
Van de Perre, M. 03258
Van de Poel, T.L. 32141
Van de Schoor, Dr. F.C.M. 32590
Van de Steeg, Dan 49003
Van de Velde, Hildegard 03313
Van de Voorde, Veronique 03895
Van de Wiele, Johan 03632
Van Delen, Wieke 38591
Van Delft, Dr. D. 32484
Van den Abeele, Jean-Pol 04036
Van den Akker, Ad 49318
Van den Akker, M. 32317
Van den Berg, S. 32672
Van den Bogaart-Vugts, Dr. H. 32612
Van den Bosch, Paula 32524
Van den Branden, Walter 04125
Van den Brink, Dr. Peter 16408
Van den Broek, J.P.G. 32047
Van den Broeke, J. 32878
Van den Bussche, Willy 03740, 03933
Van den Driessche, B. 03834
Van den Eijnde, Jerden 32915
Van den Elzen, Dr. Renate 17469
Van den Heede, Joszef 04018
Van den Hole, Edmond 03653, 03654
Van den Linden, Louis 03662
Van den Meiracker, C. 32733
Van den Steen de Jehay, Comte Guy 03743
Van den Veen, J. 32607
Van der Auwera, M. 03691
Van der Beele, P. 32415
Van der Burgh, Dr. J. 32854
Van der Coelen, P. 32729
Van der Donckt, Marie-Claire 03941
Van der Goot, Welmoed 32468
Van der Ham, Pieter 32867
Van der Heever, Chast 38722
Van Der Heijden, Merijn 49021
Van der Holst, A.P.J. 32760
Van der Horst, L. 31899
Van der Kemp, Gerald 13067
Van der Laan, Anke 32295
Van der Laar, P. 32707
Van der Lelie, H. 32483
Van der Linden, Judy 48173
Van der Lugt, Dr. P. 32338
van der Meiden, Kees 32285
Van der Meijden, Dr. Hellebora 41154
Van der Merwe, Jake 05384
Van der Merwe, T. 38597
Van der Merwe, Thijs 38591
Van der Meulen, G. 32826
Van der Meulen, G.P. 32222
Van der Mieden, O.J.E. 32908
Van der Ploeg, J. 33007
Van der Spek, Dr. T.M. 32484
Van der Veer, Hugh G. 53336
Van der Wall, Dr. Frauke 22713
Van der Werdt, E. 32844
Van der Westhuizen, V. 38714
Van der Zalm, Leo 32255
Van der Zijpp, Sue-An 32281
Van der Zwaag, M. 31855
Van Deun, Charles 04016
Van Deuren, Guy 03322
Van Deurs, Lone 09987
Van Deventer, Anriët 38710
Van Dijck, Leen 03280
Van Dijk, E.A. 33029
Van Dijk, J. 31852

Van Dongen, A.G.A. 32729
Van Dongen, P.L.F. 32488
Van Doorne, Geert 03642
Van Doorne, Veerle 03539
Van Dorst, Drs. M.H.J. 32886
van Duijn, Willem 31860
Van Duuren, D.A.P. 31852
Van Dyck, Ann 03287
Van Dyk, Pieter 01424
Van Dyk, Stephen H. 51156
Van Dyke, Dore 46463
Van Dyke, Jonathon 49182
Van Egmond, C. 32361
Van Endert, Dr. Dorothea 20331
Van Faasen, Drs. S.A.J. 32078
Van Frank, M. 49287
Van Geluwe, Johan 04131
Van Gennip, J. 32301
Van Genuchten, S. 32922
Van Gerwen, Dr. Ch. 32869
Van Gijseghem, A.M. 03640
Van Goethem, Wendy 03293
Van Gompel, F. 03876
Van Grasdorff, Maria 04015
Van Grevenstein, Alexander 32524
Van Groeningen, Drs. J.J.W.P. 32174
Van Grootheest, Joke 32083
Van Haaften, Joel 50164, 50165
van Haaren, M.A. Christiane 18151
Van Haaren, Joke 32295
Van Haarlem, Dr. W.M. 31825
Van Haecke, Bert 03536
Van Hammeren, W. 32211
Van Harinxma, Barones K.L. 32877
Van Harten-Boers, H. 32784
Van Hattem, M. 32059
Van Heerden, H. 38677
Van Heerden, Martie 31761
van Helmond, Mary 43739
Van Heugten, Sjraar 31898
Van Hoof, Werner 03319
Van Hoonacker, Albin 03782
Van Hoonacker, Marie-Henriette 03782
Van Hoorn, Dr. M.A.M. 32302
Van Horn, Walter 46288
Van Hout, I.C. 31852
Van Houten, P. 32522
Van Huis, Edwin 32389
Van Huy, Prof. Dr. Hguyen 54608
Van Impe, Dr. D. 03303
Van Kampen, Dr. Iefke Johanna 26232
Van Kerkhoven, Geertrin 03942
Van Kerkhoven, Laurence 03414
Van Klinken, M.F. 31919
Van Koetsveld, Rik 31898
Van Kooten, Drs. C. 32642
Van Krimpen, W. 32073
Van Krimpen, Wim 32071
Van Krugten, Dr. D. 19368
Van Kruistum, G. 32443
Van Laar, Barb 53120
Van Laarhoven, Dr. Jan 32345
Van Laer, Marcel 03601
Van Landschoot, G. 03261
Van Leeuwen, P.C. 31951
Van Leeuwen, P.P.H. 32978
Van Lerberghe, Anja 03495
Van Leverink, J. 32545
Van Liebergen, Dr. L.C.B.M. 32833
Van Lier, F.J. 32026
Van Lieshout, Jan 31955
Van Lingen, C. 32710
Van Loo, Rene 03691
Van Loon, V. 31903
Van Looy, Walter 03968
Van Maarseveen, Dr. M.P. 31937
Van Marion, A.J.H. 32852
Van Meerbeeck, Patrick 03764
Van Meerten, M.M. 33003, 33005
Van Melsen, Frances 38613
Van Mensch, Dr. 32388
Van Mol, Bruno 03884
Van Mourik, Justine 01327
Van Mullekom, Elisabeth 01164
Van Ness, Sally D. 48732
Van Nieuwenhuijzen, M. 31888
Van Nieuwenhyse, Ann 03638
Van Noort, Dr. S. 38592
Van Nüss, Berndt 22582
Van Oord, Aaike 32083
Van Oordt, H.W. 31809
Van Opstal, Piet 03921
Van Peteghem, Dr. Sylvia 03646
Van Pottelberghe, Michel 04133
Van Praët, Michel 14608
Van Putten, J.H.J. 40401
Van Raalte, Ronald C. 52459
Van Raay, Stefan 43830
Van Raemdonck, Mieke 03495
Van Ravensteyn, Fr. 03613
Van Regteren, R. 32595
Van Remoortel, P. 03394
Van Rhyn, Jaqueline 51874
van Rijn, Drs. Gert-Jan 32083
Van Rijnberk, Jan Willem 32841
Van Rijnswall, Dr. M. 32522
Van Rijswick, Dr. A. 32813
Van Roosmalen, Arno 32103
Van Rosmeilen 31755
Van Roy, J.P. 03454
Van Royen, Dr. P.L. 33019
Van Schaik, Prof. C. 42093
Van Schalkwyk, Frieda 38753
Van Schalkwyk, Thea 38753
Van Schie, A. 32845
Van Seagraues 49585
Van Shields 50342
Van Shields, W. 52403
Van Sinderen, Wim 32072
Van Soest, Dr. R.W.M. 31904
Van Sprang, Sabine 03462
Van Straaten, Dr. E.J. 32642
Van Stuivenberg, F. 32212
Van Suchtelen, A. 32082
Van 't Klooster, Dr. Aguita 32478
Van 't Veen, Dr. Cees 32478
Van Tichelen, Fernand 03311
Van Tiggelen, R. 03453
van Tilborg, Drs. Ben 32083
Van Tilburg, B. 32657
Van Tilburg, C. 32337
Van Tol, J. 32485
Van Tooren, M.M. 32061
Van Tuyll, Carel 14698
Van Twist, Kees 32281
Van Valkenhoef, Drs. L. 32642
Van Veen, J.W. 32286, 32287
Van Vilsteren, Dr. V.T. 31937
Van Vleet, Sue 52526
Van Vlijmen, Dr. P.M.L. 32852
Van Vollenharen, A.C. 38729
Van Volsem, Cindy 06624
Van Voorhis, Joan 46659
Van Voorst Van Beest, R. 32302
Van Wesemael, Willy 04059
Van Westerop, W.C.J. 32760
Van Wetten, Katja 21960
Van Wichelen, Peter 03362
Van Wijk, M. 32826
Van Zante, Gary 47183
Van Zijl, J.J.R. 38564
Van Zyl, S. 38712
Van Zÿverden, J. 32711
Váňa, R. 09412
Vanackere, Luc 03611
Vanadziņs, A. 29997
Vanagiene, Birute 30326
Vanausdall, John 49490
Vanautgaerden, Alexandre 03436, 03450
Vance, Alex D. 50997
Vanco, John L. 48409
Vančugov, V.P. 43163
Vandamme, Heidi 31898
Vandaule, Marianne 03509
Vande Wetering, Joe 47170
Vandecan, Suzanne 03907
Vandekerchove, Veronique 03787
Vandelannoote, Jean-André 15315
Vanden Berghe, E. 03421
Vanden Bout, Veronica 52837
Vanden Broeck, Bavo 03562
Vandenberghe, Stéphane 03414
Vandenbulcke, Anne 03439, 03451
Vandenven, Marc 03314
Vandepitte, Dr. Francisca 03460, 03463
Vandeput, Ann 03682, 03686
Vanderbeck, LeeAnne 52240
Vanderhenst, Clara 03299
Vandermaesen, M. 03426
Vanderschooten, Alain 15578
Vandersteen, Alain 03648
Vanderstraeten, Carine 04074
Vandervellen, Pascale 03477, 03495
Vanderway, Richard 46714
Vanderwerff, Laverne 46362
Vandever, William 48693
Vandevivere, I. 03834
Vándor, Dr. László 23970, 23971
Vandulek, Márta 23424
Vandura, Djuro 09011
VanDyke, Trudi 46214
Vanegas López, Walberto 08656
Vanerck, Patricia 03518
Vanhaelen, M. 03478
Vanhalewijn, Rudy 03753
Vanhemelryck, Kris 03356
Vanheusden, Margot 03308
Vanja, Prof. Dr. Christina 18870
Vanja, Prof. Dr. Konrad 17199
Vanlatum, Anne 15646
Vanloocke, Danny 03925
Vannieuwenhuyze, Sophie 54813
Vannini, Prof. Marco 26186
Vannoni, Filippo 26170
Vanns, Michael 43658, 44544
Vannson, Françoise 11568
Vannurden, James 50102
Váňová, Kristina 09845
Váňová, Dr. Libuše 09401
Vanreuerm-Rossetto, Roxane 12800
Van't Veen, Dr. Cees 32477
Vantrimpont, Jean-Denis 03773
Vaoteté, Joseph 16321, 16326
Vaoteté, Jospeh 16328
Varadinis, Mirjam 42083
Varaksina, Tamara Ivanovna 36791
Varaldo, Prof. Carlo 25176, 25178, 25497, 25498
Varani, Alberto 25321
Vardanjan, Irina Valerjevna 37506
Vardy, Anton 01488
Vare, Talis 10564
Varela, Katia 09161
Varela, Rudi 00130
Varenius, Linnéa 40614
Varga, Dr. Kálmán 23460
Varga, Tamás 23808
Varga, Vincent J. 06107
Vargas, Heriberto 08767
Vargas, Maria Cristina 08553
Vargas, Mercedes L. 34478
Vargas Cuervo, Roberto 08696
Vargas de Martínez, Evelia 08700
Vargas Guadarrama, Dr. Luis Alberto 31155
Vargha, Mihály 36599, 36601
Vargoz, Sylvine 14659
Vargyas, Herbert 02601
Varnam, Emma 43332
Vârnav, Prof. Remus 36488, 36489
Varno, Edward 47214
Varrak, Maruta 10643
Varsami, George 36309
Varsányi, Dr. András 20462
Vartiainen, Ulla 11119
Váry, Dr. László 23486
Vasconcelos, Maria João 35838, 35841
Vaseva Dimeska, Viktorija 30525
Vasher-Dean, April 51049
Vasić, Dr. Voislav 38085
Vasicek, Werner 01843
Vasile, Ioan Romulus 36405
Vasile, Prof. Radu 36283
Vasileva, Dineta 05226
Vasileva, Ljudmila 37352
Vasileva, Rumjana 05188
Vasilevič, Georgij N. 37587
Vasiliauskas, Ernestas 30231
Vasiliu, Lucian 36398, 36410, 36411, 36412, 36413, 36414, 36415, 36416, 36421, 36477
Vasiliu, Magdalena 36119
Vasilui, Lucian 36422, 36425, 36427, 36476
Vaslin, Pierre 15535
Vassallo, Dr. Adriana 27496
Vassallo de Cettour, Cristina 00315
Vasseur, Bernard 15176
Vasseur, Dominique 53236
Vasseur, Marie-Madeleine 13501
Vassiliadis, Dimitrios 23059
Vasström, Annette 10165
Vasudeva Bhatta, K. 24075
Vašut, Karel 09900
Vates-Roth, Brigitte 20763
Vatksy, Sharon 48580
Vattulainen, Päivi 11049, 11451
Vatutin, Konstantin Ivanovič 37488
Vaucher, Laurence 41649
Vaughan, Ellen 47396
Vaughan, Dr. Gerard 01288
Vaughan, James 53910
Vaughan, Robert 46491
Vaughan, Roger 43350, 43902
Vaughn, Ann L. 50869
Vaughn, Greg 47927
Vautrin, Germain 12175, 12176
Vavilova, Anna Pavlovna 37566
Vavilova, N.I. 37561
Vavilova, Natalja Veniaminovna 37320
Vavra, Jan 54139
Vázquez, Lucía 39605
Vázquez Agraz, Joaquín 30783
Vázquez Otero, Milagros 17171
Vázquez Ramírez, Nelson 30721
Vazquez Zuleta, Sixto 00387
Vazzana, Dr. Angelo 27383
Vdovin, Dr. Gennadij Viktorovič 37259
Vecarić, Nenad 08854
Vecchio, Chris 51862
Vecchio, Marji 47453
Večerková, Dr. Eva 09427
Vedaa, Lisa 50834
Vedernikova, Galina I. 37294
Vedjaeva, Natalja Anatoljevna 37534
Védrine, Laurent 15271
Védrine, Mireille 12373
Vedsted, Jakob 09994
Veen, Dr. Cees van 't 32473
Veening, Hans 51427
Veenvliet, J. 32116
Veerabhadra Rao, K. 24183
Veerman, Diederik 32083
Veerman, Hein 32902
Vees, L. 19060
Vega, Carmen L. 36071
Vegelin Van Claerbergen, Dr. Ernst 44830
Vegesack, Alexander von 22379
Vegesack, Baron Henrik von 40929
Vehma, M. 20909
Vehniä, Harri 11012, 11013
Vehns, Johannes 18974
Veisz, Ottó 23691
Veit, Johann 01699
Veit, Walter 20315
Veitch Alexander, Mary 50610
Veith, Karl-Heinrich 22249
Vejle, Jan 09988, 09989, 09990
Vekemans, B. 31999
Vela de la Torre, Antonio 39993
Velado Craña, Bernardo 38908
Velani, Dr. Livia 25261
Velaochaga Rey, Irene 34234
Velasquez, Dr. Geraldine 49906
Vélasquez, Dr. M. Manuel 35439
Velasquez Alquizaleth, Julio Cesar 04222
Velázquez, Agustín 39646
Velázquez, Teresa 40242
Velázquez Martínez del Campo, Roxana 31141
Velder, Linda 51311
Veldhuÿzen, H. A. 32272
Velers, J.B.A. 32617
Veleva, Snejana 05260
Vélez, Jesús Hernán 08669
Velez, Pedro 47517
Vélez, Pilar 38995
Velfl, Dr. Josef 09806
Velghe, Brita 03452, 03463
Velghe, Elviera 03414
Velhagen, Dr. Rudolf 41148
Velička, Gintautas 30227
Velikanova, Olga Alekseevna 37700
Velikova, Elena 05151
Velikovskaja, Galina Viktorovna 37325
Velilla Moreno, Pilar 08675
Vella, Antonino 45913
Vella, John J. 30631
Vellacott, Robert B. 01035
Vellekoop, Marije 31898
Veller, Prof. M. 38653
Velluto, Giovanna 28102
Velnic, Melchisedec 36544
Veloccia Rinaldi, Dr. Maria Luisa 26233, 27977, 27978
Veloso, Abraãg 35988
Veltheim, Mechtild von 19122
Veltmann, Dr. Claus 18891
Veltzke, Dr. Veit 20337, 22507
Venable, Charles L. 47624
Venborg Pedersen, Dr. Mikkel 10166
Venderby, Lone 10080
Vendeuvre, Elyane de 16099, 16098
Venizelos, Eleftherios 22940
Venkataraman, R. 24303
Venkatesh Jois, Dr. K.G. 24257
Venner, Peta 45341
Vento, Minna 10727
Ventosa, Silvia 39005
Ventsch, Warren 53693
Venturo Rubino, Donata 25206
Vera, Gina C. de 34347
Vera, Teresa N. de 34408
Verastegui, Maura 39605
Verbanck Piérard, Annie 03893
Verbeeck, Luc 03382
Verbrugge, Vincent 04147
Verbrugghen, Johanna M. 51709
Vercauteren, Drs. R.A.M. 32884
Verckens, Jozef 03660
Verdaasdonk, Dr. Kees 33017
Verdaasdonk, P. 32081
Verderame, Dr. Lori 46228
Verdi, Prof. Richard 43497
Verdú Alonso, María Rosa 39270
Verdugo Santos, Javier 40037
Veremeeva, Irina Lukivna 43099
Veres, Dr. László 23704
Vereščagina, Tatjana Fëdorovna 37909
Verey, A.P. 46044
Vergara, Alejandro 39573
Vergara Rodriguez, Ancízar 08529
Vergara Torres, David 07463
Verge, Laurie 47653
Verger, Sylvestre 14699
Vergho, Dr. Helge 16984, 19539
Vergne, Philippe 50743

Vergnetta, Angela 27760
Vergun, Neonila Rodionivna 43139
Verhaagh, Dr. Manfred 19449
Verhaard, A.J.G. 32618
Verheggen, Mariëtte 32128
Verheyen, Ivo 03876
Verheyen, Roland 19566
Verhoeff, R.N.S. 32305
Verhoeven, Carolina 31916
Verhoeven, P.J.C. 31928
Verhofstadt, Dr. Karl 03694
Verhulst, Sabine 04047
Verkest, Idesbalo 03262
Verlinden, Frédérique 13031
Verlingue, Bernard Jules 14968
Vermeulen, Drs. C.M.A. 31825
Vermeulen, Richard 03339
Vermeulen, S. 38566
Vermeulen, Drs. W. 32642
Vermillion, Linda Marie 54310
Vermouzek, Zdeněk 09804
Verna, Gaëtane 06157
Verner Hamilton, Elizabeth 47363
Verneuil, Yves 15997
Verney, Antoine 11873
Vernon, C.J. 38615
Vernon, G.N. 38615, 38616
Vernou, Christian 12736
Verone, Amy 51663
Verrando, Joyce 49208
Verrill, John H. 51557
Verroen, Dr. Th.L.J. 32143
Verschelden, G. 04056
Verschoor, J.C. 31822
Verschoor, M. Lynn 47059
Verschuur, P. 32180
Versluis, P. 31929
Verspeazt, V. 03438
Verster, J.L. 32700
Vertenten, A. 03575
Verthamon, Henri de 11592
Verwey, Drs. A.D.M. 31861
Verwiebe, Dr. Birgit 17032
Verževikina, Marina 37831
Verzosa, Fe Angela 34438
Verzuh, Valerie 52881
Veskovic, Esad 04284
Vesselinova, Temenuschka 25169
Vest, Ladonna 47348
Vest, Windsor 53136
Vésteinsdóttir, Auðkr{}ur 23998
Vestergaard, Ole Schou 10044
Vestergaard, Rasmus 10265
Vestermanis, Margers 30097
Vetter, Dr. Andreas W. 19282
Vetter, J.W.M. 32194
Vetter, Manfred 20673
Vetter, Maureen 50103
Vetter, Remo 41931
Veyradier, Henri 13630
Vezzosi, Alessandro 28284
Via, Marie 52396
Vial, Marie-Paule 13989, 14002
Vialla, Andre 13278
Vialla, Gérard 14964
Vialle, Christian 16218
Viana, Dr. Maria Teresa 35860
Viane, Françoise 03351
Vianin, Georges et Claire 42059
Viar, Javier 39038
Vibe, Ellen S. de 33719
Vice, Mary L. 52644
Vicini, Dr. Donata 27120, 27124, 27126
Vicini, Maria Lucrezia 27455
Vick, Daniel 49461
Vickers, Irene 44733
Vicq Carvalho, Beatriz de 04748
Victor, S. 38675
Victor, Stephanie 38676
Vida, Štefan 38322
Vidal, Juan Teodoro 39544
Vidal, Nélida 31067
Vidal, Pierre 14574
Vidal, Silvia M. 54579
Vidal de Alba, Beatriz 31136
Vidal Huguet, Carme 39481
Vidal Inglès, Anton 39293
Vidali, Dr. Gloria 28198
Vidarte, Juan Ignacio 39043
Vidović, Prof. Josip 08852
Vidulich, Tullio 28060
Viehmeier, Klaus 22568
Viehöfer, Dr. Erich 20049
Vieira, José 36025
Vienet, Bernard 13127
Viennct, Christine 11975
Vienonen, Inga 11163
Viera, Ricardo 46780
Vierck, Sigrid 22317
Vièrille, Dominique 14104
Vierk, Viola 18960
Vierlein, Ralf 20428
Vierny, Dina 14719
Vierock, Ronald 17200
Vietti, Paola 27401
Vietze, Arwed 22812
Vieux, Jacky 13072
Viéville, Dominique 14631
Viflanueva, Rudy 34367
Vigarani, Guido 26753
Vigié-Chevalier, Béatrice 14005
Vigilante, Amy 48819
Vigna Taglianti, Prof. Augusto 27502
Vignola, Dr. Chiara 26961
Vigorelli, Prof. Giancarlo 26710
Vigreux, Jean 15193
Viguier, Florence 14176
Vihovde, Anne Brit 33425
Viinikainen, Kari 11064
Vikan, Dr. Gary 46591
Vikki, Lily 33758
Vikström, Lena 40742
Viktorova, Jelena 30062
Vikulova, V.P. 37164
Vila, Dominique 12462, 12463
Vila, Emma R. 00135
Vila Álvarez, Nuria 39337
Vilà Noguera, Francesc 39614
Vilà Planas, Xavier 40291
Vilać, Pavica 08865
Vilademunt, Josep 40291
Vilain, André 03784
Vilaplana, Susana 40204
Vilas, Felipe Arias 39153
Vilchez Vilchez, Carlos 39364
Viler, Darko 38421
Viletová, Dagmar 09928, 09929
Vilhu, Raimo 10742
Viljoen, H. 38749
Viljoen, L. 38566
Vilks, Dr. Ilgonis 30067, 30068
Vilkuna, Janne 10891, 10892
Villa, Azucena 34502, 34503, 34515
Villa-Lacson, Angela de 34413, 34414
Villa Santa, Alberto 27694
Villa Santa, Emanuele F. 27694
Villa Torá, Geronimo 39181
Villadier, Francis 14103
Villagran Valdespino, Manuel Angel 31244
Villain, Agnio 15475
Villalon, Michel 12634
Villalpando, Kim 52134
Villamayor Coto, José 31099
Villamizar Duarte, Eduardo Adolfo 08705
Villamizar Lamus, Eduardo 08706
Villanueva, Gustavo M. 33280
Villarroya i Font, Joan 38937
Ville, Roy V. de 46196
Villela, Milu 04882
Villemin, Jean 30491
Villeneove, Daniel 06167
Villeon, H. de la 13702
Villiers, Jaline de 38700
Villiger, Dr. Verena 41387
Villwock, Dr. Peter 41870
Vilman, Vladimir 38467
Vilsker, Bella 47047
Vilsmeier, Cäcilie 20685
Vilus, Ljiljana 08989
Vilutienė, Aldona 30309
Vimpari, Maritta 11126
Vimpère, Jean 13329
Vincent, Ariel 15555
Vincent, Danielle 12806
Vincent, François 06195
Vincent, Glenda 53361
Vincent, Hélène 13134
Vincent, Karie C. 51013
Vincent, Patricia 51657
Vincentelli, Moira 43257, 43259
Vincenti, Dr. Patrick 49219
Vincenzi, Marco 26052
Vinduška, Jan 09724
Viner, David 43845
Vingtain, Dominique 11792
Vinje, Tor 33684
Vinke, Albert 21295
Vinogradov, Z. 24941
Vinogradova, Vira Illivna 43097
Vinsen, Bob 33155
Vinyet Estebanell, Josep 39486
Vinzens, Brigitte 42040
Vio, Ettore 28196
Virasoro, Carlos Alberto 00621
Virgili, Bianca 25673
Virgin, Louise 54302
Virieu, A. de 12955
Virole, Agnès 15289
Virtanen, Risto 11175
Vis-Best, A. 32557
Visbach, Alida 06673
Visentin, Prof. Dr. Mario 25943
Visini, Patrick 16022
Viskari, Leena 11068
Višnevskaja, Julija 37378
Vispi, Maria 26390
Visschedijk, Dr. R.A.M. 32721
Visser, A. 32086
Visser, Dr. Anna Maria 26104
Visser, H. 31888
Visser, Johann 20733
Visser, Dr. S. H. 31946
Visser, Susan R. 53159
Viswanadham, M. 24392
Vitáček, Zdeněk 09457, 09494
Vitagliano, Carol Edith 00240
Vital, Christophe 15799
Vital, M. 12969, 14166
Vitali, Christoph 41763
Vitali, Dr. des. Samuel 41210
Vitali, Tiziana 26858
Vitek, Dr. Gerhard 03028
Viti, Paolo 28206
Vitie, Heikki 10806
Vitković Zikić, Milena 38074
Vitmayer, Janet 44878
Vitt, Dr. L.J. 51376
Vitte, Bruno 14886
Vittereli, Margo 49349
Vittorelli, Dr. Pietro 25712
Vitulskis, Evaldas 30241, 30247, 30248
Vivanco Vega, Manuel 08585
Viver, F. Xavier 40149
Vivies, Bertrand de 13025, 13026, 13027
Vivirito, Jackie 47287
Vizi, Ondrej 31657
Vizzini, Salvatore 25610
Vjalikov, P.F. 37145
Vjalych, Ljudmila Makarovna 37096
Vješnic, Borko 08954
Vlack, Donald J. 51244
Vlad, Florin 36455, 36498
Vladíková, Simona 09705, 09706
Vladimirescu Băcanu, Angela 36733
Vladimirova, Galina A. 37712
Vladu, Ioana 36621
Vlaemynck, Joris 03554
Vlasáková, Beata 38365
Vlasic, Valentina 19567
Vlasova, Tetjana Mykolaïvna 43118
Vlcek, Dr. Tomas 09740
Vliegenthart, Dr. A.W. 31913
Vlieghe, Prof. Dr. Hans 03283
Vlk, Antonin 09577
Vlnas, Dr. Vít 09755, 09756
Vlnas, Dr. Vít 09795
Vlok, Rochelle 38774
Vögele, Dr. Christoph 41885
Voegele, Günter 21094, 21108
Voeks, Gretchen L. 53602
Völcker-Janssen, Dr. Wilhelm 19686
Voélin-Dubey, Monique 41691
Voelkel, David B. 48768
Völker, Susanne 17678
Völkl, Rosemarie 21215
Voelkle, William M. 51216
Vörös, Dr. Gabriella 23845, 23847, 23850
Voets, Michael 17996
Vötter-Dankl, Susanna 02431
Vogeding, Dr. Ralf 21991
Vogel, Gail 50565
Vogel, Prof. Klaus 17941
Vogel, Dr. Michael 17003
Vogel, Wilhelm Dieter 18395
Vogelaar, Dr. C.J.J. 32489
Vogelbacher, Martin 22064
Vogeler, Dr. Hildegard 20077, 20081
Vogels, J.J. 32268, 32269
Vogelsang, Harald 17399
Vogelsang, W.J. 32488
Vogelstein, John L. 51200
Vogl, Elisabeth 21985
Vogler, Franz 02956
Vogt, Barbara 30198
Vogt, Dr. Christine 20808
Vogt, Daniel 41178
Vogt, George L. 54193
Vogt, Gisela 18093
Vogt, Dr. Renate 17468
Vogt, Sharon L. 50837
Vogt, Dr. Simone 19000
Vogt, Wolfgang 20330
Vogüé, de 12593
Vogüé, Comte B. de 14505
Vogüé, Patrice de 13921
Voicu, Mariana 36530
Voigt, Dr. Kirsten 19448
Voigt, Sigrun 19238
Voigt, Stefan 17943
Voigt, Dr. Wolfgang 18361
Voigt Andersen, Mogens 23256
Voiron, Henri 11532
Voisin, Bernard 12500
Voisin-Thiberge, Claire 14370
Voithofer, Waltraud 01793
Vojtenko, Vladimir Georgievič 37470
Vokotopoulou, Dr. Julia 22985
Volker, Joye 00946
Volkert, James 51236
Volkmann, Jürgen 18776
Volkmann, Swantje 22160
Volkov, Sergej Pavlinovič 37465
Volkova, Valentina Borisovna 37001
Volkwein, Peter 19347
Vollbrecht, Dr. Jürgen 16940
Vollert, Adalbert 18374
Vollgraaff, Helene 38591
Vollmer, David 53483
Vollmer, Jennifer 48272
Vollmer, Stephen 53598
Vollnhals, Rudolf 22831
Volmar, Michael 49207
Voločkova, Olga Kuzminična 37575
Volosenkova, Nadežda Evstafievna 37773, 37777, 37778
Volosova, Galina Pavlovna 37734
Vološyna, Ljubov Vasylivna 43121
Volovenskaja, Ljudmila Vladimirovna 36948
Volpe, Jorge 00163
Volskij, Svjatoslav O. 37484
Volstorf, Bryan 53504
Volz, Günther 16687
Volz, Robert L. 54174
Vomm, Dr. Wolfgang 17014
Vomsattel, Julian 41989
von Kerssenbrock-Krosigk, Dr. Dedo 18008
von Pape, Dr. Cora 16890
Vondras, Barbara 50065
Vonhof, M.J. 32642
Vonier, Mag. (FH) Johannes 01828
Vonk-Hennig, Davien 31790
Vonscheidt, Dr. Wolfgang 17623
Vonville, Auguste 15377
Voogt, Leo 32087
Voogt, P. 32485
Vookles, Laura L. 54331
Voordendaf, A.G. 32809
Voorheis, Lynn 52374
von Voorst tot Voorst 32397
Vorachek, Pamela 52664
Vorano, Tullio 08906
Vorobjova, Solomija Oleksandrivna 43135
Voropanov, Vladimir Valentinovič 37963
Vorres, George 23129
Vorres, Ian 23129
Vorsteher, Dr. Dieter 17072
Vort-Ronald, Michael P. 01181
Vorwig, Dr. Carsten 20243
Vos, H.N. 38738
Vos, K. 32488
Vos, Margaret 52537
Vosa, Dace 30071
Voss, Dr. Björn 20513
Voss, Kurt 48851
Voß, Dr. Rolf 20592
Voss, William J. 48713
Vosseler, Peter 21979
Vossenkuhl, M.A. Katharina 20478
Vostrikov, Vladimir Ivanovič 37592
Votaw, John F. 54120
Votroubková, Dr. Iva 09473
Vouk, Dr. Morana 09013
Voute, Dr. A.M. 32857
Vovk Čepič, Taja 38453
Voyadzis, G. 22939
Voyles, Robert S. 48617
Voza, Dr. Giuseppe 05434, 27854
Voznickyj, Boris Grygorovyč 43125
Voznjuk, Dr. Volodymyr Oksentijovyč 43012
Vrabec, Vesna 09004
Vračević, Marica 38100
Vraesos, Alexis 11896
Vrat, Prof. Prem 24374
Vratislavský, Ivo 09852
Vrba, Elisabeth 51054
Vrbanus, Milan 08912
Vrčic, Vjeko 08883
Vrebos, M. 03439, 03451
Vrellis, Paul P. 23016
Vribe, Francesc 38976
Vriend, Anita 31898
Vries, D.P. de 32479
Vries, J. de 32139
Vries, Janekke de 17552
Vries, Dr. M.E.A. de 31796
Vries, P.A. de 31906
Vrieze, John 31821
Vroede, Erik de 03708
Vrousos, Alexis 13934, 15317
Vrtiska, Floyd 53404
Vrugteman, Gerda 32281
Vuadens, G. 41976
Vuille, Roger 41788

Vuilloud, R. 41635
Vujčić, Davorin 08888
Vujičić, Jasminka 08989
Vujnović, Andrej 38050
Vuk, Vili 38477
Vukadin, Hela 09004
Vukotić, Biljana 38074
Vuković, Prof. Radovan 09014
Vukovich, Franz 02361
Vulf, Galina Gerasimovna 37080, 37081, 37082
Vuori, Elina 11045, 11050
Vuorinen, Pirjo 10887, 10892
Vusoniwailala, Kate 10704, 33117
Vutela, Dr. Anneli 10775
Vuyk, Dr. K. 31894
Vydra, Dr. František 09482
Vydrina, Ljudmila Sergeevna 37689
Vykoukal, Dr. Jiří 09474
Vykydal, Vlastimil 09440
Vyntu-Terycanu, Dr. Olena Dmytrivna 43015
Vyskocil, Hubert 03066
Vysockij, Nikita Vladimirovič 37224

W

Waage, Einy 33705
Waal, M. de 38566
Wachna, Pamela 07122
Wachs, Diane C. 50048
Wachsmund, Ulrike 18160
Wachten, Dr. Johannes 18379
Wachter, Fritz 41754
Wachter, Dr. Markus 02494, 02495
Wachtmeister, Ingegerd 40782
Wackerfuß, Winfried 18322
Wacs, Darin 51689
Wada, Kunibo 29319
Wada, Norio 28904
Wadas, John W. 52738, 52749, 52755
Wade, Blake 49236, 49333
Wade, Don 53371
Wade, Hilary 43759, 43760
Wade Thomas, Randall 48784
Wadensten, Hillevi 41000
Wadsworth, David H. 47675, 47676
Wadum, Jorgen 10152
Wäckerlin, Martin 41683
Wägele, Prof. Dr. Wolfgang 17469
Wäldchen, Günter 21810
Wälder, Manfred 19982
Wærdahl, Tove 33575
Wäre, Dr. Ritva 10820
Waern, Karl Erik 40415
Wäspe, Roland 41805
Wagemakers, C. 31839
Wagemakers, T. 32829
Wageman, Patty 32281
Wagener, Danièle 30482, 30486, 30487
Waggoner, Lynda 50690
Wagini, Dr. Susanne 20487
Wagner, Angelika 19366
Wagner, Connie 52484
Wagner, Dieter 17981
Wagner, Ethel Mae 49610
Wagner, Gerhard 20212
Wagner, Gerhard A. 17513
Wagner, Günther 21043
Wagner, Prof. Herbert 01903
Wagner, Hermann 01883
Wagner, Hilke 17523
Wagner, Dr. Jens-Christian 20738
Wagner, Lynn 49726
Wagner, Prof. Marek 34861
Wagner, Patrice 46751
Wagner, Renate 19204
Wagner, René 21160
Wagner, Rita 19612
Wagner, Ruth 53063
Wagner, Dr. Siegfried 20555, 20556
Wagner, Thomas K. 48277
Wagner, Dr. Velten 18172
Wagner, Vivian 47566
Wagner-Autesserre, Pierre 13531
Wagner-Gnan, Angela 20795
Wagrodzki, Jerzy 35282
Wagstaff, Sheena 44995, 44996
Wagstaffe, Stephen 44182
Wahdan, Tracey 45616
Wahl, Hans-Rudolf 42055
Wahl, Johann 02893
Wahl, Sonja 47601
Wahl-Pletscher, Emil 41262
Wahler, Marc-Olivier 14743
Wahlers, Jack 46764
Wåhlin, Sidsel 10080
Wahlöö, Claes 40703
Wahlquist, Håkan 40892
Wahlroos, Tuija 10728
Waiboer, Adriaan 24807
Waidenschlager, Christine 17169
Wain, Lisa 01014
Waintraub, Aron 24943
Wainwright, Paige 46824
Waisbrot, Ann 50513
Waite, Dawn D. 47284
Waites, Robin 47727
Wajima, Kunio 28591
Wakae, Haruko 29621
Wakamatsu, Motoi 29522
Wake, William 44066
Wakeling, Melissa 05403
Wakili, Ousmanou-Mouhamadou 07422
Wakita, Osamu 29147
Wakitani, Masami 29338
Wakiyama, Yoshitaka 28478
Walaszczyk, Anna 35370
Walaszczyk, Antoni 34692
Walburg, Dr. Reinhold 18372
Walch, Christof 02342
Walch, Heribert 01865
Walch, Timoth 54047
Wald, Albrecht 18325
Wald, Jane 46265
Waldbott-Bassenheim, Baronin Marietheres 02041
Walde, Prof. Dr. Elisabeth 01823, 02271
Walden, Barbara B. 49802
Walden, Ian N. 44029
Walder, Emma 46079
Walder, Hans Kaspar 42012
Waldherr, Johann 02491
Waldherr, Kaspar 19955
Waldkirch, Bernhard von 42083
Waldman, Glenys A. 51851
Waldmann, Burta 41943
Waldmann-Glaser, Dr. Erich 02791
Waldmeier, Kurt 41335
Waldner, Daniel 41532
Waldren, Dr. William 39241
Waldron, Richard 51813
Waldvogel, Florian 18942
Walentowicz, T. 43803
Walgamott, Libby 52277
Walgenbach, H. 32707
Walgrave, René 03789
Waligurski, Carolyn M. 49447
Waligurski, Stephen S. 49447
Walinga, Ingeborg 32279
Walk, Stephen G.W. 51328
Walkensteiner, Wolfgang 02185
Walker, Alexandra 45444
Walker, Anita 48102
Walker, Barry 49397
Walker, Carol 45262
Walker, Catherine 50746
Walker, Celia 50963
Walker, Christopher 43673
Walker, Daniel 51212
Walker, David A. 44308
Walker, Dean 51867
Walker, Diane 45199
Walker, Edwin G. 48014
Walker, Gavin 45525
Walker, Gerald 52479
Walker, Gwendolyn 49980
Walker, Helen 44955
Walker, Jeff 50210
Walker, John 46103
Walker, June 49784
Walker, Kate 01612
Walker, Lulen 53874
Walker, Neil 45290
Walker, Nicola 45099
Walker, Pam 54265
Walker, Peter 43534
Walker, Rebecca 51694
Walker, Rob 05379
Walker, Sam 44797
Walker, Sonia 46532
Walker, Stephen 45689
Walker, Stephen D. 54114
Walker, Steward 46923
Walker, Tim 33124
Walker, Tom 53040
Walker, William 50419
Walker-Coultice, Lauren 07288
Walker Mecky, Debra 49082
Walker-Smith, G. 01159
Wall, Dora 06218
Wall, Gordon 48278
Wall, Ian 45954
Wall, Kathleen 05917, 05918
Wall, Dr. Michael 52754
Wallace, A. 07316
Wallace, Amanda 45088
Wallace, Anaret 38565
Wallace, Brian 51097, 51848
Wallace, George 51425
Wallace, Hilt 07282
Wallace, I. 44750
Wallace, J. Don 50671
Wallace, James A. 50582
Wallace, Lynda 33033
Wallace, Meghan 46273
Wallace, Mike 50722, 51780
Wallace, Dr. Patrick F. 24756, 24808, 24809, 24810
Wallace, Teresa 47768
Wallace, William D. 54301, 54303
Wallach, Nancy 49709
Wallbrecht, Dr. Andreas 17591
Wallcraft, Wilma K. 06114
Waller, Bret 49489
Waller, Josip 08912
Waller, Richard 52329
Waller, Susan R. 52680
Wallers-Bulot, Danielle 12538
Wallet, Marie 12269, 12270
Walley, G.P. 45295
Walling, Mary 51200
Wallis, Andrew 44865
Wallis, Brian 51194
Wallis, Jean 49287
Wallis, Jonathan 43965
Wallis, Simon 44819
Wallmann, Dr. Margarete 02175
Wallner, Dr. Thomas 41884
Wallner Bărbulescu, Luminiţa 36448
Wallner-Bărbulescu, Luminiţa 36450
Wallon, Jean-Marie 03847
Walloschek, Arnold 21508
Walls, Nancy 49660
Walmsley, Frances 06408
Walrath, Gary 51927
Walseman, Laurie 47885
Walser, Stefan 41966
Walseth, J. 10156
Walsh, Alice 52167
Walsh, Brian 24827
Walsh, Debbie 24765
Walsh, John 01537
Walsh, Larry 24867
Walsh, Mary 50862
Walsh, Nigel 44660
Walsh, Noa Martin 24760
Walsh-Piper, Kathleen 50057
Walt, Barbara 41823
Waltenspül, Walter 41107
Walter, A. 18104
Walter, Alfred 17873
Walter, Bakk. phil. Barbara 02625
Walter, Dr. D. 22410
Walter, Ernst-Ulrich 18668
Walter, Friederike C. 19453
Walter, H. 21859
Walter, Katharina 02136
Walter, Michael 21783
Walter, Patricia 46700
Walter, Philippe 14265
Walter, Prof. Dr. Ulrich 18310
Walter, Dipl.Biol. Wimmer 21488
Walter-Brown, Nancy F. 50745
Waltereit, Manfred 18135
Waltermire, Joan 51442
Walters, Andra 54372
Walters, Cheryl 53499
Walters, Cliff 43541
Walters, Janet 07151
Walters, John A. 53277
Walters, Margaret 54159
Walters, Nerina 38680
Walterskirchen, Prof. Dr. Gerhard 02616
Walther, Andreas 17699
Walther, Gerd 18499
Walther, Hubert 16301
Walther, Jim 46183
Walther, Romana 41901
Walton, Ann 48694
Walton, John 01059
Walton, Julia 43489
Walton, Karin 43632, 43635, 43638
Walton, Les 45449
Waltoś, Prof. Dr. Stanisław 34813
Waltz, James 47236
Walujo, Dr. Ekoparoto 24456
Walz, Dr. Alfred 17522
Walz, Carol 50879
Walz, Glenn 07003
Walz, Jack 50879
Walzer, Herwig 02411
Wamelink, Garrit 47630
Wamers, Prof. Dr. Egon 18354
Wamser, Prof. Dr. Ludwig 18818, 20422, 20586
Wamulungwe, Patrick Lisina 54656
Wan, Jamaluddin 30578
Wan, Norhiyati Ibrahim 30579
Wan, Robert 16324
Wan, Yusoff 30578
Wan Ahmad, Wan Ali Bin 30564
Wanachiwanawin, Prof. Darawan 42555
Wanamaker, Mark 50233
Wand-Seyer, Dr. Gabriele 19153, 19156
Wander, Evelyne 15225
Wandl, Monika 02956
Wandrei, Ellen A. 46691
Wandschneider, Dr. Andrea 20966, 20972, 20975, 20976
Wang, Bingsen 07860
Wang, Bingwan 08312
Wang, Bohua 08114
Wang, Chang'an 07778
Wang, Chen 07649
Wang, Chongshun 07822
Wang, Conghui 07820
Wang, Cunxiao 07900
Wang, Desong 08322
Wang, Dezhi 07927
Wang, Fengqing 08267
Wang, Fengxian 07836
Wang, Guojie 08399
Wang, Guoquan 08301
Wang, Hebing 08386
Wang, Huaqing 08120
Wang, Jiaheng 08068
Wang, Jianhua 07857, 07918
Wang, Jinbao 08179
Wang, Jinhou 08216
Wang, Jinxiang 08243
Wang, Juguo 08502
Wang, Kuiyi 07586
Wang, Kwei-lan 42443
Wang, Lanzhou 08096, 08103
Wang, Lianlan 08266
Wang, Prof. Lianmao 08130
Wang, Libin 07833, 08102
Wang, Linchang 07958
Wang, Luheng 08228
Wang, Pengfei 07650
Wang, Qi 07770
Wang, Qihui 07571
Wang, Qingshi 07651
Wang, Qinling 07594
Wang, Ragnhild 33754
Wang, Runyuan 08261
Wang, Shaoming 07819
Wang, Shenzhang 07691
Wang, Shijie 08166
Wang, Shouyin 08066
Wang, Shujin 08378
Wang, Taiming 08476
Wang, Wei-Hsien 42159
Wang, Weiyin 08276
Wang, Wenjing 08487
Wang, Wenwu 08435
Wang, Xi Rong 08173
Wang, Xiangyu 08280
Wang, Xianyou 07855
Wang, Xiaoping 07780
Wang, Xinchang 08065
Wang, Xiping 08433
Wang, Xiu 07985
Wang, Yanliang 08282
Wang, Yanqing 08477
Wang, Yonggen 07844
Wang, Yufu 07940
Wang, Yuxian 08415
Wang, Yuzhen 08375
Wang, Zheng 07976
Wang, Zhenghua 08238
Wang, Zhijie 08394
Wang, Zhiping 07515
Wang, Zhiqiong 08439
Wang, Zhiwen 07755
Wang, Zongtao 07934
Wånge, Cecilia 40586
Waning, Alexander 32496
Wankel, James 05769
Wannagat, Prof. Dr. Detlev 21386
Wanner, Prof. Dr. E. 19739
Wanner, Monika 41495
Wanriek, Benedicta 20978
Wappenschmidt, Dr. Kathrin 21367
Wappis, Erich 02186
Wappler, Dr. Friederike 17402, 17407
Warburton, Stuart 45923
Warchałowski, Walerian 34892
Ward, Alberta 46452
Ward, Barry J. 49854
Ward, Barry K. 48203
Ward, Danette 51373
Ward, Esmé 45099
Ward, James 51903
Ward, Kathy 46645
Ward, Ken 45952
Ward, Kimberley D. 48203
Ward, Dr. Lauck W. 50521
Ward, Mark 46545
Ward, Mike 43277
Ward, Nancy 49642
Ward, Philip S. 47978
Ward, Ralph E. 49190
Ward, Robert 50788
Ward, S. 43431

Ward, Scott 51718
Ward Grubb, Barbara 51818
Ward Povse, Sandra 52982
Ward Shannon, Dr. George 46666
Warda, Rebecca M. 47432
Warde, John St. Andrew 45979
Warden, Canon 43758
Warden, Donald 53099
Warden, P. 42508
Wardius, Janine 33418, 33419, 33420
Wardlaw, Alivia 49397
Wardle, Michelle 49738
Wardropper, Ian 51212
Ware, Alan 49747
Ware, Brechelle 49234
Ware, Donna M. 48288
Ware, Dr. John A. 48181
Ware, Katherine 51867
Ware, Mark 53148
Ware, Robert 46180
Warfel, Stephen G. 49181
Warger, Julia M. 53316
Wargo, Richard 46897
Warhurst, Margaret 44750
Warigai, Tsutomu 28543
Waring, H.B.H. 45074
Warinsie Kankpeyeng, Benjamin 22849
Warkiewicz Laine, Christian K. 47463
Warlitner, Crystal C. 51336
Warming, Dagmar 10177
Warne, Pam 38593
Warnecke, Jan-Christian 21950
Warner, Henry 44380
Warner, Jan 50134
Warner, Joanne 51156
Warner, Judith 52964
Warner, Malcolm 48715
Warner, R. 43448
Warnes, Dr. Carl E. 50903
Warnes, Robert 44361
Warnier, Valérie 03524
Warnke, Dr. Ursula 17569
Warp, Harold G. 50723
Warren, David B. 49397
Warren, Eleanor 47585
Warren, Elizabeth 51171
Warren, Emma 43814
Warren, George W. 50073
Warren, Harry 54133
Warren, Jack D. 53918
Warren, Sally 06595
Warstler, Pasqua 48416
Wartenberg, Dr. Ute 51116
Warth, Siegfried 20381
Warth, Werner 42017
Warther, David R. 48168
Wartke, Dr. Ralf-B. 17279
Wartna, Fred 32732
Warton, Sue 45674
Warwick Thompson, Paul 51156
Warzecha, Newton M. 48968
Waschka, Adelheid 16843
Wasescha, Elisabeth 41824
Washbrook, Harry 05735
Washington jr, W. Curtis 47377
Washington Goyaud, Rolando 00391
Waskowic, Lorraine 06655
Wasner-Peter, Mag. Isabella 03087
Wass, Glenys 45368
Wassermann, Fred 51200
Wassing, A. 32733
Wassmann, Harry 46734
Wassmer, Edouard 41390
Wastian, Mag. Josef 02691
Wastiau, Boris 41421
Wat, Kathryn 46146
Watanabe, Hisashi 28960
Watanabe, Junko 28532
Watanabe, Kiriko 29522
Watanabe, Kousho 29205
Watanabe, Osamu 29349
Watanabe, Seiichiro 28356
Watanabe, Shinsuke 29419
Watanabe, Taeko 28914
Watanabe, Takao 28860
Watanabe, Yasuhiro 29393
Watari, Shizuko 29504
Watelet, Pierre 14366
Waterlow Oam, Nick 01400
Waterman, Amy E. 51169
Waters, Dr. D. 45336
Waters, Davis 48215
Waters, Deborah D. 51231
Waters, Fergus 45253
Waters, Ruth D. 47589
Waters Yarsinske, Amy 51369
Watkin, Jeffrey 44654
Watkins, Jonathan 43505
Watkins, Kathleen 45576
Watkins, Rod 50709
Watkins, Sarah 46954
Watne, Eva 33963
Watson, Andy 47661
Watson, Bob 44330
Watson, Claudia 47986
Watson, Daryl 48827
Watson, Doreen 43437
Watson, Edwin W. 48765
Watson, Jamie 46082
Watson, Jim 06019
Watson, John 01229, 01540
Watson, Julian 44864
Watson, Julie 45229
Watson, Kristy 46888
Watson, Mark 43943
Watson, Matt 50552
Watson, Maureen 46009
Watson, Roger 44627
Watson, Ronald 48720
Watson, Sara C. 49084
Watson, Scott 07216
Watson, Sheena 45458
Watson, Dr. Sheila 44357
Watson, Wendy 53171
Watson, Wendy M. 53170
Watson, William S. 44158
Watson-Armstrong, Francis 43368
Watt, David 44120
Watt, James C.Y. 51212
Watt, Keith 44261
Watt, Marie 52081
Watt, Robert 05332
Watt, T. 44062, 44692, 44693
Wattenberg García, Eloisa 40235
Watters, Frank X. 51955
Watters, Jane 01558
Watterson, Norma 48273
Watteyne, Damien 03995, 03997
Watton, Graham C. 33171
Watton, Janet 52245
Watts, Kelly 49325
Watts, L. 01443
Watts, Michael 47354
Watts, Pete 54327
Watts, Peter 01065, 01383, 01468, 01472, 01547, 01548, 01555, 01560, 01616, 01622
Watts, Steve 44697
Watts McKinney, Sylvia 51865
Waugh, Carmen 07457
Wautier, Jacques 03736
Wawrzyk, Henryk 34954, 34955
Wawrzyniak, Andrzej 35237
Way, Catherine A. 49595
Way, Paul 44107
Way-Jones, M.F. 38630
Wayne, Cynthia 46544
Wayne, Kenneth 47114
Wayne, Ph.D. Kenneth 49430
Wayne, Nan 49372
Wcześny, Marek 34957, 34958
Weadick, Nathalie 24848
Weaner, Arthur 48900
Wear, Lisa 52680
Weatherby, Ken 06301
Weathers, Allen 50598
Weaver, Henry 05995
Weaver, Janet 49773
Weaver, Leslie J. 51902
Weaver, Nigel 45539
Weaver, Pat 52647
Weaver, R. 44530
Weaver, Susan 49251
Weaver, Suzanne 47931
Weaver, Vey O. 46121
Webb, Belinda 01564
Webb, Eddie 52112
Webb, Heather 06590
Webb, Judy 52432
Webb, Keith 45935
Webb, Lisa 43979
Webb, Vaughan 48538
Webber, Barry 51360
Webber, Chris 45028
Webber, Eddie 00879
Webber, Maurie 45312
Webel, Sophie 14600, 14776
Weber, Brigitta 22816
Weber, C.S. 19735
Weber, C. Sylvia 19736, 21657
Weber, Dr. Erich 22128, 41882
Weber, Dr. Gerhard 19514
Weber, Dr. Hans 41113
Weber, Hans Ruedi 42083
Weber, Heidi 42075
Weber, Heiko 20346
Weber, Jean 30498
Weber, Jens 18955
Weber, Johanne 21974
Weber, John 52795
Weber, Karl 16753, 16998, 22386
Weber, Dr. Klaus 17041
Weber, Dr. Marion 08684
Weber, Markus 22652
Weber, Michael 31757
Weber, Otto 20798
Weber, Dr. Peter 36469
Weber, Robin 06671
Weber, Dr. Stefan 17203
Weber, Stephan W. 01700
Weber, Timothy 53135
Weber, Ute 20858
Weber, Dr. W.I.M. 32691
Weber, Drs. W.I.M. 32900
Weber, Yvette 13949
Weber-Lehmann, Dr. Cornelia 17402
Weber Soros, Dr. Susan 51132
Weber-Stöber, Dr. Christianne 18975
Weberg, William 52155
Webersberger, Josef 02366
Webley, Dr. L. 38629
Webley, Dr. W. 38630
Webster, Barbara 47149
Webster, Harvey 47625
Webster, L. 44803
Webster, R.M.O. 46033
Webster, Robert 53522
Webster, Roger J. 43547
Webster, Susan 49062
Webster, Prof. William 01549
Wechner, Wilma 02136
Wechs, Thomas 16748
Weckbacher, Vernon 50339
Weckström, Gunnar 11445
Weddegjerde, Per 33381
Weddle, Sam 50197
Wedlake, Roger 45948
Wedler, Jacque 46667
Weeda, K.A. 32208
Weedmark, Lillian 06983
Weekley, Carolyn 54164
Weekley, Carolyn J. 54165, 54166
Weekly, Nancy 47119
Weeks, Jason 47175
Weems, Marcia 53623
Weenink, G.J. 32502
Weerdt, G.A. de 32818
Weese, Josh De 49239
Wefers, Dr. Sabine 19393
Wegener-Welte, Kathrin 16679
Wegge, Aslak 33971
Wegmann, Dr. Peter 42031, 42035
Wegner, Gene 34360
Wegner, Hartmut 17373
Wegner, Manfred 20462
Wehdorn, Armine 02991
Wehgartner, Dr. Irma 22714
Wehling, Arno 04738
Wehmeyer, Egon 16677
Wehmeyer, Pete 46509
Wehrberger, Gerhard 02371
Wehrberger, Kurt 22170
Wehrle, Karl 21244
Wehrli, Martin 41251
Wei, Biao 07584
Wei, Haibo 07579
Wei, Hsing-Ru 42360
Wei, Kangquan 07913
Wei, Mengling 07756
Wei, Zhongying 07947
Weibel, Dr. h.c. Peter 01982
Weibel, Univ. Prof. Dr. h.c. Peter 02842
Weibel, Prof. Peter 19453
Weibel, Dr. Urs 41831
Weibezahn, Dr. Ingrid 17548
Weibull, Ph.D. Dr. Nina 40943
Weichselbaum, Dr. Hans 02614
Weichselbaum, Karl 02456
Weick, Wolfgang E. 17918, 17925
Weidemann, Dr. Friedegund 17123
Weidenschläger, Christine 17168
Weidinger, Mag. Dr. Johannes T. 01951
Weidisch, Peter 16759
Weidl, Beverly 49362
Weidlich, Ariane 18809
Weidman, Mary C. 51439
Weidner, Dr. Thomas 20462
Weidner, Timothy 48934
Weidner, William 52001
Weiffenbach, Jeannie 52470
Weigel, Jay 51078
Weigel, Dr. Viola 22578
Weigl, Stephan 02310
Weigle, Petra 20777
Weigt, Karin 17831
Weil, Benjamin E. 52795
Weil, Gilbert 12135
Weil, Dr. Mark S. 52586
Weil-Seigeot, Claude 12537
Weiland, Linde 18510
Weileder, Rainer 20723
Weilepp, Bruce 53162
Weiler, Gerhard 22209, 22710, 22720
Weiler-Streichsbier, Dr. Doris 20897
Weimert, Dr. Helmut 19093
Weinberg, Adam D. 51279
Weinberg, H. Barbara 51212
Weinelt, Hannes 01996
Weinelt, Wendeline 18504
Weiner, Lutz 19933
Weinert, Renate 16731
Weingarten, Ulfried 21699
Weingartner, Heinrich 03036
Weinhäusel, Mag. Bernhard 01858
Weinhart, Dr. Martina 18391
Weinheber-Janota, Brigitte 02173
Weinig, Prof. Dr. 21931
Weinke, Jane 53993
Weinkein, John 48062
Weinland, Dr. Martina 17209
Weinmayr, M.A. Stefanje 19805
Weinshrott, William J. 48583
Weinstein, Michael 51169
Weintraub, Aviva 51200
Weinzapfel, Connie 51048
Weipert, Eve 49275
Weir, Kathryn 01512
Weir, Margaret 45148
Weir, Margaret S. 53346
Weise, Roland 19581
Weisenberg, Dr. Angelika 17032
Weiserová, Dr. Jana 09538, 09766
Weishaupt-Theopold, Kathrin 22164
Weiskircher, Arrand 15792
Weisman, Billie Milam 50224
Weisman, Bruce 49771
Weiss, Al 49875
Weiß, Bernd 21021
Weiß, Berthold 01917
Weiss, Brigitte 17692
Weiss, Christian 19541
Weiß, E. 21563
Weiß, Ferdinand 01875
Weiß, Frank 21055
Weiss, Dr. Gabriele 03038
Weiss, Ján 38238
Weiss, Dr. Judith 20058
Weiss, Lois 52955
Weiss, Martin 48578
Weiß, Peter 02225, 02226, 02227
Weiss, Prof. Dr. Rainer-Maria 18936
Weiss, Sabine 18009, 22354
Weiss, Dr. Thomas 17832, 17833, 20907, 22638, 22639
Weiss, Wolfgang 02953
Weissend, Patrick 46626
Weissengruber, Dr. Thekla 02324
Weißenseel, Günter 21862
Weisser, Dr. Bernhard 17051
Weitkamp, Janna 19550
Weitschat, Dr. W. 18930
Weitzendorfer, Dr. Alfred 02284
Weium, Frode 33744
Welch, Cheryl 53729
Welch, Richard 48517
Welfele, Odile 14728
Welin, Kjell-Arne 40587
Wellens, Luc 03321
Weller, Bernhard 16863, 16864
Weller, Dennis P. 52232
Weller, Susan J. 50735
Wellicome, Lavinia 46054
Wellington, Amy 49556
Wellington, Judith L. 47378
Wellman, Lesley 49162
Wellmann, Dr. Marc 17117
Wellmann, Nóra 23433
Wells, Carol A. 50216
Wells, Edna 48105
Wells, Gerald 45005
Wells, Keith 52199
Wells, Kenneth D. 46979
Wells, Nicole 51243
Wells, Penny 05891, 05892
Wells, Stephen D. 48647
Wells, Trevor C. 30194
Wells-Cole, Anthony 44666
Wells Fitzgerald, Barbra 49428, 49429, 49431, 49432
Wells Zaia, Kay 51440
Wellspeak, Charles 47949
Wellspeake, Raymond 54082
Wellwood, Bridget 33180
Welsh, Caroline 46876
Welsh, Dr. Peter 48927
Welsh, Tracy 44355
Welter, Barbara 41145
Welter, Matt 49041
Welti, Georges 41367
Weltmaier, Marion 20435

Welty, Claude 14326
Welu, James A. 54302
Welzig, Ingrid 20748
Wemhoff, Prof. Dr. Matthias 17206, 20973
Wemmlinger, Raymond 51190
Wen, Yanyuan 07689
Wendel, Birgit 19437
Wendel, Candice 49510
Wendel, Michael G. 46833
Wendel, Siegfried 21436, 21437
Wendl, Dr. Tobias 16953
Wendling, Britt 40799
Wendling, Pia 13176, 13177
Wendling, Urs 41565
Wendorf, Richard 46919
Wendowski-Schünemann, Andreas 17757
Wendt, M.A. Liane 19082
Wener, M.A. Markus 21950
Wenes, Bram 03867
Wenger, Rhona 05913
Wenham, Julia 46020
Weniger, Prof. Dr. Gerd-C. 20314
Weniger, Dr. Matthias 20426
Weninger, Gehard 02491
Wenn, Norbert 17611
Wentworth, Peggy 46597
Wenzel, Dr. Astrid 19437
Wenzel, Cornelia 17416
Wenzel, Gundula 22447
Wenzel, Kai 18679
Weon-Kettenhofen, Hae-Yon 17469
Wepfer, Dr. Hans-Ulrich 41516
Wepner, Dr. Friedrich 02326
Weppelmann, Dr. Stefan 17116
Werd, Drs. Guido de 19567, 19568
Were-Were, Liking 08831
Werf, Hans van der 15835
Werkmäster, Daniel 40775, 40776
Werkmäster, Lisskulla 40515
Werkner, Prof. Dr. Patrick 03065
Werlen, Franziska 41927
Werlich, Dr. Uta 21951
Wern, Christine 18379
Werneburg, Dr. Ralf 21572
Werner, Dr. Anne-Marie 21459
Werner, Bettina 51252
Werner, Constanze 20800
Werner, Dietrich 52461
Werner, Dr. Edwin 18902
Werner, Hans-Hubert 20295
Werner, Harald 02748
Werner, Heinz 19166
Werner, Jane 51957
Werner, Manuela 22323
Werner, Petra 19234
Werner, Dr. Sigrid 16926
Werner, Dr. Winfried 20443, 20472
Werner, Wolfgang 17560
Wernevik, Lena 40782
Werneyer, Rita 22665
Wernz-Kaiser, Heike 16795
Werring, Ole A. 33723
Wert, Nancy L. 50693
Werth, Linda 53976
Wertkin, Gerard C. 51113, 51171
Wéry, Benoît 04113
Wery, Bernoît 03681
Weschenfelder, Dr. Klaus 17717, 21351
Wescott, Dusty 52236
Wesle, Anna 41275
Wesling, Ferdinand 21240
Wesney, Russell 33165
Wessel, Dr. Alexandra 41169
Wessels, E.M. 38570
Wessels, Reinhard 19013
West, Charlotte 10221
West, Chris 43633
West, Gerald C. 47539
West, J. Martin 50082
West, Janis 50410
West, Dr. Patricia 49778
West, Penelope A. 50082
West, Richard 52497
West, W. Richard 51236, 53905
West, Wanda 48556
Westad, Lynn 48142
Westbrook, Nicholas 53506
Westbrook, T.G. 44031, 46072
Westby, Carole 46696
Wester la Torre, Carlos 34202
Westerlund, Jens 10277
Westerlund, Kasper 11390
Westermann, Verena 18952
Westermann-Angerhausen, Prof. Dr. Hiltrud 19625
Westers, Drs. Abraham 32576
Westfall, Deborah 46850
Westfehling, Dr. Uwe 19635
Westheider, Dr. Ortrud 18918
Westheider, Dr. Rolf 18833
Westholm, Gun 41068
Westin Berg, Elisabeth 40909
Westland, Alan 06988
Westling, Stefan 40656
Westlund, Frances 06261
Westmacott, Jean 48820, 48822
Westman, Anne 40442
Westmoreland, Debra 50850
Westmorland, Lamont 50221
Westoby, Carolyn 05339
Weston, Catherine 43342
Weston, Elizabeth 45101
Weston, Gill 45123
Weston-Lewis, Aidan 44138
Westphal, Florian 21565
Westphal, Jean 13401
Westphal, Dr. Martin 21275
Westphal, Ursula 17825
Westra, J.F. 31828
Westrop, Dr. S. 51376
Westwood, Corinna 45244
Westwood, Ros. 43705
Wetenhall, John 52914
Wetherell, Virginia 48536
Wetherington, Mark V. 50285
Wettengl, Dr. Kurt 17924
Wetter, Ellen de 06652
Wetter, Dr. Evelin 41767
Wetter, Kenneth 05578
Wetterau, Glenadore 48670
Wetton, Allan 00810
Wettre, Håkan 40547
Wetzel, Brigitte 18769
Wetzel, Dietrich 21054
Wetzel, James K. 48048
Wetzel, Roland 41173
Wetzel, Walter 41267
Weuts, Sylvain 03785
Wex, Dr. Norbert 21797, 21799
Wexell, Dale 49370
Wey-Eiholzer, Josef 41760
Weydemann, Ingrid 02436
Weyerhaeuser, Charles A. 48227
Weyers, E. 38714
Weygandt, Virginia 53235
Weyher Dawson, V. 48532
Weyns, Marie 03815
Weyns, Sara 03292
Weytjens, Laurent 03373
Whalen, Connie 51911
Whaley, Martha 49634
Whallon, Robert 46308
Whang, Prof. Won-Koo 29935
Wharton, David 53954
Whatmoor, D.J. 45326
Wheaton, Marilyn L. 53655
Wheaton, R. 01455
Wheeler, A. 44941
Wheeler, Geoffrey 50591
Wheeler, Karla 46724
Wheeler, Ken 51670
Wheeler, Lawrence J. 52232
Wheeler, Mary 49703
Wheeler, Matt 45273
Wheeler, Robert K.L. 50681
Wheeler, Verna Anne 47889
Wheeley, T.R. 44501
Wheldon, G.H. 44553
Whidden, John 07399
Whissell, Raymond 06793
Whitaker, Monique 38725
White, Adam 44666
White, Babs 47966
White, Barbara 53087
White, Barry 24783
White, Betsy K. 46126
White, Blair H. 49631
White, C.S. 45437
White, Carol 46677
White, Christopher 45891
White, David 07167, 44998, 50117
White, Derrick 46354
White, Donald 51886
White, Donald F. 53677
White, Donny 06258
White, Elisa 51110
White, Gail W. 53524
White, Sir George 44827
White, James J. 51950
White, Joe L. 49774
White, Julia 49343, 54218
White, Karen 54285
White, Kathleen 52317
White, Kathryn 43415
White, Kristy 47808
White, Marcelle 47948
White, Mark A. 53296
White, Mary 45676
White, Matthew 51390
White, Maylon 53759
White, Metthew 46678
White, Pam 45979
White, Patsy 51664
White, Richard S. 53592
White, Sally 44698
White, Susanna 47659
White, Timothy J. 47445
White, Warren O. 48306
Whitehead, F. 43708
Whitehead, James T. 53047
Whitehead, Jane 46107
Whitehead, Leah 46744
Whitehouse, Adrienne 44436
Whitehouse, C.M. 43514
Whitehouse, David B. 47810
Whitelaw, R. Lynn 53448
Whiteley, Jim 50000
Whiter, Robert 01060
Whitey, Matthew 44660
Whitfield, Carol 06197
Whiting, Dr. Cliff 33260
Whiting, Elizabeth K. 48253
Whitlock, Pamela 01111
Whitlow, Paula 06502
Whitman-Grice, Lisa 52309
Whitmore, Damien 45004
Whitmore, Dennis 46574
Whitney, Lisa 50228
Whitney, Richard 48135
Whitt, Teresa C. 53166
Whittaker, Dr. J. 43782
Whittall, Ann 44761
Whittington, Michael 47392
Whittle, Stephen 45314
Whitton, Dr. Jeannette 07199, 07199
Whitworth, Kent 48737
Whitworth, Peggy 47300
Wholey, B.J. 43624
Whorlow, David 43699
Whyel, Rosalie 46709
Whyte, Alan 07283
Wiant, Dr. Michael D. 50040
Wiblé, François 41612
Wible, Dr. John R. 51943
Wichert, Dr. Susanne 17459
Wichman, Tomasz 34784
Wichrowski, Zbigniew 34823, 34824
Wick, Susan B. 53640
Wick Reaves, Wendy 53907
Wickenden, N.P. 43803, 43806
Wickers, Tine 10152
Wickerson, David 45808
Wickham, Ann 44369
Wickham, W.C. 49163
Wicklander, Dan 53840
Wicklund, Doug 48478
Wickman, Viktor 30661
Wickramasinghe, Dr. Nanda 40370, 40374
Widder, Kenneth J. 52752
Widdowson, Prof. J.D.A. 45655
Widemann, Norbert 18436
Widén, Britt-Marie 40547
Widenheim, Cecilia 40915
Widjeskog, Susanna 11230
Widmann, Dr. Marion 17462
Widmer, Walter 41957
Widrick, Melissa 53972
Widulin, Navena 17044
Wie, Guangzhen 08116
Wie, Na 07986
Wie, Xia 08425
Wiebe, Charles M. 53044
Wiebel-Roth, Dr. Christiane 17717
Wiech, Stefanie 20075
Wiechers, Renate 18912
Wiechert, Ralph D. 52601
Wiechmann, Dr. A.H.C.A. 32484
Wieczorek, Prof. Dr. Alfried 20173
Wieczorek, Jan 35045
Wied, Fürst zu 21450
Wied, M. Prinz zu 21450
Wiedemann, M.A. Sandra 22018
Wiedenhofer, Dr. Josef 02364
Wieg, Cornelia 18904
Wiegand, Liesel 18779
Wiegand, Otto 20276
Wiegand, Tanya 05470
Wiegand, Winfried 20259, 20260, 20262
Wiegartz, Dr. Veronika 17551
Wiegmann, Dr. Karlheinz 22119
Wiehager, Dr. Renate 17063
Wiehart, Michael 03039
Wielebski, M.A. Anke 18744
Wielek, Jan 34856
Wiell, Stine 10049
Wiemann, Dr. Elsbeth 21960
Wiemer, Regine 20013
Wienand, Paul 22384
Wiench, Barbara 18008
Wiene, Inger 10135
Wienecke, Joachim 19133
Wienicke, Melanie 18952
Wieninger, Johannes 03026
Wiens, Dr. John J. 51943
Wier, Marsi H. 51791
Wier, Nan 51575
Wierbicki, Wilma 52048
Wiermann, Dr. Roland 17295
Wiermann, Susanne 16706
Wierschowski, Dr. Myriam 20010
Wiersema, J. 32696
Wierzbicka, Alina 34930
Wierzbowska, Ilona 34579
Wiese, Dr. André 41154
Wiese, Joyce 53519
Wiese, Kerstin 19898
Wiese, Larry 50609
Wiese, Prof. Dr. Rolf 20230, 20367, 20368, 21373, 22479
Wiese, Sharon A. 46362
Wiese, Dr. Vollrath 17707
Wiese, Dr. W. 21185
Wiesel, Yael 24928
Wieseman, Betsy 44937, 47555
Wiesen, Barbara 48920
Wiesen, Richard A. 47116
Wiesenhofer, Franz 02539, 02541
Wieser, Dr. Christian 02186
Wiesinger, Ingrid 01777
Wiesner, Cary 47446
Wiesner, Evelyne 03026
Wiesner, Tomáš 09892
Wiethege, Dr. Katrin 01795
Wiethoff, Charlotte 32295
Wiethorn, Hans 21839
Wieting, Margaret 49519
Wiewelhove, Dr. Hildegard 17342
Wiget, Dr. Josef 41849, 41851
Wiggins, Denise 50396
Wight, Anthony 47476
Wigmore, John R. 07243
Wihebrink, Peggy 50162
Wiig, Prof. Øystein 33737
Wijers, Gaby 31873
Wijesooriya, Bandusena 40365
Wijn, Drs. J.J.A. 32781
Wikström, Anne-Maj 11373
Wikström-Haugen, Inger 40552
Wilcox, Chris 54204
Wilcox, John 47787
Wilcox, Mike 48296
Wilcox, Richard 50493
Wilcox, Scott 51055
Wilcox, Susan 48550
Wilcox, Thomas R. 46628
Wilcox, Tim 45088
Wilczak, Susan 52558
Wild, Dr. Beate 17199
Wild, Jane 06595
Wild, Maxine F. 01069
Wild, Teresa 45877
Wilde, Cathleen 52015
Wilde, Kevan A. 33230
Wilde, Dr. Manfred 17813
Wildeman, G.J.D. 31885
Wilder, Jacalin W. 47056, 47057
Wilder, Oliver L. 52425
Wilder, Sofie 03308
Wildgans, Ilse 02391
Wildgust, Deborah 45400
Wildgust, Patrick 43910
Wildman, Janet 44337
Wildman, S.G. 44637
Wilenius-Rantala, Carita 10773
Wiley, E.O. 49991
Wiley, Florence 48152
Wiley, John 06288, 46069
Wilezek, Alexander 02841
Wilfong, Terry 46307
Wilhelm, Ada 47738
Wilhelm, Elliot 48114
Wilhelm, Franz 18069
Wilhelm, Martin 01899
Wilhelm, Robert 49223
Wilhelmer, Erika 16870
Wilhelmer, Hans-Sieghart 01723
Wilhelmson, Pia 11082
Wilhelmy, Dr. W. 20141
Wilhite, Janet 47926
Wilk, Zbigniew 34653
Wilke, Erich 17574, 22585
Wilkens, Dr. Barbara 26533
Wilker, Klaus-J. 17879
Wilkie, Jo-Anne 50871
Wilkins, Prof. David G. 51963
Wilkins, Margaret 44227
Wilkinson, Dr. D.J. 44783
Wilkinson, MaryAnn 48114
Wilkinson, Patty 46877
Wilks, Harry T. 49139
Willaert, Saskia 03495
Willard, David 48873
Willard-Wright, Rowena 44012, 44014
Willburger, Dr. Nina 21950
Wille, Cerstin 17564
Willeboordse, A. 32788

Willeboordse, A.C.J. 31787
Willeboordse, Arco 32770
Willems, Alfons 04150
Willems, Carmen 04075
Willems, Jean 03856
Willems, Roger 03342
Willemsen, M. 32059
Willer, Dr. Susanne 17462
Willer, Thérèse 15833
Willert, Dr. Barbara 22288
Willesme, Jean-Pierre 14577
Willi, Dawn 50924
Willi, Jürg 41761
Williams, Alice 48894
Williams, Barb 52277
Williams, Bart 49443
Williams, Bill 49140
Williams, Clay 49565
Williams, Dr. D. 44803
Williams, Debby 53059
Williams, Don 51181
Williams, Edna 50438
Williams, Edward 46547
Williams, Eric K. 51354
Williams, Fred 47795
Williams, Gareth H. 43709
Williams, Dr. George 54086
Williams, Dr. Gloria 52616
Williams, Hilary 43987
Williams, James 49198
Williams, Janice W. 53367
Williams, Jay 46481, 47728
Williams, Jon 54194
Williams, Judy 33255
Williams, Karen 33204
Williams, Kathy 48684
Williams, Kay P. 51020
Williams, Dr. Keith 54037
Williams, Kevin 50735
Williams, Linda 45533
Williams, Lisa 50336
Williams, Dr. Lorraine 53561
Williams, Lyle 52717
Williams, Mary-Louise 01025
Williams, Matt 43401
Williams, Matthew 43749
Williams, Michael 45289, 45298
Williams, Morag 44034
Williams, Nadezhda 51177
Williams, Nora 05708
Williams, Pamela 46970
Williams, Paul 52694
Williams, Peter J. 01651
Williams, Richard 44446
Williams, Robert J. 53048
Williams, Roger 52162
Williams, Sean 45004
Williams, Sheldon 46694
Williams, Stacy 52690
Williams, Stephen R. 52035
Williams, Teri 46422, 46438, 46449
Williams, Thomas 50459
Williams, Victoria 44415, 44416
Williams-Davies, John 45565
Williamson, Eleanor 53245
Williamson, J. Reid 49494
Williamson, Jane 48537
Williamson, Norma 48186
Williamson, Roberta 48554
Willis, Carol 51256
Willis, Kelly B. 50045
Willis, Leanne 00927, 01504
Willis, Michael 52022
Willis, Pamela 44930
Willis, Patricia 44782
Willis, William 52133
Willits, Mary Lou 49020
Willmann, Prof. Dr. R. 18702
Willmitch, Thomas R. 46877
Willmont, Joanne 50039
Willms, Dr. Christoph 18354
Willms, Karl 17575
Willomitzer, Gerry 07328
Willoughby, Alan 50833
Willoughby, Brenda 47961
Willoughby, Laura 51797
Willour, Clint 48844
Willow, Hai Chang 51150
Willows, John 45990
Willscheid, Bernd 20688
Willson, Michael 51967
Wilmes, Dr. Ulrich 19624
Wilmes, Ulrich 20445
Wilmink, Drs. G.J.G. 32039
Wilmot, Jonel 49948
Wilmot, Patricia J. 49203
Wilmot, Robert 44128
Wilms, Kurt 16688
Wilschewski, Dr. Frank 20681
Wilsdon, D.E. 01520
Wilsey, Bart 48517
Wilson, A. 45223
Wilson, Dr. Alan 43333
Wilson, Amy H. 48379
Wilson, Audrey 06209
Wilson, B. 45793
Wilson, Beth 47154
Wilson, Bob 47154
Wilson, Carolyn 54255
Wilson, Carrie 51268
Wilson, Chris 24861, 43232, 44755, 54210
Wilson, Christine 43938, 43939
Wilson, Courtney B. 46547
Wilson, Dale 48311
Wilson, Dan 49019
Wilson, David 47904
Wilson, Diane 46701
Wilson, Dorothy 43507
Wilson, Floyd S. 47385
Wilson, Gillian 50237
Wilson, Gordon 52521
Wilson, Hugh Allen 46897
Wilson, I.A. 43375
Wilson, Inge 05994
Wilson, J. Keith 50244
Wilson, Jane 51640
Wilson, Jerry 47980
Wilson, Jim 49149
Wilson, John 43448
Wilson, Jonathan 45390
Wilson, Kathy 06837
Wilson, Kay 49091
Wilson, Kevin 01579
Wilson, Lamar 53067
Wilson, Lesley 43817
Wilson, Marc F. 49687, 49695
Wilson, Mark 32281
Wilson, Martha 47073
Wilson, Nancy 50318
Wilson, Peter 44995, 44996
Wilson, Pierre 06360
Wilson, Robert 47127
Wilson, Robert C. 53321
Wilson, Robin 45376
Wilson, Dr. Rodney 33043
Wilson, Roger N. 51340
Wilson, Roy 45386
Wilson, Scott 48025
Wilson, Sophia 43810
Wilson, Steven 49191
Wilson, Dr. Thomas 50605
Wilson, Thomas H. 46725
Wilson, Tom 54250
Wilson, Viv 45221
Wilson, Wayne 06050, 06054
Wilson, William 45058
Wilson Davine, Blossom 52157
Wilson Jenkins, Kay 49090
Wilson-McKee, Marie 47448
Wilsony, Thomas A. 52883
Wiltfang, Helga 20729
Wiltzius, Ruth 51790
Wiluś, Bogusława 35001
Wilusz, Mike 53308
Wilz, Mick 47090
Wimberger, Dr. Peter 53409
Wimmer, Franz 02506
Wimsett, Brian 43657
Win, U Kyaw 31736
Winans, Cathie 53049
Winberg, Hélène 41001
Winbladh, Marie-Louise 40913
Winch, Barbara 53173
Winch, Dinah 45314
Winchell, Janey 52657
Winchester, Dr. Judi A. 47671
Wincke, Claus-Jürgen 16645
Wind, D.A. 32756
Windau, Gerda 20513
Winderickx-Joos, Myriam 03265
Winders, Dr. Bruce 52705
Windham, Michael 52696
Windholz, Sascha 02822
Winding, Dr. Norbert 02615
Windl, Dr. Helmut 02818
Windle, Ann S. 50381
Windsor, Deirdre 50303
Windsor, Noel 00864
Wine, Dr. Humphrey 44937
Winegrad, Dr. Dilys V. 51815
Wines, Claudia 48354
Wines, Darryl 06296
Wines, Richard 52366
Winfield, Rosemarie 49934
Winford, Laura 49878
Wingate, Lloyd R. 38675
Wingfield, Carolyn 45540
Winglhofer, Rudolf 01924
Wingo, Calvin 48429
Winkelmeyer, Iris 20491
Winker, Kevin 48472
Winkler, Anke 18657
Winkler, Axel 21948
Winkler, Bernhard 22312
Winkler, Dr. Ferenc 23353, 23633, 23806
Winkler, Dr. G. 01861
Winkler, Gerhard 01745
Winkler, Dr. Gerhard 01860
Winkler, Howard 46348
Winkler, Karin 50673
Winkler, Kerstin 19421
Winkler, Paul 49384
Winkler, Dr. Raffael 41175
Winkler, Steffen 18657
Winkler, Ursula 19510
Winkler-Hermaden, Burkhardt 02152
Winkworth, David R. 43868
Winn, Allan 45985
Winnekes, Dr. Katharina 19613
Winner, Scott 46875
Winnicki, Tomasz 35204
Winning, Fiona 01078
Winslow, Bruce B. 50666
Winsryg, Marian 52900
Wint, Dennis M. 51836
Winteler-Isoz, Florence 41093
Winter, Dr. Amy 48576
Winter, Dr. B. 05504
Winter, C. 33091
Winter, Cordula 16498
Winter, Dr. Eleonore 21755
Winter, Emil 19176
Winter, Gerhard 19045
Winter, Hilda 45036
Winter, Joan G. 00917
Winter, Małgorzata 34578
Winter, Petra 17283
Winter-Berger, Robert 51190
Winterhalter, Greg 46740
Winterich, Douglas E. 47139
Winters, Laurie 50711
Wintgens, Dr. D. 32489
Wintz, Klaudius 02233
Winzeler, Marius 22812
Winzingerode, Margit Freifrau von 21114
Wiotte-Franz, Dr. Claudia 21463
Wipaki, Gail 00986
Wippermann, Hartmut 20272
Wippert, Bernd 19714
Wipplinger, Mag. Hans Peter 02227, 02228
Wirenfeldt Asmussen, Marianne 10253
Wirnsberger, Mag. Karlheinz 02759, 02760
Wirrell, Wynne 01033
Wirrmann, Fanny 11706
Wirsperger, Dr. Peter 02371
Wirtanen, Lyle 47843
Wirth, Axel 20260
Wirth, Dr. Klaus 20173
Wirth, Marlies 03026
Wirth, Paul 41241
Wirth, Peter 52888
Wirthgen, G. 19894
Wirtler, Dr. Ulrike 19587
Wirtz, Gudrun 20424
Wirtz, Wes 07271
Wirtz Eybl, Dr. Irmgard 41215
Wirxel, Julia 16890
Wirz, Arno 16922
Wisbey, Peter A. 46471
Wise, Carol 49669
Wise, Philip 43872
Wise, Dr. Stephen R. 51714
Wise, Tess 50401
Wiséhn, Ian 40904
Wiseman, Christopher 49649, 49651
Wishart, E. 01159
Wismar, Franz Werner von 21266
Wismeijer, P.J. 32150
Wismer, Beat 18008
Wisowaty, Elaine 50147
Wißkirchen, Prof. Dr. Hans 20073
Wissler, Richard 47751
Wissler-Thomas, Carrie 49178
Wistar, Caroline 51846
Wistoft, Birgitte 10148
Wisuri, Marlene 47663
Wit, B.H. de 31984
Wit, Prof. Dr. G.W. de 32720
Witcherly, Zak 54126
Witcox, David E. 51357
Witenstein, Michael 50173
Witham, Jim 48423
Withrow, Sharon 46375
Witkofsky, Adrien 49080
Witmer, Linda F. 47243
Witschey, Walter R.T. 52338, 52340
Witschi, Heinz 41498
Witschi, Dr. Peter 41481
Witt, Dr. Christopher C. 46182
Witt, David 53441
Witt-Dörring, Dr. Christian 02995
Wittbold, Dr. Annette 18827
Wittboldt, Dr. Annette 18828
Witte, Amy 47105
Witte, David 53117
Witte, Hubert De 03412
Witte, Dr. Reinhard 16549
Witte de Haelen, Didier de 03666
Wittekindt, Heiner 18343
Wittenbauer, Christine 19981
Wittenzellner, Renate 21224
Witter, Thomas 19579
Wittig, Peer 16551
Wittigo Keller, Dr. 02953
Wittington, Dr. Stephen 54246
Wittkopp, Gregory M. 46855
Wittman, Jörg 22813
Wittmann, Dr. Helmut 02615
Wittmann, Dr. Reinhard G. 20456
Wittmer, Heike 21041
Wittmer, Marcilene 47802
Wittmer, Stephan 41594
Wittwer, Felix 41687
Witzke, Gitta 19154
Witzmann, Nora 03049
Wlattnig, Robert 02186, 02868
Włodarczyk, Tadeusz 35338
Woehlk Hansen, MA Lola 10096
Wöhrer, Hannes 01744
Wöhrer, Josef 02361
Wölbling, Klaus 19478
Woelk, Dr. Moritz 17970
Wörlen, Hanns Egon 20991
Wörtz, Walter 19315, 19503, 20574, 21362
Wörz, Dr. Arno 21959
Wöß, Gerhard 02837
Woestenburg, K. 32668
Wohlgemuth, Caroline 16419
Wohlgemuth, Werner 18449
Wojciechowski, Jerzy 34737
Wojciechowski, Zbigniew 34656
Wójcik, Adam 35175
Wójcik, Andrej J. 35350
Wojewoda, Maciej 35119
Wojtowicz, Wit Karol 34842
Wojtusiak, Prof. Janusz 34814
Wojtysiak, Władysława 34896
Wolanin, Dr. Barbara A. 53923
Wolde, H.B. ten 32432
Wolenitz, Dorit 25084
Wolf, Alexandra 02823
Wolf, Dr. Barbara 16611
Wolf, Dr. Beate 21960
Wolf, Connie 52771
Wolf, Dieter 17671
Wolf, Dr. Eberhard 21327
Wolf, H.-J. 21912
Wolf, Dr. Helmut 16606, 19734
Wolf, Martha Leigh 52149
Wolf, Maurice de 03937
Wolf, Reinmar 02395
Wolf, René 31841
Wolf, Dr. Rüdiger 02708
Wolf, Sara 47392
Wolf, Sarah 52414
Wolf, Sylvia 51279
Wolf, Vicki 52756
Wolf, Werner 01980
Wolfe, Adam 01107
Wolfe, Douglas 50605
Wolfe, Jeraldine J. 52270
Wolfe, L. 06434
Wolfe, Sara 51138
Wolfe, Sarah 53819
Wolfe, Susan 48933
Wolfer, Johannes 01820
Wolfer, Dr. Martin 01820
Wolff, Prof. Heinz 43682
Wolff, Thomas 18756
Wolff, Dr. Tobias 17565
Wolff, W. 19871
Wolford, Craig 47598
Wolfová, Dr. Eva 09786
Wolfs, R. 32729
Wolfs, Rein 19465
Wolfschlag, Robert 50379
Wolfson, R.D. 32553
Wolins, Inez S. 49028
Woliung, Dorothy 46710
Wolkow, Gary 48001
Woll, Gunda 22134, 22137, 22138
Wollan, Alexa 06023
Wolle, Dr. Stefan 17064
Wollert, Christian 17040
Wolleswinkel, E.J. 32682
Wollkopf, Peter 19682
Wollmann, Dr. Klaus 20974
Wollner, Dr. Bernd 20317
Wolny, Maja 03926
Wołodko, Wiesław 34645
Wolsdorff, Dr. Christian 17041
Wolsk, Nancy 50054
Wolstenholme, L. 45090

W

Wolston, Bill 53181
Wolter, Rüdiger 17550
Woltereck, Jane D. 46580
Wolters, R. 32012
Wolters, Prof. Dr. Reinhard 22125
Wolterstorff, Robert 52069
Wolther, Gian Andrea 41903
Wolthoorn, Dr. R. 32489
Wolts, Silke 18579
Wolynec, Dr. Renata 53960
Wolz, Robert J. 49761
Wolzogen, Wolf von 18377
Wong, Danny 07814
Wong, Michele 51187
Wong-Uparaj, Acharn Damrong 42534
Wonisch-Langenfelder, Dr. Renate 02625, 02627
Wooby, William 53893
Wood, Bryen 43699
Wood, Carolyn 47344
Wood, Charles A. 52208
Wood, David 47767
Wood, Dr. Donald A. 46822
Wood, Eileen 45482
Wood, Elaine 47389
Wood, James N. 47452
Wood, Jeff 50669
Wood, Jeremy 44926
Wood, Katherine 43873
Wood, Kaye 00872
Wood, Lydia 47711
Wood, Mara-Helen 45233
Wood, Margaret 51790
Wood, Margaret K. 48977
Wood, Nancy 01435
Wood, P. 43790
Wood, Ronald M. 50316
Wood, Sue 43329
Wood, Tim 01182
Wood, Virginia 53730
Wood, Willis 54016
Wood, Wilma 06162
Woodcock, Mick 52135
Woodcock, Vivi 51806
Woodings, David J. 33130
Woodrow, James 43428
Woods, Christine 45099
Woods, Claude N. 49235
Woods, Don 51926
Woods, James C. 53627
Woods, Jean 49122
Woods, John S. 51365
Woods, Margaret M. 46982
Woods, Mark H. 50655
Woods, Wally 54110
Woodside, April 53122
Woodson, Celia 48123
Woodson, Yoko 52766
Woodward, Alan 33247
Woodward, Geoff 45917
Woodward, John 43570
Woodward, Mac 49446
Woodward, Peter 43999
Woodward, Richard 52343
Woody Cywink, Linda 52877
Woof, Dr. Robert 44348
Woog, Dr. Friederike 21959
Woolcott, Isabelle S. 47841
Woolf, Derrick 45201
Woolley, Brandon 47023
Woolley, F. 45073
Woolliscroft, Pam 45763
Woolmore, J.D. 43833
Woolrich, Gill 45656
Woolsey, Sam 47945
Worcester, Mike 47679
Worden, Gretchen 51857
Worden, John 52372
Workman, Robert 48709
Workmann, Tom 49668
Worm, Dr. Heinz-Lothar 20003
Worrall, Adam 00946
Worschech, Kathrin 16498
Worschech, Thomas 18362
Worsdale, Godfrey 45140
Worsley, Lucy 44093
Worswick, Clark 52657
Worth, Claudia R. 54117
Worth, Cyril 49319
Worth, Tim 07357
Worthen jr., William B. 50139
Worthington, Gary 51983
Worthington, Robert 49098
Wortmann, Susanne 22713
Worton, Graham 44030
Worton, James 52593
Worts, Martin 33537
Woster, Don 49376
Wotovic, Audrey 12920
Wouters, Lisbeth 03315
Woyda, Stefan 35052
Woyke, Karin 17988
Woytowicz, Barbara 51194
Woźny, Dr. Alexander 34678
Wpadhyays, Dr. Chandrakant B. 24052
Wragg, Gary 43395
Wray, JaLayne 54147
Wrazidlo, Dr. Isolde 17347
Wrębiak, Celestyn 34543
Wrenn, Dr. W.J. 48990
Wrey, Dr. James D. 51762
Wright, Amina 43405
Wright, Anthony 33069
Wright, Dr. Barnaby 44830
Wright, Belinda 00923
Wright, Carolyn 47944
Wright, David M. 54270
Wright, George 44236
Wright, Geri 51904
Wright, Henry T. 46308
Wright, John 05578, 49285
Wright, Lesley 49090
Wright, Lynne 05305
Wright, Michael 49565
Wright, Raymond 54008
Wright, Rita 33037
Wright, Dr. Robin K. 52994
Wright, Dr. Sheila 45295
Wright, Sherri 50911
Wright, Simon 00898
Wright, Twyla 46627
Wright-Parsons, Ann 48037
Wrigley, John 00938
Wrobel, Horst 18643
Wroczyńska, Ewa Maria 35202
Wroz, Winfried 19493
Wu, Cheng-Shang 42410
Wu, Ching-Feng 42473
Wu, Chu Pang 08000
Wu, Douming 08480
Wu, Gangzhong 07715
Wu, Gaobin 08452
Wu, Hongzhou 08175
Wu, Hung Yuen 07656
Wu, Jin 08390
Wu, Jingchao 07645
Wu, Kai 08384
Wu, Ke Shun 08227
Wu, Li-Chu 42366
Wu, Monical 42307
Wu, Qingying 08124
Wu, Qisheng 08082
Wu, Rongguang 07680
Wu, Shengtian 07760
Wu, Shiliang 07824
Wu, Tianming 07832
Wu, Wenxin 07920
Wu, Xiaosong 07830
Wu, Xuebin 08097
Wu, Xuhui 07847
Wu, Yulu 08401
Wu, Zhengguang 07746
Wucherer, Viktoria 18196
Wudel, Tom 50762
Wüest, Charles 41517
Wünsche, Prof. Dr. Raimund 20486
Würsch, Paul 41843
Würthenberger, Ute 41589
Würzelberger, Jörg 02831
Wüthrich, Dr. Lucas 41750
Wulf, Karin 17439, 17440
Wulf, Rüdiger 17925, 17929
Wulfsohn, Iris R. 00381
Wulkan, Reba 51281
Wullen, Dr. Moritz 17165, 17202
Wullus, Linda 03495
Wunderer, Dr. Hansjörg 21230
Wunderlich, Dr. Christian-Heinrich 18900
Wunderlin, Dominik 41169
Wunderwald, Anne 19933
Wunsch, G. 18334
Wurda, Andreas 20117
Wurfel, Max 01416, 01417, 01418
Wurmfeld, Sanford 51193
Wurtele, Gail 50996
Wurzer, Günther 02186
Wurzinger, Erika 02027
Wussow, Indra 21181
Wustrack, Dr. Michael K. 18353
Wuthenau von Pietsch, Carlota von 31403
WWüst, Léonard 41924
Wyatt, Frederica 49654
Wyatt, Steve 50552
Wyckoff, Dr. D. 51376
Wyczółkowski, Mariusz 34738
Wydler, Henry 41600
Wye, Deborah 51215
Wykersloot, Baron E. de 03398
Wyld, Lesley 45114
Wyld, Martin 44937
Wylie, Charles 47931
Wyman, James 46417
Wymbs, Norm 48146
Wÿnandts, Dr. H. 32474
Wyniger, Dr. Denise 41595
Wynn, Rachel 52051
Wynne, M.V. 43837
Wyrsch, Dr. Jürg 41993
Wyse Jackson, Dr. Patrick N. 24793
Wyss, Dr. Gabriela 42109
Wyss, M. 41696
Wyss, Matthias von 41947

X

Xamena, Miguel A. 00285
Xanke, Uwe 17860
Xavier, Andrew 44817
Xerri, Alda 30658
Xi, Jiping 08169
Xi, Mude 08300
Xia, Meizhen 08428
Xia, Suixiang 08270
Xia, Xiangjun 08464
Xia, Zhaoli 07908
Xian, Shihai 07502
Xiang, Qukui 08313
Xiao, Changjin 07853
Xiao, Gaoshe 08437
Xiao, Kanghua 08329
Xiao, Qi 07973
Xiao, Runjun 08506
Xie, Daohua 07867
Xie, Daoxin 07637
Xie, Kaihua 08117
Xie, Qihuang 07530
Xie, Shanguang 07890
Xie, Suilian 08500
Xie, Ying-Cong 42371
Xie, Yuan'an 07823
Xie, Zhijie 08443
Xiejun, Chen 08184
Xin, Fa 07899
Xin, Xue Jian 47065
Xinmiao, Zheng 07559
Xiong, Bufa 08379
Xiong, Guoyao 08292
Xiong, Jianping 07906
Xiong, Keda 07903
Xiong, Qiu Shi 08211
Xu, Dali 07498
Xu, Fan 07683
Xu, Huping 08045
Xu, Qiaohua 08474
Xu, Rongmin 08303
Xu, Sanjian 07948
Xu, Xiaozhong 07827
Xu, Xinwan 08398
Xu, Xueshu 07618
Xu, Xuyao 07963
Xu, Yiqun 08419
Xu, Yiyi 08045
Xu, Yuli 08404
Xu, Zhile 07859
Xue, Deng 07617
Xue, Dongxing 08298
Xue, Fusheng 07930
Xylander, M. A. Marlen von 20735

Y

Y, Dr. Rainer 20046, 21950
Yaacobi, Effi 25073
Yabacıoğlu, Ömer 42754
Yabe, Yoshiaki 28770
Yaeger, Dan 53808
Yager, Bob 50441
Yaghobi 24695
Yagid, Meira 25122
Yajima, Loretta 49339
Yakar, Saffet 42797
Yalakki, C.S. 24402, 24403
Yaldız, Fethiye 42793
Yaldiz, Mehmet 42948
Yamada, Rikuzou 29180
Yamada, Shimako 29192
Yamada, Yasuhisa 29580
Yamada, Yasushi 29575
Yamaga, Motoko 28517
Yamagishi, Youichi 29115
Yamaguchi, Kichiko 28377
Yamaguchi, Seiichi 29003
Yamakawa, Kazutoshi 29251
Yamamoto, Akira 28357
Yamamoto, Atsuo 28952
Yamamoto, Chikyo 28773
Yamamoto, Kazumaro 28927
Yamamoto, Dr. Masao 28957
Yamamoto, Minoru 29521
Yamamoto, Shuji 29574
Yamamoto, Yasukazu 29012
Yamamoto, Yoshiko 52763
Yamanaka, Osamu 28726
Yamazaki, Taeko 29507
Yan, Hongbin 07501
Yan, Lijin 08110
Yan, Yiyuan 08154
Yanagi, Munemichi 29296
Yanagihara, Masaki 29522
Yanamoto, Akira 29368
Yanari, Sachi 48705
Yancey, Shirley S. 53998
Yande, R. 03323, 03324
Yandle, Ashley 47376
Yanega, Dr. Doug A. 52370
Yáñez Anllo, Lucila 39499
Yang, Daoyi 07713
Yang, Huifu 07874
Yang, Jianmin 08342
Yang, Jinghua 08215
Yang, Jingnan 08258
Yang, Jun 08168
Yang, Mingzhu 08472
Yang, Shi-Rong 42360
Yang, Xiaoneng 49695, 53253
Yang, Xin 08495
Yang, Yanqiang 07782
Yang, Yingji 07960
Yang, Yongfang 08411
Yang, Yuanlong 07910
Yang, Zishan 08508
Yanik, Kadir 42949
Yaniv, Shosh 25021
Yank, Paul 47308
Yankov, Dr. Angel 05126
Yano Bretón, Dr. Sabino 31478
Yao, Jingzhou 08093
Yap, Priscilla M. 34482
Yapelli, Tina 52759
Yarlow, Loretta 46271, 47082, 51249
Yarmosh, Linda 46911
Yarsinske, Amy Waters 51368
Yarzhembovska, E. 22712
Yaşar, Hasan 42888
Yashimoto, Mami 29192
Yasin, Dr. Walid 43217
Yassin, Robert A. 52242
Yasuda, Haruki 29444
Yasui, Teruo 28784
Yasumura, Toshinbu 29401
Yasuo, Hasimoto 29101
Yates, Jane M. 47360
Yates, R. 38592
Yates, Royden 38591
Yates, Sam 49813
Yates, Dr. Steve 52883
Yates, Dr. Steven A. 52880
Yatim, Prof. Dr. Othman 30582
Yatras-Dehove, Elena 22901
Yawazaki, Takashi 28959
Yazbeck, Janine 30170
Yde Iversen, Gerrud 10180
Ydstie, Ingebørg 33732
Ye, Dingxiang 08247
Ye, Faling 08389
Ye, Jianhong 08245
Ye, Qinglin 07485
Ye, Rongguang 07988
Ye, Xiangrong 08409
Yeager, Shelley 51399
Yeatts, David L. 54287
Yébenes, M. Sierra 39209
Yee, Lydia 47040
Yee de Cury, Ana 10351
Yeffeth, Laura 51277
Yelavich, Susan 51156
Yeler, Nihat 42691
Yellow Horn, Jo-Ann 05485
Yelton, Chris 52777
Yener, Emin 42721
Yerdon, Lawrence 52100
Yerdon, Lawrence J. 51968
Yerhot, Judith 46869
Yerkovich, Dr. Sally 51286
Yernaux, Danielle 03741
Yeroulanos, Marinos 22925
Yersin, Jean-Marc 41987
Yesalis, Dr. Charles 46881
Yetkin, Mehmet 42675
Yeudall, Robert 05745
Yeung, C.T. 07818
Yi, Jiasheng 08050
Yi, Shu 07557
Yi, Xuelin 08257
Yi, Zhu 07534
Yıldız, Adem 42762
Yıldız, Hacı Ahmet 42674
Yılmaz, Ayla 42825
Yin, Bilan 07962
Yin, Hexian 08499
Yin, Liming 07950
Yin, Qingsong 07919
Yin, Xueqiang 08446
Yip, Chee-Kuen 07804

W

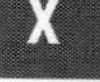

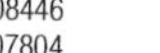

Ylioja, Merja 11286
Ylvisåker, Anne Britt 33433
Yocarini, T. 32091
Yocham, Doris R. 52910
Yochim, Jordan 49991
Yoder, Harold E. 52263
Yoel, Ben-Yosef 25037
Yokomizo, Hiroko 29392
Yokoyama, Saki 29419
Yoneya, Tetsuo 29223
Yoneyama, Torataro 29464
Yoon, Prof. Se-Young 29899
York, Beverly 54179
York, Brian 46383
Yosef, A. 25114
Yoshida, Hidekazu 28921
Yoshida, Kazuo 29216
Yoshida, Shoji 29516
Yoshihara, Lisa 49342
Yoshihisa, Suto 28374
Yoshikata, Ishihara 29361
Yoshikawa, Miho 29012
Yoshinaka, Yasumaro 28538
Yoshioka, Akemi 29444
Yoshioka, Yoji 28382
Yoshitaka, Abe 28605
Yoshiyuki, Murakawa 29589
Yotsutsuji, Hideki 29012
You, Mee Chung 29905
Youck, Ingrid 07018
Young, Anna 46798
Young, Bill 50895
Young, Bobbie 47825
Young, Brian 50138
Young, C. 44047
Young, Dan 05936
Young, David C. 46468
Young, Dede 52201
Young, Ernest 49712
Young, Gary 06914
Young, Holly 51908
Young, Jocelyn 46288
Young, John 45022
Young, John J. 07081
Young, Joyce 51551
Young, Judi Michelle 07099
Young, Karen S. 53445
Young, Kyle 50964
Young, Laurie 05823
Young, Dr. Marcia 46859
Young, Marvin R. 53086
Young, Mary Jane 46735
Young, Sir Nicholas 44806
Young, Patience 53253
Young, Richard 53175
Young, Richard L. 47271
Young, Sherilyn B. 47775
Young, Steve 48481
Young Lee, Nan 29826
Young-Sanchez, Margaret 48072
Young-Vignola, Eilee 51835
Youngken, M. Joan 51322
Youngs, Christopher 52262
Younker, Dr. Randall W. 46767
Yount, Frank 50076
Youssoufian, Jerry 06455
Yperman, Jan 03868
Ysel, H.J.D. 38718
Ysenburg und Büdingen, Elisabeth Prinzessin zu 54857
Ysenburg und Büdingen, Casimir Erbprinz zu 21368
Yu, Eric Howyi 42424
Yu, Jianling 08141
Yu, Jianyong 07768
Yu, Jun 08341, 08392
Yu, Lanshuan 08233
Yu, Louis 07811
Yu, Shenghua 08475
Yu, Sun 07547
Yu, Teng Kuei 42419
Yu, Weimin 07654
Yu, Zemin 07702
Yuan, Daojun 08501
Yuan, Hua 07661
Yuan, Zhongyi 07954
Yüce, Ahmet 42690
Yüksek, Mustafa 42869
Yuergniaux, Danièle 14928
Yūki, Ryōgo 28968
Yukio, Miyashita 28765
Yuko, Isazawa 29394
Yun, Jang Sub 29884
Yunus bin Zakaria, Dr. Mohd 30614
Yusupov, Muhammad 54441
Yut, Joyce 47229
Yutaka, Minami 28765
Yzereef, Dr. Gerard 31812

Z

Zaadnoordijk, Hans 32493
Zabarovskij, Vladimir Ivanovič 37149
Zabehlicky-Scheffenegger, Dr. Susanne 02500
Zabel, Anne 17727
Zabel, Barbara 34751
Zabel, Jürgen 22841
Zabērs, Miervaldis 30048
Zabka, Reinhard 19754
Zabrodin, A.V. 37239
Zabusch, Prof. Stephanie 02960
Zaccaguini, Dr. Marisa 27452
Zaccome, Dr. Gian Maria 28007
Žáček, PhDr. Rudolf 09692
Zacepina, Olga Vasiljevna 37755
Zacharias, Elke 21474
Zacharias, Dr. Kyllikki 17192, 17244
Zachariasse, Dr. F. 32946
Zachariasse, Dr. Fiona 32375
Zachariou, E. 09375
Zacharkiw, Irene 53821
Zacharuk, Dr. Richard 18378
Zache, Dirk 17921, 17930
Zachman, Jon 49062
Zachman, Patty 52441
Zachos, Dr. K. 23012
Zachou, Eleni 22897, 23070
Zack, Wendy 05631
Zackel, Mag. Christine 03038
Zadgorska, Valentina 05191
Zadka, Ruth 24978
Zadoropžna, Liudmila 43026
Zadorožna, Nadija Fedorivna 43175
Zäch, Benedikt 42030
Zaffarano, Antonio 28258
Zaffino, Antonio 27811
Zaforgia, Elena 26561
Zafortesza Calvet, José 39063
Zafran, Eric 49199
Zagala, Maria 00773
Žagar, Zora 38492
Zagermann, Tino 41207
Zaglmann, Klaus 20494
Zagnoli, Roberto 54501, 54506
Zagrina, Natalja Aleksandrovna 37594
Żaguń, Wanda 35059
Zahaniciuc, Marcel 36255
Žahirović, Prof. Šaban 04270
Zahn, Donald 47568
Zahn-Biemüller, Dr. Eva 22713
Zahrádka, Jiří 09438
Zahradník, Dr. Zdeněk 09534, 09917
Zaicev, Nikolai Alekseevič 29711
Zając, Zdzisława 35356
Zajączkowski, Wiesław 34630
Zajceva, Natalija Leonidovna 37041
Zajic, Václav 09584
Zajtsev, Oleg 43200
Zakatova, Anastasija Vasiljevna 36827
Zakian, Michael 50407
Zakin, Helen 51627
Zakrevskaja, G.P. 37631
Zakrzewski, R.J. 49225
Zakythinos, Prof. D. 23104
Zalba, Pat E. 47328
Zaldivar, Brenda Projas 31407
Zaleski, Pierre 14627
Zaliene, Olga 30282
Zalpienė, Danutė 30273
Zalto, Kurt 02420
Zamarchi Grassi, Dr. P. 25266
Zambrano de Gentile, Flor 54537
Zamora Bonet, Imma 00093
Zamora Canellada, Alonso 40053, 40056
Zamudio Díaz, Hermano Luis Alberto 08677
Zanche, Prof. Vittorio de 27040
Zander, Charlotte 17426
Zander, Dr. Gisela 20125
Zander-Seidel, Dr. Jutta 20764
Zandler, Richard 49983
Zanella, Prof. S. 26264
Zanella Manara, Dr. Emma 26265
Zanelli, Dr. Gianluca 26294
Zanetta, Michèle 41152
Zang, Gongmei 08406
Zangenfeind, Hermann 02637
Zankl, Mag. Eva 02887
Zankman, Patricia 48997
Zannas, D.K. 23223
Zannelli, Diletta 25896
Zanni, Dr. Annalisa 26714
Zannieri, Nina 46946
Zanola, Oscar Pablo 00660
Zanoli, Emilio 26270
Zansitis, Peter Paul 47486
Zant, Herbert 02258
Zantfrisco, Liana 03696
Žantudieva, Elizaveta Magamedovna 37438
Zapałowa, Jadwiga 34944
Zapałowa, Kazimiera 34741
Zapata, Marian 23325
Zapolis, Frank 47456
Zaporožskaja, Valentina Aleksandrovna 36858
Zappa, Giancarlo 41318
Zaracki, Kathleen 50799
Zaragüeta, Vicente 39254
Zarb, Michael 07257
Zaremba, Ewa 35056
Zaremba, Tadeusz 35014
Zarev, Dr. Kosio 05058, 05059, 05062
Zarimi, Dominique 15678
Zarina, Elena Gennadjevna 37558
Zarling, Ronald 51059
Žarmagambetov, Nurlan Kanikeevič 29684
Zarp, Frank 21566
Zarri, Dr. Enrico 26533
Zarrilli, Dr. Carla 27842
Zarzecka, Natalia 34768
Zarzuelo Villaverde, Manuel 39813
Zarzycka, Joanna 35217
Zasur, Dahas 04712
Zatloukal, Prof. Dr. Pavel 09689
Zatulovska, Olena Petrivna 43010
Zau, Daw Nu Mya 31728
Zavada, Jeanne L. 51738
Závala, Ignacio M. 38929
Zavattaro, Dr. M. 26178
Zavjalova, Irina Vasiljevna 37014
Závorková, Hana 09413
Zawacki, Andrew 50139
Zawiła, Gabriela 34712
Zawisza, Norbert 34860
Zawol, Magdalena 34630
Zawol, Zofia 35210
Zawrel, Dr. Peter 02976
Zayani, Abdelhak 31681
Zazo, Jennifer 51102
Zbarnea, Tudor 31600
Żbikowska, Janina 35053
Zbikowski, Dr. Dörte 19531
Zborowski, Janusz 34919, 34920
Zbytovský, RNDr Petr 09842
Zdesenko, Vjačeslav Iosifovič 36926
Zdravkovski, Dr. Dragiša 30519
Zea, Philip 49600, 51303
Zec, Prof. Dr. Peter 18266
Zecca, Walter 28115
Zech, Karen 47870
Zecha, Horst 21778
Zechlin, René 18997
Zeder, O. 14248
Zee, Dr. Pauline van der 03626
Zegadłowicz-Rudel, Atessa 35208
Zegg, Theo 41801
Zegher, Catherine de 51167
Zegna Rata, Olivier 11566
Zeh, Thomas 02061
Zehavi, Zvi 25053
Zehentmeier, Dr. Sabine 16578
Zehethofer, Gerhard 01886
Zehetmeier, Dr. Sabine 22731
Zehnder, Andreas 41447
Zehnder, Angela 38593
Zehnder, Walter 41975
Zeidler, Ingrid 21161
Zeiger, Antje 22629
Zeigerer, Wolfgang 19534
Zeik, Travis 54124
Zeiner, Ulrich 18892
Zeipel, Agneta von 40992
Zeisler, Stefan 03019
Zeitler, Dr. Kurt 20487
Zeitlin, Marilyn 53465
Zeitlin, Dr. Richard 50398
Zeitschel, Charlotte 19908
Zekaria, Ahmed 10678
Zekorn, Beate 18397
Zeković, Marina 38091
Zelenović, Radoslav 38068
Želev, Ivan 05097
Železnik, Adela 38454
Zelfel, Dr. Hans Peter 02395
Zelie, M. 38765
Zella, Robbin 47018
Zeller, Anselm 02872
Zeller, Peter van 45462
Zeller, Regine 17990
Zeller, Dr. Ursula 17142, 18478
Zellner, Donna 51407
Zeman, M.A. Bettina 20654
Zemanová, Jana 09837
Zembala, Dr. Dennis 48113
Zembik, Andrzej 34615
Zemer, Avshalom 24965
Zemovičs, Elmārs 30098
Zeńczak, Anna 34770
Zender, Rick 47368
Zenegas-Millan, Dorenda 51746
Zeng, Dehai 08139
Zenks, Ralph 22290
Zentgraf, Henrike 20778
Zentz d'Alnois, Philippe 12128
Zepa, Līvia 30045
Zepa, Līvija 30044
Zepeczaner, Eugen 36497
Zeppenfeld, Dr. Burkhard 20809, 20810
Zerah, Dov 14655
Zerdina, V. 30086
Zernig, Kurt 01981
Zerrudo, Eric 34472
Žertogova, Izabella Jakovlevna 36853
Zeschick, Dr. Johannes 21364
Zeskinskaja, Nelly 43159
Zettel, Ernest 01955
Zetterström, Margareta 40903
Zettl, Charleen 47004
Zettler, Richard 51886
Zeuke, Holger 17702
Zeuli, Prof. Errico 27493
Zeuli, Maria Francesca 27509
Zevallos Quinones, Dr. Jorge 34318
Zevi, Dr. Anna Gallina 27016, 27017, 27018
Ževrnja, Prof. Nediljko 08962
Zgaga, Višnja 09009
Zhai, Dang 08402
Zhai, Tunjian 07834
Zhan, Xiangsheng 08349
Zhang, Benren 08414
Zhang, Cheng Zhi 08194
Zhang, Chengzhong 08457
Zhang, Dengxian 08074
Zhang, Dianren 08277
Zhang, Guangshan 07887
Zhang, Guiyun 07698
Zhang, Guowen 08294
Zhang, Huaide 08469
Zhang, Jia 07866
Zhang, Jiaji 07732
Zhang, Jian 08293
Zhang, Jianting 07685
Zhang, Jianwen 07912
Zhang, Jidong 07595
Zhang, Jinchao 08101
Zhang, Jing 07731
Zhang, Junshao 07716
Zhang, Linxiang 08105
Zhang, Liushun 08136
Zhang, Lizhi 08362
Zhang, Lizhu 08232
Zhang, Min 08045
Zhang, Ming 08031
Zhang, Mingjun 07678
Zhang, Muga 08382
Zhang, Ning 07503, 07508
Zhang, Peilian 08471
Zhang, Qiguo 08432
Zhang, Qiqing 07846
Zhang, San 08461
Zhang, Shuliang 07666
Zhang, Songrong 07623
Zhang, Wanxing 08422
Zhang, Wei 07589
Zhang, Wen Long 08304
Zhang, Wenbao 08331
Zhang, Wenju 08085
Zhang, Wenjun 08498
Zhang, Wu'ai 08207
Zhang, Xi'ai 08087
Zhang, Xiaojun 08063
Zhang, Xinli 07907
Zhang, Xuqiu 07893
Zhang, Yingchao 08416
Zhang, Yingjun 08326
Zhang, Yongxi 08397
Zhang, Yongzhong 07771
Zhang, Yuandong 07757
Zhang, Yushan 08067
Zhang, Zeguo 08297
Zhang, Zhaosheng 07974
Zhang, Zhigang 07626
Zhang, Zhipan 07961
Zhao, Changshu 07581
Zhao, Fusheng 07573
Zhao, Genxi 08387
Zhao, Guoxiang 08353
Zhao, Mingyong 08395
Zhao, Shuzhong 08016
Zhao, Sichao 08021
Zhao, Wannian 07981
Zhao, Yunji 08310
Zhao, Yuquan 08419
Zhao, Zhiming 07616
Zhen, Qingfeng 08413
Zheng, Bingyan 08491
Zheng, Binhai 08239
Zheng, Chongjin 08347
Zheng, Junsheng 07781
Zheng, Mingjin 08268
Zheng, Peisheng 07706
Zheng, Siguang 08029
Zhi, Yunting 08217
Zhong, Chuangjian 07668
Zhong, Chuanya 07692

Zhong, Fayong 08518
Zhong, Fengcai 08033
Zhong, Guangcai 07690
Zhong, Mingqi 07938
Zhong, Yuqing 08482
Zhou, Chengshan 07896
Zhou, Chuike 08144
Zhou, Chunru 08489
Zhou, Dehua 08377
Zhou, Delin 08062
Zhou, Diguang 08013
Zhou, Diren 07657
Zhou, Enwei 07712
Zhou, Hao 07825
Zhou, Hengbin 08204
Zhou, Huajun 08012
Zhou, Jiaju 07987
Zhou, Jiakang 07681
Zhou, Jichang 07837
Zhou, Jingyi 08381
Zhou, Jinkang 07687
Zhou, Jinkui 08084
Zhou, Jisheng 07611
Zhou, Junzhi 08317
Zhou, Rongben 07750
Zhou, Ronglin 07889
Zhou, Shengrui 08022
Zhou, Shiwu 07631
Zhou, Tianlin 07872
Zhou, Tianyou 08360
Zhou, Weilin 07861
Zhou, Xinhua 08340, 08447
Zhou, Zhao 07615
Zhou, Zhongqing 08426
Zhu, Jiping 08187
Zhu, Mingye 07483
Zhu, Xianjin 08203
Zhuang, Zhongneng 08348
Zhuo, Daocheng 08368
Ziąbka, Leszek 34711
Ziauddin, Muhammad 03221
Zibelo, Fisseha 10683
Zibler, Barbara 17134
Zich, Dr. Bernd 18900
Zidić, Igor 09003
Zidlick, Pauline 46614
Zieglar, Ken 51893
Ziéglé, Anne 12066
Ziegler, Dr. Reinhard 21959
Ziegler, Dr. Susanne 17081
Ziegler, Thomas 21967
Zieglgänsberger, Dr. Roman 22557
Ziehe, Dr. Irene 17199
Zielezinski, Adam 35073
Zielinski, Melissa L. 46338
Zielonka, Mariusz 35113
Zieman, Judge Fred 49893
Ziems, Heidrun 41165
Zierden, Martha 47359
Zierke, Franciszek 35365
Ziesche, Ferdinand 18914
Zieselman, Ellen 52880
Ziesemer, Günter 18315
Ziessow, Dr. Karl-Heinz 17713
Zięzio, Ryszard 34745
Ziffer, Dr. Irit 25103
Zifferero, Dr. Andrea 27991
Žiganova, Nina Iljinična 37021
Zigmunde, Alida 30110
Zihlmann, André 41223
Zijlmans, Dr. J. 32489
Zijp, R.P. 32208
Zika, Paul 01156
Zilianti, Dr. Romualdo 25466
Žiljaeva, Natalja Evenjevna 29709
Žiljanina, Tatjana Vladimirovna 36984
Zillenbiller, Prof. Dr. Erwin 22221, 22222
Zimina, I.D. 37879
Zimmer, Georg 19973
Zimmer, Gerd 20615
Zimmer, Jim L. 50157
Zimmer, Patricia 05367
Zimmerer, Kathy 47258
Zimmerli, Martin 41682
Zimmerman, Anne 47298
Zimmerman, Bob 48157
Zimmerman, Cindy 48157
Zimmerman, Daniel W. 48614
Zimmerman, Diane 52433
Zimmerman, Ryan 47725
Zimmerman, Sara C. 51299
Zimmerman, Vickie 48205
Zimmermann, Friedrich 01869
Zimmermann, Harvey 53837
Zimmermann, Paul 19230
Zimmermann, Petra 19453
Zimmermann, Prof. Udo 17029
Zimmern, Emily F. 47390
Zimprich, Shirley 48388
Zindler, Debra 50590
Zingarelli, Thomas V. 48482
Zingg, Ernst 41709
Zink, Bob 50735
Zinke, Dr. Detlef 18428
Zinke, Olaf 19421
Zinnkann, Dr. Heidrun 18382
Zint, Günter 18958
Zintz, Daniel 15755
Ziolkowski, Anne 47873
Ziolkowski, Jan 53865
Zipfel, Dr. Bernhard 38650
Žirgulis, Rimantas 30272
Zirnitis, P. 30102
Žitko, Duška 38492
Zitko, Salvator 38434
Živankić, Radovan 38167
Živkovič, Vesna 38040
Zizman, Ron 06806
Zjeżdżałka, Ireneusz 35022
Zjumčenko, Ljudmila Petrovna 37782
Zlatar, Milena 38517
Zlatar Stambuck, José 34221
Zlobina, Nina Nikolaevna 37042
Zmrhalová, RNDr. Magda 09855
Żmuda, Anna 34895
Żmuda, Maurycy 34839
Znamenov, V.V. 37718
Žnidarič, Dr. Marjan 38476
Znojek, Jerzy 35009
Zoe, Vivian F. 51438, 54059
Zöller, Achim 21738, 21739
Zöller, Dr. Helge 22713
Zoer, J.A. 32148
Zoete, Johan de 32298
Žogina, Viktorija Stepanovna 37674
Zoli, Renato Giancarlo 27966
Zoller, Tamara 46495
Zolli, Barbara T. 51991
Zollinger Dettwiler, lic. phil. Yvonne 41561
Zolotareva, Tatjana Nikolaevna 37567
Zolotina, I.A. 37659
Zolotinkina, Dr. Larissa Igorevna 37664
Zolotova, Anastasija 37260
Zoltán, Kató 36602
Zombori, Dr. István 23843
Zombork, Márta 23944
Zomborka Szűcsné, Márta 23946
Zommer, Raya 24946
Zona, Louis A. 54353
Zong, Changhai 08069
Zonneveld-Kouters, M.C. 32654
Zopf, August 01908, 02405, 02407
Zorić, Ivanka 38074
Zorn, Dr. Bettina 03038
Zorn, Dr. Olivia 17028
Zorzi, Dr. Marino 28209
Zorzoliu, Traian 36353
Zou, Lianhe 08290
Zou, Zhifang 07714
Zoubek, Ernst 02835
Zoubek, Thomas 53248
Zrebiec, Alice 48072
Zrelov, Vladimir Andreevič 37611
Zsidi, Dr. Paula 23382, 23405
Zsigmond, Attila 23389
Zsutty, Andrea 02942
Zsutty, Dr. Gerhard 03092
Zsuzsa, Dr. Fodor 23959
Zubajdullaeva, S. 54459
Zuber, Michaela 18341
Zuberbühler, Walter 41932
Zuberer, Peter 22374
Zubiaurre, Pablo 00131
Zubov, Aleksandr Nikolaevič 37897
Zubrij, Elena Stanislavovna 36932
Zucal, Wendy R. 48058
Zuccali, Jean-Pierre 13814
Zucchi, Cristina 26726
Zuckerman, B. Michael 47235
Zuckerman Jacobson, Heidi 46397
Zuckriegl, Dr. Margit 02620
Zünd, Peter 41699
Zürcher, Barbara 41103
Züst, Ernst 42045
Zueva, Ljubov Vasiljevna 37827
Zufferey, Anne-D. 41799, 41867
Žugan, Vitalij Ivanovič 37840
Zugaza, Miguel 39573
Zugazagoitia, Julian 51220
Zuidervaart, Dr. Huib J. 32630
Zuiderwish, Dr. A.C. 31904
Žuk, Oksana Jaroslavivna 43119
Żuk, Stanisław 34566
Žukauskien£, Genovaite 30354, 30361
Žukauskienė, Genovaite 30365
Žukauskienė, Rita 30316
Zukulis, Ivars 30104
Zuleski, Jim 49556
Zuleta, Pilar 08573
Zumaeta Zúñiga, Hector 07467
Žumalieva, Tamara Kalmurzaevna 29670
Zumbach, Ueli 41802
Zumpe, Ulrike 21085, 21086, 21104, 21107
Zumwalt, May Louise 51669
Zuna-Kratky, Dr. Gabriele 03079
Zunk, Bodo 21253
Žunusov, Remistokol Tuskenovič 29736
Zuo, Zijun 08153
Zupan, Michelle 48960
Zuparko, Robert 46753
Žuravlev, M.N. 03245
Žuravleva, Tatjana Jurjevna 37443
Zurawski, Jarosław 35371
Zurbrugg-Tonezzer, Lucia 41288
Zurbuchen, Max 41555
Zurcher, Sarah 41386
Zurfluh, Dr. Anselm 41735
Zurfluh, Hubert 12468
Zuris, Donald P. 47820
Zurowski, Dr. Stanis'Elaw 35255
Zuskinová, Dr. Iveta 38285
Zutautas, Sofia 47465
Žutikov, Aleksandr Ivanovič 37509
Zuver, W. Marc 53870
Zvelebil, Miloš 09824
Zviguilsky, Dr. Alexandre 12085
Zvirin, Prof. Yoram 24966
Zwaenepoel, Dr. J. 03706
Zwahlen, Ulrich 41511
Zwaka, Petra 17148
Zwanger, Meryl 51167
Zwart, Joel 49000, 49001
Zwart-Van der Gaag, A. 32425
Zweifel, Gwen 05843
Zwez, Annelise 41954
Zwiebelhofer, Dr. Otmar 16879
Zwiers, S. 33030
Zwierzyna, John 49181
Zwink, Dr. Eberhard 21969
Zwolak, Adam 34534
Zybok, Prof. Dr. Oliver 21265
Zykaner, Michaela 03027
Zykov, Andrej Viktorovič 37030
Zyman, Daniela 03081

Z

Subject Index

List of Subjects

Agricultural Implements and Machinery
Agriculture
Amber
Anthropology → Ethnology
Apiculture → Bees
Archaeology
Archaeology, Christian and Medieval
Archaeology, Far Eastern
Archaeology, Greek and Roman
Archaeology, Ibero-American
Archaeology, Near and Middle Eastern
Architecture
Arms and Armour
Art
Art, African
Art, American
Art, Asian
Art, Australian
Art, European
Art, Greek and Roman
Art, Latin American
Art, Medieval
Art, Modern and Contemporary
Art, Oriental
Art, Russian
Astronomy
Automobiles → Vehicles
Aviation
Baking and Cookery
Balneology
Banks and Banking
Baskets
Bees
Bells
Biology
Birds
Boats and Shipping
Bones
Books, Book Art and Manuscripts
Botany
Brewing
Bronzes
Bullfights
Buttons
Cabaret → Performing Arts
Calligraphy
Cameras → Optics
Carpets
Carriages and Carts
Cartography
Carts → Carriages and Carts
Carving, Wood → Wood Carving
Castles and Fortresses
Ceramics → Porcelain and Ceramics
Chemistry
China (Porcelain) → Porcelain and Ceramics
Cigarettes → Tobacco
Cinematography
Circus → Performing Arts
Clocks and Watches → Horology
Clothing and Dress
Coaches → Carriages and Carts
Coffee → Tea and Coffee
Coins → Numismatics
Cookery → Baking and Cookery
Copper
Cosmetics
Costume → Clothing and Dress
Crafts → Handicraft
Criminology
Dairy Products
Dance → Performing Arts
Decorative and Applied Arts
Distilling → Wines and Spirits
Dolls and Puppets
Drawing
Ecology
Economics
Education
Egyptology
Electricity
Embroidery → Needlework
Enamel Works
Energy
Engraving → Graphic Arts
Entomology → Insects
Epitaphs → Inscriptions and Epitaphs
Ethnology
Expeditions → Scientific Expeditions
Explosives
Eyeglasses → Optics
Faience → Porcelain and Ceramics
Farms and Farming
Fashion → Clothing and Dress
Festivals
Fire Fighting and Rescue Work
Firearms → Arms and Armour
Fishing and Fisheries
Folk Art
Folklore
Food
Forest Products
Fortresses → Castles and Fortresses
Fossils
Fresco Painting → Mural Painting and Decoration
Fur → Hides and Skins
Furniture and Interior
Games → Toys and Games
Gas → Petroleum Industry and Trade
Gems → Precious Stones
Genealogy
Geography
Geology
Glass and Glassware
Globes → Cartography
Gold and Silver

Graphic Arts
Gravestones → Tombs
Handicapped
Handicraft
Heraldry → Genealogy
Hides and Skins
History
History, Early → Prehistory and Early History
History, Local and Regional
History, Maritime → Boats and Shipping
History, Medieval
History, Modern
History, Social → Social Conditions
Horology
Horses
Horticulture
Household Articles
Hunting
Hygiene → Medicine and Hygiene
Icons
Incunabula → Books, Book Art and Manuscripts
Indian Artifacts
Industry
Inscriptions and Epitaphs
Insects
Instruments of Torture
Ironwork → Metalwork
Ivory
Jade
Jewellery → Precious Stones
Judaica
Keys → Locks and Keys
Lace → Needlework
Lacquerwork
Lamps and Lighting
Language and Literature
Law
Lead and Tin Figures
Leather
Literature → Language and Literature
Locks and Keys
Lumber → Forest Products
Machines and Machinery
Majolica → Porcelain and Ceramics
Mammals
Man, Prehistoric
Manuscripts → Books, Book Art and Manuscripts
Maps → Cartography
Marine Archaeology
Marine Biology → Oceanography
Masks
Mass Media
Medals → Numismatics
Medicine and Hygiene
Metals and Metallurgy
Metalwork
Meteorites
Military History
Mills
Mineralogy
Miniature Painting
Mining Engineering
Mollusks
Monasteries
Money → Numismatics
Mosaics
Mummies
Mural Painting and Decoration
Museology
Music and Musical Instruments
Naive Art → Folk Art
Natural History
Needlework
Numismatics
Oceanography
Oenology → Wines and Spirits
Office Equipment
Oil → Petroleum Industry and Trade
Opera → Performing Arts
Optics
Ornithology → Birds
Painted and Stained Glass
Painting
Painting, 17th C.
Painting, 18th C.
Painting, 19th C.
Painting, 20th C.
Painting, Dutch
Painting, English
Painting, European
Painting, French
Painting, German
Painting, Italian
Painting, Latin American
Painting, Medieval
Painting, Renaissance
Painting, Spanish
Paleontology → Fossils
Paper
Peasant Life and Traditions
Performing Arts
Petroleum Industry and Trade
Pets
Pewter
Pharmacy
Philately → Postage Stamps
Philosophy
Photography
Physics
Pipes, Tobacco → Tobacco
Plaster Casts
Playing Cards
Police → Criminology
Politics and Government
Porcelain and Ceramics
Portraits
Postage Stamps
Postal Services
Posters
Pottery, Earthenware → Porcelain and Ceramics
Precious Stones
Prehistory and Early History
Press → Mass Media
Printing
Prints → Graphic Arts
Prisons → Criminology
Puppets → Dolls and Puppets
Radio → Mass Media
Railroads
Rare Books → Books, Book Art and Manuscripts
Religious Art and Symbolism
Religious History and Traditions
Reptiles
Rescue Equipment → Fire Fighting and Rescue Work
Rocks → Mineralogy
Salt
Sarcophagi → Tombs
Scientific Apparatus and Instruments
Scientific Expeditions
Sculpture
Shells
Shipping → Boats and Shipping
Shoes
Silk
Silver → Gold and Silver
Skeletons → Bones
Social Conditions
Speleology
Spirits → Wines and Spirits
Sports and Recreation
Stained Glass → Painted and Stained Glass
Stelae → Tombs
Stoves
Stucco
Sugar
Tapestries
Tea and Coffee
Technology
Telecommunications → Postal Services
Television → Mass Media
Terracotta
Textiles
Theatre → Performing Arts
Tin Figures → Lead and Tin Figures
Tobacco
Tombs
Tools
Tourism
Toys and Games
Trades and Guilds
Transport
Travel → Tourism
Treasuries
Typography → Books, Book Art and Manuscripts
Uniforms, Military
Vehicles
Veterinary Medicine
Viticulture → Wines and Spirits
Voyages of Discovery → Scientific Expeditions
Wallpaper
Water
Wax, Molded
Weapons → Arms and Armour
Weaving
Weights and Measures
Whaling → Fishing and Fisheries
Wines and Spirits
Wood → Forest Products
Wood Carving
Zoology

Agricultural Implements and Machinery

Australia

Ardrossan Historical Museum, Ardrossan 00811
Cleve National Trust Museum, Cleve 00982
Donald Agricultural Museum, Donald 01040
Doug Kerr Vintage Museum, Oaklands 01387
Farming Through the Ages, Whyte Yarcowie 01650
Mingenew Museum, Mingenew 01306
North Western Agricultural Museum, Warracknabeal 01630
Pearn's Steam World, Westbury 01645
Yankalilla District Historical Museum, Yankalilla 01677

Austria

Bäuerliches Gerätemuseum und Venezianer-Gatter, Innervillgraten 02111
Bauern- und Heimatmuseum Haigermoos, Sankt Pantaleon 02675
Dorf- und Kutschenmuseum, Eisgarn 01864
Freilichtmuseum Hayrl, Reichenthal 02571
Freilichtmuseum Stehrerhof - Dreschmaschinen-museum, Neukirchen an der Vöckla 02432
Handwerkermuseum Laufenbach, Taufkirchen an der Pram 02797
Landtechnikmuseum, Leiben 02279
Landtechnisches Museum Burgenland, Sankt Michael im Burgenland 02670
Landwirtschaftliches Museum und Pfarrer Jungblut-Museum, Prinzendorf 02533
Lokalhistorisches Pulkautalmuseum, Hadres 02034
Museum der Landtechnik, Knittelfeld 02207
Museum Schloss Peigarten, Dobersberg 01820
Der Ölschlag, Kirchberg in Tirol 02168
Schloß- und Heimatmuseum, Sigharting 02741
Traktorenmuseum, Vordernberg 02879
Troadkasten, Freinberg 01902
Troadkasten, Schardenberg 02698
Troadkasten Fornach, Fornach 01895

Belgium

Heemmuseum De Kaeck, Wommelgem 04156
La Remise Musée de la Vie Rurale, Offaing 03920
Musée de la Poterie et Musée du Jambon d'Ardenne, La Roche-en-Ardenne 03771
Musée Vivant du Cheval de Trait, Waret-l'Evêque 04132

Canada

Anderson Farm Museum, Madoc 06213
Backus Heritage Village, Simcoe 06978
British Columbia Farm Machinery and Agricultural Museum, Fort Langley 05811
Camrose and District Centennial Museum, Camrose 05544
Centre d'Interprétation de l'Agriculture et de la Ruralité, Métabetchouan-Lac-à-la-Croix 06270
Crossroads Museum, Oyen 06565
Frobisher Thresherman's Museum, Frobisher 05844
Grants Old Mill, Winnipeg 07362
Heritage Farm Village, North Battleford 06468
Keystone Pioneer Museum, Roblin 06764
Missisquoi Museum, Stanbridge East 07004
Musée des Pionniers, Grande-Digue 05898
Musée Saint-Joseph, Saint-Joseph 06858
Pembina Threshermen's Museum, Winkler 07350
Prairie Acres Museum, Picture Butte 06604
Saanich Historical Artifacts Society, Saanichton 06785
Swords and Ploughshares Museum, Kars 06044
Teeterville Pioneer Museum, Teeterville 07062
Whitemouth Municipal Museum, Whitemouth 07330

Denmark

Blaavandshuk Egnsmuseum, Oksbøl 10219

Estonia

Hurda Talumuuseum, Hurda 10571
Jõevere Talumuuseum, Lindora 10592
Sillaotsa Talumuuseum, Rapla 10613

Finland

Helsingin Yliopiston Maatalousmuseo, Helsinki 10786
Kihniön Museo, Kihniö 10955
Kodisjoen Kotiseutumuseo, Kodisjoki 10968
Loimaan Kotiseutumuseo, Loimaa 11080
Peltotyökalumuseo, Somero 11321
Sibbo Jordbruksmuseum, Nikkila 11142
Suomusjärven Kotiseutumuseo, Suomusjärvi 11339
Talomuseo, Polvijärvi 11217

France

Ecomusée Agricole du Pays de Fayence, Fayence 12911
Écomusée de l'Agriculture, Aigues-Vives 11505
Ferme-Musée de la Tonnelière, Courcerault 12651
Musée Agricole, Sainte-Geneviève (Meurthe-et-Moselle) 15561
Musée Agricole Vivant, Bissey-la-Pierre 12004
Musée Beauceron du Grand Breau, Tivernon 15916
Musée Conservatoire de l'Agriculture, Chartres 12422
Musée de la Faux, Pont-Salomon 14895
Musée de la Machine Agricole, Vroncourt 16279
Musée de la Machine Agricole Ancienne, Saint-Loup 15379
Musée de la Machine Agricole et à Vapeur, Ambert 11568
Musée de la Vie Rurale, Steenwerck 15807
Musée de l'Outil et des Vieux Métiers, La Celle-Guenand 13325
Musée de Machines Agricoles Miniatures, Guerlesquin 13161
Musée des Champs, Saint-Ségal 15509
Musée des Outils Agraires, Châteaurenard (Bouches-du-Rhône) 12455
Musée du Machinisme Agricole, Neuvic (Corrèze) 14406
Musée du Machinisme Agricole, Prissac 14949
Musée du Machinisme Agricole, Tullins-Fures 16009
Musée du Matériel Agricole, La Villeneuve-les-Convers 13456
Musée du Moteur et de l'Outil à la Ferme, Juvigné 13297
Musée du Terroir, Sisteron 15763
Musée Régional du Machinisme Agricole, La Ferté-Milon 13363
Musée Revoir le Passé, Les Loges-Marchis 13703
Musée Thimonnier Doyen, Saint-Germain-au-Mont-d'Or 15285
Musées de Vieille Forge, de la Traversée du Siècle et des Automates, Fréland 13007

Germany

Altlandtechnik-Museum, Klein Upahl 19562
Bachgau-Museum, Großostheim 18800
Bäuerliches Heimatmuseum der Landtechnik, Geiselhöring 18556
Bauern- und Waldmuseum, Haidmühle 18865
Bauerngerätemuseum Hundszell, Ingolstadt 19338
Bauernhaus-Museum, Ruhmannsfelden 21444
Bauernhausmuseum des Landkreises Erding, Erding 18193
Bauernhofmuseum Hof, Kirchanschöring 19542
Bauernmuseum Liebenau, Geising 18559
Bauernmuseum Mühlhausen, Villingen-Schwenningen 22234
Bauernmuseum Ostdorf, Balingen 16898
Bauernmuseum Pfarrscheuer, Bösingen 17430
Dinkelbergmuseum Minseln, Rheinfelden 21299
Dorfmuseum Weckbach, Weilbach 22383
Freilichtmuseum Ostenfelder Bauernhaus, Husum 19298
Freilichtmuseum Scherzenmühle Weidenberg, Weidenberg 22368
Friedrich-der-Große-Museum im Museumshof Lerchennest, Sinsheim 21784
Gerätesammlung Koch, Königsberg in Bayern 19641
Gottlieb-Häußler-Heimatmuseum, Filderstadt 18317
Heimat- und Bergbaumuseum Nentershausen, Nentershausen 20570
Heimat- und Hafnermuseum, Heidenheim 19089
Heimatmuseum, Dietenhofen 17853
Heimatmuseum, Holzgerlingen 19257
Heimatmuseum, Niederaichbach 20701
Heimatmuseum, Stadtsteinach 21861
Heimatmuseum, Waldaschaff 22281
Heimatmuseum, Weissach im Tal 22446
Heimatmuseum, Wertach 22501
Heimatmuseum Beutelsbach, Weinstadt 22436
Heimatmuseum Grimmen "Im Mühlentor", Grimmen 18766
Heimatmuseum Weiler, Schorndorf 21626
Heimatstuben im Oberen Tor, Scheinfeld 21540
Heimethüs, Denzlingen 17823
Historische Ochsentretanlage im Brunnenhausmuseum, Schillingsfürst 21552
Kreisagrarmuseum, Dorf Mecklenburg 17901
Landmaschinenmuseum - Sammlung Speer, Rimbach 21332
Landsratamt Rems-Murr-Kreis, Waiblingen 22277
Landtechnik-Museum Braunschweig Gut Steinhof, Braunschweig 17524
Mittelschwäbisches Heimatmuseum Krumbach, Krumbach 19729
Museum für bäuerliche Arbeitsgeräte des Bezirks Oberfranken, Bayreuth 16961
Museum Wildberg, Wildberg 22567
Museumsschmiede, Winnweiler-Langmeil 22604
Museumsstadl, Bernried, Kreis Deggendorf 17303
Privatmuseum Hans Klein, Prichsenstadt 21125
Privatsammlung Leo Gesell, Triefenstein 22094
Riedenburger Bauernhofmuseum, Riedenburg 21318
Sammlung landwirtschaftlicher Geräte, Unterpleichfeld 22183
Schulheimatmuseum, Asbach-Bäumenheim 16580
Sensen- und Heimatmuseum, Achern 16423
Stadtmuseum Schloss Wolfsburg, Wolfsburg 22667
Städtische Sammlungen, Adelsheim 16429
Traktoren-Museum Kempen, Horn-Bad Meinberg 19273

Hungary

Mezogazdasági Eszköz- és Gépfejlodéstörténeti Szakmúzeum, Gödöllő 23568
Mezőgazdasági Gépmúzeum, Mezőkövesd 23699

Ireland

An Dun Transport and Heritage Museum, Ballinahown 24734
Threshing Museum, Mullinahone 24879

Israel

Museum of Pioneer Settlement, Kibbutz Yifat 25059

Italy

Museo della Civiltà Contadina in Val Vibrata, Controguerra 25964
Museo della Tecnologia Contadina, Santu Lussurgiu 27735

Latvia

Latvijas Lauksaimniecības Muzejs, Talsi 30127
Latvijas Republikas Meliorācijas un Zemkopības Muzejs, Mālpils 30047
Zemgales Sētas un Lauksaimniecības Mašīnu Muzejs, Uzvara 30135

Lithuania

Žemaitijos Kaimo Muziejus, Telšiai 30380

Mexico

Museo Aperos de Labranza, Lerdo 31001

Namibia

Helmeringhausen Museum, Helmeringhausen 31741

Netherlands

Automuseum Histo-Mobil, Giethoorn 32251
Boerderij Klein Hulze, Almen 31801
Boerenwagenmuseum, Buren, Gelderland 32024
Fries Landbouw Museum, Exmorra 32228
Nationaal Museum Historisch Landbouwtechniek, Wageningen 32920
Plumershuuske, Diepenheim 32126
Streekmuseum Hoeksche Waard, Heinenoord 32349
Het Trekkermuseum, Nisse 32596

New Zealand

Pigeon Valley Steam Museum, Wakefield 33231

Norway

Auli Mølle, Årnes 33398
Egge Museum, Steinkjer 33879
Jærmuseet, Nærbø 33690

Poland

Muzeum Prasy Śląskiej, Pszczyna 35060

Portugal

Museu da Alfaia Agrícola, Estremoz 35576
Museu da Alfaia Agrícola, Mafra 35749
Museu da Alfaia Agrícola, Montemor-o-Novo 35774
Museu da Alfaia Agrícola, Santo Amaro 35913
Museu das Técnicas Rurais e Museu Municipal de Oliveira de Frades, Oliveira de Frades 35793

Romania

Muzeul Agriculturii din România, Slobozia 36629

Slovakia

Múzeum Beckov, Beckov 38198
Vlastivedné Múzeum, Trebišov 38374

Slovenia

Muzej Hmeljarstva, Žalec 38539

South Africa

Settlers Museum, Himeville 38642

Spain

Museo Etnográfico Benigno Eiriz, Alija del Infantado 38847

Sweden

Sala Traktormuseum, Sala 40827

Switzerland

Agrotechnorama, Ettenhausen 41364
Heimatmuseum, Aesch 41090
Heimatmuseum, Attiswil 41132
Musée Agricole, Coffrane 41307
Musée Romand de la Machine Agricole, Gingins 41432
Museo etnografico della Valle di Muggio, Cabbio 41279
Ortsmuseum, Nürensdorf 41688
Ortsmuseum Albisrieden, Zürich 42100
Ortsmuseum Wiesendangen, Wiesendangen 42015
Schweizerisches Agrarmuseum Burgrain, Alberswil 41099

Taiwan

Exhibition Hall of Agricultural Implements, Neipu 42284

United Kingdom

Fife Folk Museum, Ceres 43785
Greenfield Valley Museum, Greenfield 44362
Lackham Museum of Agriculture and Rural Life, Lacock 44628
Laidhay Croft Museum, Dunbeath 44043
March and District Museum, March 45103
Museum of Lincolnshire Life, Lincoln 44714
Northfield Farm Museum, New Pitsligo 45206
Sheppy's Farm and Cider Museum, Bradford-on-Tone, Taunton 43584
Sussex Farm Museum, Heathfield 44433
Walton Hall Museum, Linford 44718

United States of America

Agricultural Heritage Museum, Boerne 46888
Deere Museum, Moline 50778
Forest Park Museum, Perry 51780
Midwest Old Settlers and Threshers Association Museum, Mount Pleasant 50883
Museum of the Northern Great Plains, Fort Benton 48593
The National Agricultural Center and Hall of Fame, Bonner Springs 46901
Overstreet-Kerr Historical Farm, Keota 49747
San Joaquin County Historical Museum, Lodi 50159
Steam Engine Museum, Mabel 50338

Agriculture

Argentina

Museo de la Máquina Agrícola, Esperanza 00368
Museo Molino Nant Fach, Trevelin 00655

Australia

Avondale Discovery Farm, Beverley 00865
Balaklava Museum - Urlwin Park, Balaklava 00823
Churchill Island Agricultural Museum, Newhaven 01366
Court House Museum, Stroud 01534
Cunderdin Municipal Museum, Cunderdin 01016
D.A. Wurfel Grain Collection, Pinnaroo 01416
Farm Shed Museum, Kadina 01182
Geralka Rural Farm, Spalding 01520
Gledswood Farm Museum and Homestead, Catherine Fields 00961
Gulf Station Historic Farm, Yarra Glen 01678
Maffra Sugarbeet Museum, Maffra 01249
Maitland Museum, Maitland, South Australia 01251
National Trust Museum, Streaky Bay 01533
Old Council Chambers Museum, Cleve 00983
Old Highercombe Hotel Museum, Tea Tree Gully 01575
Old Timers Traeger Museum, Alice Springs 00801
Rosalie Shire Historical Museum, Goombungee 01119
Roseworthy Agricultural Museum, Roseworthy 01469
Rouse Hill Estate, Rouse Hill 01472
Trentham Agricultural and Railway Museum, Trentham 01602
Wyalkatchem C.B.H. Agricultural Museum, Wyalkatchem 01673

A

Austria

Agrargeschichtliche Sammlung, Wels 02911
Alpsennereimuseum, Hittisau 02083
Bauernmuseum Grottenhof, Kaindorf an der Sulm 02148
Bauernmuseum Sollingerhof, Maria Schmolln 02353
Bergbauernmuseum, Alpbach 01705
Bergbauernmuseum, Telfs 02801
Denkmalhof Gererhof, Annaberg 01722
Denkmalhof Maurerhaus, Zederhaus 03125
Freilichtmuseum, Bad Tatzmannsdorf 01766
Freilichtmuseum Mondseer Rauchhaus, Mondsee 02405
Haus des Moores, Heidenreichstein 02060
Heimatmuseum im alten Bruderhaus und in der Dechanathofscheune, Altenmarkt im Pongau 01713
Kärntner Freilichtmuseum mit Bauernmöbelsammlung, Maria Saal 02352
Keltendorf Mitterkirchen, Mitterkirchen 02388
Landwirtschaftsmuseum Schloss Ehrental, Klagenfurt 02187
Lavanttaler Obstbaumuseum, Sankt Paul 02677
Mentlhof-Bergbauernmuseum, Heiligenblut 02066
Mostviertler Bauernmuseum, Amstetten 01719
Ötztaler Heimat- und Freilichtmuseum, Längenfeld 02251

Belgium

Bardelaere Museum, Lembeke 03783
Ecomusée du Pays des Collines, Lahamaide 03773
Havesdonckhoeve, Bornem 03385
Hopmuseum De Stadsschaal, Poperinge 03953
Karrenmuseum, Essen 03587
La Remise Musée de la Vie Rurale, Offaing 03920
Landbouwmuseum Leiedal, Bissegem 03370
Maison de la Ruralité, Thuillies 04063
Musée de la Vie Rurale Li Vile Grègne, Erezée 03584
Musée du Cuir et des Industries du Péruwelzis, Péruwelz 03949
Museum Kempenland, Lommel 03829
Museum Rufferdinge, Landen 03776

Brazil

Fundação Museu do Zebu Edilson Lamartine Mendes, Uberaba 04940
Museu do Tropeiro, Castro 04408
Museu Fazenda, Nanuque 04580
Museu Municipal Visconde de Guarapuava, Guarapuava 04485

Brunei

Muzium Teknologi Melayu, Bandar Seri Begawan 04966

Bulgaria

Muzej na Bivolarstvoto i Konevädstvoto, Šumen 05217
Nacionalen Selskostopanski Muzej, Sofia 05199

Canada

1910 Boomtown, Saskatoon 06909
Agricultural Museum of New Brunswick, Sussex 07039
Anderson Farm Museum, Madoc 06213
Anderson Farm Museum, Lively 06176
Beckoning Hills Museum, Boissevain 05433
Blair House Museum, Kentville 06061
Blockhouse Museum, Merrickville 06267
Brownvale North Peace Agricultural Museum, Brownvale 05490
Canada Agriculture Museum, Ottawa 06532
Central Butte District Museum, Central Butte 05583
Centre d'Interprétation de l'Agriculture et de la Ruralité, Métabetchouan-Lac-à-la-Croix 06270
Chemainus Valley Museum, Chemainus 05606
Cole Harbour Heritage Farm Museum, Cole Harbour 05630
Comber and District Historical Society Museum, Comber 05634
Compton County Historical Museum, Eaton Corner 05734
Country Heritage Park, Milton 06288
DeBolt and District Pioneer Museum, DeBolt 05686
Écomusée du Fier Monde, Montréal 06338
Gaspesian British Heritage Village, New Richmond 06432
Hanna Pioneer Museum, Hanna 05967
Historic O'Keefe Ranch, Vernon, British Columbia 07237
International Fox Museum, Summerside 07033
John P. Fisher Memorial Museum, Kingston 06079
Keillor House Museum, Dorchester 05708
Kelowna Museum, Kelowna 06054
Kilby Store and Farm, Harrison Mills 05974
Kings Landing Historical Settlement, Prince William 06673
Lac Sainte-Anne Historical Society Pioneer Museum, Rochfort Bridge 06766
The Log Farm, Nepean 06424
Mackenzie Crossroads Museum, High Level 05984
MacPherson's Mill and Farm Homestead, New Glasgow 06430
Maison du Colon, Ville-Marie 07266
Manitoba Agricultural Hall of Fame Inc., Brandon 05464
Manitoba Agricultural Museum, Austin 05364
Maplewood Farm, North Vancouver 06479
Mayne Island Museum, Mayne Island 06254
Musée de Guérin, Guérin 05920
Musée Sainte-Martine, Sainte-Martine 06900
Museum of the Cariboo Chilcotin, Williams Lake 07337
Orwell Corner Historic Village, Orwell 06519
Peace River Centennial Museum, Peace River 06575
Pouce Coupe and District Museum, Pouce Coupe 06652
Riverview Ethnographic Museum, Bear River 05404
Ross Farm Museum, New Ross 06434
Salt Spring Island Museum, Ganges 05849
South Peace Centennial Museum, Beaverlodge 05408
Story of People, Yorkton 07416
Victoria County Archiv and Museum, Baddeck 05371

China

Anhui Agriculture Museum, Hefei 07774
China Agricultural Museum, Beijing 07526
Shihezi Junken Museum, Shihezi 08230

Cuba

Museo La Isabelica, Santiago de Cuba 09323

Czech Republic

Kotulova Dřevěnka, Havířov 09508
Lovecko Lesnicke Muzeum, Úsov 09891
Muzeum Českého Venkova, Kutná Hora 09615
Muzeum zahradnictví a životního prostředí, Valtice 09897

Denmark

Bov Museum, Padborg 10225
Dansk Landbrugsmuseum, Auning 09973
Fjordmuseet, Jyllinge 10111
Frilandsmuseet i Maribo, Maribo 10184
Den Fynske Landsby, Odense 10214
Glud Museum, Horsens 10096
Herning Museum, Herning 10072
Klosterlund Museum og Naturcenter, Engesvang 09998
Kongenhus Mindeparks Museum, Viborg 10315
Landbrugs- og Interiørmuseet, Farsø 10018
Landbrugsmuseet Melstedgård, Gudhjem 10047
Landbrugsmuseet Skarregaard, Nykøbing Mors 10202
Sydhimmerlands Museum, Hobro 10083
Vestfyns Hjemstavnsgård, Glamsbjerg 10039

Estonia

Eesti Põllumajandusmuuseum, Tõrvandi 10664
Sillaotsa Talumuuseum, Rapla 10613
Väike-Männiku Talumuuseum, Kõpu 10585

Finland

Annalan Kotiseutumuseo, Pyhäjoki 11244
Bragegården, Vaasa 11422
Etelä-Pohjanmaan Maakuntamuseo, Seinäjoki 11304
Isonkyrön Kotiseutumuseo, Isokyrö 10862
Jokioisten Pappilamuseo, Jokioinen 10876
Kalannin Kotiseutumuseo, Kalanti 10907
Karjalainen Kotitalo, Imatra 10855
Kauhajoen Museo, Kauhajoki 10927
Kaunislehdon Talomuseo, Hyrynsalmi 10845
Keuruun Kotiseutumuseo, Keuruu 10951
Kokemäen Maatalousmuseo ja Sotilaspoikamuseo, Kokemäki 10969
Laurinmäen Torpparimuseo, Janakkala 10870
Liuksialan Maatalousmuseo, Kangasala 10909
Luopioisten Museo, Luopioinen 11084
Murtovaaran Talomuseo, Valtimo 11441
Museum Jan Karlsgården och Vita Björn, Sund 11334
Mustialan Maataloushistoriallinen Museo, Mustiala 11127
Pielisen Museo, Lieksa 11067
Pienmäen Talomuseo, Niemisjärvi 11140
Pöljän Kotiseutumuseo, Pöljä 11210
Pudasjärven Kotiseutumuseo, Pudasjärvi 11235
Pyhäjärven Kotiseutumuseo, Pyhäjärvi 11243
Reipin Museo, Pirkkala 11209
Reppuniemen Ulkomuseo, Pöytyä 11213
Ruoveden Kotiseutumuseo, Ruovesi 11282
Rymättylän Kotiseutumuseo, Rymättylä 11284
Talonpoikaismuseo Yli-Kirra, Punkalaidum 11240
Tervajärven Talomuseo, Sotkamo 11328
Urjalan Museo, Urjala 11418
Vanha-Rantalan Talomuseo, Pertunmaa 11201
Viljamakasiinimuseo, Vihti 11464
Yläneen Kotiseutumuseo, Yläne 11478

France

L'Agriculture au Fil du Temps, Maniquerville 13937
Agropolis Museum, Montpellier 14236
Archéo-Site Gaulois de Saint-Julien, Saint-Julien 15343
Ecomusée Agricole du Pays de Fayence, Fayence 12911
Ecomusée de la Bresse Bourguignonne, Pierre-de-Bresse 14812
Écomusée de la Courneuve, La Courneuve 13354
Écomusée de la Forêt d'Orient, Brienne-la-Vieille 12175
Ecomusée de la Grande Lande Marquèze, Sabres 15144
Ecomusée de la Vallee d'Aspe, Accous 11489
Ecomusée de l'Ile-de-Groix, Ile-de-Groix 13237
Ecomusée de Marais Breton Vendéen, La Barre-de-Monts 13311
Ecomusée de Savigny-le-Temple, Savigny-le-Temple 15679
Ecomusée des Pipes Ropp, Baume-les-Dames 11865
Écomusée du Liège, Gonfaron 13081
Ecomusée du Pays de Rennes, Rennes (Ille-et-Vilaine) 15012
Ecomusée du Veron, Savigny-en-Veron 15678
Ecomusée Picarvie, Saint-Valéry-sur-Somme 15531
Écomusée Rural de Vigneux-La Pauelais, Vigneux-de-Bretagne 16160
Ferme-Musée de Bray, Sommery 15781
Ferme-Musée du Contentin, Sainte-Mère-Eglise 15573
Maison de la Vie Rurale, Massiac 14027
Maison des Abeilles, Brignoles 12179
Musée Accous-Fermiers Basco-Béarnais, Accous 11490
Musée Agricole, Villy-le-Maréchal 16246
Musée Agricole du Château de Didonne, Semussac 15710
Musée Artisanal et Rural d'Arts, Ars 11702
Musée Briard et de la Crypte Mérovingienne, Jouarre 13280
Musée Conservatoire de la Vie Agricole et Rurale d'Autrefois, Gimont 13062
Musée d'Art et d'Histoire de Toul, Toul 15922
Musée d'Arts et de Traditions Populaires, Mussidan 14341
Musée de la Coiffe et du Costume Oléronnais, Le Grand-Village-Plage 13588
Musée de la Dentelle, Mirecourt 14134
Musée de la Ferme et des Vieux Métiers, Bosquentin 12079
Musée de la Mémoire Paysanne, Estissac 12868
Musée de la Moisson, Sagy (Magny-en-Vexin) 15149
Musée de la Vallée à Saint-Paul-sur-Ubaye, Saint-Paul-sur-Ubaye 15447
Musée de la Vie Rurale, Desaignes 12715
Musée de la Vie Rurale et Forestière, Saint-Michel-en-Thiérache 15419
Musée de la Vigne et de la Vie Rurale, Plauzat 14834
Musée de l'Agriculture Auvergnate, Coltines 12589
Musée de l'Agriculture et de la Vie Rurale, Labrousse 13466
Musée de l'Estérel, Les Adrets-de-l'Estérel 13682
Musée de l'Habitat Rural Passé et des Traditions, La Garnache 13376
Musée de l'Outillage Artisanal Rural et Bourguignon, Cruzille-en-Mâconnais 12682
Musée de Sucy, Sucy-en-Brie 15836
Musée départemental du Revermont, Treffort 15979
Musée des Anciens Outils Agricoles, Saint-Mathurin 15408
Musée des Arts et Traditions Populaires, Valbonne 16034
Musée des Traditions Agricoles du Ségala, Pradinas 14938
Musée des Vallées Cévenoles, Saint-Jean-du-Gard 15337
Musée Domaine de Samara, La Chaussée-Tirancourt 13341
Musée du Crest-Cherel, Ugine 16015
Musée du Fromage de la Chèvrerie, Noirmoutier-en-l'Ile 14469
Musée du Mouton et de la Bergerie Nationale, Rambouillet 14982
Musée du Patrimoine Montagnard, Nages 14346
Musée du Veinazès, Lacapelle-del-Fraisse 13469
Musée Ferme de la Patte d'Oie, Le Plessis-Patte-d'Oie 13635
Musée Maison Déodat Roché, Arques (Aude) 11695
Musée Olivier-de-Serres, Mirabel (Ardèche) 14127
Musée Paysan de Bourgogne Nivernaise, La Celle-sur-Loire 13326
Musée Revoir le Passé, Les Loges-Marchis 13703
Musée Rural de l'Educations dans les Côtes d'Armor, Saint-Nicolas-du-Pelem 15429
Musée Troglodytique, Louresse 13854
Musée Vivant du Cochon, Chambonas 12377
Un Village se Raconte, Lourdios-Ichère 13853

Germany

Ackerbürgermuseum, Gartz 18543
Ackerbürgermuseum, Reichenbach, Oberlausitz 21247
Agrarhistorisches Aktionsmuseum, Trendelburg 22087
Agrarhistorisches Museum, Schlepzig 21564
Albrecht-Daniel-Thaer-Gedenkstätte-Landesausstellung, Reichenow-Möglin 21248
Bauernhaus-Museum, Lindberg 20002
Bauernhausmuseum Amerang des Bezirks Oberbayern, Amerang 16536
Bauernhausmuseum Hattingen, Hattingen 19036
Bauernhofmuseum Jexhof, Schöngeising 21609
Bauernhofmuseum Schreiner, Buchhofen 17617
Bauernmuseum, Tuttlingen 22131
Bauernmuseum Rothenburg, Rothenburg/Oberlausitz 21413
Bauernmuseum Selfklant, Selfkant 21736
Bergisches Freilichtmuseum für Ökologie und bäuerlich-handwerkliche Kultur, Lindlar 20007
Denkmalhof Rauchhaus Möllin, Gadebusch 18525
Deutsches Hirtenmuseum, Hersbruck 19164
Deutsches Landwirtschaftsmuseum, Stuttgart 21930
Deutsches Landwirtschaftsmuseum Schloss Blankenhain, Crimmitschau 17749
Deutsches Schweinemuseum Ruhlsdorf, Teltow 22024
Europäisches Spargelmuseum, Schrobenhausen 21638
F.W.Raiffeisen Museum, Hamm, Sieg 18968
Fränkisches Freilandmuseum, Bad Windsheim 16872
Freilandmuseum Ammerländer Bauernhaus, Bad Zwischenahn 16876
Freilandmuseum Grassemann Naturpark-Infostelle, Warmensteinach 22338
Freilandmuseum Lehde, Lübbenau 20070
Freilichtmuseum, Diesdorf 17850
Freilichtmuseum Alt Schwerin, Alt Schwerin 16483
Freilichtmuseum Domäne Dahlem, Berlin 17086
Freilichtmuseum Hessenpark, Neu-Anspach 20581
Freilichtmuseum Klausenhof, Herrischried 19161
Freilichtmuseum Massing, Massing 20236
Freilichtmuseum Stade, Stade 21848
Hanf-Museum, Berlin 17124
Haselünner Heimathäuser, Haselünne 19024
Heimatmuseum, Buchenberg 17616
Heimatmuseum, Venningen 22218
Heimatmuseum Hopfenhaus, Aidlingen 16448
Heimatmuseum im Hartmannhaus, Marktoberdorf 20218
Heimatmuseum im Wiggensbacher Informationszentrum, Wiggensbach 22565
Heimatmuseum im Wolfschneider-Hof, Taufkirchen, Kreis München 22014
Hopfen Erlebnis Hof, Altmannstein 16514
Isinger Dorfmuseum - Alte Kelter, Tübingen 22116
Jura-Bauernhof-Museum, Hitzhofen 19203
Kulturzentrum Sinsteden des Rhein-Kreises Neuss, Rommerskirchen 21367
Landschaftsmuseum Westerwald, Hachenburg 18847
Landwirtschaftliches Museum Wetzlar, Wetzlar 22528
Landwirtschafts- und Heimatmuseum, Karben 19431
Landwirtschaftsmuseum, Rhede 21295
Landwirtschaftsmuseum, Weil am Rhein 22375
Landwirtschaftsmuseum Reitscheid, Freisen 18447
Lebendiges Museum, Bad Wildungen 16861
LWL-Freilichtmuseum Detmold, Detmold 17840
Mühlen- und Landwirtschaftsmuseum, Fehmarn 18298
Mühlen- und Landwirtschaftsmuseum, Jever 19396
Museum Dermbach, Dermbach 17825
Museum Elbinsel Wilhelmsburg, Hamburg 18947
Museum für Bodenschätzung, Eickendorf 18087
Museum für ländliches Kulturgut, Landwirtschaft, Forsten und Jagd, Ulrichstein 22171
Museum für Landtechnik und Landarbeit, Emmerthal 18154
Museum für Landwirtschaftliche Maschinen, Erbach, Alb-Donau-Kreis 18188
Museum Grossauheim, Hanau 18980
Museum Hüsli-Sammlung Schwarzwälder Volkskunst, Grafenhausen 18730
Museum im Schlößle, Freiberg am Neckar 18419
Museum Reichenau, Reichenau 21244
Museum Wilnsdorf, Wilnsdorf 22589
Museums-Gutshof Sonnekalb, Bad Kösen 16764
Museumsdorf Volksdorf mit Spiekerhus, Hamburg 18953
Museumshof Großkochen, Großkochen 18795
Museumsscheune Fränkische Schweiz, Hollfeld 19254
Niederbayerisches Landwirtschaftsmuseum, Regen 21214
Niedersächsisches Spargelmuseum, Nienburg 20719
Njepila-Hof Rohne, Rohne 21363
Nostalgie-Museum, Wörth an der Donau 22643

Oberfränkisches Bauernhofmuseum, Zell, Oberfranken 22783
Oberpfälzer Freilandmuseum Neusath-Perschen, Nabburg 20542
Oberschwäbisches Museumsdorf Kreisfreilichtmuseum Kürnbach, Bad Schussenried 16831
Obstbaumuseum, Werder 22487
Ostfriesisches Landwirtschaftsmuseum, Krummhörn 19732
Raiffeisen-Museum, Flammersfeld 18326
Riedmuseum, Rastatt 21184
Rochow-Museum Reckahn, Kloster Lehnin 19578
Schäfertanz-Kabinett, Rothenburg ob der Tauber 21411
Schlesisch-Oberlausitzer Dorfmuseum Markersdorf, Markersdorf 20203
Schleswig-Holsteinisches Freilichtmuseum, Molfsee 20369
Schleswig-Holsteinisches Landwirtschaftsmuseum und Dithmarscher Bauernhaus, Meldorf 20270
Schwäbisches Bauernhofmuseum Illerbeuren, Kronburg 19726
Schwarzwälder Freilichtmuseum Vogtsbauernhof, Gutach 18843
Stadtmuseum Herrenmühle, Hammelburg 18973
Stadtmuseum Rastatt, Rastatt 21186
Städtische Galerie Rastatt, Rastatt 21188
Tauberländer Dorfmuseum, Weikersheim 22373
Technisches Museum Ölmühle, Pockau 21069
Vogtländisches Freilichtmuseum, Erlbach 18228
Volkskundliches Gerätemuseum, Arzberg 16578
Wein- und Heimatmuseum, Mehring 20256
Winzermuseum Bachem, Bad Neuenahr-Ahrweiler 16797

Greece

Manos Faltaits Museum, Skyros 23192

Hungary

A Badacsonyi Állami Gazdaság Üzemi Gyűjteménye, Badacsony 23342
Georgikon Majormúzeum, Keszthely 23661
Kopáncsi Tanyamúzeum, Hódmezővásárhely 23612
Magyar Fuszerpaprika Múzeum, Kalocsa 23626
Magyar Mezőgazdasági Múzeum, Budapest 23439
Magyar Mezogazdasági Múzeum Tanyamúzeuma, Lajosmizse - Bene 23688
Mezőgazdasági Gépmúzeum, Mezőkövesd 23699
Pick Szalámi és Szegedi Paprika Múzeum, Szeged 23848

India

Agriculture Museum, Imphal 24220
Meghalaya State Museum, Shillong 24386
Museum of the Agricultural College and Research Institute, Coimbatore 24122
Regional Science Centre, Guwahati 24195
Victoria and Albert Museum, Mumbai 24324

Indonesia

Museum Subak Bali, Kediri 24513
Subak Museum, Banjar Senggulan 24447

Ireland

Ballinamore Local Museum, Ballinamore 24735
Greenan Farm Museums, Greenan 24841
Irish Agricultural Museum, Wexford 24912
Roscrea Heritage Centre, Roscrea 24893

Israel

Dagon Collection, Haifa 24959
Farmer's House, Metulla 25064
Landscapes of the Holy Land Park, Tel Aviv 25114
Neot Kedumim, Lod 25061

Italy

Acquario e Museo Etnografico del Po, Monticelli d'Ongina 26859
Mostra della Civiltà Contadina, Lavello 26470
Museo Agrario Tropicale, Firenze 26166
Museo Agricolo, Ronco Briantino 27554
Museo Agricolo della Civiltà Contadina L. Carminati, San Giuliano Milanese 27649
Museo Agricolo e del Vino Ricci Curbastro, Capriolo 25668
Museo Agricolo e Museo dell'Arte Conciaria, Castelfranco Veneto 25730
Museo Agro-Forestale San Matteo, Erice 26061
Museo Agro-Pastorale dell'Area Ellenofona, Bova Marina 25520
Museo Contadino della Bassa Pavese, San Cristina e Bissone 27629
Museo del Lavoro e della Civiltà Rurale, Palaia 27056
Museo del Trattore Agricolo, Senago 27799
Museo della Bonifica, Argenta 25272
Museo della Civiltà Contadina, Aliano 25200
Museo della Civiltà Contadina, Bentivoglio 25396
Museo della Civiltà Contadina ed Artigiana, Monterosso Calabro 26848
Museo della Civiltà Contadina in Val Vibrata, Controguerra 25964
Museo della Civiltà Contadina Valle dell'Aniene, Roviano 27573
Museo della Civiltà Rurale, Altamura 25205
Museo della Frutticoltura Adofo Bonvicini, Massa Lombarda 26624
Museo della Vita e del Lavoro della Maremma Settentrionale, Cecina e Marina 25793
Museo delle Tradizioni Agroalimentari della Sardegna, Siddi 27834
Museo di Cultura Contadina, Oriolo 26994
Museo di Storia dell'Agricoltura, Cesena 25834
Museo di Storia dell'Agricoltura, Urbania 28120
Museo di Storia dell'Agricoltura e della Pastorizia, Morano Calabro 26867
Museo Enologico Grasso, Milazzo 26728
Museo Etnico della Civiltà Salentina Agrilandi Museum, Brindisi 25546
Museo Etnografico, Vione 28286
Museo Lombardo di Storia dell'Agricoltura, Sant'Angelo Lodigiano 27720

Japan

Hida Minzoku-mura, Takayama 29338
Hokkaido Daigaku Nogakubu Hakubutsukan, Sapporo 29215
Jingu Nogyokan, Ise 28582
Nihon Minka Shuraku Hakubutsukan, Toyonaka 29530
Nitobe Kinenkan, Hanamaki 28493
Sakai Greenery Museum Harvest Hill, Sakai 29198

Korea, Republic

Agricultural Museum of Jeollanam-do, Samho 29860
Jeonju National Museum, Jeonju 29842
Korean Agricultural Museum, Seoul 29894

Latvia

Burtnieku Pagasta Briedes Krogs, Burtnieki 29994
Latvijas Lauksaimniecības Universitātes Muzejs, Jelgava 30020

Lithuania

Arklio Muziejus, Anykščiai 30207

Malaysia

Agricultural Museum, Jasin 30556

Mexico

Ecomuseo Los Torres, Lerdo 31000
Museo Agrarista de Tzurúmutaro, Pátzcuaro 31254
Museo Comunitario el Asalto a las Tierras, Mexicali 31033
Museo El Remate, Comala 30789
Museo Nacional de Agricultura, Texcoco de Mora 31453

Netherlands

Agrarisch Museum Westerhem, Middenbeemster 32556
Boerderij Museum Duurswold, Slochteren 32783
Boerderijmuseum De Bovenstreek, Oldebroek 32614
Boerderijmuseum De Lebbenbrugge, Borculo 31982
Boerderijmuseum Zelhem, Zelhem 32996
Boerenbondsmuseum, Gemert 32242
Cultuurhistorisch Streekmuseum De Acht Zaligheden, Eersel 32178
De Locht - Streekmuseum, Nationaal Asperge- en Champignonmuseum, Horst 32414
Fruitteeltmuseum, Kapelle 32441
De Garstkamp, Overasselt 32662
Goemanszorg - Streek- en Landbouwmuseum, Dreischor 32156
Heemkundemuseum Paulus Van Daesdonck, Ulvenhout 32839
Historisch Museum Ede, Ede 32168
Land- en Tuinbouwmuseum, Etten Leur 32226
Landbouw- en Juttersmuseum Swartwoude, Buren, Friesland 32023
Landbouwmuseum Erve Niehof, Diepenheim 32125
Meierische Museumboerderij, Heeswijk-Dinther 32343
Museum Bebinghehoes, Exloo 32227
Museum Broeker Veiling, Broek op Langedijk 32012
Museum Saet en Cruyt, Andijk 31906
Museumboerderij de Karstenhoeve, Ruinerwold 32739
Museumboerderij de Tip, Herveld 32381
Museumboerderij de Zwemkolk, Markelo 32538
Museumboerderij Erve Hofman, Hellendoorn 32357
Museumboerderij Scholten, Den Ham 32108
Museumboerderij Wendezoele, Ambt-Delden 31805
Museumsboerderij Oud Noordwijk, Noordwijk, Zuid-Holland 32603
Nationaal Veeteelt Museum, Beers bij Cuyk 31955
Openlucht Laagveenderij Museum Damshûs, Nij Beets 32585
Openluchtmuseum Ellert en Brammert, Schoonoord 32767
Oudheidkamer Horst, Horst 32416
Peelmuseum, Ospel 32639
Pluimvee Museum, Barneveld 31950
Sallands Landbouwmuseum de Laarman, Luttenberg 32514
Stichting Zeeuwse Schaapkudde, Heinkenszand 32351
Streeklandbowmuseum Agrimunda, Sint Anna ter Muiden 32770
Streekmuseum de Meesthof, Sint Annaland 32771
Streekmuseum Het Land van Axel, Axel 31943
Streekmuseum voor de Krimpenerwaard Crimpenerhof, Krimpen aan den IJssel 32460
Veen Park, Barger Compascuum 31949
Veluws Streekmuseum Hagedoorns Plaatse, Epe 32218
Vlasbewerkingsmuseum It Braakhok, Ee 32171
Wereld Bodem Museum, Wageningen 32921
Westlands Museum voor Streek- en Tuinbouwhistorie, Honselersdijk 32403
Wieringer Boerderij, Den Oever 32113

New Zealand

Central Hawke's Bay Settlers Museum, Waipawa 33223
East Coast Museum of Technology, Gisborne 33108
Norsewood Pioneer Cottage Museum, Norsewood 33160

Norway

Ål Bygdemuseum, Ål 33377
Andøymuseet, Risøyhamn 33783
Askim Museum, Askim 33407
Aurskog-Høland Bygdetun, Hemnes i Høland 33542
Bagn Bygdesamling, Bagn 33413
Bardu Bygdetun, Salangsdalen 33806
Brudavollen Bygdetun, Ørsta 33711
Bygdetunet Jutulheimen, Vågåmo 33954
Bygland Museum, Bygland 33459
Dokken, Hol i Hallingdal 33551
Eidsvoll Bygdemuseet, Eidsvoll Verk 33477
Enebakk Museet, Enebakk 33482
Foldalbruket, Kjøllefjord 33594
Folkenborg Museum, Mysen 33687
Forsand Bygdemuseum, Forsand 33504
Fyresdal Bygdemuseet, Fyresdal 33508
Gamle Hvam Museum, Årnes 33400
Gamle Kvernes Bygdemuseum, Averøy 33412
Glomdalsmuseet, Elverum 33480
Gol Bygdemuseet, Gol 33515
Granvin Bygdamuseum, Granvin 33516
Grødaland Bygdetun, Nærbø 33689
Grytøy Bygdetun, Lundenes 33653
Havråtunet, Lonevåg 33647
Hedmarksmuseet og Domkirkeodden, Hamar 33528
Hemsedal Bygdatun, Hemsedal 33543
Hol Bygdemuseum, Hol i Hallingdal 33552
Hordamuseet, Fana 33488
Huldreheimen, Bykle 33461
Jærmuseet, Nærbø 33690
Kvæfjord Bygdetun, Harstad 33533
Kviteseid Bygdetun, Kviteseid 33621
Kystmuseet i Øygarden, Rong 33794
Kystmuseet i Sør-Trøndelag, Hitra 33548
Landbruksmuseet for Møre og Romsdal, Vikebukt 33979
Lands Museum, Dokka 33464
Lårdal Bygdemuseum, Høydalsmo 33570
Lørenskog Bygdemuseum, Skårer 33829
Målselv Bygdemuseum, Moen 33671
Mandal og Opplands Folkemuseum, Øyslebø 33764
Måsøy Museum, Havøysund 33540
Meløy Bygdemuseum, Ørnes 33710
Meråker Bygdemuseum, Meråker 33666
Mosvik Museum, Mosvik 33686
Namsdalsmuseet, Namsos 33691
Ner Hole Museum, Åndalsnes 33389
Nesna Bygdemuseum, Nesna 33699
Nordre Husan, Alvdal 33386
Norsk Hagebruks Museum, Grimstad 33520
Norsk Landbruksmuseum, Ås 33402
Odalstunet Gårdsmuseum, Skarnes 33830
Osterøy Museum, Lonevåg 33648
Rakkestad Bygdetun, Rakkestad 33775
Rennebu Bygdemuseum, Rennebu 33779
Rissa Bygdemuseum, Rissa 33784
Roan Bygdetun, Roan 33787
Sætersgårds Samlinger Dølmotunet, Tolga 33909
Sande Historielags Bygdesamlinger, Sande 33811
Sigdal og Eggedal Museum, Prestfoss 33773
Skiptvet Bygdemuseum, Skiptvet 33838
Sørli Museum, Sørli 33853
Sørlistøa Fløtermuseet, Osen 33715
Spydeberg Bygdetun, Spydeberg 33859
Stenneset Bygdetun, Mo i Rana 33668
Sunndal Bygdemuseum, Sunndalsøra 33892
Sunnfjord Museum, Førde 33503
Toten Økomuseum, Bøverbru 33451
Toten Økomuseum, Kapp 33587
Trøgstad Bygdemuseum, Trøgstad 33916
Trysil Bygdemuseum, Trysil 33934
Tylldalen Bygdetun, Tylldalen 33942
Velfjord Bygdemuseum, Hommelstø 33556
Vestfold Fylkesmuseum, Tønsberg 33911
Volda Bygdetun og Garverimuseet, Volda 33984

Pakistan

Agricultural Museum, Faisalabad 34001

Poland

Muzeum Dworu Polskiego, Płochocin 35012
Muzeum Narodowe Rolnictwa i Przemysłu Rolno-Spożywczego w Szreniawie, Komorniki 34754
Muzeum Rolnictwa im. Ks. Krzysztofa Kluka, Ciechanowiec 34608
Muzeum Wiedzy o Środowisku, Poznań 35046
Muzeum Ziemi Krzeszowickiej, Krzeszowice 34836
Muzeum Ziemi Otwockiej, Otwock 35000

Portugal

Colecção Etnográfica da Direcção-Geral Agrícola, Aljustrel 35410
Jardim-Museu Agrícola Tropical, Lisboa 35668
Museu Agrícola da Escola Profissional Agricultura de Cister, Alcobaça 35392
Museu Agrícola de Entre Douro e Minho, Vila do Conde 35988
Museu Agrícola de Riachos, Riachos 35885
Museu Agrícola e Museu de Arte Sacra, Ribeira Chã 35887
Museu da Agricultura de Fermentões, Guimarães 35638
Museu da Casa da Cultura da Orada, Orada 35794
Museu do Porco, Pampilhosa 35807
Museu Nacional de Agricultura, Coimbra 35541

Qatar

Al-Khor Museum, Al-Khor 36083

Réunion

Muséum Agricole et Industriel de Stella Matutina, Piton-Saint-Leu 36090

Romania

Casa Memorială Gheorghe Ionescu-Şişeşti, Şişeşti 36625
Expoziţia Foto-documentară Ion Ionescu de la Brad, Negri 36490
Muzeul Agriculturii din România, Slobozia 36629
Muzeul de Etnografie şi Artă Populară, Floreşti-Stoeneşti 36371
Muzeul în Aer Liber din Dumbrava Sibiului, Sibiu 36612
Muzeul Naţional al Satului Dimitrie Gusti, Bucureşti 36217

Russia

Gosudarstvennyj Muzej Životnovodstva im. Akademika E.F. Liskuna, Moskva 37206
Muzej Istorii Moskovskoj Selskochosjajstvennoj Akademii im K.A. Timirjazeva, Moskva 37293
Muzej-usadba A.T. Bolotova, Dvorjaninovo 36882
Počvenno-agronomičeskij Muzej im. Akademika V.R. Viljamsa, Moskva 37398

Slovakia

Slovenské Polnohospodárske Múzeum, Nitra 38316
Vlastivedné Múzeum, Trebišov 38374

Slovenia

Planšarski Muzej, Bohinjsko Jezero 38402

South Africa

Bathurst Agricultural Museum, Bathurst 38556
Cuyler Manor Museum, Uitenhage 38767
Howick Museum, Howick 38643
P.W. Vorster Museum, Middelburg 38689
Willem Prinsloo Museum, Rayton 38721

Spain

Museo de la Agricultura, Torremocha de Jarama 40158
Museo del Románico, Aguilar de Campoo 38791
Museo Etnológico de Horcajuleo de la Sierra, Horcajuelo de la Sierra 39392
Museu d'Eines del Camp Masia de Can Deu, Sabadell 39926
Museu del Pagès, Linyola 39467

Sweden

Åttingen Hantverk och Lantbruksmuseum, Aneby 40421
Berte Museum - Livet på Landet, Slöinge 40857
Grans Lantbruksmuseum, Piteå 40808
Hägnan Friluftsmuseet i Gammelstad, Gammelstad 40534
Halmens Hus, Bengtsfors 40456
Julita Sveriges Lantbruksmuseums, Julita 40621
Kronobergs Lantbruksmuseum, VÄXJÖ 41038
Kulturhistoriska Museet i Bunge, Fårösund 40517
Lantbruksmuseum, Ingatorp 40609
Särestad Lantbruksmuseum, Grästorp 40564

Switzerland

Alpmuseum Riederalp, Riederalp 41762
Dorfmuseum Bottmingen, Bottmingen 41247
Heimatmuseum Nutli-Hüschi, Klosters 41505
Musée Mazot de Plan-Cerisier, Martigny 41613
Museum Wasseramt, Halten 41470
Ortsgeschichtliche Sammlung, Oberrieden 41698
Ortsmuseum Altes Rathaus, Balgach 41151
Sennerei-Museum und Heimatmuseum, Unterwasser 41964
Ständerhaus, Buus 41278
Strohhaus, Muhen 41665

Taiwan

Agricultural Association Museum of Jhonghe, Jhonghe 42218
Agricultural Exhibition Hall, Taipei 42372
Egg's Farm and Museum, Ciaotou 42164
Farming and Fishing Museum of Sihu, Sihu 42328
Loquat Museum of Sinshe, Sinshe 42340
Meishan Farming Village Museum, Meishan 42274
Museum of Chinese Chive Farming, Cingshuei 42168

Turkmenistan

Lehap Regional Museum, Turkmenabat 42985

United Kingdom

Aberdeenshire Farming Museum, Mintlaw 45159
Acton Scott Historic Working Farm, Acton Scott 43263
Agricultural Museum Brook, Brook 43657
Almond Valley Heritage Centre, Livingston 44751
Ardress Farmyard Museum, Annaghmore 43304
Barleylands Farm Museum, Billericay 43485
Battle and Farm Museum, Naseby 45194
Brattle Farm Museum, Staplehurst 45740
Bygones Museum, Claydon 43850
Bygones Museum at Holkham, Wells-next-the-Sea 45959
Church Farm Museum, Skegness 45685
Cogges Manor Farm Museum, Witney 46052
Corfe Castle Museum, Corfe Castle 43892
Countryside Museum, East Budleigh 44082
Cousland Smiddy, Cousland 43898
Craigavon Museum, Craigavon 43912
Easton Farm Park, Easton 44106
Ellisland Farm, Auldgirth 43337
Farmland Museum and Denny Abbey, Waterbeach 45949
Farmlife Centre, Thornhill, Central 45853
Gower Heritage Centre, Gower 44340
Hampshire Farm Museum, Botley, Hampshire 43563
Hamptonne Country Life Museum, Saint Lawrence 45580
Heaven Farm Museum, Uckfield 45893
Hirsel Homestead Museum, Coldstream 43879
Kingsbury Watermill Museum, Saint Albans 45544
Leslie Hill Open Farm, Ballymoney 43367
Maldon and District Agricultural and Domestic Museum, Goldhanger 44324
Manor Farm, Bursledon 43684
Museum of English Rural Life, Reading 45468
Museum of Farming Life, Ellon 44169
Museum of Scottish Country Life, East Kilbride 44089
Newham Grange Leisure Farm Museum, Middlesbrough 45142
Normanby Park Farming Museum, Scunthorpe 45627
Orkney Farm and Folk Museum, Birsay 43518
Park Farm Museum, Milton Abbas 45152
Plantation of Ulster Visitor Centre, Draperstown 44022
Roots of Norfolk at Gressenhall, Gressenhall 44366
Rural Life Centre, Farnham 44219
Sacrewell Farm and Country Centre, Thornhaugh 45852
Shetland Croft House Museum, Dunrossness 44062
Somerset Rural Life Museum, Glastonbury 44308
South Molton and District Museum, South Molton 45693
Tingwall Agricultural Museum, Gott 44338
Upminster Tithe Barn Museum of Nostalgia, Upminster 45899
Upper Wharfedale Folk Museum, Grassington 44349
Warwickshire Museum of Rural Life, Moreton Morrell 45176
Wimpole Hall and Home Farm, Royston 45522
Yorkshire Museum of Farming, York 46113

United States of America

The 100th Meridian Museum, Cozad 47864
The Accokeek Foundation, Accokeek 46131
Adams County Museum, Brighton 47026
Agricultural Heritage Museum, Boerne 46888
The Agricultural Memories Museum, Penn Yan 51763
Agricultural Museum at Stone Mills, La Fargeville 49839
Alabama Agricultural Museum, Dothan 48154
Alexander and Baldwin Sugar Museum, Puunene 52205
American Maple Museum, Croghan 47885
American West Heritage Center, Wellsville 54034
The Apple Trees Museum, Burlington 47131
Arches Museum of Pioneer Life, Winona 54238
Baldwin County Heritage Museum, Elberta 48331
Banner County Historical Museum, Harrisburg 49177
Barrington Living History Farm, Washington 53945
Belair Stable Museum, Bowie 46968
Billings Farm and Museum, Woodstock 54287
Bisbee Mill Museum, Chesterfield 47435
Blackberry Farm-Pioneer Village, Aurora 46491
Browntown Museum, Lake City 49880
Calumet County Historical Society Museum, New Holstein 51059
Carroll County Farm Museum, Westminster 54096
Center for Agricultural Science and Heritage, Indianapolis 49487
Central Washington Agricultural Museum, Union Gap 53654
Chippokes Farm and Forestry Museum, Surry 53382
Chisholm Trail Heritage Center Museum, Duncan 48204
Coley Homestead and Barn Museum, Weston 54101
Collin County Farm Museum, McKinney 50358
Columbia County Museum, Bloomsburg 46871
Craftsman Farms Foundation, Parsippany 51716
Daniel Boone Homestead, Birdsboro 46818
Delaware Agricultural Museum, Dover 48162
Doric House, Flemington 48557
Drumlin Farm, Lincoln 50091
Erie Canal Village, Rome 52450
The Farm, Sturgeon Bay 53348
Farm House Museum, Ames 46257
Farmamerica, Waseca 53838
The Farmers' Museum, Cooperstown 47788
Florence Ranch Homestead, Mesquite 50612
Florida Agricultural Museum, Palm Coast 51679
Fort Walla Walla Museum, Walla Walla 53797
Fosterfields Living Historical Farm, Morristown 50851
The Fryeburg Fair Farm Museum, Fryeburg 48804
Garfield Farm Museum, Lafox 49869
Georgia Agrirama, Tifton 53508
Gibbs Farm Museum, Falcon Heights 48496
Grant-Kohrs Ranch, Deer Lodge 48028
Hadley Farm Museum, Hadley 49117
Hale Farm and Village, Bath 46630
Hallockville Museum Farm, Riverhead 52366
Hamilton Grange National Memorial, New York 51189
Harry's Truman Farm Home, Grandview 49012
Hathaway Ranch Museum, Santa Fe Springs 52893
Highland Maple Museum, Monterey 50809
Hillsboro Museums, Hillsboro 49284
Historic Schaefferstown, Schaefferstown 52956
Historic Stagville, Bahama 46529
Hofwyl-Broadfield Plantation, Brunswick 47100
Homestead National Monument of America, Beatrice 46664
Howard-Dickinson House Museum, Henderson 49252
Howell Living History Farm, Titusville 53514
Huddleston Farmhouse Inn Museum, Cambridge City 47192
Jackson's Mill Historic Area, Weston 54109
Jarrell Plantation Georgia, Juliette 49653
Jerome County Historical Society, Jerome 49623
Jewell County Historical Museum, Mankato 50445
John Brown Farm, Lake Placid 49890
Jourdan-Bachman Pioneer Farm, Austin 46512
Kiwanis Van Slyke Museum Foundation, Caldwell 47160
Kline Creek Farm, Winfield 54231
Landis Valley Museum, Lancaster 49928
Leaming's Run Garden and Colonial Farm, Cape May Court House 47237
Living History Farms Museum, Urbandale 53671
Longstreet Farm, Holmdel 49324
Loudon Heritage Farm Museums, Sterling 53287
Malabar Farm State Park, Lucas 50315
Marjorie Kinnan Rawlings Historic State Park, Cross Creek 47890
Marshall County Historical Society Museum, Warren 53823
Midwest Old Settlers and Threshers Association Museum, Mount Pleasant 50883
Miles B. Carpenter Museum, Waverly 53998
Mississippi Agriculture and Forestry/National Agricultural Aviation Museum, Jackson 49562
Mitchell County Historical Museum, Osage 51608
Montville Township Historical Museum, Montville 50832
Moriarty Historical Museum, Moriarty 50847
Mountain Farm Museum, Cherokee 47422
Mulford House and Farm, East Hampton 48247
Museum at Prophetstown, Battle Ground 46646
Museum of East Alabama, Opelika 51573
Museum of the Arkansas Grand Prairie, Stuttgart 53351
Museum of the Northern Great Plains, Fort Benton 48593
The National Agricultural Center and Hall of Fame, Bonner Springs 46901
National Apple Museum, Biglerville 46803
Neeses Farm Museum, Neeses 50999
Nelson Pioneer Farm and Museum, Oskaloosa 51617
New Hampshire Farm Museum, Milton 50701
New Jersey Museum of Agriculture, North Brunswick 51385
New Mexico Farm and Ranch Heritage Museum, Las Cruces 49960
New Sweden Farmstead Museum, Bridgeton 47020
Northeast Texas Rural Heritage Museum, Pittsburg 51937
Northeastern Montana Threshers and Antique Association, Culbertson 47901
Old Trail Museum, Choteau 47546
Oliver H. Kelley Farm, Elk River 48344
Olmstead Place State Park, Ellensburg 48360
Overstreet-Kerr Historical Farm, Keota 49747
Oxon Cove Park Museum, Oxon Hill 51659
Pawnee Bill Ranch, Pawnee 51736
Pawnee City Historical Society Museum, Pawnee City 51737
Pendleton District Agricultural Museum, Pendleton 51759
Perkins County Historical Society Museum, Grant 49015
Pioneer Farm Museum and Ohop Indian Village, Eatonville 48275
Pioneer Florida Museum, Dade City 47919
Pioneer Museum of Alabama, Troy 53571
Piqua Historical Area State Memorial, Piqua 51931
Plantation Agriculture Museum, Scott 52973
Plumb Farm Museum, Greeley 49035
Pomeroy Living History Farm, Yacolt 54318
Queens County Farm Museum, Floral Park 48562
Quiet Valley Living Historical Farm, Stroudsburg 53340
Rancho Los Alamitos, Long Beach 50177
Red Mill Museum Village, Clinton 47658
Rough and Tumble Engineers Museum, Kinzers 49798
Seward County Historical Society Museum, Goehner 48952
Shaker Village of Pleasant Hill, Harrodsburg 49190
Silvercreek Museum, Freeport 48782
Slate Run Living Historical Farm, Ashville 46395
Sotterley Plantation Museum, Hollywood 49323
South County Museum, Narragansett 50955
State Agricultural Heritage Museum, Brookings 47060
Steppingstone Museum, Havre de Grace 49220
Stonefield Historic Site, Cassville 47275
Thetford Historical Society Library and Museum, Thetford 53496
Tipton-Haynes Museum, Johnson City 49636
Tobacco Farm Life Museum, Kenly 49726
Tread of Pioneers Museum, Steamboat Springs 53280
Troy Museum and Historic Village, Troy 53572
Volkening Heritage Farm at Spring Valley, Schaumburg 52959
Ward O'Hara Agricultural Museum of Cayuga County, Auburn 46472
Waterloo Area Farm Museum, Grass Lake 49024
Watters Smith Memorial, Lost Creek 50275
West Virginia State Farm Museum, Point Pleasant 52012
Wheeler Historic Farm, Salt Lake City 52698
Wolf Point Area Historical Society, Wolf Point 54267

Amber

Denmark

Ravmuseet, Oksbøl 10220

Dominican Republic

Museo de Ámbar Dominicano, San Felipe de Puerto Plata 10334
Museo Mundo de Ambar, Santo Domingo 10356

Germany

Bernstein Museum Bad Füssing, Bad Füssing 16736
Deutsches Bernsteinmuseum, Ribnitz-Damgarten 21309

Lithuania

Gintaro Galerija-Muziejus, Nida 30319
Gintaro Galerija-Muziejus, Vilnius 30409
Palangos Gintaro Muziejus, Palanga 30329

Mexico

Museo del Ámbar de Chiapas, San Cristóbal de las Casas 31327

Poland

Muzeum Inkluzji w Bursztynie, Gdynia 34655

Romania

Colecția Muzeală Colți, Colți 36296

Russia

Muzej Jantarja, Kaliningrad 36991

Sweden

Bärnstensmuseet, Höllviken 40595

Anthropology → Ethnology

Apiculture → Bees

Archaeology

Albania

Muzeu Arkeologjik, Durrës 00013
Muzeu Arkeologjik, Tiranë 00031
Zona arkeologjike e Butrintit, Butrinti 00012

Algeria

Musée de Bordj Moussa, Béjaïa 00053
Musée de Cherchell, Cherchell 00055
Musée de Circonscription Archéologique, Sétif 00073
Musee de Plein-Air de Cherchell, Cherchell 00056
Musée d'Hippone, Annaba 00050
Musée du Temple de Minerve, Tebessa 00079
Musée National d'Archéologie de Sétif, Sétif 00075
Musée National des Antiquités, Alger 00045

Angola

Museu Nacional de Arqueologia, Benguela 00106

Argentina

Instituto de Investigaciones Arqueológicas y Museo Prof. Mariano Gambier, Albardón 00118
Museo Arqueológico, Aguilares 00117
Museo Arqueológico, Tafí del Valle 00638
Museo Arqueológico Adán Quiroga, San Fernando del Valle de Catamarca 00571
Museo Arqueológico del Instituto de Arqueología y Etnología, Mendoza 00452
Museo Arqueológico Provincial Andalgalá, Andalgalá 00123
Museo Arqueológico Provincial Condor Huasi (interim), Belén 00146
Museo Etnográfico Juán B. Ambrosetti, Buenos Aires 00230
Museo Etnográfico y Colonial Juan de Garay, Santa Fé 00615
Museo Nacional del Hombre del Instituto Nacional de Antropología y Pensamiento Latinoamericano, Buenos Aires 00263
Museo Pringles de Arqueología e Historia, Coronel Pringles 00345
Museo Privado de Arqueología Regional Julio Gironde, Carmen de Patagonés 00294
Museo Rocsen, Nono 00478

Armenia

History Museum of Armenia, Yerevan 00738

Australia

Abbey Museum of Art and Archaeology, Caboolture 00931
Antiquities Museum, Saint Lucia 01484
John Elliott Classics Museum, Sandy Bay 01494
Museum of Classical Archaeology, Adelaide 00779
Nicholson Museum, Sydney 01557

Austria

Archäologie am Landesmuseum Joanneum, Graz 01962
Archäologische Gedenkstätte (Unterkirche), Klosterneuburg 02199
Archäologische Sammlung, Wels 02912
Archäologisches Museum Innsbruck, Innsbruck 02113
Archäologisches Museum Lavant, Lavant 02271
Archäologisches Museum und Dokumentationszentrum Via Claudia, Fließ 01892
Ausgrabungen Wüstung Hard, Thaya 02809
Freilichtmuseum Römervilla in Brederis, Rankweil 02558
Heimatmuseum Traun, Traun 02824
Historisch-Archäologisches Museum, Markgrafneusiedl 02360
Keltendorf Mitterkirchen, Mitterkirchen 02388
Keltenmuseum Gracarca, Sankt Kanzian 02658
Keltenpark Schwarzenbach, Schwarzenbach 02719
Landesmuseum Burgenland, Eisenstadt 01861
Museen im Wehrturm, Perchtoldsdorf 02489
Museum der Stadt Lienz, Lienz 02301
Museum im Bürgerspital, Drosendorf an der Thaya 01830
Museum im Kastenturm, Bischofshofen 01786
Museum Schloß Lackenbach, Lackenbach 02250
Museum Wattens - Industrie Vorgeschichte, Wattens 02896
Römermuseum Favianis-Sankt Severin, Mautern 02369
Römermuseum Teurnia, Lendorf 02283
Römermuseum Wallsee-Sindelburg, Wallsee 02893
Römische Baureste Am Hof, Wien 03062
Rollettmuseum, Baden bei Wien 01776
Salzburg Museum, Salzburg 02625
Stadtmuseum Sankt Pölten, Sankt Pölten 02686
Universalmuseum Joanneum, Graz 01985
Wienerwaldmuseum, Eichgraben 01851

Azerbaijan

Artsakh State Museum, Stepanakert 03197
Raşidbey Efandiyev Historical-Regional Ethnography Museum, Şeki 03196

Bahrain

Bahrain National Museum, Manama 03212

Bangladesh

Archaeological Museum, Mahastangarh 03231
Archaeological Museum, Paharpur 03233
Bangladesh National Museum, Dhaka 03220
Dhaka City Museum, Dhaka 03222
Dinajpur Museum, Dinajpur 03230
Rangpur Archaeological Museum, Rangpur 03235
Varendra Research Museum, Rajshahi 03234

Belarus

Muzej Staražytna Belaruskaj Kultury, Minsk 03254

Belgium

Ancien Palais de Bruxelles, Bruxelles 03431
Archeologisch Museum Abdij, Affligem 03262
Archeologisch Museum Van Bogaert-Wauters, Hamme 03672
Archeologisch Site Museum, Werken 04146
Bruggemuseum - Archeologie, Brugge 03408
Espace Gallo-Romain, Ath 03328
Gallo-Romeins Museum, Tongeren 04075
Gemeentelijk Archaeologisch Museum, Grobbendonk 03663
Gemeentelijk Museum, Melle 03866
Gemeentemuseum, Temse 04053
Grotte Scladina et Centre Archéologique, Sclayn 04006
Musée Archéologique, Namur 03899
Musée Archéologique d'Autelbas, Bastogne 03346
Musée Athois, Ath 03330
Musée Communal d'Archéologie Hesbignonne, Saint-Georges-sur-Meuse 03992
Musée Communal Georges Mulpas, Elouges 03577
Musée d'Archéologie et Folklore, Verviers 04107
Musée d'Archéologie Préhistorique de l'Université de Liège, Liège 03803
Musée de la Haute Haine, Carnières 03510
Musée de la Pierre et Site des Carrières, Maffle 03840
Musée de Wanne, Trois-Ponts 04095
Musée des Fagnes, Roly 03981
Musée des Francs et de la Famenne, Marche-en-Famenne 03848
Musée d'Histoire et d'Archéologie, Warneton 04133
Musée du Malgré-Tout, Treignes 04092
Musée du Monde Souterrain, Han-sur-Lesse 03675
Musée du Vieux-Cimetière, Soignies 04036
Musée Gallo-Romain, Rumes 03989
Musée Royal de Mariemont, Morlanwelz-Mariemont 03893
Museum 't Nieuwhuys Hoegaarden, Hoegaarden 03705
Paleontologisches Museum, Burg-Reuland 03507
Rietgaverstede, Nevele 03912
Stedelijk Museum Aalst, Aalst 03259
Stedelijk Museum Brusselpoort, Mechelen 03861
Vleeshuis, Antwerpen 03318
Vleeshuismuseum, Dendermonde 03542

Bosnia and Herzegovina

Humačka Arheološko Zbirka, Ljubuški 04268
Muzej Istočne Bosnë, Tuzla 04290
Muzej Republike Srpske, Banja Luka 04257
Zavičajni Muzej Kozara, Prijedor 04271
Zemaljski Muzej Bosne i Hercegovine, Sarajevo 04284

Brazil

Centro de Arqueologia Annete Laminge Emperaire, Lagoa Santa 04527
Centro Regional de Arqueologia Ambiental, Piraju 04631
Centro Regional de Pesquisas Archeológicas Mário Neme, Piraju 04632
Museu Arqueológico, Lagoa Santa 04528
Museu Arqueológico de Sambaqui Joinville, Joinville 04514
Museu de Arqueologia e Etnologia, Salvador 04793
Museu de Arqueologia e Etnologia da Universidade de São Paulo, São Paulo 04877
Museu Municipal de Arqueologia, Monte Alto 04573
Museu Paranaense, Curitiba 04442

Bulgaria

Archaeological Preserve Nicopolis ad Istrum, Veliko Tărnovo 05249
Archeologičeska Ekspozicija v Džamia Achmed Bej, Kjustendil 05063
Archeologičeska Muzej, Silistra 05160
Archeologičeski Muzej, Burgas 04995
Archeologičeski Muzej, Plovdiv 05124
Archeologičeski Muzej, Radnevo 05142
Archeologičeski Muzej, Septemvri 05158
Archeologičeski Muzej, Sozopol 05206
Archeologičeski Muzej, Varna 05235
Archeologičeski Rezervat Starijat Grad Šumen i Šumenska Krepost, Šumen 05212
Gradski Archeologičeski Muzej, Nesebăr 05096
Gradski Istoričeski Muzej, Asenovgrad 04971
Gradski Istoričeski Muzej, Botevgrad 04992
Gradski Istoričeski Muzej, Karnobat 05051
Istoričeski Muzej, Blagoevgrad 04991
Istoričeski Muzej, Goce Delčev 05035
Istoričeski Muzej, Petrič 05113
Istoričeski Muzej, Pomorie 05133
Istoričeski Muzej, Provadia 05140
Istoričeski Muzej, Razgrad 05148
Istoričeski Muzej, Velingrad 05260
Istoričeski Muzej Targovište, Targovište 05225
Madara Nacionalen Istoriko-archeologičeski Rezervat, Madara 05089
Nacionalen Istoriko-archeologičeski Rezervat i Muzej Veliki Preslav, Veliki Preslav 05248
Nacionalen Istoriko-archeologičeski Rezervat Pliska, Pliska 05123
Okražen Istoričeski Muzej, Chaskovo 05005
Okražen Istoričeski Muzej, Sliven 05169
Peštera Rabiša Muzej, Belogradčik 04983
Regionalen Istoričeski Muzej, Stara Zagora 05211

Cambodia

Angkor Borei Museum, Angkor Borei 05282
National Museum of Cambodia, Phnom Penh 05287

Cameroon

Musée des Arts Traditionnels et de la Culture, Foumban 05297

Canada

Archaeological Survey of Alberta, Edmonton 05742
Centre d'Interprétation Archéo Topo, Bergeronnes 05421
Centre d'Interprétation du Bourg de Pabos, Pabos Mills 06566
Don Murray's Museum of History, Neepawa 06420
Kingston Archaeological Centre, Kingston 06080
Musée Canadien des Civilisations, Gatineau 05862
Musée Marguerite Bourgeoys, Montréal 06374
Museum of Antiquities, Saskatoon 06920
Museum of Archaeology and Ethnology, Burnaby 05504
Parc Archéologique de la Pointe-du-Buisson, Melocheville 06263
Pelham Historical Resource Centre, Fenwick 05793
Pointe-à-Callière - Musée d'Archéologie et d'Histoire de Montréal, Montréal 06378
Prince of Wales Northern Heritage Centre, Yellowknife 07411
Red Bay National Historic Site, Red Bay 06705
Royal British Columbia Museum, Victoria 07258
Ska-Nah-Doht Iroquoian Village and Museum, Mount Brydges 06398
Surrey Museum, Surrey 07038
The Van Egmond Foundation, Egmondville 05768
Victoria Settlement Provincial Historic Site, Saint Paul 06871
Xá:ytem Museum, Mission 06307
Yukon Beringia Interpretive Centre, Whitehorse 07328

Central African Republic

Musée Labasso, Bangassou 07420

Chile

Museo Arqueológico R.P. Gustavo Le Paige S.J., San Pedro de Atacama 07449
Museo Arqueológico San Miguel de Azapa, Arica 07430
Museo de Hualpén, Concepción 07435
Museo Regional de Atacama, Copiapó 07437
Museo Regional de Iquique, Iquique 07439

China

Beijing Ancient Architecture Museum, Beijing 07507
Guilin Stone Inscription Museum, Guilin 07742
Hong Kong Museum of History, Hong Kong 07800
Inner Mongolia Museum, Hohhot 07790
Xizhou Yandu Ruins Museum, Beijing 07573

Colombia

Casa del Escribano Real Don Juan de Vargas, Tunja 08772
Colección Arqueológica, Génova 08646
Colección Arqueológica, Pijao 08721
Colección de Minerales y Arqueología, Villanueva 08787
Monumento a Cristo Rey, Belalcázar 08541
Muestra Arqueológica, Armenia 08532
Muestra Arqueológica, Montenegro 08691
Muestra Arqueológica, Quimbaya 08735
Muestra Arqueológica Colección Rosmary, Colosó 08627
Museo Antrológico, Cereté 08623
Museo Antropológico Tairona, Santa Marta 08745
Museo Arqueológica de Mongua, Mongua 08689
Museo Arqueológico, Aranzazu 08530
Museo Arqueológico, Cartago 08621
Museo Arqueológico, Curumaní 08632
Museo Arqueológico, Inzá 08661
Museo Arqueológico, Manizales 08667
Museo Arqueológico, Oiba 08700
Museo Arqueológico, Paéz 08701
Museo Arqueológico, Puerto Berrío 08731
Museo Arqueológico, Salamina 08742
Museo Arqueológico, Sogamoso 08760
Museo Arqueológico, La Tebaida 08767
Museo Arqueológico Casa de Madame Augustín, Santa Marta 08746
Museo Arqueológico Chimica, El Copey 08638
Museo Arqueológico de Guacarí, Guacarí 08648
Museo Arqueológico del Cesar, Valledupar 08779
Museo Arqueológico del Huila, Neiva 08695
Museo Arqueológico del Sinú, Tierralta 08769
Museo Arqueológico Julio César Cubillos, San Agustín 08743
Museo Arqueológico La Bagatela, Villa del Rosario 08785
Museo Arqueológico La Merced, Cali 08607
Museo Arqueológico Marco Fidel Micolta, Guamo 08653
Museo Arqueológico Regional Guane, Floridablanca 08643
Museo Arqueológico y Paleontológico de Guane, Barichara 08536
Museo Casa de Bolívar, Bucaramanga 08598
Museo Chairama, Santa Marta 08748
Museo del Canasto Cafetero y Muestra Arqueológica de la Cultura Quimbaya, Filandia 08640
Museo del Cobre, Bogotá 08576
Museo del Oro, Ipiales 08663
Museo del Oro Calima, Cali 08612
Museo del Oro Nariño, Pasto 08708
Museo del Oro Quimbaya, Armenia 08534
Museo del Oro Quimbaya, Manizales 08670
Museo Escolar María Goretti, Pasto 08709
Museo Ethnográfico, Florencia 08641
Museo Fotográfico, Calarcá 08605
Museo Histórico de la Casa de la Cultura, Ambalema 08528
Museo Histórico y Arqueológico, Belén de Umbría 08542
Museo Histórico y Arqueológico, Charalá 08624
Museo Indígena de Guatavita, Guatavita 08654
Museo Medardo Burbano, Pasto 08714
Museo Mineralógico y Antropológico, Ibagué 08660
Museo Nacional de Colombia, Bogotá 08591
Museo Universitario, Pasto 08716
Museo y Parque Arqueológico de Obando, Obando 08698
Sala Museo de la Casa de la Cultura Martín Gamarra, Gamarra 08644

Congo, Democratic Republic

Musée de Mbandaka, Mbandaka 08802

Côte d'Ivoire

Musée des Civilisations de Côte d'Ivoire, Abidjan 08830

Croatia

Arheološka Zbirka, Stari Grad 08964
Arheološka Zbirka, Metković Split 08910
Arheološka Zbirka Osor, Nerezine 08913
Arheološki i Prirodoslovni Muzej, Brijuni 08848
Arheološki Muzej Dubrovnik, Dubrovnik 08860
Arheološki Muzej Zadar, Zadar 08981
Brački Muzej, Supetar 08966
Creski Muzej, Cres 08856
Gradski Muzej, Senj 08945
Gradski Muzej, Umag 08974
Gradski Muzej Korčula, Korčula 08894
Gradski Muzej Makarska, Makarska 08907
Gradski Muzej Sisak, Sisak 08952
Gradski Muzej Virovitica, Virovitica 08978
Hrvatska Akademija Znanosti i Umjetnosti - Gliptoteka, Zagreb 08993
Lapidarij Muzeja Grada Trogira, Trogir 08969
Muzej Franjevačkog Samostana Košljun, Punat 08936
Muzej Hvarske Baštine, Hvar 08880
Muzej Medjimurja, Čakovec 08852
Narodni Muzej Labin, Labin 08906
Zavičajni Muzej, Biograd na Moru 08844
Zbirka Umjetnina Kairos, Trogir 08973
Županijski Muzej Šibenik, Šibenik 08949

Cuba

Museo Arqueológico Guamuhaya, Trinidad 09334
Museo Chorro de Maita, Yaguajay 09353
Museo de la Provincia de Granma, Bayamo 09034
Museo Histórico Municipal de Regla, Regla 09264
Museo Memorial El Morrillo, Matanzas 09217
Museo Municipal de Rodas, Rodas 09267
Museo Provincial de Holguín La Periquera, Holguín 09184
Museo Provincial de Villa Clara y Museo Municipal, Santa Clara 09300

Cyprus

Kourion Museum, Episkopi 09363
Larnaca District Archaeological Museum, Larnaka 09367
Larnaca District Museum, Larnaka 09368
Paphos District Archaeological Museum, Paphos 09392
Pierides Museum, Larnaka 09369
Temple of Aphrodite, Kouklia 09365

Czech Republic

Archeologické Muzeum, Břeclav 09416
Chebský Hrad, Cheb 09472
Dietrichsteinský Palác, Brno 09423
Dvořákovo Muzeum Kolínska v Pravěku, Kolín 09589
Jihomoravské Muzeum ve Znojmě, Znojmo 09947
Kamenný Dům, Kutná Hora 09614
Měnínská Brána, Brno 09431
Městské Muzeum, Čelákovice 09455
Městské Muzeum, Jesenice 09556
Městské Muzeum, Mnichovo Hradiště 09660
Městské Muzeum, Počátky 09721
Městské Muzeum, Valašské Klobouky 09895
Městské Muzeum, Veselí nad Moravou 09906
Městské Muzeum, Nové Strašecí 09682
Městské Muzeum a Galerie, Břeclav 09417
Městské Muzeum a Galerie, Dačice 09489
Městské Muzeum a Galerie J. Karse, Velvary 09903
Městské Vrbasovo Muzeum, Ždánice 09932
Moravské Zemské Muzeum, Brno 09435
Muzeum Jižního Plzeňska v Blovicích, Blovice 09410
Muzeum Mohelnice, Mohelnice 09661
Muzeum Prostějovska, Prostějov 09810
Muzeum Východních Čech v Hradci Králové, Hradec Králové 09534
Muzeum Vysočiny Pelhřimov, Pelhřimov 09710
Národní Kulturní Památník - Valy, Mikulčice 09652
Ostravské Muzeum, Ostrava 09699
Regionální Muzeum a Galerie, Jičín 09559
Regionální Muzeum v Kolíně, Kolín 09591
Regionální Muzeum ve Vysokém Mýtě, Vysoké Mýto 09920
Skanzen Březno u Loun, Louny 09642

Sládečkovo Vlastivědné Muzeum v Kladně, Kladno 09583

Denmark

Den Antikvariske Samling, Ribe 10231
Arkaeologisk Museum, Korsør 10167
Blaabjerg Egnsmuseum, Nørre Nebel 10193
Danmarks Kloster Museum, Ry 10254
Djurslands Museum - Dansk Fiskerimuseum, Grenaa 10041
Hobro Museum, Hobro 10082
Klosterlund Museum og Naturcenter, Engesvang 09998
Kulturhistorisk Museum, Randers 10229
Langelands Museum, Rudkøbing 10252
Lindholm Høje Museet, Nørresundby 10196
Vendsyssel Historiske Museum, Hjørring 10080
Viborg Stiftsmuseum, Viborg 10317

Ecuador

Museo Arqueológico, Cuenca 10381
Museo Arqueológico Atahualpa, Ibarra 10411
Museo Arqueológico y Etnográfico, Ibarra 10412
Museo Augustín Landívar, Cuenca 10382
Museo Benigno Malo, Cuenca 10384
Museo de las Culturas Aborígenes, Cuenca 10390
Museo Francisco Campos, Guayaquil 10405
Museo Los Amantes de Sumpa, Guayaquil 10408
Museo Mariano Acosta, Ibarra 10415
Museo Municipal de Guayaquil, Guayaquil 10409
Museo Vásquez Fuller, Otavalo 10422
Museos Arqueológico, Portoviejo 10423
Valdivia Museo, Valdivia 10463

Egypt

Luxor Museum of Ancient Art, Al-Uqsur 10467

Eritrea

Eritrean Archaeological Museum, Asmara 10558

Ethiopia

Aksum Archaeology Museum, Aksum 10683
National Museum of Ethiopia, Addis Ababa 10680

Faroe Islands

Føroya Fornminnissavn, Hoyvík 10687

Fiji

Fiji Museum, Suva 10704

Finland

Aboa Vetus - Arkeologis-historiallinen Museo, Turku 11387
Liedon Vanhalinna, Vanhalinna 11444

France

Aître Saint-Saturnin, Blois 12030
Archéo-Musée du Kronos, Albens 11532
L'Archéodrome de Bourgogne, Meursault 14106
Archéosite Gallo-Romain des Flaviers, Mouzon 14318
Archéosite Préhistorique de Cambous, Saint-Martin-de-Londre 15399
Centre d'Etudes Archéologiques, Eyguières 12892
Château-Musée, Saumur 15661
Collections Archéologiques, Parmain 14752
Dépôt de Fouilles Gallo-Romaines et Préhistoriques de Montmaurin, Montmaurin 14224
Maison du Patrimoine, Blanquefort 12018
Musée Alfred Bonnot, Chelles 12497
Musée Antoine Vivenel, Compiègne 12597
Musée Archéologique, Brumath 12197
Musée Archéologique, Champagnole 12387
Musée Archéologique, Corseul 12633
Musée Archéologique, Dijon 12736
Musée Archéologique, Escolives-Sainte-Camille 12852
Musée Archéologique, Lectoure 13672
Musée Archéologique, Marvejols 14023
Musée Archéologique, Nîmes 14449
Musée Archéologique, Saint-Gilles-du-Gard 15300
Musée Archéologique, Soyons 15805
Musée Archéologique, Strasbourg 15820
Musée Archéologique, Toulouse 15933
Musée Archéologique, Vachères 16027
Musée Archéologique, Eauze 12804
Musée Archéologique, Saintes 15582
Musée Archéologique d'Argentomagus, Saint-Marcel (Indre) 15393
Musée Archéologique de la Porte du Croux, Nevers 14416
Musée Archéologique de la Région de Breteuil, Breteuil 12166
Musée Archéologique Départemental, Saint-Bertrand-de-Comminges 15185
Musée Archéologique Départemental du Val-d'Oise, Guiry-en-Vexin 13167
Musée Archéologique du Gâtinais, Montargis 14171
Musée Archéologique - Eglise Saint-Laurent, Grenoble 13132
Musée Archéologique et de Paléontologie, Minerve 14125
Musée Archéologique et Lapidaire, Blois 12032
Musée Archéologique Henri Prades, Lattes 13534
Musée Archéologique Municipal, Martres-Tolosane 14022
Musée Archéologique Géologique et Ethnologique des Vans, Les Vans 13720
Musée Auguste Jacquet de Beaucaire, Beaucaire 11887
Musée Baroncelli, Saintes-Maries-de-la-Mer 15588
Musée Biochet-Brechot, Caudebec-en-Caux 12322
Musée Boucher-de-Perthes, Abbeville 11486
Musée Calvet, Avignon 11789
Musée Campanaire, L'Isle-Jourdain 13782
Musée Cappon, Marans 13945
Musée Champollion, Figeac 12932
Musée Charbonneau-Lassay, Loudun 13830
Musée Charles-Portal, Cordes-sur-Ciel 12623
Musée Clément Ader, Muret 14335
Musée Crozatier, Le Puy-en-Velay 13643
Musée d'Archéologie, Izernore 13257
Musée d'Archéologie, Le Pègue 13630
Musée d'Archéologie du Jura, Lons-le-Saunier 13815
Musée d'Archéologie Méditerranéenne, Marseille 13994
Musée d'Archéologique, Saint-Raphaël 15480
Musée d'Art et d'Archéologie, Aurillac 11754
Musée d'Art et d'Archéologie, Laon 13521
Musée d'Art et d'Archéologie, Senlis 15715
Musée d'Art et d'Archéologie de Périgord, Périgueux 14778
Musée d'Art et d'Histoire, Chaumont (Haute-Marne) 12483
Musée d'Art et d'Histoire de Toul, Toul 15922
Musée de la Castre, Cannes 12269
Musée de la Société Archéologique et Historique de la Charente, Angoulême 11619
Musée de la Société d'Histoire et d'Archéologie, Briord 12182
Musée de la Tour aux Puces, Thionville 15884
Musée de l'Ancienne Abbaye de Landévennec, Landévennec 13498
Musée de l'Ardenne, Charleville-Mézières 12410
Musée de Lodève, Lodève 13800
Musée de Melun, Melun 14069
Musée de Moulages, Strasbourg 15826
Musée de Provins et du Provinois, Provins 14951
Musée de Site Galloromain Villa Loupian, Loupian 13837
Musée de Tauroentum, Saint-Cyr-les-Lecques 15227
Musée de Tautavel-Centre Européen de la Préhistoire, Tautavel 15868
Musée de Traditions Populaires et d'Archéologie, Chauvigny 12488
Musée Départemental de l'Oise, Beauvais 11910
Musée Départemental del'Ariège, Foix 12952
Musée Départemental des Antiquités de la Seine-Maritime, Rouen 15114
Musée Départemental des Hautes-Alpes, Gap 13031
Musée Départemental Jérôme-Carcopino, Aléria 11546
Musée des Antiquités Nationales, Saint-Germain-en-Laye 15291
Musée des Arts Patrimoine et des Traditions, Nanteuil-en-Vallée 14373
Musée des Beaux-Arts, Dole 12754
Musée des Beaux-Arts, Rennes (Ille-et-Vilaine) 15016
Musée des Beaux-Arts et d'Archéologie, Vienne 16152
Musée des Beaux-Arts et d'Archeologie Joseph Déchelette, Roanne 15059
Musée des Fontaines Salées, Saint-Père-sous-Vezelay 15454
Musée des Moulages, Montpellier 14243
Musée d'Evreux, Evreux 12889
Musée d'Histoire et d'Archéologie, Vannes (Morbihan) 16069
Musée d'Histoire et d'Archéologie René Bauberot, Châteauponsac 12452
Musée d'Histoire Naturelle et d'Histoire Locale, Elbeuf-sur-Seine 12815
Musée Dobrée, Nantes 14367
Musée Domaine de Gourjade, Castres 12319
Musée du Berry, Bourges 12118
Musée du Biterrois, Béziers 11977
Musée du Châtillonnais, Châtillon-sur-Seine 12474
Musée du Cloître, Tulle 16008
Musée du Colombier, Alès 11550
Musée du Louvre, Paris 14698
Musée du Nord de la Vendée, Montaigu (Vendée) 14166
Musée du Noyonnais, Noyon 14483
Musée du Patrimoine, Néris-les-Bains 14393
Musée du Pays de Sarrebourg, Sarrebourg 15643
Musée du Trophée des Alpes, La Turbie 13454
Musée du Vieux Pérouges, Pérouges 14786
Musée d'Uxellodunum et Musée de Martel, Martel 14018
Musée E. Chenon, Château-Meillant 12436
Musée Gallo-Romain de Tauroentum, Saint-Cyr-sur-Mer 15230
Musée Gassendi, Digne-les-Bains 12727
Musée Historique, Haguenau 13177
Musée Historique et Archéologique, Montségur 14265
Musée Historique et Archéologique de l'Orléanais, Orléans 14537
Musée Intercommunal d'Histoire et d'Archéologie, Louvres 13859
Musée Jeanne d'Aboville, La Fère 13358
Musée La Diana, Montbrison 14190
Musée Labenche d'Art et d'Histoire, Brive-la-Gaillarde 12189
Musée Languédocien, Montpellier 14250
Musée Le Minerve, Yzeures-sur-Creuse 16314
Musée Municipal, Châlons-en-Champagne 12364
Musée Municipal, Nuits-Saint-Georges 14485
Musée Municipal, Saint-Dizier 15244
Musée Municipal A. Danicourt, Péronne 14785
Musée Municipal Ancienne Abbaye Saint-Léger, Soissons 15774
Musée Municipal de Cambrai, Cambrai 12257
Musée Municipal de Millau, Millau 14120
Musée Municipal de Vendôme, Vendôme 16100
Musée Municipal d'Ernée, Ernée 12846
Musée Municipal des Beaux-Arts et d'Archéologie, Valence (Drôme) 16040
Musée Municipal d'Étampes, Etampes 12871
Musée Municipal Méditerranéen de Cassis, Cassis 12305
Musée Pierre Borrione, Aime-en-Tarentaise 11509
Musée Pincé, Angers 11613
Musée Rolin, Autun 11764
Musée Saint-Loup, Troyes 16001
Musée Saint-Raymond, Toulouse 15945
Musée Saint-Remi, Reims 15005
Musée Saturnin Garimond, Fons 12954
Musée Savoisien, Chambéry 12375
Musée Site Départemental d'Archéologie de Bavay, Bavay 11871
Musée Ziem, Martigues 14020
Musées de la Cour d'Or, Metz 14102
Muséum d'Histoire Naturelle, Rouen 15122
Parc Archéologique Européen de Bliesbruck-Reinheim, Bliesbruck 12027
Site Gallo-Romain du Fâ, Barzan 11855

French Polynesia

Musée de Ua Huka, Vaipaee 16328

Gambia

Gambia National Museum, Banjul 16331

Georgia

Archaeological Museum, Tbilisi 16351
Dmanisi Museum-Reserve of History and Architecture, Dmanisi 16339
Kutaisi History Museum, Kutaisi 16343
Metsamor Museum, Metsamor 16348
Mtskheta Museum, Mtskheta 16349
Otar Lordkipanidze Vani Archaeology Miseum-Reserve, Vani 16389
Tbilisi State Museum of Anthropology and Ethnography, Tbilisi 16385

Germany

Alamannenmuseum, Weingarten, Kreis Ravensburg 22425
Alamannenmuseum Ellwangen, Ellwangen 18120
Antikenmuseum der Universität Leipzig, Leipzig 19893
Antikensammlung der Friedrich-Alexander-Universität Erlangen-Nürnberg, Erlangen 18219
Archäologie-Museum Neuburg, Neuburg an der Donau 20597
Archäologiemuseum im Fränkischen Freilandmuseum, Bad Windsheim 16871
Archäologisch-Ethnographische Lehr- und Studiensammlung des Instituts für Altamerikanistik und Ethnologie der Universität Bonn, Bonn 17443
Archäologisch-Ökologisches Museum, Albersdorf 16451
Archäologische Sammlung der Universität Fachbereich III, Trier 22095
Archäologische Sammlung der Universität Rostock, Rostock 21386
Archäologische Sammlung Keltenhaus, Taufkirchen, Kreis München 22013
Archäologische Sammlung mit Heimatvertriebenenstube und Kruk-Sammlung, Königsbrunn 19644
Archäologischer Park Ostkastell, Welzheim 22471
Archäologisches Freilichtmuseum Oerlinghausen, Oerlinghausen 20874
Archäologisches Landesmuseum, Schleswig 21565
Archäologisches Landesmuseum Baden-Württemberg, Konstanz 19676
Archäologisches Landesmuseum Mecklenburg-Vorpommern, Schwerin 21691
Archäologisches Museum, Donauwörth 17894
Archäologisches Museum, Rimpar 21333
Archäologisches Museum der Stadt Kelheim, Kelheim 19499
Archäologisches Museum der Stadt Weißenhorn, Weißenhorn 22460
Archäologisches Museum Essenbach, Essenbach 18271
Archäologisches Museum Gablingen, Gablingen 18524
Archäologisches Museum in der Feldmühle, Rennertshofen 21276
Archäologisches Museum Renningen, Renningen 21277
Archäologisches Zentrum Hitzacker, Hitzacker 19202
Arche Nebra, Wangen 22323
Badisches Landesmuseum Karlsruhe, Karlsruhe 19437
Bienwaldmuseum, Kandel 19425
Braunschweigisches Landesmuseum, Braunschweig 17518
Burg Neuhaus, Wolfsburg 22660
Clemens-Sels-Museum Neuss, Neuss 20654
Dorfmuseum Dierscheid, Dierscheid 17849
Eifelmuseum mit Deutschem Schieferbergwerk, Mayen 20241
Die Eulenburg, Rinteln 21337
Federseemuseum, Bad Buchau 16706
Freilichtmuseum Groß Raden, Sternberg 21900
Glauberg-Museum, Glauburg 18655
Haus zum Stockfisch - Stadtmuseum, Erfurt 18204
Hegau-Museum, Singen 21780
Heimatmuseum, Schlitz 21575
Heimatmuseum, Wolfstein 22671
Heimatmuseum, Zusmarshausen 22828
Heimatmuseum Adlhoch-Haus, Altdorf, Kreis Landshut 16485
Heimatmuseum Sprucker Mühle, Guben 18820
Heimatmuseum Uhl'sches Haus, Göllheim 18671
Heinrich-Schliemann-Museum, Ankershagen 16549
Heuneburgmuseum, Herbertingen 19132
Historisches Museum im Marstall, Paderborn 20970
Hochburg-Museum, Emmendingen 18144
Kelten Römer Museum Manching, Manching 20159
Keltenmuseum Hochdorf/Enz, Eberdingen 18035
Kreismuseum Bitburg-Prüm, Bitburg 17368
Kur-und Stadtmuseum Bad Ems, Bad Ems 16725
Landesmuseum Birkenfeld, Birkenfeld 17359
Landesmuseum Koblenz, Koblenz 19585
Landesmuseum Mainz, Mainz 20146
LWL-Museum für Archäologie, Herne 19155
LWL-Römermuseum, Haltern am See 18912
Museen der Stadt Landshut in der Stadtresidenz, Landshut 19803
Museum Abodiacum, Denklingen 17821
Museum Bodmann, Bodman-Ludwigshafen 17415
Museum Bürgstadt, Bürgstadt 17639
Museum Burg Bederkesa, Bad Bederkesa 16676
Museum der Stadt Worms im Andreasstift, Worms 22685
Museum der Westlausitz Kamenz, Kamenz 19421
Museum für Archäologie und Volkskunde, Kösching 19665
Museum für Archäologie Westfälisches Außenstelle, Olpe 20904
Museum für Kloster- und Heimatgeschichte, Harsefeld 19016
Museum für Kunst und Kulturgeschichte Dortmund, Dortmund 17925
Museum im Graulturm, Gau-Algesheim 18546
Museum in der Kaiserpfalz, Paderborn 20973
Museum Quintana - Archäologie in Künzing, Künzing 19738
Museum Roemervilla, Bad Neuenahr-Ahrweiler 16796
Museum Römervilla, Grenzach-Wyhlen 18754
Museum Schloß Hohentübingen, Tübingen 22123
Museum Schloss Oberschwappach, Knetzgau 19582
Museum Weltkulturen, Mannheim 20170
Osterburgsammlung, Bischofsheim an der Rhön 17363
Pellenz-Museum für Archäologie, Nickenich 20691
Reinhard-Blauth-Museum, Weilerbach 22390
Reiss-Engelhorn-Museen (rem), Mannheim 20173
Rhön-Museum, Fladungen 18325
Robertinum, Halle, Saale 18903
Römer und Bajuwaren Museum Burg Kipfenberg, Kipfenberg 19541
Römerhalle, Bad Kreuznach 16767
Römerkastell Saalburg, Bad Homburg v.d.Höhe 16752
Römermuseum Kastell Boiotro, Passau 20994
Römermuseum Stettfeld, Ubstadt-Weiher 22141
Römerpark Köngen, Köngen 19636
Römische Villa Urbana Longuich, Longuich 20032
Römisches Museum und Naturkunde-Museum, Kempten 19516
Ruhr Museum, Essen 18267
Sammlung Antiker Kleinkunst, Jena 19389

Sammlungen des Winckelmann-Instituts - Lehrsammlung Klassische Archäologie, Berlin 17245
Schlossmuseum, Bad Iburg 16755
Slawenburg Raddusch, Vetschau 22223
Stadthaus Ulm, Ulm 22169
Terra Sigillata Museum, Rheinzabern 21308
Töpfereimuseum, Speicher 21832
Torfmuseum, Neustadt am Rübenberge 20665
Ur- und Frühgeschichtliche Sammlung, Erlangen 18226
Varusschlacht im Osnabrücker Land, Bramsche 17501
Victor-Schultze-Institut für christliche Archäologie und Geschichte der kirchlichen Kunst, Greifswald 18751
Vonderau Museum Fulda, Fulda 18513
Wallmuseum, Oldenburg in Holstein 20902
Wikinger Museum Haithabu, Busdorf 17668
Wittelsbachermuseum Aichach, Aichach 16446
zeittor - Museum der Stadt Neustadt in Holstein, Neustadt in Holstein 20681

Ghana

Museum of Archaeology, Legon 22861
National Museum of Ghana, Accra 22848

Gibraltar

Gibraltar Museum, Gibraltar 22864

Greece

Archaeological Museum, Delos 22981
Archaeological Museum, Mykonos 23103
Archaeological Site of Edessa, Edessa 22989
Byzantine Museum, Mytilini 23110

Greenland

Avanersuup Katersugaasivia, Qaanaaq 23264

Grenada

Grenada National Museum, Saint George's 23273

Guatemala

Museo Popol Vuh, Guatemala 23305

Hungary

Arany János Múzeum, Nagykőrös 23727
Balassa Bálint Múzeum, Esztergom 23549
Balatoni Múzeum, Keszthely 23659
Cselley Ház, Mosonmagyaróvár 23718
Déri Múzeum, Debrecen 23516
Dobó István Vármúzeum, Eger 23536
Dráva Múzeum, Barcs 23361
Erkel Ferenc Múzeum, Gyula 23594
Ferenczy Múzeum, Szentendre 23876
Göcseji Múzeum, Zalaegerszeg 23971
Hajdusági Múzeum, Hajdúböszörmény 23603
Hansági Múzeum, Mosonmagyaróvár 23720
Herman Ottó Múzeum, Miskolc 23704
Herman Ottó Múzeum - Kiállítóhelye, Miskolc 23707
Intercisa Múzeum, Dunaújváros 23533
Jász Múzeum, Jászberény 23623
Jósa András Múzeum, Nyíregyháza 23733
Kubinyi Ferenc Múzeum, Szécsény 23840
Kuny Domokos Megyei Múzeum, Tata 23918
Laczkó Dezső Múzeum, Veszprém 23959
Magyar Nemzeti Múzeum, Budapest 23442
Mátyás Király Palota Múzeum, Visegrád 23966
Móra Ferenc Múzeum, Szeged 23847
Rákóczi Múzeum, Sárospatak 23792
Régészeti Osztály és Állandó Régészeti Kiállítás, Pécs 23772
Reneszánsz Kőtár, Pécs 23773
Smidt Múzeum, Szombathely 23903
Szántó Kovács János Területi Múzeum, Orosháza 23740
Szent István Király Múzeum, Székesfehérvár 23860
Tari László Múzeum, Csongrád 23513
Tatabánya Múzeum, Tatabánya 23922
Tornyai János Múzeum, Hódmezővásárhely 23614
Tragor Ignác Múzeum, Vác 23946
Vármúzeum, Esztergom 23555
Vesztő-Mágor Történelmi Emlékhely, Vészto 23965
Wosinsky Mór Megyei Múseum, Szekszárd 23869
Xantus János Múzeum, Győr 23590

Iceland

Minjasafn Austurlands, Egilsstadir 23993
Pjóðminjasafn Islands, Reykjavik 24030

India

Akshaya Kumar Maitreya Museum, Siliguri 24394
Amreli Archaeological and Art Museum, Amreli 24064
Archaeological Museum, Bijapur 24100
Archaeological Museum, Bodhghaya 24103
Archaeological Museum, Delhi 24136
Archaeological Museum, Guntur 24183
Archaeological Museum, Guntur 24184
Archaeological Museum, Gwalior 24196
Archaeological Museum, Halebidu 24200
Archaeological Museum, Hardwar 24203
Archaeological Museum, Jaunpur 24238
Archaeological Museum, Kamalapur 24251
Archaeological Museum, Khajuraho 24258
Archaeological Museum, Nalanda 24334
Archaeological Museum, Pune 24354
Archaeological Museum, Sanchi 24375
Archaeological Museum, Sarnath 24380
Archaeological Museum, Trichur 24411
Archaeological Museum and Portrait Gallery, Goa Velha 24177
Archaeological Museum Konarak, Puri 24363
Archaeological Museum of Patiala, Patiala 24343
Archaeological Museum Red Fort Delhi, Delhi 24137
Archaeological Site Museum, Alampur 24056
Archaeology Museum, Chennai 24112
Assam State Museum, Guwahati 24188
Bhuri Singh Museum, Chamba 24108
Central Museum, Indore 24230
Central Museum, Nagpur 24332
Commercial Museum, Guwahati 24189
District Archaeological Museum, Dhar 24159
District Museum, Gulbarga 24182
District Museum, Pillalamari 24350
District Museum, Shimoga 24390
Gandhi Centenary Museum, Karimnagar 24255
Ganga Government Museum, Bikaner 24101
Government Museum, Ajmer 24054
Government Museum, Chennai 24115
Government Museum, Hassan 24206
Government Museum, Pudukottai 24353
Government Museum Jhalawar, Jhalawar 24242
Government Museum Vellore, Vellore 24428
Jeypore Branch Museum, Jaipur, Orissa 24232
Jhalawar Archaeology Museum, Jhalawar 24243
Junagadh Museum, Junagadh 24248
Kachchh Museum, Bhuj 24099
Keladi Museum, Keladi 24257
Kithur Rani Channamma Memorial Museum, Kittur 24261
Kolhapur Museum, Kolhapur 24265
Local Antiquities Museum, Chitradurga 24120
Local Museum, Bhanpura 24083
Mahant Ghasidas Memorial Museum, Raipur 24367
Manipur University Museum, Canchipur 24107
Museum of Antiquities, Jamnagar 24237
Museum of Archaeology, Vadodara 24420
Museum of Art, Archaeology and Folklore, Mysore 24330
National Museum of India, Delhi 24147
Orissa State Museum, Bhubaneshwar 24097
Palace and Museum of Antiquities, Padmanabhapuram 24339
Patna Museum, Patna 24346
Sri Pratap Singh Museum, Srinagar 24397
Sri Rallabandi Subbarao Government Museum, Rajahmundry 24369
Ramlingappa Lamture Museum, Osmanabad 24338
State Archaeological Museum, Kolkata 24282
University Museum, Vallabhvidyanagar 24423
Victoria Jubilee Museum, Vijayawada 24429
Vikram Kirti Mandir Museum, Ujjain 24417
Yeleswasam Pavilion, Hyderabad 24219

Indonesia

Archaeological Museum Blitar, Blitar 24455
Archaeological Museum Kediri, Kediri 24512
Museum Daerah Blambangan, Banyuwangi 24450
Museum Gedung Arca, Gianyar 24470
Museum Nasional, Jakarta 24492
Statue Museum Balai Arca, Nganjuk 24531

Iran

Archaeological Museum, Neishaboor 24637
Hegmataneh Museum, Samedan 24641
Khalkhal Museum, Khalkhal 24622
Khoy Museum, Khoy 24625
Miandoab Museum, Urmieh 24700
Museum Falak-ol-Aflak, Khorram-Abaad 24623
National Museum, Tehrān 24687
Qom Museum, Qom 24639
Shahrood Museum, Semnan 24644

Iraq

Mosul Museum, Mosul 24720

Ireland

Clare Museum, Ennis 24829
The Hunt Museum, Limerick 24864
Kerry County Museum, Tralee 24904
Músaem Corca Dhuibhne, Ballyferriter 24738
National Museum of Ireland - Archaeology, Dublin 24808

Israel

Antiquities Museum of Tel Aviv-Yafo, Tel Aviv 25096
Archaeological and Historical Museum, Ramat Hashofet 25088
Archaeological and Land Study Collection, Alumot 24926
Archaeological Collection, Yehi'am 25130
Archaeological Collection of Youth Village Nitzanim, Doar-Na Evtach 24943
The Archaeological Museum of Kfar Saba, Kfar Saba 25029
Archaeology Museum, Shamir 25094
Beit-Shean Museum, Beit Shean 24937
Center of Nautical and Regional Archaeology, Nahshonim 25070
Collection of Archaeology and Land Study, Nahshonim 25071
Ecce Homo, Jerusalem 24986
Genia Schreiber University Art Gallery, Ramat Aviv 25081
Greek Orthodox Museum, Jerusalem 24988
Israel Museum, Jerusalem, Jerusalem 24997
Qedem Museum, Kedumim 25027
Rockefeller Museum, Jerusalem 25011
Shephela Museum, Kibbutz Kfar Menahem 25047
Siebenberg House, Jerusalem 25017
Ussishkin House, Kibbutz Dan 25035

Italy

Antiquarium, Agropoli 25164
Antiquarium, Ariano Irpino 25274
Antiquarium, Avella 25330
Antiquarium, Borgia 25502
Antiquarium, Buccino 25558
Antiquarium, Caldarola 25596
Antiquarium, Cesenatico 25837
Antiquarium, Cimitile 25890
Antiquarium, Filadelfia 26133
Antiquarium, Golasecca 26349
Antiquarium, Loreto Aprutino 26521
Antiquarium, Lugnano in Teverina 26542
Antiquarium, Massarosa 26635
Antiquarium, Mergozzo 26665
Antiquarium, Monasterace 26772
Antiquarium, Nonantola 26939
Antiquarium, Partinico 27111
Antiquarium, San Marzano sul Sarno 27665
Antiquarium, Serravalle Scrivia 27817
Antiquarium, Trieste 28075
Antiquarium, Boscoreale 25517
Antiquarium Cantianense, San Canzian d'Isonzo 27620
Antiquarium Civico, Bagnolo San Vito 25349
Antiquarium Comunale, Colleferro 25940
Antiquarium Comunale, Contessa Entellina 25963
Antiquarium Comunale, Milena 26729
Antiquarium Comunale, Monte Romano 26802
Antiquarium Comunale, Sezze 27831
Antiquarium Comunale N. Pansoni, Cossignano 25994
Antiquarium del Castello Eurialo, Siracusa 27852
Antiquarium del Parco della Forza, Ispica 26418
Antiquarium del Seminario Vescovile, Nola 26935
Antiquarium del Varignano, Portovenere 27313
Antiquarium di Canne, Barletta 25372
Antiquarium di Megara Hyblaea, Augusta 25327
Antiquarium di Monte Cronio, Sciacca 27794
Antiquarium di Nervia, Ventimiglia 28215
Antiquarium di Poggio Civitate, Murlo 26878
Antiquarium di Sant'Appiano, Barberino Val d'Elsa 25355
Antiquarium di Tesis, Vivaro 28301
Antiquarium di Villa Romana, Patti 27115
Antiquarium e Archeologica Museo, Cuglieri 26017
Antiquarium Ipogeo dei Volumni, Perugia 27143
Antiquarium Jetino, San Cipirello 27624
Antiquarium Lucio Salvio Quintiano, Ossuccio 27015
Antiquarium Tellinum, Teglio 27951
Antiquarium - Villa Maritima, Minori 26734
Archeologica di Sovana, Sorano 27878
Castello di Giulio II, Ostia Antica 27016
Castello e Raccolta Archeologica, Montechiarugolo 26818
Castello Malaspina, Massa 26619
Centro Ambientale Archeologico, Legnano 26481
Centro di Documentazione Messapica, Oria 26989
Civica Raccolta e Ceramiche Rinascimentali, Camporgiano 25643
Civici Musei e Gallerie di Storia ed Arte, Udine 28110
Civico Museo Archeologico, Arsago Seprio 25287
Civico Museo Archeologico, Camogli 25626
Civico Museo Archeologico, Ozieri 27025
Civico Museo Archeologico L. Barni e Pinacoteca C. Ottone, Vigevano 28265
Civico Museo di Storia ed Arte e Orto Lapidario, Trieste 28082
Civico Museo Frignanese, Pavullo nel Frignano 27127
Civico Museo Ingauno, Albenga 25176
Civico Museo Storico Archeologico, Savona 27774
Collezione Martini Carissimo, Oria 26990
Collezioni Egittologiche dell'Università di Pisa, Pisa 27236
Lapidario Civico, Gradisca d'Isonzo 26357
Lapidario Comunale, Osimo 27010
Mostra Archeologica G. Venturini, San Felice sul Panaro 27632
Musei Civici, Reggio Emilia 27388
Musei Civici, Pavia 27120
Musei Civici di Padova, Padova 27034
Musei Civici e Pinacoteca L. Sturzo, Caltagirone 25605
Musei Provinciali di Gorizia, Gorizia 26351
Museo Alessi, Enna 26055
Museo Archeologica, Savignone 27770
Museo Archeologico, Acqualagna 25144
Museo Archeologico, Arborea 25257
Museo Archeologico, Artena 25288
Museo Archeologico, Atri 25321
Museo Archeologico, Bonorva 25495
Museo Archeologico, Caltanissetta 25607
Museo Archeologico, Camaiore 25612
Museo Archeologico, Castello di Godego 25741
Museo Archeologico, Dorgali 26044
Museo Archeologico, Fermo 26098
Museo Archeologico, Frosinone 26253
Museo Archeologico, Isola della Scala 26417
Museo Archeologico, Lentini 26484
Museo Archeologico, Molfetta 26763
Museo Archeologico, Ostiglia 27019
Museo Archeologico, Palazzuolo sul Senio 27061
Museo Archeologico, Povegliano Veronese 27320
Museo Archeologico, Priverno 27345
Museo Archeologico, Pula 27347
Museo Archeologico, Rocca di Mezzo 27427
Museo Archeologico, San Miniato 27671
Museo Archeologico, Santadi 27713
Museo Archeologico, Santorso 27734
Museo Archeologico, Scansano 27784
Museo Archeologico, Teramo 27955
Museo Archeologico, Ventotene 28218
Museo Archeologico, Veroli 28225
Museo Archeologico, Vetulonia 28241
Museo Archeologico, Villasimius 28281
Museo Archeologico, Cavriana 25791
Museo Archeologico, Siena 27838
Museo Archeologico, Varallo Pombia 28156
Museo Archeologico Antiquarium Arborense, Oristano 26996
Museo Archeologico Antonio Santarelli, Forlì 26223
Museo Archeologico Aquaria, Soncino 27868
Museo Archeologico Civico, Carassai 25673
Museo Archeologico Civico, Forlimpopoli 26231
Museo Archeologico Comprensoriale, Teti 27970
Museo Archeologico Comunale, Vicchio 28248
Museo Archeologico dei Campi Flegrei, Bacoli 25338
Museo Archeologico del Finale, Finale Ligure 26138
Museo Archeologico del Territorio, Cupra Marittima 26022
Museo Archeologico del Territorio, San Lorenzo in Campo 27656
Museo Archeologico del Territorio di Populonia, Piombino 27228
Museo Archeologico della Badia, Licata 26490
Museo Archeologico della Lucania Occidentale, Padula 27054
Museo Archeologico dell'Abruzzo, Crecchio 25998
Museo Archeologico dell'Alto Adige, Bolzano 25486
Museo Archeologico dell'Alto Canavese, Cuorgnè 26021
Museo Archeologico dell'Antica, Alife 25202
Museo Archeologico dell'Antica Capua, Santa Maria Capua Vetere 27705
Museo Archeologico dell'Antica Corfinio, Corfinio 25974
Museo Archeologico di Calatia, Maddaloni 26561
Museo Archeologico di Casteggio, Casteggio 25714
Museo Archeologico di Naxos, Giardini Naxos 26334
Museo Archeologico di Sepino, Sepino 27808
Museo Archeologico di Villa Sulcis, Carbonia 25675
Museo Archeologico e Antiquarium, Sala Consilina 27591
Museo Archeologico e d'Arte della Maremma, Grosseto 26371
Museo Archeologico e Paleontologico di Farneta, Cortona 25987
Museo Archeologico ed Etnografico, Ittireddu 26421
Museo Archeologico Etnografico, Paulilatino 27118
Museo Archeologico F. Milizia, Oria 26991
Museo Archeologico G. Allevi, Offida 26973
Museo Archeologico G. Rambotti, Desenzano del Garda 26034
Museo Archeologico Ibleo, Ragusa 27359
Museo Archeologico Iulium Carnicum, Zuglio 28320
Museo Archeologico Lametino, Lamezia Terme 26443
Museo Archeologico Liutprando, San Giovanni in Persiceto 27645
Museo Archeologico Lucus Feroniae, Capena 25656
Museo Archeologico Luigi Donini, San Lazzaro di Savena 27650

Museo Archeologico Michele Janora, Irsina .. 26409
Museo Archeologico Nazionale, Cassino 25711
Museo Archeologico Nazionale, Cividale del Friuli 25916
Museo Archeologico Nazionale, Eboli 26050
Museo Archeologico Nazionale, Firenze 26167
Museo Archeologico Nazionale, Manfredonia . 26587
Museo Archeologico Nazionale, Mantova 26591
Museo Archeologico Nazionale, Spoleto 27892
Museo Archeologico Nazionale d'Abruzzo, Campli 25631
Museo Archeologico Nazionale dell'Abruzzo, Chieti 25872
Museo Archeologico Nazionale delle Marche, Ancona 25227
Museo Archeologico Nazionale dell'Umbria, Perugia 27147
Museo Archeologico per la Preistoria e Protostoria del Tigullio, Chiavari 25858
Museo Archeologico Provinciale, Salerno 27594
Museo Archeologico Regionale, Aosta 25247
Museo Archeologico Regionale, Gela 26282
Museo Archeologico Regionale Paolo Orsi, Siracusa 27854
Museo Archeologico Remo Fumagalli, Broletto di Novara 25550
Museo Archeologico Sentinate, Sassoferrato .. 27761
Museo Archeologico Statale, Arcevia 25259
Museo Archeologico Statale, Teano 27947
Museo Archeologico Statale, Urbisaglia 28128
Museo Archeologico Tifernum Mataurense, Sant'Angelo in Vado 27718
Museo Archeologico Villa Abbas, Sardara 27740
Museo Berenziano, Cremona 26002
Museo Bicknell, Bordighera 25498
Museo Canonicale, Verona 28231
Museo Carlo Venturini, Massa Lombarda 26623
Museo Centrale Montemartini, Roma 27467
Museo Civico, Avola 25337
Museo Civico, Noto 26945
Museo Civico, Piedimonte Matese 27200
Museo Civico, Pizzighettone 27260
Museo Civico, Ragogna 27357
Museo Civico, Riva del Garda 27421
Museo Civico, Sanremo 27696
Museo Civico, Susa 27925
Museo Civico, Viterbo 28292
Museo Civico, Argenta 25271
Museo Civico, Nocera Umbra 26933
Museo Civico, Finale Emilia 26136
Museo Civico Albano, Albano Laziale 25175
Museo Civico Antiquarium, Nettuno 26921
Museo Civico Antiquarium Platina, Piadena .. 27190
Museo Civico Archeologica e Quadreria, Monterubbiano 26850
Museo Civico Archeologico, Angera 25235
Museo Civico Archeologico, Arona 25281
Museo Civico Archeologico, Bene Vagienna ... 25393
Museo Civico Archeologico, Bologna 25447
Museo Civico Archeologico, Castrovillari 25768
Museo Civico Archeologico, Chiomonte 25877
Museo Civico Archeologico, Cirò Marina 25899
Museo Civico Archeologico, Marciana 26602
Museo Civico Archeologico, Massa Marittima . 26628
Museo Civico Archeologico, Nicotera 26925
Museo Civico Archeologico, Padria 27053
Museo Civico Archeologico, Ramacca 27361
Museo Civico Archeologico, Remedello 27394
Museo Civico Archeologico, Rutigliano 27581
Museo Civico Archeologico, Tolentino 27984
Museo Civico Archeologico, Treia 28052
Museo Civico Archeologico della Val Tenesi, Manerba del Garda 26585
Museo Civico Archeologico Eno Bellis, Oderzo 26969
Museo Civico Archeologico F. Saverio Majellaro, Bisceglie 25424
Museo Civico Archeologico Genna Maria, Villanovaforru 28279
Museo Civico Archeologico P. Giovio, Como .. 25951
Museo Civico Archeologico Trebula Mutesca, Monteleone Sabino 26834
Museo Civico Archeologico U. Formentini, La Spezia 26433
Museo Civico di Belriguardo, Voghiera 28305
Museo Civico di Scienze Naturali ed Archeologia della Valdinievole, Pescia 27172
Museo Civico Domenico Mambrini, Galeata ... 26263
Museo Civico Don Queirolo, Vado Ligure 28131
Museo Civico e Antiquarium, Latina 26465
Museo Civico e Pinacoteca A. Vernarecci, Fossombrone 26241
Museo Civico e Risorgimentale Don Giovanni Verità, Modigliana 26760
Museo Civico F. de Rocco, San Vito al Tagliamento 27688
Museo Civico G. Sutermeister, Legnano 26482
Museo Civico Goffredo Bellini, Asola 25301
Museo Civico Pietro Lotti, Ischia di Castro 26410
Museo Civico-Pinacoteca, Vicenza 28252
Museo Civico Storico Archeologico, Leno 26483
Museo Comunale, Roggiano Gravina 27436
Museo d' Arte e Scienza, Milano 26693
Museo d'Arte e Archeologia I. Mormino, Palermo 27067
Museo dei Grandi Fiumi, Rovigo 27574
Museo del Castello, Santa Severina 27709
Museo del Sannio, Benevento 25394
Museo del Santuario di Montevergine, Mercogliano 26664
Museo del Territorio, Riccione 27400
Museo della Città, Aquino 25256
Museo della Civiltà Locale - Raccolta d'Arte di San Francesco, Trevi 28064
Museo della Civiltà Messapica, Poggiardo 27265
Museo della Civiltà Rupestre, Zungri 28321
Museo della Regina, Cattolica 25785
Museo dell'Abazzia di San Colombano, Bobbio 25433
Museo dell'Abbazia di San Nilo, Grottaferrata . 26374
Museo dell'Accademia Etrusca e della Città di Cortona, Cortona 25988
Museo dell'Arte Classica, Sezione di Archeologia, Roma 27482
Museo delle Cripta, Poggiardo 27266
Museo delle Navi Romane, Nemi 26918
Museo di Antichità Etrusche e Italiche, Roma 27490
Museo di Antropologia, Napoli 26894
Museo di Archeologia Urbana e dei Commerc Antichi in Abruzzo, Lanciano 26447
Museo di Paletnologia, Polignano a Mare 27272
Museo di Scienze Archeologiche e d'Arte, Padova 27042
Museo di Scienze Naturali ed Umane, L'Aquila 26455
Museo Diocesano, Tropea 28105
Museo e Pinacoteca Civica, Alessandria 25193
Museo e Pinacoteca Comunali, Gubbio 26390
Museo Egnazia, Monopoli 26781
Museo Etrusco Gasparri, Piombino 27231
Museo G. Whitaker, Marsala 26613
Museo Giovanni Antonio Sanna, Sassari 27755
Museo Irpino, Avellino 25331
Museo Lapidario della Canonica della Cattedrale, Novara 26952
Museo Luigi Ceselli, Subiaco 27916
Museo Memorie Pollentine, Pollenza 27274
Museo Meo-Evoli, Monopoli 26782
Museo Naturalistica-Archeologico, Vicenza ... 28254
Museo Nazionale Archeologico, Altamura 25206
Museo Nazionale Archeologico di Egnazia, Fasano-Savelletri 26090
Museo Nazionale Atestino, Este 26066
Museo Nazionale Concordiese, Portogruaro ... 27311
Museo Nazionale d'Abruzzo, L'Aquila 26457
Museo Nazionale d'Arte Orientale, Roma 27512
Museo Nazionale dell'Agro Picentino, Pontecagnano 27287
Museo Nazionale di Reggio Calabria, Reggio Calabria 27384
Museo Nazionale di San Matteo, Pisa 27245
Museo Oraziano, Licenza 26491
Museo P. Vagliasindi, Randazzo 27364
Museo Perrando, Sassello 27757
Museo Preistorico e Archeologico A.C. Blanc, Viareggio 28244
Museo Provinciale Campano, Capua 25670
Museo Provinciale Sigismondo Castromediano, Lecce 26474
Museo R. De Cesare, Spinazzola 27889
Museo Storico Archeologico G. Gabetti, Dogliani 26039
Museo Storico del Vetro e della Bottiglia J.F. Mariani, Montalcino 26796
Museo Territoriale del Lago di Bolsena, Bolsena 25481
Museo Tuscolano, Frascati 26246
Palazzo Belgiojoso, Lecco 26479
Parco Archeologico di Castel Seprio, Castelseprio 25754
Raccolta Archeologica, Terni 27965
Raccolta Archeologica, Tolentino 27990
Raccolta Archeologica Alberto Pisani Dossi, Corbetta 25969
Raccolta Archeologica e Paleontologica, Corciano 25972
Raccolta Lapidaria Capitolina, Roma 27543
Raccolta Materiale Archeologico, Bevagna 25414
Raccolta Privata Toraldo di Francia, Tropea ... 28106
Scavi Archeologici di Velia, Ascea 25289
Scavi di Ostia, Ostia Antica 27018
Scavi di Stabiae, Castellammare di Stabia 25731
Vintschger Museum, Sluderno 27858

Jamaica

Archaeological Museum, Spanish Town 28340
Museum of Historical Archaeology, Kingston . 28333

Japan

Ainu Museum, Sapporo 29213
Chiba-kenritsu Boso Fudoki-No-Oka, Sakae .. 29193
Fukahori Kaizukai Iseki Museum, Nagasaki ... 28973
Kyoto-shiritsu Kokokan, Kyoto 28848
Nagatoro Kyukokan, Nagatoro 28988
Nara-kenritsu Kashihara Kokogaku Kenkyujo Fuzoku Hakubutsukan, Kashihara 28677
Osaka Prefectural Museum of Yayoi Culture, Izumi, Osaka 28614
Saitama-kenritsu Sakitama Shiryokan, Gyoda 28460
Shibayama Haniwa Hakubutsukan, Shibayama 29255
Shiretoko Museum, Shari 29254
Tatsuuma Kokokan, Nishinomiya 29075
Tokushima-kenritsu Torii Kinen Hakubutsukan, Naruto 29045
The University Museum, Tokyo 29502

Jordan

Salt Archaeological Museum, Salt 29666

Kazakhstan

Centralnyj Gosudarstvennyj Muzej Kazachstana, Almaty 29678
Muzej Archeologii, Almaty 29685
Muzej Kazakstan Respublikasynyn Prezidenttik Mėdeniet Ortalyfy, Astana 29692
Muzej Prirody, Almaty 29690
Otrar Memlekettik Archaeologicalik Korik Muzej, Čimkent 29699
Otrarskij Gosudarstvennyj Archeologičeskij Muzej-Zapovednik, Šaulder 29726

Kenya

Fort Jesus Museum, Mombasa 29746
Gede Museum and National Monument, Watamu 29752
Hyrax Hill Museum, Hyrax Hill 29739
Kabarnet Museum, Kabarnet 29740
Kariandusi Prehistoric Site Museum, Gilgil ... 29738

Kyrgyzstan

Archeologičeskij Muzej, Biškek 29951
Muzej Architektury i Archeologii Vez Burana, Tokmok 29967

Latvia

Latvijas Vēstures Muzejs, Rīga 30093
Turaidas Muzejrezervāts, Sigulda 30124

Lithuania

Aušros Alėjos Rūmai, Šiauliai 30352
Ch. Frenkelio Rūmai, Šiauliai 30353
Lietuvos Nacionalinis Muziejus, Vilnius 30419
Senasis Arsenalas, Vilnius 30431
Šiaulių Aušros Muziejus, Šiauliai 30362
Vytauto Didžiojo Karo Muziejus, Kaunas 30269

Luxembourg

Musée Municipal, Dudelange 30467
Musée Municipal des Mosaïques Romaines, Diekirch 30464
Musée National d'Histoire et d'Art (MNHA), Luxembourg 30483

Macedonia

Makedonsko Archeološko Društvo - Zavod i Muzej Prilep, Prilep 30513
Muzej na Grad Skopje, Skopje 30518
Muzej na Makedonija - Archeološki, Etnološki i Istoriski, Skopje 30519
Naroden Muzej, Gevgelija 30501
Naroden Muzej, Veles 30533
Narodni Muzej, Štip 30526

Madagascar

Musée d'Art et d'Archéologie de l'Université d'Antananarivo, Antananarivo 30534

Malaysia

Archaeological Museum, Merbok 30608
Jabatan Muzium Sabah, Kota Kinabalu 30565
Muzium Arkeologi, Merbok 30609
Muzium Negara, Kuala Lumpur 30578
Muzium Sejarah National, Kuala Lumpur 30581

Malta

Archaeology Museum Gozo, Victoria 30650
National Museum of Archaeology, Valletta 30643

Mauritius

Historical Museum, Mahébourg 30677

Mexico

Museo de las Culturas de la Huasteca Potosina Tamuantzán, Valles 31529
Museo de las Culturas de Oaxaca, Oaxaca 31225
Museo Nacional de las Culturas, México 31138

Montenegro

Arheološka Zbirka Crne Gore, Podgorica 31653
Muzej Grada Perasta, Perast 31651

Morocco

Antiquités de Volubilis, Moulay-Idriss Zerhoun 31685
Musée Archéologique, Larache 31676
Musée Archéologique, Tétouan 31700
Musée Kasba, Tanger 31698
Museo de Arqueologia e Historia, Melilla 31683
Museo de la Basilica Tardorromana, Ceuta ... 31663

Namibia

National Museum of Namibia, Windhoek 31765

Nepal

Bronze and Brass Museum, Bhaktapur 31769

Netherlands

Archeologisch Centrum C.W. Bruinvis, Alkmaar 31792
Archeologisch en Paleontologisch Museum Hertogsgemaal, Gewande 32248
Archeologisch Museum Haarlem, Haarlem 32293
Breda's Museum, Breda 32003
Fluweelengrot Valkenburg, Valkenburg, Limburg 32862
Gemeentelijk Archeologisch Museum, Aardenburg 31787
Gemeentelijke Oudheidkamer, Ermelo 32222
Gemeentelijke Oudheidkamer, Geldrop 32239
Historisch Museum de Scheper, Eibergen 32183
Kasteelruïne Valkenburg, Valkenburg, Limburg 32864
Museum De Roode Tooren, Doesburg 32134
Museum Dorestad, Wijk bij Duurstede 32948
Museum Het Valkhof, Nijmegen 32590
Museum Oud Westdorpe, Westdorpe 32938
Museum Schokland, Ens 32209
Museum Wierdenland, Ezinge 32229
Museumkelder Derlon, Maastricht 32528
Ruïne van Brederode, Santpoort 32743
Venrays Museum, Venray 32886
Westlands Museum voor Streek- en Tuinbouwhistorie, Honselersdijk 32403

Nicaragua

Museo Nacional de Nicaragua, Managua 33290

Niger

Musée National du Niger, Niamey 33308

Nigeria

Institute of Archaeology and Museum Studies, Jos 33329
National Museum Benin, Benin City 33318
National Museum Esie, Esie 33321
National Museum Ibadan, Ibadan 33323
National Museum Ile-Ife, Ife 33325
National Museum Kaduna, Kaduna 33331

Norfolk Island

Norfolk Island Museum, Kingston 33349

Northern Cyprus

Güzelyurt Müzesi, Güzelyurt 33361
Saint Barnabas Icon and Archaeological Museum, Gazimağusa 33354

Norway

Hægebostad Bygdemuseum, Tingvatn 33905
Kon-Tiki Museum, Oslo 33729
Vitenskapsmuseet, Trondheim 33932

Oman

Muscat Gate Museum, Muscat 33989
Qurm Museum, Qurm 33995

Pakistan

Archaeological Museum, Karachi 34013
Archaeological Museum Harappa, Harappa ... 34002
Archaeological Museum Moen-Jo-Daro, Moen-Jo-Daro 34031
Archaeological Museum of Baluchistan Quetta, Quetta 34038
Archaeological Museum Peshawar, Peshawar 34032
Archaeological Museum Saidu Sharif, Saidu Sharif 34042
Bahawalpur Museum, Bhawalpur 33998
Chakdara Museum, Chakdara 33999

Museum of Archaeological Site Moenjodaro, Dokri 34000
National Museum of Pakistan, Karachi 34014
Peshawar Museum, Peshawar 34036

Palestine

Palestinian Archaeological Museum, Birzeit 34052

Peru

Complejo Arqueológico de Chavín, Huari 34183
Museo Amano, Lima 34222
Museo Arqueológical Rafael Larco Herrera, Lima 34225
Museo Arqueologico José María Morante de la Universidad Nacional de San Augustín de Arequipa, Arequipa 34109
Museo de Arqueología de la Universidad Nacional de Trujillo, Trujillo 34318
Museo de Sitio de Pachacamac, Lima 34247
Museo Regional de Ica María Reiche Gross Newman, Ica 34189
Museo Santuarios Andinos de la Universidad Católica de Santa María, Arequipa 34116

Philippines

Angono National Museum, Angono 34333
Daniel W. Tantoco jr. Collection, Manila 34438
Pambasang Museo ng Pilipinas, Manila 34450
Siliman University Anthropology Museum, Dumaguete 34386
Southwestern University Museum, Cebu 34376
Zamboanga National Museum, Zamboanga 34527

Poland

Baszta Prochowa, Chełmno 34590
Dział Archeologii i Etnografii, Bydgoszcz 34574
Muzeum, Koszalin 34761
Muzeum Archeologiczne, Gdańsk 34636
Muzeum Archeologiczne, Poznań 35030
Muzeum Archeologiczne, Wrocław 35326
Muzeum Archeologiczne i Etnograficzne, Łódź 34873
Muzeum Archeologiczne Środkowego Nadodrza, Świdnica, k Zielonej Góry 35148
Muzeum Archeologiczne w Biskupinie, Gąsawa 34630
Muzeum Archeologiczne w Krakowie, Kraków 34784
Muzeum Archeologiczne w Krakowie, Oddział w Nowej Hucie, Kraków 34785
Muzeum Archeologiczno-Historyczne, Głogów 34668
Muzeum Częstochowskie, Częstochowa 34615
Muzeum Historyczne w Sanoku, Sanok 35101
Muzeum im. Jacka Malczewskiego w Radomiu, Radom 35073
Muzeum im. Wojciecha Kętrzyńskiego, Kętrzyn 34738
Muzeum Kresów w Lubaczowie, Lubaczów 34893
Muzeum Lubelskie, Lublin 34905
Muzeum Miasta Pabianice, Pabianice 35002
Muzeum Miedzi, Legnica 34848
Muzeum Miejskie, Zabrze 35350
Muzeum Neolitycznego Górnictwa i Rezerwat Archeologiczny, Krzemionki koło Ostrowca 34834
Muzeum Okręgowe, Konin 34755
Muzeum Okręgowe, Piotrków Trybunalski 35010
Muzeum Okręgowe, Rzeszów 35094
Muzeum Okręgowe, Sandomierz 35099
Muzeum Okręgowe, Suwałki 35144
Muzeum Okręgowe im. Leona Wyczółkowskiego, Bydgoszcz 34578
Muzeum Okręgowe Ziemi Kaliskiej, Kalisz 34710
Muzeum Początków Państwa Polskiego, Gniezno 34673
Muzeum Regionalne, Cedynia 34586
Muzeum Regionalne, Szczecinek 35166
Muzeum Regionalne, Szydłów 35173
Muzeum Regionalne, Wiślica 35304
Muzeum Regionalne im. Andrzeja Kaubego, Wolin 35316
Muzeum Regionalne im. Janusza Petera, Tomaszów Lubelski 35183
Muzeum Regionalne im. Władysława Kowalskiego, Dobczyce 34622
Muzeum Regionalne PTTK, Golub-Dobrzyń 34675
Muzeum Śląska Opolskiego, Opole 34986
Muzeum Ślężańskie im. Stanisława Dunajewskiego, Sobótka 35116
Muzeum Starożytnego Hutnictwa Mazowieckiego, Pruszków 35052
Muzeum w Chrzanowie im. Ireny i Mieczysława Mazarakich, Chrzanów 34606
Muzeum w Elblągu, Elbląg 34628
Muzeum w Kwidzynie, Kwidzyń 34840
Muzeum w Lęborku, Lębork 34843
Muzeum w Łowiczu, Łowicz 34892
Muzeum w Międzyrzeczu, Międzyrzecz 34917
Muzeum w Raciborzu, Racibórz 35072
Muzeum-Zamek Górków, Szamotuły 35154
Muzeum Ziemi Kępińskiej, Kępno 34737
Muzeum Ziemi Rawickiej, Rawicz 35079
Muzeum Ziemi Rawskiej, Rawa Mazowiecka 35078
Muzeum Ziemi Wieluńskiej, Wieluń 35301
Muzeum Ziemi Zawkrzańskiej, Mława 34927
Państwowe Muzeum Archeologiczne, Warszawa 35286
Rezerwat Archeologiczny na Zawodziu, Kalisz 34711
Skansen Archeologiczny w Mrówkach, Mrówki 34931
Zamek w Gniewie - Muzeum Regionalne, Gniew 34671

Portugal

Mamoa de Lamas, Lamas 35654
Museo Arqueológico de Martins Sarmento, Guimarães 35637
Museu Arqueológico da Citânia de Sanfins, Sanfins de Ferreira 35896
Museu Arqueológico de Barcelos, Barcelos 35455
Museu Arqueológico de São Miguel de Odrinhas, Sintra 35946
Museu Arqueológico de Serpa, Serpa 35928
Museu Arqueológico do Carmo, Lisboa 35670
Museu Arqueológico e Etnográfico de Setúbal, Setúbal 35935
Museu Arqueológico e Lapidar do Infante D. Henrique, Faro 35596
Museu Arqueológico Municipal, Gouveia 35631
Museu Arqueológico Regional de Castro Marim, Castro Marim 35517
Museu da Cidade de Lisboa, Lisboa 35677
Museu da Rainha Dona Leonor, Beja 35463
Museu de Arqueologia, Montemor-o-Novo 35775
Museu de Arqueologia, Sesimbra 35931
Museu de Arqueologia, Vila Viçosa 36018
Museu de Arqueologia de Mértola, Mértola 35760
Museu de Arte Sacra e Arqueologia, Porto 35846
Museu de Arte Sacra e Arqueologia de Vila Alva, Vila Alva 35985
Museu de Setúbal - Convento de Jesus - Galeria de Arte Quinhentista, Setúbal 35936
Museu de Vila Real, Vila Real 36013
Museu Didáctico do Instituto de Arqueologia, Coimbra 35535
Museu Etnográfico e Arqueológico do Dr. Joaquim Manso, Nazaré 35780
Museu Fundação Arquivo Paes Telles, Ervedal 35561
Museu Monográfico do Bárrio e Campo Arqueológico, Bárrio 35458
Museu Municipal Abade de Pedrosa, Santo Tirso 35916
Museu Municipal Alcácer do Sal, Alcácer do Sal 35388
Museu Municipal Carlos Reis de Torres Novas, Torres Novas 35969
Museu Municipal de Arqueologia, Albufeira 35386
Museu Municipal de Arqueologia, Aljezur 35409
Museu Municipal de Arqueologia, Amadora 35431
Museu Municipal de Arqueologia de Silves, Silves 35940
Museu Municipal de Arqueologia e Etnologia, Sines 35942
Museu Municipal de Óbidos, Óbidos 35784
Museu Municipal de Santarém, Santarém 35909
Museu Municipal do Seixal, Seixal 35927
Museu Regional Abade de Baçal, Bragança 35487
Museu Regional de Arqueologia D. Diogo de Sousa, Braga 35483
Museus Municipal de Almada, Almada 35415
Núcleo Arqueológico da Rua dos Correeiros, Lisboa 35728
Núcleo Museológico da Capela do Espírito Santo dos Mareantes, Sesimbra 35933
Núcleo Museológico das Ruínas de Milreu, Faro 35599
Núcleo Museológico de S. Cucufate, Vila de Frades 35986
Sala Museu de Arqueologia, Mogadouro 35767

Puerto Rico

Museo y Centro de Estudios Humanísticos, Gurabo 36040

Romania

Cetatea de Scaun a Sucevei, Suceava 36641
Cetatea Devei, Deva 36343
Cetatea Poienari, Căpăţânenii Pământeni 36258
Cetatea Sucidava, Corabia 36310
Colecţia Arheologică, Zlatna 36744
Colecţia Maria şi Dr. George Severeanu, Bucureşti 36192
Expoziţia Arheologică Tibiscum, Caransebeş 36262
Expoziţia Foto-documentară Vasile Pârvan, Huruieşti 36406
Muzeul Arheologic Adamclisi, Adamclisi 36096
Muzeul Arheologic Sarmizegetusa, Sarmizegetusa 36588
Muzeul Castelul Corvinestilor, Hunedoara 36404
Muzeul Civilizaţiei Dacice şi Romàne, Deva 36344
Muzeul Civilizatiej Gumelniţa Olteniţa, Olteniţa 36499
Muzeul Curtea Veche, Bucureşti 36200
Muzeul de Arheologie, Săveni 36593
Muzeul de Arheologie Callatis Mangalia, Mangalia 36465
Muzeul de Arheologie şi Artă Religioasă, Băile Govora 36129
Muzeul de Arheologie şi Etnografie, Corabia 36311
Muzeul de Arheologie şi Istorie, Baia Mare 36124
Muzeul de Istorie, Târgu Jiu 36665
Muzeul de Istorie al Moldovei, Iaşi 36418
Muzeul de Istorie Naţional şi Arheologie, Constanţa 36305
Muzeul de Istorie şi Arheologie, Tulcea 36705
Muzeul de Istorie şi Artă Roman, Roman 36568
Muzeul Judeţean Argeş, Piteşti 36526
Muzeul Judeţean de Istorie, Galaţi 36380
Muzeul Judeţean de Istorie, Piatra-Neamţ 36521
Muzeul Judeţean de Istorie Braşov, Braşov 36180
Muzeul Judeţean Ialomiţa, Slobozia 36630
Muzeul Naţional al Unirii, Alba Iulia 36106
Muzeul Naţional de Istorie a României, Bucureşti 36221
Muzeul Orăşenesc de Istorie, Aiud 36103
Muzeul Sătesc, Gostavăţu 36389
Muzeul Ţării Făgăraşului Valeriu Literat, Făgăraş 36363
Rezervaţia Arheologică, Cucuteni 36334
Rezervaţia Arheologică Cotnari, Cotnari 36319
Rezervaţia Arheologică Târgşoru Vechi, Târgşoru Vechi 36662

Russia

Anapskij Archeologičeskij Muzej-zapovednik, Anapa 36758
Archeologičeskij Muzej, Čerepovec 36847
Archeologičeskij Muzej Kazanskogo Universiteta, Kazan 37004
Astrachanskij Gosudarstvennyj Obedinennyj Istoriko- architekturnyj Muzej-zapovednik, Astrachan 36778
Azovskij Kraevedčeskij Muzej, Azov 36786
Biljarskij Gosudarstvennyj Istoriko-Archeologičeskij i Prirodnyj Muzej-Zapovednik, Biljarsk 36806
Brjanskij Gosudarstvennyj Obedinennyj Kraevedčeskij Muzej, Brjansk 36824
Chabarovskij Gosudarstvennyj Muzej Dalnego Vostoka im. N.I. Grodekova, Chabarovsk 36856
Chabarovskij Muzej Archeologii im. A.P. Okladnikova, Chabarovsk 36857
Chakasskij Respublikanskij Kraevedčeskij Muzej, Abakan 36745
Chvalynskij Kraevedčeskij Muzej, Chvalynsk 36865
Čuvašskij Nacionalnyj Muzej, Čeboksary 36837
Gosudarstvennyj Istoriko-architekturnyj i Prirodno-landšaftnyj Muzej Zapovednik Izborsk, Izborsk 36953
Gosudarstvennyj Nacionalnyj Muzej Ust-Ordynskogo Burjatskogo Avtonomnogo Okruga, Ust-Ordynsk 37913
Gosudarstvennyj Okružnoj Muzej Prirody i Čeloveka Goroda Chanty-Mansijsk, Chanty-Mansijsk 36862
Irkutskij Oblastnoj Kraevedčeskij Muzej, Irkutsk 36933
Istoriko-architekturnyj Muzej Kuzneckaja Krepost', Novokuzneck 37479
Istoriko-memorialnyj Muzejnyj Kompleks Bobriki, Donskij 36878
Kaliningradskij Oblastnoj Istoriko-chudožestvennyj Muzej, Kaliningrad 36989
Kalmyckij Respublikanskij Kraevedčeskij Muzej im. Prof. N.N. Palmova, Ėlista 36913
Kamčatskij Oblastnoj Kraevedčeskij Muzej, Petropavlovsk-Kamčatskij 37553
Karelskij Gosudarstvennyj Kraevedčeskij Muzej, Petrozavodsk 37558
Kislovodskij Istoriko-kraevedčeskij Muzej Krepost', Kislovodsk 37046
Krasnodarskij Gosudarstvennyj Istoriko-archeologičeskij Muzej-zapovednik, Krasnodar 37066
Kurskij Gosudarstvennyj Oblastnoj Muzej Archeologii, Kursk 37095
Miasskij Kraevedčeskij Muzej, Miass 37130
Muzej 700 let - Landskrona, Nevskoe Ustjě, Nienšanc, Sankt-Peterburg 37668
Muzej Archeologii, Syktyvkar 37817
Muzej Archeologii i Ėtnografii, Vladivostok 37943
Muzej Archeologii i Ėtnografii, Ufa 37887
Muzej Archeologii i Ėtnografii, Syktyvkar 37818
Muzej Archeologii Moskvy, Moskva 37262
Muzej Archeologii Prikamja, Perm 37548
Muzej Archeologija, Ėtnografija i Ėkologija Sibiri Kemerovskogo Gosudarstvennogo Universiteta, Kemerovo 37033
Muzej Istorii i Kultury Srednego Prikamja, Sarapul 37746
Muzej Istorii Moskvy, Moskva 37294
Nacionalnyj Muzej Respubliki Baškortostan, Ufa 37889
Nacionalnyj Muzej Udmurtskoj Respubliki im. K. Gerda, Iževsk 36955
Narovčatskij Rajonnyj Kraevedčeskij Muzej, Narovčat 37446
Nižegorodskij Gosudarstvennyj Istoriko-architekturnyj Muzej-zapovednik, Nižnij Novgorod 37468
Novokuzneckij Kraevedčeskij Muzej, Novokuzneck 37482
Roslavlskij Istoriko-chudožstvennyj Muzej, Roslavl 37597
Šeltozerskij Vepsskij Ėtnografičeskij Muzej, Šeltozero 37758
Severo-osetinskij Gosudarstvennyj Obedinennyj Muzej Istorii, Architektury i Literatury, Vladikavkaz 37932
Tatarskoe Gorodišče - Archeologičeskij i prirodnyj Muzej-zapovednik, Stavropol 37807
Tuapsinskij Kraevedčeskij Muzej im. N.G. Poletaeva, Tuapse 37864
Volgogradskij Oblastnoj Kraevedčeskij Muzej, Volgograd 37959

Saudi Arabia

King Saud University Museum, Riyadh 38016
National Museum, Riyadh 38017

Serbia

Arheološki Muzej Đerdap, Kladovo 38103
Gradski Muzej Sombor, Sombor 38148
Gradski Muzej Vršac, Vršac 38163
Muzej Grada Beograda, Beograd 38065
Muzej Srema, Sremska Mitrovica 38151
Narodni Muzej, Bela Crkva 38033
Narodni Muzej, Kragujevac 38109
Narodni Muzej, Smederevska Palanka 38145
Narodni Muzej, Užice 38158
Narodni Muzej, Vranje 38162
Narodni Muzej, Zrenjanin 38166
Narodni Muzej Čačak, Čačak 38094
Narodni Muzej Kikinda, Kikinda 38102
Narodni Muzej Ras, Novi Pazar 38119
Narodni Muzej u Beogradu, Beograd 38082
Narodni Muzej Valjevo, Valjevo 38160
Zavičajni Muzej, Ruma 38137

Sierra Leone

Sierra Leone National Museum, Freetown 38169

Slovakia

Archeologické Múzeum, Bratislava 38204
Hrad Devín, Bratislava 38214
Kammerhof - Baníctvo na Slovensku, Banská Štiavnica 38191
Krajské Múzeum, Prešov 38335
Lesnícke a Drevárske Múzeum, Zvolen 38394
Mestské Múzeum v Pezinku, Pezinok 38325
Slovenské Národné Múzeum, Bratislava 38234
Starý Zamok Banská Štiavnica, Banská Štiavnica 38194
Vlastivedné Múzeum, Nové Zámky 38322
Západoslovenské Múzeum v Trnave, Trnava 38381

Slovenia

Arheološki Muzej, Most na Soči 38484
Dolenjski Muzej Novo Mesto, Novo Mesto 38490
Pokrajinski Muzej, Murska Sobota 38487
Pokrajinski Muzej, Maribor 38477
Pokrajinski Muzej Celje, Celje 38407
Pokrajinski Muzej Kočevje, Kočevje 38432
Pokrajinski Muzej Koper, Koper 38434
Pokrajinski Muzej Ptuj, Ptuj 38499
Posavski Muzej Brežice, Brežice 38404
Tolminski Muzej, Tolmin 38527

South Africa

Arend Dieperink Museum, Potgietersrus 38718
Mapungubwe Museum, Pretoria 38719
Museum of Anthropology and Archaeology, Tshwane 38750
National Museum Bloemfontein, Bloemfontein 38567
Schoemansdal Museum, Louis Trichardt 38683

Spain

Castell Cartoixa Vallparadis, Terrassa 40115
Clos Arqueologic Torre Llauder, Mataró 39631
Colección de la Real Academia de la Historia, Madrid 39514
Colección Museográfica Permanente de Buñol, Buñol 39062
Colección Museográfica Permanente Luís García de Fuentes, Caudete de las Fuentes 39156
Colección Museográfica Permenent Municipal de Cocentaina, Cocentaina 39196

Collecions Municipals del Foment Arqueològic Excursionista Sallentí, Sallent 39949
Conjunt Monumental de la Esglésias de Sant Pere, Terrassa 40117
Conjunto Arqueológico de Madinat Al Zahra, Córdoba 39202
Deià Museo Arqueológico, Deià 39241
Euskal Arkeologia, Etnografia eta Kondaira Museoa, Bilbao 39042
Gabinete de Antigüedades, Madrid 39521
Museo Arqueológico, Novelda 39718
Museo Arqueológico, Sagunto 39932
Museo Arqueológico, Nerja 39714
Museo Arqueológico, Santa Cruz de Tenerife 39994
Museo Arqueológico Comarcal, Orihuela 39741
Museo Arqueológico de Asturias, Oviedo 39758
Museo Arqueológico Municipal, Burriana 39075
Museo Arqueológico Municipal, Cartagena 39136
Museo Arqueológico Municipal, Crevillente 39226
Museo Arqueológico Municipal, Lorca 39492
Museo Arqueológico Municipal, Marchena 39621
Museo Arqueológico Municipal Cayetano Mergelina, Yecla 40330
Museo Arqueológico Municipal de Alhambra, Alhambra 38842
Museo Arqueológico Municipal José María Soler, Villena 40311
Museo Arqueológico Nacional, Madrid 39526
Museo Arqueológico Provincial, Alicante 38843
Museo Arqueológico Provincial, Sevilla 40064
Museo Arqueológico-Torre del Agua, Osuna 39752
Museo Arqueologico y Etnologico Municipal Dámaso Navarro, Petrer 39839
Museo Arqueolóxico e Histórico, A Coruña 39217
Museo Bonsor, Mairena del Alcor 39593
Museo de Albacete, Albacete 38803
Museo de Alcalá la Real, Alcalá la Real 38816
Museo de Almería, Almería 38861
Museo de América, Madrid 39532
Museo de Arqueología de Álava, Vitoria-Gasteiz 40321
Museo de Ávila, Ávila 38920
Museo de Burgos, Burgos 39068
Museo de Cáceres, Cáceres 39082
Museo de Ciudad Real, Ciudad Real 39187
Museo de Guadalajara, Guadalajara 39375
Museo de Huelva, Huelva 39396
Museo de los Concilios y de la Cultura Visigoda, Toledo 40133
Museo de los Orígenes, Madrid 39550
Museo de Málaga, Málaga 39599
Museo de Palencia, Palencia 39770
Museo de Pontevedra, Pontevedra 39857
Museo de Salamanca, Salamanca 39940
Museo de Santa Cruz, Toledo 40134
Museo de Segóbriga, Saelices 39930
Museo de Valladolid, Valladolid 40235
Museo del Foro de Caesaraugusta, Zaragoza 40345
Museo Etnológico y Arqueológico Municipal, Manzanares el Real 39618
Museo Histórico Municipal de Écija, Écija 39257
Museo Municipal, Consuegra 39200
Museo Municipal, Valdepeñas 40192
Museo Municipal, Zuheros 40360
Museo Municipal Jerónimo Molina, Jumilla 39434
Museo Nacional de Arte Romano, Mérida 39646
Museo Numantino, Soria 40093
Museo Provincial, Teruel 40122
Museu Arqueològic Comarcal, Banyoles 38947
Museu Arqueològic de la Ciutat de Dénia, Dénia 39244
Museu Arqueològic del Vendrell, El Vendrell 40261
Museu Arqueològic d'Ontinyent i de la Vall d'Albaida, Ontinyent 39738
Museu Arqueològic i Paleontològic, Baldomar 38943
Museu Arqueològic i Paleontològic, Moià 39656
Museu Arqueològic Municipal, Bocairent 39055
Museu Arqueològic Municipal, Els Prats de Rei 39868
Museu-Arxiu Tomàs Balvey i Bas, Cardedeu 39129
Museu d'Arqueologia de Catalunya, Olèrdola 39722
Museu d'Arqueologia de Catalunya-Empúries, L'Escala 39278
Museu d'Arqueologia de Catalunya Ullastret, Ullastret 40180
Museu d'Arqueologia Salvador Vilaseca, Reus 39894
Museu de la Conca Dellà, Isona 39415
Museu de la Noguera, Balaguer 38942
Museu de la Vall de Lord, Sant Lorenç de Morunys 39988
Museu de Vilafranca, Vilafranca del Penedès 40280
Museu d'Eines Antigues, Torregrossa 40155
Museu d'Història de Calp, Calpe 39107
Museu d'Història de Cambrils, Cambrils 39110
Museu d'Història de la Ciudad, Sant Feliu de Guíxols 39982
Museu d'Història de Manacor, Manacor 39610
Museu Diocesa de Menorca, Ciutadella de Menorca 39193
Museu Durán i Sanpere, Cervera 39166
Museu Egipci de Barcelona, Barcelona 38992
Museu Local Arqueològic, Artesa del Lleida 38903
Museu Monogràfic de Pol Lèntia, Alcúdia (Baleares) 38833
Museu Municipal Can Xifreda, Sant Feliu de Codines 39981
Museu Municipal d'Història i Arqueologia, Cullera 39238
Museu Municipal Vicenç Ros, Martorell 39627
Museu-Poblat Ibèric, Tornabous 40150
Palacio Condesa de Lebrija, Sevilla 40076
Sala d'Arqueologia i Gabinet Numismàtic, Lleida 39481
Yacimiento-Museo Arqueológico de Numancia, Garray 39326

Sri Lanka

Archaeological Museum, Colombo 40368
Archaeological Museum, Dedigama 40376
Colombo National Museum, Colombo 40370

Sudan

El-Ibied Regional Museum, Khartoum 40388
Sheikan Museum, El-Obeid 40385
Sudan National Museum, Khartoum 40392

Sweden

Eketorps Borg, Degerhamn 40485
Gene Fornby, Örnsköldsvik 40793
Göteborgs Stadsmuseum, Göteborg 40549
Härjedalens Fjällmuseum, Funäsdalen 40530
Kärnan, Helsingborg 40589
Lappstaden, Arvidsjaur 40435
Museum Gustavianum, Uppsala 41009
Museum Tre Kronor, Stockholm 40916
Regionmuseet Kristianstad, Kristianstad 40655
Såguddens Museum, Arvika 40440
Tanums Hällristningsmuseum - Underslös, Tanumshede 40968
Västmanlands Läns Museum, Västerås 41034
Vuollerim 6000 år - Arkeologiskt Museum, Vuollerim 41074

Switzerland

Archäologische Sammlung der Universität Zürich, Zürich 42064
Archäologische Sammlung des Historischen Museums, Luzern 41587
Augusta Raurica, Augst 41137
Castello di Montebello, Bellinzona 41190
Didaktische Ausstellung Urgeschichte, Chur 41301
Fondation La Tène, Hauterive 41473
Kulturama, Zürich 42080
Laténium, Hauterive 41474
Münzkabinett und Antikensammlung der Stadt Winterthur, Winterthur 42030
Musée Cantonal d'Archéologie, Sion 41873
Musée d'Art et d'Histoire, Genève 41418
Musée Gyger, Vallorbe 41977
Musée Romain de Lausanne-Vidy, Lausanne 41551
Museo Civico e Archeologico, Locarno 41569
Museum.BL, Liestal 41564
Museum für Archäologie des Kantons Thurgau, Frauenfeld 41378
Museum Schwab, Biel 41232
Ortsmuseum Wetzikon, Wetzikon 42012
Pfahlbausammlung Dr. h.c. Carl Irlet, Twann 41954
Rätisches Museum, Chur 41304
Schweizerisches Nationalmuseum, Zürich 42105
Steinzeitmuseum, Thunstetten 41943

Tajikistan

Mamnu"gohi Bostonshinosii Panjakent, Pandžakent 42504
Osorkhonai Bostonshinosiii Khujand, Chudžand 42492

Tanzania

Dar-es-Salaam National Museum, Dar-es-Salaam 42512

Thailand

Ban Kao Prehistoric Museum, Kanchanaburi 42573
Chiang Mai National Museum, Chiang Mai 42568
Phiphitthaphanthasathan Haeng Chat Ram Khamhaeng Sukhothai, Sukhothai 42603
Phipitapan Haeng Chart Phra Pathom Chedi, Nakhon Pathom 42582
Phipitapan Laeng Chart, Suphan Buri 42605
Phipittapan Tasatan Haengchart Khon Kaen, Khon Kaen 42574
Somdet Phra Narai National Museum, Lopburi 42580
Sood Sangvichien Prehistoric Museum, Bangkok 42561

Tunisia

Musée Archéologique, Gafsa 42625
Musée Archéologique de Chimtou, Jendouba 42629
Musée Archéologique de Lamta, Lamta 42635
Musée Archéologique de Sbeïtla, Sbeïtla 42649
Musée Archéologique de Sfax, Sfax 42650
Musée Archéologique de Sousse, Sousse 42656
Musée Archeologique du Cap Bon, Nabeul 42644
Musée de Bulla Regia, Jendouba 42630
Musée du Memoire de la Terre, Tataouine 42659
Musée Lella Hadria, Midoun 42641
Musée les Thermes d'Antonin, Carthage 42619
Musée Regional de Mahdia, Mahdia 42637
Le Tophet de Salammbô à Carthage, Salammbô 42648

Turkey

Adana Arkeoloji Müzesi, Adana 42670
Adıyaman Müzesi, Adıyaman 42675
Afyon Müzesi, Afyon 42678
Ahlat Müzesi, Ahlat 42680
Amasra Müzesi ve Tarihçesi, Amasra 42688
Amasya Müzesi, Amasya 42690
Antalya Müzesi, Antalya 42717
Arkeoloji Müzesi, Kütahya 42892
Bergama Arkeoloji Müzesi, Bergama 42728
Bolu Müzesi, Bolu 42734
Burdur Müzesi, Burdur 42737
Bursa Arkeoloji Müzesi, Bursa 42741
Çankırı Müzesi, Çankırı 42751
Çorum Müzesi, Çorum 42755
Diyarbakır Müzeleri ve Tarihi Kültürel Değerleri, Diyarbakır 42763
Erzurum Arkeoloji Müzesi, Erzurum 42775
Eskişehir Müzeleri, Eskişehir 42781
Fethiye Müzesi, Fethiye 42784
Gordion Müzesi, Polatlı 42925
Haçıbektaş Müzesi, Hacibektaş 42792
İzmir Arkeoloji Müzesi, İzmir 42855
İznik Müzeleri, İznik 42860
Kars Müzesi, Kars 42867
Kayseri Arkeoloji Müzesi, Kayseri 42874
Koyunoğlu Şehir Müzesi, Konya 42888
Malatya Müzesi, Malatya 42898
Milas Müzesi, Milas 42909
Nevşehir Müzesi, Nevşehir 42917
Sadberk Hanım Müzesi, İstanbul 42837
Şanliurfa Müzesi, Şanlıurfa 42935
Seyitgazi Müzesi, Seyitgazi 42938
Sinop Müzesi, Sinop 42944
Tarsus Müzesi, Tarsus 42949
Tokat Müzesi ve Örenyerleri, Tokat 42954
Truva Müzesi Örenyeri, Çanakkale 42749
Ürgüp Müzesi, Ürgüp 42958
Uşak Arkeoloji Müzesi, Uşak 42960
Van Müzesi, Van 42962
Yalvaç Müzesi ve Örenyerleri, Yalvaç 42965

Turkmenistan

Taryhy Ýer Nusaý, Nusaý 42980

Ukraine

Kerčs'kyj Deržavnyj Istoriko-kul'turnyj Zapovidnyk, Kerč 43062
Krymski Respublikanskij Kraevedčeskij Muzej, Simferopol' 43185
Odessa Archaeological Museum, Odesa 43163

United Arab Emirates

Sharjah Archaeology Museum, Sharjah 43225

United Kingdom

Abbot's House, Arbroath 43311
Anker's House Museum, Chester-le-Street 43819
Archaeology Centre, Bagshot 43352
Arnol Blackhouse Museum, Arnol 43322
Ashmolean Museum of Art and Archaeology, Oxford 45327
Basing House, Basingstoke 43394
Beaumaris Castle, Beaumaris 43425
Beeston Castle, Tarporley 45827
Bessie Surtees House, Newcastle-upon-Tyne 45224
Binchester Roman Fort, Bishop Auckland 43520
Bredy Farm Old Farming Collection, Burton Bradstock 43687
The British Museum, London 44803
Bute Museum, Rothesay 45520
Canterbury Heritage Museum, Canterbury 43742
Carnegie Museum, Inverurie 44527
Castle Museum and Art Gallery, Norwich 45276
Castleford Museum Room, Castleford 43771
Ceredigion Museum, Aberystwyth 43258
Cheddar Man the Cannibals, Cheddar 43801
City Museum and Records Office, Portsmouth 45428
Corinium Museum, Cirencester 43847
Creswell Crags Museum and Education Centre, Welbeck 45954
Criccieth Castle, Criccieth 43926
Crofton Roman Villa, Orpington 45320
Cuming Museum, London 44834
Dewa Roman Experience, Chester 43817
DIG, York 46100
Dover Museum, Dover 44013
Droitwich Spa Heritage Centre, Droitwich 44024
Egypt Centre, Swansea 45812
English Heritage South East Region, Dover 44014
Forge Mill Needle Museum and Bordesley Abbey Visitor Centre, Redditch 45475
The Garstang Museum of Archaeology, Liverpool 44734
Groam House Museum, Rosemarkie 45513
Guardhouse Museum, Tattershall 45828
Guildford Museum, Guildford 44371
Gunnersbury Park Museum, London 44867
Hull and East Riding Museum, Kingston-upon-Hull 44597
Hunterian Museum, Glasgow 44285
Ipswich Museum, Ipswich 44530
Jewry Wall Museum, Leicester 44676
Jorvik Museum, York 46103
Kildonan Museum, Kildonan, Isle of South Uist 44576
King John's Hunting Lodge, Axbridge 43340
King's Lynn Museum, King's Lynn 44587
Kings Weston Roman Villa, Bristol 43637
Kirriemuir Gateway to the Glens Museum, Kirriemuir 44618
Kyngston House Museum, Saint Albans 45545
La Hougue Bie Museum, Grouville 44368
Lancaster City Museum, Lancaster 44633
Lindisfarne Priory, Holy Island 44480
Littlehampton Museum, Littlehampton 44729
London Borough of Bromley Museum, Orpington 45321
Maison Dieu, Faversham 44222
The Manchester Museum, Manchester 45090
Manor House Art Gallery and Museum, Ilkley 44512
Marischal Museum, Aberdeen 43244
Moat Park Heritage Centre, Biggar 43484
Museum Nan Eilean, Steornabhagh, Stornoway 45767
Museum of Antiquities, Alnwick 43287
Museum of Antiquities, Newcastle-upon-Tyne 45230
Museum of Archaeology, Durham 44073
Museum of Classical Archaeology, Cambridge 43730
Museum of Sussex Archaeology, Lewes 44700
Museum of the Holmesdale Natural History Club, Reigate 45481
Newark Museum, Newark-on-Trent 45212
Norris Museum, Saint Ives, Cambridgeshire 45574
North Lincolnshire Museum, Scunthorpe 45628
Nottingham University Museum of Archaeology, Nottingham 45299
Old Rectory Museum, Loughborough 45034
Orkney Museum, Kirkwall 44615
The Oxfordshire Museum, Woodstock 46067
Penlee House Gallery and Museum, Penzance 45361
Perth Museum and Art Gallery, Perth 45366
Peterborough Museum and Art Gallery, Peterborough 45368
Pitt Rivers Museum, Oxford 45338
Powysland Museum and Montgomery Canal Centre, Welshpool 45963
Red House Museum and Art Gallery, Christchurch 43842
Rhuddlan Castle, Rhuddlan 45485
Rockbourne Roman Villa, Fordingbridge 44239
Roman Bath House, Lancaster 44636
Roman Baths Museum, Bath 43410
Roman Museum, Canterbury 43743
Royal Museum, Edinburgh 44148
Saint Bride's Crypt Museum, London 44980
Salisbury and South Wiltshire Museum, Salisbury 45606
Shepton Mallet Museum, Shepton Mallet 45658
Strathnaver Museum, Bettyhill 43465
Swansea Museum, Swansea 45816
Tees Archaeology, Hartlepool 44409
Time Machine, Weston-super-Mare 45982
Torre Abbey Historic House and Gallery, Torquay 45867
Tredegar House and Park, Newport, Gwent 45241
Ulster History Park, Omagh 45318
University Archaeology Museum, Birmingham 43515
University Museum of Archaeology and Anthropology, Cambridge 43734
Verulamium Museum, Saint Albans 45550
Warwickshire Museum, Warwick 45942
Wells Museum, Wells 45958
Welwyn Roman Baths Museum, Welwyn 45964
West Stow Anglo-Saxon Village, Bury Saint Edmunds 43697
Weston Park Museum, Sheffield 45656
Whitby Abbey Museum, Whitby 45997
Wiltshire Heritage Museum and Gallery, Devizes 43979
Winchester City Museum, Winchester 46037
World Museum Liverpool, Liverpool 44750
Yorkshire Museum, York 46112

United States of America

AAF Museum, Washington 53843
Abbe Museum at Downtown, Bar Harbor 46600
Abbe Museum at Sieur de Monts Spring, Bar Harbor 46601

Alexandria Archaeology Museum, Alexandria 46198
Alutiiq Museum and Archaeological Repository, Kodiak 49823
America's Stonehenge, North Salem 51404
Amerind Foundation, Dragoon 48181
Anasazi Heritage Center, Dolores 48150
Anasazi State Park Museum, Boulder 46962
Angel Mounds, Evansville 48455
Anthropology Museum, DeKalb 48037
Anza-Borrego Desert Park Museum, Borrego Springs 46915
Archaeological Collection, Baltimore 46546
Arizona Museum of Natural History, Mesa 50605
Arizona State Museum, Tucson 53584
Aztec Ruins, Aztec 46528
Bailey House Museum, Wailuku 53786
Bandelier National Monument, Los Alamos 50203
Biblical Resource Center Museum, Collierville 47692
Boone's Lick Site, Boonesboro 46907
Boonshoft Museum of Discovery, Dayton 47982
Brazos Valley Museum of Natural History, Bryan 47105
Brazosport Museum of Natural Science, Clute 47666
C.H. Nash Museum-Chucalissa Archaeological Museum, Memphis 50567
Cahokia Mounds, Collinsville 47694
California Museum of Ancient Art, Beverly Hills 46791
Cape Cod National Seashore Museum, Wellfleet 54025
Carlsbad Caverns National Park, Carlsbad 47248
Carlsbad Museum and Art Center, Carlsbad 47249
Casa Grande Ruins, Coolidge 47786
Catalina Island Museum, Avalon 46524
Center for American Archeology, Kampsville 49673
Cobb Institute of Archaeology Museum, Mississippi State 50750
Colonial Pemaquid Historical Site, New Harbor 51047
Cottonlandia Museum, Greenwood 49085
Crow Canyon Archaeological Center, Cortez 47827
Crystal River State Archaeological Site, Crystal River 47897
Dickson Mounds Museum, Lewistown 50040
Effigy Mounds Museum, Harpers Ferry 49173
El Paso Museum of Archaeology at Wilderness Park, El Paso 48320
Etowah Indian Mounds, Cartersville 47264
Evelyn Payne Hatcher Museum of Anthropology, Saint Cloud 52538
Favell Museum of Western Art and Indian Artifacts, Klamath Falls 49806
First National Bank of Chicago Art Collection, Chicago 47474
Florence Hawley Ellis Museum of Anthropology, Abiquiu 46128
Flowerdew Hundred Foundation, Hopewell 49363
Folsom Museum, Folsom 48582
Fort Mountain, Chattsworth 47415
Fort Robinson Museum, Crawford 47868
Frank H. McClung Museum, Knoxville 49818
Gila Cliff Dwelling Visitor Center, Silver City 53102
Gila County Historical Museum, Globe 48943
Glacier National Park Museum, West Glacier 54056
Grand Village of the Natchez Indians, Natchez 50984
Grave Creek Mound Archaeology Complex, Moundsville 50868
Gustav Jeeninga Museum of Bible and Near Eastern Studies, Anderson 46292
H. Earl Clack Museum, Havre 49218
Hampson Archeological Museum, Wilson 54208
Historic London Town, Edgewater 48288
Historic Morven, Princeton 52149
Homolovi Ruins, Winslow 54240
Indian Museum of the Carolinas, Laurinburg 49988
The Institute For American Indian Studies, Washington Green 53950
Iraan Museum, Iraan 49526
John Deere House, Dixon 48145
The Johns Hopkins University Archaeological Collection, Baltimore 46572
The Joseph A. Callaway Archaeological Museum, Louisville 50286
Journey Museum, Rapid City 52251
Kelsey Museum of Archaeology, Ann Arbor 46307
Kolomoki Mounds State Park Museum, Blakely 46849
L.C. Bates Museum, Hinckley 49295
The Lake Champlain Maritime Museum, Vergennes 53717
Lake Jackson Mounds Archaeological Park, Tallahassee 53424
Logan Museum of Anthropology, Beloit 46724
Long Island Culture History Museum, Commack 47763
Louden-Henritze Archaeology Museum, Trinidad 53568
Luzerne County Museum, Wilkes-Barre 54162
Marin Museum of the American Indian, Novato 51446
Maxwell Museum of Anthropology, Albuquerque 46181
Mesa Verde National Park Museum, Mesa Verde 50609
Mission San Juan Capistrano Museum, San Juan Capistrano 52813
Moanalua Gardens Foundation, Honolulu 49350
Museum of Anthropology, Highland Heights 49279
Museum of Anthropology, Ann Arbor 46308
Museum of Anthropology, Columbia 47717
Museum of Anthropology, Detroit 48126
Museum of Art and Archaeology, Columbia 47718
Museum of Florida History, Tallahassee 53427
Museum of Peoples and Cultures, Provo 52189
Museum of Prehistory and History, Russellville 52489
Museum of Primitive Art and Culture, Peace Dale 51742
Museum of the Red River, Idabel 49461
Native American Exhibit, Fonda 48584
New York State Museum, Albany 46161
Ocmulgee National Monument, Macon 50368
The Octagon Museum, Washington 53912
Ohio Historical Society Museum, Columbus 47757
Oriental Institute Museum, Chicago 47502
Osage Historic Village, Lamar 49911
Pawnee Indian Village, Republic 52296
Peabody Museum of Archaeology and Ethnology, Cambridge 47185
Pecos Park, Pecos 51745
Phoebe Apperson Hearst Museum of Anthropology, Berkeley 46760
Poeh Museum, Santa Fe 52886
Post-Archaeological Institute, Jackson 49567
Pueblo Grande Museum, Phoenix 51908
Puukohola Heiau Ruins, Kawaihae 49712
Robbins Museum of Archaeology, Middleborough 50649
Robert S. Peabody Museum of Archaeology, Andover 46298
Russell Cave Museum, Bridgeport 47014
Safety Harbor Museum of Regional History, Safety Harbor 52512
Saint Clements Island-Potomac River Museum, Colton Point 47711
Sanford Museum, Cherokee 47421
Saratoga Museum, Saratoga 52919
Scotts Bluff National Monument, Gering 48898
Seminole Canyon State Historical Park, Comstock 47765
Seton Hall University Museum, South Orange 53177
Siegfried H. Horn Archaeological Museum, Berrien Springs 46767
Silver River Museum, Ocala 51484
South Florida Museum of Natural History, Dania Beach 47955
Southold Indian Museum, Southold 53189
Southwest Museum of the American Indian, Los Angeles 50265
Spiro Mounds Archaeological Center, Spiro 53204
Sunwatch Indian Village - Archaeological Park, Dayton 47989
Thomas E. McMillan Museum, Brewton 47011
Thomas T. Taber Museum, Williamsport 54173
Toltec Mounds Archeological State Park, Scott 52974
Treasures of the Sea Exhibit, Georgetown 48890
University of Pennsylvania Museum of Archaeology and Anthropology, Philadelphia 51886
Van Nostrand-Starkins House, Roslyn 52464
Western Archaeological and Conservation Center, Tucson 53602
Wickliffe Mounds, Wickliffe 54156
Winnebago Area Museum, Winnebago 54235
Woolaroc Museum, Bartlesville 46620
Wormsloe State Historic Site, Savannah 52949
Wray Museum, Wray 54310
Yellowstone Gateway Museum of Park County, Livingston 50149

Uruguay

Museo Municipal de Historia y Arqueología, Rivera 54431

Uzbekistan

Surxondaryo Vilojati Arxeologiya Muzeyi, Termiz 54466

Vanuatu

Vanuatu Cultural Centre and National Museum, Port Vila 54495

Venezuela

Museo J.M. Cruxent, Los Teques 54579

Vietnam

Thua Thien Hue Museum, Thua Thien Hue 54639

Yemen

National Museum, Aden 54643
Seiyun in Wadi Hadhramaut Museum, Seiyun 54651

Zimbabwe

Great Zimbabwe Site Museum, Masvingo 54675
Khami Ruins Site Museum, Bulawayo 54663
Museum of Human Sciences, Harare 54669
Nyahokwe Ruins Site Museum, Inyanga 54672

Archaeology, Christian and Medieval

Austria

Dekanatsmuseum, Haus im Ennstal 02059
Diözesanmuseum, Klagenfurt 02182
Instituts- und Studiensammlung des Instituts für Ur- und Frühgeschichte, Wien 03006
Römermuseum Teurnia, Lendorf 02283
Zisterzienserstift Rein, Rein 02575

Belgium

Archaeological Site Walraversijde 1465, Oostende 03928
Musée des Celtes, Libramont 03792

Bulgaria

Archeologičeski Muzej, Varna 05235

Croatia

Arheološka Zbirka i Lapidarij Dr. Grga Novak, Hvar 08878
Muzej Hrvatskih Arheoloških Spomenika, Split 08961

Denmark

Vikingemuseet Moesgård, Højbjerg 10085

Egypt

Monastery of Saint Catherine, Gebel Katherîna 10525

Finland

Suomen Kansallismuseo, Helsinki 10820

France

Crypte Archéologique du Parvis de Notre-Dame, Paris 14593
Musée Archéologique de Viuz-Faverges, Faverges 12908
Musée Briard et de la Crypte Mérovingienne, Jouarre 13280
Musée d'Art et d'Histoire de Langres, Langres 13509
Musée de Notre-Dame-de-Paris, Paris 14679
Musée des Amis de Castrum Vetus, Châteauneuf-les-Martigues 12449
Musée Municipal, La Charité-sur-Loire 13339

Georgia

Archaeological Museum, Tbilisi 16351

Germany

Archäologische Staatssammlung München, München 20422
Bajuwarenmuseum Waging am See, Waging am See 22271
Bibelmuseum, Beuron 17313
Bischöfliches Dom- und Diözesanmuseum, Trier 22096
Burgmuseum Altnußberg, Geiersthal 18553
Grabungsmuseum Kirchhof, Coburg 17716
Grabungsmuseum Langenburg, Langenburg 19824
Museum Sankt Dionys, Esslingen 18276
Römisch-Germanisches Museum, Köln 19631
Torhalle Frauenchiemsee und Vikarhaus, Frauenchiemsee 18412

Hungary

Budapesti Történeti Múzeum, Budapest 23392
Középkori Romkert - Nemzeti Emlékhely, Székesfehérvár 23855
Okeresztény Mauzóleum, Pécs 23770
Régészeti Osztály és Állandó Régészeti Kiállítás, Pécs 23772
Római Kori Kotár, Sopron 23817
Scarabantia Fóruma, Régészeti Kiállítóhely, Sopron 23818

Ireland

Athlone Castle Museum, Athlone 24730
Waterford Museum of Treasures, Waterford 24909
Weingreen Museum of Biblical Antiquities, Dublin 24823

Israel

Archaeological Site Avedat, Avedat 24931
Benedictine Monastery - Church of the Multiplication of Loaves and Fish Tabgha, Tiberias 25128

Italy

Cripta e Museo di Sant'Anastasio, Asti 25312
Museo Archeologico, Montelupo Fiorentino 26837
Museo Archeologico Etneo, Adrano 25152
Museo Civico Archeologico, Verucchio 28238
Museo Diocesano, Ancona 25229
Museo Lapidario Estense, Modena 26754
Museo Paleocristiano, Aquileia 25254

Lithuania

Kernavės Archeologijos ir Istorijos Muziejus, Kernavė 30274

Netherlands

Museum Abdij van Egmond, Egmond Binnen 32182

Poland

Gród Piastowski w Gieczu, Giecz 34663
Muzeum Północno-Mazowieckie, Łomża 34890

Spain

Museo Misional de Nuestra Señora de Regla, Chipiona 39179
Museo Santa Clara de Murcia, Murcia 39703
Museu Monogràfic del Castell de Llinars, Llinars del Vallés 39482

Sweden

Lödöse Museum, Lödöse 40691
Sigtuna Museer, Sigtuna 40835

Tunisia

Musée d'Enfidha, Enfidaville 42624

Turkey

Ayasofya Müzesi, İstanbul 42805
Yerebatan Sarnıcı, İstanbul 42850

United Kingdom

Farleigh Hungerford Castle, Farleigh Hungerford 44214
Halesworth and District Museum, Halesworth 44381
Meigle Museum of Sculptured Stones, Meigle 45122
River and Rowing Museum, Henley-on-Thames 44449
Shaftesbury Abbey Museum and Garden, Shaftesbury 45644
Whithorn Priory and Museum, Whithorn 46005

United States of America

Badè Museum of Biblical Archaeology, Berkeley 46750

Archaeology, Far Eastern

Cambodia

Musée, Kampong Thom 05284

China

Arthur M. Sackler Museum of Art and Archaeology, Beijing 07505
Baoji Zhouyuan Museum, Fufeng 07694
Beijing Wabgfujing Paleolithic Museum, Beijing 07522
Chongqing Museum, Chongqing 07630
Dahe Cun Yi Zhi Guan, Zhengzhou 08497
Dingling Museum, Beijing 07536
Fuxin Chahai Ruins Museum, Fuxin 07701
Guangxi Zhuang Museum, Nanning 08055
Han Tomb Cultural Relics Exhibition Hall, Mawangdui 08010
Hemudu Ruins Museum, Yuyao 08478
Henan Museum, Zhengzhou 08498
Hunan Provincial Museum, Changsha 07597
Jiaoshan Beike Stone Museum, Zhenjiang 08501
Liangshanzhou Guan, Xichang 08383

A

Luoyang Museum, Luoyang 07985
Luoyang Tumulus Museum, Luoyang 07986
Maoling Museum, Xingping 08394
National Museum of Chinese History, Beijing 07556
Ningxia Hui Autonomous Region Museum, Yinchuan 08447
Qi Gucheng Yi Zhi Guan, Linzi 07960
Qianling Museum, Qianxian 08103
Qin Shi Huan Bingmayong Museum, Xian 08358
Qinghai Provincial Museum, Xining 08397
Shanghai Museum of Natural History, Shanghai 08187
Shangjing Ruins Museum, Ning'an 08069
Shantou Archaeology Museum, Shantou 08206
Sichuan Museum, Chengdu 07619
Suzhou Museum, Suzhou 08258
Urumchi City Museum, Urumchi 08307
Waysi Museum, Fufeng 07696
Xian Banpo Museum, Xian 08362
Xianyang Museum, Xianyang 08374
Xincheng District Museum, Nanning 08057
Xingjing Yandao Gucheng Ruins Museum, Yingjing 08449
Xinle Yi Zhi Ruins Museum, Shenyang 08218
Yangzhou Hantomb Museum, Yangzhou 08428
Yinque Shanhan Tomb Zhujian Museum, Linyi, Shandong 07956
Youzhou Youci Museum, Pingdingshan 08080
Zhaoling Museum, Liquan 07961
Zhouyuan Museum, Fufeng 07697

France

Musée Archéologique, Monségur 14153
Musée du Vieux Plassac, Plassac 14833

Germany

Museum für Asiatische Kunst, Berlin 17201

Indonesia

Museum MPU Tantular, Surabaya 24566
Museum Purbakala Trowulan, Mojokerto 24530

Israel

Burnt House Museum, Jerusalem 24982

Japan

Aikawa Kokokan, Isesaki 28584
Aomori-kenritsu Kyodokan, Aomori 28360
Asuka Shiryokan, Asuka 28379
Beppu Daigaku Fuzoku Hakubutsukan, Beppu 28388
Hida Minzoku Kokokan, Takayama 29337
Hiraide Gatokan, Shiojiri 29275
Hokkaido Home-Making School Museum, Tsugaru 29539
Ichikawa-shiritsu Koko Hakubutsukan, Ichikawa 28554
Iida-shiritsu Kokokan, Iida 28560
Kagoshima-shiritsu Kokokan, Kagoshima 28629
Kanra Ancient Art Museum, Kanra 28665
Kokugakuin University Archaeology Museum, Tokyo 29415
Korekawa Archaeological Museum, Hachinohe 28464
Kurashiki-shiritsu Kokokan, Kurashiki 28787
Kyoto Daigaku Hakubutsukan, Kyoto 28837
Machida-shiritsu Hakubutsukan, Machida 28874
Museum of Kaigarayama Shell Mound, Kiyosu 28720
Osaka-furitsu Senboku Koko Shiryokan, Sakai 29197
Sado Kokokan, Ogi 29093
Shimane Prefectural Yakumodatsu Fudoki-no-oka Museum, Matsue 28890
Shizuoka City Toro Museum, Shizuoka 29288
Tenri Daigaku Sankokan, Tenri 29358
Togari-Ishi Jomon Kokokan, Chino 28401
Wakasa Rekishi Minzoku Shiryokan, Obama 29083
Yamagata-kenritsu Hakubutsukan, Yamagata 29579
Yamanashi-kenritsu Koko Hakubutsukan, Kofu 28756
Yanaginogosho Kokokan, Hiraizumi 28523
Yatsugatake Bijutsukan, Hara 28498

Korea, Republic

Busan University Museum, Busan 29775
Buyeo National Museum, Buyeo 29783
Daegwallyeong Museum, Gangneung 29807
Dong-eui University Museum, Busan 29777
Gongju National Museum, Gongju 29817
Gwangju National Museum, Gwangju 29823
Gyeongju National Museum, Gyeongju 29826
Jinju National Museum, Jinju 29844
Koryo Taehakyo Pakmulgwan, Seoul 29899
National Museum of Korea, Seoul 29910
Seoul National University Museum, Seoul 29921
Silla University Museum, Busan 29781
Uiryeong Museum, Uiryeong 29938

Myanmar

Archaeological Museum, Myohaung 31724
Archaeological Museum, Pagan 31725
National Museum of Art and History, Yangon 31736

Nepal

Kapilavastu Museum, Lumbani 31783
National Bronze Art Museum, Lalitpur 31781

Pakistan

Archaeological Museum, Swat 34043
Archaeological Museum, Taxila 34044

Philippines

Balanghai Archaeological Site Museum, Butuan 34351
Fort San Pedro National Museum, Cebu 34370
University of San Carlos Anthropology Museum, Cebu 34377
Xavier University Museum (Museo de Oro), Cagayan de Oro City 34358

Russia

Gosudarstvennyj Istoriko-architekturnyj i Chudožestvennyj Muzej-zapovednik Drevnij Derbent, Derbent 36874

Sri Lanka

Archaeological Museum, Anuradhapura 40366

Sweden

Östasiatiska Museet, Stockholm 40924

Syria

Deir ez-Zor Museum, Deir ez-Zor 42134
Hama National Museum, Hama 42135
Musée de Palmyra, Palmyra 42137
Tartous Museum, Tartous 42139

Taiwan

Museum of the Institute of History and Philology, Taipei 42410
Shi-San-Hang Museum of Archaeology, Bali 42142

Thailand

Ancient City, Samut Prakan 42597
Chaosamphraya National Museum, Ayutthaya 42527
Kamphaeng Phet National Museum, Kamphaeng Phet 42572
Phipitapantasatan Haeng Chart Nakhon Si Thammarat, Nakhon Si Thammarat 42585

United Kingdom

Terracotta Warriors Museum, Dorchester 44003

Uzbekistan

Surxondaryo Vilojati Arxeologiya Muzeyi, Termiz 54466

Archaeology, Greek and Roman

Algeria

Musée Archéologique de Djemila, Djemila 00059
Musée de Tazoult, Tazoult 00078
Musée de Tipasa, Tipasa 00081

Australia

Hellenic Antiquities Museum, Melbourne 01283

Austria

Archäologische Sammlung der Universität Graz, Graz 01963
Archäologischer Park Carnuntum, Petronell 02494
Archäologischer Park Magdalensberg, Pischeldorf 02500
Archäologisches Museum und Dokumentationszentrum Via Claudia, Fließ 01892
Ephesos-Museum, Wien 02980
Freilichtmuseum Petronell, Petronell 02495
Freilichtmuseum Rätersiedlung Himmelreich, Volders 02871
Frühchristliche Kirchenanlage, Feistritz an der Drau 01877
Museum Aguntum, Dölsach 01823
Museum Thurnfels, Völs 02869
Römermuseum, Tulln 02833
Römermuseum, Wien 03061
Römerzeitliches Museum, Ratschendorf 02562
Tempelmuseum Frauenberg, Seggauberg 02737

Belgium

Grand Curtius, Liège 03796
Musée Archéologique, Arlon 03323
Musée Archéologique, Charleroi 03514
Musée Archéologique et Lapidaire de Montauban, Montauban-sous-Buzenol 03890
Musée de la Basilique et des Thermes Romains, Arlon 03324
Musée de la Tour Romaine, Arlon 03325
Musée Gallo-Romain, Berneau 03363
Musée Gallo-Romain, Blicquy 03374
Musée Gallo-Romain de Waudrez, Waudrez 04139
Provinciaal Archeologisch Museum van Zuid-Oost-Vlaanderen, Zottegem 04169

Bulgaria

Archeologičeski Institut s Muzej, Sofia 05172
Archeologičeski Muzej, Varna 05235
Burgaski Muzej, Burgas 04996
Gradski Archeologičeski Muzej, Chisar 05006
Gradski Archeologičeski Muzej, Sandanski 05157
Gradski Istoričeski Muzej, Balčik 04976
Gradski Istoričeski Muzej, Čirpan 05009
Istoričeski Muzej, Svištov 05222
Muzej na Mozaikite, Devnja 05012
Trakijska Grobnica, Kazanlăk 05062

Canada

Museum of Antiquities, Saskatoon 06920

Croatia

Arheološka Zbirka, Nin 08914
Arheološki Muzej Istre, Pula 08932
Arheološki Muzej u Splitu, Split 08954
Zbirka Franjevačkog Samostana, Sinj 08951

Cyprus

Archaeological Museum, Lemesos 09384
Cyprus Museum, Lefkosia 09375
Limassol District Museum, Lemesos 09387
Museum of the George and Nefeli Giabra Pierides Collection, Lefkosia 09380

Egypt

Graeco-Roman Museum, Alexandria 10471
Ismailia Museum, Ismailia 10535

France

Ancienne Mairie-Musée Annexe, Lillebonne 13760
Centre Culturel - Musée Montaut, Saint-Sever (Landes) 15513
Exposition Archéologique Gallo-Romaine, Mandeure 13935
Glanum, Saint-Rémy-de-Provence 15485
Musée Alésia, Alise-Sainte-Reine 11554
Musée Amphoralis, Sallèles-d'Aude 15601
Musée Antiquités Gallo-Romain, Aoste 11648
Musée Archéologique, Saint-Martin-de-Bromes 15397
Musée Archéologique, Saint-Paulien 15452
Musée Archéologique, Soulosse-sous-Saint-Élophe 15800
Musée Archéologique, Vaison-la-Romaine 16031
Musée Archéologique de Rom-Sainte-Soline, Rom 15082
Musée Archéologique de Viuz-Faverges, Faverges 12908
Musée Archéologique Municipal de Fréjus, Fréjus 13001
Musée Celte - Gallo Romain, Seltz 15706
Musée d'Archéologie, Antibes-Juan-les-Pins 11640
Musée de Rauranum, Rom 15083
Musée des Docks Romains, Marseille 14004
Musée d'Eysses, Villeneuve-sur-Lot 16221
Musée d'Histoire et d'Archéologie de Die et du Diois, Die 12720
Musée du Camp Romain, Saint-Jean-le-Vieux 15338
Musée et Site Archéologiques de Nice-Cimiez, Nice 14434
Musée et Site Gallo-Romains de Saint-Romain-en-Gal - Vienne, Saint-Romain-en-Gal 15502
Musée Gallo-Romain, Biesheim 11988
Musée Gallo-Romain, Petit-Bersac 14804
Musée Gallo-Romain de Lyon-Fourvière, Lyon 13896
Musée Horreum Romain, Narbonne 14379
Musée Lapidaire, Narbonne 14380
Musée Mémoires de la Terre et Horizons Cathares, Mazamet 14045
Musée Municipal-Musée d'Art et d'Archéologie, Château-Gontier 12435
Musée Villien, Avesnes-sur-Helpe 11785
Muséosite de l'Amphithéâtre Gallo-Romain, Lillebonne 13762
Nécropole de Louis Cauvin, Gareoult 13034
Site Gallo-Romaine de Sanxay, Sanxay 15627

Germany

Antikenmuseum und Abgusssammlung des Instituts für Klassische Archäologie Universität Heidelberg, Heidelberg 19069
Antikensammlung im Pergamonmuseum, Berlin 17036
Archäologische Ausgrabung Borg, Perl 21010
Archäologische Ausstellung in der Schule, Nassenfels 20548
Archäologische Sammlung der Universität, Freiburg im Breisgau 18426
Archäologische Staatssammlung München, München 20422
Archäologischer Park Cambodunum - APC, Kempten 19514
Archäologischer Park mit Museums-Turm, Kellmünz an der Iller 19503
Archäologisches Museum, Burgheim 17657
Archäologisches Museum Frankfurt, Frankfurt am Main 18354
Freilichtmuseum am Rätischen Limes, Rainau 21170
Heinrich-Schliemann-Gedenkstätte, Neubukow 20593
Kunstsammlungen der Ruhr-Universität Bochum, Bochum 17402
LVR - Archäologischer Park Xanten, Xanten 22760
Museum im Römerbad, Heidenheim an der Brenz 19092
Museum Villa Sarabodis, Gerolstein 18611
Pergamonmuseum, Berlin 17234
Rheinisches Landesmuseum Trier, Trier 22102
Römerbadmuseum, Hüfingen 19288
Römerhaus Walheim, Walheim 22306
Römermuseum, Mainhardt 20139
Römermuseum, Obernburg am Main 20817
Römermuseum Bedaium, Seebruck 21714
Römermuseum in der Schule, Ehingen 18069
Römermuseum Osterburken, Osterburken 20932
Römervilla e.V., Möckenlohe 20348
Römisch-Germanisches Museum, Köln 19631
Römischer Weinkeller Oberriexingen, Oberriexingen 20829
Römisches Freilichtmuseum, Hechingen 19063
Römisches Heiligtum für Isis und Mater Magna, Mainz 20151
Römisches Museum, Augsburg 16633
Römisches Museum Remagen, Remagen 21261
Sammlungen des Archäologischen Instituts der Universität, Göttingen 18700
Stiftung Römermuseum Homburg-Schwarzenacker, Homburg 19268
Sumelocenna - Römisches Stadtmuseum, Rottenburg am Neckar 21422
Winckelmann-Museum, Stendal 21899

Greece

Agios Nikolaos Archaeological Museum, Agios Nikolaos 22872
Amfiaraion Archaeological Museum, Kalamos 23026
Ancient Agora Museum, Athinai 22898
Antonopouleion Archaeological Museum, Pylos 23160
Archaeological and Byzantine Museum Florina, Florina 22999
Archaeological Collection, Farsala 22997
Archaeological Collection, Kozani 23066
Archaeological Collection, Loutra Aidipsou 23081
Archaeological Collection, Molyvos 23098
Archaeological Collection of Amorgos, Amorgos 22879
Archaeological Collection of Pythagorion, Pythagorio 23164
Archaeological Collection Panagia Kastrou, Platanos 23150
Archaeological Museum, Andros 22881
Archaeological Museum, Argostolion 22889
Archaeological Museum, Arhaia Olympia 22892
Archaeological Museum, Atalanti 22897
Archaeological Museum, Chalkeio 22959
Archaeological Museum, Chalkida 22960
Archaeological Museum, Corfu 22974
Archaeological Museum, Iráklion 23018
Archaeological Museum, Kavala 23046
Archaeological Museum, Kilkis 23053
Archaeological Museum, Lamia 23070
Archaeological Museum, Lárissa 23072
Archaeological Museum, Myrina 23105
Archaeological Museum, Néa Anchialos 23121
Archaeological Museum, Paros 23135
Archaeological Museum, Rethymnon 23165
Archaeological Museum, Rhodos 23170
Archaeological Museum, Vravrón 23241
Archaeological Museum Eressos, Eressos 22993
Archaeological Museum of Agrinion, Agrinion 22873
Archaeological Museum of Aiani, Aiani 22875
Archaeological Museum of Aigina, Aegina 22865
Archaeological Museum of Ancient Corinth, Ancient Corinth 22880
Archaeological Museum of Argos, Argos 22888

Archaeological Museum of Astypalaia, Astypalaia 22896
Archaeological Museum of Chios, Chios 22967
Archaeological Museum of Chora, Messenia 23091
Archaeological Museum of Corinth, Kórinthias 23062
Archaeological Museum of Delphi, Delphi 22982
Archaeological Museum of Elefsina, Elefssina 22992
Archaeological Museum of Epidaurus, Asklepieion Lygourio 22894
Archaeological Museum of Eretria, Eretria 22994
Archaeological Museum of Galaxidi, Galaxidi 23004
Archaeological Museum of Gytheion, Gythio 23011
Archaeological Museum of Ioannina, Ioannina 23012
Archaeological Museum of Isthmia, Isthmia 23022
Archaeological Museum of Kampos Ikaria, Kampos 23033
Archaeological Museum of Karystos, Karystos 23039
Archaeological Museum of Kea, Kea 23049
Archaeological Museum of Kimolos, Kimolos 23055
Archaeological Museum of Komotini, Komotini 23058
Archaeological Museum of Kos, Kos 23063
Archaeological Museum of Megara, Megara 23088
Archaeological Museum of Mytilene, Mytilini 23109
Archaeological Museum of Nafplion, Nafplion 23115
Archaeological Museum of Naxos, Naxos 23119
Archaeological Museum of Nemea, Archaia Nemea 22887
Archaeological Museum of Nicopolis, Preveza 23158
Archaeological Museum of Patras, Patras 23138
Archaeological Museum of Pella, Pella 23141
Archaeological Museum of Piraeus, Piraeus 23143
Archaeological Museum of Plaka, Milos 23097
Archaeological Museum of Polygyros, Polygyros 23152
Archaeological Museum of Samos, Samos 23175
Archaeological Museum of Samothraki, Paleopoli 23130
Archaeological Museum of Sifnos, Kastro 23044
Archaeological Museum of Siteia, Siteia 23188
Archaeological Museum of Spárti, Spárti 23194
Archaeological Museum of Stavros, Stavros Ithakis 23199
Archaeological Museum of Syros, Ermoupolis 22995
Archaeological Museum of Thassos, Thassos 23203
Archaeological Museum of Thebes, Thebes 23205
Archaeological Museum of Thera, Thera 23206
Archaeological Museum of Thermos, Thermo 23208
Archaeological Museum of Tinos, Tinos 23226
Archaeological Museum of Veroia, Veria 23234
Archaeological Museum Poros, Poros 23156
Archaeological Museum Serres, Serres 23179
Archaeological Museums of Chania and Rethymnon, Chania 22961
Archaeological Sit of Vitsa, Vitsa 23236
Archaeological Site, Paramythia 23134
Archaeological Site and Museum, Sicyon 23184
Archaeological Site of Geraki, Geraki 23006
Archaeological Site of Lindos, Lindos 23078
Archaeological site of Megalopolis, Megalopolis 23087
Archaeological Site of Perachora, Perachora 23142
Arhaiologiko Mouseio Thessaloniki, Thessaloniki 23209
Athanassakeion Archaeological Museum, Volos 23237
Benakeion Archaeological Museum, Kalamata 23023
Byzantine and Christian Museum, Athinai 22904
Byzantine Museum of Antivouniotissa, Corfu 22975
Chania Byzantine and Postbyzantine Collection, Chania 22962
Chios Museum of Byzantine and Postbyzantine Art, Chios 22968
Collection of Byzantine and Postbyzantine Art, Kythira 23069
Corfu Byzantine Collection, Corfu 22976
Dion Archaeological Museum, Dion 22985
Drama Archaeological Museum, Drama 22987
Goulandris Collection, Athinai 22911
Ionnina Byzantine Museum, Ioannina 23015
Kastoria Byzantine Museum, Kastoria 23043
Kiato Archaeological Museum, Kiato 23050
Lambropoulos Archaeological Collection, Polygyros 23154
Megisti Museum, Kastellorizo 23041
Monemvasia Archaeological Collection, Monemvasia 23100
Mouseio Akropoleos, Athinai 22917
Mouseio Ethnikon Archaiologikon, Athinai 22920
Mouseio Kerameikos, Athinai 22923
Mouseio Mpenaki, Athinai 22925
Mouseio Zakynthou, Zakynthos 23244
Museum Mistra, Mystras 23108
Museum of Casts and Archaeological Collection, Thessaloniki 23218
Museum of Kalymnos, Kalymnos 23031
Museum Papageorgiou, Limenaria 23077
New Archaeological Museum, Mytilini 23112
P. and Al. Canellopoulos Museum, Athinai 22947
Palace of the Grand Masters, Rhodos 23173
Panarcadic Archaeological Museum of Tripolis, Tripoli 23231
Pierides Museum of Ancient Cypriot Art, Athinai 22950
Sikinos Archaeological Collection, Sikinos 23187
Spetses Museum, Spetses 23198
Tower Museum of Prosphorion, Ouranoupoli 23128
Veria Byzantine Museum, Veria 23235

Hungary

Aquincumi Múzeum, Budapest 23382
Fürdő Múzeum, Budapest 23405
Görög-Római Szobormásolatok Kiállítása, Tata 23917
Gorsium Szabadtéri Múzeum, Tác 23909
Herkules Villa, Budapest 23411
Mithrász-szentély, Fertőrákos 23560
Régészeti Osztály és Állandó Régészeti Kiállítás, Pécs 23772
Román Kori Kőtár, Pécs 23774
Táborvárosi Múzeum, Budapest 23479
Villa Publica Romkert, Zalalövo 23974

Ireland

University College Dublin Classical Museum, Dublin 24819

Israel

Local Museum, Ashkelon 24930
Museum of Regional and Mediterranean Archaeology, Nir-David 25074
Wohl Archaeology Museum Herodian Mansions, Jerusalem 25024

Italy

Antiquarium, Brescello 25530
Antiquarium, Manduria 26584
Antiquarium, Numana 26963
Antiquarium, Palazzolo Acreide 27058
Antiquarium, Palmi 27084
Antiquarium, Patti 27114
Antiquarium, Porto Torres 27304
Antiquarium, Santa Flavia 27700
Antiquarium, Santa Maria Capua Vetere 27704
Antiquarium, Sant'Antioco 27724
Antiquarium Civico, Vazzano 28170
Antiquarium Comunale, Roma 27439
Antiquarium Comunale, Tiriolo 27974
Antiquarium del Serapeo, Tivoli 27977
Antiquarium delle Grotte di Catullo, Sirmione 27856
Antiquarium di Himera, Termini Imerese 27959
Antiquarium di Torre Nao, Crotone 26011
Antiquarium e Zona Archeologica, Lugagnano Val d'Arda 26541
Antiquarium Torre Cimalonga, Scalea 27781
Civico Museo Archeologico, Acqui Terme 25149
Civico Museo Archeologico, Milano 26678
Civico Museo Archeologico e di Scienze Naturali Federico Eusebio, Alba 25174
Collezione Archeologica e Foro Romano, Assisi 25304
Cripta e Museo di Sant'Anastasio, Asti 25312
Mostra Permanente di Archeologia, Rivello 27422
Musei Capitolini, Roma 27458
Musei Civici di Palazzo d'Avalos, Vasto 28169
Museo A. Versace, Bagnara Calabra 25343
Museo Antiquarium Archeologico, Falerone 26082
Museo Archeologico, Cesena 25825
Museo Archeologico, Montelupo Fiorentino 26837
Museo Archeologico, Cecina e Marina 25792
Museo Archeologico al Teatro Romano, Verona 28230
Museo Archeologico Comunale di Artimino, Carmignano 25680
Museo Archeologico dell'Agro Nocerino, Nocera Inferiore 26932
Museo Archeologico di Vallecamonica, Cividate Camuno 25918
Museo Archeologico e Antiquarium Nazionale, Formia 26233
Museo Archeologico e Paleontologico, Asti 25314
Museo Archeologico e Pinacoteca Comunale, Foligno 26214
Museo Archeologico e Storico Aristico, Bra 25523
Museo Archeologico Girolamo Rossi, Ventimiglia 28216
Museo Archeologico Nazionale, Adria 25153
Museo Archeologico Nazionale, Aquileia 25253
Museo Archeologico Nazionale, Capaccio 25652
Museo Archeologico Nazionale, Crotone 26012
Museo Archeologico Nazionale, Gioia del Colle 26341
Museo Archeologico Nazionale, Napoli 26883
Museo Archeologico Nazionale, Parma 27099
Museo Archeologico Nazionale, Quarto d'Altino 27350
Museo Archeologico Nazionale, Taranto 27941
Museo Archeologico Nazionale del Melfese, Melfi 26651
Museo Archeologico Nazionale della Sibaritide, Cassano allo Jonio 25710
Museo Archeologico Nazionale di Cagliari, Cagliari 25579
Museo Archeologico Nazionale di Cosa, Orbetello 26987
Museo Archeologico Nazionale di Locri Epizefiri, Locri 26505
Museo Archeologico Nazionale di Metaponto, Bernalda 25407
Museo Archeologico Nazionale di Palestrina, Palestrina 27081
Museo Archeologico Nazionale di Venosa, Venosa 28214
Museo Archeologico Nazionale Etrusco, Cerveteri 25819
Museo Archeologico Oliveriano, Pesaro 27154
Museo Archeologico Provinciale, Potenza 27317
Museo Archeologico Provinciale, Bari 25362
Museo Archeologico Provinciale F. Ribezzo, Brindisi 25545
Museo Archeologico Regionale, Agrigento 25161
Museo Archeologico Regionale A. Salinas, Palermo 27066
Museo Archeologico San Giovanni in Compito, Savignano sul Rubicone 27769
Museo Archeologico Statale, Cingoli 25891
Museo Archeologico Statale V. Capialbi, Vibo Valentia 28246
Museo Archeologico Statale V. Laviola, Amendolara 25222
Museo Civico, Asolo 25302
Museo Civico, Baranello 25353
Museo Civico, Cabras 25574
Museo Civico, Centuripe 25803
Museo Civico, Feltre 26094
Museo Civico, Guardiagrele 26384
Museo Civico, Iesi 26394
Museo Civico, Magliano Sabina 26566
Museo Civico, Sesto Calende 27823
Museo Civico, Tolfa 27991
Museo Civico, Trevignano Romano 28067
Museo Civico Antonino Olmo e Gipsoteca, Savigliano 27766
Museo Civico Archeologico, Bergamo 25401
Museo Civico Archeologico, Canosa di Puglia 25647
Museo Civico Archeologico, Cologna Veneta 25943
Museo Civico Archeologico, Fiesole 26127
Museo Civico Archeologico, Latronico 26468
Museo Civico Archeologico Arsenio Crespellani, Bazzano 25385
Museo Civico Archeologico C. Cellini, Ripatransone 27414
Museo Civico Archeologico O. Nardini, Velletri 28171
Museo Civico Archeologico Paleoambientale, Budrio 25559
Museo Civico Archeologico 'Romualdi', Notaresco 26944
Museo Civico Arecheologico delle Acque, Chianciano Terme 25850
Museo Civico Aufidenate, Castel di Sangro 25717
Museo Civico C. G. Nicastro, Bovino 25522
Museo Civico Castello Ursino, Catania 25773
Museo Civico di Archeologia Ligure, Genova 26296
Museo Civico e Teatro e Serbatoio Romano, Falerone 26083
Museo Civico G.B. Adriani, Cherasco 25848
Museo Civico Rogadeo, Bitonto 25430
Museo Correale di Terranova, Sorrento 27881
Museo del Opera Omnia R. Musa e Pinacoteca Parmigiani, Bedonia 25386
Museo del Patrimonium, Sutri 27930
Museo della Civitella, Chieti 25874
Museo dell'Abbazia, Casamari-Veroli 25700
Museo dell'Abbazia, Castiglione a Casauria 25759
Museo delle Grigne, Esino Lario 26065
Museo delle Mura, Roma 27485
Museo di Preistoria e del Mare, Trapani 28050
Museo Didattico Archeologico, Laterza 26463
Museo Diocesano di Arte Sacra, Sarsina 27747
Museo Etrusco, Asciano 25292
Museo Etrusco M. Guarnacci, Volterra 28311
Museo Fondazione C. Faina e Museo Civico Palazzo Faina, Orvieto 27009
Museo Lapidario, Ferrara 26117
Museo Nazionale Archeologico, Sarsina 27748
Museo Nazionale della Siritide, Policoro 27270
Museo Nazionale di Santa Maria delle Monanche e Paleolitico, Isernia 26414
Museo Nazionale di Villa Guinigi, Lucca 26536
Museo Nazionale Etrusco di Villa Giulia, Roma 27518
Museo Nazionale Etrusco Pompeo Aria, Marzabotto 26615
Museo Nazionale Jatta, Ruvo di Puglia 27582
Museo Nazionale Romano-Terme di Diocleziano, Roma 27520
Museo Nuovo - Museo del Conservatori, Roma 27522
Museo Ostiense, Ostia Antica 27017
Museo Palatino, Roma 27523
Museo Paleoveneto, Pieve di Cadore 27210
Museo Provinciale, Catanzaro 25783
Museo Romano, Tortona 28044
Museo Tolone Azzariti, Girifalco 26343
Museo Torlonia, Roma 27540
Nuovo Museo Provinciale Sannitico, Campobasso 25636
Raccolta d'Arte e Archeologia, Piazza Armerina 27195
Scavi di Ercolano, Ercolano 26060
Scavi di Oplontis, Torre Annunziata 28034
Scavi di Pompei, Pompei 27280
Villa Romana, Desenzano del Garda 26036

Macedonia

Archeološki Grad Stobi, Negotino 30509
Archeološki Muzej Cherakleja, Bitola 30499

Malta

Domus Romana, Rabat 30636

Montenegro

Arheološki Muzej, Budva 31638

Netherlands

Allard Pierson Museum Amsterdam, Amsterdam 31825
Archaeologische Verzameling van de Universiteit van Utrecht, Utrecht 32843
Kerkmuseum Elst, Elst 32203
Ouddorps Raad- en Polderhuis, Ouddorp 32650
Oudheidkundig Museum, Sint Michielsgestel 32773
Thermenmuseum, Heerlen 32339

Northern Cyprus

Sipahi Ay. Trias Bazilikası ve Kantara Kalesi, Kantara 33364
Soli Archaeological Site, Lefke 33365

Peru

Museo Nacional de Antropología, Arqueología e Historia del Perú, Lima 34264

Portugal

Museu do Teatro Romano, Lisboa 35705
Museu Monográfico de Conimbriga, Condeixa-a-Nova 35548
Museu Nacional de Arqueologia, Lisboa 35715

Romania

Edificiul Roman cu Mozaic, Constanța 36301
Muzeul Arheologic Istria, Istria 36431
Muzeul de Arheologie, Topalu 36697
Muzeul și Rezervația Arheologică de la Castrul Roman Jidava, Câmpulung 36254
Muzeul Stațiunii Băile Herculane, Băile Herculane 36130

Slovakia

Antická Gerulata v Rusovciach, Bratislava 38203

Slovenia

Muzejska Zbirka, Ajdovščina 38396

Spain

Conjunto Arqueológico de Carmona, Carmona 39131
Conjunto Arqueológico de Itálica, Santiponce 40037
Museo Arqueológico Municipal, Alcúdia (Baleares) 38832
Museo Arqueológico Municipal Alejandro Ramos Folqués, Elche 39264
Museo Arqueológico Municipal Camilo Visedo Moltó, Alcoy 38827
Museo Arqueológico Provincial, Badajoz 38933
Museo de Cádiz, Cádiz 39089
Museo de Jérica, Jérica 39429
Museo de la Catedral, Murcia 39693
Museo de la Comisión Provincial de Monumentos, León 39455
Museo Monográfico de Cástulo, Linares 39463
Museo Monográfico de Clunia, Peñalba de Castro 39828
Museo Monográfico y Necrópolis Púnica de Puig des Molins, Eivissa 39260
Museo Municipal, Antequera 38875
Museo Romano de Astorga, Astorga 38911
Museo Romano Oiasso, Irún 39413
Museu Arqueològic de l'Esquerda, Roda de Ter 39908
Museu d'Arqueologia de Catalunya, Barcelona 38970
Museu de Badalona, Badalona 38937
Museu Municipal, Tossa de Mar 40167
Museu Nacional Arqueològic de Tarragona, Tarragona 40105

A

Sweden

Antikmuseet, Göteborg ... 40543
Lunds Universitets Antikmuseum, Lund ... 40706
Medelhavsmuseet, Stockholm ... 40913

Switzerland

Historisches Museum der Stadt Baden, Baden ... 41145
Musée de la Villa Romaine de Pully, Pully ... 41738
Musée Gallo-Romain d'Octodure, Martigny ... 41612
Musée Romain, Avenches ... 41138
Musée Romain, Nyon ... 41692
Musée Romain de Vallon, Vallon ... 41973
Museo di Castelgrande, Bellinzona ... 41193
Thermen-Museum Juliomagus, Schleitheim ... 41837
Vindonissa-Museum, Brugg ... 41261

Tunisia

Musée Archéologique de Makthar, Makthar ... 42638
Musée d'Institut National de Archéologique d'El-Jem, El-Jem ... 42623
Musée National de Carthage, Carthage ... 42620
Musée National d'Utique, Utique ... 42668

Turkey

Aphrodisias Müzesi, Aydın ... 42719
Arkeoloji Müzesi, Konya ... 42882
Aydın Müzesi, Aydın ... 42721
Bodrum Sualtı Arkeolojisi Müzesi, Bodrum ... 42731
Çanakkale Müzesi, Çanakkale ... 42748
Efes Müzesi, Selçuk ... 42936
Ereğli Müzesi, Ereğli ... 42772
Hatay Müzesi, Hatay ... 42794
Hierapolis Arkeoloji Müzesi, Denizli ... 42759
Kossuth Müzesi, Kütahya ... 42894
Manisa Müzesi, Manisa ... 42903
Niğde Müzesi, Niğde ... 42918
Side Müzesi, Side ... 42939
Silifke Müzesi, Silifke ... 42942
Tire Müzesi, Tire ... 42952

Ukraine

Nacional'nyj Zapovidnyk Chersones Tavrijs'kyj, Sevastopol ... 43178

United Kingdom

Aldborough Roman Site, Aldborough ... 43266
Arbeia Roman Fort and Museum, South Shields ... 45697
Brading Roman Villa, Brading ... 43586
Castle Museum, Colchester ... 43872
Chedworth Roman Villa, Yanworth ... 46091
Chedworth Roman Villa Museum, Cheltenham 43809
Clayton Collection Museum, Chollerford ... 43839
Corbridge Roman Site and Museum, Corbridge ... 43891
Fishbourne Roman Palace and Museum, Chichester ... 43826
Hereford Museum and Art Gallery, Hereford ... 44454
Housesteads Roman Fort and Museum, Housesteads ... 44491
Lullingstone Roman Villa, Eynsford ... 44201
Lunt Roman Fort, Baginton ... 43350
Malton Museum, Malton ... 45080
National Roman Legionary Museum, Caerleon 43707
Newport Roman Villa, Newport, Isle of Wight ... 45245
Nottingham University Museum of Archaeology, Nottingham ... 45299
Ribchester Roman Museum, Ribchester ... 45487
Richborough Castle - Roman Fort, Sandwich ... 45616
Roman Army Museum, Greenhead ... 44364
Roman Painted House, Dover ... 44017
Segedunum Roman Fort, Baths and Museum, Wallsend ... 45917
Segontium Roman Museum, Caernarfon ... 43711
Senhouse Roman Museum, Maryport ... 45113
Thurrock Museum, Grays ... 44351
Ure Museum of Greek Archaeology, Reading ... 45472
Viroconium Museum, Wroxeter ... 46087
Wall Roman Site and Museum - Letocetum, Lichfield ... 44708

Vatican City

Museo Gregoriano Egizio, Città del Vaticano ... 54503

Archaeology, Ibero-American

Argentina

Museo Arqueológico, Gualeguaychú ... 00382
Museo Arqueológico de Chasicó, Chasicó ... 00301
Museo Arqueológico de Puerto Esperanza, Puerto Esperanza ... 00506
Museo Arqueológico de Tilcara, Tilcara ... 00647
Museo Arqueológico Pío P. Díaz, Cachi ... 00285
Museo Arqueológico Provincial, San Salvador de Jujuy ... 00607
Museo Arqueológico Regional Anibal Montes, Río Segundo ... 00538
Museo Arqueológico Regional Inca Huasi, La Rioja ... 00416
Museo Arqueológico y Colonial, Fuerte Quemado ... 00374
Museo de Antropología, Córdoba ... 00321
Museo de Antropología y Ciencias Naturales, Concordia ... 00315
Museo de Arqueología de la Universidad Nacional de Tucumán, San Miguel de Tucumán ... 00595
Museo de Arqueología de Salta, Salta ... 00554
Museo del Area Fundacional, Mendoza ... 00454
Museo Histórico Arqueológico Guillermo Magrassi, Mar del Plata ... 00445
Museo Jesútico Nacional de Jesús María, Jesús María ... 00393
Museo Pampeano de Chascomús, Chascomús 00300
Museo Provincial Arqueológico Eric Boman, Santa María de Catamarca ... 00623
Museo Provincial de Arqueología Wagner, Santiago del Estero ... 00630
Museo Rauzi, Palmira ... 00483
Museo Regional Dr. Adolfo Alsina, Carhué ... 00292
Museo Universitario Florentino y Carlos Ameghino, Rosario ... 00550

Belize

Altun Ha Archaeological Site, Altun Ha ... 04172
Ambergris Museum and Cultural Centre, Ambergris Caye ... 04173
Caracol Archaeological Site, Caracol ... 04177
Lamanai Archaeological Site, Lamanai ... 04178
Museum of Belize, Belize City ... 04176
Tanah Art Museum, San Antonio ... 04180

Bolivia

Museo Arqueológico, Trinidad ... 04253
Museo Arqueológico, Cochabamba ... 04206
Museo Arqueológico de Saipina, Santa Cruz ... 04239
Museo Etnoarqueológico del Beni Kenneth Lee, Trinidad ... 04254
Museo Nacional de Arqueología, La Paz ... 04222
Museo Nacional Paleontológico y Arqueológico, Tarija ... 04251
Museo Regional de la Arqueológico de Tiwanaku, Tiwanaku ... 04252

Brazil

Museu Câmara Cascudo, Natal ... 04582
Museu Comercial e Arqueológico do Sindicato dos Empregados, Maceió ... 04542
Museu Lasar Segall, São Paulo ... 04906
Museu Sitio Arqueológico Casa dos Pilões, Rio de Janeiro ... 04761

Canada

London Museum of Archaeology, London ... 06186

Chile

Museo Arqueológico de La Serena, La Serena 07440
Museo Chileno de Arte Precolombino, Santiago de Chile ... 07451
Museo de Antofagasta, Antofagasta ... 07429
Museo del Limarí, Ovalle ... 07444

Colombia

Museo Arqueológico, Pasca ... 08707
Museo Arqueologico Casa del Marqués de San Jorge, Bogotá ... 08553
Museo Arqueológico Julio César Cubillos, Cali ... 08606
Museo del Río Magdalena, Honda ... 08657
Museo Luciano Rosero, Pasto ... 08711
Museo Madre Caridad Brader, Pasto ... 08712

Costa Rica

Museo Nacional de Costa Rica, San José ... 08824
Museo Sito Arqueológico Guayabo de Turrialba, Turrialba ... 08826

Cuba

Indo-Cuban Bani Museum, Banes ... 09029
Museo Antropológico Montané, La Habana ... 09110
Museo de Arqueología, La Habana ... 09130

Dominican Republic

Museo Arqueología Regional, La Romana ... 10332
Museo Arqueológico de La Caleta, La Caleta ... 10325
Museo Cuevas de Las Maravillas, La Romana 10333
Museo de Arte Taino, San Felipe de Puerto Plata ... 10335
Museo de Sitio de La Isabela, Luperón ... 10329
Museo de Sitio de la Vega Vieja, La Vega Vieja 10361
Museo Fortaleza de San Felipe, San Felipe de Puerto Plata ... 10336
Museo Parque Los Tres Ojos, Santo Domingo ... 10359
Sala de Arte Prehispánico, Santo Domingo ... 10360

Ecuador

Museo Antropológico del Banco Central, Guayaquil ... 10400
Museo Arqueologico Carlos Aiiercado, Esmeraldas ... 10395
Museo Arqueologico del Banco Central de Bahia, Bahia de Carácluez ... 10372
Museo Banco del Pacifico, Guayaquil ... 10401
Museo Carlos Emilio Grijalva, Tulcán ... 10461
Museo de Sitio de Ingapirca, Azogues ... 10371
Museo de Sitio del Rumicucho, San Antonio ... 10459
Museo del Banco Central de Manta, Manta ... 10420
Museo del Colegio Nacional Pedro Vicente Maldonado, Riobamba ... 10456
Museo del Colegio Stella Maris, Atacames ... 10370
Museo Francisco Huerta Rendón en la Casona Universitaria, Guayaquil ... 10406
Museo Nacional de la Dirección Regional del Banco Central del Ecuador, Quito ... 10449
Museo Nahim Isaias, Guayaquil ... 10410
Museo Sala de Exposición Banco Central de Esmeraldas, Esmeraldas ... 10396
Museo Santuario Nuestra Señora Natividad del Guayco, La Magdalena ... 10419
Museo Weilbauer, Quito ... 10452

El Salvador

Museo de Antropología David J. Guzmán, San Salvador ... 10549
Museo Sitio Arqueológico de San Andrés, Valle de Zapotitán ... 10556
Ruinas del Tazumal, Chalchuapa ... 10545
Sitio Arqueológico Joya de Cerén, San Juan ... 10547

France

Musée des Jacobins, Auch ... 11735

Guatemala

Museo Nacional de Arqueología y Etnología, Guatemala ... 23301

Honduras

Museo Arqueológico de Comayagua, Comayagua ... 23325
Museo de las Esculturas de Copán, Copán ... 23328

Italy

Museo Americanistico Federico Lunardi, Genova ... 26295
Museo delle Culture del Mondo, Genova ... 26304

Mexico

Museo Arqueológico Chiametlán, Rosario ... 31307
Museo Arqueológico Cuicuilco, México ... 31051
Museo Arqueológico de Apaxco, Apaxco ... 30719
Museo Arqueológico de Campeche, Campeche ... 30739
Museo Arqueológico de Cancún, Cancún ... 30747
Museo Arqueológico de Ciudad Guzmán, Guzmán ... 30925
Museo Arqueológico de Comitán, Comitán de Domínguez ... 30792
Museo Arqueológico de la Ciudad, Emiliano Zapata ... 30851
Museo Arqueológico de la Costa Grande, Zihuatanejo ... 31596
Museo Arqueológico de Lagos de Moreno, Lagos de Moreno ... 30991
Museo Arqueológico de Mazatlán, Mazatlán ... 31020
Museo Arqueológico de Nextlalpan, Nextlalpan ... 31214
Museo Arqueológico de Tepeapulco, Tepeapulco ... 31432
Museo Arqueológico de Tizatlán, Tizatlán ... 31462
Museo Arqueológico de Tonalá, Tonalá, Chiapas ... 31495
Museo Arqueológico de Tula Jorge R. Acosta, Tula de Allende ... 31508
Museo Arqueológico de Venustiano Carranza, Venustiano Carranza ... 31534
Museo Arqueológico de Xochimilco, México ... 31052
Museo Arqueológico del Cerro de la Estrella, México ... 31053
Museo Arqueológico del Soconusco, Tapachula ... 31401
Museo Arqueológico Dr. Román Piña Chan, Tenango del Valle ... 31422
Museo Arqueológico Juan Sepúlveda y Museo Eduardo Ruiz, Uruapán ... 31520
Museo Arqueológico Luis G. Urbina, México ... 31054
Museo Arqueológico Mesoamericano, Tijuana 31455
Museo Casa del Dr. Mora, Comonfort ... 30795
Museo Comunitario Shan-Dany, Santa Ana del Valle ... 31372
Museo David Ramírez Lavoignet, Misantla ... 31169
Museo de Antropología de Xalapa, Xalapa ... 31557
Museo de Arqueología, Acapulco ... 30691
Museo de Arqueología, Ixtepec ... 30961
Museo de Arqueología de Córdoba, Córdoba ... 30800
Museo de Arqueología de León, León ... 30997
Museo de Arqueología de Occidente de México, Guadalajara ... 30891
Museo de Arqueología el Chamizal, Juárez ... 30980
Museo de Arte Prehispánico de México Rufino Tamayo, Oaxaca ... 31222
Museo de la Ciudad, Mérida ... 31027
Museo de la Cultura Huasteca, Tampico ... 31399
Museo de la Cultura Maya, Chetumal ... 30761
Museo de la Pintura Mural y Museo Manuel Gamio, Teotihuacán ... 31426
Museo de la Quemada, Guadalupe ... 30908
Museo de San Lorenzo Tenochtitlán, Texistepec ... 31454
Museo de Sitio Boca de Potrerillos, Mina ... 31167
Museo de Sitio de Cacaxtla, Nativitas ... 31210
Museo de Sitio de Cempoala, Cempoala ... 30755
Museo de Sitio de Chalcatzingo, Chalcatzingo 30757
Museo de Sitio de Chichen Itza, Timun ... 31459
Museo de Sitio de Coatetelco, Coatetelco ... 30777
Museo de Sitio de Comalcalco, Comalcalco ... 30791
Museo de Sitio de el Tajín, Papantla ... 31252
Museo de Sitio de el Zapotal, Ignacio de la Llave ... 30954
Museo de Sitio de Higueras, Vega de Alatorre ... 31532
Museo de Sitio de La Venta, Huimanguillo ... 30950
Museo de Sitio de la Zona Arqueológica de Tzintzunzan, Tzintzuntzan ... 31519
Museo de Sitio de Monte Albán, Oaxaca ... 31227
Museo de Sitio de Oculian, Ocuilan ... 31236
Museo de Sitio de Palenque Alberto Ruz Lhuillier, Palenque ... 31246
Museo de Sitio de Pomona, Tenosique de Pino Suárez ... 31424
Museo de Sitio de Querétaro, Querétaro ... 31301
Museo de Sitio de San Lorenzo Tenochtitlán, Tenochtitlán ... 31423
Museo de Sitio de San Miguel Ixtapan, San Miguel Ixtapan ... 31358
Museo de Sitio de Tenayuca, Tlalnepantla ... 31468
Museo de Sitio de Teotihuacán, Teotihuacán ... 31427
Museo de Sitio de Toniná, Ocosingo ... 31233
Museo de Sitio de Tres Zapotes, Santiago Tuxtla ... 31386
Museo de Sitio de Uxmal, Santa Elena ... 31378
Museo de Sitio de Xochicalco, Xochicalco ... 31565
Museo de Sitio de Xochitecatl, Nativitas ... 31211
Museo de Sitio Degollado Jalisco, Degollado ... 30831
Museo de Sitio El Torreoncito, Torreón ... 31498
Museo de Sitio Ocotelulco, Totolac ... 31505
Museo de Sitio San Agustín, Juárez ... 30982
Museo de Sitio Talavera, México ... 31095
Museo de Sitio Tecpan, México ... 31096
Museo del Carmen, México ... 31100
Museo del Pueblo Maya Dzibilchaltún, Chablekal ... 30756
Museo del Templo Mayor, México ... 31109
Museo Dr. Atl, Pihuamo ... 31259
Museo El Jonotal, Playa Vicente, Veracruz ... 31261
Museo El Polvorín, Monclova ... 31174
Museo Fray Bernardo Padilla, Acámbaro ... 30687
Museo Omar Huerta Escalante, Jonuta ... 30977
Museo Porfirio Corona Covarrubias, El Grullo, Jalisco ... 30847
Museo Regional de Arqueología de Tuxpan, Tuxpan de Rodríguez Cano ... 31510
Museo Regional de la Laguna, Torreón ... 31503
Museo Regional Huasteco, Valles ... 31530
Museo Sac-be, Yaxcabá ... 31572
Museo San José el Mogote, San José el Mogote ... 31334
Museo Universitario de Arqueología de Manzanillo, Manzanillo ... 31013
Museo Universitario Lariab de Historia Antigua de la Huaxteca, Tamuín ... 31400
Sala de Arqueología del Municipio de Mújica, Múgica ... 31204

Netherlands

Amerika Museum Nederland, Cuijk ... 32040

Nicaragua

Museo Arqueológico de Esteli, San Juan de Limay ... 33304
Museo Arqueológico Gregorio Aguilar Barea, Juigalpa ... 33280
Museo Chorotega-Nicarao llamado "Enrique Berio Mántica Deshon", Chinandega ... 33273
Museo del Pueblo y Museo Pre-Colombino de Chagüitillo, Chagüitillo ... 33272
Museo Prof. Ninfa María Báez, Acoyapa ... 33271

Museo Sitio Huellas de Acahualinca, Acahualinca 33270

Panama

Museo del Parque Arqueológico El Caño, Natá 34077

Peru

Muestrario Arqueológico de Amazonas, Chachapoyas 34144
Museo Arqueológico Alejandro Pezzia Aseretto, La Angostura 34198
Museo Arqueológico Comunal, Carquín 34138
Museo Arqueológico de Ancash, Huaraz 34181
Museo Arqueológico de Cajamarca y Museo Médico, Cajamarca 34126
Museo Arqueológico de Chusís, Sechura 34306
Museo Arqueológico de la Universidad Nacional del Altiplano, Pucara 34295
Museo Arqueológico del Centro de Estudios Histórico Sociales Julio Espejo Nuñez, Jauja 34196
Museo Arqueológico del Colegio Juan XXIII, Lima 34226
Museo Arqueológico e Histórico de Andahuaylas, Andahuaylas 34107
Museo Arqueológico Frederico Galvez Durand, Huancayo 34176
Museo Arqueológico Horacio Urteaga, Cajamarca 34127
Museo Arqueológico José Casinelli, Santa Inés 34305
Museo Arqueológico Municipal y Museo Municipal Miguel Justino Ramírez, Piura 34290
Museo Arqueológico Nacional Brüning, Lambayeque 34202
Museo Cabrera, Ica 34187
Museo Carlos Dreyer, Puno 34298
Museo Cassinelli, Trujillo 34317
Museo Católica de Santa María, Cercado 34141
Museo Contisuyo, Moquegua 34278
Museo de Arqueología Josefina Ramos de Cox, Lima 34229
Museo de Arqueología Samuel Humberto Espinoza Lozano, Huaytará 34185
Museo de Arqueología y Antropología, Lima 34230
Museo de Arte de Lima, Lima 34232
Museo de Arte Precolombino, Cusco 34157
Museo de Investigaciones de Zonas Aridas, Lima 34240
Museo de la Cultura Martínez de Compañon, Tambogrande 34315
Museo de la Provincia El Dorado, El Dorado 34172
Museo de Oro Vicús, Piura 34293
Museo de Sitio Arqueológico Kuntur Wasi, San Pablo 34304
Museo de Sitio Complejo Arqueológico de Sillustani, Atuncolla 34117
Museo de Sitio de Chan Chan, Trujillo 34321
Museo de Sitio de Chinchero, Urubamba 34324
Museo de Sitio de Paracas Julio C. Tello, Pisco 34287
Museo de Sitio de Puruchuco, Lima 34248
Museo de Sitio de San Isidro, Tembladera 34316
Museo de Sitio de Sipán, Zaña 34331
Museo de Sitio de Túcume, Campo 34136
Museo de Sitio de Wari, Wari 34326
Museo de Sitio El Algarrobal, El Algarrobal 34171
Museo de Sitio Huaca Pucllana, Lima 34250
Museo de Sitio Manuel Chávez Ballón, Aguas Calientes 34106
Museo de Sitio Parque Reducto 2, Lima 34251
Museo de Sitio Wari, Ayacucho 34120
Museo de Sitio Wari Willca, Huancayo 34177
Museo del Centro de Investigación Arqueológica de Ancón, Lima 34252
Museo del Instituto Nacional de Cultura, Arequipa 34112
Museo Enrico Poli, Lima 34257
Museo Inca de la Universidad Nacional San Antonio Abad, Cusco 34166
Museo Inka, Cusco 34167
Museo Lítico de Pucara, Pucara 34296
Museo Municipal de Chincha, Chincha Alta 34147
Museo Municipal Dreyer, Dreyer 34170
Museo Municipal San Juan de Iris, San Juan de Iris 34303
Museo Parque de la Muralla, Lima 34268
Museo Regional de Arqueología de Junín, Chupaca 34151
Museo Regional de Ucayali, Pucallpa 34294
Museos Oro del Perú - Armas del Mundo, Lima 34273
Museum of the Royal Tombs of Sipán, Lambayeque 34204
Sitio Arquelógico Huaca Malena, Capilla de Asia 34137
Sitio Arqueológico de Cumbemayo, Cumbemayo 34155

Puerto Rico

Museo Cemí, Jayuya 36042
Museum of History, Anthropology and Art, San Juan 36076

Spain

Colección de Arqueología y Etnografía Americana, Madrid 39511
Museo Arqueológico de Benalmádena, Benalmádena 39025

United States of America

Blackwater Draw Museum, Portales 52053
Chaco Culture National Historical Park, Nageezi 50939
El Morro National Monument, Ramah 52238
Salinas Pueblo Missions National Monument, Mountainair 50896

Uruguay

Palacio Taranco, Montevideo 54424

Archaeology, Near and Middle Eastern

Afghanistan

Balkh Provincial Museum, Mazar-i-Sharif 00008
Balkh Religious Museum, Mazar-i-Sharif 00009
Kapisa Provincial Museum and Archeological Site, Mahmud-i-Raqi 00007

France

Musée de Bible et Terre Sainte, Paris 14648

Germany

Hilprecht-Sammlung Vorderasiatischer Altertümer, Jena 19381
Pergamonmuseum, Berlin 17234
Vorderasiatisches Museum im Pergamonmuseum, Berlin 17279

Iran

Azarbayezan Museum, Tabriz 24650
Gorgan Museum, Gorgan 24605
Hakhamaneshi Museum, Shiraz 24646
Reza Abbasi Museum, Tehrān 24692
Takht-e-Jamshid, Shiraz 24649
Uroomieh Museum, Uroomieh 24702

Iraq

Al Mawsil Museum, Nineveh 24724
Archaeological Site Nimrud (Calah), Nimrud 24723
Babylon Museum, Babylon 24709
Basrah Museum, Basrah 24717
Nasiriya Museum, Nasiriya 24722
Nergal Gate Museum, Mosul 24721
Samarra Museum, Samarra 24725

Israel

Akko Municipal Museum, Akko 24923
Arad Museum, Arad 24927
Archaeological Museum, Kibbutz Ein Dor 25036
Be'eri Archaeological Collection, Be'eri 24935
Beit Miriam-Museum, Kibbutz Palmachim 25053
Bible Lands Museum Jerusalem, Jerusalem 24980
Caesarea Museum, Kibbutz Sdot Yam 25056
Ceramics Museum, Tel Aviv 25103
Golan Archaeological Museum, Qatzrin 25079
Hanita Museum, Kibbutz Hanita 25043
Hazor Museum, Ayelet-Hashahar 24933
Herbert E. Clark Collection of Near Eastern Antiquities, Jerusalem 24989
Local Museum, Kibbutz Maabarot 25049
Local Museum, Kibbutz Sasa 25055
Museum, Ancient Synagogue and Necropolis, Beit Shearim 24938
Nahsholim Museum, Hof Dor 24973
Pères Blancs - Saint Anne, Jerusalem 25008
Reuben and Edith Hecht Museum, Haifa 24967
Rockefeller Archaeological Museum, Jerusalem 25010
Saint James Museum and Library, Jerusalem 25013
Shrine of the Book, Jerusalem 25016
Shtekelis Prehistory Museum, Haifa 24968
Skirball Museum of Biblical Archaeology, Jerusalem 25019
Tell Qasile Archaeological Site, Tel Aviv 25123
Terra Sancta Museum, Nazareth 25072

Japan

Okayama-shiritsu Oriento Bijutsukan, Okayama 29102

Jordan

Irbid Archaeological Museum, Irbid 29662
Jarash Archaeological Museum, Jarash 29664
Jordan Archaeological Museum, Amman 29652
Madaba Archaeology Museum, Ajloun 29640
Museum of Aqaba Antiquities, Aqaba 29661
Umm Qais Archaeological Museum, Umm Qais 29668

Lebanon

Musée Archéologique de l'Université Americaine, Beirut 30157
Musée National, Beirut 30160

Mexico

Museo de Historia de Tabasco, Villahermosa 31551
Parque-Museo La Venta, Villahermosa 31555

Morocco

Gaselec Musée, Melilla 31682

Netherlands

Bijbels Openluchtmuseum, Heilig Landstichting 32345

Oman

Bait Muzna Gallery, Muscat 33988

Pakistan

Islamabad Museum, Islamabad 34007

Palestine

Musée de l'Ecole Biblique et Archéologique Française, Jerusalem 34057

Tajikistan

Mamnu"gohi Bostonshinosii Panjakent, Pandžakent 42504
Museum of National Antiquities, Dušanbe 42496
Osorkhonai Bostonshinosiii Khujand, Chudžand 42492

Turkey

Agora Açıkhava Müzesi, İzmir 42852
Alacahöyük Müzesi, Alaca 42686
Anadolu Medeniyetleri Müzesi, Ankara 42695
Anamur Müzesi ve Örenyerleri, Anamur 42692
Arkeoloji ve Etnografya Müzesi, Elazığ 42769
Aydın Müzesine Bağlı Örenyerleri, Aydın 42722
Bandirma Arkeoloji Müzesi, Bandirma 42725
Çorum Örenyerleri, Çorum 42756
Diyarbakır Örenyerleri, Diyarbakır 42764
Elazığ Örenyerleri, Elazığ 42770
Gaziantep Arkeoloji Müzesi, Gaziantep 42785
Harput Müzesi, Elazığ 42771
Kahramanmaraş Müzesi, Kahramanmaraş 42862
Karaman Müzesi, Karaman 42864
Karatay Çini Eserler Müzesi, Konya 42886
Karatepe-Aslantaş Açıkhava Müzesi, Osmaniye 42924
Kastamonu Arkeoloji Müzesi, Kastamonu 42868
Kuva-yi Milliye Müzesi, Balıkesir 42724
Mardin Müzesi, Mardin 42904
Marmaris Kalesi ve Arkeoloji Müzesi, Marmaris 42905
Milet Müzesi, Söke 42948
Perge Müzesi, Aksu 42685
Samsun Arkeoloji ve Etnografya Müzesi, Samsun 42931
Sebastopolis Açıkhava Müzesi, Tokat 42953
Yazılıkaya Müzesi, Çifteler 42753

Turkmenistan

Taryhy we Etnografiki Muzeý, Mary 42978
Taryhy Ýer Abiwerd, Kaka 42976
Taryhy Ýer Nusaý, Nusaý 42980

United Arab Emirates

Al-Ain Museum, Al-Ain 43217

United Kingdom

Oriental Museum Durham, Durham 44074

United States of America

Elizabeth P. Korn Gallery, Madison 50383
Harvard University Semitic Museum, Cambridge 47178
Hellenic Museum and Cultural Center, Chicago 47479

Yemen

Al-Mukalla Museum, Al-Mukalla 54645
Ataq Museum, Ataq 54646
Baihan Museum, Baihan 54647
National Museum, Sanaa 54650

Architecture

Argentina

Museo de la Santa Casa de Ejercicios Espirituales Sor María Antonia de la Paz y Figueroa, Buenos Aires 00208
Torre Monumental, Buenos Aires 00283

Armenia

Architect Alexander Tamanyan Research Institute-Museum, Yerevan 00727
National Museum-Institute of Architecture, Yerevan 00752

Australia

Burnett House and Myilly Point Heritage Precinct, Darwin 01029
City of Melbourne Collection, Melbourne 01277
Earlystreet, Norman Park 01370
Elizabeth Farm, Rosehill 01468
Eryldene Heritage House, Gordon 01120
Palma Rosa Museum, Hamilton, Brisbane 01142
Powder Magazine, Beechworth 00849
Quilpie Museum, Quilpie 01451
Rose Seidler House, Wahroonga 01622
Tin Sheds Gallery, Sydney 01562

Austria

Albertina, Wien 02929
Architekturzentrum Wien, Wien 02932
Ausstellungszentrum Heft, Hüttenberg 02101
Ausstellungszentrum Heiligenkreuzer Hof, Wien 02940
Ernst Fuchs Museum, Wien 02981
Freilichtmuseum Ensemble Gerersdorf, Gerersdorf 01930
Lehár-Schikaneder-Schlößl, Wien 03021
Minimundus, Klagenfurt 02188
muba - Museum für Baukultur, Neutal 02444
Museum Schloss Ottersbach, Großklein 02004
Österreichisches Brückenbaumuseum, Edelsbach bei Feldbach 01840
Salvatorianerkloster, Gurk 02020
Stiftsgalerie Seitenstetten, Seitenstetten 02738
Studiensammlung historischer Ziegel, Irdning 02138

Bangladesh

Varendra Research Museum, Rajshahi 03234

Belgium

Ancien Palais de Bruxelles, Bruxelles 03431
Fondation pour l'Architecture, Bruxelles 03446
Kasteel van Gaasbeek, Gaasbeek 03611
Miniatuurstad, Antwerpen 03293
Musée d'Architecture - La Loge, Bruxelles 03461
Musée du Château Comtal, Rochefort 03973

Brazil

Fundaçao Oscar Niemeyer, Rio de Janeiro 04694
Museu da Casa Brasileira, São Paulo 04867
Museu da Limpeza Urbana - Casa de Banhos de Dom João VI, Rio de Janeiro 04717

Brunei

Muzium Teknologi Melayu, Bandar Seri Begawan 04966

Bulgaria

Architekturno-muzeen Rezervat Arbanassi, Veliko Tărnovo 05250
Architekturno-muzeen Rezervat Carevec, Veliko Tărnovo 05251
Durjaven Kulturen Center Dvoreca, Balčik 04975
Muzej Kolju Fičeto, Drjanovo 05025
Nacionalen Muzej na Bălgarskata Architektura, Veliko Tărnovo 05256

Canada

Bellevue House, Kingston 06074
Brubacher House Museum, Waterloo 07284
Canadian Centre for Architecture, Montréal 06330
Jost House Museum, Sydney 07049

Musée des Hospitalières de l'Hôtel-Dieu de Montréal, Montréal 06359
Rideau Hall, Ottawa 06554
Thomas Foster Memorial, Uxbridge 07190
Westfield Heritage Village, Rockton 06768

China

Ancient Architecture Museum, Beijing 07503
Beijing Ancient Architecture Museum, Beijing 07507
Hangzhou City Construction Exhibition Hall, Shanghai 08167
Sam Tung Uk Museum, Hong Kong 07815
Shanghai Urban Planning Exhibition Hall, Shanghai 08190
Sheung Yiu Folk Museum, Hong Kong 07816

Colombia

Museo de Arquitectura Leopoldo Rother, Bogotá 08556

Croatia

Hrvatski Muzej Arhitekture, Zagreb 08994

Cuba

Casa de los Arabes, La Habana 09096
Museo Romántico, Trinidad 09338

Czech Republic

Galerie Architektury Brno, Brno 09428
Galerie Jaroslava Fragnera, Praha 09747
Krajská Galerie Výtvarného Umění ve Zlíně, Zlín 09940
Muzeum Antického Sochařstvi a Architektury, Litomyšl 09633
Muzeum Lidových Staveb, Kouřim 09600
Národopisné Muzeum Třebíz, Třebíz 09876
Rodný Dům Josefa Hoffmanna, Brtnice 09444
Vila Tugendhat, Brno 09442

Denmark

Den Gamle By, Århus 09963

Egypt

Coptic Museum, Cairo 10490

Estonia

Eesti Arhitektuurimuuseum, Tallinn 10623
Eesti Vabaõhumuuseum, Tallinn 10630

Finland

Alvar Aalto Museum, Jyväskylä 10886
Hvitträsk, Kirkkonummi 10963
Sederholmin Talo, Helsinki 10815
Suomen Rakennustaiteen Museo, Helsinki 10821

France

Abbaye de Bonnefont, Proupiary 14950
Château d'Azay-le-Ferron, Azay-le-Ferron 11808
Ecomusée des Monts d'Arrée, Saint-Rivoal 15499
Ecomusée du Pays de Rennes, Rennes (Ille-et-Vilaine) 15012
Ecomusée du Veron, Savigny-en-Veron 15678
Espace Baroque Maurienne, Lanslebourg-Mont-Cenis 13517
Espace Vauban, Besançon 11954
Fondation Le Corbusier, Paris 14601
Musée Benoît XIII, Châteaurenard (Bouches-du-Rhône) 12454
Musée Biochet-Brechot, Caudebec-en-Caux 12322
Musée des Années 30, Boulogne-Billancourt 12090
Musée Frugès - Le Corbusier, Pessac 14803
Palais des Papes, Avignon 11796
Village Miniature, Marchesieux 13952

Germany

Alt-Rothenburger Handwerkerhaus, Rothenburg ob der Tauber 21405
Architekturmuseum der Technischen Universität München, München 20423
Architekturmuseum Schwaben, Augsburg 16611
Bauernhausmuseum Gebersheim, Leonberg 19960
Bauhaus-Archiv, Berlin 17041
Befreiungshalle Kelheim, Kelheim 19500
Belvedere auf dem Klausberg, Potsdam 21080
Belvedere und Pomonatempel auf dem Pfingstberg, Potsdam 21081
Betonzeitschiene, Dresden 17938
Borstei Museum, München 20429
Chinesische Haus, Potsdam 21085
Deutsches Architekturmuseum, Frankfurt am Main 18361
Fachwerkbaumuseum im Ständerbau, Quedlinburg 21146
Fagus-Gropius-Ausstellung, Alfeld 16464
Felsengarten Sanspareil mit Morgenländischem Bau, Wonsees 22680
Henry van de Velde Museum, Chemnitz 17696
ifa-Galerie Stuttgart, Stuttgart 21945
Jan Bouman Haus, Potsdam 21092
Kunstsammlung Lorenzkapelle Rottweil, Rottweil 21425
Kunststätte Bossard, Jesteburg 19395
Miniaturpark Klein-Erzgebirge, Oederan 20860
Museum Altes Land, Jork 19403
Museum für Kunst und Technik des 19. Jahrhunderts, Baden-Baden 16888
Museumsdorf Bayerischer Wald, Tittling 22065
Museumskirche Sankt-Katharinen, Lübeck 20081
Neuer Pavillon im Schlossgarten Charlottenburg, Berlin 17227
Normannischer Turm auf dem Ruinenberg, Potsdam 21101
Regionalmuseum Wolfhager Land, Wolfhagen 22655
Residenz Kempten, Kempten 19515
Römische Bäder, Potsdam 21104
Schloß Brake - Das Weserrenaissance-Museum, Lemgo 19949
Schloss Charlottenhof, Potsdam 21107
Stiftung Bauhaus Dessau, Dessau 17834
Vitra Design Museum, Weil am Rhein 22379
Walpurgishalle Thale, Thale (Harz) 22043
Wenzel-Hablik-Museum, Itzehoe 19370

Greece

The Chicago Athenaeum - Europe, Néa Symirni 23125
Ethnological Museum of Pyrsogianni Stonemasons, Konitsa 23061

Hungary

Ernst Múzeum, Budapest 23399
Magyar Építészeti Múzeum, Budapest 23435

Iceland

Minjasafn Austurlands, Egilsstadir 23993

Ireland

Castletown, Celbridge 24758

Israel

Islamic Museum, Jerusalem 24995

Italy

Museo Civico, Busseto 25569
Museo della Linea Gotica, Montefiore Conca 26825
Museo di Architettura, Ferrara 26113
Museo di Architettura Militare, Bologna 25460
Museo di Palazzo Strozzi, Firenze 26182
Museo In Situ, Sulmona 27922
Museo Storico di Architettura Militare, Roma 27538
Oskar Schlemmer Theatre Estate and Collection, Oggebbio 26977
Palazzo Reale-Reggia di Caserta, Caserta 25705

Japan

Edo-Tokyo Tatemono-En, Koganei 28757
Hakubutsukan Meiji-Mura, Inuyama 28573
Sankei-en Garden, Yokohama 29605
Seto-Ohashi Memorial Bridge Museum, Kurashiki 28790
Yagai Hakubutsukan, Gashozukuri Minkaen, Shirakawa 29281

Kyrgyzstan

Muzej Architektury i Archeologii Vez Burana, Tokmok 29967

Lithuania

Architektūros Muziejus, Vilnius 30401
Geležinkelių Muziejus, Vilnius 30406

Mexico

Museo Municipal Los Guachimontones y Museo de Arquitectura Prehispánica, Teuchitlán 31451
Museo Nacional de Arquitectura, México 31130

Moldova

Ščusev Museum, Chişinău 31604

Morocco

Museo de Arqueologia e Historia, Melilla 31683

Netherlands

ABC Architectuurcentrum Haarlem, Haarlem 32292
ARCAM Architectuurcentrum Amsterdam, Amsterdam 31829
Madurodam, Den Haag 32081
Miniatuur Walcheren, Middelburg 32552
Nederlands Architectuurinstituut, Rotterdam 32719
Nederlands Baksteen en Dakpanmuseum de Panoven, Zevenaar 33000
Rietveld Schröderhuis, Utrecht 32856
Zuiderkerk, Amsterdam 31905

New Zealand

City Gallery Wellington, Wellington 33244
Rewa's Village, Kerikeri 33137

Nigeria

Jos National Museum and Museum of Traditional Nigerian Architecture, Jos 33330

Norway

Andøymuseet, Risøyhamn 33783
Arkitekturmuseet, Oslo 33716
Emanuel Vigeland Museum, Oslo 33721
Kvam Bygdemuseum, Norheimsund 33703
Midttunet, Skei i Jølster 33833
Nasjonalmuseet for Kunst, Arkitektur og Design, Oslo 33736
Tuddal Bygdetun, Tuddal 33935
Vennesla Bygdemuseum, Vennesla 33972
Vest-Telemark Museum, Dalen i Telemark 33463

Pakistan

Shakir Ali Museum, Lahore 34028

Peru

Museo de Miniaturas, Huaraz 34182
Museo de Suelos, Iquitos 34192

Poland

Galeria Architektury SARP, Katowice 34721
Galeria Architektury SARP, Kraków 34774
Galeria SARP, Wrocław 35324
Muzeum Architektury, Wrocław 35328
Muzeum Budownictwa Ludowego - Park Etnograficzny w Olsztynku, Olsztynek 34980
Muzeum Wsi Lubelskiej, Lublin 34909
Muzeum Wsi Opolskiej, Opole 34987

Romania

Casa Memorială Constantin Joja, Bucureşti 36185
Casa Memorială Ecaterina Teodoroiu, Târgu Jiu 36663
Muzeul Arhitecturii Populare din Gorj, Bumbeşti-Jiu 36236
Muzeul Haszmann Pál, Cernatu de Sus 36274
Muzeul Satului din Bucovina, Suceava 36647
Muzeul Satului Vâlcean, Bujoreni 36234
Rezervaţia de Arhitectură Vrânceană, Focşani 36374

Russia

Aleksandrovskaja Sloboda - Gosudarstvennyj Istoriko- architekturnyj i Chudožestvennyj Muzej-zapovednik, Aleksandrov 36751
Architekturno-chudožestvennyj Ansambl byvšego Spaso-Jakovleskogo Monastyrja, Rostov 37598
Architekturno-ėtnografičeskij Muzej Derevjannogo Zodčestva Chochlovka, Perm 37543
Astrachanskij Kreml, Astrachan 36780
Bolšoj Dvorec, Lomonosov 37106
Elabužskij Gosudarstvennyj Istoriko-architekturnyj i Chudožestvennyj Muzej-zapovednik, Elabuga 36903
Gosudarstvennaja OK Istoriko-ėtnografičeskij i Architekturnyj Muzej-zapovednik Staraja Sarepta, Volgograd 37954
Gosudarstvennyj Chudožestvenno-architekturnyj Dvorcovo-parkovyj Muzej-zapovednik Oranienbaum, Lomonosov 37108
Gosudarstvennyj Chudožestvennyj Istoriko-architekturnyj i prirodno-landšaftnyj Muzej-zapovednik Kolomenskoe, Moskva 37191
Gosudarstvennyj Istoriko-architekturnyj i Ėtnografičeskij Muzej-zapovednik Kiži, Kiži 37050
Gosudarstvennyj Istoriko-architekturnyj i Prirodno-landšaftnyj Muzej Zapovednik Izborsk, Izborsk 36953
Gosudarstvennyj Istoriko-architekturnyj i Prirodnyj Muzej-Zapovednik Monrepo, Vyborg 37981
Gosudarstvennyj Istoriko-chudožestvennyj Dvorcovo-parkovyj Muzej-zapovednik Gatčina, Gatčina 36920
Gosudarstvennyj Istoriko-kulturnyj Muzej-zapovednik "Moskovskij Kreml", Moskva 37194
Gosudarstvennyj Istoriko-memorialnyj Zapovednik Rodina V.I. Lenina, Uljanovsk 37897
Gosudarstvennyj Naučno-issledovatelskij Muzej Architektury im. A.V. Ščuseva, Moskva 37207
Gosudarstvennyj Vladimiro-Suzdalskij Istoriko-architekturnyj i Chudožestvennyj Muzej-zapovednik, Vladimir 37934
Irkutskij Architekturno-ėtnografičeskij Muzej Talcy, Irkutsk 36931
Karačaevo-Čerkesskij Istoriko-kulturnyj i Prirodnyj Muzej-Zapovednik, Čerkessk 36854
Kargopolskij Gosudarstvennyj Istoriko-Architekturnyj i Chudožestvennyj Muzej, Kargopol 37003
Kirillo-Belozerskij Istoriko-Architekturnyj i Chudožestvennyj Muzej-Zapovednik, Kirillov 37034
Kiži - Gosudarstvennyj Istoriko-architekturnyj i Ėtnografičeskij Muzej-zapovednik, Petrozavodsk 37559
Kostromskoj Gosudarstvennyj Istoriko-architekturnyj Muzej-zapovednik, Kostroma 37061
Memorialnyj Muzej N.I. Beloborodova, Tula 37867
Muzej Architektury i Byta Narodov Nižegorodskogo Povolžja, Nižnij Novgorod 37453
Muzej Architektury i Stroitelstva, Samara 37617
Muzej Derevjannogo Zodčestva Malye Karely, Archangelsk 36772
Muzej Goroda Jurjevca, Jurjevec 36984
Muzej Gradostroitelstva i Byta Taganroga, Taganrog 37825
Muzej Istorii Architektury i Architekturnogo Obrazovanija Sibiri, Novosibirsk 37485
Muzej Istorii Doma XIX-XX vekov, Tjumen 37847
Muzej Remesla, Architektury i Byta, Kaluga 37000
Muzej Semji Benua, Sankt-Peterburg 37707
Muzej Tulskij Kreml, Tula 37871
Muzej-usadba Ostafjevo "Russkij Parnas", Ostafjevo 37518
Muzej-zapovednik Dmitrovskij Kreml, Dmitrov 36876
Muzej-zapovednik Kazanskij Kreml, Kazan 37025
Muzejno-vystavočnyj Kompleks Volokolamskij Kreml, Volokolamsk 37966
Nižegorodskij Gosudarstvennyj Istoriko-architekturnyj Muzej-zapovednik, Nižnij Novgorod 37468
Novgorodskij Gosudarstvennyj Muzej-zapovednik, Velikij Novgorod 37922
Petergof - Gosudarstvennyj Muzej Zapovednik, Sankt-Peterburg 37718
Plesskij Gosudarstvennyj Istoriko-Architekturnyj i Chudožestvennyj Muzej-Zapovednik, Ples 37568
Rjazanskij Gosudarstvennyj Istoriko-architekturnyj Muzej-zapovednik, Rjazan 37595
Šlisselburgskaja Krepost Orešek, Šlisselburg 37769
Soloveckij Gosudarstvennyj Istoriko-Architekturnyj i Prirodnyj Muzej-Zapovednik, Soloveckij 37788
Staročerkasskij Istoriko-architekturnyj Muzej-zapovednik, Staročerkassk 37798
Staroladožskij Istoriko-Architekturnyj i Archeologičeskij Muzej-Zapovednik, Staraja Ladoga 37795
Svijažskij Architekturno-chudožestvennyj Muzej, Svijažsk 37812
Svjatogorskij Monastyr, Puškinskie Gory 37588
Taganrogskij Gosudarstvennyj Literaturnyj i Istoriko-architekturnyj Muzej-zapovednik, Taganrog 37828
Velikoustjugskij Gosudarstvennyj Istoriko-Architekturnyj i Chudožestvennyj Muzej-Zapovednik, Velikij Ustjug 37923
Vologodskij Gosudarstvennyj Istoriko-Architekturnyj i Chudožestvennyj Muzej-Zapovednik, Vologda 37964

Slovakia

Dom L'udového Bývania, Matúškovo 38303
Dom L'udového Bývania, Šala 38347
Expozícia Čičmany, Čičmany 38246
Expozícia L'udovej Architektúry a Bývania (Skanzen), Humenné 38261
Expozícia L'udovej Archtektúry a Bývania, Martovce 38302
Múzeum Kysuckej Dediny, Nová Bystrica 38319
Múzeum Oravskej Dediny, Zuberec 38393
Skanzen - Múzeum L'udovej Architektúry, Bardejovské Kúpele 38197

Slovenia

Arhitekturni Muzej Ljubljana, Ljubljana 38444
Jakopičeva Galerija, Ljubljana 38449
Plečnikova Zbirka, Ljubljana 38460

Spain

Casa-Museu Gaudí, Barcelona 38953
Conjunt Monumental de la Esglésias de Sant Pere, Terrassa 40117
Monasterio de Santa Clara de Tordesillas, Tordesillas 40146
Museo de la Colegiata de Santa Juliana y Claustro, Santillana del Mar 40033
Museo Diocesano de Jaca, Jaca 39418
Museo Nacional de Arquitectura, Madrid 39568
Museu d'Arquitectura de la Real Càtedra Gaudí, Barcelona 38971

Poble Espanyol, Palma de Mallorca 39794
Real Cartuja de Miraflores, Burgos 39073

Sweden

Arkitekturmuseet, Stockholm 40884
Friluftmuseet Hallandsgården, Halmstad 40577
Friluftmuseet Skansen, Stockholm 40895
Gunnebo Slott och Trädgårdar, Mölndal 40742
Hembygdsgården Rots Skans, Älvdalen 40406

Switzerland

Architekturforum Zürich, Zürich 42065
Architekturgalerie Luzern, Ausstellungsraum Partikel, Luzern 41588
Archives de la Construction Moderne, Lausanne 41534
Baugeschichtliches Archiv der Stadt Zürich, Zürich 42067
Espace des Inventions, Lausanne 41537
Grubenmann-Sammlung, Teufen 41932
Ortsmuseum Eglisau, Eglisau 41341
SAM - Schweizerisches Architekturmuseum, Basel 41180
Swissminiatur, Melide 41622

Taiwan

Ami Cultural Center, Chengkung 42161
Beitou Hot Spring Museum, Taipei 42376
Chinese Ancient Architecture Museum, Yangmei 42463
Liohuo Aboriginal Stone-roofed Hut Museum, Neipu 42285
Miniatures Museum of Taiwan, Taipei 42404

Thailand

Gallery of the Faculty of Architecture, Bangkok 42540
The Jim Thompson House, Bangkok 42543
Kamthieng House, Bangkok 42544

Tunisia

Musée de l'Architecture Traditionnelle, Sfax 42651

Ukraine

Černivec'kyj Muzej Narodnoï Architektury ta Pobutu, Černivci 43014
Muzej Narodnoï Architektury i Pobutu u Lvovi, Lviv 43134
Muzej Narodnoï Architektury ta Pobutu Ukraïny, Kyïv 43098
Zakarpats'kyj Muzej Narodnoï Architektury ta Pobutu, Užgorod 43199

United Kingdom

Ancient High House, Stafford 45728
Blackwell - The Arts and Crafts House, Bowness-on-Windermere 43571
Bramall Hall, Bramhall 43590
Building of Bath Museum, Bath 43403
Carew Manor and Dovecote, Beddington 43432
Design Museum, London 44837
Farleigh Hungerford Castle, Farleigh Hungerford 44214
Furness Abbey, Barrow-in-Furness 43389
Glasgow School of Art - Mackintosh Collection, Glasgow 44283
Guildhall, Leicester 44675
Hill House, Helensburgh 44438
House of Dun, Montrose 45171
Hughenden Manor, High Wycombe 44465
Kufa Gallery, London 44897
Loseley House, Guildford 44373
Mellerstain House, Gordon 44331
Museum of Domestic Design and Architecture, Barnet 43382
Newark Town Treasures and Art Gallery, Newark-on-Trent 45213
Old Merchant's House and Row 111 Houses, Great Yarmouth 44359
Osterley Park House, Isleworth 44542
Pitzhanger Manor-House and Gallery, London 44955
Sainsbury Centre for Visual Arts, Norwich 45285
Scotland Street School Museum, Glasgow 44302
Sir John Soane's Museum, London 44987

United States of America

Archibald Graham McIlwaine House, Petersburg 51796
Bellamy Mansion Museum of History and Design Arts, Wilmington 54199
Bronck Museum, Coxsackie 47863
C. Kermit Ewing Gallery of Art and Architecture, Knoxville 49813
Chicago Architecture Foundation, Chicago 47462
The Chicago Athenaeum - Galena, Galena 48826
Colonel Ashley House, Ashley Falls 46393
The Dana-Thomas House, Springfield 53215
Edison and Ford Winter Estates, Fort Myers 48661
Fairbanks House, Dedham 48024
Frank Lloyd Wright Museum, Richland Center 52308
Frank Lloyd Wright's Pope-Leighey House, Alexandria 46204
Frederick Law Olmsted Collection, Brookline 47062
The Gardner Museum of Architecture and Design, Quincy 52213
Glessner House Museum, Chicago 47478
Grant-Humphreys Mansion, Denver 48079
Gropius House, Lincoln 50092
Hancock Shaker Village, Pittsfield 51968
Hermann-Grima House, New Orleans 51082
Historic Lyme Village, Bellevue 46704
The House of the Seven Gables, Salem 52656
Koreshan Historic Site, Estero 48425
Lace House Museum, Black Hawk 46840
MAK Center for Art and Architecture, Los Angeles 50250
MAK Center for Art and Architecture, West Hollywood 54064
Museum of American Architecture and Decorative Arts, Houston 49396
National Building Museum, Washington 53896
The Octagon Museum, Washington 53912
Price Tower Arts Center, Bartlesville 46619
Providence Preservation Society, Providence 52175
Pullman Museum, Chicago 47509
Reshaw Exhibit, Evansville 48458
Ruthmere House Museum, Elkhart 48348
Schifferstadt Architectural Museum, Frederick 48760
Sed Gallery, Athens 46414
Skyscraper Museum, New York 51256
Spring Mill State Park Pioneer Village and Grissom Memorial, Mitchell 50760
Stone House Museum, Belchertown 46693
Strater Hotel, Durango 48213
Union of American Hebrew Congregation, New York 51275
UrbanArts Institute, Boston 46953
Van Cortlandt House Museum, Bronx 47053
William Whitley House, Stanford 53254
Wilton House Museum, Richmond 52344
The Wolfsonian, Miami Beach 50644

Venezuela

Museo de Arquitectura, Caracas 54539

Vietnam

Crazy House Museum, Đà Lat 54602
National Museum of Cham Sculpture, Da Nang 54604

Arms and Armour

Argentina

Museo de Armas de la Nación, Buenos Aires 00179
Museo de Instituto Magnasco, Gualeguaychú 00383
Museo de la Policía del Chaco, Resistencia 00523

Australia

Firearms Technology Museum, Orange 01394

Austria

Büchsenmacher- und Jagdmuseum, Ferlach 01887
Burg Kreuzenstein, Korneuburg 02217
Hexenmuseum, Riegersburg, Steiermark 02587
Hofjagd- und Rüstkammer, Wien 03004
Landeszeughaus am Landesmuseum Joanneum, Graz 01977
Museum der Stadt Steyr, Steyr 02770
Privatmuseum "Feuerwaffen einst und jetzt" im Reckturm, Wiener Neustadt 03098
Schloß Ambras, Innsbruck 02130
Schützenscheibenmuseum, Scheibbs 02703
Waffensammlung Willibald Folger, Bruck an der Mur 01806

Azerbaijan

Artsakh State Museum, Stepanakert 03197

Belgium

Le Fort et son Musée, Flémalle 03601
Moervarststede, Wachtebeke 04124
Musée d'Armes de Liège, Liège 03804
Musée d'Armes et d'Historie Militaire, Tournai 04084
Musée de la Compagnie Royale des Anciens Arquebusiers de Visé, Visé 04117
Musée de la Porte de Hal-Hallepoortmuseum, Bruxelles 03469
Musée du Fort de Huy, Huy 03724
Musée Français - 1e Armée Française Mai 1940, Cortil-Noirmont 03527
Museum van de Kruisbooggilde Sint-Joris, Brugge 03422

Brazil

Casa de Osório, Rio de Janeiro 04685
Museu Aeroespacial, Rio de Janeiro 04697
Museu de Armas Ferreira da Cunha, Petrópolis 04628
Museu do Instituto Histórico, Carpina 04405
Museu Histórico de Sergipe, São Cristóvão 04827
Museu Histórico do Exército e Forte de Copacabana, Rio de Janeiro 04750
Museu José Bonifácio, São Paulo 04905
Museu Navio Comandante Bauru, Rio de Janeiro 04758
Museu Particular de Armas de Arlindo Pedro Zatti, Porto Alegre 04653

Brunei

Royal Brunei Armed Forces Museum, Bandar Seri Begawan 04968

Canada

Army Museum, Halifax 05930
Elmont Evergreen Firearms Museum, Belmont05419
Lieutenant General Ashton Armoury Museum, Victoria 07251
Murney Tower Museum, Kingston 06086
Musée les Voltigeurs de Québec Manège Militaire, Québec 06693
Royal Military College of Canada Museum, Kingston 06090
Royal Winnipeg Rifles Museum, Winnipeg 07377

China

Chinese Museum of Military Weaponary, Tongzhou 08304
Shanghai Museum of Public Security, Shanghai 08188

Croatia

Etnografski Muzej Split, Split 08955

Cuba

Casa de los Arabes, La Habana 09096
Complejo Museos Históricos Militares, La Habana 09100
Museo Armería 9 de Abril, La Habana 09111
Museo de la Revolución, La Habana 09137

Czech Republic

Východočeské Muzeum v Pardubicích, Pardubice 09707

Denmark

Dragon- og Frihedsmuseet, Holstebro 10089
Hjemmeværnsmuseet, Padborg 10228
Krudttaarnsmuseet, Frederikshavn 10027

Dominican Republic

Museo de las Casas Reales, Santo Domingo 10351

Egypt

Military Museum, Cairo 10500

Estonia

Rakvere Linnus-Muuseum, Rakvere 10611

Ethiopia

War Museum, Addis Ababa 10681

Finland

Asesepän Paja, Ilomantsi 10852
Ilmatorjuntamuseo, Tuusula 11414

France

Château du Haut-Koenigsbourg, Orschwiller 14541
Maison de la Manufacture d'Armes Blanches, Klingenthal 13304
Musée d'Armes Anciennes, La Cluse-et-Mijoux13347
Musée de la Guerre au Moyen Âge, Castelnaud-la-Chapelle 12313
Musée de la Pierre à Fusil, Luçay-le-Male 13863
Musée du Valois et de l'Archerie, Crépy-en-Valois 12673
Musée International des Hussards, Tarbes 15863
Musée Régional d'Armes, Mutzig 14344

Germany

Bartholomäus-Schmucker-Heimatmuseum, Ruhpolding 21445
Bayerisches Armeemuseum, Ingolstadt 19339
Burg Meersburg GmbH, Meersburg 20247
Burg Neuhaus, Wolfsburg 22660
Burg Stolpen, Stolpen 21911
Burgmuseum Ortenberg, Ortenberg, Hessen 20914
Fürstlich Hohenzollernsche Sammlungen, Sigmaringen 21763
Gräfliche Sammlungen Schloss Erbach, Erbach, Odenwaldkreis 18191
Museum im Zwinger, Goslar 18713
Rüstkammer, Dresden 17967
Schloßmuseum Mespelbrunn, Mespelbrunn 20303
Schützenmuseum der Königlich privilegierten Feuerschützengesellschaft Weilheim, Weilheim in Oberbayern 22391
Schützenscheibensammlung, Mainbernheim 20135
Städtisches Museum, Aschersleben 16596
Trillarium, Cleebronn 17710
Waffenmuseum, Oberndorf am Neckar 20820
Waffenmuseum Suhl, Suhl 21977
Wehrtechnische Studiensammlung des Bundesamtes für Wehrtechnik und undBeschaffung (BWB), Koblenz 19591

Greece

Tanks Museum, Avlona 22956

India

Fort Saint George Museum, Chennai 24114
Maharaja Banaras Vidya Mandir Museum, Varanasi 24427
Victoria Memorial Hall, Kolkata 24283

Ireland

Heraldic Museum, Dublin 24795

Italy

Armeria Reale, Torino 27997
Civico Museo di Guerra per la Pace Diego de Henriquez, Trieste 28081
Collezioni d'Armi e Ceramiche della Rocca Sforzesca, Imola 26399
Museo Civico, Pizzighettone 27260
Museo Civico delle Armi L. Marzoli, Brescia 25534
Museo dei Trombini di Lessinia, San Bortolo 27617
Museo delle Armi Antiche, Martinsicuro 26614
Museo delle Armi Antiche, Offagna 26971
Museo di Rievocazione Storica, Mondavio 26776
Museo di San Michele, Sagrado 27586
Museo Fioroni, Legnago 26480
Museo Nazionale del Palazzo di Venezia, Roma 27514
Museo Nazionale dell'Arma della Cavalleria, Pinerolo 27224
Museo Nazionale di Castel Sant'Angelo, Roma 27517
Museo Stibbert, Firenze 26198
Museo Storico dela Caccia e del Territorio, Cerreto Guidi 25812
Museo Storico Italiano della Guerra, Rovereto 27571
Palazzo Ducale, Venezia 28205

Japan

Iwakuni-shi Bijutsukan no Kashiwara Shuu, Iwakuni 28608
Kurashiki-shi Nihon Token Bijutsukan, Kurashiki 28786
Museum of Antique Armour Helmets and Swords, Kyoto 28852
Oyamazumi-jinja Kokuhokan, Omishima 29116
Tanegashima Teppokan, Nishinomote 29076
Token Hakubutsukan, Tokyo 29487

Malta

The Palace Armoury, Valletta 30645

Mexico

Museo de Historia de Lampazos de Naranjo y las Armas Nacionales, Lampazos de Naranjo 30993

Morocco

Belghazi-Musée, Fès 31669
Musée d'Armes du Bordj Nord, Fès 31670

Nepal

National Museum of Nepal, Kathmandu 31776

Netherlands

Bussemakerhuis, Borne 31989

A

Nigeria

Kanta Museum Argungu, Argungu 33315

Norway

Galtebosamlingen, Spydeberg 33858
Kongsberg Våpenfabrikks Museum, Kongsberg 33601

Peru

Museos Oro del Perú - Armas del Mundo, Lima 34273

Poland

Muzeum Militariów Arsenał Miejski, Wrocław 35333
Muzeum Okręgowe, Zamość 35362
Muzeum Pierwszych Piastów na Lednicy, Lednogóra 34846
Muzeum Regionalne, Pyzdry 35069
Muzeum Wojska Polskiego, Warszawa 35280
Muzeum Zbrojownia, Liw 34858

Portugal

Museu Militar de Lisboa, Lisboa 35713
Museu Sala de Armas, Almeida 35417

Russia

Centralnyj Muzej Vnutrennych Vojsk MVD Rossii, Moskva 37150
Centralnyj Voenno-morskoj Muzej, Sankt-Peterburg 37632
Memorialnyj Kompleks 'Vorošilovskaja Batareja', Vladivostok 37940
Muzej Archeologii i Ėtnografii, Vladivostok 37943
Tulskij Gosudarstvennyj Muzej Oružija, Tula 37873
Vladivostokskaja Krepost - Voenno istoričeskij Fortifikacionnyj Muzej, Vladivostok 37951
Voenno-istoričeskij Muzej Tichookeanskogo Flota, Vladivostok 37952

San Marino

Museo delle Armi Antiche, San Marino 38004
Museo delle Armi Moderne, San Marino 38005

Serbia

Muzej Stara Livnica, Kragujevac 38107
Vojni Muzej Beograd, Beograd 38091

Slovakia

Expozícia Zbraní a Stredovekého Mestského Opevnenia, Bratislava 38208

Slovenia

Muzej Novejše Zgodovine Slovenije, Ljubljana 38455
Vojni Muzej, Logatec 38472

South Africa

Fort Beaufort Museum, Fort Beaufort 38620
Messina Museum, Messina 38688
Museum of Coast and Anti-Aircraft Artillery, Green Point 38636
R.E. Stevenson Museum, Colenso 38595

Spain

Alcázar de Segovia, Segovia 40048
Museo de Almería, Almería 38861
Museo de Armería de Álava, Vitoria-Gasteiz 40320
Museo de la Academia de Infantería, Toledo 40132
Museo de la Real Maestranza de Caballería de Ronda, Ronda 39916
Museo de Tabar, Tabar 40094
Museo Específico de la Academia de Caballería, Valladolid 40239
Museo Naval de Cartagena, Cartagena 39140
Museu Militar, Barcelona 38999

Sweden

Livrustkammaren, Stockholm 40909
Vapentekniska Samlingarna, Eskilstuna 40505

Switzerland

Château de Grandson, Grandson 41450
Kantonales Museum Altes Zeughaus, Solothurn 41883
Musée d'Histoire, La Neuveville 41684
Musée Militaire Vaudois, Morges 41644
Ortsmuseum Sust, Horgen 41488
Waffenkammer im Munot-Turm, Schaffhausen 41832

Taiwan

Chi Mei Museum, Rende 42304
Cingyun Sword Craft Museum, Shalu 42320
Dow Sheng Chinese Kuoshu Weapons Museum, Tamsui 42448
Kuo Chang-Hsi Knife and Sword Museum, Jiading 42226

Ukraine

Nacional'nyj Muzej Istorii Velykoï Vitčyznjanoï Vijny 1941-1945 Rokiv, Kyïv 43104
Nacional'nyj Muzej Zbrojnych Syl Ukraïny, Kyïv 43108

United Kingdom

Cobbaton Combat Collection, Chittlehampton 43838
Dean Castle, Kilmarnock 44578
Delgatie Castle, Turriff 45882
Firepower - Royal Artillery Museum, London 44848
Fleet Air Arm Museum - Concorde, Yeovilton 46097
Fort Brockhurst, Gosport 44333
Havant Museum, Havant 44420
Hexham Old Gaol, Hexham 44463
Infantry and Small Arms School Corps Weapons Collection, Warminster 45933
Preston Hall Museum, Stockton-on-Tees 45757
Rifles (Berkshire and Wiltshire) Museum, Salisbury 45605
Ripley Castle, Ripley, North Yorkshire 45498
Royal Armouries, London 44968
Royal Armouries at Fort Nelson, Fareham 44211
Royal Armouries Museum, Leeds 44665
Royal Marines Museum, Portsmouth 45436
Warwickshire Yeomanry Museum, Warwick 45943
West Gate Towers, Canterbury 43746

United States of America

Daisy Airgun Museum, Rogers 52439
Dane G. Hansen Memorial Museum, Logan 50160
Dixie Gun Works' Old Car Museum, Union City 53653
General Douglas L. McBride Museum, Roswell 52468
Higgins Armory Museum, Worcester 54298
J.M. Davis Arms and Historical Museum, Claremore 47578
National Firearms Museum, Fairfax 48478
Remington Firearms Museum, Ilion 49467
Saunders Memorial Museum, Berryville 46769
UDT-SEAL Museum, Fort Pierce 48669

Vatican City

Museo Storico Vaticano, Città del Vaticano 54513

Yemen

Military Museum, Aden 54642

Zimbabwe

Zimbabwe Military Museum, Gweru 54667

Art

Afghanistan

National Art Gallery, Kabul 00004

Albania

Galera Kombëtare e Arteve Figurative, Tiranë 00030

Algeria

Musée de Skikda, Skikda 00076
Musée de Timgad, Timgad-Batna 00080
Musée National Cirta de Constantine, Constantine 00057
Musée National des Beaux-Arts d'Alger, Alger 00046

American Samoa

Jean P. Haydon Museum, Pago Pago 00084

Argentina

Museo Anconetani del Acordeón, Buenos Aires 00165
Museo Casa del Teatro, Buenos Aires 00174
Museo de Arte E. Minnicelli, Río Gallegos 00530
Museo de Arte Gauchesco La Recova, San Antonio de Areco 00564
Museo de Bellas Artes Domingo Faustino Sarmiento, Mercedes 00460
Museo de Calcos y Esculturas Comparda Ernesto de La Cárcova, Buenos Aires 00188
Museo de Historia Reginal San Lorenzo y Galería del Libertador General San Martín, San Lorenzo 00590
Museo Fundación Naum Komp, Buenos Aires 00234
Museo Histórico Municipal Brigadier General Juan Martín de Pueyrredón, San Isidro 00581
Museo Judío de Buenos Aires Dr. Salvador Kibrick, Buenos Aires 00249
Museo Molino de Papel La Villa, Buenos Aires 00254
Museo Municipal de Bellas Artes, Bahía Blanca 00136
Museo Municipal de Bellas Artes, La Calera 00397
Museo Municipal de Bellas Artes, La Paz 00405
Museo Municipal de Bellas Artes, Tandil 00642
Museo Municipal de Bellas Artes Prof. Amelio Ronco Ceruti, Metán 00464
Museo Pío Collivadino, Banfield 00140
Museo Provincial de Bellas Artes, Salta 00560
Museo Provincial de Bellas Artes Dr. Juan Ramon Vidal, Corrientes 00349
Museo Provincial de Bellas Artes Emiliano Guiñazu - Casa de Fader, Luján de Cuyo 00438
Museo Provincial de Bellas Artes Ramón Gómez Cornet, Santiago del Estero 00631
Museo Provincial de Bellas Artes Timoteo Eduardo Navarro, San Miguel de Tucumán 00601
Museo Regional y de Arte Marino, Puerto San Julián 00508
Museo Universitario de Artes, Mendoza 00459
Museo y Archivo Dardo Rocha, La Plata 00415
Pinacoteca de la Escuela Provincial de Bellas Artes Dr. Figueroa Alcorta, Córdoba 00340

Armenia

H.K. Buniatyan Art Gallery, Gavar 00700
Cafesjan Museum, Yerevan 00729
Children's Art Gallery, Yerevan 00730
Echmiadzin Art Gallery, Echmiadzin 00694
National Art Gallery of Armenia - Sisian Branch, Sisian 00716
National Gallery of Armenia, Yerevan 00751
Yeghegnadzor Art Gallery, Yeghegnadzor 00723
Yegishe Charents Literature and Arts Museum, Yerevan 00760

Australia

Albury Regional Art Gallery, Albury 00792
Art Gallery of South Australia, Adelaide 00773
Art Gallery of Western Australia, Perth 01410
Artspace Mackay, Mackay 01246
Atwell House Gallery, Alfred Cove 00795
Ballarat Fine Art Gallery, Ballarat 00825
Benalla Art Gallery, Benalla 00857
Bundaberg Arts Centre, Bundaberg 00917
Cunningham Dax Collection, Parkville, Victoria 01406
Deakin University Art Gallery, Burwood 00927
Drill Hall Gallery, Acton 00768
Flinders University Art Museum, Bedford Park 00846
Flinders University City Gallery, Adelaide 00776
Fremantle Arts Centre, Fremantle 01092
Gallery M, Oaklands Park 01388
The George Adams Gallery, Melbourne 01280
George Paton Gallery, Victoria 01618
Glen Eira City Council Gallery, Caulfield South 00962
Goethe-Institut Sydney, Woollahra 01669
Hamilton Art Gallery, Hamilton, Victoria 01144
Holmes à Court Gallery, East Perth 01055
Kingaroy Shire Art Gallery, Kingaroy 01200
La Trobe University Art Museum, Bundoora 00922
May and Herbert Gibbs Art Collection, South Perth 01516
Mildura Arts Centre, Mildura 01300
Museum of Fine Arts, Hobart 01155
Narrogin Gallery, Narrogin 01353
National Gallery of Australia, Canberra 00946
National Gallery of Victoria, Melbourne 01288
Newcastle Region Art Gallery, Newcastle 01364
Plimsoll Gallery, Hobart 01156
Port Community Art Centre, Port Adelaide 01423
Queen Victoria Museum and Art Gallery at Royal Park, Launceston 01224
RMIT Gallery RMIT University's Art Collection, Melbourne 01291
Salamanca Arts Centre, Hobart 01157
Toowoomba Regional Art Gallery, Toowoomba 01594
Uleybury School Museum, One Tree Hill 01391
University of Queensland Art Museum, Brisbane 00901
University of Sydney Art Collection, Sydney 01563
Vancouver Arts Centre, Albany 00789
Victorian College of the Arts Gallery, Southbank 01519
Wagga Wagga Art Gallery, Wagga Wagga 01620
Walker Street Gallery, Dandenong 01024
Wangaratta Exhibitions Gallery, Wangaratta 01627
War Memorial Gallery of Fine Arts, Sydney 01565

Austria

Alpen-Adria-Galerie im Stadthaus, Klagenfurt 02180
Alte Galerie am Landesmuseum Joanneum, Graz 01961
Anatomisches Museum, Innsbruck 02112
Atelierhaus der Akademie der bildenden Künste Wien, Wien 02936
Bank Austria Kunstforum und Kunstsammlung, Wien 02942
Basis Wien, Wien 02943
Belvedere - Oberes Belvedere, Wien 02951
Belvedere - Unteres Belvedere, Wien 02952
Dengel-Galerie, Reutte 02582
Galerie der Stadt Vöcklabruck, Vöcklabruck 02864
Galerie im 44er Haus, Leonding 02292
Galerie im Taxispalais, Innsbruck 02117
Galerie Kulturzentrum Lerchhaus, Eibiswald 01848
Galerie Museum Rabalderhaus, Schwaz 02724
Galerie Sala Terrena, Mödling 02392
Galerie Stephanie Hollenstein, Lustenau 02341
Gemeindegalerie im Alten Rathaus, Wiener Neudorf 03094
Gironcoli Museum, Sankt Johann bei Herberstein 02655
Haenel-Pancera-Familienmuseum, Bad Ischl 01752
Hermesvilla, Wien 03003
K-HOF-Galerie, Gmunden 01950
Kaiservilla, Bad Ischl 01753
Kubin-Haus, Wernstein am Inn 02925
Kunsthalle Krems, Krems 02227
Kunsthalle Leoben, Leoben 02285
Kunsthaus Graz am Landesmuseum Joanneum, Graz 01976
Kunsthaus Köflach, Köflach 02211
Kunsthaus Mürzzuschlag, Mürzzuschlag 02413
Kunsthistorisches Museum Wien, Wien 03019
Kunstraum Sankt Virgil, Salzburg 02617
Kunstsammlungen des Stiftes Kremsmünster, Kremsmünster 02233
Moderne Galerie im Weinstadtmuseum Krems, Krems 02229
Museum der Marktgemeinde Sankt Johann in Tirol, Sankt Johann in Tirol 02657
Museum der Moderne Salzburg Mönchsberg, Salzburg 02620
Museum der Stadt Bad Ischl, Bad Ischl 01755
Museum des Nötscher Kreises, Nötsch im Gailtal 02451
Museum im Schottenstift, Wien 03039
Neidhart Fresken, Wien 03044
Niederösterreichisches Landesmuseum, Sankt Pölten 02685
Prof. Eidenberger-Museum, Niederwaldkirchen 02449
Red Bull Hangar-7, Salzburg 02623
RLB Kunstbrücke, Innsbruck 02129
Salzburg Museum, Salzburg 02625
Sammlung Friedrichshof, Zurndorf 03133
Sammlungen der Universität für angewandte Kunst Wien mit Oskar-Kokoschka-Zentrum, Wien 03065
Sammlungen Schloß Moosham, Unternberg 02841
Schloß Burgau, Burgau 01808
Schloß Farrach, Zeltweg 03130
Schloss Halbturn, Halbturn 02041
Schloß-Museum, Reidling 02574
Schubert Geburtshaus, Wien 03072
Schützenmuseum im Kanzlerturm, Eggenburg 01845
Stadtgalerie am Minoritenplatz, Wolfsberg 03119
Stadtgalerie im alten Herrenhaus, Ternitz 02804
Stadtmuseum Dornbirn, Dornbirn 01829
Stadtmuseum Graz, Graz 01983
Stadtmuseum Klosterneuburg, Klosterneuburg 02204
Stiftssammlungen Wilten, Innsbruck 02133

Bangladesh

Dhaka City Museum, Dhaka 03222

Barbados

Portobello Gallery, Saint James 03240

Belarus

Nacionalnyj Chudožestvennyj Muzej Respubliki Belarus, Minsk 03255
Rėspublikanskaja Mastackaja Galerėja, Minsk 03256

Belgium

Andre Garitte Foundation, Berchem 03359
Bruggemuseum - Stadhuis, Brugge 03413
Centre Culturel des Roches, Rochefort 03972
Collection du Château d'Attre, Attre 03334
Design Museum Gent, Gent 03625
Fondation Folon, La Hulpe 03765
Groeningemuseum, Brugge 03414
Huis Hellemans, Edegem 03574
Internationaal Exlibriscentrum, Sint-Niklaas 04020
Kortrijk 1302, Kortrijk 03760
Kunstraum Bruxelles, Bruxelles 03449
Mad Musée, Liège 03798
Musée des Francs et de la Famenne, Marche-en-Famenne 03848
Musée d'Iconographie Historique, Onoz 03923
Musée du Bois de Maredret-Galerie A la Fontaine, Maredret 03853
Musée du Petit Format, Nismes 03916
Musée Gaspar, Arlon 03326

Musée Ribauri, Lasne-Chapelle-Saint-Lambert 03778
Museum Dr. Guislain, Gent 03634
The Museum of Museums, Waregem 04131
Museum Taf Wallet, Koksijde 03754
Museum voor Schone Kunsten Gent, Gent 03639
Ric's Art Boat, Bruxelles 03502
Roger Raveelmuseum, Machelen 03839
Salons voor Schone Kunsten, Sint-Niklaas 04023
Stadsmuseum STAM, Gent 03643
Stadsmuseum 't Schippershof, Menen 03868
Stedelijk Museum Aalst, Aalst 03259
Stedelijk Museum Hof van Busleyden, Mechelen 03862
Stedelijk Museum Ieper, Ieper 03731
Stedelijke Academie voor Schone Kunsten, Roeselare 03979
Tentoonstellingzaal Hessenhuis, Antwerpen 03317
Het Van Abbemuseum, Bonheiden 03380
Vleeshuis, Antwerpen 03318
Het Zotte Kunstkabinet, Mechelen 03865

Bermuda

Bermuda National Gallery, Hamilton 04192
Bermuda Society of Arts, Hamilton 04193
The Masterworks Foundation Collection, Paget 04195

Bolivia

Fundación de Fomento a las Artes, La Paz 04211
Galería de Arte Valerio Calles, Oruro 04229
Museo Nacional de Arte, La Paz 04223

Bosnia and Herzegovina

Galerija Brčko, Brčko 04261
Galerija Likovnih Umjetnosti Republike Srpske Banja Luka, Banja Luka 04256

Brazil

Acervo Artístico Cultural dos Palácios do Governo, São Paulo 04848
Casa di Xilogravura, Campos de Jordão 04397
Ciências Educação e Artes Luiz de Queiráz, Piracibaba 04630
Coleção de Artes Visuais, São Paulo 04853
Fundação Casa de Jorge Amado, Salvador 04787
Fundação Clóuis Salgado - Palácio das Artes, Belo Horizonte 04342
Fundação Eva Klabin Rappaport, Rio de Janeiro 04693
Fundaçao Oscar Niemeyer, Rio de Janeiro 04694
Galeria de Arte UFF, Rio de Janeiro 04695
Galeria de Arte UNICAMP, Campinas 04385
Galeria Prof. Sylvio Vasconcellos, Belo Horizonte 04343
Galeria Virtual da Casa das Rosas, São Paulo 04860
Museu Arquidiocesano de Mariana, Mariana 04560
Museu Carmen Miranda, Rio de Janeiro 04702
Museu da Arte Brasileira, Maceió 04543
Museu da Arte Pierre Chalita, Maceió 04544
Museu da Gravura Brasileira, Bagé 04325
Museu da Porcelana Senhor Adelino dos Santos Gouveia, Pedreira 04619
Museu de Arte, Belém 04335
Museu de Arte, Goiânia 04473
Museu de Arte, Joinville 04516
Museu de Arte, Londrina 04538
Museu de Arte Assis Chateaubriand, Campina Grande 04383
Museu de Arte Brasileira, São Paulo 04878
Museu de Arte Contemporânea de Pernambuco, Olinda 04596
Museu de Arte Contemporônea, Curitiba 04433
Museu de Arte da Pampulha, Belo Horizonte 04346
Museu de Arte da Universidade Fedral do Ceará, Fortaleza 04463
Museu de Arte de Santa Catarina, Florianópolis 04455
Museu de Arte do Paranà, Curitiba 04434
Museu de Arte do Rio Grande do Sul Aldo Malagoli, Porto Alegre 04644
Museu de Arte e Cultura Canjira, Itaobim 04498
Museu de Arte e História, Natal 04583
Museu de Arte e Histórico, Jaboticabal 04505
Museu de Arte Moderna, Resende 04674
Museu de Arte Moderna do Rio de Janeiro, Rio de Janeiro 04722
Museu de Arte Sacra, São Cristóvão 04826
Museu de Arte Sacra, São João Del Rei 04835
Museu de Arte Sacra, São Luís 04843
Museu de Arte Sacra, São Sebastião 04913
Museu de Arte Sacra, Taubaté 04924
Museu de Arte Sacra de Angra dos Reis, Angra dos Reis 04307
Museu de Arte Sacra de Boa Morte, Goiás 04480
Museu de Arte Sacra de Iguape, Iguape 04490
Museu de Arte Sacra do Carmo / Paróquia do Pilar, Ouro Preto 04603
Museu de Arte Sacra do Estado de Alagoas, Marechal Deodoro 04559
Museu de Arte Sacra do Pará, Belém 04336
Museu de Artes Plásticas Quirino da Silva, Mococa 04568
Museu de Artes Sacras Iria Josepha da Silva, Mococa 04569
Museu de Artes Visuais, São Luís 04844
Museu de Artes Visuais Ruth Schneider, Passo Fundo 04612
Museu de Belas-Artes, Cataguases 04409
Museu de Ex-votos do Senhor do Bonfim, Salvador 04799
Museu de Images do Inconsciente, Rio de Janeiro 04728
Museu di Arte Contemporânea, Americana 04305
Museu Histórico Farroupilha, Piratini 04634
Museu Internacional de Arte Naïf do Brasil, Rio de Janeiro 04754
Museu Lasar Segall, São Paulo 04906
Museu Nacional de Belas Artes, Rio de Janeiro 04756
Museu Paranaense, Curitiba 04442
Museu Pinacoteca, Igarassu 04489
Museu Universitário de Arte, Uberlândia 04945
Pinacoteca Barão de Santo Ângelo, Porto Alegre 04654
Pinacoteca Benedicto Calixto, Santos 04822
Pinacoteca Municipal Miguel Ângelo Pucci, Franca 04469
Sítio Roberto Burle Marx, Rio de Janeiro 04771

Brunei

Bangunan Alat Kebesaran Diraja, Bandar Seri Begawan 04961

Bulgaria

Chudožestvena Galerija, Balčik 04974
Chudožestvena Galerija, Blagoevgrad 04990
Chudožestvena Galerija, Chaskovo 05003
Chudožestvena Galerija, Drjanovo 05022
Chudožestvena Galerija, Gabrovo 05030
Chudožestvena Galerija, Gălăbovo 05034
Chudožestvena Galerija, Karlovo 05048
Chudožestvena Galerija, Kavarna 05053
Chudožestvena Galerija, Kazanlăk 05056
Chudožestvena Galerija, Levski 05082
Chudožestvena Galerija, Loveč 05085
Chudožestvena Galerija, Mezdra 05093
Chudožestvena Galerija, Provadia 05138
Chudožestvena Galerija, Provadia 05139
Chudožestvena Galerija, Radnevo 05143
Chudožestvena Galerija, Silistra 05161
Chudožestvena Galerija, Sozopol 05207
Chudožestvena Galerija, Velingrad 05259
Chudožestvena Galerija Darenie Kollekcia Svetlin Russev, Pleven 05115
Chudožestvena Galerija Dobrič, Dobrič 05016
Chudožestvena Galerija Ilija Beškov, Pleven 05116
Chudožestvena Galerija Kiril Petrov, Montana 05094
Chudožestvena Galerija Nikola Marinov, Targovište 05224
Chudožestvena Galerija Nikolaj Pavlovič, Svištov 05219
Chudožestvena Galerija "Petar Persengiev", Novi Pazar 05098
Chudožestvena Galerija Stanislav Dospevski, Pazardžik 05105
Gradska Chudožestvena Galerija, Berkovica 04986
Gradska Chudožestvena Galerija, Plovdiv 05127
Gradska Chudožestvena Galerija Petko Čurčuliev, Dimitrovgrad 05014
Kăšta-muzej Ivan Lazarov, Sofia 05177
Kăšta-muzej s Dom-pametnik Jordan Jovkov, Dobrič 05020
Muzej na Chudožestvenata Akademija, Sofia 05185
Nacionalna Chudožestvena Galerija, Sofia 05200
Nacionalna Galerija za Čuždestranno Izkustvo, Sofia 05201
Okrăžna Chudožestvena Galerija Dimităr Dobrovič, Sliven 05170
Regionalen Istoričeski Muzej Varna, Varna 05247

Canada

A.K.A. Gallery, Saskatoon 06910
Academy of Spherical Arts, Toronto 07090
Agnes Etherington Art Centre, Kingston 06073
Alcan Museum and Archives, Montréal 06326
The Alfoldy Gallery, Erikson 05781
Allen Sapp Gallery, North Battleford 06465
Armstrong-Spallumcheen Museum and Art Gallery, Armstrong 05354
Art Gallery of Algoma, Sault Sainte Marie 06928
Art Gallery of Greater Victoria, Victoria 07241
Art Gallery of Hamilton, Hamilton 05951
Art Gallery of Northumberland, Cobourg 05628
Art Gallery of Nova Scotia, Halifax 05931
Art Gallery of Southwestern Manitoba, Brandon 05458
Art Gallery of Swift Current National Exhibition Centre, Swift Current 07044
Art Gallery of Windsor, Windsor, Ontario 07344
Atelier d'Estampe Imago, Moncton 06314
Beaverbrook Art Gallery, Fredericton 05828
Brackendale Art Gallery Theatre Teahouse, Brackendale 05454
Burnaby Art Gallery, Burnaby 05501
Carleton University Art Gallery, Ottawa 06541
Cedar Ridge Creative Centre, Scarborough 06936
Centre Culturel Franco-Manitobain, Winnipeg 07356
Centre d'Art, Baie-Saint-Paul 05372
Centre d'Art de Saint-Georges, Saint-Georges, Quebec 06819
Centre d'Artistes Vaste et Vague, Carleton 05563
Centre d'Exposition Art-Image, Gatineau 05856
Centre d'Exposition de Rouyon-Noranda, Rouyn-Noranda 06780
Centre d'Exposition de Saint-Hyacinthe, Saint-Hyacinthe 06820
Centre d'Exposition de Val d'Or, Val-d'Or 07193
Centre d'Exposition Mont-Laurier, Mont-Laurier 06321
Centre d'Exposition Plein Sud, Longueuil 06192
Centre Missionnaire Sainte-Therèse Musée, Montréal 06335
Centre National d'Exposition, Jonquière 06031
Charles H. Scott Gallery, Vancouver 07207
City of Calgary-Civic Art Collection, Calgary 05517
Colville Lake Museum, Colville Lake 05633
Confederation Centre Art Gallery, Charlottetown 05591
Creative Spirit Art Centre, Toronto 07103
Cultural Centre of Dorval, Dorval 05709
Dalhousie Art Gallery, Halifax 05933
Dawson Creek Art Gallery, Dawson Creek 05683
Dr. Vernon L. Watson Allied Arts Centre, Dauphin 05678
Dunlop Art Gallery, Regina 06716
Eastern Edge Art Gallery, Saint John's 06847
Écomusée de Saint-Constant, Saint-Constant 06810
Edmonton Art Gallery, Edmonton 05743
Espace Virtuel, Chicoutimi 05611
Galerie 101, Ottawa 06545
Galerie d'Art de Matane, Matane 06247
Galerie d'Art du Parc, Trois-Rivières 07172
Galerie d'Art l'Union-Vie, Drummondville 05714
Galerie d'Art Stewart Hall, Pointe-Claire 06617
Galerie d'Art Stewart Hall, Pointe-Claire 06618
Galerie Georges-Goguen, Moncton 06316
Galerie Montcalm, Gatineau 05859
Galerie Renée-Blain, Brossard 05489
Gallerie Restigouche, Campbellton 05543
Gallery Arcturus, Toronto 07108
Glenbow Museum, Calgary 05519
Glendon Gallery, Toronto 07111
Godfrey Dean Art Gallery, Yorkton 07414
Gordon Snelgrove Art Gallery, Saskatoon 06914
Grunt Gallery, Vancouver 07212
Hart House, Toronto 07113
Hector Exhibit Centre, Pictou 06601
Henry-Stuart House, Québec 06681
Homer Watson House and Gallery, Kitchener 06101
Hudson Bay Museum, Hudson Bay 05999
Huronia Museum, Midland 06277
Les Impatients, Montréal 06340
Island Mountain Arts Public Gallery, Wells 07295
The Japanese Canadian Museum, Kaslo 06045
Justina M. Barnicke Gallery, Hart House, Toronto 07117
Kenderdine Art Gallery, Saskatoon 06915
Kitchener-Waterloo Art Gallery, Kitchener 06103
The Koffler Gallery, Toronto 07118
Langage Plus, Alma 05321
Langham Cultural Centre Galleries, Kaslo 06046
Latitude 53 Contemporary Visual Culture, Edmonton 05748
Leaf Rapids National Exhibition Centre, Leaf Rapids 06153
Leonard Bina Ellen Art Gallery, Montréal 06342
Library Gallery, Cambridge 05536
Macdonald Stewart Art Centre, Guelph 05919
McIntosh Gallery, London 06188
McMichael Canadian Art Collection, Kleinburg 06107
Maison André-Benjamin-Papineau, Laval 06147
Maison de la Culture Plateau-Mont-Royal, Montréal 06347
Maltwood Art Museum and Gallery, Victoria 07253
Market Gallery, Toronto 07122
Mendel Art Gallery, Saskatoon 06918
Mount Saint Vincent University Art Gallery, Halifax 05942
Musée Communautaire des Soeurs de Sainte-Croix, Montréal 06351
Musée d'Art Contemporarin des Laurentides, Saint-Jérôme 06830
Musée d'Art de Joliette, Joliette 06030
Musée d'Art de Mont-Saint-Hilaire, Mont-Saint-Hilaire 06322
Musée des Beaux-Arts de Montréal, Montréal 06357
Musée des Beaux-Arts de Sherbrooke, Sherbrooke, Québec 06962
Musée du Centre Élisabeth Bergeron, Saint-Hyacinthe 06821
Musée International d'Art Naïf Yvon-M. Daigle, Magog 06215
Musée Louis-Hémon, Péribonka 06586
Musée Marc-Aurèle Fortin, Montréal 06373
Musée National des Beaux-Arts du Québec, Québec 06694
Musée Régional de la Côte-Nord, Sept-Îles 06945
Museum London, London 06189
Nanaimo Art Gallery, Nanaimo 06412
National Gallery of Canada, Ottawa 06551
Native Cultural Arts Museum, Grouard 05915
Nelson District Museum, Nelson 06422
Niagara Falls Art Gallery - William Kurelek Art Collection, Niagara Falls 06446
Norfolk Arts Centre, Simcoe 06980
OCAD Student Gallery, Toronto 07125
Orillia Museum of Art and History, Orillia 06514
Osoyoos Art Gallery, Osoyoos 06527
Owens Art Gallery, Sackville 06788
Peel Heritage Complex, Brampton 05456
Les Petites Franciscaines de Marie, Baie-Saint-Paul 05373
Platform - Centre for Photographic and Digital Arts, Winnipeg 07373
Prairie Panorama Museum, Czar 05670
Praxis Art Actuel, Sainte-Thérèse 06902
Prince Edward Island Museum, Charlottetown 05595
Prince of Wales Northern Heritage Centre, Yellowknife 07411
La Pulperie de Chicoutimi/Musée du Saguenay-Lac-Saint-Jean, Chicoutimi 05612
Rail's End Gallery - Haliburton Highlands Guild of Fine Arts, Haliburton 05928
Red Brick Arts Centre and Museum, Edson 05766
River Brink, Queenston 06701
Robert McLaughlin Gallery, Oshawa 06524
Rodman Hall Arts Centre, Saint Catharines 06807
Rollin Arts Centre, Port Alberni 06627
Rowsell House Interpretation Center, Harrington Harbour 05972
Ruthven Park, Cayuga 05582
Saint Thomas-Elgin Public Art Centre, Saint Thomas 06880
Saint Thomas More Art Gallery, Saskatoon 06923
Salle Alfred Pellan, Laval 06149
SAW Gallery, Ottawa 06555
Seymour Art Gallery, North Vancouver 06481
Sidney and Gertrude Zack Gallery, Vancouver 07222
Signal Hill Arts Centre, Weyburn 07315
Smithers Public Art Gallery, Smithers 06984
Sooke Region Museum, Art Gallery and Historic Moss Cottage, Sooke 06991
Southampton Art Gallery, Southampton 06998
A Space Gallery, Toronto 07139
Springbank Visual Arts Centre, Mississauga 06313
Station House Studio and Gallery Society, Williams Lake 07338
Sunbury Shores Arts and Nature Centre, Saint Andrews 06801
Temiskaming Art Gallery, Haileybury 05926
Thames Art Gallery, Chatham, Ontario 05603
Thomas Fisher Rare Book Library, Toronto 07144
Toronto Dominion Gallery of Inuit Art, Toronto 07148
University of Lethbridge Art Gallery, Lethbridge 06164
University of New Brunswick Art Centre, Fredericton 05840
University of Toronto Art Centre, Toronto 07153
Visual Arts Centre, Westmount 07308
W.K.P. Kennedy Gallery, North Bay 06471
Walter Phillips Gallery, Banff 05382
Western Front, Vancouver 07231
Whetung Craft Centre and Art Gallery, Curve Lake 05668
Whyte Museum of the Canadian Rockies, Banff 05383
Woodstock Art Gallery, Woodstock 07405
Ydessa Hendeles Art Foundation, Toronto 07154
York Quay Gallery at Harbourfront Centre, Toronto 07156
YYZ Artists' Outlet, Toronto 07157

Chile

Museo de Arte y Artesanía de Linares, Linares 07441
Museo Municipal de Bellas Artes de Valparaíso, Valparaíso 07472
Museo O'Higginiano y de Bellas Artes de Talca, Talca 07466

China

Arthur M. Sackler Museum of Art and Archaeology, Beijing 07505
Beijing World Art Museum, Beijing 07523
Galeria de Exposiçōes Temporárias da CMMP, Macau 07992
Guangzhou Art Gallery, Guangzhou 07727
Hong Kong Visual Arts Centre, Hong Kong 07806
Jincheng-gu Construction Art Museum, Jincheng 07887
Jingsha Art Museum, Jingsha 07892

A

Jintai Art Museum, Beijing 07543
Namoc-National Art Museum of China, Beijing 07554
Nanjing International Culture and Arts Exchange Center, Nanjing 08044
National Museum of Chinese History, Beijing 07556
Qin Dai Zhan Guan, Lintong 07953
Qingyunpu Museum, Nanchang 08032
Shanghai Art Museum, Shanghai 08180
Suzhou Art Gallery, Suzhou 08254
Tianjin Art Museum, Tianjin 08283
Yanhuang Art Museum, Beijing 07575

Colombia

Colección de Arte y Museo Botero, Bogotá 08548
Collectión Banco de la República, Bogotá 08552
Museo Casa Mosquera, Popayán 08724
Museo Casa Negret, Popayán 08725
Museo de Arte, Santa Marta 08749
Museo de Arte de Pereira, Pereira 08719
Museo de Arte Universidad Nacional de Colombia, Bogotá 08561
Museo de Artes y Tradiciones Populares, Espinal 08639
Museo de la Cultura de Albán, Albán 08525
Museo de los Niños, Bogotá 08568
Museo de Marionetas, Bogotá 08569
Museo del Carajo, Cali 08611
Museo del Cobre, Bogotá 08576
Museo del Siglo XIX, Bogotá 08578
Museo del Virrey Ezpeleta, Guaduas 08651
Museo Etnográfico, Albán 08526
Museo Gabriel Mateus, Sabanagrande 08741
Museo Mercedes Sierra de Pérez El Chico, Bogotá 08588
Museo Rayo, Roldanillo 08740
Museo Taminango de Artes y Tradiciones Populares de Nariño, Pasto 08715
Museo Universitario, Medellín 08686
Museo Universitario, Pasto 08716
Pinacoteca de la Facultad de Bellas Artes, Manizales 08671
Pinacoteca Eduardo Ramírez Castro, Aranzazu 08531

Costa Rica

Galeria de la Escuela de Artes Plásticas, San José 08811
Galería Nacional y Museo de los Niños, San José 08813
Museo de Arte Costarricense, San José 08815
Sala de Exhibitiones Temporales del Banco Central, San José 08825

Côte d'Ivoire

Maison du Patrimoine Culturel, Grand-Bassam 08835

Croatia

Galerija Bukovac, Cavtat 08853
Galerija Grada Krapine, Krapina 08899
Galerija Izvorne Umjetnosti, Zlatar 09018
Galerija Klovićevi Dvori, Zagreb 08990
Galerija Vjekoslav Karas, Karlovac 08885
Gradska Galerija Labin, Labin 08905
Gradski Muzej Karlovac, Karlovac 08886
Gradski Muzej Virovitica, Virovitica 08978
Kaštela Gradski Muzej, Kaštel Lukšić 08887
Memorijalna Zbirka Eugena Kumičića, Brseč 08849
Muzej za Umjetnost i Obrt, Zagreb 09008
Strossmayerova Galerija Starih Majstora, Zagreb 09011
Umjetnički Paviljon u Zagrebu, Zagreb 09014
Zbirka Umjetnina Andra Vida i Katarine Mihičić i Zbirka Starij Majstora Piperata, Mali Lošinj 08909

Cuba

Centro de Desarrollo de las Artes Visuales, La Habana 09098
La Galería de Arte, La Habana 09104
Galería de Arte Universal, Santiago de Cuba 09303
Museo Acacia, La Habana 09109
Museo de Arte Colonial de la Habana, La Habana 09131
Museo de Arte Colonial de Palacio Valle Iznaga, Sancti-Spíritus 09286
Museo des Origenes, La Habana 09142
Museo Mariano, La Habana 09149
Museo Nacional de Bellas Artes, La Habana 09162
Museo Oscar María de Rojas, Cárdenas 09056
Museo Sol y Mar, Matanzas 09219
Museo Wifredo Lam, La Habana 09176

Czech Republic

Dům Umění, České Budějovice 09462
Dům Umění, Znojmo 09946
Dům Umění Města Brna, Brno 09425
Galerie Hlavního Města Prahy - Městská knihovna, Praha 09746
Galerie Jamborův Dům, Tišnov 09870
Galerie Melantrich, Rožďalovice 09822
Galerie Města Blanska, Blansko 09408
Galerie Městského Muzea Čáslav, Čáslav 09450
Galerie pod Zámkem, Rožmitál pod Třemšínem 09825
Galerie Zázvorka, Nové Město nad Metují 09680
Horácká Galerie, Nové Město na Moravě 09678
Hrad Roštejn, Telč 09860
Jihomoravské Muzeum ve Znojmě, Znojmo 09947
Jízdárna Pražského Hradu, Praha 09753
Městská Galerie, Polička 09728
Městská Galerie Art Club, Týn nad Vltavou 09884
Městská Galerie Beroun, Beroun 09403
Městské Muzeum, Rýmařov 09833
Mičovna, Praha 09761
Moravská Galerie Brno, Brno 09432
Muzeum Telč, Telč 09861
Muzeum Umění Olomouc, Olomouc 09689
Památník Lidice, Lidice 09627
Severočeská Galerie Výtvarného Umění v Litoměřicích, Litoměřice 09631
Slezské zemské muzeum, Opava 09693
Východočeská Galerie Pardubicích, Pardubice 09705
Výstavní Síň Muzea, Ústí nad Orlicí 09894
Výstavní Síň - Synagoga, Hranice 09537
Západočeská Galeria v Plzni, Plzeň 09719
Znojemský Hrad, Znojmo 09949

Denmark

Bornholms Kunstmuseum, Gudhjem 10046
Danmarks Keramikmuseum, Middelfart 10188
Drachmanns Hus, Skagen 10259
Frederikshavn Kunstmuseum- og Exlibrissamling, Frederikshavn 10026
Fyns Kunstmuseum, Odense 10213
Grenen Kunstmuseum, Skagen 10260
Holstebro Kunstmuseum, Holstebro 10090
Horsens Kunstmuseum, Horsens 10097
Jens Nielsens Olivia Holm-Møller Museet, Holstebro 10092
Johannes Larsen Museet, Kerteminde 10116
Kunsthallen Brandts, Odense 10217
Kunstmuseet Brundlund Slot, Aabenraa 09954
Kunstmuseet Køge Skitsesamling, Køge 10160
Laugø Smedie Museum, Helsinge 10061
Museet Psykiatrisk Hospital i Århus, Risskov 10238
Odsherreds Kunstmuseum, Asnæs 09969
Ribes Vikinger Museum, Ribe 10234
Rundetaarn, København 10151
Storstrøms Kunstmuseum, Maribo 10186
Vestsjællands Kunstmuseum, Sorø 10282

Dominican Republic

Museo de Arte Moderno, Santo Domingo 10343

Ecuador

Museo de Arte Colonial, Quito 10435
Museo de Arte Moderno, Cuenca 10385
Museo de Arte Moderno de la Casa de la Cultura Ecuatoriana, Quito 10437
Museo Frey Pedro Bedón, Ibarra 10414
Museo Municipal de Guayaquil, Guayaquil 10409

Egypt

Gayer Anderson Museum, Cairo 10496

Estonia

Adamson-Eric Museum, Tallinn 10619
Eesti Kunstimuuseum, Tallinn 10624
Kadrioru Loss Väliskunsti Muuseum, Tallinn 10632
Kunstigalerii, Narva 10597
Mikkeli Muuseum, Tallinn 10636
Näitustesaal Rotermanni Soolalaos, Tallinn 10637
Niguliste Muuseum, Tallinn 10638
Palmse Mõis, Viitna 10673
Rakvere Näitustemaja-Muuseum, Rakvere 10612
Tartu Kunstimaja, Tartu 10655
Tartu Ülikooli Kunstimuuseum, Tartu 10661

Ethiopia

Ethnographic Museum, Addis Ababa 10678

Faroe Islands

Listasavn Føroya, Tórshavn 10699

Finland

Aineen Taidemuseo, Tornio 11383
Ålands Konstmuseum, Mariehamn 11097
Ålands Museum, Mariehamn 11098
Ateneumin Taidemuseo, Helsinki 10769
Etelä-Karjalan Taidemuseo, Lappeenranta 11046
Helsingin Taidehalli, Helsinki 10783
Hvitträsk, Kirkkonummi 10963
Jyväskylän Taidemuseo, Jyväskylä 10890
Kemin Taidemuseo, Kemi 10940
Keravan Taidemuseo, Kerava 10947
Kuopion taidemuseo, Kuopio 11009
Lampi-Museo, Liminka 11073
Lönnströmin Taidemuseo, Rauma 11259
Maria de Lisitzin Fine Art and Historical Museum, Porvoo 11229
Mikkelin Taidemuseo, Mikkeli 11115
Museo Villa Urpo, Siivikkala 11311
Önningebymuseet, Mariehamn 11101
Orimattilan Taidemuseo, Orimattila 11165
Pohjanmaan Museo, Vaasa 11426
Porin Taidemuseo, Pori 11220
Pyynikinlinna, Tampere 11356
Reitzin Säätiön Kokoelmat, Helsinki 10813
Riihimäen Taidemuseo, Riihimäki 11269
Rovaniemen Taidemuseo, Rovaniemi 11279
Sinebrychoffin Taidemuseo, Helsinki 10817
Tampereen Taidemuseo - Pirkanmaan Aluetaidemuseo, Tampere 11364
Turun Taidemuseo, Turku 11405
Valtion Taidemuseo, Helsinki 10835
Vantaan Kaupunginmuseo - Kuvataideasiat, Vantaa 11450
Villa Gyllenberg, Helsinki 10836
Voipaalas Taidekeskus, Sääksmäki 11288

France

Atelier Hoguet, Paris 14568
Carré d'Art, Nîmes 14446
Le Carré-Scéne Nationale, Château-Gontier 12434
Centre d'Art Contemporain, Bignan 11993
Centre d'Art Contemporain, Dijon 12731
Centre d'Art Contemporain, Pougues-les-Eaux 14928
Centre d'Art Contemporain, Tanlay 15851
Centre d'Art Contemporain le Quartier, Quimper 14966
Centre d'Art du Crestet, Crestet 12674
Centre d'Art et d'Exposition Jacques-Henri Lartigue, L'Isle-Adam 13780
Centre International d'Art-Raymond du Puy, Le Poët-Laval 13636
Le Channel - Galerie de l'Ancienne Poste, Calais 12246
Château Beaufief, Mazeray 14049
Château de Maisons-Laffitte, Maisons-Laffitte 13925
Château de Rochefort-en-Terre, Rochefort-en-Terre 15074
Château des Pêcheurs, La Bussière 13322
Château-Musée des XVIe-XVIIe s, Tanlay 15852
Le Creux de l'Enfer, Centre d'Art Contemporain, Thiers 15880
Ecomusée de Saint-Dégan, Brec'h 12151
Ecomusée du Vignoble Drouais et des Artisans, Dreux 12791
Espace Giono, Lalley 13488
Espace Joseph Besset, Vanosc 16070
Exposition du Cloître des Cordeliers, Tarascon (Bouches-du-Rhône) 15854
Exposition Permanente d'Art, Brasparts 12145
Galerie des Ponchettes, Nice 14424
Galerie d'Exposition d'Art et d'Histoire, Villaries 16168
Galerie du Vitrail, Amiens 11585
Galerie Fernand Léger - Le Crédac, Ivry-sur-Seine 13256
Galerie Municipale Edouard Manet, Gennevilliers 13043
Institut d'Art Contemporain, Villeurbanne 16236
Institut Tessin, Paris 14612
La Piscine - Musée d'Art et d'Industrie André Diligent, Roubaix 15111
Le Carmel, Tarbes 15860
Maison de l'Amandier, Saint-Rémy-de-Provence 15486
Maison dite de François 1er, Aubigny-sur-Nère 11729
Maison et Atelier de Jean-François Millet, Barbizon 11846
Maison Lansyer, Loches-en-Touraine 13794
Maison-Musée des Sciences, Lettres et Arts de Cholet et de sa Région, Cholet 12521
Monasterie Royal de Brou, Bourg-en-Bresse 12107
Musée Archéologique de Touraine, Tours 15963
Musée Arts et Histoire, Bormes-les-Mimosas 12077
Musée Barrois, Bar-le-Duc 11844
Musée Bernard Buffet, Paris 14633
Musée Bernard d'Agesci, Niort 14455
Musée Bossuet, Meaux 14056
Musée Campredon, L'Isle-sur-la-Sorgue 13783
Musée-Chapelle Bellini, Cannes 12268
Musée Charles Milcendeau-Jean Yole, Soullans 15799
Musée-Château, Grignan 13149
Musée Château d'Annecy et Observatoire Régional des Lacs Alpins, Annecy 11625
Musée-Château de Saint-Ouen, Saint-Ouen 15433
Musée Chintreuil, Pont-de-Vaux 14883
Musée Communal Robert Tatin, Cossé-le-Vivien 12637
Musée d'Art et d'Archéologie, Laon 13521
Musée d'Art et d'Archéologie de Périgord, Périgueux 14778
Musée d'Art et d'Histoire des Côtes d'Armor, Saint-Brieuc 15191
Musée d'Art et d'Histoire Louis Senlecq, L'Isle-Adam 13781
Musée d'Art et d'Industrie, Saint-Étienne 15251
Musée d'Art Religieux, Marsal 13984
Musée d'Art Roger Quilliot, Clermont-Ferrand 12549
Musée de Borda, Dax 12701
Musée de Gajac, Villeneuve-sur-Lot 16220
Musée de la Castre, Cannes 12269
Musée de la Loire, Cosne-Cours-sur-Loire 12636
Musée de la Princerie, Verdun (Meuse) 16112
Musée de la Tour des Echevins, Luxeuil-les-Bains 13872
Musée de la Vigne et du Vin-Galerie d'Art, Lods 13801
Musée de l'Art Culinaire, Villeneuve-Loubet 16217
Musée de l'Erotisme, Paris 14669
Musée de l'Holographie, Paris 14672
Musée de l'hôtel de Berny, Amiens 11587
Musée de l'Hôtel-Dieu, Mantes-La Jolie 13944
Musée de l'île de France, Sceaux 15688
Musée de Lodève, Lodève 13800
Musée de Montmartre, Paris 14678
Musée de Moulages, Strasbourg 15826
Musée de Peinture Louis Français, Plombières-les-Bains 14845
Musée de Picardie, Amiens 11588
Musée de Saint-Germain-Laval, Saint-Germain-Laval 15293
Musée Départemental des Hautes-Alpes, Gap 13031
Musée Départemental Hébert, La Tronche 13453
Musée Départemental Stéphane Mallarmé, Vulaines-sur-Seine 16280
Musée des Arts Africains, Océaniens, Amérindiens, Marseille 14001
Musée des Arts et de l'Enfance, Fécamp 12913
Musée des Arts Populaires, Reillanne 14997
Musée des Beaux-Arts, Dijon 12741
Musée des Beaux-Arts, Montbard 14180
Musée des Beaux-Arts, Vannes (Morbihan) 16068
Musée des Beaux-Arts et Arts Décoratifs Raymond Ritter, Morlanne 14278
Musée des Beaux-Arts et d'Archeologie Joseph Déchelette, Roanne 15059
Musée des Beaux-Arts et de la Dentelle, Alençon 11545
Musée des Beaux-Arts et d'Histoire, Saint-Lô 15372
Musée des Hospices Civils de Lyon, Lyon 13890
Musée des Moulages, Lyon 13891
Musée des Peintres du Pays, Saint-Cyprien 15223
Musée des Tapisseries, Aix-en-Provence 11515
Musée d'Evreux, Evreux 12889
Musée du Gemmail, Lourdes 13845
Musée du Jeu de Paume, Versailles 16127
Musée du Prieuré, Charolles 12416
Musée du Touquet, Le Touquet-Paris-Plage 13655
Musée-Fondation Louis Jou, Les Baux-de-Provence 13693
Musée François Mitterrand, Jarnac 13260
Musée Galerie Honoré Camos, Bargemon 11853
Musée Girodet, Montargis 14173
Musée Gorsline, Bussy-le-Grand 12212
Musée Henri Boez, Maubeuge 14032
Musée Henri Giron, Le Vigan (Lot) 13667
Musée Hofer-Bury, Lavérune 13552
Musée International d'Art Naïf, Vicq 16148
Musée International d'Arts Modestes, Sète 15734
Musée J.-E. Blanche, Offranville 14497
Musée Jean-Charles Cazin, Samer 15617
Musée Jean de La Fontaine, Château-Thierry 12440
Musée Jean Lurçat, Saint-Laurent-les-Tours 15363
Musée Lombard, Doullens 12782
Musée Magnelli, Vallauris 16050
Musée Mainssieux, Voiron 16267
Musée-Maison des Arts des Hauts-Cantons, Bédarieux 11917
Musée M'Am Jeanne, Fontenoy-en-Puisaye 12974
Musée Mine, Cagnac-les-Mines 12235
Musée Municipal, Hazebrouck 13197
Musée Municipal, Semur-en-Auxois 15708
Musée Municipal Auguste Grasset, Varzy 16077
Musée Municipal de Cambrai, Cambrai 12257
Musée Municipal de Saint-Maur, La Varenne-Saint-Hilaire 13455
Musée Municipal Max Claudet, Salins-les-Bains 15598
Musée Municipal Méditerranéen de Cassis, Cassis 12305
Musée Municipal Saraleguinea, Guéthary 13162
Musée Napoléonien du Château de Grosbois, Boissy-Saint-Léger 12049
Musée National des Douanes, Bordeaux 12073
Musée Nicéphore Niépce, Chalon-sur-Saône 12359
Musée Paul Dini, Villefranche-sur-Saône 16196
Musée Raymond Lafage, Lisle-sur-Tarn 13786
Musée Régional Dupuy-Méstreau, Saintes 15585
Musée René Davoine, Charolles 12417
Musée Rupert de Chièvres, Poitiers 14867
Musée Saint-Loup, Troyes 16001
Musée Tavet-Delacour, Pontoise 14909
Musée Théodore Rousseau, Barbizon 11847

Musée Villa Tamaris Pacha, La Seyne-sur-Mer 13445
Palais des Papes, Avignon 11796
Pavillon des Arts, Paris 14746
Petit Musée du Bizarre, Villeneuve-de-Berg 16210
Petit-Palais, Paris 14749

Georgia

Fine Arts Museum, Tbilisi 16356
Kakabdze Art Gallery, Kutaisi 16342
Shalva Amiranasvili Museum of Art, Tbilisi 16379

Germany

600 Jahre Kunst - Studiensammlung der Universität Leipzig, Leipzig 19890
Airport Gallery 1 2, Frankfurt am Main 18353
Akademie der Künste, Berlin 17029
Alpenländische Galerie, Kempten 19512
Alte Nationalgalerie, Berlin 17032
Arithmeum, Bonn 17445
ARKA Kulturwerkstatt e.V., Essen 18248
Arthur Boskamp-Stiftung, Hohenlockstedt 19239
Ausstellungszentrum Kloster Cismar, Grömitz 18769
Ausstellungszentrum Kroch-Haus, Leipzig 19896
Ausstellungszentrum Lokschuppen, Rosenheim 21374
Badhaus Museum und Galerie, Kulmbach 19745
Badischer Kunstverein, Karlsruhe 19436
Beate Uhse Erotik-Museum, Berlin 17042
Besenmuseum Galerie, Ehingen 18066
Bilderbogen-Galerie, Neuruppin 20648
Bildhauermuseum Prof. Wandschneider, Plau am See 21051
Brandenburgischer Kunstverein Potsdam e.V., Potsdam 21084
Brasilianisches Kulturinstitut in Deutschland - Galerie, Berlin 17054
Bröhan-Museum, Berlin 17056
Bucerius Kunst Forum, Hamburg 18918
Burg Eltz, Wierschem 22544
Caricatura Galerie, Kassel 19460
Clemens-Sels-Museum Neuss, Neuss 20654
Deutsche Guggenheim, Berlin 17067
Deutsches Märchen- und Wesersagenmuseum, Bad Oeynhausen 16798
Deutsches Museum für Schulkunst, Hagen 18854
Digital Art Museum, Berlin 17075
E-Werk Freiburg, Freiburg im Breisgau 18429
Eduard Bargheer-Haus, Hamburg 18924
Erotic Art Museum Hamburg, Hamburg 18927
Fälschermuseum, Binz, Ostseebad 17355
Familienmuseum Schloss Braunfels, Braunfels 17514
Felix-Nussbaum-Haus Osnabrück, Osnabrück 20921
Forum für Angewandte Kunst Nürnberg e.V, Nürnberg 20760
Frankfurter Goethe-Haus, Frankfurt am Main 18370
Franz Radziwill Haus, Varel 22203
Frauen Museum, Bonn 17454
Fürstlich Fürstenbergische Sammlungen, Donaueschingen 17892
Galerie Alte Reichsvogtei, Schweinfurt 21682
Galerie am Domplatz, Halle, Saale 18892
galerie.bruehl, Brühl 17605
Galerie der Hochschule für Graphik und Buchkunst Leipzig, Leipzig 19904
Galerie der Stadt, Bad Wimpfen 16865
Galerie der Stadt Backnang, Backnang 16656
Galerie der Stadt Fellbach, Fellbach 18305
Galerie der Stadt Plochingen, Plochingen 21064
Galerie der Stadt Sindelfingen, Sindelfingen 21774
Galerie des Kultur- und Festspielhauses Wittenberge, Wittenberge 22620
Galerie des Polnischen Instituts Düsseldorf, Düsseldorf 17989
Galerie Goethe 53, München 20441
Galerie Grünstraße, Berlin 17089
Galerie Haus 23, Cottbus 17731
Galerie im Amtshaus, Bad Wurzach 16875
Galerie im Fontanehaus, Berlin 17090
Galerie im Körnerpark, Berlin 17092
Galerie im Kunsthoken, Quedlinburg 21147
Galerie im Prater, Berlin 17094
Galerie im Saalbau, Berlin 17096
Galerie in der Ungarischen Botschaft, Berlin 17100
Galerie Inkatt, Bremen 17550
Galerie Kunstbrücke Osteuropa, Berlin 17101
Galerie Kunstverein Wilhelm Kimmich, Lauterbach, Schwarzwald 19863
Galerie Marschall, Bernried 17302
Galerie Münsterland, Emsdetten 18160
Galerie Palais Walderdorff, Trier 22098
Galerie Remise, Halle / Westfalen 18909
Galerie Schloss Borbeck, Essen 18253
Galerie Stihl Waiblingen, Waiblingen 22276
Galerie Waldhof in Birkenwerder, Birkenwerder 17361
Gellert-Museum Hainichen, Hainichen 18873
Gerhard-Kurt-Müller-Stiftung, Leipzig 19910
Günter Grass Haus, Forum für Literatur und bildende Kunst, Lübeck 20075
Gutshaus Steglitz, Berlin 17122
H2 - Zentrum für Gegenwartskunst im Glaspalast, Augsburg 16620
Haus am Lützowplatz, Berlin 17126
Haus Opherdicke, Holzwickede 19263
Haus Peters, Tetenbüll 22034
Heidelberger Kunstverein, Heidelberg 19073
Heiliggeistkirche, Landshut 19801
Heimatmuseum Lübbecke und Kunstgalerie im Speicher, Lübbecke 20067
Heimatstube Bühlau, Großharthau 18792
Heinrich Vogeler Stiftung Haus im Schluh, Worpswede 22693
Heinrich Zille Museum, Berlin 17136
Heussenstamm Stiftung, Frankfurt am Main 18376
Hohenhof, die Jugendstil-Villa des Osthaus Museums Hagen, Hagen 18857
Horst-Janssen-Museum, Oldenburg 20895
Institut für Auslandsbeziehungen e.V. - Abt. Kunst, Stuttgart 21946
Internationale Tage - Boehringer Ingelheim, Ingelheim am Rhein 19334
Italienisches Kulturinstitut, Berlin 17144
Jerusalemhaus mit Verwaltung der Städtischen Sammlungen Wetzlar, Wetzlar 22527
Josef Albers Museum Quadrat Bottrop, Bottrop 17488
Jüdisches Museum Berlin, Berlin 17147
Jugendkunstschule Pankow, Berlin 17149
Junge Kunsthalle, Karlsruhe 19440
Kahnweilerhaus, Rockenhausen 21339
Kelsterbacher Museum für Porzellan, Kelsterbach 19504
Kinder-Akademie Fulda Werkraummuseum, Fulda 18511
Kinder- und Jugendgalerie Sonnensegel, Brandenburg an der Havel 17509
Kirche am Hohenzollernplatz, Berlin 17156
Klostergalerie, Zehdenick 22769
Koenraad Bosman Museum, Rees 21212
Kolumba, Köln 19613
Kommunale Galerie Berlin, Berlin 17158
Kommunale Galerie Gelsenkirchen, Gelsenkirchen 18576
Kügelgenhaus, Dresden 17950
Künstlerbunker, Leverkusen 19975
Künstlerforum Bonn, Bonn 17458
Künstlerhaus Marktoberdorf, Marktoberdorf 20219
Künstlerhaus Schloß Balmoral, Bad Ems 16724
Kultur- und Museumszentrum Schloss Glatt, Sulz am Neckar 21981
Kulturamt Mitte, Berlin 17161
Kulturforum Alte Post, Neuss 20657
Kulturhistorisches Museum Franziskanerkloster, Zittau 22812
Kulturkunststatt Prora, Prora 21131
Kulturzentrum Hans-Reiffenstuel-Haus, Pfarrkirchen 21024
Kulturzentrum Schloß Bonndorf, Bonndorf im Schwarzwald 17470
Kundencenter der Stadtwerke Strausberg, Strausberg 21920
Kunst Archiv Darmstadt, Darmstadt 17791
Kunst im Turm, Sömmerda 21793
Kunst Pavillon, München 20451
Kunst:raum Sylt Quelle, Rantum 21181
Kunstforum der Berliner Volksbank, Berlin 17167
Kunstgußmuseum Lauchhammer, Lauchhammer 19845
Kunsthalle Barmen, Wuppertal 22746
Kunsthalle der Hypo-Kulturstiftung, München 20454
Kunsthalle der Sparkasse Leipzig, Leipzig 19916
Kunsthalle Dominikanerkirche, Osnabrück 20924
Kunsthalle Emden, Emden 18138
Kunsthalle Erfurt im Haus zum Roten Ochsen, Erfurt 18205
Kunsthalle Fridericianum, Kassel 19465
Kunsthalle Gießen, Gießen 18637
Kunsthalle Harry-Graf-Kessler Weimar, Weimar 22407
Kunsthalle Jesuitenkirche, Aschaffenburg 16585
Kunsthalle Villa Kobe, Halle, Saale 18898
Kunsthaus Hamburg, Hamburg 18940
Kunsthaus Kannen, Münster 20511
Kunsthaus Kaufbeuren, Kaufbeuren 19489
Kunsthaus Rehau und Archiv Eugen Gomringer, Rehau 21238
Kunstkeller Annaberg, Annaberg-Buchholz 16554
Kunstmuseen Krefeld, Krefeld 19705
Kunstmuseum Bayreuth mit Tabakhistorischer Sammlung der British American Tobacco, Bayreuth 16958
Kunstmuseum Celle mit Sammlung Robert Simon, Celle 17688
Kunstmuseum Dieselkraftwerk Cottbus, Cottbus 17732
Kunstmuseum Gelsenkirchen, Gelsenkirchen 18577
Kunstmuseum Heidenheim, Heidenheim 19090
Kunstmuseum Solingen, Solingen 21802
Kunstmuseum Walter, Augsburg 16626
Kunstraum Düsseldorf, Düsseldorf 17996
Kunstraum Kreuzberg/Bethanien, Berlin 17170
Kunstsammlung im Schloß Rheinstein, Trechtingshausen 22085
Kunstsammlung im Stadtmuseum, Jena 19383
Kunstsammlung Lausitz, Senftenberg 21743
Kunstsammlungen Böttcherstraße, Bremen 17557
Kunstsammlungen Chemnitz, Chemnitz 17697
Kunstsammlungen der Stadt Limburg, Limburg an der Lahn 19995
Kunststätte Bossard, Jesteburg 19395
Kunststiftung Erich Hauser, Rottweil 21426
Kunstverein Hannover, Hannover 18997
Kunstverein Köln rechtsrheinisch e.V., Köln 19616
Kunstverein Passau - Sankt-Anna-Kapelle, Passau 20990
Kunstverein Ruhr e.V., Essen 18260
Kunstverein Springhornhof e.V., Neuenkirchen, Kreis Soltau-Fallingbostel 20609
Kunstverein "Talstrasse" e.V, Halle, Saale 18899
Kunstverein zu Frechen e.V., Frechen 18415
Kustodie, Dresden 17955
Landesmuseum Mainz, Mainz 20146
Langen Foundation, Neuss 20658
Lettl-Atrium, Augsburg 16627
Liebermann-Villa am Wannsee, Berlin 17177
Lindenau-Museum, Altenburg 16497
Löwenpalais, Berlin 17178
Ludwig Museum im Deutschherrenhaus, Koblenz 19586
LWL-Landesmuseum für Kunst und Kulturgeschichte, Münster 20512
Marburger Universitätsmuseum für Kunst und Kulturgeschichte, Marburg 20191
Margret-Knoop-Schellbach-Museum, Körle 19663
Markgräfler Museum Müllheim, Müllheim 20417
Morat-Institut für Kunst und Kunstwissenschaft, Freiburg im Breisgau 18432
Die Mühle, Eberswalde 18046
Museum Charlotte Zander, Bönnigheim 17426
Museum der bildenden Künste Leipzig, Leipzig 19921
Museum der Stadt Ratingen, Ratingen 21191
Museum der Weltkulturen mit Galerie 37, Frankfurt am Main 18381
Museum für Islamische Kunst im Pergamonmuseum, Berlin 17203
Museum für Kunst und Gewerbe Hamburg, Hamburg 18951
Museum für Kunst und Kulturgeschichte Dortmund, Dortmund 17925
Museum Georg Schäfer, Schweinfurt 21685
Museum Giersch, Frankfurt am Main 18387
Museum Haus Cajeth, Heidelberg 19079
Museum Haus Esters, Krefeld 19707
Museum Haus Lange, Krefeld 19708
Museum Heylshof, Worms 22686
Museum im Kornhaus, Altheim 16505
Museum im Kreuzgang, Landshut 19804
Museum Insel Hombroich, Neuss 20659
Museum Kartause Astheim, Volkach 22256
Museum Katharinenhof mit Mühlenturm und Stadtscheune, Kranenburg 19699
Museum Kloster Bentlage, Rheine 21298
Museum Kulturgeschichte der Hand, Wolnzach 22678
Museum Osterzgebirgsgalerie im Schloß, Dippoldiswalde 17872
Museum Pachen, Rockenhausen 21341
Museum Pfalzgalerie, Kaiserslautern 19412
Museum Ritter, Waldenbuch 22288
Museum Sankt Ingbert, Sankt Ingbert 21500
Museum Sankt Wendel, Sankt Wendel 21514
Museum Schloss Oberschwappach, Knetzgau 19582
Museum Synagoge Gröbzig, Gröbzig 18768
Museum Synthese, Hallbergmoos 18887
Museum und Galerie im Prediger, Schwäbisch Gmünd 21650
Museum Villa Haiss, Zell am Harmersbach 22779
Museum Wurzen mit Ringelnatzsammlung, Wurzen 22751
Museumsgalerie, Altomünster 16525
Neue Galerie, Dachau 17766
Neuer Sächsischer Kunstverein, Dresden 17965
Neues Kunsthaus Ahrenshoop, Ahrenshoop, Ostseebad 16443
Neues Museum, Nürnberg 20786
Niedersächsisches Landesmuseum Hannover, Hannover 19002
Oberpfälzer Künstlerhaus, Schwandorf 21664
Oldenburg-Museum, Unna 22176
Osterzgebirgsmuseum, Lauenstein 19850
Otto-Flath-Kunsthalle, Bad Segeberg 16838
Overbeck-Museum, Bremen 17559
Palais für aktuelle Kunst, Glückstadt 18664
PAN kunstforum niederrhein, Emmerich am Rhein 18151
Pixelprojekt_Ruhrgebiet, Gelsenkirchen 18579
Pommersches Landesmuseum, Greifswald 18750
Purrmann-Haus, Speyer 21837
Rathaus-Galerie, Berlin 17237
Rathaus Galerie Balingen, Balingen 16902
Rathausgalerie, Castrop-Rauxel 17682
Rathausgalerie Munster, Munster 20532
Rathausgalerie Vellmar, Vellmar 22216
Replikate der Welt-Kunst im Schloss, Miltach 20325
Rudolf-Scharpf-Galerie, Ludwigshafen am Rhein 20054
Sammlung Crous im Alten Kurhaus, Aachen 16406
Sammlung Domnick, Nürtingen 20794
Sammlung Goetz, München 20478
Sammlung Holzinger, München 20479
Sammlung Nuss, Weinstadt 22440
Sammlung Prinzhorn, Heidelberg 19082
Sammlung und Kunstraum Grässlin, Sankt Georgen im Schwarzwald 21495
Sammlungen des Instituts für Hochschulkunde, Würzburg 22723
Schering-Museum, Berlin 17247
Schirn Kunsthalle Frankfurt, Frankfurt am Main 18391
Schloß Brake - Das Weserrenaissance-Museum, Lemgo 19949
Schloss Callenberg, Coburg 17719
Schloss Cappenberg, Selm 21741
Schloss Elisabethenburg, Meiningen 20260
Schloss Güstrow, Güstrow 18830
Schloss Hundisburg-Sammlung Apel, Haldensleben 18884
Schloss Johannisburg mit Schlossgarten und Pompejanum, Aschaffenburg 16590
Schloss Luisium, Dessau 17832
Schloss Ökonomie Gern, Eggenfelden 18063
Schloß Weikersheim, Weikersheim 22372
Schloß Wiligrad, Lübstorf 20084
Schloßmuseum, Schrozberg 21642
Schwules Museum, Berlin 17254
Simplicissimus-Haus, Renchen 21272
El Sourdog Hex, Berlin 17256
Staatsgalerie Stuttgart, Stuttgart 21960
Stadtgalerie Altena, Altena 16489
Stadtgalerie Bamberg - Villa Dessauer, Bamberg 16921
Stadtgalerie Kiel, Kiel 19534
Stadthaus-Galerie, Münster 20518
Stadtmuseum, Jena 19392
Stadtmuseum Hadamar, Hadamar 18850
Stadtmuseum Münster, Münster 20519
Städtische Galerie, Dreieich 17937
Städtische Galerie, Überlingen 22145
Städtische Galerie, Paderborn 20975
Städtische galerie ada Meiningen, Meiningen 20261
Städtische Galerie am Markt Wurzen, Wurzen 22752
Städtische Galerie Bremen, Bremen 17562
Städtische Galerie Delmenhorst - Haus Coburg - Sammlung Stuckenberg, Delmenhorst 17817
Städtische Galerie Dresden - Kunstsammlung, Dresden 17973
Städtische Galerie Filderhalle, Leinfelden-Echterdingen 19885
Städtische Galerie Grimma, Grimma 18763
Städtische Galerie Haus Seel, Siegen 21757
Städtische Galerie im Königin-Christinen-Haus, Zeven 22798
Städtische Galerie im Schloßpark Strünkede, Herne 19157
Städtische Galerie im Spital zum Heiligen Geist, Riedlingen 21320
Städtische Galerie im Theater Ingolstadt, Ingolstadt 19347
Städtische Galerie In der Badstube, Wangen im Allgäu 22327
Städtische Galerie Iserlohn, Iserlohn 19361
Städtische Galerie Kaarst, Kaarst 19409
Städtische Galerie Lehrte, Lehrte 19875
Städtische Galerie "Lovis-Kabinett", Villingen-Schwenningen 22240
Städtische Galerie Lüdenscheid, Lüdenscheid 20091
Städtische Galerie Mennonitenkirche, Neuwied 20689
Städtische Galerie Reutlingen, Reutlingen 21288
Städtische Galerie Rosenheim, Rosenheim 21378
Städtische Galerie Sonneberg, Sonneberg 21819
Städtische Galerie Villa Streccius, Landau in der Pfalz 19791
Städtische Galerie Villa Zanders, Bergisch Gladbach 17020
Städtische Galerie Wesseling, Wesseling 22513
Städtische Kunstgalerie Torhaus Rombergpark, Dortmund 17928
Städtisches Kunstmuseum Spendhaus Reutlingen, Reutlingen 21289
Städtisches Museum, Kalkar 19417
Städtisches Museum, Eisenhüttenstadt 18107
Städtisches Museum Ludwigsburg, Ludwigsburg 20048
Städtisches Museum Zirndorf, Zirndorf 22810
Steinhausen-Museum, Frankfurt am Main 18395
Stiftung Bauhaus Dessau, Dessau 17834
Stiftung für Konkrete Kunst Roland Phleps, Freiburg im Breisgau 18438
Stiftung museum kunst palast mit Sammlung Kunstakademie und Glasmuseum Hentrich, Düsseldorf 18008
Stiftung Schloss Neuhardenberg, Neuhardenberg 20614

Stiftung Simonshof, Gleisweiler 18660
Stiftung Starke, Berlin 17267
Strathwork Galerie, Ladbergen 19759
Temporäre Galerie Schloß Neersen, Willich 22583
Theo-Steinbrenner-Turm-Museum, Sommerach 21816
TVDART, Berlin 17273
Urgeschichtliches Museum und Galerie Vierzigtausend Jahre Kunst, Blaubeuren 17381
Ursula Blickle Stiftung, Kraichtal 19697
Villa Merkel und Bahnwärterhaus, Esslingen 18278
Weissenhofmuseum im Haus Le Corbusier, Stuttgart 21968
Weißenhorner Heimatmuseum, Weißenhorn 22461
Wenzel-Hablik-Museum, Itzehoe 19370
Wieland-Museum Biberach, Wieland-Gartenhaus, Biberach an der Riß 17326
Winckelmann-Museum, Stendal 21899
Zehnthaus, Jockgrim 19398
Zentral-und Landesbibliothek Berlin, Berlin 17282
Zentrale Kustodie und Universitätsmuseum, Halle, Saale 18906

Greece

Jannis Spyropoulos Museum, Ekali 22990
Marika Kotopouli Museum, Zografou 23248
Municipal Art Gallery, Molyvos 23099
Municipal Art Gallery of Xanthi, Xanthi 23243
Municipal Gallery of Corfu, Corfu 22977
Municipal Museum of Kavala, Kavala 23048
Museum-Art Collection of the National Bank, Athinai 22929
Pinakothiki Kouvoutsaki, Kifissia 23052
Polygnotos Vagis Museum, Thassos 23204
Theophilos Museum, Varia Lesbos 23232

Greenland

Maniitsoq Katersugaasivia, Maniitsoq 23257
Qalipakkanik Katersugaasivik, Ilulissat 23256

Guyana

Guyana Museum, Georgetown 23318

Haiti

Musée du Panthéon National Haïtien, Port-au-Prince 23322

Honduras

Galería Nacional de Arte, Tegucigalpa 23331

Hungary

Csepel Galéria és Helytörténeti Gyűjtemény, Budapest 23394
Csók István Képtár, Székesfehérvár 23852
Ferenczy Múzeum, Szentendre 23876
Gaál Imre Galéria, Budapest 23406
Holló László Galéria, Putnok 23787
Kassák Múzeum, Budapest 23418
Lábasház, Sopron 23813
Miskolci Galéria - Városi Művészeti Múzeum - Rákóczi-ház, Miskolc 23712
Műcsarnok, Budapest 23453
Örökség Galéria, Budapest 23460
Rátkay Endre - Átlók Galéria, Budapest 23471
Sárospataki Képtár, Sárospatak 23794
Szépművészeti Múzeum, Budapest 23477
Váczy Péter Gyűjtemény, Győr 23587
Városi Művészeti Múzeum - Városi Képtár, Győr 23589
Vidovszky Béla Helytörténeti Gyűjtemény, Gyomaendrőd 23578

Iceland

Gerdarsafn, Reykjavik 24016
Hafnarborg Art Gallery, Hafnarfjörður 23998
Hafnarhúsið, Reykjavík 24018
Kjarvalsstaðir, Reykjavík 24019
Listasafn Kópavogs - Gerðasafn, Kópavogur 24008
Listasafnið Reykjanesbæjar, Reykjanesbær 24013
Norræna Húsid, Reykjavík 24027
Safn Ásgrims Jónssonar, Reykjavík 24032

India

Allahabad Museum, Allahabad 24059
Art Museum, Thiruvananthapuram 24402
Barton Museum, Bhavnagar 24086
Bharat Kala Bhavan, Varanasi 24425
Commercial Museum, Guwahati 24189
Damerla Rama Rao Memorial Art Gallery, Rajahmundry 24368
District Museum, Barpeta 24079
Framgi Dadabhoy Alpaiwalle Museum, Mumbai 24315
Gallery of Musical Instruments, Delhi 24140
Ganga Government Museum, Bikaner 24101
Government Museum, Alwar 24063
Government Museum, Chennai 24115
Government Museum, Jodhpur 24245
Haryana Prantiya Puratatva Sangrahalaya, Jhajjar 24241
Indira Gandhi National Centre for the Arts, Delhi 24143
Kannada Research Institute Museum, Dharwad 24163
Karnataka Government Museum and Venkatappa Art Gallery, Bangalore 24073
Mahakoshal Art Gallery, Raipur 24366
Manipur State Museum, Imphal 24225
Manipur University Museum, Canchipur 24107
Maratha History Museum, Pune 24360
Marble Palace Art Gallery and Zoo, Kolkata 24277
Mehrangarh Museum, Jodhpur 24246
Rukmini Devi Museum, Tiruvanmiyur 24410
Shreemanthi Bai Memorial Government Museum, Mangalore 24308
State Museum, Hyderabad 24218

Indonesia

Agung Rai Museum of Art, Ubud 24579
Museum Kraton Yogyakarta, Yogyakarta 24589
Museum Seni, Denpasar 24467
Museum Taruna Akbari Udarat, Magelang 24520

Iran

Babol Museum, Babol 24597
Fine Arts Museum, Tehrān 24666
Museum of Contemporary Articrafts, Tehrān 24683
Sizdah Aban Museum, Tehrān 24696

Ireland

AV Gallery, Limerick 24863
Bank of Ireland Arts Centre, Dublin 24778
The Chester Beatty Library, Dublin 24781
Farmleigh House, Dublin 24788
Galway Arts Centre, Galway 24835
Garter Lane Arts Centre, Waterford 24907
The Hunt Museum, Limerick 24864
Saint John's Theatre and Arts Centre, Listowel 24868
Waterfront Gallery, Westport 24910
Wexford Arts Centre, Wexford 24914

Isle of Man

Manx Museum, Douglas 24916

Israel

Art Gallery, Haifa 24953
Beit-Hagefen Art Gallery, Haifa 24954
Beit Ha'Omanim, Jerusalem 24978
Colosseum Art Gallery, Haifa 24958
Corin Maman Ashdod Museum, Ashdod 24928
David Palombo Museum, Jerusalem 24984
Herzliya Museum of Contemporary Art, Herzliya 24972
Ilana Goor Museum, Tel Aviv 25107
Israel Museum, Jerusalem, Jerusalem 24997
Kibbutz Art Gallery, Tel Aviv 25113
Mané-Katz Museum, Haifa 24964
Negev Museum, Be'ersheva 24936
Open Museum, Migdal-Tefen 25067
Open Museum, Omer 25075
Reuben and Edith Hecht Museum, Haifa 24967
Ruth Youth Wing, Jerusalem 25012
Shephela Museum, Kibbutz Kfar Menahem 25047
Yad Labanim Museum and Memorial Center, Petah Tikva 25078
Yad-Lebanim Memorial Center, Rehovot 25091
Yad Vashem Art Museum, Jerusalem 25026

Italy

Affresco di Piero della Francesca, Monterchi 26845
Burg Taufers, Campo Tures 25633
Casa Museo del Cima, Conegliano 25961
Casa Museo F. Scaglione, Sciacca 27795
Casa Museo Ivan Bruschi, Arezzo 25264
Casa Romei, Ferrara 26105
Castello del Buonconsiglio, Trento 28054
Castello della Manta, Manta 26589
Castello di Donnafugata, Ragusa 27358
Castello Ducale, Agliè 25155
Castello Pallotta, Caldarola 25597
Castello Principesco, Merano 26655
Castello Reale, Racconigi 27353
Centro di Scultura Contemporanea, Cagli 25575
Chiostro dello Scalzo, Firenze 26149
Civica Galleria d'Arte G. Sciortino, Monreale 26783
Civica Galleria di Palazzo Rocca, Chiavari 25857
Civica Pinacoteca V. Crivelli, Sant'Elpidio a Mare 27728
Civica Raccolta d'Arte, Medole 26648
Civica Raccolta d'Arte, Roseto degli Abruzzi 27559
Civica Raccolta d'Arte B. Biancolini, Potenza Picena 27318
Civiche Raccolte d'Arte, Busto Arsizio 25571
Civici Musei di Villa Paolina, Viareggio 28243
Civico Museo d'Arte Orientale, Trieste 28076
Civico Museo G.V. Parisi-Valle, Maccagno 26551
Collezione d'Arte della Banca Carige, Genova 26291
Collezione d'Arte di Santa Maria di Piazza, Ostra 27020
Collezione d'Arte Sacra, Saludecio 27601
Collezione di Carrozze, Maser 26616
Collezione Permanente di Valentino Martinelli, Perugia 27144
Collezione Sarda L. Piloni, Cagliari 25577
Collezione Straka-Coppa, Spello 27885
Collezioni d'Arte della Cassa di Risparmio, Savona 27775
Collezioni d'Arte e di Storia della Fondazione Cassa di Risparmio, Bologna 25440
Dogana d'Arte, Atripalda 25324
Donazione Sambo, Trieste 28087
Enoteca Regionale del Monferrato, Vignale Monferrato 28268
Fondazione Antonio Mazzotta, Milano 26684
Fondazione Bandera per l'Arte, Busto Arsizio 25572
Fondazione M. Minucci, Vittorio Veneto 28297
Fondazione Merz, Torino 27999
Fondazione Palazzo Albizzini Collezione Burri, Città di Castello 25905
Fondazione Palazzo Coronini Cronberg, Gorizia 26350
Fondazione Sandretto Re Rebaudengo, Torino 28000
Fondazione T. Balestra, Longiano 26515
Galleria Aurora Pallavicini, Roma 27448
Galleria Civica d'Arte Contemporanea, Caltagirone 25603
Galleria Civica d'Arte Contemporanea, Trento 28055
Galleria Civico, Modena 26740
Galleria Comunale d'Arte, Lecco 26476
Galleria Comunale d'Arte Moderna e Contemporanea, Arezzo 25265
Galleria Comunale Santa Croce, Cattolica 25784
Galleria Corsini, Firenze 26155
Galleria d'Arte Contemporanea V. Stoppioni, Santa Sofia 27711
Galleria d'Arte Moderna A. Discovolo, Bonassola 25493
Galleria d'Arte Moderna Comunale, San Severino Marche 27683
Galleria d'Arte Moderna F. Montanari, Moncalvo 26773
Galleria d'Arte Moderna O. Marchesi, Copparo 25967
Galleria dell'Accademia di Belle Arti, Napoli 26880
Galleria di Palazzo degli Alberti, Prato 27333
Galleria Ferrari, Maranello 26600
Galleria Nazionale d'Arte Antica, Trieste 28088
Galleria Nazionale di Cosenza, Cosenza 25990
Galleria Palazzo Cini, Venezia 28183
Gallerie di Palazzo Leoni Montanari, Vicenza 28251
Giardino Storico e Villa Garzoni, Pescia 27170
Gipsoteca, Lucca 26530
Il Mauriziano, Reggio Emilia 27387
Mostra Nazionale di Pittura Contemporanea, Marsala 26610
Mostra Permanente P. Mariani, Bordighera 25497
Museo Alcamese d'Arte Sacra, Alcamo 25188
Museo Alternativo Remo Brindisi, Comacchio 25945
Museo Antoniano, Padova 27035
Museo Archeologico e Pinacoteca Comunale, Foligno 26214
Museo Artistico della Bambola, Suvereto 27931
Museo Aurelio Castelli, Siena 27839
Museo Bagatti Valsecchi, Milano 26690
Museo Capitolare, Atri 25322
Museo Capitolare, Cagliari 25580
Museo Cappuccini Emiliani, Reggio Emilia 27389
Museo Casa del Duca, Venezia 28187
Museo Casa Natale di Giotto, Vicchio 28250
Museo Civico, San Gimignano 27637
Museo Civico Archeologica e Quadreria, Monterubbiano 26850
Museo Civico Archeologico, Montalto Marche 26797
Museo Civico d'Arte, Modena 26746
Museo Civico del Sigillo, La Spezia 26434
Museo Civico e Pinacoteca, Asti 25315
Museo Civico Goffredo Bellini, Asola 25301
Museo d' Arte e Scienza, Milano 26693
Museo d'Arte d'Ammobigliamento, Stupinigi 27915
Museo d'Arte Nuoro, Nuoro 26965
Museo d'Arte Orientale, Venezia 28189
Museo d'Arte Pietro Cavoti, Galatina 26262
Museo dei Burattini e delle Figure, Cervia 25820
Museo dei Fossili, Lugo di Vicenza 26545
Museo del Bigallo, Firenze 26170
Museo del Lazio Meridionale, Anagni 25225
Museo del Paesaggio e Museo Storico Artistico, Verbania Pallanza 28219
Museo del Teatro Argentina, Roma 27475
Museo del Tesoro, Lodi 26508
Museo della Città, Ancona 25228
Museo della Congregazione Mechitarista dei Padri Armeni, Venezia 28191
Museo della Pieve di Staggia, Poggibonsi 27267
Museo della Val Gardena, Ortisei 27001
Museo dell'Abbazia, Casamari-Veroli 25700
Museo dell'Accademia di Belle Arti, Perugia 27149
Museo dell'Arte Serica e Laterizia, Malo 26579
Museo delle Tavolette di Biccherna, Siena 27842
Museo dell'Opera Metropolitana, Siena 27843
Museo di Arte Contemporanea, Stia 27903
Museo di Arte Contemporanea e del Novecento, Monsummano Terme 26791
Museo di Arte e Giacimenti Minerari, Latina 26466
Museo di Arte e Storia delle Miniere, Massa Marittima 26631
Museo di Arte Sacra del Castello, Genga 26287
Museo di Palazzo Altieri, Oriolo Romano 26995
Museo di Palazzo della Penna, Perugia 27150
Museo di Palazzo Gamilli, Nocera Umbra 26934
Museo di Palazzo Mirto, Palermo 27073
Museo di Palazzo Pepoli Campogrande, Bologna 25463
Museo di Palazzo Pretorio, Certaldo 25815
Museo di Roma, Roma 27499
Museo di San Francesco, Aversa 25333
Museo di Scienze Archeologiche e d'Arte, Padova 27042
Museo di Villa Cagnola, Gazzada 26280
Museo Diffuso, Pennabilli 27132
Museo Diocesano, Sulmona 27921
Museo elle Arti e Tradizioni Contadine, Roccasecca dei Volsci 27434
Museo Ernesto Treccani, Milano 26708
Museo Fortuny, Venezia 28201
Museo Ideale Leonardo da Vinci, Vinci 28284
Museo Laboratorio Santa Barbara, Mammola 26581
Museo Mario Praz, Roma 27510
Museo Mediceo della Petraia e Giardino, Firenze 26194
Museo Miniscalchi Erizzo, Verona 28237
Museo Piersanti, Matelica 26638
Museo Pietro della Vedova, Rima San Giuseppe 27404
Museo Pinacoteca dell'Arte dei Bambini, Smerillo 27860
Museo Pinacoteca Parrocchiale di Santa Maria, Mogliano Marche 26761
Museo Storico dela Caccia e del Territorio, Cerreto Guidi 25812
Palazzo del Principe, Genova 26321
Palazzo delle Esposizioni, Roma 27541
Palazzo Ducale, Genova 26322
Palazzo Sandretta Re Rebaudengo, Guarene 26387
Palazzo Silva, Domodossola 26043
Pinacoteca, Chiomonte 25878
Pinacoteca Andrea Alfano, Castrovillari 25770
Pinacoteca Chiesa di San Giovanni, Pieve Torina 27217
Pinacoteca Civica, Abano Terme 25136
Pinacoteca Civica, Baiardo 25351
Pinacoteca Civica, Savona 27780
Pinacoteca Civica, Montelupone 26841
Pinacoteca Civica, Vado Ligure 28132
Pinacoteca Civica Amedeo Modigliani, Follonica 26217
Pinacoteca Civica d'Arte Moderna, Latina 26467
Pinacoteca Civica e Gipsoteca U. Gera, Ripatransone 27419
Pinacoteca Civica G. Cattabriga, Bondeno 25494
Pinacoteca Comunale, Bosa 25513
Pinacoteca Comunale, Deruta 26033
Pinacoteca Comunale, Manciano 26583
Pinacoteca Comunale, Matelica 26639
Pinacoteca Comunale, Narni 26915
Pinacoteca Comunale, Ostra 27021
Pinacoteca Comunale, Quistello 27352
Pinacoteca Comunale, Ripe San Ginesio 27420
Pinacoteca Comunale, San Benedetto del Tronto 27612
Pinacoteca Comunale A. Moroni, Porto Recanati 27303
Pinacoteca Comunale D. Stefanucci, Cingoli 25894
Pinacoteca Comunale E. Giannelli, Parabita 27090
Pinacoteca Comunale e Museo A. De Felice, Terni 27964
Pinacoteca Comunale F. Galante Civera, Margherita di Savoia 26605
Pinacoteca D. Inzaghi, Budrio 25561
Pinacoteca Dantesca Fortunato Bellonzi, Torre de Passeri 28035
Pinacoteca d'Arte Antica, Gemona del Friuli 26284
Pinacoteca d'Arte Contemporanea, Smerillo 27861
Pinacoteca d'Arte Francescana, Lecce 26475
Pinacoteca Davide Bergh, Calice al Cornovoglio 25600
Pinacoteca dei Cappuccini, Voltaggio 28309
Pinacoteca della Cassa di Risparmio, Cesena 25836
Pinacoteca della Certosa, Firenze 26201
Pinacoteca della Rocca Ubaldinesca, Sassocorvaro 27759
Pinacoteca di Corinaldo, Corinaldo 25978
Pinacoteca di San Silvestro, Montecompatri 26819
Pinacoteca e Biblioteca Rambaldi, Sanremo 27697
Pinacoteca E. Notte, Ceglie Messapica 25796
Pinacoteca in Palazzo Volpi, Como 25955
Pinacoteca Internazionale dell'Età Evolutiva A. Cibaldi, Rezzato 27399

Pinacoteca-Museo Beato Sante, Mombaroccio 26770
Pinacoteca Parrocchiale, Castroreale 25767
Pinacoteca Rossetti Valentini, Santa Maria Maggiore 27707
Quadreria Comunale, Offida 26976
Quadreria Comunale, San Constanzo 27628
Quadreria della Società Economica, Chiavari 25862
Quadreria e Pinacoteca Civica, San Giovanni in Persiceto 27647
Raccolta Civica d'Arte Contemporanea, Molfetta 26766
Raccolta Comunale d'Arte, Frontone 26252
Raccolta d'Arte C. Lamberti, Codogno 25933
Raccolta d'Arte Contemporanea R. Pastori, Calice Ligure 25601
Raccolta d'Arte della Provincia, Modena 26757
Raccolta di Opere d'Arte, Solarolo 27862
Raccolta E. Guatelli, Collecchio 25939
Raccolta Privata Toraldo di Francia, Tropea 28106
Riserva Naturale Speciale del Sacro Monte di Varalio, Varallo Sesia 28159
Santa Giulia - Museo della Città, Brescia 25543
Scuderie del Quirinale, Roma 27546
Wolfsoniana, Genova 26327

Japan

The A Museum of Art, Takane 29323
Anno Bijutsukan, Tsuwano 29545
Arai Kinen Bijutsukan, Iwanai 28609
Art Gallery, Kitahiroshima 28706
Art Hall Miraku, Ikoma 28563
Chiba-kenritsu Bijutsukan, Chiba 28391
Chinchu Bijutsukan, Naoshima 29028
Daigaku Bijutsukan, Kyoto 28804
Daiichi Bijutsukan, Nagoya 28995
Former Hong Kong and Shanghai Bank Museum, Nagasaki 28972
Fuji Art Museum, Fujinomiya 28423
Fukui-kenritsu Bijutsukan, Fukui 28428
Fukushima-kenritsu Bijutsukan, Fukushima 28440
Higashi-Osaka Art Center, Higashi-Osaka 28505
Hikone Castle Museum, Hikone 28509
Hokkaidoritsu Obihiro Bijutsukan, Obihiro 29085
Imura Art Museum, Kyoto 28816
Inazawa City Oguiss Memorial Art Museum, Inazawa 28572
Ishinomaki Bunkakan, Ishinomaki 28587
Iwate-kenritsu Bijutsukan, Morioka 28934
Japan Amateur Art Museum, Ogi 29090
Kagawa-ken Bunka Kaikan, Takamatsu 29316
Kagoshima-shiritsu Bijutsukan, Kagoshima 28628
Karuizawa Wild Flower Museum of Art, Karuizawa 28669
Kasama Nichido Bijutsukan, Kasama 28675
Kawasaki-shi Hakubutsukan, Kawasaki 28693
Kazuaki Iwasaki Space Art Gallery, Ito 28596
Keio Daigaku Bijutsu Senta, Tokyo 29410
Kitakyushu-shiritsu Bijutsukan, Kitakyushu 28712
Kitamura Bijutsukan, Kyoto 28822
Kiuchi Gallery, Ichikawa 28556
Kiyoharu Shirakaba Museum, Nagasaka 28967
Kiyosato Kitazawa Bijutsukan, Takane 29324
Koganei City Hakenomori Art Museum, Koganei 28758
Komagata Jukichi Bijutsu Kinenkan, Nagaoka 28958
Komatsu Hitoshi Art Museum, Kyoto 28827
Koryo Museum of Art, Kyoto 28829
Koyama Keizo Bijutsukan, Komoro 28769
Kozu Kobunka Museum, Kyoto 28831
Kushiro Art Museum, Kushiro 28798
Kuwayama Bijutsukan, Nagoya 29000
Kyushu Sangyo Daigaku Hakubutsukan, Fukuoka 28435
Laforet Hakuba Bijutsukan, Hakuba 28481
Light and Greenery Art Museum, Sagamihara 29182
Mirasaka Peace Museum of Art, Mirasaka 28912
Miyagi-ken Bijutsukan, Sendai 29243
MOA Bijutsukan, Atami 28382
Moji-kuritsu Idemitsu Bijutsukan, Kitakyushu 28715
Munakata Shikou Art Station, Muika 28943
Musée Bernard Buffet, Shizuoka 29286
Musee d'Art Mercian Karuizawa, Miyota 28930
Musée Marie Laurencin, Chino 28400
Museum Haus Kasuya, Yokosuka 29621
Nagoya-Boston Museum of Fine Arts, Nagoya 29003
Nagoya-shi Bijutsukan, Nagoya 29005
Nitten - Japan Fine Arts Exhibition, Tokyo 29452
Ohara Art Reference Center, Kobe 28738
The Ohsha'joh Museum of Art, Hiroshima 28542
Okyo no Rosetsu Bijutsukan, Kushimoto 28797
Osaka-kenritsu Kokuritsu Bijutsukan, Osaka 29144
Otsuka Bijutsukan, Naruto 29044
Pola Museum of Art, Hakone 28480
Rokuzan Bijutsukan, Hotaka 28549
Sagamihara Citizen's Gallery, Sagamihara 29183
Saitama-kenritsu Kindai Bijutsukan, Saitama 29190
Sakaide-shi Bijutsukan, Sakaide 29202
Saku-shiritsu Bijutsukan, Saku 29208
Sapporo Geijutsu no Mori Bijutsukan, Sapporo 29222
Seiji Togo Memorial Sompo Japan Museum of Art, Tokyo 29462
Setsuryosha Bijitsukan, Niigata 29062
Shibunkaku Bijutsukan, Kyoto 28865
Shiseido Art House, Kakegawa 28638
Shizuoka Art Gallery, Shizuoka 29287
Shoundo Bijutsukan, Ube 29550
Sogo-Bunka Art Gallery, Amagasaki 28357
Sogo Museum of Art, Yokohama 29607
Striped House Museum of Art, Tokyo 29475
Sunritz Hattori Bijutsukan, Suwa 29302
Takamatsu City Museum of Art, Takamatsu 29322
Takaoka-shi Bijutsukan, Takaoka 29326
Takasaki City Gallery, Takasaki 29332
Tochigi-kenritsu Bijutsukan, Utsunomiya 29566
Tokyo Chuo-ku Bijutsukan, Tokyo 29488
Tokyo Fuji Art Museum, Hachioji 28467
Tokyo Station Gallery, Tokyo 29496
Townsend Harris Kinenkan, Shimoda 29264
Toyama Kenminkaikan Bijutsukan, Toyama 29521
Toyota Kuragaike Commemorative Hall and Art Salon, Toyota 29533
Wocoal Art Center, Tokyo 29505
Woodone Bijutsukan, Yoshiwa 29629
Yamanashi-kenritsu Bijutsukan, Kofu 28754
Yamauchi Library, Yokohama 29610
Yayoi Museum, Tokyo 29508
Yokohama Bijutsukan, Yokohama 29611
Yokohama Portside Gallery, Yokohama 29616
Yuki Bijutsukan, Osaka 29158

Kazakhstan

Karagandinskij Oblastnoj Muzej Izobrazitelnogo Iskusstva, Qaraghandy 29723
Oblastnoj Muzej Izobrazitelnych Iskusstv, Semey 29730
Oblastnoj Muzej Literatury i Iskusstva im. Buchar Žyrau, Pavlodar 29717
Pavlodarskij Oblastnoj Chudožestvennyj Muzej, Pavlodar 29718
Ščučinskij Oblastnoj Muzej Literatury i Iskusstva, Kokšetau 29704
Severo-Kazachstanskij Oblastnoj Muzej Izobrazitelnogo Iskusstva, Petropavl 29721
Vostočno-Kazachstanskij Muzej Iskusstv, Öskemen 29713

Kenya

The Museum Studio and Arts Centre, Nairobi 29748

Korea, Republic

Daejeon Museum of Art (DMA), Daejeon 29799
Dr. Park Collection and Gallery, Seoul 29873
Woo-Bong Art Museum, Daegu 29796

Latvia

Arzemju Mākslas Muzejs, Rīga 30062
Bauskas Novadpētniecības un Mākslas Muzejs, Bauska 29988
Liepājas Muzeja Izstāžu Zāles, Liepāja 30039
Naive art museum of Latvia, Rīga 30099
Raiņa Literatūras un Mākslas Vēstures Muzejs, Rīga 30102
Valsts Mākslas Muzejs, Rīga 30114

Lebanon

Musée des Beaux-Arts, Beirut 30159
Musée Nicholas Ibrahim Sursock, Beirut 30161

Liechtenstein

Vereinigung der Mund- und Fussmalenden Künstler in aller Welt e.V., Schaan 30194

Lithuania

Ch. Frenkelio Rūmai, Šiauliai 30353
Juodkrantės Parodų Salėa, Juodkrantė 30232
Kazio Varnelio Namai-Muziejus, Vilnius 30411
Klaipėdos Paveikslų Galerija, Klaipėda 30279
Lietuvos Dailės Muziejus, Vilnius 30415
Lietuvos Nacionaline Dailės Galerija, Vilnius 30418
Mykolo Žilinsko Dailės Galerija, Kaunas 30261
Nacionalinis Mikalojaus Konstantino Čiurlionio Dailės Muziejus, Kaunas 30262
Radvilų Rūmai, Vilnius 30430
Šiaulių Aušros Muziejus, Šiauliai 30362
Taikomosios Dailės Muziejus, Vilnius 30434
Venclauskių Namai, Šiauliai 30365
Vilniaus Paveikslų Galerija, Vilnius 30440

Luxembourg

Galerie d'Art d'Esch, Esch-sur-Alzette 30472
Galerie d'Art du Château, Bettembourg 30455
Galerie d'Art Municipale, Oberkorn 30489
Galerie d'Art Municipale, Rumelange 30490

Macedonia

Muzej na Makedonija - Archeološki, Etnološki i Istoriski, Skopje 30519
Muzej na Umetnosta na Makedonija, Skopje 30521

Madagascar

Musée d'Art et d'Archéologie de l'Université d'Antananarivo, Antananarivo 30534
Musée du Palais de la Reine, Antananarivo 30535

Malaysia

Balai Seni Lukis Negara, Kuala Lumpur 30573
Museum and Art Gallery, Shah Alam 30618
Pucuk Rebung-Gallery Museum, Kuala Lumpur 30585
Sabah Art Gallery, Kota Kinabalu 30566
Shah Alam Art Gallery, Shah Alam 30621

Malta

Saint James Cavalier, Valletta 30646

Mexico

Casa Siglo XIX, Chihuahua 30764
Centro Cultural Isidro Fabela, México 31037
Galería José María Velasco, México 31045
Instituto Mexicano Norteamericano de Jalisco, Guadalajara 30890
Museo Amparo, Puebla 31265
Museo Casa del Alfeñique, Puebla 31268
Museo de Arte José Luis Bello y González, Puebla 31271
Museo de la Secretaria de Hacienda y Credito Publico Antiguo Palacio del Arzobispado, México 31088
Museo Dolores Olmedo, México 31111
Museo José Luis Cuevas, México 31125
Museo Nacional de Arte Fantástico, Córdoba 30802
Museo Soumaya, México 31149
Museo Taller Luis Nishizawa, Toluca 31494
Museo Universitario de Ciencias y Artes, México 31154
Pinacoteca de Nuevo León, Monterrey 31189
Pinacoteca del Centro de la Profesa, México 31158
Pinacoteca Diego Rivera, Xalapa 31560
Robert Brady Museum, Cuernavaca 30821
San Pedro Museo de Arte, Puebla 31291

Moldova

Muzeul National de Arte Plastice, Chişinău 31600

Mongolia

Zanabazar Museum of Fine Arts, Ulaanbaatar 31634

Montenegro

Centar Savremene Umjetnosti Crne Gore, Podgorica 31654
Umjetnički Paviljon, Podgorica 31658

Morocco

Musée d'Art Islamique, Marrakech 31678
Musée de Dar Jamaï, Meknès 31681
Musée des Arts Sahariens de Laâyoune, Laâyoune 31674
Museo de los Muralles Reales, Ceuta 31665

Mozambique

Museu Nacional de Arte, Maputo 31710

Myanmar

State Museum, Mandalay 31720

Namibia

National Art Gallery of Namibia, Windhoek 31764

Netherlands

Achter de Zuilen, Overveen 32664
Armando Museum, Amersfoort 31809
Beeldentuin Belling Garde, Bellingwolde 31958
Beurs van Berlage Museum, Amsterdam 31833
De Beyerd, Breda 32000
Brink 7 Galerij, Yde 32972
Centrum Beeldende Kunst, Groningen 32279
Centrum Beeldende Kunst Emmen, Emmen 32204
Galerie Anna Paulownahuis, Soest 32792
Galerie de Lange, Emmen 32205
ID Galerie, Helmond 32364
Inter Art - Sonsbeek Art and Design, Arnhem 31927
Jachthuis Sint-Hubertus, Hoenderloo 32397
De Kabinetten van de Vleeshal, Middelburg 32551
Kasteel Heeze - Collectie H.N.C. Baron Van Tuyll Van Serooskerken Van Heeze en Leende, Heeze 32344
Kasteel Het Nijenhuis, Heino 32352
Kasteel-Museum Sypesteyn, Loosdrecht 32506
Kunstcentrum De Waagh, Oldenzaal 32616
Kunstcentrum Pand Paulus, Schiedam 32753
Kunsthal Rotterdam, Rotterdam 32710
Kunsthuis Secretarie, Meppel 32550
Kunstpaviljoen, Nieuw Roden 32572
't Kunstuus, Heinkenszand 32350
Kunstvereniging Diepenheim, Diepenheim 32124
Museum De Paviljoens, Almere 31803
Museum Jan Cunen, Oss 32640
Museum Katendrecht, Rotterdam 32713
Museum Kranenburgh, Bergen 31968
Museum Møhlman, Appingedam 31918
Museum Swaensteyn, Voorburg 32906
Museum Veluwezoom, Doorwerth 32147
Museum voor Figuratieve Kunst De Buitenplaats, Eelde 32174
MuseumgoudA, Gouda 32269
Nederlands Museum van Knipkunst, Schoonhoven 32765
Oerka Irene Verbeek Museum, Raard 32678
Orangerie, 's-Hertogenbosch 32377
Paleis aan de Blijmarkt, Zwolle 33027
Presentatieruimte de Overslag, Eindhoven 32194
Rijksmuseum Hendrik Willem Mesdag, Den Haag 32099
Stadsmuseum Woerden, Woerden 32959
Stichting tot Beheer Museum Boijmans Van Beuningen, Rotterdam 32729
Stroom Den Haag, Den Haag 32103
Tattoo Museum, Amsterdam 31893
Theo Swagemakers Museum, Haarlem 32303
De Verbeelding Kunst Landschap Natuur, Zeewolde 32993
Verpleeghuis Cornelia, Zierikzee 33006
Villa Zebra Museum voor Kinderen, Rotterdam 32732
De Vleeshal, Middelburg 32553
Voerman Museum Hattem, Hattem 32324

New Zealand

Arts Centre of Christchurch, Christchurch 33068
Cleveland Living Arts Centre, Dunedin 33089
Community Arts Centre, Wanganui 33233
Eastern Southland Gallery, Gore 33111
Franklin Arts and Cultural Centre, Pukekohe 33190
Fridge Gallery, Petone 33181
George Fraser Gallery, Auckland 33050
Hawke's Bay Museum, Napier 33150
Lake Taupo Museum and Art Gallery, Taupo 33204
Lopdell House Gallery, Waitakere 33227
Sarjeant Gallery - Te Whare o Rehua, Wanganui 33234
Southland Museum and Art Gallery, Invercargill 33130
Te Manawa Art, Palmerston North 33175
Uxbridge - Howick's Creative Centre, Auckland 33060
Waikato Museum of Art and History, Hamilton 33117
Whangarei Art Museum, Whangarei 33265

Nicaragua

Pinacoteca del Banco Central de Nicaragua, Managua 33291

Nigeria

National Museum Uyo, Uyo 33347

Northern Cyprus

Saçaklı Ev - Kültür ve Sanat Merkezi, Lefkoşa 33372

Norway

Agatunet, Nå 33688
Astruptunet, Skei i Jølster 33831
Beiarn Bygdetun, Moldfjord 33680
Bergen Kunstmuseum - Lysverket, Bergen 33418
Bergen Kunstmuseum - Rasmus Meyers Samlinger, Bergen 33419
Drammens Museum, Drammen 33466
Eikaasgalleriet, Skei i Jølster 33832
Fedrenes Minne, Hidrasund 33547
Galleri Espolin, Kabelvåg 33584
Galleri Svalbard, Longyearbyen 33649
Haugar Vestfold Kunstmuseum, Tønsberg 33910
Ingebrigt Vik's Museum, Øystese 33765
Internasjonale Barnekunstmuseet, Oslo 33728
Lillehammer Kunstmuseum, Lillehammer 33636
Lindesnes Bygdemuseum, Sør-Audnedal 33851
Magnus Dagestad Museet, Voss 33985
Mandal Bymuseum, Mandal 33658
Nasjonalmuseet for Kunst, Arkitektur og Design, Oslo 33736
Nord-Trøndelag Fylkesgalleri, Namsos 33692
Nordnorsk Kunstmuseum, Tromsø 33918
Nøstetangen Museum, Hokksund 33550
Savio-museet, Kirkenes 33592
Sørlandets Kunstmuseum, Kristiansand 33612
Stenersenmuseet, Oslo 33752
Stokke Bygdetun, Stokke 33882
Trastad Samlinger, Borkenes 33448

A

Pakistan

Aiwen e Rifat Museum and Art Gallery, Karachi 34012
Children Art Palace, Islamabad 34006
Hyderabad Museum, Hyderabad 34004

Paraguay

Centro de Artes Visuales, Asunción 34089
Museo Bernardino Caballero, Asunción 34091
Museo Nacional de Bellas Artes, Asunción 34101

Peru

Galería Piurarte, Piura 34289
Museo de Arte Colonial Pedro de Osma, Lima 34231
Pinacoteca Municipal Ignacio Merino, Lima 34274

Philippines

Ayala Museum, Makati 34414
Bulwagang Helena Benitez Gallery, Manila 34433
CAP Art Center and Don Sergio Osmeña Memorabilia, Cebu 34365
Children's Museum, Quezon City 34488
Corredor, Gallery of the College of Fine Arts, Quezon City 34489
Cultural Museum, Kabacan 34394
Pambasang Museo ng Pilipinas, Manila 34450
University of Santo Tomas Museum of Arts and Sciences, Manila 34458

Poland

Cricoteca-Galeria-Pracownia Tadeusza Kantora, Kraków 34768
Dom Jana Matejki, Kraków 34769
Dworek Jana Matejki, Krzesławice 34835
Dworek Jana Matejki w Krzesławicach, Kraków 34771
Galeria Akademii Sztuk Pięknych im. Władysława Strzemińskiego, Łódź 34861
Galeria Bałucka, Łódź 34862
Galeria Engram, Katowice 34722
Galeria Marii Ritter, Nowy Sącz 34957
Galeria Sztuki Dawna Synagoga, Nowy Sącz 34958
Galeria Sztuki im. Włodzimierza i Jerzego Kulczyckich, Zakopane 35354
Galeria Sztuki Wozownia, Toruń 35191
Galeria Sztuki Współczesnej, Katowice 34723
Galeria Sztuki Współczesnj, Łomża 34889
Galeria Władysława Hasiora, Zakopane 35355
Galeria Zamek w Reszlu, Reszel 35080
Kurza Stopa - Galeria Współczesnej Sztuki Polskiej, Chojnice 34598
Miejska Galeria Arsenał, Poznań 35027
Muzeum Akademii Sztuk Pięknych, Warszawa 35234
Muzeum Emila Zegadłowicza, Wadowice 35208
Muzeum Górnośląskie, Bytom 34584
Muzeum im. Adama Mickiewicza w Śmiełowie, Żerków 35369
Muzeum im. Anny i Jarosława Iwaszkiewiczów w Stawisku, Podkowa Leśna 35018
Muzeum im. Jerzego Dunin-Borkowskiego, Krośniewice 34828
Muzeum im. Oskara Kolberga, Przysucha 35058
Muzeum im. Stanisława Noakowskiego, Nieszawa 34946
Muzeum Łazienki Krolewskie, Warszawa 35255
Muzeum Marii Konopnickiej, Suwałki 35143
Muzeum Miedzi, Legnica 34848
Muzeum Narodowe, Szczecin 35163
Muzeum Narodowe w Krakowie, Kraków 34805
Muzeum Narodowe w Warszawie, Warszawa 35261
Muzeum Narodowe we Wrocławiu, Wrocław 35335
Muzeum Okręgowe w Toruniu, Toruń 35195
Muzeum Pałac w Wilanowie, Warszawa 35263
Muzeum Parafialne im. Jana Piechoły, Iwkowa 34693
Muzeum Rzeźby Alfonsa Karnego, Białystok 34552
Muzeum Rzeźby im. Xawerego Dunikowskiego Królikarnia, Warszawa 35269
Muzeum Śląska Opolskiego, Opole 34986
Muzeum Stanisława Wyspiańskiego w Kamienicy Szołayskich, Kraków 34809
Muzeum Sztuki Dziecka, Warszawa 35271
Muzeum Sztuki Mieszczańskiej, Wrocław 35341
Muzeum-Zamek w Gołuchowie, Gołuchów 34676
Muzeum Zamkowe, Pszczyna 35062
Muzeum Ziemi Zaborskiej i Galeria Sztuki Ludowej, Wiele 35299
Nowosądecka Mała Galeria, Nowy Sącz 34960
Ośrodek Biograficzny Komisji Turystyki Górskiej, Kraków 34816
Pałac Sztuki w Krakowie, Kraków 34817
Panorama Racławicka, Wrocław 35343
Piętro Wyżej, Katowice 34729
Rezydencja Księży Młyn, Łódź 34888
Społeczne Muzeum Konstantego Laszczki, Dobre 34624
Zamek Królewski w Warszawie, Warszawa 35291
Zbiory Sztuki, Włocławek 35312
Zwierzyniecki Salon Artystyczny, Kraków 34822

Portugal

Atelier-Museu António Duarte, Caldas da Rainha 35493
Atelier-Museu João Fragoso, Caldas da Rainha 35494
Casa Museu Abel Salazar da Universidade do Porto, São Mamede de Infesta 35922
Casa-Museu de Francisco Ernesto de Oliveira Martins, Angra do Heroísmo 35437
Casa-Museu Dr. Anastácio Gonçalves, Lisboa 35662
Casa Tait, Porto 35842
O Museu, Porto 35843
Museu da Liga dos Combatentes da Grande Guerra, Lisboa 35680
Museu de Angra do Heroísmo, Angra do Heroísmo 35439
Museu de Aveiro, Aveiro 35447
Museu de Electricidade, Fornelos 35606
Museu de José Malhoa, Caldas da Rainha 35497
Museu do Neo-Realismo, Vila Franca de Xira 35996
Museu Nacional de Soares dos Reis, Porto 35860
Museu Romântico da Quinta da Macieirinha, Porto 35861
Museu Santa Casa da Misericórdia, Coimbra 35543
Palácio dos Coruchéus, Lisboa 35731

Puerto Rico

Museo de Las Américas, San Juan 36068

Réunion

Musée Léon Dierx, Saint-Denis 36091

Romania

Muzeul Banatului. Secţia de Artă, Timişoara 36690
Muzeul Colecţiilor de Artă, Bucureşti 36198
Muzeul de Artă, Braşov 36178
Muzeul de Artă, Iaşi 36417
Muzeul de Artă Piatra-Neamţ, Piatra-Neamţ 36518
Muzeul Judeţean Argeş, Piteşti 36526
Muzeul Naţional de Artă al României, Bucureşti 36219
Muzeul Naţional de Artă Cluj-Napoca, Cluj-Napoca 36289
Muzeul Naţional Peleş, Sinaia 36620
Muzeul Ţării Făgăraşului Valeriu Literat, Făgăraş 36363
Muzeul Theodor Pallady, Bucureşti 36231

Russia

Aleksandrovskaja Sloboda - Gosudarstvennyj Istoriko- architekturnyj i Chudožestvennyj Muzej-zapovednik, Aleksandrov 36751
Aleksandrovskij Chudožestvennyj Muzej, Aleksandrov 36752
Altajskij Respublikanskij Kraevedčeskij Muzej im. A.V. Anochina, Gorno-Altajsk 36927
Archangelskij Oblastnoj Muzej Izobrazitelnych Iskusstv, Archangelsk 36768
Architekturno-chudožestvennyj Ansambl byvšego Spaso-Jakovleskogo Monastyrja, Rostov 37598
Art Centr Puškinskaja 10, Sankt-Peterburg 37624
Baškirskij Respublikanskij Chudožestvennyj Muzej im. M.V. Nesterova, Ufa 37883
Bolšoj Dvorec, Lomonosov 37106
Burjatskij Respublikanskij Chudožestvennyj Muzej im. C.S. Sampilova, Ulan-Udė 37892
Caricyno - Gosudarstvennyj Istoriko-architekturnyj, Chudožestvennyj i Landšaftnyj Muzej-Zapovednik, Moskva 37143
Centralnyj Dom Chudožnikov, Moskva 37144
Čitinskij Oblastnoj Chudožestvennyj Muzej, Čita 36868
Čuvašskij Gosudarstvennyj Chudožestvennyj Muzej, Čeboksary 36836
Dagestanskij Gosudarstvennyj Objedinennyj Istoriko-architekturnyj Muzej, Machačkala 37115
Dagestanskij Muzej Izobrazitelnych Iskusstv, Machačkala 37116
Ekaterininskij Dvorec - Gosudarstvennyj Muzej Carskoje Selo, Puškin 37579
Galereja Zarubežnogo Iskusstva im. Professora M.F. Gabyševa, Jakutsk 36961
Gosudarstvennyj Chudožestvenno-architekturnyj Dvorcovo-parkovyj Muzej-zapovednik Oranienbaum, Lomonosov 37108
Gosudarstvennyj Chudožestvenno-architekturnyj Dvorcovo-parkovyj Muzej-zapovednik Pavlovsk, Pavlovsk 37528
Gosudarstvennyj Chudožestvennyj Istoriko-architekturnyj i prirodno-landšaftnyj Muzej-zapovednik Kolomenskoe, Moskva 37191
Gosudarstvennyj Istoriko-chudožestvennyj Dvorcovo-parkovyj Muzej-zapovednik Gatčina, Gatčina 36920
Gosudarstvennyj Istoriko-kulturnyj Muzej-zapovednik "Moskovskij Kreml", Moskva 37194
Gosudarstvennyj Istoriko-literaturnyj Muzej-Zapovednik A.S. Puškina s usadbami Vjazemy i Zacharovo, Bolšie Vjazemy 36814
Gosudarstvennyj Memorialnyj Istoriko-chudožestvennyj i Prirodnyj Muzej-zapovednik V.D. Polenova, Strachovo 37808
Gosudarstvennyj Muzej Izobrazitelnych Iskusstv Respubliki Kalmykija, Ėlista 36911
Gosudarstvennyj Muzej-masterskaja Skulptora A.S. Golubkinoj, Moskva 37200
Gosudarstvennyj Muzej A.S. Puškina, Moskva 37202
Gosudarstvennyj Russkij Muzej, Sankt-Peterburg 37649
Gosudarstvennyj Vladimiro-Suzdalskij Istoriko-architekturnyj i Chudožestvennyj Muzej-zapovednik, Vladimir 37934
Irkutskij Oblastnoj Chudožestvennyj Muzej im. V.P. Sukačeva, Irkutsk 36932
Istoriko-Architekturnyj, Chudožestvennyj i Archeologičeskij Muzej "Zarajskij Kreml", Zarajsk 37986
Istoriko-Architekturnyj i Chudožestvennyj Muzej Ivangorodskaja Krepost, Ivangorod 36942
Istoriko-chudožestvennyj Muzej, Uglič 37891
Istoriko-memorialnyj i Architekturno-chudožestvennyj Muzej, Tichvin 37842
Ivanovskij Oblastnoj Chudožestvennyj Muzej, Ivanovo 36948
Jaroslavskij Chudožestvennyj Muzej - Gubernatorskij Dom, Jaroslavl 36967
Jaroslavskij Istoriko-architekturnyj i Chudožestvennyj Muzej-zapovednik, Jaroslavl 36968
Jurinskij Istoriko-chudožestvennyj Muzej, Jurino 36982
Kabardino-Balkarskij Gosudarstvennyj Muzej Izobrazitelnych Iskusstv, Nalčik 37438
Kaliningradskaja Chudožestvennaja Galereja, Kaliningrad 36988
Kaliningradskij Oblastnoj Istoriko-chudožestvennyj Muzej, Kaliningrad 36989
Kalmyckaja Gosudarstvennaja Kartinnaja Galereja, Ėlista 36912
Kalužskij Oblastnoj Chudožestvennyj Muzej, Kaluga 36997
Karačaevo-Čerkesskij Istoriko-kulturnyj i Prirodnyj Muzej-Zapovednik, Čerkessk 36854
Kartinnaja Galereja K. Vasiljeva, Kazan 37009
Kirovskij Oblastnoj Chudožestvennyj Muzej im. Viktora i Apollinarija Vasnecovych, Kirov 37038
Kostromskoj Gosudarstvennyj Istoriko-architekturnyj Muzej-zapovednik, Kostroma 37061
Kostromskoj Gosudarstvennyj Obedinennyj Chudožestvennyj Muzej, Kostroma 37062
Krasnojarskij Chudožestvennyj Muzej im. V.I. Surikova, Krasnojarsk 37069
Kulturno-vystavočnyj Centr Raduga, Uljanovsk 37900
Malojaroslaveckaja Gorodskaja Kartinnaja Galereja, Malojaroslavec 37125
Memorialnyj Muzej-Masterskaja Skulptora S.T. Konenkova, Moskva 37246
Memorialnyj Muzej-usadba Chudožnika N.A. Jarošenko, Kislovodsk 37048
Meždunarodnyj Centr-Muzej im. N.K. Rericha, Moskva 37249
Mitropoličji Palaty - Muzej Drevnerusskogo Iskusstva, Jaroslavl 36971
Mordovskij Muzej Izobrazitelnych Isskustv im. S.D. Ėrzi, Saransk 37739
Moskovskij Gosudarstvennyj Vystavočnyj Zal Novyj Manež, Moskva 37256
Muzej 'A Muzy ne Molčali', Sankt-Peterburg 37669
Muzej-institut Semji Rerichov, Sankt-Peterburg 37679
Muzej Izobrazitelnych Iskusstv im. A.S. Puškina, Moskva 37302
Muzej Izobrazitelnyh Iskusstv Respubliki Marij Ėl, Joškar-Ola 36979
Muzej Klassičeskogo i Sovremennogo Iskusstva - Dom Burganova, Moskva 37307
Muzej Krylova, Tula 37868
Muzej Nonkonformistskogo Iskusstva, Sankt-Peterburg 37703
Muzej 'Novoj Akademii Izjaščnych Iskusstv', Sankt-Peterburg 37704
Muzej Parka Iskusstv na Krymskoj Naberežnoj, Moskva 37347
Muzej Rossijskoj Akademii Chudožestv, Sankt-Peterburg 37706
Muzej Semji Benua, Sankt-Peterburg 37707
Muzej Sovremennogo Izobrazitelnogo Iskusstva im. A.A. Plastova, Uljanovsk 37908
Muzej-usadba im. N.K. Rericha, Izvara 36958
Muzej-zapovednik Dmitrovskij Kreml, Dmitrov 36876
Muzej-zapovednik Kazanskij Kreml, Kazan 37025
Muzejno-vystavočnyj Centr, Ivanovo 36952
Muzejno-vystavočnyj Centr Ego, Sankt-Peterburg 37714
Nacionalnaja Galereja Respubliki Komi, Syktyvkar 37820
Nižegorodskij Chudožestvennyj Muzej, Nižnij Novgorod 37467
Nižnetagilskij Municipalnyj Muzej Izobrazitelnych Iskusstv, Nižnij Tagil 37472
Novokuzneckij Chudožestvennyj Muzej, Novokuzneck 37481
Oblastnaja Kartinnaja Galereja Obraz, Kaluga 37001
Oloneckaja Kartinnaja Galereja, Olonec 37495
Omskij Oblastnoj Muzej Izobrazitelnych Iskusstv im. M.A. Vrubelja, Omsk 37501
Pereslavl-Zalesskij Gosudarstvennyj Istoriko-architekturnyj i Chudožestvennyj Muzej-zapovednik, Pereslavl-Zalesskij 37542
Petergof - Gosudarstvennyj Muzej Zapovednik, Sankt-Peterburg 37718
Respublikanskij Vystavočnyj Zal, Joškar-Ola 36981
Sergačskij Kraevedčeskij Muzej, Sergač 37761
Sergievo-Posadskij Gosudarstvennyj Istoriko-chudožestvennyj Muzej Zapovednik, Sergiev Posad 37764
Severo-osetinskij Gosudarstvennyj Obedinennyj Muzej Istorii, Architektury i Literatury, Vladikavkaz 37932
Smolenskaja Chudožestvennaja Galereja, Smolensk 37776
Smolenskij Muzej-Zapovednik, Smolensk 37778
Sosnovoborskij Chudožestvennyj Muzei Sovremennogo Iskusstva, Sosnovyj Bor 37791
Sovremennaja Galereja-muzej Nikor, Moskva 37406
Tarusskaja Kartinnaja Galereja, Tarusa 37835
Tatarskij Gosudarstvennyj Muzej Izobrazitelnych Iskusstv, Kazan 37027
Tjumenskij Oblastnoj Kraevedčeskij Muzej im. I.Ja. Slovcova, Tjumen 37852
Tomskij Oblastnoj Chudožestvennyij Muzej, Tomsk 37858
Učebno-chudožestvennyj Muzej Slepkov im. I.V. Cvetaeva, Moskva 37409
Udmurtskij Respublikanskij Muzej Izobrazitelnych Iskusstv, Iževsk 36956
Uljanovskij Oblastnoj Chudožestvennyj Muzej, Uljanovsk 37909
Voronežskij Chudožestvennyj Muzej im. I.N. Kramskogo, Voronež 37975
Vystavočnyj Centr Sankt-Peterburgskogo Otdelenija Sojuza Chudožnikov Rossii, Sankt-Peterburg 37733
Vystavočnyj Zal, Tula 37876
Vystavočnyj Zal Moskovskogo Sojuza Chudožnikov, Moskva 37425
Vystavočnyj Zal Sojuza Chudožnikov Rossii, Sankt-Peterburg 37735
Zvenigorodskij Istoriko-architekturnyj i Chudožestvennyj Muzej, Zvenigorod 37991

San Marino

Museo - Pinacoteca San Francesco, San Marino 38009

Serbia

Galerija Lazar Vozarević, Sremska Mitrovica 38150
Galerija Milan Konjović, Sombor 38146
Galerija Milene Pavlović Barili, Požarevac 38131
Galerija Savremene Umetnosti Likovna Jesen, Sombor 38147
Galerija Slika Sava Šumanović, Šid 38140
Galerija Umetničkih Dela, Beograd 38048
Narodni Muzej u Beogradu, Beograd 38082
Narodni Muzej Valjevo, Valjevo 38160
Zadužbina Kralja Petra I Karadjordjevića, Topola 38157

Singapore

The Art Gallery, Singapore 38170
Art Retreat, Singapore 38171

Slovakia

Dom Mikuláša Galandu, Turčianske Teplice 38382
Galéria Jozefa Kollára, Banská Štiavnica 38190
Galéria Petra Michala Bohúňa, Liptovský Mikuláš 38286
Galéria Umelcov Spiša, Spišská Nová Ves 38358
Kysucka Galéria, Čadca 38242
Martin Benka Múzeum, Martin 38295
Mestská Galéria, Rimavská Sobota 38341
Mestská Výstavná Sieň Barónka Kežmarok, Kežmarok 38266
Múzeum Arthura Fleischmanna, Bratislava 38220
Múzeum Ferdiša Kostku, Stupava 38363
Múzeum vo Svätý Antone, Svätý Anton 38364
Nitrianska Galéria, Nitra 38314
Šarišská Galéria, Prešov 38337
Slovenská Národná Galéria, Bratislava 38233
Slovenská Národná Galéria - Zvolenský Zámok, Zvolen 38395
Slovenské Banské Múzeum, Banská Štiavnica 38193
Turčianska Galéria, Martin 38301

Slovenia

Cankarjev Dom, Ljubljana 38447
Galerija Božidar Jakac, Kostanjevica na Krki ... 38435
Galerija Likovnih Samorastnikov, Trebnje 38529
Galerija Likovnih Umetnosti, Slovenj Gradec ... 38517
Galerija Mestne Hiše, Kranj 38436
Galerije Sivčeva Hiša, Radovljica 38501
Mestna Galerija, Šoštanj 38523
Obalne Galerije - Galerija Loža, Koper 38433
Obalne Galerije - Mestna Galerija, Piran 38491
Pilonova Galerija, Ajdovščina 38397
Pokrajinski Muzej, Murska Sobota 38487
Razstavni Salon Rotovž, Maribor 38478

South Africa

Anton van Wouw House, Brooklyn 38574
Caledon Museum, Caledon 38576
Jack Heath Art Gallery, Pietermaritzburg 38704
Johannes Stegmann Art Gallery, Bloemfontein 38563
Johannesburg Art Gallery, Johannesburg 38658
Mayibuye Centre Collection, Bellville 38558
Nelson Mandela Metropolitan Art Museum, Port Elizabeth 38711
Old Arts Gallery, Tshwane 38753
Oliewenhuis Art Museum, Bloemfontein 38569
Rupert Museum, Stellenbosch 38737
University Museum - Sasol Art Museum, Stellenbosch 38741
University of Johannesburg Art Gallery, Auckland Park 38554
University of Pretoria Art Collection, Tshwane 38761
Van Gybland-Oosterhoff Collection, Tshwane 38762
Van Tilburg Collection, Tshwane 38763
Works of Art Collection, Rosebank 38728

Spain

Bilboko Arte Eder Museoa / Museo de Bellas Artes de Bilbao, Bilbao 39038
Caixa Forum Madrid, Madrid 39501
CaixaForum Barcelona, Barcelona 38952
CaixaForum Palma de Mallorca, Palma de Mallorca 39776
Cartuja de la Asunción, Granada 39356
Casa de los Morlanes, Zaragoza 40337
Casa Museo de Rafael Alberti, El Puerto de Santa María 39878
Castell Cartoixa Vallparadis, Terrassa 40115
Centre de Cultura Sa Nostra, Palma de Mallorca 39779
Centro Cultural Conde Duque, Madrid 39508
Centro de Historia de Zaragoza, Zaragoza 40338
Colección de la Real Academia de la Historia, Madrid 39514
Conjunt Monumental de la Església de Sant Pere, Terrassa 40117
Fundació Antoni Tàpies, Barcelona 38956
Fundació Joan Miró, Barcelona 38957
Fundació Pilar i Joan Miró, Palma de Mallorca 39783
Fundación Carlos de Amberes, Madrid 39519
Fundación César Manrique, Teguise 40108
Fundación Museo de las Ferias, Medina del Campo 39638
Fundación Pablo o Ruiz Picasso Museo Casa Natal, Málaga 39596
Fundación Yannick y Ben Jakober, Alcúdia (Baleares) 38831
Galeria d'Art Mar, Barcelona 38961
Galeria d'Art Terra Ferma, Lleida 39474
Koldo Mitxelena Kulturunea Erakustaret, Donostia-San Sebastián 39249
Museo Artístico de Roncesvalles, Roncesvalles 39914
Museo Bonsor, Mairena del Alcor 39593
Museo Casa Natal de Isabel la Católica, Madrigal de las Altas Torres 39589
Museo Cerezo Moreno, Villatorres 40309
Museo Cerralbo, Madrid 39527
Museo de Arte Contemporáneo José María Moreno Galván, La Puebla de Cazalla 39874
Museo de Ávila, Ávila 38920
Museo de Bellas Artes, Valencia 40204
Museo de Bellas Artes de Asturias, Oviedo 39759
Museo de Bellas Artes de Castellón, Castellón de la Plana 39148
Museo de Bellas Artes de Gravina, Alicante 38844
Museo de Bellas Artes de Murcia, Murcia 39692
Museo de Bellas Artes de Sevilla, Sevilla 40069
Museo de Burgos, Burgos 39068
Museo de Ciudad Real, Ciudad Real 39187
Museo de Cuenca, Cuenca 39232
Museo de Gomellano, Gumiel de Hizán 39383
Museo de Guadalajara, Guadalajara 39375
Museo de Huelva, Huelva 39396
Museo de la Encarnación, Corella 39212
Museo de las Telecomunicaciones, Madrid 39549
Museo de las Termas Públicas de Caesaraugusta, Zaragoza 40342
Museo de Palencia, Palencia 39770
Museo de Santa Cruz, Toledo 40134
Museo de Valladolid, Valladolid 40235
Museo de Zaragoza, Zaragoza 40343
Museo del Traje, Madrid 39558
Museo Etnográfico Tanit, San Bartolomé de Lanzarote 39951
Museo Etnológico de Horcajuleo de la Sierra, Horcajuelo de la Sierra 39392
Museo-Exposición Marín Bosqued, Aguarón 38788
Museo Fundación Eugenio Granell, Santiago de Compostela 40028
Museo Fundación Jesús Otero, Santillana del Mar 40035
Museo Gregorio Prieto, Valdepeñas 40191
Museo Guerrero Malagón, Urda 40182
Museo Municipal, Requena 39892
Museo Municipal de Arte en Vidrio de Alcorcón, Alcorcón 38825
Museo Municipal Marceliano Santa María, Burgos 39072
Museo Nacional de Reproducciones Artísticas, Madrid 39572
Museo Thyssen-Bornemisza, Madrid 39583
Museu d'Art Casa Turull, Sabadell 39924
Museu d'Art Frederic Marès, Montblanc 39671
Museu d'Art Municipal, Cadaqués 39075
Museu de Granollers, Granollers 39373
Museu de la Ciutat, Valls 40250
Museu de l'Art de la Pell, Vic 40269
Museu de l'Empordà, Figueres 39298
Museu de Montserrat, Monistrol de Montserrat 39666
Museu de Rubí, Rubí 39922
Museu de Vilafranca, Vilafranca del Penedès ... 40280
Museu Deu, El Vendrell 40262
Museu Parroquial dels Sants Joans, Estivella 39291
Palacio de Viana, Córdoba 39210
Rajoloteca Salvador Miquel, Vallromanes 40249
Reial Acadèmia Catalana de Belles Arts de Sant Jordi, Barcelona 39006
Sala de Exposiciones de las Alhajas, Madrid . 39588
Sala d'Exposicions, Ponts 39858
Salas Municipales de Arte Sanz-Enea, Zarautz 40354
Sección de Bellas Artes, Ceuta 39173
Zarauzko Arte eta Historia Museoa, Zarautz .. 40355

Sudan

Art Gallery of the College of Fine and Applied Arts, Khartoum 40386

Sweden

Åmåls Konsthall, Åmål 40417
Arvika Konsthall, Arvika 40437
Axel Ebbes Konsthall, Trelleborg 40980
Båtsmanskasernen, Karlskrona 40633
Bergsjö Konsthall, Bergsjö 40457
Bollnäs Konsthallen, Bollnäs 40464
Borås Konstmuseum, Borås 40468
Bror Hjorths Hus, Uppsala 41003
Dalslands konstmuseum, Åsensbruk 40443
Dunkers Kulturhus, Helsingborg 40585
Höganäs Museum och Konsthall, Höganäs 40594
Karlskoga Konsthall, Karlskoga 40631
Kulturhuset, Stockholm 40903
Kungliga Slottet i Stockholm, Stockholm 40906
Länsmuseet Gävleborg, Gävle 40532
Landskrona Konsthall, Landskrona 40668
Lidköpings Konsthall, Lidköping 40675
Liljevalchs Konsthall, Stockholm 40908
Linköpings Konsthall, Linköping 40682
Ljungbergmuseet, Ljungby 40686
Lund University Art Collection, Lund 40704
Lunds Konsthall, Lund 40705
Malmö Konsthall, Malmö 40722
Malmö Konstmuseum, Malmö 40723
Mjellby Konstmuseum, Halmstad 40579
Museet Kvarnen, Filipstad 40520
Museum Anna Nordlander, Skellefteå 40848
Museum Lionardo da Vinci Ideale, Karlskrona 40636
Nationalmuseum Stockholm, Stockholm 40918
Örebro Konsthallen, Örebro 40788
Östergötlands Länsmuseum, Linköping 40684
Olle Olsson-Huset, Solna 40871
Rackstadmuseet, Arvika 40439
Rättviks Konstmuseum och Naturmuseum, Rättvik 40812
Regionmuseet Kristianstad, Kristianstad 40655
Skissernas Museum, Lund 40708
Skövde Konsthall och Konstmuseum, Skövde 40851
Smålands Konstarkiv, Värnamo 41029
Staffanstorps Konsthall, Staffanstorp 40878
Stockholms Universitet Schefflerska huset, Stockholm 40943
Strindbergsmuseet, Stockholm 40944
Sundsbergs Gård Museum och Konsthall, Sunne 40962
Tidaholms Konsthall, Tidaholm 40971
Tomelilla Konsthall, Tomelilla 40974
Ulricehamns Konst- och Östasiatiska Museum, Ulricehamn 40995
Uppsala Konstmuseum, Uppsala 41013
Uppsala Universitets Konstsamlingar, Uppsala 41016
Västerås Konstmuseum, Västerås 41032
Vasamuseet, Stockholm 40949
Zornmuseet, Mora 40748

Switzerland

Andrea Robbi Museum, Sils Maria 41869
Casa Anatta e Casa Selma, Ascona 41127
Castello di Sasso Corbaro, Bellinzona 41191
Cerny Inuit Collection, Bern 41204
Coninx Museum, Zürich 42068
Dichter- und Stadtmuseum Liestal (DISTL)/ Herwegh-Archiv, Liestal 41561
Espace Jean Tinguely - Niki de Saint-Phalle, Fribourg 41384
Fondation de l'Hermitage, Lausanne 41539
Fondation Louis Moret, Martigny 41607
Fondation Pierre Gianadda, Martigny 41608
Fondazione Galleria Gottardo, Lugano 41580
Galerie Gersag Emmen, Emmenbrücke 41348
Galerie Gluri-Suter-Huus, Wettingen 42008
Hans Trudel-Haus Stiftung, Baden 41144
Haus für Kunst Uri, Altdorf 41103
Heidi Weber-Museum, Zürich 42075
Kulturmühle Lützelflüh, Lützelflüh 41579
Kunst im Alten Schützenhaus, Zofingen 42060
Kunstforum Bâloise, Basel 41164
Kunsthalle Arbon, Arbon 41122
Kunsthalle Palazzo, Liestal 41563
Kunsthalle Winterthur, Winterthur 42028
Kunsthaus Baselland, Muttenz 41670
Kunsthaus Grenchen, Grenchen 41453
Kunsthof Zürich, Zürich 42084
Kunstmuseum, Sankt Gallen 41805
Kunstmuseum Bern, Bern 41210
Kunstmuseum des Kantons Thurgau, Warth .. 42002
Kunstmuseum Poschiavo, Poschiavo 41730
Kunstmuseum Thun, Thun 41937
Kunstraum Aarau, Aarau 41082
Kunstsammlung Villa Schüpbach, Steffisburg 41910
Manoir de la Ville de Martigny, Martigny 41610
Martin-Lauterburg-Stiftung, Bern 41211
Musée Barbier-Mueller, Genève 41417
Musée d'Art du Valais, Sion 41876
Musée d'art et d'histoire, Fribourg 41387
Musée d'Art et d'Histoire Neuchâtel, Neuchâtel 41680
Musée de la Science-Fiction, de l'Utopie et des Voyages Extraordinaires, Yverdon-les-Bains ... 42048
Musée Fondation Léon Perrin, Môtiers 41647
Musée Rath, Genève 41426
Museo Cantonale d'Arte, Lugano 41581
Museo d'Arte Mendrisio, Mendrisio 41625
Museo d'Arte Moderna, Lugano 41583
Museo in Erba, Bellinzona 41194
Museo Wilhelm Schmid, Brè sopra Lugano 41252
Museum im Bellpark, Kriens 41518
Museum im Lagerhaus, Sankt Gallen 41807
Museum of Porn in Art, Zürich 42096
Museum Rietberg Zürich, Zürich 42097
Museum Tinguely, Basel 41173
Museum zu Allerheiligen, Schaffhausen 41831
Nidwaldner Museum, Stans 41905
Ortsmuseum Zollikon, Zollikon 42062
Polenmuseum Rapperswil, Rapperswil 41746
Psychiatrie-Museum, Bern 41214
Sammlung Hauser Wirth, Henau 41479
Sankturbanhof Sursee, Sursee 41926
Schlößchen Vorder Bleichenberg Biberist, Biberist 41227
Verein Kunstsammlung Unterseen, Unterseen 41961

Syria

National Museum of Damascus, Damascus ... 42132

Taiwan

Accton Arts Foundation, Hsinchu 42204
Art Center, Douliou 42183
Art Gallery of Hungkuang University, Shalu ... 42319
Art Museum of Hsinchu County, Jhubei 42223
Art Site of Chai-Yi Railway Warehouse, Chai-Yi 42149
Black Woods Museum, Houbi 42203
Carrie Chang Fine Arts Center, Tamsui 42447
Cheng Ta Art Gallery, Tainan 42360
Chengshiu University Art Center, Niaosong ... 42286
Children's Art Museum, Taipei 42381
Children's Gallery, Taichung 42347
Children's Museum of Art and Creativity, Kaohsiung 42232
Chung Hua University Arts Center, Hsinchu ... 42205
Chung Yi Gallery, Danshuei 42177
Chungchern Art Museum and Jhongjheng Art Center, Pingtung 42290
CMC Art Museum, Taipei 42385
Feng Chia University Art Center, Taichung 42348
Kuandu Museum of Fine Arts, Taipei 42396
Kun-Shan University Art Center, Yongkang 42474
LHU Arts Center, Gueishan 42196
Lin Yuan Simplicity Art Museum, Yuchi 42482
Local Cultural Museum, Keelung 42248
The Mayor's Residence Arts Salon, Taipei 42403
Meidiyas Museum of Art and Papertole, Linkou 42253
National Chiao Tung University Art Center, Hsinchu 42212
National Chung Hsing University Art Center, Taichung 42351
National Taiwan Museum of Fine Arts, Taichung 42353
National Tsing Hua University Arts Center, Hsinchu 42213
Providence University Art Center, Shalu 42321
Taipei Children's Art Museum, Taipei 42427
Taiwan International Visual Arts Center, Taipei 42433
Triangle Rush Exhibition Hall, Yuanli 42479
Tsing Hua Stone Teapot Arts Center, Yungho . 42485
Tsou-Ma-Li Country Gallery, Danei 42176
Tunghai University Art Center, Taichung 42357
Wan Fang Arts Museum, Taipei 42437
Wang's Fine Art Museum, Kaohsiung 42245
Yang Ming Studio, Taipei 42440

Thailand

The Arts Center, Bangkok 42531
National Gallery, Bangkok 42551
Phiphitthaphanthasathan Haeng Chat Ram Khamhaeng Sukhothai, Sukhothai 42603

Trinidad and Tobago

National Museum and Art Gallery of Trinidad and Tobago, Port-of-Spain 42615

Tunisia

La Nationale Galerie d'Art Alyssa, Le Bardo ... 42617

Turkey

Adana Devlet Güzel Sanatlar Galerisi, Adana . 42672
Amasya Devlet Güzel Sanatlar Galerisi, Amasya 42689
Ankara Devlet Güzel Sanatlar Galerisi, Ankara 42697
Antalya Devlet Güzel Sanatlar Galerisi, Antalya 42716
Aydın Devlet Güzel Sanatlar Galerisi, Aydın ... 42720
Balıkesir Devlet Güzel Sanatlar Galerisi, Balıkesir 42723
Basın Müzesi, İstanbul 42807
Beyoglu Sanat Merkezi, İstanbul 42809
Bilecik Devlet Güzel Sanatlar Galerisi, Bilecik 42730
Bolu Devlet Güzel Sanatlar Galerisi, Bolu 42733
Burdur Devlet Güzel Sanatlar Galerisi, Burdur 42736
Bursa Devlet Güzel Sanatlar Galerisi, Bursa .. 42742
Çanakkale Devlet Güzel Sanatlar Galerisi, Çanakkale 42747
Çankırı Devlet Güzel Sanatlar Galerisi, Çankırı 42750
Çorum Devlet Güzel Sanatlar Galerisi, Çorum 42754
Diyarbakır Devlet Güzel Sanatlar Galerisi, Diyarbakır 42762
Edirne Devlet Güzel Sanatlar Galerisi, Edirne . 42767
Eskişehir Devlet Güzel Sanatlar Galerisi, Eskişehir 42780
Gaziantep Devlet Güzel Sanatlar Galerisi, Gaziantep 42786
Hatay Devlet Güzel Sanatlar Galerisi, Hatay ... 42793
İsparta Devlet Güzel Sanatlar Galerisi, İsparta 42799
İstanbul Devlet Güzel Sanatlar Galerisi, İstanbul 42825
Kahramanmaraş Devlet Güzel Sanatlar Galerisi, Kahramanmaraş 42861
Kars Devlet Güzel Sanatlar Galerisi, Kars 42866
Kastamonu Devlet Güzel Sanatlar Galerisi, Kastamonu 42869
Kayseri Devlet Güzel Sanatlar Galerisi, Kayseri 42875
Kırşehir Devlet Güzel Sanatlar Galerisi, Kırşehir 42880
Kocaeli Devlet Güzel Sanatlar Galerisi, İzmit . 42858
Konya Devlet Güzel Sanatlar Galerisi, Konya . 42887
Kütahya Devlet Güzel Sanatlar Galerisi, Kütahya 42895
Malatya Devlet Güzel Sanatlar Galerisi, Malatya 42897
Manisa Devlet Güzel Sanatlar Galerisi, Manisa 42902
Muğla Devlet Güzel Sanatlar Galerisi, Muğla . 42912
Nevşehir Devlet Güzel Sanatlar Galerisi, Nevşehir 42916
Ordu Devlet Güzel Sanatlar Galerisi, Ordu 42921
Rize Devlet Güzel Sanatlar Galerisi, Rize 42926
Sakıp Sabancı Müzesi, İstanbul 42838
Samsun Devlet Güzel Sanatlar Galerisi, Samsun 42933
Şanlıurfa Devlet Güzel Sanatlar Galerisi, Şanlıurfa 42934
Şefik Bursali Müze Evi, Ankara 42711
Sinop Devlet Güzel Sanatlar Galerisi, Sinop ... 42943
Sivas Devlet Güzel Sanatlar Galerisi, Sivas ... 42946

A

Tekirdağ Devlet Güzel Sanatlar Galerisi, Tekirdağ ... 42950
Temel Sanat Kolleksiyon, İstanbul ... 42842
Trabzon Devlet Güzel Sanatlar Galerisi, Trabzon ... 42956
Van Devlet Güzel Sanatlar Galerisi, Van ... 42961
Yozgat Devlet Güzel Sanatlar Galerisi, Yozgat 42967
Zonguldak Devlet Güzel Sanatlar Galerisi, Zonguldak ... 42968

Ukraine

Čerkas'kyj Chudožnij Muzej, Čerkasy ... 43001
Černigivs'kyj Oblasnyj Chudožnij Muzej, Černigiv ... 43004
Černivec'kyj Chudožnij Muzej, Černivci ... 43009
Charkivs'kyj Chudožnij Muzej, Charkiv ... 43017
Chmel'nyc'kyj Chudožnij Muzej, Chmel'nyc'kyj ... 43026
Chudožnij Muzej, Žytomyr ... 43211
Chudožno-memorial'nyj Muzej Ivana Truša, Lviv ... 43119
Chudožno-memorial'nyj Muzej Leopol'da Levyc'kogo, Lviv ... 43120
Chudožno-memorial'nyj Muzej Oleksi Novakivs'kogo, Lviv ... 43121
Chudožno-memorial'nyj Muzej Oleni Kul'čic'koï, Lviv ... 43122
Chudožnyj Muzej im. O. Šovkunenka, Cherson ... 43022
Deržavnyj Istoriko-kul'turnyj Zapovidnyk, Kam'janec'-Podil's'kyj ... 43051
Dnipropetrovskyj Chudožnij Muzej, Dnipropetrovs'k ... 43032
Ivano-Frankivs'kyj Chudožnij Muzej, Ivano-Frankivs'k ... 43041
Kartynna Galereja, Kam'janec'-Podil's'kyj ... 43052
Kartynna Galereja, Luc'k ... 43111
Kirovograds'kyj Oblasnyj Chudožnij Muzej, Kirovograd ... 43067
Kyïvs'kyj Literaturno-memorial'nyj Budinok-muzej Tarasa Ševčenka, Kyïv ... 43081
Lugans'kyj Chudožnij Muzej, Lugans'k ... 43114
Lvivs'ka Galereja Mystectv, Lviv ... 43125
Memorial'nyj Budinok-muzej Tarasa Ševčenka, Kyïv ... 43087
Mis'ka Chudožna Galereja im. S.I. Vasyl'kivs'kogo, Charkiv ... 43020
Muzej Narodnogo Mystectva Slobožanščyny, Charkiv ... 43021
Nacionalnij Muzej u Lvovi, Lviv ... 43137
Nacional'nyj Muzej Tarasa Ševčenka, Kyïv ... 43107
Odes'kyj Chudožnij Muzej, Odesa ... 43158
Poltavs'kyj Chudožnij Muzej, Poltava ... 43167
Rivnens'kyj Krajeznavčyj Muzej, Rivne ... 43173
Simferopolskij Chudožestvennyj Muzej, Simferopol' ... 43186
Sums'kyj Chudožnij Muzej im. N. Onac'kogo, Sumy ... 43190
Ternopil's'kyj Oblasnyj Chudožnij Muzej, Ternopil' ... 43193
Vinnyc'kyj Oblasnyj Chudožnij Muzej, Vinnycja ... 43205
Zakarpats'kyj Oblasnyj Chudožnij Muzej im. J. Bokšaja, Užgorod ... 43200
Zaporiz'kyj Chudožnij Muzej, Zaporižžja ... 43209

United Arab Emirates

Emirates Very Special Arts, Sharjah ... 43224
Sharjah Art Museum, Sharjah ... 43226

United Kingdom

Abbot Hall Art Gallery, Kendal ... 44557
Aberdeen Arts Centre, Aberdeen ... 43234
Aberdona Gallery, Alloa ... 43281
Amgueddfa Cymru - National Museum of Wales, Cardiff ... 43748
An Lanntair Art Centre, Stornoway ... 45766
Angel Row Gallery, Nottingham ... 45288
Apsley House, London ... 44784
Architecture Foundation, London ... 44785
Ards Art Centre, Newtownards ... 45260
Art Library, Leeds ... 44659
Arts Centre, Washington ... 45946
Ashmolean Museum of Art and Archaeology, Oxford ... 45327
Astley-Cheetham Art Gallery, Stalybridge ... 45733
Axiom Centre for Arts, Cheltenham ... 43807
Ballycastle Museum, Ballycastle ... 43360
Bankside Gallery, London ... 44789
Barbican Art Gallery, London ... 44791
Barreau Le Maistre Art Gallery, Saint Helier ... 45569
Bedford Central Library Gallery, Bedford ... 43433
Beecroft Art Gallery, Southend-on-Sea ... 45710
Berwick Borough Museum and Art Gallery, Berwick-upon-Tweed ... 43459
Beverley Art Gallery, Beverley ... 43468
Billingham Art Gallery, Billingham ... 43487
Blackburn Museum and Art Gallery, Blackburn, Lancashire ... 43526
Blenheim Palace, Woodstock ... 46066
Borough Museum and Art Gallery, Berwick-upon-Tweed ... 43460
Boxfield Gallery, Stevenage ... 45741
Brecknock Museum and Art Gallery, Brecon ... 43597
Bristol's City Museum and Art Gallery, Bristol 43632
Bromham Mill and Art Gallery, Bromham ... 43653
Broughton House and Garden, Kirkcudbright ... 44610
BSAT Gallery, Newmarket ... 45238
Burghley House, Stamford ... 45735
Bury Saint Edmunds Art Gallery, Bury Saint Edmunds ... 43694
Bushey Museum and Art Gallery, Bushey ... 43699
Camborne School of Mines Geological Museum and Art Gallery, Pool ... 45404
Camden Arts Centre, London ... 44812
Canada House Gallery, London ... 44813
Cannon Hall Museum, Barnsley, South Yorkshire ... 43384
Castle Museum and Art Gallery, Norwich ... 45276
Castle Park Arts Centre, Frodsham ... 44256
Cecil Higgins Art Gallery, Bedford ... 43435
Central Art Gallery, Ashton-under-Lyne ... 43331
Chapman Gallery, Salford ... 45598
Cheltenham Art Gallery and Museum, Cheltenham ... 43810
Cheshire Military Museum, Chester ... 43816
Chisenhale Gallery, London ... 44819
Christchurch Mansion, Ipswich ... 44528
City Art Centre, Edinburgh ... 44116
The City Gallery, Leicester ... 44674
City Museum and Art Gallery, Plymouth ... 45390
City Museum and Records Office, Portsmouth 45428
Cliffe Castle, Keighley ... 44550
Commonwealth Institute, London ... 44828
Corridor Gallery, Glenrothes ... 44313
Craven Museum and Gallery, Skipton ... 45688
Crawford Arts Centre, Saint Andrews ... 45553
Crescent Arts, Scarborough ... 45622
Dalkeith Arts Centre, Dalkeith ... 43945
The Danish Cultural Institute, Edinburgh ... 44118
The Dean Clough Galleries, Halifax ... 44383
Denbigh Museum and Gallery, Denbigh ... 43964
Dick Institute Museum and Art Gallery, Kilmarnock ... 44579
Doncaster Museum and Art Gallery, Doncaster ... 43992
Dudley Museum and Art Gallery, Dudley ... 44030
Durham Light Infantry Museum and Durham Art Gallery, Durham ... 44071
Ealing College Gallery, London ... 44841
Eastwood House, Glasgow ... 44279
Eden Court Art Gallery, Inverness ... 44524
Fabrica, Brighton ... 43622
Farnhem Maltings Gallery, Farnham ... 44216
The Fitzwilliam Museum, Cambridge ... 43727
Focal Point Gallery, Southend-on-Sea ... 45711
Foundation for Women's Art, London ... 44850
Foyer / James Hockey Gallery, Farnham ... 44217
Fry Art Gallery, Saffron Walden ... 45539
The Gallery, Gateshead ... 44264
Gallery II, Bradford ... 43579
Glasgow School of Art - Mackintosh Collection, Glasgow ... 44283
Gomshall Gallery, Gomshall ... 44327
Graves Art Gallery, Sheffield ... 45651
Grays School of Art Gallery and Museum, Aberdeen ... 43243
Great Yarmouth Museums, Great Yarmouth ... 44358
Guernsey Museum and Art Gallery, Saint Peter Port ... 45591
Guild Gallery, Bristol ... 43636
Guildhall Gallery, Winchester ... 46028
Hamilton's Gallery, London ... 44869
Hanbury Hall, Droitwich ... 44025
Harewood House, Harewood ... 44394
Harris Museum and Art Gallery, Preston ... 45444
Hayloft Gallery, Armagh ... 43318
Hayward Gallery, London ... 44872
Helston Folk Museum, Helston ... 44444
Hereford Museum and Art Gallery, Hereford ... 44454
Hermitage Rooms, London ... 44874
Holden Gallery, Manchester ... 45085
Horsham Arts Centre, Horsham ... 44488
Howard Gardens Gallery, Cardiff ... 43750
Hull University Art Collection, Kingston-upon-Hull ... 44598
Hunterian Art Gallery, Glasgow ... 44284
Ikon Gallery, Birmingham ... 43505
John Hansard Gallery, Southampton ... 45704
Keswick Museum and Art Gallery, Keswick ... 44564
King's Lynn Arts Centre, King's Lynn ... 44586
Kirriemuir Gateway to the Glens Museum, Kirriemuir ... 44618
Landings Gallery, Edinburgh ... 44130
Lauderdale House, London ... 44898
Leeds Metropolitan University Gallery, Leeds 44663
Leighton House Museum, London ... 44899
Letchworth Museum and Art Gallery, Letchworth ... 44695
Lewis Collection, Cambridge ... 43729
Lewis Elton Gallery, Guildford ... 44372
Llandudno Museum and Art Gallery, Llandudno ... 44759
Longleat House, Warminster ... 45934
Lyth Arts Centre, Lyth ... 45058
MAC, Birmingham ... 43507
McManus Galleries, Dundee ... 44047
Maidstone Museum and Bentlif Art Gallery, Maidstone ... 45073
Manchester Art Gallery, Manchester ... 45088
Margaret Harvey Gallery, Saint Albans ... 45546
Marischal Museum, Aberdeen ... 43244
Medway Towns Gallery, Rochester ... 45508
Melbourne Hall, Melbourne ... 45123
Metropole Gallery, Folkestone ... 44238
Milton Keynes Gallery, Milton Keynes ... 45153
Morley Gallery, London ... 44919
Museum and Art Gallery, Forfar ... 44242
Museum and Art Gallery, Newport, Gwent ... 45240
Museum and Study Collection, London ... 44920
Museum of Installation, London ... 44926
Narwhal Inuit Art Gallery, London ... 44935
Nature in Art, Gloucester ... 44318
Newark Town Treasures and Art Gallery, Newark-on-Trent ... 45213
Newry and Mourne Arts Centre and Museum, Newry ... 45250
Northampton Museum and Art Gallery, Northampton ... 45270
Norwich Gallery, Norwich ... 45282
Nottingham Castle, Nottingham ... 45298
Old Museum Arts Centre, Belfast ... 43443
Old Town Hall Arts Centre, Hemel Hempstead ... 44447
Orchard Gallery, Londonderry ... 45023
Oriel Davies Gallery, Newtown, Powys ... 45257
Oriel Mostyn Gallery, Llandudno ... 44760
Orleans House Gallery, Twickenham ... 45888
Out of the Blue Gallery, Leek ... 44671
Parc Howard Museum and Art Gallery, Llanelli ... 44762
Park Gallery, Falkirk ... 44206
Pearce Institute, Glasgow ... 44293
Penlee House Gallery and Museum, Penzance 45361
Penrhyn Castle, Bangor, Gwynedd ... 43378
Peter Scott Gallery, Lancaster ... 44635
Petworth House and Park, Petworth ... 45378
Piece Hall Art Gallery, Halifax ... 44386
Pitlochry Festival Theatre Art Gallery, Pitlochry ... 45386
Plymouth Arts Centre, Plymouth ... 45392
Portico Gallery, Helensburgh ... 44439
Portico Library and Gallery, Manchester ... 45097
Robert Gordon University Collection, Aberdeen ... 43247
Robson Gallery, Selkirk ... 45637
Rottingdean Grange Art Gallery and Museum, Brighton ... 43624
Royal Albert Memorial Museum and Art Gallery, Exeter ... 44195
Royal Glasgow Institute Kelly Gallery, Glasgow ... 44298
Royal Scottish Academy, Edinburgh ... 44151
Rozelle House Galleries, Ayr ... 43348
Ruskin Library, Lancaster ... 44637
Ruskin Museum, Coniston ... 43885
Russell-Cotes Art Gallery and Museum, Bournemouth ... 43567
Saddleworth Museum and Art Gallery, Uppermill ... 45900
Saint Barbe Museum and Art Gallery, Lymington ... 45055
Saint David's Hall, Cardiff ... 43753
Saint Ives Society of Artists Members Gallery (Norway Gallery) and Mariners Gallery, Saint Ives, Cornwall ... 45578
Salford Museum and Art Gallery, Salford ... 45601
The Scott Gallery, Hawick ... 44427
Shakespeare's Globe Exhibition and Tour, London ... 44985
Sherborne House, Sherborne ... 45660
Sir Francis Cook Gallery, Augrès ... 43336
South Hill Park Arts Centre, Bracknell ... 43574
The Stanley Picker Gallery, Kingston-upon-Thames ... 44603
Stirling Smith Art Gallery and Museum, Stirling ... 45752
Strood Library Gallery, Rochester ... 45509
Tenby Museum and Art Gallery, Tenby ... 45836
Timespan Heritage Centre and Art Gallery, Helmsdale ... 44440
UCL Art Collections, London ... 44999
Ulster Museum, Belfast ... 43448
University of Brighton Gallery, Brighton ... 43627
University of Dundee Exhibitions, Dundee ... 44051
University of Dundee Museum Collections, Dundee ... 44052
University of Liverpool Art Gallery, Liverpool ... 44747
Vennel Gallery, Irvine ... 44536
Victoria and Albert Museum, London ... 45004
Victoria Art Gallery, Bath ... 43413
Warrington Museum and Art Gallery, Warrington ... 45936
Watford Museum, Watford ... 45950
Weston Park Museum, Sheffield ... 45656
Whitstable Museum and Gallery, Whitstable ... 46007
Winchester College Treasury, Winchester ... 46038
Women's Art Library - Make, London ... 45020
Worcester City Museum and Art Gallery, Worcester ... 46073
Worthing Museum and Art Gallery, Worthing ... 46079
Yard Gallery, Nottingham ... 45303

United States of America

1890 House-Museum and Center for Arts, Cortland ... 47828
80 Washington Square East Gallery, New York 51107
A.L. Fetterman Educational Museum, Fort Valley ... 48694
Abby Aldrich Rockefeller Folk Art Museum, Williamsburg ... 54164
Abington Art Center, Jenkintown ... 49619
Academy Art Museum, Easton ... 48267
Academy Gallery of Art, New York ... 51108
Ackland Art Museum, Chapel Hill ... 47344
ACME Art Gallery, Columbus ... 47749
Actual Art Foundation, New York ... 51109
Adam East Museum Art Center, Moses Lake ... 50866
Adams County Historical Society Museum, Gettysburg ... 48900
Adell McMillan Gallery, Eugene ... 48430
Adirondack Museum, Blue Mountain Lake ... 46876
Akron Art Museum, Akron ... 46146
AKUS Gallery, Willimantic ... 54178
Alaska Heritage Museum at Wells Fargo, Anchorage ... 46286
Albany Institute of History and Art, Albany ... 46157
Albany Museum of Art, Albany ... 46155
Albrecht-Kemper Museum of Art, Saint Joseph ... 52560
Albright-Knox Art Gallery, Buffalo ... 47114
The Albuquerque Museum of Art and History, Albuquerque ... 46175
Alexander Brest Museum, Jacksonville ... 49576
Alexandria Museum of Art, Alexandria ... 46195
Alford House-Anderson Fine Arts Center, Anderson ... 46291
Alice C. Sabatini Gallery, Topeka ... 53532
Allen Memorial Art Museum, Oberlin ... 51478
Allen Priebe Gallery, Oshkosh ... 51612
Allentown Art Museum, Allentown ... 46225
Allied Arts Center and Gallery, Richland ... 52306
Alternative Museum, New York ... 51110
Alton Museum of History and Art, Alton ... 46240
Alva DeMars Megan Chapel Art Center, Manchester ... 50420
Amarillo Museum of Art, Amarillo ... 46250
American Academy of Arts and Letters Art Museum, New York ... 51111
American Advertising Museum, Portland ... 52071
American Craft Museum, New York ... 51112
American Red Cross Museum, Washington ... 53844
American Sport Art Museum, Daphne ... 47965
American Visionary Art Museum, Baltimore ... 46545
Amerind Foundation, Dragoon ... 48181
Amon Carter Museum, Fort Worth ... 48709
Anchor Graphics, Chicago ... 47451
Anchorage Museum at Rasmuson Center, Anchorage ... 46288
Ancient Spanish Monastery of Saint Bernard de Clairvaux Cloisters, Miami ... 50619
Anderson County Arts Center, Anderson ... 46293
Anderson Gallery, Richmond ... 52322
Ann Arbor Art Center, Ann Arbor ... 46302
Anna Eaton Stout Memorial Art Gallery, Owensboro ... 51637
Annmary Brown Memorial, Providence ... 52160
Apex Museum, Atlanta ... 46421
Appleton Art Center, Appleton ... 46329
Appleton Museum of Art, Ocala ... 51480
Arcadia University Art Gallery, Glenside ... 48936
Arizona Museum for Youth, Mesa ... 50604
Arkansas Arts Center, Little Rock ... 50138
Arkansas State University Art Gallery, Jonesboro ... 49646
Arlington Arts Center, Arlington ... 46355
Arnot Art Museum, Elmira ... 48378
Arrowmont School of Arts and Crafts Collection, Gatlinburg ... 48873
Art and Culture Center of Hollywood, Hollywood ... 49322
Art Association of Harrisburg, Harrisburg ... 49178
Art Center at Fuller Lodge, Los Alamos ... 50202
Art Center Gallery, Warrensburg ... 53828
Art Center In Hargate, Concord ... 47773
Art Center of Corpus Christi, Corpus Christi ... 47817
Art Center of Waco, Waco ... 53772
Art Center Sarasota, Sarasota ... 52912
Art Complex Museum, Duxbury ... 48227
Art Department Gallery, Warwick ... 53836
Art Galleries, Northridge ... 51426
Art Galleries of Ramapo College, Mahwah ... 50400
Art Gallery, Atlanta ... 46422
Art Gallery, Oneonta ... 51562
Art Gallery, San Antonio ... 52706
Art Guild of Burlington Gallery, Burlington ... 47132

The Art Institute of Boston Main Gallery, Boston 46917
The Art Institute of Chicago, Chicago 47452
Art League of Houston, Houston 49380
Art League of Manatee County, Bradenton 46984
Art Museum, Los Gatos 50272
The Art Museum at Florida International University, Miami 50621
Art Museum of Greater Lafayette, Lafayette 49864
Art Museum of South Texas, Corpus Christi 47818
Art Museum of the University of Memphis, Memphis 50566
Art Museum of Western Virginia, Roanoke 52379
Art Omi, Ghent 48905
Art Omi, New York 51124
Art Pace, San Antonio 52707
Art Sation, Stone Mountain 53309
Artemisia Gallery, Chicago 47453
Artful Deposit Galleries, Allentown 46224
Arthur A. Houghton Jr. Gallery and the Great Hall Gallery, New York 51125
Arthur M. Sackler Gallery, Washington 53849
Arthur Ross Gallery, Philadelphia 51815
Articulate Gallery, Washington 53850
Artist's Space, New York 51127
Artmobile, Newtown 51344
Artrain USA, Ann Arbor 46304
The Arts Center, Saint Petersburg 52634
Arts Council Gallery, Silver City 53100
Arts Council of Fayetteville/Cumberland County, Fayetteville 48530
Arts in the Academy, Washington 53852
Artspace, Richmond 52323
Artspace-Lima, Lima 50085
Artswatch, Louisville 50282
Artworks, Trenton 53556
Ashtabula Arts Center, Ashtabula 46394
Aspen Art Museum, Aspen 46397
Associated Artists of Pittsburgh Gallery, Pittsburgh 51940
The Athenaeum, Alexandria 46200
Athenaeum of Philadelphia, Philadelphia 51816
Atlanta Contemporary Art Center, Atlanta 46424
The Atlanta Cyclorama, Atlanta 46425
Atlanta International Museum of Art and Design, Atlanta 46427
Atwater Kent Museum - The History Museum of Philadelphia, Philadelphia 51817
Augustana College Art Gallery, Rock Island 52405
Aurora Public Art Commission, Aurora 46490
Austin Museum of Art, Austin 46503
Austin Museum of Art Laguna Gloria, Austin 46504
Ava Gallery and Art Center, Lebanon 50010
The AXA Gallery, New York 51131
Babe Ruth Birthplace Museum, Baltimore 46548
Bakersfield Museum of Art, Bakersfield 46536
Ball State University Museum of Art, Muncie 50902
The Baltimore Museum of Art, Baltimore 46551
Bank One Fort Worth Collection, Fort Worth 48711
Barker-Cawood Art Gallery, Olivet 51540
Barnard's Mill Art Museum, Glen Rose 48922
Barnsdall Art Center and Junior Arts Center, Los Angeles 50211
Barton Museum, Wilson 54210
Bass Museum of Art, Miami Beach 50641
Bates College Museum of Art, Lewiston 50037
The Bead Museum, Glendale 48926
Beatrice M. Haggerty Gallery, Irving 49532
Beaufort Museum, Beaufort 46667
Beaumont Art League, Beaumont 46673
Beeville Art Museum, Beeville 46692
The Belknap Mill Museum, Laconia 49861
Bellevue Art Museum, Bellevue 46706
Belmont, Gari Melchers Estate and Memorial Gallery, Fredericksburg 48763
Ben Maltz Gallery and Bolsky Gallery, Los Angeles 50212
The Benjamin Banneker Museum, Baltimore 46556
The Bennington Museum, Bennington 46739
The Bergen Museum of Art and Science, Paramus 51697
Berkeley Art Museum and Pacific Film Archive, Berkeley 46752
Bertha V.B. Lederer Arts Gallery, Geneseo 48876
Betty Brinn Children's Museum, Milwaukee 50705
Bicentennial Art Center and Museum, Paris 51701
Biggs Museum of American Art, Dover 48161
Bingham Gallery, Columbia 47714
Birger Sandzén Memorial Gallery, Lindsborg 50120
Birmingham Bloomfield Art Center, Birmingham 46828
Birmingham Museum of Art, Birmingham 46822
Blanden Art Museum, Fort Dodge 48615
Blount-Bridgers House/ Edgecombe County Art Museum, Tarboro 53447
B'nai B'rith Klutznick National Jewish Museum, Washington 53853
Boal Mansion Museum, Boalsburg 46880
Boca Raton Museum of Art, Boca Raton 46883
Boehm Art Gallery, San Marcos 52817
The Bohen Foundation, New York 51135
Boise Art Museum, Boise 46890
Bolinas Museum, Bolinas 46894
Bona Thompson Memorial Center, Indianapolis 49486
Boothbay Region Art Foundation, Boothbay Harbor 46911
Boston Athenæum, Boston 46919
Boston Public Library, Boston 46923
Boulder Museum of Contemporary Art, Boulder 46956
Boulevard Arts Center, Chicago 47458
Bowdoin College Museum of Art, Brunswick 47102
The Bower Gallery, San Antonio 52710
Bradford Brinton Memorial, Big Horn 46795
Braithwaite Fine Arts Gallery, Cedar City 47289
Brandywine River Museum, Chadds Ford 47322
Branigan Cultural Center, Las Cruces 49957
Brattleboro Museum and Art Center, Brattleboro 46998
Brauer Museum of Art, Valparaiso 53698
Brenau University Galleries, Gainesville 48820
Brevard Museum of Art and Science, Melbourne 50564
Brigham City Museum-Gallery, Brigham City 47024
Broadway Windows, New York 51137
Brookings Arts Council, Brookings 47058
Brooklyn Museum, Brooklyn 47069
Brown County Art Gallery and Museum, Nashville 50959
Brownsville Museum of Fine Art, Brownsville 47096
Brueckner Museum of Starr Commonwealth, Albion 46173
Brunnier Art Museum, Ames 46256
Brush Art Gallery, Lowell 50304
Bry Gallery, Monroe 50788
Buchanan Center for the Arts, Monmouth 50782
Buckley Center Gallery, Portland 52073
Buffalo Bill Historical Center, Cody 47671
Bush House Museum and Bush Barn Art Center, Salem 52665
The C.A.S.E Museum, Jersey City 49625
C.M. Russell Museum, Great Falls 49028
Callanwolde Fine Arts Center, Atlanta 46429
Camp Gallery, Sweet Briar 53390
Canadian Embassy Art Gallery, Washington 53854
Canajoharie Library and Art Gallery, Canajoharie 47211
Cape Ann Historical Museum, Gloucester 48945
Caramoor Center for Music and the Arts, Katonah 49707
Cardinal Gallery, Annapolis 46315
Carleton College Art Gallery, Northfield 51418
Carlsbad Museum and Art Center, Carlsbad 47249
Carnegie Art Center, Walla Walla 53796
Carnegie Center for Art History, New Albany 51008
Carnegie Museum of Art, Pittsburgh 51942
Carney Gallery, Weston 54104
Carroll Reece Museum, Johnson City 49631
Castellani Art Museum, Niagara University 51348
Castle Gallery, New Rochelle 51102
CCC Weeks Gallery, Jamestown 49596
Cecelia Coker Bell Gallery, Hartsville 49202
Cecille R. Hunt Gallery, Sankt Louis 52851
Cedar Rapids Museum of Art, Cedar Rapids 47301
Center Art Gallery, Grand Rapids 49001
Center Art Gallery, Athens 46409
Center for Exploratory and Perceptual Art, Buffalo 47120
Center for the Visual Arts, Philadelphia 51821
Center for the Visual Arts, Denver 48067
Central Iowa Art Museum, Marshalltown 50511
Chadron State College Main Gallery, Chadron 47324
Chaffee Art Center, Rutland 52493
The Chaim Gross Studio Museum, New York 51144
Chandler Gallery, Randolph 52245
Chapel Hill Museum, Chapel Hill 47345
Chapman Art Center Gallery, Cazenovia 47286
Chappell Art Gallery, Chappell 47347
Charles A. Wustum Museum of Fine Arts, Racine 52222
Charles Allis Art Museum, Milwaukee 50706
Charles B. Goddard Center for Visual and Performing Arts, Ardmore 46343
Charleston Heights Art Center Gallery, Las Vegas 49965
Charlotte and Philip Hanes Art Gallery, Winston-Salem 54243
Chattahoochee Valley Art Museum, La Grange 49840
Chazen Museum of Art, Madison 50388
Cheekwood Museum of Art, Nashville 50963
Chelsea Art Museum, New York 51147
Chester Art Guild, Chester 47434
Cheyenne Frontier Days Old West Museum, Cheyenne 47444
The Chicago Athenaeum, Chicago 47463
Chicago Cultural Center, Chicago 47465
Chieftains Museum, Rome 52448
Children's Art Museum, San Angelo 52700
Children's Museum of the Arts, New York 51149
Chinati Foundation, Marfa 50466
The Chrysler Museum of Art, Norfolk 51361
Chung-Cheng Art Gallery, Jamaica 49589
Cigna Museum and Art Collection, Philadelphia 51822
Cincinnati Art Museum, Cincinnati 47555
City Arts, New York 51151
City Hall Council Chamber Gallery, Charleston 47361
Clark Humanities Museum, Claremont 47571
The Cleveland Museum of Art, Cleveland 47624
Cleveland State University Art Gallery, Cleveland 47627
Clinton Academy Museum, East Hampton 48243
Clymer Museum of Art, Ellensburg 48357
Coastal Center for the Arts, Saint Simons Island 52645
Cobb Museum of Art, Marietta 50468
Coe College Collection of Art, Cedar Rapids 47302
College of Arts and Sciences Gallery, Coral Gables 47798
College of New Jersey Art Gallery, Trenton 53557
The College of Wooster Art Museum, Wooster 54294
Colorado Gallery of the Arts, Littleton 50143
Colorado Springs Fine Arts Center, Colorado Springs 47699
Colorado University Art Galleries, Boulder 46957
Columbia College Art Gallery, Chicago 47468
Columbia Museum of Art, Columbia 47723
The Columbian Museum and Art Center, Wamego 53813
Columbus Cultural Arts Center, Columbus 47751
The Columbus Museum, Columbus 47739
Columbus Museum of Art, Columbus 47752
Combat Air Museum, Topeka 53533
Commencement Art Gallery, Tacoma 53406
Communications and History Museum of Sutton, Washington 53855
Community Arts Gallery, Detroit 48111
Computing Commons Gallery, Tempe 53466
Concordia University-Wisconsin Art Gallery, Mequon 50593
Connecticut Graphic Arts Center, Norwalk 51431
Connecticut Historical Society Museum, Hartford 49193
Contemporary Museum, Baltimore 46558
Copper King Mansion, Butte 47151
Copper Village Museum and Arts Center, Anaconda 46277
Coppini Academy of Fine Arts Gallery and Museum, San Antonio 52713
Copshaholm House Museum and Historic Oliver Gardens, South Bend 53157
Corcoran Gallery of Art, Washington 53856
Cornell Museum, Delray Beach 48045
Corvallis Art Center, Corvallis 47832
Courthouse Gallery, Portsmouth 52106
Courtney and Lemmerman Galleries, Jersey City 49626
Coutts Memorial Museum of Art, El Dorado 48312
Crary Art Gallery, Warren 53826
Creative Arts Center, Kansas City 49687
Creative Arts Center, Pontiac 52020
Creative Arts Guild, Dalton 47947
Creative Growth Art Center, Oakland 51464
Cress Gallery of Art, Chattanooga 47409
Crocker Art Museum, Sacramento 52505
Crook County Museum and Art Gallery, Sundance 53370
Crowder College-Longwell Museum and Camp Crowder Collection, Neosho 51004
Crowley Art Association and Gallery, Crowley 47893
Cuban Museum of Arts and Culture, Miami 50623
Cue Art Foundation, New York 51157
The Cultural Center of Fine Arts, Parkersburg 51713
Cummer Museum of Art, Jacksonville 49577
Cuneo Museum, Vernon Hills 53725
Curfman Gallery and Duhesa Lounge, Fort Collins 48605
Currier Gallery of Art, Manchester 50421
Cy Twombly Gallery, Houston 49384
DAAP Galleries, Cincinnati 47559
Dactyl Foundation, New York 51159
Dallam-Hartley XIT Museum, Dalhart 47922
The Dallas Center for Contemporary Art, Dallas 47928
Dallas Museum of Art, Dallas 47931
Dalton Gallery, Decatur 48012
Dan Flavin Art Institute, Bridgehampton 47013
Daniel Boone Homestead, Birdsboro 46818
Danville Museum of Fine Arts and History, Danville 47963
Daura Gallery, Lynchburg 50325
Davenport Museum of Art, Davenport 47969
David L. Mason Children's Art Museum, Dunedin 48207
David Strawn Art Gallery, Jacksonville 49587
David Winton Bell Gallery, Providence 52164
Davidson County Museum of Art, Lexington 50064
Davis Art Center, Bountiful 46964
Davis Art Gallery, Columbia 47715
Davis Museum, Wellesley Hills 54023
Davison Art Center, Middletown 50657
Dawson Springs Museum and Art Center, Dawson Springs 47981
Dayton Art Institute, Dayton 47984
De Land Museum of Art, De Land 47996
De Young Museum, San Francisco 52772
DeGrazia Gallery in the Sun, Tucson 53586
Deines Cultural Center, Russell 52483
Delaware Art Museum, Wilmington 54189
Delaware Center for the Contemporary Arts, Wilmington 54190
Denison University Art Gallery, Granville 49021
Dennos Museum Center of Northwestern Michigan College, Traverse City 53553
Denver Art Museum, Denver 48072
Department of Fine Arts Gallery, Decatur 48011
DePaul University Art Gallery, Chicago 47470
Des Lee Gallery, Saint Louis 52575
Des Moines Art Center, Des Moines 48096
Desert Caballeros Western Museum, Wickenburg 54155
The Detroit Institute of Arts, Detroit 48114
Dia Center for the Arts, Quemado 52210
Diggs Gallery at Winston-Salem State University, Winston-Salem 54244
Dimock Gallery, Washington 53861
The Discovery Museum, Bridgeport 47017
Dishman Art Gallery, Beaumont 46674
District of Columbia Arts Center, Washington 53863
Dittmar Memorial Gallery, Evanston 48447
Dix Art Mix, Chicago 47471
The Dixon Gallery and Gardens, Memphis 50571
Douglas F. Cooley Memorial Art Gallery, Portland 52075
Downey Museum of Art, Downey 48175
Dubuque Museum of Art, Dubuque 48187
Duke University Union Museum and Brown Art Gallery, Durham 48217
Dumbarton Oaks Collections, Washington 53865
Dwight Frederick Boyden Gallery, Saint Mary's City 52608
East-West Center, Honolulu 49338
Eaton-Buchan Gallery and Marvin Cone Gallery, Cedar Rapids 47303
Eccles Community Arts Center, Ogden 51500
Edison Community College Gallery of Fine Art, Fort Myers 48662
Edmonds Arts Festival Museum, Edmonds 48295
Edward and Marthann Samek Art Gallery, Lewisburg 50029
Edward Hopper House Art Center, Nyack 51449
Edward M. Bannister Gallery, Providence 52165
Edward Williams Gallery, Hackensack 49110
Edwin A. Ulrich Museum of Art, Wichita 54140
Eiteljorg Museum of American Indians and Western Art, Indianapolis 49490
El Paso Museum of Art, El Paso 48321
Elder Art Gallery, Lincoln 50094
Elizabeth Myers Mitchell Art Gallery, Annapolis 46318
Elizabeth O'Neill Verner Studio Museum, Charleston 47363
Elizabeth P. Korn Gallery, Madison 50383
Elizabeth Rozier Gallery, Jefferson City 49612
Elizabeth Slocumb Galleries, Johnson City 49632
The Ellen Noel Art Museum of the Permian Basin, Odessa 51496
Elmhurst Art Museum, Elmhurst 48374
Eloise Pickard Smith Gallery, Santa Cruz 52866
Epcot, Lake Buena Vista 49875
Erie Art Museum, Erie 48409
Esther M. Klein Art Gallery, Philadelphia 51832
Evansville Museum of Arts and Science, Evansville 48456
Everhart Museum of Natural History, Science and Art, Scranton 52980
Everson Museum of Art, Syracuse 53398
The Exploratorium, San Francisco 52773
Ezra and Cecile Zilkha Gallery, Middletown 50658
The Fabric Workshop and Museum, Philadelphia 51833
Fairbanks Gallery, Corvallis 47833
Fairfield Center for Contemporary Art, Sturgeon Bay 53347
Fallingwater - Western Pennsylvania Conservancy, Mill Run 50690
Farm House Museum, Ames 46257
Farnham Galleries, Indianola 49507
Father Weyland Gallery, Epworth 48407
Faulconer Gallery at Grinnell College, Grinnell 49090
Fayette Art Museum, Fayette 48524
Fayetteville Museum of Art, Fayetteville 48531
Federal Reserve Board Art Gallery, Washington 53868
Fenimore Art Museum, Cooperstown 47789
Festival-Institute Museum, Round Top 52471
Fielding L. Wright Art Center, Cleveland 47620
Fine Art Gallery at Centro Cultural Aztlan, San Antonio 52714
Fine Arts Center for New River Valley, Pulaski 52196
Fine Arts Center Galleries, Bowling Green 46977
Fine Arts Center Gallery, Maryville 50525
Fine Arts Center of Kershaw County, Camden 47200
Fine Arts Exhibitions, Austin 46508
Fine Arts Gallery, Breckenridge 47001
Fine Arts Gallery, New Orleans 51079

A

A

Fine Arts Gallery, Los Angeles 50221
Fine Arts Gallery, Omaha 51549
Fine Arts Gallery, Great Falls 49029
Firehouse Art Center, Norman 51373
Firehouse Art Gallery, Garden City 48856
Firehouse Gallery, Oak Bluffs 51450
First National Bank of Chicago Art Collection, Chicago 47474
Fisher Gallery, Los Angeles 50222
Fisk University Galleries, Nashville 50968
Fitchburg Art Museum, Fitchburg 48548
Flamingo Gallery, Las Vegas 49967
Flanagan Valley Campus Art Gallery, Lincoln 50112
Flaten Art Museum, Northfield 51420
Flint Institute of Arts, Flint 48559
Florence Griswold Museum, Old Lyme 51536
Florence Museum of Art, Science and History, Florence 48570
The Florida Museum of Hispanic and Latin American Art, Coral Gables 47801
Florida State University Museum of Fine Arts, Tallahassee 53422
Fort Frederick, Big Pool 46796
Fort Wayne Museum of Art, Fort Wayne 48705
Foster Gallery, Eau Claire 48277
Foundation for Hellenic Culture, New York 51175
Fountain Gallery, New York 51176
Fowler Museum of Cultural History, Los Angeles 50223
The Frances Lehman Loeb Art Center, Poughkeepsie 52121
Francis Colburn Gallery, Burlington 47143
Francis McCray Gallery, Silver City 53101
Franklin G. Burroughs-Simeon B. Chapin Art Museum, Myrtle Beach 50933
Fred Harman Art Museum, Pagosa Springs 51674
Fred Wolf Jr. Gallery, Philadelphia 51837
Frederic Remington Art Museum, Ogdensburg 51506
Frederick R. Weisman Art Museum, Minneapolis 50731
Freeport Arts Center, Freeport 48781
Freer Gallery of Art, Washington 53873
French Art Colony, Gallipolis 48839
Fresno Art Museum, Fresno 48790
The Frick Art Museum-Henry Clay Frick Estate, Pittsburgh 51949
Frick Collection, New York 51178
Frist Center for the Visual Arts, Nashville 50970
Gadsden Arts Center, Quincy 52212
Gadsden Museum of Fine Arts, Gadsden 48811
Gahlberg Gallery, Glen Ellyn 48920
Galesburg Civic Art Center, Galesburg 48833
The Galleries at Moore, Philadelphia 51838
The Galleries of Texas Tech, Lubbock 50311
The Gallery, Trenton 53558
Gallery 110 North, Plymouth 52002
Gallery 181, Ames 46258
Gallery 2, Chicago 47476
Gallery 218, Milwaukee 50708
Gallery 400 at the University of Illinois at Chicago, Chicago 47477
Gallery 57, Cambridge 47175
Gallery '76, Wenatchee 54035
Gallery at Bristol-Myers Squibb, Princeton 52148
Gallery Gazelle, Philomath 51896
Gallery Lolani, Kaneohe 49675
Gallery of Art, Overland Park 51632
Gallery of the Department of Art, Portales 52054
Gallery of the National Institute of Art and Disabilities, Richmond 52310
Gallery of Visual Arts, Missoula 50752
Gallery West, Alexandria 46207
Gardner Museum, Gardner 48862
Gaston County Museum of Art and History, Dallas 47923
Gatov Gallery, Long Beach 50173
Genesee Country Village and Museum, Mumford 50901
George A. Spiva Center for the Arts, Joplin 49650
The George D. and Harriet W. Cornell Fine Arts Museum, Winter Park 54254
George L. Sturman Museum, Las Vegas 49968
George Washington's Mount Vernon, Mount Vernon 50893
Georgetown University Art Collection, Washington 53874
Georgia Museum of Art, Athens 46411
Georgia State University Art Gallery, Atlanta 46438
Gertrude Herbert Institute of Art, Augusta 46478
Gibbes Museum of Art, Charleston 47364
Gibson Barham Gallery, Lake Charles 49877
Gilcrease Museum, Tulsa 53608
Giustina Gallery, Corvallis 47834
The Glynn Art Association, Saint Simons Island 52647
Godwin-Ternbach Museum, Flushing 48576
Goethe-Institut New York - Exhibitions, New York 51185
Governor's Gallery, Santa Fe 52875
Grand Rapids Art Museum, Grand Rapids 49003
Gray Gallery, Quincy 52214
Great Lakes Lighthouse Museum, Mackinaw City 50353
Great Plains Art Collection in the Christlieb Gallery, Lincoln 50096
The Greater Reston Arts Center, Reston 52299
Green Hill Center for North Carolina Art, Greensboro 49060
Greensboro Artists' League Gallery, Greensboro 49061
Greensward Gallery, Corvallis 47835
Greenville County Museum of Art, Greenville 49079
Greenville Museum of Art, Greenville 49074
Grey Art Gallery, New York 51187
Grinnell College Art Gallery, Grinnell 49091
Grover M. Hermann Fine Arts Center, Marietta 50474
Guilford College Art Gallery, Greensboro 49063
Gulf Beach Art Center, Indian Rocks Beach 49481
Gumtree Museum of Art, Tupelo 53613
Haas Gallery of Art, Bloomsburg 46872
The Haggin Museum, Stockton 53306
Hallwalls Arts Center, Buffalo 47121
Hammer Museum, Los Angeles 50228
Hammond Museum, North Salem 51405
Hampton University Museum, Hampton 49152
Harlow Gallery, Hallowell 49129
Harris Art Gallery, James M. Lykes Maritime Gallery and Hutchings Gallery, Galveston 48846
Harris Exhibit Hall, Oxford 51652
Hartsville Museum, Hartsville 49203
Harvard University Art Museums, Cambridge 47177
Harwood Museum of Art of the University of New Mexico, Taos 53441
Hatton Gallery, Fort Collins 48609
Hawaii Pacific University Gallery, Honolulu 49341
Hawaii State Foundation on Culture and the Arts, Honolulu 49342
Heard Museum, Phoenix 51904
Hearst Art Gallery, Moraga 50838
Heckscher Museum of Art, Huntington 49430
Hefner Zoology Museum, Oxford 51653
Helen Day Art Center, Stowe 53325
Helen E. Copeland Gallery, Bozeman 46981
Helen King Kendall Memorial Art Gallery, San Angelo 52702
The Henry Sheldon Museum of Vermont Histoy, Middlebury 50652
Herbert F. Johnson Museum of Art, Ithaca 49548
The Heritage Center, Pine Ridge 51924
Heritage Center of Lancaster County, Lancaster 49923
Hershey Museum, Hershey 49263
Heuser Art Center, Peoria 51772
Hibel Museum of Art, Jupiter 49660
Hiddenite Center, Hiddenite 49272
High Museum of Art, Atlanta 46441
Highland Cultural Arts Gallery, Highland 49278
Highlands Museum of the Arts, Sebring 53021
Hill-Stead Museum, Farmington 48513
Hillwood Art Museum, Brookville 47091
Hiram Blauvelt Art Museum, Oradell 51576
Hirshhorn Museum and Sculpture Garden, Washington 53877
The Hispanic Society of America, New York 51192
Hockaday Museum of Arts, Kalispell 49671
Hofstra Museum, Hempstead 49246
Holland Tunnel Art, Brooklyn 47076
The Hollycroft Foundation, Ivoryton 49552
The Holtzman Art Gallery, Towson 53551
Honolulu Academy of Arts Museum, Honolulu 49343
Hood Museum of Art, Hanover 49162
Hoosier Salon Gallery, Indianapolis 49495
Hopkins House Gallery, Haddon Township 49113
Housatonic Museum of Art, Bridgeport 47018
Houston Harte University Center, San Angelo 52703
Howard University Gallery of Art, Washington 53879
Hoyt Institute of Fine Arts, New Castle 51042
Hoyt Sherman Place, Des Moines 48097
The Hudson River Museum, Yonkers 54331
Hunt Gallery, Staunton 53276
Hunt Institute for Botanical Documentation, Pittsburgh 51950
Hunterdon Museum of Art, Clinton 47657
The Huntington Art Collections, San Marino 52819
Huntington Museum of Art, Huntington 49435
Huntsville Museum of Art, Huntsville 49444
The Hyde Collection, Glens Falls 48935
I-5 Gallery, Los Angeles 50236
I Space, Champaign 47334
Illinois Art Gallery, Chicago 47481
Illinois State Museum, Springfield 53216
Ilwaco Heritage Museum, Ilwaco 49468
Imperial Calcasieu Museum, Lake Charles 49878
Indiana University Art Museum, Bloomington 46864
Indianapolis Art Center, Indianapolis 49499
Indianapolis Museum of Art - Columbus Gallery, Columbus 47742
International Museum of Art, El Paso 48325
International Museum of Cartoon Art, Boca Raton 46885
International Museum of the Horse, Lexington 50050
Intersection for the Arts Gallery, San Francisco 52778
Iredell Museum of Arts Heritage, Statesville 53274
Iris B. Gerald Cantor Center for Visual Arts, Stanford 53253
Irving Arts Center, Irving 49533
Isabella Stewart Gardner Museum, Boston 46934
ISE Art Foundation, New York 51197
Iupui Cultural Arts Gallery, Indianapolis 49502
J. Paul Getty Museum, Los Angeles 50237
Jack S. Blanton Museum of Art, Austin 46511
Jacqueline Casey Hudgens Center for the Arts, Duluth 48191
Jacques Marchais Museum of Tibetan Art, Staten Island 53264
Jamaica Center for Arts, Jamaica 49590
James A. Michener Art Museum, Doylestown 48179
James E. Lewis Museum of Art, Baltimore 46570
James Howe Gallery, Union 53648
The James Watrous Gallery, Madison 50391
Jane Voorhees Zimmerli Art Museum, New Brunswick 51033
Jay I. Kislak Foundation, Miami Lakes 50645
Jersey City Museum, Jersey City 49627
Jewett Hall Gallery, Augusta 46484
Jewish Institute for the Arts, Boca Raton 46886
The Jewish Museum, New York 51200
Joe and Emily Lowe Art Gallery, Syracuse 53399
Joel and Lila Harnett Museum of Art, Richmond 52329
John A. Hermann jr. Memorial Art Museum, Bellevue 46705
The John and Mable Ringling Museum of Art, Sarasota 52914
John Mariani Art Gallery, Greeley 49033
John Slade Ely House, New Haven 51051
Jonathan Hager House and Museum, Hagerstown 49120
Jordan Schnitzer Museum of Art at the University of Oregon, Eugene 48431
Joseloff Gallery, West Hartford 54057
Joseph Gross Gallery, Tucson 53593
Joslyn Art Museum, Omaha 51552
Julia C. Butridge Gallery, Austin 46513
Jundt Art Museum, Spokane 53206
Juniata College Museum of Art, Huntingdon 49422
Kalamazoo Institute of Arts, Kalamazoo 49666
Karl Drerup Fine Arts Gallery, Plymouth 52000
Karpeles Manuscript Museum, Buffalo 47122
Karpeles Manuscript Museum, Jacksonville 49581
Karpeles Manuscript Museum, Tacoma 53410
Katonah Museum of Art, Katonah 49709
Kaua'i Museum, Lihu'e 50083
Kendall Campus Art Gallery, Miami 50628
Kennedy-Douglass Center for the Arts, Florence 48563
Kennedy Museum of Art, Athens 46417
Kenosha Public Museum, Kenosha 49734
Kent Art Association, Kent 49736
Kent State University Art Galleries, Kent 49739
Kentuck Museum and Art Center, Northport 51424
Keokuk Art Center, Keokuk 49744
Key West Museum of Art and History, Key West 49763
Kiboko Projects, New York 51202
Kiehle Gallery, Saint Cloud 52539
Kimball Art Center, Park City 51706
Kimbell Art Museum, Fort Worth 48715
Kipp Gallery, Indiana 49484
Kirkland Art Center, Clinton 47660
Kirkland Fine Arts Center-Perkinson Gallery, Decatur 48015
KOA Art Gallery, Honolulu 49347
Krannert Art Museum, Champaign 47335
The Kreeger Museum, Washington 53885
Kresge Art Museum, East Lansing 48253
L.C. Bates Museum, Hinckley 49295
L.D. Brinkman Art Foundation, Kerrville 49750
La Salle University Art Museum, Philadelphia 51846
Lafayette College Art Gallery, Easton 48269
Lake George Historical Association Museum, Lake George 49884
Lakeview Museum of Arts and Sciences, Peoria 51773
Lakewood's Heritage Culture and the Arts Galleries, Lakewood 49905
Lamar Dodd Art Center, La Grange 49841
The Lamont Gallery, Exeter 48465
Lancaster Museum / Art Gallery, Lancaster 49915
Larson Gallery, Yakima 54319
Lauren Rogers Museum of Art, Laurel 49985
Leanin' Tree Museum of Western Art, Boulder 46960
Leepa-Rattner Museum, Tarpon Springs 53448
Legion of Honor, San Francisco 52781
Lehigh University Art Galleries/Museum, Bethlehem 46780
Leigh Yawkey Woodson Art Museum, Wausau 53993
The Lentz Center for Asian Culture, Lincoln 50098
Licking County Art Association Gallery, Newark 51289
List Gallery, Swarthmore 53389
The Lithuanian Museum, Chicago 47486
The Living Arts and Science Center, Lexington 50052
Lockport Gallery, Lockport 50157
Long Beach Museum of Art, Long Beach 50174
Longboat Key Center for the Arts, Longboat Key 50191
Longview Museum of Fine Art, Longview 50195
Longwood Center for the Visual Arts, Farmville 48523
Lore Degenstein Gallery, Selinsgrove 53030
Los Angeles County Museum of Art, Los Angeles 50244
Louisburg College Art Gallery, Louisburg 50278
Louisiana Arts and Science Center, Baton Rouge 46633
Louisiana State University Museum of Art, Baton Rouge 46636
Loveland Museum and Gallery, Loveland 50301
Lowe Art Museum, Coral Gables 47802
Lubeznik Center for the Arts, Michigan City 50646
Luckman Gallery, Los Angeles 50248
Luther College Fine Arts Collection, Decorah 48021
Lyndon House Art Center, Athens 46413
Mabee-Gerrer Museum of Art, Shawnee 53059
Mabel and Victor D'Amico Studio and Archive, Amagansett 46248
McAllen International Museum, McAllen 50339
Macculloch Hall Historical Museum, Morristown 50853
McMullen Museum of Art, Chestnut Hill 47441
McNay Art Museum, San Antonio 52717
Magnolia Plantation, Charleston 47372
Main Line Art Center, Haverford 49215
Maine Art Gallery, Wiscasset 54262
Maitland Art Center, Maitland 50402
MAK Center for Art and Architecture, Los Angeles 50250
MAK Center for Art and Architecture, West Hollywood 54064
Malden Public Library Art Collection, Malden 50405
Margaret Fort Trahern Gallery, Clarksville 47594
Margaret Harwell Art Museum, Poplar Bluff 52024
Margaret Hutchinson Compton Gallery, Cambridge 47181
Marias Museum of History and Art, Shelby 53066
Marine Corps Art Collection, Washington 53890
Markeim Art Center, Haddonfield 49116
Martin Art Gallery, Allentown 46228
Marxhausen Art Gallery, Seward 53043
Mary and Leigh Block Museum of Art, Evanston 48453
Mary Brogan Museum of Art and Science, Tallahassee 53426
Mary Porter Sesnon Art Gallery, Santa Cruz 52867
Mary Washington College Galleries, Fredericksburg 48769
Maryhill Museum of Art, Goldendale 48963
Maryland Art Place, Baltimore 46575
Maryland Institute Museum, Baltimore 46577
Marymount Manhattan College Museum, New York 51210
Marywood University Art Galleries, Scranton 52982
Mason Gross Art Galleries, New Brunswick 51034
Massillon Museum, Massillon 50531
Masur Museum of Art, Monroe 50789
The Mattatuck Museum of the Mattatuck Historical Society, Waterbury 53956
Maude I. Kerns Art Center, Eugene 48434
Meadow Brook Art Gallery, Rochester 52385
Meadows Museum, Dallas 47935
Meadows Museum of Art, Shreveport 53085
Mellon Financial Corporation's Collection, Pittsburgh 51954
Memorial Art Gallery of the University of Rochester, Rochester 52396
Memorial Union Art Collection, Tempe 53468
Memorial Union Art Gallery, Corvallis 47836
Memphis Brooks Museum of Art, Memphis 50577
Memphis College of Art Gallery, Memphis 50578
Mendocino Art Center Gallery, Mendocino 50587
The Menil Collection, Houston 49394
Mercer Gallery, Rochester 52397
Meridian International Center - Cafritz Galleries, Washington 53892
Meridian Museum of Art, Meridian 50600
The Merrick Art Gallery, New Brighton 51027
Mesquite Heritage Museum, Mesquite 50611
The Metropolitan Museum of Art, New York 51212
Mexican Fine Arts Center Museum, Chicago 47489
Miami Art Museum, Miami 50631
Miami University Art Museum, Oxford 51654
Michelson Museum of Art, Marshall 50510
Mid-Atlantic Center for the Arts, Cape May 47235
The Middlebury College Museum of Art, Middlebury 50653
Midwest Museum of American Art, Elkhart 48346
Millennium Arts Center International, Washington 53893
Miller Art Museum, Sturgeon Bay 53349
Millicent Rogers Museum of Northern New Mexico, Taos 53442
Milliken Gallery, Spartanburg 53193
Mills College Art Museum, Oakland 51467
Milwaukee Art Museum, Milwaukee 50711
Milwaukee Very Special Art Center, Milwaukee 50714

Mingei International Museum, Escondido 48420
Mingenback Art Center Gallery, Lindsborg 50122
Minneapolis Institute of Arts, Minneapolis 50739
Mint Museum of Art, Charlotte 47392
Miriam and Ira D. Wallach Art Gallery, New York 51214
Mission Cultural Center for Latino Arts, San Francisco 52784
Mississippi Museum of Art, Jackson 49560
Missoula Art Museum, Missoula 50754
Mitchell Museum at Cedarhurst, Mount Vernon 50892
Mobile Museum of Art, Mobile 50769
Modern Art Museum of Fort Worth, Fort Worth 48717
Montalvo Arts Center, Saratoga 52918
Montclair Art Museum, Montclair 50798
Monterey Museum of Art, Monterey 50803
Montgomery Museum and Lewis Miller Regional Art Center, Christiansburg 47547
Montgomery Museum of Fine Arts, Montgomery 50815
The Morgan Library, New York 51216
Morris Belknap Gallery, Dario Covi Gallery and SAL Gallery, Louisville 50293
Morris Museum of Art, Augusta 46481
Morristown National Historical Park, Morristown 50855
Moss-Thorns Gallery of Arts, Hays 49224
MSCC Forsyth Center Galleries, College Station 47690
The Muchnic Gallery, Atchison 46408
Muckenthaler Gallery and Mansion Museum, Fullerton 48807
Mulvane Art Museum, Topeka 53537
Munce Art Center, Zionsville 54371
Municipal Art Society, New York 51219
Munson-Williams-Proctor Arts Institute Museum of Art, Utica 53676
Muscarelle Museum of Art, Williamsburg 54170
Muscatine Art Center, Muscatine 50921
Museo de Las Americas, Denver 48083
El Museo del Barrio, New York 51220
El Museo Latino, Omaha 51553
Museum and Arts Center in the Sequim Dungeness Valley, Sequim 53037
The Museum at Drexel University, Philadelphia 51855
Museum Center, Maysville 50539
Museum in the Community, Hurricane 49452
Museum of Ancient and Modern Art, Penn Valley 51762
Museum of Art, Providence 52170
Museum of Art, Saint Louis 52589
Museum of Art, Orono 51603
Museum of Art, Ann Arbor 46309
Museum of Art, Durham 48225
Museum of Art, Pullman 52199
Museum of Art and Archaeology, Columbia 47718
Museum of Art at Brigham Young University, Provo 52188
Museum of Arts and Design, New York 51224
The Museum of Arts and Sciences, Daytona Beach 47994
Museum of Arts and Sciences, Macon 50367
Museum of Bad Art, Dedham 48025
Museum of Central Connecticut State University, New Britain 51028
Museum of Children's Art, Oakland 51468
The Museum of Contemporary Art, Atlanta 46445
The Museum of East Texas, Lufkin 50318
Museum of Fine Art, Las Cruces 49959
The Museum of Fine Arts, Houston 49397
Museum of Fine Arts, Missoula 50756
Museum of Fine Arts, Saint Petersburg 52640
Museum of Fine Arts, Santa Fe 52880
Museum of Fine Arts, Boston, Boston 46940
Museum of Geometric and Madi Art, Dallas 47936
Museum of Indian Arts and Culture, Santa Fe 52881
Museum of Nebraska Art, Kearney 49717
Museum of New Mexico, Santa Fe 52883
The Museum of Outdoor Arts, Englewood 48400
Museum of Teenage Art, Housatonic 49379
Museum of Texas Tech University, Lubbock 50312
Museum of the Dog, Saint Louis 52592
Museum of the North, Fairbanks 48472
Museum of the Southwest, Midland 50673
The Museum of Western Art, Kerrville 49751
Muskegon Museum of Art, Muskegon 50926
Mystic Art Center, Mystic 50934
Naples Museum of Art, Naples 50952
Nasher Museum of Art at Duke University, Durham 48221
Nassau County Museum of Art, Roslyn 52463
National Academy Museum, New York 51232
National Art Museum of Sport, Indianapolis 49505
National Exhibits by Blind Artists, Philadelphia 51859
National Gallery of Art, Washington 53898
National Museum of American Illustration, Newport 51324
National Museum of Wildlife Art, Jackson 49573
National Museum of Women in the Arts, Washington 53906
National Portrait Gallery, Washington 53907
NAU Art Museum, Flagstaff 48552
Nave Museum, Victoria 53741
The Nelson, Davis 47976
The Nelson-Atkins Museum of Art, Kansas City 49695
Nemours Mansion and Gardens, Wilmington 54196
Nevada Museum of Art, Reno 52289
New England College Gallery, Henniker 49255
New Hampshire Institute of Art, Manchester 50424
New Jersey Center for Visual Arts, Summit 53364
New Jersey State Museum, Trenton 53561
New Mexico State University Art Gallery, Las Cruces 49961
New Orleans Artworks Gallery, New Orleans 51089
New Orleans Museum of Art, New Orleans 51091
New York Studio School of Drawing, Painting and Sculpture - Gallery, New York 51245
New Zone Virtual Gallery, Eugene 48436
The Newark Museum, Newark 51287
Newcomb Art Gallery, New Orleans 51093
Newport Art Museum, Newport 51326
Nicholas Roerich Museum, New York 51246
The Noble Maritime Collection, Staten Island 53265
Nordic Heritage Museum, Seattle 53006
Norman R. Eppink Art Gallery, Emporia 48393
North Carolina Museum of Art, Raleigh 52232
North Carolina State University Gallery of Art and Design, Raleigh 52235
North Country Museum of Arts, Park Rapids 51709
North Dakota Museum of Art, Grand Forks 48989
North Florida Community College Art Gallery, Madison 50376
Northcutt Steele Gallery, Billings 46805
Northeastern Nevada Museum, Elko 48354
Northern Galleries, Aberdeen 46117
Northern Illinois University Art Gallery in Chicago, Chicago 47501
Northern Indiana Arts Association, Munster 50909
Northern Kentucky University Art Galleries, Highland Heights 49280
Northwest Gallery and Sinclair Gallery, Powell 52126
Norton Simon Museum of Art, Pasadena 51720
Noyes Art Gallery, Lincoln 50104
The Noyes Museum of Art, Oceanville 51493
Oak Ridge Art Center, Oak Ridge 51458
Oakland Museum of California, Oakland 51470
Oceanside Museum of Art, Oceanside 51492
The Octagon Center for the Arts, Ames 46259
Oglethorpe University Museum of Art, Atlanta 46447
Ojai Center for the Arts, Ojai 51510
Okefenokee Heritage Center, Waycross 54002
Oklahoma City Museum of Art, Oklahoma City 51521
Olana State Historic Site, Hudson 49411
Old Dominion University Gallery, Norfolk 51370
The Old Jail Art Center, Albany 46167
Old State House, Hartford 49198
Olin Hall Galleries, Salem 52671
On the Hill Cultural Arts Center, Yorktown 54345
Opelousas Museum of Art, Opelousas 51575
Orange County Museum of Art, Newport Beach 51332
Orange County Museum of Art - Orange Lounge, Costa Mesa 47840
Oriental Institute Museum, Chicago 47502
Orlando Museum of Art, Orlando 51593
Ormond Memorial Art Museum, Ormond Beach 51600
Oscar Howe Art Center, Mitchell 50764
Ossining Historical Society Museum, Ossining 51621
Ozaukee Art Center, Cedarburg 47308
Pacific Arts Center, Seattle 53009
Pacific Grove Art Center, Pacific Grove 51665
Paine Art Center, Oshkosh 51616
The Palette and Chisel Academy, Chicago 47503
Palm Beach Institute of Contemporary Art, Lake Worth 49898
Palm Springs Art Museum, Palm Springs 51681
Palmer Museum of Art, University Park 53659
Palo Alto Art Center, Palo Alto 51688
Parkland College Art Gallery, Champaign 47336
The Parrish Art Museum, Southampton 53185
Passaic County Community College Galleries, Paterson 51728
The Patrick and Beatrice Haggerty Museum of Art, Milwaukee 50718
Payne Gallery, Bethlehem 46783
Peabody Essex Museum, Salem 52657
The Peace Museum, Chicago 47505
The Pearson Museum, Springfield 53220
Pelham Art Center, Pelham 51749
Pen and Brush Museum, New York 51248
Peninsula Fine Arts Center, Newport News 51336
Pennsylvania Academy of the Fine Arts Gallery, Philadelphia 51864
Pensacola Museum of Art, Pensacola 51769
Perkins Center for the Arts, Moorestown 50833
Perkinson Gallery, Decatur 48017
Peter Paul Luce Gallery, Mount Vernon 50891
Petersburg Area Art League, Petersburg 51800
Petterson Museum of Intercultural Art, Claremont 47573
Philadelphia Art Alliance, Philadelphia 51865
Philadelphia Museum of Art, Philadelphia 51867
Philbrook Museum of Art, Tulsa 53610
Philip and Muriel Berman Museum of Art, Collegeville 47691
The Phillips Collection, Washington 53914
The Phillips Museum of Art, Lancaster 49931
The Philmont Museum, Seton Memorial Library and Kit Carson Museum, Cimarron 47551
Phoenix Art Museum, Phoenix 51906
Pickens County Museum of Art History, Pickens 51912
Piedmont Arts Museum, Martinsville 50520
Pittsburgh Center for the Arts, Pittsburgh 51956
Plains Art Museum, Fargo 48511
Plattsburgh Art Museum, Plattsburgh 51984
Polish American Museum, Port Washington 52048
Polk County Heritage Gallery, Des Moines 48098
Polk Museum of Art, Lakeland 49902
Pomona College Museum of Art, Claremont 47574
Portholes Into the Past, Medina 50558
Portland Art Museum, Portland 52086
Portland Museum of Art, Portland 52067
Portland State University Galleries, Portland 52088
Pratt Manhattan Gallery, New York 51249
Prichard Art Gallery, Moscow 50865
Princeton University Art Museum, Princeton 52151
Providence Athenaeum, Providence 52173
Provincetown Art Association and Museum, Provincetown 52183
Purdue University Galleries, West Lafayette 54067
Queens Library Gallery, Jamaica 49592
Queens Museum of Art, Flushing 48580
Quincy Art Center, Quincy 52216
R.S. Barnwell Memorial Garden and Art Center, Shreveport 53087
The R.W. Norton Art Gallery, Shreveport 53088
Radford University Art Museum, Radford 52225
Rahr West Art Museum, Manitowoc 50443
Ramsay Museum, Honolulu 49353
Rasmussen Art Gallery, Angwin 46301
The Ray Drew Gallery, Las Vegas 49964
Reader's Digest Art Gallery, Pleasantville 51990
Reading Public Museum and Art Gallery, Reading 52265
Red River Valley Museum, Vernon 53723
Reed Gallery, Presque Isle 52138
Reed Whipple Cultural Center, Las Vegas 49978
The Reeves Center, Lexington 50072
The Regina A. Quick Center for the Arts, Saint Bonaventure 52530
Rehoboth Art League, Rehoboth Beach 52284
The Renaissance Society at the University of Chicago, Chicago 47510
Renwick Gallery, Washington 53915
Reynolds Homestead, Critz 47883
Rhode Island Black Heritage Society Museum, Providence 52176
Richard E. Peeler Art Center, Greencastle 49047
Richard Gallery and Almond Tea Gallery, Cuyahoga Falls 47917
Richmond Art Museum, Richmond 52314
Ridderhof Martin Gallery, Fredericksburg 48771
Ripon College Art Gallery, Ripon 52358
River Arts Center, Clinton 47651
Robert and Mary Montgomery Armory Art Center, West Palm Beach 54074
Robert E. Wilson Art Gallery, Huntington 49424
Robert Hull Fleming Museum, Burlington 47144
Robert V. Fullerton Art Museum, San Bernardino 52734
Rockford Art Museum, Rockford 52416
Rockford College Art Gallery/Clark Arts Center, Rockford 52417
Rockland Center for the Arts, West Nyack 54070
Rocky Mount Arts Center, Rocky Mount 52436
Rocky Mountain College of Art and Design Galleries, Denver 48087
Rodin Museum, Philadelphia 51876
Roger Guffey Gallery, Kansas City 49697
Rogers House Museum and Gallery, Ellsworth 48371
Rogue Gallery and Art Center, Medford 50551
Rosenberg Gallery, Baltimore 46587
Rosenthal Gallery of Art, Caldwell 47161
Rosenwald-Wolf Gallery, Philadelphia 51878
Roswell Museum and Art Center, Roswell 52470
Round Top Center for the Arts - Arts Gallery, Damariscotta 47950
Rudolph E. Lee Gallery, Clemson 47616
Runnels Gallery, Portales 52058
Ruth Eckerd Hall, Clearwater 47611
Saco Museum, Saco 52498
Saginaw Art Museum, Saginaw 52516
Saint Gaudens National Historic Site, Cornish 47812
Saint George Art Museum, Saint George 52546
Saint Joseph College Art Gallery, West Hartford 54060
Saint Louis Art Museum, Saint Louis 52595
Saint Louis Artists' Guild Museum, Saint Louis 52596
Saint Peter's College Art Gallery, Jersey City 49628
De Saisset Museum, Santa Clara 52863
Sala Diaz, San Antonio 52722
Salisbury State University Galleries, Salisbury 52678
Salt Lake Art Center, Salt Lake City 52693
The Salt Queen Foundation, New York 51252
Samuel Cupples House, Saint Louis 52599
Samuel Dorsky Museum of Art, New Paltz 51097
Samuel P. Harn Museum of Art, Gainesville 48816
San Angelo Museum of Fine Arts, San Angelo 52704
San Antonio Museum of Art, San Antonio 52726
San Diego Art Institute, San Diego 52745
San Diego Museum of Art, San Diego 52752
San Francisco Camerawork, San Francisco 52792
The Sandor Teszler Gallery, Spartanburg 53195
Sangre de Cristo Arts Center and Buell Children's Museum, Pueblo 52194
Santa Barbara Museum of Art, Santa Barbara 52859
Santa Fe Gallery, Gainesville 48817
Santa Rosa Junior College Art Gallery, Santa Rosa 52908
Santuario de Nuestra Senora de Guadalupe, Santa Fe 52889
Sarah Campbell Blaffer Foundation, Houston 49404
Sarah Campbell Blaffer Gallery, Houston 49405
Sarah Moody Gallery of Art, Tuscaloosa 53621
Sarah Spurgeon Gallery, Ellensburg 48361
Sawhill Gallery, Duke Hall, Harrisonburg 49187
SCAD Museum of Art, Savannah 52945
Scarfone and Hartley Galleries, Tampa 53435
Schenectady Museum, Schenectady 52962
Schick Art Gallery, Saratoga Springs 52925
School of Art Gallery, Bozeman 46983
School of Fine Arts Gallery, Bloomington 46866
School of Nations Museum, Elsah 48381
Schumacher Gallery, Columbus 47760
Schweinfurth Memorial Art Center, Auburn 46470
Seattle Art Museum, Seattle 53013
Second Street Gallery, Charlottesville 47402
Sedona Arts Center, Sedona 53027
Seigfred Gallery, Athens 46418
Selby Gallery, Sarasota 52916
Senate House, Kingston 49791
SFA Galleries, Nacogdoches 50936
Sharon Arts Center, Sharon 53051
The Sheldon Art Galleries, Saint Louis 52600
Sheldon Memorial Art Gallery and Sculpture Garden, Lincoln 50105
Silvermine Guild Arts Center, New Canaan 51038
Simmons Visual Center, Gainesville 48822
Sioux City Art Center, Sioux City 53113
Skirball Cultural Center, Los Angeles 50264
Slater Memorial Museum, Norwich 51438
Slidell Art Center, Slidell 53130
Slifer House, Lewisburg 50031
Smart Museum of Art, Chicago 47513
Smith College Museum of Art, Northampton 51414
The Smithy-Pioneer Gallery, Cooperstown 47792
The Snite Museum of Art, Notre Dame 51445
Snyder Museum and Creative Arts Center, Bastrop 46624
Society for Contemporary Craft Museum, Pittsburgh 51961
The Society of the Cincinnati Museum, Washington 53918
The Society of the Four Arts Gallery, Palm Beach 51678
Solomon R. Guggenheim Museum, New York 51258
Sordoni Art Gallery, Wilkes-Barre 54163
South Arkansas Arts Center, El Dorado 48310
South Boston-Halifax County Museum of Fine Arts and History, South Boston 53165
Southeastern Center for Contemporary Art, Winston-Salem 54251
Southern Plains Indian Museum, Anadarko 46282
Southern Vermont Art Center, Manchester 50428
Southern Vermont College Art Gallery, Bennington 46740
The Southland Art Collection, Dallas 47940
Space One Eleven, Birmingham 46826
Spartanburg County Museum of Art, Spartanburg 53196
Speed Art Museum, Louisville 50297
Spelman College Museum of Fine Art, Atlanta 46454
Spencer Museum of Art, Lawrence 49992
Springfield Art Gallery, Springfield 53221
Springfield Art Museum, Springfield 53233
Springville Museum of Art, Springville 53243
Spruill Gallery, Atlanta 46455
Stables Art Gallery of Taos Art Association, Taos 53443
Stamford Museum and Nature Center, Stamford 53249
Stark Museum of Art, Orange 51578
State of the Art Gallery, Ithaca 49551
Staten Island Institute of Arts and Sciences, Staten Island 53270
Staunton Augusta Art Center, Staunton 53278
Stephens Museum, Fayette 48526

A

Sterling and Francine Clark Art Institute, Williamstown 54175
Stone House Gallery, Fredonia 48774
Stowitts Museum, Pacific Grove 51667
Stuart Collection, La Jolla 49847
Studio Gallery, Washington 53919
Studio San Giuseppe Art Gallery, Cincinnati 47564
Suffolk Museum, Suffolk 53355
Sumter Gallery of Art, Sumter 53367
Sun Valley Center for the Arts, Ketchum 49754
Sweeney Art Gallery, Riverside 52375
Sweet Briar College Art Gallery, Sweet Briar 53391
Swope Art Museum, Terre Haute 53483
Synagogue for the Arts, New York 51266
Syracuse University Art Collection, Syracuse 53403
T.C. Steele State Historic Site, Nashville 50961
T.W. Wood Gallery and Arts Center, Montpelier 50824
Tacoma Art Museum, Tacoma 53414
The Taft Museum of Art, Cincinnati 47565
Tang Teaching Museum and Art Gallery, Saratoga Springs 52926
Tarble Arts Center, Charleston 47354
Tate House, Portland 52068
Telfair Museum of Art, Savannah 52947
Tempe Center for the Arts, Tempe 53471
Tennent Art Foundation Gallery, Honolulu 49354
Tennessee Valley Art Center, Tuscumbia 53625
Texas Woman's University Art Galleries, Denton 48062
Thomas Center Galleries, Gainesville 48818
Thomas Handforth Gallery, Tacoma 53415
Thomas J. Walsh Art Gallery and Regina A. Quick Center for the Arts, Fairfield 48482
Thomasville Cultural Center, Thomasville 53501
Thorne-Sagendorph Art Gallery, Keene 49720
Thornhill Gallery, Kansas City 49702
Tibet House Museum, New York 51270
Times Square Gallery, New York 51271
Timken Museum of Art, San Diego 52758
TM Gallery, Crestview 47879
Toledo Museum of Art, Toledo 53523
Tomasulo Gallery, Cranford 47867
Torpedo Factory Art Center, Alexandria 46214
Tougaloo College Art Collection, Tougaloo 53545
Tower Fine Arts Gallery, Brockport 47035
Trianon Museum and Art Gallery, Denver 48091
Tubac Center of the Arts, Tubac 53580
Tucson Museum of Art and Historic Block, Tucson 53598
Tufts University Art Gallery, Medford 50548
Tulane University Art Collection, New Orleans 51095
Turner House Museum, Hattiesburg 49213
The Turner Museum, Sarasota 52917
Turtle Mountain Chippewa Heritage Center, Belcourt 46694
Tweed Museum of Art, Duluth 48200
Tyler Art Gallery, Oswego 51627
Tyler Museum of Art, Tyler 53635
Ucross Foundation Art Gallery, Clearmont 47609
Ukrainian Museum, Cleveland 47643
Union Art Gallery, Baton Rouge 46642
Union-Art Gallery, Milwaukee 50719
Union Street Gallery, Chicago 47521
United States Capitol Visitor Center, Washington 53923
United States Naval Academy Museum, Annapolis 46321
United States Navy Art Gallery, Washington 53927
United States Senate Commission on Art Collection, Washington 53929
University Art Galleries, Akron 46150
University Art Galleries, Murray 50918
University Art Galleries, Vermillion 53720
University Art Gallery, Bridgeport 47019
University Art Gallery, Pittsburgh 51963
University Art Gallery, Stony Brook 53315
University Art Gallery, San Diego 52759
University Art Gallery, La Jolla 49848
University Art Gallery of the University of the South, Sewanee 53041
University Art Museum, Binghamton 46817
University Art Museum, Santa Barbara 52861
University Art Museum, Lafayette 49868
University Art Museum, Albuquerque 46189
University at Buffalo Art Galleries, Buffalo 47126
University Gallery, Commerce 47764
University Gallery, Newark 51282
The University Museum, Edwardsville 48301
The University Museum, Indiana 49485
University Museums, Oxford 51650
University Museums, Newark 51283
University of Arizona Museum of Art, Tucson 53600
University of Arizona Student Union Galleries, Tucson 53601
University of Arkansas at Little Rock Art Galleries, Little Rock 50142
University of Hawaii Art Gallery, Honolulu 49355
University of Iowa Museum of Art, Iowa City 49521
University of Kentucky Art Museum, Lexington 50057
University of North Texas Art Gallery and Cora Stafford Gallery, Denton 48063
University of Rhode Island Fine Arts Center Galleries, Kingston 49795
University of South Carolina Beaufort Art Gallery, Beaufort 46668
University of Virginia Art Museum, Charlottesville 47400
University of West Florida Art Gallery, Pensacola 51770
Upper Room Chapel Museum, Nashville 50978
USM Art Gallery, Gorham 48976
Utah Museum of Fine Arts, Salt Lake City 52695
Valdosta State University Fine Arts Gallery, Valdosta 53687
Valley Art Center, Clarkston 47591
Vanderbilt University Fine Arts Gallery, Nashville 50980
Vanderbilt University Sarratt Gallery, Nashville 50981
Venango Museum of Art, Science and Industry, Oil City 51509
Vero Beach Museum of Art, Vero Beach 53728
Very Special Arts Gallery, Albuquerque 46190
Victoria Mansion, Portland 52069
Viking Union Gallery, Bellingham 46712
Villa Julie College Gallery, Stevenson 53291
Virginia Museum of Fine Arts, Richmond 52343
Visceglia Art Gallery, Caldwell 47163
Visual Aid, San Francisco 52797
Visual Art Gallery, Santa Fe 52891
Visual Arts Center of Northwest Florida, Panama City 51694
Visual Arts Gallery, Birmingham 46827
Visual Arts Gallery, Pensacola 51771
Volcano Art Center, Volcano 53766
The von Liiebig Art Center, Naples 50954
Wadsworth Atheneum, Hartford 49199
Walker Art Collection, Garnett 48866
Walter Anderson Museum of Art, Ocean Springs 51489
Walters Art Museum, Baltimore 46591
The Ward Museum of Wildfowl Art, Salisbury 52679
Warehouse Gallery, Lee 50016
Warther Museum, Dover 48168
Warwick Museum of Art, Warwick 53837
Washington County Museum of Fine Arts, Hagerstown 49122
Washington Pavillion of Arts and Science, Sioux Falls 53123
Wassenberg Art Center, Van Wert 53703
Waterloo Center for the Arts, Waterloo 53965
Waterworks Visual Arts Center, Salisbury 52682
Watts Towers Arts Center, Los Angeles 50267
Wayne Center for the Arts, Wooster 54295
Weatherspoon Art Museum, Greensboro 49068
Weber State University Art Gallery, Ogden 51504
Weir Farm, Wilton 54213
Wellington B. Gray Gallery, Greenville 49075
West Bend Art Museum, West Bend 54045
West Valley Art Museum, Surprise 53381
West Virginia University Mesaros Galleries, Morgantown 50846
Western Art Gallery and Museum, Dillon 48143
Western Colorado Center for the Arts, Grand Junction 48995
Western Gallery, Bellingham 46713
Western Illinois University Art Gallery, Macomb 50364
Western Wyoming College Art Gallery, Rock Springs 52412
Wexner Center for the Arts, Columbus 47761
Whatcom Museum of History and Art, Bellingham 46714
Whitaker Center for Science and the Arts, Harrisburg 49183
Whitney Museum of American Art at Champion, Stamford 53250
Whittier Fine Arts Gallery, Wichita 54149
Widener Gallery, Hartford 49200
Widener University Art Collection and Gallery, Chester 47432
The Wiegand Gallery, Belmont 46718
Wignall Museum of Contemporary Art, Rancho Cucamonga 52239
Wildling Art Museum, Los Olivos 50274
Wildwood Center, Nebraska City 50996
Wilkes Art Gallery, North Wilkesboro 51410
The William Benton Museum of Art, Storrs 53320
William King Regional Arts Center, Abingdon 46126
William Weston Clarke Emison Museum of Art, Greencastle 49048
William Woods University Art Gallery, Fulton 48809
Williams College Museum of Art, Williamstown 54176
Wing Luke Asian Museum, Seattle 53017
Winterset Art Center, Winterset 54257
Winthrop University Galleries, Rock Hill 52404
Wiregrass Museum of Art, Dothan 48155
Wirtz Gallery, Miami 50637
The Wise Art Gallery, Tyler 53636
Witter Gallery, Storm Lake 53317
Women and Their Work, Austin 46523
Woodmere Art Museum, Philadelphia 51889
Woodstock Artists Gallery, Woodstock 54284
Worcester Art Museum, Worcester 54302
Words and Pictures Museum, Northampton 51415
The Works, Newark 51295
The World Organization of China Painters' Museum, Oklahoma City 51527
Wright Museum of Art, Beloit 46725
Wriston Art Center Galleries, Appleton 46332
Wyoming Arts Council Gallery, Cheyenne 47447
The Yager Museum, Oneonta 51566
Yale Center for British Art, New Haven 51055
Yale University Art Gallery, New Haven 51056
Yeiser Art Center, Paducah 51672
Zanesville Art Center, Zanesville 54368

Uruguay

Centro de Artistas Plásticos, Montevideo 54389
Galería de Cinemateca, Montevideo 54391
Galería Pocitos, Montevideo 54392
Museo de Bellas Artes Agustín Araujo, Treinta y Tres 54436
Museo Juan Manuel Blanes, Montevideo 54405
Museo Municipal de Artes Plásticas, Rivera 54430
Museo Nacional de Artes Visuales, Montevideo 54411
Museo San Gregorio de Polanco, San Gregorio de Polanco 54433
Museo Virtual de Artes el Pais, Montevideo 54419
Sala de Arte Carlos F. Sáez, Montevideo 54425

Uzbekistan

Andijon Viloyat Adabiyot va San'at Muzeyi, Andijon 54438
Buxoro Qala Davlat Muzeyi Qo'riqxonasi, Buxoro 54444
Özbekiston San'at Davlat Muzeyi, Toshkent 54484
Savitsky nomidagi Qoraqalpoğiston Respublikasi Davlat San"at Muzeyi, Nukus 54453
Tasviriy San'at Muzeyi - Andijon Qal'asi, Andijon 54442

Venezuela

Ateneo de Valencia, Valencia 54596
Centro Cultural Corp Group, Caracas 54527
Centro de Arte La Estancia, Caracas 54529
Galería de Arte Nacional, Caracas 54532
Museo Arturo Michelena, Caracas 54535
Museo de Arte Colonial, Mérida 54586
Museo de Arte Moderno Jesús Soto, Ciudad Bolívar 54563
Museo de la Fundación John Boulton, Caracas 54548
Museo Pedagógico de Historia del Arte, Caracas 54555

Vietnam

Exhibition Hall, Ha Noi 54611
Saigon Open City Studio Space, Ho Chi Minh City 54626
Viet Nam National Fine Arts Museum, Ha Noi 54617

Art, African

Argentina

Museo Nacional de Arte Oriental, Buenos Aires 00258

Australia

SCA Gallery, Rozelle 01473

Belgium

Koninklijk Museum voor Midden-Afrika, Tervuren 04058

Benin

Musée d'Art Contemporarin Africain à Cotonou, Cotonou 04183

Botswana

Botswana National Museum and Art Gallery, Gaborone 04296

Burkina Faso

Centre National d'Artisanat d'Art, Ouagadougou 05277

Cameroon

Musée d'Art Nègre, Yaoundé 05301

Czech Republic

Náprstkovo Muzeum Asijských, Afrických a Amerických Kultur, Praha 09775

France

Musée Africain, Lyon 13882
Musée Charbonneau-Lassay, Loudun 13830
Musée Dapper, Paris 14643
Musée d'Art Moderne, Troyes 15998
Musée d'Arts Africains, Langonnet 13508
Musée du Quai Branly, Paris 14700
Musée National des Arts d'Afrique et d'Océanie, Paris 14726

Germany

Iwalewa Haus, Bayreuth 16953
Makonde Art Museum, Bammental 16923

Italy

Museo Africano, Roma 27460
Museo e Villaggio Africano, Basella di Urgnano 25377

Japan

Makonde Bijutsukan, Futami 28454

Morocco

Musée Dar-Si-Said, Marrakech 31677
Musée de Marrakech, Marrakech 31679
Museum of Maroccan Arts, Tanger 31699

Netherlands

Afrika Museum, Berg en Dal 31964

Nigeria

Centre for Black and African Art and Civilization, Lagos 33333
Isenbaye Art Gallery and Cultural Troupe, Oshogbo 33339

Poland

Muzeum Afrykanistyczne im. Dr. Bogdana Szczygła, Olkusz 34973

Portugal

Museu de Arte Africana da Azambuja, Azambuja 35450

Puerto Rico

Museo de Las Américas, San Juan 36068
Museo Taller Arte Africano, San Juan 36075

Sao Tome and Principe

Museu Nacional, São Tomé 38013

Senegal

Galerie Antenna, Dakar 38019
Galerie Nationale d'Art, Dakar 38020
Musée d'Art Africain, Dakar 38022

South Africa

African Art Centre, Durban 38602
Ann Bryant Art Gallery, East London 38614
Durban Art Gallery, Durban 38604
Gertrude Posel Gallery, Johannesburg 38655
Irma Stern Museum of the University of Cape Town, Rosebank 38727
Julius Gordon Africana Centre, Riversdale 38723
Lichtenburg Museums and Art Gallery, Lichtenburg 38682
Museum of Southern African Rock Art, Johannesburg 38661
Polokwane Art Museum, Polokwane 38710
Potchefstroom Museum, Potchefstroom 38714
Pretoria Art Museum, Tshwane 38755
Rust en Vreugd Museum, Cape Town 38589
South African National Gallery, Cape Town 38593
Standard Bank Gallery, Johannesburg 38665
Tatham Art Gallery, Pietermaritzburg 38707
Technikon Natal Gallery, Durban 38613
Township Museum, Lwandle 38684
UNISA Art Gallery, Tshwane 38760
University Art Gallery, Potchefstroom 38717
University of Fort Hare Museum, Alcie 38551

Spain

Museo Africano Mundo Negro, Madrid 39525

Switzerland

Asia-Africa Museum, Genève 41398

Uganda

Makerere Art Gallery, Kampala 42992

United Kingdom

Danford Collection of West African Art and Artefacts, Birmingham 43504

United States of America

African Art Museum of Maryland, Columbia 47712
African Arts Collection, San Diego 52736
African Cultural Art Forum, Philadelphia 51811
Black Arts National Diaspora, New Orleans 51076
Chattanooga African Museum, Chattanooga 47406
I.P. Stanback Museum, Orangeburg 51582
Jackson Hall Gallery, Frankfort 48736
Lois E. Woods Museum, Norfolk 51365
Muscarelle Museum of Art, Williamsburg 54170
Museum for African Art, Long Island City 50184
The Museum of African Tribal Art, Portland 52066
Museum of the National Center of Afro-American Artists, Boston 46942
National Museum of African Art, Washington 53899
Neuberger Museum of Art, Purchase 52201
New Visions Gallery, Marshfield 50513
North Carolina Central University Art Museum, Durham 48222
Northwestern Mutual Art Gallery, Milwaukee 50717
Portland Art Museum, Portland 52086
Selsor Gallery of Art, Chanute 47343
The Studio Museum in Harlem, New York 51264
Trout Gallery, Carlisle 47244
Wellington B. Gray Gallery, Greenville 49075
The Zora Neale Hurston Museum, Eatonville 48274

Zimbabwe

National Gallery of Zimbabwe, Harare 54670

Art, American

Canada

Art Gallery of Bancroft, Bancroft 05378
Art Gallery of Prince Albert, Prince Albert 06659
Atlantic Salmon Museum on the Miramichi River, Doaktown 05705
Bonavista North Regional Museum, Wesleyville 07300
Bowen Island Museum, Bowen Island 05448
Canadian Wildlife and Wilderness Art Museum, Ottawa 06540
Centre d'Exposition d'Amos, Amos 05337
City of Saint John Gallery, Saint John 06835
Cornwall Regional Art Gallery, Cornwall 05643
Court House Museum, Cardston 05561
Gallery Lambton, Sarnia 06907
Granisle Museum, Granisle 05903
Kelowna Art Gallery, Kelowna 06053
Leighton Art Centre, Calgary 05522
Maple Ridge Art Gallery, Maple Ridge 06236
Marion Scott Gallery, Vancouver 07215
Moose Jaw Museum and Art Gallery, Moose Jaw 06386
Mount Royal College Gallery, Calgary 05525
Profiles Public Art Gallery, Saint-Albert 06791
Red Deer and District Museum, Red Deer 06709
Simon Fraser Gallery, Burnaby 05505
Tom Thomson Art Gallery, Owen Sound 06563
Workers Arts and Heritage Centre, Hamilton 05964
Yorkton Arts Council, Yorkton 07417

France

Musée du Quai Branly, Paris 14700
Musée National de la Coopération Franco-Américaine, Blérancourt 12025

Lithuania

Kauno Paveikslų Galerija, Kaunas 30251

Netherlands

Museum Valse Kunst, Vledder 32896

Peru

Galería del Instituto Cultural Peruano-Norteamericano, Lima 34220

United States of America

106 South Division Gallery, Grand Rapids 49000
3M Collection of Contemporary Western and Americana Art, Long Lake 50189
A.R.C. Gallery, Chicago 47449
A.R. Mitchell Memorial Museum of Western Art, Trinidad 53565
A.Shiwi A.Wan Museum, Zuni 54375
Academy Gallery, New Orleans 51074
Addison Gallery of American Art, Andover 46296
African Arts Collection, San Diego 52736
Agua Caliente Cultural Museum, Palm Springs 51680
Alaska Centennial Center for the Arts, Fairbanks 48469
Allen R. Hite Art Institute, Louisville 50280
Amarillo Museum of Art, Amarillo 46250
Amon Carter Museum, Fort Worth 48709
Arizona State University Art Museum, Tempe 53465
Armory Art Gallery, Blacksburg 46843
The Art Center, Mount Clemens 50871
Art Center of Battle Creek, Battle Creek 46643
Art Center of South Florida, Miami Beach 50640
Art Gallery, Minneapolis 50728
Art Gallery Kendall, Miami 50620
Art Gallery Marist College, Poughkeepsie 52119
Art Museum of Southeast Texas, Beaumont 46670
Art Museum of Western Virginia, Roanoke 52379
Arts Alliance Center at Clear Lake, Nassau Bay 50983
Arts and Science Center for Southeast Arkansas, Pine Bluff 51920
Arts Club of Washington, Washington 53851
Arts Midland Galleries, Midland 50666
Ashby-Hodge Gallery of American Art, Fayette 48525
Asheville Art Museum, Asheville 46368
Associated Artists of Winston-Salem, Winston-Salem 54242
Attleboro Museum-Cerner for the Arts, Attleboro 46463
Atwood Memorial Center Art Collection, Saint Cloud 52537
Augustana College Center for Western Studies, Sioux Falls 53116
Avampato Discovery Museum, Charleston 47378
B.F. Larsen Gallery, Provo 52185
Bank of Mississippi Art Collection, Tupelo 53612
Barron Arts Center and Museum, Woodbridge 54274
Beall Center for Art and Technology, Irvine 49529
Beard Gallery, Norton 51430
Bedford Gallery, Walnut Creek 53802
Bergen Museum of Art and Science, Paramus 51696
The Berkshire Artisans, Pittsfield 51966
Biola University Art Gallery, La Mirada 49851
Bjarne Ness Gallery, Fort Ransom 48672
Black Mountain College Museum and Arts Center, Asheville 46370
Blackwell Street Center for the Arts, Denville 48092
Bloomington Art Center, Bloomington 46869
Boarman Arts Center, Martinsburg 50518
Bowery Gallery, New York 51136
Bowman, Megahan and Penelec Galleries, Meadville 50546
Braithwaite Fine Arts Gallery, Cedar City 47289
Bruce Museum of Arts and Science, Greenwich 49081
Burchfield-Penney Art Center, Buffalo 47119
California African American Museum, Los Angeles 50214
California Art Club Gallery at the Old Mill, San Marino 52818
The Canton Museum of Art, Canton 47227
Carnegie Art Center, Mankato 50447
Center Galleries, Detroit 48108
Charles H. MacNider Museum, Mason City 50529
Charles H. Taylor Arts Center, Hampton 49150
The Charles Hosmer Morse Museum of American Art, Winter Park 54253
Charles M. Schulz Museum and Research Center, Santa Rosa 52904
Chatham College Art Gallery, Pittsburgh 51946
Chazen Museum of Art, Madison 50388
Chesterwood Museum, Stockbridge 53298
Chief Oshkosh Native American Arts, Egg Harbor 48302
Clark Atlanta University Art Galleries, Atlanta 46433
Clough-Hanson Gallery, Memphis 50569
The Coburn Gallery, Ashland 46387
Colby College Museum of Art, Waterville 53977
The College Art Gallery, Ewing 48461
College of Marin Art Gallery, Kentfield 49741
College of Visual Arts Gallery, Saint Paul 52614
Colter Bay Indian Arts Museum, Colter Bay Village 47709
Community College Art Gallery, Baltimore 46557
Concord Art Association, Concord 47766
Conejo Valley Art Museum, Thousand Oaks 53502
Coral Springs Museum of Art, Coral Springs 47803
Courtyard Gallery, Washington 53946
Crooked Tree Arts Center, Petoskey 51806
Custer County Art Center, Miles City 50682
Cypress College Fine Arts Gallery, Cypress 47918
The Dalles Art Association Gallery, The Dalles 47944
Danforth Museum of Art, Framingham 48729
Decker, Meyerhoff and Pinkard Galleries, Baltimore 46560
DeCordova Museum and Sculpture Park, Lincoln 50090
DePree Art Center and Gallery, Holland 49314
DeRicci Gallery, Madison 50389
Doll Gardner Art Gallery, Portland 52074
Donna Beam Fine Art Gallery, Las Vegas 49966
Duluth Art Institute, Duluth 48193
Dupont Gallery, Lexington 50069
East Wing Art Gallery, Gardner 48861
El Camino College Art Gallery, Torrance 53542
Ellington White Project-Art Exhibitions, Detroit 48118
Emerald Art Center, Springfield 53237
Evans-Tibbs Collection, Washington 53866
Evergreen Galleries, Olympia 51543
Fairfield Art Museum, Fairfield 48484
Favell Museum of Western Art and Indian Artifacts, Klamath Falls 49806
Fawick Art Gallery, Berea 46748
Fine Arts Center of Hot Springs, Hot Springs 49372
Fine Arts Gallery, Columbus 47745
Fisk University Galleries, Nashville 50968
Foothills Art Center, Golden 48959
Ford Gallery, Ypsilanti 54356
Frame Gallery, Pittsburgh 51948
Frankenberger Art Gallery, Charleston 47379
Frankie G. Weems Gallery Rotunda Gallery, Raleigh 52229
Freedman Gallery, Reading 52262
Fresno Metropolitan Museum, Fresno 48791
The Furlong Gallery, Menomonie 50591
Gallery 825, West Hollywood 54063
Gallery A and Downstairs Gallery, San Pedro 52825
The Gallery Beyond Words Gallery, Detroit 48119
Gallery of Art, Dickinson 48139
Gallery of the Department of Art and Art History, Lincoln 50095
Gallery XII, Wichita 54142
Gavilan College Art Gallery, Gilroy 48913
George Walter Vincent Smith Art Museum, Springfield 53225
Glyndor Gallery and Wave Hill House Gallery, Bronx 47045
Goppert Gallery, Leavenworth 50009
The Grace Museum, Abilene 46124
Grants Pass Museum of Art, Grants Pass 49016
Guggenheim Museum Soho, New York 51188
Hallie Ford Museum of Art, Salem 52666
Hartnell College Art Gallery, Salinas 52676
Hera Gallery, Wakefield 53792
Heritage Conservancy, Doylestown 48178
Hickory Museum of Art, Hickory 49267
Hicks Art Center, Newtown 51345
Higgins Art Gallery, West Barnstable 54043
Holland Area Arts Council, Holland 49316
Hoxie Gallery, Westerly 54087
Hui No'eau Visual Arts Center, Makawao 50404
Hunter College Art Galleries, New York 51193
Hunter Museum of American Art, Chattanooga 47411
Huntington Museum of Art, Huntington 49435
Hutchinson Art Center, Hutchinson 49453
Hyde Art Gallery, El Cajon 48308
I.P. Stanback Museum, Orangeburg 51582
Indian Arts and Crafts Board, Washington 53882
Indian Arts Research Center, Santa Fe 52876
Irene Cullis Gallery, Greensboro 49065
Iron County Museum, Caspian 47274
Irvine Museum, Irvine 49531
Isabel Anderson Comer Museum and Arts Center, Sylacauga 53394
Jack S. Blanton Museum of Art, Austin 46511
James and Meryl Hearst Center for the Arts, Cedar Falls 47293
James J. Hill House, Saint Paul 52618
Jasper Rand Art Museum, Westfield 54092
John B. Davis Gallery of Fine Art, Pocatello 52007
John H. Vanderpoel Art Gallery, Chicago 47484
John Michael Kohler Arts Center, Sheboygan 53061
John P. Barclay Memorial Gallery, Pittsburgh 51951
Johnson Heritage Post Art Gallery, Grand Marais 48998
Jonson Gallery, Albuquerque 46180
Jordan Schnitzer Museum of Art at the University of Oregon, Eugene 48431
June Steingart Art Gallery, Oakland 51465
Junior Museum, Newark 51285
Kansas African American Museum, Wichita 54144
Katherine Nash Gallery, Minneapolis 50736
Kauffman Gallery, Shippensburg 53080
Kearny Cottage, Perth Amboy 51785
King County Arts Commission Gallery, Seattle 53001
Kirkland Museum of Fine and Decorative Art, Denver 48080
Kittredge Art Gallery, Tacoma 53411
La Raza Galeria Posada, Sacramento 52508
Lansing Art Gallery, Lansing 49942
Laure A. Sprague Art Gallery, Joliet 49644
Levy Gallery, Philadelphia 51848
Lewistown Art Center, Lewistown 50042
Lillian and Coleman Taube Museum of Art, Minot 50745
Long Island Museum of American Art, History and Carriages, Stony Brook 53313
Loop Museum and Art Center, Seagraves 52989
Los Angeles Valley College Art Gallery, Valley Glen 53696
Louisburg College Art Gallery, Louisburg 50278
Louise Wells Cameron Art Museum, Wilmington 54202
Luther W. Brady Art Gallery, Washington 53888
Lyman Allyn Museum of Art, New London 51066
Lynchburg Fine Arts Center, Lynchburg 50326
M. Christina Geis Gallery, Lakewood 49906
McPherson College Gallery, McPherson 50371
Madison Museum of Contemporary Art, Madison 50393
Maier Museum of Art, Lynchburg 50328
Manhattan Arts Center, Manhattan 50434
Mansfield Art Center, Mansfield 50451
Marianna Kistler Beach Museum of Art, Manhattan 50435
Marie Hull Gallery, Raymond 52259
Marwen Foundation, Chicago 47488
Matrixarts, Sacramento 52509
MCAD Gallery, Minneapolis 50738
Mead Art Museum, Amherst 46268
The Mexican Museum, San Francisco 52783
Michigan Chapter Gallery, Detroit 48123
Middletown Fine Arts Center, Middletown 50662
Midtown Art Center, Houston 49395
Minnesota Museum of American Art, Saint Paul 52624
Minority Arts Resource Council Studio Art Museum, Philadelphia 51852
Mississippi Museum of Art, Jackson 49560
Morris Graves Museum of Art, Eureka 48440
Moses Lake Museum and Art Center, Moses Lake 50867
Mount Holyoke College Art Museum, South Hadley 53170
Muscarelle Museum of Art, Williamsburg 54170
El Museo Francisco Oller y Diego Rivera, Buffalo 47123
Museum of African American Art, Los Angeles 50251
The Museum of Art and History, Santa Cruz 52868
Museum of Art and History, Weston 54102
Museum of Art at Brigham Young University, Provo 52188
Museum of Fine Arts, Springfield 53226
Museum of Northwest Art, La Conner 49831
Napa Valley Museum, Yountville 54355
Natalie and James Thompson Gallery, San Jose 52806
National Academy Museum, New York 51232
National Arts Club, New York 51233
National Museum of American Art, Washington 53900
New Bedford Art Museum, New Bedford 51013
New Britain Museum of American Art, New Britain 51029
Nora Eccles Harrison Museum of Art, Logan 50161
Nordstrand Visual Arts Gallery, Wayne 54006
Norman Rockwell Museum at Stockbridge, Stockbridge 53302
North Carolina Central University Art Museum, Durham 48222
North Carolina State University Gallery of Art and Design, Raleigh 52235
Northeastern Illinois University Art Gallery, Chicago 47500
Northwind Arts Center, Port Townsend 52044
Noyes and Read Gallery and Herndon Gallery, Yellow Springs 54330
NSCC Art Gallery, Seattle 53007
Ogunquit Museum of American Art, Ogunquit 51507
Oklahoma City Museum of Art, Oklahoma City 51521
Oklahoma Museum of African American Art, Oklahoma City 51523
The Old Jail Art Center, Albany 46167
Olin Art Gallery, Gambier 48852
Olive Hyde Art Gallery, Fremont 48785
Owatonna Arts Center, Owatonna 51635
Owensboro Museum of Fine Art, Owensboro 51640
Palos Verdes Art Center, Rancho Palos Verdes 52242
Paris Gibson Square Museum of Art, Great Falls 49031
Pasadena City College Art Gallery, Pasadena 51722
Pasadena Museum of California Art, Pasadena 51723
Peacock Gallery, Cochran 47667
Pennsylvania College of Art and Design Galleries, Lancaster 49930
People's Gallery, El Paso 48328
Perspective Gallery, Blacksburg 46846
Philbrook Museum of Art, Tulsa 53610
Phippen Museum of Western Art, Prescott 52134
Platt and Borstein Galleries, Los Angeles 50260
Plattsburgh Art Museum, Plattsburgh 51984
Pollock Gallery, Dallas 47938
Port Angeles Fine Arts Center, Port Angeles 52028
Portland Art Museum, Portland 52086
Price Tower Arts Center, Bartlesville 46619
Project Art-University of Iowa and Clinics, Iowa City 49520
Pump House Regional Arts Center, La Crosse 49836
QCC Art Gallery, Bayside 46657

Rahr West Art Museum, Manitowoc 50443
Ramsey Center for Arts, Shoreview 53082
Rankin Museum, Ellerbe 48363
Regina Gouger Miller Gallery, Pittsburgh 51958
Revolving Museum, Lowell 50307
Rhode Island Foundation Art Gallery, Providence 52177
Richmond Art Center, Richmond 52311
Rider University Art Gallery, Lawrenceville 49997
Riverside Art Museum, Riverside 52373
Roberson Museum and Science Center, Binghamton 46816
Robert A. Peck Gallery, Riverton 52378
Rockport Art Association Galleries, Rockport 52423
Rockwell Museum of Western Art, Corning 47811
Rothko Chapel, Houston 49403
Ryan Fine Arts Center, Abilene 46125
Sacred Circle Gallery of American Indian Art, Seattle 53011
Saint Luke's Gallery, Paxton 51740
Saint Mary's University Art Center, San Antonio 52721
Salmagundi Museum of American Art, New York 51251
San Luis Obispo Art Center, San Luis Obispo 52815
Santa Cruz Art League, Santa Cruz 52869
Sargent House Museum, Gloucester 48948
Sheldon Peck Museum, Lombard 50165
Sheppard Fine Arts Gallery, Reno 52290
Sid Richardson Collection of Western Art, Fort Worth 48719
Sioux Indian Museum, Rapid City 52254
Smart Museum of Art, Chicago 47513
Smithsonian American Art Museum, Washington 53917
South Bend Regional Museum of Art, South Bend 53159
South Dakota Art Museum, Brookings 47059
Southeast Arts Center, Atlanta 46452
Southern Alleghenies Museum of Art, Altoona 46243
Southern Alleghenies Museum of Art, Loretto 50201
Southern Alleghenies Museum of Art at Johnstown, Johnstown 49642
Southern Ohio Museum and Cultural Center, Portsmouth 52103
Southwest Museum of the American Indian, Los Angeles 50265
Spiers Gallery, Brevard 47007
Springfield Museum of Art, Springfield 53236
Steep and Brew Gallery, Madison 50394
Steffen Thomas Museum of Art, Buckhead 47109
Stephens African-American Museum, Shreveport 53091
Stubnitz Gallery, Adrian 46139
Studio Channel Islands Art Center, Camarillo 47171
The Studio Museum in Harlem, New York 51264
Swan Coach House Gallery, Atlanta 46456
Taos Center for the Arts, Taos 53444
Tekakwitha Fine Arts Center, Sisseton 53124
Terra Museum of American Art, Chicago 47518
Thea G. Korver Visual Art Center, Orange City 51581
Triton Museum of Art, Santa Clara 52864
Tryon Palace, New Bern 51020
University Art Gallery, Carson 47258
University Art Gallery, Chico 47529
University Art Gallery, Jacksonville 49586
University Art Gallery, Turlock 53615
University Art Gallery, Mount Pleasant 50885
University Art Gallery, Terre Haute 53484
University Art Gallery, Rohnert Park 52444
University Art Gallery, Florence 48565
University Art Gallery Cabinet of Natural History, New Bedford 51016
University Galleries, Edinburg 48292
University Gallery, Alexandria 46196
University Gallery, Oxford 51649
University of Hawaii Art Gallery, Honolulu 49355
University of South Carolina at Spartanburg Art Gallery, Spartanburg 53198
University of Wyoming Art Museum, Laramie 49950
University Place Art Center, Lincoln 50108
University Saint Thomas Art Collection and Exhibitions, Saint Paul 52630
Upstairs Art Gallery, Los Alamos 50206
Utah Museum of Fine Arts, Salt Lake City 52695
Ventura College Art Galleries, Ventura 53715
Ventura County Museum of History and Art, Ventura 53716
Victor Valley Museum and Art Gallery, Apple Valley 46328
Villanova University Art Gallery, Villanova 53745
Vincent Price Art Museum, Monterey Park 50810
Virginia Wright Fund, Seattle 53016
Wake Forest University Fine Arts Gallery, Winston-Salem 54252
Wally Findlay Galleries, New York 51278
Watkins Institute Art Gallery, Nashville 50982
Weatherspoon Art Museum, Greensboro 49068
Wells Fargo History Museum, Phoenix 51911
Westminster College Art Gallery, New Wilmington 51106
Westmoreland Museum of American Art, Greensburg 49070
Whistler House Museum of Art, Lowell 50308
Whitney Museum of American Art, New York 51279
Wichita Art Museum, Wichita 54150
Wichita Falls Museum and Art Center, Wichita Falls 54154
William A. Farnsworth Art Museum and Wyeth Center, Rockland 52422
William Blizard Gallery, Springfield 53230
William Bonifas Fine Arts Center, Escanaba 48416
Wisconsin Union Gallery, Madison 50397
Woodstock School of Art Gallery, Woodstock 54285
Yuma Art Center and Galleries, Yuma 54365
The Zigler Museum, Jennings 49621
The Zora Neale Hurston Museum, Eatonville 48274

Uruguay

Museo de Arte Americano de Maldonado, Maldonado 54379

Art, Asian

Argentina

Museo Nacional de Arte Oriental, Buenos Aires 00258

Australia

Asian Galleries, Sydney 01545
Queensland Art Gallery, South Brisbane 01512
Rockhampton Art Gallery, Rockhampton 01466

Austria

MAK - Österreichisches Museum für angewandte Kunst / Gegenwartskunst, Wien 03026

Azerbaijan

Rustam Mustafayev ad. Azerbaycan Dövlet İncesenet Muzeyi, Bakı 03169

Bangladesh

Bangladesh National Museum, Dhaka 03220
Institute of Arts and Crafts Gallery, Dhaka 03224
National Art Gallery, Dhaka 03227

Belgium

Hooy Kaye Museum, Bruxelles 03448
Pavillon Chinois, Tour Japonaise et Musée d'Art Japonais, Bruxelles 03501

Bhutan

National Museum of Bhutan, Paro 04201

Cambodia

Musée National de Phnom Penh, Phnom Penh 05286

Canada

Canadian Society of Asian Arts, Vancouver 07206

China

Art Museum of the Chinese University of Hong Kong, Hong Kong 07792
Beijing Art Museum, Beijing 07508
Boer Tala Zhou Museum and Gallery, Bole 07582
China Art Museum, Beijing 07527
Dalian Gallery, Dalian 07638
Doumen County Fine Art Calligraphy Society, Zhuhai 08512
Guan Return to the Ancient Dian Art Museum, Beijing 07540
Guangdong Museum of Art, Guangzhou 07725
Guangzhou Fine Arts Museum, Guangzhou 07728
He Xiang Ning Art Museum, Shenzhen 08220
Hong Kong Museum of Art, Hong Kong 07798
Jinze Xuan Art Center, Shenzhen 08221
Liaoning Art Gallery, Shenyang 08214
Liaoning Provincial Museum, Shenyang 08216
Macau Museum of Art, Macau 07997
Museum of Sacred Art, Macau 08002
Museum of Shanxi Province, Taiyuan 08273
Museum of Site, Most I, Hong Kong 07809
Museum of Site, Most II, Hong Kong 07810
Nanjing Museum, Nanjing 08045
The Palace Museum, Beijing 07559
Qing Tycoon Yashu Bowu Yuanji, Hohhot 07791
Shaanxi Chinese Painting Gallery, Xian 08359
Shandong Provincial Museum, Jinan 07884
Shenzhen Art Gallery, Shenzhen 08224
Sichuan University Museum, Chengdu 07620
Suzhou Sportqu Museum, Suzhou 08261
Tangdai Art Museum, Xian 08361
Tsui Museum of Art, Hong Kong 07817
University Museum and Art Gallery, Hong Kong 07818
Xian Forest of Stone Tablets Museum, Xian 08364
Yantai Arts Museum, Yantai 08432
Yunnan Provincial Museum, Kunming 07917
Zhangjiagang City Gallery, Zhangjiagang 08484
Zhangqiu Arts and Crafts Guan, Jinan 07886

Czech Republic

Muzeum Bedřicha Hrozného, Lysá nad Labem 09644
Náprstkovo Muzeum Asijských, Afrických a Amerických Kultur, Praha 09775
Národní Galerie v Praze, Praha 09776

Finland

Joensuun Taidemuseo, Joensuu 10872

France

Asiatica Musée d'Art Oriental, Biarritz 11981
Galeries du Panthéon Bouddhique du Japon et de la Chine, Paris 14609
Musée Cernuschi, Paris 14638
Musée d'Ennery, Paris 14681
Musée des Arts Asiatiques, Nice 14432
Musée du Quai Branly, Paris 14700
Musée Georges Labit, Toulouse 15943
Musée Municipal d'Arts Asiatiques, Toulon (Var) 15928
Musée National des Arts Asiatiques Guimet, Paris 14725

Georgia

Georgian State Museum of Oriental Art, Tbilisi 16362

Germany

Gentil-Haus, Aschaffenburg 16582
Japanisches Kulturinstitut, Köln 19609
Langen Foundation, Neuss 20658
Museum für Asiatische Kunst, Berlin 17201
Museum für Frühislamische Kunst, Bamberg 16914
Museum für Ostasiatische Kunst, Köln 19621
Museum Villa Rot - Kunstsammlung Hoenes-Stiftung, Burgrieden 17662
Siebold-Museum, Würzburg 22724

Greece

Museum of Asian Art, Corfu 22978
New Museum of European and Eastern Art, Athinai 22944

Hungary

Hopp Ferenc Kelet-Ázsiai Művészeti Múzeum, Budapest 23413
Ráth György Museum, Budapest 23470

India

Amreli Archaeological and Art Museum, Amreli 24064
Ananda Niketan Kirtishala, Bagnan 24069
Asutosh Museum of Indian Art, Kolkata 24268
India Arts Museum, Chennai 24116
Indian Academy of Fine Arts Gallery, Amritsar 24067
Jehangir Art Gallery, Mumbai 24318
K.C.S. Paniker Gallery, Thiruvananthapuram 24404
Madhavan Nair Foundation, Edapally 24167
Sri Meenakshi Sundaresvara Temple Museum, Madurai 24304
Museum and Gallery of Art and Archaeology, Dharwad 24164
Nandan Museum-Vichittra and Art Gallery, Santiniketan 24378
Prince of Wales Museum of Western India, Mumbai 24323

Indonesia

Museum Istana Mangkunegaran, Surakarta 24568
Museum Kraton Kasepuhan, Cirebon 24465

Israel

Tikotin Museum of Japanese Art, Haifa 24969
Wilfrid Israel Museum of Oriental Art and Studies, Kibbutz Hazorea 25044

Italy

Messner Mountain Museum - Juval, Castelbello 25721
Museo d'Arte Orientale, Venezia 28189
Museo d'Arte Orientale Edoardo Chiossone, Genova 26299
Museo di Arte Cinese ed Etnografico, Parma 27102
Museo Nazionale d'Arte Orientale, Roma 27512
Museo Osservanza, Bologna 25475
Museo Popoli e Culture, Milano 26715

Japan

Abashiri Kyodo Bijutsukan, Abashiri 28342
Adachi Bijutsukan, Yasugi 29588
Akamon Bijutsukan, Sendai 29241
Asago Art Village, Asago 28366
Ashikaga-shiritsu Bijutsukan, Ashikaga 28372
Ashiya-shiritsu Bijutsu oyo Rekishi Hakubutsukan, Ashiya 28375
Chikkyo Art Museum Kasaoka, Kasaoka 28676
Chinese Historical Art Museum, Nagasaki 28970
Daikaku-ji Temple, Kyoto 28806
Dewazakura Bijutsukan, Tendo 29357
Egawa Bijutsukan, Nishinomiya 29072
Eisei Bunko Museum, Tokyo 29385
Fujii Saisei-kai Yurinkan, Kyoto 28808
Fujita Bijutsukan, Fukui 28426
Fujita Bijutsukan, Osaka 29126
Fukui-shiritsu Bijutsukan, Fukui 28430
Fukuoka Ajian Bijutsukan, Fukuoka 28431
Fukuoka-kenritsu Bijutsukan, Fukuoka 28432
Fukuyama Bijutsukan, Fukuyama 28442
Fukuzawa Ichiro Memorial Museum, Nakatsu 29022
Gangidori Bijutsukan, Joetsu 28618
Geidai Bijutsukan, Tokyo 29392
Gifu-ken Bijutsukan, Gifu 28457
Goto Sumio Bijutsukan, Kamifurano 28646
Gyokudo Bijutsukan, Ome 29110
Hagi-shi Uragami Bijutsukan, Hagi 28469
Hakodate Bijutsukan, Hakodate 28472
Hakutsuru Fine Art Museum, Kobe 28726
Handa-shiritsu Hakubutsukan Niimi Nankichi Kinenkan, Handa 28495
Hara Museum ARC, Shibukawa 29256
Hasegawa Machiko Kinen Bijutsukan, Tokyo 29396
Hatakeyama Memorial Museum of Fine Art, Tokyo 29397
Hayashibara Bijutsukan, Okayama 29098
Hida Takayama Bijutsukan, Takayama 29339
Higashi-Hiroshima-shiritsu Bijutsukan, Higashi-Hiroshima 28502
Higashiyama Kaii Gallery, Nagano 28954
Himeji-shiritsu Bijutsukan, Himeji 28512
Hiratsuka Bijutsukan, Hiratsuka 28527
Hiroshima-kenritsu Bijutsukan, Hiroshima 28537
Hitomi Bijutsukan, Eigenji 28415
Hofu Tenmangu History Museum, Hofu 28545
Hokkaidoritsu Migishi Kotaro Bijutsukan, Sapporo 29219
Hokuetsu Bijutsukan, Nakajo 29016
Honma Bijutsukan, Sakata 29204
Horyu-ji Homotsukan, Tokyo 29399
Hosomi Bijutsukan, Kyoto 28813
Ibaraki-ken Tenshin Kinen Izura Bijutsukan, Kitaibaraki 28708
Iida-shi Bijutsu Hakubutsukan, Iida 28559
Ikuo Hirayama Bijutsukan, Onomichi 29118
Ippuku Bijutsukan, Asahi 28367
Ishikawa-kenritsu Bijutsukan, Kanazawa 28656
Itabashi Kuritsu Bijutsukan, Tokyo 29401
Itami-shiritsu Bijutsukan, Itami 28590
Itsuo Bijutsukan, Ikeda 28562
Iwaki-shi Bijutsukan, Iwaki 28606
Iwami Adachi Bijutsukan, Hamada 28482
Iwami Bijutsukan, Masuda 28886
Iwasaki Bijutsukan, Ibusuki 28552
Iwata Senshinkan, Inuyama 28575
Iwayama Urushi Bijutsukan, Morioka 28937
Izukogen Kirie Bijutsukan, Ito 28595
Jingu Bijutsukan, Ise 28580
K Museum, Shimizu 29261
Kaga-shi Bijutsukan, Kaga 28625
Kake Bijutsukan, Kurashiki 28782
Kakegawa-shi Ninomaru Bijutsukan, Kakegawa 28637
Kanagawa-kenritsu Kanazawabunko Museum, Yokohama 29596
Kanuma Municipal Art Museum of Kawakami Sumio, Kanuma 28666
Kanzeon-ji Temple Treasure House, Dazaifu 28408
Kariya-shiritsu Bijutsukan, Kariya 28667
Kathmandu - Hot Spring in the Art Museum, Yuzawa 29634
Katsushika Hokusai Bijutsukan, Tsuwano 29546
Kawagoe-shi Bijutsukan, Kawagoe 28683
Kawaguchiko Bijutsukan, Kawaguchiko 28688
Kawamura Memorial Museum of Art, Sakura 29209
Keiko Bijutsukan, Hirara 28526
Kitano Bijutsukan, Nagano 28955
Kitazawa Bijutsukan, Suwa 29301
Kochi-kenritsu Bijutsukan, Kochi 28743
Kodai Orient Hakubutsukan, Tokyo 29413
Komagane Kogen Bijutsukan, Komagane 28763
Kongosho-ji Treasure House, Ise 28583
Kosetsu Bijutsukan, Kobe 28736
Kotohira-gu Museum, Kotohira 28772
Kuboso Memorial Museum of Arts, Izumi, Izumi, Osaka 28613
Kuma Bijutsukan, Kuma 28774
Kumamoto-kenritsu Bijutsukan, Kumamoto 28776
Kumaya Bijutsukan, Hagi 28471
Kume Bijutsukan, Tokyo 29424

Kunaicho Sannomaru Shozokan, Tokyo 29425
Kurashiki-shi Bijutsukan, Kurashiki 28785
Kure-shiritsu Bijutsukan, Kure 28793
Kuzuu-machi Yoshizawa Kinen Bijutsukan, Kuzuu 28801
Kyoto-kenritsu Insho-Domoto Bijutsukan, Kyoto 28841
Kyoto Kokuritsu Hakubutsukan, Kyoto 28842
Kyoto-shi Bijutsukan, Kyoto 28845
Mainichi Art Collection, Nagoya 29001
Matsumoto-shiritsu Bijutsukan, Matsumoto 28894
Matsuoka Museum of Art, Tokyo 29427
Matsushita Bijutsukan, Fukuyama, Aira-gun 28451
Meguro Gajoen Bijutsukan, Tokyo 29428
Menard Art Museum, Komaki 28764
Mie-kenritsu Bijutsukan, Tsu 29536
Miho Museum, Shigaraki 29258
Mitsui Kinen Bijutsukan, Tokyo 29437
Miyama-cho Kayabuki Bijutsukan oyo Kyodo Hakubutsukan, Miyama 28925
Miyazaki-kenritsu Bijutsukan, Miyazaki 28928
Mohri Hakubutsukan, Hofu 28546
Murauchi Bijutsukan, Hachioji 28466
Musashino Art University Museum and Library, Kodaira 28752
The Museum Yamato Bunkakan, Nara 29032
Nagano-ken Shinano Bijutsukan, Nagano 28957
Nagasaki-ken Bijutsukan, Nagasaki 28979
Nakagawa Chaini-zu Kindai Bijutsukan, Fukuyama 28448
Nakamura Kinen Bijutsukan, Kanazawa 28661
Nanao-shiritsu Bijutsukan, Nanao 29024
Nara-kenritsu Bijutsukan, Nara 29035
Nara-kokuritsu Hakubutsukan, Nara 29036
Neiraku Bijutsukan, Nara 29037
Nezu Bijutsukan, Tokyo 29444
Niigata-shiritsu Bijutsukan, Niigata 29059
Niki Bijutsukan, Nasu 29051
Nomura Bijutsukan, Kyoto 28857
Northern Culture Museum, Niigata 29061
Ohara Bijutsukan, Kurashiki 28789
Oita-kenritsu Geijutsu Kaikan, Oita 29096
Okawa Bijutsukan, Kiryu 28703
Okayama-kenritsu Bijutsukan, Okayama 29100
Okayama-kenritsu Hakubutsukan, Okayama 29101
Okazaki-shiritsu Bijutsukan, Okazaki 29105
Okura Shukokan Museum of Fine Arts, Tokyo 29455
Okuuchi Bijutsukan, Toyonaka 29531
Ome-shiritsu Bijutsukan, Ome 29112
Onomichi-shiritsu Bijutsukan, Onomichi 29120
Osaka-shiritsu Bijutsukan, Osaka 29151
Ōta Kinen Bijutsukan, Tokyo 29457
Otaru-shiritsu Bijutsukan, Otaru 29163
Otsu-e Bijutsukan, Otsu 29165
Saga-kenritsu Bijutsukan, Saga 29179
Sagawa Bijutsukan, Moriyama 28940
Sakata-shiritsu Bijutsukan, Sakata 29206
Sakura-shiritsu Kampo Arai Kinenkan, Ujiie 29561
Sano Bijutsukan, Mishima 28914
Sasayama-shiritsu Bijutsu oyo Rekishi Hakubutsukan, Sasayama 29231
Sato Art Museum Toyama, Toyama 29520
Seikado Bunko Bijutsukan, Tokyo 29463
Seikado Bunko no Bijutsukan, Tokyo 29464
Seison-kaku, Kanazawa 28663
Sekiguchi Koh Museum Yuzawa, Yuzawa 29637
Sekisho Bijutsukan, Misumi 28915
Sekisui Bijutsukan, Tsu 29538
Sen-oku Hakukokan, Kyoto 28864
Setsuro Takahashi Bijutsukan, Azumino, Azumino 28387
Shimonoseki-shi Bijutsukan, Shimonoseki 29269
Shinshushin-machi Hakubutsukan, Shinshushin 29273
Shirakaba Bijutsukan, Onomichi 29121
Shizuoka-kenritsu Bijutsukan, Shizuoka 29289
Shogetsu Bijutsukan, Wakakusa 29571
The Shoto Museum of Art, Tokyo 29473
Shouhaku Bijutsukan, Nara 29039
Shunan-shi Bijutsukan nara Hakubutsukan, Shunan 29292
Sogetsu Bijutsukan, Tokyo 29474
Soun Museum, Ashikaga 28374
Sugimoto Bijutsukan, Mihama 28907
Suka Gogoda Bijutsukan, Nagasaki 28986
Sumpu Museum, Shizuoka 29290
Suntory Bijutsukan, Tokyo 29478
Suntory Hakubutsukan, Osaka 29155
Sutama-cho Bijutsukan, Sutama 29299
Suwa-shi Bitjutsukan, Suwa 29303
Tagawa Bijutsukan, Tagawa 29309
Taizi Harada Bijutsukan, Suwa 29304
Tanabe Bijutsukan, Matsue 28891
Tanaka Bijutsukan, Suzaka 29305
Tanimura Bijutsukan, Itoigawa 28602
Tatsuno Bijutsukan, Tatsuno 29355
Tekisui Bijutsukan, Ashiya-shi 28377
Tenmon Bijutsukan, Hirakata 28525
Tessai Bijutsukan, Takaraduka 29329
Toguri Bijutsukan, Tokyo 29486
Tokugawa Bijutsukan, Nagoya 29012
Tokyo Kokuritsu Hakubutsukan, Tokyo 29490
Tokyo-to Bijutsukan, Tokyo 29497
Toyama Kinenkan, Kawajima 28689
Tsukuba Bijutsukan, Tsukuba 29540
Urawa-kuritsu Bijutsukan, Saitama 29192
Utsunomiya Bijutsukan, Utsunomiya 29568
Wajima Urushi Bijutsukan, Wajima 29570
Wakayama-kenritsu Hakubutsukan, Wakayama 29574
Yamaguchi-kenritsu Bijutsukan, Yamaguchi 29581
Yamamoto Kanae Memorial Museum, Ueda 29553
Yataro Noguchi Bijutsukan, Nagasaki 28987
Yatsugatake Bijutsukan, Hara 28498
Yokohama Museum of EurAsian Cultures, Yokohama 29615
Yokoyama Taikan Kinenkan, Tokyo 29509
Yomitanson Bijutsukan, Yomitanson 29624
Yonago-shi Bijutsukan, Yonago 29626
Yugawara-machi Bijutsukan, Yugawara 29633
Yumeji Bijutsukan, Okayama 29103

Korea, Democratic People's Republic

Korean Art Gallery, Pyongyang 29756

Korea, Republic

Cheongju National Museum, Cheongju 29788
Donga University Museum, Busan 29778
Dukwon Museum of Art, Seoul 29875
Equine Museum, Gwacheon 29819
Gan-Song Art Museum, Seoul 29877
Han Kuk Art Museum, Yongin 29942
Hanwon Art Museum, Seoul 29881
Ho-Am Art Museum, Yongin 29943
Horim Museum, Seoul 29884
Hwajeong Museum, Seoul 29885
Ilmin Museum of Art, Seoul 29886
Keumho Museum of Art, Seoul 29888
Kyungin Museum of Fine Art, Seoul 29902
Munsin Museum, Masan 29851
Museum of NaJu Pears, Naju 29854
Seoul Metropolitan Museum of Art, Seoul 29919
Seoul National University Museum, Seoul 29921
Sincheonji Art Museum, Aeweol 29764
Songam Art Museum, Incheon 29834
Walker Hill Art Center, Seoul 29932
Whanki Museum, Seoul 29933

Malaysia

The Art Gallery, Kuala Lumpur 30571
Muzium Seni Asia, Kuala Lumpur 30582

Mongolia

Mongolian Art Museum, Ulaanbaatar 31622

Nepal

National Art Gallery, Bhaktapur 31770
National Birendra Art Gallery, Kathmandu 31775

Pakistan

Chughtai Museum Trust, Lahore 34018

Philippines

Artists Center and Foundation, Makati 34412
Heritage Art Center, Quezon City 34490
Metropolitan Museum of Manila, Manila 34444
Museo ng Makati, Makati 34415
Museum of Philippine Art, Manila 34448
Sining Makiling Gallery, Los Baños 34408

Poland

Galeria Azjatycka, Warszawa 35226
Galeria Nusantara, Warszawa 35229
Kamienica pod Gwiazdą, Toruń 35192
Muzeum Azji i Pacyfiku, Warszawa 35237

Russia

Chudožestvennyj Muzej, Valujki 37915
Dagestanskij Gosudarstvennyj Objedinennyj Istoriko-architekturnyj Muzej, Machačkala 37115
Gosudarstvennyj Istoriko-architekturnyj i Chudožestvennyj Muzej-zapovednik Drevnij Derbent, Derbent 36874
Gosudarstvennyj Muzej Izobrazitelnych Iskusstv Respubliki Kalmykija, Ėlista 36911
Irkutskij Oblastnoj Chudožestvennyj Muzej im. V.P. Sukačeva, Irkutsk 36932
Kitajskij Dvorec-muzej, Lomonosov 37109
Muzej Iskusstva Narodov Vostoka, Moskva 37286

Singapore

Luxe Art Museum, Singapore 38175
NUS Museum, Singapore 38178
Singapore Art Museum, Singapore 38179

South Africa

Reza Shah Museum, Johannesburg 38662

Spain

Museo de Arte Oriental, Ávila 38919
Museo Oriental, Valladolid 40241

Switzerland

Asia-Africa Museum, Genève 41398
Collections Baur-Musée des Arts d'Extrême-Orient, Genève 41411

Taiwan

Art Exhibition Room, Yilan 42464
Arts Center, Jhongli 42219
Chang Foundation Museum, Taipei 42378
Chang Gung University Arts Center, Gueishan 42195
Chuan Cheng Art Center, Taipei 42384
Creative Art Center, Jhongli 42220
Dadaocheng Art Museum, Taipei 42388
Erdai Art Hall, Magong 42267
Fo Guang Yuan Art Gallery and Museum of Treasures, Dashu 42178
Gallery of Hualien County, Hualien 42214
Gallery of National Taiwan Normal University, Taipei 42391
Huang Mun Art Center, Tongsiao 42454
Huangsi Art Exhibition, Changhua 42155
Kaohsiung Museum of Fine Arts, Kaohsiung 42237
Kwang-Ya Art Center, Zaociao 42486
Lingnan Fine Arts Museum, Taipei 42402
Local Art Museum, Pingtung 42291
Mary Leu Fine Carving Gallery and Meili Museum, Sansing 42314
Mei Ling Arts Museum, Puzi 42300
Mei-Nong Country Gallery, Qishan 42301
National Central University Art Center, Jhongli 42221
National Palace Museum, Taipei 42416
Pingtung Fo Guang Yuan Art Gallery, Pingtung 42293
Tainan National University of the Arts Museum, Kuantien 42252
Teipei Fo Guang Yuan Art Gallery, Taipei 42435
Yilan Fo Guang Yuan Art Gallery, Yilan 42471
Yuyu Yang Museum, Taipei 42441

Thailand

Art Gallery of the Faculty of Painting, Sculpture and Graphic Art, Bangkok 42530
Burapha Museum of Art, Chon Buri 42571
D.R. Visual Art Center, Nakorn-Chaisri 42587
The Jim Thompson House, Bangkok 42543
Silpakorn University Art Centre, Bangkok 42560
Srinakharinwirot Art Exhibition Hall, Bangkok 42562

Turkey

Bursa Türk-İslâm Eserleri Müzesi, Bursa 42744
Mevlânâ Müzesi, Konya 42889
Türk ve Islâm Eserleri Müzesi, Edirne 42768
Türk ve Islam Eserleri Müzesi, İstanbul 42845

Ukraine

Muzej Mystectv imeni Bogdana ta Varvari Chanenkiv, Kyïv 43097
Odes'kyj Muzej Zachidnogo ta Schidnogo Mystectva, Odesa 43162

United Kingdom

Hartlepool Art Gallery, Hartlepool 44406
Kufa Gallery, London 44897
Museum of East Asian Art, Bath 43408
Oriental Museum Durham, Durham 44074
Powis Castle, Welshpool 45962
Snowshill Manor, Broadway 43648

United States of America

American Museum of Asmat Art, Saint Paul 52613
Asian-American Arts Centre, New York 51129
Asian American Arts Centre, New York 51130
Asian Art Museum, San Francisco 52766
Asian Arts and Culture Center, Towson 53549
Asian Cultures Museum, Corpus Christi 47819
China Institute Gallery, New York 51150
Chinese Culture Center of San Francisco, San Francisco 52770
Denison University Art Gallery, Granville 49021
Gallery Korea, New York 51180
George J. Doizaki Gallery, Los Angeles 50225
George Walter Vincent Smith Art Museum, Springfield 53225
Hammond Museum, North Salem 51405
Japan Society Gallery, New York 51199
Mission Inn Museum, Riverside 52372
Monterey Museum of Art at La Mirada, Monterey 50804
Mount Holyoke College Art Museum, South Hadley 53170
Nicholas Roerich Museum, New York 51246
Pacific Asia Museum, Pasadena 51721
Peabody Place Museum, Memphis 50583
Plattsburgh Art Museum, Plattsburgh 51984
Robert W. Ryerss Museum, Philadelphia 51875
Seattle Asian Art Museum, Seattle 53014
Sheehan Gallery at Whitman College, Walla Walla 53798
Trammell and Margaret Crow Collection of Asian Art, Dallas 47942
Tremont Gallery, Boston 46952
University Art Gallery, Rohnert Park 52444
Vanderbilt University Fine Arts Gallery, Nashville 50980

Vietnam

Duc Minh Art Gallery, Ho Chi Minh City 54621
Fine Art University Exhibition Hall, Ha Noi 54612
Ho Chi Minh City Fine Arts Museum, Ho Chi Minh City 54623

Art, Australian

Australia

Adelaide Central Gallery, Norwood 01381
Art Gallery of New South Wales, Sydney 01544
Artarmon Galleries, Artarmon 00815
Bendigo Art Gallery, Bendigo 00859
Berndt Museum of Anthropology, Crawley 01008
Berrima District Art Society Gallery, Bowral 00883
Brambuk Living Cultural Centre, Halls Gap 01141
Brett Whiteley Studio Museum, Surry Hills 01539
Bribie Island Community Arts Centre, Bribie Island 00892
Brisbane City Gallery, Brisbane 00895
Bunbury Regional Art Galleries, Bunbury 00916
Burnie Regional Art Gallery, Burnie 00923
Cairns Regional Gallery, Cairns 00934
Caloundra Regional Art Gallery, Caloundra 00935
Campbelltown Art Center, Campbelltown 00940
Castlemaine Art Gallery and Historical Museum, Castlemaine 00960
Charles Sturt University Art Collection, Wagga Wagga 01619
Co-operative Storage Museum, Albany 00786
Community Gallery Spaces, Portland 01438
Cooktown School of Art Gallery, Cooktown 00997
Emerald Art Gallery, Emerald 01069
Footscray Community Arts Centre, Footscray 01083
Geelong Gallery, Geelong 01102
Geraldton Regional Art Gallery, Geraldton 01106
Gladstone Regional Art Gallery and Museum, Gladstone, Queensland 01111
Global Arts Link, Ipswich 01176
Gosford Regional Gallery and Arts Centre, Gosford 01122
Hazelhurst Regional Gallery and Arts Centre, Gymea 01137
Heide Museum of Modern Art, Bulleen 00915
Hervey Bay Regional Art Gallery, Pialba 01413
Horsham Regional Art Gallery, Horsham 01163
The Ian Potter Museum of Art, Melbourne 01284
Jane Neville-Rolfe Gallery, Alpha 00805
John Curtin Gallery, Bentley 00862
Koorie Heritage Trust Cultural Centre-Galleries, Melbourne 01286
Lady Franklin Gallery, Lenah Valley 01227
Lancefield Courthouse Museum, Lancefield 01215
Lawrence Wilson Art Gallery, Crawley 01010
Light Square Gallery, Adelaide 00777
Logan Art Gallery, Logan 01237
McClelland Gallery, Langwarrin 01217
Manly Art Gallery and Museum, Manly 01258
Metro Arts, Brisbane 00897
Moonah Arts Centre, Moonah 01313
Mornington Peninsula Regional Gallery, Mornington 01322
Museum and Art Gallery of the Northern Territory, Darwin 01030
Muswellbrook Regional Arts Centre, Muswellbrook 01341
Naracoorte Art Gallery, Naracoorte 01347
Noosa Regional Art Gallery, Tewantin 01579
Orange Regional Gallery, Orange 01395
Phillip Island Community Art and Craft Gallery, Cowes 01005
Pinnacles Gallery, Thuringowa 01586
Port Pirie Regional Art Gallery, Port Pirie 01436
Port Victor Gallery, Victor Harbor 01617
QCA Gallery, Brisbane 00898
Queensland Art Gallery, South Brisbane 01512
Queensland University of Technology Art Museum, Brisbane 00900
Quilpie Museum, Quilpie 01451
Riddoch Art Gallery, Mount Gambier 01327
S.H. Ervin Gallery, Sydney 01558
Saint Marys Community Arts Centre, Saint Marys 01487
SASA Gallery, Adelaide 00781

A

Talbot Arts and Historical Museum, Talbot 01567
Tandanya, Adelaide 00783
Tasmanian Museum and Art Gallery, Hobart 01159
Umoona Opal Mine and Museum, Coober Pedy 00996
UNSW Art Collection, Sydney 01564
Vincent Art Gallery, Townsville 01601
Warrnambool Art Gallery, Warrnambool 01634
Warwick Regional Art Gallery, Warwick 01636
Wildlife Wonderlands-Giant Earthworm Museum, Bass 00835

Netherlands

Aboriginal Art Museum, Utrecht 32841

United Kingdom

Hastings Museum and Art Gallery, Hastings 44415

United States of America

Australian Exhibition Center, Chicago 47455
New Visions Gallery, Marshfield 50513

Art, European

Armenia

National Gallery of Armenia, Yerevan 00751

Australia

Bendigo Art Gallery, Bendigo 00859
Creswick Museum and Gold Battery, Creswick 01011
Outback Regional Gallery, Winton 01657

Austria

Galerie im Traklhaus, Salzburg 02613
Grazer Kunstverein, Graz 01970
Hipp-Halle, Gmunden 01949
MAK - Österreichisches Museum für angewandte Kunst / Gegenwartskunst, Wien 03026
Museum des Stiftes Admont, Admont 01693
Museum im Ballhaus, Imst 02110
Museum Neuhaus – Sammlung Liaunig, Neuhaus 02428

Belgium

Musée d'Art Ancien, Bruxelles 03462
Musée de Louvain-la-Neuve, Louvain-la-Neuve 03834
Musée Ianchelevici, La Louvière 03768
Provinciaal Domein Rivierenhof Galerij, Deurne 03549

Bulgaria

Chudožestvena Galerija, Plovdiv 05125

Canada

McMaster Museum of Art, Hamilton 05960

Chile

Museo Ralli, Santiago de Chile 07464

Croatia

Galerija Cate Dujšin Ribar, Trogir 08968
Moderna Galerija, Zagreb 09003
Muzej Maksimilijana Vanka, Korčula 08896
Muzej Mimara, Zagreb 09006
Muzej Moslavine, Kutina 08904
Zbirci Umjetnina u Crkvi Sv. Ivana Krstitelja, Trogir 08971

Czech Republic

Alšova Jihočeská Galerie, Hluboká nad Vltavou 09515
Bílkova vila, Praha 09735
České Muzeum Výtvarných Umění, Praha 09737
Císařská Konírna, Praha 09738
Dům Františka Bílka, Chýnov 09488
Galerie Bratří Špillarů, Domažlice 09495
Obrazárna Pražského Hradu, Praha 09784
Šternberský Palác, Praha 09795
Zámek Troja, Praha 09799

Denmark

Danmarks Kunstbibliotek, København 10127
Fanø Kunstmuseum, Fanø 10013
Den Hirschsprungske Samling, København 10133
Nivågårds Malerisamling, Nivå 10192
Trapholt, Kolding 10164

Finland

Ateneumin Taidemuseo, Helsinki 10769
Hämeenlinnan Taidemuseo, Hämeenlinna 10746
Joensuun Taidemuseo, Joensuu 10872
Nykarleby Museum och Konstsamlingen Kalevala, Nykarleby 11156
Taidekoti Kirpilä, Helsinki 10827
Teresia ja Rafael Lönnströmin kotimuseo, Rauma 11262
Valtion Taidemuseo, Helsinki 10835
Villa Gyllenberg, Helsinki 10836
Yrjö A. Jäntin Taidekokoelma, Porvoo 11232

France

Musée Cognacq-Jay, Paris 14639
Musée d'Art, Toulon (Var) 15923
Musée d'Art et d'Histoire, Nice 14427
Musée d'Art et d'Histoire de Cognac, Cognac 12565
Musée d'Art et d'Histoire de Langres, Langres 13509
Musée des Années 30, Boulogne-Billancourt 12090
Musée Granet, Aix-en-Provence 11518
Musée Municipal de Draguignan, Draguignan 12790
Villa Arson, Nice 14442

Germany

Altes Schloss, Abteilung Gemäldegalerie und Kunsthandwerk, Gießen 18636
Angermuseum, Erfurt 18196
Atelierhaus Philipp Harth, Bayrischzell 16967
Bayerisches Nationalmuseum, München 20426
Bonner Kunstverein, Bonn 17450
Dieter Roth-Museum, Hamburg 18923
Galerie am Kamp, Teterow 22036
Galerie Nord-Kunstverein Tiergarten, Berlin 17104
Heinrich-Vogeler Museum, Worpswede 22692
Hessisches Landesmuseum Darmstadt, Darmstadt 17789
Hirschwirtscheuer-Museum für die Künstlerfamilie Sommer, Künzelsau 19735
Historischer Kunstbunker, Nürnberg 20769
Kunst- und Gewerbeverein Regensburg, Regensburg 21227
Kunsthaus Stade, Stade 21850
Kunstmuseum Nörvenich, Nörvenich 20728
Kunstwerk, Köln 19617
Max Liebermann Haus, Berlin 17186
Mittelrhein-Museum Koblenz, Koblenz 19587
Museum Berggruen, Berlin 17192
Museum Junge Kunst, Frankfurt (Oder) 18405
Museum Villa Rot - Kunstsammlung Hoenes-Stiftung, Burgrieden 17662
Neue Sächsische Galerie, Chemnitz 17701
Sammlung Geyer zu Lauf, Emmendingen 18148
Schlossmuseum Aulendorf-Kunst des Klassizismus-Altes Spielzeug, Aulendorf 16643
Stadtmuseum Weilheim, Weilheim in Oberbayern 22392
Städel Museum, Frankfurt am Main 18393
Städtische Galerie Karlsruhe, Karlsruhe 19451
Stiftung Schloss und Park Benrath, Düsseldorf 18009
Studio Bildende Kunst, Berlin 17269
Willi Sitte Galerie, Merseburg 20295
Worpsweder Kunsthalle, Worpswede 22699

Greece

Atelier Spyros Vassiliou, Athinai 22900
Coumantaros Art Gallery of Sparta, Spárti 23195
Mouseio Ellinikis Paidikis Tehnis, Athinai 22919
New Museum of European and Eastern Art, Athinai 22944

Hungary

Artchivum, Budapest 23384
Koszta József Múzeum, Szentes 23884
Magyar Nemzeti Galéria, Budapest 23441
Péter Pál Polgárház, Szentes 23885

Ireland

Butler Gallery, Kilkenny 24848
GAA Museum, Dublin 24790

Israel

Ralli Museum, Caesarea 24939

Italy

Collezione Contini-Bonacossi, Firenze 26150
Galleria Giorgio Franchetti alla Ca' d'Oro, Venezia 28182
Museo di Storia e Arte, Gorizia 26352
Museo Palladio, Vicenza 28255
Palazzo Rosso, Genova 26323
Palazzo Taffini d'Acceglio, Savigliano 27768
Pinacoteca Civica, Cepagatti 25804
Pinacoteca Civica, Crotone 26015
Pinacoteca Civica Tosio Martinengo, Brescia 25542

Japan

Flanders Center, Osaka 29125
Kokuritsu Seiyo Bijutsukan, Tokyo 29419
Miyazaki-kenritsu Bijutsukan, Miyazaki 28928
Nagasaki-ken Bijutsukan, Nagasaki 28979
Yokohama Museum of EurAsian Cultures, Yokohama 29615

Lithuania

Nacionalinis Mikalojaus Konstantino Čiurlionio Dailės Muziejus, Kaunas 30262

Malta

Cathedral Museum, Mdina 30630

Mexico

Museo Nacional de San Carlos, México 31141

Montenegro

Muzeji i Galerije Podgorica, Podgorica 31656
Umjetnička Galerija Nikola I i Galerija Ilija Sobajić, Nikšić 31649

Netherlands

Anton Pieck Museum, Hattem 32322
Bonnefantenmuseum, Maastricht 32524
Cultureel Centrum van Gogh, Zundert 33017
Historisch Museum Rotterdam Het Schielandshuis, Rotterdam 32707
Marie Tak van Poortvliet Museum, Domburg 32141
Museum Het Oude Raadhuis, Megen 32547
Museum Ianchelevici, Goudriaan 32272
Museum Jacques van Mourik, Mook 32561
Museum Jan van der Togt, Amstelveen 31822
Museum Kruijsenhuis, Oirschot 32613
Museum Valse Kunst, Vledder 32896
Museum van Lien, Fijnaart 32230
Stedelijk Museum Bureau Amsterdam, Amsterdam 31889

Norway

Haugesund Billedgalleri, Haugesund 33537
Jølstramuseet og Kunstgalleriet, Vassenden 33965

Poland

Gabinet Rycin Biblioteki Uniwersyteckiej w Warszawie, Warszawa 35225
Muzeum im. Juliana Fałata, Bystra Śląska 34582
Muzeum Książąt Czartoryskich, Kraków 34803
Muzeum Narodowe w Gdańsku, Gdańsk 34641
Muzeum Regionalne, Nysa 34966
Muzeum Śląskie, Katowice 34728
Muzeum Sztuki w Łodzi, Łódź 34884
Państwowe Zbiory Sztuki, Sułoszowa 35140

Portugal

Casa Museu Bissaya-Barreto, Coimbra 35526
Casa Museu da Fundação Dionísio Pinheiro e Alice Pinheiro, Águeda 35385
Casa-Museu da Fundação Medeiros e Almeida, Lisboa 35661
Casa Museu Fernando de Castro, Porto 35837
Casa Museu Marieta Solheiro Madureira, Estarreja 35571
Casa Museu Roque Gameiro, Amadora 35429
Galeria Municipal de Arte, Almada 35413
Museu Calouste Gulbenkian, Lisboa 35672
Museu da Cidade de Lisboa, Lisboa 35677
Museu da Quinta de Santiago, Leça da Palmeira 35657
Museu de Arte, Santa Maria da Feira 35904
Museu de Artes e Ofícios Francisco Távora, Escalhão 35564
Museu de Grão Vasco, Viseu 36023
Museu do Caramulo, Caramulo 35506
Museu dos Rios e das Artes Marítimas, Constância 35550
Museu Henrique e Francisco Franco, Funchal 35616
Museu Municipal Carlos Reis de Torres Novas, Torres Novas 35969
Museu Municipal de Óbidos, Óbidos 35784
Museu Municipal Polinucleado de Tomar, Tomar 35966
Sala Museu Jenny Mendes de Brito, Entroncamento 35559

Romania

Atelier Memorial Paul Erdös, Satu Mare 36589
Casa-Atelier Gheorghe Petraşcu, Târgovişte 36653
Casa Memorială Gyarfas Jenö, Sfântu Gheorghe 36599
Casa Memorială Nicu Enea, Bacău 36119
Casa-Muzeu Constantin Brâncuşi, Hobiţa 36399
Casa Personalităţilor, Lugoj 36448
Centrul de Cultură Palatele Brâncoveneşti, Mogoşoaia 36479
Colecţia de Artă Plastică Al. Bălintescu, Costeşti 36316
Colecţia Mănăstirii Zamfira, Lipăneşti 36441
Galeria de Artă, Piteşti 36525
Galeria de Artă Ioan Sima, Zalău 36741
Galeriile de Artă Emilia Dumitrescu, Brăila 36169
Galeriile de Artă Plastică Ştefan Luchian, Botoşani 36161
Muzeul Brăilei, Brăila 36170
Muzeul de Artă, Arad 36113
Muzeul de Artă, Botoşani 36162
Muzeul de Artă, Râmnicu Vâlcea 36552
Muzeul de Artă, Roman 36567
Muzeul de Artă, Târgovişte 36657
Muzeul de Artă, Tulcea 36703
Muzeul de Artă, Satu Mare 36590
Muzeul de Artă Centrul Artistic Baia Mare, Baia Mare 36125
Muzeul de Artă Comparată, Sângeorz-Băi 36584
Muzeul de Artă Constanţa, Constanţa 36302
Muzeul de Artă Craiova, Craiova 36324
Muzeul de Artă Dinu şi Sevasta Vintilă, Topalu 36698
Muzeul de Artă şi Artă Populară, Calafat 36247
Muzeul de Artă Veche Apuseană Dumitru Minovici, Bucureşti 36202
Muzeul Judeţean de Artă, Ploieşti 36532
Muzeul Judeţean de Artă şi Etnografie, Bacău 36121
Muzeul Judeţean Mureş-Secţia de Artă, Târgu Mureş 36670
Muzeul Judeţean Olt, Slatina 36628
Muzeul Krikor H. Zambaccian, Bucureşti 36205
Muzeul Memorial Gheorghe M. Tattarascu, Bucureşti 36210
Muzeul Memorial Octavian Goga, Ciucea 36283
Muzeul Naţional al Bucovinei, Suceava 36646
Muzeul Naţional Brukenthal, Sibiu 36613
Muzeul Naţional Cotroceni, Bucureşti 36218
Muzeul Orăşenesc, Lipova 36443
Muzeul Romanaţiului Secţia de Artă Plastică, Caracal 36260

Russia

Gosudarstvennyj Muzej-usadba Archangelskoe, Archangelskoe 36773
Istoriko-Architekturnyj, Chudožestvennyj i Archeologičeskij Muzej "Zarajskij Kreml'", Zarajsk 37986
Kirovskij Oblastnoj Chudožestvennyj Muzej im. Viktora i Apollinarija Vasnecovych, Kirov 37038
Samarskij Oblastnoj Chudožestvennyj Muzej, Samara 37621
Uljanovskij Oblastnoj Chudožestvennyj Muzej, Uljanovsk 37909

Serbia

Galerija ULUS, Beograd 38047
Umetnička Galerija, Kruševac 38113

Slovakia

Danubiana, Bratislava 38205
Galéria Mesta Bratislavy, Bratislava 38211
Galéria Umenia, Nové Zámky 38321
Slovenská Národná Galéria, Bratislava 38233

Spain

Monasterio de San Lorenzo de El Escorial, San Lorenzo de El Escorial 39961
Museo Camón Aznar de Ibercaja, Zaragoza 40340
Museo de Belas Artes da Coruña, A Coruña 39220
Museo de Bellas Artes de Álava, Vitoria-Gasteiz 40322
Museo de Cádiz, Cádiz 39089
Museo Nacional del Prado, Madrid 39573
Museo Pablo Gargallo, Zaragoza 40347
Museo Ralli, Marbella 39620
Museu Municipal Josep Aragay, Breda 39057
Museu Municipal Vicenç Ros, Martorell 39627

Sweden

Eskilstuna Konstmuseum, Eskilstuna 40501
Norrtälje Konsthall, Norrtälje 40776
Sörmlands Museum, Nyköping 40782
Tallbo, Järbo 40611
Ystads Konstmuseum, Ystad 41078

Switzerland

Fondation Oscar et Janine Wiggli, Muriaux 41667
Kunstmuseum Luzern, Luzern 41593
Musée des Beaux-Arts, Le Locle 41572
Musée Jurassien des Arts, Moutier 41653

Turkmenistan

Muzej Izobrazitel'nych Iskusstv, Aşgabat 42972

Ukraine

Charkivs'kyj Chudožnij Muzej, Charkiv 43017
Chudožnij Muzej, Sevastopol 43176
Mis'ka Chudožna Galereja im. S.I. Vasyl'kivs'kogo, Charkiv 43020
Muzej Mystectv imeni Bogdana ta Varvari Chanenkiv, Kyïv 43097
Muzej Narodnogo Mystectva Slobožanščyny, Charkiv 43021
Odes'kyj Muzej Zachidnogo ta Schidnogo Mystectva, Odesa 43162

United Kingdom

Art Collection, Stirling 45748
The Art Gallery, Hillsborough 44468
Barber Institute of Fine Arts, Birmingham 43497
Beldam Gallery, Uxbridge 45903
Boughton House, Kettering 44567
Bowhill Collection, Selkirk 45634
Bowood House and Gardens, Calne 43716
City Museum and Art Gallery, Gloucester 44315
Coach-House Gallery, Saint Peter Port 45590
Compton Verney, Warwickshire 45944
Ferens Art Gallery, Kingston-upon-Hull 44594
Fergusson Gallery, Perth 45365
Foundling Museum, London 44851
Government Art Collection, London 44860
Hove Museum and Art Gallery, Hove 44493
Kenwood House, London 44894
Mercantile Arts Centre Kintyre, Tarbert 45825
Museum of the Royal College of Physicians, London 44931
Newlyn Art Gallery, Penzance 45360
Osborne House, East Cowes 44084
Palace of Holyroodhouse, Edinburgh 44143
Pallant House Gallery, Chichester 43830
Polish Cultural Institute, London 44956
The Queen's Gallery, Edinburgh 44145
Royal Academy of Arts, London 44966
Royal Scottish Academy Collections, Edinburgh 44152
Royal West of England Academy New Gallery, Clifton 43853
Shire Hall Gallery, Stafford 45730
Squerryes Court, Westerham 45979
Stephen Lawrence Gallery, London 44991
The Study Gallery, Poole 45409
Torre Abbey Historic House and Gallery, Torquay 45867
University Gallery Leeds, Leeds 44670
Wyndham Lewis Society, London 45021

United States of America

3M Collection of Contemporary Western and Americana Art, Long Lake 50189
Allen R. Hite Art Institute, Louisville 50280
Beard Gallery, Norton 51430
Bob Jones University Museum Gallery, Inc., Greenville 49078
Dahesh Museum of Art, New York 51160
The Frick Art Museum-Henry Clay Frick Estate, Pittsburgh 51949
The Galeria, Orchard Lake 51583
Guggenheim Museum Soho, New York 51188
Hutchinson Art Center, Hutchinson 49453
The Martin D'Arcy Museum of Art, Chicago 47487
Mulvane Art Museum, Topeka 53537
Mulvane Art Museum, Topeka 53537
Museum of Fine Arts, Springfield 53226
Salgo Trust for Education, Port Washington 52049
The Spanish Institute, New York 51261
Wally Findlay Galleries, New York 51278
The Zigler Museum, Jennings 49621

Uruguay

Museo Ralli, Punta del Este 54429

Art, Greek and Roman

Austria

Archäologischer Park Carnuntum, Petronell 02494
Archäologischer Park Carnuntum, Bad Deutsch-Altenburg 01741
Freilichtmuseum Petronell, Petronell 02495
Museum Lauriacum, Enns 01870

Cyprus

Leventis Municipal Museum of Nicosia, Lefkosia 09379

Czech Republic

Galerie Antického Umění, Hostinné 09530

France

Musée des Beaux-Arts, Agen 11494
Musée Gallo-Romain d'Aoste, Aoste 11649
Villa Grècque Kérylos, Beaulieu-sur-Mer 11896

Germany

Antikensammlung, Kiel 19527
Antikensammlung, Kassel 19457
Archäologisches Museum der Universität Münster, Münster 20503
Dominikanermuseum Rottweil, Rottweil 21424
Humboldt-Museum Schloß Tegel, Berlin 17140
Limesmuseum Aalen, Aalen 16412
Martin-von-Wagner-Museum der Universität Würzburg, Würzburg 22714
Neues Museum, Berlin 17228
Staatliche Antikensammlungen und Glyptothek, München 20486
Städtisches Lapidarium, Stuttgart 21962

Greece

Archaeological Collection, Symi 23200
Archaeological Collection of Agios Kirikos, Agios Kirikos 22871
Art Gallery of the Society for Macedonian Studies, Thessaloniki 23210
Byzantine and Christian Museum, Athinai 22904
Coumantaros Art Gallery of Sparta, Spárti 23195
Folklore Museum and Municipal Art Gallery, Karditsa 23034
G.I. Katsigras Museum, Lárissa 23074
Gallery of Florina Artists, Florina 23002
Gallery of the Society for Epirot Studies, Ioannina 23013
Historical Museum of Arta, Arta 22893
Mouseio Vorre, Paiania 23129
Municipal Art Gallery, Thessaloniki 23216
Municipal Gallery of Herakleion, Iráklion 23021
Municipal Gallery of Lamia, Lamia 23071
Municipal Gallery of Volos, Volos 23239
Museum of Cycladic and Ancient Greek Art, Athinai 22931
Public Art Gallery of Chios, Chios 22972

Iran

Sa'dabad Museums, Tehrān 24694

Israel

Museum of the Franciscan Convent, Jerusalem 25004

Italy

Antiquarium, Tindari 27972
Museo Agricolo e Museo dell'Arte Conciaria, Castelfranco Veneto 25730
Museo Archeologico Nazionale, Sperlonga 27887
Museo Archeologico Nazionale, Venezia 28185
Museo Civico Pio Capponi, Terracina 27967
Museo Lapidario Maffeiano, Verona 28236

Russia

Ivanovskij Oblastnoj Chudožestvennyj Muzej, Ivanovo 36948

South Africa

Rhodes Museum of Classical Antiquities, Grahamstown 38633
Social History Collections, Cape Town 38591

Spain

Museo Fundación Casa de Alba, Madrid 39560

Switzerland

Antikenmuseum Basel und Sammlung Ludwig, Basel 41154

Syria

Musée de Soueida, Soueida 42138

United Kingdom

Old Speech Room Gallery, London 44949
Rugby Art Gallery and Museum, Rugby 45527
Shefton Museum of Greek Art and Archaeology, Newcastle-upon-Tyne 45231

United States of America

Kelsey Museum of Archaeology, Ann Arbor 46307

Vatican City

Museo Chiaramonti e Braccio Nuovo, Città del Vaticano 54502
Museo Pio Clementino, Città del Vaticano 54508

Art, Latin American

Argentina

Casa Museo Fernando Fader, Ischilín 00389
Centro Cultural España Córdoba, Córdoba 00317
Escuela Museo de Bellas Artes General Urquiza, Buenos Aires 00157
Galerías Pacífico de Centro Cultural Borges, Buenos Aires 00161
Malba - Fundación Costantini, Buenos Aires 00164
Museo de Arte e História de Los Toldos, Los Toldos 00431
Museo de Arte Hispanoamericano Isaac Fernández Blanco, Buenos Aires 00182
Museo de Artes Plásticas Eduardo Sívori, Buenos Aires 00185
Museo de Bellas Artes Bonaerense, La Plata 00410
Museo de Bellas Artes Ceferino Carnacini, Villa Ballester 00671
Museo de Bellas Artes Claudio León Sempere, Burzaco 00284
Museo de Bellas Artes Franklin Rawson, San Juan 00586
Museo de Bellas Artes Juan Yapari, Posadas 00497
Museo de Bellas Artes Laureano Brizuela, San Fernando del Valle de Catamarca 00572
Museo de Centro Cultural Recoleta, Buenos Aires 00189
Museo Evocativo y de Bellas Artes Osvaldo Gasparini, San Antonio de Areco 00565
Museo Florencio Molina Campos, Moreno 00473
Museo Histórico, Zapala 00687
Museo Histórico Provincial Brig. Estanislao López, Santa Fé 00617
Museo Juan Carlos Iramain, San Miguel de Tucumán 00600
Museo Municipal de Artes Plásticas Faustino Jorge Bonadeo, Rauch 00518
Museo Municipal de Artes Plásticas Pompeo Boggio, Chivilcoy 00305
Museo Municipal de Artes Visuales, Concordia 00316
Museo Municipal de Bellas Artes Dr. Genaro Perez, Córdoba 00335
Museo Municipal de Bellas Artes Dr. Genaro Pérez, Río Cuarto 00529
Museo Municipal de Bellas Artes Dr. Urbano Poggi, Rafaela 00515
Museo Municipal de Bellas Artes Juan B. Castagnino, Rosario 00547
Museo Municipal Lino Enea Spilimbergo, Unquillo 00658
Museo Municipal Primeros Pobladores, San Martín de los Andes 00593
Museo Provincial de Artes de La Pampa, Santa Rosa 00624
Museo Provincial de Bellas Artes Emilio A. Caraffa, Córdoba 00337
Museo Provincial de Bellas Artes Rene Brusau, Resistencia 00526
Museo Provincial de Bellas Artes Rosa Galisteo de Rodríguez, Santa Fé 00620
Museo Rómulo Raggio, Vicente López 00667
Museo Xul Solar, Buenos Aires 00277
Salón Electo Brizuela, Concepción del Uruguay 00314

Belize

Tanah Art Museum, San Antonio 04180

Bolivia

Casa Museo Nuñez del Prado, La Paz 04209
Casa Nacional de la Moneda, Potosí 04235
Museo de la Recoleta, Sucre 04248

Brazil

Casa Daros, Rio de Janeiro 04682
Fundação Cultural Ema Gordon Klabin, São Paulo 04857
Instituto Moreira Sales, São Paulo 04861
Museu da Abolição, Recife 04661
Museu do Índio, Florianópolis 04457
Museu Malinverni Filho, Lages 04526

Chile

Museo Artequin, Santiago de Chile 07450
Museo Ralli, Santiago de Chile 07464

Colombia

Collecíon de Antigüedades, San Andrés de Sotavento 08744
Museo Bolivariano de Arte Contemporáneo, Santa Marta 08747
Museo Casa de la Cultura de Cuacutecuta, Cuacutecuta 08629
Museo de la Caña de Azúcar, El Cerrito 08636
Museo Maestro Alfonso Zambrano, Pasto 08713

Cuba

Casa Natal de Regino Boti, Guantánamo 09090
Museo Benito Hortiz, Sancti-Spíritus 09283
Museo de Arte de Matanzas, Matanzas 09213
Museo de Arte Habana, La Habana 09132
Museo La Vigía, Matanzas 09216
Museo Servando Cabrera Moreno, La Habana 09173
Museo Teodoro Ramos Blanco, La Habana 09174
Museo Víctor Manuel, La Habana 09175

Dominican Republic

The Gallery, La Romana 10331

Ecuador

Museo Casa de Benalcazar, Quito 10430
Museo Convento de San Diego, Quito 10432
Museo Cultural del Instituto Geografico Militar, Quito 10433
Museo de Arte e Historia Alberto Mena Caamaño, Quito 10436
Museo de la Fundación Guayasamin, Quito 10442
Museo de las Artes Populares Cidap, Cuenca 10388
Museo del Banco Central, Riobamba 10455
Museo Guillermo Perez Chiriboga del Banco Central, Quito 10445
Museo Nahim Isaias, Guayaquil 10410

El Salvador

Centro Nacional de Artes, San Salvador 10548
Museo de La Palabra y La Imagen, San Salvador 10552

Haiti

Musée d'Art Haïtien, Port-au-Prince 23321

Honduras

Museo de Bellas Artes, Comayagüela 23327

Israel

Ralli Museum, Caesarea 24939

Japan

Morishita Bijutsukan, Hinase 28515

Mexico

Casa Museo Gene Byron, Guanajuato 30911
Centro Nacional de las Artes, México 31038
Colección de Roberto Montenegro, Guadalajara 30888
Galería de Arte de Tabasco, Villahermosa 31547
Galería de Arte Iztapalapa, México 31040
Galería de Escuela Nacional de Artes Plásticas, México 31041
Galeria del Sur, México 31043
Galería Dr. Miguel Angel Gómez Ventura, Villahermosa 31549
Galería IFAL, México 31044
Galería Universitaria Artistos, México 31047
Museo Casa del Maestro José Márquez Figueroa, Puebla 31270
Museo de Academia de San Carlos, México 31066
Museo de Arte Colonial, Morelia 31193
Museo de Arte de Mazatlán, Mazatlán 31021
Museo de Arte de Orizaba y del Estado de Veracruz, Orizaba 31238
Museo de Arte de Querétaro, Querétaro 31299
Museo de Arte de Sinaloa, Culiacán 30825
Museo de Arte en Ciudad Juárez, Juárez 30981
Museo de Arte Guillermo Ceniceros, Durango 30839
Museo de Arte José Luis Bello y Zetina, Puebla 31272
Museo de Arte Moderno, México 31070
Museo de Arte Popular de la Universidad, Puebla 31273
Museo de Arte Prehispánico, Mascota 31015
Museo de Arte Prehispánico Carlos Pellicer, Tepoztlán 31447
Museo de Arte Prehispánico de México Rufino Tamayo, Oaxaca 31222
Museo de Arte Virreinal de Taxco, Taxco de Alarcón 31403
Museo de Artes Visuales Aramara, Tepic 31442
Museo de Artes y Tradiciones Populares de Tlaxcala, Tlaxcala 31476
Museo de Bellas Artes de Toluca, Toluca 31485

Museo de Laca, Chiapa de Corzo 30762
Museo de las Artes, Guadalajara 30895
Museo Emilia Ortiz, Tepic 31443
Museo Francisco Goitia, Zacatecas 31579
Museo Frissel de Arte Zapoteca, Santa Ana del Valle 31374
Museo Hermila Domínguez de Castellanos, Comitán de Domínguez 30794
Museo José María Velasco, Toluca 31493
Museo Nacional de Virreinato, Tepotzotlán 31445
Museo Regional de Arte e Historia, Jerez de García Salinas 30974
Museo Sala de Arte, Tijuana 31458
Museo Universitario del Chopo, México 31156
Pinacoteca del Ateneo, Saltillo 31318
Pinacoteca del Estado Juan Gamboa Guzmán, Mérida 31030
Pinacoteca del Templo de la Compañía, Guanajuato 30922
Pinacoteca Municipal de Guadalupe, Guadalupe 30909
Pinacoteca Universitaria, Colima 30784

Nicaragua

Galería Las Pintadas, Estelí 33276
Museo Arqueológico de Solentiname, Solentiname 33307

Panama

Museo José Domingo de Obaldía, David 34072

Peru

Casa Goyeneche, Lima 34206
Casa Museo Julia Codesido, Lima 34208
Galería Cultural de la Escuela Nacional de Bellas Artes, Lima 34213
Galería de Arte Dos V's, Lima 34214
Galería de Arte UNIFE, Lima 34215
Galería de Artes Visuales, Lima 34216
Galería de la Municipalidad de Miraflores, Lima 34217
Galería del Instituto Cultural Peruano-Juan Pardo Heeren, Lima 34219
Museo de Artes y Tradiciones Populares, Lima 34236
Museo del Instituto Americano del Arte, Cusco 34163
Museo Marina Núñez del Prado, Lima 34261
Museo Universitario de la Universidad Nacional Federico Villarreal, Lima 34272
Pinacoteca Municipal Leoncio Lugo, Paucartambo 34286
Sala de Exposiciones de la Universidad del Pacífico, Lima 34275
Sala Nacional de Cultura Mariano Fuentes Lira, Cusco 34169

Puerto Rico

Ateneo Puertorriqueño, San Juan 36058
Casa Galería Marcos J. Alegría, Dorado 36037
Galería Nacional, San Juan 36064
Museo Alfredo Ramírez de Arellano Rosello, San Germán 36056
Museo de Arte de Caguas, Caguas 36031
Museo de Arte de Puerto Rico, San Juan 36067
Museo Orocoveño de Familia Avilés, Orocovis 36044
Museo Universitario Pío López Martínez, Cayey 36035

Spain

Museo de América, Madrid 39532
Museo Ralli, Marbella 39620

Switzerland

Daros-Latinamerica AG, Zürich 42070

United States of America

Casa de Unidad, Detroit 48107
Galeria de la Raza, San Francisco 52774
Guadalupe Cultural Arts Center, San Antonio 52715
Latin American Art Museum, Miami 50629
Mexic-Arte Museum, Austin 46515
Museum of Latin American Art, Long Beach 50175
Museum of Spanish Colonial Art, Santa Fe 52884
The Parthenon, Nashville 50972
Poeh Museum, Santa Fe 52886

Uruguay

Centro Municipal de Exposiciones Subste, Montevideo 54390
Museo Municipal de Historia del Arte, Montevideo 54409
Museo Municipal Precolombino y Colonial, Montevideo 54410
Museo Ralli, Punta del Este 54429
Museo Torres García, Montevideo 54418

Venezuela

Centro de Arte La Cañuela, Caracas 54528
Galería de Arte de la Escuela Armando Reverón, Barcelona 54517
Galería de la Asamblea Legislativa, Barcelona 54518
Galería del Museo Anzoátegui, Barcelona 54519
Museo de Arte Coro y Museo Alberto Henríquez, Falcón 54570
Museo El Sol de las Botellas, Barcelona 54522

Art, Medieval

Albania

Muzeu i Artit Mesjetar, Korçë 00021

Andorra

Andorra Romànica Centre d'Interpretació, La Massana 00095

Austria

Stift Herzogenburg, Herzogenburg 02076

Belgium

Musée d'Art Religieux et d'Art Mosan, Liège 03806
Musée des Arts Anciens du Namurois, Namur 03901

Bulgaria

Nacionalen Istoriko-archeologičeski Rezervat i Muzej Veliki Preslav, Veliki Preslav 05248

Cyprus

Cyprus Medieval Museum, Lemesos 09385
Leventis Municipal Museum of Nicosia, Lefkosia 09379

Czech Republic

Klášter Sv. Anežky České, Praha 09755
Klašter Sv. Jiří, Praha 09756
Sládečkovo Vlastivědné Muzeum v Kladně, Kladno 09583

Dominican Republic

Museo de las Casas Reales, Santo Domingo 10351

France

Musée d'Art et d'Histoire, Saint-Denis (Seine-Saint-Denis) 15236
Musée de Normandie, Caen 12231
Musée du Petit-Palais, Avignon 11792
Musée Municipal, Chartres 12425
Musée National du Moyen Age, Paris 14729
Musée Zervos, Vézelay 16138
Palais Jacques Coeur, Bourges 12121

Germany

Augustinermuseum, Freiburg im Breisgau 18428
Barfüßerkirche, Erfurt 18198
Diözesanmuseum Freising, Freising 18450
Diözesanmuseum Rottenburg, Rottenburg am Neckar 21420
Dominikanermuseum Rottweil, Rottweil 21424
Focke-Museum, Bremen 17549
Kunsthalle Sankt-Annen und Sankt-Annen-Museum, Lübeck 20077
Mittelrhein-Museum Koblenz, Koblenz 19587
Museum Kurhaus Kleve, Kleve 19567
Museum Schnütgen, Köln 19625
Museumskirche Sankt-Katharinen, Lübeck 20081
Staatsgalerie in der Katharinenkirche, Augsburg 16636
Strigel-Museum und Antoniter-Museum, Memmingen 20280
Ulmer Museum, Ulm 22170
Zisterzienserkloster Bebenhausen, Tübingen 22127

Greece

Museum of Byzantine Culture, Thessaloniki 23217

Italy

Galleria Regionale della Sicilia, Palermo 27065
Museo Civico Medievale, Bologna 25449
Museo dell'Alto Medioevo, Roma 27481
Museo di San Marco, Venezia 28196
Museo Diocesano San Pietro, Teggiano 27950
Museo Nazionale d'Arte Medievale e Moderna, Matera 26640
Pinacoteca Malaspina, Pavia 27126

Poland

Kamienica pod Gwiazdą, Toruń 35192
Muzeum Farmacji, Poznań 35035
Muzeum Narodowe w Poznaniu, Poznań 35041

Romania

Casa Domnească Brebu, Brebu 36183
Colecția Muzeală a Bisericii Ortodoxe Sf. Nicolae, Bereşti 36148
Colecția Muzeală a Mănăstirii Jitianu, Podari 36537
Expoziția de Artă Feudală, Domneşti 36348

Slovakia

Slovenská Národná Galéria - Zvolenský Zámok, Zvolen 38395

Slovenia

Posavski Muzej Brežice, Brežice 38404

Spain

Colleción Eugenio Fontaneda, Ampudia 38871
Instituto de Valencia de Don Juan, Madrid 39522
Museo del Real Monasterio de San Juan de la Peña, Jaca 39417
Museo Parroquial de San Cosme y San Damián, Covarrubias 39224
Museu Diocesà de Girona, Girona 39348
Museu Episcopal de Vic, Vic 40270

Sweden

Fenomenalen, Visby 41068
Historiska Museet, Stockholm 40899

Ukraine

Rivnens'kyj Krajeznavčyj Muzej, Rivne 43173

United Kingdom

Boston Guildhall Museum, Boston 43562
Norton Priory Museum, Runcorn 45530
Westminster Abbey Museum, London 45012

United States of America

The Cloisters, New York 51152
Early American Museum, Mahomet 50399
Humboldt County Old Settler's Museum, Livermore 50145
The Johns Hopkins University Archaeological Collection, Baltimore 46572
The Lane Place, Crawfordsville 47870
McAllister House Museum, Colorado Springs 47702
Michael C. Carlos Museum, Atlanta 46444
Oakley Pioneer Museum, Oakley 51475
Saunders Memorial Museum, Berryville 46769
Strater Hotel, Durango 48213

Art, Modern and Contemporary

Afghanistan

National Art Gallery, Kabul 00004

Argentina

Fundación Federico Jorge Klemm, Buenos Aires 00160
Itimuseum, Buenos Aires 00163
Museo de Arte Abstracto del Sur, Buenos Aires 00180
Museo de Arte Contemporáneo, Bahía Blanca 00133
Museo de Arte Contemporáneo Beato Angélico, La Plata 00409
Museo de Arte Moderno de Buenos Aires, Buenos Aires 00183
Museo de Artes y Artesanías Enrique Estrada Bello, Santo Tomé 00632
Museo de Bellas Artes Octavio de la Colina, La Rioja 00418
Museo Fernán Félix de Amador, Luján 00436
Museo Municipal de Arte Moderno de Mendoza, Mendoza 00458

Armenia

Ervand Kochar Museum, Yerevan 00734
Gevorg Grigoryan - Giotto - Studio-Museum, Yerevan 00736
Hayart Cultural Center, Yerevan 00737
Kapan Museum of Contemporary Art, Kapan 00713
Museum of Modern Art, Yerevan 00748

Australia

24HR Art, Parap 01404
Anne Gordon Samstag Museum of Art, Adelaide 00772
ANU School of Art Gallery, Acton 00766
Art Collection, Acton 00767
Artarmon Galleries, Artarmon 00815
Arts Project Australia, Northcote 01379
Artspace, Woolloomooloo 01670
Australian Centre for Contemporary Art, Southbank 01517
Bathurst Regional Art Gallery, Bathurst 00838
Bega Valley Regional Gallery, Bega 00854
Bribie Island Community Arts Centre, Bribie Island 00892
Broken Hill Regional Art Gallery, Broken Hill 00903
Bunbury Regional Art Galleries, Bunbury 00916
Canberra Contemporary Art Space, Braddon 00888
Canberra Contemporary Art Space Manuka, Manuka 01259
Canberra Museum and Gallery, Canberra 00944
Contemporary Art Centre of South Australia, Parkside 01405
Dubbo Regional Gallery, Dubbo 01046
Experimental Art Foundation, Adelaide 00775
Gallery of Modern Art, South Brisbane 01511
Geraldton Regional Art Gallery, Geraldton 01106
Gertrude Contemporary Art Spaces, Fitzroy 01081
Gippsland Art Gallery, Sale 01488
Gold Coast City Art Gallery, Surfers Paradise 01537
Goulburn Regional Art Gallery, Goulburn 01125
Grafton Regional Gallery, Grafton 01127
Greenaway Gallery, Camberwell 00937
Griffith Artworks, Nathan 01355
Griffith Regional Art Gallery, Griffith 01132
Heide Museum of Modern Art, Bulleen 00915
Institute Of Modern Art, Fortitude Valley 01087
Ivan Dougherty Gallery, Paddington, New South Wales 01400
Lake Macquarie City Art Gallery, Booragul 00877
Latrobe Regional Gallery, Morwell 01325
Linden - Saint Kilda Centre for Contemporary Arts, Saint Kilda, Victoria 01480
Lismore Regional Gallery, Lismore 01232
Maitland City Art Gallery, Maitland, New South Wales 01250
Manning Regional Art Gallery, Taree 01573
Metro Arts, Brisbane 00897
Monash University Museum of Art, Clayton 00980
Moree Plains Gallery, Moree 01319
Museum of Contemporary Art Sydney, Sydney 01553
National Vietnam Veterans Art Museum, Newhaven 01367
New England Regional Art Museum, Armidale 00814
Nolan Gallery, Tharwa 01581
Parliament House Art Collection, Canberra 00950
Penrith Regional Gallery and The Lewers Bequest, Emu Plains 01073
Perc Tucker Regional Gallery, Townsville 01599
Performance Space, Eveleigh 01078
Perth Institute of Contemporary Arts, Northbridge, Western Australia 01377
Port Phillip City Collection, Saint Kilda, Victoria 01481
QCA Gallery, Brisbane 00898
Queensland University of Technology Art Museum, Brisbane 00900
Rockhampton Art Gallery, Rockhampton 01466
Sir Hermann Black Gallery, Sydney 01559
Stanthorpe Regional Art Gallery, Stanthorpe 01527
Swan Hill Regional Art Gallery, Swan Hill 01541
Tin Sheds Gallery, Sydney 01562
University Fine Art Collection, Hobart 01160
University of Southern Queensland Art Gallery Art Collection, Toowoomba 01595

Austria

Alte Schmiede - Kunstverein Wien, Wien 02930
Arnulf Rainer Museum, Baden bei Wien 01772
Art Via Mondsee, Mondsee 02403
Artbox, Mattersburg 02364
Artothek-Galerie, Wien 02935
Ausstellungszentrum Heiligenkreuzer Hof, Wien 02940
BAWAG Foundation, Wien 02944
Belvedere - 20er Haus, Wien 02949
Belvedere - Augarten Contemporary, Wien 02950
Blau-gelbe Viertelsgalerie Weistrach, Weistrach 02906
Burgenländische Landesgalerie, Eisenstadt 01856
Burggalerie Laa, Laa an der Thaya 02246
Egon Schiele-Museum, Tulln 02830
Ernst Fuchs Museum, Wien 02981
Galerie 5020, Salzburg 02611
Galerie der Mitte, Linz 02314
Galerie der Stadt Traun, Traun 02823
Galerie der Stadt Wels, Wels 02915
Galerie Freihausgasse, Villach 02852
Galerie im Bürger- und Gemeindezentrum, Hofstetten-Grünau 02088

Galerie im Schloss Porcia, Spittal an der Drau 02751
Galerie im Theater am Saumarkt, Feldkirch 01880
Galerie im Troadkasten, Pram 02522
Galerie in der Hauptschule, Ulrichsberg 02837
Galerie Kunstverein Mistelbach, Mistelbach an der Zaya 02385
Galerie Stephanie Hollenstein, Lustenau 02341
Galerie Theodor von Hörmann, Imst 02108
Galerie und Oberösterreichisches Literaturhaus im Stifterhaus, Linz 02315
Galerie Weberhaus, Weiz 02910
Generali Foundation, Wien 02993
Haus Wittgenstein, Wien 02999
Kokoschka-Haus Pöchlarn, Pöchlarn 02505
Kubinkabinett im Badhaus, Leogang 02290
Kuenburg Kunststätte, Payerbach 02480
Künstlerhaus Büchsenhausen, Innsbruck 02120
Künstlerhaus Graz am Landesmuseum Joanneum, Graz 01975
Kunst Haus Wien, Wien 03014
Kunstforum Ebendorf Raumacht, Wien 03015
Kunsthalle Exnergasse, Wien 03016
Kunsthalle Krems, Krems 02227
Kunsthalle Wien, Wien 03017
Kunsthaus Bregenz, Bregenz 01795
Kunstraum Stein, Krems 02228
Landesgalerie Linz, Linz 02317
Lentos Kunstmuseum Linz, Linz 02318
Leopold Museum, Wien 03022
MAK - Österreichisches Museum für angewandte Kunst / Gegenwartskunst, Wien 03026
MUSA - Museum - Startgalerie - Artothek, Wien 03029
Museum Gugging, Maria Gugging 02350
Museum Kuenburggewölbe, Werfen 02923
Museum Moderner Kunst Kärnten MMKK, Klagenfurt 02190
Museum Moderner Kunst Stiftung Ludwig Wien, Wien 03042
MZM - Museumszentrum Mistelbach, Mistelbach an der Zaya 02386
Neue Galerie am Landesmuseum Joanneum, Graz 01982
Niederösterreichisches Dokumentationszentrum für Moderne Kunst, Sankt Pölten 02684
OK Offenes Kulturhaus Oberösterreich, Linz 02323
Phantasten Museum Wien, Wien 03056
Schloss Deutschkreutz, Deutschkreutz 01815
Schloß Farrach, Zeltweg 03130
Secession, Wien 03074
Stadtgalerie Klagenfurt, Klagenfurt 02192
Stadtgalerie Schwaz, Schwaz 02728
Stadtmuseum Bruck an der Mur, Bruck an der Mur 01805
Stadtturmgalerie, Gmünd, Kärnten 01943
Stadtturmgalerie, Innsbruck 02132
Sudhaus Sorgendorf, Bleiburg 01787
Swarovski Kristallwelten, Wattens 02898
Thomas K. Lang Gallery, Wien 03080
Tiroler Künstlerschaft - Kunstpavillon, Innsbruck 02135
Würth Art Room, Böheimkirchen 01792

Azerbaijan

Arts etc., Bakı 03141
Qiz Qalasy Gallery, Bakı 03167
Rustam Mustafayev ad. Azerbaycan Dövlet İncesenet Muzeyi, Bakı 03169

Bahrain

Bahrain Arts Society Gallery, Manama 03211
Rashid Al-Oraifi Museum, Muharraq 03213

Belgium

BPS 22 - Espace de Creation Contemporaine, Charleroi 03513
CCNOA, Bruxelles 03441
Centre d'Art Contemporain, Bruxelles 03443
Kunstforum Würth, Turnhout 04098
MAC's - Musée des Arts Contemporains, Hornu 03714
Maison Jonathas, Enghien 03581
Maison Vincent Van Gogh, Cuesmes 03531
Mu.ZEE - Kunstmuseum aan Zee, Oostende 03933
Musée d'Art Moderne et d'Art Contemporain, Liège 03805
Musée d'Art Moderne - Museum voor Moderne Kunst, Bruxelles 03463
Museum Dhondt-Dhaenens, Deurle 03546
Museum van Hedendaagse Kunst Antwerpen, Antwerpen 03301
Stedelijk Museum voor Actuele Kunst, Gent 03644
Z33, Hasselt 03687

Bosnia and Herzegovina

Centar za Sovremeni Umetnosti Sarajevo, Sarajevo 04273
Galerija na Centar za Kultura, Mostar 04269
Gradska Galerija Bihać, Bihać 04258
Umjetnička Galerija Bosne i Hercegovine, Sarajevo 04283

Brazil

Açude Museum, Rio de Janeiro 04680
Casa Daros, Rio de Janeiro 04682
Chácara do Céu Museum, Rio de Janeiro 04688
Museu de Arte Contemporâea, Jataí 04509
Museu de Arte Contemporânea, Feira de Santana 04450
Museu de Arte Contemporânea de Niterói, Niterói 04591
Museu de Arte Moderna da Bahia, Salvador 04795
Museu de Arte Moderna de São Paulo, São Paulo 04882

Bulgaria

Chudožestvena Galerija Elena Karamichajlova, Šumen 05213
Chudožestvena Galerija Georgi Papazov, Jambol 05041
Chudožestvena Galerija Prof. Ilija Petrov, Razgrad 05147
Chudožestvena Galerija Varna, Varna 05236
Galerija Centăr za Săvremenno Izkustvo, Varna 05238
Gradska Chudožestvena Galerija, Plovdiv 05127
Nacionalna Chudožestvena Galerija, Sofia 05200
Okražna Chudožestvena Galerija, Stara Zagora 05210
Okražna Chudožestvena Galerija, Vraca 05266

Cameroon

Doual'art, Douala 05294

Canada

Ace Art, Winnipeg 07352
Agnes Jamieson Gallery, Minden 06293
Allie Griffin Art Gallery, Weyburn 07314
Anna Leonowens Gallery, Halifax 05929
Art Gallery of Calgary, Calgary 05514
Art Gallery of Mississauga, Mississauga 06308
Art Gallery of Newfoundland and Labrador, Saint John's 06842
Art Gallery of North York, North York 06484
Art Gallery of Nova Scotia, Halifax 05931
Art Metropole, Toronto 07092
Artpoint Gallery and Studios Society, Calgary 05515
Artspace, Peterborough 06590
Artspeak Gallery, Vancouver 07198
Axe Néo-7 Art Contemporain, Gatineau 05855
Beaches Heritage Centre, Eastport 05733
Blackwood Gallery, Mississauga 06310
Burlington Art Centre, Burlington 05498
Campbell River and District Public Art Gallery, Campbell River 05538
Campus Gallery, Barrie 05386
Centre Culturel Ti Yarihuten, Wendake 07298
Centre d'Art de la Sarre, La Sarre 06115
Centre d'Exposition des Gouverneurs, Sorel 06992
Centre d'Exposition Raymond-Lasnier, Trois-Rivières 07170
Centre for Art Tapes, Halifax 05932
Centre International d'Art Contemporain de Montréal, Montréal 06334
City Hall Art Gallery, Ottawa 06542
Comox Valley Centre for the Arts, Courtenay 05650
Comox Valley Community Arts Council, Courtenay 05651
Contemporary Art Gallery, Vancouver 07209
Downtown Gallery, Nanaimo 06410
Durham Art Gallery, Durham 05727
Eye Level Gallery, Halifax 05934
Fine Art Collection Gallery 1C03, Winnipeg 07358
Foreman Art Gallery of Bishop's University, Lennoxville 06157
Galerie Coup d'Oeuil, Rimouski 06752
Galerie d'Art, Moncton 06315
Galerie d'Art du Centre Culturel, Sherbrooke, Québec 06959
Galerie d'Art du Vieux Presbytère, Sainte-Flavie 06891
Galerie d'Art Stewart Hall, Pointe-Claire 06618
Galerie Port-Maurice, Saint-Léonard 06862
Galerie Sans Nom, Moncton 06317
Gallery Connexion, Fredericton 05831
Gallery Gachet, Vancouver 07211
Gallery One One One, Winnipeg 07361
Gallery Stratford, Stratford 07019
Gallery TPW (Toronto Photographers Workshop), Toronto 07109
Gallery - University of Toronto, Scarborough 06937
La Gare de Rivière Blanche, Saint-Ulric-de-Rivière-Blanche 06881
Gibson Gallery, Amherstburg 05333
Grace Campbell Gallery, Prince Albert 06662
Grand Forks Art Gallery, Grand Forks 05894
Grimsby Public Art Gallery, Grimsby 05913
Hamilton Artists, Hamilton 05954
Illingworth Kerr Gallery, Calgary 05521
Kamloops Art Gallery, Kamloops 06035
Karsh-Masson Gallery, Ottawa 06548
Kelowna Art Gallery, Kelowna 06053
Khyber Digital Media Centre, Halifax 05938
Latcham Gallery, Stouffville 07016
Leamington Art Gallery, Leamington 06154
Liane and Danny Taran Gallery, Montréal 06343
MacLaren Art Centre, Barrie 05387
McMaster Museum of Art, Hamilton 05960
Maison de la Culture Ahuntsic-Cartierville, Montréal 06344
Maison de la Culture Côte-des-Neiges, Montréal 06345
Maison de la Culture Mercier, Montréal 06346
Maison de la Culture Pointe-aux-Trembles, Montréal 06348
Maison de la Culture Rosemont-Petite-Patrie, Montréal 06349
Maison du Village de Val-David, Val-David 07192
Maison Hamel-Bruneau, Sainte-Foy 06892
Maison Pierre-Lacroix, Sainte-Marie-de-Beauce 06898
Mercer Union - Centre for Contemporary Visual Art, Toronto 07123
Modern Fuel Artist - Run Centre, Kingston 06085
Morris and Helen Belkin Art Gallery, Vancouver 07216
Musée Acadien du Québec, Bonaventure 05436
Musée d'Art Contemporain de Montréal, Montréal 06352
Musée d'Art Contemporarin des Laurentides, Saint-Jérôme 06830
Musée de Lachine, Lachine 06123
Musée du Bas Saint-Laurent, Rivière-du-Loup 06761
Musée et Centre de Transmission de la Culture Daniel Weetaluktuk, Inukjuak 06015
Museum of Contemporary Canadian Art, Toronto 07124
Neutral Ground, Regina 06721
New Brunswick Art Bank, Fredericton 05834
The Nickle Arts Museum, Calgary 05529
Oakville Galleries, Centennial, Oakville 06495
Oakville Galleries, Gairloch, Oakville 06496
Open Space Gallery, Victoria 07256
Open Studio, Toronto 07127
Ottawa Art Gallery, Ottawa 06552
Penticton Art Gallery, Penticton 06583
Power Plant Contemporary Art Gallery, Toronto 07128
Presentation House Gallery, Vancouver 07219
Resource Centre for the Arts, Saint John's 06854
Richmond Art Gallery, Richmond 06744
Robert Langen Art Gallery, Waterloo 07289
Rosemont Art Gallery, Regina 06725
Russell Gallery, Peterborough 06596
Saint Lawrence College Art Gallery, Kingston 06091
Saint Mary's University Art Gallery, Halifax 05948
Saint Norbert Arts Centre, Winnipeg 07380
Salle Augustin-Chénier, Ville-Marie 07267
Saskatchewan Arts Board Collection, Regina 06728
Simon Fraser Gallery, Burnaby 05505
Sir Wilfred Grenfell College Art Gallery, Corner Brook 05642
Société Canadienne d'Éducation par l'Art, Toronto 07138
Southern Alberta Art Gallery, Lethbridge 06163
Struts Gallery and Faucet Media Arts Centre, Sackville 06789
Surrey Art Gallery, Surrey 07037
Téms Swíya Museum, Sechelt 06941
Thunder Bay Art Gallery, Thunder Bay 07080
Truck Centre, Calgary 05532
Two Rivers Gallery of Prince George and Region, Prince George 06669
Ukrainian-Canadian Art Foundation, Toronto 07152
University College of Cape Breton Art Gallery, Sydney 07051
University of Waterloo Art Gallery, Waterloo 07290
Urban Shaman, Winnipeg 07389
Varley Art Gallery of Markham, Unionville 07188
Vernon Public Art Gallery, Vernon, British Colombia 07238
Visual Arts Centre of Clarington, Bowmanville 05450
White Water Gallery, North Bay 06472
Winnipeg Art Gallery, Winnipeg 07392
Yukon Arts Centre Gallery, Whitehorse 07327

Chile

Museo de Arte Contemporáneo, Santiago de Chile 07453
Museo de Arte Contemporaneo, Valdivia 07468
Museo de Arte Moderno, Santiago de Chile 07454

China

Asia Art Archive, Hong Kong 07793
OCT-LOFT Art Terminal, Shenzhen 08223
Pao Galleries at Hong Kong Art Centre, Hong Kong 07811
Para/Site Art Space, Hong Kong 07812
Shanghai Duolun Museum of Moder Art, Shanghai 08182
Shanghai Museum of Contemporary Art, Shanghai 08185
Shanghai Zendai Museum of Modern Art, Shanghai 08191

Colombia

Museo Bolivariano de Arte Contemporáneo, Santa Marta 08747
Museo de Antioquia, Medellín 08675
Museo de Arte Contemporáneo Moderno de Ibagué, Ibagué 08659
Museo de Arte Moderno de Barranquilla, Barranquilla 08540
Museo de Arte Moderno de Bogotá, Bogotá 08559
Museo de Arte Moderno de Bogotá II, Bogotá 08560
Museo de Arte Moderno de Bucaramanga, Bucaramanga 08599
Museo de Arte Moderno de Cartagena, Cartagena 08618
Museo de Arte Moderno de Medellín, Medellín 08676
Museo de Arte Moderno La Tertulia, Cali 08608
Museo de Arte Moderno Ramírez Villamizar, Pamplona 08706
Museo de la Memoria, Cuacutecuta 08631
Museo Nacional de Colombia, Bogotá 08591

Costa Rica

Galería Nacional de Arte Contemporáneo, San José 08812
Museo de Arte y Diseño Contemporáneo, San José 08816

Côte d'Ivoire

Fondation Jean Paul Blachère, Abidjan 08829

Croatia

Branko Deškovič Art Gallery, Bol 08846
Galerija Juraj Plančić, Stari Grad 08965
Galerija Koprivnica, Koprivnica 08891
Galerija Miroslav Kraljević, Zagreb 08991
Galerija Rigo, Novigrad 08918
Galerija Umjetnina Narodnog Muzeja Zadar, Zadar 08982
Gradska Galerija Labin, Labin 08905
Hrvatski Muzej Naivne Umjetnosti, Zagreb 08995
Muzej Hvarske Baštine, Hvar 08880
Muzej Suvremene Umjetnosti, Zagreb 09007

Cuba

Museo Provincial de Villa Clara y Museo Municipal, Santa Clara 09300

Cyprus

Byzantine Museum and Art Galleries, Lefkosia 09372
Nicosia Municipal Arts Centre, Lefkosia 09382
State Gallery of Contemporary Cypriot Art, Lefkosia 09383

Czech Republic

DOX - Centrum současného umění, Praha 09739
Dům u Černé Matky Boží - Muzeum kubismu, Praha 09740
Dům u Kamenného Zvonu, Praha 09741
Dům u Zlatého Prstenu, Praha 09742
Galerie Benedikta Rejta v Lounech, Louny 09640
Galerie Moderniho Umění, Hradec Králové 09533
Galerie Rudolfinum, Praha 09749
Galerie Umění Karlovy Vary, Karlovy Vary 09573
Galerie Výtvarného Umění, Hodonín 09518
Galerie Výtvarného Umění, Náchod 09668
Krajská Galerie Výtvarného Umění ve Zlíně, Zlín 09940
Královský Letohrádek, Praha 09757
Leica Gallery Prague, Praha 09758
Moravská Galerie Brno - Pražákův Palác, Brno 09433
Moravská Galerie Brno - Uměleckoprůmyslové Muzeum, Brno 09434
Muzeum Umění Olomouc, Olomouc 09689
Obecní Dům, Praha 09783
Oblastní Galerie Vysočiny v Jihlavě, Jihlava 09562
Podještědské Muzeum, Český Dub 09466
Regionální Muzeum a Galerie, Jičín 09559
Šternberský Palác, Praha 09795
Uměleckoprůmyslové Muzeum v Praze, Praha 09797
Veletržní Palác, Praha 09798
Východočeská Galerie v Pardubicích, Pardubice 09706

Denmark

Arken Museum of Modern Art, Ishøj 10108
Aros Aarhus Kunstmuseum, Århus 09962
Esbjerg Kunstmuseum, Esbjerg 10000

A

A

Herning Kunstmuseum, Herning 10071
J.F. Willumsens Museum, Frederikssund 10028
København Museum for Moderne Kunst, København 10134
Kulturforum Würth, Kolding 10162
Kunstforeningen GL Strand, København 10137
Kunstmuseet i Tønder, Tønder 10299
Kunstmuseet Køge Skitsesamling, Køge 10160
Louisiana Museum of Modern Art, Humlebæk 10103
Museet for Samtidskunst, Roskilde 10244
Nordjyllands Kunstmuseum, Aalborg 09956
Ribe Kunstmuseum, Ribe 10232
Silkeborg Kunstmuseum, Silkeborg 10257
Skive Kunstmuseum, Skive 10265
Vendsyssel Kunstmuseum, Hjørring 10081

Dominican Republic

Museo Bellapart, Santo Domingo 10342

Ecuador

Museo Camilo Egas del Banco Central, Quito 10429
Museo Municipal de Arte Moderno, Cuenca 10394

Egypt

Museum of Modern Art, Cairo 10506

El Salvador

Museo de Arte Contemporáneo, San Salvador 10550

Finland

Amos Andersonin Taidemuseo, Helsinki 10767
Ars Nova - Nykytaiteen Museo, Turku 11389
Galleria Pinacotheca, Jyväskylä 10887
Helsingin Kaupungin Taidemuseo, Meilahti, Helsinki 10780
Helsingin Kaupungin Taidemuseo, Tennispalatsi, Helsinki 10781
Kajaanin Taidemuseo, Kajaani 10904
Kiasma Nykytaiteen Museo, Helsinki 10792
Kluuvin Galleria - Glogalleriet, Helsinki 10793
Kouvolan Taidemuseo, Kouvola 10990
Kuntsin Taidekokoelma, Vaasa 11423
NIFCA - Nordic Institute for Contemporary Art, Helsinki 10805
Oulun Taidemuseo, Oulu 11172
Taidekeskus Mältinranta, Tampere 11361
Tikanojan Taidekoti, Vaasa 11427
Vantaan Kaupunginmuseo - Kuvataideasiat, Vantaa 11450
Varkauden Taidemuseo, Varkaus 11455
Villa Gyllenberg, Helsinki 10836

France

Abbaye de Beaulieu en Rouergue, Ginals 13063
Archéo-Musée du Kronos, Albens 11532
Art en Stalles, Pouzac 14932
Artothèque, Angers 11601
L'Atelier Legault, Pouance 14925
CAPC Musée d'Art Contemporain, Bordeaux 12062
Centre Artistique et Médiéval, Richerenches 15036
Centre Culturel Ipousteguy, Doulcon 12781
Centre d'Art Contemporain, Alençon 11543
Centre d'Art Contemporain, Lacoux-Hauteville-Lompnes 13472
Centre d'Art Contemporain, Revin 15032
Centre d'Art Contemporain, Rueil-Malmaison 15137
Centre d'Art Contemporain, Marne-la-Vallée 13977
Centre d'Art Contemporain de Brétigny, Brétigny-sur-Orge 12167
Centre d'Art Contemporain de Castres, Castres 12317
Centre de Création Contemporaine, Tours 15961
Centre Départmental d'Art Contemporain Cimaise et Portique, Albi 11536
Centre International d'Art Contemporain - CIAC, Port Aven 14912
Centre International d'Art et du Paysage, Beaumont-du-Lac 11899
Centre International du Vitrail, Chartres 12421
Centre Pompidou Metz, Metz 14099
Centre Régional d'Art Contemporain, Montbéliard 14183
Centre Régional d'Art Contemporain, Sète 15733
Centre Rhenan d'Art Contemporain, Altkirch 11561
Chapelle de Capucins, Aigues-Mortes 11500
Château de l'Alba-la-Romaine, Alba-la-Romaine 11529
Château de Ratilly, Treigny 15986
Château de Villeneuve, Vence 16096
Château d'Oiron, Oiron 14503
Château-Musée de Dieppe, Dieppe 12721
Château Musée Grimaldi, Cagnes-sur-Mer 12236
Le Chemin du Fauvisme, Collioure 12573
Collection de Peintures, Issac 13249
Collection Lambert en Avignon, Avignon 11786
Le Consortium, Dijon 12733
Espace Culturel Lucien Prigent, Landivisiau 13499
Espace d'Art Contemporain, Marseille 13990
Espace d'Art Contemporain, Colmar 12580
Espace de l'Art Concret, Mouans-Sartoux 14292
Espace Mira Phalaina, Montreuil-sous-Bois 14254
Fond Régional d'Art Contemporain, Angoulême 11616
Fondation Cartier pour l'Art Contemporain, Paris 14597
Fondation Jean Paul Blachère, Apt 11652
Fondation Maeght, Saint-Paul-de-Vence 15444
Fondation pour l'Art Contemporain Claudine et Jean-Marc Salomon, Alex 11552
Fonds Régionaux d'Art Contemporain Bourgogne, Dijon 12734
FRAC du Centre, Orléans 14531
Galerie Beaubourg, Vence 16097
Galerie d'Art Contemporain, Menton 14079
Galerie de la Marine, Nice 14423
Galerie de l'Ancienne Douane, Strasbourg 15814
La Galerie de Noisy, Noisy-le-Sec 14470
Galerie de Prêt Ville d'Angers, Angers 11604
Galerie Le Carré, Bayonne 11877
Galerie Municipale, Montesquieu 14202
Gorodka Galeries, Sarlat-la-Canéda 15636
Jeu de Paume, Paris 14614
Jeu de Paume, Paris 14615
L.A.C. Lieu d'Art Contemporain, Sigean 15756
La Criée Centre d'Art Contemporain, Rennes (Ille-et-Vilaine) 15014
Le Lavoir, Mougins 14295
Les Jacobins, Morlaix 14275
MAC/VAL, Musée d'art contemporain du Val-de-Marne, Vitry-sur-Seine 16259
Maeght Musée, Paris 14616
Magasin - Centre National d'Art Contemporain, Grenoble 13130
Maison des Artisans, Angers 11605
Maison du Patrimoine, Pleyben 14843
Maison Levanneur, Chatou 12475
Maison Rouge, Paris 14624
Municipal d'Arts Plastiques, Choisy-le-Roi 12520
Musée Angladon, Avignon 11788
Musée Auguste Rodin, Paris 14631
Musée-Bibliothèque Pierre-André Benoît, Alès 11549
Musée Cantini, Marseille 13993
Musée-Centre d'Arts Plastiques Chanot, Clamart 12538
Musée-Centre de Documentation Alfred Desmasures, Hirson 13218
Musée d'Art Contemporain, Dunkerque 12797
Musée d'Art Contemporain, Marseille 13995
Musée d'Art Contemporain de Lyon, Lyon 13883
Musée d'Art Contemporain Fernet Branca, Saint-Louis-la-Chaussée 15377
Musée d'Art et d'Histoire, Livry-Gargan 13790
Musée d'Art Moderne, Céret 12340
Musée d'Art Moderne, Troyes 15998
Musée d'Art Moderne, Belfort 11923
Musée d'Art Moderne de la Ville de Paris, Paris 14646
Musée d'Art Moderne de Saint-Étienne Métropol, Saint-Étienne 15252
Musée d'Art Moderne et Contemporain, Strasbourg 15821
Musée d'Art Moderne et d'Art Contemporain, Nice 14428
Musée d'Art Moderne Jean Peské, Collioure 12575
Musée d'Art Moderne Lille Métropole, Villeneuve-d'Ascq 16202
Musée de l'Abbaye Sainte-Croix, Les Sables-d'Olonne 13715
Musée de Saint-Paul, Saint-Paul-de-Vence 15445
Musée Départemental d'Art Ancien et Contemporain, Épinal 12840
Musée Départemental d'Art Contemporain de Rochechouart, Rochechouart 15067
Musée Départemental Maurice Denis, Saint-Germain-en-Laye 15290
Musée des Beaux-Arts, Dole 12754
Musée des Beaux-Arts, Nantes 14365
Musée des Beaux-Arts de Bordeaux, Bordeaux 12069
Musée d'Orsay, Paris 14692
Musée du Lac, Nernier 14394
Musée du Petit-Palais, Avignon 11792
Musée du Valois, Vez 16135
Musée Elise Rieuf, Massiac 14028
Musée Goetz-Boumeester, Villefranche-sur-Mer 16193
Musée Hofer-Bury, Lavérune 13552
Musée Jacques Pellan, Giverny 13069
Musée Jean Aicard - Paulin Bertrand, La Garde 13373
Musée Municipal François Guiguet, Corbelin 12619
Musée Municipale, Saint-Laurent-d'Olt 15361
Musée National d'Art Moderne, Paris 14722
Musée National Fernand Léger, Biot 11998
Musée Paul Arbaud, Aix-en-Provence 11519
Musée Paul Raymond, Pont-Saint-Esprit 14894
Musée Picasso, Paris 14736
Musée Würth, Erstein 12847
Palais de Tokyo, Paris 14743
Passages Centre d'Art Contemporain, Troyes 16003
Villa du Parc Centre d'Art Contemporain, Annemasse 11627
Village d'Artistes, Rablay-sur-Layon 14978

Germany

Albert-König-Museum, Unterlüß 22180
Altana, Bad Homburg v.d.Höhe 16749
Arp Museum Bahnhof Rolandseck, Remagen 21259
Art Center Berlin Friedrichstrasse, Berlin 17039
Art Gluchowe, Glauchau 18656
artothek, Köln 19595
Artothek der Zentral- und Landesbibliothek Berlin, Berlin 17040
Asyl der Kunst, Häuslingen 18852
Atelier 13, Ellwangen 18121
AusstellungsHalle, Frankfurt am Main 18355
Ausstellungshalle Zeitgenössische Kunst Münster, Münster 20504
Bauhaus-Museum, Weimar 22396
Berlinische Galerie, Berlin 17049
Buchheim Museum, Bernried 17301
Büro Otto Koch, Dessau 17828
C/O Berlin, Berlin 17059
Caricatura Museum Frankfurt, Frankfurt am Main 18357
Cubus Kunsthalle, Duisburg 18014
DA, Hörstel 19217
Daimler Contemporary, Berlin 17063
Deichtorhallen Hamburg, Hamburg 18920
Deutsche Bank Kunstsammlung, Frankfurt am Main 18359
DLM - Deutsches Ledermuseum/ Schuhmuseum Offenbach, Offenbach am Main 20881
Edith-Ruß-Haus für Medienkunst, Oldenburg 20894
Ehemaliges Postfuhramt, Berlin 17077
EnBW-Showroom Berlin, Berlin 17078
Ernst Barlach Museum Wedel, Wedel 22353
Flottmann-Hallen, Herne 19152
Fritz-Winter-Haus, Ahlen 16436
Fünf-Giebel-Haus, Friedrichstadt 18480
Fuhrwerkswaage Kunstraum, Köln 19602
Galerie 100, Berlin 17087
Galerie am Dom, Wetzlar 22524
Galerie am Domhof, Zwickau 22832
Galerie am Domplatz, Halle, Saale 18892
Galerie am Graben, Augsburg 16618
Galerie am Ratswall, Bitterfeld 17369
Galerie an der Bleiche, Ludwigslust 20059
Galerie Bodenseekreis, Meersburg 20249
Galerie Contact, Böblingen 17418
Galerie der Künste, Berlin 17088
Galerie der Stadt Tuttlingen, Tuttlingen 22135
Galerie der Stadt Wendlingen am Neckar, Wendlingen am Neckar 22477
Galerie des Kunstvereins Neustadt an der Weinstraße, Neustadt an der Weinstraße 20670
Galerie des Landkreises Rügen, Putbus 21140
Galerie für Zeitgenössische Kunst Leipzig, Leipzig 19905
Galerie Glockengasse, Köln 19603
Galerie im Kehrwiederturm, Hildesheim 19183
Galerie im Malzhaus, Plauen 21053
Galerie im Polnischen Institut, Leipzig 19907
Galerie im Willy-Brandt-Haus, Berlin 17098
Galerie in der Klosterschule, Glonn 18661
Galerie M, Berlin 17102
Galerie Mitte, Berlin 17103
Galerie Pankow, Berlin 17105
Galeriehaus, Nürnberg 20761
Galerieverein Leonberg e.V., Leonberg 19962
Gartenatelier-Galerie-Berlin, Berlin 17108
Georg-Kolbe-Museum, Berlin 17117
Georg-Meistermann-Museum, Wittlich 22623
Gesellschaft für Aktuelle Kunst GAK, Bremen 17552
Grafik Museum Stiftung Schreiner, Bad Steben 16845
Graphothek Berlin, Berlin 17119
Gründerzeitmuseum im Gutshof Mahlsdorf, Berlin 17121
Gustav-Lübcke Museum, Hamm, Westfalen 18969
Halle für Kunst Lüneburg e.V., Lüneburg 20097
Hamburger Bahnhof - Museum für Gegenwart - Berlin, Berlin 17123
Haus am Waldsee, Berlin 17127
Haus der Kunst, München 20445
Herbert-Gerisch-Stiftung, Neumünster 20635
Historische Mühle, Potsdam 21090
Hochheimer Kunstsammlung, Hochheim am Main 19204
ifa-Galerie Berlin, Berlin 17141
ifa-Galerie Stuttgart, Stuttgart 21945
Institut für Auslandsbeziehungen e.V.- Büro Berlin, Berlin 17142
Institut für moderne Kunst im Studio des Zumikon Nürnberg, Nürnberg 20772
Institut Mathildenhöhe, Darmstadt 17790
Jagd- und Fischereimuseum Schloß Wolfstein und Galerie Wolfstein, Freyung 18461
Junge Kunst e.V., Wolfsburg 22664
Käthe Kollwitz Museum Köln, Köln 19610
Kallmann-Museum, Ismaning 19364
Kreisgalerie, Mellrichstadt 20275
Künstlerhaus Hooksiel, Hooksiel 19270
Kultur Bahnhof Eller, Düsseldorf 17994
Kunst aus Nordrhein-Westfalen, Aachen 16401
Kunst und Museum, Hollfeld 19253
Kunstbunker Tumulka, München 20452
Kunsthalle, Bremerhaven 17571
Kunsthalle Darmstadt, Darmstadt 17792
Kunsthalle Dresden im Art'otel Dresden, Dresden 17952
Kunsthalle Göppingen, Göppingen 18674
Kunsthalle Lingen, Lingen 20009
Kunsthalle Mainz, Mainz 20144
Kunsthalle Nürnberg, Nürnberg 20776
Kunsthalle Sankt-Annen und Sankt-Annen-Museum, Lübeck 20077
Kunsthalle Tübingen, Tübingen 22118
Kunsthalle Weishaupt, Ulm 22164
Kunsthaus Apolda Avantgarde, Apolda 16566
Kunsthaus Dresden, Dresden 17953
Kunsthaus Erfurt, Erfurt 18206
Kunsthaus Essen, Essen 18257
Kunstmuseum Ahlen, Ahlen 16438
Kunstmuseum Bochum, Bochum 17401
Kunstmuseum Bonn, Bonn 17460
Kunstmuseum Dr. Krupp, Weinbach 22424
Kunstmuseum Solingen, Solingen 21802
Kunstmuseum Stuttgart, Stuttgart 21948
Kunstmuseum Wolfsburg, Wolfsburg 22665
Kunstsammlung Neubrandenburg, Neubrandenburg 20591
Kunstsammlung Nordrhein-Westfalen, Düsseldorf 17997
Kunstsammlung Nordrhein-Westfalen, Düsseldorf 17998
Kunstsammlungen Chemnitz - Museum Gunzenhauser, Chemnitz 17698
Kunststiftung Baden-Württemberg, Stuttgart 21949
Kunststiftung Poll, Berlin 17171
Kunstverein Heilbronn, Heilbronn 19098
Kunstverein in Hamburg e.V, Hamburg 18942
Kunstverein Reutlingen, Reutlingen 21284
Kunstverein Ulm, Ulm 22165
Kunstverein Weiden, Weiden in der Oberpfalz 22367
Kurt-Mühlenhaupt-Museum, Zehdenick 22770
KW Institute for Contemporary Art, Berlin 17173
Lände, Kressbronn am Bodensee 19711
Landesmuseum für Kunst und Kulturgeschichte Oldenburg, Oldenburg 20897
Landschaftsmuseum der Dübener Heide, Bad Düben 16713
Leopold-Hoesch-Museum Düren, Düren 17982
Ludwig Forum für Internationale Kunst, Aachen 16402
Ludwig Galerie Schloss Oberhausen, Oberhausen 20808
Märkisches Museum, Witten 22619
Malura Museum, Unterdießen 22178
Marta Herford, Herford 19140
Max Ernst Museum Brühl des LVR, Brühl 17606
Mies van der Rohe Haus, Berlin 17187
Mönchehaus Museum Goslar, Goslar 18712
Münter-Haus, Murnau am Staffelsee 20534
Museum Abteiberg, Mönchengladbach 20357
Museum Brandhorst, München 20464
Museum Burg Abenberg, Abenberg 16418
Museum der Stadt Borna, Borna 17485
museum FLUXUS gemeinnützige GmbH, Potsdam 21096
Museum Frieder Burda, Baden-Baden 16886
Museum für bildende Kunst im Landkreis Neu-Ulm, Nersingen 20574
Museum für Gegenwartskunst Siegen, Siegen 21755
Museum für Konkrete Kunst, Ingolstadt 19344
Museum für Moderne Kunst Frankfurt am Main, Frankfurt am Main 18385
Museum für Neue Kunst, Freiburg im Breisgau 18433
Museum gegenstandsfreier Kunst, Otterndorf 20951
Museum Küppersmühle für Moderne Kunst, Duisburg 18024
Museum Kunstraum Neuss, Neuss 20660
Museum Ludwig, Köln 19624
Museum Modern Art, Hünfeld 19289
Museum Moderner Kunst - Wörlen, Passau 20991
Museum Morsbroich, Leverkusen 19976
Museum Ritter, Waldenbuch 22288
Museum Schloss Hardenberg, Velbert 22211
Museum SPUR Cham, Cham 17690
Museum Waldhof, Bielefeld 17345
Neue Galerie der Karl-Hofer-Gesellschaft, Berlin 17224
Neue Galerie im Höhmannhaus, Augsburg 16632
Neue Nationalgalerie, Berlin 17225
Neue Sammlung, Passau 20992
Neuer Berliner Kunstverein, Berlin 17226
Neues Museum Weimar, Weimar 22411
Nolde Stiftung Seebüll, Berlin 17229
Nordische Botschaften Gemeinschaftshaus/ Felleshus - Galerie, Berlin 17230
Oberschwäbische Galerie Kloster Heiligkreuztal, Altheim 16507

Orangerie, Gera 18596
Osthaus Museum Hagen, Hagen 18860
Panorama Museum, Bad Frankenhausen 16730
Parochialkirche, Berlin 17233
Pinakothek der Moderne, München 20474
Richard-Haizmann-Museum, Niebüll 20699
Robert-Sterl-Haus, Struppen 21922
Royal Art Museum, Berlin 17241
Sammlung Hoffmann, Berlin 17242
Sammlung Scharf-Gerstenberg, Berlin 17244
Schauwerk Sindelfingen, Sindelfingen 21777
Schloßgalerie Kastenscheuer, Eberdingen 18036
Schwartzsche Villa, Berlin 17253
Situation Kunst, Bochum 17407
Spital Hengersberg, Markt Hengersberg 20213
Sprengel Museum Hannover, Hannover 19005
Staatsgalerie Moderne Kunst, Augsburg 16637
Staatsgalerie Stuttgart, Stuttgart 21960
Stadtgalerie, Osnabrück 20927
Stadtgalerie im Elbeforum Brunsbüttel, Brunsbüttel 17612
Stadtmuseum Hofheim am Taunus, Hofheim am Taunus 19226
Städtische Galerie, Traunstein 22079
Städtische Galerie Böblingen, Böblingen 17421
Städtische Galerie Filderstadt, Filderstadt 18318
Städtische Galerie Fruchthalle Rastatt, Rastatt 21187
Städtische Galerie im Cordonhaus, Cham 17692
Städtische Galerie Leerer Beutel, Regensburg 21232
Städtische Galerie Sohle 1, Bergkamen 17022
Städtische Galerie Waldkraiburg, Waldkraiburg 22301
Städtische Galerie Wolfsburg, Wolfsburg 22668
Städtische Kunsthalle, Recklinghausen 21210
Städtische Kunstsammlung Schloß Salder, Salzgitter 21475
Städtische Sammlungen Freital, Freital 18453
Städtisches Kunstmuseum Singen, Singen 21781
Städtisches Museum Engen und Galerie, Engen 18172
Stiftung für Konkrete Kunst, Reutlingen 21290
Stiftung Moritzburg, Halle, Saale 18904
Stiftung Museum Schloss Moyland, Bedburg-Hau 16975
Stiftung Opelvillen, Rüsselsheim 21439
Stiftung S BC - pro arte, Biberach an der Riß .. 17321
Studio im Hochhaus, Berlin 17270
Studiogalerie Kaditzsch, Grimma 18764
Stuttgarter Kunstverein e.V., Stuttgart 21965
Synagoge Stommeln, Pulheim-Stommeln 21138
Uferhallen, Berlin 17274
Das Verborgene Museum, Berlin 17277
Villa Oppenheim, Berlin 17278
Weserburg Museum für Moderne Kunst, Bremen 17566
Wolf Kahlen Museum, Bernau bei Berlin 17290
Xylon Museum Werkstätten, Schwetzingen ... 21704
ZKM | Zentrum für Kunst- und Medientechnologie Karlsruhe, Karlsruhe 19453

Ghana

College of Art Gallery, Kumasi 22855

Greece

Averoff Gallery, Athinai 22902
Averoff Gallery, Metsovo 23094
Dimitreion Oikima House, Agios Georgios Nilias 22870
Dimitris Pierides Museum of Contemporary Art, Glyfada 23007
Florina Museum of Modern Art, Florina 23000
Gallery of Florina Artists, Florina 23002
Gallery of Modern Greek Art of Kalamata, Kalamata 23024
Gallery of the Municipality of Patras, Patras .. 23139
Giustiniani Palace, Chios 22969
Lefteris Kanakakis Gallery, Rethymnon 23168
Macedonian Museum of Contemporary Art, Thessaloniki 23215
Municipal Gallery of Athens, Athinai 22927
Municipal Gallery of Lamia, Lamia 23071
Municipal Gallery of Livadia, Livadia 23079
Municipal Gallery of Piraeus, Piraeus 23145
Municipal Gallery of Rhodes, Rhodos 23172
Museum of Contemporary Art, Andros 22883
National Gallery of Corfu, Skripero 23191
National Museum of Contemporary Art, Athinai 22943
Skironio Museum Polychronopoulos, Megara 23089
Techni-Macedonian Artistic Association Kilkis, Kilkis 23054
Tehni Art Gallery, Thessaloniki 23221
Thanassis Frissiras Museum, Athinai 22953
Venetsanos Museum of Modern Art, Piraeus .. 23149
Yannis Tsarouchis Museum, Maroussi 23084

Guatemala

Museo Nacional de Arte Moderno, Guatemala 23302

Haiti

Musée Nader, Port-au-Prince 23323

Hungary

Baktay Kortárs Képzőművészeti Múzeum, Dunaharaszti 23530
Budapest Galéria Kiállítóháza, Budapest 23389
Budapest Galéria Kiállítóterem, Budapest 23390
Budenz-Ház - Ybl Gyűjtemény, Székesfehérvár 23851
Cifrapalota, Kecskemét 23644
Czóbel Múzeum, Szentendre 23871
Ernst Múzeum, Budapest 23399
Ferenczy Museum - Barcsay Collection, Szentendre 23872
Ferenczy Museum - Imre Ámos - Margit Anna Memorial Museum, Szentendre 23873
Ferenczy Museum - Lajos Vajda Memorial Exhibition, Szentendre 23874
Hamza Gyűjtemény és Jász Galéria, Jászberény 23622
János Kmetty Memorial Museum - Kerényi Collection, Szentendre 23877
Kalocsai Érseki Kincstár, Kalocsa 23624
Ludwig Múzeum Budapest - Kortárs Művészeti Múzeum, Budapest 23432
Martyn Ferenc Múzeum, Pécs 23763
Miskolci Galéria - Városi Művészeti Múzeum - Rákóczi-ház, Miskolc 23712
Modern Magyar Képtár I, Pécs 23767
Modern Magyar Képtár II, Pécs 23768
Művészetek Kincsesháza - Vaszary Képtár, Kaposvár 23632
Népi Iparművészeti Gyűjtemény, Kecskemét .. 23652
Óbudai Pincegaléria, Budapest 23457
Óbudai Társaskör Galéria, Budapest 23458
Paksi Képtár, Paks 23747
Sárospataki Képtár, Sárospatak 23794
Schöffer Museum, Kalocsa 23627
Szentendrei Képtár, Szentendre 23881
Szolnoki Galéria, Szolnok 23898
Szombathelyi Képtár, Szombathely 23904
Új Magyar Képtár, Székesfehérvár 23863
Vasarely Múzeum, Budapest 23490
Victor Vasarely Múzeum, Pécs 23780

Iceland

Listasafn Árnesinga, Hveragerði 24005
Listasafn Así, Ásmundarsalur, Reykjavík 24020
Listasafn Íslands, Reykjavík 24022
Nýlistasafnið Nýlo, Reykjavík 24028
Safn, Reykjavík 24031

India

Museum of Fine Arts, Chandigarh 24110
National Gallery of Modern Art, Delhi 24145

Indonesia

Museum Affandi - Museum Seni, Yogyakarta . 24583
Museum Puri Lukisan, Ubud 24580

Iran

Honarhaye Moaser Tehran, Tehrān 24671

Ireland

Ashford Gallery, Dublin 24777
Butler Gallery, Kilkenny 24848
The Douglas Hyde Gallery, Dublin 24783
Gallagher Gallery, Dublin 24791
The Hugh Lane Municipal Gallery of Modern Art, Dublin 24797
Irish Museum of Modern Art, Dublin 24801

Israel

Arad Museum, Arad 24927
Art Centre, Haifa 24952
Digital Art Lab, Holon 24974
Genia Schreiber University Art Gallery, Ramat Aviv 25081
Mishkan Le'Omanut, Ein Harod 24945
Museum für verfolgte Kunst-Israel, Ashdod .. 24929
The Museum of Israeli Art, Ramat Gan 25085
Tel Aviv Museum of Art, Tel Aviv 25122
Yigal Allon Center, Ginnosar 24948

Italy

Air of Art Museum, Bologna 25437
Art Forum Würth, Capena 25655
Ca'Pesaro - Galleria Internazionale d'Arte Moderna, Venezia 28178
Castello di Rivoli, Rivoli 27423
Centro per l'Arte Contemporanea Luigi Pecci, Prato 27332
Civica Galleria d'Arte Contemporanea, Lissone 26494
Civica Galleria d'Arte Contemporanea F. Scroppo, Torre Pellice 28039
Civica Galleria d'Arte Moderna, Genova 26289
Civica Galleria d'Arte Moderna, Milano 26674
Civici Musei e Gallerie di Storia ed Arte, Udine 28110
Civico Museo d'Arte Contemporanea, Milano . 26679
Collezione Calderara, Ameno 25223
Collezioni del '900, Firenze 26151
Fondazione Prada, Milano 26685
Forum Artis Museum, Montese 26851
Galleria Civica, Valdagno 28136
Galleria Civica d'Arte Contemporanea, Suzzara 27933
Galleria Civica d'Arte Contemporanea, Termoli 27961
Galleria Civica d'Arte Moderna, Santhià 27731
Galleria Civica d'Arte Moderna e Contemporanea, Torino 28001
Galleria Credito Siciliano, Acireale 25142
Galleria Credito Valtellinese, Sondrio 27871
Galleria d'Arte Contemporanea, Ascoli Piceno 25293
Galleria d'Arte Contemporanea delle Pro Civitate Christiane, Assisi 25305
Galleria d'Arte Contemporanea L. Spazzapan, Gradisca d'Isonzo 26356
Galleria d'Arte Moderna, Udine 28111
Galleria d'Arte Moderna Arnoldo Bonzagni, Cento 25800
Galleria d'Arte Moderna Carlo Rizzarda, Feltre 26093
Galleria d'Arte Moderna Giannoni, Novara 26949
Galleria d'Arte Moderna Moretti, Civitanova Marche 25921
Galleria dell'Accademia Nazionale di San Luca, Roma 27450
Galleria di Arte Moderna Palazzo Forti, Verona 28228
Galleria di Piazza San Marco, Venezia 28181
Galleria Gruppo Credito Valtellinese, Milano .. 26688
Galleria Nazionale d'Arte Moderna e Contemporanea, Roma 27454
GAMeC - Galleria d'Arte Moderna e Contemporanea, Bergamo 25400
Hangar Bicocca - spazio d'arte contemporanea, Milano 26689
MADRE - Museo d'Arte Donnaregina, Napoli . 26881
Maxxi - Museo delle Arti dell XXI secolo, Roma 27457
Merano Arte - Kunst Meran, Merano 26656
Museion - Museo d'Arte Moderna e Contemporanea, Bolzano 25485
Museo Civico di Belriguardo, Voghiera 28305
Museo Comunale, Praia a Mare 27323
Museo Comunale d'Arte Contemporanea, Polignano a Mare 27271
Museo d'Arte Contemporanea di Roma - MACRO, Roma 27470
Museo d'Arte Contemporanea di Villa Croce, Genova 26298
Museo d'Arte Moderna di Bologna - MAMbo, Bologna 25450
Museo d'Arte Moderna e Contemporanea, Ferrara 26109
Museo d'Arte Moderna e Contemporanea di Trento e Rovereto, Rovereto 27570
Museo d'Arte Moderna e Contemporanea di Trento e Rovereto, Trento 28056
Museo d'Arte Moderna e Contemporanea - MAM, Gazoldo degli Ippoliti 26279
Museo d'Arte Moderna Mario Rimoldi, Cortina d'Ampezzo 25984
Museo d'Arte Moderna Pagani, Castellanza 25734
Museo della Villa San Carlo Borromeo, Senago 27800
Museo Marino Marini, Firenze 26193
Museo Marino Marini, Pistoia 27253
Museo Marino Marini, Milano 26711
Museo Mascagnano, Livorno 26499
Museo Nazionale d'Arte Medievale e Moderna, Matera 26640
Museo Provinciale d'Arte Contemporanea, Crotone 26014
Museo Sperimentale d'Arte Contemporanea Muspac, L'Aquila 26459
Museo Stauros di Arte Sacra Contemporanea, Isola del Gran Sasso 26416
Nuova-Icona, Venezia 28204
PAC - Padiglione d'Arte Contemporanea, Milano 26717
Palazzo Belmonte Riso, Palermo 27079
Pinacoteca Civica, Crotone 26015
Pinacoteca Comunale, Verucchio 28239
Pinacoteca Comunale Francesco Cozza, Stilo 27906
Pinacoteca d'Arte Moderna, Avezzano 25336
Pinacoteca d'Arte Moderna e Contemporanea "AM International", Bivongi 25431
Pinacoteca L. Répaci, Palmi 27087
Raccolta Agorà dell'Arte, Sersale 27820
Raccolte Frugone, Genova 26326
Rotonda di Via Besana, Milano 26722
Trevi Flash Art Museum, Trevi 28065
Villa Collezione Panza, Varese 28164

Japan

21-seiki Gendai Bijutsukan, Kanazawa 28654
Akita-kenritsu Kindai Bijutsukan, Yokote 29623
Art House Project, Naoshima 29025
Benesse House, Naoshima 29026
Bunkamura Museum, Tokyo 29375
Center for Contemporary Art Kitakyushu, Kitakyushu 28711
Chichu Art Museum, Naoshima 29027
Daimaru Museum Kobe, Kobe 28723
Daimaru Museum Kyoto, Kyoto 28807
Daimaru Museum Sapporo, Sapporo 29214
Daimaru Museum Shinsaibashi Osaka, Osaka 29123
Daimaru Museum Tokyo, Tokyo 29381
Daimaru Museum Umeda Osaka, Osaka 29124
Fukuoka Ajian Bijutsukan, Fukuoka 28431
Fukuyama Bijutsukan, Fukuyama 28442
Furukawa Bijutsukan, Nagoya 28999
Gunma-kenritsu Kindai Bijutsukan, Takasaki . 29330
Hara Gendai Bijutsukan, Tokyo 29395
Hara Museum ARC, Shibukawa 29256
Himeji-shiritsu Bijutsukan, Himeji 28512
Hiroshima-shi Gendai Bijutsukan, Hiroshima . 28538
Hokkaidoritsu Kindai Bijutsukan, Sapporo 29218
Hyogo-kenritsu Bijutsukan, Kobe 28728
Ibaraki-ken Kindai Bijutsukan, Mito 28918
Joshibi Bijutsukan, Sagamihara 29181
Kahitsukan - Kyoto Museum of Contemporary Art, Kyoto 28818
Kamakura Kindai Bijutsukan, Kamakura 28640
Kanagawa Kenritsu Kindai Bijutsukan, Hayama 28500
Kiso-ji Bijutsukan, Agematsu 28347
Kitano Bunkakan, Nagano 28956
Kokuritsu Kokusai Bijutsukan, Osaka 29130
Konoki one Artist Museum, Nasu 29047
Kumamoto-shiritsu Gendai Bijutsukan, Kumamoto 28779
Kurashiki-shi Bijutsukan, Kurashiki 28785
Kyoto Kokuritsu Kindai Bijutsukan, Kyoto 28843
Marugame Genichiro-Inokuma Gendai Bijutsukan, Marugame 28881
Marugame Hirai Museum, Marugame 28882
Meguro Museum of Art, Tokyo 29430
Mito Geijutsu-kan, Gendai Bijutsu Gallery, Mito 28921
Miyake Bijutsukan, Kagoshima 28631
Mori Bijutsukan, Tokyo 29439
Morohashi Kindai Bijutsukan, Kitashiobara 28718
Museum of Contemporary Art Sapporo, Sapporo 29220
Nagashima Museum, Kagoshima 28633
Nagi-cho Gendai Bijutsukan, Nagi 28990
Nakata Bijutsukan, Onomichi 29119
Nichido Contemporary Art, Tokyo 29446
Niigata-kenritsu Bandaijima Bijutsukan, Niigata 29055
Niimi Bijutsukan, Niimi 29066
Nishinomiya-shiritsu Otani Bijutsu Kinenkan, Nishinomiya 29074
O Bijutsukan, Tokyo 29454
Ohara Bijutsukan, Kurashiki 28789
Okinawa-kenritsu Hakubutsukan Gendai Bijutsukan, Naha 29014
Osaka-furitsu Gendai Bijutsu Senta, Osaka 29141
Osaka-shiritsu Kindai Bijutsukan Kensetsu Jumbishitsu, Osaka 29152
Sakura-shiritsu Bijutsukan, Sakura 29211
Sano Gallery, Kokubunji 28762
Satoe Kindai 21-seiki Bijutsukan, Kazo 28699
Setagaya-kuritsu Bijutsukan, Tokyo 29468
Sezon Museum of Modern Art, Karuizawa 28671
Shiga-kenritsu Kindai Bijutsukan, Otsu 29166
Shimogamo Hanki-cho Kyoto, Kyoto 28867
Sogetsu Bijutsukan, Tokyo 29474
Tamagawa Kindai Bijutsukan, Tamagawa 29349
Taro Okamoto Bijutsukan, Kawasaki 28695
Taro Okamoto Kinenkan, Tokyo 29482
Tokushima-kenritsu Kindai Bijutsukan, Tokushima 29367
Tokyo Kokuritsu Kindai Bijutsukan, Tokyo 29491
Tokyo Kokusai Bijutsukyoukai, Tokyo 29494
Tokyo Opera City Art Gallery, Tokyo 29495
Tokyo-to Gendai Bijutsukan, Tokyo 29498
Toyama-kenritsu Kindai Bijutsukan, Toyama . 29522
Toyoshina-machi Kindai Bijutsukan, Toyoshina 29532
Toyota-shiritsu Bijutsukan, Toyota 29535
Tsurui Bijutsukan, Niigata 29064
Uehara Kindai Bijutsukan, Shimoda 29265
Wakayama-kenritsu Kindai Bijutsukan, Wakayama 29575
Watari-Um, Tokyo 29504
Yorozu Tetsugoro Kinen Bijutsukan, Towa 29518
Yufuin-cho Kindai Bijutsukan, Yufuin 29632

Jordan

Jordan National Gallery of Fine Arts, Amman 29655

A

Kazakhstan

Centr Sovremennogo Iskusstva Sorosa, Almaty 29677

Korea, Republic

Artsonje Center, Seoul 29864
Artsonje Museum, Gyeongju 29825
Deosugung Art Museum, Seoul 29870
Gongpyeong Art Center, Seoul 29878
Han Kuk Art Museum, Yongin 29942
Nam June Paik Art Center, Yongin 29945
National Museum of Contemporary Art, Gwacheon 29820
Seoul Art Museum, Seoul 29916
Suka Art Space, Busan 29782
Sungkok Art Museum, Seoul 29930
Whanki Museum, Seoul 29933

Kyrgyzstan

Kyrgyzskij Nacionalnyj Muzej Izobrazitelnych Iskusstv im. Gapara Ajtijeva, Biškek 29956

Latvia

Latvian Centre for Contemporary Art, Rīga 30075
Saules muzejs, Rīga 30112
Valsts Mākslas Muzejs Izstazu Zale Arsenāls, Rīga 30115

Lebanon

Musée Dorothée Salhab Kazemi, Broumana 30165

Liechtenstein

Kunstmuseum Liechtenstein, Vaduz 30197

Lithuania

Dailininko Prano Domšaičio Galerija ir Kultūros Centras, Klaipėda 30277
Europos Parkas, Vilnius 30403
Janinos Monkutės-Marks Muziejus-Galerija, Kėdainiai 30270
Kauno Paveikslų Galerija, Kaunas 30251
Vilniaus Dailės Akademijos Muziejus, Vilnius 30439
Vytauto Kazimiero Jonyno Galerija, Druskininkai 30226

Luxembourg

Casino Luxembourg - Forum d'Art Contemporain, Luxembourg 30476
Galerie d'Art Contemporain Am Tunnel, Luxembourg 30477
Galerie de Bourglinster, Bourglinster 30457
Musée d'Art Moderne Grand-Duc Jean, Luxembourg 30478
Villa Vauban, Luxembourg 30486

Macedonia

Centar za Sovremeni Umetnosti Skopje, Skopje 30514
Muzej na Sovremenata Umetnost Skopje, Skopje 30520
Umetnička Galerija Strumica, Strumica 30529

Mexico

Instituto Cultural Cabañas, Guadalajara 30889
Museo Chihuahuense de Arte Contemporáneo, Chihuahua 30766
Museo de Arte Abstracto Manuel Felguérez, Zacatecas 31577
Museo de Arte Carrillo Gil, México 31068
Museo de Arte Contemporáneo Alfredo Zalce, Morelia 31194
Museo de Arte Contemporáneo Ángel Zárraga, Durango 30838
Museo de Arte Contemporáneo Ateneo de Yucatán, Mérida 31024
Museo de Arte Contemporáneo de Chihuahua, Chihuahua 30767
Museo de Arte Contemporáneo de Monterrey, Monterrey 31178
Museo de Arte Contemporáneo de Oaxaca, Oaxaca 31221
Museo de Arte Contemporáneo Número 8, Aguascalientes 30699
Museo de Arte Moderno de Durango, Gómez Palacio 30879
Museo de Arte Moderno del Estado de México, Toluca 31484
Museo de la Ciudad de Querétaro, Querétaro 31300
Museo de Sitio Polyforum Siqueiros, México 31094
Museo Sala de Arte Moderno, Juchitán de Zaragoza 30984
Museo Tamayo Arte Contemporáneo, México 31150
Museo Universitario Contemporáneo de Arte, México 31153
Ninart Centro de Cultura, México 31157

Monaco

Monaco Modern Art Museum, Monaco 31606

Mongolia

Mongolian National Modern Art Gallery, Ulaanbaatar 31623

Montenegro

Moderna Galerija, Podgorica 31655

Morocco

Musée d'Art Contemporain de Tanger, Tanger 31696
Villa des Arts, Casablanca 31661

Netherlands

Arti et Amicitiae, Amsterdam 31830
BAT Artventure Collectie, Amstelveen 31820
BAT Artventure Collectie, Zevenaar 32998
Centrum Beeldende Kunst Utrecht, Amersfoort 31810
Centrum Kunstlicht in de Kunst, Eindhoven 32184
Cobra Museum voor Moderne Kunst, Amstelveen 31821
CODA Museum, Apeldoorn 31911
Electric Ladyland - The First Museum of Fluorescent Art, Amsterdam 31838
Gemeentelijke Expositieruimte Kampen, Kampen 32436
Gemeentemuseum, Maassluis 32522
Gemeentemuseum Helmond, Helmond 32363
De Hallen Haarlem, Haarlem 32296
Kröller-Müller Museum, Otterlo 32642
Kunstenaarscentrum De Fabriek, Eindhoven 32189
Kunstlocatie Würth, 's-Hertogenbosch 32372
Mondriaanhuis - Museum voor Constructieve en Concrete Kunst, Amersfoort 31813
Museum Henriette Polak, Zutphen 33019
Museum Het Valkhof, Nijmegen 32590
Museum Kasteel Wijchen, Wijchen 32946
Museum Van Bommel Van Dam, Venlo 32884
Museum voor Moderne Kunst, Arnhem 31929
Nederlands Instituut voor Mediakunst - NIMk, Amsterdam 31873
Noordbrabants Museum, 's-Hertogenbosch 32375
Odapark, Venray 32885
De Pont, Tilburg 32830
Pronkkamer Uden-Museum Hedendaagse Kunst, Uden 32834
De Rietgors, Papendrecht 32667
Rietveld Schröderhuis, Utrecht 32856
Scheringa Museum voor Realisme, Spanbroek 32797
Stadsgalerij Heerlen - Museum voor Moderne en Hedendaagse Kunst, Heerlen 32338
Stedelijk Museum Het Domein, Sittard 32779
Stedelijk Museum Schiedam, Schiedam 32756
Stichting De Appel, Amsterdam 31890
Stichting De Brakke Grond, Amsterdam 31891
Van Abbemuseum, Eindhoven 32196
Witte de With, Rotterdam 32734
De Zonnehof, Centrum voor Moderne Kunst, Amersfoort 31818

New Zealand

Anderson Park Art Gallery, Invercargill 33128
Aratoi Wairarapa Museum of Art and History, Masterton 33140
Artspace, Auckland 33041
Centre of Contemporary Art, Christchurch 33070
City Gallery Wellington, Wellington 33244
Forrester Gallery, Oamaru 33161
Govett-Brewster Art Gallery, New Plymouth 33156
Left Bank Art Gallery, Greymouth 33114
Lopdell House Gallery, Waitakere 33227
New Zealand Academy of Fine Arts Gallery, Wellington 33252
Pataka, Porirua 33186

Norway

Astrup Fearnley Museet for Moderne Kunst, Oslo 33717
Bergen Kunsthall, Bergen 33417
Bergen Kunstmuseum - Stenersens Samling, Bergen 33420
Galleri Würth, Hagan 33524
Hå Gamle Prestegard, Varhaug 33963
Munch-museet, Oslo 33732
Museet for Samtidskunst, Oslo 33733
Nasjonalgalleriet, Oslo 33735
Nasjonalmuseet for Kunst, Arkitektur og Design, Oslo 33736

Pakistan

National Art Gallery, Islamabad 34008

Panama

Museo de Arte Contemporáneo, Ciudad de Panamá 34067

Papua New Guinea

Papua New Guinea National Museum and Art Gallery, Boroko 34080

Peru

Centro de Arte Visuales, Lima 34211
Galería del Banco Continental, Lima 34218
Museo de Arte Contemporáneo, Cusco 34156
Museo de la Nación, Lima 34244

Philippines

Anita Gallery, San Antonio 34497
Artists' Space, Makati 34413
Cultural Center of the Philippines Museo, Pasay 34471
GSIS Gallery, Pasay 34472
Jorge B. Vargas Museum and Filipiana Research Center, Quezon City 34492
Museo ng Kalinangang Pilipino, Pasay 34474
Museo ng Makati, Makati 34415
SM City Art Center, Cebu 34375

Poland

Centrum Rzeźby Polskiej, Orońsko 34989
Centrum Sztuki Współczesnej, Warszawa 35220
Centrum Sztuki Współczesnej Łaźnia, Gdańsk 34632
Galeria, Lublin 34897
Galeria 5, Kalisz 34708
Galeria 5, Katowice 34720
Galeria Akademii Sztuk Pięknych im. Władysława Strzemińskiego, Łódź 34861
Galeria Amfilada, Szczecin 35156
Galeria Arsenał, Białystok 34545
Galeria Biała, Lublin 34898
Galeria Biblioteka, Łódź 34863
Galeria Domek Ogrodnika, Łódź 34864
Galeria Fabryka Kestenberga, Łódź 34865
Galeria Nowa Przestrzeń, Łódź 34868
Galeria Patio, Łódź 34869
Galeria Podlaska, Biała Podlaska 34541
Galeria Sztuki Wozownia, Toruń 35191
Galeria Sztuki Współczesnej, Szczecin 35157
Galeria Sztuki Wspólczesnej Profil Galeria Fotografii pf, Poznań 35024
Galeria Sztuki Wspólczesnej Winda, Kielce 34740
Galeria w Hallu, Kalisz 34709
If Museum Inner Spaces, Poznań 35025
Muzeum, Koszalin 34761
Muzeum Farmacji, Poznań 35035
Muzeum Narodowe w Gdańsku, Gdańsk 34642
Muzeum Narodowe w Poznaniu, Poznań 35041
Muzeum Sztuki Nowoczesnej w Niepołomicach, Niepołomice 34945
Muzeum Sztuki Współczesnej, Radom 35074
Patio Centrum Sztukii, Łódź 34887
Sektor I, Katowice 34731
Zamek Książąt Pomorskich, Szczecin 35165

Portugal

Casa da Cerca, Almada 35412
Centro de Arte Moderna José de Azevedo Perdigão, Lisboa 35664
Centro de Exposições, Arte e Multimédia, Sintra 35944
Colecção Berardo, Lisboa 35665
Museu Anjos Teixeira, Sintra 35945
Museu de Arte Contemporânea, Funchal 35612
Museu de Arte Contemprânea da Bienal de Cerveira, Vila Nova de Cerveira 36002
Museu de Arte Primitiva Moderna, Guimarães 35640
Museu de Serralves, Porto 35848
Museu do Chiado, Lisboa 35698
Museu Municipal de Arte Moderna Abel Manta, Gouveia 35634
Sintra Museu de Arte Moderna, Sintra 35954

Puerto Rico

Museo de Arte Contemporáneo de Puerto Rico, San Juan 36066

Réunion

Fonds Régional d'Art Contemporain, Saint-Paul 36094

Romania

Expoziţia Permanentă de Artă Plastică Contemporană, Ştefăneşti, Botosani 36635
Galeria de Artă Lascăr Viorel, Piatra-Neamţ 36517
Muzeul de Artă, Medgidia 36468
Muzeul de Artă, Sfântu Gheorghe 36601
Muzeul de Artă, Târgu Jiu 36664
Muzeul de Artă Vizuală, Galaţi 36379
Muzeul de Etnografie şi Artă Contemporană, Comăneşti 36298
Muzeul Municipal de Artă, Turnu Măgurele 36710
Muzeul Municipal Dej şi Galeria de Artă, Dej 36341
Muzeul Naţional de Artă Contemporană al României, Bucureşti 36220

Russia

Art Centr Puškinskaja 10, Sankt-Peterburg 37624
Art Podvalčik, Chabarovsk 36855
Belgorodskij Gosudarstvennyj Chudožestvennyj Muzej, Belgorod 36793
Caricyno - Gosudarstvennyj Istoriko-architekturnyj, Chudožestvennyj i Landšaftnyj Muzej-Zapovednik, Moskva 37143
Centralnyj Vystavočnyj Zal Manež g. Sankt-Peterburga, Sankt-Peterburg 37633
Chudožestvenno-memorialnyj Muzej K.S. Petrova-Vodkina, Chvalynsk 36864
Chudožestvennyj Muzej, Staryj Oskol 37799
Chudožestvennyj Muzej, Valujki 37915
Chudožestvennyj Muzej im. M.S. Tuganova, Vladikavkaz 37928
D137, Sankt-Peterburg 37634
Galereja Forum, Sankt-Peterburg 37638
Galereja Gerceva, Moskva 37173
Galereja Iskusstv Zuraba Cereteli, Moskva 37174
Galereja-masterskaja Chudožnika G.S. Rajševa, Chanty-Mansijsk 36861
Galereja Sacha-Art, Moskva 37177
Galereja Sem Bruk, Moskva 37178
Galereja Sojuza Tvorčestvo, Moskva 37179
Galereja Sovremennogo Iskusstva Arka, Vladivostok 37935
Galereja Sovremennogo Iskusstva Artëtaž, Vladivostok 37936
Gosudarstvennoe Muzejnoe Objedinenie Chudožestvennaja Kultura Rossijskogo Severa, Archangelsk 36769
Gosudarstvennyj Centr Sovremennogo Iskusstva, Moskva 37187
Gosudarstvennyj Muzej Izobrazitelnych Iskusstv Respubliki Kalmykija, Elista 36911
Gosudarstvennyj Vystavočnyj Zal Na Kaširke, Moskva 37212
Jaroslavskij Chudožestvennyj Muzej - Gubernatorskij Dom, Jaroslavl 36967
Kaliningradskaja Chudožestvennaja Galereja, Kaliningrad 36988
Kartinnaja Galereja A.A. Plastova, Uljanovsk 37899
Kartinnaja Galereja Ministerstva Kultury Respubliki Adygeja, Majkop 37124
Krasnodarskij Chudožestvennyj Muzej im. F.A. Kovalenko, Krasnodar 37065
Krasnojarskij Gorodskoj Vystavočnyj Zal, Krasnojarsk 37070
Memorialnyj Muzej-kvartira M.S. Spiridonova, Čeboksary 36839
Moskovskij Muzej Sovremennogo Iskusstva, Moskva 37258
Most čerez Stiks - Pons per Styx, Sankt-Peterburg 37665
Municipalnaja Chudožestvennaja Galereja g. Kostromy, Kostroma 37063
Muzej Carskoselskaja Kollekcija, Puškin 37583
Muzej Drugoe Iskusstvo, Moskva 37272
Muzej-institut Semji Rerichov, Sankt-Peterburg 37679
Muzej Nonkonformistskogo Iskusstva, Sankt-Peterburg 37703
Muzej 'Novoj Akademii Izjaščnych Iskusstv', Sankt-Peterburg 37704
Muzej Sovremennogo Izobrazitelnogo Iskusstva im. A.A. Plastova, Uljanovsk 37908
Novokuzneckij Chudožestvennyj Muzej, Novokuzneck 37481
Oblastnoj Muzejnyj Centr, Sankt-Peterburg 37716
Orenburgskij Oblastnoj Muzej Izobrazitelnych Iskusstv, Orenburg 37515
Penzenskaja Kartinnaja Galerija im. K.A. Savickogo, Penza 37536
Sosnovoborskij Chudožestvennyj Muzej Sovremennogo Iskusstva, Sosnovyj Bor 37791
Tambovskaja Oblastnaja Kartinnaja Galereja, Tambov 37832
Tjumenskaja Kartinnaja Galereja, Tjumen 37851
Tulskij Muzej Izobrazitelnych Iskusstv, Tula 37874
Tverskaja Oblastnaja Kartinnaja Galereja, Tver 37880
Tvorčeskij Centr i Vystavočnyj Zal Fëdor, Sankt-Peterburg 37729
Ufimskaja Kartinnaja Galereja, Ufa 37890
Vologodskaja Oblastnaja Kartinnaja Galereja, Vologda 37963
Vystavočnyj Zal Tverskoj Oblastnoj Kartinnoj Galerei, Tver 37882
Zverevskij Centr Sovremennogo Iskusstva, Moskva 37432

San Marino

Galleria Nazionale di Arte Moderna e Contemporanea, San Marino 38002

Serbia

Artmedia Art Gallery, Beograd 38034
Galerija Doma kulture Studentski Grad, Beograd 38038
Galerija Remont, Beograd 38043
Galerija Singidunum, Beograd 38044
Galerija ULUS, Beograd 38046
Mala Galerija Udruženja Likovnih Umetnika Primenjenih Umetnosti i Dizajnera Srbije, Beograd 38057
Muzej Naivne Umetnosti, Jagodina 38100
Muzej Savremene Likovne Umetnosti, Novi Sad 38123
Muzej Savremene Umetnosti, Beograd 38076
Paviljon Cvijeta Zuzorića, Beograd 38083
Salon Muzeja Savremene Umetnosti, Beograd 38087
Spomen-Zbirka Pavla Beljanskog, Novi Sad 38125
Umetnička Galerija Nadežda Petrović, Čačak 38095

Singapore

Singapore Art Museum, Singapore 38179

Slovakia

Galéria Jána Koniarka v Trnave - Kopplov Kaštiel', Trnava 38379
Galéria KOMART, Bratislava 38210
Galéria Slovenskej Sporitel'ne, Bratislava 38212
Martin Benka Múzeum, Martin 38295
Múzeum Milana Dobeša, Bratislava 38224
Múzeum moderného umenia Andy Warhola, Medzilaborce 38304
Považská Galéria Umenia, Žilina 38390
Štátna Galéria, Banská Bystrica 38186
Synagóga - Centrum Súčasného Umenia, Trnava 38380
Východoslovenská Galéria, Košice 38274

Slovenia

Galerija Murska Sobota, Murska Sobota 38486
Galerija Sodobne Umetnosti, Celje 38405
Moderna Galerija, Ljubljana 38454

South Africa

Centre for African Studies Collection, Rondebosch 38725
Hester Rupert Art Museum, Graaff-Reinet 38627
Irma Stern Museum of the University of Cape Town, Rosebank 38727
KZNSA Gallery, Durban 38608
UNISA Art Gallery, Tshwane 38760
University Art Gallery, Stellenbosch 38740

Spain

Artium - Centro Museo Vasco de Arte Contemporáneo, Vitoria-Gasteiz 40317
Casa Aymat, Sant Cugat del Vallès 39978
Casa das Artes de Vigo, Vigo 40273
Casa Museo Antonio Padrón, Gáldar 39319
Casa-Museo Pedralba 2000, Pedralba 39820
Casa Museu Segrelles, Albaida 38805
Centre de Cultura Contemporània de Barcelona, Barcelona 38955
Centro Andaluz de Arte Contemporáneo, Sevilla 40061
Centro Atlántico de Arte Moderno, Las Palmas de Gran Canaria 39797
Centro de Arte, Almería 38857
Centro de Arte Contemporáneo de Málaga, Málaga 39594
Centro de Arte Contemporáneo Fundación Chirivella Soriano, Valencia 40198
Centro de Arte José Guerrero de la Diputación de Granada, Granada 39360
Centro de Arte Moderno y Contemporáneo Daniel Vázquez Díaz, Nerva 39715
Centro de Arte y Naturaleza. Fundación Beulas, Huesca 39397
Colección de Escultura Contemporánea, Madrid 39512
Colección Maruja Díaz, Villaviciosa de Odón 40310
Colección Municipal de Arte Contemporáneo de Alcorcón, Alcorcón 38824
Es Baluard - Museu d'Art Modern i Contemporani de Palma, Palma de Mallorca 39781
Fundación NMAC Montenmedio de Arte Contemporáneo, Vejer de la Frontera 40255
Fundación Vela Zanetti, León 39452
Guggenheim Bilbao Museoa, Bilbao 39043
IVAM Centre Julio González, Valencia 40201
Metrònom, Barcelona 38966
Museo Adolfo Lozano Sidro, Priego de Córdoba 39870
Museo Angel Orensanz y Artes de Serralbo, Sabiñánigo 39929
Museo Antonio Manuel Campoy, Cuevas del Almanzora 39237
Museo Berrocal, Vilanueva de Algaidas 40288
Museo Casa Ibáñez, Olula del Río 39730
Museo Colecciones ICO, Madrid 39530
Museo de Arte Abstracto Español, Cuenca 39231
Museo de Arte Contemporáneo, Toledo 40131
Museo de Arte Contemporáneo Ángel Miguel de Arce, Sasamón 40043
Museo de Arte Contemporáneo Esteban Vicente, Segovia 40052
Museo de Arte Contemporáneo Pajar de Agustín, Valle de Hecho 40247
Museo de Arte Contemporáneo Union Fenosa - MACUF, A Coruña 39218
Museo de Bellas Artes de Santander, Santander 40009
Museo de Cáceres, Cáceres 39082
Museo del Grabado Español Contemporáneo, Marbella 39619
Museo Delso, L'Alfàs del Pi 38838
Museo-Exposición Marín Bosqued, Aguarón 38788
Museo Extremeño e Iberoamericano de Arte Contemporáneo, Badajoz 38936
Museo Fundación Eugenio Granell, Santiago de Compostela 40028
Museo Galego de Arte Contemporánea Carlos Maside, Arès 38888
Museo Garnelo, Montilla 39679
Museo Internacional de Arte Contemporáneo, Arrecife 38896
Museo Internacional de Electrografía, Cuenca 39236
Museo Joaquín Peinado de Ronda, Ronda 39918
Museo José Ortega, Arroba de los Montes 38897
Museo Manuel López Villaseñor, Ciudad Real 39190
Museo Municipal, Valdepeñas 40192
Museo Municipal de Arte Contemporáneo de Madrid, Madrid 39564
Museo Municipal de Arte en Vidrio de Alcorcón, Alcorcón 38825
Museo Municipal López Torres, Tomelloso 40143
Museo Nacional Centro de Arte Reina Sofía, Madrid 39566
Museo Patio Herreriano, Valladolid 40242
Museo Picasso Málaga, Málaga 39605
Museo-Pinacoteca de Arte Contemporáneo, Zarzuela del Monte 40356
Museo Popular de Arte Contemporáneo, Villafamés 40299
Museo Vostell-Malpartida de Cáceres, Malpartida de Cáceres 39609
Museu Comarcal de la Garrotxa, Olot 39727
Museu d'Art, Girona 39344
Museu d'Art Contemporani de Barcelona, Barcelona 38972
Museu d'Art Contemporani d'Eivissa, Eivissa 39261
Museu d'Art Espanyol Contemporani, Palma de Mallorca 39787
Museu d'Art Modern, Tarragona 40101
Museu de la Universitat d'Alacant, Alicante 38846
Museu de Vilafamés Vicente Aguilera Cerni, Vilafamés 40279
Museu d'Història de l'Hospitalet de Llobregat, L'Hospitalet de Llobregat 39394
Museu Picasso, Barcelona 39002
Teatre-Museu Dalí, Figueres 39300

Sweden

Alingsås Kulturhus/Konsthall, Alingsås 40412
Baltic Art Centre, Visby 41067
Botkyrka Konsthall och Xet-Museet, Tumba 40986
Box Kraftverk, Bjärtrå 40459
Bruksmuseum, Hudiksvall 40603
C Konsthall, Upplands Väsby 41001
Christinehof Slott, Brösarp 40478
Eksjö Museum, Eksjö 40493
Färgfabriken, Stockholm 40893
Gävle Konstcentrum, Gävle 40531
Galeri Valfisken, Simrishamn 40838
Galleri 21, Malmö 40719
Galleri 54, Göteborg 40545
Göteborgs Konsthall, Göteborg 40546
Havets Hus, Lysekil 40711
Immigrant-institutets Museum, Borås 40470
Kabusa Konsthall, Köpingebro 40649
Kalmar Konstmuseum, Kalmar 40623
Konceptkonstmuseum, Rydboholm 40820
Konsthallen, Borgholm 40473
Konsthallen Sandviken, Sandviken 40832
Konsthallen Trollhättan, Trollhättan 40984
Kostens hus, Luleå 40697
Kristinehamns Konstmuseum, Kristinehamn 40656
Kulturcentrum i Ronneby, Ronneby 40815
Linköpings Konsthall, Linköping 40682
Lokstallet Strömstadt Konsthall, Grebbestad 40566
Magasin 3 Stockholm Konsthall, Stockholm 40910
Mannaminne, Nordingrå 40766
Moderna Museet, Stockholm 40915
Norrköpings Konstmuseum, Norrköping 40773
Rooseum, Malmö 40726
Rottneros Park, Rottneros 40817
Södertälje Konsthall, Södertälje 40863
Tensta Konsthall, Spånga 40876
Thielska Galleriet, Stockholm 40946
Växjö Konsthall, Växjö 41044
VIDA Museum, Borgholm 40474
Virserums Konsthall, Virserum 41065

Switzerland

Allschwiler Kunst-Zentrum, Allschwil 41100
Attitudes, Genève 41399
Ausstellungsraum Klingental, Basel 41156
Aux Portes de la Champagne, Bernex 41222
Bündner Kunstmuseum, Chur 41299
Centre d'art contemporain Genève, Genève 41405
Centro d'Arte Contemporanea Ticino, Bellinzona 41192
Collection de l'Art Brut, Lausanne 41535
Daros Exhibitions, Zürich 42069
Daros-Latinamerica AG, Zürich 42070
Emma Kunz-Museum, Würenlos 42047
Espace Lausannoise d'Art Contemporain, Renens 41755
F'ar, Lausanne 41538
Fond d'art contemporain de la Ville de Genève (Fmac), Genève 41412
Fondation Beyeler, Riehen 41763
Fondation Saner, Studen 41918
Forum Würth, Arlesheim 41124
Fri-Art, Fribourg 41386
Galerie art one, Zürich 42072
Hallen für neue Kunst, Schaffhausen 41828
Haus für Kunst Uri, Altdorf 41102
Helmhaus Zürich, Zürich 42077
Kulturforum Würth Chur, Chur 41303
Kunsthalle Sankt Gallen, Sankt Gallen 41804
Kunsthalle Ziegelhütte, Appenzell 41117
Kunsthaus Glarus, Glarus 41437
Kunsthaus Langenthal, Langenthal 41526
Kunsthaus Zug, Zug 42125
Kunstmuseum des Kantons Thurgau, Warth 42002
Kunstmuseum Thun, Thun 41936
Kunstpanorama Luzern, Luzern 41594
Kunstsammlung Hans und Marlis Suter, Oberhofen am Thunersee 41695
Kunst(Zeug)Haus, Rapperswil 41745
m. a. x. museo, Chiasso 41298
Mamco, Genève 41415
Migros Museum für Gegenwartskunst, Zürich 42087
Mili Weber-Haus, Sankt Moritz 41812
Musée des Beaux-Arts, La Chaux-de-Fonds 41293
Musée Fondation Deutsch, Belmont-sur-Lausanne 41196
Musée Jenisch, Vevey 41986
Museo Comunale d'Arte Moderna, Ascona 41129
Museo Epper, Ascona 41130
Museo Villa dei Cedri, Bellinzona 41195
museum franz gertsch, Burgdorf 41275
Museum für Gegenwartskunst, Basel 41170
Museum La Congiunta, Giornico 41434
Museum Liner Appenzell, Appenzell 41119
museumbickel, Walenstadt 41997
Neuer Shed im Eisenwerk, Frauenfeld 41380
Pinacoteca Comunale, Locarno 41570
Sammlung Picasso Donation Rosengart, Luzern 41598
Schaulager, Münchenstein 41659
Seedamm Kulturzentrum, Pfäffikon (Schwyz) 41721
Sotterranei dell'Arte, Monte Carasso 41634
Spontanart Artforum, Courlevon 41316
Villa Flora Winterthur, Winterthur 42041

Taiwan

Art Center, Tainan 42359
Museum of Contemporary Art, Taipei, Taipei 42407
The Pier-2, Kaohsiung 42242
Seaport Art Center, Cingshui 42169
Tainan Municipal Cultural Center, Visual Art Section, Tainan 42369
Taipei Fine Arts Museum, Taipei 42428

Thailand

Bangkok University Gallery, Bangkok 42532
Bhirasri Museum of Modern Art, Bangkok 42534
Chiang Mai Contemporary Art Museum, Chiang Mai 42567
Rama IX Art Museum, Bangkok 42557

Tunisia

Maison des Arts du Belvédère Dar El-Founoun, Tunis 42664

Turkey

Erzurum Resim ve Heykel Müzesi, Erzurum 42777
İstanbul Modern Sanat Müzesi, İstanbul 42826
Resim ve Heykel Müzesi, İzmir 42857
Turkuvaz Sanat Galerisi, Ankara 42715

Ukraine

Center for Contemporary Art, Kyiv 43076

United Kingdom

198 Gallery, London 44778
Aberdeen Art Gallery, Aberdeen 43233
Aberystwyth Arts Centre, Aberystwyth 43256
Arnolfini, Bristol 43628
Art and Design Gallery, Hatfield 44418
Batley Art Gallery, Batley 43416
Ben Uri Gallery, London 44793
Y Capel, Llangollen 44769
The Citadel, Saint Helens 45566
Collins Gallery, Glasgow 44278
County Hall Gallery, London 44829
Delfina, London 44836
Edinburgh Printmakers Workshop and Gallery, Edinburgh 44120
Estorick Collection of Modern Italian Art, London 44844
Ferens Art Gallery, Kingston-upon-Hull 44594
Firstsite at the Minories Art Gallery, Colchester 43873
Fruitmarket Gallery, Edinburgh 44124
Gallery of Modern Art, Glasgow 44281
Gallery Tresco, Tresco 45874
Harley Gallery, Worksop 46076
Himley Hall, Himley 44470
Institute of Contemporary Arts, London 44883
John Creasey Museum, Salisbury 45603
Kettle's Yard, Cambridge 43728
Kingsgate Gallery, London 44895
Lighthouse Gallery, Poole 45405
Maclaurin Art Gallery, Ayr 43347
Mall Galleries, London 44911
Matt's Gallery, London 44914
Mead Gallery, Coventry 43904
Middlesbrough Institute of Modern Art, Middlesbrough 45141
Mission Gallery, Swansea 45814
Modern Art Oxford, Oxford 45332
MOMA Wales, Machynlleth 45069
The New Art Gallery Walsall, Walsall 45920
New Hall Art Collection, Cambridge 43731
New Realms, London 44943
New Realms at Adams Street, London 44944
New Walk Museum and Art Gallery, Leicester 44681
Northern Gallery for Contemporary Art, Sunderland 45802
Ormeau Baths Gallery, Belfast 43444
Parasol Unit, London 44951
Pier Arts Centre, Stromness 45792
Pitzhanger Manor-House and Gallery, London 44955
Pump House Gallery, London 44959
Rammerscales, Lockerbie 44777
Rotherham Art Gallery, Rotherham 45518
Royal West of England Academy New Gallery, Clifton 43853
Ruchill Centre for Contemporary Arts, Glasgow 44300
Rye Art Gallery, Rye 45535
Saatchi Gallery, London 44978
Sainsbury Centre for Visual Arts, Norwich 45285
Serpentine Gallery, London 44984
South London Gallery, London 44988
Surface Gallery, Nottingham 45300
Talbot Rice Gallery, Edinburgh 44161
Tate Britain, London 44995
Tate Liverpool, Liverpool 44746
Tate Modern, London 44996
Tate Saint Ives, Saint Ives, Cornwall 45579
Tolbooth Art Centre, Kirkcudbright 44612
Turnpike Gallery, Leigh, Lancashire 44685
The University Gallery, Newcastle-upon-Tyne 45233
University of Essex Exhibition Gallery, Colchester 43877
West Wales Arts Centre, Fishguard 44231
Whitechapel Art Gallery, London 45014
Wrexham Arts Centre, Wrexham 46084
Wysing Arts Centre, Bourn 43564

United States of America

1078 Gallery, Chico 47524
ACME Art Gallery, Columbus 47749
Aldrich Museum of Contemporary Art, Ridgefield 52347
Aljira Center for Contemporary Art, Newark 51284
American Indian Contemporary Arts, San Francisco 52765
Anderson Museum of Contemporary Art, Roswell 52467
The Andy Warhol Museum, Pittsburgh 51938
apexart, New York 51121
Archer Gallery, Vancouver 53704
Arlington Museum of Art, Arlington 46350
Armory Center for the Arts, Pasadena 51718
Art Center, Flushing 48574
Art for Change, New York 51123
Art Gallery, Brooklyn 47066
The Art Gallery, College Park 47685

A

Art Gallery Marist College, Poughkeepsie 52119
The Art Gym, Marylhurst 50522
Art Museum of the Americas, Washington 53848
Art-Tech Exhibitions, San Jose 52804
Artforms Gallery Manuyunk, Philadelphia 51814
Arthouse at the Jones Center, Austin 46500
Artists Association of Nantucket Gallery, Nantucket 50942
The Arts Center of the Capital Region, Troy 53573
Arts Center of the Ozarks-Gallery, Springdale 53211
Arts Council of Wayne County - Museum, Goldsboro 48964
Artspace, Raleigh 52227
Arvada Center for the Arts and Humanities, Arvada 46367
Asian-American Arts Centre, New York 51129
Atrium Gallery, Saint Louis 52567
Baton Rouge Gallery, Baton Rouge 46631
Baum Gallery, Conway 47782
Bemis Center for Contemporary Arts, Omaha 51547
Ben Shahn Galleries, Wayne 54007
Benedicta Arts Center, Saint Joseph 52559
Berea College Doris Ulmann Galleries, Berea 46747
Berkeley Art Center, Berkeley 46751
Beverly Art Center, Chicago 47457
Black Rock Arts Foundation, San Francisco 52767
Blue Star Contemporary Art Center, San Antonio 52708
Bobbitt Visual Arts Center, Albion 46172
Bowman, Megahan and Penelec Galleries, Meadville 50546
Bridge Center for Contemporary Art, El Paso 48316
Bromfield Art Gallery, Boston 46926
Bronx Museum of the Arts, Bronx 47040
Bronx River Art Center, Bronx 47041
Bruce Gallery of Art, Edinboro 48290
Buckham Gallery, Flint 48558
Buddy Holly Center, Lubbock 50310
Bunnell Street Gallery, Homer 49329
Butler Institute of American Art, Youngstown 54353
Butte Silver Bow Arts Chateau, Butte 47150
C. Kermit Ewing Gallery of Art and Architecture, Knoxville 49813
Cabot's Pueblo Museum, Desert Hot Springs 48105
California Center for the Arts Museum, Escondido 48418
Campbell Hall Gallery, Monmouth 50784
Cape Museum of Fine Arts, Dennis 48057
Carlson Tower Gallery, Chicago 47459
Carnegie Art Museum, Oxnard 51657
Center for Contemporary Arts of Santa Fe, Santa Fe 52872
Center for Curatorial Studies, Annandale-on-Hudson 46314
Center for Maine Contemporary Art, Rockport 52425
Center for the Visual Arts, Denver 48066
Center on Contemporary Art, Seattle 52996
Centro Cultural de la Raza, San Diego 52737
City Gallery at Chastain, Atlanta 46431
City Gallery East, Atlanta 46432
Cleveland Artists Foundation at Beck Center for the Arts, Lakewood 49907
The Coburn Gallery, Ashland 46387
Commerce Bancshares Fine Art Collection, Kansas City 49686
Communications and History Museum of Sutton, Washington 53855
Community Fine Arts Center, Rock Springs 52410
Contemporary Art Center of Virginia, Virginia Beach 53755
Contemporary Art Gallery at Falkirk Cultural Center, San Rafael 52827
Contemporary Art Museum, Raleigh 52228
Contemporary Art Museum, Tampa 53431
Contemporary Art Museum of Saint Louis, Saint Louis 52573
Contemporary Art Workshop, Chicago 47469
Contemporary Arts Center, Cincinnati 47558
Contemporary Arts Center, New Orleans 51078
Contemporary Arts Museum, Houston 49383
The Contemporary Museum, Honolulu 49336
The Contemporary Museum at First Hawaiian Center, Honolulu 49337
Corita Art Center Gallery, Los Angeles 50218
Coyote Art Gallery, Oroville 51605
Craft Alliance, Saint Louis 52574
Cranbrook Art Museum, Bloomfield Hills 46855
Creative Arts Studio, Louisville 50283
Crestar Bank Art Collection, Richmond 52326
Crossman Art Gallery, Whitewater 54134
Dahl Arts Center, Rapid City 52250
The Dallas Center for Contemporary Art, Dallas 47928
Dan Flavin Art Foundation Temporary Gallery, New York 51161
Danforth Gallery, Portland 52063
Daum Museum of Contemporary Art, Sedalia 53023
Davis Art Center, Davis 47975
Dayton Visual Arts Center, Dayton 47985
Delta Axis Contemporaray Arts Center, Memphis 50570
Design Arts Gallery, Philadelphia 51826
Detroit Artists Market, Detroit 48112
DeVos Art Museum, Marquette 50498
Dia Art Foundation, New York 51163
Dia:Beacon, Riggio Galleries, Beacon 46660
Dia:Chelsea, New York 51166
Dinnerware Contemporary Art Gallery, Tucson 53587
Dudley and Mary Marks Lea Gallery, Findlay 48543
Dumbo Arts Center, Brooklyn 47072
Duncan Gallery of Art, De Land 47997
Duncan-McAshan Visual Arts Center, Ingram 49511
Dunedin Fine Arts Center, Dunedin 48208
East Campus Galleries, Orlando 51590
East Jordan Portside Art and Historical Museum, East Jordan 48252
Edison Community College Gallery of Fine Art, Fort Myers 48662
Edward and Marthann Samek Art Gallery, Lewisburg 50029
Elaine L. Jacob Gallery, Detroit 48117
Euphrat Museum of Art, Cupertino 47911
Evanston Art Center, Evanston 48449
Exhibitions @ Soap Factory, Minneapolis 50730
Eyebeam, New York 51172
Fairfield Center for Contemporary Art, Sturgeon Bay 53347
Fayetteville Museum of Art, Fayetteville 48531
Federal Reserve Bank of Boston Collection, Boston 46929
Feldman Gallery and Project Space, Portland 52076
Fondo Del Sol, Washington 53870
Fort Smith Art Center, Fort Smith 48684
Four Sisters Gallery and the Robert Lynch Collection, Rocky Mount 52435
Franklin Furnace Archive, Brooklyn 47073
Fred Jones Jr. Museum of Art, Norman 51374
Fred Jones Jr. Museum of Art, Norman 51374
Frederick R. Weisman Art Foundation, Los Angeles 50224
Frederick R. Weisman Museum of Art, Malibu 50407
Freedman Gallery, Reading 52262
Fugitive Art Center, Nashville 50971
The Furlong Gallery, Menomonie 50591
Galapagos Art Space, Brooklyn 47074
Gallery 101, River Falls 52363
Gallery 210, Saint Louis 52577
The Gallery at UTA, Arlington 46351
Gallery II, Kalamazoo 49664
Gallery North, Setauket 53038
Gallery of Art, Cedar Falls 47292
Gallery of Contemporary Art, Colorado Springs 47701
Galveston Arts Center, Galveston 48844
Gardiner Art Gallery, Stillwater 53296
Georgetown College Gallery, Georgetown 48891
Georgia O'Keeffe Museum, Santa Fe 52874
Glastonbury Gild Arts Center, New Canaan 51036
Guild Hall Museum, East Hampton 48245
Guild of Creative Art, Shrewsbury 53092
Halsey Gallery, Charleston 47365
Hamline University Galleries, Saint Paul 52617
Handwerker Gallery of Art, Ithaca 49547
Haydon Gallery, Lincoln 50097
Headlands Center for the Arts, Sausalito 52936
Helena Gallery, Helena 49240
Holter Museum of Art, Helena 49241
HUB Robeson Galleries, University Park 53658
Hunter College Art Galleries, New York 51193
Huntington Beach Art Center, Huntington Beach 49438
Hyde Park Art Center, Chicago 47480
Indian Arts and Crafts Board, Washington 53882
Indianapolis Museum of Art, Indianapolis 49501
Institute of American Indian Arts Museum, Santa Fe 52877
Institute of Contemporary Art, Boston 46933
Institute of Contemporary Art, Portland 52064
Institute of Contemporary Art, Philadelphia 51844
Institute of Visual Arts, Milwaukee 50710
Intercultural Museum and Art Gallery, Baltimore 46569
Intermedia Arts Minnesota-Galleries, Minneapolis 50734
International Images, Sewickley 53044
Iroquois Indian Museum, Howes Cave 49408
Irvine Fine Arts Center, Irvine 49530
Isamu Noguchi Garden Museum, Long Island City 50182
Islip Art Museum, East Islip 48251
J. Wayne Stark University Center Galleries, College Station 47689
Jackson Art Center, Washington 53884
Jacksonville Museum of Modern Art, Jacksonville 49580
Jean Paul Slusser and Warren Robbins Galleries, Ann Arbor 46306
John A. Logan College Museum, Carterville 47266
John J. McDonough Museum of Art, Youngstown 54354
Kala Gallery, Berkeley 46756
Karpeles Manuscript Museum, Duluth 48196
Kemper Museum of Contemporary Art, Kansas City 49691
Kendall Gallery, Grand Rapids 49005
Kent and Vicki Logan Galleries and Capp Street Project, San Francisco 52779
Kent Campus Museum and Gallery, Jacksonville 49582
Knoxville Museum of Art, Knoxville 49819
The Lab, San Francisco 52780
Laband Art Gallery, Los Angeles 50242
LACE - Los Angeles Contemporary Exhibitions, Los Angeles 50243
Lancaster Museum of Art, Lancaster 49927
Lane Community College Art Gallery, Eugene 48432
Las Vegas Art Museum, Las Vegas 49971
Lawndale Art Center, Houston 49393
Leedy-Voulke's Art Center, Kansas City 49692
Lehman College Art Gallery, Bronx 47048
LeMoyne Art Foundation, Tallahassee 53425
Lewis, James and Nellie Stratton Gallery, Columbia 47716
Lighthouse Gallery, Tequesta 53478
List Visual Arts Center, Cambridge 47179
Littman Gallery, Portland 52079
Longwood Center for the Visual Arts, Farmville 48523
Los Angeles Municipal Art Gallery, Los Angeles 50245
Louisville Visual Art Museum, Louisville 50292
Lower Columbia College, Longview 50196
Luggage Store, San Francisco 52782
Macalester College Art Gallery, Saint Paul 52620
McLean County Arts Center, Bloomington 46860
Madison Museum of Contemporary Art, Madison 50393
Main Art Gallery, Fullerton 48806
Main Gallery of Henry Street Settlement, New York 51208
Marie Walsh Sharpe Art Foundation, New York 51209
Martin Museum of Art, Waco 53776
Mass Moca, North Adams 51378
Matrixarts, Sacramento 52509
Mattress Factory Museum, Pittsburgh 51953
Memorial Union Gallery, Fargo 48510
Mesa Contemporary Arts, Mesa 50607
Miami University Hiestand Galleries, Oxford 51655
Michael C. Rockefeller Arts Center Gallery, Fredonia 48778
Mildred Lane Kemper Art Museum, Saint Louis 52586
Mobius Gallery, Boston 46938
Modified Art Gallery, Phoenix 51905
MoMA, Museum of Modern Art, New York 51215
Montana Museum of Art and Culture, Missoula 50755
Montclair State University Art Galleries, Upper Montclair 53665
Monterey Museum of Art at La Mirada, Monterey 50804
Muscarelle Museum of Art, Williamsburg 54170
Muse Art Gallery, Philadelphia 51854
Museo Italoamericano, San Francisco 52785
Museum Necca and New England Center for Contemporary Art, Brooklyn 47065
Museum of Art, Fort Lauderdale 48638
Museum of Contemporary Art, Chicago 47492
Museum of Contemporary Art, Cleveland 47634
Museum of Contemporary Art, Fort Collins 48610
Museum of Contemporary Art, Miami 50634
Museum of Contemporary Art, Washington 53894
Museum of Contemporary Art Denver, Denver 48085
Museum of Contemporary Art Los Angeles, Los Angeles 50252
Museum of Contemporary Art San Diego - Downtown, San Diego 52740
Museum of Contemporary Art San Diego - La Jolla, La Jolla 49846
Museum of Contemporary Craft, Portland 52080
Museum of Contemporary Religious Art, Saint Louis 52590
Museum of Glass, Tacoma 53412
Museum of Neon Art, Los Angeles 50254
Museum of New Art, Pontiac 52021
NAB Gallery, Chicago 47498
National Association of Women Artists Gallery, New York 51234
National Cowboy and Western Heritage Museum, Oklahoma City 51519
Neuberger Museum of Art, Purchase 52201
Neue Galerie New York, New York 51238
New Harmony Gallery of Contemporary Art, New Harmony 51049
New Image Art, West Hollywood 54065
New Museum of Contemporary Art, New York 51240
Nexus Foundation for Today's Art, Philadelphia 51862
Nicolaysen Art Museum, Casper 47272
NIU Art Museum, DeKalb 48039
The North Shore Arts Association, Gloucester 48947
North View Gallery, Portland 52081
Northern Lights Art Gallery, Mayville 50541
Northwood Art Space, Columbus 47756
Norton Museum of Art, West Palm Beach 54073
Oasis Gallery, Marquette 50500
Oberlin College Gallery, Cleveland 47636
Oklahoma City Museum of Art, Oklahoma City 51521
Olin Fine Arts Center, Washington 53942
Olive DeLuce Art Gallery, Maryville 50524
Omaha Center for Contemporary Art, Omaha 51554
Omega Gallery, Jefferson City 49615
P.S. 1 Contemporary Art Center, Long Island City 50185
Pacific Design Center, West Hollywood 54066
Paint Creek Center for the Arts Exhibitions, Rochester 52387
Painted Bride Art Center Gallery, Philadelphia 51863
Paul Whitney Larson Art Gallery, Saint Paul 52627
Pence Gallery, Davis 47977
Phoenix Art Museum, Phoenix 51906
Picker Art Gallery, Hamilton 49137
Pittsburgh Center for the Arts, Pittsburgh 51956
Playhouse Gallery, Orleans 51598
Polvo Art Studio, Chicago 47508
Pro Arts at Oakland Art Gallery, Oakland 51472
Project Art-University of Iowa and Clinics, Iowa City 49520
Rawls Museum Arts, Courtland 47856
Reese Bullen Gallery, Arcata 46339
Reinberger Galleries, Cleveland 47637
Reynolda House, Winston-Salem 54249
Reynolds Gallery, Stockton 53307
Rice University Art Gallery, Houston 49401
Rio Hondo College Art Gallery, Whittier 54137
Rochester Art Center, Rochester 52390
Roland Gibson Gallery, Potsdam 52116
Rome Art and Community Center, Rome 52452
Rose Art Museum, Waltham 53810
Rothko Chapel, Houston 49403
Rotunda Gallery, Brooklyn 47081
Rubell Family Collection, Miami Beach 50643
Ruby Green Contemporary Arts Foundations, Nashville 50973
Russell Hill Rogers Gallery, San Antonio 52720
S.C.R.A.P. Gallery, Indio 49509
Saint Joseph's University Gallery, Philadelphia 51880
Salina Art Center, Salina 52674
Salvador Dalí Museum, Saint Petersburg 52642
San Francisco Museum of Modern Art, San Francisco 52795
San Jose Institute of Contemporary Art, San Jose 52808
San Jose Museum of Art, San Jose 52809
San Marco Gallery, San Rafael 52829
Santa Barbara Contemporary Arts Forum, Santa Barbara 52857
Santa Monica Museum of Art, Santa Monica 52901
Schneider Museum of Art, Ashland 46388
School of Art and Art History Galleries, Denver 48089
School of Art Gallery, Baton Rouge 46641
Scottsdale Museum of Contemporary Art, Scottsdale 52977
Sharadin Art Gallery, Kutztown 49830
Shemer Art Center and Museum, Phoenix 51910
Sidney Mishkin Gallery of Baruch College, New York 51255
Sierra Arts Foundation, Reno 52291
Site Santa Fe, Santa Fe 52890
Slocumb Galleries, Johnson City 49635
Snug Harbor Cultural Center, Staten Island 53267
Socrates Sculpture Park, Long Island City 50188
Sol Koffler Graduate Gallery, Providence 52180
Solomon R. Guggenheim Museum, New York 51258
Sonya Art, Brooklyn 47083
Soo Visual Arts Center, Minneapolis 50742
South Shore Art Center, Cohasset 47678
Southern Arts Federation, Atlanta 46453
Space 101 Gallery, Pittsburgh 51962
Spaces, Cleveland 47640
The Stanley and Gerald Rubin Center for the Visual Arts, El Paso 48329
Stedman Art Gallery, Camden 47196
Student Center Gallery, Wilmore 54207
Sun Gallery, Hayward 49230
Sushi Performance and Visual Art, San Diego 52756
Susquehanna Art Museum, Harrisburg 49182
T.F. Chen Cultural Center, New York 51267
Tampa Museum of Art, Tampa 53437
Tang Teaching Museum and Art Gallery, Saratoga Springs 52926
Tarrant County College Southeast The Art Corridor, Arlington 46354
TBA Exhibition Space, Chicago 47517
Temple Gallery, Philadelphia 51884
Terrace Gallery, Orlando 51595
Terrain Gallery, New York 51268
This Century Art Gallery, Williamsburg 54171
Totem Heritage Center, Ketchikan 49753
Tresure Room Gallery of the Interchurch Center, New York 51272
Trova Foundation, Saint Louis 52602
Tryart Gallery, Louisville 50299
Tufts University Art Gallery, Medford 50548
Tyndale Galleria, Upland 53661

UMKC Gallery of Art, Kansas City 49704
University Art Gallery, Fort Worth 48720
University Art Gallery, Jackson 49569
University Art Museum, Long Beach 50179
University Art Museum, Albany 46165
University at Buffalo Art Galleries, Buffalo 47126
University Galleries, Clarion 47583
University Gallery, Gainesville 48819
University Gallery, Amherst 46271
University of Arizona Museum of Art, Tucson 53600
University of Hawaii Art Gallery, Honolulu 49355
University of South Carolina at Spartanburg Art Gallery, Spartanburg 53198
University of Wyoming Art Museum, Laramie 49950
Urban Institute for Contemporary Arts, Grand Rapids 49006
Ursuline Hallway Gallery, San Antonio 52729
Vanderbilt Kennedy Center Arts Gallery, Nashville 50979
Vanderbilt University Fine Arts Gallery, Nashville 50980
Visual Arts Annex Gallery, San Antonio 52730
Visual Arts Gallery, New York 51276
Vox Populi Gallery, Philadelphia 51887
Walker Art Center, Minneapolis 50743
Walker's Point Center for the Arts, Milwaukee 50721
Walter and McBean Galleries, San Francisco 52799
Watkins Gallery, Washington 53933
Wayne Art Center, Wayne 54012
Weatherspoon Art Museum, Greensboro 49068
Weil Art Gallery, Corpus Christi 47823
Weingart Gallery, Los Angeles 50268
Western Kentucky University Gallery, Bowling Green 46976
Wexner Center for the Arts, Columbus 47761
Wheeler Gallery, Providence 52181
Whitney Museum of American Art at Altria, New York 51280
Wichita Falls Museum and Art Center, Wichita Falls 54154
William Whipple Art Gallery, Marshall 50507
Wisconsin Union Gallery, Madison 50397
Wiseman and Firehouse Galleries, Grants Pass 49018
Woman Made Gallery, Chicago 47523
Wood Street Galleries, Pittsburgh 51964
Woodstock Artists Gallery, Woodstock 54284
Wright State University Art Galleries, Dayton 47990
Yellowstone Art Museum, Billings 46808
Yerba Buena Center for the Arts, San Francisco 52801
Yuma Art Center and Galleries, Yuma 54365

Uruguay

Museo de Arte Contemporáneo, Montevideo 54396

Uzbekistan

Centr Sovremennogo Iskusstva, Toshkent 54468
Gelereja Iskusstva Uzbekistana, Toshkent 54470

Venezuela

Galería Municipal de Arte Moderno, Barcelona 54520
Museo Alejandro Otero, Caracas 54533
Museo de Arte Contemporáneo de Caracas Sofía Imber, Caracas 54541
Museo de Arte Contemporáneo Francisco Narváez, Porlamar 54593
Museo de Arte Moderno, Mérida 54587
Sala Ipostel, Caracas 54559

Zimbabwe

National Gallery Bulawayo, Bulawayo 54665

Art, Oriental

Armenia

Middle East Museum, Yerevan 00743
National Gallery of Armenia, Yerevan 00751
State Museum of Wood Carving, Yerevan 00756

Belgium

Hooy Kaye Museum, Bruxelles 03448

Brazil

Açude Museum, Rio de Janeiro 04680

Cuba

Galería Oriente, Santiago de Cuba 09304

France

Musée de l'Institut du Monde Arabe, Paris 14673
Musée du Quai Branly, Paris 14700
Musée Salies, Bagnères-de-Bigorre 11820

Germany

Archäologisches Museum der Universität Münster, Münster 20503

Iran

Abkar Museum, Tehrān 24655
Behzad Museum, Tehrān 24658
Farshchian Museum, Tehrān 24664
Muséyé Honarha-ye Melli, Tehrān 24685
Reza Abbasi Museum, Tehrān 24692
Sanati Museum, Kerman 24619

Italy

Museo d'Arte Siamese Stefano Gardu, Cagliari 25582

Japan

Hagi-shi Uragami Bijutsukan, Hagi 28469
Okuuchi Bijutsukan, Toyonaka 29531

Lebanon

Musée Marie Baz, Deir-el-Qamar 30169

Pakistan

Ahmed Pervez Art Gallery, Karachi 34011

Spain

Museo de Arte Oriental, Ávila 38919
Museo Etnográfico del Oriente de Asturias, Llanes 39471

Sweden

Ulricehamns Konst- och Östasiatiska Museum, Ulricehamn 40995

Switzerland

Antikenmuseum Basel und Sammlung Ludwig, Basel 41154

Tunisia

Musée d'Art Islamique de Rakkada, Rakkada 42646

Turkey

Eksi Şark Eserleri Müzesi, İstanbul 42817
Etnografya Müzesi, Ankara 42702
İnce Minare Müzesi, Konya 42885
İstanbul Sanat Müzesi, İstanbul 42827
Resim ve Heykel Müzesi, Ankara 42710
Türk İslam Eserleri ve Etnografya Müzesi, Erzurum 42778

United Arab Emirates

Sharjah Islamic Museum, Sharjah 43228

United Kingdom

Hartlepool Art Gallery, Hartlepool 44406
Hastings Museum and Art Gallery, Hastings 44415
Maidstone Museum and Bentlif Art Gallery, Maidstone 45073

United States of America

Dane G. Hansen Memorial Museum, Logan 50160
Vanderbilt University Fine Arts Gallery, Nashville 50980

Art, Russian

Armenia

Museum of Russian Arts, Yerevan 00749
National Gallery of Armenia, Yerevan 00751

Estonia

Eestimaa Vene Muuseum, Tallinn 10631

Kazakhstan

Ķazaķstan Respublikasynyņ Memlekettīk Ävīlchan Ķasteev Atyndaghy Öner Muzejī, Almaty 29683

Poland

Muzeum Nikifora, Krynica 34833

Russia

Aleksandrovskaja Sloboda - Gosudarstvennyj Istoriko- architekturnyj i Chudožestvennyj Muzej-zapovednik, Aleksandrov 36751
Aleksandrovskij Chudožestvennyj Muzej, Aleksandrov 36752
Archangelskij Gosudarstvennyj Muzej Zodčestva i Narodnogo Ikusstva - Malye Karely, Archangelsk 36765
Archangelskij Oblastnoj Muzej Izobrazitelnych Iskusstv, Archangelsk 36768
Architekturno-chudožestvennyj Ansambl byvšego Spaso-Jakovleskogo Monastyrja, Rostov 37598
Astrachanskaja Gosudarstvennaja Kartinnaja Galereja im. B.M. Kustodieva, Astrachan 36777
Baškirskij Respublikanskij Chudožestvennyj Muzej im. M.V. Nesterova, Ufa 37883
Belgorodskij Gosudarstvennyj Chudožestvennyj Muzej, Belgorod 36793
Belozerskij Muzej Narodnogo Dekorativno-prikladnogo Iskusstva, Belozersk 36799
Bolšoj Dvorec, Lomonosov 37106
Bratskij Chudožestvennyj Vystavočnyj Zal, Bratsk 36821
Brjanskij Oblastnoj Chudožestvennyj Muzej, Brjansk 36826
Čajkovskaja Kartinnaja Galereja, Čajkovskij 36833
Caricyno - Gosudarstvennyj Istoriko-architekturnyj, Chudožestvennyj i Landšaftnyj Muzej-Zapovednik, Moskva 37143
Čeljabinskaja Oblastnaja Kartinnaja Galereja, Čeljabinsk 36842
Chudožestvenno-memorialnyj Muzej K.S. Petrova-Vodkina, Chvalynsk 36864
Chudožestvenno-vystavočnyj Kompleks, Sarapul 37744
Chudožestvennyj Muzej, Staryj Oskol 37799
Chudožestvennyj Muzej, Valujki 37915
Chudožestvennyj Muzej im. A.N. Radiščeva, Saratov 37747
Chudožestvennyj Muzej im. M.S. Tuganova, Vladikavkaz 37928
Čitinskij Oblastnoj Chudožestvennyj Muzej, Čita 36868
Dalnevostočnyj Chudožestvennyj Muzej, Chabarovsk 36858
Dom-muzej Velimira Chlebnikova, Astrachan 36781
Dom-muzej N.N. Chochrjakova, Kirov 37035
Dom-Muzej I.N. Kramskogo, Ostrogožsk 37519
Dom-muzej B.I. Prorokova, Ivanovo 36944
Dom-Muzej I.E. Repina, Žigulevsk 37987
Dom-muzej Skulptora A.S. Golubkinoj, Zarajsk 37985
Dom-muzej V.M. Vasnecova, Moskva 37169
Ekaterininskij Dvorec - Gosudarstvennyj Muzej Carskoje Selo, Puškin 37579
Galereja Forum, Sankt-Peterburg 37638
Galereja Iskusstv Zuraba Cereteli, Moskva 37174
Galereja Sojuza Tvorčestvo, Moskva 37179
Gatčinskij Literaturno-memorialnyj Muzej-usadba P.E. Ščerbova, Gatčina 36919
Gorodskoj Muzej Iskusstvo Omska, Omsk 37497
Gosudarstvennaja Tretjakovskaja Galerja, Moskva 37184
Gosudarstvennaja Tretjakovskaja Galerja na Krymskom Valu, Moskva 37185
Gosudarstvennyj Chudožestvenno-architekturnyj Dvorcovo-parkovyj Muzej-zapovednik Oranienbaum, Lomonosov 37108
Gosudarstvennyj Chudožestvenno-architekturnyj Dvorcovo-parkovyj Muzej-zapovednik Pavlovsk, Pavlovsk 37528
Gosudarstvennyj Istoriko-architekturnyj i Ėtnografičeskij Muzej-zapovednik Kiži, Kiži 37050
Gosudarstvennyj Istoriko-architekturnyj i Prirodnyj Muzej-Zapovednik Monrepo, Vyborg 37981
Gosudarstvennyj Istoriko-chudožestvennyj Dvorcovo-parkovyj Muzej-zapovednik Gatčina, Gatčina 36920
Gosudarstvennyj Istoriko-kulturnyj Muzej-zapovednik "Moskovskij Kreml", Moskva 37194
Gosudarstvennyj Istoriko-literaturnyj Muzej-Zapovednik A.S. Puškina s usadbami Vjazemy i Zacharovo, Bolšie Vjazemy 36814
Gosudarstvennyj Memorialnyj Istoriko-chudožestvennyj i Prirodnyj Muzej-zapovednik V.D. Polenova, Strachovo 37808
Gosudarstvennyj Muzej Izobrazitelnych Iskusstv Respubliki Kalmykija, Ėlista 36911
Gosudarstvennyj Russkij Muzej, Sankt-Peterburg 37649
Gosudarstvennyj Russkij Muzej, Sankt-Peterburg 37650
Gosudarstvennyj Vladimiro-Suzdalskij Istoriko-architekturnyj i Chudožestvennyj Muzej-zapovednik, Vladimir 37934
Inženernyj (Michailovskij) Zamok, Sankt-Peterburg 37651
Irkutskij Oblastnoj Chudožestvennyj Muzej im. V.P. Sukačeva, Irkutsk 36932
Irkutskij Oblastnoj Kraevedčeskij Muzej, Irkutsk 36933
Istoriko-Architekturnyj, Chudožestvennyj i Archeologičeskij Muzej "Zarajskij Kreml", Zarajsk 37986
Istoriko-memorialnyj i Architekturno-chudožestvennyj Muzej, Tichvin 37842
Ivanovskij Oblastnoj Chudožestvennyj Muzej, Ivanovo 36948
Jaroslavskij Chudožestvennyj Muzej - Gubernatorskij Dom, Jaroslavl 36967
Jaroslavskij Istoriko-architekturnyj i Chudožestvennyj Muzej-zapovednik, Jaroslavl 36968
Jurinskij Istoriko-chudožestvennyj Muzej, Jurino 36982
Jusupovskij Dvorec, Sankt-Peterburg 37653
Kabardino-Balkarskij Gosudarstvennyj Muzej Izobrazitelnych Iskusstv, Nalčik 37438
Kamčatskij Oblastnoj Kraevedčeskij Muzej, Petropavlovsk-Kamčatskij 37553
Kartinnaja Galereja A.A. Plastova, Uljanovsk 37899
Kartinnaja Galereja Ministerstva Kultury Respubliki Adygeja, Majkop 37124
Kemerovskij Oblastnoj Muzej Izobrazitelnych Iskusstv, Kemerovo 37032
Kirillo-Belozerskij Istoriko-Architekturnyj i Chudožestvennyj Muzej-Zapovednik, Kirillov 37034
Kirovskij Oblastnoj Chudožestvennyj Muzej im. Viktora i Apollinarija Vasnecovych, Kirov 37038
Kiži - Gosudarstvennyj Istoriko-architekturnyj i Ėtnografičeskij Muzej-zapovednik, Petrozavodsk 37559
Kostromskoj Gosudarstvennyj Istoriko-architekturnyj Muzej-zapovednik, Kostroma 37061
Kostromskoj Gosudarstvennyj Obedinennyj Chudožestvennyj Muzej, Kostroma 37062
Krasnodarskij Chudožestvennyj Muzej im. F.A. Kovalenko, Krasnodar 37065
Krasnojarskij Chudožestvennyj Muzej im. V.I. Surikova, Krasnojarsk 37069
Krasnojarskij Gorodskoj Vystavočnyj Zal, Krasnojarsk 37070
Krasnokamskaja Kartinnaja Galereja, Krasnokamsk 37078
Kurganskij Oblastnoj Chudožestvennyj Muzej, Kurgan 37091
Kurskaja Kartinnaja Galereja im. A.A. Dejneki, Kursk 37093
Magnitogorskaja Kartinnaja Galereja, Magnitogorsk 37121
Malojaroslaveckaja Gorodskaja Kartinnaja Galereja, Malojaroslavec 37125
Memorialno-chudožestvennyj Dom-muzej Vladimira Aleksandroviča Serova, Ėmmaus 36914
Memorialnyj Chudožestvennyj Muzej Valentina Aleksandroviča Serova v Domotkanovo, Krasnaja Nov 37064
Memorialnyj Dom-muzej I.I. Levitana, Ples 37566
Memorialnyj Dom-muzej I.P. Požalostina, Solotča 37787
Memorialnyj Dom-muzej Viktora i Apolinarija Vasnecovych, Rjabovo 37592
Memorialnyj Muzej-kvartira A.M. Vasnecova, Moskva 37245
Memorialnyj Muzej-usadba Chudožnika N.A. Jarošenko, Kislovodsk 37048
Minusinskaja Chudožestvennaja Kartinnaja Galereja, Minusinsk 37135
Mitropoličji Palaty - Muzej Drevnerusskogo Iskusstva, Jaroslavl 36971
Mramornyj Dvorec, Sankt-Peterburg 37666
Municipalnaja Chudožestvennaja Galereja g. Kostromy, Kostroma 37063
Murmanskij Oblastnoj Chudožestvennyj Muzej, Murmansk 37433
Muzej Bratjev Tkačevych, Brjansk 36827
Muzej Drugoe Iskusstvo, Moskva 37272
Muzej V.N. Gorjaeva, Moskva 37278
Muzej im. B.M. Kustodieva, Ostrovskoe 37521
Muzej Izobrazitelnych Iskusstv, Komsomolsk-na-Amure 37054
Muzej Izobrazitelnych Iskusstv, Rostov-na-Donu 37600
Muzej Krylova, Tula 37868
Muzej Moskovskogo Chudožestvenno-promyšlennogo Universiteta im. S.G. Stroganova, Moskva 37335
Muzej Narodnogo Chudožnika Rossii A.I. Morozova, Ivanovo 36951
Muzej Narodnogo Derevjannogo Zodčestva Vitoslavlicy, Velikij Novgorod 37921
Muzej Pejzaža, Ples 37567
Muzej Sovremennogo Izobrazitelnogo Iskusstva im. A.A. Plastova, Uljanovsk 37908
Muzej Tropinina i Moskovskich Chudožnikov ego Vremeni, Moskva 37375
Muzej Tverskogo Byta, Tver 37879
Muzej-usadba V.E. Borsova-Musatova, Saratov 37750
Muzej Konstantina Vasiljeva, Moskva 37382
Muzej Vereščaginych, Čerepovec 36851

A

Muzej-zapovednik Abramcevo, Abramcevo 36747
Muzej-zapovednik Dmitrovskij Kreml, Dmitrov 36876
Nacionalnyj Chudožestvennyj Muzej Respubliki Sacha (Jakutija), Jakutsk 36964
Nižnetagilskij Municipalnyj Muzej Izobrazitelnych Iskusstv, Nižnij Tagil 37472
Novgorodskij Gosudarstvennyj Muzej-zapovednik, Velikij Novgorod 37922
Novokuzneckij Chudožestvennyj Muzej, Novokuzneck 37481
Oblastnoj Chudožestvennyj Muzej, Petropavlovsk-Kamčatskij 37555
Oblastnoj Muzejnyj Centr, Sankt-Peterburg 37716
Omskij Oblastnoj Muzej Izobrazitelnych Iskusstv im. M.A. Vrubelja, Omsk 37501
Orenburgskij Oblastnoj Muzej Izobrazitelnych Iskusstv, Orenburg 37515
Orlovskij Oblastnoj Muzej Izobrazitelnych Iskusstv, Orël 37512
Ostrogožskij Rajonnyj Istoriko-chudožestvennyj Muzej im. I.N. Kramskogo, Ostrogožsk 37520
Patrijaršie Palaty, Moskva 37396
Penzenskaja Kartinnaja Galerija im. K.A. Savickogo, Penza 37536
Pereslavl-Zalesskij Gosudarstvennyj Istoriko-architekturnyj i Chudožestvennyj Muzej-zapovednik, Pereslavl-Zalesskij 37542
Permskaja Gosudarstvennaja Chudožestvennaja Galereja, Perm 37550
Petergof - Gosudarstvennyj Muzej Zapovednik, Sankt-Peterburg 37718
Plesskij Gosudarstvennyj Istoriko-Architekturnyj i Chudožestvennyj Muzej-Zapovednik, Ples 37568
Primorskaja Gosudarstvennaja Kartinnaja Galereja, Vladivostok 37948
Pskovskij Gosudarstvennyj Istoričesko-architekturnyj i Chudožestnennyj Muzej-zapovednik, Pskov 37575
Reutovskaja Gorodskaja Kartinnaja Galereja, Reutov 37591
Rjazanskij Gosudarstvennyj Istoriko-architekturnyj Muzej-zapovednik, Rjazan 37595
Rjazanskij Oblastnoj Chudožestvennyj Muzej im. I.P. Požalostina, Rjazan 37596
Roslavlskij Istoriko-chudožstvennyj Muzej, Roslavl 37597
Rostovskij Kreml - Gosudarstvennyj Muzej-zapovednik, Rostov 37599
Rybinskij Gosudarstvennyj Istoriko-architekturnyj i Chudožestvennyj Muzej-zapovednik, Rybinsk 37606
Sajanskaja Kartinnaja Galereja, Sajansk 37607
Samarskij Oblastnoj Chudožestvennyj Muzej, Samara 37621
Semenovskij Gosudarstvennyj Kraevedčeskij Istoriko-chudožestvennyj Muzej, Semenov 37759
Sergievo-Posadskij Gosudarstvennyj Istoriko-chudožestvennyj Muzej Zapovednik, Sergiev Posad 37764
Serpuchovskij Istoriko-chudožestvennyj Muzej, Serpuchov 37765
Sočinskij Chudožestvennyj Muzej, Soči 37784
Staroladožskij Istoriko-Architekturnyj i Archeologičeskij Muzej-Zapovednik, Staraja Ladoga 37795
Starorusskij Kraevedčeskij Muzej, Staraja Russa 37797
Stavropolskij Muzej Izobrazitelnych Iskusstv, Stavropol 37806
Stroganovskij Dvorec, Sankt-Peterburg 37727
Svijažskij Architekturno-chudožestvennyj Muzej, Svijažsk 37812
Taganrogskaja Kartinnaja Galereja, Taganrog 37827
Tambovskaja Oblastnaja Kartinnaja Galereja, Tambov 37832
Tarusskaja Kartinnaja Galereja, Tarusa 37835
Tjumenskij Oblastnoj Muzej Izobrazitelnych Iskusstv, Tjumen 37853
Tomskij Oblastnoj Chudožestvennyij Muzej, Tomsk 37858
Troickij Sobor, Serpuchov 37766
Tverskoj Gosudarstvennyj Obedinennyj Muzej, Tver 37881
Ufimskaja Kartinnaja Galereja, Ufa 37890
Uljanovskij Oblastnoj Chudožestvennyj Muzej, Uljanovsk 37909
Ust-Ilimskaja Kartinnaja Galereja, Ust-Ilimsk 37912
Valaamskij Naučno-issledovatelskij Cerkovno-archeologičeskij i Prirodnyj Muzej-zapovednik, Sortavala 37790
Velikoustjugskij Gosudarstvennyj Istoriko-Architekturnyj i Chudožestvennyj Muzej-Zapovednik, Velikij Ustjug 37923
Vjatskaja Kunstkamera, Kirov 37043
Volgogradskij Muzej Izobrazitelnych Iskusstv, Volgograd 37958
Vologodskaja Oblastnaja Kartinnaja Galereja, Vologda 37963
Vologodskij Gosudarstvennyj Istoriko-Architekturnyj i Chudožestvennyj Muzej-Zapovednik, Vologda 37964
Voronežskij Chudožestvennyj Muzej im. I.N. Kramskogo, Voronež 37975
Vystavočnye Zaly Rossijskoj Akademii Chudožestv, Moskva 37415
Vystavočnyj Zal, Čerepovec 36852
Vystavočnyj Zal, Kirov 37044
Vystavočnyj Zal, Tula 37876
Vystavočnyj Zal, Volgograd 37960
Vystavočnyj Zal Biblioteki im. A.A. Bloka, Sankt-Peterburg 37734
Vystavocnyj Zal Doma Učenych, Novosibirsk 37489
Vystavočnyj Zal - Russkie Chudožniki v Samarskoj Gubernii, Žigulevsk 37989
Vystavočnyj Zal Tverskoj Oblastnoj Kartinnoj Galerei, Tver 37882
Zvenigorodskij Istoriko-architekturnyj i Chudožestvennyj Muzej, Zvenigorod 37991

South Africa

Marvol Museum, Kuilsriver 38679

Ukraine

Charkivs'kyj Chudožnij Muzej, Charkiv 43017
Kyïvs'kyj Muzej Rosijs'koho Mystectva, Kyïv 43085
Mis'ka Chudožna Galereja im. S.I. Vasyl'kivs'kogo, Charkiv 43020
Muzej Narodnogo Mystectva Slobožanščyny, Charkiv 43021
Mykolaïvs'kyj Oblasnyj Chudožnij Muzej im. Vereščagina, Mykolaïv 43150
Sums'kyj Chudožnij Muzej im. N. Onac'kogo, Sumy 43190

United States of America

Chazen Museum of Art, Madison 50388
Fayetteville Museum of Art, Fayetteville 48531
Museum of Russian Art, Minneapolis 50740

Astronomy

Armenia

Outer Space Museum, Garni 00699

Australia

Sydney Observatory, Sydney 01561

Belgium

Museum voor de Geschiedenis van de Wetenschappen, Gent 03636

Brazil

Museu de Astronomia e Ciências Afins, Rio de Janeiro 04724

Canada

Brydone Jack Observatory Museum, Fredericton 05829
H.R. MacMillan Space Centre, Vancouver 07213

Denmark

Ole Rømer Museet, Taastrup 10295

Ecuador

Museo Los Amantes de Sumpa, Guayaquil 10408

France

Bibliothèque-Musée de l'Observatoire, Paris 14573
Château d'Abbadie, Hendaye 13200

Georgia

Metsamor Museum, Metsamor 16348

Germany

Arche Nebra, Wangen 22323
Archenhold Sternwarte, Berlin 17037
Berliner Planetarium und Sternwarte Wilhelm Foerster, Berlin 17045
Heimat- und Palitzsch-Museum Prohlis, Dresden 17947
Kepler-Gedächtnishaus, Regensburg 21226
Kepler-Museum, Weil der Stadt 22380
Tobias-Mayer-Museum, Marbach am Neckar 20184
Universum Bremen, Bremen 17565
Zeiss-Großplanetarium Berlin, Berlin 17281

Greece

Cultural Center Hellenic Cosmos, Athinai 22906

Hungary

Eszterházy Károly Főiskola Csillagászati Múzeuma, Eger 23538

Italy

Museo Astronomico e Copernicano, Roma 27462
Museo Astronomico e Geofisico, Modena 26744
Museo della Specola, Bologna 25455
Museo dell'Osservatorio, Monte Porzio Catone 26801
Museo dell'Osservatorio Astronomico di Capodimonte, Napoli 26891

Japan

Katsushika-ku Kyoudo to Tenmon no Hakubutsukan, Tokyo 29409

Korea, Republic

Sungshin Women's University Museum, Seoul 29931

Kuwait

Educational Science Museum, Kuwait 29948

Latvia

Frīdriha Candera Kosmosa Muzejs, Rīga 30067
Frīdriha Candera Memoriālā Māja, Rīga 30068

Lithuania

Lietuvos Etnokosmologijos Muziejus, Kulionių 30291

Mexico

Museo Planetario, Garza García 30874

Netherlands

Explorion, Heerlen 32336
Nationaal Ruimtevaart Museum, Lelystad 32494

Poland

Muzeum Mikołaja Kopernika, Frombork 34629
Muzeum Mikołaja Kopernika, Toruń 35194

Russia

Istoriko-memorialnyj Muzejnyj Kompleks Bobriki, Donskij 36878

South Africa

Observatory Museum, Grahamstown 38632

Spain

CosmoCaixa Madrid, Alcobendas 38822
Museo de Astronomía y Geodesia, Madrid 39535
Museo de la Ciencia de Valladolid, Valladolid 40231
Museo de la Ciencia y el Cosmos, La Laguna de Tenerife 39440
Museo Interactivo de la Ciencia y Tecnología - Principia, Málaga 39601

Sweden

Observatoriemuseet, Stockholm 40923

Taiwan

Kaohsiung Astronomical Museum, Kaohsiung 42236
Taipei Astronomical Museum, Taipei 42426

United Kingdom

Foredown Tower, Portslade-by-Sea 45426
The Herschel Museum of Astronomy, Bath 43404
Mills Observatory Museum, Dundee 44048
National Maritime Museum, London 44939
Royal Observatory Greenwich, London 44977

United States of America

California Academy of Sciences, San Francisco 52768
Danville Science Center, Danville 47964
Memphis Pink Palace Museum, Memphis 50579
Meteor Crater, Winslow 54241
Mitchell House, Nantucket 50945
Museum of Ancient and Modern Art, Penn Valley 51762
William McKinley Presidential Library and Museum, Canton 47229

Uzbekistan

Ulug-Beg Yodgorlik Muzeyi, Samarqand 54464

Automobiles → Vehicles

Aviation

Argentina

Museo Aeronáutico, Baradero 00141
Museo de la Aviación Naval Argentina, Bahía Blanca 00134
Museo de Tecnología Aeronáutica y Espacial, Córdoba 00327
Museo Histórico Fuerte Barragan, Ensenada 00365
Museo Histórico Sanmartíniano, Mendoza 00456
Museo Nacional de Aeronáutica, Morón 00475
Museo Tecnológico Aeroespacial, Las Higueras 00425

Australia

Aero Nautical Museum-Armadillo, Mona Vale 01312
Australian Flying Museum, Wallon 01625
Aviation Heritage Museum of Western Australia, Bull Creek 00914
Central Australian Aviation Museum, Alice Springs 00798
Fighter World, Williamtown 01653
Flypast Museum of Australian Army Flying, Oakey 01386
Hinkler House Museum, Bundaberg 00921
Moorabbin Air Museum, Cheltenham 00966
QANTAS Founders Outback Museum, Longreach 01240
Queensland Air Museum, Caloundra 00936
Royal Australian Air Force Museum, Point Cook 01420
Royal Flying Doctor Service Visitors Centre, Edge Hill 01061
South Australian Aviation Museum, Port Adelaide 01424
West Torrens Railway, Signal Telegraph and Aviation Museum, Brooklyn Park 00907

Austria

Flugmuseum Aviaticum, Wiener Neustadt 03095
Museum Fahrzeug - Technik - Luftfahrt, Lauffen 02267
Österreichisches Luftfahrtmuseum, Feldkirchen bei Graz 01883
Red Bull Hangar-7, Salzburg 02623

Belgium

First Wing Historical Center (1WHC) The Golden Falcon, Beauvechain 03351
Musée International de la Base Aérienne de Chièvres, Chievres 03522
Musée Spitfire, Florennes 03605

Brazil

Museu Aeroespacial, Rio de Janeiro 04697

Bulgaria

Muzej na Aviacijata, Krumovo 05081

Canada

17 Wing Heritage Collection, Winnipeg 07351
Aero Space Museum of Calgary, Calgary 05513
Air Force Heritage Museum and Air Park, Winnipeg 07353
Alberta Aviation Museum, Edmonton 05740
Atlantic Canada Aviation Museum, Enfield 05777
Billy Bishop Heritage Museum, Owen Sound 06560
British Columbia Aviation Museum, Sidney 06970
Canada Aviation Museum, Ottawa 06533
Canada's Aviation Hall of Fame, Wetaskiwin 07311
Canadian Museum of Flight, Langley 06136
Canadian Warplane Heritage Museum, Mount Hope 06399
Commonwealth Air Training Plan Museum, Brandon 05462
Comox Air Force Museum, Lazo 06151
Gallery of Early Canadian Flight, Wiaton 07333
Great War Flying Museum, Cheltenham 05605
Greenwood Military Aviation Museum, Greenwood, Nova Scotia 05910
Musée de la Défense Aérienne, Alouette 05326
Nanton Lancaster Air Museum, Nanton 06415
North Atlantic Aviation Museum, Gander 05848
RCAF Memorial Museum, Trenton 07161
Shearwater Aviation Museum, Shearwater 06951
Toronto Aeropspace Museum, Toronto 07146
West Coast Museum of Flying, Sidney 06974
Western Canada Aviation Museum, Winnipeg 07391

China

Aeronautics Museum, Nanjing 08037
Astronautics Museum, Beijing 07506
China Aviation Museum, Beijing 07528

A

Czech Republic

Letecké Muzeum VHÚ Praha, Praha 09759

Denmark

Dansk Veteranflysamling, Skjern 10268

Ecuador

Museo Aeronautico y del Espacio, Quito 10426

Egypt

Airport Museum, Cairo 10486

Finland

Keski-Suomen Ilmailumuseo, Tikkakoski 11376
Luotsitupa, Uusikaupunki 11419
Suomen Ilmailumuseo, Vantaa 11448

France

Aérospatiale, Toulouse 15930
Conservatoire de l'Air et de l'Espace d'Aquitaine, Mérignac (Gironde) 14088
Musée Aérorétro, Albon 11541
Musée Airborne, Sainte-Mère-Eglise 15574
Musée d'Avions Montélimar, Montélimar 14198
Musée de la Papeterie-Canson et Montgolfier, Davézieux 12700
Musée de l'Aile Delta, Athis-Mons 11716
Musée de l'Air et de l'Espace, Le Bourget (Seine-Saint-Denis) 13564
Musée de l'Aviation, Perpignan 14790
Musée de l'Aviation, Warluis 16286
Musée de l'Aviation Légère de l'Armée de Terre, Dax 12702
Musée de l'Invasion Aéroportée, Bénouville 11939
Musée des Débuts de l'Aviation, Douzy 12786
Musée Voland, Cerny 12347

Germany

Aeronauticum - Deutsches Luftschiff- und Marinefliegermuseum Nordholz, Nordholz 20741
Albatros Flugmuseum, Stuttgart 21928
Ballonmuseum Gersthofen, Gersthofen 18619
BMW Museum, München 20428
Deutsche Raumfahrtausstellung e.V., Morgenröthe-Rautenkranz 20383
Deutsches Museum Flugwerft Schleißheim, Oberschleißheim 20831
Deutsches Segelflugmuseum mit Modellflug, Gersfeld 18613
Erfatal-Museum Hardheim, Hardheim 19011
Flughafen Modellschau, Hamburg 18929
Flugpionier-Gustav-Weißkopf-Museum, Leutershausen-Wiedersbach 19969
Hans-Grade-Museum, Borkheide 17480
Hermann-Köhl-Museum, Pfaffenhofen an der Roth 21021
Hermann-Oberth-Raumfahrt-Museum e.V., Feucht 18310
Hubschraubermuseum, Bückeburg 17622
Internationales Luftfahrtmuseum, Villingen-Schwenningen 22239
Luftfahrt-Museum Laatzen-Hannover e.V., Laatzen 19755
Luftfahrt- und Technik- Museumspark Merseburg, Merseburg 20293
Luftfahrtausstellung im Alten Straßenbahndepot, Brandenburg an der Havel 17510
Luftfahrttechnisches Museum, Rothenburg/Oberlausitz 21414
Museum für Luftfahrt und Technik, Wernigerode 22498
Otto Lilienthal Gedenkstätte, Gollenberg / OT Stölln 18705
Otto-Lilienthal-Museum, Anklam 16551
Zeppelin-Museum, Neu-Isenburg 20583
Zeppelin Museum Friedrichshafen, Friedrichshafen 18478
Zeppelin-Museum Meersburg, Meersburg 20253

Hungary

Ferihegy Repülogép Emlékpark, Budapest 23403
Hadtörténeti Aero Múzeum, Szolnok 23897

Ireland

Foynes Flying Boat Museum, Foynes 24834

Italy

Museo Aeronautico Caproni di Taliedo, Roma 27459
Museo Aeronautico Caproni-Taliedo, Vizzola Ticino 28302
Museo dell'Aeronautica G. Caproni, Trento 28058
Museo dell'Aviazione, Rimini 27408
Museo Storico-Aeronautica Militare, Bracciano 25525

Japan

Kakamigahara-shiritsu Uchuu Kagakukan, Kakamigahara 28636
Tomakomai-shi Kagagukan, Tomakomai 29511

Lithuania

Lietuvos Aviacijos Muziejus, Kaunas 30254
Stasio Girėno Gimtinė, Vytogala 30449
Stepono Dariaus Gimtinė, Dariaus 30220

Malaysia

Royal Malaysian Air Force Museum, Kuala Lumpur 30586

Malta

Malta Aviation Museum, Ta'Qali 30639

Netherlands

Avog's Crash Museum, Lievelde 32501
Ballon- en Luchtvaartmuseum Zep/Allon, Lelystad 32492
Dutch Dakota Association Exhibition, Schiphol 32759
Luchtvaart Museum Texel, De Cocksdorp 32046
Luchtvaart Museum Twenthe, Enschede 32210
Militaire Luchtvaart Museum, Soesterberg 32794
Museum Vliegbasis Deelen, Deelen 32054
Nationaal Luchtvaart-Themapark Aviodome, Schiphol 32760
Old Aircraft Museum, Arnemuiden 31920
Space Expo, Noordwijk, Zuid-Holland 32604
Traditiekamer Typhoon, Volkel 32903
Vliegend Museum Lelystad, Lelystad 32496
Vliegend Museum Seppe, Bosschenhoofd 31991
Westerwolds Crashmuseum, Oude Pekela 32651

New Zealand

Air Force Museum, Christchurch 33067
Ashburton Aviation Museum, Ashburton 33036
Fleet Air Arm Museum of New Zealand, Auckland 33048
Gore Airforce Museum, Gore 33112
Museum of Aviation, Paraparaumu 33178
New Zealand Fighter Pilots Museum, Wanaka 33232
Vintage Aircraft Museum, Masterton 33143

Pakistan

PAF Museum, Karachi 34016
Pakistan Air Force Museum Peshawar, Peshawar 34035

Peru

Museo Aeronáutico del Perú, Lima 34221
Museo de la Fuerza de Aviación Naval, Callao 34131

Philippines

Philippine Air Force Aerospace Museum, Pasay 34477

Poland

Muzeum Lotnictwa Polskiego, Kraków 34804

Portugal

Museu das Tropas Aerotransportadas, Praia do Ribatejo 35876
Museu do Ar, Alverca do Ribatejo 35426

Romania

Muzeul Aviației, București 36195

Russia

Centr Istorii Aviacionnych Dvigatelej im. N.D. Kuznecova, Samara 37611
Centralnyj Muzej Aviacii i Kosmonavtiki, Moskva 37145
Centralnyj Muzej Voenno-Vozdušnych Sil Rossijskoj Federacii, Monino 37139
Dom-muzej K.É. Ciolkovskogo, Kaluga 36994
Gosudarstvennyj Muzej Istorii Aviacii, Sankt-Peterburg 37645
Gosudarstvennyj Muzej Istorii Kosmonavtiki im. K.É. Ciolkovskogo, Kaluga 36996
Memorialnyj Dom-muzej Akademika S.P. Koroleva, Moskva 37229
Memorialnyj Dom-muzej N.E. Žukovskogo, Orechovo 37502
Memorialnyj Muzej Jurija Gagarina, Gagarin 36918
Memorialnyj Muzej Kosmonavtiki, Moskva 37242
Muzej Aviacii Severa, Archangelsk 36771
Muzej K. É. Ciolkovskogo, Aviacii i Kosmonavtiki, Kirov 37040
Naučno-memorialnyj Muzej N.E. Žukovskogo, Moskva 37388

Serbia

Muzej Jugoslovenskih Vazduhoplovstva, Beograd 38069

Slovakia

Múzeum Letectva Košice, Košice 38271

South Africa

Pioneers of Aviation Museum, Kimberley 38672
SAAF Museum Swartkop, Tshwane 38756
SAAF Museum Ysterplaat, Ysterplaat 38780

Spain

Museo Aeronáutico Torreón de Gando, Telde 40113
Museo de Aeronáutica y Astronáutica, Madrid 39531
Museo de Aviones Históricos en Vuelo, Madrid 39536
Museo Nacional de Aeropuertos y Transporte Aéreo, Málaga 39604

Sweden

Aeroseum, Göteborg 40542
Flyvapenmuseum, Malmslätt 40728
Jämtlands Flyg och Lottamuseum, Östersund 40795
Landskrona Museum, Landskrona 40669
Västerås Flyg Museum, Västerås 41031

Switzerland

Clin d'Ailes, Payerne 41718
Flieger Flab Museum, Dübendorf 41335
Fliegermuseum Altenrhein, Altenrhein 41107
Musée de la Naissance de l'Aviation Suisse, Payerne 41719

Taiwan

Chung Cheng Aviation Museum, Taoyuan 42452

Turkey

Havacılık Müzesi, İstanbul 42822

Ukraine

Muzej Kosmonavtiki im. S. Koroljova, Žytomyr 43212
Poltavs'kyj Muzej Aviacii ta Kosmonavtiki, Poltava 43172

United Arab Emirates

Al-Mahatah - Sharjah Aviation Museum, Sharjah 43223

United Kingdom

448th Bomb Group Memorial Museum, Seething 45631
Brenzett Aeronautical Museum Trust, Romney Marsh 45511
Carpetbagger Aviation Museum, Harrington 44398
City of Norwich Aviation Museum, Horsham Saint Faith 44490
Dumfries and Galloway Aviation Museum, Dumfries 44035
Fenland and West Norfolk Aviation Museum, West Walton 45973
Flambards Experience, Helston 44442
Flambards Victorian Village and Gardens, Helston 44443
De Havilland Aircraft Heritage Centre, Saint Albans 45542
The Helicopter Museum, Weston-super-Mare 45980
Imperial War Museum Duxford, Duxford 44075
Lashenden Air Warfare Museum, Headcorn 44432
Midland Air Museum, Baginton 43351
Military Aviation Museum, Tangmere 45824
Montrose Air Station Museum, Montrose 45172
Museum of Army Flying, Middle Wallop 45138
Museum of Berkshire Aviation, Woodley 46065
National Museum of Flight, North Berwick 45264
Newark Air Museum, Newark-on-Trent 45211
Norfolk and Suffolk Aviation Museum, Flixton 44234
North East Aircraft Museum, Sunderland 45801
Percy Pilcher Museum, Lutterworth 45049
Royal Air Force Air Defence Radar Museum, Norwich 45283
Royal Air Force Museum, London 44967
Royal Air Force Museum, Shifnal 45664
Royal Air Force Museum 201 Squadron, Saint Peter Port 45593
Royal Air Force Museum Reserve Collection, Stafford 45729
Shoreham Aircraft Museum, Shoreham 45667
Shoreham Airport Visitor Centre, Shoreham-by-Sea 45669
Shuttleworth Collection, Old Warden 45313
Solent Sky, Southampton 45705
Solway Aviation Museum, Crosby-on-Eden 43933
Spitfire and Hurricane Memorial Museum, Ramsgate 45459
Thameside Aviation Museum, East Tilbury 44095
Wellington Aviation Museum, Moreton-in-Marsh 45175
Yorkshire Air Museum, Elvington 44171

United States of America

Air Classic Museum of Aviation, Sugar Grove 53356
Air Force Armament Museum, Eglin Air Force Base 48304
Air Force Flight Test Center Museum, Edwards 48299
Air Mobility Command Museum, Dover, Air Force Base 48170
Air Power Park and Museum, Hampton 49149
Air Victory Museum, Lumberton 50320
Air Victory Museum, Medford 50549
Airborne and Special Operations Museum, Fayetteville 48529
Airmen Memorial Museum, Suitland 53359
Airpower Museum, Ottumwa 51630
Alaska Aviation Museum, Anchorage 46285
Allied Air Force Museum, Allentown 46226
Amelia Earhart Birthplace Museum, Atchison 46406
American Airlines C.R. Smith Museum, Fort Worth 48708
American Airpower Heritage Museum, Midland 50669
American Helicopter Museum Education Center, West Chester 54048
Anderson/Abruzzo Albuquerque International Balloon Museum, Albuquerque 46176
Arizona Wing Commemorative Air Force Museum, Mesa 50606
Arkansas Air Museum, Fayetteville 48527
Aviation Museum of Kentucky, Lexington 50046
Bergen Museum of Art and Science, Paramus 51696
Carolinas Aviation Museum, Charlotte 47385
Castle Air Museum, Atwater 46464
Charles A. Lindbergh House, Little Falls 50133
Clark County Museum, Henderson 49249
College Park Aviation Museum, College Park 47686
Combat Air Museum, Topeka 53533
Cradle of Aviation Museum, Garden City 48855
Crawford Auto-Aviation Museum, Cleveland 47628
EAA AirVenture Museum, Oshkosh 51613
Edward H. White II Memorial Museum, Brooks Air Force Base 47089
Empire State Aerosciences Museum, Glenville 48940
Evergreen Aviation Museum, McMinnville 50363
Exploration Station, Bourbonnais 46965
Fargo Air Museum, Fargo 48509
Frontiers of Flight Museum, Dallas 47933
Glenn H. Curtiss Museum, Hammondsport 49143
Grissom Air Museum, Peru 51788
Grumman Memorial Park, Calverton 47170
Hill Aerospace Museum, Hill Air Force Base 49282
Hiller Aviation Institute, San Carlos 52735
Howard W. Cannon Aviation Museum, Henderson 49250
International Sport Aviation Museum, Lakeland 49900
International Women's Air and Space Museum, Cleveland 47633
Intrepid Sea-Air-Space Museum, New York 51196
Iowa Aviation Museum, Greenfield 49053
Kalamazoo Aviation History Museum, Kalamazoo 49665
Kansas Aviation Museum, Wichita 54145
Kansas Cosmosphere and Space Center, Hutchinson 49454
Kirkpatrick Science and Air Space Museum at Omniplex, Oklahoma City 51517
Lone Star Flight Museum/Texas Aviation Hall of Fame, Galveston 48848
Louisiana State Museum, Patterson 51732
March Field Air Museum, Riverside 52371
May Natural History Museum and Museum of Space Exploration, Colorado Springs 47703
Memphis Belle B17 Flying Fortress, Memphis 50576
Mid-America Air Museum, Liberal 50076
Mid Atlantic Air Museum, Reading 52264
Mighty Eighth Air Force Museum, Pooler 52023
Minnesota Air National Guard Museum, Saint Paul 52621
Minter Field Air Museum, Shafter 53047
Mitchell Gallery of Flight, Milwaukee 50715
Museum of Aviation at Robins Air Force Base, Warner Robins 53819
Museum of Flight, Seattle 53004
Museum of Flying, Santa Monica 52899
NASA Glenn Research Center's Visitor Center, Cleveland 47635
National Air and Space Museum, Washington 53895
National Model Aviation Museum, Muncie 50905
National Museum of Naval Aviation, Pensacola 51767
National Soaring Museum, Elmira 48380
National Warplane Museum, Horseheads 49369

A

B

Neil Armstrong Air and Space Museum, Wapakoneta 53815
New England Air Museum of the Connecticut Aeronautical Historical Association, Windsor Locks 54230
New Mexico Wing-Commemorative Air Force, Hobbs 49305
North Carolina Museum of Life and Science, Durham 48223
Oakland Aviation Museum, Oakland 51469
Octave Chanute Aerospace Museum, Rantoul 52249
Oregon Air and Space Museum, Eugene 48437
Pacific Coast Air Museum, Santa Rosa 52907
Parker-O'Malley Air Museum, Ghent 48906
Pearson Air Museum, Vancouver 53707
Pima Air and Space Museum, Tucson 53595
Pioneer Air Museum, Fairbanks 48473
Piper Aviation Museum, Lock Haven 50155
Portal of the Folded Wings, North Hollywood 51395
Rhinebeck Aerodrome Museum, Rhinebeck 52301
San Diego Air Space Museum, San Diego 52744
San Francisco Airport Museums, San Francisco 52791
Santa Maria Museum of Flight, Santa Maria 52895
Science Museum of Virginia, Richmond 52338
Southern Museum of Flight, Birmingham 46825
Space Center Houston, Houston 49406
Strategic Air Space Museum, Ashland 46383
Tennessee Museum of Aviation, Sevierville 53040
Texas Air Museum, Rio Hondo 52354
Tillamook Naval Air Station Museum, Tillamook 53510
United States Air Force Museum, Wright-Patterson Air Force Base 54311
United States Space and Rocket Center, Huntsville 49445
Unites States Air Force Academy Visitor Center, USAF Academy 53672
Virgil I. Grissom State Memorial, Mitchell 50761
Virginia Air and Space Center, Hampton 49154
Virginia Aviation Museum, Richmond 52340
Wings of Freedom Museum, Huntington 49427
Wings of History Air Museum, San Martin 52821
Wright Brothers National Memorial, Kill Devil Hills 49775
Yankee Air Museum, Belleville 46702

Uruguay

Museo Aeronáutico, Montevideo 54393

Venezuela

Museo Aeronáutico Colonel Luis Hernan Paredes, Maracay 54583

Vietnam

Air Force Museum, Ha Noi 54606

Baking and Cookery

Australia

Laura and Charles Ferguson Museum, Mentone 01293
Miller Bakehouse Museum, Palmyra 01402
Winns Historic Bakehouse Museum, Coromandel Valley 01002

Austria

Bäckermuseum, Wien 02941
Brot- und Mühlen-Lehrmuseum, Gloggnitz 01938
Burgenländisches Brotmuseum, Bad Tatzmannsdorf 01765
Österreichisches Gebäckmuseum, Wels 02917

Belgium

Bakkerijmuseum, Groot-Bijgaarden 03664
Bakkerijmuseum Walter Plaetinck, Veurne 04109

Canada

Maison J.A. Vachon, Sainte-Marie-de-Beauce 06897
Musée de la Cuisine, Saint-Charles-de-Drummond 06809

France

Musée de la Boulangerie, Bonnieux 12058
Musée départemental de la Bresse - Domaine des Planons, Saint-Cyr-sur-Menton 15228
Musée des Ustensiles de Cuisine Anciens, Saint-Denis-la-Chevasse 15233
Musée du Pain, Uhlwiller 16016
Musée-Maison du Pain, Commeny 12594

Germany

Badisches Bäckereimuseum und Erstes Deutsches Zuckerbäckermuseum, Kraichtal 19695
Bäckerei- und Dorfgeschichtliches Museum, Bremervörde 17577
Bäckereimuseum Poppinga's alte Bäckerei, Krummhörn 19730
Bayerisches Brauerei- und Bäckereimuseum Kulmbach, Kulmbach 19746
Europäisches Brotmuseum, Ebergötzen 18037
Glasmuseum Passau, Passau 20989
Lebzelterei- und Wachsziehereimuseum, Pfaffenhofen an der Ilm 21019
Museum Alte Pfefferküchlerei, Weißenberg 22447
Museum der Brotkultur, Ulm 22166

Italy

Museo Agricolo Brunnenburg, Tirolo di Merano 27975

Netherlands

Bakkerijmuseum, Huizen 32421
Bakkerijmuseum De Grenswachter, Luijksgestel 32511
Bakkerijmuseum de Meelzolder, Almelo 31798
Gerrit Valk's Bakkerij- en Ijsmuseum, Hellendoorn 32354
Haardplatenmuseum Klarenbeek, Klarenbeek 32448
Museum Bakkerij Mendels, Middelstum 32555
Nederlands Bakkerijmuseum Het Warme Land, Hattem 32323
Oude Bakkerij, Medemblik 32544
Schilder- en Bakkerijmuseum 't Steenhuis, Niebert 32570

Russia

Muzej Tulskie Samovary, Tula 37870

Switzerland

La Maison du Blé et du Pain, Echallens 41339
Sammlung historischer Biscuitdosen, Basel 41181
Schweizerisches Gastronomie-Museum, Thun 41940

United Kingdom

Bakelite Museum, Williton 46017
Sally Lunn's Refreshment House and Kitchen Museum, Bath 43412
Times Past - Bygones of Yester Year, Blindley Heath 43544

United States of America

Harold M. Freund American Museum of Baking, Manhattan 50432

Balneology

Austria

Kurmuseum, Bad Tatzmannsdorf 01767

Croatia

Zavizajni Muzej, Varaždinske Toplice 08976

Czech Republic

Královská Mincovna v Jáchymove, Jáchymov 09549
Městské Muzeum, Františkovy Lázně 09502
Muzeum Zlatý Klíč, Karlovy Vary 09576
Vlastivědné Muzeum Jesenicka, Jeseník 09557

France

Musée du Site Balnéaire, Dinard 12751

Germany

Historische Kuranlagen und Goethe-Theater Bad Lauchstädt, Bad Lauchstädt 16774
Kur-und Stadtmuseum Bad Ems, Bad Ems 16725
Museum Bad Orb, Bad Orb 16802
Museum im Alten Rathaus, Bad Brückenau 16705
Museum Uranbergbau, Bad Schlema 16827
Quellen Museum, Bad Wildungen 16863
Römerthermen Zülpich - Museum der Badekultur, Zülpich 22826
Römische Bäder, Potsdam 21104
Sächsisches Bademuseum Bad Elster, Bad Elster 16722
Salzmuseum, Bad Sülze 16847
Sebastian-Kneipp-Museum, Bad Wörishofen 16874
Stadt- und Bädermuseum, Bad Salzuflen 16822
Stadt- und Bädermuseum Bad Doberan, Bad Doberan 16710

Slovakia

Balneologické Múzeum, Piešt'any 38327

Switzerland

Musée Bizarre und Mäusemuseum, Rieden bei Baden 41761
Museum Altes Bad Pfäfers, Bad Ragaz 41139

Banks and Banking

Argentina

Museo Histórico del Banco Provincia Dr. Arturo Jauretche, Buenos Aires 00239
Museo Histórico y Numismático del Banco de la Nación Argentina, Buenos Aires 00246

Australia

ANZ Banking Museum, Melbourne 01273
Bank of Victoria Museum, Yackandandah 01674
Banking and Currency Museum, Kadina 01181
Westpac Museum, Sydney 01566

Austria

Geldmuseum der Oesterreichischen Nationalbank, Wien 02991
Sparkassen-Museum der Erste Bank AG, Wien 03076

Canada

Bank of Montreal Museum, Montréal 06327
Doucet House, Rustico 06783
Old Bank of New Brunswick Museum, Riverside-Albert 06759

Denmark

Sparekassemuseet, Korsør 10169

Finland

Osuuspankkimuseo, Helsinki 10806

Germany

Museum der Staatlichen Münze Berlin, Berlin 17196
Sparkassen-Museum, Greding 18743
Sparkassenmuseum Leipzig, Leipzig 19935

Haiti

Unité Musée Numismatique, Port-au-Prince 23324

Hungary

Pénzügytörténeti Gyűjtemény, Budapest 23465

Iran

Money Museum, Tehrān 24682
Sekkeh Museum, Tehrān 24695

Ireland

The House of Lords, Dublin 24796

Israel

Bank Leumi Museum, Tel Aviv 25098

Japan

Former Hong Kong and Shanghai Bank Museum, Nagasaki 28972
Money Museum of the UFJ Bank, Nagoya 29002
Sakai House, Osaka 29154

Korea, Republic

Chohung Museum of Finance, Seoul 29866

Lebanon

Audi Fondation, Saïda 30172

Malta

Bank of Valletta Museum, Sliema 30638

Mexico

Museo Túenl Casa Parker, Angangueo 30716

Mozambique

Museu Nacional da Moeda, Maputo 31709

Panama

Casa Museo del Banco Nacional, Ciudad de Panamá 34063

Peru

Museo Banco Central de Reserva del Perú, Lima 34227
Museo Numismático del Banco Wiese, Lima 34267

Philippines

Museo ng Bangko Sentral ng Pilipinas, Manila 34445

Portugal

Museu do Banco de Portugal, Lisboa 35695
Museu do Papel Moeda, Porto 35853

Romania

Muzeul Băncii Naționale a României, București 36196

South Africa

ABSA Group Museum, Johannesburg 38645

Spain

Fundación Museo de las Ferias, Medina del Campo 39638
Museo Histórico BBVA, Bilbao 39046

Sweden

Gamla Bankgården, Vrigstad 41073

Turkey

T.C. Ziraat Bankası Müzesi, Ankara 42712

United Kingdom

Savings Banks Museum, Ruthwell 45533

United States of America

Farmers Bank, Petersburg 51798
Higgins Museum, Okoboji 51529
IMF Center Exhibit, Washington 53881
The Money Museum, Kansas City 49694
Museum of American Finance, New York 51222
Old State Bank, Vincennes 53750
Wells Fargo History Exhibits, Portland 52090
Wells Fargo History Museum, Minneapolis 50744
Wells Fargo History Museum, Phoenix 51911
Wells Fargo History Museum, San Diego 52761
Wells Fargo History Museum, San Francisco 52800

Uruguay

Museos del Gaucho y de la Moneda, Montevideo 54423

Venezuela

Museo de la Moneda, Caracas 54549

Baskets

Czech Republic

Expozice Čelákovického Košikářství, Čelákovice 09454

France

Musée de la Vannerie des Compagnons du Terroir, Villaines-les-Rochers 16161

Germany

Deutsches Korbmuseum, Michelau in Oberfranken 20317
Korbmacher-Museum, Malsfeld 20158
Stadtmuseum Lichtenfels, Lichtenfels 19981

Netherlands

Nationaal Vlechtmuseum, Noordwolde 32605

Poland

Muzeum Wikliny, Olkusz 34976

Russia

Muzej Istorii Deneg i Bankovskogo Dela, Moskva 37290

United States of America

Gatekeeper's Museum and Marion Steinbach Indian Basket Museum, Tahoe City 53418
Nantucket Lightship Basket Museum, Nantucket 50948

Bees

Austria

Bienen- und Wagnereimuseum, Tollet 02814
Imkereimuseum, Pöggstall 02509
Imkereimuseum, Pramet 02527
Imkermuseum, Sitzenberg-Reidling 02745
Museum Orth, Orth an der Donau 02473

Belgium

Bijenteeltmuseum Kalmthout, Kalmthout 03747
Musée de l'Abeille, Tilff 04072
Musée du Miel, Lobbes 03826
Petit Musée des Abeilles et du Miel-De la Fleur à la Table, Alle-sur-Semois 03264

Canada

Musée de l'Abeille, Château Richer 05600

France

Ateliers de l'Abeille, Chavignon 12492
Écomusée de l'Abeille, Grateloup 13117
Écomusée de l'Abeille, Le Faou 13581
Écomusée de l'Abeille et de la Flore Locale, Faugères 12907
Ecomusée la Cité des Abeilles, Saint-Faust 15264
Ecomusée l'Abeille du Haut-Doubs, Aubonne 11733
Ferme aux Abeilles, Saint-Brévin-les-Pins 15189
Mesnil aux Abeille, Beautheil 11907
Miellerie, Villeneuve-de-Berg 16208
Musée de la Chasse et de la Nature, Baurech 11869
Musée de l'Abeille, Le Tronchet 13658
Musée de l'Abeille et du Braconnage, Salbris 15593
Musée de l'Abeille Vivante, Vitré 16254
Musée de Miellerie, Saint-Saturnin-les-Avignon 15503
Musée des Arts et Tradition Apicoles, Fontan 12967
Musée du Miel et de l'Abeille, Gramont 13095
Musée l'Abeille Vivante, Le Faouët 13583
Musée Vivant de l'Abeille, Corbeny 12620
Musée Vivant de l'Apiculture Gâtinaise, Châteaurenard (Loiret) 12457

Germany

Bayerisches Bienenmuseum, Illertissen 19315
Bienenkundemuseum, Münstertal 20522
Deutsches Bienenmuseum, Weimar 22397
Hohberger Bienenmuseum, Hohberg 19230
Imker- und Heimatmuseum Wassertrüdingen, Wassertrüdingen 22349
Imkereigeschichtliche und bienenkundliche Sammlungen, Celle 17687
Westfälisch-Niederländisches Imkereimuseum, Gescher 18624
Zeidel-Museum, Feucht 18311

Italy

Museo Apistico Didattico, Bregnano 25527
Museo dell'Apicoltura, Renon 27396
Museo dell'Apicultura, Abbiategrasso 25139
Museo di Apicoltura Guido Fregonese, Magera di Oderzo 26564
Plattner Bienenhof Museum, Soprabolzano 27875

Lithuania

Servoninė Bitininkystės Muziejus, Stripeikiai 30375

Netherlands

Bijenteeltmuseum De Bankorf, Amen 31806
Imkerijmuseum Poppendamme, Grijpskerke 32273

Norway

Norges Birøkterlags Museet, Billingstad 33436

Poland

Muzeum im. Jana Dzierżona, Kluczbork 34748
Skansen i Muzeum Pszczelarstwa im prof. Ryszarda Kosteckiego, Swarzędz 35145
Skansen Kurpiowski im. Adama Chętnika, Nowogród 34953

Russia

Muzej Pčelovodstva, Moskva 37348

Serbia

Muzej Pčelarstva Porodice Živanović, Sremski Karlovci 38153

Slovakia

Múzeum Včelárstva na Slovensku, Král'ová pri Senci 38276

Slovenia

Čebelarski Muzej, Radovljica 38500

Spain

Colección Museográfica Casa de la Apicultura, Boal 39053
Museo de la Miel, Peñalver 39830

Switzerland

Bienenmuseum des VDRB, Alberswil 41098
Imkereimuseum, Grüningen 41457
Imkereimuseum Grüningen, Wetzikon 42010

Taiwan

Bee Family Museum, Yuanshan 42481

United Kingdom

IBRA Collection, Cardiff 43751

Bells

Austria

Glockenmuseum der Glockengiesserei, Innsbruck 02118

Belgium

Klokkenmuseum van de Sint-Martinusbasiliek, Halle 03668
Musée de la Cloche et du Carillon, Tellin 04052

China

Dazhong Siguzhong Museum, Beijing 07535

Croatia

Etnografski Muzej Istre, Pazin 08926
Muzej Grada Pazina, Pazin 08927

Czech Republic

Farní muzeum Zábřeh, Zábřeh 09923

France

Cloches de Corneville, Pont-Audemer 14878
Musée Campanaire, L'Isle-Jourdain 13782

Germany

Deutsches Glockenmuseum, Greifenstein 18746
Glocken-Museum, Siegen 21754
Glockenmuseum, Apolda 16565
Glockenmuseum, Laucha an der Unstrut 19843
Westfälisches Glockenmuseum, Gescher 18625

Netherlands

Klokkengieterijmuseum, Vries 32909
Nationaal Beiaardmuseum, Asten 31941

Portugal

Museu do Chocalho, Alcáçovas 35389

Russia

Valdajskij Kraevedčeskij Muzej, Valdaj 37914

United Kingdom

John Taylor Bellfoundry Museum, Loughborough 45033

Biology

Argentina

Museo de Ciencias Naturales, Salta 00556

Armenia

Academician Norayr Sisakyan House-Museum, Ashtarak 00691

Austria

Haus der Natur, Salzburg 02615

Belgium

Museum voor de Geschiedenis van de Wetenschappen, Gent 03636
Naturzentrum Ternell, Eupen 03591

Brazil

Centro de Biologia Marinha, São Sebastião 04912
Museu de Biodiversidade do Cerrado, Uberlândia 04941
Museu de Biologia Prof. Mello Leitão, Santa Teresa 04813

Canada

Armand Frappier Museum, Laval 06145
Collection Entomologique Ouellet-Robert, Outremont 06559
Environment Canada's Biosphère, Île-Saint-Hélène 06006
Musée du Fjord, Ville de Saguenay, La Baie 07263
Thomas McCulloch Museum, Halifax 05949

Chile

Museo Nacional de Historia Natural, Santiago de Chile 07462

China

Beijing Museum of Natural History, Beijing 07518
Sun Yat-Sen University Museums, Guangzhou 07736

Colombia

Museo del Mar Mundo Marino, Santa Marta 08752

Czech Republic

Lesnické, Myslivecké a Rybářské Muzeum, Hluboká nad Vltavou 09516
Vlastivědné Muzeum, Nymburk 09686

Denmark

Naturhistorisk Museum, Århus 09965

Egypt

Hydrobiological Museum, Alexandria 10472

France

Musée de la la Faune des Pyrénées, Nébias 14388
Petit Musée Minéraux et Faune de l'Alpe, Bourg-Saint-Maurice 12111

Germany

Alfred-Erhardt-Stiftung, Berlin 17030
Biohistoricum, Neuburg an der Donau 20598
Müritzeum, Waren (Müritz) 22335
Museum der Natur und Umwelt Cottbus, Cottbus 17733
Museum der Westlausitz Kamenz, Kamenz 19421
Museum für Natur und Umwelt, Lübeck 20079
Museum für Naturkunde, Dortmund 17926
Naturkunde-Museum, Bielefeld 17347
Waldmuseum, Surwold 21990

Hungary

Újpesti Lepkemúzeum Alapítvány, Budapest 23488

Iceland

Hið Íslenzka Reðasafn, Húsavík 24003

India

Biological Museum, Imphal 24221
Zoology and Botany Museum, Ernakulam 24169

Indonesia

Museum Biologi UGM, Yogyakarta 24585

Iran

Natural History Museum, Hamedan 24608

Iraq

Natural History Museum of the University of Basrah, Basrah 24718

Israel

Bloomfield Science Museum Jerusalem, Jerusalem 24981
Man and the Living World Museum, Ramat Gan 25084
Museum of Natural History, Jerusalem 25000

Italy

Casa di Lazzaro Spallanzani, Scandiano 27783
Centro Museo Universitario di Storia Naturale e della Strumentazione Scientifica, Modena 26739
Gabinetto Geologico e Botanico Piccinini, Pergola 27138
Museo del Castagno, Pescaglia 27161
Museo della Vita, Castelmagno 25743
Museo di Biologia Marina, Porto Cesareo 27302
Museo di Biologia Marina, Fano 26085

Jordan

Biology Museum, Amman 29646

Kazakhstan

Biologičeskij Muzej, Almaty 29675

Mexico

Museo de Anatomía, México 31067
Museo Estatal de Patología, Puebla 31286

Poland

Muzeum Przyrodnicze im. Prof. Andrzeja Myrchy, Białystok 34551
Muzeum Przyrodnicze Uniwersytetu Łódzkiego, Łódź 34882
Muzeum Przyrodniczo-Leśne, Białowieża 34544

Romania

Complex Muzeal de Ştiinţele Naturii, Galaţi 36378

Russia

Gosudarstvennyj Biologičeskij Muzej im. K.A. Timirjazeva, Moskva 37186
Muzei Permskogo Gosudarstvennogo Universiteta, Perm 37547
Muzej Anatomii, Moskva 37260
Muzej Instituta Biologii Morja, Vladivostok 37944
Muzej Priroda Morja i eë Ochrana, Vladivostok 37947
Novosibirskij Muzej Prirody, Novosibirsk 37487
Zoologičeskij Muzej, Vladivostok 37953

South Africa

Sea World, Durban 38612

Sweden

Biologiska Museet, Örebro 40784
Biologiska Museet, Södertälje 40862
Biologiska Museet, Stockholm 40887

Switzerland

Kulturama, Zürich 42080
Museum der Anthropologie, Zürich 42093

United Kingdom

Biological Sciences Collection, Birmingham 43498
Hawthorns Urban Wildlife Centre, Southampton 45703
Home of Chales Darwin - Down House, Downe 44018

United States of America

The Benjamin Banneker Museum, Baltimore 46556
Chesapeake Children's Museum, Annapolis 46317
New York State Museum, Albany 46161
Regional Science and Discovery Center, Horseheads 49370

Birds

Argentina

Museo Ornitológico Patagónico, El Bolsón 00358

Australia

Murtoa Water Tower Museum, Murtoa 01339

Austria

Falknereimuseum Burg Hohenwerten, Werfen 02922
Mühlviertler Vogelkundeweg, Gutau 02024
Naturmuseum, Neuberg an der Mürz 02425

See-Vogel-Museum Herbert Vargyas, Rust 02601
Vogelmuseum, Aigen im Mühlkreis 01701

Belgium

Musée d'Histoire Naturelle de Mons, Mons 03885
Natuurpunt Museum, Turnhout 04100

Brazil

Museu Anchieta Ciencias Naturais, Porto Alegre 04639
Museu da Fauna, Rio de Janeiro 04713
Museu de Ornitologia, Goiânia 04474

Bulgaria

Prirodonaučen Muzej, Burgas 05000
Prirodonaučen Muzej, Černi Osăm 05002

Canada

Abernethy Nature Heritage Museum, Abernethy 05307
McLurg Museum, Wilkie 07335
Musée du Séminaire de Saint-Hyacinthe, Saint-Hyacinthe 06822

Colombia

Museo de Historia Natural, Popayán 08726

Cuba

Museo de Historia Natural Felipe Poey, La Habana 09135

Czech Republic

Zámek Budišov, Budišov u Třebíče 09447

Egypt

Entomological Society Museum, Cairo 10493
Ornithology and Entomology Museum, Cairo 10512

Finland

K.E. Kivirikon Lintu- ja Nisäkäskokoelma, Helsinki 10789

France

Ecomusée de la Brenne, Le Blanc 13558
Musée d'Allard, Montbrison 14188
Musée d'Histoire Naturelle, Nantes 14366
Musée d'Histoire Naturelle de Toulon et du Var, Toulon (Var) 15925
Musée d'Histore Naturelle Victor-Brun, Montauban 14175
Musée Joseph Abeilhe, Marciac 13955
Muséum d'Histoire Naturelle, Le Havre 13599

Germany

Harzer Roller-Kanarien-Museum, Sankt Andreasberg 21486
Museum für Jagdtier- und Vogelkunde des Erzgebirges, Augustusburg 16641
Museum Heineanum, Halberstadt 18878
Natureum Niederelbe, Neuhaus/Balje 20617
Naturkundliches Museum in der Harmonie, Schweinfurt 21688
Naumann-Museum Köthen, Köthen 19667
Norddeutsches Vogelmuseum, Osterholz-Scharmbeck 20936
Schloßmuseum Hohenlohe-Schillingsfürst, Schillingsfürst 21554
Staatliches Naturhistorisches Museum, Braunschweig 17529
Vogelmuseum, Waging am See 22272
Waldmuseum Wassermühle, Wingst 22598

Italy

Museo Civico di Scienze Naturali E. Caffi, Bergamo 25402
Museo Civico Ornitologico, Lonato 26512
Museo Civico Ornitologico e di Scienze Naturali Luigi Scanagatta, Varenna 28160
Museo di Scienze Naturali, Brescia 25538
Museo Geopaleontologico Naturalistico Antropico e Ornitologico Brancaleoni, Piobbico 27227
Museo Ornitologico, Poppi 27296
Museo Ornitologico, San Gimignano 27640
Museo Ornitologico C. Beni, Stia 27904
Museo Ornitologico Ferrante Foschi, Forlì 26227
Museo Ornitologico Naturalistico S. Bambini, Pietralunga 27204
Museo Provinciale di Storia Naturale, Foggia 26212
Museo Provinciale di Storia Naturale, Livorno 26500

Japan

Abiko-shiritsu Kinchou Hakubutsukan, Abiko 28346
Kure-n Hakubutsukan, Izumi 28611
Shiretoko Museum, Shari 29254

Lithuania

Kauno Tado Ivanausko Zoologijos Muziejus, Kaunas 30252

Mexico

Museo de las Aves de México, Saltillo 31313
Museo de Pericos, Culiacán 30826

Netherlands

Natuurdiorama Holterberg, Holten 32401
Natuurhistorisch en Volkenkundig Museum, Oudenbosch 32653
Natuurhistorisch Museum Rotterdam, Rotterdam 32718
Natuurmuseum Dokkum, Dokkum 32139

New Zealand

Whanganui Regional Museum, Wanganui 33235

Panama

Museo de Historia Natural, Ciudad de Panamá 34070

Poland

Muzeum im. Jacka Malczewskiego w Radomiu, Radom 35073
Muzeum Lasu i Drewna przy Lesnym Zakładzie Doświadczalnym SGGW, Rogów 35083

Romania

Colecția Oologică Kalaber Ladislau, Reghin 36558

Russia

Zoologičeskij Muzej, Vladivostok 37953
Zoologičeskij Muzej im. N.M.Kulagina, Moskva 37430

Slovenia

Prirodoslovni Muzej Slovenije, Ljubljana 38461

Sudan

Sudan Natural History Museum, Khartoum 40393

Sweden

Fågelmuseet, Jönköping 40612
Ottenby Naturum, Degerhamn 40486

Switzerland

Museum Stemmler, Schaffhausen 41830
Naturkundliche Sammlung Oberhasli, Meiringen 41620

United Kingdom

Aberdeen University Natural History Museum, Aberdeen 43236
Booth Museum of Natural History, Brighton 43619
Dunrobin Castle Museum, Golspie 44325
Natural History Museum, Tring 45876
Woodland Heritage Museum, Westbury 45976

United States of America

Birds of Vermont Museum, Huntington 49433
Carrie Weedon Natural Science Museum, Galesville 48835
Charles R. Conner Museum, Pullman 52197
Connecticut Audubon Birdcraft Museum, Fairfield 48480
International Crane Foundation Museum, Baraboo 46605
John James Audubon Museum, Henderson 49247
McLane Center at Silk Farm, Concord 47774
Murie Museum, Kelly 49722
The Museum of Natural History, Stevens Point 53289
Museum of Texas Tech University, Lubbock 50312
The Natural Science Center of Greensboro, Greensboro 49067
Northwestern Oklahoma State University Museum, Alva 46247
Pember Museum of Natural History, Granville 49019
Roberson Museum and Science Center, Binghamton 46816
Southern Oregon University Museum of Vertebrate Natural History, Ashland 46389
Trailside Nature Center and Museum, Cincinnati 47566
Trailside Nature Museum, Cross River 47891
University of North Dakota Zoology Museum, Grand Forks 48990
World of Wings Pigeon Center Museum, Oklahoma City 51526

Venezuela

Colección Ornitológica W.H. Phelps, Caracas 54530
Museo Ornitológico, Aragua 54516

Boats and Shipping

Argentina

Buque Museo Corbeta Uruguay, Buenos Aires 00150
Buque-Museo Fragata Presidente Sarmiento, Buenos Aires 00151
Museo Histórico de la Prefectura Naval Argentina, Tigre 00645
Museo Marítimo de Ushuaia y Presidio, Ushuaia 00661
Museo Marítimo y Naval de la Patagonia Austral, Río Gallegos 00533
Museo Naval de la Nación, Tigre 00646

Australia

Aero Nautical Museum-Armadillo, Mona Vale 01312
Australian National Maritime Museum, Darling Harbour 01025
Axel Stenross Maritime Museum, Port Lincoln 01431
Customs House Nautical Museum, Robe 01462
Devonport Maritime Museum, Devonport 01035
Geelong Naval and Maritime Museum, North Geelong 01372
Gladstone Maritime Museum, Gladstone, Queensland 01110
Golden City Paddle Steamer Museum, Ballarat 00828
HMAS Castlemaine - Museum Ship, Williamstown 01651
Lady Denman Maritime Museum, Huskisson 01170
Maritime Centre Newcastle, Newcastle 01363
Maritime Museum, Low Head 01242
Maritime Museum of Tasmania, Hobart 01154
Maritime Museum of Townsville, Townsville 01597
Melbourne Maritime Museum, Southbank 01518
Naval and Maritime Museum, Ballina 00830
Paddle Steamer Industry Museum, Renmark 01457
Penneshaw Maritime and Folk Museum, Penneshaw 01409
Pine Islet Lighthouse, Mackay 01247
Port Albert Maritime Museum, Port Albert 01426
Port Fairy Historic Lifeboat Station, Port Fairy 01429
Port MacDonnell and District Maritime Museum, Port MacDonnell 01433
Port Victoria Maritime Museum, Port Victoria 01437
Portland Maritime Discovery Centre, Portland 01440
Queenscliffe Maritime Museum, Queenscliff 01449
Queensland Maritime Museum, South Brisbane 01513
South Australian Maritime Museum, Port Adelaide 01425
Sydney Heritage Fleet, Pyrmont 01445
Western Australian Maritime Museum, Fremantle 01095
Western Australian Museum Geraldton, Geraldton 01107
Whyalla Maritime Museum, Whyalla 01649
Williamstown Museum, Williamstown 01652

Austria

Heimatmuseum und Apothekenmuseum, Mauthausen 02372
Schifffahrtsmuseum der Marktgemeinde Spitz an der Donau, Spitz 02753
Schifffahrtsmuseum im Schloss Greinburg, Grein an der Donau 01988
Schiffleutmuseum, Stadl-Paura 02754

Belgium

Les Ascenceurs du Canal Historique du Centre, Houdeng-Goegnies 03718
Ascenseur Funiculaire pour Bateaux de Strépy-Thieu, Thieu 04061
Ecomusée de la Batellerie et de la Voie d'Eau Le Thudo, Thuin 04064
Museum Rijn- en Binnenvaart, Antwerpen 03298
Nationaal Scheepvaartmuseum, Antwerpen 03306
Plan Incliné de Ronquières, Ronquieres 03984
Scheepvaartmuseum, Bornem 03387

Belize

Belize Maritime Terminal and Museum, Belize City 04175

Bermuda

Bermuda Maritime Museum, Mangrove Bay 04194
Bermuda National Trust Museum, Saint George's 04196

Brazil

Espaço Cultural da Marinha, Rio de Janeiro 04690
Museu da Escola Naval, Rio de Janeiro 04711
Museu de Pesca, Santos 04821
Museu Naval e Oceanográfico, Rio de Janeiro 04757

Bulgaria

Archeologičeski Muzej, Burgas 04995
Nacionalen Voenno-morski Muzej, Varna 05244

Canada

Admiralty House Museum and Archives, Mount Pearl 06400
Archelaus Smith Museum, Cape Sable Island 05553
Britannia Shipyard, Richmond 06741
Campbell River Optical Maritime Museum, Campbell River 05540
Canadian Canoe Museum, Peterborough 06591
CFB Esquimalt Naval Museum and Military Museum, Victoria 07244
Churchill House and Marine Memorial Room Museum, Hantsport 05969
Collingwood Museum, Collingwood 05632
Cowichan Bay Maritime Centre, Cowichan Bay 05654
Green Park Shipbuilding Museum and Yeo House, Port Hill 06640
Gulf Museum, Port-aux-Basque 06630
Hawthorne Cottage, Brigus 05479
HMCS Sackville, Halifax 05937
LaHave Islands Marine Museum, LaHave 06129
Lawrence House Museum, Maitland 06220
Louisbourg Marine Museum, Louisbourg 06198
Marine Museum, Toronto 07121
Marine Museum of Manitoba, Selkirk 06943
Mariners' Park Museum, Milford 06283
Maritime Command Museum, Halifax 05939
Maritime Museum of British Columbia, Victoria 07254
Maritime Museum of the Atlantic, Halifax 05940
Musée de la Mer, Îles-de-la-Madeleine 06008
Musée de la Mer de Pointe-au-Père, Rimouski 06754
Musée du Vieux-Phare, Matane 06248
Musée Maritime de Charlevoix, Saint-Joseph-de-la-Rive 06860
Musée Maritime de Tadoussac, Tadoussac 07057
Musée Maritime du Québec, L'Islet 06175
Museum Ship Norgoma, Sault Sainte Marie 06931
Muskoka Lakes Museum, Port Carling 06631
Muskoka's Steamship Museum, Gravenhurst 05908
Nancy Island Historic Site, Wasaga Beach 07280
Naval Museum, Vancouver 07218
Owen Sound Marine and Rail Museum, Owen Sound 06562
Pointe-Noire - Parc Marin du Saguenay, Baie-Sainte-Catherine 05374
Port Colborne Historical and Marine Museum, Port Colborne 06633
Port de Québec au XIXe Siècle, Saint-André 06792
Quaco Museum and Archives, Saint Martins 06867
Samson V Maritime Museum, New Westminster 06438
Site-Historique du Banc-de-Paspébiac, Paspébiac 06573
Southern Newfoundland Seamen's Museum, Grand Bank 05887
Trinity Museum, Trinity 07167
Vancouver Maritime Museum, Vancouver 07228
Vancouver Naval Museum, Qualicum Beach 06675
Voyageur Heritage Centre, Mattawa 06251
Winterton Boat Building Museum, Winterton 07397
Yarmouth County Museum, Yarmouth 07410

China

Hong Kong Maritime Museum, Hong Kong 07797
Maritime Museum of Macau, Macau 08000
PLA Naval Museum, Qingdao 08107
Quanzhou Maritime Museum, Quanzhou 08130

Croatia

Pomorski i Povijesni Muzej Hrvatskog Primorja, Rijeka 08939
Pomorski Muzej, Zadar 08984
Pomorski Muzej, Dubrovnik 08872
Pomorski muzej Orebić, Orebić 08920
Zavičajni Muzej Baška, Baška 08842

Cuba

Museo Histórico Naval, Cienfuegos 09067

Denmark

Aabenraa Museum, Aabenraa 09952
Flaske-Peters-Samling, Ærøskøbing 09958
Fregatten Jylland, Ebeltoft 09996
Handels- og Søfartsmuseet paa Kronborg, Helsingør 10064
Ladbyskibsmuseet, Kerteminde 10117
Limfjordsmuseet, Løgstør 10179
Marstal Søfartsmuseum, Marstal 10187
Orlogsmuseet, København 10147
Skagen By- og Egnsmuseum, Skagen 10262
Søfarts og Fiskerimuseet, Læsø 10173
Søfartssamlingerne i Troense, Svendborg 10292
Strandingsmuseum St. George, Ulfborg 10302
Vikingeskibsmuseet, Roskilde 10250

Egypt

National Maritime Museum, Alexandria 10477
Solar Boats Museum, Giza 10531

Estonia

Eesti Meremuuseum, Tallinn 10626

Faroe Islands

Tvøroyrar Bygda- og Sjósavn, Tvøroyri 10700

Finland

Ålands Sjöfartsmuseum, Mariehamn 11099
Ehrensvärd-Museo, Helsinki 10773
Forum Marinum, Turku 11390
Jakobstads Museum, Jakobstad 10865
K.H. Renlundin Museo ja Kamerakokoelma, Kokkola 10972
Kråkö Båtbyggarmuseum, Kråkö 10994
Kymenlaakson maakuntamuseo, Kotka 10986
Loviisan Merenkulkumuseo, Loviisa 11083
Raahen Museo, Raahe 11248
Rauman Museo, Rauma 11260
Sjöfartsmuseet, Kristiinankaupunki 10997
Sjöhistoriska Institutet vid Åbo Akademi, Turku 11398
Suomen Joutsen, Turku 11399
Suomen Merimuseo, Kotka 10988
Uudenkaupungin Kulttuurihistoriallinen Museo, Uusikaupunki 11421

France

Aquarium-Musée de la Mer, Dinard 12750
Château-Musée de Dieppe, Dieppe 12721
Croiseur Colbert, Bordeaux 12063
Ecomusée de Saint-Nazaire, Saint-Nazaire (Loire-Atlantique) 15423
Espace Maritime et Portuaire des Docks Vauban, Le Havre 13594
Maison du Patrimoine Maritime, Camaret-sur-Mer 12251
Musée à Flot de l'Escorteur d'Escadre, Nantes 14361
Musée d'Aquitaine, Bordeaux 12066
Musée de la Batellerie, Briennon 12178
Musée de la Batellerie, Conflans-Sainte-Honorine 12612
Musée de la Batellerie de l'Ouest, Redon 14991
Musée de la Construction Navale, Noirmoutier-en-l'Ile 14466
Musée de la Marine, Saint-Brévin-les-Pins 15190
Musée de la Marine, Honfleur 13221
Musée de la Marine de la Seine, Caudebec-en-Caux 12323
Musée de la Marine de Loire, Châteauneuf-sur-Loire 12450
Musée de la Marine et de l'Économie de Marseille, Marseille 13997
Musée de la Mer, Paimpol 14553
Musée de la Mer OCEAM, Les Sables-d'Olonne 13714
Musée de la Vilaine Maritime, La Roche-Bernard 13412
Musée des Mariniers du Rhône, Serrières 15729
Musée des Terre-Neuvas et de la Pêche, Fécamp 12914
Musée d'Histoire de la Ville et d'Ethnographie du Pays Malouin, Saint-Malo (Ille-et-Vilaine) 15389
Musée du Canal de Berry, Audes-Reugny 11739
Musée du Château des Ducs de Bretagne, Nantes 14368
Musée Flottant-Architecture Navale, Audierne 11740
Musée Français des Phares et Balises, Ile-d'Ouessant 13243
Musée International du Long Cours Cap Hornier, Saint-Malo (Ille-et-Vilaine) 15390
Musée la Maison du Pêcheur et "Le Hope", Saint-Gilles-Croix-de-Vie 15299
Musée Maritime, Camaret-sur-Mer 12252
Musée Maritime, Carantec 12278
Musée Maritime, Le Mont-Saint-Michel 13624
Musée Maritime l'Esvale et Ostréicole, La Tremblade 13452
Musée National de la Marine, Brest 12164
Musée National de la Marine, Paris 14724
Musée National de la Marine, Port-Louis 14919
Musée National de la Marine, Rochefort (Charente-Maritime) 15072
Musée National de la Marine, Toulon (Var) 15929
Musée Naval de Nice, Nice 14437
Musée Naval de Sarcelles, Sarcelles 15629
Musée Naval Fort Balaguier, La Seyne-sur-Mer 13444
Naviscope Alsace, Strasbourg 15835
Le Port-Musée, Douarnenez 12776

French Polynesia

Musée de Marine, Faranui 16319

Germany

Altes Hafenamt, Dortmund 17913
Altonaer Museum für Kunst und Kulturgeschichte, Hamburg 18915
Atrium an der Schleuse, Brunsbüttel 17610
Binnenschifffahrts-Museum Oderberg, Oderberg 20858
Buddelschiff-Museum, Wedel 22352
Deutsches Schiffahrtsmuseum, Bremerhaven 17569
Deutsches Sielhafenmuseum in Carolinensiel, Wittmund 22625
Donau-Schiffahrts-Museum-Regensburg, Regensburg 21220
Elbschiffahrtsmuseum mit stadtgeschichtlicher Sammlung, Lauenburg 19849
Fehn- und Schiffahrtsmuseum, Rhauderfehn 21291
Flensburger Schiffahrtsmuseum, Flensburg 18327
Heimat- und Schiffahrtsmuseum, Heinsen 19113
Heimatmuseum, Zingst, Ostseebad 22807
Heimatmuseum im Alten Leuchtturm, Wangerooge, Nordseebad 22330
Historischer Hafen Berlin, Berlin 17138
Holzkanumuseum, Stuttgart 21944
Informationszentrum am Wasserstraßenkreuz Minden, Minden 20334
Internationales Maritimes Museum Hamburg, Hamburg 18937
Kapitän Tadsen Museum, Langeneß 19831
Kieler Stadt- und Schiffahrtsmuseum, Kiel 19530
Klepper Museum, Rosenheim 21377
Küstenmuseum, Juist 19408
Modellbaumuseum, Büdingen 17629
Museum der Deutschen Binnenschifffahrt, Duisburg 18022
Museum der Seefahrt, Geiselhöring 18557
Museum für Antike Schiffahrt des Römisch-Germanischen Zentralmuseums, Mainz 20148
Museumsfeuerschiff Amrumbank/Deutsche Bucht, Emden 18139
Museumslogger AE7 Stadt Emden, Emden 18140
Museumsschiff Cap San Diego, Hamburg 18955
Museumsschiff Mannheim, Mannheim 20172
Prignitz-Museum am Dom Havelberg, Havelberg 19054
Rhein-Museum Koblenz, Koblenz 19588
Rheinmuseum, Emmerich am Rhein 18152
Schiffahrts- und Schiffbaumuseum, Wörth am Main 22640
Schiffahrtsmuseum der Oldenburgischen Weserhäfen, Brake 17498
Schiffahrtsmuseum mit Nordseeaquarium Nordseeheilbad Langeoog, Langeoog 19836
Schiffahrtsmuseum Nordfriesland, Husum 19300
Schiffbau- und Schiffahrtsmuseum, Rostock 21395
Schiffermuseum, Zehdenick 22771
Schifffahrtsmuseum, Haren (Ems) 19013
SchifffahrtsMuseum im Schlossturm, Düsseldorf 18005
Schiffsmuseum Seitenradschleppdampfer "Württemberg", Magdeburg 20129
Stadtgeschichtliches Museum, Tangermünde 22003
Technikmuseum U-Boot "Wilhelm Bauer", Bremerhaven 17574
Torfschiffwerftmuseum, Worpswede 22697
Uns lütt Schiffsmuseum, Cuxhaven 17758
Wrackmuseum, Cuxhaven 17759
Wrackmuseum Cuxhaven, Cuxhaven 17760

Greece

Aegean Maritime Museum, Mykonos 23102
Maritime Museum of Andros, Andros 22882
Maritime Museum of Crete, Chania 22964
Maritime Museum of Thera (Santorini), Oia 23126
Maritime Tradition Museum, Piraeus 23144
Museum of Nautical Art, Symi 23201
Museum of the Merchant Marine, Piraeus 23146
Nautical Museum of Galaxidi, Galaxidi 23005
Nautiko Mouseio Chiou, Chios 22970
Nautikon Mouseiontis Ellados, Piraeus 23147
Oinoussian Maritime Museum, Oinoussai 23127

Hungary

Hajózástörténeti Múzeum, Zebegény 23976

Iceland

Sjóminja- og Vélsmiðjusafn, Reykjavík 24035
Sjóminjasafn Austurlands, Eskifjörður 23994
Sjóminjasafn Islands, Reykjavík 24036
Sjóminjasafnid á Eyrarbakka, Eyrarbakki 23996

Indonesia

Museum Tni A.L. Loka Jala Carana, Surabaya 24567

Ireland

1796 Bantry French Armada Exhibition Centre, Bantry 24742
Arklow Maritime Museum, Arklow 24729
Cape Clear Heritage Centre, Skibbereen 24897
Kilmore Quay Maritime Museum, Kilmore Quay 24854
Kilmore Quay Maritime Museum, Wexford 24913
National Maritime Museum of Ireland, Dún Laoghaire 24826
Waterways Visitor Centre, Dublin 24821

Italy

Civico Museo del Mare, Trieste 28078
Galata - Museo del Mare e della Navigazione, Genova 26293
Museo Civico Marinaro Gio-Bono Ferrari, Camogli 25627
Museo dei Navigli, Milano 26696
Museo della Barca Lariana, Pianello del Lario 27193
Museo della Cultura Marinara, Tortoreto Lido 28045
Museo della Marineria dell'Alto e Medio Adriatico, Cesenatico 25839
Museo delle Navi e delle Antiche Carte Geografiche, Bologna 25457
Museo delle Navi Romane, Fiumicino 26206
Museo delle Navi Romane, Nemi 26918
Museo dello Sbarco Alleato, Nettuno 26922
Museo Navale, Napoli 26907
Museo Navale di Pegli, Genova 26317
Museo Navale Didattico, Milano 26712
Museo Navale Internazionale del Ponente Ligure, Imperia 26406
Museo Navale O. Zibetti, Caravaggio 25674
Museo Storico della Brigata Sassari, Sassari 27756
Museo Storico Navale, Venezia 28203
Museo Tecnico Navale, La Spezia 26436
Piccolo Museo Navale, Laives 26440
Raccolta di Ingegneria Navale, Genova 26325

Jamaica

Fort Charles Maritime Museum, Kingston 28325

Japan

Biwa-ko Sosui Kinenkan, Kyoto 28803
Fune no Kagakukan, Tokyo 29386
Kinenkan Mikasa, Yokosuka 29620
Kitamae-senshu-no-yakata Ukon-ke, Echizen 28414
Kobe Maritime Museum, Kobe 28732
Marine Transportation Museum, Ogi 29091
Naniwa-no Umino Jikukan, Osaka 29137
Shimonoseki-shiritsu Kaikyokan, Shimonoseki 29270
Tokai Daigaku Kaiyo Kagaku Hakubutsukan, Shizuoka 29291
Yokohama Maritime Museum and Nippon-Maru Memorial Park, Yokohama 29614

Korea, Republic

Korea Maritime University Maritime Museum, Busan 29779
National Maritime Museum, Mokpo 29853
Naval Academy Museum, Jinhae 29843

Malaysia

Maritime Archaeology Museum, Melaka 30596
Muzium Samudera, Melaka 30601
Muzium Tentera Laut Diraja Malaysia, Melaka 30604

Malta

Malta Maritime Museum, Vittoriosa 30657
Maritime Museum, Nadur, Gozo 30633
Wickman Maritime Collection, Zabbar 30661

Mexico

Museo Histórico Naval Central, México 31122
Museo Marino de Tecolutla, Tecolutla 31414
Museo Pablo Bush Romero, Puerto Aventura 31294

Monaco

Musée Naval, Monaco 31612

Montenegro

Pomorski Muzej u Kotoru, Kotor 31647

Netherlands

Bataviawerf-Nationaal Scheephistorisch Centrum, Lelystad 32493
Droogdok Jan Blanken, Hellevoetsluis 32359
Fries Scheepvaart Museum, Sneek 32789
Gemeentemuseum Het Hannemahuis, Harlingen 32316
Het Havenmuseum, Rotterdam 32705
Maas- en Scheepvaartmuseum, Maasbracht 32519
Marinemuseum, Den Helder 32110
Maritiem en Jutters Museum, Oudeschild 32657
Maritiem Museum Rotterdam, Rotterdam 32712
Maritime Museum Zierikzee, Zierikzee 33003
Minischeepvaartmuseum, Ouwerkerk 32660
Museum de Verzamelaar, Zuidhorn 33012
Museumhaven Zeeland, Zierikzee 33004
Museumwerf 't Kromhout, Amsterdam 31870
Nationaal Reddingmuseum Dorus Rijkers, Den Helder 32112
Nationaal Sleepvaart Museum, Maassluis 32523
Noordelijk Scheepvaartmuseum, Groningen 32287
Open Haven Museum, Amsterdam 31877
Oudheidkamer Nederlandse Kaap-Hoornvaarders, Hoorn 32411
Prins Hendrik de Zeevaarder Museum, Egmond aan Zee 32181
Scheepswerf De Delft, Rotterdam 32726
Het Scheepvaartmuseum, Amsterdam 31885
Scheepvaartmuseum, Gasselternijveen 32235
Woonbootmuseum, Amsterdam 31903
Zeews Maritiem MuZEEum, Vlissingen 32900
Zompenmuseum, Enter 32217

New Zealand

Bluff Maritime Museum, Bluff 33064
Museum of Wellington, City and Sea, Wellington 33251
New Zealand National Maritime Museum, Auckland 33057
Whanganui Riverboat Centre Museum, Wanganui 33236
Whangaroa County Museum, Kaeo 33131

Norway

Ålesunds Museum, Ålesund 33380
Berg-Kragerø Museum, Kragerø 33607
Bergens Sjøfartsmuseum, Bergen 33423
Frammuseet, Oslo 33725
Hurtigrutemuseet, Stokmarknes 33883
Ishavsmuseet Aarvak, Brandal 33452
Kystmuseet Hvaler, Vesterøy 33974
Kystmuseet i Sogn og Fjordane, Florø 33499
Larvik Sjøfartsmuseum, Larvik 33629
Lillesand By-og Sjøfartsmuseum, Lillesand 33640
Marinemuseet, Horten 33563
Mellemværftet, Kristiansund 33614
Museet Kystens Arv, Stadsbygd 33860
Nordnorsk Fartøyvernsenter og Båtmuseum, Gratangen 33517
Norsk Sjøfartsmuseum, Oslo 33742
Sandefjord Sjøfartsmuseum, Sandefjord 33814
Sjøfartsmuseet i Porsgrunn, Porsgrunn 33771
Skaalurensamlinga - Skipsbyggingsmuseet i Rosendal, Rosendal 33800
Stavanger Sjøfartsmuseum, Stavanger 33876
Sunnmøre Museum, Ålesund 33381
Svelvik Museum, Svelvik 33895
Vikingskipshuset, Oslo 33759

Peru

Casa Museo Miguel Grau, Lima 34209
Museo de la Fuerza de Submarinos, Callao 34132
Museo Naval, Puno 34299
Museo Naval de Ilo, Ilo 34190
Museo Naval del Perú, Callao 34135

Philippines

Philippine Navy Museum, Cavite City 34363

Poland

Centralne Muzeum Morskie, Gdańsk 34631
Muzeum Marynarki Wojennej, Gdynia 34656
Muzeum Okręt Błyskawica, Gdynia 34659
Muzeum Żuraw, Gdańsk 34649

Portugal

Ecoteca-Museu João Lúcio, Olhão 35789
Museu de Marinha, Lisboa 35693
Museu do Rio, Alcoutim 35402
Museu dos Rios e das Artes Marítimas, Constância 35550
Museu Marítima Almirante Ramalho Ortigão, Faro 35598
Museu Marítimo, Funchal 35617
Museu Marítimo e Regional de Ílhavo, Ílhavo 35648
Museu Sub-Tenente António Bernardino Apolónio Piteira, Arraiolos 35446

Núcleo Museológico da Casa do Infante, Porto ... 35862

Romania

Muzeul Mării, Constanța ... 36307
Muzeul Portului Constanța, Constanța ... 36309

Russia

Centralnyj Voenno-morskoj Muzej, Sankt-Peterburg ... 37632
Gosudarstvennyj Severnyj Morskoj Muzej, Archangelsk ... 36770
Istoričeskaja Usadba Botik Petra I, Veskovo .. 37926
Kamčatskij Voenno-istoričeskij Muzej, Petropavlovsk-Kamčatskij ... 37554
Memorialnaja Gvardejskaja Krasnoznamënnaja Podvodnaja Lodka S-56 1939 g., Vladivostok .. 37939
Morskoj Muzej, Petrozavodsk ... 37560
Muzej Istorii Černomorskogo Flota, Moskva .. 37289
Muzej Istorii Morskogo Universiteta im. G.I. Nevelskogo, Vladivostok ... 37945
Muzej-ledokol Angara, Irkutsk ... 36938
Muzej-ledokol Bajkal, Angarsk ... 36762
Muzej Morskogo Flota, Moskva ... 37333
Muzej Morskogo Flota Dalnevostočnogo Morskogo Parochodstva, Vladivostok ... 37946
Muzej Rečnogo Flota, Nižnij Novgorod ... 37465
Muzej Voenno-morskogo Flota, Moskva ... 37384
Storoževoj Korabl 'Krasnyj Vympel' 1911 g., Vladivostok ... 37950
Voenno-istoričeskij Muzej Tichookeanskogo Flota, Vladivostok ... 37952
Voenno-morskoj Muzej Severnogo Flota, Murmansk ... 37436

Serbia

Muzej Rečnog Brodarstva, Beograd ... 38075

Singapore

Maritime Museum, Singapore ... 38176

Slovenia

Pomorski Muzej Sergej Mašera, Piran ... 38492

South Africa

Bartolomeu Dias Museum Complex, Mossel Bay ... 38691
Maritime Centre, Cape Town ... 38586
Old Harbour Museum, Hermanus ... 38641
Port Natal Maritime Museum, Durban ... 38611
Shipwreck Museum, Bredasdorp ... 38573
Simon's Town Museum, Simon's Town ... 38730

Spain

Archivo-Museo Don Alvaro de Bazán, Viso del Marqués ... 40316
Museo del Mar, Santa Pola ... 40006
Museo Marítimo de Asturias, Gozón ... 39352
Museo Marítimo del Cantábrico, Santander ... 40011
Museo Marítimo Torre del Oro, Sevilla ... 40074
Museo Massó, Bueu ... 39060
Museo Naval, Madrid ... 39574
Museu de Curiositats Marineres, Vilanova i la Geltrú ... 40284
Museu del Port, Tarragona ... 40102
Museu Marítim de Barcelona, Barcelona ... 38998
Untzi Museoa - Museo Naval, Donostia-San Sebastián ... 39255

Sri Lanka

National Maritime Museum, Galle ... 40378

Sweden

Dalälvarnas Flottningsmuseum, Sandviken ... 40831
Dyvelstens Flottningsmuseum, Forshaga ... 40525
Eldnäsets Flottningsmuseum, Ånge ... 40425
Holmöns Båtmuseum, Holmön ... 40600
Kalmar Sjöfartsmuseum, Kalmar ... 40625
Marinmuseum, Karlskrona ... 40635
Maritiman, Göteborg ... 40551
Motorbåtmuseet Museihuset, Vreta Kloster ... 41072
Oskarshamns Sjöfartsmuseum, Oskarshamn . 40803
Råå Museum för Fiske och Sjöfart, Råå ... 40811
Roslagens Sjöfartsmuseum, Väddö ... 41023
Sjöfartmuseet, Oskarshamn ... 40804
Sjöfartsmuseet Akvariet, Göteborg ... 40556
Sjöhistoriska Museet, Stockholm ... 40933
Skärgårdsmuseet, Stavsnäs ... 40880
Skärhamns Sjöfartsmuseum, Skärhamn ... 40844
Sveriges Sjömanshusmuseum, Uddevalla ... 40993
Teknikens och Sjöfartens Hus, Malmö ... 40727
Torekovs Sjöfartsmuseum, Torekov ... 40975
Vasamuseet, Stockholm ... 40949
Vikens Sjöfartsmuseum, Viken ... 41062

Switzerland

Musée des Traditions et des Barques du Léman, Saint-Gingolph ... 41790
Schifffahrtsmuseum Heimethus, Männedorf .. 41601

Thailand

Royal Barges National Museum, Bangkok ... 42558

Ukraine

Muzej Sudnobuduvannja ta Flotu, Mykolaïv ... 43149

United Kingdom

Aberdeen Maritime Museum, Aberdeen ... 43235
Arctic Corsair, Kingston-upon-Hull ... 44592
Barmouth Sailors' Institute, Barmouth ... 43380
Bridlington Harbour Museum, Bridlington ... 43610
Buckler's Hard Village Maritime Museum, Beaulieu ... 43423
Caernarfon Maritime Museum, Caernarfon ... 43708
Canal Museum, Llangollen ... 44768
Canal Museum, Stoke Bruerne ... 45758
Castle Cornet Military and Maritime Museums, Saint Peter Port ... 45589
Clipper Ship Cutty Sark, London ... 44826
Clydebank Museum, Clydebank ... 43857
Clydebuilt, Glasgow ... 44277
Cowes Maritime Museum, Cowes ... 43907
Deal Maritime and Local History Museum, Deal ... 43959
Denny Ship Model Experiment Tank, Dumbarton ... 44033
Dock Museum, Barrow-in-Furness ... 43388
Dolphin Sailing Barge Museum, Sittingbourne 45682
Fort Grey and Shipwreck Museum, Saint Pierre du Bois ... 45595
Foxton Canal Museum, Foxton ... 44250
H.M. Frigate Unicorn, Dundee ... 44046
Harbour Museum, Derry ... 43973
Hartland Quay Museum, Bideford ... 43478
Hartland Quay Museum, Hartland ... 44405
Harwich Maritime Museum and Harwich Lifeboat Museum, Harwich ... 44410
Historic Dockyard Chatham, Chatham ... 43795
HMS Belfast, London ... 44875
HMS Ganges Museum, Ipswich ... 44529
HMS Trincomalee, Hartlepool ... 44407
HMS Victory, Portsmouth ... 45432
HMS Warrior 1860, Portsmouth ... 45433
Holyhead Maritime Museum, Holyhead ... 44481
House of Manannan, Peel ... 45350
Hovercraft Museum, Gosport ... 44335
Hull Maritime Museum, Hull ... 44499
Inveraray Maritime Museum, Inveraray ... 44521
John Paul Jones Birthplace Museum, Arbigland ... 43309
Lancaster Maritime Museum, Lancaster ... 44634
Linlithgow Union Canal Museum, Linlithgow .. 44722
Lleyn Historical and Maritime Museum, Nefyn ... 45197
Local and Maritime Museum, Newhaven ... 45235
London Canal Museum, London ... 44905
Lowestoft and East Suffolk Maritime Museum, Lowestoft ... 45041
McLean Museum and Art Gallery, Greenock ... 44365
Mary Rose Museum, Portsmouth ... 45434
Maryport Maritime Museum, Maryport ... 45112
Merseyside Maritime Museum, Liverpool ... 44740
The Motorboat Museum, Basildon ... 43392
Museum in Docklands, London ... 44921
Museum of Hartlepool, Hartlepool ... 44408
National Coracle Centre, Cenarth ... 43784
National Maritime Museum, London ... 44939
National Maritime Museum Cornwell, Falmouth ... 44209
National Waterfront Museum, Swansea ... 45815
National Waterways Museum, Gloucester ... 44317
National Waterways Museum - Ellesmere Port, Ellesmere Port ... 44168
Nautical Museum, Castletown ... 43776
North Carr Lightship, Dundee ... 44049
North Devon Maritime Museum, Appledore, Devon ... 43308
Old Lifeboat Museum, Poole ... 45406
Overbeck's Museum, Salcombe ... 45596
Porthmadog Maritime Museum, Porthmadog . 45422
Ramsgate Maritime Museum, Ramsgate ... 45457
Royal National Lifeboat Institution Headquarters Museum, Poole ... 45407
Royal Naval Patrol Service Association Museum, Lowestoft ... 45043
Royal Research Ship Discovery, Dundee ... 44050
Saint Margaret's Museum, Saint Margaret's Bay ... 45582
Salcombe Maritime Museum, Salcombe ... 45597
Scottish Maritime Museum, Irvine ... 44535
Shamrock and Cotehele Quay Museum, Saint Dominick, Cornwall ... 45563
Shipwreck Heritage Centre, Hastings ... 44417
Southampton Maritime Museum, Southampton ... 45707
Southend Pier Museum, Southend-on-Sea ... 45715
Spurn Lightship, Kingston-upon-Hull ... 44599
SS Great Britain Museum, Bristol ... 43639
Sunken History Exhibition, Yarmouth ... 46092
Trinity House National Lighthouse Centre, Penzance ... 45362
Valhalla Museum, Tresco ... 45875
Walton Maritime Museum, Walton-on-the-Naze ... 45926
Watchet Market House Museum, Watchet ... 45948
Waterfront Museum, Poole ... 45410
Whitstable Museum and Gallery, Whitstable .. 46007
Windermere Steamboat Museum, Bowness-on-Windermere ... 43572
Windermere Steamboat Museum, Windermere 46039
Zetland Lifeboat Museum, Redcar ... 45474

United States of America

American Merchant Marine Museum, Kings Point ... 49785
Antique Boat Museum, Clayton ... 47600
Apostles Island National Lakeshore Maritime Museum, Bayfield ... 46656
Baltimore Maritime Museum, Baltimore ... 46550
Barnegat Bay Decoy and Baymen's Museum, Tuckerton ... 53581
Battleship Cove - Maritime Heritage Museums, Fall River ... 48498
Battleship North Carolina, Wilmington ... 54198
Bremerton Naval Museum, Bremerton ... 47004
Buffalo and Erie County Naval and Military Park, Buffalo ... 47117
Cape Ann Historical Museum, Gloucester ... 48945
Captain Salem Avery House Museum, Shady Side ... 53046
Center for Wooden Boats, Seattle ... 52995
Chittenango Landing Canal Boat Museum, Chittenango ... 47545
City Island Nautical Museum, Bronx ... 47043
Coffin School Museum, Nantucket ... 50943
Columbia River Maritime Museum, Astoria ... 46401
Connecticut River Museum, Essex ... 48421
Custom House Maritime Museum, Newburyport ... 51306
Door County Maritime Museum, Gills Rock ... 48910
Door County Maritime Museum, Sturgeon Bay 53345
Dossin Great Lakes Museum, Detroit ... 48116
Eagle Harbor Lighthouse and Museums, Eagle Harbor ... 48232
East and Seaport Museum, Greenport ... 49058
Essex Shipbuilding Museum, Essex ... 48423
Fairport Marine Museum, Fairport Harbor ... 48493
Great Harbor Maritime Museum, Northeast Harbor ... 51417
Great Lakes Naval Memorial and Museum, Muskegon ... 50923
Greater Port Jefferson Museum, Port Jefferson ... 52036
Hampton Roads Naval Museum, Norfolk ... 51363
Hawaii Maritime Center, Honolulu ... 49340
Hereford Inlet Lighthouse, North Wildwood ... 51409
Herreshoff Marine Museum/America's Cup Hall of Fame, Bristol ... 47034
Howard Steamboat Museum, Jeffersonville ... 49616
Hudson River Maritime Museum, Kingston ... 49790
Independence Seaport Museum, Philadelphia 51843
Inland Seas Maritime Museum, Vermilion ... 53718
Intrepid Sea-Air-Space Museum, New York ... 51196
Iowa Great Lakes Maritime Museum, Arnolds Park ... 46363
Keewatin Maritime Museum, Douglas ... 48157
Keokuk River Museum, Keokuk ... 49745
Kittery Historical and Naval Museum, Kittery . 49805
Lake Superior Maritime Visitors Center, Duluth ... 48197
Lake Waccamaw Depot Museum, Lake Waccamaw ... 49894
Long Island Maritime Museum, West Sayville . 54081
Los Angeles Maritime Museum, San Pedro ... 52826
Maine Maritime Museum, Bath ... 46628
Marine Museum at Fall River, Fall River ... 48500
The Mariners' Museum, Newport News ... 51334
Maritime and Seafood Industry Museum, Biloxi ... 46811
Maritime Industry Museum at Fort Schuyler, Bronx ... 47049
Maritime Museum, Cohasset ... 47676
Michigan Maritime Museum, South Haven ... 53172
Mississippi River Museum, Dubuque ... 48189
Monterey Maritime and History Museum, Monterey ... 50802
Museum of Yachting, Newport ... 51323
Mystic Seaport, Mystic ... 50935
Nauticus, Norfolk ... 51367
Naval Shipyard Museum, Portsmouth ... 52110
Naval Undersea Museum, Keyport ... 49768
The Navy Museum, Washington ... 53911
New Jersey Naval Museum, Hackensack ... 49111
Newburyport Maritime Society Museum, Amesbury ... 46262
North Carolina Maritime Museum, Beaufort ... 46666
Old Lighthouse Museum, Michigan City ... 50647
The Oldest House Museum, Key West ... 49765
P.T. Boat Museum, Germantown ... 48899
Patriots Point Naval and Maritime Museum, Mount Pleasant ... 50888
Penobscot Marine Museum, Searsport ... 52992
Portsmouth Athenaeum, Portsmouth ... 52098
Queen Mary Museum, Long Beach ... 50176
River Heritage Museum, Paducah ... 51670
S.S. Meteor Maritime Museum, Barkers Island 46609
Sailor's Memorial Museum, Islesboro ... 49543
San Diego Maritime Museum, San Diego ... 52750
San Francisco Maritime National Historical Park, San Francisco ... 52794
Sandpoint Lighthouse, Escanaba ... 48415
Santa Babara Maritime Museum, Santa Barbara ... 52856
Scituate Maritime and Irish Mossing Museum, Scituate ... 52967
Ships of the Sea Maritime Museum, Savannah ... 52946
The South Street Seaport Museum, New York 51260
Steamboat Arabia Museum, Kansas City ... 49700
Steamship Collection, Baltimore ... 46589
Steamship William G. Mather Museum, Cleveland ... 47641
Submarine Force Museum and Historic Ship Nautilus, Groton ... 49095
Texas Seaport Museum, Galveston ... 48851
United States Coast Guard Museum, New London ... 51068
United States Naval Academy Museum, Annapolis ... 46321
USS Bowfin Submarine Museum, Honolulu ... 49357
USS Constitution Museum, Boston ... 46954
USS Lexington Museum on the Bay, Corpus Christi ... 47822
USS Radford National Naval Museum, Newcomerstown ... 51310
Ventura County Maritime Museum, Oxnard ... 51658
Virginia Beach Maritime Museum, Virginia Beach ... 53758
Waterfront Museum and Showboat Barge, Brooklyn ... 47084
Watermen's Museum, Yorktown ... 54346
Westport Maritime Museum, Westport ... 54111
Wisconsin Maritime Museum, Manitowoc ... 50444

Uruguay

Casa de Ximénez, Montevideo ... 54385
Museo Naval, Montevideo ... 54414

Venezuela

Museo Naval, Catia la Mar ... 54561

Bones

Argentina

Museo de Arte Popular José Hernández, Buenos Aires ... 00184

Canada

Dinosaur Provincial Park, Patricia ... 06574

Denmark

Æbelholt Klostermuseum, Hillerød ... 10075

Germany

Anatomische Sammlung, Greifswald ... 18747
Museum für Haustierkunde Julius Kühn, Halle, Saale ... 18901
Staatliches Naturhistorisches Museum, Braunschweig ... 17529
Stein- und Beinmuseum, Weiterstadt ... 22467

Greece

Museum of Anatomy of the Department of Anatomy, Athinai ... 22930

Netherlands

Gemeentelijke Oudheidkamer, Ermelo ... 32222
Schelpen Museum, Zaamslag ... 32974

Norway

Naturhistorisk museum, Oslo ... 33737

Puerto Rico

Caribbean Primate Research Center Museum, San Juan ... 36059

Switzerland

Kulturama, Zürich 42080

United Kingdom

Anatomy Museum, Aberdeen 43239

United States of America

Florence Hawley Ellis Museum of Anthropology, Abiquiu 46128

Books, Book Art and Manuscripts

Argentina

Museo Biblioteca Municipal Cesar Fernández Navarro, Santa Fé 00613
Museo del Libro, San Javier 00582
Museo Judío de Buenos Aires Dr. Salvador Kibrick, Buenos Aires 00249
Museo Notarial Argentino, Buenos Aires 00266

Armenia

Matenadaran Manuscript Museum, Yerevan 00742

Austria

Österreichische Nationalbibliothek, Wien 03045
Schrift-, Heimat- und Druckereimuseum Bartlhaus, Pettenbach 02496
Universitätsbibliothek Salzburg, Salzburg 02628
Wienbibliothek im Rathaus, Wien 03087

Belgium

Bibliothèque d'Art ENSAV de la Cambre, Bruxelles 03437
Centrum voor de Vlaamse Kunst van de 16de en de 17de Eeuw, Antwerpen 03283
Ecomusée des Techniques, Thuin 04065
Mundaneum - Centre d'Archives de la Communauté Française, Mons 03882
Musée des Imprimés en Luxembourg, Redu 03963
Musée du Livre et Cabinets de Donations de la Bibliothèque Royale de Belgique, Bruxelles 03485
Rubenianum, Antwerpen 03314

Bhutan

National Library Museum, Thimphu 04203

Brazil

Museu Casa de Rui Barbosa, Salvador 04790

Canada

Marsh Collection Society, Amherstburg 05334
Morrin Centre, Québec 06684
Royal Nova Scotia Historical Society, Halifax 05947
Thomas Fisher Rare Book Library, Toronto 07144

Cape Verde

Museu de Documentos Especiais, Praia 07419

China

Book Cultural Museum, Qujiang 08134
China First History Archive Museum, Beijing 07529
Shaanxilong Books Museum, Longxian 07973

Croatia

Zbirka Umjetnina Franjevačkog Samostana, Hvar 08881

Czech Republic

Muzeum Jana Ámoše Komenského, Uherský Brod 09888
Muzeum Knihy, Žďár nad Sázavou 09933

Denmark

Danmarks Bogmuseum, København 10126
Den Rosendahlske Bibelsamling, Esbjerg 10003

Egypt

Egypt Papyrus Museum, Haram 10532
Library of the Greek Orthodox Patriarchate of Alexandria and All Africa, Alexandria 10473
Monastery of Saint Catherine, Gebel Katherîna 10525

France

Association Mathieu Vivian, Saint-Pryvé Saint-Mesmin 15473
Bibliothèque Musée, Clermont (Oise) 12554
Bibliothèque-Musée Forney, Paris 14575
Centre de Recherches Historiques sur les Maîtres Ebénistes, Paris 14580
Centre du Livre d'Artiste Contemporain, Verderonne 16110
Fondation Custodia, Collection Frits Lugt, Paris 14598
Fondation Dosne-Thiers, Paris 14599
Maison du Livre, de l'Image et du Son, Villeurbanne 16237
Musée Bibliothèque, Anost 11632
Musée Départemental Stéphane Mallarmé, Vulaines-sur-Seine 16280
Musée des Lettres et Manuscrits, Paris 14686
Musée du Parchemin et de l'Enluminure, Duras 12801
Musée Mathurin Méheut, Lamballe 13492
Musée Michel Braibant, Montolieu 14235

Germany

Archiv Werner Kittel, Hamburg 18916
Badische Landesbibliothek, Karlsruhe 19435
Bayerische Staatsbibliothek, München 20424
Bibelmuseum Stuttgart, Stuttgart 21929
Buchmuseum der Sächsischen Landesbibliothek - Staats- und Universitätsbibliothek Dresden, Dresden 17939
Deutsches Buch- und Schriftmuseum der Deutschen Nationalbibliothek Leipzig, Leipzig 19900
Deutsches Buchbindermuseum, Mainz 20142
documenta Archiv für die Kunst des 20. und 21 Jahrhunderts, Kassel 19462
Einbecker Schreibmaschinenmuseum, Einbeck 18091
Feiligrath Museum, Rüdesheim am Rhein 21434
Fürst Thurn und Taxis Zentralarchiv und Hofbibliothek, Regensburg 21221
Göschenhaus / Seume-Gedenkstätte, Grimma 18760
Goethe- und Schiller-Archiv, Weimar 22402
Gutenberg-Museum, Mainz 20143
Herzog August Bibliothek, Wolfenbüttel 22653
Hochschul-und Landesbibliothek, Fulda 18509
Internationale Jugendbibliothek, München 20447
J. F. Schreiber-Museum, Esslingen 18275
Klingspor-Museum, Offenbach am Main 20883
Kupferstichkabinett - Sammlung der Zeichnungen und Druckgraphik, Berlin 17172
Landesbibliothek Oldenburg, Oldenburg 20896
Landesbibliothek und Murhardsche Bibliothek der Stadt Kassel, Kassel 19466
Monacensia, München 20459
Museum an der Krukenburg, Bad Karlshafen 16758
Museum Otto Schäfer, Schweinfurt 21687
Museum Schloß Burgk, Burgk 17658
Sammlung handgeschriebener Choralbücher, Kiedrich 19525
Sankt Nikolaus-Hospital Cusanusstift, Bernkastel-Kues 17299
Staatliche Bücher- und Kupferstichsammlung Greiz, Greiz 18753
Staatsbibliothek Bamberg, Bamberg 16919
Staatsbibliothek zu Berlin, Berlin 17258
Stadtarchiv und Ehemals Reichsstädtische Bibliothek Lindau, Lindau 20000
Stadtmuseum Hildburghausen, Hildburghausen 19179
Theatermuseum Zauberwelt der Kulisse, Meiningen 20262
Universitätsbibliothek der Humboldt-Universität zu Berlin, Berlin 17276
Universitätsbibliothek Johann Christian Senckenberg, Frankfurt am Main 18399
Württembergische Landesbibliothek, Stuttgart 21969

Greece

Tatarna Monastery Museum, Granitsa 23010

Guatemala

Museo del Libro Antiguo, Antigua 23292

Hungary

Főegyházmegyei Könyvtár, Eger 23539
Győr Egyházmegyei Könyvtár es Kincstár, Győr 23582
Kalocsai Foegyházmegyei Gyűjtemények - Foszékesegyházi Könyvtár, Kalocsa 23625
Magyar Ferences Könyvtár Gyöngyösi Muemlék Könyvtára, Gyöngyös 23573
Vachott Sándor Városi Könyvtár, Gyöngyös 23575

India

Asiatic Society Museum, Kolkata 24267
G.N. Jha Kendriya Sanskrit Vidyapeetha, Allahabad 24060
Museum of Gujarat Vidyasabha, Ahmedabad 24050
Museum of the Sarvajanik Wachanalaya, Nasik 24336
Sibsagar College Museum, Sibsagar 24392
Vrindavan Research Institute Museum, Vrindavan 24430

Indonesia

Museum Gedong Kirtya, Singaraja 24559

Ireland

Bolton Library, Cashel 24754
The Chester Beatty Library, Dublin 24781
Irish Architectural Archive, Dublin 24798

Israel

Berman Hall, Jerusalem 24979
Bible Museum, Tel Aviv 25102
Nahum Gutman's Museum, Tel Aviv 25118

Italy

Biblioteca Medicea Laurenziana, Firenze 26141
Casa di Pirandello, Agrigento 25160
Centro di Informazione e Documentazione Arti Visive, Prato 27330
Mostra Permanente della Biblioteca Estense, Modena 26742
Museo Carducci - Biblioteca e Casa, Bologna 25446
Museo Civico Didattico Gian Andrea Irico, Trino 28099
Museo Dantesco, Ravenna 27371
Museo del Cattedrale, Ferrara 26110
Museo dell'Illustrazione, Ferrara 26112
Museo dell'Istituto di Patologia del Libro, Roma 27487
Museo Marchigiano del Risorgimento G. e D. Spadoni, Macerata 26554
Pinacoteca Civica, Fermo 26103

Japan

Chihiro Art Museum Azumino, Nagano 28953
Erzgebirge Toys Museum Museum of Picture Books, Karuizawa 28668

Kazakhstan

Muzej Knigi, Almaty 29688

Liechtenstein

Liechtensteinische Landesbibliothek, Vaduz 30198

Lithuania

Adelės ir Pauliaus Galaunių namai, Kaunas 30238
Kazio Varnelio Namai-Muziejus, Vilnius 30411

Luxembourg

Musée de l'Abbaye, Echternach 30469

Malta

Cathedral Museum, Mdina 30630

Mexico

Museo Biblioteca Palafoxiana, Puebla 31266
Museo Biblioteca Pape, Monclova 31172

Netherlands

Museum Meermanno-Westreenianum, Den Haag 32087
Museum Oude Boekdrukkunst, Almen 31802
Schriftmuseum J.A. Dortmond, Amsterdam 31886
Scryption, Tilburg 32832

New Zealand

Renwick Museum, Renwick 33194

Philippines

Cavite City Library and Museum, Cavite City 34361
Lopez Memorial Museum, Pasig 34478

Poland

Biblioteka Sejmowa, Warszawa 35218
Dział Zbiorów Specjalnych - Muzeum, Warszawa 35222
Muzeum Sztuki Książki, Wrocław 35339
Zakład Narodowy im. Ossolińskich, Wrocław 35344

Portugal

Museu da Biblioteca Nacional, Lisboa 35675

Puerto Rico

Museo La Casa del Libro, San Juan 36073

Romania

Colecţia Bibliotecii Seminarului Teologic Ortodox, Caransebeş 36261

Russia

Muzej Knigi, Moskva 37308

Serbia

Muzej Orijentalnih Rukopisa, Prizren 38135

Slovenia

Spominski Muzej Prežihovega Voranca, Ravne na Koroškem 38504

Spain

Fundació Pública Institut d'Estudis Ilerdencs, Lleida 39473
Fundación Museo de las Ferias, Medina del Campo 39638
Museo de la Biblioteca Nacional, Madrid 39542
Museu d'Història de la Ciudad, Sant Feliu de Guíxols 39982

Swaziland

University of Swaziland Library Museums Collection, Kwaluseni 40402

Sweden

Bernadottebiblioteket, Stockholm 40886
Biblioteksmuseet, Borås 40466

Switzerland

Ausstellung in der Universitätsbibliothek Basel, Basel 41155
Fondation Martin Bodmer, Cologny 41308
Fonds ancien de la Bibliothèque cantonale jurassienne, Porrentruy 41727
Fundaziun Planta Samedan Wohnmuseum, Samedan 41800
Le Musée Gutenberg des Arts Graphiques et de la Communication, Fribourg 41389
Schweizerische Nationalbibliothek, Bern 41215

Tanzania

Zanzibar National Archives and Museums Department, Zanzibar 42525

Turkey

Hat Sanatları Müzesi, İstanbul 42821

Ukraine

Muzej Knigi i Drukarstva Ukraïny, Kyïv 43092

United Kingdom

British Library, London 44802
John Rylands University Library-Special Collections, Manchester 45087
Leadhills Miners' Library Museum, Leadhills 44653
Mappa Mundi and Chained Library, Hereford 44456
Wellcome Library, London 45008

United States of America

Armstrong Browning Library, Waco 53771
The Atlanta College of Art, Atlanta 46423
Brand Library Art Galleries, Glendale 48928
Canajoharie Library and Art Gallery, Canajoharie 47211
Center for Book Arts, New York 51141
Chancellor Robert R. Livingston Masonic Library and Museum, New York 51145
Chapin Library of Rare Books, Williamstown 54174
D.R. Barker Historical Museum, Fredonia 48776
Dwight D. Eisenhower Library-Museum, Abilene 46120
Eric Carle Museum of Picture Book Art, Amherst 46266
Fort Worth Public Library Arts and Humanities, Fort Worth 48714
Hamilton Library and Two Mile House, Carlisle 47243
Herbert Hoover Presidential Library-Museum, West Branch 54047
James Monroe Museum and Memorial Library, Fredericksburg 48768
John F. Kennedy Presidential Library-Museum, Boston 46935
Karpeles Manuscript Museum, Charleston 47370
Karpeles Manuscript Museum, Duluth 48196
Lehigh University Art Galleries/Museum, Bethlehem 46780
Living Word National Bible Museum, Aledo 46194
Lyndon Baines Johnson Museum, Austin 46514
Mary Baker Eddy Library for the Betterment of Humanity, Boston 46936
Melton Art Reference Library Museum, Oklahoma City 51518
The Morgan Library, New York 51216

B

The Museum of American Illustration at the Society of Illustrators, New York ... 51223
Museum of the Jimmy Carter Library, Atlanta 46446
Racine Art Museum, Racine ... 52223
Rosenbach Museum, Philadelphia ... 51877
Rutherford B. Hayes Presidential Center, Fremont ... 48788
State Capital Publishing Museum, Guthrie ... 49109
The Thoreau Institute at Walden Woods, Lincoln ... 50093
Tresure Room Gallery of the Interchurch Center, New York ... 51272
Wagnalls Memorial, Lithopolis ... 50131

Uzbekistan

Moyle Mubarek Library Museum, Toshkent ... 54475

Vatican City

Archivio Segreto Vaticano, Città del Vaticano .. 54496

Zambia

University of Zambia Library Museums Collection, Lusaka ... 54660

Botany

Argentina

Museo Argentino de Caza Mayor, Buenos Aires ... 00167
Museo Botánico, Córdoba ... 00320
Museo de Ciencias Naturales Dr. Amado Bonpland, Corrientes ... 00347
Museo del Colegio San José, Buenos Aires ... 00226
Museo Farmacobotánica Juan A. Domínguez de la Universidad de Buenos Aires, Buenos Aires ... 00232

Australia

Tasmanian Museum and Art Gallery, Hobart .. 01159

Austria

Biologiezentrum der Oberösterreichischen Landesmuseen, Linz ... 02310
Bonsai Museum, Seeboden ... 02732
Naturhistorisches Museum, Admont ... 01694
Naturkundemuseum, Dobersberg ... 01821
Naturwissenschaftliche Sammlungen am Landesmuseum Joanneum, Graz ... 01981

Belgium

Musée d'Histoire Naturelle et Vivarium, Tournai ... 04088
Musée du Buis, Durbuy ... 03565
Naturzentrum Ternell, Eupen ... 03591
Schoolmuseum Michel Thiery, Gent ... 03640

Brazil

Museu Botânico Dr. João Barbosa Rodrigues, São Paulo ... 04865
Museu Carpológico do Jardim Botânico do Rio de Janeiro, Rio de Janeiro ... 04703
Museu de História Natural, Belo Horizonte ... 04349
Museu do Eucalipto, Rio Claro ... 04678
Museu Regional da Fauna e Flora do Itatiaia, Itatiaia ... 04500
Museu von Martius, Teresópolis ... 04933
Sítio Roberto Burle Marx, Rio de Janeiro ... 04771

Bulgaria

Prirodonaučen Muzej, Burgas ... 05000

China

Muaseum of Nature and Agriculture, Macau . 08001

Colombia

Museo Etnobotánico, Tuluá ... 08771
Museo Mundocaña, Cali ... 08613
Museo Mundocaña, Palmira ... 08702

Czech Republic

Blatské muzeum - Rožmberský Dům, Soběslav ... 09842
Botanické Oddělení, Brno ... 09421
Botanické Oddělení Národního Muzea, Průhonice ... 09812
Dietrichsteinský Palác, Brno ... 09423
Entomologické Oddělení, Brno ... 09426
Mendelianum - Památník Gregora Mendela, Brno ... 09430
Muzeum Těšínska, Český Těšín ... 09470
Slezské zemské muzeum, Stěbořice-Nový Dvůr 09846

Denmark

Botanisk Museum, København ... 10125

Egypt

Scientific Researches Museum, Cairo ... 10518

Finland

Kasvimuseo, Turku ... 11391
Kasvimuseo, Helsinki ... 10790
Oulun Yliopiston Kasvimuseo, Oulu ... 11175

France

Écomusée de l'Abeille et de la Flore Locale, Faugères ... 12907
Galerie de Paléobotanique, Paris ... 14605
Maison de la Baie du Mont Saint-Michel, Vains ... 16029
Maison de la Flore, Longchaumois ... 13806
Maison de la Forêt, Raismes ... 14979
Musée Botanique, Angers ... 11606
Musée Conservatoire des Cépages, Saint-Ilpize ... 15314
Musée de Cerdagne d'Eyne, Eyne ... 12896
Musée d'Histoire Naturelle, Nantes ... 14366
Muséum d'Histoire Naturelle, Nice ... 14440
Muséum National d'Histoire Naturelle, Paris . 14741

Germany

Alfred-Erhardt-Stiftung, Berlin ... 17030
Botanische Staatssammlung München, München ... 20430
Botanisches Museum Berlin-Dahlem, Berlin .. 17053
Fehn- und Schiffahrtsmuseum, Rhauderfehn . 21291
Landschloss Pirna-Zuschendorf - Botanische Sammlungen der TU Dresden, Pirna ... 21044
Museum für Naturkunde der Stadt Gera, Gera 18595
Natureum am Schloß Ludwigslust, Ludwigslust ... 20061
Naturhistorisches Museum, Mainz ... 20149
Naturkundemuseum, Bayreuth ... 16963
Vonderau Museum Fulda, Fulda ... 18513

Greece

Botanical Museum of the National Gardens, Athinai ... 22903
Siatista Botanical Museum, Siatista ... 23182

Hungary

Szálkahalmi Orháza, Balmazujváros ... 23359

India

Botanical Survey Museum of India, Pune ... 24356
Botanical Survey of India, Dehradun ... 24130
Botany Museum, Faizabad ... 24171
Botany Museum, Gorakhpur ... 24179
Botany Museum, Jaunpur ... 24239
Botany Museum, Kanpur ... 24252
Botany Museum, Lucknow ... 24292
Central National Herbarium, Haora ... 24202
Government Museum, Chennai ... 24115

Indonesia

Museum Herbarium Bogoriensis, Bogor ... 24456

Ireland

Músaem Corca Dhuibhne, Ballyferriter ... 24738

Israel

Ussishkin House, Kibbutz Dan ... 25035

Italy

Museo Agrario Tropicale, Firenze ... 26166
Museo Alpino Duca degli Abruzzi, Courmayeur ... 25997
Museo Botanico Cesare Bicchi, Lucca ... 26531
Museo dei Fossili e Mostra dei Funghi, Pioraco ... 27232
Museo del Fiore, Acquapendente ... 25145
Museo del Fungo e di Scienze Naturali, Boves 25521
Museo delle Erbe, Teggiano ... 27949
Museo di Botanica, Firenze ... 26179
Museo di Storia Naturale, Genova ... 26311
Museo Erbario, Roma ... 27506
Museo Friulano di Storia Naturale, Udine ... 28114
Museo Naturalistico del Parco, San Romano di Garfagnana ... 27681

Korea, Republic

Forest Museum, Gyeonggi-do ... 29824

Latvia

Latvijas Universitātes Botānikas Muzejs, Rīga ... 30087

Mali

Musée National du Mali, Bamako ... 30624

Mexico

Museo Botánico, Tuxtla Gutiérrez ... 31511
Museo Herbario de la Facultad de Ciencias, México ... 31119

Netherlands

Breda's Begijnhof Museum, Breda ... 32002
Museum de Zwarte Tulp, Lisse ... 32503

New Zealand

Rewa's Village, Kerikeri ... 33137

Norway

Naturhistorisk museum, Oslo ... 33737

Pakistan

Botanical Museum, Rawalpindi ... 34040

Papua New Guinea

Kau Rainforest Museum Conservation Area, Madang ... 34083

Poland

Muzeum Przyrodnicze im. Prof. Andrzeja Myrchy, Białystok ... 34551

Portugal

Museu Botânico, Coimbra ... 35530
Museu Botânico da Escola Superior Agrária de Beja, Beja ... 35462
Museu do Instituto de Botânica Gonçalo Sampaio, Porto ... 35852

Romania

Muzeul Brăilei-Secţia Ştiinţele Naturii, Brăila 36172

Russia

Botaničeskij Muzej, Sankt-Peterburg ... 37625
Muzei Permskogo Gosudarstvennogo Universiteta, Perm ... 37547
Novosibirskij Muzej Prirody, Novosibirsk ... 37487

Slovakia

Múzeum Tatranského Národného Parku, Tatranská Lomnica ... 38370
Ponitrianske Múzeum, Nitra ... 38315

Slovenia

Prirodoslovni Muzej Slovenije, Ljubljana ... 38461

South Africa

Albany Museum, Grahamstown ... 38629
Life Sciences Museum, Johannesburg ... 38659
National Museum Bloemfontein, Bloemfontein 38567

Spain

Museo de Historia Natural, Santiago de Compostela ... 40021
Museu de Ciències Naturals, Onda ... 39734
Museu Salvador, Barcelona ... 39003

Sweden

Botaniska Museet, Lund ... 40701
Evolutionsmuseet, Uppsala ... 41005
Linnémuseet, Uppsala ... 41007
Linnés Hammarby, Uppsala ... 41008
Vänermuseet, Lidköping ... 40676

Switzerland

Musée de Botanique, Lausanne ... 41545
Sukkulenten-Sammlung Zürich, Zürich ... 42109

Taiwan

Lotus Explorer Museum, Baihe ... 42140

United Kingdom

Botanic Gardens Museum, Southport ... 45717
Economic Botany Collection/Museum No. 1, Richmond, Surrey ... 45491
Falconer Museum, Forres ... 44245
Kew Palace Museum Queen Charlotte's Cottage, Kew ... 44570
Margam County Park, Port Talbot ... 45417

United States of America

Botanical Museum of Harvard University, Cambridge ... 47173
Edison and Ford Winter Estates, Fort Myers ... 48661
El Paso Museum of Archaeology at Wilderness Park, El Paso ... 48320
Mount Rainier National Park Museum, Ashford ... 46378
Museum of Ancient Artifacts, Chicago ... 47490
Olympic National Park Visitor Center, Port Angeles ... 52027

Uruguay

Museo Botánico, Montevideo ... 54395
Museo y Jardin Botánico Profesor Atilio Lombardo, Montevideo ... 54421

Uzbekistan

Natural History Museum, Nukus ... 54450

Brewing

Australia

Carriage and Harness Museum Murray Breweries Historic Museum, Beechworth ... 00848

Austria

Bierarium, Oberwart ... 02464
Bindereimuseum Hofbräu Kaltenhausen, Hallein ... 02045
Braumuseum, Wieselburg an der Erlauf ... 03100
Braumuseum, Leoben ... 02284
Erstes Österreichisches Biermuseum, Laa an der Thaya ... 02247
Museum Historischer Bierkrüge, Hainfeld ... 02040

Belgium

Bocholter Brouwerij Museum, Bocholt ... 03375
Brouwershuis, Antwerpen ... 03282
Mout- en Brouwhuis De Snoek, Alveringem ... 03268
Musée Bruxellois de la Gueuze, Bruxelles ... 03454
Musée de la Bière, Sankt Vith ... 04000
Musée de la Brasserie-Museum van de Belgische Brouwers, Bruxelles ... 03467
Musée de la Cervoise, du Gruyt et des Bières Medièvales, Anthisnes ... 03276
Musée des Bières Belges, Lustin ... 03836

Czech Republic

Pivovarské Muzeum, Plzeň ... 09718

Denmark

Fjerritslev Bryggeri- og Egnsmuseum, Fjerritslev ... 10020

Finland

Kotiseutumuseo Pellava ja Sahti, Alfred Kordelinin Ajokalumuseo, Lammi ... 11042

France

Musée de la Bière, Armentières ... 11689
Musée de la Bière, Stenay ... 15808
Musée Français de la Brasserie, Saint-Nicolas-de-Port ... 15428
Musée Vosgien de la Brasserie, Ville-sur-Illon. 16177

Germany

Bayerisches Brauerei- und Bäckereimuseum Kulmbach, Kulmbach ... 19746
Bier- und Oktoberfestmuseum, München ... 20427
Brauerei-Kontor, Bochum ... 17397
Brauerei-Museum, Haßfurt ... 19030
Brauerei-Museum, Nesselwang ... 20576
Brauerei-Museum Dortmund, Dortmund ... 17916
Brauereikulturmuseum Gut Riedelsbach, Neureichenau ... 20645
Brauereimuseum, Aldersbach ... 16462
Brauereimuseum, Altomünster ... 16522
Brauereimuseum Barre's Brauwelt, Lübbecke 20066
Brauereimuseum Bräu im Moos, Tüßling ... 22130
Brauereimuseum im historischen Kronen-Brauhaus zu Lüneburg, Lüneburg ... 20095
Brauereimuseum Schöneck, Schöneck / Vogtland ... 21604
Brauhausmuseum, Stadtlauringen ... 21858
Braumuseum - Wolff Giersing's Brauerbe, Torgau ... 22069

Fahrradmuseum Zumhaus und Brauereimuseum, Feuchtwangen 18312
Felsenkeller-Labyrinth im Hirschberg - Brauereimuseum, Beilngries 16983
Fränkisches Brauereimuseum, Bamberg 16908
Hallertauer Hopfen- und Heimatmuseum, Geisenfeld 18558
Heimat- und Brauereimuseum Pleinfeld, Pleinfeld 21058
Klosterbräu Brauereimuseum, Irsee 19355
Maisel's Brauerei- und Büttnerei-Museum, Bayreuth 16959
Museum der Berliner Bürgerbräu Brauerei, Berlin 17195
Museum für Krippen, Eß- und Trinkkultur und Brauereigeschichte, Burghausen 17654
Neumarkter Brauereimuseum im Glossner-Bräu, Neumarkt in der Oberpfalz 20633
Stutzhäuser Brauereimuseum, Luisenthal 20108

Hungary
Dreher Sörmúzeum, Budapest 23395

Ireland
Guinness Storehouse, Dublin 24794

Japan
Hakutsuru Sake Brewery Museum, Kobe 28727
Kyoto Asahi Beer Oyamazaki Villa Museum, Oyamazaki 29169

Luxembourg
Musée d'Histoire de la Brasserie de Diekirch, Diekirch 30463
Musée National d'Art Brassicole et Musée de la Tannerie, Wiltz 30498

Netherlands
Heineken Experience, Amsterdam 31845
Museumbrouwerij de Hemel, Nijmegen 32591
Museumbrouwerij De Roos, Hilvarenbeek 32386
Nederlands Biermuseum de Boom, Alkmaar 31794
Villa Bergzicht/Brouwerijmuseum Raaf, Heumen 32382
Voormalig Stoombierbrouwerij De Keijzer N.A. Bosch, Maastricht 32535

Slovakia
Tribečské Múzeum Topoľčany, Topoľčany 38372

Slovenia
Pivovarski Muzej, Ljubljana 38459

Sweden
Arboga Bryggerimuseum, Arboga 40427

Switzerland
Ortsmuseum, Nürensdorf 41688

United Kingdom
Museum of Brewing, Burton-upon-Trent 43688
Stamford Brewery Museum, Stamford 45736
Tetley's Brewery Wharf, Leeds 44667
Young's Brewery Museum, London 45022

United States of America
Troy-Hayner Cultural Center, Troy 53577

Bronzes

Cambodia
National Museum of Cambodia, Phnom Penh 05287

Canada
Musée du Bronze d'Inverness, Inverness 06018

China
Capital Museum, Beijing 07524
Poly Art Museum, Beijing 07562

Italy
Museo Archeologico G. Cilnio Mecenate, Arezzo 25266

Japan
Shimane-kenritsu Hakubutsukan, Matsue 28889

Netherlands
Museum Mesdag, Den Haag 32088

Vatican City
Museo Gregoriano Etrusco, Città del Vaticano 54504

Bullfights

Canada
Stones'n Bones Museum, Sarnia 06908

Colombia
Museo Taurino, Cali 08615
Museo Taurino de Bogotá, Bogotá 08595

France
Musée des Cultures Taurines, Nîmes 14451
Musée Tauromachique, Pomarez 14871

Mexico
Museo Centro Taurino Potosino, San Luis Potosí 31344
Museo Taurino, Huamantla 30939

Peru
Museo Taurino de la Plaza de Acho, Lima 34270

Portugal
Museu Taurino, Alcochete 35400

Spain
Colección Taurina, Gerena 39331
Museo de la Real Maestranza de Caballería de Ronda, Ronda 39916
Museo Taurino, Córdoba 39208
Museo Taurino, Hellín 39386
Museo Taurino, La Línea de la Concepción 39466
Museo Taurino, Madrid 39582
Museo Taurino, Murcia 39704
Museo Taurino, Salamanca 39944
Museo Taurino, Valencia 40217
Museo Taurino de Béjar, Béjar 39015
Museo Taurino de Mijas, Mijas 39649
Museo Taurino Municipal, Antequera 38876
Museu Taurí de la Monumental, Barcelona 39004

Buttons

Germany
Deutsches Knopfmuseum, Bärnau 16895

Netherlands
Speldjesmuseum, Klaaswaal 32447

United States of America
Pearl Button Museum, Muscatine 50922

Cabaret → Performing Arts

Calligraphy

China
Art Museum of Dazu Rock Carvings in Chongqing, Chongqing 07628
Capital Museum, Beijing 07524
China Penmanship Art Museum, Xian 08357
Lin San Zhi Art Display Center, Zhujiang 08514
Luweizhao Painting and Calligraphy House, Pinghu 08083

France
Musée du Scribe, Saint-Christol-lès-Alès 15204

Germany
Graphische Sammlung der Universität, Erlangen 18221
Siebold-Museum, Würzburg 22724

Iran
Mir-Emad Calligraphy Museum, Tehrān 24681

Japan
Basho Memorial Museum, Ueno 29554
Imabari-shiritsu Kono Bijutsukan, Imabari 28568
Kanken Kanji Hakubutsukan, Kyoto 28819
Masaki Art Museum, Osaka 29133
Nara City Museum of Calligraphy in Honor of Kason Sugioka, Nara 29033
Naritasan Calligraphy Museum, Narita 29040
Nihon Shodo Bijutsukan, Tokyo 29451
Saga-kenritsu Hakubutsukan, Saga 29180
Shodo Hakubutsukan, Tokyo 29472
Ueno Royal Museum, Tokyo 29501

Korea, Republic
Namjin Art Museum, Imhoe 29831
Woljeon Art Museum, Seoul 29934

Mexico
Museo de la Pluma, México 31086

Syria
Musée de Calligraphie et Epigraphie Arabe, Damascus 42131

Turkey
Divan Edebiyatı Müzesi, İstanbul 42815
Hat Sanatları Müzesi, İstanbul 42821

United Kingdom
Ditchling Museum, Ditchling 43987

Uzbekistan
Qo'qon Viloyat Ülkashunoslik Muzeyi, Qo'qon 54458

Cameras → Optics

Carpets

Armenia
National Folk Art Museum of Armenia, Yerevan 00750

Azerbaijan
Carpet Museum, Bakı 03150
Letif Kerimov ad. Azerbaycan Xalçasi ve Xalq Tetbiqi Seneti Dövlet Muzeyi, Bakı 03159
Nakhchivan Carpet Museum, Naxçivan 03187

Belgium
Musée de la Tapiserie et des Arts du Tissu, Tournai 04085

Brazil
Fundação Eva Klabin Rappaport, Rio de Janeiro 04693

Bulgaria
Izložba na Stari Kotlenski Kilimi, Kotel 05076

Canada
Textile Museum of Canada, Toronto 07142

Germany
Orientteppich-Museum, Hannover 19004

Iran
Astan-e-Qods-e Razavi Museums, Mashad 24631
Farsh Museum, Tehrān 24663
Rassan Arabzadeh Carpet Museum, Tehrān 24691

Kuwait
Persian Carpet Museum, Raas 29950

Morocco
Belghazi-Musée, Fès 31669

Netherlands
Het Tapijtmuseum, Genemuiden 32245

Turkey
Arkoloji ve Etnografya Müzesi, Edirne 42766
Halı, Kilim ve Düz Dokuma Yaygılar Müzesi, İstanbul 42820
Yıldız Sarayı Müzesi ve Tiyatrosu ve Sahne Kostümkeri Müzesi, İstanbul 42851

Turkmenistan
Türkmenistanyt Gahrymany Gurbansoltan eje Adyndaky Türkmen Halsynyt Muzeýi, Aşgabat 42974

Carriages and Carts

Argentina
Museo del Carruaje "El Tacú", Villa Ciudad Parque 00674

Australia
Carriage and Harness Museum Murray Breweries Historic Museum, Beechworth 00848

Austria
Fiakermuseum, Wien 02987
Kutschen- und Heimatmuseum, Obertilliach 02461
Kutschen- und Schlittenmuseum, Großraming 02009
Wagenburg im Schloss Schönbrunn, Wien 03085

Barbados
Sunbury Planatation House, Saint Phillip 03244

Belgium
Karrenmuseum, Essen 03587
Musée des Chasseurs Ardennais, Marche-en-Famenne 03847
Rijtuigmuseum, Bree 03405

Bermuda
Carriage Museum of Bermuda, Saint George's 04197

Canada
Country Heritage Park, Milton 06288
Remington Carriage Museum, Cardston 05562

Czech Republic
Vagonářské Muzeum, Studénka 09854

Denmark
Slesvigske Vognsamling, Haderslev 10050
Sparresholm Gods og Vognsamling, Holme-Olstrup 10088

Egypt
Royal Carriage Museum, Bulaq-el-Dakrur 10483

France
Collection Pierre Bouton, Berneuil 11949
Musée de Voitures Hippomobiles Anciennes, Tour-en-Sologne 15949
Musée des Attelages, de la Carrosserie et du Charronnage, Vonnas 16272
Musée des Carrosses, Versailles 16126
Musée du Cheval en 1900, Cussac-Fort-Médoc 12692
Musée National de la Voiture et du Tourisme, Compiègne 12599

Germany
Hessisches Kutschen- und Wagenmuseum, Lohfelden 20022
Kutschen-, Schlitten- und Wagenmuseum Rottach-Egern, Rottach-Egern 21417
Kutschenmuseum, Augustusburg 16639
Kutschenmuseum, Ludwigslust 20060
Kutschenmuseum Hessisches Landgestüt, Dillenburg 17860
Kutschensammlung, Eigeltingen 18088
Kutschensammlung, Weimar 22408
Marstallmuseum, München 20458
Marstallmuseum, Regensburg 21229
Metz von Tessin'sche Kutschensammlung, Starzach 21865
Museum für Kutschen, Chaisen, Karren, Heidenheim an der Brenz 19091
Museum Kutschen-Wagen-Karren, Gescher 18622

Hungary
Kocsimúzeum, Bábolna 23340

C

Ireland

Vintage Car and Carriage Museum, Inishowen 24843

Italy

Eugenio Collectta Carrozze Serreria, Caronno Pertusella 25682
Mostra di Palazzo Farnese, Piacenza 27184
Museo Le Carrozze d'Epoca, Roma 27465
Museo del Biroccio Marchigiano e Museo Beltrami, Filottrano 26134
Museo della Carrozza, Macerata 26552
Museo delle Carrozze, Catanzaro 25781
Museo delle Carrozze, Napoli 26890
Museo delle Carrozze, Firenze 26173
Museo delle Carrozze dell'Ottocento, Verona 28234
Museo di Auto e Moto d'Epoca, Vigolzone 28270

Netherlands

Agrarisch en Wagenmuseum, De Waal 32052
Cultuurhistorisch Streek- en Handkarrenmuseum De Wemme, Zuidwolde, Drenthe 33014
Kinderwagens van toen, Dwingeloo 32161
Nationaal Rijtuigmuseum, Leek 32467
Zijper Museum, Schagerbrug 32750

Portugal

Museu Nacional dos Coches, Lisboa 35723

Spain

Museo de Carruajes de Sevilla, Sevilla 40070
Museu de Carrosses Fúnebres, Barcelona 38973

Sweden

Eriksborg Vagnmuseum, Ystad 41076
Livrustkammaren, Stockholm 40909

Switzerland

Kutschenmuseum, Basel 41168
Kutschensammlung Robert Sallmann, Amriswil 41111

Thailand

Thai Royal Carriages and Paraphenalia, Bangkok 42563

United Kingdom

Arlington Court, Arlington 43315
Erddig, Wrexham 46082
Tyrwhitt-Drake Museum of Carriages, Maidstone 45075
Unicorn Trust Equestrian Centre, Stow-on-the-Wold 45772

United States of America

Car and Carriage Museum, Pittsburgh 51941
Cheyenne Frontier Days Old West Museum, Cheyenne 47444
Thrasher Carriage Museum, Frostburg 48803

Vatican City

Museo Padiglione delle Carozza, Città del Vaticano 54507

Cartography

Austria

Alpinmuseum Dachstein, Ramsau am Dachstein 02557
Anich-Hueber-Museum, Oberperfuss 02457
Dokumentationsraum Georg Matthäus Vischer, Raabs an der Thaya 02545
Globenmuseum der Österreichischen Nationalbibliothek, Wien 02996

Belgium

Mercator Museum, Sint-Niklaas 04021

Brazil

Museu do Colono, Santa Leopoldina 04808

Germany

Tobias-Mayer-Museum, Marbach am Neckar 20184

Hungary

Magyar Térképészet Szakmatörténeti Múzeum, Budapest 23444

India

Survey Museum, Roorkee 24374

Italy

Mostra Cartografica, Mendatica 26652

Netherlands

Van de Poll-Stichting, Zeist 32995
Witte's Museum, Axel 31944

Norway

Norsk Kartmuseum, Hønefoss 33558

Romania

Muzeul Național al Hărților și Cărții Vechi, București 36214

Slovenia

Zemljepišni Muzej Slovenije, Ljubljana 38470

Switzerland

Alpineum, Luzern 41586
Gletschergarten Luzern, Luzern 41590

United Kingdom

American Museum in Britain, Bath 43398

United States of America

Museum of Surveying, Lansing 49944

Carts → Carriages and Carts

Carving, Wood → Wood Carving

Castles and Fortresses

Albania

Museum of Weapons, Gjirokastër 00017

Algeria

Musée du Palais du Dey, Casbah, Alger 00043

Armenia

Erebouni Museum of the Establishment of Yerevan, Yerevan 00733

Austria

Ausgrabungsstätte Schanzberg, Thunau 02813
Barockjagdschloß Eckartsau, Eckartsau 01838
Burg Greifenstein, Greifenstein 01987
Burg Hochosterwitz, Launsdorf 02268
Burgenkundliches Museum Schloss Alt-Kainach, Bärnbach 01777
Burgmuseum, Güssing 02015
Burgmuseum, Seebenstein 02731
Burgmuseum Clam, Klam 02193
Esterházy-Ausstellung, Eisenstadt 01859
Festung Kniepass, Unken 02839
Festungsmuseum Salzburg, Salzburg 02610
Franzensburg, Laxenburg 02272
Freilichtanlage Gschlößl, Leithaprodersdorf 02281
Hengistburgmuseum Volkskundliche Sammlung, Hengsberg 02070
Kaiser Maximilian von Mexiko Museum, Hardegg 02051
Museum Burg Heidenreichstein, Heidenreichstein 02062
Museum Mariastein, Kirchbichl 02169
Paul Anton Keller-Museum, Lockenhaus 02331
Renaissance Schloss Greillenstein, Röhrenbach 02591
Renaissanceschloss Rosenburg, Rosenburg am Kamp 02599
Sammlungen Burg Bernstein, Bernstein 01783
Schloss Tratzberg, Jenbach 02142
Schlossmuseum, Nauders 02423

Azerbaijan

State Historical and Architectural Museum-reserve Shirvan Shahs' Palace, Bakı 03173

Bahamas

Fort Charlotte, Nassau 03206
Fort Fincastle, Nassau 03207
Fort Montagu, Nassau 03208

Bangladesh

Ahsan Manzil Museum, Dhaka 03217
Lalbagh Fort Museum, Dhaka 03225

Belgium

Château de Beloeil, Beloeil 03358
Château de Reinhardstein, Robertville 03970
Château de Rixensart, Rixensart 03969
Château des Princes de Chimay, Chimay 03523
Château du Fosteau, Leers-et-Fosteau 03782
Fort Napoleon, Oostende 03931
Kasteel Achtendries, Gent 03631
Kasteel Beauvoorde, Wulveringem 04158
Musée Baillet-Latour, Latour 03779
Musée de la Chasse, de la Vénerie et de la Protection de la Nature, Lavaux-Sainte-Anne 03780
Musée du Château de Franchimont, Theux 04060
Musée du Château-Fort, Ecaussinnes 03571
Musée et Parc Archéologiques de Montauban, Buzenol 03508
Museum Fort Liezele, Puurs 03956
Stedelijk Archeologisch Museum, Oudenburg 03943

Bulgaria

Archeologičeski Rezervat Starijat Grad Šumen i Šumenska Krepost, Šumen 05212

Canada

Bastion, Nanaimo 06409
Castle Hill National Historic Park, Placentia Bay 06609
Château Logue, Maniwaki 06226
Craigdarroch Castle, Victoria 07245
Fort Ingall, Cabano 05508
Fort-Numéro-Un de la Pointe-de-Lévy, Lévis 06167
Fort Point Museum, LaHave 06128
Fort Rodd Hill, Victoria 07247
Fort Saint Joseph, Richard's Landing 06737
Fort Selkirk, Yukon 07418
Fortress of Louisbourg, Louisbourg 06197
Musée du Fort Saint-Jean, Saint-Jean-sur-Richelieu 06828
Old Fort Erie, Niagara Falls 06448
Old Fort William Historical Park, Thunder Bay 07078
Prince of Wales Fort, Churchill 05616

China

Chaotiangong Palace, Nanjing 08038
Museum of the Imperial Palace of Manchu State, Changchun 07589
Nan Yue Wang Palace Museum, Guangzhou 07735
Pu Le Temple, Chengde 07613
The Summer Palace, Beijing 07570

Colombia

Museo El Castillo Diego Echavarría, Medellín 08678

Cuba

Fortaleza de Nuestra Señora de los Angeles de Jagua, Jagua 09186
Museo Fuerte de la Loma, Victoria de las Tunas 09345

Czech Republic

Chebský Hrad, Cheb 09472
Expozice Oddělení Starších Českých Dějin, Vrchotovy Janovice 09916
Hrad Helfštýn, Týn nad Bečvou 09883
Státní Zámek Vizovice, Vizovice 09909
Zámecké Muzeum, Kravaře 09606
Zámek, Hradec nad Moravicí 09535
Zámek Brandýs, Brandýs nad Labem 09414
Zámek Bučovice, Bučovice 09446
Zámek Kinski, Chlumec nad Cidlinou 09475
Zámek Libochovice, Libochovice 09626
Zámek Líčkov - Galerie O. Brázda, Žatec 09927
Zámek Lysice, Lysice 09645
Zámek Slavkov - Austerlitz, Slavkov u Brna 09841

Denmark

Borgmuseum, Spøttrup 10283
Jægerspris Slot, Jægerspris 10109
Kronborg Slot, Helsingør 10066
Liselund Gamle Slot, Borre 09981
Palæsamlingerne, Roskilde 10245
Rosenholm Slot, Hornslet 10094
Voergård Slot, Dronninglund 09992

Dominican Republic

Museo Casa-Fuerte Juan de Ponce de León, Higuëy 10328
Museo de las Casas Reales, Santo Domingo 10351

Egypt

Abdeen Palace Museum, Cairo 10484
Manial Palace Museum, Cairo 10499

Estonia

Kiek in de Kök, Tallinn 10633
Rakvere Linnus-Muuseum, Rakvere 10611

Finland

Hanko Museo, Hanko 10757
Kastelholms Slott, Sund 11333
Olavinlinna Castle, Olavinlinna 11161
Suomenlinna-Museo, Helsinki 10824
Suomenlinna - Sveaborg, Helsinki 10825
Turun Kaupungin Historiallinen Museo, Turun Linna, Turku 11403

France

Château, Arlay 11681
Château Beaufief, Mazeray 14049
Château d'Angers, Angers 11602
Château d'Arcelot, Arceau 11663
Château de Bruniquel Semateg, Bruniquel 12200
Château de Fénelon, Sainte-Mondane 15576
Château de Fleurac, Nersac 14395
Château de Hautefort, Hautefort 13191
Château de Larroque-Toirac, Larroque-Toirac 13531
Château de Messilhac, Raulhac 14988
Château de Montmuran, Les Iffs 13702
Château de Rochefort-en-Terre, Rochefort-en-Terre 15074
Château de Saint-Jean-de-Beauregard, Saint-Jean-de-Beauregard 15321
Château de Talcy, Marchenoir 13950
Château-Musée, Carrouges 12300
Château-Musée, Dinan 12746
Château-Musée d'Art Populaire et Religieux, Clermont-en-Genevois 12546
Château-Musée de Courson, Courson-Monteloup 12657
ChÂteau Musée de Montmorin, Beillom 11920
Château-Musée de Terre-Neuve, Fontenay-le-Comte 12968
Château-Musée de Vollore, Vollore-Ville 16268
Château-Musée des Ducs, Duras 12800
Château-Musée et Ecuries de Chaumont-sur-Loire, Chaumont-sur-Loire 12486
Château-Musée Mansart, Sagonne 15147
Château-Musée Municipal, Nemours 14390
Collections du Château de Breil de Foin, Genneteil 13042
Collections du Château de Courances, Courances 12649
Exposition du Château Fort du Fleckenstein, Lembach 13678
Fort Carré, Antibes-Juan-les-Pins 11639
Fort de Verny, Verny 16124
Fort l'Ecluse, Leaz 13669
Mini-Château, Amboise 11577
Musée Arthur Le-Duc, Torigni-sur-Vire 15918
Musée Benoît XIII, Châteaurenard (Bouches-du-Rhône) 12454
Musée-Château de Goulaine, Haute-Goulaine 13189
Musée-Château de La Verrerie, Oizon 14505
Musée de Château Guillaume, Lignac 13745
Musée des Plans-Reliefs, Paris 14688
Musée du Château, Montbard 14181
Musée du Château, Noirmoutier-en-l'Île 14468
Museè du Château de Bourbilly, Vic-de-Chassenay 16141
Musée du Château de Flaugergues, Montpellier 14244
Musée du Château de Vincennes, Vincennes 16248
Musée du Fort Vauban, Ambleteuse 11573
Musée du Manoir d'Argentelles, Villebadin 16179
Musée du Tire-Bouchon, Ménerbes 14072
Musée Fort de Sucy, Sucy-en-Brie 15837
Musée Fort Lagarde, Prats-de-Mollo-la-Preste 14941
Musée Historial, Mazères (Gironde) 14051
Musée National des Châteaux de Versailles et de Trianon, Versailles 16129
Musée National du Château de Fontainebleau, Fontainebleau 12966
Musée National du Château de Pau, Pau 14759

Germany

Allgäuer Burgenmuseum, Kempten 19511
Bettina und Achim von Arnim Museum, Niederer Fläming 20704
Burg Gnandstein, Kohren-Sahlis 19670
Burg Lauenstein, Ludwigsstadt 20063
Burg Mildenstein, Leisnig 19943
Burg Pappenheim mit Naturmuseum und Historischem Museum, Pappenheim 20981
Burg Prunn, Riedenburg 21314
Burg Trausnitz, Landshut 19800
Burg Trifels, Annweiler am Trifels 16558
Burg- und Klosteranlage Oybin, Oybin 20965

Burg Zwernitz, Wonsees 22679
Burgenmuseum Eisenberg, Eisenberg, Kreis Ostallgäu 18101
Burgmuseum Burg Guttenberg, Haßmersheim 19032
Burgmuseum der Wasserburg Kapellendorf, Kapellendorf 19429
Burgmuseum Grünwald, Grünwald 18818
Burgmuseum Kirschau, Kirschau 19558
Burgmuseum Marksburg, Braubach 17513
Burgmuseum Ortenberg, Ortenberg, Hessen 20914
Dornburger Schlösser, Dornburg, Saale 17905
Felsenburg Neurathen, Lohmen, Sachsen 20025
Festung Königstein, Königstein 19656
Festung Marienberg mit Fürstenbaumuseum, Würzburg 22710
Die Festung Rosenberg - Deutsches Festungsmuseum, Kronach 19719
Festung Wilhelmstein im Steinhuder Meer, Wunstorf 22732
Fürstliches Schloß und Kreuzgang Sankt Emmeram, Regensburg 21222
Herzogliches Museum Gotha, Gotha 18718
Historische Wehranlage, Mühlhausen/Thüringen 20402
Historisches Kabinett Burgruine Eckartsburg, Eckartsberga 18052
Historisches Museum für Mittelanhalt und Bach-Gedenkstätte im Schloss Köthen, Köthen 19666
Jagdschoss Letzlingen, Letzlingen 19966
Kaiserburg Nürnberg, Nürnberg 20773
Kaiserpfalz Gelnhausen, Gelnhausen 18570
Königshaus am Schachen, Garmisch-Partenkirchen 18540
Kulturstiftung Schloss Britz, Berlin 17163
Kunst- und Wunderkammer Burg Trausnitz, Landshut 19802
Landesmuseum Koblenz, Koblenz 19585
Löwenburg, Kassel 19467
Museum Burg Frankenberg, Aachen 16404
Museum Burg Kriebstein, Kriebstein 19714
Museum Burg Querfurt, Querfurt 21155
Museum Burg Ramsdorf, Velen 22214
Museum Burg Stein, Hartenstein 19018
Museum Burg Stickhausen, Detern 17836
Museum für brandenburgische Kirchen- und Kulturgeschichte, Ziesar 22802
Museum im Alten Schloß, Altensteig 16503
Museum Schloss Fasanerie, Eichenzell 18079
Museum Schloss Friedrichsfelde, Berlin 17215
Museum Schloss Hardenberg, Velbert 22211
Museum Schloss Herzberg, Herzberg am Harz 19167
Museum Schloss Steinau an der Straße mit Brüder-Grimm-Gedenkstätte, Steinau an der Straße 21878
Museum Zitadelle Jülich, Jülich 19405
Neues Schloss Herrenchiemsee, Herrenchiemsee 19160
Neues Schloss Schleißheim, Oberschleißheim 20834
Privatsammlung Burg Runkel, Runkel 21450
Residenz Ansbach, Ansbach 16561
Residenzmuseum im Celler Schloss, Celle 17689
Residenzschloß Arolsen, Bad Arolsen 16673
Schloß Augustusburg, Brühl 17607
Schloß Berleburg, Bad Berleburg 16693
Schloss Burg, Solingen 21808
Schloß Egg, Bernried, Kreis Deggendorf 17304
Schloß Falkenlust, Brühl 17608
Schloß Favorite, Rastatt 21185
Schloß Hämelschenburg, Emmerthal 18155
Schloss Harburg, Harburg 19009
Schloss Hohenschwangau, Schwangau 21668
Schloß Leitzkau, Leitzkau 19944
Schloss Linderhof, Ettal 18279
Schloss Ludwigsburg, Ludwigsburg 20047
Schloss Ludwigslust, Ludwigslust 20062
Schloss Neuschwanstein, Schwangau 21669
Schloss Rosenau, Rödental 21352
Schloss Schönhausen, Berlin 17251
Schloß Schwetzingen, Schwetzingen 21703
Schloß Solitude, Stuttgart 21958
Schloss Stolzenfels, Koblenz 19590
Schloss und Hofgarten Dachau, Dachau 17767
Schloss und Park Schönbusch, Aschaffenburg 16591
Schloss und Schlossgarten Weilburg, Weilburg 22386
Schloß und Schloßpark Bad Homburg, Bad Homburg v.d.Höhe 16753
Schloß und Schloßpark Wilhelmsthal, Calden 17677
Schloss Unteraufsess, Aufseß 16610
Schloss Veitshöchheim, Veitshöchheim 22209
Schloß vor Husum, Husum 19301
Schloßmuseum, Arnstadt 16574
Schloßmuseum, Bad Bentheim 16685
Schloßmuseum, Erbach, Alb-Donau-Kreis 18189
Schloßmuseum, Kirchberg an der Jagst 19544
Schloßmuseum, Kronburg 19725
Schloßmuseum, Ortenburg 20916
Schloßmuseum Amerang, Amerang 16538
Schloßmuseum Hohenlohe-Schillingsfürst, Schillingsfürst 21554
Schloßmuseum Hubertusburg, Wermsdorf 22492
Schlossmuseum Schwerin, Schwerin 21698
Schlossmuseum Sondershausen, Sondershausen 21817
Schloßmuseum und Park Molsdorf, Erfurt 18210
Schloßparkmuseum, Bad Kreuznach 16768
Spielzeugmuseum im Alten Schloß, Sugenheim 21975
Stadt- und Kulturgeschichtliches Museum Torgau, Torgau 22073
Städtisches Heimatmuseum Landau, Landau in der Pfalz 19792
Städtisches Museum Wesel, Wesel 22508
Städtisches Museum Wesel, Wesel 22509
Toppler-Schlößchen, Rothenburg ob der Tauber 21412
Wartburg, Eisenach 18099
Wasserschloß Mitwitz, Mitwitz 20347
Willibaldsburg, Eichstätt 18085

Ghana

Manhyia Palace Museum, Kumasi 22858

Hungary

Beethoven Emlékmúzeum, Martonvásár 23691
Bodrogközi Kastélymúzeum, Pácin 23745
Diósgyőri Vármúzeum, Miskolc 23702
Esterházy Kastélymúzeum, Fertőd 23557
Festetics-Kastély, Dég 23522
Festetics-Kastély, Keszthely 23660
Gödöllői Királyi Kastélymúzeum, Gödöllő 23565
Károlyi-Kastély, Fehérvárcsurgó 23556
Károlyi-Kastély, Füzérradvány 23562
Kazinczy Ferenc Múzeum, Sátoraljaújhely 23798
Kinizsi Pál Vármúzeum, Nagyvázsony 23728
L'Huillier-Coburg-Kastély, Edelény 23535
Lónyai-Kastély, Tuzsér 23940
Monostori Erőd, Komárom 23680
Nádasdi Ferenc Múzeum, Sárvár 23796
Sümegi Vármúzeum, Sümeg 23829
Tatai Vár, Tata 23919
Vármúzeum, Dunaföldvár 23529
Vármúzeum, Pécsvárad 23784
Vármúzeum, Siklós 23800
Vármúzeum, Veszprém 23963
Vármúzeum, Simontornya 23801
Vay Adám Múzeum, Vaja 23947
Zempléni Múzeum, Szerencs 23887

Iran

Baghcheh-Jaq Palace, Urmieh 24699
Golestan Palace Museum, Tehrān 24668
Green Palace Museum, Tehrān 24669
Mianposhteh Palace Museum, Bandar-Anzali 24599
Museum Falak-ol-Aflak, Khorram-Abaad 24623
Niavaran Palace, Tehrān 24689

Ireland

Driminagh Castle, Dublin 24784
Dunguaire Castle, Kinvara 24858
King John's Castle, Limerick 24865
Malahide Castle, Malahide 24874
Ross Castle, Killarney 24852

Israel

Museum of Acre, Acre 24919

Italy

Castel Beseno, Besenello 25410
Castello Fénis, Fénis 26095
Museo del Castello di Porciano, Stia 27902
Museo del Castello di Sarre, Sarre 27745
Museo del Risorgimento Duca d'Aosta, Sanluri 27694
Museo dell'Aria dello Spazio, Due Carrare 26049
Museo di Palazzo d'Arco, Mantova 26595
Museo di Palazzo Piccolomini, Pienza 27201
Museo di Rievocazione Storica, Mondavio 26776
Museo Morando Bolognini, Sant'Angelo Lodigiano 27721
Museo Nazionale di Canossa N. Campanini, Ciano d'Enza 25884
Museo Statale, Mileto 26730
Museo Storico del Castello di Miramare, Trieste 28096
Palazzo della Ragione, Padova 27050
Rappresentazione Scenografica Gioacchino Murat, Pizzo 27262
Villa Romana del Casale, Piazza Armerina 27196

Japan

Fukuyama Castle Museum, Fukuyama 28444
Heijokyuseki Shiryokan, Nara 29029
Himeji-jo, Himeji 28510
Hiroshima Castle, Hiroshima 28533
Honjin Hakubutsukan Futagawa Syuku, Toyohashi 29527
Inuyama Kenbunkan, Inuyama 28574
Iwasaki Kinen Kenbunkan, Nisshin-city, Aichi Prefecture 29078
Katsuyama Kenbunkan, Katsuyama 28682
Kumamoto Castle, Kumamoto 28775
Odawara-jo Rekishi Kenbunkan no Tenshukaku, Odawara 29088
Okawa-suji Samurai Residence, Kochi 28745
Osaka Castle Museum, Osaka 29140
Saga-kenritsu Nagoya Jou Hakubutsukan, Chinzei 28402
Shiga-kenritsu Shiryokan, Azuchi 28386

Laos

Vat Xiengkhouane, Vientiane 29978

Latvia

Bauskas Pils Muzejs, Bauska 29989
Rundāles Pils Muzejs, Pilsrundāle 30054

Lithuania

Gedimino Pilies Bokštas, Vilnius 30405
Kauno IX Forto Muziejus, Kaunas 30249
Pilies Muzieus, Klaipėda 30284

Luxembourg

Musée de la Forteresse, Luxembourg 30479
Schloß Vianden, Vianden 30496

Malta

Palazzo Parisio, Naxxar 30634

Mexico

Museo Baluarte de Santiago, Veracruz 31535
Museo de la No Intervención Fuerte de Loreto, Puebla 31277
Museo de Sitio Castillo de Teayo, Castillo de Teayo 30751
Museo del Fuerte de San Felipe, Bacalar 30732
Museo del Fuerte de San Juan de Ulúa, Veracruz 31537
Museo Histórico Fuerte de San José el Alto, Campeche 30743

Mongolia

Bogd Khaan Winter Palace Museum, Ulaanbaatar 31617

Myanmar

Myanan Sankyaw Shwe Nandaw, Mandalay 31718

Namibia

Fort Namutoni, Tsumeb 31758

Nepal

Patan Museum, Lalitpur 31782

Netherlands

Fort Rammekens, Ritthem 32691
Fraeylemaborg, Slochteren 32784
De Groene Schuur, Culemborg 32042
Kasteel Amerongen, Amerongen 31808
Kasteel Ammersoyen, Ammerzoden 31819
Kasteel Cannenburch, Vaassen 32860
Kasteel De Doornenburg, Doornenburg 32145
Kasteel Duivenvoorde, Voorschoten 32907
Kasteel Groeneveld, Baarn 31946
Kasteel Heeswijk, Heeswijk-Dinther 32342
Kasteel Hernen, Hernen 32370
Kasteel Hoensbroek, Hoensbroek 32398
Kasteel Middachten, De Steeg 32051
Kasteel Radboud, Medemblik 32542
Kasteeltoren IJsselstein, IJsselstein 32427
Muiderslot, Muiden 32562
Museum Fort Kijkduin, Den Helder 32111
Museum Kasteel Wijchen, Wijchen 32946
Museum Menkemaborg, Uithuizen 32836
Het Nederlands Vestingmuseum, Naarden 32564
Poptaslot Heringa-State, Marssum 32541
Slot Loevestein, Poederoijen 32670
Slot Zeist, Zeist 32994
Slot Zuylen, Oud-Zuilen 32648
Stichting Huis Bergh, 's-Heerenberg 32331
Vestingmuseum, Nieuweschans 32580

Northern Cyprus

Girne Kalesi, Girne 33358
Otello Kalesi, Gazimağusa 33353
Sipahi Ay. Trias Bazilikası ve Kantara Kalesi, Kantara 33364
Vuni Sarayi, Lefke 33366

Norway

Baroniet Rosendal, Rosendal 33799
Bogstad Gård, Oslo 33718
Håkonshallen og Rosenkrantztårnet, Bergen 33427
Norges Hjemmefrontmuseum, Oslo 33738

Oman

Nahkl Fort Museum, Nahkl 33993

Poland

Muzeum Dworek Zabytkowy, Stryszów 35136
Muzeum Zamkowe w Malborku, Malbork 34914
Punkt Muzealny w Pałacu Opatów, Supraśl 35141
Twierdza Wisłoujście, Gdańsk 34652
Zamek Piastowski, Gliwice 34667
Zamek w Czersku, Czersk 34612
Zamek w Gniewie - Muzeum Regionalne, Gniew 34671

Portugal

Fortaleza de Sagres, Sagres 35894
Museu da Caça, Vila Viçosa 36017
Museu do Castelo, Ourém 35796
Museu do Castelo de Castro Marim, Castro Marim 35518
Núcleo Museológico do Castelo do Lindoso, Braga 35484
Palácio da Bolsa, Porto 35865
Palácio de Estói, Estói 35572
Palácio de Monserrate, Sintra 35950
Palácio dos Marqueses de Fronteira, Lisboa 35732
Palacio Nacional de Queluz, Queluz 35879
Palácio Nacional de Sintra, Sintra 35952
Palácio Valenças, Sintra 35953
Quinta Municipal de Nossa Senhora da Piedade, Póvoa de Santa Iria 35872

Puerto Rico

Castillo de San Felipe del Morro, San Juan 36061
Museo Fuerte Conde de Mirasol, Vieques 36082

Romania

Cetatea Braşov şi Fortificaţiile din Ţara Bârsei, Braşov 36175

Russia

Istoriko-architekturnyj Muzej Kuzneckaja Krepost', Novokuzneck 37479
Memorialnyj Kompleks 'Vorošilovskaja Batareja', Vladivostok 37940
Muzej Kronštadtskaja Krepost, Kronštadt 37086
Petropavlovskaja Krepost - Istoriko-kulturnyj Zapovednik, Sankt-Peterburg 37719
Šlisselburgskaja Krepost Orešek, Šlisselburg 37769
Vladivostokskaja Krepost - Voenno istoričeskij Fortifikacionnyj Muzej, Vladivostok 37951

San Marino

Palazzo Pubblico, San Marino 38011

Slovakia

Expozícia Zbraní a Stredovekého Mestského Opevnenia, Bratislava 38208
Hrad Devín, Bratislava 38214
Hrad Krásna Hôrka, Krásnohorské Podhradie 38278
Nový Zamok Banská Štiavnica, Banská Štiavnica 38192
Spišský Hrad, Spišské Podhradie 38360

Slovenia

Grad Bogenšperk, Šmartno pri Litiji 38522
Grad Dobrovo, Dobrovo v Brdih 38413
Grad Podsreda, Podsreda 38494
Grad pri Gradu, Grad 38419
Muzej Političnih Zapornikov, Internirancev in Izgnancev Grad Rajhenburg, Brestanica 38403
Sevniški Grad, Sevnica 38513
Zbirka v Jamarskem Domu, Domžale 38416

Spain

Castillo de Santa Catalina, Cádiz 39087
Museo del Alcázar, Toledo 40135
Museu-Monestir de Pedralbes, Barcelona 39000
Palacio Real de Aranjuez, Aranjuez 38879
Palacio Real de Madrid, Madrid 39586
Sacro Convento Castillo de Calatrava la Nueva, Aldea del Rey 38836

Sweden

Bosjökloster Slot Trädgårdar, Höör 40597
Drottningholms Slott, Drottningholm 40488
Engsö Slott, Västerås 41030
Finspångs Slott, Finspång 40521
Glimmingehus, Hammenhög 40581
Grönsöö Slott, Enköping 40498
Hallwylska Museet, Stockholm 40898
Kalmar Slott, Kalmar 40626
Kina Slott, Drottningholm 40489

C

Löfstad Slott Museum, Norrköping 40772
Mälsåkers Slott, Stallarholmen 40879
Nynäs Slott, Tystberga 40990
Rosersbergs Slott, Rosersberg 40816
Sofiero Slott, Helsingborg 40592
Strömsholms Slott, Kolbäck 40651
Teleborgs Slott, Växjö 41042
Trollenäs Slott, Eslöv 40507
Tullgarns Slott, Vagnhärad 41046
Tyresö Slott, Tyresö 40988
Vadstena Slott, Vadstena 41021
Vaxholms Fästnings Museum, Vaxholm 41054

Switzerland

Burgenkundliche Sammlung im Burgenmuseum, Gossau (Sankt Gallen) 41443
Château de Chillon, Montreux 41637
Château de Coppet, Coppet 41313
Château de Gruyères, Gruyères 41460
Château d'Oron, Oron-le-Châtel 41715
Festungsmuseum Crestawald, Filisur 41370
Museum Schloss Kyburg, Kyburg 41521
Museum Schloss Waldegg, Feldbrunnen 41369
Ortsmuseum Dietikon, Dietikon 41330
Schloß Hallwyl, Seengen 41856
Schloß Jegenstorf, Jegenstorf 41498
Schloß Spiez, Spiez 41893
Schloß Thunstetten, Thunstetten 41942
Schloss Werdenberg, Werdenberg 42007
Schloßmuseum Heidegg, Gelfingen 41397
Stadtmuseum Aarau, Aarau 41085

Thailand

Chantharakasem National Museum, Ayutthaya 42526

Turkey

Anadoluhisarı Müzesi, İstanbul 42801
Beylerbeyi Sarayı, İstanbul 42808
İshak Paşa Sarayi, Ağri 42679
Rumelihisarı Müzesi, İstanbul 42836
Tekfur Sarayi, İstanbul 42841
Yedikulehisarı Müzesi, İstanbul 42849

Turkmenistan

Taryhy Ýer Gökdepe Galasy, Gökdepe 42975

Ukraine

Istoryko-architekturna Pam'jatka-muzej Kyïvska Fortecja, Kyïv 43079

United Arab Emirates

Shalimar Palace, Dubai 43219

United Kingdom

Bamburgh Castle, Bamburgh 43368
Berkeley Castle, Berkeley 43453
Bickleigh Castle, Bickleigh 43476
Blair Castle, Blair Atholl 43537
Bodiam Castle, Robertsbridge 45503
Broughty Castle Museum, Dundee 44045
Buckingham Palace, London 44810
Caerphilly Castle, Caerphilly 43712
Caldicot Castle, Caldicot 43713
Cardiff Castle, Cardiff 43749
Castle and Regimental Museum, Monmouth 45167
Castle Cary and District Museum, Castle Cary 43767
Castle Coole, Enniskillen 44177
Castle Fraser, Sauchen 45619
Castle Menzies, Aberfeldy 43250
Castle Museum, York 46099
Castle of Saint John, Stranraer 45776
Castle Rushen, Castletown 43774
Cawdor Castle, Cawdor 43781
Compton Castle, Paignton 45342
Culross Palace, Culross 43938
Culzean Castle and Country Park, Maybole 45120
Deal Castle, Deal 43958
Dean Castle, Kilmarnock 44578
Denbigh Castle, Denbigh 43963
Dover Castle, Dover 44012
Downhill Castle, Castlerock 43772
Drum Castle, Drumoak 44028
Dunster Castle, Dunster 44066
Dunvegan Castle, Dunvegan 44067
Elizabeth Castle, Saint Helier 45570
Floors Castle, Kelso 44555
Frogmore House, Windsor 46041
Fyvie Castle, Fyvie 44258
Giant's Castle, Marazion 45102
Gilstrap Heritage Centre, Newark-on-Trent 45210
Glamis Castle, Glamis 44273
Harlech Castle, Harlech 44395
Harwich Redoubt Fort, Harwich 44411
Hever Castle and Gardens, Edenbridge 44112
Kensington Palace State Apartments, London 44893
Knaresborough Castle and Old Courthouse Museum, Knaresborough 44619
Lauriston Castle, Edinburgh 44131
Lewes Castle and Barbican House Museum, Lewes 44698
Linlithgow Palace, Linlithgow 44720
Mont Orgueil Castle, Saint Martin 45583
Muncaster Castle, Ravenglass 45460
Museum of Fulham Palace, London 44924
Museum of Local History and Industry, Llandrindod Wells 44755
Nothe Fort Museum of Coastal Defence, Weymouth 45988
Oakham Castle, Oakham 45308
Peel Castle, Peel 45351
Pendennis Castle, Falmouth 44210
Penkill Castle, Girvan 44271
Pontefract Castle Museum, Pontefract 45397
Powderham Castle, Kenton 44560
Sandal Castle, Sandal 45609
Southsea Castle, Portsmouth 45438
Stafford Castle and Visitor Centre, Stafford 45731
Story of Castle Cornet, Saint Peter Port 45594
Sudeley Castle, Winchcombe 46022
Thirlestane Castle, Lauder 44645
Urquhart Castle, Drumnadrochit 44027
Victoria Tower, Huddersfield 44498
Warwick Castle, Warwick 45940
Weoley Castle, Birmingham 43516
Windsor Castle, Windsor 46047

United States of America

Belcourt Castle, Newport 51319
Castle Clinton, New York 51139
Fort Casey Interpretive Center, Coupeville 47853
Fort Delaware, Narrowsburg 50956
Fort Inglish, Bonham 46898
Fort Leboeuf, Waterford 53960
Fort Loudoun, Chambersburg 47331
Fort Nashborough, Nashville 50969
Fort Okanogan, Pateros 51727
Fort Scott National Historic Site, Fort Scott 48680
Fort Sumner, Fort Sumner 48691
Hammond Castle Museum, Gloucester 48946
Iolani Palace, Honolulu 49344
Miramont Castle Museum, Manitou Springs 50441
Old Castle Museum, Baldwin City 46540
The Walsenburg Mining Museum and Fort Francisco Museum of La Veta, Walsenburg 53805
Wilson Castle, Proctor 52157

Uzbekistan

Tasviriy San'at Muzeyi - Andijon Qal'asi, Andijon 54442

Venezuela

Castillos de Guayana la Vieja, Guayana 54575

Vietnam

Bảo Đai Palace Museum, Đà Lat 54601

Ceramics → Porcelain and Ceramics

Chemistry

Austria

Alchimistenmuseum, Kirchberg am Wagram 02165

France

Maison de Louis Pasteur, Arbois 11657
Musée de la Chimie, Jarrie 13261

Germany

Deutsches Chemie-Museum, Merseburg 20291
Kunststoff-Museums-Verein e.V., Düsseldorf 17999
Liebig-Museum, Gießen 18639
Museum der Göttinger Chemie, Göttingen 18695
phæno gGmbH, Wolfsburg 22666
Wilhelm-Ostwald-Gedenkstätte, Großbothen 18783

Hungary

Magyar Vegyészeti Múzeum, Várpalota 23952

Italy

Museo di Chimica, Roma 27493

Lithuania

Vilniaus Universiteto Chemijos Fakulteto Muzieju, Vilnius 30443

Russia

Dom-muzej D.I. Mendeleeva, Boblovo 36811
Memorialnyj Muzej Akademikov A.E. i B.A. Arbuzovych, Kazan 37012
Muzej-Archiv D.I. Mendeleeva, Sankt-Peterburg 37671
Muzej Kazanskoj Chimičeskoj Školy, Kazan 37019

United Kingdom

Catalyst, Widnes 46012
Michael Faraday's Laboratory and Museum, London 44917

United States of America

Danville Science Center, Danville 47964

China (Porcelain) → Porcelain and Ceramics

Cigarettes → Tobacco

Cinematography

Argentina

Museo del Cine Pablo C. Ducrós Hicken, Buenos Aires 00224

Armenia

Sergei Parajanov House-Museum, Yerevan 00753

Austria

Österreichische Filmgalerie, Krems 02231
Österreichisches Filmmuseum Wien, Wien 03046
Zentralfilmarchiv Laxenburg, Laxenburg 02276

Belgium

Filmmuseum, Antwerpen 03286
Musée du Cinéma, Bruxelles 03481
Vlaams Filmmuseum en -archief, Kortrijk 03762

Canada

Cinémathèque Québécoise - Musée du Cinéma, Montréal 06336

Denmark

Det Danske Filminstitut, København 10128

France

La Cinémathèque Française, Paris 14581
Musée de l'Audiovisuel, Montarcher 14169
Musée Raimu, Cogolin 12567
Musée Vivant du Cinéma, Lyon 13902

Georgia

State Museum of Theatre, Music and Cinema, Tbilisi 16383

Germany

Deutsche Kinemathek, Berlin 17068
Deutsche Kinemathek - Museum für Film und Fernsehen, Berlin 17069
Deutsches Filmmuseum, Frankfurt am Main 18362
Filmmuseum, München 20440
Filmmuseum Landeshauptstadt Düsseldorf, Düsseldorf 17988
Filmmuseum Potsdam, Potsdam 21087
Industrie- und Filmmuseum Wolfen, Wolfen 22650
Kinemathographisches Museum, Wuppertal 22744
Münchner Stadtmuseum, München 20462
Museum für visuelle Kommunikation, Köln 19622
Stadtmuseum Tübingen, Tübingen 22124
Zeiler Foto- und Filmmuseum, Zeil am Main 22773

Hungary

Magyar Filmtörténeti Gyűjtemény, Budapest 23436

Iceland

Kvikmyndasafn Islands, Hafnarfjörður 23999

Iran

Film Museum of Iran, Tehrān 24665

Italy

Museo del Cinema, Milano 26697
Museo dell'Immagine, Cesena 25832
Museo Nazionale del Cinema, Torino 28021

Japan

Fukuoka-machi Shashinki Hakubutsukan, Fukuoka, Toyoma 28438
Ozu Yasujiro Seihunkan Museum, Matsusaka 28899
Tin Toy Museum Motion Display Museum, Yokohama 29609
Toei Movie Land Toei Uzumasa Eigamura Movie Museum, Kyoto 28871
Tokyo Kokuritsu Kindai Bijutsukan, National Film Center, Tokyo 29493

Korea, Republic

Shinyoung Cinema Museum, Namwon 29856

Latvia

Rīgas Kino Muzejs, Rīga 30107

Lithuania

Lietuvos Teatro, Muzikos ir Kino Muziejus, Vilnius 30422

Mauritius

Musée de la Photographie, Port Louis 30683

Mexico

Museo del Cine, Durango 30842

Netherlands

Filmmuseum Foundation Nederlands, Amsterdam 31841
Nederlands Instituut voor Beeld en Geluid-Colleties, Hilversum 32389

New Zealand

Museum of Audio Visual Technology, Wellington 33250

Norway

Filmmuseet, Oslo 33723

Poland

Muzeum Kinematografii w Łodzi, Łódź 34877

Portugal

Museu da Imagem em Movimento, Leiria 35658
Museu do Cinema, Lisboa 35699

Russia

Muzej Goroda Jurjevca, Jurjevec 36984
Muzej Kino, Moskva 37304
Muzej Kinostudii im. A.M. Gorkogo, Moskva 37305
Muzej Kinostudii Mosfilm, Moskva 37306
Muzej Andreja Tarkovskogo, Jurjevec 36985

Serbia

Muzej Jugoslavenska Kinoteka, Beograd 38068

Slovenia

Slovenska Kinoteka, Ljubljana 38462

South Africa

Mayibuye Centre Collection, Bellville 38558

Spain

Museu del Cinema, Girona 39345

Sweden

Biografmuséet, Säter 40822
Filmmuseet, Kristianstad 40653

Taiwan

Image Museum of Hsinchu City, Hsinchu 42209

Ukraine

Deržavnyj Muzej Teatral'noho, Muzyčnoho ta Kinomystectva Ukraïny, Kyïv 43077

United Kingdom

Cinema Museum, London 44823
East Anglian Film Archive, Norwich 45277
Laurel and Hardy Museum, Ulverston 45897

United States of America

The Academy Gallery, Beverly Hills 46790
American Advertising Museum, Portland 52071
American Film Institute, Los Angeles 50208
The Andy Warhol Museum, Pittsburgh 51938
Anthology Film Archives, New York 51120
Berkeley Art Museum and Pacific Film Archive, Berkeley 46752
Birthplace of John Wayne, Winterset 54255
Film Forum - Film Archives, Collingswood 47693
Hollywood Entertainment Museum, Los Angeles 50231
Hollywood Heritage Museum, Los Angeles 50233
The Jimmy Stewart Museum, Indiana 49483
The Light Factory, Charlotte 47391
McLarty Treasure Museum, Vero Beach 53727
Martin and Osa Johnson Safari Museum, Chanute 47342
The Movie Museum, Owosso 51643
The Museum of Films for Children, Weston 54103
The Walt Disney Family Museum, San Francisco 52798

Uzbekistan

Muzej Kinoiskusstva Uzbekistana, Toshkent 54477

Circus → Performing Arts

Clocks and Watches → Horology

Clothing and Dress

Argentina

Museo Cooperativo Eldorado, Eldorado 00361
Museo Criollo de los Corrales, Buenos Aires 00176
Museo de la Casa Rosada, Buenos Aires 00200
Museo Nacional de la Historia del Traje, Buenos Aires 00260

Australia

Benalla Costume and Pioneer Museum, Benalla 00858
Don Bank Museum, North Sydney 01373
Frances Burke Textile Resource Centre-Collections, Melbourne 01278
Scout Heritage Centre of Western Australia, West Perth 01643
Yesteryear Costume Gallery, Orroroo 01396

Austria

Heimatmuseum Traun, Traun 02824
Museum der Heimatvertriebenen, Vöcklabruck 02866

Belgium

Fashion Museum, Hasselt 03682
Modemuseum Provincie Antwerpen, Antwerpen 03294
Musée du Costume et de la Dentelle, Bruxelles 03483
Poldermuseum, Antwerpen 03310

Bolivia

Museo de Textiles Andinos Bolivianos, La Paz 04217

Brazil

Museo Municipal Barão de Santo Angelo, Rio Pardo 04776

Bulgaria

Etnografska Ekspozicija, Svištov 05220
Istoričeski Muzej, Goce Delčev 05035

Canada

Costume Museum of Canada, Dugald 05721
Ermatinger Old Stone House, Sault Sainte Marie 06930
Musée Marsil - Musée du Costume et du Textile du Québec, Saint-Lambert 06861
Ukrainian Catholic Women's League Museum, Edmonton 05757
Ukrainian Museum of Canada, Vancouver 07225
Ukrainian Museum of Canada - Manitoba Branch, Winnipeg 07388

Côte d'Ivoire

Musée National du Costume, Grand-Bassam 08836

Croatia

Etnografski Muzej, Zagreb 08989
Etnografski Muzej Split, Split 08955

Czech Republic

Muzeum Krajky, Prachatice 09732
Zámek Hrubý Rohozec, Turnov 09882

Denmark

Brede Værk, Kongens Lyngby 10165
Fanø Skibsfarts- og Dragtsamling, Fanø 10015

Ethiopia

City Museum, Makale 10685

Finland

Helsinge Hembygdsmuseum, Vantaa 11445
Kansallispukukeskus, Jyväskylä 10893
Nukke- ja Pukumuseo - Hatanpäa Mansion, Tampere 11355

France

Atelier-Musée du Chapeau, Chazelles-sur-Lyon 12494
Centre d'Exposition du Costume, Avallon 11782
La Chapeauthèque, Anduze 11598
Château Beaufief, Mazeray 14049
Ecomusée de Salazie, Salazie 15592
Ecomusée du Pays de Montfort, Montfort-sur-Meu 14210
Musée d'Allauch, Allauch 11555
Musée de la Bonneterie, Troyes 15999
Musée de la Chapellerie, Le Somail 13651
Musée de la Coiffe, Blesle 12026
Musée de la Mode, Marseille 13998
Musée de la Mode et du Textile, Paris 14654
Musée de la Mode Retrouvée, Digoin 12730
Musée de l'Automobile La Belle Époque, Pont-l'Évêque 14891
Musée de Traditions Populaires et d'Archéologie, Chauvigny 12488
Musée Départemental Breton, Quimper 14970
Musée départemental de la Bresse - Domaine des Planons, Saint-Cyr-sur-Menton 15228
Musée d'Ethnographie et d'Art Populaire, Honfleur 13222
Musée du Costume, Château-Chinon 12432
Musée du Costume Breton, Paimpol 14554
Musée du Costume Comtadin, Pernes-les-Fontaines 14783
Musée du Costume Comtois, Ornans 14539
Musée du Costume Provençal, Solliès-Ville 15777
Musée du Costume Savogard, Sévrier 15745
Musée du Folklore, Moulins (Allier) 14307
Musée du Terroir Marseillais, Marseille 14007
Musée Galliera, Paris 14708
Musée Pyrénéen, Lourdes 13852
Musée Textile de Haute-Alsace, Husseren-Wesserling 13232

Georgia

Georgian Silk Museum, Tbilisi 16361

Germany

Deutsches Fächer Museum, Bielefeld 17338
Deutsches Phonomuseum und Uhrensammlung, Sankt Georgen im Schwarzwald 21492
Feuerwehr-Helm-Museum, Stuttgart 21932
Geburtshaus Levi Strauss, Buttenheim 17670
Heimat- und Miedermuseum Heubach, Heubach 19174
Heimatmuseum, Stade 21849
Hüttenberger Heimatmuseum, Linden 20003
Modemuseum im Schloss Ludwigsburg, Ludwigsburg 20046
Museum der Deutschen Spielzeugindustrie, Neustadt bei Coburg 20675
Regionales Heimatmuseum für das Renchtal, Oppenau 20905
Rheinisches Landesmuseum für Industrie- und Sozialgeschichte, Ratingen 21193
Schwarzwälder Trachtenmuseum, Haslach im Kinzigtal 19026
Trachten- und Heimatmuseum, Weiltingen 22394
Trachten- und Mühlenmuseum, Pfullingen 21035
Trachtenmuseum, Ochsenfurt 20854
Tuchmuseum Lennep der Anna-Hardt-Stiftung, Remscheid 21266

Greece

Folklore and Ethnological Museum of Macedonia and Thrace, Thessaloniki 23212
Folklore Museum, Xanthi 23242
Folklore Museum and Municipal Art Gallery, Karditsa 23034
Nafplion Folklore Museum, Nafplion 23117

Guadeloupe

Musée Municipal de Saint-Barthélemy, Saint-Barthélemy 23280

Guatemala

Museo Ixchel del Traje Indigena, Guatemala 23300

Hungary

Mária-ruhák Múzeum, Andocs 23335
Népviseletes Babamúzeum - Történelmi Panoptikum, Keszthely 23662
Palóc Néprajzi Magángyűjtemény és Palóc Baba Kiállítás, Mátrafüred 23695
Szombathelyi Képtár, Szombathely 23904

Iceland

Heimilisidnardarsafnid Halldórustofa, Blönduós 23988

India

Bharatiya Adim Jati Sevak Sangh Museum, Delhi 24138

Iraq

Baghdad Museum, Baghdad 24713

Ireland

Pighouse Folk Museum, Cornafean 24771

Italy

Casa di Carlo Goldoni, Venezia 28180
Galleria del Costume, Firenze 26158
Mostra Permanente del Costume Arbereshe, Vaccarizzo Albanese 28130
Musei Provinciali di Gorizia, Gorizia 26351
Museo Civico, Amalfi 25214
Museo dei Costumi della Partita a Scacchi, Marostica 26609
Museo dei Magli, Ponte Nossa 27285
Museo del Cappello, Alessandria 25191
Museo del Cappello, Montappone 26800
Museo del Costume Arberesh, Frascineto 26247
Museo del Costume e della Tradizione Nostra Gente, Guardiagrele 26385
Museo del Costume Farnesiano, Gradoli 26361
Museo del Costume R. Piraino, Palermo 27068
Museo del Costume Tradizionale, Grosio 26370
Museo del Vino e della Moda, Cilíverghe di Mazzano 25888
Museo della Donna E. Ortner, Merano 26658
Museo della Sindone, Torino 28007
Museo della Valle Cannobina, Gurro 26393
Museo delle Icone e della Tradizione Bizantina, Frascineto 26248
Museo di Palazzo Mocenigo, Venezia 28195
Museo Salvatore Ferragamo, Firenze 26197
Museo Stibbert, Firenze 26198
Sala del Costume e delle Tradizioni Popolari, Corinaldo 25979

Japan

Arimatsu Narumi Shibori Kaikan, Nagoya 28993
Fukuchiyama-shi Tanba Seikatsui-kan, Fukuchiyama 28425
Fuzoku Hakubutsukan, Kyoto 28810
Hirano Image Library, Osaka 29128
Iwasaki Museum, Yokohama 29594
Kobe-shi Fukushoku Hakabutsukan, Kobe 28733
Ome Kimono Museum, Ome 29111
Oriamu Museum, Izumiotsu 28615
Sugino Gakuen Isho Hakubutsukan, Tokyo 29476
Yatsuhashi-An Shishu-Yakata Gallery, Kyoto 28873

Jordan

Jordan Folklore Museum, Amman 29653

Korea, Republic

Korean Museum of Contemporary Clothing, Seoul 29898

Macedonia

Muzej na Grad Kratovo, Kratovo 30504

Malaysia

Museum of Traditional Costumes, Alor Gajah 30548

Mexico

Museo de Cultura Populares del Estado de México, Toluca 31487
Museo de la Indumentaria Mexicana Luis Márquez Romay, México 31084
Museo de Sitio del Claustro de Sor Juana, México 31093
Museo del Traje en Tabasco, Villahermosa 31553
Museo Serfín de Indumentaria Indígena, México 31148

Morocco

Musée Ethnographique, Tétouan 31702

Netherlands

Kostuummuseum De Gouden Leeuw, Noordhorn 32602
Museum Hindeloopen, Hindeloopen 32391
Museum Scheveningen, Den Haag 32089
Museum voor Klederdracht en Oud Speelgoed, Warnsveld 32924
'T Museumke, Handel 32309
Mutsen en Poffermuseum Sint-Paulusgasthuis, Sint Oedenrode 32777
Politie-Petten Museum, Slochteren 32786
Stedelijk Museum Drenthe's Veste, Coevorden 32038
Streekmuseum Ommen, Ommen 32620
Wasch-en Strijkmuseum, Boxtel 31999
De Wascht en Strekt, Gilze 32256
Zeeuws Poppen- en Klederdrachten Museum, Zoutelande 33009

New Zealand

Gore Historical Museum and Hokonui Heritage Research Centre, Gore 33113

Norway

Konfeksjonsmuseet, Molde 33677
Norsk Trikotasjemuseum, Salhus 33807
Setesdalsmuseet, Rysstad 33804

Poland

Muzeum Kamienica Orsettich w Jarosławiu, Jarosław 34696

Portugal

Museu do Traje de Viana do Castelo, Viana do Castelo 35983
Museu Nacional do Traje, Lisboa 35722

Qatar

Al-Wakra Museum, Al-Wakra 36084

Romania

Muzeul Ţesutului şi Postăvăritului, Buhuşi 36233

Russia

Muzej Chudožestvennych Tkanej, Moskva 37267
Muzej Ėtnografičeskogo Kostjuma na Kuklach, Moskva 37274
Muzej Ivanovskogo Sitca, Ivanovo 36950
Muzej Tverskogo Byta, Tver 37879

Slovakia

Národopisná Expozícia SARM Bardejovské Kúpele, Bardejovské Kúpele 38196

South Africa

Bernberg Fashion Museum, Johannesburg 38651
Drostdy Museum, Uitenhage 38768

Spain

Colección de la Casa de Alba, Epila 39276
Colección Museográfica d'Etnografia i Costums de la Mancomunitat de Municipis de la Vall d'Albaida, Benisoda 39027
Museo de Costumbres y Artes Populares de los Montes, Guadamur 39377
Museo del Baile Flamenco, Sevilla 40073

Sweden

Livrustkammaren, Stockholm 40909

Switzerland

Musée suisse de la Mode, Yverdon-les-Bains 42050

Taiwan

Costume Museum of Shih Chien University, Taipei 42386

Turkmenistan

Museum of Turkmen Values, Aşgabat 42971

C

United Kingdom

Bexhill Museum of Costume and Social History, Bexhill-on-Sea 43472
Burton Court, Eardisland 44079
Cavalcade of Costume, Blandford Forum 43540
Chertsey Museum, Chertsey 43814
Fairlynch Museum, Budleigh Salterton 43668
Fashion and Textile Museum, London 44846
Folk and Costume Museum, Câtel Guernsey 43779
Gallery of Costume, Manchester 45082
Hollytrees Museum, Colchester 43874
Horsham Museum, Horsham 44489
Museum of Costume, Bath 43407
Museum of Costume and Textiles, Nottingham 45294
Nidderdale Museum, Pateley Bridge 45346
Paulise de Bush Costume Collection, Broadclyst 43643
Royal Ceremonial Dress Collection, London 44970
Shambellie House Museum of Costume, New Abbey 45202
Springhill Costume Museum, Moneymore 45165
St John's House, Warwick 45939
Totnes Costume Museum, Totnes 45871
Vina Cooke Museum of Dolls and Bygone Childhood, Newark-on-Trent 45217

United States of America

Alling Coverlet Museum, Palmyra 51683
Black Fashion Museum, New York 51134
Dar Museum First Ladies of Texas Historic Costumes Collection, Denton 48060
Glove Museum, New York 51184
The Hermitage, Ho-Ho-Kus 49302
Historic Costume Textiles Collection, Columbus 47754
John E. and Walter D. Webb Museum of Vintage Fashion, Island Falls 49539
Kent State University Museum, Kent 49740
Mount Mary College Costume Museum, Milwaukee 50716
Museum at the Fashion Institute of Technology, New York 51221
Owatonna Arts Center, Owatonna 51635
Perry-Downer House and Costume Gallery, Monterey 50806
Philadelphia Mummers Museum, Philadelphia 51866
Reginald F. Lewis Museum of Maryland African American History and Culture, Baltimore 46586
Yeshiva University Museum, New York 51281

Coaches → Carriages and Carts

Coffee → Tea and Coffee

Coins → Numismatics

Cookery → Baking and Cookery

Copper

Australia

Geralka Rural Farm, Spalding 01520

Germany

Kupferschmiede-Museum, Tangermünde 22002

Mexico

Museo Nacional del Cobre, Salvador Escalante 31320

Netherlands

Koperslagersmuseum Van der Beele, Horst 32415

Spain

Calcografia Nacional, Madrid 39502

Cosmetics

Andorra

Museu del Perfum, Escaldes-Engordany 00093

Bulgaria

Kazanlăška Roza, Kazanlăk 05059

Canada

Irish Moss Interpretive Centre and Museum, Miminegash 06292
Perfume Museum, Niagara-on-the-Lake 06455

China

Huangshan Beauty Spot Museum, Huangshan 07833

France

Château-Promenade des Parfums, Chilleurs-aux-Bois 12512
Musée de la Lavande, Cabrières-d'Avignon 12221
Musée de la Parfumerie, Paris 14657
Musée de la Parfumerie Cyrnarom, Bastia 11860
Musée de la Parfumerie Fragonard, Grasse 13111
Musée de la Parfumerie Fragonard, Paris 14658
Musée de la Parfumerie Galimard, Grasse 13112
Musée des Arômes et du Parfum, Graveson 13125
Musée du Savon de Marseille Marius Fabre, Salon-de-Provence 15612
Musée International de la Parfumerie, Grasse 13114
Musée Traditions Verrieres, Eu 12886

Germany

Duftmuseum im Farina Haus, Köln 19600
Wella Museum, Darmstadt 17796

Japan

Iwata-gun Kaori Hakubutsukan, Toyoda 29526
Pola Fuzoku Hakubutsukan, Tokyo 29459

Korea, Republic

Pacific Museum, Yongin 29946

Mexico

Museo de Navajas y Curiosidades del Mundo, Puebla 31282

Spain

Museo Perfumeria Gal, Alcalá de Henares 38815
Museu del Perfum, Barcelona 38985

Switzerland

Schweizer Kamm-Museum, Mümliswil 41655

United States of America

Annette Green Perfume Museum, Los Angeles 50209
Museum of Cosmetology, Chicago 47494

Costume → Clothing and Dress

Crafts → Handicraft

Criminology

Argentina

División Museo e Investigaciones Históricas de la Policia Federal, Buenos Aires 00156
Museo de la Policía del Chaco, Resistencia 00523
Museo Forense de la Morgue Judicial, Buenos Aires 00233
Museo Histórico de la Policía de la Provincia de Misiones, Posadas 00499
Museo Histórico Policial de La Plata, La Plata 00414
Museo Penitenciario Argentino Antonio Ballvé, Buenos Aires 00270

Australia

Adelaide Gaol, Thebarton 01582
Bridgetown Old Gaol Museum, Bridgetown 00893
Fremantle Prison Precinct, Fremantle 01094
Gladstone Gaol, Gladstone, South Australia 01112
Hay Gaol Museum, Hay 01150
Justice and Police Museum, Sydney 01551
Lock-up Museum and John Paynter Gallery, Newcastle 01362
Melrose Courthouse Museum, Melrose 01292
Old Dubbo Gaol, Dubbo 01047
Old Gaol Museum, Albany 00787
Old Jail House Old Court House, Croydon 01014
Old Melbourne Gaol and Penal Museum, Melbourne 01289
Old Police Station Museum, Mount Barker 01326
Police Station and Courthouse Collections, Auburn 00818
Port Arthur Historic Site, Port Arthur 01427
Queensland Police Museum, Brisbane 00899
Richmond Gaol Museum, Richmond, Tasmania 01458
Roebourne Old Goal Museum, Roebourne 01467
Silverton Gaol Museum, Silverton 01505
South Australian Police Museum, Thebarton 01583
Stuart Town Gaol Museum, Alice Springs 00802
York Courthouse Complex, York 01685

Austria

Hans Gross Kriminalmuseum, Graz 01972
Museum für Rechtsgeschichte, Pöggstall 02510
Österreichisches Kriminalmuseum, Scharnstein 02700
Wiener Kriminalmuseum, Wien 03090

Belgium

Musée de la Police et de la Gendarmerie, Floreffe 03603
Politiemuseum Oudaan, Antwerpen 03311
Service Historique de la Police, Bruxelles 03503

Bolivia

Museo del Litoral Boliviano, La Paz 04218
Museo Histórico de la Policía Nacional, La Paz 04219

Brazil

Museu da Companhia Independente do Palácio Guanabara, Rio de Janeiro 04708
Museu da Polícia Civil, Rio de Janeiro 04718

Canada

Canada's Penitentiary Museum, Kingston 06075
Fort Saskatchewan Museum, Fort Saskatchewan 05822
Halifax Police Museum, Halifax 05936
Musée de la Police de l'île de Montréal et de l'île Bizard, Montréal 06355
Old Carleton County Court House, Woodstock 07404
Ontario Police College Museum, Aylmer 05368
Ontario Provincial Police Museum, Orillia 06513
Rotary Museum of Police and Corrections, Prince Albert 06665
Royal Canadian Mounted Police Museum, Regina 06726
Royal Newfoundland Constabulary Museum, Saint John's 06855
Shoal Lake Police and Pioneer Museum, Shoal Lake 06967
Sudbury Region Police Museum, Sudbury 07030
Toronto Police Museum and Discovery Centre, Toronto 07149
Vancouver Police Centennial Museum, Vancouver 07230
Victoria Police Historical Museum, Victoria 07260
Winnipeg Police Museum, Winnipeg 07393
Wood Mountain Ranch and Rodeo Museum, Wood Mountain 07402
YouthLink Calgary, Calgary 05534

China

Police Museum, Hong Kong 07813
Public Security Museum, Gongan 07717

Colombia

Museo Histórico de la Policía National, Bogotá 08585

Cuba

Antigua Cárcel de Jaruco, Jaruco 09188
Antiguo Ayuntamiento y Cárcel de Bejucal, Bejucal 09038

Denmark

Fængselshistoriske Museum, Horsens 10095
Ribe Raadhussamling, Ribe 10233

Egypt

Criminology Museum, Abbasiya 10464
National Police Museum and Ancient Police Museum, Cairo 10511

Finland

Museum Jan Karlsgården och Vita Björn, Sund 11334

France

Musée de la Préfecture de Police, Paris 14661

Germany

Bayerisches Strafvollzugsmuseum, Kaisheim 19415
Heimatmuseum, Bernau bei Berlin 17288
Historiengewölbe, Rothenburg ob der Tauber 21406
Mittelalterliche Lochgefängnisse unter dem Alten Rathaus, Nürnberg 20779
Mittelalterliches Kriminalmuseum, Rothenburg ob der Tauber 21408
Polizeihistorische Sammlung Berlin, Berlin 17235
Sächsisches Strafvollzugsmuseum, Waldheim 22296
Strafvollzugsmuseum, Ludwigsburg 20049
Zollmuseum, Wegscheid 22360
Zollmuseum Friedrichs, Aachen 16409

India

Jail Training School Museum, Lucknow 24296

Indonesia

Museum Kepolisian Negara Republik Indonesia, Jakarta 24488

Ireland

Garda Museum, Dublin 24792
Kilmainham Gaol and Museum, Dublin 24805

Italy

Museo del Brigante, Sonnino 27874
Museo di Antropologia Criminale, Torino 28011
Museo di Criminologia Medievale, San Gimignano 27639
Museo di Criminologico, Roma 27494
Museo Malacologico, Erice 26064
Museo Storico dell'Arma dei Carabinieri, Roma 27534

Japan

Abashiri Prison Museum, Abashiri 28344
Meiji-Daigaku Hakubutsukan, Tokyo 29431

Madagascar

Musée de la Gendarmerie, Toamasina 30541

Malaysia

Briged Pasukan Polis Hutan, Kuching 30590
Muzium Polis Diraja Malaysia, Kuala Lumpur 30580

Malta

The Old Prison, Victoria 30654

Mexico

Museo Calabozo de Hidalgo, Chihuahua 30765
Museo de la Policía Preventiva de la Ciudad de México, México 31087
Museo del Policía, Monterrey 31183

Netherlands

Museum de Gevangenpoort, Den Haag 32086
Museum Historische Verzameling Nederlandse Politie, Zaandam 32977
Museum 't Gevang, Doetinchem 32136
Museum van de Koninklijke Marechaussee, Buren, Gelderland 32026
Het Nationaal Gevangenismuseum, Veenhuizen 32876
Nederlands Politie Museum, Apeldoorn 31914

New Zealand

New Zealand Police Museum, Porirua 33185

Norway

Aurland Bygdetun og Lensmannsstova, Aurland 33409
Trondheim Politimuseum, Trondheim 33931

Peru

Museo de Criminalística de la Policía Nacional del Perú, Lima 34237

Philippines

Manila Police Department Museum and Gallery, Manila 34443

Poland

Muzeum Więzienia Pawiak, Warszawa 35277

Portugal

Museu da Polícia Judiciária, Barro 35459
Museu das Alfândegas, Lisboa 35687

Romania

Muzeul Poliţiei Române, Târgovişte 36659

Russia

Muzej Istorii Milicii, Sankt-Peterburg 37680
Muzej Kriminalistiki, Barnaul 36792

South Africa

South African Police Service Museum, Tshwane 38758

Spain

Euskal Poliziaren Museoa, Arkaute 38891
Museo del Bandolero, Ronda 39917

Sweden

Långholmens Fängelsemuseum, Stockholm 40907
Polishistoriska Museet, Stockholm 40925
Polistekniska Museet, Solna 40872

Switzerland

Kriminalmuseum der Kantonspolizei Zürich, Zürich 42079
Polizeimuseum, Aarau 41084
Polizeimuseum, Basel 41178

Taiwan

World Police Museum, Gueishan 42197

United Kingdom

Beaumaris Gaol and Courthouse, Beaumaris 43426
City of London Police Museum, London 44824
Essex Police Museum, Chelmsford 43805
Fire Police Museum, Sheffield 45650
Galleries of Justice, Nottingham 45291
Greater Manchester Police Museum, Manchester 45083
Inveraray Jail, Inveraray 44520
Jedburgh Castle Jail Museum, Jedburgh 44547
Kent Police Museum, Chatham 43796
Metropolitan Police Historical Collection, London 44916
Mounted Branch Museum, East Molesey 44094
NCCL Galleries of Justice, Nottingham 45296
Ripon Prison and Police Museum, Ripon 45501
Ruthin Goal, Ruthin 45532
Sherlock Holmes Museum, London 44986
South Wales Police Museum, Bridgend 43602
Stirling Old Town Jail, Stirling 45751
Tales of the Old Gaol House, King's Lynn 44588
Tetbury Police Museum, Tetbury 45841
Thames Police Museum, London 44997
The Tolhouse, Great Yarmouth 44361
West Midlands Police Museum, Birmingham 43517

United States of America

American Police Center and Museum, Chicago 47450
Andersonville Prison, Andersonville 46295
Cleveland Police Museum, Cleveland 47626
Eastern State Penitentiary Historic Site, Philadelphia 51828
Houston Police Museum, Houston 49391
Jacinto Courthouse, Corinth 47807
Law Enforcement Museum, Roselle 52459
Miers Home Museum, Sonora 53155
Montana Auto Museum, Deer Lodge 48029
Museum of Death, Los Angeles 50253
New York City Police Museum, New York 51242
Old Fire House and Police Museum, Superior 53379
Old Gaol Museum, York 54336
Old Jail Museum, Albion 46171
The Old Jail Museum, Crawfordsville 47871
The Old Jail Museum, Warrenton 53831
Old Monterey Jail, Monterey 50805
Old Prison Museum, Deer Lodge 48030
Porter County Museum, Valparaiso 53699
Yuma Territorial Prison Museum, Yuma 54366

Dairy Products

Austria

Käsereimuseum, Lochen 02330

Belgium

Zuivelmuseum, Blankenberge 03372

Canada

Eriksdale Creamery Museum, Eriksdale 05780
Historic Markerville Creamery, Markerville 06238

Denmark

Hjedding Mejerimuseum, Ølgod 10221

Estonia

Eesti Piimandusmuuseum, Imavere 10574

Finland

Meijerimuseo, Saukkola 11300
Nakkilan Kotiseutumuseo ja Juustomeijerimuseo, Nakkila 11136
Puuhkalan Talomuseo, Ylivieska 11483

France

Moulin-Musée de la Brosserie, Saint-Félix 15265
Musée de la Meunerie, Courtelevant 12658
Musée des Tradition de l'île, Noirmoutier-en-l'Ile 14467
Musée du Bocage Normand, Saint-Lô 15373
Musée du Camembert, Vimoutiers 16247
Musée du Fromage et de la Vache Salers, Saint-Paul-de-Salers 15442
Musée du Lait, Belvedere 11935
Musée du Lait et des Traditions, Saint-Étienne-de-Tinée 15257
Musée Paysan, Viuz-en-Sallaz 16260

Germany

Carl-Hirnbein-Museum, Missen-Wilhams 20339
Heimatmuseum, Hüttenberg 19295
Historische Käsküche, Wiggensbach 22566
Molkerei-Museum, Bernbeuren 17293
Trützschler's Milch- und Reklamemuseum, Hildburghausen 19180

Hungary

Tejipari Emléktár, Székesfehérvár 23862

Italy

Museo del Profumo, Milano 26699
Museo del Sale, Paceco 27029

Netherlands

Kaasboerderijmuseum de Weistaar, Maarsbergen 32515
Kaaswaag Gouda, Gouda 32266

Russia

Muzej Tulskie Samovary, Tula 37870

South Africa

Riemland Museum, Heilbron 38640

Switzerland

La Maison du Gruyère, Pringy 41736
Milchwirtschaftliches Museum, Kiesen 41502

United Kingdom

National Dairy Museum, Ashurst, Southampton 43334

United States of America

America's Ice Cream and Dairy Museum, Medina 50556

Dance → Performing Arts

Decorative and Applied Arts

Afghanistan

Ghazni Provincial Museum, Ghazni 00001

Algeria

Musée Folklorique de Ghardaïa, Ghardaïa 00063

Argentina

Fundación Banco Francés, Buenos Aires 00159
Museo Almacén El Recreo, Chivilcoy 00304
Museo Folklórico Regional de Humahuaca, Humahuaca 00387
Museo Histórico de la Honorable Cámara de Diputados de la Nación, Buenos Aires 00238
Museo Nacional de Arte Decorativo, Buenos Aires 00257
Museo Patria Chica de Mohamed Diaz, Realicó 00520
Museo Tornambe de la Universidad Nacional de San Juan, San Juan 00589
Palais de Glace, Buenos Aires 00281

Armenia

History Museum of Armenia, Yerevan 00738
Middle East Museum, Yerevan 00743
State Museum of Wood Carving, Yerevan 00756
Wood Carving Museum, Yerevan 00758

Australia

Museum of Fine Arts, Hobart 01155
Powerhouse Museum, Ultimo 01609

Austria

Ausstellungszentrum Heiligenkreuzer Hof, Wien 02940
Heimatmuseum Egg, Egg 01842
Museum im Palais Kulturhistorische Sammlung, Graz 01979
Museum Schloss Ottersbach, Großklein 02004
Sammlung Bodingbauer, Steyr 02771
Sammlungen der Universität für angewandte Kunst Wien mit Oskar-Kokoschka-Zentrum, Wien 03065
Schloß Loosdorf mit Prunkräumen und Zinnfigurensammlung, Loosdorf, Bez. Mistelbach 02336

Azerbaijan

Letif Kerimov ad. Azerbaycan Xalçasi ve Xalq Tetbiqi Seneti Dövlet Muzeyi, Bakı 03159
Museum of National Applied Art, Şeki 03195

Belgium

Bruggemuseum - Gruuthuse, Brugge 03411
Espace des Saveurs, Herve 03702
Maagdenhuismuseum, Antwerpen 03291
Maison du Roi-Musée de la Ville, Bruxelles 03451
Musée d'Ansembourg, Liège 03802
Musée de l'Orfevrerie de la Communaute Française, Seneffe 04008
Musée du Château de Fagnolles, Fagnolles 03594
Musée du Cinquantenaire-Jubelparkmuseum, Bruxelles 03482
Musée François Duesberg des Arts Décoratifs, Mons 03888
Musées Royaux d'Art et d'Histoire, Bruxelles 03495
Romain De Saeghermuseum, Sint-Amands 04015
Stedelijk Museum Ieper, Ieper 03731

Brazil

Açude Museum, Rio de Janeiro 04680
Chácara do Céu Museum, Rio de Janeiro 04688
Museu de Arte Popular da Fortaleza dos Reis Magos, Natal 04584
Museu Henriqueta Catharino, Salvador 04804

Bulgaria

Chudožestvena Galerija, Žeravna 05268
Nacionalna Galerija za Dekorativni Izkustva, Sofia 05202

Cameroon

Musa Heritage Gallery, Kumbo-Nso 05298

Canada

Caroline Wheels of Time Museum, Caroline 05568
Centre de Design de l'UQAM, Montréal 06332
Great Hall of the Clans Museum, English Town 05779
Henry-Stuart House, Québec 06681
Joseph Schneider Haus, Kitchener 06102
Maison du Notaire, Trois-Pistoles 07169
Mary E. Black Gallery, Halifax 05941
Musée des Beaux-Arts de Montréal, Montréal 06357
Museum of Cape Breton Heritage, North East Margaree 06475
Nels Berggran Museum, Imperial 06010
New Brunswick College of Craft and Design Gallery, Fredericton 05835
Pavillon Saint-Arnaud, Trois-Rivières 07179
Port Coquitlam Heritage and Cultural Museum, Port Coquitlam 06634
Prescott House Museum, Port Williams 06650
Ross Memorial Museum, Saint Andrews 06799
Schwartz Heritage House, Altona 05327
Ukrainian Cultural Education Centre Oseredok, Winnipeg 07387
Whetung Craft Centre and Art Gallery, Curve Lake 05668

China

Chinese University of Hong Kong Art Museum, Ma Liu Shui 07989
GianFu Classic Art Museum, Beijing 07539
Handover Gifts Museum, Macau 07994
Museum of the Holy House of Mercy, Macau 08004

Colombia

Museo Patio del Moro, Guaduas 08652

Costa Rica

Museo de Arte y Diseño Contemporáneo, San José 08816

Croatia

Muzej za Umjetnost i Obrt, Zagreb 09008
Zbirka Anke Gvozdanović, Zagreb 09015

Cuba

Centro de Diseño Ambiental, La Habana 09099
Museo Artes Decorativo, Gibara 09085
Museo de Ambiente Histórico Cubano, Santiago de Cuba 09309
Museo de Artes Decorativas de Palacio Salcines, Guantánamo 09091
Museo Nacional de Artes Decorativas, La Habana 09161
Museo Nacional de Artes Decorativas de Santa Clara, Santa Clara 09299

Czech Republic

Městské Muzeum, Aš 09395
Městské Muzeum, Chotěboř 09479

Denmark

Det Danske Kunstindustrimuseum, København 10129
Davids Samling, København 10130
Georg Jensen Museum, København 10132
Michael og Anna Anchers Hus og Saxilds Gaard, Skagen 10261
Museet for Varde By og Omegn, Varde 10307
Museet på Koldinghus, Kolding 10163
Trapholt, Kolding 10164

Dominican Republic

Museo de la Familia Dominicana del Siglo XIX, Santo Domingo 10347

El Salvador

Sala Nacional de Exposiciones, San Salvador 10555

Estonia

Eesti Tarbekunsti- ja Disainimuuseum, Tallinn 10627

Finland

Alvar Aalto Museum, Jyväskylä 10886
Designmuseo, Helsinki 10771
Edelfelt-Vallgren Museo, Porvoo 11226
Hiekan Taidemuseo, Tampere 11350

France

Château de Champchevrier, Cléré-les-Pins 12545
Château-Musée, Heudicourt 13214
Château-Musée du Plessis-Bouré, Ecuillé 12810
Musée Angladon, Avignon 11788
Musée Anne de Beaujeu, Moulins (Allier) 14303
Musée Baccarat, Paris 14632
Musée Baron Gérard, Bayeux 11873
Musée Boilhet-Christofle, Saint-Denis (Seine-Saint-Denis) 15235
Musée Boucher-de-Perthes, Abbeville 11486
Musée Bourdelle, Paris 14636
Musée d'Art et d'Histoire, Belfort 11922
Musée de la Chapellerie, Espéraza 12861
Musée de la Nacre et de la Tabletterie, Méru 14091
Musée de la Vie Romantique, Paris 14665
Musée de l'Eventail, Paris 14670
Musée de Picardie, Amiens 11588
Musée départemental du Revermont, Treffort 15979
Musée des Années 30, Boulogne-Billancourt 12090
Musée des Arts Décoratifs, Lyon 13887
Musée des Arts Décoratifs, Paris 14682
Musée des Arts Décoratifs et de la Modernité, Gourdon (Alpes-Maritimes) 13088
Musée des Beaux-Arts, Blois 12035
Musée des Beaux-Arts, Reims 15001
Musée Dobrée, Nantes 14367
Musée du Château, Siorac-en-Périgord 15760
Musée École de la Perrine, Laval (Mayenne) 13543
Musée Leblanc-Duvernoy, Auxerre 11777
Musée Municipal de l'Évêché, Limoges 13770
Musée Paul Dupuy, Toulouse 15944
Musée Pierre-Loti, Rochefort (Charente-Maritime) 15073
Parcours-découverte de l'écurie n1, Le Pin-au-Haras 13632
Prieuré du Vieux Logis, Nice 14441
Villa Majorelle, Nancy 14358

D

Georgia

Folk and Applied Arts Museum, Tbilisi 16357

Germany

Augustinermuseum, Freiburg im Breisgau 18428
Auto- und Motorrad-Museum, Öhringen 20863
Bauhaus-Archiv, Berlin 17041
Bayerisches Nationalmuseum, München 20426
Belvedere auf dem Klausberg, Potsdam 21080
Das Berta-Hummel-Museum im Hummelhaus, Massing 20235
Bröhan-Museum, Berlin 17056
Chinesische Haus, Potsdam 21085
Deutsches Drachenmuseum und Stadtmuseum, Furth im Wald 18515
Deutsches Plakat Museum, Essen 18250
Die Neue Sammlung - The International Design Museum Munich, München 20436
Ehemaliges Jagdschloß Bebenhausen, Tübingen 22113
Erstes Deutsches Fliesenmuseum Boizenburg e.V, Boizenburg 17438
Forum Konkrete Kunst Erfurt, Erfurt 18202
Fürstliches Residenzschloß, Detmold 17837
Fugger-Museum, Babenhausen, Kreis Unterallgäu 16651
Gablonzer Archiv Museum, Kaufbeuren 19486
Galerie Handwerk, München 20442
Gewerbemuseum im Germanischen Nationalmuseum, Nürnberg 20765
Glasschmelzofenbau-Hütte mit Glasausstellung, Plößberg 21066
Gotisches Haus, Wörlitz 22638
Grassi-Museum für Angewandte Kunst, Leipzig 19911
Heimatmuseum, Preetz 21116
Heimatmuseum, Scheeßel 21536
Historisches Museum Hanau - Schloß Philipsruhe, Hanau 18979
Königliches Schloss Berchtesgaden und Rehmuseum, Berchtesgaden 17001
Krippenmuseum, Glattbach 18654
Krippenstube-Heimatstube, Plößberg 21067
Kunst und Museum, Hollfeld 19253
Kunstgewerbemuseum, Dresden 17951
Kunstgewerbemuseum, Berlin 17168
Kunstsammlungen der Veste Coburg, Coburg 17717
Marmorpalais, Potsdam 21094
Maximilianmuseum, Augsburg 16629
Miniaturmuseum Arikalex, Berlin 17188
Museen der Stadt Landshut in der Stadtresidenz, Landshut 19803
Museum August Kestner, Hannover 19000
Museum beim Markt, Karlsruhe 19443
Museum der Stadt Miltenberg und Porzellansammlung Kreis Dux, Miltenberg 20327
Museum der Stadt Zerbst mit Sammlung Katharina II., Zerbst 22788
Museum für Angewandte Kunst, Frankfurt am Main 18382
Museum für Angewandte Kunst, Köln 19619
Museum für Angewandte Kunst Gera, Gera 18594
Museum für Frühislamische Kunst, Bamberg 16914
Museum für schlesische Landeskunde im Haus Schlesien, Königswinter 19659
Museum Huelsmann, Bielefeld 17342
Museumsberg Flensburg, Flensburg 18328
Neue Kammern, Potsdam 21099
Neuer Pavillon im Schlossgarten Charlottenburg, Berlin 17227
Neues Palais, Potsdam 21100
OberhausMuseum, Passau 20993
red dot design museum Deutschland, Essen 18266
Residenzmuseum, München 20476
Sammlung Angewandte Kunst, Kassel 19474
Schloss Babelsberg, Potsdam 21105
Schloss Belvedere, Weimar 22418
Schloss Caputh, Schwielowsee 21708
Schloss Charlottenhof, Potsdam 21107
Schloss Glienicke, Berlin 17249
Schloß Heidelberg, Heidelberg 19083
Schloss Oranienbaum, Oranienbaum 20907
Schloss Pfaueninsel, Berlin 17250
Schloss Rheinsberg, Rheinsberg 21306
Schloss Sacrow, Potsdam 21108
Schloßmuseum der Stadt Aschaffenburg, Aschaffenburg 16592
Schlossmuseum Langenburg, Langenburg 19827
Schloßmuseum Mespelbrunn, Mespelbrunn 20303
Schlossmuseum Oranienburg, Oranienburg 20910
Schwäbisches Krippenmuseum im Jesuitenkolleg, Mindelheim 20329
Sebnitzer Kunstblumen- und Heimatmuseum Prof. Alfred Meiche, Sebnitz 21710
Spessartmuseum, Lohr am Main 20027
Stadtmuseum Fembohaus, Nürnberg 20791
Stadtmuseum Lindau, Lindau 20001
Städtisches Museum, Überlingen 22146
Städtisches Museum, Zeulenroda-Triebes 22794
Stiftung Bauhaus Dessau, Dessau 17834
Vitra Design Museum, Weil am Rhein 22379
Werdenfelser Museum, Garmisch-Partenkirchen 18542
Werkbundarchiv - Museum der Dinge, Berlin 17280

Greece

The Chicago Athenaeum - Europe, Néa Symirni 23125
Decorative Arts Museum of Rhodos, Rhodos 23171
Likeio Ton Ellinidon Collection, Rethymnon 23169

Hungary

Iparművészeti Múzeum, Budapest 23416
Patkó Imre Gyűjtemény, Győr 23585
Róth Miksa Emlékház, Budapest 23474

Iceland

Listasafn Árnesinga, Hveragerði 24005

India

Crafts Museum, Delhi 24139
Dr. Bhau Daji Lad Museum, Mumbai 24314
Government Central Museum, Jaipur, Rajasthan 24233
Maharaja Sawai Man Singh II Museum, Jaipur, Rajasthan 24234
Raja Dinkar Kelkar Museum, Pune 24362
Rukmini Devi Museum, Tiruvanmiyur 24410

Indonesia

Museum Batik, Yogyakarta 24584
Museum Radya Pustaka, Surakarta 24571

Iran

Astan-e-Qods-e Razavi Museums, Mashad 24631
Chehel-Stoun Museum, Ghazvin 24602
Chinikhaneh Museum, Ardebil 24596
Honarhaye Tazini - Rakibkhaneh, Isfahan 24612
Kandeloos Museum, Kojoor 24626
Malek Museum, Tehrān 24677
Mellat Museum, Tehrān 24678
Pars Museum, Shiraz 24648

Iraq

Mustansiriya School Collections, Baghdad 24716

Ireland

Glebe House and Gallery, Letterkenny 24861
The Hunt Museum, Limerick 24864
National Museum of Ireland - Decorative Arts History, Dublin 24809

Israel

L.A. Mayer Museum for Islamic Art, Jerusalem 24999

Italy

Collezione Permanente del Design Italiano '45-'90, Milano 26683
Collezioni Comunali d'Arte, Imola 26398
MIAAO - Museo Internazionale delle Arti Applicate oggi, Torino 28003
Musei Provinciali di Gorizia, Gorizia 26351
Museo Civico d'Arte Antica e Palazzo Madama, Torino 28004
Museo Civico Gaetano Filangieri, Napoli 26887
Museo del Corallo, Ravello 27368
Museo del Corallo, Torre del Greco 28036
Museo del Presepe, Modena 26749
Museo del Presepio, Dalmine 26031
Museo del Presepio Tipologico Internazionale, Roma 27473
Museo dell'Arte del Cappello, Ghiffa 26332
Museo di Arte Contadina, Montegallo 26833
Museo di Arti Decorative della Fondazione Pietro Accorsi, Torino 28013
Museo di Palazzo Reale Genova, Genova 26307
Museo Missionario Francescano, Fiesole 26130
Museo per l'Arte e l'Archeologia del Vastese, Gissi 26344
Oskar Schlemmer Theatre Estate and Collection, Oggebbio 26977
Palazzo Tursi, Genova 26324

Japan

21_21 Design Sight, Tokyo 29369
Araki Shuseikan Museum, Nagoya 28992
Asakura Isokichi Bijutsukan, Terai 29359
Deshine Bijutsukan, Nagoya 28997
Design Museum Design Gallery, Nagoya 28998
Fukuoka-shi Bijutsukan, Fukuoka 28433
Hida Takayama Bijutsukan, Takayama 29339
Huis Ten Bosch Museums, Sasebo 29234
Kitamura Bijutsukan, Kyoto 28822
Koishibara-mura Yakimono Bijutsukan, Koishibara 28760
Kumamoto-kenritsu Gigei Bijutsukan, Kumamoto 28777
Makuhari Kitazawa Bijutsukan, Chiba 28396
Namikawa Cloisonne Museum of Kyoto, Kyoto 28854
Nihon Kogeikan, Tokyo 29449
Sawanoi Museum of Traditional Japanese Ornaments, Ome 29114
Showa Bijutsukan, Nagoya 29010
Suntory Hakubutsukan, Osaka 29155
Tako no Hakubutsukan, Tokyo 29481
Tokyo-to Teien Bijutsukan, Tokyo 29500
Tottori-kenritsu Gigei Bijutsukan, Nakayama 29023
Toyota-shiritsu Bijutsukan, Toyota 29535
Yonezawa Municipal Uesugi Museum Uesugi Shrine Keisyo Palace Museum, Yonezawa 29627

Kazakhstan

Atyrauskij Oblastnoj Muzej Dekorativno-prikladnogo Iskusstva, Atyrau 29696
Muzej Kazakstan Respublikasynyn Prezidenttik Mēdeniet Ortalyfy, Astana 29692

Korea, Republic

Ewha Yoja Taehakkyo Pakmulgwan, Seoul 29876
Seoul Design Museum, Seoul 29917

Latvia

Dekoratīvi Lietišķās Mākslas Muzejs, Rīga 30064

Lebanon

Moussa Castle Museum, Beiteddine 30163

Lithuania

Keramikos Muziejus, Kaunas 30253
Klaipėdos Paveikslų Galerija, Klaipėda 30279
Laikrodžių Muziejus, Klaipėda 30281
Taikomosios Dailės Muziejus, Vilnius 30434

Luxembourg

Villa Vauban, Luxembourg 30486

Malaysia

Museum and Art Gallery, Shah Alam 30618
Muzium Negeri Kedah, Alor Setar 30552

Mexico

Galeria Mexicana de Diseño, México 31046
Museo Casa Ruth Lechuga, México 31061
Museo de Arte Olga Costa, Guanajuato 30913
Museo de la Casa de las Artesanías del Estado de Chihuahua, Bocoyna 30738
Museo de la Quemada, Guadalupe 30908
Museo Estudio Diego Rivera, México 31113
Museo Franz Mayer, México 31117
Museo Mural Diego Rivera, México 31127
Museo Nacional de Arte Popular, Mérida 31028
Museo Nacional de Virreinato, Tepotzotlán 31445
Museo Palacio de la Canal, San Miguel de Allende 31357
Museo Universitario Antigua Casa de los Muñecos, Puebla 31290

Montenegro

Državni Muzej, Cetinje 31640

Nepal

National Museum of Nepal, Kathmandu 31776

Netherlands

Amsterdams Historisch Museum, Amsterdam 31826
Centraal Museum, Utrecht 32844
Collectie A. Veltman, Bussum 32030
Fogelsangh State, Veenklooster 32877
Frans Hals Museum, Haarlem 32295
Galerie Numero 16, Noordbroek 32599
Het Gouverneurshuis, Heusden 32383
Hans van Riessen Museum, Vledderveen 32897
Herdenkingsbordenmuseum, Lochem 32504
Het Cuypershuis, Roermond 32698
Historisch Museum Deventer, Deventer 32120
Kasteel de Haar, Haarzuilens 32304
Merklappenmuseum, Dieteren 32128
Museum De Weem, Westeremden 32941
Museum het Tramstation, Schipluiden 32761
MuseumgoudA, Gouda 32269
Nederlands Tegelmuseum, Otterlo 32644
Oudheidkamer in het Stadhuis, Bolsward 31980
Simon van Gijn-Museum aan Huis, Dordrecht 32154
Stedelijk Museum, Alkmaar 31796
Stedelijk Museum Vianen, Vianen 32887
Stedelijk Museum Zutphen, Zutphen 33020
Stichting tot Beheer Museum Boijmans Van Beuningen, Rotterdam 32729
Streekmuseum Het Rondeel, Rhenen 32684
Het Waaierkabinet, Amsterdam 31900

New Zealand

The Dowse, Hutt City 33124

Northern Cyprus

Folk Art Institute, Lefkoşa 33369
Güzel Sanatlar Müzesi, Girne 33359

Norway

Kunstindustrimuseet, Oslo 33730
Nasjonalmuseet for Kunst, Arkitektur og Design, Oslo 33736
Nordenfjeldske Kunstindustrimuseum, Trondheim 33925
Permanenten Vestlandske Kunstindustrimuseum, Bergen 33433
Stiftsgården, Trondheim 33928

Pakistan

Bahawalpur Museum, Bhawalpur 33998

Peru

Museo de Sitio de Narihualá, Catacaos 34140
Museo Inka, Cusco 34167

Philippines

Casa Gorordo Museum, Cebu 34366
Museo ng Kalinangang Pilipino, Pasay 34474
Santo Niño Shrine and Heritage Museum, Tacloban 34511

Poland

Muzeum Kultury Szlacheckiej, Łopuszna 34891
Muzeum Okręgowe w Toruniu, Toruń 35195
Muzeum Plakatu w Wilanowie, Warszawa 35264
Muzeum Sprzętu Gospodarstwa Domowego, Ziębice 35371
Muzeum Sztuk Użytkowych, Poznań 35044
Muzeum Zamoyskich w Kozłówce, Kamionka 34713
Sztuka Dalekiego Wschodu, Toruń 35197

Portugal

Casa do Corpo Santo, Setúbal 35934
Casa-Museu da Fundação Medeiros e Almeida, Lisboa 35661
Casa Museu Frederico de Freitas, Funchal 35609
Casa-Museu José Régio, Portalegre 35832
Casa Museu Vitorino Ribeiro, Porto 35840
Centro de Artes Tradicionais, Évora 35583
Fundação Ricardo do Espírito Santo Silva, Lisboa 35667
Museu Bernardino Machado, Vila Nova de Famalicão 36004
Museu-Biblioteca da Casa de Bragança, Vila Viçosa 36016
Museu Calouste Gulbenkian, Lisboa 35672
Museu Condes de Castro Guimarães, Cascais 35512
Museu da Quinta das Cruzes, Funchal 35611
Museu da Regiao Flāviense, Chaves 35523
Museu de Artes Decorativas, Évora 35586
Museu de Design e Artes Decorativas, Lisboa 35691
Museu de Escultura Comparada, Mafra 35750
Museu dos Biscaínhos, Braga 35479
Museu-Escola de Artes Decorativas Portuguesas, Lisboa 35706
Museu José Luciano de Castro, Anadia 35434
Museu Municipal de Pinhel, Pinhel 35825
Museu Municipal de Viana do Castelo, Viana do Castelo 35984
Museu Nacional de Machado de Castro, Coimbra 35542
Museu Nacional de Soares dos Reis, Porto 35860
Museu S. Pedro da Palhaça, Palhaça 35804

Romania

Centrul de Cultură Palatele Brâncovenești, Mogoșoaia 36479
Galeriile de Artă Emilia Dumitrescu, Brăila 36169
Galeriile de Artă Plastică Ștefan Luchian, Botoșani 36161
Muzeul Casa Mureșenilor, Brașov 36177
Muzeul de Artă, Iași 36417
Muzeul de Artă, Târgoviște 36657
Muzeul de Artă Religioasă Nicolae Iorga, Vălenii de Munte 36720
Muzeul de Artă Veche Apuseană Dumitru Minovici, București 36202
Muzeul Hrandt Avakian, București 36204
Muzeul Județean Argeș, Pitești 36526
Muzeul Memorial Dimitrie Aurelia Ghiață, București 36207
Muzeul Național Brukenthal, Sibiu 36613

Muzeul Național Cotroceni, București 36218
Muzeul Orășenesc, Lipova 36443
Palatul Scroviștea, Periș 36513

Russia

Belozerskij Muzej Narodnogo Dekorativno-prikladnogo Iskusstva, Belozersk 36799
Bogorodickij Dvorec-muzej i Park, Bogorodick 36812
Dagestanskij Muzej Izobrazitelnych Iskusstv, Machačkala 37116
Dom-muzej A.A. Dydykina, Palech 37523
Dom-muzej I.I. Golikova, Palech 37524
Dom-Muzej P.D. Korina, Palech 37525
Dom-muzej N.M. Zinovjeva, Djagilevo 36875
Elaginoostrovskij Dvorec-Muzej Russkogo Dekorativno-Prikladnogo Iskusstva i Interjera XVIII-XX vv., Sankt-Peterburg 37637
Gosudarstvennyj Chudožestvennyj Muzej Altajskogo Kraja, Barnaul 36790
Gosudarstvennyj Istoričeskij Zapovednik Gorki Leninskie, Gorki Leninskie 36926
Gosudarstvennyj Muzej Istorii Sankt-Peterburga, Sankt-Peterburg 37647
Gosudarstvennyj Muzej Palechskogo Iskusstva, Palech 37526
Istoriko-Architekturnyj, Chudožestvennyj i Archeologičeskij Muzej "Zarajskij Kreml", Zarajsk 37986
Jurinskij Istoriko-chudožestvennyj Muzej, Jurino 36982
Krasnojarskij Chudožestvennyj Muzej im. V.I. Surikova, Krasnojarsk 37069
Muzej Chudožestvennych Tkanej, Moskva 37267
Muzej Dekorativno-Prikladnogo i Promyšlennogo Iskusstva, Moskva 37270
Muzej Dekorativno-prikladnogo Iskusstva, Sankt-Peterburg 37673
Muzej Dekorativno-prikladnogo Iskusstva Urala, Čeljabinsk 36844
Muzej Istorii Kamnereznogo i Juvelirnogo Iskusstva, Ekaterinburg 36894
Muzej Moskovskogo Chudožestvenno-promyšlennogo Universiteta im. S.G. Stroganova, Moskva 37335
Nižegorodskij Gosudarstvennyj Istoriko-architekturnyj Muzej-zapovednik, Nižnij Novgorod 37468
Oblastnoj Muzejnyj Centr, Sankt-Peterburg 37716
Oružejnaja Palata, Moskva 37392
Paviljon Katalnoj Gorki, Lomonosov 37111
Pereslavl-Zalesskij Gosudarstvennyj Istoriko-architekturnyj i Chudožestvennyj Muzej-zapovednik, Pereslavl-Zalesskij 37542
Petergof - Gosudarstvennyj Muzej Zapovednik, Sankt-Peterburg 37718
Rjazanskij Oblastnoj Chudožestvennyj Muzej im. I.P. Požalostina, Rjazan 37596
Samarskij Oblastnoj Chudožestvennyj Muzej, Samara 37621
Šeremetevskij Dvorec - Fontannyj Dom, Sankt-Peterburg 37726
Sergievo-Posadskij Gosudarstvennyj Istoriko-chudožestvennyj Muzej Zapovednik, Sergiev Posad 37764
Severo-osetinskij Gosudarstvennyj Obedinennyj Muzej Istorii, Architektury i Literatury, Vladikavkaz 37932
Uljanovskij Oblastnoj Chudožestvennyj Muzej, Uljanovsk 37909
Valaamskij Naučno-issledovatelskij Cerkovno-archeologičeskij i Prirodnyj Muzej-zapovednik, Sortavala 37790
Vserossijskij Muzej Dekorativno-Prikladnogo i Narodnogo Iskusstva, Moskva 37412

Serbia

Muzej Primenjene Umetnosti, Beograd 38074

Slovakia

Múzeum Kultúry Maďarov na Slovensku, Bratislava 38223

South Africa

The Foundry, Design Museum, Cape Town 38583
Melrose House, Tshwane 38749
Roodepoort Museum, Roodepoort 38726
Van Gybland-Oosterhoff Collection, Tshwane 38762

Spain

CaixaForum Palma de Mallorca, Palma de Mallorca 39776
Centro del Diseño, Bilbao 39041
Colección Museográfica Permanente de Biar, Biar 39036
Museo Art Nouveau-Art Déco, Salamanca 39936
Museo Bonsor, Mairena del Alcor 39593
Museo de la Alhambra, Granada 39367
Museo de Tabar, Tabar 40094
Museo del Alcázar, Sevilla 40072
Museo Rodera Robles, Segovia 40055
Museo Taller del Moro, Toledo 40138
Museo Valenciano de la Ilustración y la Modernidad, Valencia 40218
Museu de les Arts Decoratives, Barcelona 38982
Museu Frederic Marès, Barcelona 38995
Museu Romàntic Can Llopis, Sitges 40085
Palacio Real de Madrid, Madrid 39586
Palacio Real y Museos de Tapices, San Ildefonso 39958

Sweden

Godegårds Bruks- och Porslinsmuseum, Motala 40751
Hälsinglands Museum, Hudiksvall 40604
Jämtlands Läns Museum, Östersund 40796
Länsmuseet Halmstad, Halmstad 40578
Örebro Läns Museum, Örebro 40789
Röhsska Museet, Göteborg 40555
Skoklosters Slott, Skokloster 40854
Surahammars Bruksmuseum, Surahammar 40964

Switzerland

Abegg-Stiftung, Riggisberg 41767
Bernische Stiftung für angewandte Kunst und Gestaltung, Bern 41203
Burg Zug, Zug 42123
Froschmuseum, Münchenstein 41657
Gewerbemuseum, Winterthur 42027
Haus zum Dolder, Beromünster 41223
Haus zum Kirschgarten, Basel 41160
Historisches Museum Bern, Bern 41207
Historisches Museum Luzern, Luzern 41592
Kunstgewerbesammlung, Zürich 42081
mudac - Musée de Design et d'Arts Appliqués Contemporains, Lausanne 41540
Musée d'Histoire, La Chaux-de-Fonds 41294
Musée Eugène Burnand, Moudon 41651
Museum für Gestaltung Basel, Basel 41171
Museum für Gestaltung Zürich, Zürich 42094
Museum für Gestaltung Zürich - Sammlungen für Design, Grafik, Plakat und Kunstgewerbe, Zürich 42095
Museum Lindengut, Winterthur 42034
Rathaussammlung, Stein am Rhein 41916
Schlossmuseum Beromünster, Beromünster 41225
Schule für Gestaltung Basel, Basel 41182
Schweizerisches Nationalmuseum, Zürich 42105
Stadtmuseum Wil, Wil 42017

Taiwan

Chia Nan Museum of Culture and Art, Rende 42305
Idea Art and Design Center, Hsinchu 42208
Lee's Horsey Fine Art Studio, Kaohsiung 42239
Long-Nan Museum of Natural Lacquer Ware, Puli 42294
Mei Museum, Shueili 42327
Taiwan Museum of Art Sanri, Sanyi 42318

Thailand

Daraphirom Royal Mansion Museum, Bangkok 42537

Turkey

Galeri Suav ve Küsav, İstanbul 42819
Sadberk Hanım Müzesi, İstanbul 42837
Yıldız Sarayı Müzesi ve Tiyatrosu ve Sahne Kostümkeri Müzesi, İstanbul 42851

Ukraine

Kanivs'kyj Muzej Narodnogo Dekoratyvnogo Mystectva, Kaniv 43059
Muzej Dekoratyvno-prykladnogo Mystectva, Sumy 43189
Muzej Istoryčnych Koštovnostej Ukraïny, Kyïv 43090
Muzej Ivana Gončara, Kyïv 43091
Muzej Narodnogo Dekorativnogo Mystectva, Černigiv 43006

United Kingdom

Alfred Dunhill Museum, London 44780
American Museum in Britain, Bath 43398
Andrew Carnegie Birthplace Museum, Dunfermline 44054
Bantock House, Wolverhampton 46057
Belton House, Grantham 44344
Birmingham Institute of Art and Design, Birmingham 43499
The Bowes Museum, Barnard Castle 43381
Brighton Museum and Art Gallery, Brighton 43621
Buxton Museum and Art Gallery, Buxton 43705
Charleston Trust, Firle 44229
Cheltenham Art Gallery and Museum, Cheltenham 43810
Clocktower Museum, Elsham 44170
Clotworthy Arts Centre, Antrim 43307
Coughton Court, Alcester 43264
Courtauld Institute Gallery, London 44830
Cragside House, Rothbury 45516
Crown Liquor Saloon, Belfast 43439
Daphne du Maurier's Smugglers's at Jamaica Inn, Bolventor 43557
Design Museum, London 44837
Dimbola Lodge Museum, Freshwater Bay 44255
Doddington Hall, Lincoln 44711
The Fitzwilliam Museum, Cambridge 43727
Gawsworth Hall, Macclesfield 45063
Gilbert Collection, London 44856
Guildhall Collection, Kingston-upon-Hull 44595
Harley Gallery, Worksop 46076
Helena Thompson Museum, Workington 46075
Herbert Art Gallery and Museum, Coventry 43902
Holburne Museum of Art, Bath 43405
Kellie Castle, Pittenweem 45388
Kew Palace Museum Queen Charlotte's Cottage, Kew 44570
Laing Art Gallery, Newcastle-upon-Tyne 45229
Little Holland House, London 44902
Lotherton Hall, Aberford 43251
Manchester Art Gallery, Manchester 45088
Manor House, Sandford Orcas 45610
Merchant Adventurers' Hall, York 46104
Millennium Galleries, Sheffield 45653
Mompesson House, Salisbury 45604
Museum and Study Collection, London 44920
Museum of Domestic Design and Architecture, Barnet 43382
Museum of Farnham, Farnham 44218
Museum of the Order of Saint John, London 44930
Newby Hall, Ripon 45499
Palace of Westminster, London 44950
Peckover House, Wisbech 46048
Pickford's House Museum, Derby 43967
Polesden Lacey House, Dorking 44007
Quex Museum, House and Gardens, Birchington 43489
Ragley Hall, Alcester 43265
Royal Museum and Art Gallery, Canterbury 43744
Saltram House, Plymouth 45393
Scolton Manor Museum, Haverfordwest 44422
Speke Hall, Liverpool 44744
Sutton Park, Sutton-on-the-Forest 45807
Tatton Park, Knutsford 44625
Usher Gallery, Lincoln 44717
Victoria and Albert Museum, London 45004
Waddesdon Manor, Waddesdon 45905
Warner Archive, Milton Keynes 45155
Wernher Collection at Ranger's House, London 45011
West Berkshire Museum, Newbury 45221
West Wycombe Park House, West Wycombe 45975
William Morris Gallery and Brangwyn Gift, London 45015

United States of America

The 100th Meridian Museum, Cozad 47864
1890 House-Museum and Center for Arts, Cortland 47828
AAF Museum, Washington 53843
Adam Thoroughgood House, Virginia Beach 53752
Alleghany Highlands Arts and Crafts Center, Clifton Forge 47649
American Museum of Straw Art, Long Beach 50172
Ash Lawn-Highland, Charlottesville 47395
Asia Society Museum, New York 51128
Asian Cultures Museum, Corpus Christi 47819
Astor House Hotel Museum, Golden 48955
Atlantic Wildfowl Heritage Museum, Virginia Beach 53753
The Bard Graduate Center for Studies in the Decorative Arts, Design, and Culture, New York 51132
Barnes Foundation, Merion 50601
Batsto Village, Hammonton 49145
Bayou Bend Collection, Houston 49381
Beauport-Sleeper-McCann House, Gloucester 48944
Bellamy Mansion Museum of History and Design Arts, Wilmington 54199
The Bennington Museum, Bennington 46739
Bergstrom-Mahler Museum, Neenah 50997
Bernard Judaica Museum, New York 51133
Bidwell House Museum, Monterey 50808
Birks Museum, Decatur 48014
Boscobel Restoration, Garrison 48868
Boston Public Library, Boston 46923
Brooklyn Museum, Brooklyn 47069
Brown Mansion, Coffeyville 47673
Bruce Museum of Arts and Science, Greenwich 49081
Brunnier Art Museum, Ames 46256
Buccleuch Mansion, New Brunswick 51031
California Heritage Museum, Santa Monica 52898
Campbell House Museum, Saint Louis 52569
Campbell Whittlesey House, Rochester 52393
Captain Forbes House Museum, Milton 50699
Cass County Historical Society Museum, Plattsmouth 51985
CFCC Webber Center, Ocala 51481
Clermont State Historic Site, Clermont 47618
Clinton Academy Museum, East Hampton 48243
The Columbian Museum and Art Center, Wamego 53813
Concord Museum, Concord 47767
Cooper-Hewitt National Design Museum, New York 51156
Coral Gables Merrick House, Coral Gables 47799
Crescent Bend/Armstrong-Lockett House and William P. Toms Memorial Gardens, Knoxville 49815
The Dana-Thomas House, Springfield 53215
Daughters of the American Revolution Museum, Washington 53858
Decatur House Museum, Washington 53859
The Decorative Arts Center of Ohio, Lancaster 49919
Dedham Historical Museum, Dedham 48023
Department of the Treasury Museum, Washington 53860
Design Arts Gallery, Philadelphia 51826
DeWitt Wallace Decorative Arts Museum, Williamsburg 54168
Dezign House, Jefferson 49607
Dunham Tavern Museum, Cleveland 47630
Edsel and Eleanor Ford House, Grosse Pointe 49094
Edward-Dean Museum, Cherry Valley 47425
Elbert Hubbard-Roycroft Museum, East Aurora 48234
Elizabeth D. Walters Library, Stroudsburg 53339
Fallingwater - Western Pennsylvania Conservancy, Mill Run 50690
Fenimore Art Museum, Cooperstown 47789
Festival-Institute Museum, Round Top 52471
Flint Institute of Arts, Flint 48559
Fonthill Museum of the Bucks County Historical Society, Doylestown 48177
Forbes Collection, New York 51174
Fort Hunter Mansion, Harrisburg 49179
Fort Smith Art Center, Fort Smith 48684
Fred Wolf Jr. Gallery, Philadelphia 51837
Frederic Remington Art Museum, Ogdensburg 51506
Fuller Craft Museum, Brockton 47036
Fulton Mansion, Rockport 52426
Gallery II, Kalamazoo 49664
Gallier House, New Orleans 51081
George Washington's Mount Vernon, Mount Vernon 50893
The Goldstein Museum of Design, Saint Paul 52616
Gomez Foundation for Mill House, Marlboro 50494
Headley-Whitney Museum, Lexington 50048
Hearst Castle, San Simeon 52830
Henry B. Plant Museum, Tampa 53433
Heritage Center of Lancaster County, Lancaster 49923
Hermann-Grima House, New Orleans 51082
Hermitage Foundation Museum, Norfolk 51364
Hibel Museum of Art, Jupiter 49660
Hillwood Museum, Washington 53876
Hinckley Foundation Museum, Ithaca 49549
Historic Cherry Hill, Albany 46158
Historic Hope Plantation, Windsor 54226
The Historic Indian Agency House, Portage 52052
Historic New England, Boston 46932
Historical Society of Wells and Ogunquit Museum, Wells 54029
Horatio Colony House Museum, Keene 49719
House of Memories Museum, Wilson 54209
Houston Museum of Decorative Arts, Chattanooga 47410
Huntington Museum of Art, Huntington 49435
Hyland House, Guilford 49103
Illinois Artisans and Visitors Centers, Whittington 54138
Indianapolis Museum of Art, Indianapolis 49501
Isabella Stewart Gardner Museum, Boston 46934
Jacob Kelley House Museum, Hartsville 49204
Jamestown Settlement, Williamsburg 54169
Jeremiah Lee Mansion, Marblehead 50460
John A. Logan College Museum, Carterville 47266
The John and Mable Ringling Museum of Art, Sarasota 52914
John Brown House, Providence 52168
John Nicholas Brown Center for the Study of American Civilization, Providence 52169
Joseph Manigault House, Charleston 47369
Kansas Museum of History, Topeka 53536
Kearney Mansion Museum, Fresno 48792
Kelton House Museum, Columbus 47755
Kemerer Museum of Decorative Arts, Bethlehem 46779
Kent State University Museum, Kent 49740
Kirkland Museum of Fine and Decorative Art, Denver 48080
Ladew Manor House, Monkton 50781
Leffingwell House Museum, Norwich 51436
Leigh Yawkey Woodson Art Museum, Wausau 53993
Lemon Hill Mansion, Philadelphia 51847
Liberty Hall, Frankfort 48740
Lightner Museum, Saint Augustine 52522
Locust Lawn and Terwilliger House, Gardiner 48859
Longue Vue House, New Orleans 51085
Lower Cape Fear Historical Society Museum, Wilmington 54203
Luna Mimbres Museum, Deming 48053

McConaghy House, Hayward 49229
McMinn County Living Heritage Museum, Athens 46420
Manchester Historical Society Museum, Manchester 50419
Marblehead Museum and J.O.J. Frost Folk Art Gallery, Marblehead 50462
Maria Martinez Museum, Santa Fe 52879
Marston House, San Diego 52738
Mary Todd Lincoln House, Lexington 50053
The Mattatuck Museum of the Mattatuck Historical Society, Waterbury 53956
Maymont, Richmond 52334
Meadow Brook Hall, Rochester 52386
Mel Fisher Maritime Heritage Museum, Key West 49764
Memorial Art Gallery of the University of Rochester, Rochester 52396
Memorial Hall Museum, Deerfield 48033
Milwaukee Art Museum, Milwaukee 50711
Mint Museum of Craft and Design, Charlotte 47393
Mission Inn Museum, Riverside 52372
Montpelier Cultur Arts Center, Laurel 49983
The Moody Mansion Museum, Galveston 48849
Morlan Gallery, Lexington 50054
Morris-Butler House Museum, Indianapolis 49504
Morris-Jumel Mansion, New York 51217
Moses Myers House, Norfolk 51366
Mount Clare Museum House, Baltimore 46580
MSCC Forsyth Center Galleries, College Station 47690
Muscatine Art Center, Muscatine 50921
The Museum at Drexel University, Philadelphia 51855
Museum of American Architecture and Decorative Arts, Houston 49396
Museum of Contemporary Craft, Portland 52080
Museum of Decorative Art, Chicago 47495
Museum of Early Southern Decorative Arts, Winston-Salem 54247
Museum of Fine Arts, Springfield 53226
Museum of Glass, Tacoma 53412
Museum of New Hampshire History, Concord 47775
The Museums of Oglebay Institute, Wheeling 54124
Naomi Wood Collection at Woodford Mansion, Philadelphia 51858
National Hispanic Art Museum, Albuquerque 46184
New Orleans Artworks Gallery, New Orleans 51089
The Newark Museum, Newark 51287
Nichols House Museum, Boston 46943
Noyes Art Gallery, Lincoln 50104
The Old Stone Jail Museum, Palmyra 51687
Osborn-Jackson House, East Hampton 48248
Owensboro Museum of Fine Art, Owensboro 51640
Paris Gibson Square Museum of Art, Great Falls 49031
Parry Mansion Museum, New Hope 51062
Patterson Homestead, Dayton 47987
Peabody Essex Museum, Salem 52657
Peppers Art Gallery, Redlands 52275
Philadelphia Society for the Preservation of Landmarks, Philadelphia 51869
The Phillips Museum of Art, Lancaster 49931
Pittock Mansion, Portland 52085
Polynesian Cultural Center, Laie 49873
Potsdam Public Museum, Potsdam 52115
The Prairie Museum of Art and History, Colby 47680
Pump House Center for the Arts, Chillicothe 47536
Purdue University Calumet Library Gallery, Hammond 49142
Queen Emma Gallery, Honolulu 49351
Rahr West Art Museum, Manitowoc 50443
Raynham Hall Museum, Oyster Bay 51662
Renfrew Museum, Waynesboro 54013
Reynolda House, Winston-Salem 54249
Richard Sparrow House, Plymouth 51998
Ripley's Believe it or not!, Key West 49766
Riverview at Hobson Grove, Bowling Green 46975
Roberson Museum and Science Center, Binghamton 46816
Rockwood Museum, Wilmington 54197
Rough Point, Newport 51329
Royal Arts Foundation, Newport 51330
Rubelle and Norman Schafler Gallery, Brooklyn 47082
Ruthmere House Museum, Elkhart 48348
De Saisset Museum, Santa Clara 52863
Salgo Trust for Education, Port Washington 52049
Sam Bell Maxey House, Paris 51704
Sanford Museum, Sanford 52847
The Saratoga Springs History Museum, Saratoga Springs 52924
Sargent House Museum, Gloucester 48948
Schuyler-Hamilton House, Morristown 50856
Schuyler Mansion, Albany 46162
Shirley-Eustis House, Boston 46949
Slater Memorial Museum, Norwich 51438
Society of Arts and Crafts-Exhibition Gallery, Boston 46950
Sosa-Carillo-Fremont House Museum, Tucson 53596
Southern Ohio Museum and Cultural Center, Portsmouth 52103
Stamford Historical Society Museum, Stamford 53248
Stan Hywet Hall and Gardens, Akron 46148
Stauth Memorial Museum, Montezuma 50812
The Stephen Girard Collection, Philadelphia 51882
Sterling and Francine Clark Art Institute, Williamstown 54175
Stevens-Coolidge Place, North Andover 51380
Stone-Tolan House, Rochester 52399
Suffolk County Vanderbilt Museum, Centerport 47315
Ten Broeck Mansion, Albany 46164
Timken Museum of Art, San Diego 52758
Valentine Richmond History Center, Richmond 52339
Varner-Hogg Plantation, West Columbia 54052
The Vermont State House, Montpelier 50826
Villa Montezuma, San Diego 52760
Villa Terrace Decorative Arts Museum, Milwaukee 50720
Visual Arts Gallery, Pensacola 51771
Vizcaya Museum, Miami 50635
Warther Museum, Dover 48168
Webster House Museum, Elkhorn 48351
Wichita Center for the Arts, Wichita 54151
William Trent House, Trenton 53564
Willoughby-Baylor House, Norfolk 51371
Wilson Castle, Proctor 52157
Wilton House Museum, Richmond 52344
Winterthur Museum, Winterthur 54258
The Wolfsonian, Miami Beach 50644

Uruguay

Museo del Azulejo Francés, Punta Ballena 54428
Palacio Taranco, Montevideo 54424

Uzbekistan

Buxoro Qala Davlat Muzeyi Qo'riqxonasi, Buxoro 54444
Muzej Prikladnogo Iskusstva Uzbekistana, Toshkent 54480

Venezuela

Museo de la Estampa y del Diseño Carlos Cruz Diez, Caracas 54547
Museo de la Fundación John Boulton, Caracas 54548

Distilling → Wines and Spirits

Dolls and Puppets

Australia

Jerilderie Doll World, Jerilderie 01179

Austria

Badener Puppen- und Spielzeugmuseum, Baden bei Wien 01773
Elli Riehl Puppenwelt, Einöde 01852
Puppenmuseum, Villach 02861
Puppenmuseum Hanisch, Waldkirchen an der Thaya 02892
Puppenmuseum Kärntner Eisenwurzen, Hüttenberg 02104
Welser Puppenweltmuseum, Wels 02918

Belgium

Centre de la Marionnette de la Communauté Française, Tournai 04081
Maison des Géants, Ath 03329
Musée du Jouet, Bruxelles 03484
Musée International de la Marionnette, Bruxelles 03489
Musée Marial de Beauraing, Beauraing 03350
Musée Tchantchès - Musée de la République Libre d'Outre-Meuse, Liège 03815
Poppenmuseum Christus Koning, Wetteren 04150
Poppentheatermuseum, Tienen 04069

Canada

Mildred M. Mahoney Silver Jubilee Dolls House Gallery, Fort Erie 05808
Puppet Centre, North York 06488
V. Gross Doll Museum, Warren 07279

China

Weifang Kite Museum, Weifang 08314

Czech Republic

Muzeum Loutek Milana Knížáka, Děčín 09490
Muzeum Loutkářských Kultur, Chrudim 09485

Estonia

Nukumuuseum, Tallinn 10639

Finland

Nukke- ja Pukumuseo - Hatanpäa Mansion, Tampere 11355
Nukkemuseo Suruton, Savonlinna 11302
Suomenlinnan Lelumuseo, Helsinki 10826

France

Collection Historial Cognacq-Jay, Paris 14584
Collections Internationales de Marionnettes, Charleville-Mézières 12408
Musée-Atelier de Poupées Francépoque, Nances-Lac d'Aiguebelette 14348
Musée de Jouet et da la Poupée Ancienne, L'Isle-sur-la-Sorgue 13784
Musée de la Poupée, Aubeterre-sur-Dronne 11725
Musée de la Poupée, La Côte-d'Aime 13349
Musée de la Poupée, Le Beausset 13557
Musée de la Poupée et du Jouet Ancien, Wambrechies 16283
Musée des Poupées, Josselin 13279
Musée des Poupées d'Antan, Malicorne-sur-Sarthe 13931
Musée des Poupées d'Antan et de la Tonnellerie, Maisons-lès-Chaource 13926
Musée du Folklore, Moulins (Allier) 14307
Musée Gadagne, Lyon 13895
Musée La Pouponnière, Bourg-Charente 12105
Musée Miniatures et Poupées, Gréoux-les-Bains 13143

Georgia

T. Tumanishvili Children Toy Museum, Tbilisi 16384

Germany

Coburger Puppen-Museum, Coburg 17714
Deutsch-Französisches Puppen und Spielzeugmuseum, Schweigen-Rechtenbach 21681
Hessisches Puppenmuseum, Hanau 18978
Historisches Puppen-, Spielzeug- und Ikonenmuseum, Bernkastel-Kues 17297
Käthe-Kruse-Puppen-Museum, Donauwörth 17897
Katharinas Puppenhaus, Pleystein 21061
Kindergartenmuseum, Bruchsal 17597
Kindheitsmuseum, Marburg 20190
Das Kleine Museum, Hagnau am Bodensee 18863
Münchner Puppenmuseum, München 20461
Museen der Stadt Bad Kösen, Bad Kösen 16763
Museum Aschenbrenner, Garmisch-Partenkirchen 18541
Museum Burg Wendelstein, Vacha 22195
Museum der Deutschen Spielzeugindustrie, Neustadt bei Coburg 20675
Museum der Puppengeschichte - Joyce Merlet, Köln 19618
Museum für Puppen, Bären und Spielzeug, Solingen 21804
Musikinstrumenten- und Puppenmuseum, Goslar 18715
Nostalgie-Museum, Wörth an der Donau 22643
Oberrheinische Narrenschau, Kenzingen 19518
PuK- Museum für PuppentheaterKultur, Bad Kreuznach 16766
Puppen-Stuben-Museum, Zimmerschied 22803
Puppen- und Spielzeugmuseum, Lichtenstein/Sachsen 19983
Puppen- und Spielzeugmuseum mit Zinfigurenkabinett, Dornstetten 17907
Puppen- und Spielzeugmuseum Sammlung Katharina Engels, Rothenburg ob der Tauber 21409
Puppenmuseum, Bad Breisig 16704
Puppenmuseum, Bad Schönborn 16829
Puppenmuseum, Nordrach 20745
Puppenmuseum Falkenstein, Hamburg 18956
Puppenmuseum Laufeld, Laufeld 19853
Puppentheatermuseum, Kaufbeuren 19491
Rügener Puppen- und Spielzeugmuseum, Putbus 21141
Schlossmuseum, Ellwangen 18124
Spielmuseum Soltau und Stiftung Spiel, Soltau 21815
Spielzeugmuseum, Bad Lauterberg 16776
Stadtmuseum Borken, Westfalen, Borken, Westfalen 17479
Steinhuder Spielzeugmuseum, Wunstorf 22735
Teddy Museum Hof, Hof 19221
Teddymuseum, Klingenberg am Main 19569
Zum Puppenhaus, Immenstaad am Bodensee 19327

Hungary

Babamúzeum, Tihany 23924
Bábkiállítás, Bosta 23378
Fehérvári Babaház, Székesfehérvár 23853
Népművészetek Háza, Szentendre 23878
Palóc Néprajzi Magángyűjtemény és Palóc Baba Kiállítás, Mátrafüred 23695

India

Shankar's International Dolls Museum, Delhi 24154

Indonesia

Museum Wayang, Jakarta 24506

Italy

Collezione di Marionette Ferrari, Parma 27095
Museo dei Burattini, Mantova 26593
Museo della Marionetta, Torino 28006
Museo Internazionale delle Marionette A. Pasqualino, Palermo 27078

Japan

Echizen Take-ningyo-no-sato, Maruoka 28883
Japan Clay Doll Museum, Nakano 29020
Kobe Doll Museum, Kobe 28729
Miharu-machi Ningyo Hakubutsukan, Miharu 28908
Nippon Kokeshikan, Naruko 29042
Omoide Museum, Kyoto 28858
Tekiho Memorial Museum of Dolls, Ogose 29095
Yokohama Doll Museum, Yokohama 29612

Korea, Republic

Teddy Bear Museum, Seogwipo 29862

Latvia

Leļļu Mākslas muzeja, Rīga 30094

Luxembourg

Musée Dicks, Vianden 30494

Mexico

Museo Casa de Carranza, México 31057
Museo Nacional del Títere, Huamantla 30938

Monaco

Musée National de Monaco, Monaco 31611

Netherlands

Klederdracht en Visserijmuseum, Bunschoten Spakenburg 32020
Museum de Verzamelaar, Zuidhorn 33012
Poppen- en Speelgoedmuseum, Tilburg 32831
Poppenhuis Carmen, Utrecht 32855
Poppenhuismuseum, Heesch 32341
Poppenmuseum, Ter Apel 31910
Poppenmuseum, Maastricht 32530
Poppenspe(e)lmuseum, Vorchten 32908
Popppenhuismuseum Alida's Kleine Wereldje, Veendam 32871
Speelgoedmuseum De Brug, Eerbeek 32177
Speelgoedmuseum Op Stelten, Oosterhout 32629
Zeeuws Poppen- en Klederdrachten Museum, Zoutelande 33009

Russia

Chudožestvennaja Galereja - Dom Kukly Tatjany Kalininoj, Petrozavodsk 37556
Muzej Kukol i Detskoj Knigi "Strana Čudes", Ekaterinburg 36895
Muzej Kukol 'Kukolnyj Dom', Moskva 37312
Muzej Miniaturnych Kukol Malenkij Mir, Semipalatinsk 37760
Muzej Teatralnoj Kukly A.A Veselova, Voronež 37974
Muzej Teatralnych Kukol, Moskva 37374
Muzej Unikalnych Kukol, Moskva 37378

Slovakia

Múzeum Bábkarských Kultúr a Hračiek Hrad Modry Kameň, Modrý Kameň 38307

Spain

El Mundo de Muñecas, Icod de los Vinos 39406
Museu Internacional de Titelles d'Albaida, Albaida 38806

Sweden

Carlotte Weibulls Foklivscenter, Åkarp 40411
Dock och Leksaksmuseum, Landsbro 40666
Dock- och Textilmuseum, Katrineholm 40640
Hobby- och Leksaksmuseum, Stockholm 40900

Switzerland

Krippen- und Spielzeugmuseum, Bubendorf 41262
Musée Suisse de la Marionnette, Fribourg 41391

Puppen- und Spielzeugmuseum, Solothurn ... 41887
Puppenmuseum Jeannine, Kreuzlingen ... 41515
Spielzeugmuseum, Davos Platz ... 41322
Teddybär Museum, Luzern ... 41599

Taiwan

Li Tien-lu Hand Puppet Historical Museum, Sanjhi ... 42310
Lin Liu-Hsin Puppet Theatre Museum, Taipei ... 42400
Museum of Shadow Puppet, Gangshan ... 42191
Puppetry Art Center of Taipei, Taipei ... 42420
Taishan Doll Museum, Taishan ... 42443

United Kingdom

Abbey House Museum, Leeds ... 44658
Bear Museum, Petersfield ... 45372
Childhood and Costume Museum, Bridgnorth ... 43604
Coventry Toy Museum, Coventry ... 43899
The Dolly Mixture, Langbank ... 44638
Dorset Teddy Bear Museum, Dorchester ... 44000
Ethnic Doll and Toy Museum, Whitstable ... 46006
Lilliput Antique Doll and Toy Museum, Brading ... 43588
Mechanical Music and Doll Collection, Chichester ... 43829
Mellerstain House, Gordon ... 44331
Museum of Childhood, Sudbury, Derbyshire ... 45796
The Teddy Bear Museum, Stratford-upon-Avon ... 45786
Teddy Bear Shop and Museum, Ironbridge ... 44533
The Toy Museum, Stow-on-the-Wold ... 45771
VA Museum of Childhood, Bethnal Green, London ... 45001
Vina Cooke Museum of Dolls and Bygone Childhood, Newark-on-Trent ... 45217
Warwick Doll Museum, Warwick ... 45941

United States of America

American Museum of the Miniature Arts, Dallas ... 47926
Angels Attic Museum, Santa Monica ... 52897
Arizona Doll and Toy Museum, Phoenix ... 51897
Bread Puppet Museum, Glover ... 48950
Creatabilitoys! - Museum of Advertising Icons, Coral Gables ... 47800
Denver Museum of Miniatures, Dolls and Toys, Denver ... 48074
The Doll and Toy Museum of NYC, Brooklyn ... 47071
Eliza Cruce Hall Doll Museum, Ardmore ... 46344
Enchanted Mansion Doll Museum, Baton Rouge ... 46632
Enchanted World Doll Museum, Mitchell ... 50762
Franks Antique Doll Museum, Marshall ... 50508
Gay Nineties Button and Doll Museum, Eureka Springs ... 48444
Heritage Museum of Fine Arts for Youth, Detroit ... 48120
House of a Thousand Dolls, Loma ... 50163
Milan Historical Museum, Milan ... 50680
Museum of East Alabama, Opelika ... 51573
Rosalie Whyel Museum of Doll Art, Bellevue ... 46709
Strong National Museum of Play, Rochester ... 52400
Teddy Bear Museum of Naples, Naples ... 50953
Toy and Miniature Museum of Kansas City, Kansas City ... 49703
Washington Dolls House and Toy Museum, Washington ... 53931
Wax Museum, Santa Claus ... 52865
Yesterday's Children Antique Doll Toy Museum, Vicksburg ... 53736
Yesteryears Doll and Toy Museum, Sandwich ... 52843

Venezuela

Museo de las Muñecas, Píritu ... 54592

Drawing

Argentina

Museo de Bellas Artes de la Boca Quinquela Martín, Buenos Aires ... 00187
Museo de la Caricatura Severo Vaccaro, Buenos Aires ... 00199
Museo Itinerante de Arte Contemporáneo del Mercosur, Buenos Aires ... 00248

Armenia

Ara Sargsyan and Hakob Kojoyan House-Museum, Yerevan ... 00725

Austria

Albertina, Wien ... 02929
Kupferstichkabinett, Wien ... 03020
Nordico - Museum der Stadt Linz, Linz ... 02320

Belgium

Alfred Ost Museum, Zwijndrecht ... 04171
Centre Belge de la Bande Dessinée, Bruxelles ... 03442
Musée Constantin Meunier, Bruxelles ... 03460
Musée de la Ville d'Eaux, Spa ... 04043
Musée de l'Imprimerie-Museum van de Boekdrukkunst, Bruxelles ... 03473
Museum Plantin-Moretus / Prentenkabinet, Antwerpen ... 03297

Brazil

Museu de Arte Contemporânea da Universidade de São Paulo, São Paulo ... 04879

Canada

Acadia University Art Gallery, Wolfville ... 07398

China

Ningde Zone Shezu Museum, Ningde ... 08074

Croatia

Ivan Meštrović Gallery, Split ... 08957

Czech Republic

Letohrádek Ostrov, Ostrov nad Ohří ... 09703
Muzeum umění a designu, Benešov ... 09402
Národní Galerie v Praze - Palác Kinských, Praha ... 09777

Denmark

Museet for Dansk Bladtegning, København ... 10140
Vejen Kunstmuseum, Vejen ... 10309
Vejle Kunstmuseum, Vejle ... 10310

Finland

Tampereen Taidemuseon Muumilaakso, Tampere ... 11365

France

Collections de l'Abbaye Bénédictine de Saint-Louis-du-Temple, Vauhallan ... 16084
Collections de l'Ecole Nationale Supérieure des Beaux-Arts, Paris ... 14590
Fondation Custodia, Collection Frits Lugt, Paris ... 14598
Fondation Le Corbusier, Paris ... 14601
Musée Atger, Montpellier ... 14238
Musée Calvet, Avignon ... 11789
Musée-Chapelle Saint-Blaise-des-Simples, Milly-la-Forêt ... 14122
Musée du Luxembourg, Paris ... 14699
Musée Jean Cocteau, Menton ... 14083
Musée Magnin, Dijon ... 12742
Musée Maison Robert Doisneau, Gentilly ... 13044
Musée Municipal de l'Évêché, Limoges ... 13770
Musée National Ernest Hébert, Paris ... 14731
Musée Raymond Peynet, Brassac-les-Mines ... 12147

Germany

1822-Forum Frankfurter Sparkasse, Frankfurt am Main ... 18352
Edwin Scharff Museum, Neu-Ulm ... 20587
Ernst Barlach Atelierhaus und Ausstellungsforum-Graphikkabinett, Güstrow ... 18827
Galerie Parterre, Berlin ... 17106
Gertrudenkapelle, Güstrow ... 18828
Grafische Sammlung, Augsburg ... 16619
Kunsthalle Bremen, Bremen ... 17556
Kupferstich-Kabinett, Dresden ... 17954
Kupferstichkabinett - Sammlung der Zeichnungen und Druckgraphik, Berlin ... 17172
Lechner Museum, Ingolstadt ... 19343
Luza Acht, Lübben ... 20068
Olaf-Gulbransson-Museum für Graphik und Karikatur, Tegernsee ... 22018
Wilhelm-Busch-Museum Hannover, Hannover ... 19008

Greece

Museum Alex Mylona, Athinai ... 22928
N. Hadjikyriakos-Ghikas Gallery, Athinai ... 22941

Iraq

Al Shaheed Monument and Museum, Baghdad ... 24712

Italy

Casa Buonarroti, Firenze ... 26143
Civico Gabinetto dei Disegni, Milano ... 26677
Fondazione Europea del Disegno, Meina ... 26649
Gabinetto Disegni e Stampe degli Uffizi, Firenze ... 26153
Museo Civico, Bassano del Grappa ... 25378
Museo dell'Accademia Ligustica di Belle Arti, Genova ... 26303
Museo Internazionale della Caricatura, Tolentino ... 27988
Museo Storico Carlo Levi, Aliano ... 25201
Raccolta d'Arte Pagliara, Napoli ... 26912
Raccolte d'Arte dell'Ospedale Maggiore di Milano, Milano ... 26721

Japan

Masaki Art Museum, Osaka ... 29133
Saitama-shiritsu Manga Bijutsukan, Saitama ... 29191
Shinano Drawing Museum, Ueda ... 29552
Yokoyama Memorial Manga Museum, Kochi ... 28750
Yumiko Igarashi Bijutsukan, Kurashiki ... 28791

Mexico

Museo de la Caricatura, México ... 31078
Museo Francisco Sarabia, Lerdo ... 31005

Netherlands

Abraham's Mosterdmakerij, Eenrum ... 32176
Teylers Museum, Haarlem ... 32302
Ton Smits Huis, Eindhoven ... 32195

Poland

Centrum Rysunku i Grafiki im. Tadeuza Kulisiewicza, Kalisz ... 34707
Galeria Foksal, Warszawa ... 35227
Galeria Krytyków Pokaz, Warszawa ... 35228
Galeria Test, Warszawa ... 35230
Galeria XXI, Warszawa ... 35231
Kordegarda, Warszawa ... 35232
Zachęta - Narodowa Galeria Sztuki, Warszawa ... 35290

Portugal

Museu Rafael Bordalo Pinheiro, Lisboa ... 35726

Serbia

Galerija Srpske Akademije Nauka i Umetnosti, Beograd ... 38045

Slovakia

Považská Galéria Umenia, Žilina ... 38390

Spain

Museo de Bellas Artes, Córdoba ... 39205
Museo de Dibujo "Julio Gavín-Castillo de Larrés", Larrés (Huesca) ... 39444
Museu Nacional d'Art de Catalunya (MNAC), Barcelona ... 39001

Sweden

Albert Engström-Museerna, Grisslehamn ... 40568
Museum Kasper, Norrtälje ... 40775
Seriemuseet Comicland, Ullared ... 40994

Switzerland

Kunstmuseum Solothurn, Solothurn ... 41885
Kunstmuseum Winterthur, Winterthur ... 42029
Kupferstichkabinett, Basel ... 41167
Museum Sursilvan, Trun ... 41953

Taiwan

Yang Ming Oceanic Culture and Art Museum, Keelung ... 42250

Turkey

İstanbul Büyükşehir Belediyesi Karikatür ve Mizah Müzesi, İstanbul ... 42824

United Kingdom

Beatrix Potter Gallery, Hawkshead ... 44429
Courtauld Institute Gallery, London ... 44830
Guildford House Gallery, Guildford ... 44370
Linley Sambourne House, London ... 44901
The Lowry Gallery, Salford ... 45599
National Fairground Archive, Sheffield ... 45654
RIBA Drawings and Archives Collections, London ... 44964

United States of America

Americas Society Art Gallery, New York ... 51117
Atrium Gallery, Lewiston ... 50036
Avampato Discovery Museum, Charleston ... 47378
Cartoon Art Museum, San Francisco ... 52769
Center Gallery, Demarest ... 48052
Drawing Center, New York ... 51167
Eiteljorg Museum of American Indians and Western Art, Indianapolis ... 49490
Gallery of Art, Saint Louis ... 52578
Hughes Fine Arts Center, Grand Forks ... 48988
International Museum of Cartoon Art, Boca Raton ... 46885
Joseph A. Cain Memorial Art Gallery, Corpus Christi ... 47821
Mabel Larson Fine Arts Gallery, Clarksville ... 47593
Merwin and Wakeley Galleries, Bloomington ... 46862
Muse Gallery, Jackson ... 49572
Pete and Susan Barrett Art Gallery, Santa Monica ... 52900
Putney School Gallery, Putney ... 52204
Santa Ana College Art Galleries, Santa Ana ... 52854
The Studio Annex, New York ... 51263
Synagogue for the Arts, New York ... 51266

Ecology

Argentina

Ecomuseo Regional Maipú, Maipú ... 00441

Australia

Western Power World of Energy, Fremantle ... 01096

Austria

Alpinarium Galtür, Galtür ... 01920
UnterWasserReich, Schrems ... 02715

Belgium

La Maison du Parc Naturel des Plaines de l'Escaut, Bon-Secours ... 03379

Bolivia

Museo de Historia Natural Noel Kempf Mercado, Santa Cruz ... 04240

Brazil

Espaço Museu da Vida, Rio de Janeiro ... 04691

Canada

Écomusée de Hearst Ecomuseum, Hearst ... 05981
Écomusée des Deux-Rives, Salaberry-de-Valleyfield ... 06903
Lynn Canyon Ecology Centre, North Vancouver ... 06478
Musée Écologique Vanier, Laval ... 06148

Denmark

Naturhistorisk Museum, Århus ... 09965
Strandgaarden Museum, Ulfborg ... 10301

France

Conservatoire de la Faune Sauvage de Sologne, Ménestreau-en-Villette ... 14074
Eco-Musée de la Noix, Castelnaud-la-Chapelle ... 12312
Ecomusée de la Lomagne, Miradoux ... 14128
Ecomusée de Saint-Joseph, Fargues-sur-Ourbise ... 12906
Mémoire du Rhin, Saint-Louis-la-Chaussée ... 15376
Musée-Maison de la Nature et de l'Estuaire de l'Orne, Sallenelles ... 15603

Germany

Archäologisch-Ökologisches Museum, Albersdorf ... 16451
Elbschloss Bleckede, Bleckede ... 17384
Haus des Waldes, Köln ... 19606
Naturmuseum Lüneburg, Lüneburg ... 20101
Naturschutzausstellung der Naturschutzstation, Neschwitz ... 20575
Niedersächsisches Deichmuseum, Dorum ... 17931
Stadtmuseum Schlüsselfeld, Schlüsselfeld ... 21578

Italy

Civico Museo Naturalistico Ferruccio Lombardi, Stradella ... 27910
Museo di Ecologia e Storia Naturale, Marano sul Panaro ... 26601

Japan

Fuji Visitor Center, Kawaguchiko ... 28687

Kazakhstan

Karaganda Ecological Museum, Karaganda ... 29702

Mexico

Museo Na-Bolom, San Cristóbal de las Casas ... 31329

Netherlands

Afval-Museum, Zwolle ... 33025
Bezoekerscentrum Oortjespad, Kamerik ... 32434

Milieu Educatie Centrum, Eindhoven ... 32190
De Verbeelding Kunst Landschap Natuur, Zeewolde ... 32993

Peru

Museo Multidisciplinario del Colegio La Salle, Lima ... 34262

Philippines

Dinosaur Museum, Malabon ... 34418

Poland

Muzeum Wiedzy o Środowisku, Poznań ... 35046

Portugal

Museu de São Miguel de Entre-Ambos-os-Rios, Ponte da Barca ... 35830

Puerto Rico

Casa Ramón Power y Giralt, San Juan ... 36060

Russia

Muzej Archeologija, Ėtnografija i Ėkologija Sibiri Kemerovskogo Gosudarstvennogo Universiteta, Kemerovo ... 37033

South Africa

Sea World, Durban ... 38612

Spain

Centro Visitantes Fabrica de Hielo, Sanlúcar de Barrameda ... 39974
Ingurugiro Etxea, Azpeitia ... 38930
Museo de la Ciencia de Valladolid, Valladolid ... 40231

Sweden

Aquaria Vattenmuseum, Stockholm ... 40883
Vänermuseet, Lidköping ... 40676

Taiwan

Hanbisuan Mikado Pheasant Ecology Museum, Su-Ao ... 42344

Tanzania

Shinyanga Mazingira Museum, Shinyanga ... 42519

Tunisia

Ecomusée d'Ichkeul, Bizerte ... 42618

United Kingdom

Harestanes Countryside Visitor Centre, Jedburgh ... 44546
River and Rowing Museum, Henley-on-Thames ... 44449

United States of America

Asri Environmental Education Center, Bristol ... 47030
Austin Nature and Science Center, Austin ... 46505
Chesapeake Children's Museum, Annapolis ... 46317
Coyote Point Museum for Environmental Education, San Mateo ... 52822
Eavanston Ecology Center, Evanston ... 48448
EcoTarium, Worcester ... 54297
Koke'e Natural History Museum, Kauai ... 49710
Robert A. Vines Environmental Science Center, Houston ... 49402
Weymouth Woods-Sandhills Nature Preserve Museum, Southern Pines ... 53187

Zambia

Copperbelt Museum, Ndola ... 54662

Economics

Austria

Österreichisches Gesellschafts- und Wirtschaftsmuseum, Wien ... 03048
Zoll- und Finanzgeschichtliche Sammlung, Linz ... 02327
Zoll- und Heimatmuseum, Perwang ... 02493

Belgium

Musée de la Banque Nationale de Belgique, Bruxelles ... 03466

Canada

Lester-Garland Premises, Trinity ... 07165

Finland

Suur-Miehikkälän Kylämuseo, Suur-Miehikkälä ... 11341

France

Musée de la Douane et des Frontières, Hestrud ... 13211
Musée des Commerces Anciens, Doué-la-Fontaine ... 12780
Musée National des Douanes, Bordeaux ... 12073

Germany

Deutsches Zollmuseum, Hamburg ... 18922
Erstes Deutsches Historic-Actien-Museum, Kürnbach ... 19739
Heimat- und Naturkunde-Museum Wanne-Eickel, Herne ... 19153
NRW-Forum Kultur und Wirtschaft, Düsseldorf ... 18003
Wirtschaftsgeschichtliches Museum Villa Grün, Dillenburg ... 17862

Hungary

Magyar Kereskedelmi és Vendéglátóipari Múzeum, Budapest ... 23438

India

Commercial and Industrial Museum, Kanpur ... 24253
Government Industrial and Commercial Museum, Kolkata ... 24273

Japan

Takayama Jinya, Takayama ... 29344

Mexico

Museo Histórico Ex Aduana de Ciudad Juárez, Juárez ... 30983

Netherlands

Belasting en Douane Museum, Rotterdam ... 32702
Grenslandmuseum, Dinxperlo ... 32131

Norway

Bankmuseet, Stavanger ... 33864
Bergen Skolemuseum, Bergen ... 33422
Norsk Tollmuseum, Oslo ... 33745

Pakistan

Industrial and Commercial Museum, Lahore ... 34020

Portugal

Museu das Alfândegas, Lisboa ... 35687

Russia

Muzej Istorii Rossijskogo Predprinimatelstva, Moskva ... 37297
Muzej Predprinimatelej, Mecenatov i Blagotvoritelej, Moskva ... 37349

Spain

Museo Histórico BBVA, Bilbao ... 39046

Sweden

Tullmuseum, Stockholm ... 40948

Switzerland

Museo Doganale Svizzero, Gandria ... 41396

United States of America

Discovery World - The James Lovell Museum of Science, Economics and Technology, Milwaukee ... 50707
Hagley Museum and Library, Wilmington ... 54194

Education

Albania

Muzeu i Arsimit, Korçë ... 00020

Algeria

Musée du Mont Riant, Alger ... 00042

Argentina

Eureka - Parque de la Ciencia, Mendoza ... 00450
Museo de Escuela de Educación Técnica Número 1, Dolores ... 00354
Museo de la E.E.M. Número 3, Dolores ... 00355
Museo de las Campanas, Mina Clavero ... 00466
Museo de las Escuelas, Buenos Aires ... 00211
Museo de los Niños Barrilete, Córdoba ... 00325
Museo y Centro Estudios Históricos de la Facultad de Odontología, Buenos Aires ... 00279

Australia

Alumny Creek Museum, Grafton ... 01126
Andrew Ross Museum, Kangaroo Ground ... 01187
Frankston Primary Old School Museum, Frankston ... 01090
Hale School Museum, Wembley Downs ... 01641
Old School Museum, Merimbula ... 01295
Past Students Association Museum, Gatton ... 01099
Ripley's Believe It or Not! Museum, Surfers Paradise ... 01538
Royal Prince Alfred Hospital Museum, Camperdown ... 00943
Saint John's Schoolhouse Museum, Reid ... 01456
Sir Edgeworth David Memorial Museum, Kurri Kurri ... 01207
Uleybury School Museum, One Tree Hill ... 01391

Austria

Archiv der Universität Wien, Schausammlung, Wien ... 02933
Frida Fred - Grazer Kindermuseum, Graz ... 01968
Kinderwelt Walchen, Vöcklamarkt ... 02867
Michelstettner Schule, Asparn an der Zaya ... 01733
Schulmuseum, Lockenhaus ... 02332
Schulmuseum Schulstubn Glocknhäusl, Sankt Peter am Wimberg ... 02679
Zoom Kindermuseum, Wien ... 03093

Azerbaijan

Baku Museum of Education, Bakı ... 03147

Belgium

Ardoisalle, Alle-sur-Semois ... 03263
De School van Toen, Gent ... 03624
Instituut voor Sociale Geschiedenis, Gent ... 03630
Le Petit Musée, Wavre ... 04141
Mu-Zee-Um, Oostende ... 03934
Provinciaal Suske en Wiske Kindermuseum, Kalmthout ... 03748
Schoolarchief en Museum, Sint-Katelijne-Waver ... 04017
Stedelijk Onderwijsmuseum Ieper, Ieper ... 03732

Bolivia

Museo Kusillo, La Paz ... 04221
Museo Universitario Dr. Ricardo Bohorquez R., Potosí ... 04238
Museo Universitario La Memoria Colectiva, Llallagia ... 04228
Museos Universitarios de Bolivia, La Paz ... 04227

Brazil

Museu da UFRGS, Porto Alegre ... 04643
Museu Histórico e Pedagógico, Mococa ... 04570
Museu Histórico e Pedagógico Conselheiro Rodrigues Alves, Guaratinguetá ... 04487
Museu Histórico e Pedagógico Dr. Washington Luís, Batatais ... 04330
Museu Histórico e Pedagógico Fernão Dias Paes, Penápolis ... 04623
Museu Histórico e Pedagógico Marechal Cândido Rondon, Araçatuba ... 04315
Museu Histórico e Pedagógico Monteiro Lobato, Taubaté ... 04926
Museu Histórico e Pedagógico Pe. Manoel da Nóbrega, São Manuel ... 04846
Museu Histórico e Pedagógico Prof. Lourenço Filho, Porto Ferreira ... 04656
Museu Histórico e Pedagógico Zequinha de Abreu, Santa Rita do Passa Quatro ... 04812
Museu Histórico Pedagógico Conselheiro Dr. João da Silva Carrão, Americana ... 04306
Museu Histórico Pedagógico Índia Vanuíre, Tupã ... 04938

Bulgaria

Nacionalen Muzej na Obrazovanieto, Gabrovo ... 05033
Školoto Muzej, Trjavna ... 05231

Canada

Alex Youck School Museum, Regina ... 06714
Algonquin College Museum, Nepean ... 06423
Amos Seaman School Museum, Minudie ... 06300
Athol Murray College of Notre Dame Museum, Wilcox ... 07334
Cape Breton Centre for Heritage and Science, Sydney ... 07047
Comox Archives Museum, Comox ... 05637
East Coulee School Museum, East Coulee ... 05731
École du Rang II d'Authier, Authier ... 05365
Edmonton Public Schools Museum, Edmonton ... 05744
Enoch Turner Schoolhouse, Toronto ... 07105
Evolution of Education Museum, Prince Albert ... 06661
The Hamilton Children's Museum, Hamilton ... 05955
Historic Zion Schoolhouse, Toronto ... 07115
Little White Schoolhouse, Truro ... 07181
McCulloch House Museum and Hector Exhibit Centre, Pictou ... 06602
Manitoba Children's Museum, Winnipeg ... 07367
Metchosin School Museum, Victoria ... 07255
Mohawk Trail School Museum, Hamilton ... 05961
Morrison Museum of the Country School, Islay ... 06025
Musée Acadien du Québec, Bonaventure ... 05436
Musée du Collège de Lévis, Lévis ... 06168
Musée National de l'Éducation, Montréal ... 06375
Old Crofton School Museum, Crofton ... 05660
Quinte Educational Museum, Ameliasburgh ... 05330
Redpath Museum, Montréal ... 06380
River Valley School Museum, Virden ... 07269
Rocky Lane School Museum, Fort Vermilion ... 05825
Ronathahon:ni Cultural Centre, Cornwall ... 05644
School Days' Museum, Fredericton ... 05839
School House Museum, Deep River ... 05688
Science North, Sudbury ... 07029
Star Mound School Museum, Snowflake ... 06989
Thames Valley Museum School, Burgessville ... 05495
Toronto Board of Education, Toronto ... 07147
Tupperville School Museum, Tupperville ... 07182
University of Alberta Museums, Edmonton ... 05761
Victoria School Museum, Carleton Place ... 05564
Victoria School Museum and Archives, Edmonton ... 05762
Wilson MacDonald Memorial School Museum, Selkirk ... 06944
York Region District School Board Heritage Schoolhouse, Markham ... 06240

Chile

Museo Pedagógico de Chile, Santiago de Chile ... 07463

China

Central Nation University Museum, Beijing ... 07525
Children's Museum, Shanghai ... 08158
Jilin University Museum, Changchun ... 07588
Ripley's Believe It or Not Museum, Hong Kong ... 07814
University History Museum, Guangzhou ... 07737

Costa Rica

Galería Nacional y Museo de los Niños, San José ... 08813

Croatia

Hrvatski Školski Muzej, Zagreb ... 08999
Hrvatski Sportski Muzej, Zagreb ... 09000

Cuba

Museo Abel Santamaria, Santa Clara ... 09294
Museo Nacional de la Alfabetización, La Habana ... 09164

Czech Republic

Městské Muzeum, Příbor ... 09805
Muzeum Komenského v Přerově, Přerov ... 09804
Památník Zapadlých Vlastenců, Paseky nad Jizerou ... 09708
Pedagogické Muzeum Jana Ámoše Komenského, Praha ... 09788
Stará Polenská Škola, Polná ... 09730
Stará Škola, Police nad Metují ... 09726

Denmark

Børnenens Museum, København ... 10124
Flakkebjerg Skolemuseum, Slagelse ... 10274
Gærup Skolemuseum og Brahetrolleborg Sogns Folkemindesamling, Faaborg ... 10009
Ripley's Believe It or Not Museum, København ... 10149

Dominican Republic

Museo del Niño - Trampolín, Santo Domingo ... 10354

Egypt

Museum of Education and Documents Library, Cairo ... 10502

El Salvador

Museo de los Niños Tin Marin, San Salvador ... 10553

Estonia

Antsla Gümnaasiumi Muuseum, Antsla 10560
Obinitsa Põhikooli Muuseum, Obinitsa 10599
Õisu Tehnikumi Muuseum, Halliste 10567
Parksepa Keskkooli Muuseum, Parksepa 10607
Pikakannu Põhikooli Muuseum, Pikakannu 10608
Tartu Ülikooli Ajaloomuuseum, Tartu 10659
Vastseliina Gümnaasiumi Muuseum, Vastseliina 10669
Viljandimaa Gümnaasiumi Muuseum, Viljandi 10675

Finland

Elimäen Koulumuseo, Elimäki 10725
Finlands Svenska Skolmuseum, Munsala 11124
Helsingin Yliopistomuseo Arppeanum, Helsinki 10785
Himangan Kotiseutumuseo, Himanka 10840
Jyväskylän Lyseon Museo, Jyväskylä 10888
Jyväskylän Näkövammaisten Koulun Museo, Jyväskylä 10889
Jyväskylän Yliopiston Museo - Kulttuurihistoriallinen Osasto, Jyväskylä 10892
Kaurilan Koulumuseo, Tohmajärvi 11378
Koulumuseo, Somero 11320
Koulumuseo, Helsinki 10794
Lastenmuseo, Helsinki 10798
Lastentarhamuseo, Helsinki 10799
Näkövammaismuseo, Jiris 10871
Saukkojärven Kotiseutu- ja Koulumuseo, Ranua 11256
Suomen Koulumuseo, Tampere 11359
Yttilän Museokoulu, Yttilä 11485

France

Écomusée des Boutères-Pilat, Riotord 15048
Musée Aubois d'Histoire de l'Éducation, Troyes 15997
Musée Bernard d'Agesci, Niort 14455
Musée de Jouet et da la Poupée Ancienne, L'Isle-sur-la-Sorgue 13784
Musée de la Mine et de l'École, Harnes 13181
Musée de la Vallée à Pontis, Pontis 14905
Musée de La Vieille École, Les Valagans 13719
Musée de l'École, Carcassonne 12284
Musée de l'Ecole, Chartres 12423
Musée de l'Ecole, Echery 12806
Musée de l'École, Saint-Remy 15484
Musée de l'Ecole, Villeneuve-d'Ascq 16203
Musée de l'Ecole 1900, Saint-Martin-des-Olmes 15402
Musée de l'École Jules Ferry, Fraize 12995
Musée de l'Ecole Rurale, Trégarvan 15981
Musée de l'École Rurale d'Auvergne, Messeix 14097
Musée Départemental de l'Ecole Publique, Vergne 16116
Musée Départemental de l'Éducation, Saint-Ouen-l'Aumône 15435
Musée du Jeu de l'Oie, Rambouillet 14981
Musée-École, Saint-Christophe-en-Brionnais 15207
Musée Ecole Publique en Charente, Saint-Fraigne 15274
Musée Jean Macé, Beblenheim 11915
Musée la Vie des Jouets, Mauleon 14035
Musée-Maison d'Ecole, Montceau-les-Mines 14192
Musée National de l'Éducation, Rouen 15121
Musée Nivernais de l'Education, Nevers 14418
Musée Rural de l'Educations dans les Côtes d'Armor, Saint-Nicolas-du-Pelem 15429
Musée Valentin Haüy, Paris 14739

Germany

Bayerisches Schulmuseum, Ichenhausen 19305
Bergringmuseum, Teterow 22035
Deutsches Museum für Schulkunst, Hagen 18854
Dorfschulmuseum, Ködnitz 19593
Ebersdorfer Schulmuseum, Chemnitz 17694
Erstes Allgäu-Schwäbisches Dorfschulmuseum, Erkheim 18214
Erstes Bayerisches Schulmuseum, Sulzbach-Rosenberg 21983
Exploratorium-Kindermuseum Stuttgart und Region, Stuttgart 21931
Faszinosum, Borna 17484
Die Franckeschen Stiftungen zu Halle, Halle, Saale 18891
Friedrich-Fröbel-Museum, Bad Blankenburg 16698
Hamburger Schulmuseum, Hamburg 18933
Heimat-und Handwerkermuseum, Leutershausen-Wiedersbach 19970
Heimatmuseum, Hallstadt 18911
Heimatstube Wendhausen, Schellerten 21542
Johann-Baptist-Graser-Schulmuseum, Bayreuth 16956
Jugend Museum Schöneberg, Berlin 17148
Junge Kunsthalle, Karlsruhe 19440
kinder museum frankfurt, Frankfurt am Main 18380
Kinder- und Jugendmuseum München, München 20450
Kindermuseum, Wuppertal 22743
Kleines Museum - Kinder der Welt, Frankfurt (Oder) 18403
Klick Kindermuseum Hamburg, Hamburg 18939
Labyrinth Kindermuseum Berlin, Berlin 17174
Landschulmuseum Göldenitz, Dummerstorf 18028
LIPSIKUS Kindermuseum, Leipzig 19918
MACHmit! Museum für Kinder, Berlin 17180
Memorialmuseum Friedrich Fröbel, Oberweißbach 20849
Mobiles Kindermuseum im Freizeitheim Vahrenwald, Hannover 18998
Museum im Koffer, Nürnberg 20782
Museum Kindheit und Jugend, Berlin 17209
Museum zur Geschichte Hohenheims, Stuttgart 21955
Nordwestdeutsches Schulmuseum Friesland, Zetel 22791
Ostfriesisches Schulmuseum Folmhusen, Westoverledingen 22518
Rochow-Museum Reckahn, Kloster Lehnin 19578
Sammlungen des Instituts für Hochschulkunde, Würzburg 22723
Schulgeschichtliche Sammlung, Magdeburg 20130
Schulgeschichtliche Sammlung im Main-Taunus-Kreis, Kriftel 19715
Schulhausmuseum, Hollingstedt 19255
Schulheimatmuseum, Asbach-Bäumenheim 16580
Schulhistorische Sammlung Cruismannschule, Bochum 17406
Schulmuseum, Haldensleben 18885
Schulmuseum, Neumark 20632
Schulmuseum Bremen e.V., Bremen 17561
Schulmuseum des Bezirks Unterfranken, Bad Bocklet 16700
Schulmuseum Friedrichshafen am Bodensee, Friedrichshafen 18477
Schulmuseum Fronau, Roding 21348
Schulmuseum Lilienthal, Lilienthal 19991
Schulmuseum Mozartschule Rheingönheim, Ludwigshafen am Rhein 20056
Schulmuseum Nordwürttemberg in Kornwestheim, Kornwestheim 19691
Schulmuseum Nürnberg, Nürnberg 20789
Schulmuseum Steinhorst, Steinhorst 21893
Schulmuseum und Hallenhaus, Middelhagen 20322
Schulmuseum - Werkstatt für Schulgeschichte Leipzig, Leipzig 19933
Städtisches Schulmuseum, Lohr am Main 20028
Theodor-Mommsen-Gedächtnisstätte, Garding 18539
Universitätsmuseum, Altdorf bei Nürnberg 16484
Westfälisches Schulmuseum, Dortmund 17929

Greece

Athens University Museum, Athinai 22901
Hellenic Children's Museum-The Station, Athinai 22912

Guatemala

Museo Universitario de San Carlos, Guatemala 23306

Hungary

Angyalföldi Helytörténeti Gyűjtemény, Budapest 23381
Eötvös József Emlékmúzeum, Ercsi 23546
Iskolamúzeum, Békés 23363
Iskolatörténeti Kiállítás, Kecskemét 23646
Kőbányai Helytörténeti Gyűjtemény, Budapest 23423
Országos Pedagógiai Könyvtár és Múzeum, Budapest 23462
Óvodamúzeum, Martonvásár 23692
Városi Könyvtár és Múzeum, Tapolca 23914
XVIII. Kerületi Pedagógiai és Helytörténeti Gyűjtemény, Budapest 23492

India

Bal Sangrahalaya, Lucknow 24290
Carey Museum, Serampore 24383
Children's Museum, Bhavnagar 24087
Government Educational Museum, Etawah 24170
National Children's Museum, Delhi 24144
Nehru Children's Museum, Kolkata 24278
Regional Science Centre, Bhopal 24095
Regional Science Centre and Planetarium, Kozhikode 24286
Regional Science Centre Bhubaneswar, Bhubaneshwar 24098
Shri Girdharbhai Sangrahalaya, Amreli 24065

Ireland

Plunket Museum of Irish Education, Dublin 24816

Israel

Havayeda Museum, Holon 24976
Israel Children's Museum, Holon 24977

Italy

Museo dei Bambini Explora, Roma 27472
Museo della Scuola, Bolzano 25488
Museo della Scuola, Pramollo 27326
Museo Didattico, Niscemi 26928
Museo Didattico della Scuola Medica Salernitana, Salerno 27596
Museo per la Storia dell'Università di Pavia, Pavia 27125
Museo Scuola Odin-Bertot, Angrogna 25240
Museo Storico della Didattica Mauro Laeng, Roma 27529

Japan

Children's Museum Big Bang, Sakai 29195
H.C.Andersen-koen Kodomotachi Hakubutsukan, Funabashi 28452
Historical Museum of Private Schools at the Foreign Settlement, Nagasaki 28975
Iwata-shi Kyu Mitsukegakko, Iwata, Shizuoka 28610
Kanazawa Daigaku Hakubutsukan, Kanazawa 28660
Kyoto-shiritsu Gakkou Rekishi Hakubutsukan, Kyoto 28847
Muroran-shi Seishonen Kagakukan, Muroran 28947
Nagoya-shi Daigaku Hakubutsukan, Nagoya 29006
Saigo Kinenkan, Kagoshima 28634
Watashi no Shigotokan, Seika 29239

Jordan

Children's Heritage and Science Museum, Amman 29647

Kazakhstan

Respublikanskij Pedagogičeskij Muzej, Almaty 29691

Korea, Republic

Chonbuk University Museum, Jeonju 29840
Chonnam University Museum, Gwangju 29821
Daegu National University of Education Museum, Daegu 29793
Dan Kook University Museum, Seoul 29869
Deokpojin Educational Museum, Gimpo 29815
Dongguk University Museum, Seoul 29872
Gyeongju University Museum, Gyeongju 29827
Hanbat Education Museum, Daejeon 29800
Hansong Womens College Museum, Seoul 29880
Jeju Education Museum, Jeju 29837
Joongang University Museum, Seoul 29887
Kookmin College Museum, Seoul 29891
Kwangdong College Museum, Gangneung 29809
Kyongju University Museum, Kyongju 29849
Kyongnam College Museum, Masan 29850
Myong Ji University Museum, Seoul 29908
Samsung Children's Museum, Seoul 29913
Sangji Vocational College Museum, Andong 29767
Sejong College Museum, Seoul 29915
Seoul Education Museum, Seoul 29918
Sokang University Museum, Seoul 29924
Sung Kyun Kwan University Museum, Seoul 29929
Sungshin Women's University Museum, Seoul 29931
Won Kwang College Museum, Iri 29835
Yeungnam University Museum, Gyeongsan 29828

Latvia

Daugavpils Universitātes Muzejs, Daugavpils 29999
Latvijas Universitātes Pedagoģijas Muzejs, Rīga 30089
Rīgas Tehniskās Universitātes Muzejs, Rīga 30110
Valkas Novadpētniecības Muzejs, Valka 30137

Lebanon

Planète de la Découverte, Beirut 30162

Lithuania

Pedagoginis Muziejus, Kaunas 30264

Malaysia

Malaysian Youth Museum, Melaka 30595
Ripley's Believe It or Not Museum, Shah Alam 30620

Mexico

La Burbuja-Museo del Niño, Hermosillo 30928
Centro de Investigaciones Culturales-Museo, Mexicali 31031
Museo Antiguo Colegio de San Ildefonso, México 31049
Museo del Niño, Santa Ana del Valle 31373
Museo del Niño Globo mi Ciudad Pequeña, Guadalajara 30901
Museo del Niño Papalote, México 31105
Museo del Recuerdo, México 31107
Museo Escolar de Sitio de la Matamba, Jamapa 30971
Museo Escolar y Comunitario de Acamalín, Xico 31562
Museo Histórico de la Educación Tecnológica en México, Chilpancingo 30774
Museo Interactivo de Ciencia, Tecnología y Medio Ambiente Sol del Niño, Mexicali 31035
Museo Interactivo del Centro de Ciencias de Sinaloa, Culiacán 30828
Museo Interactivo del Medio Ambiente, México 31124
Museo Interactivo El Rehilete, Pachuca 31244
Museo Interactivo La Avispa, Chilpancingo 30775
Ripley's Believe It or Not Museum, Guadalajara 30905
Ripley's Believe It or Not Museum, México 31160

Netherlands

Academisch Historisch Museum van de Universiteit Leiden, Leiden 32481
Comenius Museum, Naarden 32563
Historisch Documentatiecentrum voor het Nederlands Protestantisme 1800-Heden, Amsterdam 31846
Joodse Schooltje, Leek 32466
Museum de Burghse Schoole, Burgh-Haamstede 32027
Museum Katendrecht, Rotterdam 32713
Nationaal Schoolmuseum, Rotterdam 32717
Onderwijsmuseum Educatorium, Ootmarsum 32633
De Ontdekhoek Kindermuseum, Rotterdam 32722
Schooltijd Schoolmuseum, Terneuzen 32816
Scouting Museum de Ducdalf, Rotterdam 32727
Scouting Museum Haagse Randstad, Den Haag 32101
Universiteitsmuseum, Groningen 32288
Universiteitsmuseum De Agnietenkapel, Amsterdam 31897
Van 't Lindenhoutmuseum, Nijmegen 32595

New Zealand

Saint Andrews College Museum, Christchurch 33077

Norway

Averøy Skolemuseum, Averøy 33411
Faleide Skulemuseum, Stryn 33889
Flatanger Bygdemuseum, Flatanger 33496
Gulli Skoletun, Andebu 33391
Holt Skolemuseum, Tvedestrand 33937
Lindås Skulemuseum, Isdalstø 33577
Meløy Bygdemuseum, Ørnes 33710
Modalen Skulemuseum, Modalen 33669
Norsk Barnemuseum, Stavanger 33868
Saltstraumen Museum, Saltstraumen 33808
Sleire Skulemuseum, Hosteland 33566
Trosterud Skolemuseum, Rømskog 33793
Vestlandske Skolemuseum, Stavanger 33878

Oman

Children's Museum, Qurm 33994

Pakistan

Education Museum, Hyderabad 34003

Peru

Museo de la Universidad Nacional Hermilio Valdizán, Huánuco 34180
Museo Didáctico Antonini, Nasca 34280

Philippines

Brillantes Ancestral House, Tayum 34517
Central Luzon State University Museum, Muñoz 34463
Centro Escolar University Museum and Archives, Manila 34437
Colegio de Santa Isabel Museum, Naga 34465
Museo de la Salle, Dasmariñas 34384
Museo Pambata, Manila 34447
Philippine Science Centrum, Marikina 34461
University Historical Museum, Musuan 34464
University of Southern Philippines Museum, Cebu 34379
Urios College Museum, Butuan 34356

Poland

Muzeum Oświatowe, Puławy 35064
Muzeum Oświaty, Bydgoszcz 34579
Muzeum Pedagogiczne im. Prof. K. Sośnickiego, Gdańsk 34644
Muzeum Uniwersytetu im. Ks. Jana Władzińskiego, Lublin 34907
Muzeum Uniwersytetu Warszawskiego, Warszawa 35274
Muzeum Uniwersytetu Wrocławskiego, Wrocław 35342
Oświatowe Muzeum, Szczecin 35164

Portugal

Museu Académico de Coimbra, Coimbra ... 35528
Museu Escolar de Marrazes, Leiria ... 35659
Núcleo Museológico dos Escuteiros, Torres Vedras ... 35973
Pavilhão do Conhecimento, Lisboa ... 35733

Puerto Rico

Museo del Niño, San Juan ... 36071

Romania

Muzeul Primei Şcoli Româneşti din Scheii Braşovului, Braşov ... 36181
Muzeul Universităţii Bucureşti, Bucureşti ... 36232

Russia

Memorialnyj Kabinet A.V. Lunačarskogo, Moskva ... 37233
Memorialnyj Muzej-kvartira E.F. Gnesinoj, Moskva ... 37243
Muzej Istorii Prosveščenija Komi Kraja, Syktyvkar ... 37819
Muzej Istorii Rossijskogo Gosudarstvennogo Pedagogičeskogo Universiteta im A.I. Gercena, Sankt-Peterburg ... 37685
Muzej Istorii Sankt-Peterburgskogo Universiteta, Sankt-Peterburg ... 37686
Muzej Narodnogo Obrazovanija Oblasti, Penza ... 37533
Muzej Narodnoje Obrazovanie Simbirskoj Gubernii v 70-80 gody 19 veka, Uljanovsk ... 37905
Muzej Simbirskaja Čuvašskaja Škola - Kvartira I.Ja. Jakovleva, Uljanovsk ... 37906
Muzej Simbirskaja Klassičeskaja Gimnazija, Uljanovsk ... 37907
Muzj Istorii Rossijskich Voennych Učilišč, Sankt-Peterburg ... 37715
Ogni Moskvy - Muzej Istorii Gorodskogo Osveščenija, Moskva ... 37391
Pedagogičeskij Muzej A.S. Makarenko, Moskva ... 37397
Učebno-chudožestvennyj Muzej Slepkov im. I.V. Cvetaeva, Moskva ... 37409

Serbia

Pedagoški Muzej, Beograd ... 38084

Slovakia

Múzeum Školstva a Pedagogiky Bratislava, Bratislava ... 38227
Múzeum Špecialneho Školstva, Levoča ... 38283

Slovenia

Slovenski Šolski Muzej, Ljubljana ... 38465

South Africa

Edoardo Villa Museum, Tshwane ... 38745

Spain

Centro de Historia de Zaragoza, Zaragoza ... 40338
Museo de la Academia de Infantería, Toledo ... 40132
Museo de la Ciencia de Valladolid, Valladolid ... 40231
Museo del Niño, Albacete ... 38804
Museo del Real Seminario de Bergara, Bergara ... 39029
Museo Etnográfico Comarcal de Azuaga, Azuaga ... 38932
Museo Exposición Permanente de Pedagogía, Huesca ... 39401
Museo Pedagóxico de Galicia, Santiago de Compostela ... 40029
Museo Scout-Guía, Madrid ... 39580
Museu de l'Escola de Gràmatica Ramon Llull, Algaida ... 38839

Sweden

Årbols Skolmuseum, Åmål ... 40418
Friluftmuseet Hallandsgården, Halmstad ... 40577
Helsingborgs Skolmuseum, Helsingborg ... 40588
Leksbergs Skolmuseum, Mariestad ... 40736
Moheda Skolmuseum, Moheda ... 40745
Skolmuseet, Fjälkinge ... 40522
Skolmuseum Bunge, Fårösund ... 40518
Stockholms Skolmuseum, Stockholm ... 40941
Universeum, Göteborg ... 40557
Västerås Skolmuseum, Västerås ... 41033
Vetlanda Skolmuseum, Vetlanda ... 41059

Switzerland

Musée d'Yverdon-les-Bains et sa Région, Yverdon-les-Bains ... 42049
Schulmuseum Mühlebach, Amriswil ... 41113
Schweizer Kindermuseum, Baden ... 41149

Taiwan

Museum of the Taipei Municipal Cheng Kung Senior High School, Taipei ... 42411
National Taiwan Science Education Center, Taipei ... 42418

Thailand

Ripley's Believe It or Not Museum, Pattaya ... 42589

Turkey

Cumhuriyet Egitim Müzesi, İstanbul ... 42813
Cumhuriyet Eğitim Müzesi, Karataş ... 42865
Eğitim Özel Müzesi, Ankara ... 42701
Eğitim ve Kültür Vakfı Özel Müzesi, Malkara ... 42901
Kız Teknik Öğretim Müzesi, Ankara ... 42704
ODTÜ Müzesi, Ankara ... 42707
Zonguldak Ereğli Müzesi, Zonguldak Ereğli ... 42969

United Kingdom

Andrew Carnegie Birthplace Museum, Dunfermline ... 44054
Auld Sköll, Fair Isle ... 44202
Böd of Gremista Museum, Lerwick ... 44692
British Schools Museums, Hitchin ... 44473
Captain Cook Schoolroom Museum, Great Ayton ... 44352
Causeway School Museum, Bushmills ... 43701
Eureka! The Museum for Children, Halifax ... 44384
Hands on History, Kingston-upon-Hull ... 44596
Haslemere Educational Museum, Haslemere ... 44413
Heriot-Watt University Museum, Edinburgh ... 44127
Islington Education Artefacts Library, London 44884
Livesey Museum for Children, London ... 44903
Museum of Childhood, Edinburgh ... 44132
Museum of Eton Life, Windsor ... 46043
Old Grammar School, Castletown ... 43777
Ragged School Museum, London ... 44963
Ripley's Believe It or Not Museum, Blackpool ... 43530
Rugby School Museum, Rugby ... 45528
Scaplen's Court Museum, Poole ... 45408
Scotland Street School Museum, Glasgow ... 44302
Tom Brown's School Museum, Uffington ... 45895
UCL Museums and Collections, London ... 45000

United States of America

4-H Schoolhouse Museum, Clarion ... 47581
A.C. Gilbert's Discovery Village, Salem ... 52664
A.D. Buck Museum of Natural History and Science, Tonkawa ... 53530
Adams State College Luther Bean Museum, Alamosa ... 46153
Arkansas State University Museum, Jonesboro ... 49647
Austin Children's Museum, Austin ... 46501
Bay Area Discovery Museum, Sausalito ... 52935
Black Mountain College Museum and Arts Center, Asheville ... 46370
Boston Children's Museum, Boston ... 46920
Brooklyn Children's Museum, Brooklyn ... 47067
Buffalo and Erie County Naval and Military Park, Buffalo ... 47117
Chicago Children's Museum, Chicago ... 47464
Chief John Ross House, Rossville ... 52465
Children's Art Museum, San Angelo ... 52700
Children's Discovery Museum of North San Diego, Carlsbad ... 47246
Children's Discovery Museum of San Jose, San Jose ... 52805
Children's Discovery Museum of the Desert, Rancho Mirage ... 52240
The Children's Discovery Museum of Vermont, Burlington ... 47142
Children's Hands-On Museum, Tuscaloosa ... 53617
The Children's Metamorphosis, Londonderry ... 50169
Children's Museum, Detroit ... 48110
Children's Museum, Holyoke ... 49327
Children's Museum, La Habra ... 49845
Children's Museum, Saint Petersburg ... 52635
The Children's Museum, Seattle ... 52997
The Children's Museum at Saratoga, Saratoga Springs ... 52920
Children's Museum at Yunker Farm, Fargo ... 48508
Children's Museum in Dartmouth, South Dartmouth ... 53167
Children's Museum in Easton, North Easton ... 51392
Children's Museum of Boca Raton, Boca Raton ... 46884
The Children's Museum of Cleveland, Cleveland ... 47622
Children's Museum of Denver, Denver ... 48068
The Children's Museum of Green Bay, Green Bay ... 49036
Children's Museum of History, Natural History and Science at Utica, New York, Utica ... 53675
The Children's Museum of Houston, Houston ... 49382
The Children's Museum of Indianapolis, Indianapolis ... 49488
The Children's Museum of Kansas City, Kansas City ... 49679
Children's Museum of Lake Charles, Lake Charles ... 49876
Children's Museum of Los Angeles, Los Angeles ... 50217
Children's Museum of Maine, Portland ... 52062
Children's Museum of Manhattan, New York ... 51148
The Children's Museum of Memphis, Memphis ... 50568
Children's Museum of Oak Ridge, Oak Ridge ... 51457
Children's Museum of Portsmouth, Portsmouth ... 52093
Children's Museum of Richmond, Richmond ... 52325
The Children's Museum of Rose Hill Manor Park, Frederick ... 48755
Children's Museum of South Carolina, Myrtle Beach ... 50932
Children's Museum of Spokane, Spokane ... 53205
Children's Museum of Stockton, Stockton ... 53305
Children's Museum of Tampa - Kid City, Tampa ... 53430
The Children's Museum of Utah, Salt Lake City 52689
Children's Museum of Virginia, Portsmouth ... 52105
The Children's Science Center, Cape Coral ... 47232
Childventure Museum, Fort Washington ... 48699
Cincinnati Museum Center, Cincinnati ... 47557
Cobb County Youth Museum, Marietta ... 50467
Coyote Point Museum for Environmental Education, San Mateo ... 52822
Creative Discovery Museum, Chattanooga ... 47408
The Curious Kids' Museum, Saint Joseph ... 52557
Discovery Center Museum, Rockford ... 52414
Discovery Center of the Southern Tier, Binghamton ... 46815
Discovery Creek Children's Museum of Washington DC, Washington ... 53862
The Discovery Museums, Acton ... 46133
Doozoo Children's Museum, Grand Junction ... 48993
Duke University Union Museum and Brown Art Gallery, Durham ... 48217
Duluth Children's Museum, Duluth ... 48194
Edventure, Columbia ... 47724
Elder Art Gallery, Lincoln ... 50094
Experience Children's Museum, Erie ... 48411
Exploration Place, Wichita ... 54141
Explore More ... a Children's Museum, East Aurora ... 48235
Explorium of Lexington, Lexington ... 50047
Family Museum of Arts and Science, Bettendorf ... 46784
Goodnow Museum, Manhattan ... 50431
Grand Rapids Children's Museum, Grand Rapids ... 49004
Great Explorations, Saint Petersburg ... 52639
Grout Museum of History and Science, Waterloo ... 53961
Hands On! Regional Museum, Johnson City ... 49633
Hannah Lindahl Children's Museum, Mishawaka ... 50748
Hawaii Children's Discovery Center, Honolulu 49339
Historic Jefferson College, Washington ... 53940
Hudson Valley Children's Museum, Nanuet ... 50949
Imaginarium Hands-On Museum, Fort Myers ... 48663
Joel and Lila Harnett Museum of Art, Richmond ... 52329
The Junior Museum, Troy ... 53574
The Junior Museum of Bay County, Panama City ... 51692
Kaleidoscope, Kansas City ... 49688
Kansas Health Museum, Halstead ... 49130
The Kansas Teachers' Hall of Fame, Dodge City ... 48149
Kauai Children's Discovery Museum, Kapaa ... 49706
Kearney Area Children's Museum, Kearney ... 49716
Kidspace Children's Museum, Pasadena ... 51719
Kidsquest Children's Museum, Bellevue ... 46708
Kohl Children's Museum of Greater Chicago, Glenview ... 48939
Lanesfield School, Edgerton ... 48286
Laredo Children's Museum, Laredo ... 49952
Las Vegas International Scouting Museum, Las Vegas ... 49972
Lied Discovery Children's Museum, Las Vegas ... 49975
Lincoln Children's Museum, Lincoln ... 50099
Little Red School House Museum, Weyauwega 54115
Little Red Schoolhouse-Living Library, Beloit ... 46722
The Living Arts and Science Center, Lexington 50052
Long Island Children's Museum, Garden City ... 48857
Louisiana Children's Museum, New Orleans ... 51086
Lutz Children's Museum, Manchester ... 50417
McKenna Children's Museum, New Braunfels ... 51022
Madison Children's Museum, Madison ... 50392
Magic House-Saint Louis Children's Museum, Saint Louis ... 52585
Marbles Kids Museum, Raleigh ... 52230
Melvin B. Tolson Black Heritage Center, Langston ... 49937
Miami Children's Museum, Miami ... 50632
Mid-Hudson Children's Museum, Poughkeepsie ... 52123
Minnesota Children's Museum, Saint Paul ... 52622
Mississippi University for Women Museum, Columbus ... 47747
Muncie Children's Museum, Muncie ... 50904
Museum and Archives of Georgia Education, Milledgeville ... 50691
Museum in the Community, Hurricane ... 49452
Museum of Anthropology, Columbia ... 47717
Museum of Art and Archaeology, Columbia ... 47718
Museum of Sex, New York ... 51228
National Children's Museum, Washington ... 53897
New Britain Youth Museum, New Britain ... 51030
New Britain Youth Museum at Hungerford Park, Kensington ... 49735
The New Children's Museum, San Diego ... 52742
New Jersey Children's Museum, Paramus ... 51698
New Mexico State University Museum, Las Cruces ... 49962
North Star Museum of Boy Scouting and Girl Scouting, North Saint Paul ... 51403
Northern Indiana Center for History, South Bend ... 53158
Old School House, York ... 54337
Omaha Children's Museum, Omaha ... 51555
Ottawa Scouting Museum, Ottawa ... 51628
Pittsburgh Children's Museum, Pittsburgh ... 51957
Please Touch Museum, Philadelphia ... 51871
Port Discovery - Children's Museum in Baltimore, Baltimore ... 46585
Portland Children's Museum, Portland ... 52087
President Andrew Johnson Museum, Greeneville ... 49052
Princeton University Museum of Natural History, Princeton ... 52152
Providence Children's Museum, Providence ... 52174
Randall Museum, San Francisco ... 52790
Rocky Mount Children's Museum, Rocky Mount ... 52437
Sam Houston Historical Schoolhouse, Maryville ... 50526
San Antonio Children's Museum, San Antonio 52723
Sangre de Cristo Arts Center and Buell Children's Museum, Pueblo ... 52194
Santa Fe Children's Museum, Santa Fe ... 52888
Science Learning Center, Upton ... 53668
Scitech, Aurora ... 46494
Scotia-Glenville Children's Museum, Scotia ... 52971
Seminole Nation Museum, Wewoka ... 54114
Sound School House Museum, Mount Desert ... 50875
Southeast Missouri State University Museum, Cape Girardeau ... 47234
Spertus Museum, Chicago ... 47514
Staten Island Children's Museum, Staten Island ... 53268
Stephens Museum, Fayette ... 48526
Stepping Stones Museum for Children, Norwalk ... 51433
Tougaloo College Art Collection, Tougaloo ... 53545
Town House, East Hampton ... 48249
Tucson Children's Museum, Tucson ... 53597
The University Museum, Edwardsville ... 48301
University Museum, Carbondale ... 47241
University Museums, Oxford ... 51650
University of Northern Iowa Museum Museums Collections, Cedar Falls ... 47294
The Virginia Discovery Museum, Charlottesville ... 47401
Wake Forest College Bithplace Society Museum, Wake Forest ... 53789
Wylie House Museum, Bloomington ... 46868
Young at Art Children's Museum, Davie ... 47974
Youth Museum of Southern West Virginia Beckley Exhibition Coal Mine, Beckley ... 46685

Uruguay

Museo Didáctico Artiguista, Maldonado ... 54380
Museo Pedagógico José Pedro Varela, Montevideo ... 54416

Venezuela

Museo de Los Niños de Caracas, Caracas ... 54550

Egyptology

Austria

Papyrusmuseum und Papyrussammlung, Wien ... 03054

Belgium

Musée Vivant de l'Egypte pour Enfants, Jambes ... 03741

Egypt

Egyptian Museum, Cairo ... 10492
Elephantine Island's Museum, Aswan ... 10480

Luxor Museum of Ancient Art, Al-Uqsur 10467
Nubia Museum, Aswan 10481
The Pharaonic Village, Giza 10529

France

Musée d'Egyptologie, Villeneuve-d'Ascq 16204
Musée Municipal de l'Évêché, Limoges 13770

Germany

Ägyptisches Museum der Universität Leipzig - Georg Steindorff, Leipzig 19891
Ägyptisches Museum und Papyrussammlung, Berlin 17028
Gustav-Lübcke Museum, Hamm, Westfalen 18969
Museum August Kestner, Hannover 19000
Museum Schloß Hohentübingen, Tübingen 22123
Museum Schloss Seefeld, Seefeld 21715
Neues Museum, Berlin 17228
Roemer- und Pelizaeus-Museum, Hildesheim 19185
Sammlung des Ägyptologischen Instituts, Heidelberg 19081
Staatliches Museum Ägyptischer Kunst, München 20489

Israel

Pontifical Biblical Institute Museum, Jerusalem 25009

Italy

Museo Egizio, Torino 28018

Japan

Miho Museum, Shigaraki 29258

Netherlands

Allard Pierson Museum Amsterdam, Amsterdam 31825
Rijksmuseum van Oudheden, Leiden 32487

Spain

Museu Egipci de Barcelona, Barcelona 38992

Sweden

Medelhavsmuseet, Stockholm 40913
Museum Gustavianum, Uppsala 41009

Switzerland

Bibel Orient Museum, Fribourg 41383

United Kingdom

Blackburn Museum and Art Gallery, Blackburn, Lancashire 43526
Bolton Museum and Art Gallery, Bolton 43553
Egypt Centre, Swansea 45812
Hancock Museum, Newcastle-upon-Tyne 45227
Myers Museum of Egyptian and Classical Art, Eton 44190
Old Speech Room Gallery, London 44949
Petrie Museum of Egyptian Archaeology, London 44953
Tutankhamun Exhibition, Dorchester 44004
Ure Museum of Greek Archaeology, Reading 45472
West Park Museum, Macclesfield 45067

United States of America

California Museum of Ancient Art, Beverly Hills 46791
Rosicrucian Egyptian Museum, San Jose 52807

Uruguay

Museo Egipcio, Montevideo 54402

Vatican City

Museo Gregoriano Egizio, Città del Vaticano 54503

Electricity

Argentina

Museo de la Energía, Embalse 00363
Museo Participativo de Ciencias, Buenos Aires 00269

Australia

Powerhouse House and Car Museum, Portland 01441
Powerhouse Museum Yanco, Yanco 01676

Austria

Vorarlberger Museumswelt, Frastanz 01900

Belgium

Centrale Hydroélectrique du Val de Poix, Saint-Hubert 03994

Brazil

Museu da Eletricidade, Rio de Janeiro 04709

Canada

Cité de l'Énergie, Shawinigan 06950
Electrical Engineering Museum, Fredericton 05830
Électrium, Sainte-Julie 06896
Manitoba Electrical Museum Education Centre, Winnipeg 07369

Czech Republic

Památník Dr. Františka Křižíka, Plánice 09715

Finland

Voimalamuseo, Helsinki 10838

France

Musée André-Marie Ampère, Poleymieux-au-Mont-d'Or 14869
Musée de l'Électricité, Dijon 12739
Musée de l'Électricité, Saint-Aignan 15154
Musée EDF Electropolis, Mulhouse 14326

Germany

Elektrizitätszähler-Kabinett, Braunschweig 17519
Elektro-Museum Rendsburg-E.ON Hanse, Rendsburg 21273
Energiemuseum, Fürstenfeldbruck 18492
Energiemuseum, Rickenbach 21312
Fachausstellung Arbeitsschutz/ Sicherheitstechnik, München 20438
LVR-Industriemuseum, Engelskirchen 18171
Museum der Stadtwerke, Pirmasens 21042
Museum für Energiegeschichte(n), Hannover 19001
Museum im Trafohaus, Leipzig 19926
SiemensForum, München 20483
VSE Elektro-Museum, Illingen, Kreis Neunkirchen 19319

Greece

Open-Air Water-Power Museum, Dimitsana 22984

Hungary

Magyar Elektrotechnikai Múzeum, Budapest 23434

Iceland

Safn Rafmagnsveitu Reykjavíkur, Reykjavík 24033

Iran

The Electrical Industry Museum, Tehrān 24660

Italy

Museo dell'Energia Elettrica, Brescia 25537
Museo dell'Energia Elettrica, Brescia 25537

Japan

Asahikawa Youth Science Museum, Asahikawa 28370
Denki no Kagakukan, Nagoya 28996
Kobe Lamp Museum, Kobe 28730
Okukiyotsu Electric Power Museum, Yuzawa 29635
Tepco Denryokukan, Yokohama 29608

Mexico

Museo Tecnológico de la Comisión Federal de Electricidad, México 31151

Netherlands

Bedrijfsmuseum NUON-ENW Amsterdam, Amsterdam 31832
Blikmuseum De Blikvanger, Renesse 32681
Elektriciteits- en Techniekmuseum, Hoenderloo 32396
EnergeticA, Museum voor Energietechniek, Amsterdam 31839
Museum Stoomgemaal Winschoten, Winschoten 32952
Het Philips Gloeilampenfabriekje anno 1891, Eindhoven 32192
Philips Museum Eindhoven, Eindhoven 32193
Zaltbommels Stoom- en Energiemuseum, Zaltbommel 32988

Norway

Theta-Museet, Bergen 33434
Vassdragsmuseet Labro, Skollenborg 33840

Peru

Museo de la Electricidad, Lima 34242

Poland

Muzeum Energetyki Jeleniogórskiej, Szklarska Poręba 35169

Portugal

Museu da Electricidade, Lisboa 35678
Museu de Electricidade, Fornelos 35606

Romania

Expoziția Hidrocentrală Porțile de Fier, Drobeta-Turnu Severin 36358

Serbia

Muzej Nikole Tesle, Beograd 38071

Spain

Museo Exposición Permanente de la Electricidad, Murillo de Gállego 39706

Sweden

Industrimuseet Tollereds Öfvre Kraftstation, Tollered 40973

Switzerland

Electrobroc, Broc-Fabrique 41259
Elektrizitätsmuseum, Münchenstein 41656
Elektro Museum, Baden 41143
Les Roues de L'Areuse, Saint-Sulpice (Neuchâtel) 41794

Taiwan

Northern Taiwan Visitors Center, Wanli 42458

United Kingdom

Museum of Bath at Work, Bath 43406
Museum of Electricity, Christchurch 43841

United States of America

A.W.A. Electronic-Communication Museum, Bloomfield 46853
American Museum of Radio and Eletricity, Bellingham 46711
Con Edison Energy Museum, New York 51154
Hearthstone Historic House Museum, Appleton 46330
Niagara Power Project Visitors' Center, Lewiston 50039
Real World Computer Museum, Boothwyn 46912
Robinson Visitors Center, Hartsville 49205
World of Energy at Keowee-Toxaway, Seneca 53034

Venezuela

Museo de la Electricidad, Caracas 54546

Embroidery → Needlework

Enamel Works

Austria

Emailmuseum Gertrude Stöhr, Vorchdorf 02874

Canada

Canadian Clay and Glass Gallery, Waterloo 07285

France

Maison de l'Émail, Morez 14271
Musée d'Art et d'Histoire Alfred Douët, Saint-Flour (Cantal) 15270

Netherlands

Kachelmuseum De Drie Kronen, Boekel 31977

Switzerland

Musée de l'Horlogerie et de l'Emaillerie, Genève 41420

United Kingdom

Bilston Craft Gallery, Wolverhampton 46058

United States of America

Wichita Center for the Arts, Wichita 54151

Energy

Argentina

Museo Participativo de Ciencias, Buenos Aires 00269

Australia

Longreach Power House Museum, Longreach 01239
Western Power World of Energy, Fremantle 01096

Canada

Cité de l'Énergie, Shawinigan 06950
Hamilton Museum of Steam and Technology, Hamilton 05957
Musée de la Gare, Temiscaming 07065
Pump House Steam Museum, Kingston 06089

Czech Republic

Technické Muzeum v Brně, Brno 09440

Finland

Korvensuun Voimalaitos- ja Konepajamuseo, Mynämäki 11131

Germany

Atom Museum, Haigerloch 18868
Energie-Museum Berlin, Berlin 17079
Technisches Landesmuseum Mecklenburg-Vorpommern, Wismar 22615
Wasserkraftmuseum, Ziegenrück 22799

Hungary

Gázmúzeum, Budapest 23407

India

Bardhaman Science Centre, Bardhaman 24076

Italy

Museo Italgas, Torino 28020

Japan

Fu Hakubutsukan, Hijikawa 28508
Gasu no Kagakukan, Tokyo 29391
The Great Hanshin-Awaji Earthquake Memorial, Kobe 28725
Kitai Hakubutsukan, Kodaira 28751
Mitsubishi Minatomirai Indasutoriaru Hakubutsukan, Yokohama 29601
Tepco Denryokukan, Tokyo 29484

Latvia

Dienvidu Elektrisko Tīlu Muzejs, Jelgava 30016

Lithuania

Lietuvos Energetikos Muziejus, Vilnius 30416

Madagascar

Musée de l'Université de Tuléar, Tuléar 30543

Netherlands

Stoomhoutzagerij, Groenlo 32276

New Zealand

Engine House, Dunedin 33091

Norway

Fotland Kraftverk, Bryne 33457
Industrimuseet, Kopperå 33606
Norsk Oljemuseum, Stavanger 33871

Russia

Muzej Istorii OAO Gaz, Nižnij Novgorod 37457
Muzej Istorii Razvitija Mosenergo, Moskva 37296

Sweden

Umeå Energicentrum med Kraftverksmuséet, Umeå 40999

Switzerland

EKZ Stromhaus Burenwisen, Glattfelden ... 41438
Kraftwerkmuseum Löntsch, Netstal ... 41675

United Kingdom

Electric Mountain - Museum of North Wales, Llanberis ... 44752
Ellenroad Engine House, Milnrow ... 45151
Fakenham Museum of Gas and Local History, Fakenham ... 44203
Flame!, Carrickfergus ... 43766
Markfield Beam Engine and Museum, London 44913

United States of America

Coolspring Power Museum, Coolspring ... 47787
Edison Museum, Beaumont ... 46675
National Atomic Museum, Albuquerque ... 46183
North Anna Nuclear Information Center, Mineral ... 50725
Texas Energy Museum, Beaumont ... 46680
World of Energy at Keowee-Toxaway, Seneca 53034

Engraving → Graphic Arts

Entomology → Insects

Epitaphs → Inscriptions and Epitaphs

Ethnology

Albania

Muzeu Etnografik, Berat ... 00010
Muzeu Etnografik, Elbasan ... 00015
Muzeu Etnografik, Gjirokastër ... 00018

Algeria

Musée National de Préhistoire et d'Ethnographie du Bardo, Alger ... 00044

Andorra

Museu Etnogràfic Casa Cristo, Encamp ... 00089

Angola

Museu Nacional de Antropologia, Luanda ... 00112
Museu Regional de Etnografia do Lobito, Lobito ... 00109
Museu Regional do Huambo, Huambo ... 00108

Argentina

Centro Mundo Aborigen, Córdoba ... 00318
Museo Cacique Balata, La Cumbre ... 00400
Museo de Arqueología de Salta, Salta ... 00554
Museo de Ciencias Antropológicas y Naturales, Santiago del Estero ... 00628
Museo de Ciencias Naturales y Antropológicas Juan Cornelio Moyano, Mendoza ... 00453
Museo de Ciencias Naturales y Misional del Colegio San José, Esperanza ... 00366
Museo del Indio, Los Toldos ... 00432
Museo del Mundo, Buenos Aires ... 00227
Museo Ethnográfico de las Obras Misionales Pontificias, Buenos Aires ... 00229
Museo Etnográfico Juán B. Ambrosetti, Buenos Aires ... 00230
Museo Etnográfico Municipal Dámaso Arce, Olavarría ... 00480
Museo Etnográfico y Archivo Histórico Enrique Squirru, Azul ... 00132
Museo Histórico Municipal y Colección de Arte Precolombino Arminio Weiss, Rafaela ... 00514
Museo Independencia, Córdoba ... 00331
Museo Indigenista, Esquel ... 00370
Museo Nacional del Hombre del Instituto Nacional de Antropología y Pensamiento Latinoamericano, Buenos Aires ... 00263
Museo Privado de Arqueología Regional Julio Gironde, Carmen de Patagonés ... 00294
Museo Regional de Antropología, Resistencia 00527
Museo Rocsen, Nono ... 00478
Museo Runa de Antropología, Tafí del Valle ... 00640

Armenia

History Museum of Armenia, Yerevan ... 00738
Ijevan Local Lore Museum, Ijevan ... 00710

Australia

Adelaide Lithuanian Museum, Norwood ... 01382
Alpenrail Swiss Model Village and Railway, Claremont, Tasmania ... 00975
Anthropology Museum, Saint Lucia ... 01483
Australian Museum, Sydney ... 01546
Berndt Museum of Anthropology, Crawley ... 01008
Boandik Cultural Museum, Beachport ... 00842
Chinese Museum, Melbourne ... 01276
Golden Dragon Museum, Bendigo ... 00860
J.L. Shellshear Museum of Physical Anthropology and Comparative Anatomy, Sydney ... 01549
J.T. Wilson Museum of Human Anatomy, Sydney ... 01550
Koorie Heritage Trust Cultural Centre-Galleries, Melbourne ... 01286
Macleay Museum, Sydney ... 01552
Miegunyah Pioneer Women's Memorial House Museum, Bowen Hills ... 00882
Philippine House Museum, Shepparton ... 01503
Queensland Museum, South Bank ... 01510
School of Political and Social Inquiry Museum, Clayton ... 00981
South Australian Museum, Adelaide ... 00782
Tandanya, Adelaide ... 00783
Templin Historical Village, Boonah ... 00876
Tineriba Tribal Gallery and Museum, Hahndorf 01140
Western Australian Museum Perth, Perth ... 01412

Austria

Afrikamuseum, Gallneukirchen ... 01918
Ethnographisches Museum Schloss Kittsee, Kittsee ... 02175
Fasnachtsmuseum, Imst ... 02107
Haus der Völker, Schwaz ... 02725
Heimathaus Mörbisch, Mörbisch ... 02401
Heimathaus Neudorf - Seoski muzej, Neudorf bei Parndorf ... 02426
Heimatmuseum im Amonhaus, Lunz ... 02340
Missions-Ethnographisches Museum St. Gabriel, Mödling ... 02393
Museum - Begegnung der Kulturen, Krenglbach ... 02235
Museum der Stadt Bad Ischl, Bad Ischl ... 01755
Museum für Völkerkunde, Wien ... 03038
Museum Humanum, Waldkirchen an der Thaya 02891
Museum Oberschützen, Oberschützen ... 02459
Richard-Simoncic-Museum, Rabensburg ... 02547
Volkskundemuseum am Landesmuseum Joanneum, Graz ... 01986
Volkskundliche Sammlung alter bäuerlicher Geräte, Ludesch ... 02339

Azerbaijan

Raşidbey Efandiyev Historical-Regional Ethnography Museum, Şeki ... 03196

Bangladesh

Bangla Academy Folklore Museum, Dhaka ... 03219
Chittagong University Museum, Chittagong ... 03214
Ethnological Museum, Chittagong ... 03215
Varendra Research Museum, Rajshahi ... 03234

Belarus

Muzej Staražytna Belaruskaj Kultury, Minsk ... 03254

Belgium

Etnografische Verzamelingen van de Universiteit Gent, Gent ... 03626
Koninklijk Museum voor Midden-Afrika, Tervuren ... 04058
Musée Africain de Namur, Namur ... 03898
Musée de Chine, Bruxelles ... 03465
Musée de la Vie Wallonne, Liège ... 03807
Museum Aan de Stroom, Antwerpen ... 03295
Museum voor Volkskunde, Brugge ... 03425
Volkskundemuseum, Antwerpen ... 03319

Benin

Musée de Natitingou, Natitingou ... 04184
Musée d'Homme Palais Royaux de Porto-Novo, Porto-Novo ... 04188
Musée Ethnographique Alexandre Senou Adande, Porto-Novo ... 04189

Bolivia

Museo Antropológico, Sucre ... 04243
Museo Nacional de Etnografía y Folklore, La Paz ... 04224

Bosnia and Herzegovina

Zemaljski Muzej Bosne i Hercegovine, Sarajevo ... 04284

Botswana

Kgosi Bathoen II (Segopotso) Museum, Kanye04298
Nhabe Museum, Nhabe ... 04300
Supa-Ngwao Museum Centre, Francistown ... 04295

Brazil

Museu da Cidade, São Paulo ... 04868
Museu de Arqueologia e Etnologia da Universidade de São Paulo, São Paulo ... 04877
Museu de Folclore Édison Carneiro, Rio de Janeiro ... 04727
Museu do Homem, Curitiba ... 04437
Museu do Homem do Nordeste, Recife ... 04668
Museu do Índio, Manaus ... 04555
Museu do Indio, Rio de Janeiro ... 04737
Museu do Instituto de Antropologia Câmara Cascudo, Natal ... 04586
Museu-Escola Esacro do Estado da Paraiba, João Pessoa ... 04513
Museu Paranaense, Curitiba ... 04442
Museu Rondon, Cuiabá ... 04424
Museus de História Natural e Etnologia Indígena, Juiz de Fora ... 04523

Bulgaria

Ekspozicija Kërpeeva Kašta, Kotel ... 05075
Etnografska Ekspozicija, Pazardžik ... 05107
Etnografska Ekspozicija, Svištov ... 05220
Etnografski Muzej, Berkovica ... 04985
Etnografski Muzej, Dobrič ... 05018
Etnografski Muzej, Elchovo ... 05026
Etnografski Muzej, Plovdiv ... 05126
Etnografski Muzej, Silistra ... 05162
Etnografski Muzej, Varna ... 05237
Etnografski muzej, Burgas ... 04998
Etnografski Muzej Dunavski Ribolov i Lodkostroenie, Tutrakan ... 05233
Istoričeski Muzej, Dimitrovgrad ... 05015
Istoričeski Muzej, Razgrad ... 05148
Istoričeski Muzej Batak, Batak ... 04981
Istoričeski Muzej Targovište, Targovište ... 05225

Burkina Faso

Musée National de Burkina Faso, Ouagadougou ... 05279
Musée Provincial du Houet, Bobo-Dioulasso ... 05272

Burundi

Musée Vivant de Bujumbura, Bujumbura ... 05280

Cameroon

Musée de Diamare, Maroua ... 05299
Musée International de Akum, Bamenda ... 05293
Musée Municipal de Mokolo, Mokolo ... 05300

Canada

Arkona Lion's Museum, Arkona ... 05353
Canadian Museum of Hindu Civilization, Richmond Hills ... 06748
Centre d'Interprétation du Moulin Bernier, Courcelles ... 05649
Eskimo Museum, Churchill ... 05614
Glenbow Museum, Calgary ... 05519
Haida Gwaii Museum at Qay'llnagaay, Skidegate ... 06982
Institut Culturel Avataq, Westmount ... 07307
Inuit Heritage Center, Baker Lake ... 05375
Ksan Historical Village and Museum, Hazelton ... 05979
Manitoba Museum, Winnipeg ... 07370
Musée Canadien des Civilisations, Gatineau ... 05862
Musée de Charlevoix, La Malbaie ... 06111
Musée de la Civilisation, Québec ... 06687
Musée des Abénakis, Odanak ... 06499
Musée d'Héritage de Notukeu, Ponteix ... 06622
Musée du Bas Saint-Laurent, Rivière-du-Loup 06761
Musée du Château Ramezay, Montréal ... 06366
Musée et Centre de Transmission de la Culture Daniel Weetaluktuk, Inukjuak ... 06015
Musée François Pilote, La Pocatière ... 06113
Musée Kateri Tekakwitha, Kahnawake ... 06034
Musée Kio-Warini, Loretteville ... 06195
Musée McCord, Montréal ... 06371
Musée Québécois de Culture Populaire, Trois-Rivières ... 07178
Musée Ukraina, Saskatoon ... 06919
Museum of Anthropology, Vancouver ... 07217
Museum of Archaeology and Ethnology, Burnaby ... 05504
Nebenaigoching Heritage, Batchawana Bay ... 05394
North American Black Historical Museum, Amherstburg ... 05335
Northern Life Museum and National Exhibition Centre, Fort Smith ... 05823
Prince of Wales Northern Heritage Centre, Yellowknife ... 07411
La Pulperie de Chicoutimi/Musée du Saguenay-Lac-Saint-Jean, Chicoutimi ... 05612
Rainy River District Women's Institute Museum, Emo ... 05775
Royal British Columbia Museum, Victoria ... 07258
Surrey Museum, Surrey ... 07038
Transcona Historical Museum, Winnipeg ... 07386
Ukrainian Canadian Archives and Museum of Alberta, Edmonton ... 05756
Ukrainian Cultural Heritage Village, Edmonton05758
Ukrainian Museum, Gardenton ... 05850
Ukrainian Museum of Canada, Saskatoon ... 06925
Ukrainian Museum of Canada - Alberta Branch, Edmonton ... 05759

Central African Republic

Musée Ethnographique Regional, Bouar ... 07422
Musée National Barthélémy Boganda, Bangui 07421

Chile

Museo Antropológico de Antofagasta, Antofagasta ... 07428
Museo Antropológico de Iquique, Iquique ... 07438
Museo Arqueológico San Miguel de Azapa, Arica ... 07430
Museo Regional Salesiano Maggiorino Borgatello, Punta Arenas ... 07447

China

Anthropological Museum, Xiamen ... 08354
Beijing China Nationality Museum, Beijing ... 07510
Hainan Nationality Museum, Tongshi ... 08301
Hainanzhou Nationality Museum, Gonghe ... 07718
Haixizhou Nationality Museum, Delingha ... 07659
Hebei Provincial Museum, Shijiazhuang ... 08232
Hunsu Nationality Museum, Xingyi ... 08395
Jiangxi Provincial Museum, Nanchang ... 08028
Jilin Wenmiao Museum, Jilin ... 07879
Liangshan Yizu Thrall Society Museum, Xichang ... 08382
Liannan Nationality Museum, Qingyuan ... 08115
Lianshan Chuang Yaozu Self-Rule Nationality Museum, Qingyuan ... 08116
Milkyuan Yaozu Self-Rule Nationality Museum, Ruyuan ... 08146
Museum of the Cultural Palace of National Minorities, Beijing ... 07553
Nanzhou Nationality Museum, Kaili ... 07910
Qiannan Zhousandu Nationality Museum, Sandu ... 08147
Qijinshang Culture Museum, Qixian ... 08128
Yaozu Self-Rule Nationality Museum, Jinxiu ... 07898
Youjiang Nationality Museum, Baise ... 07496
Youli Zhouyi Museum, Tangyin ... 08278
Yunnan Nationality Museum, Kunming ... 07916

Colombia

Exposición Permanente de Cerámica Indígena, Manizales ... 08666
Museo Chairama, Santa Marta ... 08748
Museo de Artesanías del Mundo, Silvia ... 08755
Museo de Historia Natural, Bogotá ... 08566
Museo de la Cultura de Albán, Albán ... 08525
Museo de la Diversidad Biológica y Cultural del Pacífico, Quibdó ... 08734
Museo de Trajes Regionales de Colombia, Bogotá ... 08575
Museo Etnográfico, Albán ... 08526
Museo Etnográfico, Inzá ... 08662
Museo Etnográfico, Santa Marta ... 08753
Museo Etnográfico Madre Laura, Medellín ... 08679
Museo Etnográfico Miguel Angel Builes, Medellín ... 08680
Museo Histórico de la Casa de la Cultura, Ambalema ... 08528
Museo Indígena Regional Enin, Mitú ... 08687
Museo Inter-étnico de la Amazonía, Florencia 08642
Museo Nacional de Antropología, Bogotá ... 08590
Museo Nacional de Colombia, Bogotá ... 08591
Museo Paisa, Caicedonia ... 08604

Congo, Democratic Republic

Musée du Roi de Cuba, Mushenge ... 08803
Musée National de Kananga, Kananga ... 08794
Musée National de Lubumbashi, Lumumbashi 08801
Musée Regional, Kisangani ... 08800

Congo, Republic

Musée Régional André Grenard Matsoua, Kinkala ... 08806
Musée Régional Ma-Loango Diosso, Pointe-Noire ... 08807

Côte d'Ivoire

Musée des Civilisations de Côte d'Ivoire, Abidjan ... 08830
Musée Théâtre Ki-Yi, Abidjan ... 08831

Croatia

Creski Muzej, Cres 08856
Etnografski Muzej Rupe, Dubrovnik 08862
Gradski Muzej, Bjelovar 08845
Gradski muzej Čazma, Čazma 08855
Gradski Muzej Karlovac, Karlovac 08886
Gradski Muzej Makarska, Makarska 08907
Gradski Muzej Virovitica, Virovitica 08978
Memorijalna Zbirka Eugena Kumičića, Brseč 08849
Narodni Muzej Labin, Labin 08906
Staro Selo Kumrovec, Kumrovec 08903
Zbirka Baltazara Bogišića, Cavtat 08854

Cuba

Museo Antropológico Montané, La Habana 09110
Museo Casa de Africa, La Habana 09114
Museo Casa de Asia, La Habana 09115
Museo Ciudades del Mundo, La Habana 09128
Museo Etnográfico, Madruga 09199
Museo Etnográfico, Santa Lucía 09302

Curaçao

Museum Kurá Hulanda, Willemstad 09359

Cyprus

Ethnographic Museum, Avgorou 09362
Ethnographical Museum, Paphos 09391

Czech Republic

Dietrichsteinský Palác, Brno 09423
Etnografický Ústav, Brno 09427
Hrdličkovo Muzeum Člověka, Praha 09752
Lesnické, Myslivecké a Rybářské Muzeum, Hluboká nad Vltavou 09516
Městské Muzeum, Valašské Klobouky 09895
Městské Muzeum a Galerie Julie W. Mezerové, Úpice 09890
Moravské Zemské Muzeum, Brno 09435
Muzeum Dr. Aleše Hrdličky, Humpolec 09541
Muzeum Jindřicha Jindřicha, Domažlice 09497
Muzeum Vysočiny Jihlava, Jihlava 09561
Náprstkovo Muzeum Asijských, Afrických a Amerických Kultur, Praha 09775
Národopisné Muzeum Plzeňska, Plzeň 09717
Sládečkovo Vlastivědné Muzeum v Kladně, Kladno 09583
Slovácké Muzeum, Uherské Hradiště 09887
Zámecké Muzeum, Tovačov 09872

Denmark

Djurslands Museum - Dansk Fiskerimuseum, Grenaa 10041
Grindsted Museum, Grindsted 10044
Kulturhistorisk Museum, Randers 10229
Nordsjaellandsk Folkemuseum, Hillerød 10077
Vendsyssel Historiske Museum, Hjørring 10080

Dominican Republic

Museo del Hombre Dominicano, Santo Domingo 10352

East Timor

Museum Siwa Lima, Ambon 10363

Ecuador

Museo Antropológico Shuar, Quito 10427
Museo Azuayo del Folklore, Cuenca 10383
Museo de Etnografía, Quito 10439
Museo Etnografico del Cologio Nacional Mejia, Quito 10444

Egypt

Ethnological Museum, Cairo 10494
Nubia Museum, Aswan 10481

El Salvador

Museo de Antropología David J. Guzmán, San Salvador 10549

Estonia

Eesti Põllumajandusmuuseum, Tõrvandi 10664
Eesti Rahva Muuseum, Tartu 10647
Rannarootsi Muuseum, Haapsalu 10565

Ethiopia

City Museum, Makale 10685
City Museum, Yirgalem 10686
Ethnographic Museum, Addis Ababa 10678

Fiji

Fiji Museum, Suva 10704

Finland

Ethnographisches Museum Pöykkölä, Rovaniemi 11275
Helinä Rautavaaran museo, Espoo 10730
Kulttuurien Museo, Helsinki 10795
Lapin Maakuntamuseo, Rovaniemi 11276
Liedon Vanhalinna, Vanhalinna 11444
Mäntyharjun Museo, Mäntyharju 11094
Perniön Museo, Perniö 11200
Rautalammin Museo, Rautalampi 11263
Ruokolahden Kotiseutumuseo, Ruokolahti 11280
Suomen Kansallismuseo, Helsinki 10820
Tornionlaakson Maakuntamuseo, Tornio 11385
Vanajan Kotiseutumuseo, Hämeenlinna 10749

France

Ecomusée de la Brenne, Le Blanc 13558
Ecomusée de la Margeride, Loubaresse 13827
Ecomusée de la Margeride, Ruynes-en-Margeride 15142
Ecomusée de l'Ile-de-Groix, Ile-de-Groix 13237
Ecomusée de Plein Air du Quercy, Sauliac-sur-Célé 15654
Ecomusée du Pays de Rennes, Rennes (Ille-et-Vilaine) 15012
Musée Africain, Ile-d'Aix 13235
Musée Alsacien, Haguenau 13176
Musée Arménien de France, Paris 14629
Musée d'Art et d'Histoire de Lisieux, Lisieux 13778
Musée Dauphinois, Grenoble 13133
Musée de la Camargue, Arles 11683
Musée de la Haute-Auvergne, Saint-Flour (Cantal) 15271
Musée de la Loire, Cosne-Cours-sur-Loire 12636
Musée de la Poupée, Paris 14660
Musée de l'Algérie Française, Perpignan 14789
Musée de Plein Air du Quercy de Cuzals, Sauliac-sur-Célé 15655
Musée de Traditions Populaires et d'Archéologie, Chauvigny 12488
Musée Départemental des Hautes-Alpes, Gap 13031
Musée des Arts Africains, Océaniens, Amérindiens, Marseille 14001
Musée des Beaux-Arts, Angoulême 11620
Musée d'Ethnographie de l'Université de Bordeaux II, Bordeaux 12071
Musée d'Histoire Naturelle et d'Ethnographie, Colmar 12584
Musée du Biterrois, Béziers 11977
Musée du Château Saint-Jean, Nogent-le-Rotrou 14462
Musée du Papillon et des Arts Africains, Aubeterre-sur-Dronne 11727
Musée du Patrimoine Rural, Dompierre (Orne) 12763
Musée du Pays de Luchon, Bagnères-de-Luchon 11822
Musée du Pays de Retz, Bourgneuf-en-Retz 12122
Musée Ethnographique de l'Olivier, Cagnes-sur-Mer 12238
Musée Ethnographique du Donjon, Niort 14458
Musée-Fondation Alexandra David Néel, Digne-les-Bains 12726
Musée Hospitalier, Charlieu 12412
Musée Joseph Vaylet, Espalion 12857
Musée Missionnaire, Pontmain 14907
Musée Napoléonien, Ile-d'Aix 13236
Musée National de la Marine, Paris 14724
Musée National des Arts et Traditions Populaires, Paris 14727
Musée Portuaire, Dunkerque 12799
Musée Provençal des Arts et Traditions Poplaires, Donzère 12770
Musée Régional d'Ethnologie, Béthune 11967
Museon Arlaten, Arles 11686
Muséum d'Histoire Naturelle, Rouen 15122
Muséum d'Histoire Naturelle de Toulouse, Toulouse 15946

French Polynesia

Musée de Tahiti et des Iles - Te Fare Manaha, Punaauia 16327

Georgia

Giorgi Chitaia Ethnography Museum, Tbilisi 16363
Ivane Javakhishvili Samtskhe-Javaxeti History Museum, Akhaltsikhe 16332
Kutaisskij Gosudarstvennyj Muzej Istorii i Etnografii, Kutaisi 16344
Svaneti History-Ethnography Museum, Mestia 16347

Germany

Abenteuermuseum Saarbrücken - Sammlung Heinz Rox-Schulz, Saarbrücken 21453
Afrika-Museum, Bergen, Kreis Celle 17008
Afrikahaus Museum und Nold-Namibia-Bibliothek, Sebnitz 21709
Anthropologische Sammlung der Universität Göttingen, Göttingen 18689
Archäologisch-Ethnographische Lehr- und Studiensammlung des Instituts für Altamerikanistik und Ethnologie der Universität Bonn, Bonn 17443
Brasilienmuseum im Kloster Bardel, Bad Bentheim 16680
Darß-Museum, Prerow, Ostseebad 21119
Deutsches Albert-Schweitzer-Zentrum, Frankfurt am Main 18360
Deutsches Landwirtschaftsmuseum Schloss Blankenhain, Crimmitschau 17749
Deutsches Märchen- und Wesersagenmuseum, Bad Oeynhausen 16798
Druckereimuseum und Schaudepot der Museen der Stadt Erfurt im Benary-Speicher, Erfurt 18200
Ethnologische Sammlung der Universität Göttingen, Göttingen 18690
Ethnologisches Museum, Berlin 17081
GRASSI-Museum für Völkerkunde zu Leipzig, Leipzig 19913
Hamaland-Museum, Vreden 22258
Haus Völker und Kulturen - Völkerkundliches Museum, Sankt Augustin 21489
Iwalewa Haus, Bayreuth 16953
JuniorMuseum im Ethnologischen Museum, Berlin 17150
Karl-Otto-Braun-Museum, Ludwigshafen am Rhein 20052
Kindheitsmuseum, Schönberg, Holstein 21598
Linden-Museum Stuttgart, Stuttgart 21951
Lippisches Landesmuseum, Detmold 17839
Ludwig-Harms-Haus, Hermannsburg 19146
Missionsmuseum, Sankt Ottilien 21508
Missionsmuseum der Mariannhiller Missionare, Würzburg 22716
Museum der deutschen Sprachinselorte bei Brünn, Erbach, Alb-Donau-Kreis 18187
Museum Europäischer Kulturen, Berlin 17199
Museum Forum der Völker, Werl 22490
Museum für Regionalgeschichte und Volkskunde, Gotha 18720
Museum für Stadtgeschichte, Naturkunde und Völkerkunde Julius Riemer, Lutherstadt Wittenberg 20117
Museum für Stadtgeschichte und Volkskunde, Heppenheim 19129
Museum für Völkerkunde Dresden, Dresden 17964
Museum für Völkerkunde Hamburg, Hamburg 18952
Museum Schloß Hohentübingen, Tübingen 22123
Museum Wilnsdorf, Wilnsdorf 22589
Naturhistorisches Museum, Nürnberg 20785
Naturmuseum, Freiburg im Breisgau 18436
Niedersächsisches Landesmuseum Hannover, Hannover 19002
Rautenstrauch-Joest-Museum, Köln 19629
Reiss-Engelhorn-Museen (rem), Mannheim 20173
Ruhr Museum, Essen 18267
Schatzhaus in der Lausitz-Museum für Morgenlandfahrer, Göda 18668
Schwarz-Afrika-Museum, Vilshofen 22246
Staatliches Museum für Völkerkunde München, München 20490
Übersee Museum Bremen, Bremen 17564
Ungarndeutsches Heimatmuseum, Backnang 16664
Völkerkundemuseum der von Portheim-Stiftung, Heidelberg 19087
Völkerkundemuseum Herrnhut, Herrnhut 19163
Völkerkundesammlung, Lübeck 20083
Völkerkundliche Sammlung der Philipps-Universität, Marburg 20195
Völkerkundliches Museum, Witzenhausen 22634
Volkskundliche Sammlungen, Witzenhausen 22635
Wallenfels'sches Haus, Abteilung Vor- und Frühgeschichte, Archäologie und Völkerkunde, Gießen 18641
Wieland-Museum Biberach, Wieland-Gartenhaus, Biberach an der Riß 17326
Zweigmuseum des Staatlichen Museum für Völkerkunde München, Oettingen in Bayern 20880

Ghana

Museum of Ethnography, Accra 22846
Upper East Region Museum, Bolgatanga 22849

Greece

Ethnographical Historical Museum of Lárissa, Lárissa 23073
Mouseio Mpenaki, Athinai 22925
Museum of Anthropology, Goudi 23008
Museum of Cretan Ethnology, Vori 23240

Guatemala

Museo Nacional de Arqueología y Etnología, Guatemala 23301

Guinea

Musée National de Conakry, Conakry 23310
Musée Préfectoral de Kissidougou, Kissidougou 23313
Musée Préfectoral de Koundara, Koundara 23314
Petit Musée, Conakry 23312

Guinea-Bissau

Musée da Guine Portugesa, Bissau 23316
Museu Etnografico Nacional, Bissau 23317

Haiti

Musée du Peuple Haïtien, Fermath 23319

Honduras

Museo Antropologia e Historia Valle de Sula, San Pedro Sula 23330

Hungary

Déri Múzeum, Debrecen 23516
Dráva Múzeum, Barcs 23361
Erdős Kamill Cigány Múzeum, Pécs 23760
Erkel Ferenc Múzeum, Gyula 23594
Ferenczy Múzeum, Szentendre 23876
Györffy István Nagykun Múzeum, Karcag 23638
Ifj. Lele J. Néprajzi Gyűjteménye, Szeged 23844
Intercisa Múzeum, Dunaújváros 23533
Kanizsai Dorottya Múzeum, Mohács 23716
Kiskun Múzeum, Kiskunfélegyháza 23666
Kiss Pál Múzeum, Tiszafüred 23930
Kossuth Múzeum, Cegléd 23503
Laczkó Dezső Múzeum, Veszprém 23959
Matyó Múzeum, Mezőkövesd 23698
Munkácsy Mihály Múzeum, Békéscsaba 23371
Néprajzi Kiállítás, Pécs 23769
Néprajzi Kiállítás, Sopron 23814
Néprajzi Múzeum, Budapest 23455
Ópusztaszeri Nemzeti Történeti Emlékpark, Opusztaszer 23736
Rábaközi Muzeális Gyűjtemény, Kapuvár 23637
Rákóczi Múzeum, Sárospatak 23792
Szántó Kovács János Területi Múzeum, Orosháza 23740
Szatmári Múzeum, Mátészalka 23693
Tari László Múzeum, Csongrád 23513
Thorma János Múzeum, Kiskunhalas 23671
Thury György Múzeum, Nagykanizsa 23725
Türr István Múzeum, Baja 23347
Viski Károly Múzeum, Kalocsa 23628

Iceland

Pjóðminjasafn Islands, Reykjavík 24030

India

Anthropological Museum, Guwahati 24186
Anthropology Museum, Delhi 24135
Anthropology Museum, Ranchi 24371
Anthropology Museum, Lucknow 24288
Bharat Itihas Samshodhak Mandal, Pune 24355
Bihar Tribal Welfare Research Institute Museum, Ranchi 24372
Cultural Research Institute, Kolkata 24272
Ethnological Museum, Pune 24357
Government Museum, Pudukottai 24353
Jawaharlal Nehru State Museum, Itanagar 24231
Kurukshetra Panorama Science Centre, Kurukshetra 24287
Prabhas Patan Museum, Prabhas Patan 24352
Ragailong Museum, Imphal 24228
State Museum, Kohima 24264
Tibet House Museum, Delhi 24156
Tribal Museum, Ahmedabad 24052
Victoria and Albert Museum, Mumbai 24324
Zonal Anthropological Museum, Shillong 24388
Zonal Museum, Dehradun 24132

Indonesia

Museum Loka Budaya, Jayapura 24509
Museum Negeri Propinsi Nusa Tenggara Timur, Kupang 24517
Museum Negeri Sumatera Utara, Medan 24526
Museum Pemda Balige, Balige 24437
Museum Wanua Paksinanta, Manado 24523
Subak Museum, Banjar Senggulan 24447

Iran

Ethnological Museum, Tehrān 24661
Ethnological Museum, Tehrān 24662
Ghoochan Museum, Ghoochan 24604
Kandelous Museum, Kandelous 24614
Khoy Museum, Khoy 24625
Mardom-shenasi-ye Kavir, Nayin 24636
Moaven Al- Molk Museum, Kermanshah 24621
Muséyé Kerman, Kerman 24618
Muséyé Mardom Shenassi, Tehrān 24686
Rasht Museum, Rasht 24640

Sa'dabad Museums, Tehrān 24694
Zabol Museum, Zabol 24705

Israel

Albadia Museum, Osafia 25077
Babylonian Jewry Museum, Or Yehuda 25076
Museum for Beduin Culture, Kibbutz Lahav 25048

Italy

Acquario e Museo Etnografico del Po, Monticelli d'Ongina 26859
Casa della Nonna, Cairate 25590
Collezione Indiana e dell' Estremo Oriente F. L. Pullè, Bologna 25439
Museo Africano, Verona 28229
Museo Archeologico Etnografico, Paulilatino 27118
Museo Calabrese di Etnografia e Folklore Raffaele Corso, Palmi 27085
Museo Civico Archeologico Etnologico, Modena 26745
Museo Civico Etnografico, Atri 25323
Museo Civico Etnografico C.G. Franchini, Oleggio 26978
Museo Civico Etnografico M.G. Fontana, Sappada 27737
Museo dei Ferri Taglienti, Scarperia 27787
Museo del Bosco, Sovicille 27884
Museo del Sud-Ovest Americano, Cuveglio 26030
Museo della Cultura Arbereshe, San Paolo Albanese 27675
Museo della Gente dell'Appennino Pistoiese, Cutigliano 26027
Museo della Vita Contadina in Romagnolo, Russi 27579
Museo delle Culture del Mondo, Genova 26304
Museo delle Culture Extraeuropee Dinz Rialto, Rimini 27409
Museo delle Genti d'Abruzzo, Pescara 27163
Museo delle Terre Marchigiane, San Lorenzo in Campo 27657
Museo delle Valli Valdesi, Torre Pellice 28040
Museo di Antropologia, Bologna 25459
Museo di Antropologia, Napoli 26894
Museo di Antropologia ed Etnografia, Torino 28012
Museo di Antropologia ed Etnologia, Firenze 26178
Museo di Arte Cinese ed Etnografico, Parma 27102
Museo di Etnopreistoria, Napoli 26896
Museo Ethnografico Don L. Pellegrini delle Civiltà Contadina, San Pellegrino in Alpe 27676
Museo Etiopico G. Massaia, Frascati 26245
Museo Etnico Arbereshe, Civita 25919
Museo Etno-Antropologico, Calatafimi 25592
Museo Etno-Antropologico, San Cipirello 27625
Museo Etno-Antropologico, San Mauro Castelverde 27667
Museo Etno-Antropologico, Scaletta Zanclea 27782
Museo Etno-Antropologico della Terra di Zabut, Sambuca di Sicilia 27605
Museo Etno-Antropologico della Valle del Belice, Gibellina 26339
Museo Etno-Antropologico delle Madonie, Geraci Siculo 26330
Museo Etnografico, Aprica 25252
Museo Etnografico, Aritzo 25279
Museo Etnografico, Bomba 25492
Museo Etnografico, Collinas 25941
Museo Etnografico, Faeto 26080
Museo Etnografico, Fossalta di Portogruaro 26237
Museo Etnografico, Fratta Polesine 26250
Museo Etnografico, Lusevera 26549
Museo Etnografico, Malborghetto Valbruna 26573
Museo Etnografico, Montodine 26861
Museo Etnografico, Novara di Sicilia 26956
Museo Etnografico, Oneta 26983
Museo Etnografico, Ortonovo 27006
Museo Etnografico, Ossimo Superiore 27014
Museo Etnografico, Parma 27107
Museo Etnografico, Ponte in Valtellina 27283
Museo Etnografico, Roccalbegna 27431
Museo Etnografico, Roseto Capo Spulico 27558
Museo Etnografico, Schilpario 27791
Museo Etnografico, Vione 28286
Museo Etnografico Africo-Mozambico, Bari 25368
Museo Etnografico Antico Mulino ad Acqua Licheri, Fluminimaggiore 26207
Museo Etnografico Cerginolano, Cerignola 25810
Museo Etnografico del Pinerolese e Museo del Legno, Pinerolo 27223
Museo Etnografico del Ponente Ligure, Cervo 25823
Museo Etnografico della Cultura Contadina, Morigerati 26869
Museo Etnografico della Lunigiana, Villafranca in Lunigiana 28277
Museo Etnografico della Trinità, Botticino 25518
Museo Etnografico della Valle, Ultimo 28117
Museo Etnografico e della Stregoneria, Triora 28100
Museo Etnografico e di Scienze Naturali, Torino 28019
Museo Etnografico Francesco Bande, Sassari 27754
Museo Etnografico G. Carpani, Lizzano in Belvedere 26503
Museo Etnografico Gino di Benedetto, Torricella Sicura 28043
Museo Etnografico Il Cicle della Vita, Quartu Sant'Elena 27351
Museo Etnografico La Steiva, Piverone 27259
Museo Etnografico San Domu de is Ainas, Armungia 25280
Museo Etnografico sulla Lavorazione del Legno, San Vito di Leguzzano 27691
Museo Etnografico Tiranese, Madonna di Tirano 26563
Museo Etnografico Tiranese, Tirano 27973
Museo Etnografico U. Ferrandi, Novara 26951
Museo Etnografico Vallivo, Val Masino 28134
Museo Etnologico della Apuane, Massa 26621
Museo Geopaleontologico Naturalistico Antropico e Ornitologico Brancaleoni, Piobbico 27227
Museo Missionario delle Grazie, Rimini 27410
Museo Nazionale Preistorico ed Etnografico Luigi Pigorini, Roma 27519
Museo Polare Etnografico Silvio Zavatti, Fermo 26102
Museo Provinciale degli Usi e Costumi, Brunico 25556
Museo Storico Etnografico, Sampeyre 27607
Museo Storico Etnografico della Bassa Valsesia, Romagnano Sesia 27549
Museo Storico Etnografico Naturalistico, Chiesa in Valmalenco 25870
Museum Zeitreise Mensch, Cortaccia 25983
Nest Ume Tzimbari Gabizza, Giazza-Selva di Progno 26336

Japan

Hokkaido-kenritsu Kita-hitobito Hakubutsukan, Abashiri 28345
Imaizumi Hakubutsukan, Shiozawa 29277
Kokuritsu Minzokugaku Hakubutsukan, Minpaku, Suita 29295
Man'yo Bunkakan, Wakayama 29573
Namban Bunkakan, Osaka 29136
Nara-kenritsu Man'yo Bunkakan, Asuka 28380
Yokota Hakubutsukan, Tsukude 29541

Jordan

The Anthropological National Folklore Museum, Amman 29644

Kazakhstan

Centralnyj Gosudarstvennyj Muzej Kazachstana, Almaty 29678
Muzej Kazakstan Respublikasynyn Prezidenttik Mėdeniet Ortalyfy, Astana 29692
Vostočno-Kazachstanskij Etnografičeskij Muzej, Özkemen 29711

Kenya

Kabarnet Museum, Kabarnet 29740
Kitale Museum, Kitale 29743
Meru Museum, Meru 29745

Korea, Democratic People's Republic

Korean Central Ethnographic Museum, Pyongyang 29757

Korea, Republic

Africa Museum, Seoul 29863
Latin American Cultural Center Museum, Koyang 29847
Pacific Museum, Yongin 29946
Sookmyung Women's University Museum, Seoul 29925

Kyrgyzstan

Gumbez Manace Literary Ethnographic Museum, Taš-Aryk 29965

Laos

Parc National de la Culture des Ethnies, Vientiane 29976

Latvia

Turaidas Muzejrezervāts, Sigulda 30124

Lebanon

Cilicia Museum, Antélias 30154

Lesotho

Lesotho National Museum, Maseru 30179
Morija Museum, Morija 30180

Liberia

Africana Museum, Monrovia 30181
National Museum, Monrovia 30182

Lithuania

Antano ir Jono Juškų Etninės Kultūros Muziejus, Vilkija 30399
Aušros Alėjos Rūmai, Šiauliai 30352
Kėdainių Krašto Muziejus, Kėdainiai 30272
Kelmės Krašto Muziejus, Kelmė 30273
Kretingos Muziejus, Kretinga 30288
Lietuvos Nacionalinis Muziejus, Vilnius 30419
Šiaulių Aušros Muziejus, Šiauliai 30362
Vaclovo ir Elenos Venskų Etnografinė Sodyba, Tytuvėnai 30387
Žeimelio Žiemgalos Muziejus, Žeimelis 30451

Luxembourg

Musée d'Art Rustique, Vianden 30493

Macedonia

Muzej Bitola, Bitola 30500
Muzej na Makedonija - Archeološki, Etnološki i Istoriski, Skopje 30519

Malawi

Museum of Malawi, Blantyre 30544

Malaysia

Jabatan Muzium Sabah, Kota Kinabalu 30565
Malay Ethnographic Museum, Kuala Lumpur 30576
Museum of Aboriginal Affairs, Gombak 30553
Muzium Kecantikan, Melaka 30598
Muzium Negara, Kuala Lumpur 30578
Muzium Negeri Sarawak, Kuching 30591
Muzium Negeri Sembilan, Seremban 30616
Muzium Rakyat, Melaka 30600

Mali

Musée National du Mali, Bamako 30624

Malta

Inquisitor's Palace, Vittoriosa 30655
Ta' Kola Windmill, Xaghra 30660

Mexico

Museo Antropología e Historia, Victoria 31543
Museo Centro de Cultura Mayo de el Júpare Blas Mazo, Huatabampo 30943
Museo Coahuila y Texas, Monclova 31173
Museo Comunitario Raramuri, Guachochi 30882
Museo de Antropología de Xalapa, Xalapa 31557
Museo de Artes Populares, Tepic 31441
Museo de Cocula es el Mariachi, Cocula 30778
Museo de Culturas Populares de San Luis Potosí, San Luis Potosí 31345
Museo de Culturas Populares e Indígenas de Sonora, Hermosillo 30929
Museo de la Cultura Maya, Chetumal 30761
Museo de la Huaxteca, Amatlán de los Reyes 30714
Museo de la Soledad, Oaxaca 31224
Museo de las Californias, Los Cabos 31008
Museo de las Californias, Tijuana 31457
Museo de las Culturas, Saltillo 31314
Museo de las Culturas Afromestizas, Cuajinicuilapa 30810
Museo de las Culturas de la Huasteca Potosina Tamuantzán, Valles 31529
Museo de las Culturas de Occidente, Colima 30780
Museo de las Culturas Prehispánicas, Puerto Vallarta 31295
Museo de las Culturas Prehispánicas de Durango, Durango 30841
Museo de las Miniaturas Poblanas, Palmar de Bravo 31247
Museo de Sitio de Cholula, San Pedro Cholula 31368
Museo del Centro Cultural Universitario Quinta Gameros, Chihuahua 30770
Museo del Instituto de Investigaciones Antropológicas, México 31103
Museo del Origen, Santiago Ixcuintla 31383
Museo Étnico de los Seris, Hermosillo 30930
Museo Étnico de los Yaquis, Obregón 31230
Museo Etnográfico, Tuxtla Gutiérrez 31515
Museo Etnográfico Huichol Wixarica, Zapopan 31590
Museo Nacional de Antropología, México 31129
Museo Nacional de las Culturas, México 31138
Museo Naucalpan de la Cultura Tlatilca, Naucalpan de Juárez 31212
Museo Nuestro Pueblo y Museo Etnográfico de Tecpatán, Tecpatán 31416
Museo Regional de Actopán, Actopán 30696
Museo Regional de Antropología Carlos Pellicer Cámara, Villahermosa 31554
Museo Regional de Antropología e Historia de Baja California Sur, La Paz 30989
Museo Regional de Guadalajara, Guadalajara 30904
Museo Regional de Guerrero, Chilpancingo 30776
Museo Regional de Nayarit, Tepic 31444
Museo Regional de Yucatán, Mérida 31029
Museo Renacimiento Indígena, Huamuxtitlán 30940
Museo Sala, San Juan Bautista Tuxtepec 31336
Museo Tradicional Ik Al Ojov, Zinacantán 31599

Moldova

Muzeul National de Etnografie si Istorie Naturala, Chişinău 31601

Mongolia

Museum of Mongolian Ethnography, Ulaanbaatar 31626

Montenegro

Muzeji i Galerije Podgorica, Podgorica 31656

Morocco

Musée Ethnographique, Tétouan 31701
Musée Ethnographique, Tétouan 31702

Myanmar

University of Mandalay Collections, Mandalay 31721

Namibia

National Museum of Namibia, Windhoek 31765

Netherlands

Afrika Centrum, Cadier en Keer 32032
Afrika Museum, Berg en Dal 31964
Educatieve Boerderij en Zijdemuseum de Schans, Oud-Gastel 32647
Expositie Versteend Leven, Drouwen 32159
Gemeentegrot Valkenburg, Valkenburg, Limburg 32863
KIT Tropenmuseum, Amsterdam 31852
Moluks Historisch Museum, Utrecht 32848
Museum Kennemerland, Beverwijk 31972
Museum Nusantara, Delft 32064
Museum Stapelen, Boxtel 31997
Museum Suriname, Amsterdam 31865
Natuurhistorisch en Volkenkundig Museum, Oudenbosch 32653
Nederlands Openluchtmuseum, Arnhem 31930
Nijmeegs Volkenkundig Museum, Nijmegen 32594
Rijksmuseum voor Volkenkunde, Leiden 32488
Tropenmuseum Junior, Amsterdam 31896
Wereldmuseum Rotterdam, Rotterdam 32733

New Zealand

Okains Bay Maori and Colonial Museum, Okains Bay 33163
Otago Museum, Dunedin 33095
Te Awamutu Museum, Te Awamutu 33207
Waikato Museum of Art and History, Hamilton 33117
Whanganui Regional Museum, Wanganui 33235

Nicaragua

Museo ADIAC, León 33283
Museo del Arbol El Genízaro, Nagarote 33299

Niger

Musée National du Niger, Niamey 33308

Nigeria

Kanta Museum Argungu, Argungu 33315
Museum of the Institute of African Studies, Ibadan 33322
Museum of the Institute of African Studies, Oshogbo 33340
National Museum, Oshogbo 33341
National Museum Gidan Makama, Kano, Kano 33332

Northern Cyprus

Canbulat Türbe ve Müzesi, Gazimağusa 33350
Dervish Pasha Mansion Ethnographical Museum, Lefkoşa 33368
Mevlevi Tekke Müzesi, Lefkoşa 33371

Norway

Etnografisk Museum, Oslo 33722
Finnetunet, Grue Finnskog 33523
Kon-Tiki Museum, Oslo 33729
Østsamisk Museum, Neiden 33697
Varjjat Sámi Musea, Varangerbotn 33960

Pakistan

Bahawalpur Museum, Bhawalpur 33998

Panama

Museo Antropológico Reina Torres de Araúz, Ciudad de Panamá 34065

Papua New Guinea

Papua New Guinea National Museum and Art Gallery, Boroko 34080

Paraguay

Museo Etnográfico Andrés Barbero, Asunción 34097

Peru

Museo de la Cultura José Arens Berg, Sullana 34308
Museo del Centro de Investigación Arqueológica de Ancón, Lima 34252
Museo del Colegio Nacional Amauta, Chupaca 34150
Museo Etnográfico Santa Rosa de Ocopa, Concepción 34153
Museo Etnohistórico de Ayabacara, Ayabacara 34118
Museo Etnológico de Sechura, Sechura 34307
Museo Nacional de la Cultura Peruana, Lima 34266
Museo Schaferrer, Oxapampa 34284
Museo Yanque, Yanque 34329

Philippines

Aga Khan Museum of Islamic Arts, Marawi 34459
Ayala Museum, Makati 34414
Baguio-Mountain Provinces Museum, Baguio 34340
Bahay Tsinoy Museum, Manila 34431
Barangay Museum, Mandaluyong 34424
Cotabato National Museum, Cotabato 34381
Cultural Center of the Philippines Museo, Pasay 34471
Divine Word University Museum, Tacloban 34509
Museo Pambata, Manila 34447
Nayong Pilipino Museum, Pasay 34475
Notre Dame of Jolo College Museum, Jolo 34392
Palawan State University Museum, Puerto Princesa 34483
Pambasang Museo ng Pilipinas, Manila 34450
Panaddaman-Cagayan State University Museum, Tuguegarao 34519
Quijano Museum, Mandaue 34428
Saint Louis University Museum of Arts and Culture, Baguio 34342
Saint Theresa's College Folklife Museum, Cebu 34372
Siliman University Anthropology Museum, Dumaguete 34386
Sulu National Museum, Jolo 34393
Tawi-Tawi Ethnological Museum, Bongao 34348
University Museum of Anthropology, Quezon City 34495
University of San Carlos Anthropology Museum, Cebu 34377
University of Santo Tomas Museum of Arts and Sciences, Manila 34458
Zamboanga National Museum, Zamboanga 34527

Poland

Dział Archeologii i Etnografii, Bydgoszcz 34574
Górnośląski Park Etnograficzny, Chorzów 34603
Muzeum, Koszalin 34761
Muzeum Etnograficzne, Wrocław 35329
Muzeum Etnograficzne im. Franciszka Kotuli, Rzeszów 35092
Muzeum Etnograficzne im. M. Znamierowskiej-Prüfferowej, Toruń 35193
Muzeum Etnograficzne im. Seweryna Udzieli, Kraków 34792
Muzeum Etnograficzne w Zielonej Górze z Siedzibą w Ochli, Ochla 34968
Muzeum Historii Katowic - Dzial Etnologii Miasta, Katowice 34726
Muzeum Historyczno-Etnograficzne, Chojnice 34599
Muzeum im. Jana Nikodema Jaronia, Olesno 34972
Muzeum-Kaszubski Park Etnograficzny w Wdzydzach Kiszewskich, Wdzydze Kiszewskie 35294
Muzeum Kaszubskie, Kartuzy 34718
Muzeum Kultury Ludowej, Kolbuszowa 34749
Muzeum Kultury Ludowej, Węgorzewo 35295
Muzeum Mazurskie, Szczytno 35168
Muzeum Misyjne Ojców Sercanów, Polanica Zdrój 35020
Muzeum Misyjno-Etnograficzne, Pieniężno 35007
Muzeum Nadwiślański Park Etnograficzny, Babice 34532
Muzeum Narodowe w Gdańsku, Gdańsk 34640
Muzeum Okręgowe - Muzeum Etnograficzny, Tarnów 35178
Muzeum Podróżników, Toruń 35196
Muzeum Pojezierza Łęczyńsko-Włodawskiego, Włodawa 35313
Muzeum Regionalne, Jarocin 34695
Muzeum Regionalne Dom Grecki, Myślenice 34934
Muzeum Regionalne im. Hieronima Ławniczaka, Krotoszyn 34832
Muzeum Regionalne im. Marcina Rożka, Wolsztyn 35318
Muzeum Rzemiosła Tkackiego, Turek 35201
Muzeum Śląska Opolskiego, Opole 34986
Muzeum Tatrzańskie im. Tytusa Chałubińskiego, Zakopane 35360
Muzeum w Raciborzu, Racibórz 35072
Muzeum w Tomaszowie Mazowieckim, Tomaszów Mazowiecki 35184
Muzeum Wisły, Tczew 35182
Muzeum Wsi Kieleckiej, Kielce 34744
Muzeum Wsi Mazowieckiej, Sierpc 35107
Muzeum Zachodnio-Kaszubskie, Bytów 34585
Muzeum Ziemi Puckiej, Puck 35063
Orawski Park Etnograficzny, Zubrzyca Górna 35379
Pałucka Izba Muzealna, Kcynia 34736
Państwowe Muzeum Etnograficzne, Warszawa 35287
Prywatne Muzeum Etnograficzne, Góra Kalwaria 34677
Punkt Etnograficzny w Rogierówku, Rogierówko 35082
Sądecki Park Etnograficzny, Nowy Sącz 34961

Portugal

Museu Antropológico, Coimbra 35529
Museu Arqueológico e Etnográfico de Setúbal, Setúbal 35935
Museu da Terra de Miranda, Miranda do Douro 35764
Museu de Arte Sacra e Etnologia de Fátima, Fátima 35602
Museu de Olaria, Barcelos 35456
Museu do Centro Científico e Cultural de Macau, Lisboa 35697
Museu do Mar, Sesimbra 35932
Museu Etnográfico, Santa Cruz das Flores 35899
Museu Etnográfico da Freguesia de Almoster, Almoster 35422
Museu Etnográfico da Madeira, Ribeira Brava 35886
Museu Etnográfico da Sociedade de Geografia de Lisboa, Lisboa 35707
Museu Etnográfico de Espiche, Luz 35744
Museu Etnográfico de Serpa, Serpa 35930
Museu Etnográfico de Vila Franca, Lisboa 35708
Museu Etnográfico de Vila Franca de Xira, Vila Franca de Xira 35997
Museu Etnográfico de Vilarinho da Furna, Terras de Bouro 35962
Museu Etnográfico do Espírito Santo, Vila do Porto 35994
Museu Etnográfico do Grupo Típico de Ançã, Ançã 35436
Museu Etnográfico e Arqueológico do Dr. Joaquim Manso, Nazaré 35780
Museu Etnográfico e Arte Sacra de São Torcato, São Torcato 35926
Museu Etnográfico Regional do Algarve, Faro 35597
Museu Municipal da Azambuja, Azambuja 35451
Museu Municipal de Arqueologia e Etnologia, Sines 35942
Museu Nacional de Etnologia, Lisboa 35717
Museu Vivo de Vilar do Pinheiro, Vilar do Pinheiro 36021

Puerto Rico

Museo y Centro de Estudios Humanísticos, Gurabo 36040

Qatar

Ethnographical Museum, Doha 36086
Qatar National Museum, Doha 36088

Romania

Casa Memorială Simion Florea Marian, Suceava 36640
Colecţia Etnografică, Poiana Sibiului 36538
Colecţia Etnografică, Răşinari 36554
Colecţia Etnografică Zahacinshi, Mihăileni 36474
Colecţia Eugenia şi Costache Burada, Dăneşti 36339
Colecţia Vasile Heisu, Răcăciuni 36545
Complexul Muzeal Măldăreşti, Măldăreşti 36457
Complexul Naţional Muzeal Astra, Sibiu 36604
Expoziţia Etnografică, Horezu 36402
Muzeul Conacul Bellu, Urlaţi 36715
Muzeul de Arheologie şi Etnografie, Corabia 36311
Muzeul de Artă Populară şi Etnografie, Tulcea 36704
Muzeul de Etnografie, Botoşani 36163
Muzeul de Etnografie Braşov, Braşov 36179
Muzeul de Etnografie Piatra-Neamţ, Piatra-Neamţ 36519
Muzeul de Etnografie şi a Regimentului de Graniţă, Caransebeş 36263
Muzeul de Etnografie şi Artă Populară, Reghin 36559
Muzeul de Etnografie şi Artă Populară Orăştie, Orăştie 36506
Muzeul de Etnografie Universală Franz Binder, Sibiu 36607
Muzeul Etnografic, Vatra Dornei 36728
Muzeul Etnografic al Moldovei, Iaşi 36420
Muzeul Etnografic al Transilvaniei, Cluj-Napoca 36287
Muzeul Etnografic Lupşa, Lupşa 36454
Muzeul Etnografic Tehnici Populare Bucovinene, Rădăuţi 36547
Muzeul Judeţean de Etnografie, Slatina 36627
Muzeul Judeţean Mureş-Secţia de Etnografie şi Artă Populară, Târgu Mureş 36671
Muzeul Municipal Câmpulung-Secţia de Etnografie şi, Câmpulung 36253
Muzeul Şvăbesc, Petreşti 36514
Muzeul Tarii Beiusului, Beiuş 36146
Muzeul Ţării Făgăraşului Valeriu Literat, Făgăraş 36363
Muzeul Tării Oaşului, Negreşti-Oaş 36489
Muzeul Tarisznyás Márton, Gheorghieni 36382
Muzeul Vrancei, Focşani 36373
Punct Muzeal, Moroeni 36482
Secţia de Etnografie-Bastion, Timişoara 36693

Russia

Adygejskij Respublikanskij Kraevedčeskij Muzej, Majkop 37123
Altajskij Respublikanskij Kraevedčeskij Muzej im. A.V. Anochina, Gorno-Altajsk 36927
Amurskij Gorodskoj Kraevedčeskij Muzej, Amursk 36756
Architekturno-ètnografičeskij Muzej Angarskaja Derevnja, Bratsk 36820
Astrachanskij Gosudarstvennyj Obedinennyj Istoriko- architekturnyj Muzej-zapovednik, Astrachan 36778
Astrachanskij Kraevedčeskij Muzej, Astrachan 36779
Bratskij Gorodskoj Objedinennyj Muzej Istorii Osvoenija Angary, Bratsk 36822
Chakasskij Respublikanskij Kraevedčeskij Muzej, Abakan 36745
Čuvašskij Nacionalnyj Muzej, Čeboksary 36837
Dagestanskij Gosudarstvennyj Objedinennyj Istoriko-architekturnyj Muzej, Machačkala 37115
Dom Poèta A.N. Širjaevca - Muzej Krestjanskogo Byta, Žigulevsk 37988
Enisejskij Kraevedčeskij Muzej, Enisejsk 36915
Ètnografičeskij Muzej Kazanskogo Universiteta, Kazan 37006
Ètnografičeskij Muzej Narodov Zabajkalja, Ulan-Udè 37893
Ètnografičeskij Muzej pod Otkrytym Nebom Torum Maa, Chanty-Mansijsk 36860
Èvenkijskij Okružnoj Kraevedčeskij Muzej, Tura 37877
Gosudarstvennaja OK Istoriko-ètnografičeskij i Architekturnyj Muzej-zapovednik Staraja Sarepta, Volgograd 37954
Gosudarstvennyj Istoriko-ètnografičeskij Muzej-zapovednik Šušenskoe, Šušenskoe 37811
Gosudarstvennyj Nacionalnyj Muzej Ust-Ordynskogo Burjatskogo Avtonomnogo Okruga, Ust-Ordynsk 37913
Gosudarstvennyj Okružnoj Muzej Prirody i Čeloveka Goroda Chanty-Mansijsk, Chanty-Mansijsk 36862
Ibresinskij Ètnografičeskij Muzej pod Otkryrym Nebom, Ibresi 36929
Ingušskij Gosudarstvennyj Muzej Kraevedenija im. T.Ch. Malsagova, Nazran 37448
Irkutskij Architekturno-ètnografičeskij Muzej Talcy, Irkutsk 36931
Irkutskij Oblastnoj Kraevedčeskij Muzej, Irkutsk 36933
Istoriko-ètnografičeskij Muzej Usadba Galskich, Čerepovec 36849
Jakutskij Gosudarstvennyj Muzej Istorii i Kultury Narodov Severa im. Jaroslavskogo, Jakutsk 36962
Jamalo-Neneckij Okružnoj Kraevedčeskij Muzej im. I.S. Šemanovskogo, Salechard 37608
Kalmyckij Respublikanskij Kraevedčeskij Muzej im. Prof. N.N. Palmova, Èlista 36913
Kamčatskij Oblastnoj Kraevedčeskij Muzej, Petropavlovsk-Kamčatskij 37553
Kislovodskij Istoriko-kraevedčeskij Muzej Krepost', Kislovodsk 37046
Komi Respublikanskij Istoriko-Kraevedčeskij Muzej, Syktyvkar 37815
Krasnodarskij Gosudarstvennyj Istoriko-archeologičeskij Muzej-zapovednik, Krasnodar 37066
Literaturno-memorialnyj Dom-muzej Sulejmana Stalskogo, Ašaga-Stalsk 36776
Muzej Antropologii i Ètnografii im. Petra Velikogo (Kunstkamera), Sankt-Peterburg 37670
Muzej Antropologii im. D.N. Anučina Moskovskogo Gosudarstvennogo Universiteta im M.V. Lomonosova, Moskva 37261
Muzej Archeologii i Ètnografii, Vladivostok 37943
Muzej Archeologii i Ètnografii, Ufa 37887
Muzej Archeologii i Ètnografii, Syktyvkar 37818
Muzej Archeologija, Ètnografija i Èkologija Sibiri Kemerovskogo Gosudarstvennogo Universiteta, Kemerovo 37033
Muzej Ètnografii, Istorii i Kultury Narodov Baškortostana, Ufa 37888
Muzej Istorii Aksaja, Aksaj 36749
Muzej Istorii i Kultury Srednego Prikamja, Sarapul 37746
Muzej Istorii Kolskich Saamov, Lovozero 37113
Nacionalnyj Muzej Respubliki Baškortostan, Ufa 37889
Nacionalnyj Muzej Respubliki Komi, Syktyvkar 37821
Nacionalnyj Muzej Respubliki Marij Èl im. T. Evseeva, Joškar-Ola 36980
Nacionalnyj Muzej Udmurtskoj Respubliki im. K. Gerda, Iževsk 36955
Neneckij Okružnoj Kraevedčeskij Muzej, Narjan-Mar 37443
Novokuzneckij Kraevedeskij Muzej, Novokuzneck 37482
Oloneckij Nacionalnyj Muzej Karelov-Livvikov im. N.T.Prilukin, Olonec 37496
Omskij Gosudarstvennyj Istoriko-kraevedčeskij Muzej, Omsk 37499
Primorskij Gosudarstvennyj Muzej im. V.K. Arseneva, Vladivostok 37949
Pskovskij Gosudarstvennyj Istoričesko-architekturnyj i Chudožestnennyj Muzej-zapovednik, Pskov 37575
Roslavlskij Istoriko-chudožstvennyj Muzej, Roslavl 37597
Rossijskij Ètnografičeskij muzej, Sankt-Peterburg 37722
Saratovskij Ètnografičeskij Muzej, Saratov 37752
Šeltozerskij Vepsskij Ètnografičeskij Muzej, Šeltozero 37758
Severo-osetinskij Gosudarstvennyj Obedinennyj Muzej Istorii, Architektury i Literatury, Vladikavkaz 37932
Taganrogskij Kraevedčeskij Muzej, Taganrog 37829
Tajmyrskij Okružnoj Kraevedčeskij Muzej, Dudinka 36880
Tarskij Istoriko-kraevedčeskij Muzej, Tara 37834
Tjumenskij Oblastnoj Kraevedčeskij Muzej im. I.Ja. Slovcova, Tjumen 37852
Tuapsinskij Kraevedčeskij Muzej im. N.G. Poletaeva, Tuapse 37864
Tuvinskij Respublikanskij Kraevedčeskij Muzej im. Aldan Maadyr, Kyzyl 37099
Vserossijskij Istoriko-ètnografičeskij Muzej, Toržok 37862

Rwanda

Musée de Kabgayi, Kabgayi 37993
Musée National du Rwanda, Butare 37992

Serbia

Etnografski Muzej Srbije, Beograd 38035
Galerija Naivnih Slikara i Etnološka Zbirka, Kovačica 38106
Kompleksi Monumental i Lidhjes Shqiptare të Prizrenit, Prizren 38134
Manakova Kuća, Beograd 38058
Muzej Afričke Umetnosti, Beograd 38060
Muzej Ponišavlja Pirot, Pirot 38130
Muzej Vojvodine, Novi Sad 38124
Narodni Muzej, Bački Petrovac 38031
Narodni Muzej, Kragujevac 38109
Narodni Muzej Kraljevo, Kraljevo 38111
Narodni Muzej Valjevo, Valjevo 38160
Zavičajni Muzej, Aleksinac 38029

Singapore

Asian Civilisations Museum, Singapore 38172

Slovakia

Dokumentačné Centrum Českej Kultúry Martin, Martin 38293
Etnografické Múzeum, Martin 38294
Historické a Národopisné Múzeum Želovce, Želovce 38389
Horehronské Múzeum, Brezno 38238
Karol Plicka Múzeum, Blatnica 38200
Krajské Múzeum, Prešov 38335
Múzeum Kultúry Karpatských Nemcov, Bratislava 38222
Múzeum Kultúry Maďarov na Slovensku, Bratislava 38223
Múzeum Kultúry Rómov na Slovensku, Martin 38297
Múzeum Samuela Jurkoviča Sobotište-Nyaryiovský kaštiel', Sobotište 38354
Národopisná Expozícia SARM Bardejovské Kúpele, Bardejovské Kúpele 38196
Skanzen - Múzeum L'udovej Architektúry, Bardejovské Kúpele 38197
Vajnorský L'udový Dom - Etnografická Expozícia, Bratislava 38235
Vihorlatské Múzeum, Humenné 38262
Západoslovenské Múzeum v Trnave, Trnava 38381

Slovenia

Belokranjski Muzej, Metlika 38481
Grad Bogenšperk, Šmartno pri Litiji 38522
Groharjeva Hiša v Sorici, Spodnja Sorica 38525

E

Kurnikova Hiša, Tržič 38530
Loški Muzej, Škofja Loka 38514
Mestni Muzej Idrija, Idrija 38421
Muzej Političnih Zapornikov, Internirancev in Izgnancev Grad Rajhenburg, Brestanica 38403
Pokrajinski Muzej, Murska Sobota 38487
Pokrajinski Muzej, Maribor 38477
Pokrajinski Muzej Kočevje, Kočevje 38432
Posavski Muzej Brežice, Brežice 38404
Slovenski Etnografski Muzej, Ljubljana 38463
Tolminski Muzej, Tolmin 38527
Tržiški Muzej z Galerijo, Tržič 38532

Somalia

Somali National Museum, Mogadishu 38549

South Africa

Africana Museum, Mariannhill 38686
Anthropology Museum, Johannesburg 38647
Barberton Museum, Barberton 38555
Duggan-Cronin Gallery, Kimberley 38667
Museum of Anthropology and Archaeology, Tshwane 38750
National Cultural History Museum, Tshwane 38752
University Museum - Sasol Art Museum, Stellenbosch 38741
Zululand Historical Museum, Eshowe 38618

Spain

Colección Museográfica d'Etnografia i Costums de la Mancomunitat de Municipis de la Vall d'Albaida, Benisoda 39027
Colección Museográfica Permanente de la Font de la Figuera, La Font de la Figuera 39303
Ecomuseu de les Valls d'Aneu, Esterri d'Aneu 39290
Etnografi Museoa, Arcienaga 38883
Etnografikoa Museoa Artziniega, Artziniega 38904
Euskal Arkeologia, Etnografia eta Kondaira Museoa, Bilbao 39042
Museo Africano Mundo Negro, Madrid 39525
Museo Agrícola El Patio, Teguise 40109
Museo Angel Orensanz y Artes de Serralbo, Sabiñánigo 39929
Museo Arqueologico y Etnologico Municipal Dámaso Navarro, Petrer 39839
Museo Canario, Las Palmas de Gran Canaria 39798
Museo de América, Madrid 39532
Museo de Antropología de Tenerife, La Laguna de Tenerife 39438
Museo de Artes y Costumbres Populares, Pedrafita do Cebrero 39819
Museo de Artes y Costumbres Populares e internacional de Arte Naif, Jaén 39421
Museo de Artes y Tradiciones Populares, San Sebastián de los Reyes 39973
Museo de Ávila, Ávila 38920
Museo de Cáceres, Cáceres 39082
Museo de Guadalajara, Guadalajara 39375
Museo de Orozko, Orozko 39750
Museo de Usos y Costumbres Etnológicas, Paracuellos de Jarama 39810
Museo de Zaragoza, Zaragoza 40344
Museo del Carro y Aperos de Labranza, Tomelloso 40142
Museo del Labrador, Humanes de Madrid 39404
Museo del Pueblo de Asturias, Gijón 39338
Museo do Pobo Galego, Santiago de Compostela 40026
Museo Etnográfico Comarcal de Azuaga, Azuaga 38932
Museo Etnográfico de Autilla, Autilla del Pino 38917
Museo Etnográfico de Cantabria, Muriedas 39705
Museo Etnográfico de Oyón-Oion, Oyón 39762
Museo Etnográfico del Reino de Pamplona, Ollo 39725
Museo Etnográfico Tanit, San Bartolomé de Lanzarote 39951
Museo Etnográfico Textil Pérez Enciso, Plasencia 39844
Museo Etnológica del Valle Medio del Ebro, Belchite 39017
Museo Etnológico, Cehegín 39159
Museo Etnológico de la Diputación, Castellón de la Plana 39149
Museo Etnológico de la Huerta de Murcia, Alcantarilla 38817
Museo Etnológico de Morella y del Meastrazgo, Morella 39685
Museo Etnológico de Ribadavia, Ribadavia 39897
Museo Etnológico la Posada, Chinchón 39178
Museo-Exposición de Creencias y Religiosidad Popular, Abizanda 38782
Museo-Exposición de Oficios y Artes Tradicionales, Aínsa 38798
Museo Histório de Aspe, Aspe 38907
Museo Molino de Teodoro, Cieza 39182
Museo Municipal de San Telmo, Donostia-San Sebastián 39253
Museo Nacional de Antropología, Madrid 39567
Museo Parroquial de Arte Sacro y Etnología, Anso 38873
Museo Parroquial de Frómista, Frómista 39311
Museo Provincial, Teruel 40122
Museo Taurino, Salamanca 39944
Museu de Cabrera, Palma de Mallorca 39788
Museu de la Fusta, Areu 38889
Museu de l'Acordió, Arsèguel 38898
Museu de l'Horta d'Almàssera, Almàssera 38853
Museu de Vilafranca, Vilafranca del Penedès 40280
Museu del Blat i la Pagesia, Cervera 39165
Museu del Pagès, Castelldans 39146
Museu d'Etnografia d'Eivissa, Santa Eulària del Río 40000
Museu d'Etnologia i Arqueologia, Callosa d'En Sarrià 39105
Museu d'Història de Cambrils, Cambrils 39110
Museu Etnogràfic de Ripoll, Ripoll 39904
Museu Etnològic, Benaguasil 39023
Museu Etnològic de Barcelona, Barcelona 38994
Museu Etnologic de Guadalest, Guadalest 39376
Museu Etnològic de l'Alcúdia, L'Alcúdia (Valencia) 38834
Museu Etnològic del Montseny La Gabella, Arbúcies 38882
Museu Etnològic i Arqueològic, Juneda 39435
Museu Marès de la Punta, Arenys de Mar 38886
Museu Valencià d'Etnologia, Valencia 40223

Sri Lanka

Colombo National Museum, Colombo 40370
Museum of Ethnology and Folk Art, Ratnapura 40381
Ratnapura National Museum, Ratnapura 40383

Sudan

Ethnographical Museum, Khartoum 40389
Halfa Museum, Wadi Halfa 40398

Swaziland

Swaziland National Museum, Lobamba 40403

Sweden

Ájtte - Svenskt Fjäll- och Samemuseum, Jokkmokk 40618
Arboga Museum, Arboga 40428
Disagården, Uppsala 41004
Etnografiska Museet, Stockholm 40892
Fotevikens Museum, Höllviken 40596
Grekiskt Kulturcentrum, Stockholm 40896
Hantverks- och Sjörfartsmuseét på Norra Berget, Sundsvall 40959
Jamtli, Östersund 40797
Jukkasjärvi Museet, Jukkasjärvi 40620
Trelleborgs Museum, Trelleborg 40981
Världskulturmuseet, Göteborg 40558

Switzerland

Afrika-Museum, Zug 42122
Cerny Inuit Collection, Bern 41204
Ethnographische Sammlung, Fribourg 41385
Historisches und Völkerkundemuseum, Sankt Gallen 41803
Kulturama, Zürich 42080
Musée Cantonal d'Histoire, Sion 41874
Musée d'Ethnographie, Genève 41421
Musée d'Ethnographie, Neuchâtel 41681
Musée d'Ethnographie de Genève - Annexe de Conches, Conches 41312
Museo delle Culture, Castagnola 41283
Museo di San Martino, Olivone 41705
Museum der Kulturen Basel, Basel 41169
Museum für Völkerkunde, Burgdorf 41276
Museum Vitznau-Rigi, Vitznau 41991
Tibet Songtsen House, Zürich 42112
Völkerkundemuseum der Universität Zürich, Zürich 42116
Völkerkundliche Sammlung, Winterthur 42042

Taiwan

Atayal Museum, Ren-Ai 42302
Hong-Gah Museum, Taipei 42392
Jibei Cultural Relics Exhibition Hall, Baisha 42141
Museum of the Institute of Ethnology, Taipei 42409
Museum of the Institute of History and Philology, Taipei 42410
National Chengchi University Museum of Ethnology, Taipei 42412
Pingtung Hakka Museum, Jhutian 42225
Sakalhe's Cultural Art Museum, Majia 42269
Shangrila Hakka Museum, Zaociao 42487
Shung Ye Museum of Formosan Aborigines, Taipei 42424
Taitung Aboriginal Exhibition Room, Taitung 42445
Taiwan Aboriginal Culture Park, Majia 42270

Togo

Musée d'Histoire et d'Ethnographie, Aneho 42610
Musée National du Togo, Lomé 42612

Tunisia

Musée Dar Sghir, Mahdia 42636
Musée des Arts et Traditions Populaires, Djerba 42622
Musée des Arts et Traditions Populaires 'Dar Ben Abdallah' de Tunis, Tunis 42666
Musée du Village, Moknine 42642
Musée Sidi Ben-Aissa, Le Kef 42632

Turkey

Afyon Etnografya Müzesi, Afyon 42677
Etnografya Müzesi, Ankara 42702
Hasan Süzer Etnografya Müzesi, Gaziantep 42787
İzmir Arkeoloji Müzesi, İzmir 42855
İzmir Etnografya Müzesi, İzmir 42856
Kız Teknik Öğretim Müzesi, Ankara 42704
Koyunoğlu Şehir Müzesi, Konya 42888
Tarsus Müzesi, Tarsus 42949

Turkmenistan

Lehap Regional Museum, Turkmenabat 42985
Taryhy we Etnografiki Muzeý, Türkmenabat 42982
Türkmenistanyn Milli Muzeýi, Aşgabat 42973

Uganda

Folk Museum, Mbarara 42995
Folk Museum, Soroti 42997

Ukraine

Deržavnyj Istoriko-kul'turnyj Zapovidnyk, Kam'janec'-Podil's'kyj 43051
Etnografičeskij Muzej Tavrika, Simferopol' 43184
Muzej Istoriï ta Etnografiï Grekiv Priazov'ja, Mariupol' 43145
State Museum of Ethnography, Arts and Crafts, Lviv 43141

United Kingdom

Brunei Gallery, London 44808
Dunrobin Castle Museum, Golspie 44325
The Eastern Museum, Derby 43966
Economic Botany Collection/Museum No. 1, Richmond, Surrey 45491
Gurkha Museum, Winchester 46029
Hancock Museum, Newcastle-upon-Tyne 45227
Hastings Museum and Art Gallery, Hastings 44415
Horniman Museum, London 44878
Marischal Museum, Aberdeen 43244
Pitt Rivers Museum, Oxford 45338
Quex Museum, House and Gardens, Birchington 43489
Rhodes Memorial Museum and Commonwealth Centre, Bishop's Stortford 43524
Royal Engineers Museum, Gillingham, Kent 44269
Russell-Cotes Art Gallery and Museum, Bournemouth 43567
Saffron Walden Museum, Saffron Walden 45540
Sainsbury Centre for Visual Arts, Norwich 45285
University Museum of Archaeology and Anthropology, Cambridge 43734
World Museum Liverpool, Liverpool 44750

United States of America

Abbe Museum at Downtown, Bar Harbor 46600
Abbe Museum at Sieur de Monts Spring, Bar Harbor 46601
Adán E. Treganza Anthropology Museum, San Francisco 52763
African Art Museum of the S.M.A. Fathers, Tenafly 53476
African Cultural Art Forum, Philadelphia 51811
Afro-American Cultural Center, Charlotte 47384
Alabama-Coushatta Indian Museum, Livingston 50150
Alabama Museum of Natural History, Tuscaloosa 53616
Alaska Heritage Museum at Wells Fargo, Anchorage 46286
Alaska Indian Arts, Haines 49124
Alutiiq Museum and Archaeological Repository, Kodiak 49823
American Swedish Institute, Minneapolis 50727
Amerind Foundation, Dragoon 48181
Anasazi Heritage Center, Dolores 48150
Anthropology Museum, DeKalb 48037
Appalachian Center for Crafts, Smithville 53135
The Appaloosa Museum and Heritage Center, Moscow 50863
Arizona State Museum, Tucson 53584
Armenian Museum of America, Watertown 53970
The Balch Institute for Ethnic Studies, Philadelphia 51818
Baltimore's Black American Museum, Baltimore 46555
Balzekas Museum of Lithuanian Culture, Chicago 47456
Bernice Pauahi Bishop Museum, Honolulu 49335
Buechel Memorial Lakota Museum, Saint Francis 52542
Burke Museum of Natural History and Culture, Seattle 52994
C.E. Smith Museum of Anthropology, Hayward 49227
C.H. Nash Museum-Chucalissa Archaeological Museum, Memphis 50567
Cahokia Mounds, Collinsville 47694
California State Indian Museum, Sacramento 52502
Capitol Reef National Park Visitor Center, Torrey 53543
Carnegie Museum of Natural History, Pittsburgh 51943
Carrie M. McLain Memorial Museum, Nome 51359
Carter County Museum, Ekalaka 48306
Castine Scientific Society Museum, Castine 47278
Center for Intuitive and Outsider Art, Chicago 47460
Chattanooga African Museum, Chattanooga 47406
Cherokee National Museum, Tahlequah 53417
College of Eastern Utah Prehistoric Museum, Price 52140
Colorado River Indian Tribes Museum, Parker 51712
Croatian Heritage Museum, Eastlake 48266
Cultural Heritage Center Museum, Toppenish 53538
Cultural Museum at Scarritt-Bennett Center's, Nashville 50965
Cultural Rights and Protection/Ute Indian Tribe, Fort Duchesne 48618
CWU Anthropology Department Collection, Ellensburg 48358
Danish Immigrant Museum, Elk Horn 48343
DePauw University Anthropology Museum, Greencastle 49046
Douglas County Museum of History and Natural History, Roseburg 52457
Eastern Arizona Museum, Pima 51919
Eastern California Museum, Independence 49469
Edge of the Cedars State Park Museum, Blanding 46850
Elgin Public Museum, Elgin 48335
Esther Prangley Rice Gallery, Westminster 54097
Etowah Indian Mounds, Cartersville 47264
Evelyn Payne Hatcher Museum of Anthropology, Saint Cloud 52538
Exhibition of the Norwegian-American Historical Association, Northfield 51419
Favell Museum of Western Art and Indian Artifacts, Klamath Falls 49806
Five Civilized Tribes Museum, Muskogee 50928
Florence Hawley Ellis Museum of Anthropology, Abiquiu 46128
Fort Ancient Museum, Oregonia 51586
Fort Fetterman, Douglas 48158
Fort Recovery Museum, Fort Recovery 48674
Frisco Native American Museum and Natural History Center, Frisco 48798
Grace Hudson Museum and The Sun House, Ukiah 53638
Grand Village of the Natchez Indians, Natchez 50984
Haffenreffer Museum of Anthropology, Bristol 47033
Hawaii's Plantation Village, Waipahu 53787
Heard Museum, Phoenix 51904
The Heritage Center, Pine Ridge 51924
Heritage Hjemkomst Interpretive Center, Moorhead 50836
Herrett Center for Arts and Science, Twin Falls 53627
Heym-Oliver House, Russell 52486
High Desert Museum, Bend 46733
Hopewell Culture National Historic Park, Chillicothe 47534
Horry County Museum, Conway 47784
Hudson Museum, Orono 51602
Imperato Collection of West African Artifacts, Chanute 47340
Indian City U.S.A., Anadarko 46280
Indian Museum of Lake County-Ohio, Painesville 51675
Indian Museum of North America, Crazy Horse 47873
Indian Museum of the Carolinas, Laurinburg 49988
Indian Temple Mound Museum, Fort Walton Beach 48696
The Institute For American Indian Studies, Washington Green 53950
Institute of Texan Cultures, San Antonio 52716
International Museum of Cultures, Dallas 47934
Iroquois Indian Museum, Howes Cave 49408
Jacques Marchais Museum of Tibetan Art, Staten Island 53264
Jean Lafitte National Historical Park and Preserve, New Orleans 51084
Jensen Arctic Museum, Monmouth 50785
Jesse Peter Museum, Santa Rosa 52905
Journey Museum, Rapid City 52251
Ka-Do-Ha Indian Village Museum, Murfreesboro 50911

Kaw Mission, Council Grove 47852
The Kendall College Mitchell Museum of the American Indian, Evanston 48451
Koshare Indian Museum, La Junta 49850
The Kurdish Museum, Brooklyn 47077
Latvian Museum, Rockville 52430
Lesbian Herstory Educational Foundation, Brooklyn 47079
Logan Museum of Anthropology, Beloit 46724
Lompoc Museum, Lompoc 50166
Lucy Craft Laney Museum of Black History, Augusta 46479
MacCallum More Museum, Chase City 47403
Malki Museum, Banning 46599
Maxwell Museum of Anthropology, Albuquerque 46181
Merritt Museum of Anthropology, Oakland 51466
Miami County Museum, Peru 51789
Michel Brouillet House and Museum, Vincennes 53749
Mid-America All-Indian Center Museum, Wichita 54146
Middle Border Museum of American Indian and Pioneer Life, Mitchell 50763
Millicent Rogers Museum of Northern New Mexico, Taos 53442
Milwaukee Public Museum, Milwaukee 50713
The Morikami Museum and Japanese Gardens, Delray Beach 48046
Moundbuilders State Memorial and Museum, Newark 51291
Museum for African Art, Long Island City 50184
Museum of Afro-American History, Boston 46939
Museum of Anthropology, Denver 48084
Museum of Anthropology, Lawrence 49990
Museum of Anthropology, Pullman 52198
Museum of Anthropology, Tempe 53469
Museum of Anthropology, Winston-Salem 54246
Museum of Anthropology, Fullerton 48808
Museum of Anthropology, Chico 47528
Museum of Anthropology, Highland Heights 49279
Museum of Anthropology, Ann Arbor 46308
Museum of Anthropology, Columbia 47717
Museum of Indian Arts and Culture, Santa Fe 52881
Museum of Indian Culture, Allentown 46229
Museum of Natural and Cultural History, Eugene 48435
Museum of Peoples and Cultures, Provo 52189
Museum of Primitive Art and Culture, Peace Dale 51742
Museum of Russian Culture, San Francisco 52787
Museum of the Cherokee Indian, Cherokee 47423
Museum of the Red River, Idabel 49461
Nance Museum, Lone Jack 50171
National Hall of Fame for Famous American Indians, Anadarko 46281
Native American Museum, Terre Haute 53481
Native American Resource Center, Pembroke 51755
New Mexico State University Museum, Las Cruces 49962
Nez Perce National Historical Park, Spalding 53191
Old World Wisconsin, Eagle 48230
Oneida Nation Museum, De Pere 47999
Pacific Asia Museum, Pasadena 51721
Pawnee Indian Village, Republic 52296
Peabody Museum of Archaeology and Ethnology, Cambridge 47185
The Peary-MacMillan Arctic Museum, Brunswick 47103
Pella Historical Village, Pella 51750
Pennsylvania German Cultural Heritage Center, Kutztown 49829
Phoebe Apperson Hearst Museum of Anthropology, Berkeley 46760
Plains Indians and Pioneers Museum, Woodward 54293
Polynesian Cultural Center, Laie 49873
Red Rock Museum, Church Rock 47549
Redwood Museum, Crescent City 47875
Reuel B. Pritchett Museum, Bridgewater 47022
Riley House Museum of African American History and Culture, Tallahassee 53428
Robert Hull Fleming Museum, Burlington 47144
Romanian Ethnic Art Museum, Cleveland 47639
Roy Rogers-Dale Evans Museum, Branson 46997
Saint Joseph Museum, Saint Joseph 52565
Samuel K. Fox Museum, Dillingham 48141
San Diego Museum of Man, San Diego 52753
Schingoethe Center for Native American Cultures, Aurora 46493
Schomburg Center for Research in Black Culture, New York 51254
Seminole Nation Museum, Wewoka 54114
Serpent Mound Museum, Peebles 51747
Sheldon Jackson Museum, Sitka 53125
Simon Paneak Memorial Museum, Anaktuvuk Pass 46284
Smoki Museum, Prescott 52136
Southern Plains Indian Museum, Anadarko 46282
Southern Ute Museum and Cultural Center, Ignacio 49466
Southwest Museum of the American Indian, Los Angeles 50265
Spurlock Museum, Urbana 53669
Stewart Indian Cultural Center, Carson City 47262
Suquamish Museum, Poulsbo 52124
Suquamish Museum, Suquamish 53380
Texas Memorial Museum, Austin 46519
Teysen's Woodland Indian Museum, Mackinaw City 50356
Town Creek Indian Mound Historic Site, Mount Gilead 50877
University of Arkansas Collections, Fayetteville 48528
University of Northern Iowa Museum Museums Collections, Cedar Falls 47294
University of Pennsylvania Museum of Archaeology and Anthropology, Philadelphia 51886
University of Wyoming Anthropology Museum, Laramie 49949
Ute Indian Museum, Montrose 50830
Vesterheim Norwegian-American Museum, Decorah 48022
Vincent Price Art Museum, Monterey Park 50810
Western Archaeological and Conservation Center, Tucson 53602
Western New Mexico University Museum, Silver City 53104
Wheelwright Museum of the American Indian, Santa Fe 52892
William Hammond Mathers Museum, Bloomington 46867
William Pryor Letchworth Museum, Castile 47277
Wing Luke Asian Museum, Seattle 53017
Winnebago Area Museum, Winnebago 54235
Woolaroc Museum, Bartlesville 46620

Uruguay

Casa de Giró, Montevideo 54383
Museo del Gaucho y de la Moneda, Montevideo 54401

Uzbekistan

Termuridi Tarikhi Muzeyi, Toshkent 54489

Vanuatu

Vanuatu Cultural Centre and National Museum, Port Vila 54495

Vatican City

Museo Missionario Etnologico, Città del Vaticano 54506

Venezuela

Museo de Antropología e Historia, Maracay 54584
Museo Etnológico de Amazonas, Puerto Ayacucho 54594
Museo J.M. Cruxent, Los Teques 54579
Museo Nueva Cádiz, La Asunción 54577

Vietnam

Bao Tàng Dân Tôc Hoc Viêt Nam, Ha Noi 54608
Bao Tàng Văn Hóa Các Dân Tôc Viêt Nam, Thai Nguyen 54636
Viet Nam Women's Museum, Ha Noi 54619
Women's Museum of South Vietnam, Ho Chi Minh City 54632

Yemen

Ethnographical Museum, Aden 54641

Zambia

Choma Museum and Crafts Centre, Choma 54653
Copperbelt Museum, Ndola 54662
Lusaka National Museum, Lusaka 54658
Moto Moto Museum, Mbala 54661

Zimbabwe

Museum of Human Sciences, Harare 54669

Expeditions → Scientific Expeditions

Explosives

Brazil

Museu Sitio Arqueológico Casa dos Pilões, Rio de Janeiro 04761

Germany

Fachausstellung Arbeitsschutz/ Sicherheitstechnik, München 20438

Poland

Muzeum Filumenistyczne, Bystrzyca Kłodzka 34583

United Kingdom

Royal Gunpowder Mills, Waltham Abbey 45925

Eyeglasses → Optics

Faience → Porcelain and Ceramics

Farms and Farming

Belgium

Ecomusée du Pays des Collines, Lahamaide 03773

Canada

Musée Historique des Soeurs de l'Assomption, Nicolet 06458
Pinhey's Point, Ottawa 06553
Spruce Lane Farm, Oakville 06498
Spruce Row Museum, Waterford 07283
Stony Plain and District Pioneer Museum, Stony Plain 07015

China

Muaseum of Nature and Agriculture, Macau 08001

France

Ferme-Musée du Léon, Tréflaouenan 15980

Germany

Museumsbauernhof Wennerstorf, Wennerstorf 22479

Hungary

Csabai Tanya- és Gabonatörténeti Kiállítóhely, Békéscsaba 23368
Magyar Mezogazdasági Múzeum Tanyamúzeuma, Lajosmizse - Bene 23688
Szabadtéri Néprajzi Múzeum, Nagyvázsony 23730

Switzerland

Bauernmuseum Althuus, Gurbrü 41463

United Kingdom

Leslie Hill Open Farm, Ballymoney 43367

United States of America

Hurst Ranch, West Covina 54053

Fashion → Clothing and Dress

Festivals

Austria

Faschings- und Brauchtumsmuseum, Knittelfeld 02206
Fasnacht- und Heimatmuseum Noaflhaus, Telfs 02802
Matschgerermuseum, Absam 01689

Belgium

Animalaine, Bastogne 03343
Folk Experience Museum, Dranouter 03562

Brazil

Museu do Carnaval, Rio de Janeiro 04735

China

China Caideng Museum, Zigong 08517

Cuba

Museo del Carnaval, Santiago de Cuba 09317

France

L'Amusant Musée, Juigne-sur-Sarthe 13290
Maison Jean-Vilar, Avignon 11787
Musée Spectacle de Musiques du Monde, Montoire-sur-le-Loir 14233

Germany

Georg Papendicks Faschings- und Karnevalsordenmuseum, Bad Reichenhall 16807
Internationales Artistenmuseum in Deutschland, Klosterfelde 19581
Zunftstube im Torli, Endingen am Kaiserstuhl 18169

Japan

Hino-shiritsu Shinsengumino Furusato Hakubutsukan, Hino 28516

Fire Fighting and Rescue Work

Australia

Fire Services Museum of Victoria, East Melbourne 01053
Queensland Fire Museum, Kurwongbah 01209

Austria

Burgenländisches Feuerwehrmuseum, Eisenstadt 01857
Erlauftaler Feuerwehrmuseum, Purgstall an der Erlauf 02539
Erstes Österreichisches Rettungsmuseum, Hohenems 02091
Erstes Tiroler Feuerwehrmuseum, Schwaz 02723
Feuerwehrmuseum, Dobersberg 01819
Feuerwehrmuseum, Perchtoldsdorf 02485
Feuerwehrmuseum, Sankt Leonhard am Forst 02660
Feuerwehrmuseum, Steyrermühl 02778
Feuerwehrmuseum, Türnitz 02828
Museum der Freiwilligen Feuerwehr, Laxenburg 02274
Niederösterreichisches Feuerwehrmuseum, Tulln 02831
Oberösterreichisches Feuerwehrmuseum St. Florian, Sankt Florian 02638
Steirisches Feuerwehrmuseum, Gross Sankt Florian 01996
Vorarlberger Museumswelt, Frastanz 01900
Wiener Feuerwehrmuseum, Wien 03088

Belarus

Fire Museum, Minsk 03248

Belgium

Musée des Sapeurs Pompiers-L'Univers des Pompiers, Athus 03333

Brazil

Museu Histórico do Corpo de Bombeiros Militar do Estado do Rio de Janeiro, Rio de Janeiro 04749

Canada

Canadian FireFighter's Museum, Port Hope 06642
City of Kingston Fire Department Museum, Kingston 06076
Dory Shop Museum, Shelburne 06953
Fire Fighters Museum of Winnipeg, Winnipeg 07359
Firefighters Museum of Nova Scotia, Yarmouth 07409
Musée des Pompiers Auxiliaires de Mont-Royal, Montréal 06362
Niagara Fire Museum, Niagara-on-the-Lake 06453
Prince Rupert Fire Museum, Prince Rupert 06672
Regina Firefighters Museum, Regina 06722

Cuba

Museo de los Bomberos, Matanzas 09214
Sala de los Bomberos, La Habana 09177

Czech Republic

Centrum Hasičského Hnuti, Přibyslav 09807
Hasičské Muzeum, Bechyně 09396

Denmark

Blåvand Redningsbådsmuseum, Blåvand 09978
Nymindegab Redningsbådsmuseum, Nørre Nebel 10195

Ecuador

Museo Coronel Félix Luque Plata, Guayaquil 10402
Museo de los Bomberos, Guayaquil 10403

Estonia

Järva-Jaani Tultõrjeseltsi Muuseum, Järva-Jaani 10576
Vanatehnikaklubi "Retrom" Tuletorjemuuseum, Kärevere 10579

Finland

Palomuseo, Helsinki 10807
Palomuseo, Sairaalamäki 11294

France

Musée Centre Historique des Sapeurs-Pompiers, Firminy 12939
Musée de la Mine et de Meunerie-Moulin de Marcy, Le Molay-Littry 13616
Musée de La Remise, Gex 13052
Musée de la Tradition de la Brigade des Sapeurs-Pompiers de Paris, Paris 14664
Musée de l'Histoire du Feu, Campuzan 12259
Musée Départemantal des Sapeurs-Pompiers de l'Orne, Bagnoles-de-l'Orne 11823
Musée Départemental des Sapeurs-Pompiers, Sciez 15692
Musée Départemental des Sapeurs-Pompiers du Val-d'Oise, Osny 14543
Musée des Pompiers, Wasquehal 16287
Musée des Pompiers, Fontainebleau 12963
Musée des Sapeurs Pompiers, Bièvres 11990
Musée des Sapeurs-Pompiers, Fessenheim 12929
Musée des Sapeurs-Pompiers, Thônes 15892
Musée des Sapeurs-Pompiers de France, Montville 14267
Musée des Sapeurs-Pompiers du Grand Lyon, Lyon 13892
Musée des Soldats du Feu, Wambrechies 16284
Musée du Sapeur-Pompier, Bouzigues 12137
Musée du Sapeur Pompier, Saint-Hippolyte-du-Fort 15311
Musée du Sapeurs-Pompiers, Ille-sur-Têt 13245
Musée Le Patrimoine du Sapeur-Pompier, Beaune 11906

Germany

Deutsches Feuerwehr-Museum, Fulda 18505
Fachausstellung Arbeitsschutz/Sicherheitstechnik, München 20438
Feuerwehr-Helm-Museum, Stuttgart 21932
Feuerwehr-Museum, Stuttgart 21933
Feuerwehr-Museum Schloß Waldmannshofen, Creglingen 17739
Feuerwehr- und Technikmuseum, Eisenhüttenstadt 18106
Feuerwehrmuseum, Kaufbeuren 19485
Feuerwehrmuseum, Lengenfeld 19952
Feuerwehrmuseum, Niederwiesa 20715
Feuerwehrmuseum, Rehau 21237
Feuerwehrmuseum, Roding 21347
Feuerwehrmuseum, Salem, Baden 21471
Feuerwehrmuseum, Sayda 21526
Feuerwehrmuseum, Spalt 21827
Feuerwehrmuseum, Stadtprozelten 21860
Feuerwehrmuseum, Wasserburg am Inn 22344
Feuerwehrmuseum Berlin, Berlin 17082
Feuerwehrmuseum des Landkreises Harburg, Marxen 20229
Feuerwehrmuseum Ellricher Spritzenhaus, Ellrich 18118
Feuerwehrmuseum Grethen, Parthenstein 20986
Feuerwehrmuseum Hannover, Hannover 18989
Feuerwehrmuseum Kalmbach, Riedbach 21313
Feuerwehrmuseum Marxen, Marxen 20230
Feuerwehrmuseum Musberg, Leinfelden-Echterdingen 19882
Feuerwehrmuseum Nürnberg, Nürnberg 20759
Feuerwehrmuseum Ravensburg, Ravensburg 21203
Feuerwehrmuseum Salzbergen, Salzbergen 21473
Feuerwehrmuseum Schleswig-Holstein, Norderstedt 20734
Feuerwehrmuseum Schröttinghausen, Preußisch Oldendorf 21124
Feuerwehrmuseum Winnenden, Winnenden 22600
Feuerwehrmuseum Zeven, Zeven 22796
Haller Feuerwehrmuseum, Schwäbisch Hall 21653
Heimat- und Feuerwehrmuseum, Schauenstein 21534
Historische Lehrsammlung der Feuerwehr Köln, Köln 19607
Historisches Feuerwehrmuseum Lüchow-Dannenberg, Dannenberg 17779
Kreis Feuerwehrmuseum Lohmen, Lohmen, Kreis Güstrow 20024
Münchner Feuerwehrmuseum, München 20460
Museum für Seenotrettungsgeräte, Neuharlingersiel 20616
Niederbayerisches Feuerwehrmuseum, Eggenfelden 18062

Hungary

Gyor-Moson-Sopron Megyei Tűzoltó Kiállítás, Mosonmagyaróvár 23719
Huszita-Ház - Helytörténeti Gyűjtemény, Gönc 23570
Tűzoltástörténeti kiállítás, Barcs 23362
Tűzoltó Múzeum, Bonyhád 23375
Tűzoltó Múzeum, Budapest 23486

Lithuania

Priešgaisrinės Apsaugos Muziejus, Šiauliai 30360

Netherlands

Brandweermuseum, Wouwse Plantage 32971
Brandweermuseum en Stormrampmuseum, Borculo 31983
Brandweermuseum Wassenaar, Wassenaar 32926
Nationaal Brandweermuseum, Hellevoetsluis 32361
Nationaal Reddingmuseum Dorus Rijkers, Den Helder 32112
Reddingsmuseum Abraham Fock, Hollum 32400

New Zealand

Fire Brigadesmen's Museum, Wellington 33246
Southland Fire Service Museum, Invercargill 33129

Norway

Ringsaker Vekter- og Brannhistoriske Museum, Moelv 33670
Stavanger Brannmuseum, Stavanger 33874

Peru

Museo de la Compañía de Bomberos Italia No. 5, Bellavista 34123

Poland

Muzeum Pożarnictwa, Alwernia 34528

Portugal

Museu do Batalhão de Sapadores Bombeiros, Porto 35849
Museu do Bombeiro, Lisboa 35696
Museu do Bombeiro Manuel Valdés Sobral, Valença 35978
Museu dos Bombeiros, Vila do Conde 35992
Museu dos Bombeiros Voluntários de Bucelas, Bucelas 35489
Museu Renato Lobo Garcia, Colares 35545

Puerto Rico

Museo del Antiquo Parque de Bombas, Ponce 36052

Romania

Muzeul Național al Pompierilor, București 36216
Muzeul Pompierilor, Teremia Mare 36686

Russia

Postojannaja Požarno-techničeskaja Vystavka, Sankt-Peterburg 37721

Slovenia

Muzej Gasilstva Dravinjske doline, Slovenske Konjice 38520
Sevniški Grad, Sevnica 38513
Slovenski Gasilski Muzej Dr. Branka Božiča, Metlika 38482

Spain

Museo de Bomberos, Madrid 39537
Museo del Ferrocarril, Ponferrada 39852

Sweden

Brandkårsmuseet, Simonstorp 40837
Motala Brandförsvarsmuseum, Motala 40753

Switzerland

Feuerwehr- und Handwerkermuseum, Endingen 41350
Feuerwehrmuseum, Adligenswil 41089
Feuerwehrmuseum, Kreuzlingen 41512
Feuerwehrmuseum Kradolf-Schönenberg, Kradolf 41510
Schweizerisches Feuerwehrmuseum Basel, Basel 41183

Taiwan

Fire Fly Museum of Dongshi Forest Farm, Dongshi 42182
Fire Safety Museum of Taipei Fire Department, Taipei 42390
Hsinchu City Fire Museum, Hsinchu 42207

Turkey

İtfaiye Müzesi, İstanbul 42829

United Kingdom

Balfour Museum of Hampshire Red Cross History, Winchester 46027
Chartered Insurance Institute's Museum, London 44818
Fire Police Museum, Sheffield 45650
Henry Blogg Lifeboat Museum, Cromer 43930
Lifeboat House, Margate 45105
Lifeboat Museum, Eastbourne 44099
London Fire Brigade Museum, London 44906
Lytham Lifeboat Museum, Lytham Saint Anne's 45061
Museum of Fire, Edinburgh 44134
Royal National Lifeboat Institution Headquarters Museum, Poole 45407
Selsey Lifeboat Museum, Selsey 45639
Uppark Exhibition, Petersfield 45375
Wiltshire Fire Defence and Brigades Museum, Potterne 45440

United States of America

Alma Firehouse and Mining Museum, Alma 46234
American Heritage "Big Red" Fire Museum, Louisville 50300
American Museum of Fire Fighting, Hudson 49410
Atlantic 1 Museum, Swampscott 53387
Bare Cove Fire Museum, Hingham 49297
Boston Fire Museum, Boston 46921
Cincinnati Fire Museum, Cincinnati 47556
Columbia Fire Department Museum, Columbia 47722
Connecticut Fire Museum, East Windsor 48263
Conway Fire Museum Aka Vintage Fire Engines, New Albany 51009
County of Los Angeles Fire Museum, Los Angeles 50219
Denver Firefighters Museum, Denver 48073
Eureka Fire Museum, Milltown 50695
Fire Museum of Maryland, Lutherville 50323
Fire Museum of Memphis, Memphis 50572
Fire Museum of Texas, Beaumont 46676
Fire Museum of York County, York 54340
Firehouse Museum, Ellicott City 48366
Firehouse No. 1 Museum, Nevada City 51007
Fireman's Hall Museum, Philadelphia 51834
Fort Wayne Firefighters Museum, Fort Wayne 48704
Friends of Firemen's Museum, New Bern 51019
Friendship Firehouse, Alexandria 46205
Hall of Flame - Museum of Firefighting, Phoenix 51903
Hinckley Fire Museum, Hinckley 49296
Homeville Antique Fire Department, Homer 49331
Houston Fire Museum, Houston 49389
Hull Lifesaving Museum, Hull 49417
Jacksonville Fire Museum, Jacksonville 49579
John J. Harvey Fireboat Collection, New York 51201
Lexington Historical Society, Lexington 50060
Maple Valley Historical Museum, Maple Valley 50456
Nantucket Life-Saving Museum, Nantucket 50947
Nevada State Fire Museum and Comstock Firemen's Museum, Virginia City 53761
New England Fire and History Museum, Brewster 47009
New Orleans Fire Department Museum, New Orleans 51090
New York City Fire Museum, New York 51241
North Plainfield Exempt Firemen's Museum, North Plainfield 51400
Oklahoma Firefighters Museum, Oklahoma City 51522
Old Fire House and Police Museum, Superior 53379
Old Firehouse Museum, Greenville 49073
Plantation Historical Museum, Plantation 51978
Robert A. Bogan Fire Museum, Baton Rouge 46639
San Francisco Fire Department Museum, San Francisco 52793
Uppertown Firefighters Museum, Astoria 46404
Volunteer Firemen's Mall and Museum of Kingston, Kingston 49793
Woodbine International Fire Museum, Woodbine 54272

Firearms → Arms and Armour

Fishing and Fisheries

Australia

Eden Killer Whale Museum, Eden 01060
Fisherman's Cottage Museum, Tooradin 01591
Old Wool and Grain Store Museum, Beachport 00843
Whale World, Albany 00790

Austria

Kärntner Fischereimuseum, Seeboden 02734
Kunstsammlungen des Stiftes Kremsmünster, Kremsmünster 02233
Museum Orth, Orth an der Donau 02473
Österreichisches Donau- und Fischereimuseum, Orth an der Donau 02474
Schopper- und Fischermuseum, Aschach an der Donau 01731

Belgium

Amandine - d'Ostensjhe leslandvisjherie, Oostende 03927
Heemkundig Museum Sincfala, Knokke 03753
Musée de la Pêche et des Poissons, Amberloup 03271
Nationaal Visserijmuseum van Oostduinkerke, Oostduinkerke 03926

Bolivia

Museo Icticola del Beni, Trinidad 04255

Brazil

Museu de Pesca, Santos 04821

Brunei

Muzium Teknologi Melayu, Bandar Seri Begawan 04966

Canada

Alert Bay Museum, Alert Bay 05314
Basin Head Fisheries Museum, Souris 06994
Britannia Shipyard, Richmond 06741
Durrell Museum, Durrell 05728
Fisheries Museum of the Atlantic, Lunenburg 06202
Fishermen's Museum, Port-de-Grave 06635
Fishermen's Museum, Saint Vincent's 06884
Flatrock Museum, Flatrock 05798
Fox Harbour Community Museum, Fox Harbour 05826
Gulf of Georgia Cannery, Richmond 06743
Huntsman Marine Science Centre Aquarium Museum, Saint Andrews 06797
Lundar Museum, Lundar 06201
Margaree Salmon Museum, North East Margaree 06474
Musée de Kent, Bouctouche 05446
Musée du Centre Acadien de l'Université Sainte-Anne, Pointe de l'Eglise 06619
North Pacific Cannery Village Museum, Port Edward 06637
Northumberland Fisheries Museum, Pictou 06603
La Pirogue Museum, Cheticamp 05609
Port Dover Harbour Museum, Port Dover 06636
Prince Edward Island Shellfish Museum, Bidford 05424
Saint Lewis Heritage Museum, Saint Lewis 06863
Salvage Fisherman's Museum, Salvage 06906
Site Historique Matamajaw, Causapscal 05580
Tehkummah Township Little Schoolhause and Museum, Tehkummah 07063

Czech Republic

Lesnické, Myslivecké a Rybářské Muzeum, Hluboká nad Vltavou 09516
Lovecko Lesnicke Muzeum, Úsov 09891

Denmark

Blåvand Museum, Blåvand 09977
Fiskeri- og Søfartsmuseet, Esbjerg 10002
Fiskeriets Hus - Vestkyst Akvariet, Hvide Sande 10107
Fjordmuseet, Jyllinge 10111
Gilleleje Museum, Gilleleje 10036
Museumscenter Hanstholm, Hanstholm 10054
Nordsømuseet, Hirtshals 10079
Nymindegab Museum, Nørre Nebel 10194
Søfarts og Fiskerimuseet, Læsø 10173

Egypt

Aquatics Museum, Suez 10541

Estonia

Karepa Kalame Talumuuseum, Vihula 10672

Finland

Ålands Jakt och Fiskemuseum, Eckerö 10720
Kalajoen Kalastusmuseo, Kalajoki 10905
Langinkoski Imperial Fishing Lodge, Kotka 10987
Merikarvian Kalastusmuseo, Merikarvia 11109
Runebergs stugan, Jakobstad 10867
Skärgårdsmuseet, Lappoby-Åland 11052

France

Ecomusée de la Pêche, Thonon-les-Bains 15895
Ecomusée de Montjean, Montjean-sur-Loire 14218
Écomusée du Pêrche de Saint-Cyr-la-Rosière, Bellême 11928
Maison de la Rivière et du Pêcheur, Saint-Georges-de-Montaigu 15279

Maison de l'Eau et de la Pêche, Besse-en-Chandesse 11961
Maison de l'Eau et de la Pêche, Neuvic (Corrèze) 14404
Maison des Etangs, Saint-Viâtre 15535
Maison du Patrimoine Maritime, Camaret-sur-Mer 12251
Maison Nationale de la Pêche et de l'Eau, Ornans 14538
Mémoire de Saumon, Saint-Louis-la-Chaussée 15375
Musée de la Conserverie Le Gall, Loctudy 13798
Musée de la Mer, Paimpol 14553
Musée de la Mer OCEAM, Les Sables-d'Olonne 13714
Musée de la Mytiliculture, Esnandes 12853
Musée de la Pêche, Concarneau 12602
Musée de la Vilaine Maritime, La Roche-Bernard 13412
Musée la Maison du Pêcheur et "Le Hope", Saint-Gilles-Croix-de-Vie 15299
Musée-Maison du Saumon et de la Rivière, Brioude 12186

Germany

Deutsches Jagd- und Fischereimuseum, München 20432
Deutsches Meeresmuseum, Stralsund 21913
Fischer- und Webermuseum Steinhude, Wunstorf 22733
Fischereimuseum, Wassertrüdingen 22348
Jagd- und Fischereimuseum, Adelsdorf 16427
Jagd- und Fischereimuseum Schloss Tambach, Weitramsdorf 22468
Jagd- und Fischereimuseum Schloß Wolfstein und Galerie Wolfstein, Freyung 18461
Museum für Wattenfischerei, Wremen 22702
Museum im Malhaus, Wasserburg, Bodensee 22347
Museumslogger AE7 Stadt Emden, Emden 18140
Nationalpark-Haus, Butjadingen 17669
Norderneyer Fischerhausmuseum, Norderney 20733
Oberpfälzer Fischereimuseum, Tirschenreuth 22062
Schleimuseum, Kappeln 19430

Greenland

Cold Museum, Ilulissat 23252
Hunting Fishing Museum, Ilulissat 23254

Iceland

Síldarminjasafnið a Siglufirdi, Siglufjörður 24039
Sjóminjasafn Islands, Reykjavík 24036

India

Meghalaya State Museum, Shillong 24386
Museum of the Central Marine Fisheries Research Station, Mandapam Camp 24306

Israel

House of the Anchors, Kibbutz Ein Gev 25037

Italy

Museo della Pesca, Magione 26565
Museo della Tonnara, Milazzo 26727
Museo della Tonnara, Stintino 27907
Museo Ittico, Pescara 27165
Museo Provinciale della Caccia e della Pesca, Racines 27355

Japan

Chiba-kenritsu Awa Hakubutsukan, Tateyama, Chiba 29353
Echizen Crab Museum, Echizen 28411
Iyoboya Salmon Museum, Murakami 28944
Koto-ku Nakagawa Funabansho Museum, Tokyo 29421
Seto Rekishi Mingeikan, Takamatsu 29320
Tokai Daigaku Kaiyo Kagaku Hakubutsukan, Shizuoka 29291

Korea, Republic

Bukyoung University Museum, Busan 29773

Latvia

Rojas Jūras un Zvejniecības Muzejs, Roja 30117
Vecdaugavas Novadpētniecības un Zvejniecības Muzejs, Rīga 30116
Zvejniecības Brīvdabas Muzejs, Jūrmala 30027
Zvejniecības un Kugniecības Vestures Muzejs, Liepāja 30041

Macedonia

Ribarsko-Biološki Muzej, Nov Dojran 30511

Netherlands

It Earmhus en Oud Friese Greidboerderij, Warten 32925
Gemeentemuseum Elburg, Elburg 32197
Jan Vissermuseum, Helmond 32365
Klederdracht en Visserijmuseum, Bunschoten Spakenburg 32020
Museum 't Fiskershúske, Moddergat 32558
Museum In 't Houten Huis, De Rijp 32049
Museum Scheveningen, Den Haag 32089
Museum van Egmond, Egmond aan Zee 32180
Oudheidkundig Museum Arnemuiden, Arnemuiden 31921
Visafslag Elburg, Elburg 32200
Visserij- en Cultuurhistorisch Museum, Woudrichem 32969
Visserijmuseum, Breskens 32009
Visserijmuseum, Bruinisse 32016
Visserijmuseum, Vlaardingen 32892
Visserijmuseum Zoutkamp, Zoutkamp 33011
Zee- en Havenmuseum de Visserijschool, IJmuiden 32425

Norway

Andøymuseet, Risøyhamn 33783
Berlevåg Havnemuseum, Berlevåg 33435
Fiskerimuseet på Hjertøya, Molde 33676
Fiskerimuseum i Måløy, Måløy 33657
Gamvik Museum, Gamvik 33509
Grytøy Bygdetun, Lundenes 33653
Ishavsmuseet Aarvak, Brandal 33452
Karmøy Fiskerimuseum, Vedavågen 33967
Kommandør Chr. Christensen's Hvalfangst Museum, Sandefjord 33812
Kråksundet notnaust Sjøbruksmuseum, Tustna 33936
Kystmuseet i Sogn og Fjordane, Florø 33499
Leka Bygdemuseum, Leka 33631
Lenvik Bygdemuseet, Finnsnes 33492
Lofotmuseet, Kabelvåg 33585
Meløy Bygdemuseum, Ørnes 33710
Nordkappmuseet, Honningsvåg 33560
Nordlandsmuseet, Bodø 33445
Norges Fiskerimuseet, Bergen 33432
Norsk Fiskeindustrimuseum, Melbu 33663
Norsk Fiskerværsmuseet, Sørvågen 33855
Norsk Hermetikkmuseum, Stavanger 33870
Norsk Klippfiskmuseum, Kristiansund 33616
Norsk Skogmuseum, Elverum 33481
Rogaland Fiskerimuseum, Åkrehamn 33376
Skrolsvik Fiskebruksmuseum, Moen 33674
Sund Fiskerimuseum, Sund i Lofoten 33891
Sunnhordland Museum, Stord 33885

Poland

Muzeum Rybackie, Jastarnia 34698
Muzeum Rybołówstwa, Hel 34687
Skansen Rybołowstwa Morskiego, Gdynia 34661

Portugal

Museu do Pescador, Montijo 35777
Museu dos Baleeiros, Lajes do Pico 35653
Museu Oceanográfico Prof. Luiz Saldanha, Azeitão 35453
Museu Sobre a Pesca do Atum, Tavira 35960

Russia

Morskoj Muzej-Akvarium, Vladivostok 37942
Muzej Istorii Rybnoj Promyšlennosti Dagestana, Machačkala 37117
Muzej Ochoty i Rybolovstva, Moskva 37342

Senegal

Musée de la Mer, Gorée 38025

South Africa

Old Harbour Museum, Hermanus 38641
Sea World, Durban 38612

Spain

Arrantzaleen Museoa, Bermeo 39030
Museu de la Pesca, Palamós 39768
Museu Marítim de Barcelona, Barcelona 38998

Sri Lanka

National Maritime Museum, Galle 40378

Sweden

Hälleviks Fiskemuseum, Sölvesborg 40866
Holmöns Båtmuseum, Holmön 40600
Kukkola Fiskemuseum, Haparanda 40582
Nordvärmlands Jakt och Fiskemuseum, Sysslebäck 40967
Råå Museum för Fiske och Sjöfart, Råå 40811

Switzerland

Fischerei-Museum, Zug 42124
Kantonales Fischereimuseum Mettlen, Netstal 41674
Musée du Léman, Nyon 41690
Museo della Pesca, Caslano 41281
Seemuseum in der Kornschütte, Kreuzlingen 41516

Syria

Fish Museum Golden Angel, Damascus 42130

Taiwan

Farming and Fishing Museum of Sihu, Sihu 42328
Fishery Exhibition Hall, Lugang 42256
Pei Kuan Crab Museum, Toucheng 42456

United Kingdom

Arbuthnot Museum, Peterhead 45371
Brighton Fishing Museum, Brighton 43620
Brixham Heritage Museum, Brixham 43641
Buckhaven Museum, Buckhaven 43662
Buckie Drifter Fishing Heritage Centre, Buckie 43663
Eyemouth Museum, Eyemouth 44200
Fishermen's Museum, Hastings 44414
Forge Mill Needle Museum and Bordesley Abbey Visitor Centre, Redditch 45475
Golden Hind Museum, Brixham 43642
Hull Maritime Museum, Hull 44499
Izaak Walton's Cottage, Shallowford 45646
Lleyn Historical and Maritime Museum, Nefyn 45197
Lossiemouth Fisheries and Community Museum, Lossiemouth 45029
National Fishing Heritage Centre, Great Grimsby 44355
Scottish Fisheries Museum, Anstruther 43306
Time Tide - Museum of Great Yarmouth Life, Great Yarmouth 44360
True's Yard Fishing Museum, King's Lynn 44590
Whitstable Oyser and Fishery Exhibition, Whitstable 46008

United States of America

The American Museum of Fly Fishing, Manchester 50426
Catskill Fly Fishing Center and Museum, Livingston Manor 50152
Clausen Museum, Petersburg 51793
Cold Spring Harbor Whaling Museum, Cold Spring Harbor 47682
D.C. Booth Historic National Fish Hatchery, Spearfish 53199
Fishermen's Museum, Pemaquid Point 51752
National Fresh Water Fishing Hall of Fame, Hayward 49231
New Bedford Whaling Museum, New Bedford 51014
Reedville Fishermen's Museum, Reedville 52281
Sag Harbor Whaling and Historical Museum, Sag Harbor 52514
Virginia Institute of Marine Science, Fish Collection, Gloucester Point 48949
Whale Museum, Friday Harbor 48796

Folk Art

Algeria

Musée des Arts Populaires et Traditions, Alger 00041

Argentina

Museo Cooperativo Eldorado, Eldorado 00361
Museo de Artes y Artesanías Enrique Estrada Bello, Santo Tomé 00632
Museo Monseñor José Fagnano, Río Grande 00536

Australia

Bega Valley Regional Gallery, Bega 00854
Latvian Museum, Wayville 01639
Man From Snowy River Folk Museum, Corryong 01004
Tiagarra Aboriginal Museum, Devonport 01037

Austria

Bienen- und Wagnereimuseum, Tollet 02814
Fasnacht- und Heimatmuseum Noaflhaus, Telfs 02802
Haus der Völker, Schwaz 02725
Krippenmuseum, Vösendorf 02870
Museum für Volkskultur, Spittal an der Drau 02752
Museum für Volkskunst, Haslach 02056
Niederösterreichisches Museum für Volkskultur, Groß Schweinbarth 01997
Österreichisches Museum für Volkskunde, Wien 03049
Salzburg Museum, Salzburg 02625
Sammlungen Schloß Moosham, Unternberg 02841
Tiroler Volkskunstmuseum, Innsbruck 02137
Volkskundemuseum Salzburg, Salzburg 02629
Walsermuseum Lech-Tannberg, Lech 02278

Bangladesh

Sonargaon Folk Art and Craft Museum, Sonargaon 03236

Belarus

Gallery of Traditional Ornamental Art, Minsk 03249
Grodnenski Dziarzhauny Hystorika-Archealagičny Muziej, Grodno 03246

Belgium

Folk Experience Museum, Dranouter 03562
Museum voor Volkskunde, Brugge 03425
Rariteiten- en Ambachtenmuseum 't Krekelhof, Koksijde 03755

Benin

Musée da Silva des Arts et de la Culture, Porto-Novo 04187

Bolivia

Museo Colonial Charcas, Sucre 04245

Brazil

Museu Casa do Pontal, Rio de Janeiro 04706
Museu da Cultura e Arte Popular, João Pessoa 04512
Museu da Inconfidência, Ouro Preto 04601
Museu de Arte e Cultura Popular da UFMT, Cuiabá 04422

Bulgaria

Etnografska Ekspozicija Stojova Kăšta, Radomir 05144
Nacionalen Etnografski Muzej, Sofia 05190
Nacionalna Izložba na Narodnite Chudožestveni Zanajati i Priložni Izkustva, Orešak 05100

Canada

Court House Museum, Cardston 05561
Heritage Models Centre, River Herbert 06756
Japanese Canadian National Museum, Burnaby 05503
Lower Fort Garry, Saint Andrews 06798
Mennonite Historical Society of British Columbia, Abbotsford 05306
Musée de Charlevoix, La Malbaie 06111
Musée Québécois de Culture Populaire, Trois-Rivières 07178
Prince Albert Arts Centre, Prince Albert 06663
Yesterday Museum, Bay Roberts 05402

Chile

Museo de Arte Popular Americano, Santiago de Chile 07455
Museo Regional de la Araucania, Temuco 07467

China

Anyang Glaswegianjian Art Museum, Anyang 07487
Guangdong Folk Arts and Crafts Museum, Guangzhou 07723
Nanjing Museum, Nanjing 08045
Zhangqiu Arts and Crafts Guan, Jinan 07886

Colombia

Museo de Arte y Tradiciones Populares, Bogotá 08562

Costa Rica

Museo Indigeno, San José 08823

Croatia

Galerija Naivne Umetnosti - Donacija Ivana Generalića, Hlebine 08877
Galerija Zavjetnih Slika Brodova, Dubrovnik 08864
Kulturno Povijesni Muzej, Dubrovnik 08865
Narodni Muzej, Zadar 08983
Zavičajna Zbirka Đelekovec, Đelekovec 08859

Cuba

Museo del Danza, La Habana 09140

Cyprus

Cyprus Folk Art Museum, Lefkosia 09373
Ethnographic Museum, Avgorou 09362
Folk Art Museum, Lemesos 09386
Folk Art Museum, Yeroskipos 09393
Pierides Museum, Larnaka 09369

Czech Republic

Krkonošské Muzeum - Tri historické domy, Vrchlabí 09915
Městské Muzeum, Česká Třebová 09461
Muzeum Beskyd Frýdek-Místek, Frýdek-Místek 09504
Muzeum Krajky, Prachatice 09732
Muzeum Krajky, Vamberk 09898
Muzeum Luhačovického Zálesí, Luhačovice 09643
Muzeum Romské Kultury, Brno 09437
Slezské zemské muzeum, Opava 09693

Denmark

Lolland-Falsters Stiftsmuseum, Maribo 10185

Dominican Republic

Museo Cándido Bido, Bonao 10324

F

Ecuador

Museo de Artesanias Cefa, Quito 10438
Museo de la Inmaculada Concepción, Cuenca 10386
Museo de las Artes Populares Cidap, Cuenca 10388

Estonia

Hanila Muuseum, Hanila 10568

Finland

Emil Cedercreutzin Museo, Harjavalta 10758
Ilmajoen Museo, Ilmajoki 10850
Muurlan Kotiseutumuseo, Muurla 11128

France

Maison d'Offwiller, Offwiller 14498
Musée Abbaye d'Airvault, Airvault 11512
Musée Alsacien, Strasbourg 15819
Musée Catalan des Arts et Traditions Populaires, Perpignan 14788
Musée Cévenol, Le Vigan (Gard) 13666
Musée d'Art et d'Histoire de Puisaye, Villiers-Saint-Benoît 16244
Musée d'Art et Traditions Populaires, Nancy 14352
Musée d'Arts et Traditions Populaires, La Guérinière 13383
Musée de Cerdagne, Sainte-Léocadie 15562
Musée de Dol, Dol-de-Bretagne 12753
Musée de la Crèche Provençale, Cavaillon 12330
Musée de la Fève et de la Crèche, Blain 12008
Musée de l'Amour et des ses Traditions, Oger 14499
Musée Départemental des Arts et Traditions Populaires du Perche, Saint-Cyr-la-Rosière 15225
Musée des Arts et Métiers, Saint-Gildas-de-Rhuys 15298
Musée des Arts et Traditions Populaires, La Petite-Pierre 13402
Musée d'Histoire et d'Archéologie René Bauberot, Châteauponsac 12452
Musée du Charroi Rural, Salmiech 15608
Musée du Rouergue Arts Métiers, Salles-la-Source 15607
Musée du Vieux-Château, Laval (Mayenne) 13542
Musée Folklorique et Archéologique, Châtelus-le-Marcheix 12465
Musée Municipal de Pithiviers, Pithiviers-le-Vieil 14828
Musée National des Arts et Traditions Populaires, Paris 14727
Musée Régional d'Arts et de Traditions Popoulaires, Grimaud 13150
Musée Vivant de la Passementerie, Jonzieux 13278

Gabon

Musée National des Arts et Traditions, Libreville 16329

Georgia

Folk and Applied Arts Museum, Tbilisi 16357

Germany

Altonaer Museum für Kunst und Kulturgeschichte, Hamburg 18915
Augustinermuseum, Freiburg im Breisgau 18428
Bauernhaus-Museum, Bielefeld 17337
Buddelschiff-Museum, Neuharlingersiel 20615
Buddelschiff-Museum, Tangerhütte 22000
Erzgebirgischer Spielzeugwinkel, Eisenheim 18104
Haus der erzgebirgischen Tradition, Deutschneudorf 17844
Heimatmuseum im Unteren Turm, Leutershausen-Wiedersbach 19971
Heimatstube Freest, Kröslin 19718
Heimatstube für Volkskunst, Zschadraß 22823
Heimatstube Podersam-Jechnitz, Kronach 19722
Manufaktur der Träume, Annaberg-Buchholz 16555
Missionsmuseum des päpstlichen Missionswerks der Kinder in Deutschland, Aachen 16403
Museum Erzgebirgische Volkskunst "Buntes Holz" im Postgut am Altmarkt, Hohenstein-Ernstthal 19245
Museum für bergmännische Volkskunst Schneeberg, Schneeberg 21591
Museum für Sächsische Volkskunst mit Puppentheatersammlung, Dresden 17963
Museum Knochenstampfe, Zwönitz 22841
Museum Otzberg, Otzberg 20963
Museum Rade am Schloß Reinbek, Reinbek 21251
Museum Schloß Adelsheim, Berchtesgaden 17002
Museum Steinau...das Museum an der Straße, Steinau an der Straße 21879
Museumsdorf Bayerischer Wald, Tittling 22065
Nussknackermuseum, Neuhausen / Erzgebirge 20620
Oderlandmuseum, Bad Freienwalde 16733
Schloßmuseum Murnau, Murnau am Staffelsee 20535
Siegerlandmuseum mit Ausstellungsforum Haus Oranienstraße, Siegen 21756
Spielzeugmuseum, Görlitz 18682
Theodor-Zink-Museum und Wadgasserhof, Kaiserslautern 19414
Volkskundemuseum des Bezirks Unterfranken, Bad Bocklet 16701

Greece

Archipelagos Cultural Center, Athinai 22899
Folk Art Museum, Pyrgi 23161
Folklore Museum, Thessaloniki 23213
Granitsa Folklore Museum, Granitsa 23009
Historical and Folk Art Museum of Gavalohori, Chania 22963
Mouseio Ellinikis Laïkis Technis, Athinai 22918
Mouseio Vorre, Paiania 23129
Museum of Greek Folk Art, Athinai 22934
Museum of Greek Folk Art-Pottery Collection, Athinai 22935
National Folk Museum of Lesbos, Mytilini 23111
National Historical Museum, Athinai 22942

Hungary

Erdős Kamill Cigány Múzeum, Pécs 23760
Kis Géza Ormánság Múzeum és Skanzen, Sellye 23799
Magyar Naiv Művészek Múzeuma, Kecskemét 23651
Német Nemzetiségi Gyűjtemény, Gyula 23599
Népi Iparművészeti Gyűjtemény, Kecskemét 23652
Palóc Múzeum - Palóc Ház, Balassagyarmat 23350
Palóc Népművészet, Eger 23543
Szabadtéri Néprajzi Múzeum, Szentendre 23880
Tájház, Balatonszentgyörgy 23357

India

Acharya Jogesh Chandra Purakirti Bhavan, Bishnupur 24102
Ananda Niketan Kirtishala, Bagnan 24069
Gurusaday Museum, Kolkata 24274
Kachchh Museum, Bhuj 24099
Lady Wilson Museum, Dharampur 24161
Madhya Pradesh Tribal Research and Development Institute, Bhopal 24092
Museum and Art Gallery, Burdwan 24105
Purnananda Memorial Museum, Hazarikapra 24207
State Tribal Museum, Chhindwada 24119

Indonesia

Museum Negeri Sumatera Barat, Padang 24533

Iran

Sanandaj Museum, Sanandaj 24642

Italy

Casa Natale di Pascoli, San Mauro Pascoli 27668
Museo Arti e Tradizioni Popolari del Gargano Giovanni Tancredi, Monte Sant'Angelo 26806
Museo Carnico delle Arti Popolari Michele Gortani, Tolmezzo 27992
Museo Civico delle Arti e Tradizioni Popolari, Micigliano 26671
Museo della Cultura Popolare e Contadina, Carrega Ligure 25692
Museo della Vita e delle Tradizioni Popolari Sarde, Nuoro 26967
Museo delle Tradizioni ed Arti Contadine, Picciano 27198
Museo e Pinacoteca della Comunità di Fiemme, Cavalese 25788
Museo Etnografico Siciliano Giuseppe Pitrè, Palermo 27076
Museo Ibleo delle Arti e Tradizioni Popolari - Guastella, Modica 26759
Museo Nazionale delle Arti e Tradizioni Popolari, Roma 27515
Museo Nazionale delle Arti Naïves C. Zavattini, Luzzara 26550

Japan

Ehime Mingeikan, Saijo 29185
Fujii Bijutsu Mingeikan, Takayama 29335
Fukushima-ken Bunka Centre, Fukushima 28439
Gotoh Bijutsukan, Tokyo 29394
ICU Hachiro Yuasa Memorial Museum, Mitaka 28916
Ikebana Bijutsukan, Kyoto 28815
Ishikawa-kenritsu Dento Sangyo Kogeikan, Kanazawa 28657
Iwayama Urushi Bijutsukan, Morioka 28937
Kaishu Ichizo Museum, Aizuwakamatsu 28351
Kochi-ken Kaitokukan, Kochi 28742
Kodai Yuzen-en Bijutsukan, Kyoto 28826
Kumamoto Kokusai Mingeikan, Kumamoto 28778
Kurashiki Mingeikan, Kurashiki 28784
Kyu Ikegamike, Takatomachi Inashi 29334
Matsumoto Museum of Folk and Arts, Matsumoto 28893
Middle Eastern Culture Centre in Japan, Mitaka 28917
Nihon Mingeikan, Tokyo 29450
Osaka Nippon Mingeikan, Suita 29296
Seikado Bunko no Bijutsukan, Tokyo 29464
Suifu-Meitoku Shokoh-kan Tokugawa Museum, Mito 28922
Takasawa Folk Craft Museum, Niigata 29063
Tottori-kenritsu Hakubutsukan, Tottori 29515
Tottori Mingeikan, Tottori 29516

Kazakhstan

Atyrauskij Oblastnoj Muzej Dekorativno-prikladnogo Iskusstva, Atyrau 29696

Korea, Republic

Duksung Women's University Art Museum, Seoul 29874
Gwangju Municipal Folk Museum, Gwangju 29822
Jeju-do Folklore and Natural History Museum, Jeju 29836
Jeonju National Museum, Jeonju 29842
Kuknip Minsok Pakmulgwan, Seoul 29900
Suk Joo-sun Memorial Museum of Korean Folk Arts, Seoul 29927

Kyrgyzstan

Issyk-Kulskij Gosudarstvennyj Istoriko-Kulturnyj Muzej-Zapovednik, Čolpan-Ata 29961

Latvia

Latvijas Etnogrāfiskais Brīvdabas Muzejs, Rīga 30079

Lithuania

Antano Baranausko ir Antano Vienuolio-Žukausko Memorialinis Muziejus, Anykščiai 30206
Antano ir Jono Juškų Etninės Kultūros Muziejus, Vilkija 30399
Antano Žmuidzinavičiaus Memorialinis Muziejus, Kaunas 30239
Daugyvenės Kultūros Istorijos Muziejus-draustinis, Radviliškis 30341
Karaimų Etnografinė Paroda, Trakai 30381
Kretingos Muziejus, Kretinga 30288
Lietuvos Liaudis Buities Muziejus, Rumšiškės 30347
Marijampolės Kraštotyros Muziejus, Marijampolė 30308
Nacionalinis Mikalojaus Konstantino Čiurlionio Dailės Muziejus, Kaunas 30262
Šeduvos Kraštotyros Ekspozicija, Seduva 30351
Šilutės Muziejus, Šilutė 30369
Vilniaus Krašto Etnografijos Muziejus, Nemenčinės 30318
Žemaičių Dailės Muziejus, Plungė 30338

Mexico

Museo Casa Ruth Lechuga, México 31061
Museo de Arte Popular de San Bartolo de Coyotepec, San Baltazar Chichicapan 31322
Museo de Arte Popular Poblano y Museo de Artesanías, Puebla 31274
Museo de Arte Popular Tlajomulco, Tlajomulco de Zúñiga 31467
Museo de Artes e Industrias Populares, Pátzcuaro 31255
Museo de Artes Populares, Tepic 31441
Museo de Artes y Tradiciones Populares de Tlaxcala, Tlaxcala 31476
Museo de Culturas Populares, Durango 30840
Museo de la Casa de las Artesanías del Estado de Chihuahua, Bocoyna 30738
Museo de la Platería, Taxco de Alarcón 31406
Museo de las Artes Populares de Jalisco, Guadalajara 30896
Museo de las Culturas, Saltillo 31314
Museo Michoacano de las Artesanías, Morelia 31200
Museo Nacional de Artes e Industrias Populares, México 31132
Museo Regional de Arte Popular La Huatápera, Uruapán 31521
Museo Sala de Arte Público David Alfaro Siqueiros, México 31146
Museo Universitario de Artes Populares, Colima 30783

Morocco

Musée du Patrimoine Amazigh, Agadir 31660
Musée Ethnographique, Chefchaouen 31666
Musée Sidi Mohamed Ben Abdellah, Essaouira 31668

Namibia

Shambyu Museum, Shambyu 31755

Netherlands

Creativiteitscentrum Gouden Handen, 's-Heerenberg 32330
Flessenscheepjesmuseum, Enkhuizen 32207
Huize Nijenstede, Hardenberg 32312
Katwijks Museum, Katwijk aan Zee 32443
Museum De Vier Quartieren, Oirschot 32612

New Zealand

Rotorua Museum of Art and History, Rotorua 33196
Whakatane District Museum and Gallery, Whakatane 33263

Nigeria

Isenbaye Art Gallery and Cultural Troupe, Oshogbo 33339

Northern Cyprus

Halk Sanatları Müzesi, Girne 33360

Norway

Bu Museum, Utne 33948
Hufthamartunet, Storebø 33886
Lågdalsmuseet, Kongsberg 33602
Maihaugen, Lillehammer 33637
Nordfjord Folkemuseum, Sandane 33810
Norsk Husflidsmuseum, Moen 33672
Oscarsborg Festningsmuseum, Oscarsborg 33713
Telemark og Grenland Museum, Skien 33836
Tinn Museum, Rjukan 33786
Vest-Telemark Museum, Dalen i Telemark 33463
Vikedal Bygdemuseum, Vikedal 33980

Pakistan

National Museum of Ethnology, Islamabad 34009

Palestine

Maha Al Saqqa Center - Palestinian Folk Museum, Bethlehem 34051

Peru

Museo Comunitario Inti Raymi, Lima 34228
Museo de Arte Popular de la Universidad Nacional San Cristóbal de Huamanga, Ayacucho 34119
Museo de Arte Popular de Quinua, Quinua 34300
Museo de Artes y Tradiciones Populares, Lima 34236
Museo de Filigrana de Cobre, La Esperanza 34199
Museo Histórico Hipólito Unanue, Ayacucho 34122

Philippines

Bahay Laguna Museum, Liliw 34403
Balay Negrense Museum, Silay 34505
Bayanihan Folk Arts Museum, Manila 34432
Bontoc Museum, Bontoc, Mountain Province 34350
Museo It Akean, Kalibo 34396

Poland

Dom Rodzinny Jana Kasprowicza, Inowrocław 34690
Muzeum Beskidzkie im. Andrzeja Podżorskiego, Wisła 35303
Muzeum Budownictwa Ludowego, Sanok 35100
Muzeum Etnograficzne, Poznań 35034
Muzeum Etnograficzne, Włocławek 35309
Muzeum Kultury Ludowej, Osiek nad Notecią 34990
Muzeum Lachów Sądeckich im. Zofii i Stanisława Chrząstowskich, Podegrodzie 35017
Muzeum Pomorza Środkowego, Słupsk 35112
Muzeum Regionalne, Łuków 34913
Muzeum Regionalne, Radomsko 35076
Muzeum Regionalne im. Seweryna Udzieli, Stary Sącz 35132
Muzeum Regionalne Ziemi Zbąszyńskiej PTTK, Zbąszyń 35365
Muzeum Rzemiosła, Krosno 34831

Muzeum Stylu Zakopiańskiego im. Stanisława Witkiewicza - Willa Koliba, Zakopane 35359

Portugal

Museu da Escola Regional de Artesanato de Santo Amaro, Santo Amaro 35914
Museu de Arte Popular, Lisboa 35690
Museu Dr. Marcelino Moules, Cinco Ribeiras 35525
Museu João de Deus, Lisboa 35711
Museu Rural de Villa-Lobos, Borba 35473
Núcleo Museológico da Quinta do Prado, Celorico de Basto 35521

Puerto Rico

Museo de Las Américas, San Juan 36068
Taller Tanamá y Casa Jesús T. Piñero, Canóvanas 36034

Romania

Colecția de Etnografie și Artă Populară, Chilia 36276
Colecția Muzeală, Colacu 36294
Complexul Național Muzeal Astra, Sibiu 36604
Expoziția de Etnografie și Artă Populară, Ciacova 36277
Expoziția de Etnografie și Artă Populară, Pucioasa 36543
Muzeul Arta lemnului, Câmpulung Moldovenesc 36255
Muzeul Brăilei-Secția de Etnografie și Artă Populară, Brăila 36171
Muzeul de Artă Comparată, Sângeorz-Băi 36584
Muzeul de Artă Populară, Constanța 36303
Muzeul de Artă Populară Prof. Dr. Nicolae Minovici, București 36201
Muzeul de Artă Populară și Etnografie, Tulcea 36704
Muzeul de Artă și Artă Populară, Calafat 36247
Muzeul de Etnografie și Artă Populară, Baia Mare 36126
Muzeul de Etnografie și Artă Populară, Reghin 36559
Muzeul de Etnografie și Artă Populară Orăștie, Orăștie 36506
Muzeul Județean Mureș-Secția de Etnografie și Artă Populară, Târgu Mureș 36671
Muzeul Maramureșului, Sighetu Marmației 36615
Muzeul Municipal Câmpulung-Secția de Etnografie și, Câmpulung 36253
Muzeul Național al Satului Dimitrie Gusti, București 36217
Muzeul Obiceiurilor Populare din Bucovina, Gura Humorului 36394
Muzeul Sătesc, Botoșana 36159
Muzeul Țăranului Român, Hotarele 36403

Russia

Adygejskij Respublikanskij Kraevedčeskij Muzej, Majkop 37123
Amurskij Gorodskoj Kraevedčeskij Muzej, Amursk 36756
Archangelskij Gosudarstvennyj Muzej Zodčestva i Narodnogo Ikusstva - Malye Karely, Archangelsk 36765
Architekturno-ėtnografičeskij Muzej Derevjannogo Zodčestva Chochlovka, Perm 37543
Chabarovskij Gosudarstvennyj Muzej Dalnego Vostoka im. N.I. Grodekova, Chabarovsk 36856
Chudožestvennyj Muzej im. M.S. Tuganova, Vladikavkaz 37928
Dom-Muzej I.E. Repina, Žigulevsk 37987
Ėtnografičeskij Muzej Narodov Zabajkalja, Ulan-Udė 37893
Ėtnografičeskij Muzej pod Otkrytym Nebom Torum Maa, Chanty-Mansijsk 36860
Ėvenkijskij Okružnoj Kraevedčeskij Muzej, Tura 37877
Gosudarstvennyj Istoriko-architekturnyj i Chudožestvennyj Muzej-zapovednik Drevnij Derbent, Derbent 36874
Gosudarstvennyj Moskovskij Muzej Naivnogo Iskusstva, Moskva 37196
Gosudarstvennyj Nacionalnyj Muzej Ust-Ordynskogo Burjatskogo Avtonomnogo Okruga, Ust-Ordynsk 37913
Jakutskij Gosudarstvennyj Muzej Istorii i Kultury Narodov Severa im. Jaroslavskogo, Jakutsk 36962
Kargopolskij Gosudarstvennyj Istoriko-Architekturnyj i Chudožestvennyj Muzej, Kargopol 37003
Kirillo-Belozerskij Istoriko-Architekturnyj i Chudožestvennyj Muzej-Zapovednik, Kirillov 37034
Kulturno-Vystavočnyj Kompleks Sokolniki, Moskva 37223
Moskovskij Muzej-usad'ba Ostankino, Moskva 37259
Muzej Istorii Narodnych Chudožestvennych Promyslov, Nižnij Novgorod 37456
Muzej Narodnogo Iskusstva, Moskva 37338
Muzej Narodnogo Tvorčestva, Penza 37534
Muzej Narodnogo Tvorčestva, Uljanovsk 37904
Muzej Remesla, Architektury i Byta, Kaluga 37000
Muzej Vjatskie Narodnye Chudožestvennye Promysly, Kirov 37041
Nacionalnyj Muzej Respubliki Baškortostan, Ufa 37889
Rybinskij Gosudarstvennyj Istoriko-architekturnyj i Chudožestvennyj Muzej-zapovednik, Rybinsk 37606
Saratovskij Ėtnografičeskij Muzej, Saratov 37752
Tjumenskij Oblastnoj Muzej Izobrazitelnych Iskusstv, Tjumen 37853
Tuvinskij Respublikanskij Kraevedčeskij Muzej im. Aldan Maadyr, Kyzyl 37099
Udmurtskij Respublikanskij Muzej Izobrazitelnych Iskusstv, Iževsk 36956
Vserossijskij Muzej Dekorativno-Prikladnogo i Narodnogo Iskusstva, Moskva 37412
Vystavočnyj Zal Narodnaja Galereja, Moskva 37426
Zvenigorodskij Istoriko-architekturnyj i Chudožestvennyj Muzej, Zvenigorod 37991

Serbia

Galerija Naivnih Slikara i Etnološka Zbirka, Kovačica 38106
Muzej Naivne Umetnosti Ilianum, Šid 38141

Slovakia

Historické Múzeum, Bratislava 38213
Kammerhof - Baníctvo na Slovensku, Banská Štiavnica 38191
Múzeum Kultúry Karpatských Nemcov, Bratislava 38222
Múzeum Kultúry Maďarov na Slovensku, Bratislava 38223
Múzeum Ždiarsky Dom, Ždiar 38388

Spain

Museo Parroquial, Écija 39258
Museo Taurino, Córdoba 39208
Museu d'Art Popular, Horta de Sant Joan 39393
Museu de l'Empordà, Figueres 39298
Museu Municipal de Molins de Rei, Molins de Rei 39658

Sri Lanka

Folk Museum, Anuradhapura 40367

Sweden

Dalarnas Museum, Falun 40515

Switzerland

Musée du Vieux-Pays d'Enhaut, Château-d'Oex 41292
Museo di Vallemaggia, Giumaglio 41435
Museum im Lagerhaus, Sankt Gallen 41807
Museum Murten, Murten 41668
Trummlehus, Langenthal 41528

Syria

Museé d'Art Populaire du Hauran, Bosra 42129
Popular Traditions Museum Qasrelazem, Damascus 42133

Taiwan

Anli Art and Culture Museum, Shengang 42324
Dr. Lin Sian Tang Gallery, Wufeng 42459
Hengchuen Peninsula Aboriginal Arts and Crafts Museum, Mudan 42278
Huang Guei-li Memorial Hall, Banciao 42144
Lanyang Museum of Yilan County, Yilan 42467
Lugang Folk Arts Museum, Lugang 42257
Taiwan Folk Arts Museum, Taipei 42431
Ying-Min JHS Folk Art Museum, Kaohsiung 42246

Tunisia

Musée des Arts Populaires et Traditions Dar Jallouli, Sfax 42652
Musée El Kobba des Arts et Traditions, Sousse 42657

Turkey

1997 Avrupa'da Yılın Müzesi, Atpazarı 42718
Adana Etnografya Müzesi, Adana 42673

Uganda

Makerere Art Gallery, Kampala 42992

Ukraine

Černivec'kyj Muzej Narodnoï Architektury ta Pobutu, Černivci 43014
Deržavnyj Muzej Ukraïns'koho Narodnoho Dekoratyvnoho Mystectva, Kyïv 43078
Kanivs'kyj Muzej Narodnoho Dekoratyvnogo Mystectva, Kaniv 43059
Muzej Ivana Gončara, Kyïv 43091
Muzej Narodnogo Dekorativnogo Mystectva, Černigiv 43006
Muzej Narodnoho Mystectva Gucul'ščini ta Pokuttja im. J. Kobryns'koho, Kolomija 43072
Muzej Narodnoï Architektury i Pobutu u Lvovi, Lviv 43134
Muzej Narodnoï Architektury ta Pobutu Ukraïny, Kyïv 43098
Nacionalnij Muzej u Lvovi, Lviv 43137
Nacionalnij Muzej u Lvovi - Viddil Narodnogo Mystectva, Lviv 43138
Zakarpats'kyj Muzej Narodnoï Architektury ta Pobutu, Užgorod 43199

United Kingdom

Polish Cultural Institute, London 44956

United States of America

A:Shiwi A:Wan Museum, Zuni 54375
Abby Aldrich Rockefeller Folk Art Museum, Williamsburg 54164
Academy Art Museum, Easton 48267
African American Historical and Cultural Society Museum, San Francisco 52764
American Folk Art Museum, New York 51113
American Museum of Straw Art, Long Beach 50172
Americana Museum, El Paso 48315
Art and Culture Center of Hollywood, Hollywood 49322
The Arts Center of the Capital Region, Troy 53573
Asian-American Arts Centre, New York 51129
Beck Cultural Exchange Center, Knoxville 49811
Bedford Gallery, Walnut Creek 53802
Belair Mansion, Bowie 46967
Berlin Art and Historical Collections, Berlin 46763
Birmingham Bloomfield Art Center, Birmingham 46828
Blue Ridge Institute and Museum, Ferrum 48538
Boone County Historical Center, Boone 46903
Bowers Museum of Cultural Art, Santa Ana 52852
Branigan Cultural Center, Las Cruces 49957
Brattleboro Museum and Art Center, Brattleboro 46998
Brookings Arts Council, Brookings 47058
Brooklyn Museum, Brooklyn 47069
Brueckner Museum of Starr Commonwealth, Albion 46173
Bulloch Hall, Roswell 52466
CAFAM, Los Angeles 50213
Center for Intuitive and Outsider Art, Chicago 47460
CFCC Webber Center, Ocala 51481
Chase Home Museum of Utah Folk Art, Salt Lake City 52688
Chattahoochee Valley Art Museum, La Grange 49840
Columbus Cultural Arts Center, Columbus 47751
Cuban Museum of Arts and Culture, Miami 50623
Delta Cultural Center, Helena 49237
Denver Art Museum, Denver 48072
Durham Center Museum, East Durham 48238
East Martello Museum, Key West 49758
Eide-Dalrymple Gallery, Sioux Falls 53119
Emerson Gallery, Clinton 47659
Eva and Morris Feld Gallery, New York 51171
Fayette Art Museum, Fayette 48524
Fenimore Art Museum, Cooperstown 47789
Folk Art and Photography Galleries, Atlanta 46436
Folk Art Society of America, Richmond 55134
Four Rivers Cultural Center, Ontario 51570
Fowler Museum of Cultural History, Los Angeles 50223
Fresno Metropolitan Museum, Fresno 48791
Goschenhoppen Folklife Museum, Green Lane 49042
The Greater Reston Arts Center, Reston 52299
Hambidge Center for the Creative Arts and Sciences, Rabun Gap 52221
Havre de Grace Decoy Museum, Havre de Grace 49219
Hellenic Museum and Cultural Center, Chicago 47479
Henry County Museum and Cultural Arts Center, Clinton 47655
Heritage Center, Dayton 47991
Hiwan Homestead Museum, Evergreen 48460
Hopi Cultural Center Museum, Second Mesa 53022
Hungarian Folk-Art Museum, Port Orange 52039
Huntington Museum of Art, Huntington 49435
International Institute of Metropolitan Detroit, Detroit 48122
Jaffrey Civic Center, Jaffrey 49588
Jamaica Center for Arts, Jamaica 49590
Jimmie Rodgers Museum, Meridian 50599
Kentucky Folk Art Center, Morehead 50841
Kentucky Museum of Arts and Design, Louisville 50288
Krikorian Gallery, Worcester 54300
The Kurdish Museum, Brooklyn 47077
La Raza Galeria Posada, Sacramento 52508
Larson Gallery, Yakima 54319
Liberty Village Arts Center and Gallery, Chester 47431
Lladro Museum, New York 51204
Lyndon House Art Center, Athens 46413
McKissick Museum, Columbia 47728
Mattye Reed African Heritage Center, Greensboro 49066
Meadow Farm Museum, Glen Allen 48917
Mennello Museum of American Folk Art, Orlando 51591
Mercer Museum of the Bucks County Historical Society, Doylestown 48180
Michigan Historical Museum, Lansing 49943
Miles B. Carpenter Museum, Waverly 53998
Mingei International Museum, San Diego 52739
Mingei International Museum, Escondido 48420
Mission Cultural Center for Latino Arts, San Francisco 52784
Mount Dora Center for the Arts, Mount Dora 50876
El Museo Latino, Omaha 51553
Museum in the Community, Hurricane 49452
Museum of Craft and Folk Art, San Francisco 52786
Museum of Cultural and Natural History, Mount Pleasant 50884
Museum of International Folk Art, Santa Fe 52882
Museum of the American Hungarian Foundation, New Brunswick 51035
National Hall of Fame for Famous American Indians, Anadarko 46281
National Hispanic Art Museum, Albuquerque 46184
Native American Museum, Terre Haute 53481
New England Carousel Museum, Bristol 47028
New Visions Gallery, Marshfield 50513
The Octagon Center for the Arts, Ames 46259
Old West Wax Museum, Thermopolis 53493
Oscar Howe Art Center, Mitchell 50764
Ozark Folk Center, Mountain View 50895
Parson Fisher House, Blue Hill 46874
Pelham Art Center, Pelham 51749
Pence Gallery, Davis 47977
Pennsylvania Dutch Folk Culture Society Museum, Lenhartsville 50023
Petterson Museum of Intercultural Art, Claremont 47573
Philbrook Museum of Art, Tulsa 53610
Plaza de la Raza - Boathouse Gallery, Los Angeles 50261
Polish American Museum, Port Washington 52048
Polish Museum of America, Chicago 47507
Ralph Milliken Museum, Los Banos 50271
Reginald F. Lewis Museum of Maryland African American History and Culture, Baltimore 46586
Rialto Museum, Rialto 52303
Richland County Museum, Lexington 50066
Rockford Art Museum, Rockford 52416
Romanian Ethnic Art Museum, Cleveland 47639
Rose Center Museum, Morristown 50859
Sautee-Nacoochee Museum, Sautee 52937
Schweinfurth Memorial Art Center, Auburn 46470
Scottsdale Museum of Contemporary Art, Scottsdale 52977
Sedona Arts Center, Sedona 53027
Serra Museum, San Diego 52755
The Shaker Museum, Old Chatham 51534
Shoshone Tribal Cultural Center, Fort Washakie 48697
Sioux Indian Museum, Rapid City 52254
Smithsonian American Art Museum, Washington 53917
Snug Harbor Cultural Center, Staten Island 53267
South Arkansas Arts Center, El Dorado 48310
South Carolina Artisans Center, Walterboro 53807
Southern Highland Craft Guild at the Folk Art Center, Asheville 46376
Southern Ohio Museum and Cultural Center, Portsmouth 52103
Southern Vermont Art Center, Manchester 50428
Southwest Museum of the American Indian, Los Angeles 50265
Stephanie Ann Roger Gallery, Frostburg 48802
Stewart Indian Cultural Center, Carson City 47262
The Studio Museum in Harlem, New York 51264
Swedish American Museum Association of Chicago Museum, Chicago 47516
Tennessee Valley Art Center, Tuscumbia 53625
Torpedo Factory Art Center, Alexandria 46214
Tucumcari Historical Research Institute Museum, Tucumcari 53604
Ukrainian-American Museum, Hamtramck 49155
Ukrainian Museum, Cleveland 47643
The Ukrainian Museum, New York 51274
Ukrainian National Museum, Chicago 47520
Vermont Folklife Center, Middlebury 50654
Virginia Museum of Fine Arts, Richmond 52343
Washington Pavillion of Arts and Science, Sioux Falls 53123
Wendell Gilley Museum, Southwest Harbor 53190
Wheelwright Museum of the American Indian, Santa Fe 52892
White River Museum, Meeker 50560

Uruguay

Museo del Indio y del Gaucho, Tacuarembó 54435

Venezuela

Museo de Arte Popular Venezolano, Juangriego 54576

Vietnam

Viet Nam National Fine Arts Museum, Ha Noi 54617

Folklore

Albania

Muzeu i Filmit Shkodër, Shkodër 00029

American Samoa

Jean P. Haydon Museum, Pago Pago 00084

Argentina

Centro Mundo Aborigen, Córdoba 00318
Museo Centro Cultural Pachamama, Amaicha del Valle 00122
Museo de Artesanías Tradicionales Folklóricas, Corrientes 00346
Museo de Esculturas Luis Perlotti, Buenos Aires 00192
Museo de la Artesanía Tradicional Juan Alfonso Carrizo, La Plata 00412
Museo del Traje, La Rioja 00419
Museo Etnográfico y Colonial Juan de Garay, Santa Fé 00615
Museo Folklórico, La Rioja 00420
Museo Gauchesco Ricardo Güiraldes, San Antonio de Areco 00566
Museo Histórico y de la Tradiciones Populares de la Boca, Buenos Aires 00244
Museo La Cinacina, San Antonio de Areco 00567
Museo Manoblanca, Buenos Aires 00250
Museo Mundo Yámana, Ushuaia 00662
Museo Rafael Escriña, Santa Clara de Saguier 00612
Museo Rauzi, Palmira 00483
El Museo Viajero, Buenos Aires 00276

Armenia

Dilijan Historic Center, Dilijan 00693
Gavar Local Tradition Museum, Gavar 00701
Goris Local Tradition Museum, Goris 00703
Kapan Local Tradition Museum, Kapan 00712
Lori Local Tradition Museum, Vanadzor 00719
Regional Museum of Yeghegnadzor, Yeghegnadzor 00722

Australia

Albany Residency Museum, Albany 00785
Australian Children's Folklore Collection, Melbourne 01274
Balaklava Museum - Centenary Hall, Balaklava 00822
Barossa Valley Historical Museum, Tanunda 01570
Bowraville Folk and Military Museum, Bowraville 00884
Brambuk Living Cultural Centre, Halls Gap 01141
Carbethon Folk Museum and Pioneer Village, Crows Nest, Queensland 01013
Channel Folk Museum, Lower Snug 01243
City of Armadale History House Museum, Armadale, Western Australia 00812
Clarendon House, Evandale 01077
Dawson Folk Museum, Theodore 01584
Daylesford and District Historical Museum, Daylesford 01031
Deloraine Museum, Deloraine 01032
Dowerin District Museum, Dowerin 01042
Dreamtime Cultural Centre, Rockhampton 01464
Franklin House, Launceston 01221
Gawler Museum, Gawler 01100
Gippsland Heritage Park, Moe 01309
Kempsey Museum, Kempsey 01193
Latrobe Court House Museum, Latrobe 01219
Mingenew Museum, Mingenew 01306
Mount Laura Homestead Museum, Whyalla 01648
Newcastle Gaol Museum, Toodyay 01590
Nor West Bend Museum, Morgan 01320
Old Mornington Post Office Museum, Mornington 01323
Old Wool and Grain Store Museum, Beachport 00843
Penneshaw Maritime and Folk Museum, Penneshaw 01409
Pioneer Settlement Museum, Swan Hill 01540
Port Pirie National Trust Museum, Port Pirie 01435
Rockcavern, Beechworth 00851
Settlers Museum, Brewarrina 00891
Tibooburra Local Aboriginal Land Council Keeping Place, Tibooburra 01588
Wellington Courthouse Museum, Wellington 01640
Wonnerup House and Old School, Busselton 00928
Woodloes Homestead Folk Museum, Cannington 00952
Zara Clark Folk Museum, Charters Towers 00964

Austria

Dr. Eduard Benninger Heimathaus, Bad Wimsbach-Neydharting 01770
Faschings- und Brauchtumsmuseum, Knittelfeld 02206
Gailtaler Heimatmuseum-Sammlung Georg Essl, Hermagor 02071
Galerie Volkskundesammlung Piaty, Waidhofen an der Ybbs 02888
Handwerkskunst-Museum, Rattenberg, Inn 02565
Heimat- und Landlermuseum, Bad Goisern 01748
Heimathaus mit Photographiemuseum, Mariazell 02357
Heimatkundliche Sammlung Strick, Bad Mitterndorf 01760
Himmelkeller, Kronberg im Weinviertel 02239
Krahuletz-Museum, Eggenburg 01843
Krippenmuseum, Fulpmes 01911
Mährisch-Schlesisches Heimatmuseum, Klosterneuburg 02201
Mühlviertler Schlossmuseum Freistadt, Freistadt 01904
Museum der Heimatvertriebenen, Vöcklabruck 02866
Museum im Rathaus, Gleisdorf 01934
Museum Lauriacum, Enns 01870
Museum Schloss Bruck, Lienz 02302
Österreichisches Museum für Volkskunde, Wien 03049
Ortskundliches Museum Jois, Jois 02145
Peter Rosegger-Geburtshaus, Alpl 01707
Schloßmuseum Landeck, Landeck 02254
Stadtmuseum Judenburg, Judenburg 02146
Urgeschichtssammlung und Heimatmuseum, Jedenspeigen 02140
Volkskundemuseum, Mödling 02398
Volkskundliches Museum, Deutschlandsberg 01817
Volkskundliches Museum Alois Alphons, Hirschwang 02082
Weinlandmuseum, Asparn an der Zaya 01735

Barbados

Heritage Park Folk Museum, Saint Phillip 03243

Belgium

De Botermolen, Keerbergen 03750
Domein De Locht, Duffel 03564
Gemeentelijk Museum, Melle 03866
Geschiedkundig Museum Mesen, Mesen 03871
Heemkundig Museum Tempelhof, Beerse 03354
Heemkundig Museum Wetteren, Wetteren 04149
Heemmuseum Die Swane, Heist-op-den-Berg 03691
Heemmuseum Eeklo, Eeklo 03575
Hopmuseum De Stadsschaal, Poperinge 03953
Het Huis van Alijn, Gent 03628
La Maison Tournaisienne - Musée de Folklore, Tournai 04082
Musée Communal de Folklore Léon Maes, Mouscron 03895
Musée Communal Herstalien d'Archéologie et de Folklore, Herstal 03701
Musée de Folklore, Nismes 03913
Musée de la Porte, Tubize 04097
Musée du Folklore et de la Vie Montoise, Mons 03886
Musée du Vieux Nimy, Mons 03887
Musée Rue Etienne Lenoir, Mussy-la-Ville 03897
Museum De Kijkuit, Merksem 03870
Museum voor Folklore, Zottegem 04168
Museum voor Heem- en Oudheidkunde, Kontich 03757
Museum voor Volkskunde, Dendermonde 03541
Openluchtmuseum Bachten de Kupe, Alveringem 03269
Provinciaal Openluchtmuseum Bokrijk, Genk 03621
Rietgaverstede, Nevele 03912
Stedelijk Museum Aarschot, Aarschot 03260
Stedelijk Museum Het Toreke, Tienen 04070
Stedelijk Museum voor Folklore en Regionale Geschiedenis, Ronse 03986
Vi Gurnis-Vieux Greniers, Bertrix 03366

Bolivia

Museo Casa de la Cultura Simón Patiño, Oruro 04230
Museo de Textiles Andinos Bolivianos, La Paz 04217
Museo Nacional de Etnografía y Folklore, La Paz 04224

Brazil

Museu da Cultura e Arte Popular, João Pessoa 04512
Museu de Arte e Cultura Canjira, Itaobim 04498
Museu Folclorico da Divisão de Discoteca e Biblioteca de Música de São Paulo, São Paulo 04900
Museu Histórico e de Artes, Santana do Ipanema 04815
Museu Municipal do Folclore, Penápolis 04624

Canada

Basilian Fathers Museum, Mundare 06402
Bathurst Heritage Museum Cultural Centre / Musée de Bathurst/Musée Centre Culturel, Bathurst 05397
Black Creek Pioneer Village, North York 06486
Black Cultural Centre for Nova Scotia, Dartmouth 05674
Centre Culturel et Patrimonial la Poudrière de Windsor, Windsor, Québec 07348
Centre Franco-Ontarien de Folklore, Sudbury 07027
Chinese Cultural Centre of Vancouver, Vancouver 07208
Doon Heritage Crossroads, Kitchener 06100
Dr. Henry N. Payne Community Museum, Cow Head 05653
Great Hall of the Clans Museum, English Town 05779
Heritage Park - Historical Village, Calgary 05520
Highland Village Museum, Iona 06019
Kinistino and District Pioneer Museum, Kinistino 06094
Kitimat Centennial Museum, Kitimat 06105
Langley Centennial Museum and National Exhibition Centre, Fort Langley 05813
Latvian History Museum, Toronto 07119
Lougheed House Conservation Society, Calgary 05524
McCord and District Museum, McCord 06209
Maison Cyr - Musée de Chapelle, Saint-Basile 06803
Mountain View Doukhobor Museum, Grand Forks 05895
Le Musée Acadien, Caraquet 05554
Ukrainian Museum, Gardenton 05850
Ukrainian Peoples Home of Ivan Franco, Angusville 05341
U'mista Cultural Centre, Alert Bay 05315

Central African Republic

Musée Ethnographique Regional, Bouar 07422

Chile

Museo Folklórico Araucano de Cañete Juan A. Ríos M., Cañete 07431

China

Baimasi Han Wei Gucheng Display Center, Luoyang 07982
Beijing Changcheng Hua Ren Huai, Beijing 07509
Beijing Cultural Relic Protection Foundation, Beijing 07511
Beijing Folklore Museum, Beijing 07513
Dunhuang Glaswegiansu Guan, Dunhuang 07676
Folk Custom Museum, Xiangfen 08369
Fuzhou City Hualin Si Museum, Fuzhou 07708
Huangping Nationality Holiday Museum, Huangping 07832
Law Uk Folk Museum, Hong Kong 07807
Li Zong Ren Display Center, Guilin 07743
Liangzhu Culture Museum, Hangzhou 07763
Longgang Ke Jia Folklorish Museum, Shenzhen 08222
Longjing Koreazu Glaswegiansu Guan, Longjing 07970
Longquanyi Mingshu Wangling Museum, Chengdu 07618
Luoyang Glaswegiansu Guan, Luoyang 07984
Museum of Dr. Sun Yat-Sen, Zhongshan 08506
Nanjing Folklorish Museum, Nanjing 08043
Ping Jin Zhan Yi, Old Zhi Display Center, Yangliuqing 08423
Pinggu Shangzhai Cultural Display Center, Beijing 07561
Qi County Folkways Museum, Qixian 08127
Qingdao City Folklorish Museum, Qingdao 08109
Shan Rong Cultural Display Center, Beijing 07566
Shanghai Museum of Folk Collectibles, Shanghai 08186
Sui Tangyao Zhi Display Center, Chengdu 07621
Suzhou Folkways Museum, Suzhou 08256
Tangjia Glaswegiansu Guan, Xunyi 08415
Tianjin Glaswegiansu Guan, Tianjin 08284
Wangmiao Display Center, Dujiangyan 07675
Wuxi Glaswegianjian Guan, Wuxi 08343
Xijiashan Glaswegiansu Guan, Jiang'an 07855
Yanbian Museum, Yanji 08430
Yingde Cultureal Museum, Qingyuan 08119
Yuzhang Glaswegiansu Bowu Yuan, Nanchang 08033

Colombia

Casa de los Abuelos, Sonsón 08761
Museo de Antropología, Barranquilla 08539
Museo de Trajes Regionales de Colombia, Bogotá 08575
Museo Etnográfico de Colombia, Bogotá 08579
Museo Histórico de Cartagena Casa de la Inquisición, Cartagena 08619

Costa Rica

Museo de Cultura Popular Costa Rica, Heredia 08809

Croatia

Kulturno Povijesni Muzej, Dubrovnik 08865
Zavičajna Zbirka Đelekovec, Đelekovec 08859

Cuba

Galería Bienes Culturales, Victoria de las Tunas 09344
Museo de las Parrandas de Remedios, Remedios 09265
Museo de los Orishas, La Habana 09138
Museo Etnográfico, Madruga 09199
Museo Histórico Municipal de Regla, Regla 09264

Czech Republic

Kotulova Dřevěnka, Havířov 09508
Krkonošské Muzeum - Tri historické domy, Vrchlabí 09915
Muzeum Českého Lesa, Tachov 09859
Muzeum Mladaboleslavska, Mladá Boleslav 09657
Muzeum Romské Kultury, Brno 09437
Národopisné Oddělení, Praha 09782

Denmark

Haastrup Folkemindesamling, Faaborg 10011
Hjerl Hedes Frilandsmuseum, Vinderup 10318
Museet for Holbæk og Omegn, Holbæk 10087
Odsherreds Kulturhistoriske Museum, Nykøbing Sjælland 10206

Dominican Republic

Museo Tomás Morel, Santiago de Los Caballeros 10340

East Timor

East Timorese Cultural Centre Museum Collection, Dili 10364

Ecuador

Museo Azuayo del Folklore, Cuenca 10383
Museo de Arte Moderno de la Casa de la Cultura Ecuatoriana, Quito 10437

Egypt

Folklore Museum, Cairo 10495

Estonia

Eesti Kirjandusmuuseum, Tartu 10645
Hanila Muuseum, Hanila 10568

Finland

Riuttalan Talonpoikaismuseo, Karttula 10922
Tenala Hembygdsmuseum, Tenala 11373
Varpaisjärven Kotiseutumuseo, Varpaisjärvi 11456

France

Écomusée d'Alzen, Alzen 11563
Ecomusée de la Ferme et des Vieux Métiers, Lizio 13791
Ecomusée de Savigny-le-Temple, Savigny-le-Temple 15679
Ecomusée des Pays de l'Oise, Beauvais 11908
Maison des Archers, Quimperlé 14972
Maison des Traditions, Chassignolles 12431
Maison du Platin, La Flotte 13370
Musée Agathois, Agde 11491
Musée Basque et de l'Histoire de Bayonne, Bayonne 11878
Musée Bourguignon Perrin de Puycousin, Tournus 15958
Musée Comtadin, Carpentras 12295
Musée de la Corse, Corte 12635
Musée de l'Abbaye, Gaillac 13025
Musée de l'Image Populaire, Pfaffenhoffen 14810
Musée de Plein Air du Quercy de Cuzals, Sauliac-sur-Célé 15655
Musée de Sologne, Romorantin-Lanthenay 15091
Musée Départemental de Flandre, Cassel 12304
Musée des Arts et Traditions Populaires, Le Sel-de-Bretagne 13650
Musée des Tradition de l'île, Noirmoutier-en-l'Ile 14467
Musée des Traditions et Arts Normands, Martainville-Epreville 14017
Musée d'Ethnographie et d'Histoire de Conflans, Albertville 11535
Musée du Berry, Bourges 12118
Musée du Chablais, Thonon-les-Bains 15896

Musée du Rouergue, Espalion 12855
Musée du Santon et des Traditions de Provence, Fontaine-de-Vaucluse 12960
Musée du Vieil Aix, Aix-en-Provence 11517
Musée Émile Guillaumin, Ygrande 16307
Musée Municipal Méditerranéen de Cassis, Cassis 12305
Musée Provençal et Folklorique Taradeau, Les Arcs 13685
Musée Traditions et Vie, Châtillon-sur-Chalaronne 12470
Musées Départementaux Albert Demard, Champlitte 12393

French Guiana

Musée des Cultures Guyanaises, Cayenne 16317

Germany

Bauernhofmuseum Jexhof, Schöngeising 21609
Bergbaumuseum, Aldenhoven 16461
Bienwaldmuseum, Kandel 19425
Brauchtums- und Trachtenpuppenmuseum der Banater Schwaben e.V., Würzburg 22707
Burgmuseum Schönfels, Schönfels 21608
Donauschwäbisches Zentralmuseum, Ulm 22160
Dorfmuseum mit Pfarrer-Lehmann Stube im Bauernkrieghaus, Landau in der Pfalz 19788
Druckereimuseum und Schaudepot der Museen der Stadt Erfurt im Benary-Speicher, Erfurt 18200
Erstes Niederrheinisches Karneval-Museum, Duisburg 18015
Erzgebirgisches Freilichtmuseum, Seiffen / Erzgebirge Kurort 21729
Fastnachtsmuseum, Sankt Ingbert 21499
Fastnachtsmuseum Fasenickl, Kipfenberg 19540
Heesberg Museum, Gevensleben 18629
Heimatmuseum, Duderstadt 17981
Heimatmuseum Haus der Südostdeutschen, Böchingen 17422
Heimatmuseum Warnemünde, Rostock 21392
Josefine-Weihrauch-Heimatmuseum Neudenau, Neudenau 20603
Lahnsteiner Fastnachtsmuseum, Lahnstein 19766
Landesmuseum für Kunst und Kulturgeschichte, Schleswig 21566
Landschaftsmuseum Angeln, Langballig 19807
Landschaftsmuseum Schönhengstgau, Göppingen 18675
Loreley-Museum im historischen Stadtturm, Sankt Goarshausen 21497
Museum der Alltagskultur - Schloss Waldenbuch, Waldenbuch 22287
Museum der Stadt Hagenow, Hagenow 18862
Museum für ländliches Kulturgut im Dorfstadel Leupolz, Wangen im Allgäu 22326
Museum für Thüringer Volkskunde Erfurt, Erfurt 18207
Museum im Alten Schloß, Altensteig 16503
Museum im Bierlinghaus, Bad Bayersoien 16675
Museum im Graulturm, Gau-Algesheim 18546
Museum im Pflegschloss, Schrobenhausen 21640
Museum Der Kunststall, Dahlem 17769
Museum Zunftturm, Bacharach 16654
Niederrheinisches Museum für Volkskunde und Kulturgeschichte, Kevelaer 19524
Njepila-Hof Rohne, Rohne 21363
Potsdam-Museum, Potsdam 21103
Privatsammlung Burg Runkel, Runkel 21450
Reinhard-Blauth-Museum, Weilerbach 22390
Rheinisches Fastnachtsmuseum Koblenz, Koblenz 19589
Riedmuseum, Rastatt 21184
Riesengebirgsmuseum, Marktoberdorf 20221
Sammler- und Glücksschweinmuseum, Bad Wimpfen 16870
Sammlung Volkskunde, Kassel 19475
Schifffahrtsmuseum, Nierstein 20721
Schwäbisches Volkskundemuseum Oberschönenfeld, Gessertshausen 18627
Stadtmuseum Rastatt, Rastatt 21186
Städtische Galerie Rastatt, Rastatt 21188
Städtisches Museum, Aschersleben 16596
Trachtenhaus Jatzwauk, Hoyerswerda 19283
Töötten-Museum, Mettingen 20311
Volkskunde Museum Schleswig, Schleswig 21570
Volkskundemuseum, Ostrach 20944
Volkskundemuseum Treuchtlingen, Treuchtlingen 22091
Volkskundemuseum Wyhra, Borna 17486
Volkskundliches Freilichtmuseum im Stadtpark Speckenbüttel, Bremerhaven 17575
Waldarbeits-Museum und Historische Wappenschmiede, Elmstein 18127

Greece

Folklore Museum, Pogoniani 23151
Folklore Museum at Drama, Drama 22988
Folklore Museum of Folegandros, Folegandros 23003
Folklore Museum of Polygyros, Polygyros 23153
Folklore Museum of the Culture Club Florina, Florina 23001
Hellenic Folklore Research Centre, Athinai 22913
Historical Museum of Crete, Iráklion 23019
Mouseio Isrorias tis Ellinikis Endymasias, Athinai 22921
Museum of Folklore and Popular Art, Sifnos 23186
Museum of Folklore Arts and Culture, Salamina 23174
Mykonos Folklore Museum, Mykonos 23104
Sarakatsani Folklore Museum, Serres 23181

Guam

Faninadahen Kosas Guahan, Adelup 23284

Guatemala

Centro de Estudios Folklóricos, Guatemala 23297

Hungary

Bosnyák Helytörténeti és Néprajzi Szoba, Áta 23339
Erdős Kamill Cigány Múzeum, Pécs 23760
Lábasház, Sopron 23813
Magyar Fuszerpaprika Múzeum, Kalocsa 23626
Mária-ruhák Múzeum, Andocs 23335
Meseház, Békéscsaba 23369
Német Nemzetiségi Gyűjtemény, Gyula 23599
Palóc Néprajzi Magángyűjtemény és Palóc Baba Kiállítás, Mátrafüred 23695
Szlovák Emlékház, Tótkomlós 23936
Szlovák Tájház, Tótkomlós 23937

Iceland

Byggðasafn Húnvetninga og Strandarmanna, Brú 23990
Pjóðmenningarhúsið, Reykjavik 24029
Stofnun Arna Magnússonar, Reykjavik 24037

India

Heras Institute of Indian History and Culture, Mumbai 24317
Museum of Art, Archaeology and Folklore, Mysore 24330

Iran

Babol Museum, Babol 24597
Hazrati Museum, Semnan 24643
Kandelous Museum, Kandelous 24614
Shahrood Museum, Semnan 24644

Ireland

Knock Museum, Knock 24860
Muckross House Gardens and Traditional Farms, Killarney 24850
National Museum of Ireland - Country Life, Castlebar 24756
Noreview Folk and Heritage Museum, Bennettsbridge 24744

Israel

Beit Hameiri, Zefat 25132
Kibbutz Negba Museum, Kibbutz Negba 25052

Italy

Civico Museo del Merletto al Tombolo, Rapallo 27365
Museo Civico delle Arti e Tradizioni Popolari, Micigliano 26671
Museo degli Usi e Costumi della Gente di Romagna, Santarcangelo di Romagna 27725
Museo degli Usi e Costumi della Valle di Goima, Zoldo Alto 28319
Museo degli Usi e Costumi delle Genti dell'Etna, Giarre 26335
Museo del Folklore e delle Tradizioni Popolari, Caltanissetta 25608
Museo della Civiltà Contadina Valle dell'Aniene, Roviano 27573
Museo delle Arti e delle Tradizioni Popolari, Sassoferrato 27764
Museo delle Tradizioni Popolari, Città di Castello 25908
Museo di Roma in Trastevere, Roma 27500
Museo Etnografico e del Folklore Valsesiano, Borgosesia 25509
Museo Etnografico Siciliano Giuseppe Pitrè, Palermo 27076
Museo Fittile, Ripatransone 27418
Museo Internazionale del Folklore e Civiltà Contadina, Atina 25320
Museo Statale, Anghiari 25238
Ultner Talmuseum, Ulten 28116

Japan

Ajia-no Mingeikan, Noda 29079
Arita-machi Minzoku Hakubutsukan, Arita 28363
Atsukeshi-machi Mingeikan, Atsukeshi 28383
Chigira Mingeikan, Katashina 28679
Chita-shi Rekishi Minzoku Hakubutsukan, Chita 28405
Edo-Tokyo Museum, Tokyo 29384
Ehime Bunkakan, Imabari 28567
Fukuchiyama Shato-Mingeikan, Fukuchiyama 28424
Gifu-ken Hakubutsukan, Seki 29240
Hida Minzoku Kokokan, Takayama 29337
Hondo-shiritsu Mingeikan, Hondo 28548
Imari-shi Rekishi Minzoku Shiryokan, Imari 28569
Itayanagi-machi Mingeikan, Itayanagi 28592
Itsukaichi Folklore Museum, Akiruno 28352
Izumi-mura Rekishi no Sato, Izumi, Fukui 28612
Izumo Mingeikan, Izumo 28616
Kaminokuni Mingeikan, Kaminokumi 28648
Kawasaki-shiritsu Nihon Minkaen, Kawasaki 28694
Kunisaki-machi Rekishi Minzoku Shiryokan, Kunisaki 28781
Kyoto-kenritsu Tango Minzoku Shiryokan, Miyau 28926
Kyoto-kenritsu Yamashiro Minzoku Shiryokan, Yamashiro 29583
Matsumoto-shiritsu Hakubutsukan, Matsumoto 28895
Minzoku Shiryokan, Miyama 28924
Moromi Mingeikan, Okinawa 29106
Murakami Folk Museum, Murakami 28945
Nagasaki Kite Museum, Nagasaki 28980
Nagasaki Rekishi no Bunka Hakubutsukan, Nagasaki 28981
Nakama-shiritsu Minzoku Hakubutsukan, Nakama 29017
Nara-kenritsu Mingeikan, Yamato-koriyama 29586
Nichi-Mongoru Mingeikan, Tantou 29351
Nihon Mingeikan, Osaka 29139
Nihon Oni Hakubutsukan, Oe 29089
Obu-shiritsu Mingeikan, Obu 29086
Takayama-shi Kyoudokan, Takayama 29345
Toyama Municipal Folkcraft Village, Toyama 29523
Yamagata-kenritsu Hakubutsukan, Yamagata 29579

Jordan

The Anthropological National Folklore Museum, Amman 29644
Jordan Museum of Popular Tradition, Amman 29654
Salt Folklore Museum, Salt 29667

Kenya

Lamu Museum, Lamu 29744
Narok Museum, Narok 29750

Korea, Republic

Andong Folk Museum, Andong 29765
Cheongju University Museum, Cheongju 29789
Gongju Folk Drama Museum, Gongju 29816
Hangguk Minsok-chon, Kihung 29845
Kuknip Minsok Pakmulgwan, Seoul 29900
Kyungsung Museum, Busan 29780
National Folk Museum, Seoul 29909
Onyang Minsok Pakmulgwan, Asan 29770
Samsung Foundation of Culture, Seoul 29914
Yeongil Folk Museum, Pohang 29859

Kyrgyzstan

Issyk-Kulskij Gosudarstvennyj Istoriko-Kulturnyj Muzej-Zapovednik, Čolpan-Ata 29961

Laos

Muang Sing Exhibitions Museum, Muang Sing 29969
Parc National de la Culture des Ethnies, Vientiane 29976

Lebanon

Moussa Castle Museum, Beiteddine 30163
Qozhaya Museum, Zgharta 30177

Lithuania

Antano ir Jono Juškų Etninėes Kultūros Muziejus, Vilkija 30399
Daugyvenės Kultūros Istorijos Muziejus-draustinis, Radviliškis 30341
Etnografinė Prano Dzūko Sodyba-Muziejus, Delnicai 30221
Karaimų Etnografinė Paroda, Trakai 30381
Kėdainių Krašto Muziejus, Kėdainiai 30272
Kelmės Krašto Muziejus, Kelmė 30273
Kražių Kraštotyros Muziejus, Kražiai 30287
Kretingos Muziejus, Kretinga 30288
Kupiškio Etnografijos Muziejus, Kupiškis 30292
Lietuvos Liaudis Buities Muziejus, Rumšiškės 30347
P. Stulgos Lietuvių Tautinės Muzikos Muziejus, Kaunas 30263
Šeduvos Kraštotyros Ekspozicija, Seduva 30351
Simono Stanevičiaus Memorialinė Sodyba-muziejus, Kanopenai 30237
Skriaudžių Buities Muziejus, Skriaudžiai 30370
Smilgių Etnografinė Sodyba, Smilgiai 30374
Vilniaus Krašto Etnografijos Muziejus, Nemenčinės 30318
Žvejo Etnografinė Sodyba, Nida 30323

Macedonia

Muzej na Grad Kratovo, Kratovo 30504

Madagascar

Museum du Parc Tsimbazaza, Antananarivo 30538

Malta

Folklore Museum, Victoria 30651

Mauritius

Maison Creole Eureka, Moka 30679

Mexico

Casa Museo Leonardo Valdez Esquer, Etchojoa 30865
Museo Casa de Cultura de Ixtlán, Ixtlán del Río 30963
Museo Costumbrista de Sonora, Alamos 30709
Museo Cultura y Tradición de Jalcomulco, Jalcomulco 30966
Museo de Artes y Tradiciones Populares de Tlaxcala, Tlaxcala 31476
Museo de Cultura Popular Angel Gil Hermidas, Villahermosa 31550
Museo de Culturas Populares de Chiapas, San Cristóbal de las Casas 31324
Museo de Culturas Populares e Indígenas, Obregón 31229
Museo de Culturas Populares e Indígenas de Sonora, Hermosillo 30929
Museo de la Casa de Cultura de Tepeapulco, Tepeapulco Hidalgo 31434
Museo de la Charrería, México 31081
Museo de la Cultura Potosina, San Luis Potosí 31347
Museo de las Culturas del Norte, Monte Verde 31175
Museo de las Tradiciones Potosinas, San Luis Potosí 31349
Museo del Centro Cultural Patio de los Ángeles, Guadalajara 30898
Museo del Centro Cultural Zacualpan, Comala 30788
Museo Diego Rivera Anahuacalli, México 31110
Museo Estatal de Culturas Populares, Monterrey 31185
Museo Fomento Cultural Banamex, México 31116
Museo Instituto Cultural México Israel, México 31123
Museo Juárez del Jardín Borda, Cuernavaca 30818

Namibia

Tsumeb Cultural Village, Tsumeb 31759

Netherlands

Cultureel Maçonniek Centrum Prins Frederik, Den Haag 32070
Educatieve Boerderij en Zijdemuseum de Schans, Oud-Gastel 32647
Folkloristisch Museum, Swalmen 32813
Goemanszorg - Streek- en Landbouwmuseum, Dreischor 32156
Holland Experience, Amsterdam 31847
Limburgs Museum, Venlo 32883
Museum Kameleon Dorp en Museum Hans Brinker, Terherne 32815
Nederlands Openluchtmuseum, Arnhem 31930
Het Regthuijs, Abbekerk 31789
Streekmuseum Jan Anderson, Vlaardingen 32891
Volksbuurtmuseum, Den Haag 32104
Volksbuurtmuseum Wijk, Utrecht 32858

New Zealand

Waikawa District Museum, Tokanui 33214

Nigeria

Jos National Museum and Museum of Traditional Nigerian Architecture, Jos 33330

Norway

Drangedal Bygdetun, Drangedal 33468
Fauske Museum, Fauske 33490
Finnetunet, Grue Finnskog 33523
Follo Museum, Drøbak 33469
Fossesholm Herregård, Vestfossen 33975
Hallingdal Folkemuseum, Nesbyen 33698
Hedrum Bygdetun, Kvelde 33619
Hordamuseet, Fana 33488
Jutulheimen Bygdemuseum, Vaagaa 33952
Karmsund Folkemuseum, Haugesund 33538
Kittilbu Utmarksmuseum, Østre Gausdal 33761
Kvæven Bygdetun, Tjørhom 33907

Mattisrud Småbruksmuseum, Løten 33652
Meldal Bygdemuseum, Meldal 33665
Norsk Folkemuseum, Oslo 33741
Norsk Utvandrermuseum, Ottestad 33762
Ringerikes Museum, Hønefoss 33559
Sjukehusmuseet i Molde, Molde 33679
Stalheim Folkemuseum, Stalheim 33861
Tromsø Folkemuseum, Folkeparken, Tromsø 33920
Trøndelag Folkemuseum, Trondheim 33929
Voss Folkemuseum, Skulestadmo 33842

Palestine

Artas Folklore Museum, Bethlehem 34048
Museum of Palestinian Popular Heritage, Al-Bireh 34046

Panama

Museo Afro Antillano de Panamá, Ciudad de Panamá 34064
Museo Manuel F. Zárate, Guarraré 34073

Papua New Guinea

J.K. MacCarthy Museum, Goroka 34081

Paraguay

Centro Cultural Paraguayo-Americano, Asunción 34088
Centro Cultural Paraguayo-Argentino, Concepción 34102

Philippines

Bayanihan Folk Arts Museum, Manila 34432
Museo de la Salle, Dasmariñas 34384
Museo ng Buhay Pilipino, Pasay 34473
Xavier University Museum (Museo de Oro), Cagayan de Oro City 34358

Poland

Dział Archeologii i Etnografii, Bydgoszcz 34574
Muzeum Kultury Ludowej, Węgorzewo 35295
Muzeum Ludowych Instrumentów Muzycznych, Szydłowiec 35174
Muzeum Regionalne PTTK, Muszyna 34932
Muzeum Wsi Słowińskiej w Klukach, Smołdzino 35114
Skansen Budownictwa Ludowego Zachodniej Wielkopolski, Wolsztyn 35320
Skansen Kultury Ludowej Pogórza Sudeckiego, Kudowa Zdrój 34837

Portugal

Museu do Grupo Folclórico, Santa Eulália de Barrosas 35901
Museu do Rancho Folclórico das Papoilas do Campo, Cela 35520
Museu do Rancho Folclórico de Alenquer, Alenquer 35403
Museu do Rancho Folclórico de São Miguel do Milharado, Milharado 35762
Museu do Rancho Folclórico do Calvário, Estômbar 35574
Museu do Trajo, São Brás de Alportel 35917
Museu Etnográfico da Lousã, Lousã 35741
Museu Etnográfico e Museu Casa Tradicional, Glória do Ribatejo 35626
Museu João de Deus, Lisboa 35711
Museu Popular Beatriz Costa, Malveira 35754
Núcleo Museológico da Cultura do Pastel dos Açores, Ribeira Chã 35889

Puerto Rico

Centro Ceremonial Indígena de Caguana, Utuado 36079

Romania

Expoziția de Artă Populară Iulia Folea Troceanu, Buziaș 36245
Muzeul Arta Lemnului, Câmpulung Moldovenesc 36255
Muzeul Brăilei-Secția de Etnografie și Artă Populară, Brăila 36171
Muzeul de Artă Populară, Ploiești 36530
Muzeul de Etnografie Universală Franz Binder, Sibiu 36607
Muzeul Maghiar, Bogdand 36157
Muzeul Memorial Ion Pop Reteganul, Reteag 36564
Muzeul Satului Bănățean-Pădurea Verde, Timișoara 36692

Russia

Adygejskij Respublikanskij Kraevedčeskij Muzej, Majkop 37123
Amurskij Gorodskoj Kraevedčeskij Muzej, Amursk 36756
Archangelskij Gosudarstvennyj Muzej Zodčestva i Narodnogo Ikusstva - Malye Karely, Archangelsk 36765
Architekturno-ėtnografičeskij Muzej Angarskaja Derevnja, Bratsk 36820
Bogorodickij Dvorec-muzej i Park, Bogorodick 36812
Chabarovskij Gosudarstvennyj Muzej Dalnego Vostoka im. N.I. Grodekova, Chabarovsk 36856
Detskij Muzej Dom-Skazok Žili-byli, Moskva 37159
Dom-muzej Semji Cvetaevych, Novo-Talicy 37476
Enisejskij Kraevedčeskij Muzej, Enisejsk 36915
Ėtnografičeskij Muzej Narodov Zabajkalja, Ulan-Udė 37893
Ėtnografičeskij Muzej pod Otkrytym Nebom Torum Maa, Chanty-Mansijsk 36860
Ėvenkijskij Okružnoj Kraevedčeskij Muzej, Tura 37877
Gosudarstvennyj Istoriko-ėtnografičeskij Muzej-zapovednik Šušenskoe, Šušenskoe 37811
Gosudarstvennyj Nacionalnyj Muzej Ust-Ordynskogo Burjatskogo Avtonomnogo Okruga, Ust-Ordynsk 37913
Ibresinskij Ėtnografičeskij Muzej pod Otkryrym Nebom, Ibresi 36929
Istoki, Belozersk 36800
Jakutskij Gosudarstvennyj Muzej Istorii i Kultury Narodov Severa im. Jaroslavskogo, Jakutsk 36962
Jamalo-Neneckij Okružnoj Kraevedčeskij Muzej im. I.S. Šemanovskogo, Salechard 37608
Krasnodarskij Gosudarstvennyj Istoriko-archeologičeskij Muzej-zapovednik, Krasnodar 37066
Muzej Archeologii i Ėtnografii, Ufa 37887
Muzej Tulskie Samovary, Tula 37870
Muzej "V Mire Skazok", Smolensk 37774
Nacionalnyj Muzej Respubliki Baškortostan, Ufa 37889
Tuvinskij Respublikanskij Kraevedčeskij Muzej im. Aldan Maadyr, Kyzyl 37099

Serbia

Muzej Staro Selo, Sirogojno 38142

Slovakia

Banícky Dom, Vyšna Boca 38387
Dom L'udového Bývania, Matúškovo 38303
Dom L'udového Bývania, Šala 38347
Expozícia Čičmany, Čičmany 38246
Expozícia L'udovej Architektúry a Bývania (Skanzen), Humenné 38261
Expozícia L'udovej Archtektúry a Bývania, Martovce 38302
Múzeum Kultúry Karpatských Nemcov, Bratislava 38222
Múzeum Kultúry Maďarov na Slovensku, Bratislava 38223
Múzeum Kysuckej Dediny, Nová Bystrica 38319
Múzeum Ludovita Štúra, Modra 38306
Múzeum Oravskej Dediny, Zuberec 38393
Múzeum Samuela Jurkoviča Sobotište-Nyaryiovský kaštiel', Sobotište 38354
Múzeum Ukrajinsko-Rusínskej Kultúry, Svidník 38367
Národopisná Expozícia SARM Bardejovské Kúpele, Bardejovské Kúpele 38196
Národopisné Múzeum Liptova, Liptovský Hrádok 38285
Obecné múzeum a Galéria Jana Hálu, Važec 38385

Slovenia

Muzej na Prostem, Rogatec 38508
Muzej na Prostem pri Kartuziji Pleterje, Drča 38417
Rudarska in Etnologoška Zbirka v Črni na Koroškem, Črna na Koroškem 38409

South Africa

Pioneer Museum, Silverton 38729

Spain

Museo de Artes y Costumbres Populares, Sevilla 40068
Museo de Artes y Tradiciones Populares, Madrid 39534
Museu d'Història dels Jueui, Girona 39347

Sweden

Burlövs Gamla Prästgård, Arlöv 40432
Friluftsmuseet Gammelgården, Köping 40646
Holmöns Båtmuseum, Holmön 40600
Kulturhistoriska Museet 'Kulturen i Lund', Lund 40703
Mångkulturellt Centrum, Tumba 40987
Nordiska Museet, Stockholm 40922
Norrbottens Museum, Luleå 40699
Torsby Finnkulturcentrum, Torsby 40976
Tunabygdens Gammelgård, Borlänge 40477

Switzerland

Appenzeller Brauchtumsmuseum, Urnäsch 41966
Appenzeller Volkskunde-Museum, Stein 41912
Museum Sursilvan, Trun 41953
Museum Vitznau-Rigi, Vitznau 41991
Museum Wasseramt, Halten 41470
Walserhaus Gurin, Bosco 41246
Wiggertaler Museum, Schötz 41844

Taiwan

Anli Art and Culture Museum, Shengang 42324
Bunun Culture Exhibition Center, Taoyuan 42450
Caifeng Culture Museum, Danei 42175
Children's Folklore and Folkgame Museum, Sikou 42329
Chingsi Culture Museum, Kaohsiung 42233
Folk Cultural Collection Museum, Jhongpu 42222
Foot-Binding Culture Museum, Tucheng 42457
Hakka Cultural Museum of Shigang, Shigang 42325
Hakka Culture and History Museum, Kaohsiung 42234
Hakka Relic and Folk Custom Museum, Meinong 42272
Hojing Cultural Museum, Kaohsiung 42235
Huafan Cultural Gallery, Shihtin 42326
Kaohsiung County Museum of Natural History, Lujhu 42263
Keelung Mid-Summer Ghost Festival Museum, Keelung 42247
Mazu Folklore Museum of Tianhou Temple, Lugang 42258
Mongolian and Tibetan Cultural Center, Taipei 42405
Museum of Old Western Time, Sindian 42335
Museum of Sisiyat Culture, Nanjhuang 42279
Museum of the Institute of Ethnology, Taipei 42409
Spectacular Folk Cultural Collection Museum, Longtan 42254
Taiwan Folk Culture Museum, Bali 42143
Taiwan Folklore Museum, Taichung 42355
Wulai Atayal Museum, Wulai 42462

Tanzania

Sukuma Museum, Mwanza 42518
Village Museum, Dar-es-Salaam 42513

Thailand

Hill Tribes Museum, Bangkok 42542

Tunisia

Musée Dar Hammamet, Hammamet 42627
Musée d'Art et Traditions Populaires, Tunis 42665
Musée des Arts et Traditions Populaires, Djerba 42622
Musée Sidi Ben-Aissa, Le Kef 42632

Turkey

Ereğli Müzesi, Ereğli 42772
Etnografya Müzesi, Konya 42884
Rıfat Ilgaz Kültür Merkezi, Kastamonu 42871
Yapı Kredi Vedat Nedim Tör Müzesi, İstanbul 42848

Ukraine

Černivec'kyj Muzej Narodnoï Architektury ta Pobutu, Černivci 43014
Muzej Ivana Gončara, Kyïv 43091
Muzej Narodnogo Dekorativnogo Mystectva, Černigiv 43006
Muzej Narodnogo Pobutu, Mariupol' 43146
Muzej Narodnoï Architektury i Pobutu u Lvovi, Lviv 43134
Muzej Narodnoï Architektury ta Pobutu, Perejaslav-Chmel'nyc'kyj 43164
Muzej Narodnoï Architektury ta Pobutu Ukraïny, Kyïv 43098
Zakarpats'kyj Muzej Narodnoï Architektury ta Pobutu, Užgorod 43199

United Kingdom

Breadalbane Folklore Centre, Killin 44577
Calverton Folk Museum, Calverton 43717
Carmarthenshire County Museum, Carmarthen 43762
Cregneash Folk Village, Cregneash 43920
Cregneash Village Folk Museum, Port Saint Mary 45413
Curtis Museum, Alton 43293
Dalgarven Mill, Kilwinning 44583
Folk and Costume Museum, Câtel Guernsey 43779
Folk Museum, Mevagissey 45135
Folk Museum, Millom 45147
Jorvik Museum, York 46103
Kingussie Museum, Kingussie 44605
Moravian Museum, Pudsey 45450
Museum of Welsh Life, Saint Fagans 45565
Newtonmore Museum, Newtonmore 45255
Ordsall Hall Museum, Salford 45600
Strathnaver Museum, Bettyhill 43465
Swaledale Museum, Reeth 45479
Tolson Memorial Museum, Huddersfield 44497
Ulster-American Folk Park, Omagh 45317
Ulster Folk and Transport Museum, Holywood 44482
World of Country Life, Exmouth 44198

United States of America

African American Museum, Cleveland 47621
African American Museum, Dallas 47924
Afro-American Cultural Center, Charlotte 47384
Akta Lakota Museum, Chamberlain 47329
Alaska State Museum, Juneau 49656
Alexandria Black History Resource Center, Alexandria 46199
American Swedish Historical Museum, Philadelphia 51813
American Swedish Institute, Minneapolis 50727
Amherst County Museum, Amherst 46273
Anderson County Historical Museum, Garnett 48865
Atrium Gallery, Saint Louis 52567
Avery County Historical Museum, Newland 51315
Balzekas Museum of Lithuanian Culture, Chicago 47456
Beau Fort Plantation Home, Bermuda 46765
Beck Cultural Exchange Center, Knoxville 49811
Black American West Museum and Heritage Center, Denver 48064
Brick Store Museum, Kennebunk 49727
Bridgton Historical Museum, Bridgton 47023
Buechel Memorial Lakota Museum, Saint Francis 52542
Burke Museum of Natural History and Culture, Seattle 52994
Carter County Museum, Ekalaka 48306
Centro Cultural de la Raza, San Diego 52737
Charles A. Grignon Mansion, Kaukauna 49711
Charles H. Wright Museum of African American History, Detroit 48109
Chisholm Trail Museum, Kingfisher 49781
Connecticut River Museum, Essex 48421
Croatian Heritage Museum, Eastlake 48266
Czech Center New York, New York 51158
Daniel Boone Homestead, Birdsboro 46818
Delta Cultural Center, Helena 49237
Dubuque Museum of Art, Dubuque 48187
East-West Center, Honolulu 49338
Eastern Oregon Museum on the Old Oregon Trail, Haines 49126
Edwards Memorial Museum, Chesterfield 47436
El Rancho de Las Golondrinas Museum, Santa Fe 52873
Eldridge Street Project, New York 51169
Five Civilized Tribes Museum, Muskogee 50928
Four Rivers Cultural Center, Ontario 51570
Frankenmuth Historical Museum, Frankenmuth 48732
Frontier Culture Museum, Staunton 53275
General Adam Stephen House, Martinsburg 50519
George W. Brown jr. Ojibwe Museum, Lac du Flambeau 49857
Gibbs Farm Museum, Falcon Heights 48496
Goliad State Historical Park, Goliad 48967
Hammond-Harwood House Museum, Annapolis 46319
Hampton County Historical Museum, Hampton 49148
Historic Crab Orchard Museum and Pioneer Park, Tazewell 53457
Historic Saint Mary's City, Saint Mary's City 52609
Horry County Museum, Conway 47784
Hoskins House, Burlington 47141
House of Memories Museum, Wilson 54209
The Hudson River Museum, Yonkers 54331
Hulihee Palace, Kailua-Kona 49662
Irish American Heritage Museum, East Durham 48239
Jensen Arctic Museum, Monmouth 50785
Kanabec History Center, Mora 50837
Korean American Museum, Los Angeles 50240
Korean Cultural Center Art Gallery, Los Angeles 50241
The Kurdish Museum, Brooklyn 47077
The Lacrosse Museum and National Hall of Fame, Baltimore 46573
Lahaina Restoration Foundation, Lahaina 49872
Latvian Museum, Rockville 52430
Linn County Historical Museum and Moyer House, Brownsville 47095
Malki Museum, Banning 46599
Maramec Museum, Saint James 52552
Mattye Reed African Heritage Center, Greensboro 49066
Melrose Plantation Home Complex, Melrose 50565
Meriden Historical Society Museum, Meriden 50598
Meridian International Center - Cafritz Galleries, Washington 53892
Mexican Fine Arts Center Museum, Chicago 47489
The Mexican Museum, San Francisco 52783
Michigan State University Museum, East Lansing 48254

Mid-America All-Indian Center Museum, Wichita 54146
Mountain Heritage Center, Cullowhee 47903
Museo Italoamericano, San Francisco 52785
The Museum at Warm Springs, Warm Springs 53817
Museum of Afro-American History, Boston 46939
Museum of Amana History, Amana 46249
Museum of Appalachia, Norris 51377
Museum of Chinese in the Americas, New York 51226
Museum of Indian Arts and Culture, Santa Fe 52881
Museum of Russian Culture, San Francisco 52787
Museum of the American Hungarian Foundation, New Brunswick 51035
National Afro-American Museum and Cultural Center, Wilberforce 54158
National Museum of Women in the Arts, Washington 53906
Native American Exhibit, Fonda 48584
Native American Museum, Terre Haute 53481
Nevada State Fire Museum and Comstock Firemen's Museum, Virginia City 53761
Nordic Heritage Museum, Seattle 53006
Northeast Louisiana Delta African American Heritage Museum, Monroe 50790
Northwest Museum of Arts and Culture, Spokane 53207
Old West Wax Museum, Thermopolis 53493
The Oldest House Museum, Key West 49765
Orcas Island Historical Museum, Eastsound 48272
P.H. Sullivan Museum and Genealogy Library, Zionsville 54372
Pacific University Museum, Forest Grove 48585
Paynes Creek Historic State Park, Bowling Green 46972
Pendarvis Historic Site, Mineral Point 50726
Pennsbury Manor, Morrisville 50860
Pennsylvania German Cultural Heritage Center, Kutztown 49829
Piatt Castles, West Liberty 54068
Point of Honor, Lynchburg 50329
Pro Rodeo Hall of Fame and Museum of the American Cowboy, Colorado Springs 47706
Putney Historical Society, Putney 52203
Richland County Museum, Lexington 50066
S. Ray Miller Auto Museum, Elkhart 48350
Schingoethe Center for Native American Cultures, Aurora 46493
Schwenkfelder Library and Heritage Center, Pennsburg 51765
Seabrook Village, Midway 50678
Shakowi Cultural Center, Oneida 51561
Shoal Creek Living History Museum, Kansas City 49698
Shoshone Tribal Cultural Center, Fort Washakie 48697
Sioux Indian Museum, Rapid City 52254
Skagit County Historical Museum, La Conner 49832
Southern Plains Indian Museum, Anadarko 46282
Spurlock Museum, Urbana 53669
Stephen Foster Folk Culture Center, White Springs 54129
Stewart M. Lord Memorial Museum, Burlington 47137
Strasburg Museum, Strasburg 53334
Swedish American Museum Association of Chicago Museum, Chicago 47516
Terrebonne Museum, Houma 49378
Teysen's Woodland Indian Museum, Mackinaw City 50356
Thousand Islands Museum of Clayton, Clayton 47602
Tubman African-American Museum, Macon 50370
Ukrainian-American Museum, Hamtramck 49155
Ukrainian Museum, Cleveland 47643
The Ukrainian Museum, New York 51274
Volcano Art Center, Volcano 53766
Wanapum Dam Heritage Center, Beverly 46788
Whitingham Historical Museum, Whitingham 54136
Wilber Czech Museum, Wilber 54157
Yeshiva University Museum, New York 51281
Zoar State Memorial, Zoar 54373

Uruguay

Museo Parque de Esculturas, Montevideo 54415
Museo Romántico - Casa de Montero, Montevideo 54417
Museo y Centro Cultural AGADU, Montevideo 54420

Uzbekistan

Akmal Ikramov nomidagi Tarikhi Kültüri va San"atshunoslik Özbekiston Khalqlari Davlat Müzeyi, Samarqand 54460

Venezuela

Museo de Arte Popular Petare, Caracas 54542
Museo de Arte Popular Venezolano, Juangriego 54576

Food

Australia

Old Council Chambers Museum, Norton Summit 01380

Belgium

Frietkotmuseum, Antwerpen 03288
Le Petit Musée du Pain, Chievres 03521
Musée Brasserie des Fagnes, Mariembourg 03854
Musée de la Gourmandise, Hermalle-sous-Huy 03700
Musée du Champignon, Tillet-Saint-Ode 04073
Musée Vivant de la Brasserie à Vapeur, Pipaix 03952
Musée Vivant de la Pomme de Terre, Genappe 03618
Musée Vivant du Jambon - La Ferme des Sanglochons, Verlaine 04105

Canada

Choco-Musée Erico, Québec 06679
Ingersoll Cheese Factory Museum and Sports Hall of Fame, Ingersoll 06012
Maple Syrup Museum, Saint Jacobs 06823
Musée du Fromage Cheddar, Saint-Prime 06876
Prince Edward Island Potato Museum, O'Leary 06510

China

Nanji Guan, Qingdao 08106

Czech Republic

Jan Becher Muzeum, Karlovy Vary 09574
Malé Máslovické Muzeum Másla, Vodochody 09912

France

Cassissium, Nuits-Saint-Georges 14484
Hameau du Fromage, Cléron 12556
Maison de la Pomme et de la Poire, Barenton 11851
Maison de l'Abeille et du Miel, Riez 15039
Maison des Fromages, Les Moussières 13707
Maison des Fromages d'Auvergne, Egliseneuve-d'Entraigues 12812
Musée Amora, Dijon 12735
Musée Art du Chocolat, Lisle-sur-Tarn 13785
Musée Conservatoire des Productions Fruitières, Saint-Jean-d'Arvey 15320
Musée de la Casse, Orcières 14524
Musée de la Mytiliculture, Esnandes 12853
Musée de la Pomme, Bourg-Saint-Maurice 12109
Musée de la Pomme et du Cidre et des Métiers Traditionnels, Rosay 15101
Musée de l'Amande et du Macaron, Montmorillon 14231
Musée de l'Art Culinaire, Villeneuve-Loubet 16217
Musée de l'Olivier, Les Baux-de-Provence 13689
Musée des Techniques Fromagères, Saint-Pierre-sur-Dives 15463
Musée du Bonbon Haribo, Uzès 16024
Musée du Chocolat, Barcelonnette 11849
Musée du Foie Gras, Frespech 13011
Musée du Foie Gras, Thiviers 15887
Musée du Gâteau Basque, Sare 15633
Musée du Jus de Fruit, Saint-Andiol 15165
Musée du Lait, Montebourg 14195
Musée et Plaisir du Chocolat, Biarritz 11984
Musée Vivant de l'Abeille, Valensole 16044
Le Palais du Chocolat, La Côte-Saint-André 13352

Germany

Deutsches Bananenmuseum, Sierksdorf 21760
1. Deutsches Bratwurstmuseum, Wachsenburggemeinde 22260
Deutsches Currywurst Museum, Berlin 17071
Deutsches Fleischermuseum, Böblingen 17417
Deutsches Kartoffelmuseum Fußgönheim, Fußgönheim 18522
Deutsches Kochbuchmuseum, Dortmund 17918
Europäisches Spargelmuseum, Schrobenhausen 21638
Historisches Käsereimuseum, Altusried 16527
Das Kartoffelmuseum, München 20449
Museum der Brotkultur, Ulm 22166
Museum für Krippen, Eß- und Trinkkultur und Brauereigeschichte, Burghausen 17654
Museum im Bock, Leutkirch im Allgäu 19973
Neudorfer Suppenmuseum, Sehmatal 21727
Schokoladenmuseum Köln, Köln 19632
Spicy's Gewürzmuseum, Hamburg 18960

Greece

Museum of the Olive and Greek Olive Oil, Spárti 23197

Hungary

Húsipari Üzemtörténeti Gyűjtemény, Gyula 23595
Komiszkenyér Kiállítás, Komáron 23679
Méhészeti Gyűjtemény, Gödöllő 23567

Ireland

Cork Butter Museum, Cork 24766

Israel

Israel Oil Industry Museum, Haifa 24962

Italy

Museo Agrumario, Reggio Calabria 27380
Museo del Cioccolato, Norma 26943
Museo del Pane, Sant'Angelo Lodigiano 27719
Museo del Peperoncino, Maierà 26568
Museo dell'Arte e della Tecnologia Confetteria, Sulmona 27919
Museo dell'Olivo, Imperia 26405
Museo dell'Olivo e dell'Olio, Torgiano 27994
Museo Nazionale delle Paste Alimentari, Roma 27516

Japan

Sanuki Museum of Soy Sauce Art, Sakaide 29203
Su-no-sato Hakubutsukan, Handa 28497
Yokohama-shi Kare-Hakubutsukan, Yokohama 29618

Korea, Republic

Kimchi Field Museum, Seoul 29889

Netherlands

Bakkerij Museum, Oosterhout 32627
Culinair Historisch Kookmuseum de Vleer, Appelscha 31916
Culinair Museum, Amersfoort 31811
Flipje en Jam Museum Tiel, Tiel 32823
Het Hollands Kaasmuseum, Alkmaar 31793
Kaasmuseum Bodegraven, Bodegraven 31976
Museum in den Halven Maen, Rockanje 32692

Norway

Meierimuseet, Ås 33401

Peru

Coca Museum, Puno 34297

Philippines

Grains Industry Museum, Cabanatuan 34357
Ricewolrd Museum, Los Baños 34407

Portugal

Museu do Arroz, Comporta 35546

South Africa

Elgin Apple Museum, Grabouw 38628

Spain

Euskal Gastronomi Museoa, Llodio 39487
Gastronomiaren Euskal Museoa, Laudio 39446
Museo de la Miel, Peñalver 39830
Museo del Arroz, Valencia 40207
Museo del Chocolate, Astorga 38910
Museo del Queso, Villarejo-Periesteban 40307
Museu de la Taronja, Burriana 39076

Switzerland

Alimentarium, Vevey 41983
Appenzeller Schaukäserei, Stein 41911
Metzgermuseum Zunft zum Widder, Zürich 42086
Schokoland Alprose, Caslano 41282

Taiwan

Jierjia Sweet Potato Museum, Taichung 42349
Kuo Yuan Ye Museum of Cake and Pastry, Taipei 42397
Mango Explorer Museum, Yujing 42483
Onion Museum, Checheng 42160
Rice Museum, Dajia 42174
Strawberry Culture Museum, Dahu 42172
Taiwan Mushroom Museum, Wufeng 42460

United Kingdom

Mustard Shop Museum, Norwich 45281
Radbrook Culinary Museum, Shrewsbury 45674

United States of America

Beverage Containers Museum, Millersville 50694
Culinary Archives and Museum, Providence 52163
George Washington Carver House, Diamond 48137
New England Maple Museum, Pittsford 51970
The Rice Museum, Crowley 47894
SPAM Museum, Austin 46499
Yankee Candlemaking Museum, South Deerfield 53168

Forest Products

Argentina

Museo Histórico La Gallareta Forestal, La Gallareta 00403

Australia

Forest and Heritage Centre, Geeveston 01105
Manjimup Timber Museum, Manjimup 01257
Timbertown Museum, Wauchope 01637
Wirrabara Forestry Museum, Wirrabara 01659

Austria

Erstes Tiroler Holzmuseum, Wildschönau 03105
Forstmuseum, Reichraming 02573
Hacken- und Holzmuseum, Köflach 02210
Heimathaus, Steinbach am Attersee 02764
Huebmer-Gedächtnisstätte, Naßwald 02422
Lignorama - Holz- und Werkzeugmuseum, Riedau 02585
Montanmuseum, Hieflau 02077
Mühlviertler Waldhaus, Windhaag bei Freistadt 03112
Museum Alte Brettersäge, Rappottenstein 02561
Museum Schloß Lackenbach, Lackenbach 02250
Österreichisches Forstmuseum Silvanum, Großreifling 02010
Schau- und Erlebnissägewerk Mühlauersäge, Fusch 01914
Steirisches Holzmuseum, Sankt Ruprecht 02688
Stierhübelteichhaus, Karlstift 02155
Waldbauernmuseum, Gutenstein 02027

Belgium

Bosmuseum Pijnven, Hechtel 03689
Bosmuseum Wildert, Essen 03585
Musée Provincial de la Forêt, Namur 03905

Brazil

Laboratório de Anatomia e Identificação de Madeiras, Manaus 04553
Museu da Secção de Tecnologia do Serviço Florestal, Rio de Janeiro 04720
Museu Florestal Octávio Vecchi, São Paulo 04899

Brunei

Brunei Forestry Museum, Sungai Liang 04970

Canada

A Logger's Life Provincial Museum, Grand Falls-Windsor 05890
Alberta Forestry Service Museum, Hinton 05990
British Columbia Forest Discovery Centre, Duncan 05722
British Columbia Forest Service Museum, Victoria 07242
Logging Museum, Peachland 06576
Penetanguishene Centennial Museum, Penetanguishene 06582
Prince Georg Railway and Forestry Museum, Prince George 06668
Riding Mountain National Park Interpretive Centre, Wasagaming 07281
Sandilands Forest Centre and Museum, Hadashville 05923
Sherbrooke Village, Sherbrooke, Nova Scotia 06957

China

Guifei Muwen Guan (Museum), Xinping 08401

Czech Republic

Lovecko Lesnicke Muzeum, Úsov 09891

Denmark

Dansk Jagt- og Skovbrugsmuseum, Hørsholm 10100
Vorbasse Museum, Vorbasse 10319

Finland

Lapin Metsämuseo, Rovaniemi 11277
Lusto-Suomen Metsämuseo, Punkaharju 11238
Pielisen Museo, Lieksa 11067
Viitasaaren Metsätyömuseo, Viitasaari 11466

France

Château de Roquedols, Meyrueis 14112
Maison des Tresses et des Lacets, La Terrasse-sur-Dorlay 13446
Musée de la Schlitte et des Métiers du Bois, Muhlbach-sur-Munster 14322
Musée de la Vie Rurale et Forestière, Saint-Michel-en-Thiérache 15419
Musée des Métiers du Bois, Labaroche 13459
Musée du Bois et de la Forêt, La Chaise-Dieu 13328
Musée-Maison des Amis de la Forêt, Mervent 14092

French Polynesia

Musée du Bois, Papuakeikaha 16326

Germany

Brandenburgisches Forstmuseum Fürstenberg/Havel, Fürstenberg, Havel 18488
Forst- und Köhlerhof, Rostock 21390
Forstbotanisches Museum, Tharandt 22049
Forstmuseum Heringen, Heringen (Werra) 19142
Forstmuseum im Hochwildpark Karlsberg, Weikersheim 22371
Gemeinde- und Forstmuseum, Oftersheim 20889
Holzknechtmuseum Ruhpolding, Ruhpolding 21447
Holztechnisches Museum, Rosenheim 21375
Köhlereimuseum, Hasselfelde 19029
Museum für ländliches Kulturgut, Landwirtschaft, Forsten und Jagd, Ulrichstein 22171
Museum Wald und Umwelt, Ebersberg 18045
Sägmühlmuseum Marhördt, Oberrot 20830
Schüttesägemuseum, Schiltach 21557
Technisches Museum Holzschleiferei Weigel, Breitenbrunn 17542
Wald- und Lönsmuseum, Heinade 19110
Wald- und Moormuseum Berumerfehn, Großheide 18793
Waldgeschichtliches Museum Sankt Oswald, Sankt Oswald-Riedlhütte 21507
Waldmuseum, Münstertal 20524
Waldmuseum, Zwiesel 22839
Waldmuseum im Wildpark Neuhaus, Holzminden 19262
Waldmuseum Stendenitz, Neuruppin 20652
Waldmuseum Wassermühle, Wingst 22598
Waldmuseum Watterbacher Haus, Kirchzell 19556

Guatemala

Centro de Estudios Conservacionistas, Guatemala 23296

Hungary

Bakonybéli Erdok Háza, Bakonybél 23348
Erdészeti-, Faipari- és Földméréstörténeti Gyűjtemény, Sopron 23808
Erdészeti Szabadtéri Erdészeti Múzeum, Szilvásvárad 23893
Nyírségi Erdészeti Zrt. Debreceni Erdészet - Bánki Tájház, Debrecen 23519
Orbán Ház, Szilvásvárad 23895
Stájer-házi Erdészeti Múzeum, Kőszeg 23684

India

Assam Forest Museum, Guwahati 24187
F.R.I. Museums, Dehradun 24131
Gass Forest Museum, Coimbatore 24121

Indonesia

Museum Manggala Wanabakti, Jakarta 24490

Italy

Museo Agro-Forestale San Matteo, Erice 26061
Museo del Legno, Cantù 25650

Japan

Kitayama Cedar Museum, Kyoto 28824
Museum for Wood Culture, Muraoka 28946
Shimazu Foundation Memorial Hall, Kyoto 28866
Woody Keihoku, Keihoku 28701

Korea, Republic

Forest Museum, Gyeonggi-do 29824
Tamyang Bamboo Museum, Tamyang 29937

Malaysia

Forestry Research Institute Malaysia Museum, Kepong 30562
Penang Forestry Museum, Pulau Pinang 30614

Myanmar

Forest Museum, Taungdwingyi 31729

Netherlands

Boomkwekerijmuseum, Boskoop 31990

New Zealand

Timber Museum of New Zealand, Putaruru 33191

Norway

Fetsund Lenser Fløtingsmuseum, Fetsund 33491
Norsk Skogmuseum, Elverum 33481
Sunnhordland Museum, Stord 33885

Pakistan

Forest Museum, Peshawar 34033

Poland

Muzeum Lasu i Drewna przy Lesnym Zakładzie Doświadczalnym SGGW, Rogów 35083

Russia

Muzej Lesovodstva, Moskva 37328
Rossijskij Muzej Lesa, Moskva 37404

Slovakia

Lesnícke a Drevárske Múzeum, Zvolen 38394

Slovenia

Muzej Vrbovec - Muzej Gozdarstva in Lesarstva, Nazarje 38488
Tehniški Muzej Slovenije, Bistra pri Vrhniki 38399
Tehniški Muzej Slovenije, Ljubljana 38467

Sweden

Finnskogsmuseet, Viksjöfors 41063
Siljansfors Skogsmuseum, Mora 40746
Skärgårdens Sågverksmuseum, Holmsund 40601
Skogsmuseet i Lycksele, Lycksele 40710
Skogsmuseum, Åsele 40442
Skogsmuseum, Sunne 40961
Tändsticksmuseet, Jönköping 40616

Switzerland

Gemossenschaft Pro Sagi Samstagern, Samstagern 41802
Musée du Bois, Aubonne 41136
Museum Internationales Baum-Archiv, Winterthur 42033

Taiwan

Bamboo Museum, Lugu 42259
Forestry Exhibition Hall, Fengyuan 42189
Taiwan Forestry Exhibition Hall, Taipei 42432

Uganda

Forest Department Utilisation Division Museum, Kampala 42990

United Kingdom

Dean Heritage Museum, Soudley 45692
New Forest Museum and Visitor Centre, Lyndhurst 45056
Queen Elizabeth's Hunting Lodge, London 44961

United States of America

Big Timbers Museum, Lamar 49909
Camp 6 Logging Museum, Tacoma 53405
Camp Five Museum, Laona 49946
Chippokes Farm and Forestry Museum, Surry 53382
Fisher Museum of Forestry, Petersham 51805
Forest Capital State Museum, Perry 51779
Forest Discovery Center, Portland 52077
Forest History Center, Grand Rapids 49008
Hoo-Hoo International Forestry Museum, Gurdon 49107
Iron County Museum, Caspian 47274
Klamath National Forest Interpretive Museum, Yreka 54358
Mississippi Agriculture and Forestry/National Agricultural Aviation Museum, Jackson 49562
New England Maple Museum, Rutland 52494
North Carolina Museum of Forestry, Whiteville 54133
Oklahoma Forest Heritage Center Forestry Museum, Broken Bow 47038
Paul Bunyan Logging Camp, Eau Claire 48278
Pennsylvania Lumber Museum, Galeton 48836
Rhinelander Logging Museum, Rhinelander 52302
Rim Country Museum, Payson 51741
Scotia Museum, Scotia 52969
Southern Forest World Museum, Waycross 54003
Texas Forestry Museum, Lufkin 50319

Fortresses → Castles and Fortresses

Fossils

Argentina

Complejo Museológico del Instituto Félix Bernasconi, Buenos Aires 00155
Museo de Ciencias Naturales Luma, Tanti 00643
Museo de Geología y Paleontología, Neuquén 00477
Museo de História y Ciencias Naturales, Lobería 00429
Museo Municipal Carmen Funes, Plaza Huincul 00495
Museo Municipal de Ciencias Naturales Lorenzo Scaglia, Mar del Plata 00447
Museo Paleontológico Egidio Feruglio, Trelew 00651
Museo Paleontológico y Petrolero Astra, Comodoro Rivadavia 00311
Museo Prof. Dr. Juan Augusto Olsacher, Zapala 00688
Museo Provincial de Historia Natural de La Pampa, Santa Rosa 00625
Museo Universitario Florentino y Carlos Ameghino, Rosario 00550
Paleorama - Museo Itinerante, Don Torcuato 00357

Australia

Boulia Stone House Museum, Boulia 00881
Museum of Tropical Queensland, Townsville 01598
National Dinosaur Museum, Nicholls 01368
Queensland Museum, South Bank 01510

Austria

Erstes österreichisches Bodenschätzungsmuseum, Hollabrunn 02095
Museum Steinmetzhaus, Eggenburg 01844
Paläontologisches Museum Dr. Herbert Schaffer, Offenhausen 02470

Belgium

Mineralogisch Museum, Merksem 03869
Musée de la Préhistoire en Wallonie, Flémalle 03602
Musée de l'Iguanodon, Bernissart 03364
Muséum des Sciences Naturelles de Belgique, Bruxelles 03497
Museum Ons Mijnverleden, Houthalen 03719
Schoolmuseum Michel Thiery, Gent 03640

Bolivia

Museo Nacional Paleontológico y Arqueológico, Tarija 04251

Brazil

Museu de Ciências da Terra - DNPM, Rio de Janeiro 04725
Museu de Ciências Naturais PUC Minas, Belo Horizonte 04348
Museu de Paleontologia, Monte Alto 04572

Bulgaria

Paleontologičen Muzej, Asenovgrad 04972

Canada

Centre d'Interprétation des Mammifères Marins (CIMM), Tadoussac 07055
Centre Thématique Fossilifère, Notre-Dame-du-Nord 06492
Courtenay and District Museum and Paleontology Centre, Courtenay 05652
Devil's Coulee Dinosaur Heritage Museum, Warner 07278
Dinosaur Provincial Park, Patricia 06574
Garden Park Farm Museum, Alberta Beach 05312
Geological Museum, Saskatoon 06913
Grande Prairie Museum, Grande Prairie 05900
Logan Hall, Ottawa 06550
London Regional Children's Museum, London 06187
Morden and District Museum, Morden 06391
Musée de la Paléontologie et de l'Évolution, Montréal 06354
Musée du Squelette, île-Verte 06007
Museum of Natural Sciences, Saskatoon 06921
Princeton and District Museum, Princeton 06674
Royal Tyrrell Museum of Palaeontology, Drumheller 05713
Yukon Beringia Interpretive Centre, Whitehorse 07328

Chile

Museo de Historia Natural de Valparaíso, Valparaíso 07471
Museo Histórico de Osorno, Osorno 07443

China

Dalian Natural History Museum, Dalian 07639
Lufeng Saurian Museum, Lufeng 07978
Nanjing Museum of Palaeontology, Nanjing 08046
Sanying Pagoda and Museum of Dinosaur, Shaoguan 08208
Zigong Saurian Museum, Zigong 08519

Colombia

Museo Geológico José Royo y Gómez, Bogotá 08584
Museo Paleontológico, Mariquita 08674

Croatia

Gradski Muzej, Požega 08930
Muzej Krapinskog Pračovjeka, Krapina 08900

Cuba

Museo Paleontológico, Pinar del Río 09252

Cyprus

Tornaritis Pierides Municipal Museum of Paloentology, Larnaka 09371

Czech Republic

Muzeum a Galerie Litomyšl, Litomyšl 09632
Muzeum Českého Krasu v Berouně, Beroun 09404
Vlastivědné Muzeum, Slany 09838
Zámek Budišov, Budišov u Třebíče 09447

Denmark

Fur Museum, Fur 10032
Molermuseet, Nykøbing Mors 10203
Statens Naturhistoriske Museum (SNM), København 10153

Ecuador

Museo David Paltan, Tulcán 10462

Egypt

The Egyptian Geological Museum, Cairo 10491

Finland

Kivimuseo, Tampere 11351

France

Collection Archéologique, Villarzel-Cabardes 16172
Écomusée de la Cévenne, Saint-Laurent-de-Trèves 15358
Espace Pierres Folles, Saint-Jean-des-Vignes 15335
Galeries de Paléontologie et d'Anatomie Comparée, Paris 14608
Musée de la Pierre, Trambly 15977
Musée de la Thiérache, Vervins 16131
Musée de Paléontologie, La Voulte-sur-Rhône 13457
Musée de Paléontologie, Villers-sur-Mer 16233
Musée de Paléontologie, Villeurbanne 16238
Musée de Paléontologie Christian Guth, Chilhac 12511
Musée de Préhistoire Amédée Lemozi, Cabrerets 12219
Musée des Fossiles, Montceau-les-Mines 14191
Musée des Minéraux et Fossiles des Ardennes, Bogny-sur-Meuse 12044
Musée d'Histoire Naturelle Philadelphe Thomas, Gaillac 13027
Musée du Cloître, Tulle 16008
Musée du Fossile, Lion-sur-Mer 13774
Musée du Savignéen, Savigné-sur-Lathan 15677
Musée Municipal d'Archéologique, Murviel-les-Montpellier 14340
Musée Municipal de Préhistoire, Fismes 12940
Musée Municipal des Fossiles, Tourtour 15972
Musée Théophile Jouglet, Anzin 11647
Muséum d'Histoire Naturelle, Le Havre 13599
Muséum d'Histoire Naturelle, Dijon 12744
Muséum National d'Histoire Naturelle, Paris 14741
Parc Préhistorique Imaginaire, Chamoux 12382
Rhinopolis, Gannat 13029

Germany

Archäologische Ausstellung in der Schule, Nassenfels 20548
Bürgermeister-Müller-Museum, Solnhofen 21813
Dinosaurier-Freilichtmuseum Münchehagen, Rehburg-Loccum 21240
Dobergmuseum/Geologisches Museum Ostwestfalen-Lippe, Bünde 17635
Eifelmuseum Blankenheim, Blankenheim 17375
Erdgeschichtliches Museum, Lauffen am Neckar 19855
Fossilien-Museum, Fürth 18497
Fossilien- und Heimatmuseum, Messel 20304

Fossilien- und Steindruck-Museum, Gunzenhausen 18838
Fossilienmuseum im Werkforum Rohrbach Zement, Dotternhausen 17933
Geiseltalmuseum, Halle, Saale 18895
Geomuseum der Universität Münster, Münster 20509
Geowissenschaftliche Sammlungen, Freiberg, Sachsen 18421
Geowissenschaftliche Sammlungen und Geologischer Garten, Halle, Saale 18896
Heimatmuseum, Günzburg 18825
Heimatmuseum, Niederaichbach 20701
Hohenloher Urweltmuseum, Waldenburg, Hohenlohekreis 22289
Muschelkalkmuseum Hagdorn, Ingelfingen 19333
Museum Alzey, Alzey 16530
Museum Bergér, Eichstätt 18083
Museum für Naturkunde, Berlin 17205
Museum im Hollerhaus, Dietfurt an der Altmühl 17855
Museum Natur und Mensch, Greding 18742
Naturhistorisches Museum, Mainz 20149
Naturkunde-Museum Bamberg, Bamberg 16915
Naturkundliche Sammlung, Königsbrunn 19646
Odenwald- Spielzeugmuseum, Michelstadt 20321
Paläontologisches Museum München, München 20472
Paläontologisches Museum Nierstein, Nierstein 20720
Petrefaktensammlung, Bad Staffelstein 16844
Privatmuseum Hans Klein, Prichsenstadt 21125
Sammlung Fossilien des Jura, Lichtenfels 19980
Sieblos-Museum Poppenhausen, Poppenhausen 21076
Stadtmuseum Stadtoldendorf, Stadtoldendorf 21859
Stein- und Fossiliensammlung Albert, Sulzdorf an der Lederhecke 21989
Südostbayerisches Naturkunde- und Mammut-Museum, Siegsdorf 21759
Trias-Museum, Ochsenfurt 20855
Turmuhrenmuseum Rammensee, Gräfenberg 18725
Urwelt-Museum Hauff, Holzmaden 19260
Urwelt-Museum Oberfranken, Bayreuth 16966
Urweltmuseum Aalen, Aalen 16415
Zeittunnel Wülfrath, Wülfrath 22704

Greece

Natural History Museum of the Aegean, Samos 23178
Natural History Museum of Tilos, Tilos 23224
Paleontological and Geological Museum of the University of Athens, Athinai 22949
Siatista Paleontological Museum, Siatista 23183

Hungary

Encsy György Tokaj-Hegyalja Földtani Gyűjteménye, Tállya 23910
Komlói Természettudományi Gyűjtemény, Komló 23681
Mátra Múzeum, Gyöngyös 23574

Ireland

James Mitchell Museum, Galway 24836

Israel

House of the Scribe, Almog 24925

Italy

Civico Museo Naturalistico Ferruccio Lombardi, Stradella 27910
Mostra Permanente di Paleontologia, Terni 27962
Museo Archeologico della Valle Sabbia, Gavardo 26276
Museo Civico Antiquarium, Nettuno 26921
Museo Civico D. dal Lago, Valdagno 28137
Museo Civico dei Fossili di Besano, Besano 25409
Museo Civico di Paleontologia, Empoli 26053
Museo Civico di Scienze Naturali, Faenza 26074
Museo Civico di Scienze Naturali, Voghera 28303
Museo Civico di Scienze Naturali ed Archeologia della Valdinievole, Pescia 27172
Museo Civico Paleontologico, Macerata Feltria 26557
Museo dei Fossili, Vestenanova 28240
Museo dei Fossili, Zogno 28316
Museo dei Fossili Don Giuseppe Mattiacci, Serra San Quirico 27812
Museo dei Fossili e dei Minerali, Smerillo 27859
Museo dei Fossili e Minerali del Monte Nerone, Apecchio 25249
Museo dei Fossili e Mostra dei Funghi, Pioraco 27232
Museo del Fossile del Monte Baldo, Brentonico 25529
Museo di Antropologia G. Sergi, Roma 27491
Museo di Fossili, Ronca 27553
Museo di Geologia e Paleontologia, Torino 28014
Museo di Geologia e Paleontologia, Padova 27040
Museo di Minerali e Fossili, Montefiore Conca 26826
Museo di Mineralogia e Paleontologia, Saint Vincent 27589
Museo di Mineralogia e Petrologia, Padova 27041
Museo di Paleontologia, Milano 26705
Museo di Paleontologia, Modena 26752
Museo di Paleontologia, Napoli 26901
Museo di Paleontologia, Roma 27498
Museo di Paleontologia dell'Accademia Federiciana, Catania 25774
Museo di Paleontologia e Mineralogia, Campomorone 25642
Museo di Paleontologia e Speleologia E.A. Martel, Carbonia 25676
Museo di Paleontologia G. Buriani, San Benedetto del Tronto 27610
Museo di Scienze Naturali Don Bosco, Alassio 25171
Museo di Storia Naturale, Parma 27105
Museo Friulano di Storia Naturale, Udine 28114
Museo Geo-Paleontologico, Ferrara 26115
Museo Geologico e Paleontologico G. Capellini, Bologna 25472
Museo Geologico G.G. Gemmellaro, Palermo 27077
Museo Geopaleontologico Alto Aventino, Palena 27063
Museo Geopaleontologico dei Fossili della Lessinia, Velo Veronese 28173
Museo Paleontologico, Amandola 25218
Museo Paleontologico, Bova 25519
Museo Paleontologico, Fluminimaggiore 26208
Museo Paleontologico, L'Aquila 26458
Museo Paleontologico, Montevarchi 26858
Museo Paleontologico, Mondaino 26774
Museo Paleontologico della Rocca di Monfalcone, Monfalcone 26779
Museo Paleontologico M. Gortani, Portogruaro 27312
Museo Paleontologico Parmense, Parma 27109
Museo Paleontologico Rinaldo Zardini, Cortina d'Ampezzo 25986
Museo Paleontologico S. Lai, Ceriale 25808
Museo Provinciale di Storia Naturale, Foggia 26212
Raccolta Archeologica e Paleontologica, Corciano 25972
Raccolta di Fossili Francesco Angellotti, Ostra 27022

Japan

Fukui-kenritsu Bunka Hakubutsukan, Fukui 28429
Fukui-kenritsu Kyoryu Hakabutsukan, Katsuyama 28681
Hayashibara Shizenkagaku Hakubutsukan, Okayama 29099
Mizunami-shi Kaseki Hakubutsukan, Mizunami 28932
Osaka Museum of Natural History, Osaka 29149
Shinshushin-machi Hakubutsukan, Shinshushin 29273

Korea, Republic

Gyeongbo Museum of Paleontology, Namjeong 29855

Latvia

Latvijas Universitātes Ģeologijas Muzejs, Rīga 30088

Lebanon

Hakel Fossils Museum, Byblos 30167

Luxembourg

Musée Collection Eugène Pesch, Differdange 30466

Mexico

Museo Casa de la Plomada, Vallecillo 31528
Museo de Paleontología, Delicias 30832
Museo de Paleontología de Guadalajara, Guadalajara 30897
Museo de Paleontología de la Preparatoria, Apaxtla de Castrejón 30720
Museo de Paleontología de Rincón Gallardo, General Cepeda 30875
Museo de Paleontología Eliseo Palacios Aguilera, Tuxtla Gutiérrez 31514
Museo José de Jesús Almaza, Xicoténcatl 31563
Museo Palenteológico de la Laguna, Torreón 31502
Museo Universitario de Paleontología, México 31155

Morocco

Musée for Fossils Minerals, Erfoud 31667

Netherlands

Natuurhistorisch Museum Maastricht, Maastricht 32529
Palaeobotanisch Museum, Utrecht 32854
Provincial Natuurhistorisch Museum Natura Docet, Denekamp 32116
Verzameling Steenbakker Ten Bokkel Huinink, Neede 32567

New Zealand

Geology Museum, Auckland 33049

Nicaragua

Museo Comunitario de San Rafael del Sur, San Rafael del Sur 33306

Philippines

Dinosaur Museum, Malabon 34418

Portugal

Museu GEAL, Lourinhã 35740

Romania

Muzeul de Ştiinte Naturale, Piatra-Neamţ 36520

Russia

Gosudarstvennyj Geologičeskij Muzej im V.I. Vernadskogo, Moskva 37192
Paleontologičeskij Muzej im. Ju.A. Orlova, Moskva 37395

Slovenia

Zbirka v Jamarskem Domu, Domžale 38416

South Africa

Bernard Price Institute Paleontology Museum, Johannesburg 38650
South African Museum, Cape Town 38592

Spain

Museo de Ciencias Naturales, Santa Cruz de Tenerife 39995
Museo de Dinosaurios, Salas de los Infantes 39945
Museo Geominero, Madrid 39561
Museo Paleontológico de Elche, Elche 39269
Museo Paleontológico - Sala Longinos Navás, Zaragoza 40349
Museo Paleontológico - Sala Lucas Mallada, Zaragoza 40350
Museu de la Conca Dellà, Isona 39415
Museu de l'Institut Paleontològic Miquel Crusafont, Sabadell 39925
Museu Paleontològic, Valencia 40222

Sweden

Evolutionsmuseet, Paleontologi, Uppsala 41006

Switzerland

focusTerra, Zürich 42071
Mammutmuseum, Niederweningen 41687
Museo dei Fossili, Meride 41629
Paläontologisches Museum der Universität Zürich, Zürich 42103
Saurier-Museum, Frick 41393
Sauriermuseum Aathal, Aathal-Seegräben 41087

Taiwan

Chia Hsien Fossil Museum, Jiasian 42229
Kiko's Fossil and Mineral Museum, Banciao 42145
Land Fossils and Minerals Museum, Yongkang 42475
Ts'ai-Liao Fossil Museum, Zuojhen 42490

United Kingdom

Bolton Museum and Art Gallery, Bolton 43553
Booth Museum of Natural History, Brighton 43619
Cockburn Museum, Edinburgh 44117
Creetown Gem Rock Museum, Creetown 43918
Dinosaur Isle, Sandown 45613
Dinosaur Museum, Dorchester 43998
Dinosaurland, Lyme Regis 45053
Dorking and District Museum, Dorking 44006
Elgin Museum, Elgin 44166
Planet Earth Museum, Newhaven 45237
Portland Museum, Portland 45425
Staffin Museum, Staffin 45727
Swanage Museum, Swanage 45810

United States of America

Alaska Museum of Natural History, Anchorage 46287
Bergen Museum of Art and Science, Paramus 51696
Big Bone Lick State Park Museum, Union 53646
Brigham Young University Earth Science Museum, Provo 52186
Dakota Dinosaur Museum, Dickinson 48138
Dinosaur Gardens Museum, Ossineke 51620
Dinosaur Museum, Tucumcari 53603
Dinosaur Valley State Park, Glen Rose 48923
Dunn-Seiler Museum, Mississippi State 50751
Fernbank Museum of Natural History, Atlanta 46434
Fick Fossil and History Museum, Oakley 51476
Fort Peck Museum, Fort Peck 48666
Frontier Gateway Museum, Glendive 48931
Greene Memorial Museum, Milwaukee 50709
John Day Fossil Beds National Monument, Kimberly 49776
Mulberry Phospate Museum, Mulberry 50899
Museum of Long Island Natural Sciences, Stony Brook 53314
The Museum of Natural History, Stevens Point 53289
Museum of Paleontology, Berkeley 46758
Northwestern Oklahoma State University Museum, Alva 46247
Orton Geological Museum, Columbus 47758
Petrified Creatures Museum of Natural History, Richfield Springs 52305
Ruth Hall Museum of Paleontology, Abiquiu 46130
Stone Museum, Monroe Township 50794
Trailside Nature Center and Museum, Cincinnati 47566
Wyoming Dinosaur Center, Thermopolis 53494

F

Fresco Painting → Mural Painting and Decoration

Fur → Hides and Skins

Furniture and Interior

Andorra

Museu Casa d'Areny-Plandolit, Ordino 00100

Argentina

Fundación Alfredo L. Palacios, Buenos Aires 00158
Museo Municipal de Arte Decorativo Firma y Odilo Estévez, Rosario 00546

Australia

Carrick Hill Museum, Springfield 01522
Clarendon House, Evandale 01077
Cooma Cottage, Yass 01681
Dowerin District Museum, Dowerin 01042
Eryldene Heritage House, Gordon 01120
Franklin House, Launceston 01221
Gledswood Farm Museum and Homestead, Catherine Fields 00961
Old Blythewood, Pinjarra 01415
Rouse Hill Estate, Rouse Hill 01472
White House, Westbury 01646

Austria

Barockschloß Riegersburg, Riegersburg, Niederösterreich 02586
Bauernmöbelmuseum Hirschbach, Hirschbach 02080
Bauernmöbelmuseum Propstkeusche, Malta 02345
Erzherzog Franz Ferdinand Museum, Artstetten 01728
Europäisches Museum für Frieden, Stadtschlaining 02756
Freilichtmuseum Sumerauerhof, Sankt Florian 02637
Geymüller Schlössel, Wien 02995
Handwerkskunst-Museum, Rattenberg, Inn 02565
Hofburg Innsbruck, Innsbruck 02119
Hofmobiliendepot, Wien 03005
Kaiserappartements, Sisi Museum, Silberkammer, Wien 03009
Kunstsammlungen und Graphische Sammlung, Furth bei Göttweig 01913
Lehár-Schikaneder-Schlößl, Wien 03021
Museum des Zisterzienserstifts mit Margret Bilger-Galerie, Schlierbach 02707
Peter Rosegger-Museum, Krieglach 02237
Sammlungen des Augustiner Chorherrenstiftes, Sankt Florian 02640
Schloß Ambras, Innsbruck 02130
Schloss Greinburg, Grein an der Donau 01989
Schloss Hof, Schlosshof 02709
Schloß Schönbrunn, Wien 03070
Schloss Tratzberg, Jenbach 02142
Thonet-Museum, Friedberg 01906
Volkskundliche Sammlungen, Straßburg 02787

Bahamas

Balcony House, Nassau 03205

Barbados

Sunbury Planatation House, Saint Phillip 03244

Belgium

Bormshuis, Antwerpen 03281
Design Museum Gent, Gent 03625
Gemeentelijk Museum van Evere, Bruxelles 03447
Hôpital Notre-Dame de la Rose, Lessines 03786
Hospitaalmuseum Onze-Lieve-Vrouw ter Potterie, Brugge 03416
Hotel-Museum Arthur Merghelynck, Ieper 03727
Jeneverhuis Van Hoorebeke, Eeklo 03576
Kantmuseum Olsene, Olsene 03922
Musée de Groesbeeck de Croix, Namur 03900
Musée de la Haute Sûre, Martelange 03855
Musée de la Vie Régionale des Rièzes et des Sarts, Cul-des-Sarts 03532
Musée du Château des Comtes de Marchin, Modave 03874
Musée Horta, Bruxelles 03488
Museum Smidt van Gelder, Antwerpen 03299
Museumkerk Sint Pieter, Rekem 03964
Provinciaal Museum Stijn Streuvels, Ingooigem 03733
Rockoxhuis, Antwerpen 03313
Taxandriamuseum, Turnhout 04102

Bermuda

Saint George's Historical Society Museum, Saint George's 04198
Tucker House Museum, Saint George's 04199
Verdmont Historic House Museum, Smith's Parish 04200

Brazil

Casa dos Sete Candeeiros, Salvador 04786
Fundação Eva Klabin Rappaport, Rio de Janeiro 04693
Museu Capela Santa Luzia, Vitória 04956
Museu Carlos Costa Pinto, Salvador 04789
Museu Casa de Rui Barbosa, Rio de Janeiro 04705
Museu da República, Rio de Janeiro 04719
Museu de Arte da Bahia, Salvador 04794
Museu Dom José, Sobral 04918
Museu Mariano Procópio, Juiz de Fora 04522
Museu Municipal, São João Del Rei 04836
Museu Solar Monjardim, Vitória 04958

Brunei

Bumbongan Dua Belas, Bandar Seri Begawan 04962

Canada

Allan Macpherson House, Napanee 06416
Annandale, Tillsonburg 07083
Anse-Blanchette, Gaspé 05852
Barbour's General Store Museum, Saint John 06834
Barnum House Museum, Grafton 05886
Battlefield House Museum, Stoney Creek 07011
Billings Estate Museum, Ottawa 06529
Brubacher House Museum, Waterloo 07284
Campobello Island Museum, Welshpool 07297
Captain Kennedy Museum and Tea House, Saint Andrews 06796
Casa Loma, Toronto 07100
Centre des Dossiers, Québec 06677
Colborne Lodge, Toronto 07102
Dalnavert Museum, Winnipeg 07357
Daly House Museum, Brandon 05463
Écomusée d'Anticosti, Port-Menier 06645
Eldon House, London 06183
Erland Lee Museum, Stoney Creek 07012
Ermatinger Old Stone House, Sault Sainte Marie 06930
Eva Brook Donly Museum, Simcoe 06979
Fort La Reine Museum and Pioneer Village, Portage-la-Prairie 06651
Green Gables House, Charlottetown 05593
Green Park Shipbuilding Museum and Yeo House, Port Hill 06640
Helmcken House, Victoria 07250
Hillcrest Museum, Souris 06995
Hiscock House, Trinity 07164
Lieu Historique National du Manoir Papineau, Montebello 06324
Little Stone School, Saskatoon 06916
Lucy Maud Montgomery Birthplace, Charlottetown 05594
Maison Centrale des Filles de la Charité du Sacré-Coeur de Jésus, Sherbrooke, Québec 06960
Maison Lamontagne, Rimouski 06753
Maison Louis-Hippolyte Lafontaine, Boucherville 05445
Maison Trestler, Vaudreuil-Dorion 07233
Manoir Mauvide-Jenest, Saint-Jean 06826
Maud's Place - The Putnam House, Brookfield 05487
North Hills Museum, Granville Ferry 05904
Peterborough Centennial Museum and Archives, Peterborough 06595
Point Ellice House, Victoria 07257
Proctor House Museum, Brighton 05478
Quaker William Ray House, Dartmouth 05677
Randall House Museum, Wolfville 07399
Roedde House Museum, Vancouver 07220
Saint Marys Museum, Saint Marys 06869
Seven Oaks House Museum, Winnipeg 07384
Twillingate Museum, Twillingate 07185
La Vieille Maison, Belliveau Cove 05417
Vieux Presbytère de Batiscan, Batiscan 05398
Whitehern Historic House and Garden, Hamilton 05963
Willistead Manor, Windsor, Ontario 07345
Woodchester Villa, Bracebridge 05453
Woodside National Historic Site, Kitchener 06104

Chile

Casa-Museo La Sebastiana, Valparaíso 07470

China

Li Xian Jun Former Residence, Nanjing 08041
Mao Dun Former Residence, Beijing 07549
Song Qingling Former Residence, Shanghai 08195
Song Qingling Tongzhi Former Residence, Beijing 07567
Taipa House Museum, Macau 08008
Zhan Tianyou Former Residence, Wuhan 08333

Croatia

Dvor Trakošćan, Trakošćan 08967
Ivan Meštrović Gallery, Split 08957

Cuba

Castillo de los Tres Reyes del Morro, La Habana 09097
Museo Ambiente Cubano, Gibara 09084
Museo del Ambiente Histórico Cubano, Santiago de Cuba 09316
Museo Doña Rosario, Santiago de Cuba 09321
Museo La Isabelica, Santiago de Cuba 09323
Museo Romántico, Trinidad 09338

Cyprus

House of Hadjigeorgakis Kornesios, Lefkosia 09378

Czech Republic

Městské Muzeum, Holešov 09521
Muzeum města Duchcova, Duchcov 09498
Starý Královský Palác, Praha 09794
Státní Hrad Bouzov, Bouzov 09412
Zámek Hrubý Rohozec, Turnov 09882

Denmark

Egnsmuseet Jacob Michelsens Gård, Aabenraa 09953
Gammel Estrup - Herregårdsmuseet, Auning 09974
KulturhistorieTønder, Tønder 10298
Michael og Anna Anchers Hus og Saxilds Gaard, Skagen 10261
Reventlow-Museet Pederstrup, Horslunde 10102

Dominican Republic

Museo de la Familia Dominicana del Siglo XIX, Santo Domingo 10347

Egypt

Qasr Al-Gawharah, Cairo 10514

Estonia

Memme-Taadi Kamber, Karksi 10582

Finland

Alikartano, Numminen 11151
Anjalan Kartanomuseo, Anjalankoski 10709
Askon Museo, Lahti 11025
Hankasalmen Kotiseutumuseo, Hankasalmi 10755
Hartmanin Mökki, Oripää 11166
Hvitträsk, Kirkkonummi 10963
Kotkaniemi - Presidentti P.E. Svinhufvudin koti, Luumäki 11086
Lebell Residence, Kristiinankaupunki 10996
Liedenpohjan Kylämuseo, Liedenpohja 11066
Louhisaaren Kartanolinna, Askainen 10715
Museet Ett Hem, Turku 11395
Nivalan Museo Katvala, Nivala 11145
Nokian Kotiseutumuseo Hinttala, Nokia 11147
Paikkarin Torppa, Sammatti 11298
Pukkilan Kartanomuseo, Piikkiö 11207
Ruiskumestarin Talo, Helsinki 10814
Söderlångvikin Museo, Dragsfjärd 10719
Taivassalon Museo, Taivassalo 11345
Turkansaaren Ulkomuseo, Oulu 11178
Urajärven Kartanomuseo, Urajärvi 11417
Vanha-Rantalan Talomuseo, Pertunmaa 11201
Yli-Lauroselan Talomuseo, Ilmajoki 10851

France

Château de Bouges, Bouges-le-Château 12084
Château de Lourmarin, Lourmarin 13855
Château de Saint-Aubin-sur-Loire, Saint-Aubin-sur-Loire 15178
Château du Bousquet, Laguiole 13479
Château du Haut-Koenigsbourg, Orschwiller 14541
Château du XVIe s, Ancy-le-Franc 11595
Château-Musée de la Grange, Manom 13939
Château Musée Grimaldi, Cagnes-sur-Mer 12236
Château-Musée Renaissance, Azay-le-Rideau 11809
Domaine National de Chambord, Chambord 12378
La Maison de la Dordogne Quercynoise, Souillac 15793
Musée-Château de Saint-Germain-de-Livet, Saint-Germain-de-Livet 15286
Musée Comtois, Besançon 11955
Musée d'Armes Anciennes, La Cluse-et-Mijoux 13347
Musée d'Art et d'Histoire de Provence, Grasse 13109
Musée d'Art et d'Histoire Marcel Dessal, Dreux 12792
Musée de Jules Gounon-Loubens, Loubens-Lauragais 13828
Musée de la Faucillonnaie, Vitré 16253
Musée de la Folie Marco, Barr 11854
Musée de l'Hôtel-Dieu, Beaune 11902
Musée de l'Hôtel Sandelin, Saint-Omer 15431
Musée départemental de la Bresse - Domaine des Planons, Saint-Cyr-sur-Menton 15228
Musée des Arts et des Meubles, Poilley 14865
Musée Jacquemart-André, Fontaine-Chaalis 12955
Musée Jules Garinet, Châlons-en-Champagne 12363
Musée Lamartine, Mâcon 13908
Musée Louis Philippe, Eu 12885
Musée Magnin, Dijon 12742
Musée National du Château de Compiègne - Musée du Second Empire, Compiègne 12600
Musée Palais Lascaris, Nice 14438
Parc Château de Valencay, Valençay 16037
Pavillon de Vendôme, Aix-en-Provence 11520

Germany

50er Jahre Erlebniswelt, Zusmarshausen 22827
Alt-Rothenburger Handwerkerhaus, Rothenburg ob der Tauber 21405
Alt-Segeberger Bürgerhaus, Bad Segeberg 16836
Altes und Neues Schloss Eremitage, Bayreuth 16944
Altfriesisches Haus, Sylt 21993
Antikes Wohnmuseum, Weinheim 22430
Bauernhausmuseum, Grafenau, Kreis Freyung-Grafenau 18727
Benedikt-Nimser-Haus, Wilhelmsdorf, Kreis Ravensburg 22572
Burg Falkenberg, Falkenberg, Kreis Tirschenreuth 18289
Burg Kronberg, Kronberg im Taunus 19723
Burg Lauenstein, Ludwigsstadt 20063
Burg Meersburg GmbH, Meersburg 20247
Burg zu Burghausen, Burghausen 17651
Coburger Puppen-Museum, Coburg 17714
Couven Museum, Aachen 16395
Flatowturm, Potsdam 21088
Friesenstube Honkenswarf, Langeneß 19830
Fürstlich Ysenburg- und Büdingensches Schloßmuseum, Büdingen 17627
Gentil-Haus, Aschaffenburg 16582
Haus Hohe Pappeln, Weimar 22404
Heimat- und Uhrenmuseum, Villingen-Schwenningen 22236
Heimatmuseum, Spangenberg 21829
Heimatmuseum, Wertach 22501
Heimatmuseum Beutelsbach, Weinstadt 22436
Heimatmuseum der Stadt Herrnhut, Herrnhut 19162
Heimatmuseum Strümpfelbach, Weinstadt 22437
Historische Räume des Stadtschlosses, Fulda 18508
Jagdschloss Grunewald, Berlin 17145
Jagdschloss Stern, Potsdam 21091
Jenisch Haus, Hamburg 18938
Kavalierhaus Gifhorn - Museum für bürgerliche Wohnkultur, Gifhorn 18644
Kunstgewerbemuseum im Schloss Köpenick, Berlin 17169
Kunsthaus Meyenburg, Nordhausen 20737
Marmorpalais, Potsdam 21094
Michaels Privatmuseum, Nellingen 20569
Münchner Stadtmuseum, München 20462
Museum Altes Bürgerhaus, Stolberg (Harz) Kurort 21908
Museum Burg Falkenstein, Falkenstein, Harz 18292
Museum der Stadt Boppard, Boppard 17475
Museum Hohenzollern in Franken, Kulmbach 19749
Museum Knoblauchhaus, Berlin 17210
Museum Nostalgie der 50er Jahre, Burgpreppach 17661
Museum Schloß Burgk, Burgk 17658
Museum Schloss Lembeck, Dorsten 17911
Museum Schloss Neuenburg, Freyburg (Unstrut) 18459
Museum Schloß Rochsburg, Lunzenau 20109
Museum Schloß Weesenstein, Müglitztal 20396
Museum Thonet, Frankenberg, Eder 18344
Museum und Kunstsammlung Schloß Hinterglauchau, Glauchau 18657
Museumswohnung WBS 70, Berlin 17220
Neue Kammern, Potsdam 21099
Neue Residenz Bamberg, Bamberg 16916
Neues Palais, Potsdam 21100
Neues Schloss Bayreuth, Bayreuth 16964
Neues Schloss Wallerstein, Wallerstein 22314
Nietzsche-Archiv, Weimar 22412
Plassenburg, Kulmbach 19750
Pompejanum, Aschaffenburg 16589
Residenz und Hofgarten Würzburg, Würzburg 22720
Residenzschloß Oettingen, Oettingen in Bayern 20879
Römisches Haus, Weimar 22416
Roentgen-Museum, Neuwied 20688
Sammlung Dr. Irmgard von Lemmers-Danforth, Wetzlar 22531
Schloss Babelsberg, Potsdam 21105
Schloss Belvedere, Weimar 22418
Schloss Caputh, Schwielowsee 21708
Schloss Cecilienhof, Potsdam 21106
Schloss Charlottenburg, Berlin 17248
Schloss Charlottenhof, Potsdam 21107
Schloss Drachenburg, Königswinter 19660
Schloss Glienicke, Berlin 17249
Schloss Glücksburg, Glücksburg 54857
Schloss Kirchheim, Kirchheim unter Teck 19550
Schloss Königs Wusterhausen, Königs Wusterhausen 19638
Schloss Luisium, Dessau 17832
Schloss Lustheim, Oberschleißheim 20837
Schloss Oranienbaum, Oranienbaum 20907
Schloss Paretz, Ketzin 19523
Schloss Pfaueninsel, Berlin 17250
Schloss Rheinsberg, Rheinsberg 21306
Schloss Sacrow, Potsdam 21108
Schloß Sanssouci, Potsdam 21109
Schloss Tiefurt, Weimar 22419
Schloss und Schlossgarten Weilburg, Weilburg 22386
Schloß Wernigerode, Wernigerode 22500
Schloß Wilflingen, Langenenslingen 19829
Schloss Wilhelmshöhe-Weißensteinflügel, Kassel 19476
Schloss Wörlitz, Wörlitz 22639
Schlossmuseum, Kirchentellinsfurt 19546
Schlossmuseum Darmstadt, Darmstadt 17794
Schlossmuseum Jever, Jever 19397
Schlossmuseum Oranienburg, Oranienburg 20910
Schloßmuseum und Park Molsdorf, Erfurt 18210
Staatliche Münzsammlung, München 20488
Stadtgeschichtliches Museum, Tangermünde 22003
Städtisches Kramer-Museum, Kempen 19509
Stiftung Schloss Ahrensburg, Ahrensburg 16441
Stuck-Museum, Freiburg im Breisgau 18439
Stuhlbau- und Heimatmuseum, Rabenau 21157
Stuhlmuseum Burg Beverungen, Beverungen 17316
Technisches Museum Frohnauer Hammer, Annaberg-Buchholz 16556
Thüringer Landesmuseum Heidecksburg, Rudolstadt 21431
Tucherschloss mit Hirsvogelsaal, Nürnberg 20792
Vogtlandmuseum Plauen, Plauen 21055
Vonderau Museum Fulda, Fulda 18513
Weingutmuseum Hoflößnitz, Radebeul 21161
Wittumspalais, Weimar 22423

Greece

National Resistance Museum, Karpenissi 23038

Guatemala

Museo de Arte Colonial, Antigua 23289

Hungary

Bodrogközi Kastélymúzeum, Pácin 23745
Endrődi Tájház, Gyomaendrőd 23576
Fabricius-Ház, Sopron 23809
Ladics-ház, Gyula 23597
Mudrány Kúria, Szabolcs 23832
Nagytétényi Kastélymúzeum, Budapest 23454

India

Darbarhall Museum, Junagadh 24247

Indonesia

Museum Asserajah El Hasyimiah Palace, Bengkalis 24451

Ireland

Bantry House, Bantry 24743
Bunratty Castle and Folkpark, Bunratty 24748
Castletown, Celbridge 24758
Clonalis House, Castlerea 24757
Rothe House Museum, Kilkenny 24849

Tecck an Phiarsaigh, Rosmuck 24894
Westport House, Westport 24911

Israel

Liebermann House, Nahariya 25069

Italy

Casa Da Noal, Treviso 28068
Casa Natale di A. Rosmini, Rovereto 27568
Castello Malaspina, Bobbio 25432
Civica Raccolta d'Arte Applicata, Milano 26675
Civico Museo del Castello di San Giusto, Trieste 28077
Civico Museo Revoltella e Galleria d'Arte Moderna, Trieste 28084
Civico Museo Sartorio, Trieste 28085
Fondazione Magnani Rocca, Mamiano di Traversetolo 26580
Museo Bagatti Valsecchi, Milano 26690
Museo Civico Casa Cavassa, Saluzzo 27602
Museo d'Arte e Ammobiliamento, Nichelino 26924
Museo della Fondazione Horne, Firenze 26171
Museo dell'Arredo Contemporaneo, Russi 27580
Museo di Palazzo Reale Genova, Genova 26307
Museo e Pinacoteca Nazionale di Palazzo Mansi, Lucca 26535
Museo Mario Praz, Roma 27510
Palazzina Marfisa, Ferrara 26120
Palazzo Pfanner, Lucca 26537
Palazzo Reale, Torino 28029
Rocca Sanvitale, Fontanellato 26221
Villa San Michele, Anacapri 25224
Walsermuseum, Alagna Valsesia 25170

Jamaica

Greenwood Great House Museum, Greenwood 28323

Japan

Kagu no Hakubutsukan, Tokyo 29405
Kanemori Museum with Baccarat Collection, Hakodate 28475
Nasu Queen's Museum, Nasu 29050
Niitsu House, Niigata 29060

Korea, Republic

Korean Furniture Museum, Seoul 29896

Lithuania

Žemaitijos Kaimo Muziejus, Telšiai 30380

Luxembourg

Palais Grand Ducal, Luxembourg 30485

Malta

Palazzo Armeria, Zurrieq 30663

Mexico

Casa Luis Barragán, México 31036
Museo Casa de León Trotsky, México 31059
Museo Casa del Alfeñique, Puebla 31268
Museo Casa del Marques de Aguayo y Museo Municipal, Mazapil Zacatecas 31019
Museo Colonial del Conde de Sierra Gorda, Santander Jiménez 31382
Museo de las Leyendas, Guanajuato 30915
Museo Ex Hacienda San Gabriel de Barrera, Guanajuato 30919
Museo Jarocho Salvador Ferrando, Tlacotalpan 31464
Museo José María Morelos y Pavón, Cuautla 30814
Museo Palacio de la Canal, San Miguel de Allende 31357
Museo Timoteo L. Hernández, Villaldama 31556

Moldova

Ščusev Museum, Chişinău 31604

Monaco

Palais Princier, Monaco 31614

Morocco

Nejarine-Musée, Fès 31672

Namibia

Duwisib Castle, Maltahohe 31746

Netherlands

Borg Verhildersum, Leens 32468
Dekema State, Jelsum 32431
Heemkundemuseum Paulus Van Daesdonck, Ulvenhout 32839
Heemkundig Streekmuseum Jan Uten Houte, Etten Leur 32225
Hilleshuis Museum, Rotterdam 32706
Huis Doorn, Doorn 32143
Huis Verwolde, Laren, Gelderland 32463
Huis Zypendaal, Arnhem 31926
Huize Betje Wolff, Middenbeemster 32557
Jachthuis Sint-Hubertus, Hoenderloo 32397
Kasteel Middachten, De Steeg 32051
Kasteel Rosendael, Rozendaal 32737
Koninklijk Paleis te Amsterdam, Amsterdam 31855
Muiderslot, Muiden 32562
Museum Ald Slot, Wergea 32936
Museum Beeckestijn, Velsen Zuid 32882
Museum Geelvinck Hinlopen Huis, Amsterdam 31861
Museum Hindeloopen, Hindeloopen 32391
Museum It Kokelhûs van Jan en Sjut, Earnewald 32164
Museum Spaans Gouvernement, Maastricht 32527
Museum van de Stichting Bisdom van Vliet, Haastrecht 32305
Museum Van Loon, Amsterdam 31866
Museum Willet-Holthuysen, Amsterdam 31869
Museumboerderij Gilde Koat, Zeddam 32992
Ons Museum, Wintelre 32955
Saksische en Museumboerderij Erve Brooks Niehof, Gelselaar 32241
Schepenzaal Stadhuis, Zwolle 33028
Schultehuis, Diever 32130
Schutterskamer, Sneek 32791
Sint-Pieters Museum op de Lichtenberg, Maastricht 32533
Slot Loevestein, Poederoijen 32670
Stedelijk Museum Zwolle, Zwolle 33029
Wiechers Woon Oase, Dwingeloo 32163
't Wienkeltje van Wullempje, Hoedekenskerke 32394

New Zealand

Alberton, Auckland 33040
Ewelme Cottage, Auckland 33047
Theomin Gallery, Dunedin 33099

Northern Cyprus

Lüzinyan Evi, Lefkoşa 33370

Norway

Asker Museum, Hvalstad 33572
Flekkefjord Museum, Flekkefjord 33497
Gamle Bergen Museum, Bergen 33426
Grindheim Bygdemuseum, Kollungtveit 33597
Herregården Museum, Larvik 33628
Merdøgård, Arendal 33397
Olav Holmegaards Samlinger, Mandal 33659
Roparshaugsamlinga, Isdalstø 33578
Selbu Bygdemuseum, Selbu 33823

Oman

Bait Al Zubair, Muscat 33987

Paraguay

Museo Doctor Francia, Yaguarón 34104

Peru

Museo de Arte Colonial Pedro de Osma, Lima 34231

Philippines

Bernandino Jalandoni Ancestral House, Silay 34506
CAP Art Center and Don Sergio Osmeña Memorabilia, Cebu 34365
Casa Manila Museum, Manila 34436

Poland

Muzeum Narodowe w Poznaniu, Oddział w Rogalinie, Rogalin 35081
Muzeum Wnętrz Dworskich, Ożarów 35001
Muzeum Wnętrz Pałacowych w Choroszczy, Choroszcz 34602
Muzeum Wnętrz Stylowych, Oporów 34988
Muzeum Wsi Lubelskiej, Lublin 34909
Muzeum Zespoł Zamkowy w Niedzicy, Niedzica 34943
Muzeum Ziemi Średzkiej Dwór w Koszutach, Koszuty 34762
Pałac Radziwiłłów w Nieborowie, Nieborów 34942
Zamek Królewski na Wawelu - Państwowe Zbiory Sztuki, Kraków 34821
Zamek w Gniewie - Muzeum Regionalne, Gniew 34671

Portugal

Casa-Museu Egas Moniz, Estarreja 35570
Casa Museu Manuel Nunes Corrêa, Pedrógão Grande 35812
Casa Museu Mestre João da Silva, Lisboa 35663
Casa Museu Vieira Natividade, Alcobaça 35391
Museu Biblioteca Anselmo Braamcamp Freire, Santarém 35908
Museu da Casa Cadaval, Évora 35584
Museu da Fundação da Casa de Mateus, Vila Real 36011
Museu Palácio Nacional da Ajuda, Lisboa 35725
Paço dos Duques de Bragança, Guimarães 35642
Palácio Nacional da Pena, Sintra 35951
Solar Condes de Resende, Canelas 35504
Solar Museu Quinta dos Cónegos, Maia 35753

Puerto Rico

Casa Dr. José Celso Barbosa, Bayamón 36028
Museo Casa Blanca, San Juan 36065
Museo Castillo Serrallés, Ponce 36048
Museum of the Puerto Rican Family of the 19th Century, San Juan 36077

Romania

Hanul Domnesc, Suceava 36643
Muzeul Casa Mureşenilor, Braşov 36177
Muzeul Nicolae Iorga, Vălenii de Munte 36722

Russia

Dom-muzej Gannibalov v Petrovskom, Puškinskie Gory 37585
Dom-muzej E.M. Jaroslavskogo, Jakutsk 36959
Dom-muzej Pavla Kuznecova, Saratov 37748
Dom-muzej Osipovych-Vulf v Trigorskom, Puškinskie Gory 37586
Dom-muzej V.M. Vasnecova, Moskva 37169
Dvorec-Muzej Petra I, Sankt-Peterburg 37636
Dvorec-muzej Petra III, Lomonosov 37107
Gosudarstvennyj Istoričeskij Zapovednik Gorki Leninskie, Gorki Leninskie 36926
Jurinskij Istoriko-chudožestvennyj Muzej, Jurino 36982
Muzej Mebeli v Moskve, Moskva 37331

Serbia

Dvorac u Čelarevu, Čelarevo 38097
Njegošev Muzej, Cetinje 38098

Slovakia

Expozícia Meštianskej Bytovej Kultúry, Kežmarok 38265
Múzeum Červený Kameň, Častá 38244
Nábytkové Múzeum, Markušovce 38292
Slovenské Národné Múzeum - Múzeum Betliar, Betliar 38199

Slovenia

Dvorec Štatenberg, Makole 38474
Mestni Muzej Ljubljana, Ljubljana 38453

South Africa

Bertram House Museum, Cape Town 38578
Caledon Museum, Caledon 38576
Campbell Collections, Durban 38603
Drostdy Museum, Swellendam 38744
Dunluce House Museum, Kimberley 38668
Gately House, East London 38616
Graaf-Reinet Museum, Graaff-Reinet 38626
Great Fish River Museum, Cradock 38599
Koopmans-De Wet House Museum, Cape Town 38585
Macrorie House Museum, Pietermaritzburg 38705
Marie Rawdon Museum, Matjiesfontein 38687
Montagu Museum, Montagu 38690
Natale Labia Museum, Muizenberg 38692
Old House Museum, Durban 38610
De Oude Drostdy, Tulbagh 38765
Oude Kerk Volksmuseum, Tulbagh 38766
Our Heritage Museum, Adelaide 38550
Richmond Museum, Richmond 38722
Rudd House Museum, Kimberley 38673
Somerset East Museum, Somerset East 38733
Stellenbosch Museum, Stellenbosch 38738
Temlett House, Grahamstown 38634
Worcester Museum, Worcester 38779

Spain

Casa de Pilatos, Sevilla 40060
L'Enrajolada Casa-Museu Santacana, Martorell 39626
Museo Rodera Robles, Segovia 40055
Museo Taller del Moro, Toledo 40138
Museu Romantic Can Papiol, Vilanova i la Geltrú 40286

Sweden

Charlotte Berlins Museum, Ystad 41075
Ebbas Hus, Malmö 40718
Ekenäs Slott, Linköping 40680
Friluftmuseet Skansen, Stockholm 40895
Gustav III:s Paviljong, Solna 40870
Julita Sveriges Lantbruksmuseums, Julita 40621
Kattlunds Museigård, Havdhem 40583
Läckö Slott, Lidköping 40674
Lundströmska Gården, Sigtuna 40834
Nääs Slott Slöjdseminarium, Floda 40524
Nobynäs Säteri, Aneby 40423
Petes Museigård, Havdhem 40584
Rosendals Slott, Stockholm 40931
Steninge Slott Kulturcenter, Märsta 40712
Stjernsunds Slott, Askersund 40445
Svaneholms Slotts Museum, Skurup 40856
Tjolöholms Slott, Fjärås 40523
Ulriksdals Slott samt Orangerimuseet, Solna 40875
Virserums Möbelindustrimuseum, Virserum 41066

Switzerland

Château, La Sarraz 41822
Château de Grandson, Grandson 41450
Château et Musée de Valangin, Valangin 41971
Collection Fondation Zoubov, Genève 41410
Engadiner Museum, Sankt Moritz 41811
Fundaziun Planta Samedan Wohnmuseum, Samedan 41800
Heimatmuseum Allschwil, Allschwil 41101
Heimatmuseum Schanfigg, Arosa 41126
Maison Michel, Coppet 41314
Museum Bärengasse, Zürich 42091
Museum Engiadine Bassa, Scuol 41852
Museum für Wohnkultur des Historismus und des Jugendstils, Hilterfingen 41482
Museum Langmatt, Baden 41148
Palazzo Castelmur, Stampa 41903
Schloß Tarasp, Tarasp 41928
Schlossmuseum, Oberhofen am Thunersee 41697

Taiwan

Chinese Furniture Museum, Taoyuan 42451
Furniture Manufacturing Museum in Tainan, Rende 42306
Museum of Bamboo Art, Nantou 42280

Turkey

Atatürkün Mekanı Müze Köşk, Ankara 42699
İnönü Evi Müzesi, İzmir 42854
On Yedinci Yüzyıl Osmalı Evi, Bursa 42745
Şemaki Evi Müzesi, Yenisehir 42966
Tahir Pasa Konagi, Mudanya 42911

Ukraine

Alupka Deržavnyj Palac ta Park Zapovidnyk, Alupka 42998
Municypal'nyj Muzej Pryvatnych Kollekcij im. O. Bleščunova, Odesa 43156

United Kingdom

Anne Hathaway's Cottage, Stratford-upon-Avon 45779
Ardress Farmyard Museum, Annaghmore 43304
The Argory, Dungannon 44057
Ascott House, Leighton Buzzard 44687
Astley Hall Museum and Art Gallery, Chorley 43840
Aston Hall, Birmingham 43495
Athelhampton House, Dorchester 43997
Attingham Park, Shrewsbury 45671
Audley End House, Saffron Walden 45538
Banqueting House, London 44790
Belgrave Hall Museums and Gardens, Leicester 44673
Belton House, Grantham 44344
Bishops' House Museum, Sheffield 45649
Blackwell - The Arts and Crafts House, Bowness-on-Windermere 43571
Bolling Hall Museum, Bradford 43575
Brodsworth Hall Gardens, Doncaster 43991
Buckland Abbey, Yelverton 46094
Building of Bath Museum, Bath 43403
Burton Constable Hall, Kingston-upon-Hull 44593
Castle Ward, Downpatrick 44019
Chatsworth House, Bakewell 43354
Chethams's Hospital and Library, Manchester 45081
Chiswick House, London 44820
Church Farmhouse Museum, London 44821
Clarke Hall, Wakefield 45907
Claydon House, Middle Claydon 45137
Clergy House, Alfriston 43279
Cotehele House, Saint Dominick, Cornwall 45562
Cottage Museum, Lancaster 44630
Crathes Castle, Banchory 43373
Croxteth Hall, Liverpool 44733
Dalemain Historic House, Penrith 45356
Dan Winters House - Ancestral Home, Loughgall 45036
Drumlanrig Castle, Thornhill, Dumfriesshire 45855
Dudmaston House, Quatt 45454
Dunham Massey Hall, Altrincham 43294
East Riddlesden Hall, Keighley 44551
Eltham Palace, London 44843
English Heritage South East Region, Dover 44014
Epworth Old Rectory, Epworth 44184
Erddig, Wrexham 46082
Fairfax House Museum, York 46102
Felbrigg Hall, Norwich 45278

F

Fenton House, London 44847
Firle Place, Firle 44230
Florence Court House, Enniskillen 44179
Ford Green Hall, Stoke-on-Trent 45760
Gainsborough Old Hall, Gainsborough 44259
Geffrye Museum, London 44854
Georgian House, Bristol 43635
The Georgian House, Edinburgh 44125
Glamis Castle, Glamis 44273
Goodwood House, Chichester 43827
Haddo House, Methlick 45134
Haden Hall House, Cradley Heath 43911
Hall i'th' Wood Museum, Bolton 43555
Hall's Croft, Stratford-upon-Avon 45780
Hampton Court Palace, East Molesey 44093
Hardwick Hall, Chesterfield 43821
Haremere Hall, Etchingham 44188
Harewood House, Harewood 44394
Hauteville House, Saint Peter Port 45592
Heaton Hall, Manchester 45084
Hellens Manor, Much Marcle 45184
Hezlett House, Coleraine 43880
Hill House, Helensburgh 44438
Hill of Tarvit Mansion House, Cupar 43942
Hinchingbrooke House, Huntingdon 44501
Hopetoun House, South Queensferry 45695
Ickworth House, Horringer 44485
Ingatestone Hall, Ingatestone 44515
Inveraray Castle, Inveraray 44519
Judges' Lodgings, Lancaster 44631
Kelmscott Manor, Kelmscott 44554
Kirkham House, Paignton 45343
Knebworth House, Knebworth 44621
Knole, Sevenoaks 45641
Lennoxlove House, Haddington 44376
Lydiard House, Swindon 45817
Lytham Hall, Lytham Saint Anne's 45059
Manor House, Donington-le-Heath 43996
Mary Arden's House and the Countryside Museum, Stratford-upon-Avon 45782
Melford Hall, Long Melford, Sudbury, Suffolk 45027
Michelham Priory, Hailsham 44380
Mount Edgcumbe House, Cremyll 43921
Mount Stewart House, Newtownards 45261
Nash's House and New Place, Stratford-upon-Avon 45783
National Portrait Gallery, Montacute 45169
Normanby Hall, Scunthorpe 45626
Norton Conyers, Ripon 45500
Nostell Priory, Wakefield 45911
Number 1 Royal Crescent Museum, Bath 43409
Oak House Museum, West Bromwich 45967
Oakwell Hall Country Park, Birstall 43519
Old Bridge House Museum, Dumfries 44038
Old House, Hereford 44457
Osborne House, East Cowes 44084
Packwood House, Lapworth 44641
Plas Mawr, Conwy 43887
Plas Newydd, Llangollen 44772
Preston Manor, Brighton 43623
Provan Hall House, Glasgow 44296
Provost Skene's House, Aberdeen 43246
Rainham Hall, Rainham 45456
Red Lodge, Bristol 43638
Revolution House, Chesterfield 43823
Royal Pavilion, Brighton 43625
Rufford Old Hall, Rufford 45526
Salomons Museum, Southborough 45709
Scone Palace, Perth 45367
Seaton Delaval Hall, Seaton 45630
Selly Manor Museum, Birmingham 43511
Shakespeare Birthplace Trust, Stratford-upon-Avon 45785
Shibden Hall Museum and West Yorkshire Folk Museum, Halifax 44388
Shrewsbury Castle, Shrewsbury 45675
Smithills Hall Museum, Bolton 43556
Southchurch Hall Museum, Southend-on-Sea 45713
Southside House, London 44989
Spencer House, London 44990
Springhill House, Magherafelt 45071
Stourhead House, Stourton 45770
Strangers Hall Museum of Domestic Life, Norwich 45287
Stratfield Saye House and Wellington Exhibition, Stratfield Saye 45778
Strathnaver Museum, Bettyhill 43465
Syon House, London 44994
Tenement House, Glasgow 44305
Town House Museum of Lynn Life, King's Lynn 44589
Treasurer's House, York 46107
Trevithick Cottage, Camborne 43722
Tudor House Museum, Southampton 45708
Turton Tower, Turton Bottoms 45885
Ugbrooke House, Chudleigh 43844
The Vyne, Basingstoke 43396
Wallington, Cambo 43721
Walmer Castle, Walmer 45918
Wightwick Manor, Wolverhampton 46059
Wilton House, Wilton, Salisbury 46019
Woburn Abbey, Woburn 46054
Wythenshawe Hall, Manchester 45100

United States of America

Ace of Clubs House, Texarkana 53488
Adsmore Museum, Princeton 52145
African Art Museum of the S.M.A. Fathers, Tenafly 53476
Alexander Ramsey House, Saint Paul 52612
Amos Herr House, Landisville 49935
The Apple Trees Museum, Burlington 47131
Ashtabula Plantation, Pendleton 51758
Ashtabula Plantation, Pendleton 51758
Barbara Fritchie House Museum, Frederick 48754
Baxter House Museum, Gorham 48975
Beauchamp Newman Museum, Elizabeth 48337
Bloom Mansion, Trinidad 53567
Bluff Hall Museum, Demopolis 48054
Boscobel Restoration, Garrison 48868
Burnside Plantation, Bethlehem 46777
Byers-Evans House Museum, Denver 48065
Campbell Whittlesey House, Rochester 52393
Camron-Stanford House, Oakland 51463
Captain George Flavel House Museum, Astoria 46400
Carpenter Home Museum, Cle Elum 47603
Casa Navarro, San Antonio 52712
Cedar Grove Mansion, Philadelphia 51820
Chief John Ross House, Rossville 52465
Chinqua-Penn Plantation, Reidsville 52285
Codman Estate - The Grange, Lincoln 50089
Collinsville Depot Museum, Collinsville 47695
Colony House, Newport 51320
Copper King Mansion, Butte 47151
Cranbrook House, Bloomfield Hills 46856
Cuneo Museum, Vernon Hills 53725
Cupola House, Egg Harbor 48303
Darnall's Chance House Museum, Upper Marlboro 53664
David Conklin Farmhouse, Huntington 49428
Dinsmore Homestead, Burlington 47135
Donkey Milk House, Key West 49757
Donkey Milk House, Key West 49757
Dumbarton House, Washington 53864
Dyckman Farmhouse Museum, New York 51168
E. St. Julien Cox House, Saint Peter 52632
Edison and Ford Winter Estates, Fort Myers 48661
Elisabet Ney Museum, Austin 46507
Elizabeth Perkins House, York 54333
Excelsior House, Jefferson 49608
Ezekiel Harris House, Augusta 46476
Fabyan Villa Museum and Dutch Windmill, Geneva 48878
Fairbanks House, Dedham 48024
Fairlawn Mansion and Museum, Superior 53378
Fall River County Historical Museum, Hot Springs 49374
Frances Willard House Museum, Evanston 48450
Gaineswood Mansion, Demopolis 48055
General Adam Stephen House, Martinsburg 50519
General Wait House, Waitsfield 53788
Gertrude Smith House, Mount Airy 50869
Gilbert House, Atlanta 46439
Glenn House, Cape Girardeau 47233
Glensheen-Historic Congdon Estate, Duluth 48195
Gore Place, Waltham 53809
Gracie Mansion, New York 51186
Granger House, Marion 50480
Groton Historical Museum, Groton 49096
Haas-Lilienthal House, San Francisco 52776
Hackley Hume Houses, Muskegon 50924
Hamilton House, South Berwick 53163
Hanley House, Clayton 47599
Harvey House Museum, Florence 48569
Helena Modjeska Historic House, Modjeska Canyon 50776
Heyward-Washington House, Charleston 47366
Hill-Hold Museum, Montgomery 50817
The Hill House, Portsmouth 52107
Hillforest House Museum, Aurora 46495
Historic Deepwood Estate, Salem 52667
Historic General Dodge House, Council Bluffs 47850
Historic Newton Home, Decatur 48019
Historic Rosedale, Charlotte 47389
Hixon House, La Crosse 49835
Home Sweet Home Museum, East Hampton 48246
Howard-Dickinson House Museum, Henderson 49252
Howland House, Plymouth 51993
Hyland House, Guilford 49103
Jacob Kelley House Museum, Hartsville 49204
James Dunklin House, Laurens 49987
James J. Hill House, Saint Paul 52618
Jeremiah Lee Mansion, Marblehead 50460
John Jay Homestead, Katonah 49708
John Wornall House Museum, Independence 49477
Josiah Quincy House, Quincy 52219
Julian H. Sleeper House, Saint Paul 52619
Kelly-Griggs House Museum, Red Bluff 52270
Kentucky State Capitol, Frankfort 48739
Kimball House Museum, Battle Creek 46644
King Hooper Mansion, Marblehead 50461
Lexington County Museum, Lexington 50067
Locust Grove, Poughkeepsie 52122
Locust Lawn and Terwilliger House, Gardiner 48859
Longfellow-Evangeline State Historic Site, Saint Martinville 52604
Loudoun Mansion, Philadelphia 51849
McConaghy House, Hayward 49229
McFaddin-Ward House, Beaumont 46678
Mallory-Neely House, Memphis 50575
Mary Todd Lincoln House, Lexington 50053
Merchant's House Museum, New York 51211
Merwin House, Stockbridge 53299
Molly Brown House Museum, Denver 48082
Montpelier Cultur Arts Center, Laurel 49983
Montpellier Mansion, Laurel 49984
Moss Mansion Museum, Billings 46804
Mount Clare Museum House, Baltimore 46580
Mount Pleasant, Philadelphia 51853
National Trust for Historic Preservation, Washington 53910
Neill-Cochran Museum House, Austin 46516
Old Capitol, Iowa City 49518
Old Constitution House, Windsor 54229
Old Hoxie House, Sandwich 52840
Old Ordinary House Museum, Hingham 49298
Old Westbury Gardens, Old Westbury 51539
Orlando Brown House, Frankfort 48742
Parry Mansion Museum, New Hope 51062
Pearl S. Buck House, Perkasie 51776
Pena-Peck House, Saint Augustine 52524
Phelps House, Burlington 47134
Philadelphia Society for the Preservation of Landmarks, Philadelphia 51869
Philipse Manor Hall, Yonkers 54332
Pointe Coupee Museum, New Roads 51101
Pownalborough Court House, Dresden 48182
Propst House and Marple Grove, Hickory 49268
Rancho Los Alamitos, Long Beach 50177
Rancho Los Cerritos, Long Beach 50178
Rebecca Nurse Homestead, Danvers 47957
Restoration House, Mantorville 50455
Riordan Mansion, Flagstaff 48553
Rising Sun Tavern, Fredericksburg 48772
Robert Fulton Birthplace, Quarryville 52209
Rosemont Plantation, Woodville 54289
Rotch-Jones-Duff House and Garden Museum, New Bedford 51015
Rothschild House, Port Townsend 52047
Rough Point, Newport 51329
Sagamore Hill National Historic Site, Oyster Bay 51663
Salem Maritime Houses, Salem 52658
Sam Brown Log House, Browns Valley 47094
Sayward-Wheeler House, York Harbor 54343
Schmidt House Museum, Grants Pass 49017
Scotchtown, Beaverdam 46683
Seguine-Burke Plantation, Staten Island 53266
The Shadows-on-the-Teche, New Iberia 51064
Sonnenberg Mansion, Canandaigua 47215
Spencer-Peirce-Little Farm, Newbury 51303
The Stephen Girard Collection, Philadelphia 51882
Stephen Phillips Memorial House, Salem 52661
Stone-Tolan House, Rochester 52399
Suffolk County Vanderbilt Museum, Centerport 47315
The Taft Museum of Art, Cincinnati 47565
Taylor Brown and Sarah Dorsey House, Midland 50675
Texas Governor's Mansion, Austin 46518
Thomas Griswold House and Museum, Guilford 49104
Thomas Whaley House Museum, San Diego 52757
Tinker Swiss Cottage Museum, Rockford 52418
Travellers Rest Plantation and Museum, Nashville 50977
Tullis-Toledano Manor, Biloxi 46813
Turner House Museum, Hattiesburg 49213
United States Department of the Interior Museum, Washington 53924
Victoria Mansion, Portland 52069
Vikingsholm Museum, Tahoma 53420
Villa Montezuma, San Diego 52760
Ward Hall, Georgetown 48892
Waveland Museum, Lexington 50058
Wentworth Gardner and Tobias Lear Houses, Portsmouth 52102
Westover House, Charles City 47353
Wharton Esherick Museum, Malvern 50411
White Hall Historic Site, Richmond 52317
William Heath Davis House Museum, San Diego 52762
Winslow Crocker House, Yarmouth Port 54326
Woodburn Plantation, Pendleton 51760
Wyck Museum, Philadelphia 51890
The Wyckoff Farmhouse Museum, Brooklyn 47085

Venezuela

Casa San Isidro, Ciudad Bolívar 54562
Museo Biblioteca Rosauro Rosa Acosta, Pampatar 54590
Museo Inés Mercedes Gómez Álvarez, Guanare 54574

Games → Toys and Games

Gas → Petroleum Industry and Trade

Gems → Precious Stones

Genealogy

Australia

Australian Institute of Genealogical Studies, Blackburn 00873
Hawkesbury Regional Museum, Windsor, New South Wales 01656
Hope Cottage, Kingscote 01202

Austria

Esterházy-Ausstellung, Eisenstadt 01859
Großer Wappensaal im Landhaus, Klagenfurt 02183
K-HOF Kammerhof Museen Gmunden, Gmunden 01951
Liechtenstein Schloss Wilfersdorf, Wilfersdorf 03107

Canada

Alberton Museum, Alberton 05313
Beaconsfield Historic House, Charlottetown 05589
Cape Sable Historical Society Centre, Barrington 05390
Musée Acadien de Pubnico-Ouest, West Pubnico 07303

China

Fan Tower Museum, Fanshi 07685

Germany

Museum The Kennedys, Berlin 17216

Ireland

Nenagh Heritage Centre, Nenagh 24883

Korea, Republic

Royal Museum, Seoul 29912

Latvia

Kurzemes Hercogu Kapenes, Jelgava 30018

Mexico

Museo Norawa, Guachochi 30883

Netherlands

Huygensmuseum Hofwijck, Voorburg 32905
Museum Buren en Oranje, Buren, Gelderland 32025

New Zealand

Albertland and Districts Museum, Wellsford 33262
Waipu Heritage Centre, Waipu 33224

Philippines

Balay Negrense Museum, Silay 34505
Clarin Ancestral House, Loay 34406
Teodoro P. Resurreccion Memorial Museum, Luna 34409

South Africa

Huguenot Memorial Museum, Franschhoek 38621

Spain

Museo de Heráldica Alavesa, Mendoza 39644

United Kingdom

Clare Ancient House Museum, Clare 43848
Greenwich Local History Centre, London 44864
Mountbatten Exhibition, Romsey 45512

United States of America

Adams County Historical Society Museum, Gettysburg 48900
Anoka County Museum, Anoka 46323
Bellflower Genealogical and Historical Society Museum, Bellflower 46710
Bronck Museum, Coxsackie 47863
Camden Archives and Museum, Camden 47199
Cayuga-Owasco Lakes Historical Society Museum, Moravia 50839
Durham Center Museum, East Durham 48238
Dyer Memorial Library, Abington 46127

Fryer Memorial Museum, Munnsville 50908
Genoa Courthouse Museum, Genoa 48885
Hagadorn House Museum, Almond 46236
Hamilton Library and Two Mile House, Carlisle 47243
Held-Poage Memorial Home Research Library, Ukiah 53639
Historical Society of Middletown and the Wallkill Precinct, Middletown 50661
House of the Temple, Washington 53878
Interlaken Historical Society Museum, Interlaken 49512
McClurg Mansion, Westfield 54094
Mayville Area Museum of History and Genealogy, Mayville 50540
Minisink Valley Historical Society Museum, Port Jervis 52037
Monroe County Local History Room and Library, Sparta 53192
Northampton County Historical and Genealogical Society Museum, Easton 48271
Old Brutus Historical Society Museum, Weedsport 54020
Oliver House Museum, Penn Yan 51764
Ontario County Historical Society Museum, Canandaigua 47214
Platt R. Spencer Memorial Archives and Special Collections Area, Geneva 48882
Rensselaer County Historical Society Museum, Troy 53575
Rome Historical Society Museum, Rome 52453
Saint Marys and Benzinger Township Museum, Saint Marys 52607
Sidney Historical Association Museum, Sidney 53098
Terwilliger Museum, Waterloo 53968
Town of Clarence Museum, Clarence 47580
Union County Museum, Lewisburg 50032

Geography

Argentina

Complejo Museológico del Instituto Félix Bernasconi, Buenos Aires 00155

Australia

Edward de Courcy Clarke Earth Science Museum, Crawley 01009

Austria

Lungauer Landschaftsmuseum, Mauterndorf 02371

Brazil

Museu Histórico e Geográfico de Poços de Caldas, Poços de Caldas 04637

Dominican Republic

Museo de Historia y Geografía, Santo Domingo 10345

Egypt

The Egyptian Geological Museum, Cairo 10491

France

Maison des Sirènes et des Siréniens, Castellane 12307

Germany

Universum Bremen, Bremen 17565

Hungary

Erdészeti-, Faipari- és Földméréstörténeti Gyűjtemény, Sopron 23808
Magyar Földrajzi Gyűjtemény, Érd 23548

Italy

Società Geografica Italiana, Roma 27547

Netherlands

Abel Tasman Kabinet, Lutjegast 32513

Poland

Dworek Wincentego Pola, Lublin 34896

Portugal

Casa do Corpo Santo, Setúbal 35934
Museu das Ciências da Terra, Estremoz 35577
Museu do Instituto Geográfico Português, Lisboa 35702

Russia

Muzej Geodezii i Kartografii, Moskva 37277
Russkij Gosudarstvennyj Muzej Arktiki i Antarktiki, Sankt-Peterburg 37723

Switzerland

Museum Langenthal, Langenthal 41527

United Kingdom

National Maritime Museum Cornwell, Falmouth 44209
Tower Bridge Exhibition, London 44998

United States of America

Explorers Hall, Washington 53867
Turquoise Museum, Albuquerque 46188

Geology

Andorra

Museu obert de geologia 'Roques al Carrer', Andorra la Vella 00086

Argentina

Museo Mineralógico Prof. Manuel Tellechea, Mendoza 00457

Armenia

Geological Museum, Yerevan 00735

Australia

Albert Kersten GeoCentre, Broken Hill 00902
Norseman Historical and Geological Collection, Norseman 01371
School of Earth Sciences Museum, Townsville 01600
Tasmanian Museum and Art Gallery, Hobart 01159

Austria

Alpineum, Hinterstoder 02079
Alpinmuseum Dachstein, Ramsau am Dachstein 02557
Amethystwelt Maissau, Maissau 02344
Erstes österreichisches Bodenschätzungsmuseum, Hollabrunn 02095
Geologische Sammlung Schloß Kapfenstein, Kapfenstein 02152
Geologisches Freilichtmuseum, Gmünd, Niederösterreich 01944
GeoZentrum Gams, Gams bei Hieflau 01923
Granatium, Radenthein 02549
Haus der Natur, Salzburg 02615
K-HOF Kammerhof Museen Gmunden, Gmunden 01951
Landesmuseum Burgenland, Eisenstadt 01861
Mineralien-Fossilien-Bergbau, Langenlois 02259
Museum Steinmetzhaus, Eggenburg 01844
Naturwissenschaftliche Sammlungen am Landesmuseum Joanneum, Graz 01981
SBW Terra Mystica Montana, Bad Bleiberg-Kreuth 01740
Städtisches Museum, Schärding 02696
Wienerwaldmuseum, Eichgraben 01851

Bangladesh

Geological Survey Museum, Dhaka 03223

Belgium

Archéoscope du Pays de Salm, Vielsalm 04110
Jardin Géologique d'Obourg, Obourg 03919
Musée de Minéralogie et de Géologie, Bruxelles 03474
Musée de Wanne, Trois-Ponts 04095
Musée du Monde Souterrain, Han-sur-Lesse 03675
Musée Hydrogéologique, Chaudfontaine 03519

Brazil

Museu de Geociências, São Paulo 04887
Museu de Geociências da Universidade de Brasília, Brasília 04361
Museu de Geologia, Maringá 04564
Museu Geológico da Bahia, Salvador 04803
Museu Geológico Valdemar Lefèvre, São Paulo 04901

Canada

Banff Park Museum, Banff 05380
Earth Sciences Museum, Waterloo 07286
Geological Museum, Saskatoon 06913
Logan Hall, Ottawa 06550
M.W. Williams Geological Museum, Vancouver 07214
Maison du Granit, Lac-Drolet 06117
Miller Museum of Geology, Kingston 06084
Moose River Gold Mines Museum, Middle Musquodoboit 06274
Musée de géologie René-Bureau, Sainte-Foy 06894
Museum of Geology, Edmonton 05750

China

Exhibition Hall of Geology, Beijing 07537
Geological Museum of China, Beijing 07538
Keshike Tengqi Museum, Chifeng 07627
Sun Yat-Sen University Museums, Guangzhou 07736

Colombia

Museo Casa Colonial, Pamplona 08705
Museo de Geología y Suelos, Armenia 08533
Museo de Historia Natural, Alta Suiza 08527
Museo Geológico José Royo y Gómez, Bogotá 08584
Museo Geológico Marino Arce Herrera, Bucaramanga 08601

Congo, Democratic Republic

Musée Geologique Sengier-Cousin, Jadotville 08793

Cuba

Museo del Transporte Terrestre, Santiago de Cuba 09320

Czech Republic

Blatské muzeum - Rožmberský Dům, Soběslav 09842
Krkonošské Muzeum, Jilemnice 09563
Městské Muzeum, Čáslav 09451
Mistni Vlastivědné Muzeum, Železnice 09936
Muzeum Těšínska, Český Těšín 09470
Podhorácké Muzeum - Klášter Porta Coeli, Předklášteří 09802
Sládečkovo Vlastivědné Muzeum v Kladně, Kladno 09583
Vlastivědné Muzeum a Galerie, Česká Lípa 09457

Denmark

Midtsønderjyllands Museum, Gram 10040
Naturhistorisk Museum, Århus 09965

Egypt

The Egyptian Geological Museum, Cairo 10491

Estonia

Tartu Ülikooli Geoloogiamuuseum, Tartu 10660

Finland

Kivimuseo, Tampere 11351

France

Les Fosses d'Enfer, Saint-Rémy-sur-Orne 15495
Géothermia-Musée de la Géothermie et du Thermalisme, Chaudes-Aigues 12478
Musée de la Vallée à Jausiers, Jausiers 13266
Musée d'Histoire Naturelle de Toulon et du Var, Toulon (Var) 15925
Musée d'Histoire Naturelle et de Géologie, Lille 13758
Musée Geologique, Vernet-les-Bains 16119
Musée Géologique et Salle Archéologique, Saint-Avit-Senieur 15180
Musée Régional de Géologie Pierre Vetter, Decazeville 12704
Museum de Géologie Provençale, La Roque-d'Anthéron 13439
Muséum d'Histoire Naturelle, Nice 14440
Volcan de Lemptégy, Saint-Ours-les-Roches 15437

Germany

Anton-Buchner-Heimatmuseum, Betzenstein 17311
Audorfer Museum im Burgtor, Oberaudorf 20801
Bachmann-Museum Bremervörde, Bremervörde 17576
Bergbaumuseum Hückelhoven, Hückelhoven 19286
Deutsches Erdölmuseum Wietze, Wietze 22564
Deutsches Goldmuseum, Theuern 22051
Deutsches Schiefermuseum Steinach/Thür., Steinach, Kreis Sonneberg 21875
Dobergmuseum/Geologisches Museum Ostwestfalen-Lippe, Bünde 17635
Dr.-Carl-Haeberlin-Friesen-Museum, Wyk auf Föhr 22759
Eifelmuseum mit Deutschem Schieferbergwerk, Mayen 20241
Erlebniswelt Grundwasser, Langenau 19817
Fossilien- und Heimatmuseum, Messel 20304
Geiseltalmuseum, Halle, Saale 18895
Geologische Landessammlung von Vorpommern, Greifswald 18748
Geologische Schausammlung Eselsmühle, Leinfelden-Echterdingen 19883
Geologisches Museum, Kamp-Lintfort 19423
Geologisches Museum München, München 20443
Geologisches und Mineralogisches Museum der Christian-Albrechts-Universität, Kiel 19528
Geomuseum der Universität Münster, Münster 20509
GeoMuseum der Universität zu Köln, Köln 19605
Geosammlung der Technischen Universität Clausthal, Clausthal-Zellerfeld 17708
Geowissenschaftliche Sammlungen, Freiberg, Sachsen 18421
Geowissenschaftliche Sammlungen und Geologischer Garten, Halle, Saale 18896
Geowissenschaftliches Museum am Geowissenschaftlichen Zentrum der Universität Göttingen, Göttingen 18691
Haus des Gastes mit heimatkundlicher Sammlung, Gößweinstein 18688
Heimatecke am Seifenbach, Beierfeld 16981
Heimatmuseum, Bad Laer 16772
Heimatmuseum, Camburg 17681
Heimatmuseum, Günzburg 18825
Heimatmuseum, Mügeln 20395
Heimatmuseum, Spangenberg 21829
Heimatmuseum im Wiggensbacher Informationszentrum, Wiggensbach 22565
Heimatstuben im Oberen Tor, Scheinfeld 21540
Höhlenkundemuseum Dechenhöhle, Iserlohn 19358
Höhlenmuseum Kubacher Kristallhöhle und Freilicht-Steinemuseum, Weilburg 22385
Informationsstelle des Naturparks Fichtelgebirge, Zell, Oberfranken 22782
Jura-Museum Eichstätt, Eichstätt 18082
Klever Schwanenturm, Kleve 19565
Kreidemuseum Gummanz, Sagard 21469
Kreisheimatmuseum Weißes Haus, Rotenburg an der Fulda 21399
Kreismuseum Bad Liebenwerda, Bad Liebenwerda 16778
Landschaftsinformationszentrum (LIZ) Hessisches Kegelspiel Rasdorf, Rasdorf 21182
Müritzeum, Waren (Müritz) 22335
Muschelkalkmuseum Hagdorn, Ingelfingen 19333
Museum am Lindenbühl, Mühlhausen/Thüringen 20403
Museum der Natur, Gotha 18719
Museum der Natur und Umwelt Cottbus, Cottbus 17733
Museum der Westlausitz Kamenz, Kamenz 19421
Museum Dermbach, Dermbach 17825
Museum für Archäologie und Volkskunde, Kösching 19665
Museum für Bodenschätzung, Eickendorf 18087
Museum für Geologie und Paläontologie, Tübingen 22122
Museum für Natur und Umwelt, Lübeck 20079
Museum für Naturkunde, Dortmund 17926
Museum für Naturkunde der Stadt Gera, Gera 18595
Museum Geologie/Paläontologie, Heidelberg 19078
Museum Schloß Bernburg, Bernburg 17295
Museum Villa Erckens, Grevenbroich 18756
Natureum am Schloß Ludwigslust, Ludwigslust 20061
Naturhistorisches Museum, Mainz 20149
Naturkunde-Museum, Bielefeld 17347
Naturkundemuseum Erfurt, Erfurt 18209
Naturkundemuseum Ostbayern, Regensburg 21230
Naturkundliche Sammlung, Königsbrunn 19646
Paläontologisches Museum München, München 20472
Parkhöhle Weimar, Weimar 22414
Peter-Wiepert-Museum, Fehmarn 18299
Regionalmuseum Wolfhager Land, Wolfhagen 22655
Rieskrater-Museum, Nördlingen 20725
Ruhr Museum, Essen 18267
Salinenmuseum im Gradierbau, Bad Dürkheim 16717
Senckenberg Naturhistorische Sammlungen Dresden, Dresden 17969
Sieblos-Museum Poppenhausen, Poppenhausen 21076
Stadtmuseum Coesfeld, Coesfeld 17723
Stadtmuseum Schelklingen, Schelklingen 21541
Stadtmuseum Schlüsselfeld, Schlüsselfeld 21578
Stadtmuseum Stadtoldendorf, Stadtoldendorf 21859
Städtisches Museum, Aschersleben 16596
Südmährisches Landschaftsmuseum, Geislingen an der Steige 18563
Südostbayerisches Naturkunde- und Mammut-Museum, Siegsdorf 21759
Urwelt-Museum Oberfranken, Bayreuth 16966
Vonderau Museum Fulda, Fulda 18513
Waldnaabtal-Museum in der Burg Neuhaus, Windischeschenbach 22597

Ghana

Geological Survey Department Museum, Accra 22845
Geology Museum, Legon 22860

Greece

Museum of Mineralogy, Petrology and Geology, Zografou ... 23249
Natural History Museum of the Aegean, Samos ... 23178
Paleontological and Geological Museum of the University of Athens, Athinai ... 22949

Guinea

Musée Géologique, Conakry ... 23309

Hungary

Geológiai Múzeum, Tata ... 23916
Magyarországi Kalcitkristályok Színes Világa, Fertőrákos ... 23559
Országos Földtani Múzeum, Budapest ... 23461

India

Geological Museum, Varanasi ... 24426
Geology and Geophysics Museum, Roorkee ... 24373
Geology Museum, Lucknow ... 24295
Government Museum, Chennai ... 24115

Indonesia

Museum Geologi, Bandung ... 24442

Iran

Geological Museum of Iran, Tehrān ... 24667

Ireland

Geological Museum, Dublin ... 24793

Israel

Geological Museum, Mizpe Ramon ... 25068

Italy

Civico Museo Naturalistico Ferruccio Lombardi, Stradella ... 27910
Museo Astronomico e Geofisico, Modena ... 26744
Museo Civico Craveri di Storia Naturale, Bra ... 25524
Museo Civico della Val Fiorentina Vittorino Cazzetta, Selva di Cadore ... 27798
Museo Civico di Scienze Naturali ed Archeologia della Valdinievole, Pescia ... 27172
Museo Civico di Storia Naturale, Ferrara ... 26108
Museo Civico Geologia e Etnografia, Predazzo 27341
Museo di Geologia, Camerino ... 25619
Museo di Geologia, Napoli ... 26898
Museo di Geologia, Roma ... 27495
Museo di Geologia e Paleontologia, Firenze ... 26180
Museo di Geologia e Paleontologia, Padova ... 27040
Museo di Scienza dela Terra, San Gemini ... 27634
Museo di Scienze della Terra U. Baroli, Crodo 26010
Museo di Storia Naturale dell'Accademia dei Fisocritici, Siena ... 27844
Museo Geo-Paleontologico, Ferrara ... 26115
Museo Geologico, Castell'Arquato ... 25736
Museo Geopaleontologico Alto Aventino, Palena ... 27063
Museo Geopaleontologico del Castello, Lerici 26486
Museo Geopaleontologico Naturalistico Antropico e Ornitologico Brancaleoni, Piobbico ... 27227
Museo Paleontologico Rinaldo Zardini, Cortina d'Ampezzo ... 25986
Museo Sardo di Geologia e Paleontologia D. Lovisato, Cagliari ... 25588
Museo Vesuviano G. B. Alfano, Pompei ... 27279

Jamaica

Geology Museum, Kingston ... 28326

Japan

Aso-gun Kazan Hakubutsukan-Orugoru Kyouwakoku, Aso ... 28378
Fossa Magna Museum, Itoigawa ... 28599
Omi Natural History Museum, Omi ... 29115

Jordan

Geology Museum, Amman ... 29649

Kazakhstan

Memorialnyj Muzej Akademika K.I. Satpaeva, Almaty ... 29684

Kyrgyzstan

Geologičeskij i Mineralogičeskij Muzej, Biškek ... 29953

Lithuania

Geologijos ir Mineralogijos Muziejus, Vilnius ... 30408
Lietuvos Geologijos Muziejus, Vievis ... 30397
Mineralų Muziejus, Vilnius ... 30425
Respublikinis Unikalių Amenų Muziejus, Mosėdis ... 30315

Madagascar

Musée National de Géologie, Antananarivo ... 30537

Malaysia

Geological Survey Museum, Ipoh ... 30554

Mexico

Museo de Geología de la UNAM, México ... 31074
Museo de Geología del IPN, México ... 31075
Museo de Geología Dr. Jenaro González Reyna, Morelia ... 31195
Museo de Mineralogía y Geología Ing. Luis Silva Ruelas, Morelia ... 31198

Mozambique

Museu Nacional de Geologia, Maputo ... 31711

Namibia

Geological Survey Museum, Windhoek ... 31763

Netherlands

Artis Geologisch Museum, Amsterdam ... 31831
Gelders Geologisch Museum, Velp ... 32881
Geologisch Museum Hofland, Laren, Noord-Holland ... 32464
Geologisch Streek-Museum de IJsselvallei, Olst ... 32618
Kristalmuseum 't Los Hoes, Borculo ... 31984
Museum Freriks, Winterswijk ... 32957
Natuurmuseum Groningen, Groningen ... 32285
Oudheidkamer Heerde, Heerde ... 32329
Stenenexpositie de Molen - Museum Batjuchin, Borculo ... 31986

New Zealand

Alexander McKay Geological Museum, Wellington ... 33242
Geology Museum of the University of Otago, Dunedin ... 33092
Institute of Geological and Nuclear Sciences, Hutt City ... 33125

Nicaragua

Museo Comunitario de San Rafael del Sur, San Rafael del Sur ... 33306
Museo de Geología, Managua ... 33289
Museo del Volcán Masaya, Masaya ... 33294

Norway

Naturhistorisk museum, Oslo ... 33737
Norsk Bremuseum, Fjærland ... 33493

Pakistan

Museum of Historical Geology and Geological Survey of Pakistan, Quetta ... 34039

Peru

Museo Geológico de la Universidad Nacional de Ingeniería del Perú, Lima ... 34259

Poland

Muzeum Geologiczne, Kraków ... 34794
Muzeum Geologiczne, Wrocław ... 35330
Muzeum Geologiczne Państwowego Instytutu Geologicznego, Warszawa ... 35244
Muzeum Geologiczne Uniwersytetu Szczecińskiego, Szczecin ... 35159
Muzeum Geologii Złóż im. Czesława Poborskiego, Gliwice ... 34664
Muzeum Górnicze, Bogatynia ... 34566
Muzeum Okręgowe, Konin ... 34755
Muzeum Puszczy Kampinoskiej, Kampinos ... 34714
Muzeum Regionalne w Sosnowcu, Sosnowiec 35123
Muzeum Zbiorów Geologicznych, Kielce ... 34746
Muzeum Ziemi Polskiej Akademii Nauk, Warszawa ... 35284

Portugal

Museu das Ciências da Terra, Estremoz ... 35577
Museu de Geologia, Vila Real ... 36012
Museu de Jazigos Minerais Portugueses, São Mamede de Infesta ... 35923
Museu Geológico, Lisboa ... 35710
Museu Laboratório Mineralógico e Geológico, Porto ... 35856
Museu Mineralógico e Geológico, Lisboa ... 35714

Romania

Casa Memorială Ion Popescu, Voitești ... 36734
Muzeul Național Geologic, București ... 36223
Muzeul Sării, Slănic Prahova ... 36626

Russia

Centralnyj Geologorazvedočnyj Muzej im. Akademika F.N. Černyševa, Sankt-Peterburg ... 37627
Geologičeskij Muzej, Vorkuta ... 37967
Geologičeskij Muzej im. A.A. Černova, Syktyvkar ... 37814
Geologičeskij Muzej im. Professora V.E. Eršova MGGU, Moskva ... 37180
Geologičeskij Muzej Vserossijskogo NII Mineralnogo Syrja im. N.M. Fedorovskogo, Moskva ... 37181
Geologo-mineralogičeskij Muzej, Kazan ... 37007
Geologo-mineralogičeskij Muzej, Moskva ... 37182
Geologo-mineralogičeskij Muzej im. A.I. Kozlova, Vladivostok ... 37937
Gornyj Muzej, Sankt-Peterburg ... 37639
Gosudarstvennyj Geologičeskij Muzej im V.I. Vernadskogo, Moskva ... 37192
Mineralogičeskij Muzej Moskovskoj Geologorazvedyvatelnoj Akademii, Moskva ... 37251
Muzei Permskogo Gosudarstvennogo Universiteta, Perm ... 37547
Muzej Geologii, Čeljabinsk ... 36845
Muzej Geologii Centralnoj Sibirii, Krasnojarsk 37075
Muzej Geologii, Nefti i Gaza, Tjumen ... 37845
Muzej Mineralov, Rud, Samocvetov - Udivitelnoe v Kamne, Tereda ... 37840
Narodnyj Geologičeskij Muzej Volgageologija, Nižnij Novgorod ... 37466
Paleontologo-stratigrafičeskij Muzej pri Kafedre Istorii Geologii, Sankt-Peterburg ... 37717
Regionalnyj Muzej Severnogo Priladožja, Sortavala ... 37789
Rudno-petrografičeskij Muzej, Moskva ... 37405
Uralskij Geologičeskij Muzej V.V. Vachrušev, Ekaterinburg ... 36901

Rwanda

Musée Géologique du Rwanda, Kigali ... 37994

St. Lucia

Pigeon Island, Castries ... 37995

Saudi Arabia

Geological Museum, Riyadh ... 38015

Senegal

Musée Géologique, Dakar ... 38023

Slovakia

Berggericht - Mineralogická Expozícia, Banská Štiavnica ... 38189
Liptovské Múzeum, Ružomberok ... 38343
Slovenské Národné Múzeum - Múzeum Bojnice, Bojnice ... 38201

Slovenia

Prirodoslovni Muzej Slovenije, Ljubljana ... 38461
Zbirka v Jamarskem Domu, Domžale ... 38416

South Africa

Barberton Museum, Barberton ... 38555
Bleloch Museum, Johannesburg ... 38652
Geological Museum, Johannesburg ... 38654
Geology Education Museum, Durban ... 38606
Geology Museum of the University of Stellenbosch, Stellenbosch ... 38735
Geoscience Museum, Tshwane ... 38747

Spain

Museo de Astronomía y Geodesia, Madrid ... 39535
Museo de Ciencias Naturales, Pamplona ... 39807
Museo de Historia Natural, Santiago de Compostela ... 40021
Museo de los Minerales y Rocas, Porriño ... 39861
Museo Geológico del Seminari, Barcelona ... 38969
Museo Nacional de Ciencias Naturales, Madrid ... 39571
Museo Paleontológico - Sala Longinos Navás, Zaragoza ... 40349
Museu d'Arqueologia Salvador Vilaseca, Reus ... 39894
Museu de Geologia, Barcelona ... 38977
Museu de Geologia de la Universitat de València, Burjassot ... 39074
Museu de Geologia Valenti Masachs, Manresa ... 39615

Sudan

Geological Museum, Khartoum ... 40390

Switzerland

focusTerra, Zürich ... 42071
Gletschergarten Luzern, Luzern ... 41590
Musée Cantonal de Géologie, Lausanne ... 41542
Musée des Glaciers, Lourtier ... 41576
Naturmuseum Olten, Olten ... 41708

Taiwan

Black Woods Museum, Houbi ... 42203
Chai-Yi Museum of Stone and Art, Chai-Yi ... 42151
Geology Museum, Tainan ... 42361
Xiaomen Geology Hall, Siyu ... 42342

Thailand

Geological Museum, Bangkok ... 42541

Tunisia

Musée Géologique, La Charguia ... 42634

Uganda

Geological Survey and Mines Museum, Entebbe ... 42987
Geology Museum, Kampala ... 42991

United Kingdom

Amberley Museum and Heritage Centre, Amberley ... 43297
Bolton Museum and Art Gallery, Bolton ... 43553
Castle Museum, Clitheroe ... 43854
Dewey Museum, Warminster ... 45932
Dick Institute Museum and Art Gallery, Kilmarnock ... 44579
Dudley Museum and Art Gallery, Dudley ... 44030
Dunrobin Castle Museum, Golspie ... 44325
Galloway Deer Museum, Clatteringshaws ... 43849
Geology Collections, London ... 44855
Geology Department Museum, Aberdeen ... 43241
Gosport Museum, Gosport ... 44334
Halesworth and District Museum, Halesworth 44381
Hugh Miller's Cottage, Cromarty ... 43928
Hunterian Museum, Glasgow ... 44285
La Hougue Bie Museum, Grouville ... 44368
Lapworth Museum of Geology, Birmingham ... 43506
Ludlow Museum, Ludlow ... 45045
Much Wenlock Museum, Much Wenlock ... 45185
Natural History Museum, Belfast ... 43442
Natural History Society and Folk Museum, Bacup ... 43349
Oxford University Museum of Natural History, Oxford ... 45336
Peterborough Museum and Art Gallery, Peterborough ... 45368
Quex Museum, House and Gardens, Birchington ... 43489
Ruskin Museum, Coniston ... 43885
Sedgwick Museum of Earth Sciences, Cambridge ... 43733
Treasures of the Earth, Fort William ... 44248
Tweeddale Museum, Peebles ... 45349
Warwickshire Museum, Warwick ... 45942

United States of America

Anza-Borrego Desert Park Museum, Borrego Springs ... 46915
Arizona Museum of Natural History, Mesa ... 50605
Berea College Burroughs Geological Museum, Berea ... 46746
Bob Campbell Geology Museum, Clemson ... 47613
Burpee Museum of Natural History, Rockford ... 52413
Carter County Museum, Ekalaka ... 48306
Chesapeake Children's Museum, Annapolis ... 46317
Colburn Gem and Mineral Museum, Asheville 46371
Colorado School of Mines Geology Museum, Golden ... 48958
Dunn-Seiler Museum, Mississippi State ... 50751
Ed Clark Museum of Missouri Geology, Rolla ... 52445
Eleanor Barbour Cook Museum of Geology, Chadron ... 47326
Emporia State University Geology Museum, Emporia ... 48391
Fiedler Memorial Museum, Seguin ... 53028
Franklin Mineral Museum, Franklin ... 48748
Fryxell Geology Museum, Rock Island ... 52408
The Geological Museum, Laramie ... 49947
Geology Museum, Huntington ... 49434
Grand Canyon Museum Collection, Grand Canyon ... 48985
Greene Memorial Museum, Milwaukee ... 50709
Henry S. Reuss Ice Age Visitor Center, Campbellsport ... 47209
Lora Robins Gallery of Design from Nature, Richmond ... 52332
Luzerne County Museum, Wilkes-Barre ... 54162

Mammoth Site of Hot Springs, Hot Springs 49375
Mississippi Petrified Forest, Flora 48561
Mount Rainier National Park Museum, Ashford 46378
Museum of Ancient Artifacts, Chicago 47490
Museum of Geology, Rapid City 52253
Museum of the Geological Sciences, Blacksburg 46845
The Natural Science Center of Greensboro, Greensboro 49067
New York State Museum, Albany 46161
North Carolina Museum of Natural Sciences, Raleigh 52234
Northwestern Oklahoma State University Museum, Alva 46247
Olympic National Park Visitor Center, Port Angeles 52027
Orton Geological Museum, Columbus 47758
Petrified Forest of the Black Hills, Piedmont 51914
Rolla Minerals Museum, Rolla 52447
Ruth Hall Museum of Paleontology, Abiquiu 46130
Searchlight Historic Museum and Mining Park, Searchlight 52991
Silver River Museum, Ocala 51484
Springfield Science Museum, Springfield 53229
Tate Geological Museum, Casper 47273
Vermont Marble Museum, Proctor 52156
Voas Museum, Minburn 50722
W.M. Keck Museum, Reno 52292

Uruguay

Museo Geológico del Uruguay, Montevideo 54403

Uzbekistan

Muzej Geologii, Toshkent 54476

Vietnam

Geology Museum, Ha Noi 54613

Glass and Glassware

Armenia

Cafesjan Museum, Yerevan 00729

Australia

Colac Otway Shire Hammerton Bottle Collection, Colac 00991
Devonport Gallery and Arts Centre, Devonport 01034
Fusions Gallery, Fortitude Valley 01086
House of Bottles, Kinglake 01201
House of Bottles and Bottle Museum, Tewantin 01578
Wagga Wagga Art Gallery, Wagga Wagga 01620

Austria

Gläsermuseum, Gerersdorf 01931
Glas- und Steinmuseum, Gmünd, Niederösterreich 01945
Glasmuseum Zalto, Nagelberg 02420
Heimathaus, Ulrichsberg 02838
Heimatstube, Schwarzenberg am Böhmerwald 02720
Hinterglasmuseum, Sandl 02631
Museum Freudenthaler Glassammlung, Weißenkirchen im Attergau 02903
Schnapsglas-Museum, Echsenbach 01837
Steirisches Glaskunstzentrum und Glasmuseum Bärnbach, Bärnbach 01778
Wiener Glasmuseum, Wien 03089

Belgium

Artonivo, Brugge 03407
Château du Val-Saint-Lambert, Seraing 04010
Musée d'Histoire et des Arts Décoratifs, Tournai 04087
Musée du Verre, Marcinelle 03851
Musée du Verre, Liège 03812

Canada

Canadian Clay and Glass Gallery, Waterloo 07285
Verrerie La Mailloche, Québec 06698

China

Liuligongfang Museum, Shanghai 08171

Czech Republic

Muzeum Skla, Harrachov 09507
Muzeum Skla a Bižuterie, Jablonec nad Nisou 09545
Muzeum Šumavy, Kašperské Hory 09581
Severočeské Muzeum v Liberci, Liberec 09625
Sklářské Muzeum, Kamenický Šenov 09571
Sklářské Muzeum Moser, Karlovy Vary 09577
Sklářské Muzeum Nový Bor, Nový Bor 09683
Sklenka Muzeum v Železném Brodě, Železný Brod 09938
Uměleckoprůmyslové Muzeum v Praze, Praha 09797
Východočeské Muzeum v Pardubicích, Pardubice 09707

Denmark

Anneberg-Samlingerne, Nykøbing Sjælland 10205
Den Gamle Gård, Faaborg 10010
Glasmuseet Ebeltoft, Ebeltoft 09997

Finland

Iittalan lasimuseo, Iittala 10849
Pykäri Nuutajärven Lasimuseo, Nuutajärvi 11154
Suomen Lasimuseo, Riihimäki 11270

France

Atelier Musée du Verre, Trélon 15988
Centre International du Vitrail, Chartres 12421
Cristalleries Royale de Saint-Louis, Saint-Louis-les-Bitche 15378
Galerie d'Art Moderne, Passavant-la-Rochère 14754
Maison du Verre, Puy-Guillaume 14960
Musée Baccarat, Paris 14632
Musée Calvet, Avignon 11789
Musée d'Art Moderne, Troyes 15998
Musée de La Glacerie, La Glacerie 13378
Musée de la Poterie et de la Céramique, Sadirac 15146
Musée de la Verrerie, Blangy-sur-Bresle 12015
Musée de l'Ecole de Nancy, Nancy 14353
Musée de l'Homme et de l'Industrie, Le Creusot 13577
Musée de Traditions Populaires et d'Archéologie, Chauvigny 12488
Musée Despiau Wlérick, Mont-de-Marsan 14158
Musée du Verre, Blaye-les-Mines 12024
Musée du Verre, Sars-Poteries 15646
Musée du Vin, Cissac-Médoc 12531
Musée du Vitrail et du Verre, Gordes 13084
Musée International de la Parfumerie, Grasse 13114
Musée-Maison du Verre, Meisenthal 14063
Musée National Adrien Dubouché, Limoges 13771

Germany

Erich Mäder-Glasmuseum, Grünenplan 18813
Europäisches Museum für Modernes Glas, Rödental 21351
Glas- Heimat-Museum Warndt, Völklingen 22248
Glas- und Keramikmuseum, Großalmerode 18779
Glashütte Annenwalde, Templin 22027
Glasmuseum, Frauenau 18410
Glasmuseum, Immenhausen 19325
Glasmuseum, Warmensteinach 22339
Glasmuseum Boffzen, Boffzen 17431
Glasmuseum der Dorotheenhütte, Wolfach 22647
Glasmuseum Grünenplan, Grünenplan 18814
Glasmuseum Hadamar, Hadamar 18849
Glasmuseum in der Glashütte, Monschau 20374
Glasmuseum Passau, Passau 20989
Glasmuseum Rheinbach, Rheinbach 21297
Glasmuseum Steina, Bad Sachsa 16811
Glasmuseum Weisswasser, Weißwasser 22465
Glasmuseum Wertheim, Wertheim 22502
Goethe-Museum Stützerbach mit Museum zur Geschichte des technischen Glases, Stützerbach 21924
Gralglasmuseum, Dürnau 17984
Heimatmuseum Uelzen, Uelzen 22150
Heimattreff Gersweiler, Saarbrücken 21455
Historische Fraunhofer-Glashütte, Benediktbeuern 16991
Kunstmuseum Walter, Augsburg 16626
LWL-Industriemuseum, Glashütte Gernheim, Petershagen 21016
Märkisches Glasmuseum, Stechlin 21871
Museum des Kreises Plön mit norddeutscher Glassammlung, Plön 21065
Museum für Glaskunst, Lauscha 19861
Museum für modernes Glas, Öhringen 20865
Museumsdorf Baruther Glashütte, Baruth 16932
Neues Schloss Wallerstein, Wallerstein 22314
Schott GlasMuseum, Jena 19391
Stadtgeschichtliche Sammlung, Gaggenau 18528
Stadtmuseum Neustadt an der Waldnaab, Neustadt an der Waldnaab 20668
Stadtmuseum Waldkraiburg, Waldkraiburg 22300
Stiftung museum kunst palast mit Sammlung Kunstakademie und Glasmuseum Hentrich, Düsseldorf 18008
Teplitz-Schönauer Heimatmuseum, Frankfurt am Main 18398
Theresienthaler Glasmuseum, Zwiesel 22838
Waldmuseum, Zwiesel 22839
Wüstenroter Glas- und Heimatmuseum, Wüstenrot 22729

Hungary

Báthory Júlia Üveggyűjtemény, Dömsöd 23526
Bükki Üveghuták Ipartörténeti Múzeuma, Bükkszentkereszt 23496
Róth Miksa Emlékház, Budapest 23474

Iran

Abgineh va Sofalineh, Tehrān 24654

Italy

Museo del Cristallo, Colle di Val d'Elsa 25938
Museo del Vetro, Murano 26877

Japan

Gunze Museum, Ayabe 28385
Notojima Glass Art Museum, Notojima 29082

Mexico

Museo del Vidrio, Monterrey 31184

Netherlands

De Drentse Glasbloazer, Dalen 32044
Glasmuseum Veenhuizen, Veenhuizen 32875
Museum Hedendaagse Glaskunst, Vledder 32894
Nationaal Glasmuseum, Leerdam 32471
Ontmoetingscentrum de Veldkei, Expositie, Havelte 32325

Norway

Hadeland Glassverks Museum, Jevnaker 33581

Philippines

MPG Town Gallery, Odiongan 34468

Poland

Muzeum Karkonoskie, Jelenia Góra 34704

Portugal

Mosteiro de Santa Clara-a-Velha, Coimbra 35527
Museu do Vidro, Marinha Grande 35756
Museu Santos Barosa da Fábrica do Vidro, Marinha Grande 35757

Romania

Muzeul de Artă Decorativă, Dorohoi 36349

Russia

Gosudarstvennyj Chudožestvenno-architekturnyj Dvorcovo-parkovyj Muzej-zapovednik Pavlovsk, Pavlovsk 37528

Slovakia

Slovenské Sklárske Múzeum, Lednické Rovne 38281
Tribečské Múzeum Topoľčany, Topoľčany 38372

South Africa

Our Heritage Museum, Adelaide 38550

Spain

Museo del Castillo de Peralada, Peralada 39836
Museo del Vidrio, San Ildefonso 39957
Museu del Vidre, Algaida 38840

Sweden

Glasbruksmuseet i Surte, Surte 40965
Smålands Museum, Växjö 41041

Switzerland

Glasmuseum, Hergiswil 41480
Historisches Museum Wiedlisbach, Wiedlisbach 42013
mudac - Musée de Design et d'Arts Appliqués Contemporains, Lausanne 41540
Musée Ariana, Genève 41416
Musée du Verrier, Saint-Prex 41793
Museum Bellerive, Zürich 42092
Schauglashütte, Uetendorf 41955

Taiwan

Glass Museum of Hsinchu City, Hsinchu 42206
Tittot Glass Art Museum, Taipei 42436

United Kingdom

Bennie Museum, Bathgate 43414
Broadfield House Glass Museum, Kingswinford 44604
Clevedon Court, Clevedon 43852
Dartington Crystal Glass Museum, Torrington 45868
Guildford House Gallery, Guildford 44370
Haworth Art Gallery, Accrington 43262
Himley Hall, Himley 44470
National Glass Centre, Sunderland 45800
Red House Glass Cone Museum, Stourbridge 45769
Royal Brierley Crystal Museum, Dudley 44031
Shipley Art Gallery, Gateshead 44265
South Shields Museum and Art Gallery, South Shields 45698
Stained Glass Museum, Ely 44174
World of Glass, Saint Helens 45568
Yelverton Paperweight Centre, Yelverton 46095

United States of America

Blenko Museum of Seattle, Seattle 52993
Cambridge Glass Museum, Cambridge 47189
Corning Museum of Glass, Corning 47810
Degenhart Paperweight and Glass Museum, Cambridge 47190
Fenton Glass Museum, Williamstown 54177
The Glass Museum, Dunkirk 48210
Greentown Glass Museum, Greentown 49071
The Jones Museum of Glass and Ceramics, Sebago 53018
Museum of American Glass at WheatonArts, Millville 50696
Museum of Glass, Tacoma 53412
Museum of Glass and Ceramics, South Portland 53179
National Heisey Glass Museum, Newark 51292
National Museum of Ceramic Art and Glass, Baltimore 46583
Neustadt Museum of Tiffany Art, New York 51239
Racine Art Museum, Racine 52223
Sandwich Glass Museum, Sandwich 52841

Globes → Cartography

Gold and Silver

Australia

Creswick Museum and Gold Battery, Creswick 01011
Gold Museum, Ballarat 00827
Gold Treasury Museum, Melbourne 01281
Warrandyte Old Post Office Museum, Warrandyte 01633

Austria

Gold- und Silberschmiedemuseum, Wien 02997
Rauriser Talmuseum, Rauris 02566
Schatzkammer, Wien 03068
Schloß- und Goldbergbaumuseum, Großkirchheim 02002

Belgium

Kasteelmuseum Slot van Laarne, Laarne 03772
Musée de l'Orfevrerie de la Communaute Française, Seneffe 04008
Musée du Chapitre, Soignies 04035
Stedelijk Museum, Oudenaarde 03942
Trésor de la Collégiale, Huy 03725
Trésor de la Collegiale Sainte-Waudru, Mons 03889
Trésor du Frère Hugo d'Oignies, Namur 03907
Zilvermuseum Sterckshof Provincie Antwerpen, Antwerpen 03321

Bermuda

Tucker House Museum, Saint George's 04199

Bolivia

Museo de Metales Preciosos Precolombinos o Museo del Oro, La Paz 04216

Bulgaria

Etnografiski Kompleks na Istoričeski Muzej, Vraca 05264

Canada

Historic Hat Creek Ranch, Cache Creek 05510
Klondike National Historic Sites, Dawson City 05682
Sherbrooke Village, Sherbrooke, Nova Scotia 06957
Silvery Slocan Museum, New Denver 06428
Soo Line Historical Museum, Weyburn 07316

Colombia

Museo del Oro, Bogotá 08577

Costa Rica

Museo de Oro Precolombino, San José 08819

Croatia

Riznica Crkvene Umjetnosti - Zlato i Srebro Nina, Nin 08915

Cuba

Museo Casa de la Orfebrería, La Habana 09118

Czech Republic

Regionální Muzeum v Jílovém u Prahy, Jílové u Prahy 09564

Denmark

KulturhistorieTønder, Tønder 10298

France

Collection Charles de l'Escalopier, Amiens 11583
Musée de l'Or, Jumilhac-le-Grand 13293
Petit Musée de l'Argenterie, Paris 14748

Germany

G

Bergbau- und Heimatmuseum, Goldkronach 18704
Deutsches Goldmuseum, Schalkau 21529
Deutsches Goldmuseum, Theuern 22051
Deutsches Goldschmiedehaus, Hanau 18975
Diözesanmuseum Freising, Freising 18450
Glashüttenmuseum des Erzgebirges, Neuhausen / Erzgebirge 20619
Historische Grünes Gewölbe und Neues Grünes Gewölbe, Dresden 17948
Historisches Silbererzbergwerk Grube Samson und Heimatmuseum Sankt Andreasberg, Sankt Andreasberg 21487
Ikonenmuseum Schloß Autenried, Ichenhausen 19307
Museum im Goldschmiedehaus, Ahlen 16439
Städtisches Museum Wesel, Wesel 22510
Technisches Museum Silberwäsche Antonsthal, Breitenbrunn 17543
Zunftstube, Gernsbach 18609

Italy

Museo Alessi, Enna 26055
Museo Civico, Lucignano 26539
Museo d'Arte e Storia Antica Ebraica, Casale Monferrato 25696
Museo del Duomo, Città di Castello 25906
Museo della Propositura di San Pietro, Montecatini Val di Nievole 26814
Museo e Pinacoteca della Città, Todi 27981
Museo Ebraico di Roma, Roma 27505
Museo Sacro, Pompei 27278
Museo Storico dell'Oro Italiano, Predosa 27342
Tesoro del Duomo, Milano 26725
Tesoro della Cattedrale, Palermo 27080

Korea, Republic

Ewha Yoja Taehakkyo Pakmulgwan, Seoul 29876
National Museum of Korea, Seoul 29910

Latvia

Rīgas Vēstures un Kuģniecības muzejs, Rīga 30111

Mexico

Museo Guillermo Spratling, Taxco de Alarcón 31407

Netherlands

Nederlands Goud-, Zilver- en Klokkenmuseum, Schoonhoven 32764
Oudheidkamer Beilen, Beilen 31957
Schoonhovens Edelambachtshuys, Schoonhoven 32766

New Zealand

Clyde Historical Museum, Clyde 33081
Maniototo Early Settlers Museum, Naseby 33151

Norway

Norsk Bergverksmuseum, Kongsberg 33603

Peru

Museo de Arte Precolombino, Cusco 34157

Portugal

Museu de Alberto Sampaio, Guimarães 35639
Museu do Ouro de Travassos, Travassos - Póvoa de Lanhoso 35974

Romania

Colecţia Muzeală de Argintărie, Braşov 36176

Russia

Almaznyj Fond, Moskva 37140

Saudi Arabia

Silver Museum, Al-Khobar 38014

Spain

Cámara Santa de la Catedral de Oviedo, Oviedo 39757
Museo de Arte Sacra de Vilanova, Lourenzá 39495

Sweden

Gruvmuseet, Sala 40826
Silvermuseet, Arjeplog 40431
Skattkammaren - Uppsala Domkyrkas Museum, Uppsala 41011

Switzerland

Helvetisches Goldmuseum, Burgdorf 41274

Turkmenistan

Museum of Turkmen Values, Aşgabat 42971

United Kingdom

Brodick Castle, Brodick 43651
Goldsmiths' Hall, London 44858
Ickworth House, Horringer 44485
Nelson Collection at Lloyd's, London 44942
Treadgolds Museum, Portsmouth 45439

United States of America

Argo Gold Mine, Idaho Springs 49463
Big Thunder Gold Mine, Keystone 49769
Custer Museum, Clayton 47598
Gold Discovery Museum, Coloma 47696
Reed Gold Mine, Midland 50668

Vatican City

Museo Storico Artistico, Città del Vaticano 54512

Graphic Arts

Argentina

Museo Itinerante de Arte Contemporáneo del Mercosur, Buenos Aires 00248
Museo Municipal de Bellas Artes Lucas Braulio Areco, Posadas 00500
Museo Nacional del Grabado, Buenos Aires 00262

Armenia

Ara Sargsyan and Hakob Kojoyan House-Museum, Yerevan 00725
Dom Muzej M. Saryana, Yerevan 00732

Austria

Albertina, Wien 02929
Franz Traunfellner-Dokumentation, Pöggstall 02507
Galerie Adolf Blaim, Messern 02377
Himmelkeller, Kronberg im Weinviertel 02239
Karikaturmuseum Krems, Krems 02226
Kunstsammlungen und Graphische Sammlung, Furth bei Göttweig 01913
Kupferstichkabinett, Wien 03020
Werner Berg Museum, Bleiburg 01788

Belgium

Cabinet des Estampes et des Dessins, Liège 03794
Centre de la Gravure et de l'Image Imprimée de la Communauté Française de Belgique, La Louvière 03766
Collections Artistiques de l'Université de Liège, Liège 03795
Museum Plantin-Moretus / Prentenkabinet, Antwerpen 03297

Bulgaria

Chudožestvena Galerija, Žeravna 05268
Sofijska Gradska Chudožestvena Galerija, Sofia 05203

Canada

Durham Art Gallery, Durham 05727
Galerie de l'UQAM, Montréal 06339
Glenhyrst Art Gallery of Brant, Brantford 05469
McMaster Museum of Art, Hamilton 05960
The Station Gallery, Whitby 07319

China

Changshu Beike Museum, Changshu 07599
Hong Kong Space Museum, Hong Kong 07805
Nianhua Museum, Mianzhu 08017
Qijiang Stone Engraving Museum, Gunan 07747
Shandong Stone Inscription Art Museum, Jinan 07885
Stone Inscription Art Museum, Beijing 07569
Suzhou Beike Guan, Suzhou 08255
Wuxi Beike Guan, Wuxi 08341

Colombia

Museo de Museos Colsubsidio, Bogotá 08572
Museo Luis Alberto Acuña, Villa de Leyva 08784

Croatia

Grafička Zbirka Nacionalne i Sveučilišne Knjižnice, Zagreb 08992
Kabinet Grafike, Zagreb 09001

Cuba

Antigua Casa de los Marqueses de Campo Florido, San Antonio de los Baños 09272

Czech Republic

Galerie Výtvarného Umění, Havlíčkův Brod 09511
Národní Galerie v Praze - Palác Kinských, Praha 09777

Denmark

Kastrupgårdsamlingen, Kastrup 10114
Kunsthallen Brandts, Odense 10217
Storm P.-Museet, Frederiksberg 10023
Vejle Kunstmuseum, Vejle 10310

Finland

Jyväskylän Taidemuseo, Jyväskylä 10890

France

Cabinet des Estampes et des Dessins, Strasbourg 15809
Château de Lourmarin, Lourmarin 13855
Dalí - Espace Montmartre, Paris 14594
Musée Adzak, Paris 14628
Musée de Nogent, Nogent-sur-Marne 14463
Musée des Beaux-Arts, Marseille 14002
Musée du Dessin et de l'Estampe Originale, Gravelines 13122
Musée Toulouse-Lautrec, Naucelle 14384

Germany

1822-Forum Frankfurter Sparkasse, Frankfurt am Main 18352
A. Paul Weber-Museum, Ratzeburg 21195
Albrecht-Dürer-Gesellschaft, Nürnberg 20754
Artothek Worpswede, Worpswede 22690
Bauernfeind-Museum, Sulz am Neckar 21979
Bayerisches Armeemuseum, Ingolstadt 19339
Ernst Barlach Atelierhaus und Ausstellungsforum-Graphikkabinett, Güstrow 18827
Erwin von Kreibig-Museum, München 20437
Feld-Haus - Museum für populäre Druckgrafik, Neuss 20655
Forum Konkrete Kunst Erfurt, Erfurt 18202
Fossilien- und Steindruck-Museum, Gunzenhausen 18838
Galerie Albstadt, Albstadt 16453
Galerie Parterre, Berlin 17106
Gertrudenkapelle, Güstrow 18828
Grafik Museum Stiftung Schreiner, Bad Steben 16845
Grafische Sammlung, Augsburg 16619
Grafische Sammlung, Nürnberg 20766
Graphik-Kabinett der Stadt Backnang, Backnang 16658
Graphikmuseum Pablo Picasso Münster, Münster 20510
Graphische Sammlung, Kassel 19464
Graphische Sammlung der Universität, Erlangen 18221
Heimatmuseum, Scheeßel 21536
Heinrich Zille Museum, Berlin 17136
Holzschnitt-Museum Klaus Herzer, Mössingen 20366
Hudetz-Turm, Wiesent 22561
Käthe-Kollwitz-Museum Berlin, Berlin 17151
Kunstbibliothek, Berlin 17165
Kunstforum Ostdeutsche Galerie, Regensburg 21228
Kunsthalle Bremen, Bremen 17556
Kunsthalle Mannheim, Mannheim 20166
Kunsthalle Wilhelmshaven, Wilhelmshaven 22578
Kunsthaus Meyenburg, Nordhausen 20737
Kunstpalais, Erlangen 18222
Kupferstich-Kabinett, Dresden 17954
Lichtdruck-Kunst Leipzig e.V., Leipzig 19917
Ludwig-Doerfler-Galerie, Schillingsfürst 21553
Luza Acht, Lübben 20068
LWL-Landesmuseum für Kunst und Kulturgeschichte, Münster 20512
Lyonel-Feininger-Galerie, Quedlinburg 21151
Martin-von-Wagner-Museum der Universität Würzburg, Würzburg 22714
Museum Folkwang, Essen 18263
Museum Mitterteich, Mitterteich 20344
Museum Otto Schäfer, Schweinfurt 21687
Museum Rolf Werner, Bansin, Seebad 16926
Museum Schloß Moritzburg, Zeitz 22777
Museum und Stiftung Anton Geiselhart, Münsingen 20501
Otto-Dix-Haus Hemmenhofen, Gaienhofen 18532
Otto Pankok Museum, Hünxe 19293
Sammlungen des Kunstvereins, Ibbenbüren 19304
Siegerlandmuseum mit Ausstellungsforum Haus Oranienstraße, Siegen 21756
Staatliche Graphische Sammlung München, München 20487
Staatsgalerie Stuttgart, Stuttgart 21960
Stadtmuseum Bocholt, Bocholt 17396
Städtische Galerie, Bietigheim-Bissingen 17350
Städtische Galerie im Park, Viersen 22232
Städtische Wessenberg-Galerie Konstanz, Konstanz 19683
Städtisches Museum, Aschersleben 16596
Stiftung Käthe Kollwitz Haus, Moritzburg 20387
Vonderau Museum Fulda, Fulda 18513
Wilhelm-Hack-Museum Ludwigshafen am Rhein, Ludwigshafen am Rhein 20058

Greece

Museum of Engravings and Graphic Arts, Athinai 22933

Guam

Isla Center for the Arts at the University of Guam, Mangilao 23287

Hungary

Bodrogközi Kastélymúzeum, Pácin 23745
Kass Galéria, Szeged 23845
Lajos Ferenc és Kortársai Grafikai Gyűjtemény, Gyula 23598

Italy

Calcografia, Roma 27441
Civica Raccolta delle Stampe Achille Bertarelli, Milano 26676
Gabinetto dei Disegni e delle Stampe, Bologna 25442
Gabinetto Disegni e Stampe, Pisa 27239
Gabinetto Disegni e Stampe degli Uffizi, Firenze 26153
Gabinetto Nazionale delle Stampe, Roma 27447
Museo Casabianca, Malo 26577
Museo di Milano, Milano 26704
Pinacoteca Ambrosiana, Milano 26719

Japan

Center for Contemporary Graphic Art and Tyler Graphics Archive Collection, Sukagawa 29297
Chihiro Art Museum Tokyo, Tokyo 29377
Ehagaki Shiryokan, Kobe 28724
Ehime-ken Bijutsukan, Matsuyama 28904
Engraving Print Village Art Museum, Sado 29177
Kagawa-kenritsu Higashiyama Kaii Setouchi Bijutsukan, Sakaide 29201
Kamigata Ukiyo-e Bijutsukan, Osaka 29129
Katsushika Hokusai Bijutsukan, Tsuwano 29546
Machida-shiritsu Kokusai Hanga Bijutsukan, Machida 28875
Munakata Prints Museum, Kamakura 28645
Munakata Shiko Kinenkan, Aomori 28361
Nihon Ukiyo-e Hakubutsukan, Matsumoto 28896
Ukiyo-e and Pottery Museum, Osaka 29156

Lithuania

Rokiškio Krašto Muziejus, Rokiškis 30346
Žemaičių Dailės Muziejus, Plungė 30338

Mexico

Casa Museo Vladimir Cora, Acaponeta 30689
Instituto de Artes Gráficas de Oaxaca, Oaxaca 31220
Museo de Arte Gráficas Juan Pablos, México 31069
Museo de la Estampa, Cuitzeo 30823
Museo de la Estampa del Estado de México, Toluca 31490
Museo de la Gráfica de la Historia Social de Taxco Siglo XX, Taxco de Alarcón 31404
Museo del Periodismo y las Artes Gráficas, Guadalajara 30902
Museo Franco Lázaro, Chiapa de Corzo 30763
Museo Gráfico de la Mujer Guerrenese, Chilpancingo 30773

Museo José Guadalupe Posada, Aguascalientes ... 30702
Museo Nacional de la Estampa, México ... 31136

Netherlands

Affiche Museum Hoorn, Hoorn ... 32408
Atelier Kea Homan, Assen ... 31935
Bierreclamemuseum, Breda ... 32001
Escher in Het Paleis, Den Haag ... 32071
Grafisch Museum Groningen, Groningen ... 32280
Jelly van den Bosch Museum, Roden ... 32694
Museum Hedendaagse Grafiek, Vledder ... 32895
Museum Het Rembrandthuis, Amsterdam ... 31862
Museum Jan Heestershuis, Schijndel ... 32757
Prentenkabinet Universiteit Leiden, Leiden ... 32486
Rijksprentenkabinet, Amsterdam ... 31884
Zandvoorts Museum, Zandvoort ... 32989

Norway

Nesch-Museet, Ål ... 33378

Pakistan

Tasneem Arts Gallery, Peshawar ... 34037

Peru

Buque Museo Ex BAP América, Iquitos ... 34191

Philippines

Samoy Art Museum, Bacoor ... 34338

Poland

Biblioteka Narodowa, Warszawa ... 35217
Centrum Rysunku i Grafiki im. Tadeuza Kulisiewicza, Kalisz ... 34707
Gabinet Rycin Polskiej Akademii Umiejętności w Krakowie, Kraków ... 34773
Galeria Willa, Łódź ... 34871
Muzeum Karykatury, Warszawa ... 35250
Oddział Zbiorów Graficznych i Kartograficznych, Kraków ... 34815
Ośrodek Propagandy Sztuki, Łódź ... 34886

Portugal

Museu Galeria Manuel Cabanas, Vila Real de Santo António ... 36014

Puerto Rico

Museo del Cartel José P.H. Hernández, Río Grande ... 36054
Popular Arts Museum, San Juan ... 36078

Romania

Casa-Atelier Gabriel Popescu, Vulcana Pandele ... 36736

Russia

Dom-Muzej N.N. Žukova, Elec ... 36908
Gosudarstvennyj Memorialnyj Istoriko-chudožestvennyj i Prirodnyj Muzej-zapovednik V.D. Polenova, Strachovo ... 37808
Gosudarstvennyj Muzej Vadima Sidura, Moskva ... 37203
Muzej Knigi, Moskva ... 37308
Muzej Narodnoj Grafiki, Moskva ... 37339
Novosibirskaja Kartinnaja Galereja, Novosibirsk ... 37486
Regionalnyj Muzej Severnogo Priladožja, Sortavala ... 37789
Rybinskij Gosudarstvennyj Istoriko-architekturnyj i Chudožestvennyj Muzej-zapovednik, Rybinsk ... 37606

Serbia

Galerija Matice Srpske, Novi Sad ... 38120
Galerija Srpske Akademije Nauka i Umetnosti, Beograd ... 38045
Savremena Galerija Umetničke Kolonije Ečka, Zrenjanin ... 38167

Slovakia

Novohradská Galéria, Lučenec ... 38289

Slovenia

Grafični Muzej, Rogaska Slatina ... 38507
Grafični Muzej Rogoška Slatina, Rogoška Slatina ... 38509
Mednjarodni Grafični Likovni Centar, Ljubljana ... 38451

Spain

Gabinet de les Arts Gràfiques, Barcelona ... 38959
Museo de Dibujo "Julio Gavín-Castillo de Larrés", Larrés (Huesca) ... 39444
Museo Taller Litográfico, Cádiz ... 39092

Sweden

Åbergs Museum, Bålsta ... 40451
Grafikens Hus, Mariefred ... 40730
Techningsmuseet, Laholm ... 40665

Switzerland

Bourbaki-Panorama Luzern, Luzern ... 41589
Cabinet des Estampes du Musée d'Art et d'Histoire, Genève ... 41403
Graphische Sammlung der Eidgenössischen Technischen Hochschule Zürich, Zürich ... 42073
Herrliberger-Sammlung, Maur ... 41616
Karikatur und Cartoon Museum Basel, Basel ... 41163
Kirchner Museum Davos, Davos Platz ... 41321
Kunsthaus Zürich, Zürich ... 42083
Kupferstichkabinett, Basel ... 41167
Musée Alexis Forel, Morges ... 41643
Musée de l'Hôtel-Dieu Porrentruy (MHDP), Porrentruy ... 41728
Museum für Gestaltung Zürich, Zürich ... 42094

Taiwan

Chen Feng-Hsien Miniature Art Museum, Sindian ... 42333
Dong Zuo Inkslab Art Museum, Ershuei ... 42185

Turkey

Ayşe ve Ercümend Kalmık Vakfı-Müzesi, Gümüşsuyu ... 42791
Karikatür ve Mizah Müzesi, İstanbul ... 42830

Ukraine

Deržavnyj Ajvazovskyj Kartynna Galereja, Feodosija ... 43039

United Kingdom

Bank of England Museum, London ... 44788
Beatrix Potter Gallery, Hawkshead ... 44429
Buddle Arts Centre, Wallsend ... 45916
Glasgow Print Studio, Glasgow ... 44282
Hogarth's House, London ... 44876
The John Southern Gallery, Dobwalls ... 43988
School of Art Gallery and Museum, Aberystwyth ... 43259
Strang Print Room, London ... 44993

United States of America

Anchor Graphics, Chicago ... 47451
Art Galleries at UAH, Huntsville ... 49441
Audubon State Historic Site, Saint Francisville ... 52543
Bobbitt Visual Arts Center, Albion ... 46172
Brand Library Art Galleries, Glendale ... 48928
Centrum Arts and Creative Education, Port Townsend ... 52042
Chapin Library of Rare Books, Williamstown ... 54174
Claypool-Young Art Gallery, Morehead ... 50840
Dowd Fine Arts Gallery, Cortland ... 47831
Emerson Gallery, McLean ... 50362
Emerson Gallery, Clinton ... 47659
Esther Prangley Rice Gallery, Westminster ... 54097
Ford Gallery, Ypsilanti ... 54356
Gallery of Art, Fort Valley ... 48695
Grier Musser Museum, Los Angeles ... 50226
Grunwald Center for the Graphic Arts, Los Angeles ... 50227
Harlow Ceramics Collection Kleven Print Collection, Bemidji ... 46731
Harry D. Hendren Gallery, Saint Charles ... 52533
Hughes Fine Arts Center, Grand Forks ... 48988
Janet Turner Print Museum, Chico ... 47527
Joel and Lila Harnett Print Study Center, Richmond ... 52330
Lassen Museum Waikiki, Honolulu ... 49348
Martin Art Gallery, Allentown ... 46228
Museum of Antropology, History and Art, Rio Piedras ... 52355
Museum of Prints and Printmaking, Albany ... 46160
Myra Powell Art Gallery and Gallery at the Station, Ogden ... 51502
Piedra Lumbre Visitors Center, Abiquiu ... 46129
Print and Picture Collection, Philadelphia ... 51873
The Print Center, Philadelphia ... 51874
Print Consortium, Kansas City ... 49696
Richard F. Brush Art Gallery and Permanent Collection, Canton ... 47224
Rubelle and Norman Schafler Gallery, Brooklyn ... 47082
Self Help Graphics and Art Gallery, Los Angeles ... 50263
SFA Galleries, Nacogdoches ... 50936
Trout Gallery, Carlisle ... 47244
University of Nebraska Art Gallery at Omaha, Omaha ... 51556
Weyers-Sampson Art Gallery, Greenville ... 49077
William H. Van Every jr. and Edward M. Smith Galleries, Davidson ... 47973

Vatican City

Museo Profano, Città del Vaticano ... 54510

Gravestones → Tombs

Handicapped

Austria

Museum des Blindenwesens, Wien ... 03033

Belgium

Mad Musée, Liège ... 03798

Canada

Show Gallery, Toronto ... 07137

Croatia

Tiflološki Muzej, Zagreb ... 09013

Denmark

Blindehistorisk Museum, Hellerup ... 10056

Finland

Jyväskylän Näkövammaisten Koulun Museo, Jyväskylä ... 10889
Näkövammaismuseo, Jiris ... 10871

Germany

Deutsches Blinden-Museum (DBM), Berlin ... 17070
Geschichtliche Sammlung der Schule-Ausbildung für Blinde, Hannover ... 18990

Greece

Tactual Museum, Kallithea ... 23029

Russia

Centralnyj Muzej Vserossijskogo Obščestva Slepych, Moskva ... 37152
Literaturno-memorialnyj Muzej Vasilija Erošenko, Staryj Oskol ... 37800

United States of America

Perkins History Museum, Watertown ... 53971

Handicraft

Argentina

Casa de Manuel Mujica Lainez, La Cumbre ... 00399
Museo de Arqueología Colonial Rodolfo I. Bravo, Cafayate ... 00286

Australia

Phillip Island Community Art and Craft Gallery, Cowes ... 01005
Tweed River Art Gallery, Murwillumbah ... 01340

Austria

Buchdruckmuseum, Rohrbach/OÖ ... 02596
Handwerksmuseum, Sankt Leonhard am Hornerwald ... 02662
Kaiser Franz-Josef-Museum für Handwerk und Volkskunst, Baden bei Wien ... 01775
Maurermuseum und Dachdeckermuseum, Langenlois ... 02258
Messerermuseum, Steinbach an der Steyr ... 02767
Museum Stoffels Säge-Mühle, Hohenems ... 02093
Österreichisches Papiermacher- und Druckereimuseum, Steyrermühl ... 02779
Pechermuseum, Hernstein ... 02072
Rauchfangkehrermuseum, Wien ... 03060
Schmiedemuseum, Fulpmes ... 01912
Sensen-Schmiede-Museum/ Klangwelten, Micheldorf ... 02379
Steinmetzmuseum, Burgschleinitz ... 01809
Wäschepflegemuseum, Rainbach im Mühlkreis ... 02555
Wiener Ziegelmuseum, Wien ... 03092

Belgium

Bocholter Handwerksmuseum, Bocholt ... 03376
Folkloremuseum Florishof, Oostduinkerke ... 03924
Heemkundig Museum Boekhoute, Boekhoute ... 03378
Kempisch Museum, Brecht ... 03400
Musée de la Vie Régionale des Rièzes et des Sarts, Cul-des-Sarts ... 03532
Musée des Métiers Anciens et de la Tradition Rurale, Estinnes-au-Mont ... 03590
Musée d'Histoire Locale, Ecaussinnes ... 03570
Musée du Sabot, Porcheresse ... 03955
Musée Piperie Léonard, Andenelle ... 03273
Museum 't Nieuwhuys Hoegaarden, Hoegaarden ... 03705
Nationaal Borstelmuseum, Izegem ... 03737
Regionale Heemmusea Bachten de Kupe, Izenberge ... 03739
Streekmuseum van Klein-Brabant De Zilverreiger, Weert ... 04142

Brazil

Museu Vivo da Memória Candanga, Brasília ... 04370

Brunei

Muzium Teknologi Melayu, Bandar Seri Begawan ... 04966
Pusat Latihan Kesenian Pertukangan Tangan Brunei, Bandar Seri Begawan ... 04967

Bulgaria

Architekturno-etnografski Muzej Etăr, Gabrovo ... 05029
Nacionalna Izložba na Narodnite Chudožestveni Zanajati i Priložni Izkustva, Orešak ... 05100

Canada

Alcan Museum and Archives, Montréal ... 06326
Carman House Museum, Iroquois ... 06022
Centre de Design de l'UQAM, Montréal ... 06332
Design Exchange, Toronto ... 07104
Huron County Museum, Goderich ... 05879
Lenoir Forge Museum, Arichat ... 05352
Manitoba Crafts Museum and Library, Winnipeg ... 07368
Milton Blacksmith Shop Museum, Milton ... 06290
Musée Acadien, Cheticamp ... 05607
Musée des Maîtres et Artisans, Montréal ... 06360
Musée Maritime du Québec, L'Islet ... 06175
Saskatchewan Craft Council Gallery, Saskatoon ... 06924

Chile

Museo de Arte y Artesanía de Linares, Linares ... 07441

China

China Silk Museum, Hangzhou ... 07759
Handover Gifts Museum, Macau ... 07994
Shanghai Arts and Crafts Museum, Shanghai ... 08181
Yanshi Marketplace Museum, Yanshi ... 08431

Congo, Democratic Republic

Musée d'Anthropologie, Kinshasa ... 08795

Cuba

Galería Los Oficios, La Habana ... 09105

Czech Republic

Hrad Helfštýn, Týn nad Bečvou ... 09883
Muzeum Krajky, Prachatice ... 09732
Muzeum Krajky, Vamberk ... 09898
Muzeum Prostějovska, Prostějov ... 09810
Muzeum Šumavy, Sušice ... 09856
Technické Muzeum v Brně, Brno ... 09440
Vlastivědné Muzeum pro Vysoké nad Jizerou a Okolí, Vysoké nad Jizerou ... 09921

Denmark

Holstebro Museum, Holstebro ... 10091
Museet for Thy og Vester Hanherred, Thisted ... 10296
Næstved Museum, Næstved ... 10190
Slagelse Museum for Handel, Håndværk og Industri, Slagelse ... 10276
Søfartssamlingerne i Troense, Svendborg ... 10292
Viborg Stiftsmuseum, Viborg ... 10317

Finland

Fransu-tupa, Himanka ... 10839
Gallen-Kallelan Museo, Espoo ... 10728
Halikkon Museo, Halikko ... 10752
Kemin Työläismuseo ja Kähertäjämuseo, Kemi ... 10941
Luostarinmäen Käsityöläismuseo, Turku ... 11394

Lusto-Suomen Metsämuseo, Punkaharju 11238
Naantalin Museo, Naantali 11133
Porvoon Museo, Porvoo 11230
Satakunnan Museo, Pori 11223
Sepänmäen Käsityömuseo, Mäntsälä 11092
Suomen Käsityön Museo, Jyväskylä 10897
Työväenasuntomuseo ja Kotiseututalo, Kuusankoski 11018
Väinontalo - Järviseudun Museo, Vasikka-Aho 11457
Valtakunnallinen Työväentalomuseo, Riihimäki 11273
Virkki Käsityömuseo, Helsinki 10837

France

Ecomusée de la Ferme et des Vieux Métiers, Lizio 13791
Musée de la Coutellerie, Thiers 15881
Musée de Sucy, Sucy-en-Brie 15836
Musée du Compagnonnage, Tours 15966
Musée du Feutre, Mouzon 14319
Musée Paul Dupuy, Toulouse 15944
Musée Régional des Métiers, La Cheze 13343

Germany

Alte Wache, Groitzsch 18770
Bauernhofmuseum Schreiner, Buchhofen 17617
Bergisches Museum für Bergbau, Handwerk und Gewerbe, Bergisch Gladbach 17014
Breuberg-Museum, Breuberg 17587
Burg Stollberg, Stolberg, Rheinland 21909
Deutsches Maler- und Lackierer-Museum, Hamburg 18921
Dorfmuseum Dietersweiler, Freudenstadt 18455
Erstes Deutsches Türmermuseum, Vilseck 22244
Erzgebirgsmuseum mit Besucherbergwerk, Annaberg-Buchholz 16553
Europäisches Klempner- und Kupferschmiede-Museum, Karlstadt 19454
Familienmuseum Schloss Braunfels, Braunfels 17514
Fellenbergmühle Feinmechanisches Museum, Merzig 20300
Flößermuseum Unterrodach, Marktrodach 20223
Freilicht- und Heimatmuseum Donaumoos, Karlshuld 19434
Freilichtmuseum Glentleiten, Großweil 18809
Gedenkstätte Ehemalige Synagoge, Adelsheim 16428
Gewerbe-Museum, Spaichingen 21826
Gilde-Museum, Oldenburg in Holstein 20901
Grönegau-Museum, Melle 20272
Hammer- und Waffenschmiede-Museum Hexenagger, Altmannstein 16513
Handwerkerstuben, Feuchtwangen 18314
Handwerksmuseum, Deggendorf 17805
Handwerksmuseum Ovelgönne, Ovelgönne 20964
Haselünner Heimathäuser, Haselünne 19024
Heimat- und Feuerwehrmuseum, Schauenstein 21534
Heimat-und Handwerkermuseum, Leutershausen-Wiedersbach 19970
Heimathaus-Handwerkermuseum, Scheidegg 21539
Heimatmuseum, Brackenheim 17494
Heimatmuseum, Maßbach 20234
Heimatmuseum, Müllrose 20418
Heimatmuseum, Oederan 20859
Heimatmuseum, Roßwein 21385
Heimatmuseum, Stadtilm 21857
Heimatmuseum, Wertingen 22505
Heimatmuseum und Naturalienkabinett, Waldenburg, Kreis Chemnitzer Land 22290
Heimatmuseum Vilsbiburg - Kröninger Hafnermuseum, Vilsbiburg 22243
Heimatstuben der Stadt, Titisee-Neustadt 22064
Heimschneidermuseum, Großwallstadt 18808
Historische Handwerkerstuben, Gingst 18647
Historische Spinnerei Gartetal, Gleichen 18659
Kolpings- und Handwerksmuseum im Faltertor, Dettelbach 17841
Kreisheimatmuseum Demmin, Demmin 17818
Kreismuseum Oberhavel, Oranienburg 20909
Kulturhistorisches Museum der Hansestadt Stralsund, Stralsund 21914
Kulturhistorisches Museum Rostock, Rostock 21393
Kunstgewerbemuseum im Schloss Köpenick, Berlin 17169
Landesmuseum Mainz, Mainz 20146
Lohgerber-, Stadt- und Kreismuseum, Dippoldiswalde 17871
Ludwig-Gebhard-Museum, Tiefenbach, Oberpfalz 22059
LWL-Freilichtmuseum Hagen, Hagen 18858
Mainfränkisches Museum Würzburg, Würzburg 22713
Museum Alte Münze, Stolberg (Harz) Kurort 21907
Museum Altes Rathaus, Leingarten 19886
Museum Dermbach, Dermbach 17825
Museum des Kreises Plön mit norddeutscher Glassammlung, Plön 21065
Museum ehemalige Klöppelschule Tiefenbach, Tiefenbach, Oberpfalz 22060
Museum für bäuerliches Handwerk und Kultur, Wilhelmsdorf, Kreis Ravensburg 22573
Museum für Holzhandwerk, Sinzig 21788
Museum Göltzsch, Rodewisch 21346
Museum im Alten Schloß, Altensteig 16503
Museum Kloster Asbach, Asbach 16579
Museum Schloß Wilhelmsburg, Schmalkalden, Kurort 21580
Natur-Museum mit Bauerngarten, Goldberg 18703
Oberpfälzer Handwerksmuseum, Rötz 21361
Potsdam-Museum, Potsdam 21103
Regionalmuseum Alsfeld, Alsfeld 16481
Schefflenztal-Sammlungen, Schefflenz 21538
Schiefer- und Ziegelmuseum Dörfles-Esbach, Dörfles-Esbach 17884
Schloss Moritzburg und Fasanenschlösschen, Moritzburg 20386
Schwäbisches Handwerkermuseum, Augsburg 16634
Stadt- und Bergbaumuseum Freiberg, Freiberg, Sachsen 18425
Stadtmuseum Burghausen, Burghausen 17656
Stadtmuseum Gardelegen, Gardelegen 18538
Städtisches Hutmuseum, Lindenberg im Allgäu 20005
Städtisches Museum Braunschweig, Braunschweig 17530
Stiftung Moritzburg, Halle, Saale 18904
Südhessisches Handwerksmuseum, Roßdorf 21382
Tabakspeicher, Nordhausen 20740
Technik anno dazumal - Museum Kratzmühle, Kinding 19539
Thüringer Museum Eisenach, Eisenach 18098
Voithenberghammer, Furth im Wald 18517
Volkskundliche Sammlung des Fichtelgebirgsvereins, Weidenberg 22369
Wachsstöcklkabinett - Wachszieher- und Lebzelter-Museum, Viechtach 22227
Waldnaabtal-Museum in der Burg Neuhaus, Windischeschenbach 22597
Wannenmacher-Museum, Emsdetten 18161
Wegmachermuseum, Wasserburg am Inn 22346
Zeiselmairhaus, Schrobenhausen 21641

Greece

Ethnological Museum of Pyrsogianni Stonemasons, Konitsa 23061
Museum of Kalymnos, Kalymnos 23031

Hungary

Barlanglakás Emlékkiállítás, Budapest 23386
Bozsó Gyűjtemény, Kecskemét 23643
Ercsi Helytörténeti Gyűjtemény, Ercsi 23547
Kékfestő Múzeum, Pápa 23753
Kis Géza Ormánság Múzeum és Skanzen, Sellye 23799
Korom Gyula Fodrászati Eszközök Gyűjtemény, Budapest 23427
Néprajzi Kiállítás, Sopron 23814
Somogy Kereskedelme Anno..., Kaposvár 23634

Iceland

Heimilisidnardarsafnid Halldórustofa, Blönduós 23988
Minjasafnid Burstarfelli, Vopnafjörður 24044

India

Arts and Crafts Museum, Bhavnagar 24085
Magan Sangrahalaya Samiti, Wardha 24431
Mahatma Phule Vastu Sangrahalaya, Pune 24359
Sree Moolam Shastyabdapurti Memorial Institute, Thiruvananthapuram 24405
Museum of Arts and Crafts, Lucknow 24297

Italy

Mostra degli Antichi Mestieri di Calabria, Tropea 28104
Museo del Falegname T. Sana-Onlus, Almenno San Bartolomeo 25204
Museo della Civiltà Contadina ed Artigiana, Monterosso Calabro 26848
Museo della Paglia e dell'Intreccio D. Michalacci, Signa 27850
Museo dello Spazzacamino, Santa Maria Maggiore 27706
Museo dell'Ombrello e del Parasole, Gignese 26340

Japan

Bekko Crafts Museum, Nagasaki 28969
Floral Museum, Chiba 28394
Fureaikan - Kyoto Museum of Traditional Crafts, Kyoto 28809
Gotoh Bijutsukan, Tokyo 29394
Iwate-kenritsu Meisho Hakubutsukan, Morioka 28936
Maizuru-shiritsu Sekei Renga Hakubutsukan, Maizuru 28879
Meiji-Daigaku Shouhin Hakubutsukan, Tokyo 29433
Nagasaki-shiritsu Hakabutsukan, Nagasaki 28983
Oushuu-shokoku Mingeikan, Teshikaga 29360
Sapporo Geijutsu no Mori Bijutsukan, Sapporo 29222
Token Hakubutsukan, Tokyo 29487
Tokyo Chuo-ku Bijutsukan, Tokyo 29488
Tokyo Kokuritsu Kindai Bijutsukan Kogeikan, Crafts Gallery, Tokyo 29492

Jordan

Jordan Museum of Popular Tradition, Amman 29654

Korea, Republic

Straw and Grass Handicraft Museum, Seoul 29926

Kuwait

The Tareq Rajab Museum, Hawelli 29947

Latvia

Kandavas Novada Muzejs, Kandava 30029
Salacgrīvas Muzejs, Salacgrīva 30121

Malaysia

Heritage Museum, Kuah 30567
Kemajuan Kraftangan Malaysia, Rawang 30615
Kuala Lumpur Craft Museum, Kuala Lumpur 30575

Morocco

Musée de Dar Jamaï, Meknès 31681
Nejarine-Musée, Fès 31672

Netherlands

Afdeling Nuttige en Fraaie Handwerken, Uithuizermeeden 32837
Atelier Na-iem, Gees 32238
Craftselijke Zadelmakerij Museum, Bellingwolde 31959
De Dubbelde Palmboom, Rotterdam 32704
Kantmuseum, Uithuizen 32835
Museum De Grutterswinkel, Leeuwarden 32475
Museum de Koperen Knop, Hardinxveld-Giessendam 32314
Museum de Timmerwerf, De Lier 32048
Museum de Valkhof, Hellendoorn 32356
Museum Valkenheide, Maarsbergen 32516
Museum van het Ambacht, Den Haag 32090
Museumboerderij, Paasloo 32665
Nederlands Baksteen en Dakpanmuseum de Panoven, Zevenaar 33000
Nederlands Centrum voor Handwerken, Breda 32006
Oude Ambachten en Speelgoedmuseum, Terschuur 32819
Stadsboerderij Het Wevershuisje, Almelo 31800
Streekmuseum Opsterland, Gorredijk 32264

Niue

Huanaki Cultural Centre, Alofi 33348

Norway

Bygdetunet Jutulheimen, Vågåmo 33954
Hardanger Folkemuseum, Utne 33949
Stjørdal Museum KF, Stjørdal 33881
Trastad Samlinger, Borkenes 33448
Vest-Telemark Museum, Dalen i Telemark 33463

Philippines

Museo ng Kalinangang Pilipino, Pasay 34474

Poland

Ekspozycja Rzemiosło w Białym Spichrzu, Bydgoszcz 34575
Muzeum Dawnego Kupiectwa, Świdnica 35147
Muzeum Miedzi, Legnica 34848
Muzeum Regionalne, Chojnów 34600
Muzeum Regionalne, Krasnystaw 34826
Muzeum Zegarów, Rzemiosł Artystycznych i Precyzyjnych, Warszawa 35283
Państwowe Muzeum Etnograficzne, Warszawa 35287
Skansen Budownictwa Ludowego Zachodniej Wielkopolski, Wolsztyn 35320

Portugal

Museu da Latoaria, Tomar 35963
Museu de Artes e Ofícios Francisco Távora, Escalhão 35564

Romania

Muzeul de Etnografie Universală Franz Binder, Sibiu 36607

Russia

Gorodeckij Kraevedčeskij Muzej, Gorodec 36928
Muzej Istorii Narodnych Chudožestvennych Promyslov, Nižnij Novgorod 37456
Muzej Remesla, Architektury i Byta, Kaluga 37000
Muzej Vjatskie Narodnye Chudožestvennye Promysly, Kirov 37041
Novgorodskij Gosudarstvennyj Muzej-zapovednik, Velikij Novgorod 37922

Serbia

Gradski Muzej Sombor, Sombor 38148
Muzej Staro Selo, Sirogojno 38142

Slovakia

Historické a Národopisné Múzeum Želovce, Želovce 38389
Mestský Vlastivedný Dom, Šamorín 38348
Múzeum Mošovských Remeselníkov, Mošovce 38309
Múzeum Umeleckých Remesiel, Bratislava 38229
Múzeum v Kežmarku a Mestský Hrad, Kežmarok 38267
Národopisná Expozícia SARM Bardejovské Kúpele, Bardejovské Kúpele 38196

Slovenia

Mestni Muzej Ljubljana, Ljubljana 38453

South Africa

Vukani Zulu Cultural Museum, Eshowe 38617
Zululand Historical Museum, Eshowe 38618

Spain

Colección Municipal, Granada 39361
Museo Etnológico de Navarra Julio Caro Baroja, Ayegui 38925

Sweden

Åttingen Hantverk och Lantbruksmuseum, Aneby 40421
Färgargården, Norrköping 40770
Falkenbergs Museum, Falkenberg 40508
Hantverks- och Sjörfartsmuseét på Norra Berget, Sundsvall 40959
Hantverksloftet, Simrishamn 40840
Hantverksmuseet, Ängelholm 40408
Läckö Slott, Lidköping 40674
Länsmuseet Gävleborg, Gävle 40532
Länsmuseet Varberg, Varberg 41048
Nora Museum, Gyttorp 40570
Sadelmakare Öbergs Hus, Solna 40873
Skellefteå Museum, Skellefteå 40849
Slöjdmuseet, Ånge 40426

Switzerland

Feuerwehr- und Handwerkermuseum, Endingen 41350
Gewerbemuseum, Winterthur 42027
Heimatmuseum, Reinach (Basel-Land) 41754
Historisches Museum Blumenstein, Solothurn 41882
Historisches Museum Schloss Arbon, Arbon 41121
Malermuseum Wetzikon, Wetzikon 42011
Musée des Arts et des Sciences, Sainte-Croix 41797
Musée du Vieux-Moudon, Moudon 41650
Musée Paysan et Artisanal, La Chaux-de-Fonds 41297
Ortsmuseum, Kaltbrunn 41500
Stiftung Ziegelei-Museum, Cham 41288

Taiwan

Ami Cultural Center, Chengkung 42161
Exhibition Center, Caotun 42148

United Kingdom

Abingdon Museum, Abingdon 43260
Bewdley Museum, Bewdley 43469
Bilston Craft Gallery, Wolverhampton 46058
Castle Park Arts Centre, Frodsham 44256
Cousland Smiddy, Cousland 43898
Crafts Council Gallery, London 44831
Cumberland Pencil Museum, Keswick 44562
Devon Guild of Craftsmen Gallery, Bovey Tracey 43569
Fordyce Joiner's Visitor Centre, Fordyce 44240
Friary Art Gallery, Lichfield 44703
Gordon Brown Collection, Ravenshead 45463
Hat Works, Museum of Hatting, Stockport 45753
Jersey Battle of Flowers Museum, Saint Ouen 45588
Museum of Local History, Burnley 43677
The Petworth Cottage Museum, Petworth 45377
Saint Ives Museum, Saint Ives, Cornwall 45577
Shire Hall Gallery, Stafford 45730
Somerset Rural Life Museum, Glastonbury 44308

Stockwood Park Museum, Luton ... 45047
Strachur Smiddy Museum, Strachur ... 45775

United States of America

Alleghany Highlands Arts and Crafts Center, Clifton Forge ... 47649
American Craft Museum, New York ... 51112
Arrowmont School of Arts and Crafts Collection, Gatlinburg ... 48873
Arts Center of the Ozarks-Gallery, Springdale 53211
Brookfield Craft Center, Brookfield ... 47055
Center for Book Arts, New York ... 51141
Chimneyville Crafts Gallery, Jackson ... 49557
Craft Alliance, Saint Louis ... 52574
Crafts Museum, Mequon ... 50594
Craftsman Farms Foundation, Parsippany ... 51716
Crowley Art Association and Gallery, Crowley 47893
Cultural Museum at Scarritt-Bennett Center's, Nashville ... 50965
Dawson Springs Museum and Art Center, Dawson Springs ... 47981
Fowler Museum of Cultural History, Los Angeles ... 50223
Frog Hollow Vermont State Craft Center, Middlebury ... 50651
Georgetown College Gallery, Georgetown ... 48891
Hambidge Center for the Creative Arts and Sciences, Rabun Gap ... 52221
Hammond Museum, North Salem ... 51405
Hand Workshop Art Center, Richmond ... 52328
Haystack Mountain School of Crafts Gallery, Deer Isle ... 48027
Historic Sauder Village, Archbold ... 46340
Kentucky Museum of Arts and Design, Louisville ... 50288
Krikorian Gallery, Worcester ... 54300
The London Brass Rubbing Centre in Washington D.C., Gaithersburg ... 48824
Lyndon House Art Center, Athens ... 46413
Manchester Craftsmen's Guild, Pittsburgh ... 51952
Mattye Reed African Heritage Center, Greensboro ... 49066
Mint Museum of Craft and Design, Charlotte ... 47393
Mississippi Crafts Center, Ridgeland ... 52349
Museum and Arts Center in the Sequim Dungeness Valley, Sequim ... 53037
Museum of Contemporary Craft, Portland ... 52080
Museum of the American Quilter's Society, Paducah ... 51669
Native American Resource Center, Pembroke 51755
Old Depot Museum, Ottawa ... 51629
Pioneer Log Cabin, Manhattan ... 50436
Pope County Historical Museum, Glenwood ... 48942
Red Rock Museum, Church Rock ... 47549
Renwick Gallery, Washington ... 53915
Richard Sparrow House, Plymouth ... 51998
Schingoethe Center for Native American Cultures, Aurora ... 46493
Snug Harbor Cultural Center, Staten Island ... 53267
Society for Contemporary Craft Museum, Pittsburgh ... 51961
South Carolina Artisans Center, Walterboro ... 53807
Southern Highland Craft Guild at the Folk Art Center, Asheville ... 46376
Steppingstone Museum, Havre de Grace ... 49220
Tama County Historical Museum, Toledo ... 53519
Thomas E. McMillan Museum, Brewton ... 47011
Western Colorado Center for the Arts, Grand Junction ... 48995
Westville Historic Handicrafts Museum, Lumpkin ... 50322
Wharton Esherick Museum, Malvern ... 50411
World Kite Museum and Hall of Fame, Long Beach ... 50180

Yemen

Seiyun in Wadi Hadhramaut Museum, Seiyun 54651

Heraldry → Genealogy

Hides and Skins

Canada

Canal-de-Lachine - Lieu Historique National du Canada, Lachine ... 06121
Ermatinger Old Stone House, Sault Sainte Marie ... 06930
Fort Carlton, Regina ... 06717
Nor' Wester and Loyalist Museum, Williamstown ... 07339
Sturgeon River House, Sturgeon Falls ... 07024

Germany

Leder- und Gerbermuseum, Mülheim an der Ruhr ... 20415

United States of America

Fort Clark Trading Post, Center ... 47313
Fort Uncompahgre History Museum, Delta ... 48049
Gingras Trading Post, Walhalla ... 53794
Stuart House City Museum, Mackinac Island ... 50352

History

Afghanistan

National Museum Kabul, Kabul ... 00005

Albania

Muzeu Historik Fier, Fier ... 00016

Algeria

History and Natural History Museum, Alger ... 00038
Musée Emir Abdelkader, Miliana ... 00069

Andorra

Museu Casa Rull, La Massana ... 00097

Angola

Museu da Escravatura, Luanda ... 00111

Argentina

Buque Museo Corbeta Uruguay, Buenos Aires 00150
Casa Natal de Sarmiento, San Juan ... 00585
Fuerte de la Punta del Sauce, La Carlota ... 00398
Galería Monumento Histórico Nacional a la Bandera, Rosario ... 00541
Museo Americanista, Lomas de Zamora ... 00430
Museo de la Administración Federal de Ingresos Públicos, Buenos Aires ... 00197
Museo de la Casa Histórica de la Independencia Nacional, San Miguel de Tucumán ... 00597
Museo de los Corrales Viejos de Parque de los Patricios, Buenos Aires ... 00212
Museo de los Pioneros, Río Gallegos ... 00531
Museo General Belgrano, Buenos Aires ... 00235
Museo Hermanos Nacif Weiss, Rivadavia ... 00539
Museo Histórico Conventual San Carlos, San Lorenzo ... 00591
Museo Histórico de la Honorable Cámara de Diputados de la Nación, Buenos Aires ... 00238
Museo Histórico de la Provincia de Catamarca, San Fernando del Valle de Catamarca ... 00573
Museo Histórico Municipal Brigadier General Juan Martín de Pueyrredón, San Isidro ... 00581
Museo Histórico Nacional, Buenos Aires ... 00241
Museo Histórico Provincial Colonel Manuel José Olascoaga, Chos Malal ... 00306
Museo La Cinacina, San Antonio de Areco ... 00567
Museo Malvinas Argentinas, Río Gallegos ... 00532
Museo Metropolitano, Buenos Aires ... 00252
Museo Nacional Justo José de Urquiza, Concepción del Uruguay ... 00313
Museo Notario Roberto Mario Arata, Buenos Aires ... 00267
Museo Provincial Mario Brozoski, Puerto Deseado ... 00505
Museo Ramón Pérez Fernández, Rivadavia ... 00540
Museo Regional Aníbal Cambas, Posadas ... 00501
Museo Regional Carlos Funes Derieul, Coronel Dorrego ... 00342
Museo Regional Dr. José Luro, Pedro Luro ... 00489
Museo Regional José G. Brochero, Santa Rosa de Río Primero ... 00626
Museo Rural de la Posta de Sinsacate, Jesús María ... 00394
Museo Senador Domingo Faustino Sarmiento, Buenos Aires ... 00273
Museos Integrales de Antofagasta de la Sierra y Laguna Blanca, San Fernando del Valle de Catamarca ... 00575
Sala Josefa Rodriguez del Fresno, Santa Fé ... 00622

Armenia

Echmiadzin Cathedral Complex and Museum, Echmiadzin ... 00695
History Museum of Armenia, Yerevan ... 00738
Mother Armenia Military Museum, Yerevan ... 00746
Museum-Institute of the Armenian Genocide, Yerevan ... 00747
Sisian Museum of History N. Adonts, Sisian ... 00717

Australia

Adelaide City Council Civic Collection, Adelaide ... 00770
Adelaide Lithuanian Museum, Norwood ... 01382
Adelaide Masonic Centre Museum, Adelaide ... 00771
Australian Stockman's Hall of Fame and Outback Heritage Centre, Longreach ... 01238
Commissariat Store Museum, Brisbane ... 00896
Conservation Resource Centre, Glebe ... 01113
Garrison Gallery Museum, Millers Point ... 01304
Griffith Pioneer Park Museum, Griffith ... 01131
Hyde Park Barracks Museum, Sydney ... 01548
Melbourne Museum, Carlton South ... 00956
Migration Museum, Adelaide ... 00778
Museum of Childhood, Edith Cowan University, Perth ... 01411
Museum of Sydney, Sydney ... 01555
Museum Victoria, Carlton South ... 00957
National Museum of Australia, Canberra ... 00947
Powder Magazine, Beechworth ... 00849
Queen Victoria Museum and Art Gallery at Royal Park, Launceston ... 01224
Royal Western Australian Historical Society Museum, Nedlands ... 01357
Tasmanian Museum and Art Gallery, Hobart ... 01159
Torres Strait Museum, Thursday Island ... 01587
Ukrainian Museum, Torrens Park ... 01596
Western Australian Museum Perth, Perth ... 01412

Austria

Cowboy-Museum Fatsy, Linz ... 02311
Dorfmuseum, Kritzendorf ... 02238
Erzherzog Franz Ferdinand Museum, Artstetten ... 01728
Erzherzog Johann-Dokumentation Thernberg, Thernberg ... 02811
Fürstlich Esterházy'sche Sammlungen Burg Forchtenstein, Forchtenstein ... 01894
Gemeindemuseum, Hittisau ... 02085
K K-Museum, Türnitz ... 02829
K.u.k. Hausmuseum, Weyregg am Attersee ... 02928
Kunstraum Stein, Krems ... 02228
Landesmuseum Kärnten, Klagenfurt ... 02186
Mährisch-Schlesisches Heimatmuseum, Klosterneuburg ... 02201
Malteser-Museum, Mailberg ... 02343
Museum der Mechitharisten-Congregation, Wien ... 03031
Museum für Österreichische Zeitgeschichte, Scharnstein ... 02699
Museum Goldenes Dachl, Innsbruck ... 02124
Museum Laxenburg, Laxenburg ... 02275
Museum Orth, Orth an der Donau ... 02473
Napoleon- und Heimatmuseum, Deutsch Wagram ... 01812
Oberösterreichisches Schulmuseum, Bad Leonfelden ... 01759
Otto Wagner Hofpavillion Hietzing, Wien ... 03052
Otto Wagner Pavillion Karlsplatz, Wien ... 03053
Rogendorfer-Ausstellung, Pöggstall ... 02511
Salzburg Museum, Salzburg ... 02625
Schlossmuseum, Peuerbach ... 02497
Schlossmuseum, Linz ... 02324
Sigmund Freud Museum, Wien ... 03075
Stadtmuseum, Gmünd, Niederösterreich ... 01947
Stadtmuseum Innsbruck, Innsbruck ... 02131
Das Tirol Panorama, Innsbruck ... 02134
Wienerwaldmuseum, Eichgraben ... 01851

Azerbaijan

Azerbaycan Tarixi Muzeyi, Bakı ... 03144
State Museum of the History of Religion, Bakı 03175

Bahamas

Bahamas Historical Society Museum, Nassau 03203
Pompey Museum of Slavery and Emancipation, Nassau ... 03210

Bahrain

Bahrain National Museum, Manama ... 03212

Bangladesh

Bangabandhu Shek Mujibur Rahman Museum, Dhaka ... 03218
Chittagong University Museum, Chittagong ... 03214
Mukti Juddha Museum, Dhaka ... 03226

Barbados

Barbados Museum, Saint Michael ... 03242

Belarus

Gosudarstvennyj Muzej Istorii i Kultury Belarusa, Minsk ... 03250
Grodnenski Dziarzhauny Hystorika-Archealagičny Muziej, Grodno ... 03246

Belgium

Archeologisch Museum Van Bogaert-Wauters, Hamme ... 03672
Athus et l'Acier, Athus ... 03332
Ecomusée de la Batellerie et de la Voie d'Eau Le Thudo, Thuin ... 04064
Gallo-Romeins Museum, Tongeren ... 04075
Kortrijk 1302, Kortrijk ... 03760
Landcommanderij Alden Biesen, Bilzen ... 03367
Le Musée Archéologique de Montaigle, Falaën ... 03596
Mercator Museum, Sint-Niklaas ... 04021
Musée Archéologique, Arlon ... 03323
Musée Communal d'Archéologie, d'Art et d'Histoire, Nivelles ... 03917
Musée Communal de Braine-l'Alleud, Braine-l'Alleud ... 03395
Musée de la Vie Locale de Walhain, Walhain ... 04129
Musée de la Vie Populaire, Soumagne ... 04039
Musée de Waterloo, Waterloo ... 04137
Musée des Métiers Anciens et de la Tradition Rurale, Estinnes-au-Mont ... 03590
Musée d'Histoire et d'Archéologie, Marcinelle 03850
Musée d'Histoire et d'Archéologie, Warneton ... 04133
Musée d'Histoire et de Vie Locale, Genappe ... 03617
Musées Royaux d'Art et d'Histoire, Bruxelles ... 03495
Muséobus, Naninne ... 03909
Pauselijk Zouavenmuseum, Roeselare ... 03977
Rietgaverstede, Nevele ... 03912
Stadhuis, Gent ... 03642
Stadsmuseum STAM, Gent ... 03643
Stedelijk Museum Aalst, Aalst ... 03259
Torenmuseum, Mol ... 03877
Vredesmuseum IJzertoren, Diksmuide ... 03555

Belize

Museum of Belize, Belize City ... 04176

Benin

Musée d'Histoire, Ouidah ... 04185
Musée Historique d'Abomey, Abomey ... 04182

Bermuda

Bermuda Maritime Museum, Mangrove Bay ... 04194

Bhutan

National Museum of Bhutan, Paro ... 04201

Bolivia

Casa Pedro Domingo Murillo, La Paz ... 04210
Museo Martín Cárdenas, Cochabamba ... 04208

Bosnia and Herzegovina

Muzej Hercegovine, Mostar ... 04270
Spomen Kuća Bitke na Sutjesci, Tjentište ... 04285

Botswana

Khama III Memorial Museum, Serowe ... 04301

Brazil

Divisão de Museus, Patrimônio e Arquivo Histórico, Taubaté ... 04922
Museu Afro Brasileiro de Sergipe, Laranjeiras 04533
Museu Arquivo Histórico da Santa Casa, Belém ... 04332
Museu Banespa, São Paulo ... 04864
Museu Bi Moreira, Lavras ... 04534
Museu Casa de Portinari, Brodowski ... 04373
Museu da Associação Nacional dos Veteranos, São João Del Rei ... 04834
Museu da Família Colonial, Blumenau ... 04353
Museu de Arte e Histórico, Jaboticabal ... 04505
Museu de Folclore Saul Martins, Vespasiano ... 04950
Museu de História Natural, Crato ... 04421
Museu de Instrumentos de Cálculo Numérico, São Carlos ... 04825
Museu de Pedras Ramis Bucair, Cuiabá ... 04423
Museu de Tradições do Nordeste, Brasília ... 04362
Museu do Ceará, Fortaleza ... 04464
Museu do Estado da Bahia, Salvador ... 04800
Museu do Instituto Arqueológico, Histórico e Geográfico Pernam Bucano, Recife ... 04669
Museu do Instituto Geografico e Historico da Bahia, Salvador ... 04801
Museu do Instituto Histórico e Geográfico Brasileiro, Rio de Janeiro ... 04738
Museu do Instituto Histórico e Geográfico de Alagoas, Maceió ... 04547
Museu do Instituto Histórico e Geográfico do Rio Grande do Norte, Natal ... 04587
Museu do Minstério do Trabalho, Brasília ... 04364
Museu do Primeiro Reinado, Rio de Janeiro ... 04741
Museu do STF, Brasília ... 04365
Museu Educativo Gama d'Eça, Santa Maria ... 04810
Museu Guido Straube, Curitiba ... 04438
Museu Gustavo Barroso, Fortaleza ... 04465

Museu Histórico, Cambé 04381
Museu Histórico, Divinópolis 04445
Museu Histórico, Londrina 04539
Museu Histórico, Nova Ponte 04593
Museu Histórico, Rio Pomba 04777
Museu Histórico, São Francisco do Sul 04829
Museu Histórico Aurélio Dolabella, Santa Luzia 04809
Museu Histórico Bárbara Heliodora, São Gonçalo do Sapucaí 04832
Museu Histórico Beato José de Anchieta, Pedreira 04620
Museu Histórico Casa do Imigrante, Bento Gonçalves 04352
Museu Histórico Corália Venites Maluf, Sacramento 04785
Museu Histórico Cultural, Rio do Sul 04774
Museu Histórico da Cidade, Brasília 04367
Museu Histórico da Universidade Federal, Viçoasa 04952
Museu Histórico de Araxá - Don Beja, Araxá 04319
Museu Histórico de Jataí Francisco Honório de Campos, Jataí 04510
Museu Histórico do Senado Federal, Brasília 04368
Museu Histórico Dona Mariana Joaquina da Costa, Vespasiano 04951
Museu Histórico e de Ordem Geral Plínio Travassos dos Santos, Ribeirão Preto 04660
Museu Histórico e Diplomático do Itamaraty, Rio de Janeiro 04751
Museu Histórico Municipal, Carangola 04402
Museu Histórico Municipal e da Imigração Italiana Oswaldo Samuel Massei, São Caetano do Sul 04824
Museu Histórico Municipal Tuany Toledo, Pouso Alegre 04658
Museu Histórico Nacional, Rio de Janeiro 04752
Museu Histórico Regional, Apucarana 04310
Museu Histórico Regional, Itambacuri 04496
Museu Histórico Tenente Coronel PM Augusto de Almeida Garrett, Curitiba 04439
Museu Histórico Visconde de São Leopoldo, São Leopoldo 04842
Museu Joaquim José Felizardo, Porto Alegre 04650
Museu Julio de Castilhos, Porto Alegre 04651
Museu Leprológico, Curitiba 04440
Museu Meridional, Porto Alegre 04652
Museu Metropolitano de Arte, Curitiba 04441
Museu Municipal, Conquista 04414
Museu Nacional do Mar - Embarações Brasileiras, São Francisco do Sul 04830
Museu Patríco Corrêa da Câmara, Bagé 04327
Museu Paulista da Universidade de São Paulo, São Paulo 04908
Museu Regional D. Bosco, Campo Grande 04394
Museu Regional de Olinda, Olinda 04598
Museu Republicano Convenção de Itu, Itu 04504
Museu Sargento Edésio de Carvalho, Sousa 04920
Museu Theodormiro Carneiro Santiago, Itajubá 04495
Sala de Memórias de Chapada dos Guimarães, Chapada dos Guimarães 04412

Brunei

Constitutional History Gallery, Bandar Seri Begawan 04963
Muzium Brunei, Bandar Seri Begawan 04965

Bulgaria

Istoričeski Muzej, Drjanovo 05024
Istoričeski Muzej, Pravec 05137
Istoričeski Muzej Batak, Batak 04981
Istoričeski Muzej - Preminavane na Ruskite Vojski pri Svištov, Svištov 05223
Istoričeski Muzej Targovište, Targovište 05225
Istoričeskij Muzej, Tutrakan 05234
Kăšta-muzej Christo G. Danov, Plovdiv 05128
Kăšta-muzej Dimităr Poljanov, Karnobat 05052
Kăšta-muzej Dobri Voinikov, Šumen 05214
Kăšta-muzej Georgi Benkovski, Koprivštica 05071
Kăšta-muzej Neofit Rilski, Bansko 04979
Kăšta-muzej Panaët Volov, Šumen 05215
Kăšta-muzej Pančo Vladigerov, Šumen 05216
Kăšta-muzej Svetoslav Obretenov, Provadia 05141
Kăšta-muzej Zachari Stojanov, Medven 05090
Kăšta-muzej Zachari Stojanov, Ruse 05152
Muzej Kăkrinsko Chanče, Kăkrina 05044
Muzej na Kotlenskite Vazroždenci i Panteon na Georgi S. Rakovski, Kotel 05077
Muzej na Văzraždeneto i Nacionalno-osvoboditelnite Borbi, Elena 05027
Muzej Vojniško Văstanie 1918, Radomir 05146
Nacionalen Istoričeski Muzej, Sofia 05191
Nacionalen Muzej Parachod Radecki, Kozloduj 05080
Nacionalen Muzej Roženski Manastir, Melnik 05092
Nacionalen Park-muzej Samuilova krepost, Petrič 05114
Okražen Istoričeski Muzej, Veliko Tărnovo 05257
Panorama Plevenska Epopeja 1877, Pleven 05119
Park-muzej General V.Z. Lavrov, Gorni Dabnik 05038
Park-Muzej Vladislav Varnenčik, Varna 05245
Regionalen Istoričeski Muzej, Pleven 05120
Sarafkina Kašta Gradski Bit, Veliko Tărnovo 05258
Skobelev Park-muzej, Pleven 05121
Voenno-istoričeski Muzej, Pleven 05122
Voenno-istoričeski Muzej Osvoboditelna Vojna 1877-1878, Bjala 04988

Canada

100 Mile District Historical Society, 100 Mile House 05304
Admiral Digby Museum, Digby 05703
Avon River Heritage Society Museum, Hants County 05968
Campbell River Museum, Campbell River 05539
Cape Spear Lighthouse National Historic Site, Saint John's 06844
Centre Culturel et d'Interprétation de Havre Saint-Pierre, Havre-Saint-Pierre 05977
Centre d'Interprétation du Patrimoine de Plaisance, Plaisance 06610
Centre National d'Exposition, Jonquière 06031
Chatham-Kent Museum, Chatham, Ontario 05602
Cossit House Museum, Sydney 07048
Deep Cove Heritage Society, North Vancouver 06477
Diefenbaker Canada Centre, Saskatoon 06912
Dorothy's House Museum, Port Hope 06643
Dundas Historical Society Museum, Dundas 05724
Etobicoke's Museum - Montgomery's Inn, Etobicoke 05787
Farmers' Bank of Rustico, Hunter River 06002
Foreman Art Gallery of Bishop's University, Lennoxville 06157
Fort Anne, Annapolis Royal 05342
Fort Beauséjour, Aulac 05362
Fort Langley, Fort Langley 05812
Fort Malden, Amherstburg 05332
Fort Ostell Museum, Ponoka 06621
Fort Saint John-North Peace Museum, Fort Saint John 05821
Fultz House Museum, Lower Sackville 06200
Gaspesian British Heritage Village, New Richmond 06432
Gatineau Valley Historical Society, Chelsea 05604
Grand-Pré National Historic Site, Grand-Pré 05897
The Grange, Toronto 07112
Greenwood Centre for Living History, Hudson 05998
Head-Smashed-In Buffalo Jump, Fort Macleod 05815
Hector Exhibit Centre, Pictou 06601
Henry-Stuart House, Québec 06681
Heritage Sutton, Sutton, Québec 07041
Hermitage And Gatehouse Museum, Ancaster 05338
Homewood Museum, Prescott 06658
Hutchison House Museum, Peterborough 06592
Ireland House at Oakridge Farm, Burlington 05499
James House Museum, Bridgetown 05473
Keir Memorial Museum, Malpeque 06224
Lieu Historique National du Canada de la Bataille-de-la-Ristigouche, Pointe-à-la-Croix 06616
Lithuanian Museum-Archives of Canada, Mississauga 06312
Log Cabin Museum, Murray Harbour 06403
MacDonald House Museum, East Lake Ainslie 05732
McFarland House, Niagara Falls 06444
Maison Dumulon, Rouyn-Noranda 06781
Maison LePailleur, Châteauguay 05601
Maison Louis Bertrand, L'Isle-Verte 06174
Maison Nationale des Patriotes, Saint-Denis-sur-Richelieu 06814
Manoir Leboutillier, Anse-au-Griffon 05347
Market Gallery, Toronto 07122
Monument Lefebvre, Memramcook 06266
Musée Canadien des Civilisations, Gatineau 05862
Musée Colby-Curtis, Stanstead 07005
Musée de l'Amérique Française, Québec 06688
Musée de Saint-Boniface, Saint-Boniface 06804
Musée des Pioneers, Saint-André-Avellin 06793
Musée des Ursulines, Trois-Rivières 07175
Musée du Château Dufresne, Montréal 06365
Musée du Château Ramezay, Montréal 06366
Musée Laurier, Victoriaville 07261
Musée Stewart au Fort de île Sainte-Hélène, Montréal 06376
Museum London, London 06189
Myrtleville House Museum, Brantford 05470
Nanton Lancaster Air Museum, Nanton 06415
New Iceland Heritage Museum, Gimli 05866
North American Black Historical Museum, Amherstburg 05335
Northern Lights Centre, Watson Lake 07291
Norwegian Laft Hus, Red Deer 06708
Palais Montcalm, Québec 06696
Perkins Bull Collection, Peterborough 06594
Pier 21, Halifax 05945
Pioneer Historical Connors Museum, Connors 05639
Pioneer Museum, Sayward 06935
Port Royal, Annapolis Royal 05344
Prairie Mountain Regional Museums Collection, Shoal Lake 06966
Prince Edward Island Museum, Charlottetown 05595
Prince Township Museum, Sault Sainte Marie 06932
Province House, Charlottetown 05597
Riel House, Saint-Vital 06885
Rocky Mountain House Museum, Rocky Mountain House 06769
Rowsell House Interpretation Center, Harrington Harbour 05972
Royal Alberta Museum, Edmonton 05752
Ruthven Park, Cayuga 05582
Saint Michael's Museum, Miramichi 06303
Secwepemc Museum and Heritage Park, Kamloops 06038
Signal Hill National Historic Site, Saint John's 06857
Sod House Museum, Arctic Bay 05351
Sodbuster Archives Museum, Strome 07023
South Peace Historical Museum, Dawson Creek 05685
Spadina Museum, Toronto 07140
Sukanen Ship Pioneer Village and Museum of Saskatchewan, Moose Jaw 06387
Thomas Foster Memorial, Uxbridge 07190
Uncle Tom's Cabin Historic Site, Dresden 05711
UVAN Historical Museum, Winnipeg 07390
Whitman House Museum, Canso 05548
Whyte Museum of the Canadian Rockies, Banff 05383
Wilno Heritage Society Museum, Wilno 07341
Wiltondale Pioneer Village, Bonne Bay 05440
Wood Mountain Post, Wood Mountain 07401
York Museum, Toronto 07155

Chile

Museo de la Solidaridad Salvador Allende, Santiago de Chile 07457
Museo del Carmen de Maipú, Santiago de Chile 07458
Museo Histórico Regional de Magallanes, Punta Arenas 07446
Museo Martín Gusinde, Puerto Williams 07445
Museo O'Higginiano y de Bellas Artes de Talca, Talca 07466

China

Anhui Museum, Hefei 07775
Anhui Sheng Bo Wu Guan, Hefei 07776
Beijing China Nationality Museum, Beijing 07510
Beijing Dabao Taixi Western Han Tomb Museum, Beijing 07512
Bund Museum, Shanghai 08155
Changyangtu Sept Self-Rule Museum, Yidu 08444
China Great Wall Museum, Badaling 07492
China Nationality Museum, Beijing 07530
Dan Estuary Museum, Shiyan 08237
Dayi Liu's Manor-House Museum, Chengdu 07615
Elm Museum, Yushu 08477
Eternity Museum, Wannian 08312
Guangdong Museum, Guangzhou 07724
Hainan Museum, Haikou 07752
Hebei Museum, Shijiazhuang 08231
Hebei Provincial Museum, Shijiazhuang 08232
Hebei Provincial Museum, Wuchang 08328
Heilongjiang Nationality Museum, Harbin 07771
History Museum of the Taiping Heavenly Kingdom in Nanjing, Nanjing 08039
Hunan Provincial Museum, Changsha 07597
Jiangnan Gongyuan History Display Center, Nanjing 08040
Jiangxi Provincial Museum, Nanchang 08028
Jilin Provincial Museum, Changchun 07585
Jin Ancestral Hall Museum, Taiyuan 08272
Jinggangshan Revolution Museum, Jinggangshan 07890
Leadshan Museum, Qianshan, Jiangxi 08102
LiaoJin Cheng Yuan, Beijing 07546
Liaoning Provincial Museum, Shenyang 08216
Longhua Mausoleum of Fallen Heroes, Shanghai 08172
Macao Museum, Macau 07995
Mountain Resort Museum, Chengde 07612
Museum of the Mausoleum of King Nanyue, Guangzhou 07734
Nation Museum, Beijing 07555
National Museum of Chinese History, Beijing 07556
Qin Shi Huang Terracotta Warriors and Horses Museum, Lingtong 07945
Qinghai Provincial Museum, Xining 08397
Quanzhou Overseas-Relations History Museum, Quanzhou 08132
Quaternary Glacial Vestige of China Exhibition Hall, Beijing 07563
Santiaoshi History Museum, Tianjin 08282
Shaanxi History Museum, Xian 08360
Shandong Provincial Museum, Jinan 07884
Shanghai History Museum, Shanghai 08183
Shenyang Palace Museum, Shenyang 08217
Suzhou Museum, Shanghai 08197
Xin'an Qiantang Zhizhai Museum, Xin'an 08387
Xinjiang Uighur Autonomous Region Museum, Urumchi 08309
Yunnan Provincial Museum, Kunming 07917
Zhejiang Provincial Museum, Hangzhou 07768
Zhuhai Museum, Changning 07594

Colombia

Casa Museo 20 de Julio de 1810, Bogotá 08544
Casa Museo Quinta de Bolívar, Bogotá 08546
Cuadros Vivos de la Sabana, Corozal 08628
Museo Antonio Ricaurte, Villa de Leyva 08782
Museo Casa de la Cultura de Pensilvania, Pensilvania 08717
Museo Casa de la Cultura "Luis Camacho Rueda", Socorro 08759
Museo Casa del Fundador Ponzalo Suárez Rendón, Tunja 08774
Museo Casa Natal General Santander, Villa del Rosario 08786
Museo de Artes y Tradiciones Populares de Ráquira, Ráquira 08737
Museo de la Ferrería de La Pradera, Subachoque 08764
Museo del Hombre en el Universo, Sopó 08763
Museo Fotográfico, Calarcá 08605
Museo Histórico Casa de la Convención de Rionegro, Rionegro 08739
Museo Histórico y Arqueológico, Charalá 08624
Museo Luis A. Calvo, Agua de Dios 08522
Museo Quinta de San Pedro Alejandrino, Santa Marta 08754
Museo Taminango de Artes y Tradiciones Populares de Nariño, Pasto 08715
Museo Univérsitas, El Cerrito 08637

Comoros

Musée National, Moroni 08792

Congo, Republic

Musée National du Congo, Brazzaville 08805

Costa Rica

Museo Histórico Cultural Juan Santamaría, Alajuela 08808

Côte d'Ivoire

Musée des Civilisations de Côte d'Ivoire, Abidjan 08830

Croatia

Creski Muzej, Cres 08856
Galerija Sebastian, Dubrovnik 08863
Gradski Muzej Virovitica, Virovitica 08978
Hrvatski Povijesni Muzej, Zagreb 08997
Memorijalna Zbirka Eugena Kumičića, Brseč 08849
Muzej Slavonije Osijek, Osijek 08923
Pomorski i Povijesni Muzej Hrvatskog Primorja, Rijeka 08939

Cuba

Museo 13 de Marzo, Pinar del Río 09246
Museo Casa Natal de Calixto Garcia, Holguín 09178
Museo Casa Natal de Carlos Manuel de Céspedes, Bayamo 09033
Museo Comandancia La Plata, Santo Domingo 09327
Museo de Ambiente Histórico Cubano, Santiago de Cuba 09309
Museo de Baracoa, Baracoa 09030
Museo de la Guerra Hispano-Cubano-Norteamericana, Santiago de Cuba 09313
Museo Historico de Ciego de Avila, Ciego de Avila 09062
Museo Histórico de Palma Soriano, Palma Soriano 09240
Museo Histórico Municipal de Guanabacoa, Guanabaoa 09088
Museo Máximo Gómez, La Habana 09150
Museo Memorial 26 de Julio, Victoria de las Tunas 09346
Museo Memorial El Morrillo, Matanzas 09217
Museo Memorial José Martí, La Habana 09152
Museo Memorial La Demajagua, Manzanillo 09208
Museo Memorial Los Malagones, Moncada 09226
Museo Memorial Vicente García, Victoria de las Tunas 09348
Museo Municipal del III Frente, Santiago de Cuba 09324
Museo Nacional de Camilo Cienfuegos, Sancti-Spíritus 09289
Museo Provincial de la Historia de Pinar del Río, Pinar del Río 09253

Cyprus

Cyprus Historical Museum, Lefkosia 09374
Museum of the History of Cypriot Coinage, Lefkosia 09381

Czech Republic

Historické Muzeum, Praha 09751
Jihomoravské Muzeum ve Znojmě, Znojmo 09947
Lobkovicky Palác, Praha 09760
Masarykovo Muzeum, Hodonín 09519

Muzeum v Bruntále, Bruntál 09445
Muzeum Východních Čech v Hradci Králové, Hradec Králové 09534
Muzeum Židovské Komunity, Tovačov 09871
Národní Muzeum, Praha 09779
Národopisná Pobočka, Železný Brod 09937
Památník Krále Jiřího z Poděbrad a Lapidarium, Poděbrady 09723
Památník Lidice, Lidice 09627
Slezské zemské muzeum, Opava 09692

Denmark

Aalborg Historiske Museum, Aalborg 09955
Amalienborg, København 10122
Ballerup Egnsmuseum, Ballerup 09975
Blaabjerg Egnsmuseum, Nørre Nebel 10193
Danmarks Kloster Museum, Ry 10254
Ebeltoft Museum, Ebeltoft 09994
Faaborg Arrest, Faaborg 10006
Farvergården, Ebeltoft 09995
Frøslevlejrens Museum, Padborg 10227
Furesø Museer, Farum 10019
Det Gamle Fiskerhus og Skibshallerne, Gilleleje 10035
Grundtvigs Mindestuer i Udby, Lundby 10182
Helgenæs Præstegård, Knebel 10121
Hjemsted Oldtidspark, Skærbæk 10258
Horsens Museum, Horsens 10098
Kulturhistorie Sønderborg Slot, Sønderborg 10279
Lejre Forsøgscenter, Lejre 10174
Museet ved Trelleborg, Slagelse 10275
Nationalmuseet, København 10144
Ribes Vikinger Museum, Ribe 10234
Silkeborg Kulturhistoriske Museum, Silkeborg 10256

Dominican Republic

Museo Casa Generalísimo Máximo Gómez, Montecristi 10330
Museo de Historia y Geografía, Santo Domingo 10345
Museo de los Héroes de la Restauración, Santiago de los Caballeros 10339
Museo Faro a Colón, Santo Domingo 10355
Museo Hermanas Mirabal, Conuco 10326

East Timor

East Timorese Cultural Centre Museum Collection, Dili 10364
Museum Sonyine Malige, Halmahera Tengah 10366

Ecuador

Museo Casa de Sucre, Quito 10431
Museo Cultural del Instituto Geografico Militar, Quito 10433
Museo Historico Casa de los Tratados, Cuenca 10393

Egypt

Museum of the Faculty of Arts, Alexandria 10476

Estonia

Eesti Ajaloomuuseum, Tallinn 10622
Eesti Põllumajandusmuuseum, Tõrvandi 10664
Maarjamäe Loss, Tallinn 10635
Peeter I Maja, Tallinn 10641
Rakvere Linnakodaniku Muuseum, Rakvere 10610

Ethiopia

National Museum of Ethiopia, Addis Ababa 10680

Faroe Islands

Føroya Fornminnissavn, Hoyvík 10687

Finland

Aboa Vetus - Arkeologis-historiallinen Museo, Turku 11387
Hämeen Museo, Tampere 11349
Historiallinen Museo, Kokkola 10971
Lahden Historiallinen Museo, Lahti 11028
Mannerheim museo, Helsinki 10804
Maria de Lisitzin Fine Art and Historical Museum, Porvoo 11229
Museokeskus Vapriikki, Tampere 11354
Olkijoen Rauhanpirtti, Pattijoki 11194
Suomen Kansallismuseo, Helsinki 10820

France

Centre Jeanne d'Arc, Orléans 14530
Château de Gratot, Gratot 13118
Château et Musée de Filain-Fort, Filain 12934
Château-Musée d'Aulteribe, Sermentizon 15726
Collections de l'Abbaye Bénédictine de Saint-Louis-du-Temple, Vauhallan 16084
Domaine National de Chambord, Chambord 12378
Fondation Dosne-Thiers, Paris 14599
Galerie d'Exposition d'Art et d'Histoire, Villaries 16168
Historium de Sedan, Sedan 15696
La Chaumière de Granièr, Thiézac 15882
Maison de Jean Monnet, Bazoches-sur-Guyonne 11884
Maison dite de la Duchesse Anne, Morlaix 14277
Maison du Bâti, Sansac-de-Marmiesse 15625
Maison du Forgeron et Maréchal Ferrand, Carnusse 12293
Maison du Journalier, Pleaux 14835
Maison du Patrimoine, Argentat 11673
Le Mémorial de Caen, Caen 12229
Mémorial des Soldats de La Nouvelle France, Le Château-d'Oléron 13570
Musée America-Gold Beach, Ver-sur-Mer 16108
Musée Arthur Batut, Labruguiere 13467
Musee Bajén-Vega, Monesties 14147
Musée Bourbonnais, Moulins (Allier) 14304
Musée Centre Culturel, Vichy 16146
Musée Charles VII, Mehun-sur-Yèvre 14061
Musée Conservation, Saint-Paul-Trois-Châteaux 15448
Musée de Bretagne, Rennes (Ille-et-Vilaine) 15015
Musée de Clichy et Société Historique, Clichy 12559
Musée de la Bataille de Tilly sur Seulles 1944, Tilly-sur-Seulles 15913
Musée de La Bouteille, Saint-Émilion 15246
Musée de la Curiosité et de la Magie, Paris 14652
Musée de la Déportation et de la Résistance, Tarbes 15862
Musée de la Guerre 1939-45, Pourrain 14931
Musée de la Monnaie de Paris, Paris 14655
Musée de la Résistance, Clavières 12540
Musée de la Résistance et Déportation, Blois 12034
Musée de la Résistance Henri Queuille, Neuvic (Corrèze) 14405
Musée de la Vie d'Autrefois, Vivoin 16263
Musée de la Vie Romantique, Paris 14665
Musée de la Vie Rurale, Desaignes 12715
Musée de l'Ancien Havre, Le Havre 13595
Musée de l'Histoire de France, Paris 14671
Musée de l'Histoire de la Médecine, Vandœuvre-les-Nancy 16064
Musée de l'Hôtel de Barral, Soissons 15773
Musée de l'Île de France, Sceaux 15688
Musée de l'Œuvre, Avignon 11790
Musée de Notre-Dame-de-Paris, Paris 14679
Musée de Tourisme et d'Artisanat, Lys-Saint-Georges 13905
Musée Départemental Albert Kahn, Boulogne-Billancourt 12089
Musée départemental de la Bresse - Domaine des Planons, Saint-Cyr-sur-Menton 15228
Musée Départemental de la Résistance et de la Déportation, Lorris 13826
Musée Départemental de la Résistance et de la Déportation, Tulle 16006
Musée Départemental de la Résistance et de la Déportation Jean Philippe, Toulouse 15938
Musée départemental du Revermont, Treffort 15979
Musée des Amis du Vieux Corbie, Corbie 12621
Musée des Traditions Populaires Marius Audin, Beaujeu 11894
Musée d'Histoire Locale, Mougins 14298
Musée d'Histoire Locale Pierre Salvi, Viarmes 16140
Musée d'Izieu Mémorial des Enfants Juifs Exterminés, Izieu 13258
Musée d'Orange, Orange 14518
Musée du Biterrois, Béziers 11977
Musée du Café Gondrée, Bénouville 11940
Musée du Capitellu, Ajaccio 11526
Musée du Palais des Evêques, Saint-Lizier 15371
Musée du Patrimoine Religieux et des Croyances Populaires, Moustey 14313
Musée du Pays Rabastinois, Rabastens 14976
Musée du Plan Relief, Bitche 12006
Musée du Prieuré, Perrecy-les-Forges 14795
Musée du Septennat, Château-Chinon 12433
Musée du Souvenir du Combattant, Chalon-sur-Saône 12358
Musée et du Vieux Chinon, Chinon 12516
Musée Historique, Mulhouse 14328
Musée Historique de Graffiti Anciens et d'Archéologie, Marsilly 14015
Musée Historique du Domaine National, Saint-Cloud 15219
Musée Hospitalier, Charlieu 12412
Musée Intercommunal d'Histoire et d'Archéologie, Louvres 13859
Musée Lacoune, Lacaune 13470
Musée Lapérouse, Albi 11538
Musée Lautrecois, Lautrec 13539
Musée Lorrain, Nancy 14356
Musée Maison Déodat Roché, Arques (Aude) 11695
Musée Municipal, Saint-Jean-d'Angély 15319
Musée Napoléonien, Antibes-Juan-les-Pins 11642
Musée Napoléonien, Ile-d'Aix 13236
Musée National de la Légion d'Honneur et des Ordres de Chevalerie, Paris 14723
Musée National des Douanes, Bordeaux 12073
Musée National du Château de Fontainebleau, Fontainebleau 12966
Musée Remenber 39-45, Léhon 13677
Musée Une Halte sur le Chemin de Saint-Jacques, Borce 12060
Musée Vie et Métiers d'Autrefois dans le Haut-Jura, La Pesse 13400
Musée Vivant de la Laine et du Mouton, Saint-Pierreville 15464
Petit Musée des Silos de Jouques, Jouques 13284
Refuge Fortifié Muséalisé, Dossenheim-sur-Zinsel 12773

Gabon

Musée National du Gabon, Libreville 16330

Gambia

Gambia National Museum, Banjul 16331

Georgia

D. Baazov Georgian Jewish Historic and Ethnographic Museum, Tbilisi 16353
Dadiani Museum, Dadiani 16338
Giorgi Chitaia Museum of Architecture, Tbilisi 16364
Ioseb Grishashvili Tbilisi History Museum, Tbilisi 16368
Ivane Javakhishvili Samtskhe-Javaxeti History Museum, Akhaltsikhe 16332
Open Air Museum of Ethnography, Tbilisi 16378
Simon Janashia Museum of Georgia, Tbilisi 16381
Svaneti History-Ethnography Museum, Mestia 16347
Tskhinvali Museum, Tskhinvali 16387

Germany

600 Jahre Kunst - Studiensammlung der Universität Leipzig, Leipzig 19890
Adolf-Bender-Zentrum, Sankt Wendel 21510
Albert-Sammt-Zeppelin-Museum, Niederstetten 20709
Alpinmuseum, Kempten 19513
Antifaschistische Mahn- und Gedenkstätte, Lieberose 19985
Ausstellungszentrum Kroch-Haus, Leipzig 19896
Bach-Stammhaus Wechmar, Günthersleben-Wechmar 18823
Bachritterburg Kanzach, Kanzach 19428
Bandwebermuseum, Wuppertal 22736
Bauspar-Museum im Georg-Kropp-Haus, Wüstenrot 22728
Begegnungsstätte Alte Synagoge, Wuppertal 22737
Bergbaumuseum BUV Kleinzeche, Dortmund 17915
Bismarck-Museum, Schönhausen 21610
Breslauer Sammlung Köln, Köln 19596
Bunkermuseum Oberhausen, Oberhausen 20807
Burg Gößweinstein, Gößweinstein 18686
Burg Hohenzollern, Burg Hohenzollern 17645
Carl Bosch Museum, Heidelberg 19070
DDR-Museum im Filmpalast, Malchow 20154
Deutsches Bauernkriegsmuseum, Böblingen 17416
Deutsches Historisches Museum, Berlin 17072
Dokumentationszentrum Alltagskultur der DDR, Eisenhüttenstadt 18105
Dokumentationszentrum Oberer Kuhberg-Ulm KZ Gedenkstätte, Ulm 22159
Domschatzkammer, Essen 18251
Dorfschmiede Neckargröningen, Remseck am Neckar 21267
Ehemalige Synagoge, Freudental 18457
Ehemalige Synagoge Ahrweiler, Bad Neuenahr-Ahrweiler 16794
Emslandmuseum Schloss Clemenswerth, Sögel 21791
Erinnerungsstätte an die Zwangsarbeit auf dem Gelände des Volkswagenwerk, Wolfsburg 22661
Erinnerungsstätte Baltringer Haufen - Bauernkrieg in Oberschwaben, Mietingen 20324
Erlebniswelt Grundwasser, Langenau 19817
Familienmuseum, Schwerin 21693
Familienmuseum Bad Homburg, Bad Homburg v.d.Höhe 16750
Feuchtmayer Museum, Salem, Baden 21470
Frank-Loebsches Haus, Landau in der Pfalz 19789
Franz-Stock-Museum, Arnsberg 16571
Frauen Museum, Bonn 17454
Freilichtmuseum, Herbertingen 19131
Gartenkunst-Museum Schloss Fantasie, Eckersdorf 18054
Gedenkturm Großbeeren 1813, Großbeeren 18780
Gelsenkirchen im Nationalsozialismus, Gelsenkirchen 18573
Gerichtsmuseum Bad Fredeburg, Schmallenberg 21583
Germanisches Nationalmuseum, Nürnberg 20764
Gottliebin-Dittus-Haus, Bad Liebenzell 16779
Gottscheer Stube, Bad Wildbad 16858
Graf-Luxburg-Museum, Bad Bocklet 16699
Gustav-Adolf-Museum, Geleitshaus, Weißenfels 22455
Hammerschmiede mit Heimatmuseum und Bienengarten, Lahr 19772
Handwerker- und Heimatstube Gernsbach, Gernsbach 18607
Handwerksmuseum Kork, Kehl 19496
Heimatgeschichtliche Sammlung des Kirchspiels, Schwabstedt 21647
Heimatmuseum, Bad Teinach-Zavelstein 16850
Heimatmuseum, Eppelborn 18179
Heimatmuseum, Müllrose 20418
Heimatmuseum der Deutschen aus Bessarabien, Stuttgart 21940
Heimatstube Rautheim, Braunschweig 17521
Henkerhaus, Bernau bei Berlin 17289
Historische Dauerausstellung Schloss Borbeck, Essen 18255
Historisches Museum im Marstall, Paderborn 20970
Historisches Museum Regensburg, Regensburg 21224
Historisches Museum Schloß Urach, Bad Urach 16853
Hofgarten an der Residenz, München 20446
Idstedt-Gedächtnishalle, Idstedt 19310
John-Wagener-Haus Sievern, Langen, Kreis Cuxhaven 19815
Judenbad, Friedberg, Hessen 18467
Jüdisches Museum, Creglingen 17742
Jüdisches Museum Berlin, Berlin 17147
Jüdisches Museum Westfalen, Dorsten 17910
Käseglocke, Museum für Worpsweder Möbel und Kunsthandwerk, Worpswede 22694
Kaiserpfalz mit Sankt Ulrichskapelle, Goslar 18711
Karnevalsmuseum, Blankenheim 17376
Kettenschmiedemuseum, Fröndenberg 18486
Klosterruine Memleben, Memleben 20277
Königspesel, Hooge 19269
Kreismuseum Bitburg-Prüm, Bitburg 17368
Küferei-Museum, Eberbach 18032
Küstenmuseum Wilhelmshaven, Wilhelmshaven 22577
Kulturhistorisches Museum, Magdeburg 20123
Kyffhäuser-Denkmal, Bad Frankenhausen 16729
KZ Gedenkstätte Moringen, Moringen 20385
KZ-Gedenkstätte Neckarelz, Mosbach 20389
KZ-Gedenkstätte Sandhofen, Mannheim 20167
Landesmuseum Birkenfeld, Birkenfeld 17359
Landesmuseum Württemberg, Stuttgart 21950
Louise-Henriette-Stift, Kloster Lehnin 19576
Die Männer von Brettheim, Rot am See 21397
Marmorbad, Kassel 19469
Melanchthonhaus Wittenberg, Lutherstadt Wittenberg 20116
Menzer Heimatstube, Stechlin 21872
Mindener Museum für Geschichte, Landes- und Volkskunde, Minden 20336
Missionsmuseum der Liebenzeller Mission, Bad Liebenzell 16780
Mithrasheiligtum, Mundelsheim 20526
Modemuseum Feigel, Ehingen 18067
Museum Abtsküche, Heiligenhaus 19105
Museum Burg Abenberg, Abenberg 16418
Museum der 50er Jahre, Bad Bevensen 16696
Museum der 50er und 60er Jahre, Vaihingen an der Enz 22198
Museum der Freiwilligen Feuerwehr, Heiligenhaus 19106
Museum der Landsmannschaft Ostpreußen, Gelsenkirchen 18578
Museum der Stadt Bad Schwartau, Bad Schwartau 16835
Museum für ehemals Großherzoglich Badische Schulen, Offenburg 20886
Museum für Kommunikation Nürnberg (Verkehrsmuseum), Nürnberg 20781
Museum für schlesische Landeskunde im Haus Schlesien, Königswinter 19659
Museum Geiserschmiede, Bühlertal 17633
Museum Hoffmann'sche Sammlung, Kohren-Sahlis 19672
Museum im ehemaligen Augustiner-Chorherrenstift, Herrenchiemsee 19159
Museum im Kinzigtorturm, Gengenbach 18584
Museum Priesterhäuser, Zwickau 22834
Museum Schloß Lichtenburg, Prettin 21122
Museum Stadt Königsberg, Duisburg 18025
Museum Steinau...das Museum an der Straße, Steinau an der Straße 21879
Museum Theulegium, Tholey 22056
Museum unter der Y-Burg, Kernen im Remstal 19519
Museum zur brandenburg-preußischen Geschichte - Zinnfigurenmuseum, Gusow-Platkow 18842
Museum zur Geschichte der Herrnhuter Brüdergemeinde, Königsfeld im Schwarzwald 19650
Museum zur Geschichte der Juden im Ostalbkreis, Bopfingen 17473
Museumsverein Dokumentation 2. Weltkrieg Hürtgenwald e.V, Hürtgenwald 19294
Nostalgie-Museum, Wörth an der Donau 22643
NS-Dokumentationszentrum der Stadt Köln, Köln 19626
Oberer Torturm-Ausstellung, Marbach am Neckar 20180

Ofenmuseum und Baumgalerie, Allmersbach im Tal 16474
Ostdeutsche Heimatstube, Fellbach 18306
Ostdeutsche Heimatstube, Zweibrücken 22829
Otto-König von Griechenland-Museum, Ottobrunn 20957
Pfeifermuseum im alten Rathaus, Werbach 22482
Philipp-Reis-Haus und Hugenottenmuseum, Friedrichsdorf 18476
Piratenamüseum, Wilhelmshaven 22579
Pommersches Landesmuseum, Greifswald 18750
Preußen-Museum Nordrhein-Westfalen, Wesel 22507
Preußen-Museum Nordrhein-Westfalen, Minden 20337
Reichskammergerichtsmuseum, Wetzlar 22530
Römerkeller Ottmarsheimer Höhe, Mundelsheim 20527
Römermuseum, Heitersheim 19115
Römermuseum, Mengen 20283
Römische Badruine, Badenweiler 16893
Das Römische Kastell in Waldmüssingen, Schramberg 21633
Römische Villa Nennig, Perl 21012
Rößler Museum, Untermünkheim 22182
Rommel Museum, Blaustein 17383
Ruhr Museum, Essen 18267
Rumänisches Institut, Freiburg im Breisgau 18437
Sammlung zur Landeskunde Ost- und Westpreußens "Es war ein Land...", Oberschleißheim 20836
Schloß Brake - Das Weserrenaissance-Museum, Lemgo 19949
Schloss Freienwalde mit Walther-Rathenau-Gedenkstätte, Bad Freienwalde 16734
Schloss Rheda-Wiedenbrück, Rheda-Wiedenbrück 21293
Schloss Wolkenburg, Limbach-Oberfrohna 19994
Schraube Museum, Halberstadt 18880
Schwedenspeicher-Museum Stade, Stade 21852
Shloss Nöthnitz, Bannewitz 16924
Staatspark Fürstenlager, Bensheim 16998
Staatspark Wilhelmsbad, Hanau 18983
Stadtmuseum Borken, Westfalen, Borken, Westfalen 17479
Stadtmuseum Meißen, Meißen 20267
Staufergedächtnisstätte und Museum Wäscherschloß, Wäschenbeuren 22267
Stauffenberg-Gedenkstätte, Albstadt 16460
Stiftung Gerhart-Hauptmann-Haus, Deutsch-osteuropäisches Forum, Düsseldorf 18007
Störtebeker Park, Wilhelmshaven 22580
Synagoge, Worms 22689
Techniksammlung Backnang, Backnang 16663
Turenne-Museum Sasbach, Sasbach 21516
Umuzi-Afrika-Haus, Freiberg am Neckar 18420
Villa Hügel, Essen 18270
Völkerkundliche Sammlung der Steyler Missionare, Aulendorf 16644
Westpreussisches Landesmuseum, Münster 20521
Westwall-Bunker, Beckingen 16973
Westwallmuseum, Irrel 19353
Widukind-Museum, Enger 18173
Wilhelm Busch Zimmer, Dassel 17799
Wilhelm-Fabry-Museum, Hilden 19181
Zinnfiguren-Museum Goslar, Goslar 18716

Ghana

Cape Coast Castle Museum, Cape Coast 22850

Greece

Cultural Center Hellenic Cosmos, Athinai 22906
Eleftherios K. Venizelos Museum, Athinai 22907
Historical and Ethnological Museum of Patras, Patras 23140
Historical Museum of Crete, Iráklion 23019
Mouseio Mpenaki, Athinai 22925
Pavlos Vrellis Museum of Greek History, Ioannina 23016
Therisso Museum of Eleftherios Venizelos, Chania 22966
Vouros Eftaxias Museum, Athinai 22954

Guatemala

Museo de Santiago de los Caballeros, Antigua 23291
Museo Nacional de Historia, Guatemala 23303

Guinea

Musée National de Guinée, Conakry 23311

Haiti

Musée du Panthéon National Haïtien, Port-au-Prince 23322

Honduras

Museo Antropologia e Historia Valle de Sula, San Pedro Sula 23330
Museo Nacional de Historia Colonial, Omoa 23329

Hungary

Bihari Múzeum, Berettyóujfalu 23373
Blaskovich Múzeum, Tápiószele 23911
Bocskai István Emlékkiállítás, Nagykereki 23726
Budavári Mátyás-templom Egyházművészeti Gyűjteménye, Budapest 23393
Erdős Kamill Cigány Múzeum, Pécs 23760
Ferenczy Múzeum, Szentendre 23876
Gödöllői Királyi Kastélymúzeum, Gödöllő 23565
Határőrség Központi Múzeuma, Körmend 23675
Jantyik Mátyás Múzeum, Békés 23364
Laczkó Dezső Múzeum, Veszprém 23959
Magyar Nemzeti Múzeum, Budapest 23442
Magyar Zsidó Múzeum és Levéltár, Budapest 23447
Noszlopy Kúria, Újvárfalva 23941
Rákóczi Múzeum, Sárospatak 23792
Rétközi Múzeum, Kisvárda 23673
Várostörténeti Múzeum, Pécs 23779
Zrinyi Miklós Vármúzeum, Szigetvár 23892

Iceland

Byggðasafn Dalamanna, Búdardalur 23991

India

Assam State Museum, Guwahati 24188
Bhagavan Mahavir Government Museum, Cuddapah 24123
Childrens Museum, Imphal 24222
Gandhi National Memorial, Pune 24358
Government Museum Vellore, Vellore 24428
H.H. Maharaja Jayaji Rao Scindia Museum, Gwalior 24197
Heras Institute of Indian History and Culture, Mumbai 24317
Leimarel Museum and Research Centre, Imphal 24224
Police Museum, Imphal 24227
Shillong Tribal Research Institute, Shillong 24387
Shri Chhatrapati Shivaji Maharaj Museum, Satara 24381
Tipu Sahib Museum, Srirangapatna 24398

Indonesia

Monumen Nasional, Jakarta 24476
Museum Asia Africa, Bandung 24441
Museum Dewantara Kirti Griya, Yogyakarta 24586
Museum Fatahillah, Jakarta 24483
Museum Joang '45, Medan 24525
Museum Kebangkitan Nasional, Jakarta 24486
Museum Nasional, Jakarta 24492
Museum Negeri Kalimantan Timur Mulawarman, Tenggarong 24577
Museum Negeri of Aceh, Banda Aceh 24438
Museum Negeri Propinsi Nusa Tenggara Timur, Kupang 24517
Museum Pancacsila Sakti, Jakarta 24493
Museum Pangeran Diponegoro, Yogyakarta 24590
Museum Pendidikan Islam, Yogyakarta 24591
Museum Pers, Surakarta 24569
Museum Pugung Ulago Sembah, Donggala 24469
Museum Sasmita Loka A. Yani, Jakarta 24496
Museum Sono Budoyo, Yogyakarta 24594
Museum Sudirman, Magelang 24519
Museum Sumpah Pemuda, Jakarta 24501
Museum Tugu Nasional, Jakarta 24505
Museum Wanua Paksinanta, Manado 24523
Sultan Mahmud Badaruddin II Museum, Palembang 24538

Iran

Golestan Palace Museum, Tehrān 24668
Khorassan Natural History Museum, Mashad 24632
Melli Museum, Tehrān 24679
Nader Mausoleum, Mashad 24633
National Museum, Tehrān 24687
National Museum of Kashan, Kashan 24616
Rais Ali Delvary Museum, Bushehr 24601
Sa'dabad Museums, Tehrān 24694

Iraq

Iraqi National Museum, Baghdad 24715
Kirkuk Museum, Kirkuk 24719

Ireland

Bank of Ireland Arts Centre, Dublin 24778
DeValera Museum and Bruree Heritage Centre, Bruree 24747
Famine Museum, Strokestown 24902
Kerry County Museum, Tralee 24904
Michael Davitt National Memorial Museum, Straide 24901
National 1798 Visitor Centre, Enniscorthy 24831
The Queenstown Story, Cobh 24765
Wild Geese Heritage Museum, Portumna 24887

Israel

Beit Abraham Krinizi, Ramat Gan 25082
Beit Hameiri, Zefat 25132
Bet Hashomer Museum, Kibbutz Kfar Giladi 25046
Edward and Helen Mardigian Museum, Jerusalem 24987
Eretz-Israel Museum Tel Aviv, Tel Aviv 25104
Hamizgaga Museum, Kibbutz Nachsholim 25051
Man in the Galilee Museum, Kibbutz Ginosar 25042
Museum of Taxes, Jerusalem 25003
Museum of Yarmukian, Kibbutz Shaar Hagolan 25058
Resistance Museum, Akko 24924
Tower of David, Jerusalem 25021
Weizmann Archives and House, Rehovot 25090
Yad Labanim Museum and Memorial Center, Petah Tikva 25078
Yad Yaari Museum of Jewish Youth Movement, Givat Haviva 24949

Italy

Castello di Masnago, Varese 28161
Museo Archeologico, Venafro 28175
Museo Casa Walser, Macugnaga 26558
Museo Centrale dell'Istituto Storico della Resistenza, Imperia 26404
Museo Civico, Barga 25360
Museo Civico Archeologico, Anzio 25244
Museo Civico Archeologico, Castelleone di Suasa 25739
Museo Civico Archeologico, Castro dei Volsci 25765
Museo Civico Archeologico, Cavaion Veronese 25787
Museo Civico Archeologico, Concordia Sagittaria 25960
Museo Civico Archeologico, Gazzo Veronese 26281
Museo Civico Archeologico, Gottolengo 26354
Museo Civico Archeologico, Grotte di Castro 26377
Museo Civico Archeologico, Mel 26650
Museo Civico Archeologico, Mondragone 26778
Museo Civico Archeologico, Nepi 26919
Museo Civico Archeologico, Norma 26942
Museo Civico Archeologico, Pitigliano 27255
Museo Civico Archeologico, Roccagloriosa 27429
Museo Civico Archeologico, Sarteano 27749
Museo Civico Archeologico, Portoferraio 27306
Museo Civico Archeologico di Fregellae, Ceprano 25805
Museo Civico Archeologico e Paleobotanico, Perfugas 27137
Museo Civico Archeologico L. Fantini, Monterenzio 26847
Museo Civico Archeologico M. Petrone, Vieste 28263
Museo Civico Archeologico Sa Domu Nosta, Senorbì 27807
Museo Civico dei Villaggi Scomparsi, Villa Estense 28272
Museo Civico del Marmo, Carrara 25689
Museo Civico della Val Fiorentina Vittorino Cazzetta, Selva di Cadore 27798
Museo Civico delle Centuriazioni Romane Padovana, Borgoricco 25507
Museo Civico Goffredo Bellini, Asola 25301
Museo Civico Messapico, Alezio 25194
Museo Civico Storico-Etnografico, Ripatransone 27415
Museo Comunale, Montaione 26794
Museo Comunale Archeologico, Viddalba 28262
Museo Comunale della Valle dei Nuraghi, Torralba 28033
Museo degli Scavi, Piuro 27257
Museo del Brigante, Sonnino 27874
Museo del Lazio Meridionale, Anagni 25225
Museo del Medioevo e del Rinascimento, Sorano 27879
Museo del Paesaggio, Castelnuovo Berardenga 25746
Museo del Risorgimento, Catanzaro 25779
Museo del Vicino Oriente, Roma 27476
Museo della Battaglia di San Martino, Desenzano del Garda 26035
Museo della Civiltà Messapica, Poggiardo 27265
Museo della Guerra Bianca, Temù 27953
Museo della Via Ostiense, Roma 27480
Museo dello Sbarco, Anzio 25245
Museo di Palazzo Reale, Napoli 26899
Museo di Storia, d'Arte e d'Antichità Don Florindo Piolo, Serravalle Sesia 27818
Museo Didattico d'Arte e Vita Preistorica, Capo di Ponte 25658
Museo Didattico della Riserva Regionale Incisioni Rupestri, Ceto 25841
Museo Documentario della Città, Gradisca d'Isonzo 26358
Museo Etnografico della Valle Brembana, Zogno 28318
Museo Garibaldino della Campagna dell'Agro Romano per la Liberazione di Roma 1867, Mentana 26654
Museo Glauco Lombardi, Parma 27108
Museo Laboratorio di Archeologia, Monsampolo del Tronto 26787
Museo Marsiliano, Bologna 25473
Museo Muratoriano, Modena 26755
Museo Napoleonico, Rivoli Veronese 27424
Museo Nazionale Archeologico di Vulci, Canino 25645
Museo Paleontologico Archeologico V. Caccia, San Colombano al Lambro 27626
Museo Preistorico, Paceco 27030
Museo Preistorico dell'Isolino, Biandronno 25415
Museo Storico, Voghera 28304
Museo Storico Archeologico, Cicagna 25887
Museo Storico Archeologico dell'Antica Nola, Nola 26938
Museo Storico Badogliano, Bari 25370
Museo Storico del Risorgimento della Società Economica, Chiavari 25861
Museo Villa Medicea, Poggio a Caiano 27268
Nest Ume Tzimbari Gabizza, Giazza-Selva di Progno 26336
Rappresentazione Scenografica Gioacchino Murat, Pizzo 27262
Südtiroler Landesmuseum für Kultur- und Landesgeschichte Schloss Tyrol, Tirolo di Merano 27976
Villa Pisani, Strá 27909

Jamaica

Institute of Jamaica, Kingston 28328

Japan

Aka-Renga Kokusaikan, Yokohama 29592
Dewa San-Zan History Museum, Tsuruoka 29543
Former Hong Kong and Shanghai Bank Museum, Nagasaki 28972
Fukuyama-renritsu Rekishi Hakubutsukan, Fukuyama 28446
Gunma-kenritsu Rekishi Hakubutukan, Takasaki 29331
Hakubutsukan Meiji-Mura, Inuyama 28573
Hirado Kanko Shiryokan, Hirado 28517
Historical Museum Chokokan, Saga 29178
Hokkaido Kaitaku Kinenkan, Sapporo 29216
Hyogo-kenritsu Rekishi Hakubutsukan, Himeji 28514
Ibaraki-kenritsu Rekishikan, Mito 28919
Iga-ryu Ninja Yashiki, Ueno 29556
Kagawa-kenritsu Rekishi Hakubutsukan, Takamatsu 29317
Kanagawa-kenritsu Kanazawabunko Museum, Yokohama 29596
Kirishitan Shiryokan, Hirado 28518
Kochi Liberty and People's Rights Museum, Kochi 28744
Kokuritsu Rekishi Minzoku Hakubutsukan, Sakura 29210
Kunozan-Toshogu-ji Homotsukan, Shizuoka 29284
Kyushu Rekishi Shiryokan, Dazaifu 28409
Museum of the Meiji Restoration, Kagoshima 28632
Nagano-kenritsu Rekishi Hakubutsukan, Chikuma 28398
Nagoya City Hideyoshi and Kiyomasa Memorial Museum, Nagoya 29004
Nara-kokuritsu Hakubutsukan, Nara 29036
Oita-kenritsu Rekishi Hakubutsukan, Usa 29564
Osaka-kenritsu Chikatsu-Asuka Hakubutsukan, Kanan 28653
Rekishi Shiryokan, Ranzan, Ranzan 29170
Ryosan-cho Rekishi Hakubutsukan, Kyoto 28863
Sakamoto Ryoma Kinenkan, Kochi 28747
Takaoka Shiritsu Hakubutsukan, Takaoka 29327
Takenouchi Kaido Rekishi Hakubutsukan, Taishi 29310
Tohoku Rekishi Shiryokan, Tagajou 29308
Tokushima-kenritsu Hakubutsukan, Tokushima 29366
Tokyo Kokuritsu Hakubutsukan, Tokyo 29490
Uwajima-shiritsu Date Hakubutsukan, Uwajima 29569
Yamanashi-kenritsu Hakubutsukan, Fuefuki 28419
Yokohama Kaiko Shiryokan, Yokohama 29613
Yokohama Rekishi Hakabutsukan, Yokohama 29617

Jordan

Archaeological Museum, Amman 29645

Kazakhstan

Centralnyj Gosudarstvennyj Muzej Kazachstana, Almaty 29678
Gosudarstvennyj Muzej-Reservat Pamjatniki Drevnego Taraza, Taraz 29732
Pugachev Museum, Uralsk 29735

Kenya

Fort Jesus Museum, Mombasa 29746

Korea, Democratic People's Republic

Haeju Historical Museum, Haeju 29753
Hamhung Historical Museum, Hamhung 29754
Korean Central Historical Museum, Pyongyang 29758
Shinuiju Historical Museum, Shinuiju 29762
Wonsan Historical Museum, Wonsan 29763

Korea, Republic

Bokcheon Municipal Museum, Busan 29772
Hyeonchungsa Shrine, Asan 29769
Independence Hall of Korea, Cheonan 29784
King Sejong The Great Memorial Exhibition, Seoul 29890
National Museum of Korea, Seoul 29910
Naval Academy Museum, Jinhae 29843
Royal Museum, Seoul 29912

Kuwait

Kuwait National Museum, Kuwait 29949

Kyrgyzstan

Gosudarstvennyj Istoričeskij Muzej Kyrgyzskoj Respubliki, Biškek 29954
Issyk-Kulskij Gosudarstvennyj Istoriko-Kulturnyj Muzej-Zapovednik, Čolpan-Ata 29961
Memorialnyj Dom-Muzej Aaly Tokombaeva, Biškek 29957

Laos

Lao National Museum, Vientiane 29974
Royal Place Museum, Luang Prabang 29968

Latvia

Cēsu Vēstures un Mākslas Muzejs, Cēsis 29997
Latvijas Kara Muzejs, Rīga 30081
Oskara Kalpaka Muzejs Airīšu Piemiņas Vietā, Zirņi 30152

Lesotho

Lesotho National Museum, Maseru 30179

Lithuania

Aušros Alėjos Rūmai, Šiauliai 30352
Dionizo Poškos Baubliai, Bijotai 30213
Lietuvos Nacionalinis Muziejus, Vilnius 30419
Mažosios Lietuvos Istorijos Muzieus, Klaipėda 30283
Pilies Muzieus, Klaipėda 30284
Varpo - Viurinės Mokyklos Istorijos Muzieus, Klaipėda 30285

Luxembourg

Musée National d'Histoire Militaire, Diekirch 30465

Macedonia

Istoriski Muzej, Kruševo 30505
Muzej na Makedonija - Archeološki, Etnološki i Istoriski, Skopje 30519
Muzej na Tetovskiot Kraj, Tetovo 30532
Muzej Zapadna Makedonija vo NOV, Kičevo 30503

Madagascar

Musée Historique, Antananarivo 30536

Malawi

Museum of Malawi, Blantyre 30544

Malaysia

History and Legendary Museum, Kuah 30568
Jabatan Muzium Sabah, Kota Kinabalu 30565
Kedah Royal Museum, Alor Setar 30550
Memorial Pengisytiharan Kemerdekaan, Melaka 30597
Muzium dan Galeri Seni, Minden 30610
Muzium Diraja Abu Bakar, Johor Bahru 30558
Muzium Diraja Perak, Kuala Kangsar 30570
Muzium Negeri Kelantan, Kota Bharu 30564
Muzium Negeri Sembilan, Seremban 30616
Muzium Sejarah National, Kuala Lumpur 30581
Perbadanan Muzium Melaka, Melaka 30607
Sultan Abdul Aziz Royal Gallery, Kelang 30560
Tun Abdul Razak Memorial, Kuala Lumpur 30587

Mauritania

Musée National de Nouakchott, Nouakchott 30675

Mauritius

B. Bissoondoyal Memorial Centre, Tyack 30686
Historical Museum, Mahébourg 30677
Sir Seewoosagar Ramgoolam Memorial Centre, Plaine Verte 30681

Mexico

Casa de la Constitución de 1814, Apatzingán 30717
Casa Museo Cadete Juan Escutia, Tepic 31440
Galería de Historia o Museo del Caracol, México 31042
Museo Antiguo Colegio de San Ildefonso, México 31049
Museo Antropología e Historia, Victoria 31543
Museo Casa de Hidalgo la Francia Chiquita, San Felipe 31330
Museo Casa del General Alvaro Obregón, Huatabampo 30942
Museo Casa Dr. Belisario Domínguez, Comitán de Domínguez 30793
Museo Casa Natal de Morelos, Morelia 31192
Museo Colonial de Acolman, Acolman 30695
Museo Comunitario el Asalto a las Tierras, Mexicali 31033
Museo Comunitario General Francisco Villa, Durango 30837
Museo Comunitario Unidad Indígena Emiliano Zapata, Hueyapan de Ocampo 30947
Museo Cristero Ing. Efrén Quezada, Encarnación de Díaz, Jalisco 30854
Museo de Historia de Tabasco, Villahermosa 31551
Museo de Historia Mexicana, Monterrey 31180
Museo de la Amistad México-Cuba, Tuxpan de Rodríguez Cano 31509
Museo de la Independencia, Dolores Hidalgo 30835
Museo de la Insurgencia, Rincón de Romos 31305
Museo de la Lealtad Republicana, Chihuahua 30769
Museo de la Memoria, Tlaxcala 31477
Museo de la Revolución, Puebla 31279
Museo de la Revolución de San Pedro, San Pedro 31367
Museo de las Californias, Tijuana 31457
Museo de las Estelas Mayas, Campeche 30742
Museo de las Revoluciones Mariano Jiménez, San Luis Potosí 31348
Museo del Camino Real de Hecelchakan, Hecelchakán 30927
Museo del Centro de Estudios de la Revolución Mexicana, Jiquilpan 30975
Museo del General Toribio Ortega, Coyame 30806
Museo División del Norte, Canutillo 30749
Museo División del Norte, Lerdo 31004
Museo General Francisco Villa, San Juan del Río 31340
Museo Hermanos López Rayón, Tlalpujahua 31469
Museo Hidalgo, Salamanca 31311
Museo Histórico de Acapulco, Acapulco 30692
Museo Histórico de la Revolución Mexicana, Chihuahua 30771
Museo Histórico de la Sierra Gorda, Jalpan 30969
Museo Histórico del Oriente de Morelos, Cuautla 30813
Museo Histórico Ex Aduana de Ciudad Juárez, Juárez 30983
Museo Histórico Mexicano de Cananea, Cananea 30746
Museo José Luis Cuevas, Colima 30781
Museo Nacional de Historia, México 31134
Museo Regional de Antropología e Historia de Baja California Sur, La Paz 30989
Museo Regional de Guadalajara, Guadalajara 30904
Museo Regional de Historia, Hermosillo 30931
Museo Regional de Historia de Aguascalientes, Aguascalientes 30703
Museo Regional de Historia de Tamaulipas, Victoria 31544
Museo-Sala de los Símbolos Patrios, Hermosillo 30933
Sala Homenaje a Juárez, Guelatao de Juárez 30924

Moldova

Muzeul National de Istorie a Moldovei, Chişinău 31602

Mongolia

National Museum of Mongolian History, Ulaanbaatar 31628

Montenegro

Istorijski Muzej Crne Gore, Cetinje 31641

Myanmar

Bogyoke Aung San Museum, Yangon 31732

Namibia

Alte Feste, Windhoek 31762
National Museum of Namibia, Windhoek 31765
Owela Display Centre, Windhoek 31766

Nepal

Birendra Museum, Kathmandu 31773
Mahendra Museum, Kathmandu 31774
National Museum of Nepal, Kathmandu 31776

Netherlands

Bezoekerscentrum Binnenhof, Den Haag 32069
De Brede, Maasbree 32520
Centraal Museum, Utrecht 32844
Dief- en Duifhuisje, Capelle aan den IJssel 32035
Drents Museum, Assen 31937
Fries Museum, Leeuwarden 32473
Haags Historisch Museum, Den Haag 32074
Historisch Museum, Capelle aan den IJssel 32036
Historisch Museum de Scheper, Eibergen 32183
Historisch Museum Haarlemmermeer, Hoofddorp 32404
Historisch Museum Rotterdam Het Schielandshuis, Rotterdam 32707
Historisch Museum Warsenhoeck, Nieuwegein 32577
Historisch Museum Zuid-Kennemerland, Haarlem 32297
Koninklijk Oudheidkundig Genootschap, Amsterdam 31854
Koninklijk Paleis te Amsterdam, Amsterdam 31855
Kralingsmuseum, Rotterdam 32709
Madurodam, Den Haag 32081
Mannenzaal van het Sint Pieters en Bloklands Gasthuis, Amersfoort 31812
Museon, Den Haag 32083
Museum Buren en Oranje, Buren, Gelderland 32025
Museum De Helpoort, Maastricht 32525
Nationaal Monument Kamp Vught, Vught 32915
Oud Amelisweerd Museum, Bunnik 32018
Oudheidskamer Vreeswijk, Nieuwegein 32578
Paleis Het Loo - Nationaal Museum, Apeldoorn 31915
Stichting Atlas Van Stolk, Rotterdam 32728
Titus Brandsma Museum, Bolsward 31981
Universiteitsmuseum, Groningen 32288
Van-der-Werf's Wedgwoodmuseum, Zuidwolde, Groningen 33016
Van 't Lindenhoutmuseum, Nijmegen 32595
Venrays Museum, Venray 32886
Zeeuws Museum, Middelburg 32554

New Zealand

Canterbury Museum, Christchurch 33069
Colonial Cottage Museum, Wellington 33245
Lake Taupo Museum and Art Gallery, Taupo 33204
South Otago Historical Museum, Balclutha 33062
Treaty House, Waitangi, Auckland 33228
Waipara County Historical Society Museum, Hawarden 33120

Nicaragua

Museo de Inmigrantes de Alemania, Mataglpa 33295
Museo Nacional de Nicaragua, Managua 33290

Niger

Musée National du Niger, Niamey 33308

Nigeria

National Museum of Colonial History, Aba 33310

Northern Cyprus

Barbarlık Müzesi, Lefkoşa 33367

Norway

Akershus Fylkesmuseum, Strømmen 33887
Bergen Museum, Bergen 33421
Emanuel Vigeland Museum, Oslo 33721
Halden Historiske Samlinger, Halden 33525
Herøy Kystmuseum, Herøy 33546
Historisk Museum, Oslo 33726
Hovden Jernvinnemuseum, Bykle 33460
Kaperdalen Samemuseum, Senja 33825
Kongsvinger Festningsmuseum, Kongsvinger 33604
Lørenskog Bygdemuseum, Skårer 33829
Museet i Mælandsgården, Skudeneshavn 33841
Nord-Jarlsbergmuseene, Holmestrand 33554
Norsk Seminmuseum, Stange 33862
Norske Grenselosers Museum, Bjørkelangen 33441
Ny Ålesund By- og Gruvemuseum, Ny-Ålesund 33706
Osen Bygdemuseum, Osen 33714
Polarmuseet i Tromsø, Tromsø 33919
Risør Museum, Risør 33782
Rogaland Krigshistorisk Museum, Sola 33849
Tromsø Universitets Museum, Tromsø 33921
Vardømuseene, Vardø 33961
Vevelstad Bygdetun, Vevelstad 33976
Vikingmuseet på Borg, Bøstad 33450

Oman

National Museum at Ruwi, Ruwi 33996
Oman Museum, Muscat 33990
Omani-French Museum Bait Fransa, Muscat 33992

Pakistan

Archaeological Museum Umarkot, Umarkot 34045
Punjab Archives Museum, Lahore 34027

Panama

Museo Bolivar, Ciudad de Panamá 34066
Museo de Historia de Panamá, Ciudad de Panamá 34069
Museo de la Nacionalidad, La Villa de Los Santos 34074
Museo Ricardo J. Alfaro, Bella Vista 34060

Papua New Guinea

Papua New Guinea National Museum and Art Gallery, Boroko 34080

Peru

Casa Museo Almirante Miguel Grau, Piura 34288
Museo Bolívar Periodista, Piura 34291
Museo Bolivariano, Pativilca 34285
Museo Casa de Zela, Tacna 34310
Museo de Arte e Historia, Lima 34233
Museo del Virreinato, Lima 34256
Museo Histórico Andrés Avelino Cáceres, Huamanga 34173
Museo Histórico Regional - Casa Garcilaso Cusco, Cusco 34165
Museo Provincial José María Morante, Arequipa 34115
Museo Universitario y Museo de Anatomía General, Yanacancha 34327
Museos de la Universidad Nacional de Cajamarca, Cajamarca 34130

Philippines

Ayala Museum, Makati 34414
Baldomero Aguinaldo House, Kawit 34397
Bibak Museum, Baguio 34341
Carlos P. Romulo Collection, Manila 34435
Casa Real Shrine, Malolos 34421
Emilio Aguinaldo Shrine Museum, Kawit 34398
Geronimo Berenguer de los Reyes jr. Museum, Cavite City 34362
José Rizal Shrine Museum, Manila 34441
Leon Apacible Historical Landmark Museum, Taal 34507
Malacañan Palace Presidential Museum, Manila 34442
Marcela M. Agoncillo Historical Museum, Taal 34508
Marcelo H. del Pilar Museum, Malolos 34423
Munting Museo ng Namayan, Mandaluyong 34426
Museo ng Rebolusyon, San Juan 34498
Museo Valenzuela, Valenzuela 34520
Museums of Filipinana and Rizaliana, Sampaloc 34496
Pambasang Museo ng Pilipinas, Manila 34450
Sulu National Museum, Jolo 34393
University of Nueva Caceres Museum, Naga 34467
Veterans Federation of the Philippines Museum, Taguig 34514

Poland

Centralne Muzeum Pożarnictwa, Mysłowice 34936
Centrum Pamięci Gen. Józefa Hallera, Władysławowo 35305
Dom Rodzinny Ojca Świętego Jana Pawła II, Wadowice 35207
Dwór Artusa, Gdańsk 34634
Ekspozycija Rzemiosło w Białym Spichrzu, Bydgoszcz 34575
Izba Pamięci Żydów Lublina, Lublin 34899
Muzeum Dom Wincentego Witosa, Wierzchosławice 35302
Muzeum Dworu Polskiego, Płochocin 35012
Muzeum Gross-Rosen, Wałbrzych 35211
Muzeum Gross-Rosen w Rogoźnicy, Rogoźnica 35084
Muzeum Historii Miasta, Szczecin 35161
Muzeum Historii Polskiego Rucha Ludowego, Piaseczno koło Gniewa 35006
Muzeum Historii Polskiego Ruchu Ludowego w Warszawie, Oddział w Sandomierzu, Sandomierz 35097
Muzeum Historii Żydów Polskich, Warszawa 35246
Muzeum Historyczne, Białystok 34549
Muzeum Historyczne Miasta Krakowa, Kraków 34798
Muzeum Historyczno-Etnograficzne, Chojnice 34599
Muzeum Ignacego Jana Paderewskiego i Wychodźstwa Polskiego w Ameryce, Warszawa 35249

H

Muzeum im. Jana Kasprowicza, Inowrocław .. 34691
Muzeum Małego Miasta, Bieżuń .. 34563
Muzeum Martyrologii Pod Zegarem, Lublin .. 34906
Muzeum Miedzi, Legnica .. 34848
Muzeum Pamiątek po Janie Matejce "Koryznówka", Nowy Wiśnicz .. 34965
Muzeum Podlaskie w Białymstoku, Białystok. 34550
Muzeum Pojezierza Łęczyńsko-Włodawskiego, Włodawa .. 35313
Muzeum Powstania Poznańskiego - Czerwiec 1956, Poznań .. 35042
Muzeum Powstania Wielkopolskiego 1918-1919, Poznań .. 35043
Muzeum w Raciborzu, Racibórz .. 35072
Muzeum - Zamek Książąt Pomorskich, Darłowo .. 34620

Portugal

Casa-Museu de Francisco Ernesto de Oliveira Martins, Angra do Heroísmo .. 35437
Museu Arqueológico do Carmo, Lisboa .. 35670
Museu da Emigração, das Comunidades e da Luso-Descendência, Fafe .. 35591
Museu da República e Resistência, Lisboa .. 35684
Museu Maçonico Português, Lisboa .. 35712
Museu Pio XII, Braga .. 35482
Palácio Nacional da Pena, Sintra .. 35951

H

Puerto Rico

Fundación Luis Muñoz Marín, San Juan .. 36063
Museo Casa Roig, Humacao .. 36041

Romania

Casa Martirilor de la Moisei, Moisei .. 36480
Casa Memorială Aloisie Tăutu, Valea Vinului .. 36717
Casa Memorială Emil Racoviţă, Dăneşti .. 36338
Casa Memorială Moş Dumitru, Baldovineşti .. 36133
Complexul Muzeal Naţional Goleşti, Ştefaneşti 36597
Expoziţia Curtea Domnească, Piatra-Neamţ .. 36516
Expoziţia Memorială Silviu Dragomir, Gurasada .. 36396
Memorial Museum B.P. Hasdeu, Câmpina .. 36249
Muzeul Badea Cârţan, Cârţişoara .. 36267
Muzeul Brăilei, Brăila .. 36170
Muzeul de Istorie, Sibiu .. 36609
Muzeul de Istorie al Moldovei, Iaşi .. 36418
Muzeul de Istorie Naţional şi Arheologie, Constanţa .. 36305
Muzeul de Istorie şi Artă Roman, Roman .. 36568
Muzeul Judeţean de Istorie, Piatra-Neamţ .. 36521
Muzeul Judeţean de Istorie Braşov, Braşov .. 36180
Muzeul Judetean de Istorie şi Arheologie Prahova, Ploieşti .. 36533
Muzeul Naţional al Unirii, Alba Iulia .. 36106
Muzeul Naţional de Istorie a României, Bucureşti .. 36221
Muzeul Naţional de Istorie a Transilvaniei, Cluj-Napoca .. 36290
Muzeul Naţional Peleş, Sinaia .. 36620
Muzeul Ţării Făgăraşului Valeriu Literat, Făgăraş .. 36363
Muzeul Unirii, Iaşi .. 36428
Muzeul Universitatii Babeş-Bolyai, Cluj-Napoca .. 36291

Russia

Aleksandrovskij Dvorec - Gosudarstvennyj Muzej Carskoje Selo, Puškin .. 37577
Astrachanskij Gosudarstvennyj Obedinennyj Istoriko- architekturnyj Muzej-zapovednik, Astrachan .. 36778
Astrachanskij Kraevedčeskij Muzej, Astrachan .. 36779
Belgorodskij Gosudarstvennyj Istoriko-Kraevedčeskij Muzej, Belgorod .. 36794
Biljarskij Gosudarstvennyj Istoriko-Archeologičeskij i Prirodnyj Muzej-Zapovednik, Biljarsk .. 36806
Bogorodickij Dvorec-muzej i Park, Bogorodick36812
Bratskij Gorodskoj Objedinennyj Muzej Istorii Osvoenija Angary, Bratsk .. 36822
Brjanskij Gosudarstvennyj Obedinennyj Kraevedčeskij Muzej, Brjansk .. 36824
Bunker Stalina, Samara .. 37610
Chabarovskij Gosudarstvennyj Muzej Dalnego Vostoka im. N.I. Grodekova, Chabarovsk .. 36856
Čuvašskij Nacionalnyj Muzej, Čeboksary .. 36837
Dagestanskij Gosudarstvennyj Objedinennyj Istoriko-architekturnyj Muzej, Machačkala .. 37115
Diorama Kurskaja Bitva - Belgorodskoe Napravlenie, Belgorod .. 36795
Dom-muzej Istorii Molodežnogo Dviženija, Rjazan .. 37593
Dom-muzej Petra Pervogo (Petrovskij Domik), Vologda .. 37961
Elabužskij Gosudarstvennyj Istoriko-architekturnyj i Chudožestvennyj Muzej-zapovednik, Elabuga .. 36903
Fort N5 - Muzej Istorii Velikoj Otečestvennoj Vojny, Kaliningrad .. 36987
Gorodskoj Kraevedčeskij Muzej, Elec .. 36910
Gosudarstvennaja OK Istoriko-ėtnografičeskij i Architekturnyj Muzej-zapovednik Staraja Sarepta, Volgograd .. 37954
Gosudarstvennyj Borodinskij Voenno-istoričeskij Muzej-zapovednik, Borodino .. 36819
Gosudarstvennyj Chudožestvennyj Istoriko-architekturnyj i prirodno-landšaftnyj Muzej-zapovednik Kolomenskoe, Moskva .. 37191
Gosudarstvennyj Istoričeskij Muzej, Moskva . 37193
Gosudarstvennyj Istoričeskij Zapovednik Gorki Leninskie, Gorki Leninskie .. 36926
Gosudarstvennyj Istoriko-memorialnyj Sankt-Peterburgskij Muzej Smolnyj, Sankt-Peterburg 37641
Gosudarstvennyj Istoriko-memorialnyj Zapovednik Rodina V.I. Lenina, Uljanovsk .. 37897
Gosudarstvennyj Muzej-usadba Archangelskoe, Archangelskoe .. 36773
Gosudarstvennyj Muzej Vyborgskij Zamok, Vyborg .. 37982
Gosudarstvennyj Voenno-istoričeskij i Prirodnyj Muzej-zapovednik Kulikovo Pole, Tula .. 37866
Gosudarstvennyj Voenno-istoričeskij Muzej-zapovednik Prochorovskoe Pole, Prochorovka 37571
Istoričeskaja Usadba Botik Petra I, Veskovo .. 37926
Istoriko-memorialnyj Kompleks Gerojam Stalingradskoj Bitvy na Mamaevom Kurgane, Volgograd .. 37956
Istoriko-memorialnyj Muzej M.V. Lomonosova, Lomonosovo .. 37112
Istoriko-memorialnyj Muzejnyj Kompleks Bobriki, Donskij .. 36878
Jakutskij Gosudarstvennyj Muzej Istorii i Kultury Narodov Severa im. Jaroslavskogo, Jakutsk .. 36962
Jamalo-Neneckij Okružnoj Kraevedčeskij Muzej im. I.S. Šemanovskogo, Salechard .. 37608
Kaliningradskij Oblastnoj Istoriko-chudožestvennyj Muzej, Kaliningrad .. 36989
Kalmyckij Respublikanskij Kraevedčeskij Muzej im. Prof. N.N. Palmova, Ėlista .. 36913
Karačaevo-Čerkesskij Istoriko-kulturnyj i Prirodnyj Muzej-zapovednik, Čerkessk .. 36854
Kirillo-Belozerskij Istoriko-Architekturnyj i Chudožestvennyj Muzej-Zapovednik, Kirillov . 37034
Komi Respublikanskij Istoriko-Kraevedčeskij Muzej, Syktyvkar .. 37815
Kostromskoj Gosudarstvennyj Istoriko-architekturnyj Muzej-zapovednik, Kostroma .. 37061
Kraevedčeskij Muzej Evrejskoj Avtonomnoj Oblasti, Birobidžan .. 36807
Krasnojarskij Istoričeskij Muzej Memorial Pobedy, Krasnojarsk .. 37071
Krasnojarskij Kraevoj Kraevedčeskij Muzej, Krasnojarsk .. 37072
Memorialnyj Dom-muzej Gercena, Moskva .. 37231
Memorialnyj Dom-muzej G.V. Plechanova, Lipeck .. 37105
Memorialnyj Dom-muzej Uljanovych, Astrachan .. 36782
Memorialnyj Muzej-Kvartira S.M. Kirova, Vladikavkaz .. 37930
Mordovskij Respublikanskij Kraevedčeskij Muzej, Saransk .. 37740
Muzej 700 let - Landskrona, Nevskoe Ustjė, Nienšanc, Sankt-Peterburg .. 37668
Muzej Archeologii Prikamja, Perm .. 37548
Muzej Arsenal - Muzej Velikoj Otečestvennoj Vojny, Voronež .. 37971
Muzej Avtogravov imeni Imperatora Nikolaja Vtorogo, Novosibirsk .. 37484
Muzej Boevogo i Trudovogo Podviga 1941-1945, Saransk .. 37741
Muzej Boevoj Slavy, Astrachan .. 36783
Muzej Cerkov Dekabristov, Čita .. 36870
Muzej Dekabristov, Minusinsk .. 37137
Muzej Dekabristov - Dom Trubeckogo, Irkutsk36935
Muzej Dekabristov - Dom Volkonskich, Irkutsk .. 36936
Muzej Diplomatičeskogo Korpusa, Vologda .. 37962
Muzej Ėtnografii, Istorii i Kultury Narodov Baškortostana, Ufa .. 37888
Muzej Istorii Kolskich Saamov, Lovozero .. 37113
Muzej Istorii Političeskoj Policii Rossii - Gorochovaja 2, Sankt-Peterburg .. 37683
Muzej Istorii Političeskoj Ssylki, Bratsk .. 36823
Muzej Istorii Revoljucionno-demokratičeskogo Dviženija 1880-1890 gg, Sankt-Peterburg .. 37684
Muzej Konspiritavnaja Kvartira Simbirskoj Gruppy RSDRP 1904-1906, Uljanovsk .. 37902
Muzej Kutuzovskaja Izba, Moskva .. 37313
Muzej-panorama Borodinskaja Bitva, Moskva 37345
Muzej Podviga Ivana Susanina, Susanino .. 37810
Muzej Vladimira Raevskogo, Bogoslovka .. 36813
Muzej Smolenščina v Gody Velikoj Otečestvennoj Vojny 1941 -1945, Smolensk .. 37772
Muzej Štab-kvartira V.K. Bljuchera, Tjumen .. 37850
Muzej-zapovednik Dmitrovskij Kreml, Dmitrov .. 36876
Muzejno-vystavočnyj Centr - Diorama, Kirov . 37042
Muzejno-vystavočnyj Centr Tulskie Drevnosti, Tula .. 37872
Muzejno-vystavočnyj Kompleks Volokolamskij Kreml, Volokolamsk .. 37966
Nacionalnyj Muzej Respubliki Baškortostan, Ufa .. 37889
Nacionalnyj Muzej Respubliki Komi, Syktyvkar .. 37821
Nacionalnyj Muzej Udmurtskoj Respubliki im. K. Gerda, Iževsk .. 36955
Novgorodskij Gosudarstvennyj Muzej-zapovednik, Velikij Novgorod .. 37922
Novočerkasskij Muzej Istorii Donskich Kazakov, Novočerkassk .. 37478
Oloneckij Nacionalnyj Muzej Karelov-Livvikov im. N.T.Prilukin, Olonec .. 37496
Omskij Gosudarstvennyj Istoriko-kraevedčeskij Muzej, Omsk .. 37499
Osinskij Kraevedčeskij Muzej, Osa .. 37517
Pavlovskij Istoričskij Muzej, Pavlovo .. 37527
Penzenskij Gosudarstvennyj Obedinennyj Kraevedčeskij Muzej, Penza .. 37537
Permskij Oblastnoj Kraevedčeskij Muzej, Perm .. 37551
Piskarevskoe Memorialnoe Kladbišče - Muzej, Sankt-Peterburg .. 37720
Plesskij Gosudarstvennyj Istoriko-Architekturnyj i Chudožestvennyj Muzej-Zapovednik, Ples .. 37568
Podpolnaja Tipografija CK RSDRP 1905-1906, Moskva .. 37399
Presnja Istoriko-memorialnyj Muzej, Moskva 37402
Primorskij Gosudarstvennyj Muzej im. V.K. Arseneva, Vladivostok .. 37949
Pskovskij Gosudarstvennyj Istoričesko-architekturnyj i Chudožestnennyj Muzej-zapovednik, Pskov .. 37575
Pskovskij Memorialnyj Muzej V.I. Lenina, Pskov .. 37576
Rjazanskij Gosudarstvennyj Istoriko-architekturnyj Muzej-zapovednik, Rjazan .. 37595
Roslavlskij Istoriko-chudožstvennyj Muzej, Roslavl .. 37597
Rostovskij Oblastnoj Muzej Kraevedenija, Rostov-na-Donu .. 37602
Saratovskij Gosudarstvennyj Muzej Boevoj Slavy, Saratov .. 37753
Saratovskij Kraevedčeskij Muzej, Saratov .. 37754
Šlisselburgskaja Krepost Orešek, Šlisselburg. 37769
Soloveckij Gosudarstvennyj Istoriko-Architekturnyj i Prirodnyj Muzej-Zapovednik, Soloveckij .. 37788
Staročerkasskij Istoriko-architekturnyj Muzej-zapovednik, Staročerkassk .. 37798
Staroladožskij Istoriko-Architekturnyj i Archeologičeskij Muzej-Zapovednik, Staraja Ladoga .. 37795
Stavropolskij Kraevedčeskij Muzej im. G.N. Prozriteleva i G.K. Prave, Stavropol .. 37805
Taganrogskij Gosudarstvennyj Literaturnyj i Istoriko-architekturnyj Muzej-zapovednik, Taganrog .. 37828
Tambovskij Oblastnoj Kraevedčeskij Muzej, Tambov .. 37833
Tobolskij Gosudarstvennyj Istoriko-architekturnyj Muzej-zapovednik, Tobolsk .. 37855
Tomskij Oblastnoj Kraevedčeskij Muzej, Tomsk .. 37859
Uljanovskij Oblastnoj Kraevedčskij Muzej im. I.A. Gončarova, Uljanovsk .. 37910
Velikoustjugskij Gosudarstvennyj Istoriko-Architekturnyj i Chudožestvennyj Muzej-Zapovednik, Velikij Ustjug .. 37923
Volgogradskij Oblastnoj Kraevedčeskij Muzej, Volgograd .. 37959
Vologodskij Gosudarstvennyj Istoriko-Architekturnyj i Chudožestvennyj Muzej-Zapovednik, Vologda .. 37964
Voronežskij Oblastnoj Kraevedčeskij Muzej, Voronež .. 37976
Vserossijskij Istoriko-ėtnografičeskij Muzej, Toržok .. 37862

Rwanda

Musée National du Rwanda, Butare .. 37992

St. Lucia

Pigeon Island, Castries .. 37995

St. Pierre and Miquelon

Musée Héritage, Saint-Pierre .. 37999

Samoa

Museum of Samoa, Apia .. 38000

San Marino

Museo delle Cere - Strumenti di Tortura, San Marino .. 38006
I Torre Guaita, San Marino .. 38012

Senegal

Musée Historique de Gorée, Gorée .. 38026

Serbia

Jevrejski Istorijski Muzej, Beograd .. 38051
Kompleksi Monumental i Lidhjes Shqiptare të Prizrenit, Prizren .. 38134
Kuća Cveća, Beograd .. 38055
Muzej Istorije Jugoslavije, Beograd .. 38066
Muzej Vuka i Dositeja, Beograd .. 38080
Muzej Žrtava Genocida, Beograd .. 38081
Narodni Muzej, Kragujevac .. 38109
Narodni Muzej, Leskovac .. 38114
Narodni Muzej u Beogradu, Beograd .. 38082

Sierra Leone

Sierra Leone National Museum, Freetown .. 38169

Singapore

National Museum of Singapore, Singapore .. 38177

Slovakia

Historické a Národopisné Múzeum Želovce, Želovce .. 38389
Múzeum Kultúry Karpatských Nemcov, Bratislava .. 38222
Múzeum Kultúry Maďarov na Slovensku, Bratislava .. 38223
Múzeum Slovenských Národných Rád, Myjava .. 38311
Východoslovenské Múzeum, Košice .. 38275
Západoslovenské Múzeum v Trnave, Trnava .. 38381

Slovenia

Kobariški Muzej, Kobarid .. 38431
Narodni Muzej Slovenije, Ljubljana .. 38458
Predjamski Grad, Postojna .. 38497

South Africa

Arbeidsgenot, Oudtshoorn .. 38696
Bo-Kaap Museum, Cape Town .. 38579
Great Fish River Museum, Cradock .. 38599
Museum Connection Die, Cape Town .. 38588
National Cultural History Museum, Tshwane .. 38752
National Cultural History Museum, Tshwane .. 38752
Port Elizabeth Museum, Port Elizabeth .. 38712
South African Jewish Museum, Gardens .. 38622
Voortrekker Museum, Pietermaritzburg .. 38708

Spain

Casa del Cordón, Vitoria-Gasteiz .. 40318
Colegio del Rey, Alcalá de Henares .. 38811
Fundación Carlos de Amberes, Madrid .. 39519
Museo-Archivo de Falset y Comarca, Falset .. 39293
Museo Arqueolóxico e Histórico, A Coruña .. 39217
Museo de Bardados del Paso Azul, Lorca .. 39493
Museo de la Ciudad, Madrid .. 39543
Museo de la Vida Rural, L'Espluga de Francolí. 39282
Museo de las Cortes de Cádiz, Cádiz .. 39090
Museo Municipal, Ayllón .. 38926
Museo Sinagoga de Santa María la Blanca, Toledo .. 40137
Museo Thermalia, Caldes de Montbui .. 39099
Museu de Formentera, Formentera .. 39306
Museu de la Ciutat Casa de Polo, Villarreal .. 40308
Museu de la Història de Fontcoberta, Fontcoberta .. 39304
Museu d'Història de Catalunya, Barcelona .. 38988
Museu d'Història de Manacor, Manacor .. 39610
Sala de Exposiciones Edificio Historico, Oviedo .. 39761
Zarauzko Arte eta Historia Museoa, Zarautz .. 40355

Sudan

Khalifa's House Museum, Omdurman .. 40396

Swaziland

Swaziland National Museum, Lobamba .. 40403

Sweden

Årsunda Viking Museum, Årsunda .. 40433
Aschanska Gården, Eksjö .. 40492
Birka Vikingastaden, Adelsö .. 40404
Ekehagens Forntidsby, Åsarp .. 40441
Ekomuseum Bergslagen, Smedjebacken .. 40859
Fågelsjö Gammelgård Bortom Åa, Los .. 40694
Kullängsstugan, Askersund .. 40444
Kulturhistoriska Museet 'Kulturen i Lund', Lund .. 40703
Kungsstugan och Cajsa Wargs Hus, Örebro .. 40787
Östergötlands Länsmuseum, Linköping .. 40684
Riddarholmskyrkan, Stockholm .. 40928
Riddarhuset, Stockholm .. 40929

Stiftelsen Militärfordons Museum, Malmköping 40717
Stora Nyckelvikens Gård, Nacka 40762
Utvandrarnas Hus, Växjö 41043
Vasamuseet, Stockholm 40949

Switzerland

Barfüsserkirche, Basel 41157
Bundesbriefmuseum, Schwyz 41849
Maison du Vieux-Zinal, Zinal 42059
Mazzini-Gedenkstätte, Grenchen 41454
Moulagensammlung des Universitätsspitals und der Universität Zürich, Zürich 42089
Musée Cantonal d'Histoire, Sion 41874
Musée d'art et d'histoire, Fribourg 41387
Musée des Suisses dans le Monde, Pregny-Chambésy 41735
Musée International de la Croix-Rouge et du Croissant-Rouge, Genève 41423
Musée National Suisse - Château de Prangins, Prangins 41732
Polenmuseum Rapperswil, Rapperswil 41746
Schweizerisches Nationalmuseum, Zürich 42105
Suworow-Museum, Schwanden 41847
Tell-Museum, Bürglen 41270
Thomas-Legler-Haus, Diesbach 41326

Syria

National Museum of Damascus, Damascus 42132

Taiwan

Chi Mei Museum, Rende 42304
Fengshan Dr. Sun Yat-sen Memorial Hall, Fengshan 42187
The Forbidden City Museum, Lujhou 42262
Jheng Cheng-Gong Memorial Hall, Sikou 42330
Koxinga Museum, Tainan 42362
Lin An-Tai Historical Home, Taipei 42399
Lion's Kingdom Museum, Toucheng 42455
Museum of the Institute of History and Philology, Taipei 42410
National Cheng Kung University Museum, Tainan 42364
National Dr. Sun Yat-sen Memorial Hall, Taipei 42414
National Museum of History, Taipei 42415
National Museum of Taiwanese History, Tainan 42365
National Palace Museum, Taipei 42416
Tainan County History Museum, Syuejia 42346
Taiwan Historica, Nantou 42283

Tajikistan

Osorkhonai Akademiiai Ulumi Jumhurii Tojikiston, Dušanbe 42497
Osorkhonai Davlatii Ta"rikhi - Kishvarshinosi va Sa"nati Tasvirii Jumhurii Tojikiston ba nomi Kamoliddin Behzod, Dušanbe 42498

Tanzania

Beit al-Ajaib Museum with House of Wonders Museum, Zanzibar 42522
Livingstone and Stanley Memorial, Tabora 42521
Palace Museum, Zanzibar 42523
Pemba Museum, Chake Chake 42511
Zanzibar National Archives and Museums Department, Zanzibar 42525

Tonga

Tupou College Museum, Nuku'alofa 42614

Trinidad and Tobago

National Museum and Art Gallery of Trinidad and Tobago, Port-of-Spain 42615

Turkey

I. Türkiye Büyük Millet Meclisi Binası, Ankara 42693
II. Türkiye Büyük Millet Meclisi Binası, Ankara 42694
Anıtkabir-Atatürk ve Kurtuluş Müzesi, Ankara 42696
Cumhuriyet Devri Müzesi, Ankara 42700
Gazi Müzesi, Samsun 42930
Mudanya Mutareke Evi Müzesi, Mudanya 42910
Osman Hamdi Bey Müzesi, Gebze 42788
Tanzimat Müzesi, İstanbul 42840

Turkmenistan

Taryhy Ýer Abiwerd, Kaka 42976
Taryhy Ýer Gadymy Merw, Mary 42979
Taryhy Ýer Gökdepe Galasy, Gökdepe 42975
Taryhy Ýer Saragt, Saragt 42981
Türkmenistanyn Milli Muzeýi, Aşgabat 42973

Uganda

Uganda Museum, Kampala 42993

Ukraine

Baturins'kyj Deržavnyj Istoričnyj Zapovidnyk Get'mans'ka Stolicja, Baturin 43000
Budinok-muzej Sevastopol's'ke Komunistične Pidpillja 1942-1944, Sevastopol 43175
Deržavnyj Istoriko-kul'turnyj Zapovidnyk, Kam'janec'-Podil's'kyj 43051
Deržavnyj Istoryko-kul'turnyj Zapovidnyk Pole Poltavskoï Bytvy, Poltava 43166
Diorama "Bitva za Dnepr", Dnipropetrovs'k 43031
Istoričnij Muzej, Dnipropetrovs'k 43033
Lvivs'kyj Istoryčnyj Muzej, Lviv 43126
Memorial Geroïčnoï Oboroni Odesi - 411 Beregova Linija, Odesa 43155
Muzei Geroïčnoï Oborony i Vyzvolennja Sevastopolja, Sevastopol 43177
Muzej Istoriï Mista Dniprodzeržins'ka, Dniprodzeržins'k 43030
Muzej Istoriï Mista Kyjeva, Kyïv 43089
Muzej Istoriï ta Etnografiï Grekiv Priazov'ja, Mariupol' 43145
Nacional'nyj Istoriko-etnografičnyj Zapovidnyk Perejaslav, Perejaslav-Chmel'nyc'kyj 43165
Nacional'nyj Muzej Istoriï Ukraïny, Kyïv 43103
Nacional'nyj Muzej Istoriï Velykoï Vitčyznjanoï Vijny 1941-1945 Rokiv, Kyïv 43104
Novgorod-Siverskyj Istoriko-kulturnyj Muzej-zapovidnyk Slovo o Polku Igorevim, Novgorod-Sivers'kyj 43153
Oblasnyj Istoryčnyj Muzej im. V. Tarnovs'kogo, Černigiv 43007
Oboronna Bašta Malachova Kurgana, Sevastopol 43179
Panorama Oborona Sevastopolja 1854-1855 rr., Sevastopol 43180
Panorama Šturm Sapun-gori 7 Travnja 1944 r., Sevastopol 43181

United Arab Emirates

National Museum, Ras-Al-Khaimah 43222
Sheikh Saeed's House, Dubai 43220

United Kingdom

Anne Hathaway's Cottage, Stratford-upon-Avon 45779
Bannockburn Heritage Centre, Stirling 45749
Black Cultural Archives, London 44797
Blackburn Museum and Art Gallery, Blackburn, Lancashire 43526
Blakesley Hall, Birmingham 43501
British Empire and Commonwealth Museum, Bristol 43634
British Red Cross Museum and Archives, London 44806
Brontë Parsonage Museum, Haworth 44430
Burston Strike School, Burston 43685
Captain Cook Memorial Museum, Whitby 45994
Castle Museum, Colchester 43872
A Celebration of Immigration, London 44815
Chiddingstone Castle, Chiddingstone 43834
Childhood and Costume Museum, Bridgnorth 43604
City and County Museum, Lincoln 44710
Clan Armstrong Museum, Langholm 44639
Clan Cameron Museum, Spean Bridge 45725
Clan Macpherson House and Museum, Newtonmore 45254
Cleethorpes Discovery Centre, Cleethorpes 43851
Commandery Museum, Worcester 46070
Crom Estate, Newtownbutler 45263
Cromarty Courthouse Museum, Cromarty 43927
Culloden Battlefield and Visitor Centre, Inverness 44523
Darby Houses, Coalbrookdale 43858
Devon and Cornwall Constabulary Museum, Exeter 44193
East Ham Nature Reserve, London 44842
Edinburgh Scout Museum, Edinburgh 44121
Elliot Colliery Winding House, New Tredegar 45208
Golden Hinde Educational Trust, London 44857
Her Majesty Tower of London, London 44873
Historic Scotland, Edinburgh 44128
Holst Birthplace Museum, Cheltenham 43811
Jewish Museum, London 44889
John Muir's Birthplace, Dunbar 44042
Kent and Medway Museum, West Malling 45971
Magdalen College Museum, Wainfleet All Saints 45906
Marischal Museum, Aberdeen 43244
Mary Queen of Scots' House, Jedburgh 44548
Milestones: Hampshire's Living History Museum, Basingstoke 43395
Museum of Scotland, Edinburgh 44136
Museum - Treasures from the National Archives, Kew 44571
National Football Museum, Preston 45446
National Wallace Monument, Stirling 45750
Nelson Collection at Lloyd's, London 44942
New Lanark Visitor Centre, New Lanark 45203
Newark Town Treasures and Art Gallery, Newark-on-Trent 45213
The Oxford Story, Oxford 45335
Pass of Killiecrankie Visitor Centre, Pitlochry 45385
People's Palace Museum, Glasgow 44294
Piece Hall Art Gallery, Halifax 44386
POSK Gallery, London 44958
Ripon Workhouse Museum, Ripon 45502
Rob Roy and Trossachs Victor Centre, Callander 43714
Royal Borough Museum Collection, Windsor 46045
Royal Museum, Edinburgh 44148
Royal Naval Museum Portsmouth, Portsmouth 45437
SEARCH-Centre, Gosport 44337
Social History Curators Group, Croydon 55093
Strathnaver Museum, Bettyhill 43465
Tolpuddle Martyrs Museum, Tolpuddle 45863
Tower Bridge Exhibition, London 44998
Tudor House, Weymouth 45989
Ulster History Park, Omagh 45318
Ulster Museum, Belfast 43448
United States Grant Ancestral Homestead, Dungannon 44058
Wellesbourne Wartime Museum, Wellesbourne 45955
Wellington Arch, London 45010
Western Approaches, Liverpool 44749
Wilberforce House, Kingston-upon-Hull 44601
Workhouse Museum, Derry 43976

United States of America

19th Century Willowbrook Village, Newfield 51312
A.D. Buck Museum of Natural History and Science, Tonkawa 53530
Abraham Lincoln Museum, Harrogate 49191
Adams Museum and Historic Adams House, Deadwood 48003
African American Historical and Cultural Society Museum, San Francisco 52764
African American Historical Museum, Cedar Rapids 47299
African American Museum, Dallas 47924
African American Museum at Oakland, Oakland 51462
Afro-American Historical Society Museum, Jersey City 49624
Akta Lakota Museum, Chamberlain 47329
Alabama Department of Archives and History Museum, Montgomery 50813
Albany Institute of History and Art, Albany 46157
The Albuquerque Museum of Art and History, Albuquerque 46175
Alton Museum of History and Art, Alton 46240
American Airpower Heritage Museum, Midland 50669
American Catholic Historical Society Museum, Philadelphia 51812
The American Historical Foundation Museum, Richmond 52321
American Independence Museum, Exeter 48463
American Irish Historical Society Museum, New York 51114
American Merchant Marine Museum, Kings Point 49785
American Printing House for the Blind, Callahan Museum, Louisville 50281
The American Revolution Center, Wayne 54010
American Saddlebred Horse Museum, Mexico 50617
Americana Museum, El Paso 48315
America's Black Holocaust Museum, Milwaukee 50704
Anacostia Museum, Washington 53845
Anchorage Museum at Rasmuson Center, Anchorage 46288
Anna Bemis Palmer Museum, York 54339
Antietam National Battlefield, Sharpsburg 53053
Antiquarian and Landmarks Society Museum, Hartford 49192
Antique Gas and Steam Engine Museum, Vista 53764
Apex Museum, Atlanta 46421
Aptucxet Trading Post Museum, Bourne 46966
Arizona Historical Museum, Tucson 53582
Arkansas State University Museum, Jonesboro 49647
Armenian Museum of America, Watertown 53970
Arrow Rock State Historic Site, Arrow Rock 46364
Arts Club of Washington, Washington 53851
Ashland - Henry Clay Estate, Lexington 50045
The Atlanta Cyclorama, Atlanta 46425
Audubon House, Key West 49756
Avery Research Center for African American History and Culture, Charleston 47357
Aztec Museum and Pioneer Village, Aztec 46527
Baca House, Trinidad 53566
Barbed Wire Museum, La Crosse 49833
Barnacle Historic State Park, Coconut Grove 47670
Barrett House, Forest Hall, New Ipswich 51065
Bates-Scofield Homestead, Darien 47966
Battle of Lexington State Historic Site, Lexington 50062
Beauregard-Keyes House, New Orleans 51075
Beauvoir, Biloxi 46809
Beaverhead County Museum, Dillon 48142
Bellamy Mansion Museum of History and Design Arts, Wilmington 54199
Belmont County Victorian Mansion, Barnesville 46610
Bennington Battle Monument, Old Bennington 51531
Bennington Battlefield State Historic Site, Walloomsac 53801
Bent's Old Fort, La Junta 49849
Betsy Ross House, Philadelphia 51819
Black Hawk State Historic Site, Rock Island 52406
Black Heritage Museum, Miami 50622
Black Hills Mining Museum, Lead 50004
Boal Mansion Museum, Boalsburg 46880
Bodie Park Museum, Bridgeport 47015
Boone County Historical Center, Boone 46903
Boonsborough Museum of History, Boonsboro 46908
Boot Hill Museum, Dodge City 48147
Boston National Historical Park, Boston 46922
Brazos Valley Museum of Natural History, Bryan 47105
Brooklyn Historical Society Museum, Brooklyn 47068
Brown House Museum, Sandersville 52832
Brucemore, Cedar Rapids 47300
Buffalo Bill Historical Center, Cody 47671
Buffalo Bill Memorial Museum, Golden 48956
Buffalo Bill Museum of Le Claire, Iowa, Le Claire 50001
Buffalo Bill Ranch, North Platte 51401
Bunker Hill Museum, Charlestown 47381
Butte County Pioneer Memorial Museum, Oroville 51604
Butterfield Trail Historical Museum, Russell Springs 52488
C.H. Nash Museum-Chucalissa Archaeological Museum, Memphis 50567
Cahokia Courthouse, Cahokia 47158
Cairo Museum, Vicksburg 53732
California African American Museum, Los Angeles 50214
California Heritage Museum, Santa Monica 52898
California Museum, Sacramento 52500
Cape Cod National Seashore Museum, Wellfleet 54025
Cape Fear Museum, Wilmington 54201
Cappon House Museum, Holland 49313
Captain Forbes House Museum, Milton 50699
Captain Phillips' Rangers Memorial, Saxton 52951
Carillon Historical Park, Dayton 47983
Carpinteria Valley Museum of History, Carpinteria 47254
Carroll County Farm Museum, Westminster 54096
Carroll Reece Museum, Johnson City 49631
Carter's Grove Visitor Center, Williamsburg 54166
Carver County Historical Society Museum, Waconia 53782
Castillo de San Marcos Museum, Saint Augustine 52520
Castle Tucker, Wiscasset 54261
Cedar Grove Mansion Inn, Vicksburg 53733
Centennial Village Museum, Greeley 49032
Center for American History, Austin 46506
Center for Jewish History, New York 51143
Century Village Museum, Burton 47147
Chancellor Robert R. Livingston Masonic Library and Museum, New York 51145
Charles H. Wright Museum of African American History, Detroit 48109
Charles Towne Landing 1670, Charleston 47358
Cherokee Courthouse Museum, Gore 48974
Cherokee National Museum, Tahlequah 53417
Cherokee Strip Museum and Henry S. Johnston Library, Perry 51781
Cheyenne Frontier Days Old West Museum, Cheyenne 47444
Chicago Historical Society, Chicago 47466
Chickasaw Council House Museum, Tishomingo 53512
Children's Museum in Easton, North Easton 51392
Children's Museum of Southeastern Connecticut, Niantic 51349
Chisholm Trail Museum, Wellington 54027
Cigna Museum and Art Collection, Philadelphia 51822
The Citadel Archives and Museum, Charleston 47360
City of Las Vegas and Rough Riders Memorial Museum, Las Vegas 49963
Civil War Museum, Philadelphia 51823
Civil War Museum, Carrollton 47256
Civilian Conservation Corps Museum, Sebring 53020
Clark County Museum, Henderson 49249
Clemenceau Heritage Museum, Cottonwood 47842
Coastal Discovery Museum, Hilton Head Island 49294
Coffin House, Newbury 51302
Colesville and Windsor Museum at Saint Luke's Church, Harpursville 49176
Collier Memorial State Park and Logging Museum, Chiloquin 47538
Colorado Historical Society Museum, Denver 48069
Columbia River Maritime Museum, Astoria 46401

H

Columbus-Belmont Civil War Museum, Columbus 47743
The Columbus Museum, Columbus 47739
Commonwealth Museum, Boston 46927
Confederate Museum, Crawfordville 47872
Constitution Square, Danville 47961
Coppini Academy of Fine Arts Gallery and Museum, San Antonio 52713
Coronado National Memorial Museum, Hereford 49256
Cottonwood County Historical Society, Windom 54224
Creek Council House Museum, Okmulgee 51528
Cu Heritage Center, Boulder 46959
Cultural Heritage Center Museum, Toppenish 53538
Cultural Rights and Protection/Ute Indian Tribe, Fort Duchesne 48618
Cumberland Gap National Historical Park, Middlesboro 50655
Custer Museum, Clayton 47598
Czech Heritage Museum and Genealogy Center, Temple 53473
Dade Battlefield, Bushnell 47149
Dahlonega Courthouse Gold Museum, Dahlonega 47920
Dalton Defenders Museum, Coffeyville 47674
Daughters of the American Revolution Museum, Washington 53858
Dawson Springs Museum and Art Center, Dawson Springs 47981
Delaware and Hudson Canal Museum, High Falls 49274
Depot Museum, Mammoth Spring 50413
Depreciation Lands Museum, Allison Park 46233
Desert Caballeros Western Museum, Wickenburg 54155
Desoto National Memorial, Bradenton 46985
Deutschheim State Historic Site, Hermann 49260
Discovery Creek Children's Museum of Washington DC, Washington 53862
Discovery Museum, Sacramento 52506
Donner Memorial and Emigrant Trail Museum, Truckee 53578
Door County Maritime Museum, Sturgeon Bay 53345
Douglass Historical Museum, Douglass 48160
Downtown History Museum, Tucson 53588
Dr. Pepper Museum and Free Enterprise Institute, Waco 53773
Drum Barracks Civil War Museum, Wilmington 54188
Du Page County Historical Museum, Wheaton 54119
Duke Homestead, Durham 48216
Dusable Museum of African-American History, Chicago 47472
E.M. Violette Museum, Kirksville 49800
Early American Museum, Mahomet 50399
East-West Center, Honolulu 49338
Edison National Historic Site, West Orange 54071
Eisenhower National Historic Site, Gettysburg 48901
El Morro National Monument, Ramah 52238
El Paso Museum of History, El Paso 48322
Elijah Clark Memorial Museum, Lincolnton 50114
Ephrata Cloister, Ephrata 48404
The Ernest W. Michel Historical Judaica Collection, New York 51170
The Esther Thomas Atkinson Museum, Hampden-Sydney 49146
Eugene and Marilyn Glick Indiana History Center, Indianapolis 49492
Eugene V. Debs Home, Terre Haute 53480
Eustis Historical Museum, Eustis 48445
Exchange Hotel Civil War Museum, Gordonsville 48973
Exhibition of the Norwegian-American Historical Association, Northfield 51419
Fairfax Museum and Visitor Center, Fairfax 48477
Farmington Historic Home, Louisville 50284
Favell Museum of Western Art and Indian Artifacts, Klamath Falls 49806
Federal Hall National Memorial, New York 51173
Fick Fossil and History Museum, Oakley 51476
First Missouri State Capitol, Saint Charles 52532
First White House of the Confederacy, Montgomery 50814
Five Civilized Tribes Museum, Muskogee 50928
Florence McLeod Hazard Museum, Columbus 47746
Florida History Museum, Jupiter 49659
Florissant Valley Historical Society, Florissant 48571
Folsom House Museum, Taylors Falls 53455
Forest Lawn Museum, Glendale 48930
Fort Boonesborough Museum, Richmond 52316
Fort Caroline Memorial Museum, Jacksonville 49578
Fort Clatsop National Memorial, Astoria 46402
Fort Clinch, Fernandina Beach 48536
Fort de Chartres Museum, Prairie du Rocher 52130
Fort Fisher, Kure Beach 49828
Fort Frederica, Saint Simons Island 52646
Fort Huachuca Museum, Fort Huachuca 48626
Fort King George, Darien 47967
Fort Ligonier Museum, Ligonier 50082
Fort McHenry, Baltimore 46565
Fort Macon, Atlantic Beach 46459
Fort Massac, Metropolis 50616
Fort Matanzas, Saint Augustine 52521
Fort Saint Jean Baptiste, Natchitoches 50987
The Fort Smith, Fort Smith 48683
Fort Union Trading Post, Williston 54181
Fort Vasquez Museum, Platteville 51979
Fort Worth Museum of Science and History, Fort Worth 48713
Fort Yargo, Winder 54223
Frank Lloyd Wright Home and Studio, Oak Park 51453
Franklin Pierce Homestead, Hillsborough 49292
Frederick C. Robie House, Chicago 47475
Frederick Douglass National Historic Site, Washington 53872
The French Legation Museum, Austin 46509
Fresno Flats Museum, Oakhurst 51461
Frontier Times Museum, Bandera 46593
Gadsden Museum, Mesilla 50610
Ganondagan State Historic Site, Victor 53738
General Grant National Memorial, New York 51183
General Henry Knox Museum, Thomaston 53497
General John J. Pershing Boyhood Home, Laclede 49860
General Lew Wallace Museum, Crawfordsville 47869
George Rogers Clark Park Museum, Vincennes 53747
George W. Carver Museum, Tuskegee Institute 53626
George W. Somerville Historical Library, Chillicothe 47531
George Washington Masonic National Memorial, Alexandria 46208
George Washington's Headquarters, Cumberland 47907
George Washington's Mount Vernon, Mount Vernon 50893
Georgia Mountains History Museum, Gainesville 48821
Gerald R. Ford Museum, Grand Rapids 49002
Geronimo Springs Museum, Truth or Consequences 53579
Gettysburg National Military Park, Gettysburg 48902
Gilcrease Museum, Tulsa 53608
Glendower State Memorial, Lebanon 50011
Governor Bent Museum, Taos 53440
Grand River Historical Society Museum, Chillicothe 47532
Great Blacks in Wax Museum, Baltimore 46566
Great Plains Black Museum, Omaha 51551
Green River Museum, Woodbury 54276
Grout Museum of History and Science, Waterloo 53961
The Hagley Museum, Wilmington 54193
Hampton History Museum, Hampton 49151
Hampton National Historic Site Museum, Towson 53550
Hampton University Museum, Hampton 49152
Hancock House, Hancock's Bridge 49158
Harn Homestead and 1889er Museum, Oklahoma City 51514
Harriet Beecher Stowe Center, Hartford 49194
Harry S. Truman Birthplace, Lamar 49910
Harry S. Truman Home, Independence 49473
Harry S. Truman Presidential Library and Museum, Independence 49475
Harvard Historical Society Collection, Harvard 49208
Hearthstone Historic House Museum, Appleton 46330
Helena Modjeska Historic House, Modjeska Canyon 50776
Hellenic Museum and Cultural Center, Chicago 47479
Hennepin History Museum, Minneapolis 50732
Henry Ford Estate, Dearborn 48008
Henry Whitfield State Historical Museum, Guilford 49102
Herbert Hoover National Historic Site, West Branch 54046
Hereford Inlet Lighthouse, North Wildwood 51409
Heritage Museum, Baltimore 46567
Heritage Square Museum, Los Angeles 50229
Herreshoff Marine Museum/America's Cup Hall of Fame, Bristol 47034
Hershey Museum, Hershey 49263
Hickories Museum, Elyria 48386
High Desert Museum, Bend 46733
Hinckley Foundation Museum, Ithaca 49549
Historic Bethany, Bethany 46772
Historic Columbus Foundation, Columbus 55137
Historic Edenton, Edenton 48284
Historic Landmarks Foundation of Indiana, Indianapolis 49494
Historic Langhorne Museum, Langhorne 49936
Historic Latta Plantation, Huntersville 49420
Historic Lyme Village, Bellevue 46704
Historic Mann House, Concord 47772
Historic Museum of Arkansas, Little Rock 50139
Historic Spanish Point, Osprey 51619
Historical Center for Southeast New Mexico, Roswell 52469
Historical Museum of Southern Florida, Miami 50625
Historical Society of Quaker Hill and Pawling, Pawling 51735
History Museum Historical Society of Western Virginia, Roanoke 52381
Homewood House Museum, Baltimore 46568
Horry County Museum, Conway 47784
House of Roses, Deadwood 48004
House of the Temple, Washington 53878
Howard County Center of African American Culture, Columbia 47713
Howard University Museum, Washington 53880
Huddleston Farmhouse Inn Museum, Cambridge City 47192
Huguenot Historical Society Museum, New Paltz 51096
Humboldt County Old Settler's Museum, Livermore 50145
Idaho State Historical Museum, Boise 46893
Ilwaco Heritage Museum, Ilwaco 49468
Indian City U.S.A., Anadarko 46280
Indian Mill Museum State Memorial, Upper Sandusky 53666
Indian Pueblo Cultural Center, Albuquerque 46178
Indian Temple Mound Museum, Fort Walton Beach 48696
Indiana State Museum, Indianapolis 49497
Indiana Territory Capitol, Vincennes 53748
International Museum of Women, San Francisco 52777
International Spy Museum, Washington 53883
Iraan Museum, Iraan 49526
Irish American Museum, Albany 46159
Iron Mission Museum, Cedar City 47290
Italian American Museum, New York 51198
J.M. Davis Arms and Historical Museum, Claremore 47578
James K. Polk Ancestral Home, Columbia 47735
James Whitcomb Riley Birthplace and Museum, Greenfield 49054
James Whitcomb Riley Museum, Indianapolis 49503
Jamestown Settlement, Williamsburg 54169
Japanese American National Museum, Los Angeles 50238
Japanese Cultural Center of Hawaii, Honolulu 49345
Jefferson County Historical Society Museum, Port Townsend 52043
Jefferson National Expansion Memorial, Saint Louis 52583
The Jewish Museum, New York 51200
Jim Crow Museum, Big Rapids 46797
John Adams and John Quincy Adams Birthplaces, Quincy 52218
John G. Neihardt Center, Bancroft 46592
Joseph Priestley House, Northumberland 51427
Joseph Smith Historic Center, Nauvoo 50991
Kalaupapa Historical Park, Kalaupapa 49668
Kansas Museum of History, Topeka 53536
Kaua'i Museum, Lihu'e 50083
The Kendall College Mitchell Museum of the American Indian, Evanston 48451
Kennesaw Civil War Museum, Kennesaw 49730
Kennesaw Mountain National Battlefield Park, Kennesaw 49731
Kentucky Historical Society, Frankfort 48737
Knights of Columbus Museum, New Haven 51052
Knox County Historical Society Museum, Edina 48289
Lane County Historical Museum, Eugene 48433
Laura Ingalls Wilder-Rose Wilder Lane Historic Home and Museum, Mansfield 50450
Le Sault de Sainte Marie Historical Sites, Sault Sainte Marie 52931
Lea County Cowboy Hall of Fame and Western Heritage Center, Hobbs 49304
Legends of New Mexico Museum, Algodones 46217
Levere Memorial Temple, Evanston 48452
Levine Museum of the New South, Charlotte 47390
Liberty Hall Museum, Union 53649
Liberty Memorial Museum of World War One, Kansas City 49693
Lincoln Homestead, Springfield 53223
Lincoln Memorial Shrine, Redlands 52274
The Lincoln Museum, Fort Wayne 48707
Lincoln Tomb, Springfield 53218
The Lithuanian Museum, Chicago 47486
Little Norway, Blue Mounds 46875
Locust Grove, Louisville 50289
Long Island Culture History Museum, Commack 47763
Longfellow House, Cambridge 47180
Longhorn Museum, Pleasanton 51989
Los Angeles Police Museum, Los Angeles 50247
Lost City Museum, Overton 51634
Louis E. May Museum, Fremont 48787
Louisiana Old State Capitol, Center for Political and Governmental History, Baton Rouge 46635
Louisiana State Museum, New Orleans 51087
Lower East Side Tenement Museum, New York 51206
Luzerne County Museum, Wilkes-Barre 54162
The Lyceum - Alexandria's History Museum, Alexandria 46211
Lyndon Baines Johnson Museum, Austin 46514
McCollum-Chidester House, Camden 47193
Macculloch Hall Historical Museum, Morristown 50853
McLarty Treasure Museum, Vero Beach 53727
Mahaffie Stagecoach Stop and Farm, Olathe 51530
Major John Bradford House, Kingston 49787
Mansfield State Historc Site, Mansfield 50449
Manship House Museum, Jackson 49558
Marais Des Cygnes Memorial Historic Site, Pleasanton 51988
Maramec Museum, Saint James 52552
Margaret Mitchell House and Museum, Atlanta 46442
The Mariners' Museum, Newport News 51334
Marjorie Kinnan Rawlings Historic State Park, Cross Creek 47890
Mark Twain Birthplace, Stoutsville 53323
Mark Twain Home and Museum, Hannibal 49159
Mark Twain House, Hartford 49195
Marland's Grand Home, Ponca City 52017
Marrett House, Standish 53251
Marshall County Historical Museum, Holly Springs 49321
Marshall County Historical Society Museum, Warren 53822
Martin Van Buren Historic Site, Kinderhook 49778
Masonic Grand Lodge Library and Museum, Waco 53777
Massachusetts Historical Society, Boston 46937
Massillon Museum, Massillon 50531
The Mattatuck Museum of the Mattatuck Historical Society, Waterbury 53956
Mattye Reed African Heritage Center, Greensboro 49066
Melrose Plantation Home Complex, Melrose 50565
Melvin B. Tolson Black Heritage Center, Langston 49937
Mennonite Heritage Museum, Goessel 48953
Mercer Museum of the Bucks County Historical Society, Doylestown 48180
Meriwether Lewis National Monument, Hohenwald 49309
Meux Home Museum, Fresno 48793
Michigan State University Museum, East Lansing 48254
Middleton Place Foundation, Charleston 47373
Milton House Museum, Milton 50703
Minnesota Historical Society Museum, Saint Paul 52623
Minnesota Pioneer Park Museum, Annandale 46313
Minnesota State Capitol Historic Site, Saint Paul 52625
Miracle of America Museum, Polson 52015
Mission San Carlos Borromeo del Rio Carmelo, Carmel 47251
Mississippi County Historical Society, Charleston 47355
Mississippi Governor's Mansion, Jackson 49559
Missouri Historical Society, Saint Louis 52587
Missouri State Museum, Jefferson City 49614
Missouri Town 1855, Lees Summit 50017
Monocacy National Battlefield, Frederick 48757
Monticello, Home of Thomas Jefferson, Charlottesville 47398
The Moody Mansion Museum, Galveston 48849
Moravian Historical Society Museum, Nazareth 50993
Mordecai Historic Park, Raleigh 52231
Morningside Nature Center, Gainesville 48815
Morristown National Historical Park, Morristown 50855
Mount Rushmore National Memorial, Keystone 49771
Mountain Heritage Center, Cullowhee 47903
Muscatine Art Center, Muscatine 50921
Museo de Las Americas, Denver 48083
El Museo Latino, Omaha 51553
Museum and Arts Center in the Sequim Dungeness Valley, Sequim 53037
Museum at Southwestern Michigan College, Dowagiac 48173
Museum of American Frontier Culture, Staunton 53277
Museum of American Historical Society of Germans from Russia, Lincoln 50100
Museum of Chinese in the Americas, New York 51226
Museum of Connecticut History, Hartford 49197
Museum of Discovery, Little Rock 50140
Museum of East Alabama, Opelika 51573
The Museum of East Texas, Lufkin 50318
Museum of Florida History, Tallahassee 53427
Museum of History and Industry, Seattle 53005
Museum of Man in the Sea, Panama City 51693
Museum of Nebraska History, Lincoln 50101
Museum of New Mexico, Santa Fe 52883
Museum of Prehistory and History, Russellville 52489
Museum of Rural Life, Denton 48059
Museum of Science and History of Jacksonville, Jacksonville 49584

Museum of Southern History, Jacksonville 49585
Museum of the Albemarle, Elizabeth City 48338
Museum of the American West, Los Angeles 50255
Museum of the Cape Fear Historical Complex, Fayetteville 48532
Museum of the Cherokee Indian, Cherokee 47423
Museum of the Fur Trade, Chadron 47327
Museum of the Mountain Man, Pinedale 51925
Napoleonic Society of America, Clearwater 47610
National Afro-American Museum and Cultural Center, Wilberforce 54158
National Border Patrol Museum and Memorial Library, El Paso 48327
National Center for the American Revolution, Valley Forge 53694
National Cowboy and Western Heritage Museum, Oklahoma City 51519
National Cowgirl Museum and Hall of Fame, Fort Worth 48718
National Czech and Slovak Museum, Cedar Rapids 47306
National Frontier Trails Center, Independence 49479
National Hall of Fame for Famous American Indians, Anadarko 46281
National Heritage Museum, Lexington 50061
National Museum of American History, Washington 53901
National Museum of American Jewish History, Philadelphia 51861
National Museum of Funeral History, Houston 49400
National Museum of Industrial History, Bethlehem 46782
National Museum of the American Indian, Washington 53905
National Museum of the American Indian, New York 51236
National Society of the Children of the American Revolution Museum, Washington 53909
National Trust for Historic Preservation, Washington 53910
National Yiddish Book Center, Amherst 46270
Native American Exhibit, Fonda 48584
Native American Heritage Museum at Highland Mission, Highland 49277
Native American Museum, Terre Haute 53481
Naumkeag House, Stockbridge 53301
Nevada Historical Society Museum, Reno 52288
Nevada State Fire Museum and Comstock Firemen's Museum, Virginia City 53761
Nevada State Museum, Carson City 47260
New Echota, Calhoun 47167
New Jersey Historical Society Museum, Newark 51286
New Jersey State Museum, Trenton 53561
New Mexico Farm and Ranch Heritage Museum, Las Cruces 49960
Newseum, Arlington 46358
Nickels-Sortwell House, Wiscasset 54264
Nobles County Historical Society, Worthington 54306
North Carolina Museum of History, Raleigh 52233
Northeast Louisiana Delta African American Heritage Museum, Monroe 50790
Northeastern Nevada Museum, Elko 48354
Northern Indiana Center for History, South Bend 53158
Norwich University Museum, Northfield 51423
Oakland Museum of California, Oakland 51470
Oaks House Museum, Jackson 49564
Ocean City Historical Museum, Ocean City 51487
O'Fallon Historical Museum, Baker 46533
Ohio Historical Society Museum, Columbus 47757
Oklahoma Museum of History, Oklahoma City 51524
Oklahoma Territorial Museum, Guthrie 49108
Oktibbeha County Heritage Museum, Starkville 53258
Old Capitol Museum of Mississippi History, Jackson 49565
Old Fort Garland, Fort Garland 48621
Old Fort Western, Augusta 46486
The Old Manse, Concord 47769
Old Spanish Fort Museum, Pascagoula 51725
Old State Capitol, Springfield 53219
Old State House, Boston 46945
Old State House, Hartford 49198
Old State House Museum, Little Rock 50141
Old World Wisconsin, Eagle 48230
Olde Colonial Courthouse, Barnstable 46613
Oldest House Museum Complex, Saint Augustine 52523
Oliver House Museum, Saint Martinville 52605
Olustee Battlefield, Olustee 51541
Oregon Coast History Center, Newport 51316
Oregon Trail Regional Museum, Baker City 46535
Our House State Memorial, Gallipolis 48840
Owens-Thomas House, Savannah 52944
Pacific University Museum, Forest Grove 48585
Palmer Museum, Jewell 49629
Pardee Home Museum, Oakland 51471
The Park-McCullough House, North Bennington 51383
Parker Tavern Museum, Reading 52260
Paul H. Karshner Memorial Museum, Puyallup 52207
Paul Revere House, Boston 46946
Pawnee Bill Ranch, Pawnee 51736
Paynes Creek Historic State Park, Bowling Green 46972
The Peace Museum, Chicago 47505
Pembina State Museum, Pembina 51753
Pennsylvania Anthracite Heritage Museum, Scranton 52983
Pennsylvania Lumber Museum, Galeton 48836
Peoria Historical Society, Peoria 51774
Perry's Victory and International Peace Memorial, Put in Bay 52202
The Petroleum Museum, Midland 50674
Pierce Manse, Concord 47776
Pioneer Arizona Living History Village, Phoenix 51907
Pioneer Florida Museum, Dade City 47919
Pioneer Heritage Center, Shreveport 53086
Pioneer Museum, Fairbanks 48474
Pioneer Museum, Watford City 53980
Pioneer Woman Statue and Museum, Ponca City 52018
Pioneers, Trail and Texas Rangers Memorial Museum, San Antonio 52719
Plum Grove Historic Home, Iowa City 49519
Plymoth Plantation Museum, Plymouth 51996
Polish American Museum, Port Washington 52048
Polish Museum of America, Chicago 47507
Port Discovery Children's Museum, Baltimore 46584
Porter-Phelps-Huntington House Museum, Hadley 49118
Portholes Into the Past, Medina 50558
Postville Courthouse Museum, Lincoln 50087
Pratt Building, Cohasset 47677
President Andrew Johnson Museum, Greeneville 49052
President Benjamin Harrison Home, Indianapolis 49506
The Presidential Museum, Odessa 51497
El Pueblo de Los Angeles, Los Angeles 50262
Pullman Museum, Chicago 47509
Putnam Museum of History and Natural Science, Davenport 47970
Railroad and Heritage Museum, Temple 53474
Red River Valley Museum, Vernon 53723
Rensselaer Russell House Museum, Waterloo 53962
The Rice Museum, Crowley 47894
The Rice Museum, Georgetown 48895
Roanoke Island Festival Park, Manteo 50453
Robert Louis Stevenson House, Monterey 50807
Rochester Museum, Rochester 52398
Rockingham, Princeton 52153
Ronald Reagan Presidential Library and Museum, Simi Valley 53110
Rosalie House Museum, Natchez 50985
Roscoe Village Foundation, Coshocton 47839
Rose Hill Manor Park Children's Museum, Frederick 48759
Rose Hill Plantation State Historic Site, Union 53651
Rowan Museum, Salisbury 52681
Royall House, Medford 50547
Ruggles House, Columbia Falls 47737
Rumford Historical Museum, North Woburn 51411
Rutherford B. Hayes Presidential Center, Fremont 48788
Safe Haven Museum Education Center, Oswego 51626
Sag Harbor Whaling and Historical Museum, Sag Harbor 52514
Sagamore Hill National Historic Site, Oyster Bay 51663
Saint Augustine Lighthouse and Museum, Saint Augustine 52526
Saint Clements Island-Potomac River Museum, Colton Point 47711
Salem Witch Museum, Salem 52660
Salter Museum, Argonia 46346
Sam Houston Memorial Museum, Huntsville 49446
San Jacinto Museum of History, La Porte 49855
San Marcos de Apalache Historic State Park, Saint Marks 52603
Santa Fe Trail Museum, Trinidad 53569
Schomburg Center for Research in Black Culture, New York 51254
Seminole Nation Museum, Wewoka 54114
Senate House, Kingston 49791
Shaker Heritage Society Museum, Albany 46163
Shaker Museum, New Gloucester 51046
Shaker Museum at South Union, South Union 53183
Shawneetown Historic Site, Old Shawneetown 51537
Sheldon Museum, Haines 49125
Sibley House, Mendota 50588
Skirball Cultural Center, Los Angeles 50264
Skowhegan History House, Skowhegan 53129
Smith Robertson Museum, Jackson 49566
Smith-Zimmermann Museum, Madison 50387
Smoky Hill Museum, Salina 52675
Snowden House, Waterloo 53964
The Society of the Cincinnati Museum, Washington 53918
Sons of the American Revolution Museum, Louisville 50296
Sotterley Plantation Museum, Hollywood 49323
South Carolina Confederate Relic Room and Museum, Columbia 47731
South Carolina Historical Society Museum, Charleston 47376
South County Museum, Narragansett 50955
South Dakota Hall of Fame, Chamberlain 47330
Southern Ute Museum and Cultural Center, Ignacio 49466
Southwest Virginia Museum, Big Stone Gap 46800
Spirit of '76 Museum, Wellington 54028
Spurlock Museum, Urbana 53669
Star of the Republic Museum, Washington 53947
Star-Spangled Banner Flag House and 1812 Museum, Baltimore 46588
State Museum of Pennsylvania, Harrisburg 49181
Statue of Liberty National Monument and Ellis Island Immigration Museum, New York 51262
Steamboat Bertrand Museum, Missouri Valley 50758
Stephen A. Douglas Tomb, Chicago 47515
Stephens Museum, Fayette 48526
Stevens County Gas and Historical Museum, Hugoton 49416
Stewart M. Lord Memorial Museum, Burlington 47137
Stockbridge Library Historical Collection, Stockbridge 53303
Storrowton Village Museum, West Springfield 54083
Strecker Museum's Collection, Waco 53779
Sunwatch Indian Village - Archaeological Park, Dayton 47989
Surratt House Museum, Clinton 47653
Swindler's Ridge Museum, Benton 46742
Tate House, Portland 52068
Tennessee Historical Society Museum, Nashville 50975
Tennessee State Museum, Nashville 50976
Texas City Museum, Texas City 53490
Texas Memorial Museum, Austin 46519
Texas Ranger Hall of Fame and Museum, Waco 53780
Theodore Roosevelt Birthplace, New York 51269
Theodore Roosevelt Inaugural National Historic Site, Buffalo 47124
Thomas Edison Birthplace Museum, Milan 50681
Thomas P. Kennard House, Lincoln 50106
Thousand Islands Museum of Clayton, Clayton 47602
Tibet House Museum, New York 51270
Tippecanoe Battlefield, Battle Ground 46647
Town of Yorktown Museum, Yorktown Heights 54349
Trail End State Historic Museum, Sheridan 53074
Tryon Palace, New Bern 51020
Tubman African-American Museum, Macon 50370
Turtle Mountain Chippewa Heritage Center, Belcourt 46694
Ukrainian-American Museum, Hamtramck 49155
Ukrainian Museum, Cleveland 47643
Uncle Remus Museum, Eatonton 48273
United States Army Museum of Hawaii, Fort DeRussy 48613
United States Marine Corps Museum, Washington 53926
Utah State Historical Society, Salt Lake City 52697
Valley Forge National Historical Park, Valley Forge 53695
Vandalia State House, Vandalia 53709
Virginia Historical Museum, Richmond 52342
The Walker African American Museum, Las Vegas 49980
Walter F. George Law Office Museum, Vienna 53744
Washington Crossing Historic Park, Washington Crossing 53949
Weir Farm, Wilton 54213
Wells Fargo History Museum, Los Angeles 50269
Wentworth-Coolidge Mansion, Portsmouth 52101
Western Archaeological and Conservation Center, Tucson 53602
Western Hotel Museum, Lancaster 49916
Wickersham House Museum, Fairbanks 48475
Wilber Czech Museum, Wilber 54157
Will Rogers Memorial Museums, Claremore 47579
William Hammond Mathers Museum, Bloomington 46867
William Howard Taft National Historic Site, Cincinnati 47567
William M. Colmer Visitor Center, Ocean Springs 51490
William McKinley Presidential Library and Museum, Canton 47229
Windham Textile and History Museum, Willimantic 54179
Wing Luke Asian Museum, Seattle 53017
Winnebago Area Museum, Winnebago 54235
Winnie Davis Museum of History, Gaffney 48812
Winona County Historical Museum, Winona 54239
Winston Churchill Memorial and Library in the United States, Fulton 48810
Winterthur Museum, Winterthur 54258
Witte Museum, San Antonio 52731
Women's History Museum, West Liberty 54069
Workman and Temple Family Homestead Museum, City of Industry 47569
The Works, Newark 51295
Wright Museum, Wolfeboro 54270
Wyoming State Museum, Cheyenne 47448
Yazoo Historical Museum, Yazoo City 54328
Yorba and Slaughter Families Adobe Museum, Chino 47540
Yorktown Battlefield Visitor Center Museum, Yorktown 54347
Yorktown Colonial Park, Yorktown 54348
Yuma Territorial Prison Museum, Yuma 54366

Uruguay

Casa de Garibaldi, Montevideo 54382
Casa de Lavalleja, Montevideo 54384
Casa Quinta de Luis Alberto de Herrera, Montevideo 54388
Museo Histórico Nacional - Casa Rivera, Montevideo 54404

Uzbekistan

Akmal Ikramov nomidagi Tarikhi Kŭltŭri va San"atshunoslik Ŏzbekiston Khalqlari Davlat Mŭzeyi, Samarqand 54460
Andijon Viloyat Adabiyot va San'at Muzeyi, Andijon 54438
History Museum of the People of Uzbekistan, Toshkent 54471
M.T. Oibek nomidagi Ŏzbekiston Khalqlari Tarikhi Muzyi, Toshkent 54472
Muzej Pamjati Žertv Repressij, Toshkent 54479
Ŏzbekiston Tarikhi Davlat Muzeyi, Toshkent 54485
Qo'qon Viloyat Ŭlkashunoslik Muzeyi, Qo'qon 54458
Qoraqalpoğiston Tarikhi Muzeyi, Nukus 54452
Sergej Borodin Muzeyi, Toshkent 54486
Tasviriy San'at Muzeyi - Andijon Qal'asi, Andijon 54442
Termuridi Tarikhi Muzeyi, Toshkent 54489
Yoldosh Oxunboboev Yodgorlik Muzeyi, Marg'ilon 54449

Venezuela

Casa Natal de José María España, La Guaira 54578
Casa Natal del Libertador Simón Bolívar, Caracas 54526
Museo Bolívariano, Caracas 54537
Museo Cristóbal Mendoza, Trujillo 54595
Museo de Historia y Artesanía de la Colonial Tovar, Colonia Tovar 54565
Museo Gran Mariscal de Ayacucho, Cumaná 54568
Museo José Antonio Páez, Acarigua 54515

Vietnam

Chu Dong Tu Temple, Binh Minh 54600
Vien Bao Tang Lich Sa Viet Nam, Ha Noi 54616
Viet Nam History Museum Ho Chi Minh City, Ho Chi Minh City 54630

Zambia

Lusaka National Museum, Lusaka 54658
Moto Moto Museum, Mbala 54661

Zimbabwe

Queen Victoria Museum, Harare 54671

History, Early → Prehistory and Early History

History, Local and Regional

Afghanistan

Ghazni Provincial Museum, Ghazni 00001
Herat Provincial Museum, Herat 00003
Kandahar Provincial Museum, Kandahar 00006

Albania

Muzeu Gjergj Kastrioti Skënderbeu, Krujë 00026
Muzeu Historik i Vlorës, Vlorë 00035
Muzeu Historik Kombëtar, Tiranë 00032
Muzeu Historik Shkodër, Shkodër 00028

Algeria

Musée Archéologique du Théâtre de Guelma, Guelma 00065
Musée de Béjaïa, Béjaïa 00052
Musée de Temouchent, Ain Temouchent 00037
Musée de Tlemcen, Tlemcen 00083
Musée d'El-Oued, El-Oued 00062
Musée Marsa El-Kharez, El-Kala 00060

Musée National Ahmed Zabana - Demaeght Museum, Oran 00071
Musée National de Préhistoire et d'Ethnographie du Bardo, Alger 00044
Musée Saharien de Ouargla, Ouargla 00072

American Samoa

Jean P. Haydon Museum, Pago Pago 00084

Andorra

Casa de la Vall, Andorra la Vella 00085

Angola

Museu Regional da Huila, Lumbango 00114

Argentina

Cartref Taid, Trevelin 00654
Casa de la Cultura, Curuzú Cuatiá 00351
Centro de Exposiciones La Casona de los Olivera, Buenos Aires 00153
Complejo Histórico Chivilcoy, Chivilcoy 00303
Complejo Museológico La Estación, Cruz Alta 00350
Exposición de Gregorio de Laferrere, Gregorio de Laferrere 00380
Fundación Archivo Gráfico y Museo Histórico de la Ciudad de San Francisco y la Región, San Francisco 00576
Museo Activo del Pago de los Arroyos, Rosario 00542
Museo Capitán Juan de Zevallos, Valle Hermoso 00665
Museo Casa Taller Celia Chevalier, Buenos Aires 00175
Museo Clarisse Coulombie de Goyaud, Ituzaingó 00391
Museo Colonial e Histórico, Luján 00433
Museo Comunal de Chañar Ladeado, Chañar Ladeado 00299
Museo Comunal Pedro Bargero, Emilio Bunge 00364
Museo Comunal Regional de San Guillermo, San Guillermo 00579
Museo Coyug-Curá, Pigüé 00492
Museo de Antropología e Historia Natural Los Desmochados, Casilda 00296
Museo de Arte e História de Los Toldos, Los Toldos 00431
Museo de Chilecito, Chilecito 00302
Museo de Ciencias Naturales e Historia, Posadas 00498
Museo de Firmat, Firmat 00372
Museo de Historia Reginal San Lorenzo y Galeria del Libertador General San Martín, San Lorenzo 00590
Museo de Historia Regional de Tristán Suárez, Trisán Suárez 00657
Museo de la Casa Histórica de la Independencia Nacional, San Miguel de Tucumán 00597
Museo de la Ciudad, Paraná 00485
Museo de la Ciudad, Salta 00557
Museo de la Ciudad, Victoria 00668
Museo de la Ciudad de Colón, Colón 00308
Museo de la Ciudad de San Francisco, San Francisco 00577
Museo de la Colonización, Esperanza 00367
Museo de la Patagonia Dr. Francisco P. Moreno, Bariloche 00144
Museo de La Plata, La Plata 00413
Museo de los Asentamientos, Federación 00371
Museo de Mitos y Leyendas, Tafí del Valle 00639
Museo del Este, Monte Caseros 00471
Museo del Fin del Mundo, Ushuaia 00660
Museo del Hombre y su Entorno, Caleta Olivia 00287
Museo del Periodismo Bonaerense, Capilla del Señor 00290
Museo del Pueblo, Pirané 00494
Museo del Puerto de Ingeniero White, Ingeniero White 00388
Museo El Reencuentro, Darregueira 00352
Museo General de Arequito, Arequito 00126
Museo Germán Guglielmetti, Benito Juárez 00147
Museo Histórico, Arqueológico y de Arte Pedro Balduín, San Pedro de Jujuy 00605
Museo Histórico, Colonial y de Bellas Artes y Museo de Ciencias Naturales Regional Mesopotámico, Mercedes 00461
Museo Histórico Comunal, Villa Trinidad 00685
Museo Histórico Comunal de San Genaro Norte, San Genaro Norte 00578
Museo Histórico de Arenaza, Arenaza 00125
Museo Histórico de Arrecifes, Arrecifes 00129
Museo Histórico de Cañuelas, Cañuelas 00289
Museo Histórico de Entre Ríos Martiniano Leguizamón, Paraná 00486
Museo Histórico de Guatraché, Guatraché 00385
Museo Histórico de la Colonia San Carlos, San Carlos Centro 00569
Museo Histórico de la Provincia Dr. Orestes di Lullo, Santiago del Estero 00629
Museo Histórico de Laborde, Laborde 00421
Museo Histórico de Navarro, Navarro 00476
Museo Histórico de Ranchos, Ranchos 00517
Museo Histórico General Julio de Vedia, 9 de Julio 00115
Museo Histórico "Juan Bautista Ambrosetti", Gualeguay 00381
Museo Histórico La Gallareta Forestal, La Gallareta 00403
Museo Histórico Manuel A. Moreira, Laboulaye 00422
Museo Historico Municipal, Bahía Blanca 00135
Museo Histórico Municipal Alfredo E. Múlgura, General Belgrano 00377
Museo Histórico Municipal Andrés A. Roverano, Santo Tomé 00633
Museo Histórico Municipal de Armstrong, Armstrong 00128
Museo Histórico Municipal de Cañada de Gómez, Cañada de Gómez 00288
Museo Histórico Municipal de Lincoln, Lincoln 00428
Museo Histórico Municipal de Luque, Luque 00439
Museo Histórico Municipal de Pergamino, Pergamino 00490
Museo Histórico Municipal de Villa Gesell, Villa Gesell 00680
Museo Histórico Municipal General Levalle, General Levalle 00378
Museo Histórico Municipal Hércules J. Rabagliati, Ramallo 00516
Museo Histórico Municipal Juan Lavalle, Baradero 00142
Museo Histórico Municipal La Para, La Para 00404
Museo Histórico Municipal Monte Grande, Monte Grande 00472
Museo Histórico Municipal Víctor E. Miguez, Mercedes 00462
Museo Histórico Municipal Villa Clara, Villa Clara 00675
Museo Histórico Municipal Villa del Rosario, Villa del Rosario 00678
Museo Histórico Municipal y Colección de Arte Precolombino Arminio Weiss, Rafaela 00514
Museo Histórico Provincial Agustín V. Gnecco, San Juan 00588
Museo Histórico Provincial de Rosario Dr. Julio Marc, Rosario 00545
Museo Histórico Provincial Marqués de Sobremonte, Córdoba 00330
Museo Histórico Regional, Alpachiri 00119
Museo Histórico Regional de Ayacucho, Ayacucho 00131
Museo Histórico Regional de Gaiman, Gaiman00375
Museo Histórico Regional de la Colonia San José, San José 00583
Museo Histórico Regional De la Isla, Isla del Cerrito 00390
Museo Histórico Regional de Río Cuarto, Río Cuarto 00528
Museo Histórico Regional de Urdinarrain, Urdinarrain 00659
Museo Histórico Regional de Villa La Angostura, Villa La Angostura 00681
Museo Histórico Regional Gabriel Campomar Cervera, Salliqueló 00553
Museo Histórico Regional Ichoalay, Resistencia 00525
Museo Histórico Regional Municipal de Magdalena, Magdalena 00440
Museo Histórico Regional Pablo Argilaga, Santo Tomé 00634
Museo Histórico Regional Padre Francisco Cremasco, Villa Concepción del Tío 00676
Museo Histórico San Ignacio de Loyola, San José 00584
Museo Histórico y Natural de Lavalle, Lavalle Villa Tulumaya 00427
Museo Histórico y Regional Pico Truncado, Pico Truncado 00491
Museo Histórico y Tradicional de Barracas al Sud, Avellaneda 00130
Museo Irureta, Tilcara 00648
Museo José Manuel Estrada, Suipacha 00637
Museo López Claro, Santa Fé 00618
Museo Municipal Carmen Funes, Plaza Huincul 00495
Museo Municipal de El Trébol, El Trébol 00360
Museo Municipal de Eldorado, Eldorado 00362
Museo Municipal de la Ciudad de Carcarañá, Carcarañá 00291
Museo Municipal de la Ciudad de Rosario, Rosario 00548
Museo Municipal de Porteña, Porteña 00496
Museo Municipal Dr. Rodolfo Doval Fermi, Sastre 00635
Museo Municipal Histórico Regional Santiago Lischetti, Villa Constitución 00677
Museo Municipal Ignacio Balvidares, Puan 00503
Museo Municipal José A. Mulazzi, Tres Arroyos 00653
Museo Municipal José Manuel Maciel, Coronda 00341
Museo Municipal Punta Hermengo, Miramar 00469
Museo Municpal Dr. Santos F. Tosticarelli, Casilda 00297
Museo Oncativo, Oncativo 00482
Museo Orlando Binaghi, Moreno 00474
Museo Paleontológico, Arqueológico e Histórico de la Ciudad de Deán Funes, Deán Funes 00353
Museo Particular Los Abuelos, Franck 00373
Museo Patria Chica de Mohamed Diaz, Realicó 00520
Museo Polifacético Regional Vottero, Laguna Larga 00423
Museo Pringles de Arqueología e Historia, Coronel Pringles 00345
Museo Provincial Dora Ochoa de Masramón, San Luis 00592
Museo Pueblo de Luis, Trelew 00652
Museo Regional Castelli, Castelli 00298
Museo Regional Cayetano Alberto Silva, Venado Tuerto 00666
Museo Regional de Cinco Saltos, Cinco Saltos 00307
Museo Regional de Claromecó, Balneario Claromecó 00139
Museo Regional de la Ciudad de La Paz, La Paz 00406
Museo Regional de Pigüé, Pigüé 00493
Museo Regional de Sacanta, Sacanta 00552
Museo Regional Dr. Federico Escalada, Río Mayo 00537
Museo Regional Las Varillas, Las Varillas 00426
Museo Regional Malargüe, Malargüe 00442
Museo Regional Maracó, General Pico 00379
Museo Regional Municipal de El Calafate, El Calafate 00359
Museo Regional Particular Epifanio Saravia, Santa Catalina 00611
Museo Regional Patagónico, Comodoro Rivadavia 00312
Museo Regional Provincial Padre Manuel Jesús Molina, Río Gallegos 00534
Museo Regional Rincón de Vida, Villa Mirasol 00683
Museo Regional Rumi Mayu (Río de Piedra), Sanagasta 00610
Museo Regional Salesiano de Rawson, Rawson 00519
Museo Regional Trevelin, Trevelin 00656
Museo Regional Valcheta, Valcheta 00664
Museo Regional y de Arte Marino, Puerto San Julián 00508
Museo Rocsen, Nono 00478
Museo San Antonio de Padua, Córdoba 00338
Museo Ucraniano, Apóstoles 00124
Museo Usina Vieja, San Antonio de Areco 00568
Museo y Monumento Histórico Las Bóvedas, Uspallata 00663
Museo y Parque Libres del Sur, Dolores 00356
Rincón de Historia, Salto 00563
Unidad Museológica Municipal de La Banda, La Banda 00396

Armenia

Gavar Local Tradition Museum, Gavar 00701
Goris Local Tradition Museum, Goris 00703
Kapan Local Tradition Museum, Kapan 00712
Lori Local Tradition Museum, Vanadzor 00719
Museum of National Architecture and Urban Life of Gyumri, Gyumri 00707
Regional Museum of Yeghegnadzor, Yeghegnadzor 00722
Sisian Museum of History N. Adonts, Sisian 00717
Yerevan History Museum, Yerevan 00761

Australia

Addington - Ryde House of Heritage, Ryde 01475
Adelaide House Museum, Alice Springs 00796
Albury Regional Museum, Albury 00793
Allora Historical Museum, Allora 00803
Allwood House, Hurstbridge 01168
Artspace Mackay, Mackay 01246
Augusta Historical Museum, Augusta 00819
Australian Freethought Heritage Library, Balwyn 00831
Ballan Shire Historical Museum, Ballan 00824
Bank of Victoria Museum, Yackandandah 01674
Barcaldine and District Historical Museum, Barcaldine 00833
Bathurst Museum, Bathurst 00837
Beaudesert Museum, Beaudesert 00845
Beenleigh and District Museum, Beenleigh 00852
Benalla Costume and Pioneer Museum, Benalla 00858
Benga Oral Historic Centre, Dandenong 01022
Berkshire Mill Museum, Moora 01316
Berrima Museum, Berrima 00863
Berry Museum, Berry 00864
Berwick Pakenham Historical Museum, Pakenham 01401
Bicentennial Historical Museum, Cunnamulla 01017
Biggenden Museum, Biggenden 00866
Bingara Museum, Bingara 00868
Birchip Local History Museum, Birchip 00869
Birdsville Working Museum, Birdsville 00870
Bland District Historical Museum, West Wyalong 01644
Bond Store Museum, Maryborough, Queensland 01266
Boondooma Homestead - Museum and Heritage Complex, Mundubbera 01335
Boulia Stone House Museum, Boulia 00881
Boyne Valley Historical Cottage, Ubobo 01608
Braidwood Museum, Braidwood 00889
Brewarrina Aboriginal Cultural Museum, Brewarrina 00890
Bridgetown Old Gaol Museum, Bridgetown 00893
Broome Museum, Broome 00909
Broomehill Museum, Broomehill 00910
Brunswick Valley Museum, Mullumbimby 01334
Buderim Pioneer Cottage, Buderim 00913
Bundaberg Historical Museum, Bundaberg 00918
Burnett House and Myilly Point Heritage Precinct, Darwin 01029
Byfield and District Museum, Byfield 00929
C.L. Alexander Museum, Tumby Bay 01604
Cairns Museum, Cairns 00933
Calala Cottage, Tamworth 01568
Camden Haven Museum, Laurieton 01225
Camden Museum, Camden 00938
Camperdown and District Museum, Camperdown 00941
Canberra Museum and Gallery, Canberra 00944
Captain's Cottage Museum, Murray Bridge 01336
Carbethon Folk Museum and Pioneer Village, Crows Nest, Queensland 01013
Carcoar Historic Village, Carcoar 00954
Cardwell Shire Museum, Tully 01603
Carss Cottage Museum, Blakehurst 00874
Charleville Historic House Museum, Charlesville 00963
Chillagoe Historical Centre, Chillagoe 00969
Chiltern Athenaeum, Chiltern 00970
Chinchilla and District Historical Society Museum, Chinchilla 00973
City of Belmont Museum, Belmont 00856
City of Gosnells Museum, Gosnells 01124
City of Unley Museum, Unley 01612
Clare National Trust Museum, Clare 00974
Claremont Museum, Claremont, Western Australia 00976
Clarencetown and District Historical Museum, Clarencetown 00978
Cloncurry and District Museum, Cloncurry 00986
Clunes Museum, Clunes 00987
Coal Creek Heritage Village, Korumburra 01206
Coalfields Museum, Collie 00993
Cohuna Museum, Cohuna 00990
Coleraine Local History Museum, Coleraine 00992
Concord Heritage Museum, Concord 00995
Corrigin Pioneer Museum, Corrigin 01003
Court House Museum, Stroud 01534
Cowra Museums, Cowra 01006
Creswick Museum and Gold Battery, Creswick 01011
Crystal Brook Heritage Centre, Crystal Brook 01015
Dance Cottage Museum, Ravensthorpe 01453
Dartmoor District Coach House Museum, Dartmoor 01027
Denmark Historical Museum, Denmark 01033
Dr. Arratta Memorial Museum, Muttaburra 01343
Dromana and District Museum, Dromana 01043
Dungog Historical Museum, Dungog 01048
Dunolly Museum, Dunolly 01049
East Gippsland Museum, Bairnsdale 00821
Echuca Museum, Echuca 01057
Edithburgh Museum, Edithburgh 01062
Eidsvold Historical Society Museum, Eidsvold 01063
El Dorado Museum, Eldorado 01064
Elizabeth Bay House, Elizabeth Bay 01065
Elmore Progress Association Station Museum, Elmore 01066
Emu Park Historical Museum, Emu Park 01072
Endeavour Museum, Wollombi 01661
Eskbank House, Lithgow 01234
Esperance Municipal Museum, Esperance 01074
Eumundi Museum, Eumundi 01075
Fairfield City Museum and Gallery, Smithfield 01508
Farm Shed Museum, Kadina 01182
Farmer's Arms Hotel Museum, Euroa 01076
Forbes Museum, Forbes 01085
Foster and District Historical Society Museum, Foster 01089
Fremantle History Museum, Fremantle 01093
Furneaux Museum, Flinders Island 01082
Gascoyne Junction Museum, Gascoyne Junction 01098
Gayndah Museum, Gayndah 01101
Geelong Heritage Centre, Geelong 01103
Gerringong and District Museum, Gerringong 01108
Gilgandra Museum, Gilgandra 01109
Gladstone Regional Art Gallery and Museum, Gladstone, Queensland 01111
Glen Innes Museum, Glen Innes 01114
Glenreagh Memorial Museum, Glenreagh 01116
Gloucester Historic Museum, Gloucester 01117

Gold Museum, Ballarat 00827
Gold Treasury Museum, Melbourne 01281
Goolwa National Trust Museum, Goolwa 01118
Great Cobar Heritage Centre, Cobar 00988
Great Lakes Historical Museum, Tuncurry 01607
Grenfell and District History Museum, Grenfell 01130
Gumeracha and District History Centre, Gumeracha 01134
Gundagai Historical Museum, Gundagai 01135
Gympie and District Historical and Gold Mining Museum, Gympie 01138
Hamilton Heritage Museum, Hamilton, Tasmania 01143
Hamilton Hume Museum, Yass 01682
Hastings-Western Port Historical Museum, Hastings 01147
Hawker Museum, Hawker, South Australia 01148
Hawkesbury Regional Museum, Windsor, New South Wales 01656
Headland Historical Museum, Nambucca Heads 01345
Henry Kendall Cottage and Museum, Gosford 01123
Heritage Hill Museum, Dandenong 01023
Hervey Bay Historical Village and Museum, Scarness 01495
Historical Society Museum of Saint Kilda, Saint Kilda, Victoria 01478
Hobart Heritage Museum, Hobart 01153
Homestead Park Pioneer Museum, Port Augusta 01428
Hope Cottage, Kingscote 01202
Horsham Museum, Horsham 01162
Hub Local History Collection, Aberfoyle Park 00764
Hunters Hill Historical Museum, Hunters Hill 01165
Huntly and Districts Historical Museum, Huntly 01166
Huon Valley Apple and Heritage Museum, Huonville 01167
Ilfracombe Machinery and Heritage Museum, Ilfracombe 01171
Illawarra Museum, Wollongong 01662
Inglewood District Historical Museum, Inglewood 01173
Innisfail and District Historical Museum, Innisfail 01174
Isis District Historical Complex, Childers 00968
James Cook Museum, Cooktown 00998
Jamestown Museum, Jamestown 01177
Junee Historical Museum, Junee 01180
Kandos Bicentennial Industrial Museum, Kandos 01186
Kapunda Museum, Kapunda 01188
Katamatite Museum, Katamatite 01189
Katanning Historical Museum, Katanning 01190
Katherine Museum, Katherine 01191
Keith Museum, Keith 01192
Kempsey Museum, Kempsey 01193
Kew Historical Museum, Kew 01195
Kilkivan Historical Museum, Kilkivan 01197
Kimba and Gawler Ranges Historical Society Museum, Kimba 01198
King Island Historical Museum, Currie 01019
Kingston Pioneer Museum, Kingston 01204
Koorda Museum, Koorda 01205
Kyneton Museum, Kyneton 01210
Laidley District Historical Village, Laidley 01211
Lake Tabourie Museum, Lake Tabourie 01213
Lambing Flat Museum, Young 01686
Lancefield Courthouse Museum, Lancefield 01215
Landsborough Shire Historical Museum, Landsborough 01216
Langi Morgala Museum, Ararat 00810
Laura and Charles Ferguson Museum, Mentone 01293
Learmonth Museum, Learmonth 01226
Lighthouse Keepers Cottage, Carnarvon 00959
Linton and District Museum, Linton 01231
Local Studies Collection @ Library, Marrickville 01265
Longreach Power House Museum, Longreach 01239
Lord Howe Island Museum, Lord Howe Island 01241
Loxton Historical Village, Loxton 01244
McCrossin's Mill Museum, Uralla 01614
Maldon Museum, Maldon 01252
Mallala and Districts Historical Museum, Mallala 01253
Mandurah Community Museum, Mandurah 01255
Manilla Historical Royce Cottage Museum, Manilla 01256
Mareeba District Rodeo Museum, Mareeba 01262
Mareeba Heritage Museum, Mareeba 01263
May and Herbert Gibbs Art Collection, South Perth 01516
Melbourne's Living Museum of the West, Maribyrnong 01264
Melville Discovery Centre, Applecross 00807
Merredin Museum, Merredin 01297
Merriwa Colonial Museum, Merriwa 01298
Mid-Richmond Historical Museum, Coraki 01000
Miles and District Historical Village Museum, Miles 01302
Millicent National Trust Museum, Millicent 01305
Minlaton National Trust Museum, Minlaton 01307
Moe Historical Museum, Moe 01310
Molong Historical Museum, Molong 01311
Monarch Historical Museum, Williamtown 01654
Mont De Lancey Historical Museum, Wandin 01626
Moonta Mines Museum, Moonta 01314
Moonta National Trust Museum, Moonta 01315
Mooroopna Hospital and District Museum, Mooroopna 01317
Morawa Old Police Station Museum, Morawa 01318
Mordialloc Historical Society Museum, Mentone 01294
Morven Historical Museum, Morven 01324
Mount Morgan Historical Museum, Mount Morgan 01329
Mount Victoria and District Historical Museum, Mount Victoria 01330
Mudgee Colonial Inn Museum, Mudgee 01331
Murrumburrah-Harden Historical Museum, Murrumburrah 01337
Murrurundi Museum, Murrurundi 01338
Murtoa Water Tower Museum, Murtoa 01339
Museum of Wonders, Imbil 01172
Nandewar Historical Museum, Barraba 00834
Narembeen Historical Museum, Narembeen 01348
Narrabri Old Gaol Museum, Narrabri 01350
Narrogin Old Courthouse Museum, Narrogin 01354
Narryna Heritage Museum, Battery Point 00841
National Trust Museum, Streaky Bay 01533
Nepean Historical Museum, Sorrento 01509
New Norcia Museum and Art Gallery, New Norcia 01360
Newcastle Regional Museum, Newcastle 01365
Nimbin Museum, Nimbin 01369
Noosa Shire Museum, Pomona 01421
Nor West Bend Museum, Morgan 01320
North Stradbroke Island Historical Museum, Dunwich 01050
Northcliffe Pioneer Museum, Northcliffe 01378
Nowra Museum, Nowra 01384
Nundah and Districts Historical Museum, Nundah 01385
Oakleigh and District Museum, Oakleigh 01389
Old Cable Station Museum, Apollo Bay 00806
Old Council Chambers Museum, Norton Summit 01380
Old Court House Museum, Drysdale 01044
Old Highercombe Hotel Museum, Tea Tree Gully 01575
Old Police Station Museum, Mount Barker 01326
Old Police Station Museum Brookton, Brookton 00908
Old School Museum, Merimbula 01295
Old Timers Traeger Museum, Alice Springs 00801
Ongerup and Needilup District Museum, Ongerup 01392
Ouyen Local History Resource Centre, Ouyen 01397
Paterson Court House, Adamstown 00769
Patrick Taylor Cottage Folk Museum, Albany 00788
Pilots Cottage Museum, Kiama 01196
Pine Creek Museum, Pine Creek 01414
Pinnaroo Heritage Museum, Pinnaroo 01417
Pioneer Village Museum, Burnie 00924
Pittsworth Pioneer Historical Museum, Pittsworth 01419
Port Fairy History Centre, Port Fairy 01430
Port MacDonnell and District Maritime Museum, Port MacDonnell 01433
Port Macquarie Historic Museum, Port Macquarie 01434
Port Phillip City Collection, Saint Kilda, Victoria 01481
Pringle Cottage Museum, Warwick 01635
Proserpine Historical Museum, Proserpine 01443
Prospect Hill Historical Museum, Meadows 01272
Pyramid Hill and District Historical Museum, Pyramid Hill 01444
Queanbeyan and District Historical Museum, Queanbeyan 01446
Queenscliffe Historical Museum, Queenscliff 01448
Quirindi and District Historical Cottage and Museum, Quirindi 01452
Railway Station Museum, Wonthaggi 01666
Redcliffe Museum, Redcliffe 01454
Redhill Museum, Redhill 01455
Redland Museum, Cleveland 00984
Residency Museum, York 01684
Richmond River Historical Museum, Lismore 01233
Ringwood Miners Cottage, Ringwood 01460
Rippon Lea Estate, Elsternwick 01068
Robert O'Hara Burke Memorial Museum, Beechworth 00850
Rochester Historical and Pioneer Museum, Rochester 01463
Rockhampton and District Historical Museum, Rockhampton 01465
Roebourne Old Goal Museum, Roebourne 01467
Rosalie Shire Historical Museum, Goombungee 01119
Rosewood Scrub Historical Museum, Marburg 01261
Rottnest Island Museum, Rottnest Island 01471
Rushworth Museum, Rushworth 01474
Saint Georges Regional Museum, Hurstville 01169
Saint Helens History Room, Saint Helens 01476
Sale Historical Museum, Sale 01489
Salisbury Folk Museum, Salisbury 01490
Samford and District Historical Museum, Samford 01491
Sandgate and District Museum, Sandgate 01493
Schaeffer House Museum, Grafton 01128
Schwerkolt Cottage Museum, Mitcham 01308
Scone and Upper Hunter Museum, Scone 01498
Scout Heritage Centre of Western Australia, West Perth 01643
Serviceton Historic Railway Station, Serviceton 01501
Settlers Museum, Brewarrina 00891
Sexton's Cottage Museum, Crows Nest, New South Wales 01012
Seymour and District Historical Museum, Seymour 01502
Singleton Historical Museum, Singleton 01507
Sir Edgeworth David Memorial Museum, Kurri Kurri 01207
Springvale and District Historical Museum, Springvale 01525
Stansbury Museum, Stansbury 01526
Stawell Historical Society Museum, Stawell 01529
Steiglitz Court House, Steiglitz 01530
Stirk Cottage, Kalamunda 01184
Stonnington Local History Collection - Malvern, Malvern, Victoria 01254
Stonnington Local History Collection - Prahran, Prahran 01442
Stratford and District Museum, Stratford 01531
Strathalbyn National Trust Museum, Strathalbyn 01532
Subiaco Museum, Subiaco 01535
Susannah Place Museum, Sydney 01560
Talbot Arts and Historical Museum, Talbot 01567
Tamborine Mountain Heritage Centre, Eagle Heights 01051
Tara and District Pioneer Memorial Museum, Tara 01571
Taralga Historical Museum, Taralga 01572
Tatura Irrigation and Wartime Camps Museum, Tatura 01574
Tenterfield Centenary Cottage, Tenterfield 01576
Texas Historical Museum and Tobacco Museum, Texas 01580
Timber Creek Police Station Museum, Timber Creek 01589
Tongarra Museum, Albion Park 00791
Townsville Museum, Currajong 01018
Tranby House, Maylands, Western Australia 01271
Tumut and District Historical Museum, Tumut 01606
Ulverstone Local History Museum, Ulverstone 01610
Ungarie Museum, Ungarie 01611
Upper Yarra Valley Museum, Yarra Junction 01679
Wagin Historical Village Museum, Wagin 01621
Walkaway Station Museum, Walkaway 01623
Wallaroo Heritage and Nautical Museum, Wallaroo 01624
Waltzing Matilda Centre and Qantilda Museum, Winton 01658
Wangaratta Museum, Wangaratta 01628
Warooka and District Museum, Warooka 01629
Warracknabeal Historical Centre, Warracknabeal 01631
Warragul and District Historical Museum, Warragul 01632
Warrandyte Old Post Office Museum, Warrandyte 01633
Water Tower Museum, Gunnedah 01136
Wauchope District Museum, Wauchope 01638
Wayville Latvian Museum, Brooklyn Park 00906
Weidmann Cottage Heritage Centre, Muswellbrook 01342
Werribee and District Historical Museum, Werribee 01642
Wide Bay and Burnett Historical Museum, Maryborough, Queensland 01269
Wimmera Mallee Pioneers Museum, Jeparit 01178
Winchelsea and District Historical Records Centre, Winchelsea 01655
Winns Historic Bakehouse Museum, Coromandel Valley 01002
Wollondilly Heritage Centre, The Oaks 01390
Wondai Heritage Museum, Wondai 01664
Wongan Ballidu Museum, Wongan Hills 01665
Woocoo Museum, Brooweena 00911
Woodend History Resource Centre, Woodend 01668
Woomera Heritage Centre, Woomera 01671
Wowan and District Museum, Wowan 01672
Wyalong Park Private Museum, Maddington 01248
Yamba Museum, Yamba 01675
Yankalilla District Historical Museum, Yankalilla 01677
Yarraman Heritage Centre, Yarraman 01680
Yesteryear Museum, Captains Flat 00953

Austria

5 Elemente Museum, Waidhofen an der Ybbs 02887
Aberseer Lipphaus, Strobl 02789
Achenseer Museumswelt, Maurach am Achensee 02368
Alpin- und Heimatmuseum, Hohe Wand-Stollhof 02089
Angelika Kauffmann Museum - Gemeindemuseum, Schwarzenberg in Vorarlberg 02722
Aussiedlermuseum im Schüttkasten, Allentsteig 01704
Bad Gasteiner Museum, Bad Gastein 01746
Bauern-Technik-Museum, Steyr-Gleink 02774
Bauern- und Heimatmuseum Haigermoos, Sankt Pantaleon 02675
Bauernmuseum Bundwerkstadel, Haigermoos 02037
Bergbau- und Heimatmuseum, Jochberg 02143
Bezirksheimatmuseum, Völkermarkt 02868
Bezirksheimatmuseum mit Zdarsky-Skimuseum, Lilienfeld 02305
Bezirksmuseum Alsergrund, Wien 02954
Bezirksmuseum Brigittenau, Wien 02955
Bezirksmuseum Donaustadt, Wien 02957
Bezirksmuseum Favoriten, Wien 02958
Bezirksmuseum Floridsdorf, Wien 02959
Bezirksmuseum Hernals, Wien 02960
Bezirksmuseum Hietzing, Wien 02961
Bezirksmuseum Innere Stadt, Wien 02962
Bezirksmuseum Josefstadt, Wien 02963
Bezirksmuseum Landstraße, Wien 02964
Bezirksmuseum Leopoldstadt, Wien 02965
Bezirksmuseum Liesing, Wien 02966
Bezirksmuseum Margareten, Wien 02967
Bezirksmuseum Mariahilf, Wien 02968
Bezirksmuseum Meidling, Wien 02969
Bezirksmuseum Neubau, Wien 02970
Bezirksmuseum Ottakring, Wien 02971
Bezirksmuseum Penzing, Wien 02972
Bezirksmuseum Rudolfsheim-Fünfhaus, Wien 02973
Bezirksmuseum Simmering, Wien 02974
Bezirksmuseum Stockerau, Stockerau 02782
Bezirksmuseum Währing, Wien 02975
Bezirksmuseum Wieden, Wien 02976
Bild- und Tonarchiv, Graz 01965
Brunner Heimathaus mit Rudolf-Steiner-Gedenkstätte, Brunn am Gebirge 01807
Carl Goldmark-Gedenkhaus, Deutschkreutz 01814
Denkmalhof Kösslerhäusl, Großarl 01999
Dokumentationsarchiv des österreichischen Widerstandes, Wien 02978
Dorfmuseum Herrnbaumgarten, Herrnbaumgarten 02073
Dorfmuseum Mönchhof, Mönchhof 02399
Dorfmuseum Nikitsch, Nikitsch 02450
Dorfmuseum Pöttelsdorf, Pöttelsdorf 02513
Dorfmuseum Roiten, Rappottenstein 02560
Dorfmuseum Weinburg, Weinburg 02901
Dorfmuseum Zwingendorf, Zwingendorf 03138
Ehrwalder Heimatmuseum, Ehrwald 01846
Ennsmuseum Weyer, Weyer 02926
Erstes Österreichisches Friedensmuseum und Heimatstube, Wolfsegg am Hausruck 03120
Eva Faschaunerin Heimatmuseum, Gmünd, Kärnten 01940
Falger-Museum, Elbigenalp 01865
Fastnachtmuseum, Nassereith 02421
Felberturmmuseum, Mittersill 02390
Ferrum - Museum Kremayr, Ybbsitz 03124
Festungs- und Heimatmuseum, Kufstein 02243
Forum Hall, Bad Hall 01751
Frankenburger Heimatstube, Frankenburg 01896
Franz Michael Felder-Stube, Schoppernau 02712
Freilichtmuseum Oberlienz, Lienz 02300
Freilichtmuseum Pelmberg, Hellmonsödt 02069
Freilichtmuseum Venetianersäge, Windhaag bei Freistadt 03110
Freilichtmuseum Vorau, Vorau 02873
Ganghofer-Museum, Leutasch 02296
Gedächtnisstätte Mida Huber, Landsee 02255
Gemeindemuseum, Schweiggers 02730
Gemeindemuseum Absam, Absam 01688
Geschichtliches Museum Enns-Donauwinkel, Sankt Valentin 02690
Großmitterberger Troadkasten, Laussa 02269
Haager Heimatstuben, Haag am Hausruck 02028
Handwerkermuseum Laufenbach, Taufkirchen an der Pram 02797
Heimat- und Krippenmuseum, Zirl 03131
Heimat- und Montanmuseum in Schloß Oberkindberg, Kindberg 02163
Heimat- und Pfarrmuseum, Wildalpen 03102
Heimathaus, Königswiesen 02212
Heimathaus, Neufelden 02427
Heimathaus, Obernberg am Inn 02454
Heimathaus, Pregarten 02530
Heimathaus, Raab 02544
Heimathaus, Sankt Peter im Sulmtal 02680
Heimathaus, Schörfling am Attersee 02711
Heimathaus, Ulrichsberg 02838

Heimathaus Beandhaus, Sankt Johann am Walde 02653
Heimathaus Ebensee, Ebensee 01835
Heimathaus Gallneukirchen, Gallneukirchen 01919
Heimathaus im Alten Turm, Haslach 02054
Heimathaus Julbach, Julbach 02147
Heimathaus-Stadtmuseum Vöcklabruck, Vöcklabruck 02865
Heimathaus und Schimuseum, Saalbach 02605
Heimathaus und Stadtmuseum, Perg 02490
Heimathaus Wartberg an der Krems, Wartberg an der Krems 02894
Heimatkundehaus und Münzkabinett, Sankt Marien 02665
Heimatkundesammlung Inzersdorf, Herzogenburg 02075
Heimatkundliche Sammlung Strick, Bad Mitterndorf 01760
Heimatkundliches Museum Medaria, Matrei in Osttirol 02363
Heimatkundliches Museum Wetzlhäusl, Sankt Gilgen 02647
Heimatmuseum, Altlichtenwarth 01716
Heimatmuseum, Kaprun 02153
Heimatmuseum, Millstatt 02382
Heimatmuseum, Radstadt 02551
Heimatmuseum, Rietz 02589
Heimatmuseum, Sankt Aegyd am Neuwalde 02632
Heimatmuseum, Sankt Martin am Wöllmißberg 02668
Heimatmuseum, Schwanenstadt 02717
Heimatmuseum, Wilfersdorf 03106
Heimatmuseum Achental, Achenkirch 01692
Heimatmuseum am Kobernaußerwald, Lohnsburg am Kobernaußerwald 02334
Heimatmuseum Bad Großpertholz, Bad Großpertholz 01750
Heimatmuseum Bernhardsthal, Bernhardsthal 01781
Heimatmuseum Bezau, Bezau 01784
Heimatmuseum der Gemeinde Weißenkirchen, Perschling 02491
Heimatmuseum der Hauptschule Taufkirchen, Taufkirchen an der Pram 02798
Heimatmuseum Elsbethen, Elsbethen 01866
Heimatmuseum Fügen, Fügen 01909
Heimatmuseum Gaaden, Gaaden 01915
Heimatmuseum Gablitz, Gablitz 01917
Heimatmuseum Gnas, Gnas 01953
Heimatmuseum Gröbming, Gröbming 01992
Heimatmuseum Groß-Enzersdorf, Groß-Enzersdorf 01995
Heimatmuseum Großengersdorf, Großengersdorf 02000
Heimatmuseum Großes Walsertal, Sonntag 02747
Heimatmuseum Großschönau, Großschönau 02011
Heimatmuseum Guntramsdorf, Guntramsdorf 02018
Heimatmuseum Hadersdorf, Hadersdorf 02031
Heimatmuseum Holzgau, Holzgau 02098
Heimatmuseum Ilz, Ilz 02106
Heimatmuseum im alten Marktturm, Fischamend 01890
Heimatmuseum im alten Zollhaus, Kobersdorf 02208
Heimatmuseum im Einlegerhaus, Obertrum am See 02463
Heimatmuseum im Studierstüberl, Krieglach 02236
Heimatmuseum Kalchofengut, Unken 02840
Heimatmuseum Kals, Kals am Großglockner 02150
Heimatmuseum Katsdorf, Katsdorf 02156
Heimatmuseum Kaumberg, Kaumberg 02158
Heimatmuseum Kautzen, Kautzen 02160
Heimatmuseum Kematen, Kematen an der Ybbs 02161
Heimatmuseum Kirchham, Kirchham 02171
Heimatmuseum Kirchschlag, Kirchschlag 02172
Heimatmuseum Kohbauernhaus, Köstendorf 02213
Heimatmuseum Kuchl, Kuchl 02242
Heimatmuseum Langenlois, Langenlois 02256
Heimatmuseum Langenzersdorf, Langenzersdorf 02261
Heimatmuseum Lembach, Lembach 02282
Heimatmuseum Litschau, Litschau 02328
Heimatmuseum Losenstein, Losenstein 02337
Heimatmuseum Mank, Mank 02346
Heimatmuseum Marchegg, Marchegg 02349
Heimatmuseum Markt Piesting, Markt Piesting 02361
Heimatmuseum Mautern, Mautern in Steiermark 02370
Heimatmuseum Metzgerhaus, Kirchdorf 02170
Heimatmuseum Michelhausen, Michelhausen 02380
Heimatmuseum mit Folterkammer, Pöggstall 02508
Heimatmuseum mit Raimund von Montecuccoli-Gedenkraum, Hafnerbach 02036
Heimatmuseum Mureck, Mureck 02419
Heimatmuseum Neunkirchen, Neunkirchen 02440
Heimatmuseum Niederleis, Niederleis 02445
Heimatmuseum Niederndorf, Niederndorf 02447
Heimatmuseum Oberbichl, Prägraten 02521
Heimatmuseum Oberhofen, Oberhofen im Inntal 02453
Heimatmuseum Oberwölz und Österreichisches Blasmusikmuseum, Oberwölz 02466
Heimatmuseum Pabneukirchen, Pabneukirchen 02478
Heimatmuseum Persenbeug-Gottsdorf, Persenbeug 02492
Heimatmuseum Pfunds, Pfunds 02498
Heimatmuseum Pressbaum, Pressbaum 02532
Heimatmuseum Pulkau, Pulkau 02537
Heimatmuseum Rabenstein, Rabenstein an der Pielach 02548
Heimatmuseum Rauchstubenhaus Edelschachen, Anger 01721
Heimatmuseum Reith, Reith im Alpbachtal 02577
Heimatmuseum Rother Hof, Pottendorf 02514
Heimatmuseum Sammlung Fuchs, Kirchbach in Steiermark 02164
Heimatmuseum Schloß Fels, Fels am Wagram 01885
Heimatmuseum Schloß Hochhaus, Vorchdorf 02875
Heimatmuseum Schönau, Günselsdorf 02014
Heimatmuseum Schrems, Schrems 02714
Heimatmuseum Tannheimer Tal, Tannheim 02794
Heimatmuseum Tarrenz, Tarrenz 02795
Heimatmuseum Thaya, Thaya 02810
Heimatmuseum und Keltenhaus, Ligist 02304
Heimatmuseum und Klöpferhaus, Eibiswald 01849
Heimatmuseum und landwirtschaftliche Sammlung, Alberschwende 01702
Heimatmuseum und Schaubergwerk, Arzberg 01729
Heimatmuseum und Schiele Museum, Neulengbach 02434
Heimatmuseum Wartberg, Wartberg im Mürztal 02895
Heimatmuseum Weng, Weng im Innkreis 02920
Heimatmuseum Windischgarsten, Windischgarsten 03116
Heimatmuseum Wörgl, Wörgl 03117
Heimatstube, Großraming 02007
Heimatvertriebenen-Stuben, Traun 02825
Hundsmarktmühle, Thalgau 02808
Irrseer Heimathaus, Zell am Moos 03126
Jenbacher Museum, Jenbach 02141
Joslowitzer Heimatstube, Zwingendorf 03139
K-HOF Kammerhof Museen Gmunden, Gmunden 01951
Kammerhofmuseum, Bad Aussee 01739
Karl Heinrich Waggerl-Haus, Wagrain 02882
Katschtaler Heimatmuseum, Rennweg 02578
Kellermuseum Preßhaus, Großkrut 02006
Kellerviertel Heiligenbrunn, Heiligenbrunn 02067
Klostertal Museum, Dalaas 01810
Koschatmuseum, Klagenfurt 02184
Kreuzstadelmuseum, Mogersdorf 02402
Kultur-Gut-Museum, Aigen im Mühlkreis 01699
Kutschen- und Heimatmuseum, Obertilliach 02461
Landesmuseum Kärnten, Klagenfurt 02186
Landschaftsmuseum der Kulmregion, Pischelsdorf 02501
Landschaftsmuseum im Schloß Trautenfels, Trautenfels 02827
Lavanttaler Heimatmuseum, Wolfsberg 03118
Leopold Figl-Museum, Rust im Tullnerfeld 02603
Lichtentaler Pfarrmuseum, Wien 03023
LinzGenesis, Linz 02319
Literaturmuseum Altaussee, Altaussee 01708
Lungauer Heimatmuseum Tamsweg, Tamsweg 02793
Marchfeldmuseum, Weikendorf 02899
Montafoner Heimatmuseum Schruns, Schruns 02716
Mostmuseum und Heimathaus, Sankt Marienkirchen an der Polsenz 02666
Mostviertelmuseum Haag, Haag, Niederösterreich 02029
Mühlviertler Kulturgütermuseum, Sankt Johann am Wimberg 02654
Mühlviertler Schlossmuseum Freistadt, Freistadt 01904
Museum Burg Golling, Golling an der Salzach 01959
Museum der Heimatvertriebenen, Wels 02916
Museum der Marktgemeinde Sankt Johann in Tirol, Sankt Johann in Tirol 02657
Museum der Stadt Bludenz, Bludenz 01790
Museum der Stadt Kapfenberg, Kapfenberg 02151
Museum der Stadt Korneuburg, Korneuburg 02218
Museum der Stadt Schwaz auf Burg Freundsberg, Schwaz 02726
Museum der Stadt Villach, Villach 02859
Museum der Stadt Vils, Vils 02863
Museum Frankomarchia, Frankenmarkt 01899
Museum Herzogsburg, Braunau am Inn 01794
Museum Hohenau an der March, Hohenau an der March 02090
Museum im alten Zeughaus, Bad Radkersburg 01764
Museum im Bürgerspital, Drosendorf an der Thaya 01830
Museum im Fürstenstöckl, Ebenau 01833
Museum im Grünen Haus, Reutte 02583
Museum im Ledererhaus, Purgstall an der Erlauf 02540
Museum im Rathaus, Gleisdorf 01934
Museum im Stadtturm, Waidhofen an der Ybbs 02889
Museum im Troadkostn zu Giem, Feldbach 01879
Museum im Wasserschloß Kottingbrunn, Kottingbrunn 02220
Museum im Zeughaus, Innsbruck 02125
Museum Kierling, Kierling 02162
Museum Kitzbühel, Kitzbühel 02177
Museum Kuenburggewölbe, Werfen 02923
Museum Lauriacum, Enns 01870
Museum Mannersdorf, Mannersdorf 02348
Museum Mödling, Mödling 02395
Museum Sankt Veit, Sankt Veit an der Glan 02691
Museum Schöngrabern, Hollabrunn 02096
Museum Wattens - Industrie Vorgeschichte, Wattens 02896
Museumcenter Leoben, Leoben 02287
Museumsdorf Krumbach, Krumbach 02241
MZM - Museumszentrum Mistelbach, Mistelbach an der Zaya 02386
Neubistritzer Heimatstube, Reingers 02576
Neues Museum Schwechat, Schwechat 02729
Niederösterreichisches Freilichtmuseum, Haag, Niederösterreich 02030
Niederösterreichisches Museum für Volkskultur, Groß Schweinbarth 01997
Nordico - Museum der Stadt Linz, Linz 02320
Oberkärntner Brauchtums- und Maskenmuseum, Oberdrauburg 02452
Österreichisches Freilichtmuseum Stübing, Stübing bei Graz 02790
Österreichisches Pfahlbaummuseum und Museum Mondseeland, Mondsee 02406
Österreichisches Wandermuseum und Roseggers Waldschule, Alpl 01706
Ortskundliches Museum Jois, Jois 02145
Ortsmuseum, Weikertschlag 02900
Pannonisches Heimatmuseum, Neusiedl am See 02443
Pfeilburg Museum, Fürstenfeld 01910
Pinzgauer Heimatmuseum, Saalfelden am Steinernen Meer 02607
Pongauer Heimatmuseum im Schloß Goldegg, Goldegg 01957
Pratermuseum, Wien 03058
Privatmuseum "Feuerwaffen einst und jetzt" im Reckturm, Wiener Neustadt 03098
Proviant-Eisen Museum, Gresten 01990
Rauriser Talmuseum, Rauris 02566
Regional- und Telegrafenmuseum, Stegersbach 02762
Salzburger Freilichtmuseum, Großgmain 02001
Sammlungen der Stadt Amstetten auf Schloß Ulmerfeld, Ulmerfeld-Hausmening 02836
Sauwald-Heimathaus, Sankt Roman 02687
Schiffleutmuseum, Stadl-Paura 02754
Schloß- und Heimatmuseum, Sigharting 02741
Schloßmuseum, Texing 02807
Schneebergmuseum, Puchberg 02536
Schützenmuseum im Kanzlerturm, Eggenburg 01845
Seelackenmuseum, Sankt Veit im Pongau 02693
Sensenmuseum Geyerhammer, Scharnstein 02701
Siglhaus, Sankt Georgen bei Salzburg 02644
Silomuseum, Waidhofen an der Thaya 02885
Stadt- und Heimatmuseum, Heidenreichstein 02063
Stadtmuseum, Murau 02418
Stadtmuseum, Stadtschlaining 02757
Stadtmuseum, Traiskirchen 02817
Stadtmuseum, Waidhofen an der Thaya 02886
Stadtmuseum Alte Hofmühle, Hollabrunn 02097
Stadtmuseum Arelape-Bechelaren-Pöchlarn, Pöchlarn 02506
Stadtmuseum Bruck an der Leitha, Bruck an der Leitha 01803
Stadtmuseum Dornbirn, Dornbirn 01829
Stadtmuseum Friesach, Friesach 01907
Stadtmuseum Graz, Graz 01983
Stadtmuseum Hall in Tirol, Hall in Tirol 02044
Stadtmuseum Hartberg, Hartberg 02053
Stadtmuseum Judenburg, Judenburg 02146
Stadtmuseum Klosterneuburg, Klosterneuburg 02204
Stadtmuseum Kremayr, Rust 02602
Stadtmuseum Melk, Melk 02376
Stadtmuseum mit Weinmuseum, Bad Vöslau 01769
Stadtmuseum Purkersdorf, Purkersdorf 02542
Stadtmuseum Schladming, Schladming 02706
Stadtmuseum Traismauer, Traismauer 02820
Stadtmuseum und Fürstlich Starhembergisches Familienmuseum, Eferding 01841
Stadtmuseum und Tuchmachermuseum Pinkafeld, Pinkafeld 02499
Stadtmuseum Vogtturm, Zell am See 03127
Stadtmuseum Wienertor Hainburg, Hainburg an der Donau 02039
Stadtmuseum Zistersdorf, Zistersdorf 03132
Stadtmuseum Zwettl, Zwettl 03136
Städtisches Museum, Schärding 02696
Stefan Fadinger Museum, Sankt Agatha 02634
Stein- und Bauernmuseum Großdöllnerhof, Rechberg 02567
Stelzhamer-Gedenkstätte, Pramet 02528
Stiftsmuseum, Vomp 02872
Stille-Nacht- und Heimatmuseum Bruckmannhaus, Oberndorf bei Salzburg 02455
Südmährer Heimatmuseum, Laa an der Thaya 02249
Talmuseum Hüttschlag, Hüttschlag 02105
Talmuseum Kaunertal, Feichten 01876
Talmuseum Lachitzhof, Klein Sankt Paul 02194
Tauriska-Galerie, Neukirchen am Großvenediger 02431
Tiroler Kaiserjägermuseum Obergricht, Serfaus 02739
Triestingtaler Heimatmuseum, Weissenbach an der Triesting 02902
Tullner Museen im Minoritenkloster, Tulln 02834
Turm 9 - Stadtmuseum Leonding, Leonding 02293
Turmmuseum, Breitenbrunn, Neusiedlersee 01799
Turmmuseum Oetz, Oetz 02469
Unterwarter Heimathaus, Unterwart 02843
Urgeschichtssammlung und Heimatmuseum, Jedenspeigen 02140
Viechtauer Heimathaus Neukirchen, Neukirchen bei Altmünster 02433
Vorarlberger Landesmuseum, Bregenz 01796
Walsermuseum, Riezlern 02590
Weinstadtmuseum Krems, Krems 02232
Weinstadtmuseum Poysdorf, Poysdorf 02519
Weinviertler Museumsdorf, Niedersulz 02448
Wien Museum Karlsplatz, Wien 03086
Würfelspielhaus - Mensch.Macht.Leben, Frankenburg 01897
Zeitspurenmuseum Ochzethaus, Altheim 01714
Zimmermannshaus Lackinger, Windhaag bei Freistadt 03114
Zoll- und Heimatmuseum, Perwang 02493

Azerbaijan

Azerbaijan State Museum of Independence, Bakı 03143
Djamshid Nakhchivansky Home Museum, Naxçivan 03182
Jafar Jabarly State Museum of Theatre, Bakı 03158
Julfa Regional History Museum, Julfa 03178
Memorial Museum, Naxçivan 03185
Nakhchivan State History Museum, Naxçivan 03188
Ordubad Regional History Studies, Ordubad 03191
Şahbuz District Regional History Museum, Şahbuz 03193
Şarur District Regional History Museum, Şarur 03194

Bahamas

Bahamas Museum, Nassau 03204
Long Island Museum, Long Island 03202
Nassau Public Library and Museum, Nassau 03209
Wyannie Malone Historical Museum, Hope Town 03201

Bangladesh

Ahsan Manzil Museum, Dhaka 03217
Bangladesh Shamorik Jadughar, Dhaka 03221
Osmani Museum, Sylhet 03237
Ram Mala Museum, Comilla 03216
Sooner Ga Museum, Dhaka 03229

Belgium

Archeologisch Museum Abdij, Affligem 03262
Begijnhofmuseum, Dendermonde 03540
Bruggemuseum - Belfort, Brugge 03409
Druivenmuseum, Overijse 03945
Gemeentelijk Heemkundig Museum, Hoeilaart 03707
Gemeentelijk Museum Gevaert-Minne, Sint-Martens-Latem 04018
Gemeentelijk Museum voor Geschiedenis en Heemkunde, Destelbergen 03544
Gemeentemuseum, Temse 04053
Gohltalmuseum, Neu-Moresnet 03910
Grevenbroek Museum, Hamont-Achel 03673
Heemhuis en Sigarenmuseum, Arendonk 03322
Heemkundig Hagelands Museum, Tielt-Winge 04068
Heemkundig Museum, Asse 03327
Heemkundig Museum, Borgerhout 03383
Heemkundig Museum, Bree 03402
Heemkundig Museum, Moen 03875
Heemkundig Museum, Tervuren 04057
Heemkundig Museum, Waasmunster 04122
Heemkundig Museum, Wilrijk 04154
Heemkundig Museum, Zolder 04164
Heemkundig Museum De Drie Rozen, Gravenwezel 03660
Heemkundig Museum Gerard Meeusen, Essen 03586
Heemkundig Museum Wissekerke, Kruibeke 03763
Heemmuseum De Pastorij, Edegem 03573
Heemmuseum van de Zuiderkempen, Westerlo 04148
La Ferme Musée de Marcq, Enghien 03580
Maison du Roi-Musée de la Ville, Bruxelles 03451
Musée Charlier, Bruxelles 03457
Musée Communal de Huy, Huy 03723
Musée Communal de la Ville de Braine-le-Comte, Braine-le-Comte 03399
Musée Communal de Woluwe-Saint-Lambert, Bruxelles 03458

Musée d'Archéologie, Tournai 04083
Musée d'Arenberg, Rebecq 03962
Musée de Folklore et d'Histoire Armand Pellegrin, Opheylissem 03937
Musée de la Tour Salamandre, Beaumont 03349
Musée de la Vie Régionale, Cerfontaine 03512
Musée de la Vie Tihangeoise, Tihange 04071
Musée de l'Entité de Walcourt, Walcourt 04126
Musée de l'Histoire et de la Vie Salmiennes, Vielsalm 04111
Musée de Warcoing, Warcoing 04130
Musée de Waterloo, Waterloo 04137
Musée des Archives, Saint-Hubert 03996
Musée d'Histoire Locale, Ecaussinnes 03570
Musée du Bailles, Nismes 03914
Musée Ducal, Bouillon 03393
Musée Gaspar, Arlon 03326
Musée Historique de la Principauté de Stavelot-Malmedy, Stavelot 04050
Musée National de la Résistance, Bruxelles 03492
Museum De Bres, Ruisbroek, Antwerpen 03988
Museum Het Gebied van Staden, Staden 04047
Museum Meerhem, Gent 03635
Museum Stellingwerff-Waerdenhof, Hasselt 03685
Museum van de Willebroekse Vaart, Willebroek 04152
Museum van Deinze en de Leiestreek, Deinze 03539
Museum Vleeshuis, Antwerpen 03302
Museum voor Heem- en Volkskunde, Wenduine 04143
De Notelaer, Hingene 03703
Oostends Historisch Museum De Plate, Oostende 03935
Oudheidkundig Museum Sint-Janshospitaal, Damme 03535
Priester Poppe Museum, Temse 04055
Stadsmuseum 't Schippershof, Menen 03868
Stedelijk Museum Diest, Diest 03552
Stedelijk Museum Diksmuide, Diksmuide 03554
Stedelijk Museum Hof van Busleyden, Mechelen 03862
Stedelijk Museum Hoogstraten, Hoogstraten 03711
Stedelijk Museum Ieper, Ieper 03731
Stedelijk Museum Lokeren, Lokeren 03827
Stedelijk Museum Schepenhuis, Mechelen 03863
Stedelijke Musea, Torhout 04080
Taxandriamuseum, Turnhout 04102
Ten Duinen 1138, Koksijde 03756
Torenmuseum, Mol 03877
Volkskundemuseum Deurne, Deurne 03550
Volksmuseum Deurne, Antwerpen 03320
Zuidwestbrabants Museum, Halle 03670
Zwijvekemuseum, Dendermonde 03543

Belize

Ambergris Museum and Cultural Centre, Ambergis Caye 04173

Benin

Musée de Natitingou, Natitingou 04184
Musée de Plein Air, Parakou 04186

Bermuda

Bermuda Historical Society Museum, Hamilton 04191
Tucker House Museum, Saint George's 04199
Verdmont Historic House Museum, Smith's Parish 04200

Bolivia

Museo Casa de la Libertad, Sucre 04244
Museo Casa Dorada-Maison d'Or, Tarija 04250
Museo Costumbristá Juan de Vargas, La Paz 04212
Museo Historico Okinawa Bolivia, Santa Cruz 04242
Museo Norte de Potosí, Potosí 04237

Bosnia and Herzegovina

Muzej Grada Sarajeva, Sarajevo 04277
Muzej Grada Zenice, Zenica 04294
Muzej Herzegovine-Trebinje, Trebinje 04287
Muzej Semberija, Bijeljina 04260
Muzej Unsko-Sanskog Kantona, Bihać 04259
Muzej Visoko, Visoko 04293
Regionalni Muzej Doboj, Doboj 04262
Zavičajni Muzej, Travnik 04286

Botswana

Botswana National Museum and Art Gallery, Gaborone 04296
Kgosi Bathoen II (Segopotso) Museum, Kanye 04298
Phuthadikobo Museum, Mochudi 04299
Supa-Ngwao Museum Centre, Francistown 04295

Brazil

Arquivo Histórico e Museu de Canonas Dr. Sezefredo Azambuja Vieira, Canoas 04401
Casa de Cultura Afrónio Peixoto, Lençóis 04536
Casa de Deodoro, Rio de Janeiro 04683
Casa de José Bonifácio, Rio de Janeiro 04684
Casa de Oliveira Vianna, Niterói 04589
Casa do Escultor Ervin Curt Teichmann, Pomerode 04638
Casa Roche Pombo, Morretes 04578
Centro de Estudos Murilo Mendes, Juiz de Fora 04519
Espaço Lucio Costa, Brasília 04357
Fundação Clóuis Salgado - Palácio das Artes, Belo Horizonte 04342
Museu Alfredo Varela, Jaguarão 04507
Museu Anibal Ribeiro Filho, Paranaguá 04610
Museu Antonio Granemann de Souza, Curitibanos 04443
Museu Arquivo Histórico Florense, Flores da Cunha 04451
Museu Barão de Mauá, Mauá 04565
Museu Casa Cel. Joaquim Lacerda, Lapa 04530
Museu Casa da Hera, Vassouras 04948
Museu Casa do Anhanguera, Santana de Parnaíba 04814
Museu Casa do Sertão, Feira de Santana 04449
Museu Casa dos Contos, Ouro Preto 04599
Museu Casa Fritz Alt, Joinville 04515
Museu Casa Pedro Américo, Areia 04322
Museu Comendador Ananias Arruda, Baturit 04331
Museu Corina Novelino, Sacramento 04784
Museu D. João VI, Rio de Janeiro 04707
Museu da Borracha, Rio Branco 04677
Museu da Casa da Cultura de Teresina, Teresina 04929
Museu da Cidade, Campinas 04386
Museu da Cidade de Recife, Recife 04662
Museu da Companhia Paulista, Jundiá 04525
Museu da Estrada de Ferro Sorocabana, Sorocaba 04919
Museu da Imigração, Santa Bárbara d'Oeste 04806
Museu da Inconfidência, Ouro Preto 04601
Museu da Limpeza Urbana - Casa de Banhos de Dom João VI, Rio de Janeiro 04717
Museu das Bandeiras, Goiás 04479
Museu das Monções, Porto Feliz 04655
Museu de Corrente, Corrente 04419
Museu de Rua Alto Paraná Ontem e Hoje, Alto Paran 04304
Museu de Santo André, Santo André 04817
Museu de Sertão, Petrolina 04627
Museu de Venâncio Aires, Venâncio Aires 04949
Museu do Automóvel, Arte História, São Francisco de Paula 04828
Museu do Cangaço, Triunfo 04936
Museu do Colégio Mauá, Santa Cruz do Sul 04807
Museu do Estado de Pernambuco, Recife 04667
Museu do Mamulengo, Olinda 04597
Museu do Piauí, Teresina 04930
Museu do Pio XII, Novo Hamburgo 04595
Museu do Porto do Rio, Rio de Janeiro 04740
Museu do Seridó, Caicó 04380
Museu do Sol, Penápolis 04622
Museu do Vaqueiro, Morada Nova 04577
Museu do Vinho, Videira 04954
Museu Dom Avelar Brandão Vilela, Teresina 04931
Museu dos Inhamuns, Tauá 04921
Museu e Arquivo Histórico, Itu 04503
Museu e Arquivo Histórico Municipal, Guapor 04483
Museu e Arquivo Público Municipal de Campo Belo, Campo Belo 04391
Museu e Biblioteca Pública Pelotense, Pelotas 04621
Museu Estadual do Carvão, Arroio dos Ratos 04323
Museu Farroupilha, Triunfo 04937
Museu Ferroviário de Pires do Rio, Pires do Rio 04635
Museu Getúlio Vargas, São Borja 04823
Museu Goiano Zoroastro Artiaga, Goiânia 04475
Museu Histórico, Igarassu 04488
Museu Histórico, Lavras 04535
Museu Histórico, Sertanópolis 04916
Museu Histórico, Siqueira Campos 04917
Museu Histórico Abílio Barreto, Belo Horizonte 04350
Museu Histórico da Cidade do Rio de Janeiro, Rio de Janeiro 04748
Museu Histórico de Brejo Madre de Deus, Brejo Madre de Deus 04372
Museu Histórico de Witmarsum, Palmeira 04607
Museu Histórico Desembargador Edmundo Mercer Júnior, Tibagi 04934
Museu Histórico e Artístico, Quitandinha 04659
Museu Histórico e Geográfico de Monte Sião, Monte Sião 04575
Museu Histórico e Geral da Cidade de São Vicente, São Vicente 04915
Museu Histórico Kiyoshi Takano, Ura 04946
Museu Histórico Muncipal José Chiachiri, Franca 04468
Museu Histórico Municipal, Dois Irmãos 04446
Museu Histórico Municipal, Montenegro 04576
Museu Histórico Municipal, Telêmaco Borba 04928
Museu Histórico Professor José Alexandre Vieira, Palmas 04606
Museu Integrado de Roraima, Boa Vista 04354
Museu Jaguaribano, Aracati 04314
Museu João Pedro Nunes, São Gabriel 04831
Museu Municipal, Barbacena 04329
Museu Municipal, Caxias do Sul 04411
Museu Municipal, Garibaldi 04471
Museu Municipal, Itambaracá 04497
Museu Municipal, Missal 04567
Museu Municipal, Paulínia 04617
Museu Municipal, Pedras Grandes 04618
Museu Municipal, Rolândia 04779
Museu Municipal, Rosário de Oeste 04780
Museu Municipal, Uberlândia 04944
Museu Municipal, Visconde do Rio Branco 04955
Museu Municipal Adolfo Eurich, Turvo 04939
Museu Municipal Atílio Rocco, São José dos Pinhais 04840
Museu Municipal Capitão Henrique José Barbosa, Canguçu 04400
Museu Municipal David Canabarro, Sant'Anna do Livramento 04816
Museu Municipal de Antônio Prado, Antônio Prado 04308
Museu Municipal de São José dos Campos, São José dos Campos 04839
Museu Municipal Deolindo Mendes Pereira, Campo Mourão 04396
Museu Municipal Domingos Battistel, Nova Prata 04594
Museu Municipal Elisabeth Aytai, Monte Mor 04574
Museu Municipal Embaixador Hélio A. Scarabôtolo, Marília 04562
Museu Municipal Francisco Manoel Franco, Itaúna 04501
Museu Municipal Francisco Veloso, Cunha 04425
Museu Municipal Joaquim de Bastos Bandeira, Perdões 04626
Museu Municipal Karl Ramminger, Mondaí 04571
Museu Municipal Oswaldo Russomano, Bragança Paulista 04356
Museu Municipal Pedro Laragnoit, Miracatu 04566
Museu Municipal Prof. Hugo Daros, Gramado 04482
Museu Municipal Wenceslau Braz, Itajubá 04494
Museu Municipal Wülson Jehovah Lütz Farias, Frederico Westphalen 04470
Museu Nossa Senhora Aparecida, Aparecida 04309
Museu Oswaldo Aranha, Alegrete 04302
Museu Padre Júlio Maria, Russas 04782
Museu Paranaense, Curitiba 04442
Museu Paulo Firpo, Dom Pedrito 04447
Museu Regional, Vitória da Conquista 04959
Museu Regional de São João Del Rei, São João Del Rei 04837
Museu Regional do Alto Uruguai, Erechim 04448
Museu Regional do Sul de Minas, Campanha 04382
Museu Regional Dr. Alípio de Araújo Silva, Rio Preto 04778
Museu Regional Olívio Otto, Carazinho 04404
Museu Salles Cunha, Rio de Janeiro 04760
Museu São Norberto, Pirapora do Bom Jesus 04633
Museu Tempostal, Salvador 04805
Museu Teresa Bernardes Adami de Carvalho, Santo Antônio do Monte 04819
Museu Territorial do Amapá, Macapá 04540
Museu Tinguí-Cuera, Araucária 04317
Museu Victor Bersani, Santa Maria 04811
Museu Vivo da Memória Candanga, Brasília 04370
Museu Zé Didor, Campo Maior 04395
O Museu do Marajó, Cachoeira do Arari 04377
Paço Imperial, Rio de Janeiro 04766
Palácio das Laranjeiras, Rio de Janeiro 04767
Palácio Guanabara, Rio de Janeiro 04768
Palácio Gustavo Capanema, Rio de Janeiro 04769
Setor de Malacologia, Juiz de Fora 04524
Solar de Dom João VI, Rio de Janeiro 04772
Solar Grandjean de Montigny, Rio de Janeiro 04773

Brunei

Taman Warisan Tasek Merimbun, Mukim Rambai 04969

Bulgaria

Architekturno-istoričski Muzej-rezervat, Boženci 04993
Drjanovski Manastir Muzej, Drjanovo 05023
Etnografska Ekspozicija, Svištov 05220
Etnografski Kompleks Starijat Dobrič, Dobrič 05017
Etnografski-vázroždenski Kompleks, Kavarna 05054
Gradski Istoričeski Muzej, Čirpan 05009
Gradski Istoričeski Muzej, Dalgopol 05011
Gradski Istoričeski Muzej, Kavarna 05055
Gradski Istoričeski Muzej, Klisura 05069
Gradski Istoričeski Muzej, Lom 05084
Gradski Istoričeski Muzej, Melnik 05091
Gradski Istoričeski Muzej, Orjachovo 05101
Gradski Istoričeski Muzej, Panagjurište 05102
Gradski Istoričeski Muzej, Samakov 05156
Gradski Istoričeski Muzej, Sevlievo 05159
Gradski Istoričeski Muzej, Teteven 05226
Gradski Istoričeski Muzej Aleko Konstantinov, Svištov 05221
Gradski Istoričeski Muzej Iskra, Kazanlăk 05058
Istoričeski Muzej, Bansko 04978
Istoričeski Muzej, Bracigovo 04994
Istoričeski Muzej, Burgas 04999
Istoričeski Muzej, Čiprovci 05007
Istoričeski Muzej, Dobrič 05019
Istoričeski Muzej, Gabrovo 05032
Istoričeski Muzej, Goce Delčev 05035
Istoričeski Muzej, Gorna Orjachovica 05036
Istoričeski Muzej, Ichtiman 05039
Istoričeski Muzej, Isperich 05040
Istoričeski Muzej, Karlovo 05049
Istoričeski Muzej, Loveč 05086
Istoričeski Muzej, Omurtag 05099
Istoričeski Muzej, Petrič 05113
Istoričeski Muzej, Pomorie 05133
Istoričeski Muzej, Popovo 05134
Istoričeski Muzej, Provadia 05140
Istoričeski Muzej, Razlog 05149
Istoričeski Muzej Pavlikeni, Pavlikeni 05104
Istoričeski Muzej Silistra, Silistra 05163
Kăšta-muzej Angel Kănčev, Trjavna 05228
Kăšta-muzej Asen Razcvetnikov, Draganovo 05021
Kăšta-muzej Canko Cerkovski, Bjala Čerkva 04989
Kăšta-muzej Dimitar Pešev, Kjustendil 05064
Kăšta-muzej Iljo Vojvoda, Kjustendil 05065
Kăšta-muzej Neofit Rilski, Bansko 04979
Kăšta-muzej Pejo K. Javorov, Čirpan 05010
Kăšta-muzej Rusi Čorbadži, Žeravna 05270
Kăšta-muzej Sava Filaretov, Žeravna 05271
Muzej Aleksandăr Stambolijski, Slavovica 05164
Muzej Emfiedžieva Kăšta, Kjustendil 05066
Muzej-krepost Baba Vida, Vidin 05262
Muzej na Gradskij Bit Oslekova Kăšta, Koprivštica 05074
Muzej na Gradskija Bit, Ruse 05153
Muzej Rumănski Voin, Pordim 05136
Muzej Văzraždane i Učreditelno Săbranie, Veliko Tărnovo 05255
Muzej za Istorija na Varna, Varna 05243
Nacionalen Istoriko-archeologičeski Rezervat i Muzej Veliki Preslav, Veliki Preslav 05248
Nacionalen Park-muzej Šipka-Buzludža, Kazanlăk 05060
Oblasten Istoričeski Muzej, Jambol 05043
Obščinski Istoričeski Muzej, Etropole 05028
Okražen Istoričeski Muzej, Montana 05095
Okražen Istoričeski Muzej, Smoljan 05171
Okražen Istoričeski Muzej, Vidin 05263
Okražen Istoričeski Muzej, Vraca 05265
Regionalen Istoričeski Muzej, Kardžali 05047
Regionalen Istoričeski Muzej, Kjustendil 05068
Regionalen Istoričeski Muzej, Pazardžik 05109
Regionalen Istoričeski Muzej, Šumen 05218
Regionalen Istoričeski Muzej Varna, Varna 05247
Školoto Muzej, Trjavna 05231
Sofijski Istoričeski Muzej, Sofia 05204

Burkina Faso

Musée Pobe Mengao, Mengao 05276
Musée Provincial du Houet, Bobo-Dioulasso 05272
Musée Provincial du Poni, Gaoua 05273
Musée Provincial du Sanmatenga, Kaya 05274

Burundi

Musée National de Gitega, Gitega 05281
Musée Vivant de Bujumbura, Bujumbura 05280

Cameroon

Musée National de Yaoundé, Yaoundé 05302
Musée Provincial de Douala, Douala 05296

Canada

1910 Boomtown, Saskatoon 06909
Agassiz-Harrison Historical Society Museum, Agassiz 05309
Albert County Museum, Hopewell Cape 05996
Alert Bay Museum, Alert Bay 05314
Algonquin Logging Museum, Whitney 07332
Alix Wagon Wheel Museum, Alix 05317
Alliance and District Museum Society, Alliance 05318
Ameliasburgh Historical Museum, Ameliasburgh 05329
Ancien Palais de Justice - Centre d'Art et d'Histoire, Kamouraska 06040
Anderson Farm Museum, Lively 06176
Andrew and District Local History Museum, Andrew 05340
Annandale, Tillsonburg 07083
Annapolis Valley Macdonald Museum, Middleton 06276
Anola and District Museum, Anola 05346
L'Anse aux Meadows National Historic Park, Saint-Lunaire-Griquet 06865
Anthony Henday Museum, Delburne 05690
Antigonish Heritage Museum, Antigonish 05348
Antler River Historical Society Museum, Melita 06262
Archibald Historical Museum, La Rivière 06114

Arcola Museum, Arcola 05349
Arnprior and District Museum, Arnprior 05355
Ashcroft Museum, Ashcroft 05357
Ashern Pioneer Museum, Ashern 05358
Assiginack Historical Museum and Norisle Heritage Park, Manitowaning 06225
Assiniboia and District Museum, Assiniboia 05359
Atikokan Centennial Museum, Atikokan 05360
Aurora Museum, Aurora 05363
Avonlea and District Museum, Avonlea 05366
Aylmer and District Museum, Aylmer 05367
Ball's Falls Historical Park, Jordan 06032
Bancroft Historical Museum, Bancroft 05379
Barkerville Historic Town, Barkerville 05384
Barr Colony Heritage Cultural Centre, Lloydminster 06180
Barrhead Centennial Museum, Barrhead 05385
Bateman Historical Museum, Bateman 05395
Batoche, Batoche 05399
Beachville District Museum, Beachville 05403
Beau Village Museum, Saint Victor 06882
Beautiful Plains Museum, Neepawa 06419
Beaver River Museum, Beaverton 05409
Beckoning Hills Museum, Boissevain 05433
Bell Island Community Museum, Bell Island 05413
Belmont and District Museum, Belmont 05418
Bengough and District Museum, Bengough 05420
Beothuck Village, Grand Falls-Windsor 05891
Bethune Memorial House, Gravenhurst 05907
Big Bear Trails Museum, Loon Lake 06194
Big Beaver Museum and Nature Centre, Big Beaver 05425
Big River Memorial Museum, Big River 05426
Biggar Museum and Gallery, Biggar 05427
Binscarth and District Gordon Orr Memorial Museum, Binscarth 05429
Black Creek Pioneer Village, North York 06486
Blaine Lake Museum, Blaine Lake 05430
Blockhouse Museum, Merrickville 06267
Bonavista Museum, Bonavista 05437
Bonavista North Regional Museum, Wesleyville 07300
Bonnechere Museum, Eganville 05767
Botwood Heritage Centre, Botwood 05444
Boundary Museum, Grand Forks 05893
Bowden Pioneer Museum, Bowden 05447
Bowen Island Museum, Bowen Island 05448
Bowmanville Museum, Bowmanville 05449
Brant Museum, Brantford 05467
Bresaylor Heritage Museum, Bresaylor 05472
Briercrest and District Museum, Briercrest 05476
Broadview Historical Museum, Broadview 05484
Brockville Museum, Brockville 05486
Broken-Beau Pioneer Village Museum, Beauséjour 05406
Brome County Historical Society Museum, Knowlton 06108
Brooks and District Museum, Brooks 05488
Bruce County Museum, Southampton 06997
Bruce Groner Museum, Groundbirch 05916
Bruce Mines Museum, Bruce Mines 05491
Bulkley Valley Museum, Smithers 06983
Burgeo Centennial Memorial Museum, Burgeo 05494
Burin Heritage House, Burin 05496
Burk's Falls and District Museum, Burk's Falls 05497
Burnaby Village Museum, Burnaby 05502
Burrough of the Gleann Museum, Glenboro 05872
Buxton Museum, North Buxton 06473
C.O. Card Home, Cardston 05560
Cabri and District Museum, Cabri 05509
Cadillac Museum, Cadillac 05511
Callander Bay Heritage Museum, Callander 05535
Campbell House Museum, Toronto 07096
Campbellford-Seymour Heritage Centre, Campbellford 05542
Camrose and District Centennial Museum, Camrose 05544
Canadian Military Heritage Museum, Brantford 05468
Canadian Transportation Museum and Historic Village, Kingsville 06092
Canmore Museum, Canmore 05545
Cannington Centennial Museum, Cannington 05546
Cannington Manor, Kenosee Lake 06058
Canso Museum, Canso 05547
Canwood Museum, Canwood 05549
Cape Breton Centre for Heritage and Science, Sydney 07047
Cape Ray Lightkeeper's House, Cape Ray 05552
Cape Sable Historical Society Centre, Barrington 05390
Carberry Plains Museum, Carberry 05556
Carmicheal-Stewart House, New Glasgow 06429
Caroline Wheels of Time Museum, Caroline 05568
Carscadden's Museum, Plenty 06612
Castor and District Museum Society, Castor 05578
Centennial Museum of Sheguiandah, Sheguiandah 06952
Centennial Park 1910 Logging Museum, Thunder Bay 07076
Centennial Park Museum, Moose Factory 06384
Central Butte District Museum, Central Butte 05583
Centre d'Exposition d'Amos, Amos 05337
Centre d'Histoire de Montréal, Montréal 06333
Centre d'Historie et d'Archéologie de la Métabetchouane, Desbiens 05697
Centre d'Interpretation de l'Ardoise, Richmond 06742
Centre d'Interpretation de l'Histoire de Sherbrooke, Sherbrooke, Québec 06958
Cereal Prairie Pioneer Museum, Cereal 05584
Chapel Hill Museum, Shag Harbour 06946
Chapleau Centennial Museum, Chapleau 05586
Chapman Museum, Brandon 05461
Charlotte County Museum, Saint Stephen 06878
Chase and District Museum, Chase 05599
Chemainus Valley Museum, Chemainus 05606
Chestico Museum, Port Hood 06641
Chilliwack Museum, Chilliwack 05613
Christ Church Community Museum, Lakefield 06131
Claresholm Museum, Claresholm 05619
Clarke Museum, Kirby 06096
Claude Crayston Museum, Glenora 05873
Clayton McLain Memorial Museum, Cut Knife 05669
Climax Community Museum, Climax 05623
Coaticook Historical Society Museum, Coaticook 05625
Colchester Historical Museum, Truro 07180
Coldwater Canadiana Heritage Museum, Coldwater 05629
Cole Harbour Heritage Farm Museum, Cole Harbour 05630
Colville Lake Museum, Colville Lake 05633
Compton County Historical Museum, Eaton Corner 05734
Conception Bay Museum, Harbour Grace 05971
Cook's Creek Heritage Museum, Dugald 05720
Copper Cliff Museum, Copper Cliff 05640
Corner Brook Museum Archives, Corner Brook 05641
Coronach District Museum, Coronach 05647
Cossit House Museum, Sydney 07048
Cowichan Valley Museum, Duncan 05723
Craig Heritage Park Museum, Parksville 06568
Creamery Square Museum, Tatamagouche 07060
Creston and District Museum, Creston 05659
Crowsnest Museum, Coleman 05631
Crowsnest Pass Ecomuseum, Bellevue 05416
Crystal City Community Museum, Crystal City 05663
Cumberland County Museum, Amherst 05331
Cumberland Heritage Village Museum, Cumberland, Ontario 05665
Cumberland Museum and Archives, Cumberland 05664
Cumberland Township Historical Society, Ottawa 06543
Cupar and District Heritage Museum, Cupar 05666
Cupids Museum, Cupids 05667
Darlingford School Heritage Museum, Darlingford 05673
Dartmouth Heritage Museum, Dartmouth 05675
Dawson City Museum, Dawson City 05681
Dawson Creek Station Museum, Dawson Creek 05684
DeBolt and District Pioneer Museum, DeBolt 05686
Deep South Pioneer Museum, Ogema 06501
Delhi Ontario Tobacco Museum Heritage Centre, Delhi 05691
Delta Museum, Delta 05692
Denman Island Museum, Denman Island 05695
DesBrisay Museum and Exhibition Centre, Bridgewater 05474
Dewberry Valley Museum, Dewberry 05699
Didsbury and District Museum, Didsbury 05702
Dionne Quints Museum, North Bay 06469
Discovery North Bay Museum, North Bay 06470
Dodsland and District Museum, Dodsland 05706
Donalda and District Museum, Donalda 05707
Drayton Valley Museum, Drayton Valley 05710
Dryden and District Museum, Dryden 05716
Duck Lake Regional Interpretive Centre, Duck Lake 05717
Duff Community Heritage Museum, Duff 05718
Dufferin County Museum, Rosemont 06774
Dufferin Historical Museum, Carman 05566
Dundurn Castle, Hamilton 05953
Dunwell and Community Museum, Weekes 07293
Eagle Valley Museum, Sicamous 06968
Ear Falls District Museum, Ear Falls 05729
East Hants Historical Museum, Maitland 06219
Écomusée de Hull, Gatineau 05858
Écomusée de Saint-Constant, Saint-Constant 06810
Edgerton and District Museum, Edgerton 05737
Edinburgh Square Heritage and Cultural Centre, Caledonia 05512
Eildon Hall Sibbald Memorial Museum, Sutton, Ontario 07040
Elbow Museum, Elbow 05769
Elman W. Campbell Museum, Newmarket 06439
Elrose Heritage Museum, Elrose 05774
Enderby and District Museum, Enderby 05776
Englehart and Area Historical Museum, Englehart 05778
Eptek Art Culture Centre, Summerside 07032
Erland Lee Museum, Stoney Creek 07012
Esterhazy Community Museum, Esterhazy 05782
Estevan Art Gallery and Museum, Estevan 05784
Etzikom Museum, Etzikom 05788
Evergreen House, Dartmouth 05676
Evolution of Education Museum, Prince Albert 06661
Fairbridge Chapel Museum, Mill Bay 06284
Fairfield Museum, Bothwell 05443
Fairview R.C.M.P. Centennial Museum, Fairview 05791
Fanshawe Pioneer Village, London 06184
Fenelon Falls Museum, Fenelon Falls 05792
Fernie and District Historical Society Museum, Fernie 05795
Fishermen's Museum, Saint Vincent's 06884
Flambro Head Heritage Museum, Job's Cove 06029
Flatrock Museum, Flatrock 05798
Flin Flon Museum, Flin Flon 05800
Foam Lake Museum, Foam Lake 05801
Forest-Lambton Museum, Forest 05802
Forestburg and District Museum, Forestburg 05803
Fort Calgary, Calgary 05518
Fort Chipewyan Bicentennial Museum, Fort Chipewyan 05806
Fort Dauphin Museum, Dauphin 05679
Fort Edmonton Park, Edmonton 05746
Fort Erie Historical Museum, Ridgeway 06750
Fort Frances Museum, Fort Frances 05809
Fort Lennox, Saint-Paul-de-l'Île-aux-Noix 06872
Fort Museumof the North West Mounted Police, Fort Macleod 05814
Fort Nelson Heritage Museum, Fort Nelson 05818
Fort Pelley Livingston Museum, Pelly 06578
Fort Qu'appelle Museum, Fort Qu'appelle 05819
Fort Sackville Foundation, Bedford 05411
Fort Saint James, Fort Saint James 05820
Fort Saskatchewan Museum, Fort Saskatchewan 05822
Fort Steele Heritage Town, Fort Steele 05824
Fort Témiscamingue, Ville-Marie 07265
Fort Walsh, Maple Creek 06232
Fortifications-de-Québec, Québec 06680
Fortress of Louisbourg, Louisbourg 06197
Forwarders' Museum, Prescott 06657
Founders' Hall, Charlottetown 05592
Fox Harbour Community Museum, Fox Harbour 05826
Frank Cameron Museum of Wynyard, Wynyard 07407
Frank Slide Interpretive Centre, Crowsnest Pass 05661
Frazer's Museum, Beauval 05407
Fred Light Museum, Battleford 05400
Frenchman Butte Museum, Frenchman Butte 05843
Gabriola Museum, Gabriola 05845
Galloway Station Museum, Edson 05765
Gananoque Historical Museum, Gananoque 05847
George Johnston Tlingit Indian Museum, Teslin 07069
Georgina Pioneer Village, Keswick 06066
Gibson House Museum, North York 06487
Gillam Community Museum, Gillam 05865
Glace Bay Heritage Museum, Glace Bay 05869
Gladstone and District Museum, Gladstone 05870
Glanmore National Historic Site of Canada, Belleville 05415
Glengarry Pioneer Museum, Dunvegan 05726
Glentworth Museum, Glentworth 05874
Gloucester Museum, Gloucester 05875
Glovertown Heritage Museum, Glovertown 05876
Golden and District Museum, Golden 05880
Goldstream Region Museum, Victoria 07248
Goodsoil Historical Museum, Goodsoil 05882
Gowganda and Area Museum, Gowganda 05885
Grand Falls Museum, Grand Falls 05889
Grand Manan Museum and Walter B. McLaughlin Marine Gallery, Grand Manan 05896
Granisle Museum, Granisle 05903
Gravelbourg and District Museum, Gravelbourg 05906
Great Sandhills Museum, Sceptre 06939
Greater Vernon Museum, Vernon, British Colombia 07236
Greenwood Museum, Greenwood, British Columbia 05909
Grenfell Community Museum, Grenfell 05911
Grey Roots Heritage, Owen Sound 06561
Grimsby Museum, Grimsby 05912
Guelph Civic Museum, Guelph 05917
Gut of Canso Museum, Port Hastings 06639
Haileybury Heritage Museum, Haileybury 05925
Haldimand County Museum, Cayuga 05581
Haliburton Highlands Museum, Haliburton 05927
Hallmark Society, Victoria 07249
Halton Region Museum, Milton 06289
Hamiota Pioneer Club Museum, Hamiota 05965
Haney House Museum, Maple Ridge 06235
Harris Heritage and Museum, Harris 05973
Harry S. Washbrook Museum, Edam 05735
Hart-Cam Museum, Hartney 05976
Harvey Grant Heritage Centre, Springdale 07001
Heritage Hazenmore Museum, Hazenmore 05980
Heritage House Museum, Smiths Falls 06986
Heritage Museum, Deer Lake 05689
Heritage Park, Fort McMurray 05816
Heritage Park Museum, Terrace 07066
High Prairie and District Museum, High Prairie 05985
Highland Pioneers Museum, Baddeck 05370
Historic Ferryland Museum, Ferryland 05797
Historic Fort York, Toronto 07114
Historic Yale Museum, Yale 07408
Historical Heritage Complex in Old Harry - CAMI, Old Harry 06506
Historical Museum and Pioneer Village, Sundre 07035
Hodgeville Community Museum, Hodgeville 05991
Holden Historical Society Museum, Holden 05992
Homestead Antique Museum, Drumheller 05712
Homestead Museum, Biggar 05428
Hope Museum, Hope 05994
House of Memories, Latchford 06144
Howard House of Artifacts, Old Perlican 06508
Huble Homestead, Prince George 06667
Hudson Bay Museum, Hudson Bay 05999
Hudson's Hope Histrical Society Museum, Hudson's Hope 06000
Humboldt and District Museum and Gallery, Humboldt 06001
Huronia Museum, Midland 06277
Hymers Museum, Hymers 06004
Îlot des Palais, Québec 06682
Imperial and District Museum, Imperial 06009
Indian Head Museum, Indian Head 06011
Ingersoll Cheese Factory Museum and Sports Hall of Fame, Ingersoll 06012
Ingledale, Stoney Creek 07013
Innisfail and District Historical Village, Innisfail 06013
Innisville and District Museum, Innisville 06014
Inverness Miners Museum, Inverness 06017
Iron Creek Museum, Lougheed 06196
Iroquois Falls Pioneer Museum, Iroquois Falls 06023
Irving House Historic Centre and New Westminster Museum, New Westminster 06436
Islands Museum, Tiverton 07085
Ituna Cultural and Historical Museum, Ituna 06026
J.A.V. David Museum, Killarney 06067
Jasper Cultural and Historical Centre, Maple Creek 06233
John N. Leamon Museum, Brigus 05480
Jordan Historical Museum of the Twenty, Jordan 06033
Jost House Museum, Sydney 07049
Kamloops Museum Archives, Kamloops 06036
Kamsack and District Museum, Kamsack 06042
Kaposvar Historic Site, Esterhazy 05783
Keillor House Museum, Dorchester 05708
Kelliher and District Heritage Museum, Kelliher 06049
Kellross Heritage Museum, Leross 06159
Kelowna Museum, Kelowna 06054
Keno City Mining Museum, Keno City 06056
Kerrobert and District Museum, Kerrobert 06065
Killarney Centennial Museum, Killarney 06068
Kimberley District Heritage Museum, Kimberley 06069
Kincaid Museum, Kincaid 06070
Kindersley and District Plains Museum, Kindersley 06071
King Township Historical Museum, King City 06072
Kings County Museum, Hampton 05966
Kings County Museum, Kentville 06062
Kings Landing Historical Settlement, Prince William 06673
Kipling District Museum, Kipling 06095
Kisbey Museum, Kisbey 06099
Klondike National Historic Sites, Dawson City 05682
Knaut Rhuland House Museum, Lunenburg 06203
Knee Hill Historical Society Museum, Three Hills 07075
Kootenay Gallery of Art, History and Science, Castlegar 05575
Labrador Interpretation Centre, North West River 06483
Labrador Straits Museum, L'Anse-au-Loup 06139
Lac Cardinal Regional Pioneer Village Museum, Grimshaw 05914
Lac-Du-Bonnet and District Historical Society Museum, Lac-Du-Bonnet 06118
Lac-La-Hache Museum, Lac-La-Hache 06119
Lake Country Museum, Okanagan 06504
Lake of the Woods Museum, Kenora 06057
Lakes District Museum Society, Burns Lake 05506
Lakeside Museum, Bulyea 05493
Lambton Heritage Museum, Grand Bend 05888
Lanark and District Museum, Lanark 06132
Lancer Centennial Museum, Lancer 06134
Lang Pioneer Village, Keene 06048
Langham and District Heritage Village Museum, Langham 06135
Lanigan and District Heritage Centre, Lanigan 06137

Lashburn Centennial Museum, Lashburn 06143
Laurier House, Ottawa 06549
Leading Tickles Heritage Centre, Leading Tickles 06152
Lennox and Addington County Museum, Napanee 06417
Leroy and District Heritage Museum, Leroy 06160
Lieu Historique National de Côteau-du-Lac, Côteau-du-Lac 05648
Lillooet District Museum, Lillooet 06170
Little Prairie Heritage Society, Chetwynd 05610
Lockmaster's House Museum, Elgin 05771
Log Cabin Museum, Saanichton 06784
London Regional Children's Museum, London 06187
Londonderry Mines Museum, Londonderry 06191
Loyalist Cultural Centre, Bath 05396
Lundar Museum, Lundar 06201
Lundy's Lane Historical Museum, Niagara Falls 06443
Luseland and Districts Museum, Luseland 06205
Lutz Mountain Meeting House and Heritage Museum, Moncton 06318
Lytton Museum, Lytton 06207
Macaulay Heritage Park, Picton 06600
MacBride Museum, Whitehorse 07325
McCrae House, Guelph 05918
Mackenzie and District Museum, Mackenzie 06210
Mackenzie Crossroads Museum, High Level 05984
Macklin and District Museum, Macklin 06211
Macrorie Museum, Macrorie 06212
Madonna House Museum, Combermere 05635
Magnetawan Historical Museum, Magnetawan 06214
Maidstone and District Museum, Maidstone 06217
Main Centre Heritage Museum, Main Centre 06218
Maison Chevalier, Québec 06683
Maison de Nos Aïeux, Sainte-Famille 06890
Maison des Chapais, Saint-Denis-de-la-Bouteillerie 06813
Mankota Museum, Mankota 06227
Manor Museum, Manor 06229
Maple Ridge Museum, Maple Ridge 06237
Marble Mountain Library and Museum, West Bay 07301
Markham Museum and Historic Village, Markham 06239
Marringhurst Pioneer Park Museum, Pilot Mound 06605
Marten River Logging Museum, Marten River 06241
Mary March Regional Museum and Logging Exhibition, Grand Falls-Windsor 05892
Maryfield Museum, Maryfield 06243
Marystown Heritage Museum, Marystown 06244
Massey Area Museum, Massey 06246
Matsqui Sumas Abbotsford Museum, Abbotsford 05305
Mattawa and District Museum, Mattawa 06250
Meadow Lake Museum, Meadow Lake 06255
Meaford Museum, Meaford 06256
Meanskinisht Museum, Kitwanga 06106
Medicine Hat Museum and Art Gallery, Medicine Hat 06258
Melfort and District Museum, Melfort 06261
Melville Heritage Museum, Melville 06264
Mennonite Heritage Village, Steinbach 07007
Miami Museum, Miami 06271
Middle Lake Museum, Middle Lake 06273
Middleville and District Museum, Lanark 06133
Milden Community Museum, Milden 06282
Millertown Museum, Millertown 06285
Millet and District Museum, Millet 06286
Milltown-Head Bay d'Espoir Museum, Milltown 06287
Milton Blacksmith Shop Museum, Milton 06290
Miniota Municipal Museum, Miniota 06297
Minnedosa and District Museum, Minnedosa 06298
Minto Museum, Minto 06299
Mirror and District Museum, Mirror 06304
Mission Museum, Mission 06306
Mistasinihk Place Interpretive Centre, LaRonge 06140
Mockbeggar Plantation, Bonavista 05439
Moncton Museum, Moncton 06319
Montgomery House Museum, Frankville 05827
Moore Museum, Mooretown 06383
Moose Factory Centennial Museum, Kirkland Lake 06097
Moose Jaw Museum and Art Gallery, Moose Jaw 06386
Moosehorn Heritage Museum, Moosehorn 06388
Moreton's Harbour Museum, Moreton's Harbour 06392
Morris and District Centennial Museum, Morris 06394
Morse Museum, Morse 06396
Mossbank and District Museum, Mossbank 06397
Multicultural Heritage Centre, Stony Plain 07014
Musée Acadien, Miscouche 06305
Musée Acadien, Moncton 06320
Musée Beaulne, Coaticook 05626
Musée de Kamouraska, Kamouraska 06041
Musée de Kent, Bouctouche 05446
Musée de la Gaspésie, Gaspé 05854
Musée de Lachine, Lachine 06123
Musée de l'Auberge Symmes, Gatineau 05863
Musée de Saint-Georges, Saint-Georges, Manitoba 06818
Musée de Saint-Pierre-Jolys, Saint-Pierre-Jolys 06875
Musée d'Histoire et du Patrimoine de Dorval, Montréal 06364
Musée du Fjord, Ville de Saguenay, La Baie 07263
Musée du Haut-Richelieu, Saint-Jean-sur-Richelieu 06829
Musée du Nouveau-Brunswick, Saint John 06838
Musée du Père Gédéon, Sainte-Marie-de-Beauce 06899
Musée François Pilote, La Pocatière 06113
Musée Girouxville, Girouxville 05867
Musée Héritage, Saint-Albert 06790
Musée Historique de Saint-Paul, Saint Paul 06870
Musée Historique du Madawaska, Edmundston 05764
Musée Joseph Filion, Sainte-Thérèse 06901
Musée Le Pionnier, Saint-Malo 06866
Musée les Trois Pignons, Cheticamp 05608
Musée Maison Saint-Gabriel, Montréal 06372
Musée Marguerite Bourgeoys, Montréal 06374
Musée Namesokanjic, Lac-Mégantic 06120
Musée Regional d'Argenteuil, Saint-André-d'Argenteuil 06795
Musée Régional de la Côte-Nord, Sept-Îles 06945
Musée Régional de Rimouski, Rimouski 06755
Musée Régional de Vaudreuil-Soulanges, Vaudreuil-Dorion 07234
Musée Saint-Brieux, Saint-Brieux 06805
Musée Saint-Joachim, La Broquerie 06109
Musée Sainte-Martine, Sainte-Martine 06900
Muséocabinet, Saint-Jérôme 06831
Museum by the Bay, Lewisporte 06169
Museum of Northern British Columbia, Prince Rupert 06671
Museum of Northern History at the Sir Harry Oakes Château, Kirkland Lake 06098
Museum of the Highwood, High River 05986
Museum of the Rural Municipality of Roland, Roland 06771
Museum on the Boyne, Alliston 05319
Muskoka Heritage Place, Huntsville 06003
Naicam Museum, Naicam 06406
Nain Museum, Nain 06407
Nakusp District Museum, Nakusp 06408
Nanaimo District Museum, Nanaimo 06413
Naramata Heritage Museum Society, Naramata 06418
Ned Shed Museum, Meldrum Bay 06260
Nelson District Museum, Nelson 06422
Nepean Museum, Nepean 06425
Niagara Historical Museum, Niagara-on-the-Lake 06454
Nicola Valley Museum, Merritt 06269
Nipawin Living Forestry Museum, Nipawin 06459
Nipigon Museum, Nipigon 06460
Nipissing Township Museum, Nipissing 06461
Nokomis and District Museum, Nokomis 06462
Nor' Wester and Loyalist Museum, Williamstown 07339
Norman Wells Historical Centre, Norman Wells 06463
Norris Arm Heritage Museum, Norris Arm 06464
North Atlantic Aviation Museum, Gander 05848
North Highland Community Museum, Cape North 05551
North Huron District Museum, Wingham 07349
North Lanark Regional Museum, Almonte 05325
North Pacific Cannery Village Museum, Port Edward 06637
North Sydney Museum, North Sydney 06476
North Thompson Museum, Barriere 05388
North Vancouver Museum, North Vancouver 06480
Northern Gateway Museum, Denare Beach 05694
Nose Creek Valley Museum, Airdrie 05311
Nova Scotia Museum, Halifax 05943
Nunatta Sunakkutaangit Museum, Iqaluit 06020
Oakville Museum at Erchless Estate, Oakville 06497
O'Dell House Museum, Annapolis Royal 05343
Ogniwo Polish Museum, Winnipeg 07372
Old Massett Village Council's Heritage, Old Massett 06507
Old Meeting House, Barrington 05391
Old Town Hall Museum, Thamesville 07071
Oldman River Cultural Centre, Brocket 05485
Oliver Heritage Society Museum, Oliver 06511
Orillia Museum of Art and History, Orillia 06514
Osgoode Township Museum, Vernon, Ontario 07239
Oshawa Community Museum and Archives, Oshawa 06522
Oshawa Sydenham Museum, Oshawa 06523
Osler Historical Museum, Osler 06525
Osoyoos Museum, Osoyoos 06528
Ottawa House By-The-Sea Museum, Parrsboro 06571
Our Lady of Mercy Museum, Port-au-Port 06629
Outlook and District Heritage Museum and Gallery, Outlook 06558
Paipoonge Historical Museum, Thunder Bay 07079
Park House Museum, Amherstburg 05336
Parkdale-Maplewood Community Museum, Barss Corner 05393
Parks Canada Visitor Reception Centre, Churchill 05615
Peel Heritage Complex, Brampton 05456
Pelee Island Heritage Centre, Pelee Island 06577
Pemberton District Museum, Pemberton 06579
Pembina Lobstick Historical Museum, Evansburg 05789
Penetanguishene Centennial Museum, Penetanguishene 06582
Peninsula and Saint Edmunds Township Museum, Tobermory 07086
Perth Museum, Perth 06587
Petit Musée, Clair 05617
La Petite Anglicane, Forestville 05805
Petrolia Discovery, Petrolia 06598
Pickering Museum Village, Pickering 06599
Pincher Creek District Museum, Pincher Creek 06606
Pinhey's Point, Ottawa 06553
Pioneer Museum Society of Grande Prairie and District, Grande Prairie 05901
Pitt Meadows Museum, Pitt Meadows 06607
Placentia Area Museum, Placentia 06608
Plamondon and District Museum, Plamondon 06611
Plenty and District Museum, Plenty 06613
Plum Coulee and District Museum, Plum Coulee 06614
Port Clements Museum, Port Clements 06632
Port Colborne Historical and Marine Museum, Port Colborne 06633
Port Hardy Museum, Port Hardy 06638
Port-la-Joye-Fort Amherst, Rocky Point 06770
Port Moody Station Museum, Port Moody 06646
Port Union Museum, Port Union South 06649
Poste de Traite Chauvin, Tadoussac 07058
Pouch Cove Museum, Pouch Cove 06653
Prairie Memories Museum, Irvine 06024
Prairie Panorama Museum, Czar 05670
Prairie Pioneer Museum, Craik 05656
Prairie River Museum, Prairie River 06655
Prairie West Historical Centre, Eston 05785
Presqu'île Provincial Park Museum, Brighton 05477
Prince Albert Historical Museum, Prince Albert 06664
Princeton and District Museum, Princeton 06674
Provincial Museum of Newfoundland and Labrador at The Rooms, Saint John's 06851
Quaco Museum and Archives, Saint Martins 06867
Queens County Museum, Liverpool 06178
Queens County Museum, Tilley House and Court House, Gagetown 05846
R.N. Atkinson Museum, Penticton 06584
Railway Station Museum, Castlegar 05576
Rainy Hills Pioneer Exhibits, Iddesleigh 06005
Rankin House Museum, Miramichi 06302
Rapid City Museum, Rapid City 06703
Raymore Pioneer Museum, Raymore 06704
RCMP Centennial Celebration Museum, Fairfield 05790
Red Lake Regional Heritage Centre, Red Lake 06711
Redcliff Museum, Redcliff 06712
Redwater and District Museum, Redwater 06713
Regina Plains Museum, Regina 06724
Restigouche Regional Museum, Dalhousie 05671
Reston and District Historical Museum, Reston 06734
Revelstoke Museum, Revelstoke 06735
Reynold Rapp Museum, Spalding 06999
Richibucto River Museum, Richibucto 06739
Richmond County Historical Society Museum, Melbourne 06259
Richmond Museum, Richmond 06745
Rideau District Museum, Westport 07309
Ridge House Museum, Ridgetown 06749
Riding Mountain National Park Interpretive Centre, Wasagaming 07281
Riverton Transportation Heritage Centre, Riverton 06760
Rocanville and District Museum, Rocanville 06765
Rockglen Rolling Hills Museum, Rockglen 06767
Rose Valley and District Heritage Museum, Rose Valley 06772
Rosebud Centennial Museum, Rosebud 06773
Rosetown District Museum, Rosetown 06775
Ross Museum, Foresters Falls 05804
Rossburn Museum, Rossburn 06776
Rouleau and District Museum, Rouleau 06779
Roulston Museum, Carstairs 05570
Royal Ontario Museum, Toronto 07133
Rusty Relics Museum, Carlyle 05565
S.S. Moyie National Historic Site, Kaslo 06047
Sachs Harbour Museum, Sachs Harbour 06787
Saint Catharines Museum, Saint Catharines 06808
Saint Claude Museum, Claude 05621
Saint Joseph Island Museum Village, Richard's Landing 06738
Saint Lewis Heritage Museum, Saint Lewis 06863
Saint Mary's Museum of Maxstone, Maxstone 06252
Saint Norbert Provincial Heritage Park, Winnipeg 07381
Saint Patrick's Church Museum, Sydney 07050
Saint Walburg and District Historical Museum, Saint Walburg 06887
Salmo Museum, Salmo 06904
Saskatchewan River Valley Museum, Hague 05924
Saskatchewan Western Development Museum, Yorkton 07415
Sault Sainte Marie Museum, Sault Sainte Marie 06934
Scarborough Historical Museum, Scarborough 06938
Scugog Shores Historical Museum Village, Port Perry 06647
Seal Island Light Museum, Barrington 05392
Seba Beach Heritage Museum, Seba Beach 06940
Sedgewick Archives Museum and Gallery, Sedgewick 06942
Selkirk College Mansbridge Kootenay Collection, Castlegar 05577
Shades of the Past, Carbonear 05559
Shamrock Museum, Shamrock 06947
Sharon Temple Museum Historic Site, Sharon 06948
Shelburne County Museum, Shelburne 06955
Shell Lake Museum, Shell Lake 06956
Shoal Lake Police and Pioneer Museum, Shoal Lake 06967
Sicamous and District Museum, Sicamous 06969
Sidney Historical Museum, Sidney 06973
Siksika Nation Museum, Siksika 06975
Simcoe County Museum, Minesing 06296
Simeon Perkins House, Liverpool 06179
Sipalaseequtt Museum Society, Pangnirtung 06567
Sir Alexander Galt Museum, Lethbridge 06162
Sir Andrew MacPhail Foundation, Orwell 06520
Sir William F. Coaker Heritage Foundation, Port Union 06648
Smith's Cove Historical Society Museum, Smith's Cove 06985
Sooke Region Museum, Art Gallery and Historic Moss Cottage, Sooke 06991
South Cariboo Historical Museum, Clinton 05624
South Grey Museum and Historical Library, Flesherton 05799
South Similkameen Museum, Keremeos 06064
Southwestern Saskatchewan Oldtimer's Museum, Maple Creek 06234
Spirit River District Museum, Spirit River 07000
Spruce Row Museum, Waterford 07283
Squamish Valley Museum, Garibaldi Heights 05851
Star City Heritage Museum, Star City 07006
Stettler Town and Country Museum, Stettler 07009
Stony Plain and District Pioneer Museum, Stony Plain 07015
Story of People, Yorkton 07416
Stoughton and District Museum, Stoughton 07017
Strasbourg and District Museum, Strasbourg 07018
Stratford-Perth Museum, Stratford 07020
Strathclair Museum, Strathclair 07021
Strathcona County Heritage Museum, Sherwood Park 06964
Strathroy Middlesex Museum, Strathroy 07022
Sturgis Station House Museum, Sturgis 07025
Sulphur Springs Station, Ancaster 05339
Summerland Museum, Summerland 07031
Sunrise Trail Museum, Tatamagouche 07061
Sunshine Coast Museum, Gibsons 05864
Surrey Museum, Surrey 07038
Swan Valley Historical Museum, Swan River 07043
Swift Current Museum, Swift Current 07045
Taber and District Museum, Taber 07053
Tabusintac Centennial Museum, Tabusintac 07054
Tahsis Museum, Tahsis 07059
Tehkummah Township Little Schoolhause and Museum, Tehkummah 07063
Telkwa Museum, Telkwa 07064
Telling the Menonite Story - Visitor Centre, Saint Jacobs 06824
Teulon and District Museum, Teulon 07070
Thelma Miles Museum, Matheson 06249
Thunder Bay Museum, Thunder Bay 07082
Timber Village Museum, Blind River 05431
Timmins Museum, Timmins 07084
Tofield Historical Museum, Tofield 07087
Torbay Museum, Torbay 07088
Trail Museum and Sports Hall of Memories, Trail 07159
Transcona Historical Museum, Winnipeg 07386
Treherne Museum, Treherne 07160
Trepassey Area Museum, Trepassey 07162
Trinity Interpretation Centre, Trinity 07166
Trochu and District Museum, Trochu 07168
Tweed and Area Heritage Centre, Tweed 07184
Two Hills and District Museum, Two Hills 07186
Ukrainian Homestead Museum, Winnipeg Beach 07395
United Counties Museum in the Wood House, Cornwall 05646
Unity and District Heritage Museum, Unity 07189
Upper Canada Village, Morrisburg 06395

Uxbridge Historical Centre Museum, Uxbridge 07191
Valley Museum, McBride 06208
Vancouver Museum, Vancouver 07229
Vanguard Centennial Museum, Vanguard 07232
Vegreville Regional Museum, Vegreville 07235
Verwood Community Museum, Verwood 07240
Victoria County Historical Society Museum, Lindsay 06173
Le Vieux Moulin, Sainte-Jeanne-d'Arc 06895
Viking Historical Museum, Viking 07262
Village Historique Acadien, Caraquet 05555
Village Québécois d'Antan, Drummondville 05715
Waba Cottage Museum, White Lake 07322
Wabowden Historical Museum, Wabowden 07270
Wadena and District Museum and Nature Centre, Wadena 07271
Wainwright Museum, Wainwright 07272
Wakaw Heritage Museum, Wakaw 07274
Wallace and Area Museum, Wallace 07275
Wallaceburg and District Museum, Wallaceburg 07276
Wanuskewin Heritage Park, Saskatoon 06926
Waskada Museum, Waskada 07282
Watson Crossley Community Museum, Grandview 05902
Wawota and District Museum, Wawota 07292
Welland Historical Museum, Welland 07294
Wellington County Museum, Fergus 05794
Wells Museum, Wells 07296
West Hants Historical Society Museum, Windsor, Nova Scotia 07343
West Parry Sound District Museum, Parry Sound 06572
West Vancouver Museum, West Vancouver 07304
Western Development Museum, Saskatoon 06927
Western Manitoulin Island Historical Society Museum, Gore Bay 05883
Wetaskiwin and District Museum, Wetaskiwin 07313
Whitbourne Museum, Whitbourne 07318
White Fox Museum, White Fox 07321
White Rock Museum, White Rock 07323
Whiteley Museum, Saint Paul's River 06873
Whitewood Historical Museum, Whitewood 07331
Whitney Pier Historical Museum, Sydney 07052
Wilkie and District Museum, Wilkie 07336
Willoughby Historical Museum, Niagara Falls 06450
Willow Bunch Museum, Willow Bunch 07340
Windermere Valley Museum, Invermere 06016
Windsor's Community Museum, Windsor, Ontario 07347
Winnipegosis Museum, Winnipegosis 07396
Wolseley and District Museum, Wolseley 07400
Wolverine Hobby and Historical Society Museum, Spy Hill 07003
Woodlands Pioneer Museum, Woodlands 07403
Woodstock Museum, Woodstock 07406
Workers Arts and Heritage Centre, Hamilton 05964
Yester-Year Artifacts Museum, Rowley 06782
Yester-Years Community Museum, Dinsmore 05704
Ymir Museum, Ymir 07412
York Sunbury Museum, Fredericton 05842

Chad

Musée d'Abéché, Abéché 07423
Musée National de Sarh, Sarh 07426
Musée National N'Djamena, N'Djamena 07425

Chile

Museo Casa de Isla Negra, San Felipe 07448
Museo de Antofagasta, Antofagasta 07429
Museo Regional de Iquique, Iquique 07439

China

A City Jinshangjing History Museum, Acheng 07476
Anji Museum, Anji 07478
Anlu Museum, Anlu 07480
Anqing Museum, Anqing 07481
Anqiu Museum, Anqiu 07482
Anshan Museum, Anshan 07483
Anshun Zone Museum, Anshun 07484
Anxi Museum, Anxi, Fujian 07485
Anxi Museum, Anxi, Gansu 07486
Anyang Museum, Anyang 07488
Anyuan Museum, Anyuan 07490
Aohan Qi Museum, Aohan Qi 07491
Badong Museum, Badong 07493
Baicheng Museum, Baicheng 07494
Bairin Youqi Museum, Balin Youqi 07497
Bairin Zuoqi Museum, Bairin Zuoqi 07495
Bangbu Museum, Bangbu 07498
Baoji Museum, Baoji 07500
Baoji Xiàn Museum, Baoji 07501
Baoshan Museum, Baoshan 07502
Bayin Guoleng Zhou Museum, Kuerle 07915
Beijing Museum, Beijing 07517
Beiliu Museum, Beiliu 07578
Benxi Museum, Benxi 07579
Bo'ai Museum, Bo'ai 07581
Boer Tala Zhou Museum and Gallery, Bole 07582
Boyang Museum, Boyang 07583
Bozhou Museum, Bozhou 07584
Caidian District Museum, Wuhan 08330
Changdao Museum, Changdao 07590
Changji Mulei County Museum, Changji 07592
Changjizhou Museum, Changji 07593
Changqing Museum, Changqing 07595
Changsha Museum, Changsha 07596
Changshu Museum, Changshu 07600
Changtai Museum, Changtai 07601
Changting Museum, Changting 07602
Changwu Museum, Changwu 07603
Changxing Museum, Changxing 07604
Changzhi Museum, Changzhi 07605
Changzhou Museum, Changzhou 07606
Chaotiangong Palace, Nanjing 08038
Chaoyang Museum, Chaoyang 07608
Chaoyang Museum, Chaoyang, Liaoning 07610
Chaozhou Museum, Chaozhou 07611
Chengdu Museum, Chengdu 07614
Chenghai Museum, Chenghai 07624
Chi City Museum, Chicheng 07625
Chifeng Museum, Chifeng 07626
China Suxiu Art Museum, Suzhou 08253
Chongming Museum, Shanghai 08163
Chuxiongzhou Museum, Chuxiong 07633
Da'an Museum, Da'an 07634
Dagu Paotai Yi Zhi Museum, Tianjin 08281
Dai County Museum, Daixian 07635
Dali Museum, Dali 07636
Dalizhou Museum, Dali 07637
Dan City Museum, Dancheng 07645
Dangyang Museum, Dangyang 07647
Dapu Museum, Dapu 07648
Daqing Dumeng Museum, Daqing 07649
Datian Museum, Datian 07650
Datong Museum, Datong 07651
Dawu Museum, Dawu 07652
Daxin Museum, Daxin 07653
Daye Museum, Daye 07654
De'an Museum, De'an 07657
Dengzhou Guchuan Museum, Penglai 08078
Deqing Museum, Deqing, Guangdong 07660
Deqing Museum, Deqing, Zhejiang 07661
Dexing Museum, Dexing 07662
Deyang Museum, Deyang 07663
Dianbai Museum, Dianbai 07664
Dingxi Zone Museum, Dingxi 07665
Dingzhou Museum, Dingzhou 07666
Donggang Museum, Donggou 07667
Dongguan Museum, Dongguan 07668
Dongshan Museum, Dongshan 07670
Dongtai Museum, Dongtai 07671
Dongying History Museum, Guangrao 07722
Doumen Museum, Doumen 07673
Duchang Museum, Douchang 07672
Dunhuang Museum, Dunhuang 07677
Emeishan Museum, Emei 07678
Engping Museum, Enping 07680
Erlianhaote Saurian Museum, Erlianhaote 07681
Ezhou Museum, Ezhou 07682
Fanchang Museum, Fanchang 07683
Fang County Museum, Fangxian 07684
Feng City Museum, Fengcheng 07686
Fenghua Xikou Museum, Fenghua 07687
Fengkai Museum, Fengkai 07688
Fengshun Museum, Fengshun 07689
Fengxian Museum, Shanghai 08166
Fengxiang Museum, Fengxiang 07690
Fenxi Museum, Linfen 07940
Fenyang Museum, Fenyang 07691
Fenyi Museum, Fenyi 07692
Foshan Museum, Foshan 07693
Fufeng Museum, Fufeng 07695
Fugou Museum, Fugou 07698
Fujian Museum, Fuzhou 07706
Fuling Museum, Fuling 07699
Fullyjie Zone Museum, Bijie 07580
Fushun Museum, Fushun 07700
Fuxin Museum, Fuxin 07702
Fuyang Museum, Fuyang, Anhui 07703
Fuyang Museum, Fuyang, Zhejiang 07704
Fuyu Museum, Fuyu 07705
Fuzhou Cangshan Museum, Fuzhou 07707
Gan County Museum, Ganxian 07709
Gannan Storezu Zone Museum, Xiahe 08352
Gansu Museum, Lanzhou 07921
Ganyu Museum, Ganyu 07710
Ganzhou Museum, Ganzhou 07711
Ganzi Storezu Museum, Kangding 07913
Gao'an Museum, Gao'an 07713
Gaomi Museum, Gaomi 07714
Gaoming Museum, Gaoming 07715
Gaowish Museum, Zhaoqing 08493
Gaozhou Museum, Gaozhou 07716
Gu City Museum, Gucheng, Hubei 07739
Guandu Museum, Guandu 07719
Guangdong Beginxing Museum, Shixing 08236
Guangning Museum, Guangning 07721
Guangshui Museum, Xiaogan 08377
Guangzhou Liwan District Museum, Guangzhou 07730
Guangzhou Museum, Guangzhou 07732
Gucity Museum, Gucheng, Jiangsu 07740
Guilin Museum, Guilin 07741
Guiping Museum, Guiping 07745
Guizhou Provincial Museum, Guiyang 07746
Gutian Museum, Gutian 07748
Guyuan Museum, Guyuan 07749
Haian Museum, Haian 07750
Haikou Museum, Haikou 07751
Haining Museum, Haining 07753
Haiyan Museum, Haiyan 07754
Haiyang Museum, Haiyang 07755
Hancheng Museum, Hancheng 07756
Hanchuan Museum, Hanchuan 07757
Handan Museum, Handan 07758
Harbin Museum, Harbin 07769
Harbin Museum, Harbin 07770
He County Museum, Hexian 07785
Hebi Museum, Hebi 07773
Hefeng Museum, Hefeng 07777
Heihe Museum, Heihe 07778
Hengfeng Museum, Hengfeng 07780
Hengyang Museum, Hengyang 07781
Hepu Museum, Hepu 07783
Heshan Museum, Heshan 07784
Heyuan Museum, Heyuan 07786
Heze Museum, Heze 07787
Hohhot City Museum, Hohhot 07789
Hong Kong Heritage Museum, Hong Kong 07796
Hong Kong Museum of History, Hong Kong 07800
Hongdong Museum, Hongdong 07819
Honghu Revolution History Museum, Honghu 07820
Huai'an Museum, Huai'an 07821
Huaibei Museum, Huaibei 07824
Huaihua Zone Museum, Huaihua 07825
Huaiji Museum, Huaiji 07826
Huainan Museum, Huainan 07827
Huaiyang Museum, Huaiyang 07828
Huaiyin Museum, Huai'an 07822
Huanggang Museum, Huanggang 07830
Huangmei Museum, Huangmei 07831
Huangshan Museum, Huangshan 07834
Huangshi Museum, Huangshi 07838
Huangyan Museum, Huangyan 07839
Huangzhong Museum, Huangzhong 07840
Huangzhou Museum, Huangzhou 07841
Huating Museum, Huating 07842
Huazhou Museum, Huazhou 07843
Huguan Museum, Huguan 07844
Hui'an Museum, Hui'an 07845
Huichang Museum, Huichang 07846
Huidong Museum, Huidong 07847
Huilai Museum, Huilai 07848
Huizhou Museum, Huizhou 07849
Huludao Museum, Huludao 07850
Huzhou Museum, Huzhou 07851
Inner Mongolia Museum, Hohhot 07790
Jiading District Museum, Shanghai 08168
Jian City Museum, Jian, Jilin 07854
Jian Museum, Jian, Jiangxi 07853
Jiangdu Museum, Jiangdu 07857
Jiangmen Museum, Jiangmen 07860
Jiangning Museum, Jiangning 07861
Jiangshan Museum, Jiangshan 07862
Jiangxian Museum, Jiangxian 07863
Jiangyin Museum, Jiangyin 07864
Jian'ou Museum, Jian'ou 07866
Jianyang Museum, Jianyang 07867
Jiaoling Museum, Jiaoling 07868
Jiaonan Museum, Jiaonan 07869
Jiaozhou Museum, Jiaozhou 07870
Jiaozuo Museum, Jiaozuo 07871
Jiashan Museum, Jiashan 07872
Jiaxing Museum, Jiaxing 07873
Jiayuguan Changcity Museum, Jiayuguan 07874
Jiexi Museum, Jiexi 07875
Jiexiu Museum, Jiexiu 07876
Jieyang Museum, Jieyang 07877
Jilin City Museum, Jilin 07878
Jimo Museum, Jimo 07880
Jinan Museum, Jinan 07881
Jingchuan Museum, Jingchuan 07888
Jingmen Museum, Jingmen 07891
Jingsha Jingzhou Museum, Jingsha 07893
Jingxi Chuang Museum, Jingxi 07894
Jingyang Museum, Jingyang 07895
Jining Museum, Jining 07896
Jinjiang Museum, Jinjiang 07897
Jinshan Museum, Shanghai 08169
Jinzhou District Museum, Dalian 07640
Jinzhou Museum, Jinzhou 07899
Jishan Museum, Jishan 07900
Jishui Museum, Jishui 07902
Jiujiang Museum, Jiujiang 07903
Jixian Display Center, Jixian 07905
Jiyang Museum, Jiyang 07906
Jiyuan Museum, Jiyuan 07907
Juxian Museum, Juxian 07908
Kaifeng Museum, Kaifeng 07909
Kaiping Huaqiao Museum, Kaiping 07912
Kang Xicao Museum, Beijing 07544
Kazuo Museum, Chaoyang 07609
Laiyang Museum, Laiyang 07918
Laizhou Museum, Laizhou 07919
Landlord's Manor House Museum, Dayi 07656
Lanxi Museum, Lanxi 07920
Lanzhou Museum, Lanzhou 07922
Laohekou Museum, Laohekou 07923
Le'an Museum, Le'an 07924
Lechang Museum, Lechang 07925
Leping Museum, Leping 07926
Leye Museum, Leye 07927
Li County Museum, Lixian 07967
Lianjiang Museum, Lianjiang, Fujian 07930
Lianjiang Museum, Lianjiang, Guangdong 07931
Lianping Museum, Lianping 07932
Lianshan Chuang Yaozu Self-Rule Nationality Museum, Qingyuan 08116
Lianyungang Museum, Lianyungang 07933
Lianzhou Museum, Qingyuan 08117
Liaoning Jin Modern City Museum, Shenyang 08215
Liaoyang Museum, Liaoyang 07935
Licheng District Museum, Licheng 07936
Lijiang Museum, Lijiang 07937
Liling Museum, Liling 07938
Linan Museum, Hangzhou 07764
Linchuan Museum, Linchuan 07939
Linfen Museum, Linfen 07941
Lingao Museum, Lingao 07943
Lingtai Museum, Lingtai 07944
Lingyuan Museum, Lingyuan 07946
Lingyun Museum, Lingyun 07947
Linhai Museum, Linhai 07948
Linli Museum, Linli 07949
Linqing Museum, Linqing 07950
Linquan Museum, Linquan 07951
Lintong County Museum, Lintong 07952
Linxiazhou Museum, Tiyuchang 08295
Linyi Museum, Linyi, Shandong 07955
Linyi Museum, Linyi, Shanxi 07957
Linyou Museum, Linyou 07958
Lishui Museum, Lishui, Jiangsu 07962
Lishui Museum, Lishui, Zhejiang 07963
Liuhe County Display Center, Liucheng 07964
Liuzhou Museum, Liuzhou 07966
Longchuan Museum, Longchuan 07968
Longhua Museum, Longhua 07969
Longkou Museum, Longkou 07971
Longquan Museum, Longquan 07972
Longquanyi District Museum, Chengdu 07617
Longrock Museum, Longyan 07974
Longyou Museum, Longyou 07975
Lu Xun and Qiu Jin Museum, Shaoxing 08211
Luanping Museum, Luanping 07977
Lugouqiao Display Center, Beijing 07548
Luochuan Museum, Luochuan 07979
Luoding Museum, Luoding 07980
Luohe Wieldyang Museum, Wuyang 08345
Luonan Museum, Luonan 07981
Luoyang Museum, Luoyang 07985
Lushan Museum, Lushan 07987
Lushun Museum, Dalian 07641
Ma'anshan Museum, Ma'anshan 07990
Maoming Museum, Maoming 08009
Mayang Miaozu Self-Rule Museum, Mayang 08011
Mei Museum, Meixian 08013
Mengzhou Museum, Mengzhou 08014
MenTouGou Museum, Beijing 07550
Mian County Museum, Mianxian 08015
Mianyang Museum, Mianyang 08016
Mingcheng Yuan Shi Museum, Nanjing 08042
Mingxi Museum, Mingxi 08019
Minhe Hui Tuzu Self-Rule Museum, Minhe 08020
Minle Museum, Minle 08021
Minqin Museum, Minqin 08022
Minqing Museum, Minqing 08023
Mirouge Museum, Mizhi 08025
Museum of Shanxi Province, Taiyuan 08273
Museum of Taipa Coloane History, Macau 08003
Museum of the Holy House of Mercy, Macau 08004
Nan'ao Coast Defence City Museum, Nan'ao 08027
Nanchang Museum, Nanchang 08029
Nancheng Museum, Nancheng 08034
Nanhai Museum, Nanhai 08035
Nanjiang Museum, Nanjiang 08036
Nanjing Museum, Nanjing, Zhangzhou 08052
Nankang County Museum, Nankang 08053
Nanning Museum, Nanning 08056
Nanping Museum, Nanping 08058
Nantong Bowu House, Nantong 08059
Nantong Museum, Nantong 08060
Nanxiong Museum, Nanxiong 08062
Nanyang Guan, Nanyang 08063
Nanzhang Museum, Nanzhang 08066
Nei Mongol Zhelimu Meng Guan, Tongliao 08300
Neixiang Museum, Neixiang 08068
Ningbo Baoguo Si Display Center, Ningbo 08070
Ningbo Museum, Ningbo 08071
Ningde Museum, Ningde 08073
Ningdu Museum, Ningdu 08075
Ningxia Hui Autonomous Region Museum, Yinchuan 08447
Ningyang Museum, Ningyang 08076
Pawnshop Museum, Macau 08005
Pei County Museum, Peixian 08077

Pianguan Museum, Pianguan 08079
Piaohao Museum, Pingyao 08090
Pingdu Museum, Pingdu 08081
Pinghe Museum, Pinghe 08082
Pinghu Museum, Pinghu 08084
Pingliang Museum, Pingliang 08085
Pingliang Zone Museum, Pingliang 08086
Pinglu Museum, Pinglu 08087
Pingshun Museum, Pingshun 08088
Pingxiang Museum, Pingxiang 08089
Pingyuan Museum, Pingyuan 08092
Pixian Museum, Pixian, Sichuan 08094
Pizhou Museum, Pixian, Jiangsu 08093
Pouryun Museum, Guanyun 07738
Pubei Museum, Pubei 08095
Pucheng Museum, Pucheng 08096
Pulandian Museum, Dalian 07644
Puning Museum, Puning 08097
Putian Museum, Putian 08098
Puyang Museum, Puyang 08099
Qianjiang Museum, Qianjiang 08100
Qiankou Minzai Museum, Huangshan 07835
Qiannanzhou Museum, Douyun 07674
Qianshan Museum, Qianshan, Anhui 08101
Qimen Museum, Qimen 08104
Qinan Museum, Qinan 08105
Qingdao Municipal Museum, Qingdao 08110
Qinghai Logroll Museum, Huzhu 07852
Qingliu Museum, Qingliu 08112
Qingpu Museum, Shanghai 08179
Qingyang Museum, Qingyang 08114
Qingyang Zone Museum, Xifeng 08384
Qingyuan Museum, Qingyuan 08118
Qingzhou Museum, Qingzhou 08120
Qinyang Museum, Qinyang 08121
Qinzhou Museum, Qinzhou 08123
Qiongshan Museum, Qiongshan 08124
Qishan Museum, Qishan 08125
Quanzhou Museum, Quanzhou 08131
Qujiang Museum, Qujiang 08135
Quwo Museum, Quwo 08136
Quzhou Museum, Quzhou 08137
Raoping Museum, Raoping 08138
Renhua Museum, Renhua 08139
Rinsing Museum, Qingshui 08113
Rizhao Museum, Rizhao 08140
Rong County Museum, Rongxian 08142
Rongshui Museum, Rongshui 08141
Rugao Museum, Rugao 08143
Ruichang Museum, Ruichang 08144
Ruicheng Museum, Ruicheng 08145
Saltyning Zone Museum, Xianning 08373
Sanmen Museum, Sanmen 08148
Sanmenxia Museum, Sanmenxia 08149
Sanming Museum, Sanming 08150
Sanshui Museum, Sanshui 08151
Sanxingdui Museum, Guanghan 07720
Sanya Museum, Sanya 08152
Sha Museum, Shaxian 08213
Shanggao Museum, Shanggao 08154
Shanghai Museum, Shanghai 08184
Shanghang Museum, Shanghang 08200
Shangqiu Museum, Shangqiu 08201
Shangrao County Museum, Shangrao 08203
Shangzhou Museum, Shangzhou 08205
Shantou Museum, Shantou 08207
Shaoguan Museum, Shaoguan 08209
Shaowu Museum, Shaowu 08210
Shaoxing Museum, Shaoxing 08212
Shenzhen Museum, Shenzhen 08225
Sheqi Museum, Sheqi 08228
Shi Liao Display Center, Nanjing 08049
Shicheng Museum, Shicheng 08229
Shijiazhuang Museum, Shijiazhuang 08233
Shikumen Open House Museum, Shanghai 08192
Shimen Museum, Shimen 08234
Shishou Museum, Shishou 08235
Shou Museum, Shouxian 08240
Shouguang Museum, Shouguang 08239
Shunchang Museum, Shunchang 08242
Sihui Museum, Sihui 08243
Siping Museum, Siping 08244
Site of the 1st National Congress of the CCP, Shanghai 08193
Songjiang Museum, Shanghai 08196
Songyang Museum, Songyang 08245
Spring Museum, Yangchun 08421
Startqing Museum, Zhaoqing 08494
Suichuan Museum, Suichuan 08247
Suide Museum, Suide 08248
Suining Museum, Suining, Jiangsu 08249
Suining Museum, Suining, Sichuan 08250
Suixi Museum, Suixi 08251
Sumove Museum, Suqian 08252
Suzhou Museum, Suzhou 08258
Tai Hao Ling Museum, Huaiyang 07829
Tai'an Museum, Tai'an 08263
Taicang Museum, Taicang 08264
Taihe Museum, Taihe 08266
Tailai Museum, Tailai 08267
Taining Museum, Taining 08268
Taishan Museum, Taishan 08269
Taishun Museum, Taishun 08270
Taizhou Museum, Taizhou 08276
Tangshan Museum, Tangshan 08277
Teng County Museum, Tengxian 08279
Tengzhou Museum, Tengzhou 08280
Term City Museum, Xiangcheng 08367
Tianjin History Museum, Tianjin 08285
Tianshui Museum, Tianshui 08292
Tiantai Museum, Tiantai 08293
Tibet Museum, Lhasa 07929
Tieling Museum, Tieling 08294
Tongan Museum, Tongan 08296
Tongcheng Museum, Tongcheng 08297
Tongzhou Museum, Beijing 07571
Tumen Museum, Tumen 08305
Tungxiang Museum, Tongxiang 08303
Tunxi Museum, Huangshan 07836
Tuoketuo Museum, Tuoketuo 08306
Wafangdian Museum, Wafangdian 08310
Wallqu Museum, Yuanqu 08466
Wan Zhou District Museum, Chongqing 07632
Wangfu Museum, Naiman Qi 08026
Wangjiang Museum, Wangjiang 08311
Wansanxia Museum, Wanxian 08313
Wanxi Museum, Lu'an 07976
Wei County Museum, Weixian 08321
Weifang Museum, Weifang 08315
Weihai Museum, Weihai 08317
Weihui Museum, Weihui 08318
Weinan Center Museum, Weinan 08320
Wendeng Museum, Wendeng 08322
Wengyuan Museum, Wengyuan 08323
Wenshanzhou Museum, Wenshan 08324
Wenshui Museum, Wenshui 08325
Wenxi Museum, Wenxi 08326
Wenzhou Museum, Wenzhou 08327
Wouldle Museum, Jiangle 07858
Wuchuan Museum, Wuchuan 08329
Wuhan Museum, Wuhan 08331
Wuhole Museum, Xiaogan 08378
Wuhou Museum, Chengdu 07623
Wuhua Museum, Wuhua 08334
Wujiang Museum, Wujiang 08335
Wujin Museum, Wujin 08336
Wuping Museum, Wuping 08338
Wuwei Museum, Wuwei 08339
Wuwei Zone Museum, Wuwei 08340
Wuxi City Museum, Wuxi 08342
Wuyi Museum, Wuyi 08346
Wuyishan Museum, Wuyishan 08347
Wuyuan Museum, Wuyuan 08349
Wuzhangyuan-Zhu Geliangmiao Museum, Qishan 08126
Wuzhi Museum, Wuzhi 08350
Wuzhou Museum, Wuzhou 08351
Xi County Museum, Huangshan 07837
Xiajiang Museum, Xiajiang 08353
Xiamen Museum, Xiamen 08356
Xian Beilin Museum, Xian 08363
Xiangfan Museum, Xiangfan 08368
Xiangfen Museum, Xiangfen 08370
Xiangtan Museum, Xiangtan 08371
Xiangwall Museum, Xianyuan 08376
Xiangxi Zone Museum, Jishou 07901
Xiangxiang Museum, Xiangxiang 08372
Xianyang Museum, Xianyang 08374
Xianyou Museum, Xianyou 08375
Xiaogan Museum, Xiaogan 08379
Xiaoshan Museum, Xiaoshan 08380
Xiaoyi Museum, Xiaoyi 08381
Xifeng Museum, Xifeng 08385
Xilin Museum, Xilin 08386
Xincity Tusi Museum, Xincheng 08388
Xinfeng Museum, Xinfeng, Guangdong 08389
Xinfeng Museum, Xinfeng, Jiangxi 08390
Xingan-meng Museum, Wulanhoate 08337
Xingan Museum, Xingan 08391
Xinghua Museum, Xinghua 08392
Xingning Museum, Xingning 08393
Xinjian Museum, Xinjian 08398
Xinjiang Xian Museum, Xinjiang 08399
Xinmi Museum, Xinmi 08400
Xintai Museum, Xintai 08403
Xinxiang Museum, Xinxiang 08404
Xinyi Museum, Xinyi, Guangdong 08405
Xinyi Museum, Xinyi, Jiangsu 08406
Xinyu Museum, Xinyu 08407
Xishui Museum, Xishui 08409
Xiuning Museum, Xiuning 08410
Xuanzhou Museum, Xuanzhou 08412
Xuchang Museum, Xuchang 08413
Xunyang Museum, Xunyang 08414
Xunyi Museum, Xunyi 08416
Xuyi Museum, Huai'an 07823
Xuzhou Museum, Xuzhou 08418
Yancheng Museum, Yancheng 08419
Yanchi Museum, Yanchi 08420
Yang County Antiquity Museum, Yangxian 08426
Yangjiang Museum, Yangjiang 08422
Yangliuqing Museum, Tianjin 08291
Yangshan Museum, Yangshan 08424
Yangxin Museum, Yangxin 08427
Yangzhou Museum, Yangzhou 08429
Yantai Museum, Yantai 08433
Yanzhou Museum, Yanzhou 08434
Yao County Museum, Yaoxian 08436
Yao'an Museum, Yao'an 08435
Yaoyao Wangshan Museum, Yaoxian 08437
Yi City Museum, Yicheng, Hubei 08440
Yibin Museum, Yibin 08438
Yichang Museum, Yichang 08439
Yicheng Museum, Yicheng, Shandong 08441
Yichun Museum, Yichun, Jiangxi 08443
Yifeng Museum, Yifeng 08445
Ying City Museum, Yingcheng 08448
Yingkou Museum, Yingkou 08450
Yingtan Museum, Yingtan 08451
Yiwu Museum, Yiwu 08452
Yiyang Museum, Yiyang, Hunan 08454
Yiyang Museum, Yiyang, Jiangxi 08455
Yizheng Museum, Yizheng 08456
Yongan Museum, Yongan 08457
Yongchun Museum, Yongchun 08458
Yongding Museum, Yongding 08459
Yongfeng County Museum, Enjiang 07679
Yongji Museum, Yongji 08460
Yongtai Museum, Yongtai 08461
Youxi Museum, Youxi 08462
Yuanling Museum, Yuanling 08464
Yuanping Museum, Yuanping 08465
Yudu Museum, Yudu 08467
Yulin Museum, Yulin 08468
Yumen Museum, Yumen 08469
Yun County Museum, Shiyanyun 08238
Yunan Museum, Yunan 08470
Yuncheng Museum, Yuncheng 08471
Yunfu Museum, Yunfu 08473
Yunmeng Museum, Yunmeng 08474
Yushan Museum, Yushan 08475
Yuzhong Museum, Yuzhong 08479
Zaoyang Museum, Zaoyang 08480
Zaozhuang Museum, Zaozhuang 08481
Zengcheng Museum, Zengcheng 08482
Zhangheng Museum, Nanyang 08064
Zhangjiachuan Museum, Zhangjiachuan 08483
Zhangjiakou Museum, Zhangjiakou 08485
Zhangping Museum, Zhangping 08486
Zhangpu Museum, Zhangpu 08487
Zhangqiu Museum, Zhangqiu 08488
Zhangshu Museum, Zhangshu 08489
Zhangye Museum, Zhangye 08490
Zhangzhou Museum, Zhangzhou 08491
Zhanjiang Museum, Zhanjiang 08492
Zhaotong Museum, Zhaotong 08495
Zhengding County Display Center, Zhengding 08496
Zhengjiatun Museum, Shuangliao 08241
Zhengzhou Museum, Zhengzhou 08500
Zhenjiang Museum, Zhenjiang 08502
Zhi City Museum, Zhicheng 08504
Zhijiang Museum, Zhijiang 08505
Zhongshan Museum, Zhongshan 08507
Zhou Hejing County Museum, Hejing 07779
Zhoukou Zone Museum, Zhoukou 08508
Zhoushan Museum, Zhoushan 08509
Zhuanglang Museum, Zhuanglang 08510
Zhucheng Museum, Zhucheng 08511
Zhuhai Museum, Zhuhai 08513
Zhuzhou Museum, Zhuzhou 08515
Zibo Museum, Zibo 08516
Zijin Museum, Zijin 08520

Colombia

Casa-Museo Jorge Eliécer Gaitán, Bogotá 08545
Casa Museo Quevedo Zornoza, Zipaquirá 08790
Casa Natal de Aquileo Parra, Barichara 08535
Museo Academia de Historia Leonardo Tascón, Buga 08602
Museo Anzoátegui, Pamplona 08703
Museo Arqueológico de Muzo, Muzo 08693
Museo de Historia Natural, Manizales 08669
Museo de la Casa de la Cultura, Pinchote 08722
Museo de la Sal, Zipaquirá 08791
Museo de los Andes, Socha 08758
Museo de Tradiciones Populares de Aguadas, Aguadas 08523
Museo del Centro de Historia de Tuluá, Tuluá 08770
Museo del Disco, Zipacón 08789
Museo del Samán, Guacarí 08649
Museo Ethnohistórico, Soatá 08757
Museo Histórico, El Carmen de Bolívar 08634
Museo Histórico Anton García de Bonilla, Ocaña 08699
Museo Histórico de Antioquia, Medellín 08683
Museo Histórico y Arqueológico, Belén de Umbría 08542
Museo Hotel Marquetá, Mariquita 08673
Museo Jorge Eliécer Gaitán, Puerto Gaitán 08732
Museo Municipal de Historia, Necolí 08694
Museo Nacional del Sombrero, Aguadas 08524
Museo Regional Guayupe de Puerto Santander, Puerto Santander 08733

Congo, Democratic Republic

Musée de Mayombe, Kinshasa 08796
Musée Ethnolopique Provincial, Tshikappa 08804
Musée National de Kananga, Kananga 08794
Musée National de Lubumbashi, Lumumbashi 08801
Musées Universitaire, Kinshasa 08799

Costa Rica

Museo Histórico La Casona de Santa Rosa, Liberia 08810

Côte d'Ivoire

Maison du Patrimoine Culturel, Grand-Bassam 08835
Musée Adja Swa, Yamoussoukro 08840
Musée Adjaman Mihiin Raphaël, Vavoua 08839
Musée Binger de Zaranou, Abengourou 08827
Musée Charles et Marguerithe Bieth, Abengourou 08828
Musée de Bonoua, Bonoua 08832
Musée Don Bosco, Doékoué 08834
Musée du Parc M'Ploussoue, Bonoua 08833
Musée Peleforo Gbon Coulibaly, Korhogo 08837

Croatia

Arheološki i Prirodoslovni Muzej, Brijuni 08848
Etnografski Muzej Istre, Pazin 08926
Gradski Muzej, Križevci 08902
Gradski Muzej Bakar, Bakar 08841
Gradski Muzej Jastrebarsko i Gradska Galerija, Jastrebarsko 08884
Gradski Muzej Karlovac, Karlovac 08886
Gradski Muzej Korčula, Korčula 08894
Gradski muzej Nova Gradiška, Nova Gradiška 08916
Gradski Muzej Varaždin, Varaždin 08975
Kaštela Gradski Muzej, Kaštel Lukšić 08887
Kninski Muzej, Knin 08889
Muzej Belišće, Belišće 08843
Muzej Brdovec, Prigorje Brdověko 08931
Muzej Cetinjske Krajine, Sinj 08950
Muzej Đakovština, Đakovo 08857
Muzej Gacka, Otočac 08924
Muzej Grada Koprivnice, Koprivnica 08892
Muzej Grada Pazina, Pazin 08927
Muzej Grada Trogira, Trogir 08970
Muzej Grada Zagreba, Zagreb 09004
Muzej Ljudevit Gaj, Krapina 08901
Muzej Moslavine, Kutina 08904
Muzej Prigorja, Sesvete 08947
Muzej Turopolja, Velika Gorica 08977
Muzejska Zbirka, Imotski 08882
Narodni Muzej, Zadar 08983
Narodni Muzej Novi Vinodolski, Novi Vinodolski 08917
Samoborski Muzej, Samobor 08944
Stjepan Gruber Museum, Županja 09019
Zavičajni Muzej, Buzet 08851
Zavičajni Muzej, Ogulin 08919
Zavičajni Muzej Grada Rovinja, Rovinj 08943
Zavičajni Muzej Našice, Našice 08912
Zavičajni Muzej Ozlja, Ozalj 08925
Zavičajni Muzej Poreštine, Poreč 08929
Zbirka Baltazara Bogišića, Cavtat 08854

Cuba

Antiguo Ayuntamiento y Cárcel de Bejucal, Bejucal 09038
Museo Casa Natal de Antonio Maceo, Santiago de Cuba 09306
Museo Comunidad Las Terrazas, San Cristobal 09276
Museo de Bahía de Cochinos, Playa Girón 09256
Museo de Frank País, Holguín 09181
Museo de la Ciudad de La Habana, La Habana 09136
Museo de la Constitución, Guáimaro 09087
Museo de la Piratería, Santiago de Cuba 09315
Museo de la Provincia de Granma, Bayamo 09034
Museo de los Combates contra los Bandidos, Trinidad 09336
Museo Historia Municipal, Antilla 09026
Museo Histórico, Pinar del Río 09249
Museo Historico, San Antonio de los Baños 09274
Museo Histórico Municipal de Regla, Regla 09264
Museo Historico Provincial, Cienfuegos 09068
Museo Histórico Provincial, Sancti-Spíritus 09287
Museo Histórico Provincial de Guantánamo, Guantánamo 09092
Museo Isabel Rubio, Isabel Rubio 09185
Museo Melia Cayo Coco y Museo Tryp Colonial, Cayo Coco 09059
Museo Municipal, Ciego de Avila 09063
Museo Municipal, Holguín 09183
Museo Municipal, Sancti-Spíritus 09288
Museo Municipal Alex Urquiola, El Salvador 09270
Museo Municipal de 10 de Octubre, La Habana 09153
Museo Municipal de Abreus, Abreus 09021

Museo Municipal de Aguada de Pasajeros, Aguada de Pasajeros 09022
Museo Municipal de Alto Songo y La Maya, Alto Songo 09023
Museo Municipal de Amancio, Amancio 09025
Museo Municipal de Báguano, Báguano 09027
Museo Municipal de Bahía Honda, Bahía Honda 09028
Museo Municipal de Baraguá, Baraguá 09031
Museo Municipal de Bartolomé Masó, Bartolomé Masó 09032
Museo Municipal de Bayamo, Bayamo 09036
Museo Municipal de Bolivia, Bolivia 09040
Museo Municipal de Buey Arriba, Buey Arriba 09041
Museo Municipal de Cabaiguán, Cabaiguán 09042
Museo Municipal de Cacomun, Cacomun 09043
Museo Municipal de Caibarién, Caibarién 09044
Museo Municipal de Caimanera, Caimanera 09045
Museo Municipal de Calimete, Calimete 09047
Museo Municipal de Camajuaní Hermanos Vidal Caro, Camajuaní 09053
Museo Municipal de Campechuela, Campechuela 09054
Museo Municipal de Candelaria, Candelaria 09055
Museo Municipal de Carlos Manuel de Céspedes, Carlos Manuel de Céspedes 09057
Museo Municipal de Cauto Cristo, Cauto Cristo 09058
Museo Municipal de Chambas, Chambas 09060
Museo Municipal de Cifuentes, Cifuentes 09069
Museo Municipal de Ciro Redondo, Ciro Redondo 09070
Museo Municipal de Colón José R. Zulueta, Colón 09072
Museo Municipal de Consolación del Sur, Consolación del Sur 09073
Museo Municipal de Contramaestre, Contramaestre 09074
Museo Municipal de Corralillo, Corralillo 09075
Museo Municipal de Cruces, Cruces 09076
Museo Municipal de Cueto, Cueto 09078
Museo Municipal de Cumanayagua, Cumanayagua 09079
Museo Municipal de Esmeralda, Esmeralda 09080
Museo Municipal de Florencia, Florencia 09081
Museo Municipal de Florida, Florida 09082
Museo Municipal de Fomento, Fomento 09083
Museo Municipal de Gibara, Gibara 09086
Museo Municipal de Guamá, La Plata 09195
Museo Municipal de Guane, Guane 09089
Museo Municipal de Guantánamo - Antigua Cárcel, Guantánamo 09093
Museo Municipal de Guisa, Guisa 09094
Museo Municipal de Habana del Este, La Habana 09155
Museo Municipal de Imías, Playitas y Cajobabo, Cajobabo 09046
Museo Municipal de Jagüey Grande, Jagüey Grande 09187
Museo Municipal de Jesús Menéndez, Jesús Menéndez 09190
Museo Municipal de Jiguaní, Jiguaní 09191
Museo Municipal de Jovellanos Domingo Mujica, Jovellanos 09194
Museo Municipal de la Lisa, La Habana 09156
Museo Municipal de La Sierpe, La Sierpe 09331
Museo Municipal de La Yaya, La Yaya 09357
Museo Municipal de Lajas, Lajas 09196
Museo Municipal de Los Arabos Clotilde García, Los Arabos 09197
Museo Municipal de Los Palacios, Los Palacios 09198
Museo Municipal de Maisí, Maisí 09200
Museo Municipal de Majagua, Majagua 09201
Museo Municipal de Majibacoa, Majibacoa 09202
Museo Municipal de Manatí Jesús Suárez Gayol, Manatí 09203
Museo Municipal de Manicaragua, Manicaragua 09204
Museo Municipal de Mantua, Mantua 09205
Museo Municipal de Manuel Tames, Manuel Tames 09206
Museo Municipal de Manzanillo, Manzanillo 09209
Museo Municipal de Marianao, La Habana 09157
Museo Municipal de Martí, Martí 09210
Museo Municipal de Mayarí, Mayarí 09220
Museo Municipal de Media Luna, Media Luna 09222
Museo Municipal de Minas, Minas 09223
Museo Municipal de Minas de Matahambre, Minas de Matahambre 09224
Museo Municipal de Moa, Moa 09225
Museo Municipal de Morón, Morón 09227
Museo Municipal de Najasa, Sibanicú 09329
Museo Municipal de Niceto Pérez, Niceto Pérez 09228
Museo Municipal de Niquero, Niquero 09229
Museo Municipal de Nuevitas, Nuevitas 09239
Museo Municipal de Palma Soriano, Palma Soriano 09241
Museo Municipal de Pedro Betancourt, Pedro Betancourt 09243
Museo Municipal de Perico, Perico 09244
Museo Municipal de Pilón, Pilón 09245
Museo Municipal de Placetas, Placetas 09255
Museo Municipal de Poblado de Jimaguayú, Jimaguayú 09192
Museo Municipal de Primero de Enero, Primero de Enero 09257
Museo Municipal de Quemado de Güines, Quemado de Güines 09261
Museo Municipal de Rafael Freyre, Rafael Freyre 09262
Museo Municipal de Río Cauto, Río Cauto 09266
Museo Municipal de Rodas, Rodas 09267
Museo Municipal de Sagua de Tánamo, Sagua de Tánamo 09268
Museo Municipal de Sagua la Grande, Sagua la Grande 09269
Museo Municipal de San Cristobal, San Cristobal 09277
Museo Municipal de San Juan y Martínez, San Juan y Martínez 09279
Museo Municipal de San Luis 29 de Abril, San Luis 09280
Museo Municipal de San Miguel del Padrón, La Habana 09158
Museo Municipal de Sandino, Sandino 09291
Museo Municipal de Santa Cruz del Sur, Santa Cruz del Sur 09301
Museo Municipal de Santo Domingo, Santo Domingo 09328
Museo Municipal de Sibanicú, Sibanicú 09330
Museo Municipal de Sierra de Cubitas, Cubitas 09077
Museo Municipal de Taguasco, Taguasco 09333
Museo Municipal de Trinidad, Trinidad 09337
Museo Municipal de Unión de Reyes Juan G. Gómez, Unión de Reyes 09339
Museo Municipal de Urbano Noris, Urbano Noris 09340
Museo Municipal de Varadero, Varadero 09341
Museo Municipal de Vedado, La Habana 09159
Museo Municipal de Venezuela, Venezuela 09342
Museo Municipal de Vertientes, Vertientes 09343
Museo Municipal de Viñales Adela Azcuy Labrador, Pinar del Río 09251
Museo Municipal de Yaguajay, Yaguajay 09354
Museo Municipal de Yara, Yara 09355
Museo Municipal de Yateras, Yateras 09356
Museo Municipal del Cerro, La Habana 09160
Museo Municipal Emilio Daudinot, San Antonio del Sur 09275
Museo Municipal Fernando García Grave de Peralta, Puerto Padre 09260
Museo Municipal Francisco Javier Balmaseda, Santa Clara 09298
Museo Municipal Jatibonico, Jatibonico 09189
Museo Municipal Juan M. Ameijeiras, El Salvador 09271
Museo Municipal Nueva Gerona, Nueva Gerona 09237
Museo Municipal Roberto Rojas Tamayo, Colombia 09071
Museo Municipal Rosendo Arteaga, Jobabo Dos 09193
Museo Oscar María de Rojas, Cárdenas 09056
Museo Pedrín Troya, San Nicolás de Bari 09281
Museo Polivalente de Palmira, Palmira 09242
Museo Presidio Modelo, Nueva Gerona 09238
Museo Provincial, Ciego de Avila 09064
Museo Provincial de Holguín La Periquera, Holguín 09184
Museo Provincial de Matanzas, Matanzas 09218
Museo Provincial de Villa Clara y Museo Municipal, Santa Clara 09300
Museo Provincial Don Emilio Bacardí Moreau, Santiago de Cuba 09326
Museo Provincial General Vicente García, Victoria de las Tunas 09349
Museo Provincial Ignacio Agramonte, La Vigía 09351
Museo San Miguel del Padrón, La Habana 09172

Curaçao

Curaçao Museum, Willemstad 09358

Cyprus

Palaepaphos Museum, Kouklia 09364
Thalassa - Aghia Napa Municipal Museum, Aghia Napa 09361

Czech Republic

Brdský Památník - Městské Muzeum, Rožmitál pod Třemšínem 09824
Bredovský Letohrádek na Lemberku, Jablonné v Podještědí 09546
Centrum Bohuslava Martinů v Poličce, Polička 09727
Chebský Hrad, Cheb 09472
Expozice Městského Vlastivědného Mudruňkovo Muzeuma na Městském úřadu, Říčany u Prahy 09816
Fričovo Muzeum, Lázně Bělohrad 09619
Galerie Ludvika Kuby a Městské Muzeum, Březnice 09418
Horácké Muzeum, Nové Město na Moravě 09679
Hrad Roštejn, Telč 09860
Jandovo Muzeum - Vodní hrad, Budyně nad Ohří 09448
Jihočeské Muzeum v Českých Budějovicích, České Budějovice 09464
Karlovarské Muzeum, Karlovy Vary 09575
Karlovské Muzeum, Velké Karlovice 09900
Kotulova Dřevěnka, Havířov 09508
Krajské Muzeum Cheb, Cheb 09473
Krkonošské Muzeum - Augustiniánský Klášter, Vrchlabí 09914
Masarykovo Muzeum, Hodonín 09519
Měnínská Brána, Brno 09431
Městská Galerie Josefa Lieslerova, Kadaň 09567
Městská Galerie Karla Havlíčka, Kadaň 09568
Městské Muzeum, Bystřice nad Pernštejnem 09449
Městské Muzeum, Chrast u Chrudimě 09480
Městské Muzeum, Chrastava 09482
Městské Muzeum, Dobruška 09493
Městské Muzeum, Frenštát pod Radhoštěm 09503
Městské Muzeum, Horažďovice 09523
Městské muzeum, Kamenice nad Lipou 09570
Městské Muzeum, Klobouky u Brna 09588
Městské Muzeum, Králíky 09603
Městské Muzeum, Ledeč nad Sázavou 09621
Městské Muzeum, Letohrad 09623
Městské Muzeum, Nepomuk 09673
Městské Muzeum, Nové Město nad Metují 09681
Městské Muzeum, Nový Bydžov 09684
Městské Muzeum, Přelouč 09803
Meštské Muzeum, Přibyslav 09808
Městské Muzeum, Protivín 09811
Městské Muzeum, Rýmařov 09833
Městské Muzeum, Sadská 09834
Městské Muzeum, Sedlčany 09835
Městské Muzeum, Skuteč 09837
Městské Muzeum, Strážnice 09852
Městské Muzeum, Valašské Klobouky 09895
Městské Muzeum, Velká Bíteš 09899
Městské Muzeum, Vimperk 09907
Městské Muzeum, Žamberk 09925
Městské Muzeum, Zbiroh 09928
Městské Muzeum, Moravský Krumlov 09664
Městské Muzeum, Krnov 09609
Městské Muzeum, Napajedla 09670
Městské Muzeum, Bílovec 09407
Městské Muzeum a Galerie, Břeclav 09417
Městské Muzeum a Galerie, Dvůr Králové nad Labem 09500
Městské Muzeum a Galerie, Lomnice nad Popelkou 09637
Městské Muzeum a Galerie, Svitavy 09857
Městské Muzeum a Galerie, Vodňany 09911
Městské Muzeum a Galerie, Hranice 09536
Městské Muzeum - Františkánský klášter, Kadaň 09569
Městské Muzeum v Kralupech nad Vltavou, Kralupy nad Vltavou 09604
Městské Vlastivědné Mudruňkovo Muzeum, Říčany u Prahy 09817
Milevské Muzeum, Milevsko 09655
Místní Muzeum, Mladá Vožice 09659
Místní Muzeum, Netolice 09674
Místní Vlastividni Muzeum, Kopidlno 09594
Muzeum a Galerie Orlických Hor, Rychnov nad Kněžnou 09832
Muzeum a Galerie Severního Plzeňska, Mariánská Tynice - Kralovice 09648
Muzeum Benátky nad Jizerou, Benátky nad Jizerou 09400
Muzeum Blansko, Blansko 09409
Muzeum Boskovicka, Boskovice 09411
Muzeum Brňenska, Šlapanice 09839
Muzeum Broumovska, Broumov 09443
Muzeum Českého Krasu v Berouně - pobačka muzeum v Žebráku, Žebrák 09935
Muzeum Chodska, Domažlice 09496
Muzeum Dr. Bohuslava Horáka, Rokycany 09818
Muzeum Fojtství, Kopřivnice 09595
Muzeum Hlavního Města Prahy, Praha 09769
Muzeum Jihovýchodní Moravy, Zlín 09941
Muzeum Jindřicha Šimona Baara, Klenčí pod Čerchovem 09586
Muzeum Jindřichohradecka, Jindřichův Hradec 09565
Muzeum Josefov, Jaroměř 09551
Muzeum Klimkovice, Klimkovice 09587
Muzeum Kouřimska, Kouřim 09599
Muzeum Krupka, Krupka 09611
Muzeum Litovel, Litovel 09635
Muzeum Města Brna, Brno 09436
Muzeum Města Ústí nad Labem, Ústí nad Labem 09892
Muzeum Městec Králové, Městec Králové 09651
Muzeum Okresu Benešov, Benešov 09401
Muzeum Podkrkonoší, Trutnov 09880
Muzeum Polná, Polná 09729
Muzeum Prostějovska, Prostějov 09810
Muzeum Rumburk, Rumburk 09831
Muzeum Středního Pootaví, Strakonice 09849
Muzeum Strojíren Poldi, Kladno 09582
Muzeum T.G. Masaryka, Rakovník 09814
Muzeum Telč, Telč 09861
Muzeum Těšínska, Český Těšín 09470
Muzeum Třešť, Třešť 09878
Muzeum Týn nad Vltavou, Týn nad Vltavou 09885
Muzeum v Ivančicích, Ivančice 09543
Muzeum ve Štramberku, Štramberk 09850
Muzeum ve Žluticích, Žlutice 09945
Muzeum Velké Meziříčí, Velké Meziříčí 09902
Muzeum Vysočiny, Třebíč 09875
Muzeum Vysočiny Jihlava, Jihlava 09561
Muzeum Židovské Komunity, Tovačov 09871
Národopisné Muzeum, Nejdek 09671
Oblastní Muzeum v Lounech, Louny 09641
Okresní Muzeum, Děčín 09491
Okresní Muzeum, Litoměřice 09630
Okresní Muzeum, Vlašim 09910
Okresní Muzeum Praha-východ, Brandýs nad Labem 09413
Orlické Muzeum, Choceň 09476
Památník Bible Kralické, Kralice nad Oslavou 09602
Památník Města, Police nad Metují 09725
Památník Životické Tragedie, Havířov 09509
Pevnost Josefov - Podsemní Chodby, Jaroměř 09552
Podještědské Muzeum, Český Dub 09466
Podlipanské muzeum v Českém Brodě, Český Brod 09465
Prachatické Muzeum, Prachatice 09733
Prácheňské Muzeum, Písek 09714
Regionální Muzeum, Český Krumlov 09468
Regionální Muzeum, Mělník 09650
Regionální Muzeum, Mikulov na Moravě 09653
Regionální Muzeum, Náchod 09669
Regionální Muzeum, Žďár nad Sázavou 09934
Regionální Muzeum a Galerie, Jičín 09559
Regionální Muzeum K.A. Polánka, Žatec 09926
Regionální muzeum v Chrudimi, Chrudim 09486
Regionální Muzeum v Teplicích, Teplice 09863
Smrčkův Dům, Soběslav 09843
Stará Škola, Police nad Metují 09726
Starý Zámek, Jevišovice 09558
Státní Hrad a Zámek, Frýdlant v Čechách 09505
Státní Hrad a Zámek, Horšovský Týn 09529
Sucharduv Dům, Nová Paka 09677
Tvrz Žumberk, Žumberk 09951
Vlastivědné Muzeum, Bělá pod Bezdězem 09399
Vlastivědné Muzeum, Kyjov 09616
Vlastivědné Muzeum, Šumperk 09855
Vlastivědné Muzeum, Žirovnice 09939
Vlastivědné Muzeum Dr. Hostaše v Klatovech, Klatovy 09585
Vlastivědné Muzeum v Olomouci, Olomouc 09690
Výstavní síň Muzeum Těšínska, Jablunkov 09548
Wax Museum Český Krumlov, Český Krumlov 09469
Weisův Dům, Veselí nad Lužnicí 09904
Zámek Kinských, Valašské Meziříčí 09896
Zámek Manětín, Manětín 09647
Zámek Vsetín, Vsetín 09918
Západomoravské Muzeum v Třebíči - Muzeum Jemnice, Jemnice 09555
Znojemský Hrad, Znojmo 09949

Denmark

Amagermuseet, Dragør 09988
Anne Hvides Gård, Svendborg 10288
Århus Bymuseum, Århus 09961
Ballerup Egnsmuseum, Ballerup 09975
Bangsbo Museum, Frederikshavn 10024
Blicheregnens Museum, Kjellerup 10119
Blichermuseet på Herningsholm, Herning 10068
Brande Museum, Brande 09982
Brønshøj Museum, Brønshøj 09984
Bymuseet, Skjern 10267
Bymuseet Møntergården, Odense 10209
Dragør Museum, Dragør 09989
Ernsts Samlinger, Assens 09971
Faaborg Kulturhistoriske Museer, Faaborg 10007
Fanø Museum, Fanø 10014
Farvergården, Kerteminde 10115
Frederikssværk Bymuseum, Frederiksværk 10029
Gærup Skolemuseum og Brahetrolleborg Sogns Folkemindesamling, Faaborg 10009
Gilleleje Museum, Gilleleje 10036
Goldborgsund Museum, Nykøbing Falster 10200
Greve Museum, Greve 10042
Haderslev Museum, Haderslev 10049
Hadsund Egns Museum, Hadsund 10053
Haslev Museum, Haslev 10055
Heltborg Museum, Hurup Thy 10105
Historisk Museum, Slangerup 10277
Hørsholm Egns Museum, Hørsholm 10101
HTS-Museum, Rødovre 10239
Kalundborg og Omegns Museum, Kalundborg 10113
Karlebo Museum, Kokkedal 10161
Københavns Bymuseum, København 10135
Køge Museum, Køge 10159
Korsør By- og Overfartsmuseum, Korsør 10168
Krudtværksmuseet, Frederiksværk 10030
Læsø Museum, Læsø 10171
Lemvig Museum Vesterhus, Lemvig 10176

Løkken Museum, Løkken 10181
Middelfart Museum, Middelfart 10189
Morslands Historiske Museum, Nykøbing Mors 10204
Museerne i Fredericia, Fredericia 10021
Museet Færgegården, Jægerspris 10110
Museet for Varde By og Omegn, Varde 10307
Museet Holmen, Løgumkloster 10180
Museet på Sønderskov, Brørup 09985
Museet Tikøb Frysehus, Tikøb 10297
Næstved Museum, Næstved 10190
Nexø Museum, Nexø 10191
Nielstrup Museum, Hadsten 10051
Nordfyns Museum, Bogense 09980
Nordsjaellandsk Folkemuseum, Hillerød 10077
Nyborg og Omegns Museer, Nyborg 10199
Odder Museum, Odder 10208
Odsherreds Museum, Asnæs 09970
Ølgod Museum, Ølgod 10222
Øregaard-Museum, Hellerup 10058
Østsjællands Museum, Stevns Museum, Store Heddinge 10285
Otterup Museum, Otterup 10224
Ringe Museum, Ringe 10235
Ringkøbing Museum, Ringkøbing 10236
Ringsted Museum, Ringsted 10237
Roskilde Museum, Roskilde 10247
Sæby Museum and Sæbygaard Manor Museum, Sæby 10255
Skagen By- og Egnsmuseum, Skagen 10262
Skanderborg Museum, Skanderborg 10264
Sorø Amts Museum, Sorø 10281
Sporvejsmuseet Skjoldenæsholm, Jystrup Midtsj 10112
Sundby Samling Bryggergården, Nørresundby 10197
Sundby Samling Raschgården, Nørresundby 10198
Svendborg og Omegns Museum, Svendborg 10293
Sydsjællands Museum, Vordingborg 10320
Tåsinge Skipperhjem og Folkemindesamling, Svendborg 10294
Try Museum, Dronninglund 09991
Værløse Museum, Værløse 10304
Vejle Museum, Vejle 10311

Dominican Republic

Centro Cultural Eduardo León Jimenes, Santiago de los Caballeros 10338
Museo Alcázar de Colón, Santo Domingo 10341
Museo de Duarte, Santo Domingo 10344
Museo Histórico de Baní, Bani 10323
Museo Histórico y Arqueológico Dr. Estrada Torres, Azua 10322
Museo Imágenes de la Frontera, Dajabón 10327

East Timor

Museum Negeri Timor-Timur, Dili 10365

Ecuador

Casa de la Cultura Benjamín Carrion, Cuenca 10380
Casa Museo María Augusta Urrutia, Quito 10425
Museo Aurelio Espinosa Polit, Cotocollao 10379
Museo Aurelio Espinoza Pólit, Quito 10428
Museo Balseros del Mar del Sur, Salango 10458
Museo Cultural Iijdio Guaranga, Guaranga 10399
Museo David Paltán, San Gabriel 10460
Museo de Arte Alberto Mena Caamaño, Quito 10434
Museo de Arte e Historia Alberto Mena Caamaño, Quito 10436
Museo de la Ciudad, Quito 10441
Museo de la Comunidad de Agua Blanca, Puerto López 10424
Museo de la Comunidad de Chordeleg, Chordeleg 10377
Museo del Banco Central, Cuenca 10392
Museo del Banco Central, Loja 10417
Museo Historico, Riobamba 10457
Museo Historico Bae Calderon, Guayaquil 10407
Museo Municipal Casa de Alfaro, Montecristi 10421
Museo Quinta Casa de Juan Montalvo, Ambato 10368
Museo Quinta Juan Leon Mera, Ambato 10369
Museo Víctor Alejandro Jaramillo, Ibarra 10416

Egypt

Alexandria National Museum, Alexandria 10469
Antiquities Museum, Alexandria 10470
Densheway National Museum, Densheway Village 10523
Hadiqat al-Asmak Museum, Cairo 10498
Mallawy Museum, Mallawy 10536
Port-Said Museum, Port Said 10538
Qasr Ali Ibrahim, Cairo 10516
Saad Zaghlul Museum, Munira 10537
Sharia Museum, Cairo 10519
Tanta Museum, Tanta 10542

Estonia

Aavikute Majamuuseum, Kuressaare 10586
Dr. Fr.R. Kreutzwaldi Memoriaalmuuseum, Võru 10676
Eesti Vabaõhumuuseum, Tallinn 10630
Eestimaa Vene Muuseum, Tallinn 10631
Hanila Muuseum, Hanila 10568
Harjumaa Muuseum, Keila 10583
Heimtali Koduloomuuseum, Heimtali 10569
Heliloojate Kappide Majamuuseum, Suure-Jaani 10617
Helme Kihelkonnamuuseum, Helme 10570
Hiiumaa Muuseum, Kärdla 10578
Iisaku Muuseum, Iisaku 10573
Järvamaa Muuseum, Paide 10605
KGB Kongide Muuseum, Tartu 10652
Läänemaa Muuseum, Haapsalu 10564
Lehtse Koduloomuuseum, Lehtse 10590
Mahtra Talurahvamuuseum, Juuru 10577
Mart Saare Majamuuseum, Suure-Jaani 10618
Mihkli Talumuuseum, Kihelkonna 10584
Mõniste Muuseum, Mõniste 10595
Mõisaküla Muuseum, Mõisaküla 10594
Mõniste Muuseum, Kuutsi 10588
Muuseum, Tartu 10653
Narva Muuseum, Narva 10598
Oisu Valla Muuseum, Oisu 10601
Olustvere Muuseum, Olustvere 10602
Pärnu Muuseum, Pärnu 10604
Palamuse O. Lutsu Kihelkonnakoolimuuseum, Palamuse 10606
Põltsamaa Muuseum, Põltsamaa 10609
Põlva Talurahvamuuseum, Karilatsi 10580
Rakvere Linnus-Muuseum, Rakvere 10611
Ruhnu Muuseum, Ruhnu 10614
Ruusmäe Koduloomuuseum, Ruusmäe 10615
Saaremaa Muuseum, Kuressaare 10587
Tallinna Linnamuuseum, Tallinn 10643
Tartu Linnamuuseum, Tartu 10657
Tartumaa Muuseum, Elva 10562
Tarvastu Kihelkonnamuuseum, Tarvastu 10663
Türi Muuseum, Türi 10666
Türi Valla ja Türi Tehnika- ja Maamajanduskooli Muuseum, Särevere 10616
Valga Muuseum, Valga 10667
Vastseliina Koduloomuuseum, Vastseliina 10670
Viljandi Muuseum, Viljandi 10674
Võrumaa Muuseum, Võru 10677

Ethiopia

Castle Museum, Gonder 10684
City Museum, Yirgalem 10686

Faroe Islands

Blásastova, Norðagøta 10694
Brunn Museum, Nólsoy 10693
Bygdarsavnið í Vestmanna, Vestmanna 10703
Gamli Prestagarðurin og Bygdarsavnið, Nes, Eysturoy 10692
Málinga- og Bátasavnið, Leirvík 10691
Norðoya Fornminnissavn, Klaksvík 10690
Porkeris Bygdasavn, Porkeri 10695
Roykstovan, Kirkjubøur 10689
Sands Bygdasavn, Sandur 10696
Sørvágs Bygdasavn, Sørvágur 10697
Tvøroyrar Bygda- og Sjósavn, Tvøroyri 10700
Útisavnið í Hoyvík, Hoyvík 10688
Vágs Bygdasavn, Vágur 10701

Finland

Abraham Ojanperän Museo, Liminka 11072
Äänekosken Kaupunginmuseo, Äänekoski 10705
Ähtärin Kotiseutumuseo, Ähtäri 10706
Aholan Talomuseo, Keitele 10936
Aleksis Kiven Syntymäkoti, Palojoki 11185
Alisgårdenin Kotiseutumuseo, Lapinkylä 11043
Artjärven Kotiseutumuseo, Artjärvi 10713
Asikkalan Kotiseutumuseo, Asikkala 10714
Ekenäs Museum, Ekenäs 10721
Elimäen Kotiseutumuseo ja Koulumuseo, Elimäki 10724
Eloaitta-Kaavin Kotiseutumuseo, Kaavi 10900
Esinemuseo ja Koulumuseo, Polvijärvi 11216
Espoon Kaupunginmuseo, Espoo 10727
Etelä-Karjalan museo, Lappeenranta 11045
Gillesgården, Pohja 11214
Glims Talomuseo, Espoo 10729
Häkkilän Museo, Toholampi 11380
Hämeenlinnan Kaupungin Historiallinen Museo, Hämeenlinna 10744
Härkätien Museo, Renko 11266
Halosenniemi Museo, Tuusula 11413
Haminan Kaupunginmuseo, Hamina 10753
Hauhon Esinemuseo, Hauho 10761
Hausjärven Kotiseutumuseo, Hausjärvi 10763
Heinolan Kaupunginmuseo, Heinola 10765
Helsingin Kaupunginmuseo, Helsinki 10782
Himangan Kotiseutumuseo, Himanka 10840
Itä-Hämeen Museo, Hartola 10760
Jakobstads Museum, Jakobstad 10865
Jalasjärven Museo, Jalasjärvi 10869
Johannes Lohilammen Museo, Sammatti 11297
Joroisten Kotiseutumuseo, Joroinen 10877
Kaarlelan Kotiseutumuseo, Kokkola 10973
Kaivolan Museo, Laitila 11038
Kalajoen Kotiseutumuseo, Kalajoki 10906
Kallenaution Kievari ja Kotiseutumuseo, Juupajoki 10883
Kanavamuseo, Varkaus 11452
Kangasalan Kotiseutumuseo, Kangasala 10908
Kangaslammin Kotiseutumuseo, Kangaslampi 10911
Kangasniemen Kotiseutumuseo, Kangasniemi 10912
Kannuksen Museo, Kannus 10915
Karjalan Kotiseutumuseo, Mynämäki 11130
Karkkilan Ruukkimuseo, Karkkila 10918
Karttulan Kotiseutumuseo, Karttula 10921
Karvian Museo ja Ulkomuseoalue, Karvia 10923
Kaskö Hembygdsmuseum, Kaskinen 10925
Kaurilan Koulumuseo, Tohmajärvi 11378
Kaustisen Kotiseutumuseo, Kaustinen 10932
Keikyän Kotiseutumuseo, Keikyä 10935
Kemijärven Kotiseutumuseo, Kemijärvi 10943
Kemin Kultuurihistoriallinen Museo, Kemi 10939
Keminmaan Kotiseutumuseo, Keminmaa 10944
Kempeleen Kotiseutumuseo, Kempele 10945
Keravan Museo, Kerava 10946
Kesälahden Museo, Kesälahti 10949
Keski-Suomen Museo, Jyväskylä 10894
Kiihtelysvaaran Museo, Heinävaara 10764
Kiimingin Museo, Kiiminki 10958
Kinnulan Kotiseutumuseo, Kinnula 10960
Kiskon Kotiseutumuseo, Kisko 10964
Kivirannan Kotimuseo, Säkylä 11289
Klaavola Museo, Tuusula 11415
Komperon Kotiseutumuseo, Viinijärvi 11465
Konneveden Kotiseutumuseo, Konnevesi 10978
Kontiolahden Museo, Kontiolahti 10979
Korpilahden Kotiseutumuseo, Korpilahti 10980
Korppoon Kotiseutumuseo, Korppoo 10981
Korsnäs Hembygdsmuseum, Korsnäs 10982
Kotiseutumuseo Lukkarin Puustelli, Loppi 11081
Kovjoki Museo, Kovjoki 10993
Krookilan Kotiseutukeskus, Raisio 11251
Kuddnäs Museum, Nykarleby 11155
Kuhmalahden Kotiseutumuseo, Kuhmalahti 10998
Kuhmoisten Kotiseutumuseo, Kuhmoinen 11001
Kullaan Kotiseutumuseo, Kullaa 11004
Kuokkalan Museoraitti, Lempäälä 11059
Kuopion Korttelimuseo, Kuopio 11007
Kurun Ulkomuseo, Kuru 11015
Kustavin Kotiseutumuseo, Kustavi 11016
Kuusiston Kartano, Kuusisto 11019
Kvevlax Hembygdsmuseum, Kvevlax 11020
Kyyjärven Kotiseutumuseo, Kyyjärvi 11021
Lääninkivalteri Aschanin Talo, Heinola 10766
Längelmäen Korsumuseo, Länkipohja 11023
Lahden Kaupunginmuseo, Lahti 11029
Laihian Museo, Laihia 11037
Larsmo Hembygdsmuseum, Larsmo 11054
Lemin Kotiseutumuseo, Lemi 11058
Lemun Kotiseutumuseo, Lemu 11062
Liljendal Hembygdsmuseum, Liljendal 11070
Lohjan Museo, Lohja 11077
Loviisan Kaupunginmuseo, Loviisa 11082
Luopioisten Torpparimuseo, Luopioinen 11085
Luusuan Kylämuseo, Luusua 11088
Luvian Kotiseutumuseo, Luvia 11089
Mäntsälän Kotiseutumuseo, Mäntsälä 11091
Malmska Gården, Jakobstad 10866
Marttilan Kotiseutumuseo, Marttila 11103
Maskun Museo, Masku 11104
Matilda Roslin-Kalliolan Kirjailijakoti, Merikarvia 11108
Mellilän Museoriihi, Mellilä 11105
Merijärven Kotiseutumuseo, Merijärvi 11107
Merimiehenkotimuseo, Oulu 11170
Möhkön Ruukki, Möhkö 11119
Muhoksen Kotiseutumuseo, Muhos 11121
Munsala Hembygdsmuseum, Munsala 11125
Muonion Kotiseutumuseo, Muonio 11126
Museo Eerola, Tuulos 11410
Muuruveden Kotiseutumuseo, Muuruvesi 11129
Myllysaaren Museo, Valkeakoski 11437
Myrbergsgården Local Museum, Vörå 11475
Myrskylän Kotiseutumuseo, Myrskylä 11132
Närpes Hembygdsmuseum, Närpes 11134
Nakkilan Kotiseutumuseo ja Juustomeijerimuseo, Nakkila 11136
Nedervetil Hembygdsgård och -museum, Nedervetil 11139
Nilsiän Kotiseutumuseo, Nilsiä 11143
Nousiaisten Kotiseutumuseo ja Topoisten Ulkomuseoalue, Nousiainen 11149
Nurmeksen Museo, Nurmes 11152
Nurmon Museo, Nurmo 11153
Nykarleby Museum och Konstsamlingen Kalevala, Nykarleby 11156
Nyslotts Landskapsmuseum, Nyslott 11157
Oripään Kotiseutumuseo, Oripää 11167
Oulaisten Kotiseutumuseo, Oulainen 11169
Padasjoen Kotiseutumuseo, Padasjoki 11183
Padasjoen Museo Nyystölä, Nyystölä 11158
Padasjoen Museo Torittu, Torittu 11382
Paltamon Kotiseutumuseo, Paltamo 11186
Parainen District Museum, Parainen 11187
Parainen Industrial Museum, Parainen 11188
Parikkalan Kotiseutumuseo, Parikkala 11190
Pattijoen Kotiseutumuseo Ojala, Pattijoki 11195
Pelkosenniemen Kotiseutumuseo, Pelkosenniemi 11196
Peltolan Mäkitupalaismuseo, Tammijärvi 11347
Perhon Kotiseutumuseo, Perho 11199
Perttelin Kotiseutumuseo, Inkere 10861
Petalax Hembygdsmuseum, Petalax 11202
Pihlajaveden Kotiseutumuseo, Keuruu 10953
Pitkänpellon Talomuseo, Kangasniemi 10913
Pörtom Hembygdsmuseum, Pörtom 11211
Pöytyän Kotiseutumuseo, Pöytyä 11212
Pohjanmaan Museo, Vaasa 11426
Pohjois-Karjalan Museo, Joensuu 10873
Pohjois-Pohjanmaan Museo, Oulu 11177
Pomarkun Kotiseutumuseo, Pomarkku 11218
Posion Kotiseutumuseo, Posio 11234
Pudasjärven Kotiseutumuseo, Pudasjärvi 11235
Pukkilan Kotiseutumuseo, Pukkila 11236
Pulkkilan Kotiseutumuseo, Pulkkila 11237
Punkaharjun Kotiseutumuseo, Punkaharju 11239
Pyhämaan Kotiseutumuseo, Pyhämaa 11245
Rääkkylän Kotiseutumuseo, Rääkkylä 11250
Rantasalmen Museo, Rantasalmi 11253
Rantsilan Kotiseutumuseo, Rantsila 11254
Rauhaniemen Kotiseutumuseo, Sulkava 11331
Reisjärven Kotiseutumuseo, Reisjärvi 11265
Replot Hembygdsmuseum, Replot 11267
Riihimäen Kaupunginmuseo, Riihimäki 11268
Ruskon Kotiseutumuseo ja Rindellin Mäkitupamuseo, Rusko 11283
Saarijärven Museo, Saarijärvi 11286
Säkylän Kotiseutumuseo, Säkylä 11291
Säräisniemen Kotiseutumuseo, Säräisniemi 11292
Sagalunds Museum, Kimito 10959
Sauhutupamuseo, Somero 11322
Saukkojärven Kotiseutu- ja Koulumuseo, Ranua 11256
Sauvon Kotiseutumuseo, Sauvo 11301
Savonlinnan Maakuntamuseo, Savonlinna 11303
Sibbesgården Museo, Sipoo 11313
Siikaisten Kotiseutumuseo, Siikainen 11309
Siuntion Kotiseutumuseo, Siuntio 11315
Skyttalan Museo, Parainen 11189
Sodankylän Kotiseutumuseo, Sodankylä 11317
Sotkamon Makasiinimuseo, Sotkamo 11327
Suodenniemi Kotiseutumuseo, Suodenniemi 11336
Suolahden Museo, Suolahti 11338
Suomusjärven Kotiseutumuseo, Suomusjärvi 11339
Suonenjoen Kotiseutumuseo, Suonenjoki 11340
Suur-Savon Museo, Mikkeli 11117
Sysmäen Kotiseutumuseo, Sysmä 11342
Taalintehtaan Ruukinmuseo, Dalsbruk 10718
Tarvasjoen Kotiseutumuseo, Tarvasjoki 11371
Tavastilan Kotiseutumuseo, Mietoinen 11111
Teiskon Kotiseutumuseo, Tampere 11366
Temmeksen Kotiseutumuseo, Temmes 11372
Teuvan Museo, Teuva 11374
Töysän Museo, Töysä 11377
Turun Maakuntamuseo, Turku 11404
Tuusniemen Kotiseutumuseo, Tuusniemi 11412
Untamalan Kotiseutumuseo, Laitila 11041
Vähänkyrön Museo, Vähäkyrö 11432
Vantaan Kaupunginmuseo, Vantaa 11449
Vehmaan Kotiseutumuseo ja Kivityömuseo, Vehmaa 11458
Vekkilän Kotiseutumuseo, Tammela 11346
Vetelin Museo, Veteli 11460
Vieremän Kotiseutumuseo, Vieremä 11461
Vihdin Museo, Vihti 11463
Viirilän Kotiseutumuseo, Kuninkaankylä 11005
Viitainahon Kotiseutumuseo, Piippola 11208
Virtasalmen Kotiseutumuseo, Virtasalmi 11473
Wolkoffin talomuseo, Lappeenranta 11050
Ylä-Savon Kotiseutumuseo, Iisalmi 10848
Ylikiimingin Kotiseutumuseo, Ylikiiminki 11481
Ylistaron Kotiseutumuseo, Ylistaro 11482

France

Abbaye du Val Richer, Saint-Ouen-le-Pin 15436
L'Archéodrome de Bourgogne, Meursault 14106
Catacombes, Paris 14577
Château, Duilhac-sous-Peyrepertuse 12794
Château de Tarascon, Tarascon (Bouches-du-Rhône) 15853
Château-Musée de Lantheuil, Lantheuil 13518
Château-Musée du Rhône, Tournon-sur-Rhône 15956
Cité des Métiers de Tradition, Saint-Laurent-de-la-Plaine 15357
Collection Archéologique de la Maison des Fouilles et Maison du Memontois, Malain 13927
Conservation Départementale Musées des Pays de l'Ain, Bourg-en-Bresse 12106
Ecomusée de Hauteluce, Hauteluce 13192
Ecomusée de la Brenne, Le Blanc 13558
Ecomusée de la Communauté Le Creusot Montceau-les-Mines, Le Creusot 13576

Ecomusée de la Forêt Méditerranéenne, Gardanne 13032
Ecomusée de la Sainte-Baume, Plan-d'Aups 14830
Ecomusée de l'Ile-de-Groix, Ile-de-Groix 13237
Ecomusée des Monts d'Arrée, Commana 12592
Ecomusée des Monts d'Arrée, Saint-Rivoal 15499
Ecomusée du Mont-Lozère, Le Pont-de-Montvert 13640
Ecomusée du Pays de Rennes, Rennes (Ille-et-Vilaine) 15012
Écomusée du Val de Saône, Seurre 15738
Écomusée Sauvegarde du Vieil Auzon, Auzon 11781
Espace Culturel La Douve, Langeais 13505
Espace Patrimoine le Cœur, Tignes-le-Lac 15910
Faïences et Emaux, Longwy 13812
Familistère de Guise, Guise 13169
Galerie Heitz, Strasbourg 15817
Hôtel Dieu - Musée Greuze, Tournus 15957
Maison à Pondalez, Morlaix 14276
Maison de Pays, Pont-sur-Sambre 14896
Maison du Patrimoine, Bourguignon-les-Morey 12127
Maison du Patrimoine, Sarcelles 15628
Maison du Pays Roussillonnais, Roussillon (Isère) 15128
Maison du Platin, La Flotte 13370
Maison du Vieill Alby, Albi 11537
Musée, Frontignan-la-Peyrade 13016
Musée, Quiberon 14965
Musée Adrien Mentienne, Bry-sur-Marne 12203
Musée Agathois, Agde 11491
Musée Agro-Pastoral d'Aussois, Aussois 11759
Musée Alpin, Chamonix-Mont-Blanc 12381
Musée André Dunoyer-de-Ségonzac, Boussy-Saint-Antoine 12131
Musée Archéologique, Mormoiron 14280
Musée Archéologique Hôtel Dieu, Cavaillon 12329
Musée Au fil du papier, Pont-à-Mousson 14877
Musée Basque et de l'Histoire de Bayonne, Bayonne 11878
Musée Bigouden, Pont-l'Abbé 14888
Musée Biochet-Brechot, Caudebec-en-Caux 12322
Musée Bonaparte, Auxonne 11780
Musée Bonnevallais, Bonneval 12055
Musée Bourguignon Perrin de Puycousin, Tournus 15958
Musée Briard et de la Crypte Mérovingienne, Jouarre 13280
Musée Calbet, Grisolles 13151
Musée Carnavalet - Histoire de Paris, Paris 14637
Musée-Centre de Documentation Alfred Desmasures, Hirson 13218
Musée Charles de Bruyères, Remiremont 15008
Musée Château du Bucheneck, Soultz-Haut-Rhin 15801
Musée Ciotaden, La Ciotat 13344
Musée Clément Ader, Muret 14335
Musée Communal, Clérac 12544
Musée d'Archéologie, Séverac-le-Château 15741
Musée d'Archéologie Bargoin, Clermont-Ferrand 12548
Musée d'Art Breton, Saint-Vougay 15541
Musée d'Art et d'Archéologie, Hyères 13233
Musée d'Art et d'Histoire, Cholet 12522
Musée d'Art et d'Histoire, Kaysersberg 13301
Musée d'Art et d'Histoire, Livry-Gargan 13790
Musée d'Art et d'Histoire, Nice 14427
Musée d'Art et d'Histoire, Rueil-Malmaison 15138
Musée d'Art et d'Histoire, Saint-Julien-du-Sault 15347
Musée d'Art et d'Histoire, Rochefort (Charente-Maritime) 15071
Musée d'Art et d'Histoire de Cognac, Cognac 12565
Musée d'Art et d'Histoire de Toul, Toul 15922
Musée d'Art et d'Histoire des Côtes d'Armor, Saint-Brieuc 15191
Musée d'Art et d'Histoire du Pays Blayais, Blaye 12023
Musée d'Art et d'Histoire Local, Pezens 14809
Musée d'Art et d'Histoire Louis Senlecq, L'Isle-Adam 13781
Musée d'Art et d'Histoire Roger Rodière, Montreuil-sur-Mer 14256
Musée d'Art et Historique, Montbéliard 14184
Musée d'Art et Traditions Populaires, Binic 11995
Musée d'Art Local, Oisemont 14504
Musée d'Arts et de Traditions Populaires, Palinges 14558
Musée d'Arts et de Traditions Populaires, Saint-Paul-de-Fenouillet 15441
Musée d'Assier, Feurs 12931
Musée de Bastia, Bastia 11859
Musée de Biot, Biot 11997
Musée de Borda, Dax 12701
Musée de Bouxwiller et du Pays de Hanau, Bouxwiller 12134
Musée de Cervières, Cervières 12349
Musée de Charroux et de son Canton, Charroux-d'Allier 12420
Musée de Dol, Dol-de-Bretagne 12753
Musée de France de Berck-sur-Mer, Berck-sur-Mer 11941
Musée de Jouques, Jouques 13283
Musée de la Bande Dessinée et Musée Municipal, Angoulême 11617
Musée de la Bonneterie, Troyes 15999
Musée de la Corbillière, Mer 14084
Musée de la Correrie, Saint-Pierre-de-Chartreuse 15456
Musée de la Loire, Cosne-Cours-sur-Loire 12636
Musée de la Mairie de Crécy, Crécy-la-Chapelle 12670
Musée de la Résistance en Zone Interdite, Denain 12709
Musée de la Roue Tourne, La Mothe-Achard 13394
Musée de la Tour aux Puces, Thionville 15884
Musée de la Tour de l'Abbaye, Jouarre 13281
Musée de la Vallée à Barcelonnette, Barcelonnette 11848
Musée de la Vie Bourguignonne Perrin-de-Puycousin, Dijon 12738
Musée de la Ville de Saint-Quentin-en-Yvelines, Montigny-le-Bretonneux 14217
Musée de la Ville de Saverne, Saverne 15675
Musée de l'Alta Rocca, Lévie 13732
Musée de l'Ancien Évêché, Grenoble 13136
Musée de l'Ancien Havre, Le Havre 13595
Musée de l'Auditoire, Sainte-Suzanne 15581
Musée de l'Eau de Vie et des Vieux Métiers, Valognes 16057
Musée de l'Empéri de Salon et de la Crau, Salon-de-Provence 15611
Musée de l'Habitat Ancien, Bessans 11960
Musée de l'Histoire Vivante Parc Montreau, Montreuil-sous-Bois 14255
Musée de l'Historire de Rosny, Rosny-sous-Bois 15109
Musée de l'Hôpital, Yssingeaux 16309
Musée de l'Hôtel de Vermandois, Senlis 15717
Musée de l'Hôtel de Ville, Amboise 11578
Musée de l'île d'Oleron Aliénor d'Aquitaine, Saint-Pierre-d'Oléron 15462
Musée de l'Ostrévant, Bouchain 12081
Musée de l'Outil, Magny-en-Vexin 13913
Musée de Maison-Alfort, Maisons-Alfort 13923
Musée de Montmartre, Paris 14678
Musée de Nogent, Nogent-sur-Marne 14463
Musée de Saint-Germain-Laval, Saint-Germain-Laval 15293
Musée de Salers, Salers 15595
Musée de Sarreguemines, Sarreguemines 15644
Musée de Suresnes René Sordes, Suresnes 15841
Musée de Verrières-le-Buisson, Verrières-le-Buisson 16125
Musée de Vieux Sallèles, Sallèles-d'Aude 15602
Musée d'Entrecasteaux, Entrecasteaux 12825
Musée Départemental de Flandre, Cassel 12304
Musée Départemental Ignon-Fabre, Mende 14071
Musée des Allues, Meribel 14086
Musée des Alpilles, Saint-Rémy-de-Provence 15490
Musée des Amis de Thann, Thann 15874
Musée des Amis du Vieux Saint-Étienne, Saint-Étienne 15254
Musée des Arts et Traditions Populaires, Bressuire 12155
Musée des Canonniers Sédentaires de Lille, Lille 13757
Musée des Mines d'Argent des Gorges du Fournel, L'Argentière-la-Bessée 13526
Musée des Traditions, Les Avanchers-Valmorel 13687
Musée des Traditions et de l'Histoire de Biscarrosse, Biscarrosse 12000
Musée des Traditions Locales, Sainte-Maxime 15570
Musée des Traditions Populaires, Escaudain 12849
Musée d'Ethnographie et d'Art Populaire, Honfleur 13222
Musée d'Histoire de la Ville et d'Ethnographie du Pays Malouin, Saint-Malo (Ille-et-Vilaine) 15389
Musée d'Histoire de Marseille, Marseille 14005
Musée d'Histoire et d'Archéologie, Moutiers (Savoie) 14317
Musée d'Histoire et d'Archéologie du Pays d'Apt, Apt 11655
Musée d'Histoire et d'Art, Mirande 14131
Musée d'Histoire Local, Maubourguet 14033
Musée d'Histoire Locale, Saint-Paul-de-Vence 15446
Musée d'Histoire Locale, Saint-Renan 15496
Musée d'Histoire Locale et des Vieux Mètiers, Saint-Ciers-sur-Gironde 15209
Musée d'Histoire Naturelle et d'Histoire Locale, Elbeuf-sur-Seine 12815
Musée d'Orbigny-Bernon, La Rochelle 13428
Musée du Château, Flers 12944
Musée du Château, Fougères (Ille-et-Vilaine) 12986
Musée du Château de Dourdan, Dourdan 12783
Musée du Château de Vitré, Vitré 16256
Musée du Château des Baux, Les Baux-de-Provence 13692
Musée du Château des Ducs de Bretagne, Nantes 14368
Musée du Château d'Henri IV, Nérac 14392
Musée du Château Paul Pastre, Marsillargues 14013
Musée du Château Saint-Jean, Nogent-le-Rotrou 14462
Musée du Châtillonnais, Châtillon-sur-Seine 12474
Musée du Compagnonnage, Paris 14695
Musée du Débarquement à Utah Beach, Sainte-Marie-du-Mont 15568
Musée du Dessin et de l'Estampe Originale, Gravelines 13122
Musée du Dolder, Riquewihr 15052
Musée du Donjon de la Toque, Huriel 13231
Musée du Florentinois, Saint-Florentin 15269
Musée du Léon, Lesneven 13723
Musée du Noyonnais, Noyon 14483
Musée du Patrimoine, Roquebrune-sur-Argens 15096
Musée du Patrimoine, Six-Fours-les-Plages 15768
Musée du Patrimoine des Pères de Bétharram, Lestelle-Bétharram 13725
Musée du Patrimoine et de la Dentelle, Chantilly 12400
Musée du Paye de Guérande, Guérande 13157
Musée du Pays Brignolais, Brignoles 12180
Musée du Pays de Thônes, Thônes 15893
Musée du Pays Vaurais, Lavaur 13549
Musée du Pilori, Niort 14457
Musée du Présidial, Castelnaudary 12315
Musée du Ranquet, Clermont-Ferrand 12550
Musée du Saut du Tarn, Saint-Juéry 15342
Musée du Sceau Alsacien, La Petite-Pierre 13403
Musée du Site Balnéaire, Dinard 12751
Musée du Terroir, Villeneuve-d'Ascq 16206
Musée du Terroir et Musée de la Volaille, Romenay 15089
Musée du Terroir Hérissonnais, Hérisson 13205
Musée du Textile et de la Vie Sociale, Fourmies 12992
Musée du Trièves, Mens 14076
Musée du Val-de-Cher, Selles-sur-Cher 15705
Musée du Valois, Vez 16135
Musée du Vieil Argenteuil, Argenteuil 11674
Musée du Vieux Brest, Brest 12162
Musée du Vieux Clairac, Clairac 12535
Musée du Vieux Figeac, Figeac 12933
Musée du Vieux Granville, Granville 13105
Musée du Vieux Lambesc, Lambesc 13493
Musée du Vieux Mazères, Mazères (Ariège) 14050
Musée du Vieux Nîmes, Nîmes 14453
Musée du Vieux Sisteron, Sisteron 15764
Musée du Vieux Toulon, Toulon (Var) 15926
Musée du Vieux Toulouse, Toulouse 15941
Musée du Vieux-Tréport, Le Tréport 13657
Musée du Vieux Warcq, Warcq 16285
Musée du Village, Saint-Sorlin-d'Arves 15516
Musée du Vivarais Protestant, Pranles 14940
Musée d'Uxellodunum et Musée de Martel, Martel 14018
Musée Ernest Cognacq, Saint-Martin-de-Ré 15400
Musée Ethnologique Intercommunal du Vermandois, Vermand 16118
Musée Eugène Farcot, Sainville 15590
Musée Friry, Remiremont 15009
Musée Gadagne, Lyon 13895
Musée Galerie Honoré Camos, Bargemon 11853
Musée Gallé-Juillet, Creil 12671
Musée Gardanne Autrefois, Gardanne 13033
Musée Georges Borias, Uzès 16025
Musée Guillion, Romanèche-Thorins 15086
Musée Henri-Barre, Thouars 15902
Musée Henri Boez, Maubeuge 14032
Musée Henri Dupuis, Saint-Omer 15432
Musée Historial, Grézolles 13147
Musée Historique, Gourdon (Alpes-Maritimes) 13089
Musée Historique, Haguenau 13177
Musée Historique de la Ville de Strasbourg, Strasbourg 15831
Musée Historique de Troyes et de la Champagne, Troyes 16000
Musée Historique du Bailliage de Rouffach, Rouffach 15123
Musée Historique du Centre-Var, Le Luc 13602
Musée Historique et Militaire, Huningue 13227
Musée-Hôtel Le Vergeur, Reims 15004
Musée Huppy Autrefois, Huppy 13229
Musée J.B. Mathon et André Durand, Neufchâtel-en-Bray 14398
Musée Jeanne d'Arc, Chinon 12517
Musée Johannique, Vaucouleurs 16082
Musée Joseph Denais, Beaufort-en-Vallée 11892
Musée la Muse, Mérindol 14089
Musée Labenche d'Art et d'Histoire, Brive-la-Gaillarde 12189
Musée Lapidaire, Morlaàs 14274
Musée Louis Braille, Coupvray 12648
Musée Maçonnique de la Grande Loge de France, Paris 14718
Musée Marc Deydier, Cucuron 12684
Musée Maurice Poignant, Bourg-sur-Gironde 12113
Musée Montebello, Trouville-sur-Mer 15993
Musée Municipal, Châtellerault 12463
Musée Municipal, Cusset 12693
Musée Municipal, Forcalquier 12978
Musée Municipal, Mazan 14046
Musée Municipal, Saint-Amand-les-Eaux 15160
Musée Municipal, Saint-Antonin-Noble-Val 15172
Musée Municipal, Saint-Calais 15195
Musée Municipal, Villandraut 16162
Musée Municipal A.G. Poulain, Vernon 16123
Musée Municipal Albert Marzelles, Marmande 13972
Musée Municipal Auguste Grasset, Varzy 16077
Musée Municipal Bourbon-Lancy, Bourbon-Lancy 12099
Musée Municipal de Cambrai, Cambrai 12257
Musée Municipal de Civray, Civray 12533
Musée Municipal de la Vie Sauvage, Le Bugue 13566
Musée Municipal de Montmorillon, Montmorillon 14232
Musée Municipal de Saint-Gaudens, Saint-Gaudens 15276
Musée Municipal de Sault, Sault (Vaucluse) 15659
Musée Municipal des Beaux Arts et d'Histoire Naturelle, Châteaudun 12444
Musée Municipal des Capucins, Coulommiers 12645
Musée Municipal d'Histoire et d'Archéologie, Harnes 13182
Musée Municipal d'Orbec, Orbec 14520
Musée Municipal du Château, Gannat 13028
Musée Municipal du Présidial, Bazas 11881
Musée Municipal Gautron du Coudray, Marzy 14025
Musée Municipal Georges Turpin, Parthenay 14753
Musée Municipal Paul Lafran, Saint-Chamas 15200
Musée Municipal Robert Dubois-Corneau, Brunoy 12201
Musée Municipal Urbain-Cabrol, Villefranche-de-Rouergue 16189
Musée Municipale, Saint-Laurent-d'Olt 15361
Musée National d'Enéerune, Nissan-lez-Ensérune 14460
Musée Oberlin, Waldersbach 16281
Musée-Observatoire de la Vallée d'Anjou, Saint-Mathurin-sur-Loire 15409
Musée Patrimoine et Traditions, Mirepoix 14136
Musée Percheron, Mortagne-au-Perche 14287
Musée Pierre Gaudin, Puteaux 14959
Musée Pillon, Chaumont-en-Vexin 12481
Musée Quentovic, Etaples 12875
Musée Raymond Peynet, Brassac-les-Mines 12147
Musée Reflet Brionnais, Iguerande 13234
Musée Régional d'Auvergne, Riom 15046
Musée Régional de Fouras, Fouras 12990
Musée Régional de l'Orléanais, Beaugency 11893
Musée Régional Dupuy-Méstreau, Saintes 15585
Musée Roybet Fould, Courbevoie 12650
Musée Saint-Remi, Reims 15005
Musée Sainte-Croix, Poitiers 14868
Musée Serret, Saint-Amarin 15162
Musée Suffren et du Vieux Saint-Cannat, Saint-Cannat 15197
Musée Vauban, Neuf-Brisach 14396
Musée Vendéen, Fontenay-le-Comte 12969
Musée Victor Aubert, Maule 14034
Musée Villageois d'Orgon, Orgon 14528
Musée Villeneuvien, Villeneuve-sur-Yonne 16223
Musées de Sens, Sens 15719
Musées des Colombiers Cauchois, Oherville 14500
Museo Départemental A. Demard, Le Haut-du-Them 13593
Palais Bénédictine, Fécamp 12915
Parc Château de Valencay, Valençay 16037
Réserve Géologique de Haute-Provence, Digne-les-Bains 12728
Sainte Maison de la Flocellière, La Flocellière 13369
Site Gallo-Romain de Grand, Grand 13096
Village-Musée, Fos-sur-Mer 12982

Georgia

Dmanisi Museum-Reserve of History and Architecture, Dmanisi 16339
State Museum of Abkhasia, Suchumi 16350

Germany

50er-Jahre-Museum, Büdingen 17626
Abteilung Handwerk und dörfliches Leben des Heimatmuseums Neu-Ulm, Neu-Ulm 20585
Ahler Kråm - Volkskundliche Sammlung, Partenstein 20985
Aktivmuseum Ländliches Brauchtum, Breitungen an der Werra 17545
Alexander-Mack-Museum, Bad Berleburg 16691
Allgäu-Museum, Kempten 19510
Alstertal-Museum, Hamburg 18914
Alt-Arnstorf-Haus, Arnstorf 16576
Alt-Freden-Sammlung, Freden 18417
Alt-Stade im Baumhaus, Stade 21847
Alte Apotheke, Ottweiler 20958
Alte Arche, Furtwangen im Schwarzwald 18519
Alte Dorfmühle, Monzelfeld 20377
Alte Dorfschmiede, Waldbreitbach 22282
Alte Oberamtei, Gerabronn 18600
Alte Schmiede, Bennhausen 16995
Alte Wassermühle, Splietsdorf 21841
Alte Welt Museum, Nußbach 20796
Das Alte Zollhaus Heimatmuseum, Hitzacker 19201
Altes Amtshaus, Werne 22493
Altes Rathaus, Lüneburg 20094
Altonaer Museum für Kunst und Kulturgeschichte, Hamburg 18915

Ammerländer Schinken-Museum F. W. Meyer, Apen 16563
Amrumer Museum und Windmühle, Nebel 20558
Amtsturm-Museum, Lüchow 20086
Anton-Buchner-Heimatmuseum, Betzenstein 17311
Archäologische Sammlung mit Heimatvertriebenenstube und Kruk-Sammlung, Königsbrunn 19644
Archäologisches Museum, Oldendorf 20903
Auberlehaus, Trossingen 22109
August-Becker-Museum, Klingenmünster 19571
August-Gottschalk-Haus, Esens 18238
August-Holländer-Museum, Emsdetten 18159
Aule Mettmanner, Mettmann 20313
Ausstellung 'Juden in Buttenhausen', Münsingen 20499
Auswanderermuseum, Oberalben 20799
Bachgau-Museum, Großostheim 18800
Backnanger Stadtturm mit Gotischem Chor St. Michael, Backnang 16655
Bad Schwalbacher Kur-Stadt-Apotheken-Museum, Bad Schwalbach 16833
Badhaus Museum und Galerie, Kulmbach 19745
Badisches Landesmuseum Karlsruhe, Karlsruhe 19437
Bartholomäus-Schmucker-Heimatmuseum, Ruhpolding 21445
Bauernhaus Habach, Eppelborn 18178
Bauernhaus-Museum Bortfeld, Wendeburg 22474
Bauernhausmuseum, Steinen 21886
Bauernhausmuseum Hof Haina, Biebertal 17332
Bauernhausmuseum Knechtenhofen, Oberstaufen 20840
Bauernmuseum, Simmerath 21769
Bauernmuseum Schwarzbach, Schwarzbach 21671
Bauernstube Pfahlheim, Ellwangen 18122
Bayerisches Bienenmuseum, Illertissen 19315
Beerewein Museum, Eulenbis 18282
Berg- und Stadtmuseum Obernkirchen, Obernkirchen 20822
Bergbau- und Heimatmuseum, Goldkronach 18704
Bergbaumuseum, Aldenhoven 16461
Bergbaumuseum, Mechernich 20242
Bergbaumuseum Grube Anna II, Alsdorf 16479
Bergmanns-Bauernmuseum, Breitenbach, Kreis Kusel 17538
Bergwerkmuseum Grube Glasebach, Straßberg 21917
Bergwinkelmuseum, Schlüchtern 21576
Berliner Unterwelten-Museum, Berlin 17047
Besucherbergwerk Vereinigt Zwitterfeld zu Zinnwald, Altenberg 16494
Bezirksmuseum Buchen, Buchen 17615
Bezirksmuseum Dachau, Dachau 17763
Bezirksmuseum Marzahn-Hellersdorf, Berlin 17050
Biebergrund-Museum, Biebergemünd 17330
Blaues Schloss mit Museumspädagogischer Werkstatt Rotes Schloss, Obernzenn 20825
Bodenheimer Heimatmuseum, Bodenheim 17412
Das Bördemuseum, Lamstedt 19784
Bördenheimatmuseum, Heeslingen 19064
Bomann-Museum Celle, Celle 17683
Borstei Museum, München 20429
Braith-Mali-Museum, Biberach an der Riß 17319
Brandenburg-Preußen Museum, Wustrau 22755
Brandenburgisches Textilmuseum Forst, Forst, Lausitz 18341
Braunauer Heimatmuseum, Forchheim, Oberfranken 18336
Braunschweigisches Landesmuseum, Braunschweig 17518
Brensbach Museum und Galerie im alten Rathaus, Brensbach 17581
Breuberg-Museum, Breuberg 17587
Brünner Heimatmuseum, Schwäbisch Gmünd 21648
Brüxer und Komotauer Heimatstuben, Erlangen 18220
Brunnenmuseum, Bad Vilbel 16855
s'Buchholze Hisli, Schuttertal 21643
Buchholzer Heimatmuseum, Buchholz in der Nordheide 17618
Bürgermeister-Ringenhain-Haus, Torgau 22070
Burg Berwartstein, Erlenbach bei Dahn 18229
Burg Eltz, Wierschem 22544
Burg Hagen, Hagen im Bremischen 18861
Burg Herzberg, Breitenbach am Herzberg 17537
Burg-Museum, Parsberg 20984
Burg Neustadt-Glewe, Neustadt-Glewe 20679
Burg Reichenstein, Trechtingshausen 22083
Burg Sooneck, Niederheimbach 20706
Burg Thurant, Alken 16470
Burg- und Schloßmuseum Allstedt, Allstedt 16475
Burg- und Schloßmuseum Jägersburg, Homburg 19267
Burg- und Stadtmuseum Königstein, Königstein im Taunus 19657
Burg Vischering, Lüdinghausen 20092
Burgenmuseum, Dahn 17775
Burghofmuseum, Soest 21795
Burgmuseum Bad Bodenteich, Bad Bodenteich 16702
Burgmuseum Greifenstein, Greifenstein 18745
Burgmuseum Krautheim, Krautheim 19701
Burgmuseum Lenzen, Lenzen 19957
Burgmuseum Nideggen, Nideggen 20695
Burgmuseum Runneburg, Weißensee 22462
Burgmuseum Waldeck, Waldeck 22285
Burgmuseum Wolfsegg, Wolfsegg 22669
Burgruine Montclair, Merzig 20298
Burgturm Davert, Ascheberg 16595
Buxtehude Museum für Regionalgeschichte und Kunst, Buxtehude 17673
Christian-Wolff-Haus, Halle, Saale 18890
Clemens-Sels-Museum Neuss, Neuss 20654
Cryptoneum, Rostock 21387
Danevirke Museum, Dannewerk 17782
Dat lütt Museum, Vielank 22228
Dauerausstellung im Stadtarchiv, Heilbronn 19096
Denkmalhof Bechelsdorfer Schulzenhaus, Schönberg, Mecklenburg 21600
Denkmalhof Rauchhaus Möllin, Gadebusch 18525
Detlefsen-Museum, Glückstadt 18663
Deusserhaus, Monheim am Rhein 20371
Dithmarscher Landesmuseum, Meldorf 20269
Dörpmuseum Münkeboe, Südbrookmerland 21973
Dokumentation Obersalzberg - Orts- und Zeitgeschichte, Berchtesgaden 17000
Dokumentations- und Informationszentrum, Stadtallendorf 21855
Dokumentationszentrum NS-Zwangsarbeit Berlin-Schöneweide, Berlin 17076
Domowniski muzej Dešno, Dissen-Striesow 17876
Dorf- und Fahrradmuseum, Pflückuff 21025
Dorf- und Heimatmuseum, Winterbach, Rems-Murr-Kreis 22607
Dorf- und Uhrenmuseum, Gütenbach 18831
Dorfgemeinschaftshaus Heimatmuseum Dörrenbach, Sankt Wendel 21511
Dorfmuseum, Günzach 18824
Dorfmuseum, Mertingen 20296
Dorfmuseum Alter Forsthof, Wetter 22521
Dorfmuseum Altkirchen, Altkirchen 16511
Dorfmuseum Bad Alexandersbad, Bad Alexandersbad 16667
Dorfmuseum Daniel-Martin-Haus, Rauschenberg 21201
Dorfmuseum Delligsen, Delligsen 17814
Dorfmuseum Dierscheid, Dierscheid 17849
Dorfmuseum Dobin am See, Dobin am See 17881
Dorfmuseum Düppenweiler, Beckingen 16971
Dorfmuseum Eichstetten am Kaiserstuhl, Eichstetten am Kaiserstuhl 18086
Dorfmuseum Emmingen-Liptingen, Emmingen-Liptingen 18156
Dorfmuseum Gahlenz, Gahlenz 18529
Dorfmuseum Gebhardshain, Gebhardshain 18547
Dorfmuseum Görzig, Görzig bei Köthen 18684
Dorfmuseum Großgörschen, Großgörschen 18791
Dorfmuseum Grube, Grube 18811
Dorfmuseum Holzburg, Schrecksbach 21635
Dorfmuseum im Greifenhaus, Hausen, Oberfranken 19051
Dorfmuseum Jockenhof, Simonswald 21772
Dorfmuseum Kelter, Marbach am Neckar 20178
Dorfmuseum Kienberg, Kienberg 19537
Dorfmuseum Leo's Tenne, Schweisweiler 21689
Dorfmuseum Leutesdorf, Leutesdorf 19972
Dorfmuseum mit Pfarrer-Lehmann Stube im Bauernkrieghaus, Landau in der Pfalz 19788
Dorfmuseum Niederbreitbach, Niederbreitbach 20702
Dorfmuseum Niederlinxweiler, Sankt Wendel 21512
Dorfmuseum Ostheim, Nidderau 20693
Dorfmuseum Ringenwalde, Temmen-Ringenwalde 22026
Dorfmuseum Rohrbach, Rohrbach, Pfalz 21365
Dorfmuseum Schönbach, Großbothen 18782
Dorfmuseum Tündern, Hameln 18966
Dorfmuseum Weckbach, Weilbach 22383
Dorfschmiede- und Heimatmuseum, Empfingen 18157
Dorfstube Münchholzhausen, Wetzlar 22523
Dotzheimer Museum, Wiesbaden 22547
Dr.-Bauer-Heimatmuseum, Bad Rothenfelde 16810
Dr. Berger Heimatmuseum, Forchtenberg 18338
Dr. Eisenbarth- und Heimatmuseum, Oberviechtach 20848
Dreieich-Museum, Dreieich 17936
Drilandmuseum Gronau, Gronau, Westfalen 18773
Dümmer-Museum, Lembruch 19946
Ebinger Heimatmuseum, Albstadt 16452
Egerländer-Elbogner Heimatstuben, Illertissen 19316
Egerland-Museum, Marktredwitz 20222
Eichsfelder Heimatmuseum, Heiligenstadt, Heilbad 19107
Eidelstedter Heimatmuseum, Hamburg 18925
Eifel-Heimatmuseum Bergweiler, Bergweiler 17025
Eifeler Bauernhausmuseum, Adenau 16430
Eifelmuseum Blankenheim, Blankenheim 17375
Einricher Heimatmuseum, Katzenelnbogen 19481
Elbschiffahrtsmuseum mit stadtgeschichtlicher Sammlung, Lauenburg 19849
Elsenzer Heimatstuben, Eppingen 18182
Emslandmuseum Lingen, Lingen 20008
Emslandmuseum Schloss Clemenswerth, Sögel 21791
Erlebniswelt Renaissance - Schloß Bevern, Bevern, Kreis Holzminden 17314
Ernst-Moritz-Arndt-Museum, Garz 18545
Erzbergbaumuseum und Besucherbergwerk, Bestwig 17309
Erzgebirgisches Heimatmuseum Hospital Sankt Johannis, Sayda 21525
Die Eulenburg, Rinteln 21337
Eulenthurm-Museum, Weißenthurm 22464
Fasnets-Museum am Vorderochsen, Weingarten, Kreis Ravensburg 22426
Fastnachtsmuseum, Bräunlingen 17496
Fehnmuseum Eiland, Großefehn 18786
Feringa Sach - Ortsgeschichte und heimatkundliche Sammlung, Unterföhring 22179
Festung Marienberg mit Fürstenbaumuseum, Würzburg 22710
Feuerwehrmuseum Gahlenz, Gahlenz 18530
Fichtelgebirgsmuseum, Wunsiedel 22731
Findorffhof, Grasberg 18738
Fischer- und Webermuseum Steinhude, Wunstorf 22733
Fischerstube, Mönkebude 20359
Fischlandhaus, Wustrow, Ostseebad 22756
Flachsbrechhütte, Creglingen 17741
Flößer- und Heimatmuseum, Wolfach 22645
Focke-Museum, Bremen 17549
Fossilien- und Heimatmuseum, Messel 20304
Fränkische Schweiz Museum, Pottenstein 21113
Fränkisches Bauern- und Handwerkermuseum, Iphofen 19350
Frankenwaldmuseum, Kronach 19721
Franziskanermuseum, Villingen-Schwenningen 22235
Frau Holle Expreß, Hessisch Lichtenau 19170
Frauen Museum Wiesbaden, Wiesbaden 22548
Freilichtmuseum, Wolfach 22646
Freilichtmuseum am Kiekeberg, Rosengarten, Kreis Harburg 21373
Freilichtmuseum "Frelsdorfer Brink", Frelsdorf 18454
Freilichtmuseum "Mittelalterliche Bergstadt Bleiberg", Frankenberg, Sachsen 18346
Freilichtmuseum Rhöner Museumsdorf, Tann 22004
Friesisches Heimatmuseum, Niebüll 20697
Fruchtkasten, Tuttlingen 22134
Fürstlich Leiningensche Sammlungen - Heimatmuseum Amorbach, Amorbach 16541
Gäubodenmuseum, Straubing 21918
Galerie Biebertal, Biebertal 17333
Gedenkstätte und Museum Trutzhain, Schwalmstadt 21661
Gemeinde-Heimatmuseum Harsum, Harsum 19017
Gemeindemuseum, Bad Krozingen 16769
Gemeindemuseum im Historischen Rathaus, Seeheim-Jugenheim 21717
Geroldsecker-Museum im Storchenturm, Lahr 19771
Geschichtliche Begegnungstätte Hachborn, Ebsdorfergrund 18049
Geschichtsmuseum Lüdenscheid, Lüdenscheid 20088
Glas- Heimat-Museum Warndt, Völklingen 22248
Glashaus-Derneburg, Holle 19251
Glasmuseum Weisswasser, Weißwasser 22465
Glauberg-Museum, Glauburg 18655
GoetheStadtMuseum Ilmenau, Ilmenau 19320
Goslarer Museum, Goslar 18708
Grafschafter Museum im Moerser Schloss, Moers 20362
Grafschaftsmuseum, Wertheim 22503
Gramzower Heimatstuben, Gramzow 18736
Graslitzer Gedenk- und Informationsraum, Aschaffenburg 16583
Grenzland- und Trenckmuseum, Waldmünchen 22302
Grenzlandheimatstuben des Heimatkreises Marienbad, Neualbenreuth 20590
Grenzsteinmuseum Ostrach, Ostrach 20942
Hackenberger Mühle, Saarburg 21462
Hällisch-Fränkisches Museum, Schwäbisch Hall 21652
Härtsfeld-Museum, Neresheim 20571
Haidhausen-Museum, München 20444
Hallertauer Heimat- und Hopfenmuseum, Mainburg 20136
Hallertauer Hopfen- und Heimatmuseum, Geisenfeld 18558
Hamaland-Museum, Vreden 22258
Handwebmuseum Rupperath, Bad Münstereifel 16788
Handwerkermuseum, Freinsheim 18446
Handwerkermuseum, Schönecken 21606
Handwerksmuseum Groß-Gerau, Groß-Gerau 18775
Haniel Museum, Duisburg 18018
Haus am Kleistpark, Berlin 17125
Haus der badisch-pfälzischen Fasnacht, Speyer 21833
Haus der Brandenburgisch-Preußischen Geschichte, Potsdam 21089
Haus der Donauschwaben, Sindelfingen 21775
Haus der Dorfgeschichte Bliesmengen-Bolchen, Mandelbachtal 20160
Haus der Geschichte Baden-Württemberg, Stuttgart 21938
Haus der Heimat, Olbernhau 20892
Haus der Ortsgeschichte, Schwäbisch Hall 21654
Haus der Stadtgeschichte, Offenbach am Main 20882
Haus der Stadtgeschichte, Donauwörth 17895
Haus des Gastes mit heimatkundlicher Sammlung, Gößweinstein 18688
Haus Kickelhain, Mosbach 20388
Haus Opherdicke, Holzwickede 19263
Haus Rottels, Neuss 20656
Haus Stöcken, Haan 18846
Haus zum Stockfisch - Stadtmuseum, Erfurt 18204
Hausenhäusl, Reit im Winkl 21256
Heidemuseum Dat ole Huus, Bispingen 17365
Heidemuseum Rischmannshof Walsrode, Walsrode 22316
Heiliggeistkirche, Landshut 19801
Heimat-Haus Großderschau, Großderschau 18785
Heimat-Museum Maintal, Maintal 20140
Heimat- und Bauernmuseum Bruck, Bruck in der Oberpfalz 17599
Heimat- und Bauernmuseum Sulz, Sulz am Neckar 21980
Heimat- und Bergbaumuseum Nentershausen, Nentershausen 20570
Heimat- und Bergbaumuseum Reinsdorf, Reinsdorf bei Zwickau 21255
Heimat- und Brauereimuseum Pleinfeld, Pleinfeld 21058
Heimat- und Braunkohlemuseum Steinberg, Steinberg, Kreis Schwandorf 21883
Heimat- und Buddelmuseum Osten, Osten 20930
Heimat- und Bürstenbindermuseum, Ramberg 21173
Heimat- und Burgmuseum, Kirkel 19557
Heimat- und Grimmelshausenmuseum, Oberkirch 20812
Heimat- und Humboldtmuseum, Eibau 18072
Heimat- und Industriemuseum Kolbermoor, Kolbermoor 19674
Heimat- und Industriemuseum Wackersdorf, Wackersdorf 22261
Heimat- und Kleinbrennermuseum, Steinach, Ortenaukreis 21876
Heimat- und Schiffahrtsmuseum, Heinsen 19113
Heimat- und Ski-Museum, Braunlage 17516
Heimat- und Torfmuseum, Gröbenzell 18767
Heimat- und Uhrenmuseum, Villingen-Schwenningen 22236
Heimat- und Zunftmuseum, Adenau 16431
Heimatarchiv Zwittau, Esslingen 18273
Heimatdiele Kutenholz, Kutenholz 19753
Heimatecke am Seifenbach, Beierfeld 16981
Heimathaus Alte Mühle, Schladen 21561
Heimathaus Aying, Aying 16650
Heimathaus De Theeshof, Schneverdingen 21593
Heimathaus der Stadt Lauingen, Lauingen 19857
Heimathaus der Stadt Warendorf, Warendorf 22337
Heimathaus des Rupertiwinkels und Heimatstube der Sudetendeutschen, Tittmoning 22066
Heimathaus Dingden, Hamminkeln 18974
Heimathaus Greven Worth, Selsingen 21742
Heimathaus Mehedorf, Bremervörde 17578
Heimathaus Nesselwang, Nesselwang 20577
Heimathaus Neuchl-Anwesen, Hohenschäftlarn 19242
Heimathaus Pfarrkirchen, Pfarrkirchen 21023
Heimathaus Plettenberg, Plettenberg 21060
Heimathaus Sittensen, Sittensen 21790
Heimathaus Sonthofen, Sonthofen 21823
Heimathaus Welver, Welver 22470
Heimathausanlage Schafstall, Bremervörde 17579
Heimathof Hüll, Drochtersen 17978
Heimatkabinett Westerholt, Herten 19166
Heimatkundliche Sammlung Bergendorf, Holzheim 19259
Heimatkundliche Sammlung Blaustein, Blaustein 17382
Heimatkundliche Sammlung Elzach, Elzach 18133
Heimatkundliche Sammlung Heideck, Heideck 19068
Heimatkundliche Sammlung Heroldsbach, Heroldsbach 19158
Heimatkundliche Sammlung Isen, Isen 19356
Heimatkundliche Sammlung, Mühle Oberkleinich, Kleinich 19563
Heimatkundliche Sammlung Pfronten, Pfronten 21032
Heimatkundliche Sammlung und Ausstellung, Hermannsburg 19145
Heimatkundliches Museum, Friedeburg 18470
Heimatmuseum, Abstatt 16420
Heimatmuseum, Adelsdorf 16426
Heimatmuseum, Affalterbach 16434
Heimatmuseum, Aken 16450
Heimatmuseum, Alfdorf 16463
Heimatmuseum, Algermissen 16469
Heimatmuseum, Allendorf 16472
Heimatmuseum, Altenkirchen, Kreis Kusel 16500

H

Heimatmuseum, Althütte 16509
Heimatmuseum, Amöneburg 16540
Heimatmuseum, Annaburg 16557
Heimatmuseum, Arendsee 16568
Heimatmuseum, Arneburg 16570
Heimatmuseum, Aßlar 16602
Heimatmuseum, Auetal 16609
Heimatmuseum, Bad Aibling 16666
Heimatmuseum, Bad Dürrheim 16719
Heimatmuseum, Bad Eilsen 16721
Heimatmuseum, Bad Endbach 16726
Heimatmuseum, Bad König 16760
Heimatmuseum, Bad Laer 16772
Heimatmuseum, Bad Lauterberg 16775
Heimatmuseum, Bad Oldesloe 16801
Heimatmuseum, Bad Rodach 16809
Heimatmuseum, Bad Schandau 16826
Heimatmuseum, Bad Überkingen 16852
Heimatmuseum, Bad Vilbel 16856
Heimatmuseum, Balingen 16900
Heimatmuseum, Bammental 16922
Heimatmuseum, Baunach 16936
Heimatmuseum, Beelitz 16977
Heimatmuseum, Bergneustadt 17024
Heimatmuseum, Berkatal 17026
Heimatmuseum, Bermel 17286
Heimatmuseum, Biebesheim am Rhein 17334
Heimatmuseum, Bleicherode 17387
Heimatmuseum, Blindheim 17389
Heimatmuseum, Bobenheim-Roxheim 17392
Heimatmuseum, Breckerfeld 17533
Heimatmuseum, Brigachtal 17589
Heimatmuseum, Brüssow 17609
Heimatmuseum, Brunsbüttel 17611
Heimatmuseum, Buchenberg 17616
Heimatmuseum, Buchloe 17619
Heimatmuseum, Büddenstedt 17624
Heimatmuseum, Camburg 17681
Heimatmuseum, Crivitz 17751
Heimatmuseum, Dahlen 17772
Heimatmuseum, Dahme 17774
Heimatmuseum, Darlingerode 17785
Heimatmuseum, Denkendorf 17819
Heimatmuseum, Dienheim 17847
Heimatmuseum, Dietenhofen 17853
Heimatmuseum, Dietzenbach 17857
Heimatmuseum, Dirlewang 17873
Heimatmuseum, Dohna 17887
Heimatmuseum, Domersleben 17888
Heimatmuseum, Dossenheim 17932
Heimatmuseum, Duderstadt 17981
Heimatmuseum, Dunningen 18029
Heimatmuseum, Ebermannstadt 18039
Heimatmuseum, Ebern 18040
Heimatmuseum, Echzell 18051
Heimatmuseum, Eggenstein-Leopoldshafen 18064
Heimatmuseum, Eibelstadt 18073
Heimatmuseum, Eich 18075
Heimatmuseum, Eichenzell 18078
Heimatmuseum, Eisenbach 18100
Heimatmuseum, Ellrich 18119
Heimatmuseum, Elsenfeld 18128
Heimatmuseum, Eltmann 18130
Heimatmuseum, Elze 18134
Heimatmuseum, Emskirchen 18162
Heimatmuseum, Epfenbach 18177
Heimatmuseum, Ergoldsbach 18211
Heimatmuseum, Eschenburg 18234
Heimatmuseum, Eschwege 18237
Heimatmuseum, Falkensee 18291
Heimatmuseum, Falkenstein, Vogtland 18296
Heimatmuseum, Finsterbergen 18320
Heimatmuseum, Föhren 18333
Heimatmuseum, Forchheim, Breisgau 18335
Heimatmuseum, Freiamt 18418
Heimatmuseum, Freigericht 18443
Heimatmuseum, Freudenstadt 18456
Heimatmuseum, Friedewald 18471
Heimatmuseum, Friedland 18472
Heimatmuseum, Friesenheim 18481
Heimatmuseum, Fußgönheim 18523
Heimatmuseum, Gadebusch 18526
Heimatmuseum, Garbsen 18535
Heimatmuseum, Geisa 18555
Heimatmuseum, Geislingen an der Steige 18561
Heimatmuseum, Geretsried 18601
Heimatmuseum, Gersfeld 18614
Heimatmuseum, Gerstungen 18620
Heimatmuseum, Geversdorf 18630
Heimatmuseum, Giesen 18634
Heimatmuseum, Gimbsheim 18645
Heimatmuseum, Ginsheim-Gustavsburg 18648
Heimatmuseum, Glonn 18662
Heimatmuseum, Görwihl 18683
Heimatmuseum, Gößnitz 18685
Heimatmuseum, Grafing bei München 18734
Heimatmuseum, Gransee 18737
Heimatmuseum, Greußen 18755
Heimatmuseum, Großenlüder 18790
Heimatmuseum, Großröhrsdorf 18802
Heimatmuseum, Grünstadt 18817
Heimatmuseum, Gschwend 18819
Heimatmuseum, Haiger 18866
Heimatmuseum, Haimhausen 18869
Heimatmuseum, Hallstadt 18911
Heimatmuseum, Haßloch 19031
Heimatmuseum, Haynsburg 19058
Heimatmuseum, Hebertsfelden 19059
Heimatmuseum, Heiligenhafen 19104
Heimatmuseum, Hettenleidelheim 19172
Heimatmuseum, Heuchelheim bei Frankenthal 19177
Heimatmuseum, Heusenstamm 19178
Heimatmuseum, Hillscheid 19188
Heimatmuseum, Himmelpforten 19193
Heimatmuseum, Hohberg 19229
Heimatmuseum, Hohenroda 19241
Heimatmuseum, Hohenwestedt 19247
Heimatmuseum, Holle 19252
Heimatmuseum, Hornburg 19276
Heimatmuseum, Hoya / Weser 19280
Heimatmuseum, Igel 19312
Heimatmuseum, Ilvesheim 19323
Heimatmuseum, Immenstaad am Bodensee 19326
Heimatmuseum, Karlstein am Main 19456
Heimatmuseum, Kefenrod 19494
Heimatmuseum, Kehl 19497
Heimatmuseum, Keltern 19506
Heimatmuseum, Kenn 19517
Heimatmuseum, Kirchen 19545
Heimatmuseum, Kirchhain 19547
Heimatmuseum, Kölleda 19594
Heimatmuseum, Königs Wusterhausen 19637
Heimatmuseum, Königsdorf 19647
Heimatmuseum, Kraiburg am Inn 19694
Heimatmuseum, Kroppenstedt 19728
Heimatmuseum, Kuhardt 19742
Heimatmuseum, Ladbergen 19757
Heimatmuseum, Landau an der Isar 19786
Heimatmuseum, Langenzenn 19835
Heimatmuseum, Leimersheim 19879
Heimatmuseum, Leipheim 19889
Heimatmuseum, Lengenfeld 19953
Heimatmuseum, Lenggries 19954
Heimatmuseum, Leutenbach 19968
Heimatmuseum, Lorch, Württemberg 20034
Heimatmuseum, Luckenwalde 20041
Heimatmuseum, Lügde 20093
Heimatmuseum, Magstadt 20132
Heimatmuseum, Mainz-Kostheim 20153
Heimatmuseum, March 20196
Heimatmuseum, Markgröningen 20204
Heimatmuseum, Markranstädt 20210
Heimatmuseum, Marktl 20217
Heimatmuseum, Maßbach 20234
Heimatmuseum, Meerane 20245
Heimatmuseum, Meinhard 20258
Heimatmuseum, Mellrichstadt 20274
Heimatmuseum, Melsungen 20276
Heimatmuseum, Merchweiler 20288
Heimatmuseum, Mering 20289
Heimatmuseum, Merkendorf 20290
Heimatmuseum, Merzenich 20297
Heimatmuseum, Miesbach 20323
Heimatmuseum, Mindelheim 20328
Heimatmuseum, Mitterfels 20343
Heimatmuseum, Möckmühl 20350
Heimatmuseum, Moosburg an der Isar 20380
Heimatmuseum, Murg 20533
Heimatmuseum, Nauheim 20552
Heimatmuseum, Naumburg, Hessen 20553
Heimatmuseum, Neidenstein 20567
Heimatmuseum, Nellingen 20568
Heimatmuseum, Neu-Ulm 20588
Heimatmuseum, Neukirchen, Schwalm-Eder-Kreis 20626
Heimatmuseum, Neumagen-Dhron 20631
Heimatmuseum, Neunkirchen, Neckar-Odenwald-Kreis 20641
Heimatmuseum, Neupotz 20644
Heimatmuseum, Neureichenau 20646
Heimatmuseum, Neuried, Ortenaukreis 20647
Heimatmuseum, Neustadt in Sachsen 20682
Heimatmuseum, Neustadt, Wied 20683
Heimatmuseum, Nidda 20692
Heimatmuseum, Nideggen 20696
Heimatmuseum, Nieder-Olm 20700
Heimatmuseum, Niedernhall 20708
Heimatmuseum, Niederstetten 20710
Heimatmuseum, Northeim 20747
Heimatmuseum, Nüdlingen 20751
Heimatmuseum, Ober-Flörsheim 20797
Heimatmuseum, Oberndorf am Lech 20818
Heimatmuseum, Oberndorf am Neckar 20819
Heimatmuseum, Obernfeld 20821
Heimatmuseum, Oberstdorf 20842
Heimatmuseum, Ockenheim 20856
Heimatmuseum, Oettingen in Bayern 20878
Heimatmuseum, Osterwieck 20938
Heimatmuseum, Ostrach 20943
Heimatmuseum, Otterberg 20949
Heimatmuseum, Papenburg 20980
Heimatmuseum, Perl 21011
Heimatmuseum, Perleberg 21013
Heimatmuseum, Pfaffenhausen 21017
Heimatmuseum, Pfullendorf 21033
Heimatmuseum, Philippsburg 21037
Heimatmuseum, Plankstadt 21049
Heimatmuseum, Polch 21072
Heimatmuseum, Polling 21074
Heimatmuseum, Postbauer-Heng 21077
Heimatmuseum, Pressath 21120
Heimatmuseum, Pretzsch 21123
Heimatmuseum, Prien am Chiemsee 21127
Heimatmuseum, Queidersbach 21153
Heimatmuseum, Quierschied 21156
Heimatmuseum, Radeburg 21162
Heimatmuseum, Radevormwald 21164
Heimatmuseum, Rattenberg 21194
Heimatmuseum, Raunheim 21200
Heimatmuseum, Rauschenberg 21202
Heimatmuseum, Rehlingen-Siersburg 21241
Heimatmuseum, Reilingen 21250
Heimatmuseum, Reinheim 21254
Heimatmuseum, Renningen 21278
Heimatmuseum, Rheda-Wiedenbrück 21292
Heimatmuseum, Rieneck 21328
Heimatmuseum, Rodenberg 21344
Heimatmuseum, Rodewald 21345
Heimatmuseum, Rötha 21358
Heimatmuseum, Rötz 21360
Heimatmuseum, Ronnenberg 21370
Heimatmuseum, Rotenburg, Wümme 21400
Heimatmuseum, Sandhausen 21483
Heimatmuseum, Schalkau 21530
Heimatmuseum, Schermbeck 21544
Heimatmuseum, Schifferstadt 21547
Heimatmuseum, Schliersee 21574
Heimatmuseum, Schlitz 21575
Heimatmuseum, Schnaittach 21589
Heimatmuseum, Schöneck / Vogtland 21605
Heimatmuseum, Schöningen 21612
Heimatmuseum, Schwanfeld 21666
Heimatmuseum, Seeg 21716
Heimatmuseum, Seelze 21721
Heimatmuseum, Sersheim 21746
Heimatmuseum, Siegbach 21750
Heimatmuseum, Simbach am Inn 21767
Heimatmuseum, Sinzig 21787
Heimatmuseum, Spangenberg 21829
Heimatmuseum, Speicher 21831
Heimatmuseum, Sprendlingen 21844
Heimatmuseum, Stade 21849
Heimatmuseum, Stadt Wehlen 21854
Heimatmuseum, Stadtsteinach 21861
Heimatmuseum, Staufenberg 21869
Heimatmuseum, Steinbach, Taunus 21882
Heimatmuseum, Steinwiesen 21894
Heimatmuseum, Sternberg 21901
Heimatmuseum, Stockstadt am Main 21905
Heimatmuseum, Strausberg 21919
Heimatmuseum, Talheim 21998
Heimatmuseum, Todtmoos 22067
Heimatmuseum, Trebur 22082
Heimatmuseum, Trechtingshausen 22084
Heimatmuseum, Treffurt 22086
Heimatmuseum, Triptis 22107
Heimatmuseum, Undenheim 22173
Heimatmuseum, Unterschneidheim 22184
Heimatmuseum, Vallendar 22201
Heimatmuseum, Varel 22204
Heimatmuseum, Velburg 22213
Heimatmuseum, Venningen 22218
Heimatmuseum, Viernheim 22230
Heimatmuseum, Wadern 22262
Heimatmuseum, Wächtersbach 22266
Heimatmuseum, Waldfischbach-Burgalben 22294
Heimatmuseum, Waldheim 22295
Heimatmuseum, Waldsee 22304
Heimatmuseum, Waldthurn 22305
Heimatmuseum, Wanna 22331
Heimatmuseum, Wedel 22354
Heimatmuseum, Weisendorf 22442
Heimatmuseum, Weißenborn 22448
Heimatmuseum, Wemding 22473
Heimatmuseum, Wennigsen 22480
Heimatmuseum, Wertach 22501
Heimatmuseum, Westerhausen 22514
Heimatmuseum, Westheim 22516
Heimatmuseum, Windeck 22594
Heimatmuseum, Winden 22595
Heimatmuseum, Wippra 22610
Heimatmuseum, Wolfratshausen 22657
Heimatmuseum, Wolfstein 22671
Heimatmuseum, Zarrentin am Schaalsee 22767
Heimatmuseum, Zemmer 22787
Heimatmuseum, Zerf 22789
Heimatmuseum, Zorge 22818
Heimatmuseum, Zusmarshausen 22828
Heimatmuseum, Hüttenberg 19295
Heimatmuseum, Marne 20227
Heimatmuseum, Grafenau, Kreis Böblingen 18726
Heimatmuseum, Mülheim an der Ruhr 20413
Heimatmuseum Adolphshöhe, Wallerfangen 22313
Heimatmuseum Ahlem, Hannover 18992
Heimatmuseum Ahrbergen, Giesen 18635
Heimatmuseum Alte Windmühle Groß Reken, Reken 21257
Heimatmuseum Alter Pfarrhof, Raisting 21171
Heimatmuseum Altes Rathaus, Böhl-Iggelheim 17423
Heimatmuseum Altes Schulhaus, Böhl-Iggelheim 17424
Heimatmuseum Amt Blankenstein, Gladenbach 18652
Heimatmuseum Aschen, Diepholz 17848
Heimatmuseum Barbaramühle, Longkamp 20031
Heimatmuseum Berching mit Ritter-v.-Gluck-Archiv, Berching 16999
Heimatmuseum Bergen-Enkheim, Frankfurt am Main 18373
Heimatmuseum Bierstadt, Wiesbaden 22549
Heimatmuseum Bislich, Wesel 22506
Heimatmuseum Blankenau, Hosenfeld 19279
Heimatmuseum Blankenloch, Stutensee 21927
Heimatmuseum Bockwindmühle, Lebusa 19866
Heimatmuseum Börgerende, Börgerende-Rethwisch 17429
Heimatmuseum Boizenburg, Boizenburg 17439
Heimatmuseum Boxberg, Boxberg 17491
Heimatmuseum Brueckenhof, Königswinter 19658
Heimatmuseum Bühl, Baden, Bühl 17630
Heimatmuseum Carl Swoboda, Schirgiswalde 21558
Heimatmuseum Crumstadt, Riedstadt 21323
Heimatmuseum Dachtel, Aidlingen 16447
Heimatmuseum Debstedt, Langen, Kreis Cuxhaven 19813
Heimatmuseum Delkenheim, Wiesbaden 22550
Heimatmuseum der Parabutscher Donauschwaben, Bad Schönborn 16828
Heimatmuseum der Stadt Bad Tölz, Bad Tölz 16851
Heimatmuseum der Stadt Gernsheim, Gernsheim 18610
Heimatmuseum der Stadt Herrnhut, Herrnhut 19162
Heimatmuseum der Stadt Rheinau in Freistett, Rheinau 21296
Heimatmuseum der Stadt Wilsdruff, Wilsdruff 22590
Heimatmuseum der Verbandsgemeinde Wallhalben, Schmitshausen 21587
Heimatmuseum des Daadener Landes, Daaden 17761
Heimatmuseum des Heimatbundes Jauernig, Vaihingen an der Enz 22197
Heimatmuseum des Landkreises Regensburg, Altenthann 16504
Heimatmuseum des Nordböhmischen Niederlandes, Böblingen 17419
Heimatmuseum Dietingen, Dietingen 17856
Heimatmuseum Dykhus, Borkum 17481
Heimatmuseum Ebenhausen, Oerlenbach 20873
Heimatmuseum Eisenbach, Obernburg am Main 20816
Heimatmuseum Erbenheim, Wiesbaden 22551
Heimatmuseum Esterau, Holzappel 19256
Heimatmuseum Eversberg, Meschede 20302
Heimatmuseum Flacht, Weissach 22445
Heimatmuseum Frankenberg, Frankenberg, Sachsen 18347
Heimatmuseum Freudenthal, Memmingen 20279
Heimatmuseum Friedrichshain, Berlin 17130
Heimatmuseum für Stadt und Kreis Gelnhausen, Gelnhausen 18568
Heimatmuseum für Stadt- und Landkreis, Kirchheimbolanden 19553
Heimatmuseum Gadernheim, Lautertal 19864
Heimatmuseum Garbenheim, Wetzlar 22525
Heimatmuseum Grabow, Grabow 18724
Heimatmuseum Grafling, Oberschneiding 20839
Heimatmuseum "Grafschaft Molbach", Kreuzau 19713
Heimatmuseum Greene, Kreiensen 19709
Heimatmuseum Grimmen "Im Mühlentor", Grimmen 18766
Heimatmuseum Großseelheim, Kirchhain 19548
Heimatmuseum Großzschepa, Hohburg 19231
Heimatmuseum Harthausen, Winterlingen 22608
Heimatmuseum Haueneberstein, Baden-Baden 16880
Heimatmuseum Haus Horn, Niederzier 20716
Heimatmuseum Heidelsheim, Bruchsal 17595
Heimatmuseum Heinrich Zoller, Wittelshofen 22616
Heimatmuseum Helfs Hof, Bochum 17400
Heimatmuseum Herzogenweiler, Villingen-Schwenningen 22237
Heimatmuseum Heuchelheim-Kinzenbach, Heuchelheim 19176
Heimatmuseum Hochzeitshaus, Homberg (Efze) 19264
Heimatmuseum Höchstädt, Höchstädt an der Donau 19210
Heimatmuseum Hoheneggelsen, Söhlde 21792
Heimatmuseum Hus tu Löwenberg, Löwenberger Land 20017
Heimatmuseum im Ackerbürgerhaus, Vohenstrauß 22253
Heimatmuseum im Alten Kloster, Erbendorf 18192
Heimatmuseum im Alten Leuchtturm, Wangerooge, Nordseebad 22330

Heimatmuseum im alten Pfarrhaus, Schwieberdingen 21705
Heimatmuseum im Alten Rathaus, Bruchköbel 17593
Heimatmuseum im Berchtoldshof, Uhingen 22157
Heimatmuseum im Fünfeckigen Turm, Neckarbischofsheim 20560
Heimatmuseum im Haus Catoir, Bad Dürkheim 16714
Heimatmuseum im Herrenhaus, Burghaun 17650
Heimatmuseum im Hintermeierhaus, Donauwörth 17896
Heimatmuseum im Kolb'schen Haus, Bickenbach 17328
Heimatmuseum im Oberen Torturm, Lauchheim 19846
Heimatmuseum im Schelfenhaus, Volkach 22255
Heimatmuseum im Schillerhaus, Oberkochen 20814
Heimatmuseum im Schlösschen, Ochsenfurt 20852
Heimatmuseum im Schloß, Zörbig 22816
Heimatmuseum im Tuchmacherhaus, Thannhausen 22048
Heimatmuseum im Unteren Turm, Leutershausen-Wiedersbach 19971
Heimatmuseum im Wachtturm, Geyer 18631
Heimatmuseum im Wiggensbacher Informationszentrum, Wiggensbach 22565
Heimatmuseum in der Mühle Gettorf, Gettorf 18628
Heimatmuseum Ittersbach, Karlsbad 19432
Heimatmuseum Kappe, Zehdenick 22768
Heimatmuseum Karlsdorf, Karlsdorf-Neuthard 19433
Heimatmuseum Kiechlinsbergen, Endingen am Kaiserstuhl 18164
Heimatmuseum Kirberg, Hünfelden 19291
Heimatmuseum Kirchdorf, Insel Poel 19348
Heimatmuseum Kirchheim, Heidelberg 19074
Heimatmuseum Klosterlangheim, Lichtenfels 19979
Heimatmuseum Köpenick, Berlin 17131
Heimatmuseum Kornburg, Nürnberg 20767
Heimatmuseum Lahnau-Waldgirmes, Lahnau 19765
Heimatmuseum Laisa, Battenberg 16933
Heimatmuseum Lamspringe, Lamspringe 19783
Heimatmuseum Langelsheim, Langelsheim 19809
Heimatmuseum Langenau, Langenau 19818
Heimatmuseum Langenseifen, Bad Schwalbach 16834
Heimatmuseum Langenselbold, Langenselbold 19834
Heimatmuseum Laubach, Laubach 19842
Heimatmuseum Lauchhammer, Lauchhammer 19844
Heimatmuseum Lauda-Königshofen, Lauda-Königshofen 19847
Heimatmuseum Lauenau, Lauenau 19848
Heimatmuseum Leeheim, Riedstadt 21324
Heimatmuseum Leer, Leer 19870
Heimatmuseum Lette, Coesfeld 17722
Heimatmuseum Lich, Lich 19977
Heimatmuseum Lichtenberg, Berlin 17132
Heimatmuseum Liebenwalde im historischen Stadtgefängnis, Liebenwalde 19984
Heimatmuseum Lübbecke und Kunstgalerie im Speicher, Lübbecke 20067
Heimatmuseum Lütgendortmund, Dortmund 17920
Heimatmuseum Medenbach, Wiesbaden 22552
Heimatmuseum Meerholz, Gelnhausen 18569
Heimatmuseum-Menton, Teningen 22031
Heimatmuseum Michelbacher Schlößchen, Alzenau in Unterfranken 16529
Heimatmuseum mit ostdeutscher Heimatstube, Hanerau-Hademarschen 18985
Heimatmuseum Mittelbuchen, Hanau 18976
Heimatmuseum Möhringen, Stuttgart 21941
Heimatmuseum Münchingen, Korntal-Münchingen 19688
Heimatmuseum Naunheim, Wetzlar 22526
Heimatmuseum Naurod, Wiesbaden 22553
Heimatmuseum Neckarau, Mannheim 20164
Heimatmuseum Neuenwalde, Langen, Kreis Cuxhaven 19814
Heimatmuseum Neustadtgödens, Sande 21482
Heimatmuseum Nied, Frankfurt am Main 18374
Heimatmuseum Niemes und Prachatitz, Ingolstadt 19341
Heimatmuseum Nürnberg, Nürnberg 20768
Heimatmuseum Obbornhofen, Hungen 19297
Heimatmuseum Obergünzburg mit Südseesammlung, Obergünzburg 20806
Heimatmuseum Peldemühle, Wittmund 22626
Heimatmuseum Pfarrscheuer, Römerstein 21355
Heimatmuseum Plieningen, Stuttgart 21942
Heimatmuseum Rain am Lech, Rain 21169
Heimatmuseum Ratzenried, Argenbühl 16569
Heimatmuseum Rauchkate, Zetel 22790
Heimatmuseum Reinickendorf, Berlin 17133
Heimatmuseum Reischenau, Dinkelscherben 17869
Heimatmuseum Reiterhaus, Neusalza-Spremberg 20653
Heimatmuseum Reutlingen, Reutlingen 21283
Heimatmuseum Rheiderland, Weener 22357
Heimatmuseum Rhüden, Seesen 21723
Heimatmuseum Rodenstein, Fränkisch-Crumbach 18342
Heimatmuseum Römstedthaus, Bergen, Kreis Celle 17009
Heimatmuseum Rohrbach, Heidelberg 19075
Heimatmuseum Rohrhirschmühle, Bühl 17631
Heimatmuseum Runder Turm, Sigmaringen 21764
Heimatmuseum Sandweier, Baden-Baden 16881
Heimatmuseum Sanhofen, Mannheim 20165
Heimatmuseum Sankt Laurentius, Dahlenburg 17773
Heimatmuseum Schierstein, Wiesbaden 22554
Heimatmuseum Schloß Brenz, Sontheim an der Brenz 21821
Heimatmuseum Schloß Schönebeck, Bremen 17554
Heimatmuseum Schloß Wolkenstein, Wolkenstein 22675
Heimatmuseum Schwanheim, Frankfurt am Main 18375
Heimatmuseum Schwarzes Tor, Sankt Georgen im Schwarzwald 21493
Heimatmuseum Seulberg, Friedrichsdorf 18475
Heimatmuseum Sprucker Mühle, Guben 18820
Heimatmuseum Stadt Bernstadt, Bernstadt auf dem Eigen 17307
Heimatmuseum Stadt Nagold, Nagold 20545
Heimatmuseum Stadt und Landkreis Neudek im Erzgebirge, Augsburg 16622
Heimatmuseum Stammheim, Stuttgart 21943
Heimatmuseum Steinbach, Ottweiler 20959
Heimatmuseum Steinfischbach, Waldems 22286
Heimatmuseum Tieringen, Meßstetten 20309
Heimatmuseum - Traktoren- und Landmaschinenmuseum, Großengottern 18788
Heimatmuseum Treptow, Berlin 17134
Heimatmuseum und Dokumentationszentrum zur Deutschen Nachkriegsgeschichte, Wanfried 22322
Heimatmuseum und Galerie Neue Diele, Jork 19402
Heimatmuseum und Gedenkstätte Bützow, Bützow 17641
Heimatmuseum und Hedwig Courths-Mahler Archiv, Nebra 20559
Heimatmuseum und Textil-Schauwerkstatt, Greiz 18752
Heimatmuseum Usseln, Willingen 22585
Heimatmuseum Vietze, Höhbeck 19213
Heimatmuseum Vilsbiburg - Kröninger Hafnermuseum, Vilsbiburg 22243
Heimatmuseum Walldorf, Mörfelden-Walldorf 20360
Heimatmuseum Waltersweier, Offenburg 20885
Heimatmuseum Wandsbek, Hamburg 18934
Heimatmuseum Wiesenbach, Wiesenbach 22559
Heimatmuseum Winterberg im Böhmerwald, Freyung 18460
Heimatmuseum Wißmar, Wettenberg 22520
Heimatmuseum Wolfskehlen, Riedstadt 21325
Heimatmuseum Wommelshausen, Bad Endbach 16727
Heimatmuseum Worms Abenheim, Worms 22681
Heimatmuseum Worms Horchheim, Worms 22682
Heimatmuseum Worms-Weinsheim, Worms 22683
Heimatmuseum Zehlendorf, Berlin 17135
Heimatscheune, Zschepplin 22824
Heimatstätte der Kuhländer, Ludwigsburg 20044
Heimatstube, Altenbrak 16496
Heimatstube, Altenstadt an der Waldnaab 16502
Heimatstube, Benneckenstein 16993
Heimatstube, Brüel 17603
Heimatstube, Dabel 17762
Heimatstube, Derenburg 17824
Heimatstube, Ditfurt 17877
Heimatstube, Eggesin 18065
Heimatstube, Elsteraue 18129
Heimatstube, Enkirch 18176
Heimatstube, Esslingen 18274
Heimatstube, Fehrbellin 18300
Heimatstube, Gittelde 18649
Heimatstube, Hasselfelde 19027
Heimatstube, Hasselfelde 19028
Heimatstube, Hinternah 19194
Heimatstube, Lenzkirch 19958
Heimatstube, Liepgarten 19987
Heimatstube, Marlow 20226
Heimatstube, Matzlow-Garwitz 20237
Heimatstube, Neulußheim 20629
Heimatstube, Neuruppin 20649
Heimatstube, Quellendorf 21154
Heimatstube, Remseck am Neckar 21268
Heimatstube, Schäplitz 21528
Heimatstube, Schochwitz 21595
Heimatstube, Schonach im Schwarzwald 21619
Heimatstube, Schwedt / Oder 21677
Heimatstube, Tanne 22007
Heimatstube, Zörbig 22817
Heimatstube, Stapelburg 21863
Heimatstube 1813 Möckern, Möckern 20349
Heimatstube Abbenrode, Abbenrode 16416
Heimatstube Adlergebirge, Waldkraiburg 22299
Heimatstube Altenau, Altenau 16490
Heimatstube Arenborn, Oberweser 20850
Heimatstube Batzsch-Brestowatz, Königsbach-Stein 19640
Heimatstube Bliedersdorf, Bliedersdorf 17388
Heimatstube Börnecke, Blankenburg 17371
Heimatstube Bredenbeck, Wennigsen 22481
Heimatstube Brunkensen, Alfeld 16465
Heimatstube der Gemeinde und des Harzklub Zweigvereins Bad Suderode, Bad Suderode 16846
Heimatstube "Der Große Hof", Wienrode 22543
Heimatstube der Stadt Saaz, Roth 21402
Heimatstube des Heimatkreises Reichenberg, Augsburg 16623
Heimatstube Eime, Eime 18090
Heimatstube Elbingerode, Elbingerode, Harz 18112
Heimatstube Eltze, Uetze 22153
Heimatstube Endersbach, Weinstadt 22438
Heimatstube Frauenwald und Bunkermuseum, Frauenwald 18414
Heimatstube Gadenstedt, Lahstedt 19775
Heimatstube Gräfenhain/ Nauendorf, Georgenthal 18590
Heimatstube Groß Denkte, Denkte 17822
Heimatstube Groß Gleidinggen, Vechelde 22206
Heimatstube Groß Lafferde, Lahstedt 19776
Heimatstube Groß Nemerow, Groß Nemerow 18777
Heimatstube Großbodungen, Großbodungen 18781
Heimatstube Gutenberg, Oberostendorf 20826
Heimatstube Hänigsen, Uetze 22154
Heimatstube Hetzdorf, Uckerland 22143
Heimatstube Hohenhameln, Hohenhameln 19237
Heimatstube im alten Rathaus, Hessigheim 19169
Heimatstube im Burgturm, Plau am See 21052
Heimatstube im Elbtor, Werben 22484
Heimatstube im "Spieker Anno 1754", Tarmstedt 22008
Heimatstube in der Wasserburg, Gerswalde 18621
Heimatstube Kappel, Lenzkirch 19959
Heimatstube Lebus, Lebus 19865
Heimatstube Lichtenstadt, Zirndorf 22809
Heimatstube Mederitzsch, Pflückuff 21026
Heimatstube Münstedt, Lahstedt 19777
Heimatstube Neckarrems, Remseck am Neckar 21269
Heimatstube Nendingen, Tuttlingen 22136
Heimatstube Oberg, Lahstedt 19778
Heimatstube Petznick, Templin 22028
Heimatstube Radegast, Radegast 21163
Heimatstube Röckwitz, Röckwitz 21350
Heimatstube Röhrnbach-Kaltenbach, Röhrnbach 21354
Heimatstube Schandelah, Cremlingen 17746
Heimatstube Schlesien, Kaufbeuren 19487
Heimatstube Sitzenroda, Schildau, Gneisenaustadt 21549
Heimatstube Sperenberg, Am Mellensee 16531
Heimatstube Stadt und Landkreis Saaz, Georgensgmünd 18587
Heimatstube Tribsees, Tribsees 22093
Heimatstube Trochtelfingen, Bopfingen 17471
Heimatstube Untere Baranya, Gingen an der Fils 18646
Heimatstube Usedom, Usedom 22190
Heimatstube Vietmannsdorf, Templin 22029
Heimatstube Wahrenbrück, Uebigau-Wahrenbrück 22148
Heimatstube Warthe, Boitzenburger Land 17435
Heimatstuben im Arresturm, Spalt 21828
Heimatstuben im Oberen Tor, Scheinfeld 21540
Heimatstuben Weipert und Erzgebirgsschau, Gunzenhausen 18839
Heimatsube, Uchtspringe 22142
Heimatsube Klaushagen, Boitzenburger Land 17436
Heimatsube Staupitz, Pflückuff 21027
Heimatverein Räntzelstecher Güsten Osmarsleben e.V, Güsten 18826
Hein's Mühle, Bendorf 16988
Hellweg-Museum Unna, Unna 22175
Hennebergisches Museum, Kloster Veßra 19579
Hermann-Grochtmann-Museum Datteln, Datteln 17800
Hermann-Hesse-Höri-Museum, Gaienhofen 18531
Herzogkasten, Abensberg 16419
Herz'sche Heimatstiftung Hilgerhof, Pittenhart 21047
Hessisches Forst-Kulturhistorisches Museum, Biebergemünd 17331
Heuson-Museum im Rathaus, Büdingen 17628
Hexenbürgermeisterhaus, Lemgo 19947
Hinterlandmuseum Schloss Biedenkopf, Biedenkopf 17335
Hirschberger Heimatstuben, Alfeld 16466
Historia Elstal, Wustermark 22754
Historiengewölbe, Rothenburg ob der Tauber 21406
Historische Einungsmeisterstube, Dogern 17886
Historische Gerberei, Burg bei Magdeburg 17643
Historische Sammlung Eltville, Eltville am Rhein 18132
Historische Sammlungen im Gesundheitspark, Bad Gottleuba 16740
Historische Turmstuben, Winnenden 22601
Historische Werkstatt Nordenstadt, Wiesbaden 22555
Historisches Centrum Hagen, Hagen 18856
Historisches Heimatmuseum Östringen, Östringen 20876
Historisches Museum am Hohen Ufer, Hannover 18993
Historisches Museum am Strom - Hildegard von Bingen, Bingen am Rhein 17353
Historisches Museum Aurich, Aurich 16647
Historisches Museum Bayreuth, Bayreuth 16952
Historisches Museum Bremerhaven, Bremerhaven 17570
Historisches Museum Frankfurt am Main, Frankfurt am Main 18377
Historisches Museum für Mittelanhalt und Bach-Gedenkstätte im Schloss Köthen, Köthen 19666
Historisches Museum im Spital, Dinkelsbühl 17866
Historisches Museum Saar, Saarbrücken 21456
Historisches Museum Schloß Gifhorn, Gifhorn 18642
Historisches Museum Verden, Verden 22220
Hiwwelhaus, Marpingen 20228
Hobelmuseum Langenfeld, Langenfeld 19832
Hochrheinmuseum, Bad Säckingen 16814
Hofmarkmuseum, Eggenfelden 18061
Hofmarkmuseum Schloß Eggersberg, Riedenburg 21315
Hohenlohe-Museum, Neuenstein, Württemberg 20613
Hohenzollerisches Landesmuseum, Hechingen 19062
Hohhaus-Museum, Lauterbach, Hessen 19862
Hufschmiede Museum Frehrking, Neustadt am Rübenberge 20663
Huldigungssaal und Rathaus, Goslar 18710
Hunsrück-Museum, Simmern 21771
Huttermuseum, Erdweg 18195
Imker- und Heimatmuseum Wassertrüdingen, Wassertrüdingen 22349
Industriemuseum Elmshorn, Elmshorn 18126
Industriemuseum Freudenthaler Sensenhammer, Leverkusen 19974
Inselmuseum Spiekeroog, Spiekeroog 21839
Isergebirgs Museum Neugablonz, Kaufbeuren 19488
Jägerndorfer Heimatstuben, Ansbach 16559
Jagdschloß, Gelbensande 18565
Jerusalemhaus mit Verwaltung der Städtischen Sammlungen Wetzlar, Wetzlar 22527
Johann-Friedrich-Danneil-Museum, Salzwedel 21480
John-Wagener-Haus Sievern, Langen, Kreis Cuxhaven 19815
Josef Albers Museum Quadrat Bottrop, Bottrop 17488
Jüdisches Museum der Nordpfalz, Winnweiler 22603
Jüdisches Museum Georgensgmünd, Georgensgmünd 18588
Jüdisches Museum Shalom Europa, Würzburg 22712
Kaisersaal Schloß Schwarzburg, Schwarzburg 21672
Kaiserstühler Heimatmuseum, Endingen am Kaiserstuhl 18166
Kampffmeyersche Holzlöffelsammlung, Bad Dürkheim 16715
Karasek-Museum, Seifhennersdorf 21732
Karl-Wörn-Haus, Schwetzingen 21702
Karrasburg Museum Coswig, Coswig 17727
Kelnhof-Museum, Bräunlingen 17497
Kettenschmiedemuseum, Fröndenberg 18486
Ketterer-Haus-Museum, Biberach, Ortenaukreis 17327
Kienzler-Schmiede, Kirchzarten 19555
Kirchengeschichtliches Museum in der Pfalzkapelle, Bad Wimpfen 16866
Kirschenmuseum, Endingen am Kaiserstuhl 18167
Klaus-Groth-Museum und Neue Museumsinsel Lüttenheid, Heide 19066
Das Kleine Museum, Gelsenkirchen 18575
Kleines Heimatmuseum Windmühle, Werdum 22489
Kloster Hadmersleben, Hadmersleben 18851
Kloster-Museum, Sankt Märgen 21505
Kobisenmühle, Sankt Georgen im Schwarzwald 21494
Kölnisches Stadtmuseum, Köln 19612
Koenraad Bosman Museum, Rees 21212
Korallen- und Heimatmuseum, Nattheim 20550
Korbinian-Brodmann-Museum, Hohenfels 19236
Kornhaus-Museum, Weiler-Simmerberg 22388
Das Kranichhaus, Otterndorf 20950
Kreis- und Heimatmuseum, Wittstock/Dosse 22628
Kreis- und Heimatmuseum Bogen, Bogen 17432
Kreis- und Stadtmuseum Dieburg, Dieburg 17846
Kreis- und Universitätsmuseum Helmstedt, Helmstedt 19123
Kreisheimatmuseum Frankenberg, Frankenberg, Eder 18343
Kreisheimatmuseum Striedieck's Hof, Bünde 17636
Kreisheimatstube, Ellzee 18125
Kreismuseum, Osterburg 20931
Kreismuseum Bad Liebenwerda, Bad Liebenwerda 16778
Kreismuseum Bitterfeld, Bitterfeld 17370
Kreismuseum der Heimatvertriebenen, Seligenstadt 21738
Kreismuseum Finsterwalde, Finsterwalde 18321
Kreismuseum Geilenkirchen, Geilenkirchen 18554
Kreismuseum Heinsberg, Heinsberg 19112
Kreismuseum Herzogtum Lauenburg, Ratzeburg 21197
Kreismuseum Jerichower Land, Genthin 18586
Kreismuseum Mühldorf, Mühldorf am Inn 20399
Kreismuseum Oberhavel, Oranienburg 20909
Kreismuseum Peine, Peine 21001

Kreismuseum Prinzeßhof, Itzehoe 19369
Kreismuseum Schönebeck, Schönebeck 21602
Kreismuseum Syke, Syke 21991
Kreismuseum Walderbach, Walderbach 22291
Kreuzberg Museum, Berlin 17160
Krippenmuseum, Telgte 22022
Krönungssaal des Aachener Rathauses, Aachen 16400
Kubiner Heimatstube, Weinsberg 22433
Kultur- und Heimattreff, Tholey 22055
Kultur- und Stadthistorisches Museum Duisburg, Duisburg 18020
Kulturelles Forum im Freiherr-vom-Stein-Haus, Langenfeld 19833
Kulturgeschichtliches Museum Bügeleisenhaus, Hattingen 19037
Kulturgeschichtliches Museum Osnabrück, Osnabrück 20923
Kulturhistorische Sammlung, Iphofen 19352
Kulturhistorisches Karnevalsmuseum des Regionalverband Düren, Niederzier 20717
Kulturhistorisches Museum, Görlitz 18679
Kulturhistorisches Museum Franziskanerkloster, Zittau 22812
Kulturhistorisches Museum Prenzlau im Dominikanerkloster, Prenzlau 21118
Kulturhistorisches Museum Rostock, Rostock 21393
Kulturkunststatt Prora, Prora 21131
Kulturzentrum Ostpreußen, Ellingen 18116
Kunst- und Heimathaus, Paderborn 20971
Kunsthof Bad Salzelmen, Schönebeck 21603
Kurfürstliche Kanzlei, Torgau 22072
Ländliche Bildergalerie, Dischingen 17875
Lambertsmühle, Burscheid 17667
Landesmuseum für Kunst und Kulturgeschichte, Schleswig 21566
Landesmuseum für schaumburg-lippische Geschichte, Landes- und Volkskunde, Bückeburg 17623
Landlmuseum Sulzbürg, Mühlhausen, Oberpfalz 20400
Landschaftsmuseum, Seligenstadt 21739
Landschaftsmuseum der Dübener Heide, Bad Düben 16713
Landschaftsmuseum Obermain, Kulmbach 19748
Landschaftsmuseum Westerwald, Hachenburg 18847
Landwirtschaftsmuseum Lüneburger Heide, Suderburg 21972
Langbeinmuseum, Hirschhorn 19199
Langobardenwerkstatt Zethlingen, Zethlingen 22792
Langsdorfer Heimatmuseum, Lich 19978
Lechfeldmuseum, Königsbrunn 19645
Lehmmuseum, Buchberg 17614
Lehrerhaus-Friemersheim, Duisburg 18021
Leib'sches Haus, Abteilung Stadtgeschichte und Volkskunde, Gießen 18638
Leinfelder Haus, Leinfelden-Echterdingen 19884
Letschiner Heimatstuben, Letschin 19965
Liebenberger Geschichtsmuseum, Löwenberger Land 20018
Lieblinger Heimatstube, Willstätt 22588
Lilienhof, Lilienthal 19989
Lindenfelser Museum, Lindenfels 20006
Lippisches Landesmuseum, Detmold 17839
Lotschnauer Heimatstube, Eislingen 18110
Lottehaus, Wetzlar 22529
Lovriner Stube, Donauwörth 17898
LVR-Landesmuseum Bonn, Bonn 17462
LWL-Landesmuseum für Kunst und Kulturgeschichte, Münster 20512
Märkisches Museum, Berlin 17181
Manfreds Haus- und Hofmuseum, Weisweil 22466
Marine-Museum Limburg, Limburg an der Lahn 19996
Markgräfler Museum Müllheim, Müllheim 20417
Markgrafenmuseum Ansbach, Ansbach 16560
Markt Einersheimer Heimatstuben, Markt Einersheim 20211
Maschinen-u.Heimatmuseum, Eslohe 18244
Meyerhof Scheeßel, Scheeßel 21537
Mies-Pilsner-Heimatmuseum, Dinkelsbühl 17867
Miletitscher Heimatstube, Schönaich 21596
Millpetersch Haus Güdesweiler, Oberthal 20845
Mineralien- und Heimatmuseum Pleystein, Pleystein 21062
Mitte Museum, Berlin 17189
Mittelschwäbisches Heimatmuseum Krumbach, Krumbach 19729
Möllner Museum Historisches Rathaus, Mölln 20352
Mörfelden Museum, Mörfelden-Walldorf 20361
Mörsbacher Museum, Mörsbach 20364
Moltke-Gedächtnisstätte, Parchim 20982
Montfort-Museum, Tettnang 22040
Mühlenhof-Freilichtmuseum Münster, Münster 20515
Mühlenmuseum Mitling-Mark mit Sammlung Omas Küche, Westoverledingen 22517
Münchhausen-Museum, Bodenwerder 17414
Münchner Stadtmuseum, München 20462
Multivision "Worpswede und das Teufelsmoor", Worpswede 22695
Murgtal-Museum, Forbach 18334
MUSE-O, Stuttgart 21953
Museen Alte Bischofsburg, Wittstock/Dosse 22629
Museen der Stadt Bad Kösen, Bad Kösen 16763
Museen im Alten Schloss, Neustadt an der Aisch 20666
Museen im Kulturzentrum, Rendsburg 21275
Museen im Markgrafen-Schloss, Emmendingen 18146
Museum, Coswig 17728
Museum Abodiacum, Denklingen 17821
Museum Adlerturm, Dortmund 17923
Museum Adorf, Adorf 16433
Museum Alexandrowka, Potsdam 21095
Museum alte Ziegelei Westerholt, Wardenburg 22334
Museum Altes Gymnasium, Schweinfurt 21684
Museum Altes Land, Jork 19403
Museum Altomünster, Altomünster 16524
Museum Alzey, Alzey 16530
Museum am Lindenbühl, Mühlhausen/ Thüringen 20403
Museum am Lindenplatz, Weil am Rhein 22376
Museum am Markt, Schiltach 21556
Museum am Maxplatz, Rehau 21239
Museum am Thie e.V., Göttingen 18694
Museum Amelinghausen mit Zinnfigurenausstellung, Amelinghausen 16535
Museum Amtspforte Stadthagen, Stadthagen 21856
Museum Appeleshof, Gechingen 18548
Museum Auerbach, Auerbach, Vogtland 16608
Museum Bad Abbach, Bad Abbach 16665
Museum Bargteheide, Bargteheide 16928
Museum Barockschloß Delitzsch, Delitzsch 17812
Museum Bautzen, Bautzen 16940
Museum Bayerisches Vogtland, Hof 19220
Museum Bernhard Brühl, Landsberg, Saalekreis 19798
Museum Bischofsheim, Bischofsheim 17362
Museum Blauer Anger, Bergen, Kreis Traunstein 17011
Museum Burg Brome, Brome 17591
Museum Burg Creuzburg, Creuzburg 17748
Museum Burg Frankenberg, Aachen 16404
Museum Burg Hornberg, Neckarzimmern 20566
Museum Burg Mylau, Mylau 20540
Museum Burg Posterstein, Posterstein 21078
Museum Burg Pottenstein, Pottenstein 21114
Museum Burg Ronneburg, Ronneburg, Hessen 21368
Museum Castellum, Mainz-Kastel 20152
Museum Charlottenburg-Wilmersdorf, Berlin 17194
Museum der Barbarossastadt Gelnhausen, Gelnhausen 18571
Museum der Donauschwaben, Herbrechtingen 19135
Museum der Grafschaft Rantzau, Barmstedt 16929
Museum der Landschaft Eiderstedt, Sankt Peter-Ording 21509
Museum der Oberzent, Beerfelden 16978
Museum der Schildbürger, Schildau, Gneisenaustadt 21550
Museum der Stadt Alfeld (Leine), Alfeld 16467
Museum der Stadt Aue, Aue 16605
Museum der Stadt Bad Berleburg, Bad Berleburg 16692
Museum der Stadt Bad Gandersheim, Bad Gandersheim 16738
Museum der Stadt Bad Neuenahr-Ahrweiler, Bad Neuenahr-Ahrweiler 16795
Museum der Stadt Bad Staffelstein, Bad Staffelstein 16843
Museum der Stadt Bensheim, Bensheim 16997
Museum der Stadt Boppard, Boppard 17475
Museum der Stadt Borna, Borna 17485
Museum der Stadt Butzbach im Solms-Braunfelser Hof, Butzbach 17671
Museum der Stadt Calw, Calw 17680
Museum der Stadt Dommitzsch, Dommitzsch 17889
Museum der Stadt Eberbach, Eberbach 18033
Museum der Stadt Eschborn, Eschborn 18233
Museum der Stadt Ettlingen, Ettlingen 18280
Museum der Stadt Gladbeck, Gladbeck 18650
Museum der Stadt Gronau (Leine), Gronau, Leine 18772
Museum der Stadt Güstrow, Güstrow 18829
Museum der Stadt Hagenow, Hagenow 18862
Museum der Stadt Kraichtal, Kraichtal 19696
Museum der Stadt Lahnstein im Hexenturm, Lahnstein 19767
Museum der Stadt Lahr, Lahr 19773
Museum der Stadt Langen, Langen, Hessen 19811
Museum der Stadt Lünen, Lünen 20103
Museum der Stadt Nauen, Nauen 20551
Museum der Stadt Neu-Isenburg Haus zum Löwen, Neu-Isenburg 20582
Museum der Stadt Neustadt an der Weinstraße, Neustadt an der Weinstraße 20672
Museum der Stadt Neutraubling, Neutraubling 20685
Museum der Stadt Parchim, Parchim 20983
Museum der Stadt Pegau, Pegau 20999
Museum der Stadt Ratingen, Ratingen 21191
Museum der Stadt Schopfheim, Schopfheim 21622
Museum der Stadt Troisdorf, Troisdorf 22108
Museum der Stadt Waiblingen, Waiblingen 22278
Museum der Stadt Weinheim, Weinheim 22431
Museum der Stadt Worms im Andreasstift, Worms 22685
Museum der Westlausitz Kamenz, Kamenz 19421
Museum des ehemaligen Kreises Stuhm/Westpr., Bremervörde 17580
Museum des Haager Landes, Haag in Oberbayern 18845
Museum des Handwerks, Bad Bederkesa 16677
Museum des Heimatkundlichen Vereins, Scheßlitz 21545
Museum des Historischen Vereins Freising, Freising 18451
Museum des Teltow, Zossen 22820
Museum Dingolfing-Herzogsburg, Dingolfing 17865
Museum Eckernförde, Eckernförde 18053
Museum Ehingen im Heilig-Geist-Spital, Ehingen 18068
Museum Elbinsel Wilhelmsburg, Hamburg 18947
Museum Erding, Erding 18194
Museum Flederwisch, Furth im Wald 18516
Museum-Freundschaften in der Kulturkunststatt Prora, Prora 21133
Museum Friesenheim, Ludwigshafen am Rhein 20053
Museum für das Fürstentum Lüneburg, Lüneburg 20100
Museum für dörfliche Alltagskultur Rubenheim, Gersheim 18615
Museum für Gerberei- und Stadtgechichte, Hirschberg, Saale-Orla-Kreis 19198
Museum für Hamburgische Geschichte, Hamburg 18949
Museum für Heimatgeschichte und Nordpfalzgalerie, Alsenz 16480
Museum für Kloster- und Heimatgeschichte, Harsefeld 19016
Museum für Kunst und Kulturgeschichte Dortmund, Dortmund 17925
Museum für Ortsgeschichte, Mutterstadt 20538
Museum für Ortsgeschichte und Weinbau, Zellertal 22786
Museum für Schnitzkunst und Kulturgeschichte, Oberammergau 20800
Museum für Stadtgeschichte, Alpirsbach 16478
Museum für Stadtgeschichte, Breisach am Rhein 17536
Museum für Stadtgeschichte, Höhr-Grenzhausen 19215
Museum für Stadtgeschichte, Neuenburg am Rhein 20605
Museum für Stadtgeschichte Dessau, Dessau 17831
Museum für Stadtgeschichte im Adam-und-Eva-Haus, Paderborn 20972
Museum für Stadtgeschichte und Volkskunde, Heppenheim 19129
Museum für Weinbau und Stadtgeschichte, Edenkoben 18055
Museum Garching an der Alz, Garching an der Alz 18536
Museum Gensungen, Felsberg 18309
Museum Goch, Goch 18666
Museum Goldener Steig, Waldkirchen 22298
Museum Griesheim, Griesheim 18759
Museum Grossauheim, Hanau 18980
Museum Großkrotzenburg, Großkrotzenburg 18796
Museum Gülden Creutz, Leinefelde-Worbis 19880
Museum Haldensleben, Haldensleben 18883
Museum Hameln, Hameln 18967
Museum Haus Kast, Gaggenau 18527
Museum Heimathaus Irmintraut, Ottersberg 20953
Museum Heimathaus Münsterland, Telgte 22023
Museum Hessenstube, Baunatal 16937
Museum Hofmühle, Immenstadt im Allgäu 19328
Museum Hohenzollern in Franken, Kulmbach 19749
Museum Holstentor, Lübeck 20080
Museum Illingen, Illingen, Kreis Neunkirchen 19318
Museum im alten Brauhaus, Homberg (Ohm) 19265
Museum im alten Pfarrhaus, Wiedensahl 22539
Museum im Alten Rathaus, Mönchberg im Spessart 20355
Museum im Alten Rathaus, Neckargemünd 20562
Museum im Alten Rathaus, Waghäusel 22270
Museum im Alten Schloß, Neckarbischofsheim 20561
Museum im Ammes Haus, Hatzfeld 19044
Museum im Amtshaus, Grünsfeld 18816
Museum im Amtshausschüpfla, Erlangen 18223
Museum im Astorhaus, Walldorf 22308
Museum im Blaahaus am Unteren Römerweg, Kiefersfelden 19526
Museum im Boyneburgischen Schloß, Sontra-Wichmannshausen 21824
Museum im 'Fressenden Haus' - Burgkasten Weißenstein, Regen 21213
Museum im Gotischen Haus, Bad Homburg v.d.Höhe 16751
Museum im Heimathaus, Traunstein 22078
Museum im Herrenhaus, Hausach 19048
Museum im Hexenturm, Markdorf 20202
Museum im Kloster - Museum des Landkreises Osnabrück, Bersenbrück 17308
Museum im Kreuzgang, Landshut 19804
Museum im Malhaus, Wasserburg, Bodensee 22347
Museum im Marstall, Winsen, Luhe 22606
Museum im Mesnerhaus, Pfaffenhofen an der Ilm 21020
Museum im Mönchenkloster, Jüterbog 19407
Museum im Pflegschloss, Schrobenhausen 21640
Museum im Prediger, Schwäbisch Gmünd 21649
Museum im Rathaus Möhringen, Tuttlingen 22137
Museum im Ritterhaus, Offenburg 20887
Museum im Ritterhaus, Osterode am Harz 20937
Museum im Schäferhaus, Wahlsburg 22273
Museum im Schafstall, Neuenstadt am Kocher 20612
Museum im Schloss, Bernstadt, Alb-Donau-Kreis 17306
Museum im Schloß Bad Pyrmont, Bad Pyrmont 16804
Museum im Schloss Wolfenbüttel, Wolfenbüttel 22654
Museum im Schweizer Hof, Bretten 17585
Museum im Stadtturm, Groitzsch 18771
Museum im Steinhaus, Bad Wimpfen 16867
Museum im Steintor, Anklam 16550
Museum im Steintorturm, Brandenburg an der Havel 17512
Museum im Stern, Warburg 22332
Museum im Strumpferhaus, Oberstaufen 20841
Museum im Turm, Neuleiningen 20628
Museum im Wächterhaus, Nürnberg 20783
Museum im Wehener Schloss, Taunusstein 22015
Museum im Westrich, Ramstein-Miesenbach 21176
Museum im Wetterbergschen Adelshof, Bad Münder am Deister 16785
Museum im Wittelsbacher Schloss Friedberg, Friedberg, Kreis Aichach-Friedberg 18469
Museum im Zeughaus, Vechta 22207
Museum in den Berglen, Berglen 17023
Museum in der Adler-Apotheke, Eberswalde 18047
Museum in der Alten Schule, Efringen-Kirchen 18058
Museum in der Burg, Coppenbrügge 17725
Museum in der Osterburg, Weida 22365
Museum in der Präparandenschule, Höchberg 19208
Museum in der Remise, Bad Harzburg 16745
Museum in der Scheuergasse, Zwingenberg, Kreis Bergstraße 22840
Museum in der Stadthalle Stern (Interim), Riesa 21329
Museum "Kaffeemühle", Wolgast 22673
Museum Kleines Schloß, Blankenburg 17373
Museum Kloster Bentlage, Rheine 21298
Museum Kloster Zeven, Zeven 22797
Museum Leben am Meer, Esens 18242
Museum Leuchtenburg, Seitenroda 21733
Museum Lichtenberg im Stadthaus, Berlin 17211
Museum Löffingen, Löffingen 20015
Museum Malerwinkelhaus, Marktbreit 20215
Museum Mengeringhausen, Bad Arolsen 16672
Museum Naila im Schusterhof, Naila 20546
Museum Neukirchen-Vluyn, Neukirchen-Vluyn 20627
Museum Neukölln, Berlin 17212
Museum Neuruppin, Neuruppin 20650
Museum Neustadt an der Orla, Neustadt an der Orla 20667
Museum Nicolaihaus, Berlin 17213
Museum Nienburg, Nienburg 20718
Museum Niesky, Niesky 20722
Museum Nikolaikirche, Berlin 17214
Museum Nordenham, Nordenham 20732
Museum NordJura, Weismain 22444
Museum Ober-Ramstadt, Ober-Ramstadt 20798
Museum Obere Saline mit Bismarck-Museum, Bad Kissingen 16759
Museum Oberes Donautal, Fridingen an der Donau 18466
Museum Otto Ludwig, Eisfeld 18108
Museum Petersberg, Petersberg bei Halle, Saale 21014
Museum Prüm, Prüm 21136
Museum Reichenau, Reichenau 21244
Museum Sankt Leon-Rot, Sankt Leon-Rot 21504
Museum Sankt Wendel, Sankt Wendel 21514
Museum Scheuerle, Backnang 16660
Museum Schloss Colditz, Colditz 17724
Museum Schloß Ehrenstein, Ohrdruf 20890
Museum Schloss Fellenberg, Merzig 20301
Museum Schloss Hellenstein, Heidenheim an der Brenz 19093
Museum Schloß Kuckuckstein, Liebstadt 19986
Museum Schloss Netzschkau, Netzschkau 20580
Museum Schloß Ratibor, Roth 21404
Museum Schloß Rochlitz, Rochlitz 21338
Museum Schloß Steinheim, Hanau 18981
Museum Schloss und Gut Liebenberg, Löwenberger Land 20019
Museum Schöne Stiege, Riedlingen 21319
Museum Schwarzes Ross, Hilpoltstein 19189
Museum Seitingen-Oberflacht, Seitingen-Oberflacht 21734
Museum Soltau, Soltau 21814
Museum Stadt Bad Hersfeld, Bad Hersfeld 16747
Museum Starnberger See, Starnberg 21864
Museum Steinarbeiterhaus, Hohburg 19232
Museum Tegernseer Tal, Tegernsee 22017

Museum Theulegium, Tholey 22056
Museum und Begegnungsstätte Alter Krug, Zossen 22821
Museum und Galerie der Stadt Schwabmünchen, Schwabmünchen 21646
Museum Uslar, Uslar 22193
Museum Viadrina, Frankfurt (Oder) 18406
Museum Viechtach, Viechtach 22226
Museum Villa Stahmer, Georgsmarienhütte 18592
Museum Voswinckelshof, Dinslaken 17870
Museum Wasserburg, Wasserburg am Inn 22345
Museum Wasserburg Anholt, Isselburg 19368
Museum Weißenfels Schloß Neu-Augustusburg, Weißenfels 22457
Museum Wendener Hütte, Wenden 22476
Museum Wiesbaden, Wiesbaden 22557
Museum Wildberg, Wildberg 22567
Museum Wilmersdorf - Forum für Stadtkultur und Geschichte, Berlin 17217
Museum Wirtles Haus, Rottenacker 21418
Museum Wurzen mit Ringelnatzsammlung, Wurzen 22751
Museum Wustrow, Wustrow, Wendland 22757
Museum Zella-Mehlis, Zella-Mehlis 22785
Museum zur Geschichte von Christen und Juden, Laupheim 19859
Museum zur Stadtgeschichte, Königslutter am Elm 19654
Museumsanlage Kulturstiftung Landkreis Osterholz, Osterholz-Scharmbeck 20935
Museumsdorf Cloppenburg, Cloppenburg 17713
Museumshof, Neuruppin 20651
Museumshof, Roßtal 21384
Museumshof am Sonnenluch, Erkner 18216
Museumshof Emmerstedt, Helmstedt 19124
Museumshof Historischer Moorhof Augustendorf, Gnarrenburg 18665
Museumshof-Senne, Bielefeld 17346
Museumshof und Heimatmuseum, Falkenstein, Harz 18293
Museumshof Zirkow, Zirkow 22808
Museumslandschaft Deilbachtal, Essen 18264
Museumsscheune, Helmenzen 19120
Museumsscheune Fränkische Schweiz, Hollfeld 19254
Museumsscheune Kremmen, Kremmen 19710
Museumsverbund Pankow - Prenzlauer Allee, Berlin 54851
Museumszentrum Lorsch, Lorsch 20035
Narrenmuseum, Rheinfelden 21300
Natur-Heimat-Museum, Langenburg 19826
Natur- und Heimatmuseum Vellberg, Vellberg 22215
Naturhistorisches Museum Schloss Bertholdsburg, Schleusingen 21572
Neisser Haus - Heimathaus und Archiv, Hildesheim 19184
Neudorfer Suppenmuseum, Sehmatal 21727
Neues Schloß, Schloßmuseum, Tettnang 22041
Neues Stadtmuseum Landsberg am Lech, Landsberg am Lech 19796
Niederbergisches Museum Wülfrath, Wülfrath 22703
Niederebersdorfer Heimat- und Archivstuben, Tutzing 22140
Niederlausitz-Museum Luckau, Luckau, Niederlausitz 20040
Niederlausitzer Sorbisches Dorfmuseum Bloischdorf, Felixsee 18303
Nordfriesisches Museum Ludwig-Nissen-Haus, Husum 19299
Nordhannoversches Bauernhaus-Museum, Isernhagen 19363
Nordpfälzer Heimatmuseum, Rockenhausen 21342
OberhausMuseum, Passau 20993
Oberlausitzer Sechsstädtebund- und Handwerksmuseum Löbau, Löbau 20013
Oberrheinisches Bäder- und Heimatmuseum, Bad Bellingen 16679
Oberschlesisches Landesmuseum, Ratingen 21192
Oma-Freese-Huus, Dornum 17909
Oranien-Nassau-Museum, Diez 17858
Oranien-Nassauisches Museum, Dillenburg 17861
Ortsgeschichtliche Sammlung, Bovenden 17490
Ortsgeschichtliches Museum, Siegenburg 21758
Ortsmuseum, Brachwitz-Saalkreis 17493
Ortsmuseum, Nackenheim 20544
Ortsmuseum, Neuburg am Rhein 20596
Ortssammlung Wettelsheim, Treuchtlingen 22090
Ost-, Südostdeutsche Heimatstuben, Ulm 22168
Ostdeutsche Heimatsammlung, Ravensburg 21205
Ostdeutsche Heimatstube, Bad Zwischenahn 16878
Ostdeutsche Heimatstube, Weinstadt 22439
Ostdeutsche Kultur- und Heimatstuben mit Schönbacher Stube, Heppenheim 19130
Osterzgebirgsmuseum, Lauenstein 19850
Das Ostfriesische Teemuseum und Norder Heimatmuseum, Norden 20730
Ostfriesisches Landesmuseum, Emden 18141
Ostholstein-Museum, Eutin 18285
Ostpreußisches Landesmuseum, Lüneburg 20102
Otto-Ditscher-Galerie, Neuhofen 20622
Otto-Schwabe-Museum, Hochheim am Main 19205
Otto-Ubbelohde-Haus, Lahntal 19768
Otto-von-Guericke-Museum in der Lukasklause, Magdeburg 20128
Paffenroter Dorfmuseum, Marxzell 20232
Panke Museum, Berlin 17231
Papiermacher- und Heimatmuseum, Frankeneck 18350
Patenschaftsmuseum Goldap in Ostpreußen, Stade 21851
Peter-Wiepert-Museum, Fehmarn 18299
Peter-Zirbes-Haus Niederkail, Landscheid 19799
Pfalzmuseum, Forchheim, Oberfranken 18337
Pfingstritt-Museum Kötzting, Kötzting 19668
Philipp Schäfer II Museum, Riedstadt 21326
Pilzmuseum Taura, Schildau, Gneisenaustadt 21551
Pomázer Heimatstube, Sinsheim 21785
Poppelsdorfer Heimatsammlung Stöcker, Bonn 17464
Priental-Museum, Aschau im Chiemgau 16594
Probstei-Museum, Schönberg, Holstein 21599
Prof. Bertlein Schulmuseum, Worms 22688
Puppenmuseum Laufeld, Laufeld 19853
Queichtalmuseum, Offenbach an der Queich 20884
Randeck-Museum, Mannweiler-Cölln 20176
Rangau-Handwerkermuseum, Markt Erlbach 20212
Rathausmuseum Nieder-Ohmen, Mücke 20394
Reblandmuseum, Baden-Baden 16889
Regionales Heimatmuseum für das Renchtal, Oppenau 20905
Regionalgeschichtliches Museum, Lutherstadt Eisleben 20112
Regionalmuseum Alte Schule, Kaufungen 19493
Regionalmuseum Bad Frankenhausen, Bad Frankenhausen 16731
Regionalmuseum Burg Beeskow, Beeskow 16979
Regionalmuseum des Amtes Malliß, Neu Kaliß 20584
Regionalmuseum Fritzlar, Fritzlar 18485
Regionalmuseum Harkerode, Harkerode 19014
Regionalmuseum Leben und Arbeiten im Blauen Ländchen, Nastätten 20549
Regionalmuseum Neubrandenburg, Neubrandenburg 20592
Regionalmuseum Reichelsheim Odenwald, Reichelsheim 21243
Regionalmuseum Stockstadt, Stockstadt am Rhein 21906
Reichsdorf-Museum Gochsheim, Gochsheim 18667
Reichsstadtmuseum, Rothenburg ob der Tauber 21410
Reichsstadtmuseum im Ochsenhof, Bad Windsheim 16873
Reichsstadtmuseum Weißenburg, Weißenburg in Bayern 22450
Reichsstädtisches Archiv, Rathaushalle und Ratsstube, Mühlhausen/Thüringen 20405
Reiss-Engelhorn-Museen (rem), Mannheim 20173
Renaissance-Schloss Demerthin, Gumtow 18834
Rhein-Museum Koblenz, Koblenz 19588
Rhön-Museum, Fladungen 18325
Rieck-Haus, Hamburg 18957
Riesengebirgsmuseum, Marktoberdorf 20221
Riesengebirgsstube, Würzburg 22721
Rittergut Haus Laer, Bochum 17405
Robert-Struppmann-Museum, Lorch, Rheingau 20033
Römer und Bajuwaren Museum Burg Kipfenberg, Kipfenberg 19541
Römerbad, Schwangau 21667
Römermuseum im Haus Isenburg, Eisenberg, Pfalz 18102
Römermuseum Weißenburg, Weißenburg in Bayern 22451
Römermuseum Xanten, Xanten 22762
Römische Kalkbrennerei Iversheim, Bad Münstereifel 16790
Römische Thermen, Weißenburg in Bayern 22452
Roentgen-Museum, Neuwied 20688
Rohnstädter Heimatstube, Weilmünster 22393
Rosgartenmuseum, Konstanz 19682
Rossfelder Heimatstuben, Crailsheim 17737
Roter Haubarg, Witzwort 22636
Rügen-Museum in der Kulturkunststatt Prora, Binz, Ostseebad 17358
Ruhrtalmuseum, Schwerte 21700
Rundlingsmuseum Wendlandhof-Lübeln, Küsten 19741
Sächsisches Feuerwehrmuseum Zeithain, Zeithain 22774
Salinenmuseum im Gradierbau, Bad Dürkheim 16717
Samson-Haus, Leer 19871
Sandauer Heimatstube, Arzberg 16577
Sandelsches Museum, Kirchberg an der Jagst 19543
Sangerberger Heimatstube, Regenstauf 21236
Sankt Pauli-Museum, Hamburg 18958
Sauerland-Museum, Arnsberg 16573
Schachmuseum Ströbeck, Halberstadt 18879
Schaumanufaktur Ostsee-Schmuck, Leopoldshagen 19964
Schifffahrtsmuseum, Nierstein 20721
Schleifermuseum, Solingen 21807
Schleipehammer, Kierspe 19538
Schlesisches Museum zu Görlitz, Görlitz 18680
Schliekau-Museum, Bad Bevensen 16697
Schloss Agathenburg, Agathenburg 16435
Schloß Bürgeln, Schliengen 21573
Schloss Bürresheim, Sankt Johann bei Mayen 21501
Schloss Ehrenburg, Coburg 17720
Schloss Elisabethenburg, Meiningen 20260
Schloss Kirchheim, Kirchheim unter Teck 19550
Schloß Marienburg, Pattensen 20995
Schloss Neuburg, Neuburg an der Donau 20599
Schloss Sachsenburg, Frankenberg, Sachsen 18349
Schloss Schlettau, Schlettau 21571
Schloß Strünkede, Herne 19156
Schloß- und Spielkartenmuseum, Altenburg 16499
Schloss Wiederau, Pegau 21000
Schlossbergmuseum Chemnitz, Chemnitz 17704
Schlossmuseum, Höchstädt im Fichtelgebirge 19212
Schloßmuseum, Ortenburg 20916
Schloßmuseum, Schrozberg 21642
Schloßmuseum, Quedlinburg 21152
Schloßmuseum Ismaning, Ismaning 19365
Schloßmuseum Jagsthausen, Jagsthausen 19373
Schloßmuseum Rimpar, Rimpar 21335
Schmiede- und Schlossermuseum Schwalbach, Schwalbach, Saar 21660
Schmiedemuseum, Sersheim 21747
Schul- und Heimatmuseum, Mark Landin 20201
Schullandheim Sassen, Sassen 21517
Schulmuseum, Immerath 19329
Schulmuseum, Königsfeld im Schwarzwald 19651
Schulmuseum, Obersulm 20844
Schulmuseum Fronau, Roding 21348
Schwarzachtaler Heimatmuseum, Neunburg vorm Wald 20637
Sebnitzer Kunstblumen- und Heimatmuseum Prof. Alfred Meiche, Sebnitz 21710
Sensen- und Heimatmuseum, Achern 16423
Serbski muzej / Wendisches Museum, Cottbus 17734
Serpentinsteinmuseum Zöblitz, Zöblitz 22815
Sickingen-Museum, Landstuhl 19806
Sieben-Schwaben-Museum, Türkheim 22129
Siebenbürgisches Museum Gundelsheim, Gundelsheim, Württemberg 18836
Siebengebirgsmuseum, Königswinter 19661
Sindringer Heimatmuseum - Stadtmühle, Forchtenberg 18340
Singhofener Heimatstuben, Singhofen 21782
Smiterlöwsche Sammlung, Franzburg 18408
Spengler-Museum, Sangerhausen 21485
Staatlicher Schloßbetrieb Schloß Nossen / Kloster Altzella, Nossen 20748
Stadt-und Bergbaumuseum der Stadt Staßfurt, Staßfurt 21866
Stadt- und Brauereimuseum Pritzwalk, Pritzwalk 21128
Stadt- und Burgmuseum, Eppstein 18186
Stadt- und Dampfmaschinenmuseum, Werdau 22485
Stadt- und Fachwerkmuseum Alte Universität, Eppingen 18183
Stadt- und Freilichtmuseum Sinsheim, Sinsheim 21786
Stadt- und Heimatmuseum, Malchow 20156
Stadt- und Heimatmuseum der Stadt Marl, Marl, Westfalen 20225
Stadt- und Heimatmuseum Kusel, Kusel 19752
Stadt- und Heimatmuseum mit Galerie, Waischenfeld 22279
Stadt- und Hochstiftmuseum, Dillingen an der Donau 17863
Stadt- und Industriemuseum, Wetzlar 22533
Stadt- und Kreisgeschichtliches Museum, Hünfeld 19290
Stadt- und Kulturgeschichtliches Museum Torgau, Torgau 22073
Stadt- und Manfred-Kyber-Museum, Löwenstein 20020
Stadt- und Regionalmuseum Lübben, Lübben 20069
Stadt- und Schulmuseum, Ronneburg, Thüringen 21369
Stadt- und Turmmuseum, Bad Camberg 16708
Stadt- und Wallfahrtsmuseum, Walldürn 22312
Stadtarchiv mit städtischen Sammlungen, Lauf an der Pegnitz 19852
Stadtgeschichtliche Sammlung, Velbert 22212
Stadtgeschichtliches Museum, Kehl 19498
Stadtgeschichtliches Museum, Ottweiler 20962
Stadtgeschichtliches Museum, Tangermünde 22003
Stadtgeschichtliches Museum, Waren (Müritz) 22336
Stadtgeschichtliches Museum, Pfullingen 21034
Stadtgeschichtliches Museum im Landrichterhaus, Karlstadt 19455
Stadtgeschichtliches Museum Leipzig, Leipzig 19937
Stadtgeschichtliches Museum Leipzig - Altes Rathaus, Leipzig 19938
Stadtgeschichtliches Museum Leipzig - Neubau, Leipzig 19939
Stadtgeschichtliches Museum Schabbellhaus, Wismar 22613
Stadtgeschichtliches Museum Spandau, Berlin 17259
Stadtgeschichts- und Schradenmuseum, Ortrand 20917
Stadthaus Ulm, Ulm 22169
Stadtmuseum, Bad Bergzabern 16687
Stadtmuseum, Hornberg 19275
Stadtmuseum, Riesa 21330
Stadtmuseum, Jena 19392
Stadtmuseum Aichach, Aichach 16445
Stadtmuseum "Alte Burg" Wittenberge, Wittenberge 22621
Stadtmuseum Alte Post, Ebersbach an der Fils 18042
Stadtmuseum Amberg, Amberg 16533
Stadtmuseum Amtsturm, Lübz 20085
Stadtmuseum Andernach, Andernach 16545
Stadtmuseum Apolda, Apolda 16567
Stadtmuseum Bad Cannstatt, Stuttgart 21961
Stadtmuseum Bad Soden, Bad Soden am Taunus 16841
Stadtmuseum Bad Wildungen, Bad Wildungen 16864
Stadtmuseum Baden-Baden, Baden-Baden 16891
Stadtmuseum Battenberg, Battenberg 16934
Stadtmuseum Baunatal, Baunatal 16938
Stadtmuseum Beckum, Beckum 16974
Stadtmuseum Bergen, Bergen auf Rügen 17007
Stadtmuseum Bergkamen, Bergkamen 17021
StadtMuseum Bonn, Bonn 17467
Stadtmuseum Brakel, Brakel 17499
Stadtmuseum Breuberg-Neustadt, Breuberg 17588
Stadtmuseum Brilon, Brilon 17590
Stadtmuseum Burgdorf, Burgdorf 17649
Stadtmuseum Burgstädt, Burgstädt 17664
Stadtmuseum Cottbus, Cottbus 17735
Stadtmuseum Cuxhaven, Cuxhaven 17757
Stadtmuseum Damme, Damme 17778
Stadtmuseum Deggendorf, Deggendorf 17806
Stadtmuseum Ditzingen, Ditzingen 17878
Stadtmuseum Döbeln/Kleine Galerie, Döbeln 17882
Stadtmuseum Dresden, Dresden 17972
Stadtmuseum Eilenburg, Eilenburg 18089
Stadtmuseum Einbeck mit Radhaus im Stadtmuseum, Einbeck 18092
Stadtmuseum Eisenberg, Eisenberg, Saale-Holzland-Kreis 18103
Stadtmuseum Erlangen, Erlangen 18225
Stadtmuseum Fellbach, Fellbach 18307
Stadtmuseum Freilassing, Freilassing 18445
Stadtmuseum Fürstenfeldbruck, Fürstenfeldbruck 18493
Stadtmuseum Gehrden, Gehrden bei Hannover 18552
Stadtmuseum Geithain, Geithain 18564
Stadtmuseum Gera, Gera 18598
Stadtmuseum Gerlingen - Gerlinger Heimatmuseum - Museum der Deutschen aus Ungarn, Gerlingen 18603
Stadtmuseum Gerolzhofen, Gerolzhofen 18612
Stadtmuseum Giengen, Giengen an der Brenz 18633
Stadtmuseum Grafenau, Grafenau, Kreis Freyung-Grafenau 18729
Stadtmuseum Groß-Gerau, Groß-Gerau 18776
Stadtmuseum Gütersloh, Gütersloh 18833
Stadtmuseum Gunzenhausen, Gunzenhausen 18841
Stadtmuseum Hattingen, Hattingen 19041
Stadtmuseum Haus Kupferhammer, Warstein 22341
Stadtmuseum Herzogenaurach, Herzogenaurach 19168
Stadtmuseum Hildburghausen, Hildburghausen 19179
Stadtmuseum Hofgeismar, Hofgeismar 19224
Stadtmuseum Hofheim am Taunus, Hofheim am Taunus 19226
Stadtmuseum Hohe Lilie, Naumburg, Saale 20556
Stadtmuseum Holzminden, Holzminden 19261
Stadtmuseum Horb, Horb am Neckar 19272
Stadtmuseum Hornmoldhaus, Bietigheim-Bissingen 17349
Stadtmuseum im Alten Forstamt, Stockach 21904
Stadtmuseum im Augustinerkloster Bad Langensalza, Bad Langensalza 16773
Stadtmuseum im Gelben Haus, Esslingen 18277
Stadtmuseum im Historischen Rathaus, Staufen im Breisgau 21868
Stadtmuseum im Knochenhauer-Amtshaus, Hildesheim 19186
Stadtmuseum im Schlössle, Weingarten, Kreis Ravensburg 22429
Stadtmuseum im Schloss, Münsingen 20502
Stadtmuseum im Spital, Crailsheim 17738
Stadtmuseum Ingolstadt im Kavalier Hepp, Ingolstadt 19346
Stadtmuseum Iserlohn, Iserlohn 19360
Stadtmuseum Karlsruhe, Karlsruhe 19450
Stadtmuseum Kassel, Kassel 19478
Stadtmuseum Kaufbeuren, Kaufbeuren 19492
Stadtmuseum Kelsterbach, Kelsterbach 19505
Stadtmuseum Killingerhaus, Idstein 19311
Stadtmuseum Klostermühle, Bad Urach 16854
Stadtmuseum Landeshauptstadt Düsseldorf, Düsseldorf 18006
Stadtmuseum Leimen - Sankt Ilgen, Leimen 19878
Stadtmuseum Leonberg, Leonberg 19963
Stadtmuseum Leun, Leun 19967
Stadtmuseum Lichtenfels, Lichtenfels 19981

Stadtmuseum Ludwig Erhard, Fürth 18500
Stadtmuseum Ludwigshafen, Ludwigshafen am Rhein 20057
Stadtmuseum Meersburg, Meersburg 20251
Stadtmuseum Meppen, Meppen 20287
Stadtmuseum Mosbach, Mosbach 20392
Stadtmuseum Mühlberg, Elbe, Mühlberg 20398
Stadtmuseum Mühlheim, Mühlheim am Main 20407
Stadtmuseum Mülheim-Kärlich, Mülheim-Kärlich 20416
Stadtmuseum Münster, Münster 20519
Stadtmuseum Mutzschen, Mutzschen 20539
Stadtmuseum Neumarkt, Neumarkt in der Oberpfalz 20634
Stadtmuseum Neuötting, Neuötting 20643
Stadtmuseum Nittenau, Nittenau 20723
Stadtmuseum Nördlingen, Nördlingen 20727
Stadtmuseum Norderstedt, Norderstedt 20735
Stadtmuseum Nürtingen, Nürtingen 20795
Stadtmuseum Obermühle, Braunfels 17515
Stadtmuseum Oldenburg, Oldenburg 20900
Stadtmuseum Penzberg, Penzberg 21008
Stadtmuseum Pößneck, Pößneck 21071
Stadtmuseum Pulsnitz, Pulsnitz 21139
Stadtmuseum Quakenbrück, Quakenbrück 21144
Stadtmuseum Rheinfelden, Rheinfelden 21301
Stadtmuseum Rothenburg, Rothenburg/ Oberlausitz 21415
Stadtmuseum Saalfeld im Franziskanerkloster, Saalfeld/Saale 21452
Stadtmuseum Sachsenheim, Sachsenheim 21466
Stadtmuseum Schleswig, Schleswig 21568
Stadtmuseum Schloß Hoyerswerda, Hoyerswerda 19282
Stadtmuseum Schloss Wolfsburg, Wolfsburg 22667
Stadtmuseum Schongau, Schongau 21620
Stadtmuseum Schramberg, Schramberg 21634
Stadtmuseum Schwabach, Schwabach 21645
Stadtmuseum Schwandorf mit Falkenauer Heimatstube, Schwandorf 21665
Stadtmuseum Schwedt, Schwedt / Oder 21678
Stadtmuseum Siegburg, Siegburg 21752
Stadtmuseum Simeonstift Trier, Trier 22105
Stadtmuseum Sindelfingen, Sindelfingen 21778
Stadtmuseum Stadtoldendorf, Stadtoldendorf 21859
Stadtmuseum Steinfurt, Steinfurt 21887
Stadtmuseum Sulzbach-Rosenberg, Sulzbach-Rosenberg 21985
Stadtmuseum Teterow, Teterow 22037
Stadtmuseum Trostberg, Trostberg 22111
Stadtmuseum Tübingen, Tübingen 22124
Stadtmuseum Waldkraiburg, Waldkraiburg 22300
Stadtmuseum Wasungen, Wasungen 22350
Stadtmuseum Wehr, Wehr 22361
Stadtmuseum Weimar, Weimar 22421
Stadtmuseum Wendlingen, Wendlingen am Neckar 22478
Stadtmuseum Werne, Werne 22494
Stadtmuseum Zweibrücken, Zweibrücken 22830
Stadtteilmuseum Nowaweser Weberstube, Potsdam 21110
Stadttormuseum, Wehrheim 22363
Städtische Museen Heilbronn, Heilbronn 19100
Städtisches Heimatmuseum, Meßkirch 20307
Städtisches Heimatmuseum Ballenstedt, Ballenstedt 16903
Städtisches Heimatmuseum Höchstadt, Höchstadt an der Aisch 19209
Städtisches Heimatmuseum im Waldemarturm, Dannenberg 17780
Städtisches Heimatmuseum Mengen, Mengen 20284
Städtisches Heimatmuseum Reinfeld, Reinfeld 21253
Städtisches Hellweg-Museum, Geseke 18626
Städtisches Museum, Bad Reichenhall 16808
Städtisches Museum, Bruchsal 17598
Städtisches Museum, Herborn 19133
Städtisches Museum, Kitzingen 19561
Städtisches Museum, Menden 20281
Städtisches Museum, Pfungstadt 21036
Städtisches Museum, Salzgitter 21476
Städtisches Museum, Überlingen 22146
Städtisches Museum, Welzheim 22472
Städtisches Museum, Werl 22491
Städtisches Museum, Wiesloch 22563
Städtisches Museum, Eisenhüttenstadt 18107
Städtisches Museum Engen und Galerie, Engen 18172
Städtisches Museum Fürstenwalde, Fürstenwalde 18495
Städtisches Museum Göppingen im Storchen, Göppingen 18677
Städtisches Museum Göttingen, Göttingen 18701
Städtisches Museum Grevesmühlen, Grevesmühlen 18758
Städtisches Museum Hann. Münden, Hann. Münden 18988
Städtisches Museum Haus Letmathe, Iserlohn 19362
Städtisches Museum im ehemaligen Heiliggeistspital, Munderkingen 20529
Städtisches Museum im Hospital, Nidderau 20694
Städtisches Museum im Kornhaus, Kirchheim unter Teck 19552
Städtisches Museum im Kornhaus Bad Waldsee, Bad Waldsee 16857
Städtisches Museum im Spiegelschen Palais, Halberstadt 18881
Städtisches Museum Kamen, Kamen 19419
Städtisches Museum Ludwigsburg, Ludwigsburg 20048
Städtisches Museum Peterskirche, Vaihingen an der Enz 22199
Städtisches Museum Rosenheim, Rosenheim 21379
Städtisches Museum Saarlouis, Saarlouis 21464
Städtisches Museum Seesen, Seesen 21724
Städtisches Museum Wesel, Wesel 22508
Städtisches Museum Wesel, Wesel 22509
STATT-Museum, Herbstein 19137
Steglitz Museum, Berlin 17263
Stickereimuseum Eibenstock, Eibenstock 18074
Stiftlandmuseum, Waldsassen 22303
Stiftung Historische Museen Hamburg, Hamburg 18961
Stiftung Scheibler-Museum Rotes Haus, Monschau 20376
Stiftung Schleswig-Holsteinische Landesmuseen Schloß Gottorf, Schleswig 21569
Stiftung Stadtmuseum Berlin, Landesmuseum für Kultur und Geschichte Berlins, Berlin 17266
Stoltze-Museum der Frankfurter Sparkasse, Frankfurt am Main 18396
Storchenturm-Museum, Zell am Harmersbach 22780
Stuhlbau- und Heimatmuseum, Rabenau 21157
Südhessisches Handwerksmuseum, Roßdorf 21382
Südsauerlandmuseum, Attendorn 16603
Sülchgau-Museum, Rottenburg am Neckar 21421
Sylter Heimatmuseum, Sylt 21995
Tabakpfeifenmuseum und Museum für Stadtgeschichte, Ruhla 21442
Tagelöhner- und Brennereimuseum Faißt, Bad Peterstal-Griesbach 16803
Tauberfränkisches Landschaftsmuseum, Tauberbischofsheim 22010
Technisches Museum Zechau, Rositz 21380
Tempelhof Museum, Berlin 17271
Teplitz-Schönauer Heimatmuseum, Frankfurt am Main 18398
Theatermuseum Zauberwelt der Kulisse, Meiningen 20262
Theresienthaler Heimatstube, Gaildorf 18534
Thüringer Korbmachermuseum, Bad Berka 16690
Thüringer Wald - Kreativ Museum mit 1. Deutschem Kloßpressenmuseum, Großbreitenbach 18784
Töpfereimuseum, Speicher 21832
Tonbergbaumuseum Westerwald, Siershahn 21761
Torhaus Dölitz, Leipzig 19940
Trachten- und Heimatmuseum, Weiltingen 22394
Troadkasten, Sauerlach 21524
Tüshaus-Mühle, Dorsten 17912
Tuffstein-Museum "Steinmetzbahnhof", Weibern 22364
Turmmuseum, Öhringen 20867
Turmmuseum Schloß Mengerskirchen, Mengerskirchen 20285
Turmmuseum Stadt Blankenberg, Hennef 19128
Turmuhren- und Heimatmuseum Bockenem, Bockenem 17409
Tuttlinger Haus, Tuttlingen 22138
Uckermärkisches Volkskundemuseum Templin, Templin 22030
Übersee Museum Bremen, Bremen 17564
Überwälder Heimatmuseum, Wald-Michelbach 22280
Uffenheimer Gollachgaumuseum, Uffenheim 22155
Uniseum, Freiburg im Breisgau 18440
Universitätsmuseum Heidelberg, Heidelberg 19086
Uns lütt Museum, Dargun 17784
Ur-Donautal-Museum, Wellheim 22469
Verein "Chemnitzer Gewölbegänge" e. V., Chemnitz 17705
Verein für Heimatpflege, Viersen 22233
Vestisches Museum, Recklinghausen 21211
Viehstrichmuseum, Wörth am Rhein 22642
Vielstedter Bauernhaus, Hude 19285
Völkerschlachtdenkmal / Forum 1813, Leipzig 19941
Vogelsberger Heimatmuseum, Schotten 21628
Vogteimuseum mit Blumenauer Heimatstube, Aurach 16646
Vogtländisches Dorfmuseum, Erlbach 18227
Vogtländisches Freilichtmuseum, Erlbach 18228
Volkskulturmuseum Raußmühle, Eppingen 18185
Volkskundemuseum, Schönberg, Mecklenburg 21601
Volkskundliche Sammlung, Mömbris 20354
Volkskundliche Sammlung des Fichtelgebirgsvereins, Weidenberg 22369
Vonderau Museum, Fulda 18512
Vorderösterreichisches Museum, Endingen am Kaiserstuhl 18168
Waldmuseum Göhrde, Göhrde 18669
Waldnaabtal-Museum in der Burg Neuhaus, Windischeschenbach 22597
Wallfahrts- und Heimatmuseum, Altötting 16521
Wassermühle Kuchelmiß, Kuchelmiß 19733
Weberhaus Marlesreuth, Naila 20547
Wehrturm, Niederstetten 20714
Weibertreu-Museum, Weinsberg 22434
Weimar-Haus, Weimar 22422
Wein-und Heimatmuseum, Leiwen 19945
Wein- und Heimatmuseum, Mehring 20256
Wein- und Heimatmuseum, Winningen 22602
Weinbau- und Heimatmuseum Klingenberg, Klingenberg am Main 19570
Weißenhorner Heimatmuseum, Weißenhorn 22461
Weißgerberwalk, Biberach an der Riß 17324
Wentzingerhaus, Freiburg im Breisgau 18441
Werberger Stuben, Motten 20393
Westallgäuer Heimatmuseum, Weiler-Simmerberg 22389
Wetterau-Museum, Friedberg, Hessen 18468
Wieland-Museum Biberach, Wieland-Gartenhaus, Biberach an der Riß 17326
Wilhelm-Fabry-Museum, Hilden 19181
Willinger Milchmuhseum, Willingen 22586
Windmühle Waldfeucht, Waldfeucht 22293
Winser Museumshof, Winsen, Aller 22605
Wipperkotten, Solingen 21809
Wittenberger Freilichtmuseum, Edewecht 18057
Wixhäuser Dorfmuseum, Darmstadt 17797
Wolfsteiner Heimatmuseum im Schramlhaus, Freyung 18462
Worpsweder fotografische Werkstätten, Worpswede 22698
zeittor - Museum der Stadt Neustadt in Holstein, Neustadt in Holstein 20681
Zinnfiguren-Museum Goslar, Goslar 18716
Zinnfigurenkabinett im Haspelturm, Eberbach 18034
Zuckmantler Heimatstube, Bietigheim-Bissingen 17351

Ghana

Elmina Castle Museum, Cape Coast 22851
Upper East Region Museum, Bolgatanga 22849
Volta Regional Museum, Ho 22853

Greece

Agia Triada Folklore Museum, Agia Triada 22867
Archipelagos Cultural Center, Athinai 22899
Domnista Folklore and History Museum, Domnista 22986
Ecclesiastical Museum of Alexandroupolis, Alexandroupolis 22876
Folk Art Museum, Pyrgi 23161
Folklore Museum of Polygyros, Polygyros 23153
Gallery of the Society for Epirot Studies, Ioannina 23013
Historical and Ethnological Museum of Cappadocia, Kavala 23047
Historical and Ethnological Museum of the Mani, Kranae 23068
Historical and Folk Art Museum of Gavalohori, Chania 22963
Historical, Folk Art and Natural History Museum of Kozani, Kozani 23067
Historical Museum of Arta, Arta 22893
Manos Faltaits Museum, Skyros 23192
Megalo Chorio Folklore Museum, Megalo Chorio 23085
Municipal Gallery of Samos, Samos 23176
Municipal Museum of Kavala, Kavala 23048
Museum of Komotini, Komotini 23060
Museum of Postindependence Athens, Athinai 22936
Museum of the City of Athens, Athinai 22939
Museum of the History and Art of the Holy City of Missolonghi, Mesolongion 23090
Museum of the Statesman Eleftherios Venizelos and the Corresponding Historical Period, Athinai 22940
Nature-Study and Folk Art Museum of Loutra Almopias, Loutraki 23082
Palio Mikro Chorio Museum, Mikro Chorio 23096
Spetses Museum, Spetses 23198
Thessaloniki Museum for the Macedonian Struggle, Thessaloniki 23223

Greenland

Aasiaat Katersugaasiviat Museum, Aasiaat 23251
Avanersuup Katersugaasivia, Qaanaaq 23264
Ilulissani Katersugaasivik, Ilulissat 23255
Maniitsoq Katersugaasivia, Maniitsoq 23257
Nanortallip Katersugaasivia, Nanortalik 23258
Narsaq Museum, Narsaq 23259
Narsarsuaq Museum, Narsarsuaq 23260
Nunatta Katersugaasivia Allagaateqarfialu, Nuuk 23262
Paamiune Katersugausivik, Paamiut 23263
Qaqortup Katersugaasivia, Qaqortoq 23265
Qasigiannguit Katersugaasiviat, Qasigiannguit 23267
Qeqertarsuaq Museum, Qeqertasuaq 23268
Sisimiut Katersugaasiviat, Sisimiut 23269
Tasiilap Katersugaasivia, Tasiilaq 23270
Upernavik Museum, Upernavik 23271
Uummannap Katersugaasivia, Uummannaq 23272

Guadeloupe

Ecomusée de Marie Galante, Grand Bourg 23274
Musée Municipal de Saint-Barthélemy, Saint-Barthélemy 23280

Guam

Faninadahen Kosas Guahan, Adelup 23284
Guam Museum, Agana 23285
War in the Pacific National Historical Park, Piti 23288

Guatemala

Museo de Casa Popenoe, Antigua 23290
Museo del Palacio Nacional, Guatemala 23299
Museo Quezaltenango, Quezaltenango 23307
Museo Regional Arqueológico Rossbach, Chichicastenango 23294
Museo Silvanus G. Morley y de las Estelas, El Petén 23295

Guinea

Musée Préfectoral de Boké, Boké 23308
Musée Préfectoral de Kissidougou, Kissidougou 23313
Musée Préfectoral de Koundara, Koundara 23314
Musée Préfectoral de N'Zerekore, N'Zerekore 23315

Hungary

Abony Lajos Falumúzeum, Abony 23334
Albertfalvai Helytörténeti Gyűjtemény és Iskolamúzeum, Budapest 23380
Andor Semsey Múzeum, Balmazujváros 23358
Angyalföldi Helytörténeti Gyűjtemény, Budapest 23381
Békés Megyei Múzeumok Igazgatósága, Békéscsaba 23367
Beregi Múzeum, Vásárosnamény 23955
Bodrogközi Kastélymúzeum, Pácin 23745
Csepel Galéria és Helytörténeti Gyűjtemény, Budapest 23394
Cukormúzeum, Szerencs 23886
Damjanich János Múzeum, Szolnok 23896
Dr. Batthyány-Strattmann László Múzeum, Körmend 23674
Dráva Múzeum, Barcs 23361
Eötvös Kúria, Vásárosnamény 23956
Ercsi Helytörténeti Gyűjtemény, Ercsi 23547
Espersit Ház, Makó 23689
Etele XI. Kerületi Helytörténeti Kör Kiállítóhelye, Budapest 23400
Fekete-Ház, Szeged 23843
Ferencvárosi Helytörténeti Gyűjtemény, Budapest 23402
Fővárosi Képtár, Budapest 23404
Gödöllői Városi Múzeum, Gödöllő 23566
Gömöri Múzeum, Putnok 23786
Gróf Esterházy Károly Kastély- és Tájmúzeum, Pápa 23752
Gyermek- és ifjúságszervezet-történeti Múzeum, Zánka 23975
Hegyvidéki Helytörténeti Gyűjtemény, Budapest 23410
Helytörténeti Gyűjtemény, Tiszaújváros 23932
Helytörténeti Gyűjtemény és Tájház, Várpalota 23951
Helytörténeti Múzeum, Csempeszkopács 23505
Herman Ottó Múzeum - Kiállítóhelye, Miskolc 23707
Hortobágyi Pásztormúzeum, Hortobágy 23619
Intercisa Múzeum, Dunaújváros 23533
Iskolamúzeum, Békés 23363
Iskolatörténeti és Egyházművészeti Múzeum, Cegléd 23502
Janus Pannonius Múzeum, Pécs 23762
Jókai Mór Emlékszoba és Tájház, Tardona 23915
Kiscelli Múzeum, Budapest 23421
Kispesti Helytörténeti Gyűjtemény, Budapest 23422
Klapka György Múzeum, Komáron 23678
Könyvtár és Múzeum - Borsodi Tájház, Edelény 23534
Lábasház, Sopron 23813
Megyeháza Börtönépület, Eger 23541
Múzeum-Ház, Csongrád 23512
Népi Műemlékház, Zalalövo 23973
Ózdi Városi Múzeum, Ózd 23744
Pákozdi Csata Emlékműve, Pákozd 23746
Pékmúzeum, Sopron 23816
Pesterzsébeti Múzeum, Budapest 23466
Rákospalotai Múzeum, Budapest 23469
Sinka István Emlékház - Helytörténeti kiállítás, Vészto 23964
Somogyi Megyei Múzeumok Igazgatósága, Kaposvár 23635
Szatmári Múzeum, Mátészalka 23693
Szlovák Tájház, Szarvas 23835
Szlovák Tájház, Kiskőrös 23665
Szőnyi István Emlékmúzeum, Zebegény 23977
Tabán Múzeum, Budapest 23478

Tari László Múzeum, Csongrád 23513
Tatabánya Múzeum, Tatabánya 23922
Tessedik Sámuel Múzeum, Szarvas 23836
Tímárház - Kézmuvesek Háza, Debrecen 23521
Török Ház, Szigetvár 23891
Tokaji Múzeum, Tokaji 23934
Újpesti Helytörténeti Gyűjtemény, Budapest 23487
Vármúzeum, Esztergom 23555
Vármúzeum - Kotár, Szeged 23850
Városi Helytörténeti Múzeum, Komló 23682
Városi Könyvtár és Múzeum, Tapolca 23914
Városi Múzeum, Kőszeg 23685
Városi Múzeum, Paks 23748
Vay Adám Múzeum, Vaja 23947
Völgységi Múzeum, Bonyhád 23376
XVIII. Kerületi Pedagógiai és Helytörténeti Gyűjtemény, Budapest 23492

Iceland

Arbæjarsafn, Reykjavík 24014
Bókasafn Menntaskólans á Ísafirði, Ísafirði 24007
Byggðasafn, Vestmannaeyjar 24042
Byggðasafn Akraness og Nærsveita, Akranes 23980
Byggðasafn Árnesinga, Eyrarbakki 23995
Byggðasafn Austur-Skaftafellssýslu, Höfn 24001
Byggðasafn Borgarfjardar, Borgarnes 23989
Byggðasafn Dalvíkurbyggdar, Dalvík 23992
Byggðasafn Hafnarfjarðar, Hafnarfjörður 23997
Byggðasafn Reykjanesbaejar, Reykjanesbaer 24012
Byggðasafn Skagfirdinga, Varmahlid 24041
Byggðasafn Snæfellinga og Hnappdæla, Stykkishólmur 24040
Byggðasafnið Grenjadarstad, Húsavík 24002
Byggðasafnið í Skógum, Hvolsvöllur 24006
David Stefánsson Memorial Museum, Akureyri 23981
Minjasafn Austurlands, Egilsstadir 23993
Minjasafn Egils Ólafssonar, Patreksfjörður 24011
Minjasafnida å Akureyri, Akureyri 23983
Nonnahûs, Akureyri 23985
Pjóðminjasafn Islands, Reykjavík 24030
Safnahusid Húsavík, Húsavík 24004
Sigurhaedir - Hús Skáldsins Museum, Akureyri 23987

India

Akshaya Kumar Maitreya Museum, Siliguri 24394
City Museum, Ahmedabad 24047
Dakshina Chitra Museum, Muttukadu 24325
District Museum, Darrang 24128
Gaya Museum, Gaya 24176
Geological Museum, Guwahati 24191
Manjusha Museum, Dharmasthala 24162
Municipal Museum Gwalior, Gwalior 24198
Museum and Art Gallery, Hooghly 24209
Museum of Kerala History, Edapally 24168
Pondicherry Museum, Pondicherry 24351
Prabhas Patan Museum, Prabhas Patan 24352
Sangli State Museum, Sangli 24377
Sardar Patel University Museum, Vallabhvidyanagar 24422
Uttar Pradesh State Museum, Lucknow 24301

Indonesia

Museum Adam Malik, Jakarta 24478
Museum Bahari, Jakarta 24482
Museum Balanga, Palangkaraya 24535
Museum Bangga Lore, Poso 24548
Museum Brawijaya, Malang 24521
Museum Bundo Kandung, Bukittinggi 24462
Museum Daerah Bangkalan, Bangkalan 24446
Museum Daerah Sumenep, Sumenep 24563
Museum Diponegoro, Magelang 24518
Museum Gayo, Aceh Tengah 24432
Museum Gedung Joang '45, Jakarta 24484
Museum Huta Bolon Simanindo, Samosir 24551
Museum Indonesia, Jakarta 24485
Museum Istant Siak Sri Indapura, Bengkalis 24452
Museum Kalimantan Tengah, Palangkaraya 24536
Museum Kartini, Jepara 24511
Museum Kartini Rembang, Rembang 24550
Museum Loka Budaya, Jayapura 24509
Museum Malikusaleh, Aceh Utara 24434
Museum Memorial Sultan Ternate, Ternate 24578
Museum Negeri Bali, Denpasar 24466
Museum Negeri Jambi, Jambi 24508
Museum Negeri Jawa Barat, Bandung 24444
Museum Negeri Kalimantan Barat, Pontianak 24547
Museum Negeri Lampung, Bandarlampung 24439
Museum Negeri Nusa Tenggara Barat, Mataram 24524
Museum Negeri of Bengkulu, Bengkulu 24453
Museum Negeri of Irian Jaya, Jayapura 24510
Museum Negeri Propinsi Sumatera Selatan Balaputra Dewa, Palembang 24537
Museum Negeri Sulawesi Tenggara, Kendari 24514
Museum Negeri Sulawesi Utara, Manado 24522
Museum Pemirintah Daerah Grobogan, Purwodadi Grobogan 24549
Museum Perjuangan, Yogyakarta 24592
Museum Perjuangan Bukit Barisan, Medan 24527
Museum Prabu Geusan Ulun, Sumedang 24562
Museum Purbakala Mojokerto, Mojokerto 24529
Museum Sepakat Segenap, Aceh Tenggara 24433
Museum Simalungun, Pematangsiantar 24545
Museum Suaka Budaya, Surakarta 24572
Museum Sulawesi Tengah, Palu 24540
Puri Gamelan Suar Agung, Denpasar 24468
Tridaya Eka Dharma Museum, Bukittinggi 24463

Iran

Azarbayezan Museum, Tabriz 24650
Birjand Museum, Birjand 24600
Chehel-Sotoon Museum, Qazvin 24638
Chehel-Sotoon Palace Museum, Isfahan 24611
Contemporary History Museum, Tehrān 24659
Gorgan Museum, Gorgan 24605
Haft Tappeh Museum, Haft Tappeh 24607
Iran Bastan Museum, Tehrān 24672
Islamic Museum, Tehrān 24674
Maragheh Museum, Maragheh 24629
Miyandoab Museum, Miyandoab 24635
Muséyé Kerman, Kerman 24618
Rakhtshuy-Khaneh Museum, Zanjan 24707
Shohada Museum, Ghazvin 24603
Zahedan Museum, Zahedan 24706

Ireland

Ballybunion Heritage Museum, Ballybunion 24736
Belcarra Eviction Cottage, Castlebar 24755
Bray Heritage Centre and Museum, Bray 24746
Burren Centre, Kilfenora 24847
Cape Clear Heritage Centre, Skibbereen 24897
Carlow County Museum, Carlow 24751
Cavan County Museum, Ballyjamesduff 24740
Clare Heritage Centre, Corofin 24772
Cobh Museum, Cobh 24764
Connemara History and Heritage Centre, Clifden 24760
Cork Public Museum, Cork 24767
County Museum, Sligo 24899
County Museum Dundalk, Dundalk 24827
De Valera Library and Museum, Ennis 24830
Donegal Historical Society Museum, Rossnowlagh 24895
Dublin Civic Museum, Dublin 24786
Dungarvan Museum, Dungarvan 24828
Glencolmcille Folk Village Museum, Glencolmcille 24839
Glor na nGael Museum, Carrickmacross 24752
Hillview Museum, Bagenalstown 24732
Irish Palatine Association Museum, Rathkeale 24891
Kilmallock Museum, Kilmallock 24853
Kinlough Folk Museum, Kinlough 24855
Kinsale Regional Museum, Kinsale 24857
Knappogue Castle, Quin 24888
Limerick Museum, Limerick 24867
Mayo County Museum, Ballina 24733
Millmount Museum and Tower, Drogheda 24776
Millstreet Local Museum, Millstreet 24876
Monaghan County Museum, Monaghan 24878
Musaem Chontae Dhún na nGall, Letterkenny 24862
Nenagh Heritage Centre, Nenagh 24883
North Kerry Museum, Ballyduff 24737
Old Barracks Heritage Centre, Cahersiveen 24750
Quaker Museum, Ballytore 24741
Reginalds Tower, Waterford 24908
Roscommon County Museum, Roscommon 24892
Roscrea Heritage Centre, Roscrea 24893
Rothe House Museum, Kilkenny 24849
Saint Catherine's National School Museum, Aughrim 24731
Saint Mullins Heritage Centre, Saint Mullins 24896
South Tipperary County Museum, Clonmel 24763
Tullow Museum, Tullow 24906
Valentia Island Heritage Centre, Knightstown 24859
Waterford Museum of Treasures, Waterford 24909
West Cork Regional Museum, Clonakilty 24762

Isle of Man

Manx Museum, Douglas 24916

Israel

Antiquities Museum of Tel Aviv-Yafo, Tel Aviv 25096
Beit Hankin, Kfar Yehoshua 25031
Beit Shturman Museum, Kibbutz Ein Harod 25038
Bialik-Museum, Tel Aviv 25101
Corin Maman Ashdod Museum, Ashdod 24928
Ekron - The Philistine City and it's Culture, Shiqmim 25095
Founders Room, Raanana 25080
Gush Ezyon Museum, Gush Ezyon 24950
Haifa City Museum, Haifa 24960
Liebermann House, Nahariya 25069
Meir Dizengoff Museum, Tel Aviv 25116
Museum of Acre, Acre 24919
Museum of the History of Tiberias and the Lake Kinnereth, Tiberias 25129
Old Courtyard Museum, Kibbutz Ein Shemer 25039
Ramla Museum, Ramla 25089
Sharon Museum, Emek Hefer 24947
Shephela Museum, Kibbutz Kfar Menahem 25047
Sturman Institute Museum of Regional Science, Afula 24921
Tel Hai Museum, Tel Hai 25126
Yigal Allon Center, Ginnosar 24948

Italy

Antiquarium, Ravello 27367
Antiquarium Civico, Vico del Gargano 28258
Casa Contadina, Forenza 26222
Casa Museo Francesco Baracca, Lugo 26543
Casa Museo R. Bendandi, Faenza 26072
Civico Museo Dianese, Diano Marina 26038
Civico Museo e Antiquarium Comunale, Lanuvio 26451
Civico Museo Flaminio Massetano, Massa Martana 26634
Civico Museo G. Galletti, Domodossola 26042
Civico Museo Morpurgo, Trieste 28083
Civico Museo Storico, Palmanova 27083
Civico Studio Museo Francesco Messina, Milano 26681
Collezione Civica d'Arte, Pinerolo 27219
Dolomythos, San Candido 27618
Enoteca Regionale del Monferrato, Vignale Monferrato 28268
Mostra della Civiltà Contadina e Pastorale, Avezzano 25334
Mostra Permanente della Cultura Materiale, Levanto 26489
Mostra Permanente della Resistenza, Lugo 26544
Musei della Cultura Popolare Radana, San Benedetto Po 27613
Museo Agro-Pastorale dell'Area Ellenofona, Bova Marina 25520
Museo Archeologico Civico-Diocesano G. B. Leopardi, Penne 27134
Museo Archeologico di Morgantina, Aidone 25165
Museo Archeologico Nazionale dell'Abruzzo, Chieti 25871
Museo Bernardo de Muro, Tempio Pausania 27952
Museo Biblioteca Comunale, Etroubles 26067
Museo Camillo Leone, Vercelli 28220
Museo Camuno e Biblioteca Civica, Breno 25528
Museo Casa Galimberti, Cuneo 26018
Museo Castello Masegra, Sondrio 27872
Museo Cervi, Gattatico 26275
Museo Civico, Aci Castello 25141
Museo Civico, Agrigento 25162
Museo Civico, Airola 25167
Museo Civico, Alatri 25172
Museo Civico, Atina 25319
Museo Civico, Bardonecchia 25357
Museo Civico, Belluno 25389
Museo Civico, Bisacquino 25423
Museo Civico, Borgo Velino 25506
Museo Civico, Bormio 25510
Museo Civico, Caprino Veronese 25667
Museo Civico, Casamicciola Terme 25701
Museo Civico, Castel Bolognese 25715
Museo Civico, Castelleone 25738
Museo Civico, Castelnovo Bariano 25744
Museo Civico, Castelnuovo Scrivia 25750
Museo Civico, Cerchio 25806
Museo Civico, Cittadella 25911
Museo Civico, Conversano BA 25965
Museo Civico, Cori 25975
Museo Civico, Crema 25999
Museo Civico, Crotone 26013
Museo Civico, Cuneo 26019
Museo Civico, Ferentino 26097
Museo Civico, Foggia 26209
Museo Civico, Fondi 26219
Museo Civico, Fornovo San Giovanni 26236
Museo Civico, Fratta Polesine 26249
Museo Civico, Fucecchio 26255
Museo Civico, Guardiagrele 26384
Museo Civico, Larciano 26460
Museo Civico, Lodi 26507
Museo Civico, Luino 26546
Museo Civico, Maddaloni 26562
Museo Civico, Manerbio 26586
Museo Civico, Mattinata 26642
Museo Civico, Medicina 26647
Museo Civico, Mirandola 26737
Museo Civico, Osimo 27011
Museo Civico, Pegognaga 27130
Museo Civico, Pieve di Cento 27212
Museo Civico, Pontecorvo 27289
Museo Civico, Putignano 27348
Museo Civico, Rende 27395
Museo Civico, San Cesario di Lecce 27622
Museo Civico, San Ferdinando di Puglia 27633
Museo Civico, San Severo 27687
Museo Civico, Sulmona 27917
Museo Civico, Taverna 27946
Museo Civico, Trinitapoli 28098
Museo Civico, Vasanello 28168
Museo Civico, Gangi 26271
Museo Civico, Canneto sull'Oglio 25646
Museo Civico A.E. Baruffaldi, Badia Polesine 25339
Museo Civico A. Giacomelli, Montagnana 26793
Museo Civico A. Tubino, Masone 26618
Museo Civico Alta Val Brembana, Valtorta 28153
Museo Civico Archeologico, Barbarano Romano 25354
Museo Civico Archeologico, Montalto Marche 26797
Museo Civico Archeologico, Rosignano Marittimo 27561
Museo Civico Archeologico, Salò 27599
Museo Civico Archeologico Trebula Mutesca, Monteleone Sabino 26834
Museo Civico C. Verri, Biassono 25416
Museo Civico Chiusa, Chiusa 25880
Museo Civico del Castello, Conegliano 25962
Museo Civico del Risorgimento, Modena 26747
Museo Civico della Cultura Materiale, Roseto degli Abruzzi 27560
Museo Civico di Merano - Stadtmuseum Meran, Merano 26657
Museo Civico di Palazzo Te, Mantova 26592
Museo Civico Didattico Gian Andrea Irico, Trino 28099
Museo Civico Diocesano La Castellina, Norcia 26941
Museo Civico e d'Arte Sacra, Colle di Val d'Elsa 25937
Museo Civico E. Nardi, Poggio Mirteto 27269
Museo Civico Ernesto e Teresa della Torre, Treviglio 28066
Museo Civico F. L. Belgiorno, Modica 26758
Museo Civico F. Rittatore Von Willer, Farnese 26087
Museo Civico Filippo Meli, Ciminna 25889
Museo Civico G. Ugonia, Brisighella 25547
Museo Civico-Galleria di Palazzo Pretorio, Prato 27334
Museo Civico Giacomo Rodolfo, Carignano 25678
Museo Civico Gonzaga, Novellara 26959
Museo Civico Luigi Dalla Laita, Ala 25168
Museo Civico P. A. Garda, Ivrea 26422
Museo Civico P. Rosario, Ascoli Satriano 25299
Museo Civico Padre Michele Jacobelli, Casalvieri 25699
Museo Civico Preistorico, Pofi 27264
Museo Civico Storico ed Etnografico, Primaluna 27344
Museo Communale Antiquarium, Atena Lucana 25318
Museo Complesso della Collegiata, Castiglione Olona 25763
Museo Comunale, Praia a Mare 27323
Museo Comunale della Valle del Sarno, Sarno 27744
Museo Comunale E. Durio, Civiasco 25915
Museo Correr e Biblioteca d'Arte e Storia Veneziana, Venezia 28188
Museo degli Alpini, Bassano del Grappa 25379
Museo dei Grandi Fiumi, Rovigo 27574
Museo del Borgo, Amato 25220
Museo del Camoscio, Opi 26984
Museo del Centro Caprense Ignazio Cerio, Capri 25665
Museo del Lavoro Contadino, Brisighella 25548
Museo del Lavoro Povero e della Civiltà Contadina, Livraga 26502
Museo del Lino, Pescarolo 27167
Museo del Modellismo Storico, Voghiera 28306
Museo del Risorgimento, Castelfidardo 25723
Museo del Risorgimento, Imola 26400
Museo del Risorgimento, Lucca 26534
Museo del Risorgimento, Villafranca di Verona 28276
Museo del Risorgimento, Piacenza 27186
Museo del Risorgimento Aurelio Saffi, Forlì 26224
Museo del Risorgimento F. Tanara, Langhirano 26449
Museo del Territorio, Foggia 26210
Museo del Territorio, Longiano 26516
Museo del Territorio, San Severino Marche 27685
Museo del Territorio, Finale Emilia 26137
Museo del Territorio dell'Alta Valle dell'Aulella, Casola in Lunigiana 25706
Museo del Villaggio, Chiusa 25881
Museo dell' Antica Civiltà Locale, Bronte 25551
Museo della Biblioteca Maldotti, Guastalla 26388
Museo della Casa Carsica, Monrupino 26785
Museo della Casa Trevigiana, Treviso 28071
Museo della Città, Aquino 25256
Museo della Città, Rimini 27407
Museo della Città e del Territorio, Monsummano Terme 26790
Museo della Civiltà Contadina, Acri 25151
Museo della Civiltà Contadina, Ateleta 25317
Museo della Civiltà Contadina, Barolo 25375
Museo della Civiltà Contadina, Bastiglia 25383
Museo della Civiltà Contadina, Guardia Piemontese 26382
Museo della Civiltà Contadina, Loreto Aprutino 26522
Museo della Civiltà Contadina, Maierato 26569
Museo della Civiltà Contadina, Mombaroccio 26769
Museo della Civiltà Contadina D. Bianco, Sammichele di Bari 27606
Museo della Civiltà Romana, Roma 27479
Museo della Civiltà Solandra, Malé 26575

H

Museo della Figurina di Gesso e dell'Emigrazione, Coreglia Antelminelli 25973
Museo della Memoria, Sant'Andrea Apostolo 27717
Museo della Misericordia, San Miniato 27672
Museo della Nostra Terra, Pieve Torina 27216
Museo della Resistenza e della Civiltà Contadina, Pietrabruna 27203
Museo della Resistenza e del Folclore Valsabbino, Pertica Bassa 27141
Museo della Società di Studi Patri, Gallarate 26266
Museo della Stampa, Jesi 26425
Museo della Val Gardena, Ortisei 27001
Museo della Valchiavenna, Chiavenna 25863
Museo delle Arti e delle Tradizioni di Puglia, Latiano 26464
Museo delle Genti d'Abruzzo, Pescara 27163
Museo delle Genti delle Valli di Lanzo, Ceres 25807
Museo delle Rimembranze, Cosenza 25991
Museo dell'Intreccio Mediterraneo, Castelsardo 25753
Museo Demologico dell'Economia, del Lavoro e della Storia Sociale Silana, San Giovanni in Fiore 27644
Museo di Archeologia Urbana e dei Commerc Antichi in Abruzzo, Lanciano 26447
Museo di Archeologia Urbana Fiorelli, Lucera 26538
Museo di Criminologia Medievale, San Gimignano 27639
Museo di Prali e della Val Germanasca, Prali 27324
Museo di Rocca Fregoso, Sant'Agata Feltria 27715
Museo di Rodoretto, Prali 27325
Museo di Storia dell'Agricoltura e della Pastorizia, Morano Calabro 26867
Museo di Storia delle Mezzadria, Senigallia 27803
Museo di Storia e Arte, Gorizia 26352
Museo di Storia e Cultura della Val del Biois, Vallada Agordina 28143
Museo di Storia Locale, Velturno 28174
Museo di Storia Quarnese, Quarna Sotto 27349
Museo di Tossicia, Tossicia 28046
Museo Diefenbach, Capri 25666
Museo e Certosa di San Martino, Napoli 26906
Museo elle Arti e Tradizioni Contadine, Roccasecca dei Volsci 27434
Museo Emidiano, Agnone 25157
Museo Eritreo Bottego, Parma 27106
Museo Etnografico, Bomba 25492
Museo Etnografico, Cortina d'Ampezzo 25985
Museo Etnografico di Servola, Trieste 28093
Museo Etnografico Don Luigi Pellegrini, Castiglione di Garfagnana 25762
Museo F. Renzi, Borghi 25501
Museo Illichiano, Castell'Arquato 25737
Museo Ladino di Fassa, Pozza di Fassa 27321
Museo Lapidario, Urbino 28127
Museo Lapidario del Duomo di Sant'Eufemia, Grado 26359
Museo Leoniano, Carpineto Romano 25688
Museo Luigi Varoli, Cotignola 25996
Museo Manzoniano, Lesa 26487
Museo Marinaro Gio Bono Gerrari, Cairo Montenotte 25591
Museo Medievale Borbonico, Salle 27598
Museo Memorie Pollentine, Pollenza 27274
Museo Nazionale Archeologico, Civitavecchia 25924
Museo Nazionale della Residenza Napoleonica dei Mulini, Portoferraio 27308
Museo Nazionale di Casa Giusti, Monsummano Terme 26792
Museo Nazionale Garibaldino, La Maddalena 26429
Museo Parrocchiale e Museo P. Mascagni, Bagnara di Romagna 25344
Museo Pietro Micca e dell'Assedio, Torino 28024
Museo Pomposiano, Codigoro 25931
Museo Preistorico N. Lamboglia e Val Varatello, Toirano 27983
Museo Quattroruote, Rozzano 27577
Museo Regionale di Camarina, Ragusa 27360
Museo Risorgimentale, Montichiari 26860
Museo Risorgimentale, Pieve di Cadore 27211
Museo Rudolf Stolz, Sesto in Pusteria 27826
Museo Sacrario Galileo Ferraris, Livorno Ferraris 26501
Museo Sardo di Antropologia ed Etnografia, Cagliari 25587
Museo Storico, Montese 26853
Museo Storico Alpino, Antrodoco 25243
Museo Storico C. Musazzi, Parabiago 27089
Museo Storico Civico Cuggionese, Cuggiono 26016
Museo Storico Dittico Alto Livenza, Sacile 27585
Museo Storico Etnografico, Sampeyre 27607
Museo Storico Michele Ghislieri, Vicoforte 28261
Museo Storico Navale, Venezia 28203
Museo Storico Novarese Aldo Rossini, Novara 26955
Museo Sveviano, Trieste 28097
Museo Territoriale del Lago di Bolsena, Bolsena 25481
Museo Vallivo, Valfurva 28140
Museo Valtellinese di Storia e Arte, Sondrio 27873
Museo Virgiliano, Virgilio 28289
Museum Passeier - Andreas Hofer, San Leonardo in Passiria 27654
Parco Nazionale delle Incisioni Rupestri, Capo di Ponte 25659
Pinacoteca Civica, Imola 26403
Pinacoteca e Musei Comunali, Macerata 26556
Raccolta Archeologica Comunale, Celenza Valfortore 25799
Raccolta Civica, Città di Castello 25910
Raccolta Comunale, Sigillo 27849
Sala A. De Carolis, Montefiore dell'Aso 26828
Santa Giulia - Museo della Città, Brescia 25543
Stadt- und Multschermuseum, Vipiteno 28288

Jamaica

Jamaica Folk Museum, Kingston 28329
Jamaican People's Museum of Craft and Technology, Spanish Town 28341
Seville Great House and Heritage Park, Ocho Rios 28339

Japan

Abashiri Kyodo Hakubutsukan, Abashiri 28343
Aikawa Folk Museum and Exhibition Hall of Folk Crafts, Sado 29175
Akadomari-mura Folk Museum, Sado 29176
Akita-kenritsu Hakubutsukan, Akita 28355
Akizuki Hakubutsukan, Amagi 28358
Aomori-kenritsu Kyodokan, Aomori 28360
Arai-shiritsu Hakubutsukan, Arai 28362
Asahikawa Hakabutsukan, Asahikawa 28369
Ashiya-shiritsu Bijutsu oyo Rekishi Hakubutsukan, Ashiya 28375
Ayabe-shiritsu Hakubutsukan, Ayabe 28384
Chiba-kenritsu Boso-no-Mura Hakubutsukan, Sakae 29194
Chiba-kenritsu Kazusa Hakubutsukan, Kisarazu 28704
Chido Hakubutsukan, Tsuruoka 29542
Chihaya-Akasaka Kyodo-shiryokan, Chihaya-Akasaka 28397
Chofu Museum, Shimonoseki 29268
Chojagahara Kokokan, Itoigawa 28598
Chokaido Hakubutsukan, Yokkaichi 29591
Daikan Yashiki oyo Setagaya Rekishi Hakubutsukan, Tokyo 29380
Dejima-machi Hakabutsukan, Nagasaki 28971
Fujimi-machi Rekishi Hakubutsukan, Fujimi 28421
Fujiyama Museum of Historical Data, Oonishi 29122
Fukui-kenritsu Bunka Hakubutsukan, Fukui 28429
Fukuoka-shi Hakubutsukan, Fukuoka 28434
Fukushima-kenritsu Hakubutsukan, Aizuwakamatsu 28350
Gifu-ken Hakubutsukan, Seki 29240
Hachinohe-shi Bijutsukan, Hachinohe 28461
Hachinohe-shi Hakubutsukan, Hachinohe 28462
Hachinohe-shiritsu Rekishi Hakubutsukan, Hachinohe 28463
Hachioji-shi Rekishi Hakubutsukan, Hachioji 28465
Hagi Hakubutsukan, Hagi 28468
Hakodate-shiritsu Hakubutsukan, Hakodate 28474
Hamamatsu-shi Hakubutsukan, Hamamatsu 28484
Hamura-shiritsu Hakubutsukan, Hamura 28489
Hanamaki-shiritsu Hakubutsukan, Hanamaki 28490
Higashi-Osaka-shiritsu Kyodo Hakubutsukan, Higashi-Osaka 28506
Higashi-yamate District Historic Preservation Center, Nagasaki 28974
Higashimurayama Mingeikan, Higashimurayama 28507
Hino-shiritsu Shinsengumino Furusato Hakubutsukan, Hino 28516
Hiraizumi Historical Artifacts Museum, Hiraizumi 28521
Hirata Kinenkan, Takayama 29341
Hiratsuka-shi Hakubutsukan, Hiratsuka 28528
Hirosaki-shiritsu Hakubutsukan, Hirosaki 28529
Hirose-machi Rekishi oyo Minzoku Hakubutsukan, Hirose 28530
Hitachi Kyodo Hakubutsukan, Hitachi 28543
Ichinomiya-shiritsu Hakubutsukan, Ichinomiya 28557
Ichinoseki-shiritsu Hakubutsukan, Ichinoseki 28558
Iida-shi Bijutsu Hakubutsukan, Iida 28559
Iruma-shi Hakubutsukan Alit, Iruma 28579
Ishikawa-kenritsu Rekishi Hakubutsukan, Kanazawa 28658
Itami-shiritsu Hakubutsukan, Itami 28591
Itsutsubashikan, Sendai 29242
Iwakuni Historical Museum Kikkawa Family Museum, Iwakuni 28607
Iwate-kenritsu Hakubutsukan, Morioka 28935
Izumi-mura Rekishi no Sato, Izumi, Fukui 28612
Jingu Hakubutsukan, Ise 28581
Joetsu-shiritsu Hakubutsukan, Joetsu 28620
Kagoshima-kenritsu Hakubutsukan, Kagoshima 28627
Kanagawa-kenritsu Rekishi Hakabutsukan, Yokohama 29597
Kanai-machi Hakabutsukan, Kanai 28652
Katsushika-ku Kyoudo to Tenmon no Hakubutsukan, Tokyo 29409
Kawagoe-shiritsu Hakubutsukan, Kawagoe 28684
Kawasaki-shi Hakubutsukan, Kawasaki 28693
Kazuno Cultural Property Preservation Center, Kazuno 28700
Kensei Kinenkan, Niigata 29054
Kibiji Kyodo Hakubutsukan, Soja 29294
Kii-Fudoki-no-oka Museum of History, Wakayama 29572
Kinpo-cho Rekishi Akusha, Kinpo 28702
Kishu-furitsu Hakubutsukan, Shirahama 29280
Kita-kamakura Museum, Kamakura 28644
Kitakami-shiritsu Hakubutsukan, Kitakami 28709
Kitami-Abashiri-furitsu Bunkakan, Kitami 28716
Kiyose-shi Kyodo Hakubutsukan, Kiyose 28719
Kobe-shiritsu Hakubutsukan, Kobe 28735
Komatsu Shiritsu Hakubutsukan, Komatsu 28765
Kosanji Temple Museum, Setoda 29253
Koto-ku Nakagawa Funabansho Museum, Tokyo 29421
Kumamoto-shiritsu Hakubutsukan, Kumamoto 28780
Kunisaki-machi Rekishi Minzoku Shiryokan, Kunisaki 28781
Kuramayama Museum, Kyoto 28832
Kurayoshi Hakubutsukan, Kurayoshi 28792
Kushiro City Museum, Kushiro 28799
Kuwana-shiritsu Hakubutsukan, Kuwana 28800
Kyoto-fu Kyoto Bunka Hakubutsukan, Kyoto 28838
Machida-shiritsu Hakubutsukan, Machida 28874
Matsudo Museum, Matsudo 28887
Matsumae-cho Bunkakan, Matsumae 28892
Matsumoto-shiritsu Hakubutsukan, Matsumoto 28895
Matsushima Historical Museum, Matsushima 28901
Matsushita Hakubutsukan, Aira 28349
Mie-kenritsu Hakubutsukan, Tsu 29537
Minokamo-shiritsu Hakubutsukan, Minokamo 28911
Miyama-cho Kayabuki Bijutsukan oyo Kyodo Hakubutsukan, Miyama 28925
Miyazaki-ken Sogo Hakubutsukan, Miyazaki 28927
Miyazawa Kenji Kinenkan, Hanamaki 28491
Mizuhomachi Kyodo Hakubutsukan, Mizuho 28931
Morioka City Local Hall, Morioka 28938
Morita-mura Rekishi Mingeikan, Morita 28939
Motobu-cho Hakubutsukan, Motobu 28942
Muroran-shi Tontenkan, Muroran 28948
Musashi-murayama-shi Rekishi Minzoku Shiryokan, Musashi-murayama 28949
Museum of Yokohama Urban History, Yokohama 29602
Nagaoka-shiritsu Hakabutsukan, Nagaoka 28961
Nagasaki Rekishi no Bunka Hakubutsukan, Nagasaki 28981
Nagoya-shi Hakubutsukan, Nagoya 29007
Nakano Historical Museum, Tokyo 29441
Nakano-shi Rekishi Minzoku Shiryokan, Nakano 29021
Niigata-kenritsu Rekishi Hakubutsukan, Nagaoka 28964
Niigata-shi Hakubutsukan, Niigata 29057
Ninohe-shiritsu Rekishi Mingeikan, Ninohe 29071
Okaya Museum, Okaya 29097
Okayama-kenritsu Hakubutsukan, Okayama 29101
Okinawa-kenritsu Hakubutsukan Gendai Bijutsukan, Naha 29014
Oku Kyodokan, Oku 29107
Ome-shiritsu Hakubutsukan, Ome 29113
Osaka Museum of History, Osaka 29147
Osaka Museum of Housing and Living, Osaka 29148
Ota-kuritsu Mingeikan, Tokyo 29458
Otaru-shi Hakubutsukan, Otaru 29161
Rekishi Minzoku Shiryokan, Shimosuwa 29271
Rekishi no Bunka Hakubutsukan, Matsusaka 28900
Rikuzentakata-shi Hakubutsukan, Rikuzentakata 29171
Ryoma's Hometown Museum, Kochi 28746
Ryotsu Folk Museum, Ryotsu 29173
Sado Hakabutsukan, Sawata 29237
Sado-koku Ogi Folk Museum, Ogi 29094
Saga-kenritsu Hakubutsukan, Saga 29180
Sagamihara-shi Hakubutsukan, Sagamihara 29184
Saijo-shi Minzokugaku Hakubutsukan, Saijo 29186
Saitama-kenritsu Hakubutsukan, Saitama 29189
Sakai-shi Hakubutsukan, Sakai 29199
Sapporo Clock Tower Exhibition, Sapporo 29221
Sapporo-shiritsu Hakubutsukan, Sapporo 29225
Sasayama-shiritsu Bijutsu oyo Rekishi Hakubutsukan, Sasayama 29231
Sendai-shi Hakubutsukan, Sendai 29246
Senshu Bunko Museum, Tokyo 29466
Shikoku-mura Minka Hakubutsukan no Bijutsukan, Takamatsu 29321
Shinagawa Historical Museum, Tokyo 29469
Shinshushin-machi Hakubutsukan, Shinshushin 29273
Shitamachi Museum, Tokyo 29470
Shizukuishi-machi Rekishi Mingeikan, Shizukuishi 29283
Shoko Shuseikan Historical Museum, Kagoshima 28635
Shunan-shi Bijutsukan nara Hakubutsukan, Shunan 29292
Soma Gyofu Kinenkan, Itoigawa 28601
Sukagawa-shiritsu Hakubutsukan, Sukagawa 29298
Sutama-cho Rekishi Hakubutsukan, Sutama 29300
Takachiho-cho Rekishi oyo Minzoku Hakubutsukan, Takachiho 29313
Tateyama Hakubutsukan anna Toyama-ken, Tateyama, Toyoma 29354
Tobinodai Shiseki Koen Hakubutsukan, Funabashi 28453
Tochigi-kenritsu Hakubutsukan, Utsunomiya 29567
Tokushima Prefectural Local Culture Hall Awa Deko-puppet Museum, Tokushima 29368
Tokyo Kasei Daigaku Hakubutsukan, Tokyo 29489
Tomakomai-shi Hakubutsukan, Tomakomai 29510
Tomo-no-Ura Rekishi oyo Minzoku Hakubutsukan, Fukuyama 28450
Tono Kyodo Hakubutsukan, Tono 29513
Tottori-kenritsu Hakubutsukan, Tottori 29515
Tsuyama-furitsu Rekishi Hakubutsukan, Tsuyama 29548
Uchiko Hakubutsukan, Uchiko 29551
Uozu-shi Rekishi no Mingei Hakubutsukan, Uozo 29562
Urahoro-cho Kyodo Hakubutsukan, Urahoro 29563
Wakayama-kenritsu Hakubutsukan, Wakayama 29574
Yakage Museum, Yakage 29577
Yamaguchi Hakabutsukan, Yamaguchi 29580
Yamato Hakubutsukan, Kure 28794
Yanaga Hokkaido Museums, Sapporo 29229
Yatsushiro Kyodo Hakubutsukan, Yashiro 29587
Yomitanson Rekishi oyo Minzoku Hakubutsukan, Yomitanson 29625
Yufuin-cho Hakubutsukan, Yufuin 29631
Yuzawa History and Folk Museum, Yuzawa 29639

Jordan

Al-Karak Museum for Archaeology and Folklore, Al-Karak 29641
Madaba Archaeology Museum, Ajloun 29640
Petra Museum, Petra 29665

Kazakhstan

Almaty City History Museum Museum of Repression, Almaty 29674
Atyrauskij Oblastnoj Istoriko-Kraevedčeskij Muzej, Atyrau 29695
Balchašskij Gorodskoj Istoriko-Kraevedčeskij Muzej, Balchaš 29697
Dom-muzej M. Auèzova, Almaty 29680
Gorodskoj Istoriko-Kraevedčeskij Muzej, Ridder 29724
History Local Studies Museum, Semey 29727
Istoriko-Kraevedčeskij Muzej, Merke 29710
Istoriko-Kraevedčeskij Muzej, Rudnyj 29725
Istoriko-Kulturnyj Muzej-zapovednik Azret Sultan, Turkestan 29734
Južno-Kazachstanskij Istoriko-Kraevedčeskij Muzej, Čimkent 29698
Karagandinskij Oblastnoj Istoriko-Kraevedčeskij Muzej, Qaraghandy 29722
Karkaralinskij Rajonnyj Istoriko-Kraevedčeskij Muzej, Karkaralinsk 29703
Kostanajskij Oblastnoj Istoriko-Kraevedčeskij Muzej, Kostanaj 29706
Kurchatov Museum, Kurchatov 29707
Kyzylordynskij Istoriko-Kraevedčeskij Muzej, Kyzylorda 29708
Lisakovskij Muzej Istorii i Kultury Verchnego Pritobolja, Lisakovsk 29709
Local Museum, Fort Shevchenko 29700
Mangystauskij Oblastnoj Istoriko-Kraevedčeskij Muzej, Aktau 29670
Memorialnyj Muzej Akademika K.I. Satpaeva, Almaty 29684
Memorialnyj Muzej Alii Moldagulovoj, Aktöbe 29672
Muzej Kazakstan Respublikasynyn Prezidenttik Mèdeniet Ortalyfy, Astana 29692
Oblastnoj Istoriko-Kraevedčeskij Muzej, Aktöbe 29673
Osakarovskij Rajonnyj Istoriko-Kraevedčeskij Muzej, Osakarovka 29715
Pavlodarskij Oblastnoj Istoriko-Kraevedčeskij Muzej im. G.N. Potanina, Pavlodar 29719
Regional Museum, Taraz 29733
Severo-Kazachstanskij Oblastnoj Istoriko-Kraevedčeskij Muzej, Petropavl 29720
Taldykorganskij Istoriko-Kraevedčeskij Muzej, Taldykorgan 29731
Vostočno-Kazachstanskij Istoriko-Kraevedčeskij Muzej, Özkemen 29712
Zajsanskij Gorodskoj Istoriko-Kraevedčeskij Muzej, Zajsan 29736
Zapadno-Kazachstanskij Oblastnoj Istoriko-Kraevedčeskij Muzej, Oral 29714
Žezkazganskij Istoriko-Kraevedčeskij Muzej, Žezkazgan 29737

Kenya

Kisumu Museum, Kisumu 29742
Lamu Museum, Lamu 29744
Meru Museum, Meru 29745

Korea, Democratic People's Republic

Myohyang-san Museum, Hyangsan 29755
Shinchon Museum, Shinchon 29761

Korea, Republic

Busan Museum, Busan 29774
Cheongju National Museum, Cheongju 29788
Chungju Museum, Chungju 29790
Chungnam University Museum, Daejeon 29798
Daegu National Museum, Daegu 29792
Dongguk University Museum, Seoul 29872
Gangneung Municipal Museum, Gangneung 29808
Gangwon Folk Museum, Cheongil 29786
Geochang Museum, Geochang 29811
Geoje Museum, Geoje 29812
Gwangju Municipal Folk Museum, Gwangju 29822
Gwangju National Museum, Gwangju 29823
Gyemyung University Museum, Daegu 29794
Gyeongbuk National University Museum, Daegu 29795
Gyeonggi Provincial Museum, Yongin 29941
Gyeongju National Museum, Gyeongju 29826
Hong-ik University Museum, Seoul 29882
Incheon Metroploitan City Museum, Incheon 29833
Jeonju Municipal Museum, Jeonju 29841
Jeonju National Museum, Jeonju 29842
Korea University Museum, Seoul 29892
Korean Christian Museum, Seoul 29895
Kyunghee University Museum, Seoul 29901
Mireuksa Temple Museum, Iksan 29830
Miryang Municipal Museum, Miryang 29852
Museum of Daejeon History, Daejeon 29801
Yeongil Folk Museum, Pohang 29859

Kyrgyzstan

Frunze House Museum, Biškek 29952
Istoriko-Kraevedceskij Muzej, Tokmok 29966
Karacolskij Istoričeskij Kraevedčeskij Muzej, Karacol 29962
Oškij Objedinennyj Istoriko-Kulturnyj Muzej-Zapovednik, Oš 29963

Laos

Champasak Historical Heritage Museum, Vientiane 29972
Musée Kaysone Phomvihane, Vientiane 29975
Pha That Luang, Vientiane 29977

Latvia

Aizkraukles Vēstures un Mākslas Muzejs, Aizkraukle 29982
Bauskas Novadpētniecības un Mākslas Muzejs, Bauska 29988
Jāņa Akuratera Muzejs, Rīga 30070
Kandavas Novada Muzejs, Kandava 30029
Liepāja Okupaciju Rezimos, Liepāja 30038
Liepājas Muzejs, Liepāja 30040
Naujenes Novadpētniecības Muzejs, Naujene 30049
Rīgas Vēstures un Kuģniecības muzejs, Rīga 30111
Valkas Novadpētniecības Muzejs, Valka 30137

Lebanon

Musée de Baalbeck, Baalbeck 30155
Musée Terbol, Zahlé 30175

Liechtenstein

Bäuerliches Wohnmuseum, Schellenberg, Schellenberg 30195
DoMuS - Museum und Galerie der Gemeinde Schaan, Schaan 30192
Liechtensteinisches Landesmuseum, Vaduz 30199
Walser Heimatmuseum, Triesenberg 30196

Lithuania

Joniškio Istorijos ir Kultūros Muziejus, Joniškis 30231
Kernavės Archeologijos ir Istorijos Muziejus, Kernavė 30274
Kražių Kraštotyros Muziejus, Kražiai 30287
Laukminiškių Kaimo Muziejus, Laukminiškes 30296
Lazdijų Krašto Muziejus, Lazdijai 30300
Lietuvos Liaudis Buities Muziejus, Rumšiškės 30347
Mažosios Lietuvos Istorijos Muzieus, Klaipėda 30283
Motiejaus Gustaičio Memorialinis Muziejus, Lazdijai 30301
Naujasis Arsenalas, Vilnius 30427
Neringos Istorijos Muziejus, Nida 30321
Panevėžio Kraštotyros Muziejus, Panevėžys 30332
Pasvalio Krašto Muziejus, Pasvalys 30335
Pilies Muzieus, Klaipėda 30284
Prienų Krašto Muziejus, Prienai 30340
Raseinių Krašto Istorijos Muziejus, Raseiniai 30342
Seniausias Panevėžio Pastatas, Panevėžys 30333
Šiaulių Aušros Muziejus, Šiauliai 30362
Trakų Istorijos Muziejus, Trakai 30382
Žeimelio Žiemgalos Muziejus, Žeimelis 30451
Žemaičių Dailės Muziejus, Plungė 30338

Luxembourg

A. Schiiwesch Musée Rural, Binsfeld 30456
Musée Aal Braanebuerg, Brandenbourg 30458
Musée d'Histoire de la Ville de Luxembourg, Luxembourg 30482
Musée Municipal, Dudelange 30467

Macedonia

Muzej Bitola, Bitola 30500
Muzej-Galerija Kavadarci, Kavadarci 30502
Muzej na Grad Kratovo, Kratovo 30504
Muzej na Grad Negotino, Negotino 30510
Muzej na Tetovskiot Kraj, Tetovo 30532
Muzej Zapadna Makedonija vo NOV, Kičevo 30503
Naroden Muzej, Ohrid 30512
Naroden Muzej, Sveti Nikole 30531
Naroden Muzej D-r Nikola Nezlobinski, Struga 30527
Naroden Muzej Kumanovo, Kumanovo 30507
Pokrajinski Muzej, Strumica 30528
Zavod za Zaštita na Spomenicite na Kulturata i Muzej - Strumica, Strumica 30530

Madagascar

Musée Faniahy, Fianarantsoa 30539
Musée Régional du Cerel, Toamasina 30542

Malawi

Lake Malawi Museum, Mangochi 30545
Mzuzu Regional Museum, Mzuzu 30546

Malaysia

Istana Batu, Kota Bharu 30563
Lembaga Muzium Negeri, Pulau Pinang 30613
Muzium Alor Gajah, Alor Gajah 30549
Muzium Darul Ridzwan, Ipoh 30555
Muzium Jasin, Jasin 30557
Muzium Negeri Perlis, Kangar 30559
Muzium Negeri Selangor, Shah Alam 30619
Muzium Negeri Trengganu, Kuala Terengganu 30589
Muzium Sejarah Etnografi, Melaka 30603
P. Ramlee Memorial, Kuala Lumpur 30584
Royal Museum Sri Menanti, Seremban 30617
Sultan Abu Bakar Museum Pahang, Pekan 30611
Tin Museum - Gedung Raja Abdullah, Kelang 30561

Maldives

National Museum, Malé 30623

Mali

Musée Régional de Sikasso, Sikasso 30626
Musée Régional du Sahel, Gao 30625

Malta

Casa Rocca Piccola, Valletta 30640

Mauritania

Musée de Ouadane, Nouakchott 30674

Mauritius

Vieux Grand Port Archaeological Site, Old Grand Port 30680

Mexico

Casa de Hidalgo, Dolores Hidalgo 30834
Ex Convento de Santo Domingo de Guzman, Oaxtepec 31228
Museo Anenecuilco, Ayala 30729
Museo Antzetik Ta Jtelum, Zinacantán 31598
Museo Bahía de los Angeles, Ensenada 30855
Museo Balaa Xte Guech Gulal, Teotitlán del Valle 31429
Museo Benito Juárez, José Azueta 30978
Museo Bernabé de las Casas, Mina 31166
Museo Campamento Irritila, San Pedro 31365
Museo Casa de la Cultura, Montemorelos 31176
Museo Casa de Morelos, Ecatepec de Morelos 30845
Museo Casa de Morelos, Morelia 31191
Museo Casa del Deán, Puebla 31269
Museo Casa del Diego, Celaya 30752
Museo Casa del Dr. Mora, Comonfort 30795
Museo Casa Francisco Villa, Hidalgo del Parral 30935
Museo Casa Mata, Matamoros 31017
Museo Casa Maya, Cancún 30748
Museo Cerro de la Campana, Santiago Suchilquitongo 31385
Museo Cerro de los Huizaches, San Pablo Huixtepec 31364
Museo Ciénaga de San Pedro, Madera 31009
Museo Coahuila y Texas, Monclova 31173
Museo Comunitario, Mineral del Monte 31168
Museo Comunitario, Puente de Ixtla 31293
Museo Comunitario Altagracia de Arauz, Ensenada 30856
Museo Comunitario Amuzgo de Xochistlahuaca, Xochistlahuaca 31567
Museo Comunitario Colonia Vicente Guerrero, Colonia Vicente Guerrero 30787
Museo Comunitario Concuemitl, Cuencamé 30815
Museo Comunitario Coronela Amelia La Güera Robles, Xochipala 31566
Museo Comunitario Crónica y Arte Agropecuario de la Sierra Gorda, Landa de Matamoros 30994
Museo Comunitario Cuarenta Casas, Valentín Gómez Farías 31523
Museo Comunitario de Antropología e Historia, Apatzingán 30718
Museo Comunitario de Atoyac, Atoyac 30727
Museo Comunitario de Carrillo Puerto, Alvaro Obregón 30713
Museo Comunitario de Ciudad Lerdo, Lerdo 31002
Museo Comunitario de Historia de San José de Gracia, San José de Gracia 31333
Museo Comunitario de Huitzapula, Huitzapula 30951
Museo Comunitario de Ixtlahuacán, Ixtlahuacán 30962
Museo Comunitario de la Ballena, La Paz 30986
Museo Comunitario de las Varas, Compostela 30796
Museo Comunitario de Pejelagartero, Huimanguillo 30949
Museo Comunitario de Pipillola, Españita 30864
Museo Comunitario de San Andrés Mixquic, México 31064
Museo Comunitario de San Nicolás Zoyatlán, San Nicolás Zoyatlán 31362
Museo Comunitario de Santa Eulalia, Santa Eulalia 31379
Museo Comunitario de Santa Martha, Mulegé 31205
Museo Comunitario de Tototepec, Tototepec 31507
Museo Comunitario de Xalisco, Xalisco 31561
Museo Comunitario de Xolalpan, Ixtacamaxtitlán 30960
Museo Comunitario del Valle de Guadalupe, Ensenada 30857
Museo Comunitario El Rosario, El Rosario 30850
Museo Comunitario Felipe Carrillo Puerto, Motul 31203
Museo Comunitario Francisco I. Madero, Francisco I. Madero, Coahuila 30867
Museo Comunitario Hicupa, Tehuacán 31418
Museo Comunitario Hitalulu, San Martín Huamelulpan 31354
Museo Comunitario Huitza Chilin, San Pedro 31366
Museo Comunitario Iluikatlachiyalistli, Yahualica 31570
Museo Comunitario Ismael Girón González, Pánuco, Zacatecas 31250
Museo Comunitario Itzmal Kauil, Izamal 30964
Museo Comunitario Ji, Ocosingo 31232
Museo Comunitario Jna Niingui, San Miguel Tequixtepec 31359
Museo Comunitario Joyonaque, Tuxtla Gutiérrez 31512
Museo Comunitario Juan García Aldama de la Comunidad Cucapá, El Mayor Indígena 30848
Museo Comunitario Kumkuy Ys Untzi, Copainalá 30799
Museo Comunitario Maika, Poanas 31263
Museo Comunitario Note Ujía, Santa María Asunción Tlaxiaco 31380
Museo Comunitario Nu-kuiñe, Cuquila 30830
Museo Comunitario Rubén Jaramillo, Tlaquiltenango 31475
Museo Comunitario Serafín Olarte, Papantla 31251
Museo Comunitario Shan-Dany, Santa Ana del Valle 31372
Museo Comunitario sin Paredes Colonia Orizaba, Mexicali 31034
Museo Comunitario Snuuvico, San Juan Mixtepec Juxtlahuaca 31342
Museo Comunitario Tejamen, Nuevo Ideal 31218
Museo Comunitario Tzakualli, Zacualpan 31587
Museo Comunitario Unión y Progreso, Cuatrocienegas 30812
Museo Comunitario Xiximes, Gómez Palacio 30878
Museo Comunitario Xolalpan Calli, Hueyapan 30946
Museo Comunitario Ya Nfădi Yu Nohño Los Conocimientos de los Otomíes, Tolimán 31483
Museo Comunitario Yucu-Iti, Santa María Yucu-Iti 31381
Museo Comunitario Ze Acatl Topiltzin Quetzalcoatl, Tepoztlán 31446
Museo Contalpa de Nombre de Dios, Nombre de Dios 31217
Museo Costumbrista de Sonora, Alamos 30709
Museo de Acaponeta, Acaponeta 30690
Museo de Aguascalientes, Aguascalientes 30698
Museo de Ahumada, Villa Ahumada y Anexas 31545
Museo de Akil Uyotoch Cah, Akil 30708
Museo de Ameca, Vetagrande 31538
Museo de Antropología e Historia de Allende, Allende 30711
Museo de Antropología e Historia de Santiago Papasquiaro, Santiago Papasquiaro 31384
Museo de Atoyac, Atoyac de Alvarez 30728
Museo de Ayutla, Ayutla de los Libres 30730
Museo de Batopilas, Francisco I. Madero, Coahuila 30868
Museo de Bermejillo, Bermejillo 30736
Museo de Buenos Aires, Tacotalpa 31394
Museo de Campeche, Campeche 30740
Museo de Ciudades Hermanas, Guadalajara 30893
Museo de Cochoapa, Ometepec 31237
Museo de Colón, Colón 30785
Museo de Conkal, Conkal 30798
Museo de Cosautlán de Carvajal, Cosautlán de Carvajal 30804
Museo de Coscomatepec, Coscomatepec de Bravo 30805
Museo de Cuajinicuilapa, Cuajinicuilapa 30809
Museo de Emiliano Zapata, Emiliano Zapata 30852
Museo de General Zaragoza, General Zaragoza 30876
Museo de Guadalupe, Guadalupe 30906
Museo de Guamúchil de Salvador Alvarado, Guamúchil 30910
Museo de Historia de Ensenada, Ensenada 30859
Museo de Historia de Higueras, Higueras 30936
Museo de Historia de Lampazos de Naranjo y las Armas Nacionales, Lampazos de Naranjo 30993
Museo de Historia de Linares, Linares 31006
Museo de Historia Regional de Pánuco, Pánuco, Veracruz 31249
Museo de Historia Regional de Sabinas Hidalgo, Sabinas Hidalgo 31309
Museo de Hostotipaquillo, Hostotipaquillo 30937
Museo de Jala, Jala 30965
Museo de Jamapa, Jamapa 30970
Museo de la Casa de Cultura de Zacapú, Zacapu 31575
Museo de la Casa de la Cultura, San Luis Potosí 31346
Museo de la Casa del Risco, México 31080
Museo de La Cerería, Tlayacapan 31480
Museo de la Ciudad, Campeche 30741
Museo de la Ciudad, Guadalajara 30894
Museo de la Ciudad, Mérida 31027
Museo de la Ciudad, Tecate 31412
Museo de la Ciudad de Chetumal, Chetumal 30760
Museo de la Ciudad de Córdoba, Córdoba 30801
Museo de la Ciudad de Irapuato, Irapuato 30957
Museo de la Ciudad de León, León 30999
Museo de la Ciudad de México, México 31082
Museo de la Ciudad de Querétaro, Querétaro 31300
Museo de la Ciudad de Sombrerete, Sombrerete 31392
Museo de la Ciudad de Tepatitlán de Morelos, Tepatitlán de Morelos 31430
Museo de la Ciudad de Veracruz, Veracruz 31536
Museo de la Ciudad Tuxtlán, Tuxtla Gutiérrez 31513
Museo de la Colonia Siete de Marzo, Juan Aldama 30979
Museo de la Isla de Cozumel, Cozumel 30808
Museo de la Labranza, Francisco I. Madero, Hidalgo 30871
Museo de la Tortuga, Aquila 30722
Museo de la Villa de Ahome, Ahome 30704
Museo de Lagos de Moreno, Lagos de Moreno 30992
Museo de las Culturas de la Huasteca Potosina Tamuantzán, Valles 31529
Museo de los Altos de Chiapas, San Cristóbal de las Casas 31325
Museo de Maní y Museo del Convento San Miguel Arcángel, Maní 31011
Museo de Mexticacan, Mexticacán 31163
Museo de Mier y Noriega, Mier y Noriega 31165
Museo de Minería e Historia de Cosalá, Cosalá 30803
Museo de Paracho, Paracho 31253
Museo de Pungarabato, Pungarabato 31298
Museo de Ruiz, Ruiz 31308
Museo de San Juan Raya, Zapotitlán Salinas 31592
Museo de San Juan Tlacotenco, San Juan Tlacotenco 31343
Museo de San Miguel, Tlaxiaco 31479
Museo de San Pablo, San Pablo Huitzo 31363
Museo de San Pedro, Navolato 31213
Museo de San Pedro, San Pedro Lagunillas 31369
Museo de Sayula Juan Rulfo, Sayula 31389
Museo de Susticacán, Susticacán 31393
Museo de Tacuichamona, Culiacán 30827
Museo de Tahdziú, Tahdziú 31396
Museo de Tapalpa, Tapalpa 31402
Museo de Taximaroa, Hidalgo 30934
Museo de Tepeticpa, Tepeticpa 31436
Museo de Texcalyacac, Texcalyacac 31452
Museo de Tlahualilo, Tlahualilo de Zaragoza 31465
Museo de Tlalancaleca, San Matías Tlalancaleca 31355
Museo de Tlapa, Tlapa de Comonfort 31471
Museo de Tlayacapan, Tlayacapan 31481

Museo de Totolapan, Totolapan 31506
Museo de Velardeña, Velardeña 31533
Museo de Xiutetelco, Xiutetelco 31564
Museo de Xochitepec, Xochitepec 31568
Museo de Xolalpancalli, Zacapoaxtla 31574
Museo de Yautepec, Yautepec 31571
Museo del Centro Cultural Joaquín Arcadio Pagaza, Valle de Bravo 31525
Museo del Centro Cultural La Cantera, Jonacatepec 30976
Museo del Centro Cultural Sor Juana Inés de la Cruz, Tepetlixpa 31437
Museo del Centro Histórico, San Juan del Río 31339
Museo del Cuale, Puerto Vallarta 31296
Museo del Ejidatario, Guasave 30923
Museo del Estado, Morelia 31199
Museo del Pueblo, Guanajuato 30918
Museo del Tequila, Tequila 31449
Museo del Valle de Tehuacán, Tehuacán 31420
Museo del Valle de Xico, Valle de Chalco 31526
Museo Dr. José Gómez Panaco, Balancán 30734
Museo el Centenario, Garza García 30873
Museo El Pariancito, San Luis Potosí 31350
Museo Escolar y Comunitario de Acamalín, Xico 31562
Museo Filemón Gutiérrez Ramírez, Ameca 30715
Museo Fin de la Tierra, Francisco I. Madero, Coahuila 30869
Museo Fuentes de Loreto y Guadelupe, Palmar de Bravo 31248
Museo Fundación Estero Beach, Ensenada 30860
Museo General Porfirio Rubio, Landa de Matamoros 30995
Museo Gonzalo Carrasco, Otumba 31239
Museo Hacienda San Pedro, General Zuazua 30877
Museo Hermenegildo Galeana, Tecpan de Galeana 31415
Museo Histórica Ex Convento de Tepoztlán, Tepoztlán 31448
Museo Histórico de Acapulco, Acapulco 30692
Museo Histórico de Ayotzinapa, Tixtla de Guerrero 31461
Museo Histórico de Guachinango, Guachinango 30881
Museo Histórico de la Ciudad, Torreón 31500
Museo Histórico de Múzquiz, Múzquiz 31207
Museo Histórico de Reynosa, Reynosa 31304
Museo Histórico de San Miguel de Allende, San Miguel de Allende 31356
Museo Histórico de San Nicolás de los Garza, San Nicolás de los Garza 31361
Museo Histórico Municipal, Ocotlán 31234
Museo Histórico Municipal, Taxco de Alarcón 31408
Museo Histórico Regional Ex Cuartel de la Compañía Fija, Ensenada 30861
Museo Huehuetla-tolli, Comala 30790
Museo Hunucmá de Sisal, Hunucmá 30952
Museo Irineo Germán, Azoyú 30731
Museo José María Luis Mora, Ocoyoacac 31235
Museo José y Tomás Chávez Morado, Silao 31391
Museo Juárez e Historia de Mapimí y Ojuela, Mapimí 31014
Museo Juarista de Congregación Hidalgo, Entronque Congregación Hidalgo 30862
Museo La Chole, Petatlán 31257
Museo La Flor de Jimulco, Torreón 31501
Museo La Pesca, Benito Juárez 30735
Museo Local de Acámbaro, Dolores Hidalgo 30836
Museo Local de Antropología e Historia de Compostela, Compostela 30797
Museo Local de Valle de Santiago, Valle de Santiago 31527
Museo Local Tuxteco, Santiago Tuxtla 31387
Museo Lucio Balderas Márquez, Landa de Matamoros 30996
Museo Manuel Ojinaga, Manuel Ojinaga 31012
Museo Metropolitano de Monterrey, Monterrey 31186
Museo Mohicca, Mohicca 31171
Museo Morelos, Carácuaro 30750
Museo Municipal, Mier 31164
Museo Municipal de Historia Malitzín de Huiloapan, Huiloapan de Cuauhtémoc 30948
Museo Municipal de Historia y Arqueología, González 30880
Museo Municipal IV Centenario, Pinos 31260
Museo Municipal Jesús González Herrera, Matamoros 31018
Museo Municipal Jesús H. Salgado, Teloloapan 31421
Museo Municipal José Reyes Meza, Altamira 30712
Museo Municipal Los Guachimontones y Museo de Arquitectura Prehispánica, Teuchitlán 31451
Museo Municipal Quetzalpapalotl, Teotihuacán 31428
Museo Nezahualcóyotl Acomiztli, Nezahualcóyotl 31215
Museo Ñiace, Tepelmeme de Morelos 31435
Museo Nuestro Pueblo y Museo Etnográfico de Tecpatán, Tecpatán 31416
Museo Parroquial, San Sebastián del Oeste 31371
Museo Peñón Blanco, Peñón Blanco 31256
Museo Petatlán, Petatlán 31258
Museo Peten Ak, Valladolid 31524
Museo Prof. Arturo Reyes Viramontes, Jalpa 30967
Museo Prof. Moisés Sáenz Garza, Apodaca 30721
Museo Prof. Ricardo Vega Noriega, San Ignacio 31332
Museo Raíces de Satevó, Satevó 31388
Museo Regional, San Juan Bautista Tuxtepec 31335
Museo Regional Comunitario Altepepialcalli, México 31144
Museo Regional Cuauhnáhuac, Cuernavaca 30819
Museo Regional de Acambay, Acambay 30688
Museo Regional de Aldama, Aldama 30710
Museo Regional de Arte e Historia, Jerez de García Salinas 30974
Museo Regional de Campeche, Campeche 30744
Museo Regional de Chiapas, Tuxtla Gutiérrez 31517
Museo Regional de Cuetzalan, Cuetzalán 30822
Museo Regional de Durango y Museo del Niño, Durango 30844
Museo Regional de Guanajuato Alhóndiga de Granaditas, Guanajuato 30921
Museo Regional de Guerrero, Chilpancingo 30776
Museo Regional de Hidalgo, Pachuca 31245
Museo Regional de Historia de Colima, Colima 30782
Museo Regional de Iguala, Iguala de la Independencia 30956
Museo Regional de la Laguna, Torreón 31503
Museo Regional de las Culturas Occidente, Guzmán 30926
Museo Regional de Mascota, Mascota 31016
Museo Regional de Nayarit, Tepic 31444
Museo Regional de Nuevo León, Monterrey 31187
Museo Regional de Puebla, Puebla 31289
Museo Regional de Querétaro, Querétaro 31302
Museo Regional de Río Verde, Río Verde 31306
Museo Regional de Sinaloa, Culiacán 30829
Museo Regional de Sonora, Hermosillo 30932
Museo Regional de Tierra Caliente, Coyuca de Catalán 30807
Museo Regional de Tlaxcala, Tlaxcala 31478
Museo Regional de Yucatán, Mérida 31029
Museo Regional del Valle del Fuerte, Ahome 30705
Museo Regional El Fuerte, El Fuerte 30846
Museo Regional Huasteco, Valles 31530
Museo Regional Lázaro Cárdenas, Tlapehuala 31472
Museo Regional Luciano Márquez, Zacatlán 31586
Museo Regional Michoacano, Morelia 31201
Museo Regional Potosino, San Luis Potosí 31353
Museo Regional Rescate de Raíces, Ahualulco de Mercado 30707
Museo Regional Rufino Muñiz Torres, Ocampo 31231
Museo Regional Valparaíso, Valparaíso 31531
Museo San José el Mogote, San José el Mogote 31334
Museo San Juan en la Historia, San Juan del Río 31341
Museo San Miguel Teotongo, México 31147
Museo Senen Mexic, Acatlán de Osorio 30694
Museo Tarike, Chihuahua 30772
Museo Taurino, Huamantla 30939
Museo Tepeapulco, Tepeapulco 31433
Museo Tetetzontlilco, Tizayuca 31463
Museo Tlallán, Tala 31397
Museo Tohue, Vicente Guerrero, Durango 31542
Museo Tomaskitla, Epazoyucan 30863
Museo Tomochic, Vicente Guerrero 31541
Museo Tonallán, Tonalá, Jalisco 31497
Museo Tonatiuh, Zempoala 31595
Museo Towi, Guachochi 30884
Museo Wakuatay, Playas de Rosarito 31262
Museo Xochiltepec, Magdalena 31010
Museo Xólotl, San Bartolo Tenayuca 31323
Museo Yoloxúchitl, Cuernavaca 30820
Museo Zacatecano, Zacatecas 31584
Museo Zoloxúchitl, Zacualpan 31588
Recinto Casa Benito Juárez, Saltillo 31319

Mongolia

Ulaanbaatar City Museum, Ulaanbaatar 31631

Montenegro

Etnološki Muzej, Budva 31639
J.U. Polimski Muzej, Berane 31637
Muzeji i Galerije Podgorica, Podgorica 31656
Narodni Muzej Crne Gore, Cetinje 31643
Zavičajni Muzej, Bar 31636
Zavičajni Muzej, Danilovgrad 31644
Zavičajni Muzej, Herceg-Novi 31646
Zavičajni Muzej, Nikšić 31650
Zavičajni Muzej, Pljevlja 31652
Zavičajni Muzej, Ulcinj 31659

Morocco

Musée Dar Belghazi, Km 17, Route de Kénitra 31673
Musée Saint-Euxpéry, Laâyoune 31675
Musée Sidi Mohamed Ben Abdellah, Essaouira 31668
Museo de Ceuta, Ceuta 31662

Mozambique

Fondaçao-Museu Chissano, Matola 31713
Museu de Nampula, Nampula 31714
Museum Municipal, Beira 31703

Myanmar

National Museum of Mandalay, Mandalay 31719
Pathein Museum, Pathein 31727
Rakhine State Museum, Sitture 31728
Shan State Museum, Taunggyi 31730
State Library and Museum, Mawlamyine 31723

Namibia

Gobabis Museum, Gobabis 31739
Grootfontein Museum Das Alte Fort, Grootfontein 31740
Helmeringhausen Museum, Helmeringhausen 31741
Keetmanshoop Museum, Keetmanshoop 31742
Kolmanskop Museum, Kolmanskop 31743
Lizauli Cultural Village, Lizauli 31744
Lüderitz Museum, Lüderitz 31745
Nakambale Museum, Ondangwa 31751
Omaruru Museum, Omaruru 31749
Ombalantu Baobab Tree, Ombalantu 31750
Outjo Museum, Outjo 31753
Rehoboth Museum, Rehoboth 31754
Sperrgebiet Museum, Oranjemund 31752
Swakopmund Museum, Swakopmund 31757
Tsumeb Museum, Tsumeb 31760
Walvis Bay Museum, Walvis Bay 31761

Netherlands

Aldfaers Erf Route, Allingawier 31797
Aldheidskeamer Uldrik Bottema, Aldeboarn 31791
Amerongs Historisch Museum/ Tabaksmuseum, Amerongen 31807
Amsterdams Historisch Museum, Amsterdam 31826
Archeologisch en Paleontologisch Museum Hertogsgemaal, Gewande 32248
Archeologisch Museum Haarlem, Haarlem 32293
De Baracquen Museum voor Bourtanger Bodemvondsten, Bourtange 31992
Biesbosch Museum, Werkendam 32937
Brabants Museum Oud Oosterhout, Oosterhout 32628
Breda's Museum, Breda 32003
Brouws Museum, Brouwershaven 32014
CODA Museum, Apeldoorn 31911
Cultuur-Historisch Museum Sorgdrager, Hollum 32399
Cultuurhistorisch Museum Valkenswaard, Valkenswaard 32868
Dongha Museum, Dongen 32142
Dorpsmuseum de Kleuskes, Liempde 32500
Dorpsmuseum De Kluis, Eext 32179
Dorpsmuseum in de Drie Snoeken, Oud-Gastel 32646
Driels Oudheidkundig Museum, Kerkdriel 32445
Edams Museum, Edam 32167
Enkhuizer Almanak Museum, Enkhuizen 32206
Epema-State, IJsbrechtum 32426
Gemeentelijk Historisch Museum Ouder-Amstel, Ouderkerk aan de Amstel 32655
Gemeentelijke Expositieruimte Kampen, Kampen 32436
Gemeentemuseum, Maassluis 32522
Gemeentemuseum, Weesp 32934
Gemeentemuseum Het Markiezenhof, Bergen op Zoom 31969
Gemeentemuseum 't Behouden Huijs, Terschelling-West 32818
Gemeentemuseum 't Sterkenhuis, Bergen 31967
Gemeentemuseum Weert - De Tiendschuur, Weert 32932
Het Gildenhuys, Blokzijl 31975
Goois Museum, Hilversum 32388
Goors Historisch Museum, Goor 32261
Gorcums Museum, Gorinchem 32263
Heemerf De Schutsboom, Goirle 32260
Heemkundemuseum Beek, Beek 31952
Heemkundig Museum, Ommeren 32621
Heemschuur Maasland, Heesch 32340
De Heksenwaag, Oudewater 32658
Historisch Museum Arnhem, Arnhem 31923
Historisch Museum De Bevelanden, Goes 32258
Historisch Museum De Tien Malen, Putten 32673
Historisch Museum Den Briel, Brielle 32011
Historisch Museum Deventer, Deventer 32120
Historisch Museum Hedel, Hedel 32327
Historisch Museum Hengelo, Hengelo, Overijssel 32367
Historisch Museum Het Kleine Veenlo, Veenendaal 32873
Historisch Museum Het Palthe Huis, Oldenzaal 32615
Historisch Museum Piet Dorenbosch, Boxtel 31996
Historisch Museum Ter Aar, Ter Aar 31786
Historisch Museum Tweestromenland, Beneden Leeuwen 31961
Historisch Museum Wolfheze, Wolfheze 32960
Historisch Muzeum Hazerswoude, Hazerswoude Dorp 32326
Historisch Openluchtmuseum Eindhoven, Eindhoven 32187
Hof van Hessen, Huissen 32420
Home of History, Rotterdam 32708
Huis Verwolde, Laren, Gelderland 32463
Huize Keizer, Denekamp 32115
Huizer Museum Het Schoutenhuis, Huizen 32422
Kapiteinshuis Pekela, Nieuwe Pekela 32576
Katwijks Museum, Katwijk aan Zee 32443
't Kiekhuus, Wolvega 32961
Kijk en Luister Museum, Bennekom 31962
De Klomphoek, Beilen 31956
Koornmarktspoort, Kampen 32438
Het Land van Strijen, Strijen 32809
Het Land van Thorn/ Panorama Thorn, Thorn 32820
Liemers Museum, Zevenaar 32999
Limburgs Miniatuurmuseum, Sint Geertruid 32772
Limburgs Museum, Venlo 32883
Limburgs Openluchtmuseum Eynderhoof, Nederweert Eind 32566
Loosduins Museum De Korenschuur, Den Haag 32079
Maarten van Rossum Museum, Zaltbommel 32987
Marker Museum, Marken 32540
Martinustoren, Losser 32510
Middendorpshuis, Den Ham 32107
Minimuseum, Groningen 32283
Molen van Frans Expositie, Mander (Tubbergen) 32537
Museum Ceuclum, Cuijk 32041
Museum de 5000 Morgen, Hoogeveen 32405
Museum De Koloniehof, Frederiksoord 32234
Museum De Ronde Venen, Vinkeveen 32889
Museum De Schilpen, Maasland 32521
Museum de Speeltoren, Monnickendam 32559
Museum de Stratemakerstoren, Nijmegen 32589
Museum de Striid tsjin it Wetter, Wommels 32963
Museum de Trije Gritenijen, Grou 32290
Museum Delfzijl, Delfzijl 32067
Museum Dorestad, Wijk bij Duurstede 32948
Museum Dorpsbehoud Papendrecht, Papendrecht 32666
Museum Echt, Echt 32165
Museum Elisabeth Weeshuis, Culemborg 32043
Museum Flehite Amersfoort, Amersfoort 31814
Museum Freriks, Winterswijk 32957
Museum Frerikshuus, Aalten 31785
Museum Gesigt van't Dok, Hellevoetsluis 32360
Museum Havezate Mensinge, Roden 32695
Museum Het Groot Graffel, Warnsveld 32923
Museum Het Petershuis, Gennep 32246
Museum Het Verscholen Dorp, Vierhouten 32888
Museum In 't Houten Huis, De Rijp 32049
Museum Kempenland, Eindhoven 32191
Museum Maarssen, Maarssen 32517
Museum Nagele, Nagele 32565
Museum Orvelte, Orvelte 32636
Museum Oud-Asperen, Asperen 31933
Museum Oud Overschie, Rotterdam 32714
Museum Oud-Rijnsburg, Rijnsburg 32687
Museum Oud Soest, Soest 32793
Museum Palthehof, Nieuwleusen 32582
Museum Rijswijk Het Tollenshuis, Rijswijk, Zuid-Holland 32690
Museum Schietkamp-Harskamp, Harskamp 32320
Museum Slag bij Heiligerlee, Heiligerlee 32347
Museum Stad Appingedam, Appingedam 31919
Museum Stedhûs Sleat, Sloten 32787
Museum Swaensteyn, Voorburg 32906
Museum 't Coopmanshûs, Franeker 32232
Museum van de Vrouw, Echt 32166
Museum voor Heemkunde Almelo, Almelo 31799
Museum Wierdenland, Ezinge 32229
Museum Willem van Haren, Heerenveen 32333
Museum Yerseke, Yerseke 32973
Museum Zwaantje Hans Stockman's Hof, Schoonebeek 32763
Museumboerderij, Leens 32469
Museumboerderij Eungs Schöppe, Markelo 32539
Museumboerderij Tante Jaantje, Callantsoog 32034
Museumboerderij Vreeburg, Schagen 32747
Museumboerderij Westfrisia, Hoogwoud 32407
Museumsboerderij Oud Noordwijk, Noordwijk, Zuid-Holland 32603
Museumwinkel Albert Heijn, Zaandam 32979
Muzee Scheveningen, Den Haag 32093
Nationaal Coöperatie Museum, Schiedam 32755
Het Nationaal Gevangenismuseum, Veenhuizen 32876
Natuur Historisch Museum en Heemkunde Centrum, Meerssen 32546
Noordbrabants Museum, 's-Hertogenbosch 32375
Het Noorderhuis, Zaandam 32981
Openluchtmuseum De Duinhuisjes, Rockanje 32693
De Oude Aarde, Giethoorn 32255

Het Oude Raadhuis, Oud-Beijerland 32645
Het Oude Raadhuis, Urk 32840
Het Oude Stadhuis, Hasselt 32321
Oudheidkamer, Den Burg 32068
Oudheidkamer, Koudekerk 32458
Oudheidkamer, Steenwijk 32803
Oudheidkamer, Wijngaarden 32949
Oudheidkamer Bleiswijk, Bleiswijk 31973
Oudheidkamer Boerderij Strunk, Raalte 32676
Oudheidkamer Brederwiede, Vollenhove 32904
Oudheidkamer Buisjan Enter, Enter 32216
Oudheidkamer Dantumadeel, Zwaagwesteinde 33021
Oudheidkamer de Oude Pastory, Heiloo 32348
Oudheidkamer Doorn, Doorn 32144
Oudheidkamer Eemnes, Eemnes 32175
Oudheidkamer Geervliet, Geervliet 32237
Oudheidkamer Hellevoetsluis, Hellevoetsluis 32362
Oudheidkamer Hoolt'n, Holten 32402
Oudheidkamer IJsselstreek, Wijhe 32947
Oudheidkamer Leiderdorp, Leiderdorp 32491
Oudheidkamer Lemster Fiifgea, Lemmer 32498
Oudheidkamer Leshuis, Dedemsvaart 32053
Oudheidkamer Lunteren, Lunteren 32512
Oudheidkamer/ Museum Mr. Andreae, Kollum 32454
Oudheidkamer Nunspeet, Nunspeet 32609
Oudheidkamer Renswoude, Renswoude 32682
Oudheidkamer Ridderkerk, Ridderkerk 32686
Oudheidkamer Rozenburg, Rozenburg 32736
Die Oudheidkamer tot Medenblick, Medemblik 32545
Oudheidkamer Wassenaar, Wassenaar 32928
Oudheidkamer Wieringerwaard het Polderhuis, Wieringerwaard 32945
Oudheidkundeige Verzameling, Schinveld 32758
Oudheidkundig Museum Arnemuiden, Arnemuiden 31921
Pasmans Huus, Ruinen 32738
Purmerends Museum, Purmerend 32672
Regthuis Oudkarspel, Oudkarspel 32659
't Rieuw, Nuis 32607
Rijssens Museum, Rijssen 32689
Roerstreekmuseum, Sint Odiliënberg 32774
De Schotse Huizen, Veere 32878
Slaait'n Hoes, Onstwedde 32623
Slag van de Somme Museum, Schagen 32749
Sliedrechts Museum, Sliedrecht 32782
Spakenburgs Museum 't Vurhuus, Bunschoten Spakenburg 32021
Stadsmuseum Doetinchem, Doetinchem 32138
Stadsmuseum Groenlo, Groenlo 32275
Stadsmuseum IJsselstein, IJsselstein 32428
Stadsmuseum Woerden, Woerden 32959
Stadsmuseum Zoetermeer, Zoetermeer 33007
Stedelijk Museum Amsterdam, Amsterdam 31888
Stedelijk Museum Drenthe's Veste, Coevorden 32038
Stedelijk Museum Het Domein, Sittard 32779
Stedelijk Museum Kampen, Kampen 32440
Stedelijk Museum Zutphen, Zutphen 33020
Stedelijke Oudheidkamer De Roos, Geertruidenberg 32236
Streekheemmuseum den Aanwas, Ossendrecht 32641
Streekhistorisch Centrum, Stadskanaal 32800
Streekmuseum De Groote Sociëteit, Tiel 32825
Streekmuseum de Moennik, Helden 32353
Streekmuseum De Vier Ambachten, Hulst 32423
Streekmuseum Goeree en Overflakkee, Sommelsdijk 32796
Streekmuseum Het Admiraliteitshuis, Dokkum 32140
Streekmuseum Het Dorp van Bartje, Rolde 32699
Streekmuseum Het Land van Axel, Axel 31943
Streekmuseum Jan Anderson, Vlaardingen 32891
Streekmuseum Land van Valkenburg, Valkenburg, Limburg 32866
Streekmuseum Leudal, Haelen 32307
Streekmuseum Oudheidkamer Reeuwijk, Reeuwijk 32679
Streekmuseum Stevensweert/ Ohé en Laak, Stevensweert 32808
Streekmuseum/Volkssterrenwacht Burgum, Burgum 32029
Streekmuseum West Zeeuws-Vlaanderen, IJzendijke 32429
Themapark Openluchtmuseum De Spitkeet, Harkema 32315
Torenmuseum, Goedereede 32257
De Turfschuur, Kolhorn 32453
Het Veenkoloniaal Museum, Veendam 32872
Veenmuseum het Vriezenveense Veld, Vriezenveen 32912
Veluws Museum Nairac, Barneveld 31951
Vestingmuseum Oudeschans, Oudeschans 32656
Visserij- en Cultuurhistorisch Museum, Woudrichem 32969
Het Vlaemsche Erfgoed, Groede 32274
Voerman Museum Hattem, Hattem 32324
Volendams Museum, Volendam 32902
Volksbuurtmuseum Wijk, Utrecht 32858
Vughts Historisch Museum, Vught 32916
Het Waalres Museum, Waalre 32917
Waterland Neeltje Jans, Vrouwenpolder 32913
Werkspoor Museum, Amsterdam 31901
Westfries Museum, Hoorn 32413
Zaanlandse Oudheidkamer Honig Breet Huis, Zaandijk 32986
Zaans Museum, Zaandam 32984
De Zaanse Schans, Zaandam 32985
Zandvoorts Museum, Zandvoort 32989
Zuiderzeemuseum, Enkhuizen 32208

New Caledonia

Musée Historique, Bourail 33031

New Zealand

Alexandra Museum, Alexandra 33034
Amuri Historical Museum, Waihu 33219
Aratoi Wairarapa Museum of Art and History, Masterton 33140
Ashburton Museum, Ashburton 33037
Auckland Museum and Auckland War Memorial Museum, Auckland 33043
Brain-Watkins House, Tauranga 33205
Cambridge Museum, Cambridge 33065
Catlins Historical Museum, Owaka 33167
Central Hawke's Bay Settlers Museum, Waipawa 33223
Chatham Islands Museum, Waitangi, Chatham Islands 33229
Cheviot Museum, Cheviot 33066
Clyde Historical Museum, Clyde 33081
Cobblestones Museum, Greytown 33116
Coromandel School of Mines Museum, Coromandel 33082
Cust Museum, Cust 33083
Dargaville Museum, Dargaville 33085
Eketahuna and Districts Museum, Eketahuna 33101
Far North Regional Museum, Kaitaia 33135
Featherston Heritage Museum, Featherston 33102
Featherston Memorabilia Museum, Featherston 33103
Firth Tower Historical Reserve, Matamata 33145
Flaxbourne Settlers Museum, Ward 33237
Forrester Gallery, Oamaru 33161
Founders Historic Park, Nelson 33152
Foxton Museum, Foxton 33107
Golden Bay Museum and Gallery, Golden Bay 33110
Gore Historical Museum and Hokonui Heritage Research Centre, Gore 33113
Havelock Museum, Havelock 33119
Hawke's Bay Museum, Napier 33150
Helensville and District Pioneer Museum, Helensville 33122
Hororata Historic Museum, Darfield 33084
Howick Historical Village, Pakuranga 33173
Hurworth Cottage, New Plymouth 33157
Journey's End Cottage and Laishley House, Auckland 33052
Kaikohe Pioneer Village and Museum, Kaikohe 33133
Kaikoura District Museum, Kaikoura 33134
Kapiti Coast Museum Waikanae, Wellington 33247
The Kauri Museum, Matakohe 33144
Kawhia Regional Museum Gallery, Kawhia 33136
Lyttelton Historical Museum, Christchurch 33074
Marlborough Provincial Museum, Blenheim 33063
Marton Historic Village Museum, Marton 33139
Mercury Bay and District Museum, Whitianga 33267
Morrin Museum, Morrinsville 33147
Motueka District Museum, Motueka 33148
Mount Bruce Pioneer Museum, Masterton 33141
Murchison Museum, Murchison 33149
Museum of Wellington, City and Sea, Wellington 33251
Nelson Provincial Museum, Nelson 33153
North Otago Museum, Oamaru 33162
Opotiki Museum, Opotiki 33164
Otago Settlers Museum, Dunedin 33096
Otakou Marae Museum, Dunedin 33098
Otautau and District Local History Museum, Otautau 33165
Oxford Museum, Oxford 33168
Paeroa and District Historical Society Museum, Paeroa 33171
Pahiatua and District Museum, Pahiatua 33172
Papakura and District Historical Society Museum, Papakura 33177
Pataka, Porirua 33186
Petone Settlers Museum, Wellington 33258
Picton Community Museum, Picton 33182
Pioneer Village, Silverdale 33201
Piopio and District Museum, Piopio 33183
Plains Vintage Railway and Historical Museum, Ashburton 33039
Pleasant Point Railway and Historical Society Museum, Pleasant Point 33184
Port Chalmers Museum, Port Chalmers 33188
Puhoi Historical Society Museum, Puhoi 33189
Raglan and District Museum, Raglan 33192
Rakiura Museum, Stewart Island 33202
Rangiora Museum, Rangiora 33193
Renwick Museum, Renwick 33194
Russell Museum, Russell 33200
South Canterbury Museum, Timaru 33212
South Taranaki District Museum, Patea 33180
Southland Museum and Art Gallery, Invercargill 33130
Sumner/Redcliffs Historical Museum, Christchurch 33078
Taieri Historical Museum, Outram 33166
Taihape and District Museum, Taihape 33203
Tairawhiti Museum, Gisborne 33109
Tauranga Museum, Tauranga 33206
Tawhiti Museum, Hawera 33121
Te Amorangi Trust Museum, Rotorua 33197
Te Awamutu Museum, Te Awamutu 33207
Te Kuiti and District Historical Society Museum, Te Kuiti 33208
Te Manawa Life Mind, Palmerston North 33176
Te Whare Taonga O Akaroa, Akaroa 33033
Te Whare Whakaaro o Pito-one, Hutt City 33127
Temuka Courthouse Museum, Temuka 33209
Teviot District Museum, Roxburgh 33198
Tokomairiro Historical Society Museum, Milton 33146
Tuatapere Bushmans Museum, Tuatapere 33216
Upper Waitaki Pioneer Museum and Art Gallery, Kurow 33138
Waiheke Island Historical Society Museum, Waiheke Island 33217
Waikato Museum of Art and History, Hamilton 33117
Waikouaiti Museum, Waikouaiti 33220
Waipukurau Museum, Waipukurau 33225
Wairoa Museum Kopututanga, Wairoa 33226
Wallace Early Settlers Museum, Riverton 33195
Warkworth and District Museum, Warkworth 33239
West Coast Historical Museum, Hokitika 33123
Whanganui Regional Museum, Wanganui 33235
Whangarei Museum, Whangarei 33266
Whangaroa County Museum, Kaeo 33131
Woodville Pioneer Museum, Woodville 33268
Wyndham Museum, Wyndham 33269

Nicaragua

Museo de Antropología e Historia de Rivas, Rivas 33301
Museo de Antropología y Municipal de Boaco, Juigalpa 33281
Museo de la Diócesis de Matagalpa, Mataglpa 33296
Museo del Archivo Histórico de Granada, Granada 33279
Museo del Pueblo y Museo Pre-Colombino de Chagüitillo, Chagüitillo 33272
Museo Histórico y Antropológico de Nindirí, Nindirí 33300

Niger

Musée Régional de Zinder, Zinder 33309

Nigeria

Aragon Museum, Ilupeju 33328
Borno State Museum, Maiduguri 33334
Didi Museum, Abuja 33312
Ethnographic Museum, Port Harcourt 33344
National Museum, Abuja, Abuja 33313
National Museum Bauchi, Bauchi 33317
National Museum Ibadan, Ibadan 33323
National Museum, Imo State, Owerri 33342
National Museum, Lagos, Onikan 33337
National Museum Makurdi, Makurdi 33335
National Museum Minna, Minna 33336
National Museum of Calabar, Calabar 33319
National Museum of Enugu, Enugu 33320
National Museum of Ilorin, Ilorin 33327
National Museum of the Ogun State, Abeokuta 33311
National Museum of the Ondo State, Akure 33314
National Museum Uyo, Uyo 33347
Odinani Museum, Awka 33316
Owo Museum, Owo 33343
University Museum, Port Harcourt 33345

Norfolk Island

Norfolk Island Museum, Kingston 33349

Norway

Ålen Bygdetun, Ålen 33379
Alta Museum, Alta 33384
Alvdal Museum, Alvdal 33385
Andrøymuseet, Andenes 33392
Arendal Bymuseum, Arendal 33394
Atnabrufossen Vannbruksmuseum, Atna 33408
Aust-Agder Museet, Arendal 33395
Austefjord Museum, Austefjorden 33410
Austråttborgen, Trondheim 33922
Ballangen Bygdemuseet, Ballangen 33415
Bamble Museum, Langesund 33624
Bautahaugen Samlinger, Hedalen 33541
Berg-Kragerø Museum, Kragerø 33607
Bjarkøy Bygdemuseum, Bjarkøy 33438
Bjugn Bygdatun, Bjugn 33442
Blakstad Sykehus Museum, Asker 33405
Blokkodden Villmarksmuseum, Engerdal 33483
Bokn Bygdemuseet, Bokn 33447
Bommen Elvemuseum, Vennesla 33970
Bomsholmen Museum, Arendal 33396
Borgarsyssel Museum, Sarpsborg 33818
Brandval Museum, Brandval 33453
Breidablikk, Stavanger 33865
Brevik Bymuseum, Brevik 33455
Burøsund Bygdemuseum, Vannareid 33958
Bymodellen, Oslo 33719
Bymuseet, Oslo 33720
Cudrio Kystmuseum, Langesund 33625
Dagali Museum, Geilo 33511
Dalane Folkemuseum, Egersund 33470
Dåsettunet, Lyngdal 33654
Dikemark Sykehus Museum, Asker 33406
Dølmo Bygdetun, Tolga 33908
Dønna Bygdesamling, Dønna 33465
Drammens Museum for Kunst og Kulturhistorie, Drammen 33467
Dyrøy Bygdemuseum, Tennevoll 33902
Eidskog Museum, Eidskog 33475
Eidsvoll Bygdemuseet, Eidsvoll Verk 33477
Eiktunet Kulturhistorisk Museum, Gjøvik 33513
Elingaard Museum, Gressvik 33518
Elvarheim Museum, Åmli 33387
Enebakk Bygdetun, Trysil 33933
Evje og Hornnes Museum, Evje 33485
Fedrenes Minne, Hidrasund 33547
Finnkroken Bygdemuseum, Tromsdalen 33917
Fjordane Forsvarsmuseum, Nordfjordeid 33700
Fjøsangersamlingene, Fjøsanger 33495
Flatanger Bygdemuseum, Flatanger 33496
Folldal Bygdetun, Folldal 33500
Forsand Bygdemuseum, Forsand 33504
Fosnes Bygdemuseum, Jøa 33583
Fredrikstad Museum, Fredrikstad 33505
Frosta Bygdemuseet, Frosta 33507
Frøya Bygdemuseum, Sistranda 33827
Gildeskål Bygdesamling, Gildeskål 33512
Gjesdal Bygdemuseum, Ålgård 33382
Godøy Kystmuseum, Godøy 33514
Grong Bygdamuseum, Grong 33521
Gruetunet Museum, Kirkenær i Solør 33591
Guovdageainnu Gilisillju, Kautokeino 33590
Hå Bygdemuseum, Varhaug 33962
Hamarøy Bygdetun, Harmarøy 33532
Hardanger Folkemuseum, Utne 33950
De Heibergske Samlinger - Sogn Folkemuseum, Kaupanger 33589
Hemne Bygdemuseum, Kyrksæterøra 33622
Herdla Museum, Herdla 33544
Herøy Bygdesamling, Herøy 33545
Holdhus Skulemuseum, Eikelandsosen 33479
Holmestova, Frekhaug 33506
Holmestrand Museum, Holmestrand 33553
Høvåg Museum, Høvåg 33567
Hurdal Bygdetun, Hurdal 33571
Indre Sør-Troms Distriktsmuseum, Tennevoll 33903
Iveland og Vegusdal Bygdemuseum, Vatnestrom 33966
Karmsund Folkemuseum, Haugesund 33538
Klæbu Bygdemuseum, Klæbu 33595
Klepp Bygdemuseum, Kleppe 33596
Klokkergården Bygdetun Lyngdal Misjonsmuseum, Lyngdal 33655
Krambuvika Bygdemuseum, Tennevoll 33904
Kystmuseet i Nord-Trøndelag, Rørvik 33797
Laudal Museum, Øyslebø 33763
Ledaal Museum, Stavanger 33866
Leirfjord Bygdesamlinger, Leirfjord 33630
Leksvik Bygdesamling, Leksvik 33632
Levanger Museum, Levanger 33635
Lier Bygdetun, Tranby 33914
Lista Museum, Vanse 33959
Lom Bygdamuseum, Lom 33645
Lund Bygdemuseum, Moi 33675
Måbødalen Kulturlandskapsmuseum, Eidfjord 33474
Marnardal Museum, Marnardal 33661
Meløy Bygdemuseum, Ørnes 33710
Midttunet, Skei i Jølster 33833
Miljømuseet Dokken, Haugesund 33539
Namsskogan Bygdatun, Namsskogan 33693
Nes Lensemuseum, Skiptvet 33837
Nord-Troms Museum, Sørkjosen 33852
Norddal Museum, Valldal 33957
Nordli Bygdemuseum, Nordli 33702
Nordmøre Museum, Kristiansund 33615
Nore og Uvdal Bygdetun, Uvdal i Numedal 33951
Norsk Kjøretøyhistorisk Museum, Lillehammer 33639
Norsk Tindemuseum, Åndalsnes 33390
Oddentunet, Os i Østerdalen 33712
Ofoten Museum - Museum Nord, Narvik 33696
Økomuseum Grenseland, Sarpsborg 33819
Øksnes Bygdemuseum, Alsvåg 33383
Orkdal Bygdemuseum, Svorkmo 33897
Osbane Museet, Fana 33489
Osen Bygdetun, Steinsdalen 33880
Petran Museum, Haltdalen 33527

Porsanger Museum, Lakselv 33623
Porsgrunn Bymuseum, Porsgrunn 33770
Rabekk Museum, Moss 33684
Rælingen Bygdetun, Fjerdingby 33494
Rana Museum, Mo i Rana 33667
Randaberg Bygdemuseum, Randaberg 33776
Rennesøy Bygdemuseum, Rennesøy 33780
Rindal Bygdemuseum, Rindal 33781
Rød Bygdetunet, Uskedalen 33947
Røldal Bygdemuseum, Røldal 33790
Rørosmuseet, Røros 33796
Røyrvik Bygdatun, Limingen 33641
Ryfylkemuseet, Sand 33809
Rygnestadtunet, Rysstad 33803
Salangen Bygdetun, Sjøvegan 33828
Saltdal Museum, Rognan 33789
Sandefjord Bymuseum, Sandefjord 33813
Sandnes Museum, Sandnes 33815
Sandtorg Bygdetun, Sørvik 33857
Sannidal Bygdetun, Sannidal 33817
Sauda Museum, Sauda 33821
Sekken Museum, Sekken 33822
Selbu Strikkemuseum, Selbu 33824
Singsås Museum, Singsås 33826
Sjukehusmuseet i Skien, Skien 33835
Skaun Bygdamuseum, Børsa 33449
Skiptvet Bygdemuseum, Skiptvet 33838
Smøla Museum, Smøla 33844
Snåsa Bygdamuseum, Snåsa 33845
Soknedal Bygdemuseum, Soknedal 33847
Sømna Bygdetun, Sømna 33850
Sør-Senja Museum, Stonglandseidet 33884
Sør-Troms Museum, Evenskjer 33484
Stavanger Museum, Stavanger 33875
Stokke Bygdetun, Stokke 33882
Straumsnes Bygdemuseum, Kanestraum 33586
Sund Bygdemuseum, Skogsvåg 33839
Sunndal Bygdemuseum, Sunndalsøra 33892
Sunnmøre Museum, Ålesund 33381
Surnadal Bygdemuseum, Surnadal 33893
Svalbard-Museum, Longyearbyen 33650
Svelvik Museum, Svelvik 33895
Tana Museum, Tana 33900
Time Bygdemuseum, Undheim 33946
Tørvikbygd Bygdemuseum, Tørvikbygd 33913
Toten Økomuseum, Kapp 33587
Træe, Bryne 33458
Tresfjord Museum, Tresfjord 33915
Trondarnes Distriktmuseum, Harstad 33535
Tune Bygdemuseum, Sarpsborg 33820
Tvedestrand Museum, Tvedestrand 33939
Tveit Bygdemuseum, Tveit 33940
Tveitens Samlinger, Eggedal 33473
Tydal Museum, Tydal 33941
Ullensaker Bygdemuseum, Jessheim 33580
Vadsø Museum - Ruija Kvenmuseum, Vadsø 33953
Varanger Museum Sør, Kirkenes 33593
Varden Redningsmuseum, Narbø 33694
Varteig Bygdemuseum, Varteig 33964
Vefsn Museum, Mosjøen 33683
Vega Bygdemuseum, Vega 33968
Vegårshei Bygdetun, Vegårshei 33969
Verran Museum, Malm 33656
Vingelen Kirke- og Skolemuseum, Vingelen 33981
Woksens Samlinger, Rørvik 33798
Yrjar Heimbygdslag, Opphaug 33708

Pakistan

Lahore Fort Museum, Lahore 34022
Lahore Museum, Lahore 34023
Peshawar Museum, Peshawar 34036
Sind Provincial Museum, Hyderabad 34005

Palestine

Armenian Museum, Jerusalem 34054
Darl At Tifl Museum, Jerusalem 34055
Municipality Museum, Hebron 34053

Panama

Museo de Herrera, Chitré-Herrera 34062
Museo de Penonomé, Penonomé 34078
Museo del Cristo Negro de Portobelo, Portobelo 34079
Museo José Domingo de Obaldía, David 34072

Papua New Guinea

Madang Museum, Madang 34084

Peru

Centro Andino de Tecnología Tradicional y Cultura de la Comunidades, Ollantaytambo 34282
Museo Antonio Raimondi, La Merced 34201
Museo Arqueológico e Histórico de Andahuaylas, Andahuaylas 34107
Museo Arqueológico Municipal y Museo Municipal Miguel Justino Ramírez, Piura 34290
Museo Carlos Dreyer, Puno 34298
Museo Catcco, Ollantaytambo 34283
Museo Comunitario de Villa El Salvador, Villa El Salvador 34325
Museo de Abancay, Abancay 34105
Museo de Arqueología e Historia Natural de Yungay, Yungay 34330
Museo de Etnohistoria de Suyo, Suyo 34309
Museo de la Reincorporación, Tacna 34311
Museo de Sitio de Huallamarca, Lima 34246
Museo de Sitio de la Pampa Galeras, Quinua 34301
Museo de Sitio del Mirador del Cerro San Cristóbal, Lima 34249
Museo Etnográfico, Cajamarca 34129
Museo General, Yanahuanca 34328
Museo Histórico Andrés Avelino Cáceres, Ayacucho 34121
Museo Histórico de Chacamarka, Chacamarka 34143
Museo Histórico Municipal Guillermo Zegarra Meneses, Arequipa 34113
Museo Histórico Regional de Tacna, Tacna 34314
Museo Monumental de la Municipalidad Distrital de Huaura, Huaura 34184
Museo Municipal, Iquitos 34194
Museo Municipal de Cabana, Cabana 34124
Museo Municipal de Chancay, Chancay 34145
Museo Municipal de Huaytará, Huaytará 34186
Museo Municipal de Nazca y Casa Museo María Reiche, Nazca 34281
Museo Orve, Chaupimascos 34146
Museo Parque de la Muralla, Lima 34268
Museo Regional Amazónico, Iquitos 34195
Museo Regional Daniel Hernández Morillo, Huancavelica 34175
Museo Regional de Arqueología, Huamanga 34174
Museo Regional de Casma Max Uhle, Casma 34139
Museo Regional de Ica María Reiche Gross Newman, Ica 34189
Museo Regional de San Martín, Moyobamba 34279
Museo Salesiano, Huancayo 34178
Museo Técnico Pedagógico de la Región de Pasco, San Juan 34302

Philippines

Aera Memorial Museum, San Pablo 34499
Asian Center Museum, Quezon City 34486
Aurelio Sevilla Alvero Library and Museum, Pasay 34470
Baguio-Mountain Provinces Museum, Baguio 34340
Benguet Museum, La Trinidad 34400
Bohol Museum, Tagbilaran 34513
Bolinao National Museum, Bolinao 34347
Bontoc Museum, Bontoc, Ifugao 34349
Butuan City Museum, Butuan 34352
Butuan National Museum, Butuan 34354
Cagayan Museum, Tuguegarao 34518
Casagwa National Museum, Daraga 34383
Cebu City Museum, Cebu 34368
Davao Museum, Davao 34385
Ford Museum, Dumalag 34387
Hiyas ng Bulakan Museum, Malolos 34422
Iligan Museum, Iligan 34388
Jabonga Parish Museum, Jabonga 34391
Kabayan National Museum, Kabayan 34395
Kiangan National Museum, Kiangan 34399
Laoag Museum, Vigan 34522
Legaspi City Museum, Legaspi 34402
Magsingal National Museum, Magsingal 34411
Museo de Tarlac, Tarlac 34516
Museo del Seminario Conciliar de Nueva Caceres, Naga 34466
Museo Iloilo, Iloilo 34389
Museo ng Batangas, Lipa 34404
Museo ng Katipunan, Lipa 34405
Museo ng Maynila, Manila 34446
The Negros Museum, Bacolod 34337
Padre Burgos National Museum, Vigan 34523
Palawan Museum, Puerto Princesa 34482
Palawan National Museum, Quezon 34484
Philippine Presidential Museum, Malacañang 34419
Pila Museum, Pila 34480
Quezonia Museum, Manila 34455
Santa Fe Community Museum, Santa Fe 34501
Southwestern University Museum, Cebu 34376
Teodoro P. Resurreccion Memorial Museum, Luna 34409
University of Northern Philippines Museum, Vigan 34524
Vargas Museum and Art Gallery, Mandaluyong 34427
Vigan House Museum, Vigan 34525
Xavier University Museum (Museo de Oro), Cagayan de Oro City 34358

Poland

Baszta Pranieńska, Chełmno 34589
Dom Eskenów, Toruń 35186
Dom Rodzinny Jana Kasprowicza, Inowrocław 34690
Dwór w Dołędze, Zaborów 35348
Historyczny Oddział Muzeum w Grudziądzu, Grudziądz 34683
Izba Muzealna im. Antoniego Krajewskiego, Lanckorona 34841
Izba Pamięci Żydów Lublina, Lublin 34899
Jan Kochanowski Muzeum, Czarnolas 34611
Klasztorna, Grudziądz 34684
Kolekcija Historyczno-regionalna Rodziny Makowskich, Chojnice 34597
Muzeum Biblioteki Kórnickiej, Kórnik 34756
Muzeum Bitwy nad Bzurą, Kutno 34838
Muzeum Chełmskie, Chełm 34587
Muzeum Dawnego Kupiectwa, Świdnica 35147
Muzeum Dom Urbańczyka, Chrzanów 34605
Muzeum Filumenistyczne, Bystrzyca Kłodzka 34583
Muzeum Historii Katowic, Katowice 34725
Muzeum Historii Katowic - Dzial Etnologii Miasta, Katowice 34726
Muzeum Historii Miasta, Zduńska Wola 35366
Muzeum Historii Miasta Łodzi, Łódź 34876
Muzeum Historii Miasta Lublina, Lublin 34902
Muzeum Historii Miasta Poznania, Poznań 35036
Muzeum Historii Przemysłu, Opatówek 34981
Muzeum Historii Ratusza i Trybunału Koronnego, Lublin 34903
Muzeum Historii Włocławka, Włocławek 35310
Muzeum Historyczne, Przasnysz 35053
Muzeum Historyczne Miasta Gdańska, Gdańsk 34638
Muzeum Historyczne Miasta Stołecznego Warszawy, Warszawa 35247
Muzeum Historyczne Miasta Tarnobrzega, Tarnobrzeg 35175
Muzeum Historyczne w Wrocławiu, Wrocław 35331
Muzeum Historyczno-Archeologiczne, Ostrowiec Świętokrzyski 34995
Muzeum im. Aleksandra Kłosińskiego, Kęty 34739
Muzeum im. Edmunda Bojanowskiego w Grabonogu, Grabonóg 34682
Muzeum im. Michała Kajki, Ogródek 34969
Muzeum im. Stanisława Staszica, Hrubieszów 34688
Muzeum im. Władysława St. Reymonta, Lipce Reymonotowskie 34857
Muzeum-Kaszubski Park Etnograficzny w Wdzydzach Kiszewskich, Wdzydze Kiszewskie 35294
Muzeum Kaszubskie, Kościerzyna 34758
Muzeum Kościuszkowskie Dworek Zacisze, Miechów 34915
Muzeum Kultury Łemkowskiej w Zyndranowej, Tylawa 35203
Muzeum Martyrologiczno-Historyczny, Biała Podlaska 34542
Muzeum Martyrologii, Bielsk Podlaski 34555
Muzeum Mazurskie, Ostrzeszów 34996
Muzeum Miasta Gdyni, Gdynia 34657
Muzeum Miasta i Rzeki Warty, Warta 35292
Muzeum Miasta Ostrowa Wielkopolskiego, Ostrów Wielkopolski 34994
Muzeum Miasta Zgierza, Zgierz 35370
Muzeum Miejskie, Świętochłowice 35150
Muzeum Miejskie im. Maksymiliana Chroboka, Ruda Śląska 35087
Muzeum Miejskie w Wadowicach, Wadowice 35209
Muzeum Miejskie Wrocławia, Wrocław 35332
Muzeum Nadwiślańskie, Kazimierz Dolny 34733
Muzeum Narodowe Ziemi Przemyskiej, Przemyśl 35055
Muzeum Okręgowe, Konin 34755
Muzeum Okręgowe, Krosno 34830
Muzeum Okręgowe, Nowy Sącz 34959
Muzeum Okręgowe, Ostrołęka 34993
Muzeum Okręgowe, Tarnów 35177
Muzeum Okręgowe w Lesznie, Leszno 34850
Muzeum Okręgowe w Sieradzu, Sieradz 35105
Muzeum Okręgowe w Tarnowie, Tarnów 35179
Muzeum Okręgowe w Toruniu, Toruń 35195
Muzeum Oręża Polskiego w Kołobrzegu, Kołobrzeg 34753
Muzeum Oświaty Ziemi Łódzkiej, Łódź 34880
Muzeum Pamięci Narodowej lata 1939 -1956, Kielce 34743
Muzeum Pienińskie im. Józefa Szalaya, Szczawnica 35155
Muzeum Podhalańskie PTTK, Nowy Targ 34963
Muzeum Południowego Podlasia, Biała Podlaska 34543
Muzeum Pomorza Środkowego, Słupsk 35112
Muzeum Pożarnictwa, Lidzbark 34854
Muzeum Puszczy Kampinoskiej, Kampinos 34714
Muzeum Regionalne, Barlinek 34537
Muzeum Regionalne, Brodnica 34570
Muzeum Regionalne, Brzeziny 34572
Muzeum Regionalne, Człuchów 34618
Muzeum Regionalne, Głogówek 34669
Muzeum Regionalne, Główno 34670
Muzeum Regionalne, Gostyń 34681
Muzeum Regionalne, Iłża 34689
Muzeum Regionalne, Jasło 34697
Muzeum Regionalne, Jawor 34699
Muzeum Regionalne, Kościan 34757
Muzeum Regionalne, Kozienice 34763
Muzeum Regionalne, Kutno 34839
Muzeum Regionalne, Lubań 34894
Muzeum Regionalne, Międzychód 34916
Muzeum Regionalne, Mielec 34920
Muzeum Regionalne, Nowe Miasto nad Pilicą 34952
Muzeum Regionalne, Nysa 34966
Muzeum Regionalne, Ojców 34971
Muzeum Regionalne, Opoczno 34983
Muzeum Regionalne, Pińczów 35009
Muzeum Regionalne, Pułtusk 35066
Muzeum Regionalne, Siedlce 35103
Muzeum Regionalne, Sławków 35109
Muzeum Regionalne, Środa Śląska 35126
Muzeum Regionalne, Wągrowiec 35210
Muzeum Regionalne, Wodzisław Śląski 35314
Muzeum Regionalne, Wronki 35345
Muzeum Regionalne, Słupca 35111
Muzeum Regionalne, Kraśnik 34824
Muzeum Regionalne, Łęczna 34844
Muzeum Regionalne, Myślibórz 34935
Muzeum Regionalne Dom Szewca, Pszczew 35059
Muzeum Regionalne im. Adama Fastnachta, Brzozów 34573
Muzeum Regionalne im. Albina Nowickiego, Ryn 35090
Muzeum Regionalne im. Antoniego Minkiewicza, Olkusz 34975
Muzeum Regionalne im. Dzieci Wrzesińskich, Września 35346
Muzeum Regionalne im. Hieronima Ławinczaka, Krotoszyn 34832
Muzeum Regionalne im. Marcina Rożka, Wolsztyn 35318
Muzeum Regionalne im. Władysława Kowalskiego, Dobczyce 34622
Muzeum Regionalne im Wojciechy Dutkiewicz, Rogoźno 35085
Muzeum Regionalne Kromerówka - Baszta Kowalska, Biecz 34554
Muzeum Regionalne PTTK, Puławy 35065
Muzeum Regionalne PTTK, Starachowice 35128
Muzeum Regionalne PTTK, Zagórze Śląskie 35353
Muzeum Regionalne w Sosnowcu, Sosnowiec 35123
Muzeum Regionalne w Stęszewie, Stęszew 35135
Muzeum Regionalne w Świebodzinie, Świebodzin 35149
Muzeum Regionalne Ziemi Limanowskiej, Limanowa 34856
Muzeum Regionalne Ziemi Sadowieńskiej, Sadowne 35095
Muzeum Regionalne Ziemi Sulmierzyckiej, Sulmierzyce 35139
Muzeum Rybołówstwa Morskiego, Świnoujście 35153
Muzeum Rzemiosł Ludowych - Zagroda Sitarska, Biłgoraj 34564
Muzeum Skansen Kolejnictwa, Kościerzyna 34759
Muzeum Skansen-Zagroda Pszczyńska, Pszczyna 35061
Muzeum Śląska Cieszyńskiego w Cieszynie, Cieszyn 34610
Muzeum Śląska Opolskiego, Opole 34986
Muzeum Śląskie, Katowice 34728
Muzeum Sopotu, Sopot 35120
Muzeum Społeczne, Strzyżów 35137
Muzeum Sportu i Turystyki Regionu Karkonoszy, Karpacz 34715
Muzeum Śremskie, Śrem 35125
Muzeum Szlachty Mazowieckiej, Ciechanów 34607
Muzeum Tatrzańskie im. Tytusa Chałubińskiego, Zakopane 35360
Muzeum Tkactwa Dolnośląskiego w Kamiennej Górze, Kamienna Góra 34712
Muzeum w Bielsku Podlaskim, Bielsk Podlaski 34556
Muzeum w Chorzowie, Chorzów 34604
Muzeum w Dobrzycy, Dobrzyca 34625
Muzeum w Elblągu, Elbląg 34628
Muzeum w Gliwicach, Gliwice 34666
Muzeum w Grudziądzu, Grudziądz 34685
Muzeum w Łęczycy, Łęczyca 34845
Muzeum w Rybniku, Rybnik 35089
Muzeum w Stargardzie, Stargard 35129
Muzeum w Szklarskiej Porębie Hoffmanna, Szklarska Poręba 35171
Muzeum w Tarnowskich Górach, Tarnowskie Góry 35180
Muzeum w Tykocinie, Tykocin 35202
Muzeum w Żywcu, Żywiec 35382
Muzeum Walk o Wał Pomorski przy Miejsko-Gminnym Ośrodku Kultury, Mirosławiec 34925
Muzeum Warmii i Mazur, Olsztyn 34978
Muzeum Warmińskie, Lidzbark Warmiński 34855
Muzeum Wartownia nr 1 na Westerplatte Sucharskiego, Gdańsk 34647
Muzeum Wnętrz Zabytkowych, Dębno 34621
Muzeum Woli, Warszawa 35281
Muzeum Zagłębia, Będzin 34538
Muzeum Zamek, Baranów Sandomierski 34534
Muzeum-Zamek w Bolkowie, Bolków 34569
Muzeum Zamek w Łańcucie, Łańcut 34842
Muzeum Ziemi Augustowskiej, Augustów 34531
Muzeum Ziemi Chełminskiej, Chełmno 34591
Muzeum Ziemi Dobrzyńskiej, Rypin 35091
Muzeum Ziemi Kłodzkiej, Kłodzko 34747
Muzeum Ziemi Kociewskiej, Starogard Gdański 35130
Muzeum Ziemi Krajeńskiej, Nakło nad Notecią 34937

Muzeum Ziemi Lubawskiej, Nowe Miasto Lubawskie ... 34951
Muzeum Ziemi Mińskiej, Mińsk Mazowiecki ... 34924
Muzeum Ziemi Mrągowskiej, Mrągowo ... 34930
Muzeum Ziemi Nadnoteckiej im. Wiktora Stachowiaka, Trzcianka ... 35199
Muzeum Ziemi Pałuckiej, Żnin ... 35378
Muzeum Ziemi Prudnickiej, Prudnik ... 35051
Muzeum Ziemi Puckiej, Puck ... 35063
Muzeum Ziemi Sochaczewskiej i Pola Bitwy nad Bzurą, Sochaczew ... 35119
Muzeum Ziemi Suskiej, Sucha Beskidzka ... 35138
Muzeum Ziemi Wałeckiej, Wałcz ... 35214
Muzeum Ziemi Wschowskiej, Wschowa ... 35347
Muzeum Ziemi Zaborskiej i Galeria Sztuki Ludowej, Wiele ... 35299
Muzeum Ziemi Złotowskiej, Złotów ... 35374
Muzeum Zofii i Wacława Nałkowskich, Wołomin ... 35317
Pałac w Rogalinie, Świątniki ... 35146
Pomorskie Muzeum Wojskowe, Bydgoszcz ... 34581
Prywatne Muzeum Etnograficzno-Historyczne, Bielsk Podlaski ... 34557
Regionalna Izba Muzealna, Poddębice ... 35016
Regionalne Muzeum Młodej Polski - Rydlówka, Kraków ... 34819
Stała Ekspozycja Muzealna Ziemi Sierakowickiej, Sierakowice ... 35106
Wielkopolskie Muzeum Pożarnictwa PTTK, Rakoniewice ... 35077
Zamek Piastowski, Gliwice ... 34667
Zbiory Historyczno-Etnograficzne, Oświęcim ... 34998

Portugal

Casa Memória d' Alte, Alte ... 35424
Casa Museu de Aljustrel, Fátima ... 35600
Casa Museu de Penacova, Penacova ... 35816
Casa Museu do Lorvão, Lorvão ... 35736
Casa-Museu Teixeira Lopes, Vila Nova de Gaia 36009
Centro de Exposição Rural da Urzelina, Urzelina ... 35976
Colecção Etnográfica da Câmara Municipal de Sobral de Monte Agraço, Sobral de Monte Agraço ... 35955
Cronomuseu de Queiriga, Queiriga ... 35877
Ecomuseu de Góis, Góis ... 35627
Museu a Casa do Povo, Ferro ... 35604
Museu Arqueológico e Etnológico António Tomás Pires, Elvas ... 35556
Museu Casa do Assento, Friande ... 35608
Museu Convento dos Lóios, Santa Maria da Feira ... 35902
Museu da Casa do Povo do Raminho, Raminho ... 35880
Museu da Casa Grande, Freixo de Numão ... 35607
Museu da Cidade, Almada ... 35414
Museu da Cidade, Coimbra ... 35531
Museu da Freguesia de Estômbar, Estômbar ... 35573
Museu da Graciosa, Santa Cruz da Graciosa ... 35897
Museu da Guarda, Guarda ... 35636
Museu da Margalha, Gavião ... 35625
Museu da Matriz, Ponta Delgada ... 35827
Museu das Flores, Santa Cruz das Flores ... 35898
Museu das Técnicas Rurais e Museu Municipal de Oliveira de Frades, Oliveira de Frades ... 35793
Museu de Afife, Afife ... 35384
Museu de Alhandra, Alhandra ... 35405
Museu de Angra do Heroísmo, Angra do Heroísmo ... 35439
Museu de Arte e Etnografia da Ribeira Chã, Ribeira Chã ... 35888
Museu de Aveiro, Aveiro ... 35447
Museu de Etnografia e História de Póvoa de Varzim, Póvoa de Varzim ... 35873
Museu de Francisco Tavares Proença Júnior, Castelo Branco ... 35515
Museu de Janes e Malveira, Alcabideche ... 35387
Museu de Lamego, Lamego ... 35656
Museu de Monforte, Monforte ... 35770
Museu de Monte Redondo, Monte Redondo ... 35773
Museu de Ovar, Ovar ... 35799
Museu de Santa Maria de Lamas, Santa Maria de Lamas ... 35907
Museu de São Jorge, Calheta de São Jorge ... 35499
Museu de Vila Verde de Ficalho, Vila Verde de Ficalho ... 36015
Museu do Barreiro e Museu Industrial da Quimiparque, Barreiro ... 35457
Museu do Campo da Lacam, Carapinheira ... 35507
Museu do Douro, Peso da Régua ... 35821
Museu do Ferro e da Região de Moncorvo, Torre de Moncorvo ... 35967
Museu do Forte de Santiago, Funchal ... 35614
Museu do Monte da Guerreira, Tavira ... 35959
Museu do Nordeste, Nordeste ... 35783
Museu e Arquivo Municipal, Lagoa ... 35649
Museu Etnogáfico e Arqueológico de Ourém, Ourém ... 35797
Museu Etnográfico, Santa Maria da Feira ... 35905
Museu Etnográfico da Arada, Arada ... 35441
Museu Etnográfico da Casa do Povo, Almeirim ... 35419
Museu Etnográfico da Pedreira, Pedreira ... 35811
Museu Etnográfico da Povoação, Povoação ... 35874
Museu Etnográfico da Praia de Mira, Praia de Mira ... 35875
Museu Etnográfico da Região do Vouga, Trofa 35975
Museu Etnográfico de Campelos, Campelos ... 35501
Museu Etnográfico do Cabouco, Cabouco ... 35490
Museu Etnográfico do Espinheiro, Espinheiro ... 35565
Museu Local do Pico da Pedra, Pico da Pedra 35824
Museu Marítimo e Regional de Ílhavo, Ílhavo ... 35648
Museu Monográfico do Bárrio e Campo Arqueológico, Bárrio ... 35458
Museu Monográfico do Fundão, Fundão ... 35622
Museu Municipal, Amadora ... 35430
Museu Municipal, Ferreira do Alentejo ... 35603
Museu Municipal, Sobralinho ... 35956
Museu Municipal Armindo Teixeira Lopes, Mirandela ... 35765
Museu Municipal Carmen Miranda, Marco de Canaveses ... 35755
Museu Municipal da Azambuja, Azambuja ... 35451
Museu Municipal da Chamusca, Chamusca ... 35522
Museu Municipal da Moita, Moita ... 35768
Museu Municipal da Ribeira Grande, Ribeira Grande ... 35890
Museu Municipal de Alcochete - Núcleo Sede, Alcochete ... 35399
Museu Municipal de Alvaiázere, Alvaiázere ... 35425
Museu Municipal de Avis, Avis ... 35449
Museu Municipal de Baião, Baião ... 35454
Museu Municipal de Bobadela, Bobadela ... 35469
Museu Municipal de Bombarral, Bombarral ... 35470
Museu Municipal de Caminha, Caminha ... 35500
Museu Municipal de Campo Maior, Campo Maior ... 35503
Museu Municipal de Castro Daire, Castro Daire ... 35516
Museu Municipal de Esposende, Esposende ... 35569
Museu Municipal de Estremoz, Estremoz ... 35578
Museu Municipal de Grândola, Grândola ... 35635
Museu Municipal de Loulé, Loulé ... 35737
Museu Municipal de Loures, Loures ... 35738
Museu Municipal de Mafra, Mafra ... 35751
Museu Municipal de Marvão, Marvão ... 35758
Museu Municipal de Montijo, Montijo ... 35778
Museu Municipal de Moura, Moura ... 35779
Museu Municipal de Olhão, Olhão ... 35790
Museu Municipal de Paços de Ferreira, Paços de Ferreira ... 35803
Museu Municipal de Palmela, Palmela ... 35806
Museu Municipal de Paredes, Paredes ... 35808
Museu Municipal de Paredes de Coura, Paredes de Coura ... 35809
Museu Municipal de Penafiel, Penafiel ... 35817
Museu Municipal de Penamacor, Penamacor ... 35819
Museu Municipal de Peniche, Peniche ... 35820
Museu Municipal de Pinhel, Pinhel ... 35825
Museu Municipal de Portalegre, Portalegre ... 35834
Museu Municipal de Portimão, Portimão ... 35835
Museu Municipal de Porto de Mós, Porto de Mós ... 35869
Museu Municipal de Santarém, Santarém ... 35909
Museu Municipal de Torres Vedras, Torres Vedras ... 35972
Museu Municipal de Vale de Cambra, Vale de Cambra ... 35977
Museu Municipal de Viana do Castelo, Viana do Castelo ... 35984
Museu Municipal de Vila Flor, Vila Flor ... 35995
Museu Municipal de Vila Franca de Xira, Vila Franca de Xira ... 35998
Museu Municipal de Vila Franca do Campo, Vila Franca do Campo ... 35999
Museu Municipal de Vouzela, Vouzela ... 36026
Museu Municipal Dias de Oliveira, Valongo ... 35980
Museu Municipal do Cadaval, Cadaval ... 35492
Museu Municipal do Crato, Crato ... 35554
Museu Municipal do Fundão, Fundão ... 35623
Museu Municipal do Sabugal, Sabugal ... 35892
Museu Municipal do Seixal, Seixal ... 35927
Museu Municipal Dr. António Gabriel Ferreira Lourenço, Benavente ... 35466
Museu Municipal Dr. João Calado Rodrigues, Mação ... 35745
Museu Municipal Dr. José Formosinho, Lagos 35651
Museu Municipal Dr. Santos Rocha, Figueira da Foz ... 35605
Museu Municipal Hipólito Cabaço, Alenquer ... 35404
Museu Municipal Prof. Álvaro Viana de Lemos, Lousã ... 35742
Museu Municipal Sobral Mendes, Penafiel ... 35818
Museu Regional Carlos Machado, Ponta Delgada ... 35829
Museu Regional de Arqueologia e Etnografia de Arganil, Arganil ... 35444
Museu Regional de Leiria, Leiria ... 35660
Museu Regional de Oliveira de Azeméis, Oliveira de Azeméis ... 35792
Museu Regional de Paredes de Coura, Paredes de Coura ... 35810
Museu Regional de Sintra, Sintra ... 35949
Museu Regional Manuel Nobre, Azambujeira ... 35452
Museu Restauraçao, Lisboa ... 35727
Museu Rural do Grupo dos Amigos de Estremoz, Estremoz ... 35580
Museus Municipal de Almada, Almada ... 35415
Núcleo Etnográfico de Cedros, Cedros ... 35519
Núcleo Museológico da Bemposta, Bemposta 35465
Núcleo Museológico da Junta de Freguesia da Orada, Borba ... 35474
Núcleo Museológico de Alverca, Alverca do Ribatejo ... 35427
Núcleo Museológico de Amieira do Tejo, Nisa 35781
Núcleo Museológico de Cachopo, Cachopo ... 35491
Núcleo Museológico de Espinho, Espinho ... 35566
Núcleo Museológico do Centro Cultural de Borba, Borba ... 35475
Núcleo Museológico do Lagido, São João ... 35919
Núcleo Museológico do Tempo, Santarém ... 35910
Núcleo Museológico Etnográfico José Ramires Saraiva, Ervidel ... 35562
Núcleo Museológico Medieval / Moderno, Almada ... 35416
Núcleo Rural de Ervidel, Ervidel ... 35563
Núcleos Museológicos da Lagoa, Lagoa ... 35650
Sala Museu Fialho de Almeida, Cuba ... 35555
Santiago do Cacém Museu Municipal, Santiago do Cacém ... 35912

Puerto Rico

Caparra Ruins Historical Museum, Guaynabo 36039
Casa Dr. José Celso Barbosa, Bayamón ... 36028
Fideicomisio de Conservación e Historia de Vieques, Vieques ... 36081
Fortaleza San Cristóbal, San Juan ... 36062
Hacienda Buena Vista, Ponce ... 36045
Museo Casa Blanca, San Juan ... 36065
Museo Casa Rosada Abelardo Díaz Alfaro, Caguas ... 36030
Museo de Arte, Historia y Cultura Casa Alonso, Vega Baja ... 36080
Museo de Coamo, Coamo ... 36036
Museo de la Historia de Ponce, Ponce ... 36050
Museo de San Juan, San Juan ... 36070
Museum of the Puerto Rican Family of the 19th Century, San Juan ... 36077

Qatar

Al Zubara Fort, Doha ... 36085

Réunion

Musée Historique de Saint-Gilles-les-Hauts, Saint-Paul ... 36095

Romania

Casa Memorială Crişan, Crişan ... 36330
Casa Memorială Elena Cuza, Soleşti ... 36633
Casa Memorială George Pop de Băseşti, Băseşti ... 36143
Casa Memorială Moş Ion Roată, Câmpuri ... 36257
Casa Memorială Nicolae Iorga, Botoşani ... 36160
Casa Memorială Theodor Bălăşel, Ştefăneşti, Vâlcea ... 36636
Casa-Muzeu Cristian, Cristian ... 36331
Casa-Muzeu Solca, Solca ... 36632
Castelul Regal Săvârşin, Săvârşin ... 36592
Cetatea Neamţului, Târgu Neamţ ... 36675
Cetatea Ţărănească Râşnov, Râşnov ... 36556
Colecţia Boiţa, Tălmaciu ... 36651
Colecţia de Etnografie şi Artă Populară, Buzău ... 36242
Colecţia de Etnografie şi Artă Populară, Chilia 36276
Colecţia de Istorie şi Arheologie, Piatra Olt ... 36523
Colecţia Etnografică, Bărbăteşti ... 36138
Colecţia Etnografică, Irimeşti ... 36430
Colecţia Muzeală, Beliu ... 36147
Colecţia Muzeală, Cotorca ... 36320
Colecţia Muzeală, Holod ... 36400
Colecţia Muzeală, Rimetea ... 36565
Colecţia Muzeală, Smeeni ... 36631
Colecţia Muzeală, Tărcaia ... 36652
Colecţia Muzeală Birchiş, Birchiş ... 36151
Colecţia Muzeală de Istorie, Săcueni ... 36578
Colecţia Muzeală Sătească, Cârligele ... 36265
Colecţia Muzeală Sătească, Coteşti ... 36318
Colecţia Muzeală Sătească, Gugeşti ... 36393
Colecţia Muzeală Sătească, Paltin ... 36510
Colecţia Muzeală Sătească, Vulcăneasa ... 36737
Colecţia Sătească, Atid ... 36115
Colecţia Sătească, Breaza ... 36182
Colecţia Sătească, Cârţa ... 36266
Colecţia Sătească, Ciumani ... 36284
Colecţia Sătească, Corund ... 36313
Colecţia Sătească, Dăneşti ... 36340
Colecţia Sătească, Gura Râului ... 36395
Colecţia Sătească, Joseni ... 36433
Colecţia Sătească, Lăpuş ... 36435
Colecţia Sătească, Lăzarea ... 36436
Colecţia Sătească, Lunca de Sus ... 36451
Colecţia Sătească, Lupeni ... 36453
Colecţia Sătească, Mereşti ... 36471
Colecţia Sătească, Muşăteşti ... 36485
Colecţia Sătească, Ocland ... 36494
Colecţia Sătească, Plăieşii de Jos ... 36527
Colecţia Sătească, Praid ... 36541
Colecţia Sătească, Remetea ... 36560
Colecţia Sătească, Sâncrăieni ... 36582
Colecţia Sătească, Sândominic ... 36583
Colecţia Sătească, Sărmaş ... 36587
Colecţia Sătească, Subcetate ... 36639
Colecţia Sătească, Suseni ... 36649
Colecţia Sătească, Tulgheş ... 36708
Colecţia Sătească, Voşlăbeni ... 36735
Colecţia Sătească, Zetea ... 36743
Complexul Muzeal Arad, Arad ... 36111
Complexul Muzeal Bucovina, Suceava ... 36642
Complexul Naţional Muzeal Curtea Domnească, Târgovişte ... 36656
Expoziţia de Arheologie şi Istorie Locală, Deta ... 36342
Expoziţia de Istorie a Culturii Nămăeşti, Câmpulung ... 36251
Expoziţia de Istorie Locală, Ciacova ... 36278
Expoziţia Muzeală, Turnu Roşu ... 36712
Expoziţia Muzeală de Etnografie, Cisnădie ... 36281
Expoziţia Permanentă de Etnografie, Pietroşiţa ... 36524
Expoziţia Sătescă, Stejaru ... 36637
Expoziţie Sătească, Lisa ... 36444
Expoziţia Sătească, Sericu ... 36598
Galeria Oamenilor de Seamă, Fălticeni ... 36365
Muzeul Banatului, Timişoara ... 36689
Muzeul Banatului Montan, Reşiţa ... 36562
Muzeul Brăilei-Secţia de Etnografie şi Artă Populară, Brăila ... 36171
Muzeul Bran, Bran ... 36173
Muzeul Câmpiei Băileştilor, Băileşti ... 36131
Muzeul Câmpiei Boianului, Drăgăneşti-Olt ... 36353
Muzeul Carpaţilor Răsăriteni, Sfântu Gheorghe 36600
Muzeul Casa Săsească, Livezile ... 36446
Muzeul Comunei Buteni, Buteni ... 36239
Muzeul de Artă şi Artă Populară, Calafat ... 36247
Muzeul de Etnografie, Botoşani ... 36163
Muzeul de Etnografie, Comarnic ... 36299
Muzeul de Etnografie, Topliţa ... 36700
Muzeul de Etnografie al Văii Teleajenului, Vălenii de Munte ... 36721
Muzeul de Etnografie Casa Băniei, Craiova ... 36325
Muzeul de Etnografie Gheorghe Cernea, Rupea ... 36576
Muzeul de Etnografie Săsească Emil Sigerus, Sibiu ... 36606
Muzeul de Etnografie şi Artă, Drăgăşani ... 36354
Muzeul de Etnografie şi Artă Bisericească, Fântânele ... 36369
Muzeul de Etnografie şi Artă Contemporană, Comăneşti ... 36298
Muzeul de Etnografie şi Artă Populară, Baia Mare ... 36126
Muzeul de Etnografie şi Artă Populară, Floreşti-Stoeneşti ... 36371
Muzeul de Istorie, Bicaz ... 36150
Muzeul de Istorie, Oneşti ... 36500
Muzeul de Istorie, Scorniceşti ... 36595
Muzeul de Istorie, Siret ... 36622
Muzeul de Istorie, Târgovişte ... 36658
Muzeul de Istorie, Târgu Jiu ... 36665
Muzeul de Istorie, Teremia Mare ... 36685
Muzeul de Istorie a Clisurii Dunării, Moldova Nouă ... 36481
Muzeul de Istorie a Oraşului Hârlău, Hârlău ... 36397
Muzeul de Istorie Augustin Bunea, Blaj ... 36155
Muzeul de Istorie, Etnografie şi Artă Plastică, Lugoj ... 36450
Muzeul de Istorie Gherla, Gherla ... 36383
Muzeul de Istorie Locală şi Etnografie, Brad ... 36166
Muzeul de Istorie şi Arheologie, Tulcea ... 36705
Muzeul de Istorie şi Etnografie, Târgu Neamţ ... 36676
Muzeul de Istorie Sighişoara, Sighişoara ... 36617
Muzeul de Istorie Turda, Turda ... 36709
Muzeul de Istorie Valea Hârtibaciului, Agnita 36101
Muzeul Dunării de Jos, Călăraşi ... 36248
Muzeul Etnografic, Săcele ... 36577
Muzeul Etnografic al Bucovinei, Suceava ... 36645
Muzeul Etnografic Ioan Tugui, Câmpulung Moldovenesc ... 36256
Muzeul Grăniceresc Năsăudean, Năsăud ... 36486
Muzeul Haáz Rezsö, Odorheiu Secuiesc ... 36497
Muzeul în Aer Liber, Negreşti-Oaş ... 36488
Muzeul Judeţean Bistriţa-Năsăud, Bistriţa, Năsăud ... 36152
Muzeul Judeţean Botoşani, Botoşani ... 36164
Muzeul Judeţean Buzău, Buzău ... 36244
Muzeul Judeţean de Artă şi Etnografie, Bacău ... 36121
Muzeul Judeţean de Istorie, Galaţi ... 36380
Muzeul Judeţean de Istorie şi Arheologie Iulian Antonescu Bacău, Bacău ... 36122
Muzeul Judeţean de Istorie şi Artă Zalău, Zalău ... 36742
Muzeul Judeţean de Istorie Teleorman, Alexandria ... 36107
Muzeul Judeţean Ialomiţa, Slobozia ... 36630

Muzeul Județean Maramureș, Baia Mare 36128
Muzeul Județean Mureș, Târgu Mureș 36669
Muzeul Județean Olt, Slatina 36628
Muzeul Județean Satu Mare, Satu Mare 36591
Muzeul Județean Ștefan cel Mare, Vaslui 36726
Muzeul Județean Teohari Antonescu, Giurgiu 36386
Muzeul Județean Vâlcea, Râmnicu Vâlcea 36553
Muzeul Maghiar, Bogdand 36157
Muzeul Maramureșului, Sighetu Marmației 36615
Muzeul Memorial Alexandru Ioan Cuza, Ruginoasa 36575
Muzeul Municipal, Huși 36408
Muzeul Municipal, Râmnicu Sărat 36549
Muzeul Municipal Câmpulung, Câmpulung 36252
Muzeul Municipal Curtea de Argeș, Curtea de Argeș 36336
Muzeul Municipal de Istorie, Roșiorii de Vede 36573
Muzeul Municipal Dej și Galeria de Artă, Dej 36341
Muzeul Municipal Ioan Raica Sebeș, Sebeș 36596
Muzeul Municipal Medias, Mediaș 36469
Muzeul Municipiului București, București 36213
Muzeul National Secuisec, Sfântu Gheorghe 36602
Muzeul Olteniei, Craiova 36328
Muzeul Orășanesc Fălticeni, Fălticeni 36368
Muzeul Orășenesc, Adjud 36097
Muzeul Orășenesc, Baraolt 36137
Muzeul Orășenesc, Carei 36264
Muzeul Orășenesc, Târnăveni 36678
Muzeul Orășenesc, Tășnad 36679
Muzeul Orășenesc, Tecuci 36680
Muzeul Orășenesc de Istorie, Aiud 36103
Muzeul Orășenesc Molnar Istvan, Cristuru Secuiesc 36332
Muzeul Orașului Avrig, Avrig 36117
Muzeul Regiunii Porților de Fier, Drobeta-Turnu Severin 36360
Muzeul Romanațiului, Caracal 36259
Muzeul Sătesc, Arcani 36114
Muzeul Sătesc, Bănia 36136
Muzeul Sătesc, Bărbătești 36139
Muzeul Sătesc, Bârca 36140
Muzeul Sătesc, Bechet 36145
Muzeul Sătesc, Castranova 36268
Muzeul Sătesc, Celaru 36271
Muzeul Sătesc, Cernătești 36273
Muzeul Sătesc, Cornu 36312
Muzeul Sătesc, Dănciulești 36337
Muzeul Sătesc, Drăcșenei 36352
Muzeul Sătesc, Dranic 36357
Muzeul Sătesc, Frătești 36375
Muzeul Sătesc, Ghighera 36385
Muzeul Sătesc, Goicea 36387
Muzeul Sătesc, Gornea 36388
Muzeul Sătesc, Grumezoaia 36392
Muzeul Sătesc, Lelești 36437
Muzeul Sătesc, Leu 36440
Muzeul Sătesc, Mehadica 36470
Muzeul Sătesc, Săliște 36579
Muzeul Sătesc, Tăcuta 36650
Muzeul Sătesc, Teiu 36681
Muzeul Sătesc, Valea Stanciului 36716
Muzeul Sătesc, Vulturești 36738
Muzeul Sătesc, Vutcani 36739
Muzeul Sătesc Constantin T. Cumpănașu, Orlea 36509
Muzeul Sătesc Prăjești, Traian 36701
Muzeul Satului Bănățean-Pădurea Verde, Timișoara 36692
Muzeul Satului din Bucovina, Suceava 36647
Muzeul Secuiesc al Ciucului, Miercurea-Ciuc 36472
Muzeul Tarii Beiusului, Beiuș 36146
Muzeul Țării Codrului, Valea Vinului 36718
Muzeul Țării Crișurilor, Oradea 36505
Muzeul Tării Oașului, Negrești-Oaș 36489
Muzeul Tarisznyás Márton, Gheorghieni 36382
Muzeul Vasile Pârvan, Bârlad 36142
Muzeul Viei și Vinului, Miniș 36475
Muzeul Viticol, Odobești 36496
Muzeul Vrancei, Focșani 36373
Parcul Etnografic Romulus Vuia, Cluj-Napoca 36293
Punct Muzeal, Balș 36135
Punct Muzeal, Bezdead 36149
Punct Muzeal, Bunești 36237
Punct Muzeal, Cața 36269
Punct Muzeal, Comana de Jos 36297
Punct Muzeal, Cuciulata 36333
Punct Muzeal, Fundata 36376
Punct Muzeal, Gostavățu 36390
Punct Muzeal, Maieru 36456
Punct Muzeal, Malu cu Flori 36459
Punct Muzeal, Mateiaș 36466
Punct Muzeal, Moțăieni 36483
Punct Muzeal, Pucheni 36542
Punct Muzeal, Râu Alb 36557
Punct Muzeal, Romoș 36571
Punct Muzeal, Rotbav 36574
Punct Muzeal, Sângeorz-Băi 36585
Punct Muzeal, Tărlungeni 36677
Punct Muzeal, Ulmi 36714
Punct Muzeal, Văleni 36719
Punct Muzeal, Viștea 36731
Punct Muzeal, Vlădeni 36732
Punct Muzeal Turnu Măgurele, Turnu Măgurele 36711

Russia

Adlerskij Kraevedčeskij Muzej - Muzej Istorii Adlerskogo Rajona, Soči 37781
Adygejskij Respublikanskij Kraevedčeskij Muzej, Majkop 37123
Aginskij Okružnoj Kraevedčeskij Muzej im. G. Cybikova, Aginskoe 36748
Altajskij Gosudarstvennyj Kraevedčeskij Muzej, Barnaul 36789
Altajskij Respublikanskij Kraevedčeskij Muzej im. A.V. Anochina, Gorno-Altajsk 36927
Amurskij Gorodskoj Kraevedčeskij Muzej, Amursk 36756
Amurskij Oblastnoj Kraevedčeskij Muzej im. G.S. Novikova-Daurskogo, Blagoveščensk 36809
Archangelskij Literaturnyj Muzej, Archangelsk 36766
Archangelskij Oblastnoj Kraevedčeskij Muzej, Archangelsk 36767
Architekturno-ètnografičeskij Muzej Angarskaja Derevnja, Bratsk 36820
Arzamasskij Istoriko-chudožestvennyj Muzej, Arzamas 36775
Astrachanskij Gosudarstvennyj Obedinennyj Istoriko- architekturnyj Muzej-zapovednik, Astrachan 36778
Astrachanskij Kraevedčeskij Muzej, Astrachan 36779
Astrachanskij Kreml, Astrachan 36780
Azovskij Kraevedčeskij Muzej, Azov 36786
Balachninskij Kraevedčeskij Muzej, Balachna 36787
Barabinskij Kraevedčeskij Muzej, Barabinsk 36788
Belgorodskij Gosudarstvennyj Istoriko-Kraevedčeskij Muzej, Belgorod 36794
Belinskij Rajonnyj Kraevedčeskij Muzej, Belinskij 36796
Belozerskij Istoriko-chudožestvennyj Muzej, Belozersk 36798
Bereznikovskij Istoriko-kraevedčeskij Muzej, Berezniki 36803
Bijskij Kraevedčeskij Muzej V.V. Bianki, Bijsk 36805
Bolšemuraškinskij Gosudarstvennyj Istoriko-chudožestvennyj Muzej, Bolšoe Muraškino 36815
Borisoglebskij Kraevedčeskij Muzej, Borisoglebsk 36816
Bratskij Gorodskoj Objedinennyj Muzej Istorii Osvoenija Angary, Bratsk 36822
Brjanskij Gosudarstvennyj Obedinennyj Kraevedčeskij Muzej, Brjansk 36824
Buguruslanskij Istoriko-kraevedčeskij Muzej, Buguruslan 36830
Bujnakskij Istoriko-kraevedčeskij Muzej, Bujnaksk 36831
Buzulukskij Gorodskoj Istoriko-kraevedčeskij Muzej, Buzuluk 36832
Čajkovskij Kraevedčeskij Muzej, Čajkovskij 36834
Čeljabinskij Oblastnoj Kraevedčeskij Muzej, Čeljabinsk 36843
Čerdynskij Kraevedčeskij Muzej A.S. Puškina, Čerdyn 36846
Čerkechskij Muzej Jakutskaja Političeskaja Ssylka, Čerkech 36853
Chabarovskij Gosudarstvennyj Muzej Dalnego Vostoka im. N.I. Grodekova, Chabarovsk 36856
Chakasskij Respublikanskij Kraevedčeskij Muzej, Abakan 36745
Chvalynskij Kraevedčeskij Muzej, Chvalynsk 36865
Čitinskij Oblastnoj Kraevedčeskij Muzej im A.K. Kuznecova, Čita 36869
Čuvašskij Nacionalnyj Muzej, Čeboksary 36837
Dagestanskij Gosudarstvennyj Objedinennyj Istoriko-architekturnyj Muzej, Machačkala 37115
Diorama Kurskaja Bitva - Belgorodskoe Napravlenie, Belgorod 36795
Dobrjanskij Istoriko-kraevedčeskij Muzej, Dobrjanka 36877
Dom-muzej N.N. Chochrjakova, Kirov 37035
Dom-muzej Dekabristov, Kurgan 37090
Dom-muzej M.B. Grekova, Novočerkassk 37477
Dom-muzej im. T.N. Granovskogo, Orël 37504
Dom-muzej E.M. Jaroslavskogo, Jakutsk 36959
Dom-muzej V.I. Lenina, Kazan 37005
Dom-muzej V.I. Lenina, Samara 37614
Dom-muzej M.K. Ammosova, Jakutsk 36960
Dom-muzej N.S. Muchina, Joškar-Ola 36976
Dom-muzej V.A. Rusanova, Orël 37506
Dom-muzej Semji Lopatinych, Stavropol 37802
Domik Petra I, Sankt-Peterburg 37635
Donskoj Istoriko-kraevedčeskij Muzej, Donskoe 36879
Dzeržinskij Gorodskoj Kraevedčeskij Muzej, Dzeržinsk 36883
Èkspozicija "Ivanovskaja starina", Ivanovo 36946
Elabužskij Gosudarstvennyj Istoriko-architekturnyj i Chudožestvennnyj Muzej-zapovednik, Elabuga 36903
Enisejskij Kraevedčeskij Muzej, Enisejsk 36915
Èssentukskij Gorodskoj Kraevedčeskij Muzej im. V.P. Špakovskogo, Èssentuki 36917
Ètnografičeskij Muzej pod Otkrytym Nebom Torum Maa, Chanty-Mansijsk 36860
Gatčinskij Literaturno-memorialnyj Muzej-usadba P.E. Ščerbova, Gatčina 36919
Glazovskij Kraevedčeskij Muzej, Glazov 36924
Gorodskoj Kraevedčeskij Muzej, Elec 36910
Gorodskoj Muzej Boevoj Slavy, Kaluga 36995
Gosudarstvennaja OK Istoriko-ètnografičeskij i Architekturnyj Muzej-zapovednik Staraja Sarepta, Volgograd 37954
Gosudarstvennyj Istoričeskij Zapovednik Gorki Leninskie, Gorki Leninskie 36926
Gosudarstvennyj Istoriko-architekturnyj i Chudožestvennyj Muzej-zapovednik Drevnij Derbent, Derbent 36874
Gosudarstvennyj Istoriko-architekturnyj i Prirodno-landšaftnyj Muzej Zapovednik Izborsk, Izborsk 36953
Gosudarstvennyj Istoriko-ètnografičeskij Muzej-zapovednik Šušenskoe, Šušenskoe 37811
Gosudarstvennyj Istoriko-memorialnyj Zapovednik Rodina V.I. Lenina, Uljanovsk 37897
Gosudarstvennyj Muzej Istorii Sankt-Peterburga, Sankt-Peterburg 37647
Gosudarstvennyj Muzej Vyborgskij Zamok, Vyborg 37982
Gosudarstvennyj Muzej-zapovednik S.A. Esenina, Konstantinovo 37056
Gosudarstvennyj Okružnoj Muzej Prirody i Čeloveka Goroda Chanty-Mansijsk, Chanty-Mansijsk 36862
Igarskij Kraevedčeskij Kompleks Muzej Večnoj Merzloty, Igarka 36930
Ingušskij Gosudarstvennyj Muzej Kraevedenija im. T.Ch. Malsagova, Nazran 37448
Irkutskij Oblastnoj Kraevedčeskij Muzej, Irkutsk 36933
Iskitimskij Kraevedčeskij Muzej, Iskitim 36941
Istoričeskij Muzej Dom na Naberežnoj, Moskva 37218
Istoriko-chudožestvennyj Muzej, Uglič 37891
Istoriko-ètnografičeskij Muzej Usadba Galskich, Čerepovec 36849
Istoriko-Kraevedčeskij Muzej, Ivanovo 36947
Istoriko-kraevedčeskij Muzej g. Kronštadta, Kronštadt 37084
Istoriko-kulturnyj Muzej-zapovednik Idnakar, Glazov 36925
Istoriko-memorialnyj i Architekturno-chudožestvennyj Muzej, Tichvin 37842
Jakutskij Gosudarstvennyj Muzej Istorii i Kultury Narodov Severa im. Jaroslavskogo, Jakutsk 36962
Jamalo-Neneckij Okružnoj Kraevedčeskij Muzej im. I.S. Šemanovskogo, Salechard 37608
Jaroslavskij Istoriko-architekturnyj i Chudožestvennyj Muzej-zapovednik, Jaroslavl 36968
Jaroslavskij Muzej Boevoj Slavy, Jaroslavl 36969
Jurjeveckij Kraevedčeskij Muzej, Jurjevec 36983
Kaliningradskij Oblastnoj Istoriko-chudožestvennyj Muzej, Kaliningrad 36989
Kaliningradskij Oblastnoj Kraevedčeskij Muzej, Kaliningrad 36990
Kalužskij Oblastnoj Kraevedčeskij Muzej, Kaluga 36998
Kamčatskij Oblastnoj Kraevedčeskij Muzej, Petropavlovsk-Kamčatskij 37553
Kamčatskij Voenno-istoričeskij Muzej, Petropavlovsk-Kamčatskij 37554
Kargopolskij Gosudarstvennyj Istoriko-Architekturnyj i Chudožestvennyj Muzej, Kargopol 37003
Kemerovskij Istoriko-architekturnyi Muzej Krasnaja Gorka, Kemerovo 37030
Kemerovskij Oblastnoj Kraevedčeskij Muzej, Kemerovo 37031
Kirovskij Oblastnoj Kraevedčeskij Muzej, Kirov 37039
Kislovodskij Istoriko-kraevedčeskij Muzej Krepost', Kislovodsk 37046
Kizljarskij Kraevedčeskij Muzej im. P.I. Bagrationa, Kizljar 37051
Klinskoe Muzejnoe Obedinenie, Klin 37053
Komi-Permjackij Okružnoj Kraevedčeskij Muzej im. Subbotina-Permjaka, Kudymkar 37088
Komi Respublikanskij Istoriko-Kraevedčeskij Muzej, Syktyvkar 37815
Korablinkskij Kraevedčeskij Muzej, Korablino 37057
Kraevedčeskij Muzej Evrejskoj Avtonomnoj Oblasti, Birobidžan 36807
Kraevedčeskij Muzej g. Lomonosova, Lomonosov 37110
Kraevedčeskij Muzej g. Puškina, Puškin 37580
Kraevedčeskij Muzej g. Puškino, Puškino 37584
Kraevedčeskij Muzej Novgorodskogo Rajona, Velikij Novgorod 37919
Kraevoj Muzej Istorii Literatury, Iskusstva i Kultury Altaja, Barnaul 36791
Krasnojarskij Istoričeskij Muzej Memorial Pobedy, Krasnojarsk 37071
Krasnojarskij Kraevoj Kraevedčeskij Muzej, Krasnojarsk 37072
Krasnojarskij Kulturno-istoričeskij Muzejnyj Kompleks, Krasnojarsk 37073
Krasnokamskij Kraevedčeskij Muzej, Krasnokamsk 37079
Krasnoturinskij Kraevedčeskij Muzej, Krasnoturinsk 37080
Kungurskij Kraevedčeskij Muzej, Kungur 37089
Kurganskij Oblastnoj Kraevedčeskij Muzej, Kurgan 37092
Kurskij Gosudarstvennyj Oblastnoj Kraevedčeskij Muzej, Kursk 37094
Lgovskij Kraevedčeskij Muzej, Lgov 37101
Lipeckij Oblastnoj Kraevedčeskij Muzej, Lipeck 37104
Literaturnyj Muzej im. K.V. Ivanova, Čeboksary 36838
Magnitogorskij Kraevedčeskij Muzej, Magnitogorsk 37122
Malojaroslaveckij Istoriko-kraevedčeskij Muzej, Malojaroslavec 37126
Malojaroslaveckij Muzej Voennoj Istorii 1812 Goda, Malojaroslavec 37127
Memorialnyj Dom-muzej Podpolnaja Tipografija Permskogo Komiteta RSDRP 1906 g., Perm 37545
Memorialnyj Dom-Muzej I.I. Šiškina, Elabuga 36904
Memorialnyj Dom-muzej Uljanovych, Astrachan 36782
Memorialnyj Muzej Jurija Gagarina, Gagarin 36918
Memorialnyj Muzej-kabinet M.V. Frunze, Ivanovo 36949
Memorialnyj Muzej-kvartira M.S. Spiridonova, Čeboksary 36839
Michajlovskij Istoričeskij Muzej, Michajlov 37131
Mineralovodskij Kraevedčeskij Muzej, Mineralnye Vody 37132
Minusinskij Regionalnyj Kraevedčeskij Muzej im. N.M. Martjanova, Minusinsk 37136
Murmanskij Oblastnoj Kraevedčeskij Muzej, Murmansk 37434
Muzej 'A Muzy ne Molčali', Sankt-Peterburg 37669
Muzej A.V. Suvorova, Velikij Novgorod 37920
Muzej Alpinizma, Turizma i Istorii Kurorta Teberda, Teberda 37839
Muzej Arsenal - Muzej Velikoj Otečestvennoj Vojny, Voronež 37971
Muzej Boevogo i Trudovogo Podviga 1941-1945, Saransk 37741
Muzej Boevoj Slavy, Astrachan 36783
Muzej Butyrskoj Tjurmy, Moskva 37263
Muzej V.I. Čapaeva, Čeboksary 36840
Muzej Cerkov Dekabristov, Čita 36870
Muzej Dekabristov, Minusinsk 37137
Muzej Dekabristov - Dom Trubeckogo, Irkutsk 36935
Muzej Ètnografii, Istorii i Kultury Narodov Baškortostana, Ufa 37888
Muzej Goroda Jurjevca, Jurjevec 36984
Muzej Goroda Penzy, Penza 37531
Muzej Gorodskaja Duma, Tjumen 37846
Muzej Gradostroitelstva i Byta Taganroga, Taganrog 37825
Muzej Istorii Aksaja, Aksaj 36749
Muzej Istorii Ekaterinburga, Ekaterinburg 36893
Muzej Istorii Goroda Irkutska, Irkutsk 36937
Muzej Istorii Goroda Joškar-Oly, Joškar-Ola 36978
Muzej Istorii Goroda-Kurorta Soči, Soči 37783
Muzej Istorii Goroda Naberežnye Čelny, Naberežnye Čelny 37437
Muzej Istorii Goroda Šlisselburga, Šlisselburg 37768
Muzej Istorii Goroda Stavropolja, Stavropol 37804
Muzej Istorii Goroda Tjumeni, Tjumen 37848
Muzej Istorii i Kultury Goroda Votkinska, Votkinsk 37978
Muzej Istorii i Kultury Moskovskogo Rajona, Nižnij Novgorod 37455
Muzej Istorii i Kultury Srednego Prikamja, Sarapul 37746
Muzej Istorii Izučenija i Osvoenija Evropejskogo Severa, Apatity 36764
Muzej Istorii Jaroslavlja, Jaroslavl 36972
Muzej Istorii Kolskich Saamov, Lovozero 37113
Muzej Istorii Kraja, Belozersk 36801
Muzej Istorii Moskovskogo Gosudarstvennogo Universiteta im. M.V. Lomonosova, Moskva 37291
Muzej Istorii Moskvy, Moskva 37294
Muzej Istorii Orenburga, Orenburg 37513
Muzej Istorii Permskogo Rajona, Perm 37549
Muzej Istorii Političeskoj Ssylki, Bratsk 36823
Muzej Istorii Politssylki v Jakutii, Jakutsk 36963
Muzej Istorii Priokskogo Rajona, Nižnij Novgorod 37459
Muzej Istorii Rossijskogo Gosudarstvennogo Pedagogičeskogo Universiteta im A.I. Gercena, Sankt-Peterburg 37685
Muzej Istorii Rossijskogo Predprinimatelstva, Moskva 37297
Muzej Istorii Teatrov Dagestana, Machačkala 37118
Muzej Konspiritavnaja Kvartira Simbirskoj Gruppy RSDRP 1904-1906, Uljanovsk 37902
Muzej Kraevedenija g. Elabuga, Elabuga 36905
Muzej Kronštadtskaja Krepost, Kronštadt 37086
Muzej Kultury Astrachani, Astrachan 36785
Muzej Kupečeskogo Byta, Ekaterinburg 36896

Muzej-kvartira A.M. Gorkogo, Nižnij Novgorod 37460
Muzej Lefortovo, Moskva 37326
Muzej Literatury Burjatii im. Choca Namsaraeva, Ulan-Udė 37895
Muzej Meterologičeskaja Stancija Simbirska, Uljanovsk 37903
Muzej Mologskogo Kraja, Rybinsk 37605
Muzej Narodnoje Obrazovanie Simbirskoj Gubernii v 70-80 gody 19 veka, Uljanovsk 37905
Muzej Nekropol Demidovych i Vystavočnyj Zal Tulskij Metall, Tula 37869
Muzej Nižegorodskij Ostrog, Nižnij Novgorod 37463
Muzej Nižegorodskoj Intelligencii, Nižnij Novgorod 37464
Muzej Oktjabrskogo Vooružennogo Vosstanija, Voronež 37973
Muzej Predprinimatelej, Mecenatov i Blagotvoritelej, Moskva 37349
Muzej Prioratskij Dvorec, Gatčina 36921
Muzej Vladimira Raevskogo, Bogoslovka 36813
Muzej Russko-Armjanskoj Družby, Rostov-na-Donu 37601
Muzej Simbirskaja Čuvašskaja Škola - Kvartira I.Ja. Jakovleva, Uljanovsk 37906
Muzej Simbirskaja Klassičeskaja Gimnazija, Uljanovsk 37907
Muzej Smolenščina v Gody Velikoj Otečestvennoj Vojny 1941 -1945, Smolensk 37772
Muzej Sovremennoj Istorii i Kultury Severnoj Osetii, Vladikavkaz 37931
Muzej Uezdnogo Goroda, Čistopol 36867
Muzej-usadba N.A. Durovoj, Elabuga 36906
Muzej-zapovednik Dmitrovskij Kreml, Dmitrov 36876
Muzejno-vystavočnyj Centr, Ivanovo 36952
Muzejno-vystavočnyj Centr - Diorama, Kirov 37042
Muzejno-vystavočnyj Centr Tulskie Drevnosti, Tula 37872
Muzejno-vystavočnyj Kompleks Volokolamskij Kreml, Volokolamsk 37966
Muzejnyj Centr Nasledie Čukotki, Anadyr 36757
Nacionalnyj Muzej Kabardino-Balkarskoj Respubliki, Nalčik 37442
Nacionalnyj Muzej Respubliki Marij Ėl im. T. Evseeva, Joškar-Ola 36980
Nacionalnyj Muzej Respubliki Tatarstan, Kazan 37026
Nacionalnyj Muzej Udmurtskoj Respubliki im. K. Gerda, Iževsk 36955
Narodnyj Istoriko-kraevedčeskij Muzej Tatarska, Tatarsk 37838
Narovčatskij Rajonnyj Kraevedčeskij Muzej, Narovčat 37446
Neneckij Okružnoj Kraevedčeskij Muzej, Narjan-Mar 37443
Nižegorodskij Gosudarstvennyj Istoriko-architekturnyj Muzej-zapovednik, Nižnij Novgorod 37468
Nižnetagilskij Muzej-zapovednik Gornozavodskoj Ural, Nižnij Tagil 37473
Novokuzneckij Kraevedčeskij Muzej, Novokuzneck 37482
Novorossijskij Gosudarstvennyj Istoričeskij Muzej Zapovednik, Novorossijsk 37483
Novotroickij Istoriko-kraevedčeskij Muzej, Novotroick 37491
Oblastnoj Kraevedčeskij Muzej, Ekaterinburg 36900
Oloneckij Nacionalnyj Muzej Karelov-Livvikov im. N.T.Prilukin, Olonec 37496
Omskij Gosudarstvennyj Istoriko-kraevedčeskij Muzej, Omsk 37499
Orenburgskij Oblastnoj Kraevedčeskij Muzej, Orenburg 37514
Orlovskij Oblastnoj Kraevedčeskij Muzej, Orël 37511
Orskij Istoriko-kraevedčeskij Muzej, Orsk 37516
Osinskij Kraevedčeskij Muzej, Osa 37517
Penzenskij Gosudarstvennyj Obedinennyj Kraevedčeskij Muzej, Penza 37537
Pjatigorskij Kraevedčeskij Muzej, Pjatigorsk 37565
Pskovskij Gosudarstvennyj Istoričesko-architekturnyj i Chudožestnennyj Muzej-zapovednik, Pskov 37575
Pskovskij Memorialnyj Muzej V.I. Lenina, Pskov 37576
Pustozerskij Kompleksnyj Istoriko-prirodnyj Muzej, Narjan-Mar 37444
Roslavlskij Istoriko-chudožstvennyj Muzej, Roslavl 37597
Samarskij Oblastnoj Istoriko-kraevedčeskij Muzej im. P.V. Alabina, Samara 37622
Saratovskij Ėtnografičeskij Muzej, Saratov 37752
Saratovskij Gosudarstvennyj Muzej Boevoj Slavy, Saratov 37753
Saratovskij Kraevedčeskij Muzej, Saratov 37754
Šeltozerskij Vepsskij Ėtnografičeskij Muzej, Šeltozero 37758
Semenovskij Gosudarstvennyj Kraevedčeskij Istoriko-chudožestvennyj Muzej, Semenov 37759
Sergačskij Kraevedčeskij Muzej, Sergač 37761
Severo-osetinskij Gosudarstvennyj Obedinennyj Muzej Istorii, Architektury i Literatury, Vladikavkaz 37932
Šlisselburgskaja Krepost Orešek, Šlisselburg 37769
Smolenskij Istoričeskij Muzej, Smolensk 37777
Smolenskij Muzej-Zapovednik, Smolensk 37778
Solikamskij Gorodskoj Kraevedčeskij Muzej, Solikamsk 37785
Staroladožskij Istoriko-Architekturnyj i Archeologičeskij Muzej-Zapovednik, Staraja Ladoga 37795
Starooskolskij Kraevedčeskij Muzej, Staryj Oskol 37801
Starorusskij Kraevedčeskij Muzej, Staraja Russa 37797
Stavropolskij Kraevedčeskij Muzej im. G.N. Prozriteleva i G.K. Prave, Stavropol 37805
Syzranskij Gorodskoj Kraevedčeskij Muzej, Syzran 37823
Taganrogskij Kraevedčeskij Muzej, Taganrog 37829
Tajmyrskij Okružnoj Kraevedčeskij Muzej, Dudinka 36880
Tambovskij Oblastnoj Kraevedčeskij Muzej, Tambov 37833
Tarskij Istoriko-kraevedčeskij Muzej, Tara 37834
Tarusskij Kraevedčeskij Muzej, Tarusa 37836
Troickij Sobor, Serpuchov 37766
Tuapsinskij Kraevedčeskij Muzej im. N.G. Poletaeva, Tuapse 37864
Tulskij Oblastnoj Kraevedčeskij Muzej, Tula 37875
Tuvinskij Respublikanskij Kraevedčeskij Muzej im. Aldan Maadyr, Kyzyl 37099
Tverskoj Gosudarstvennyj Obedinennyj Muzej, Tver 37881
Uljanovskij Oblastnoj Kraevedčeskij Muzej im. I.A. Gončarova, Uljanovsk 37910
Valaamskij Naučno-issledovatelskij Cerkovno-archeologičeskij i Prirodnyj Muzej-zapovednik, Sortavala 37790
Velikolukskij Kraevedčeskij Muzej, Velikie Luki 37918
Vetlužskij Kraevedčeskij Muzej, Vetluga 37927
Vjatskaja Kunstkamera, Kirov 37043
Volgogradskij Memorialno-istoričeskij Muzej, Volgograd 37957
Volgogradskij Oblastnoj Kraevedčeskij Muzej, Volgograd 37959
Vorkutinskij Mežrajonnyj Kraevedčeskij Muzej, Vorkuta 37968
Voronežskij Oblastnoj Kraevedčeskij Muzej, Voronež 37976
Vserossijskij Istoriko-ėtnografičeskij Muzej, Toržok 37862
Vsevoložskij Gosudarstvennyj Istoriko-kraevedčeskij Muzej, Vsevoložsk 37980
Vystavocnyj Zal, Vologda 37965
Vystavočnyj Zal Spasskaja Cerkov, Irkutsk 36940

St. Pierre and Miquelon

Musée-Archives de Saint-Pierre, Saint-Pierre 37998
Musée de Miquelon, Miquelon 37996

San Marino

Museo di Stato, San Marino 38008

Senegal

Musée Régional Ex-Dixième Riaom, Thiès 38028

Serbia

Gradski Muzej, Bečej 38032
Gradski Muzej, Kosovska Mitrovica 38105
Gradski Muzej, Sremski Karlovci 38152
Gradski Muzej Kruševac, Kruševac 38112
Gradski Muzej - Városi Muzeum, Subotica 38156
Muzej Grada Novog Sada, Novi Sad 38121
Muzej Jovana Jovanovića Zmaja, Sremska Kamenica 38149
Muzej Krajine Negotin, Negotin 38116
Muzej Srema, Sremska Mitrovica 38151
Muzej u Majdanpeku, Majdanpek 38115
Muzej u Smederevu, Smederevo 38144
Muzej Vojvodine, Novi Sad 38124
Muzejska Zbirka, Knić 38104
Muzeu i Kosoves / Muzej Kosova i Metohije, Priština 38133
Narodni Muzej, Bela Crkva 38033
Narodni Muzej, Pančevo 38128
Narodni Muzej, Požarevac 38132
Narodni Muzej, Zaječar 38164
Narodni Muzej Kikinda, Kikinda 38102
Narodni Muzej Niš, Niš 38118
Narodni Muzej Sabac, Sabac 38138
Narodni Muzej Toplice, Prokuplje 38136
Narodni Muzej Valjevo, Valjevo 38160
Regionalni Muzej, Djakovica 38099
Senćanski Gradski Muzej, Senta 38139
Spomen-Park Kragujevački Oktobar, Kragujevac 38110
Zavičajna Muzejska Zbirka, Ćajetina 38096
Zavičajni Muzej u Jagodini, Jagodina 38101
Zavicajni Muzej Zemuna, Zemun 38165

Singapore

Fort Siloso, Singapore 38174

Slovakia

Gemersko-Malohontské Múzeum, Rimavská Sobota 38340
Hrad Devín, Bratislava 38214
Hutnícke Múzeum Železiarne Podbrezová, Podbrezová 38329
Jozef Gregor Tajovský Pamätný Dom, Tajov 38368
Krajanské Múzeum MS Bratislava, Bratislava 38217
Krajské Múzeum, Prešov 38335
Kysucke Múzeum, Čadca 38243
Literárne a Vlastivedné Múzeum, Svätý Jur 38365
Lubovnianske Múzeum - Hrad, Stará Lubovňa 38361
Malokarpatské Múzeum v Pezinku, Pezinok 38324
Městské Múzeum, Ilava 38263
Městské Múzeum, Zlaté Moravce 38392
Městské Múzeum Fiľakovo, Fiľakovo 38254
Městské Múzeum Michala Tillnera, Malacky 38291
Městské Múzeum Poltár, Poltár 38330
Městské Múzeum Rajec, Rajec 38339
Mestské Múzeum v Bratislave, Bratislava 38219
Mestský Vlastivedný Dom, Šamorín 38348
Misijné Múzeum Nitra, Nitra 38313
Múzeum Andreja Sládkoviča Krupina, Krupina 38280
Múzeum Bábkarských Kultúr a Hračiek Hrad Modry Kameň, Modrý Kameň 38307
Múzeum Beckov, Beckov 38198
Múzeum Čachtice, Čachtice 38241
Múzeum Ludovita Štúra, Modra 38306
Múzeum Mad'arskej Kultúry a Podunajska, Komárno 38269
Múzeum Michala Greisigera, Spišská Béla 38356
Múzeum Samuela Jurkoviča Sobotište-Nyaryiovský kaštiel', Sobotište 38354
Múzeum Spiša, Spišská Nová Ves 38359
Múzeum v Kežmarku a Mestský Hrad, Kežmarok 38267
Múzeum vo Svätý Antone, Svätý Anton 38364
Novohradské Múzeum, Lučenec 38290
Nový Zamok Banská Štiavnica, Banská Štiavnica 38192
Podjavorinské Múzeum, Nové Mesto nad Váhom 38320
Pohronské Múzeum v Novej Bani, Nová Baňa 38318
Ponitrianske Múzeum, Nitra 38315
Považské Múzeum, Žilina 38391
Šarišské Múzeum, Bardejov 38195
Slovenské Banské Múzeum, Banská Štiavnica 38193
Spišské Múzeum, Levoča 38284
Stredoslovenské Múzeum, Banská Bystrica 38187
Tekovské Múzeum v Leviciach, Levice 38282
Trenčianske Múzeum, Trenčín 38376
Vlastivedné Múzeum, Hanušovce nad Topl'ou 38257
Vlastivedné Múzeum, Trebišov 38374
Vlastivedné Múzeum v Galante, Galanta 38255
Zemplínske Múzeum, Michalovce 38305
Žitnoostrovné Múzeum, Dunajská Streda 38253

Slovenia

Cerkljanski Muzej, Cerkno 38408
Dolenjski Muzej Novo Mesto, Novo Mesto 38490
Dvorec-Muzej Betnava, Maribor 38475
Etnološki Muzej, Ribnica na Pohorju 38506
Galerija-Muzej Lendava, Lendava 38443
Goriški Muzej, Nova Gorica 38489
Grad Bogenšperk, Šmartno pri Litiji 38522
Kamniški Muzej, Kamnik 38427
Koroški Pokrajinski Muzej, Slovenj Gradec 38518
Kosova Graščina, Jesenice 38426
Loški Muzej, Škofja Loka 38514
Mestni Muzej Idrija, Idrija 38421
Mestni Muzej Radovljica, Radovljica 38502
Miklova Hiša, Ribnica 38505
Muzej na Blejskem Gradu, Bled 38400
Muzej Političnih Zapornikov, Internirancev in Izgnancev Grad Rajhenburg, Brestanica 38403
Muzej Velenje na Velenjskem Gradu, Velenje 38534
Muzejska Zbirka Laško, Laško 38442
Nacetova Hiša, Škofja Loka 38516
Partizanska Bolnica Franja, Dolenji Novaki 38415
Pokrajinski Muzej, Maribor 38477
Rojstna Hiša Matije Čopa, Žirovnica 38543
Sokličev Muzej, Slovenj Gradec 38519
Tolminski Muzej, Tolmin 38527
Triglavska Muzejska Zbirka, Mojstrana 38483
Tržiški Muzej z Galerijo, Tržič 38532
Zbirka v Jamarskem Domu, Domžale 38416

Solomon Islands

National Museum Cultural Centre, Honiara 38547

South Africa

Albany Museum, Grahamstown 38629
Aliwal North Museum and Church Plein Museum, Aliwal North 38552
Arend Dieperink Museum, Potgietersrus 38718
Bartolomeu Dias Museum Complex, Mossel Bay 38691
Beaufort West Museum, Beaufort West 38557
Bergtheil Museum, Westville 38777
Bethlehem Museum, Bethlehem 38559
Burgerhuis, Stellenbosch 38734
Burgersdorp Cultural Historical Museum, Burgersdorp 38575
C.P. Nel Museum, Oudtshoorn 38697
Calvinia Regional Museum, Calvinia 38577
Colesberg/Kemper Museum, Colesberg 38596
District Six Museum, Cape Town 38582
Drostdy Museum, Uitenhage 38768
East London Museum, East London 38615
Elgin Apple Museum, Grabouw 38628
First Raadsaal Museum, Bloemfontein 38561
Freshford House Museum, Bloemfontein 38562
George Museum, George 38624
Graaf-Reinet Museum, Graaff-Reinet 38626
Great Brak River Museum, Great Brak River 38635
Hartenbos Museum, Hartenbos 38639
Heritage Collection Catherina Brand, Ladybrand 38681
History Museum, Grahamstown 38630
Howick Museum, Howick 38643
Julius Gordon Africana Centre, Riversdale 38723
Kalahari-Oranje Museum, Upington 38771
Klerksdorp Museum, Klerksdorp 38677
Knysna Museum, Knysna 38678
Kwa Muhle Museum, Durban 38607
Lichtenburg Museums and Art Gallery, Lichtenburg 38682
McGregor Museum and Duggan Cronin Gallery, Kimberley 38670
Mafikeng Museum, Mafikeng 38685
Mary Moffat Museum, Griquatown 38638
Messina Museum, Messina 38688
Michaelis Collection, Cape Town 38587
Museum Africa, Johannesburg 38660
Museum Caledon River, Smithfield 38732
Museum Chris Van Niekerk, Boshof 38572
Old Court House Museum, Durban 38609
P.W. Vorster Museum, Middelburg 38689
Paarl Museum, Paarl 38700
Parys Museum, Parys 38701
Pellisier House Museum, Bethulie 38560
Pilgrim's Rest Museum, Pilgrim's Rest 38709
Rhodes Cottage Museum, Muizenberg 38693
Robben Island Museum, Robben Island 38724
Roodepoort Museum, Roodepoort 38726
Saint Vincent Church Museum, Isandlwana 38644
Schoemansdal Museum, Louis Trichardt 38683
Simon's Town Museum, Simon's Town 38730
South African Police Service Museum, Tshwane 38758
Sterkstroom Museum, Sterkstroom 38743
Talana Museum and Heritage Park, Dundee 38601
Township Museum, Lwandle 38684
Vaal Teknorama Cultural Museum, Vereeniging 38772
Village Museum, Stellenbosch 38742
Voortrekker Monument Heritage Site, Tshwane 38764
Welkom Museum, Welkom 38776

Spain

Casa Alegre de Sagrera, Terrassa 40114
Casa Cercantes, Vélez Málaga 40256
Casa Museo Ignacio Zuloaga, Zumaia 40362
Casa Museo José María de Cossío, Tudanca 40172
Casa-Museu Torres Amat, Sallent 39947
Casa Natal de Sant Antoni Maria Claret, Sallent 39948
Castell Cartoixa Vallparadis, Terrassa 40115
Centre d'Interpretació del Castell de Gilida, Gelida 39330
Centro de Historia de Zaragoza, Zaragoza 40338
Colleció d'Autòbils de Salvador Claret Sargatal, Sils 40081
El Castell-Ecomuseu Urbà, Rubí 39921
Etnografikoa Museoa Artziniega, Artziniega 38904
Larraulgo Museoa, Larraul 39443
Museo Arqueológico de Úbeda, Úbeda 40178
Museo Arqueológico Etnológico Municipal, Guardamar de Segura 39379
Museo Ciudad de Baeza, Baeza 38941
Museo de Albacete, Albacete 38803
Museo de Cáceres, Cáceres 39082
Museo de Cádiz, Cádiz 39089
Museo de Cuenca, Cuenca 39232
Museo de Geologia, Paleontologia i Prehistòria del Bruc, El Bruc 39058
Museo de Historia de Agüimes, Agüimes 38789
Museo de Historia de Tenerife, La Laguna de Tenerife 39439

H

Museo de Historia y Cultura Casa Pedrilla, Cáceres 39083
Museo de la Ciudad, Melilla 39642
Museo de la Ciudad, Murcia 39695
Museo de la Ciudad, Vinaròs 40315
Museo de la Guardia Civil, Madrid 39546
Museo de la Paeria, Lleida 39475
Museo de la Rioja, Logroño 39491
Museo de la Sociedad Amigos de Laguardia, Laguardia 39437
Museo de León - Edificio Pallares, León 39457
Museo de los Orígenes, Madrid 39550
Museo de Menorca, Mahón 39591
Museo de Segovia, Segovia 40053
Museo de Zamora, Zamora 40335
Museo del Almudín, Xàtiva 40327
Museo Etnografico Extremeño González Santana, Olivenza 39724
Museo Etnológico de Canet d'En Berenguer, Canet d'En Berenguer 39119
Museo Histórico de la Ciudad, Valencia 40212
Museo Histórico de San Roque, San Roque 39970
Museo Histórico Municipal, Baena 38939
Museo Historico Municipal, Montilla 39680
Museo Histórico Municipal de Écija, Écija 39257
Museo Histórico Municipal de Teba, Teba 40107
Museo Lara, Ronda 39919
Museo Militar Regional de Burgos, Burgos 39071
Museo Municipal, Algeciras 38841
Museo Municipal, Alzira 38867
Museo Municipal, Calatayud 39098
Museo Municipal, Nerva 39716
Museo Municipal, Puertollano 39882
Museo Municipal, Requena 39892
Museo Municipal, Carmona 39132
Museo Municipal de Daimiel, Daimiel 39239
Museo Municipal de Madrid, Madrid 39565
Museo Municipal de Santiponce, Santiponce 40038
Museo Municipal de Vigo Quiñones de León, Vigo 40276
Museo Municipal Eduard Camps Cava, Guissona 39382
Museo Municipal Elisa Cendrero, Ciudad Real 39191
Museo Municipal Estados del Duque, Malagón 39608
Museo Municipal Ulpiano Checa, Colmenar de Oreja 39199
Museo Nájerillense, Nájera 39710
Museo-Palacio Spínola, Teguise 40111
Museo Provincial, Teruel 40122
Museográfica de la Villa, Camporrobles 39113
Museu, Pallejà 39774
Museu Arxiu Comarcal, Moià 39657
Museu-Arxiu de Santa María de Mataró, Mataró 39632
Museu Arxiu de Vilassar de Dalt, Vilassar de Dalt 40291
Museu-Arxiu Municipal de Calella, Calella 39102
Museu Biblioteca Fundació Mauri, La Garriga 39327
Museu Cerdà, Puigcerdà 39884
Museu Comarcal de Berga, Berga 39028
Museu Comarcal de la Conca de Barberà, Montblanc 39670
Museu Comarcal de l'Anoia-Museu de la Pell, Igualada 39408
Museu Comarcal de l'Urgell, Tárrega 40106
Museu Comarcal de Manresa, Manresa 39614
Museu d'Art i Història de Reus, Reus 39895
Museu de Badalona, Badalona 38937
Museu de Coses del Poble, Collbató 39198
Museu de Gavà, Gavà 39329
Museu de Granollers, Granollers 39373
Museu de la Mar, Peñíscola 39834
Museu de la Vila, Vilar-Rodona 40290
Museu de l'Adet, Torelló 40149
Museu de l'Aigua Colleció Gavarró, Vilanova del Camí 40283
Museu de l'Estable i Troballes, Guimerà 39381
Museu de Llavaneres, Sant Andreu de Llavaneres 39975
Museu de Lloret de Mar, Lloret de Mar 39488
Museu de Mallorca, Palma de Mallorca 39789
Museu de Mataró, Mataró 39633
Museu de Montgrí i del Baix Ter, Torroella de Montgrí 40162
Museu de Terrassa, Terrassa 40119
Museu del Càntir, Argentona 38890
Museu del Castell de Onda, Onda 39735
Museu del Montsec, Artesa de Segre 38902
Museu del Montsià, Amposta 38870
Museu del Pagès, Torroella de Fluvià 40161
Musèu d'Era Val d'Aran, Vielha e Mijaran 40272
Museu d'Història, Sabadell 39927
Museu d'Història de la Ciutat, Girona 39346
Museu d'Història de la Ciutat de Barcelona, Barcelona 38989
Museu d'Història de la Ciutat de Palma, Palma de Mallorca 39790
Museu d'Historia de Tarragona, Tarragona 40103
Museu Els Cups, Montbrió del Camp 39672
Museu Històric Municipal, Riudoms 39905
Museu Historic Municipal de Polinyà, Polinyà 39849
Museu la Granja, Santa Perpètua de Mogoda 40005
Museu Manolo Hugué, Caldes de Montbui 39100
Museu Municipal, Alcover 38826
Museu Municipal, Llívia 39486
Museu Municipal, Manlleu 39613
Museu Municipal, Tortosa 40166
Museu Municipal Can Xifreda, Sant Feliu de Codines 39981
Museu Municipal d'Agramunt, Agramunt 38786
Museu Municipal de la Pagesia, Castellbisbal 39145
Museu Municipal de Nàutica, El Masnou 39630
Museu Municipal de Sant Pol de Mar, Sant Pol de Mar 39989
Museu Municipal Joan Pla i Gras, Llinars del Vallés 39483
Museu Municipal Miquel Soldevila, Prats de Lluçanès 39867
Museu Particular del Pastor Joaquim Gassó, Fornells de la Muntanya 39307
Museu Rocaguinarda de Terrissa dels Països Catalans, Oristà 39746
Naturhiscope, Oropesa del Mar 39748
Palau Reial de l'Almudaina, Palma de Mallorca 39793
Parque Cultural, Zerain 40357

Sri Lanka

Galle National Museum, Galle 40377
Jaffna National Museum, Jaffna 40379
Kandy National Museum, Kandy 40380
Ratnapura National Museum, Ratnapura 40383

Sudan

El-Barkai Museum, Khartoum 40387
Khalif's House Museum Abdallah at-Ta'ayushi, Khartoum 40391
Sultan Ali Dianr Museum, El-Fasher 40384

Suriname

Openluchtmuseum Nieuw Amsterdam, Nieuw Amsterdam 40399
Surinaams Museum, Paramaribo 40401

Sweden

Åhus Museum, Åhus 40410
Åkers Hembygdsmuseum, Strängnäs 40951
Alingsås Museum, Alingsås 40413
Åmåls Hembygdsmuseum, Åmål 40415
Arboga Museum, Arboga 40428
Arvidsjaurs Hembygdsmuseum, Arvidsjaur 40434
Åsavallen, Edsbro 40491
Aspö-Tosterö Hembygdsmuseum, Strängnäs 40952
Åtvidabergs Bruks- och Facitmuseum, Åtvidaberg 40447
Berte Museum - Livet på Landet, Slöinge 40857
Bjärnums Museum, Bjärnum 40458
Björklidens Gårdsmuseum, Ljusterö 40689
Blekinge Museum, Karlskrona 40634
Bohusläns Museum, Uddevalla 40992
Bollnäs Konsthallen, Bollnäs 40464
Borås Museum, Borås 40469
Borgsjö Hembygdsgård, Erikslund 40500
Bruksmuseet Smedsgården, Sandviken 40830
Brunskogs Hembygdsgård, Brunskog 40480
Christinehof Slott, Brösarp 40478
Dalslands konstmuseum, Åsensbruk 40443
Ekomuseum Gränsland, Strømstad 40954
Eksjö Museum, Eksjö 40493
Enköpings Museum, Enköping 40497
Eriksbergs Museum, Tranås 40978
Eslövs Stadsmuseum, Eslöv 40506
Falbygdens Museum, Falköping 40512
Falsterbo Museum, Falsterbo 40513
Fredriksdal Museer och Trädgårdar, Helsingborg 40586
Frejamuseét, Kil 40642
Friluftsmuseet i Apladalen, Värnamo 41028
Frösö Hembygdsmuseum, Frösön 40528
Gamla Linköping, Linköping 40681
Gamla Skeninge, Skänninge 40842
Gammelgården Friluftsmuseet, Bengtsfors 40455
Glommersträsks Hembygdsmuseum, Glommersträsk 40538
Göteborgs Stadsmuseum, Göteborg 40549
Grangärde Hembygdsgården, Grangärde 40565
Grassagården Miljömuseum, Strängnäs 40953
Hembygdsgården, Bjurholm 40460
Hembygdsgården i Kristvalla, Nybro 40779
Hembygdsmuseet i Korsberga, Korsberga 40652
Höganäs Museum och Konsthall, Höganäs 40594
Högarps Bymuseum, Vetlanda 41055
Hörby Museum, Hörby 40598
Huskvarna Stadsmuseum, Huskvarna 40607
Jämtlands Läns Museum, Östersund 40796
Jönköpings Läns Museum, Jönköping 40614
Kalix Flottningsmuseum, Kalix 40622
Kalmar Läns Museum, Kalmar 40624
Karlshamns Museum, Karlshamn 40628
Katrineholms Hembygdsmuseet, Kristineholm 40657
Köpings Museum, Köping 40647
Kortedala Museum, Göteborg 40550
Kungsudden Local Museum, Kungsör 40661
Länsmuseet Västernorrland, Härnösand 40572
Länsmuseet Varberg, Varberg 41048
Landskrona Museum, Landskrona 40669
Leksands Hembygdsmuseum, Leksand 40670
Lesjöforsmuseum, Lesjöfors 40671
Lima Hembygdsgård, Lima 40677
Lindesbergs Museum, Lindesberg 40679
Ljungby Museum, Ljungby 40687
Ljusdalsbygdens Museum, Ljusdal 40688
Ljusterö Hembygdsmuseum, Ljusterö 40690
Måleriyrkets Museum, Stockholm 40911
Malmö Museer, Malmö 40724
Malungs Hembygdsgård, Malung 40729
Mariefreds Hembygdsmuseum, Mariefred 40732
Markarydsortens Hembygdsmuseum, Markaryd 40739
Mjölby Hembygdsgård, Mjölby 40740
Mölndals museum, Mölndal 40743
Motala Museum - Charlottenborgs Slott, Motala 40756
Museet Näktergalen, Vimmerby 41064
Njudungs Hembygdsmuseum, Vetlanda 41056
Nordhallands Hembygdsförening, Kungsbacka 40659
Österlens Museum, Simrishamn 40841
Övertorneå Hembygdsmuseum, Övertorneå 40800
Olofsfors Bruksmuseum, Nordmaling 40767
Ornunga Museum, Vårgårda 41051
Pengsjö Nybyggarmuseum, Umeå 40998
Piteå Museum, Piteå 40809
Regionmuseet Kristianstad, Kristianstad 40655
Regionmuseum Västra Götaland, Vänersborg 41025
Rörbäcksnäs Hembygdsgård, Lima 40678
Roslagsmuseet, Norrtälje 40778
Saab Museum, Trollhättan 40985
Säters Hembygdsmuseum, Säter 40824
Sigtuna Rådhus, Sigtuna 40836
Skövde Stadsmuseum, Skövde 40852
Söderbärke Hembygdsgård, Söderbärke 40860
Söderhamns Stadsmuseum, Söderhamn 40861
Sölje Bygdegård, Glava 40536
Sölvesborgs Museum, Sölvesborg 40867
Sörmlands Museum, Nyköping 40782
Sörmlandsgården, Eskilstuna 40504
Sollefteå Museum, Sollefteå 40868
Solna Hembygdsmuseum, Solna 40874
Stadsmuseet och Ahlbergshallen, Östersund 40798
Stafsjö Bruksmuseum, Falkenberg 40510
Stockholms Läns Museum, Stockholm 40939
Stockholms Stadsmuseum, Stockholm 40942
Strømstads Museum, Strømstad 40955
Sundbybergs Museum, Sundbyberg 40957
Sundsbergs Gård Museum och Konsthall, Sunne 40962
Sundsvalls Museum, Sundsvall 40960
Swensbylijda, Svensbyn 40966
Tibro Museum, Tibro 40970
Tidaholms Museum, Tidaholm 40972
Torekällbergets Friluftsmuseum, Södertälje 40865
Tumbergs Skol- och Hembygdsmuseum, Vårgårda 41052
Tunabygdens Gammelgård, Borlänge 40477
Ulrika museum, Ulrika 40996
Upplandsmuseet, Uppsala 41012
Vadsbo Museum, Mariestad 40738
Vadstena Stadsmuseum, Vadstena 41022
Vämöparken, Karlskrona 40637
Vänersborgs Museum, Vänersborg 41026
Värmlands Museum, Karlstad 40639
Väsby Kungsgård, Sala 40828
Västerbottens Museum, Umeå 41000
Västergötlands Museum, Skara 40846
Västerviks Museum, Västervik 41037
Vetlanda Museum, Vetlanda 41058
Viktor Rydbergmuseet, Jönköping 40617

Switzerland

Bergsturz-Museum Goldau, Goldau 41441
Bibliothèque de Genève, Genève 41401
Birsfelder Museum, Birsfelden 41238
Büecheler-Hus, Kloten 41506
Burg Zug, Zug 42123
Ciäsa Granda, Stampa 41901
Dichter- und Stadtmuseum Liestal (DISTL)/ Herwegh-Archiv, Liestal 41561
Dörflihaus-Museum, Spiringen 41894
Dorfmuseum Bellach, Bellach 41189
Dorfmuseum Bennwil, Bennwil 41198
Dorfmuseum Birr, Birr 41237
Dorfmuseum Bönigen, Bönigen 41243
Dorfmuseum Feldbrunnen, Feldbrunnen 41368
Dorfmuseum Gontenschwil, Gontenschwil 41442
Dorfmuseum Graberhaus, Strengelbach 41917
Dorfmuseum Hüntwangen, Hüntwangen 41490
Dorfmuseum im Guggerhuus, Ettingen 41365
Dorfmuseum im Kulturzentrum, Fislisbach 41371
Dorfmuseum im Pulverturm, Schwanden 41846
Dorfmuseum Kirchbözberg, Unterbözberg 41957
Dorfmuseum Konolfingen, Konolfingen 41509
Dorfmuseum Langendorf, Langendorf 41525
Dorfmuseum Linthal, Linthal 41567
Dorfmuseum Maschwanden, Maschwanden 41614
Dorfmuseum Melihus, Möhlin 41631
Dorfmuseum Nunningen, Nunningen 41689
Dorfmuseum Rupperswil, Rupperswil 41783
Dorfmuseum Schlossweid, Ringgenberg 41768
Dorfmuseum Seon, Seon 41864
Dorfmuseum Stricklerhuus, Hombrechtikon 41486
Dorfmuseum Therwil, Therwil 41935
Dorfmuseum Wiesen, Wiesen 42014
Dorfmuseum Zeihen, Zeihen 42051
Dorfmuseum Ziefen, Ziefen 42055
Drachenloch Museum, Vättis 41970
Dürstelerhaus, Ottikon 41717
Ecomuseum Simplon, Simplon Dorf 41871
Entlebucher Museum, Schüpfheim 41845
Forum Stadtscheune-Ortsmuseum und Raum für Kunst, Mellingen 41623
Freilichtmuseum Säge Buch, Buch 41265
Fricktaler Museum, Rheinfelden 41757
Les Galeries de l'Historie, Neuchâtel 41679
Gandahus, Vals 41978
Das Gelbe Haus, Flims Dorf 41374
Gemeindemuseum, Krauchthal 41511
Gemeindemuseum, Regensdorf 41750
Gemeindemuseum Rothus, Oberriet 41699
Heimat- und Ortsmuseum, Zürich 42076
Heimatkundliche Sammlung, Richterswil 41758
Heimatmuseum, Donzhausen 41332
Heimatmuseum, Grindelwald 41455
Heimatmuseum, Grüningen 41456
Heimatmuseum, Hallau 41468
Heimatmuseum, Küssnacht am Rigi 41520
Heimatmuseum, Rothrist 41777
Heimatmuseum, Wald 41996
Heimatmuseum, Waltensburg 42000
Heimatmuseum, Brugg 41260
Heimatmuseum Aarburg, Aarburg 41086
Heimatmuseum Adelboden, Adelboden 41088
Heimatmuseum Alt-Falkenstein, Klus 41507
Heimatmuseum am Pfäffikersee, Pfäffikon (Zürich) 41723
Heimatmuseum Arnold Bärtschi, Dulliken 41336
Heimatmuseum der Talschaft Lauterbrunnen, Lauterbrunnen 41552
Heimatmuseum Elgg, Elgg 41347
Heimatmuseum Leuk Stadt, Leuk Stadt 41557
Heimatmuseum Oltingen-Wenslingen-Anwil, Oltingen 41711
Heimatmuseum Prättigau, Grüsch 41459
Heimatmuseum Prestegg, Altstätten 41108
Heimatmuseum Rheinwald, Splügen 41895
Heimatmuseum Schinznach-Dorf, Schinznach Dorf 41834
Heimatmuseum Schwarzbubenland, Dornach 41333
Heimatmuseum Seehof, Buonas 41273
Heimatmuseum Sissach, Sissach 41879
Heimatmuseum Spittel, Büren an der Aare 41269
Heimatmuseum Spycher, Rickenbach (Luzern) 41760
Heimatmuseum Stein, Stein am Rhein 41913
Heimatmuseum Suhr, Suhr 41920
Heimatmuseum Trimmis, Trimmis 41950
Heimatmuseum Trubschachen, Trubschachen 41952
Heimatmuseum Weesen, Weesen 42003
Historisches Museum, Bischofszell 41239
Historisches Museum der Stadt Baden, Baden 41145
Historisches Museum Heiden, Heiden 41476
Historisches Museum Obwalden, Sarnen 41820
Historisches Museum Olten, Olten 41706
Historisches Museum Thurgau, Frauenfeld 41377
Historisches Museum Uri, Altdorf 41104
Historisches Museum Wiedlisbach, Wiedlisbach 42013
Ital Reding-Haus, Schwyz 41851
Klostermuseum Müstair, Müstair 41664
Kultur-Historisches Museum Grenchen, Grenchen 41452
Kulturpfad Root, Root 41774
Lötschentaler Museum, Kippel 41504
Maison Buttin-de Loës, Grandvaux 41451
La Maison du Blé et du Pain, Echallens 41339
Maison Tavel, Genève 41414
March-Museum, Vorderthal 41993
Musée, Evolène 41367
Musée, Le Landeron 41523
Musée d'Art et d'Histoire Neuchâtel, Neuchâtel 41680
Musée de Carouge, Carouge 41280
Musée de la Confrérie des Vignerons, Vevey 41984
Musée de l'Areuse, Boudry 41249
Musée de l'Evéché, Sion 41877
Musée de Sonvilier, Sonvilier 41890
Musée de Vieux-Lavaux, Cully 41317
Musée d'Estavayer, Estavayer-le-Lac 41363
Musée d'Orbe, Orbe 41713
Musée du Grand-Lens, Lens 41553
Musée du Mont-Repais, Montmelon 41636
Musée du Montreux, Montreux 41639
Musée du Pays et Val de Charmey, Charmey 41291
Musée du Tour Automatique et d'Histoire de Moutier, Moutier 41652

Musée du Vieux-Baulmes, Baulmes 41187
Musée du Vieux-Monthey, Monthey 41635
Musée du Vieux-Pays, Le Châble 41287
Musée du Vieux-Saxon, Saxon 41826
Musée Gruérien, Bulle 41272
Musée Historique de Lausanne, Lausanne 41548
Musée Historique de Vevey, Vevey 41985
Musée Isérables, Isérables 41494
Musée Régional de la Sagne, La Sagne 41788
Musée Rural Jurassien, Les Genevez 41430
Musée Vissoie, Vissoie 41990
Museo Casa del Padre, Orselina 41716
Museo del Malcantone, Curio 41318
Museo di Blenio, Lottigna 41575
Museo di Leventina, Giornico 41433
Museo di Val Verzasca, Sonogno 41889
Museo Elisarion, Minusio 41630
Museo Nazionale del San Gottardo, Airolo 41097
Museo Onsernonese, Loco 41574
Museo Poschiavino, Poschiavo 41731
Museo Regionale delle Centovalli e del Pedemonte, Intragna 41493
Museo Vallerano Bregagliotto, Stampa 41902
Museum Aargau, Lenzburg 41554
Museum Alpin Pontresina, Pontresina 41726
Museum Amden, Amden 41110
Museum auf der Burg, Raron 41747
Museum.BL, Liestal 41564
Museum Burghalde, Lenzburg 41555
Museum Chasa Jaura, Valchava 41972
Museum Chornhuus, Gränichen 41447
Museum der Landschaft, Saanen 41784
Museum der Landschaft Hasli, Meiringen 41619
Museum der Winterthur-Versicherungen, Winterthur 42032
Museum des Landes Glarus, Näfels 41672
Museum Herisau, Herisau 41481
Museum im Bellpark, Kriens 41518
Museum im Schulhaus, Niederrohrdorf 41686
Museum im Stockalperschloß, Brig 41256
Museum im Zendenrathaus, Ernen 41359
Museum Langenthal, Langenthal 41527
Museum Laufental, Laufen 41531
Museum Lindengut, Winterthur 42034
Museum Lindwurm, Stein am Rhein 41915
Museum Löwenburg, Ederswiler 41340
Museum Neuhaus, Biel 41231
Museum Regiunal, Savognin 41824
Museum Regiunal Surselva Casa Carniec, Ilanz 41492
Museum Rosenegg, Kreuzlingen 41514
Museum Salzbütte, Huttwil 41491
Museum Sarganserland, Sargans 41818
Museum Schloss Münsingen, Münsingen 41660
Museum Wolfhalden, Wolfhalden 42045
Museum zu Allerheiligen, Schaffhausen 41831
Museum zum Schiff, Laufenburg 41533
Nidwaldner Museum, Stans 41905
Obersimmentaler Heimatmuseum, Zweisimmen 42127
Ortsgeschichtliche Sammlung, Oberrieden 41698
Ortsgeschichtliche Sammlung Rebstein, Rebstein 41748
Ortsmuseum, Amriswil 41112
Ortsmuseum, Andwil 41116
Ortsmuseum, Bergün 41199
Ortsmuseum, Binningen 41235
Ortsmuseum, Bülach 41268
Ortsmuseum, Bütschwil 41271
Ortsmuseum, Häggenschwil 41467
Ortsmuseum, Jona 41499
Ortsmuseum, Laax 41522
Ortsmuseum, Liesberg 41560
Ortsmuseum, Meilen 41618
Ortsmuseum, Merenschwand 41628
Ortsmuseum, Oftringen 41704
Ortsmuseum, Opfikon 41712
Ortsmuseum, Rafz 41740
Ortsmuseum, Roggwil 41771
Ortsmuseum, Rüti 41782
Ortsmuseum, Sankt Stephan 41815
Ortsmuseum, Schänis 41827
Ortsmuseum, Thalwil 41933
Ortsmuseum, Unterengstringen 41958
Ortsmuseum, Untersiggenthal 41962
Ortsmuseum, Vaz 41979
Ortsmuseum, Vnà 41992
Ortsmuseum, Wängi 41995
Ortsmuseum, Walenstadt 41998
Ortsmuseum Altstetten, Altstetten 41109
Ortsmuseum Beringen, Beringen 41200
Ortsmuseum Brittnau, Brittnau 41258
Ortsmuseum Dietikon, Dietikon 41330
Ortsmuseum Erlenbach, Erlenbach 41355
Ortsmuseum Frenkendorf, Frenkendorf 41382
Ortsmuseum für Heimatkunde, Schlieren 41838
Ortsmuseum Grächen, Grächen 41446
Ortsmuseum Kefiturm, Belp 41197
Ortsmuseum Kilchberg, Kilchberg (Zürich) 41503
Ortsmuseum Küsnacht, Küsnacht 41519
Ortsmuseum Mollis, Mollis 41632
Ortsmuseum Mühle, Maur 41617
Ortsmuseum Neunkirch, Neunkirch 41683
Ortsmuseum Oberes Bad, Marbach 41605
Ortsmuseum Oberuzwil, Oberuzwil 41700
Ortsmuseum Oetwil, Oetwil am See 41702
Ortsmuseum Rüschlikon, Rüschlikon 41781
Ortsmuseum Schleitheim, Schleitheim 41836
Ortsmuseum Schmitten, Schmitten, Albula 41840
Ortsmuseum Spreitenbach, Spreitenbach 41896
Ortsmuseum Steinmaur, Sünikon 41919
Ortsmuseum Trotte, Arlesheim 41125
Ortsmuseum Urdorf, Urdorf 41965
Ortsmuseum Vaz, Lenzerheide 41556
Ortsmuseum Vechigen, Boll 41244
Ortsmuseum Wallisellen, Wallisellen 41999
Ortsmuseum Weiach, Weiach 42004
Ortsmuseum Wila, Wila 42018
Ortsmuseum Wilchingen, Wilchingen 42019
Ortsmuseum Wilderswil, Wilderswil 42021
Ortsmuseum Wollishofen, Zürich 42102
Ortsmuseum zur Farb, Stäfa 41900
Ortsstube, Bolligen 41245
Privatmuseum Im Blauen Haus, Appenzell 41120
Rathausmuseum, Sempach 41860
Regionalmuseum Binn, Binn 41233
Regionalmuseum Langnau im Chüechlihus, Langnau im Emmental 41529
Regionalmuseum Schlangenhaus, Werdenberg 42006
Regionalmuseum Schwarzwasser, Schwarzenburg 41848
Reiat-Museum, Thayngen 41934
Saaser Museum, Saas Fee 41785
Safier Heimatmuseum, Safien 41787
Sankturbanhof Sursee, Sursee 41926
Schamser Talmuseum, Zillis 42057
Schloß Mörsburg, Stadel, Winterthur 41899
Schloss Sargans - Schlossmuseum und Schlossgalerie, Sargans 41819
Schlossmuseum, Burgdorf 41277
Sensler Museum, Tafers 41927
Simon-Gfeller-Gedenkstube, Heimisbach 41478
Sontg Hippolytus, Veulden 41982
Städtisches Ortsmuseum Schwamendingen, Zürich 42107
Staufferhaus - Sammlung Alt Unterentfelden, Unterentfelden 41959
Tal Museum Engelberg, Engelberg 41352
Talmuseum Ursern, Andermatt 41115
Textil- und Heimatmuseum, Sennwald 41862
Toggenburger Museum, Lichtensteig 41559
Uetiker Museum, Uetikon am See 41956
Weberei- und Heimatmuseum Ruedertal, Schmiedrued 41839
Weissenstein-Museum, Weissenstein bei Solothurn 42005

Syria

Aleppo National Museum, Aleppo 42128
Hama National Museum, Hama 42135

Taiwan

Cultural Museum of Hemei, Hemei 42199
Debao Museum, Guansi 42194
Desing Artifact Museum, Sinpu 42339
Discovery Center of Taipei, Taipei 42389
Farming Village Museum, Cingshuei 42167
Historical Museum of Sansia Town, Sansia 42311
Jian Museum, Jian 42227
Jiaosi Museum, Jiaosi 42228
Kaohsiung Museum of History, Kaohsiung 42238
Kunying Art and Culture Museum, Beimen 42146
Local Cultural Museum, Lujhu 42264
Local Cultural Museum, Keelung 42248
Lugu Museum, Lugu 42260
Mei-Nong Country Gallery, Qishan 42301
Meinong Hakka Museum, Meinong 42273
Meishan Farming Village Museum, Meishan 42274
Museum for Local Culture, Beipu 42147
Museum of Changhua, Changhua 42157
Museum of Chia-Yi City, Chai-Yi 42154
Nantou History Museum, Nantou 42282
Northern Paiwan Art and Culture Museum, Sandimen 42309
Phoenix Cultural Artifacts Center, Tainan 42367
Taipei Story House, Taipei 42429
Yilan County Historical Museum, Yilan 42470

Tajikistan

Historical Museum of Sughd Province, Khojand 42501
Museum of Archaeology and Fortifications, Khojand 42502
Osorkhonai Jumkhuriiavii Ta"rikhi - Kishvarshinosii ba nomi Abuabdullo Rudaki, Pandžakent 42505
Osorkhonai Ta"rikhi - Kishvarshinosii Isfara, Isfara 42499
Osorkhonai Ta"rikhi - Kishvarshinosii Shahri Uro-Teppa, Uro-Teppa 42506
Osorkhonai Ta"rikhi - Kishvarshinosii Viloiati Khujand, Chudžand 42493
Osorkhonai Ta"rikhi - Kishvarshinosii Viloiatii Kŭlob, Kŭlob 42503
Osorkhonai Viloiatii Ta"rikhi - Kishvarshinosii, Chorog 42491

Tanzania

Peace Memorial Museum, Zanzibar 42524
Shinyanga Mazingira Museum, Shinyanga 42519
Singida Museum, Singida 42520

Thailand

Benchamabophit National Museum, Bangkok 42533
Chaiya National Museum, Surat Thani 42606
Chiang Saen National Museum, Chiang Rai 42570
Chumbhot-Punthip Museum, Bangkok 42535
Hariphunchai National Museum, Lampun 42578
In Buri National Museum, Sing Buri 42600
Kamphaeng Phet National Museum, Kamphaeng Phet 42572
Kao Chong, Trang 42608
Kao Luang, Nakhon Si Thammarat 42584
Kao Ta Phet, Sukhothai 42602
Kao Yai, Saraburi 42599
Lan Sang, Tak 42607
Maha Wirawong National Museum, Nakhon Ratchasima 42583
Matchimawas National Museum, Songkhla 42601
Nam Nao, Petchabun 42590
National Museum, Bangkok 42552
Ob Luang, Chiang Mai 42569
Phipittapan Tasatan Haengchart Khon Kaen, Khon Kaen 42574
Phra Borommathat National Museum, Nakhon Si Thammarat 42586
Phra Chetuponwimonmangkhalaram National Museum, Bangkok 42556
Phra Phuttachinnarat National Museum, Phitsanulok 42595
Phukradung, Loei 42579
Sam Roi Yod, Phachuab Khiri Khan 42592
Sawanworanayok National Museum, Sukhothai 42604
Thrai Kao, Battani 42566
Ton Sak Yai, Uttaradit 42609
Tung Salaeng Luang, Phitsanulok 42596
Wachiraprasat Museum, Phetchaburi 42593
Wat Ko Museum, Phetchaburi 42594
Wat Phra Thart Lampang Luang Museum, Lampang 42577

Togo

Musée d'Histoire et d'Ethnographie, Aneho 42610
Musée Régional de Sokadé, Sokadé 42613
Musée Régional des Savanes, Dapaong 42611

Tunisia

Musée Dar Cheraït, Tozeur 42661
Musée de Tabarka, Tabarka 42658
Musée de Zarzis, Zarzis 42669
Musée du Patrimoine, Guellala 42626
Musée Regional de Mahdia, Mahdia 42637
Musée Sidi Bou Saïd, Sidi Bou Saïd 42655
Musée Sidi Zitouni, Houmt Souk 42628

Turkey

Adana Etnografya Müzesi, Adana 42673
Adıyaman Müzesi, Adıyaman 42675
Afyon Müzesi, Afyon 42678
Aksaray Müzesi ve Bağli Örenyerleri, Aksaray 42681
Akşehir Müzesi, Akşehir 42682
Alanya Müzeleri ve Örenyerleri, Alanya 42687
Amasra Müzesi ve Tarihçesi, Amasra 42688
Amasya Müzesi, Amasya 42690
Anamur Müzesi ve Örenyerleri, Anamur 42692
Antalya Müzesi, Antalya 42717
Arkoloji ve Etnografya Müzesi, Edirne 42766
Atatürk ve Etnografya Müzesi, Akşehir 42683
Atatürk ve Etnografya Müzesi, Denizli 42757
Atatürk ve Etnografya Müzesi, Uşak 42959
Aydın Müzesi, Aydın 42721
Beypazarı Kültür ve Tarih Müzesi, Beypazarı 42729
Boğazköy Müzesi, Boğazkale 42732
Bolu Müzesi, Bolu 42734
Bursa Kent Müzesi, Bursa 42743
Çankırı Müzesi, Çankırı 42751
Çeşme Müzesi, Çeşme 42752
Giresun Müzesi, Giresun 42790
İsparta Müzesi, İsparta 42800
İznik Müzesi, İznik 42860
Kars Müzesi, Kars 42867
Kayseri Etnografya Müzesi, Kayseri 42876
Kırklareli Müzesi, Kırklareli 42879
Kırşehir Müzesi, Kırşehir 42881
Kocaeli Müzeleri ve Örenyerleri, İzmit 42859
Kütahya Müzeleri ve Örenyerleri, Kütahya 42896
Kuva-yi Milliye Müzesi, Balıkesir 42724
Livapaşa Konağı Etnografya Müzesi, Kastamonu 42870
Manisa Müzesi, Manisa 42903
Mersin Müzesi, Mersin 42907
Muğla Müzesi, Muğla 42914
Nevşehir Müzesi, Nevşehir 42917
Ocak Köyü Özel Müzesi, Kemaliye 42877
Ödemiş Müzesi, Ödemiş 42920
Ordu Etnografya Müzesi, Ordu 42922
Sakarya Müzesi, Sakarya 42928
Samsun Arkeoloji ve Etnografya Müzesi, Samsun 42931
Şehir Müzesi, İstanbul 42839
Silifke Müzesi, Silifke 42942
Sinop Müzesi, Sinop 42944
Söğüt Müzesi, Söğüt 42947
Tas Oda Etnografya Müzesi, Burdur 42738
Tekirdağ Müzesi, Tekirdağ 42951
Trabzon Müzesi, Trabzon 42957
TRT Müzesi ve Ulusal Kurtulus Sergisi, Ankara 42713
Türk İslam Eserleri ve Etnografya Müzesi, Erzurum 42778
Van Müzesi, Van 42962
Yalvaç Müzesi ve Örenyerleri, Yalvaç 42965

Turkmenistan

Earthquake Museum, Aşgabat 42970
Etnografiki Muzeý, Türkmenbaşi 42983
Museum of Regional History, Türkmenbaşi 42984
Taryhy we Etnografiki Muzeý, Türkmenabat 42982
Taryhy Ýer Könürgenç, Könürgenç 42977

Ukraine

Baturins'kyj Deržavnyj Istoričnyj Zapovidnyk Get'mans'ka Stolicja, Baturin 43000
Čerkas'kyj Oblasnyj Krajeznavčyj Muzej, Čerkasy 43002
Černivec'kyj Krajeznavčyj Muzej, Černivci 43010
Černivec'kyj Muzej Bukovins'koï Diaspory, Černivci 43013
Chmel'nyc'kyj Oblasnyj Krajeznavčyj Muzej, Chmel'nyc'kyj 43027
Diorama Šturm Izmaïla, Izmaïl 43045
Donec'kij Oblasnyj Krajeznavčyj Muzej, Donec'k 43038
Istoriko-memorial'nyj Muzej Polityčnych V'jazniv, Ternopil' 43192
Istoryko-architekturna Pam'jatka-muzej Kyïvska Fortecja, Kyïv 43079
Izmaïl's'kyj Krajeznavčyj Muzej Pridunav'ja, Izmaïl 43047
Izmaïl's'kyj Muzej O.V. Suvorova, Izmaïl 43048
Kam'jans'kyj Istoryčnyj Muzej, Kam'janka 43055
Kanivs'kyj Istoryčnyj Muzej, Kaniv 43058
Kirovograds'kyj Oblasnyj Krajeznavčyj Muzej, Kirovograd 43068
Krivoriz'kyj Krajeznavčyj Muzej, Kryvyj Rig 43075
Krymski Respublikanskij Kraevedčeskij Muzej, Simferopol' 43185
Literaturno-mystec'kyj Muzej, Chmel'nyc'kyj 43028
Lugans'kyj Oblasnyj Krajeznavčyj Muzej, Lugans'k 43115
Mariupol's'kij Krajeznavčyj Muzej, Mariupol' 43144
Muzej Get'manstva, Kyïv 43088
Muzej Istoriï Mista Chmel'nyc'kogo, Chmel'nyc'kyj 43029
Muzej Istoriï Mista Kolomyï, Kolomija 43071
Muzej Istoriï ta Kul'tury Mista Lugans'ka, Lugans'k 43117
Muzej Pidpil'nyj ta Partizans'kyj Ruch na Mykolaïvščyni u Roki Velikoï Vitčyznjanoï Vijny 1941-1945 rr, Mykolaïv 43148
Muzej Vyzvol'nych Zmagan' Prikarpats'koho Kraju, Ivano-Frankivs'k 43044
Mykolaïvs'kyj Oblasnyj Krajeznavčyj Muzej, Mykolaïv 43151
Oblasnyj Krajeznavčyj Muzej, Žytomyr 43214
Oblasnyyj Krajeznavčyj Muzej, Cherson 43024
Odes'kyj Istoriko-krajeznavčyj Muzej, Odesa 43159
Odes'kyj Literaturnyj Muzej, Odesa 43161
Rivnens'kyj Krajeznavčyj Muzej, Rivne 43173
Sums'kyj Oblasnyj Krajeznavčyj Muzej, Sumy 43191
Ternopil's'kyj Oblasnyj Krajeznavčyj Muzej, Ternopil' 43194
Vijs'kovo-istoryčnyj Muzej, Černigiv 43008
Vinnyc'kyj Oblasnyj Krajeznavčyj Muzej, Vinnycja 43206
Volodymyr-Volyns'kij Istoryčnyj Muzej, Volodymyr-Volyns'kij 43207
Volyns'kyj Krajeznavčyj Muzej, Luc'k 43113
Zaporiz'kyj Krajeznavčyj Muzej, Zaporižžja 43210

United Arab Emirates

Dubai Museum, Dubai 43218
Fujairah Museum, Fujairah 43221
Sharjah Heritage Museum, Sharjah 43227

United Kingdom

Abbey House Museum, Leeds 44658
Abbeydale Industrial Hamlet, Sheffield 45648
Abbotsford House, Melrose 45124
Aberconwy House, Conwy 43886
Abergavenny Museum, Abergavenny 43252
Abernethy Museum, Abernethy 43254
Abertillery Museum, Abertillery 43255
Abingdon Museum, Abingdon 43260
Abington Park Museum, Northampton 45268
Aldeburgh Museum, Aldeburgh 43267
Alderney Society Museum, Alderney 43269
Aldershot Military Museum and Rushmoor Local History Gallery, Aldershot 43271
Alford Heritage Centre, Alford, Aberdeenshire 43276
Alford Manor House Museum, Alford, Lincolnshire 43278
Allhallows Museum of Lace and Antiquities, Honiton 44483
Almond Valley Heritage Centre, Livingston 44751
Almonry Heritage Centre, Evesham 44191
Alyth Museum, Alyth 43296
Amersham Museum, Amersham 43301
Amgueddfa Pontypool Museums, Pontypool 45400
Ancient House Museum, Thetford 45846
Andover Museum, Andover 43302
Anne of Cleves House Museum, Lewes 44697
Arbroath Museum, Arbroath 43313
Arbuthnot Museum, Peterhead 45371
Armadale Community Museum, Armadale 43316
Armagh County Museum, Armagh 43317
Arran Heritage Museum, Brodick 43650
Arthur Cottage, Ballymena 43363
Arundel Museum and Heritage Centre, Arundel 43323
Ashburton Museum, Ashburton 43325
Ashby-de-la-Zouch Museum, Ashby-de-la-Zouch 43326
Ashford Museum, Ashford 43327
Ashton Court Visitor Centre, Bristol 43629
Ashwell Village Museum, Ashwell 43335
Athelstan Museum, Malmesbury 45078
Atholl Country Life Museum, Blair Atholl 43536
Auld Kirk Museum, Kirkintilloch 44613
Avoncroft Museum of Historic Buildings, Bromsgrove 43654
Axminster Museum, Axminster 43341
Ayscoughfee Hall Museum, Spalding 45721
Bagshaw Museum, Batley 43415
Baird Institute Museum, Cumnock 43941
Ballycastle Museum, Ballycastle 43360
Ballymena Museum, Ballymena 43364
Ballymoney Museum, Ballymoney 43366
Banbury Museum, Banbury 43370
Banchory Museum, Banchory 43372
Banff Museum, Banff 43374
Bantock House, Wolverhampton 46057
Barnet Museum, London 44792
Bassetlaw Museum and Percy Laws Memorial Gallery, Retford 45484
Battle Museum of Local History, Battle 43418
Bayle Museum, Bridlington 43609
Baysgarth House Museum, Barton-upon-Humber 43391
The Beacon, Whitehaven, Whitehaven 46000
Beaminster Museum, Beaminster 43421
Beamish, the North of England Open Air Museum, Beamish 43422
Beccles and District Museum, Beccles 43428
Beck Isle Museum of Rural Life, Pickering 45381
Beckford's Tower and Museum, Bath 43402
Bedale Museum, Bedale 43430
Bedford Museum, Bedford 43434
Bellaghy Bawn, Magherafelt 45070
Bennie Museum, Bathgate 43414
Berkswell Village Museum, Berkswell 43456
Bersham Heritage Centre and Ironworks, Wrexham 46081
Berwick Borough Museum and Art Gallery, Berwick-upon-Tweed 43459
Bexhill Museum, Bexhill-on-Sea 43471
Bexley Museum, London 44795
Bishop Bonner's Cottage Museum, Dereham 43971
Bishop's Stortford Museum, Bishop's Stortford 43523
Bishops Waltham Museum, Bishop's Waltham 43525
Bitterne Local History Centre, Southampton 45702
Blackridge Community Museum, Blackridge 43531
Blackwater Valley Museum, Benburb 43450
Blake Museum, Bridgwater 43606
Blandford Forum Museum, Blandford Forum 43539
Blickling Hall, Norwich 45274
Bloxham Village Museum, Bloxham 43545
Bodmin Town Museum, Bodmin 43548
Bognor Regis Museum, Bognor Regis 43552
Bo'ness Heritage Area, Bo'ness 43558
Borough Museum and Art Gallery, Berwick-upon-Tweed 43460
Boscobel House, Brewood 43601
Bourne Hall Museum, Ewell 44192
Bradford-on-Avon Museum, Bradford-on-Avon 43583
Braintree District Museum, Braintree 43589
Brander Museum, Huntly 44503
Brandon Heritage Museum, Brandon 43591
Braunton and District Museum, Braunton 43592
Brechin Museum, Brechin 43595
Brecknock Museum and Art Gallery, Brecon 43597
Bredy Farm Old Farming Collection, Burton Bradstock 43687
Brent Museum, London 44799
Brentwood Museum, Brentwood 43600
Brewery Chapel Museum, Halstead 44390
Brewhouse Yard Museum, Nottingham 45289
Bridewell Museum, Norwich 45275
Bridgnorth Northgate Museum, Bridgnorth 43603
Bridport Museum, Bridport 43612
Brightlingsea Museum, Brightlingsea 43617
Brighton Museum and Art Gallery, Brighton 43621
Bristol's City Museum and Art Gallery, Bristol 43632
Bromsgrove Museum, Bromsgrove 43655
Broughton House and Garden, Kirkcudbright 44610
Broughty Castle Museum, Dundee 44045
Bruce Castle Museum, London 44807
Buckfast Abbey, Buckfastleigh 43659
Buckinghamshire County Museum, Aylesbury 43342
Bude-Stratton Museum, Bude 43667
Bungay Museum, Bungay 43672
Burghead Museum, Burghead 43674
Burnham on Crouch and District Museum, Burnham-on-Crouch 43675
Burntisland Edwardian Fair Museum, Burntisland 43681
Burton Court, Eardisland 44079
Burwell Museum, Burwell 43690
Bury Art Gallery and Museum, Bury, Lancashire 43691
Bushey Museum and Art Gallery, Bushey 43699
Buxton Museum and Art Gallery, Buxton 43705
Caernarfon Maritime Museum, Caernarfon 43708
Callendar House, Falkirk 44205
Cambridge and County Folk Museum, Cambridge 43723
Campbeltown Museum, Campbeltown 43739
Canolfan Y Plase, Bala 43357
Canolfan Y Plase, Y Bala 46090
Canterbury Heritage Museum, Canterbury 43742
Cardigan Heritage Centre, Cardigan 43756
Carew Manor and Dovecote, Beddington 43432
Carisbrooke Castle Museum, Newport, Isle of Wight 45243
Carlisle Collection of Miniature Rooms, Nunnington 45306
Castle Combe Museum, Castle Combe 43768
Castle House Museum, Dunoon 44061
Castle Museum, Clitheroe 43854
Castleton Village Museum, Castleton 43773
Castletown Museum, Gerston 44267
Cater Museum, Billericay 43486
Champs Chapel Museum, East Hendred 44086
Chard and District Museum, Chard 43790
Charlbury Museum, Charlbury 43793
Charnwood Museum, Loughborough 45031
Charterhouse School Museum, Godalming 44321
Chatelherault, Ferniegair 44223
Chelmsford Museum, Chelmsford 43803
Chepstow Museum, Chepstow 43812
Chertsey Museum, Chertsey 43814
Chesham Museum-The Stables, Chesham 43815
Chesterfield Museum and Art Gallery, Chesterfield 43820
Chichester District Museum, Chichester 43825
Chiltern Open Air Museum, Chalfont Saint Giles 43786
Chippenham Museum and Heritage Centre, Chippenham 43835
Chipping Norton Museum of Local History, Chipping Norton 43836
Church Farmhouse Museum, London 44821
Cirencester Lock-Up, Cirencester 43846
City Museum and Art Gallery, Gloucester 44315
City Museum and Art Gallery, Plymouth 45390
City Museum and Records Office, Portsmouth 45428
Clackmannshire Council Museum, Alloa 43282
Clan Donnachaidh Museum, Pitlochry 45384
Clan Gunn Museum, Latheron 44644
Clare Ancient House Museum, Clare 43848
Clink Prison Museum, London 44825
Clive House Museum, Shrewsbury 45672
Clun Local History Museum, Clun 43856
Coggeshall Heritage Centre and Museum, Coggeshall 43871
Coldstream Museum, Coldstream 43878
Colne Valley Museum, Golcar 44323
Colzium Museum, Kilsyth 44581
Combe Martin Museum, Combe Martin 43881
Commandery Museum, Worcester 46070
Corinium Museum, Cirencester 43847
Cothey Bottom Heritage Centre, Ryde, Isle of Wight 45534
Countryside Museum, East Budleigh 44082
County Museum Technical Centre, Halton 44391
Court Hall Museum, Sittingbourne 45681
Court House and Museum, Pevensey 45379
Cowbridge and District Museum, Cowbridge 43906
Crail Museum, Crail 43913
Cranbrook Museum, Cranbrook 43914
Craven Museum and Gallery, Skipton 45688
Crawfordjohn Heritage Venture, Crawfordjohn 43915
Crawley Museum, Crawley 43916
Creetown Heritage Museum, Creetown 43919
Crewkerne and District Museum and Heritage Centre, Crewkerne 43924
Cromer Museum, Cromer 43929
Croydon Museum, Croydon 43934
Croydon Natural History and Scientific Society, Croydon 43935
Crystal Palace Museum, London 44833
Cuckfield Museum, Cuckfield 43936
Cumbernauld Museum, Cumbernauld 43940
Cynon Valley Museum and Gallery, Aberdare 43232
Dacorum Heritage, Berkhamsted 43455
Dales Countryside Museum, Hawes 44424
Dartford Borough Museum, Dartford 43951
Dartmouth Museum, Dartmouth 43953
Daventry Museum, Daventry 43956
Dawlish Museum, Dawlish 43957
Deal Maritime and Local History Museum, Deal 43959
Dean Heritage Museum, Soudley 45692
Denbigh Museum and Gallery, Denbigh 43964
Derby Museum and Art Gallery, Derby 43965
Derrymore House, Bessbrook 43464
Dewey Museum, Warminster 45932
Dewsbury Museum, Dewsbury 43980
Dingwall Museum, Dingwall 43982
Discovery Museum, Newcastle-upon-Tyne 45226
Diss Museum, Diss 43986
Ditchling Museum, Ditchling 43987
Dollar Museum, Dollar 43989
Doncaster Museum and Art Gallery, Doncaster 43992
Doon Valley Museum, Dalmellington 43946
Dorchester Abbey Museum, Dorchester-on-Thames 44005
Dorman Memorial Museum, Middlesbrough 45140
Dorney Court, Windsor 46040
Dornoch Heritage Museum, Dornoch 44009
Dorset County Museum, Dorchester 43999
Douglas Heritage Museum, Douglas, Lanarkshire 44010
Dover Museum, Dover 44013
Down County Museum, Downpatrick 44020
Drenewydd Museum, Bute Town 43704
Drumlanrig's Tower, Hawick 44425
Dualchas-Museum Bharraigh Agus Bhatarsaidh, Isle-of-Barra 44537
Dumfries Museum Camera Obscura, Dumfries 44036
Dunbar Town House Museum, Dunbar 44041
Dunblane Museum, Dunblane 44044
Dunfermline Museum, Dunfermline 44055
Dunkeld Cathedral Chapter House Museum, Dunkeld 44059
Dunrobin Castle Museum, Golspie 44325
Duns Area Museum, Duns 44064
Durham Heritage Centre and Museum, Durham 44070
Earls Barton Museum of Local Life, Earls Barton 44080
Easdale Island Folk Museum, Easdale Island 44081
East Surrey Museum, Caterham 43780
Eastbourne Heritage Centre, Eastbourne 44097
Eastleigh Museum, Eastleigh 44104
Eden Valley Museum, Edenbridge 44111
Egham Museum, Egham 44163
Elmbridge Museum, Weybridge 45986
Ely Museum, Ely 44172
Emsworth Museum, Emsworth 44175
Erewash Museum, Ilkeston 44511
Erith Museum, Erith 44185
Exmouth Museum, Exmouth 44197
Eyam Museum, Eyam 44199
Fairlynch Museum, Budleigh Salterton 43668
Fakenham Museum of Gas and Local History, Fakenham 44203
Feering and Kelvedon Local History Museum, Kelevedon 44553
Fermanagh County Museum at Enniskillen Castle, Enniskillen 44178
Fernhill House, Belfast 43440
Fetlar Interpretive Centre, Fetlar 44225
Fife Folk Museum, Ceres 43785
Filey Museum, Filey 44227
First Garden City Heritage Museum, Letchworth 44694
Fleur de Lis Heritage Centre, Faversham 44221
Flintham Museum, Flintham 44233
Fochabers Folk Museum, Fochabers 44235
Folkestone Museum and Art Gallery, Folkestone 44237
Folklife Display, Lisnaskea 44726
Forty Hall Museum, Enfield 44176
Fossil Grove, Glasgow 44280
Fraserburgh Museum, Fraserburgh 44253
Gairloch Heritage Museum, Gairloch 44260
Gallery Oldham, Oldham 45314
Gillingham Museum, Gillingham, Dorset 44268
Gladstone Court Museum, Biggar 43481
Glastonbury Lake Village Museum, Glastonbury 44307
Glencoe and North Lorn Folk Museum, Glencoe 44310
Glenesk Folk Museum, Glenesk 44311
Glossop Heritage Centre, Glossop 44314
Gloucester Folk Museum, Gloucester 44316
Godalming Museum, Godalming 44322
Goole Museum and Art Gallery, Goole 44329
Gosport Museum, Gosport 44334
Gower Heritage Centre, Gower 44340
Grace Darling Museum, Bamburgh 43369
Grangemouth Museum, Grangemouth 44342
Grantham Museum, Grantham 44345
Grantown Museum, Grantown-on-Spey 44347
Gravesham Museum, Gravesend, Kent 44350
Great Bardfield Cage, Great Bardfield 44353
Great Bardfield Cottage Museum, Great Bardfield 44354
Greenwich Heritage Centre, London 44863
Greenwich Local History Centre, London 44864
Grosvenor Museum, Chester 43818
The Guards Museum, London 44865
Guernsey Museum and Art Gallery, Saint Peter Port 45591
Guildford Museum, Guildford 44371
Guildhall Museum, Carlisle 43759
Guildhall Museum, Chichester 43828
Guildhall Museum, Sandwich 45615
Gwynedd Education and Culture, Caernarfon 43709
Gwynedd Museum and Art Gallery, Bangor, Gwynedd 43376
Hackney Museum, London 44868
Hailsham Heritage Centre, Hailsham 44379
Hall Place, Bexley 43473
Hallaton Museum, Hallaton 44389
Halliwell's House Museum, Selkirk 45635
Ham House, Richmond, Surrey 45492
Hamilton Low Parks Museum, Hamilton 44392
Hampstead Museum, London 44870
Handsworth Saint Mary's Museum, Handsworth 44393
Hanseatic Booth or The Pier House, Symbister 45821
Harborough Museum, Market Harborough 45109
Harbour Life Exhibition, Bridport 43613
Harbour Museum, Derry 43973
Harris Museum and Art Gallery, Preston 45444
Harrow Museum, Harrow 44404
Hartland Quay Museum, Bideford 43478
Hartland Quay Museum, Hartland 44405
Havant Museum, Havant 44420
Haverhill and District Local History Centre, Haverhill 44423
Hawick Museum, Hawick 44426
Hedon Museum, Hedon 44437
Helena Thompson Museum, Workington 46075
Helston Folk Museum, Helston 44444
Henfield Museum, Henfield 44448
Heptonstall Museum, Heptonstall 44452
Herbert Art Gallery and Museum, Coventry 43902
Herne Bay Museum and Gallery, Herne Bay 44461
Hertford Museum, Hertford 44462
Hinckley and District Museum, Hinckley 44471
Hindley Museum, Hindley 44472
Historic Resources Centre, Annan 43305
History Shop, Wigan 46013
Hitchin Museum and Art Gallery, Hitchin 44474
Holsworthy Museum, Holsworthy 44478
Honeywood Heritage Centre, London 44877
Hornsea Museum of Village Life, Hornsea 44484
Horsforth Village Museum, Horsforth 44486
House on Crutches Museum, Bishop's Castle 43522
Howell Harris Museum, Brecon 43598
Hull and East Riding Museum, Kingston-upon-Hull 44597
Hythe Local History Room, Hythe, Kent 44507
Iceni Village and Museums, Cockley Cley 43870
Ilchester Museum, Ilchester 44508
Ilfracombe Museum, Ilfracombe 44510
Inverkeithing Museum, Inverkeithing 44522
Inverness Museum and Art Gallery, Inverness 44525
Iona Heritage Centre, Isle-of-Iona 44539
Irish Linen Centre and Lisburn Museum, Lisburn 44725
Ironbridge Gorge Museum Trust, Coalbrookdale 55069
Isles of Scilly Museum, Saint Mary's 45584
Islington Museum, London 44885
Jane Welsh Carlyle's House, Haddington 44375
Jersey Museum, Saint Helier 45571
Jewel Tower, London 44887
John Hastie Museum, Strathaven 45787
John McDouall Stuart Museum, Kirkcaldy 44608
Judge's Lodging, Presteigne 45443
Kegworth Museum, Kegworth 44549
Kelham Island Museum, Sheffield 45652
Kendal Museum, Kendal 44558
Keswick Museum and Art Gallery, Keswick 44564
Kilsyth's Heritage, Kilsyth 44582
Kilwinning Abbey Tower, Kilwinning 44584

King John's Hunting Lodge, Axbridge 43340
Kingston Museum, Kingston-upon-Thames 44602
Kinross Museum, Kinross 44607
Kirkcaldy Museum and Art Gallery, Kirkcaldy 44609
Kirkleatham Museum, Redcar 45473
Kirriemuir Gateway to the Glens Museum, Kirriemuir 44618
Knaresborough Castle and Old Courthouse Museum, Knaresborough 44619
Knutsford Heritage Centre, Knutsford 44623
Kyle and Carrick District Library and Museum, Ayr 43346
Kyngston House Museum, Saint Albans 45545
Laing Museum, Newburgh 45218
Lanark Museum, Lanark 44629
Lancaster City Museum, Lancaster 44633
Lanman Museum, Framlingham 44251
Largs Museum, Largs 44642
Larne Museum, Larne 44643
Larrybane and Carrick-A-Rede, Ballycastle 43361
Lauderdale House, London 44898
Lawrence House Museum, Launceston 44648
Laxfield and District Museum, Laxfield 44652
Leamington Spa Art Gallery and Museum, Leamington Spa 44654
Leatherhead Museum of Local History, Leatherhead 44655
Ledbury Heritage Centre, Ledbury 44657
Leeds Museum, Leeds 44664
Leighton Hall, Carnforth 43763
Leominster Folk Museum, Leominster 44691
Letchworth Museum and Art Gallery, Letchworth 44695
Lichfield Heritage Centre, Lichfield 44705
The Lightbox, Woking 46055
Linlithgow Story, Linlithgow 44721
Litcham Village Museum, Litcham 44727
Little Houses, Dunkeld 44060
Littlehampton Museum, Littlehampton 44729
Llancaiach Fawr Living History Museum, Nelson 45199
Llandudno Museum and Art Gallery, Llandudno 44759
Llanidloes Museum, Llanidloes 44773
Lleyn Historical and Maritime Museum, Nefyn 45197
London Borough of Bromley Museum, Orpington 45321
Long Eaton Town Hall, Long Eaton 45024
Lossiemouth Fisheries and Community Museum, Lossiemouth 45029
Lostwithiel Museum, Lostwithiel 45030
Louth Museum, Louth 45037
Lower Methil Heritage Centre, Lower Methil 45039
Lowestoft Museum, Lowestoft 45042
Lowewood Museum, Hoddesdon 44476
Ludlow Museum, Ludlow 45045
Lydd Town Museum, Lydd 45051
Lyme Regis Philpot Museum, Lyme Regis 45054
Lyn and Exmoor Museum, Lynton 45057
Lytham Heritage Centre, Lytham Saint Anne's 45060
Macclesfield Silk Museum, Macclesfield 45065
McKechnie Institute, Girvan 44270
McManus Galleries, Dundee 44047
Maidstone Museum and Bentlif Art Gallery, Maidstone 45073
Maldon District Museum, Maldon 45076
Mallaig Heritage Centre, Mallaig 45077
Malvern Museum, Great Malvern 44356
Manor Cottage Heritage Centre, Southwick 45719
Manor House Art Gallery and Museum, Ilkley 44512
Manor House Museum, Kettering 44568
Mansfield Museum and Art Gallery, Mansfield 45101
Margate Old Town Hall Local History Museum, Margate 45106
Margrove - South Cleveland Heritage Centre, Boosbeck 43561
Market Lavington Village Museum, Market Lavington 45111
Marlipins Museum, Shoreham-by-Sea 45668
Measham Museum, Measham 45121
Medway Heritage Centre, Chatham 43797
Mellerstain House, Gordon 44331
Melton Carnegie Museum, Melton Mowbray 45126
Merchant's House Museum, Plymouth 45391
Mere Museum, Mere 45128
Mersea Island Museum, West Mersea 45972
Methil Heritage Centre, Methil 45133
Mildenhall and District Museum, Mildenhall, Suffolk 45144
Milford Haven Museum, Milford Haven 45145
Milton Keynes Museum, Milton Keynes 45154
Minster Abbey Gatehouse Museum, Sheerness 45647
Moat Park Heritage Centre, Biggar 43484
Moffat Museum, Moffat 45161
Mold Museum, Mold 45163
Motherwell Heritage Centre, Motherwell 45179
Moyse's Hall Museum, Bury Saint Edmunds 43695
Much Wenlock Museum, Much Wenlock 45185
Mull Museum, Tobermory 45861
The Museum, Dunwich 44068
The Museum, Newton Stewart 45252
Museum and Art Gallery, Forfar 44242
Museum and Art Gallery, Newport, Gwent 45240
Museum and Art Gallery, Nuneaton 45305
The Museum in the Park, Stroud 45794
Museum Nan Eilean, Lionacleit 44723
Museum Nan Eilean, Steornabhagh, Stornoway 45767
Museum of Ambleside, Ambleside 43299
Museum of Barnstaple and North Devon, Barnstaple 43387
Museum of Cannock Chase, Hednesford 44436
Museum of Dartmoor Life, Okehampton 45312
Museum of East Anglian Life, Stowmarket 45773
The Museum of Edinburgh, Edinburgh 44133
Museum of Harlow, Harlow 44396
Museum of Island History, Newport, Isle of Wight 45244
Museum of Islay Life, Port Charlotte 45411
Museum of Lakeland Life, Kendal 44559
Museum of Lancashire, Preston 45445
Museum of Lincolnshire Life, Lincoln 44714
Museum of Local History, Hastings 44416
Museum of London, London 44928
Museum of North Craven Life, Settle 45640
Museum of Oxford, Oxford 45333
Museum of Reading, Reading 45469
Museum of Richmond, Richmond, Surrey 45493
Museum of Saint Albans, Saint Albans 45547
Museum of South Somerset, Yeovil 46096
Museum of South Yorkshire Life, Doncaster 43994
Museum of Speed, Pendine 45355
Museum of the Cumbraes, Millport 45148
Museum of the Holmesdale Natural History Club, Reigate 45481
Museum of the Isles, Isle-of-Skeye 44541
Museum of the Manchester Regiment, Ashton-under-Lyne 43332
Museum of Victorian Whitby, Whitby 45995
Nairn Fishertown Museum, Nairn 45189
Nantwich Museum, Nantwich 45191
Narberth Museum, Narberth 45193
National Horseracing Museum, Newmarket 45239
Navan Centre, Armagh 43319
Neath Museum, Neath 45196
Ness Historical Society Museum, Port-of-Ness 45412
New Hall, Dymchurch 44077
New Mills Heritage and Information Centre, New Mills 45204
Newham Heritage Centre, London 44945
Newhaven Heritage Museum, Edinburgh 44142
Newport Pagnell Historical Society Museum, Newport Pagnell 45247
Newry and Mourne Arts Centre and Museum, Newry 45250
Newton Abbot Town and Great Western Railway Museum, Newton Abbot 45251
Nidderdale Museum, Pateley Bridge 45346
Normanton Church Museum, Oakham 45307
North Ayrshire Museum, Saltcoats 45608
North Berwick Museum, North Berwick 45265
North Cornwall Museum and Gallery, Camelford 43738
North Down Heritage Centre, Bangor, County Down 43375
North Lincolnshire Museum, Scunthorpe 45628
North Somerset Museum, Weston-super-Mare 45981
Norwood Museum, Holm 44477
Oban War and Peace Museum, Oban 45310
Old Bell Museum, Montgomery 45170
The Old Byre Heritage Centre, Dervaig 43977
Old Gaol Museum, Buckingham 43666
Old Guildhall Museum, East Looe 44091
Old Haa, Yell 46093
Old House Museum, Bakewell 43355
Old Post Office Museum, Turriff 45883
Oriel Ynys Môn, Llangefni 44766
Otley Museum, Otley 45324
Oundle Museum, Oundle 45325
The Oxfordshire Museum, Woodstock 46067
Padstow Museum, Padstow 45341
Paisley Museum and Art Galleries, Paisley 45345
Parc Howard Museum and Art Gallery, Llanelli 44762
Peacock Heritage Centre, Chesterfield 43822
Pendle Heritage Centre, Barrowford 43390
Penlee House Gallery and Museum, Penzance 45361
Penrith Museum, Penrith 45357
The People's Museum, Belfast 43445
Perranzabuloe Folk Museum, Perranporth 45363
Petersfield Museum, Petersfield 45374
Pewsey Heritage Centre, Pewsey 45380
Pittencrieff House Museum, Dunfermline 44056
The Police Museum, Belfast 43446
Pontefract Museum, Pontefract 45398
Pontypridd Museum, Pontypridd 45403
Port Sunlight Heritage Centre, Port Sunlight 45416
Porthcawl Museum, Porthcawl 45420
Portland Basin Museum, Ashton-under-Lyne 43333
Potland Museum, Basildon 43393
Potters Bar Museum, Potters Bar 45441
Prescot Museum, Prescot 45442
Priest's House, Easton-on-the-Hill 44107
Priest's House Museum, Wimborne Minster 46020
Prittlewell Priory Museum, Southend-on-Sea 45712
Pumphouse Educational Museum, London 44960
Purton Museum, Purton 45452
Queensferry Museum, South Queensferry 45696
Raasay Heritage Museum, Kyle 44626
Radnorshire Museum, Llandrindod Wells 44757
Radstock Museum, Radstock 45455
Ramsgate Museum, Ramsgate 45458
Red House Museum, Gomersal 44326
Redbridge Museum, Ilford 44509
Reigate Priory Museum, Reigate 45482
Rhondda Heritage Park, Trehafod 45873
Rhondda Museum, Llwynypia 44776
Rhyl Library, Museum and Arts Centre, Rhyl 45486
Richmondshire Museum, Richmond, North Yorkshire 45490
River and Rowing Museum, Henley-on-Thames 44449
Robin Hood's Bay and Fylingdale Museum, Robin Hood's Bay 45504
Rochdale Art and Heritage Centre, Rochdale 45505
Rossendale Museum, Rossendale 45515
Rottingdean Grange Art Gallery and Museum, Brighton 43624
Royal Albert Memorial Museum and Art Gallery, Exeter 44195
Royal Pump Room Museum, Harrogate 44402
Royston and District Museum, Royston 45521
Rozelle House Galleries, Ayr 43348
Ruddington Village Museum, Ruddington 45525
Rustington Heritage Exhibition Centre, Rustington 45531
Rutland County Museum, Oakham 45309
Rydal Mount, Ambleside 43300
Rye Castle Museum, Rye 45536
Rye Heritage Centre Town Model, Rye 45537
Saffron Walden Museum, Saffron Walden 45540
Saint Andrews Museum, Saint Andrews 45555
Saint Andrews Preservation Museum, Saint Andrews 45556
Saint Barbe Museum and Art Gallery, Lymington 45055
Saint Margaret's Museum, Saint Margaret's Bay 45582
Saint Neots Museum, Saint Neots 45585
Saint Patrick's Trian, Armagh 43321
Salford Museum and Art Gallery, Salford 45601
Saltash Heritage Centre, Saltash 45607
Sanquhar Tolbooth Museum, Sanquhar 45618
Scalloway Museum, Scalloway 45621
Scolton Manor Museum, Haverfordwest 44422
Seaford Museum of Local History, Seaford 45629
Segedunum Roman Fort, Baths and Museum, Wallsend 45917
Sevenoaks Museum and Gallery, Sevenoaks 45642
Shaftesbury Town Museum, Shaftesbury 45645
Shambles Museum, Newent 45234
Sherborne Museum, Sherborne 45661
Sheringham Museum, Sheringham 45663
Sherrier Resources Centre, Lutterworth 45050
Shetland Museum, Lerwick 44693
Shirehall Museum, Little Walsingham 44728
Shotts Heritage Centre, Shotts 45670
Shrewsbury Museum and Art Gallery, Shrewsbury 45676
Shugborough Estate Museum, Milford, Staffordshire 45146
Sidmouth Museum, Sidmouth 45679
Sir Henry Jones Museum, Llangernyw 44767
Sir Walter Scott's Courtroom, Selkirk 45638
Sittingbourne Heritage Museum, Sittingbourne 45684
Sledmere House, Driffield 44023
Slough Museum, Slough 45690
Somerset County Museum, Taunton 45829
South Ribble Museum and Exhibition Centre, Leyland 44702
South Shields Museum and Art Gallery, South Shields 45698
Southend Central Museum, Southend-on-Sea 45714
Southwick Hall, Peterborough 45370
Southwold Museum, Southwold 45720
Spalding Gentlemen's Society Museum, Spalding 45723
Spelthorne Museum, Staines 45732
Sperrin Heritage Centre, Plumridge 45389
Stamford Museum, Stamford 45737
Stevenage Museum, Stevenage 45742
Stewart Collection, Pocklington 45394
Stewarton and District Museum, Stewarton 45744
The Stewartry Museum, Kirkcudbright 44611
Steyning Museum, Steyning 45745
Stirling Smith Art Gallery and Museum, Stirling 45752
Stockport Museum, Stockport 45755
Stranraer Museum, Stranraer 45777
Stromness Museum, Stromness 45793
Summerlee Heritage Park, Coatbridge 43864
Sunderland Museum, Sunderland 45805
Surrey Heath Museum, Camberley 43720
Sutton Windmill and Broads Museum, Sutton, Norwich 45806
Swaffham Museum, Swaffham, Norfolk 45808
Swalcliffe Barn, Swalcliffe 45809
Swansea Museum, Swansea 45816
Swindon Museum and Art Gallery, Swindon 45820
Tamworth Castle Museum, Tamworth 45823
Teignmouth and Shaldon Museum, Teignmouth 45832
Tenbury and District Museum, Tenbury Wells 45835
Tenby Museum and Art Gallery, Tenby 45836
Tenterden and District Museum, Tenterden 45840
Tewkesbury Museum, Tewkesbury 45844
The Tolbooth Museum, Aberdeen 43249
Thetford Collection, Thetford 45848
Thirsk Museum, Thirsk 45849
Thornbury and District Museum, Thornbury, South Gloucestershire 45850
Thorney Heritage Museum, Thorney 45851
Three Rivers Museum of Local History, Rickmansworth 45495
Thurrock Museum, Grays 44351
Thurso Heritage Museum, Thurso 45857
Time Machine, Weston-super-Mare 45982
Timespan Heritage Centre and Art Gallery, Helmsdale 44440
Tiptree Museum, Tiptree 45859
Tiverton Museum of Mid Devon Life, Tiverton 45860
Toad Hole Cottage Museum, Ludham 45044
Tolbooth Museum, Stonehaven 45765
The Tolhouse, Great Yarmouth 44361
Tolsey Museum, Burford 43673
Tolson Memorial Museum, Huddersfield 44497
Tomintoul Museum, Tomintoul 45864
Topsham Museum, Topsham 45865
Torfaen Museum, Pontypool 45402
Torquay Museum, Torquay 45866
Torrington Museum, Torrington 45869
Totnes Elizabethan House Museum, Totnes 45872
Tower Museum, Derry 43974
Town and Crown Exhibition, Windsor 46046
Town House, Culross 43939
Town Museum, East Grinstead 44085
Towneley Hall Art Gallery and Museums, Burnley 43679
Traditional Heritage Museum, Sheffield 45655
Tudor House, Margate 45107
Tudor House Museum, Southampton 45708
Tullie House - City Museum and Art Gallery, Carlisle 43760
Tunbridge Wells Museum and Art Gallery, Tunbridge Wells 45881
Tweeddale Museum, Peebles 45349
Twickenham Museum, Twickenham 45889
Ullapool Museum, Ullapool 45896
Upminster Tithe Barn Museum of Nostalgia, Upminster 45899
Upperlands Eco-Museum, Derry 43975
Uttoxeter Heritage Centre, Uttoxeter 45902
Vale and Downland Museum, Wantage 45929
Valence House Museum and Art Gallery, Dagenham 43943
The Valiant Soldier, Buckfastleigh 43661
Ventnor Heritage Museum, Ventnor 45904
Vestry House Museum, London 45002
Victoria Jubilee Museum, Cawthorne 43783
Vindolanda Fort and Museum Chesterholm, Bardon Mill 43379
Wakefield Museum, Wakefield 45914
Wallingford Museum, Wallingford 45915
Walsall Local History Centre, Walsall 45922
Walsall Museum, Walsall 45923
Walton Hall Heritage Centre, Warrington 45935
Walton Maritime Museum, Walton-on-the-Naze 45926
Wandsworth Museum, London 45007
Ware Museum, Ware 45930
Wareham Town Museum, Wareham 45931
Warrington Museum and Art Gallery, Warrington 45936
Watford Museum, Watford 45950
Weardale Museum, Weardale 45952
Weaver's Cottage, Kilbarchan 44575
Wednesbury Museum and Art Gallery, Wednesbury 45953
Welholme Galleries, Grimsby 44367
Wellingborough Heritage Centre, Wellingborough 45957
Wellington Aviation Museum, Moreton-in-Marsh 45175
West Berkshire Museum, Newbury 45221
West Highland Museum, Fort William 44249
West Kilbride Museum, West Kilbride 45970
West Somerset Rural Life Museum, Allerford 43280
Westbury Manor Museum, Fareham 44212
Weymouth Museum, Weymouth 45991
Whitburn Community Museum, Whitburn 45993
Whitby Museum, Whitby 45998
Whitehall, Cheam 43800
Whittlesey Museum, Whittlesey 46009
Whitworth Historical Society Museum, Whitworth 46010

Wick Heritage Centre, Wick 46011
Wigan Pier, Wigan 46014
Wightwick Manor, Wolverhampton 46059
Willenhall Museum, Willenhall 46016
Willis Museum, Basingstoke 43397
Wiltshire Heritage Museum and Gallery, Devizes 43979
Wimbledon Society Museum of Local History, London 45018
Wincanton Museum, Wincanton 46021
Winchcombe Folk and Police Museum, Winchcombe 46023
Winchelsea Museum, Winchelsea 46025
Winchester City Museum, Winchester 46037
Wirral Museum, Birkenhead 43493
Wisbech and Fenland Museum, Wisbech 46049
Witney and District Museum and Art Gallery, Witney 46053
Wollaston Museum, Wollaston 46056
Woodbridge Museum, Woodbridge 46062
Woodchurch Village Life Museum, Woodchurch 46063
Woodhall Spa Cottage Museum, Woodhall Spa 46064
Woodhorn, Ashington 43329
Woolpit and District Museum, Woolpit 46069
Worcester City Museum and Art Gallery, Worcester 46073
Worcestershire County Museum, Kidderminster 44573
Worksop Museum, Worksop 46078
World of Glass, Saint Helens 45568
Worthing Museum and Art Gallery, Worthing 46079
Wotton Heritage Centre, Wotton-under-Edge 46080
Wrexham County Borough Museum, Wrexham 46085
Wycombe Museum, High Wycombe 44466
Wymondham Heritage Museum, Wymondham 46089
York Story, York 46111

United States of America

The 1811 Old Lincoln County Jail and 1839 Jailer's House Museum, Wiscasset 54260
1859 Jail-Marshal's Home and Museum, Independence 49471
1914 Plant City High School Community Exhibition, Plant City 51977
A.R. Bowman Memorial Museum, Prineville 52155
Academy Hall Museum of the Rocky Hill Historical Society, Rocky Hill 52434
Acorn Hall House Museum, Morristown 50850
Adair County Historical Society Museum, Kirksville 49799
Adam Thoroughgood House, Virginia Beach 53752
Adams County Historical Museum, Decatur 48018
Adams County Historical Society Museum, Lind 50116
Adams County Museum, Brighton 47026
Adams Museum, Deadwood 48002
Adamson House and Malibu Lagoon Museum, Malibu 50406
Adena State Memorial, Chillicothe 47533
Adirondack History Center, Elizabethtown 48340
Adirondack Museum, Blue Mountain Lake 46876
Agecroft Hall, Richmond 52320
Aiken County Historical Museum, Aiken 46140
Aitkin County Museum, Aitkin 46143
Alamo Township Museum - John E. Gray Memorial Museum, Kalamazoo 49663
Alaska State Museum, Juneau 49656
Albany Historical Society, Sabetha 52496
Albany Regional Museum, Albany 46166
Albion Academy Historical Museum, Edgerton 48287
Alden Historical Society Museum, Alden 46192
Alden House, Duxbury 48226
Alexandria Black History Resource Center, Alexandria 46199
Alger County Heritage Center, Munising 50906
Alice Austen House Museum, Staten Island 53260
Alice T. Miner Colonial Collection, Chazy 47416
Allan Shivers Museum, Woodville 54291
Allegan County Historical and Old Jail Museum, Allegan 46222
Allegany County Historical Museum, Cumberland 47905
Allegany County Museum, Belmont 46719
Allen County-Fort Wayne Historical Society Museum, Fort Wayne 48703
Allen County Historical Society, Iola 49515
Allen County Museum, Lima 50084
Amador County Museum, Jackson 49555
Amelia Island Museum of History, Fernandina Beach 48535
American Labor Museum, Haledon 49127
American Swedish Historical Museum, Philadelphia 51813
American West Heritage Center, Wellsville 54034
Americana Manse, Whitney-Halsey Home, Belmont 46720
Americana Museum, Terra Alta 53479
Amherst County Museum, Amherst 46273
Amherst History Museum, Amherst 46264
Amherst Museum, Amherst 46272
Amity and Woodbridge Historical Society Museum, Woodbridge 54273
Amory Regional Museum, Amory 46275
Anacortes Museum, Anacortes 46278
Anadarko Philomathic Museum, Anadarko 46279
Anaheim Museum, Anaheim 46283
Anderson County Historical Museum, Garnett 48865
Anderson County Museum, Anderson 46294
Andover Historical Society Museum, Andover 46297
Andres Pico Adobe, Mission Hills 50749
Andrew County Museum, Savannah 52950
Androscoggin Historical Society Museum, Auburn 46468
Animas Museum, Durango 48212
Anna Miller Museum, Newcastle 51308
The Anne Spencer Memorial Foundation, Lynchburg 50324
Annie E. Woodman Institute, Dover 48166
Annie Riggs Memorial Museum, Fort Stockton 48688
Anoka County Museum, Anoka 46323
Anson County Historical Society Museum, Wadesboro 53783
Antelope County Historical Museum, Neligh 51000
The Appaloosa Museum and Heritage Center, Moscow 50863
Appomattox Court House, Appomattox 46333
Arbor Lodge, Nebraska City 50995
Arcade Historical Museum, Arcade 46335
Arcadia Historical Museum, Arcadia 46337
Archer County Museum, Archer City 46341
Arches Museum of Pioneer Life, Winona 54238
Archibald Graham McIlwaine House, Petersburg 51796
Arizona Historical Society Museum at Papago Park, Tempe 53464
Arizona Historical Society Pioneer Museum, Flagstaff 48550
Arizona Women Hall of Fame, Phoenix 51902
Arkansas Post Museum, Gillett 48908
Arlington Antebellum Museum, Birmingham 46820
Arlington Heights Historical Museum, Arlington Heights 46360
Arlington Historical Museum, Arlington 46356
Arlington Historical Society Collection, Arlington 46348
Arlington House - The Robert E. Lee Memorial, Arlington 46357
Arms Family Museum of Local History, Youngstown 54352
Arrowhead Museum, Pittsfield 51965
Artesia Historical Museum and Art Center, Artesia 46365
Arthurdale Heritage Museum, Arthurdale 46366
Asa Packer Mansion, Jim Thorpe 49630
Ash Lawn-Highland, Charlottesville 47395
Ashland Historical Society Museum, Ashland 46381
Ashland Historical Society Museum, Ashland 46391
Ashtabula County Historical Society Museum, Geneva-on-the-Lake 48884
Ashton Villa, Galveston 48842
Aspen Historical Society Museum, Aspen 46398
Astors' Beechwood-Victorian Living History Museum, Newport 51318
Atascadero Historical Society Museum, Atascadero 46405
Atchison County Historical Society, Atchison 46407
Atlanta History Museum, Atlanta 46426
Atlanta Museum, Atlanta 46428
Atlantic City Historical Museum, Atlantic City 46460
Atlantic County Historical Society Museum, Somers Point 53146
Attmore-Oliver House, New Bern 51018
Atwood House Museum, Chatham 47404
Au Glaize Village, Defiance 48036
Audrain Historical Museum, Graceland, Mexico 50618
Auglaize County Historical Society Museum, Saint Marys 52606
Augusta Historical Museum, Augusta 46482
Augusta Museum of History, Augusta 46475
Aurora Historical Museum, Aurora 46489
Aurora Historical Museum, East Aurora 48233
Aurora Historical Society Museum, Aurora 46498
Aurora History Museum, Aurora 46488
Austin History Center, Austin 46502
Avery County Historical Museum, Newland 51315
Avon Historical Society Museum, Avon 46526
Aztalan Museum, Jefferson 49610
Bacon's Castle, Surry County 53384
Badger Mine and Museum, Shullsburg 53093
Bailey House Museum, Wailuku 53786
Baker-Cederberg Museum, Rochester 52392
Baldwin County Heritage Museum, Elberta 48331
Baldwin Historical Society Museum, Baldwin 46539
Baldwin-Reynolds House Museum, Meadville 50545
Baltimore County Historical Museum, Cockeysville 47668
Bangor Historical Society Museum, Bangor 46595
Bank Museum, Meeteetse 50561
Banner County Historical Museum, Harrisburg 49177
The Banning Museum, Wilmington 54187
Bannock County Historical Museum, Pocatello 52005
Bar Harbor Historical Society Museum, Bar Harbor 46602
Baranov Museum, Kodiak 49824
Barn Museum and Rail Depot Museum, Troutdale 53570
Barnes County Historical Museum, Valley City 53693
Barnet Historical Society, Barnet 46611
Barnwell County Museum, Barnwell 46614
Barre Historical Museum, Barre 46615
Barrington Area Historical Museum, Barrington 46616
Barron County Historical Society's Pioneer Village Museum, Cameron 47204
Bartholomew County Historical Museum, Columbus 47741
Bartlesville Area History Museum, Bartlesville 46617
Bartlett Museum, Amesbury 46260
Barton County Historical Society Village and Museum, Great Bend 49027
Bartow History Center, Cartersville 47263
Bartow-Pell Mansion Museum, Bronx 47039
Batavia Depot Museum, Batavia 46625
Batsto Village, Hammonton 49145
Baxter House Museum, Gorham 48975
Baxter Springs Heritage Center and Museum, Baxter Springs 46649
Bay Area Museum, Seabrook 52986
Bay View Historical Museum, Bay View 46652
Beacon Historical Society Museum, Beacon 46659
Bear Butte State Park Visitors Center, Sturgis 53350
Beaufort Historic Site, Beaufort 46665
Beaufort Museum, Beaufort 46667
Beauvais Heritage Center, Clark 47586
Beaver Island Historical Museum, Beaver Island 46682
Beck Cultural Exchange Center, Knoxville 49811
Becker County Historical Museum, Detroit Lakes 48131
Bedford City/County Museum, Bedford 46691
Bedford Historical Society Museum, Bedford 46689
Behringer-Crawford Museum, Covington 47858
Belhaven Memorial Museum, Belhaven 46695
The Belknap Mill Museum, Laconia 49861
Bell County Museum, Belton 46727
Bella Vista Historical Museum, Bella Vista 46696
Belle Grove Plantation, Middletown 50663
Belle Meade Plantation, Nashville 50962
Belleville Area Museum, Belleville 46701
Bellflower Genealogical and Historical Society Museum, Bellflower 46710
Bellport-Brookhaven Historical Society Museum, Bellport 46717
Beloit Historical Society Museum, Beloit 46723
Beltrami County Historical Center, Bemidji 46730
Bement-Billings Farmstead, Newark Valley 51296
Benicia Historical Museum, Benicia 46734
Benjamin Patterson Inn Museum Complex, Corning 47809
Bennett Place, Durham 48215
The Bennington Museum, Bennington 46739
Benton County Historical Museum, Philomath 51895
Benton County Historical Museum, Prosser 52159
The Benton Homestead, Tolland 53524
Benzie Area Historical Museum, Benzonia 46745
Bergen County Historical Society Museum, River Edge 52361
Bergen Museum of Local History, Bergen 46749
Berkeley Plantation, Charles City 47350
Berkshire Museum, Pittsfield 51967
Berlin Art and Historical Collections, Berlin 46763
Berlin Historical Society Museum of Local History, Berlin 46764
Bernard Historical Museum, Delton 48051
Bernice Pauahi Bishop Museum, Honolulu 49335
Bessemer Hall of History, Bessemer 46771
Bethel Historical Society's Regional History Center, Bethel 46774
Bethlehem Historical Association Museum, Selkirk 53031
Betsey Williams Cottage, Providence 52162
Betts House Research Center, Cincinnati 47553
Beverly Historical Museum, Beverly 46785
Bexley Historical Society Museum, Bexley 46793
Bidwell House Museum, Monterey 50808
Big Horn County Historical Museum, Hardin 49165
Big Springs Museum, Caledonia 47165
Bigelow House Museum, Olympia 51542
Billings Farm and Museum, Woodstock 54287
Billy The Kid Museum, Fort Sumner 48690
Biltmore Estate, Asheville 46369
Birmingham Historical Museum Park, Birmingham 46829
Bisbee Mining and Historical Museum, Bisbee 46830
Bishop Hill Colony, Bishop Hill 46832
Bishop Hill Heritage Museum, Bishop Hill 46833
The Bishop's Palace, Galveston 48843
Black Kettle Museum, Cheyenne 47443
Black River Academy Museum, Ludlow 50317
Black River Historical Society of Lorain Museum, Lorain 50199
Blackberry Farm-Pioneer Village, Aurora 46491
Blackman Museum, Snohomish 53136
Blackstone Valley Historical Society Museum, Lincoln 50111
Blaine County Historical Museum, Hailey 49123
Blaine County Museum, Chinook 47541
Blaine House, Augusta 46483
Blewitt-Harrison-Lee Museum, Columbus 47744
Bloomington Historical Museum, Bloomington 46870
Blount Mansion, Knoxville 49812
Blue Earth County Museums, Mankato 50446
Blue Licks Battlefield Museum, Mount Olivet 50881
Bob Burns Museum, Van Buren 53700
Boca Raton Historical Society Museum, Boca Raton 46882
Bolduc House Museum, Sainte Genevieve 52649
Bolinas Museum, Bolinas 46894
Bolivar Hall, San Antonio 52709
Bolton Historical Museum, Bolton 46896
Bonner County Historical Museum, Sandpoint 52835
Bonnet House, Fort Lauderdale 48635
Booker T. Washington National Monument, Hardy 49167
Boone County Historical Society Museum, Belvidere 46728
The Boorman House, Mauston 50537
Boothbay Railway Village, Boothbay 46910
Bordentown Historical Society Museum, Bordentown 46913
Boulder History Museum, Boulder 46955
Bowen House / Roseland Cottage, Woodstock 54281
Bowers Mansion, Washoe Valley 53952
Bowie Railroad Station and Huntington Museum, Bowie 46969
Bowie's Heritage and Children's Museum, Bowie 46970
Bowne House, Flushing 48575
Boyer Museum, Wildwood 54161
The Boyle House, Malvern 50410
Boys Town Hall of History Father Flanagan House, Boys Town 46980
Bradford County Historical Society Museum, Towanda 53546
Bradford Historical Society, Bradford 46990
Bradford Ohio Railroad Museum, Bradford 46988
Bragg-Mitchell Mansion, Mobile 50766
Braintree Historical Society Museum, Braintree 46992
Brattleboro Museum and Art Center, Brattleboro 46998
Brazoria County Historical Museum, Angleton 46299
Brevard Museum, Cocoa 47669
Brick House, Montgomery 50816
Brick Store Museum, Kennebunk 49727
Bridgehampton Historical Society Museum, Bridgehampton 47012
Bridgton Historical Museum, Bridgton 47023
Brigham Young's Winter Home, Saint George 52544
Brimstone Museum, Sulphur 53361
Bristol Historical and Preservation Society Museum, Bristol 47031
Bronck Museum, Coxsackie 47863
Brookings County Museum, Volga 53767
Brookline Historical Society Museum, Brookline 47061
Brooklyn Historical Society Museum, Brooklyn 47086
Brooks Academy Museum, Harwich 49209
Brookside Museum, Ballston Spa 46542
Broome County Historical Society Museum, Binghamton 46814
Broward County Historical Museum, Fort Lauderdale 48636
Brown County Historical Society Pioneer Museum, Nashville 50960
Brown County Museum, New Ulm 51105
Brownella Cottage, Galion 48837
Brownville Historical Society Museum, Brownville 47097
Brunswick Town/Fort Anderson State Historic Site, Winnabow 54234
Brush Country Museum, Cotulla 47847
Buccleuch Mansion, New Brunswick 51031
Bucksport Historical Museum, Bucksport 47110
Bucyrus Historical Society Museum, Bucyrus 47111
Buena Park Historical Society Museum, Buena Park 47112
Buffalo and Erie County Historical Society Museum, Buffalo 47116
Buffalo Bill Cody Homestead, Princeton 52143
Buffalo Trails Museum, Epping 48406
Bureau County Historical Society Museum, Princeton 52144
Burgwin-Wright Museum, Wilmington 54200
Burke County Historical Powers Lake Complex, Powers Lake 52127
Burleson County Historical Museum, Caldwell 47164
Burlington County Historical Society Museum, Burlington 47139
Burlington Historical Museum, Burlington 47136

Burnham Tavern, Machias 50349
Burritt on the Mountain, Huntsville 49442
Burrowes Mansion Museum, Matawan 50532
Burt County Museum, Tekamah 53460
Bush House Museum and Bush Barn Art Center, Salem 52665
Bushwhacker Museum, Nevada 51006
Butler County Historical Museum, Allison 46232
Butler County Museum, Hamilton 49138
Butler County Society Museum and Kansas Oil Museum, El Dorado 48311
The Buttonwoods Museum, Haverhill 49216
Byron Historical Museum, Byron 47154
Cabot Historical Society, Cabot 47156
Cades Cove Open-Air Museum, Townsend 53548
Calaveras County Museum, San Andreas 52699
Caleb Pusey House, Upland 53662
Calhoun County Museum, Port Lavaca 52038
Calhoun County Museum, Rockwell City 52432
Calhoun County Museum, Saint Matthews 52610
California Route 66 Museum, Victorville 53742
Callahan County Pioneer Museum, Baird 46532
Calumet County Historical Society Museum, New Holstein 51059
Cambria County Historical Society Museum, Ebensburg 48279
Cambria Historical Society Museum, New Providence 51100
Cambridge Historical Museum, Cambridge 47174
Cambridge Museum, Cambridge 47188
Camden Archives and Museum, Camden 47199
Camden County Historical Society Museum, Camden 47195
Camp Floyd, Fairfield 48486
Campbell County Rockpile Museum, Gillette 48909
Campbell Historical Museum, Campbell 47208
Campbell House Museum, Saint Louis 52569
Campus Martius Museum, Marietta 50472
Canaan Historical Society, Falls Village 48503
Canaan Historical Society Museum, Canaan 47210
Canadian County Historical Museum, El Reno 48330
Canal Fulton Heritage Society Museum, Canal Fulton 47212
Canby Depot Museum, Canby 47216
Caney Valley Historical Society, Caney 47217
Cannon Falls Area Historical Museum, Cannon Falls 47218
Cannon Village Visitor Center, Kannapolis 49677
Canon City Municipal Museum, Canon City 47219
Canterbury Shaker Village, Canterbury 47221
Canton Historical Museum, Canton 47222
Cape May County Historical Museum, Cape May Court House 47236
Cape Vincent Historical Museum, Cape Vincent 47238
Capitola Historical Museum, Capitola 47239
Captain George Flavel House Museum, Astoria 46400
Captain John Wilson House, Cohasset 47675
Captain Nathaniel B. Palmer House, Stonington 53311
Carbon County Museum, Rawlins 52258
Carlsbad Museum and Art Center, Carlsbad 47249
Carlton County Historical Museum and Heritage Center, Cloquet 47663
Carlyle House, Alexandria 46201
Carnegie Center for Art History, New Albany 51008
Carnegie Historical Museum, Fairfield 48483
Carnifex Ferry Battlefield State Park and Museum, Summersville 53362
Carriage House Museum, Camden 47198
Carroll Mansion Home - Leavenworth County Historical Society Museum, Leavenworth 50008
Carson County Square House Museum, Panhandle 51695
Carson Valley Museum, Gardnerville 48863
Carter County Museum, Ekalaka 48306
The Carter House, Franklin 48751
Caryl House and Fisher Barn, Dover 48164
Casa Adobe de San Rafael, Glendale 48929
Casa Grande Valley History Museum, Casa Grande 47269
Casey Farm, Saunderstown 52933
Cashmere Museum Pioneer Village, Cashmere 47270
Cass County Historical Society Museum, Logansport 50162
Cass County Historical Society Museum, Plattsmouth 51985
Cass County History Museum at Bonanzaville, West Fargo 54054
Cassia County Museum, Burley 47130
Casteel-Linn House and Museum, Princeton 52146
The Castle, Marietta 50473
Castle Museum of Saginaw County History, Saginaw 52515
Castleton Historical Society, Castleton 47279
Catalina Island Museum, Avalon 46524
Catawba County Museum of History, Newton 51342
The Catlin House Museum, Scranton 52979
Cattaraugus Area Historical Center, Cattaraugus 47284
Cattle Raisers Museum, Fort Worth 48712
Cayuga Museum, Auburn 46469
Cayuga-Owasco Lakes Historical Society Museum, Moravia 50839
Cedar Falls Historical Museum, Cedar Falls 47291
Cedar Grove Historical Society Museum, Cedar Grove 47295
Cedar Key Historical Society Museum, Cedar Key 47296
Cedar Key State Park Museum, Cedar Key 47297
Cedar Valley Memories, Osage 51607
Center for Sacramento History, Sacramento 52504
Centerville Historical Museum, Centerville 47316
Central Missouri State University Archives and Museum, Warrensburg 53829
Central Montana Historical Association, Lewistown 50041
Central Museum, Central 47318
Central Nevada Museum, Tonopah 53531
Central Texas Area Museum, Salado 52652
Centre County Historical Society Museum, State College 53259
Centre County Library Historical Museum, Bellefonte 46698
Centre Hill Museum, Petersburg 51797
Centura House Historical Society, Rosendale 52461
Chagrin Falls Historical Society Museum, Chagrin Falls 47328
Chalet of the Golden Fleece, New Glarus 51044
Chamizal National Memorial, El Paso 48318
Champaign County Historical Museum, Champaign 47333
Champaign County Historical Museum, Urbana 53670
Chandler Museum, Chandler 47338
Chapman Historical Museum, Glens Falls 48934
Chappell Hill Historical Society Museum, Chappell Hill 47348
Charles A. Grignon Mansion, Kaukauna 49711
Charles A. Weyerhaeuser Memorial Museum, Little Falls 50134
Charles B. Aycock Birthplace, Fremont 48786
Charles Carroll House of Annapolis, Annapolis 46316
Charles Gates Dawes House, Evanston 48446
Charles J. Beldon Museum of Western Photography, Meeteetse 50562
The Charleston Museum, Charleston 47359
Charlotte County Historical Center, Charlotte Harbor 47394
Charlotte Museum of History and Hezekiah Alexander Homesite, Charlotte 47386
Charnley-Persky House Museum, Chicago 47461
Chase County Historical Society, Cottonwood Falls 47844
Chateau de Mores State Historic Site, Medora 50559
Chatillon-DeMenil Mansion, Saint Louis 52570
Chattanooga Regional History Museum, Chattanooga 47407
Chelmsford Historical Society Museum, Chelmsford 47418
Chemung Valley History Museum, Elmira 48379
Chenango County Historical Society Museum, Norwich 51439
Cheney Homestead, Manchester 50416
Cheraw Lyceum Museum, Cheraw 47419
Cherokee County Historical Museum, Murphy 50917
Cherokee Strip Land Rush Museum, Arkansas City 46347
Cherokee Strip Museum, Alva 46246
Cherry County Historical Society Museum, Valentine 53689
Cherry Valley Museum, Cherry Valley 47426
Chesapeake Beach Railway Museum, Chesapeake Beach 47428
Cheshire County Museum, Keene 49718
Chester County Historical Society Museum, West Chester 54049
Chester County Museum, Chester 47433
Chesterfield Historical Society of Virginia, Chesterfield 47438
Chetopa Historical Museum, Chetopa 47442
Chickasaw County Historical Society Museum, Bradford Village, Nashua 50957
Chico Museum, Chico 47526
Childress County Heritage Museum, Childress 47530
Chimmey Point, Addison 46136
Chippewa County Museum, Montevideo 50811
Chippewa Valley Museum, Eau Claire 48276
Chisholm Trail Historical Museum, Waurika 53992
Chisholm Trail Museum, Kingfisher 49781
Christian C. Sanderson Museum, Chadds Ford 47323
Christiansted National Historic Site, Saint Croix 52541
CHTJ Southard House Museum, Richmond 52318
Churchill County Museum and Archives, Fallon 48502
Cincinnati Museum Center, Cincinnati 47557
City County Pioneer Museum, Sweetwater 53393
City Island Nautical Museum, Bronx 47043
City Museum, Saint Louis 52571
City of Gilroy Museum, Gilroy 48912
City of Las Vegas and Rough Riders Memorial Museum, Las Vegas 49963
City of Le Sueur Museum, Le Sueur 50003
City of Wayne Historical Museum, Wayne 54005
Claremont Museum, Claremont 47577
Clark County Historical Museum, Vancouver 53705
Clark County Museum, Marshall 50501
Clark Hotel Museum, Van Horn 53701
Clark House Museum Complex, Wolfeboro 54268
Clarke County Historical Museum, Berryville 46770
Clarke Historical Museum, Eureka 48439
Clarke House Museum, Chicago 47467
Clarkson Historical Museum, Clarkson 47589
Clarksville-Montgomery County Museum-Customs House Museum, Clarksville 47592
Clausen Museum, Petersburg 51793
Clawson Historical Museum, Clawson 47596
Clay County Historical Museum, Liberty 50077
Clay County Museum, Moorhead 50834
Clear Lake Area Historical Museum, Clear Lake 47606
Clearfield County Historical Society Museum, Clearfield 47607
Clearwater County Museum, Shevlin 53077
Clearwater Historical Museum, Orofino 51601
Cleo Redd Fisher Museum, Loudonville 50276
Clermont State Historic Site, Clermont 47618
Cleveland County Historical Museum, Shelby 53067
Clewiston Museum, Clewiston 47646
Clifton Community Historical Society, Clifton 47647
Clinton County Historical Museum, Plattsburgh 51982
Clinton County Museum, Clinton 47650
Clinton County Museum, Frankfort 48733
Cliveden House, Philadelphia 51825
Cloud County Historical Society Museum, Concordia 47777
Coachella Valley Museum, Indio 49508
Coastal Heritage Museum, Crystal River 47896
Cocalico Valley Museum, Ephrata 48403
Codington County Heritage Museum, Watertown 53974
Coe Hall, Oyster Bay 51660
Coffin School Museum, Nantucket 50943
Cokato Museum and Akerlund Photography Studio, Cokato 47679
Cole County Historical Museum, Jefferson City 49611
Coleman House Museum, Ainsworth 46142
Coley Homestead and Barn Museum, Weston 54101
Colleton Museum, Walterboro 53806
Collier County Museum, Naples 50951
Colonel Davenport Historical Foundation, Rock Island 52407
Colonial Burlington Foundation, Burlington 47140
Colonial Industrial Quarter, Bethlehem 46778
Colonial Pennsylvania Plantation, Media 50553
Colonial Williamsburg, Williamsburg 54167
Colorado Springs Museum, Colorado Springs 47700
Colorado University Heritage Center, Boulder 46958
Colton Area Museum, Colton 47710
Columbia County Historical Museum, Lake City 49879
Columbia County Historical Society Museum, Kinderhook 49777
Columbia County Historical Society Museum, Saint Helens 52549
Columbia County Museum, Bloomsburg 46871
Columbia George Interpretive Center, Stevenson 53292
Comanche Crossing Museum, Strasburg 53327
Community Historical Museum, Maxwell 50538
Community Memorial Museum of Sutter County, Yuba City 54360
Comstock House, Moorhead 50835
Concord Museum, Concord 47767
Confederate Memorial Hall-Bleak House, Knoxville 49814
Confederate Memorial State Historic Site, Higginsville 49273
The Conference House, Staten Island 53261
Conklin Reed Organ and History Museum, Hanover 49161
Connecticut Historical Society Museum, Hartford 49193
Connecticut Valley Historical Museum, Springfield 53224
Conrad Mansion Museum, Kalispell 49670
Conrad Weiser Homestead, Womelsdorf 54271
Constitution Convention Museum, Port Saint Joe 52040
Constitution Island - Warner House, West Point 54077
Cook County Historical Museum, Grand Marais 48997
Cooper Regional History Museum, Upland 53660
Coopersville Area Historical Society Museum, Coopersville 47793
Coos County Historical Society Museum, North Bend 51381
Copper Village Museum and Arts Center, Anaconda 46277
Copshaholm House Museum and Historic Oliver Gardens, South Bend 53157
Coquille River Museum, Brandon 46994
Coquille River Museum, Bandon 46594
Cordova Historical Museum, Cordova 47806
Cornelius Low House/Middlesex County Museum, Piscataway 51932
Cornwall Historical Museum, Cornwall 47813
Coronado-Quivira Museum, Lyons 50335
Corpus Christi Museum of Science and History, Corpus Christi 47820
Corry Area Historical Society Museum, Corry 47824
Cortland County Historical Society Museum, Cortland 47830
Cottage Grove Museum, Cottage Grove 47841
Cottage Lawn, Oneida 51559
Courthouse Museum, Berrien Springs 46766
Courthouse Museum, Exira 48466
Courthouse Museum, Jacksonport 49574
Courthouse Square Museum, Charlotte 47383
The Cowley County Historical Museum, Winfield 54232
Cowlitz County Historical Museum, Kelso 49723
Cracker Country Museum, Tampa 53432
Crailo State Historic Site, Rensselaer 52294
Cranbury Historical and Preservation Society Museum, Cranbury 47866
Cravens House, Lookout Mountain 50197
Crawford County Historical Museum, Pittsburg 51934
Crazy Mountain Museum, Big Timber 46801
Crestline Shunk Museum, Crestline 47877
Cripple Creek District Museum, Cripple Creek 47882
Crocker House, Mount Clemens 50872
Crockett County Museum, Ozona 51664
Crockett Tavern Museum, Morristown 50858
Crook County Museum and Art Gallery, Sundance 53370
Crosby County Pioneer Memorial Museum, Crosbyton 47889
Crow Wing County Historical Museum, Brainerd 46991
Crown Gardens Museum, Dalton 47948
Crystal Caverns at Hupp's Hill, Strasburg 53330
Crystal Lake Falls Historical Association, Barton 46621
Culbertson Museum, Culbertson 47900
Cullman County Museum, Cullman 47902
Cumberland County Historical Society Museum, Greenwich 49084
Cupertino Historical Museum, Cupertino 47910
Curry Historical Society Museum, Gold Beach 48954
Cushing House Museum, Newburyport 51305
Custer County 1881 Courthouse Museum, Custer 47915
Custer County Historical Society Museum, Broken Bow 47037
Cuyuna Range Museum, Crosby 47887
Cyrus H. McCormick Memorial Museum, Steeles Tavern 53281
Dacotah Prairie Museum and Lamont Art Gallery, Aberdeen 46116
Dakota County Museum, South Saint Paul 53181
Dakota Territorial Museum, Yankton 54323
Dakotaland Museum, Huron 49450
Dal-Paso Museum, Lamesa 49912
Dallam-Hartley XIT Museum, Dalhart 47922
Dallas Historical Society Museum, Dallas 47929
Dan O'Laurie Canyon Country Museum, Moab 50765
Danbury Museum, Danbury 47952
Daniels County Museum and Pioneer Town, Scobey 52968
Danvers Historical Society Exhibition, Danvers 47956
Dar-Hervey Ely House, Rochester 52394
Darnall's Chance House Museum, Upper Marlboro 53664
Daughters of Utah Pioneers Pioneer Memorial Museum, Salt Lake City 52691
David Bradford House, Washington 53941
David Crawford House, Newburgh 51300
Davidson County Historical Museum, Lexington 50063
Dawes County Historical Society Museum, Chadron 47325
Dawson County Historical Museum, Lexington 50065
Dayton Historical Depot Society Museum, Dayton 47992
De Mores State Historic Site, Bismarck 46835
De Smet Depot Museum, De Smet 48001
Deaf Smith County Museum, Hereford 49257
Dearborn Historical Museum, Dearborn 48007
Death Valley National Park Museum, Death Valley 48010
Dedham Historical Museum, Dedham 48023
Deer Isle-Stonington Historical Society Museum, Deer Isle 48026
Deerfield Beach Historical Society Museum, Deerfield Beach 48034
DeKalb Historical Society Museum, Decatur 48013
Delano Heritage Park Museum, Delano 48043
Delaware Agricultural Museum, Dover 48162

Delaware Museum, Wilmington 54191
Delaware State Museums, Dover 48163
Delta County Historical Society Museum, Escanaba 48414
Delta County Museum, Delta 48048
Dennison Railroad Depot Museum, Dennison 48058
Denton County Historical Museum, Denton 48061
Depot Museum and Fort Sedgwick Museum, Julesburg 49652
The Depot Museum Complex, Henderson 49251
Depot Museum Complex, Condon 47778
Depot Park Museum, Sonoma 53154
Des Plaines Historical Museum, Des Plaines 48104
Deschutes County Historical Society Museum, Bend 46732
Desha County Museum, Dumas 48201
Detroit Historical Museum, Detroit 48113
Devils Tower Visitor Center, Devils Tower 48132
Dewey Hotel, Dewey 48133
DeWitt County Historical Museum, Cuero 47899
DeWitt Historical Society Museum, Ithaca 49546
Dexter Area Museum, Dexter 48136
Dexter Historical Society Museum, Dexter 48135
Dickinson County Heritage Center, Abilene 46119
Dillaway-Thomas House, Boston 46928
Discover Houston County Visitors Center-Museum, Crockett 47884
Discovery Center, Fresno 48789
Divide County Historical Society Museum, Crosby 47888
Doak House Museum, Greeneville 49051
Dodge County Historical Museum, Mantorville 50454
Dodge County Historical Society Museum, Beaver Dam 46681
Donald G. Trayser Memorial Museum, Barnstable 46612
Door County Museum, Sturgeon Bay 53346
Dorchester County Historical Society Museum, Cambridge 47186
Doric House, Flemington 48557
Dorothea B. Hoover Historical Museum, Joplin 49649
Dorothy G. Page Museum, Wasilla 53953
Dossin Great Lakes Museum, Detroit 48116
Douglas County Historical Society Museum, Superior 53377
Douglas County Museum Complex, Armour 46362
Douglas County Museum of History and Natural History, Roseburg 52457
The Downers Grove Park District Museum, Downers Grove 48174
Downieville Museum, Downieville 48176
Dr. Increase Mathews House, Zanesville 54367
Dr. William Robinson Plantation, Clark 47585
Drake House Museum, Plainfield 51973
Drake Well Museum, Titusville 53516
Drayton Hall, Charleston 47362
Drew County Historical Museum, Monticello 50818
Driebe Freight Station, Stroudsburg 53338
Drummond Home, Hominy 49333
Drummond Island Historical Museum, Drummond Island 48183
Dry Falls Interpretive Center, Coulee City 47849
Dublin-Laurens Museum, Dublin 48185
Dubois Museum, Dubois 48186
Duluth Children's Museum, Duluth 48194
Duncan Cottage Museum, Metlakatla 50615
Dundee Township Historical Society Museum, Dundee 48206
Dundy County Historical Society Museum, Benkelman 46736
Dunedin Historical Society Museum, Dunedin 48209
Dunham Tavern Museum, Cleveland 47630
Dunklin County Museum, Kennett 49732
DuPont Historical Museum, DuPont 48211
Durham Center Museum, East Durham 48238
Durham Historic Association Museum, Durham 48224
Durham Western Heritage Museum, Omaha 51548
Dutchess County Historical Society, Poughkeepsie 52120
Duxbury Rural and Historical Society, Duxbury 48228
Dyer Memorial Library, Abington 46127
Eagle Harbor Lighthouse and Museums, Eagle Harbor 48232
Eagle Historical Society and Museums, Eagle City 48231
The Earle-Harrison House, Waco 53774
Earle Wightman Museum, Oyster Bay 51661
Earlyworks Museums, Huntsville 49443
East Brunswick Museum, East Brunswick 48237
East Hampton Historical Society Museum, East Hampton 48244
East Jersey Olde Towne, Piscataway 51933
East Jordan Portside Art and Historical Museum, East Jordan 48252
East Poultney Museum, East Poultney 48258
East Tennessee Historical Society Museum, Knoxville 49817
Eastchester Historical Society Museum, Bronxville 47054
Eastern Arizona Museum, Pima 51919
Eastern California Museum, Independence 49469
Eastern Oregon Museum on the Old Oregon Trail, Haines 49126
Eastside Heritage Center, Bellevue 46707
Ebenezer Maxwell Mansion, Philadelphia 51829
Echo Historical Museum, Echo 48281
Eckley Miners' Village, Eckley 48283
Edgar Allan Poe Cottage, Bronx 47044
Edgar Allan Poe Museum, Richmond 52327
Edgar County Historical Museum, Paris 51702
Edison Museum, Beaumont 46675
Edmond Historical Society Museum, Edmond 48294
Edmonds South Snohomish County Historical Society Museum, Edmonds 48296
Edna Historical Museum, Edna 48297
Edsel and Eleanor Ford House, Grosse Pointe 49094
Edwards County Historical Museum, Kinsley 49797
Edwin Wolters Memorial Museum, Shiner 53079
Eells-Stow House, Milford 50683
Eisenhower Birthplace, Denison 48056
El Dorado County Historical Museum, Placerville 51972
El Monte Historical Society Museum, El Monte 48313
El Pueblo History Museum, Pueblo 52191
El Rancho de Las Golondrinas Museum, Santa Fe 52873
Elberton Granite Museum, Elberton 48332
Eleanor Roosevelt National Historic Site, Hyde Park 49457
Elgin Area Museum, Elgin 48334
Elk Grove Farmhouse Museum, Elk Grove Village 48342
Elkhart County Historical Museum, Bristol 47029
Ella Sharp Museum, Jackson 49556
Elliott Museum, Stuart 53342
Ellis County Historical Society Museum, Hays 49222
Ellis County Museum, Waxahachie 53999
Elmhurst Historical Museum, Elmhurst 48375
Emmett Kelly Historical Museum, Sedan 53025
Empire Area Heritage Group Museum, Empire 48390
Enfield Historical Society Museum, Enfield 48397
Enfield Shaker Museum, Enfield 48398
Ephraim Foundation Museums, Ephraim 48402
Ephrata Cloister, Ephrata 48404
Erie Canal Museum, Syracuse 53397
Erie Canal Village, Rome 52450
Erie History Center, Erie 48410
Erwin Museum, Painted Post 51676
Essex Historical Museum, Essex 48422
Essley-Noble Museum, Aledo 46193
Estes Park Area Historical Museum, Estes Park 48426
Eugene Field House and Saint Louis Toy Museum, Saint Louis 52576
Eureka Pioneer Museum of McPherson County, Eureka 48442
Eureka School House, Springfield 53239
Evansville Museum of Arts and Science, Evansville 48456
Excelsior-Lake Minnetonka Historical Museum, Excelsior 48462
The Executive Mansion, Frankfort 48734
Ezra Meeker Mansion, Puyallup 52206
Fairbanks Community Museum, Fairbanks 48471
Fairbanks Museum and Planetarium, Saint Johnsbury 52554
Fairfield County Museum, Winnsboro 54237
Fairport Historical Museum, Fairport 48492
Fairview Museum of History and Art, Fairview 48494
Faith Trumbull Chapter Museum, Norwich 51435
Fall River County Historical Museum, Hot Springs 49374
Fall River Historical Society Museum, Fall River 48499
Falmouth Historical Museum, Falmouth 48505
The Farmers' Museum, Cooperstown 47788
Farmington Museum, Farmington 48517
Farrar-Mansur House and Old Mill Museum, Weston 54108
Faulkner County Museum, Conway 47783
Fayette County Historical Museum, West Union 54084
Fayette County Museum, Washington Court House 53948
Fayette Heritage Museum, La Grange 49843
Felix Valle State Historic Site, Sainte Genevieve 52650
Fenton History Center-Museum and Library, Jamestown 49597
Fillmore County Historical Center, Fountain 48724
Fillmore Historical Museum, Fillmore 48541
The Filson Historical Society, Louisville 50285
The Finley House, Wayne 54011
Finney County Kansas Historical Museum, Garden City 48854
Firelands Historical Society Museum, Norwalk 51434
Fisher Grove Country School, Frankfort 48744
Fitchburg Historical Society Museum, Fitchburg 48549
Fleming Historical Museum, Fleming 48556
Flint House, Scotia 52970
Flood Museum, Johnstown 49640
Florence Griswold Museum, Old Lyme 51536
Florence Price Pioneer Museum, Florence 48568
Florewood State Park, Greenwood 49086
Florida Adventure Museum, Punta Gorda 52200
Florida International Museum, Saint Petersburg 52638
Floyd County Historical Museum, Charles City 47349
Follett House Museum, Sandusky 52836
Folsom History Museum, Folsom 48581
Ford County Historical Society Museum, Paxton 51739
The Forges and Manor of Ringwood, Ringwood 52353
Former Governors' Mansion, Bismarck 46837
Fort Abercrombie Historic Site, Abercrombie 46115
Fort Bedford Museum, Bedford 46690
Fort Bend Museum, Richmond 52319
Fort Benton Museum of the Upper Missouri, Fort Benton 48592
Fort Bridger State Museum, Fort Bridger 48600
Fort Buenaventura, Ogden 51501
Fort Collins Museum, Fort Collins 48607
Fort Crawford Museum, Prairie du Chien 52128
Fort Croghan Museum, Burnet 47145
Fort Crook Museum, Fall River Mills 48501
Fort Dodge Historical Museum, Fort Dodge 48616
Fort Henrietta, Echo 48282
Fort Hill Museum, Hillsboro 49286
Fort Hill - The John C. Calhoun House, Clemson 47614
Fort Hunter Mansion, Harrisburg 49179
Fort Jones Museum, Fort Jones 48632
Fort Klock Historic Restoration, Saint Johnsville 52556
Fort Leaton, Presidio 52137
Fort Mackinac, Mackinac Island 50351
Fort Morgan Museum, Fort Morgan 48659
Fort Nisqually Living History Museum, Tacoma 53407
Fort Osage, Sibley 53095
Fort Plain Museum, Fort Plain 48670
Fort Richardson, Jacksboro 49554
Fort Saint Joseph Museum, Niles 51351
Fort Simcoe, White Swan 54132
Fort Sisseton, Lake City 49881
Fort Smith Museum of History, Fort Smith 48685
Fort Stanwix, Rome 52451
Fort Totten State Historic Museum, Fort Totten 48692
Fort Vancouver, Vancouver 53706
Fort Walla Walla Museum, Walla Walla 53797
Fort Washington Museum, Fort Washington 48700
Fort Wilkins Historic Complex, Copper Harbor 47796
Fort Winnebago Surgeons Quarters, Portage 52051
Fossil Country Museum, Kemmerer 49724
Fossil Museum, Fossil 48722
Fossil Station Museum, Russell 52484
Fostoria Area Historical Museum, Fostoria 48723
Foundation for Historic Christ Church, Irvington 49537
Four Rivers Cultural Center, Ontario 51570
Fox Island Historical Society Museum, Fox Island 48727
Fox Lake Historical Museum, Fox Lake 48728
Framingham Historical Museum, Framingham 48730
Francis Land House, Virginia Beach 53756
Francisco Fort Museum, La Veta 49856
Frank H. McClung Museum, Knoxville 49818
Frank Lloyd Wright's Pope-Leighey House, Alexandria 46204
Frank Phillips Home, Bartlesville 46618
Frankenmuth Historical Museum, Frankenmuth 48732
Frankfort Area Historical Museum, West Frankfort 54055
Franklin County Historical Museum, Pasco 51726
Franklin County Museum, Brookville 47090
Franklin County Old Jail Museum, Winchester 54221
Franklin Historical Society Museum, Franklin 48747
Fraunces Tavern Museum, New York 51177
Fred J. Johnston House Museum, Kingston 49789
Fredericksburg Area Museum and Cultural Center, Fredericksburg 48765
Freeborn County Museum, Albert Lea 46170
Freeport Historical Society Museum, Freeport 48784
Freestone County Historical Museum, Fairfield 48485
Fremont County Pioneer Museum, Lander 49933
French Azilum, Towanda 53547
Friendship Hill, Point Marion 52009
Frisco Historical Society Museum, Frisco 48797
Front Street Museum, Ogallala 51498
Frontier Gateway Museum, Glendive 48931
Frontier Museum, Williston 54182
Frontier Village Museum, Jamestown 49593
Fruitlands Museums, Harvard 49207
Fullerton Museum Center, Fullerton 48805
Fulton County Historical Museum, Wauseon 53995
Fulton County Historical Society Museum, Rochester 52384
Fulton County Museum, Groversville 49100
Furnas-Gosper Historical Society and Museum, Arapahoe 46334
Gadsby's Tavern Museum, Alexandria 46206
Gage County Historical Museum, Beatrice 46663
Galena-Jo Daviess County Historical Museum, Galena 48827
Galion Historical Museum, Galion 48838
Galloway House and Village-Blakely Museum, Fond du Lac 48583
Galveston County Historical Museum, Galveston 48845
Garden County Museum, Oshkosh 51611
Gardner House Museum, Albion 46174
Gardner Museum, Gardner 48862
Garibaldi-Meucci Museum, Staten Island 53262
Garnavillo Historical Museum, Garnavillo 48864
Garner Memorial Museum, Uvalde 53678
Garrett County Historical Museum, Oakland 51474
Garrett Historical Museum, Garrett 48867
Garst Museum, Greenville 49076
Gaston County Museum of Art and History, Dallas 47923
Gatekeeper's Museum and Marion Steinbach Indian Basket Museum, Tahoe City 53418
Gates House, Machiasport 50350
Gates Mills Historical Society Museum, Gates Mills 48872
Gateway to the Panhandle, Gate 48871
Geddes Historic District Village, Geddes 48875
General Adam Stephen House, Martinsburg 50519
General Crook House Museum, Omaha 51550
General Daniel Bissell House, Saint Louis 52579
General Sterling Price Museum, Keytesville 49773
Genesee Country Village and Museum, Mumford 50901
Genoa Courthouse Museum, Genoa 48885
Geographical Center Historical Museum, Rugby 52478
George I. Ashby Memorial Museum, Copper Center 47795
George Taylor House, Catasauqua 47280
Georgetown Museum, Columbus 47748
Georgia Historical Society Museum, Savannah 52941
Georgia Salzburger Society Museum, Rincon 52352
The Georgian, Lancaster 49920
Georgia's Stone Mountain Park, Stone Mountain 53310
Germantown Historical Society Museum, Philadelphia 51839
Gernon House and Blacksmith Shop, Russell 52485
Gig Harbor Peninsula Historical Society Museum, Gig Harbor 48907
Gila County Historical Museum, Globe 48943
Gilman Garrison House, Exeter 48464
Gilman Museum, Hellertown 49243
Gilman Town Hall Museum, Issaquah 49544
Gilpin History Museum, Central City 47320
Glastonbury Museum, Glastonbury 48916
The Glebe House Museum, Woodbury 54275
Glenn H. Curtiss Museum, Hammondsport 49143
Glenview Area Historical Museum, Glenview 48937
Gloucester County Historical Society Museum, Woodbury 54277
Gnadenhutten Historical Park and Museum, Gnadenhutten 48951
Golden Ball Tavern Museum, Weston 54105
Golden Pioneer Museum, Golden 48960
Golden Spike National Historic Site, Brigham City 47025
Goliad State Historical Park, Goliad 48967
Gomez Foundation for Mill House, Marlboro 50494
Gonzales Memorial Museum, Gonzales 48969
Goochland County Museum, Goochland 48970
Goodhue County Historical Society Museum, Red Wing 52273
Gorgas House Museum, Tuscaloosa 53618
Goschenhoppen Folklife Museum, Green Lane 49042
Goshen Historical Society Museum, Goshen 48977
Gouverneur Historical Association Museum, Gouverneur 48980
Governor Henry Lippitt House Museum, Providence 52166
Governor John Langdon House, Portsmouth 52094
Governor Printz Park, Essington 48424
Governor Ross Plantation, Seaford 52987
Governor Seay Mansion, Kingfisher 49782
Governor Stephen Hopkins House, Providence 52167
Governor's Mansion, Columbia 47725
The Governor's Mansion, Frankfort 48735
Governor's Mansion, Sacramento 52507
Grace Hudson Museum and The Sun House, Ukiah 53638
The Grace Museum, Abilene 46124
Graeme Park/Keith Mansion, Horsham 49371
Grafton Historical Society, Grafton 48982
Graham County Historical Museum, Thatcher 53491
Grand County Museum, Hot Sulphur Springs 49376
Grand Forks County Historical Society Museum, Grand Forks 48987

H

Grand Lake Area Historical Museum, Grand Lake 48996
Grand Traverse Heritage Center, Traverse City 53554
Granger Homestead Society Museum, Canandaigua 47213
Grant County Historical Museum, Canyon City 47231
Grant County Historical Museum, Elbow Lake 48333
Grant County Historical Museum, Ephrata 48405
Grant County Museum, Elgin 48336
Grant County Museum, Hyannis 49456
Grant County Museum, Medford 50550
Grant County Museum, Sheridan 53073
Grant County Museum, Ulysses 53641
Grant's Birthplace State Memorial, Point Pleasant 52010
Granville County Historical Society Museum, Oxford 51651
Granville Historical Museum, Granville 49022
Graue Mill and Museum, Oak Brook 51451
Great Basin Historical Society Museum, Delta 48050
Greater Harvard Area Historical Society, Harvard 49206
Greater Hazleton Historical Society Museum, Hazleton 49233
Greater Loveland Historical Society Museum, Loveland 50302
Greater Port Jefferson Museum, Port Jefferson 52036
Green Mountain Club, Waterbury Center 53958
Greenbelt Museum, Greenbelt 49044
Greene County Historical Museum, Waynesburg 54015
Greene County Historical Society Museum, Xenia 54317
Greenfield Historical Society Museum, Greenfield 49057
Greenfield Museum, Greenfield 49056
Greensboro Historical Museum, Greensboro 49062
Greenwood County Historical Society, Eureka 48441
Gregg County Historical Museum, Longview 50194
Gresham History Museum, Gresham 49088
Greybull Museum, Greybull 49089
Grinnell Historical Museum, Grinnell 49092
Grinter Place, Kansas City 49681
Grinter Place State Historical Site, Fairway 48495
Grosse Ile Historical Museum, Grosse Ile 49093
Groveland Yosemite Gateway Museum, Groveland 49098
Grover Cleveland Birthplace, Caldwell 47162
Guernsey County Museum, Cambridge 47191
Gulf Shores Museum, Gulf Shores 49105
Gunn Memorial Library and Museum, Washington 53842
Gunnison County Pioneer and Historical Museum, Gunnison 49106
Gunston Hall Plantation, Mason Neck 50530
Gwinnett History Museum, Lawrenceville 49996
Haakú Museum, Pueblo of Acoma 52195
Hackettstown Historical Society Museum, Hackettstown 49112
Haddonfield Museum, Haddonfield 49114
Hagadorn House Museum, Almond 46236
The Haggin Museum, Stockton 53306
Hahns Peak Area Historical Museum, Clark 47584
HAHS Museum, Hattiesburg 49212
Haines Museum, Waukegan 53989
Hale Farm and Village, Bath 46630
Halifax Historical Museum, Daytona Beach 47993
Hall Museum, Meeteetse 50563
Hall of Fame for Great Americans, Bronx 47046
Hamilton van Wogener Museum, Clifton 47648
Hamlin Garland Homestead, West Salem 54079
Hammond Historical Museum, Hammond 49141
Hampton County Historical Museum, Hampton 49148
Hampton Plantation, McClellanville 50341
Hana Cultural Center, Hana, Maui 49156
Hanby House, Westerville 54090
Hancock County Historical Museum, Carthage 47267
Hancock Historical Museum, Findlay 48544
Hancock Historical Society Museum, Hancock 49157
Hannah Lindahl Children's Museum, Mishawaka 50748
Hanover Historical Society Museum, Hanover 49163
Hanover House, Clemson 47615
Har-Ber Village, Grove 49097
Hardin County Historical Museum, Kenton 49742
Harding Home and Museum, Marion 50487
Harding Museum, Franklin 48749
Hardwick Historical Museum, Hardwick 49166
Harney County Historical Museum, Burns 47146
Harold Warp Pioneer Village Foundation, Minden 50723
Harper City Historical Museum, Harper 49172
Harpers Ferry National Historical Park, Harpers Ferry 49174
Harrison County Historical Museum, Marshall 50509
Harrison House, Branford 46995
Hartsville Museum, Hartsville 49203
Harvard Historical Society Museum, Still River 53293
Harvey County Historical Society, Newton 51340
The Haskell County Historical Museum, Sublette 53352
Hastings Museum, Hastings 49211
Hatton-Eielson Museum, Hatton 49214
Haverford Township Historical Society Museum, Havertown 49217
Hawks Inn, Delafield 48041
Hawks Nest State Park, Ansted 46324
Hayward Area Museum, Hayward 49228
Headlund Museum, Hugo 49414
Headwaters Heritage Museum, Three Forks 53503
Healy House and Dexter Cabin, Leadville 50005
Heart of West Texas Museum, Colorado City 47697
Heisey Museum, Lock Haven 50154
Helvetia Museum, Helvetia 49244
Hemet Museum, Hemet 49245
Hendrickson House Museum and Old Swedes Church, Wilmington 54195
Hennepin History Museum, Minneapolis 50732
Henry County Historical Museum, New Castle 51040
Henry County Museum and Cultural Arts Center, Clinton 47655
Henry Guest House, New Brunswick 51032
The Henry Sheldon Museum of Vermont Histoy, Middlebury 50652
Heritage Center, Dayton 47991
Heritage Center Museum, Berryville 46768
Heritage Center of Clark County, Springfield 53235
Heritage Center of Lancaster County, Lancaster 49923
Heritage Farmstead Museum, Plano 51976
Heritage Hall Museum and Archives, Freeman 48780
Heritage Hill State Park, Green Bay 49038
Heritage House Museum of Montgomery County, Mount Ida 50880
Heritage House of Orange County Museum, Orange 51577
Heritage Museum, Idaho Springs 49464
The Heritage Museum, Astoria 46403
Heritage Museum and Potton House, Big Spring 46799
The Heritage Museum at Falfurrias, Falfurrias 48497
Heritage Museum Foundation of Tate County, Senatobia 53033
Heritage Museum of Northwest Florida, Valparaiso 53697
Heritage Museums, Sandwich 52839
The Heritage Society Museum, Houston 49386
Heritage Square Museum, Ontario 51569
Heritage Village, Mountain Lake 50894
Heritage Village, Sharonville 53052
Heritage Village Museum, Woodville 54292
Heritage Village - Pinellas County Historical Museum, Largo 49953
Heritage Walk Museum, Escondido 48419
Herkimer County Historical Society Museum, Herkimer 49259
Herkimer Home, Little Falls 50136
The Hermitage - Home of President Andrew Jackson, Hermitage 49262
Hettinger County Historical Society Museum, Regent 52283
Heurich House Museum, Washington 53875
Hibbing Historical Museum, Hibbing 49265
Hickory Grove Rural School Museum, Ogden 51499
Hicksville Gregory Museum, Hicksville 49270
Hiddenite Center, Hiddenite 49272
High Cliff General Store Museum, Sherwood 53076
High Plains Heritage Center, Great Falls 49030
High Plains Museum, McCook 50343
High Point Museum, High Point 49275
Highland House Museum, Hillsboro 49287
Highland Park Historical Museum, Highland Park 49281
Highlands, Fort Washington 48701
Highlands Museum and Discovery Center, Ashland 46380
Hildene, Manchester 50427
Hill Country Museum, Kerrville 49749
Hill-Hold Museum, Montgomery 50817
Hillsboro Area Historical Society Museum, Hillsboro 49289
Hillsboro Museums, Hillsboro 49284
Hinsdale Historical Society Museum, Hinsdale 49299
Historic Allaire Village, Allaire 46221
Historic Annapolis Foundation, Annapolis 46320
Historic Bath, Bath 46629
Historic Bethabara Park, Winston-Salem 54245
Historic Bowens Mills and Pioneer Park, Middleville 50664
Historic Brattonsville, McConnells 50342
Historic Burlington County Prison Museum, Mount Holly 50878
Historic Camden Revolutionary War Site, Camden 47201
Historic Carson House, Marion 50486
Historic Charleston Foundation, Charleston 47367
Historic Charlton Park Museum, Hastings 49210
Historic Columbia Foundation, Columbia 47727
Historic Courthouse Museum, Lakeport 49903
Historic Crab Orchard Museum and Pioneer Park, Tazewell 53457
Historic Cragfont, Castalian Springs 47276
Historic Daniel Boone Home and Boonesfield Village, Defiance 48035
Historic Deerfield, Deerfield 48032
Historic Fallsington, Fallsington 48504
Historic Gardner's Basin, Atlantic City 46461
Historic General Dodge House, Council Bluffs 47850
Historic Georgetown, Georgetown 48888
Historic Governors' Mansion, Cheyenne 47445
Historic Halifax, Halifax 49128
Historic Hermann Museum, Hermann 49261
Historic Hope Plantation, Windsor 54226
Historic Houses of Odessa, Delaware, Odessa 51495
Historic Hudson Valley, Tarrytown 53450
The Historic Indian Agency House, Portage 52052
Historic Kenmore, Fredericksburg 48766
Historic Lincoln, Lincoln 50109
Historic New Harmony, New Harmony 51048
The Historic New Orleans Collection, New Orleans 51083
Historic Northampton, Northampton 51413
Historic Occoquan, Occoquan 51485
Historic Palmyra, Palmyra 51685
Historic Pensacola Village, Pensacola 51766
Historic Pittsford, Pittsford 51969
Historic Preservation Association of Bourbon County, Fort Scott 48681
Historic Richmond Town, Staten Island 53263
Historic Rock Ford, Lancaster 49924
Historic Rugby, Rugby 52479
Historic Schaefferstown, Schaefferstown 52956
Historic Shepherdstown Museum, Shepherdstown 53072
Historic Smithfield, Blacksburg 46844
Historic Speedwell, Morristown 50852
Historic Stagville, Durham 48219
Historic Ward-Meade Park Museum, Topeka 53535
Historic White Pine Village, Ludington 50316
Historical and Genealogical Society of Indiana County, Indiana 49482
Historical and Natural History Museum of Natick, Natick 50989
Historical Museum at Fort Missoula, Missoula 50753
The Historical Museum at Saint Gertrude, Cottonwood 47843
Historical Museum of Bay County, Bay City 46650
Historical Museum of Cecil County, Elkton 48355
Historical Museum of Frederick County, Frederick 48756
Historical Museum of Old Randolph, Randolph 52244
Historical Museum of Palm Beach County, West Palm Beach 54072
Historical Museum of Talbot County, Easton 48268
Historical Society Museum, Hudson 49412
Historical Society Museum of Carroll County, Westminster 54098
Historical Society Museum of Ocean Grove, New Jersey, Ocean Grove 51488
Historical Society Museum of the Town of Greenwich, Greenwich 49082
Historical Society of Berks County Museum, Reading 52263
Historical Society of Bloomfield Museum, Bloomfield 46851
Historical Society of Brookfield, Brookfield 47056
Historical Society of Kent County Museum, Chestertown 47439
Historical Society of Middletown and the Wallkill Precinct, Middletown 50661
Historical Society of Oak Park and River Forest, Oak Park 51455
Historical Society of Old Yarmouth Museum, Yarmouth Port 54325
Historical Society of Pennsylvania, Philadelphia 51841
Historical Society of Wells and Ogunquit Museum, Wells 54029
History Museum, Los Gatos 50273
History Museum for Springfield-Greene County, Springfield 53232
The History Place, Morehead City 50842
Hiwan Homestead Museum, Evergreen 48460
Hoard Historical Museum, Fort Atkinson 48590
Hobart Historical Society Museum, Hobart 49303
Hodgen House Museum Complex, Ellsworth 48370
Hoiles-Davis Museum, Greenville 49072
Holland Historical Society Museum, Holland 49319
Holland Land Office Museum, Batavia 46626
Holland Museum, Holland 49317
Holmes County Historical Society Museum, Millersburg 50692
Home of Franklin D. Roosevelt, Hyde Park 49459
Home of Stone - a Ford County Museum, Dodge City 48148
Home Sweet Home Museum, East Hampton 48246
Homeville Museum, Homer 49332
Honolulu House Museum, Marshall 50503
Hood River County Historical Museum, Hood River 49359
Hope Historical Society Museum, Hope 49361
Hope Lodge and Mather Mill, Fort Washington 48702
Hopewell Museum, Hopewell 49362
Hopewell Museum, Paris 51703
Hopsewee Plantation, Georgetown 48893
Horace Greeley House, Chappaqua 47346
Horizons Unlimited Supplementary Educational Center, Salisbury 52680
Hornby Museum, Hornby 49368
Hoskins House, Burlington 47141
Hot Springs County Museum, Thermopolis 53492
Houghton County Historical Museum, Lake Linden 49887
House in the Horseshoe, Sanford 52848
House of History, Malone 50408
The House of the Seven Gables, Salem 52656
House of Wickersham, Juneau 49657
The House on the Rock, Spring Green 53209
Howard County Historical Museum, Kokomo 49826
Howard County Historical Society Museum, Ellicott City 48367
Howell Historical Society Committee Museum, Howell 49407
Hubbard Museum of the American West, Ruidoso Downs 52480
The Hudson River Museum, Yonkers 54331
Humboldt County Historical Association Museum, Dakota City 47921
Humboldt County Historical Association Museum, Humboldt 49418
Humboldt Historical Museum, Humboldt 49419
Humpback Rocks Mountain Farm Visitor Center, Waynesboro 54014
Hunt-Morgan House, Lexington 50049
Hunter-Dawson State Historic Site, New Madrid 51071
Huntingdon County Museum, Huntingdon 49421
Huntington County Historical Society Museum, Huntington 49426
Huntington Trade School, Huntington 49431
Hurley Patentee Manor, Hurley 49447
Huron City Museums, Port Austin 52030
Hurricane Valley Heritage Park Museum, Hurricane 49451
Hurst Ranch, West Covina 54053
Hutchinson County Museum, Borger 46914
Hutchinson House Museum, Waupaca 53991
Hyde Hall, Cooperstown 47790
Hyde Hall, East Springfield 48260
Illinois and Michigan Canal Museum, Lockport 50156
Illinois State Museum, Springfield 53216
Immigrant City Museum, Lawrence 49995
Imogene Herbert Historical Museum, Manistique 50440
Imperial Calcasieu Museum, Lake Charles 49878
Independence Historical Museum, Independence 49470
Independence National Historical Park, Philadelphia 51842
Indian Hill Historical Society Museum, Cincinnati 47562
Indian King Tavern Museum, Haddonfield 49115
Interlaken Historical Society Museum, Interlaken 49512
Iosco County Historical Museum, East Tawas 48261
Iowa Masonic Library and Museum, Cedar Rapids 47304
Ipswich Historical Society, Ipswich 49524
Iredell Museum of Arts Heritage, Statesville 53274
Ironwood Area Historical Museum, Ironwood 49528
Ironworld Discovery Center, Chisholm 47544
Iroquois County Historical Society Museum, Watseka 53985
Isaac Farrar Mansion, Bangor 46597
Isanti County Museums, Cambridge 47187
Ischua Valley Historical Society, Franklinville 48753
Island County Historical Society Museum, Coupeville 47855
Isle La Motte Historical Society, Isle La Motte 49540
Isle of Wight Courthouse, Smithfield 53133
Isleford Historical Museum, Isleford 49541
Islesboro Historical Museum, Islesboro 49542
Issaquah Depot Museum, Issaquah 49545
Itasca Heritage Center Museum, Grand Rapids 49009
J.E. Reeves Home and Museum, Dover 48167
J. Evetts Haley History Center, Midland 50670
J.W. Parmely Historical Home Museum, Ipswich 49525
Jack Jackson Memorial Museum, Weaverville 54017
Jackson County Historical Museum, Lakefield 49899
Jackson County Historical Museum, Maquoketa 50457
Jackson County Historical Society Museum, Black River Falls 46841
Jackson Hole Historical Museum, Jackson 49571
Jackson-Washabaugh Historical Museum, Kadoka 49661
Jackson's Mill Historic Area, Weston 54109
Jacob Walden House, Walden 53793
Jaffrey Civic Center, Jaffrey 49588

James A. Garfield National Historic Site, Mentor 50592
The James Madison Museum, Orange 51580
James Mitchell Varnum House and Museum, East Greenwich 48240
James Monroe Museum and Memorial Library, Fredericksburg 48768
James W. Dillon House Museum, Dillon 48144
Jamestown Museum, Jamestown 49599
Jamestown Visitor Center Museum, Jamestown 49601
Jefferson County Historical Association Museum, Steubenville 53288
Jefferson County Historical Museum, Madison 50379
Jefferson County Historical Museum, Pine Bluff 51922
Jefferson County Historical Society Museum, Watertown 53972
Jefferson Historical Museum, Jefferson 49609
Jefferson Landing State Historic Site, Jefferson City 49613
Jefferson Patterson Museum, Saint Leonard 52566
Jekyll Island Museum, Jekyll Island 49617
Jemez State Monument, Jemez Springs 49618
Jennings-Brown House Female Academy, Bennettsville 46737
Jericho Historical Society, Jericho Corners 49622
Jerome County Historical Society, Jerome 49623
Jersey City Museum, Jersey City 49627
Jesse Besser Museum, Alpena 46237
Jesse James Bank Museum, Liberty 50079
Jesse James Farm and Museum, Kearney 49713
Jesse James Home Museum, Saint Joseph 52562
Jewell County Historical Museum, Mankato 50445
Jim Gatchell Memorial Museum, Buffalo 47127
Jim Thorpe Home, Yale 54321
John Abbott II House, Hamilton 49135
John E. Conner Museum, Kingsville 49796
John G. Voigt House, Plymouth 52003
John Harris/Simon Cameron Mansion, Harrisburg 49180
John Humphrey House, Swampscott 53388
John Jay French House, Beaumont 46677
John Marshall House, Richmond 52331
John Paul Jones House Museum, Portsmouth 52095
John R. Jackson House, Winlock 54233
John Ralston Museum, Mendham 50585
John S. Barry Historical Society Museum, Constantine 47781
John Smart House, Medina 50557
John Stark Edwards House, Warren 53824
John Strong Mansion, Addison 46137
John Sydnor's 1847 Powhatan House, Galveston 48847
Johnson County Historical Society Museum, Coralville 47804
Johnson County Historical Society Museum, Tecumseh 53459
Johnson County Historical Society Museum, Warrensburg 53830
Johnson County History Museum, Franklin 48746
Johnson County Museums, Shawnee 53057
Johnson Ferry House Museum, Titusville 53515
Johnson Hall, Johnstown 49638
Johnson-Humrickhouse Museum, Coshocton 47838
Johnstown Area Heritage Association, Johnstown 49641
Johnstown Historical Society Museum, Johnstown 49639
Joliet Area Historical Museum, Joliet 49643
Jonesborough-Washington County History Museum, Jonesborough 49648
Josiah Day House, West Springfield 54082
Judge Roy Bean Visitor Center, Langtry 49938
Judith Basin County Museum, Stanford 53255
Julia A. Purnell Museum, Snow Hill 53138
Junction City Historical Society Museum, Junction City 49655
Juneau-Douglas City Museum, Juneau 49658
Justin Smith Morrill Homestead, Strafford 53326
Kalamazoo Valley Museum, Kalamazoo 49667
Kalkaska County Historical Museum, Kalkaska 49672
Kaminski House Museum, Georgetown 48894
Kanabec History Center, Mora 50837
Kandiyohi County Historical Society Museum, Willmar 54184
Kankakee County Historical Society Museum, Kankakee 49676
Kansas City Museum, Kansas City 49690
Kauffman Museum, North Newton 51397
Kearney County Historical Museum, Minden 50724
Kearney Mansion Museum, Fresno 48792
Keeler Tavern Museum, Ridgefield 52348
Kell House Museum, Wichita Falls 54153
Kelley House Museum, Mendocino 50586
Kellogg Historical Society Museum, Kellogg 49721
The Kelso House Museum, Kent 49738
Kelton House Museum, Columbus 47755
Kemp-McCarthy Memorial Museum, Rowe 52474
Kenilworth Historical Society Museum, Kenilworth 49725
Kennebunkport Historical Museum, Kennebunkport 49728
Kenosha County Museum, Kenosha 49733
Kent-Delord House Museum, Plattsburgh 51983
Kentuck Museum and Art Center, Northport 51424
The Kentucky Museum, Bowling Green 46973
Kerbyville Museum, Kerby 49748
Kern County Museum, Bakersfield 46537
Kerr Place, Onancock 51557
Kershaw County Historical Society Museum, Camden 47202
Kewaunee County Historical Museum, Kewaunee 49755
Key West Lighthouse Museum, Key West 49762
Key West Museum of Art and History, Key West 49763
Keystone Area Museum, Keystone 49770
Kimball House Museum, Battle Creek 46644
Kimble County Historical Museum, Junction 49654
Kingman County Historical Museum, Kingman 49784
Kingman Tavern Historical Museum, Cummington 47908
Kingsland Homestead, Flushing 48577
Kinney Pioneer Museum, Clear Lake 47605
Kit Carson Historical Society Museum, Kit Carson 49804
Kit Carson Museum, Las Animas 49956
Kitsap County Historical Museum, Bremerton 47005
Kittery Historical and Naval Museum, Kittery 49805
Kittson County History Center Museum, Lake Bronson 49874
Kiwanis Van Slyke Museum Foundation, Caldwell 47160
Klamath County Museum, Klamath Falls 49807
Klein Museum, Mobridge 50773
Klickitat County Historical Society Museum, Goldendale 48962
Klyne Esopus Museum, Ulster Park 53640
Knife River Indian Villages National Historic Site, Stanton 53256
Knight Museum, Alliance 46230
Knox County Museum, Benjamin 46735
Knox County Museum, Knoxville 49810
Koochiching County Historical Museum Bronko Nagurski Museum, International Falls 49513
Kruse House Museum, West Chicago 54050
Kuhlmann King Historical House and Museum, Boerne 46889
Kuser Farm Mansion, Hamilton 49136
Lac Qui Parle County Historical Museum, Madison 50382
Lacey Historical Society Museum, Forked River 48586
Lafayette Museum - Alexandre Mouton House, Lafayette 49866
LaGuardia and Wagner Archives, Long Island City 50183
Lake County Discovery Museum, Wauconda 53987
Lake County Historical Museum, Tavares 53454
Lake County Historical Society Museum, Kirtland Hills 49803
Lake County Museum, Two Harbors 53630
Lake Forest-Lake Bluff Historical Society Museum, Lake Forest 49882
Lake George Historical Association Museum, Lake George 49884
Lake Guernsey Museum, Guernsey 49101
Lake Hopatcong Historical Museum, Landing 49934
Lake of the Red Cedars Museum, Cedar Lake 47298
Lake of the Woods County Museum, Baudette 46648
Lake Placid-North Elba Historical Society Museum, Lake Placid 49891
Lake Wales Depot Museum, Lake Wales 49895
Lakewood's Heritage Culture and the Arts Galleries, Lakewood 49905
Lancaster County Museum, Lancaster 49926
Lancaster Historical Society Museum, Lancaster 49918
Lancaster Museum / Art Gallery, Lancaster 49915
Land of the Yankee Fork Historical Museum, Nampa 50941
Landis Valley Museum, Lancaster 49928
Lane County Historical Museum, Dighton 48140
Lane House, Roseburg 52458
Langlade County Historical Society Museum, Antigo 46325
Lansing Manor House Museum, North Blenheim 51384
Laramie Plains Museum, Laramie 49948
Larchmont Historical Society Museum, Mamaroneck 50412
Lars Noak Blacksmith Shop, Larsson/Ostlund Log Home One-Room Capitol School, New Sweden 51103
LaSalle County Historical Society Museum, Utica 53673
Lassen Historical Museum, Susanville 53385
Lauder Museum, Amityville 46274
Laura Ingalls Wilder Museum, Walnut Grove 53804
The Laurel Museum, Laurel 49982
Laurence C. Jones Museum, Piney Woods 51928
Lawrence County Gray House Museum, Ironton 49527
Lawrence County Historical Museum, Bedford 46687
Layland Museum, Cleburne 47612
Le Roy House and Jell-o Gallery, Le Roy 50002
Le Sueur County Historical Museum, Elysian 48388
Lee Chapel and Museum, Lexington 50071
Lee County Historical Society Museum, Loachapoka 50153
Lee-Fendall House Museum, Alexandria 46210
Leelanau Historical Museum, Leland 50022
Lefferts Historic House, Brooklyn 47078
Lehigh County Historical Society Museum, Allentown 46227
Lemee House, Natchitoches 50988
Lemhi County Historical Museum, Salmon 52685
Leverett Historical Museum, Leverett 50027
Lewes Historical Society Museum, Lewes 50028
Lewis County Historical Museum, Chehalis 47417
Lewis County Historical Society Museum, Lowville 50309
Lewis County Historical Society Museum, Lyons Falls 50337
Lexington County Museum, Lexington 50067
Lexington Historical Society, Lexington 50060
Lexington History Museum, Lexington 50051
Libby Museum, Wolfeboro 54269
Liberty County Museum, Chester 47430
Liberty Hall Historic Center, Lamoni 49913
Libertyville-Mundelein Historical Society Museum, Libertyville 50080
Licking County Historical Society, Newark 51290
Lightner Museum, Saint Augustine 52522
Limon Heritage Museum, Limon 50086
Lincoln County Historical Museum, Davenport 47972
Lincoln County Historical Society, Lincoln 50088
Lincoln County Historical Society Museum, North Platte 51402
Lincoln County Historical Society Museum of Pioneer History, Chandler 47339
Lincoln County Pioneer Museum, Hendricks 49254
Lincoln Parish Museum, Ruston 52490
Lincoln Park Historical Museum, Lincoln Park 50113
The Lincoln-Tallman Restorations, Janesville 49602
Lincoln's New Salem Historic Site, Petersburg 51794
Linn County Historical Museum and Moyer House, Brownsville 47095
Linn County Historical Society Museum, Cedar Rapids 47305
Linn County Museum, Pleasanton 51987
Lisbon Historical Society Museum, Lisbon 50125
Lisle Station Park, Lisle 50127
Litchfield Historical Society Museum, Litchfield 50128
Little Compton Historical Society Museum, Little Compton 50132
Little Cowboy Bar and Museum, Fromberg 48800
Little Falls Historical Museum, Little Falls 50137
Little Traverse Historical Museum, Petoskey 51807
Littleton Historical Museum, Littleton 50144
Livingston County Historical Society Museum, Geneseo 48877
Livingston Depot Center, Livingston 50148
Lockehaven Schoolhouse Museum, Enfield 48399
Lockhouse-Friends of the Delaware Canal, New Hope 51061
Lockwood-Mathews Mansion Museum, Norwalk 51432
Log Cabin Village, Fort Worth 48716
Log House Museum, Columbiana 47738
Logan County Historical Society Museum, Bellefontaine 46697
Logan County Museum, Paris 51700
Lombard Historical Museum, Lombard 50164
Lompoc Museum, Lompoc 50166
Lon C. Hill Home, Harlingen 49169
Long Branch Historical Museum, Long Branch 50181
Longfellow's Wayside Inn Museum, Sudbury 53353
Longmont Museum, Longmont 50193
Loop Museum and Art Center, Seagraves 52989
Lopez Island Historical Museum, Lopez Island 50198
Lorenzo State Historic Site, Cazenovia 47287
Los Alamos County Historical Museum, Los Alamos 50205
Los Altos History Museum, Los Altos 50207
Los Nogales Museum, Seguin 53029
Lost City Museum, Overton 51634
Lou Holtz and Upper Ohio Valley Hall of Fame, East Liverpool 48255
Loudoun Museum, Leesburg 50018
Louis H. and Lena Firn Grover Museum, Shelbyville 53068
Louisa County Historical Society, Louisa 50277
Louisiana State Exhibit Museum, Shreveport 53084
Loveland Museum and Gallery, Loveland 50301
Lowell Damon House, Wauwatosa 53996
Lower Cape Fear Historical Society Museum, Wilmington 54203
Lower White River Museum, Des Arc 48095
Lowndes County Museum, Valdosta 53686
Lucretia Vaile Museum, Palmer Lake 51682
Lucy Hayes Heritage Center, Chillicothe 47535
Lummis Home 'El Alisal', Los Angeles 50249
Luna Mimbres Museum, Deming 48053
Lura Watkins Museum, Middleton 50656
Lyman House Memorial Museum, Hilo 49293
Lynchburg Museum System, Lynchburg 50327
Lynden Pioneer Museum, Lynden 50331
Lyndhurst, Tarrytown 53451
Lyndon B. Johnson National Historical Park, Johnson City 49637
Lynn Museum, Lynn 50333
Lynnhaven House, Virginia Beach 53757
Lyon County Historical Society and Museum, Emporia 48392
Lyon County Historical Society Museum, Marshall 50505
The Lyons Redstone Museum, Lyons 50334
Mabry-Hazen House, Knoxville 49820
McCallum Museum, Sibley 53094
McClurg Mansion, Westfield 54094
McCone County Museum, Circle 47568
McConnell Mansion, Moscow 50864
McCoy House, Lewistown 50043
McCutchen Overland Inn, McCutchenville 50345
McDade Museum, McDade 50346
Macedon Historical Society Museum, Macedon 50347
McFaddin-Ward House, Beaumont 46678
McFarland Historical Society Museum, McFarland 50348
Machan Museum, Lagrange 49870
McHenry County Historical Society Museum, Union 53645
McHenry Museum, Modesto 50775
McIntosh County Historical Society Museum, Ashley 46392
Macktown Living History Site and Whitman Trading Post, Rockton 52428
McLean-Alanreed Area Museum, McLean 50360
McLean County Historical Society Museum, Washburn 53840
McLean County Museum of History, Bloomington 46861
McMinn County Living Heritage Museum, Athens 46420
Macon County Museum Complex, Decatur 48016
MacPheadris/Warner House, Portsmouth 52096
McPherson County Old Mill Museum, Lindsborg 50121
McPherson Museum, McPherson 50372
The Madam Brett Homestead, Beacon 46661
Madeline Island Historical Museum, La Pointe 49852
Madera County Museum, Madera 50374
Madison County Historical Museum, Edwardsville 48300
Madison County Historical Society Museum, Winterset 54256
Madison Historical Society Museum, Madison 50375
Madison-Morgan Cultural Center Collection, Madison 50377
Magevney House, Memphis 50574
Magnolia Grove House Museum, Greensboro 49059
Magnolia Manor Museum, Cairo 47159
Magnolia Plantation, Charleston 47372
Magoffin Home, El Paso 48326
Main Museum, Crescent City 47874
Maine Historical Society Museum, Portland 52065
Maine State Museum, Augusta 46485
Maitland Historical Museums, Maitland 50403
The Major General Frederick Funston Boyhood Home and Museum, Iola 49516
Manassas Museum System, Manassas 50414
Manatee Village Historical Park Museum, Bradenton 46986
Manchester Historic Association Museum, Manchester 50423
Manchester Historical Society Museum, Manchester 50419
Manistee County Historical Museum, Manistee 50439
Manitowoc County History Museum, Manitowoc 50442
Mansfield Historical Society Museum, Storrs 53319
Maple Valley Historical Museum, Maple Valley 50456
Marathon County Historical Society Museum, Wausau 53994
Marble Historical Museum, Marble 50459
Marble Springs State Historic Farmstead Governor John Sevier Memorial, Knoxville 49821
Marblehead Museum and J.O.J. Frost Folk Art Gallery, Marblehead 50462
Marcellus Historical Society Museum, Marcellus 50465
Marias Museum of History and Art, Shelby 53066
Marietta House Museum, Glenn Dale 48933
Marietta Museum of History, Marietta 50469
Marilla Historical Society Museum, Marilla 50477
Marin History Museum, San Rafael 52828
Marinette County Historical Museum, Marinette 50479

Marion County Historical Society Museum, Marion 50488
Marion County Historical Society Museum, Salem 52668
Marion Heritage Center, Marion 50481
Marion Public Library Museum, Marion 50483
Marlboro County Historical Museum, Bennettsville 46738
Marlboro Historical Society, Marlboro 50495
Marquette County History Museum, Marquette 50499
Marquette County Museum, Westfield 54095
Marshall County Historical Museum, Plymouth 51992
Marshall County Historical Society Museum, Warren 53823
Marshall County Museum, Marshalltown 50512
Martha Berry Museum, Rome 52449
Martha's Vineyard Exhibition, Edgartown 48285
Martin County Historical Museum, Stanton 53257
Martin County Pioneer Museum, Fairmont 48489
Martinez Museum, Martinez 50517
Marvin Newton House, Brookfield 47057
Mary Ball Washington Museum, Lancaster 49932
Mary McLeod Bethune House, Washington 53891
Mary Washington House, Fredericksburg 48770
Maryland Historical Society Museum, Baltimore 46576
Marymoor Museum of Eastside History, Redmond 52277
Mashantucket Pequot Museum, Mashantucket 50527
Mason County Museum, Mason 50528
The Masonic Library and Museum of Pennsylvania, Philadelphia 51851
Matagorda County Museum, Bay City 46651
Mather Homestead, Wellsville 54032
Matheson Museum, Gainesville 48814
Mathias Ham House, Dubuque 48188
Mattapoisett Museum and Carriage House, Mattapoisett 50535
Matthews Museum of Maine Heritage, Union 53647
Matt's Museum, Lawtell 49999
Maturango Museum of the Indian Wells Valley, Ridgecrest 52346
May Museum and Park, Farmville 48522
Mayflower Society Museum, Plymouth 51994
Maymont, Richmond 52334
Mayville Area Museum of History and Genealogy, Mayville 50540
Mayville Historical Society Museum, Mayville 50542
Mazomanie Historical Society Museum, Mazomanie 50543
Meade County Historical Society Museum, Meade 50544
Meadowcroft Museum of Rural Life, Avella 46525
Meadowlands Museum, Rutherford 52492
Meagher County Historical Association Castle Museum, White Sulphur Springs 54130
Mecosta County Historical Museum, Big Rapids 46798
Meeker County Museum and G.A.R. Hall, Litchfield 50130
Meeker Home Museum, Greeley 49034
Meigs County Museum, Pomeroy 52016
Mellette House, Watertown 53975
Mem-Erie Historical Museum, Erie 48408
Memorial Hall Museum, Deerfield 48033
Memorial Museum Pioneer Museum, Ada 46134
Memory Lane Museum at Seattle Goodwill, Seattle 53003
Mendocino County Museum, Willits 54183
Mennonite Heritage Center, Harleysville 49168
Menominee County Heritage Museum, Menominee 50589
Mentor Graham Museum, Blunt 46879
Merced County Courthouse Museum, Merced 50595
Mercer County Genealogical and Historical Society, Princeton 52147
Mercer County Historical Museum, Celina 47310
Mercer County Museum, Mercer 50596
Meriden Historical Society Museum, Meriden 50598
Meriwether Lewis Dredge Museum, Brownville 47098
Merrill Historical Museum, Merrill 50602
Merrill Museum, Carrollton 47255
Mesa Historical Museum, Mesa 50608
Miami County Museum, Peru 51789
Michel Brouillet House and Museum, Vincennes 53749
Michie Tavern Ca. 1784, Charlottesville 47397
Michigan Historical Museum, Lansing 49943
Middle Border Museum of American Indian and Pioneer Life, Mitchell 50763
Middleborough Historical Museum, Middleborough 50648
Middlebury Historical Society Museum, Wyoming 54314
Middlesex Canal Collection, Lowell 50305
Middlesex County Historical Society Museum, Middletown 50659
Middletown Valley Historical Society Museum, Middletown 50660
Midland County Historical Museum, Midland 50667
Midland County Historical Museum, Midland 50672
Midway Museum, Midway 50677
Miers Home Museum, Sonora 53155
Milam County Museum and Annex, Cameron 47203
Milan Historical Museum, Milan 50680
Milford Historical Museum, Milford 50686
Milford Historical Society Museum, Milford 50684
Milford Museum, Milford 50685
Mill Race Historical Village, Northville 51428
Millard Fillmore House, East Aurora 48236
Miller Art Center, Springfield 53240
Miller-Cory House Museum, Westfield 54093
Miller House Museum, Hagerstown 49121
Mills County Museum, Glenwood 48941
Milton Historical Museum, Milton 50702
Milwaukee County Historical Society Museum, Milwaukee 50712
Mineral County Museum, Superior 53376
Minisink Valley Historical Society Museum, Port Jervis 52037
Minnilusa Pioneer Museum, Rapid City 52252
Mission Bell Museum, Coweta 47862
Mission Mill Museum, Salem 52669
Mission San Juan Capistrano Museum, San Juan Capistrano 52813
Mission San Luis Rey Museum, Oceanside 51491
Missisquoi Valley Historical Society, North Troy 51408
Mississippi River Museum at Mud Island River Park, Memphis 50580
Mitchell County Historical Museum, Osage 51608
Modoc County Historical Museum, Alturas 46244
Moffatt-Ladd House, Portsmouth 52097
Mohave Museum of History and Arts, Kingman 49783
Mohonk Preserve, Gardiner 48860
Mondak Heritage Center, Sidney 53096
The Monhegan Museum, Monhegan 50780
Monmouth County Historical Museum, Freehold 48779
Monmouth Museum, Lincroft 50115
Monroe County Heritage Museum, Monroeville 50795
Monroe County Historical Museum, Monroe 50791
Monroe County Historical Society, Bloomington 46865
Monroe County Local History Room and Library, Sparta 53192
Monroe Historical Society Museum, Monroe 50786
Montague Museum, Montague 50796
Montana Historical Society, Helena 49242
Montclair Historical Society Museum, Montclair 50799
Montgomery County Historical Society Center, Dayton 47986
Montgomery County Historical Society Museum, Rockville 52431
Montgomery Museum and Lewis Miller Regional Art Center, Christiansburg 47547
Montrose County Historical Museum, Montrose 50829
Montville Township Historical Museum, Montville 50832
Moody County Museum, Flandreau 48554
Moore County Historical Museum, Dumas 48202
Morgan Row Museum, Harrodsburg 49188
Mormon Station Museum, Genoa 48886
The Morris Museum, Morristown 50854
Morton Grove Historical Museum, Morton Grove 50862
Morton Homestead, Prospect Park 52158
Morton House Museum, Benton Harbor 46743
Morton Museum of Cooke County, Gainesville 48823
Moses Lake Museum and Art Center, Moses Lake 50867
Moses Myers House, Norfolk 51366
Mount Airy Museum of Regional History, Mount Airy 50870
Mount Desert Island Historical Museum, Somesville 53153
Mount Holly Community Historical Museum, Belmont 46721
Mount Pleasant Historical Society Museum, Mount Pleasant 50886
Mount Prospect Historical Society Museums, Mount Prospect 50889
Mount Vernon Hotel Museum, New York 51218
Mountain Gateway Museum, Old Fort 51535
Murray-Lindsay Mansion, Lindsay 50119
The Museum, Greenwood 49087
Museum at Prophetstown, Battle Ground 46646
Museum Center, Maysville 50539
Museum of Amana History, Amana 46249
Museum of Anthracite Mining, Ashland 46390
Museum of Appalachia, Norris 51377
The Museum of Art and History, Santa Cruz 52868
The Museum of Arts and Sciences, Daytona Beach 47994
Museum of Bronx History, Bronx 47050
Museum of Central Connecticut State University, New Britain 51028
Museum of Chicot County Arkansas, Lake Village 49893
The Museum of Clallam County Historical Society, Port Angeles 52026
Museum of Early Trades and Crafts, Madison 50385
Museum of History and Art, Ontario 51568
Museum of Idaho, Idaho Falls 49462
Museum of Maryland History, Baltimore 46582
Museum of Migrating People, Bronx 47051
Museum of Mobile, Mobile 50770
Museum of Natural and Cultural History, Eugene 48435
Museum of New Hampshire History, Concord 47775
The Museum of Newport History, Newport 51322
Museum of North Idaho, Coeur d'Alene 47672
Museum of Northern Arizona, Flagstaff 48551
Museum of Northwest Colorado, Craig 47865
Museum of Orient and East Marion History, Orient 51587
Museum of Seminole County History, Sanford 52846
Museum of South Texas History, Edinburg 48291
The Museum of Southern History, Sugarland 53357
Museum of Texas Tech University, Lubbock 50312
Museum of the Bedford Historical Society, Bedford 46688
Museum of the Berkeley Springs, Berkeley Springs 46762
Museum of the Big Bend, Alpine 46238
Museum of the Cherokee Strip, Enid 48401
Museum of the City of Lake Worth, Lake Worth 49896
Museum of the City of New York, New York 51231
Museum of the Great Plains, Lawton 50000
Museum of the Gulf Coast, Port Arthur 52029
Museum of the Historical Society of Trappe, Collegeville, Perkiomen Valley, Trappe 53552
Museum of the Llano Estacado, Plainview 51975
Museum of the Rockies, Bozeman 46982
The Museum of the South Dakota State Historical Society, Pierre 51915
Museum of the Southwest, Midland 50673
Museum of the Waxhaws and Andrew Jackson Memorial, Waxhaw 54001
Museum of the Weathersfield Historical Society, Weathersfield 54016
Museum of the Western Prairie, Altus 46245
Museum of the Weston Historical Society, Weston 54106
Museum of Western Colorado, Grand Junction 48994
Museum of Yarmouth History, Yarmouth 54324
Museum of York County, Rock Hill 52403
Museum on Main Street, Pleasanton 51986
Museum Village, Monroe 50793
The Museums of Oglebay Institute, Wheeling 54124
Muskegon County Museum, Muskegon 50925
Musselshell Valley Historical Museum, Roundup 52473
Nantucket Historical Association Museum, Nantucket 50946
Napa Valley Museum, Yountville 54355
Narcissa Prentiss House, Prattsburgh 52133
Narragansett Historical Society Museum, Templeton 53475
Nathan Denison House, Forty Fort 48721
Nathaniel W. Faison Home and Museum, La Grange 49844
National Historic Oregon Trail Interpretive Center, Baker City 46534
National McKinley Birthplace Memorial, Niles 51352
National Route 66 Museum and Old Town Museum, Elk City 48341
National Women's Hall of Fame, Seneca Falls 53035
Nauvoo Historical Society Museum, Nauvoo 50992
Navarro County Historical Society Museum, Corsicana 47825
Nebraska Prairie Museum, Holdrege 49312
Neill Museum, Fort Davis 48612
Nemours Mansion and Gardens, Wilmington 54196
Neptune Township Historical Museum, Neptune 51005
Netherland Inn House Museum, Kingsport 49786
Nevada State Museum, Las Vegas 49977
Neversink Valley Area Museum, Cuddebackville 47898
Neville Public Museum of Brown County, Green Bay 49041
New Bedford Whaling Museum, New Bedford 51014
The New Berlin Historical Society Museum, New Berlin 51017
New Braunfels Conservation Society Museum, New Braunfels 51023
New Canaan Historical Society Museum, New Canaan 51037
New Castle Historical Society Museum, New Castle 51039
New Hampshire Antiquarian Society Museum, Hopkinton 49365
New Haven Colony Historical Society Museum, New Haven 51053
New London County Historical Society Museum, New London 51067
New London Historical Society Museum, New London 51069
New London Public Museum, New London 51070
New Milford Historical Society Museum, New Milford 51073
New Sweden Historical Museum, New Sweden 51104
New Windsor Cantonment State Historic Site and National Purple Heart Hall of Honor, Vails Gate 53683
The New York Historical Society Museum, New York 51243
Newaygo County Museum, Newaygo 51298
Newell Museum, Newell 51311
Newport Mansions, Newport 51327
Newsome House Museum, Newport News 51335
Newton History Museum at Jackson Homestead, Newton 51341
Newton Museum, Newton 51339
Newtown Historic Museum, Newtown 51346
Nez Perce County Museum, Lewiston 50035
Niagara County Historical Center, Lockport 50158
Nici Self Museum, Centennial 47312
Ninety-Six National Historic Site, Ninety-Six 51354
Nishna Heritage Museum, Oakland 51473
No Man's Land Historical Museum, Goodwell 48972
Noah Webster House - Museum of West Hartford History, West Hartford 54059
Noank Museum, Noank 51355
Nobleboro Historical Society Museum, Nobleboro 51356
Nohwike' Bagowa, Fort Apache 48589
Nordic Heritage Museum, Seattle 53006
Nordica Homestead Museum, Farmington 48516
Norfolk Historical Museum, Norfolk 51360
Norfolk Historical Society, Norfolk 51368
Norman Cleveland County Historical Museum, Norman 51375
North Andover Historical Museum, North Andover 51379
North Castle Historical Society Museum, Armonk 46361
North Haven Historical Society Museum, North Haven 51394
North House Museum, Lewisburg 50034
North Lee County Historic Center and Santa Fe Depot Museum Complex, Fort Madison 48652
North Platte Valley Museum, Gering 48896
North Tonawanda History Museum, North Tonawanda 51407
Northampton County Historical and Genealogical Society Museum, Easton 48271
Northborough Historical Museum, Northborough 51416
Northeast Mississippi Museum, Corinth 47808
Northeast Oakland Historical Museum, Oxford 51647
Northfield Historical Society Museum, Northfield 51421
Northland Historical Society Museum, Lake Tomahawk 49892
Northport Historical Society Museum, Northport 51425
The Northumberland County Historical Society Museum, Sunbury 53369
Northwest Arctic Heritage Center, Kotzebue 49827
Northwest Museum of Arts and Culture, Spokane 53207
Norwich Historical Society, Norwich 51443
Norwood Historical Association Museum, Norwood 51444
Novato History Museum, Novato 51447
Nowata County Historical Society Museum, Nowata 51448
O. Henry Museum, Austin 46517
Oak Creek Pioneer Village, Oak Creek 51452
Oakdale Museum and History Center, Oakdale 51459
Oaklands Historic House Museum, Murfreesboro 50914
Oakleigh House Museum, Mobile 50771
Oakley Pioneer Museum, Oakley 51475
Oatlands, Leesburg 50020
Oberlin Heritage Center, Oberlin 51479
Ocean County Historical Museum, Toms River 53528
Oconto County Historical Society Museum, Oconto 51494
The Octagon House, Hudson 49413
Octagon House, Watertown 53976
Odyssey Maritima Discovery Museum, Seattle 53008
Ogden Historical Society Museum, Spencerport 53202
Ogden House, Fairfield 48481
Ogden Union Station Museums, Ogden 51503
Ohio River Museum, Marietta 50476
Ojai Valley Museum, Ojai 51511
Okefenokee Heritage Center, Waycross 54002
Oklahoma Museum of History, Oklahoma City 51524
Oklahoma Route 66 Museum, Clinton 47661
Olana State Historic Site, Hudson 49411
Old Bethpage Village Restoration, Old Bethpage 51532
Old Bohemia Historical Museum, Warwick 53833

Old Brown's Mill School, Chambersburg 47332
Old Brutus Historical Society Museum, Weedsport 54020
Old City Park - The Historical Village of Dallas, Dallas 47937
Old Colony Historical Museum, Taunton 53453
Old Conway Homestead Museum and Mary Meeker Cramer Museum, Camden 47194
Old Court House Museum-Eva Whitaker Davis Memorial, Vicksburg 53734
Old Courthouse Heritage Museum, Inverness 49514
Old Courthouse Museum, Santa Ana 52853
Old Cowtown Museum, Wichita 54147
Old Dorchester State Historic Site, Summerville 53363
Old Dutch Parsonage, Somerville 53150
Old Economy Village Museum, Ambridge 46254
Old Exchange and Provost Dungeon, Charleston 47374
Old Falls Village, Menomonee Falls 50590
Old Fort Bissell, Phillipsburg 51894
Old Fort Harrod Museum, Harrodsburg 49189
Old Fort House Museum, Fort Edward 48619
Old Fort Johnson, Fort Johnson 48631
Old Fort Lauderdale Museum of History, Fort Lauderdale 48640
Old Fort Number 4 Associates, Charlestown 47382
The Old Governor's Mansion, Frankfort 48741
Old Greer County Museum and Hall of Fame, Mangum 50429
Old Independence Regional Museum, Batesville 46627
Old Iron County Courthouse Museum, Hurley 49448
The Old Jail Museum, Warrenton 53831
Old Jefferson Town, Oskaloosa 51618
Old Lighthouse Museum, Stonington 53312
Old Lincoln County Courthouse Museum, Lincoln 50110
Old Manchester Museum, Manchester 50418
Old Mill Museum, Cimarron 47550
Old Mill Museum, San Marino 52820
Old Salem, Winston-Salem 54248
Old Shawnee Town, Shawnee 53058
Old Stone Fort Museum Complex, Schoharie 52963
The Old Stone House, Washington 53913
Old Stone House Museum, Windsor 54227
The Old Stone House Museum, Brownington 47093
The Old Stone Jail Museum, Palmyra 51687
Old Sturbridge Village, Sturbridge 53344
Old Tavern Museum, Tuscaloosa 53619
Old Town Hall Museum, Fifield 48540
Old Town Museum, Old Town 51538
Old Trail Museum, Choteau 47546
Old Village Hall Museum, Lindenhurst 50118
Old West Museum, Sunset 53375
Old Westbury Gardens, Old Westbury 51539
Old York Historical Museum, York 54338
Old York Road Historical Society, Jenkintown 49620
Oldest Stone House Museum, Lakewood 49908
Oldham County History Center, La Grange 49842
Oliver House Museum, Penn Yan 51764
Oliver Tucker Historic Museum, Beverly 46787
Olmstead Place State Park, Ellensburg 48360
Olmsted County Museum, Rochester 52389
Oneida Community Mansion House, Oneida 51560
Oneida County Historical Society Museum, Utica 53677
Onondaga Historical Association Museum, Syracuse 53402
Onslow County Museum, Richlands 52309
Ontario County Historical Society Museum, Canandaigua 47214
Ontonagon County Historical Society Museum, Ontonagon 51571
Opelousas Museum, Opelousas 51574
Opus 40 and the Quarryman's Museum, Saugerties 52927
Orange County Historical Museum, Hillsborough 49291
Orange County Regional History Center, Orlando 51592
Orangetown Historical Museum, Pearl River 51744
Orcas Island Historical Museum, Eastsound 48272
Orchard Park Historical Society Museum, Orchard Park 51584
Oregon Historical Society, Portland 52082
Oregon-Jerusalem Historical Museum, Oregon 51585
Oregon Trail Museum, Gering 48897
Orland Historical Museum, Orland 51589
Osage County Historical Society, Lyndon 50332
Osage County Historical Society Museum, Pawhuska 51733
Osborn-Jackson House, East Hampton 48248
Oscar Anderson House Museum, Anchorage 46290
Oshkosh Public Museum, Oshkosh 51615
Ossining Historical Society Museum, Ossining 51621
Osterville Historical Museum, Osterville 51622
Oswego Historical Museum, Oswego 51623
Otoe County Museum of Memories, Syracuse 53396
Ottawa County Historical Museum, Port Clinton 52031
Otter Tail County Museum, Fergus Falls 48534
Outagamie Museum, Appleton 46331
Overfield Tavern Museum, Troy 53576
Overholser Mansion, Oklahoma City 51525
Overland Trail Museum, Sterling 53284
Overstreet-Kerr Historical Farm, Keota 49747
Owensboro Area Museum of Science and History, Owensboro 51639
Owyhee County Historical Museum, Murphy 50916
Oxford Library Museum, Oxford 51645
Oxford Museum, Oxford 51646
Ozaukee County Historical Society Pioneer Village, Saukville 52930
P.H. Sullivan Museum and Genealogy Library, Zionsville 54372
Pacem in Terris, Warwick 53834
Pacific County Historical Museum, South Bend 53162
Packwood House Museum, Lewisburg 50030
Paine-Gillam-Scott Museum, Saint Johns 52553
Pajaro Valley Historical Museum, Watsonville 53986
Palace of the Governors, Santa Fe 52885
Palmer/Gullickson Octagon Home, West Salem 54080
Palmer House, Northfield 51422
Panhandle-Plains Historical Museum, Canyon 47230
Parishville Museum, Parishville 51705
Park City Historical Society Museum, Park City 51707
Parris Island Museum, Parris Island 51714
Parson Historical Commission Museum, Parsons 51717
Parsonsfield-Porter History House, Porter 52059
Pasadena Museum of History, Pasadena 51724
Pascack Historical Society Museum, Park Ridge 51711
Passaic County Historical Society Museum, Paterson 51729
Paterson Museum, Paterson 51730
Paul Laurence Dunbar State Memorial, Dayton 47988
Pawnee City Historical Society Museum, Pawnee City 51737
Pearce Collections Museum, Corsicana 47826
Pebble Hill Plantation, Thomasville 53500
Peekskill Museum, Peekskill 51748
Peel Mansion Museum, Bentonville 46744
Pejepscot Historical Society Museum, Brunswick 47104
Pella Historical Village, Pella 51750
Pembroke Historical Society Museum, Pembroke 51754
Pendarvis Historic Site, Mineral Point 50726
Pennsbury Manor, Morrisville 50860
Pennyroyal Area Museum, Hopkinsville 49364
Pensacola Historical Museum, Pensacola 51768
Perkins County Historical Society Museum, Grant 49015
Perry-Downer House and Costume Gallery, Monterey 50806
Peshtigo Fire Museum, Peshtigo 51790
Peter Conser House, Heavener 49236
Peter Rice Homestead, Marlboro 50493
Peter Wentz Farmstead, Worcester 54304
Peter Whitmer Farm, Waterloo 53967
Peter Yegen Jr. Yellowstone County Museum, Billings 46806
Peterborough Historical Society Museum, Peterborough 51792
The Petersburg Museums, Petersburg 51801
Pettaquamscutt Museum, Kingston 49794
Pettis County Historical Society, Sedalia 53024
Phillips County Museum, Helena 49238
Phillips County Museum, Holyoke 49326
Phillips Historical Society Museum, Phillips 51893
The Philmont Museum, Seton Memorial Library and Kit Carson Museum, Cimarron 47551
Piatt Castles, West Liberty 54068
Piatt County Museum, Monticello 50820
Pickens County Museum of Art History, Pickens 51912
Pierrepont Museum, Canton 47223
Pike County Museum, Milford 50689
Pilgrim Hall Museum, Plymouth 51995
Pilgrim Monument and Provincetown Museum, Provincetown 52182
Pimeria Alta Historical Society Museum, Nogales 51358
Pinal County Historical Museum, Florence 48567
Pine County Historical Society Museum, Askov 46396
Pine Grove Historic Museum, Pontiac 52022
Pioneer Corner Museum, New Holstein 51060
Pioneer Farm Museum and Ohop Indian Village, Eatonville 48275
Pioneer Heritage Center, Cavalier 47285
Pioneer-Krier Museum, Ashland 46379
Pioneer Memorial Association of Fenton and Mounty Townships, Fenton 48533
Pioneer Museum, Fabius 48468
Pioneer Museum, Wild Rose 54159
Pioneer Museum and Vereins Kirche, Fredericksburg 48762
Pioneer Museum of Alabama, Troy 53571
Pioneer Town, Wimberley 54217
Pioneer Village, Farmington 48520
Pipestone County Historical Museum, Pipestone 51930
Pittock Mansion, Portland 52085
Pittsford Historical Society Museum, Pittsford 51971
Plains Indians and Pioneers Museum, Woodward 54293
Plainsman Museum, Aurora 46497
Plantation Agriculture Museum, Scott 52973
Plantation Historical Museum, Plantation 51978
Platt R. Spencer Memorial Archives and Special Collections Area, Geneva 48882
Plumas County Museum, Quincy 52211
Plymouth Antiquarian Society Museum, Plymouth 51997
Plymouth Historical Museum, Plymouth 51999
Pocahontas County Iowa Historical Society Museum, Laurens 49986
Pocahontas County Museum, Marlinton 50497
Point of Honor, Lynchburg 50329
Polk County Historical Museum, Bartow 46622
Polk County Historical Museum, Cedartown 47309
Polk County Historical Museum, Osceola 51610
Polk County Historical Society, Des Moines 48099
Polk County Memorial Museum, Livingston 50151
Polk County Museum, Balsam Lake 46543
Polk County Museum, Crookston 47886
Polson Park and Museum, Hoquiam 49366
Pony Express Museum, Saint Joseph 52564
Pope County Historical Museum, Glenwood 48942
Pope's Tavern Museum, Florence 48564
Poplar Grove Historic Plantation, Wilmington 54204
Port Gamble Historic Museum, Port Gamble 52032
Port Huron Museum, Port Huron 52035
Portage County Historical Society Museum, Ravenna 52256
Portage County Museum, Stevens Point 53290
La Porte County Historical Society Museum, La Porte 49853
Porter County Museum, Valparaiso 53699
Portland Children's Museum, Portland 52087
Portland Museum, Louisville 50295
Portsmouth Museum, Portsmouth 52104
Post Museum, Fort Leavenworth 48643
Potsdam Public Museum, Potsdam 52115
Potter County Historical Society Museum, Coudersport 47848
Potts Inn Museum, Pottsville 52118
Pottsgrove Manor, Pottstown 52117
Pound Ridge Museum, Pound Ridge 52125
Powder Magazine, Charleston 47375
Powell County Museum, Deer Lodge 48031
Powers Museum, Carthage 47268
Prairie Homestead, Philip 51891
The Prairie Museum of Art and History, Colby 47680
Prairie Trails Museum of Wayne County, Corydon 47837
Prairie Village, Madison 50386
Pratt County Historical Society Museum, Pratt 52132
Pratt Museum, Homer 49330
President Chester A. Arthur Historic Site, Fairfield 48487
President's Cottage Museum, White Sulphur Springs 54131
Presque Isle County Historical Museum, Rogers City 52442
Prestwould Foundation, Clarksville 47595
Pricketts Fort, Fairmont 48490
Princeton Museum, Princeton 52150
Promont House Museum, Milford 50688
Prouty-Chew Museum, Geneva 48880
Providence Preservation Society, Providence 52175
Provincetown Heritage Museum, Provincetown 52184
Pueblo County Historical Society Museum, Pueblo 52192
Putnam Cottage, Greenwich 49083
Putnam County Historical Society and Foundry School Museum, Cold Spring 47681
Putnam County Historical Society Museum, Kalida 49669
Queens Historical Society, Flushing 48579
Quincy and Adams County Museum, Quincy 52215
Quincy Historical Museum, Quincy 52220
The Quincy Museum, Quincy 52217
Rachel Carson Homestead, Springdale 53213
Railroad House Historical Museum, Sanford 52849
Raleigh City Museum, Raleigh 52236
Ralls Historical Museum, Ralls 52237
Ralph Foster Museum, Point Lookout 52008
Ramsay House, Alexandria 46212
Ramsey House Museum Plantation, Knoxville 49822
Rancho Los Cerritos, Long Beach 50178
Randolph County Historical Museum, Winchester 54219
Randolph County Museum, Beverly 46789
Randolph Historical Society, Randolph 52246
Rankin House State Memorial, Ripley 52356
Rankin Museum, Rankin 52247
Ransom County Historical Society Museum, Fort Ransom 48673
Raton Museum, Raton 52255
Ravalli County Museum, Hamilton 49132
Rawhide Old West Museum, Scottsdale 52976
Ray E. Powell Museum, Grandview 49013
Raynham Hall Museum, Oyster Bay 51662
Reading Historical Society, Reading 52267
Reading Historical Society Museum, Reading 52261
Readsboro Historical Society, Readsboro 52269
Red Barn Museum, Morristown 50857
Red Clay State Historical Park, Cleveland 47645
Red Hill-Patrick Henry National Memorial, Brookneal 47087
Red Mill Museum Village, Clinton 47658
Red River Historical Society Museum, Clay City 47597
Redington Museum, Waterville 53978
Redwood County Museum, Redwood Falls 52279
Refugio County Museum, Refugio 52282
Regional Museum of Spartanburg County, Spartanburg 53194
Reitz Home Museum, Evansville 48457
Remick Country Doctor Museum and Farm, Tamworth 53439
Reno County Museum, Hutchinson 49455
Rensselaer County Historical Society Museum, Troy 53575
Renton Museum, Renton 52295
Renville County Historical Society Museum, Mohall 50777
Republic County Historical Society Museum, Belleville 46700
Reynolds Homestead, Critz 47883
Rhode Island Black Heritage Society Museum, Providence 52176
Rhode Island Historical Society Exhibition, Providence 52178
Rialto Museum, Rialto 52303
Rice County Museum of History, Faribault 48512
Richard Salter Storrs House, Longmeadow 50192
Richardson-Bates House Museum, Oswego 51625
Richey Historical Society, Richey 52304
Richland County Historical Museum, Wahpeton 53785
Richland County Museum, Lexington 50066
Richmond Museum of History, Richmond 52312
Rifle Creek Museum, Rifle 52351
Riley County Historical Museum, Manhattan 50437
Rimrocker Historical Museum of West Montrose County, Naturita 50990
Rio Grande Valley Museum, Harlingen 49170
Riordan Mansion, Flagstaff 48553
Ripley County Historical Society Museum, Versailles 53729
Ripon Historical Society Museum, Ripon 52359
Rittman Historical Society Museum, Rittman 52360
River Heritage Museum, Paducah 51670
River Museum, Wellsville 54033
Riverdale Mansion, Riverdale Park 52365
Riverside Municipal Museum, Riverside 52374
Riverside Museum, La Crosse 49837
Riverton Museum, Riverton 52377
Riverview at Hobson Grove, Bowling Green 46975
Robbins Hunter Museum, Granville 49023
Robert Crown Center for Health Education, Hinsdale 49300
Robert E. Lee Memorial Association, Stratford 53336
Robert Mills Historic House and Park, Columbia 47730
Robert Newell House, Saint Paul 52631
Robert S. Kerr Museum, Poteau 52114
Roberts County Museum, Miami 50639
Rochester Hills Museum at Van Hoosen Farm, Rochester 52388
Rock County Historical Society Museum, Janesville 49603
Rock Creek Station State Historic Park, Fairbury 48476
Rock House Museum, Wytheville 54315
Rock Island County Historical Museum, Moline 50779
Rock Springs Historical Museum, Rock Springs 52411
Rockbridge Historical Society, Lexington 50073
Rockingham Free Public Library and Museum, Bellows Falls 46716
Rockland County Museum, New City 51043
Rockwood Museum, Wilmington 54197
Rocky Ford Historical Museum, Rocky Ford 52433
Rocky Mount Museum, Piney Flats 51927
Rocky Reach Dam Museum, Wenatchee 54036
Roger Williams National Memorial, Providence 52179
Rogers Historical Museum, Rogers 52440
Rokeby Museum, Ferrisburgh 48537
Rollo Jamison Museum, Platteville 51981
Rome Historical Society Museum, Rome 52453
Roosevelt County Museum, Portales 52057
Root House Museum, Marietta 50471

Rose Center Museum, Morristown 50859
Rose Hill Mansion, Geneva 48881
Rose Hill Museum, Bay Village 46655
Roseau County Historical Museum, Roseau 52456
Rosebud County Pioneer Museum, Forsyth 48588
Rosemount Museum, Pueblo 52193
Ross County Historical Society Museum, Chillicothe 47537
Roswell Museum and Art Center, Roswell 52470
The Rotunda, Charlottesville 47399
Rowley Historical Museum, Rowley 52475
Royal Arts Foundation, Newport 51330
The Royal Governor's Mansion, Perth Amboy 51786
Royalton Historical Society, Royalton 52477
Ruddick's Folly, Suffolk 53354
Rundlet-May House, Portsmouth 52099
Runestone Museum, Alexandria 46197
Rural Life Museum and Windrush Gardens, Baton Rouge 46640
Rush County Historical Society Museum, Rushville 52482
Rusk County Historical Society Museum, Ladysmith 49862
Ruth and Charles Gilb Arcadia Historical Museum, Arcadia 46336
The Rye Historical Society and Square House Museum, Rye 52495
Sachem Historical Society Museum, Holbrook 49310
Sacramento Mountains Historical Museum, Cloudcroft 47664
Safety Harbor Museum of Regional History, Safety Harbor 52512
Saguache County Museum, Saguache 52517
Saint Albans Historical Museum, Saint Albans 52519
Saint Augustine Historical Society Museum, Saint Augustine 52525
Saint Charles County Historical Society, Saint Charles 52535
Saint Charles Heritage Center, Saint Charles 52531
Saint Clair County Historical Museum, Belleville 46699
Saint Francis County Museum, Forrest City 48587
Saint James' House, Fredericksburg 48773
Saint John's Church and Parish Museum, Hampton 49153
Saint Louis County Historical Museum, Duluth 48199
Saint Lucie County Historical Museum, Fort Pierce 48668
Saint Marys and Benzinger Township Museum, Saint Marys 52607
Saint Paul Museum, Randleman 52243
Saint Petersburg Museum of History, Saint Petersburg 52641
Saint Simons Island Lighthouse Museum, Saint Simons Island 52648
Sainte Genevieve Museum, Sainte Genevieve 52651
Sainte Marie among the Iroquois, Liverpool 50146
Salem County Historical Society Museum, Salem 52662
Salem Historical Society Museum, Salem 52663
Salem Museum, Salem 52672
Salem Pioneer Village, Salem 52659
Salida Museum, Salida 52673
Saline County Historical Society Museum, Dorchester 48153
Salisbury Historical Society Museum, Salisbury 52683
Salisbury Mansion, Worcester 54301
Salmon Brook Historical Society Museum, Granby 48984
Sam Houston Historical Schoolhouse, Maryville 50526
Sam Rayburn House Museum, Bonham 46899
The Sam Rayburn Museum, Bonham 46900
San Antonio Missions Visitor Center, San Antonio 52725
San Benito County Historical Museum, Hollister 49320
San Buenaventura Mission Museum, Ventura 53714
San Diego Historical Society Museum, San Diego 52749
San Dieguito Heritage Museum, Encinitas 48396
San Jacinto Museum, San Jacinto 52803
San Joaquin County Historical Museum, Lodi 50159
San Juan County Historical Museum, Silverton 53109
San Juan County Museum, Bloomfield 46852
San Juan Historical Society Museum, Friday Harbor 48794
San Luis Obispo County Historical Museum, San Luis Obispo 52816
San Mateo County History Museum, Redwood City 52278
Sanchez Adobe, Pacifica 51668
Sand Springs Cultural and Historical Museum, Sand Springs 52831
Sandown Historical Museum, Sandown 52834
Sands-Willets House, Port Washington 52050
Sandwich Historical Society Museum, Center Sandwich 47314
Sandwich Historical Society Museum, Sandwich 52838
Sandy Bay Museums, Rockport 52424
Sandy Spring Museum, Sandy Spring 52845
Sanford Museum, Cherokee 47421
Sanford Museum, Sanford 52847
Sanilac County Historical Museum, Port Sanilac 52041
Santa Barbara Historical Museum, Santa Barbara 52858
Santa Cruz Museum of Natural History, Santa Cruz 52870
Santa Fe Trail Museum, Larned 49955
Santa Fe Trail Museums of Gray County, Ingalls 49510
Santa Maria Valley Historical Society Museum, Santa Maria 52896
Sapulpa Historical Museum, Sapulpa 52910
Saratoga Museum, Saratoga 52919
The Saratoga Springs History Museum, Saratoga Springs 52924
Sarpy County Historical Museum, Bellevue 46703
Satterlee Clark House, Horicon 49367
Sauk County Historical Museum, Baraboo 46606
Sauk Prairie Area Historical Society Museum, Prairie du Sac 52131
Saunders County Historical Museum, Wahoo 53784
Sawin Museum, Dover 48165
Sawyer's Sandhills Museum, Valentine 53690
Sayville Historical Society Museum, Sayville 52952
Scantic Academy Museum, East Windsor 48265
Scarborough Historical Museum, Scarborough 52953
Scarsdale Historical Society Museum, Scarsdale 52954
Schenectady County Historical Society Museum, Schenectady 52961
Schenectady Museum, Schenectady 52962
Schminck Memorial Museum, Lakeview 49904
Schoenbrunn Village State Memorial, New Philadelphia 51098
Schoharie Colonial Heritage Association Museum, Schoharie 52964
Schoolhouse Museum, Ridgewood 52350
Schuyler County Historical Society Museum, Montour Falls 50822
Schuyler Jail Museum, Rushville 52481
Schuyler Mansion, Albany 46162
Schwenkfelder Library and Heritage Center, Pennsburg 51765
Scituate Historic Sites, Scituate 52966
Scotland Heritage Chapel and Museum, Scotland 52972
Scurry County Museum, Snyder 53139
Sea Cliff Village Museum, Sea Cliff 52985
Seaford Historical Museum, Seaford 52988
Sedgwick-Brooklin Historical Museum, Sedgwick 53026
Senator George Baldwin Hotel, Klamath Falls 49808
Senator George Norris State Historic Site, McCook 50344
Senator John Heinz Pittsburgh Regional History Center, Pittsburgh 51959
Seneca County Museum, Tiffin 53507
Seneca Falls Historical Society Museum, Seneca Falls 53036
Sequoyah Cabin, Sallisaw 52684
Seward County Historical Society Museum, Goehner 48952
Seward House, Auburn 46471
Seward Museum, Seward 53042
Seymour Community Museum, Seymour 53045
Shaftsbury Historical Society, Shaftsbury 53048
Shaker Historical Society Museum, Shaker Heights 53049
The Shaker Museum, Old Chatham 51534
Shaker Village of Pleasant Hill, Harrodsburg 49190
Shandy Hall, Geneva 48883
Sharlot Hall Museum, Prescott 52135
Sharpsteen Museum, Calistoga 47168
Shaw Island Historical Society Museum, Shaw Island 53055
Sheboygan County Historical Museum, Sheboygan 53062
Shekspear Ghost Town Museum, Lordsburg 50200
Shelburne Museum, Shelburne 53065
Shelter House, Emmaus 48389
Shelter Island Historical Society Museum, Shelter Island 53071
Sherburne County Historical Museum, Becker 46684
The Sherman House, Lancaster 49921
Sherwood-Davidson House, Newark 51293
Sherwood Forest Plantation, Charles City 47351
Shiloh Museum of Ozark History, Springdale 53212
Shippensburg Historical Society Museum, Shippensburg 53081
Shirley Plantation, Charles City 47352
Shoreline Historical Museum, Seattle 53015
Shorter Mansion Museum, Eufaula 48429
Sibley County Historical Museum, Henderson 49248
Sidney Historical Association Museum, Sidney 53098
Sidney Lanier Cottage, Macon 50369
Silas Wright House, Canton 47225
Siloam Springs Museum, Siloam Springs 53099
Silver City Museum, Silver City 53103
Silver Cliff Museum, Silver Cliff 53105
Silver River Museum, Ocala 51484
Silvercreek Museum, Freeport 48782
Simon Paneak Memorial Museum, Anaktuvuk Pass 46284
Simsbury Historical Society Museum, Simsbury 53112
Sioux City Public Museum, Sioux City 53114
Siouxland Heritage Museums, Sioux Falls 53122
Siskiyou County Museum, Yreka 54359
Sitka Historical Society Museum, Sitka 53126
Skagit County Historical Museum, La Conner 49832
Skagway Museum, Skagway 53128
Skinner Museum, South Hadley 53171
Slate Run Living Historical Farm, Ashville 46395
Slate Valley Museum, Granville 49020
Slaughter Ranch Museum, Douglas 48156
Slidell Museum, Slidell 53131
Sloan Museum Longway Planetarium, Flint 48560
Smith County Historical Society, Tyler 53634
Smith-McDowell House Museum, Asheville 46375
Smith's Fort Plantation, Surry 53383
Smithtown Historical Society Museum, Smithtown 53134
Smyth County Museum, Marion 50490
Snake River Heritage Center, Weiser 54022
Snoqualmie Valley Historical Museum, North Bend 51382
Snug Harbor Cultural Center, Staten Island 53267
The Snyder County Historical Society Museum, Middleburg 50650
Snyder Museum and Creative Arts Center, Bastrop 46624
Sod House Museum, Aline 46220
Sod House Museum, Gothenburg 48979
Sodus Bay Historical Society Museum, Sodus Point 53141
Soldiers and Sailors Memorial Building, Huntington 49432
Somers Historical Society Museum and Museum of the Early American Circus, Somers 53145
Somers Mansion, Somers Point 53147
Somerset Historical Center, Somerset 53148
Somerset Place, Creswell 47881
Somervell County Museum, Glen Rose 48924
Somerville Museum, Somerville 53149
Somesville Museum, Mount Desert 50874
Sonoma County Museum, Santa Rosa 52909
Sophienburg Museum, New Braunfels 51025
South Bannock County Historical Center, Lava Hot Springs 49989
South Boston-Halifax County Museum of Fine Arts and History, South Boston 53165
South Carolina State Museum, Columbia 47734
South Charleston Museum, South Charleston 53166
South Florida Museum, Bradenton 46987
South Hero Bicentennial Museum, South Hero 53173
South Holland Historical Museum, South Holland 53174
South Lake Tahoe Museum, South Lake Tahoe 53175
South Milwaukee Historical Society Museum, South Milwaukee 53176
South Park City Museum, Fairplay 48491
South Pass City State Historic Site, South Pass City 53178
South River Meeting House, Lynchburg 50330
South Texas Museum, Alice 46218
South Wood County Historical Museum, Wisconsin Rapids 54265
Southampton Historical Museum, Southampton 53186
Southeast Alaska Indian Cultural Center, Sitka 53127
Southeast Museum, Brewster 47010
Southern Oregon Historical Society Museum, Medford 50552
Southold Historical Society Museum, Southold 53188
Southwest Florida Museum of History, Fort Myers 48664
Space Farms Zoological Park and Museum, Sussex 53386
Spanish Governor's Palace, San Antonio 52727
Spindletop and Gladys City Boomtown Museum, Beaumont 46679
SPLIA Gallery, Cold Spring Harbor 47683
Spotsylvania Historical Association and Museum, Spotsylvania 53208
Spring Mill State Park Pioneer Village and Grissom Memorial, Mitchell 50760
Spring Street Historical Museum, Shreveport 53090
Spring Valley Community Museums, Spring Valley 53210
Springfield Historical Society Museum, Springfield 53234
Springfield Museum, Springfield 53238
Springfield Museum of Old Domestic Life, High Point 49276
Springs Museum, Springs 53241
Staatsburgh State Historic Site, Staatsburg 53245
Stacy's Tavern Museum, Glen Ellyn 48921
Stagecoach Inn Museum, Newbury Park 51304
Stamford Historical Society Museum, Stamford 53248
Stanley-Whitman House, Farmington 48515
Stanly County Historic Museum, Albemarle 46169
Stanton County Museum, Pilger 51917
Starved Rock State Park, Utica 53674
State Historical Society of Iowa Museum, Des Moines 48102
State Historical Society of North Dakota Museum, Bismarck 46839
Staten Island Historical Society Museum, Staten Island 53269
Staten Island Institute of Arts and Sciences, Staten Island 53270
Stauth Memorial Museum, Montezuma 50812
Steamboat Dock Museum, Keyport 49767
Stearns History Museum, Saint Cloud 52540
Steele County Museum, Hope 49360
Steilacoom Historical Museum, Steilacoom 53282
Stengel-True Museum, Marion 50489
Stephens County Historical Society Museum, Duncan 48205
Stephenson County Historical Society Museum, Freeport 48783
Stephenville Museum, Stephenville 53283
Sterling Historical Society and Little Red Schoolhouse Museum, Sterling 53286
Sterling-Rock Falls Historical Society Museum, Sterling 53285
Sterne-Hoya Museum, Nacogdoches 50937
Steuben House Museum, River Edge 52362
Stevens County Historical Society Museum, Morris 50849
Stevens Museum, Salem 52655
Steves Homestead Museum, San Antonio 52728
Stockholm Historical Society Museum, Stockholm 53304
Stone Fort Museum, Nacogdoches 50938
Stonehouse Museum, Vale 53688
Stonewall Jackson House, Lexington 50074
Stoughton Historical Society Museum, Stoughton 53322
Stow House, Goleta 48966
Stow West 1825 School Museum, Stow 53324
The Stoy Museum of the Lebanon County Historical Society, Lebanon 50013
Stranahan House, Fort Lauderdale 48641
Strasburg Museum, Strasburg 53334
The Stratford Historical Society and Catherine B. Mitchell Museum, Stratford 53335
Strathearn Museum, Simi Valley 53111
Strawbery Banke Museum, Portsmouth 52100
Stroud Mansion, Stroudsburg 53341
Stuart House City Museum, Mackinac Island 50352
Stuhr Museum of the Prairie Pioneer, Grand Island 48992
Sturdivant Hall, Selma 53032
Stutsman County Memorial Museum, Jamestown 49594
Suffolk County Historical Society Museum, Riverhead 52368
Suffolk Resolves House, Milton 50700
Sullivan County Historical Society Museum, Hurleyville 49449
Summit County Historical Society Museum, Akron 46149
The Sumter County Museum, Sumter 53366
Sun Prairie Historical Museum, Sun Prairie 53368
Sunnyside Historical Museum, Sunnyside 53371
Sunnyvale Heritage Park Museum, Sunnyvale 53373
Susan B. Anthony House, Rochester 52401
Susquehanna County Historical Society, Montrose 50831
Susquehanna Museum of Havre de Grace, Havre de Grace 49221
Sussex County Historical Society Museum, Newton 51343
Sutton-Ditz House Museum, Clarion 47582
Swarthout Memorial Museum, La Crosse 49838
Sweet Briar Museum, Sweet Briar 53392
Sweetwater County Historical Museum, Green River 49043
Swenson Memorial Museum of Stephens County, Breckenridge 47003
Swift County Historical Museum, Benson 46741
Swisher County Archives and Museum Association, Tulia 53606
Swiss Historical Village, New Glarus 51045
Switzerland County Historical Society Museum, Vevay 53731
T.B. Ferguson Home, Watonga 53983
Table Rock Historical Society Museum, Table Rock 53404
Talkeetna Historical Society Museum, Talkeetna 53421
Tallahassee Museum of History and Natural Science, Tallahassee 53429
Tama County Historical Museum, Toledo 53519
Tamastslikt Cultural Institute, Pendleton 51756
Tampa Bay History Center, Tampa 53436
Tante Blanche Museum, Madawaska 50373

Taos Historic Museums, Taos 53445
Tappantown Historical Society Museum, Tappan 53446
Tarpon Springs Cultural Center, Tarpon Springs 53449
Tarrytowns Museum, Tarrytown 53452
Teackle Mansion, Princess Anne 52142
Telluride Historical Museum, Telluride 53461
Temecula Valley Museum, Temecula 53462
Tempe Historical Museum, Tempe 53472
Temperance Tavern, Newcomerstown 51309
Ten Broeck Mansion, Albany 46164
Tenino Depot Museum, Tenino 53477
Terrace Hill Historic Site and Governor's Mansion, Des Moines 48103
Territorial Capital-Lane Museum, Lecompton 50015
Territorial Statehouse State Museum, Fillmore 48542
Territory Town USA, Old West Museum, Okemah 51512
Terwilliger Museum, Waterloo 53968
Teutopolis Monastery Museum, Teutopolis 53487
Texana Museum, Edna 48298
Texarkana Museums System, Texarkana 53489
Texas Governor's Mansion, Austin 46518
Thetford Historical Society Library and Museum, Thetford 53496
This is the Place Heritage Park, Salt Lake City 52694
Thomas Clarke House-Princeton Battlefield, Princeton 52154
Thomas-Foreman Home, Muskogee 50930
Thomas Hart Benton Home and Studio, Kansas City 49701
Thomas J. Boyd Museum, Wytheville 54316
Thomas Price House, Woodruff 54279
Thomas T. Taber Museum, Williamsport 54173
Thomas Warne Historical Museum, Old Bridge Township 51533
Thomas Warne Historical Museum, Matawan 50533
Thomaston Historical Museum, Thomaston 53498
Thornhill Historic Site and 19th Century Village, Chesterfield 47437
Three Rivers Museum, Muskogee 50931
Three Village Historical Society Museum, Setauket 53039
Thronateeska Heritage Foundation, Albany 46156
Thunderbird Museum, Merrillan 50603
Tillamook County Pioneer Museum, Tillamook 53509
Tinley Park Historical Society Museum, Tinley Park 53511
Tintic Mining Museum, Eureka 48443
Tioga County Historical Society Museum, Owego 51636
Tioga Point Museum, Athens 46419
Tippecanoe County Historical Museum, Lafayette 49865
Tipton County Museum, Covington 47861
Tipton-Haynes Museum, Johnson City 49636
Tobacco Farm Life Museum, Kenly 49726
Tobias Community Historical Society Museum, Tobias 53517
Tolland County Jail and Warden's Home Museum, Tolland 53525
Tom Mix Museum, Dewey 48134
Tomball Community Museum Center, Tomball 53526
Tonawandas Museum, Tonawanda 53529
Tongass Historical Museum, Ketchikan 49752
Top of Oklahoma Historical Museum, Blackwell 46847
Toppenish Museum, Toppenish 53540
Topsfield Historical Museum, Topsfield 53541
Torrington Historical Society Museum, Torrington 53544
Totem Heritage Center, Ketchikan 49753
Town of Clarence Museum, Clarence 47580
Town of Manlius Museum, Manlius 50448
Town of North Hempstead Museum, Westbury 54086
Town of Warwick Museum, Warwick 53835
Traill County Historical Society Museum, Hillsboro 49285
Trapezium House, Petersburg 51804
Travellers Rest Plantation and Museum, Nashville 50977
Traverse County Historical Society Museum, Wheaton 54123
Tread of Pioneers Museum, Steamboat Springs 53280
Treaty Site History Center, Saint Peter 52633
Trenton City Museum, Trenton 53563
Trenton Historical Museum, Trenton 53555
Tri-Cities Museum, Grand Haven 48991
Tri-County Historical Society and Museum, Herington 49258
Tribal Museum, Grand Ronde 49011
Troxell-Steckel House and Farm Museum, Egypt 48305
Troy Museum and Historic Village, Troy 53572
Tsa Mo Ga Memorial Museum, Plains 51974
Tu-Endie-Wei State Park, Point Pleasant 52011
Tuck Museum, Hampton 49147
Tucumcari Historical Research Institute Museum, Tucumcari 53604
Tulare County Museum, Visalia 53763
Tularosa Basin Historical Society Museum, Alamogordo 46152
Turquoise Museum, Albuquerque 46188
Twin Falls County Museum, Twin Falls 53628
Twin Falls Museum, Mullens 50900
Twinsburg Historical Society Museum, Twinsburg 53629
Tybee Museum and Tybee Island Light Station, Tybee Island 53632
Ulster County Historical Society Museum, Marbletown 50463
Ulysses S. Grant Cottage State Historic Site, Wilton 54216
Umatilla County Historical Society Museum, Pendleton 51757
Umpqua Discovery Center, Reedsport 52280
Underground Gold Miners Museum, Alleghany 46223
Union County Heritage Museum, New Albany 51011
Union County Historical Complex, Creston 47878
Union County Historical Foundation Museum, Union 53652
Union County Historical Society Museum, Blairsville 46848
Union County Historical Society Museum, Marysville 50523
Union County Museum, Lewisburg 50032
Union County Museum, Union 53650
Union Station Kansas City, Kansas City 49705
University Museum, Martin 50516
University of Colorado Museum, Boulder 46961
Upham Mansion, Marshfield 50514
Upper Musselshell Museum, Harlowtown 49171
Upper Paxton Township Museum, Millersburg 50693
Upper Snake River Valley Historical Museum, Rexburg 52300
Utica Museum, Hobson 49306
Vacaville Museum, Vacaville 53679
Vaile Mansion - Dewitt Museum, Independence 49480
Valdez Museum, Valdez 53685
Valentown Museum, Victor 53739
Vallejo Naval and Historical Museum, Vallejo 53691
Valley Community Historical Society Museum, Valley 53692
Valley County Pioneer Museum, Glasgow 48915
Van Andel Museum Center, Grand Rapids 49007
Van Buren County Historical Society Museum, Keosauqua 49746
Van Nostrand-Starkins House, Roslyn 52464
Van Riper-Hopper House, Wayne 54009
Van Wyck Homestead Museum, Fishkill 48547
Vanderbilt Mansion National Museum, Hyde Park 49460
Vashon Maury Island Heritage Museum, Vashon 53712
Ventura County Museum of History and Art, Ventura 53716
The Verdier House, Beaufort 46669
Vermilion County Museum, Danville 47960
Vermont Historical Society Museum, Montpelier 50825
The Vermont State House, Montpelier 50826
Vernon County Historical Museum, Viroqua 53762
Vernon Historians, Vernon 53724
Vernonia Historical Museum, Vernonia 53726
Vest-Lindsey House, Frankfort 48743
Vestal Museum, Vestal 53730
Victor Heritage Museum, Victor 53737
Victor Valley Museum and Art Gallery, Apple Valley 46328
Vigo County Historical Museum, Terre Haute 53485
Villa Louis Historic Site, Prairie du Chien 52129
Village Historical Society Museum, Harrison 49185
Village of Elsah Museum, Elsah 48382
The Vinalhaven Historical Society Museum, Vinalhaven 53746
Virginia Baptist Historical Society, Richmond 52341
Virginia City Madison County Historical Museum, Virginia City 53760
Vista Historical Museum, Vista 53765
Volendam Windmill Museum, Milford 50687
W.H. Over Museum, Vermillion 53721
The W.H. Stark House, Orange 51579
Wabasha County Museum, Reads Landing 52268
Wabasso Historical Society Museum, Wabasso 53770
Wahkiakum County Historical Museum, Cathlamet 47282
Wakefield Historical Museum, Wakefield 53791
Wakefield House Museum, Blue Earth 46873
Wakefield Museum Association, Wakefield 53790
Wallace House, Somerville 53152
Wallingford Historical Society Museum, Wallingford 53800
Wallis Museum at Connors State College, Warner 53818
Walt Whitman Birthplace, Huntington Station 49440
Walt Whitman House, Camden 47197
Walter Elwood Museum, Amsterdam 46276
Waltham Historical Society Museum, Waltham 53811
The Waltham Museum, Waltham 53812
Walton House Museum, Centerville 47317
Wanapum Dam Heritage Center, Beverly 46788
Wantagh Historical Preservation Society, Wantagh 53814
Wapello County Historical Museum, Ottumwa 51631
Ward County Historical Society Museum, Minot 50747
Warden's House Museum, Stillwater 53294
Warner Museum, Springville 53242
Warren County Historical Society Museum, Lebanon 50012
Warren County Museum, Warren 53827
Warren Historical Museum, Warren 53820
Warrick County Museum, Boonville 46909
Warsaw Historical Museum, Warsaw 53832
Wasco County Historical Museum, The Dalles 47946
Waseca County Museum, Waseca 53839
Washakie Museum, Worland 54305
Washburn County Historical Society Museum, Shell Lake 53070
Washburn Historical Museum, Washburn 53841
Washington County Historical Association Museum, Fort Calhoun 48602
Washington County Historical Society, Washington 53938
Washington County Historical Society Museum, Portland 52089
Washington County Historical Society Museum, Washington 53944
Washington County Historical Society Museum, West Bend 54044
Washington County Museum, Akron 46145
Washington County Museum, Sandersville 52833
Washington Historical Museum, Washington 53937
Washington Museum, Washington 53939
Washington State Capital Museum, Olympia 51545
Washington State History Museum, Tacoma 53416
Washington's Lands Museum and Sayre Log House, Ravenswood 52257
Water Mill Museum, Water Mill 53955
Waterford Historical Museum, Waterford 53959
Watertown Historical Society Museum, Watertown 53969
Watkins Community Museum of History, Lawrence 49994
Watkins Museum, Taylorsville 53456
Watson-Curtze Mansion, Erie 48413
Watters Smith Memorial, Lost Creek 50275
Wauconda Township Museum, Wauconda 53988
Waukesha County Museum, Waukesha 53990
Wayland Historical Museum, Wayland 54004
Wayne County Historical Museum, Richmond 52315
Wayne County Historical Society Museum, Lyons 50336
Wayne County Historical Society Museum, Wooster 54296
Wayne County Museum, Honesdale 49334
The Webb-Deane-Stevens Museum, Wethersfield 54112
Webb House Museum, Newark 51294
Webster County Historical Museum, Red Cloud 52271
Webster House Museum, Elkhorn 48351
Weems-Botts Museum, Dumfries 48203
Weeping Water Valley Historical Society Museum, Weeping Water 54021
Wellesley Historical Society, Wellesley Hills 54024
Wellfleet Historical Society Museum, Wellfleet 54026
Wells County Historical Museum, Bluffton 46878
Wells County Museum, Fessenden 48539
Wenatchee Valley Museum, Wenatchee 54037
Weslaco Museum, Weslaco 54040
West Allis Historical Society Museum, West Allis 54042
West Baton Rouge Museum, Port Allen 52025
West Chicago City Museum, West Chicago 54051
West of the Pecos Museum, Pecos 51746
West Overton Museums, Scottdale 52975
West Pasco Historical Society Museum, New Port Richey 51099
West River Museum, Philip 51892
West Virginia Independence Hall, Wheeling 54125
West Virginia Northern Community College Alumni Association Museum, Wheeling 54126
West Virginia State Museum, Charleston 47380
Western Hennepin County Pioneers Museum, Long Lake 50190
Western Heritage Center, Billings 46807
Western Museum, Abilene 46123
Western Reserve Historical Society Museum, Cleveland 47644
Western Rhode Island Civic Museum, Coventry 47857
Western Springs Museum, Western Springs 54088
Westminster Historical Society, Westminster 54100
Westport Historical Museum, Westport 54110
Westport Maritime Museum, Westport 54111
Westville Historic Handicrafts Museum, Lumpkin 50322
Wethersfield Museum, Wethersfield 54113
Wexford County Historical Museum, Cadillac 47157
Wharton County Historical Museum, Wharton 54116
Whatcom Museum of History and Art, Bellingham 46714
Wheat Ridge Sod House and Museum, Log Cabin, First Post Office and Tool Shed, Wheat Ridge 54117
Wheaton History Center, Wheaton 54122
Wheeler Historic Farm, Salt Lake City 52698
Whipple House Museum, Ashland 46386
White County Historical Museums, Carmi 47253
White County Historical Society Museum, Cleveland 47619
White County Historical Society Museum, Monticello 50821
White Deer Land Museum, Pampa 51690
White Pillars Museum, De Pere 48000
White Pine Public Museum, Ely 48385
White River Museum, Meeker 50560
White River Valley Museum, Auburn 46473
Whitefield House Museum, Nazareth 50994
Whitehead Memorial Museum, Del Rio 48040
Whitewater Historical Museum, Whitewater 54135
Whitingham Historical Museum, Whitingham 54136
Whitley County Historical Museum, Columbia City 47736
Wichita-Sedgwick County Historical Museum Association, Wichita 54152
Wilbur D. May Museum, Reno 52293
Wilder Memorial Museum, Strawberry Point 53337
Wilderness Road Regional Museum, Newbern 51299
Wildwood Center, Nebraska City 50996
Wilkin County Historical Museum, Breckenridge 47002
Wilkinson County Museum, Woodville 54290
Willa Cather State Historic Site, Red Cloud 52272
William Clark Market House Museum, Paducah 51671
William H. Harrison Museum/ Grouseland, Vincennes 53751
William Holmes McGuffey Museum, Oxford 51656
William Pryor Letchworth Museum, Castile 47277
William Rea Museum, Murfreesboro 50912
William S. Hart Museum, Newhall 51314
William Trent House, Trenton 53564
Williams County Historical Museum, Montpelier 50823
Williamson County Historical Society Museum, Marion 50482
Willoughby-Baylor House, Norfolk 51371
Wilmette Historical Museum, Wilmette 54186
Wilson County Historical Society Museum, Fredonia 48775
Wilson Historical Museum, Wilson 54212
Wilton Historical Museums, Wilton 54214
Wilton House Museum, Richmond 52344
Winchendon Historical Museum, Winchendon 54218
Winchester-Frederick County Historical Society, Winchester 54222
Windmill Island Municipal Park Museum, Holland 49318
Winedale Historical Center, Round Top 52472
Winnetka Historical Museum, Winnetka 54236
Wisconsin Historical Museum, Madison 50396
Wise County Heritage Museum, Decatur 48020
Wistariahurst Museum, Holyoke 49328
Wolcott House Museum Complex, Maumee 50536
Wolf Point Area Historical Society, Wolf Point 54267
Wood County Historical Center, Bowling Green 46978
Woodlawn Museum, Ellsworth 48373
Woodlawn Plantation, Alexandria 46215
Woodside House, Rochester 52402
Woodson County Historical Society, Yates Center 54327
Woodstock Historical Society, Woodstock 54288
Woodstock Historical Society Museum, Bryant Pond 47106
Woodstock Historical Society Museum, Woodstock 54282
Woodstock Museum of Shenandoah County, Woodstock 54286
Worcester Historical Museum, Worcester 54303
Worthington Historical Society, Worthington 54308
Wrangell Museum, Wrangell 54309
Wrather West Kentucky Museum, Murray 50919
Wray Museum, Wray 54310
Wright County Historical Museum, Buffalo 47113
Wright's Ferry Mansion, Columbia 47721
Wyandot County Historical Society Museum, Upper Sandusky 53667
Wyandotte County Museum, Bonner Springs 46902
Wyandotte Museum, Wyandotte 54312
Wyatt Earp Birthplace, Monmouth 50783
Wyoming Pioneer Home, Thermopolis 53495
Wyoming Pioneer Memorial Museum, Douglas 48159
Wyoming Territorial Prison, Laramie 49951
Yakima Valley Museum, Yakima 54320
Ybor City State Museum, Tampa 53438
Yellow Medicine County Historical Museum, Granite Falls 49014
Yellowstone Gateway Museum of Park County, Livingston 50149
Yolo County Historical Museum, Woodland 54278
York County Museum, York 54342

Yorktown Historical Museum, Yorktown 54344
The Yosemite Museum, Yosemite National Park 54350
Ypsilanti Historical Museum, Ypsilanti 54357
Yturri-Edmunds Historic Site, San Antonio 52733
Yucaipa Adobe, Yucaipa 54362
Yupiit Piciryarait Museum, Bethel 46773
Zalud House, Porterville 52061
Zelienople Historical Museum, Zelienople 54369
Zion Historical Museum, Zion 54370
Zoar State Memorial, Zoar 54373

Uruguay

Museo de Piria, Piriápolis 54426
Museo Martin Perez, Montevideo 54406
Museo Municipal Cabildo de Montevideo, Montevideo 54407
Museo Municipal de Historia y Arqueología, Rivera 54431
Museo Regional de San Carlos, San Carlos 54432
Museo Regional Francisco Mazzoni, Maldonado 54381

Uzbekistan

Afrosiyob - Samarqand Thariki Muzeyi, Samarqand 54459
Amir Timur Museum, Shakhrisabz 54465
Andijon Viloyat Ülkashunoslik Muzeyi, Andijon 54439
Angren Viloyat Ülkashunoslik Muzeyi, Angren 54443
Buxoro Qala Davlat Muzeyi Qo'riqxonasi, Buxoro 54444
Farg'ona Viloyat Ülkashunoslik Muzeyi, Farg'ona 54446
Hamza Hakim-Jade Niazy Muzeyi, Farg'ona 54447
Ichan Qala Davlat Muzej Qòriqxonasi, Xiva 54492
Jizzax Viloyat Ülkashunoslik Muzeyi, Jizzax 54448
Qashqadaryo Viloyat Ülkashunoslik Muzeyi, Qarshi 54454
Qoraqalpoġiston Respublikasi Ülkashunoslik Muzeyi, Nukus 54451
Sadriddin Aini Yodgorlik Muzeyi, Samarqand 54462
Samarqand Ülkashunoslik Muzeyi, Samarqand 54463

Venezuela

Casa Museo Páez, Valencia 54597
Museo Biblioteca Nueva Cádiz y Casa Natal de Juan Bautista Arismendi, Nueva Cádiz 54589
Museo Colonial, El Tocuyo 54569
Museo Cuadra de Bolívar, Caracas 54538
Museo de Anzoátegui, Barcelona 54521
Museo de Barquisimeto, Barquisimeto 54524
Museo de Caracas, Caracas 54544
Museo de Ciudad Bolívar, Ciudad Bolívar 54564
Museo de la Ciudad de Píritu, Píritu 54591
Museo de los Llanos, Guanare 54573
Museo del Estado de Mérida, Mérida 54588
Museo Jacobo Borges, Caracas 54554

Vietnam

Bao Tàng Hai Phong, Hai Phong 54620
Ha Noi Museum, Ha Noi 54614
Hung-Yen Museum, Hung-Yen 54633
Khanh Hoa Provincial Museum, Nha Trang 54634
Lâm Dông Museum, Đà Lat 54603
Quàng Ninh Museum, Ha Long 54605
Ton Duc Thang Museum, Ho Chi Minh City 54629

Yemen

Al-Mukalla Museum, Al-Mukalla 54645

Zambia

Livingstone Museum, Livingstone 54655
National Archives of Zambia, Lusaka 54659
Nayuma Museum, Limulunga 54654

Zimbabwe

Mutare Museum, Mutare 54676

History, Maritime → Boats and Shipping

History, Medieval

Austria

Ateliersammlung Herbert Hiesmayr, Sankt Thomas am Blasenstein 02689
Frühmittelaltermuseum Carantana, Rothenthurn 02600
Kaiser Maximilian von Mexiko Museum, Hardegg 02051
Ostarrichi-Kulturhof, Neuhofen an der Ybbs 02429

Belgium

Archéoscope Godefroid de Bouillon, Bouillon 03391

Bulgaria

Regionalen Istoričeski Muzej, Stara Zagora 05211

Canada

Port McNeill Museum, Port McNeill 06644

Czech Republic

Muzeum historie, Kolín 09590

France

Centre Histoirque Médiéval, Azincourt 11814
Château de Châteaudun, Châteaudun 12443
Château Guillaume le Conquérant, Falaise 12902
Donjon, Ballon 11833
Exposition Jadis, La Roche-Derrien 13414
Maison des Châteaux Forts, Obersteinbach 14494
Maison Natale de Jeanne d'Arc et Centre d'Interprètation, Domremy-la-Pucelle 12766
Musée, Joinville (Haute-Marne) 13274
Musèe Benoît XIII, Châteaurenard (Bouches-du-Rhône) 12454
Musée Colette, Saint-Sauveur-en-Puisaye 15506
Musée d'Art Sacré, Saint-Béat 15182
Musée des Temps Barbares, Marle 13970
Musée du Papogay, Rieux-Volvestre 15038
Musée du Vieux-Château, Laval (Mayenne) 13542
Musée Municipal d'Avranches, Avranches 11802

Germany

Ausstellung Haus Wysburg und Ruine Wysburg, Remptendorf 21262
Burgmuseum Schlossfreiheit, Tangermünde 22001
Burgmuseum Sulzberg, Sulzberg 21986
Dokumentationsraum für staufische Geschichte, Göppingen 18672
Fuggerei-Museum, Augsburg 16617
Gildehaus, Bardowick 16927
Hexenmuseum, Ringelai 21336
Historisches Museum der Pfalz Speyer, Speyer 21834
Mittelalterliches Foltermuseum, Rüdesheim am Rhein 21435
Museum bei der Kaiserpfalz, Ingelheim am Rhein 19335
Museum Burg Scharfenstein, Scharfenstein 21533
Museumsdorf Düppel, Berlin 17218
Naturhistorisches Museum, Nürnberg 20785
Rheinisches Landesmuseum Trier, Trier 22102
Stadtmuseum Coesfeld, Coesfeld 17723

Iran

Azadi Museum, Tehrān 24657

Italy

Palazzo del Senato, Pinerolo 27226

Kazakhstan

Otrarskij Gosudarstvennyj Archeologičeskij Muzej-Zapovednik, Šaulder 29726

Lithuania

Naujasis Arsenalas, Vilnius 30427

Netherlands

Museum Catharijneconvent, Utrecht 32849

Norway

Arkeologisk Museum i Stavanger, Stavanger 33863
Bryggens Museum, Bergen 33425
Erkebispegården, Trondheim 33923
Hedmarksmuseet og Domkirkeodden, Hamar 33528

Poland

Muzeum Mikołaja Kopernika, Toruń 35194
Muzeum Piastów Śląskich, Brzeg 34571

Romania

Complexul Monumental Curtea Domnească, Târgovişte 36654
Muzeul Judeţean Alexandru Ştefulescu, Târgu Jiu 36666

Russia

Aleksandrovskaja Sloboda - Gosudarstvennyj Istoriko- architekturnyj i Chudožestvennyj Muzej-zapovednik, Aleksandrov 36751
Gosudarstvennyj Istoriko-architekturnyj i Chudožestvennyj Muzej-zapovednik Drevnij Derbent, Derbent 36874
Gosudarstvennyj Voenno-istoričeskij i Prirodnyj Muzej-zapovednik Kulikovo Pole, Tula 37866
Gosudarstvennyj Voenno-istoričeskij Muzej-zapovednik Prochorovskoe Pole, Prochorovka 37571
Muzej Podviga Ivana Susanina, Susanino 37810

Slovakia

Nový Zamok Banská Štiavnica, Banská Štiavnica 38192

Spain

Monasterio Santa María la Real de las Huelgas, Burgos 39066
Museo Arqueológico y Etnológico Córdoba, Córdoba 39204
Museo de Prehistoria y Arqueología de Cantabria, Santander 40010
Museo del Románico, Aguilar de Campoo 38791
Museo Vivo de al-Andalus, Córdoba 39209

Sweden

Historiska Museet, Stockholm 40899
Historiska Museet, Lund 40702
Linköpings Slotts- och Domkyrkomuseum, Linköping 40683
Statens Historiska Museum, Stockholm 40937
Stockholms Medeltidsmuseum, Stockholm 40940

Switzerland

Forum der Schweizer Geschichte, Schwyz, Schwyz 41850
Musée Jurassien d'Art et d'Histoire, Delémont 41324
Museum Kleines Klingental, Basel 41172

Turkmenistan

Taryhy we Etnografiki Muzeý, Mary 42978

United Kingdom

Bishop of Winchester's Palace Site, Witney 46051
Cistercian Museum, Hailes 44378
Old Gala House, Galashiels 44261
Provand's Lordship, Glasgow 44297
Saint John and Coningsby Medieval Museum, Hereford 44459
Saint Robert's Cave, Knaresborough 44620

United States of America

Swords Into Plowshares Peace Center and Gallery, Detroit 48130

History, Modern

Algeria

Musee d'Ifri, Ifri 00066
Musée National du Moudjahid, El Madania 00061

Argentina

Museo Casa de Ricardo Rojas, Buenos Aires 00172
Museo de Aduanas y Puerto, Buenos Aires 00177
Museo de la Casa de Acuerdo de San Nicolás, San Nicolás de los Arroyos 00604
Museo de la Palabra, Villa Adelina 00670
Museo de la Reconquista, Tigre 00644
Museo de la Shoá- Fundación Memoria del Holocausto, Buenos Aires 00209
Museo Fuerte Independencia, Tandil 00641
Museo Histórico Brigadier General Cornelio de Saavedra, Buenos Aires 00236
Museo Historico del Norte, Salta 00558
Museo Histórico Nacional Casa del Virrey Liniers, Alta Gracia 00120
Museo Histórico Nacional del Cabildo de Buenos Aires y de la Revolucion de Mayo, Buenos Aires 00242
Museo Nacional de la Inmigración, Buenos Aires 00261
Museo Presidente José Evaristo Uriburu, Salta 00559
Museo Roca, Buenos Aires 00272
Museo y Archivo Histórico Localista Almirante Guillermo Brown, Adrogué 00116

Armenia

Mother Armenia Military Museum, Yerevan 00746

Australia

Defence Services Museum, Adelaide 00774
Jewish Holocaust Centre, Elsternwick 01067
Parkside Cottage Museum, Narrandera 01351
Sydney Jewish Museum, Darlinghurst 01026
Vietnam Veterans Museum, San Remo 01492

Austria

Einsiedelei Erzherzog Maximilians des Deutschmeisters, Innsbruck 02115
Gedenkräume Schloss Hartheim, Alkoven 01703
Gedenkraum 1945, Hochwolkersdorf 02087
Heimathaus, Sankt Georgen an der Gusen 02641
Heimatvertriebenen-Stuben, Traun 02825
KZ-Gedenkstätte Mauthausen, Mauthausen 02373
KZ-Gedenkstätte und Zeitgeschichte-Museum Ebensee, Ebensee 01836
Museum Aspern-Essling 1809, Wien 03030
Museum der Heimatvertriebenen, Wels 02916
Österreichisches Freimaurermuseum, Schloß Rosenau 02708
Weg des Friedens, Purgstall an der Erlauf 02541

Azerbaijan

Azerbaijan State Museum of Independence, Bakı 03143
Memorial Museum, Naxçivan 03185
Nakhchivan State History Museum, Naxçivan 03188

Belgium

Boerenkrijgmuseum, Overmere 03947
In Flanders Fields Museum, Ieper 03728
Joods Museum van Deportatie en Verzet, Mechelen 03858
Les Territoires de la Mémoire, Liège 03797
Memorial Prins Karel, Oostende 03932
Musée Ducal, Bouillon 03393
Musée Napoléon, Ligny 03823
Museum voor Sleutel en Slot, Oostduinkerke 03925
National Memorial of the Fortress of Breendonk, Willebroek 04153
Sanctuary Wood Museum - Hill 62, Zillebeke 04163
Talbot House, Poperinge 03954

Bolivia

Museo Casa de la Libertad, Sucre 04244
Museo Dr. Alfredo Gutierrez Valenzuela, Sucre 04249

Brazil

Memorial do Imigrante, São Paulo 04863
Museu do Monumento Nacional aos Mortos da Segunda Guerra Mundial, Rio de Janeiro 04739
Museu Imperial, Petrópolis 04629
Museu José Bonifácio, São Paulo 04905

Bulgaria

Gradski Istoričeski Muzej, Nova Zagora 05097
Gradski Istoričeski Muzej, Peruštica 05112
Istoričeski Muzej, Dimitrovgrad 05015
Istoričeski Muzej Belogradčik, Belogradčik 04982
Kăšta-muzej Chadži Dimităr, Sliven 05166
Kăšta-muzej Georgi Dimitrov, Kovačevci 05079
Kăšta-muzej Panaĕt Chitov, Sliven 05167
Kăšta-muzej Rajna Knjaginja, Panagjurište 05103
Kăšta-muzej Todor Kableškov, Koprivštica 05073
Muzej Glavna Kvartira na Ruskata Armija 1877-1878, Pordim 05135
Muzej Osvoboždenieto na Pleven 1877, Pleven 05118
Muzej Vasil Levski, Karlovo 05050
Muzej Vasil Levski, Loveč 05087
Okražen Istoričeski Muzej, Pernik 05111
Okražen Muzej na Văzraždaneto i Nacionalno-osvoboditelnite Borbi, Plovdiv 05131
Regionalen Istoričeski Muzej, Ruse 05155
Regionalen Istoričeski Muzej, Stara Zagora 05211

Cambodia

Documentation Center of Cambodia, Phnom Penh 05285
Land Mine Museum, Siem Reap 05291
Tuol Sleng Museum, Phnom Penh 05289
War Museum, Siem Reap 05292

Canada

Adelaide Hunter Hoodless Homestead, Saint George 06817
Bamfield Community Museum, Bamfield 05377
Bathurst Heritage Museum Cultural Centre / Musée de Bathurst/Musée Centre Culturel, Bathurst 05397
Belfast Historical Society, Belfast 05412
Bytown Museum, Ottawa 06530
Centre Commémoratif de l'Holocauste à Montréal, Montréal 06331

Elgin County Pioneer Museum, Saint Thomas 06879
Grenfell House Museum, Saint Anthony 06802
Haig-Brown House, Campbell River 05541
The Japanese Canadian Museum, Kaslo 06045
Kitikmeot Heritage Society, Cambridge Bay 05537
Mackenzie House, Toronto 07120
Mahone Bay Settlers Museum, Mahone Bay 06216
Memorial Library and Art Gallery, Saskatoon 06917
Nicolas Denys Museum, Saint Peters 06874
Nurse Bennett Heritage House, Daniel's Harbour 05672
Point Leamington Heritage Interpretation Centre, Point Leamington 06615
Ross-Thomson House Museum, Shelburne 06954
Salmon Arm Museum, Salmon Arm 06905
Sheriff Andrews House, Saint Andrews 06800
Town of Okotoks Museum and Archives, Okotoks 06505
Trembowla Cross of Freedom Museum, Dauphin 05680
Whistler Museum and Archives Society, Whistler 07317
William Henry Steeves Museum, Hillsborough 05988
Wright Historical Museum, Swift Current 07046

China

Beijing Police Museum, Beijing 07519
Changning District Revolutionary Display Center, Shanghai 08157
China Revolution Museum, Beijing 07532
Dr. Sun Yat Sen Memorial Museum, Macau 07991
Guangdong Revolutionary History Museum, Guangzhou 07726
Hunan Revolutionary Cemetery, Changsha 07598
Jilin Provincial Museum of Revolution, Changchun 07587
Jinan Revolutionary Mausoleum of Fallen Heroes, Jinan 07882
Lhasa Museum, Lhasa 07928
Lushun Su Jun Mausoleum of Fallen Heroes, Dalian 07643
Ma City Revolution Museum, Jiangling 07859
Minxi Revolution History Museum, Minxi 08024
Museum of International Friendship, Beijing 07552
Nanchang Uprising Museum, Nanchang 08030
Peace Museum, Heping 07782
Revolution Museum, Harbin 07772
Taiping Kingdom History Museum, Nanjing 08050
Wuxi Revolutionary Display Center, Wuxi 08344
Xin Hai Shouyi Mausoleum of Fallen Heroes, Wuhan 08332
Xin Metropolis Revolution Museum, Weihui 08319
Zhabei Revolutionary Shi Liao Display Center, Shanghai 08199
Zhenjiang Revolution History Museum, Zhenjiang 08503

Croatia

Gradski Muzej Makarska, Makarska 08907
Muzej Brodskog Posavlja, Slavonski Brod 08953
Muzej Grada Rijeke, Rijeka 08937
Muzej Seljačkih Buna, Gornja Stubica 08876
Muzej Suvremene Povijesti, Dubrovnik 08871
Povijesni Muzej Istre, Pula 08935

Cuba

Casa Natal de Jesús Montané Oropesa, Nueva Gerona 09230
Granma Memorial, La Habana 09106
Museo 9 de Abril, Santa Clara 09293
Museo Antimperialista, Nueva Gerona 09231
Museo Antonio Maceo, San Pedro 09282
Museo Benito Juárez, La Habana 09112
Museo Casa de los Mártires, La Habana 09120
Museo Casa del Libertador Simón Bolívar, La Habana 09122
Museo Casa Natal de Camilo Cienfuegos, La Habana 09125
Museo Casa Natal de Frank País, Santiago de Cuba 09307
Museo Casa Natal de Ignacio Agramonte, Camagüey 09050
Museo Casa Natal de Pedro Martínez Brito, Ciego de Avila 09061
Museo Casa Natal de Serafín Sánchez, Sancti-Spíritus 09284
Museo de Granjita Siboney, Santiago de Cuba 09311
Museo de la Clandestinidad, Nueva Gerona 09233
Museo de la Lucha Clandestina, Nueva Gerona 09234
Museo de la Lucha Clandestina, Santiago de Cuba 09314
Museo de la Lucha Estudiantil Jesús Suárez Gayol, Camagüey 09051
Museo de la Revolución, La Habana 09137
Museo de la Revolución, Nueva Gerona 09235
Museo de las Luchas Obreras, Manzanillo 09207
Museo El Cañón, Puerto Boniato 09259
Museo Finca El Abra, Siguanea 09332
Museo Fragua Martiana, La Habana 09145
Museo Hermanas Giralt, Cienfuegos 09066
Museo Histórico de 26 de Julio, Santiago de Cuba 09322
Museo Marcha del Pueblo Combatiente, La Habana 09148
Museo Memorial 12 de Septiembre, La Habana 09151
Museo Memorial Antonio Guiteras Holmes, Pinar del Río 09250
Museo Memorial Ernesto Che Guevara, Santa Clara 09297
Museo Memorial Mártires de Barbados, Victoria de las Tunas 09347
Museo Napoleónico, La Habana 09168
Museo Ñico Lopez, Bayamo 09037
Museo Plaza de Revolución General Antonio Maceo, Santiago de Cuba 09325

Cyprus

Cyprus National Struggle Museum, Lefkosia 09376

Czech Republic

Muzeum Komunismu, Praha 09771
Památník Františka Palackého, Hodslavice 09520
Památník II. Světové Války, Hrabyně 09531
Památník Mohyla Míru, Prace u Brna 09731
Památník Terezín, Terezín 09864
Památník Terezín - Bývalé Krematorium na Židovském Hřbitově, Terezín 09865
Památník Terezín - Kolumbárium, Obřadní Místnosti a Ústřední Márnice, Terezín 09866
Památník Terezín - Magdeburská Kasárna, Terezín 09867
Památník Terezín - Malá Pevnost, Terezín 09868
Památník Terezín - Muzeum Ghetta, Terezín 09869
Památník Životické Tragedie, Havířov 09509
Zámek, Jirkov 09566
Zámek Slavkov - Austerlitz, Slavkov u Brna 09841

Denmark

Besættelsessamlingen 1940-45, Grindsted 10043
Bunker Museet, Frederikshavn 10025
FN Museet, Padborg 10226
Frihedsmuseet, København 10131
Den Gamle By, Århus 09963

Ecuador

Museo Jacinto Jijón y Caamaño, Quito 10446

El Salvador

Museo de la Revolución Salvadoreña, Perquín 10546

Estonia

Kaitseväe Ühendatud Õppeasutuste Muuseum, Tartu 10649
Okupatsioonimuuseum, Tallinn 10640

Finland

Lenin-Museo, Tampere 11353

France

Centre de la Mémoire d'Oradour, Oradour-sur-Glane 14515
Centre européen du résistant déporté, Natzwiller 14383
Centre National et Musée Jean Jaurès, Castres 12318
Château-Musée, Collonges-lès-Bévy 12579
Historial de la Grande Guerre, Péronne 14784
Maison des Jardies-Musée Léon Gambetta, Sèvres 15742
Maison Natale du Maréchal Foch, Tarbes 15861
Maison Nationale Bonaparte, Ajaccio 11524
Mémorial de la Shoah, Paris 14625
Mémorial de Vendée, Saint-Sulpice-le-Verdon 15518
Mémorial du Maréchal Leclerc de Hauteclocque et de la Libération de Paris et Musée Jean Moulin, Paris 14626
Musée Charles-Louis Philippe, Cérilly 12342
Musée Communal de la Résistance et de la Déportation, Vénissieux 16103
Musée d'Argonne, Varennes-en-Argonne 16075
Musée de Cahors Henri-Martin, Cahors 12242
Musée de la Franc-Maçonnerie, Paris 14653
Musée de la Guerre, Fromelles 13015
Musée de la Résistance, La Balme-de-Thuy 13307
Musée de la Résistance, Limoges 13766
Musée de la Résistance Bretonne, Saint-Marcel (Morbihan) 15394
Musée de la Résistance et de la Déportation, Agen 11493
Musée de la Résistance et de la Déportation, Angoulême 11618
Musée de la Résistance et de la Déportation, Besançon 11956
Musée de la Résistance et de la Déportation, Montauban 14174
Musée de la Résistance et de la Déportation, Castelnau-le-Lez 12309
Musée de la Résistance Nationale, Champigny-sur-Marne 12390
Musée de la Révolution Française, Vizille 16264
Musée de l'Arc de Triomphe, Paris 14666
Musée de l'Ordre de la Libération, Paris 14675
Musée Departemental de la Résistance et de la Déportation, Forges-les-Eaux 12980
Musée des Troupes de Marine, Fréjus 13002
Musée d'Histoire Contemporaine, Paris 14690
Musée d'Histoire de la Résistance et de la Déportation de l'Ain et du Haut-Jura, Nantua 14374
Musée d'Histoire et de Tambow, Amnéville-les-Thermes 11590
Musée du Camp d'Internement du Vernet d'Ariège, Le Vernet d'Ariège 13665
Musée du Château du Marais, Val Saint-Germain 16033
Musée du Corps Franc Pommiès, Castelnau-Magnoac 12311
Musée du Groupe Jovinien de Résistance Bayard, Joigny (Yonne) 13273
Musée du Maréchal Murat, Labastide-Murat 13462
Musée du Mur de l'Atlantique, Ouistreham 14544
Musée Émile Jean, Villiers-sur-Marne 16245
Musée Haut Savoyard de la Résistance, Bonneville 12057
Musée-Hôtel Bertrand, Châteauroux 12459
Musée Juin 1944, L'Aigle 13485
Musée Lénine, Paris 14717
Musée-Maison du Petit Poitou, Chaillé-les-Marais 12354
Musée Mémorial de la Bataille de Normandie, Bayeux 11875
Musée Mémorial de la Ligne Maginot du Rhin, Marckolsheim 13958
Musée Municipal, Richelieu 15035
Musée Napoléon Ier et Tresors des Eglises, Brienne-le-Château 12177
Musée National de la Coopération Franco-Américaine, Blérancourt 12025
Musée National des Châteaux de Malmaison et de Bois-Préau, Rueil-Malmaison 15139
Musée Raymond Poincaré, Sampigny 15619
Musée Rosine Perrier, Villargondran 16167
Musexpo, Clairfayts 12536
Le Refuge de Grasla, Les Brouzils 13695

Georgia

Museum of Soviet Repression, Tbilisi 16374

Germany

Aktives Museum Spiegelgasse für Deutsch-Jüdische Geschichte in Wiesbaden, Wiesbaden 22545
Alte Schule mit Clara-Zetkin-Gedächtnisstätte, Königshain-Wiederau 19653
Anna-Seghers-Gedenkstätte, Berlin 17033
Anti-Kriegs-Museum, Berlin 17035
Archiv Bürgerbewegung Leipzig e.V., Leipzig 19895
Archiv und Museum des Heimatkreises Leitmeritz, Fulda 18504
Bauernkriegsmuseum, Weinstadt 22435
Befreiungshalle Kelheim, Kelheim 19500
Bismarck-Ausstellung, Friedrichsruh 18479
Bundesarchiv, Rastatt 21183
Clara-Zetkin-Museum, Birkenwerder 17360
DDR Museum, Berlin 17064
Denkmal für die ermordeten Juden Europas, Berlin 17065
DenkStätte Weiße Rose, Weiße Rose Stiftung, München 20431
Deutsch-Russisches Museum Berlin-Karlshorst, Berlin 17066
Deutsches Auswandererhaus Bremerhaven, Bremerhaven 17568
Deutsches Hugenotten-Museum, Bad Karlshafen 16757
Dokumentation Obersalzberg - Orts- und Zeitgeschichte, Berchtesgaden 17000
Dokumentations- und Informationszentrum Emslandlager (DIZ), Papenburg 20979
Dokumentationszentrum des Landes für die Opfer der Diktaturen in Deutschland, Schwerin 21692
Dokumentationszentrum ehemaliges KZ Osthofen, Osthofen 20941
Dokumentationszentrum Reichsparteitagsgelände, Nürnberg 20758
Erich Maria Remarque-Friedenszentrum, Osnabrück 20920
Erinnerungsstätte Notaufnahmelager Marienfelde, Berlin 17080
Erinnerungsstätte Ständehaus, Karlsruhe 19438
Freiland-Grenzmuseum Sorge, Sorge 21825
Friedenshistorisches Museum, Bad Hindelang 16748
Friedensmuseum, Meeder 20244
Friedrich-der-Große-Museum im Museumshof Lerchennest, Sinsheim 21784
Gedenk- und Dokumentationsstätte Opfer politischer Gewaltherrschaft, Frankfurt (Oder) 18402
Gedenkstätte Bergen-Belsen, Lohheide 20023
Gedenkstätte Berliner Mauer, Berlin 17112
Gedenkstätte Breitenau, Guxhagen 18844
Gedenkstätte Buchenwald, Weimar 22399
Gedenkstätte "Erinnern für die Zukunft", Leipzig 19908
Gedenkstätte Goldener Löwe, Eisenach 18094
Gedenkstätte Hadamar, Hadamar 18848
Gedenkstätte in der JVA Wolfenbüttel, Wolfenbüttel 22652
Gedenkstätte Köpenicker Blutwoche Juni 1933, Berlin 17114
Gedenkstätte Moritzplatz Magdeburg für die Opfer politischer Gewaltherrschaft 1945-1989, Magdeburg 20122
Gedenkstätte Museum in der "Runden Ecke", Leipzig 19909
Gedenkstätte/Museum Seelower Höhen, Seelow 21720
Gedenkstätte Plötzensee für die Opfer des Nationalsozialismus, Berlin 17115
Gedenkstätte und Museum Sachsenhausen, Oranienburg 20908
Grenzhuus, Schlagsdorf 21562
Grenzland-Museum Bad Sachsa, Bad Sachsa 16812
Grenzlandmuseum, Schnackenburg 21588
Grenzlandmuseum Eichsfeld, Teistungen 22020
Grenzlandmuseum Swinmark, Schnega 21592
Grenzmuseum Philippsthal (Werra), Philippsthal 21038
Grenzmuseum Schifflersgrund, Asbach-Sickenberg 16581
Gründerzeitmuseum im Gutshof Mahlsdorf, Berlin 17121
Gustav-Adolf-Gedenkstätte, Lützen 20105
Haus der Geschichte der Bundesrepublik Deutschland, Bonn 17455
Haus der Ost- und Westpreußen, Oberschleißheim 20832
Haus der Wannsee-Konferenz, Berlin 17129
Haus Mährisch-Schönberg, Bad Hersfeld 16746
Heimat und Bildhauer Kern Museum, Forchtenberg 18339
Heimatmuseum, Geisa 18555
Heimatmuseum, Günzburg 18825
Heimatmuseum Boizenburg, Außenstelle am Elbberg, Boizenburg 17440
Heimatmuseum Freudenthal, Memmingen 20279
Heimatmuseum und Gedenkstätte Bützow, Bützow 17641
Historisch-Technisches Informationszentrum, Peenemünde 20996
Jüdisches Museum Georgensgmünd, Georgensgmünd 18588
KdF-Museum in der Kulturkunststatt Prora, Prora 21130
Kindheitsmuseum, Marburg 20190
Kreismuseum Wewelsburg, Büren 17637
KZ-Gedenk- und Begegnungsstätte, Ladelund 19760
KZ-Gedenkstätte Dachau, Dachau 17765
KZ-Gedenkstätte Flossenbürg, Flossenbürg 18332
KZ-Gedenkstätte Husum-Schwesing, Schwesing 21701
KZ-Gedenkstätte Mittelbau-Dora, Nordhausen 20738
KZ-Gedenkstätte Neuengamme, Hamburg 18943
Mahn- und Gedenkstätte Düsseldorf, Düsseldorf 18001
Mahn- und Gedenkstätte Isenschnibber-Feldscheune, Gardelegen 18537
Mahn- und Gedenkstätte Ravensbrück, Fürstenberg, Havel 18490
Mahn- und Gedenkstätte Steinwache, Dortmund 17922
Mahn- und Gedenkstätte Wernigerode, Wernigerode 22497
Mahn- und Gedenkstätten Wöbbelin, Wöbbelin 22637
Mauermuseum - Museum Haus am Checkpoint Charlie, Berlin 17184
Memorium Nürnberger Prozesse, Nürnberg 20778
Museen Alte Bischofsburg, Wittstock/Dosse 22629
Museum Blindenwerkstatt Otto Weidt, Berlin 17193
Museum Burg Hohnstein, Hohnstein 19249
Museum des Todesmarsches im Belower Wald, Wittstock/Dosse 22630
Museum im Schloß Lützen, Lützen 20106
Museum im Stasi-Bunker, Machern 20119
Museum "Währungsreform 1948" und Konklave in Rothwesten, Fuldatal 18514
Museumsverbund Pankow - Heynstraße, Berlin 17219
Nostalgie Museum, Beetzseeheide 16980
NS-Dokumentationszentrum der Stadt Köln, Köln 19626
Parlamentsausstellung des Deutschen Bundestages "Wege, Irrwege, Umwege", Berlin 17232
Pfinzgaumuseum, Karlsruhe 19446
Plan-Weseritzer Heimatstuben mit Archiv, Tirschenreuth 22063
Reichspräsident-Friedrich-Ebert-Gedenkstätte, Heidelberg 19080

Reichstagsmuseum, Regensburg 21231
Ruhmeshalle und Bavaria, München 20477
Schloss Cecilienhof, Potsdam 21106
Schwules Museum, Berlin 17254
Stasimuseum, Berlin 17261
Stauffenberg-Erinnerungsstätte, Stuttgart 21963
Stauffenberg-Gedenkstätte, Albstadt 16460
Steinscher Hof Kirberg, Hünfelden 19292
Stiftung Bundeskanzler-Adenauer-Haus, Bad Honnef 16754
Templerhaus, Amorbach 16543
Theodor-Heuss-Haus, Stuttgart 21966
Thomas-Müntzer-Gedenkstätte, Heldrungen 19116
Topographie des Terrors, Berlin 17272
Torhaus Markkleeberg, Markkleeberg 20205
Unterfränkisches Grenzmuseum, Bad Königshofen 16761
Walhalla, Donaustauf 17893
Zeitgeschichtliches Forum Leipzig, Leipzig 19942

Greece

Historical Museum of Azogire, Azogire 22957
Historical Museum of the National Resistance of Rentina, Agrapha, Karditsa 23037
Museum of the Sacrifice of the People of Kalavryta, Kalavryta 23028
National Historical Museum, Athinai 22942
Viniani Museum, Palia Viniani 23131

Honduras

Museo Nacional de Historia Colonial, Omoa 23329

Hungary

Bajcsy-Zsilinszky Endre Emlékmúzeum, Kővágóörs-Pálköve 23686
Emlékpont, Hódmezővásárhely 23611
Holokauszt Emlékközpont, Budapest 23412
József Attila Múzeum, Makó 23690
Kossuth Lajos Emlékmúzeum, Monok 23717
Mindszenty Emlékhely, Esztergom 23554
Ópusztaszeri Nemzeti Történeti Emlékpark, Opusztaszer 23736
Terror Háza Múzeum, Budapest 23482

India

Gandhi Sangrahalaya, Patna 24345
Gandhi Smarak Sangrahalaya, Ahmedabad 24048
Gandhi Smarak Sangrahalaya, Sevagram 24384
National Gandhi Museum and Library, Delhi 24146
Nehru Memorial Museum and Library, Delhi 24152
Sarat Smriti Granthagar, Panitras 24342
Shaheed-e-Azam Bhagat Singh Museum, Khatkar Kalan 24259

Indonesia

Museum Joang '45, Surabaya 24565

Ireland

Pearse Museum, Dublin 24815
Thurles Famine Museum, Thurles 24903
War Memorial, Dublin 24820

Israel

Aaronson House, Zichron Yaakov 25134
Beit Lohamei Haghetaot-Yad Layeled, Lohamei-Hagetaot 25062
Dubrovin Farm Museum, Yisod Hama'ala 25131
From Holocaust to Revival Museum, Chof Ashkelon 24940
Hana Senesh House, Kibbutz Sdot Yam 25057
Historical Museum, Jerusalem 24990
Jabotinsky Museum, Tel Aviv 25110
Jewish National Fund House, Tel Aviv 25111
The Khan Museum, Hadera 24951
Massua-Educational Museum on the Holocaust, Tel Yitzhak 25127
Mitzpe Revivim Museum, Kibbutz Revivim 25054
Museum of the Underground Prisoners-Acre, Acre 24920
Rishon Le-Zion Museum, Rishon Le-Zion 25092
Shoah Museum, Jerusalem 25015
Underground Prisoners Museum, Jerusalem 25023
Village Museum, Kfar Tavor 25030

Italy

Civico Museo del Risorgimento e Sacrario Oberdan, Trieste 28079
Civico Museo della Risiera di San Sabba, Trieste 28080
Istituto per la Storia della Resistenza e della Società Contemporanea nelle Province di Biella e Vercelli, Varallo 28155
Mostra Permanente della Resistenzo, Massa Marittima 26627
Musei Civici, Pavia 27120
Museo Cavour, Santena 27730
Museo Centrale del Risorgimento, Roma 27466
Museo Civico, Salemi 27593
Museo Civico del Risorgimento, Bologna 25448
Museo del Risorgimento, Brescia 25536
Museo del Risorgimento, Massa Marittima 26629
Museo del Risorgimento, Milano 26700
Museo del Risorgimento e della Resistenza, Ferrara 26111
Museo del Risorgimento e della Resistenza, Vicenza 28253
Museo del Risorgimento L. Musini, Fidenza 26123
Museo della Repubblica Partigiana, Montefiorino 26829
Museo della Resistenza Cà Malanca, Brisighella 25549
Museo di Storia Contemporanea, Milano 26706
Museo e Biblioteca della Resistenza, Sansepolcro 27699
Museo Garibaldino, Caprera 25663
Museo Internazionale della Croce Rossa, Castiglione delle Stiviere 25760
Museo Napoleonico, Roma 27511
Museo Napoleonico, Tolentino 27989
Museo Napoleonico G. Antonelli, Arcole 25260
Museo Nazionale del Risorgimento Italiano, Torino 28022
Museo Nazionale di Villa San Martino, Portoferraio 27309
Museo Palazzina Storica, Peschiera del Garda 27169
Museo Storico dei Granatieri di Sardegna, Roma 27528
Museo Storico del Nastro Azzurro, Salò 27600
Museo Storico della Città, Bergamo 25405
Museo Storico della Guardia di Finanza, Roma 27531
Museo Storico della Liberazione, Roma 27532
Museo Storico della Resistanza S. Maneschi, Neviano degli Arduini 26923
Museo Storico della Resistenza, Stazzema 27900
Museo Storico G. Garibaldi, Como 25954
Museo Storico in Trento Onlus, Trento 28061

Japan

Chiran Museum, Chiran 28403
Daigo Fukuryu Maru Exhibition Hall, Tokyo 29379
Fukuyama City Human Rights and Peace Museum, Fukuyama 28445
Hiroshima City Museum of History and Traditional Ctrafts, Hiroshima 28535
Hiroshima Heiwa Kinen Shiryokan, Hiroshima 28536
Kawasaki Peace Museum, Kawasaki 28692
Kyoto Museum for World Peace, Kyoto 28844
Maizuru Museum of Salvage, Maizuru 28878
Memorial Museum of the Kanto Earthquake Disaster, Tokyo 29436
Nagasaki Atomic Bomb Museum, Nagasaki 28978
Okinawa-ken Heiwa Kinen Shiryokan, Itoman 28603
Osaka International Peace Center, Osaka 29143

Kazakhstan

Almaty City History Museum Museum of Repression, Almaty 29674

Kenya

Kapenguria Museum, Kapenguria 29741

Korea, Democratic People's Republic

Korean Revolutionary Museum, Pyongyang 29759
Memorial Museum of the War of Liberation, Pyongyang 29760
Shinchon Museum, Shinchon 29761

Korea, Republic

Independence Hall of Korea, Cheonan 29784
Korea War Memorial, Seoul 29893

Latvia

Bārricdes 1991, Rīga 30063

Lebanon

Musée Al Mathaf El Lubnani, Beiteddine 30164

Lithuania

Dr. Jono Šliūpo Memorialinė Sodyba, Palanga 30328
Druskininkų Ž. Lipšico Žydų Muziejus, Druskininkai 30223
Genocido Aukų Muziejus, Vilnius 30407
Holokausto Ekspozicija, Vilnius 30410
Kauno IX Forto Muziejus, Kaunas 30249
Laisvės Kovų Muziejus Krauji Takais, Lazdijai 30299
Lietuvos Laisvės Kovų ir Kančių Muziejus, Merkinė 30312
Lietuvos Prezidento K. Griniaus Memorialinis Muziejus, Marijampolė 30307
Lietuvos Šaulių Sąjungos Muziejus, Kaunas 30256
Lietuvos Savanorių Muziejus, Vilnius 30421
Macikų Lagerio Karceris-muziejus, Macikai 30305
Marijampolės Kraštotyros Muziejus, Marijampolė 30308
Marijos ir Jurgio Šlapelių Namas-muziejus, Vilnius 30424
Mutinės Muziejus, Vilnius 30426
Naujasis Arsenalas, Vilnius 30427
Panerių Memorialinis Muziejus, Vilnius 30428
Signatarų Namai, Vilnius 30432
Tauro Apygardos Partizanu ir Tremties Muziejus, Marijampolė 30309
Tolerancijos Centras, Vilnius 30435
Valstybinis Vilniaus Gaono Žydų Muziejus, Vilnius 30436

Luxembourg

Musée Historique, Clervaux 30462
Musée National de la Résistance, Esch-sur-Alzette 30473

Malaysia

Cultural Museum, Melaka 30594

Malta

The Malta at War Museum, Vittoriosa 30656

Martinique

Musée Historique de Saint-Pierre, Saint-Pierre 30671

Mexico

Museo de la Lucha Obrera, Cananea 30745
Museo de la Resistencia Indígena, Ixcateopan de Cuauhtémoc 30959
Museo Mural Homenaje a Benito Juárez, México 31128
Museo Nacional de la Revolución, México 31137
Museo Nacional de las Intervenciones, México 31139

Monaco

Musée des Souvenirs Napoléoniens et des Archives du Palais, Monaco 31609

Mongolia

Friendship Museum, Darhan 31616
Museum for the Victims of Political Repression, Ulaanbaatar 31625
Muzej Revoljucionnogo Dviženija, Altanbulag 31615

Mozambique

Museu da Revolução, Maputo 31706

Netherlands

Aaltense Oorlogs en Verzetscollectie 1940-1945, Aalten 31784
Achterhoeks Museum 1940-1945, Hengelo, Gelderland 32366
Anne Frank Huis, Amsterdam 31828
Arnhems Oorlogsmuseum 1940-45, Arnhem 31922
Corpsmuseum van het Utrechtsch Studenten Corps, Utrecht 32845
Corrie ten Boomhuis, Haarlem 32294
Czaar Peterhuisje, Zaandam 32975
Expositie 40-45, Blitterswijk 31974
Ferdinand Domela Nieuwenhuis Museum, Heerenveen 32332
Herinneringscentrum Kamp Westerbork, Hooghalen 32406
De Hollandsche Schouwburg, Amsterdam 31848
Maas en Waals Museum 1939-1945, Winssen 32954
Museum 1940-1945, Dordrecht 32151
Museum '40-'45, Slochteren 32785
Museum Bevrijdende Vleugels, Best 31971
Museum de Maurits 1940-1945, Doesburg 32133
Museum Oorlog en Vrede, Breda 32005
Museum van de Twintigste Eeuw, Hoorn 32409
Museum voor Vrede en Geweldloosheid, Amsterdam 31867
Nationaal Oorlogs- en Verzetsmuseum, Overloon 32663
Oorlogs- en Verzetsmuseum Johannes Post, Ridderkerk 32685
Regionaal Museum 1940-1945 Schagen en Omstreken, Schagen 32748
Ridderhofstad Gunterstein, Breukelen 32010
Verzets Museum Zuid-Holland, Gouda 32271
Verzetsmuseum Amsterdam, Amsterdam 31899
Verzetsmuseum Friesland, Leeuwarden 32480
Watersnoodmuseum, Ouwerkerk 32661
Yi Jun Peace Museum, Den Haag 32106

Nicaragua

Museo Camilo Ortega de la Ciudad Masaya, Masaya 33293
Museo General Augusto César Sandino, San Rafael del Norte 33305
Museo Heroes y Martires Revolución Sandinista en Estelí, Estelí 33277

Norway

Eidsvoll 1814 - Rikspolitisk Senter, Eidsvoll Verk 33476
Gamlehaugen, Paradis 33767
Gjenreisningsmuseet, Hammerfest 33531
Det Hanseatiske Museum og Schøtstuene, Bergen 33428
Jacob Breda Bullmuseet, Rendalen 33778
Nordland Røde Kors Krigsminnemuseum, Narvik 33695

Pakistan

Quaid-i-Azam Birthplace and Museum, Karachi 34017

Panama

Museo Belisario Porras, Las Tablas 34075
Museo del Canal Interoceanico de Panamá, Ciudad de Panamá 34071

Papua New Guinea

East New Britain Historical Cultural Center, Kokopo 34082
Rabaul Museum, Rabaul 34086

Paraguay

Museo de la Casa de la Independencia, Asunción 34095

Peru

Museo Nacional de Antropología, Arqueología e Historia del Perú, Lima 34264

Philippines

Apolinario Mabini Birthplace Museum, Tanauan 34515
Apolinario Mabini Shrine and Museum, Manila 34429
Cebu City State College Museum, Cebu 34369
José Rizal Shrine Museum, Calamba 34359
José Rizal Shrine Museum, Dapitan 34382
Wenceslao Vinzons Historical Landmark Museum, Vinzons 34526

Poland

Centralne Muzeum Jeńców Wojennych w Łambinowicach -Opolu, Opole 34984
Galicja Muzeum, Kraków 34781
Miejsce Pamięci i Muzeum Auschwitz-Birkenau, Oświęcim 34997
Muzeum Armii Krajowej im. Generala Emila Fieldorfa "Nila", Kraków 34787
Muzeum Byłego Obozu Zagłady w Chełmnie nad Narem, Chełmno nad Nerem 34592
Muzeum Czynu Niepodległościowego, Kraków 34788
Muzeum Czynu Partyzanckiego, Polichno 35021
Muzeum Gross-Rosen, Wałbrzych 35211
Muzeum Gross-Rosen w Rogoźnicy, Rogoźnica 35084
Muzeum Historii Polskiego Ruchu Ludowego, Warszawa 35245
Muzeum Historyczne Miasta Krakowa - Oddział Pomorska, Kraków 34799
Muzeum im. Kazimierza Pułaskiego, Warka-Winiary 35215
Muzeum Martyrologii i Walki Radogoszcz, Łódź 34879
Muzeum Martyrologii Wielkopolan Fort VII, Poznań 35039
Muzeum Niepodległości, Warszawa 35262
Muzeum Powstania Poznańskiego - Czerwiec 1956, Poznań 35042
Muzeum Powstania Warszawskiego, Warszawa 35266
Muzeum Powstania Wielkopolskiego 1918-1919, Poznań 35043
Muzeum Regionalne PTTK im. Ignacego Łukasiewicza, Gorlice 34679
Muzeum Ubezpieczeń, Kraków 34812
Muzeum Uzbrojenia - Cytadela Poznańska, Poznań 35045
Muzeum Walki i Męczeństwa 1939-1945, Warszawa 35275
Muzeum Walki i Męczeństwa w Palmirach, Warszawa 35276
Muzeum Walki i Męczeństwa w Treblince, Kosów Lacki 34760
Muzeum X Pawilonu Cytadeli Warszawskiej, Warszawa 35282
Państwowe Muzeum na Majdanku, Lublin 34911
Państwowe Muzeum Stutthof, Sztutowo 35172

H

Państwowe Muzeum Stutthof w Sztutowie, Sopot 35122
Wielkopolskie Muzeum Walk Niepodległościowych w Poznaniu, Poznań 35048

Puerto Rico

Museo Casa de la Masacre de Ponce, Ponce 36046

Romania

Memorialul Victimelor Comunismului şi al Rezistenţei, Sighetu Marmaţiei 36614
Muzeul Doftana, Doftana 36347
Muzeul Memorial Avram Iancu, Avram Iancu 36116

Russia

Blindaž gde podpisan akt o kapituljacii, Kaliningrad 36986
Centralnyj Muzej Velikoj Otečestvennoj Vojny, Moskva 37149
Gosudarstvennyj Centralnyj Muzej Sovremennoj Istorii Rossii, Moskva 37189
Gosudarstvennyj Istoriko-memorialnyj Sankt-Peterburgskij Muzej Smolnyj, Sankt-Peterburg 37641
Gosudarstvennyj Memorialnyj Muzej Oborony i Blokady Leningrada, Sankt-Peterburg 37644
Gosudarstvennyj Muzej Oborony Moskvy, Moskva 37201
Gosudarstvennyj Muzej-panorama Stalingradskaja Bitva, Volgograd 37955
Gosudarstvennyj Muzej Političeskoj Istorii Rossii, Sankt-Peterburg 37648
Kraevedčeskij Muzej Evrejskoj Avtonomnoj Oblasti, Birobidžan 36807
Muzej 700 let - Landskrona, Nevskoe Ustjě, Nienšanc, Sankt-Peterburg 37668
Muzej 'A Muzy ne Molčali', Sankt-Peterburg 37669
Muzej Diplomatičeskogo Korpusa, Vologda 37962
Muzej i Obščestvennyj Centr Mir, Progress, Prava Čeloveka im. Andreja Sacharova, Moskva 37284
Muzej im. S.M. Kirova, Sankt-Peterburg 37678
Muzej Krejsera Avrora, Sankt-Peterburg 37688
Muzej V.I. Lenina, Moskva 37327
Muzej Sovremennoj Istorii i Kultury Severnoj Osetii, Vladikavkaz 37931
Muzej Tvorčestva i Byta v GULAGe, Moskva 37376

Serbia

Muzej Radničkog Pokreta i Narodne Revolucije, Novi Sad 38122
Muzej u Arandjelovcu, Arandjelovac 38030
Zadužbina Kralja Petra I Karadjordjevića, Topola 38157

Slovakia

Historické Múzeum, Bratislava 38213
Muzeum Slovenského Narodného Povstania, Banská Bystrica 38184
Podpolianske Múzeum Detva, Detva 38247

Slovenia

Koroški Pokrajinski Muzej, Slovenj Gradec 38518
Muzej Narodne Osvobodilve Maribor, Maribor 38476
Muzej NOB - Pustova Hiša, Slovenske Konjice 38521
Muzej Novejše Zgodovine Celje, Celje 38406
Muzej Novejše Zgodovine Slovenije, Ljubljana 38455
Muzej Talcev, Begunje na Gorenjskem 38398
Muzejska Zbirka NOB, Zagorje ob Savi 38537
Partizanska Bolnica Franja, Dolenji Novaki 38415
Pokrajinski Muzej, Murska Sobota 38487

Solomon Islands

World War II Museum, Betikama 38546

South Africa

Apartheid Museum, Johannesburg 38648
Cape Town Holocaust Centre, Cape Town 38580
Kruger Museum, Tshwane 38748
Mahatma Gandhi Museum, Verulam 38773
Nelson Mandela Museum, Umtata 38770
Nieuwe Republiek Museums - Vryheid, Vryheid 38775

Spain

Casa Natal de Legazpi, Zumárraga 40363
Euskal Abertzaletasunaren Museoa, Artea 38901
Museo de la Paz de Gernika, Gernika-Lumo 39332
Museo de Tabar, Navarra 39712
Museo Histórico-Artístico, Cuacos de Yuste 39228
Museo Simón Bolívar, Ziortza-Bolibar 40358

Sri Lanka

Dutch Period Museum, Colombo 40371

Switzerland

Kosciuszko-Museum, Solothurn 41884
Napoleonmuseum Thurgau, Salenstein 41798

Taiwan

2-28 Memorial Museum, Taipei 42371
Chai-Yi 2-28 Memorial Park Museum, Chai-Yi 42150
National Chiang Kai-shek Memorial Hall, Taipei 42413
Taiwan Democratic Movement Museum, Wujie 42461

Tanzania

Arusha Declaration Museum, Arusha 42507

Tunisia

Dar Maâkal Az-Zaïm, Tunis 42663

Turkey

Atatürk Kongre ve Etnografya Müzesi, Sivas 42945

Ukraine

Charkivs'kyj Istoryčnyj Muzej, Charkiv 43018

United Kingdom

Althorp Museum, Althorp 43291
Baden-Powell House, London 44787
Bletchley Park, Bletchley 43543
Brenzett Aeronautical Museum Trust, Romney Marsh 45511
Britain at War Experience, London 44800
British in India Museum, Nelson 45198
Channel Islands Military Museum, Saint Ouen 45587
Eden Camp, Malton 45079
Flambards Experience, Helston 44442
German Occupation Museum, Forest 44241
The Grand Lodge of Scotland Museum, Edinburgh 44126
Imperial War Museum, London 44881
Jersey War Tunnels, Saint Lawrence 45581
John Knox's House, Edinburgh 44129
Kelvedon Hatch Secret Nuclear Bunker, Kelvedon Hatch 44556
Noirmont Command Bunker, Saint Brelade 45561
Oban War and Peace Museum, Oban 45310
Peace Museum, Bradford 43582
Pitstone Green Farm Museum, Pitstone 45387
Robert Owen Memorial Museum, Newtown, Powys 45258
Rochdale Pioneers Museum, Rochdale 45506
Scotland's Secret Bunker, Saint Andrews 45558
Second World War Experience Centre, Horsforth 44487
War Room and Motor House Collection, Harrogate 44403

United States of America

Abraham Lincoln Birthplace, Hodgenville 49307
Ainsley House, Campbell 47207
America's Black Holocaust Museum, Milwaukee 50704
Andrew Johnson National Historic Site, Greeneville 49050
Anne Frank Center USA, New York 51119
Arizona State Capitol Museum, Phoenix 51901
Atwater Kent Museum - The History Museum of Philadelphia, Philadelphia 51817
Bedingfield Inn Museum, Lumpkin 50321
Bidwell Mansion, Chico 47525
Black American West Museum and Heritage Center, Denver 48064
Bloomfield Academy Museum, Bloomfield 46854
Boscobel Restoration, Garrison 48868
Bowers Mansion, Carson City 47259
Bryant Cottage, Bement 46729
California State Capitol Museum, Sacramento 52501
Calvin Coolidge Presidential Library and Museum, Northampton 51412
Carl Sandburg State Historic Site, Galesburg 48832
Carry A. Nation Home Memorial, Medicine Lodge 50554
Chieftains Museum, Rome 52448
Church-Waddel-Brumby House Museum, Athens 46410
City Hall Council Chamber Gallery, Charleston 47361
Clark County Museum, Henderson 49249
Colonel William Jones House, Gentryville 48887
Columbia County Historical Society Museum, Kinderhook 49777
Columbus Cultural Arts Center, Columbus 47751
Conner Prairie Living History Museum, Fishers 48546
Crowley Museum, Sarasota 52913
Culbertson Mansion, New Albany 51010
Dallas Holocaust Museum, Dallas 47930
Danville Museum of Fine Arts and History, Danville 47963
Davenport House Museum, Savannah 52939
The David Davis Mansion, Bloomington 46859
East Martello Museum, Key West 49758
Edward H. White II Memorial Museum, Brooks Air Force Base 47089
El Paso Holocaust Museum, El Paso 48319
Elfreth's Alley Museum, Philadelphia 51831
Ellwood House Museum, DeKalb 48038
Ernie Pyle House, Dana 47951
Evergreen House, Baltimore 46564
First Consolidated School in Minnesota, Saum 52932
Five Civilized Tribes Museum, Muskogee 50928
Flagler Museum, Palm Beach 51677
Florida Holocaust Museum, Saint Petersburg 52637
Fonthill Museum of the Bucks County Historical Society, Doylestown 48177
Fort Concho, San Angelo 52701
Franklin D. Roosevelt Presidential Library-Museum, Hyde Park 49458
Gene Stratton-Porter House, Rome City 52454
George C. Marshall Museum, Lexington 50070
Gibson House Museum, Boston 46931
Gilbert's Bar House of Refuge, Stuart 53343
Gunderson House, Kenyon 49743
Hampton-Preston Mansion, Columbia 47726
Harlan-Lincoln House, Mount Pleasant 50882
Hartford House Museum, Manhattan 50433
Hay House Museum, Macon 50366
Henry B. Plant Museum, Tampa 53433
Historic Cherry Hill, Albany 46158
Historic Madison House, Madison 50378
Holocaust Documentation and Education Center, Miami 50626
Holocaust Memorial Center of Central Florida, Maitland 50401
Holocaust Museum Houston, Houston 49387
Iredell Museum of Arts Heritage, Statesville 53274
James Buchanan Foundation for the Preservation of Wheatland, Lancaster 49925
Jane Addams' Hull-House Museum, Chicago 47483
John Fitzgerald Kennedy House, Brookline 47063
John Wesley Powell Memorial Museum, Page 51673
Juliette Gordon Low Birthplace, Savannah 52942
Kingsley Plantation, Jacksonville 49583
Klondike Gold Rush National Historical Park, Seattle 53002
The Lane Place, Crawfordsville 47870
Lanier Mansion, Madison 50380
Lapham-Patterson House, Thomasville 53499
Levi Coffin House, Fountain City 48725
Lincoln Home, Springfield 53217
Lincoln Log Cabin, Lerna 50025
The Little Brick House, Vandalia 53708
Little White House, Warm Springs 53816
Little White Schoolhouse, Ripon 52357
Los Angeles Museum of the Holocaust, Los Angeles 50246
McNamara House Museum, Victoria 53740
Magnolia Mound Plantation, Baton Rouge 46637
Malabar Farm State Park, Lucas 50315
Mamie Doud Eisenhower Birthplace, Boone 46906
Mann-Simons Cottage, Columbia 47729
Mardi Gras Museum, Biloxi 46810
Martin Luther King jr. Center and Preservation District, Atlanta 46443
Metamora Courthouse, Metamora 50613
Mississippi University for Women Museum, Columbus 47747
Montauk, Clermont 47617
Moore Home, Lerna 50026
Morris-Jumel Mansion, New York 51217
Museum of Jewish Heritage - A Living Memorial to the Holocaust, New York 51227
Museum of Tolerance, Los Angeles 50256
National Civil Rights Museum, Memphis 50581
Nelson Pioneer Farm and Museum, Oskaloosa 51617
New Harmony State Historic Site, New Harmony 51050
Patterson Homestead, Dayton 47987
Paul Dresser Memorial Birthplace, Terre Haute 53482
Pierre Menard Home, Ellis Grove 48369
Pitot House Museum, New Orleans 51094
Porterville Historical Museum, Porterville 52060
Queen Emma Summer Palace, Honolulu 49352
Roosevelt Campobello, Lubec 50314
Saint Joseph Museum, Saint Joseph 52565
Salisbury House, Des Moines 48100
Sarguinetti House Museum, Yuma 54364
Shrewsbury Windle House, Madison 50381
Sixth Floor Museum at Dealey Plaza, Dallas 47939
Spirit of '76 Museum, Elyria 48387
Stan Hywet Hall and Gardens, Akron 46148
The Supreme Court of the United States Museum, Washington 53920
Taylor-Grady House, Athens 46415
Thomas Sappington House Museum, Crestwood 47880
Traveler's Rest State Historic Site, Toccoa 53518
United States Grant's Home, Galena 48830
United States Holocaust Memorial Museum, Washington 53925
Valentine Richmond History Center, Richmond 52339
Vann House, Chatsworth 47405
Vizcaya Museum, Miami 50635
W.C. Handy Home Museum and Library, Florence 48566
Walnut Grove Plantation, Roebuck 52438
Wayne County Museum, Honesdale 49334
The White House, Washington 53934
Wilbur Wright Birthplace and Interpretive Center, Hagerstown 49119
Wisconsin Maritime Museum, Manitowoc 50444
Wolf House Museum, Manhattan 50438
Woodrow Wilson House, Washington 53935
Worker's Home Museum, South Bend 53161
Zekelman Family Holocaust Memorial Center, Farmington Hills 48521

Uzbekistan

International Museum of Peace and Solidarity, Samarqand 54461

Vietnam

Ho Chi Minh Memorial House, Thua Thien Hue 54637
Ho Chi Minh Museum, Ha Noi 54615
Ho Chi Minh Museum, Ho Chi Minh City 54624
Ho Chi Minh Museum, Thua Thien Hue 54638
Nghe-Tinh Museum, Vinh 54640
Revolutionary Museum Ho Chi Minh City, Ho Chi Minh City 54625
Viet Nam Revolution Museum, Ha Noi 54618
War Remnants Museum, Ho Chi Minh City 54631

Yemen

Taizz National Museum, Taizz 54652

History, Social → Social Conditions

Horology

Australia

Melbourne Clocks Museum, Hampton 01145

Austria

Eisenuhren- und Antikuhrenmuseum, Steyr 02768
Uhrenmuseum, Wien 03082
Uhrenstube Aschau, Oberschützen 02460

Belgium

Astronomisch Compensatieuurwerk Kamiel Festraets, Sint-Truiden 04026
Horlogeriemuseum, Mechelen 03857
Klokkenmuseum, Hechtel 03690
Musée de la Cloche et du Carillon, Tellin 04052
Musée des Traditions Namuroises, Namur 03902
Musées de l'Horlogerie Pater, Longueville 03831
Stadtmuseum Eupen, Eupen 03593
Zimmertoren, Lier 03821

Czech Republic

Hrad a Muzeum Hodin, Šternberk 09848

Finland

Suomen Kellomuseo, Espoo 10733

France

Musée de la Montre Droz-Flores, Villers-le-Lac 16230
Musée de l'Horlogerie, Pesmes 14801
Musée de l'Horlogerie Ancienne, Fougères (Ille-et-Vilaine) 12985
Musée de l'Horlogerie du Haut-Doubs, Morteau 14289
Musée Frédéric Japy, Beaucourt 11890
Musée Horlogerie Automates Yves Cupillard, Morteau 14290
Musée Royaume de l'Horloge, Villedieu-les-Poêles 16184

Germany

Babls Uhrensammlung, Waffenbrunn 22268
Deutsches Uhrenmuseum, Furtwangen im Schwarzwald 18520
Deutsches Uhrenmuseum Glashütte, Glashütte 18653
ErfinderZeiten, Schramberg 21630
Fränkisches Turmuhrenmuseum, Mistelbach 20340
Grüttert Uhrenmuseum Bremen, Bremen 17553

Heimat- und Uhrenmuseum, Villingen-Schwenningen 22236
Heimatmuseum, Kastl, Kreis Amberg-Sulzbach 19480
Heimatstuben der Stadt, Titisee-Neustadt 22064
Mathematisch-Physikalischer Salon, Dresden 17959
Museum für Uhren und Schmuck, Frankfurt am Main 18386
Museum für Zeit - Pfälzisches Turmuhrenmuseum, Rockenhausen 21340
Ruhlaer Uhrenmuseum, Ruhla 21441
Saarländisches Uhrenmuseum, Püttlingen 21137
Schwäbisches Turmuhrenmuseum, Mindelheim 20330
Stadtmuseum Schramberg, Schramberg 21634
Turmuhrenmuseum, Naunhof 20557
Turmuhrenmuseum Rammensee, Gräfenberg 18725
Uhrenindustriemuseum, Villingen-Schwenningen 22241
Uhrenmuseum, Bad Grund 16744
Uhrenmuseum, Bad Iburg 16756
Uhrenmuseum, Regensburg 21234
Uhrentechnische Lehrschau Hennig, Tharandt 22050
Wuppertaler Uhrenmuseum, Wuppertal 22749

Italy

Museo dell'Orologio da Torre G.B. Bergallo, Tovo San Giacomo 28047
Torre dell'Orologio, Venezia 28213

Japan

Daimyo Tokei Hakubutsukan, Tokyo 29382
Fukuyama Auto and Clock Museum, Fukuyama 28441
Marie Sekei Tokei oyo Orugoru Hakubutsukan, Inuyama 28577

Netherlands

Klokkengieterij Museum, Heiligerlee 32346
Klokkenmuseum, Frederiksoord 32233
Museum van het Nederlandse Uurwerk, Zaandam 32978
Nederlands Goud-, Zilver- en Klokkenmuseum, Schoonhoven 32764
Zandlopermuseum Glanerbrug, Enschede 32214

New Zealand

Claphams Clocks-The National Clock Museum, Whangarei 33264

Poland

Muzeum Zegarów, Rzemiosł Artystycznych i Precyzyjnych, Warszawa 35283
Muzeum Zegarów Wieżowych, Gdańsk 34648
Państwowe Muzeum im. Przypkowskich, Jędrzejów 34702

Romania

Muzeul Ceasului Nicolae Simache, Ploiești 36529

Slovakia

Expozícia Historických Hodín, Bratislava 38207

Spain

Colección de Relojes Populares, Salamanca 39935
Museo Castillo del Macho, Peñiscola 39833
Museo de Relojes, A Coruña 39221
Museos de la Atalaya, Jerez de la Frontera 39427

Switzerland

Espace Horloger de la Vallée de Joux, Le Sentier 41863
Musée Audemars Piguet, Le Brassus 41251
Musée de l'Horlogerie et de l'Emaillerie, Genève 41420
Musée d'Horlogerie du Locle, Le Locle 41573
Musée International d'Horlogerie, La Chaux-de-Fonds 41296
Musée Longines, Saint-Imier 41791
Musée Omega, Biel 41230
Musée Régional d'Histoire et d'Artisanat du Val-de-Travers, Môtiers 41649
Museum für Uhren und mechanische Musikinstrumente, Oberhofen am Thunersee 41696
Patek Philippe Museum, Genève 41428
Uhrenmuseum Beyer Zürich, Zürich 42114
Uhrensammlung Kellenberger, Winterthur 42040

United Kingdom

Belmont, Faversham 44220
Clockmakers' Museum, London 44827
Coventry Watch Museum, Coventry 43901
Keith Harding's World of Mechanical Music, Northleach 45271
Mill House Cider Museum and Dorset Collection of Clocks, Owermoigne 45326
Prescot Museum, Prescot 45442
The Time Museum, Newark-on-Trent 45216
Tymperleys Clock Museum, Colchester 43876

United States of America

American Clock and Watch Museum, Bristol 47027
American Watchmakers-Clockmakers Institute, Harrison 49184
Bily Clock Museum and Antonin Dvorak Exhibition, Spillville 53203
The National Time Museum, Chicago 47499
National Watch and Clock Museum, Columbia 47720
Old Clock Museum, Pharr 51808
Timexpo Museum, Waterbury 53957
Willard House and Clock Museum, Grafton 48981

Horses

Australia

Australian Racing Museum and Hall of Fame, Melbourne 01275

Austria

Museum für Beschirrung und Besattelung, Hufbeschlag und Veterinär-Orthopädie, Wien 03035

Belgium

Musée Spadois du Cheval, Spa 04044
Musée Vivant du Cheval de Trait, Waret-l'Evêque 04132

Bulgaria

Muzej na Bivolarstvoto i Konevădstvoto, Šumen 05217

Canada

Appaloosa Horse Club of Canada Museum, Claresholm 05618
Fort Garry Horse Regimental Museum, Winnipeg 07360

China

Hong Kong Racing Museum, Hong Kong 07802

Czech Republic

Hippologické Muzeum, Slatiňany 09840
Muzeum Koněspřežné Cráhy v Bujanově, Kaplice 09572

France

Musée de l'Horlogerie, Saint-Nicolas-d'Aliermont 15426
Musée du Cheval en 1900, Cussac-Fort-Médoc 12692
Musée Vivant du Cheval, Chantilly 12401
Parcours-découverte de l'écurie n1, Le Pin-au-Haras 13632

Germany

Deutsches Pferdemuseum, Verden 22219
Gestütsmuseum Offenhausen, Gomadingen 18706

Hungary

Kincsem Múzeum, Tápiószentmárton 23913
Lipicai Múzeum, Szilvásvárad 23894
Lószerszám Múzeum, Sümeg 23826
Lovasmúzeum, Bábolna 23341
Mezohegyesi Állami Ménes Kocsimúzeuma, Mezőhegyes 23696

Ireland

Irish Horse Museum, Kildare 24846

Japan

Equine Museum of Japan JRA Racing Museum, Yokohama 29593

Korea, Republic

Equine Museum, Gwacheon 29819

Latvia

Burtnieku Pagasta Briedes Krogs, Burtnieki 29994

Lithuania

Arklio Muziejus, Anykščiai 30207

Poland

Muzeum Łowiectwa i Jeździectwa, Warszawa 35257

Russia

Muzej Konevodstva, Moskva 37309

Slovakia

Hipologické Múzeum, Topoľčianky 38373

Sweden

Nordiska Travmuseet i Årjäng, Årjäng 40430

Switzerland

Armeepferdemuseum, Scherz 41833
Musée du Cheval, La Sarraz 41823

United Kingdom

National Horseracing Museum, Newmarket 45239
The Royal Mews, London 44976
Suffolk Punch Heavy Horse Museum, Woodbridge 46061
York Racing Museum, York 46110

United States of America

Aiken Thoroughbred Racing Hall of Fame and Museum, Aiken 46141
American Royal Museum, Kansas City 49683
American Saddlebred Museum, Lexington 50044
American Work Horse Museum, Lexington 50068
International Museum of the Horse, Lexington 50050
National Museum of the Morgan Horse, Shelburne 53064

Horticulture

Australia

Emerald Museum, Emerald 01070

Austria

Haus des Apfels, Puch bei Weiz 02535
Mohnmuseum, Ottenschlag 02477
Österreichisches Gartenbaumuseum, Wien 03047

Brazil

Sítio Roberto Burle Marx, Rio de Janeiro 04771

Canada

Lawn Heritage Museum, Lawn 06150
Maison Léon-Provancher, Cap-Rouge 05550

China

Suzhou Gardens Museum, Suzhou 08257

Czech Republic

Muzeum zahradnictví a životního prostředí, Valtice 09897

Denmark

Flynderupgård Museet, Espergærde 10005

Egypt

Agricultural Museum and Cotton Museum, Cairo 10485

Germany

Deutsches Gartenbaumuseum Erfurt, Erfurt 18199
Deutsches Kleingärtnermuseum, Leipzig 19901
Gärtner- und Häckermuseum, Bamberg 16909
Obstmuseum, Sörup 21794
Orangerieschloss und Turm, Potsdam 21102
Rosenmuseum Steinfurth, Bad Nauheim 16791

Japan

Sankei-en Garden, Yokohama 29605

Latvia

Pētera Upīša Piemiņas Muzejs, Dobele 30001

Netherlands

Betuws Fruitteelt Museum, Erichem 32220
Land- en Tuinbouwmuseum, Etten Leur 32226
Museum Lammert Boerma, Borgercompagnie 31988
'T Olde Ras, Doesburg 32135

Norway

Gamle Hvam Museum, Årnes 33400
Våler Torvdriftsmuseum, Våler i Solør 33955
Vigatunet, Hjelmeland 33549

United Kingdom

Botanic Gardens Museum, Southport 45717
British Lawnmower Museum, Southport 45718
Harlow Carr Museum of Gardening, Harrogate 44400
Museum of Garden History, London 44925
National Museum of Gardening, Helston 44445
Rievaulx Abbay, Rievaulx 45496

United States of America

Barnes Foundation, Merion 50601
Loudon Heritage Farm Museums, Sterling 53287
Luther Burbank Home, Santa Rosa 52906

Household Articles

Australia

Cleve National Trust Museum, Cleve 00982

Austria

Europäisches Museum für Frieden, Stadtschlaining 02756
Privatsammlung Hermine Brandstetter, Ostermiething 02475

Belgium

Musée Communal, Xhoris 04159

Canada

Badger Creek Museum, Cartwright 05571
Heritage Farm Village, North Battleford 06468
New Denmark Memorial Museum, New Denmark 06426
Sombra Township Museum, Sombra 06990

China

Shikumen Open House Museum, Shanghai 08192

Denmark

L. Lange's Ovnmuseum, Svendborg 10290

Finland

Lounais-Hämeen Museo, Forssa 10737
Tuomarinkylän Museo, Helsinki 10832
Varkauden Museot, Varkaus 11454

France

Musée d'Art et d'Histoire de Provence, Grasse 13109

Germany

Besenmuseum Galerie, Ehingen 18066
Bürsten- und Heimatmuseum, Schönheide 21611
Deutsches Pinsel- und Bürstenmuseum, Bechhofen 16970
Gerbermuseum, Bretten 17583
Geschichtsmuseum Lüdenscheid, Lüdenscheid 20088
Miele-Museum, Gütersloh 18832
Museum Bayerische Metallwarenfabrik (BMF), Nürnberg 20780
Ortsmuseum Ursulastift, Gerstetten 18617
Schloß Neunhof, Nürnberg 20788
Uckermärkische Heimatstuben Fürstenwerder, Fürstenwerder 18496
Wäscherei-Museum Omas Waschküche, Eberswalde 18048

Greece

Historical and Folklore Museum of Rethymnon, Rethymnon 23167

Ireland

Castle Brand Visitor Centre, Nenagh 24882

Italy

Museo del Coltello Sardo, Arbus 25258
Museo della Bilancia, Campogalliano 25639
Museo della Civiltà Contadina ed Artigiana, Ripatransone 27416
Museo della Fondazione Ugo da Como, Lonato 26513

Japan

Sanuki Folk Art Museum, Takamatsu 29319

Korea, Republic

Jeju Folklore Museum, Jeju 29838

Lebanon

Beiteddine Museum, Shouf 30174

Netherlands

Boerderij Klein Hulze, Almen 31801
Gemeentelijk Museum 't Oude Slot, Veldhoven 32880
Grietje Tump Museum, Landsmeer 32462
Grootmoeders Keukenmuseum, Nijmegen 32588
Museum - Herberg de Ar, Westerbork 32939
Museum voor het Kruideniersbedrijf, Utrecht 32850
Nederlands Strijkijzer-Museum, Noordbroek .. 32601
Oudheidkamer, Bruinisse 32015
Oudheidkamer Willem Van Strijen, Zevenbergen 33001

New Zealand

North Otago Museum, Oamaru 33162
Waimate Historical Museum, Waimate 33221

Norway

Sandnes Museum, Sandnes 33815

Philippines

Zaldivar Museum, Albuera 34332

Russia

Muzej Tulskie Samovary, Tula 37870

South Africa

Free State Voortrekker Museum, Winburg 38778
Sammy Marks Museum, Kungwini 38680

Sweden

Kronbloms Kök, Örebro 40786

Switzerland

Collection d'Etains, Sierre 41865
Sammlung historischer Biscuitdosen, Basel .. 41181

United Kingdom

Biscuit Tin Museum, Duns 44063
Bygones Museum at Holkham, Wells-next-the-Sea 45959
Dalmeny House, South Queensferry 45694
Fan Museum, London 44845
Hornsea Museum of Village Life, Hornsea 44484
Priest House, West Hoathly 45969
Session Cottage Museum, Turriff 45884
Strathnaver Museum, Bettyhill 43465
Sussex Farm Museum, Heathfield 44433
Swinford Museum, Filkins 44228

Hunting

Argentina

Museo Argentino de Caza Mayor, Buenos Aires 00167

Austria

Jagdmuseum Gußwerk, Gußwerk 02021
Jagdmuseum Schloss Stainz, Stainz 02759
Museum Schloß Lackenbach, Lackenbach 02250
Oberösterreichisches Jagdmuseum, Sankt Florian 02639
Trophäensaal, Altenfelden 01711

Croatia

Lovački Muzej, Zagreb 09002

Czech Republic

Domeček T.G.M-muzeum, Čejkovice 09453
Lesnické, Myslivecké a Rybářské Muzeum, Hluboká nad Vltavou 09516
Zámek Budišov, Budišov u Třebíče 09447

Denmark

Dansk Jagt- og Skovbrugsmuseum, Hørsholm 10100
Fjordmuseet, Jyllinge 10111
Spillemands-Jagt og Skovbrugsmuseet i Rebild, Skørping 10272

Finland

Ålands Jakt och Fiskemuseum, Eckerö 10720
Tyrvään Seudun Museo, Vammala 11442

France

Centre Jean Rostand, Roquefort 15097
Musée de la Chasse et de la Nature, Paris 14649
Musée de la Vénerie et des Spahis, Senlis 15716
Musée International de la Chasse, Gien 13055

Germany

Deutsches Jagd- und Fischereimuseum, München 20432
Gräfliche Sammlungen Schloss Erbach, Erbach, Odenwaldkreis 18191
Jagd- und Falknereimuseum, Riedenburg 21316
Jagd- und Fischereimuseum, Adelsdorf 16427
Jagd- und Fischereimuseum Schloss Tambach, Weitramsdorf 22468
Jagd- und Fischereimuseum Schloß Wolfstein und Galerie Wolfstein, Freyung 18461
Jagd- und Naturkunde Museum, Niederstetten 20711
Jagd- und Schloßmuseum Spangenberg, Spangenberg 21830
Jagdhaus Gabelbach, Ilmenau 19321
Jagdmuseum Pfahlheim, Ellwangen 18123
Jagdschau im Jagdschloss Springe, Springe 21845
Jagdschloß Granitz, Binz, Ostseebad 17357
Jagdschloss Grunewald, Berlin 17145
Jagdschloß Paulinzella, Rottenbach 21419
Jagdschloss Stern, Potsdam 21091
Museum Burg Falkenstein, Falkenstein, Harz .. 18292
Museum für Jagdtier- und Vogelkunde des Erzgebirges, Augustusburg 16641
Museum für ländliches Kulturgut, Landwirtschaft, Forsten und Jagd, Ulrichstein 22171
Museum Jagd und Wild auf Burg Falkenstein, Falkenstein, Kreis Cham 18295
Museum Jagdschloß Kranichstein, Darmstadt 17793
Museum Schloss Homburg, Nümbrecht 20753
Schloss Ludwigslust, Ludwigslust 20062
Schloss Moritzburg und Fasanenschlösschen, Moritzburg 20386

Greenland

Hunting Fishing Museum, Ilulissat 23254

Hungary

Ceglédi Természetrajzi Vadászati Múzeum, Cegléd 23501
Mátra Múzeum, Gyöngyös 23574

Italy

Museo Provinciale della Caccia e della Pesca, Racines 27355

Latvia

Jaunmoku Pils Muzejs, Tukums 30131

Netherlands

Landbouw- en Juttersmuseum Swartwoude, Buren, Friesland 32023
Museum voor Natuur- en Wildbeheer, Doorwerth 32148
Stinzenmuseum, Beers 31954

Norway

Norsk Skogmuseum, Elverum 33481

Poland

Muzeum Łowiectwa i Jeździectwa, Warszawa 35257

Portugal

Museu da Caça, Vila Viçosa 36017

Romania

Muzeul de Vânătoare August von Spiess, Sibiu 36611

Russia

Muzej Ochotovedenija, Irkutsk 36939
Muzej Ochoty i Rybolovstva, Moskva 37342

Slovakia

Múzeum vo Svätý Antone, Svätý Anton 38364

Slovenia

Tehniški Muzej Slovenije, Bistra pri Vrhniki 38399
Tehniški Muzej Slovenije, Ljubljana 38467

Spain

Museo Nacional de Caza, Riofrío 39902

Sweden

Jaktvårdsmuseum, Eksjö 40495
Kungajaktmuseet, Vargön 41053
Nordvärmlands Jakt och Fiskemuseum, Sysslebäck 40967

Switzerland

Museum Alpin Pontresina, Pontresina 41726
Schweizer Museum für Wild und Jagd, Utzenstorf 41968

United Kingdom

Dog Collar Museum, Maidstone 45072
Dunrobin Castle Museum, Golspie 44325

United States of America

Atlantic Wildfowl Heritage Museum, Virginia Beach 53753
Morven Park Museum, Leesburg 50019

Hygiene → Medicine and Hygiene

Icons

Albania

Muzeu Ikonografik Onufri, Berat 00011

Andorra

Museu Iconogràfic, Ordino 00102

Armenia

Echmiadzin Cathedral Complex and Museum, Echmiadzin 00695

Austria

Ikonenmuseum, Potzneusiedl 02516

Bosnia and Herzegovina

Muzej Stare Srpske-Pravoslavne Crkve, Sarajevo 04281
Umjetnička Galerija Bosne i Hercegovine, Sarajevo 04283

Bulgaria

Chudožestvena Galerija, Čirpan 05008
Chudožestvena Galerija Prof. Ilija Petrov, Razgrad 05147
Durjaven Kulturen Center Dvoreca, Balčik 04975
Muzej-kripta Aleksandăr Nevski, Sofia 05184
Muzej Văzraždenski Ikoni, Radomir 05145

Canada

Musée Itinérant d'Art Byzantin, Montréal 06369

Croatia

Muzej Ikona, Korčula 08895
Muzej Srpske Pravoslavne Crkve, Dubrovnik . 08870

Cyprus

Byzantine Museum, Paphos 09390
Byzantine Museum and Art Galleries, Lefkosia 09372

Egypt

Coptic Museum, Cairo 10490

Finland

Ortodoksinen Kirkkomuseo, Kuopio 11010

Germany

Diözesanmuseum Freising, Freising 18450
Historisches Puppen-, Spielzeug- und Ikonenmuseum, Bernkastel-Kues 17297
Ikonen-Museum, Recklinghausen 21208
Ikonen-Museum der Stadt Frankfurt am Main, Frankfurt am Main 18378
Ikonenmuseum, Autenried 16649
Ikonenmuseum, Bad Breisig 16703
Ikonenmuseum Schloß Autenried, Ichenhausen 19307
Ikonenzentrum Alexej Saweljew, Traben-Trarbach 22074
Museum Gunnar-Wester-Haus, Schweinfurt ... 21686
Schenkbarsches Haus Museum, Biedenkopf .. 17336

Greece

Archaeological Collection, Filiatra 22998
Byzantine Museum, Tinos 23227
Byzantine Museum of Ekatontapyliani, Paros 23136
Collection of Icons and Church Heirlooms at Pyrgos, Thera 23207
Collection of the Evangelismos tis Theotokou, Skiathos 23189
Collection of the Mega Spileon Monastery, Kalavryta 23027
Collection of the Saint John the Divine Monastery, Patmos 23137
Corgialenios Historical and Cultural Museum, Argostolion 22890
Ecclesiastical Museum, Komotini 23059
Monastery of Aghiou Tessarakonta, Spárti ... 23196
Monastery of Leimon Collection, Kalloni 23030
Monastery of the Zoodochos Pighi, Poros 23157
Museum of Ecclesiastical Art, Samos 23177
Museum of Ecclesiastical Art of the Metropolis of Trikki and Stagoi, Trikala 23229
Museum of Epirot Folk Art, Metsovo 23095
Museum of Sacred Icons, Athinai 22937
Museum of the Metropolis Square, Naxos 23120
Preveli Monastery, Myrthios 23107

Hungary

Görögkatolikus Egyházművészeti Gyűjtemény, Nyíregyháza 23732

India

Namgyal Institute of Tibetology, Gangtok 24175

Italy

Museo delle Icone e della Tradizione Bizantina, Frascineto 26248
Museo di Dipinti Sacri Ponte dei Greci, Venezia 28194
Piccolo Museo di San Paolo, Reggio Calabria .. 27385

Macedonia

Galerija na Ikoni, Skopje 30517
Muzej na Makedonija - Archeološki, Etnološki i Istoriski, Skopje 30519

Mexico

Museo Iconográfico del Quijote, Guanajuato .. 30920

Netherlands

Odigia Ikonen-Museum, Den Haag 32097

Northern Cyprus

Archangelos Mihail İkon Müzesi, Girne 33355
İskele İon Müzesi, İskele 33363
Maraş Ikon Müzesi, Gazimağusa 33351
Saint Barnabas Icon and Archaeological Museum, Gazimağusa 33354
Saint Mamas Manastırı İkon Müzesi, Güzelyurt 33362

Romania

Colecţia Muzeală a Mănăstirii Agapia, Agapia 36099
Muzeul de Artă, Iaşi 36417

Russia

Aleksandrovskaja Sloboda - Gosudarstvennyj Istoriko- architekturnyj i Chudožestvennyj Muzej-zapovednik, Aleksandrov 36751
Belgorodskij Gosudarstvennyj Chudožestvennyj Muzej, Belgorod 36793
Bereznikovskij Istoriko-kraevedčeskij Muzej, Berezniki 36803
Centralnyj Muzej Drevnerusskoj Kultury i Isskustva im. Andreja Rubleva, Moskva 37146
Chudožestvenno-memorialnyj Muzej K.S. Petrova-Vodkina, Chvalynsk 36864
Chudožestvenno-vystavočnyj Kompleks, Sarapul 37744
Čitinskij Oblastnoj Chudožestvennyj Muzej, Čita 36868
Gosudarstvennaja Tretjakovskaja Galerja, Moskva 37184
Gosudarstvennyj Muzej Palechskogo Iskusstva, Palech 37526
Gosudarstvennyj Russkij Muzej, Sankt-Peterburg 37649
Gosudarstvennyj Vladimiro-Suzdalskij Istoriko-architekturnyj i Chudožestvennyj Muzej-zapovednik, Vladimir 37934
Irkutskij Oblastnoj Chudožestvennyj Muzej im. V.P. Sukačeva, Irkutsk 36932
Jaroslavskij Chudožestvennyj Muzej - Gubernatorskij Dom, Jaroslavl 36967

Kargopolskij Gosudarstvennyj Istoriko-Architekturnyj i Chudožestvennyj Muzej, Kargopol 37003
Kirillo-Belozerskij Istoriko-Architekturnyj i Chudožestvennyj Muzej-Zapovednik, Kirillov 37034
Kostromskoj Gosudarstvennyj Istoriko-architekturnyj Muzej-zapovednik, Kostroma 37061
Kostromskoj Gosudarstvennyj Obedinennyj Chudožestvennyj Muzej, Kostroma 37062
Krasnojarskij Chudožestvennyj Muzej im. V.I. Surikova, Krasnojarsk 37069
Mitropoličji Palaty - Muzej Drevnerusskogo Iskusstva, Jaroslavl 36971
Muzej Istorii i Kultury Srednego Prikamja, Sarapul 37746
Muzej Izobrazitelnych Iskusstv Karelii, Petrozavodsk 37561
Muzej Tulskij Kreml, Tula 37871
Novokuzneckij Chudožestvennyj Muzej, Novokuzneck 37481
Pereslavl-Zalesskij Gosudarstvennyj Istoriko-architekturnyj i Chudožestvennyj Muzej-zapovednik, Pereslavl-Zalesskij 37542
Permskaja Gosudarstvennaja Chudožestvennaja Galereja, Perm 37550
Pokrovskij Sobor-chram Vasilija Blažennogo, Moskva 37400
Rostovskij Kreml - Gosudarstvennyj Muzej-zapovednik, Rostov 37599
Samarskij Eparchialnyj Cerkovno-istoričeskij Muzej, Samara 37620
Sergievo-Posadskij Gosudarstvennyj Istoriko-chudožestvennyj Muzej Zapovednik, Sergiev Posad 37764
Solikamskij Gorodskoj Kraevedčeskij Muzej, Solikamsk 37785
Staročerkasskij Istoriko-architekturnyj Muzej-zapovednik, Staročerkassk 37798
Staroladožskij Istoriko-Architekturnyj i Archeologičeskij Muzej-Zapovednik, Staraja Ladoga 37795
Stavropolskij Muzej Izobrazitelnych Iskusstv, Stavropol 37806
Svijažskij Architekturno-chudožestvennyj Muzej, Svijažsk 37812
Tjumenskij Oblastnoj Muzej Izobrazitelnych Iskusstv, Tjumen 37853
Tomskij Oblastnoj Chudožestvennyij Muzej, Tomsk 37858
Udmurtskij Respublikanskij Muzej Izobrazitelnych Iskusstv, Iževsk 36956
Uspenskij Sobor, Moskva 37410

Serbia

Muzej Srpske Prvoslavne Crkve, Beograd 38078

Ukraine

Municypal'nyj Muzej Pryvatnych Kollekcij im. O. Bleščunova, Odesa 43156
Muzej Volyns'koj Ikony, Luc'k 43112
Nacional'nyj Chudožnij Muzej Ukraïny, Kyïv 43100
Nacional'nyj Kyjvo-Pečerskyj Istoryko-kul'turnyj Zapovidnyk, Kyïv 43101

United States of America

Harris Art Gallery, James M. Lykes Maritime Gallery and Hutchings Gallery, Galveston 48846
Jordan Schnitzer Museum of Art at the University of Oregon, Eugene 48431

Incunabula → Books, Book Art and Manuscripts

Indian Artifacts

Australia

Museum of Tropical Queensland, Townsville 01598
Tiagarra Aboriginal Museum, Devonport 01037

Brazil

Museu de Índio, Uberlândia 04942
Museu do Índio, Lagoa Seca 04529
Museu do Índio, Manaus 04555

Canada

Alberni Valley Museum, Port Alberni 06624
Algonquin Culture and Heritage Centre, Golden Lake 05881
Beothuk Interpretation Centre, Boyd's Cove 05451
Broadview Historical Museum, Broadview 05484
Buffalo Nations Luxton Museum, Banff 05381
Campbell River Museum, Campbell River 05539
Clayton McLain Memorial Museum, Cut Knife 05669
Colville Lake Museum, Colville Lake 05633
Dufferin Historical Museum, Carman 05566
Ear Falls District Museum, Ear Falls 05729
Eskimo Museum, Churchill 05614
Fort Frances Museum, Fort Frances 05809
Fort Nelson Heritage Museum, Fort Nelson 05818
Fort Qu'appelle Museum, Fort Qu'appelle 05819
Fort Témiscamingue, Ville-Marie 07265
Fort Whoop-Up Centre, Lethbridge 06161
Fraser-Fort George Regional Museum, Prince George 06666
George Johnston Tlingit Indian Museum, Teslin 07069
Grey Roots Heritage, Owen Sound 06561
High Prairie and District Museum, High Prairie 05985
Historic Hat Creek Ranch, Cache Creek 05510
Historical Museum of Saint James-Assiniboia, Winnipeg 07363
Hudson Bay Museum, Hudson Bay 05999
Irving House Historic Centre and New Westminster Museum, New Westminster 06436
Joseph Brant Museum, Burlington 05500
Lake of the Woods Museum, Kenora 06057
Lillooet District Museum, Lillooet 06170
McGillis Pioneer Home, Saint Victor 06883
Miniota Municipal Museum, Miniota 06297
Moncur Gallery, Boissevain 05435
Musée Amérindien de Mashteuiatsh, Mashteuiatsh 06245
Musée Amérindien et Inuit de Godbout, Godbout 05878
Le Musée Historique de Tracadie, Tracadie-Sheila 07158
Nakusp District Museum, Nakusp 06408
Nk'Mip Desert Heritage Centre, Osoyoos 06526
Porcupine Plain and District Museum, Porcupine Plain 06623
Port-au-Choix, Port-au-Choix 06628
Riverhurst F.T. Hill Museum, Riverhurst 06757
Ronathahon:ni Cultural Centre, Cornwall 05644
Salmon Arm Museum, Salmon Arm 06905
Sam Waller Museum, The Pas 07072
Strasbourg and District Museum, Strasbourg 07018
Sulphur Springs Station, Ancaster 05339
Thunder Bay Museum, Thunder Bay 07082
Tofield Historical Museum, Tofield 07087
Windermere Valley Museum, Invermere 06016

Malaysia

Orang Asli Museum, Melaka 30606

Sweden

Dalarnas Indianmuseum, Dala Husby 40482

Switzerland

Indian Land Museum, Gossau (Zürich) 41445
Nordamerika Native Museum-NONAM, Zürich 42099

United States of America

Adair County Historical Society Museum, Kirksville 49799
Adams County Historical Society Museum, Gettysburg 48900
Agua Caliente Cultural Museum, Palm Springs 51680
Ah-Tah-Thi-Ki Museum, Big Cypress 46794
Akwesasne Museum, Hogansburg 49308
Amador County Museum, Jackson 49555
Amory Regional Museum, Amory 46275
Andrew J. Blackbird Museum, Harbor Springs 49164
Angel Mounds, Evansville 48455
Antelope County Historical Museum, Neligh 51000
Antelope Valley Indian Museum, Lancaster 49914
Appalachian Center for Crafts, Smithville 53135
Ataloa Lodge Museum, Muskogee 50927
Audrain Historical Museum, Graceland, Mexico 50618
Big Horn County Historical Museum, Hardin 49165
Black Hawk State Historic Site, Rock Island 52406
Blackwater Draw Museum, Portales 52053
Buffalo Bill Historical Center, Cody 47671
Buffalo Bill Museum of Le Claire, Iowa, Le Claire 50001
Bushwhacker Museum, Nevada 51006
The California Indian Museum, Santa Rosa 52903
California State Indian Museum, Sacramento 52502
Campbell County Rockpile Museum, Gillette 48909
Carson Valley Museum, Gardnerville 48863
Cass County Museum, Walker 53795
Chemung Valley History Museum, Elmira 48379
Cherokee Courthouse Museum, Gore 48974
Cherokee National Museum, Tahlequah 53417
Chickasaw Council House Museum, Tishomingo 53512
Chief Plenty Coups Museum, Pryor 52190
Chimneyville Crafts Gallery, Jackson 49557
Coachella Valley Museum, Indio 49508
Colesville and Windsor Museum at Saint Luke's Church, Harpursville 49176
Collingwood Library and Museum on Americanism, Alexandria 46202
Colorado River Indian Tribes Museum, Parker 51712
Colter Bay Indian Arts Museum, Colter Bay Village 47709
Coronado-Quivira Museum, Lyons 50335
Creek Council House Museum, Okmulgee 51528
Devils Tower Visitor Center, Devils Tower 48132
Douglas County Historical Society Museum, Superior 53377
Edwards County Historical Museum, Kinsley 49797
Effigy Mounds Museum, Harpers Ferry 49173
Elizabeth D. Walters Library, Stroudsburg 53339
Favell Museum of Western Art and Indian Artifacts, Klamath Falls 49806
First Territorial Capitol of Kansas, Fort Riley 48675
Five Civilized Tribes Museum, Muskogee 50928
Flathead Indian Museum, Saint Ignatius 52551
Fort Harker Museum Complex, Kanopolis 49678
Fort Hays State Historic Site, Hays 49223
Fort Phil Kearny, Story 53321
Fort Plain Museum, Fort Plain 48670
Fort Recovery Museum, Fort Recovery 48674
Frisco Native American Museum and Natural History Center, Frisco 48798
Gadsden Museum, Mesilla 50610
Grant County Museum, Ulysses 53641
H. Earl Clack Museum, Havre 49218
The Haskell County Historical Museum, Sublette 53352
Hicksville Gregory Museum, Hicksville 49270
High Plains Museum, Goodland 48971
Historic Hermann Museum, Hermann 49261
Hopewell Culture National Historic Park, Chillicothe 47534
Indian Museum of Lake County-Ohio, Painesville 51675
Indian Museum of the Carolinas, Laurinburg 49988
Indian Pueblo Cultural Center, Albuquerque 46178
Iroquois Indian Museum, Howes Cave 49408
Isabel Anderson Comer Museum and Arts Center, Sylacauga 53394
J.M. Davis Arms and Historical Museum, Claremore 47578
Jeff Matthews Memorial Museum, Galax 48825
Jeffers Petroglyphs Historic Site, Comfrey 47762
Ka-Do-Ha Indian Village Museum, Murfreesboro 50911
Kateri Galleries, Auriesville 46487
Kit Carson Museum, Las Animas 49956
Knight Museum, Alliance 46230
Lake Perris Regional Indian Museum, Perris 51777
Last Indian Raid Museum, Oberlin 51477
Lewis and Clark Boat House and Nature Center, Saint Charles 52534
Liberty County Museum, Chester 47430
MacCallum More Museum, Chase City 47403
Marias Museum of History and Art, Shelby 53066
Marin Museum of the American Indian, Novato 51446
Mashantucket Pequot Museum, Mashantucket 50527
Meade County Historical Society Museum, Meade 50544
Mid-America All-Indian Center Museum, Wichita 54146
Mississippi County Historical Society, Charleston 47355
Mitchell Museum of the American Indian, Evanston 48454
Montville Township Historical Museum, Montville 50832
Moundbuilders State Memorial and Museum, Newark 51291
Mount Rainier National Park Museum, Ashford 46378
Mountain Heritage Center, Cullowhee 47903
Mountain Life Museum, London 50168
Murrell Home, Park Hill 51708
Museum of the Aleutians, Unalaska 53642
Museum of the Cherokee Indian, Cherokee 47423
Museum of the Mountain Man, Pinedale 51925
Museum of the Plains Indian, Browning 47092
Museum of the Red River, Idabel 49461
Musselshell Valley Historical Museum, Roundup 52473
National Hall of Fame for Famous American Indians, Anadarko 46281
National Museum of the American Indian, New York 51236
Native American Exhibit, Fonda 48584
Native American Heritage Museum at Highland Mission, Highland 49277
Native American Resource Center, Pembroke 51755
Navajo Nation Museum, Window Rock 54225
New London Public Museum, New London 51070
Nohwike' Bagowa, Fort Apache 48589
North Platte Valley Museum, Gering 48896
Oconto County Historical Society Museum, Oconto 51494
O'Fallon Historical Museum, Baker 46533
Old Court House Museum-Eva Whitaker Davis Memorial, Vicksburg 53734
Old Log Jail and Chapel Museums, Greenfield 49055
Old Trail Museum, Choteau 47546
Olde Colonial Courthouse, Barnstable 46613
Peter Yegen Jr. Yellowstone County Museum, Billings 46806
Putney Historical Society, Putney 52203
Rawls Museum Arts, Courtland 47856
Rock Island County Historical Museum, Moline 50779
Roniger Memorial Museum, Cottonwood Falls 47845
Saint Charles County Historical Society, Saint Charles 52535
Sainte Genevieve Museum, Sainte Genevieve 52651
Salem County Historical Society Museum, Salem 52662
School of Nations Museum, Elsah 48381
Seminole Nation Museum, Wewoka 54114
Seneca-Iroquois National Museum, Salamanca 52654
Serpent Mound Museum, Peebles 51747
Sheldon Museum, Haines 49125
Six Nations Indian Museum, Onchiota 51558
Smith's Fort Plantation, Surry 53383
South Lake Tahoe Museum, South Lake Tahoe 53175
Southern Plains Indian Museum, Anadarko 46282
Southold Indian Museum, Southold 53189
Spokane House Interpretive Center, Nine Mile Falls 51353
Starved Rock State Park, Utica 53674
Stewart Indian Cultural Center, Carson City 47262
Sussex County Historical Society Museum, Newton 51343
Tamastslikt Cultural Institute, Pendleton 51756
Tantaquidgeon Indian Museum, Uncasville 53643
Town Creek Indian Mound Historic Site, Mount Gilead 50877
Trailside Nature and Science Center, Mountainside 50897
Tribal Museum, Grand Ronde 49011
Tularosa Basin Historical Society Museum, Alamogordo 46152
Ute Indian Museum, Montrose 50830
Van Meter State Park, Miami 50638
Van Riper-Hopper House, Wayne 54009
Vermont Historical Society Museum, Montpelier 50825
Vista Historical Museum, Vista 53765
Wabaunsee County Historical Museum, Alma 46235
Wichita-Sedgwick County Historical Museum Association, Wichita 54152
Wolf Point Area Historical Society, Wolf Point 54267
Woodruff Museum of Indian Artifacts, Bridgeton 47021
Woolaroc Museum, Bartlesville 46620

Industry

Australia

Historical Woolscour Blackall, Blackall 00872
Kandos Bicentennial Industrial Museum, Kandos 01186
Museum of Transportation and Rural Industries, Boyanup 00886

Austria

Montanmuseum, Hieflau 02077
Museum der Stadt Kapfenberg, Kapfenberg 02151
Nadelburgmuseum, Lichtenwörth 02298
Ternitzer Stahlstadtmuseum, Ternitz 02805

Belgium

Musée Bruxellois de l'Industrie et du Travail, Bruxelles 03455
Musée du Coticule, Vielsalm 04112
Site du Bois du Cazier, Marcinelle 03852

Canada

British Columbia Orchard Industry Museum, Kelowna 06050
Campbell Carriage Factory Museum, Middle Sackville 06275
Écomusée du Fier Monde, Montréal 06338
Industrial Heritage Complex - Merrickville Lockstation, Merrickville 06268
Irish Moss Interpretive Centre and Museum, Miminegash 06292
John Walter Museum, Edmonton 05747
Musée de la Gare, Temiscaming 07065
Musée du Centre Acadien de l'Université Sainte-Anne, Pointe de l'Eglise 06619
Musée J. Armand Bombardier, Valcourt 07196
Museum of Industry, Stellarton 07008
Shand House Museum, Ferry Hill 05796

I

Cuba

Museo de la Industria Azucarera, Abel Santamaría ... 09020

Denmark

Brede Værk, Kongens Lyngby ... 10165
Cathrinesminde Teglværk, Broager ... 09983

Finland

Askon Museo, Lahti ... 11025
Parainen Industrial Museum, Parainen ... 11188
Rosenlew-museo, Pori ... 11222

France

Atelier des Produits Résineux Vidal, Luxey ... 13874
Cité des Sciences et de l'Industrie, Paris ... 14582
Ecomusée du Pays de Rennes, Rennes (Ille-et-Vilaine) ... 15012
Musée Campanaire, L'Isle-Jourdain ... 13782
Musée d'Art et d'Industrie, Saint-Étienne ... 15251
Musée de la Charronnerie, Brienne-la-Vieille ... 12176
Musée de la Contrefaçon, Paris ... 14651
Musée de la Houille Blanche et de ses Industries, Lancey ... 13496
Musée de l'Aventure Industrielle du Pays d'Apt, Apt ... 11654
Musée des Industries du Viemeu, Friville-Escarbotin ... 13013
Musée du Papier Le Nil, Angoulême ... 11621
Musée Régional de l'Alsace Bossue, Sarre-Union ... 15642

Germany

Bayerisches Moor- und Torfmuseum, Grassau 18739
Bergbaumuseum, Deuben ... 17843
Deutsches Schuhmuseum Hauenstein, Hauenstein ... 19045
Fellenbergmühle Feinmechanisches Museum, Merzig ... 20300
Förderkreis Industrie- und Technikgeschichte, Frankfurt am Main ... 18366
Heimat- und Industriemuseum Kolbermoor, Kolbermoor ... 19674
Heimat- und Industriemuseum Wackersdorf, Wackersdorf ... 22261
Historisches Zentrum, Wuppertal ... 22741
Industrie- und Filmmuseum Wolfen, Wolfen ... 22650
Industriemuseum Elmshorn, Elmshorn ... 18126
Industriemuseum Hartha, Hartha ... 19019
Industriemuseum Lauf, Lauf an der Pegnitz ... 19851
Industriemuseum Lohne, Lohne ... 20026
Johann Wülfing Sohn Museum, Radevormwald ... 21165
LVR-Industriemuseum, Engelskirchen ... 18171
LVR-Industriemuseum, Rheinisches Landesmuseum für Industrie- und Sozialgeschichte, Oberhausen ... 20809
LWL-Industriemuseum, Glashütte Gernheim, Petershagen ... 21016
LWL-Industriemuseum, Henrichshütte Hattingen, Hattingen ... 19038
LWL-Industriemuseum, Schiffshebewerk Henrichenburg, Waltrop ... 22319
LWL-Industriemuseum, Textilmuseum in Bocholt, Bocholt ... 17395
LWL-Industriemuseum, Westfälisches Landesmuseum für Industriekultur, Dortmund 17921
LWL-Industriemuseum, Zeche Hannover, Bochum ... 17403
LWL-Industriemuseum, Zeche Nachtigall, Witten ... 22618
LWL-Industriemuseum, Ziegeleimuseum in Lage, Lage ... 19764
Museum der Arbeit, Hann. Münden ... 18987
Museum "Werkstatt Plaumer", Öhringen ... 20866
Neugablonzer Industrie- und Schmuckmuseum im Isergebirgs-Museum, Kaufbeuren ... 19490
Nordwolle Delmenhorst, Delmenhorst ... 17816
Sammlung industrielle Gestaltung, Berlin ... 17243
Stadt- und Dampfmaschinenmuseum, Werdau ... 22485
Stadt- und Festungsmuseum im Ludwigstor, Germersheim ... 18605
Stadt- und Industriemuseum, Wetzlar ... 22533
Stadtmuseum Povelturm, Nordhorn ... 20743
Tabakspeicher, Nordhausen ... 20740
tim - Bayerisches Textil- und Industriemuseum, Augsburg ... 16638
Weltkulturerbe Völklinger Hütte, Völklingen ... 22249
Zeche Zollern, Dortmund ... 17930
Zeittunnel Wülfrath, Wülfrath ... 22704
Ziegel- und Kalkmuseum, Winzer ... 22609
Ziegelei Benzin, Kritzow ... 19716
Ziegeleipark Mildenberg, Zehdenick ... 22772
Zinkhütter Hof, Stolberg, Rheinland ... 21910

Hungary

Húsipari Múzeum, Budapest ... 23415
Magyar Alumíniumipari Múzeum, Székesfehérvár ... 23856

Italy

Ecomuseo dell'Archeologia Industriale, Schio 27792
Ecomuseo di Archeologia Industriale E. Crumi, Villar Pellice ... 28280
Museo del Patrimonio Industriale, Bologna ... 25452
Museo del Tessile e della Tradizione Industriale, Busto Arsizio ... 25573
Museo della Plastica, Pont Canavese ... 27281
Museo dell'Arte e della Tecnologia Confetteria, Sulmona ... 27919
Museo delle Ferrovie, Monserrato ... 26788
Museo delle Ferrovie in Sardegna, Cagliari ... 25584
Museo dell'Ingegno e della Tecnologia Preindustriale, Colorno ... 25944
Museo Laboratorio del Tessile, Soveria Mannelli ... 27883

Japan

Chiba-kenritsu Indasutori oyo Kagakukan, Ichikawa ... 28553
Hiroshima City Museum of History and Traditional Crafts, Hiroshima ... 28535
Kouseiroudoushou Sangyo Anzen Gijutsukan, Tokyo ... 29422
Mitsubishi Minatomirai Indasutoriaru Hakubutsukan, Yokohama ... 29601
Museum of the Mitsubishi Heavy Industries, Nagasaki ... 28977
Nagaoka-shiritsu Kogyo Hakurankai, Nagaoka ... 28962
Omuta-shi Tanko oyo Kagakukan, Omuta ... 29117
Toyota Commemorative Museum of Industry and Technology, Nagoya ... 29013

Mexico

Museo Industrial de Santa Catarina El Blanqueo, Santa Catarina ... 31376
Museo Textil La Trinidad, Santa Cruz Tlaxcala ... 31377

Netherlands

Ons Museum, Giessenburg ... 32249
Techniekmuseum Heim, Hengelo, Overijssel ... 32368
Verfmolen de Kat, Zaandam ... 32983

Norway

Birkenes Bygdemuseum, Birkeland ... 33437
Hydro Industripark-Museum, Porsgrunn ... 33769
Kistefos-museet, Jevnaker ... 33582
Klevfos Industrimuseum, Ådalsbruk ... 33375
Norsk Fiskeindustrimuseum, Melbu ... 33663
Norsk Hermetikkmuseum, Stavanger ... 33870
Norsk Vasskraft- og Industristadmuseum, Tyssedal ... 33944

Pakistan

Mughal Museum, Lahore ... 34024

Philippines

The Coca-Cola Pavillion, Santa Cruz ... 34500

Poland

Muzeum Przemysłu, Warszawa ... 35268
Muzeum Przemysłu i Techniki, Wałbrzych ... 35212

Portugal

Museu da Cerâmica de Sacavém, Sacavém ... 35893
Museu da Ciência e Indústria, Porto ... 35844
Museu da Cortiça da Fábrica do Inglês, Silves 35939
Museu da Fábrica de Cimentos Lis, Maceira ... 35746
Museu da Indústria Baleeira, São Roque do Pico ... 35925
Museu da Indústria da Cortiça, Santa Maria da Feira ... 35903
Museu da Indústria Têxtil, Calendário ... 35498
Museu de Etnografia e Arqueologia Industrial, Torres Novas ... 35968
Museu do Barreiro e Museu Industrial da Quimiparque, Barreiro ... 35457
Museu do Vidro, Marinha Grande ... 35756
Núcleo Museológico da Nacional, Lisboa ... 35729

Réunion

Muséum Agricole et Industriel de Stella Matutina, Piton-Saint-Leu ... 36090

Russia

Kemerovskij Istoriko-architekturnyi Muzej Krasnaja Gorka, Kemerovo ... 37030
Muzej Istorii Architektury i Promyšlennoj Techniki Urala, Ekaterinburg ... 36892
Muzej Kronštadtskogo Morskogo Zavoda, Kronštadt ... 37087
Sankt-Peterburgskij Muzej Chleba, Sankt-Peterburg ... 37724

Slovakia

Plynárenské Múzeum, Bratislava ... 38231

Spain

Museo Cemento Rezola, Donostia-San Sebastián ... 39251
Museo Etnográfico y de Arqueología Industrial, Chillarón de Cuenca ... 39174
Museo Preindustrial de la Caña de Azúcar, Motril ... 39687
Museu de la Rajoleria de Paiporta, Paiporta ... 39766

Sweden

Arbetets Museum, Norrköping ... 40769
Åssamuséet, Åtvidaberg ... 40446
Bofors Industrimuseum, Karlskoga ... 40629
Forsviks Bruk, Forsvik ... 40526
Gåröströms Industrimuseum, Gnosjö ... 40539
Gislaveds Industrimuseum, Gislaved ... 40535
Huseby Bruk, Grimslöv ... 40567
Husqvarna Fabriksmuseum, Huskvarna ... 40608
Hylténs Industrimuseum, Gnosjö ... 40540
Ifőverkens Industrimuseum, Bromölla ... 40479
Industrimuséet C.W. Thorstensons Mekaniska Verkstad, Åmål ... 40419
Industrimuseet Tollereds Öfvre Kraftstation, Tollered ... 40973
Järnvägs- och Industrimuseum, Hagfors ... 40575
Mariestads Industrimuseum, Mariestad ... 40737
Nobelmuseet och Bofors Industrimuseum, Karlskoga ... 40632
Norrahammars Industrimuseum, Norrahammar ... 40768
Repslagarmuseet, Älvängen ... 40405
Stickmaskinsmuseum, Glemminebro ... 40537
Töllstorps Industrimuseum, Gnosjö ... 40541

Switzerland

Gewerbemuseum, Winterthur ... 42027
Museum im Greuterhof, Islikon ... 41495

Taiwan

Sisal Industry Historical Exhibition, Hengchun 42201
Sun-Kong Factory Museum, Su-Ao ... 42345

Turkey

Rahmi M. Koç Müzesi, İstanbul ... 42834

United Kingdom

Abbeydale Industrial Hamlet, Sheffield ... 45648
Amberley Museum and Heritage Centre, Amberley ... 43297
Bersham Heritage Centre and Ironworks, Wrexham ... 46081
Black Country Living Museum, Dudley ... 44029
Bradford Industrial Museum and Horses at work, Bradford ... 43576
Bridewell Museum, Norwich ... 45275
Bristol's Industrial Museum, Bristol ... 43633
China Clay Country Park Mining and Heritage Centre, Saint Austell ... 45560
Corfe Castle Museum, Corfe Castle ... 43892
Dalbeattie Museum, Dalbeattie ... 43944
Engine House Project, Chelmsford ... 43804
Etruria Industrial Museum, Stoke-on-Trent ... 45759
Frome Museum, Frome ... 44257
Green Dragon Museum, Stockton-on-Tees ... 45756
Industrial Museum, Nottingham ... 45293
Kelham Island Museum, Sheffield ... 45652
Kidwelly Industrial Museum, Kidwelly ... 44574
Langton Matravers Museum, Langton Matravers ... 44640
Leeds Industrial Museum, Leeds ... 44662
Llywernog Silver-Lead Mine Museum, Ponterwyd ... 45399
Long Warehouse, Coalbrookdale ... 43859
Museum of Dartmoor Life, Okehampton ... 45312
Museum of Local History, Burnley ... 43677
Museum of Science and Industry in Manchester, Manchester ... 45092
Museum Workshop and Stores, Grangemouth 44343
Riverside Museum at Blake's Lock, Reading ... 45471
Silk Mill - Derby's Museum of Industry and History, Derby ... 43970
Somerset Brick and Tile Museum, Bridgwater 43607
Wandle Industrial Museum, Mitcham ... 45160

United States of America

Alexander and Baldwin Sugar Museum, Puunene ... 52205
American Precision Museum, Windsor ... 54228
American Textile History Museum, Lowell ... 50303
Antique Gas and Steam Engine Museum, Vista ... 53764
Arkansas Museum of Natural Resources, Smackover ... 53132
Attleboro Area Industrial Museum, Attleboro ... 46462
Baltimore Museum of Industry, Baltimore ... 46552
Bisbee Mill Museum, Chesterfield ... 47435
Charles River Museum of Industry, Waltham ... 53808
Colonial Industrial Quarter, Bethlehem ... 46778
Cornwall Iron Furnace, Cornwall ... 47814
Cornwall Iron Furnace, Cornwall ... 47814
Crane Museum of Papermaking, Dalton ... 47949
Crystal Lake Falls Historical Association, Barton ... 46621
Dr. Pepper Museum and Free Enterprise Institute, Waco ... 53773
Drake Well Museum, Titusville ... 53516
East Ely Railroad Depot Museum, Ely ... 48384
Fossil Station Museum, Russell ... 52484
Frye's Measure Mill, Wilton ... 54215
Grand Portage National Monument, Grand Portage ... 48999
Hamburg State Park Museum, Mitchell ... 50759
Heddon Museum, Dowagiac ... 48172
Herschell Carrousel Factory Museum, North Tonawanda ... 51406
Historic Madison House, Madison ... 50378
Hopewell Furnace Hopewell Furnace National Historic Site, Elverson ... 48383
Industrial and Agricultural Museum, York ... 54341
The Invention Factory, Trenton ... 53559
Lightner Museum, Saint Augustine ... 52522
Liquid Paper Correction Fluid Museum, Santa Fe ... 52878
Maramec Museum, Saint James ... 52552
Maritime and Seafood Industry Museum, Biloxi ... 46811
Mifflinburg Buggy Museum, Mifflinburg ... 50679
Mission Mill Museum, Salem ... 52669
Museum of Alaska Transportation and Industry, Wasilla ... 53954
Museum of History and Industry, Seattle ... 53005
National Museum of Industrial History, Bethlehem ... 46782
Racine Heritage Museum, Racine ... 52224
Rough and Tumble Engineers Museum, Kinzers ... 49798
RV/MH Heritage Foundation, Elkhart ... 48349
Scotia Museum, Scotia ... 52969
Slater Mill Historic Site, Pawtucket ... 51738
Sloss Furnaces, Birmingham ... 46824
Texas City Museum, Texas City ... 53490
Venango Museum of Art, Science and Industry, Oil City ... 51509
Watermen's Museum, Yorktown ... 54346
Watkins Woolen Mill, Lawson ... 49998
Windham Textile and History Museum, Willimantic ... 54179

Inscriptions and Epitaphs

Austria

Museumsfriedhof Tirol, Kramsach ... 02222

France

Musée Lapidaire, Carpentras ... 12297

Germany

Studentenkarzer, Heidelberg ... 19084

Greece

Epigraphical Museum, Athinai ... 22908

India

Rajputana Museum, Ajmer ... 24055

Italy

Antiquarium, Fondi ... 26218
Antiquarium del Teatro Greco-Romano, Taormina ... 27940
Museo Civico Archeologico, Camerino ... 25617
Museo di Santo Stefano, Bologna ... 25466
Museo R. Campelli, Pievebovigliana ... 27218

Philippines

Divine Word University Museum, Tacloban ... 34509

Portugal

Museu Epigráfico Egitanense, Idanha-a-Nova ... 35646

Turkey

Medrese Müzesi, Akşehir 42684

Vatican City

Galleria Lapidaria, Città del Vaticano 54500
Museo Pio Cristiano, Città del Vaticano 54509

Insects

Argentina

Museo Entomológco Mariposas del Mundo, San Miguel 00594

Australia

Carnaby Collection of Beetles and Butterflies, Boyup Brook 00887

Austria

Naturmuseum, Neuberg an der Mürz 02425

Belgium

Bosmuseum Pijnven, Hechtel 03689
Natuurpunt Museum, Turnhout 04100

Bulgaria

Prirodonaučen Muzej, Černi Osăm 05002

Canada

Insectarium de Montréal, Montréal 06341
J.B. Wallis Museum of Entomology, Winnipeg 07365
Maison des Insectes, Charlesbourg 05587
Spencer Entomological Museum, Vancouver 07223
Stones'n Bones Museum, Sarnia 06908

Chile

Museo Entomológico Luis E. Peña Guzman, Santiago de Chile 07459

Costa Rica

Museo de Insectos, San José 08817

Czech Republic

Entomologické Oddělení, Praha 09744

France

La Serre aux Papillons, La Queue-lèz-Yvelines 13404
Musée d'Entomologique, Chaveyriat 12491
Musée des Papillons, Fuveau 13021
Musée des Papillons, Saint-Chély-d'Apcher 15203
Musée Entomologique, Saint-Quentin (Aisne) 15475

Germany

Naturkundemuseum Erfurt, Erfurt 18209
Staatliches Naturhistorisches Museum, Braunschweig 17529

Hungary

Mátra Múzeum, Gyöngyös 23574
Újpesti Lepkemúzeum Alapítvány, Budapest 23488

Israel

Sharon Museum Emek Hefer, Midreshet Ruppin 25066

Italy

Collezioni del Dipartimento di Entomologia e Zoologia Agraria, Portici 27300
Museo Civico di Storia Naturale, Carmagnola 25679
Museo degli Insetti, Bisegna 25426

Japan

Osaka-kenritsu Chu Hakabutsukan, Mino 28910

Jordan

Insects Museum, Amman 29650

Lithuania

Kauno Tado Ivanausko Zoologijos Muziejus, Kaunas 30252

Macedonia

Prirodonaučen Muzej na Makedonija, Skopje 30523

Nicaragua

Museo Entomológico, León 33286

Oman

Oman Natural History Museum, Muscat 33991

Peru

Museo Entomológico, Lima 34258

Philippines

Jumalon Museum and Art Gallery, Cebu 34371

Poland

Muzeum Regionalne, Lubartów 34895

Romania

Muzeul Aristide Caradja, Grumăzeşti 36391
Muzeul Brăilei-Secţia Ştiinţele Naturii, Brăila 36172

Russia

Zoologičeskij Muzej im. N.M.Kulagina, Moskva 37430

Slovenia

Prirodoslovni Muzej Slovenije, Ljubljana 38461

South Africa

Queenstown and Frontier Museum, Queenstown 38720

Switzerland

Tierwelt-Panorama, Ebikon 41337

Taiwan

Insect Collections at the Entomology Department, Taipei 42395
Muh Sheng Museum of Entomology, Puli 42295
Wood Carving Museum and Taiwan Sericultural Museum, Zaociao 42488

Thailand

Forest Entomology Museum, Bangkok 42539

United Kingdom

Booth Museum of Natural History, Brighton 43619
National Dragonfly Biomuseum, Ashton, Oundle 43330
Oxford University Museum of Natural History, Oxford 45336

United States of America

Entomology Research Museum, Riverside 52370
Essig Museum of Entomology, Berkeley 46753
The Frost Entomological Museum, University Park 53657
May Natural History Museum and Museum of Space Exploration, Colorado Springs 47703
Pember Museum of Natural History, Granville 49019
R.M. Bohart Museum of Entomology, Davis 47978
Rosenbruch Wildlife Museum, Saint George 52545
Trailside Nature Center and Museum, Cincinnati 47566
Trailside Nature Museum, Cross River 47891
University of North Dakota Zoology Museum, Grand Forks 48990

Instruments of Torture

Austria

Foltermuseum, Seeboden 02733

Belgium

Gravensteen Museum voor Gerechtsvoorwerpen, Gent 03627

Brazil

Museu Solar Monjardim, Vitória 04958

Czech Republic

Museum of Medieval Torture Instruments, Praha 09765

France

Tour des Voleurs, Riquewihr 15054

Germany

Burg Stolpen, Stolpen 21911
Foltermuseum, Burghausen 17652
Hexenmuseum, Ringelai 21336
Museum im Zwinger, Goslar 18713

Italy

Museo delle Armi, Armature e Strumenti di Tortura, Ischia Ponte 26412

Spain

Museu de l'Institut de Criminologia, Barcelona 38984

United Kingdom

Clink Prison Museum, London 44825

Ironwork → Metalwork

Ivory

Belgium

Trésor de la Cathédrale Notre-Dame, Tournai 04089

France

Musée de la Céramique et de l'Ivoire, Commercy 12595
Musée des Ivoires, Yvetot 16310

Germany

Deutsches Elfenbeinmuseum Erbach, Erbach, Odenwaldkreis 18190
Domschatz, Trier 22097
Elfenbein-Museum, Michelstadt 20318
Elfenbeinmuseum, Walldürn 22309

Iran

Golestan Palace Museum, Tehrān 24668

Italy

Civica Raccolta d'Arte Applicata, Milano 26675

Portugal

Museu do Scrimshaw, Horta 35644

Jade

China

Capital Museum, Beijing 07524

New Zealand

Left Bank Art Gallery, Greymouth 33114

Jewellery → Precious Stones

Judaica

Argentina

Museo Judío de Buenos Aires Dr. Salvador Kibrick, Buenos Aires 00249

Australia

Jewish Holocaust Centre, Elsternwick 01067
Jewish Museum of Australia, Saint Kilda, Victoria 01479
Sydney Jewish Museum, Darlinghurst 01026

Austria

Jüdisches Museum der Stadt Wien, Wien 03008
Jüdisches Museum Hohenems, Hohenems 02092
Museum Judenplatz Wien, Wien 03041
Österreichisches Jüdisches Museum, Eisenstadt 01862

Belgium

Musée Juif de Belgique-Joods Museum van België, Bruxelles 03490

Bosnia and Herzegovina

Muzej Jevreja Bosne i Hercegovine, Sarajevo 04278

Canada

Aron Museum, Westmount 07305
Baycrest Heritage Museum, Toronto 07094
Beth Tzedec Reuben and Helene Dennis Museum, Toronto 07095
Congregation Shaar Hashomayim Museum, Westmount 07306
Saint John Jewish Historical Museum, Saint John 06839

Czech Republic

Muzeum Židovské Komunity, Tovačov 09871
Židovské Muzeum v Praze, Praha 09801

France

Musée d'Art et d'Histoire du Judaïsme, Paris 14644
Musée d'Art Juif, Paris 14645
Musée d'Arts et Traditions Populaires, Marmoutier 13975
Musée Judéo-Alsacien, Bouxwiller 12135
Musée Juif Comtadin, Cavaillon 12331

Georgia

D. Baazov Georgian Jewish Historic and Ethnographic Museum, Tbilisi 16353

Germany

Aktives Museum Südwestfalen, Siegen 21753
Alte Synagoge, Essen 18247
Alte Synagoge, Hechingen 19060
August-Gottschalk-Haus, Esens 18238
August-Gottschalk-Haus, Esens 18239
Berend Lehmann Museum, Halberstadt 18874
Braunschweigisches Landesmuseum, Braunschweig 17518
Ehemalige Synagoge Erfelden, Riedstadt 21322
Ehemalige Synagoge mit Ausstellung Juden auf dem Lande, Ichenhausen 19306
Gedenkstätte Ehemalige Synagoge, Adelsheim 16428
Gedenkstätte Synagoge, Dornum 17908
Jüdische Gedenkstätte und ehemalige Synagoge, Wallhausen 22315
Jüdisches Kulturmuseum Augsburg-Schwaben, Augsburg 16624
Jüdisches Kulturzentrum, Fulda 18510
Jüdisches Museum, Frankfurt am Main 18379
Jüdisches Museum, Göppingen 18673
Jüdisches Museum, München 20448
Jüdisches Museum, Steinbach am Glan 21880
Jüdisches Museum Berlin, Berlin 17147
Jüdisches Museum der Nordpfalz, Winnweiler 22603
Jüdisches Museum Emmendingen, Emmendingen 18145
Jüdisches Museum Franken in Fürth, Fürth 18498
Jüdisches Museum Franken in Schnaittach, Schnaittach 21590
Jüdisches Museum im Raschi-Haus, Worms 22684
Jüdisches Museum Rendsburg, Rendsburg 21274
Jüdisches Museum Shalom Europa, Würzburg 22712
Jüdisches Museum Westfalen, Dorsten 17910
Kunstamt Tempelhof-Schöneberg, Berlin 17164
Landesrabbiner Dr. I.E. Lichtigfeld-Museum, Michelstadt 20319
Museum in der Präparandenschule, Höchberg 19208
Museum Judengasse, Frankfurt am Main 18388
Museum jüdischer Geschichte und Kultur, Aschaffenburg 16587
Museum Synagoge Gröbzig, Gröbzig 18768
Museum zur Geschichte der Juden in Kreis und Stadt Heilbronn, Obersulm 20843
Stiftung Neue Synagoge Berlin - Centrum Judaicum, Berlin 17265
Synagoge Saffig, Saffig 21467
Synagoge und Jüdisches Museum Ermreuth, Neunkirchen am Brand 20639
Umstädter Museum Gruberhof, Groß-Umstadt 18778

Greece

Evraiko Mouseio tis Ellados, Athinai 22910
Jewish Museum of Thessaloniki, Thessaloniki 23214

Hungary

Középkori O-Zsinagóga, Sopron 23811
Középkori Zsidó Imaház, Budapest 23425
Magyar Zsidó Múzeum és Levéltár, Budapest 23447

Ireland

Irish Jewish Museum, Dublin 24799

Israel

Babylonian Jewry Museum, Or Yehuda 25076
Bar David Museum, Kibbutz Bar Am 25033
Beit-Alfa Ancient Synagogue, Kibbutz Chefzi-bah 25034
Beit Hatefutsoth, Tel Aviv 25099
Capharnaum Ancient Synagogue, Kfar Nahum 25028
Central Archives for the History of the Jewish People, Jerusalem 24983
Institute for Jewish Studies Gallery, Jerusalem 24992
Isaak Kaplan Old Yishuv Coourt Museum, Jerusalem 24994
Israel Museum, Jerusalem, Jerusalem 24997
Kook Museum, Jerusalem 24998
Mishkan Le'Omanut, Ein Harod 24945
Museum of Ethnography and Folklore, Tel Aviv 25117
Schocken Institute of Jewish Research, Jerusalem 25014
Sholem Asch House, Bat Yam 24934
Sir Isaac and Lady Edith Wolfson Museum, Jerusalem 25018
Ticho House, Jerusalem 25020
The U. Nahon Museum of Italian Jewish Art, Jerusalem 25022

Italy

Museo della Comunità Ebraica, Venezia 28190
Museo Ebraico - Jüdisches Museum, Merano 26659

Latvia

Muzejs "Ebreji Latvijā", Rīga 30097

Mexico

Museo Histórico Judío y del Holocausto Tuvie Maizel, México 31121

Netherlands

Joods Historisch Museum, Amsterdam 31850
Synagoge, Bourtange 31993

Poland

Galicja Muzeum, Kraków 34781
Muzeum Dzieje i Kultura Żydów, Kraków 34790
Muzeum Okręgowe w Lesznie Dział Judaistyczny, Leszno 34851
Muzeum Sztuki Cmentarnej, Wrocław 35338
Muzeum Żydowskiego Instytutu Historycznego w Polsce, Warszawa 35285

Portugal

Museu Luso-Hebraico Abraham Zacut, Tomar 35965

Romania

Muzeul de Istorie a Evreilor din România, București 36203

Serbia

Jevrejski Istorijski Muzej, Beograd 38051

Slovakia

Múzeum Židovskej Kultúry, Bratislava 38230
Židovské Múzeum, Prešov 38338

South Africa

South African Jewish Museum, Gardens 38622

Spain

Museo Sefardí, Toledo 40136
Museu de l'Ebre, Tortosa 40165
Museu d'Història dels Jueui, Girona 39347

Suriname

Stichting Joden Savanna, Paramaribo 40400

Sweden

Judiska Museet i Stockholm, Stockholm 40902

Switzerland

Jüdisches Museum der Schweiz, Basel 41162

United Kingdom

Jewish Museum, London 44889
Jewish Museum Finchley, London 44890
Manchester Jewish Museum, Manchester 45089
Scottish Jewish Museum, Glasgow 44304

United States of America

Arizona Jewish Historical Exhibitions, Phoenix 51898
Associated American Jewish Museum, Pittsburg 51936
Benjamin and Dr. Edgar R. Cofeld Judaic Museum of Temple Beth Zion, Buffalo 47115
Bernard Judaica Museum, New York 51133
Beth Ahabah Museum and Archives, Richmond 52324
Center for Jewish History, New York 51143
Contemporary Jewish Museum, San Francisco 52771
The Ernest W. Michel Historical Judaica Collection, New York 51170
Hillel Jewish Student Center Gallery, Cincinnati 47561
Jane L. and Robert H. Weiner Judaic Museum, Rockville 52429
The Jewish Museum, New York 51200
Jewish Museum of Florida, Miami Beach 50642
Jewish Museum of Maryland, Baltimore 46571
Judah L. Magnes Museum, Berkeley 46755
The Judaica Museum, Bronx 47047
Judaica Museum of Central Synagogue, Irvington 49536
Lillian and Albert Small Jewish Museum, Washington 53887
Los Angeles Museum of the Holocaust, Los Angeles 50246
Museum of Jewish Heritage - A Living Memorial to the Holocaust, New York 51227
Museum of the Southern Jewish Experience, Jackson 49563
National Museum of American Jewish Military History, Washington 53902
Philadelphia Museum of Judaica, Philadelphia 51868
Sherwin Miller Museum of Jewish Art, Tulsa 53611
Sylvia Plotkin Judaica Museum, Scottsdale ... 52978
Temple Judea Museum, Elkins Park 48353
The Temple Museum of Religious Art, Cleveland 47642
William Breman Jewish Heritage Museum, Atlanta 46457
Zimmer Children's Museum, Los Angeles 50270

Keys → Locks and Keys

Lace → Needlework

Lacquerwork

France

Musée de La Glacerie, La Glacerie 13378
Musée des Beaux-Arts et de la Dentelle, Alençon 11545

Germany

Museum für Lackkunst, Münster 20516

Japan

Hachiga Minzoku Bijutsukan, Takayama 29336
Hida Takayama Shunkei Kaikan, Takayama ... 29340
Kaisendo Museum, Kaminoyama 28649

Mexico

Museo de Laca, Chiapa de Corzo 30762

Myanmar

Lacquerware Museum, Pagan 31726

Russia

Dom-muzej A.A. Dydykina, Palech 37523
Dom-muzej I.I. Golikova, Palech 37524
Dom-Muzej P.D. Korina, Palech 37525
Dom-muzej N.M. Zinovjeva, Djagilevo 36875
Gosudarstvennyj Muzej Palechskogo Iskusstva, Palech 37526

Spain

Museo y Centro Didáctico del Encaje de Castilla y León, Tordesillas 40148

Taiwan

Fengyuan Lacquer Art Museum, Fengyuan 42188
Taiwan Nature Lacquer Culture Museum, Taichung 42356

Lamps and Lighting

Australia

Firelight Museum, Leongatha 01228
Narooma Lighthouse Museum, Narooma 01349

Belgium

Lampenmuseum, Wezemaal 04151
Musée d'Histoire et de l'Éclairage, Liège 03811

Canada

Cape Bonavista Lighthouse, Bonavista 05438
Donalda and District Museum, Donalda 05707
Musée des Phares, La Martre 06112
Northern Lights Centre, Watson Lake 07291
Point Amour Lighthouse, L'Anse Amour 06138
West Point Lighthouse Museum, West Point ... 07302

Germany

Gaslaternen-Freilichtmuseum Berlin, Berlin .. 17109
Museum Gunnar-Wester-Haus, Schweinfurt ... 21686
Zentrum für Internationale Lichtkunst, Unna 22177

Iran

Aineh va Roshananal Museum, Yazd 24703

Ireland

Wexford County Museum, Enniscorthy 24832

Japan

Himezaki Lighthouse Museum, Ryotsu 29172
Kobe Lamp Museum, Kobe 28730
Kobe Lampwork Glass Museum, Kobe 28731
Kyoto Gion Oil Lamp Museum, Kyoto 28840

Korea, Republic

Janggigot Lighthouse Museum, Pohang 29858

Mexico

Museo de la Luz, México 31085

Poland

Muzeum Latarnia Morska, Władysławowo 35306

Spain

Casita del Príncipe, Madrid 39506

United Kingdom

Museum of Lighting, Edinburgh 44135
Museum of Scottish Lighthouses, Fraserburgh 44254
Pendeen Lighthouse, Pendeen 45354
Skerryvore Museum, Hynish, Isle of Tiree 44506
Withernsea Lighthouse Museum, Withernsea 46050

United States of America

Great Lakes Lighthouse Museum, Mackinaw City 50353
Key West Lighthouse Museum, Key West 49762
Lightship Portsmouth Museum, Portsmouth ... 52109
Magic Lantern Castle Museum, San Antonio .. 52718
Montauk Point Lighthouse Museum, Montauk 50797
National Packard Museum, Warren 53825
Neustadt Museum of Tiffany Art, New York ... 51239
Old Mackinac Point Lighthouse Colonial Michilimackinac, Mackinaw City 50355
Ponce DeLeon Inlet Lighthouse, Ponce Inlet ... 52019
Split Rock Lighthouse, Two Harbors 53631

Language and Literature

Argentina

Museo Almafuerte, La Plata 00407
Museo Histórico José Hernández-Chacra Pueyrredón, Villa Ballester 00672
Museo y Centro Cultural Victoria Ocampo, Mar del Plata 00448

Armenia

Aksel Bakuntz House-Museum, Goris 00702
Avetik Issahakyan House-Museum, Yerevan ... 00728
Derenik Demirchyan House-Museum, Yerevan 00731
Hovhannes Hovhanisyan House-Museum, Echmiadzin 00696
Hovhannes Tumanyan Museum, Yerevan 00739
Isahakyan House Museum, Gyumri 00705
Khachatour Abovyan House-Museum, Yerevan 00740
Perch Proshyan House-Museum, Ashtarak ... 00692
Shiraz House Museum, Gyumri 00709
Stepan Zoryan House-Museum, Vanadzor 00720
Yegishe Charents Literary House-Museum, Yerevan 00759
Yegishe Charents Literature and Arts Museum, Yerevan 00760

Australia

Henry Kendall Cottage and Museum, Gosford 01123
Henry Lawson Centre, Gulgong 01133
Nutcote Museum, Neutral Bay 01358
Palma Rosa Museum, Hamilton, Brisbane 01142
Tom Collins House, Swanbourne 01542

Austria

Adalbert-Stifter-Haus, Linz 02307
Anton Wildgans-Haus, Mödling 02391
Bernhard-Haus, Ohlsdorf 02471
Bezirksmuseum Alsergrund, Wien 02954
Bezirksmuseum Döbling, Wien 02956
Esperanto-Museum der Österreichischen Nationalbibliothek und Sammlung für Plansprachen, Wien 02984
Franz Xaver Gruber-Gedächtnishaus, Ach 01690
Ganghofer-Museum, Leutasch 02296
Georg Rendl-Museum, Sankt Georgen bei Salzburg 02643
Georg Trakl Forschungs- und Gedenkstätte, Salzburg 02614
Grillparzer-Gedenkzimmer, Wien 02998
Hanrieder-Gedenkraum, Putzleinsdorf 02543
Heimathaus und Schimuseum, Saalbach 02605
Heimatmuseum und Klöpferhaus, Eibiswald .. 01849
Hermann Broch-Museum, Teesdorf 02800
Josef Weinheber-Museum, Kirchstetten 02173
Joseph Misson-Gedenkstätte, Mühlbach am Mannhartsberg 02409
Karl Bacher-Stüberl, Zellerndorf 03129
Karl Heinrich Waggerl-Haus, Wagrain 02882
Kernstock-Museum, Bruck an der Lafnitz 01802
Österreichische Nationalbibliothek, Wien 03045
Österreichisches Sprachinselmuseum, Wien. 03050
Paul Ernst-Gedenkstätte, Sankt Georgen an der Stiefing 02642
Peter Rosegger-Geburtshaus, Alpl 01707
Raimund-Gedenkstätte, Gutenstein 02025
Richard Billinger-Gedenkraum, Sankt Marienkirchen bei Schärding 02667
Robert Hamerling - Museum, Kirchberg am Walde 02166
Robert Musil-Literatur-Museum, Klagenfurt ... 02191
Sterbehaus Ferdinand Raimunds, Pottenstein 02515
Strindberg-Museum Saxen, Saxen 02695
Wystan Hugh Auden-Dokumentationsräume, Kirchstetten 02174

Azerbaijan

Cefer Cabbarlının Ev-Muzeyi, Bakı 03151
Celil Memmedquluzadenin Ev-Muzeyi, Bakı .. 03152
Djalil Mammadkulizade Home Museum, Naxçivan 03181
Djalil Mammadkulizade Memorial Museum, Nehram 03189
Hüseyn Cavidin Ev-Muzeyi, Bakı 03157
Hüseyn Cavidin Ev-Muzeyi, Naxçivan 03184
Mammad Seid Ordubad Home Museum, Ordubad 03190
Memmed Seid Ordubadinin Xatire Muzeyi, Bakı 03160
Miniature Books Museum, Bakı 03161
Museum-Appartment of Abdulla Shaig, Bakı . 03162
Nakhchivan Autonomous Republic Djalil Mammadkulizade Literature Museum, Naxçivan 03186
Neriman Nerimanovun Xatire Muzeyi, Bakı ... 03165
Nizami ad Dövlet Azerbaycan Edebiyyati Muzeyi, Bakı 03166
Semed Vurğunun Ev-Muzeyi, Bakı 03171

Belarus

Litaraturny Muzej Janki Kupaly, Minsk 03251
Literary History Museum, Minsk 03252
Literaturnyj Muzej Maksima Bogdanoviča, Minsk 03253

Belgium

AMVC-Letterenhuis, Antwerpen 03280
Archives et Musée de la Littérature, Bruxelles 03433
Centre Provincial d'Hébergement Le Caillou, Roisin 03980
Cyriel Verschaeve Museum, Alveringem 03267
Guido Gezellemuseum, Brugge 03415
Literair Museum, Hasselt 03683
Maison Adolphe Hardy, Dison 03560
Musée Camille Lemonnier, Bruxelles 03456
Musée Émile Verhaeren, Honnelles 03710

Musée Guillaume Apollinaire, Stavelot 04049
Museum Arnold Vander Haeghen, Gent 03633
Museum René de Clercq, Deerlijk 03537
Protestants Historisch Museum Abraham Hans, Horebeke 03712
Provinciaal Museum Emile Verhaeren, Sint-Amands 04014
Stedelijk Museum Huize Ernest Claes, Scherpenheuvel 04004
Timmermans-Opsomerhuis, Lier 03820
Trignolles - Espace Arthur Masson, Treignes 04094
Uilenspiegel Museum, Damme 03536

Bolivia

Museo de Escritores Bolivianos, La Paz 04213

Bosnia and Herzegovina

Muzej Književnosti i Pozorišne Umjetnosti Bosne i Hercegovine i Galerija MAK, Sarajevo 04279

Brazil

Academia Brasileira de Letras, Rio de Janeiro 04679
Museu Casa Guimarães Rosa, Cordisburgo 04417
Museu Castro Alves, Salvador 04791

Bulgaria

Dom-muzej Čudomir, Kazanlăk 05057
Dom-muzej Elin Pelin, Bajlovo 04973
Dom-muzej Ivan Vazov, Sopot 05205
Dom-muzej Penjo Penev, Dimitrovgrad 05013
Dom-muzej Konstantin Veličkov, Pazardžik 05106
Istoričeski Muzej, Bansko 04978
Kăšta-muzej Angel Karalijčev, Sofia 05173
Kăšta-muzej Christo Botev, Kalofer 05045
Kăšta-muzej Christo Smirnenski, Sofia 05174
Kăšta-muzej Dimcho Debeljanov, Koprivštica 05070
Kăšta-muzej Dimităr Dimov, Sofia 05176
Kăšta-muzej Dobri Voinikov, Šumen 05214
Kăšta-muzej Emilijan Stanev, Veliko Tărnovo 05254
Kăšta-muzej Geo Milev, Stara Zagora 05209
Kašta-muzej Ivan Vazov, Berkovica 04987
Kăšta-muzej Ivan Vazov, Sofia 05178
Kăšta-muzej Jordan, Jovkov, Žeravna 05269
Kăšta-muzej Lamartine, Plovdiv 05129
Kăšta-muzej Ljuben Karavelov, Koprivštica 05072
Kăšta-muzej Nikola Ěnkov Vapcarov, Bansko 04980
Kăšta-muzej Nikola Vapcarov, Sofia 05179
Kăšta-muzej Pejo Javorov, Sofia 05180
Kăšta-muzej Pejo K. Javorov, Čirpan 05010
Kăšta-muzej Petko i Pencho Slavejkovi, Sofia 05181
Kăšta-muzej s Dom-pametnik Jordan Jovkov, Dobrič 05020
Kăšta-muzej Slavejkovi, Trjavna 05229
Nacionalen Literaturen Muzej, Sofia 05192

Canada

Anne of Green Gables Museum, Kensington 06059
Benares Historic House, Mississauga 06309
Centre d'Exposition de la Bibliothèque Gabrielle Roy, Québec 06678
Chiefswood Museum, Ohsweken 06502
Craik Community Archives Oral History Society, Craik 05655
Gitxsan Treaty Office, Hazelton 05978
Green Gables House, Charlottetown 05593
Haig-Brown House, Campbell River 05541
Ivan Franko Museum, Winnipeg 07364
Lashburn Centennial Museum, Lashburn 06143
McCrae House, Guelph 05918
McPherson Library Special Collections, Victoria 07252
Margaret Laurence Home, Neepawa 06421
Morrin Centre, Québec 06684
Stephansson House, Edmonton 05754
Stephen Leacock Museum, Orillia 06515
Taras H. Shevchenko Museum Memorial Park Foundation, Toronto 07141

Chile

Museo Gabriela Mistral de Vicuña, Vicuña 07473

China

Guangzhou Lu Xun Museum, Guangzhou 07731
Lao She Museum, Beijing 07545
Lu Xun Former Residence, Shanghai 08173
Lu Xun Memorial Hall, Shanghai 08174
Lu Xun Museum, Beijing 07547
National Museum of Modern Chinese Literature, Beijing 07557

Colombia

Colección del Poeta Baudilio Montoya, Clarcá 08626
Museo Eduardo Carranza, Villavicencio 08788
Museo Luis Alberto Acuña, Villa de Leyva 08784

Croatia

Muzej Hrvatske Književnosti i Kazališne Umjetnosti, Zagreb 09005
Muzej Ljudevit Gaj, Krapina 08901

Cuba

Casa de Dulce María Loynaz, La Habana 09095
Casa de José Jacinto Milanés, Matanzas 09211
Casa Natal de Regino Boti, Guantánamo 09090
Museo Alejo Carpentier, Camagüey 09048
Museo Andrés Cué, Amancio 09024
Museo Casa de la Poesía, La Habana 09119
Museo Casa Natal de José María Heredia, Santiago de Cuba 09308
Museo Casa Natal de José Martí, La Habana 09126
Museo del Humor, San Antonio de los Baños 09273
Museo Ernest Hemingway, San Francisco de Paula 09278
Museo José Martí, La Habana 09147
Museo Quinta Amalia Simoni, Camagüey 09052

Czech Republic

Barunčina Škola, Česká Skalice 09458
Domek Boženy Němcové, Červený Kostelec 09456
Franz Kafka museum, Praha 09745
Literární Oddělení, Brno 09429
Městské Muzeum Antonína Sovy, Pacov 09704
Muzeum Aloise Jiráska, Hronov 09538
Muzeum Aloise Jiráska a Mikoláše Alše, Praha 09766
Muzeum Boženy Němcové, Česká Skalice 09459
Muzeum Bratří Čapků, Malé Svatoňovice 09646
Muzeum Josefa Dobrovského, Chudenice 09487
Muzeum Otokara Březiny, Jaroměř nad Rokytnou 09554
Muzeum Prostějovska, Prostějov 09810
Muzeum J.V. Sládka, Zbiroh 09929
Muzeum Svatopluka Čecha, Liteň 09628
Památník Bedřicha Smetany, Obříství 09687
Památník Bratří Mrštíků, Diváky 09492
Památník Karla Čapka, Stará Hut u Dobříše 09845
Památník Karla Havlíčka Borovského, Havlíčkův Brod 09512
Památník Karla Hynka Máchy, Doksy 09494
Památník Karla Jaromíra Erbena, Miletín 09654
Památník Karla Václava Raise, Lázně Bělohrad 09620
Památník Národního Písemnictví, Praha 09786
Památník Petra Bezruče, Kostelec na Hané 09597
Památník Svatopluka Čecha, Obříství 09688
Památník Svatopluka Čecha, Ostředek 09702
Památník Terezy Novákové, Proseč u Skutče 09809
Památník Václava Beneše Třebízského, Třebíz 09877
Rodný Domek Aloise Jiráska, Hronov 09539
Rodný Domek Otokara Březiny, Počátky 09722
Rodný dům Adalberta Stiftera, Horní Planá 09526
Slezské zemské muzeum, Opava 09694
Slezské zemské muzeum, Ostravice 09701

Denmark

Bakkemuseet, Frederiksberg 10022
Digterhuset, Farsø 10017
Drachmanns Hus, Skagen 10259
E. Bindstouw, Viborg 10314
Hans Christian Andersen Museum, Odense 10215
Hans Christian Andersens Barndomshjem, Odense 10216
Jenle Museum, Roslev 10251
Johannes Jørgensens Mindestuer, Svendborg 10289
Karen Blixen Museet, Rungsted Kyst 10253
Skjoldborgs Hus, Vesløs 10312
Struer Museum og Johs. Buchholtz Hus, Struer 10286
Tersløsegaard, Dianalund 09987
Wesselstuerne, København 10157

Egypt

Manuscript Museum, Alexandria 10474
Shawqi Museum, Cairo 10520

El Salvador

Museo de La Palabra y La Imagen, San Salvador 10552

Estonia

A.H. Tammsaare Muuseum, Vetepere 10671
Eesti Kirjandusmuuseum, Tartu 10645
Karl Ristikivi Majamuuseum, Tartu 10651
Oskar Luts Majamuuseum, Tartu 10654

Finland

Danielson-Kalmarin Huvila, Vääksy 11430
Einari Vuorelan Kirjailijakoti, Keuruu 10950
J.L. Runebergin Koti, Porvoo 11227
Juhani Ahon Museo, Iisalmi 10847
Kangasalan Kotiseutumuseo, Kangasala 10908
Kuddnäs Museum, Nykarleby 11155
Paikkarin Torppa, Sammatti 11298
Westmansmors stugan, Jakobstad 10868

France

Bibliothèque-Musée Valéry Larbaud, Vichy 16144
Centre d'Études Edmond Michelet, Brive-la-Gaillarde 12188
Château de Monte-Cristo, Le Port-Marly 13641
Exposition Permanente sur François Pétrarque, Lombez 13805
Galerie Bovary, Ry 15143
Maison de Balzac, Paris 14618
Maison de George Sand, Nohant-Vic 14465
Maison de Jules Verne, Amiens 11586
Maison de Victor Hugo, Paris 14621
Maison d'Elsa Triolet et Louis Aragon, Saint-Arnoult-en-Yvelines 15176
Maison d'Émile Zola, Medan 14057
Maison Francis Jammes, Orthez 14542
Maison Jacques Prévert, Omonville-la-Petite 14513
Maison Littéraire de Victor Hugo, Bièvres 11989
Maison Natale de Jean Giraudoux, Bellac 11925
Maison Stendhal, Grenoble 13131
Musée Adam Mickiewicz, Paris 14627
Musée Alexandre Dumas, Villers-Cotterêts 16229
Musée Alfred de Vigny, Champagne-Vigny 12385
Musée Alphonse Daudet, Fontvieille 12977
Musée Arthur Rimbaud, Charleville-Mézières 12409
Musée Balzac, Saché 15145
Musée Barbey d'Aurévilly, Saint-Sauveur-le-Vicomte 15507
Musée-Bibliothèque François Pétrarque, Fontaine-de-Vaucluse 12957
Musée Charles Milcendeau-Jean Yole, Soullans 15799
Musée d'Art et d'Histoire, Saint-Denis (Seine-Saint-Denis) 15236
Musée de la Musique Mécanique, Les Gets 13700
Musée de l'Imprimerie de Lyon, Lyon 13885
Musée de Pierre Corneille, Le Petit-Couronne 13631
Musée de Vulliod Saint-Germain, Pézenas 14808
Musée des Charmettes, Chambéry 12373
Musée-École du Grand Meaulnes, Épineuil-le-Fleuriel 12841
Musée Edmond Rostand, Cambo-les-Bains 12254
Musée Émile Guillaumin, Ygrande 16307
Musée Espéranto a Gray, Gray 13128
Musée Flaubert et d'Histoire de la Médecine, Rouen 15116
Musée Frédéric Mistral, Maillane 13917
Musée George Sand et de la Vallée Noire, La Châtre 13340
Musée Goethe, Sessenheim 15732
Musée Henri Barbusse, Aumont-en-Halatte 11747
Musée Historique et Erckmann-Chatrian, Phalsbourg 14811
Musée J.-E. Blanche, Offranville 14497
Musée Jean de La Fontaine, Château-Thierry 12440
Musée Jean-Jacques Rousseau, Montmorency 14229
Musée Joachim du Bellay, Liré 13775
Musée Jules Verne, Nantes 14369
Musée Lamartine, Mâcon 13908
Musée Maison de Clemenceau, Saint-Vincent-sur-Jard 15539
Musée-Maison Marcel Proust, Illiers-Combray 13246
Musée Marc Leclerc, La Ménitré 13392
Musée Maurice Dufresne, Azay-le-Rideau 11811
Musée Maurice et Eugénie de Guérin, Andillac 11597
Musée Memorial Alain, Mortagne-au-Perche 14286
Musée Natale de Pierre Corneille, Rouen 15120
Musée Paul Valéry, Sète 15735
Musée Pierre-Loti, Rochefort (Charente-Maritime) 15073
Musée Rabelais, Seuilly 15737
Musée Schiller et Goethe, Châlons-en-Champagne 12365
Musée Verlaine, Juniville 13294
Musée Victor Hugo, Villequier 16225
Pavillon Flaubert, Canteleu 12271

Georgia

G. Leonidze State Literature Museum of Georgia, Tbilisi 16358
Ilja Chavchavadze House Museum, Tbilisi 16366
Ilja Chavchavadze Literature-Memorial Museum, Tbilisi 16367

Germany

Agnes-Miegel-Haus, Bad Nenndorf 16793
Anna-Seghers-Gedenkstätte, Berlin 17033
Ausstellung Fontane und Hankels Ablage, Zeuthen 22795
Ausstellung Max-Kommerell, Münsingen 20500
Berthold-Auerbach-Museum, Horb am Neckar 19271
Brecht-Weigel-Gedenkstätte, Berlin 17055
Brechthaus, Augsburg 16613
Brehm-Gedenkstätte, Renthendorf 21279
Brüder Grimm-Haus Steinau, Steinau an der Straße 21877
Brüder Grimm-Museum Kassel, Kassel 19459
Buddenbrookhaus, Heinrich-und-Thomas-Mann-Zentrum, Lübeck 20073
Büchnerhaus, Riedstadt 21321
Deutsches Eichendorff-Museum, Wangen im Allgäu 22324
Deutsches Märchen- und Wesersagenmuseum, Bad Oeynhausen 16798
Dichtermuseum Joseph Maria Lutz, Pfaffenhofen an der Ilm 21018
Dorfmuseum Hausen im Wiesental, Hausen im Wiesental 19049
Droste-Museum, Münster 20508
Droste-Museum, Havixbeck 19056
Droste-Museum im Fürstenhäusle, Meersburg 20248
E.T.A. Hoffmann-Haus, Bamberg 16907
Ehm Welk-Haus, Bad Doberan 16709
Ehm Welk- und Heimatmuseum, Angermünde 16548
Erich Kästner Museum Dresden, Dresden 17942
Ernst-Moritz-Arndt-Haus, Bonn 17453
Faust-Museum, Knittlingen 19583
Frankfurter Goethe-Haus, Frankfurt am Main 18370
Friedrich-Gerstäcker-Museum, Braunschweig 17520
Friedrich-Rückert-Gedächtnisstätte, Coburg 17715
Friedrich-Wilhelm-Weber-Museum, Bad Driburg 16711
Fritz-Reuter-Literaturmuseum, Stavenhagen 21870
Gellert-Museum Hainichen, Hainichen 18873
Gerhart-Hauptmann-Haus, Kloster 19574
Gerhart-Hauptmann-Museum, Erkner 18215
Das Gleimhaus, Halberstadt 18877
Goethe-Gedenkstätte, Jena 19380
Goethe-Museum Düsseldorf, Düsseldorf 17990
Goethe-Museum Stützerbach mit Museum zur Geschichte des technischen Glases, Stützerbach 21924
Goethe-Nationalmuseum mit Goethes Wohnhaus, Weimar 22401
Goethes Gartenhaus, Weimar 22403
Gottfried-August-Bürger-Museum, Molmerswende 20370
Günter Grass Haus, Forum für Literatur und bildende Kunst, Lübeck 20075
Gustav-Freytag-Archiv und Museum, Wangen im Allgäu 22325
Hans-Fallada-Haus, Feldberger Seenlandschaft 18301
Hans Werner Richter Haus, Bansin, Seebad 16925
Hansjakobmuseum im Freihof, Haslach im Kinzigtal 19025
Hauffs Märchen-Museum, Baiersbronn 16896
Hebbel-Museum, Wesselburen 22511
Heimatmuseum und Hedwig Courths-Mahler Archiv, Nebra 20559
Heinrich-Hansjakob-Gedenkstätte, Freiburg im Breisgau 18430
Heinrich-Heine-Institut-Museum, Düsseldorf 17991
Hermann-Allmers-Haus, Sandstedt 21484
Hermann-Hesse-Museum, Calw 17678
Hölderlinturm, Tübingen 22115
Hoffmann-von-Fallersleben-Museum, Wolfsburg 22663
Internationale Jugendbibliothek, München 20447
Jagdhaus Gabelbach, Ilmenau 19321
Jean-Paul-Museum der Stadt Bayreuth, Bayreuth 16954
Jean-Paul-Zimmer in der Rollwenzelei, Bayreuth 16955
Joachim Ringelnatz Museum, Cuxhaven 17755
Justinus-Kerner-Haus, Weinsberg 22432
Karl-May-Haus Hohenstein-Ernstthal, Hohenstein-Ernstthal 19244
Karl-May-Museum, Radebeul 21160
Kirms-Krackow-Haus, Weimar 22405
Klaus-Groth-Museum und Neue Museumsinsel Lüttenheid, Heide 19066
Kleist-Museum, Frankfurt (Oder) 18404
Klopstockhaus, Quedlinburg 21150
Kolbenheyer-Archiv und Gedenkstätte, Geretsried 18602
Kraszewski-Museum, Dresden 17949
Kügelgenhaus, Dresden 17950
Kunstmuseum Solingen, Solingen 21802
Kurt Tucholsky Literaturmuseum, Rheinsberg 21304
Lessing-Museum, Kamenz 19420
Literarisches Museum im Max-Eyth-Haus, Kirchheim unter Teck 19549
Literaturarchiv Sulzbach-Rosenberg, Sulzbach-Rosenberg 21984
Literaturhaus, Magdeburg 20125
Literaturmuseum Baumbachhaus, Meiningen 20259
Literaturmuseum Brecht-Weigel-Haus, Buckow 17620
Literaturmuseum der Moderne, Marbach am Neckar 20179
Literaturmuseum Theodor Storm, Heiligenstadt, Heilbad 19108
Ludwig-Thoma-Haus, Tegernsee 22016
Mahn- und Gedenkstätten Wöbbelin, Wöbbelin 22637

Maxim-Gorki-Gedächtnisstätte, Heringsdorf, Seebad ... 19144
Mörike Museum im alten Schulhaus, Neuenstadt am Kocher ... 20611
Mörikehaus Ochsenwang, Bissingen an der Teck ... 17366
Mori-Ōgai-Gedenkstätte, Berlin ... 17190
Münchhausen-Museum, Bodenwerder ... 17414
Museum Friedrichshagener Dichterkreis, Berlin ... 17200
Museum für Literatur am Oberrhein, Karlsruhe ... 19444
Museum Neuruppin, Neuruppin ... 20650
Museum Raabe-Haus, Eschershausen ... 18235
Museum Schillerhaus, Mannheim ... 20169
Museum Wolfram von Eschenbach, Wolframs-Eschenbach ... 22656
Museum Wurzen mit Ringelnatzsammlung, Wurzen ... 22751
Nietzsche-Archiv, Weimar ... 22412
Nietzsche-Haus, Naumburg, Saale ... 20555
Novalis-Museum, Wiederstedt ... 22541
Pater Haw-Museum, Schweich an der römischen Weinstraße ... 21680
Platenhäuschen, Erlangen ... 18224
Polenz-Museum, Cunewalde ... 17753
Profanierte Marienkapelle, Ludwigsstadt ... 20064
Raabe-Haus, Braunschweig ... 17528
Reuter-Villa mit Wagner-Sammlung, Eisenach 18096
Romantikerhaus, Jena ... 19388
Sammlung Crous im Alten Kurhaus, Aachen ... 16406
Schiller-Museum Bauerbach, Bauerbach ... 16935
Schiller-Nationalmuseum, Marbach am Neckar 20181
Schillerhäuschen, Dresden ... 17968
Schillerhaus, Ludwigshafen am Rhein ... 20055
Schillerhaus, Leipzig ... 19932
Schillers Gartenhaus der Friedrich-Schiller-Universität, Jena ... 19390
Schillers Geburtshaus, Marbach am Neckar ... 20182
Schillers Wohnhaus, Weimar ... 22417
Schloss Kochberg mit Park und Liebhabertheater, Großkochberg ... 18794
Schubartstube, Blaubeuren ... 17380
Serbski muzej - Sorbisches Museum, Bautzen ... 16941
Stefan-George-Museum im Stefan-George-Haus, Bingen am Rhein ... 17354
Stiftung für Konkrete Kunst, Reutlingen ... 21290
Struwwelpeter-Museum, Frankfurt am Main ... 18397
Theatermuseum Zauberwelt der Kulisse, Meiningen ... 20262
Theodor-Fontane-Archiv, Potsdam ... 21111
Theodor-Storm-Haus, Husum ... 19302
Till-Eulenspiegel-Museum, Schöppenstedt ... 21618
Waldschmidt-Ausstellung, Eschlkam ... 18236
Wieland-Archiv, Biberach an der Riß ... 17325
Wieland-Gedenkzimmer, Achstetten ... 16424
Wieland-Museum Biberach, Wieland-Gartenhaus, Biberach an der Riß ... 17326
Wielandgut Oßmannstedt mit Wieland-Museum, Oßmannstedt ... 20928
Wilhelm-Busch-Geburtshaus, Wiedensahl ... 22540
Wilhelm-Busch-Gedenkstätte, Seesen ... 21725

Greece

Kazantzakis Museum, Iráklion ... 23020
Museum of Dionysios Solomos and Eminent Zakynthians, Zakynthos ... 23245
Palamas Museum, Athinai ... 22948
Solomos Museum, Corfu ... 22980

Guadeloupe

Musée Municipal Saint-John-Perse, Pointe-à-Pitre ... 23277

Hungary

Ady Endre Emlékmúzeum, Budapest ... 23379
Arany János Irodalmi Emlékház, Geszt ... 23564
Babits Mihály Emlékház, Szekszárd ... 23865
Balassa Bálint Múzeum - Babits Mihály Emlékház, Esztergom ... 23550
Berzsenyi Dániel Emlékmúzeum, Egyházashetye ... 23545
Berzsenyi Dániel Emlékmúzeum, Nikla ... 23731
Cserépfalvi Imre Emlékszoba, Cserépfalu ... 23506
Darvas József Irodalmi Emlékház és Kutatóépület, Orosháza ... 23738
Deák Ferenc Szülőháza, Söjtör ... 23805
Debreceni Irodalmi Múzeum, Debrecen ... 23514
Fekete István Emlékház, Gölle ... 23569
Gárdonyi Géza Emlékmúzeum, Eger ... 23540
Irodalom Háza - Mészöly Miklós Múzeum, Szekszárd ... 23867
Jókai Mór Emlékmúzeum, Balatonfüred ... 23352
József Attila Emlékház, Öcsöd ... 23735
József Attila Emlékmúzeum, Balatonszárszó ... 23353
József Attila Emlékszoba, Budapest ... 23417
Katona József Emlékház, Kecskemét ... 23647
Kazinczy Ferenc Emlékcsarnok, Sátoraljaújhely ... 23797
Kazinczy Ferenc Múzeum, Sátoraljaújhely ... 23798
Kisfaludy Károly Emlékkiállítás, Tét ... 23923
Kisfaludy Sándor Emlékház, Sümeg ... 23825
Kodolányi Emlékmúzeum, Vajszló ... 23948
Kölcsey Ferenc Emlékszoba, Szatmárcseke ... 23837
Madách Imre Emlékmúzeum, Csesztve ... 23510
Móricz Zsigmond Emlékház, Prügy ... 23785
Móricz Zsigmond Emlékház, Tiszacsécse ... 23929
Nagy Lajos Emlékkiállítás, Apostag ... 23336
Nagy László Emlékház, Iszkáz ... 23621
Németh László Emlékkiállítás, Hódmezővásárhely ... 23613
Petőfi Emlékmúzeum, Kiskőrös ... 23664
Petőfi Irodalmi Múzeum, Budapest ... 23467
Petőfi Múzeum, Aszód ... 23338
Petőfi Sándor Emlékház, Dömsöd ... 23527
Petőfi Sándor Emlékmúzeum, Szalkszentmárton ... 23834
Petőfi Sándor Emlékszoba, Kiskunfélegyháza ... 23668
Remenyik Zsigmond Emlékszoba, Dormánd ... 23528
Szántó Kovács János Területi Múzeum, Orosháza ... 23740
Szegedy Róza Ház Irodalmi Emlékmúzeum, Badacsony ... 23344
Tűzoltó- és Községi Emlékház, Vál ... 23949
Vajda János Emlékház, Vál ... 23950
Vörösmarty Mihály Emlékmúzeum, Kápolnásnyék ... 23629

Iceland

Sigurhaedir - Hús Skáldsins Museum, Akureyri ... 23987
Stofnun Arna Magnússonar, Reykjavík ... 24037

India

Ghalib Museum, Delhi ... 24141
Rabindra Bharati Museum, Kolkata ... 24280
Rabindra Bhavana, Santiniketan ... 24379

Indonesia

Literary Documentation Centre H.B. Yasin, Jakarta ... 24475

Iran

Hessaby Museum, Tehrān ... 24670
Shahryar Museum, Tabriz ... 24653

Ireland

James Joyce Centre, Dublin ... 24803
James Joyce Museum, Dublin ... 24804
Munster Literature Centre, Cork ... 24769
Newman House, Dublin ... 24813
Nora Barnacle House Museum, Galway ... 24837
Writers Museum, Dublin ... 24824
Yeats Tower - Thoor Ballylee, Gort ... 24840

Israel

Nahum Gutman's Museum, Tel Aviv ... 25118
Sholem Asch House, Bat Yam ... 24934
Sholom Aleichem Museum, Tel Aviv ... 25121

Italy

Casa del Boccaccio, Certaldo ... 25814
Casa dell'Ariosto, Ferrara ... 26104
Casa di Goethe, Roma ... 27442
Casa di Luigi Pirandello, Roma ... 27443
Casa di Petrarca, Arquà Petrarca ... 25286
Casa di Pirandello, Agrigento ... 25160
Casa Leopardi, Recanati ... 27375
Casa Museo Giovanni Verga, Catania ... 25771
Casa Natale di Giosuè Carducci, Pietrasanta ... 27205
Centro Studi I. Silone, Pescina ... 27175
Collezioni dell'Accademia Nazionale Virgiliana, Mantova ... 26590
Keats-Shelley House, Roma ... 27456
Manzoni Villa, Lecco ... 26477
Museo Alfieriano, Asti ... 25313
Museo Carducci - Biblioteca e Casa, Bologna 25446
Museo Casa di Dante, Firenze ... 26168
Museo Casa Natale G. D'Annunzio, Pescara ... 27162
Museo Cerlogne, Saint Nicolas ... 27587
Museo Dantesco, Ravenna ... 27371
Museo d'Arte dello Splendore, Giulianova ... 26346
Museo del Risorgimento, Genova ... 26301
Museo del Tasso, Roma ... 27474
Museo Deleddiano, Nuoro ... 26966
Museo della Guerra, Gardone Riviera ... 26272
Museo della Tipografia, Città di Castello ... 25907
Museo Manzoniano, Milano ... 26710
Pinacoteca Civica, Cepagatti ... 25804

Japan

Basho Museum Koto, Tokyo ... 29371
Fukuyama Bungakukan, Fukuyama ... 28443
Hermann Hesse Museum, Hiroshima ... 28531
Himeji-shiritsu Bungaku Hakubutsukan, Himeji ... 28513
Hokkaido-ritsu Bungakukan, Sapporo ... 29217
Ishikawa Takuboku Memorial Hall, Tamayama 29350
Jiro Osaragi Kinenkan, Yokohama ... 29595
Kagoshima City Modern Literature Museum Marchen Fairy Tale Museum, Kagoshima ... 28626
Kamakura-shi Bungaku, Kamakura ... 28642
Kan Kikuchi Kinenkan, Takamatsu ... 29318
Kanagawa Kindai Bungakukan, Yokohama ... 29598
Kochi Bungakukan, Kochi ... 28741
Koizumi Yakumo Kinenkan, Matsue ... 28888
Kyoto Daigaku Bungakukan, Kyoto ... 28836
Matsuyama Municipal Shiki-Kinen Museum, Matsuyama ... 28906
Mokichi Saito Museum, Kaminoyama ... 28650
Motoori Norinaga Kinenkan, Matsusaka ... 28898
Museum of Contemporary Japanese Poetry, Tanka and Haiku, Kitakami ... 28710
Nihon Kindai Bungakukan, Tokyo ... 29448
Niigata-shiritsu Aizu Yaichi Kinenkan, Niigata 29058
Setagaya-kuritsi Riterashi Hakubutsukan, Tokyo ... 29467
Shiojiri Tankakan, Shiojiri ... 29276
Suishouen Toji Shiryokan, Nagoya ... 29011
Suzuki Bokushi Kinenkan, Shiozawa ... 29279
Tachihara Michizo Kinenkan, Tokyo ... 29480
Takamura's Cottage Memorial Museum, Hanamaki ... 28494
Takehisa Yumeji Ikaho Kinenkan, Ikaho ... 28561
Tale of Genji Museum, Uji, Kyoto ... 29559
Torahiko Terada Kinenkan, Kochi ... 28748
Toyohashi-shiritsu Bijutsukan, Toyohashi ... 29529

Kazakhstan

Dom-muzej M. Auėzova, Almaty ... 29680
Gosudarstvennyj Literaturno-memorialnyj Muzejnyj Kompleks Sabita Mukanova i Gabita Musrepova, Almaty ... 29681
Literaturno-memorialnyj Muzej F.M. Dostoevskogo, Semey ... 29728
Literaturnyj Memorialnyj Dom-muzej Abaja, Semey ... 29729
Memorialnyj Muzej Akademika K.I. Satpaeva, Almaty ... 29684
Muzej Abaja, Abaj ... 29669
Muzej Knigi, Almaty ... 29688
Muzej Sakena Sejfullina, Astana ... 29693
Oblastnoj Muzej Literatury i Iskusstva im. Buchar Žyrau, Pavlodar ... 29717
Ščučinskij Oblastnoj Muzej Literatury i Iskusstva, Kokšetau ... 29704

Kenya

Karen Blixen Museum, Nairobi ... 29747

Korea, Republic

Sung-Am Archives of Classical Literature, Seoul ... 29928

Kyrgyzstan

Gumbez Manace Literary Ethnographic Museum, Taš-Aryk ... 29965
Kyrgyzskij Gosudarstvennyj Muzej Literatury i Iskusstva im. Toktogula Satylganova, Biškek 29955
Memorialnyj Dom-Muzej Aaly Tokombaeva, Biškek ... 29957

Latvia

Aleksandra Čaka Memoriālais Dzīvoklis, Rīga 30060
Andreja Upīša Memoriālais Muzejs, Rīga ... 30061
Annas Brigaderes Memoriālais Muzejs, Tērvete ... 30129
Bārdu Dzimtas Muzejs Rumbiņi, Katvaru ... 30030
Brāļu Kaudzīšu Memoriālais Muzejs, Kaudzīši 30031
Ernesta Birznieka-Upīša Muzejs Bisnieki, Zentene ... 30151
Jāņa Akuratera Muzejs, Rīga ... 30070
Jāņa Jaunsudrabiņa Muzejs Riekstiņi, Nereta 30050
Jaņa Rozentāla un Rūdolfa Blaumaņa Muzejs, Rīga ... 30071
Kārļa Skalbes Memoriālais Muzejs Saulrieti, Vecpiebalga ... 30142
Ojāra Vācieša Memoriālais Muzejs, Rīga ... 30100
Raiņa Literatūras un Mākslas Vēstures Muzejs, Rīga ... 30102
Raiņa un Aspazijas Memoriālā Vasarnīca, Jūrmala ... 30026
Rakstniecības un Mūzikas Muzejs, Rīga ... 30104
Rakstnieka Herberta Dorbes Māja-Muzejs Senču Putekļi, Ventspils ... 30143

Lithuania

Anselmo Matučio Memorialinis Muziejus, Alytus ... 30204
Antano Baranausko ir Antano Vienuolio-Žukausko Memorialinis Muziejus, Anykščiai ... 30206
Antano Jonyno Memorialinis Muziejus, Alytus 30205
Balio ir Vandos Sruogų Namai-Muziejus, Kaunas ... 30240
Bronės Buivydaitės Memorialinis Muziejus, Anykščiai ... 30208
Didžiulių Muziejus, Anykščiai ... 30209
Dr. Vinco Kudirkos Muziejus, Kudirkos Naumiestis ... 30290
Eugenijaus ir Leonardo Matuzevičių Memorialinis Muziejus, Krinčinas ... 30289
Gabrielės Petkevičaitės-Bitės Literatūrinis Muziejus, Smilgiai ... 30373
Gabrielės Petkevičaitės Memorialinė Ekspozicija, Panevėžys ... 30331
Hermano Zudermano Memorialinis Muziejus, Macikai ... 30304
Ievos Simonaitytės Memorialinis Muziejus, Priekulė ... 30339
Jono Biliūno Memorialinis Muziejus, Anykščiai ... 30210
Juozo Grušo Memorialinis Muziejus, Kaunas ... 30244
Juozo Paukštelio Memorialinis Muziejus, Kėdainiai ... 30271
Juozo Tumo-Vaižganto Memorialinis Muziejus, Kaunas ... 30245
Kalbininko Kazimiero Būgos Memorialinis Muziejus, Paziegė ... 30336
Konstantinas Sankenis Memorialinis Muziejus, Veleniskiai ... 30392
Lazdijų Krašto Muziejus, Lazdijai ... 30300
Literatūrinis Aleksandro Puškino Muziejus, Vilnius ... 30423
Maironio Lietuvių Literatūros Muziejus, Kaunas ... 30259
Marijos ir Jurgio Šlapelių Namas-muziejus, Vilnius ... 30424
Motiejaus Gustaičio Memorialinis Muziejus, Lazdijai ... 30301
Muziejus, Kintai ... 30275
Poeto Jovaro Namas, Šiauliai ... 30359
Professoriaus Adomo Hrebnickio Memorialinis Muziejus, Dūkšto ... 30227
Rašytojo Tomo Mano Memorialinis Muziejus, Nida ... 30322
Rašytojos Žemaitės Memorialinis Muziejus, Bukante ... 30218
Rumšiškių Jono Aisčio (Jono Aleksandravičiaus) Muziejus, Rumšiškės ... 30348
Salomėjos Nėries Memorialinis Muziejus, Kaunas ... 30266
Šiaulių Rajono Literatūros Muziejus, Naisiai ... 30316
Simono Stanevičiaus Memorialinė Sodyba-muziejus, Kanopenai ... 30237
Vaikų Literatūros Muziejus, Kaunas ... 30267
Vilniaus Universiteto Adomo Mickevičiaus Muziejus, Vilnius ... 30442
Vinco Krėvės-Mickevičiaus Memorialinis Muziejus, Vilnius ... 30447
Vinco Mykolaičio-Putino Memorialinis Butas-muziejus, Vilnius ... 30448
Vinco Mykolaičio-Putino Memorialinis Muziejus, Marijampolė ... 30310
Vinco Mykolaičio-Putino Memorialinis Muziejus, Skriaudžiai ... 30371

Luxembourg

Musée Littéraire Victor Hugo, Vianden ... 30495

Malaysia

Muzium Sastera, Melaka ... 30602

Mauritius

Robert Edward Hart Memorial Museum, Souillac ... 30685

Mexico

Casa Museo Amado Nervo, Tepic ... 31439
Casa Museo Carlos Pellicer Cámara, Villahermosa ... 31546
Casa Museo Maestro Manuel Altamirano, Tixtla de Guerrero ... 31460
Museo Casa de Ramón López Velarde, Jerez de García Salinas ... 30973
Museo del Recuerdo, México ... 31107
Museo Elías Nandino, Cocula ... 30779
Museo José Fernando Ramírez, Durango ... 30843
Museo Othoniano, San Luis Potosí ... 31352
Museo Ramón López Velarde, México ... 31143

Mongolia

Natsagdorj Museum, Ulaanbaatar ... 31629

Netherlands

A.M. de Jonghuis, Nieuw Vossemeer ... 32573
Bilderdijkmuseum, Amsterdam ... 31835
It Gysbert Japicxhûs Museum, Bolsward ... 31979
Letterkundig Museum en Kinderboekenmuseum, Den Haag ... 32078
Multatuli Museum, Amsterdam ... 31860
Theo Thijssen Museum, Amsterdam ... 31895

L

Tresoar, Frysk Histoarysk en Letterkundich Sintrum, Leeuwarden 32479

Nicaragua

Museo Archivo Rubén Darío, León 33284
Museo Casa Natal Rubén Darío, Darío 33275

Northern Cyprus

Namık Kemal Zindanı ve Müzesi, Gazimağusa 33352

Norway

Aulestad, Follebu 33502
Bjørgan Prestegård, Kvikne 33620
Henrik Ibsen Museum, Skien 33834
Ibsen-museet, Oslo 33727
Ibsenhuset og Grimstad Bymuseet, Grimstad 33519
Ivar Aasen-tunet, Hovdebygda 33568
Kittelsenhuset, Kragerø 33608
Knudaheio - Garborgheimen, Undheim 33945
Petter Dass-Museet, Sandnessjøen 33816
Ratvolden, Røros 33795
Sagstua Skolemuseum, Sagstua 33805
Vinjestoga, Vinje 33983

Pakistan

Iqbal Museum, Lahore 34021

Panama

Museo el Pausílipo, Las Tablas 34076

Peru

Casa Museo Ricardo Palma, Lima 34210

Philippines

Ayala Museum, Vigan 34521
University of Nueva Caceres Museum, Naga 34467

Poland

Dom Gerharta Hauptmanna, Jelenia Góra 34703
Dworek Wincentego Pola, Lublin 34896
Gabinet Pisarza Pawła Hulki-Lasowskiego, Żyradów 35380
Jan Kochanowski Muzeum, Czarnolas 34611
Muzeum Andrzeja Struga, Warszawa 35235
Muzeum Biograficzne Władysława Orkana, Niedźwiedź 34944
Muzeum Bolesława Prusa, Nałęczów 34938
Muzeum Emila Zegadłowicza, Wadowice 35208
Muzeum Henryka Sienkiewicza, Wola Okrzejska 35315
Muzeum Henryka Sienkiewicza, Oblęgorek 34967
Muzeum im. Adama Mickiewicza w Śmiełowie, Żerków 35369
Muzeum im. J.G. Herdera w Morągu, Morąg 34928
Muzeum Jana Kasprowicza, Zakopane 35356
Muzeum Józefa Ignacego Kraszewskiego, Romanów 35086
Muzeum Kornela Makuszyńskiego, Zakopane 35358
Muzeum Lat Szkolnych Stefana Żeromskiego, Kielce 34741
Muzeum Literackie Henryka Sienkiewicza, Poznań 35038
Muzeum Literackie im. Józefa Czechowicza, Lublin 34904
Muzeum Literatury im. Adama Mickiewicza, Warszawa 35256
Muzeum Literatury im. Jarosława Iwaszkiewicza, Sandomierz 35098
Muzeum Marii Konopnickiej w Żarnowcu, Jedlicze 34701
Muzeum Okręgowe, Suwałki 35144
Muzeum Piśmiennictwa i Muzyki Kaszubsko-Pomorskiej, Wejherowo 35296
Muzeum Pozytywizmu im. Aleksandra Świętochowskiego, Gołotczyzna 34674
Muzeum-Pracownia Literacka Arkadego Fiedlera, Puszczykowo 35067
Muzeum Romantyzmu, Opinogóra 34982
Muzeum Stanisława Staszica, Piła 35008
Muzeum Stefana Żeromskiego, Nałęczów 34940
Muzeum Władysława Broniewskiego, Warszawa 35278
Oddział Literacki im. Marii Dąbrowskiej w Russowie, Russów 35088
Regionalne Muzeum Młodej Polski - Rydlówka, Kraków 34819

Portugal

Casa-Musei de Aquilino Ribeiro, Peva 35822
Casa-Museu de Camilo, Vila Nova de Famalicão 36003
Casa Museu de Sarah Beirão, Tábua 35958
Casa Museu Fernando de Castro, Porto 35837
Casa Museu Fernando Namora, Condeixa-a-Nova 35547
Casa-Museu Ferreira de Castro, Ossela 35795
Casa Museu Guerra Junqueiro, Porto 35838
Casa-Museu José Régio, Portalegre 35832
Casa Museu José Régio, Vila do Conde 35987
Casa-Museu Maria da Fontinha, Gafanhão 35624
Casa Museu Vasco Lima Couto, Constância 35549
Fundação Eça de Queiroz, Santa Cruz do Douro 35900
Museu Ferreira de Castro, Sintra 35948
Museu Júlio Dinis, Ovar 35800
Museu Nacional da Literatura, Porto 35859
Museu Rodrigues Lapa, Anadia 35435
Museu Sebastião da Gama, Vila Nogueira de Azeitão 36001

Romania

Casa Memorială a Poetului Ion Şiugariu, Băiţa 36132
Casa Memorială Alexandru Sahia, Mănăstirea 36460
Casa Memorială Alexandru Vlahuţă, Agapia 36098
Casa Memorială Anton Pann, Râmnicu Vâlcea 36550
Casa Memorială Benedek Elek, Bătanii Mici 36144
Casa Memorială Cezar Petrescu, Buşteni 36238
Casa Memorială Costache Negruzzi, Hermeziu 36398
Casa Memorială D.P. Perpessicius, Brăila 36167
Casa Memorială Elena Farago, Craiova 36321
Casa Memorială Eusebiu Camilar, Udeşti 36713
Casa Memorială Gala Galaction, Bucureşti 36186
Casa Memorială George Agatha Bacovia, Bucureşti 36187
Casa Memorială George Bacovia, Bacău 36118
Casa Memorială George Topârceanu, Iaşi 36410
Casa Memorială Gheorghe Rădulescu-Motru, Butoieşti 36240
Casa Memorială I.C. Vissarion, Costeştii din Vale 36317
Casa Memorială Ion Creangă, Târgu Neamţ 36673
Casa Memorială Ion Minulescu Claudia Millian, Bucureşti 36188
Casa Memorială Liviu Fanny Rebreanu, Bucureşti 36189
Casa Memorială Liviu Rebreanu, Ştefăneşti 36634
Casa Memorială Lucian Blaga, Lancrăm 36434
Casa Memorială Mihai Codreanu, Iaşi 36411
Casa Memorială Mihail Sadoveanu, Fălticeni 36364
Casa Memorială Mihail Sadoveanu, Iaşi 36412
Casa Memorială Mihail Sadoveanu, Vânători-Neamţ 36723
Casa Memorială N.D. Cocea, Sighişoara 36616
Casa Memorială Nichita Stănescu, Ploieşti 36528
Casa Memorială Nicolae Gane, Iaşi 36413
Casa Memorială Nicolae Labiş, Mălini 36458
Casa Memorială Nicu Enea, Bacău 36119
Casa Memorială Nikolaus Lenau, Lenauheim 36438
Casa Memorială Otilia Cazimir, Iaşi 36414
Casa Memorială Panait Cerna, Cerna 36272
Casa Memorială Panait Istrati, Brăila 36168
Casa Memorială Tamasi Aron, Lupeni 36452
Casa Memorială Traian Demetrescu, Craiova 36322
Casa Memorială Tudor Arghezi, Bucureşti 36190
Casa Memorială Vasile Alecsandri, Mireşti 36476
Casa Memorială Vasile Pogor, Iaşi 36415
Casa Memorială Vasile Voiculescu, Pârscov 36511
Casa Memorială Veronica Micle, Târgu Neamţ 36674
Casa Mitropolitului Dosoftei, Iaşi 36416
Colecţia Memorială Octavian Goga, Răşinari 36555
Colecţia Muzeală a Mănăstirii Ortodoxe Suzana, Măneciu 36462
Expoziţia de Istorie a Culturii Nămăeşti, Câmpulung 36251
Expoziţia Memorială Adam Müller-Guttenbrunn, Zăbrani 36740
Expoziţia Muzeală I.L. Caragiale, Ghirdoveni 36384
Expoziţia Nichita Stănescu, Teremia Mare 36684
Memorialul Ipoteşti-Centrul Naţional de Studii Mihai Eminescu, Ipoteşti 36429
Muzeul Ion Luca Caragiale, Ploieşti 36531
Muzeul Literaturii Române, Iaşi 36421
Muzeul Memorial Ady Endre, Căuaş 36270
Muzeul Memorial Ady Endre, Oradea 36503
Muzeul Memorial Alexandru Vlahuţă, Dumbrăveni 36361
Muzeul Memorial Bojdeuca Ion Creangă, Iaşi 36422
Muzeul Memorial Calistrat Hogaş, Piatra-Neamţ 36522
Muzeul Memorial Costache Negri, Costache Negri 36315
Muzeul Memorial Emil Isac, Cluj-Napoca 36288
Muzeul Memorial George Călinescu, Bucureşti 36209
Muzeul Memorial George Coşbuc, Coşbuc 36314
Muzeul Memorial Ioan Slavici şi Emil Monţia, Şiria 36623
Muzeul Memorial Iosif Vulcan, Oradea 36504
Muzeul Memorial János Arany, Salonta 36580
Muzeul Memorial Liviu Rebreanu, Năsăud 36487
Muzeul Memorial Sándor Petőfi, Coltău 36295
Muzeul Mihai Eminescu, Iaşi 36425
Muzeul National al Literaturii Române, Bucureşti 36215
Muzeul Scriitorilor Târgovişteni, Târgovişte 36660
Punct Muzeal Dimitrie Anghel, Miroslava 36477

Russia

Aleksandrovskij Literaturno-chudožestvennyj Muzej Mariny i Anastasii Cvetaevych, Aleksandrov 36753
Archangelskij Literaturnyj Muzej, Archangelsk 36766
Arzamasskij Gosudarstvennyj Literaturno-Memorialnyj Muzej A.P. Gajdara, Arzamas 36774
Brjanskij Literaturnyj Muzej, Brjansk 36825
Detskij Muzej Buratino-Pinokkio, Moskva 37158
Detskij Muzej Dom-Skazok Žili-byli, Moskva 37159
Dom I.S. Ostrouchova, Moskva 37160
Dom-muzej L.N. Andreeva, Orël 37503
Dom-muzej A.P. Čechova, Moskva 37161
Dom-muzej Velimira Chlebnikova, Astrachan 36781
Dom-muzej K.I. Čukovskogo, Peredelkino 37539
Dom-muzej M.I. Cvetaevoj, Moskva 37162
Dom-muzej F.M. Dostoevskogo, Staraja Russa 37796
Dom-muzej N.V. Gogolja, Moskva 37164
Dom-muzej A.S. Grina, Kirov 37036
Dom-muzej V.G. Korolenko, Gelendžik 36922
Dom-muzej M.Ju. Lermontova, Moskva 37165
Dom-muzej N.S. Leskova, Orël 37505
Dom-muzej N.S. Muchina, Olikjal 37494
Dom-muzej N.S. Muchina, Joškar-Ola 36976
Dom-muzej N.A. Nekrasova, Čudovo 36871
Dom-muzej I.S. Nikitina, Voronež 37970
Dom-muzej A.N. Ostrovskogo, Moskva 37166
Dom-muzej B.L. Pasternaka, Peredelkino 37540
Dom-muzej M.M. Prišvina, Dunino 36881
Dom-muzej M.E. Saltykova-Ščedrina, Kirov 37037
Dom-muzej Semji Cvetaevych, Novo-Talicy 37476
Dom-muzej Tukaevych, Košlauč 37060
Dom-muzej G.I. Uspenskogo, Sjabrenicy 37767
Dom-Muzej V.V. Veresaeva, Tula 37865
Dom Poėta A.N. Širjaevca - Muzej Krestjanskogo Byta, Žigulevsk 37988
Eleckij Literaturno-memorialnyj Muzej Pisatelja I.A. Bunina, Elec 36909
Ėžvinskij Literaturnyj Muzej, Syktyvkar 37813
Gatčinskij Literaturno-memorialnyj Muzej-usadba P.E. Ščerbova, Gatčina 36919
Gosudarstvennyj Istoriko-kulturnyj i Prirodnyj Muzej-zapovednik A.S. Griboedova - Chmelita, Chmelita 36863
Gosudarstvennyj Istoriko-literaturnyj i Prirodnyj Muzej-zapovednik A.A. Bloka, Solnečnogorsk 37786
Gosudarstvennyj Istoriko-literaturnyj Muzej-Zapovednik A.S. Puškina s usadbami Vjazemy i Zacharovo, Bolšie Vjazemy 36814
Gosudarstvennyj Literaturno-memorialnyj Muzej Anny Achmatovoj v Fontannom Dome, Sankt-Peterburg 37642
Gosudarstvennyj Literaturno-memorialnyj Muzej N.A. Dobroljubova, Nižnij Novgorod 37451
Gosudarstvennyj Literaturno-memorialnyj Muzej im. M.Ju. Lermontova, Pjatigorsk 37564
Gosudarstvennyj Literaturno-memorialnyj Muzej N. Ostrovskogo, Soči 37782
Gosudarstvennyj Literaturno-memorialnyj Muzej A.N. Radiščeva, Radiščevo 37589
Gosudarstvennyj Literaturno-memorialnyj Muzej-zapovednik A.P. Čechova, Melichovo 37129
Gosudarstvennyj Literaturno-memorialnyj Muzej-zapovednik N.A. Nekrasova "Karabicha", Karabicha 37002
Gosudarstvennyj Literaturnyj Muzej, Moskva 37195
Gosudarstvennyj Literaturnyj Muzej A.M. Gorkogo, Nižnij Novgorod 37452
Gosudarstvennyj Literaturnyj Muzej im. I.S. Turgeneva, Orël 37507
Gosudarstvennyj Memorialnyj Dom-muzej Bulata Okudžavy, Peredelkino 37541
Gosudarstvennyj Memorialnyj i Prirodnyj Muzej-zapovednik A.N. Ostrovskogo - Ščelykovo, Ščelykovo 37756
Gosudarstvennyj Memorialnyj i Prirodnyj Muzej-zapovednik I.S. Turgeneva, Spasskoe-Lutovinovo 37792
Gosudarstvennyj Memorialnyj Muzej-usadba V.G. Belinskogo, Belinskij 36797
Gosudarstvennyj Muzej K.A. Fedina, Saratov 37749
Gosudarstvennyj Muzej - Gumanitarnyj Centr 'Preodolenie' im. N.A. Ostrovskogo, Moskva 37198
Gosudarstvennyj Muzej A.I. Kuprina, Narovčat 37445
Gosudarstvennyj Muzej V.V. Majakovskogo, Moskva 37199
Gosudarstvennyj Muzej A.S. Puškina, Moskva 37202
Gosudarstvennyj Muzej L.N. Tolstogo, Moskva 37204
Gosudarstvennyj Muzej L.N. Tolstogo, Moskva 37205
Gosudarstvennyj Muzej L.N. Tolstogo "Jasnaja Poljana", Jasnaja Poljana (Tula) 36975
Gosudarstvennyj Muzej-zapovednik S.A. Esenina, Konstantinovo 37056
Gosudarstvennyj Muzej-zapovednik M.A. Šolochova, Vešenskaja 37925
Istoriko-literaturnyj Muzej I.A. Gončarova, Uljanovsk 37898
Istoriko-memorialnyj Muzej M.V. Lomonosova, Lomonosovo 37112
Jaroslavskij Istoriko-architekturnyj i Chudožestvennyj Muzej-zapovednik, Jaroslavl 36968
Kirovskij Oblastnoj Kraevedčeskij Muzej, Kirov 37039
Kraevoj Muzej Istorii Literatury, Iskusstva i Kultury Altaja, Barnaul 36791
Kulturno-obrazovatelnyj centr Škola L.N. Tolstogo, Moskva 37222
Kulturnyj Centr-muzej V.S. Vysockogo, Moskva 37224
Lermontovskij Gosudarstvennyj Muzej-zapovednik Tarchany, Lermontovo 37100
Lgovskij Literaturno-memorialnyj Muzej N.N. Aseeva, Lgov 37102
Lgovskij Literaturno-memorialnyj Muzej A.P. Gajdara, Lgov 37103
Literaturno-chudožestvennyj Muzej-usadba Prijutino, Vsevoložsk 37979
Literaturno-memorialnyj Dom-Muzej D.N. Mamina-Sibirjaka, Ekaterinburg 36887
Literaturno-memorialnyj Dom-Muzej F.M. Rešetnikova, Ekaterinburg 36888
Literaturno-memorialnyj Dom-muzej Sulejmana Stalskogo, Ašaga-Stalsk 36776
Literaturno-memorialnyj Muzej F.A. Abramova, Verkola 37924
Literaturno-memorialnyj Muzej F.M. Dostoevskogo, Novokuzneck 37480
Literaturno-memorialnyj Muzej F.M. Dostoevskogo, Sankt-Peterburg 37656
Literaturno-memorialnyj Muzej Vasilija Erošenko, Staryj Oskol 37800
Literaturno-memorialnyj Muzej M. Gorkogo, Samara 37616
Literaturno-memorialnyj Muzej A.M. Gorkogo, Kazan 37011
Literaturno-memorialnyj Muzej Stancionnyj Smotritel, Vyra 37983
Literaturno-memorialnyj Muzej M.M. Zoščenko, Sankt-Peterburg 37657
Literaturno-teatralnyj Muzej N.M. Djakonova, Syktyvkar 37816
Literaturnyj Muzej, Krasnojarsk 37074
Literaturnyj Muzej, Penza 37529
Literaturnyj Muzej Abakumcevo, Abakumcevo 36746
Literaturnyj Muzej Goroda Nalčik, Nalčik 37439
Literaturnyj Muzej im. K.V. Ivanova, Čeboksary 36838
Literaturnyj Muzej Instituta Russkoj Literatury, Sankt-Peterburg 37658
Literaturnyj Muzej F.I. Tjutčeva, Ovstug 37522
Memorialnaja Kvartira Andreja Belogo, Moskva 37227
Memorialno-literaturnyj Muzej A.P. Bondina, Ekaterinburg 36889
Memorialnyj Dom-muzej S.T. Aksakova, Ufa 37884
Memorialnyj Dom-muzej P.P. Bažova, Ekaterinburg 36890
Memorialnyj Dom-muzej Gercena, Moskva 37231
Memorialnyj Dom-muzej G.D. Krasilnikova, Alnaši 36755
Memorialnyj Dom-muzej Mažita Gafuri, Ufa 37885
Memorialnyj Dom-muzej Marko Vovčok, Nalčik 37440
Memorialnyj Kabinet A.V. Lunačarskogo, Moskva 37233
Memorialnyj Kompleks G. Tukaja, Novyj Kyrlaj 37493
Memorialnyj Muzej-dača A.S. Puškina, Puškin 37581
Memorialnyj Muzej-kvartira N.A. Nekrasova, Sankt-Peterburg 37661
Memorialnyj Muzej-kvartira A.S. Puškina, Moskva 37244
Memorialnyj Muzej-kvartira A.A. Šogencukova, Nalčik 37441
Memorialnyj Muzej Licej, Puškin 37582
Memorialnyj Muzej A.G. Malyškina, Mokšan 37138
Memorialnyj Muzej B.L. Pasternaka, Čistopol 36866
Michajlovskoe - Gosudarstvennyj Memorialnyj Istoriko-literaturnyj i Prirodno-landšaftnyj Muzej-Zapovednik A.S. Puškina, Puškinskie Gory 37587
Moskovskij Gosudarstvennyj Muzej S.A. Esenina, Moskva 37255
Moskovskij Literaturnyj Muzej-centr K.G. Paustovskogo, Moskva 37257
Municipalnyj Objedinennyj Muzej Pisatelej Urala, Ekaterinburg 36891
Municipalnyj Muzej Anna Achmatova - Serebrjannyj vek, Sankt-Peterburg 37667
Muzej E.A. Boratynskogo, Kazan 37014
Muzej I.A. Bunina, Orël 37508
Muzej-čitalnja N.V. Fëdorova, Moskva 37268
Muzej-čitalnja na Nacionalnych Jazykach, Penza 37530
Muzej V.I. Dalja, Moskva 37269
Muzej Detstva A.M. Gorkogo - Domik Kaširina, Nižnij Novgorod 37454
Muzej Ėkslibrisa, Moskva 37273
Muzej Gabdulla Tukaja, Kazan 37015
Muzej A.M. Gorkogo, Moskva 37279
Muzej Istorii Teatrov Dagestana, Machačkala 37118

Muzej Kajuma Nasyri, Kazan ... 37018
Muzej A. Kolcova, Voronež ... 37972
Muzej Kukol i Detskoj Knigi "Strana Čudes", Ekaterinburg ... 36895
Muzej Kultury Astrachani, Astrachan ... 36785
Muzej-kvartira M.I. Cvetaevoj v Bolševo, Korolev ... 37059
Muzej-kvartira F.M. Dostoevskogo, Moskva ... 37316
Muzej-kvartira A.M. Gorkogo, Moskva ... 37318
Muzej-kvartira A.M. Gorkogo, Nižnij Novgorod 37460
Muzej-kvartira G.D. Krasilnikova, Iževsk ... 36954
Muzej-kvartira V.S. Mejercholda, Moskva ... 37319
Muzej-kvartira Musa Džalilja, Kazan ... 37020
Muzej-kvartira A. Puškina, Sankt-Peterburg ... 37694
Muzej-kvartira T.G. Ševčenko, Sankt-Peterburg ... 37696
Muzej-kvartira I.D. Sytina, Moskva ... 37323
Muzej-kvartira A.N. Tolstogo, Moskva ... 37325
Muzej Literaturnaja Žizn' Urala XIX Veka, Ekaterinburg ... 36897
Muzej Literaturnoj Žizni Jaroslavskogo Kraja, Jaroslavl ... 36973
Muzej Literatury Burjatii im. Choca Namsaraeva, Ulan-Udė ... 37895
Muzej Literatury XVII - XIX vekov, Moskva ... 37330
Muzej V.V. Nabokova, Sankt-Peterburg ... 37701
Muzej-nekropol Literatorskie Mostki, Sankt-Peterburg ... 37702
Muzej Pisatelej-orlovcev, Orël ... 37510
Muzej Pisatelja A.P. Bibika, Mineralnye Vody ... 37133
Muzej Poėta S. Orlova, Belozersk ... 36802
Muzej A.S. Puškina, Toržok ... 37861
Muzej A.S. Puškina v Bernove, Bernov ... 36804
Muzej Vladimira Raevskogo, Bogoslovka ... 36813
Muzej Nikolaja Rubcova, Moskva ... 37353
Muzej Russkaja Literatura XX Veka, Moskva ... 37355
Muzej M.E. Saltykova-Ščedrina, Tver ... 37878
Muzej Šarifa Kamala, Kazan ... 37024
Muzej Sceničeskogo Iskusstva im. V.Ė. Mejercholda, Penza ... 37535
Muzej Serebrjanogo Veka - Dom V.Ja. Brjusova, Moskva ... 37361
Muzej A.K. Tolstogo, Krasnyj Rog ... 37083
Muzej L.N. Tolstogo na stancii Lev Tolstoj, Stancija Lev Tolstoj ... 37794
Muzej-usadba N.G. Černyševskogo, Saratov ... 37751
Muzej- Usadba F.M. Dostoevskogo Darovoe, Darovoe ... 36873
Muzej-usadba N.A. Durovoj, Elabuga ... 36906
Muzej-usadba "Muranovo" im. F.I. Tjutčeva, Lugovskoe ... 37114
Muzej-usadba Ostafjevo "Russkij Parnas", Ostafjevo ... 37518
Muzej-usadba poėta D. Venevitinova, Novoživotinnoe ... 37492
Muzej-usadba Roždestveno, Roždestveno ... 37603
Muzej-usadba A.N. Tolstogo, Samara ... 37618
Muzej-usadba L.N. Tolstogo Nikolskoe Vjazemskoe, Nikolskoe-Vjazemskoe ... 37449
Muzej-Usadba L.N. Tolstogo v Chamovnikach, Moskva ... 37380
Muzej-zapovednik Abramcevo, Abramcevo ... 36747
Muzej-zapovednik Pisatelja S.T. Aksakova, Aksakovo ... 36750
Omskij Gosudarstvennyj Literaturnyj Muzej im. F.M. Dostoevskogo, Omsk ... 37500
Polibinskij Memoralnyj Muzej-usadba S.V. Kovalevskoj, Polibino ... 37570
Severo-osetinskij Literaturnyj Muzej im. K.L. Četagurova, Vladikavkaz ... 37933
Taganrogskij Gosudarstvennyj Literaturnyj i Istoriko-architekturnyj Muzej-zapovednik, Taganrog ... 37828
Taganrogskij Muzej im. A.P. Čechova, Taganrog ... 37830
Tarusskij Muzej Semji Cvetaevych, Tarusa ... 37837
Vserossijskij Memorialnyj Muzej-zapovednik V.M. Šukšina, Srostki ... 37793
Vserossijskij Muzej Puškina, Sankt-Peterburg 37732
Vystavočnye Zaly na Starom Arbate, Moskva 37414
Vystavočnye Zaly v Dome Aksakovych, Moskva ... 37416

Samoa

Robert Louis Stevenson Museum, Vailima ... 38001

Serbia

Memorijalni Muzej Džuro Jakšić, Srpska Crnja 38154
Muzej Kuca Bore Stankovica, Vranje ... 38161
Spomen Muzej Ive Andrića, Beograd ... 38089

Slovakia

Jozef Gregor Tajovský Pamätný Dom, Tajov ... 38368
Kaštieľ Imre Madácha a Literárne osobnosti, Dolná Strehová ... 38249
Literárna Expozícia - Múzeum Janka Jesenského, Bratislava ... 38218
Literárne a Hudobné Múzeum, Banská Bystrica 38183
Literárne a Vlastivedné Múzeum, Svätý Jur ... 38365
Literárne Múzeum Pavla Horova, Bánovce nad Ondavou ... 38182
Múzeum Beckov, Beckov ... 38198
Múzeum Čachtice, Čachtice ... 38241
Múzeum Jána Kollára, Mošovce ... 38308
Múzeum Janka Kráľa, Liptovský Mikuláš ... 38287
Múzeum Ludovíta Štúra, Modra ... 38306
Oravské Múzeum Pavla Országha Hviezdoslava, Dolný Kubín ... 38251
Pamätná Izba Sama Chalupku, Horná Lehota ... 38260
Pamätný Dom Kálmána Mikszátha, Sklabiná ... 38351
Pamätný Dom Pavla Jozefa Šafárika, Kobeliarovo ... 38268
Rodný Dom Jána Hollého, Borský Mikuláš ... 38202
Rodny Dom Pavla E. Dobšinského, Slavošovce 38352
Slovenské Múzeum A.S. Puškina, Brodzany ... 38239
Slovenské Národné Literárne Múzeum, Martin ... 38300

Slovenia

Kurnikova Hiša, Tržič ... 38530
Muzej Josipa Jurčiča, Muljava ... 38485
Prešernov Spominski Muzej, Kranj ... 38438
Rojstna Hiša Matije Čopa, Žirovnica ... 38543
Rojstna Hiša Pesnika Dr. Franceta Prešerna, Žirovnica ... 38544
Rojstna Hiša Pisatelja Frana Saleškega Finžgarja, Žirovnica ... 38545
Rojstna Hiša Simona Gregorčiča, Vrsno ... 38536
Spominska Soba Dragotina Ketteja na Premu, Prem ... 38498
Spominska Zbirka Pisatelja Ivana Cankarja, Vrhnika ... 38535

South Africa

Afrikaanse Taalmuseum, Paarl ... 38699
Laurens Van der Post Museum, Philippolis ... 38702
Nasionale Afrikaanse Letterkundige Museum en Navorsingsentrum, Bloemfontein ... 38566
National English Literary Museum, Grahamstown ... 38631
Olive Schreiner House, Cradock ... 38600

Spain

Casa de Cervantes, Esquivias ... 39284
Casa de la Panadería, Madrid ... 39503
Casa-Museo Antonio Machado, Segovia ... 40049
Casa-Museo Azorín, Monóvar ... 39667
Casa Museo Blasco Ibáñez, Valencia ... 40195
Casa Museo Curros Enríquez, Celanova ... 39160
Casa Museo de Cervantes, Valladolid ... 40227
Casa-Museo de Lope de Vega, Madrid ... 39504
Casa Museo de Unamuno, Salamanca ... 39934
Casa Museo de Zorrilla, Valladolid ... 40229
Casa Museo Federico García Lorca, Fuente Vaqueros ... 39317
Casa-Museo Menéndez Pelayo, Santander ... 40008
Casa Museo Rosalía de Castro, Padrón ... 39764
Casa-Museo Tomás Morales, Moya ... 39688
Casa Museo Zenobia y Juan Ramón Jiménez, Moguer ... 39653
Casa Museu Àngel Guimerà, El Vendrell ... 40259
Casa-Museu del Poeta Verdaguer, Folgueroles 39301
Museo Casa Natal de Cervantes, Alcalá de Henares ... 38813
Museo del Esparto, Cieza ... 39181
Museo del Quijote, Ciudad Real ... 39188
Museu-Casa Verdaguer, Vallvidrera ... 40251
Museu Víctor Balaguer, Vilanova i la Geltrú ... 40287

Sweden

Alsters Herrgård, Karlstad ... 40638
C.M. Bellman Museum, Stockholm ... 40888
Dan Andersson Museum, Ludvika ... 40695
Engströmsgården, Hult ... 40605
Eyvind Johnson-Stugan, Boden ... 40462
Ivar Lo-Museet, Stockholm ... 40901
Mårbacka Manor, Östra Ämtervik ... 40799
Orgelmuseet i Fläckebo, Salbohed ... 40829
Sandellmuseet, Ramkvilla ... 40813
Strindbergsmuseet, Stockholm ... 40944
Vättehult - Pälle Nävers Diktarstuga, Vetlanda ... 41057
Verner von Heidenstams Hem, Motala ... 40760

Switzerland

Bibliothèque Publique et Universitaire, Salle Rousseau, Neuchâtel ... 41676
Casa Anatta e Casa Selma, Ascona ... 41127
Centre Dürrenmatt, Neuchâtel ... 41678
Dichter- und Stadtmuseum Liestal (DISTL)/ Herwegh-Archiv, Liestal ... 41561
Fondation Martin Bodmer, Cologny ... 41308
Fondation Rainer Maria Rilke, Sierre ... 41866
Gottfried-Keller-Zentrum, Glattfelden ... 41439
Gotthelf-Stube, Lützelflüh ... 41578
Institut et Musée Voltaire, Genève ... 41413
Johanna-Spyri-Museum im Alten Schulhaus, Hirzel ... 41484
Musée de la Science-Fiction, de l'Utopie et des Voyages Extraordinaires, Yverdon-les-Bains ... 42048
Musée Jean-Jacques Rousseau, Genève ... 41425
Musée Jean-Jacques Rousseau, Môtiers ... 41648
Musée Sherlock Holmes, Lucens ... 41577
Museo Hermann Hesse, Montagnola ... 41633
Museo Rainis e Aspazija, Castagnola ... 41284
Museum Strauhof, Zürich ... 42098
Schloß Jegenstorf, Jegenstorf ... 41498
Sherlock Holmes-Museum, Meiringen ... 41621
Thomas-Mann-Archiv, Zürich ... 42111

Taiwan

Ch'ien Mu House, Taipei ... 42380
Chung Li-Ho Memorial Museum, Meinong ... 42271
Laiho Museum, Changhua ... 42156
Lin Yutang House, Taipei ... 42401
National Museum of Taiwanese Literature, Tainan ... 42366

Turkey

Aşiyan Müzesi, İstanbul ... 42802
Cahit Sıtkı Tarancı Müze-Evi, Diyarbakır ... 42761
Divan Edebiyatı Müzesi, İstanbul ... 42815
Mehmet Akif Ersoy Evi, Ankara ... 42705
Orhan Kemal Müzesi, İstanbul ... 42833
Sait Faik Müzesi, Burgaz Adası ... 42739
Yunus Emre Müzesi, Eskişehir ... 42782

Ukraine

Biblioteka-muzej A.P. Gajdara, Kaniv ... 43057
Budinok-muzej A. Čechova, Sumy ... 43188
Černigivs'kyjy Literaturno-memorial'nyj Muzej-zapovidnik Mychajlo Kocjubynskogo, Černigiv ... 43005
Černivec'kyj Literaturno-Memorial'nyj Muzej im. Ju.A. Fedkoviča, Černivci ... 43011
Černivec'kyj Literaturno-Memorial'nyj Muzej O. Kobyljans'koï, Černivci ... 43012
Charkivs'kyj Literaturnyj Muzej, Charkiv ... 43019
Deržavnyj Istoriko-kul'turnyj Zapovidnyk Bat'kivščyna Tarasa Ševčenka, Ševčenkove ... 43182
Dom-muzej A.P. Čechova, Jalta ... 43049
Istoriko-memorialnyj Muzej Semji Fadeevyh-Chan-Blavackij, Dnipropetrovs'k ... 43034
Kirovograds'kyj Literaturno-memorial'nyj Muzej M.L. Kropyvnyc'kogo, Kirovograd ... 43064
Kirovograds'kyj Literaturnyj Muzej, Kirovograd ... 43065
Kyïvs'kyj Literaturno-memorial'nyj Budinok-muzej M. Ryl's'kogo, Kyïv ... 43080
Kyïvs'kyj Literaturno-memorial'nyj Budinok-muzej Tarasa Ševčenka, Kyïv ... 43081
Kyïvs'kyj Literaturno-memorial'nyj Muzej-kvartyra P. Tyčyny, Kyïv ... 43082
Kyïvskyj Muzej Puškina, Kyïv ... 43084
Literaturno-Memoral'nyj Muzej N.A Ostrovskogo, Šepetovka ... 43174
Literaturno-Memorialny Muzej A.A. Achmatovoï, Slobidka Šelechivs'ka ... 43187
Literaturno-memorial'nyj Budinok-muzej M. Bulgakova, Kyïv ... 43086
Literaturno-memorialnyj Muzei O.S. Puškina ta P.I. Čajkovskogo, Kam'janka ... 43056
Literaturno-memorialnyj Muzej Ivana Franko, Lviv ... 43123
Literaturno-Memorial'nyj Muzej Lesi Ukrayinky, Kolodjažne ... 43070
Literaturno-Memorial'nyj Muzej M. Eminesku, Černivci ... 43015
Literaturno-memorial'nyj Muzej Puškina, Odesa ... 43154
Literaturno-memorial'nyj Muzej T.G. Ševčenka, Ševčenkove ... 43183
Literaturno-mystec'kyj Muzej, Chmel'nyc'kyj ... 43028
Literaturnyj Lviv Peršoï Polovyny XXst, Lviv ... 43124
Literaturnyj Muzej Literaturna Chersonščina, Cherson ... 43023
Manujlivs'kyj Literaturno-memorialnyj Muzej O.M. Gor'kogo, Verchnja Manujlivka ... 43202
Memorial'nyj Budinok-muzej V. Dalja, Lugans'k ... 43116
Memorial'nyj Budinok-muzej Tarasa Ševčenka, Kyïv ... 43087
Memorijalnyj Budinok-muzej D. Javornic'kogo, Dnipropetrovs'k ... 43035
Muzej I. Karpenka-Karogo, Kirovograd ... 43069
Muzej "Kobzarja" T.G. Ševčenka, Čerkasy ... 43003
Muzej Lesi Ukraïnki, Kyïv ... 43094
Muzej Vydatnych Dijačiv Ukraïns'koï Kul'tury, Kyïv ... 43099
Nacional'nyj Muzej Literatury Ukraïny, Kyïv ... 43105
Nacional'nyj Muzej Tarasa Ševčenka, Kyïv ... 43107
Novgorod-Siverskyj Istoriko-kulturnyj Muzej-zapovidnyk Slovo o Polku Igorevim, Novgorod-Sivers'kyj ... 43153
Odes'kyj Literaturno-memorial'nyj Muzej K. Paustovskogo, Odesa ... 43160
Odes'kyj Literaturnyj Muzej, Odesa ... 43161
Poltavs'kyj Krajeznavčyj Muzej, Poltava ... 43168
Poltavs'kyj Literaturno-memorial'nyj Muzej I.P. Kotljarevskogo, Poltava ... 43169
Poltavs'kyj Literaturno-memorial'nyj Muzej Panasa Myrnogo, Poltava ... 43170
Poltavs'kyj Literaturno-memorial'nyj Muzej V.G. Korolenka, Poltava ... 43171
Ševčenkivskyj Nacional'nyj Zapovidnyk, Kaniv ... 43060
Velykosoročynskyj Literaturno-memorialnyj Muzej Gogolja, Velyki Soročynci ... 43201
Vinnyc'kyj Literaturno-memorial'nyj Muzej M.M. Kocjubyns'koho, Vinnycja ... 43203
Zapovidnyk-muzej Gogolja, Gogoleve ... 43040
Žytomyr Oblasnyi Literaturnyj Muzej, Žytomyr 43215
Žytomyrs'kyj Literaturno-memorial'nyj Muzej V.G. Korolenka, Žytomyr ... 43216

United Kingdom

Abbotsford House, Melrose ... 45124
Bachelor's Club, Tarbolton ... 45826
Barrie's Birthplace Camera Obscura, Kirriemuir ... 44617
Bateman's, Etchingham ... 44187
Bateman's Kipling's House, Burwash ... 43689
Bleak House, Broadstairs ... 43644
Brontë Parsonage Museum, Haworth ... 44430
Broughton House and Garden, Kirkcudbright ... 44610
Brownsbank Cottage, Biggar ... 43480
Burns Cottage Museum, Alloway ... 43283
Burns House Museum, Mauchline ... 45118
Carlyle's House, London ... 44814
Ceiriog Memorial Institute, Glyn Ceiriog ... 44320
Charles Dickens Birthplace Museum, Portsmouth ... 45427
The Charles Dickens Museum, London ... 44817
Coleridge Cottage, Nether Stowey ... 45201
Cowper and Newton Museum, Olney ... 45316
D.H. Lawrence Birthplace Museum, Eastwood 44108
Dickens House Museum Broadstairs, Broadstairs ... 43646
Dove Cottage and the Wordsworth Museum, Grasmere ... 44348
Dr. Johnson's House, London ... 44839
Dylan Thomas Boathouse, Laugharne ... 44646
Ellisland Farm, Auldgirth ... 43337
Elstow Moot Hall, Bedford ... 43436
Hardy's Cottage, Higher Bockhampton ... 44467
Hauteville House, Saint Peter Port ... 45592
Hill Top, Ambleside ... 43298
Izaak Walton's Cottage, Shallowford ... 45646
James Hogg Exhibition, Selkirk ... 45636
Jane Austen's House, Chawton ... 43799
Jerome K. Jerome Birthplace Museum, Walsall ... 45919
John Buchan Centre, Biggar ... 43483
John Bunyan Museum, Bedford ... 43437
John Creasey Museum, Salisbury ... 45603
John Moore Countryside Museum, Tewkesbury ... 45842
Keats House, London ... 44892
Michael Bruce Cottage Museum, Kinnesswood 44606
Milton's Cottage, Chalfont Saint Giles ... 43787
Newstead Abbey Historic House and Gardens, Nottingham ... 45297
Richard Jefferies Museum, Coate ... 43865
Robert Burns Centre, Dumfries ... 44039
Robert Burns House, Dumfries ... 44040
Ruskin Library, Lancaster ... 44637
Samuel Johnson Birthplace Museum, Lichfield ... 44706
Shandy Hall, Coxwold ... 43910
Shaw's Corner, Ayot Saint-Lawrence ... 43343
Swaffham Museum, Swaffham, Norfolk ... 45808
Thomas Carlyle's Birthplace, Ecclefechan ... 44110
Tobias Smollett Museum, Alexandria ... 43275
Ty Mawr Wybrnant, Dolwyddelan ... 43990
Vennel Gallery, Irvine ... 44536
Wordsworth House, Cockermouth ... 43869
The Writers' Museum, Edinburgh ... 44162

United States of America

Arrowhead Museum, Pittsfield ... 51965
Boston Public Library, Boston ... 46923
Carl Sandburg Home, Flat Rock ... 48555
Cary Cottage, Cincinnati ... 47554
Curwood Castle, Owosso ... 51642
Edgar Allan Poe Cottage, Bronx ... 47044
Edgar Allan Poe House, Philadelphia ... 51830
Edgar Allan Poe House and Museum, Baltimore ... 46562
Edgar Allan Poe Museum, Richmond ... 52327
Emily Dickinson Museum, Amherst ... 46265
Ernest Hemingway House Museum, Key West 49759
Eugene O'Neill Home, Danville ... 47959
Harry Ransom Humanities Research Center, Austin ... 46510
Hemingway Museum, Oak Park ... 51454
House of Happy Walls, Glen Ellen ... 48919
John G. Neihardt Center, Bancroft ... 46592
John Greenleaf Whittier Home, Amesbury ... 46261

The Kate Chopin House and Bayou Folk Museum, Cloutierville 47665
Knott House Museum, Tallahassee 53423
Laura Ingalls Wilder-Rose Wilder Lane Historic Home and Museum, Mansfield 50450
Limberlost State Historic Site, Geneva 48879
Longfellow House, Cambridge 47180
Minute Man, Concord 47768
Museum of the Alphabet, Waxhaw 54000
My Old Kentucky Home, Bardstown 46607
National Steinbeck Center, Salinas 52677
O. Henry Museum, Austin 46517
Orchard House, Concord 47770
Pearl S. Buck Birthplace, Hillsboro 49290
Ralph Waldo Emerson House, Concord 47771
Robert Louis Stevenson Memorial Cottage, Saranac Lake 52911
Robert Louis Stevenson Silverado Museum, Saint Helena 52548
Rowan Oak, William Faulkner's Home, Oxford 51648
Sarah Orne Jewett House, South Berwick 53164
Scholte House Museum, Pella 51751
Sinclair Lewis Museum, Sauk Centre 52929
Slabsides, West Park 54076
Thomas Wolfe Memorial, Asheville 46377
Thornton W. Burgess Museum, Sandwich 52842
Vachel Lindsay Home, Springfield 53222
Volcano Art Center, Volcano 53766
Wadsworth-Longfellow House, Portland 52070
William Cullen Bryant Homestead, Cummington 47909
Wren's Nest House Museum, Atlanta 46458
Zane Grey Museum, Lackawaxen 49858

Uruguay

Casa Museo de Juan Zorrilla de San Martín, Montevideo 54386

Uzbekistan

Alisher Navoii nomidagi Davlat Adabiyot Muzeyi, Toshkent 54467
Andijon Viloyat Adabiyot va San'at Muzeyi, Andijon 54438
Babur Yodgorlik Muzeyi, Andijon 54440
Chulpan Yodgorlik Muzeyi, Andijon 54441
Hamza Hakim-zade Niazy Yodgorlik Muzeyi, Qo'qon 54455
M.T. Oibek Yodgorlik Muzeyi, Toshkent 54473
Sadriddin Aini Yodgorlik Muzeyi, Samarqand 54462
Sergej Esenin Muzeyi, Toshkent 54487

Venezuela

Casa de Andrés Eloy Blanco, Cumaná 54566
Casa Natal de José Antonio Ramos Sucre, Cumaná 54567
Museo Alberto Arvelo Torrealba, Barinas 54523

Law

Argentina

Museo Notarial Argentino, Buenos Aires 00266
Museo Notario Roberto Mario Arata, Buenos Aires 00267

Australia

Beechworth Historic Court House, Beechworth 00847
Chelsea Court House Museum, Chelsea 00965
Justice and Police Museum, Sydney 01551
Melrose Courthouse Museum, Melrose 01292
Old Borroloola Police Station Museum, Borroloola 00879
South Australian Police Museum, Thebarton 01583
Stuart Town Gaol Museum, Alice Springs 00802
York Courthouse Complex, York 01685

Belgium

Nationaal Museum en Archief van Douane en Accijnzen, Antwerpen 03305

Brazil

Museu da Faculdade de Direito, São Paulo 04870

Canada

Campbell House Museum, Toronto 07096
Old Court House Museum, Guysborough 05922

Finland

Rikosmuseo, Vantaa 11447

France

Musée du Barreau de Paris, Paris 14693

Germany

Bayerisches Strafvollzugsmuseum, Kaisheim 19415
Rechtshistorisches Museum, Karlsruhe 19447
Wieland-Museum Biberach, Wieland-Gartenhaus, Biberach an der Riß 17326

Mexico

Museo del Primer Supremo Tribunal de Justicia de la Nación, Ario de Rosales 30723

Netherlands

Vredespaleis, Den Haag 32105

Peru

Museo Tribunal de la Santa Inquisición y del Congreso, Lima 34271

Philippines

Kalantiaw Shrine Museum, Batan 34344

Poland

Muzeum Adwokatury Polskiej, Warszawa 35233

Slovakia

Múzeum Polície Slovenskej Republiky, Bratislava 38226

Sweden

Långelanda Tingshus, Årjäng 40429

United Kingdom

Galleries of Justice, Nottingham 45291
Judge's Lodging, Presteigne 45443
Old Crown Court and Cells, Dorchester 44002
The Tolhouse, Great Yarmouth 44361

United States of America

Birmingham Civil Rights Institute Gallery, Birmingham 46821
Harry S. Truman Office and Courtroom, Independence 49474
Held-Poage Memorial Home Research Library, Ukiah 53639
King Kamehameha V - Judiciary History Center, Honolulu 49346
Law Enforcement Museum, Roselle 52459
Mount Pulaski Courthouse, Mount Pulaski 50890
National Border Patrol Museum and Memorial Library, El Paso 48327
South Carolina Law Enforcement Officers Hall of Fame, Columbia 47732
US Patent and Trademark Office - Museum, Arlington 46359
The Women's Museum, Dallas 47943

Lead and Tin Figures

Austria

Zinnfigurenwelt, Katzelsdorf, Leitha 02157
Zinngießerhaus, Mattighofen 02366

Belgium

Tinelmuseum, Sinaai-Waas 04012

Germany

Deutsches Zinnfigurenmuseum, Kulmbach 19747
Museum zur brandenburg-preußischen Geschichte - Zinnfigurenmuseum, Gusow-Platkow 18842
Zinnfiguren-Klause im Schwabentor, Freiburg im Breisgau 18442
Zinnfigurenmuseum Aach, Aach 16390
Zinnkeller, Weidenberg 22370

Netherlands

Nationaal Tinnen Figuren Museum, Ommen 32619
Speelgoedmuseum Deventer, Deventer 32121

Spain

Museo de Miniaturas Militares Ciudadela de Jaca, Jaca 39416
Museo L'Iber, Valencia 40214

Switzerland

Zinnfigurenmuseum, Grüningen 41458
Zinnfigurenmuseum, Zürich 42117

Leather

Argentina

Museo de Arte Popular José Hernández, Buenos Aires 00184

Egypt

Leather Museum, Haram 10533

France

Maison des Métiers du Cuir, Graulhet 13120

Germany

DLM - Deutsches Ledermuseum/ Schuhmuseum Offenbach, Offenbach am Main 20881
Weißgerbermuseum, Doberlug-Kirchhain 17880

Namibia

Leather Museum, Swakopmund 31756

Poland

Muzeum Cechu Rzemiosł Skórzanych im. Jana Kilińskiego, Warszawa 35238

Spain

Museu de l'Art de la Pell, Vic 40269

United Kingdom

Northampton Museum and Art Gallery, Northampton 45270
Walsall Leather Museum, Walsall 45921

United States of America

Leather Museum, Chicago 47485

Literature → Language and Literature

Locks and Keys

Belgium

Provinciaal Veiligheidsinstituut, Antwerpen 03312

France

Musée Le Secq des Tournelles - Musée de la Ferronnerie, Rouen 15118

Germany

Deutsches Schloss-und Beschlägemuseum, Velbert 22210

Netherlands

LIPS Slotenmuseum, Dordrecht 32150

United Kingdom

Locksmith's House, Willenhall 46015
Neville Blakey Museum of Locks, Keys and Safes, Brierfield 43615

United States of America

Lock Museum of America, Terryville 53486

Lumber → Forest Products

Machines and Machinery

Australia

Boiler House Steam and Engine Museum, Kurwongbah 01208
Booleroo Steam and Traction Museum, Booleroo Centre 00875
Chinchilla and District Historical Society Museum, Chinchilla 00973
Cobdogla Irrigation and Steam Museum, Cobdogla 00989
Corrigin Pioneer Museum, Corrigin 01003
Manjimup Timber Museum, Manjimup 01257
Morphett's Enginehouse Museum, Burra 00926
National Motor Museum, Birdwood 00871
Rushworth Museum, Rushworth 01474

Belgium

Nationaal Borstelmuseum, Izegem 03737
Nationaal Schoeiselmuseum, Izegem 03738
Nationaal Vlas-, Kant- en Linnenmuseum, Kortrijk 03761

Canada

Antique Tractor Museum and Frontier Village, Maple Creek 06231
Mankota Museum, Mankota 06227
Prairie Acres Museum, Picture Butte 06604
Rocanville and District Museum, Rocanville 06765
Village Site Museum, Eddystone 05736

Czech Republic

Technické Muzeum v Brně, Brno 09440

Denmark

Brede Værk, Kongens Lyngby 10165
Industrimuseet, Horsens 10099

Finland

Rosenlew-museo, Pori 11222

France

Ecomusée la Cité des Abeilles, Saint-Faust 15264
Galerie Bovary, Ry 15143
Le Village des Automates, Saint-Cannat 15196
Musée Barthélemy Thimonnier de la Machine à Coudre et du Cycle, Amplepuis 11593
Musée Charolais du Machinisme Agricole, Neuvy-Grandchamp 14412
Musée de la Motocyclette, Entrevaux 12827
Musée des Automates, La Rochelle 13425
Musée des Automates, Sauclières 15649
Musée du Moteur, Saumur 15666
Musée Industriel Corderie Vallois, Notre-Dame-de-Bondeville 14474

L

M

Germany

Bayerisches Moor- und Torfmuseum, Grassau 18739
Dampfmaschinenhaus-Moschee, Potsdam 21086
Deutsches Baumaschinen-Modellmuseum, Weilburg-Gaudernbach 22387
Dieselmuseum, Schramberg 21629
Druckereimuseum und Schaudepot der Museen der Stadt Erfurt im Benary-Speicher, Erfurt 18200
Fachausstellung Arbeitsschutz/ Sicherheitstechnik, München 20438
Gerätemuseum des Coburger Landes, Ahorn 16440
Landesmuseum Koblenz, Koblenz 19585
MAN-Museum, Augsburg 16628
Molkerei-Museum, Bernbeuren 17293
Museum im Trafohaus, Leipzig 19926
Museum Industriekultur mit Motorradmuseum, Nürnberg 20784
Sächsisches Industriemuseum Chemnitz, Chemnitz 17703
Sammlung Gauselmann, Espelkamp 18245
Sammlung historischer Maschinen und Geräte, Calau 17675
Stadtmuseum Apolda, Apolda 16567
Technisches Museum der Pforzheimer Schmuck- und Uhrenindustrie, Pforzheim 21031
Weltkulturerbe Völklinger Hütte, Völklingen 22249
Werksmuseum der MTU, München 20495

Ireland

The Steam Museum, Straffan 24900
Steam Museum Strabally, Port Laoise 24886

Italy

Museo All'Aperto di Storia dell'Agricultura, Sassari 27752
Museo Archeologico dell'Arte della Lana, Arpino 25283
Museo dei Tasso e della Storia Postale, Camerata Cornello 25615

Japan

Kawaguchi-ko Motor Museum, Narusawa 29043
Nosaka O-tomatto Hakubutsukan, Ito 28597

Netherlands

De Brede, Maasbree 32520
Museumwerf 't Kromhout, Amsterdam 31870
Naaimachinemuseum Dordrecht, Dordrecht 32152
Nederlands Stoommachinemuseum, Medemblik 32543
Noord-Hollands Motoren Museum, Nieuwe Niedorp 32575
Poldermuseum, Puttershoek 32674

Poldermuseum de Hooge Boezem Achter Haastrecht, Haastrecht 32306
Racemotormuseum Lexmond, Lexmond 32499
Stoomhoutzagerij, Groenlo 32276

New Zealand

Otago Vintage Machinery Club Museum, Dunedin 33097
Pigeon Valley Steam Museum, Wakefield 33231
Tokomaru Steam Engine Museum, Tokomaru 33215

Norway

Bergens Tekniske Museum, Bergen 33424

Russia

Muzej Istorii Architektury i Promyšlennoj Techniki Urala, Ekaterinburg 36892

Slovakia

Hutnícke Múzeum Železiarne Podbrezová, Podbrezová 38329

Spain

Makina Erremintaren Museoa, Elgoibar 39273
Museu dels Autòmats, Barcelona 38987

Sweden

Eskilstuna Stadsmuseum, Eskilstuna 40502
Fornminnesgården, Eksjö 40494
Motala Industrimuseum, Motala 40754
Rubens Maskinhistoriska Samlingar, Götene 40559

Switzerland

Maschinenmuseum-Musée de la Machine-Centre Müller, Biel 41229
Musée Suisse de la Machine à Coudre, Fribourg 41390
Nähmaschinen-Museum, Steckborn 41909
Schweizerisches Dampfmaschinenmuseum Vaporama, Thun 41939

United Kingdom

Bolton Steam Museum, Bolton 43554
The Brunel Museum, London 44809
Cheddleton Flint Mill, Cheddleton 43802
Coleham Pumping Station, Shrewsbury 45673
Combe Mill Beam Engine and Working Museum, Long Hanborough 45025
Cornish Mines, Engines and Cornwall Industrial Discovery Centre, Redruth 45476
Corris Railway Museum, Machynlleth 45068
The Crossness Engines, London 44832
Dingles Steam Village - Fairground Museum, Lifton 44709
Eastney Industrial Museum, Portsmouth 45430
Etruria Industrial Museum, Stoke-on-Trent 45759
Forncett Industrial Steam Museum, Forncett Saint Mary 44243
Garlogie Mill Power House Museum, Garlogie 44262
Haynes International Motor Museum, Sparkford 45724
Hobbies Museum of Fretwork, Dereham 43972
Hollycombe Steam Collection, Liphook 44724
House Mill, London 44879
Kew Bridge Steam Museum, Kew 44569
Lark Lane Motor Museum, Liverpool 44736
Lewis Textile Museum, Blackburn, Lancashire 43527
Middleton Top Engine House, Middleton-by-Wirksworth 45143
Midland Motor Museum, Bridgnorth 43605
Museum of Bath at Work, Bath 43406
Museum Workshop and Stores, Grangemouth 44343
National Tramway Museum, Matlock 45115
National Wool Museum, Llandysul 44761
Newcomen Engine House, Dartmouth 43954
Papplewick Pumping Station, Ravenshead 45465
Pinchbeck Marsh Engine and Land Drainage Museum, Spalding 45722
Ruddington Framework Knitters' Museum, Ruddington 45524
Ryhope Engines Museum, Sunderland 45803
Shore Road Pumping Station, Birkenhead 43491
Stretham Old Engine, Stretham 45791
Waterworks Museum - Hereford, Hereford 44460
West Wycombe Motor Museum, West Wycombe 45974
Westonzoyland Pumping Station Museum, Bridgwater 43608

United States of America

Carpenter Museum of Antique Outboard Motors, Gilmanton 48911
Hidalgo Pumphouse Heritage, Hidalgo 49271
Museo Abarth, Marietta 50470
Steam Engine Museum, Mabel 50338

Majolica → Porcelain and Ceramics

Mammals

Australia

Whale World, Albany 00790

Belgium

Natuurpunt Museum, Turnhout 04100

Bulgaria

Prirodonaučen Muzej, Černi Osăm 05002

Canada

B.J. Hales Museum of Natural History, Brandon 05459
Centre d'Interprétation des Mammifères Marins (CIMM), Tadoussac 07055
New Marine Centre, Sidney 06971
Shaw Ocean Discovery Centre, Sidney 06972
Terra Nova National Park Museum, Glovertown 05877

Czech Republic

Zámek Budišov, Budišov u Třebíče 09447

Finland

K.E. Kivirikon Lintu- ja Nisäkäskokoelma, Helsinki 10789

India

Zoology Museum, Lucknow 24302

Italy

Museo di Storia Naturale della Maremma, Grosseto 26372
Museo Provinciale di Storia Naturale, Foggia 26212

Japan

Shiretoko Museum, Shari 29254

Lithuania

Kauno Tado Ivanausko Zoologijos Muziejus, Kaunas 30252

Netherlands

Natuurhistorisch Museum Rotterdam, Rotterdam 32718

Poland

Muzeum Lasu i Drewna przy Lesnym Zakładzie Doświadczalnym SGGW, Rogów 35083

Puerto Rico

Caribbean Primate Research Center Museum, San Juan 36059

Slovenia

Prirodoslovni Muzej Slovenije, Ljubljana 38461

South Africa

Amathole Museum, King William's Town 38675

Switzerland

Museum Stemmler, Schaffhausen 41830

United States of America

Charles R. Conner Museum, Pullman 52197
Murie Museum, Kelly 49722
The Museum of Natural History, Stevens Point 53289
Museum of Texas Tech University, Lubbock 50312
The Natural Science Center of Greensboro, Greensboro 49067
Northwestern Oklahoma State University Museum, Alva 46247
Pember Museum of Natural History, Granville 49019
Roberson Museum and Science Center, Binghamton 46816
Southern Oregon University Museum of Vertebrate Natural History, Ashland 46389
Trailside Nature Museum, Cross River 47891
University of North Dakota Zoology Museum, Grand Forks 48990

Man, Prehistoric

Argentina

Museo de Antropología, Córdoba 00321

Belgium

Archéo- et Géo-Espaces Moulin du Broukay, Eben-Emael 03567

Bolivia

Museo Nacional Antropológico Eduardo López Rivas, Oruro 04234

Brazil

Museu de Ciências Morfológicas, Pampulha 04608
Museu Nacional, Rio de Janeiro 04755

Canada

Anthropology Museum, Winnipeg 07354

Chile

Museo de Arte y Artesanía de Linares, Linares 07441

China

Beijing Man Zhoukoudian Museum, Beijing 07516
Nanzhao Pithecanthrope Museum, Nanzhao 08067

Colombia

Museo Universitario, Medellín 08686

Croatia

Hrvatski Prirodoslovni Muzej, Zagreb 08998

Ecuador

Museo de Antropología, Ibarra 10413
Museo de Sitio Arqueologico el Mongote Real Alto, Guayaquil 10404

Germany

Neanderthal Museum, Mettmann 20314
Urmensch-Museum, Steinheim an der Murr 21892

Greece

Museum of Anthropology, Goudi 23008

Iran

Mardom-shenasi-ye Kavir, Nayin 24636

Italy

Museo Preistorico Balzi Rossi, Ventimiglia 28217

Japan

Kitakyushu-shiritsu Shizenshi Hakubutsukan oyo Ningenteki Rekishi, Kitakyushu 28713

Mexico

Museo de Antropología, Tecolotlán 31413
Museo de Antropología de Xalapa, Xalapa 31557
Museo Regional de Antropología J. Jesús Figueroa Torres, Sayula 31390

Peru

Museo Universitario de la Universidad Nacional Federico Villarreal, Lima 34272

Portugal

Sala de Arqueologia e Pré-Historia, Porto 35866

Spain

Museo de Antropolocía Forense, Paleopatología y Criminalística Profesor Reverte, Madrid 39533

Taiwan

Museum of Anthropology, Taipei 42406

United States of America

Anthropology Museum, Marshall 50504
The Bead Museum, Glendale 48926
Heritage of the Americas Museum, El Cajon 48307
Museum of Anthropology, Detroit 48126
San Diego Museum of Man, San Diego 52753
Science Museum of Minnesota, Saint Paul 52629
William S. Webb Museum of Anthropology, Lexington 50059
Zion Human History Museum, Springdale 53214

Manuscripts → Books, Book Art and Manuscripts

Maps → Cartography

Marine Archaeology

Australia

Museum of Tropical Queensland, Townsville 01598
Western Australian Maritime Museum, Fremantle 01095

Brazil

Espaço Cultural da Marinha, Rio de Janeiro 04690

Bulgaria

Archeologičeski Muzej, Burgas 04995

Cameroon

Musée Maritime Cameroun, Douala 05295

Canada

Admiralty House Museum and Archives, Mount Pearl 06400
Maritime Heritage Society of Vancouver, Britannia Beach 05483
Musée Navale du Québec, Québec 06695
New Marine Centre, Sidney 06971
Port de Québec au XIXe Siècle, Saint-André 06792
Shaw Ocean Discovery Centre, Sidney 06972

Dominican Republic

Museo de las Atarazanas Reales, Santo Domingo 10350

France

Itsas Begia, Ciboure 12528
Musée Conservatoire Maritime et Rural, Narbonne 14376
Musée de la Mer, Cannes 12270
Musée de la Mer OCEAM, Les Sables-d'Olonne 13714
Musée d'Istres, Istres 13254
Musée Maritime de l'île de Tatihou, Saint-Vaast-la-Hougue 15528
Musée Maritime, Fluvial et Portuaire, Rouen 15119

Germany

Museum für Unterwasserarchäologie, Sassnitz 21520
Nationalpark-Haus, Butjadingen 17669

Iran

Anzali Military Museum, Bandar-Anzali 24598

Israel

Clandestine Immigration and Naval Museum, Haifa 24957
National Maritime Museum, Haifa 24965

Italy

Museo della Cultura Marinara, Tortoreto Lido 28045
Museo della Marineria dell'Alto e Medio Adriatico, Cesenatico 25839
Museo Navale Romano, Albenga 25178

Malaysia

Maritime Archaeology Museum, Melaka 30596

Northern Cyprus

Batik Gemi Müzesi, Girne 33356

Norway

Stavanger Sjøfartsmuseum, Stavanger 33876

Qatar

Al-Khor Museum, Al-Khor 36083
Al-Wakra Museum, Al-Wakra 36084

Slovenia

Pomorski Muzej Sergej Mašera, Piran 38492

South Africa
Port Elizabeth Museum, Port Elizabeth 38712

Spain
Bilboko Itsasadarra Itsas Museoa, Bilbao 39039
Museo Militar Regional de Sevilla, Sevilla 40075
Museo Naval de Cartagena, Cartagena 39140
Museu d'Arqueològia Submarina, Pals 39805

Sweden
Vasamuseet, Stockholm 40949

Turkey
Bodrum Sualtı Arkeolojisi Müzesi, Bodrum 42731

United Kingdom
Dartmouth Museum, Dartmouth 43953
National Maritime Museum, London 44939

Marine Biology → Oceanography

Masks

Armenia
S. Merkurov House-Museum, Gyumri 00708

Belgium
Centre d'Étude du Masque et Mouvement, Walhain 04128
Musée International du Carnaval et du Masque, Binche 03368

Burkina Faso
Musée de la Bendrologie, Manéga 05275

Cameroon
Musée d'Art Nègre, Yaoundé 05301
Musée des Arts Traditionnels et de la Culture, Foumban 05297

Denmark
Radio- og Motorcykel Museet, Stubbekøbing 10287

Germany
Deutsches Fastnachtmuseum, Kitzingen 19560
Narrenschopf, Bad Dürrheim 16720

Guatemala
Museo de las Máscaras Ceremoniales, Chichicastenango 23293

Japan
Shishi Kaikan, Takayama 29343

Korea, Republic
Galchon Mask Museum, Goseong 29818
Hahoedong Mask Museum, Andong 29766

Mexico
Museo de la Máscara, Morelia 31197
Museo Nacional de la Máscara, San Luis Potosí 31351

Papua New Guinea
Cultural Center and Museum, Wabag 34087

Sri Lanka
Ambalangoda Mask Museum, Ambalangoda 40365

Switzerland
Musée Suisse de la Marionnette, Fribourg 41391
Theater-Atelier, Zürich 42110

Taiwan
Face Painting Culture Museum, Sanyi 42316

United States of America
New Visions Gallery, Marshfield 50513

Mass Media

Argentina
Museo de Medios de Comunicación, Resistencia 00524
Museo del las Comunicaciones de Mar del Plata, Mar del Plata 00444
Museo Radiofónico de la Ciudad de Boulogne, Boulogne 00149

Australia
City Park Radio Museum, Launceston 01220

Austria
Ars Electronica Center, Linz 02308
Erdefunkstelle Aflenz, Au bei Aflenz 01737
Erstes Österreichisches Funk- und Radiomuseum, Wien 02982
Museum in Progress, Wien 03040
Museum Sender Dobl, Dobl 01822
Radiomuseum, Bad Tatzmannsdorf 01768

Belgium
Ecomusée des Techniques, Thuin 04065
Economuseum voor Papier, Karton en Aandrijftechnieken, Alsemberg 03265
Musée de la Radio, Mons 03883
Olens Radiomuseum, Olen 03921

Brazil
Museu da Imprensa, Brasília 04360
Museu do Rádio, Juiz de Fora 04521

Canada
CBC Museum, Toronto 07101
CBC Radio Museum, Saint John's 06845
Manitoba Amateur Radio Museum, Brandon 05465

Dominican Republic
Museo de la Radio y TV - Lulio Moscoso, Santo Domingo 10349

Estonia
Eesti Ringhäälingumuuseum, Türi 10665

Finland
Keskisuomalaisen Museo, Jyväskylä 10896
Mellilän Radio- ja Puhelinmuseo, Mellilä 11106
Radio- ja TV-Museo, Lahti 11032

France
Musée de la Publicité, Paris 14662
Musée de la Radio, Villers-le-Lac 16231
Musée de la Radio Galletti, Saint-Maurice-de-Rotherens 15410
Musée de Radio France, Paris 14680
Musée Maison de la Radio TSF, Saint-Appolinard 15174
Musée Municipal, Louhans 13834

Germany
Bremer Rundfunkmuseum, Bremen 17547
Deutsches Film- und Fototechnik Museum, Deidesheim 17807
Internationales Radiomuseum, Bad Laasphe 16771
Internationales Zeitungsmuseum der Stadt Aachen, Aachen 16399
Radiomuseum, Rottenburg an der Laaber 21423
Radiomuseum, Waldbronn 22284
Rundfunkmuseum der Stadt Fürth, Fürth 18499
Rundfunkmuseum Schloß Brunn, Emskirchen 18163

Hungary
Rádiós Hírközlési Múzeum, Siófok 23804
Televízió Technikatörténeti és Műsor Múzeum, Budapest 23481

Italy
Museo Civico della Media Valle del Liri, Sora 27876
Museo della Guerra Adamellina, Spiazzo 27888

Japan
Koto-ku Fukagawa Edo Siryokan, Tokyo 29420
Newspark - The Japan Newspaper Museum, Yokohama 29603
Newsstand Museum, Osaka 29138
NHK Hoso Hakubutsukan, Tokyo 29445
Teishin Sogo Hakubutsukan, Tokyo 29483

Korea, Republic
Korean Magazine Museum, Seoul 29897

Lithuania
Lietuvos Radijo ir Televizijos Muziejus, Vilnius 30420
Radijo ir Televizijos Muziejus, Šiauliai 30361

Mexico
Museo Casa de León Trotsky, México 31059
Museo del Retrato Hablado, México 31108

Netherlands
Museum Radio-Wereld, Diever 32129
Museum voor het Radiozendamateurisme Jan Corver, Budel 32017
Nederlands Instituut voor Beeld en Geluid-Colleties, Hilversum 32389
Persmuseum, Amsterdam 31878
Radio Amateur Museum, Reusel 32683
Radio- en Speelgoed Museum, Onstwedde 32622
Rotterdams Radio Museum, Rotterdam 32725

Portugal
Museu da Rádio, Lisboa 35683
Museu da Televisão, Lisboa 35685
Museu Nacional da Imprensa, Jornais e Artes Gráficas, Porto 35858

Russia
Centralnyj Muzej Svjazi im. A.S. Popova, Sankt-Peterburg 37630
Muzej-studija Radioteatra, Moskva 37365

Spain
Museo Verbum - Casa das Palabras, Vigo 40277

Sweden
Radiomuseet i Göteborg, Göteborg 40554
Regementsmuseet med 91:an, Halmstad 40580
Rolf Bergendorffs Radio Museum, Lessebo 40672
Sveriges Rundradiomuseum, Motala 40759

Switzerland
Fond d'art contemporain de la Ville de Genève (Fmac), Genève 41412
Museum für Kommunikation, Bern 41212
Radiomuseum, Zürich 42104

Taiwan
Children's Museum of Transportation and Communication, Taipei 42382
National Broadcasting Museum, Minsiong 42277
Taiwan Television Exposition, Taipei 42434

Turkey
Basın Müzesi, İstanbul 42807

United Kingdom
BFI Southbank Gallery, London 44796
Museum of Entertainment, Whaplode 45992
National Arts Education Archive, Wakefield 45909
National Vintage Wireless and Television Museum, Harwich 44412
Orkney Wireless Museum, Kirkwall 44616
Vintage Wireless Museum, London 45005

United States of America
American Museum of the Moving Image, Astoria 46399
California Museum, Sacramento 52500
French Cable Station Museum, Orleans 51597
Jasper's Antique Radio Museum, Saint Louis 52581
Lum and Abner Museum, Pine Ridge 51923
The Museum of Broadcast Communications, Chicago 47491
The Museum of Television and Radio, New York 51229
Newseum, Arlington 46358
The Paley Center for Media, Beverly Hills 46792
Radio-Television Museum, Bowie 46971
Robert R. McCormick Museum at Cantigny, Wheaton 54121

Venezuela
Museo Audiovisual, Caracas 54536

Medals → Numismatics

Medicine and Hygiene

Argentina
Museo Anatómico Pedro Ara, Córdoba 00319
Museo Casa de Bernardo A. Houssay, Buenos Aires 00171
Museo de Ciencias Morfológicas, Rosario 00543
Museo de Historia de la Medicina y de la Cirugía Vicente A. Risolía, Buenos Aires 00194
Museo de la Psicología Experimental Argentina Horacio G. Piñero, Buenos Aires 00207
Museo de Patología, Buenos Aires 00216
Museo de Urología de Buenos Aires, Buenos Aires 00219
Museo del Médico, Santa Fé 00614
Museo Houssay de Historia de la Ciencia y la Técnologia, Buenos Aires 00247
Museum Oculorum, Buenos Aires 00280

Australia
Anatomy Museum, Saint Lucia 01482
Dr. Arratta Memorial Museum, Muttaburra 01343
Geoffrey Kaye Museum of Anaesthetic History, Melbourne 01279
Marks-Hirshfeld Museum of Medical History, Herston 01151
Medical History Museum, Melbourne 01287
Mooroopna Hospital and District Museum, Mooroopna 01317
Museum of Pathology, Sydney 01554
Old Springsure Hospital Museum, Springsure 01524
Pathology Museum, Herston 01152
Royal Flying Doctor Service Visitors Centre, Edge Hill 01061
Royal Prince Alfred Hospital Museum, Camperdown 00943
Tasmanian Medical History Museum, South Hobart 01514
Tracy Maund Historical Collection, Carlton South 00958
Western Australian Medical Museum, Subiaco 01536
Wide Bay Hospital Museum, Maryborough, Queensland 01270

Austria
Anatomisches Museum, Innsbruck 02112
Dokumentation des ehemaligen Heilbades, Bad Pirawarth 01761
Erstes Österreichisches Rettungsmuseum, Hohenems 02091
Konventmuseum der Barmherzigen Brüder, Wien 03012
Medizinhistorisches Museum, Wien 03027
Museum für Medizin-Meteorologie, Zwettl 03135
Pathologisch-anatomisches Bundesmuseum, Wien 03055
Sanitärhistorisches Museum, Wien 03067
Sigmund Freud Museum, Wien 03075
Zahnmuseum Linz, Linz 02326

Azerbaijan
Museum of the History of Medicine, Bakı 03164

Belgium
Gasthuismuseum Sint-Dimpna, Geel 03613
Hôpital Notre-Dame de la Rose, Lessines 03786
Musée Belge de Radiologie-Belgisch Museum voor Radiologie, Bruxelles 03453
Musée de la Médecine, Bruxelles 03468
Museum Dr. Guislain, Gent 03634
Museum voor Anesthesie Verantare, Antwerpen 03303
Museum voor Geschiedenis van de Geneeskunde, Gent 03637

Bolivia
Museo de la Facultad de Ciencias de la Salud, Sucre 04247

Brazil
Museu da Faculdade de Odontologia, Araraquara 04316
Museu de Images do Inconsciente, Rio de Janeiro 04728
Museu de Medicina, Curitiba 04435
Museu de Odontologia Professor Salles Cunha, Rio de Janeiro 04730
Museu de Patologia do Instituto Oswaldo Cruz, Rio de Janeiro 04731
Museu Inaldo de Lyra Neves-Manta, Rio de Janeiro 04753

M

Bulgaria

Muzej Čovek i negovoto Zdrave, Varna 05239
Muzej za Istorija na Medicinata, Varna 05242

Canada

Alberta Association of Registered Nurses Museum, Edmonton 05739
Alberta Hospital Museum, Ponoka 06620
Banting House, London 06181
Banting Interpretation Centre, Musgrave Harbour 06404
Brandon Mental Health Centre Historical Society Museum and Archives, Brandon 05460
British Columbia Congenital Heart Museum, Vancouver 07201
British Columbia Museum of Medicine, Vancouver 07203
Canadian Medical Hall of Fame, London 06182
Hamilton Psychiatric Hospital Museum, Hamilton 05958
Lillian Stevenson Nursing Archives Museum, Saint John's 06849
Maison Docteur Joseph-Frenette, Causapscal 05579
Musée des Augustines de l'Hôtel Dieu de Québec, Québec 06690
Musée Eudore-Dubeau, Montréal 06368
Museum of Health Care, Kingston 06087
NANB Nursing History Centre, Fredericton 05833
New Brunswick Healthcare Museum, Fredericton 05836
Regina General Hospital Museum, Regina 06723
University of Alberta Dental Museum, Edmonton 05760

China

China Tradition Medicine Museum, Chongqing07629
Chinese Medicine Museum, Hangzhou 07762
Chinese Medicine Museum, Shanghai 08162
Gutao Ure Museum, Beijing 07541
Hong Kong Museum of Medical Sciences, Hong Kong 07801
Museum of Medical History, Shanghai 08175
Zhangzhongjing Museum, Nanyang 08065

Colombia

Museo Bernardo Samper Sordo, Bogotá 08554
Museo de Historia de la Medicina, Bogotá 08564

Cuba

Museo Casa Alejandro de Humboldt, La Habana 09113
Museo Casa de Sais Montes de Oca Hermanos, Pinar del Río 09248
Museo Casa Natal de Hermanos Ameijeiras, Pueblo Viejo 09258
Museo de Medicina Tropical Carlos J. Finlay, La Habana 09139

Czech Republic

Muzeum historie, Kolín 09590
U Zlaté Koruny, Brno 09441
Zdravotnické Muzeum, Praha 09800

Denmark

Medicinsk-Historisk Museum, København 10139
Museet Psykiatrisk Hospital i Århus, Risskov 10238

Ecuador

Museo de la Medicine, Cuenca 10387
Museo Nacional de Historia de la Medicina Eduardo Estrella, Quito 10448

Egypt

Museum of Hygiene, Cairo 10503

Estonia

Eesti Tervishoiu Muuseum, Tallinn 10629

Finland

Hammaslääketieteen Museo, Helsinki 10777
Helsingin Yliopistomuseo Arppeanum, Helsinki 10785
Kellokosken Sairaalan Museo, Kellokoski 10937
Keski-Suomen Sairaanhoitopiirin Sairaalamuseo, Jyväskylä 10895
Sairaalamuseo, Lahti 11034
Sotilaslääketieteen Museo, Lahti 11035
Turun Terveydenhuoltomuseo, Turku 11406

France

Ancienne École de Médecine Navale, Rochefort (Charente-Maritime) 15069
Musée Albert Schweitzer, Kaysersberg 13300
Musée Claude Bernard, Saint-Julien-en-Beaujolais 15348
Musée d'Anatomie Delmas-Orfila-Rouvière, Paris 14642
Musée de la Faculté de Médecine, Strasbourg 15823
Musée de l'Assistance Publique-Hôpitaux de Paris, Paris 14668
Musée de l'Ecorché d'Anatomie, Le Neubourg 13627
Musée de l'Histoire de la Médecine, Vandœuvre-les-Nancy 16064
Musée de l'Histoire de la Médecine de Toulouse, Toulouse 15936
Musée de l'Institut Pasteur de Lille, Lille 13755
Musée de Matière Médicale, Paris 14676
Musée Dentaire de Lyon, Lyon 13886
Musée des Hospices Civils de Lyon, Lyon 13890
Musée d'Histoire de la Médecine, Paris 14691
Musée Dupuytren, Paris 14704
Musée Flaubert et d'Histoire de la Médecine, Rouen 15116
Musée Larrey, Beaudean 11891
Musée National de l'Assurance Maladie, Lormont 13824
Musée Ollier, Les Vans 13721
Musée Pasteur, Dole 12755
Musée Testut-Latarjet de Médecine et d'Anatomie, Lyon 13900

Germany

Albert-Schweitzer-Gedenk- und Begegnungsstätte, Weimar 22395
Anatomische Sammlung, Erlangen 18218
Anatomische Sammlungen des Institut für Anatomie und Zellbiologie, Halle, Saale 18888
Arzney Küche, Bönnigheim 17425
Berliner Medizinhistorisches Museum der Charité, Berlin 17044
Chirurgiemuseum, Tuttlingen 22132
DASA, Dortmund 17917
Dentalhistorisches Museum, Zschadraß 22822
Deutsches Epilepsiemuseum Kork, Kehl 19495
Deutsches Hygiene-Museum Dresden, Dresden 17941
Deutsches Medizinhistorisches Museum, Ingolstadt 19340
Deutsches Orthopädisches Geschichts- und Forschungsmuseum, Frankfurt am Main 18363
Deutsches Röntgen-Museum, Remscheid 21263
Drogerie-Museum, Weißenstadt 22463
Emil-von-Behring-Ausstellung, Marburg 20189
Fränkisches Spitalmuseum Aub, Aub 16604
Historische Arztpraxis, Burgstädt 17663
Institut für Geschichte der Medizin der Robert Bosch Stiftung, Stuttgart 21947
Johann-Peter-Frank Museum, Rodalben 21343
Johann-Wunter-Museum Andernach, Andernach 16544
Karthaus-Prüll, Regensburg 21225
Kneipp-Museum, Bad Endbach 16728
Krankenhaus-Museum am Klinikum Bremen-Ost, Bremen 17555
Krankenhausmuseum, Nürnberg 20775
Medizin- und Pharmaziehistorische Sammlung Kiel der Christian-Albrechts-Universität, Kiel 19533
Medizinhistorische Sammlung der Ruhr-Universität Bochum, Bochum 17404
Medizinhistorische Sammlungen, Chemnitz 17699
Museum Anatomicum, Marburg 20193
Narkose Schmerztherapie Museum, Potsdam 21097
Psychiatrie-Museum, Emmendingen 18147
Psychiatrie Museum im Philippshospital, Riedstadt 21327
Psychiatriemuseum Haina, Haina (Kloster) 18870
Robert-Koch-Museum, Berlin 17239
Rot-Kreuz-Museum, Regenstauf 21235
Rotkreuz-Museum, Nürnberg 20787
Rotkreuz-Museum Berlin, Berlin 17240
Rotkreuzmuseum, Hofheim in Unterfranken 19228
Sächsisches Apothekenmuseum Leipzig, Leipzig 19930
Sächsisches Psychiatriemuseum, Leipzig 19931
Sammlung zur Geschichte der Geburtshilfe, Göttingen 18699
Stadtmuseum Gütersloh, Gütersloh 18833
Strigel-Museum und Antoniter-Museum, Memmingen 20280
Thermometermuseum, Geraberg 18599
Wilhelm-Fabry-Museum, Hilden 19181
Zahnärztliches Museum, Tübingen 22126

Greece

International Hippocratic Foundation of Kos, Kos 23064

Hungary

Kresz Géza Mentomúzeum, Budapest 23429
Orvos- és Gyógyszerészettörténeti Kiállítás, Kecskemét 23654
Richter Múzeum, Budapest 23473
Semmelweis Orvostörténeti Múzeum, Budapest 23475

Iceland

Nesstofusafn, Seltjarnarnes 24038

India

Anatomy Museum, Chennai 24111
Anatomy Museum, Mumbai 24313
Grant Medical College Museum, Mumbai 24316
Guwahati Medical Museum, Guwahati 24192
Health Museum, Hyderabad 24212
Museum of the Medical College, Mysore 24331
Pathology and Bacteriology Museum, Lucknow 24298
Pathology Museum, Mumbai 24322
State Ayurvedic College Museum, Hardwar 24204
State Health Institute Museum, Lucknow 24300
Sulabh International Museum of Toilets, Delhi 24155

Indonesia

Museum Anatomy FK UI, Jakarta 24479
Museum Kedokteran, Jakarta 24487

Ireland

Royal College of Surgeons in Ireland Museum, Dublin 24818

Italy

Donazione Putti Biblioteca, Bologna 25441
Musei Anatomici, Modena 26743
Museo Anatomia Veterinaria, Napoli 26882
Museo Anatomico, Siena 27837
Museo Anatomico G. Tumiati, Ferrara 26106
Museo Comunale La Malaria e la sua Storia, Pontinia 27292
Museo dei Ferri Chirurgici, Pistoia 27251
Museo della Croce Rossa Italiana, Campomorone 25640
Museo delle Cere Anatomiche Luigi Cattaneo, Bologna 25456
Museo dell'Istituto di Anatomia Patologica, Padova 27038
Museo di Anatomia Comparata Battista Grassi, Roma 27488
Museo di Anatomia e Ostetricia, Bologna 25458
Museo di Anatomia Patologica, Roma 27489
Museo di Anatomia Umana, Torino 28009
Museo di Anatomia Umana, Napoli 26892
Museo di Anatomia Umana Normale, Parma 27101
Museo di Anatomia Veterinaria, Napoli 26893
Museo di Antropologia Criminale, Torino 28011
Museo di Etnomedicina A. Scarpa, Piazzola sul Brenta 27197
Museo di Scienze Naturali del Collegio, Lodi 26509
Museo di Storia della Medicina, Roma 27501
Museo Ostetrico, Padova 27047
Museo Storico dell'Arte Sanitaria, Roma 27536
Raccolte del Dipartimento di Scienze Radiologiche, Roma 27544
Raccolte dell'Istituto di Clinica Otorinolaringoiatrica, Roma 27545

Japan

Chiba-kenritsu Sonan Hakubutsukan, Isumi 28589
Inro Bijitsukan, Takayama 29342
Museumsfarm Tomita, Nakafurano 29015
Noguchi Hideyo Kinenkan, Inawashiro 28571
Toyama Museum, Toyama 29524

Jordan

Medical Museum, Amman 29657

Korea, Republic

Gacheon Museum, Incheon 29832

Latvia

Paula Stradiņa Medicīnas Vēstures Muzejs, Rīga 30101

Lebanon

Soap Museum, Saïda 30173

Lithuania

Lietuvos Medicinos ir Farmacijos Istorijos Muziejus, Kaunas 30255
Vilniaus Universiteto Medicinos Fakulteto Medicinos Istorijos Muziejus, Vilnius 30444

Malaysia

Biomedical Museum, Kuala Lumpur 30574

Mexico

Museo Antiguo Palacio de la Medicina Mexicana, México 31050
Museo de Medicina Maya, San Cristóbal de las Casas 31326
Museo de Medicina Tradicional y Herbolaria, Cuernavaca 30817
Museo de Odontología, México 31092
Museo del Pueblo-Hospital, Quiroga 31303
Museo Estatal de Patología, Puebla 31286
Sala Histórica de la Facultad de Medicina, Mitras 31170

Myanmar

Health Museum, Yangon 31735

Netherlands

Anatomisch Museum, Groningen 32278
Anatomisch Museum, Utrecht 32842
Anatomisch Museum Nijmegen, Nijmegen 32587
Anatomisch Museum van de Rijksuniversiteit Leiden, Leiden 32482
Historische Collectie UMCG, Groningen 32282
Historische Verzameling Militair Geneeskundige Dienst, Loosdrecht 32505
Medisch Farmaceutisch Museum de Griffioen, Delft 32060
Museum Boerhaave, Leiden 32484
Museum De Kluis, Boekel 31978
Museum GGZ-Drenthe, Assen 31938
Nationaal Ambulance- en Eerste Hulpmuseum, Kampen 32439
Oorlogs- en Verzetsmuseum Johannes Post, Ridderkerk 32685
Parkzicht, Ermelo 32223
Tuberculosemuseum Beatrixoord, Appelscha 31917
Verpleegkundig Historisch Bezit, Amersfoort 31817
Ziekenhuis- en Verpleegkundig Museum, Deventer 32122

New Zealand

Ernest and Marion Davis Medical History Museum, Auckland 33046
Order of Saint John Museum, Christchurch 33075
Porirua Hospital Museum, Porirua 33187
Royal New Zealand Army Medical Corps Museum, Christchurch 33076
W.D. Trotter Anatomy Museum, Dunedin 33100

Norway

Lepramuseet Sant Jørgens Hospital, Bergen 33430
Redningsselskapets Museum, Horten 33565
Sand Bakkens Medisinske Museum, Flekkefjord 33498
Sykepleiemuseum, Oslo 33753
Trastad Samlinger, Borkenes 33448
Ullevål Sykehus Museum, Oslo 33756

Papua New Guinea

Kau Rainforest Museum Conservation Area, Madang 34083

Peru

Museo Arqueológico de Cajamarca y Museo Médico, Cajamarca 34126
Museo del Hospital Santo Toribio de Mogrovejo, Lima 34255
Museo Yanque, Yanque 34329

Philippines

Philippine Radiation Science Museum, Manila 34451

Poland

Muzeum Akademii Medycznej, Poznań 35029
Muzeum Historii Medycyny i Farmacji, Szczecin 35160
Muzeum Pamięci Narodowej, Kraków 34807
Muzeum Polskiej Wojskowej Służby Zdrowia, Łódź 34881
Muzeum Roberta Kocha, Wolsztyn 35319

Portugal

Museu da História da Medicina Maximiano Lemos, Porto 35845
Museu do Instituto de Anatomia, Porto 35851
Museu do Instituto de Anatomia Patológica, Coimbra 35536
Museu do Instituto de Medicina Legal, Lisboa 35701
Núcleo Museológico do Hospital de Santo António dos Capuchos, Lisboa 35730

Romania

Muzeul Memorial Dr. Victor Babeş, Bucureşti 36208
Muzeul Universitar de Istoria Medicinei şi Farmaciei Victor Gomoiu, Târgu Jiu 36667

Russia

Dom-muzej Akademika E.N. Pavlovskogo, Borisoglebsk 36817
Medicinskij Muzej, Moskva 37226
Memorialnyj Muzej-usadba Akademika I.P. Pavlova, Rjazan 37594
Muzej Evoljucii Mozga, Moskva 37275
Muzej Istorii Bolnicy im. N.A. Alekseeva, Moskva 37288
Muzej Istorii Preobraženskoj Bolnicy im. V.A. Giljarovskogo, Moskva 37295
Muzej Istorii Razvitija Mediciny, Astrachan 36784
Muzej Rossijskogo Krasnogo Kresta, Moskva 37352
Muzej Zdravoochranenija, Sankt-Peterburg 37712
Voenno-medicinskij Muzej, Sankt-Peterburg 37731

Serbia

Medicinski Muzej, Beograd 38059

Slovakia

Múzeum Slovenského Červeného Kríža, Martin 38298

Slovenia

Farmacevtsko-Medicinska Zbirka Bohuslava Lavičke, Ljubljana 38448

South Africa

Adler Museum of Medicine, Johannesburg 38646
Brebner Surgical Museum, Johannesburg 38653
Pathology Museum, Tshwane 38754

Spain

Colección Anatomía, Madrid 39509
Museo de la Escuela Militar de Montaña y Operaciones Especiales, Huesca 39399
Museo de Odontología, Madrid 39551
Museo de Oftalmología Doctor Saracíbar, Valladolid 40234
Museo de Sanidad e Higiene Pública, Madrid 39552
Museo del Colegio Médico, Santiago de Compostela 40024

Sweden

Ersta Diakonimuseum, Stockholm 40891
Hospitalsmuseet, Vadstena 41019
Målilla Sanatoriums Museum, Målilla 40715
Medicinhistoriska Museet, Göteborg 40552
Medicinhistoriska Museet, Lund 40707
Medicinhistoriska Museet, Vänersborg 41024
Medicinhistoriska Museet Eugenia, Stockholm 40914
Mentalvårdsmuseet Säter, Säter 40823
Psykiatrihistoriska Museet, Uppsala 41010
Psykiatrihistoriskt Muséum, Växjö 41040
Psykiatrimuseet, Hisings Backa 40593
Sveriges VVS-Museum, Katrineholm 40641
Uppsala Medicinhistoriska Museum, Uppsala 41014

Switzerland

Anatomisches Museum Basel, Basel 41153
Henry-Dunant-Museum, Heiden 41475
Medizinhistorisches Museum, Zürich 42085
Moulagensammlung des Universitätsspitals und der Universität Zürich, Zürich 42089
Psychiatrie-Museum, Bern 41214

Taiwan

Brand's Health Museum, Lugang 42255
Lifu Museum of Chinese Medicine, Taichung 42350
Museum of Medical Missionary in Remote Areas, Puli 42296
Taiwan Medical Museum, Kaohsiung 42244

Tajikistan

Osorkhonai Tibb va Giëhhoi Shifobakhshi Tojikiston, Isfara 42500

Thailand

Congdom Anatomical Museum, Bangkok 42536
Forensic Medicine Museum, Bangkok 42538
Museum of Medical Equipment, Bangkok 42547
Museum of Pediatric Surgery, Bangkok 42548
Orthopedics Museum, Bangkok 42554
Wajawithayawat Museum, Bangkok 42565

Ukraine

Nacional'nyj Muzej Medycyny Ukraïny, Kyïv 43106

United Kingdom

Aberdeen University Pathology and Forensic Medicine Collection, Aberdeen 43238
Alexander Fleming Laboratory Museum, London 44779
The Anaesthesia Heritage Centre, London 44782
Anaesthetic Museum, London 44783
Army Medical Services Museum, Mytchett 45187
Bethlem Royal Hospital Archives and Museum, Beckenham 43429
British Dental Association Museum, London 44801
British Red Cross Museum and Archives, London 44806
Chamberlain Museum of Pathology, Birmingham 43503
Crichton Museum, Dumfries 44034
The Crossness Engines, London 44832
Florence Nightingale Museum, London 44849
Freud Museum, London 44852
George Eliot Hospital Museum, Nuneaton 45304
Gordon Museum, London 44859
Great Ormond Street Hospital for Children Museum, London 44862
Hunter House, East Kilbride 44088
Hunterian Museum, London 44880
Jenner Museum, Berkeley 43454
Mervyn Quinlan Museum, Benenden 43451
Museum of Anatomy, Glasgow 44290
Museum of Dentistry, Liverpool 44741
Museum of the Royal College of Surgeons of England, London 44932
Museum of the Royal Pharmaceutical Society of Great Britain, London 44933
National Hearing Aid Museum, London 44938
Old Operating Theatre, Museum and Herb Garret, London 44948
Royal Army Dental Corps Historical Museum, Aldershot 43273
Royal College of Obstetricians and Gynaecologists Collection, London 44972
Royal London Hospital Archives and Museum, London 44975
Saint Bartholomew's Hospital Museum, London 44979
Saint Thomas' Hospital Mini Museum, London 44981
Stephen G. Beaumont Museum, Wakefield 45912
Sunnyside Museum, Hillside 44469
Surgeons' Hall Museum, Edinburgh 44160
Thackray Museum, Leeds 44668
University of Dundee Museum Collections, Dundee 44052
University of Manchester Medical School Museum, Manchester 45098

United States of America

Alabama Museum of the Health Science, Birmingham 46819
Alumni Museum, Catonsville 47283
American Museum of Nursing, Tempe 53463
Das Anwesen and Pharmacy Museum, New Braunfels 51021
Arkansas County Doctor Museum, Lioncoln 50124
Carter-Coile Country Doctors Museum, Winterville 54259
Children's Health Museum, Charlottesville 47396
Clara Barton Birthplace, North Oxford 51399
Clara Barton National Historic Site, Glen Echo 48918
Clendening History of Medicine Library and Museum, Kansas City 49680
The Country Doctor Museum, Bailey 46530
Crawford W. Long Museum, Jefferson 49605
DeWitt Stetten Jr., Bethesda 46775
Dittrick Museum of Medical History, Cleveland 47629
Dr. Daniel W. Kissam House, Huntington 49429
Dr. Francis Medical and Apothecary Museum, Jacksonville 49575
Dr. John Harris Dental Museum, Bainbridge 46531
Dr. Samuel D. Harris National Museum of Dentistry, Baltimore 46561
East Tennessee Discovery Center, Knoxville 49816
Feet First: The Scholl Story, North Chicago 51387
Fort Winnebago Surgeons Quarters, Portage 52051
Global Health Odyssey, Atlanta 46440
Glore Psychiatric Museum, Saint Joseph 52561
Hall of Health, Berkeley 46754
Halle Heart Center, Tempe 53467
Health Adventure, Asheville 46374
Healthspace Cleveland, Cleveland 47632
Heartland -California Museum of the Heart, Rancho Mirage 52241
History of Medicine Collections, Durham 48220
Homeopathy Works, Berkeley Springs 46761
Hospitals and Clinics Medical Museum, Iowa City 49517
Indiana Medical History Museum, Indianapolis 49496
International Museum of Surgical Science, Chicago 47482
Ivy Green, Tuscumbia 53624
John P. McGovern Museum of Health and Medical Science, Houston 49392
The John Q. Adams Center for the History of Otolaryngology - Head and Neck Surgery, Alexandria 46209
Joseph A. Tallman Museum, Cherokee 47420
Kansas Health Museum, Halstead 49130
Macaulay Museum of Dental History, Charleston 47371
McDowell House Museum, Danville 47962
Marian Koshland Museum, Washington 53889
Menczer Museum of Medicine and Dentistry, Hartford 49196
Mobile Medical Museum, Mobile 50768
Montgomery County Historical Society Museum, Rockville 52431
Museum at Mountain Home, Johnson City 49634
Museum of Chicot County Arkansas, Lake Village 49893
The Museum of Nursing History, Philadelphia 51856
Museum of the University of Connecticut Health Center, Farmington 48514
Museum of Vision, San Francisco 52788
Mutter Museum, Philadelphia 51857
National Museum of Civil War Medicine, Frederick 48758
National Museum of Health and Medicine, Washington 53903
New Orleans Pharmacy Museum, New Orleans 51092
The Pearson Museum, Springfield 53220
Sioux Empire Medical Museum, Sioux Falls 53121
Still National Osteopathic Museum, Kirksville 49801
Transylvania Museum, Lexington 50056
United States Army Medical Department Museum, Fort Sam Houston 48679
University Medical Museum, Charleston 47377
Wood Library-Museum of Anesthesiology, Park Ridge 51710

Uzbekistan

Muzej Zdravoochranenija, Toshkent 54481

Metals and Metallurgy

Argentina

Museo de Bellas Artes de la Boca Quinquela Martín, Buenos Aires 00187

Australia

New England Brass and Iron Lace Foundry Museum, Uralla 01615

Austria

Fahrngruber Hammer mit Köhlereimuseum, Ybbsitz 03123
Hochofen-Museum Bundschuh, Thomatal 02812
Hochofenmuseum Radwerk IV, Vordernberg 02878
Huf - und Hackenschmiede Lindermayr, Spital am Pyhrn 02749
Metallmuseum, Tullnerbach-Lawies 02835
Montanmuseum, Gußwerk 02022
Nagelschmiedemuseum, Losenstein 02338
Schmiedemuseum, Lölling 02333
Stadtmuseum im Kammerhof, Eisenerz 01855
Volkskundemuseum Spiralschmiede, Lasberg 02266

Belgium

Le Puits-Marie, Blegny 03373
Maison de la Metallurgie et de l'Industrie de Liège, Liège 03799
Musée de la Bourrellerie et de la Ferronnerie, Lahamaide 03774
Musée du Fer et de la Métallurgie Ancienne, Saint-Hubert 03997

China

Lijiashan Bronze Museum, Jiangchuan 07856
Tonglushan Ancient Metallurgy Museum, Daye 07655

Czech Republic

Museum Vítkovických Železáren, Ostrava 09698
Muzeum Strojíren Poldi, Kladno 09582
Muzeum Třineckých Železáren a Města Třince, Třinec 09879
Stará Hut' v Josefovském Údolí, Adamov 09394
Technické Muzeum v Brně, Brno 09440

Finland

Suomen Valimomuseo ja Högforsin Masuuni, Karkkila 10919

France

Écomusée de la Métairie, Plouigneau 14860
Ecomusée Industriel de Lochrist-Inzinzac, Lochrist-Inzinzac 13796
Musée de la Grosse Forge, Aube 11722
Musée de la Métallurgie, Saint-Chély-d'Apcher 15202
Musée de l'Histoire du Fer, Jarville 13262
Musée des Forges, Brocas-les-Forges 12190

Germany

Aurnhammer-Sammlung, Treuchtlingen 22089
Deutsches Kaltwalzmuseum, Hagen 18853
Eisenhammer Dorfchemnitz, Dorfchemnitz 17902
Gesenkschmiede Lubenbach, Zella-Mehlis 22784
Hammer- und Waffenschmiede-Museum Hexenagger, Altmannstein 16513
Historische Ausstellung Krupp, Essen 18254
Historische Hammerschmiede, Blaubeuren 17378
Industriemuseum Brandenburg, Brandenburg an der Havel 17508
Mansfeld-Museum Hettstedt, Hettstedt 19173
Maxhütten-Museum, Bergen, Kreis Traunstein 17010
Museum Eisenhüttenwerk Peitz, Peitz 21005
Museum Saigerhütte Olbernhau- Kupferhammer, Olbernhau 20893
Museum Wasseralfingen, Aalen 16413
Schmiedemuseum Bremecker Hammer, Lüdenscheid 20090
Technisches Denkmal Kupferhammer, Thießen 22053
Technisches Denkmal Neue Hütte, Schmalkalden, Kurort 21581
Technisches Schaudenkmal Gießerei Heinrichshütte, Wurzbach 22750

Hungary

Központi Kohászati Múzeum, Miskolc 23708
Massa Múzeum, Miskolc 23710
Öntödei Múzeum, Budapest 23459

Italy

Museo Italiano della Ghisa, Longiano 26518

Japan

Wakoh Hakubutsukan, Yasugi 29589

Korea, Republic

Seyun Iron Museum, Eumseong 29803

Netherlands

Het Behouden Blik, Uithuizermeeden 32838

Poland

Muzeum Starożytnego Hutnictwa Świętokrzyskiego im. prof. M. Radwana, Nowa Słupia 34949
Zabytkowa Huta Żelaza, Chlewiska 34593
Zabytkowa Kuźnia Wodna w Starej Kuźnicy, Stara Kuźnica 35127

Portugal

Museu do Ferro e da Região de Moncorvo, Torre de Moncorvo 35967

Russia

Muzej Istorii Architektury i Promyšlennoj Techniki Urala, Ekaterinburg 36892
Muzej Razvitija Metallurgii, Moskva 37351

Serbia

Muzej Crne Metalurgije Sartid A.D., Smederevo 38143
Muzej Rudarstva i Metalurgije, Bor 38093

Slovakia

Hutnícke Múzeum Železiarne Podbrezová, Podbrezová 38329

Slovenia

Muzej Železniki, Železniki 38541

Sweden

Bruksmuseum, Hudiksvall 40603
Hälleforsnäs Gjuterimuseum, Hälleforsnäs 40571
Rademachersmedjorna, Eskilstuna 40503

Switzerland

Toggenburger Schmiede- und Werkzeug-Museum, Bazenheid 41188

Turkey

Kardemir Demir Çelik Müzesi, Karabük 42863

United Kingdom

Brass Rubbing Centre, Edinburgh 44114
London Brass Rubbing Centre, London 44904

United States of America

Buckeye Furnace State Memorial, Wellston ... 54031
Iron and Steel Museum of Alabama, McCalla. 50340
Putnam County Historical Society and Foundry School Museum, Cold Spring 47681

Metalwork

Austria

Alte Marktschmiede, Lasberg 02264
Freilichtmuseum Fürstenhammer, Lasberg.... 02265
Freilichtmuseum Hackenschmiede, Bad Wimsbach-Neydharting 01771
Karlinger-Schmiede, Unterweissenbach 02844
Montanmuseum, Gußwerk 02022
Prof. Hans Gerstmayr - Museum, Mauthausen 02374
Schaumuseum Alte Schmiede, Gmünd, Niederösterreich 01946

Belgium

Musée de la Forge, Romedenne 03982

Canada

Forge-Menuiserie Cauchon, Rivière-Malbaie... 06762
Green Family Forge - A Blacksmith Museum, Trinity 07163
Lieu Historique National des Forges du Saint-Maurice, Trois-Rivières 07173

Czech Republic

Slevarenske Muzeum Komarov, Komárov 09593

Finland

Fiskars Museum, Fiskars 10734
Iloniemen Pajamuseo, Raatala 11249
Kaarlo Kangasniemen Urheilumuseo, Kullaa.. 11003

France

Musée-Atelier du Cuivre et de l'Argent, Paris 14630
Musée de La Ferblanterie, La Tour-Blanche 13448
Musée de la Grosse Forge, Aube 11722
Musée du Couteau de Laguiole, Laguiole 13480
Musées de Vieille Forge, de la Traversée du Siècle et des Automates, Fréland 13007

Germany

Blankschmiede Neimke, Dassel 17798
Deutsches Drahtmuseum, Altena 16486
Deutsches Klingenmuseum Solingen, Solingen 21801
Domschatzkammer, Essen 18251
Eisenkunstgussmuseum, Büdelsdorf 17625
Fabrikmuseum der Leonischen Industrie, Roth 21401
Glockenschmiede, Ruhpolding 21446
Gold- und Silberschmiedemuseum, Mülsen.... 20419
Hammerschmiede und Stockerhof, Neuburg an der Kammel 20602
Heinrich-Blickle-Museum, Sammlung Gußeiserner Ofenplatten, Rosenfeld 21372
Historische Schlosserei Pelzer, Jena 19382
Historischer Eisenhammer, Roth 21403
Hüttenmuseum Thale, Thale (Harz) 22042
Museum Bayerische Metallwarenfabrik (BMF), Nürnberg 20780
Museum der Grafschaft Mark, Altena 16487
Museum für Eisenkunstguß, Hirzenhain 19200
Museum Haus Martfeld, Schwelm 21690
Ofenplattensammlung der Schwäbische Hüttenwerke, Aalen 16414
Rheinisches Eisenkunstguss Museum, Bendorf 16989
Rheinisches Industriemuseum, Solingen 21806
Sayner Gießhalle, Bendorf 16990
Schaudenkmal Gaszentrale, Unterwellenborn . 22185
Schmiedemuseum Kirchlauter, Kirchlauter 19554
Südharzer Eisenhüttenmuseum, Bad Lauterberg 16777
Voithenberghammer, Furth im Wald 18517
Zentrale Kustodie und Universitätsmuseum, Halle, Saale 18906

Iceland

Sjóminja- og Vélsmiðjusafn, Reykjavik 24035

Italy

Museo Davia Bargellini, Bologna 25451
Museo degli Usi e Costumi della Gente Trentina, San Michele all'Adige 27669
Museo del Ferro e della Ghisa, Follonica 26215
Museo Etnografico Comunale, Premana 27343

Lithuania

Kalvystės Muziejus, Klaipėda 30278

Netherlands

H.J. van de Kamp Museum, Sint Oedenrode ... 32776
Oale Smederie, Hellendoorn 32358
Smederij Museum Wijlen dhr Verkley, Nieuwkoop 32581

Norway

Næs Jernverksmuseum, Tvedestrand 33938

Peru

Museo de Arte Colonial Pedro de Osma, Lima 34231

Poland

Muzeum Kowalstwa w Warszawie, Warszawa 35254

Romania

Muzeul Fierului, Hunedoara 36405

Slovenia

Kovaski Muzej Kropa, Kropa 38441

Sweden

Gislöfs Smidesmuseum, Simrishamn 40839

Switzerland

Musée du Fer et du Chemin de Fer, Vallorbe .. 41976

Turkey

Vakıf Inşaat ve Sanat Eserleri Müzesi, İstanbul 42847

United Kingdom

East Carlton Steel Heritage Centre, East Carlton 44083
Museum of Iron and Darby Furnace, Coalbrookdale 43860
Wortley Top Forge, Thurgoland 45856

United States of America

National Ornamental Metal Museum, Memphis 50582
World O' Tools Museum, Waverly 53997

Meteorites

Argentina

Museo de Mineralogía y Geología Dr. Alfredo Stelzner, Córdoba 00326
Museo del Cielo, Villa Carlos Paz 00673

Denmark

Statens Naturhistoriske Museum (SNM), København 10153

France

Musée de la Météorite, L'Aigle 13483

Germany

Heimatmuseum auf dem Klosterhof, Steinheim am Albuch 21889
Meteorkrater-Museum, Steinheim am Albuch.. 21890
Mineralogische Schau- und Lehrsammlung, Tübingen 22120

Italy

Museo di Scienze Naturali, Brescia 25538

Japan

Museum of Old Photographs and Museum of Unearthed Artifacts, Nagasaki 28976

Lithuania

Mineralų Muziejus, Vilnius 30425

United States of America

Big Well, Greensburg 49069
Cernan Earth and Space Center, River Grove .. 52364
Craters of the Moon, Arco 46342
Institute of Meteoritics Meteorite Museum, Albuquerque 46179
Museum of the Center for Meteorite Studies, Tempe 53470
Orton Geological Museum, Columbus 47758

Military History

Afghanistan

Herat Provincial Military Museum, Herat 00002

Algeria

Musée de l'Armée Nationale Populaire, Alger 00039
Musée National du Jihad, Alger 00048

Argentina

Museo de la Policía Federal Argentina, Buenos Aires 00206
Museo de los Patricios de Buenos Ayres, Buenos Aires 00215
Museo Historiador Hernández Galiano, Tostado 00650
Museo Histórico del Regimiento de Granaderos a Caballo General San Martín, Buenos Aires 00240
Museo Municipal Histórico Regional Almirante Brown, Bernal 00148
Museo Naval Puerto Belgrano, Puerto Belgrano 00504
Sala Histórica del Regimiento de Infantería Mecanizado 24, Río Gallegos 00535
Sala Historica General Savio, Buenos Aires.... 00282

Armenia

Military Museum, Yerevan 00744

Australia

Alice Springs RSL War Museum, Alice Springs 00797
Anzac Cottage, Mount Hawthorn 01328
ANZAC Memorial, Sydney 01543
Army Museum Bandiana, Bandiana 00832
Army Museum of New South Wales, Paddington, New South Wales 01398
Army Museum of South Australia, Keswick 01194
Army Museum of Western Australia, Fremantle 01091
Australian War Memorial, Campbell 00939
Bowraville Folk and Military Museum, Bowraville 00884
Dubbo Military Museum, Dubbo 01045
Fort Glanville, Semaphore Park 01499
Fort Queenscliff Museum, Queenscliff 01447
Fort Scratchley, Newcastle 01361
Frank Partridge V.C. Military Museum, Bowraville 00885
Garrison Gallery Museum, Millers Point 01304
Hassett's Uralla Military Museum, Uralla 01613
HMAS Castlemaine - Museum Ship, Williamstown 01651
Melbourne Tank Museum, Narre Warren North 01352
Merredin Military Museum, Merredin 01296
Milne Bay Military Museum, Toowoomba 01593
New South Wales Lancers Memorial Museum, Parramatta 01407
Queensland Military Memorial Museum, Fortitude Valley 01088
Returned Services League War Museum, Stawell 01528
Royal Australian Infantry Corps Museum, Singleton 01506
Tatura Irrigation and Wartime Camps Museum, Tatura 01574

Austria

Bunkermuseum Wurzenpass, Riegersdorf 02588
Dragonermuseum, Wels 02914
Freilichtmuseum des Gebirgskrieges, Kötschach-Mauthen 02214
Heeresgeschichtliches Museum HGM, Wien .. 03001
Militärgeschichtliche Ortssammlung, Schnifis 02710
Militärmuseum Rosenau, Sonntagberg 02748
Museen im Rathaus, Perchtoldsdorf 02488
Museum 1915-1918, Kötschach-Mauthen 02215
Radetzky-Gedenkstätte Heldenberg, Kleinwetzdorf 02196
Rainermuseum, Salzburg 02622
Stadtmuseum und Tuchmachermuseum Pinkafeld, Pinkafeld 02499
Truppenmuseum, Straß in Steiermark 02786
Vorarlberger Militärmuseum, Bregenz 01797
Waffensammlung Schloss Ebelsberg, Linz 02325
Wehrgeschichtliches Museum, Salzburg 02630

Belarus

Belorusskij Gosudarstvennyj Muzej Istorii Velikoj Otečestvennoj Vojny, Minsk 03247

Belgium

Ardennen Poteau '44 Museum, Sankt Vith 03998
Atlantic Wall Museum, Oostende 03929
Boerenkrijgmuseum, Overmere 03947
Canada War Museum, Adegem 03261
Ensival Historical Museum, Ensival 03583
Mémorial Musée du Régiment, Spa 04040
Memorial Museum Passchendaele 1917, Zonnebeke 04166
Musée de la Bataille des Ardennes, La Roche-en-Ardenne 03769
Musée de la Hesbaye, Remicourt 03965
Musée de la Ligne Koningshooikt-Wavre, Chaumont 03520
Musée des 18 Jours-Mémoires de Guerre, Pipaix 03951
Musée des Chasseurs à Pied, Charleroi 03517
Musée des Commandos, Flawinne 03600
Musée du 3è Lanciers, Namur 03904
Musée du Cimetière Américain Henri-Chapelle, Hombourg 03709
Musée du Fort de Battice, Battice 03348
Musée du Fusil Rouillé, Bioul 03369
Musée du Souvenir des Guerres 1914-18 et 1940-45, Ecaussinnes 03572
Musée Royal de l'Armée et d'Histoire Militaire, Bruxelles 03494
Musée Truschbaum, Elsenborn 03578
Musée Wellington, Waterloo 04138
Museum Slag der Zilveren Helmen, Halen 03666
Museum van het Kamp van Beverlo, Leopoldsburg 03785
Panorama de la Bataille de Waterloo, Braine-l'Alleud 03396
Remember Museum 39-45, Thimister-Clermont04062
Rietgaverstede, Nevele 03912

Bermuda

Bermuda National Trust Museum, Saint George's 04196

Bolivia

Museo Histórico Militar, La Paz 04220

Brazil

Army Museum and Fort Copacabana, Rio de Janeiro 04681
Museu da Brigada Militar, Porto Alegre 04641
Museu da Força Expedicionária Brasileira, Rio de Janeiro 04715
Museu da Polícia Militar de Pernambuco, Recife 04663
Museu de Armas Major Lara Ribas, Florianópolis 04454
Museu de Polícia Militar, São Paulo 04888
Museu Militar do Forte do Brum, Recife 04672

Bulgaria

Gradski Istoričeski Muzej, Klisura 05069
Istoričeski Muzej - Preminavane na Ruskite Vojski pri Svištov, Svištov 05223
Muzej Rumănski Voin, Pordim 05136
Muzej Vojniško Văstanie 1918, Radomir 05146
Panorama Plevenska Epopeja 1877, Pleven... 05119
Voenno-istoričeski Muzej, Gorna Studena 05037
Voenno-istoričeski Muzej, Pleven 05122
Voenno-istoričeski Muzej Osvoboditelna Vojna 1877-1878, Bjala 04988

Canada

12 Vancouver Service Battalion Museum, Richmond 06740
15th Field Artillery Regiment Museum, Vancouver 07197
17 Wing Heritage Collection, Winnipeg 07351
26th Field Regiment and XII Manitoba Dragoons Museum, Brandon 05457
48th Highlands Museum, Toronto 07089
Base Borden Military Museum, Borden 05441
Beaver Valley Military Museum, Clarksburg ... 05620
British Columbia Regiment Museum, Vancouver 07204
Cameron Highlanders of Ottawa Regimental Museum, Ottawa 06531
Canadian Military Engineers Museum, Oromocto 06518

Canadian Military Heritage Museum, Saskatoon 06911
Canadian Military Studies Museum, Limehouse 06171
Canadian Scottish Regimental Museum, Victoria 07243
Canadian War Museum, Ottawa 06539
CFB Esquimalt Naval Museum and Military Museum, Victoria 07244
CFB Petawawa Military Museum, Petawawa 06588
Currahee Military Museum, Virden 07268
Diefenbunker-Canada's Cold War Museum, Carp 05569
Discovery Harbour, Penetanguishene 06581
First Hussars Citizen Soldiers Museum, London 06185
Fort Beauséjour, Aulac 05362
Fort-Chambly, Chambly 05585
Fort Garry Horse Regimental Museum, Winnipeg 07360
Fort George, Niagara-on-the-Lake 06451
Fort Henry, Kingston 06077
Fort Normandeau, Red Deer 06707
Fort Saint Joseph, Richard's Landing 06737
Fort Wellington, Prescott 06656
Fort York Armoury, Toronto 07106
Governor General's Foot Guards Museum, Ottawa 06546
Guard House and Soldiers Barracks, Fredericton 05832
Halifax Citadel, Halifax 05935
Hamilton Military Museum, Hamilton 05956
HMCS Haida, Hamilton 05959
Lieu Historique National du Canada de la Bataille-de-la-Châteauguay, Ormstown 06516
Lorne Scots Regimental Museum, Brampton 05455
Loyal Edmonton Regiment Military Museum, Edmonton 05749
Musée de la Défense Aérienne, Alouette 05326
Musée des Fusiliers Mont-Royal, Montréal 06358
Musée du Corp des Magasins Militaires, Montréal 06367
Musée du Royal 22e Régiment, Québec 06692
Musée Militaire Baie-des-Chaleurs, New Richmond 06433
Musée Militaire du 12e Régiment Blindé du Canada, Trois-Rivières 07176
Musée Navale du Québec, Québec 06695
Musée Régimentaire les Fusiliers de Sherbrooke, Sherbrooke, Québec 06963
Musée Stewart au Fort de île Sainte-Hélène, Montréal 06376
Museum of the Regiments, Calgary 05526
Museum of The Royal Westminster Regiment Historical Society, New Westminster 06437
Nancy Island Historic Site, Wasaga Beach 07280
Naval Museum of Alberta, Calgary 05528
Northern Lights Military Museum, Happy Valley 05970
Okanagan Military Museum, Kelowna 06055
Parc de l'Artillerie, Québec 06697
The Prince Edward Island Regiment Museum, Charlottetown 05596
Prince of Wales Martello Tower, Halifax 05946
Princess of Wales' Own Regiment Military Museum, Kingston 06088
Princess Patricia's Canadian Light Infantry Museum, Calgary 05531
Queen's Own Rifles of Canada Regimental Museum, Toronto 07129
Queen's York Rangers Regimental Museum, Toronto 07130
Quidi Vidi Battery, Saint John's 06852
Rocky Mountain Rangers Museum, Kamloops 06037
Royal Canadian Artillery Museum, Shilo 06965
Royal Canadian Military Institute Museum, Toronto 07132
Royal Hamilton Light Infantry Heritage Museum, Hamilton 05962
Royals Museum, Toronto 07134
Salvation Army George Scott Railton Heritage Centre, Toronto 07135
Saskatchewan Military Museum, Regina 06729
Seaforth Highlanders of Canada Regimental Museum, Vancouver 07221
Stormont, Dundas and Glengarry Highlanders Regimental Museum, Cornwall 05645
Thunder Bay Military Museum, Thunder Bay 07081
Uniacke Estate Museum Park, Mount Uniacke 06401
Vancouver Island Military Museum, Nanaimo 06414
Vancouver Naval Museum, Qualicum Beach 06675
Veterans' Memorial Military Museum, Kensington 06060
York Sunbury Museum, Fredericton 05842

Chile

Museo Histórico Nacional, Santiago de Chile 07460

China

China Jiawu Zhan Zhen Guan, Weihai 08316
China People's Revolution Military Museum, Beijing 07531
Dandong Museum on the War to resist U.S. Aggression and Aid Korea, Dandong 07646
Guangzhou Huangpu War Exhibition Center, Guangzhou 07729
Han Military Forces Yong Museum, Xuzhou 08417
Macau Security Forces Museum, Macau 07998
Nan'ao Coast Defence City Museum, Nan'ao 08027
Old Zhi Display Center, Nanchang 08031
Qinshihuang Military Forces Yong Museum, Lintong 07954

Colombia

Museo Militar de Colombia, Bogotá 08589

Cuba

Museo Antiguo Cuartel de Caballería del Ejército Español, Camagüey 09049
Museo de Ejército Oriental, Holguín 09180
Museo Fortaleza de San Carlos de la Cabaña, La Habana 09144
Museo Máximo Gómez, La Habana 09150

Czech Republic

Armádní Muzeum Žižkov, Praha 09734
Památník Bitvy 1866 na Chlumu, Všestary 09917
Slezské zemské muzeum, Opava 09692
Slezské zemské muzeum, Hlučín-Darkovičky 09517
Slezské zemské muzeum, Hrabyně 09532
Vojenské technické museum Lešany, Krhanice 09608

Denmark

Blaavandshuk Egnsmuseum, Oksbøl 10219
Tirpitz-Stillingen, Blåvand 09979
Varde Artillerimuseum, Varde 10308

Ecuador

La Casa de los Tratados, Girón 10397
Museo Templete de los Heros Nacionales, Quito 10451

Egypt

Military Museum, El-Alamein 10524
Port-Said Museum, Port Said 10538

Finland

Artillery Museum of Finland, Hämeenlinna 10742
Hanko Front Museum, Hanko 10756
Ilmavoimien Viestikoulun Perinnehuone, Tikkakoski 11375
Jalkaväkimuseo, Mikkeli 11113
Kokemäen Maatalousmuseo ja Sotilaspoikamuseo, Kokemäki 10969
Lylyn Viestivarikon Museo, Lyly 11090
Maneesi Sotamuseo, Helsinki 10803
Päämajamuseo, Mikkeli 11116
Panssarimuseo, Parola 11193
Pioneerimuseo, Ala-Pihlaja 10707
Porilaismuseo, Säkylä 11290
Rajamuseo, Imatra 10856
Rannikkotykistömuseo, Helsinki 10812
Reserviupseerikoulun Museo, Hamina 10754
Sotamuseo, Helsinki 10818
Sukellusvene Vesikko, Helsinki 10819
Suomen Viestimuseo, Riihimäki 11272

France

L'Argonaute, Paris 14565
Blockhaus d'Éperlecques, Éperlecques 12828
Centre Juno Beach, Courseulles-sur-Mer 12653
Château, Duilhac-sous-Peyrepertuse 12794
Des Hommes dans la Guerre de 1870, Bitche 12005
Espace Historique de Sword Beach, Hermanville-sur-Mer 13206
Le Point Fortifié Hillmann, Colleville-Montgomery 12571
Mémorial Fort Montbarey, Brest 12160
Militarial Musée Mémorial pour la Paix, Boissezon 12048
Musée 2ème Guerre Mondiale, Ambleteuse 11572
Musée 39-45, Roscoff 15103
Musée à Flot de l'Escorteur d'Escadre, Nantes 14361
Musée Août 1944, Falaise 12903
Musée Bernadotte, Pau 14755
Musée d'Arromanches, Arromanches-les-Bains 11701
Musée d'Art Militaire, Vincey 16249
Musée de la Bataille, Fontenoy-en-Puisaye 12973
Musée de la Bataille du 6 Août 1870, Woerth 16301
Musée de la Batterie de Merville, Merville-Franceville-Plage 14094
Musée de la Figurine, Toulon (Var) 15924
Musée de la Figurine Historique, Compiègne 12598
Musée de la Légion Étrangère, Aubagne 11719
Musée de la Libération, Cherbourg-Octeville 12503
Musée de la Ligne Maginot, Hunspach 13228
Musée de la Ligne Maginot, Lembach 13679
Musée de la Mémoire et de la Paix, Clerval 12558
Musée de la Première Guerre Mondiale, Villeneuve-les-Convers 16215
Musée de la Résistance du Mont Mouchet, Auvers (Haute-Loire) 11767
Musée de la Seconde Guerre Mondiale, Digne-les-Bains 12724
Musée de la Vallée à Meyronnes, Meyronnes 14111
Musée de l'Armée, Paris 14667
Musée de l'Empéri, Salon-de-Provence 15610
Musée de l'Infanterie, Montpellier 14241
Musée des Canonniers Sédentaires de Lille, Lille 13757
Musée des Troupes de Marine, Fréjus 13002
Musée d'Histoire et de Tambow, Amnéville-les-Thermes 11590
Musée d'Histoire Militaire à Bandera, Ajaccio 11525
Musée du Fort de la Pompelle, Puisieulx 14957
Musée du Lycée Militaire de Saint-Cyr, Saint-Cyr-l'École 15226
Musée du Service de Santé des Armées, Paris 14701
Musée du Souvenir, Guer 13155
Musée du Souvenir du Génie Militaire, Angers 11611
Musée-Fort de Leveau, Feignies 12917
Musée Fort de Sucy, Sucy-en-Brie 15837
Musée Guerre et Paix en Ardennes, Novion-Porcien 14477
Musée Historique et Erckmann-Chatrian, Phalsbourg 14811
Musée Historique et Militaire, Huningue 13227
Musée la Perçée du Bocage 1944, Saint-Martin-des-Besaces 15401
Musée Libertador du Général San Martin, Boulogne-sur-Mer 12094
Musée Mémorial des Combat de la Poche de Colmar, Turckheim 16010
Musée Mémorial du Front d'Orient 1915-1918, Troissereux 15991
Musée-Mémorial National du Débarquement en Provence (1944), Toulon (Var) 15927
Musée Militaire des Gloires et Souvenirs du Périgord, Périgueux 14780
Musée Militaire Faller, Mars-la-Tour 13982
Musée Municipal, Cusset 12693
Musée National des Deux Victoires, Mouilleron-en-Pareds 14301
Musée No.4 Commando, Ouistreham 14545
Musée Pierre Noël, Saint-Dié-des-Vosges 15241
Musée Poche de Royan, Le Gua 13591
Musée Serret, Saint-Amarin 15162
Musée Servie de Santé des Armées au Val-du-Grâce, Paris 14738
Musée Somme 1916, Albert 11533
Musée sur l'Artillerie, la Cavalerie et l'Infanterie, Seclin 15695
Musée Vauban, Neuf-Brisach 14396
Musée Wagon de l'Armistice, Compiègne 12601
Musée Westercamp, Wissembourg 16300
Salle de Traditions de la Garde Républicaine, Paris 14750

Germany

Armeemuseum 'Friedrich der Große', Kulmbach 19744
Blüchermuseum, Kaub 19482
Das Bunkermuseum, Emden 18136
Celler Garnison-Museum, Celle 17684
Celler Schützenmuseum, Celle 17685
Deutsches Marinemuseum, Wilhelmshaven 22575
Deutsches Panzermuseum Munster, Munster 20530
Garnisonmuseum Nürnberg, Nürnberg 20762
Gebirgsjägermuseum, Sonthofen 21822
Geroldsecker Waffenschmiede, Seelbach 21719
Gneisenau Museum, Schildau, Gneisenaustadt 21548
Heimatmuseum, Friedland 18472
Heimatmuseum, Karlstein am Main 19456
Historisches Museum im Marstall, Paderborn 20970
Kleines Militärmuseum Burg Laurenburg, Laurenburg 19860
Krell'sche Schmiede, Wernigerode 22496
Kultur- und Militärmuseum, Grafenwöhr 18733
Luftwaffenmuseum der Bundeswehr, Berlin 17179
Marinemuseum Stralsund, Stralsund 21915
Militär-Museum Kossa, Kossa 19693
Militär- und Jagdgeschichtliche Sammlung, Bad Wildungen 16862
Militärhistorisches Museum der Bundeswehr, Dresden 17960
Motorradwelt der DDR, Prora 21132
Museum 1806, Jena 19384
Museum für historische Wehrtechnik, Röthenbach an der Pegnitz 21359
Museum für Militär- und Zeitgeschichte, Kolitzheim 19675
Museum für Waffentechnik, Ausrüstung, Auszeichnungswesen, Vilshofen 22245
Museum im Zeughaus, Vechta 22207
Museum Zitadelle Jülich, Jülich 19405
Osthofentor-Museum, Soest 21797
Rheinisches Schützenmuseum, Neuss 20661
Rottauer Museum für Fahrzeuge, Wehrtechnik und Zeitgeschichte, Pocking 21070
Schloss Höchstädt, Höchstädt an der Donau 19211
Stadt- und Festungsmuseum im Ludwigstor, Germersheim 18605
Stadtmauermuseum Nördlingen, Nördlingen 20726
Torhaus Markkleeberg, Markkleeberg 20205
Völkerschlachtdenkmal / Forum 1813, Leipzig 19941
Wehrgeschichtliches Museum Rastatt, Rastatt 21189
Westwallmuseum Gerstfeldhöhe, Pirmasens 21043

Greece

Historical and Ethnological Museum of Patras, Patras 23140
Polemiko Mouseio, Athinai 22951

Guadeloupe

Fort Fleur d'Epée, Le Gosier 23275

Guam

War in the Pacific National Historical Park, Asan 23286
War in the Pacific National Historical Park, Piti 23288

Hungary

Hadtörténeti Múzeum, Budapest 23408
Huszármúzeum, Balatonszentgyörgy 23356
Lovasmúzeum, Bábolna 23341
Pintér Muvek Hadtörténeti Múzeum, Hadtörténeti Park, Kecel 23642
Római Kötár, Szentendre 23879

India

Bishnu Museum, Shillong 24385
Indian War Memorial Museum, Delhi 24142

Indonesia

Museum Dirgantara Mandala, Yogyakarta 24588
Museum Mandala Wangsit Siliwangi, Bandung 24443
Museum Militer, Jakarta 24491
Museum Palagan, Ambarawa 24436
Museum Tni A.L. Loka Jala Carana, Surabaya 24567

Iran

Afif Abad Military Museum, Shiraz 24645
Anzali Military Museum, Bandar-Anzali 24598
Military Museum, Tehrān 24680

Ireland

Dublin Castle, Dublin 24785
Military Heritage Tours, Longford 24869
National Museum of Ireland - Decorative Arts History, Dublin 24809
Saint Catherine's National School Museum, Aughrim 24731

Israel

Battle Museum, Kibbutz Gesher 25041
Haghna Museum, Tel Aviv 25106
Israel Defense Forces History Museum, Tel Aviv 25109
Israeli Air Force Museum, Hatzerim 24971
Jewish Battalions Museum, Ahivil 24922
Museum Zahal, Hakirya 24970

Italy

Museo dei Combattenti, Soncino 27869
Museo dei Reperti Bellici e Storici della Guerra 1915-1918, Capovalle 25661
Museo dei Soldati del Montello, Nervesa della Battaglia 26920
Museo dei Soldatino e della Figurina Storica, Calenzano 25599
Museo del Combattente, Modena 26748
Museo del Reggimento "Giovani Fascisti", Ponti sul Mincio 27290
Museo del Risorgimento, Catanzaro 25779
Museo del Risorgimento, Solferino 27865
Museo della Battaglia, Vittorio Veneto 28299
Museo della Battaglia di Marengo, Alessandria 25192
Museo delle Armi Antiche e Fortezza Medievale, Acquaviva Picena 25148
Museo delle Truppe Alpine Mario Balocco, Biella 25419
Museo di Architettura Militare, Bologna 25460
Museo Ferrucciano, Gavinana 26277

Museo Sacrario delle Bandiere, Roma 27526
Museo Storico degli Truppe Alpini, Trento 28060
Museo Storico dei Bersaglieri, Roma 27527
Museo Storico della Motorizzazione Militare, Roma 27533
Museo Storico delle Armi, Civitella del Tronto 25926
Museo Storico Militare, Comun Nuovo 25958
Museo Storico Militare e Sacrario di Redipuglia, Fogliano Redipuglia 26213
Museo Storico Militare Umberto I, Turate 28108
Museo Storico Nazionale di Artiglieria, Torino 28026

Jamaica

Military Museum, Kingston 28331

Japan

Chiran Tokko Heiwa Kaikan, Chiran 28404
Himeji-kenritsu Budokan, Himeji 28511
Maritime Self-Defense Force Sasebo Museum, Sasebo 29235
Yamamoto Isoroku Kinenkan, Nagaoka 28965

Jordan

Martyr's Memorial at Sports City, Amman 29656

Korea, Republic

Military Academy Museum, Seoul 29903

Kyrgyzstan

Frunze House Museum, Biškek 29952

Laos

Savannakhet provincial Museum, Savannakhet 29971

Latvia

Latvijas Kara Muzejs, Rīga 30081
Oskara Kalpaka Muzejs Airīšu Piemiņas Vietā, Zirņi 30152

Lebanon

Moussa Castle Museum, Beiteddine 30163

Lithuania

Vytauto Didžiojo Karo Muziejus, Kaunas 30269

Luxembourg

General Patton Memorial Museum, Ettelbruck 30474
Musée de la Bataille des Ardennes, Wiltz 30497
Musée de la Guerre, Clervaux 30459
Musée National d'Histoire Militaire, Diekirch 30465
Musée Régional des Enrôles de Force, Dudelange 30468

Malaysia

Muzium Angkatan Tentera, Kuala Lumpur 30577
Muzium Tentera Laut Diraja Malaysia, Melaka 30604
Royal Malaysian Navy Museum, Lumut 30592

Mexico

Casa Museo Cadete Juan Escutia, Tepic 31440
Casa Museo Coronel Gregorio Méndez Magaña, Jalpa de Méndez 30968
Museo Casa del General Alvaro Obregón, Huatabampo 30942
Museo Cuartel de Emiliano Zapata, Tlaltizapán 31470
Museo Cuartel Zapatista, México 31065
Museo de la Guerra de Castas, Felipe Carrillo Puerto 30866
Museo de la No Intervención Fuerte de Loreto, Puebla 31277
Museo del Ejército y Fuerza Aérea Bethlemitas, México 31101
Museo del Ejército y Fuerza Aérea Mexicanos Cuartel Colorado, Guadalajara 30899
Museo General Ignacio Zaragoza, Puebla 31287
Museo Histórico Naval Central, México 31122
Museo Teniente José Azuela, Zihuatanejo 31597
Museo Toma de Zacatecas o Cerro de la Bufa, Zacatecas 31582

Mongolia

Military Museum, Ulaanbaatar 31620

Morocco

Ameziane Museum, Nador 31686
Museo Militar, Melilla 31684

Mozambique

Museu Histórico Militar, Maputo 31708

Myanmar

War Museum, Yangon 31738

Netherlands

Airborne Museum Hartenstein, Oosterbeek 32624
Gelders Schuttersmuseum, Didam 32123
Generaal Maczek Museum, Breda 32004
Historische Genie Verzameling, Vught 32914
Historische Verzameling Aan- en Afvoertroepen, Stroe 32810
Historische Verzameling Intendance, Bussum 32031
Historische Verzameling Korps Nationale Reserve, Harskamp 32318
Historische Verzameling Luchtdoelartillerie, Ede 32169
Historische Verzameling Regiment van Heutsz, Arnhem 31925
Infanterie Museum, Harskamp 32319
Kazemattenmuseum, Oosterend 32626
Kazemattenmuseum Kornwerderzand, Kornwerderzand 32457
Koninklijk Militair Historisch Museum, Delft 32059
Koninklijk Tehuis voor Oud-Militairen en Museum Bronbeek, Arnhem 31928
Mariniersmuseum, Rotterdam 32711
Militair Historisch Museum De Veteraan, Eefde 32172
Museum Brigade en Garde Prinses Irene, Oirschot 32611
Museum Nederlandse Cavalerie, Amersfoort 31816
Museum 't Schilderhuis, Driebergen-Rijsenburg 32157
Museum Verbindingsdienst, Ede 32170
Museumschip Mercuur, Den Haag 32092
Nationaal Bevrijdingsmuseum 1944-1945, Groesbeek 32277
Nederlands Artillerie Museum, 't Harde 32310
Nederlands Militair Kustverdedigingsmuseum, Hoek van Holland 32395
Oorlogsverzetsmuseum Rotterdam, Rotterdam 32723
Traditiekamer Regiment Stoottroepen, Assen 31940
Vesting Bourtange, Bourtange 31994
Vischpoort en Kazematten, Elburg 32201

New Zealand

Air Force Museum, Christchurch 33067
Auckland Museum and Auckland War Memorial Museum, Auckland 33043
Otago Military Museum, Dunedin 33094
Royal New Zealand Army Medical Corps Museum, Christchurch 33076
Royal New Zealand Navy Museum, Devonport 33086

Nicaragua

Museo Casa Hacienda San Jacinto, San Jacinto 33303

Nigeria

National War Museum of Umauhia, Umauhia 33346

Norway

Arquebus Krigshistorisk Museum, Haugesund 33536
Austrått Fort, Brekstad 33454
Bjørn West-Museet, Matredal 33662
Blodveimuseet, Rognan 33788
Falstadminnet, Levanger 33634
Flyhistorisk Museum, Sola 33848
Forsvarsmuseet, Oslo 33724
Gudbrandsdal Krigsminnesamling, Kvam 33618
Kristiansand Kanonmuseum, Kristiansand 33610
Lofoten Krigsminnemuseum, Svolvær 33896
Marinemuseet, Horten 33563
Nordland Røde Kors Krigsminnemuseum, Narvik 33695
Nordsjøfartmuseet, Tælavåg 33899
Norges Hjemmefrontmuseum, Oslo 33738
Okkupasjonsmuseet, Eidsvoll Verk 33478
Senjehesten Kystforsvarsmuseum, Moen 33673

Oman

Sultan's Armed Forces Museum, Ruwi 33997

Pakistan

PAF Museum, Karachi 34016
Pakistan Air Force Museum Peshawar, Peshawar 34035

Paraguay

Colección Carlos Alberto Pusineri Scala, Asunción 34090
Museo Histórico Militar, Asunción 34098

Peru

Museo Andrés Avelino Cáceres, Lima 34223
Museo de los Combatientes del Morro de Arica, Lima 34245
Museo de Sitio de la Pampa Galeras, Quinua 34301
Museo de Sitio del Alto de la Alianza, Tacna 34312
Museo del Ejército, Callao 34133
Museo Memoria Coronel Leoncio Prado, Callao 34134

Philippines

Armed Forces of the Philippines Museum, Quezon City 34485
Miguel Malvar Museum, Santo Tomas 34503
Philippine Army Museum, Makati 34416
Philippine Navy Museum, Cavite City 34363

Poland

Muzeum 24. Pułku Ułanów, Kraśnik 34823
Muzeum 7 Pułku Ułanów Lubelskich, Mińsk Mazowiecki 34923
Muzeum Armii Krajowej im. Generala Emila Fieldorfa "Nila", Kraków 34787
Muzeum Armii Poznań, Poznań 35032
Muzeum Bitwy Legnickiej, Legnickie Pole 34849
Muzeum Broni Pancernej, Poznań 35033
Muzeum Czynu Niepodległościowego, Kraków 34788
Muzeum Czynu Powstańczego, Góra św. Anny 34678
Muzeum Czynu Zbrojnego Pracowników Huty im. Tadeusza Sendzimira, Kraków 34789
Muzeum Historyczne - Pałac w Dukli, Dukla 34626
Muzeum im. Orła Białego, Skarżysko-Kamienna 35108
Muzeum Martyrologii Alianckich Jeńców Wojennych, Żagań 35352
Muzeum Oręża Polskiego, Kołobrzeg 34751
Muzeum Wojska, Białystok 34553
Pomorskie Muzeum Wojskowe, Bydgoszcz 34581
Wielkopolskie Muzeum Wojskowe, Poznań 35049

Portugal

Centro de Interpretação da Batalha de Aljubarrota, Porto de Mós 35868
Museu da Liga dos Combatentes da Grande Guerra, Lisboa 35680
Museu de Escola Prática de Artilharia, Vendas Novas 35981
Museu do Fuzileiro, Palhais 35805
Museu do Regimento de Infantaria 14, Viseu 36024
Museu Militar, Ponta Delgada 35828
Museu Militar de Bragança, Bragança 35486
Museu Militar de Coimbra, Coimbra 35538
Museu Militar do Baixo Alentejo, Beja 35464
Museu Militar do Buçaco, Buçaco 35488
Museu Militar do Porto, Porto 35857
Núcleo Museológico do Lagar de Azeite, Mafra 35752

Puerto Rico

Fortaleza San Cristóbal, San Juan 36062

Romania

Muzeul de Etnografie şi a Regimentului de Graniţă, Caransebeş 36263
Muzeul de Istorie Militară, Constanţa 36304
Muzeul Marinei Române, Constanţa 36308
Muzeul Militar Dolj, Craiova 36327
Muzeul Militar Naţional, Bucureşti 36211

Russia

Centralnyj Muzej Federalnoj Pograničnoj Služby RF, Moskva 37147
Centralnyj Muzej Vnutrennych Vojsk MVD Rossii, Moskva 37150
Centralnyj Muzej Vooružennych Sil, Moskva 37151
Diorama Kurskaja Bitva - Belgorodskoe Napravlenie, Belgorod 36795
Dom-muzej S.M. Budënnogo, Proletarsk 37572
Fort N5 - Muzej Istorii Velikoj Otečestvennoj Vojny, Kaliningrad 36987
Gorodskoj Muzej Boevoj Slavy, Kaluga 36995
Gosudarstvennyj Borodinskij Voenno-istoričeskij Muzej-zapovednik, Borodino 36819
Gosudarstvennyj Memorialnyj Muzej im. A.V. Suvorova, Sankt-Peterburg 37643
Gosudarstvennyj Muzej-panorama Stalingradskaja Bitva, Volgograd 37955
Gosudarstvennyj Voenno-istoričeskij i Prirodnyj Muzej-zapovednik Kulikovo Pole, Tula 37866
Gosudarstvennyj Voenno-istoričeskij Muzej-zapovednik Prochorovskoe Pole, Prochorovka 37571
Istoričeskaja Usadba Botik Petra I, Veskovo 37926
Istoriko-memorialnyj Kompleks Gerojam Stalingradskoj Bitvy na Mamaevom Kurgane, Volgograd 37956
Jaroslavskij Muzej Boevoj Slavy, Jaroslavl 36969
Kamčatskij Voenno-istoričeskij Muzej, Petropavlovsk-Kamčatskij 37554
Krasnojarskij Istoričeskij Muzej Memorial Pobedy, Krasnojarsk 37071
Malojaroslaveckij Muzej Voennoj Istorii 1812 Goda, Malojaroslavec 37127
Memorialnaja Gvardejskaja Krasnoznamënnaja Podvodnaja Lodka S-56 1939 g., Vladivostok 37939
Memorialnyj Muzej-kabinet M.V. Frunze, Ivanovo 36949
Muzej Arsenal - Muzej Velikoj Otečestvennoj Vojny, Voronež 37971
Muzej Boevoj Slavy, Astrachan 36783
Muzej V.I. Čapaeva, Čeboksary 36840
Muzej-Diorama Orlovskaja Nastupatelnaja Operacija, Orël 37509
Muzej Istorii Vojsk Moskovskogo Voennogo Okruga, Moskva 37300
Muzej Kutuzovskaja Izba, Moskva 37313
Muzej-panorama Borodinskaja Bitva, Moskva 37345
Muzej Štab-kvartira V.K. Bljuchera, Tjumen 37850
Muzej-usadba A.V. Suvorova, Končanskoe-Suvorovskoe 37055
Muzej Voenno-morskogo Flota, Moskva 37384
Muzejno-vystavočnyj Kompleks Volokolamskij Kreml, Volokolamsk 37966
Muzj Istorii Rossijskich Voennych Učilišč, Sankt-Peterburg 37715
Saratovskij Gosudarstvennyj Muzej Boevoj Slavy, Saratov 37753
Storoževoj Korabl 'Krasnyj Vympel' 1911 g., Vladivostok 37950
Voenno-istoričeskij Muzej Artillerii, Inženernych Vojsk i Vojsk Svjazi, Sankt-Peterburg 37730
Voenno-istoričeskij Muzej Tichookeanskogo Flota, Vladivostok 37952
Volgogradskij Memorialno-istoričeskij Muzej, Volgograd 37957

Slovakia

Dukelské Múzeum - Vojenské Múzeum Svidník, Svidník 38366
Nový Zamok Banská Štiavnica, Banská Štiavnica 38192

Slovenia

Partizanska Bolnica Franja, Dolenji Novaki 38415
Vojaški Muzej Tabor, Lokev 38473

Solomon Islands

Vilu Open-Air Museum, Vilu 38548

South Africa

Castle Military Museum, Cape Town 38581
Magersfontein Battlefields Museum, Kimberley 38671
Military Museum Fort, Bloemfontein 38564
Oorlogsmuseum, Bloemfontein 38570
South African National Museum of Military History, Johannesburg 38664
Talana Museum and Heritage Park, Dundee 38601

Spain

Casa Museo General San Martín, Cervatos de la Cueza 39164
Museo de Unidades Acorazadas, Madrid 39553
Museo del Ejército, Madrid 39555
Museo Específico de la Legión, Ceuta 39171
Museo Histórico Militar de Valencia, Valencia 40213
Museo Militar de Menorca, Es Castell 39143
Museo Militar Regional de Canarias, Santa Cruz de Tenerife 39996
Sala de Recuerdos de la Escuela Militar de Montaña, Jaca 39419

Sweden

Armémuseum, Stockholm 40885
Beredskapsmuseet, Viken 41060
Eda Skans Museum, Åmotfors 40420
Eldnäsets Flottningsmuseum, Ånge 40425
Fästningsmuseet Karlsborg, Karlsborg 40627
Femörefortet, Oxelösund 40805
Garnisonsmuseet, Boden 40463
Garnisonsmuseet i Skövde, Skövde 40850
Hässleholms Museum, Hässleholm 40574
Marinmuseum, Karlskrona 40635
Pansarmuséet, Axvall 40450
Smålands Militärhistoriska Museum, Eksjö 40496
Stiftelsen Militärfordons Museum, Malmköping 40717
Ystads Militärmuseum, Ystad 41079

Switzerland

Armeepferdemuseum, Scherz 41833
Bloodhound-Lenkwaffenstellung, Menzingen 41627
Bourbaki-Panorama Luzern, Luzern 41589
Clin d'Ailes, Payerne 41718

M

Espace Guisan, Pully 41737
Festungsmuseum Crestawald, Filisur 41370
Festungsmuseum Heldsberg, St. Margrethen 41897
Festungsmuseum Reuenthal, Reuenthal 41756
Fort d'Artillerie A46, Champex 41290
Fort de Vallorbe Pré-Giroud, Vallorbe 41974
Fort Découvertes, Martigny 41609
Militärmuseum Luzern, Kriens 41517
Militärmuseum St. Luzisteig, Maienfeld 41603
Musée Militaire, Colombier (Neuchâtel) 41311
Musée Militaire, Sonvilier 41891
Museum im Zeughaus, Schaffhausen 41829
Schloßmuseum Thun, Thun 41938
Schweizerisches Militärmuseum Full, Full 41394

Taiwan

Armed Forces Museum, Taipei 42374
Military Dependants Village Museum, Hsinchu 42211

Thailand

Armed Forces Survey Department Museum, Bangkok 42529
Army Museum of Infantry Center, Phachuab Khiri Khan 42591
Museum of the Royal Thai Air Force, Bangkok 42550

Tunisia

Musée de Téboursouk, Téboursouk 42660
Musée Militaire Ligne Mareth, Mareth 42640
Musée Militaire National, Manouba 42639

Turkey

Askeri Müze ve Kültür Sitesi Komtuanlığı, İstanbul 42803
Çanakkale Anıtlari ve Savaş Müzesi, Çanakkale 42746
Deniz Müzesi, İstanbul 42814

Ukraine

Budinok-muzej Sevastopol's'ke Komunistične Pidpillja 1942-1944, Sevastopol 43175
Deržavnyj Istoryko-kul'turnyj Zapovidnyk Pole Poltavskoï Bytvy, Poltava 43166
Diorama Šturm Izmaïla, Izmaïl 43045
Istoryko-architekturna Pam'jatka-muzej Kyïvska Fortecja, Kyïv 43079
Izmaïl's'kyj Muzej O.V. Suvorova, Izmaïl 43048
Memorial Geroïčnoï Oboroni Odesi - 411 Beregova Linija, Odesa 43155
Muzej Pidpil'nyj ta Partizans'kyj Ruch na Mykolaïvščyni u Roki Velikoï Vitčyznjanoï Vijny 1941-1945 rr, Mykolaïv 43148
Nacional'nyj Muzej Istoriï Velykoï Vitčyznjanoï Vijny 1941-1945 Rokiv, Kyïv 43104
Nacional'nyj Muzej Zbrojnych Syl Ukraïny, Kyïv 43108
Oboronna Bašta Malachova Kurgana, Sevastopol 43179
Panorama Oborona Sevastopolja 1854-1855 rr., Sevastopol 43180
Panorama Šturm Sapun-gori 7 Travnja 1944 r., Sevastopol 43181

United Kingdom

100th Bomb Group Memorial Museum, Diss 43984
448th Bomb Group Memorial Museum, Seething 45631
A Soldier's Life, Newcastle-upon-Tyne 45223
Adjutant General's Corps Museum, Winchester 46026
Airborne Forces Museum, Aldershot 43270
Army Physical Training Corps Museum, Aldershot 43272
Aylmer Military Collection, Brading 43585
Ayrshire Yeomanry Museum, Ayr 43344
Beatty Museum, Newport Pagnell 45246
Bedfordshire and Hertfordshire Regiment Association Museum Collection, Luton 45046
Berwick Barracks, Berwick-upon-Tweed 43458
The Black Watch Castle Museum, Perth 45364
Bosworth Battlefield Heritage Centre and Country Park, Market Bosworth 45108
Britain at War Experience, London 44800
Buffs Museum, Canterbury 43741
Castle and Regimental Museum, Monmouth 45167
Castle Cornet Military and Maritime Museums, Saint Peter Port 45589
Channel Islands Military Museum, Saint Ouen 45587
D-Day Museum and Overlord Embroidery, Portsmouth 45429
Dartmouth Castle, Dartmouth 43952
Divisional Kohima Museum, York 46101
East Essex Aviation Society and Museum of the 40's, Saint Osyth 45586
Essex Regiment Museum, Chelmsford 43806
Fort Amherst Heritage Museum, Chatham 43794
Fort Grey and Shipwreck Museum, Saint Pierre du Bois 45595
Fort Widley, Portsmouth 45431
Fusiliers London Volunteers' Museum, London 44853
Fusiliers Museum of Northumberland, Alnwick 43286
Gordon Highlanders Museum, Aberdeen 43242
Grand Shaft, Dover 44015
Green Howards Regimental Museum, Richmond, North Yorkshire 45489
Gurkha Museum, Winchester 46029
Hamilton Low Parks Museum, Hamilton 44392
Herefordshire Regiment Museum, Hereford 44455
HMS Belfast, London 44875
Houghton Hall Soldier Museum, King's Lynn 44585
Imperial War Museum Duxford, Duxford 44075
Inns of Court and City Yeomanry Museum, London 44882
Jewish Military Museum, London 44888
Kent Battle of Britain Museum, Hawkinge 44428
King's Own Royal Regiment Museum, Lancaster 44632
King's Own Yorkshire Light Infantry Regimental Museum, Doncaster 43993
The King's Royal Hussars Museum, Winchester 46030
Leith Hall, Huntly 44504
Light Infantry Museum, Winchester 46031
Liverpool Scottish Regimental Museum, Liverpool 44738
London Scottish Regimental Museum, London 44907
Military Heritage Museum, Lewes 44699
Military Museum, Bodmin 43550
Military Vehicle Museum, Sunderland 45798
Muckleburgh Collection, Holt 44479
Museum Military Intelligence, Chicksands 43833
Museum of the 40's, Point Clear Bay 45395
Museum of The Duke of Wellington's Regiment, Halifax 44385
Museum of the Manchester Regiment, Ashton-under-Lyne 43332
Museum of the Northamptonshire Regiment, Northampton 45269
Museum of the Percy Tenantry Volunteers 1798-1814, Alnwick 43288
Museum of the Royal Leicestershire Regiment, Leicester 44678
Museum of the Worcestershire Regiment, Worcester 46071
National Army Museum, London 44936
National Army Museum, Camberley 43718
National Army Museum Sandhurst Departments, Sandhurst, Berkshire 45611
Newhaven Fort, Newhaven 45236
Noirmont Command Bunker, Saint Brelade 45561
North Weald Airfield Museum and Memorial, Epping 44182
Nothe Fort Museum of Coastal Defence, Weymouth 45988
Parham Airfield Museum and Museum of the Resistance Organisation Museum, Framlingham 44252
The Police Museum, Belfast 43446
Princess of Wales's Royal Regiment and Queen's Regiment Museum, Dover 44016
Quebec House, Westerham 45978
Queen's Lancashire Regiment Museum, Preston 45447
Queen's Own Royal West Kent Regimental Museum, Maidstone 45074
Queen's Royal Irish Hussars Museum, Eastbourne 44100
The Queen's Royal Lancers Regimental Museum, Grantham 44346
Queen's Royal Surrey Regiment Museum, Guildford 44374
Queen's Royal Surrey Regiment Museum, West Clandon 45968
Redoubt Fortress - Military Museum of Sussex, Eastbourne 44101
Regimental Museum of 1st The Queen's Dragoon Guards, Cardiff 43752
Regimental Museum of the 13th/18th Royal Hussars and Kent Dragoons, Cawthorne 43782
Regimental Museum of the 9th/12th Royal Lancers, Derby 43968
Regimental Museum of the Oxfordshire and Buckinghamshire Light Infantry, Oxford 45339
Regimental Museum of the Royal Dragoon Guards and the Prince of Wales's Own Regiment of Yorkshire, York 46106
Regimental Museum of the Royal Scots Dragoon Guards, Edinburgh 44146
Regimental Museum of The Royal Welsh (Brecon), Brecon 43599
REME Museum of Technology, Arborfield 43310
Royal Anglian Regiment Museum, Duxford 44076
Royal Berkshire Yeomanry Cavalry Museum, Windsor 46044
Royal Fusiliers Museum, London 44973
Royal Green Jackets Museum, Winchester 46032
Royal Hampshire Regiment Museum, Winchester 46033
Royal Highland Fusiliers Regimental Museum, Glasgow 44299
Royal Inniskilling Fusiliers Regimental Museum, Enniskillen 44180
Royal Irish Regiment Museum, Ballymena 43365
Royal Lincolnshire Regiment Museum, Lincoln 44715
Royal Logistic Corps Museum, Deepcut 43962
Royal Marines Museum, Portsmouth 45436
Royal Military Academy Sandhurst Collection, Camberley 43719
Royal Military Police Museum, Chichester 43831
Royal Navy Submarine Museum, Gosport 44336
Royal Norfolk Regimental Museum, Norwich 45284
Royal Observer Corps Museum, Eastleigh 44105
Royal Regiment of Fusiliers Museum, Warwick 45938
Royal Welch Fusiliers Regimental Museum, Caernarfon 43710
Royal Wiltshire Yeomanry Museum, Swindon 45818
Scapa Flow Museum, Hoy 44495
Shropshire Regimental Museum, Shrewsbury 45677
Soldiers of Gloucestershire Museum, Gloucester 44319
Somme Heritage Centre, Newtownards 45262
South Notts Hussars Museum, Bulwell 43670
Staffordshire Regiment Museum, Lichfield 44707
Suffolk Regiment Museum, Bury Saint Edmunds 43696
Sussex Combined Services Museum, Eastbourne 44102
Tank Museum, Bovington 43570
Tilbury Fort, Tilbury 45858
Ugbrooke House, Chudleigh 43844
War Memorial Carillon Tower and Military Museum, Loughborough 45035
The Welch Regiment Museum (41st/ 69th Foot) of the Royal Regiment of Wales, Cardiff 43755
Westminster Dragoons Collection, London 45013
The Worcestershire and Sherwood Foresters Regimental Museum, Nottingham 45302
York and Lancaster Regimental Museum, Rotherham 45519

United States of America

103rd Ohio Volunteer Infantry Civil War Museum, Sheffield Lake 53063
45th Infantry Division Museum, Oklahoma City 51513
4th Infantry Division Museum, Fort Hood 48624
82nd Airborne Division War Memorial Museum, Fort Bragg 48598
Air Force Armament Museum, Eglin Air Force Base 48304
Air Force Flight Test Center Museum, Edwards 48299
Air Power Park and Museum, Hampton 49149
Airborne and Special Operations Museum, Fayetteville 48529
Airmen Memorial Museum, Suitland 53359
Alamance Battleground, Burlington 47138
The Alamo, San Antonio 52705
Aliceville Museum, Aliceville 46219
Allied Air Force Museum, Allentown 46226
American Airpower Heritage Museum, Midland 50669
The American Historical Foundation Museum, Richmond 52321
American Military Museum, Charleston 47356
Ancient and Honorable Artillery Company Museum, Boston 46916
Antietam National Battlefield, Sharpsburg 53053
Arizona Wing Commemorative Air Force Museum, Mesa 50606
Arkansas Air Museum, Fayetteville 48527
Arkansas National Guard Museum, North Little Rock 51396
Artillery Company of Newport Military Museum, Newport 51317
Baltimore Civil War Museum, Baltimore 46549
Battleship North Carolina, Wilmington 54198
Battleship South Dakota Memorial, Sioux Falls 53117
Battleship Texas, La Porte 49854
Baxter Springs Heritage Center and Museum, Baxter Springs 46649
Beauvoir, Biloxi 46809
Benicia Historical Museum, Benicia 46734
Bentonville Battleground, Four Oaks 48726
Big Hole National Battlefield, Wisdom 54266
Brandywine Battlefield, Chadds Ford 47321
Bremerton Naval Museum, Bremerton 47004
Bushy Run Battlefield Museum, Jeannette 49604
Cairo Museum, Vicksburg 53732
Cape May County Historical Museum, Cape May Court House 47236
Casemate Museum, Fort Monroe 48658
Castle Air Museum, Atwater 46464
Chickamauga-Chattanooga National Military Park, Fort Oglethorpe 48665
Civil War Museum of Lone Jack, Jackson County, Lone Jack 50170
Columbus-Belmont Civil War Museum, Columbus 47743
Combat Air Museum, Topeka 53533
Confederate Museum, New Orleans 51077
Cowpens National Battlefield, Chesnee 47429
Crown Point State Historic Site, Crown Point 47895
Custer Battlefield Museum, Garryowen 48869
Desoto National Memorial, Bradenton 46985
Dey Mansion, Wayne 54008
Don F. Pratt Memorial Museum, Fort Campbell 48603
Downtown History Museum, Tucson 53588
Drum Barracks Civil War Museum, Wilmington 54188
Emil A. Blackmore Museum of the American Legion, Indianapolis 49491
Exchange Hotel Civil War Museum, Gordonsville 48973
Fifth Maine Regiment Center, Peaks Island 51743
First Cavalry Division Museum, Fort Hood 48625
The First Division Museum at Cantigny, Wheaton 54120
First Michigan Museum of Military History, Holland 49315
Fort Atkinson State Historical Park, Fort Calhoun 48601
Fort Belknap Museum, Newcastle 51307
Fort Bliss Museum, Fort Bliss 48594
Fort Bridger State Museum, Fort Bridger 48600
Fort Buford, Williston 54180
Fort Caspar Museum, Casper 47271
Fort Christmas, Christmas 47548
Fort Clatsop National Memorial, Astoria 46402
Fort Columbia House Museum, Chinook 47542
Fort Dalles Museum, The Dalles 47945
Fort Davidson, Pilot Knob 51918
Fort Davis Museum, Fort Davis 48611
Fort de Chartres Museum, Prairie du Rocher 52130
Fort Delaware Society Museum, Delaware City 48044
Fort Dix Museum, Fort Dix 48614
Fort Dobbs State Historic Site, Statesville 53273
Fort Dodge Historical Museum, Fort Dodge 48616
Fort Donelson National Battlefield Museum, Dover 48169
Fort Douglas Military Museum, Fort Douglas 48617
Fort Fetterman, Douglas 48158
Fort Frederick, Big Pool 46796
Fort George G. Meade Museum, Fort Meade 48653
Fort Gibson Historic Site, Fort Gibson 48622
Fort Harker Museum Complex, Kanopolis 49678
Fort Hartsuff, Burwell 47148
Fort Hays State Historic Site, Hays 49223
Fort Huachuca Museum, Fort Huachuca 48626
Fort Inglish, Bonham 46898
Fort Jackson Museum, Fort Jackson 48628
Fort Kearney Museum, Kearney 49714
Fort Kearny, Kearney 49715
Fort King George, Darien 47967
Fort Knox State Historic Site, Stockton Springs 53308
Fort Laramie, Fort Laramie 48634
Fort Larned, Larned 49954
Fort Laurens State Memorial, Bolivar 46895
Fort Lee Historic Park and Museum, Fort Lee 48644
Fort Lewis Military Museum, Fort Lewis 48649
Fort Lowell Museum, Tucson 53590
Fort McAllister, Richmond Hill 52345
Fort MacArthur Museum, San Pedro 52824
Fort McKavett State Historic Park, Fort McKavett 48650
Fort Mackinac, Mackinac Island 50351
Fort Meigs State Memorial, Perrysburg 51782
Fort Mifflin, Philadelphia 51835
Fort Morris, Midway 50676
Fort Necessity National Battlefield, Farmington 48518
Fort Ontario, Oswego 51624
Fort Phil Kearny, Story 53321
Fort Pike, New Orleans 51080
Fort Pitt Museum, Pittsburgh 51947
Fort Polk Military Museum, Fort Polk 48671
Fort Pulaski, Savannah 52940
Fort Recovery Museum, Fort Recovery 48674
Fort Robinson Museum, Crawford 47868
Fort Sam Houston Museum, Fort Sam Houston 48678
Fort Savannah Museum, Lewisburg 50033
Fort Selden State Monument, Radium Springs 52226
Fort Sidney Museum and Post Commander's Home, Sidney 53097
Fort Sill Museum, Fort Sill 48682
Fort Snelling, Saint Paul 52615
Fort Spokane Visitor Center and Museum, Davenport 47971
Fort Stewart Museum, Fort Stewart 48687
Fort Sumter National Monument, Sullivan's Island 53360
Fort Tejon, Lebec 50014
Fort Ticonderoga, Ticonderoga 53506
Fort Towson Military Park, Fort Towson 48693
Fort Union, Watrous 53984

Fort Verde, Camp Verde ... 47206
Fort Ward Museum and Historic Site, Alexandria ... 46203
Fort Washington, Fort Washington ... 48698
Fort Washita, Durant ... 48214
Fort William Henry Museum, Lake George ... 49883
Fort Zachary Taylor, Key West ... 49760
Fredericksburg and Spotsylvania National Military Park, Fredericksburg ... 48764
Frontier Army Museum, Fort Leavenworth ... 48642
General Crook House Museum, Omaha ... 51550
General Douglas L. McBride Museum, Roswell ... 52468
General Douglas MacArthur Memorial, Norfolk ... 51362
General Jacob Brown Historical Society Museum, Brownville ... 47099
General Lewis B. Hershey Museum, Angola ... 46300
General Patton Memorial Museum, Chiriaco Summit ... 47543
General Sweeney's Museum, Republic ... 52297
Georgia Veterans Memorial Museum, Cordele 47805
Gettysburg National Military Park, Gettysburg 48902
Grand Army of the Republic Civil War Museum, Philadelphia ... 51840
Grand Army of the Republic Memorial and Veteran's Military Museum, Aurora ... 46492
Grand Gulf Military State Park Museum, Port Gibson ... 52033
Great Lakes Naval Memorial and Museum, Muskegon ... 50923
Grissom Air Museum, Peru ... 51788
Guilford Courthouse National Military Park, Greensboro ... 49064
Halifax Historical Museum, Daytona Beach ... 47993
Hampton Roads Naval Museum, Norfolk ... 51363
Harbor Defense Museum of New York City, Brooklyn ... 47075
Hill Aerospace Museum, Hill Air Force Base ... 49282
Historic Fort Steilacoom, Tacoma ... 53408
Historic Fort Stockton, Fort Stockton ... 48689
History and Traditions Museum, Lackland Air Force Base ... 49859
Homeville Museum, Homer ... 49332
Horseshoe Bend National Military Park, Daviston ... 47979
Hubbardton Battlefield Museum, Hubbardton ... 49409
Illinois Citizen Soldier Museum, Galesburg ... 48834
Indiana War Memorials Museum, Indianapolis 49498
Jean Lafitte National Historical Park and Preserve, New Orleans ... 51084
Jefferson Barracks, Saint Louis ... 52582
JFK Special Warfare Museum, Fort Bragg ... 48599
John Brown House, Osawatomie ... 51609
Kalamazoo Aviation History Museum, Kalamazoo ... 49665
Kennesaw Mountain National Battlefield Park, Kennesaw ... 49731
Kentucky Military History Museum, Frankfort 48738
Kings Mountain National Military Park, Blacksburg ... 46842
Knox's Headquarters State Historic Site, Vails Gate ... 53682
Korean War Veterans National Museum, Rantoul ... 52248
Korean War Veterans National Museum, Tuscola ... 53622
Lansing Veterans Memorial Museum, Lansing 49940
Little Bighorn Battlefield Museum, Crow Agency ... 47892
Lone Star Flight Museum/Texas Aviation Hall of Fame, Galveston ... 48848
Louisiana Naval War Memorial U.S.S. Kidd, Baton Rouge ... 46634
Manassas National Battlefield Park, Manassas ... 50415
March Field Air Museum, Riverside ... 52371
Meeker County Museum and G.A.R. Hall, Litchfield ... 50130
Memorial Day Museum, Waterloo ... 53966
Military Historians Museum, Westbrook ... 54085
Military Museum of Southern New England, Danbury ... 47953
Military Veterans Museum, Oshkosh ... 51614
Minter Field Air Museum, Shafter ... 53047
Minute Man, Concord ... 47768
Moores Creek National Battlefield, Currie ... 47913
Morristown National Historical Park, Morristown ... 50855
Motts Military Museum, Groveport ... 49099
Museum at Mountain Home, Johnson City ... 49634
Museum of Flying, Santa Monica ... 52899
The Museum of the Confederacy, Richmond ... 52335
Muskogee War Memorial Park and Military Museum, Muskogee ... 50929
National Cryptologic Museum, Fort Meade ... 48654
National D-Day Museum, New Orleans ... 51088
National Infantry Museum, Fort Benning ... 48591
National Medal of Honor Museum of Military History, Chattanooga ... 47412
National Museum of American Jewish Military History, Washington ... 53902
National Museum of Naval Aviation, Pensacola ... 51767
Naval War College Museum, Newport ... 51325
The Navy Museum, Washington ... 53911
New Jersey Naval Museum, Hackensack ... 49111
New Market Battlefield State Historical Park, New Market ... 51072
New Mexico Wing-Commemorative Air Force, Hobbs ... 49305
New York State Military Museum, Saratoga Springs ... 52923
Old Barracks Museum, Trenton ... 53562
Old Fort Garland, Fort Garland ... 48621
Old Fort Jackson, Savannah ... 52943
Old Fort Meade Museum, Fort Meade ... 48655
Old Fort Niagara, Youngstown ... 54351
The Old Guard Museum, Fort Myer ... 48660
The Old Manse, Concord ... 47769
Oldest House Museum Complex, Saint Augustine ... 52523
Olustee Battlefield, Olustee ... 51541
Oregon Military Museum, Clackamas ... 47570
Oriskany Battlefield, Oriskany ... 51588
Pamplin Historical Park and the National Museum of the Civil War Soldier, Petersburg ... 51799
Patriots Point Naval and Maritime Museum, Mount Pleasant ... 50888
Patton Museum of Cavalry and Armor, Fort Knox ... 48633
Pennsylvania Military Museum and 28th Division Shrine, Boalsburg ... 46881
Pennsylvania Military Museum and 28th Division Shrine, Boalsburg ... 46881
Perryville Battlefield Museum, Perryville ... 51783
Petersburg National Battlefield, Petersburg ... 51802
Phillips County Museum, Helena ... 49238
Port Columbus National Civil War Naval Center, Columbus ... 47740
Port Townsend Marine Science Center, Port Townsend ... 52045
Portsmouth Museum of Military History, Portsmouth ... 52111
Pownalborough Court House, Dresden ... 48182
Presidio La Bahia, Goliad ... 48968
Puget Sound Coast Artillery Museum at Fort Worden, Port Townsend ... 52046
Rankin Museum, Ellerbe ... 48363
Richmond National Battlefield Park, Richmond ... 52337
Rock Island Arsenal Museum, Rock Island ... 52409
Sackets Harbor Battlefield, Sackets Harbor ... 52497
Saint John's Northwestern Military Academy Museum, Delafield ... 48042
San Diego Command Museum, San Diego ... 52747
San Juan Island National Historical Park, Friday Harbor ... 48795
Saratoga National Historical Park, Stillwater ... 53295
Seabee Museum, Port Hueneme ... 52034
Shiloh National Military Park and Cemetery, Shiloh ... 53078
Siege Museum, Petersburg ... 51803
Smallwood's Retreat, Marbury ... 50464
Soldiers' Memorial Military Museum, Saint Louis ... 52601
Soldiers National Museum, Gettysburg ... 48904
Sons of the American Revolution, Roselle ... 52460
South Carolina Military Museum, Columbia ... 47733
South Carolina National Guard Museum, Sumter ... 53365
Springfield Armory Museum, Springfield ... 53228
Steuben Memorial State Historic Site, Remsen ... 52286
Stones River National Battlefield, Murfreesboro ... 50915
Stonewall Jackson Museum at Hupp's Hill, Strasburg ... 53333
Stony Point Battlefield State Historic Site, Stony Point ... 53316
Strategic Air Space Museum, Ashland ... 46383
Submarine Force Museum and Historic Ship Nautilus, Groton ... 49095
Texas Heritage Museum, Hillsboro ... 49288
Texas Military Forces Museum, Austin ... 46520
Third Cavalry Museum, Fort Carson ... 48604
Thomas Clarke House-Princeton Battlefield, Princeton ... 52154
UDT-SEAL Museum, Fort Pierce ... 48669
Ukrainian-American Museum, Warren ... 53821
United States Air Force Academy Museum, El Paso ... 48314
United States Air Force Museum, Wright-Patterson Air Force Base ... 54311
United States Army Air Defense Artillery Museum, Fort Bliss ... 48595
United States Army Aviation Museum, Fort Rucker ... 48677
United States Army Center of Military History, Museum Division, Fort McNair ... 48651
United States Army Chaplain Museum, Fort Jackson ... 48629
United States Army Chaplain Museum, Fort Jackson ... 48629
United States Army Communications-Electronics Museum, Fort Monmouth ... 48657
United States Army Engineer Museum, Fort Leonard Wood ... 48647
United States Army Finance Corps Museum, Fort Jackson ... 48630
United States Army Heritage and Education Center, Carlisle ... 47245
United States Army Heritage and Education Center, Carlisle ... 47245
United States Army Military Police Corps Museum, Fort Leonard Wood ... 48648
United States Army Museum of Hawaii, Fort DeRussy ... 48613
United States Army Ordnance Museum, Aberdeen Proving Ground ... 46118
The United States Army Quartermaster Museum, Fort Lee ... 48645
United States Army Signal Corps and Fort Gordon Museum, Fort Gordon ... 48623
United States Army Transportation Museum, Fort Eustis ... 48620
United States Army Women's Museum, Fort Lee ... 48646
United States Brig Niagara, Homeport Erie Maritime Museum, Erie ... 48412
United States Cavalry Museum, Fort Riley ... 48676
United States Marine Corps Air-Ground Museum, Quantico ... 52208
United States Marine Corps Museum, Washington ... 53926
United States Navy Memorial Foundation and Naval Heritage, Washington ... 53928
United States Navy Supply Corps Museum, Athens ... 46416
Unites States Air Force Academy Visitor Center, USAF Academy ... 53672
USS Alabama Battleship Memorial Park, Mobile ... 50772
USS Arizona Memorial, Honolulu ... 49356
USS Bowfin Submarine Museum, Honolulu ... 49357
USS Constellation Museum, Baltimore ... 46590
USS Constitution Museum, Boston ... 46954
USS Lexington Museum on the Bay, Corpus Christi ... 47822
Varnum Memorial Armory and Military Museum, East Greenwich ... 48242
Vicksburg National Military Park, Vicksburg ... 53735
Vietnam Era Educational Center, Holmdel ... 49325
Virginia Military Institute Museum, Lexington 50075
The Virginia War Museum, Newport News ... 51338
Warren ICBM and Heritage Museum, Cheyenne ... 47446
Warren Rifles Confederate Museum, Front Royal ... 48801
Washington's Headquarters, Newburgh ... 51301
Watervliet Arsenal's Museum of the Big Guns, Watervliet ... 53979
West Point Museum, West Point ... 54078
Wilson's Creek National Battlefield, Republic 52298
Wings of Freedom Museum, Huntington ... 49427
Wisconsin National Guard Memorial Library and Museum, Camp Douglas ... 47205
Wisconsin Veterans Museum King, King ... 49779
Wisconsin Veterans Museum Madison, Madison ... 50398
Wright Museum, Wolfeboro ... 54270
Yankee Air Museum, Belleville ... 46702

Uzbekistan

Centralnyj Muzej Vooružennych Sil Uzbekistana, Toshkent ... 54469

Venezuela

Museo Urdaneta Histórico Militar, Maracaibo . 54582

Vietnam

Anti-Aircraft Museum, Ha Noi ... 54607
Bao Tàng Quan Doi, Ha Noi ... 54609
South-East Armed Force Museum, Ho Chi Minh City ... 54627

Yemen

Military Museum, Sanaa ... 54649

Mills

Andorra

Mola e Serradora de Cal Pal, Ordino ... 00099

Argentina

Museo Molino Nant Fach, Trevelin ... 00655

Australia

Berkshire Mill Museum, Moora ... 01316
Chinchilla and District Historical Society Museum, Chinchilla ... 00973

Austria

Alte Mühle Schrattenberg, Schrattenberg ... 02713
Freilichtmuseum Apriacher Stockmühlen, Heiligenblut ... 02065
Freilichtmuseum Katzensteiner Mühle, Weyer 02927
Freilichtmuseum Ledermühle, Sankt Oswald bei Freistadt ... 02672
Freilichtmuseum Oberlienz, Lienz ... 02300
Heimathaus Wenigzell, Wenigzell ... 02921
Heimatmuseum Obermühle, Ach ... 01691
Hundsmarktmühle, Thalgau ... 02808
Kürbismühlenmuseum, Preding ... 02529
Leithenmühle, Windhaag bei Freistadt ... 03111
Museum Furthmühle, Pram ... 02523
Museum Stoffels Säge-Mühle, Hohenems ... 02093
Schattleitenmühle, Garsten ... 01925
Windmühle, Retz ... 02581

Belgium

Mercators Schrijn Graventoren, Rupelmonde . 03991
Mola- het Provinciaal Molenmuseum, Wachtebeke ... 04125
Molenmuseum, Sint-Amands ... 04013
Musée de la Meunerie et de la Boulangerie, Harzé ... 03681
Musée de la Meunerie-Moulin de la Petite Strument, La Roche-en-Ardenne ... 03770
Musée Moulin Faber, Hotton ... 03715
Sint-Janshuismolen Koeleweimolen, Brugge 03428

Canada

Balmoral Grist Mill Museum, Balmoral Mills ... 05376
Bell Rock Mill Museum, Bethany ... 05423
Etzikom Museum, Etzikom ... 05788
Flour Mill Museum, Sudbury ... 07028
Grants Old Mill, Winnipeg ... 07362
Grist Mill, Keremeos ... 06063
Historic Babcock Mill, Odessa ... 06500
Kingston Mills Blockhouse, Elgin ... 05770
Lang Water Powered Grist Mill, Peterborough. 06593
McDougall Mill Museum, Renfrew ... 06733
McLean Mill, Port Alberni ... 06626
MacPherson's Mill and Farm Homestead, New Glasgow ... 06430
Moulin de Beaumont, Beaumont ... 05405
Moulin de la Chevrotière, Deschambault ... 05698
Moulin des Jésuites, Charlesbourg ... 05588
Moulin Fleming, LaSalle ... 06141
Moulin Légaré, Saint-Eustache ... 06815
Moulin Michel de Gentilly, Bécancour ... 05410
Les Moulins de l'Isle-aux-Coudres, Saint-Louis-de-Île-aux-Coudres ... 06864
Mountain Mills Museum, Saint Catharines ... 06806
Old Stone Mill Museum, Delta ... 05693
La Pulperie de Chicoutimi/Musée du Saguenay-Lac-Saint-Jean, Chicoutimi ... 05612
Site du Moulin des Pionniers de la Doré, La Doré ... 06110
Site Historique de l'Île-des-Moulins, Terrebonne ... 07068
Sutherland Steam Mill Museum, Denmark ... 05696
Todmorden Mills Heritage Museum and Arts Centre, Toronto ... 07145
Ulverton Wool Mill, Ulverton ... 07187
Le Vieux Moulin, Sainte-Jeanne-d'Arc ... 06895
Watson's Mill, Manotick ... 06230
Wile Carding Mill Museum, Bridgewater ... 05475

Czech Republic

Větrný Mlýn, Rudice ... 09830

Denmark

Højer Mølle- og Marskmuseum, Højer ... 10086
Kaleko Mølle, Faaborg ... 10012
Pibe Mølle, Helsinge ... 10062
Skjern Vindmølle, Skjern ... 10271
Tadre Mølle, Hvalsø ... 10106

Estonia

Eemu Tulik, Linnuse ... 10593

Finland

Harjukosken Vesimylly, Mikkeli ... 11112
Harrströmin Myllymäkimuseo, Harrström ... 10759
Sotavallan Tuulimylly, Lempäälä ... 11060
Värtsilän Myllymuseo, Värtsilä ... 11433

France

Maison de la Meunerie, Nieul-sur-l'Autise ... 14444
Moulin à Eau Maître Marcel, Sainte-Agathe-en-Donzy ... 15545

Moulin à Papier Vallis-Clausa, Fontaine-de-Vaucluse 12956
Moulin à Vent Gaillardin, Chapelon 12403
Moulin de la Herpinière, Turquant 16012
Moulin de Malet, Clermont-Soubiran 12555
Moulin de Pierre, Hauville 13196
Moulin des Grands Vignes, Périssac 14781
Moulin d'Eschviller, Eschviller 12851
Moulin du Bois-Landon, Beaumont-sur-Sarthe 11901
Les Moulins, Sagy 15148
Musée de la Meunerie, Bannegon 11840
Musée du Moulin, Saint-Jean-Saint-Nicolas 15340
Musée du Moulin à Huile, Rustrel 15141
Musée du Moulins des Bouillons, Gordes 13083
Musée-Moulin de l'Espine, Fougax-et-Barrineuf 12984
Musée Moulin de Maupertuis, Donzy 12771

Germany

Altmühltaler Mühlenmuseum, Dietfurt an der Altmühl 17854
Bockwindmühlenmuseum, Doberlug-Kirchhain 17879
Buchholzer Heimatmuseum, Buchholz in der Nordheide 17618
Freilichtmuseum Scherzenmühle Weidenberg, Weidenberg 22368
Furthmühle, Egenhofen 18060
Galtze Mühle, Seelbach 21718
Haarener Mühle, Waldfeucht 22292
Heimatmuseum, Elze 18134
Heimatmuseum Bockwindmühle, Lebusa 19866
Heimatmuseum Sprucker Mühle, Guben 18820
Heimatmuseum Vollmers Mühle, Seebach 21712
Historische Mühle, Potsdam 21090
Historische Ölmühle, Simonswald 21773
Historische Senfmühle, Monschau 20375
Historische Wassermühle Dickendorfer Mühle, Dickendorf 17845
Internationales Wind- und Wassermühlenmuseum, Gifhorn 18643
Klostermühle, Boitzenburger Land 17437
Klostermühlenmuseum, Thierhaupten 22052
Lassaner Mühle, Lassan 19841
Lindigtmühle am Lindenvorwerk, Kohren-Sahlis 19671
Linnenschmidt'sche Vormals Landesherrliche Mühle zu Venne, Ostercappeln 20933
Moisburger Mühlenmuseum, Moisburg 20367
Mühlen- und Landwirtschaftsmuseum, Fehmarn 18298
Mühlen- und Landwirtschaftsmuseum, Jever 19396
Mühlen- und Mennonitenmuseum Bettinger Mühle, Schmelz 21586
Das Mühlendorf, Ottenhöfen im Schwarzwald 20946
Mühlenfachmuseum, Aurich 16648
Mühlenhof Altkalen, Altkalen 16510
Mühlenmuseum, Haren (Ems) 19012
Mühlenmuseum, Oberkrämer-Vehlefanz 20815
Mühlenmuseum Katzbrui, Apfeltrach 16564
Mühlenmuseum Mitling-Mark mit Sammlung Omas Küche, Westoverledingen 22517
Mühlenmuseum Moisburg, Moisburg 20368
Mühlenmuseum Teichmühle, Steinwiesen 21895
Museum Bockwindmühle, Falkenhain 18290
Museum für Mühlenbautechnik, Syrau 21997
Museum Moorseer Mühle, Nordenham 20731
Museum Neue Mühle, Erfurt 18208
Museumsmühle, Woldegk 22644
Museumsmühle Hasbergen, Delmenhorst 17815
Museumsmühle mit heimatkundlicher Sammlung, Varel 22205
Ölmühle Dörzbach, Dörzbach 17885
Ölmühle Michelau, Rudersberg 21430
Ölmühlenmuseum, Ottweiler 20960
Paltrockwindmühle Brehna, Brehna 17534
Schiffmühle Höfgen, Grimma 18762
Schleifmühle Schwerin, Schwerin 21697
Schwarzwälder Mühlenmuseum, Grafenhausen 18731
Strötzbacher Mühle, Mömbris 20353
Technisches Kulturdenkmal Ölmühle, Marbach am Neckar 20183
Tessiner Mühlenmuseum, Tessin 22033
Trachten- und Mühlenmuseum, Pfullingen 21035
Volkskunde- und Mühlenmuseum, Großschönau 18807
Wassermühle Höfgen, Grimma 18765
Wassermühle Kuchelmiß, Kuchelmiß 19733
Weicheltmühle, Hartmannsdorf-Reichenau 19021
Wilhelm-Busch-Mühle, Ebergötzen 18038

Greece

Mykonos Folklore Museum, Mykonos 23104

Hungary

Csabai Tanya- és Gabonatörténeti Kiállítóhely, Békéscsaba 23368
Malomipari Múzeum, Budapest 23448

Ireland

Tuam Mill Museum, Tuam 24905

Lithuania

Ginučių Vandens Malūno Ekspozicija, Ginučiai 30228
Žaliūkių Vėjo Malūnas, Šiauliai 30366

Netherlands

Gitstappermolen, Vlodrop 32901
De Grafelijke Torenmolen, Zeddam 32990
Koren en Pelmolen De Noordstar, Noordbroek 32600
Korenmolen de Hoop, Hellendoorn 32355
Korenmolen de Hoop, Klarenbeek 32449
Korenmolen de Phenix, Nes 32568
Korenmolen de Zandhaas, Santpoort 32742
Molen van Sloten, Amsterdam 31859
Molenmuseum, Koog aan de Zaan 32455
Molenmuseum de Assumburg, Nieuw Vossemeer 32574
Molenmuseum De Valk, Leiden 32483
Molenmuseum De Wachter, Zuidlaren 33013
Museum De Doornboom, Hilvarenbeek 32385
Museum De Eendracht, Anjum 31908
Museum-Gemaal de Hoogte, Nieuwolda 32584
Museum van Zwerfstenen in de Rosmolen, Zeddam 32991
Museumgemaal Caners, 's-Hertogenbosch 32374
Museummolen, Schermerhorn 32751
Museummolen de Nieuwe Palmboom, Schiedam 32754
Museummolen Jan Pol, Dalen 32045
Nederlands Graanmuseum/Olie- en Korenmolen Woldzigt, Roderwolde 32697
Oliemolen de Zoeker, Zaandam 32982
Oliemolen Het Pink, Koog aan de Zaan 32456
Oudheidkamer Weststellingwerf, Wolvega 32962
Poldermuseum Het Grootslag - Nationaal Saet en Cruytmuseum, Andijk 31907
Stoomgemaal Halfweg, Halfweg 32308
Stoomgemaal Hertog Reijnout en Bezoekerscentrum Arkemheen, Nijkerk 32586
Stoomgemaal Mastenbroek, Genemuiden 32244

Norway

Aremark Historielag, Ostfold 33760
Fotland Bygdemølle, Bryne 33456

Poland

Muzeum Historii Młynarstwa i Wodnych Urządzeń Przemysłu Wiejskiego w Jaraczu, Jaracz 34694
Muzeum Młynarstwa Powietrznego, Bęsia 34540

Portugal

Museu Municipal de Montijo, Montijo 35778
Núcleo do Moinho de Maré de Corroios, Corroios 35551
Núcleo Museológico dos Moinhos de Carreço, Carreço 35508

Slovakia

Múzeum Oravskej Dediny, Zuberec 38393

Spain

Conjunto de la Ferrería y Molinos de Agorregi, Aia 38797
Ecomuseo - Molino de Zubieta, Zubieta 40359
Museo de los Molinos de Gregorio Prieto, Valdepeñas 40189
Museo de los Molinos de Viento, Campo de Criptana 39112

Sweden

Frövifors Pappersbruksmuseum, Frövi 40529
Kulla Gunnarstorps Mölla, Helsingborg 40590
Oderljunga Mölla, Perstorp 40807
Pålsjö Kvarn, Helsingborg 40591

Switzerland

Galerie Rainmühle, Emmenbrücke 41349
Moulins Souterrains, Le Locle 41571
Mühlenmuseum Brüglingen, Münchenstein 41658
Mühlerama, Zürich 42090

United Kingdom

Arlington Mill Museum, Bibury 43474
Bancroft Mill Engine Trust, Barnoldswick 43383
Bembridge Windmill Museum, Mottistone 45180
Blackgang Sawmill and Saint Catherine's Quay, Blackgang 43528
Blackpool Mill, Narberth 45192
Bradford Industrial Museum and Horses at work, Bradford 43576
Bromham Mill and Art Gallery, Bromham 43653
Bunbury Watermill, Bunbury 43671
Bursledon Windmill, Bursledon 43683
Calbourne Water Mill and Rural Museum, Newport, Isle of Wight 45242
Caudwell's Mill Craft Centre, Matlock 45114
Coldharbour Mill, Uffculme 45894
Crabble Corn Mill, Dover 44011
Cromford Mill, Cromford 43932
Dalgarven Mill, Kilwinning 44583
Dell Mill, Isle-of-Lewis 44540
Ford End Watermill, Ivinghoe 44543
Green's Windmill and Science Centre, Nottingham 45292
Heatherslaw Mill, Cornhill-on-Tweed 43893
Heckington Windmill, Heckington 44435
Heron Corn Mill and Museum of Papermaking, Beetham 43438
Hornsbury Mill, Chard 43792
Ifield Watermill, Crawley 43917
King's Mill Visitor Centre, Wrexham 46083
Lytham Windmill Museum, Lytham Saint Anne's 45062
Meopham Windmill, Meopham 45127
Mill Green Museum and Mill, Hatfield 44419
Muncaster Watermill, Ravenglass 45461
North Leverton Windmill, North Leverton 45266
Otterton Mill Centre and Working Museum, Budleigh Salterton 43669
Patterson's Spade Mill, Templepatrick 45834
Quarry Bank Mill Styal Estate, Styal 45795
Queen Street Mill Textile Museum, Burnley 43678
Sarehole Mill, Birmingham 43510
Saxtead Green Post Mill, Saxtead 45620
Skidby Windmill and Museum of East Riding Rural Life, Skidby 45686
Stansted Mountfitchet Windmill, Stansted Mountfitchet 45739
Stevington Windmill, Stevington 45743
Stott Park Bobbin Mill, Newby Bridge 45222
Stretton Water Mill, Farndon 44215
Swanton Mill, Ashford 43328
Thwaite Mills Watermill, Leeds 44669
Trader Windmill, Sibsey 45678
Upminster Mill, Upminster 45898
Wellbrook Beetling Mill, Cookstown 43890
West Blatchington Windmill, Hove 44494
White Mill Rural Heritage Centre, Sandwich 45617
Wimbledon Windmill Museum, London 45019
Windmill and Milling Museum, Polegate 45396
Worsbrough Mill Museum, Barnsley, South Yorkshire 43386

United States of America

Adams Old Stone Grist Mill, Bellows Falls 46715
Bollinger Mill, Burfordville 47129
Burwell-Morgan Mill, Millwood 50697
Champion Mill, Champion 47337
Crane Museum of Papermaking, Dalton 47949
Dillard Mill State Historic Site, Davisville 47980
Fabyan Villa Museum and Dutch Windmill, Geneva 48878
Frye's Measure Mill, Wilton 54215
Graue Mill and Museum, Oak Brook 51451
Guest House Museum, Fort Bragg 48596
Hamburg State Park Museum, Mitchell 50759
Hanford Mills Museum, East Meredith 48257
Historic Bowens Mills and Pioneer Park, Middleville 50664
Historic Mill Creek, Mackinaw City 50354
History Museum, Los Gatos 50273
McPherson County Old Mill Museum, Lindsborg 50121
Mayflower Gold Mill, Silverton 53108
Neligh Mills, Neligh 51001
Old Schwamb Mill, Arlington 46349
Pennypacker Mills, Schwenksville 52965
Praters Mill, Varnell 53710
Saugus Iron Works, Saugus 52928
Union Mills Homestead and Grist Mill, Westminster 54099
Youth Museum of Southern West Virginia Beckley Exhibition Coal Mine, Beckley 46685

Mineralogy

Argentina

Centro de Exposiciónes de Geologia y Mineria, Buenos Aires 00152
Museo de Ciencias Naturales Luma, Tanti 00643
Museo de Mineralogía y Geología Dr. Alfredo Stelzner, Córdoba 00326
Museo Municipal de Historia Natural, General Alvear 00376
Museo Provincial de Historia Natural de La Pampa, Santa Rosa 00625

Australia

Ashworths Treasures of the Earth, Home Hill 01161
Banking and Currency Museum, Kadina 01181
Crystal Kingdom, Coonabarabran 00999
Echuca Gem Club Collection, Echuca 01056
House of Bottles, Kinglake 01201
School of Earth Sciences Museum, Townsville 01600
West Coast Pioneers Museum, Zeehan 01687

Austria

alpinartgallery - Schatzhaus der Natur, Ebene Reichenau 01834
Bergbaumuseum, Hall in Tirol 02042
Bergbaumuseum mit Mineralienschau und Schaubergwerk, Hüttenberg 02102
Felsenmuseum, Bernstein 01782
Heimathaus Richard Eichinger und Steinmuseum, Enzenkirchen 01872
Hochschwabmuseum, Sankt Ilgen 02651
Kalkofenmuseum, Steinbach am Ziehberg 02766
Magnesitmuseum, Radenthein 02550
Marmormuseum, Adnet 01695
Mineraliensammlung Steger, Wiener Neustadt 03097
Mineralienzentrum Steinstadel, Aggsbach Dorf 01698
Museum Bramberg "Wilhelmgut", Bramberg 01793
Naturhistorisches Museum, Admont 01694
Naturwissenschaftliche Sammlungen am Landesmuseum Joanneum, Graz 01981
Paläontologisches Museum Dr. Herbert Schaffer, Offenhausen 02470
Privatsammlung Bergkristallmuseum, Saalbach 02606
Schaubergwerk Arzberg, Arzberg 01730
Stadtmuseum Schladming, Schladming 02706

Belgium

Mineralogisch Museum, Merksem 03869
Musée Communal de la Pierre de Sprimont, Sprimont 04045
Musée de Minéralogie et de Géologie, Bruxelles 03474
Museum Ons Mijnverleden, Houthalen 03719
Schoolmuseum Michel Thiery, Gent 03640

Bolivia

Museo Mineralogico y Geológico, Oruro 04233

Bosnia and Herzegovina

Zavičajni Muzej, Travnik 04286

Brazil

Museu Amsterdam Sauer de Pedras Preciosas, Rio de Janeiro 04698
Museu de Ciências da Terra - DNPM, Rio de Janeiro 04725
Museu de Geologia, Porto Alegre 04647
Museu de Minerais e Rochas, Uberlândia 04943
Museu de Mineralogia, Congonhas 04413
Museu de Mineralogia Luiz Englert, Porto Alegre 04648
Museu H. Stern, Rio de Janeiro 04747

Bulgaria

Prirodonaučen Muzej, Burgas 05000

Canada

Centre d'Interprétation des Mines Minerais, Petit-Rocher 06597
Fundy Geological Museum, Parrsboro 06569
Logan Hall, Ottawa 06550
Mineral and Gem Geological Museum, Parrsboro 06570
Musée Minéralogique d'Asbestos, Asbestos 05356
Musée Minéralogique et Minier de Thetford Mines, Thetford Mines 07073
Princeton and District Museum, Princeton 06674
Robert B. Ferguson Museum of Mineralogy, Winnipeg 07375

Chile

Museo de Historia Natural de Valparaíso, Valparaíso 07471

China

Xiu Rock Museum, Xiuyan 08411

Colombia

Colección de Minerales y Arqueología, Villanueva 08787
Museo de Historia Natural, Alta Suiza 08527
Museo de Minerales del Instituto Geofísico, Bogotá 08571

Museo Mineralógico Salón Tulio Ospina, Medellín 08684
Museo Mineralógico y Antropológico, Ibagué 08660
Museo Vitaliano Zuccardi, Sincelejo 08756

Croatia

Hrvatski Prirodoslovni Muzej, Zagreb 08998
Muzej Krapinskog Pračovjeka, Krapina 08900

Czech Republic

Blatské muzeum - Rožmberský Dům, Soběslav 09842
Dietrichsteinský Palác, Brno 09423
Krajské Muzeum Sokolov, Sokolov 09844
Městské Muzeum, Radnice u Rokycan 09813
Městské Muzeum, Stříbro 09853
Městské Muzeum, Volyně 09913
Městské Muzeum v Mariánských Lázních, Mariánské Lázně 09649
Muzeum Českého Ráje, Turnov 09881
Muzeum Novojičínska, Nový Jičín 09685
Oblastní Muzeum, Most 09667
Ostravské Muzeum, Ostrava 09699
Sládečkovo Vlastivědné Muzeum v Kladně, Kladno 09583

Denmark

Statens Naturhistoriske Museum (SNM), København 10153

Dominican Republic

Museo del Larimar, Santo Domingo 10353

Ecuador

Museo del Colegio Nacional Pedro Vicente Maldonado, Riobamba 10456
Museo Petrográfico del Servicio Nacional de Geología y Minería, Quito 10450
Museo Vásquez Fuller, Otavalo 10422

Egypt

The Egyptian Geological Museum, Cairo 10491

Finland

Geologian museo, Helsinki 10775
Kieppi-Luontomuseo ja Mineraalikokoelma, Kokkola 10974
Kivimuseo, Tampere 11351
Kultamuseo Tankavaara, Tankavaara 11370

France

Anicennes Mines de Grandfontaine, Grandfontaine 13102
Espace Minéralogique Minéraux Couleur Nature, Eymoutiers 12894
Expositions Pipe et Diamant, Saint-Claude 15215
Galerie de Minéralogie, Paris 14603
Maison de la Pierre de Volvic, Volvic 16269
Maison de la Vie Rurale, Massiac 14027
Maison des Minéraux, Crozon 12680
Musée de la Mine, Grande-Synthe 13101
Musée de la Mine de Cap Garonne, Le Pradet 13642
Musée de la Pierre, Trambly 15977
Musée de Minéralogie de l'Ecole des Mines de Paris, Paris 14677
Musée de Minéralogie et du Pétrographie, Ambazac 11565
Musée des Minéralogie, Strasbourg 15830
Musée des Minéraux et Fossiles des Ardennes, Bogny-sur-Meuse 12044
Musée d'Histoire Naturelle, Nantes 14366
Musée du Dolon, Chanas 12396
Musée Minéralogique, Nîmes 14454
Musée Minéralogique de la Société Industrielle, Mulhouse 14329
Musée Minéralogique de l'École des Mines d'Alès, Alès 11551
Musée Municipal, Saint-Dizier 15244
Musée Pierre Borrione, Aime-en-Tarentaise 11509
Muséum d'Histoire Naturelle, Le Havre 13599
Petit Musée Minéraux et Faune de l'Alpe, Bourg-Saint-Maurice 12111
Le Roc des Harmonies: Palais de Mineralogique, Granville 13107

Germany

Anton-Buchner-Heimatmuseum, Betzenstein 17311
Bartholomäus-Schmucker-Heimatmuseum, Ruhpolding 21445
Basaltpark Bad Marienberg, Bad Marienberg 16782
Bergbau- und Mineralienmuseum, Oberwolfach 20851
Bergbaumuseum Hückelhoven, Hückelhoven 19286
Besucherbergwerk Finstergrund, Wieden 22538
Besucherbergwerk "Mühlenberger Stollen", Bleialf 17386
Deutsches Naturstein-Archiv (DNSA), Wunsiedel 22730
Emser Bergbaumuseum, Bad Ems 16723
Geologisches Freimuseum, Bettendorf 17310
Geologisches und Mineralogisches Museum der Christian-Albrechts-Universität, Kiel 19528
GeoMuseum der Universität zu Köln, Köln 19605
Geosammlung der Technischen Universität Clausthal, Clausthal-Zellerfeld 17708
Geowissenschaftliche Sammlungen, Freiberg, Sachsen 18421
Geowissenschaftliche Sammlungen und Geologischer Garten, Halle, Saale 18896
Geowissenschaftliches Museum am Geowissenschaftlichen Zentrum der Universität Göttingen, Göttingen 18691
Hammerschmiede mit Heimatmuseum und Bienengarten, Lahr 19772
Heimatmuseum im Alten Kloster, Erbendorf 18192
Höhlenmuseum Kubacher Kristallhöhle und Freilicht-Steinemuseum, Weilburg 22385
Informationszentrum Donnersberghaus, Dannenfels 17781
Kali-Bergbaumuseum Volpriehausen, Uslar 22192
Kreidemuseum Gummanz, Sagard 21469
Kreisbergbaumuseum Sassenroth, Herdorf 19138
Kristallmuseum, Viechtach 22225
Marienglashöhle Friedrichroda, Friedrichroda 18474
Mineralien-Kabinett des Mineralogisch-Petrografischen Museums der TU Braunschweig, Braunschweig 17526
Mineralien- und Heimatmuseum Pleystein, Pleystein 21062
Mineralienmuseum, Bad Säckingen 16816
Mineralienmuseum, Freisen 18448
Mineralienmuseum, Neubulach 20595
Mineralienmuseum Essen, Essen 18262
Mineralienmuseum Pforzheim-Dillweißenstein, Pforzheim 21028
Mineraliensammlung Lauingen, Lauingen 19853
Mineralienschau, Vohenstrauß 22254
Mineralogisch-Geologische Gesteins-Sammlung, Schwarzenfeld 21676
Mineralogisch-petrographische Sammlung der Universität Leipzig, Leipzig 19919
Mineralogische Sammlung, Oberthal 20846
Mineralogische Schau- und Lehrsammlung, Tübingen 22120
Mineralogisches Museum der Philipps-Universität Marburg, Marburg 20192
Mineralogisches Museum der Universität, Würzburg 22715
Mineralogisches Museum der Universität Bonn, Bonn 17463
Mineralogisches Museum der Universität Hamburg, Hamburg 18944
Mineralogisches Museum (Kristalle und Gesteine), Münster 20514
Mineralogisches Museum Oberkirchen, Freisen 18449
Motorrad-Träume, Zschopau 22825
Museum für Naturkunde, Berlin 17205
Museum im Hollerhaus, Dietfurt an der Altmühl 17855
Museum Reich der Kristalle, München 20468
Naturkunde-Museum, Bielefeld 17347
Naturkunde-Museum Bamberg, Bamberg 16915
Privatmuseum Hans Klein, Prichsenstadt 21125
Senckenberg Naturhistorische Sammlungen Dresden, Dresden 17969
Stadthistorisches Sole, Salz und Kali-Bergbaumuseum, Bad Salzdetfurth 16820
Stadtmuseum Nittenau, Nittenau 20723
Stadtmuseum Stadtoldendorf, Stadtoldendorf 21859
Stein- und Fossiliensammlung Albert, Sulzdorf an der Lederhecke 21989
Urwelt-Museum Oberfranken, Bayreuth 16966
Urweltmuseum Aalen, Aalen 16415
Waldmuseum, Mehlmeisel 20255
Zinngrube Ehrenfriedersdorf, Ehrenfriedersdorf 18071

Greece

Mineralogical Museum of Kamarzia, Attica 22955
Mineralogical Museum of Lavrion, Lavrion 23075
Museum of Mineralogy, Petrology and Geology, Zografou 23249

Hungary

Ásványkiállítás, Kaposvár 23630
Ásványmúzeum, Siófok 23802
Encsy György Tokaj-Hegyalja Földtani Gyűjteménye, Tállya 23910
Herman Ottó Múzeum Ásványtára, Miskolc 23705
Herman Ottó Múzeum - Kiállítóhelye, Miskolc 23707
Komlói Természettudományi Gyűjtemény, Komló 23681

India

Central College Museum, Bangalore 24072
Geography Museum, Faizabad 24172
Geological Museum, Lucknow 24294
Our India Project Museum, Narendrapur 24335

Ireland

James Mitchell Museum, Galway 24836

Italy

Frasassi-Le Grotte, Genga 26286
Mostra Permanente della Tradizione Mineraria, Tarvisio 27944
Museo A. Ricci dei Minerali Elbani, Rio nell'Elba 27413
Museo Carlo Venturini, Massa Lombarda 26623
Museo Civico, Finale Emilia 26136
Museo Civico A. Klitsche de la Grange, Allumiere 25203
Museo Civico D. dal Lago, Valdagno 28137
Museo Civico di Scienze Naturali, Voghera 28303
Museo Civico di Storia Naturale, Ferrara 26108
Museo Comunale A. Mendola, Favara 26092
Museo dei Fossili e dei Minerali, Smerillo 27859
Museo dei Fossili e di Storia Naturale, Montefalcone Appennino 26822
Museo dei Fossili e Minerali del Monte Nerone, Apecchio 25249
Museo dei Minerali e della Miniera, Oltre il Colle 26981
Museo dei Minerali Elbani, Rio Marina 27412
Museo dell'Ardesia, Cicagna 25886
Museo di Geologia e Paleontologia, Pavia 27121
Museo di Minerali e Fossili, Montefiore Conca 26826
Museo di Mineralogia, Massa Marittima 26632
Museo di Mineralogia, Palermo 27072
Museo di Mineralogia, Roma 27497
Museo di Mineralogia, di Petrologia e Geologia, Modena 26751
Museo di Mineralogia e Paleontologia, Saint Vincent 27589
Museo di Mineralogia e Petrografia, Parma 27103
Museo di Mineralogia e Petrografia, Pavia 27122
Museo di Mineralogia e Petrografia L. Bombicci, Bologna 25462
Museo di Mineralogia L. De Prunner, Cagliari 25585
Museo di Paleontologia e Mineralogia, Campomorone 25642
Museo di Storia Naturale, Cremona 26004
Museo di Storia Naturale A. Stoppani, Venegono Inferiore 28177
Museo di Storia Naturale - Sezione Mineralogia, Firenze 26185
Museo Mineralogico, Bortigiadas 25512
Museo Mineralogico, Valle Aurina 28145
Museo Mineralogico Campano, Vico Equense 28260
Museo Mineralogico e Naturalistico, Bormio 25511
Museo Mineralogico e Paleontologico delle Zolfare, Caltanissetta 25610
Museo Mineralogico Permanente, Carro 25693
Museo Mineralogico Sardo, Iglesias 26397
Museo Minerario Alpino, Cogne 25935
Museo P. Calderini, Varallo Sesia 28157
Museo Storico Minerario, Perticara 27142
Raccolte di Geologia, Mineralogia e Vulcanologia, Catania 25777
Real Museo Mineralogico, Napoli 26913

Japan

Ikuno Kobutsukan no Kogyo Hakubutsukan, Ikuno 28566
Tanakami Mineral Museum, Otsu 29167

Kazakhstan

Memorialnyj Muzej Akademika K.I. Satpaeva, Almaty 29684
Muzej Geologii, Almaty 29686

Kyrgyzstan

Geologičeskij i Mineralogičeskij Muzej, Biškek 29953

Latvia

Latvijas Universitātes Ģeologijas Muzejs, Rīga 30088

Lithuania

Geologijos ir Mineralogijos Muziejus, Vilnius 30408
Mineralų Muziejus, Vilnius 30425

Luxembourg

Musée Collection Eugène Pesch, Differdange 30466

Mexico

Museo de Mineralogía, Pachuca 31242
Museo de Mineralogía, Guanajuato 30917
Museo de Mineralogía Dr Miguel Romero Sánchez, Tehuacán 31419
Museo de Mineralogía y Geología Ing. Luis Silva Ruelas, Morelia 31198
Museo Ora Ton, Chamula 30758

Morocco

Musée for Fossils Minerals, Erfoud 31667

Mozambique

Museu Nacional de Geologia, Maputo 31711

Netherlands

Geo Inn, Oosterbeek 32625
Mineralogisch-Geologisch Museum, Delft 32061
Mineralogisch Museum, Grou 32289

New Zealand

Crystal Mountain - Crystal Mineral Gallery Museum, Auckland 33045
Geology Museum, Auckland 33049
Thames School of Mines and Mineralogical Museum, Thames 33210

Nicaragua

Museo de Geología, Managua 33289

Peru

Museo Geológico Georg Petersen, Lima 34260

Poland

Muzeum Mineralogiczne, Szklarska Poręba 35170
Muzeum Mineralogiczne, Wrocław 35334

Portugal

Museu de Jazigos Minerais Portugueses, São Mamede de Infesta 35923
Museu Laboratório Mineralógico e Geológico, Porto 35856
Museu Mineralógico e Geológico, Coimbra 35539
Museu Mineralógico e Geológico, Lisboa 35714

Romania

Colecția de Mineralogie Estetică a Fierului Constantin Gruescu, Ocna de Fier 36495
Muzeul de Mineralogie Baia Mare, Baia Mare 36127

Russia

Angarskij Muzej Mineralov, Angarsk 36761
Geologičeskij Muzej, Vorkuta 37967
Geologičeskij Muzej im. Professora V.E. Ersova MGGU, Moskva 37180
Geologičeskij Muzej Vserossijskogo NII Mineralnogo Syrja im. N.M. Fedorovskogo, Moskva 37181
Geologo-mineralogičeskij Muzej, Kazan 37007
Geologo-mineralogičeskij Muzej, Moskva 37182
Geologo-mineralogičeskij Muzej im. A.I. Kozlova, Vladivostok 37937
Gornyj Muzej, Sankt-Peterburg 37639
Gosudarstvennyj Geologičeskij Muzej im V.I. Vernadskogo, Moskva 37192
Mineralogičeskij Muzej, Apatity 36763
Mineralogičeskij Muzej im. A.E. Fersmana Rossijskoj Akademii Nauk, Moskva 37250
Mineralogičeskij Muzej im. A.V. Sidorova, Irkutsk 36934
Mineralogičeskij Muzej Moskovskoj Geologorazvedyvatelnoj Akademii, Moskva 37251
Muzei Permskogo Gosudarstvennogo Universiteta, Perm 37547
Muzej Geologii, Čeljabinsk 36845
Muzej Mineralov, Rud, Samocvetov - Udivitelnoe v Kamne, Teberda 37840
Muzej Samocvety, Moskva 37360
Muzejno-vystavočnyj Centr, Dalnegorsk 36872

Slovakia

Berggericht - Mineralogická Expozícia, Banská Štiavnica 38189

Spain

Museo de la Real Academia Nacional de Farmacia, Madrid 39548
Museo de los Minerales y Rocas, Porriño 39861
Museo Geominero, Madrid 39561
Museo Histórico Minero Francisco Pablo Holgado, Almadén 38851
Museu Mollfulleda de Mineralogía, Arenys de Mar 38887

Sri Lanka

Museum of Gems and Minerals, Ratnapura 40382

M

Sweden

Evolutionsmuseet, Paleontologi, Uppsala 41006
Jokkmokks Stencenter, Jokkmokk 40619
Mineral- och Gruvmuseet, Holsbybrunn 40602
Skottvångs Gruva, Mariefred 40734

Switzerland

Collezione di Minerali e Fossili, Semione 41859
focusTerra, Zürich 42071
Kristallmuseum, Guttannen 41466
Mineralien Museum, Seedorf 41855
Musée des Minéraux et des Fossiles, Rougemont 41778
Museum Alpin Pontresina, Pontresina 41726
Museum La Truaisch, Sedrun 41854
Steinmuseum, Solothurn 41888

Taiwan

Kiko's Fossil and Mineral Museum, Banciao 42145
Kung Long Mineral Museum, Sincheng 42331
Land Fossils and Minerals Museum, Yongkang 42475
Yang Ming Oceanic Culture and Art Museum, Keelung 42250

Tanzania

Mineral Resources Divisional Museum, Dodoma 42514

Thailand

Mineralogy Museum, Bangkok 42545

United Kingdom

Abbey and Stones Museum, Margam 45104
Camborne School of Mines Geological Museum and Art Gallery, Pool 45404
Cornwall Geological Museum, Penzance 45359
Creetown Gem Rock Museum, Creetown 43918
Geological Museum and Art Gallery, Redruth 45477
Killhope, the North of England Lead Mining Museum, Weardale 45951
National Slate Museum, Llanberis 44754
Planet Earth Museum, Newhaven 45237
Royal Cornwall Museum, Truro 45879
Royal Pump Room Museum, Harrogate 44402

United States of America

A.E. Seaman Mineral Museum, Houghton 49377
Alaska Museum of Natural History, Anchorage 46287
Annie E. Woodman Institute, Dover 48166
Arizona Mining and Mineral Museum, Phoenix 51899
California State Mining and Mineral Museum, Mariposa 50491
Cass County Historical Society Museum, Plattsmouth 51985
Colburn Gem and Mineral Museum, Asheville 46371
Colorado School of Mines Geology Museum, Golden 48958
Dakota Dinosaur Museum, Dickinson 48138
Danville Science Center, Danville 47964
East Kentucky Center for Sciece, Mathematics and Technology, Prestonsburg 52139
The Geological Museum, Laramie 49947
Gillespie Museum of Minerals, De Land 47998
Greene Memorial Museum, Milwaukee 50709
Idaho Museum of Mining and Geology, Boise 46892
Land of the Yankee Fork Historical Museum, Nampa 50941
Lizzadro Museum of Lapidary Art, Elmhurst 48376
Meadowlands Museum, Rutherford 52492
Miles Mineral Museum, Portales 52055
Mineral and Lapidary Museum of Henderson County, Hendersonville 49253
Mineral Museum, Butte 47152
Mineralogical Museum of Harvard University, Cambridge 47182
Mousley Museum of Natural History, Yucaipa 54361
Museum of Comparative Zoology, Cambridge 47184
Museum of Long Island Natural Sciences, Stony Brook 53314
Museum of North Carolina Minerals, Spruce Pine 53244
New Mexico Bureau of Geology Mineral Museum, Socorro 53140
Northwestern Oklahoma State University Museum, Alva 46247
Oregon Museum of Science and Industry, Portland 52083
Orton Geological Museum, Columbus 47758
Post Rock Museum, La Crosse 49834
Stone Museum, Monroe Township 50794
Trailside Nature Center and Museum, Cincinnati 47566
Tri-State Mineral Museum, Joplin 49651
University of Arizona Mineral Museum, Tucson 53599
W.M. Keck Museum, Reno 52292
Wallis Museum at Connors State College, Warner 53818
Weinman Mineral Museum, White 54127

Vietnam

Geological Museum, Ho Chi Minh City 54622

Zimbabwe

Macgregor Museum, Harare 54668

Miniature Painting

France

Collections de l'Ecole Nationale Supérieure des Beaux-Arts, Paris 14590

India

Government Museum and Art Gallery, Chandigarh 24109
M.C. Mehta Gallery, Ahmedabad 24049
Maharaja Sawai Man Singh II Museum, Jaipur, Rajasthan 24234

Netherlands

Reflex Miniatuur Museum voor Hedendaagse Kunst, Amsterdam 31881

Portugal

Museu de Miniaturas de Arte e Oficios, Cartaxo 35510

Switzerland

Patek Philippe Museum, Genève 41428

Taiwan

Egg Artistry Museum, Cyonglin 42170

United Kingdom

Sherborne Castle, Sherborne 45659

Mining Engineering

Argentina

Museo del Cielo, Villa Carlos Paz 00673

Australia

Acland Coal Mine Museum, Acland 00765
Australian Prospectors and Miners Hall of Fame, Kalgoorlie 01185
Burra Mine Open Air Museum, Burra 00925
Emmaville Mining Museum, Emmaville 01071
Farm Shed Museum, Kadina 01182
Grubb Shaft Gold and Heritage Museum, Beaconsfield 00844
Gympie and District Historical and Gold Mining Museum, Gympie 01138
Herons Reef Historic Gold Diggings, Fryerstown 01097
Leonora Gwalia Historical Museum, Leonora 01229
Morphett's Enginehouse Museum, Burra 00926
Richmond Main Mining Museum, Pelaw Main 01408
Sovereign Hill, Ballarat 00829
State Coal Mine Museum, Wonthaggi 01667
Umoona Opal Mine and Museum, Coober Pedy 00996

Austria

Ausstellungszentrum Heft, Hüttenberg 02101
Bergbau- und Gotikmuseum Leogang, Leogang 02289
Bergbau- und Heimatmuseum, Gloggnitz 01937
Bergbaumuseum, Grünbach am Schneeberg 02013
Bergbaumuseum, Mühlbach am Hochkönig 02408
Bergbaumuseum Goberling, Stadtschlaining 02755
Bergbaumuseum Klagenfurt, Klagenfurt 02181
Bergbaumuseum Knappenhaus, Altenmarkt bei Sankt Gallen 01712
Bergbaumuseum mit Mineralienschau und Schaubergwerk, Hüttenberg 02102
Bergbaumuseum Pöllau, Neumarkt in Steiermark 02438
Bergbaumuseum und Schaustollen, Fohnsdorf 01893
Bergbauschaustollen, Pölfing-Brunn 02512
Böcksteiner Montanmuseum Hohe Tauern, Böckstein 01791
Granatium, Radenthein 02549
Montafoner Bergbaumuseum Silbertal, Silbertal 02743
Museum Bramberg "Wilhelmgut", Bramberg 01793
Museum und Kupferzeche Hüttau, Hüttau 02100
Rauriser Talmuseum, Rauris 02566
SBW Terra Mystica Montana, Bad Bleiberg-Kreuth 01740
Schaubergwerk, Leogang 02291
Schaubergwerk Arzberg, Arzberg 01730
Schaubergwerk Grillenberg, Payerbach 02481
Schaubergwerk - Historische Silbergruben, Oberzeiring 02467
Schaubergwerk Hochfeld, Neukirchen am Großvenediger 02430
Schaubergwerk Kupferplatte, Jochberg 02144
Schaubergwerk Seegrotte, Hinterbrühl 02078
Schloß Murau / Eisenmuseum, Murau 02417
Schloß- und Goldbergbaumuseum, Großkirchheim 02002
Schwazer Silberbergwerk, Schwaz 02727
Silberbergwerk, Ramingstein 02556
Tiroler Bergbau- und Hüttenmuseum, Brixlegg 01801
Untersbergmuseum Fürstenbrunn, Grödig 01994
Zillertaler Goldschaubergwerk, Zell am Ziller 03128

Belgium

Au Coeur de l'Ardoise, Bertrix 03365
Ecomusée Régional du Centre, Houdeng-Aimeries 03716
Le Puits-Marie, Blegny 03373
Musée de la Mine-Mémoire Ouvrière, Harchies 03676
Musée de la Mine, Musée de la Clouterie, Musée du Gazomètre, Fontaine-l'Evêque 03606
Musée Minier d'Halanzy, Halanzy 03665
Museum Ons Mijnverleden, Houthalen 03719
Site du Bois du Cazier, Marcinelle 03852
Vlaams Mijnmuseum, Beringen 03361

Bolivia

Museo Etnográfico Minero, Oruro 04232

Brazil

Museu do Ouro, Sabará 04783
Museu Mineralógico da Escola de Minas, Ouro Preto 04605

Canada

Atlas Coal Mine, East Coulee 05730
Barkerville Historic Town, Barkerville 05384
Bell Island Community Museum, Bell Island 05413
Black Nugget Museum, Ladysmith 06126
British Columbia Museum of Mining, Britannia Beach 05482
Buchans Miners Museum, Buchans 05492
Cape Breton Miners' Museum, Glace Bay 05868
Centre d'Interprétation des Mines Minerais, Petit-Rocher 06597
Cobalt Mining Museum, Cobalt 05627
Copper Belt Museum, Whitehorse 07324
Crowsnest Pass Ecomuseum, Bellevue 05416
Elliot Lake Nuclear and Mining Museum, Elliot Lake 05773
Fernie and District Historical Society Museum, Fernie 05795
Inverness Miners Museum, Inverness 06017
Jack Lynn Memorial Museum, Horsefly 05997
Keno City Mining Museum, Keno City 06056
Leitch Collieries, Crowsnest Pass 05662
Lynn Lake Mining Town Museum, Lynn Lake 06206
Malagash Area Heritage Museum, Malagash 06222
Marble Mountain Library and Museum, West Bay 07301
Moose River Gold Mines Museum, Middle Musquodoboit 06274
Musée Régional des Mines de Malartic, Malartic 06223
Ron Morel Memorial Museum, Kapuskasing 06043
Rossland Historical Museum, Rossland 06777
Sandon Museum, New Denver 06427
The Silver Ledge Hotel, Ainsworth Hot Springs 05310
Silverton Outdoor Mining Exhibit, Silverton 06977
Sioux Lookout Museum, Sioux Lookout 06981
Springhill Miner's Museum, Springhill 07002
Stewart Historical Museum, Stewart 07010
Village Minier de Bourlamaque, Val-d'Or 07194

China

China Coal Museum, Taiyuan 08271

Croatia

Narodni Muzej Labin, Labin 08906
Tehnički Muzej, Zagreb 09012

Czech Republic

České muzeum stříbra - Hrádek, Kutná Hora 09612
Hornické Muzeum OKD, Ostrava 09697
Hornické Muzeum Příbram, Příbram 09806
Hornický Skanzen důl Mayrau, Vinařice u Kladna 09908
Královská Mincovna v Jáchymove, Jáchymov 09549
Muzeum Českého granátu, Třebenice 09874
Muzeum Sokolov, Krásno 09605

Denmark

Søby Brunkulslejer og Brunkulsmuseet, Herning 10073

Estonia

Kohtla Kaevanduspark-muuseum, Ida 10572

Finland

Leppävirran Kotiseutumuseo, Leppävirta 11064
Outokummun Kaivosmuseo, Outokumpu 11180
Tytyrin Kalkkikaivosmuseo, Lohja 11078

France

Anicennes Mines de Grandfontaine, Grandfontaine 13102
Caves Muséalisées Byrrh, Thuir 15905
Centre de Culture Scientifique la Rotonde Ecole des Mines, Saint-Étienne 15249
Centre Historique Minier, Lewarde 13735
Centre Minier de Faymoreau, Faymoreau-les-Mines 12912
Ecomusée des Bruneaux, Firminy 12937
Les Hautes-Mynes, Le Thillot 13653
Le Grand Filon, Saint-Georges-d'Hurtieres 15282
Maison de Pays, Sainte-Marie-aux-Mines 15565
La Mine-Image, La Motte-d'Aveillans 13395
Mine Témoin d'Alès, Alès 11548
Musée Ardoisier, Saint-Julien-Mont-Denis 15351
Musée de la Machine à Vapeur, Oignies 14501
Musée de la Mine, Brassac-les-Mines 12146
Musée de la Mine, Bruay-la-Buissière 12193
Musée de la Mine, Champagnac-les-Mines 12384
Musée de la Mine, La Machine 13391
Musée de la Mine, La Ricamaire 13408
Musée de la Mine, Saint-Étienne 15253
Musée de la Mine, Grande-Synthe 13101
Musée de la Mine, Gréasque 13129
Musée de la Mine, de la Verrerie et du Chemin de Fer, Epinac 12838
Musée de la Mine et de l'École, Harnes 13181
Musée de la Mine et de Meunerie-Moulin de Marcy, Le Molay-Littry 13616
Musée de la Mine et des Hommes, Blanzy 12019
Musée de la Mine Lucien-Mazars, Aubin 11732
Musée de la Mine Marcel Maulini, Ronchamp 15093
Musée de l'Ardoise, Renazé 15010
Musée des Mines d'Argent des Gorges du Fournel, L'Argentière-la-Bessée 13526
Musée des Mines d'Aumetz, Aumetz 11745
Musée des Mines de Fer, Aumetz 11746
Musée des Mines de Fer de Lorraine, Neufchef 14400
Musée du Fer et du Fil, Dompierre (Orne) 12762
Musée du Mineur, La Grand-Combe 13379
Musée Régional du Chemin de Fer Nord-Pas-de-Calais, Denain 12710

Germany

Baumberger-Sandstein-Museum, Havixbeck 19055
Bergbau- und Greifenstein-Museum, Ehrenfriedersdorf 18070
Bergbau- und Heimatmuseum im Paulushof, Essen 18249
Bergbau- und Industriemuseum Ostbayern, Kümmersbruck 19734
Bergbau- und Stadtmuseum Weilburg, Weilburg 22384
Bergbaumuseum, Altenberg 16493
Bergbaumuseum, Hausham 19052
Bergbaumuseum, Langelsheim 19808
Bergbaumuseum, Niedermoschel 20707
Bergbaumuseum, Oelsnitz 20870
Bergbaumuseum Achthal, Teisendorf 22019
Bergbaumuseum Grube Anna II, Alsdorf 16479
Bergbaumuseum Hückelhoven, Hückelhoven 19286
Bergbaumuseum im Besucherbergwerk Tiefer Stollen, Aalen 16410
Bergbaumuseum Peißenberg, Peißenberg 21003
Bergbaumuseum Schachtanlage Knesebeck, Bad Grund 16743
Bergbaumuseum und Schaubergwerk Röhrigschacht, Wettelrode 22519
Bergisches Museum für Bergbau, Handwerk und Gewerbe, Bergisch Gladbach 17014
Bergmännisches Traditionskabinett, Breitenbrunn 17540
Bergwerksmuseum, Penzberg 21007

Bergwerksmuseum Schacht Mehren, Käbschütztal 19410
Besucher-Bergwerk Hüttenstollen, Salzhemmendorf 21478
Besucherbergwerk, Nothweiler 20749
Besucherbergwerk "Drei Kronen und Ehrt", Elbingerode, Harz 18111
Besucherbergwerk Finstertal, Schmalkalden, Kurort 21579
Besucherbergwerk Frischglück, Neuenbürg 20604
Besucherbergwerk Grube Fortuna, Solms 21810
Besucherbergwerk Grube Gustav, Meißner 20268
Besucherbergwerk Schiefergrube Christine, Willingen 22584
Deutsches Bergbau-Museum, Bochum 17398
Energiefabrik Knappenrode, Hoyerswerda 19281
Erzstollen Silbergründle, Seebach 21711
Feld- und Grubenbahnmuseum, Solms 21811
Forum Fränkischer Hof, Bad Rappenau 16805
Fossilien- und Heimatmuseum, Messel 20304
Granitabbaumuseum, Königshain 19652
Grube Alte Hoffnung Erbstolln, Mittweida 20345
Haniel Museum, Duisburg 18018
Heimat- und Bergbaumuseum Lugau, Lugau 20107
Heimat- und Bergbaumuseum Nentershausen, Nentershausen 20570
Heimat- und Braunkohlemuseum Steinberg, Steinberg, Kreis Schwandorf 21883
Heimatmuseum, Friedewald 18471
Heimatmuseum, Klipphausen 19573
Heimatmuseum, Markranstädt 20210
Hessisches Braunkohle Bergbaumuseum, Borken, Hessen 17478
Historicum - Sammlung für Bergbaukunde, Modellsammlung, Winkler-Gedenkstätte und Karzer, Freiberg, Sachsen 18422
Historischer Schieferbergbau Lehesten, Lehesten 19873
Historisches Bergamt Bad Wildungen-Bergfreiheit, Bad Wildungen 16860
Historisches Erzbergwerk im Silberbergwerk Bodenmais, Bodenmais 17413
Historisches Schmucksteinbergwerk Silberschacht, Bach an der Donau 16653
Historisches Silberbergwerk "Hella-Glücksstollen", Neubulach 20594
Informationszentrum Schloss Paffendorf, Bergheim (Erft) 17012
Kali-Bergbaumuseum Volpriehausen, Uslar 22192
Das Kleine Museum, Gelsenkirchen 18575
Kunstsammlungen Zwickau, Zwickau 22833
Kupferbergwerk, Beckingen 16972
Landesbergbaumuseum Baden-Württemberg, Sulzburg 21988
Lehr- und Schaubergwerk Frisch Glück, Johanngeorgenstadt 19400
Lehr- und Schaubergwerk Herkules Frisch Glück, Beierfeld 16982
Luisenhütte, Balve 16905
Maffeischächte der Grube Auerbach-Nitzlbuch, Auerbach, Kreis Deggendorf 16606
Mansfeld-Museum Hettstedt, Hettstedt 19173
Mansfelder Bergwerksbahn e.V., Benndorf 16992
Mineralien-Museum Andreas Gabrys, Lam 19781
Mineralienmuseum, Neubulach 20595
Museum der Stadt Aue, Aue 16605
Museum der Stadt Gladbeck, Gladbeck 18650
Museum Huthaus Einigkeit, Brand-Erbisdorf 17504
Museum Huthaus Zinnwald, Altenberg 16495
Museum Industriekultur, Osnabrück 20926
Museum Schloß Schwarzenberg, Schwarzenberg 21675
Museum und Besucherbergwerk, Goslar 18714
Museum und Besucherbergwerk der Graphit Kropfmühl AG, Hauzenberg 19053
Museum Uranbergbau, Bad Schlema 16827
Museum Wilnsdorf, Wilnsdorf 22589
Museumspark, Rüdersdorf bei Berlin 21433
Niedersächsisches Museum für Kali- und Salzbergbau, Ronnenberg 21371
Oberharzer Bergwerksmuseum, Clausthal-Zellerfeld 17709
Oberpfälzer Flußspat-Besucherbergwerk Reichhart-Schacht, Stulln 21925
Saarländisches Bergbaumuseum, Bexbach 17318
Sächsisches Lehr- Besucherbergwerk-Himmelfahrt-Fundgrube, Freiberg, Sachsen 18424
Sandsteinmuseum Kloster Cornberg, Cornberg 17726
Schaubergwerk Büchenberg, Elbingerode, Harz 18114
Schaubergwerk Teufelsgrund, Münstertal 20523
Schaubergwerk Zum Tiefen Molchner Stolln, Pobershau 21068
Schiefermuseum, Ludwigsstadt 20065
Silbereisenbergwerk Gleißinger Fels, Fichtelberg 18316
Silberstollen Geising, Geising 18560
Stadt- und Bergbaumuseum Freiberg, Freiberg, Sachsen 18425
Städtische Sammlungen Freital, Freital 18453
Steinkohlen-Besucherbergwerk Rabensteiner Stollen, Ilfeld 19314
Technisches Denkmal Museum Kalkwerk Lengefeld, Lengefeld 19951
Technisches Denkmal Tobiashammer, Ohrdruf 20891
Werra-Kalibergbau-Museum, Heringen (Werra) 19143
Westfälisches Schieferbergbau- und Heimatmuseum Holthausen, Schmallenberg 21585
Zinngrube Ehrenfriedersdorf, Ehrenfriedersdorf 18071

Hungary

Alapítvány Érc és Ásványbányászati Múzeum, Rudabánya 23789
Balás Jenő Bauxitbányászati Múzeum, Gánt 23563
Bányászati Kiállítás, Gyöngyös 23572
Bányászati Kiállítóhelye, Salgótarján 23790
Brennbergbányai Bányászati Emlékház, Sopron 23807
Központi Bányászati Múzeum, Sopron 23812
Komlói Természettudományi Gyűjtemény, Komló 23681
Mecseki Banyászati Múzeum, Pécs 23764
Mecseki Bányászati Múzeum, Pécs 23765
Mecseki Bányászati Múzeum - Uránérc Bányászati Kiállítás, Kővágószőlős 23687
Oroszlányi Bányász Múzeum Alapítvány, Oroszlány 23743
Szabadtéri Bányászati Múzeuma, Tatabánya 23921
Tatabánya Múzeum, Tatabánya 23922
Városi Helytörténeti Múzeum, Komló 23682
Várpalotai Bányászattörténeti Gyűjtemény, Várpalota 23954

Italy

Collezione P. Mariani, Desio 26037
Museo Antica Miniera di Talco Brunetta, Cantoira 25649
Museo dei Minerali e della Miniera, Oltre il Colle 26981
Museo della Miniera, Cesena 25829
Museo della Miniera, Massa Marittima 26630
Museo della Miniera, Prata di Pordenone 27328
Museo della Miniera, Schilpario 27790
Museo della Miniera Aurifera della Guia, Macugnaga 26559
Museo della Miniera di Zolfo di Cabernardi, Sassoferrato 27763
Museo dell'Arte Mineraria, Iglesias 26396
Museo delle Miniere, Valle Aurina 28144
Museo di Arte e Giacimenti Minerari, Latina 26466
Museo di Arte e Storia delle Miniere, Massa Marittima 26631
Museo Provinciale delle Miniere, Racines 27356
Museo Provinciale delle Miniere, Vipiteno 28287
Museo Storico dell'Oro Italiano, Predosa 27342
Museo Storico Minerario di Perticara, Novafeltria 26947
Parco Museo Minerario, Abbadia San Salvatore 25138

Japan

Akita Daigaku Kogyo Hakubutsukan, Akita 28353
Ikuno Kobutsukan no Kogyo Hakubutsukan, Ikuno 28566
Nogata-shi Sekitan Kinenkan, Nogata 29081
Omuta-shi Tanko oyo Kagakukan, Omuta 29117
Sekitan Kogyo Hakubutsukan, Ube 29549

Korea, Republic

Boryeong Coal Museum, Boryeong 29771

Luxembourg

Musée du Cuivre à Stolzembourg, Stolzembourg 30492
Musée National des Mines de Fer Luxembourgeoises, Rumelange 30491

Mexico

Museo de Historia de la Minería de Santa Rosalía, Mulegé 31206
Museo de la Minería, San Juan de Guadalupe 31337
Museo de la Minería del Estado de México, El Oro 30849
Museo de la Minería José de la Borda, Taxco de Alarcón 31405
Museo de Minería de la Compañía Real del Monte, Pachuca 31243
Museo de Minería e Historia de Cosalá, Cosalá 30803
Museo Dr. Atl, Pihuamo 31259
Museo El Minero, Santa Bárbara 31375

Namibia

Kolmanskop Museum, Kolmanskop 31743

Netherlands

Industrion Museum for Industry and Society, Kerkrade 32446
Steenkolenmijn Valkenburg, Valkenburg aan de Geul 32861

New Zealand

Alexandra Museum, Alexandra 33034
Brunner Mine Site, Dobson 33087
Coromandel School of Mines Museum, Coromandel 33082
Lakes District Museum, Arrowtown 33035
Waihi Arts Centre and Museum Association, Waihi 33218

Norway

Folldal Gruver, Folldal 33501
Gammelgruva, Løkken Verk 33643
Hadeland Bergverksmuseum, Grua 33522
Museum Blaafarveværket, Åmot i Modum 33388
Norsk Bergverksmuseum, Kongsberg 33603
Sulitjelma Gruvemuseum, Sulitjelma 33890
Sunnhordland Museum, Stord 33885

Peru

Museo Geológico Georg Petersen, Lima 34260

Poland

Muzeum Górnictwa i Hutnictwa Kopalina Złota, Złoty Stok 35375
Muzeum Górnictwa Podziemnego, Nowa Ruda 34947
Muzeum Górnictwa Węglowego, Zabrze 35349
Muzeum Górnicze, Bogatynia 34566
Muzeum Historii Akademii Górniczo-Hutniczej im. S. Staszica, Kraków 34795
Muzeum Hutnictwa i Górnictwa Rud Żelaza, Częstochowa 34616
Muzeum Przemysłu i Techniki, Wałbrzych 35212
Muzeum Żup Krakowskich Wieliczka, Wieliczka 35300
Skansen Górniczy Królowa Luiza, Zabrze 35351
Zabytkowa Kopalnia, Tarnowskie Góry 35181

Portugal

Museu da Mina de Pirites, Aljustrel 35411
Museu Mineiro, São Pedro da Cova 35924

Romania

Muzeul Aurului, Brad 36165
Muzeul Mineritului, Petroșani 36515
Muzeul Mineritului Avrifer Roșia Montană, Roșia Montană 36572

Russia

Gornyj Muzej, Sankt-Peterburg 37639
Kemerovskij Istoriko-architekturnyj Muzej Krasnaja Gorka, Kemerovo 37030
Memorialnyj Istoriko-biografičeskij Dom-muzej Akademika N.V. Melnikova, Sarapul 37745

Serbia

Muzej Rudarstva i Metalurgije, Bor 38093

Slovakia

Banícke Múzeum, Gelnica 38256
Banicke Múzeum, Rožňava 38342
Banské Múzeum v Prírode, Banská Štiavnica 38188
Kammerhof - Baníctvo na Slovensku, Banská Štiavnica 38191
Slovenské Banské Múzeum, Banská Štiavnica 38193

Slovenia

Antonijev Rov - Rudnik Živega Srebra, Idrija 38420
Gornjesavski Muzej Jesenice, Jesenice 38425
Koroški Muzej Ravne na Koroškem, Ravne na Koroškem 38503
Muzej Premogovništva Slovenije, Velenje 38533
Muzej Železniki, Železniki 38541
Rudarska in Etnologoška Zbirka v Črni na Koroškem, Črna na Koroškem 38409

South Africa

Barberton Museum, Barberton 38555
Gold of Africa Museum, Cape Town 38584
Pilgrim's Rest Museum, Pilgrim's Rest 38709
Talana Museum and Heritage Park, Dundee 38601

Spain

Museo Histórico Minero Don Felipe de Borbón y Grecia, Madrid 39562

Sweden

Bruzaholms Bruksmuseum, Bruzaholm 40481
Ebbamåla Bruk, Kyrkhult 40663
Falu Gruva Gruvmuseum, Falun 40516
Gamla Brukets Muséer, Munkfors 40761
Gruvmuseet, Sala 40826
LKAB Gruvmuseum, Kiruna 40643
Loos Koboltgruva, Loos 40693
Ludvika Gammelgård och Gruvmuseum, Ludvika 40696
Mineral- och Gruvmuseet, Holsbybrunn 40602
Siggebohyttans Bergmansgård, Nora 40765

Switzerland

Bergbaumuseum, Horgen 41487
Bergbaumuseum Graubünden mit Schaubergwerk, Davos Platz 41320
Eisenbergwerk Gonzen, Sargans 41816

Taiwan

Jingtong Mining Museum, Pingsi 42288
Taiwan Coal Mine Museum, Pingsi 42289

United Kingdom

Big Pit National Coal Museum of Wales, Blaenavon 43534
Cefn Coed Colliery Museum, Neath 45195
Cleveland Ironstone Mining Museum, Skinningrove 45687
Dunaskin Open Air Museum, Ayr 43345
Durban House Heritage Centre, Eastwood 44109
Elliot Colliery Winding House, New Tredegar 45208
Florence Mine Heritage Centre, Egremont 44165
Geevor Tin Mining Museum, Pendeen 45353
Gloddfa Ganol Slate Mine, Blaenau Ffestiniog 43532
Great Laxey Wheel and Mines Trail, Laxey 44650
Keswick Mining Museum, Keswick 44563
Llywernog Silver-Lead Mine Museum, Ponterwyd 45399
Minera Lead Mines, Minera 45157
Morwellham Quay Historic Port Copper Mine, Morwellham 45178
Museum of Lead Mining, Wanlockhead 45927
Museum of Yorkshire Dales Lead Mining, Earby 44078
National Coal Mining Museum for England, Wakefield 45910
National Slate Museum, Llanberis 44754
Nenthead Mines Heritage Centre, Nenthead 45200
Peak District Mining Museum, Matlock 45116
Poldark Mine and Heritage Complex, Helston 44446
Prestongrange Museum, Prestongrange 45448
Scottish Mining Museum, Newtongrange 45253
South Wales Miner's Museum, Port Talbot 45418
Sygun Museum of Wales, Beddgelert 43431
Tar Tunnel, Coalport 43862
Tolgus Tin Mill and Streamworks, Redruth 45478
Woodhorn, Ashington 43329

United States of America

Alabama Mining Museum, Dora 48152
Argo Gold Mine, Idaho Springs 49463
Arizona Mining and Mineral Museum, Phoenix 51899
Big Thunder Gold Mine, Keystone 49769
Bisbee Mining and Historical Museum, Bisbee 46830
Black Hills Mining Museum, Lead 50004
California State Mining and Mineral Museum, Mariposa 50491
Coal Miners' Museum, Van Lear 53702
Comer Museum, Morgantown 50844
Coppertown U.S.A. Mining Museum, Calumet 47169
Dahlonega Courthouse Gold Museum, Dahlonega 47920
Dugger Coal Museum, Dugger 48190
Fort Wilkins Historic Complex, Copper Harbor 47796
Franklin Mineral Museum, Franklin 48748
Galena Mining and Historical Museum, Galena 48831
Grand Encampment Museum, Encampment 48395
Heritage House Museum of Montgomery County, Mount Ida 50880
Idaho Museum of Mining and Geology, Boise 46892
Iron County Museum, Caspian 47274
Iron Mountain Iron Mine Museum, Vulcan 53769
Ironworld Discovery Center, Chisholm 47544
Lafayette Miners Museum, Lafayette 49863
The Mining Museum, Platteville 51980
Museum of Anthracite Mining, Ashland 46390
National Mining Hall of Fame and Museum, Leadville 50006
North Star Mining Museum, Grass Valley 49025
Reed Gold Mine, Midland 50668
Reed Gold Mine State Historic Site, Stanfield 53252
Searchlight Historic Museum and Mining Park, Searchlight 52991
Slate Valley Museum, Granville 49020

Sterling Hill Mining Museum, Ogdensburg 51505
Tintic Mining Museum, Eureka 48443
Underground Gold Miners Museum, Alleghany 46223
Underhill Museum, Idaho Springs 49465
Vermont Marble Museum, Proctor 52156
Wallace District Mining Museum, Wallace 53799
The Walsenburg Mining Museum and Fort Francisco Museum of La Veta, Walsenburg 53805
Western Museum of Mining and Industry, Colorado Springs 47707
World Museum of Mining, Butte 47153
Yesteryear House-Central Mine, Clarkston 47590
Youth Museum of Southern West Virginia Beckley Exhibition Coal Mine, Beckley 46685

Uruguay

Museo Geológico del Uruguay, Montevideo 54403

Zambia

Copperbelt Museum, Ndola 54662

Zimbabwe

National Mining Museum, Kwekwe 54673

Mollusks

Australia

Sea Shell Museum, Scarness 01496

Croatia

Hrvatski Prirodoslovni Muzej, Zagreb 08998

France

Maison de l'Amiral, Pouldreuzic 14929
Musée de la Prehistoire, Rânes 14984
Musée de l'Ostréiculture, Marennes 13966
Musée du Coquillage, Le Lavandou 13601
Musée du Jamais Vu, Puymaurin 14962
Musée Maison de l'Huître, Gujan-Mestras 13171

Germany

Haus der Natur - Cismar, Cismar 17707

Greece

Moschakeion, Moschato 23101

Italy

Museo Zoologico Cambria, Messina 26669

Netherlands

Natuurhistorisch Museum Rotterdam, Rotterdam 32718

Russia

Muzej Instituta Biologii Morja, Vladivostok 37944
Zoologičeskij Muzej, Vladivostok 37953

Spain

Museo de Ciencias Naturales, Santa Cruz de Tenerife 39995
Museo Nacional de Ciencias Naturales, Madrid 39571
Museu de Malacología Cau del Cargol, Vilassar de Dalt 40292

Monasteries

Argentina

Museo de la Santa Casa de Ejercicios Espirituales Sor María Antonia de la Paz y Figueroa, Buenos Aires 00208

Austria

Kartause Gaming, Gaming 01921

Bangladesh

Archaeological Museum, Mainamati 03232

Belgium

Archief en Museum Zusters van Liefde van Jezus en Maria, Gent 03623
Chapelle Saint-Calixte - Musée du Château Comtal, Mons 03880
Monasterium De Wijngaard, Brugge 03420
Musée Communal d'Archéologie et d'Art Religieux, Amay 03270
Musée de l'Abbaye d'Orval, Villers-devant-Orval 04114
Musée de l'Eglise Abbatiale de Floreffe, Floreffe 03604
Museum van de Abdij van Roosenberg, Waasmunster 04123

Bosnia and Herzegovina

Franjevački Samostan Gorica, Livno 04267
Franjevački Samostan Jajce, Jajce 04264
Franjevački Samostan Kraljeva Sutjeska, Kraljeva Sutjeska 04265
Franjevački Samostan Krešovo, Krešovo 04266
Muzejske Zbirke Fojničkog Samostana, Fojnica 04263

Bulgaria

Drjanovski Manastir Muzej, Drjanovo 05023
Nacionalen Muzej Rilski Manastir, Rilski Manastir 05150
Nacionalen Muzej Roženski Manastir, Melnik 05092

Cambodia

The Angkor Conservation, Siem Reap 05290

Croatia

Etnografska Zbírka Františkánský Klášter, Zaostrog 09016
Muzej Dominikanskog Samostana, Bol 08847
Muzej Franjevačkog Samostana, Rovinj 08942
Muzej Franjevačkog Samostana Košljun, Punat 08936
Muzejska Zbirka Franjevačkog Samostana, Imotski 08883
Zbirka Franjevacki Samostan, Orebić 08921
Zbirka Franjevačkog Samostana, Sinj 08951
Zbirka Franjevačkog Samostana Visovac, Visovac 08979
Zbirka Umjetnina Franjevačkog Samostana, Hvar 08881

Denmark

Danmarks Kloster Museum, Ry 10254

Egypt

Monastery of Saint Catherine, Gebel Katherîna 10525

Estonia

Dominiiklaste Kloostri Muuseum, Tallinn 10621

France

Cloître de Cadouin, Le Buisson-de-Cadouin 13567
Musée-Abbaye Saint-Germain, Auxerre 11776
Musée de la Vie Monastique, Saint-Hilaire-du-Harcouët 15308

Germany

Deutsches Kartausen-Museum, Buxheim 17672
Dominikanerkloster Prenzlau, Prenzlau 21117
Kloster Irsee, Irsee 19354
Kloster Wiblingen, Ulm 22163
Kloster Zinna, Kloster Zinna 19580
Klosteranlage mit Klostermuseum, Maulbronn 20239
Klostermuseum Heggbach, Maselheim 20233
Klostermuseum Hirsau, Calw 17679
Klostermuseum Jerichow, Jerichow 19394
Kornhaus Georgenthal-Klosterruinen, Georgenthal 18591
Missionsmuseum, Schwarzach am Main 21670
Museum Abtei Liesborn, Wadersloh 22263
Museum Höxter-Corvey, Höxter 19219
Museum Sankt Blasien, Sankt Blasien 21490
Museum zur Kloster- und Stadtgeschichte, Steinheim an der Murr 21891
Prignitz-Museum am Dom Havelberg, Havelberg 19054
Residenz Kempten, Kempten 19515
Sammlungen der Benediktiner-Abtei Braunau, Rohr in Niederbayern 21364
Schloss Salem, Salem, Baden 21472
Torbogenmuseum, Königsbronn 19643
ZisterzienserMuseum Kloster Walkenried, Walkenried 22307

Greece

Ano Moni Panagias Xenias, Almyros 22877
Arkadi Monastery Collection, Rethymnon 23166
Karyes-Protaton, Agion Oros 22868
Monastery of Aghiou Tessarakonta, Spárti 23196
Monastery of Evangelistria Daponte, Skopelos 23190
Monastery of Petra, Kataphygi 23045
Monastery of the Archangel Michael at Panormitis, Panormitis 23132
Monastery of the Taxiarchs, Megalo Livadi 23086
Monastery of the Zoodochos Pighi, Poros 23157
Museum of Agios Stephanos, Meteora 23093
Museum of the Gonia Monastery, Kissamos 23056
Proussos Monastery Museum, Prousos 23159
Refectory of the Hosios Loukas Monastery, Boeotia 22958
Tatarna Monastery Museum, Granitsa 23010

Ireland

Adare Heritage Centre, Adare 24727

Israel

Benedictine Monastery - Church of the Multiplication of Loaves and Fish Tabgha, Tiberias 25128

Italy

Museo dei Benedettini, Pontida 27291
Museo della Collegiata, Castell'Arquato 25735
Museo dell'Abbazia, San Benedetto Po 27614
Museo dell'Opera di Santa Croce, Firenze 26175
Museo Paleocristiano, Aquileia 25254
Museo Parrocchiale, Brescello 25533

Japan

Treasure House of the Eiheiji Temple, Eiheiji 28416

Laos

Haw Pha Kaeo, Vientiane 29973
Wat Si Saket, Vientiane 29979

Mongolia

Chojin Lama Monastery Museum, Ulaanbaatar 31618

Montenegro

Muzej Cetinjskog Manastira, Cetinje 31642

Netherlands

Klooster Ter Apel, Ter Apel 31909

Norway

Klostertunet, Halsnøy Kloster 33526
Rissa Bygdemuseum, Rissa 33784
Utstein Kloster, Mosterøy 33685

Peru

Monasterio de Santa Catalina, Arequipa 34108

Poland

Muzeum Klasztorne na Jasnej Górze, Częstochowa 34617

Portugal

Mosteiro de S. Martinho de Tibães, Mire de Tibães 35766
Mosteiro de Santa Clara-a-Velha, Coimbra 35527
Mosteiro de São João de Tarouca, São João de Tarouca 35921
Museu das Oferendas ao Soldado Desconhecido, Batalha 35460
Museu do Mosteiro de Alcobaça, Alcobaça 35394
Museu do Mosteiro de São Martinho de Tibães, Braga 35478
Museu do Mosteiro dos Jerónimos, Lisboa 35703

Romania

Colecţia Muzeală a Mănăstirii Adormirea Maicii Domnului, Bodrogu Vechi 36156
Colecţia Muzeală a Mănăstirii Bistriţa, Bistriţa, Vâlcea 36154
Colecţia Muzeală a Mănăstirii Cozia, Căciulata 36246
Colecţia Muzeală a Mănăstirii Dintr-un Lemn, Dezrobiţi 36345
Colecţia Muzeală a Mănăstirii Dragomirna, Dragomirna 36356
Colecţia Muzeală a Mănăstirii Govora, Mihăeşti 36473
Colecţia Muzeală a Mănăstirii Gura Motrului, Butoieşti 36241
Colecţia Muzeală a Mănăstirii Hurezi, Românii de Jos 36570
Colecţia Muzeală a Mănăstirii Jitianu, Podari 36537
Colecţia Muzeală a Mănăstirii Moldoviţa, Vatra Moldoviţei 36729
Colecţia Muzeală a Mănăstirii Neamţ, Vânători-Neamţ 36724
Colecţia Muzeală a Mănăstirii Nicula, Fizeşu Gherlii 36370
Colecţia Muzeală a Mănăstirii Ortodoxe Celic-Dere, Poşta 36540
Colecţia Muzeală a Mănăstirii Ortodoxe Cernica, Cernica 36275
Colecţia Muzeală a Mănăstirii Ortodoxe Ciolanu, Tisău 36695
Colecţia Muzeală a Mănăstirii Ortodoxe Cocoş, Niculiţel 36493
Colecţia Muzeală a Mănăstirii Ortodoxe Curtea de Argeş, Curtea de Argeş 36335
Colecţia Muzeală a Mănăstirii Ortodoxe Dealu, Aninoasa 36108
Colecţia Muzeală a Mănăstirii Ortodoxe Pasărea, Pasărea 36512
Colecţia Muzeală a Mănăstirii Ortodoxe Suzana, Măneciu 36462
Colecţia Muzeală a Mănăstirii Ortodoxe Viforâta, Viforâta 36730
Colecţia Muzeală a Mănăstirii Polovragi, Polovragi 36539
Colecţia Muzeală a Mănăstirii Putna, Putna 36544
Colecţia Muzeală a Mănăstirii Râmeţ, Râmeţ 36548
Colecţia Muzeală a Mănăstirii Rohia, Târgu Lăpuş 36668
Colecţia Muzeală a Mănăstirii Secu, Mănăstirea Neamţ 36461
Colecţia Muzeală a Mănăstirii Sf. Ilie, Topliţa 36699
Colecţia Muzeală a Mănăstirii Sf. Simion-Stâlpnicul, Arad 36110
Colecţia Muzeală a Mănăstirii Strehaia, Strehaia 36638
Colecţia Muzeală a Mănăstirii Suceviţa, Suceviţa 36648
Colecţia Muzeală a Mănăstirii Tismana, Tismana 36696
Colecţia Muzeală a Mănăstirii Văratec, Văratec 36725
Colecţias Muzeală a Mănăstirii Ortodoxe Căldăruşani, Moara Vlăsiei 36478
Muzeul Mănăstirii Bistriţa, Bistriţa, Neamţ 36153

Russia

Novodevičij Monastyr, Moskva 37389
Svjatogorskij Monastyr, Puškinskie Gory 37588

Serbia

Narodni Muzej, Užice 38158

Slovakia

Múzeum Červený Kláštor, Červený Kláštor 38245

Spain

Colección Museográfica Permanente Arte Bizantino, Tángel 40096
Madres Dominicanas - Monasterio Sancti Spiritus, Toro 40152
Monasterio Cisterciense de Santa María, Sobrado de los Monjes 40086
Monasterio de la Cartuja, Sevilla 40063
Monasterio de la Encarnación, Madrid 39523
Monasterio de la Encarnación, Osuna 39751
Monasterio de las Descalzas Reales, Madrid 39524
Monasterio de San Juan de los Reyes, Toledo 40130
Monasterio de San Lorenzo de El Escorial, San Lorenzo de El Escorial 39961
Monasterio de Santa Clara de Tordesillas, Tordesillas 40146
Monasterio de Santa María de la Rábida, Palos de la Frontera 39804
Monasterio de Santa María San Salvador, Cañas 39115
Monasterio San Juan de Duero, Soria 40091
Monasterio Santa María la Real de las Huelgas, Burgos 39066
Monestero de Santes Creus, Santes Creus 40013
Museo Conventual de las Carmelitas Descalzas de Antequera, Antequera 38874
Museo de Arte Sacro, Santiago de Compostela 40019
Museo del Monasterio de San Joaquín y Santa Ana, Valladolid 40236
Museo del Monasterio de Santa Isabel, Valladolid 40237
Museo del Monasterio de Santa María de la Vid, La Vid 40271
Museo del Monasterio de Tulebras, Tulebras 40177
Museo del Real Monasterio de San Juan de la Peña, Jaca 39417
Museo del Románico, Aguilar de Campoo 38791
Museo Diocesano Regina Coeli, Santillana del Mar 40034
Museo do Mosteiro de San Paio de Antealtares, Santiago de Compostela 40025
Museo Histórico-Artístico, Cuacos de Yuste 39228
Museo Parroquial, Celanova 39161
Museo Teresiano Monasterio de la Encarnación, Ávila 38924
Museu de Poblet, Monestir de Poblet 39663
Museu de Sant Cugat, Sant Cugat del Vallès 39979
Museu del Monestir de Sant Bartomeu, Inca 39409

Museu del Monestir de Santa Maria de L'Estany, L'Estany 39285
Museu Monàstic, Vallbona de les Monges 40244
Museu-Monestir de Pedralbes, Barcelona 39000
Real Monasterio de San Jerónimo, Granada 39371
Real Monasterio de Santa María de El Puig, El Puig de Santa María 39883

Sweden

Stadsmuseum i Gråbrödraklostret, Ystad 41077

Switzerland

Hausmuseum Sankt Katharinental, Diessenhofen 41327
Klostermuseum St. Georgen, Stein am Rhein 41914
Musée Lapidaire, Saint-Ursanne 41795
Museum Klösterli, Ettiswil 41366
Trésor et Fouilles Archéologiques de l'Abbaye, Saint-Maurice 41792

Ukraine

Nacional'nyj Kyjvo-Pečerskyj Istoryko-kul'turnyj Zapovidnyk, Kyïv 43101

United Kingdom

Beaulieu Abbey Exhibition of Monastic Life, Brockenhurst 43649
Bede's World, Jarrow 44545
Glastonbury Abbey Museum, Glastonbury 44306

United States of America

Teutopolis Monastery Museum, Teutopolis 53487

Money → Numismatics

Mosaics

France

Musée de la Mosaïque et des Émaux, Briare 12172

Greece

Refectory of the Hosios Loukas Monastery, Boeotia 22958

Italy

Antiquarium e Mosaico Romano, Bevagna 25412
Musei Civici di Padova, Padova 27034
Museo di Torcello, Venezia 28198
Pinacoteca Comunale, Faenza 26079

Portugal

Museu Monográfico de Conimbriga, Condeixa-a-Nova 35548

Spain

Museo Municipal, Alcázar de San Juan 38819

Switzerland

Pro Urba, Orbe 41714

Tunisia

Musée Archéologique de Sousse, Sousse 42656
Musée National du Bardo, Le Bardo 42616

Turkey

Büyük Saray Mozaik Müzesi, İstanbul 42810
Istanbul Arkeoloji Müzeleri, İstanbul 42823
Misis Mozaik Müzesi, Yakapınar 42964
Mozaik Müzesi, İstanbul 42832
Narlıkuyu Mozaik Müzesi, Silifke 42940

Ukraine

Saint Sophia of Kiev, Kyïv 43109

United Kingdom

Aldborough Roman Site, Aldborough 43266
Roman Villa Museum at Bignor, Pulborough 45451

United States of America

Tile Heritage Foundation, Healdsburg 49234

Mummies

Egypt

Elephantine Island's Museum, Aswan 10480

Germany

Bleikeller, Bremen 17546
Petrefaktensammlung, Bad Staffelstein 16844

Italy

Museo delle Mummie, Ferentillo 26096

Mexico

Museo de las Momias, Guanajuato 30916
Museo de Momias de Celaya, Celaya 30754

Poland

Muzeum im. Władysława Orkana, Rabka 35070

Yemen

Taweelah Museum of Mummies, Mahweet 54648

Mural Painting and Decoration

Austria

Panoramamuseum, Salzburg 02621

Belgium

Sint-Baafskathedraal Gent, Gent 03641

Bulgaria

Muzej na Rezbarskoto i Zografsko Izkustvo, Trjavna 05230

France

Musée Palais Lascaris, Nice 14438

Germany

Fürstlich Ysenburg- und Büdingensches Schloßmuseum, Büdingen 17627
Leitheimer Schloß-Museum, Kaisheim 19416
Museum für Vor- und Frühgeschichte, Saarbrücken 21457
Torhalle Frauenchiemsee und Vikarhaus, Frauenchiemsee 18412

Greece

Kitsos Makris Folk Art Center, Volos 23238

Italy

Cappella degli Scrovegni, Padova 27032
Cenacolo di Sant'Apollonia, Firenze 26147
Chiesa di Santa Maria Maddalena de Pazzi, Firenze 26148
Chiostro dello Scalzo, Firenze 26149
Complesso Museale di Santa Maria della Scala, Siena 27836
Fondazione Romano nel Cenacolo di Santo Spirito, Firenze 26152
Museo Benedettino Nonantolano e Diocesano di Arte Sacra, Nonantola 26940
Museo Civico, Sansepolcro 27698
Museo Civico A. Parazzi, Viadana 28242
Museo Civico di Schifanoia, Ferrara 26107
Museo degli Argenti, Firenze 26169
Museo del Camposanto Vecchio e dell'Opera, Pisa 27242
Museo del Castello Colleoni Porto Thiene, Thiene 27971
Museo della Villa Bettoni, Bogliaco 25434
Museo dell'Opera del Duomo, Prato 27337
Museo di Palazzo Vecchio, Firenze 26183
Museo di Pittura Murale, Prato 27338
Museo di San Marco, Firenze 26184
Museo Diocesano, Gaeta 26261
Museo Diocesano Albani, Urbino 28126
Museo e Chiostri Monumentali di Santa Maria Novella, Firenze 26188
Palazzo Medici Riccardi, Firenze 26200
Pinacoteca Civica, Spello 27886
Pinacoteca e Museo Civici, Camerino 25623
Refettorio dell'ex-Convento di Sant'Onofrio detto di Fuligno, Firenze 26202
Rocca Borromeo, Angera 25236
Sale Monumentali della Biblioteca Marciana, Venezia 28209
Villa Farnesina, Roma 27548

Japan

Horyuji Kondo Hekiga Mosha Tenjikan, Nagakute 28951
Takamatsuzuka-Hekigakan, Asuka 28381

Mexico

Casa Museo La Moreña, La Barca 30985
Museo de la Pintura Mural y Museo Manuel Gamio, Teotihuacán 31426
Museo de la Sep, México 31089

Serbia

Galerija Fresaka, Beograd 38040

Spain

Museo de Navarra, Pamplona 39808
Museo Real Colegiata de San Isidoro, León 39459
Museu Diocesá d'Urgell, La Seu d'Urgell 40059

Turkey

Göreme Açıkhava Müzesi, Nevşehir 42915

Ukraine

Saint Sophia of Kiev, Kyïv 43109

United Kingdom

Lady Waterford Hall, Berwick-upon-Tweed 43462
Plas Newydd, Llanfairpwll 44765

United States of America

Gallery of Prehistoric Paintings, New York 51182

Museology

Chile

Museo Histórico y Antropológico Mauricio Van de Maele, Valdivia 07469

South Africa

Iziko Museums of Cape Town, Cape Town 55017

United Kingdom

Conservation Centre Presentation, Liverpool 44732

Venezuela

Museo Armando Reverón, Macuto 54580

Music and Musical Instruments

Argentina

Museo de Instrumentos Musicales, Buenos Aires 00196
Museo de la Partitura Historica, Rosario 00544
Museo de los Instrumentos del Teatro Colón, Buenos Aires 00213
Museo de SADAIC Vincente López y Planes, Buenos Aires 00217
Museo del Tango Roberto Firpo, Salto 00561
Museo Manoblanca, Buenos Aires 00250
Museo Manuel de Falla, Alta Gracia 00121
Museo Mundial del Tango, Buenos Aires 00256

Armenia

Alexander Spendiaryan House Museum, Yerevan 00724
Aram Khachaturyan Museum, Yerevan 00726

Australia

Grainger Museum, Melbourne 01282
Historic and Ethnic Instrument Collection, Nedlands 01356
McCrae Homestead Museum, McCrae 01245
Museum of Lillydale, Lillydale 01230
Das Neumann Haus Museum, Laidley 01212
Waltzing Matilda Centre and Qantilda Museum, Winton 01658

Austria

Anton Bruckner-Museum, Ansfelden 01724
Arnold Schönberg Center, Wien 02934
Beethoven Eroicahaus, Wien 02945
Beethoven-Gedenkstätte Floridsdorf, Wien 02946
Beethoven Pasqualatihaus, Wien 02947
Beethoven Wohnung Heiligenstadt, Wien 02948
Beethovenhaus, Krems 02224
Beethovenhaus "Haus der Neunten", Baden bei Wien 01774
Blasmusik-Museum, Ratten 02563
Brucknerzimmer, Kronstorf 02240
Franz Liszt Zentrum, Raiding 02553
Franz Schmidt Gedenkraum, Perchtoldsdorf 02486
Franz Xaver Gruber-Museum, Lamprechtshausen 02253
Gemeindemuseum Absam, Absam 01688
Gesellschaft der Musikfreunde in Wien, Wien 02994
Harmonikazentrum, Graz 01974
Haydn-Geburtshaus, Rohrau 02594
Haydn-Haus Eisenstadt, Eisenstadt 01860
Haydnhaus, Wien 03000
Heimatmuseum Oberwölz und Österreichisches Blasmusikmuseum, Oberwölz 02466
Hugo Wolf-Haus, Perchtoldsdorf 02487
Ignaz J. Pleyel Museum, Großweikersdorf 02012
Johann Michael Haydn-Museum, Salzburg 02616
Johann Strauß Wohnung, Wien 03007
Johannes Brahms-Museum, Mürzzuschlag 02412
K-HOF Kammerhof Museen Gmunden, Gmunden 01951
Komponierstube Gustav Mahlers, Steinbach am Attersee 02765
Lehár-Villa, Bad Ischl 01754
Mozart Wohnhaus, Salzburg 02618
Mozarthaus St.Gilgen, Sankt Gilgen 02648
Mozarthaus Vienna, Wien 03028
Mozarts Geburtshaus, Salzburg 02619
Museum des Wiener Männergesang-Vereins, Wien 03034
Museum Franz Schubert und sein Freundeskreis, Atzenbrugg 01736
Museum mechanische Klangfabrik, Haslach 02057
Musikinstrumente-Museum der Völker, Sankt Gilgen 02650
Musikinstrumentenmuseum Schloss Kremsegg, Kremsmünster 02234
Rudolf Weinwurm-Museum, Göpfritz 01954
Sammlung alter Musikinstrumente, Wien 03063
Schönberg-Haus in Mödling, Mödling 02397
Schubert Geburtshaus, Wien 03072
Schubert Sterbewohnung, Wien 03073
Stille Nacht-Museum, Hallein 02047
Stoani Haus der Musik, Gasen 01928
Wilhelm-Kienzl-Museum, Paudorf 02479
Wilhelm Kienzl-Stüberl, Waizenkirchen 02890

Azerbaijan

Bul-bul Memorial Museum, Bakı 03148
Bul-bul Museum-Flat, Bakı 03149
House-Museum of Uzeyir Hajibeyov, Bakı 03156
Museum-Appartment of Niyazi, Bakı 03163
State Museum of Azerbaijan Musical Culture, Bakı 03174

Belgium

AMVC-Letterenhuis, Antwerpen 03280
Harmonium-Art-Museum, Schelle 04001
Jukeboxmuseum, Menen 03867
Maison de la Pataphonie, Dinant 03557
Maison du Jazz de Liège de la Communauté Française, Liège 03801
Musée des Instruments de Musique/Muziekinstrumentenmuseum, Bruxelles 03477
Musée Grétry, Liège 03814
Musée Monopoli, Barsy-Flostoy 03340
Muziekinstrumenten-Museum, Peer 03948
Peter Benoitmuseum, Harelbeke 03679
Stedelijk Museum Aarschot, Aarschot 03260
Studio Eugène Ysaye, Liège 03817
Vleeshuis, Antwerpen 03318

Bolivia

Museo de Instrumentos Musicales de Bolivia, La Paz 04214

Brazil

Carlos Gomes Museum, Campinas 04384
Centro de Pesquisas Folclóricas, Rio de Janeiro 04687
Museu Clara Nunes, Caetanópolis 04379
Museu da Imagem e do Som, Rio de Janeiro 04716
Museu Villa-Lobos, Rio de Janeiro 04764

Bulgaria

Etnografska Ekspozicija, Pazardžik 05107

Burkina Faso

Musée de la Musique de Ouagadougou, Ouagadougou 05278

M

Canada

Bathurst Heritage Museum Cultural Centre / Musée de Bathurst/Musée Centre Culturel, Bathurst 05397
Hank Snow Country Music Centre, Liverpool 06177
Musée de l'Accordéon, Montmagny 06325
Site Mary-Travers dite - La Bolduc, Newport 06440

China

Kûnqû Bówùguǎn, Shanghai 08170
National Museum of Music, Beijing 07558
Sound of the Century Museum, Macau 08007

Colombia

Colleccíon de Instrumentos Musicales "Monsenór José Ignacio Perdomo Escobar", Bogotá 08551
Museo de la Parroquia de Gámbita, Gámbita 08645
Museo Organológico Folklórico Musical, Bogotá 08593

Croatia

Zbirka Umjetnina Kairos, Trogir 08973

Cuba

Casa de Miguel Failde Pérez, Matanzas 09212
Museo de la Música Alejandro García Caturla, Santa Clara 09295
Museo de la Música Rodrigo Prats, Santa Clara 09296
Museo Nacional de la Música, La Habana 09166

Czech Republic

České Muzeum Hudby - Oddělení Hudebních Nástrojů a Hudebně Historické Oddělení, Praha 09736
České Muzeum Hudby - Zámek Hořovice, Hořovice 09527
Dům Gustava Mahlera, Jihlava 09560
Dům Pánů z Kunštátu, Brno 09424
Muzeum Antonína Dvořáka, Praha 09767
Muzeum Bedřicha Smetany, Praha 09768
Muzeum Středniho Pootavi, Strakonice 09849
Památník Antonína Dvořáka, Rožmitál pod Třemšínem 09826
Památník Antonína Dvořáka, Zlonice 09943
Památník Antonína Dvořáka, Nelahozeves 09672
Památník Bedřicha Smetany, Jabkenice 09544
Památník Josefa Suka, Křečovice 09607
Památník Leoše Janáčka, Hukvaldy 09540
Památník Leoše Janáčka, Brno 09438
Památník W.A. Mozarta a Manželů Duškových, Praha 09787
Starý Zámek, Jevišovice 09558
Zámek Chropyně, Chropyně 09484

Denmark

Carl Nielsen Museet, Odense 10210
Carl Nielsens Barndomshjem, Årslev 09968
Musikhistorisk Museum og Carl Claudius' Samling, København 10142

Ecuador

Museo de Instrumentos Musicales Pedro Pablo Traversari, Quito 10440

Estonia

Andres Särevi Kortermuuseum, Tallinn 10620
August Kitzbergi Tubamuuseum, Karksi 10581
Eesti Teatri- ja Muusikamuuseum, Tallinn 10628

Finland

Ainola, Järvenpää 10863
Heureka - The Finnish Science Centre, Vantaa 11446
Sibelius Museum, Turku 11397
Suomen Kansansoitinmuseo, Kaustinen 10933

France

Centre de Patrimoine de Facture Instrumentale, Le Mans 13605
Maison de la Musique Mécanique, Mirecourt 14133
Maison du Luthier-Musée, Jenzat 13269
Maison Satie, Honfleur 13220
Moulin à Musique Mécanique, Mormoiron 14279
Musée Atelier d'Automates, Lusigny-sur-Barse 13869
Musée Claude Debussy, Saint-Germain-en-Laye 15289
Musée de la Musique, Anduze 11599
Musée de la Musique, Paris 14656
Musée de la Musique et des Instruments, Blangy-sur-Bresle 12014
Musée de la Musique Mécanique, Haubourdin 13187
Musée de la Musique Mécanique, Mormoiron 14281
Musée de l'Accordéon, Siran 15761
Musée de l'Art Forain et de la Musique Mécanique, Conflans-en-Jarnisy 12611
Musée de Rouget de Lisle, Lons-le-Saunier 13816
Musée des Automates de Grenoble, Grenoble 13137
Musée des Flûtes du Monde, Bollène 12050
Musée des Instruments de Musique à Vent, La Couture-Boussey 13356
Musée des Musiques Populaires, Montluçon 14221
Musée du Phonographe et de la Musique Mécanique, Sainte-Maxime 15571
Musée du Phonographe et de Musique Mécanique, Limoges 13768
Musée Edith Piaf, Paris 14705
Musée Hector Berlioz, La Côte-Saint-André 13351
Musée Mado Robin, Yzeures-sur-Creuse 16315
Musée Magic de la Musique Mécanique et de la Machine Parlante, Mesnil-Raoul 14096
Musée Maurice Ravel, Montfort-l'Amaury 14209
Musée-Placard d'Erik Satie, Paris 14737
Musée Pyrénéen, Lourdes 13852

Georgia

State Museum of Georgian National Music Instruments, Tbilisi 16382
State Museum of Theatre, Music and Cinema, Tbilisi 16383

Germany

Albert-Schweitzer-Gedenk- und Begegnungsstätte, Weimar 22395
Archiv und Kollektion Paul Kaiser-Reka, Brandenburg an der Havel 17506
Bach-Museum, Leipzig 19898
Bachhaus Eisenach, Eisenach 18093
Beatles Museum, Halle, Saale 18889
Beatles-Museum Little Cavern, Biebelnheim 17329
Beatles Museum Remember, Gemmerich 18581
Beethoven-Haus Bonn, Bonn 17448
Brahms-Museum, Hamburg 18917
Brahmshaus, Heide 19065
Carl-Maria-von-Weber-Museum, Dresden 17940
Carl Orff Museum, Dießen am Ammersee 17851
Deutsches Harmonikamuseum, Trossingen 22110
Deutsches Musikautomaten Museum, Bruchsal 17594
Deutsches Phonomuseum und Uhrensammlung, Sankt Georgen im Schwarzwald 21492
Ellermeiers Burgmannshof, Hardegsen 19010
Elztalmuseum - Regionalgeschichte und Orgelbau, Waldkirch 22297
Erlebniswelt Blockflöte, Fulda 18507
Franz-Liszt-Museum der Stadt Bayreuth, Bayreuth 16951
Gebrüder-Lachner-Museum, Rain 21168
Geigenbau-Museum, Bubenreuth 17613
Geigenbaumuseum, Mittenwald 20342
Grassi-Museum für Musikinstrumente der Universität Leipzig, Leipzig 19912
Harmonikamuseum Zwota, Zwota 22844
Heinrich-Schütz-Haus, Bad Köstritz 16765
Heinrich-Schütz-Haus Weißenfels, Weißenfels 22456
Historisches Museum für Mittelanhalt und Bach-Gedenkstätte im Schloss Köthen, Köthen 19666
Internationales Artistenmuseum in Deutschland, Klosterfelde 19581
Kultur- und Orgelzentrum, Valley 22202
Liszt-Haus, Weimar 22409
Louis Spohr-Gedenk- und Forschungsstätte, Kassel 19468
Mauersberger-Museum, Großrückerswalde 18804
Max-Reger-Gedächtniszimmer, Brand 17503
Mechanischer Musiksalon, Weiskirchen 22443
Mecklenburgisches Orgelmuseum, Malchow 20155
Mozarthaus, Augsburg 16630
Museum Bürgstadt, Bürgstadt 17639
Museum der Stadt Füssen, Füssen 18501
Museum für Design der 60er 70er Jahre im PopDom, Köln 19620
Museum für Musikinstrumente, Kürnbach 19740
Museum Gottfried Silbermann, Frauenstein 18413
Museum im Mendelssohn-Haus, Leipzig 19924
Musik- und Wintersportmuseum, Klingenthal 19572
Musikhistorische Sammlung Jehle, Albstadt 16456
Musikinstrumenten-Ausstellung des Händel-Hauses, Halle, Saale 18902
Musikinstrumenten-Museum, Markneukirchen 20207
Musikinstrumenten-Museum, Berlin 17221
Musikinstrumenten- und Puppenmuseum, Goslar 18715
Musikinstrumentenmuseum Lißberg, Ortenberg-Lißberg 20915
Musikinstrumentensammlung, Würzburg 22719
Musikinstrumentensammlung der Universität, Göttingen 18696
Musikinstrumentensammlung Hans und Hede Grumbt, Hattingen 19039
Orgelbaumuseum, Ostheim vor der Rhön 20940
Orgelbaumuseum, Weißenfels 22459
Orgelmuseum Borgentreich, Borgentreich 17477
Orgelmuseum Kelheim, Kelheim 19501
Pianoforte-Museum, Meißen 20266
Private Sammlung Mechanischer Musikinstrumente, Markneukirchen 20208
Reuter-Villa mit Wagner-Sammlung, Eisenach 18096
Richard-Wagner-Museum, Bayreuth 16965
Richard-Wagner-Museum Graupa, Pirna 21045
Robert-Schumann-Haus Zwickau, Zwickau 22835
rock'n'pop museum, Gronau, Westfalen 18774
Sängermuseum, Feuchtwangen 18315
Sammlung historischer Tasteninstrumente, Wassenach 22342
Sammlung historischer Tasteninstrumente Neumeyer-Junghanns-Tracey, Bad Krozingen 16770
Schlossmuseum Arnstadt, Arnstadt 16575
Schumann-Haus, Leipzig 19934
Schwarzwald-Museum, Triberg im Schwarzwald 22092
Siegfrieds Mechanisches Musikkabinett, Rüdesheim am Rhein 21437
Silcher-Museum des Schwäbischen Chorverbandes, Weinstadt 22441
Stadt- und Heimatmuseum Kusel, Kusel 19752
Stadtmuseum im Spital, Crailsheim 17738
Stadtmuseum Lindau, Lindau 20001
Stiftung Kloster Michaelstein-Museum, Blankenburg 17374
Technik Museum Speyer, Speyer 21838
Theatermuseum Zauberwelt der Kulisse, Meiningen 20262
Trompetenmuseum, Bad Säckingen 16818
Werner-Egk-Begegnungsstätte, Donauwörth 17900
Westpfälzer Musikantenmuseum, Mackenbach 20120

Greece

Mouseio Laikon Organon, Athinai 22924
Museum of Greek Musical Instruments, Thessaloniki 23219

Hungary

Bartók Béla Emlékház, Budapest 23387
Beethoven Emlékmúzeum, Martonvásár 23691
Erkel Ferenc Emlékház, Gyula 23593
Erkel Ferenc Múzeum, Gyula 23594
Helytörténeti Kiállítás és J. Haydn Emlékszoba, Fertőd 23558
Kálmán Imre Emlékmúzeum, Siófok 23803
Leskowsky Hangszergyűjtemény Közalapítvány, Kecskemét 23649
Liszt Ferenc Emlékmúzeum és Kutatóközpont, Budapest 23431
Magyar Állami Operaház Múzeuma, Budapest 23433
Zenetörténeti Múzeum, Budapest 23494

India

TTD Museum, Tirupati 24409

Iran

Sabā's House, Tehrān 24693

Ireland

Ceol - Irish Traditional Music Centre, Dublin 24780
Irish Music Hall of Fame, Dublin 24802

Israel

Music Box of Zami Museum, Metulla 25065
Musical Instruments Museum, Jerusalem 25006

Italy

Biblioteca-Museo Alessandro Longo, Amantea 25219
Casa Museo dei Puccini, Pescaglia 27160
Casa Natale di Giuseppe Verdi, Busseto 25568
Casa Natale di Toscanini, Parma 27092
Casa Natale Gaetano Donizetti, Bergamo 25399
Civico Museo Bibliografico Musicale, Bologna 25438
Civico Museo degli Strumenti Musicali, Milano 26680
Collezione di Violini, Cremona 26001
Donazione Raffaele Vaccari, Brescello 25531
Galleria dell'Accademia, Firenze 26159
House of Music Parma, Parma 27097
Museo Alfieriano, Asti 25313
Museo B. Gigli, Recanati 27376
Museo Casa Busoni, Empoli 26052
Museo Casa Natale Giacomo Puccini, Lucca 26532
Museo Casa Pascoli, Barga 25359
Museo Casa Rossini, Pesaro 27155
Museo Civico Belliniano, Catania 25772
Museo degli Strumenti Musica Populare Sardi, Tadasuni 27934
Museo degli Strumenti Musicali, Cesena 25826
Museo degli Strumenti Musicali Meccanici, Savio 27772
Museo degli Strumenti Musicali Meccanici, Sestola 27827
Museo del Conservatorio die Napoli, Napoli 26888
Museo del Jazz G. Dagnino, Genova 26300
Museo del Pianoforte Antico, Ala 25169
Museo della Canzone e della Riproduzione Sonora, Vallecrosia 28147
Museo della Liuteria, Arpino 25284
Museo della Pipa, Gavirate 26278
Museo della Zampogna, Scapoli 27785
Museo delle Musiche dei Popoli, Genova 26305
Museo dello Strumento Musicale, Reggio Calabria 27382
Museo di Strumenti del Conservatorio Statale di Musica Giuseppe Verdi, Torino 28016
Museo Donizettiano, Bergamo 25404
Museo Etnografico Francesco Bande, Sassari 27754
Museo Etnomusicale I Gigli di Nola, Nola 26937
Museo Francesco Cilea e Nichola Antonio Manfroce, Palmi 27086
Museo G. Spontini, Maiolati Spontini 26570
Museo Internazionale della Fisarmonica, Castelfidardo 25724
Museo Musicale d'Abruzzo, Ortona 27003
Museo Nazionale degli Strumenti Musicali, Roma 27513
Museo Organologico Didattico, Cremona 26005
Museo Pucciniano, Torre del Lago 28037
Museo Storico della Filarmonica Sestrese, Genova 26319
Museo Stradivariano, Cremona 26006
Raccolta di Fisarmoniche d'Epoca, Camerano 25614
Tempietto Rossiniano della Fondazione Rossini, Pesaro 27158
Villa Museo Puccini, Viareggio 28245
Villa Verdi, Villanova sull'Arda 28278

Jamaica

Bob Marley Museum, Kingston 28324

Japan

Arsbel Tendo O-tomachikku Gakki Hakubutsukan, Tendo 29356
Aso-gun Kazan Hakubutsukan-Orugoru Kyouwakoku, Aso 28378
Drum Museum Taikokan, Tokyo 29383
Gakki Hakubutsukan, Tokyo 29388
Gakkigaku Shiryôkan, Tachikawa 29307
Hakodate-shi Meijikan, Hakodate 28473
Hamamatsu-shi Gakki Hakubutukan, Hamamatsu 28483
Hamana-ko Orugorukan, Hamamatsu 28487
Horie Orugorukan, Nishinomiya 29073
Izu Orugorukan, Ito 28594
Izumi-mura Rekishi no Sato, Izumi, Fukui 28612
John Lennon Museum, Saitama 29187
Koga Masao Museum of Music, Tokyo 29414
Kyoto Arashiyama Orugoru Hakubutsukan oyo Guido Reuge Hakubutsukan, Kyoto 28834
Marie Sekei Tokei oyo Orugoru Hakubutsukan, Inuyama 28577
Matsushima Orugoru Hakubutsukan, Matsushima 28902
Miyagi Michio Kinenkan, Tokyo 29438
Moji-ko Orugorukan, Kitakyushu 28714
Mori-no-Naka no Chiisana Ongakukan, Nikko 29067
Museum of Contemporary Japanese Poetry, Tanka and Haiku, Kitakami 28710
Musical Boxes Museum, Kobe 28737
Nasu Orugoru Hakubutsukan, Nasu 29049
Oikawa Museum of Sound Instruments, Higashi-kurume 28504
Orugoru no Chiisana Hakubutukan, Tokyo 29456
Otaru Music Boxes Museum Antique Museum, Otaru 29160
Sendai Music Box Museum, Sendai 29245
Suwa-ko Somei-kan (Orugoru Hakubutsukan), Shimosuwa 29272
Takehisa Yumeji Ikaho Kinenkan, Ikaho 28561
Ukai Museum, Kawaguchi-ko 28686

Kazakhstan

Muzej Narodnych Muzykalnych Instrumentov Kazachstana im. Ychlasa, Almaty 29689

Korea, Republic

Chamsori Gramophone and Audio Science Museum, Gangneung 29806
Museum of Korean Traditional Music, Seoul 29907

Latvia

Brāļu Jurjānu Memoriālais Muzejs Meņģeļi, Ērgļi 30007
Krišjāņa Barona Muzejs, Rīga 30073
Mūzikas Muzejs, Rīga 30098
Rakstniecības un Mūzikas Muzejs, Rīga 30104

Lithuania

Beatričės Grincevičiūtės Memorialinis Muziejus, Beatričės Namai, Vilnius 30402
Juozo Gruodžio Memorialinis Muziejus, Kaunas 30243

Juozo Naujalio Memorialinis Muziejus, Raudondvaris 30343
Kompozitoriaus Antano Budriūno Kambarys, Vilnius 30413
Lietuvos Teatro, Muzikos ir Kino Muziejus, Vilnius 30422
Miko ir Kipro Petrauskų Lietuvių Muzikos Muziejos, Kaunas 30260
Muzikų Brolių Motiejaus, Broniaus ir Antano Budriūnų Gimtinė, Pabiržė 30324
P. Stulgos Lietuvių Tautinės Muzikos Muziejus, Kaunas 30263
Stasio Vainiūno Namai-Muziejus, Vilnius 30433

Mexico

Guitar Museo Ex Teresa Arte Actual, México 31048
Museo Casita Blanca de Agustín Lara, Boca del Río 30737
Museo de la Canción Yucateca, Mérida 31026
Museo de la Casa de la Música Mexicana, México 31079
Museo de la Música Grabada, Zapotlanejo 31594
Museo Silvestre Vargas, Tecalitlán 31411

Morocco

Musée des Arts Sahariens de Laâyoune, Laâyoune 31674
Nejarine-Musée, Fès 31672

Netherlands

Accordeon en Harmonica Museum, Malden 32536
Christ Boelens Jukeboxen Museum, Sint Oedenrode 32775
Draaiorgelmuseum, Assen 31936
Kijk en Luister Museum, Bennekom 31962
Museum Musica, Stadskanaal 32798
Muziekinformatie- en documentatiecentrum Ton Stolk, Vlaardingen 32890
Muziekinstrumentenmakersmuseum, Tilburg 32827
Nationaal Historisch Orgelmuseum Elburg, Elburg 32199
Nationaal Museum van Speelklok tot Pierement, Utrecht 32851
Pianola Museum, Amsterdam 31879

Nigeria

National Museum Ibadan, Ibadan 33323

Norway

Edvard Grieg Museum, Paradis 33766
Jacob Breda Bullmuseet, Rendalen 33778
Kirsten Flagstad Museum, Hamar 33529
Myllarheimen, Rauland 33777
Prøysenhuset, Rudshøgda 33801
Ringve Museum, Trondheim 33927
Siljustøl Museum, Paradis 33768

Philippines

Sala Piano Museum, Cebu 34373

Poland

Biblioteka, Muzeum i Archiwum Warszawskiego Towarzystwa Muzycznego im. Stanisława Moniuszki, Warszawa 35216
Dom Urodzenia Fryderyka Chopina, Sochaczew 35117
Muzeum Feliksa Nowowiejskiego, Barczewo 34535
Muzeum Fryderyka Chopina, Warszawa 35242
Muzeum Instrumentów Muzycznych, Poznań 35037
Muzeum Ludowych Instrumentów Muzycznych, Szydłowiec 35174
Muzeum Piśmiennictwa i Muzyki Kaszubsko-Pomorskiej, Wejherowo 35296
Salon Muzyczny im. Fryderyka Chopina, Antonin 34529
Salonik Chopinów, Warszawa 35288

Portugal

Museu da Música, Lisboa 35682
Museu da Música Portuguesa, Estoril 35575
Museu da Sociedade Imparcial 15 de Janeiro de 1898, Alcochete 35396
Museu de Instrumentos Musicais do Conservatório Nacional, Lisboa 35692
Museu do Fado e da Guitarra Portuguesa, Lisboa 35700
Museu dos Cordofones, Tebosa 35961
Museu Fundação Amalia Rodrigues, Lisboa 35709

Puerto Rico

Museo Casa Paoli, Ponce 36047
Museo de la Música Puertorriqueña, Ponce 36051
Museo Pablo Casals, San Juan 36074

Romania

Casa Memorială Dinu Lipatti, Leordeni 36439
Casa Memorială George Enescu, Liveni 36445
Casa Memorială George Enescu, Sinaia 36618
Casa Memorială Gheorghe Dima, Braşov 36174
Casa Memorială Ionel Perlea, Ograda 36498
Casa Muzicii, Lugoj 36447
Complexul Muzeistic Ciprian Porumbescu, Ciprian Porumbescu 36280
Expoziţia Memorială Béla Bartók, Sânnicolau Mare 36586
Expoziţia Memorială Dimitrie Cantemir, Dimitrie Cantemir 36346
Muzeul Memorial George Enescu, Dorohoi 36351
Muzeul Memorial Paul Constantinescu, Ploieşti 36535
Muzeul Naţional George Enescu, Bucureşti 36224

Russia

Častnyj Muzej Grammofonov i Fonografov, Sankt-Peterburg 37626
Dom-Muzej T.N. Chrenikova, Elec 36907
Dom-Muzej S.V. Rachmaninova, Ivanovka 36943
Dom-muzej F.I. Šaljapina, Moskva 37167
Gosudarstvennaja Kollekcija Unikalnych Muzykalnych Instrumentov, Moskva 37183
Gosudarstvennyj Centralnyj Muzej Muzykalnoj Kultury im. M.I. Glinki, Moskva 37188
Gosudarstvennyj Dom-muzej P.I. Čajkovskogo, Klin 37052
Gosudarstvennyj Memorialnyj Dom-muzej N.A. Rimskogo-Korsakova, Tichvin 37841
Gosudarstvennyj Muzej-usadba P.I. Čajkovskogo, Votkinsk 37977
Kamernyj Šaljapinskij Zal, Kazan 37008
Kraevoj Muzej Istorii Literatury, Iskusstva i Kultury Altaja, Barnaul 36791
Literaturno-muzykalnyj Muzej Dača Šaljapina, Kislovodsk 37047
Memorialnyj Dom-muzej L.V. Sobinova, Jaroslavl 36970
Memorialnyj Muzej I.S. Ključnikova-Palantaja, Joškar-Ola 36977
Memorialnyj Muzej-kvartira E.F. Gnesinoj, Moskva 37243
Memorialnyj Muzej-kvartira Vadima Alekseeviča Kozina, Magadan 37120
Memorialnyj Muzej-kvartira Rimskogo-Korsakova, Sankt-Peterburg 37663
Memorialnyj Muzej A.N. Skrjabina, Moskva 37247
Memorialnyj Muzej-usadba M.P. Musorgskogo, Naumovo 37447
Memorialnyj Muzej-zapovednik N.A. Rimskogo-Korsakova Ljubensk-Večaša, Pljussa 37569
Muzej Igorja Talkova, Moskva 37285
Muzej Istorii Peterburgskoj Konservatorii, Sankt-Peterburg 37682
Muzej-kvartira Dirižëra N.S. Golovanov, Moskva 37315
Muzej-kvartira Naziba Žiganova, Kazan 37021
Muzej-kvartira Svjatoslava Richtera, Moskva 37322
Muzej-kvartira F.I. Šaljapina, Sankt-Peterburg 37695
Muzej Muzykalnoj i Teatralnoj Kultury na Kavkazskich Mineralnyh Vodach, Kislovodsk 37049
Muzej Muzykalnych Instrumentov, Sankt-Peterburg 37700
Muzej S.S. Prokofjeva, Moskva 37350
Muzej N.G. Rubinštejna - Muzej Moskovskoj Konservatorii, Moskva 37354
Muzej Russkoj Garmoniki Alfred Mireka, Moskva 37357
Muzej Russkoj Pesni im. A. Averkina, Sasovo 37755
Muzej Salicha Sajdaševa, Kazan 37023
Muzej S.I. Taneeva v Djutkovo, Zvenigorod 37990
Muzej-usadba M.I. Glinki, Novospasskoe 37490
Muzyka i Vremja - Častnyj Muzej Džona Mostoslavskogo, Jaroslavl 36974
Sankt-Peterburgskij Muzej Teatralnogo i Musykalnogo Iskusstva, Sankt-Peterburg 37725
Šeremetevskij Dvorec - Fontannyj Dom, Sankt-Peterburg 37726

Slovakia

Dom Hudby Mikuláša Schneidera-Trnavského, Trnava 38378
Hudobná Expozícia - Rodný Dom J.N. Hummela, Bratislava 38215
Hudobné Múzeum, Bratislava 38216
Literárne a Hudobné Múzeum, Banská Bystrica 38183
Nábytkové Múzeum, Markušovce 38292
Pamätná Izba Franza Schuberta, Želiezovce 38371
Pamätná Izba Frica Kafendu, Mošovce 38310
Pamätník Ludwiga van Beethovena, Dolná Krupá 38248

Slovenia

Muzej Ipavčeva Hiša, Šentjur 38510
Spominska Soba Giuseppe Tartini, Piran 38493

South Africa

National Music Museum, Bloemfontein 38568

Spain

Casa Museo Julián Gayarre, Roncal 39913
Casa-Museo Manuel de Falla, Granada 39359
Casa Nadiua de Pau Casals, El Vendrell 40260
Colección de la Biblioteca Musical, Madrid 39513
Museo de la Gaita, Gijón 39336
Museo del Baile Flamenco, Sevilla 40073
Museo Interactivo de la Música, Málaga 39602
Museu de la Música, Barcelona 38980
Museu Pau Casals, El Vendrell 40263

Sweden

C.M. Bellman Museum, Stockholm 40888
Carl Jularbo Museum, Avesta 40449
Hobby- och Leksaksmuseum, Stockholm 40900
Jazzens Museum, Kolbäck 40650
Jussi Björlingmuseet, Borlänge 40476
Musikmuseet, Stockholm 40917
Smålands Bil, Musik och Leksaksmuseum, Rydaholm 40818
Stiftelsen Musikkulturens Främjande, Stockholm 40938

Switzerland

Ausstellung der Geigenbauschule Brienz, Brienz (Bern) 41254
Blasinstrumenten-Sammlung, Zimmerwald 42058
Fredy's Mechanisches Musikmuseum, Lichtensteig 41558
Harmonium-Museum, Liestal 41562
Mechanisches Musikmuseum, Sursee 41924
Musée Baud, L'Auberson 41134
Musée de Boîtes à Musique et Automates, Sainte-Croix 41796
Musée de Musiques Méchanique, L'Auberson 41135
Musée Paderewski, Morges 41645
Musée Suisse de l'Orgue, Roche 41769
Museo Leoncavallo, Brissago 41257
Museo Rainis e Aspazija, Castagnola 41284
Museum für Musikautomaten, Seewen 41857
Museum für Uhren und mechanische Musikinstrumente, Oberhofen am Thunersee 41696
Musikmuseum, Basel 41174
Nostalgisches Musikparadies, Oberhasli 41694
Richard Wagner-Museum, Luzern 41597
Schweizer Jazzmuseum, Uster 41967
Schweizer Zither-Kultur-Zentrum, Trachselwald 41946
Streichinstrumentensammlung, Einsiedeln 41346
Trummlehus, Langenthal 41528

Taiwan

Music Information Museum, Kaohsiung 42240
Wei Chin-Yuan Musical Instruments Museum, Taipei 42439

Tajikistan

Museum od Musical Instruments, Dušanbe 42495

Tunisia

Musée des Musiques Arabes et Méditerranéennes Ennejma Ezzahra, Sidi Bou Saïd 42654

Turkey

Aynalıkavak Kasrı, İstanbul 42806
Divan Edebiyatı Müzesi, İstanbul 42815

Ukraine

Deržavnyj Muzej Teatral'noho, Muzyčnoho ta Kinomystectva Ukraïny, Kyïv 43077
Kirovograds'kyj Muzej Musyčnoï Kul'tury im. Karolja Šymanovs'kogo, Kirovograd 43066
Literaturno-memorialnyj Muzei O.S. Puškina ta P.I. Čajkovskogo, Kam'janka 43056
Muzej Igora Stravinskogo, Ustyloog 43195
Muzej M.V. Lisenka, Kyïv 43095
Muzej S.S. Prokofjeva, Krasne 43074
Muzej Vydatnych Dijačiv Ukraïns'koï Kul'tury, Kyïv 43099
Muzyčno-memorialnyj Muzej Solomiï Krušel'nic'koï, Lviv 43136

United Kingdom

Bate Collection of Musical Instruments, Oxford 45328
The Beatles Story, Liverpool 44731
Dean Castle, Kilmarnock 44578
The Elgar Birthplace Museum, Lower Broadheath 45038
Fenton House, London 44847
Finchcocks, Goudhurst 44339
Handel House Museum, London 44871
The Herschel Museum of Astronomy, Bath 43404
Holst Birthplace Museum, Cheltenham 43811
Horniman Museum, London 44878
Joseph Parry's Ironworkers Cottage, Merthyr Tydfil 45131
Keith Harding's World of Mechanical Music, Northleach 45271
Lennon House, Liverpool 44737
McCartney House, Liverpool 44739
Mechanical Music and Doll Collection, Chichester 43829
Mechanical Music Museum and Bygones, Cotton 43897
Morpeth Chantry Bagpipe Museum, Morpeth 45177
Museum of Instruments, London 44927
Museum of Piping, Glasgow 44291
Museum of the Royal Military School of Music, Twickenham 45887
Museum of Victorian Reed Organs and Harmoniums, Shipley 45666
Musical Museum, London 44934
Piano Collection, Hereford 44458
Reid Concert Hall Museum of Instruments, Edinburgh 44147
Saint Albans Organ Museum, Saint Albans 45549
Saint Cecilia's Hall Museum of Instruments, Edinburgh 44153

United States of America

Alabama Music Hall of Fame, Tuscumbia 53623
American Classical Music Hall of Fame and Museum, Cincinnati 47552
American Jazz Museum, Kansas City 49682
America's National Music Museum, Vermillion 53719
Bagpipe Music Museum, Ellicott City 48365
The Band Museum, Pine Bluff 51921
Barrett House, Forest Hall, New Ipswich 51065
Boston Public Library, Boston 46923
Boswell Museum of Music, East Springfield 48259
Buddy Holly Center, Lubbock 50310
Center for American Music, Stephen Foster Memorial, Pittsburgh 51945
Cincinnati Art Museum, Cincinnati 47555
Conklin Reed Organ and History Museum, Hanover 49161
Country Music Hall of Fame and Museum, Nashville 50964
Deer Valley Rock Art Center, Glendale 48927
Delta Blues Museum, Clarksdale 47587
Edward Hopper House Art Center, Nyack 51449
Eubie Blake National Jazz Museum and Cultural Center, Baltimore 46563
Experience Music Project, Seattle 52998
Fender Museum of Music and the Arts, Corona 47816
Fort Morgan Museum, Fort Morgan 48659
Georgia Music Hall of Fame, Macon 50365
Graceland, Memphis 50573
Hollywood Bowl Museum, Los Angeles 50230
International Bluegrass Music Museum, Owensboro 51638
International Gospel Music Hall of Fame and Museum, Detroit 48121
Jimmie Rodgers Museum, Meridian 50599
Kenneth G. Fiske Museum, Claremont 47572
Liberace Museum, Las Vegas 49974
Lockwood-Mathews Mansion Museum, Norwalk 51432
Louisiana Country Music Museum, Marthaville 50515
Mario Lanza Museum, Philadelphia 51850
Maxwell Museum of Anthropology, Albuquerque 46181
Motown Historical Museum, Detroit 48125
Museum of Fife and Drum, Ivoryton 49553
Museum of Making Music, Carlsbad 47247
Museum of the American Piano, New York 51230
The Music House Museum, Acme 46132
Musical Box Society International Museum, Houston 49399
Musical Wonder House, Wiscasset 54263
New Braunfels Museum of Art Music, New Braunfels 51024
Rock 'n Roll Blues Heritage Museum, Clarksdale 47588
Rock 'n' Roll Hall of Fame and Museum, Cleveland 47638
Sarasota Classic Car Museum, Sarasota 52915
Schubert Club Museum of Musical Instruments, Saint Paul 52628
Stearns Collection of Musical Instruments, Ann Arbor 46311
Stephen Foster Folk Culture Center, White Springs 54129
Texas Music Museum, Austin 46521
W.C. Handy Home Museum and Library, Florence 48566
Yale University Collection of Musical Instruments, New Haven 51057
Yesteryear Museum, Pamplin 51691

Uruguay

Museo Romántico - Casa de Montero, Montevideo 54417

Uzbekistan

Mouhtar Ashrafi Yodgorlik Muzeyi, Toshkent 54474

Naive Art → Folk Art

Natural History

Albania

Muzeu i Shkencave Natyrore, Tiranë 00033

Algeria

History and Natural History Museum, Alger 00038

Angola

Museu Nacional de História Natural, Luanda 00113

Argentina

Museo Argentino de Ciencias Naturales Bernardino Rivadavia, Buenos Aires 00168
Museo de Antropología y Ciencias Naturales, Concordia 00315
Museo de Ciencias Antropológicas y Naturales, Santiago del Estero 00628
Museo de Ciencias Naturales, Coronel Pringles 00343
Museo de Ciencias Naturales Augusto G. Schulz, Resistencia 00522
Museo de Ciencias Naturales Bartolomé Mitre, Córdoba 00324
Museo de Ciencias Naturales de Oberá, Oberá 00479
Museo de Ciencias Naturales del Departamento San Cristóbal, San Cristóbal 00570
Museo de Ciencias Naturales e Historia, Posadas 00498
Museo de Ciencias Naturales Florentino Ameghino, Las Flores 00424
Museo de Ciencias Naturales Rvdo. P. Antonio Scasso, San Nicolás de los Arroyos 00603
Museo de Ciencias Naturales y Antropológicas Juan Cornelio Moyano, Mendoza 00453
Museo de Ciencias Naturales y Antropológicas Prof. Antonio Serrano, Paraná 00484
Museo de Ciencias Naturales y Misional del Colegio San José, Esperanza 00366
Museo de Historia Natural Dr. Ricardo S. Vadell, Suipacha 00636
Museo de la Facultad de Ciencias Naturales y del Instituto Miguel Lillo, San Miguel de Tucumán 00598
Museo de la Región de Ansenuza Aníbal Montes, Miramar 00467
Museo de la Selva Juan Forster, Miramar 00468
Museo del Agua y del Suelo, Viedma 00669
Museo del Parque Nacional Los Alerces, Villa Futalaufquen 00679
Museo El Hombre y la Naturaleza, San Juan 00587
Museo Histórico y Natural de Lavalle, Lavalle Villa Tulumaya 00427
Museo Interactivo de Ciencias Naturales, Ituzaingó 00392
Museo Jorge Pasquini López, San Salvador de Jujuy 00609
Museo Marcelo López del Instituto Hellen Keller, Córdoba 00333
Museo Municipal de Ciencias, Bahía Blanca 00137
Museo Municipal de Ciencias Naturales Carlos Ameghino, Mercedes 00463
Museo Municipal de Ciencias Naturales Carlos Darwin, Punta Alta 00509
Museo Municipal de Historia Natural, San Rafael 00606
Museo Paleontológico Egidio Feruglio, Trelew 00651
Museo Provincial de Ciencias Naturales Dr. Angelo Gallardo, Rosario 00549
Museo Provincial de Ciencias Naturales Florentino Ameghino, Santa Fé 00621
Museo Provincial de Ciencias Naturales y Oceanográfico, Puerto Madryn 00507
Museo Provincial de Historia Natural de La Pampa, Santa Rosa 00625

Armenia

State Museum of Nature, Yerevan 00755

Australia

Albany Residency Museum, Albany 00785
Australian Museum, Sydney 01546
Lake Broadwater Natural History Museum, Dalby 01021
Macleay Museum, Sydney 01552
Museum of Central Australia, Alice Springs 00799
Museum of Tropical Queensland, Townsville 01598
Papua New Guinea Display Centre, Crafers 01007
Queen Victoria Museum and Art Gallery at Royal Park, Launceston 01224
Rottnest Island Museum, Rottnest Island 01471
South Australian Museum, Adelaide 00782
Western Australian Museum Geraldton, Geraldton 01107

Austria

Burgmuseum Waldaist-Stöckl, Tragwein 02816
Hochschwabmuseum, Sankt Ilgen 02651
Inatura - Erlebnis Naturschau Dornbirn, Dornbirn 01825
Lobaumuseum, Wien 03025
Museum Schloß Lackenbach, Lackenbach 02250
Naturhistorisches Museum, Wien 03043
Naturwissenschaftliche Sammlungen, Innsbruck 02126
Niederösterreichisches Landesmuseum, Sankt Pölten 02685
Oberösterreichische Landesmuseen, Linz 02322

Azerbaijan

Hassanbey Zardabi Museum of Natural History, Bakı 03154

Bahrain

Bahrain National Museum, Manama 03212

Belarus

Beloveżskaja Pušča Muzej, Beloveżskaja Pušča 03245

Belgium

Animalaine, Bastogne 03343
Musée Centre Nature de Botrange, Robertville 03971
Musée de Hottemme, Barvaux 03341
Musée de la Montagne Saint-Pierre, Lanaye 03775
Musée d'Histoire Naturelle de Mons, Mons 03885
Natuurwetenschappelijk Museum, Antwerpen 03307

Belize

Belize Audubon Society, Belize City 04174
Chaa Creek Natural History Centre, San Ignacio 04181
Natural History Museum, Las Cuevas 04179

Bermuda

Bermuda Natural History Museum, Flatts 04190

Bolivia

Museo de Historia Natural, Sucre 04246
Museo de Historia Natural Alcide D'Orbigni, Cochabamba 04207
Museo de Historia Natural Noel Kempf Mercado, Santa Cruz 04240
Museo de Historia Natural Pedro Villalobos, Cobija 04205
Museo Guaraní, Santa Cruz 04241
Museo Nacional de Historia Natural, La Paz 04225

Bosnia and Herzegovina

Zemaljski Muzej Bosne i Hercegovine, Sarajevo 04284

Botswana

Nhabe Museum, Nhabe 04300

Brazil

Centro de Educação Ambiental Gralha Azul e Museu de História Natural, Cascavel 04406
Ecomuseu, Rio de Janeiro 04689
Ecomuseu da Fazenda Boa Vista, Roseira 04781
Ecomuseu do Rabeirão do Ilha, Florianópolis 04452
Laboratário de Ensino de Ciências, Ribeirão Preto 04675
Museu Arquediocesano Dom Joaquim, Brusque 04374
Museu Civico-Religioso Padre Cicero, Juazeiro do Norte 04518
Museu de Ciência e Técnica da Escola de Minas, Ouro Preto 04604
Museu de Ciências Naturais, Guarapuava 04484
Museu de Ciências Naturais, Porto Alegre 04645
Museu de Ciências Naturais, Recife 04665
Museu de Ciências Naturais PUC Minas, Belo Horizonte 04348
Museu de História e Ciências Naturais, Além Paraíba 04303
Museu de História Natural, Belo Horizonte 04349
Museu de História Natural, Campinas 04389
Museu de História Natural Prof. Antonio Pergola, Atibaia 04324
Museu de Paleontologia, Monte Alto 04572
Museu e Biblioteca do Instituto Geográfico e Histórico do Amazonas, Manaus 04556
Museu Nacional, Rio de Janeiro 04755
Museu Paraense Emílio Goeldi, Belém 04341
Museu Parque Malwee, Jaraguá do Sul 04508
Museu Zoobotânico Augusto Ruschi, Passo Fundo 04614
Museus de História Natural e Etnologia Indígena, Juiz de Fora 04523

Brunei

Muzium Brunei, Bandar Seri Begawan 04965
Taman Warisan Tasek Merimbun, Mukim Rambai 04969

Bulgaria

Nacionalen Prirodonaučen Muzej, Sofia 05198
Prirodonaučen Muzej, Burgas 05000
Prirodonaučen Muzej, Kotel 05078
Prirodonaučen Muzej, Plovdiv 05132
Prirodonaučen Muzej, Belogradčik 04984

Canada

B.J. Hales Museum of Natural History, Brandon 05459
Banff Park Museum, Banff 05380
Beaty Biodiversity Museum, Vancouver 07199
Big Beaver Museum and Nature Centre, Big Beaver 05425
Biodôme de Montréal, Montréal 06328
Canadian Museum of Nature, Ottawa 06536
Cape Breton Centre for Heritage and Science, Sydney 07047
Centre d'Interpretation du Parc de l'île-Bonaventure-et-du Rocher-Percé, Percé 06585
Centre d'Interprétation du Patrimoine de Sorel, Sorel 06993
Chatham-Kent Museum, Chatham, Ontario 05602
Écomusée de Hull, Gatineau 05858
Ecomuséum, Sainte-Anne-de-Bellvue 06889
Garden of the Gulf Museum, Montague 06323
George Hooey Wildlife Collection, North Battleford 06467
Grand Coteau Heritage and Cultural Centre, Shaunavon 06949
Henry-Stuart House, Québec 06681
Huntsman Marine Science Centre Aquarium Museum, Saint Andrews 06797
Irvin Goodon International Wildlife Museum, Boissevain 05434
Kluane Museum of Natural History, Burwash Landing 05507
Leslie M. Frost Natural Resources Centre, Minden 06294
Living Prairie Museum, Winnipeg 07366
MacBride Museum, Whitehorse 07325
McLurg Museum, Wilkie 07335
Manitoba Museum, Winnipeg 07370
Manning Park Visitor Centre, Manning Park 06228
Miramichi Natural History Museum, Miramichi 06301
Musée Canadien de la Nature, Gatineau 05860
Musée de la Nature et des Sciences, Sherbrooke, Québec 06961
Musée d'Histoire Naturelle du Parc de Miguasha, Nouvelle 06493
Museum of Natural History, Nanaimo 06411
New Iceland Heritage Museum, Gimli 05866
Nova Scotia Museum, Halifax 05943
Nova Scotia Museum of Natural History, Halifax 05944
Ontario Science Centre, Toronto 07126
Ovens Natural Park and Museum, Riverport 06758
Point Pelee Natural Park Visitor Centre, Leamington 06155
Presqu'île Provincial Park Museum, Brighton 05477
Quetico Provincial Park Heritage Pavilion, Atikokan 05361
Riding Mountain National Park Interpretive Centre, Wasagaming 07281
Rondeau Provincial Park Visitor Centre, Morpeth 06393
Royal Alberta Museum, Edmonton 05752
Royal British Columbia Museum, Victoria 07258
Royal Saskatchewan Museum, Regina 06727
Sam Waller Museum, The Pas 07072
Shuswap Lake Provincial Park Nature House, Kamloops 06039
Terra Nova National Park Museum, Glovertown 05877
Transcona Historical Museum, Winnipeg 07386
Woodland Cultural Centre, Brantford 05471
Wye Marsh Wildlife Centre, Midland 06280

Chad

Musée National, N'Djamena 07424

Chile

Museo de Historia Natural de Concepción, Concepción 07434
Museo de Historia Natural de San Pedro Nolasco, Santiago de Chile 07456
Museo de Historia Natural de Valparaíso, Valparaíso 07471
Museo Dillmann S. Bullock, Angol 07427
Museo Regional Salesiano Maggiorino Borgatello, Punta Arenas 07447

China

Anyang Yinxu Bowu Yuan, Anyang 07489
Bailian Dongdongkong Science Museum, Liuzhou 07965
Chongqing Natural History Museum, Chongqing 07631
Dalian Natural History Museum, Dalian 07639
Dongtinghu Guan, Yuanjiang 08463
Guangxi Natural Museum, Nanning 08054
Jiangxi Provincial Museum, Nanchang 08028
Jilin Provincial Museum of Natural History, Changchun 07586
Lingshan Museum, Qinzhou 08122
Luzhou Bowu Guan, Luzhou 07988
Museum of Natural Resources in Xiaoxinganling, Yichun, Heilongjiang 08442
Sanyuan Bowu Guan, Sanyuan 08153
Science Technology Museum, Beijing 07565
Shanghai Museum of Natural History, Shanghai 08187
Shanghai Science Technology Museum, Shanghai 08189
Tianjin Museum of Natural History, Tianjin 08286
Wuyishan Nature Museum, Wuyishan 08348
Xinping Bowuguan, Xinping 08402
Zhejiang Natural Museum, Hangzhou 07767

Colombia

Museo Casa de la Cultura de Barichara, Barichara 08537
Museo de Ciencias Naturales del Colegio de San José de La Salle, Medellín 08677
Museo de Ciencias Naturales Federico Carlos Lehmann, Cali 08610
Museo de Geología, Manizales 08668
Museo de Geología y Suelos, Armenia 08533
Museo de Historia Natural, Popayán 08726
Museo de la Diversidad Biológica y Cultural del Pacífico, Quibdó 08734
Museo de Suelos de Colombia, Bogotá 08574
Museo del Carbón, Lenguazaque 08665
Museo del Mar El Rodadero, Santa Marta 08751
Museo Gabriel Mateus, Sabanagrande 08741
Museo Gemológico, Bogotá 08583
Museo Geológico Petrolero, Neiva 08696
Museo Nacional del Petróleo Samuel Schneider Uribe, Barrancabermeja 08538
Museo Natural, Pensilvania 08718
Museo Universitario, Medellín 08686
Museo y Parque Nacional, Montenegro 08692

Congo, Republic

Musée National du Congo, Brazzaville 08805

Croatia

Arheološki i Prirodoslovni Muzej, Brijuni 08848
Pomorski Muzej, Zadar 08984
Prirodoslovni Muzej, Rijeka 08940
Prirodoslovni Muzej, Split 08962

Cuba

Museo Ciencias Naturales, Nueva Gerona 09232
Museo Ciencias Naturales Carlos de la Torre, Sancti-Spíritus 09285
Museo de Ciencias Naturales Carlos de la Torre, Holguín 09179
Museo de Ciencias Naturales Tomás Romay, Santiago de Cuba 09310
Museo de Historia Natural Felipe Poey, La Habana 09135
Museo de Historia Natural Valle de la Prehistoria, Santiago de Cuba 09312
Museo Historia de las Ciencias Carlos J. Finlay, La Habana 09146
Museo Nacional de Historia Natural, La Habana 09163
Tranquilino Sandalio de Noda Museo de Ciencias Naturales, Pinar del Río 09254

Czech Republic

Hrad Roštejn, Telč 09860
Jihomoravské Muzeum ve Znojmě, Znojmo 09947

Městské Muzeum a Galerie, Hlinsko v Čechách 09514
Moravské Zemské Muzeum, Brno 09435
Muzeum Českého Krasu, Hořovice 09528
Muzeum v Bruntále, Bruntál 09445
Muzeum Východních Čech v Hradci Králové, Hradec Králové 09534
Muzeum Vysočiny Jihlava, Jihlava 09561
Národní Muzeum, Praha 09779
Památník Prokopa Diviše, Znojmo 09948
Polabské Muzeum, Poděbrady 09724
Přírodovědecké Muzeum, Praha 09791
Slezské zemské muzeum, Opava 09692
Slezské zemské muzeum, Opava 09693
Středočeské Muzeum, Roztoky u Prahy 09828
Východočeské Muzeum v Pardubicích, Pardubice 09707
Zámek Kinských, Valašské Meziříčí 09896
Zámek Vsetín, Vsetín 09918
Západočeské Muzeum, Plzeň 09720

Denmark

Økomuseum Samsø, Tranebjerg 10300
Salling-Fjends Museum, Fur 10034
Wellings Landsbymuseum, Lintrup 10178

Dominican Republic

Museo Nacional de Historia Natural, Santo Domingo 10357

Ecuador

Museo del Santuario de la Virgen de Agua Santa, Baños 10375
Museo Ecuatoriano de Ciencias Naturales, Quito 10443
Museo Farallón Dillon, Ballenita 10373
Vivarium, Quito 10453

El Salvador

Museo de Historia Natural de El Salvador, San Salvador 10551
Museo y Biblioteca de Ciencias Stephen Hawking, San Salvador 10554

Estonia

Eesti Loodusmuuseum, Tallinn 10625
Karl Ernst von Baeri Muuseum, Tartu 10650

Faroe Islands

N

Føroya Náttúrugripasavn, Tórshavn 10698

Finland

Forssan Luonnonhistoriallinen Museo, Forssa 10736
Jyväskylän Yliopiston Museo - Keski-Suomen Luontomuseo, Jyväskylä 10891
Kieppi-Luontomuseo ja Mineraalikokoelma, Kokkola 10974
Kortesjärven Luontomuseo, Kortesjärvi 10983
Kuopion Luonnontieteellinen Museo, Kuopio 11008
Luonnonhistoriallinen Museo, Porvoo 11228
Luonnontieteellinen museo, Helsinki 10801
Pohjanmaan Museo, Vaasa 11426
Seinäjoen Luonto-Museo, Seinäjoki 11305
Tampereen Luonnontieteelinen Museo, Tampere 11362
Tieteelliset Kokoelmat, Forssa 10738

France

Centre de Découverte de la Nature et du Patrimoine Cynégètique, Sault (Vaucluse) 15658
Centre d'Hébergement, d'Étude sur la Nature et l'Environnement, Allouville-Bellefosse 11558
Château-Musée, Tourrette-Levens 15960
Éco Musée Vivant de Provence, La Gaude 13377
Ecomusée de la Crau, Saint-Martin-de-Crau 15398
Écomusée de Saint-Léger, Saint-Léger-les-Mélèzes 15365
Ecomusée des Appeaux et de la Faune, Saint-Didier 15238
Maison de la Loire, Montlouis-sur-Loire 14220
Maison de la Nature, Eze 12898
Maison de la Réserve Naturelle du Lac de Remoray, Labergement-Sainte-Marie 13464
Maison de la Rivière, de l'Eau et de la Pêche, Sizun 15770
Maison du Parc Naturel Régional du Luberon, Apt 11653
Maison-Musée des Sciences, Lettres et Arts de Cholet et de sa Région, Cholet 12521
Musée Bernard d'Agesci, Niort 14455
Musée d'Allard, Montbrison 14188
Musée de Cerdagne d'Eyne, Eyne 12896
Musée de la Faune et de la Nature, Séez 15700
Musée de la Faune et la Flore, Etaples 12873
Musée de la Vallée à Lauzet, La Lauzet 13388
Musée des Sciences Naturelles, Orléans 14536
Musée d'Histoire Naturelle, Aix-en-Provence 11516
Musée d'Histoire Naturelle, Angers 11610
Musée d'Histoire Naturelle, Chambéry 12374
Musée d'Histoire Naturelle, Gray 13127
Musée d'Histoire Naturelle, La Rochelle 13427
Musée d'Histoire Naturelle, Amnéville-les-Thermes 11591
Musée d'Histoire Naturelle de Toulon et du Var, Toulon (Var) 15925
Musée d'Histoire Naturelle et de Géologie, Lille 13758
Musée d'Histoire Naturelle et de la Préhistoire, Nîmes 14452
Musée d'Histoire Naturelle et d'Ethnographie, Colmar 12584
Musée d'Histoire Naturelle et d'Histoire Locale, Elbeuf-sur-Seine 12815
Musée d'Histoire Naturelle et Régionale, Gruffy 13152
Musée d'Histoire Naturelle Gabriel Foucher, Bourges 12117
Musée d'Histoires Naturelles, Boulogne-sur-Mer 12093
Musée d'Initiation à la Nature de Normandie, Caen 12234
Musée du Biterrois, Béziers 11977
Musée du Château des Ducs de Wurtemberg, Montbéliard 14185
Musée du Château d'Oigny-en-Valois, Oigny-en-Valois 14502
Musée du Granit, Saint-Michel-de-Montjoie 15416
Musée Harmas Jean-Henri Fabre, Sérignan-du-Comtat 15724
Musée Joseph Durand, Saint-Clément-des-Baleines 15216
Musée-Maison de la Nature, Samer 15618
Musée Municipal des Beaux Arts et d'Histoire Naturelle, Châteaudun 12444
Musée Nature en Provence, Riez 15040
Musée Océanographique, L'Ile-Rousse 13751
Musée Pasteur, Paris 14735
Musée Requien, Avignon 11795
Musée Vert, muséum d'histoire naturelle du Mans, Le Mans 13609
Muséum des Volcans, Aurillac 11756
Muséum d'Histoire Naturell, Auxerre 11778
Muséum d'Histoire Naturelle, Autun 11766
Museum d'Histoire Naturelle, Blois 12036
Muséum d'Histoire Naturelle, Chartres 12426
Muséum d'Histoire Naturelle, Grenoble 13139
Muséum d'Histoire Naturelle, Lyon 13903
Muséum d'Histoire Naturelle, Marseille 14011
Muséum d'Histoire Naturelle, Nice 14440
Muséum d'Histoire Naturelle, Perpignan 14793
Muséum d'Histoire Naturelle, Rouen 15122
Muséum d'Histoire Naturelle, Troyes 16002
Muséum d'Histoire Naturelle, Gras 13108
Muséum d'Histoire-Naturelle, Cherbourg-Octeville 12506
Muséum d'Histoire Naturelle, Dijon 12744
Muséum d'Histoire Naturelle de Bordeaux, Bordeaux 12074
Muséum d'Histoire Naturelle de la Ville de Bayonne, Bayonne 11880
Muséum d'Histoire Naturelle de Toulouse, Toulouse 15946
Muséum d'Histoire Naturelle de Tours, Tours 15971
Muséum Municipal d'Histoire Naturelle Henri-Lecoq, Clermont-Ferrand 12552
Rhinopolis, Gannat 13029

French Guiana

Musée Départemental Alexandre Franconie, Cayenne 16316

French Polynesia

Musée du Coquillage, Papara 16322

Georgia

Khariton Akhvlediani State Museum, Batumi 16334

Germany

Alpines Museum des Deutschen Alpenvereins, München 20420
Aquazoo-Löbbecke Museum, Düsseldorf 17985
Bayerisches Moor- und Torfmuseum, Grassau 18739
Bergbaumuseum Erzpoche, Hausach 19046
Berliner Waldmuseum mit Waldschule Grunewald, Berlin 17048
Besucherbergwerk, Nothweiler 20749
Bienenmuseum, Duisburg 18013
Bienenmuseum, Plau am See 21050
Blumberger Mühle, Angermünde 16547
Bodensee-Naturmuseum, Konstanz 19677
Borner Forst- und Jagdmuseum, Born am Darß 17482
Carl-Schweizer-Museum, Murrhardt 20536
Darß-Museum, Prerow, Ostseebad 21119
Darßer Arche, Wieck am Darß 22537
Deutsches Vulkanmuseum, Mendig 20282
Eifel-Vulkanmuseum, Daun 17801
Elbschloss Bleckede, Bleckede 17384
Emsland-Moormuseum, Geeste 18549
Erlebnismuseum "Mensch und Landschaft", Hermeskeil 19148
Felsendome Rabenstein, Chemnitz 17695
Freilichtmuseum, Herbertingen 19131
Fürstlich Fürstenbergische Sammlungen, Donaueschingen 17892
Fuhlrott-Museum, Wuppertal 22740
Gaytal-Park, Körperich 19664
Geologisch-Mineralogische Sammlung, Hillesheim 19187
Geologisches Freimuseum, Bettendorf 17310
Geoskop-Urweltmuseum, Thallichtenberg 22044
Harzmuseum, Wernigerode 22495
Haus der Natur, Eichendorf 18077
Haus Ruhrnatur, Mülheim an der Ruhr 20412
Heimat- und Bauernmuseum Bruck, Bruck in der Oberpfalz 17599
Heimat- und Naturkunde-Museum Wanne-Eickel, Herne 19153
Heimat- und Waldmuseum, Eppenbrunn 18181
Heimatmuseum, Gaildorf 18533
Heimatmuseum, Immendingen 19324
Heimatmuseum Barbaramühle, Longkamp 20031
Heimatmuseum und Naturalienkabinett, Waldenburg, Kreis Chemnitzer Land 22290
Hessisches Forstmuseum Alte Fasanerie, Hanau 18977
Hochwaldmuseum, Hermeskeil 19151
Informationsstelle des Naturparks Fichtelgebirge, Zell, Oberfranken 22782
Insects World, Wunstorf 22734
Jagd und Naturkundemuseum, Hahn am See 18864
Jagdschloß, Gelbensande 18565
Karl-Otto-Braun-Museum, Ludwigshafen am Rhein 20052
Kinder-Akademie Fulda Werkraummuseum, Fulda 18511
Königliches Schloss Berchtesgaden und Rehmuseum, Berchtesgaden 17001
Korallen- und Heimatmuseum, Nattheim 20550
Kultur- und Heimattreff, Tholey 22055
KunstNaturLandschaft, Heideblick 19067
Kurioses Muschel-Museum, Nordholz 20742
Kurioses Muschel-Museum, Spiekeroog 21840
Lahn-Marmor-Museum, Villmar 22242
Landesmuseum für Natur und Mensch, Oldenburg 20898
Landschaftsmuseum Obermain, Kulmbach 19748
Landschaftsmuseum Westerwald, Hachenburg 18847
Lippisches Landesmuseum, Detmold 17839
LWL-Museum für Naturkunde, Münster 20513
Mauritianum, Altenburg 16498
Meeresmuseum, Öhringen 20864
Moor- und Fehnmuseum, Barßel 16930
Moormuseum Moordorf, Südbrookmerland 21974
Müritzeum, Waren (Müritz) 22335
Museum am Schölerberg, Osnabrück 20925
Museum Arbeit am Rhein, Elchesheim-Illingen 18115
Museum Bayerisches Vogtland, Hof 19220
Museum der Stadt Gladbeck, Gladbeck 18650
Museum Fischerhaus, Öhningen 20862
Museum für Naturkunde, Chemnitz 17700
Museum für Naturkunde, Magdeburg 20127
Museum für Naturkunde und Vorgeschichte, Dessau 17830
Museum für Stadtgeschichte, Naturkunde und Völkerkunde Julius Riemer, Lutherstadt Wittenberg 20117
Museum im Deutschhof, Heilbronn 19099
Museum im Umweltschutz-Informationszentrum Oberfranken mit Kinder-Erlebnis-Museum, Bayreuth 16962
Museum Mensch und Natur, München 20466
Museum Weltkulturen, Mannheim 20170
Nationalpark-Besucherzentrum TorfHaus, Altenau 16491
Nationalpark-Haus Berchtesgaden, Berchtesgaden 17003
Nationalparkhaus Sankt Andreasberg, Sankt Andreasberg 21488
Natur-Heimat-Museum, Langenburg 19826
Natureum am Schloß Ludwigslust, Ludwigslust 20061
Natureum Niederelbe, Neuhaus/Balje 20617
Naturhistorisches Museum Schloss Bertholdsburg, Schleusingen 21572
Naturkunde-Museum, Coburg 17718
Naturkundemuseum, Glees 18658
Naturkundemuseum, Kassel 19472
Naturkundemuseum, Potsdam 21098
Naturkundemuseum, Reutlingen 21286
Naturkundemuseum Freiberg, Freiberg, Sachsen 18423
Naturkundemuseum im Marstall, Paderborn 20974
Naturkundemuseum Leipzig, Leipzig 19928
Naturkundemuseum Niebüll, Niebüll 20698
Naturkundliche Bildungsstätte Nordeifel, Simmerath 21770
Naturkundliches Bildungszentrum Stadt Ulm, Ulm 22167
Naturkundliches Museum, Wiesenfelden 22560
Naturkundliches Museum und Schulungsstätte "Alte Schmiede", Handeloh 18984
Naturmuseum, Augsburg 16631
Naturmuseum, Freiburg im Breisgau 18436
Naturparkhaus Stechlin, Stechlin 21873
Naturschau in der Zehntscheune, Thallichtenberg 22045
Naturschutzzentrum, Nettersheim 20578
Naturwissenschaftliche Sammlungen, Berlin 17223
Naturwissenschaftliches Museum, Aschaffenburg 16588
Naturwissenschaftliches Museum der Stadt Flensburg, Flensburg 18329
Naturwissenschaftliches Museum Duisburg, Duisburg 18026
Niedersächsisches Landesmuseum Hannover, Hannover 19002
Noctalis-Welt der Fledermäuse, Bad Segeberg 16837
Otto-Klages-Sammlung, Königslutter am Elm 19655
Pfalzmuseum für Naturkunde, Bad Dürkheim 16716
Phänomenta Lüdenscheid, Lüdenscheid 20089
Pilz-Insekten-Vogel-Museum, Ribnitz-Damgarten 21311
Regionalmuseum Bad Frankenhausen, Bad Frankenhausen 16731
Reiss-Engelhorn-Museen (rem), Mannheim 20173
Riffmuseum im Bahnhof, Gerstetten 18618
Römersiedlung, Friesenheim 18482
Römischer Keller in Sulz am Neckar, Sulz am Neckar 21982
Römisches Museum und Naturkunde-Museum, Kempten 19516
Schloss Drachenburg, Königswinter 19660
Senckenberg-Forschungsinstitut und Naturmuseum, Frankfurt am Main 18392
Senckenberg Museum für Naturkunde Görlitz, Görlitz 18681
Staatliches Museum für Naturkunde Karlsruhe, Karlsruhe 19449
Staatliches Museum für Naturkunde Stuttgart, Stuttgart 21959
Stadtmuseum Schloß Hoyerswerda, Hoyerswerda 19282
Stadtmuseum Zehentstadel, Nabburg 20543
Städtisches Naturkundliches Museum, Göppingen 18678
Stein und Wein, Aspach 16599
Stiftung Naturschutzgeschichte, Königswinter 19662
Stiftung Schloss und Park Benrath, Düsseldorf 18009
Thüringer Wald - Kreativ Museum mit 1. Deutschem Kloßpressenmuseum, Großbreitenbach 18784
Torfmuseum, Neustadt am Rübenberge 20665
Übersee Museum Bremen, Bremen 17564
Vylym-Hütte, Userin 22191
Wald- und Forstmuseum, Herschberg 19165
Waldmuseum, Burg, Dithmarschen 17644
Waldmuseum, Furth im Wald 18518
Waldmuseum Reichental, Gernsbach 18608
Waldmuseum Wassermühle, Wingst 22598
Wasser- und Fischkundemuseum, Kuhlen-Wendorf 19743
Das Wattenmeerhaus, Wilhelmshaven 22581
Wiesenmeisterei, Neustadt-Glewe 20680
Wilhelm-Panetzky-Museum, Rammelsbach 21174

Greece

Mouseion Goulandri Fysikis Istorias, Kifissia 23051
Museum of Natural History and Historical Photography, Polygyros 23155
Museum of the White Mountains of Samaria, Chania 22965
Natural History Museum of Amarousio, Amarousio 22878
Natural History Museum of Cephalonia and Ithaca, Argostolion 22891
Natural History Museum of Elassona, Elassona 22991
Natural History Museum of Oeta, Ipati 23017
Zakynthos Museum of the Natural Heritage, Zakynthos 23246

Guatemala

Museo de Historia Natural, Guatemala 23298
Museo Quezaltenango, Quezaltenango 23307

Hungary

Agrármûszaki Emlékek Gyûjteménye, Zirc 23978
Bakonyi Természettudományi Múzeum, Zirc 23979
Hungarian Museum for Science, Technology and Transport - Industrial Collection, Budapest 23414
Járdányi Paulovics István Romkert, Szombathely 23901
Magyar Természettudományi Múzeum, Budapest 23445
Savaria Múzeum, Szombathely 23902

Természet Háza, Kecskemét 23656
Természettudományi Osztály, Pécs 23778

Iceland

Byggðasafn Austur-Skaftafellssýslu, Höfn 24001
Fiska- og Náttúrngripasafn Vestmannaeyja, Vestmannaeyjar 24043
Náttúrufræðaistofnun Islands, Akureyrarsetur, Akureyri 23984
Natturufræðistofa Kopavogs, Kópavogur 24009
Náttúrufræðlistofnun Íslands, Reykjavík 24026
Náttúrugripasafnid i Neskaupstad, Neskaupstadur 24010

India

Baroda Museum and Picture Gallery, Vadodara 24418
Bengal Natural History Museum, Darjeeling 24126
Birbal Sahni Institute of Palaeobotany Museum, Lucknow 24291
Goa Science Centre, Panaji 24340
Government Museum, Udaipur 24414
Manipur University Museum, Canchipur 24107
Municipal Museum Gwalior, Gwalior 24198
National Museum of Natural History, Delhi 24148
Natural History Museum of Bombay Natural History Society, Mumbai 24320
North Bengal Science Centre, Siliguri 24395
Regional Museum of Natural History, Bhopal 24094
Regional Science Centre, Guwahati 24195
Saint Joseph's College Museum, Tiruchirappalli 24406
State Museum, Trichur 24412
University Museum of Science and Culture, Aligarh 24057

Iran

Eastern Azarbaijan Province History Museum, Tabriz 24652
Museum of Natural History and Technology, Shiraz 24647
Natural History Museum, Gorgan 24606
Natural History Museum, Ilam 24609
Natural History Museum, Tehrān 24688
Natural History Museum, Urmieh 24701
Natural History Museum, Yazd 24704
Natural History Museum, Khorram-Abaad 24624
Taymori Hall, Isfahan 24613
Zoological Museum, Tehrān 24698

Iraq

Iraq Natural History Research Centre and Museum, Baghdad 24714

Ireland

Ballyheigue Maritime Centre, Ballyheigue 24739

Israel

Bet Pinhas Museum of Nature, Haifa 24955
Natural Sciences Museum, Kibbutz Hulata 25045
Nature Park and Galleries, Jerusalem 25007
Neot Kedumim, Lod 25061
Sharon Museum, Emek Hefer 24947
Sturman Institute Museum of Regional Science, Afula 24921

Italy

Centro di Scienze Naturali e Raccolta Natura ed Arte, Prato 27331
Centro Studi Naturalistici del Pollino Il Nibbio, Morano Calabro 26866
Civico Museo di Scienze Naturali, Domodossola 26041
Civico Museo di Storia Naturale della Lombardia, Jerago con Orago 26423
Civico Museo Insubrico di Storia Naturale, Induno Olona 26408
Civico Museo Naturalistico, Ovada 27024
Collezione di Farfalle, Guardia Sanframondi 26383
Musei Civici, Reggio Emilia 27388
Museo Aldrovandiano, Bologna 25443
Museo Berenziano, Cremona 26002
Museo Brandolini e Giol, Oderzo 26968
Museo Brembano di Scienze Naturali, San Pellegrino Terme 27677
Museo Civico, Ragogna 27357
Museo Civico, Finale Emilia 26136
Museo Civico A. Klitsche de la Grange, Allumiere 25203
Museo Civico delle Scienze, Pordenone 27298
Museo Civico di Scienze Naturali, Faenza 26074
Museo Civico di Storia Naturale, Casalnuovo Monterotaro 25698
Museo Civico di Storia Naturale, Chies d'Alpago 25869
Museo Civico di Storia Naturale, Cittanova 25913
Museo Civico di Storia Naturale, Crocetta del Montello 26009
Museo Civico di Storia Naturale, Milano 26691
Museo Civico di Storia Naturale, Morbegno 26868
Museo Civico di Storia Naturale, Trieste 28089
Museo Civico di Storia Naturale, Verona 28232
Museo Civico di Storia Naturale del Liceo Glassico N. Machiavelli, Lucca 26533
Museo Civico Didattico di Scienze Naturali, Pinerolo 27220
Museo Civico E. Barba, Gallipoli 26268
Museo Civico Naturalistico, Maserada sul Piave 26617
Museo Civico Naturalistico dei Monti Prenestini, Capranica Prenestina 25662
Museo Civico Naturalistico Severo Sini, Villa d'Almè 28271
Museo Civico P. A. Garda, Ivrea 26422
Museo Civico per la Preistoria del Monte Cetona, Cetona 25842
Museo Civicodi Scienze Naturali, Randazzo 27363
Museo Civicodi Storia Naturale, Stazzano 27899
Museo d'Arte Naturale, Gioiosa Ionica 26342
Museo degli Alpini, Savignone 27771
Museo dei Fossili, Lugo di Vicenza 26545
Museo dei Fossili e di Storia Naturale, Montefalcone Appennino 26822
Museo del Capriolo, Bisegna 25427
Museo del Fungo e di Scienze Naturali, Boves 25521
Museo del Lupo Appenninico, Civitella Alfedena 25925
Museo della Flora, Fauna e Mineralogia, Auronzo di Cadore 25329
Museo dell'Avifauna Appenninica, Sarnano 27741
Museo delle Scienze della Terra e del Lavoro Contadino, Piandimeleto 27191
Museo delle Valli, Argenta 25273
Museo di Ecologia e Storia Naturale, Marano sul Panaro 26601
Museo di Preistoria e Protostoria della Valle del Fiume Fiora, Manciano 26582
Museo di Scienze Naturali, Belvi 25392
Museo di Scienze Naturali, Bolzano 25489
Museo di Scienze Naturali, Cesena 25833
Museo di Scienze Naturali, Città della Pieve 25902
Museo di Scienze Naturali, Malnate 26576
Museo di Scienze Naturali, Pavia 27123
Museo di Scienze Naturali, Camerino 25620
Museo di Scienze Naturali A. Brandolini, Ravenna 27373
Museo di Scienze Naturali ed Umane, L'Aquila 26455
Museo di Scienze Naturali L. Paolucci, Offagna 26972
Museo di Scienze Naturali Tommaso Salvadori, Fermo 26100
Museo di Storia Naturale, Follonica 26216
Museo di Storia Naturale, Macerata 26553
Museo di Storia Naturale, Merate 26661
Museo di Storia Naturale, Perugia 27151
Museo di Storia Naturale, Piacenza 27189
Museo di Storia Naturale, Senna Lodigiana 27806
Museo di Storia Naturale, Stroncone 27914
Museo di Storia Naturale, Sulmona 27920
Museo di Storia Naturale, Venezia 28197
Museo di Storia Naturale, Parma 27104
Museo di Storia Naturale, Parma 27105
Museo di Storia Naturale A. Orsini, Ascoli Piceno 25295
Museo di Storia Naturale A. Stoppani, Venegono Inferiore 28177
Museo di Storia Naturale Aquilegia, Assemini 25303
Museo di Storia Naturale del Cilento, Laureana Cilento 26469
Museo di Storia Naturale Don Bosco, Torino 28015
Museo di Storia Naturale e del Territorio, Calci 25593
Museo di Storia Naturale e dell'Uomo, Carrara 25690
Museo di Storia Naturale Faraggiana Ferrandi, Novara 26950
Museo di Storia Naturale S. Ferrari, Bedonia 25387
Museo di Storia Naturali e Archeologia, Montebelluna 26809
Museo Didattico, Niscemi 26928
Museo Etnografico e di Scienze Naturali, Torino 28019
Museo Friulano di Storia Naturale, Udine 28114
Museo Giuseppe Scarabelli, Imola 26402
Museo Marinaro Tommasino-Andreatta, San Colombano Certenoli 27627
Museo Mineralogico e Naturalistico, Bormio 25511
Museo Missionario Cinese e di Storia Naturale, Lecce 26472
Museo Naturali e Appennino, Revi nel Lazio 27398
Museo Naturalistica-Archeologico, Vicenza 28254
Museo Naturalistico, Champdepraz 25846
Museo Naturalistico, Lesina 26488
Museo Naturalistico, Morino 26871
Museo Naturalistico, Picinisco 27199
Museo Naturalistico, Portoferraio 27307
Museo Naturalistico, Chieri 25867
Museo Naturalistico Archeologico M. Locati, Lama dei Peligni 26441
Museo Naturalistico Archeologico Paolo Barrasso, Caramanico Terme 25672
Museo Naturalistico Centro Natura, Pescasseroli 27168
Museo Naturalistico del Frignano, Pavullo nel Frignano 27128
Museo Naturalistico e della Civiltà Contadina, Berceto 25398
Museo Naturalistico F. Minà Palumbo, Castelbuono 25722
Museo Naturalistico G. Zanardo, Tambre 27939
Museo Naturalistico Libero Gatti, Copanello 25966
Museo Naturalistico Missionario, Chiusa Pesio 25882
Museo Naturalistico N. De Leone, Penne 27135
Museo Naturalistico O. Perini, Borca di Cadore 25496
Museo Pavese di Scienze Naturali, Pavia 27124
Museo Regionale di Scienze Naturali, Saint Pierre 27588
Museo Regionale di Scienze Naturali, Torino 28025
Museo Storico dela Caccia e del Territorio, Cerreto Guidi 25812
Museo Tridentino di Scienze Naturali, Trento 28062
Raccolta dell'Avifauna delle Marche, Montefortino 26831
Raccolta Naturalistica, Tagliolo Monferrato 27937
Raccolta Paleontologica e Preistorica, Narni 26917

Jamaica

Institute of Jamaica, Kingston 28328
Natural History Museum, Kingston 28335

Japan

Ainu Museum, Sapporo 29213
Akiyoshi-dai Kagaku Hakubutsukan, Shuuhou 29293
Biwa-ko Bunkakan, Kusatsu 28796
Chiba-kenritsu Chuo Hakubutsukan, Chiba 28392
Ehime-kenritsu Hakubutsukan, Matsuyama 28905
Fukui City Museum of Natural History, Fukui 28427
Gakumon Kagaku Hakubutsukan, Tokyo 29389
Gifu-ken Hakubutsukan, Seki 29240
Gunma-kenritsu Shizenshi Hakubutsukan, Tomioka 29512
Hamamatsu-shiritsu Kagakukan, Hamamatsu 28486
Hayashibara Shizenkagaku Hakubutsukan, Okayama 29099
Hiwa Kagaku Hakubutsukan, Hiwa 28544
Horaiji-san Shizenkshi Hakubutsukan, Hourai 28550
Hyogo kenritsu Hito to Shizen no Hakubutsukan, Sanda 29212
Ibaraki-kenritsu Shizenshi Hakubutsukan, Iwai 28604
Ichikawa-shiritsu Shizenshi Hakubutsukan, Ichikawa 28555
Ishikawa -kenritsu Shizenshi Shiryokan, Kanazawa 28659
Kagoshima-kenritsu Hakubutsukan, Kagoshima 28627
Kanagawa-kenritsu Seimei-no-Hoshi Chikyu Hakubutsukan, Odawara 29087
Kenritsu Kagakukan Joetsu, Joetsu 28621
Kitakyushu-shiritsu Shizenshi Hakubutsukan oyo Ningenteki Rekishi, Kitakyushu 28713
Kokuritsu Kagaku Hakubutsukan, Tokyo 29416
Kyoto-shi Seishonen Kagaku, Kyoto 28846
Miyazaki-ken Sogo Hakubutsukan, Miyazaki 28927
Mountaineering Museum, Hanamaki 28492
Nagaoka Municipal Science Museum, Nagaoka 28960
Nogyo Hakubutsukan, Miyazaki 28929
Okinawa-kenritsu Hakubutsukan Gendai Bijutsukan, Naha 29014
Omachi Alpine Museum, Omachi 29109
Ryuga-Do Cave Museun, Tosayamada 29514
Saga-kenritsu Hakubutsukan, Saga 29180
Saitama-kenritsu Shizenshi Hakubutsukan, Nagatoro 28989
Saito Ho-on Kai Shizenshi Hakubutsukan, Sendai 29244
Sapporo-shiritsu Hakubutsukan, Sapporo 29225
Shizen Kyoiku En, Tokyo 29471
Shizenshi Hakubutsukan, Sendai 29247
Shizenshi Hakubutsukan, Shimizu 29262
Tohoku Daigaku Hakabutsukan, Sendai 29248
Towada Kyodo Hakubutsukan, Towada 29519
Toyohashi-shi Shizenshi Hakubutsukan, Toyohashi 29528
The University Museum, Tokyo 29502
Yaeyama Museum, Ishigaki 28586
Yamaguchi Hakabutsukan, Yamaguchi 29580
Yamanashi-kenritsu Kagaku Senta, Kofu 28755
Yokosuka-shi Hakubutsukan, Yokosuka 29622

Jordan

Children's Heritage and Science Museum, Amman 29647
Jordan Natural History Museum, Irbid 29663

Korea, Republic

Jeju-do Folklore and Natural History Museum, Jeju 29836
Kyunghee University Museum, Seoul 29901
National Science Museum of Korea, Daejeon 29802
Seoul National Science Museum, Seoul 29920

Kuwait

Educational Science Museum, Kuwait 29948

Latvia

Latvijas Dabas Muzejs, Rīga 30077

Lesotho

Lesotho National Museum, Maseru 30179

Lithuania

Panevėžio Kraštotyros Muziejus, Panevėžys 30332

Luxembourg

Musée National d'Histoire Naturelle, Luxembourg 30484

Malawi

Museum of Malawi, Blantyre 30544

Malaysia

Jabatan Muzium Sabah, Kota Kinabalu 30565
Muzium Negara, Kuala Lumpur 30578

Malta

Ghar Dalam Cave and Museum, Birzebbuga 30627
National Museum of Natural History, Mdina 30631
Natural Science Museum, Victoria 30653

Mauritius

Natural History Museum, Port Louis 30684

Mexico

Museo de Ciencia, Cuernavaca 30816
Museo de Ciencias de Ensenada, Ensenada 30858
Museo de Ciencias Explora, León 30998
Museo de Ciencias Naturales, Toluca 31486
Museo de Historia Natural, Mérida 31025
Museo de Historia Natural, México 31076
Museo de Historia Natural, Villahermosa 31552
Museo de Historia Natural, Zapotlán del Rey 31593
Museo de Historia Natural, La Paz 30987
Museo de Historia Natural Alfredo Dugès, Guanajuato 30914
Museo de Historia Natural del Ateneo Fuente, Saltillo 31312
Museo de Historia Natural Dr. Manuel M. Villada, Toluca 31488
Museo de Historia Natural Dr. Manuel Martínez Solórzano, Morelia 31196
Museo de Historia Natural José Vilet Brullet, Mexquitic 31162
Museo de Historia Natural Regional, Lerdo 31003
Museo de la Fauna y Ciencias Naturales, Monterrey 31181
Museo de los Altos de Chiapas, San Cristóbal de las Casas 31325
Museo del Desierto, Saltillo 31315
Museo Interactivo de Ciencia, Tecnología y Medio Ambiente Sol del Niño, Mexicali 31035
Museo Laguna del Caimán, Tlahualilo de Zaragoza 31466
Museo Miguel Álvarez del Toro, Tuxtla Gutiérrez 31516
Museo Parque de la Cristianía, Chapala 30759
Museo Universitario de Ciencias y Artes, México 31154

Moldova

Muzeul National de Etnografie si Istorie Naturala, Chişinău 31601

Mongolia

Museum of Natural History, Ulaanbaatar 31627
Ulsyn Tōw Muzei, Ulaanbaatar 31632

Montenegro

Prirodnjački Muzej Crne Gore, Podgorica 31657

Morocco

Museum National d'Histoire Naturelle, Rabat 31692

Mozambique

Museu de História Natural, Maputo 31707
Museu Monstruário de Manica, Manica 31704

Myanmar

Natural History Museum, Yangon 31737

Namibia

Möwe Bay Museum, Möwe Bay 31747
National Museum of Namibia, Windhoek 31765
Okaukuejo Museum, Okaukuejo 31748
Swakopmund Museum, Swakopmund 31757

Nepal

Natural History Museum, Kathmandu 31777

Netherlands

Bezoekerscentrum de Meinweg, Herkenbosch 32369
Bezoekerscentrum Mijl Op Zeven, Ospel 32637
Centrum voor Natuur en Landschap, Terschelling-West 32817
Collectie Anderson, Losser 32509
Ecodrome Zwolle, Zwolle 33026
Ecomare, De Koog 32047
Fort Kijkduin, Castricum 32037
Fries Natuurmuseum, Leeuwarden 32474
Missiemuseum Steijl, Steijl 32804
Museon, Den Haag 32083
Natuurhistorisch en Volkenkundig Museum, Oudenbosch 32653
Natuurhistorisch Museum de Peel, Asten 31942
Natuurhistorisch Museum het Diorama, Nunspeet 32608
Natuurmuseum Ameland, Nes 32569
Natuurmuseum Brabant, Tilburg 32828
Natuurmuseum de Wielewaal, Lopik 32507
Natuurmuseum E. Heimans, Zaandam 32980
Natuurmuseum Enschede, Enschede 32212
Natuurmuseum het Drents-Friese Woud, Wateren 32929
Natuurmuseum Mar en Klif, Oudemirdum 32652
Natuurmuseum Nijmegen, Nijmegen 32593
Nederlands Centrum voor Biodiversiteit Naturalis, Leiden 32485
De Noordwester, Vlieland 32898
Pieter Vermeulen Museum, IJmuiden 32424
Regionaal Natuurmuseum Westflinge, Sint Pancras 32778
Tute-Natura, Bennekom 31963
Zeeuws Biologisch Museum, Oostkapelle 32630
Zomerkoestal uut 't Wold, Ruinerwold 32741

New Zealand

Auckland Museum and Auckland War Memorial Museum, Auckland 33043
Canterbury Museum, Christchurch 33069
South Canterbury Museum, Timaru 33212
Te Papa, Wellington 33260

Nicaragua

Museo Arqueológico de Solentiname, Solentiname 33307
Museo de Ciencias Naturales, Managua 33288

Niger

Musée National du Niger, Niamey 33308

Nigeria

National Museum, Oshogbo 33341
Natural History Museum, Ife 33326

Northern Cyprus

Güzelyurt Müzesi, Güzelyurt 33361

Norway

Agder Naturmuseum, Kristiansand 33609
Bergen Museum, Bergen 33421
Massnes Villmarksmuseum, Bjordal 33440
Naturhistoriske Samlinger, Bergen 33431
Norsk Fjellmuseum, Lom 33646
Stabbursnes Naturhus og Museum, Indre Billefjord 33576
Sykkylven Naturhistorisk Museum, Sykkylven 33898
Vitenskapsmuseet, Trondheim 33932

Oman

Oman Natural History Museum, Muscat 33991

Pakistan

Dam Site Museum, Mangla 34030
Natural History Museum, Karachi 34015
Natural History Museum, Lahore 34026
Pakistan Museum of Natural History, Islamabad 34010

Papua New Guinea

Papua New Guinea National Museum and Art Gallery, Boroko 34080

Paraguay

Museo de Ciencias Naturales, Asunción 34093
Museo de Historia Natural del Paraquay, Asunción 34094

Peru

Museo de Arqueología e Historia Natural de Yungay, Yungay 34330
Museo de Ciencias, Huánuco 34179
Museo de Ciencias Naturales, Cusco 34159
Museo de Historia de Ciencias Naturales, Chuquimarca 34152
Museo de Historia Natural de la Universidad Antenor Orrego, Trujillo 34319
Museo de Historia Natural de la Universidad Ricardo Palma, Lima 34238
Museo de Historia Natural Javier Prado, Lima 34239
Museo de Historia Natural Víctor Vaca Aguinaga, Lambayeque 34203
Museo Multidisciplinario del Colegio La Salle, Lima 34262
Museo Municipal de Nazca y Casa Museo María Reiche, Nazca 34281
Museo Universitario y Museo de Anatomía General, Yanacancha 34327
Museos de la Universidad Nacional de Cajamarca, Cajamarca 34130

Philippines

Museum of Natural History, College Los Baños 34380
Natural Science Museum, Marawi 34460
University of Nueva Caceres Museum, Naga 34467
University of San Carlos Biological Museum, Cebu 34378
University of Santo Tomas Museum of Arts and Sciences, Manila 34458
University of Southern Philippines Museum, Cebu 34379

Poland

Muzeum Borów Tucholskich, Tuchola 35200
Muzeum Częstochowskie, Częstochowa 34615
Muzeum Ewolucji Instytutu Paleobiologii PAN, Warszawa 35240
Muzeum im. Jacka Malczewskiego w Radomiu, Radom 35073
Muzeum im. prof. Władysława Szafera, Ojców 34970
Muzeum Karkonoskie w Jeleniej Górze, Jelenia Góra 34705
Muzeum Łowiectwa i Jeździectwa, Warszawa 35257
Muzeum Miejskie, Nowa Sól 34950
Muzeum Przyrodnicze, Jelenia Góra 34706
Muzeum Przyrodnicze, Kazimierz Dolny 34734
Muzeum przyrodnicze, Mosina 34929
Muzeum Przyrodnicze Babiogórskiego Parku Narodowego, Zawoja 35364
Muzeum Przyrodnicze Bieszczadzkiego Parku Narodowego, Ustrzyki Dolne 35204
Muzeum Przyrodnicze Uniwersytetu Łódzkiego, Łódź 34882
Muzeum Przyrodnicze Uniwersytetu Wrocławskiego, Wrocław 35337
Muzeum Przyrodnicze Wolińskiego Parku Narodowego im. Prof. Adama Wodziczki, Międzyzdroje 34918
Muzeum Przyrodniczo-Leśne Słowińskiego Parku Narodowego, Smołdzino 35113
Muzeum Regionalne w Świebodzinie, Świebodzin 35149
Muzeum w Gorzowie Wielkopolskim, Gorzów Wielkopolski 34680
Muzeum w Wałbrzychu, Wałbrzych 35213
Muzeum Wiedzy o Środowisku, Poznań 35046
Muzeum Ziemi Piskiej, Pisz 35011
Muzeum Zoologiczne Instytutu Zoologii Uniwersytetu Jagiellońskiego, Kraków 34814

Portugal

Museu da Companhia das Lezírias, Salvaterra de Magos 35895
Museu de História Natural, Coimbra 35534
Museu de História Natural de Sines, Sines 35941
Museu Municipal do Funchal - História Natural, Funchal 35618
Museu Nacional de História Natural e Museu de Ciência, Lisboa 35718

Réunion

Muséum d'Histoire Naturelle, Saint-Denis 36092

Romania

Complexul Muzeal de Stiințe ale Naturii Constanța, Constanța 36300
Complexul Muzeal de Stiintele Naturii Ion Borcea, Bacău 36120
Muzeul Banatului. Secția de Ştiințele Naturii, Timişoara 36691
Muzeul Brăilei, Brăila 36170
Muzeul de Istorie Naturală din Sibiu, Sibiu 36610
Muzeul de Istorie Naturală Iaşi, Iaşi 36419
Muzeul de Ştiințele Naturii, Craiova 36326
Muzeul de Ştiințele Naturii, Dorohoi 36350
Muzeul de Ştiințele Naturii, Focşani 36372
Muzeul de Ştiințele Naturii, Suceava 36644
Muzeul de Stiințele Naturii, Roman 36569
Muzeul de Ştiințele Naturii Aiud, Aiud 36102
Muzeul de Ştiințele Naturii şi Cinegetică, Vatra Dornei 36727
Muzeul de Ştiinte Naturale, Tulcea 36706
Muzeul Județean Argeş, Piteşti 36526
Muzeul Județean de Ştiințele Naturii Prahova, Ploieşti 36534
Muzeul Județean Mureş-Secția de Sțiințele Naturii, Târgu Mureş 36672
Muzeul Național de Istorie Naturala Grigore Antipa, Bucureşti 36222
Muzeul Orăşenesc, Adjud 36097
Muzeul Rezervației Bucegi, Sinaia 36621
Muzeul Văii Superioare a Teleajenului, Măneciu 36463
Muzeul Vivariul Bacău, Bacău 36123

Russia

Brjanskij Gosudarstvennyj Obedinennyj Kraevedčeskij Muzej, Brjansk 36824
Chvalynskij Kraevedčeskij Muzej, Chvalynsk 36865
Darvinovskij Muzej, Moskva 37157
Kurskij Gosudarstvennyj Oblastnoj Kraevedčeskij Muzej, Kursk 37094
Lipeckij Oblastnoj Kraevedčeskij Muzej, Lipeck 37104
Miasskij Kraevedčeskij Muzej, Miass 37130
Muzej Brjanskij Les, Brjansk 36828
Muzej Gosudarstvennogo Prirodnogo Nacionalnogo Parka Losinogo Ostrova, Moskva 37282
Muzej-kvartira K.A. Timirjazeva, Moskva 37324
Muzej Prirody Burjatii, Ulan-Udė 37896
Muzej-teatr Lednikovyj period, Moskva 37366
Muzej Zemlevedenija Moskovskogo Gosudarstvennogo Universiteta im. M.V. Lomonosova, Moskva 37386
Nižnetagilskij Muzej-zapovednik Gornozavodskoj Ural, Nižnij Tagil 37473
Novosibirskij Oblastnoj Kraevedčeskij Muzej, Novosibirsk 37488
Osinskij Kraevedčeskij Muzej, Osa 37517
Pjatigorskij Kraevedčeskij Muzej, Pjatigorsk 37565
Primorskij Gosudarstvennyj Muzej im. V.K. Arseneva, Vladivostok 37949
Pustozerskij Kompleksnyj Istoriko-prirodnyj Muzej, Narjan-Mar 37444
Rossijskij Muzej Lesa, Moskva 37404
Sebežskij Kraevedčeskij Muzej, Sebež 37757
Tajmyrskij Okružnoj Kraevedčeskij Muzej, Dudinka 36880

St. Pierre and Miquelon

Musée-Archives de Saint-Pierre, Saint-Pierre 37998

Senegal

Musée du Centre de Recherches et de Documentation du Sénégal, Saint-Louis 38027

Serbia

Gradski Muzej Vršac, Vršac 38163
Prirodnjački Muzej, Beograd 38085

Seychelles

National Museum, Victoria 38168

Slovakia

Městské Múzeum, Sabinov 38346
Městské Múzeum, Zlaté Moravce 38392
Múzeum Andreja Kmet'a, Martin 38296
Múzeum Mad'arskej Kultúry a Podunajska, Komárno 38269
Múzeum Spiša, Spišská Nová Ves 38359
Podtatranské Múzeum, Poprad 38332
Prírodovedné Múzeum, Bratislava 38232
Vlastivedné Múzeum v Hlohovci, Hlohovec 38258
Záhorské Múzeum, Skalica 38350

South Africa

Durban Natural Science Museum, Durban 38605
East London Museum, East London 38615
Natal Museum, Pietermaritzburg 38706
National Museum Bloemfontein, Bloemfontein 38567
Stevenson-Hamilton Memorial Information Centre, Skukuza 38731

Spain

Centre d'Interpretació p.n. d'Aigüestortes i Estany de Sant Maurici, Espot 39283
Hontza Natur Zientzien Museoa, Mañaria 39611
Museo de Ciencias Naturales, Santa Cruz de Tenerife 39995
Museo de Ciencias Naturales, Pamplona 39807
Museo de Ciencias Naturales de Álava, Vitoria-Gasteiz 40323
Museo de Ciencias Naturales de Atarfe, Atarfe 38913
Museo del CosmoCaixa Barcelona, Barcelona 38968
Museo Interactivo de la Ciencia y Tecnología - Principia, Málaga 39601
Museo Paleontológico, Murcia 39700
Museu Darder, Banyoles 38948
Museu de Ciències Naturals de Barcelona, Barcelona 38976
Museu de Granollers-Ciències Naturals, Granollers 39374
Museu de Vilafranca, Vilafranca del Penedès 40280
Museu Municipal Guilleries, Sant Hilari Sacalm 39984
Museu Regional d'Artá, Artá 38899
Museu València d'Història Natural, Valencia 40224

Sri Lanka

National Museum of Natural History, Colombo 40374

Sudan

Natural History Museum, Medani 40394
Natural History Museum, Port Sudan 40397

Sweden

Ájtte - Svenskt Fjäll- och Samemuseum, Jokkmokk 40618
Fjärilshuset i Hagaparken, Stockholm 40894
Göteborgs Naturhistoriska Museum, Göteborg 40548
Gotlands Naturmuseum, Visby 41070
Naturens Hus, Stockholm 40919
Naturhistoriska Riksmuseet, Stockholm 40920
Naturum Stendörren, Tystberga 40989
Nordens Ark, Hunnebostrand 40606
Rättviks Konstmuseum och Naturmuseum, Rättvik 40812
Regionmuseet Kristianstad, Kristianstad 40655

Switzerland

Bündner Naturmuseum, Chur 41300
Engadiner Museum, Sankt Moritz 41811
Glarner Naturmuseum, Engi 41353
Mammutmuseum, Niederweningen 41687
Musée Cantonal d'Histoire Naturelle, Sion 41875
Musée de la Banderette, Travers 41948
Musée d'Histoire Naturelle, La Chaux-de-Fonds 41295
Musée d'Histoire Naturelle, Fribourg 41388
Musée Jurassien des Sciences Naturelles et Jardin botanique, Porrentruy 41729
Museo Cantonale di Storia Naturale, Lugano 41582
Museum, Zofingen 42061
Muséum d'Histoire Naturelle, Neuchâtel 41682
Muséum d'Histoire Naturelle de la Ville de Genève, Genève 41427
Nationalparkzentrum, Zernez 42054
Natur-Museum Luzern, Luzern 41595
Naturama Aargau, Aarau 41083
Naturhistorisches Museum, Heiden 41477
Naturhistorisches Museum Basel, Basel 41175
Naturhistorisches Museum der Burgergemeinde Bern, Bern 41213
Naturkundemuseum, Altdorf 41106
Naturmuseum, Sankt Gallen 41808
Naturmuseum Solothurn, Solothurn 41886
Naturmuseum Thurgau, Frauenfeld 41379
Naturmuseum Winterthur, Winterthur 42036
Naturwissenschaftliche Sammlung, Stans 41904
Schweizerisches Alpines Museum, Bern 41217
Wildnispark Zürich, Sihlwald 41868

Taiwan

Chiayi Natural History Museum, Singang 42336
Kaohsiung County Museum of Natural History, Lujhu 42263
Nantou County Natural History Museum, Ren-Ai 42303
National Museum of Natural Science, Taichung 42352
National Science and Technology Museum, Kaohsiung 42241
National Taiwan Museum, Taipei 42417
Pingtung Museum of Natural History, Hengchun 42200

Rueylong Museum, Cyonglin 42171
Science Hall of Penghu County, Magong 42268
Tainan Natural History Museum, Zuojhen 42489
Taitung Museum of Natural History, Chengkung 42162
Taoyuan County Museum of Natural History, Taoyuan 42453
Yilan County Natural History Museum, Luodong 42266

Tanzania

Manyara Museum, Manyara 42515
Mikumi Museum, Mikumi 42517
Ngurdoto Gate Museum, Arusha 42508
Serengeti Museum, Arusha 42509

Thailand

Natural History Museum, Bangkok 42553
Nonthaburi Natural History Museum, Nonthaburi 42588
Parasite Museum, Bangkok 42555

Trinidad and Tobago

National Museum and Art Gallery of Trinidad and Tobago, Port-of-Spain 42615

Tunisia

Musée Naturel du Sud, Chenini Gabes 42621

Turkey

MTA Tabiat Tarihi Müzesi, Ankara 42706

Uganda

Game and Fisheries Museum, Entebbe 42986
Uganda Museum, Kampala 42993

Ukraine

Chmel'nyc'kyj Oblasnyj Krajeznavčyj Muzej, Chmel'nyc'kyj 43027
Ivano-Frankivs'kyj Krajeznavčyj Muzej, Ivano-Frankivs'k 43042
Kirovograds'kyj Oblasnyj Krajeznavčyj Muzej, Kirovograd 43068
Muzej Pryrody, Žytomyr 43213
Prirodnyčyj Muzej, Cherson 43025
State Museum of Natural History of NAS of Ukraine, Lviv 43142
Vinnyc'kyj Oblasnyj Krajeznavčyj Muzej, Vinnycja 43206
Zakarpats'kyj Kraeznavčyj Muzej, Užgorod 43198

United Arab Emirates

Sharjah Natural History Museum, Sharjah 43229

United Kingdom

At-Bristol, Bristol 43630
Bath Royal Literary and Scientific Institution, Bath 43401
Bournemouth Natural Science Society Museum, Bournemouth 43565
Bracken Hall Countryside Centre, Baildon 43353
City Museum and Art Gallery, Plymouth 45390
Cliffe Castle, Keighley 44550
College Museum, Epsom 44183
Dorman Memorial Museum, Middlesbrough 45140
Dunnet Pavilion, Castletown 43775
Edinburgh University Natural History Collections, Edinburgh 44123
Eton College Natural History Museum, Eton 44189
Fleetwood Museum, Fleetwood 44232
Giant's Causeway Visitor Centre, Bushmills 43703
Guernsey Museum and Art Gallery, Saint Peter Port 45591
Hancock Museum, Newcastle-upon-Tyne 45227
Home of Chales Darwin - Down House, Downe 44018
Horniman Museum, London 44878
Ipswich Museum, Ipswich 44530
John Dony Field Centre, Bushmead 43700
John Moore Countryside Museum, Tewkesbury 45842
King's Lynn Museum, King's Lynn 44587
Kinneil Museum and Roman Fortlet, Bo'ness 43559
Landmark Forest Heritage Park, Carrbridge 43765
Maidstone Museum and Bentlif Art Gallery, Maidstone 45073
The Manchester Museum, Manchester 45090
Museum of Lancashire, Preston 45445
Museum of the Holmesdale Natural History Club, Reigate 45481
Natural History Museum, Colchester 43875
Natural History Museum, London 44941
Natural History Museum, Nottingham 45295
Natural History Museum, Portsmouth 45435
New Walk Museum and Art Gallery, Leicester 44681
North Lincolnshire Museum, Scunthorpe 45628
The Old Byre Heritage Centre, Dervaig 43977
The Oxfordshire Museum, Woodstock 46067
Perth Museum and Art Gallery, Perth 45366
Phillips Countryside Museum, Brokerswood 43652
Royal Naval Museum Portsmouth, Portsmouth 45437
SEARCH-Centre, Gosport 44337
Sound Visitor Centre, Port Saint Mary 45414
University of Dundee Museum Collections, Dundee 44052
Warwickshire Museum, Warwick 45942
Wiltshire Heritage Museum and Gallery, Devizes 43979
Wood End Museum, Scarborough 45625
Yorkshire Museum, York 46112

United States of America

A.D. Buck Museum of Natural History and Science, Tonkawa 53530
Academy of Natural Sciences of Philadelphia Museum, Philadelphia 51809
Alabama Museum of Natural History, Tuscaloosa 53616
Alabama River Museum, Franklin 48745
American Museum of Natural History, New York 51115
Amherst College Museum of Natural History, Amherst 46263
Anniston Museum of Natural History, Anniston 46322
Arizona-Sonora Desert Museum, Tucson 53583
Arts and Science Center for Southeast Arkansas, Pine Bluff 51920
Austin Nature and Science Center, Austin 46505
Aztec Museum and Pioneer Village, Aztec 46527
Bailey-Matthews Shell Museum, Sanibel 52850
Bear Mountain Trailside Museums Wildlife Center, Bear Mountain 46662
The Bergen Museum of Art and Science, Paramus 51697
Bernice Pauahi Bishop Museum, Honolulu 49335
Biology Department Teaching Museum, Muncie 50903
Black Hills Museum of Natural History, Hill City 49283
Blue Hills Trailside Museum, Milton 50698
Bluestone Museum, Hinton 49301
Bolin Wildlife Exhibit and Antique Collection, McKinney 50357
Boone County Historical Center, Boone 46903
Boonshoft Museum of Discovery, Dayton 47982
Bowers Science Museum, Cortland 47829
Bradbury Science Museum, Los Alamos 50204
Branigan Cultural Center, Las Cruces 49957
Brazos Valley Museum of Natural History, Bryan 47105
Brazosport Museum of Natural Science, Brazosport 46999
Brazosport Museum of Natural Science, Clute 47666
Brevard Museum, Cocoa 47669
Buckhorn Saloon Museum, San Antonio 52711
Buffalo Museum of Science, Buffalo 47118
Burke Museum of Natural History and Culture, Seattle 52994
Burpee Museum of Natural History, Rockford 52413
Burritt on the Mountain, Huntsville 49442
Cable Natural History Museum, Cable 47155
Cairo Museum, Vicksburg 53732
California Academy of Sciences, San Francisco 52768
Calkins Field Museum, Iowa Falls 49523
Call of the Wild - Wildlife Museum, Gaylord 48874
Calvert Marine Museum, Solomons 53142
Cape Cod Museum of Natural History, Brewster 47008
Cape Cod National Seashore Museum, Wellfleet 54025
Cape Fear Museum, Wilmington 54201
Capitol Reef National Park Visitor Center, Torrey 53543
Carlsbad Caverns National Park, Carlsbad 47248
Carnegie Museum of Natural History, Pittsburgh 51943
Carrie Weedon Natural Science Museum, Galesville 48835
Cass County Museum, Walker 53795
Castine Scientific Society Museum, Castine 47278
Cedar Key State Park Museum, Cedar Key 47297
The Centennial Museum at the University of Texas at El Paso, El Paso 48317
Central Missouri State University Archives and Museum, Warrensburg 53829
Charles R. Conner Museum, Pullman 52197
Charlotte Nature Museum, Charlotte 47387
Children's Discovery Museum of Central Illinois, Bloomington 46858
Children's Museum of History, Natural History and Science at Utica, New York, Utica 53675
Cincinnati Museum Center, Cincinnati 47557
The Cleveland Museum of Natural History, Cleveland 47625
College of Eastern Utah Prehistoric Museum, Price 52140
Connecticut State Museum of Natural History, Storrs 53318
Conner Prairie Living History Museum, Fishers 48546
Cosi Toledo, Toledo 53521
Cottonlandia Museum, Greenwood 49085
Cracker Trail Museum, Zolfo Springs 54374
Cranbrook Institute of Science - Natural History Collections, Bloomfield Hills 46857
Crane Point Hammock Museum, Marathon 50458
Crowley Museum, Sarasota 52913
Cumberland County Historical Society Museum, Greenwich 49084
Cumberland Science Museum, Nashville 50966
The Curious Kids' Museum, Saint Joseph 52557
Cylburn Nature Museum, Baltimore 46559
D. Ralph Hostetter Museum of Natural History, Harrisonburg 49186
Dallas Museum of Natural History, Dallas 47932
Death Valley National Park Museum, Death Valley 48010
Delano Heritage Park Museum, Delano 48043
Delaware Museum of Natural History, Wilmington 54192
Delbridge Museum, Sioux Falls 53118
Denver Museum of Nature and Science, Denver 48075
The Discovery Museum, Bridgeport 47017
Discovery Place, Charlotte 47388
Dubois Museum, Dubois 48186
East Tennessee Discovery Center, Knoxville 49816
East-West Center, Honolulu 49338
Eastern California Museum, Independence 49469
EcoTarium, Worcester 54297
Ehrhart Museum, Antwerp 46326
Ehrman Mansion, Tahoma 53419
El Campo Museum of Natural History, El Campo 48309
Elgin Public Museum, Elgin 48335
Environmental Education Center, Basking Ridge 46623
Everhart Museum of Natural History, Science and Art, Scranton 52980
Exhibit Museum of Natural History, Ann Arbor 46305
Fairbanks Museum and Planetarium, Saint Johnsbury 52554
Fairfax Museum and Visitor Center, Fairfax 48477
Fernbank Museum of Natural History, Atlanta 46434
Fernbank Science Center, Atlanta 46435
Field Museum of Natural History, Chicago 47473
Flint Ridge State Memorial Museum, Glenford 48932
Florida Museum of Natural History, Gainesville 48813
Forest History Center, Grand Rapids 49008
Fort Ancient Museum, Oregonia 51586
Fort Caroline Memorial Museum, Jacksonville 49578
Fort Hancock Museum, Sandy Hook 52844
Fort Raleigh, Manteo 50452
Frank H. McClung Museum, Knoxville 49818
George B. Dorr Museum of Natural History, Bar Harbor 46603
George W. Carver Museum, Tuskegee Institute 53626
Georgia Museum of Natural History, Athens 46412
Georgia Southern University Museum, Statesboro 53272
Georgia's Stone Mountain Park, Stone Mountain 53310
Gilman Museum, Hellertown 49243
Glacier National Park Museum, West Glacier 54056
Glen Helen Ecology Institute Trailside Museum, Yellow Springs 54329
Governor Bent Museum, Taos 53440
Great Lakes Science Center, Cleveland 47631
Great Valley Museum of Natural History, Modesto 50774
Gunnison Museum of Natural History, Pawling 51734
Hands on Children's Museum, Olympia 51544
Hands-on House, Lancaster 49922
Harbor Branch Oceanographic Institution, Fort Pierce 48667
Harvard Museum of Natural History, Cambridge 47176
Hastings Museum, Hastings 49211
Heard Natural Science Museum, McKinney 50359
Heritage of the Americas Museum, El Cajon 48307
Herrett Center for Arts and Science, Twin Falls 53627
Hi-Desert Nature Museum, Yucca Valley 54363
High Desert Museum, Bend 46733
Hinchman House Natural Science Museum, Nantucket 50944
Historic Bethabara Park, Winston-Salem 54245
Historical and Natural History Museum of Natick, Natick 50989
Horizons Unlimited Supplementary Educational Center, Salisbury 52680
Houston Museum of Natural Science, Houston 49390
Howard B. Owens Science Center, Lanham-Seabrook 49939
Idaho Museum of Natural History, Pocatello 52006
Illinois State Museum, Springfield 53216
Imagination Station Science Museum, Wilson 54211
Indian City U.S.A., Anadarko 46280
Indian Springs State Park Museum, Flovilla 48573
Indiana State Museum, Indianapolis 49497
International Wildlife Museum, Tucson 53592
Iredell Museum of Arts Heritage, Statesville 53274
James A. McFaul Environmental Center of Bergen County, Wyckoff 54313
James Ford Bell Museum of Natural History, Minneapolis 50735
James R. Slater Museum of Natural History, Tacoma 53409
Jean Lafitte National Historical Park and Preserve, New Orleans 51084
John Burroughs Memorial, Roxbury 52476
John Hutchings Museum of Natural History, Lehi 50021
Joseph Moore Museum, Richmond 52313
Jurica Nature Museum, Lisle 50126
Kenosha Public Museum, Kenosha 49734
Kingman Museum of Natural History, Battle Creek 46645
Kirkpatrick Science and Air Space Museum at Omniplex, Oklahoma City 51517
L.C. Bates Museum, Hinckley 49295
Lafayette Natural History Museum, Lafayette 49867
Lake Erie Nature and Science Center, Bay Village 46654
Lake Meredith Aquatic and Wildlife Museum, Fritch 48799
Lanier Museum of Natural History, Buford 47128
Las Vegas Natural History Museum, Las Vegas 49973
Lawrence Hall of Science, Berkeley 46757
Libby Museum, Wolfeboro 54269
Limberlost State Historic Site, Geneva 48879
Lindsay Wildlife Museum, Walnut Creek 53803
Lora Robins Gallery of Design from Nature, Richmond 52332
Mann Museum, Opelika 51572
Mary Brogan Museum of Art and Science, Tallahassee 53426
Massacre Rocks State Park Museum, American Falls 46255
May Natural History Museum and Museum of Space Exploration, Colorado Springs 47703
Mayborn Museum Complex, Waco 53778
Memphis Pink Palace Museum, Memphis 50579
Michigan State University Museum, East Lansing 48254
Milton J. Rubenstein Museum of Science and Technology, Syracuse 53400
Milwaukee Public Museum, Milwaukee 50713
Mississippi Museum of Natural Science, Jackson 49561
Mississippi River Museum at Mud Island River Park, Memphis 50580
Mohonk Preserve, Gardiner 48860
Monte L. Bean Life Science Museum, Provo 52187
Montshire Museum of Science, Norwich 51442
Morningside Nature Center, Gainesville 48815
Morro Bay State Park Museum of Natural History, Morro Bay 50861
Morrow Mountain State Park Museum, Albemarle 46168
Mount Washington Museum, North Conway 51389
Mousley Museum of Natural History, Yucaipa 54361
Murie Museum, Kelly 49722
Museum of Cultural and Natural History, Mount Pleasant 50884
Museum of Long Island Natural Sciences, Stony Brook 53314
Museum of Natural and Cultural History, Eugene 48435
Museum of Natural History, Providence 52171
The Museum of Natural History, Stevens Point 53289
Museum of Natural History, Marshall 50506
Museum of Natural History, Amherst 46269
Museum of Natural History, Detroit 48127
Museum of Natural History and Primitive Technology, Somers 53144
Museum of Natural Science, Baton Rouge 46638
Museum of Science and Industry, Chicago 47497
Museum of Southwestern Biology, Albuquerque 46182
Museum of Texas Tech University, Lubbock 50312
Museum of the Hudson Highlands, Cornwall-on-Hudson 47815
Museum of the North, Fairbanks 48472
Museum of Western Colorado, Grand Junction 48994
Natchez Trace Parkway, Tupelo 53614
National Museum of Natural History, Washington 53904
Natural History Museum, Arcata 46338
Natural History Museum, Santa Maria 52894
Natural History Museum, Portales 52056
Natural History Museum, Lawrence 49991
Natural History Museum of Los Angeles County, Los Angeles 50257

N

The Natural Science Center of Greensboro, Greensboro 49067
Nature Museum at Grafton, Grafton 48983
Nevada State Museum, Carson City 47260
Neversink Valley Area Museum, Cuddebackville 47898
New Jersey State Museum, Trenton 53561
New Mexico Museum of Natural History and Science, Albuquerque 46185
North Carolina Museum of Life and Science, Durham 48223
North Carolina Museum of Natural Sciences, Raleigh 52234
The North Museum of Natural History and Science, Lancaster 49929
Northeastern Nevada Museum, Elko 48354
Nylander Museum, Caribou 47242
Oakland Museum of California, Oakland 51470
Ojai Valley Museum, Ojai 51511
Oregon Museum of Science and Industry, Portland 52083
Organ Pipe Cactus Museum, Ajo 46144
The Oyster and Maritime Museum of Chincoteague, Chincoteague 47539
Pacific Grove Museum of Natural History, Pacific Grove 51666
Palo Alto Junior Museum, Palo Alto 51689
Paterson Museum, Paterson 51730
Peabody Museum of Natural History, New Haven 51054
Peggy Notebaert Nature Museum, Chicago 47506
Pember Museum of Natural History, Granville 49019
Petaluma Wildlife and Natural Science Museum, Petaluma 51791
Pittsburg State University Natural History Museum, Pittsburg 51935
Portland Children's Museum, Portland 52087
Princeton University Museum of Natural History, Princeton 52152
Putnam Museum of History and Natural Science, Davenport 47970
The Quincy Museum, Quincy 52217
Randall Museum, San Francisco 52790
Raymond M. Alf Museum of Paleontology, Claremont 47575
Reading Public Museum and Art Gallery, Reading 52265
Red Rock Museum, Church Rock 47549
Redwood Museum, Crescent City 47875
Rhodes College, Memphis 50584
Richard H. Schmidt Museum of Natural History, Emporia 48394
River Hills Park Museum, Wildwood 54160
River Legacy Living Science Center, Arlington 46353
Robert A. Vines Environmental Science Center, Houston 49402
Robert Toombs House, Washington 53936
Rochester Museum, Rochester 52398
Rocky Mountain National Park Museum, Estes Park 48427
Roger Tory Peterson Institute of Natural History Museum, Jamestown 49598
Saint Joseph Museum, Saint Joseph 52565
Sam Noble Oklahoma Museum of Natural History, Norman 51376
San Bernardino County Museum, Redlands 52276
San Diego Natural History Museum, San Diego 52754
Santa Barbara Museum of Natural History, Santa Barbara 52860
Santa Cruz Museum of Natural History, Santa Cruz 52870
Schiele Museum of Natural History, Gastonia 48870
Science and Technology Museum of Atlanta, Atlanta 46451
Science Center of Pinellas County, Saint Petersburg 52643
Science Center of West Virginia, Bluefield 46877
Science Museum of Minnesota, Saint Paul 52629
Science Museum of Western Virginia, Roanoke 52382
Sciworks, Winston-Salem 54250
South Florida Museum, Bradenton 46987
South Florida Museum of Natural History, Dania Beach 47955
Southern Oregon University Museum of Vertebrate Natural History, Ashland 46389
Southern Vermont Natural History Museum, Marlboro 50496
Spartanburg Science Center, Spartanburg 53197
Springfield Science Museum, Springfield 53229
Squam Lakes Natural Science Center, Holderness 49311
Stamford Museum and Nature Center, Stamford 53249
Stephen C. Foster State Park, Fargo 48507
Stephens Museum, Fayette 48526
Sternberg Museum of Natural History, Hays 49225
Strecker Museum's Collection, Waco 53779
Tallahassee Museum of History and Natural Science, Tallahassee 53429
Texas Memorial Museum, Austin 46519
Thames Museum, Newport 51331
Thornton W. Burgess Museum, Sandwich 52842
Trailside Nature and Science Center, Mountainside 50897
Trailside Nature Center and Museum, Cincinnati 47566
Trailside Nature Museum, Cross River 47891
Tucker Tower Nature Center Park Museum, Ardmore 46345
Union Station Kansas City, Kansas City 49705
University Art Gallery Cabinet of Natural History, New Bedford 51016
University of Colorado Museum, Boulder 46961
University of Iowa Museum of Natural History, Iowa City 49522
University of Nebraska State Museum, Lincoln 50107
UNLV Barrick Museum, Las Vegas 49979
Utah Field House of Natural History State Park, Vernal 53722
Utah Museum of Natural History, Salt Lake City 52696
Virginia Living Museum, Newport News 51337
Virginia Museum of Natural History, Martinsville 50521
W.H. Over Museum, Vermillion 53721
Wagner Free Institute of Science, Philadelphia 51888
Weinberg Nature Center, Scarsdale 52955
Weymouth Woods-Sandhills Nature Preserve Museum, Southern Pines 53187
Whitaker Center for Science and the Arts, Harrisburg 49183
White Memorial Conservation Center, Litchfield 50129
White Pine Public Museum, Ely 48385
William McKinley Presidential Library and Museum, Canton 47229
World Museum of Natural History, Riverside 52376
Wrangell Museum, Wrangell 54309
The Yosemite Museum, Yosemite National Park 54350

Uruguay

Museo Antártico, Montevideo 54394
Museo Antonio Lussich, Punta Ballena 54427
Museo Fernando García, Canelones 54376
Museo Nacional de Historia Natural y Antropología, Montevideo 54413

Uzbekistan

Natural History Museum, Nukus 54450

Venezuela

Museo de Ciencias, Caracas 54545

Needlework

Armenia

National Folk Art Museum of Armenia, Yerevan 00750

Australia

Embroiderers Guild Museum, Mile End 01301

Austria

Handarbeitsmuseum, Traunkirchen 02826

Belgium

Musée de la Dentelle, Marche-en-Famenne 03846

Czech Republic

Městské Muzeum a Galerie, Dvůr Králové nad Labem 09500
Muzeum Krajky, Prachatice 09732
Muzeum Krajky, Vamberk 09898
Oblastní Muzeum v Chomutově, Chomutov 09478

Finland

Rauman Museo, Rauma 11260

France

150 ans de Couture, Aix-les-Bains 11521
Maison de la Dentelle, Urçay 16018
Maison du Boutis Les Cordelles, Calvisson 12249
Maison du Tisserand, Quintin 14974
Musée de la Coutellerie, Nogent-en-Bassigny 14461
Musée de la Dentelle, Caudry 12324
Musée de la Dentelle, Chamalières-sur-Loire 12370
Musée de la Dentelle à la Main, Arlanc-en-Livradois 11680
Musée de La Glacerie, La Glacerie 13378
Musée des Coiffes et des Traditions, Les Ponts-de-Cé 13708
Musée des Manufactures de Dentelles, Retournac 15023
Musée-Maison des Dentelles et du Point d'Argentan, Argentan 11672

Germany

Das Deutsche Stickmuster-Museum Celle, Celle 17686
Fingerhutmuseum, Creglingen 17740
Klöppelmuseum, Abenberg 16417
Museum der Schwalm, Schwalmstadt 21662
Museum Schloß Schwarzenberg, Schwarzenberg 21675
Nadlerhaus, Stützengrün 21923
Plauener Spitzenmuseum, Plauen 21054

Hungary

Halasi Csipke Közalapítvány - Csipkeház, Kiskunhalas 23670
Kisjankó Bori Emlékház, Mezőkövesd 23697
Tájház, Buzsák 23499
Városi Galéria, Mezőkövesd 23700

Italy

Museo del Merletto, Offida 26974
Museo del Merletto, Burano 25565

Korea, Republic

Museum of Korean Embroidery, Seoul 29906

Lithuania

Kuršių Nerijos Nacionalino Parko Gamtos Muziejus, Klaipėda 30280

Netherlands

Museum voor Naaldkunst, Wedde 32931
Nieuweschans Naaimachinemuseum, Nieuweschans 32579
Vingerhoedmuseum, Veenendaal 32874

Portugal

Museu das Rendas de Bilros, Vila do Conde 35989
Núcleo Museológico do IBTAM, Funchal 35619

Réunion

Musée-Ecole de Broderie, Cilaos 36089

Spain

Museo de Usos y Costumbres Etnológicas, Paracuellos de Jarama 39810
Museo Regional de Bolillo, Cartagena 39141
Museu Marès de la Punta, Arenys de Mar 38886

Sweden

Irma Dahn-Anderssons Nävermuseum, Bollnäs 40465

United Kingdom

Allhallows Museum of Lace and Antiquities, Honiton 44483
Cowper and Newton Museum, Olney 45316
Embroiderers' Guild Museum Collection, East Molesey 44092
Forge Mill Needle Museum and Bordesley Abbey Visitor Centre, Redditch 45475
Gawthorpe Hall, Padiham 45340
Guildford Museum, Guildford 44371
Hardwick Hall, Chesterfield 43821
London Sewing Machine Museum, London 44908
Manor Cottage Heritage Centre, Southwick 45719
Moravian Museum, Pudsey 45450
Museum of Costume and Textiles, Nottingham 45294
Sheelin Antiques and Irish Lace Museum, Enniskillen 44181
Traquair House, Innerleithen 44517
Wardown Park Museum, Luton 45048

United States of America

The Lace Museum, Sunnyvale 53372

Numismatics

Argentina

Complejo Museológico del Instituto Félix Bernasconi, Buenos Aires 00155
Museo del Banco Provincia de Córdoba, Córdoba 00329
Museo del Pasado Cuyano, Mendoza 00455
Museo Histórico y Numismático del Banco Central de la República Argentina "Dr. José Evaristo Uriburu", Buenos Aires 00245

Australia

Banking and Currency Museum, Kadina 01181

Austria

Heimatkundehaus und Münzkabinett, Sankt Marien 02665
Münze Hall und Münzerturm, Hall in Tirol 02043
Münzkabinett am Landesmuseum Joanneum, Graz 01978
Universalmuseum Joanneum, Graz 01985

Belgium

Cabinet des Médailles, Bruxelles 03440
Musée de la Banque Nationale de Belgique, Bruxelles 03466

Bolivia

Casa Nacional de la Moneda, Potosí 04235

Brazil

Centro Cultural Banco do Brasil, Rio de Janeiro 04686
Museu da Fazenda Federal, Rio de Janeiro 04714
Museu de Valores, Rio de Janeiro 04733
Museu de Valores do Banco Central do Brasil, Brasília 04363
Museu do Banco de Crédito Real de Minas Gerais, Juiz de Fora 04520
Museu e Arquivo Histórico do Centro Cultural Banco do Brasil, Rio de Janeiro 04746
Museu Histórico do Piaú, Teresina 04932
Museu Numismático e Filatélico, Rio de Janeiro 04759
Museu Regional do Sul de Minas, Campanha 04382

Canada

Currency Museum of the Bank of Canada, Ottawa 06544

China

Ancient Coins Museum, Beijing 07504
Nansong Qian Currency Museum, Hangzhou 07766
Suzhou National Treasure Numismatics Museum, Suzhou 08259
Zhongguo Qianbi Bowuguan, Beijing 07577

Colombia

Colección Numismática, Bogotá 08549
Museo Casa de la Moneda, Mariquita 08672
Museo de Historia Económica, Bogotá 08565

Costa Rica

Museo de Numismática, San José 08818

Croatia

Arheološki Muzej Zagreb, Zagreb 08986

Cuba

Museo Numismático, La Habana 09169

Cyprus

Museum of the History of Cypriot Coinage, Lefkosia 09381

Czech Republic

Muzeum Prostějovska, Prostějov 09810
Východočeské Muzeum v Pardubicích, Pardubice 09707

Denmark

Den Kongelige Mønt- og Medaillesamling, København 10136

Dominican Republic

Museo Numismático y Filatélico del Banco Central, Santo Domingo 10358

Ecuador

Museo Manuela Sáenz, Quito 10447

Ethiopia

Ethnographic Museum, Addis Ababa 10678

Finland

Raha- ja Mitalikokoelma, Helsinki 10810

France

Cabinet des Medailles et Antiques, Paris 14576
Cabinet des Monnais et Médailles du Revest-les-Eaux, Le Revest-les-Eaux 13646
Musée Archéologique, Lectoure 13672
Musée de la Monnaie de Paris, Paris 14655
Musée des Monnais et Médailles J. Puig, Perpignan 14792
Musée Gallo-Romain d'Aoste, Aoste 11649
Musée Oscar Roty, Jargeau 13259

Germany

Akademisches Münzkabinett, Jena 19375
Allgäu-Museum, Kempten 19510
Bode-Museum - Münzkabinett, Berlin 17051
Burg Neuhaus, Wolfsburg 22660
Finanzgeschichtliche Sammlung der Bundesfinanzakademie, Brühl 17604
Geldgeschichtliches Museum, Köln 19604
Geldmuseum der Deutschen Bundesbank, Frankfurt am Main 18372
Hellweg-Museum Unna, Unna 22175
Motorrad-Träume, Zschopau 22825
Münzkabinett, Dresden 17961
Münzkabinett der TUI AG, Hannover 18999
Münzkabinett im Pergamonmuseum, Berlin 17191
Münzkabinett im Stadtarchiv Mainz, Mainz 20147
Museum August Kestner, Hannover 19000
Museum Schloß Hohentübingen, Tübingen 22123
Niedersächsisches Münzkabinett der Deutschen Bank, Hannover 19003
Orientalisches Münzkabinett Jena (OMJ), Jena 19386
Pergamonmuseum, Berlin 17234
Schlossmuseum, Bad Iburg 16755
Staatliche Münzsammlung, München 20488
Stadtmuseum Schongau, Schongau 21620
Universitätssammlung antiker und nachantiker Münzen und Medaillen, Tübingen 22125

Greece

Museum of Banknotes of the Ionian Bank, Corfu 22979
Numismatic Museum, Athens, Athinai 22946

Haiti

Unité Musée Numismatique, Port-au-Prince 23324

Hungary

Magyar Nemzeti Bank Látogatóközpont, Budapest 23440
Pénzügytörténeti Gyűjtemény, Budapest 23465

Iceland

Myntsafn Sedlabanka og Thjódminjasafns, Reykjavík 24025

India

Aitihasic Puratatva Sangrahalaya, Delhi 24134
Archaeological Museum, Gorakhpur 24178
Archaeological Museum, Pune 24354
Chattishgarh Mudra Sangrahalaya, Katora Talab 24256
Governmental Educational Museum, Muzaffarnagar 24326
Rajkiya Sangrahalaya, Udaipur 24415

Iran

Sekkeh Museum, Kerman 24620

Israel

Kadman Numismatic Pavilion, Tel Aviv 25112

Italy

Musée de l'Académie Saint-Anselme, Aosta 25246
Museo Alessi, Enna 26055
Museo Berenziano, Cremona 26002
Museo Bottacin, Padova 27036
Museo Carlo Venturini, Massa Lombarda 26623
Museo della Fondazione Mandralisca, Cefalù 25794
Museo Numismatico della Zecca, Roma 27521
Palazzo Tursi, Genova 26324
Piccolo Museo della Moneta, Vicenza 28256
Pinacoteca e Museo Civico, Volterra 28312

Jamaica

Money Museum, Kingston 28332

Japan

Kahei Hakubutsukan, Tokyo 29406
Mint Museum, Osaka 29134

Jordan

Coins Museum, Amman 29648

Lithuania

Lietuvos Banko Muziejus, Vilnius 30414

Malaysia

Bak Negara Muzium Matawang, Kuala Lumpur 30572
Muzium Numismatik Maybank, Kuala Lumpur 30579

Malta

Cathedral Museum, Mdina 30630

Mexico

Museo de Numismática del Estado de México, Toluca 31491
Museo Pedro Coronel, Zacatecas 31580

Monaco

Musée des Timbres et des Monnaies, Monaco 31610

Morocco

Museo de Arqueologia e Historia, Melilla 31683

Myanmar

State Library and Museum, Kyaukpyu 31717

Nepal

Numismatic Museum, Kathmandu 31778

Netherlands

Het Geld-en Bankmuseum, Utrecht 32846
Munt- en Penningkabinet van de Spaar- en Voorschotbank, Surhuisterveen 32811
Museum van de Kanselarij der Nederlandse Orden, Apeldoorn 31913
Nederlands Economisch Penningkabinet, Rotterdam 32720
Veluws Museum van Oudheden - Stadsmuseum, Harderwijk 32313

Nicaragua

Sala Numismática del Banco Central de Nicaragua, Managua 33292

Norway

Den Kongelige Mynts Museum, Kongsberg 33599
Løten Bankmuseum, Løten 33651
Myntkabinett, Oslo 33734
Sparebankmuseet, Oslo 33750

Peru

Museo Numismático del Banco Continental, Arequipa 34114
Sala Numismática del Banco de Crédito del Peru y Sala de Acuarelas Pancho Fierro, Lima 34276

Philippines

Museo ng Bangko Sentral ng Pilipinas, Manila 34445

Poland

Gabinet Numizmatyczny Mennicy Państwowej, Warszawa 35224
Muzeum Emeryka Hutten-Czapskiego w Kamienicy Łozińskych, Kraków 34791
Muzeum Sztuki Medalierskiej, Wrocław 35340
Muzeum w Chorzowie, Chorzów 34604
Zakład Narodowy im. Ossolińskich, Wrocław 35344

Portugal

Museu Eng. António de Almeida, Porto 35855
Museu Numismático Português, Lisboa 35724

Romania

Muzeul Băncii Naţionale a României, Bucureşti 36196

Russia

Chakasskij Respublikanskij Kraevedčeskij Muzej, Abakan 36745
Gosudarstvennyj Okružnoj Muzej Prirody i Čeloveka Goroda Chanty-Mansijsk, Chanty-Mansijsk 36862
Muzej Istorii Deneg i Bankovskogo Dela, Moskva 37290
Regionalnyj Muzej Severnogo Priladožja, Sortavala 37789

Slovakia

Hutnícke Múzeum Železiarne Podbrezová, Podbrezová 38329
NBS-Múzeum Mincí a Medailí, Kremnica 38279
Tribečské Múzeum Topoľčany, Topoľčany 38372

South Africa

ABSA Group Museum, Johannesburg 38645
South African Mint Museum-Coin World, Tshwane 38757

Spain

Gabinete Numismático de Cataluña, Barcelona 38960
Museo de la Fabrica Nacional de Moneda y Timbre, Madrid 39544
Museo de Palencia, Palencia 39770
Museu Episcopal de Vic, Vic 40270
Museu Nacional d'Art de Catalunya (MNAC), Barcelona 39001
Sala d'Arqueologia i Gabinet Numismàtic, Lleida 39481
Yacimiento-Museo Arqueológico de Numancia, Garray 39326

Sweden

Kungl. Myntkabinettet, Stockholm 40904
Uppsala Universitet Myntkabinett, Uppsala 41015

Switzerland

Bibel Orient Museum, Fribourg 41383
MoneyMuseum, Zürich 42088
Münzkabinett und Antikensammlung der Stadt Winterthur, Winterthur 42030
Musée Casino de Sa, Saxon 41825
Musée de la Fausse Monnaie, Saillon 41789
Musée Monétaire Cantonal, Lausanne 41549
Musée Romain de Lausanne-Vidy, Lausanne 41551

United Kingdom

Bank of England Museum, London 44788
The British Museum, London 44803
The Fitzwilliam Museum, Cambridge 43727
Museum on the Mound, Edinburgh 44137
Royal Mint Museum, Llantrisant 44775

United States of America

American Numismatic Society Museum, New York 51116
Franklin Mint Museum, Franklin Center 48752
The Money Museum, Kansas City 49694
Museum of the American Numismatic Association, Colorado Springs 47705
United States Army Finance Corps Museum, Fort Jackson 48630
United States Mint-Philadelphia, Philadelphia 51885

Uruguay

Museo del Gaucho y de la Moneda, Montevideo 54401

Uzbekistan

Qo'qon Viloyat Ülkashunoslik Muzeyi, Qo'qon 54458
Surxondaryo Vilojati Arxeologiya Muzeyi, Termiz 54466

Oceanography

Argentina

Museo del Lago Gutierrez Dr. Rosendo Pascual, Bariloche 00145
Museo Provincial de Ciencias Naturales y Oceanográfico, Puerto Madryn 00507

Brazil

Museu Oceanográfico do Valo do Itajaí, Itajaí 04493

Bulgaria

Prirodonaučen Muzej, Varna 05246

Canada

Centre d'interprétation du Phoque, Grande-Entrée 05899
Marine Museum of the Great Lakes, Kingston 06082
New Marine Centre, Sidney 06971
Shaw Ocean Discovery Centre, Sidney 06972

Chile

Museo Comparativo de Biología Marina, Viña del Mar 07474

Cyprus

Museum of Marine Life of Cyprus, Aghia Napa 09360

Egypt

Aquatics Museum, Suez 10541

Estonia

Eesti Meremuuseum, Tallinn 10626

France

Aquarium-Musée-Institut Paul Ricard, Six-Fours-les-Plages 15767
Centre de la Mer, Paris 14579
Musée Aquarium, Bergerac 11942
Musée-Aquarium d'Arcachon, Arcachon 11662
Musée Aquarium de la Rivière Dordogne, Creysse 12678
Musée-Aquarium du Laboratoire Arago, Banyuls-sur-Mer 11842
Musée de la Mer, Biarritz 11982
Musée-Marinarium du Haut Lavédan, Pierrefitte-Nestalas 14818
Musée Mer et Désert, Villerville 16235
Musée Océanographique, Bonifacio 12052
Musée Océanographique, Kingersheim 13303
Musée Océanographique, Saint-Macaire 15381
Musée Océanographique de l'Odet, Ergué-Gabéric 12843
Nausicaa, Boulogne-sur-Mer 12095
Océanopolis, Brest 12165
Océarium du Croisic, Le Croisic 13578

French Polynesia

Musée de la Mer, Hane 16320

Germany

Das Wattenmeerhaus, Wilhelmshaven 22574
Deutsches Meeresmuseum, Stralsund 21913
Internationales Muschelmuseum, Wangerland 22329
Muschel- und Schneckenmuseum, Norden 20729
Nationalpark-Haus, Butjadingen 17669
Nordseemuseum, Bremerhaven 17573

Guadeloupe

Musée International du Coquillage, Saint-Barthélemy 23279

Iceland

Listasafnið Reykjanesbæjar, Reykjanesbær 24013

India

District Science Centre, Tirunelveli 24407

Israel

Red Sea Maritime Museum, Eilat 24944

Italy

Galata - Museo del Mare e della Navigazione, Genova 26293
Museo del Mare, Furci Siculo 26259
Museo del Mare, Pesaro 27156
Museo del Mare, Piombino 27230
Museo del Mare, Pizzo 27261
Museo del Mare, Ischia Ponte 26411
Museo del Mare, Pollica 27275
Museo di Biologia Marina, Porto Cesareo 27302
Museo di Biologia Marina, Taranto 27942
Museo di Biologia Marina, Fano 26085
Museo di Biologia Marina Paleontologia, Reggio Calabria 27383
Museo Navale Archeologico N. Lamboglia, La Maddalena 26428
Museo Nazionale della Attività Subacquee, Marina di Ravenna 26608

Japan

Fukushima Kaiyo Kagakukan, Iwaki 28605
Marine Science Museum, Shizuoka 29285
NYK Kaiun Hakubutsukan, Yokohama 29604
Shima Marineland, Ago 28348
Umi-no Hakubutsukan, Toba 29361

Korea, Republic

National Maritime Museum, Mokpo 29853

Lebanon

Les Merveilles de la Mer, Jdeidet El-Metn 30170

Lithuania

Lietuvos Jūrų Muziejus, Klaipėda 30282

Macedonia

Ribarsko-Biološki Muzej, Nov Dojran 30511

Madagascar

Musée de la Mer, Nosy-Bé 30540

Mexico

Museo del Mar, Mazatlán 31022

Monaco

Musée Océanographique de Monaco, Monaco 31613

Mozambique

Museum of Inhaca Island, Maputo 31712

Netherlands

Brouws Museum, Brouwershaven 32014
Miramar Zeemuseum, Vledder 32893
Muzee Scheveningen, Den Haag 32093
Sealife Scheveningen, Den Haag 32102

New Zealand

New Zealand Marine Studies Centre, Dunedin 33093

Panama

Centro de Exhibiciones Marinas, Calzada de Amador 34061

Poland

Muzeum Morskie, Szczecin 35162
Muzeum Oceanograficzne Mir, Gdynia 34658

Portugal

Museu da Baleia, Caniçal 35505
Museu da Vida Submarina e da História Submersa, Lisboa 35686
Museu do Mar, Sesimbra 35932
Museu do Mar, Vila do Conde 35991
Museu do Mar - Rei D. Carlos, Cascais 35513
Museu Municipal do Funchal - História Natural, Funchal 35618
Museu Oceanográfico, Setúbal 35938
Museu Oceanográfico Prof. Luiz Saldanha, Azeitão 35453

Puerto Rico

Marine Sciences Museum, Mayagüez 36043

Romania

Muzeul Apelor Mihai Băcescu, Fălticeni 36366
Muzeul Mării, Constanța 36307

Russia

Gosudarstvennyj Severnyj Morskoj Muzej, Archangelsk 36770
Morskoj Muzej, Petrozavodsk 37560
Morskoj Muzej-Akvarium, Vladivostok 37942
Muzej Mirovogo Okeana, Kaliningrad 36992
Muzej Priroda Morja i eë Ochrana, Vladivostok 37947
Naučnyj Morskoj Muzej, Kaliningrad 36993

St. Pierre and Miquelon

Musée Archipélitude, Saint-Pierre 37997

Senegal

Musée de la Mer, Gorée 38025

Slovenia

Pomorski Muzej Sergej Mašera, Piran 38492

Spain

Museo del Mar y de la Sal, Torrevieja 40160
Museo Etnológico de la Laguna Grande, Beteta 39034
Museo Provincial do Mar, Cervo 39169
Palacio del Mar, Donostia-San Sebastián 39254

Sweden

Aquaria Vattenmuseum, Stockholm 40883

Taiwan

Coral Creation, the Cosmos of Buddha Coral Museum, Su-Ao 42343
National Museum of Marine Biology, Checheng 42159
National Museum of Marine Science and Technology, Keelung 42249
Tamkang University Maritime Museum, Tamsui 42449

Tunisia

Musée Océanographique Dar-El-Hout de Salammbô, Salammbô 42647

United Kingdom

Robertson Museum, Millport 45149
Scottish Maritime Museum, Irvine 44535

United States of America

Calvert Marine Museum, Solomons 53142
Chesapeake Bay Maritime Museum, Saint Michaels 52611
Custom House Maritime Museum, Newburyport 51306
Kendall Whaling Museum, Sharon 53050
Maine Maritime Museum, Bath 46628
Marine Museum at Fall River, Fall River 48500
Marineland Ocean Museum, Marineland 50478
Maritime Industry Museum at Fort Schuyler, Bronx 47049
Mel Fisher Maritime Heritage Museum, Key West 49764
MIT Museum, Cambridge 47183
Museum of Yarmouth History, Yarmouth 54324
Niagara Gorge Discovery Center, Niagara Falls 51347
North Carolina Maritime Museum, Beaufort 46666
North Wind Undersea Institute, Bronx 47052
Penobscot Marine Museum, Searsport 52992
Ponce DeLeon Inlet Lighthouse, Ponce Inlet 52019
Shore Village Museum, Rockland 52421
The South Street Seaport Museum, New York 51260
Suffolk County Vanderbilt Museum, Centerport 47315
Texas Maritime Museum, Rockport 52427
Treasures of the Sea Exhibit, Georgetown 48890
Ventura County Maritime Museum, Oxnard 51658
Virginia Marine Science Museum, Virginia Beach 53759
William Scarbrough House, Savannah 52948
Woods Hole Oceanographic Institution Ocean Science Exhibit Center, Woods Hole 54280

Uruguay

Museo del Mar, La Barra 54377

Venezuela

Museo del Mar, Boca de Río 54525

Vietnam

Oceanographic Museum, Nha Trang 54635

Oenology → Wines and Spirits

Office Equipment

Argentina

Museo de la Casa Rosada, Buenos Aires 00200

Austria

Kopiergerätemuseum, Wien 03013
Schreibmaschinenmuseum, Wattens 02897

Finland

Suomen Tietojenkäsittelymuseo, Jyväskylä 10898

Germany

Büromuseum, Mülheim an der Ruhr 20410
Deutsches Schreibmaschinenmuseum, Bayreuth 16950
Museum der Kommunikationstechnik und Bürogeschichte, Bamberg 16913
Telefonmuseum Hittfelder Bahnhof, Seevetal 21726

Mexico

Museo de Telefonía, Delicias 30833

Peru

Museo Nacional de Informática, Lima 34265

Switzerland

Beck's Schreib- und Büromaschinenmuseum, Pfäffikon (Zürich) 41722

Oil → Petroleum Industry and Trade

Opera → Performing Arts

Optics

Belgium

Het Illuseum, Gent 03629

Canada

Museum of Vision Science and Optometry, Waterloo 07288

France

Musée de la Lunette, Morez 14272
Musée de l'Optique, Biesheim 11987

Germany

Brillen-Museum, Homburg 19266
Camera Obscura, Mülheim an der Ruhr 20411
Carl Zeiss Ausstellungszentrum, Oberkochen 20813
Holowood - Holographiemuseum Bamberg, Bamberg 16911
Leica Stammbaum und Museum, Solms 21812
Museum für Optik, Meßtechnik und mathematische Geräte, Nothweiler 20750
Optik Industrie Museum, Rathenow 21190
Optisches Museum der Ernst-Abbe-Stiftung Jena, Jena 19385
Sammlung historischer Mikroskope von Ernst Leitz, Wetzlar 22532
Zeiler Foto- und Filmmuseum, Zeil am Main 22773

Japan

Anekojikan - Kyoto Mangekyo Hakabutsukan, Kyoto 28802

Netherlands

Nationaal Brilmuseum Amsterdam, Amsterdam 31871
Universiteitsmuseum Utrecht, Utrecht 32857

Slovakia

Múzeum J.M. Petzvala, Spišská Béla 38355

Spain

Museo-Colección del Observatorio Astronómico Nacional, Madrid 39529

Switzerland

Illusoria-Land, Ittigen 41497

United Kingdom

British Optical Association Museum, London 44804
Buckingham Movie Museum, Buckingham 43665
Camera Obscura and World of Illusions, Edinburgh 44115

Ornithology → Birds

Painted and Stained Glass

Argentina

Museo Nacional de Arte Decorativo, Buenos Aires 00257

France

Musée de l'Hospice Saint-Roch, Issoudun 13251

Germany

Deutsches Glasmalerei-Museum Linnich, Linnich 20010
Museum für bäuerliche und sakrale Kunst, Ruhpolding 21448
Museum für Zeitgenössische Glasmalerei, Langen, Hessen 19812

Japan

Hamamatsu-shiritsu Bijutsukan, Hamamatsu 28485

Switzerland

Vitromusée Romont, Romont 41773

Painting

Argentina

Casa Bruzzone, Mar del Plata 00443
Estudio Museo Ramoneda, Humahuaca 00386
El Fogon de los Arrieros, Resistencia 00521
Museo Atelier Antonio Ortiz Echagüe, Carro Quemado 00295
Museo Austral de Pintura Primitiva Moderna Naif, Esquel 00369
Museo Casa Taller Celia Chevalier, Buenos Aires 00175
Museo de Bellas Artes de la Boca Quinquela Martín, Buenos Aires 00187
Museo Maurice Minkowski, Buenos Aires 00251
Museo Municipal de Arte Angel María de Rosa, Junín 00395
Museo Municipal de Arte Juan Carlos Castagnino, Mar del Plata 00446
Museo Municipal de Artes Visuales, Quilmes 00511
Museo Municipal de Artes Visuales Sor Josefa Díaz y Clucellas, Santa Fé 00619
Museo Nacional de Bellas Artes, Buenos Aires 00259
Museo Provincial de Bellas Artes Dr. Pedro E. Martínez, Paraná 00488
Palacio del Mate, Posadas 00502
Pinacoteca, Córdoba 00339

Armenia

Dom Muzej M. Saryana, Yerevan 00732
Gevorg Grigoryan - Giotto - Studio-Museum, Yerevan 00736
Khoren Ter-Harutyan Museum, Echmiadzin 00697
Mher Abeghyan Museum, Echmiadzin 00698
Minas Avetisyan Museum, Yerevan 00745
National Gallery of Armenia, Yerevan 00751

Australia

Brauhinia Shire Bicentennial Art Gallery, Springsure 01523
Dalby Regional Gallery, Dalby 01020
Harvery Art Gallery, Harvey 01146
McCrae Homestead Museum, McCrae 01245
National Gallery of Victoria, Melbourne 01288
Old Ware House Gallery, Maryborough, Queensland 01268
Tamworth Regional Gallery, Tamworth 01569
Teys Art Gallery, Miles 01303
Tweed River Art Gallery, Murwillumbah 01340
Wollongong City Gallery, Wollongong 01663

Austria

Alfons Graber-Museum, Steinach 02763
Artothek, Krems 02223
Bilger Breustedt Haus, Taufkirchen an der Pram 02796
Burg Kreuzenstein, Korneuburg 02217
Essl Museum - Kunst der Gegenwart, Klosterneuburg 02200
Europäisches Museum für Frieden, Stadtschlaining 02756
Galerie Adolf Blaim, Messern 02377
Galerie im Teisenhoferhof der Malschule Motiv Wachau, Weißenkirchen in der Wachau 02904
Galerie Ludwig Bürgel, Seekirchen am Wallersee 02735
Gauermann-Museum, Miesenbach 02381
Georg Rendl-Museum, Sankt Georgen bei Salzburg 02643
Liechtenstein Museum, Wien 03024
Museum Atelier Franz Zeh, Heidenreichstein 02061
Museum der Stadt Lienz, Lienz 02301
Museum des Augustiner Chorherrenstifts, Reichersberg 02572
Museum Zinkenbacher Malerkolonie, Sankt Gilgen 02649
Otto Wagner Pavillion Karlsplatz, Wien 03053
Phantasten Museum Wien, Wien 03056
Prof. Eidenberger-Museum, Niederwaldkirchen 02449
Riesenrundgemälde Schlacht am Bergisel, Innsbruck 02128
Schloß Ambras, Innsbruck 02130
Schloss Greinburg, Grein an der Donau 01989
Schloßmuseum Niederleis, Niederleis 02446
Stiftsgalerie Seitenstetten, Seitenstetten 02738
Stiftsmuseum Praemonstratenser-Chorherren Stift Schlägl, Aigen im Mühlkreis 01700
Stiftung Hubert Aratym, Gutenstein 02026

Azerbaijan

Bakhruz Kangarli Museum, Naxçivan 03180
Kartinnaya Gallery, Stepanakert 03198
Rustam Mustafayev ad. Azerbaycan Dövlet İncesenet Muzeyi, Bakı 03169
State Art Museum, Bakı 03172

Belarus

Chagall Museum, Vitebsk 03257

Belgium

BAM - Beaux Arts Mons, Mons 03879
Broelmuseum, Kortrijk 03759
Bruggemuseum - Onthaalkerk van Onze-Lieve-Vrouw, Brugge 03412
Château feodal de Corroy-le-Château, Corroy-le-Château 03526
Collections Artistiques de l'Université de Liège, Liège 03795
Eugeen van Mieghem Museum, Antwerpen 03285
Hôpital Notre-Dame de la Rose, Lessines 03786
Hospitaalmuseum Onze-Lieve-Vrouw ter Potterie, Brugge 03416
Keizerskapel, Antwerpen 03289
Koninklijk Museum voor Schone Kunsten Antwerpen, Antwerpen 03290
Maagdenhuismuseum, Antwerpen 03291
Musée Alexandre-Louis Martin, Carnières 03509
Musée Constantin Meunier, Bruxelles 03460
Musée de l'Art Wallon, Liège 03808
Musée de l'Assistance Publique, Bruxelles 03470
Musée des Beaux-Arts Charleroi, Charleroi 03516
Musée d'Ixelles-Museum van Elsene, Bruxelles 03479
Musée du Vieux-Cimetière, Soignies 04036
Musée Ducal, Bouillon 03393
Musée Leon de Smet, Deurle 03545
Museum Charles Delporte, Damme 03534
Museum Constant Permeke, Jabbeke 03740
Museum Félix de Boeck, Drogenbos 03563
Museum Gust de Smet, Deurle 03547
Museum Jakob Smits, Mol 03876
Museum Mayer van den Bergh, Antwerpen 03296
Museum Modest Huys, Zulte 04170
Museum Smidt van Gelder, Antwerpen 03299
Museum Tysmans, Houthalen 03720
Provinciaal Museum Stijn Streuvels, Ingooigem 03733
Rockoxhuis, Antwerpen 03313
Stadhuis van Antwerpen, Antwerpen 03316
Timmermans-Opsomerhuis, Lier 03820

Brazil

Coleção do Palácio do Governo de Campos do Jordão, Campos de Jordão 04398
Museu Alfredo Andersen, Curitiba 04428
Museu de Arte Contemporânea da Universidade de São Paulo, São Paulo 04879
Museu de Arte Sacra Dom Ranulfo, Maceió 04546
Museu Internacional de Arte Naïf do Brasil, Rio de Janeiro 04754
Pinacoteca do Estado de São Paulo, São Paulo 04911

Bulgaria

Chudožestvena Galerija, Burgas 04997
Chudožestvena Galerija, Čirpan 05008
Chudožestvena Galerija, Vidin 05261
Chudožestvena Galerija, Žeravna 05268
Chudožestvena Galerija Boris Denev, Veliko Tărnovo 05252
Chudožestvena Galerija Ruse, Ruse 05151
Kăšta-muzej Stanislav Dospevski, Pazardžik 05108
Muzej Georgi Velčev, Varna 05240
Okrăžna Chudožestvena Galerija Vladimir Dimitrov-Majstora, Kjustendil 05067
Sofijska Gradska Chudožestvena Galerija, Sofia 05203

Canada

Art Gallery of Ontario, Toronto 07091
Art Gallery of Peterborough, Peterborough 06589
Bibliothèque Centrale de Prêt du Saguenay Lac-Saint-Jean, Alma 05320
Centre Culturel de Lachute, Lachute 06124
Centre d'Exposition l'Imagier, Gatineau 05857
Chapel Gallery, North Battleford 06466
Estevan Art Gallery and Museum, Estevan 05784
Galerie Colline, Edmundston 05763
Legislative Building Art Galleries, Regina 06719
Lindsay Gallery, Lindsay 06172
Maison Rodolphe-Duguay, Nicolet 06456
Michener House Museum, Lacombe 06125
Mind and Matter Gallery, South Surrey 06996
Musée du Témiscouata, Notre-Dame-du-Lac 06491
New Brunswick Art Bank, Fredericton 05834
Oak Hall, Niagara Falls 06447
Perrault's Museum, Val Marie 07195
Vancouver Art Gallery, Vancouver 07226
Wulastook Museums, Fredericton 05841

Chile

Casa del Arte de la Universidad de Concepción, Concepción 07432
Museo de Bellas Artes, Viña del Mar 07475
Museo Nacional de Bellas Artes de Santiago de Chile, Santiago de Chile 07461
Pinacoteca de la Universidad de Concepción, Concepción 07436

China

Lin San Zhi Art Display Center, Zhujiang 08514
Luweizhao Painting and Calligraphy House, Pinghu 08083
Museum of Fine Arts, Beijing 07551
Qingdao Painting Gallery, Qingdao 08111
Quanzhou Painting Gallery, Quanzhou 08133
Shaanxi Chinese Painting Gallery, Xian 08359
Shandong Painting Gallery, Jinan 07883
Shanxi Painting Gallery, Taiyuan 08274
Shenzhen Painting Gallery, Shenzhen 08226
Taiyuan Painting Gallery, Taiyuan 08275
Tianjin Painting Gallery, Tianjin 08287
Xinjiang Painting Gallery, Urumchi 08308

Colombia

Museo de Arte Colonial, Bogotá 08557
Museo de Arte Contemporáneo, Bogotá 08558
Museo Efrain Martínez Zambrano, Popayán 08728
Museo Pedro Nel Gomez, Medellín 08685

Croatia

Ca' Bon Gallery, Motovun 08911
Galerija Likovnih Umjetnosti, Osijek 08922
Galerija Umjetnina, Split 08956
Ivan Meštrović Gallery, Split 08957
Kaštelet Crikvine, Split 08958
Muzej za Umjetnost i Obrt, Zagreb 09008
Zavičajni Muzej Grada Rovinja, Rovinj 08943

Cuba

Casa Natal del Pintor Wifredo Lam, Santa Clara 09292
Mariano Rodríguez Gallery, La Habana 09107
Museo Casa de Carmen Montilla, La Habana 09116
Museo Concha Ferrant, La Habana 09129
Museo de La Imagen, Vista Alegre 09352
Museo El Hurón Azul, La Habana 09143

Cyprus

Byzantine Museum and Art Galleries, Lefkosia 09372
Limassol Municipal Art Gallery, Lemesos 09388

Czech Republic

Galerie Antonína Chittussiho, Ronov nad Doubravou 09819
Galerie Felixe Jeneweina, Kutná Hora 09613
Muzeum Aloise Jiráska, Hronov 09538
Muzeum umění a designu, Benešov 09402
Památník Adolfa Kašpara, Loštice 09639

Denmark

Ballerup Egnsmuseum, Ballerup 09975
Faaborg Museum for Fynsk Malerkunst, Faaborg 10008
Jugendhuset, Varde 10306
Michael og Anna Anchers Hus og Saxilds Gaard, Skagen 10261
Mølsteds Museum, Dragør 09990
Museet på Koldinghus, Kolding 10163
Randers Kunstmuseum, Randers 10230
Skagens Museum, Skagen 10263
Skovgaard Museet i Viborg, Viborg 10316
Statens Museum for Kunst, København 10152
Storm P.-Museet, Frederiksberg 10023
Vejen Kunstmuseum, Vejen 10309
Vejle Kunstmuseum, Vejle 10310

Egypt

Museum of Fine Arts and Cultural Center, Alexandria 10475

Estonia

Johann Köleri Muuseum, Vastemõisa 10668
Kadrioru Loss Väliskunsti Muuseum, Tallinn 10632
Kristjan Raua Majamuuseum, Tallinn 10634
Mikkeli Muuseum, Tallinn 10636

Finland

Nelimarkka Museo . Etelä-Pohjanmaan Aluetaidemuseo ja Nelimarkka-resedenssi, Alajärvi 10708
Särestöniemi-Museo, Kaukonen 10931
Taidemuseo, Kokkola 10975
Tampereen Taidemuseon Muumilaakso, Tampere 11365

France

Centre Artistique Jean Baffier, Sancoins 15622
Centre d'Art Contemporain, Istres 13253
Collection Historial Cognacq-Jay, Paris 14584
Collections de la Fondation Roger-van-Rogger, Bandol 11836
Collections de l'Ecole Nationale Supérieure des Beaux-Arts, Paris 14590
Exposition Permanente Pierre Miquel, Maurs 14042
Fondation Custodia, Collection Frits Lugt, Paris 14598
Fondation Le Corbusier, Paris 14601
Fondation Van-Gogh Arles, Arles 11682
Galerie de Nesle - Exposition, Paris 14604
Maison André Derain, Chambourcy 12380
Maison d'Artiste de la Grande Vigne, Dinan 12747
Maison Natale du Maréchal Foch, Tarbes 15861
Musée Adzak, Paris 14628
Musée Antoine Lécuyer, Saint-Quentin (Aisne) 15474
Musée Arteum, Châteauneuf-le-Rouge 12448
Musée Atelier du Trompe l'Oeil et du Décor Peint, Périgueux 14777
Musée Baron Gérard, Bayeux 11873
Musée Bourdelle, Paris 14636
Musée Calvet, Avignon 11789
Musée Charles Léandre, Domfront 12758
Musée Comtadin-Duplessis, Carpentras 12296
Musée Condé, Chantilly 12399
Musée d'Art et d'Archéologie, Aurillac 11754
Musée d'Art et d'Archéologie, Senlis 15715
Musée d'Art et d'Histoire Marcel Dessal, Dreux 12792
Musée d'Art et d'Histoire Romain Rolland, Clamecy 12539
Musée d'Art Naïf, Nice 14429
Musée d'Art Religieux, Castelnau-Magnoac 12310
Musée d'Art Sacré de Rostilas Loukine, Arsonval 11704
Musée Daubigny, Auvers-sur-Oise 11770
Musée de Grenoble, Grenoble 13134
Musée de la Tour du Moulin, Marcigny 13957
Musée de l'Image, Épinal 12839
Musée de l'Œuvre Viollet-le-Duc, Vézelay 16137
Musée de Région Auguste Chabaud, Graveson 13124
Musée de Vulliod Saint-Germain, Pézenas 14808
Musée Départemental d'Art Ancien et Contemporain, Épinal 12840
Musée des Arts et d'Histoire, Locronan 13797
Musée des Augustins, Toulouse 15939
Musée des Beaux-Arts, Arras 11698
Musée des Beaux-Arts, Béziers 11976
Musée des Beaux-Arts, Blois 12035
Musée des Beaux-Arts, Brest 12161
Musée des Beaux-Arts, Carcassonne 12285
Musée des Beaux-Arts, Chambéry 12364
Musée des Beaux-Arts, Gaillac 13026
Musée des Beaux-Arts, Libourne 13741
Musée des Beaux Arts, Lons-le-Saunier 13817
Musée des Beaux-Arts, Marseille 14002
Musée des Beaux-Arts, Menton 14082
Musée des Beaux-Arts, Mirande 14130
Musée des Beaux-Arts, Quimper 14971
Musée des Beaux-Arts, Rennes (Ille-et-Vilaine) 15016
Musée des Beaux-Arts, Rouen 15115
Musée des Beaux-Arts, Strasbourg 15829
Musée des Beaux-Arts, Tourcoing 15953
Musée des Beaux-Arts, Saintes 15584
Musée des Beaux-Arts, Mulhouse 14325
Musée des Beaux-Arts de Dunkerque, Dunkerque 12798
Musée des Beaux-Arts et d'Archéologie, Besançon 11957
Musée des Beaux-Arts et de la Dentelle, Calais 12248
Musée des Beaux-Arts Hyacinthe Rigaud, Perpignan 14791
Musée du Centre de Recherches sur les Monuments Historiques, Paris 14694
Musée du Château, Flers 12944
Musée du Faouët, Le Faouët 13582
Musée du Louvre, Paris 14698
Musée du Luxembourg, Paris 14699
Musée du Pastel, Magrin 13916
Musée du Terroir Hérissonnais, Hérisson 13205
Musée Église Catholique de l'Assomption, Rosenwiller 15105
Musée Elise Rieuf, Massiac 14028
Musée en Herbe, Paris 14707
Musée Estrine, Saint-Rémy-de-Provence 15492
Musée Eugène Boudin, Honfleur 13223
Musée Fabre, Montpellier 14248
Musée Fesch, Ajaccio 11527
Musée Francisque Mandet, Riom 15045
Musée François Desnoyer, Saint-Cyprien 15224
Musée Georges Garret, Vesoul 16133
Musée Grobet-Labadié, Marseille 14009
Musée Gustave Stoskopf, Brumath 12199
Musée Hardy, Clécy 12542
Musée Historique, Mont-Saint-Michel 14162
Musee International de la Miniature, Lyon 13897
Musée Jeanne d'Aboville, La Fère 13358
Musée Jules Garinet, Châlons-en-Champagne 12363
Musée Langlois, Beaumont-en-Auge 11900
Musée Le Petit Louvre, La Pacaudière 13399
Musée Louis Cornu, Wassigny 16289
Musée Louis Jourdan, Saint-Paul-de-Varax 15443
Musée Magnin, Dijon 12742
Musée Malraux, Le Havre 13598
Musée Marguerite Vidal des Arts et Traditions Populaires, Moissac 14140
Musée Maurice Gottlob, Mougins 14299
Musée Moralès, Port-de-Bouc 14914
Musée Morice Lipsi, Rosey 15106
Musée Municipal, Bernay 11948
Musée Municipal, Coutances 12661
Musée Municipal, La Roche-sur-Yon 13419
Musée Municipal, Vire 16251
Musée Municipal de la Chartreuse, Douai 12775
Musée Municipal des Beaux-Arts et d'Archéologie, Valence (Drôme) 16040
Musée Municipal Paul Lafran, Saint-Chamas 15200
Musée National Ernest Hébert, Paris 14731
Musée Paul-José Gosselin, Saint-Vaast-la-Hougue 15529
Musée Paul Surtel, Reuilly (Indre) 15026
Musée Raoul Dastrac, Aiguillon 11506
Musée Sahut, Simiane-Collongue 15759
Musée Saint-Vic, Saint-Amand-Montrond 15161
Musée Simon-Segal, Aups 11749
Musée Terrus, Elne 12818
Musée Thomas-Henry, Cherbourg-Octeville 12505
Musée Vivant Denon, Chalon-sur-Saône 12360
Musées de la Cour d'Or, Metz 14102

Georgia

Georgian National Gallery, Tbilisi 16360

Germany

1822-Forum Frankfurter Sparkasse, Frankfurt am Main 18352
Allgäu-Museum, Kempten 19510
Anhaltische Gemäldegalerie, Dessau 17826
August Macke Haus, Bonn 17447
Bayerisches Armeemuseum, Ingolstadt 19339
Braith-Mali-Museum, Biberach an der Riß 17319
C.O. Müller-Galerie, Eichstätt 18080
Edwin Scharff Museum, Neu-Ulm 20587
Erstes Imaginäres Museum, Wasserburg am Inn 22343
Felix-Müller-Museum im Zehntspeicher, Neunkirchen am Brand 20638
Forum Konkrete Kunst Erfurt, Erfurt 18202
Friedrich-Eckenfelder-Galerie, Balingen 16899
Friese-Galerie, Lüchow 20087
Fürstliches Schloß Mausoleum, Bückeburg 17621
Gablonzer Archiv Museum, Kaufbeuren 19486
Galerie am Alten Markt, Rostock 21391
Galerie der Fekete-Stiftung, Roßdorf 21381
Galerie im Alten Rathaus, Prien am Chiemsee 21126
Galerie im Angelburger Kunst- und Kulturhaus, Angelburg 16546
Galerie im Kulturhaus Spandau, Berlin 17093
Galerie in der Brasilianischen Botschaft, Berlin 17099
Galerie Parterre, Berlin 17106
Gaudnek-Museum, Altomünster 16523
Gemälde- und Skulpturensammlung, Nürnberg 20763
Gemäldegalerie, Berlin 17116
Gemäldegalerie Alte Meister, Kassel 19463
Gentil-Haus, Aschaffenburg 16582
Große Kunstschau im Roselius Museum, Worpswede 22691
Heimatmuseum, Scheeßel 21536
Heinrich-von-Zügel-Gedächtnisgalerie, Wörth am Rhein 22641
Herkomer-Museum am Mutterturm, Landsberg am Lech 19794
Historisches Museum Bamberg, Bamberg 16910
Hudetz-Turm, Wiesent 22561
Joachim Ringelnatz Museum, Cuxhaven 17755
Künstlerhaus Walter Helm, Aschaffenburg 16584
Kulturhaus der Otto-Hellmeier-Stiftung, Raisting 21172
Kulturhistorisches Museum Prenzlau im Dominikanerkloster, Prenzlau 21118
Kunsthalle im Schloss, Isny im Allgäu 19366
Kunsthalle Schweinfurt, Schweinfurt 21683
Kunstmühle Schwaan, Schwaan 21644
Kunststätte Bossard, Jesteburg 19395
Kunstverein Konstanz e.V., Konstanz 19680
Ludwig-Doerfler-Galerie, Schillingsfürst 21553
Luza Acht, Lübben 20068
Mainfränkisches Museum Würzburg, Würzburg 22713
Malerstübchen Willingshausen, Willingshausen 22587
Marie-Hager-Haus, Burg Stargard 17647
Max Böhlen Museum, Kandern 19427
Museum Altes Rathaus, Pirmasens 21041
Museum am Burghof, Lörrach 20016

P

Museum am Modersohn-Haus, Worpswede ... 22696
Museum Bautzen, Bautzen ... 16940
Museum Burg Falkenstein, Falkenstein, Harz .. 18292
Museum Folkwang, Essen ... 18263
Museum Künstlerhaus Gasteiger, Utting am Ammersee ... 22194
Museum Rolf Werner, Bansin, Seebad ... 16926
Museum Starnberger See, Starnberg ... 21864
Museum und Galerie in Uigendorf, Unlingen .. 22174
Museum und Stiftung Anton Geiselhart, Münsingen ... 20501
Museum Wilnsdorf, Wilnsdorf ... 22589
Neue Residenz Bamberg, Bamberg ... 16916
Olaf-Gulbransson-Museum für Graphik und Karikatur, Tegernsee ... 22018
Otto Dill-Museum, Neustadt an der Weinstraße 20673
Otto-Dix-Haus, Gera ... 18597
Paul-Röder-Museum, Marktoberdorf ... 20220
Podium Kunst, Schramberg ... 21632
Potsdam-Museum, Potsdam ... 21103
Sammlung Crous im Alten Kurhaus, Aachen .. 16406
Schiffahrtsmuseum Nordfriesland, Husum ... 19300
Schloss Königs Wusterhausen, Königs Wusterhausen ... 19638
Schloß Villa Ludwigshöhe, Edenkoben ... 18056
Schloss Weissenstein, Pommersfelden ... 21075
Schloss Wilhelmshöhe-Weißensteinflügel, Kassel ... 19476
Schloßmuseum Murnau, Murnau am Staffelsee 20535
Siegerlandmuseum mit Ausstellungsforum Haus Oranienstraße, Siegen ... 21756
Staatliche Kunsthalle Karlsruhe, Karlsruhe ... 19448
Staatsgalerie in der Benediktiner-Abtei, Ottobeuren ... 20956
Städtische Galerie, Schieder-Schwalenberg ... 21546
Städtische Gemäldegalerie, Füssen ... 18503
Städtische Kunstgalerie im Deutschordenshaus, Donauwörth ... 17899
Städtische Wessenberg-Galerie Konstanz, Konstanz ... 19683
Stiftung Schloss Ahrensburg, Ahrensburg ... 16441
Strieffler-Haus, Landau in der Pfalz ... 19793
Strübhaus-Haus der Malkunst, Veringenstadt . 22222
Suermondt-Ludwig-Museum, Aachen ... 16408
Syker Vorwerk, Syke ... 21992
Vonderau Museum, Fulda ... 18512
Vonderau Museum Fulda, Fulda ... 18513
Wallraf-Richartz-Museum Fondation Corboud, Köln ... 19635
Weibertreu-Museum, Weinsberg ... 22434
Werner Weckwerth Museum, Friedrich-Wilhelm-Lübke-Koog ... 18473
Wilhelm-Hack-Museum Ludwigshafen am Rhein, Ludwigshafen am Rhein ... 20058
Wilhelm-Ostwald-Gedenkstätte, Großbothen .. 18783

Greece

Angelos and Angeliki Giallina Gallery, Corfu .. 22973
Gounaropoulos Museum of Zografou, Zografou ... 23247
Museum Alex Mylona, Athinai ... 22928
Museum of Tinian Artists, Pyrgos ... 23163
N. Hadjikyriakos-Ghikas Gallery, Athinai ... 22941
Teloglion Foundation of Art, Thessaloniki ... 23222
Teriade Museum, Mytilini ... 23113
Theofilos Museum, Mytilini ... 23114

Guam

Isla Center for the Arts at the University of Guam, Mangilao ... 23287

Hungary

Alföldi Galéria, Hódmezővásárhely ... 23609
Benedek Péter Emlékház, Uszód ... 23942
Bozsó Gyűjtemény, Kecskemét ... 23643
Cselley Ház, Mosonmagyaróvár ... 23718
Egry József Emlékmúzeum, Badacsony ... 23343
Gyűjtemények Háza, Kiskunhalas ... 23669
Hincz Gyula Állandó Gyűjtemény, Vác ... 23944
Horváth József Festőművész Állandó Kiállítása, Sopron ... 23810
Káplár Miklós Gyűjtemény, Hajdúböszörmény . 23604
Kohán György Múzeum, Gyula ... 23596
Kunffy Lajos Emlékmúzeum, Somogytúr ... 23806
Molnár C. Pál Műterem-múzeum, Budapest ... 23452
Munkácsy Mihály Emlékház, Békéscsaba ... 23370
Munkácsy Mihály Múzeum, Békéscsaba ... 23371
Művészetek Kincsesháza - Vaszary Emlékház, Kaposvár ... 23631
Rippl-Rónai Emlékház, Kaposvár ... 23633
Szántó Piroska Emlékmúzeum, Budapest ... 23476
Vachott Sándor Városi Könyvtár, Gyöngyös ... 23575
Zettl Langer Gyűjtemény, Sopron ... 23823
Zichy Mihály Emlékmúzeum, Zala ... 23969

Iceland

Listasafn Einars Jónssonar, Reykjavík ... 24021

India

Academy of Fine Arts Museum, Kolkata ... 24266
Central Museum, Bhopal ... 24091
Sri Chitra Art Gallery, Thiruvananthapuram ... 24403
Dogra Art Museum, Jammu ... 24236
Folklore Museum, Mysore ... 24328
Government Museum, Udaipur ... 24414
Government Museum Vellore, Vellore ... 24428
Indian Museum, Kolkata ... 24275
Institute Menezes Braganza, Panaji ... 24341
Sri Jayachamarajendra Art Gallery, Mysore ... 24329
Maharaja Fate Singh Museum, Vadodara ... 24419
Mutua Museum, Imphal ... 24226
Salar Jung Museum, Hyderabad ... 24217
Sangli State Museum, Sangli ... 24377
TTD Museum, Tirupati ... 24409
Victoria Memorial Hall, Kolkata ... 24283

Indonesia

Gunarsa Museum, Klungkung ... 24516
Museum Le Mayeur, Sanur ... 24552
Neka Museum and Gallery, Gianyar ... 24472

Ireland

Bantry House, Bantry ... 24743
Glebe House and Gallery, Letterkenny ... 24861
Limerick City Gallery of Art, Limerick ... 24866
The Maritime Museum, Dublin ... 24806

Israel

Uri and Rami Nechushtan, Kibbutz Ashdot Yaakov ... 25032
Zetlin Museum, Ramat Gan ... 25087

Italy

Casa Museo Signorini-Corsi, L'Aquila ... 26454
Casa Museo V. Cermignani, Giulianova ... 26345
Casa Natale di Leonardo da Vinci, Vinci ... 28283
Cenacolo di Sant'Apollonia, Firenze ... 26147
Ciruito Museale Urbano, Cascia ... 25703
Civico Museo d'Arte Moderne, Anticoli Corrado ... 25241
Complesso Museale di Santa Maria della Scala, Siena ... 27836
Galleria Civica, Valdagno ... 28136
Galleria Regionale d'Arte Moderna e Contemporanea, Sassoferrato ... 27760
Musei Civici di Palazzo d'Avalos, Vasto ... 28169
Musei Provinciali di Gorizia, Gorizia ... 26351
Museo Alessi, Enna ... 26055
Museo Archeologico e Pinacoteca, Fano ... 26084
Museo Archeologico e Pinacoteca, Piacenza .. 27185
Museo Archeologico G. Moretti, San Severino Marche ... 27684
Museo Baroffio e del Santuario del Sacro Monte sopra Varese, Varese ... 28162
Museo Berenziano, Cremona ... 26002
Museo Borromeo, Isola Bella ... 26415
Museo Capitolare, Ortona ... 27002
Museo Carlo Venturini, Massa Lombarda ... 26623
Museo Civico, Casale Monferrato ... 25695
Museo Civico, Rieti ... 27402
Museo Civico Ala Ponzone, Cremona ... 26003
Museo Civico d'Arte, Pordenone ... 27297
Museo Civico e d'Arte Sacra, Colle di Val d'Elsa ... 25937
Museo Civico Il Correggio, Correggio ... 25980
Museo Civico Rocca Flea, Gualdo Tadino ... 26380
Museo d'Arte Antica, Milano ... 26694
Museo d'Arte della Città di Ravenna, Ravenna 27372
Museo d'Arte Sacra, San Leo ... 27653
Museo degli Affreschi G.B. Cavalcaselle, Verona ... 28233
Museo del Cenedese, Vittorio Veneto ... 28298
Museo della Civiltà Contadina, Sassocorvaro .. 27758
Museo della Fondazione Giovanni Scaramangà di Altomonte, Trieste ... 28092
Museo di Capodimonte, Napoli ... 26895
Museo di Castelvecchio, Verona ... 28235
Museo di Palazzo Ducale, Mantova ... 26596
Museo di Sant'Agostino, Genova ... 26309
Museo Domenicani, Taggia ... 27935
Museo e Pinacoteca Basilio Cascella, Pescara ... 27164
Museo e Pinacoteca Civica, Alessandria ... 25193
Museo e Pinacoteca della Basilica di San Paolo, Roma ... 27504
Museo e Pinacoteca S. Gentili, San Ginesio ... 27642
Museo Francesco Borgogna, Vercelli ... 28221
Museo Franco Azzinari, Altomonte ... 25210
Museo Lodovico Pogliaghi, Varese ... 28163
Museo Missionario delle Grazie, Rimini ... 27410
Museo Pinacoteca A. Salvucci, Molfetta ... 26765
Museo Poldi Pezzoli, Milano ... 26714
Museo Sant'Andrea, Clusone ... 25928
Museo Storico Aloisiano, Castiglione delle Stiviere ... 25761
Palazzo Attems-Petzenstein, Gorizia ... 26353
Palazzo Zabarella, Padova ... 27051
Pinacoteca, Teramo ... 27956
Pinacoteca Cascella, Ortona ... 27004
Pinacoteca Chiesa di San Tomaso Becket, Padova ... 27052
Pinacoteca Civica, Cento ... 25802
Pinacoteca Civica, Fermo ... 26103
Pinacoteca Civica, Imperia ... 26407
Pinacoteca Civica, Monza ... 26865
Pinacoteca Civica, Pieve di Cento ... 27214
Pinacoteca Civica A. Ricci, Monte San Martino 26804
Pinacoteca Civica Melozzo degli Ambrogi, Forlì ... 26230
Pinacoteca Comunale, Faenza ... 26079
Pinacoteca Comunale, Massa Marittima ... 26633
Pinacoteca Comunale, Spoleto ... 27895
Pinacoteca Comunale Alberto Martini, Oderzo 26970
Pinacoteca Comunale Foresiana, Portoferraio 27310
Pinacoteca del Duomo, Cittadella ... 25912
Pinacoteca dell' Accademia Albertina di Belle Arti, Torino ... 28030
Pinacoteca dell'Accademia dei Concordi e del Seminario, Rovigo ... 27575
Pinacoteca e Musei Civici, Jesi ... 26427
Pinacoteca e Musei Comunali, Macerata ... 26556
Pinacoteca G. A. Levis, Racconigi ... 27354
Pinacoteca G. Spano, Ploaghe ... 27263
Pinacoteca Giovanni e Marella Agnelli, Torino 28031
Pinacoteca Giovanni Morscio, Dolceacqua ... 26040
Pinacoteca Manfrediana, Venezia ... 28208
Pinacoteca Nazionale, Cagliari ... 25589
Pinacoteca Nazionale, Siena ... 27846
Pinacoteca Parrocchiale, Corridonia ... 25982
Pinacoteca Repossi, Chiari ... 25856
Quadreria Cesarini, Fossombrone ... 26242
Raccolta O. Recchione, Palena ... 27064
Refettorio dell'ex-Convento di Sant'Onofrio detto di Fuligno, Firenze ... 26202
Villa Carlotta, Tremezzo ... 28053

Jamaica

Institute of Jamaica, Kingston ... 28328

Japan

Akita-kenritsu Bijutsukan, Akita ... 28354
Asahikawa Bijutsukan, Asahikawa ... 28368
Bridgestone Bijutsukan, Tokyo ... 29373
Chiba-shi Bijutsukan, Chiba ... 28393
Eisei Bunko Museum, Tokyo ... 29385
Fujiyama Museum, Fuji-Yoshida ... 28420
Hashimoto Kansetsu Kinenkan, Kyoto ... 28812
Hazama Inosuke Bijutsukan, Kaga ... 28624
Hiroshima Bijutsukan, Hiroshima ... 28532
Ibaraki Daigaku Izura Bijutsu Bunka Kenkyujo, Kitaibaraki ... 28707
Kagawa-kenritsu Higashiyama Kaii Setouchi Bijutsukan, Sakaide ... 29201
Kansetsu Hashimoto Bijutsukan, Kyoto ... 28820
Kawamura Bijutsukan, Yuda ... 29630
Kawanabe Kyosai Memorial Museum, Warabi 29576
Kodama Museum, Kagoshima ... 28630
Kojima Torajiro Kinenkan, Kurashiki ... 28783
Kumagai Morikazu Museum, Tokyo ... 29423
Kuroda Kinenshitsu, Tokyo ... 29426
Kushiro Art Museum, Kushiro ... 28798
Marugame Bijutsukan, Marugame ... 28880
Meiji Memorial Picture Gallery, Tokyo ... 29435
Mukai Junkichi Fuzoku Bijutsukan, Tokyo ... 29440
Narukawa Bijutsukan, Hakone ... 28479
Ryushi Kinenkan, Tokyo ... 29461
Shibuya Bijutsukan, Fukuyama ... 28449
Takasaki-shi Bijutsukan, Takasaki ... 29333
Tasaki Bijutsukan, Karuizawa ... 28672
Tenshi Bijutsukan, Nasu ... 29052
Trick Bijutsukan, Nasu ... 29053
Tsuchida Bakusen Bijutsukan, Sawata ... 29238
Ueno Royal Museum, Tokyo ... 29501
Yamagata Bijutsukan, Yamagata ... 29578
Yonezawa Municipal Uesugi Museum Uesugi Shrine Keisyo Palace Museum, Yonezawa ... 29627

Korea, Republic

Onggi Folk Museum and Institute, Seoul ... 29911
Woljeon Art Museum, Seoul ... 29934

Kyrgyzstan

Muzej-masterskaja Gapara Ajtieva, Biškek ... 29958

Latvia

Jaņa Rozentāla un Rūdolfa Blaumaņa Muzejs, Rīga ... 30071

Lebanon

Daheshite Museum, Beirut ... 30156
Gibran Museum, Bsharri ... 30166

Lithuania

Antano Žmuidzinavičiaus Memorialinis Muziejus, Kaunas ... 30239
Dailininkės Veronikos Šleivytės Paveikslų Galerija, Viktariskes ... 30398
Kretingos Muziejus, Kretinga ... 30288
Nacionalinis Mikalojaus Konstantino Čiurlionio Dailės Muziejus, Kaunas ... 30262
Rokiškio Krašto Muziejus, Rokiškis ... 30346
Žemaičių Dailės Muziejus, Plungė ... 30338

Luxembourg

Villa Vauban, Luxembourg ... 30486

Malaysia

Ibrahim Hussein Foundation Museum, Pulau Langkawi ... 30612
Kedah State Art Gallery, Alor Setar ... 30551

Malta

National Museum of Fine Arts, Valletta ... 30644
Zabbar Sanctuary Museum, Zabbar ... 30662

Martinique

Musée Paul Gauguin, Le Carbet ... 30668

Mexico

Galeria de Arte del Aeropuerto, México ... 31039
Galería de Arte Jaguar Despertado, Villahermosa ... 31548
Museo Casa de Agustín Arrieta, Puebla ... 31267
Museo Centro de la Imagen, México ... 31062
Museo de la Acuarela del Estado de México, Toluca ... 31489
Museo de las Pinturas Rupestres de San Ignacio, Pueblo de San Ignacio ... 31292
Museo Franco Lázaro, Chiapa de Corzo ... 30763
Museo Galería Manuel Lepe, Puerto Vallarta ... 31297
Museo Nacional de la Acuarela, México ... 31135
Museo Rubén Herrera, Saltillo ... 31317

Montenegro

Umetnička Galerija Velimir A. Leković, Bar ... 31635
Umjetnička Galerija Josip-Bepo Benkovic, Herceg-Novi ... 31645

Morocco

American Legation Museum, Tanger ... 31694

Myanmar

Mon State Museum, Mawlamyine ... 31722

Nepal

Picture Gallery, Bhaktapur ... 31772

Netherlands

Diem Gallery, Amsterdam ... 31837
Familie Vrouwenhofje - Hofje van Aerden, Leerdam ... 32470
Frans Hals Museum, Haarlem ... 32295
Kijk- en Luistermuseum, Bant ... 31948
Koninklijk Paleis te Amsterdam, Amsterdam .. 31855
Marc Chagall Kunsthuis, Amsterdam ... 31857
Museum De Weem, Westeremden ... 32941
Museum Paul Tétar Van Elven, Delft ... 32065
Museum Slager, 's-Hertogenbosch ... 32373
Museum Ton Schulten, Ootmarsum ... 32632
Museum Veluwezoom, Doorwerth ... 32147
Museum voor Moderne Kunst, Arnhem ... 31929
Oudheidkamer Heerde, Heerde ... 32329
Panorama Mesdag, Den Haag ... 32098
Ruurd Wiersma Hûs, Burdaard ... 32022
Schilder- en Bakkerijmuseum 't Steenhuis, Niebert ... 32570
Sikkens Schildersmuseum, Sassenheim ... 32745
Tentoonstellingsruimte 't Oute Hus, Burgh-Haamstede ... 32028
Het Tromp's Huys, Vlieland ... 32899
Het Vrouwenhuis, Zwolle ... 33030

New Zealand

Auckland Art Gallery, Auckland ... 33042
Christchurch Art Gallery Te Puna o Waiwhetu, Christchurch ... 33071

Northern Cyprus

Güzel Sanatlar Müzesi, Girne ... 33359

Pakistan

Shakir Ali Museum, Lahore ... 34028
Tasneem Arts Gallery, Peshawar ... 34037

Peru

Casa Museo Mario Urteaga, Cajamarca 34125
Museo de Arte e Historia, Lima 34233
Museo de Arte Virreinal de Santa Teresa, Arequipa 34110
Museo de San Jerónimo, Cusco 34161
Museo de Sitio de Chinchero, Urubamba 34324
Museo San Juan de Letrán, Chucuito 34149
Pinacoteca Municipal Ignacio Merino, Lima 34274
Sala Numismática del Banco de Crédito del Peru y Sala de Acuarelas Pancho Fierro, Lima 34276

Philippines

Blanco Family Museum, Angono 34334
Jorge B. Vargas Museum and Filipiana Research Center, Quezon City 34492
Juan Luna Shrine Museum, Badoc 34339
Jumalon Museum and Art Gallery, Cebu 34371
Lopez Memorial Museum, Pasig 34478
Vicente Manansala Museum, Binangonan 34346

Poland

Dom Józefa Mehoffera, Kraków 34770
Galeria Foksal, Warszawa 35227
Galeria Krytyków Pokaz, Warszawa 35228
Galeria Rynek, Olsztyn 34977
Galeria Sleńdzińskich, Białystok 34547
Galeria Test, Warszawa 35230
Galeria Willa, Łódź 34871
Galeria XXI, Warszawa 35231
Kordegarda, Warszawa 35232
Kurza Stopa - Galeria Współczesnej Sztuki Polskiej, Chojnice 34598
Miejska Galeria Sztuki, Częstochowa 34614
Muzeum Mazowieckie w Płocku, Płock 35014
Muzeum Narodowe w Kielcach, Kielce 34742
Muzeum Narodowe w Poznaniu, Oddział w Rogalinie, Rogalin 35081
Muzeum Uniwersytetu Jagiellońskiego - Collegium Maius, Kraków 34813
Muzeum w Grudziądzu, Grudziądz 34685
Ośrodek Propagandy Sztuki, Łódź 34886
Płocka Galeria Sztuki, Płock 35015
Spichlerz, Olsztyn 34979
Zachęta - Narodowa Galeria Sztuki, Warszawa 35290

Portugal

Casa-Museu de Almeida Moreira, Viseu 36022
Casa Museu Marta Ortigão Sampaio, Porto 35839
Casa Museu Pintor José Cercas, Aljezur 35406
Casa-Oficina de António Carneiro, Porto 35841
Fundação Arpad Szenes e Vieira da Silva, Lisboa 35666
Museu de Evora, Évora 35587
Museu de Pintura, Lourinhã 35739
Museu de Pintura e Escultura Martins Correia, Golegã 35628
Museu de Roque Gameiro, Minde 35763
Museu-Escola de Artes Decorativas Portuguesas, Lisboa 35706
Museu Galeria Manuel Cabanas, Vila Real de Santo António 36014
Museu Municipal Severo Portela, Almodôvar 35421
Museu Nogueira da Silva, Braga 35481

Puerto Rico

Museo de Arte de Ponce, Ponce 36049
Museo de Arte Francisco Oller, Bayamón 36029
Museum of History, Anthropology and Art, San Juan 36076

Romania

Muzeul Hrandt Avakian, București 36204
Muzeul Memorial Nicolae Grigorescu, Câmpina 36250
Muzeul Theodor Aman, București 36230

Russia

Belozerskij Muzej Narodnogo Dekorativno-prikladnogo Iskusstva, Belozersk 36799
Bolšoj Dvorec, Lomonosov 37106
Chudožestvenno-memorialnyj Muzej K.S. Petrova-Vodkina, Chvalynsk 36864
Chudožestvennyj Muzej im. M.S. Tuganova, Vladikavkaz 37928
Dom-Muzej I.N. Kramskogo, Ostrogožsk 37519
Dom-muzej Narodnogo Chudožnika V.A. Igoševa, Chanty-Mansijsk 36859
Dom-muzej Osipovych-Vulf v Trigorskom, Puškinskie Gory 37586
Dom-muzej V.M. Vasnecova, Moskva 37169
Ekaterinburgskij Muzej Izobrazitelnych Iskusstv, Ekaterinburg 36884
Ekaterininskij Dvorec - Gosudarstvennyj Muzej Carskoje Selo, Puškin 37579
Gosudarstvennaja Tretjakovskaja Galerja, Moskva 37184
Gosudarstvennyj Chudožestvenno-architekturnyj Dvorcovo-parkovyj Muzej-zapovednik Oranienbaum, Lomonosov 37108
Gosudarstvennyj Chudožestvenno-architekturnyj Dvorcovo-parkovyj Muzej-zapovednik Pavlovsk, Pavlovsk 37528
Gosudarstvennyj Chudožestvennyj Istoriko-architekturnyj i prirodno-landšaftnyj Muzej-zapovednik Kolomenskoe, Moskva 37191
Gosudarstvennyj Memorialnyj Istoriko-chudožestvennyj i Prirodnyj Muzej-zapovednik V.D. Polenova, Strachovo 37808
Gosudarstvennyj Muzej Vadima Sidura, Moskva 37203
Gosudarstvennyj Muzej-usadba Archangelskoe, Archangelskoe 36773
Gosudarstvennyj Russkij Muzej, Sankt-Peterburg 37649
Gosudarstvennyj Russkij Muzej, Sankt-Peterburg 37650
Gumanitarnyj Centr-muzej Tvorčestva Autsajderov, Moskva 37217
Jaroslavskij Chudožestvennyj Muzej - Gubernatorskij Dom, Jaroslavl 36967
Kartinnaja Galerija im. P.I. Šolochova, Borisoglebsk 36818
Memorialnyj Dom-muzej P.D. Korina, Moskva 37232
Meždunarodnyj Centr-Muzej im. N.K. Rericha, Moskva 37249
Moskovskaja Gosudarstvennaja Kartinnaja Galereja A. Šilova, Moskva 37252
Moskovskij Gosudarstvennyj Vystavočnyj Zal Novyj Manež, Moskva 37256
Muzej Istorii Moskvy, Moskva 37294
Muzej-kvartira A.I. Kuindži, Sankt-Peterburg 37693
Muzej-usadba im. N.K. Rericha, Izvara 36958
Novosibirskaja Kartinnaja Galereja, Novosibirsk 37486
Orlovskij Oblastnoj Muzej Izobrazitelnych Iskusstv, Orël 37512
Ostrogožskij Rajonnyj Istoriko-chudožestvennyj Muzej im. I.N. Kramskogo, Ostrogožsk 37520
Petergof - Gosudarstvennyj Muzej Zapovednik, Sankt-Peterburg 37718
Primorskaja Gosudarstvennaja Kartinnaja Galereja, Vladivostok 37948
Regionalnyj Muzej Severnogo Priladožja, Sortavala 37789
Rostovskij Kreml - Gosudarstvennyj Muzej-zapovednik, Rostov 37599
Rybinskij Gosudarstvennyj Istoriko-architekturnyj i Chudožestvennyj Muzej-zapovednik, Rybinsk 37606
Tobolskaja Kartinnaja Galereja, Tobolsk 37854
Tomskij Oblastnoj Chudožestvennyij Muzej, Tomsk 37858
Uljanovskij Oblastnoj Chudožestvennyj Muzej, Uljanovsk 37909
Vystavočnyj Zal-masterskaja D.I. Nalbandjana, Moskva 37423

Serbia

Galerija Matice Srpske, Novi Sad 38120
Galerija Srpske Akademije Nauka i Umetnosti, Beograd 38045
Gradski Muzej, Kosovska Mitrovica 38105
Memorijalni Muzej Džuro Jakšić, Srpska Crnja 38154

Singapore

Dr. Kui-Fang Lai Art Museum, Singapore 38173
Tan Oe Pang Art Studio, Singapore 38181

Slovakia

Andrašiovská Obrazáreň, Krásnohorské Podhradie 38277
Galéria Jozefa Kollára, Banská Štiavnica 38190
Gáléria Márie Medveckej, Tvrdošín 38383
Galéria Mesta Bratislavy, Bratislava 38211
Novohradská Galéria, Lučenec 38289
Obecné múzeum a Galéria Jana Hálu, Važec 38385
Oravská Galéria, Dolný Kubín 38250
Slovenská Národná Galéria - Zvolenský Zámok, Zvolen 38395
Tatranská Galéria, Poprad 38333

Slovenia

Groharjeva Hiša v Sorici, Spodnja Sorica 38525
Sadnikarjeva Muzejska Zbirka v Kamniku, Kamnik 38428

South Africa

William Fehr Collection, Cape Town 38594
William Humphreys Art Gallery, Kimberley 38674

Spain

Basílica de Nuestra Señora del Pilar, Zaragoza 40336
Carmen Museo Max Moreau, Granada 39355
Casa de los Tiros, Granada 39357
Casa-Museu Castell Gala Dalí, La Pera 39835
Casa-Museu de Salvador Dalí de Cadaqués, Cadaqués 39085
Colección de Pintura de Olmeda de las Fuentes, Olmeda de las Fuentes 39726
Convento de San Antonio el Real, Segovia 40050
Fundación Luis Seoane, A Coruña 39216
Hospital de San Juan Bautista, Toledo 40129
MUSAC, León 39453
Museo de la Colegiata San Luis, Villagarcía de Campos 40301
Museo de la Real Academia de Bellas Artes de San Fernando, Madrid 39547
Museo de la Real Academia Nacional de Farmacia, Madrid 39548
Museo de Málaga, Málaga 39600
Museo del Monasterio de San Joaquín y Santa Ana, Valladolid 40236
Museo del Patriarca, Valencia 40208
Museo Diocesà, Tarragona 40100
Museo Lázaro Galdiano, Madrid 39563
Museo Municipal, Ayllón 38926
Museo Municipal de Bellas Artes, Santa Cruz de Tenerife 39997
Museo Municipal de San Telmo, Donostia-San Sebastián 39253
Museo Municipal de Vigo Quiñones de León, Vigo 40276
Museo Municipal Manuel Torres, Marín 39625
Museo Naval de Cartagena, Cartagena 39140
Museo Néstor, Las Palmas de Gran Canaria 39802
Museo Parroquial, Paradas 39811
Museo Pecharromán, Pasarón de la Vera 39814
Museu Cau Ferrat, Sitges 40083
Museu Krekovic, Palma de Mallorca 39792
Museu Maricel de Mar, Sitges 40084
Museu Nacional d'Art de Catalunya (MNAC), Barcelona 39001
Museu Víctor Balaguer, Vilanova i la Geltrú 40287
Palacio Real de El Pardo, El Pardo 39812

Sri Lanka

Lionel Wendt Art Gallery, Colombo 40372
National Art Gallery, Colombo 40373
Sapumal Foundation Gallery, Colombo 40375

Sweden

Aguélimuseet, Sala 40825
Bildmuseet, Umeå 40997
Internationella Barnkonstmuseet, Vårby 41049
Måleriyrkets Museum, Stockholm 40911
Nordiska Akvarellmuseet, Skärhamn 40843

Switzerland

Alexej von Jawlensky-Archiv, Locarno 41568
Centre Dürrenmatt, Neuchâtel 41678
Galleria Matasci, Tenero 41930
Hans Erni Museum, Luzern 41591
Kunsthaus Zürich, Zürich 42083
Kunstmuseum Olten, Olten 41707
Kunstmuseum Solothurn, Solothurn 41885
Kunstmuseum Winterthur, Winterthur 42029
Museo Hermann Hesse, Montagnola 41633
Museum H.R. Giger, Gruyères 41461
Schweizerisches Alpines Museum, Bern 41217
Schweizerisches Nationalmuseum, Zürich 42105
Wocher-Panorama der Stadt Thun um 1810, Thun 41941

Taiwan

Chang Da-Chien's Residence, Taipei 42377
Chen Chung-Tsang Museum, Dongshan 42180
Hwa Kang Museum, Taipei 42394
Jeoufen Art Gallery, Rueifang 42307
Lee Tze-Fan Memorial Art Gallery, Hsinchu 42210
Li Mei-Shu Memorial Gallery, Sansia 42312
Lotus Image Museum, Luodong 42265
Professor Li Mei-Shu Memorial Museum, Sansia 42313
Ran-In-Ting Museum, Taipei 42421

Tanzania

Kibo Art Gallery, Marangu 42516

Turkey

Resim ve Heykel Müzesi, İstanbul 42835

Ukraine

Deržavnyj Ajvazovskyj Kartynna Galereja, Feodosija 43039
Ivano-Frankivs'kyj Chudožnij Muzej, Ivano-Frankivs'k 43041
Izmaïl's'ka Kartynna Galereja, Izmaïl 43046
Kam'jans'ka Kartynna Galereja, Kam'janka 43053
Municypal'nyj Muzej Pryvatnych Kollekcij im. O. Bleščunova, Odesa 43156
Nacional'nyj Chudožnij Muzej Ukraïny, Kyïv 43100
Ukraine National Gallery, Kyïv 43110
Vinnyc'kyj Oblasnyj Chudožnij Muzej, Vinnycja 43205

United Kingdom

Arbroath Art Gallery, Arbroath 43312
Atkinson Art Gallery, Southport 45716
Brantwood, Coniston 43884
Brighton Museum and Art Gallery, Brighton 43621
Burrell Collection, Glasgow 44275
Buscot House, Buscot 43698
Y Capel, Llangollen 44769
Castell y Waun, Chirk 43837
Castle Douglas Art Gallery, Castle Douglas 43770
CCA-Centre for Contemporary Arts, Glasgow 44276
Christ Church Picture Gallery, Oxford 45330
Cirencester Art Local Studies Collection, Cirencester 43845
Corsham Court, Corsham 43895
Courtauld Institute Gallery, London 44830
Cyfarthfa Castle Museum and Art Gallery, Merthyr Tydfil 45130
De Morgan Centre, London 44835
Diorama Arts, London 44838
Drumlanrig's Tower, Hawick 44425
Edwin Young Collection, Salisbury 45602
Gascoigne Gallery, Harrogate 44399
Glynn Vivian Art Gallery, Swansea 45813
Gracefield Arts Centre, Dumfries 44037
Hagley Hall, Hagley 44377
Hastings Museum and Art Gallery, Hastings 44415
Hawick Museum, Hawick 44426
Heaton Hall, Manchester 45084
Hitchin Museum and Art Gallery, Hitchin 44474
Hopetoun House, South Queensferry 45695
J.M.W. Turner House, London 44886
The John Southern Gallery, Dobwalls 43988
Kirkcaldy Museum and Art Gallery, Kirkcaldy 44609
Knole, Sevenoaks 45641
Lanhydrock, Bodmin 43549
Leamington Spa Art Gallery and Museum, Leamington Spa 44654
Leighton Hall, Carnforth 43763
Lillie Art Gallery, Milngavie 45150
Long Eaton Town Hall, Long Eaton 45024
Loseley House, Guildford 44373
The Lowry Gallery, Salford 45599
Lytham Hall, Lytham Saint Anne's 45059
McKechnie Institute, Girvan 44270
McLean Museum and Art Gallery, Greenock 44365
Marble Hill House, Twickenham 45886
Mellerstain House, Gordon 44331
Mercer Art Gallery, Harrogate 44401
Mirehouse, Keswick 44565
Mitchell Library Exhibition, Glasgow 44289
Penwith Galleries, Saint Ives, Cornwall 45576
Perth Museum and Art Gallery, Perth 45366
Peter Anson Gallery, Buckie 43664
POSK Gallery, London 44958
The Queen's Gallery, London 44962
Rockingham Castle, Market Harborough 45110
Royal Cambrian Academy of Art, Conwy 43888
Royal Cornwall Museum, Truro 45879
Royal Museum and Art Gallery, Canterbury 43744
Scarborough Art Gallery, Scarborough 45624
Sir Alfred Munnings Art Museum, Dedham 43961
Sir Max Aitken Museum, Cowes 43908
Smith Art Gallery, Brighouse 43616
Southside House, London 44989
Stockport Art Gallery, Stockport 45754
Sutton Park, Sutton-on-the-Forest 45807
Talbot Rice Gallery, Edinburgh 44161
Thetford Collection, Thetford 45848
Towneley Hall Art Gallery and Museums, Burnley 43679
Trevithick Cottage, Camborne 43722
Upton House and Gardens, Banbury 43371
Usher Gallery, Lincoln 44717
Westgate Museum, Winchester 46035
Williamson Art Gallery and Museum, Birkenhead 43492
Woburn Abbey, Woburn 46054

United States of America

Americas Society Art Gallery, New York 51117
Artists' Cooperative Gallery, Omaha 51546
The Artists Museum, New York 51126
Atrium Gallery, Fort Worth 48710
Atrium Gallery, Lewiston 50036
Avampato Discovery Museum, Charleston 47378
Barnard's Mill Art Museum, Glen Rose 48922
Barnes Foundation, Merion 50601
Bass Manor Museum, Malta 50409
Baycrafters, Bay Village 46653

Blair Museum of Lithophanes, Toledo 53520
Boatmen's National Bank Art Collection, Saint Louis 52568
Boston University Art Gallery, Boston 46925
Bradley Gallery of Art, Plymouth 52001
Cahoon Museum of American Art, Cotuit 47846
Center for Figurative Painting, New York 51142
Center Gallery, Demarest 48052
Central Washington Gallery, Ellensburg 48356
Coos Art Museum, Coos Bay 47794
Dowd Fine Arts Gallery, Cortland 47831
Eastern Shore Art Center, Fairhope 48488
Eiteljorg Museum of American Indians and Western Art, Indianapolis 49490
Emerson Gallery, McLean 50362
Emerson Gallery, Clinton 47659
Free Public Library Special Collections, New Bedford 51012
Gallery at the Wharf, Coupeville 47854
Gallery Nature and Temptation, Boston 46930
Gallery of Art, Fort Valley 48695
Gallery One, Ellensburg 48359
Haas-Lilienthal House, San Francisco 52776
Henry Art Gallery, Seattle 53000
Herter Art Gallery, Amherst 46267
John G. Johnson Collection, Philadelphia 51845
José Drudis-Biada Art Gallery, Los Angeles 50239
Klemm Gallery, Adrian 46138
Laguna Art Museum, Laguna Beach 49871
Lehigh University Art Galleries/Museum, Bethlehem 46780
Leslie-Lohman Gay Art Foundation, New York 51203
Louis Bernal Gallery, Tucson 53594
Merwin and Wakeley Galleries, Bloomington 46862
Miller Art Center, Springfield 53240
Morton J. May Foundation Gallery, Saint Louis 52588
Muse Gallery, Jackson 49572
Museum of History and Art, Ontario 51568
Nicholas Roerich Museum, New York 51246
North Lee County Historic Center and Santa Fe Depot Museum Complex, Fort Madison 48652
Ocean City Art Center, Ocean City 51486
Orlando Brown House, Frankfort 48742
Orono Museum of Art, Bangor 46598
The Painting Center, New York 51247
The Parthenon, Nashville 50972
The Patrick and Beatrice Haggerty Museum of Art, Milwaukee 50718
Pharmaka Gallery, Los Angeles 50259
The Phillips Museum of Art, Lancaster 49931
Piedra Lumbre Visitors Center, Abiquiu 46129
Pittock Mansion, Portland 52085
Plastic Club, Philadelphia 51870
Print and Picture Collection, Philadelphia 51873
Putney School Gallery, Putney 52204
Richard F. Brush Art Gallery and Permanent Collection, Canton 47224
Robert E. Wilson Art Gallery, Huntington 49424
Rubelle and Norman Schafler Gallery, Brooklyn 47082
Runnels Gallery, Portales 52058
Ruth Chandler Williamson Gallery, Claremont 47576
Saint Johnsbury Athenaeum, Saint Johnsbury 52555
Santa Ana College Art Galleries, Santa Ana 52854
Southwestern Utah Art Gallery, Saint George 52547
Stowitts Museum, Pacific Grove 51667
The Studio Annex, New York 51263
The Taft Museum of Art, Cincinnati 47565
Union County Public Library Union Room, Monroe 50792
University at Buffalo Anderson Art Gallery, Buffalo 47125
University College Gallery, College Park 47687
University Galleries, Normal 51372
University Gallery, Boca Raton 46887
University Saint Thomas Art Collection and Exhibitions, Saint Paul 52630
Weyers-Sampson Art Gallery, Greenville 49077
Woolaroc Museum, Bartlesville 46620
Wynn Collection, Las Vegas 49981

Uruguay

Museo de Bellas Artes Departamental de San José, San José de Mayo 54434
Museo Municipal de Bellas Artes Juan Manuel Blanes, Montevideo 54408
Museo Nacional de Bellas Artes, Montevideo 54412

Uzbekistan

Tasviriy San'at Muzeyi - Andijon Qal'asi, Andijon 54442

Vanuatu

Michoutouchkine Pilioko Foundation, Port Vila 54494

Vatican City

Pinacoteca Vaticana, Città del Vaticano 54514

Venezuela

Museo de Arte Colonial, Caracas 54540

Painting, 17th C.

Austria

Salzburger Barockmuseum, Salzburg 02626

France

Maison Natale de Claude Gellée, Chamagne 12367
Musée des Beaux-Arts, Dole 12754

Germany

Deutsche Barockgalerie, Augsburg 16614

Italy

Casino dell'Aurora Pallavicini, Roma 27445
Museo Didattico, Tivoli 27978
Palazzo Borromeo, Stresa 27912
Pinacoteca Civica F. Duranti, Montefortino 26830
Quadreria dei Girolamini, Napoli 26911
Quadreria della Cassa Depositi e Prestiti, Roma 27542

Netherlands

Museum Het Rembrandthuis, Amsterdam 31862

Russia

Cerkov Troicy v Nikitnikach, Moskva 37156

South Africa

Michaelis Collection, Cape Town 38587

Sweden

Strömsholms Slott, Kolbäck 40651

United Kingdom

Brodie Castle, Forres 44244
Dulwich Picture Gallery, London 44840

Painting, 18th C.

Austria

Salzburger Barockmuseum, Salzburg 02626

Belgium

Château d'Aigremont, Seraing 04009

Georgia

Adjara Arts Museum, Batumi 16333

Germany

Bildergalerie, Potsdam 21083
Deutsche Barockgalerie, Augsburg 16614
Gotisches Haus, Wörlitz 22638
Neues Schloss Meersburg, Meersburg 20250
Schloss Charlottenburg, Berlin 17248
Schloss Mosigkau, Dessau 17833
Schloss Wörlitz, Wörlitz 22639
Staatsgalerie im Neuen Schloß, Oberschleißheim 20838

Italy

Museo Diego Aragona Pignatelli Cortes, Napoli 26904
Museo Glauco Lombardi, Parma 27108

Japan

Ike Taiga Bijutsukan, Kyoto 28814

Russia

Gosudarstvennyj Vladimiro-Suzdalskij Istoriko-architekturnyj i Chudožestvennyj Muzej-zapovednik, Vladimir 37934

Switzerland

Klostermuseum und Caspar Wolf-Kabinett, Muri 41666

United Kingdom

Dulwich Picture Gallery, London 44840

Painting, 19th C.

Argentina

Museo de Arte Español Enrique Larreta, Buenos Aires 00181

Austria

Kunstkammer des Innsbrucker Servitenklosters, Innsbruck 02121
Mathias Schmid-Museum, Ischgl 02139
Wachaumuseum, Weißenkirchen in der Wachau 02905

Belgium

Fondation Félicien Rops, Mettet 03873
Musée Antoine Wiertz, Bruxelles 03452
Musée Provincial Félicien Rops, Namur 03906

Croatia

Muzej Moderne i Suvremene Umjetnosti, Rijeka 08938
Umjetnička Galerija, Dubrovnik 08874

Czech Republic

Památník Mikoláše Alše a Matěje Kopeckého, Mirotice 09656
Zámek - Pamětní Síň Josefa Mánesa, Čechy pod Košířem 09452

Denmark

Ordrupgaard, Charlottenlund 09986

Finland

Cygnaeuksen Galleria, Helsinki 10770
Gösta Serlachiuksen Taidemuseo, Mänttä 11093
Reitzin Säätiön Kokoelmat, Helsinki 10813

France

Atelier Paul Cézanne, Aix-en-Provence 11513
Espace d'Exposition Minal, Héricourt 13204
Musée-Atelier Werner Lichtner-Aix, Sérignan-du-Comtat 15723
Musée Claude Monet, Giverny 13067
Musée d'Art Américain, Giverny 13068
Musée d'Art et d'Histoire de Lisieux, Lisieux 13778
Musée de Melun, Melun 14069
Musée du Docteur Faure, Aix-les-Bains 11522
Musée Ingres, Montauban 14176
Musée Marmottan Monet, Paris 14720
Musée National Eugène Delacroix, Paris 14732
Musée Paul Gauguin, Pont-Aven 14881
Musée Petiet, Limoux 13773
Musée Renoir, Cagnes-sur-Mer 12239
Musée Toulouse-Lautrec, Albi 11540

French Polynesia

Musée Paul Gauguin, Papeete 16325

Georgia

Adjara Arts Museum, Batumi 16333

Germany

Augustinermuseum, Freiburg im Breisgau 18428
Behnhaus/ Drägerhaus, Lübeck 20072
Gemäldegalerie Dachau, Dachau 17764
Hans-Thoma-Gedächtnisstätte und Vortaunusmuseum, Oberursel 20847
Hans-Thoma-Museum, Bernau im Schwarzwald 17291
Jakob-Grünenwald-Gedächtnisstätte, Ebersbach an der Fils 18041
Kunsthaus Meyenburg, Nordhausen 20737
Kunstmuseum Hohenkarpfen, Hausen ob Verena 19050
Lenbachmuseum, Schrobenhausen 21639
Leonhardi-Museum, Dresden 17957
Museum Bad Arolsen, Bad Arolsen 16670
Museum Bad Arolsen, Bad Arolsen 16671
Museum der Havelländischen Malerkolonie, Schwielowsee 21707
Museum Feuerbachhaus Speyer, Speyer 21836
Museum Kunst der Westküste, Alkersum 16471
Museum Schlösschen im Hofgarten, Wertheim 22504
Museum Villa Stuck, München 20469
Neue Pinakothek, München 20471
Orangerieschloss und Turm, Potsdam 21102
Schack-Galerie, München 20481
Schloss Charlottenburg, Berlin 17248
Slevogthof Neukastel, Leinsweiler 19887
Stiftung B.C. Koekkoek-Haus, Kleve 19568
Stiftung Fürst-Pückler-Museum, Cottbus 17736

Italy

Museo Bottega della Tarsia Lignea, Sorrento 27880
Museo Civico Giovanni Fattori, Livorno 26497
Museo Glauco Lombardi, Parma 27108
Pinacoteca Comunale Vincenzo Bindi, Giulianova 26348

Japan

Ehime-ken Bijutsukan, Matsuyama 28904
Ishibashi Museum of Art and Asian Gallery, Kurume 28795
Yamatane Bijutsukan, Tokyo 29507

Kazakhstan

Shevchenko Museum, Fort Shevchenko 29701

Netherlands

Museum Mesdag, Den Haag 32088
Van Gogh Documentatiecentrum, Nuenen 32606
Van Gogh Museum, Amsterdam 31898

New Zealand

Aigantighe Art Museum, Timaru 33211
The Suter, Te Aratoi o Whakatu, Nelson 33154

Norway

Oscarshall Slott, Oslo 33747

Russia

Bogorodickij Dvorec-muzej i Park, Bogorodick 36812
Chudožestvenno-memorialnyj Muzej K.S. Petrova-Vodkina, Chvalynsk 36864
Dom-muzej P.P. Čistjakova, Puškin 37578
Muzej-kvartira A.I. Kuindži, Sankt-Peterburg 37693
Muzej Tropinina i Moskovskich Chudožnikov ego Vremeni, Moskva 37375

Spain

Museu d'Art Modern Jaume Morera, Lleida 39478

Sweden

Carl Larsson-Gården, Sundborn 40956
Prins Eugens Waldemarsudde, Stockholm 40927

Switzerland

Aargauer Kunsthaus, Aarau 41080
Kunstmuseum, Sankt Gallen 41805
Kunstmuseum Poschiavo, Poschiavo 41730
Museum Oskar Reinhart am Stadtgarten, Winterthur 42035
Ortsmuseum mit Karl-Jauslin-Sammlung, Muttenz 41671
Pinacoteca Cantonale Giovanni Züst, Rancate 41741
Segantini-Museum, Sankt Moritz 41813
Stiftung Sammlung E.G. Bührle, Zürich 42108

United Kingdom

Birmingham Museum and Art Gallery, Birmingham 43500
Hartlepool Art Gallery, Hartlepool 44406
Haworth Art Gallery, Accrington 43262

United States of America

Art Gallery, Jamestown 49595
Fred Jones Jr. Museum of Art, Norman 51374
Frye Art Museum, Seattle 52999
Gallery of Art, Saint Louis 52578

Painting, 20th C.

Algeria

Musée Emile Aubry, Béjaïa 00054

Argentina

Museo de Arte Español Enrique Larreta, Buenos Aires 00181

Armenia

Martiros Sarian Museum, Yerevan 00741

Australia

Nolan Gallery, Tharwa 01581

Austria

Dokumentationszentrum Prof. Sepp Mayrhuber, Golling an der Erlauf 01958

Gedächtnisstätte Galerie Maria v. Ohmeyer, Poysdorf 02518

Belgium

Ensorhuis, Oostende 03930
Musée de Louvain-la-Neuve, Louvain-la-Neuve 03834
Museum Jan Vaerten, Beerse 03355
Museum Paul Delvaux, Sint-Idesbald 04016

Bulgaria

Chudožestvena Galerija, Pernik 05110

Canada

Art Gallery of Sudbury, Sudbury 07026
Emily Carr House, Victoria 07246
Memorial Library and Art Gallery, Saskatoon 06917

China

Xu Bei-Hong Museum, Beijing 07574

Croatia

Muzej Gacka, Otočac 08924
Muzej Moderne i Suvremene Umjetnosti, Rijeka 08938
Umjetnička Galerija, Dubrovnik 08874

Czech Republic

Egon Schiele Art Centrum, Český Krumlov 09467
Galerie Výtvarného Umění, Roudnice nad Labem 09820
Zámek Moravský Krumlov - Galerie Slovanská Epopej Alfonse Muchy, Moravský Krumlov 09665

Denmark

Carl-Henning Pedersen og Else Alfelts Museum, Herning 10069
Jens Søndergaard Museet, Lemvig 10175

Finland

Didrichsenin Taidemuseo, Helsinki 10772
Gallen-Kallelan Museo, Espoo 10728
Gösta Serlachiuksen Taidemuseo, Mänttä 11093
Imatran Taidemuseo, Imatra 10854
Lahden Taidemuseo, Lahti 11030
Sara Hildénin Taidemuseo, Tampere 11357
Tampereen Nykytaiteen Museo, Tampere 11363
Wäinö Aaltonen Museo, Turku 11408

France

Centre Artistique Jean Baffier, Sancoins 15622
Halle Saint-Pierre Museum, Paris 14611
Musée d'Art Moderne Richard Anacréon, Granville 13104
Musée de Peinture J. Drevon, Saint-Jean-de-Bournay 15322
Musée de Peinture Mario Prassinos, Saint-Rémy-de-Provence 15489
Musée des Années 30, Boulogne-Billancourt 12090
Musée des Beaux-Arts, Dole 12754
Musée des Beaux-Arts Denys Puech, Rodez 15079
Musée Emmanuel de la Villéon, Fougères (Ille-et-Vilaine) 12987
Musée Mathurin Méheut, Lamballe 13492
Musée Matisse, Le Cateau-Cambrésis 13569
Musée Municipal d'Art Naïf, Noyers-sur-Serein 14481
Musée Pol Mara, Gordes 13085
Musée Toulouse-Lautrec, Albi 11540
Musée Utrillo-Valadon, Sannois 15624

French Polynesia

Musée Paul Gauguin, Papeete 16325

Germany

Bauhaus-Archiv, Berlin 17041
Brücke Museum, Berlin 17058
Diözesanmuseum Freising, Freising 18450
Emil Schumacher-Museum, Hagen 18855
Franz Marc Museum, Kochel am See 19592
Fritz-Best-Museum, Kronberg im Taunus 19724
Galerie Am Markt, Bad Saulgau 16823
Galerie Neue Meister, Dresden 17944
Gemäldegalerie Dachau, Dachau 17764
Kunsthalle Bielefeld, Bielefeld 17341
Kunsthalle Mannheim, Mannheim 20166
Kunsthalle Würth, Schwäbisch Hall 21657
Kunsthaus Meyenburg, Nordhausen 20737
Kunstmuseum Hohenkarpfen, Hausen ob Verena 19050
Kunstsammlungen der Ruhr-Universität Bochum, Bochum 17402
Molerhiisli, Hausach 19047
Museum am Ostwall, Dortmund 17924
Museum im Kulturspeicher, Würzburg 22718
Museum Kunst der Westküste, Alkersum 16471
Museum Langenargen/ Bodensee, Langenargen 19816
Museum Theo Kerg, Schriesheim 21637
Museum Voswinckelshof, Dinslaken 17870
Museum Wiesbaden, Wiesbaden 22557
Museum Würth, Künzelsau 19736
Neues Schloss Meersburg, Meersburg 20250
Nolde Stiftung Seebüll, Neukirchen, Kreis Nordfriesland 20625
Otto-Dix-Haus Hemmenhofen, Gaienhofen 18532
Otto Modersohn Museum, Ottersberg 20954
Saarlandmuseum, Saarbrücken 21459
Sammlung Moderne Kunst, München 20480
Sammlung Scharf-Gerstenberg, Berlin 17244
Städtische Galerie, Ettlingen 18281
Städtische Galerie im Rathauspark, Gladbeck 18651
Städtische Kunstsammlung, Murrhardt 20537
Wilhelm-Morgner-Haus, Soest 21799

Greece

Papadopoulos Picture Gallery, Tinos 23228

Greenland

Emanuel A Petersen Art Museum, Ilulissat 23253

Hungary

Endre Nemes Múzeum, Pécs 23759
Martyn Ferenc Múzeum, Pécs 23763
Modern Magyar Képtár I, Pécs 23767
Modern Magyar Képtár II, Pécs 23768
Victor Vasarely Múzeum, Pécs 23780

India

Baroda Museum and Picture Gallery, Vadodara 24418

Ireland

Crawford Municipal Art Gallery, Cork 24768

Israel

Chagall House, Haifa 24956
Haifa Museum of Art, Haifa 24961
Janco-Dada Museum, Ein Hod 24946
Municipal Museum, Dimona 24942
Rubin Museum, Tel Aviv 25120
Ticho House, Jerusalem 25020

Italy

Casa Museo Depero, Rovereto 27567
Civica Galleria d'Arte Moderna, Gallarate 26264
Galleria Civica d'Arte Moderna, Spoleto 27891
Galleria d'Arte Moderna e Contemporanea R. Guttuso, Bagheria 25340
Museo d'Arte Costantino Barbella, Chieti 25873
Peggy Guggenheim Collection, Venezia 28207

Japan

Abashiri Kyodo Bijutsukan, Abashiri 28342
Aichi-ken Bijutsukan, Nagoya 28991
Ehime-ken Bijutsukan, Matsuyama 28904
Ikeda 20-seiki Bijutsukan, Ito 28593
Niigata-kenritsu Kindai Bijutsukan, Nagaoka 28963
Yamatane Bijutsukan, Tokyo 29507

Lithuania

Dailininko Prano Domšaičio Galerija ir Kultūros Centras, Klaipėda 30277
Kauno Paveikslų Galerija, Kaunas 30251
Kazio Varnelio Namai-Muziejus, Vilnius 30411

Netherlands

Chabot Museum, Rotterdam 32703
Gemeentemuseum Den Haag, Den Haag 32073
Museum De Wieger, Deurne 32117
Stedelijk Museum Amsterdam, Amsterdam 31888

New Zealand

Aigantighe Art Museum, Timaru 33211

Norway

Henie Onstad Kunstsenter, Høvikodden 33569
Mons Breidvik Galleries, Norheimsund 33704

Philippines

Ateneo Art Gallery, Quezon City 34487

Portugal

Casa Museu Atelier Henrique Medina, Esposende 35567
Museu Medina, Braga 35480
Museu Municipal Amadeo de Souza-Cardoso, Amarante 35433

Romania

Expoziţia de Pictură Românească Contemporană, Teremia Mare 36682

Russia

Galereja Forum, Sankt-Peterburg 37638
Kartinnaja Galereja Pejzažej P.M. Grečiškina, Stavropol 37803
Memorialnyj Dom-muzej P.D. Korina, Moskva 37232
Muzej-kvartira I.I. Brodskogo, Sankt-Peterburg 37690
Muzej-usadba I.E. Repina, Repino 37590
Muzejnyj - Vystavočnyj Kompleks Školy Akvareli Sergeja Andrijaki, Moskva 37387
Udmurtskij Respublikanskij Muzej Izobrazitelnych Iskusstv, Iževsk 36956

Serbia

Savremena Galerija Umetničke Kolonije Ečka, Zrenjanin 38167

Slovakia

Galéria Slovenskej Sporitel'ne, Bratislava 38212

Slovenia

Grad Dobrovo, Dobrovo v Brdih 38413
Mestna Galerija Ljubljana, Ljubljana 38452
Napotnikova Galerija, Šoštanj 38524
Umetnostna Galerija Maribor, Maribor 38479

Spain

Museu d'Art Modern Jaume Morera, Lleida 39478

Switzerland

Aargauer Kunsthaus, Aarau 41080
Adolf Dietrich-Haus, Berlingen 41201
Fondazione Rolf Gérard, Ascona 41128
Kirchner Museum Davos, Davos Platz 41321
Kunstsammlung Robert Spreng, Reiden 41751
Musée de Pully, Pully 41739
Picasso-Museum, Luzern 41596
Zentrum Paul Klee, Bern 41220

Turkey

Ayşe ve Ercümend Kalmık Vakfı-Müzesi, Gümüşsuyu 42791

United Kingdom

Dean Gallery, Edinburgh 44119
Grundy Art Gallery, Blackpool 43529
Hartlepool Art Gallery, Hartlepool 44406
Scottish National Gallery of Modern Art Gallery, Edinburgh 44156

United States of America

Frye Art Museum, Seattle 52999
Gallery of Art, Saint Louis 52578

Uzbekistan

Qo'qon Viloyat Ülkashunoslik Muzeyi, Qo'qon 54458

Venezuela

Museo de Bellas Artes de Caracas, Caracas 54543

Painting, Dutch

Belgium

Emile van Doren Museum, Genk 03619
Hospitaalmuseum Sint-Janshospitaal, Brugge 03417
M - Museum Leuven, Leuven 03787
Musée des Beaux-Arts, Tournai 04086
Musée des Beaux-Arts et Céramique, Verviers 04108
Museum Albert van Dyck, Schilde 04005
Museum van het Heilig Bloed, Brugge 03424
Museum Wuyts-Van Campen en Baron Caroly, Lier 03819
Rubenshuis, Antwerpen 03315
Schatkamer van de Sint-Salvatorskathedraal, Brugge 03427
Stedelijk Museum, Oudenaarde 03942

Germany

Staatsgalerie in der Residenz, Ansbach 16562
Staatsgalerie Neuburg - Flämische Barockmalerei, Neuburg an der Donau 20600

Italy

Galleria Sabauda, Torino 28002
Palazzo Bianco, Genova 26320

Netherlands

Dordrechts Museum, Dordrecht 32149
Frans Walkate Archief/ SNS Historisch Archief, Kampen 32435
Groninger Museum, Groningen 32281
Jopie Huisman Museum, Workum 32964
Mauritshuis, Den Haag 32082
Museum Bredius, Den Haag 32085
Museum Het Prinsenhof, Delft 32062
Museum Kempenland, Eindhoven 32191
Museum voor Moderne Kunst, Arnhem 31929
Noordbrabants Museum, 's-Hertogenbosch 32375
Oudheidkundige Verzameling, Sluis 32788
Peter Van den Braken Centrum, Sterksel 32807
Rijksmuseum, Amsterdam 31883
Schilderijenzaal Prins Willem V, Den Haag 32100
Singer Laren, Laren, Noord-Holland 32465
Stedelijk Museum De Lakenhal, Leiden 32489
Ton Smits Huis, Eindhoven 32195

South Africa

Groot Constantia Manor House and Wine Museum, Constantia 38597

United Kingdom

Breamore House Museum, Breamore 43594
Raby Castle, Darlington 43950
Shipley Art Gallery, Gateshead 44265

Painting, English

Belgium

Arentshuis, Brugge 03406

India

Birla Academy of Art and Culture Museum, Kolkata 24270

Ireland

Fota House, Carrigtwohill 24753

New Zealand

Dunedin Public Art Gallery, Dunedin 33090

United Kingdom

Alfred East Art Gallery, Kettering 44566
Burton Agnes Hall, Burton Agnes 43686
Burton Art Gallery and Museum, Bideford 43477
Bury Art Gallery and Museum, Bury, Lancashire 43691
Buscot Park House, Faringdon 44213
Cartwright Hall Art Gallery, Bradford 43577
Derby Museum and Art Gallery, Derby 43965
Falkland Palace and Garden, Falkland 44207
Falmouth Art Gallery, Falmouth 44208
Gainsborough's House, Sudbury, Suffolk 45797
Gallery Oldham, Oldham 45314
Guildhall Art Gallery, London 44866
Huddersfield Art Gallery, Huddersfield 44496
Lady Lever Art Gallery, Port Sunlight 45415
Laing Art Gallery, Newcastle-upon-Tyne 45229
Lotherton Hall, Aberford 43251
Pencarrow House, Bodmin 43551
Pollok House, Glasgow 44295
Rochdale Art and Heritage Centre, Rochdale 45505
Royal Holloway College Picture Gallery, Egham 44164
Sewerby Hall Art Gallery and Museum, Bridlington 43611
Shipley Art Gallery, Gateshead 44265
Southampton City Art Gallery, Southampton 45706
Stanley Spencer Gallery, Cookham 43889
Sudley House, Liverpool 44745
Tate Britain, London 44995
Temple Newsam House, Leeds 44666
Towner Art Gallery and Local Museum, Eastbourne 44103
Tunbridge Wells Museum and Art Gallery, Tunbridge Wells 45881
Wakefield Art Gallery, Wakefield 45913
Watts Gallery, Compton, Surrey 43882
Wednesbury Museum and Art Gallery, Wednesbury 45953
Wolverhampton Art Gallery, Wolverhampton 46060

Painting, European

Austria

Edmund Adler-Galerie, Mannersdorf 02347
Gemäldegalerie der Akademie der Bildenden Künste, Wien 02992
Residenzgalerie Salzburg, Salzburg 02624
Tiroler Landesmuseum Ferdinandeum, Innsbruck 02136

Belgium

Musée David et Alice Van Buuren, Bruxelles 03464

Brazil

Museu de Arte de São Paulo Assis Chateaubriand, São Paulo 04880

Canada

MacKenzie Art Gallery, Regina 06720
Outlook and District Heritage Museum and Gallery, Outlook 06558

Czech Republic

Galerie Výtvarného Umění, Ostrava 09696
Klašter Sv. Jiří, Praha 09756
Muzeum Kroměřížska, Kroměříž 09610
Oblastní Galerie, Liberec 09624
Památník Otakara Štáfla, Havlíčkův Brod 09513
Strahovská Obrazárna, Praha 09796

Estonia

Tartu Kunstimuuseum, Tartu 10656

France

Musée Boleslaw Biegas, Paris 14634
Musée Bonnat, Bayonne 11879
Musée d'Art et d'Histoire, Narbonne 14377
Musée des Beaux-Arts, Angers 11609
Musée des Beaux-Arts, Caen 12233
Musée des Beaux-Arts, Lille 13756
Musée des Beaux-Arts, Lyon 13889
Musée des Beaux-Arts, Nancy 14354
Musée des Beaux-Arts, Orléans 14535
Musée des Beaux-Arts, Pau 14757
Musée des Beaux-Arts, Tours 15964
Musée des Beaux-Arts Jules Chéret, Nice 14433
Musée des Ursulines, Mâcon 13907
Musée Jacquemart-André, Paris 14714
Musée Manoir de Kerazan, Loctudy 13799
Musée Municipal, Bergues 11947
Musée Municipal de l'Évêché, Limoges 13770

Germany

Alte Pinakothek, München 20421
Gemäldegalerie Alte Meister, Dresden 17946
Hamburger Kunsthalle, Hamburg 18932
Herzog Anton Ulrich-Museum, Braunschweig 17522
Kunsthalle Bremen, Bremen 17556
Kunsthalle zu Kiel, Christian-Albrechts-Universität, Kiel 19531
Kunstsammlung der Universität Göttingen, Göttingen 18693
Kunstsammlungen, Schwerin 21694
Kurpfälzisches Museum der Stadt Heidelberg, Heidelberg 19076
Martin-von-Wagner-Museum der Universität Würzburg, Würzburg 22714
Neue Galerie - Malerei und Skulptur ab 1750, Kassel 19473
Schloss und Schlosspark Charlottenburg mit Schinkelpavillon, Belvedere und Mausoleum, Berlin 17252
Schlossmuseum, Weimar 22420
Städtische Galerie im Lenbachhaus und Kunstbau München, München 20491
Von der Heydt-Museum, Wuppertal 22748

Greece

Ethniki Pinakothiki Mouseio Alexander Soytzoy, Athinai 22909

Ireland

National Gallery of Ireland, Dublin 24807

Italy

Fondazione Magnani Rocca, Mamiano di Traversetolo 26580
Galleria Civica Anna e Luigi Parmiggiani, Reggio Emilia 27386
Galleria Nazionale d'Arte Antica Palazzo Corsini, Roma 27453
Galleria Spada, Roma 27455

Luxembourg

Villa Vauban, Luxembourg 30487

Netherlands

Huis Singraven, Denekamp 32114
Rijksmuseum Twenthe, Enschede 32213

Norway

Hagan, Eggedal 33472
Holmsbu Billedgalleri, Holmsbu 33555
Munchs Hus, Åsgårdstrand 33403
Rogaland Kunstmuseum, Stavanger 33873
Trondheim Kunstmuseum, Trondheim 33930

Poland

Galeria Sztuki Polskiej XIX w. w Sukiennicach, Kraków 34779
Muzeum im. Jacka Malczewskiego w Radomiu, Radom 35073

Portugal

Casa dos Patudos - Museu de Alpiarça, Alpiarça 35423
Museu Nacional de Arte Antiga, Lisboa 35716

Romania

Expoziția Memorială Ştefan Jäger, Jimbolia 36432
Expoziția Muzeală Vasile Dobrian, Rod 36566
Muzeul de Artă, Sfântu Gheorghe 36601
Muzeul Memorial Dimitrie Aurelia Ghiață, Bucureşti 36207
Muzeul Memorial Henri Catargi, Scăeşti 36594

Russia

Dalnevostočnyj Chudožestvennyj Muzej, Chabarovsk 36858
Gosudarstvennyj Ėrmitaž, Sankt-Peterburg 37640
Muzej Dvorec A.D. Menšikova, Sankt-Peterburg 37675
Smolenskij Muzej-Zapovednik, Smolensk 37778
Voronežskij Chudožestvennyj Muzej im. I.N. Kramskogo, Voronež 37975

Slovenia

Narodna Galerija, Ljubljana 38457

Spain

Museo de la Capilla Real, Granada 39368

Sweden

Göteborgs Konstmuseum, Göteborg 40547
Kungliga Akademien för de Fria Konsterna, Stockholm 40905

Switzerland

Kunstmuseum Basel, Basel 41166
Musée Cantonal des Beaux-Arts, Lausanne 41544
Museo Storico Villa Ciani, Lugano 41585
Museum Briner und Kern, Winterthur 42031
Sammlung Oskar Reinhart 'Am Römerholz', Winterthur 42037

Turkey

Erzurum Resim ve Heykel Müzesi, Erzurum 42777
Resim ve Heykel Müzesi, İzmir 42857

United Kingdom

The Bowes Museum, Barnard Castle 43381
Buscot Park House, Faringdon 44213
Cooper Gallery, Barnsley, South Yorkshire 43385
Hampton Court Palace, East Molesey 44093
Hatton Gallery, Newcastle-upon-Tyne 45228
Kelvingrove Art Gallery and Museum, Glasgow 44287
National Gallery, London 44937
National Gallery of Scotland, Edinburgh 44138
The New Art Gallery Walsall, Walsall 45920
Tabley House, Knutsford 44624
The Walker Art Gallery, Liverpool 44748
The Wallace Collection, London 45006
Weston Park, Weston-under-Lizard 45983
York Art Gallery, York 46108

Painting, French

Belgium

Musée d'Art Moderne et d'Art Contemporain, Liège 03805
Musée des Beaux-Arts, Tournai 04086

France

L'Annonciade, Saint-Tropez 15525
Auberge Ganne, Barbizon 11845
Musée Albert André, Bagnols-sur-Cèze 11825
Musée Baron Martin, Gray 13126
Musée Camille Pissarro, Pontoise 14908
Musée de la Grenouillère, Croissy-sur-Seine 12679
Musée de la Vie Romantique, Paris 14665
Musée de l'Orangerie, Paris 14674
Musée de Pont-Aven, Pont-Aven 14880
Musée Départemental de l'Oise, Beauvais 11910
Musée des Beaux-Arts, Beaune 11903
Musée des Beaux-Arts, La Rochelle 13426
Musée des Beaux-Arts, Valenciennes 16043
Musée des Peintres de l'Ecole de Murol, Murol 14337
Musée Fournaise, Chatou 12476
Musée Gustave Moreau, Paris 14711
Musée Jean-Honoré Fragonard, Grasse 13115
Musée-Maison Natale Gustave Courbet, Ornans 14540
Musée Matisse, Nice 14435
Musée Municipal François Guiguet, Corbelin 12619
Musée Municipal Gautron du Coudray, Marzy 14025
Musée National Eugène Delacroix, Paris 14732
Musée National Jean-Jacques Henner, Paris 14733
Musée National Message Biblique Marc Chagall, Nice 14436
Musée Picasso, Antibes 11638
Musée Toulouse-Lautrec, Naucelle 14384
Musée Utrillo-Valadon, Sannois 15624
Musée Yves Brayer, Cordes-sur-Ciel 12625

Germany

Schloss Charlottenburg, Berlin 17248
Staatsgalerie in der Residenz, Ansbach 16562

Switzerland

Museum Langmatt, Baden 41148
Petit Palais, Genève 41429

Painting, German

Germany

Atelier Otto Niemeyer-Holstein, Koserow 19692
Conzen-Sammlung, Düsseldorf 17987
Emil Schumacher-Museum, Hagen 18855
Fritz-von-Wille-Museum im Haus Beda, Bitburg 17367
Johanniterhalle, Schwäbisch Hall 21656
Künstlerhaus Exter, Übersee 22147
Kunstforum Ostdeutsche Galerie, Regensburg 21228
Kunsthalle Rostock, Rostock 21394
Museum der Stadt Pasewalk und Künstlergedenkstätte Paul Holz, Pasewalk 20987
Museum Zitadelle Jülich, Jülich 19405
Nolde Stiftung Seebüll, Berlin 17229
Pahl Museum, Mainhardt 20138
Staatsgalerie im Hohen Schloß, Füssen 18502
Toni-Merz-Museum, Sasbach 21515

Italy

Stadt- und Multschermuseum, Vipiteno 28288

Painting, Italian

Austria

Graf Harrach'sche Familiensammlung, Rohrau 02593

France

Musée Départemental de l'Oise, Beauvais 11910
Musée des Beaux-Arts Denys Puech, Rodez 15079

Germany

Staatsgalerie in der Residenz Würzburg, Würzburg 22726

Italy

Ca'Rezzonico - Museo del Settecento Veneziano, Venezia 28179
Casa Buonarroti, Firenze 26143
Casa di Giorgione, Castelfranco Veneto 25729
Casa Natale di Raffaello, Urbino 28122
Cenacolo del Ghirlandaio, Firenze 26145
Cenacolo Vinciano, Milano 26673
Centro Culturale Polivalente, Bagnacavallo 25342
Galleria Alberoni, Piacenza 27182
Galleria Colonna, Roma 27449
Galleria Comunale d'Arte, Cagliari 25578
Galleria d'Arte Moderna, Firenze 26156
Galleria d'Arte Moderna Ricci-Oddi, Piacenza 27183
Galleria degli Uffizi, Firenze 26157
Galleria dell'Accademia, Firenze 26159
Galleria dell'Accademia Tadini, Lovere 26528
Galleria Doria Pamphilj, Roma 27451
Galleria Estense, Modena 26741
Galleria M. Rizzi, Sestri Levante 27829
Galleria Municipale, Francavilla al Mare 26243
Galleria Nazionale, Parma 27096
Galleria Nazionale d'Arte Antica Palazzo Barberini, Roma 27452
Galleria Nazionale dell'Umbria, Perugia 27145
Galleria Nazionale di Palazzo Spinola, Genova 26294
Galleria Palatina e Appartamenti Reali, Firenze 26161
Galleria Regionale di Palazzo Bellomo, Siracusa 27853
Galleria Sabauda, Torino 28002
Gallerie dell'Accademia, Venezia 28184
Musei Civici di Padova, Padova 27034
Musei Civici di Pesaro, Pesaro 27153
Museo Bandini, Fiesole 26126
Museo Canonicale, Verona 28231
Museo Civico, Pescia 27171
Museo Civico, Argenta 25271
Museo Civico di San Francesco, Montefalco 26821
Museo Civico di Santa Caterina, Treviso 28069
Museo Civico e Diocesano d' Arte Sacra, Montalcino 26795
Museo Civico e Pinacoteca P.F. Crociani, Montepulciano 26844
Museo Civico Giulio Ferrari, Carpi 25685
Museo Correr e Biblioteca d'Arte e Storia Veneziana, Venezia 28188
Museo del Risorgimento Vittorio Emanuele Orlando, Palermo 27070
Museo della Civiltà Locale - Raccolta d'Arte di San Francesco, Trevi 28064
Museo della Collegiata, Chianciano Terme 25851
Museo della Collegiata di Sant'Andrea, Empoli 26054
Museo della Fondazione Querini Stampalia, Venezia 28192
Museo della Magnifica Comunità di Cadore, Pieve di Cadore 27208
Museo della Scuola Grande dei Carmini, Venezia 28193
Museo della Società di Esecutori di Pie Disposizioni, Siena 27841
Museo dell'Arte Contemporanea e dei Pittori dell'Emigrazione, Recanati 27378
Museo dell'Opera e del Territorio, Caserta 25704
Museo di Arte Medioevale e Moderna, Arezzo 25267
Museo di Casa Vasari, Arezzo 25268
Museo di Santa Maria delle Grazie, San Giovanni Valdarno 27648
Museo Didattico Giuseppe Pellizza, Volpedo 28307
Museo Diocesano, Cortona 25989
Museo Diocesano, Pienza 27202
Museo e Casa Morandi, Bologna 25469
Museo Giannettino Luxoro, Genova 26316
Museo Giovanni Boldini, Ferrara 26116
Museo Novarese d'Arte e Storia R. Fumagalli, Novara 26953
Museo Pietro Canonica, Roma 27525
Museo Pinacoteca Comunale G. De Nittisu, Barletta 25374
Museo Regionale, Messina 26668
Museo Renato Brozzi, Traversetolo 28051
Museo Storico e Topografico, Firenze 26199
Museo Villa Colloredo Mels, Recanati 27379
Palazzo Bianco, Genova 26320
Palazzo Ducale, Venezia 28205
Pinacoteca, Castelfiorentino 25726
Pinacoteca Civica, Ascoli Piceno 25298
Pinacoteca Civica, Iesi 26395
Pinacoteca Civica, San Gimignano 27641
Pinacoteca Civica Francesco Podesti, Ancona 25231
Pinacoteca Comunale, Assisi 25311
Pinacoteca Comunale, Bevagna 25413
Pinacoteca Comunale, Cesena 25835
Pinacoteca Comunale, Città di Castello 25909
Pinacoteca Comunale, Massa Fermana 26622
Pinacoteca del Pio Monte della Misericordia, Napoli 26910
Pinacoteca dell'Accademia Carrara, Bergamo 25406
Pinacoteca di Brera, Milano 26720
Pinacoteca e Museo Civico, Bettona 25411

Pinacoteca e Museo Civico, Volterra 28312
Pinacoteca e Museo de Napoli, Terlizzi 27957
Pinacoteca Giuseppe Stuard, Parma 27110
Pinacoteca Nazionale, Bologna 25478
Pinacoteca Nazionale, Ferrara 26121
Pinacoteca Provinciale Corrado Giaquinto, Bari 25371
Pinacoteca Zelantea, Acireale 25143
Scuola Dalmata dei Santi Giorgio e Trifone, Venezia 28210
Scuola Grande Arciconfraternità di San Rocco, Venezia 28211
Scuola Grande di San Giovanni Evangelista, Venezia 28212
Studio-Museo di Giuseppe Pellizza, Volpedo 28308

Peru

Museo de Arte Italiano, Lima 34234

United Kingdom

Estorick Collection of Modern Italian Art, London 44844

United States of America

Clowes Fund Collection, Indianapolis 49489

Painting, Latin American

Argentina

Museo Municipal de Artes Plásticas, Olavarría 00481
Museo Municipal de Bellas Artes Lucas Braulio Areco, Posadas 00500
Museo Regional de Pintura José Antonio Terry, Tilcara 00649

Brazil

Museu Antonio Parreiras, Niterói 04590
Museu Dom João VI, Rio de Janeiro 04744

Cuba

Museo Wifredo Lam, La Habana 09176

Mexico

Casa Museo José Clemente Orozco, Guadalajara 30885
Museo Casa Diego Rivera, Guanajuato 30912
Museo Casa Estudio Diego Rivera y Frida Kahlo, México 31060
Museo del Palacio de Bellas Artes, México 31106
Museo Frida Kahlo, México 31118
Museo Nacional de Arte, México 31131
Pinacoteca Virreinal de San Diego, México 31159

Puerto Rico

Popular Arts Museum, San Juan 36078

United States of America

Museum of Antropology, History and Art, Rio Piedras 52355

Painting, Medieval

Austria

Alte Galerie am Landesmuseum Joanneum, Graz 01961
Museum des Chorherrenstifts Klosterneuburg, Klosterneuburg 02202

Czech Republic

Klašter Sv. Jiří, Praha 09756

France

Musée d'Unterlinden, Colmar 12585

Germany

Johanniterhalle, Schwäbisch Hall 21656
Staatsgalerie in der Burg, Burghausen 17655
Staatsgalerie in der Neuen Residenz, Bamberg 16920

Hungary

Keresztény Múzeum, Esztergom 23553

India

Birla Academy of Art and Culture Museum, Kolkata 24270

Italy

Museo Civico, Pistoia 27249
Museo Emilio Greco, Orvieto 27008

Macedonia

Umjetnička Galerija, Skopje 30525

Russia

Centralnyj Muzej Drevnerusskoj Kultury i Isskustva im. Andreja Rubleva, Moskva 37146
Rjazanskij Gosudarstvennyj Istoriko-architekturnyj Muzej-zapovednik, Rjazan 37595

Spain

Instituto Amatller de Arte Hispánico, Barcelona 38965
Museo Diocesano de Huesca, Huesca 39400

Painting, Renaissance

Belgium

Da Vinci-Museum, Tongerlo 04076

Canada

Imhoff Art Gallery, Saint Walburg 06886

France

Château du Clos-Lucé, Amboise 11575
Musée Pierre de Luxembourg, Villeneuve-lès-Avignon 16214

Germany

Albrecht-Dürer-Haus, Nürnberg 20755
Burg Trausnitz, Landshut 19800
Kunst- und Wunderkammer Burg Trausnitz, Landshut 19802
Städtisches Museum Schloss Rheydt, Mönchengladbach 20358

Italy

Castello d'Issogne, Issogne 26419
Cenacolo di Andrea del Sarto a San Salvi, Firenze 26146
Galleria Nazionale delle Marche, Urbino 28123
Museo Casa Natale di Tiziano Vecellio, Pieve di Cadore 27207
Pinacoteca Comunale, Sarnano 27743

Portugal

Museu de Setúbal - Convento de Jesus - Galeria de Arte Quinhentista, Setúbal 35936

Painting, Spanish

Austria

Graf Harrach'sche Familiensammlung, Rohrau 02593

France

Musée Goya, Castres 12320

Spain

Casa Museo de Goya, Fuendetodos 39313
Casa-Museo José Benlliure, Valencia 40197
Casa y Museo de El Greco, Toledo 40127
Colección del Banco de España, Madrid 39517
Ermita de San Antonio de la Florida y Museo Panteón de Goya, Madrid 39518
Museo Casa Natal de Jovellanos, Gijón 39335
Museo de Bellas Artes, Badajoz 38935
Museo de Bellas Artes, Córdoba 39205
Museo de Bellas Artes de Granada, Granada 39365
Museo de Bellas Artes de Santander, Santander 40009
Museo Evaristo Valle, Gijón 39339
Museo Gustavo de Maeztu, Estella 39286
Museo Julio Romero de Torres, Córdoba 39207
Museo Municipal Mariano Benlliure, Crevillente 39227
Museo Nicanor Piñole, Gijón 39340
Museo Rafael Zabaleta, Quesada 39885
Museo Romántico, Madrid 39579
Museo Sorolla, Madrid 39581
Museu Barrau, Santa Eulària del Río 39999
Museu Nacional d'Art de Catalunya (MNAC), Barcelona 39001

United Kingdom

Pollok House, Glasgow 44295

Paleontology → Fossils

Paper

Australia

Tamworth Regional Gallery, Tamworth 01569

Austria

Österreichisches Papiermacher- und Druckereimuseum, Steyrermühl 02779

Belgium

Economuseum voor Papier, Karton en Aandrijftechnieken, Alsemberg 03265
Musée National du Papier, Malmedy 03844

Canada

Centre d'Exposition sur l'Industrie des Pâtes et Papiers, Trois-Rivières 07171

Czech Republic

Ruční Papírna - Muzeum Papíru, Velké Losiny 09901

Denmark

Bruunshåb Gamle Papfabrik, Viborg 10313

Finland

Myllysaaren Museo, Valkeakoski 11437
Paperimuseo, Espoo 10731
Verlan Tehdasmuseo, Verla 11459

France

Musée Au fil du papier, Pont-à-Mousson 14877
Musée de la Bande Dessinée et Musée Municipal, Angoulême 11617
Musée de la Papeterie-Canson et Montgolfier, Davézieux 12700
Musée du Papier Chiffon, Muzillac 14345
Musée du Papier Le Nil, Angoulême 11621
Musée Historique du Papier, Ambert 11570
Musée Historique du Papier, Couze-et-Saint-Front 12664
Musée Rizla Croix, Mazères-sur-Salat 14053

Germany

Firmenmuseum Novatech, Reutlingen 21282
LVR-Industriemuseum, Bergisch Gladbach 17017
Museum Papiermühle Homburg, Markt Triefenstein 20214
Papiergeschichtliche Sammlung, Bergisch Gladbach 17018
Papiermühle Weddersleben, Weddersleben 22351
Papiermuseum, Düren 17983
Sammlung für Papier- und Druckgeschichte, Frankenberg, Sachsen 18348

Hungary

Diósgyőri Papírgyár Zrt. Papíripari Múzeuma, Miskolc 23701
Dunaújvárosi Papírgyár Üzemtörténeti Gyűjteménye, Dunaújváros 23532

Italy

Museo del Papiro, Siracusa 27855
Museo della Carta, Amalfi 25215
Museo della Carta e della Filigrana, Fabriano 26069
Museo della Carta e della Filigrana, Pioraco 27233
Museo della Cartolina S. Nuvoli, Isera 26413
Museo della Filigrana, Campo Ligure 25632

Japan

Kami no Hakubutsukan, Tokyo 29407
Obara-mura Kami no Bijutsukan, Obara 29084

Malaysia

Muzium Layang-Layang, Melaka 30599

Mexico

Museo del Molino de Papel, México 31104

Netherlands

Museum van Knipkunst, Westerbork 32940
Papiermolen de Schoolmeester, Westzaan 32943

Norway

Norsk Emballasje Museum, Oslo 33739

Poland

Muzeum Papiernictwa w Dusznikach Zdroju, Duszniki Zdrój 34627

Portugal

Museu do Papel Moeda, Porto 35853
Museu do Papel Terras de Santa Maria, Paços de Brandão 35802

Spain

Museu Molí Paperer de Banyeres de Mariola, Banyeres de Mariola 38946
Museu Molí Paperer de Capellades, Capellades 39124

Switzerland

Basler Papiermühle, Basel 41158

Taiwan

Su Ho Paper Museum, Taipei 42425

United Kingdom

Heron Corn Mill and Museum of Papermaking, Beetham 43438
The Whitworth Art Gallery, Manchester 45099

United States of America

Liquid Paper Museum, Nambe 50940
Petersen Automotive Museum, Los Angeles 50258
Racine Art Museum, Racine 52223
Robert C. Williams American Museum of Papermaking, Atlanta 46449

Peasant Life and Traditions

Argentina

Museo Posta de Yatasto, Metán 00465

Australia

Armidale Folk Museum, Armidale 00813
Blacksmith's Cottage, Bacchus Marsh 00820
Caboolture Historical Village and Museum, Caboolture 00932
Chiverton House Museum, Northampton 01376
City of Gosnells Museum, Gosnells 01124
Clifton and District Historical Museum, Clifton 00985
Cottage Blundell's, Canberra 00945
Hunter Valley Museum of Rural Life, Scone 01497
Kalamunda History Village, Kalamunda 01183
Murchison Settlement Museum, Mullewa 01333

Austria

Bauernhofmuseum, Wagrain 02881
Bauernmuseum Lanzenkirchen, Lanzenkirchen 02263
Bauernmuseum Osternach, Ort im Innkreis 02472
Bergbauernmuseum z' Bach, Wildschönau 03104
Dorfmuseum, Finkenstein 01889
Erlebnismuseum Anzenaumühle, Bad Goisern 01747
Familienmuseum im Altbauernhaus, Göstling an der Ybbs 01955
Freilichtmuseum Kalte Kuchl, Rohr im Gebirge 02592
Freilichtmuseum Sumerauerhof, Sankt Florian 02637
Freilichtmuseum Tiroler Bauernhöfe, Kramsach 02221
Heimathaus, Litzelsdorf 02329
Holzknechtmuseum im Salzkammergut, Bad Goisern 01749
Huberhus Museum, Lech 02277
Menthof-Bergbauernmuseum, Heiligenblut 02066
Mühlenmuseum und Mühlenweg, Maria Luggau 02351
Museum Breitenfurt, Breitenfurt 01800
Museum für bäuerliche Arbeitsgeräte, Imst 02109
Museum im Tabor, Feldbach 01878
Paznauner Bauernmuseum, Mathon 02362
Tiroler Bauernhausmuseum, Kitzbühel 02178

P

Waldviertler Bauernhaus-Museum, Weitra 02909
Waldviertler Bauernhof-Museum, Gföhl 01933

Belgium

Domein d'Oude Smelterij - Landbouwmuseum, Torhout 04077
Heimatmuseum Zwischen Venn und Schneifel, Sankt Vith 03999
La Remise Musée de la Vie Rurale, Offaing 03920
Maison de la Vie Paysanne, Han-sur-Lesse 03674
Maison Vinette, Malmedy 03842
Musée de la Fourche et de la Vie Rurale, Mortier 03894
Musée de la Vie Paysanne, Montquintin 03892
Musée de la Vie Rurale en Wallonie, Saint-Hubert 03995
Musée de Racour, Racour 03958
Musée du Cuir et des Industries du Péruwelzis, Péruwelz 03949
Stedelijk Museum-Hoevemuseum, Sint-Truiden 04031

Bhutan

Folk Heritage Museum, Thimphu 04202

Brazil

Museu de Comunicação Social Hipólito José da Costa, Porto Alegre 04646

Canada

Danish Canadian National Museum, Dickson 05700
Doukhobor Village Museum, Castlegar 05574
Fishermen's Life Museum, Jeddore Oyster Ponds 06028
La Grande Ferme, Saint-Joachim 06833
Lieu Historique National Sir Wilfrid Laurier, Ville-des-Laurentides 07264
Lougheed House Conservation Society, Calgary 05524
Maison Tsawenhohi, Wendake 07299
Maplewood Farm, North Vancouver 06479
Parc Historique Pointe-du-Moulin, Notre-Dame-de-l'île-Perrot 06490
Red Deer and District Museum, Red Deer 06709
Sunnybrook Farm Museum and Interpretive Centre, Red Deer 06710
Whitchurch-Stouffville Museum, Gormley 05884
White Elephant Museum, Makkovik 06221

Cuba

Museo El Piñero, Nueva Gerona 09236
Museo Municipal de Boyeros, La Habana 09154

Czech Republic

Valašské Muzeum v Přírodě, Rožnov pod Radhoštěm 09827

P

Denmark

Ærø Museum, Ærøskøbing 09957
Als Hjemstavnsmuseum, Hadsund 10052
Frilandsmuseet, Kongens Lyngby 10166
Glud Museum, Horsens 10096

Egypt

Nubia Museum, Aswan 10481

Estonia

A.H. Tammsaare Muuseum, Vetepere 10671
Aravete Külamuuseum, Aravete 10561
Karepa Kalame Talumuuseum, Vihula 10672
Mulgi Külamuuseum, Halliste 10566

Finland

Alatornion Pitäjämuseo, Tornio 11384
Herttoniemen Museo, Helsinki 10787
Juvan Museo, Juva 10884
Kärsämäen Kotiseutumuseo, Kärsämäki 10902
Kainuun Museo, Kajaani 10903
Kannonkosken Kotiseutukokoelma, Kannonkoski 10914
Kokemäen Ulkomuseo, Kokemäki 10970
Kustavin Kotiseutumuseo, Kustavi 11016
Lehtimäen Kotiseutumuseo, Lehtimäki 11057
Merimiehen Koti, Uusikaupunki 11420
Museokylä Kalluntalo, Laukaa 11055
Nivalan Museo Katvala, Nivala 11145
Nolhin Torppa, Laitila 11040
Orimattilan Kotiseutumuseo, Orimattila 11164
Oriveden Paltanmäen Museo, Orivesi 11168
Oulunsalon Kotiseutumuseo, Oulunsalo 11179
Pellonpään Talonpoikaistalo, Inkere 10860
Pielaveden Kotiseutumuseo, Pielavesi 11205
Porlammin Kotiseutumuseo, Porlammi 11224
Riuttalan Talonpoikaismuseo, Karttula 10922
Rovaniemen Kotiseutumuseo, Rovaniemi 11278
Saamelaismuseo Siida, Inari 10859
Saloisten Kotiseutumuseo, Arkkukari 10712
Seurasaaren Ulkomuseo, Helsinki 10816
Talomuseo, Kuortane 11013
Torppamuseo, Somero 11324
Työväenasuntomuseo, Helsinki 10833
Tyrnävän Kotiseutumuseo, Tyrnävä 11416
Vihannin Kotiseutumuseo, Vihanti 11462
Ypäjän Kotiseutumuseo, Ypäjä 11484

France

Archéo-Site Gaulois de Saint-Julien, Saint-Julien 15343
Ecomusée d'Alsace, Ungersheim 16017
Ecomusée d'Arts et Traditions Populaires, Ligny-le-Ribault 13749
Écomusée de la Bourrine du Bois Juquaud, Saint-Hilaire-de-Riez 15307
Ecomusée des Pays de l'Oise, Beauvais 11908
Ecomusée d'Hannonville, Hannonville-sous-les-Côtes 13179
Ferme d'Antan, Plédéliac 14837
Maison de la Vie Rurale, Massiac 14027
Maison de l'Eclusier, Saint-Malo-de-Guersac 15386
Musée, Saint-Hilaire-en-Lignière 15309
Musée à la Ferme de Rome, Bézu-la-Forêt 11980
Musée Alice Taverne, Ambierle 11571
Musée Conservatoire Maritime et Rural, Narbonne 14376
Musée d'Arts et Traditions Populaires, Marmoutier 13975
Musée d'Arts et Traditions Populaires, Yssingeaux 16308
Musée de Kerhinet, Saint-Lyphard 15380
Musée de la Goubaudière, Cholet 12523
Musée de la Paysannerie, Baguer-Morvan 11827
Musée de la Paysannerie et des Vieux Métiers, Valentigney 16045
Musée de la Vallée d'Aure, Ancizan 11594
Musée de la Vie Rurale, Steenwerck 15807
Musée de l'Habitat Rural en Pays d'Aigues, La Tour-d'Aigues 13450
Musée Départemental des Pays de Seine-et-Marne, Saint-Cyr-sur-Morin 15231
Musée départemental du Bugey-Valmorey, Lochieu 13795
Musée des 4 Saisons, Generagues 13040
Musée des Outils du Bois et de la Vie Paysanne, Bligny-sur-Ouche 12028
Musée des Petits Métiers de la Ferme, Wervicq-Sud 16294
Musée des Tradition de l'Île, Noirmoutier-en-l'Ile 14467
Musée des Vieux Métiers, Saissac 15591
Musée du Charroi Rural, Salmiech 15608
Musée du Felicien, Argentine 11675
Musée du Monde Rural, Rettel 15024
Musée du Pays Welche, Fréland 13006
Musée du Veinazès, Lacapelle-del-Fraisse 13469
Musée Ferme d'Autrefois et Matériel Artisanal, Thourie 15903
Musée Ferme de la Patte d'Oie, Le Plessis-Patte-d'Oie 13635
Musée-Maison de la Vie Rurale, La Flocellière 13368
Musée Mnemosina, Prunelli-di-Fium'Orbu 14955
Musée Pastoral le Monde Paysan d'Autrefois, Saint-Pé-de-Bigorre 15453
Musée Rural, Foucherans 12983
Musée Rural, Saint-André-d'Embrun 15167
Musée Rural de la Sologne Bourbonnaise, Beaulon 11897
Musée Rural du Porzay, Plomodiern 14846
Musée Villageois d'Orgon, Orgon 14528

Germany

Ackerbürgermuseum Haus Leck, Grebenstein 18741
Agrarmuseum Wandlitz, Wandlitz 22321
Bäuerliches Museum im Achentaler Heimathaus, Rohrdorf 21366
Bauern- und Handwerker-Museum, Malgersdorf 20157
Bauernhaus-Museum, Wolfegg 22649
Bauernhaus-Museum Bortfeld, Wendeburg 22474
Bauernhausmuseum Hohenstein, Hohenstein 19243
Bauernhausmuseum Schniderlihof, Oberried 20828
Bauernmuseum, Zahna 22766
Bauernmuseum Blankensee, Trebbin 22081
Bauernmuseum Inzigkofen, Inzigkofen 19349
Bauernmuseum Schwarzbach, Schwarzbach 21671
Bauernmuseum Zabeltitz, Zabeltitz 22765
Börde-Museum Burg Ummendorf, Ummendorf Bördekreis 22172
Brandenburgisches Freilichtmuseum Altranft, Bad Freienwalde 16732
Bürger- und Bauernmuseum, Hilzingen 19191
Burg- und Mühlenmuseum Pewsum, Krummhörn 19731
Dorfmuseum Ahnenhaus, Pliezhausen 21063
Dorfmuseum Buchenberg, Königsfeld im Schwarzwald 19649
Dorfmuseum Deckenpfronn, Deckenpfronn 17802
Dorfmuseum Dettingen und Illertal, Dettingen an der Iller 17842
Dorfmuseum Poppenweiler, Ludwigsburg 20043
Dorfstube Ötlingen, Weil am Rhein 22374
Fränkische Hopfenscheune mit heimatkundlicher Sammlung, Neunkirchen am Sand 20640
Fränkisches Freilandmuseum Fladungen mit dem Rhön-Zügle, Fladungen 18324
Freilichtmuseum Beuren, Beuren 17312
Freilichtmuseum Neuhausen ob Eck, Neuhausen ob Eck 20621
Fürstenberger Hof, Zell am Harmersbach 22778
Heimatmuseum, Ihringen 19313
Heimatmuseum, Mehrstetten 20257
Heimatmuseum, Neckargerach 20564
Heimatmuseum, Neukirch/Lausitz 20623
Heimatmuseum, Schöneiche bei Berlin 21607
Heimatmuseum, Seßlach 21749
Heimatmuseum im Vogteigebäude, Hüttlingen 19296
Heimatmuseum Obere Mühle, Loßburg 20039
Heimatmuseum Untergrombach, Bruchsal 17596
Heimatstube Fischbach, Niedereschach 20705
Heimatstube Mühlhausen, Villingen-Schwenningen 22238
Heimatstube Neuhaus-Schierschnitz, Neuhaus-Schierschnitz 20618
Heimatstuben, Waldbronn 22283
Hohenloher Freilandmuseum, Schwäbisch Hall 21655
Ignaz Günther Museum, Altmannstein 16515
Inselmuseum, Pellworm 21006
Karpatendeutsches Museum, Karlsruhe 19441
Kreisheimatmuseum, Grimma 18761
Lohrbacher Heimatstuben, Mosbach 20390
LWL-Freilichtmuseum Detmold, Detmold 17840
Mönchguter Museen im Ostseebad Göhren, Göhren, Ostseebad 18670
Museum, Mühlacker 20397
Museum Alte Lateinschule Großenhain, Großenhain 18789
Museum für ländliches Kulturgut im Dorfstadel Leupolz, Wangen im Allgäu 22326
Museum im Adler, Benningen am Neckar 16996
Museum Im Dorf, Reutlingen 21285
Museum im Seelhaus, Bopfingen 17472
Museum Kalt-Heiß, Deckenpfronn 17803
Museumsdorf Bayerischer Wald, Tittling 22065
Museumsdorf Cloppenburg, Cloppenburg 17713
Museumshof, Bad Oeynhausen 16800
Museumshof auf dem Braem, Gescher 18623
Niederlausitzer Heidemuseum, Spremberg 21843
Niederrheinisches Freilichtmuseum, Grefrath 18744
Oberschwäbisches Museumsdorf Kreisfreilichtmuseum Kürnbach, Bad Schussenried 16831
Odenwälder Freilandmuseum, Walldürn 22311
Reubacher Heimatmuseum, Rot am See 21398
Rheinisches Freilichtmuseum und Landesmuseum für Volkskunde Kommern, Mechernich 20243
Rhön-Museum, Fladungen 18325
Rieser Bauernmuseum, Maihingen 20134
Schwäbisches Bauern- und Technikmuseum, Eschach 18232
Schwarzwälder Freilichtmuseum Vogtsbauernhof, Gutach 18843
Stadt-und Wagenmuseum Oschatz, Oschatz 20918
Städtische Museen Wangen im Allgäu, Wangen im Allgäu 22328
Stormarnsches Dorfmuseum, Altes Bauernhaus am Thie, Hoisdorf 19250
Umstädter Museum Gruberhof, Groß-Umstadt 18778
Ungarndeutsche Heimatstuben, Langenau 19821
Vogtländisches Freilichtmuseum Landwüst, Markneukirchen 20209
Volkskunde-Museum, Rudolstadt 21432
Volkskunde- und Freilichtmuseum, Konz 19684
Waldhufen-Heimatmuseum Salmbach, Engelsbrand 18170
Wolfgang-Bonhage-Museum, Korbach 19686

Hungary

Bakonyi Ház, Veszprém 23958
Endrődi Tájház, Gyomaendrőd 23576
Gazdaház, Cserépfalu 23507
Göcseji Falumúzeum, Zalaegerszeg 23970
Kisjankó Bori Emlékház, Mezőkövesd 23697
Kopáncsi Tanyamúzeum, Hódmezővásárhely 23612
Magyar Mezogazdasági Múzeum Tanyamúzeuma, Lajosmizse - Bene 23688
Mezőföldi Tájház, Dég 23524
Nagykunsági Tájház, Karcag 23640
Néprajzi Ház, Tápiószele 23912
Sóstói Múzeumfalu, Sóstógyógyfürdő 23824
Szabadtéri Néprajzi Gyűjtemény, Szenna 23870
Szabadtéri Néprajzi Múzeum, Szentendre 23880
Szabadtéri Néprajzi Múzeum, Tihany 23927
Szabadtéri Néprajzi Múzeum - Tájház, Bakonybél 23349
Szatmárcsekei Tájház, Szatmárcseke 23838
Tabáni Tájház, Szolnok 23899
Tanyamúzeum, Tótkomlós 23938
Vasi Múzeumfalu, Szombathely 23906

Ireland

Inniskeen Folk Museum, Inniskeen 24844

Isle of Man

Grove House Museum, Ramsey 24917

Italy

Casa Museo A. Uccello, Palazzolo Acreide 27059
Cascina Museo della Civiltà Contadina, Cremona 26000
Castello di Costigliole d'Asti - Museo Nazionale del Territorio, Costigliole d'Asti 25995
Ecomuseo della Val Nerina, Cerreto di Spoleto 25811
Ecomuseo-Luogo della Memoria, Lamezia Terme 26442
Galleria del Maggio, Villa Minozzo 28273
Mostra della Civiltà Contadina, Lavello 26470
Mostra della Civiltà Contadina, Massa Marittima 26626
Museo Contadino, Varese Ligure 28166
Museo del Lavoro Contadino, Campobello di Mazara 25637
Museo del Lavoro e della Tradizioni Popolari della Versilia Storica, Seravezza 27809
Museo della Civiltà Contadina, Moio della Civitella 26762
Museo della Civiltà Contadina, Novellara 26960
Museo della Civiltà Contadina del Cilento, Montecorice 26820
Museo della Civiltà Contadina dell'Alto Monferrato e della Bassa Langa, Castagnole delle Lanze 25713
Museo della Civiltà Rurale del Vicentino, Malo 26578
Museo della Rocca, Dozza 26046
Museo della Transumanza e del Costume Abruzzese Molisano, Sulmona 27918
Museo dell'Agricoltura e del Mondo Rurale, San Martino in Rio 27663
Museo di Storia e Cultura Contadina, Genova 26310
Museo Etnografico Coumboscuro della Civiltà Provenzale in Italia, Sancto Lucia de Coumboscuro 27692
Museo Etnografico Romagnolo Benedetto Pergoli, Forlì 26226
Museo Ettore Pomarici Santomasi, Gravina di Puglia 26367
Museo Minimo, Corigliano Calabro 25977
Museo Provinciale della Vita Contadina, San Vito al Tagliamento 27689
Raccolta d'Arte Ubaldiana, Piandimeleto 27192

Korea, Republic

Jungmun Folklore Museum, Seogwipo 29861

Latvia

Kalncempju Pagasta Novadpētniecības Muzejs Ates Dzirnavās, Kalncempji 30028
Latvijas Lauksaimniecības Muzejs, Talsi 30127
Rūdolfa Blaumaņa Memoriālais Muzejs Braki, Ērgļi 30008

Lebanon

Saint Sharbel Museum, Annaya 30153

Lithuania

Etnografinė Prano Dzūko Sodyba-Muziejus, Delnicai 30221
Klebonišklų Kaimo Buities Ekspozicija, Kleboniškes 30286
Lietuvos Liaudis Buities Muziejus, Rumšiškės 30347
Smilgių Etnografinė Sodyba, Smilgiai 30374

Malaysia

Wedding and Custom Museum, Kuah 30569

Mexico

Museo del Campesino, Xochitepec 31569

Netherlands

Allerhande, Hippolytushoef 32392
Het Oale Meestershuus, Slagharen 32780
Openluchtmuseum Het Hoogeland, Warffum 32922
De Sfeer van Weleer, Rekken 32680
Themapark Openluchtmuseum De Spitkeet, Harkema 32315

Nicaragua

Museo Tradiciones y Leyendas del Coronel Arechavala, León 33287

Norway

Agatunet, Nå 33688
Bø Museum, Bø i Telemark 33443

Hemnes Museum, Bjerka 33439
Inderøy Museums- og Historielag, Inderøy 33574
Lesja Bygdatun, Lesja 33633
Oppdal Bygdemuseum, Oppdal 33707
Rollag Bygdetun, Rollag 33791
De Samiske Samlinger, Karasjok 33588
Søgne Bygdemuseum, Søgne 33846
Stor-Elvdal Museum, Koppang 33605
Sunnhordland Museum, Stord 33885
Tingvoll Museum, Tingvoll 33906
Tynset Bygdemuseum, Tynset 33943
Valdres Folkemuseum, Fagernes 33487

Poland

Muzeum Przyrodniczo-Leśne Świętokrzyskiego Parku Narodowego, Święty Krzyż 35152
Muzeum Wsi Lubelskiej, Lublin 34909
Skansen Budownictwa Ludowego Zachodniej Wielkopolski, Wolsztyn 35320

Portugal

Museu de Etnografia e Arqueologia Industrial, Torres Novas 35968
Museu Etnográfico de Carregueiros, Carregueiros 35509
Museu Rural, Alcáçovas 35390
Museu Rural - Centro de Arte e Cultura, Reguengo Grande 35883
Museu Rural de Beringel, Beringel 35467
Museu Rural de Boticas, Boticas 35476
Museu Rural de Vila Franca do Campo, Vila Franca do Campo 36000
Museu Rural e Etnográfico, São João da Ribeira 35920

Romania

Muzeul Țăranului Român, București 36227

Russia

Gosudarstvennyj Istoriko-ėtnografičeskij Muzej-zapovednik Šušenskoe, Šušenskoe 37811
Ibresinskij Ėtnografičeskij Muzej pod Otkryrym Nebom, Ibresi 36929
Oloneckij Nacionalnyj Muzej Karelov-Livvikov im. N.T.Prilukin, Olonec 37496
Osinskij Kraevedčeskij Muzej, Osa 37517
Šeltozerskij Vepsskij Ėtnografičeskij Muzej, Šeltozero 37758

Slovenia

Breginjski Muzej, Kobarid 38430
Liznjekova Hiša, Kranjska Gora 38439
Muzej na Prostem, Rogatec 38508
Muzej na Prostem pri Kartuziji Pleterje, Drča 38417
Pocarjeva Domačija v dolini Radovne, Kranjska Gora 38440
Tonina Hiša v Sveti Petru, Portorož 38495

Spain

Casa de la Ribera, Peñafiel 39825
Ecomuseo, Tordehumos 40145
Ecomuseo del Caserío Vasco, Artea 38900
Museu dels Raiers, El Pont de Claverol 39854
Pipaongo Etnografia Museoa, Pipaón 39842

Sweden

Berte Museum - Livet på Landet, Slöinge 40857
Bruksmuseum, Hudiksvall 40603
Ekehagens Forntidsby, Åsarp 40441
Hägnan Friluftsmuseet i Gammelstad, Gammelstad 40534
Himmelsberga Museum - Ölands Museum för Konst och Kulturhistoria, Borgholm 40472
Ingeborrarps Friluftsmuseum, Örkelljunga 40792
Örnsköldsviks Museum Konsthall, Örnsköldsvik 40794

Switzerland

Bauernhaus-Museum, Muttenz 41669
Bauernmuseum Althuus, Gurbrü 41464
Heimatmuseum, Oberweningen 41701
Heimatmuseum Bucheggberg, Buchegg 41266
Heimatmuseum der Albert-Edelmann-Stiftung, Ebnat-Kappel 41338
Musée Montagnard et Pictural, Les Haudères 41471
Musée Paysan et Artisanal, La Chaux-de-Fonds 41297
Museo della Civiltà Contadina del Mendrisiotto, Stabio 41898
Museo etnografico della Valle di Muggio, Cabbio 41279
Museum Stammertal, Unterstammheim 41963
Museum zur Ronmühle, Schötz 41843
Orts- und Wohnmuseum, Marthalen 41606
Ortsmuseum Hinwil, Hinwil 41483
Ortsmuseum zur Hohlen Eich, Wädenswil 41994
Schweizerisches Freilichtmuseum für ländliche Kultur, Hofstetten 41485

Tunisia

Musée Dar Taieb, Nabeul 42645
Musée Documentaire Dar Ayed, Ksar Hellal 42633

United Kingdom

Angus Folk Museum, Glamis 44272
Auchindrain Museum, Inveraray 44518
Blists Hill Victorian Town, Telford 45833
Breamore Countryside Museum, Breamore 43593
Calbourne Water Mill and Rural Museum, Newport, Isle of Wight 45242
Folk and Costume Museum, Câtel Guernsey 43779
Glenesk Folk Museum, Brechin 43596
Lackham Museum of Agriculture and Rural Life, Lacock 44628
Lloyd George Museum, Cricciert 43925
Mesolithic Museum, Abinger Common 43261
Mullach Ban Folk Museum, Mullach Ban 45186
Museum of Kent Life, Sandling 45612
Perry's Cider Mills, Ilminster 44513
Ramsey Rural Museum, Huntingdon 44502
Roots of Norfolk at Gressenhall, Gressenhall 44366
Rural Life Centre, Farnham 44219
Stanmer Rural Museum, Brighton 43626
Usk Rural Life Museum, Usk 45901
Wayside Folk Museum, Zennor 46114

United States of America

Ashland Logging Museum, Ashland 46382
Battles Museums of Rural Life, Girard 48914
Blue Ridge Institute and Museum, Ferrum 48538
Calvin Coolidge State Historic Site, Plymouth Notch 52004
Coggeshall Farm Museum, Bristol 47032
Forest Capital State Museum, Perry 51779
Fred Dana Marsh Museum, Ormond Beach 51599
Frontier Culture Museum, Staunton 53275
Georgia Agrirama, Tifton 53508
Hans Herr House and Museum, Willow Street 54185
Independence Visitors Center, Independence 49476
Lum and Abner Museum, Pine Ridge 51923
Lumberman's Museum, Patten 51731
Midway Village and Museum, Rockford 52415
Mountain Life Museum, London 50168
Museum of the Arkansas Grand Prairie, Stuttgart 53351
Naper Settlement Museum, Naperville 50950
The Spanish Quarter Museum, Saint Augustine 52528
Storrowton Village Museum, West Springfield 54083
Volo Antique Auto Museum, Volo 53768
Watson Farm, Jamestown 49600

Performing Arts

Argentina

Museo de Instituto Nacional de Estudios de Teatro, Buenos Aires 00195
Museo Nacional del Teatro, Buenos Aires 00264

Armenia

Mher Mkrtchyan House Museum, Gyumri 00706

Australia

Performance Space, Eveleigh 01078
Performing Arts Collection, Melbourne 01290
Performing Arts Collection of South Australia, Adelaide 00780

Austria

Museum für Unterhaltungskunst, Wien 03037
Museum Sensenwerk, Deutschfeistritz 01813
MUWA - Museum der Wahrnehmung, Graz 01980
Österreichisches Theatermuseum, Wien 03051
Puppentheatermuseum, Mistelbach an der Zaya 02387
Staatsopernmuseum, Wien 03078
Villa Sinnenreich, Rohrbach/OÖ 02597

Azerbaijan

Jafar Jabarly State Museum of Theatre, Bakı 03158

Belgium

AMVC-Letterenhuis, Antwerpen 03280
Archives et Musée de la Littérature, Bruxelles 03433
Musée du Théâtre Royal de Toone, Bruxelles 03486

Brazil

Museu de Arte Mágica e Ilusionismo João Peixoto dos Santos, São Paulo 04881
Museu de Bonecos e Ventrílocos, São Paulo 04884
Museu de Cartão, São Paulo 04885
Museu dos Teatros do Rio de Janeiro, Rio de Janeiro 04745
Museu Teatro, Rio de Janeiro 04762

Bulgaria

Muzej na Nacionalnata Opera i Balet, Sofia 05187
Muzej na Naroden Teatar Ivan Vazov, Sofia 05188

Canada

Beaches Heritage Centre, Eastport 05733
Theatre Museum Canada, Toronto 07143

China

Beijing Katong Art Museum, Beijing 07515
Beijing Theater Museum, Beijing 07521
Kūnqǔ Bówùguǎn, Shanghai 08170
Píngtán Bówùguǎn, Shanghai 08176
Sanying Pagoda and Museum of Dinosaur, Shaoguan 08208
Tianjin Theatre Museum, Tianjin 08289

Costa Rica

Galeria Teatro Nacional Enrique Echandi y Joaquín García Monge, San José 08814

Côte d'Ivoire

Musée Théâtre Ki-Yi, Abidjan 08831

Croatia

Dom Marina Držića, Dubrovnik 08861
Muzej Hrvatske Književnosti i Kazališne Umjetnosti, Zagreb 09005

Cyprus

Nikos Nikolaides Theatrical Archives, Lemesos 09389

Czech Republic

Měnínská Brána, Brno 09431
Moravské Zemské Muzeum, Brno 09435
Muzeum Loutek Milana Knížáka, Děčín 09490

Denmark

Teatermuseet i Hofteatret, København 10154

Estonia

Andres Särevi Kortermuuseum, Tallinn 10620
Eesti Teatri- ja Muusikamuuseum, Tallinn 10628

Finland

Nukketeatteri- ja Lelumuseo Unelma, Pieksämäki 11203
Teatterimuseo, Helsinki 10828

France

Bibliothèque-Musée de la Comédie-Française, Paris 14572
Bibliothèque-Musée de l'Opéra, Paris 14574
Centre Pompidou Metz, Metz 14099
Manoir de Veygoux, Scénomusée des Combrailles, Charbonnières-les-Varennes 12407
Musée du Cirque, Vatan 16081
Musée du Cirque, Wasquehal 16288
Musée du Théâtre, Couilly-Pont-aux-Dames 12641
Musée Humoristique Albert Dubout, Palavas-les-Flots 14557
Musée Jean Gabin, Mériel 14087
Musée Kwok-On, Paris 14716
Musée Raimu, Cogolin 12567

Georgia

State Museum of Theatre, Music and Cinema, Tbilisi 16383

Germany

Artistenmuseum, Marburg 20186
Augsburger Puppentheatermuseum Die Kiste, Augsburg 16612
Brecht-Weigel-Gedenkstätte, Berlin 17055
Chaplin-Archiv, Frankfurt am Main 18358
Dance Collection - Stiftung John Neumeier, Hamburg 18919
Deutsches Tanzarchiv Köln, Köln 19598
Deutsches Theatermuseum, München 20435
Ekhof-Theater, Gotha 18717
Erstes Circusmuseum in Deutschland, Preetz 21115
Fastnachtsmuseum, Herbstein 19136
Filmmuseum, Bendestorf 16987
Historische Kuranlagen und Goethe-Theater Bad Lauchstädt, Bad Lauchstädt 16774
Karnevalmuseum, Mönchengladbach 20356
Markt- und Schaustellermuseum, Essen 18261
Museum für Sächsische Volkskunst mit Puppentheatersammlung, Dresden 17963
Neuberin-Museum, Reichenbach im Vogtland 21246
Dat Otto Huus, Emden 18142
Puppentheater Museum Berlin, Berlin 17236
Stadtmuseum Rottweil, Rottweil 21429
Theaterfigurenmuseum, Lübeck 20082
Theatergeschichtliche Sammlung und Hebbel-Sammlung, Kiel 19535
Theatermuseum der Landeshauptstadt Düsseldorf, Düsseldorf 18010
Theatermuseum Hannover, Hannover 19006
Theatermuseum Zauberwelt der Kulisse, Meiningen 20262
Theaterwissenschaftliche Sammlung, Köln 19634
Valentin-Karlstadt-Musäum, München 20493
Zirkus und Varieté Archivsammlung Reinhard Tetzlaff, Hamburg 18964

Greece

Hellenic Theatre Museum and Study Center, Athinai 22914
Katina Paxinou Museum, Athinai 22916
Marika Kotopouli Museum, Zografou 23248
Museum of Delphic Celebrations of Angelos and Eva Sikelianou, Athinai 22932
Panos Aravantinos Theatrical Museum of Painting and Stage Design, Piraeus 23148
Spathareion Museum of the Shadow Theatre, Maroussi 23083

Hungary

Bajor Gizi Emlékmúzeum, Budapest 23385
Jászai Mari Emlékház, Ászár 23337
Jászai Mari Emlékszoba, Rábatamási 23788
Latinovits Zoltán Emlékmúzeum, Balatonszemes 23354
Magyar Állami Operaház Múzeuma, Budapest 23433
Országos Széchényi Könyvtár, Budapest 23463
Országos Színháztörténeti Múzeum és Intézet, Budapest 23464
Színháztörténeti és Színészmúzeum, Miskolc 23715
Varga Mátyás Színháztörténeti Kiállítása, Szeged 23849

Israel

Israel Children's Museum, Holon 24977
Israel Goor Theater Archive and Museum, Jerusalem 24996

Italy

Civico Museo Biblioteca dell'Attore del Teatro, Genova 26290
Civico Museo Setificio Monti, Abbadia Lariana 25137
Civico Museo Teatrale di Fondazione Carlo Schmidl, Trieste 28086
Collezione Minici Zotti-Un Museo di Magiche Visioni, Padova 27033
Collezione Titta Ruffo, Pisa 27235
Mostra dei Pupi Siciliani, Caltagirone 25604
Museo del Teatro, Spoleto 27893
Museo del Teatro Argentina, Roma 27475
Museo del Teatro dell'Aquila, Fermo 26099
Museo del Teatro Municipale, Piacenza 27187
Museo dell'Attore Napoletano, Napoli 26889
Museo delle Marionette, Campomorone 25641
Museo dell'Opera di Santa Niccolao, Buggiano 25562
Museo di Peppone e Don Camillo, Brescello 25532
Museo Rodolfo Valentino, Castellaneta 25733
Museo Romagnolo del Teatro, Forlì 26228
Museo Teatrale alla Scala, Milano 26716
Museo Teatrale del Burcardo, Roma 27539

Japan

Misora Hibari Kinenkan, Kyoto 28850
Osaka-kenritsu Waha Kamigata Bungeikan, Osaka 29145
Sado Nohgaku-no-sato, Ryotsu 29174
Sumiya Motenashi Museum of Art and Culture, Kyoto 28868
Waseda Daigaku Tsubouchi Hakushi Kinen Engeki Hakubutsukan, Tokyo 29503
Yoshimoto Bungeikan, Osaka 29157

Kazakhstan

Muzej im. Auėzova Kzachskogo Dramatičeskogo Teatra, Almaty 29687

Latvia

Ādolfa Alunāna Memoriālais Muzejs, Jelgava 30015
Eduarda Smiļga Teātra Muzejs, Rīga 30065
Rakstniecības un Mūzikas Muzejs, Rīga 30104

Lithuania

Kauno Lėlių Muziejus, Kaunas 30250

P

Konstantino Glinskio Sodyba-Muziejus, Jurbarka ... 30233
Lietuvos Teatro, Muzikos ir Kino Muziejus, Vilnius ... 30422
Liudo Truikio ir Marijonos Rakauskaitės Memorialinis Muziejus, Kaunas ... 30258

Malta

Manoel Theatre and Museum, Valletta ... 30641

Mexico

Museo Teatro La Casa de los Títeres, Monterrey ... 31188

Mongolia

Mongolian Theatre Museum, Ulaanbaatar ... 31624

Netherlands

Dickens Museumtheater, Bronkhorst ... 32013
Hot Poppentheatermuseum, Gouda ... 32265
Kermis- en Circusmuseum Steenwijk, Steenwijk ... 32802
Oeteldonks Gemintemuzejum, 's-Hertogenbosch ... 32376
Poppen en Poppenhuis Museum 't Duvelke, Breda ... 32007
Theater Museum, Amsterdam ... 31894

Norway

Fritzøe Museum, Larvik ... 33627
Teatermuseet i Oslo, Oslo ... 33754

Pakistan

Faqir Khana Museum, Lahore ... 34019

Peru

Museo Municipal del Teatro, Lima ... 34263

Poland

Centrum Scenografii Polskiej - Galerija Anrzeja Majewskiego, Katowice ... 34719
Muzeum Starego Teatru, Kraków ... 34810
Muzeum Teatralne, Warszawa ... 35272
Muzeum Teatralne im. Stanisława Wyspiańskiego, Kraków ... 34811

Portugal

Museu da Marioneta, Lisboa ... 35681
Museu do Teatro Nacional de São Carlos, Lisboa ... 35704
Museu Nacional do Teatro, Lisboa ... 35721

Romania

Colecția Memorială Maria Filotti, București ... 36193
Expoziția Teatrului, Arad ... 36112
Muzeul de Istorie a Teatrului și a Orașului Oravița, Oravița ... 36508
Muzeul Memorial C.I. C.C. Nottara, București 36206
Muzeul Operei Române, București ... 36225
Muzeul Teatrului, Iași ... 36427
Muzeul Teatrului Național, București ... 36228
Muzeul Teatrului Național, Craiova ... 36329

Russia

Detskij Muzej Buratino-Pinokkio, Moskva ... 37158
Dom-Muzej A.L. Durova, Voronež ... 37969
Dom-muzej M.N. Ermolovoj, Moskva ... 37163
Dom-muzej F.I. Šaljapina, Moskva ... 37167
Dom-muzej K.S. Stanislavskogo, Moskva ... 37168
Gorodskoj Muzej Teatralnogo Iskusstva, Omsk ... 37498
Gosudarstvennyj Centralnyj Teatralnyj Muzej im. Bachrušina, Moskva ... 37190
Gosudarstvennyj Muzej Detskich Teatrov, Moskva ... 37197
Kulturnyj Centr-muzej V.S. Vysockogo, Moskva ... 37224
Literaturno-teatralnyj Muzej N.M. Djakonova, Syktyvkar ... 37816
Memorialnyj Muzej-kvartira Aktërov Samojlovych, Sankt-Peterburg ... 37659
Muzej 'A Muzy ne Molčali', Sankt-Peterburg ... 37669
Muzej Cirkovogo Iskusstva, Sankt-Peterburg ... 37672
Muzej Detskogo Muzykalnogo Teatra im. N. Sats, Moskva ... 37271
Muzej Ėtnografičeskogo Kostjuma na Kuklach, Moskva ... 37274
Muzej Gosudarstvennogo Akademičeskogo Bolšogo Teatra Rossii, Moskva ... 37280
Muzej Gosudarstvennogo Akademičeskogo Malogo Teatra, Moskva ... 37281
Muzej Gosudarstvennogo Akademičeskogo Marijnskogo Teatra Opery i Baleta, Sankt-Peterburg ... 37676
Muzej Gosudarstvennogo Akademičeskogo Teatra Opery i Baleta imeni Mussorgskogo, Sankt-Peterburg ... 37677
Muzej Istorii Teatrov Dagestana, Machačkala 37118
Muzej-kvartira Aktërskoj Semi M.V. i A.A. Mironovych i A.S. Menakera, Moskva ... 37314
Muzej-kvartira Diriżëra N.S. Golovanov, Moskva ... 37315
Muzej-kvartira V.S. Mejercholda, Moskva ... 37319
Muzej-kvartira V.I. Nemiroviča-Dančenko, Moskva ... 37320
Muzej-kvartira F.I. Šaljapina, Sankt-Peterburg 37695
Muzej Moskovskogo Chudožestvennogo Akademičeskogo Teatra, Moskva ... 37336
Muzej Muzykalnoj i Teatralnoj Kultury na Kavkazskich Mineralnyh Vodach, Kislovodsk 37049
Muzej N.G. Rubinštejna - Muzej Moskovskoj Konservatorii, Moskva ... 37354
Muzej Sceničeskogo Iskusstva im. V.Ė. Mejercholda, Penza ... 37535
Muzej Teatra, Sankt-Peterburg ... 37710
Muzej Teatra im. Mossoveta, Moskva ... 37367
Muzej Teatra im. K.S. Stanislavskogo, Moskva ... 37368
Muzej Teatra im. Evg. Vachtangova, Moskva ... 37369
Muzej Teatra Junogo Zritelja, Moskva ... 37370
Muzej Teatra Operetty, Moskva ... 37371
Muzej Teatra Rossijskoj Armii, Moskva ... 37372
Muzej Teatra Satiry, Moskva ... 37373
Muzej Teatralnoj Kukly A.A Veselova, Voronež 37974
Muzej-zapovednik Abramcevo, Abramcevo ... 36747
Sankt-Peterburgskij Muzej Teatralnogo i Musykalnogo Iskusstva, Sankt-Peterburg ... 37725
Teatralnyj Muzej, Kazan ... 37028

Serbia

Muzej Pozorišne Umetnosti Srbije, Beograd ... 38073

Slovenia

Slovenski Gledališki Muzej, Ljubljana ... 38464

Spain

Museo del Baile Flamenco, Sevilla ... 40073
Museo Nacional del Teatro, Almagro ... 38852
Museu de les Arts Escèniques, Barcelona ... 38983

Sweden

Dansmuseet, Stockholm ... 40890
Gripes Modellteatermuseum, Nyköping ... 40781
Marionettmuseet, Stockholm ... 40912
Sveriges Teatermuseum, Nacka ... 40763

Switzerland

Museo Comico Verscio, Verscio ... 41981
Schweizerische Theatersammlung, Bern ... 41216

Taiwan

Chinese Opera Museum, Taipei ... 42383
Museum of Traditional Nan Bei Music and Theater, Changhua ... 42158
Taiwan Theater Museum, Yilan ... 42469

Ukraine

Deržavnyj Muzej Teatral'noho, Muzyčnoho ta Kinomystectva Ukraïny, Kyïv ... 43077
Memorial'nyj Muzej V.I. Nemiroviča-Dančenka, Neskučne ... 43152

United Kingdom

Ellen Terry Memorial Museum, Tenterden ... 45838
Georgian Theatre Royal Museum, Richmond, North Yorkshire ... 45488
Mander and Mitchenson Theatre Collection, London ... 44912
Museum of Entertainment, Whaplode ... 45992
Renishaw Hall Museum and Art Gallery, Renishaw ... 45483
Royal Shakespeare Company Collection, Stratford-upon-Avon ... 45784
University of Bristol Theatre Collection, Bristol ... 43640
Victoria and Albert Museum, London ... 45004

United States of America

American Museum of Magic, Marshall ... 50502
Barnum Museum, Bridgeport ... 47016
Bob Burns Museum, Van Buren ... 53700
Center for Puppetry Arts, Atlanta ... 46430
Central City Opera House Museum, Central City ... 47319
Charles B. Goddard Center for Visual and Performing Arts, Ardmore ... 46343
Circus City Festival Museum, Peru ... 51787
Circus World Museum, Baraboo ... 46604
Cumberland Theatre Arts Gallery, Cumberland 47906
Detroit Repertory Theatre Gallery, Detroit ... 48115
Diverse Works, Houston ... 49385
Edward Hopper House Art Center, Nyack ... 51449
Emmett Kelly Historical Museum, Sedan ... 53025
Folger Shakespeare Library, Washington ... 53869
Ford's Theatre, Washington ... 53871
Hampden-Booth Theatre Library at the Players, New York ... 51190
Hocking Valley Museum of Theatrical History, Nelsonville ... 51002
Hollywood Bowl Museum, Los Angeles ... 50230
Houdini Museum, Scranton ... 52981
International Clown Hall of Fame and Research Center, West Allis ... 54041
The John and Mable Ringling Museum of Art, Sarasota ... 52914
Laura and Paul Mesaros Galleries, Morgantown ... 50845
Marcella Sembrich Opera Museum, Bolton Landing ... 46897
National Museum of Dance and Hall of Fame, Saratoga Springs ... 52921
New York Public Library for the Performing Arts, New York ... 51244
Rose Museum at Carnegie Hall, New York ... 51250
Roy Rogers-Dale Evans Museum, Branson ... 46997
Tabor Opera House Museum, Leadville ... 50007
Vent Haven Museum, Fort Mitchell ... 48656
Visual Arts Center, Portsmouth ... 52113
Western Gallery, Bellingham ... 46713
Wexner Center for the Arts, Columbus ... 47761

Petroleum Industry and Trade

Argentina

Museo Nacional del Petróleo, Comodoro Rivadavia ... 00310

Australia

Sagasco Historical Group Museum, Brompton 00905

Austria

Freilichtmuseum Erdöl- und Erdgaslehrpfad, Prottes ... 02534

Canada

Bothwell-Zone Oil Museum, Bothwell ... 05442
Oil Museum of Canada, Oil Springs ... 06503
Oil Sands Discovery Centre, Fort McMurray ... 05817
Petrolia Discovery, Petrolia ... 06598
Ralph Allen Memorial Museum, Oxbow ... 06564
Turner Valley Gas Plant, Turner Valley ... 07183

France

Musée du Peigne et de la Plasturgie, Oyonnax 14550
Musée du Pétrole, Merkwiller-Pechelbronn ... 14090

Germany

Deutsches Erdölmuseum Wietze, Wietze ... 22564
Heimatstube Hänigsen, Uetze ... 22154

Hungary

Magyar Olajipari Múzeum, Zalaegerszeg ... 23972

India

Regional Science Centre, Guwahati ... 24195

Japan

Gasu no Kagakukan, Takaishi ... 29315

Oman

Oil and Gas Exhibition Centre, Mina Al-Fahal ... 33986

Peru

Expopetro - Museo del Petróleo, Lima ... 34212

Poland

Muzeum Przemysłu Naftowego w Bóbrce, Chorkówka ... 34601

Romania

Muzeul Național al Petrolului, Ploiești ... 36536

Switzerland

Öle-Museum, Münsingen ... 41661
Schweizerisches Tanksäulenmuseum, Gänsbrunnen ... 41395

United Kingdom

Aberdeen Maritime Museum, Aberdeen ... 43235
Biggar Gasworks Museum, Biggar ... 43479
The National Gas Museum, Leicester ... 44679

United States of America

Butler County Society Museum and Kansas Oil Museum, El Dorado ... 48311
Caddo-Pine Island Oil and Historical Society Museum, Oil City ... 51508
California Oil Museum, Santa Paula ... 52902
East Texas Oil Museum at Kilgore College, Kilgore ... 49774
Historic Pithole City, Pleasantville ... 51991
Norman No. 1 Oil Well Museum, Neodesha ... 51003
Oil Patch Museum, Russell ... 52487
Spindletop and Gladys City Boomtown Museum, Beaumont ... 46679

Pets

Austria

Erstes Europäisches Hundemuseum, Klostermarienberg ... 02197

France

Musée du Chien, Villeneuve-les-Convers ... 16216

Netherlands

Hondenmuseum, Eindhoven ... 32188
Katten Kabinet, Amsterdam ... 31851

United Kingdom

Dog Collar Museum, Maidstone ... 45072

United States of America

Presidential Pet Museum, Williamsburg ... 54172

Pewter

Czech Republic

Muzeum Mladaboleslavska, Mladá Boleslav ... 09657

France

Musée du Terroir Marseillais, Marseille ... 14007

Germany

Heimatmuseum im Wachtturm, Geyer ... 18631
Kreismuseum Zons, Dormagen ... 17904
Torhaus Dölitz, Leipzig ... 19940
Völkerschlachtdenkmal / Forum 1813, Leipzig ... 19941
Weygang-Museum, Öhringen ... 20868

Netherlands

Het Zakkendragershuisje, Rotterdam ... 32735

Pharmacy

Argentina

Museo de la Farmacia Dra. D'Alessio Bonino, Buenos Aires ... 00203

Australia

Childers Pharmaceutical Museum, Childers ... 00967
Medical History Museum, Melbourne ... 01287

Austria

Pharma- und Drogistenmuseum, Wien ... 03057
Stadtmuseum Graz, Graz ... 01983

Belgium

Hôpital Notre-Dame de la Rose, Lessines ... 03786
Museactron, Maaseik ... 03837

P

Brazil
Museu Antonio Lago, Rio de Janeiro 04700
Museu da Farmáciada Santa Casa da Misericórdia do Rio de Janeiro, Rio de Janeiro 04712
Museu de Farmácia, São Paulo 04886

Canada
James J. O'Mara Pharmacy Museum, Saint John's 06848
Le Musée Historique de Tracadie, Tracadie-Sheila 07158
Niagara Apothecary, Niagara-on-the-Lake 06452
Saskatchewan Pharmacy Museum, Regina 06730

Cuba
Farmacia Museo Taquechel, La Habana 09102
Museo Farmacéutico, Matanzas 09215

Finland
Apteekkimuseo ja Qwenselin Talo, Turku 11388
Öjskogsparkens Museiapoteket, Närpes 11135
Puuhkalan Talomuseo, Ylivieska 11483

France
Ancienne École de Médecine Navale, Rochefort (Charente-Maritime) 15069
Apothicairerie, Pont-de-Veyle 14884
Apothicairerie de l'Hôtel Dieu-le-Comte, Troyes 15994
Musée Augustin Bernard, Bourbon-l'Archambault 12101
Musée de la Pharmacie, Tence 15869
Musée de la Pharmacie Albert Ciurana, Montpellier 14239
Musée de l'Hôpital Général de Dijon, Dijon 12740
Musée Moissan, Paris 14721
Musée Testut-Latarjet de Médecine et d'Anatomie, Lyon 13900

Germany
Apotheken-Museum, Schiltach 21555
Apothekenmuseum, Hofgeismar 19223
Bad Schwalbacher Kur-Stadt-Apotheken-Museum, Bad Schwalbach 16833
Brandeburgisches Apothkenmuseum, Cottbus 17729
Brandenburgisches Apothekenmuseum, Cottbus 17730
Deutsches Apotheken-Museum im Heidelberger Schloß, Heidelberg 19071
Krankenhausmuseum, Nürnberg 20775
Löwen-Drogerie und Museum, Oelsnitz 20872
Medizin- und Pharmaziehistorische Sammlung Kiel der Christian-Albrechts-Universität, Kiel 19533
Stiftung Kohl'sche Einhorn-Apotheke, Weißenburg in Bayern 22454

Hungary
Arany Egyszarvú Patikamúzeum, Kőszeg 23683
Arany Sas Patika - Könyvtár és Levéltár, Budapest 23383
Fekete Sas Patikamúzeum, Székesfehérvár 23854
Orvostörténeti és Patikakiállítás, Karcag 23641
Patika-ház, Sopron 23815
Széchenyi Patika, Győr 23586
Szerecsen Patikamúzeum, Pécs 23776
Telekessy Patikamúzeum, Eger 23544

Italy
Farmamuseo Sa Potecaria, Villacidro 28275
Museo della Farmacia Picciola, Trieste 28091
Museo Farmacia Cinquecentesca, Roccavaldina 27435
Museo Farmacia Conventuale, Venezia 28200

Japan
Kenkoukanri Hakubutsukan, Hiroshima 28539
Naito Kinen Kusuri Hakubutsukan, Kawashima 28697

Netherlands
Kruidentuin de Groene Kruidhof, Elburg 32198
Medisch Farmaceutisch Museum de Griffioen, Delft 32060
Nederlands Drogisterij Museum, Maarssen 32518

New Zealand
Ernest and Marion Davis Medical History Museum, Auckland 33046

Norway
Ibsenhuset og Grimstad Bymuseet, Grimstad 33519
Norsk Farmasihistorisk Museum, Oslo 33740

Poland
Muzeum Farmacji, Kraków 34793
Muzeum Farmacji im. Antoniny Leśniewskiej, Warszawa 35241
Muzeum Farmacji przy Aptece pod Łabędziem, Bydgoszcz 34577
Muzeum Historii Medycyny i Farmacji, Szczecin 35160
Muzeum Roberta Kocha, Wolsztyn 35319
Muzeum Zakładu Historii Farmacji, Lublin 34910

Portugal
Museu da Farmácia, Lisboa 35679
Núcleo Museológico da Faculdade de Farmácia, Porto 35864

Romania
Colecţia de Istorie a Farmaciei Cluj-Napoca, Cluj-Napoca 36285
Expoziţia de Istoria Farmaciei, Sibiu 36605
Expoziţia Permanentă de Istoria Farmaciei Montanistice Bănăţene, Oraviţa 36507
Muzeul Universitar de Istoria Medicinei şi Farmaciei Victor Gomoiu, Târgu Jiu 36667

Slovakia
Farmaceutická Expozícia, Bratislava 38209

Slovenia
Farmacevtsko-Medicinska Zbirka Bohuslava Lavičke, Ljubljana 38448

South Africa
Adler Museum of Medicine, Johannesburg 38646

Spain
Museo de Farmacia, Llívia 39485
Museo de Farmacia, Peñaranda de Duero 39832
Museo de la Farmacia Hispana, Madrid 39545
Museo de la Real Academia Nacional de Farmacia, Madrid 39548
Museo Monástico Medival y de Farmacia, Santo Domingo de Silos 40041
Real Cartuja, Valldemosa 40245

Sweden
Stranda Hembygdsförening, Mönsterås 40744

Switzerland
Pharmazie-Historisches Museum der Universität Basel, Basel 41176

Tajikistan
Osorkhonai Tibb va Giëhhoi Shifobakhshi Tojikiston, Isfara 42500

Thailand
Museum of Thai Herbal Medicine and Thai Pharmaceutical History, Bangkok 42549

United Kingdom
Museum of the Royal Pharmaceutical Society of Great Britain, London 44933
Royal Pharmaceutical Society Late Victorian Pharmacy, Edinburgh 44149

United States of America
Hugh Mercer Apothecary Shop, Fredericksburg 48767
New Orleans Pharmacy Museum, New Orleans 51092
Stabler-Leadbeater Apothecary Museum, Alexandria 46213
Texas Pharmacy Museum, Amarillo 46253

Philately → **Postage Stamps**

Philosophy

Austria
Gsellmanns Weltmaschine, Edelsbach bei Feldbach 01839
Ludwig Wittgenstein - Gedenkraum, Trattenbach 02822
Ludwig Wittgenstein - Wirklichkeit und Mythos, Kirchberg am Wechsel 02167
Sigmund Freud Museum, Wien 03075

Belgium
Maison d'Erasme, Bruxelles 03450
Museum Donkmeer, Berlare 03362

Czech Republic
Muzeum a Památník Jana Ámoše Komenského, Horní Branná 09525
Památník Jana Ámoše Komenského, Bílá Třemešná 09406
Památník Jana Ámoše Komenského, Fulnek 09506
Pamětní Síň Jana Ámoše Komenského, Brandýs nad Orlicí 09415

France
Maison Musée René Descartes, Descartes 12716
Musée Jean-Jacques Rousseau, Montmorency 14229

Germany
Albert-Schweitzer-Gedenk- und Begegnungsstätte, Weimar 22395
Deutsches Freimaurermuseum, Bayreuth 16949
Ernst-Bloch-Zentrum, Ludwigshafen am Rhein 20051
Freimaurermuseum der Großen Landesloge der Freimaurer von Deutschland, Sankt Michaelisdonn 21506
Hegelhaus, Stuttgart 21939
Historisches Zentrum Engelshaus, Wuppertal 22742
Museum Barockschloß Rammenau, Rammenau 21175
Museum Karl-Marx-Haus, Trier 22100
Nietzsche-Archiv, Weimar 22412
Nietzsche-Haus, Naumburg, Saale 20555
Novalis-Museum, Wiederstedt 22541
Wilhelm-Ostwald-Gedenkstätte, Großbothen 18783

Greece
Theophilos Kairis Museum, Andros 22885

Italy
Collezioni dell'Accademia Nazionale Virgiliana, Mantova 26590

Lebanon
Gibran Museum, Bsharri 30166

Netherlands
Het Spinozahuis, Rijnsburg 32688

Peru
Casa Museo José Carlos Mariátegui, Lima 34207

Russia
Ékspozicija "Čelovek. Duša. Duchovnost.", Ivanovo 36945
Muzej-usadba poéta D. Venevitinova, Novoživotinnoe 37492

Switzerland
Casa Anatta e Casa Selma, Ascona 41127
Nietzsche-Haus, Sils Maria 41870

Turkey
Ziya Gökalp Müzesi, Diyarbakır 42765

Ukraine
Istoriko-memorialnyj Muzej Semji Fadeevyh-Chan-Blavackij, Dnipropetrovs'k 43034

United Kingdom
Museum of Freemasonry, London 44923

United States of America
Orchard House, Concord 47770
Ralph Waldo Emerson House, Concord 47771

Uzbekistan
Abu Ali Ibn Sina Yodgorlik Muzeyi, Afchona 54437

Photography

Argentina
Museo Cooperativo Eldorado, Eldorado 00361
Museo de la Fotografía, Rafaela 00513
Museo Itinerante de Arte Contemporáneo del Mercosur, Buenos Aires 00248
Museo Municipal Histórico Fotográfico de Quilmes, Quilmes 00512
Torre Monumental, Buenos Aires 00283

Australia
Australian Centre for Photography, Paddington, New South Wales 01399
Beatties Historic Photograph Museum, Dodges Ferry 01039
Centre for Contemporary Photography, Fitzroy 01080
Eric Thomas Galley Museum, Queenstown 01450
Horsham Regional Art Gallery, Horsham 01163
Latrobe Court House Museum, Latrobe 01219
Monash Gallery of Art, Wheelers Hill 01647

Austria
Albertina, Wien 02929
Bank Austria Kunstforum und Kunstsammlung, Wien 02942
Bild- und Tonarchiv, Graz 01965
Essl Museum - Kunst der Gegenwart, Klosterneuburg 02200
Fotoforum West, Innsbruck 02116
Galerie im Rathaus, Bad Aussee 01738
Museum der Photographie, Fischamend 01891
Oberösterreichische Fotogalerie, Linz 02321
Photomuseum des Landes Oberösterreich, Bad Ischl 01757
Vorarlberger Museumswelt, Frastanz 01900

Azerbaijan
History Museum, Martakert 03179
Museum of Fallen Soldiers, Stepanakert 03199

Belgium
Archives et Musée de la Littérature, Bruxelles 03433
Fotomuseum Provincie Antwerpen, Antwerpen 03287
Musée de la Photographie, Charleroi 03515

Brazil
Museu Histórico, Pará de Minas 04609

Canada
Canadian Museum of Contemporary Photography, Ottawa 06535
Dazibao, Montréal 06337
Gallery 44, Toronto 07107
Paved Arts, Saskatoon 06922
Presentation House Gallery, Vancouver 07219
Saint Marys Museum, Saint Marys 06869
South Sask Photo Museum, Arcola 05350
VU Centre de Diffusion et de Production de la Photographie, Québec 06699

Chile
Sala Sergio Larrain, Santiago de Chile 07465

Cuba
Fototeca de Cuba, La Habana 09103

Czech Republic
Galerie Josefa Sudka, Praha 09748
Moravská Galerie Brno - Uměleckoprůmyslové Muzeum, Brno 09434
Muzeum umění a designu, Benešov 09402
Uměleckoprůmyslové Muzeum v Praze, Praha 09797

Denmark
Danmarks Fotomuseum, Herning 10070
Museet for Fotokunst, Odense 10218
Det Nationale Fotomuseum, København 10143

Estonia
Raevangla Fotomuuseum, Tallinn 10642

Finland
Helsingin Kaupungin Taidemuseo, Meilahti, Helsinki 10780
K.H. Renlundin Museo ja Kamerakokoelma, Kokkola 10972
Suomen Valokuvataiteen Museo, Helsinki 10823

France
Centre Atlantique de la Photographie, Brest 12156
Centre de Photographie de Lectoure, Lectoure 13671
Centre Photographique d'Ile-de-France, Pontault-Combault 14900
Centre Régional de la Photographie Nord-Pas-de-Calais, Douchy-les-Mines 12778

P

Le Château d'Eau, Toulouse 15931
Fondation Van-Gogh Arles, Arles 11682
Jeu de Paume, Paris 14614
Jeu de Paume, Paris 14615
Maison Européenne de la Photo, Paris 14623
Mémoire de la Photo, Donzère 12768
Musée Adzak, Paris 14628
Musée Charles Cros, Fabrezan 12900
Musée de la Photo, Gracay 13091
Musée de la Photographie, Mougins 14296
Musée des Appareils Photographiques et de Cinéma d'Amateurs, Marsilly 14014
Musée du Centre de Recherches sur les Monuments Historiques, Paris 14694
Musée Français de la Photographie, Bièvres 11991
Musée Gorsline, Bussy-le-Grand 12212
Musée Municipal, La Roche-sur-Yon 13419
Musée Nicéphore Niépce, Chalon-sur-Saône 12359

Georgia

Ilia Chavchavadze's house museum, Kvareli 16346

Germany

Alfred-Erhardt-Stiftung, Berlin 17030
Alphons-Stübel-Sammlung früher Orientphotographien, Jena 19376
Bildarchiv Foto Marburg, Marburg 20187
Brotfabrik Galerie, Berlin 17057
Camera Obscura, Hainichen 18872
Camera Obscura, Mülheim an der Ruhr 20411
Deichtorhallen Hamburg, Hamburg 18920
Deutsches Film- und Fototechnik Museum, Deidesheim 17807
Forum für Fotografie, Köln 19601
Forum Internationale Photographie, Mannheim 20163
Fotografie Forum International, Frankfurt am Main 18367
Fotohistorische Sammlung G. und W. Pabst, Uhingen 22156
Freiburger Photohistorische Sammlung, Schallstadt 21532
Galerie Heinz Teufel Multimedia, Zingst, Ostseebad 22804
Galerie Heinz Teufel Schule des Sehens, Zingst, Ostseebad 22805
Galerie Heinz Teufel Villa Ruh, Zingst, Ostseebad 22806
Haus der Fotografie - Dr. Robert-Gerlich-Museum, Burghausen 17653
Heimatmuseum, Glonn 18662
Helmut Newton Stiftung im Museum für Fotografie, Berlin 17137
Holowood - Holographiemuseum Bamberg, Bamberg 16911
Industrie- und Filmmuseum Wolfen, Wolfen 22650
Kamera- und Fotomuseum Leipzig, Leipzig 19914
Kunsthalle Mannheim, Mannheim 20166
Kunstsammlungen der Ruhr-Universität Bochum, Bochum 17402
Kupferstich-Kabinett, Dresden 17954
Münchner Stadtmuseum, München 20462
Museen im Markgrafen-Schloss, Emmendingen 18146
Museum Folkwang, Essen 18263
Museum für Fotografie - Sammlung Fotografie, Kunstbibliothek, Berlin 17202
Museum für Gegenwartskunst Siegen, Siegen 21755
Museum für Photographie, Braunschweig 17527
Die Photographische Sammlung/SK Stiftung Kultur, Köln 19627
Reiss-Engelhorn-Museen (rem), Mannheim 20173
Ruhr Museum, Essen 18267
Städtische Galerie Iserlohn, Iserlohn 19361
Stiftung Bauhaus Dessau, Dessau 17834
Weltkulturerbe Völklinger Hütte, Völklingen 22249
ZEPHYR - Raum für Fotografie!, Mannheim 20175

Greece

National Resistance Museum, Karpenissi 23038

Hungary

André Kertész Fotómúzeum, Szigetbecse 23888
Fridrich János Fényírdája, Szentes 23883
Magyar Fotográfiai Múzeum, Kecskemét 23650

Iceland

Ljósmyndasafn Reykjavíkur Grófarhús, Reykjavík 24024

India

Gandhi Museum, Bhavnagar 24088

Ireland

Irish Architectural Archive, Dublin 24798

Israel

Israeli Museum of Photography, Tel Hai 25125

Italy

Civica Raccolta delle Stampe Achille Bertarelli, Milano 26676
Forma-Centro Internazionale di Fotografia, Milano 26686
Gabinetto Fotografico, Firenze 26154
Istituto di Fotografia Alpina Vittorio Sella, Biella 25417
Museo-Archivio di Fotografia Storica, Roma 27461
Museo del Foto - Gabinetto di Fisica, Urbino 28124
Museo della Fotografia Storica e Contemporanea, Torino 28005
Museo dell'Immagine, Cesena 25832
Museo di Fotografia Contemporanea, Cinisello Balsamo 25896
Museo Ken Damy di Fotografia Contemporanea, Brescia 25540
Museo Nazionale Alinari della Fotografia - MNAF, Firenze 26195
Museo Nazionale del Cinema, Torino 28021
Museo Nazionale della Fotografia, Brescia 25541
Spazio Oberdan, Milano 26724

Japan

Hirano Image Library, Osaka 29128
Kaigan Bijutsukan, Chikura 28399
Ken Domon Museum of Photography, Sakata 29205
Kiyosato Museum of Photographic Arts, Takane 29325
Kushiro Art Museum, Kushiro 28798
Machida-shiritsu Shashin Bijutsukan, Machida 28876
Museum of Old Photographs and Museum of Unearthed Artifacts, Nagasaki 28976
Nakagawa Photo Gallery, Kyoto 28853
Namikawa Banri Bukkyo Bijutsu Shashin Gallery, Hiroshima 28541
Nara City Museum of Photography, Nara 29034
Nihon Camera Hakubutsukan, Tokyo 29447
Shirahata Shiro World Mountains Photograph Collection, Yuzawa 29638
Shoji Ueda Museum of Photography, Kishimoto 28705
Tokyo-to Shashin Bijutsukan, Tokyo 29499
Tsuwano Gendai Photo Gallery, Tsuwano 29547

Kazakhstan

Bogaev Museum, Pavlodar 29716

Latvia

Latvijas Fotogrāfijas Muzejs, Rīga 30080

Lithuania

Fotografijos Muziejus, Šiauliai 30356

Luxembourg

Musée Municipal, Dudelange 30467
Villa Vauban, Luxembourg 30486

Mauritius

Musée de la Photographie, Port Louis 30683

Mexico

Centro Fotográfico Álvarez Bravo, Oaxaca 31219
Museo de la Fotografía, Pachuca 31241
Museo del Cine y la Fotografía, Puebla 31285
Museo Fotográfico de Nanchital, Nanchital de Lázaro Cárdenas del Río 31209

Morocco

Musée de la Fondation Lorin, Tanger 31697

Netherlands

Ter Beziens, Kolhorn 32452
Foam Fotografiemuseum Amsterdam, Amsterdam 31842
Fotomuseum Den Haag, Den Haag 32072
Huis Marseille, Amsterdam 31849
Kunsthal Rotterdam, Rotterdam 32710
Museum Kempenland, Eindhoven 32191
Nederlands Fotomuseum, Rotterdam 32721
Nederlands Instituut voor Beeld en Geluid-Colleties, Hilversum 32389
Oudheidkamer/ Museum Mr. Andreae, Kollum 32454
Prentenkabinet Universiteit Leiden, Leiden 32486
Spaarnestad Fotoarchief, Haarlem 32301
Stedelijk Museum Het Domein, Sittard 32779

New Zealand

Albertland and Districts Museum, Wellsford 33262
City Gallery Wellington, Wellington 33244
Nelson Provincial Museum, Nelson 33153
New Zealand Centre for Photography, Wellington 33253

Norway

Levanger Museum, Levanger 33635
Preus Museum, Horten 33564

Poland

Dolnośląskie Centrum Fotografii Domek Romański-Galeria, Wrocław 35323
Fotogaleria 1/3, Toruń 35188
Fotograficzny Muzeum, Mielec 34919
Fotoplastikon, Warszawa 35223
Galeria Art-Foto, Częstochowa 34613
Galeria Forum Fotografii, Łódź 34866
Galeria Fotografii Instytutu Sztuki Akademii Pedagogicznej, Kraków 34775
Galeria Sztuki Współczesnej Profil Galeria Fotografii pf, Poznań 35024
Muzeum Historii Fotografii, Kraków 34796
Muzeum Historii Fotografii im. Prof. Władisława Bogackiego, Kraków 34797
Muzeum Narodowe w Gdańsku, Gdańsk 34639
Pusta, Katowice 34730
Stara Galeria ZPAF, Warszawa 35289

Portugal

Museu da Imagem em Movimento, Leiria 35658
Museu Municipal da Fotografia João Carpinteiro, Elvas 35557
Museu Municipal de Fotografia Carlos Relvas, Golegã 35629
Photographia-Museu Vicentes, Funchal 35620

Romania

Muzeul Memorial Poni-Cernătescu, Iaşi 36424

Russia

Fotoarchiv Sojuza Fotochudožnikov Rossii, Moskva 37171
Fotogalereja im. Bratjev Ljumjer, Moskva 37172
Fotografičeskij Muzej Dom Metenkova, Ekaterinburg 36885
Moskovskij Dom Fotografii, Moskva 37254
Objedinenie Fotocentr Sojuza Žurnalistov, Moskva 37390
Russkij Muzej Fotografii, Nižnij Novgorod 37470

Slovakia

Dom Fotografie v Poprade, Poprad 38331

Slovenia

Fotomuzej Vlastja, Gorenja Vas 38418

South Africa

Bensusan Museum of Photography, Johannesburg 38649

Spain

Centro Andaluz de la Fotografía, Almería 38856
Centro de Fotografía, Santa Cruz de Tenerife 39993
Centro Ordóñez Falcón de Fotografía, Donostia-San Sebastián 39248
Fototeca Municipal de Reus, Reus 39893
Metrònom, Barcelona 38966
Museo de Reproducciones Artísticas, Bilbao 39044
Museo del Pueblo de Asturias, Gijón 39338
Museu Nacional d'Art de Catalunya (MNAC), Barcelona 39001

Sweden

Alingsås Museum, Alingsås 40413
Erna och Victor Hasselblads Fotografiska Centrum, Göteborg 40544
Foto Museet Sundvall - Bildens Hus, Sundsvall 40958
Fotomuséet i Osby, Osby 40801
Fotomuseet Olympia, Falkenberg 40509
Fotomuseum, Motala 40750
Moderna Museet, Stockholm 40915

Switzerland

Centre de la Photographie Genève, Genève 41406
Coalmine Fotogalerie, Winterthur 42024
Fondation Herzog, Basel 41159
Fotomuseum Winterthur, Winterthur 42025
Fotostiftung Schweiz, Winterthur 42026
Musée de l'Elysée, Lausanne 41547
Museum für Kommunikation, Bern 41212
Museum im Bellpark, Kriens 41518
Tal Museum Engelberg, Engelberg 41352
Uetiker Museum, Uetikon am See 41956

United Kingdom

British Photographic Museum, Totnes 45870
Cambridge University Collection of Air Photographs, Cambridge 43725
Dimbola Lodge Museum, Freshwater Bay 44255
Focal Point Gallery, Southend-on-Sea 45711
Fox Talbot Museum, Lacock 44627
HMS Ganges Museum, Ipswich 44529
Impressions Gallery of Photography, Bradford 43580
Jersey Photographic Museum, Saint Helier 45572
Kildonan Museum, Kildonan, Isle of South Uist 44576
Leigh Heritage Centre, Leigh-on-Sea 44686
National Fairground Archive, Sheffield 45654
National Media Museum, Bradford 43581
Nelson Tower, Forres 44246
Open Eye Gallery, Liverpool 44743
Peacock Visual Arts, Aberdeen 43245
The Photographers' Gallery, London 44954
The Royal Photographic Society, Fenton House, Bath 43411
Sutcliffe Gallery, Whitby 45996
Turner House Gallery, Penarth 45352

United States of America

Adell McMillan Gallery, Eugene 48430
The African American Museum in Philadelphia, Philadelphia 51810
Albany Civil Rights Movement Museum, Albany 46154
The Albin O. Kuhn Gallery, Baltimore 46544
Arvada Center for the Arts and Humanities, Arvada 46367
Baldwin Photo Gallery, Murfreesboro 50913
Bass Manor Museum, Malta 50409
Blue Sky, Portland 52072
Boston Public Library, Boston 46923
Burden Gallery, New York 51138
California Museum of Photography, Riverside 52369
Center for Creative Photography, Tucson 53585
Center for Photographic Art, Carmel 47250
Center for Photography at Woodstock, Woodstock 54283
Center for Visual Communication, Coral Gables 47797
Center Gallery, Demarest 48052
Central Washington Gallery, Ellensburg 48356
Cokato Museum and Akerlund Photography Studio, Cokato 47679
Colorado Photographic Arts Center, Denver 48070
Columbus Museum of Art, Columbus 47752
Connecticut Graphic Arts Center, Norwalk 51431
Dorothea C. Witham Gallery, Lewiston 50038
Emerson Gallery, McLean 50362
Folk Art and Photography Galleries, Atlanta 46436
Gallery at the Wharf, Coupeville 47854
Gallery of Art, Fort Valley 48695
Gallery of Henry Street Settlement, New York 51181
Griffin Museum of Photography, Winchester 54220
Historical Society of Oak Park and River Forest, Oak Park 51455
Houston Center for Photography, Houston 49388
International Center of Photography, New York 51194
International Museum of Photography and Film, Rochester 52395
Iris and B. Gerald Cantor Art Gallery, Worcester 54299
J. Paul Getty Museum, Los Angeles 50237
Japanese American National Museum, Los Angeles 50238
Kiboko Projects, New York 51202
Leslie-Lohman Gay Art Foundation, New York 51203
The Lyons Redstone Museum, Lyons 50334
Main Gallery of Henry Street Settlement, New York 51208
Merwin and Wakeley Galleries, Bloomington 46862
Museum of Antropology, History and Art, Rio Piedras 52355
The Museum of Contemporary Photography, Chicago 47493
Museum of Photographic Arts, San Diego 52741
New Braunfels Museum of Art Music, New Braunfels 51024
Palm Beach Photographic Center, Delray Beach 48047
Photo Antiquities, Pittsburgh 51955
Photographic Archives, Lexington 50055
Photographic Investments Gallery, Atlanta 46448
Photographic Resource Center, Boston 46947
Putney School Gallery, Putney 52204
Racine Art Museum, Racine 52223
Richard F. Brush Art Gallery and Permanent Collection, Canton 47224
Ruddell Gallery, Spearfish 53200
Runnels Gallery, Portales 52058
Silver Eye-Center for Photography, Pittsburgh 51960

Society for Contemporary Photography, Kansas City 49699
Soho Photo Gallery, New York 51257
Southeast Museum of Photography, Daytona Beach 47995
State of the Art Gallery, Ithaca 49551
Sterling Historical Society and Little Red Schoolhouse Museum, Sterling 53286
The Studio Annex, New York 51263
Synagogue for the Arts, New York 51266
Tang Teaching Museum and Art Gallery, Saratoga Springs 52926
University College Gallery, College Park 47687
University Museums, Newark 51283
Video History Museum, Grass Valley 49026
Visual Arts Museum, New York 51277
Washington Center for Photography, Washington 53930
White Gallery, Portland 52091
Wings of Love, Clinton 47654

Venezuela

Roberto Mata Taller de Fotografía, Caracas 54557

Physics

Argentina

Museo de Física de la Universidad Nacional de La Plata, La Plata 00411
Museo Participativo de Ciencias, Buenos Aires 00269

Belgium

Museum voor de Geschiedenis van de Wetenschappen, Gent 03636

Bulgaria

Palata na Fizikata, Kazanläk 05061

Canada

TELUS World of Science, Edmonton 05755

Germany

Faszinosum, Borna 17484
Naturkundliche Sammlung, Altusried 16528
phæno gGmbH, Wolfsburg 22666
Phänomenta Lüdenscheid, Lüdenscheid 20089
Röntgen-Gedächtnisstätte, Würzburg 22722
Wilhelm-Ostwald-Gedenkstätte, Großbothen 18783

Hungary

Jedlik Ányos Emlékkiállítás, Győr 23583

Israel

Bloomfield Science Museum Jerusalem, Jerusalem 24981

Italy

Domus Galilaeana, Pisa 27237
Museo di Fisica, Bologna 25461
Museo di Storia della Fisica, Padova 27043
Museo Torricelliano, Faenza 26078
Tempio Voltiano, Como 25956

Japan

Kids' Science Museum of Photons, Kizu 28721

Lithuania

Fizikos Muziejus, Vilnius 30404

Mexico

Museo de Ciencias Explora, León 30998
Museo de Física Recreativa, Puebla 31276
Museo de la Radiodifusión, Puebla 31278

Philippines

Philippine Radiation Science Museum, Manila 34451

Poland

Muzeum Marii Skłodowskiej Curie, Warszawa 35259

Portugal

Museu de Física, Coimbra 35533

Spain

Museo del CosmoCaixa Barcelona, Barcelona 38968

Sweden

Nora Museum, Gyttorp 40570
Observatoriemuseet, Stockholm 40923

Switzerland

CERN - Microcosm, Genève 41409

United Kingdom

Aberdeen University Natural Philosophy Museum, Aberdeen 43237
Fox Talbot Museum, Lacock 44627
Michael Faraday's Laboratory and Museum, London 44917
World Museum Liverpool, Liverpool 44750

United States of America

Danville Science Center, Danville 47964
Louisville Science Center, Louisville 50290
Regional Science and Discovery Center, Horseheads 49370
Science Discovery Center of Oneonta, Oneonta 51565
Science Imaginarium, Waterloo 53963
Whitaker Center for Science and the Arts, Harrisburg 49183

Pipes, Tobacco → Tobacco

Plaster Casts

Germany

Abguss Sammlung antiker Plastik, Berlin 17027
Gipsformerei, Berlin 17118
Knauf-Museum, Iphofen 19351
Museum Kloster Asbach, Asbach 16579

India

Government Educational Museum, Deoria 24158

Italy

Gipsoteca Canoviana e Casa del Canova, Possagno 27316
Gipsoteca del Castello Normanno Svevo, Bari 25361
Gipsoteca Istituto d'Arte, Firenze 26164
Museo del Palazzo del Podestà e Gipsoteca Libero Andreotti, Pescia 27173

Spain

Museo del Azulejo, Onda 39733
Museu del Castell de Onda, Onda 39735

Switzerland

Gipsmuseum Oberwiesen, Schleitheim 41835

United Kingdom

House of the Binns, Linlithgow 44719

Uruguay

Museo del Azulejo, Montevideo 54399

Playing Cards

Belgium

Nationaal Museum van de Speelkaart, Turnhout 04099

France

Musée du Vieux Marseille, Marseille 14008

Germany

Deutsches Spielkartenmuseum, Leinfelden-Echterdingen 19881
Schloß- und Spielkartenmuseum, Altenburg 16499

Spain

Museo Fournier de Naipes de Álava, Vitoria-Gasteiz 40324

Police → Criminology

Politics and Government

Argentina

Museo de la Casa de Gobierno, San Miguel de Tucumán 00596
Museo del Antiguo Recinto del Congreso Nacional, Buenos Aires 00220
Museo Nuestra Señora de Carrodilla, Luján de Cuyo 00437
Museo Senador Domingo Faustino Sarmiento, Buenos Aires 00273

Australia

Chifley Home, Bathurst 00839
Government House, Sydney 01547
Hawke House, Bordertown 00878
Museum of Sydney, Sydney 01555
Old Government House, Paramatta 01403
Old Parliament House, Canberra 00949
Vaucluse House, Vaucluse 01616

Austria

Burgenländisches Geschichte(n)haus, Bildein 01785
Erzherzog Franz Ferdinand Museum, Artstetten 01728
Heimatmuseum mit Franz Jonas-Gedenkraum, Teesdorf 02799
Museum Arnoldstein, Arnoldstein 01727
Zeitreise - Renner Villa, Gloggnitz 01939

Azerbaijan

Azerbaijan State Museum of Independence, Bakı 03143
Memorial Museum, Naxçivan 03185

Benin

Musée d'Histoire, Ouidah 04185

Bulgaria

Kašta-muzej Dimitâr Blagoev, Bankja 04977
Kăšta-muzej Dimitâr Blagoev, Sofia 05175
Kăšta-muzej Dimitar Pešev, Kjustendil 05064
Muzej na Ministerstvoto na Vnatrešnite raboti, Sofia 05186
Nacionalen Muzej Georgi Dimitrov, Sofia 05194

Canada

Bellevue House, Kingston 06074
Commissariat House, Saint John's 06846
Cossit House Museum, Sydney 07048
Diefenbaker House Museum, Prince Albert 06660
Government House Museum and Heritage Property, Regina 06718
Haliburton House Museum, Windsor, Nova Scotia 07342
House of Commons, Ottawa 06547
John G. Diefenbaker Replica Law Office, Wakaw 07273
Louis-S.-Saint-Laurent, Compton 05638
Loyalist House Museum, Saint John 06837
Motherwell Homestead, Abernethy 05308
Musée Laurier, Victoriaville 07261
Old Government House, Fredericton 05838
Rutherford House, Edmonton 05753
Sinclair Inn, Annapolis Royal 05345
Vancouver Holocaust Education Centre, Vancouver 07227
William Henry Steeves Museum, Hillsborough 05988

China

Beijing Police Museum, Beijing 07519
Luoyang Ducheng Guan, Luoyang 07983
Political Museum, Shanghai 08177
Rongyu Guan, Jiangyin 07865
Site of the 1st National Congress of the CCP, Shanghai 08193
Site of the First National Congress of the Communist Party of China, Shanghai 08194

Colombia

Museo Casa Rafael Núñez, Cartagena 08617

Costa Rica

Museo Historico Dr. Rafael Angel Calderón Guardia, San José 08822

Cuba

Museo Casa Natal de Celia Sanchez Manduley, Media Luna 09221
Museo Nacional del Ministerio del Interior, La Habana 09167

Denmark

Reventlow-Museet Pederstrup, Horslunde 10102

Egypt

Museum of the Egyptian People's Assembly, Cairo 10507

Finland

Eduskunnan Talvisotamuseo, Kauhajoki 10926
Lepikon Torppa, Pielavesi 11204
Urho Kekkonen Museo Tamminiemi, Helsinki. 10834

France

Maison Natale de Jean Giraudoux, Bellac 11925
Musée Georges Clemenceau, Paris 14709
Musée Georges Pompidou, Montboudif 14187

Germany

Bismarck-Museum, Schönhausen 21610
Deutsch-Deutsches Museum Mödlareuth, Töpen 22068
Deutsch-Russisches Museum Berlin-Karlshorst, Berlin 17066
Gedenkstätte Berlin-Hohenschönhausen, Berlin 17111
Grenzland-Museum Bad Sachsa, Bad Sachsa 16812
Reichstagsmuseum, Regensburg 21231
Residenz Ellingen, Ellingen 18117
Stasi Die Ausstellung zur DDR-Staatssicherheit, Berlin 17260
Unterfränkisches Grenzmuseum, Bad Königshofen 16761

Greece

Alexandros Papanastassiou Museum, Levidion 23076
Capodistria Museum, Koukouritsa, Evropouli 23065
Typaldos-Iakobatos Museum, Lixouri 23080

Honduras

Museo de Historia Republicana, Tegucigalpa 23332

Hungary

Memento Park Budapest, Budapest 23449
Széchenyi István Emlékmúzeum, Nagycenk 23723
Vay Ádám Múzeum, Vaja 23947

India

Chandradhari Museum, Darbhanga 24125
Gandhi Memorial Museum, Madurai 24303
Gandhi Museum, Bhavnagar 24088
Gandhi Museum, Lucknow 24293
Gandhi Smarak Sangrahalaya, Barrackpore 24080
Krishna Menon Museum and Art Gallery, Kozhikode 24285
National Gandhi Museum and Library, Delhi 24146
Netaji Museum, Kolkata 24279

Iran

Constitution Museum, Tabriz 24651
Presidency Museum, Tehrān 24690

Ireland

Avondale House, Rathdrum 24889
Custom House Visitor Centre, Dublin 24782
Derrynane House, Caherdaniel 24749
Parnell Museum, Rathdrum 24890

Israel

Ben-Gurion House, Tel Aviv 25100

Italy

Casa Museo A. Gramsci, Ghilarza 26333
Domus Mazziniana, Pisa 27238
Museo G. Marconi, Pontecchio Marconi 27288

Japan

Gaimusho, Gaiko Shiryokan, Tokyo 29387
Museum of the Democratic Movements, Machida 28877
Osaka Human Rights Museum, Osaka 29142
Yoshizawa Kinenkan, Joetsu 28623

Jordan

Political Life Museum, Amman 29659

Korea, Republic

Diplomatic History Museum, Seoul 29871

Lebanon

Musée Michel Aboujaoudé, Zalka ... 30176

Lithuania

Dr. Jono Šliūpo Memorialinė Sodyba, Palanga 30328
Lietuvos Prezidento K. Griniaus Memorialinis Muziejus, Marijampolė ... 30307
Signatarų Namai, Vilnius ... 30432

Madagascar

Musée du Palais de la Reine, Antananarivo ... 30535

Malaysia

Muzium Tuan Yang Terutama Melaka, Melaka 30605
Tunku Abdul Rahman Putra-Memorial, Kuala Lumpur ... 30588

Mexico

Casa Museo López Portillo, Guadalajara ... 30886
Casa Museo Silvestre Rodríguez, Nacozari de García ... 31208
Museo Abraham González, Vicente Guerrero ... 31540
Museo Adolfo López Mateos, Atizapán de Zaragoza ... 30725
Museo Casa de Carranza, Cuatrocienegas ... 30811
Museo Casa del General Alvaro Obregón, Huatabampo ... 30942
Museo de Sitio Casa de Juárez, Oaxaca ... 31226
Museo Isidro Fabela, Atlacomulco ... 30726
Museo Legislativo los Sentimientos de la Nación, México ... 31126
Museo Luis Donaldo Colosio, Francisco I. Madero, Coahuila ... 30870
Museo-Sala de los Símbolos Patrios, Hermosillo ... 30933
Sala Melchor Ocampo, Morelia ... 31202

Nepal

Birendra Museum, Kathmandu ... 31773
Mahendra Museum, Kathmandu ... 31774
Tribhuvan Memorial Museum, Kathmandu ... 31780

Netherlands

't Oude Raadhuis, Beek en Donk ... 31953
Theo Thijssen Museum, Amsterdam ... 31895

Norway

Johan Nygaardsvold Museum, Hommelvik ... 33557
Politimuseet i Oslo, Oslo ... 33748

Peru

Museo Tribunal de la Santa Inquisición y del Congreso, Lima ... 34271

Philippines

Carlos P. Garcia Memorabilia, Manila ... 34434
Earist Museum, Manila ... 34439
Jose P. Laurel Memorial Museum, Manila ... 34440
José P. Laurel Monument Museum, Santo Tomas ... 34502
Kaban ng Hiyas ng Lungsod ng Mandaluyong, Mandaluyong ... 34425
Manuel L. Quezon Memorial Shrine Museum, Quezon City ... 34493
Marcos Museum, Batac ... 34343
Museum of Philippine Political History, Manila ... 34449
Ramón Magsaysay Memorabilia Museum, Manila ... 34456
Redoña Residence Museum, Tacloban ... 34510
Sarrat Museum, Sarrat ... 34504
Sen. MacArthur Memorabilia, Tacloban ... 34512
Vigan House Museum, Vigan ... 34525

Poland

Muzeum Dyplomacji i Uchodźstwa Polskiego, Bydgoszcz ... 34576
Muzeum Hymnu Narodowego, Nowy Karczma 34955

Portugal

Casa-Museu João Soares, Cortes ... 35552
Casa Museu Rafael Bordalo Pinheiro, Caldas da Raínha ... 35495

Puerto Rico

Casa Jesús T. Piñero, Canóvanas ... 36033
Luis Muñoz Rivera Museum, Barranquitas ... 36027
Museo Casa Blanca, San Juan ... 36065
Museo Felisa Rincón de Gautier, San Juan ... 36072

Romania

Casa Memorială I.G. Duca, Horezu ... 36401
Complexul Memorial Nicolae Titulescu, Nicolae Titulescu ... 36492
Complexul Muzeistic Gheorghe Magheru, Râmnicu Vâlcea ... 36551
Muzeul Memorial Mihail Kogălniceanu, Iași ... 36423
Muzeul Memorial Nicolae Bălcescu, Nicolae Bălcescu ... 36491
Muzeul Memorial Octavian Goga, Ciucea ... 36283

Russia

Aleksandrovskij Dvorec - Gosudarstvennyj Muzej Carskoje Selo, Puškin ... 37577
Bunker Stalina, Samara ... 37610
Centralnyj Muzej MVD Rossii, Moskva ... 37148
Čerkechskij Muzej Jakutskaja Političeskaja Ssylka, Čerkech ... 36853
Dom-muzej Istorii Molodėžnogo Dviženija, Rjazan ... 37593
Dom-muzej M.K. Ammosova, Jakutsk ... 36960
Gosudarstvennyj Istoriko-memorialnyj Zapovednik Rodina V.I. Lenina, Uljanovsk ... 37897
Kraevedčeskij Muzej Evrejskoj Avtonomnoj Oblasti, Birobidžan ... 36807
Memorialnyj Dom-muzej G.V. Plechanova, Lipeck ... 37105
Memorialnyj Kabinet A.V. Lunačarskogo, Moskva ... 37233
Memorialnyj Muzej Karima Chakimova, Ufa ... 37886
Memorialnyj Muzej-Kvartira S.M. Kirova, Vladikavkaz ... 37930
Muzej Dekabristov - Dom Trubeckogo, Irkutsk 36935
Muzej Dekabristov - Dom Volkonskich, Irkutsk ... 36936
Muzej Diplomatičeskogo Korpusa, Vologda ... 37962
Muzej i Obščestvennyj Centr Mir, Progress, Prava Čeloveka im. Andreja Sacharova, Moskva ... 37284
Muzej Istorii Političeskoj Policii Rossii - Gorochovaja 2, Sankt-Peterburg ... 37683
Muzej Istorii Političeskoj Ssylki, Bratsk ... 36823
Muzej Istorii Politssylki v Jakutii, Jakutsk ... 36963
Muzej Istorii Revoljucionno-demokratičeskogo Dviženija 1880-1890 gg, Sankt-Peterburg ... 37684
Muzej Istorii UVD Kalužskoj Oblasti, Kaluga ... 36999
Muzej-kvartira Alliluevych, Sankt-Peterburg ... 37689
Muzej-kvartira A.D. Sacharova, Nižnij Novgorod ... 37461
Podpolnaja Tipografija CK RSDRP 1905-1906, Moskva ... 37399
Pskovskij Memorialnyj Muzej V.I. Lenina, Pskov ... 37576

Serbia

Kuća Cveća, Beograd ... 38055
Muzej Istorije Jugoslavije, Beograd ... 38066

Slovenia

Muzej Policijske Akademije, Ljubljana ... 38456

South Africa

Goetz/Fleischack Museum, Potchefstroom ... 38713
Kruger Museum, Tshwane ... 38748
Museum Presidency, Bloemfontein ... 38565
President Pretorius Museum, Potchefstroom ... 38715
Rhodes Cottage Museum, Muizenberg ... 38693

Spain

Casa-Museo León y Castillo, Telde ... 40112
Museo Euskal Herria, Gernika-Lumo ... 39333
Museo Fundación Antonio Maura, Madrid ... 39559

Sri Lanka

Bandaranaike Museum, Colombo ... 40369

Sweden

Skånska Lasses Hus, Mjölby ... 40741

Switzerland

Villa Kruger, Clarens ... 41306

Taiwan

Chiang Kai-shek's Shihlin Residence, Taipei ... 42379
Hu Shih Memorial Hall, Taipei ... 42393
Land Reform Museum, Taipei ... 42398
Museum of Yilan Government History, Yilan ... 42468

Turkey

Adana Atatürk Müzesi, Adana ... 42671
Anıtkabir-Atatürk ve Kurtuluş Müzesi, Ankara 42696
Atatürk Evi, Ankara ... 42698
Atatürk Evi Müzesi, Erzurum ... 42774
Atatürk Evi Müzesi, Mersin ... 42906
Atatürk Evi ve Müzesi, Sakarya ... 42927
Atatürk Köşkü, Diyarbakır ... 42760
Atatürk Müzesi, Bursa ... 42740
Atatürk Müzesi, İzmir ... 42853
Atatürk Müzesi, Konya ... 42883
Atatürk Müzesi, Kayseri ... 42872
Atatürk ve Etnografya Müzesi, Akşehir ... 42683
Dolmabahçe Sarayı Müzesi, İstanbul ... 42816
Eskişehir Atatürk ve Kültür Müzesi, Eskişehir 42779
Havza Atatürk Evi, Havza ... 42795
İstanbul Şişli'deki Ataürk Evi, İstanbul ... 42828
Samsun Atatürk Müzesi, Samsun ... 42932
Silifke Atatürk Evi, Silifke ... 42941

United Kingdom

Apsley House, London ... 44784
Ballance House, Glenavy ... 44309
Benjamin Franklin House, London ... 44794
Chartwell, Westerham ... 45977
Churchill Museum and Cabinet War Rooms, London ... 44822
Commonwealth Institute, London ... 44828
Cromwell Museum, Huntingdon ... 44500
Greenhill Covenanter's House, Biggar ... 43482
Hughenden Manor, High Wycombe ... 44465
Lloyd George Museum, Cricciert ... 43925
Milton's Cottage, Chalfont Saint Giles ... 43787
Old House of Keys, Castletown ... 43778
Pankhurst Centre, Manchester ... 45095
Rhodes Memorial Museum and Commonwealth Centre, Bishop's Stortford ... 43524
Thomas Muir Museum, Bishopbriggs ... 43521
Washington Old Hall, Washington ... 45947

United States of America

Allison-Antrim Museum, Greencastle ... 49049
Andrew Johnson National Historic Site, Greeneville ... 49050
Anti-Saloon League Museum, Westerville ... 54089
CIA Museum, McLean ... 50361
Colton Hall Museum, Monterey ... 50800
Dan Quayle Center and Museum, Huntington ... 49425
Edmonston House, Vails Gate ... 53681
Federal Hall National Memorial, New York ... 51173
Franklin D. Roosevelt Presidential Library-Museum, Hyde Park ... 49458
George Bush Presidential Library and Museum, College Station ... 47688
George Washington Birthplace National Monument, Washington's Birthplace ... 53951
George Washington's Ferry Farm, Stafford County ... 53246
Georgia Capitol Museum, Atlanta ... 46437
Harry S. Truman Little White House Museum, Key West ... 49761
Harry's Truman Farm Home, Grandview ... 49012
Held-Poage Memorial Home Research Library, Ukiah ... 53639
Hiram Smith Octagon House, Neenah ... 50998
Humphrey Forum, Minneapolis ... 50733
International Spy Museum, Washington ... 53883
James K. Polk Memorial, Pineville ... 51926
James Madison's Montpelier, Montpelier Station ... 50827
Jeane Dixon Museum and Library, Strasburg ... 53331
Jim Crow Museum, Big Rapids ... 46797
King Manor, Jamaica ... 49591
McConnell Mansion, Moscow ... 50864
Meadow Garden Museum, Augusta ... 46480
Monticello, Home of Thomas Jefferson, Charlottesville ... 47398
Museum of American Political Life, West Hartford ... 54058
Museum of American Presidents, Strasburg ... 53332
Museum of the Waxhaws and Andrew Jackson Memorial, Waxhaw ... 54001
National Liberty Museum, Philadelphia ... 51860
New Jersey State House, Trenton ... 53560
Old Constitution House, Windsor ... 54229
Old South Meeting House, Boston ... 46944
Quincy and Adams County Museum, Quincy ... 52215
Ronald Reagan Boyhood Home, Dixon ... 48146
Sewall-Belmont House, Washington ... 53916
Tudor Place Museum, Washington ... 53922
Varner-Hogg Plantation, West Columbia ... 54052
The White House, Washington ... 53934
Woodrow Wilson Presidential Library at his Birthplace, Staunton ... 53279
Zebulon B. Vance Birthplace, Weaverville ... 54018

Uruguay

Casa Quinta de José Battle y Ordóñez, Montevideo ... 54387

Uzbekistan

Babur Yodgorlik Muzeyi, Andijon ... 54440
Muzej Pamjati Žertv Repressij, Toshkent ... 54479

Porcelain and Ceramics

Australia

Colac Otway Shire Hammerton Bottle Collection, Colac ... 00991
Devonport Gallery and Arts Centre, Devonport 01034
Fusions Gallery, Fortitude Valley ... 01086
National Museum of Australian Pottery, Wodonga ... 01660
Northcote Pottery Gallery, Thornbury ... 01585
Orange Regional Gallery, Orange ... 01395
Shepparton Art Gallery, Shepparton ... 01504

Austria

Galerie Schloß Weyer, Gmunden ... 01948
Geschirr-Museum, Wilhelmsburg an der Traisen 03108
K-HOF Kammerhof Museen Gmunden, Gmunden ... 01951
KK Keramikmuseum, Bad Radkersburg ... 01763
Keramikmuseum Scheibbs, Scheibbs ... 02702
Mühlviertler Keramikwerkstätte Hafnerhaus, Leopoldschlag ... 02294
Novum Forum - Schule der Sinne, Neumarkt in Steiermark ... 02439
Sommerhuber-Kachelofenmuseum, Steyr ... 02772
Szolnay-Museum, Potzneusiedl ... 02517
Töpfermuseum, Stoob ... 02784
Virgilkapelle, Wien ... 03084

Belgium

Broelmuseum, Kortrijk ... 03759
Espace Terre et Materiaux de la Faculté Polytechnique, Mons ... 03881
Hôpital Notre-Dame de la Rose, Lessines ... 03786
Hotel-Museum Arthur Merghelynck, Ieper ... 03727
Maagdenhuismuseum, Antwerpen ... 03291
Maison de la Poterie, Bouffioulx ... 03389
Musée de la Céramique d'Andenne, Andenne 03274
Musée de la Faïence de la Manufacture Royal Boch, La Louvière ... 03767
Musée de la Poterie et Musée du Jambon d'Ardenne, La Roche-en-Ardenne ... 03771
Musée d'Histoire et des Arts Décoratifs, Tournai ... 04087
Musée du Vieux Nimy, Mons ... 03887
Museum van het Abtenhuis, Geraardsbergen ... 03652
Pavillon Chinois, Tour Japonaise et Musée d'Art Japonais, Bruxelles ... 03501
Poterie Dubois-Musée la Grange aux Potiers, Bouffioulx ... 03390
Roelantsmuseum Oude Abdij Heemmuseum, Hemiksem ... 03694
Stadtmuseum Eupen, Eupen ... 03593
Töpfereimuseum Raeren, Raeren ... 03960

Bermuda

Verdmont Historic House Museum, Smith's Parish ... 04200

Brazil

Museu Barão de Mauá, Mauá ... 04565

Bulgaria

Muzej na Chudožestvenite Zanajati i Priložnite Izkustva, Trojan ... 05232

Cambodia

National Museum of Cambodia, Phnom Penh ... 05287

Canada

Burlington Art Centre, Burlington ... 05498
Canadian Clay and Glass Gallery, Waterloo ... 07285
Community Arts Council of the Alberni Valley, Port Alberni ... 06625
Gallery of British Columbia Ceramics, Granville Island ... 05905
Gardiner Museum of Ceramic Art, Toronto ... 07110
Hayward China Museum, Saint John ... 06836
Medicine Hat Clay Industries National Historic District, Medicine Hat ... 06257
Nanaimo Art Gallery, Nanaimo ... 06412
The Van Egmond Foundation, Egmondville ... 05768

Chile

Museo Histórico de Osorno, Osorno ... 07443

China

Dehua Ceramic Museum, Dehua ... 07658
Flagstaff House Museum of Tea Ware, Hong Kong ... 07794
Jingdezhen Ceramic History Museum, Jingdezhen ... 07889

Jun Pottery Shu House, Tongxiang 08302
Qin Shi Huang Terracotta Warriors and Horses Museum, Lingtong 07945
University Museum and Art Gallery, Hong Kong 07818
Xi Bao Celadon Museum, Shenzhen 08227
Yixing Ceramic Museum, Yixing 08453

Colombia

Museo del Oro, Bogotá 08577

Cuba

Farmacia Museo Taquechel, La Habana 09102
Museo Nacional de la Cerámica, La Habana 09165

Czech Republic

Městské Muzeum, Netvořice 09675
Městské Muzeum Dr. E. Beneše, Kožlany 09601
Mezinárodní Muzeum Keramiky, Bechyně 09398
Muzeum Českého Porcelánu, Klášterec nad Ohří 09584
Muzeum Hrnčířství, Kostelec nad Černými Lesy 09598
Muzeum Mladaboleslavska, Mladá Boleslav 09657
Muzeum Olomučanské Keramiky, Olomučany 09691
Muzeum Porcelánu Pirkenhammer, Březová u Karlových Var 09419
Muzeum Tynecké Keramiky, Týnec nad Sázavou 09886
Muzeum Vyškovska, Vyškov 09919

Denmark

Bornholms Museum, Rønne 10241
Den Gamle Gård, Faaborg 10010
Hammerichs Hus, Ærøskøbing 09959
Mariager Museum, Mariager 10183
Vejen Kunstmuseum, Vejen 10309

Dominican Republic

Museo de la Cerámica Contemporánea, Santo Domingo 10346
Museo de la Porcelana, Santo Domingo 10348

Egypt

Museum of Islamic Ceramic, Cairo 10505

Finland

Arabia Museum-Gallery, Helsinki 10768
Savenvalajamuseo, Somero 11323

France

Ancien Couvent, Treigny 15985
Ani C, Longfosse 13807
Château de Ratilly, Treigny 15986
Espace Culturel La Douve, Langeais 13505
Galerie Terra Viva, Saint-Quentin-la-Poterie 15478
Maison de la Faïence, Badonviller 11815
Maison de la Faïence, Desvres 12717
Musée Agathois, Agde 11491
Musée Archéologique, Lectoure 13672
Musée Atelier de Poterie, Neuvic-sur-l'Isle 14407
Musée Benoît de Puydt, Bailleul 11829
Musée Charles VII, Mehun-sur-Yèvre 14061
Musée Comtois, Besançon 11955
Musée d'Allauch, Allauch 11555
Musée d'Art et d'Histoire, Narbonne 14377
Musée d'Art et d'Histoire de Puisaye, Villiers-Saint-Benoît 16244
Musée d'Art et d'Histoire de Toul, Toul 15922
Musée de Bible et Terre Sainte, Paris 14648
Musée de Biot, Biot 11997
Musée de Faïences Christine Viennet, Béziers 11975
Musée de la Céramique, Digoin 12729
Musée de la Céramique, Lezoux 13737
Musée de la Céramique, Rouen 15113
Musée de la Céramique et de l'Ivoire, Commercy 12595
Musée de la Faïence, Forges-les-Eaux 12979
Musée de la Faïence, Marseille 13996
Musée de la Faïence, Montereau-Fault-Yonne 14201
Musée de la Faïence et de la Poterie, Ferriére-la-Petite 12924
Musée de la Faïence et des Arts de la Table, Samadet 15615
Musée de la Faïence Jules Verlingue, Quimper 14968
Musée de la Porcelaine, Couleuvre 12643
Musée de la Porcelaine, Saint-Yrieix-la-Perche 15543
Musée de la Poterie de Grès, Betschdorf 11968
Musée de la Poterie du Sartre, Saint-Alban-Auriolles 15157
Musée de la Poterie et de la Céramique, Sadirac 15146
Musée de la Reine Bérengère, Le Mans 13607
Musée de la Tour du Moulin, Marcigny 13957
Musée de l'Hospice Comtesse, Lille 13754
Musée de Poteries de Terre Vernissée, Saint-Jean-de-Fos 15324
Musée de Traditions Populaires et d'Archéologie, Chauvigny 12488
Musée Départemental Breton, Quimper 14970
Musée Départemental de l'Oise, Beauvais 11910
Musée Départemental des Hautes-Alpes, Gap 13031
Musée des Arts Décoratifs, Bordeaux 12068
Musée des Arts Décoratifs, Strasbourg 15828
Musée des Beaux-Arts et d'Archeologie Joseph Déchelette, Roanne 15059
Musée des Faïences Anciennes de la Tour d'Aigues, La Tour-d'Aigues 13451
Musée des Faïences de Thiviers, Thiviers 15886
Musée des Faïences de Varages, Varages 16071
Musée des Ivoires, Yvetot 16310
Musée Gallé-Juillet, Creil 12671
Musée Gallo-Romain, Chanaz 12397
Musée Haviland, Limoges 13769
Musée Historique de la Faïence, Moustiers-Sainte-Marie 14314
Musée Hospitalier et Apothicairerie, Louhans 13833
Musée la Faïencerie, Ancy-le-Franc 11596
Musée Léon Marès, Lovagny 13860
Musée Magnelli, Vallauris 16050
Musée Maison du Potier, Le Fuilet 13585
Musée Marguerite Vidal des Arts et Traditions Populaires, Moissac 14140
Musée Mathurin Méheut, Lamballe 13492
Musée Municipal de la Poterie, La Chapelle-aux-Pots 13330
Musée Municipal de Louviers, Louviers 13858
Musée Municipal de Saint-Gaudens, Saint-Gaudens 15276
Musée Municipal Frédéric Blandin, Nevers 14417
Musée National Adrien Dubouché, Limoges 13771
Musée National de Céramique, Sèvres 15743
Musée National de la Renaissance, Ecouen 12809
Musée National des Arts Asiatiques Guimet, Paris 14725
Musée Paul Charnoz, Paray-le-Monial 14562
Musée Privé Saint-Jean l'Aigle, Herserange 13208
Musée Sarret de Grozon, Arbois 11659
Musée souterrain de la Poterie, Saint-Émilion 15247
Musée Vivant de la Porcelaine, Foëcy 12949
Palais Ducal, Nevers 14419

Germany

Antikensammlung der Friedrich-Alexander-Universität Erlangen-Nürnberg, Erlangen 18219
Belvedere im Schlossgarten Charlottenburg, Berlin 17043
Coburger Puppen-Museum, Coburg 17714
Deutsches Porzellanmuseum, Hohenberg an der Eger 19234
Emslandmuseum Schloss Clemenswerth, Sögel 21791
Erkenbert-Museum, Frankenthal, Pfalz 18351
Europäisches Industriemuseum für Porzellan und Europäisches Museum für Technische Keramik, Rosenthal Museum, Selb 21735
Großherzoglich-Hessische Porzellansammlung, Darmstadt 17788
Heimat- und Hafnermuseum, Heidenheim 19089
Heimat- und Keramikmuseum, Kandern 19426
Heimatmuseum, Büddenstedt 17624
Heimatmuseum, Oettingen in Bayern 20878
Heimatmuseum, Wächtersbach 22266
Heimatmuseum Vilsbiburg - Kröninger Hafnermuseum, Vilsbiburg 22243
Henneberg-Museum, Münnerstadt 20498
Hetjens-Museum, Düsseldorf 17992
Internationales Keramik-Museum, Weiden in der Oberpfalz 22366
Kaffeekannenmuseum, Schöppenstedt 21617
Kelsterbacher Museum für Porzellan, Kelsterbach 19504
Keramik-Museum Berlin, Berlin 17152
Keramik-Museum Bürgel, Bürgel 17638
Keramik Museum Mettlach, Mettlach 20312
Keramikmuseum Ludwigsburg, Ludwigsburg 20045
Keramikmuseum Schloss Obernzell, Obernzell 20824
Keramikmuseum Staufen, Staufen im Breisgau 21867
Keramikmuseum Westerwald, Höhr-Grenzhausen 19214
KPM Welt, Berlin 17159
Krügemuseum, Creußen 17747
Kunsthalle im Schloss, Isny im Allgäu 19366
Lotte Reimers-Stiftung, Deidesheim 17808
Märkisches Ziegelei-Museum Glindow, Werder 22486
Meissener Porzellan-Sammlung Sigrid Pless, Bad Soden am Taunus 16840
Meißener Porzellan-Sammlung Stiftung Ernst Schneider, Oberschleißheim 20833
Museum, Coswig 17728
Museum am Maxplatz, Rehau 21239
Museum Aschenbrenner, Garmisch-Partenkirchen 18541
Museum der Porzellan-Manufaktur Meissen, Meißen 20265
Museum im Hirsch, Remshalden 21271
Museum im Kleihues-Bau, Kornwestheim 19690
Museum im Schloß, Frohburg 18487
Museum im Schloss, Fürstenberg, Kreis Holzminden 18491
Museum in der Majolika, Karlsruhe 19445
Museum Kellinghusen, Kellinghusen 19502
Museum Leuchtenburg, Seitenroda 21733
Museum Nymphenburger Porzellan - Sammlung Bäuml, München 20467
Museum Otto Ludwig, Eisfeld 18108
Museum Schlösschen im Hofgarten, Wertheim 22504
Museum Schloß Ratibor, Roth 21404
Neue Kammern, Potsdam 21099
Neues Schloss Bayreuth, Bayreuth 16964
Neues Schloss Wallerstein, Wallerstein 22314
Ofen- und Keramikmuseum Velten, Velten 22217
Ortssammlung Wettelsheim, Treuchtlingen 22090
Das Ostfriesische Teemuseum und Norder Heimatmuseum, Norden 20730
Porzellan-Museum Höchst, Frankfurt am Main 18390
Porzellanmuseum Reichmannsdorf, Reichmannsdorf 21249
Porzellansammlung, Dresden 17966
Sammlung Ernst Schneider, Düsseldorf 18004
Sammlung Ludwig in Bamberg, Bamberg 16917
Schloss Belvedere, Weimar 22418
Schlossmuseum, Ellwangen 18124
Schloßmuseum Mespelbrunn, Mespelbrunn 20303
Schlossmuseum Oranienburg, Oranienburg 20910
Stadtmuseum Nittenau, Nittenau 20723
Stadtmuseum Schorndorf, Schorndorf 21627
Stiftung Keramion, Frechen 18416
Stiftung Schloss Ahrensburg, Ahrensburg 16441
Teplitz-Schönauer Heimatmuseum, Frankfurt am Main 18398
Töpferei und Museum im Kannenofen, Höhr-Grenzhausen 19216
Töpfereimuseum Langerwehe, Langerwehe 19837
Töpfermuseum, Duingen 18011
Töpfermuseum, Kohren-Sahlis 19673
Töpfermuseum, Rödermark 21353
Töpfermuseum Im alten Kannenofen, Ransbach-Baumbach 21180
Töpfermuseum Thurnau, Thurnau 22058
Werksmuseum der Waechtersbacher Keramik, Brachttal 17492

Ghana

Prempeh Jubilee Museum, Kumasi 22859

Greece

Agios Nikolaos Archaeological Museum, Agios Nikolaos 22872
Archaeological Museum of Tegea, Tegea 23202
Centre for the Study of Greek Traditional Pottery, Athinai 22905

Hungary

Csúcsi Fazekasház, Hódmezővásárhely 23610
Ferenczy Museum - Margit Kovács Ceramics Exhibition, Szentendre 23875
Herendi Porcelánművészeti, Herend 23605
Kántor Sándor Fazekasház, Karcag 23639
Kovács Margit Állandó Kerámia Kiállítás, Győr 23584
Meseház, Békéscsaba 23369
Nyúzó Gáspár Fazekas Tájház, Tiszafüred 23931
Patonai Fazekasház, Sümeg 23827
Porcelánmúzeum, Hollóháza 23615
Zsolnay Kerámia Kiállítás, Pécs 23781

India

Abhai Smarak Panchal Sangrahalaya, Bareilly 24077
Government Educational Museum, Bulandshahar 24104
Tamralipta Museum and Research Center, Tamluk 24400

Indonesia

Museum Asserajah El Hasyimiah Palace, Bengkalis 24451
Museum Seni Rupa dan Keramik, Jakarta 24500

Iran

Abgineh va Sofalineh, Tehrān 24654
Golestan Palace Museum, Tehrān 24668

Italy

Antiquarium, Prato 27329
Civica Raccolta d'Arte Applicata, Milano 26675
Civica Raccolta di Terraglia-Museo Internazionale Design Cermico, Laveno Mombello 26471
Collezioni d'Armi e Ceramiche della Rocca Sforzesca, Imola 26399
Musei Civici di Pesaro, Pesaro 27153
Museo Archaeologico di Santa Restituta, Lacco Ameno 26437
Museo Artistico Industriale F. Palizzi, Napoli 26884
Museo Carlo Venturini, Massa Lombarda 26623
Museo Carlo Zauli, Faenza 26073
Museo Civico della Ceramica, Nove 26958
Museo Civico e della Ceramica, Cerreto Sannita 25813
Museo Civico e Pinacoteca, Urbania 28119
Museo Comunale della Ceramica, Cutrofiano 26028
Museo della Ceramica, Bassano del Grappa 25381
Museo della Ceramica, Castellamonte 25732
Museo della Ceramica, Fiorano Modenese 26140
Museo della Ceramica, Forlì 26225
Museo della Ceramica, Montelupo Fiorentino 26839
Museo della Ceramica, Patti 27116
Museo della Ceramica del Castello, Salerno 27595
Museo della Ceramica di Castelli, Castelli 25740
Museo dell'Agro Falisco, Civita Castellana 25920
Museo delle Ceramiche, Tolentino 27986
Museo delle Ceramiche Acerbo, Loreto Aprutino 26523
Museo delle Porcellane, Firenze 26174
Museo dell'Opera del Santuario, Tolentino 27987
Museo Didattico delle Maioliche, Grottaglie 26375
Museo Duca di Martina, Napoli 26905
Museo e Pinacoteca Basilio Cascella, Pescara 27164
Museo e Pinacoteca Comunali, Gubbio 26390
Museo Internazionale delle Ceramiche, Faenza 26076
Museo Nazionale Archeologico, Ferrara 26119
Museo Provinciale della Ceramica, Vietri 28264
Museo Regionale della Ceramica, Caltagirone 25606
Museo Regionale della Ceramica, Deruta 26032
Museo Richard Ginori della Manifattura di Doccia, Sesto Fiorentino 27825
Museo Selinuntino, Castelvetrano 25757
Museo Villa Urania-Antiche Maioliche di Castelli, Pescara 27166
Pinacoteca Civica, Fermo 26103
Raccolta delle Piastrelle di Ceramica, Sassuolo 27765

Japan

Aichi-ken Toji Siryokan, Seto 29250
Arita Toji Bijutsukan, Arita 28364
Asahi-yaki Kiln Museum, Uji, Kyoto 29557
Ceramic Art Center, Mashiko 28884
Chaini-zu Toji Bijutsukan, Usuki 29565
Chuuko Tamba Yakimono Hakubutsukan, Sasayama 29230
Ebetsu-shi Bijutsu Togeikan, Ebetsu 28410
Echizen Pottery Gallery, Echizen 28412
Fukui-ken Togeikan, Echizen 28413
Gifu-ken Gendai Togei Bijutsukan, Tajimi 29311
Gifu-ken Toji Shiryokan, Tajimi 29312
Hachiga Minzoku Bijutsukan, Takayama 29336
Hakone Bijutsukan, Hakone 28477
Hayashibara Bijutsukan, Okayama 29098
Hyogo-ken Togeikan, Sasayamashi 29233
Ibaraki-kenritsu Tougeikan, Kasama 28673
Idemitsu Bijutsukan, Tokyo 29400
Iga and Shigaraki Pottery Museum, Ueno 29555
Inax Tile Museum, Tokoname 29364
Ishii Chawan Bijutsukan, Hagi 28470
Itaya Hazan Kinenkan, Shimodate 29266
Jomon Bijutsukan, Nakaniida 29018
Jomon-Kan, Kizukuri 28722
Kamogawa-shiritsu Tougeikan, Kamogawa 28651
Kasama Inari Bijutsukan, Kasama 28674
Kawai Kanjiro's House, Kyoto 28821
Kei Fujiwara Bijutsukan, Bizen 28389
Kibi Kokoshizou, Yamateson 29584
Kikusui Handicraft Museum, Kawanishi 28690
Kiso-ji Bijutsukan, Agematsu 28347
Kiyomizu Sannenzaka Museum, Kyoto 28825
Kodama Museum, Kagoshima 28630
Koishibara-mura Yakimono Bijutsukan, Koishibara 28760
Kondo Yuzo Kinenkan, Kyoto 28828
Kuma Bijutsukan, Kuma 28774
Kurita Hakabutsukan, Ashikaga 28373
Kyocera Museum of Fine Ceramics, Kyoto 28833
Kyoto Gigei Bijutsukan, Kyoto 28839
Marugame Bijutsukan, Marugame 28880
Marusu Museum, Ishinomaki 28588
Masaki Art Museum, Osaka 29133
Mashiko Sankokan, Mashiko 28885
Matsuo Museum, Takigawa 29347
Miharu-machi Ningyo Hakubutsukan, Miharu 28908
Miyake Bijutsukan, Kagoshima 28631
Mizunami Toji Shiryokan, Mizunami 28933
Naeshirogawa Minkan-yaki Bijutsukan, Higashi-Ichiki 28503
Nippon Doreikan, Shirotori 29282
Ohi Museum 10th Ohi Chozaemon Ware Gallery, Kanazawa 28662
Okayama-kenritsu Bizen-yaki Hakubutsukan, Bizen 28390
Osaka-shiritsu Toyo Toji Bijutsukan, Osaka 29153
Otaru Venice Museum, Otaru 29164
Paramita Museum, Komono 28767
Raku Museum, Kyoto 28861

P

Rakushu Rakuchu Gallery, Kyoto 28862
Saga-kenritsu Kyushu Toji Bunkakan, Arita 28365
Saitama-kenritsu Sakitama Shiryokan, Gyoda 28460
Sekido Bijutsukan, Tokyo 29465
Seto-shi Rekishi Minzoku Shiryokan, Seto 29251
Shigaraki-cho Dentouteki Bijutsu Senta, Shigaraki 29259
Shigaraki-yaki Bijutsukan, Shigaraki 29260
Suishouen Toji Shiryokan, Nagoya 29011
Suzu-yaki Kyodo Bijutsukan, Suzu 29306
Takahama-shiritsu Kawara Bijutsukan, Takahama 29314
Tamba Tachikui Traditional Ceramic Center, Sasayama 29232
Tanakamaru Collection, Fukuoka 28437
Tohoku Kindai Toji oyo Yaki Bijutsukan, Nakaniida 29019
Toki-shiritsu Mino-yaki Rekishi Hakubutsukan, Toki 29363
Tokoname-shiritsu Mingeikan oyo Toji Rekishi, Tokoname 29365
Tomimoto Kenkichi Kinenkan, Ando 28359
Tottori Mingeikan, Tottori 29516
Toyama-shiritsu Yakimono Bijutsukan, Toyama 29525
Toyohashi-shiritsu Bijutsukan, Toyohashi 29529
Toyota-shi Mingei oyo Toji Shiryokan, Toyota 29534
Toyozo Shiryoukan, Kani 28664
Ukiyo-e and Pottery Museum, Osaka 29156
Umeyama Furui Yakimono Bijutsukan, Tobe 29362
Zeze-yaki Bijutsukan, Otsu 29168

Korea, Republic

Celamuse - Porcelain Museum, Seoul 29865
Ewha Yoja Taehakkyo Pakmulgwan, Seoul 29876
Haegang Ceramics Museum, Icheon 29829
Hongsan Museum, Seoul 29883
Horim Museum, Seoul 29884
Onggi Folk Museum and Institute, Seoul 29911
Sojeon Art Museum, Siheung 29936

Latvia

Rīgas Porcelana Muzejs, Rīga 30109
Rīgas Vēstures un Kugniecības muzejs, Rīga 30111

Lithuania

Bronės ir Bronius Radeckų Keramikos Muziejus, Kuršėnai 30294
Keramikos Muziejus, Kaunas 30253
Vytauto Valiušio Keramikos Muziejus, Leliūnai 30302

Luxembourg

Musée de la Poterie, Nospelt 30488

Mexico

P

Museo Nacional de la Cerámica, Tonalá, Jalisco 31496
Museo Pantaleón Panduro, Tlaquepaque 31473
Museo Regional de la Cerámica, Tlaquepaque 31474

Morocco

Musée du Batha, Fès 31671
Musée National de la Céramique, Safi 31693

Netherlands

Aardewerkvermuseum Petrus Pegout, Noordbroek 32598
Fini's Hoeve, Vriezenveen 32910
Harlinger Aardewerk Museum, Harlingen 32317
Keramiekcentrum Tiendschuur, Tegelen 32814
Keramisch Museum Goedewaagen, Nieuw Buinen 32571
Mobach Keramiek Museum, Utrecht 32847
Museum De Ghulden Roos, Roosendaal 32700
Museum De Moriaan, Gouda 32268
Museum Het Prinsenhof, Delft 32062
Museum Lambert Van Meerten, Delft 32063
Museum Mesdag, Den Haag 32088
Museum Paul Tétar Van Elven, Delft 32065
Museum voor Keramiek Pablo Rueda Lara, Rotterdam 32716
Nederlands Kachelmuseum, Alkmaar 31795
Oudheidkamer Beilen, Beilen 31957
Porselein Dierenpark, Wildervank 32950
Pottenbakkerij Het Ovenhuis, Ruinerwold 32740
Pottenbakkerij Museum, Kattendijke 32442
Princessehof Leeuwarden, Leeuwarden 32478
Purmerends Museum, Purmerend 32672
SM's - Stedelijk Museum's-Hertogenbosch, 's-Hertogenbosch 32379
Tassenmuseum Hendrikje, Amstelveen 31823
Warkums Erfskip, Workum 32967

New Zealand

The Suter, Te Aratoi o Whakatu, Nelson 33154

Nicaragua

Museo de Condega, Condega 33274
Museo de Moyogalpa, Moyogalpa 33298

Norway

Egersund Fayancemuseum, Egersund 33471

Paraguay

Museo de Cerámica y Bellas Artes Julián de la Herreria, Asunción 34092
Museo del Barro, Asunción 34096

Peru

Museo de Arte Precolombino, Cusco 34157
Museo Nacional de Antropología, Arqueología e Historia del Perú, Lima 34264

Philippines

MPG Town Gallery, Odiongan 34468
Puerto Galera National Museum, Puerto Galera 34481

Poland

Muzeum Ceramiki, Bolesławiec 34568
Muzeum Ceramiki Kaszubskiej Neclów, Chmielno 34594
Muzeum Technik Ceramicznych, Koło 34750
Muzeum Ziemi Kujawskiej i Dobrzyńskiej, Włocławek 35311
Pałac Radziwiłłów w Nieborowie, Nieborów 34942

Portugal

Casa Museu Rafael Bordalo Pinheiro, Caldas da Rainha 35495
Mosteiro de Santa Clara-a-Velha, Coimbra 35527
Museu da Cerâmica, Caldas da Rainha 35496
Museu da Cerâmica de Sacavém, Sacavém 35893
Museu da Fundação Castro Alves, Vila Nova de Famalicão 36005
Museu de Barroso, Montalegre 35772
Museu de Cerâmica, Borba 35471
Museu de Olaria, Barcelos 35456
Museu do Barro, Redondo 35882
Museu Histórico da Vista Alegre, Ílhavo 35647
Museu Monográfico de Conimbriga, Condeixa-a-Nova 35548
Museu Municipal de Arqueologia de Silves, Silves 35940
Museu Nacional do Azulejo, Lisboa 35719
Museu Nogueira da Silva, Braga 35481

Romania

Colecția de Ceramică Populară de pe Valea Olteţului, Balş 36134
Muzeul de Artă Decorativă, Dorohoi 36349

Russia

Dalnevostočnyj Chudožestvennyj Muzej, Chabarovsk 36858
Gosudarstvennyj Chudožestvenno-architekturnyj Dvorcovo-parkovyj Muzej-zapovednik Pavlovsk, Pavlovsk 37528
Muzej Keramiki i Usadba Kuskovo XVIII Veka, Moskva 37303
Omskij Oblastnoj Muzej Izobrazitelnych Iskusstv im. M.A. Vrubelja, Omsk 37501
Rybinskij Gosudarstvennyj Istoriko-architekturnyj i Chudožestvennyj Muzej-zapovednik, Rybinsk 37606

Slovakia

Múzeum Keramiky Mestské Múzeum, Holíč 38259
Múzeum Ľudovíta Štúra, Modra 38306
Novohradská Galéria, Lučenec 38289

Slovenia

Grad Dobrovo, Dobrovo v Brdih 38413

South Africa

Engelenburg House Art Collection, Tshwane 38746
Koopmans-De Wet House Museum, Cape Town 38585
Van Tilburg Collection, Tshwane 38763

Spain

Centro Agost, Museo de Alfarería, Agost 38784
Colección Museográfica d'Cerámica de Potríes, Potríes 39862
Museo de Almería, Almería 38861
Museo de Cerámica de la Cartuja-Pickman, Salteras 39950
Museo de Cerámica Ruíz de Luna, Talavera de la Reina 40095
Museo de Farmacia Militar, Madrid 39540
Museo Municipal de Cerámica, Paterna 39816
Museo Nacional de Artes Decorativas, Madrid 39569
Museo Nacional de Cerámica y de las Artes Suntuarias González Martí, Valencia 40216
Museo Zuloaga, Segovia 40056
Museu de Cerámica, Barcelona 38975
Museu de Ceràmica de l'Alcora, Alcora 38823
Museu de Ceràmica de Manises, Manises 39612
Palacio Real de El Pardo, El Pardo 39812

Sweden

Gamla Krukmakarverkstaden, Arvika 40438
Godegårds Bruks- och Porslinsmuseum, Motala 40751
Gustavsbergs Porslinmuseum, Gustavsberg 40569
Höganäs Museum och Konsthall, Höganäs 40596
Kungliga Slottet i Stockholm, Stockholm 40906
Rörstrand Museum, Linköping 40685
Törngrens Krukmakeri, Falkenberg 40511

Switzerland

Historisches Museum Wiedlisbach, Wiedlisbach 42013
Keramik-Museum, Triengen 41949
Keramikmuseum und Maria-Felchin-Sammlung, Matzendorf 41615
Musée Ariana, Genève 41416
Musée Historique et des Porcelaines, Nyon 41691
Ortsmuseum Kilchberg, Kilchberg (Zürich) 41503
Regionalmuseum Langnau im Chüechlihus, Langnau im Emmental 41529
Zunfthaus zur Meisen, Zürich 42121

Taiwan

Aurora Art Museum, Taipei 42375
Cochin Ceramics Museum, Chai-Yi 42152
Hsin Wang Ceramics Memorial Gallery, Yingge 42472
Hwa Kang Museum, Taipei 42394
Hwataoyao - Kiln Exhibition Gallery, Yuanli 42477
Koji Pottery Museum, Sindian 42334
Miaoli Ceramics Museum, Gongguan 42192
Nantou Ceramic Museum, Nantou 42281
San Cozy Ceramic Museum, Puli 42298
Ten Sing Ceramic Art Village, Jiji 42231
Wan Li Brick Museum, Yuanli 42480
Yingge Ceramics Museum, Yingge 42473

Tunisia

Centre National de la Céramique d'Art de Sidi Kacem Jelizi, Tunis 42662

Turkey

Çamlı Köşk, İstanbul 42811
Çini Müzesi, Kütahya 42893
Çinili Köşk Müzesi, İstanbul 42812

Ukraine

Municypal'nyj Muzej Pryvatnych Kollekcij im. O. Bleščunova, Odesa 43156

United Kingdom

Allen Gallery, Alton 43292
Barlow Collection of Chinese Ceramics, Bronzes and Jades, Brighton 43618
Burton Art Gallery and Museum, Bideford 43477
Castle Museum and Art Gallery, Norwich 45276
Ceramics Gallery, Aberystwyth 43257
China Clay Country Park Mining and Heritage Centre, Saint Austell 45560
Clevedon Court, Clevedon 43852
Clifton Park Museum, Rotherham 45517
Coalport China Museum, Coalport 43861
Collection of Martinware Pottery, Southall 45700
Croydon Museum, Croydon 43934
De Morgan Centre, London 44835
Gladstone Pottery Museum, Stoke-on-Trent 45761
Glynn Vivian Art Gallery, Swansea 45813
Guildhall Museum, Carlisle 43759
Hastings Museum and Art Gallery, Hastings 44415
Hull University Art Collection, Kingston-upon-Hull 44598
Jackfield Tile Museum, Jackfield 44544
Kirkcaldy Museum and Art Gallery, Kirkcaldy 44609
Middlesbrough Institute of Modern Art, Middlesbrough 45141
Mill Green Museum and Mill, Hatfield 44419
Museum of Harlow, Harlow 44396
Museum of the Royal Pharmaceutical Society of Great Britain, London 44933
Museum of Worcester Porcelain, Worcester 46072
Nantgarw China Works Museum, Nantgarw 45190
Penwith Galleries, Saint Ives, Cornwall 45576
Peter Scott Gallery, Lancaster 44635
Pitzhanger Manor-House and Gallery, London 44955
The Potteries Museum and Art Gallery, Stoke-on-Trent 45762
Raby Castle, Darlington 43950
Roman Villa Museum at Bignor, Pulborough 45451
Royal Crown Derby Museum, Derby 43969
Rye Castle Museum, Rye 45536
School of Art Gallery and Museum, Aberystwyth 43259
Sherborne Castle, Sherborne 45659
Spode Museum, Stoke-on-Trent 45763
Sunderland Museum, Sunderland 45805
Upton House and Gardens, Banbury 43371
Wallington, Cambo 43721
Wedgwood Museum, Stoke-on-Trent 45764
Wisbech and Fenland Museum, Wisbech 46049

United States of America

A.L. Fetterman Educational Museum, Fort Valley 48694
The Art Studio, Beaumont 46671
The Bradford Museum of Collector's Plates, Niles 51350
Butter Pat Museum, Bigfork 46802
Central Iowa Art Museum, Marshalltown 50511
China Institute Gallery, New York 51150
Clay Studio, Philadelphia 51824
Concordia University-Wisconsin Art Gallery, Mequon 50593
Goodman Museum, Tyler 53633
Harlow Ceramics Collection Kleven Print Collection, Bemidji 46731
Home Sweet Home Museum, East Hampton 48246
Houston Harte University Center, San Angelo 52703
The Jones Museum of Glass and Ceramics, Sebago 53018
Julian H. Sleeper House, Saint Paul 52619
Lladro Museum, New York 51204
Macalester College Art Gallery, Saint Paul 52620
Museum of Ceramics, East Liverpool 48256
National Museum of Ceramic Art and Glass, Baltimore 46583
North Carolina Pottery Center, Seagrove 52990
Northern Clay Center Galleries, Minneapolis 50741
Ohio Ceramic Center, Roseville 52462
The Ohr-O'Keefe Museum of Art, Biloxi 46812
Pewabic Pottery Museum, Detroit 48129
The Ross C. Purdy Museum of Ceramics, Westerville 54091
Ruth Chandler Williamson Gallery, Claremont 47576
The Schein-Joseph International Museum of Ceramic Art, Alfred 46216
Sheehan Gallery at Whitman College, Walla Walla 53798
Stan Hywet Hall and Gardens, Akron 46148
Touchstone Center for Crafts, Farmington 48519
The World Organization of China Painters' Museum, Oklahoma City 51527

Venezuela

Museo de Cerámica y Loza Popular, Falcón 54571

Portraits

Australia

National Portrait Gallery, Canberra 00948

Bosnia and Herzegovina

Galerija Portreta Tuzla, Tuzla 04289

Colombia

Museo de Arte Religioso Presbítero Antonio María Franco, Supía 08765

Denmark

Det Nationalhistoriske Museum på Frederiksborg, Hillerød 10076

Germany

Burg Hohenzollern, Burg Hohenzollern 17645
Stiftung Schloss Eutin, Eutin 18287

Greece

Municipal Gallery of Samos, Samos 23176

Hungary

Székesfehérvári Egyházmegyei Múzeum, Székesfehérvár 23859

India

Archaeological Museum and Portrait Gallery, Goa Velha 24177

Italy

Quadreria Comunale, Prato 27339

Netherlands

Ter Beziens, Kolhorn 32452
Dekema State, Jelsum 32431
Hofje Van Gratie, Delft 32058

New Zealand

New Zealand Portrait Gallery, Wellington 33255

Spain

Galería de Catalanes Ilustres, Barcelona 38962

Sweden

Gripsholms Slott, Mariefred 40731
Svenska Statens Porträttsamling, Mariefred 40735

United Kingdom

Beningbrough Hall, York 46098
The Blairs Museum, Aberdeen 43240
Bodelwyddan Castle Museum, Bodelwyddan 43546
Djanogly Art Gallery, Nottingham 45290
Fyvie Castle, Fyvie 44258
Goodwood House, Chichester 43827
National Portrait Gallery, London 44940
Parham House Gardens, Storrington 45768
RIBA Drawings and Archives Collections, London 44964
Scottish National Portrait Gallery, Edinburgh 44157
Ugbrooke House, Chudleigh 43844

United States of America

The Gilbert Stuart Museum, Saunderstown 52934
John Jay Homestead, Katonah 49708
Marion Art Center, Marion 50485
Redwood Library and Athenaeum, Newport 51328

Postage Stamps

Brazil

Museu Numismático e Filatélico, Rio de Janeiro 04759

Canada

Centre d'Interprétation et de la Recherche Philatélique du Canada, Terrebonne 07067
Musée Canadien de la Poste, Gatineau 05861

China

China Stamp Museum, Beijing 07534

Cuba

Museo Postal Cubano, La Habana 09170

Egypt

Postal Museum, Cairo 10513

France

Musée de la Poste, Paris 14659
Musée Régional du Timbre et de la Philatélie, Le Luc 13603

French Polynesia

Centre Philatélique de Polynésie Française, Papeete 16323

Germany

Archiv für Philatelie der Museumsstiftung Post und Telekommunikation, Bonn 17444
Briefmarkenmuseum im Kloster Bardel, Bad Bentheim 16681

Greece

Tahydromiko Moyseio, Athinai 22952

Hungary

Bélyegmúzeum, Budapest 23388
Delizsánsz Kiállítóterem, Debrecen 23515

India

National Philatelic Museum, Delhi 24149

Iran

Museum of Post and Telecomunication and Mokhaberat Museum, Tehrān 24684

Israel

Postal and Philatelic Museum, Tel Aviv 25119

Italy

Gabinetto Comunale delle Stampe, Roma 27446
Museo della Stampa, Soncino 27870

Japan

Kitte no Hakubutsukan, Tokyo 29412

Jordan

Postal Museum, Amman 29660

Liechtenstein

Postmuseum des Fürstentums Liechtenstein, Vaduz 30200

Mexico

Museo de Filatelia de Oaxaca, Oaxaca 31223

Monaco

Musée des Timbres et des Monnaies, Monaco31610

Peru

Museo Postal y Filatélico del Perú, Lima 34269

Philippines

Postal Museum and Philatelic Library, Manila 34453

San Marino

Museo Postale, Filatelico-Numismatico, San Marino 38010

Slovakia

Poštové Múzeum Banská Bystrica, Banská Bystrica 38185

Spain

Museo Postal y Telegráfico, Madrid 39578

United States of America

The Collectors Club, New York 51153
National Postal Museum, Washington 53908
Spellman Museum of Stamps and Postal History, Weston 54107

Zambia

National Archives of Zambia, Lusaka 54659

Postal Services

Andorra

Museu Postal, Ordino 00104

Argentina

Museo de Telecomunicaciones, Buenos Aires. 00218

Australia

Telstra Museum Brisbane, Clayfield 00979
Wireless Hill Telecommunications Museum, Applecross 00808

Austria

Museum Sankt Veit, Sankt Veit an der Glan 02691

Belgium

Musée du Téléphone, Warneton 04134

Botswana

Postal Museum Botswana, Gaborone 04297

Brazil

Museu de Telecomunicações, Rio de Janeiro 04732
Museu do Telephone, Rio de Janeiro 04742
Museu Postal e Telegráfico da ECT, Brasília 04369

Canada

Edmonton Telephone Historical Centre, Edmonton 05745
Mather Post Office Museum, Cartwright 05572
Musée Canadien de la Poste, Gatineau 05861
Ross House Museum, Winnipeg 07376
Toronto's First Post Office, Toronto 07151

China

Beijing Telephone History Museum, Beijing 07520
Postal Museum, Shanghai 08178

Colombia

Museo Filatélico del Banco de la República, Medellín 08681
Museo Postal Eduardo Santos, Bogotá 08594

Costa Rica

Museo Filatélico, San José 08821

Croatia

Hrvatski Poštansko-telekomunikacijski Muzej, Zagreb 08996

Czech Republic

Poštovní Muzeum, Praha 09790
Poštovní Muzeum, Vyšší Brod 09922

Denmark

Post Tele Museum, København 10148
Telefonmuseet, Hellerup 10059

Estonia

Eesti Postimuuseum, Tartu 10646

Finland

Postimuseo, Helsinki 10809

France

Musée de la Communication en Alsace, Riquewihr 15051
Musée de la Poste, Amélie-les-Bains-Palalda 11582
Musée de la Poste, Nantes 14362
Musée de la Poste, Paris 14659
Musée de la Poste, Prunelli-di-Fium'Orbu 14954
Musée de la Poste, Saint-Flour (Cantal) 15272
Musée du Téléphone, Nancy 14355
Musée du Téléphone Narbonne, Narbonne 14378
Musée Postal d'Aquitaine, Saint-Macaire 15382
Musée Régional des Télécommunications en Flandres, Marcq-en-Barœul 13960

Germany

Deutsches Telefon-Museum, Morbach 20381
Fürstliches Schloß und Kreuzgang Sankt Emmeram, Regensburg 21222
Museum für Kommunikation, Hamburg 18950
Museum für Kommunikation Berlin, Berlin 17204
Museum für Kommunikation Frankfurt, Frankfurt am Main 18383
Museum für Kommunikation Nürnberg (Verkehrsmuseum), Nürnberg 20781
Postmuseum Mettingen, Mettingen 20310
Postmuseum Rheinhausen, Oberhausen-Rheinhausen 20811
Telefonmuseum Hittfelder Bahnhof, Seevetal. 21726

Greece

Museum of Telecommunications, Néa Kifissia 23122

Hungary

Delizsánsz Kiállítóterem, Debrecen 23515
Postakürt Galéria, Miskolc 23713
Postamúzeum, Balatonszemes 23355
Postamúzeum, Budapest 23468
Postamúzeum, Opusztaszer 23737
Postapalota, Pécs 23771
Telefónia Múzeum, Budapest 23480
Telegaléria, Pécs 23777

Iceland

Póstminjasafn Íslands, Hafnarfjörður 24000
Símaminjasafnið, Reykjavík 24034

Indonesia

Museum Pos Giro, Bandung 24445

Iran

Museum of Post and Telecomunication and Mokhaberat Museum, Tehrān 24684

Israel

Postal and Philatelic Museum, Tel Aviv 25119

Italy

Museo dei Tasso e della Storia Postale, Camerata Cornello 25615
Museo Internazionale dell'Immagine Postale, Belvedere Ostrense 25391
Museo Storico delle Poste e Telecomunicazioni, Roma 27537

Japan

Maejima Memorial Museum, Joetsu 28622

Korea, Republic

Chungnam Telecommunication Museum, Daejeon 29797
Communications Monument, Seoul 29868
Postal Museum, Cheonan 29785

Latvia

Lattelekom Telekomunikciju Muzejs, Rīga 30074

Liechtenstein

Postmuseum des Fürstentums Liechtenstein, Vaduz 30200

Lithuania

Ryšių Istorijos Muziejus, Kaunas 30265
Vilniaus Apskrities Centrinio Pašto Muziejus, Vilnius 30438

Luxembourg

Musée des Postes et Telecommunications, Luxembourg 30480

Malawi

Mtengatenga Postal Museum, Namaka 30547

Mauritius

Mauritius Postal Museum, Port Louis 30682

Mexico

Museo Biblioteca Palacio Postal, México 31055
Museo de las Telecomunicationes, México 31090
Museo Regional de Telecomunicaciones Tomás Guzmán Cantú, La Paz 30990
Museo Telefónico Victoria, México 31152

Morocco

Musée de Maroc Telecom, Rabat 31689
Musée Postal, Rabat 31691

Norway

Kjerringøy Gamle Handelssted, Bodø 33444

Peru

Museo Postal y Filatélico del Perú, Lima 34269

Poland

Muzeum Poczty i Telekomunikacji, Wrocław 35336
Muzeum Poczty i Telekomunikacji Oddział w Gdańsku, Gdańsk 34645

Portugal

Museu CTT/TLP, Lisboa 35673
Museu das Comunicações, Lisboa 35688
Museu Vivo da Comutação Manual, Vilar 36020

Russia

Centralnyj Muzej Svjazi im. A.S. Popova, Sankt-Peterburg 37630
Literaturno-memorialnyj Dom-Muzej F.M. Rešetnikova, Ekaterinburg 36888

Serbia

PTT Muzej Zajednice Jugoslovenskih Pošta, Telegrafa i Telefona, Beograd 38086

Slovakia

Poštové Múzeum Banská Bystrica, Banská Bystrica 38185

Slovenia

Muzej Pošte in Telekomunikacij, Škofja Loka 38515

Spain

Museo de las Telecomunicaciones, Madrid 39549
Museo Postal y Telegráfico, Madrid 39578

Sweden

Bodaskogs Affärs- och Postmuseum, Aneby 40422
Postmuseum, Stockholm 40926

Switzerland

Musée des Téléphones de Genève, Plan-les-Ouates 41725
Museum für Kommunikation, Bern 41212
Museum Telefonzentrale Magden, Magden 41602
Telefonmuseum im Greuterhof, Islikon 41496

Taiwan

Postal Museum, Taipei 42419

Tunisia

Musée Postal de Tunis, Tunis 42667

United Kingdom

Bath Postal Museum, Bath 43400
The British Postal Museum Archive, London 44805
Timeball Tower, Deal 43960

United States of America

Maitland Historical Museums, Maitland 50403
National Postal Museum, Washington 53908
Post Museum, Fort Leavenworth 48643
Telephone Museum, Ellsworth 48372

Posters

Finland

Julistemuseo, Lahti 11027

Germany

Kleines Plakatmuseum, Bayreuth 16957
Puppentheatermuseum, Kaufbeuren 19491

Spain

Museu Nacional d'Art de Catalunya (MNAC), Barcelona 39001

Switzerland

Museum für Gestaltung Zürich, Zürich 42094
Museum für Gestaltung Zürich - Sammlungen für Design, Grafik, Plakat und Kunstgewerbe, Zürich 42095

United Kingdom

Centre for the Magic Arts, London 44816

United States of America

Center for the Study of Political Graphics, Los Angeles 50216
Galerie at Lincoln Center, New York 51179

Pottery, Earthenware → Porcelain and Ceramics

Precious Stones

Argentina

Museo de Mineralogía y Geología Dr. Alfredo Stelzner, Córdoba 00326

Australia

Australian Pearling Exhibition, Darwin 01028
Berlins Gem and Historical Museum, Kingaroy 01199
Griffith Regional Art Gallery, Griffith 01132
Kealley's Gemstone Museum, Nannup 01346
Port Stephens Shell Museum, Corlette 01001

Austria

alpinartgallery - Schatzhaus der Natur, Ebene Reichenau 01834
Gablonzer Schmuckmuseum, Enns 01869
Granatium, Radenthein 02549
Schatzkammer, Wien 03068
Swarovski Kristallwelten, Wattens 02898

Belgium

Diamantmuseum, Antwerpen 03284
Diamantmuseum, Grobbendonk 03662
Grand Curtius, Liège 03796
Musée du Diamant, Durbuy 03566

Brazil

Museu Amsterdam Sauer de Pedras Preciosas, Rio de Janeiro 04698
Museu de Geologia, Porto Alegre 04647
Museu do Diamante, Diamantina 04444
Museu H. Stern, Rio de Janeiro 04747

Costa Rica

Museo del Jade, San José 08820

Croatia

Etnografski Muzej Split, Split 08955

Cyprus

Cyprus Jewellers Museum, Laiki Geitonia 09366

Czech Republic

Městské Muzeum Klenotnice, Nová Paka 09676
Muzeum Českého Ráje, Turnov 09881

Egypt

Palace of Royal Jewelleries, Zizinia 10544

Finland

Jalokivi Galleria, Kemi 10938
Ylämaan Jalokivimuseo, Ylämaa 11477

France

Espace d'Exposition Minal, Héricourt 13204
Musée de l'Ardoise, Trélazé 15987

Germany

Bernsteinmuseum, Sellin, Ostseebad 21740
Deutsches Edelsteinmuseum, Idar-Oberstein 19308
Fossilien-Museum, Fürth 18497
Historische Grünes Gewölbe und Neues Grünes Gewölbe, Dresden 17948
Kristallmuseum Riedenburg, Riedenburg 21317
Museum Idar-Oberstein, Idar-Oberstein 19309

Greece

Ilias Lalaounis Jewelry Museum, Athinai 22915
Mouseio Kompologiou, Nafplion 23116

Iran

Javaherat, Tehrān 24675

Israel

Harry Oppenheimer Diamond Museum, Ramat Gan 25083

Italy

Museo del Bijou, Casalmaggiore 25697
Museo della Pietra Serena, Firenzuola 26205

Japan

Akiba Anti-ku Okimono Bijutsukan, Nasu 29046

Jordan

Jordan Folklore Museum, Amman 29653

Kazakhstan

Muzej Geologii, Almaty 29686

Lithuania

Brangakmenių Muziejus, Kaunas 30242
Mineralų Muziejus, Vilnius 30425

Morocco

Musée Ethnographique des Oudaïa, Rabat 31690

Netherlands

Edelsteenslijperij de Steenarend, Eindhoven 32186
Museum voor Moderne Kunst, Arnhem 31929
SM's - Stedelijk Museum's-Hertogenbosch, 's-Hertogenbosch 32379

Poland

Muzeum Kamieni Szlachetnych, Polanica Zdrój 35019

Russia

Almaznyj Fond, Moskva 37140
Geologo-mineralogičeskij Muzej im. A.I. Kozlova, Vladivostok 37937
Mineralogičeskij Muzej im. A.V. Sidorova, Irkutsk 36934
Muzej Prirody Burjatii, Ulan-Udė 37896
Muzej Samocvety, Moskva 37360

Sweden

Porfyrmuseet, Älvdalen 40407

Switzerland

Musée de l'Horlogerie et de l'Emaillerie, Genève 41420

Ukraine

Muzej Istoryčnych Koštovnostej Ukraïny, Kyïv 43090

United Kingdom

The Crypt - Town Hall Plate Room, Oxford 45331
Her Majesty Tower of London, London 44873
Museum of the Jewellery Quarter, Birmingham 43508
Town and Crown Exhibition, Windsor 46046

United States of America

Touchstone Center for Crafts, Farmington 48519

Prehistory and Early History

Argentina

Museo Municipal C.I.P.A.S., Salto 00562
Museo Paleontológico, Arqueológico e Histórico de la Ciudad de Deán Funes, Deán Funes 00353

Australia

Hawkesbury Regional Museum, Windsor, New South Wales 01656
National Dinosaur Museum, Nicholls 01368

Austria

Archäologisches Museum und Dokumentations-zentrum Via Claudia, Fließ 01892
Archäologisches Pilgermuseum Hemmaberg-Juenna, Globasnitz 01935
Ateliersammlung Herbert Hiesmayr, Sankt Thomas am Blasenstein 02689
Ausgrabungsdokumentation 6000 Jahre Wohnberg Oberleis, Ernstbrunn 01874
Burgmuseum Archeo Norico, Deutschlandsberg 01816
Dr. Eduard Benninger Heimathaus, Bad Wimsbach-Neydharting 01770
Eisenverhüttungs-Schauraum, Oberpullendorf 02458
Erstes Urgeschichtliches Freilichtmuseum der Steiermark, Kulm bei Weiz 02245
Grenzlandmuseum Raabs an der Thaya, Raabs an der Thaya 02546
Heimatmuseum, Schwanenstadt 02717
Heimatmuseum Steyregg, Steyregg 02777
Höbarth- und Madermuseum der Stadt Horn, Horn 02099
Keltenmuseum, Hallein 02046
Museum des Geschichtsvereins Keltisches Noreia, Mühlen 02410
Museum für Frühgeschichte des Landes Niederösterreich, Traismauer 02818
Museum für Ur- und Frühgeschichte, Stillfried 02780
Museum für Ur- und Frühgeschichte, Wieselburg an der Erlauf 03101
Museum für Urgeschichte, Koblach 02209
Museum Großklein, Großklein 02003
Museum Thurnfels, Völs 02869
Museum Wattens - Industrie Vorgeschichte, Wattens 02896
Österreichisches Felsbildermuseum, Spital am Pyhrn 02750
Prähistorische Privatsammlung Walzer, Mühlen 02411
Stadtmuseum Wiener Neustadt, Wiener Neustadt 03099
Stadtmuseum Wienertor Hainburg, Hainburg an der Donau 02039
Tiroler Landesmuseum Ferdinandeum, Innsbruck 02136
Urgeschichtemuseum Niederösterreich, Asparn an der Zaya 01734
Urgeschichtliche Sammlung, Engerwitzdorf 01868
Urgeschichtliches Museum, Peggau 02483
Urzeitmuseum Nußdorf ob der Traisen, Traismauer 02821
Welterbemuseum Hallstatt, Hallstatt 02049

Belgium

Archéo- et Géo-Espaces Moulin du Broukay, Eben-Emael 03567
Archeologisch Museum van de Universiteit, Gent 03622
Domus Romana Archéosite, Aubechies 03335
Heemkundig Museum Slag van Lafelt, Vlijtingen 04121
Musée Archéologique Régional d'Orp-le-Grand, Orp-le-Grand 03939
Musée Archéologique Régional d'Oupeye-(M.A.R.O.), Vivegnis 04120
Musée de la Préhistoire, Jemeppe-sur-Sambre 03745
Musée de la Préhistoire, Goyet-Gesves 03658
Musée de la Préhistoire en Wallonie, Flémalle 03602
Musée des Mégalithes et Centre d'Exposition de Wéris, Wéris 04145
Musée d'Ourthe-Ambléve, Comblain-au-Pont 03524
Musée Historique et Préhistorique Van den Steen, Jehay-Amay 03743
Musée Régional d'Histoire et d'Archéologie, Visé 04118
Ondergronds Museum, Kanne 03749
Voorhistorisch Museum, Zonhoven 04165

Brazil

Museu de Arqueologia e Artes Populares, Paranaguá 04611
Museu de Pré-História Professora Márcia Angelina Alves, Perdizes 04625

Bulgaria

Nacionalen Muzej Zemjata i Chorata, Sofia 05196

Canada

Powell River Historical Museum, Powell River 06654

China

Xinhui Museum, Xinhui 08396

Congo, Democratic Republic

Musée de Préhistoire, Kinshasa 08797

Croatia

Arheološki Muzej Zagreb, Zagreb 08986

Cuba

Museo de Historia Natural Valle de la Prehistoria, Santiago de Cuba 09312

Czech Republic

Melicharovo Městské Muzeum, Unhošt 09889
Muzeum historie, Kolín 09590
Regionální Muzeum K.A. Polánka, Žatec 09926

Denmark

Esbjerg Museum, Esbjerg 10001
Haderslev Museum, Haderslev 10049
Moesgård Museum, Højbjerg 10084
Ringkøbing Museum, Ringkøbing 10236
Skive Museum, Skive 10266
Skjern-Egvad Museum, Skjern 10270
Søllerød Museum, Holte 10093
Vesthimmerlands Museum, Års 09967

Egypt

National Museum for Egyptian Civilization, Cairo 10510

Equatorial Guinea

Museo Etnológico Misional C.M.F., Santa Isabel 10557

Finland

Ålands Museum, Mariehamn 11098

France

Collection Préhistorique Pagès, Le Moustier 13625
Dinosauria - Musée des Dinosaures, Espéraza 12860
Ecomusée de la Vallée du Galeizon, Cendras 12335
Exposition de la Préhistoire, Mazières-en-Mauges 14055
Expostion Grotte Chauvet Pont d'Arc, Vallon-Pont-d'Arc 16054
Grotte de Pair non Pair, Prignac-et-Marcamps 14947
Les Grottes de Gargas, Aventignan 11784
Mas de la Pyramide, Saint-Rémy-de-Provence 15487
Musée Archéologique, Martizay 14021

P

Musée Archéologique, Simiane-Collongue 15758
Musée Archéologique, Sainte-Bazeille 15551
Musée Archéologique de Site, Liffol-le-Grand 13744
Musée-Château des Adhémar, Montélimar 14197
Musée d'Archéologieque, Saint-Ciers-sur-Gironde 15208
Musée de la Grotte de Cougnac, Payrignac 14767
Musée de la Pierre à Fusil, Meusnes 14107
Musée de la Préhistoire, Le Mas-d'Azil 13611
Musée de la Prehistoire, Rânes 14984
Musée de la Préhistoire, Saint-Flovier 15273
Musée de la Préhistoire, Saint-Porchaire 15467
Musée de la Préhistoire-Le Cairn, Saint-Hilaire-la-Forêt 15310
Musée de la Préhistoire/Paléontologie, Peyzac-le-Moustier 14807
Musée de l'Avallonnais, Avallon 11783
Musée de l'Homme de Néandertal, La Chapelle-aux-Saints 13331
Musée de l'Ours des Cavernes, Entremont-le-Vieux 12826
Musée de Paléontologie, Menat 14070
Musée de Paléontologie Humaine de Terra Amata, Nice 14431
Musée de Préhistoire, Aurignac 11753
Musée de Préhistoire Amédée Lemozi, Cabrerets 12219
Musée de Préhistoire des Gorges du Verdon, Quinson 14973
Musée de Préhistoire d'île-de-France, Nemours 14391
Musée de Préhistoire James Miln-Zacharie le Rouzic, Carnac 12292
Musée de Préhistoire Régionale, Menton 14081
Musée de Ventabren, Ventabren 16106
Musée Départemental Breton, Quimper 14970
Musée Départemental de la Préhistoire, Le Grand-Pressigny 13587
Musée Départemental de Préhistoire, Solutré-Pouilly 15779
Musée Départemental de Préhistoire Corse, Sartène 15647
Musée des Arts et Traditions Populaires, Esse 12863
Musée des Bujoliers, Saint-Cezaire 15198
Musée des Corbières, Sigean 15757
Musée du Vieil Auvillar, Auvillar 11772
Musée Municipal de Préhistoire, Les Matelles 13706
Musée Municipal de Préhistoire Régionale, Saint-Pons-de-Thomières 15466
Musée Municipal des Beaux Arts et d'Histoire Naturelle, Châteaudun 12444
Musée National de Préhistoire, Les Eyzies-de-Tayac-Sireuil 13699
Musée Préhistorique Charles Nardin, Sainte-Foy-la-Grande 15559
Musée Préhistorique Finistérien, Penmarch 14773
Musée Régional de Préhistoire, Orgnac-l'Aven 14526
Musée Sundgauvien, Altkirch 11562
Parc de la Préhistoire, Tarascon-sur-Ariège 15858
Parc de Préhistoire de Bretagne, Malansac 13929
Parc Préhistorique Imaginaire, Chamoux 12382
Plaine des Dinosaures, Mèze 14113
Le Préhistorama, Rousson 15130
Tour-Musée de Velaux, Velaux 16093

Georgia

Metsamor Museum, Metsamor 16348

Germany

Alamannenmuseum Ellwangen, Ellwangen 18120
Altmärkisches Museum, Stendal 21897
Archäologische Staatssammlung München, München 20422
Archäologisches Landesmuseum Brandenburg, Brandenburg an der Havel 17505
Archäologisches Museum Amberg, Amberg 16532
Archäologisches Museum Colombischlössle, Freiburg im Breisgau 18427
Archäologisches Museum des Historischen Vereins für Oberfranken, Bayreuth 16945
Archäologisches Museum Frankfurt, Frankfurt am Main 18354
Archäologisches Museum Neu-Ulm, Neu-Ulm 20586
Archäologisches Zentrum Hitzacker, Hitzacker 19202
Audorfer Museum im Burgtor, Oberaudorf 20801
Ausstellung zur Erd- und Vorgeschichte, Breitscheid, Lahn-Dill-Kreis 17544
Bachmann-Museum Bremervörde, Bremervörde 17576
Braith-Mali-Museum, Biberach an der Riß 17319
Braunschweigisches Landesmuseum, Wolfenbüttel 22651
Dischinger Heimatmuseum, Dischingen 17874
Dr.-Carl-Haeberlin-Friesen-Museum, Wyk auf Föhr 22759
Geschichtlich-heimatkundliche Sammlung, Aschheim 16597
Goldberg-Museum, Riesbürg 21331
Haffmuseum Ueckermünde, Ueckermünde 22149
Heimat- und Handfeuerwaffenmuseum, Kemnath 19507
Heimat- und Torfmuseum, Gröbenzell 18767
Heimatmuseum, Allensbach 16473
Heimatmuseum, Blaubeuren 17377
Heimatmuseum, Camburg 17681
Heimatmuseum, Garbsen 18535
Heimatmuseum, Maßbach 20234
Heimatmuseum, Osterhofen 20934
Heimatmuseum der Stadt Rerik, Rerik, Ostseebad 21280
Heimatmuseum Ebersbach in der Humboldt-Baude, Ebersbach, Sachsen 18043
Heimatmuseum Prieros, Heidesee 19094
Heimatmuseum Uelzen, Uelzen 22150
Helms-Museum, Hamburg 18936
Hürten-Heimatmuseum, Bad Münstereifel 16789
Josef Albers Museum Quadrat Bottrop, Bottrop 17488
Karl-Wagenplast-Museum, Schwaigern 21658
Kreisheimatmuseum Weißes Haus, Rotenburg an der Fulda 21399
Kulturgeschichtliches Museum Osnabrück, Osnabrück 20923
Kulturhistorisches Museum Schloß Merseburg, Merseburg 20292
Landesmuseum für Natur und Mensch, Oldenburg 20898
Landesmuseum für Vorgeschichte, Dresden 17956
Landesmuseum für Vorgeschichte Sachsen-Anhalt, Halle, Saale 18900
Lobdengau-Museum, Ladenburg 19763
Museum Alzey, Alzey 16530
Museum Burg Eisenhardt, Belzig 16986
Museum Burg Ranis, Ranis 21179
Museum der Stadt Lauffen am Neckar, Lauffen am Neckar 19856
Museum der Stadt Schkeuditz, Schkeuditz 21559
Museum des Ohrekreises, Wolmirstedt 22676
Museum für die Archäologie des Eiszeitalters, Neuwied 20687
Museum für Naturkunde und Vorgeschichte, Dessau 17830
Museum für Ur- und Frühgeschichte, Eichstätt 18084
Museum für Ur- und Frühgeschichte Thüringens, Weimar 22410
Museum für Vor- und Frühgeschichte, Gunzenhausen 18840
Museum für Vor- und Frühgeschichte, Langenau 19820
Museum für Vor- und Frühgeschichte, Berlin 17206
Museum für Vor- und Frühgeschichte, Saarbrücken 21457
Museum für Vor- und Frühgeschichte sowie Stadtgeschichte, Egeln 18059
Museum Garching an der Alz, Garching an der Alz 18536
Museum Gensungen, Felsberg 18309
Museum im Kräuterkasten, Albstadt 16455
Museum im Marstall, Winsen, Luhe 22606
Museum im Vorderen Schloss, Mühlheim an der Donau 20408
Museum "Kaffeemühle", Wolgast 22673
Museum Reichenfels, Hohenleuben 19238
Museum Schloss Hellenstein, Heidenheim an der Brenz 19093
Museum Schloß Hohentübingen, Tübingen 22123
Museum Schloß Klippenstein, Radeberg 21158
Museum Schloß Moritzburg, Zeitz 22777
Neues Museum, Berlin 17228
Niederbayerisches Vorgeschichtsmuseum, Landau an der Isar 19787
Niedersächsisches Landesmuseum Hannover, Hannover 19002
Oderlandmuseum, Bad Freienwalde 16733
Peter-Wiepert-Museum, Fehmarn 18299
Pfahlbaumuseum Unteruhldingen, Uhldingen-Mühlhofen 22158
Prähistorische Siedlung Pestenacker, Weil, Kreis Landsberg am Lech 22382
Prähistorisches Heimatmuseum, Veringenstadt 22221
Rheinisches Landesmuseum Trier, Trier 22102
Roemer- und Pelizaeus-Museum, Hildesheim 19185
Römermuseum Kastell Boiotro, Passau 20994
Römisch-Germanisches Zentralmuseum, Mainz 20150
Sammlung Irmgard Friedl, Bad Griesbach 16742
Schloss Neuburg, Neuburg an der Donau 20599
Schloß und Festung Senftenberg, Senftenberg 21744
Schloßmuseum, Quedlinburg 21152
Siegerlandmuseum mit Ausstellungsforum Haus Oranienstraße, Siegen 21756
Spengler-Museum, Sangerhausen 21485
Spreewald-Museum Lübbenau, Lübbenau 20071
Stadtmuseum, Riesa 21330
Stadtmuseum Erlangen, Erlangen 18225
Stadtmuseum Mutzschen, Mutzschen 20539
Städtische Museen Heilbronn/Archäologie, Heilbronn 19101
Städtisches Museum, Bad Reichenhall 16808
Städtisches Museum, Herborn 19133
Städtisches Museum Ludwigsburg, Ludwigsburg 20048
Steinsburg-Museum, Römhild 21357
Steinzeitmuseum, Korb 19685
Südschwäbisches Archäologiemuseum im Jesuitenkolleg, Mindelheim 20331
Uffenheimer Gollachgaumuseum, Uffenheim 22155
Ulmer Museum, Ulm 22170
Ur- und Frühgeschichtliche Sammlung, Erlangen 18226
Urgeschichtliches Museum und Galerie Vierzigtausend Jahre Kunst, Blaubeuren 17381
Vor- und Frühgeschichtliche Sammlung, Kassel 19479
Vor- und Frühgeschichtliches Museum, Thalmässing 22047
Vorgeschichtsmuseum im Grabfeldgau, Bad Königshofen 16762
Wallenfels'sches Haus, Abteilung Vor- und Frühgeschichte, Archäologie und Völkerkunde, Gießen 18641
Zeittunnel Wülfrath, Wülfrath 22704

Greece

P. and Al. Canellopoulos Museum, Athinai 22947

Guinea

Musée National de Conakry, Conakry 23310

Hungary

Vértesszőlősi Bemutatóhely, Vértesszőllős 23957

India

Archaeological Museum, Pune 24354
Archaeological Museum, Udaipur 24413
Baripada Museum, Baripada 24078
Museum and Gallery of Art and Archaeology, Dharwad 24164
Prof. G.R. Sharma Memorial Archaeology Museum, Allahabad 24062
University Museum of Science and Culture, Aligarh 24057

Israel

Bet Gordon, Deganya Aleph 24941
Huleh Valley Regional Prehistoric Museum, Kibbutz Maayan Baruch 25050
Museum of Prehistory, Jerusalem 25001
Museum of Prehistory, Shaar Hagolan 25093

Italy

Antiquarium Forense, Roma 27440
Centro Camuno di Studi Preistorici, Capo di Ponte 25657
Civico Museo Archeologico e di Scienze Naturali Federico Eusebio, Alba 25174
Museo Archeologico, Montelupo Fiorentino 26837
Museo Archeologico Eoliano, Lipari 26493
Museo Archeologico G. Moretti, San Severino Marche 27684
Museo Archeologico Statale, Ascoli Piceno 25294
Museo Civico, Erba 26058
Museo Civico, Mazara del Vallo 26643
Museo Civico, Rovereto 27569
Museo Civico A. Klitsche de la Grange, Allumiere 25203
Museo Civico Archeologico Etnologico, Modena 26745
Museo Civico Archeologico U. Granafei, Mesagne 26666
Museo Civico B. Romano, Termini Imerese 27960
Museo Civico della Val Fiorentina Vittorino Cazzetta, Selva di Cadore 27798
Museo Civico di Archeologia Ligure, Genova 26296
Museo Civico di Paleontologia e Archeologia, Ugento 28115
Museo Civico di Paleontologia e Paletnologia, Borgosesia 25508
Museo Civico e Pinacoteca, Sassoferrato 27762
Museo Civico per la Preistoria del Monte Cetona, Cetona 25842
Museo d'Arte Preistorica, Pinerolo 27221
Museo della Preistoria, Celano 25798
Museo della Preistoria e Museo Egizio, Milano 26703
Museo delle Origini, Roma 27486
Museo delle Palafitte, Molina di Ledro 26767
Museo di Antichita, Torino 28010
Museo di Paleontologia e Paletnologia, Maglie 26567
Museo di Preistorico e Paleontologico, Sant'Anna d'Alfaedo 27723
Museo Fiorentino di Preistoria, Firenze 26191
Museo Nazionale dell'Alta Val d'Agri, Grumento Nova 26379
Museo Nazionale di Santa Maria delle Monanche e Paleolitico, Isernia 26414
Museo Nazionale Domenico Ridola, Matera 26641
Museo Nazionale Preistorico ed Etnografico Luigi Pigorini, Roma 27519
Museo Sella e Mosca, Alghero 25199
Villa Mirabello, Varese 28165

Japan

Gifu-ken Hakubutsukan, Seki 29240
Idojiri Kokokan, Fujimi 28422
Kokubunji-shi Bunkazai Hozonkan, Kokubunji 28761
Meiji-Daigaku Kouko Hakubutsukan, Tokyo 29432

Kazakhstan

Kazakstan Paleolit Muzeji, Almaty 29682
Muzej Archeologii, Almaty 29685
Muzej Prirody, Almaty 29690

Kenya

Kariandusi Prehistoric Site Museum, Gilgil 29738
National Museum Nairobi, Nairobi 29749
Olorgesailie Prehistoric Site Museum, Olorgesailie 29751

Korea, Republic

Yonsei University Museum, Seoul 29935

Lebanon

Musée de Préhistoire Libanaise, Beirut 30158

Lithuania

Senasis Arsenalas, Vilnius 30431

Luxembourg

Musée de Préhistoire, Echternach 30470

Malta

Ggantija Prehistoric Temples, Xaghra 30658

Mexico

Museo Prehistórico de Tepexpan, Tepexpan 31438
Museo Regional Doctor Leonardo Oliva, Ahualulco de Mercado 30706

Monaco

Musée d'Anthropologie Préhistorique, Monaco 31607

Morocco

Musée Archéologique, Rabat 31688

Netherlands

Archeologiemuseum, Stein 32805
Eicha-Museum, Bergeijk 31966
Historiekamer Hardenberg, Hardenberg 32311
Hunebedcentrum, Borger 31987
Museum Oud Vriezenveen, Vriezenveen 32911
Museum Smallingerland, Drachten 32155
Noordbrabants Museum, 's-Hertogenbosch 32375
Openluchtmuseum Erve Kots, Lievelde 32502
Streekmuseum De Oude Wolden, Bellingwolde 31960
Streekmuseum Schippersbeurs, Elsloo 32202
Veluws Museum Nairac, Barneveld 31951

Norway

Arkeologisk Museum i Stavanger, Stavanger 33863
Universitetets Oldsaksamling, Oslo 33757

Portugal

Museu Nacional de Arqueologia, Lisboa 35715

Romania

Muzeul de Arheologie şi Istorie, Baia Mare 36124
Muzeul de Istorie, Sibiu 36609
Rezervaţia Arheologică Târgşoru Vechi, Târgşoru Vechi 36662

Russia

Muzej Archeologii, Syktyvkar 37817

Serbia

Muzejska Jedinica Odžaci, Odžaci 38127

Slovenia

Muzej Evolucije i Nalazište Pračovjeka Hušnjakovo, Klanjec 38429

P

South Africa

Stevenson-Hamilton Memorial Information Centre, Skukuza 38731
Victoria West Regional Museum, Victoria West 38774

Spain

Museo Arqueológico d'Eivissa i Formentera, Eivissa 39259
Museo Arqueológico Municipal, Jerez de la Frontera 39426
Museo Arqueológico Provincial, Huesca 39398
Museo Arqueológico Provincial, Ourense 39754
Museo Arqueológico y Etnológico Córdoba, Córdoba 39204
Museo Arqueológico y Etnológico de Granada, Granada 39364
Museo de Jaén, Jaén 39422
Museo de León, León 39456
Museo de Lluc, Escorca 39279
Museo de Murcia, Murcia 39696
Museo de Navarra, Pamplona 39808
Museo de Prehistoria y Arqueología de Cantabria, Santander 40010
Museo de Prehistoria y Servicio de Investigación Prehistórica, Valencia 40206
Museo Didáctico de Prehistoria, Alba de Tormes 38801
Museo Fundación Díaz Caneja de Arte Contemporaneo, Palencia 39772
Museo Municipal, Aroche 38895
Museo Municipal de Calahorra, Calahorra 39095
Museo Nacional y Centro de Investigación de Altamira, Santillana del Mar 40036
Museo Provincial, Lugo 39499
Museu d'Arqueologia de Catalunya Girona, Girona 39343

Sweden

Hällristningsmuseet, Norrköping 40771
Historiska Museet, Lund 40702

Switzerland

Heimatmuseum Rapperswil, Rapperswil 41743
Historische Ortssammlung, Reigoldswil 41752
Historisches und Völkerkundemuseum, Sankt Gallen 41803
Musée Cantonal d'Archéologie et d'Histoire, Lausanne 41541
Museum, Zofingen 42061
Museum Appenzell, Appenzell 41118
Museum für Urgeschichte(n), Zug 42126
Museum im Kornhaus, Rorschach 41776
Préhisto-Parc et Grottes, Réclère 41749
Village Lacustre Gletterens, Gletterens 41440

Syria

Homs Museum, Homs 42136

Taiwan

National Museum of Prehistory, Taitung 42444

United Kingdom

Alexander Keiller Museum, Avebury 43338
Jewry Wall Museum, Leicester 44676
Museum of Ancient Culture, Kilmartin 44580
Museum of the Iron Age, Andover 43303
Poldowrian Museum of Prehistory, Coverack 43905
Rotunda Museum of Archaeology and Local History, Scarborough 45623

United States of America

Edisto Island Historic Preservation Society Museum, Edisto Island 48293
Museum of Jurassic Technology, Culver City 47904
Pinson Mounds State Archaeological Area, Pinson 51929
Wyoming Dinosaur Center, Thermopolis 53494

Venezuela

Museo de Arte e Historia Casa de Los Celis, Valencia 54598

Zambia

Victoria Falls Information Centre, Livingstone 54657

Zimbabwe

Children's Library Museum, Marondera 54674
Matopos National Park Site Museums, Bulawayo 54664

Press → Mass Media

Printing

Australia

Melbourne Museum of Printing, Footscray 01084
Pinnaroo Printing Museum, Pinnaroo 01418

Austria

Buchdruckmuseum, Rohrbach/OÖ 02596
Buchdruckmuseum Druckwerk, Dornbirn 01824
Druckmuseum Graz, Graz 01967
Museum für Druckgrafik, Rankweil 02559
Österreichisches Papiermacher- und Druckereimuseum, Steyrermühl 02779

Belgium

Dagbladmuseum Abraham Verhoevenhuis, Hoboken 03704
Economuseum voor Papier, Karton en Aandrijftechnieken, Alsemberg 03265
Nationaal Museum van de Speelkaart, Turnhout 04099

Brazil

Museu de Arte Contemporânea da Universidade de São Paulo, São Paulo 04879

Canada

Mackenzie Printery and Newspaper Museum, Queenston 06700
Marsh Collection Society, Amherstburg 05334

Czech Republic

Muzeum Knihy, Žďár nad Sázavou 09933
Památník Jiřího Melantricha z Aventýna, Roždalovice 09823

Denmark

Bogtrykmuseet, Esbjerg 09999
Danmarks Mediemuseum, Odense 10212

Finland

Keskisuomalaisen Museo, Jyväskylä 10896

France

Atelier-Musée Livre et Typographie, Grignan 13148
Centre d'Art Graphique, Parly 14751
Musée Champenois de l'Imprimerie, Épernay 12831
Musée de l'Imprimerie, Nantes 14363
Musée de l'Imprimerie de Lyon, Lyon 13885
Musée du Luxembourg, Paris 14699
Musée Goupil, Bordeaux 12072
Musée Réattu, Arles 11685
Musée Renaudot, Loudun 13831

Germany

Deutsches Zeitungsmuseum, Wadgassen 22264
Druckereimuseum, Hatten 19034
Druckereimuseum, Wildeshausen 22569
Gutenberg-Gedenkstätte, Eltville am Rhein 18131
Gutenberg-Museum, Mainz 20143
Heimatmuseum, Buchloe 17619
Kupferstichkabinett - Sammlung der Zeichnungen und Druckgraphik, Berlin 17172
Lichtdruck, Dresden 17958
MAN-Museum, Augsburg 16628
Museen im Kulturzentrum, Rendsburg 21275
Museum für Druck, Mosbach 20391
Museum für Druckkunst Leipzig, Leipzig 19922
Museum für visuelle Kommunikation, Köln 19622
Ottmar-Mergenthaler-Museum, Bad Mergentheim 16784
Pavillon-Presse Druckgrafisches Museum Weimar, Weimar 22415
Strich-Chapell-Zimmer, Sersheim 21748
Vermessen in Bayern, München 20494
Werksmuseum Koenig Bauer AG, Würzburg 22727

Greenland

Nitz Gallery, Nuuk 23261

Hungary

Kner Nyomdaipari Múzeum, Gyomaendrőd 23577

Ireland

Dublin Print Museum, Dublin 24787
National Print Museum, Dublin 24811

Italy

Gabinetto Disegni e Stampe degli Uffizi, Firenze 26153
Museo Bodoniano, Parma 27100
Museo della Stampa e Museum Graphia, Urbino 28125

Japan

Kappan Hakubutsukan, Tokyo 29408

Kazakhstan

Muzej Knigi, Almaty 29688

Korea, Republic

Cheongju Early Printing Museum, Cheongju 29787

Lebanon

Qozhaya Museum, Zgharta 30177

Netherlands

Drukkerijmuseum Meppel, Meppel 32549
Grafisch Historisch Centrum - Drukkerijmuseum, Etten Leur 32224
Grafisch Musem-Atelier in den Groenen Zonck, Wouw 32970
Grafisch Museum Zutphen, Zutphen 33018
Museum de Historische Drukkerij, Maastricht 32526
Museum Enschedé, Haarlem 32298
Museum Lettergieten 1983, Westzaan 32942
Museum Oude Boekdrukkunst, Almen 31802
Nederlands Steendrukmuseum, Valkenswaard 32870

New Zealand

Yaldhurst Museum of Transport and Science, Christchurch 33079

Norway

Norsk Grafisk Museum, Stavanger 33869

Peru

Museo Gráfico El Peruano, Cercado 34142

Philippines

Museo Rosendo Mejica, Iloilo 34390

Portugal

Museu da Imprensa de Fafe, Fafe 35592
Museu Nacional da Imprensa, Jornais e Artes Gráficas, Porto 35858

Romania

Muzeul Tiparului şi al Cǫrţii Vechi Româneşti, Târgovişte 36661

Russia

Muzej Istorii Pečati Sankt-Peterburga, Sankt-Peterburg 37681
Muzej-kvartira I.D. Sytina, Moskva 37323

Spain

Fundación Museo de las Ferias, Medina del Campo 39638
Museu Nacional d'Art de Catalunya (MNAC), Barcelona 39001

Sweden

Rosenlöfs Tryckerimuseum, Kungsgården 40660

Switzerland

Centre d'Edition Contemporaine, Genève 41407
Druckereimuseum, Freienbach 41381
Schlossmuseum Beromünster, Beromünster 41225
Typorama, Bischofszell 41241

Ukraine

Muzej Knigi i Drukarstva Ukraïny, Kyïv 43092
Nacionalnij Muzej u Lvovi - Viddil Starodrukiv ta Rukopysiv, Lviv 43139

United Kingdom

Gray's Printing Press, Strabane 45774
Guildford House Gallery, Guildford 44370
Guildford House Gallery, Guildford 44370
John Jarrold Printing Museum, Norwich 45280
Oxford University Press Museum, Oxford 45337
Peacock Visual Arts, Aberdeen 43245
Printing House Museum, Cockermouth 43868
Thomas Bewick's Birthplace, Mickley 45136

United States of America

American Printing House for the Blind, Callahan Museum, Louisville 50281
Avampato Discovery Museum, Charleston 47378
Galerie at Lincoln Center, New York 51179
International Print Center New York, New York 51195
International Printing Museum, Carson 47257
The Mezzanine Gallery, New York 51213
Muse Gallery, Jackson 49572
Museum of Printing History, Houston 49398
Orono Museum of Art, Bangor 46598
Printmaking Council of New Jersey, Somerville 53151
Putney School Gallery, Putney 52204
State Capital Publishing Museum, Guthrie 49109
University of Kansas Ryther Printing Museum, Lawrence 49993
Wooden Nickel Historical Museum, San Antonio 52732

Prints → Graphic Arts

Prisons → Criminology

Puppets → Dolls and Puppets

Radio → Mass Media

Railroads

Argentina

Museo de Ferrocarriles en Miniatura Werner Dura, La Falda 00401
Museo Ferroviario de Gualeguaychú, Gualeguaychú 00384
Museo Histórico Ferroviario Escribano Alfredo Rueda, Rueda 00551
Museo La Vieja Estación, María Teresa 00449
Museo Nacional Ferroviario, Buenos Aires 00265

Australia

Adelaide River Railway Heritage Precinct Motorcycle Haven, Adelaide River 00784
Alexandra Timber Tramway and Museum, Alexandra 00794
Alpenrail Swiss Model Village and Railway, Claremont, Tasmania 00975
Ballarat Tramway Museum, Ballarat 00826
Beta Railway Station Historic Museum, Alpha 00804
Bundaberg Railway Museum, Bundaberg 00919
Don River Railway Museum, Devonport 01036
Dorrigo Steam Railway and Museum, Dorrigo 01041
Gisborne Steam Park, New Gisborne 01359
Jamestown Museum, Jamestown 01177
National Railway Museum, Port Adelaide 01422
Old Railway Station Museum, Northam 01375
Onslow Goods Shed Museum, Onslow 01393
Pinnaroo Heritage Museum, Pinnaroo 01417
Platform 1 Heritage Farm Railway, Littlehampton 01235
Puffing Billy Railway Museum, Belgrave 00855
Queen Victoria Museum and Art Gallery at Inveresk, Launceston 01223
Rail Transport Museum, Bassendean 00836
Railway Station Museum, Wonthaggi 01666
Richmond Vale Railway Museum, Richmond Vale 01459
Serviceton Historic Railway Station, Serviceton 01501
Sulphide Street Station Railway and Historical Museum, Broken Hill 00904
Sydney Tramway Museum, Loftus 01236
Tramway Museum, Bendigo 00861
Trentham Agricultural and Railway Museum, Trentham 01602
Walkaway Station Museum, Walkaway 01623
West Coast Pioneers Museum, Zeehan 01687
West Torrens Railway, Signal Telegraph and Aviation Museum, Brooklyn Park 00907
Yass Railway Museum, Yass 01683

Austria

Bahnzeit, Wels 02913
Eisenbahnmuseum, Deutsch Wagram 01811
Eisenbahnmuseum, Knittelfeld 02205
Eisenbahnmuseum Strasshof - Das Heizhaus, Strasshof 02788
Erzbergbahn- und Wintersportmuseum, Vordernberg 02877

Feld- und Industriebahnmuseum (FIM), Freiland 01901
Florianerbahn, Sankt Florian 02636
Historama - Museum für Technik und Verkehr, Ferlach 01888
Localbahnmuseum, Innsbruck 02123
Lokpark Ampflwang, Ampflwang im Hausruckwald 01718
Museum Sankt Veit, Sankt Veit an der Glan 02691
Museum Tauernbahn, Schwarzach 02718
Museumseisenbahn Payerbach-Hirschwang, Hirschwang 02081
Pferde-Eisenbahn Museum, Rainbach im Mühlkreis 02554
Salzkammergut-Lokalbahn-Museum, Mondsee 02407
Schmalspurbahn-Museum, Frojach 01908
Steyrtal-Museumsbahn, Steyr 02773
Südbahn, Mürzzuschlag 02414
Technisches Eisenbahnmuseum, Lieboch 02299
Tramway-Museum Graz, Graz 01984
Waldviertler Eisenbahnmuseum, Sigmundsherberg 02742

Belgium

Musée des Chemins de Fer Belges-Museum van de Belgische Spoorwegen, Bruxelles 03475
Musée Ferroviaire de Treignes, Treignes 04093
Musée Vivant d'Archéologie Industrielle Ferroviaire, Sprimont 04046
Musée Vivant du Chemin de Fer Vicinal, Thuin 04066
Spoorwegmuseum van Miniatuurmodellen, Heist-op-den-Berg 03693
Stoomcentrum, Maldegem 03841
Trammuseum, Schepdaal 04003

Brazil

Centro Ferroviário de Cultura, Volta Grande 04960
Museu da Estrada de Ferro Madeira-Mamoré, Porto Velho 04657
Museu do Trem, Rio de Janeiro 04743

Canada

Alberta Central Railway Museum, Wetaskiwin 07310
Alberta Railway Museum, Edmonton 05741
British Columbia Forest Discovery Centre, Duncan 05722
Campbell Carriage Factory Museum, Middle Sackville 06275
Canadian Museum of Rail Travel, Cranbrook 05658
Carbonear Museum, Carbonear 05558
Copper Belt Museum, Whitehorse 07324
Cottonwood House, Quesnel 06702
End of Steel Heritage Museum, Hines Creek 05989
Exporail, Saint-Constant 06811
Fort Erie Railroad Museum, Fort Erie 05807
Herbert CPR Train Station Museum, Herbert 05983
Hillsborough Railway Museum, Hillsborough 05987
Kaatza Station Museum, Lake Cowichan 06130
Kettle River Museum, Midway 06281
Kwinitsa Railway Station Museum, Prince Rupert 06670
Ladysmith Railway Museum, Ladysmith 06127
Melville Railway Museum, Melville 06265
Miami Station Museum, Miami 06272
Musquodoboit Railway Museum, Musquodoboit 06405
Ontario Electric Railway Museum, Milton 06291
Orangedale Railway Museum, Orangedale 06512
Owen Sound Marine and Rail Museum, Owen Sound 06562
Prince Georg Railway and Forestry Museum, Prince George 06668
Railway Coastal Museum, Saint John's 06853
Ralph Allen Memorial Museum, Oxbow 06564
Revelstoke Railway Museum, Revelstoke 06736
Riverton Transportation Heritage Centre, Riverton 06760
Sioux Lookout Museum, Sioux Lookout 06981
Smiths Falls Railway Museum of Eastern Ontario, Smiths Falls 06988
Station Museum, North Vancouver 06482
Sydney and Louisburg Railway Museum, Louisbourg 06199
Winnipeg Railway Museum, Winnipeg 07394

China

Beijing Gu Guanxiang Station, Beijing 07514
Hong Kong Railway Museum, Hong Kong 07803
Jiaozhuang Hu Didao Station, Beijing 07542

Colombia

Museo Historia del Ferrocarril de Antioquia, Medellín 08682

Cuba

Estación Ferrocarril Bejucal, Bejucal 09039

Czech Republic

Městské Muzeum, Česká Třebová 09461
Muzeum těžby a dopravy vápence v Českém krasu, Beroun 09405
Železniční Muzeum Zlonice, Zlonice 09944
Zubrnická Museální Železnice, Zubrnice 09950

Denmark

Danmarks Jernbanemuseum, Odense 10211
Museet på Gudhjem Station, Gudhjem 10048

Egypt

Rail Museum, Cairo 10517

Estonia

Eesti Muuseumraudtee, Lavassaare 10589
Eesti Raudteemuuseum, Haapsalu 10563

Finland

Jokioisten Museorautatie ja Kapearaidemuseo, Jokioinen 10874
Keitele-Museo, Suolahti 11337
Pietarin Radan Museo, Lahti 11031
Suomen Rautatiemuseo, Hyvinkää 10846
Veturimuseo, Toijala 11381

France

Collection de Locomotives à Vapeur, Obermodern 14490
Collection de Matériel Ferroviaire, Burnhaupt-le-Haut 12209
Espace du Cheminot, Nîmes 14447
Musée-Animé du Jouet et des Petits Trains, Colmar 12581
Musée Collection Ferroviaire, Miramas 14129
Musée de la Locomotion et de l'Attelage de l'Ile de France, Coignières 12570
Musée de la Machine Agricole et à Vapeur, Ambert 11568
Musée de la Mine, de la Verrerie et du Chemin de Fer, Epinac 12838
Musée des Chemins de Fer Militaires et Industriels, La Neuville-les-Bray 13398
Musée des Technologies Ferroviaires, La Seyne-sur-Mer 13443
Musée des Tramways à Vapeur et des Chemins de Fer Secondaires Français, Butry-sur-Oise 12213
Musée des Transports - Amitram, Marquette-Lez-Lille 13980
Musée des Transports de Pithiviers, Bouzonville-en-Beauce 12138
Musée du Cheminot, Ambérieu-en-Bugey 11567
Musée du Rail, Cajarc 12245
Musée du Rail, Dinan 12748
Musée du Train, Clairac 12534
Musée du Train et du Jouet, Arpaillargues-et-Aureillac 11693
Musée Ferroviaire, Saint-Sulpice-de-Pommeray 15517
Musée Ferroviaire de Saint-Géry, Saint-Géry 15297
Musée Ferroviaire et des Vieilles Gares, Guîtres 13170
Musée Français du Chemin de Fer, Mulhouse 14327
Musée Laumonier de la Locomotive à Vapeur, Vierzon 16155
Musée Lorrain des Cheminots, Rettel 15025
Musée Régional du Chemin de Fer, Rosny-sous-Bois 15110

Germany

Arbeitsgemeinschaft "Historische Nahverkehrsmittel Leipzig" e.V., Leipzig 19894
Bayerisches Eisenbahnmuseum, Nördlingen 20724
Berliner S-Bahn-Museum, Potsdam 21082
Berliner U-Bahn-Museum, Berlin 17046
Dampflok-Museum, Hermeskeil 19147
Dampflok- und Modelleisenbahnmuseum, Tuttlingen/Möhringen 22139
DB Museum, Nürnberg 20756
Deutsches Dampflokomotiv-Museum, Neuenmarkt 20610
Deutsches Feld- und Kleinbahnmuseum, Deinste 17810
Deutsches Museum - Verkehrszentrum, München 20434
DGEG Eisenbahnmuseum Würzburg, Würzburg 22708
Eisenbahn-Museum Lokschuppen Aumühle, Aumühle 16645
Eisenbahn- und Technikmuseum Rügen, Prora 21129
Eisenbahnmuseum, Losheim am See 20038
Eisenbahnmuseum Bayerischer Bahnhof zu Leipzig e.V., Leipzig 19902
Eisenbahnmuseum Bochum-Dahlhausen, Bochum 17399
Eisenbahnmuseum Darmstadt-Kranichstein, Darmstadt 17787
Eisenbahnmuseum Historischer Wasserturm, Bebra 16968
Eisenbahnmuseum Neustadt an der Weinstraße, Neustadt an der Weinstraße 20669
Eisenbahnmuseum - Sammlung Frey, Seifhennersdorf 21731
Eisenbahnmuseum Schwarzenberg, Schwarzenberg 21674
Eisenbahnmuseum Vienenburg, Vienenburg 22229
Feld- und Grubenbahnmuseum, Solms 21811
Fränkisches Freilandmuseum Fladungen mit dem Rhön-Zügle, Fladungen 18324
Frankfurter Feldbahnmuseum, Frankfurt am Main 18369
Frau Holle Expreß, Hessisch Lichtenau 19170
Härtsfeldbahn-Museum, Neresheim 20572
Hammerschmiede Gröningen, Satteldorf 21523
Localbahnmuseum, Bayerisch Eisenstein 16943
Lokwelt Freilassing, Freilassing 18444
Miniatur Wunderland Hamburg, Hamburg 18945
Museumsbahn Bremerhaven-Bederkesa, Bad Bederkesa 16678
Museumseisenbahn, Hamm, Westfalen 18971
Museumseisenbahn Schwalm-Knüll, Fritzlar 18484
MVG-Museum, München 20470
Niedersächsisches Kleinbahn-Museum Bruchhausen-Vilsen, Bruchhausen-Vilsen 17592
Oberweißbacher Bergbahn, Mellenbach-Glasbach 20273
Sächsisches Eisenbahnmuseum, Chemnitz 17702
Sächsisches Schmalspurbahn-Museum Rittersgrün, Breitenbrunn 17541
Verkehrsmuseum Dresden GmbH, Dresden 17976

Greece

Mouseio Trenon, Athinai 22926
Railway Museum, Kalamata 23025

Hungary

Közlekedési Múzeum, Budapest 23426
Magyar Vasúttörténeti Park, Budapest 23446
Vasúti Helytörténeti Gyűjtemény, Mátészalka 23694
Vasúti Múzeum, Paks 23749

India

National Rail Museum, Delhi 24150

Indonesia

Museum Kereta, Ambarawa 24435

Ireland

County Donegal Railway Heritage Centre, Donegal 24775
Fry Model Railway, Demesne 24773
Fry Model Railway Museum, Malahide 24873
The Steam Museum, Straffan 24900
West Cork Model Railway Village, Clonakilty 24761

Israel

Israel Railway Museum, Haifa 24963

Italy

Museo Ferrovario, Trieste 28094
Museo Ferroviario, Cuneo 26020
Museo Ferroviario del Verbano, Luino 26547
Museo Ferroviario Piemontese, Savigliano 27767
Museo Nazionale Ferroviario, Napoli 26908

Japan

Kaya Steam Locomotive Square, Kaya 28698
Railway Museum, Saitama 29188

Korea, Republic

BuGok Railroad Exhibition Hall, Uiwang 29939
Seoul Station Hall, Seoul 29923

Latvia

Latvijas Dzelzceļa Vēstures Muzeja Jelgavas Nodaļa, Jelgava 30019
Latvijas Dzelzceļa Vēstures Muzeja Rīgas Nodaļa, Rīga 30078

Mexico

Museo Comunitario de Estación Coahuila, Mexicali 31032
Museo de la Antigua Estación Ferroviaría, Teocelo 31425
Museo del Ferrocarril, Aguascalientes 30701
Museo del Ferrocarril, Guadalajara 30900
Museo del Ferrocarril, Monterrey 31182
Museo del Ferrocarril, Torreón 31499
Museo Ferrocarrilero, Empalme 30853
Museo la Antigua Casa del Agua, Hunucmá 30953
Museo Nacional de Ferrocarriles Mexicanos, Puebla 31288
Museo Nacional de los Ferrocarriles Mexicanos, México 31140
Museo Toma de Zacatecas o Cerro de la Bufa, Zacatecas 31582

Namibia

TransNamib Museum, Windhoek 31768

Netherlands

Blokwachterswoning Waterhuizen, Waterhuizen 32930
Gelderse Smalspoor Museum, Erlecom 32221
Het Gols-Station, Winterswijk 32956
Industrieel Smalspoor Museum, Erica 32219
Museum Buurt Spoorweg, Haaksbergen 32291
Museum Stoomdepot, Rotterdam 32715
Museumspoorlijn Star, Stadskanaal 32799
Museumstoomtram Hoorn-Medemblik, Hoorn 32410
Nationaal Smalspoormuseum - Stoomtrein Valkenburgse Meer, Valkenburg, Zuid-Holland 32867
Nederlands Spoorwegmuseum, Utrecht 32852
NZH Vervoers Museum, Haarlem 32299
Rijdend Electrisch Tram Museum, Amsterdam 31882
Spoor- en Tramweg Verzameling, Loppersum 32508
Stoomtrein Goes-Borsele, Goes 32259
Streekmuseum Alphen, Alphen, Noord-Brabant 31804
Zuid-Limburgse Stoomtreinmaatschappij, Simpelveld 32769

New Zealand

The Engine Shed, Paekakariki 33169
Fell Locomotive Museum, Featherston 33104
Founders Historic Park, Nelson 33152
Paekakariki Rail and Heritage Museum, Paekakariki 33170
Plains Vintage Railway and Historical Museum, Ashburton 33039
Pleasant Point Railway and Historical Society Museum, Pleasant Point 33184
Silver Stream Railway Museum, Hutt City 33126

Norway

Bergens Tekniske Museum, Bergen 33424
Krøderbanen, Krøderen 33617
Lommedalsbanen, Rykkinn 33802
Norsk Jernbanemuseum, Hamar 33530
Osbane Museet, Fana 33489
Setesdalsbanen Stiftelsen, Vennesla 33971
Sporveismuseet Vognhall 5, Oslo 33751
Thamshavnbanen, Løkken Verk 33644
Urskog-Hølandsbanen, Sørumsand 33854

Peru

Museo Ferroviario, Tacna 34313

Poland

Muzeum Kolei Wąskotorowej, Sochaczew 35118
Muzeum Kolejnictwa, Warszawa 35252
Osrodek Muzealny Bydgoskiego Wezla Kolejowego, Bydgoszcz 34580
Skansen Lokomotyw, Zduńska Wola 35368

Portugal

Museu dos Caminhos de Ferro de Lousado, Lousado 35743
Museu Ferroviário de Arco de Baúlhe, Arco de Baúlhe 35442
Museu Ferroviário de Chaves, Chaves 35524
Museu Nacional Ferroviário, Entroncamento 35558
Secção Museológica da CP de Estremoz, Estremoz 35582
Secção Museológica da CP de Lagos, Lagos 35652
Secção Museológica da CP de Macinhata do Vouga, Macinhata do Vouga 35747
Secção Museológica da CP de Santarém, Santarém 35911
Secção Museológica da CP de Valença, Valença 35979

Romania

Muzeul Locomotivelor cu Abur, Reşiţa 36563

Russia

Centralnyj Muzej Oktjabrskoj Železnoj Dorogi, Sankt-Peterburg 37628
Centralnyj Muzej Železnodorožnogo Transporta, Sankt-Peterburg 37631
Muzej Železnodorožnoj Techniki, Sankt-Peterburg 37713

Serbia

Železnički Muzej, Beograd 38092

Slovenia

Koroški Muzej Ravne na Koroškem, Ravne na Koroškem ... 38503
Muzej Južne Železnice, Šentjur ... 38511
Muzej Ozkotirne Železnice Parenzana, Izola ... 38423
Železniški Muzej, Ljubljana ... 38469

South Africa

Railway Station Museum, Uitenhage ... 38769
South African National Railway Steam Museum, Auckland Park ... 38553

Spain

Ferrería de Mirandaola, Legazpia ... 39449
Museo del Ferrocarril, Águilas ... 38794
Museo del Ferrocarril, Castejón ... 39142
Museo del Ferrocarril, Mérida ... 39645
Museo del Ferrocarril de Asturias, Gijón ... 39337
Museo del Ferrocarril de Madrid, Madrid ... 39556
Museo del Tren, Murcia ... 39697
Museo Ferroviario, Alcázar de San Juan ... 38818
Museo Vasco del Ferrocarril, Azpeitia ... 38931
Museu del Ferrocarril, Vilanova i la Geltrú ... 40285

Sweden

Åmåls Järnvägsmuseum, Åmål ... 40416
Anten-Gräfsnäs Järnväg, Alingsås ... 40414
Grängesbergsbanornas Järnvägsmuseum, Grängesberg ... 40560
Järnvägs- och Industrimuseum, Hagfors ... 40575
Järnvägsmuseet, Kristianstad ... 40654
Museisällskapet Jädraås-Tallås Järnväg, Jädraås ... 40610
Museispårvägen Malmköping, Malmköping ... 40716
Norrbottens Järnvägsmuseum, Luleå ... 40698
Nynäshamns Järnvägsmuseum, Nynäshamn ... 40783
Östra Södermanlands Museijärnväg, Mariefred ... 40733
Ohsabanan Museijärnväg, Växjö ... 41039
Rallarmuseet, Moskosel ... 40749
Risten-Lakviks Järnväg, Åtvidaberg ... 40448
Skara Järnvägsmuseum, Skara ... 40845
Sveriges Järnvägsmuseum, Gävle ... 40533
Sveriges Jörnvögsmuseum, Ängelholm ... 40409

Switzerland

Chemin de Fer - Musée Blonay-Chamby, Blonay ... 41242
Galleria Baumgartner, Mendrisio ... 41624
Musée du Fer et du Chemin de Fer, Vallorbe ... 41976
Tram Museum Zürich, Zürich ... 42113

Taiwan

Hualien Railway Museum, Hualien ... 42216
Jiji Railway Museum, Jiji ... 42230
Miaoli Railway Museum, Miaoli ... 42275
Taiwan High Speed Rail Exhibition Center, Jhubei ... 42224

United Kingdom

Amberley Museum and Heritage Centre, Amberley ... 43297
Bala Lake Railway, Bala ... 43356
Bluebell Railway, Uckfield ... 45892
Bodmin and Wenford Railway, Bodmin ... 43547
Bowes Railway Heritage Museum, Gateshead ... 44263
Brecon Mountain Railway, Merthyr Tydfil ... 45129
Buckinghamshire Railway Centre, Quainton ... 45453
Colonel Stephens Railway Museum, Tenterden ... 45837
Conwy Valley Railway Museum, Betws-y-Coed ... 43467
Crampton Tower Museum, Broadstairs ... 43645
Darlington Railway Centre and Museum, Darlington ... 43949
Dean Forest Railway Museum, Lydney ... 45052
Didcot Railway Centre, Didcot ... 43981
Downpatrick Railway Museum, Downpatrick ... 44021
East Anglian Railway Museum, Chappel ... 43789
East Somerset Railway, Shepton Mallet ... 45657
Elham Valley Railway Museum, Folkestone ... 44236
Embsay Bolton Abbey Steam Railway, Skipton ... 45689
Errol Station Railway Heritage Centre, Blairgowrie ... 43538
Giant's Causeway and Bushmills Railway, Bushmills ... 43702
Glenfinnan Station Museum, Fort William ... 44247
Gloucestershire Warwickshire Railway, Toddington ... 45862
Great Central Railway Museum, Loughborough ... 45032
Groudle Glen Railway, Onchan ... 45319
Gwili Steam Railway, Bronwydd ... 43656
Haig Colliery Mining Museum, Whitehaven ... 46001
Immingham Museum, Immingham ... 44514
Ingrow Loco Museum, Keighley ... 44552
Irchester Narrow Gauge Railway Museum, Wellingborough ... 45956
Isle of Wight Steam Railway and Isle of Wight Railway Heritage Museum, Havenstreet ... 44421
Keighley and Worth Valley Railway Museum, Haworth ... 44431
Kent and East Sussex Railway, Tenterden ... 45839
Kidderminster Railway Museum, Kidderminster ... 44572
Lappa Valley Steam Railway, Newquay ... 45248
Ledbury Heritage Centre, Ledbury ... 44657
Leighton Buzzard Railway, Leighton Buzzard ... 44688
Llanberis Lake Railway, Llanberis ... 44753
Locomotion - The National Railway Museum at Shildon, Shildon ... 45665
Mangapps Railway Museum, Burnham-on-Crouch ... 43676
Maud Railway Museum, Maud ... 45119
Mid Hants Railway 'Watercress Line', Alresford ... 43289
Midland Railway Butterley, Ripley, Derbyshire ... 45497
Monkwearmouth Station Museum, Sunderland ... 45799
Museum of Rail Travel, Ingrow ... 44516
Museum of the Mid-Suffolk Light Railway, Wetheringsett ... 45984
Narrow Gauge Railway Museum, Tywyn ... 45891
National Railway Museum, York ... 46105
Nene Valley Railway, Stibbington ... 45746
Newton Abbot Town and Great Western Railway Museum, Newton Abbot ... 45251
North Norfolk Railway Museum, Sheringham ... 45662
North Woolwich Old Station Museum, London ... 44946
North Yorkshire Moors Railway, Pickering ... 45382
Northampton and Lamport Railway, Chapel Brampton ... 43788
Paignton and Dartmouth Steam Railway, Paignton ... 45344
Peak Rail Museum, Matlock ... 45117
Penrhyn Castle, Bangor, Gwynedd ... 43378
Pontypool and Blaenavon Railway, Blaenavon ... 43535
Railway Age, Crewe ... 43923
Railway Museum, Ravenglass ... 45462
Railway Museum, Wylam ... 46088
Railworld, Peterborough ... 45369
Romney Toy and Model Museum, New Romney ... 45207
Rutland Railway Museum, Cottesmore ... 43896
Scottish Industrial Railway Centre, Dalmellington ... 43947
Scottish Railway Museum, Bo'ness ... 43560
Severn Valley Railway, Bewdley ... 43470
Sittingbourne and Kemsley Light Railway, Sittingbourne ... 45683
Somerset and Dorset Railway Trust Museum, Washford ... 45945
South Devon Railway, Buckfastleigh ... 43660
South Tynedale Railway Preservation Society, Alston ... 43290
Steam Museum of the Great Western Railway, Swindon ... 45819
Steamtown Railway Museum, Carnforth ... 43764
Stephenson Railway Museum, North Shields ... 45267
Strathspey Railway, Aviemore ... 43339
Swanage Railway, Swanage ... 45811
Teifi Valley Railway, Henllan ... 44451
Thursford Collection, Fakenham ... 44204
Tiverton Museum of Mid Devon Life, Tiverton ... 45860
Tyseley Loco Works Visitor Centre, Birmingham ... 43514
Wells Walsingham Light Railway, Wells-next-the-Sea ... 45960
Welsh Highland Railway Museum, Porthmadog ... 45423
Welshpool and Llanfair Light Railway, Llanfair Caereinion ... 44764
West Somerset Railway, Minehead ... 45156
Winchcombe Railway Museum, Winchcombe ... 46024

United States of America

Age of Steam Railroad Museum, Dallas ... 47925
Alabama Mining Museum, Dora ... 48152
Allegheny Portage Railroad, Gallitzin ... 48841
Ashland Railroad Station Museum, Ashland ... 46384
The B O Railroad Museum, Baltimore ... 46547
B O Railroad Station Museum, Ellicott City ... 48364
Baltimore Streetcar Museum, Baltimore ... 46554
Barn Museum and Rail Depot Museum, Troutdale ... 53570
Baycrafters, Bay Village ... 46653
Berkshire Scenic Railway Museum, Lenox ... 50024
Bicycle Museum, Three Oaks ... 53504
Boothbay Railway Village, Boothbay ... 46910
Bowie Railroad Station and Huntington Museum, Bowie ... 46969
Branchville Railroad Shrine and Museum, Branchville ... 46993
Brunswick Railroad Museum, Brunswick ... 47101
Burlington-Rock Island Railroad and Historical Museum, Teague ... 53458
California State Railroad Museum, Sacramento ... 52503
Casey Jones Home and Railroad Museum, Jackson ... 49568
Casey Jones Museum, Vaughan ... 53713
Catawissa Railroad Museum, Catawissa ... 47281
Chesapeake Beach Railway Museum, Chesapeake Beach ... 47428
Choctaw Museum, Hugo ... 49415
Cimarron Valley Railroad Museum, Cushing ... 47914
Colorado Railroad Museum, Golden ... 48957
Comanche Crossing Museum, Strasburg ... 53327
Conneaut Railroad Museum, Conneaut ... 47779
Conway Scenic Railroad, North Conway ... 51388
Cookeville Depot Museum, Cookeville ... 47785
Dennison Railroad Depot Museum, Dennison ... 48058
East Ely Railroad Depot Museum, Ely ... 48384
East Troy Electric Railroad Museum, East Troy ... 48262
End-O-Line Railroad Museum, Currie ... 47912
Florida Gulf Coast Railroad Museum, Parrish ... 51715
Fox Lake Historical Museum, Fox Lake ... 48728
Fox River Trolley Museum, South Elgin ... 53169
Garrett Historical Museum, Garrett ... 48867
Gold Coast Railroad Museum, Miami ... 50624
Golden Gate Railroad Museum, Sunol ... 53374
Golden State Model Railroad Museum, Point Richmond ... 52013
Harmar Station, Marietta ... 50475
Heart of Dixie Railroad Museum, Calera ... 47166
Huntington Railroad Museum, Huntington ... 49436
Illinois Railway Museum, Union ... 53644
Indiana Transportation Museum, Noblesville ... 51357
Iowa Railroad Historical Museum, Boone ... 46904
Kate Shelley Railroad Museum, Boone ... 46905
Kennesaw Civil War Museum, Kennesaw ... 49730
Kentucky Railway Museum, New Haven ... 51058
Lake Shore Railway Museum, North East ... 51391
Lake Superior Railroad Museum, Duluth ... 48198
Laws Railroad Museum, Bishop ... 46831
Lincoln Train Museum, Gettysburg ... 48903
Linden Railroad Museum, Linden ... 50117
Livingston Depot Center, Livingston ... 50148
Medina Railroad Museum, Medina ... 50555
Mid-Continent Railway Museum, North Freedom ... 51393
Monticello Railway Museum, Monticello ... 50819
National Capital Trolley Museum, Silver Spring ... 53107
National New York Central Railroad Museum, Elkhart ... 48347
National Railroad Museum, Green Bay ... 49040
National Railroad Museum and Hall of Fame, Hamlet ... 49140
Nevada State Railroad Museum, Carson City ... 47261
Newark Valley Depot Museum, Newark Valley ... 51297
North Carolina Transportation Museum, Spencer ... 53201
Northern Pacific Railway Museum, Toppenish ... 53539
Northwest Railway Museum, Snoqualmie ... 53137
Ohio Railway Museum, Worthington ... 54307
Old Dominion Railway Museum, Richmond ... 52336
Oregon Electric Railway Historical Society Museum, Lake Oswego ... 49889
Oregon Electric Railway Museum, Salem ... 52670
Pennsylvania Trolley Museum, Washington ... 53943
Railroad and Heritage Museum, Temple ... 53474
Railroad Museum of Long Island, Riverhead ... 52367
Railroad Museum of Pennsylvania, Strasburg ... 53329
Railroader's Memorial Museum, Altoona ... 46242
The Railway Exposition, Covington ... 47859
Reading Railroad Heritage Museum, Reading ... 52266
Rockhill Trolley Museum, Rockhill Furnace ... 52420
Salamanca Rail Museum, Salamanca ... 52653
San Diego Model Railroad Museum, San Diego ... 52751
Shore Line Trolley Museum, East Haven ... 48250
Smoky Mountain Trains, Bryson City ... 47108
South Coast Railroad Museum, Goleta ... 48965
Southeast Museum, Brewster ... 47010
Southeastern Railway Museum, Duluth ... 48192
Tennessee Central Railway Museum, Nashville ... 50974
Tennessee Valley Railroad Museum, Chattanooga ... 47414
Tri-County Historical Society and Museum, Herington ... 49258
Trolley Museum of New York, Kingston ... 49792
Union Pacific Railroad Museum, Council Bluffs ... 47851
Wabash Frisco and Pacific Association, Glencoe ... 48925
Western Pacific Railroad Museum, Portola ... 52092
The Western Railway Museum, Suisun City ... 53358
Wilmington Railroad Museum, Wilmington ... 54205

Uruguay

Museo del Círculo de Estudios Ferroviarios del Uruguay, Montevideo ... 54400

Uzbekistan

Muzej Železnodorožnoj Techniki, Toshkent ... 54482

Zambia

Railway Museum, Livingstone ... 54656

Zimbabwe

National Railways of Zimbabwe Museum, Raylton ... 54677

Rare Books → Books, Book Art and Manuscripts

Religious Art and Symbolism

Andorra

Museu d'Art Sacre de Santa Eulàlia, Encamp ... 00088

Argentina

Claustros del Pilar, Buenos Aires ... 00154
Museo de Arte Religioso Juan de Tejeda, Córdoba ... 00323
Museo de Arte Sacro de La Rioja, La Rioja ... 00417
Museo de Arte Sacro San Francisco, Salta ... 00555
Museo Histórico de Corrientes Manuel Cabral de Mayo y Alpoín, Corrientes ... 00348
Museo Histórico de San Francisco, Santa Fé ... 00616
Museo Jesútico Nacional de Jesús María, Jesús María ... 00393
Museo Sacro de la Catedral, San Miguel de Tucumán ... 00602

Armenia

Echmiadzin Cathedral Complex and Museum, Echmiadzin ... 00695

Austria

Augustinermuseum, Rattenberg, Inn ... 02564
Benediktinerstift und Stiftssammlungen, Altenburg ... 01710
Diözesanmuseum Graz, Graz ... 01966
Diözesanmuseum Sankt Pölten, Sankt Pölten ... 02681
Dom- und Diözesanmuseum Wien, Wien ... 02979
Dommuseum zu Salzburg, Salzburg ... 02609
Eisenerzer Krippenhaus, Eisenerz ... 01854
Klostersammlung der Benediktinerabtei, Michaelbeuern ... 02378
Krippenmuseum, Dornbirn ... 01826
Kunstkammer des Innsbrucker Servitenklosters, Innsbruck ... 02121
Museum des Zisterzienserstifts mit Margret Bilger-Galerie, Schlierbach ... 02707
Museum Innviertler Volkskundehaus und Galerie der Stadt Ried, Ried im Innkreis ... 02584
Museum Stift Melk, Melk ... 02375
Prämonstratenser-Chorherrenstift mit Stiftssammlungen, Geras ... 01929
Salvatorianerkloster, Gurk ... 02020
Sammlungen des Stiftes Sankt Lambrecht, Sankt Lambrecht ... 02659
Sammlungen Zisterzienserstift Stams, Stams ... 02761
Schatzkammer der Wallfahrtskirche Maria Schutz, Maria Schutz am Semmering ... 02354
Stiftsmuseum, Millstatt ... 02383
Stiftsmuseum Schatzhaus Kärntens, Sankt Paul ... 02678
Stiftsmuseum Steyr-Gleink, Steyr-Gleink ... 02775
Votivsammlung der Wallfahrtskirche Maria Kirchental, Sankt Martin bei Lofer ... 02669
Wallfahrtsbasilika Mariazell, Mariazell ... 02359
Wallfahrtsmuseum, Maria Taferl ... 02355

Belarus

Muzej Staražytna Belaruskaj Kultury, Minsk ... 03254

Belgium

Béguinage d'Anderlecht, Bruxelles ... 03436
Heemmuseum Bystervelt, Schelle ... 04002
Keizerskapel, Antwerpen ... 03289
Kunsthal Sint-Pietersabdij, Gent ... 03632
Musée d'Art Religieux et d'Art Mosan, Liège ... 03806
Musée Diocésain et Trésor de la Cathédrale Saint-Aubain, Namur ... 03903
Musée Marial de Beauraing, Beauraing ... 03350
Trésor de la Collegiale Sainte-Waudru, Mons ... 03889
Trésor de l'Eglise Primaire Saint-Sebastien, Stavelot ... 04051

Bolivia

Museo de Arte Sacro San Miguel, Oruro ... 04231
Museo de la Recoleta, Sucre ... 04248

Bosnia and Herzegovina

Muzej Stare Srpske-Pravoslavne Crkve, Sarajevo 04281
Muzejske Zbirke Fojničkog Samostana, Fojnica 04263

Brazil

Museu da Curia Metropolitana, Porto Alegre 04642
Museu da Imagem, Campinas 04387
Museu da Imagem e do Som de Alagoas, Maceió 04545
Museu da Imagem e do Som de Campinas, Campinas 04388
Museu da Imagem e do Som de Campos, Campos dos Goitacazes 04399
Museu da Imagem e do Som de Cascavel, Cascavel 04407
Museu da Imagem e do Som de Curitiba, Curitiba 04431
Museu da Imagem e do Som de e da Imprensa da Bahia, Salvador 04792
Museu da Imagem e do Som de Goiás, Goiás 04478
Museu da Imagem e do Som de Piracicaba, São Paulo 04871
Museu da Imagem e do Som de Resende, Resende 04673
Museu da Imagem e do Som de Santa Catarina, Florianópolis 04453
Museu da Imagem e do Som de São Paulo, São Paulo 04872
Museu da Imagem e do Som de Taubaté, Taubaté 04923
Museu da Imagem e do Som do Ceará, Fortaleza 04462
Museu da Imagem e do Som do Pará, Belém 04333
Museu da Imagem e do Som do Paraná, Curitiba 04432
Museu da Venrável Ordem Terceira de São Francisco da Penitência, Rio de Janeiro 04721
Museu das Alfaias, Cachoeira 04376
Museu de Arte Religiosa, Vitória 04957
Museu de Arte Sacra, Rio de Janeiro 04723
Museu de Arte Sacra, Salvador 04796
Museu de Arte Sacra, São Paulo 04883
Museu de Arte Sacra da Catedral-Basílica, Salvador 04797
Museu de Arte Sacra de Santos, Santos 04820
Museu de Arte Sacra Padre Roberto Barbalho, Recife 04664
Museu de Artes e Ofícios, Belo Horizonte 04347
Museu de Imagem e do Som de Pernambuco, Recife 04666
Museu de Nossa Senhora da Glória do Outeiro, Rio de Janeiro 04729
Museu Dom Vital, Recife 04671
Museu Henriqueta Catharino, Salvador 04804
Museu Municipal de Arte e História, Nova Era 04592

Bulgaria

Muzej-kripta Aleksandăr Nevski, Sofia 05184
Muzej na Văzraždaneto, Varna 05241
Nacionalen Muzej Bojanska Crkva, Sofia 05193

Cambodia

Battambang Museum, Battambang 05283

Cameroon

Petit Musée d'Art Camerounais, Yaoundé 05303

Canada

Augustines de la Miséricorde de Jésus de Roberval, Roberval 06763
Chapelle des Indiens, Tadoussac 07056
Mennonite Heritage Museum, Rosthern 06778
Musée de l'Oratoire Saint-Joseph, Montréal 06356
Musée de Sainte-Anne-de-Beaupré, Sainte-Anne-de-Beaupré 06888
Musée des Augustines de l'Hôpital Général de Québec, Québec 06689
Musée des Maîtres et Artisans, Montréal 06360
Musée Marguerite Bourgeoys, Montréal 06374
Musée Pierre Boucher, Trois-Rivières 07177
Museum of Religious Art, Summerside 07034
Old Saint Edward's Anglican Church, Clementsport 05622
Saint John's College Museum, Winnipeg 07379
Saint Volodymyr Ukrainian Catholic Centre Museum, Winnipeg 07383

Chile

Museo de Arte Colonial de San Francisco, Santiago de Chile 07452

China

Capital Museum, Beijing 07524
Museum of Sacred Art, Macau 08002
Museum of the Holy House of Mercy, Macau 08004
Poly Art Museum, Beijing 07562
Saint Dominic Church Museum, Macau 08006

Colombia

Biblioteca Luis-Angel Arango, Bogotá 08543
Museo Arquidiocesano de Arte Religioso, Pamplona 08704
Museo Arquidiocesano de Arte Religioso de Popayán, Popayán 08723
Museo Convento del Desierto de la Candelaria, Ráquira 08736
Museo Convento Santo Ecce Homo, Sutamarchán 08766
Museo de Arte Religioso Cardenal Luque, Tenjo 08768
Museo de Arte Religioso de Girón, Girón 08647
Museo de Arte Religioso de la Catedral de San Nicolás, Rionegro 08738
Museo de Arte Religioso de Monguí, Monguí 08690
Museo de Arte Religioso Julio Franco Arango, Duitama 08633
Museo de Arte Religioso Tiberio de J. Salazar y Herrera, Sonsón 08762
Museo de Arte Religioso y Colonial, Cali 08609
Museo de la Parroquia de Gámbita, Gámbita 08645
Museo de Santa Clara, Bogotá 08573
Museo del Carmen, Villa de Leyva 08783
Museo del Milagroso, Buga 08603
Museo Iberoamericano de Arte Moderno de Popayán, Popayán 08729
Museo Juan Lorenzo Lucero, Pasto 08710
Museo Medardo Burbano, Pasto 08714
Museo Nacional Mariano, Chiquinquira 08625
Museo Parroquial de Arte Religioso, Turmequé 08778
Museo Raíces, Jesús María 08664
Museo Religioso de Guatavita, Guatavita 08655
Museo Religioso y Colonial de San Francisco, Cali 08614
Museo Santa Clara la Real, Tunja 08777
Museo-Santuario de San Pedro Claver, Cartagena 08620
Sala de Memorias Culturales, Mogotes 08688

Croatia

Biskupski Muzej i Katedralna Zbirka, Hvar 08879
Dijecezanski Muzej Zagrebačke Nadbiskupije, Zagreb 08988
Galerija Zavjetnih Slika Brodova, Dubrovnik 08864
Katedralna Riznica, Split 08959
Museum Collection and Old Pharmacy of the Friars Minor, Dubrovnik 08866
Muzej Dominikanskog Samostana, Bol 08847
Muzej Dominikanskog Samostana, Dubrovnik 08867
Muzej Franjevačkog Samostana, Dubrovnik 08868
Muzej Franjevačkog Samostana, Rovinj 08942
Muzej Samostana Sestara od Sigurate, Dubrovnik 08869
Muzejska Zbirka Franjevačkog Samostana, Imotski 08883
Opatska Riznica Sv. Marka, Korčula 08897
Riznica Crkvene Umjetnosti - Zlato i Srebro Nina, Nin 08915
Riznica i Galerija Svetista Gospe Trsatske, Rijeka 08941
Riznica Samostana Svetog Frane, Zadar 08985
Riznica Zagrebačke Katedrala, Zagreb 09010
Sakralna Baština Muzej, Senj 08946
Zbirka Crkvene Umjetnosti, Šibenik 08948
Zbirka Franjevacki Samostan, Orebić 08921
Zbirka Franjevačkog Samostana Visovac, Visovac 08979
Zbirka Klašter Sv. Nikolaus, Korčula 08898
Zbirka Sakralne Umjetnini Župe Sveti Blaža, Vodnjan 08980
Zbirka Umjetnina Crkve Sv. Domenika, Trogir 08972
Zbirka Umjetnina Franjevačkog Samostana, Hvar 08881
Zbirka Umjetnina Kairos, Trogir 08973

Cuba

Museo de Arte Religioso, La Habana 09133

Czech Republic

Farní muzeum Zábřeh, Zábřeh 09923
Kaple Sv. Kříže, Praha 09754
Muzeum Betlémů, Karlštejn 09578
Třebechovické Muzeum Betlémů, Třebechovice pod Orebem 09873

Denmark

Davids Samling, København 10130
Møns Museum, Stege 10284
Museet for Religiøs Kunst, Lemvig 10177
Roskilde Domkirkemuseum, Roskilde 10246

Ecuador

Museo de Arte Religioso Concepción, Riobamba 10454
Museo de Arte Religioso la Concepción, Baños 10374
Muséo de Las Conceptas, Cuenca 10389
Museo Diocesano de Arte Religioso Nuestra Señora del Cisne, Loja 10418
Museo Manuela Sáenz, Quito 10447
Museo Mariano Acosta, Ibarra 10415
Museo Santuario Nuestra Señora Natividad del Guayco, La Magdalena 10419

Ethiopia

Museum of the Holy Trinity Church of Ethiopia, Addis Ababa 10679

Finland

Kivisakasti, Mikkeli 11114
Lähetysmuseo, Helsinki 10797
Rautalammin Museo, Rautalampi 11263

France

Collection d'Art Religieux Ancien, Prunay-en-Yvelines 14953
Dépôt-Musée Diocésain, Metz 14100
Exposition Historique du Monastère, Sauxillanges 15673
Galeries du Panthéon Bouddhique du Japon et de la Chine, Paris 14609
Maison Natale de Saint-Vincent-de-Paul, Saint-Vincent-de-Paul 15537
Musée d'Allauch, Allauch 11555
Musée d'Art Religieux, Blois 12033
Musée d'Art Religieux, Fréland 13005
Musée d'Art Sacré, Dijon 12737
Musée d'Art Sacré de Fourvière, Lyon 13884
Musée d'Art Sacré du Gard, Pont-Saint-Esprit 14893
Musée de l'Abbaye de Port-Royal, Magny-lès-Hameaux 13914
Musée de l'Abbaye Saint-Vigor, Cérisy-la-Forêt 12344
Musée de l'Œuvre Notre-Dame, Strasbourg 15824
Musée Départemental d'Art Religieux, Sées 15697
Musée Départemental d'Art Sacré Contemporain, Saint-Pierre-de-Chartreuse 15457
Musée des Santons des Baux, Les Baux-de-Provence 13690
Musée du Santon, Generagues 13041
Musée du Santon, Le Val 13662
Musée du Santon, Marseille 14006
Musée du Santon et des Traditions de Provence, Fontaine-de-Vaucluse 12960
Musée du Vieux Marseille, Marseille 14008
Musée François Pompon, Saulieu 15656
Musée Lambinet, Versailles 16128
Musée Pierre de Luxembourg, Villeneuve-lès-Avignon 16214
Musée Saint-Jacques, Moissac 14141
Musée Saint-Nicolas, Vitré 16257

Georgia

Ethnographic Museum, Ushguli 16388

Germany

Bischöfliches Diözesanmuseum für christliche Kunst, Münster 20506
Bischöfliches Dom- und Diözesanmuseum Mainz, Mainz 20141
Buxtehude Museum für Regionalgeschichte und Kunst, Buxtehude 17673
Cryptoneum, Rostock 21387
Diözesanmuseum, Bamberg 16906
Diözesanmuseum Freising, Freising 18450
Diözesanmuseum Sankt Afra, Augsburg 16615
Diözesanmuseum Sankt Ulrich, Regensburg 21216
Dom-Museum, Fulda 18506
Dom-Museum Bremen, Bremen 17548
Dom-Museum Hildesheim, Hildesheim 19182
Dom St. Peter, Fritzlar 18483
Dom und Domschatz Halberstadt, Halberstadt 18875
Domkammer der Kathedralkirche Sankt Paulus zu Münster, Münster 20507
Dommuseum, Brandenburg an der Havel 17507
Dommuseum, Frankfurt am Main 18364
Domschatz, Würzburg 22709
Domschatz der Stiftskirche St.Servatii, Quedlinburg 21145
Domschatz- und Diözesanmuseum, Passau 20988
Domschatz- und Diözesanmuseum Eichstätt, Eichstätt 18081
Domschatzkammer, Aachen 16396
Domschatzkammer Sankt Petri, Bautzen 16939
Domschatzkammer und Diözesanmuseum, Osnabrück 20919
Domvorhalle, Goslar 18707
Erzbischöfliches Diözesanmuseum und Domschatzkammer, Paderborn 20968
Europäische Glasmalerei- und Krippenausstellung, Schramberg 21631
Fränkische Galerie, Kronach 19720
Haus Papst Benedikt XVI. - neue Schatzkammer und Wallfahrtsmuseum, Altötting 16518
Helmstedter Freundeskreis für Paramentik und christliche Kunst, Helmstedt 19121
Henneberg-Museum, Münnerstadt 20498
Kloster Ebstorf, Ebstorf 18050
Kloster Isenhagen, Hankensbüttel 18986
Kloster Lüne mit Textilmuseum, Lüneburg 20098
Kloster Mariensee, Neustadt am Rübenberge 20664
Kloster Medingen, Bad Bevensen 16695
Kloster Sankt Marienberg, Helmstedt 19122
Kloster Walsrode, Walsrode 22317
Kloster Wienhausen, Wienhausen 22542
Klostermuseum, Roggenburg 21362
Klostermuseum der Benediktiner-Abtei, Ottobeuren 20955
Klostermuseum der Sankt Josefskongregation, Ursberg 22188
Klostermuseum in der Pfarrkirche Sankt Magnus, Bad Schussenried 16830
Kunstkammer der Pfarrkirche Sankt Georg, Bocholt 17394
Kunstsammlung des Herzoglichen Georgianums, München 20455
Kunstsammlung Lorenzkapelle Rottweil, Rottweil 21425
Laufener Stiftsschatz, Laufen 19854
Museum Abtei Liesborn, Wadersloh 22263
Museum am Dom, Würzburg 22717
Museum der Cistercienserinnen-Abtei Lichtenthal, Baden-Baden 16885
Museum für bäuerliche und sakrale Kunst, Ruhpolding 21448
Museum für Niederrheinische Sakralkunst, Kempen 19508
Museum für Sakrale Kunst und Liturgie, Heidelberg 19077
Museum Haus Löwenberg, Gengenbach 18583
Museum in der Bruderkirche, Altheim 16506
Museum Kirche zum Heiligen Kreuz, Zittau 22813
Pergamonmuseum, Berlin 17234
Pfarrmuseum, Dießen am Ammersee 17852
Pfarrmuseum, Flintsbach am Inn 18330
Schatzkammer der Basilika St. Ludgerus, Essen 18268
Schatzkammer der Evangelisch-Lutherischen Sankt Andreaskirche, Weißenburg in Bayern 22453
Schatzkammer des Münsters, Reichenau 21245
Schatzkammer Kloster Sankt Marienstern, Panschwitz-Kuckau 20978
Schloss Johannisburg mit Schlossgarten und Pompejanum, Aschaffenburg 16590
Schloß Lichtenstein, Lichtenstein 19982
Schloss Neuburg, Neuburg an der Donau 20599
Schlossmuseum, Ellwangen 18124
St. Johann Baptist, Aachen 16407
Stadtmuseum Aichach, Aichach 16445
Städtisches Kramer-Museum, Kempen 19509
Staurothek, Domschatz und Diözesanmuseum Limburg, Limburg an der Lahn 19998
Stift Fischbeck, Hessisch Oldendorf 19171
Stiftskirchenmuseum, Himmelkron 19192
Stiftsmuseum, Bad Buchau 16707
Turmmuseum Sankt Magnus, Esens 18243
Vom Kloster zum Dorf, Creglingen 17744
Wallfahrtsmuseum, Neukirchen beim Heiligen Blut 20624
Wallfahrtsmuseum, Bruckmühl 17601
Wallfahrtsmuseum Maria Steinbach, Legau 19872

Greece

Agios Andreas Convent, Kammeni Chora 23032
Agios Stephanos Basilica, Kastoria 23042
Arkadi Monastery Collection, Rethymnon 23166
Metamorphosis Church Museum, Meteora 23092
Museum of Ecclesiastical Art of Serrai and Nigrita, Serres 23180
Néa Moni Collection, Chios 22971
Panachrantos Monastery, Andros 22884
Proussos Monastery Museum, Prousos 23159

Honduras

Museo de Arte Religioso Colonial, Comayagua 23326

Hungary

Budavári Mátyás-templom Egyházművészeti Gyűjteménye, Budapest 23393
Dunamelléki Református Egyházkerület Ráday, Budapest 23396
Dunamelléki Református Egyházkerület Ráday Múzeuma, Kecskemét 23645
Dunántúli Református Egyházkerület Tudományos Gyűjteményei Múzeum, Pápa 23751
Egyházmegyei Múzeum és Kincstár, Szeged 23842
Evangélikus Országos Múzeum, Budapest 23401
Főszékesegyházi Kincstár és Könyvtár, Esztergom 23552

R

Görögkatolikus Egyházművészeti Gyűjtemény, Nyíregyháza 23732
Középkori O-Zsinagóga, Sopron 23811
Magyar Ortodox Egyházművészeti Múzeum, Miskolc 23709
Ortodox Egyházművészeti Gyűjtemény, Kecskemét 23653
Pannonhalmi Főapátság Gyűjteménye, Pannonhalma 23750
Református Kollégiumi és Egyházművészeti Múzeum, Debrecen 23520
Római Katolikus Egyházi Gyűjtemény, Sárospatak 23793
Sárospataki Református Kollégium Tudomanyos Gyűjteményei Múzeum, Sárospatak 23795
Soproni Római Katolikus Egyházművészeti Gyűjtemény, Sopron 23821
Székesfehérvári Egyházmegyei Múzeum, Székesfehérvár 23859
Szerb Egyházy Múzeum, Szentendre 23882
Szuz Mária Kegy- és Emléktárgyak Gyűjteménye, Gyula 23600

India

Hindi Sangrahalaya, Allahabad 24061
Phulsanda Religious Arts Museum, Phulsanda24348

Iran

Holy Defense Museum, Kerman 24617

Ireland

Ecclesiastical Museum, Mullingar 24880
Saint Mel's Diocesan Museum, Longford 24870

Israel

Crusader Church, Abu Ghosh 24918
Israel Bible Museum, Zefat 25133
Museum of Psalms, Jerusalem 25002
Wolfson Museum, Jerusalem 25025

Italy

Basilica Collegiata di Santa Maria Assunta, San Gimignano 27635
Capella Brancacci, Firenze 26142
Complesso Museale di Santa Chiara, Napoli 26879
Galleria d'Arte Sacra dei Contemporanei, Milano 26687
Galleria degli ex-voto del Santuario di Montenero, Livorno 26496
Galleria dello Spedale degli Innocenti, Firenze26160
Musei Sistini del Piceno, San Benedetto del Tronto 27608
Museo Alessi, Enna 26055
Museo Alfonsiano, Pagani 27055
Museo Archeologico Attrezzato e Cattedrale di San Zeno, Pistoia 27247
Museo Aurelio Marena, Bitonto 25429
Museo Badia di San Gemolo, Valganna 28141
Museo Beato Angelico, Vicchio 28249
Museo Canonicale, Verona 28231
Museo Capitolare, Assisi 25306
Museo Christiano, Cividale del Friuli 25917
Museo d'Arte Sacra, Camaiore 25613
Museo d'Arte Sacra, Ponzone 27294
Museo d'Arte Sacra, San Gimignano 27638
Museo d'Arte Sacra, Asciano 25291
Museo d'Arte Sacra Cardinal Agnifili, Rocca di Mezzo 27428
Museo d'Arte Sacra della Insigne Collegiata di San Lorenzo, Montevarchi 26856
Museo d'Arte Sacra della Marsica, Celano 25797
Museo d'Arte Sacra della Santa Maria Assunta, Gallarate 26265
Museo d'Arte Sacra della Val d'Arbia, Buonconvento 25564
Museo d'Arte Sacra San Marco, Mombaroccio 26768
Museo d'Arte Sacra Silvestro Frangipane, Zagarise 28314
Museo d'Arte Sacro San Martino, Alzano Lombardo 25213
Museo degli Arazzi delle Madrice, Marsala 26612
Museo degli Arredi Sacri, Castel Sant'Elia 25719
Museo del Capitolo di San Lorenzo, Perugia 27148
Museo del Duomo, Massa 26620
Museo del Duomo, Ravello 27369
Museo del Int. Presepio in Miniatura Guido Colitti, Campobasso 25635
Museo del Merletto a Tombolo, Pescocostanzo27176
Museo del Santuario di Santa Maria Santissima della Fontenuova, Monsummano Terme 26789
Museo della Basilica, Gandino 26270
Museo della Basilica di San Michele Arcangelo, Monte Sant'Angelo 26807
Museo della Basilica di San Nicola, Bari 25363
Museo della Basilica di Santa Maria della Passione, Milano 26701
Museo della Basilica Patriarcale di Santa Maria degli Angeli, Porziuncola 27315
Museo della Basilica Patriarcale Santa Maria degli Angeli, Assisi 25308
Museo della Cattedrale, Barletta 25373
Museo della Cattedrale, Chiusi Città 25883
Museo della Certosa, Certosa di Pavia 25816
Museo della Pieve di San Pietro, Prato 27336
Museo di Arte Religiosa, Oleggio 26979
Museo di Arte Sacra, Castrovillari 25769
Museo di Arte Sacra, Vibo Valentia 28247
Museo di Arte Sacra San Giuseppe, Rogliano 27437
Museo di San Domenico, Bologna 25464
Museo di San Petronio, Bologna 25465
Museo di Santa Maria di Castello, Genova 26308
Museo di Sant'Antonino, Piacenza 27188
Museo Diocesano, Agrigento 25163
Museo Diocesano, Albenga 25177
Museo Diocesano, Andria 25234
Museo Diocesano, Ascoli Piceno 25296
Museo Diocesano, Bari 25367
Museo Diocesano, Castelvecchio Subequo 25755
Museo Diocesano, Gubbio 26389
Museo Diocesano, Imola 26401
Museo Diocesano, Lanciano 26448
Museo Diocesano, Mantova 26597
Museo Diocesano, Mazara del Vallo 26644
Museo Diocesano, Monopoli 26780
Museo Diocesano, Osimo 27012
Museo Diocesano, Piazza Armerina 27194
Museo Diocesano, Pistoia 27252
Museo Diocesano, San Miniato 27673
Museo Diocesano, Trani 28049
Museo Diocesano, Velletri 28172
Museo Diocesano A. Bernareggi, Bergamo 25403
Museo Diocesano Antonio Bergamaschi, Pennabilli 27133
Museo Diocesano d'Arte Sacra, Brescia 25539
Museo Diocesano d'Arte Sacra, Nicotera 26926
Museo Diocesano d'Arte Sacra, Orte 27000
Museo Diocesano d'Arte Sacra, Rossano 27565
Museo Diocesano di Arte Sacra, Lodi 26510
Museo Diocesano di Arte Sacra, Palermo 27075
Museo Diocesano di Arte Sacra, Volterra 28310
Museo Diocesano di Arte Sacra del Duomo, Arezzo 25269
Museo Diocesano di Arte Sacra San Gregorio Barbarigo, Padova 27044
Museo Diocesano e Gallerie del Tiepolo, Udine 28113
Museo Diocesano Intercomunale di Arte Sacra, Castignano 25764
Museo Diocesano Intercomunale di Arte Sacra, Comunanza 25959
Museo Diocesano Intercomunale di Arte Sacra, Grottammare 26376
Museo Diocesano Intercomunale di Arte Sacra, Monteprandone 26843
Museo Diocesano Intercomunale di Arte Sacra, Ripatransone 27417
Museo Diocesano Intercomunale di Arte Sacra, Rotella 27566
Museo Diocesano Teatino, Chieti 25875
Museo Diocesano Tridentino, Trento 28059
Museo e Chiostri Monumentali di Santa Maria Novella, Firenze 26188
Museo e Pinacoteca Diocesana, Camerino 25621
Museo Ecclesiastico, Caorle 25651
Museo Francescano, Roma 27507
Museo Francescano - Raccolta d'Arte, Gubbio 26391
Museo Orientale, Tagliacozzo 27936
Museo Parrocchiale della Chiesa Collegiata, Mercatello sul Metauro 26663
Museo Parrocchiale di Arte Sacra, Ponte in Valtellina 27284
Museo Parrocchiale e Museo P. Mascagni, Bagnara di Romagna 25344
Museo Pinacoteca della Santa Casa, Loreto 26519
Museo Pio IX, Senigallia 27804
Museo Santa Cabrini, Sant'Angelo Lodigiano 27722
Museo Stauros di Arte Sacra Contemporanea, Isola del Gran Sasso 26416
Museo Tesoro Basilica di San Francesco e Collezione F. M. Perkins, Assisi 25310
Museo Tipologico del Presepe, Macerata 26555
Palazzo Foro Boario, Modena 26756
Pinacoteca Comunale, Castiglion Fiorentino 25758
Pinacoteca Diocesana di Arte Sacra, Senigallia 27805
Raccolta d'Arte, Città della Pieve 25903
Raccolta d'Arte, Roccalbegna 27432
Raccolta dell'Opera del Duomo, Oristano 26997

Japan

Atsuta-Jingu Museum, Nagoya 28994
Chikurin-ji Homotsukan, Kochi 28740
Choraku Homotsukan, Shimoda 29263
Chusonji Sankozo - Fujiwara Kinenkan, Hiraizumi 28520
Daigoji Reihokan, Kyoto 28805
Daihozoden Homotsukan, Ikoma 28564
Dazaifu Tenman-gu Hômotsuden, Dazaifu 28407
Hakone-Jinja Homotsuden, Hakone 28478
Hamana-ko Tateyama-ji Bijutsukan, Hamamatsu 28488
Hosyokan Byodoin Museum, Uji, Kyoto 29558
Itsukushima-jinja Homotsukan, Hatsukaichi 28499
Izumo Taisha Homutsuden, Izumo 28617
Jotenkaku Hakubutsukan, Kyoto 28817
Kakurin-ji Treasure House, Katsuura 28680
Kamakura Kokuhokan, Kamakura 28641
Kamakuragu Homotsukan, Kamakura 28643
Kashima-jingu Homotsukan, Kashima, Ibaraki 28678
Kasuga Taisha Homostuden, Nara 29030
Katori-jingu Homotsukan, Sawara 29236
Kitano Temman-gu Treasure House, Kyoto 28823
Kofukuji Kokuhokan, Nara 29031
Koryu-ji Reihoden, Kyoto 28830
Koyasan Reihokan, Koya 28773
Kyoto Kokuritsu Hakubutsukan, Kyoto 28842
Mishima Taisha Treasure House, Mishima 28913
Munakata Taisha Shinpōkan, Genkai 28456
Naritasan Reikokan and Reikoden Museum, Narita 29041
Natadera Fumon-kaku Homotsukan, Komatsu28766
Nikko Futarasan-jinja Hakubutsukan, Nikko 29068
Nikko Toshogu Treasure Museum, Nikko 29069
Nyoirin-ji Homotsukan, Yoshino 29628
Oyamazumi-jinja Kokuhokan, Omishima 29116
Rinno-ji Jokodo Homotsukan, Nikko 29070
Seiryuden Bijutsukan, Matsushima 28903
Shido-ji Homotsukan, Shido 29257
Shingenkou Treasure Hall Historical Museum, Enzan 28417
Shiogama-jingu Homotsukan, Shiogama 29274
Shosoin Treasure Repository, Nara 29038
To-ji Homotsukan, Kyoto 28870
Yamaguchi Hoshun Memorial Hall, Hayama 28501
Yamauchi Collection of Art Museum, Kochi 28749

Korea, Republic

Moka Buddhist Museum, Gangcheon 29804
Museum of Korean Buddhist Art, Seoul 29905

Laos

Haw Pha Kaeo, Vientiane 29973
Wat Si Saket, Vientiane 29979

Lebanon

Qozhaya Museum, Zgharta 30177

Lithuania

Baznytinio Meno Skyrius, Kaunas 30241
Birštono Sakralinis Muziejus, Birštonas 30215
Kauno Arkivyskupijos Muziejus, Kaunas 30248
Žemaičių Vyskupystės Muziejus, Varniai 30391

Malta

Saint John's Co-Cathedral and Museum, Valletta 30647
Zabbar Sanctuary Museum, Zabbar 30662

Mexico

Museo de Arte Religioso, Huaquechula 30941
Museo de Arte Religioso de Santa Mónica, Puebla 31275
Museo de Arte Religioso Ex Convento Franciscano, Huejotzingo 30944
Museo de Arte Sacro, Chihuahua 30768
Museo Ex Convento Agustino de San Pablo, Yuriria 31573
Museo Virreinal de Zinacantepec, San Miguel Zinacantepec 31360

Morocco

Nejarine-Musée, Fès 31672

Myanmar

The Buddhist Art Museum, Yangon 31733
Museum of the Shwenawdaw Pagoda, Bago 31716

Nepal

Swayambunath Museum and Art Gallery, Kathmandu 31779

Netherlands

Breda's Museum, Breda 32003
Centrum voor Heiligenbeelden, Kranenburg 32459
Gaudete, Museum voor Volksdevotie, Boxtel 31995
Gemeentemuseum Weert - Jacob van Horne, Weert 32933
Kerkmuseum Janum, Janum 32430
Museum in de Stiftskerk, Thorn 32822
Museum Kerkelijke Kunst, Workum 32965
Museum Ons' Lieve Heer op Solder, Amsterdam 31864
Museum Sacrum, Tiel 32824
Museum Van Gerwen-Lemmens, Valkenswaard 32869
Museum voor Religieuze Kunst, Uden 32833
Schatkamer Sint Walburgis Basiliek, Arnhem 31932
Schatkamer van de Basiliek van Onze Lieve Vrouwe, Maastricht 32531
Schatkamer van de Sint-Servaasbasiliek, Maastricht 32532
Sint Jansmuseum De Bouwloods, 's-Hertogenbosch 32378
Sint-Plechelmusbasiliek met Schatkamer, Oldenzaal 32617
Stichting Santjes en Kantjes, Maastricht 32534
't Stift, Susteren 32812
Wilhelmietenmuseum, Huijbergen 32419
Zwanenbroedershuis, 's-Hertogenbosch 32380

Nicaragua

Centro de Arte, León 33282
Museo de Arte Sacro, León 33285

Norway

Heddal Bygdetun, Notodden 33705
Stiklestad Nasjonal Kultursenter, Verdal 33973

Palestine

Holyland Arts Museum, Bethlehem 34050

Panama

Museo de Arte Religioso Colonial, Ciudad de Panamá 34068

Paraguay

Museo Iconográfico, Itaugua 34103
Museo Monseñor Agustín Blujaki, Asunción 34099
Museo Monseñor Juan Sinforiano Bogarin, Asunción 34100

Peru

Museo de Arte Religioso Convento de San Francisco, Cajamarca 34128
Museo de Arte Religioso de la Catedral, Lima 34235
Museo de Arte Religioso de Piura, Piura 34292
Museo de Arte Virreinal de Santa Teresa, Arequipa 34110
Museo del Convento de La Merced, Mantas 34277
Museo del Convento de los Descalzos, Lima 34253
Museo del Convento de San Francisco de Asís, Lima 34254
Museo del Convento de Santa Catalina, Cusco 34162
Museo Inka, Cusco 34167
Museo Lítico de Pucara, Pucara 34296
Palacio Arzobispal, Cusco 34168
Pinacoteca Carmelita del Convento del Carmen, Trujillo 34323

Philippines

Barasoain Church Museum, Malolos 34420
San Agustin Museum Intramuros, Manila 34457
University of San Carlos Anthropology Museum, Cebu 34377
Xavier University Museum (Museo de Oro), Cagayan de Oro City 34358

Poland

Galicja Muzeum, Kraków 34781
Muzeum Archidiecezjalne, Katowice 34724
Muzeum Archidiecezjalne, Poznań 35031
Muzeum Archidiecezjalne, Przemyśl 35054
Muzeum Archidiecezjalne, Wrocław 35327
Muzeum Archidiecezjalne Lubelskie, Lublin 34901
Muzeum Archidiecezji Gnieźnieńskiej, Gniezno 34672
Muzeum Diecezjalne, Tarnów 35176
Muzeum Diecezjalne Sztuki Kościelnej, Sandomierz 35096
Muzeum Diecezji Pelplińskiej, Pelplin 35005
Muzeum Katedralne Jana Pawła II, Kraków 34802
Muzeum Klasztorne na Jasnej Górze, Częstochowa 34617
Muzeum Klasztorne OO Cystersów, Szczyrzyc 35167
Muzeum Kolekcji im. Jana Pawła II, Warszawa 35253
Muzeum Parafialne, Dobra 34623
Muzeum Sztuki Cmentarnej, Wrocław 35338
Muzeum Sztuki Sakralnej, Bardo 34536
Muzeum Sztuki Sakralnej, Żnin 35377
Parafia Ewangelicko - Augsburska Wang, Karpacz 34717

Portugal

Museu Agrícola e Museu de Arte Sacra, Ribeira Chã 35887
Museu Antonino, Faro 35595
Museu da Igreja Matriz, Gondomar 35630

Museu de Arte Sacra, Aljezur 35408
Museu de Arte Sacra, Amarante 35432
Museu de Arte Sacra, Castelo Branco 35514
Museu de Arte Sacra, Esposende 35568
Museu de Arte Sacra, Évora 35585
Museu de Arte Sacra, Funchal 35613
Museu de Arte Sacra, Gouveia 35632
Museu de Arte Sacra, Mértola 35761
Museu de Arte Sacra, Pedrógão Grande 35813
Museu de Arte Sacra, Vila do Conde 35990
Museu de Arte Sacra, Vila Viçosa 36019
Museu de Arte Sacra da Misericórdia do Fundão, Fundão 35621
Museu de Arte Sacra da Sé de Portalegre, Portalegre 35833
Museu de Arte Sacra da Universidade, Coimbra 35532
Museu de Arte Sacra da Universidade Lusíada, Vila Nova de Famalicão 36007
Museu de Arte Sacra de Alcoutim, Alcoutim 35401
Museu de Arte Sacra de Arouca, Arouca 35445
Museu de Arte Sacra de Monsaraz, Monsaraz 35771
Museu de Arte Sacra e Arqueologia, Porto 35846
Museu de Arte Sacra e Arqueologia de Vila Alva, Vila Alva 35985
Museu de Arte Sacra e Etnografia Religiosa, Horta 35643
Museu de Arte Sacra e Etnologia de Fátima, Fátima 35602
Museu de Aveiro, Aveiro 35447
Museu de Escultura Comparada, Mafra 35750
Museu de São Roque, Lisboa 35694
Museu Etnográfico e Arte Sacra de São Torcato, São Torcato 35926
Museu Municipal de Alcochete-Núcleo de Arte Sacra, Alcochete 35397
Museu Nacional de Machado de Castro, Coimbra 35542
Museu Paroquial de Óbidos, Óbidos 35785
Museu S. Pedro da Palhaça, Palhaça 35804
Núcleo de Arte Sacra, Évora 35589
Núcleo Museológico da Misericórdia de Aveiro, Aveiro 35448
Núcleo Museológico de Arte Sacra, Nisa 35782
Núcleo Museológico do Santuário do Senhor Jesus da Pedra, Óbidos 35786
Tesouro da Sé Patriarcal de Lisboa, Lisboa 35735
Tesouro-Museu da Catedral, Viseu 36025
Tesouro-Museu da Sé de Braga, Braga 35485

Romania

Aşezământul de Artă şi Cultură Religioasă, Maia 36455
Colcţia Muzeală a Mănăstirii Lainici, Bumbeşti-Jiu 36235
Colecţia de Artă Veche Bisericească, Reşiţa 36561
Colecţia de Carte Veche şi Icoane Ştefan Crişan, Lipova 36442
Colecţia de Obiecte Bisericeşti, Agnita 36100
Colecţia Mănăstirii Zamfira, Lipăneşti 36441
Colecţia Mitropolit Antonie Plămădeală, Sâmbăta de Sus 36581
Colecţia Muzeală a Arhiepiscopiei Ortodoxe Cluj, Cluj-Napoca 36286
Colecţia Muzeală a Arhiepiscopiei Ortodoxe Galaţi, Galaţi 36377
Colecţia Muzeală a Episcopiei Armeano-Gregoriene, Bucureşti 36194
Colecţia Muzeală a Episcopiei Ortodoxe, Alba Iulia 36104
Colecţia Muzeală a Episcopiei Ortodoxe, Buzău 36243
Colecţia Muzeală a Episcopiei Ortodoxe, Oradea 36501
Colecţia Muzeală a Mănăstirii Gura Motrului, Butoieşti 36241
Colecţia Muzeală a Mănăstirii Moldoviţa, Vatra Moldoviţei 36729
Colecţia Muzeală a Mănăstirii Nicula, Fizeşu Gherlii 36370
Colecţia Muzeală a Mănăstirii Ortodoxe Dealu, Aninoasa 36108
Colecţia Muzeală a Mănăstirii Ortodoxe Ghighiu, Bărcăneşti 36141
Colecţia Muzeală a Mănăstirii Ortodoxe Sinaia, Sinaia 36619
Colecţia Muzeală a Mănăstirii Ortodoxe Suzana, Măneciu 36462
Colecţia Muzeală a Mănăstirii Ortodoxe Ţigăneşti, Ciolpani 36279
Colecţia Muzeală a Mănăstirii Ortodoxe Viforâta, Viforâta 36730
Colecţia Muzeală a Mănăstirii Rohia, Târgu Lăpuş 36668
Colecţia Muzeală a Mănăstirii Strehaia, Strehaia 36638
Colecţia Muzeală a Mitropoliei Ortodoxe a Olteniei, Craiova 36323
Colecţia Muzeală de Artă Bisericească Veche a Mitropoliei Banatului, Timişoara 36688
Colecţia Muzeală Geamia Medgidia, Medgidia 36467
Colecţias Muzeală a Mănăstirii Ortodoxe Căldăruşani, Moara Vlăsiei 36478
Complexul Muzeal Stelea, Târgovişte 36655
Geamia Esman Sultan, Mangalia 36464
Muzeul de Artă Religioasă al Mănăstirii Sfinţii Petru şi Pavel, Huşi 36407
Muzeul de Artă Religioasă Nicolae Iorga, Vălenii de Munte 36720
Muzeul de Cultură şi Artă Relgioasă, Borzeşti 36158
Muzeul de Icoane şi Obiecte Bisericesti, Sibiu 36608
Muzeul Episcopiei Romano-Catolice, Oradea 36502
Muzeul Mitropolit Antim Ivireanul, Bucureşti 36212
Muzeul Parohial de Icoane pe Sticlă, Sibiel 36603

Russia

Archangelskij Sobor, Moskva 37141
Blagoveščenskij Sobor, Moskva 37142
Centralnyj Muzej Drevnerusskoj Kultury i Isskustva im. Andreja Rubleva, Moskva 37146
Cerkov Pokrova v Filjach, Moskva 37154
Cerkov Položenija riz Presvjatoj Bogorodicy, Moskva 37155
Gosudarstvennaja Tretjakovskaja Galerja, Moskva 37184
Isaakevskij Sobor - Memorialnyj Muzej-pamjatnik, Sankt-Peterburg 37652
Mitropoličji Palaty - Muzej Drevnerusskogo Iskusstva, Jaroslavl 36971
Muzej Cerkov Petra i Pavla, Tjumen 37843
Muzej-chram Svjatitelja Nikolaja v Tolmačach, Moskva 37265
Muzej Chrama Christa Spasitelja, Moskva 37266
Muzej Spasskaja Cerkov, Tjumen 37849
Novodevičij Monastyr, Moskva 37389
Patrijaršie Palaty, Moskva 37396
Svjato-Troickij Sobor, Sankt-Peterburg 37728
Svjatogorskij Monastyr, Puškinskie Gory 37588
Vystavočnyj Zal Spasskaja Cerkov, Irkutsk 36940

Serbia

Muzej Srpske Pravoslavne Crkve, Beograd 38077
Muzej Srpske Prvoslavne Crkve, Beograd 38078

Slovakia

Cirkevno-archeologické Múzeum, Nizná Mysla 38317

Slovenia

Slovenski Verski Muzej, Ivančna Gorica 38422

Spain

Arxiu del Monestir, Sant Joan de les Abadess 39985
Casa Natal de Santa Teresa de Jesús, Ávila 38918
Colección Museo Parroquial del Santuari de la Font de la Salut, Traiguera 40168
Colección Museográfica Permanente Arte Bizantino, Tángel 40096
Elizbarrutiko Arte Sakratuaren Museoa/ Museo Diocesano de Arte Sacro, Vitoria-Gasteiz 40319
Mezquita Catedral de Córdoba, Córdoba 39203
Museo Archiconfradía de la Sangre, Murcia 39691
Museo Arxiu Capitular, Tortosa 40164
Museo Catedral, Palma de Mallorca 39785
Museo Catedralicio, Almería 38860
Museo Catedralicio, Badajoz 38934
Museo Catedralicio, Burgo de Osma 39064
Museo Catedralicio, Burgos 39067
Museo Catedralicio, Cádiz 39088
Museo Catedralicio, Ciudad Real 39186
Museo Catedralicio, Guadix 39378
Museo Catedralicio, Jaén 39420
Museo Catedralicio, Málaga 39597
Museo Catedralicio, Ourense 39755
Museo Catedralicio, Santo Domingo de la Calzada 40040
Museo Catedralicio, Segovia 40051
Museo Catedralicio, Sigüenza 40078
Museo Catedralicio, Palencia 39769
Museo Catedralicio de Tui, Tui 40175
Museo Catedralicio-Diocesano, León 39454
Museo Catedralicio Diocesano, Pamplona 39806
Museo das Peregrinacións, Santiago de Compostela 40018
Museo de Arte Religioso de San Gil, Atienza 38915
Museo de Arte Sacra das Clarisas, Monforte de Lemos 39664
Museo de Arte Sacra de Iria Flavia, Padrón 39765
Museo de Arte Sacro, Osuna 39753
Museo de la Catedral de Santiago de Compostela, Santiago de Compostela 40022
Museo de la Colegiata, Medinaceli 39639
Museo de la Colegiata, Belmonte 39020
Museo de la Santa Iglesia Catedral, Ávila 38921
Museo de los Caminos, Astorga 38909
Museo de los Condestables de Castilla, Medina de Pomar 39634
Museo de los Corporales, Daroca 39240
Museo de Roda de Isábena, Roda de Isábena 39907
Museo de Semana Santa, Zamora 40334
Museo Diocesà i Comarcal, Lleida 39476
Museo Diocesano, Cuenca 39235
Museo Diocesano, Tudela 40174
Museo Diocesano-Catedralicio, Segorbe 40046
Museo Diocesano-Catedralicio, Valladolid 40238
Museo Diocesano-Catedralicio, Barbastro 38950
Museo Diocesano Catedralicio de Arte Sacro, Orihuela 39744
Museo Diocesano de Arte Antiguo, Sigüenza 40079
Museo Diocesano de Arte Sacro, Las Palmas de Gran Canaria 39800
Museo Diocesano de Arte Sacro, Teruel 40121
Museo Diocesano de Tui, Tui 40176
Museo Diocesano e Catedralicio de Lugo, Lugo 39498
Museo Interparroquial de Arte Sacro, Medina de Rioseco 39636
Museo Municipal, Antequera 38875
Museo Parroquial, Alquezar 38865
Museo Parroquial, El Barco de Ávila 39008
Museo Parroquial, Bocairent 39054
Museo Parroquial, Santa María del Campo 40003
Museo Parroquial de Santa Eulalia, Paredes de Nava 39813
Museo Parroquial de Zurbarán, Marchena 39624
Museo Santa Iglesia Catedral, Plasencia 39845
Museo Sor María de Jesús de Agreda, Agreda 38787
Museo-Tesoro Catedralicio, Toledo 40139
Museo-Tesoro de la Cofradía de la Expiración, Málaga 39606
Museu-Arxiu de Santa María de Mataró, Mataró 39632
Museu Capitular de la Seu Nova, Lleida 39477
Museu d'Art, Girona 39344
Museu de l'Arxiprestal, Morella 39686
Museu Diocesà de Barcelona, Barcelona 38991
Museu Diocesà de Mallorca, Palma de Mallorca 39791
Museu Diocesà Santa Maria, Eivissa 39262
Museu Històric de la Seu, Manresa 39616
Museu Parroquial, Castelló d'Empúries 39147
Museu Parroquial, Santa Pau 40004
Palau Ducal dels Borja, Gandía 39325
Parroquial Arxiprestal, San Mateo 39967
Real Casa del Labrador, Aranjuez 38880
Valle de los Caídos, Valle de Cuelgamuros 40246

Switzerland

Domschatz Sankt Ursen Kathedrale, Solothurn 41881
Domschatzmuseum, Chur 41302
Ittinger Museum, Warth 42001
Kirchenschatz, Bremgarten (Aargau) 41253
Kirchenschatz, Glarus 41436
Kirchenschatz Sankt Pelagius Kirche, Bischofszell 41240
Klostermuseum Disentis, Disentis 41331
Musée de l'Ordre de Malte, Bardonnex 41152
Musée de Payerne et Abbatiale, Payerne 41720
Musée du Grand Saint-Bernard, Le Grand-Saint-Bernard 41449
Museo di Santa Maria degli Angioli, Lugano 41584
Ortsmuseum Albisrieden, Zürich 42100

Taiwan

Rosily Arts Museum, Taipei 42423
Taiwan Temple and Shrine Art Museum, Douliou 42184

Tunisia

Musée Bourguiba de Art Islamique, Monastir 42643
Musée d'Art Islamique Raqqada, Kairouan 42631

Ukraine

Nacional'nyj Kyjvo-Pečerskyj Istoryko-kul'turnyj Zapovidnyk, Kyïv 43101

United Arab Emirates

Sharjah Islamic Museum, Sharjah 43228

United Kingdom

The Blairs Museum, Aberdeen 43240
Cathedral Treasury Museum, Carlisle 43758
Christ Church Cathedral Treasury, Oxford 45329
The John Wesley House and The Museum of Methodism, London 44891
Lincoln Cathedral Treasury, Lincoln 44712
Rievaulx Abbay, Rievaulx 45496
Saint Andrews Cathedral Museum, Saint Andrews 45554
Saint Augustine's Abbey, Canterbury 43745
Saint Mungo Museum of Religious Life and Art, Glasgow 44301
Saint Peter Hungate Church Museum, Norwich 45286
Saint Vigeans Museum, Arbroath 43314
Westminster Abbey Museum, London 45012
York Minster Undercroft Treasury and Crypt, York 46109

United States of America

Badè Museum of Biblical Archaeology, Berkeley 46750
Bernard Judaica Museum, New York 51133
Biblical Arts Center, Dallas 47927
B'nai B'rith Klutznick National Jewish Museum, Washington 53853
The Chassidic Art Institute, Brooklyn 47070
Concordia University-Wisconsin Art Gallery, Mequon 50593
Contemporary Jewish Museum, San Francisco 52771
Dadian Gallery, Washington 53857
Glencairn Museum, Bryn Athyn 47107
Jane L. and Robert H. Weiner Judaic Museum, Rockville 52429
The Jewish Museum, New York 51200
Judah L. Magnes Museum, Berkeley 46755
Lutheran Brotherhood Gallery, Minneapolis 50737
Mizel Museum, Denver 48081
Mormon Visitors Center, Independence 49478
Museum of Biblical Art, New York 51225
Museum of Church History and Art, Salt Lake City 52692
Museum of Contemporary Religious Art, Saint Louis 52590
National Museum of Catholic Art and History, New York 51235
Palace of the Governors, Santa Fe 52885
Rothko Chapel, Houston 49403
Samuel S. Fleisher Art Memorial, Philadelphia 51881
Sherwin Miller Museum of Jewish Art, Tulsa 53611
Siskin Museum of Religious and Ceremonial Art, Chattanooga 47413
Temple Museum of Religious Art, Beachwood 46658
Union of American Hebrew Congregation, New York 51275
Upper Room Chapel Museum, Nashville 50978
Yeshiva University Museum, New York 51281

Vatican City

Cappella Sistina, Sala e Gallerie Affrescate, Città del Vaticano 54497
Collezione d'Arte Religiosa Moderna, Città del Vaticano 54498
Museo Pio Cristiano, Città del Vaticano 54509

Venezuela

Museo Diocesano Lucas Guillermo Castillo, Falcón 54572

Religious History and Traditions

Argentina

Museo Catedral de La Plata, La Plata 00408
Museo Histórico Franciscano, San Salvador de Jujuy 00608
Museo Jesuítico de Sitio, Yapeyú 00686
Museo Jesuitico Nacional de Jesus María, Córdoba 00332
Museo Jesútico Nacional de Jesús María, Jesús María 00393
Museo Monseñor José Fagnano, Río Grande 00536
Museo Obispo Fray José Antonio de San Alberto, Córdoba 00336
Museo Padre Coll, Buenos Aires 00268

Armenia

Echmiadzin Cathedral Complex and Museum, Echmiadzin 00695

Australia

Bishop's Lodge Museum, Hay 01149
Christmas Museum Gallery, Horsley Park 01164
Goold Catholic Museum, East Melbourne 01054
Linton and District Museum, Linton 01231
Lit Sing Kwang Chinese Temple, Innisfail 01175
Mary Mackillop Place Museum, North Sydney 01374
Monsignor Hawes Priest House Museum, Mullewa 01332
Steiglitz Court House, Steiglitz 01530

Austria

Diözesanmuseum Eisenstadt, Eisenstadt 01858
Diözesanmuseum Graz, Graz 01966
Diözesanmuseum Linz, Linz 02312
Evangelisches Diözesanmuseum, Fresach 01905
Evangelisches Diözesanmuseum, Murau 02416
Evangelisches Museum in Wien, Wien 02985
Evangelisches Museum Oberösterreich, Rutzenmoos 02604

Kaplanstöckl, Hohenzell 02094
Kartause Aggsbach, Aggsbach Dorf 01697
Kirchenhäusl, Sankt Oswald bei Freistadt 02673
Krippenmuseum, Fulpmes 01911
Krippensammlung, Großraming 02008
Museum der Votivkirche, Wien 03032
Museum im Kastenturm, Bischofshofen 01786
MZM - Museumszentrum Mistelbach, Mistelbach an der Zaya 02386
Notburga-Museum, Eben am Achensee 01831
Pfarrmuseum, Sankt Georgen im Attergau 02646
Prämonstratenser-Chorherrenstift mit Stiftssammlungen, Geras 01929
Salvatorianerkloster, Gurk 02020
Schatzkammer - Turmmuseum der Pfarrkirche, Sankt Wolfgang im Salzkammergut 02694
Stadtmuseum und Tuchmachermuseum Pinkafeld, Pinkafeld 02499
Stiftsmuseum, Garsten 01926
Stiftsmuseum, Zwettl 03137
Stiftsmuseum Mattsee, Mattsee 02367
Stiftssammlungen Lambach, Lambach 02252
Zisterzienserabtei Mehrerau, Bregenz 01798
Zisterzienserstift, Lilienfeld 02306

Azerbaijan

State Museum of the History of Religion, Bakı03175

Belgium

Archief en Museum Zusters van Liefde van Jezus en Maria, Gent 03623
Begijnhofkerk, Sint-Truiden 04027
Eglise Saint-Rémy de Cuesmes, Cuesmes 03530
Heemmuseum De Pastorij, Edegem 03573
Keizerskapel, Antwerpen 03289
Musée des Processions, Hyon-Ciply 03726
Musée du Frère Mutien-Marie, Malonne 03845
Musée Gaspar, Arlon 03326
Museum Abdij Mariënlof, Kerniel 03752
Museum Godshuis Belle, Ieper 03729
Museum Grootseminarie, Brugge 03421
Museum Pater Valentinus Paquay-Heilig Paterke, Hasselt 03684
Museum van de Onze-Lieve-Vrouwkerk, Brugge 03423
Museum Vlaamse Minderbroeders, Sint-Truiden 04029
Oudheidkundig Museum van het Begijnhof, Turnhout 04101
Protestants Historisch Museum Abraham Hans, Horebeke 03712
Sint-Godelievemuseum, Gistel 03657
Ten Duinen 1138, Koksijde 03756
Trésor de la Basilique Saint-Materne, Walcourt 04127

Bolivia

Museo de San Francisco, Potosí 04236
Museo San Francisco, La Paz 04226

Brazil

Espaço dos Anjos, Leopoldina 04537
Museu e Arquivo Municipal, Bom Jesus 04355
Museu Municipal Padre Antônio Ribeiro Pinto, Urucânia 04947
Museu Sacro São José de Ribamar, Aquiraz 04311

Brunei

Exhibition Gallery of the Islamic Dáwah Centre, Bandar Seri Begawan 04964

Bulgaria

Nacionalen Cărkoven Istoriko-archeologičeski Muzej, Sofia 05189
Pametnik na Kulturata Zemenski Manastir, Zemen 05267

Canada

Anglican Cathedral Museum, Saint John's 06841
Basilian Fathers Museum, Mundare 06402
Biblical Museum of Canada, Vancouver 07200
Cartier-Brébeuf, Québec 06676
Centre historique des Soeurs de Sainte-Anne, Lachine 06122
Father Pandosy Mission, Kelowna 06052
Maison des Jésuites de Sillery, Sillery 06976
Martyrs' Shrine, Midland 06278
Michif Historical and Cultural Preservation Society Museum, Surrey 07036
Musée Bon-Pasteur, Québec 06685
Musée des Filles de Jésus, Trois-Rivières 07174
Musée des Religions, Nicolet 06457
Musée des Soeurs de Miséricorde, Montréal 06363
Musée des Ursulines, Québec 06691
Musée Marie-Rose Durocher, Longueuil 06193
Norwich and District Museum, Norwich 06489
Old Log Church Museum, Whitehorse 07326
Prêtres de Saint-Sulpice de Montréal, Montréal 06379
Saint-Boniface Museum, Winnipeg 07378
Saint George's Heritage Museum, Brigus 05481
Saint Michael's Museum, Miramichi 06303
Sainte-Marie Among the Hurons, Midland 06279
The Salvation Army Museum, Toronto 07136
Sharon Temple Museum Historic Site, Sharon 06948
Trembowla Cross of Freedom Museum, Dauphin 05680

China

Kulunqi Religion Museum, Kailu 07911
Wen Tianxiang Temple, Beijing 07572
Wenmiao Museum, Tianjin 08290
Yushe Reliquary Museum, Yushe 08476

Côte d'Ivoire

Musée Privee de N'Douci, N'Douci 08838

Croatia

Etnografska Zbírka Františkánský Klášter, Zaostrog 09016
Spomen Muzej Biskupa Josipa Jurja Strossmayera, Đakovo 08858

Cuba

Museo Abel Santamaría, La Habana 09108
Museo Arquidiocesano, Santiago de Cuba 09305
Museo del Parque Abel Santamaría Cuadrado, Santiago de Cuba 09318

Czech Republic

Expozice Dominikánský Klášter, Jablonné v Podještědí 09547
Galerie a Muzeum Litoměřické Dieceze, Litoměřice 09629
Husitské Muzeum v Táboře, Tábor 09858
Památník Bible Kralické, Kralice nad Oslavou 09602
Památník Mistra Jana Husa, Husinec 09542

Denmark

Museerne i Fredericia, Fredericia 10021
Museet ved Sct. Ansgar Kirke, København 10141

Dominican Republic

Museo de la Cofradía del Espíritu Santo de los Congos, Villa Mella 10362
Museo Señorita Elupina Cordero, Santa Teresa 10337

Ecuador

Museo Diocesano de Arte Religioso Nuestra Señora del Cisne, Loja 10418
Museo Religioso Santa Domingo, Bolivar 10376

Egypt

Islamic Museum, Abbasiya 10465
Museum of Islamic Art, Cairo 10504

Estonia

Dominiiklaste Kloostri Muuseum, Tallinn 10621

Finland

Kirkkomuseo, Nokia 11146
Längelmäen Kirkkomuseo, Längelmäki 11022
Laihian Kirkkomuseo, Laihia 11036
Marttilan Kirkkomuseo, Marttila 11102
Ranuan Pappila- ja Pitäjämuseo, Ranua 11255
Tapulimuseo, Lempäälä 11061
Turun Tuomiokirkkomuseo, Turku 11407

France

Abbaye de Flaran, Valence-sur-Baïse 16041
Basilique Royale Saint-Denis, Saint-Denis (Seine-Saint-Denis) 15234
Chapelle de Pénitents Blanc, Aigues-Mortes 11501
Domaine de l'Abbé Saunière, Rennes-le-Château 15017
Maison d'Abraham Mazel à Falguières, Mialet 14115
Maison du Protestantisme, Nîmes 14448
Maison du Protestantisme Poitevin Beaussais, La Couarde 13353
Maison-Musée de Jeanne Jugan, Cancale 12261
Musée Archéologique, Lectoure 13672
Musée Archéologique de Saint-Pierre, Vienne 16149
Musée-Bibliothèque de l'Histoire du Protestantisme, Joigny (Yonne) 13272
Musée Calvin, Noyon 14482
Musée Cloître Roman Saint-André-le-Bas, Vienne 16150
Musée Comtois, Besançon 11955
Musée d'Art Sacré, Hambye 13178
Musée d'Art Sacré, Le Val 13660
Musée d'Art Sacré, Lectoure 13673
Musée d'Art Sacré, Lorgues 13820
Musée d'Art Sacré, Saint-Béat 15182
Musée d'Art Sacré, Viviers-en-Ardèche 16262
Musée d'Art Sacré Copte et Byzantin Scete, Le Revest-les-Eaux 13647
Musée d'Art Sacré - Francis Poulenc, Rocamadour 15061
Musée d'Art Sacré Occidental, Le Revest-les-Eaux 13648
Musée de Germigny et Grange Germignonne, Germigny-des-Prés 13049
Musée de la France Protestante de l'Ouest, Monsireigne 14154
Musée de la Société de l'Histoire du Protestantisme Français, Paris 14663
Musée de l'Abbaye de Saint-Mathieu, Plougonvelin 14854
Musée de l'Histoire de l'Abbaye, Saint-Sauveur-le-Vicomte 15508
Musée de Notre-Dame-de-Paris, Paris 14679
Musée Départemental d'Art Religieux, Digne-les-Bains 12725
Musée des Patenôtriers, Brain-sur-Allonnes 12140
Musée des Pénitents Blancs, Marsac-en-Livradois 13983
Musée d'Sacré, Mours-Saint-Eusèbe 14312
Musée du Désert, Mialet 14116
Musée du Patrimoine des Pères de Bétharram, Lestelle-Bétharram 13725
Musée du Protestantisme Dauphinois, Le Poët-Laval 13637
Musée du Protestantisme en Haut Languedoc, Ferrières (Tarn) 12927
Musée Église Catholique de l'Assomption, Rosenwiller 15105
Musée Ernest Renan, Tréguier 15983
Musée Notre Dame de la Pierre, Sarrance 15641
Musée Rochelais d'Histoire Protestante, La Rochelle 13435
Musée Saint-François-Régis, La Louvesc 13390
Petit Musée de la Chartreuse Saint-Hugon, Arvillard 11710

Georgia

D. Baazov Georgian Jewish Historic and Ethnographic Museum, Tbilisi 16353

Germany

Abtei Rommersdorf, Neuwied 20686
Anne Frank Zentrum, Berlin 17034
Backnanger Stadtturm mit Gotischem Chor St. Michael, Backnang 16655
Bibelgalerie Meersburg, Meersburg 20246
Bibelhaus Erlebnismuseum, Frankfurt am Main 18356
Bibelmuseum, Münster 20505
Bruder-Konrad-Museum, Altötting 16516
Crescentia-Gedenkstätte, Kaufbeuren 19484
Deutschordensmuseum Bad Mergentheim, Bad Mergentheim 16783
Dioramenschau Altötting, Altötting 16517
Domschatzkammer, Köln 19599
Doppelkapelle Sankt Crucis, Landsberg, Saalekreis 19797
Ehemalige Benediktinerabtei Seligenstadt, Seligenstadt 21737
EinBlick-Dauerausstellung von Mission EineWelt, Neuendettelsau 20606
Evangelisches Bibelwerk im Rheinland - Bibelzentrum, Wuppertal 22739
Evangelisches Kirchenmuseum Pleidelsheim, Pleidelsheim 21056
Die Franckeschen Stiftungen zu Halle, Halle, Saale 18891
Geburtshaus des heiligen Konrad von Parzham, Bad Griesbach 16741
Gemeindegalerie der Evangelischen Brüdergemeinde Korntal, Korntal-Münchingen 19687
Heimatmuseum, Neupotz 20644
Heimatmuseum Lamspringe, Lamspringe 19783
Henri-Arnaud-Haus - Waldensermuseum, Ötisheim 20877
Historama Kloster Hornbach, Hornbach 19274
Hochstift Meissen, Meißen 20264
Hugenotten-Museum, Berlin 17139
Hus-Museum, Konstanz 19679
Jagdschloß Paulinzella, Rottenbach 21419
Johanniter-Museum, Krautheim 19702
Johanniter- und Maltesermuseum, Heitersheim 19114
Kardinal-Bea-Museum, Blumberg 17391
Kartäusermuseum Tückelhausen, Ochsenfurt 20853
Katharina-Luther-Stube, Torgau 22071
Kirchenmuseum, Marienhafe 20200
Kirchliches Heimatmuseum Bärenthal, Bärenthal 16894
Kleines Kirchengeschichtsmuseum Sankt Gumbertus, Illesheim 19317
Kloster Alpirsbach, Alpirsbach 16477
Kloster Mariawald, Heimbach 19109
Kloster Mariensee, Neustadt am Rübenberge 20664
Kloster Schöntal, Schöntal 21614
Klostermuseum Wittichen, Schenkenzell 21543
Klostervilla Adelberg, Adelberg 16425
Koptisches Museum, Borgentreich 17476
Kornhaus Georgenthal-Klosterruinen, Georgenthal 18591
Krippenmuseum, Glattbach 18654
Leonhardi-Museum Aigen, Bad Füssing 16737
Luther-Stube Mansfeld, Mansfeld 20177
Lutherhaus mit Bibel-Café, Eisenach 18095
Lutherhaus Wittenberg, Lutherstadt Wittenberg 20115
Lutherkirche und Lutherzimmer im Dorfgemeinschaftshaus, Moorgrund 20379
Martin Luthers Geburtshaus, Lutherstadt Eisleben 20110
Melanchthonhaus Bretten, Bretten 17584
Melanchthonhaus Wittenberg, Lutherstadt Wittenberg 20116
Missionsmuseum Bug, Bamberg 16912
Missionsmuseum der Pallottiner Limburg, Limburg an der Lahn 19997
Museum Alte Kirche, Reken 21258
Museum Altomünster, Altomünster 16524
Museum für klösterliche Kultur, Weingarten, Kreis Ravensburg 22427
Museum für Klosterkultur, Weingarten, Kreis Ravensburg 22428
Museum im Kloster Grafschaft, Schmallenberg 21584
Museum im Zisterzienserkloster Lehnin, Kloster Lehnin 19577
Museum Jerusalem Panorama Kreuzigung Christi, Altötting 16519
Museum Kartause Astheim, Volkach 22256
Museum Kloster Hude, Hude 19284
Museum Kloster Kamp, Kamp-Lintfort 19424
Museum Kloster Zeven, Zeven 22797
Museum Nikolaikirche, Berlin 17214
Museum Stift Börstel, Berge 17006
Museum zur Stadtgeschichte, Königslutter am Elm 19654
Museumszentrum Lorsch, Lorsch 20035
Oekumenische Sammlung Gertrud Weinhold 'Das Gottesjahr und seine Feste', Oberschleißheim 20835
Peterstormuseum, Zwiefalten 22836
Praedicantenbibliothek der Nikolaikirche, Isny im Allgäu 19367
Religionskundliche Sammlung der Philipps-Universität, Marburg 20194
Sakrale Kunst in der Sankt-Anna-Kapelle, Cham 17691
Sammlung zur Geschichte der Benediktinerabtei Tholey, Tholey 22057
Sankt Matthäus-Kirche im Kulturforum, Berlin 17246
Schatzkammer der Katholischen Pfarrkirche Sankt-Servatius, Siegburg 21751
Schatzkammer der Wallfahrtskirche, Hohenpeißenberg 19240
Schatzkammer Grafenrheinfeld, Grafenrheinfeld 18732
Stadt- und Wallfahrtsmuseum, Walldürn 22312
Stift Fischbeck, Hessisch Oldendorf 19171
Stiftsmuseum, Xanten 22763
Stiftung Stift Neuzelle, Neuzelle 20690
Synagogen-Museum, Urspringen 22189
Wallfahrtsmuseum, Steingaden 21888
Wallfahrtsmuseum, Bruckmühl 17601
Wallfahrtsmuseum Inchenhofen, Inchenhofen 19332
ZisterzienserMuseum Kloster Walkenried, Walkenried 22307
Zisterziensermuseum Riddagshausen, Braunschweig 17531

Greece

Ayios Dimitrios Crypt Museum, Thessaloniki 23211
Bible Museum, Trilofo 23230
Byzantine Museum, Mytilini 23110
Byzantine Museum of Naousa, Naousa 23118
Church Museum, Dimitsana 22983
Ecclesiastical Museum, Sifnos 23185
Ecclesiastical Museum of Alexandroupolis, Alexandroupolis 22876
Historical Museum of Anthiros, Karditsa 23035
Historical Museum of Portitsa, Karditsa 23036
Museum of the Convent of Agios Andreas, Agios Andreas Arkadias 22869
Religious Museum, Ano Mera 22886

Hungary

Egyházmegyei Múzeum és Kincstár, Szeged 23842
Érseki Gyűjteményi Központ, Eger 23537
Jákováli Hasszán Múzeum, Pécs 23761
Károli Gáspár Múzeum és Bibliakiállítás, Gönc 23571
Középkori O-Zsinagóga, Sopron 23811
Középkori Zsidó Imaház, Budapest 23425
Magyar Zsidó Múzeum és Levéltár, Budapest 23447
Mindszenty Emlékhely, Esztergom 23554

R

Okeresztény Mauzóleum, Pécs 23770
Püspöki Kastély, Fertőrákos 23561
Püspöki Palota, Sümeg 23828
Soproni Evangélikus Egyházközség Múzeuma, Sopron 23819
Soproni Római Katolikus Egyházművészeti Gyűjtemény, Sopron 23821
Székesegyházi Kincstár és Egyházmegyei Gyűjtemény, Vác 23945
Székesfehérvári Egyházmegyei Múzeum, Székesfehérvár 23859

India

Central Sikh Museum, Amritsar 24066

Iran

Armenian Cathedral and Museum, Isfahan 24610
Astan-e-Qods-e Razavi Museums, Mashad 24631
Holy Defense Museum, Kerman 24617

Ireland

Heritage Centre of Saint Louis Convent, Monaghan 24877
Knock Museum, Knock 24860

Israel

Islamic Museum, Jerusalem 24995
Neot Kedumim, Lod 25061

Italy

Augustiner Chorherrenstift Neustift, Varna 28167
Casa Museo G. Mazzarino, Pescina 27174
Conservatorio di Santa Chiara, San Miniato 27670
Mostra Permanente della Giudaica e Raccolta di Minerali, Laino Borgo 26439
Museo A. Versace, Bagnara Calabra 25343
Museo Archeologico Civico-Diocesano G. B. Leopardi, Penne 27134
Museo Beato Angelo, Acri 25150
Museo Casa Natale di San Pio X, Riese Pio X 27401
Museo Civico di Santa Maria della Consolazione, Altomonte 25208
Museo del Duomo, Guardiagrele 26386
Museo del Duomo, Udine 28112
Museo del Santuario di Crea, Serralunga di Crea 27813
Museo della Certosa, Serra San Bruno 27811
Museo della Civiltà Rupestre, Zungri 28321
Museo della Collegiata, San Candido 27619
Museo dell'Abbazia di Santa Maria in Sylvis, Sesto al Reghena 27822
Museo delle Anime dei Defunti, Roma 27483
Museo delle Sinopie, Pisa 27243
Museo dell'Opera del Santuario, Tolentino 27987
Museo di Santa Cecilia, San Lazzaro di Savena 27651
Museo di Scienze Naturali ed Umane, L'Aquila 26455
Museo Diocesano, Catanzaro 25782
Museo Diocesano, Gerace 26328
Museo Diocesano, Lamezia Terme 26444
Museo Diocesano, Spoleto 27894
Museo Diocesano, Squillace 27896
Museo Diocesano, Sulmona 27921
Museo Diocesano, Tropea 28105
Museo Diocesano d'Arte Sacra, Scaria di Lonzo 27786
Museo Diocesano d'Arte Sacra, Treviso 28073
Museo Diocesano d'Arte Sacra Santa Apollonia, Venezia 28199
Museo Diocesano di Arte Sacra, Santa Severina 27710
Museo Diocesano e Tesoro della Cattedrale, Troia 28103
Museo Missionario, Padova 27046
Museo Missionario Cinese e di Storia Naturale, Lecce 26472
Museo Parrocchiale del Santuario del Santissimo Crocifisso, Buggiano 25563
Museo Parrocchiale Madonna dei Raccomandati, Gessopalena 26331
Museo Provinciale Campano, Capua 25670
Museo Storico Cappuccino, Camerino 25622
Museo Valdese della Balziglia, Massello 26636
Piccolo Museo Parrocchiale, Torgnon 27995
Piccolo Museo Parrocchiale, Valsavarenche 28150
Roccolta d'Arte Sacra, San Piero a Sieve 27678
Santuario Casa Santa Caterina, Siena 27847

Japan

26 Martyrs Museum, Nagasaki 28968
Akama-jingu Homotsukan, Shimonoseki 29267
Hondo-shi Amakusa Kirishitankan, Hondo 28547
Kaieji Buddhist Temple Site Museum, Sennan 29249
Konan-ji Homotsukan, Setoda 29252
Motsuji Homotsukan, Hiraizumi 28522
Museum of the Grove of the Village Shrine, Osaka 29135
Rengebu-ji Temple, Ogi 29092
Secret Christians Museum, Nagoya 29009
Seinan Gakuin Daigaku Hakubutsukan, Fukuoka 28436

Jordan

Islamic Museum, Amman 29651
Mazar Islamic Museum, Al-Mazar 29642

Korea, Republic

Jikjiseongbo Museum, Gimcheon 29813
Tongdosa Museum, Yangsan 29940

Laos

Haw Pha Kaeo, Vientiane 29973
Pha That Luang, Vientiane 29977
Wat Si Saket, Vientiane 29979

Lebanon

Cilicia Museum, Antélias 30154

Lithuania

Baznytinio Meno Skyrius, Kaunas 30241
Birštono Sakralinis Muziejus, Birštonas 30215
Kauno Arkivyskupijos Muziejus, Kaunas 30248
Žemaičių Vyskupystės Muziejus, Varniai 30391

Malta

Gozo Cathedral Museum, Victoria 30652

Mexico

Galería Episcopal, Zacatecas 31576
Museo a la Bandera y Santuario de la Patria, Iguala de la Independencia 30955
Museo Altar de la Patria, Ixcateopan de Cuauhtémoc 30958
Museo Católico Peninsular, Mérida 31023
Museo Coltzin, Culiacán 30824
Museo Cristero Señor Cura Cristóbal Magallanes, Totatiche 31504
Museo de la Basílica de Guadalupe, México 31077
Museo de la Evangelización, Huejotzingo 30945
Museo de la Purísima Concepción, Celaya 30753
Museo de las Misiones Jesuitas, Loreto 31007
Museo de Maní y Museo del Convento San Miguel Arcángel, Maní 31011
Museo de Nuestra Señora del Rosario, Talpa de Allende 31398
Museo de Oxolotán, Tacotalpa 31395
Museo de San Esteban Tetelpa, Zacatepec de Hidalgo 31585
Museo de San José María de Yermo y Parres, Puebla 31283
Museo de Todos los Santos, La Paz 30988
Museo del Ex Convento de Culhuacán, México 31102
Museo Ex Convento Agustino Siglo XVI, México 31114
Museo Ex-Convento de San Juan Bautista, Tlayacapan 31482
Museo Ex Convento del Carmen, Guadalajara 30903
Museo Ex Templo de San Agustín, Zacatecas 31578
Museo Ex Votos Sr. de la Misericordia, Tepatitlán de Morelos 31431
Museo Félix de Jesús, México 31115
Museo Kan Pepen, Teabo 31409
Museo Mariano Matamoros, Jantetelco 30972
Museo Rafael Coronel, Zacatecas 31581
Museo Regional Doctor Leonardo Oliva, Ahualulco de Mercado 30706
Museo Virgen de Zapopan, Zapopan 31591

Mongolia

Palace Museum, Ulaanbaatar 31630

Namibia

Schmelenhaus Museum, Windhoek 31767

Netherlands

Baarle's Museum, Baarle-Hertog 31945
Bijbels Museum, Amsterdam 31834
Gaudete, Museum voor Volksdevotie, Boxtel 31995
Heemkamer Barthold Van Heessel, Aarle Rixtel 31788
In de Zevende Hemel, Apeldoorn 31912
Kapucijnenmuseum, 's-Hertogenbosch 32371
Missiemuseum Steijl, Steijl 32804
Museum Abdijkerk, Thorn 32821
Museum Catharijneconvent, Utrecht 32849
Museum der Katakombenstichting, Valkenburg, Limburg 32865
Museum Sint Bernardushof, Aduard 31790
Museum Sint Sellian, Sellingen 32768
Museum Stapelen, Boxtel 31997
Nederlands Zouavenmuseum, Oudenbosch 32654
Religieus Museum Kijk-je Kerk-Kunst, Gennep 32247
Schatkamer Sint Lambertuskerk, Horst 32417

New Zealand

The Salvation Army Territorial Archives and Museum, Wellington 33259

Nicaragua

Museo de la Diócesis de Matagalpa, Mataglpa 33296
Museo de Sor de María Romero, Granada 33278

Northern Cyprus

Bellapais Manastiri, Girne 33357

Norway

Hans Nielsens Hauges Minne, Rolvsøy 33792
Jødisk Museum, Trondheim 33924
Misjonsmuseet, Stavanger 33867
Misjonsmuseet på Fjellhaug, Oslo 33731
Rygge Museum, Larkollen 33626

Palestine

Islamic Museum, Jerusalem 34056

Peru

Museo de Arte Religoso, Cusco 34158
Museo de la Biblia, Lima 34241
Museo de la Catedral, Cusco 34160
Museo de la Catedral, Trujillo 34320
Museo de la Iglesia de la Compañía de Jesús, Ica 34188
Museo de la Iglesia de San Pedro, Chucuito 34148
Museo de la Recoleta, Arequipa 34111
Museo de Sitio Arqueológico Kuntur Wasi, San Pablo 34304
Museo de Sitio Huaca El Dragón, La Esperanza 34200
Museo del Sitio de Qorikancha, Cusco 34164
Museo Etnográfico Santa Rosa de Ocopa, Concepción 34153
Museo Santuarios Andinos de la Universidad Católica de Santa María, Arequipa 34116

Philippines

Archdiocesan Museum of Manila, Manila 34430
Baclayon Church Museum, Baclayon 34336
Basilica del Santo Niño Museum, Cebu 34364
Butuan Diocesan Museum, Butuan 34353
Casa Gorordo Museum, Cebu 34366
Cebu Archdiodesan Museum, Cebu 34367
Christ the King College Museum, Calbayog 34360
Iglesia ni Cristo Museum and Gallery, Quezon City 34491
Jorge Barlin National Monument, Baao 34335
Monsignor Yatco Ecclesiastical Museum, Batangas 34345
Museo de Santa Monica, Panay 34469
Museo del Seminario Conciliar de Nueva Caceres, Naga 34466
Museo It Akean, Kalibo 34396
Museo Recoleto, Quezon City 34494
Puerta de Isabel II Gallery, Manila 34454
Santo Niño Shrine and Heritage Museum, Tacloban 34511

Poland

Galicja Muzeum, Kraków 34781
Muzeum Archidiecezjalne, Szczecin 35158
Muzeum Archidiecezjalne w Krakowie, Kraków 34786
Muzeum Archidiecezji Gnieźnieńskiej, Gniezno 34672
Muzeum Archidiecezji Łódzkiej, Łódź 34874
Muzeum Archidiecezji Warszawskiej, Warszawa 35236
Muzeum Cystersów, Wąchock 35206
Muzeum Diecezjalne, Płock 35013
Muzeum Diecezjalne, Siedlce 35102
Muzeum Diecezjalne, Włocławek 35308
Muzeum Diecezjalne w Opolu, Opole 34985
Muzeum Martyrologii (Sonnenburg), Słońsk 35110
Muzeum Misji Karmelitańskich w Afryce, Krceszowice 34827
Muzeum Misyjne, Krosno 34829
Muzeum Misyjne Misjonarzy Oblatów MN, Święty Krzyż 35151
Muzeum Parafialne, Widawa 35298
Muzeum Parafialne, Tropie 35198
Muzeum Parafialne, Grybów 34686
Muzeum Parafialne, Złota 35373
Muzeum Parafialne Sztuki Ludowej im. Ks. Edwarda Nitki, Paszyn 35004
Muzeum Prowincji Ojców Bernardynów, Leżajsk 34853
Muzeum Reformacji Polskiej, Mikołajki 34921
Parafialne Muzeum Regionalne, Krasnobród 34825

Portugal

Casa Museu da Freira, Penacova 35815
Casa Museu da Ordem Terceira de São Francisco, Ovar 35798
Casa Museu Dom Hugo, Porto 35836
Colecção Etnográfica da Paróquia de São Mamede da Ventosa, Torres Vedras 35970
Museu Antoniano, Aljezur 35407
Museu Antoniano, Lisboa 35669
Museu da Capela de Sã Frutuoso de Montélios, São Jerónimo de Real 35918
Museu da Paróquia, Redondo 35881
Museu da Paróquia da Benedita, Alcobaça 35393
Museu da Santa Casa da Misericórdia da Vila da Ericeira, Ericeira 35560
Museu da Santa Casa da Misericórdia de Alvor, Alvor 35428
Museu da Sé de Angra do Heroísmo, Angra do Heroísmo 35438
Museu de D. Lopo de Almeida, Abrantes 35383
Museu de Santa Maria, Vila do Porto 35993
Museu de Santa Maria Adelaide, Arcozelo 35443
Museu do Convento de São Domingos, Montemor-o-Novo 35776
Museu do Santuário de Nossa Senhora da Abadia, Bouro 35477
Museu dos Cristos, Borba 35472
Museu dos Cristos, Sousel 35957
Museu dos Relógios Antigos, Serpa 35929
Museu dos Terceiros, Ponte de Lima 35831
Museu Municipal de Alcochete - Núcleo Sede, Alcochete 35399
Museu Nossa Senhora d'Aires, Viana do Alentejo 35982
Museu Paroquial, Moncaparracho 35769
Museu Paroquial, Queiriga 35878
Museu Paroquial de Oliveira de Azeméis, Oliveira de Azeméis 35791
Museu Paroquial de São Sebastião, Guimarães 35641
Museu Paroquial de Vila Praia de Âncora, Vila Praia de Âncora 36010
Museu Pedro Cruz, Pedrógão Grande 35814
Museu Rural da Casa do Povo de Santa Maria, Estremoz 35579
Núcleo Museológico da Capela do Espírito Santo dos Mareantes, Sesimbra 35933
Núcleo Museológico da Paróquia da Beirã, Beirã 35461
Núcleo Museológico da Roda dos Expostos, Almeida 35418
Tesouros de Santa Maria e São Pedro de Óbidos, Óbidos 35787

Puerto Rico

Museo de Arte Religioso Santo Domingo de Porta Coeli, San Germán 36057

Romania

Casa Memorială Aloisie Tăutu, Valea Vinului 36717
Colecţia Bibliotecii Seminarului Teologic Ortodox, Caransebeş 36261
Colecţia de Obiecte Bisericeşti, Movile 36484
Colecţia Muzeală a Bisericii Ortodoxe Sf. Nicolae, Bereşti 36148
Colecţia Muzeală a Episcopiei Ortodoxe Sârbe, Timişoara 36687
Colecţia Muzeală a Mănăstirii Neamţ, Vânători-Neamţ 36724
Colecţia Muzeală a Protoieriei Ortodoxe Lugoj, Lugoj 36449
Colecţia Muzeală Batthyaneum, Alba Iulia 36105
Muzeul de Etnografie, Comarnic 36299
Muzeul Sătesc, Broşteni 36184

Russia

Cerkov Pokrova v Filjach, Moskva 37154
Dom-muzej Tukaevych, Košlauč 37060
Ėkspozicija "Čelovek. Duša. Duchovnost.", Ivanovo 36945
Gosudarstvennyj Muzej Istorii Religii, Sankt-Peterburg 37646
Kraevedčeskij Muzej Evrejskoj Avtonomnoj Oblasti, Birobidžan 36807
Muzej Cerkov Petra i Pavla, Tjumen 37843
Muzej-chram Svjatitelja Nikolaja v Tolmačach, Moskva 37265
Muzej Chrama Christa Spasitelja, Moskva 37266
Muzej Palomničestva po Svjatym Mestam, Moskva 37344
Oloneckij Nacionalnyj Muzej Karelov-Livvikov im. N.T.Prilukin, Olonec 37496
Samarskij Eparchialnyj Cerkovno-istoričeskij Muzej, Samara 37620
Troickij Sobor, Serpuchov 37766

R

Valaamskij Naučno-issledovatelskij Cerkovno-archeologičeskij i Prirodnyj Muzej-zapovednik, Sortavala ... 37790
Vystavočnyj Zal Spasskaja Cerkov, Irkutsk ... 36940

Slovenia

Rojstna Hiša Blaženega Antona Martina Slomška, Šentjur ... 38512
Slovenski Verski Muzej, Ivančna Gorica ... 38422

South Africa

Africana Museum, Mariannhill ... 38686
Genadendal Museum, Genadendal ... 38623
Macrorie House Museum, Pietermaritzburg ... 38705
Missionary Museum, King William's Town ... 38676
Missionary Museum of the N.G. Kerk, Stellenbosch ... 38736
Transgariep Museum, Philippolis ... 38703
Zululand Mission Museum Chapel, Eshowe ... 38619

Spain

Archivo-Museo San Juan de Dios-Casa de los Pisas, Granada ... 39354
Casa Natal de San Ignacio, Azpeitia ... 38928
Casa Natal del Hermano Francisco Gárate, Azpeitia ... 38929
Casa Santuario de San Pedro Claver, Verdú ... 40265
Castillo de Javier, Javier ... 39424
Colección Museográfica Permanente Parroquial de Jérica, Jérica ... 39428
Colección Parroquial de Villahermosa del Río, Villahermosa del Río ... 40302
Museo Arqueológico y Etnológico de Granada, Granada ... 39364
Museo Catedralicio de Tui, Tui ... 40175
Museo Catedralicio-Diocesano, Valencia ... 40203
Museo Catedralico y Giralda, Sevilla ... 40066
Museo Comarcal de Arte Sacro de Peñafiel, Peñafiel ... 39826
Museo Convento de Santa Clara, Salamanca ... 39937
Museo de la Concatedral de San Pedro, Soria ... 40092
Museo de la Coria, Trujillo ... 40171
Museo de la Iglesia, Oviedo ... 39760
Museo de la Iglesia de Santa María la Real, Sasamón ... 40044
Museo de la Real Capilla, Arenas de San Pedro ... 38885
Museo de la Semana Santa de Cuenca, Cuenca ... 39233
Museo de la Semana Santa Marinera de Valencia Salvador Caurín Alarcón, Valencia ... 40205
Museo de San Esteban, Salamanca ... 39941
Museo de Semana Santa, Medina de Rioseco ... 39635
Museo de Terra Santa, Santiago de Compostela ... 40023
Museo del Convento de Santa Úrsula, Salamanca ... 39942
Museo del Santo Rosario, Aroche ... 38894
Museo Diocesano de Jaca, Jaca ... 39418
Museo Diocesano de San Sebastián, Donostia-San Sebastián ... 39252
Museo Diocesano de Tui, Tui ... 40176
Museo Diocesano Regina Coeli, Santillana del Mar ... 40034
Museo Diocesano y Catedralicio, Salamanca ... 39943
Museo Iglesia San Juan de Dios, Murcia ... 39699
Museo Madres Benedictinas, Sahagún ... 39933
Museo Parroquial, Chillón ... 39175
Museo Parroquial, Espejo ... 39281
Museo Parroquial, Valdemoro ... 40188
Museo Parroquial de Calatañazor, Calatañazor ... 39097
Museo Parroquial de la Asunción de Nuestra Señora de Navarrés, Navarrés ... 39713
Museo Parroquial de Santa María, Becerril de Campos ... 39012
Museo Parroquial de Yanguas, Yanguas ... 40329
Museo Parroquial Nuestra Sacra de la Asunción, Palomares del Campo ... 39803
Museo Santa Clara de Murcia, Murcia ... 39703
Museo Teresiano, Alba de Tormes ... 38802
Museu-Arxiu Parroquial, Canet de Mar ... 39118
Museu de la Catedral, Barcelona ... 38978
Museu del Convent de la Beata sor Fracinaina de la Mare de Déu dels Dolors, Sencelles ... 40057
Museu del Temple Expiatori de la Sagrada Familia, Barcelona ... 38986
Museu Parroquial, Altafulla ... 38866
Petit Museu Parroquial, Calonge de Mar ... 39106

Sweden

Missionsmuséet Nårunga, Vårgårda ... 41050
Sancta Birgitta Klostermuseum, Vadstena ... 41020
Sankta Karins Kyrkoruin, Visby ... 41071

Switzerland

Hausmuseum Kloster Kreuzlingen, Kreuzlingen ... 41513
Musée International de la Reforme, Genève ... 41424
Museum Bruder Klaus, Sachseln ... 41786
Museum der Schweizer Kapuzinerprovinz, Sursee ... 41925
Ritterhaus Bubikon, Bubikon ... 41264
Zwinglegeburtshaus, Wildhaus ... 42022

Taiwan

Bisia Temple Museum, Yilan ... 42465
Chinese Catholic Museum and University History Exhibition Gallery, Sinjhuang ... 42338
Culture Exhibition Hall of Singang Fengtian Temple, Singang ... 42337
Hosanna Museum, Chai-Yi ... 42153
Jhenlangong Mazu Exhibition Hall, Dajia ... 42173
Museum of World Religions, Yungho ... 42484
Shaolin Temple Religion Museum, Yuanli ... 42478
Wanho Shrine Museum, Taichung ... 42358
Zambala Art Museum, Shanhua ... 42322

Turkey

Aya İrini Kilisesi Müzesi, İstanbul ... 42804
Ayasofya Müzesi, Trabzon ... 42955
Fethiye Cami Müzesi, İstanbul ... 42818
Hacı Bayram Camii, Ankara ... 42703
Kariye Müzesi, İstanbul ... 42831
Meryem Ana Evi, Selçuk ... 42937

Turkmenistan

Taryhy Ýer Saragt, Saragt ... 42981

Ukraine

Lvivs'kyj Muzej Istorii Religii, Lviv ... 43127

United Kingdom

All Hallows Undercroft Museum, London ... 44781
Ancient Order of Foresters, Southampton ... 45701
Bath Abbey Heritage Vaults, Bath ... 43399
Birkenhead Priory and Saint Mary's Tower, Birkenhead ... 43490
Byland Abbey, Coxwold ... 43909
Cardinal Ó Fiaich Heritage Centre, Cullyhanna ... 43937
Cathedral Treasury, Chichester ... 43824
Forde Abbey, Chard ... 43791
Iona Abbey and Nunnery, Isle-of-Iona ... 44538
Melrose Abbey Museum, Melrose ... 45125
Monks' Dormitory Durham Cathedral, Durham ... 44072
Museum of Methodism, London ... 44929
Museum of Primitive Methodism, Englesea Brook, Crewe ... 43922
Museum of the Order of Saint John, London ... 44930
Peel Castle, Peel ... 45351
Reading Abbay Gateway, Reading ... 45470
Rosslyn Chapel, Roslin ... 45514
Rushen Abbey, Ballasalla ... 43358
Saint Michael's Loft Museum, Christchurch ... 43843
Saint Peter's Church, Sunderland ... 45804
Saint Robert's Cave, Knaresborough ... 44620
Tintern Abbey, Chepstow ... 43813
Whithorn - Cradle of Christianity, Whithorn ... 46004
William Carey Museum, Leicester ... 44684

United States of America

American Baptist - Samuel Colgate Historical Library, Rochester ... 52391
American Catholic Historical Society Museum, Philadelphia ... 51812
Archives Museum - Temple Mickve Israel, Savannah ... 52938
The Arthur J. Moore Methodist Museum, Saint Simons Island ... 52644
Beehive House, Salt Lake City ... 52687
Bernard Judaica Museum, New York ... 51133
Biblical Resource Center Museum, Collierville ... 47692
Biedenharn Museum and Gardens, Monroe ... 50787
Billy Graham Center Museum, Wheaton ... 54118
Bishop Martin Museum, Natchitoches ... 50986
Catholic Historical Society of the Roanoke Valley, Roanoke ... 52380
Center for Jewish History, New York ... 51143
Chinese House Museum, Echo ... 48280
Community of Christ, Independence ... 49472
Concordia Historical Institute, Saint Louis ... 52572
Creation Museum, Petersburg ... 51795
Disciples of Christ Historical Society Museum, Nashville ... 50967
Eldridge Street Project, New York ... 51169
The Ernest W. Michel Historical Judaica Collection, New York ... 51170
Father Marquette National Memorial and Museum, Saint Ignace ... 52550
Fifth Meeting House, Lancaster ... 49917
Foundation for Historic Christ Church, Irvington ... 49537
Gustav Jeeninga Museum of Bible and Near Eastern Studies, Anderson ... 46292
Hill Cumorah Visitors Center, Palmyra ... 51684
Historic Liberty Jail Visitors Center, Liberty ... 50078
History Center of the United Methodist Church, Madison ... 50384
Holyland Exhibition, Los Angeles ... 50235
The Jewish Museum, New York ... 51200
The Joseph A. Callaway Archaeological Museum, Louisville ... 50286
The Judaica Museum, Bronx ... 47047
Judaica Museum of Central Synagogue, Irvington ... 49536
Kateri Galleries, Auriesville ... 46487
Kirtland Temple Historic Center, Kirtland ... 49802
Knights of Columbus Museum, New Haven ... 51052
Living Word National Bible Museum, Aledo ... 46194
Longyear Museum, Chestnut Hill ... 47440
Lovely Lane Museum, Baltimore ... 46574
Luckhard Museum, Sebewaing ... 53019
Mennonite Heritage Center, Harleysville ... 49168
Mennonite Library and Archives Museum, North Newton ... 51398
The Mission House, Stockbridge ... 53300
Mission Houses Museum, Honolulu ... 49349
Mission San Carlos Borromeo del Rio Carmelo, Carmel ... 47251
Mission San Luis Obispo de Tolosa, San Luis Obispo ... 52814
Mission San Luis Rey Museum, Oceanside ... 51491
Mission San Miguel, San Miguel ... 52823
Mississippi Baptist Historical Commission, Clinton ... 47656
Moravian Historical Society Museum, Nazareth ... 50993
Moravian Museum of Bethlehem, Bethlehem ... 46781
Museum of Church History and Art, Salt Lake City ... 52692
Museum of the Southern Jewish Experience, Jackson ... 49563
Museum of the Western Jesuit Missions, Hazelwood ... 49232
Museum of Waldensian History, Valdese ... 53684
National Museum of American Jewish History, Philadelphia ... 51861
National Museum of Catholic Art and History, New York ... 51235
National Yiddish Book Center, Amherst ... 46270
Native American Exhibit, Fonda ... 48584
Nebraska Conference United Methodist Historical Center, Lincoln ... 50103
Ohef Sholom Temple Archives, Norfolk ... 51369
Old Bohemia Historical Museum, Warwick ... 53833
Old Cathedral Museum, Saint Louis ... 52594
Old Mission Santa Ines Collection, Solvang ... 53143
Old Saint Ferdinand's Shrine, Florissant ... 48572
Oroville Chinese Temple, Oroville ... 51606
Perry County Lutheran Historical Society, Altenburg ... 46239
Presbyterian Historical Society Museum, Montreat ... 50828
Presbyterian Historical Society Museum, Philadelphia ... 51872
La Purisima Mission, Lompoc ... 50167
Quaker Meeting House, Mount Pleasant ... 50887
Rose Lawn Museum, Cartersville ... 47265
Saint George's United Methodist Church Museum, Philadelphia ... 51879
Saint Photios Greek Orthodox National Shrine, Saint Augustine ... 52527
Salvation Army Southern Historical Museum, Atlanta ... 46450
San Antonio Missions Museum, Jolon ... 49645
San Buenaventura Mission Museum, Ventura ... 53714
San Gabriel Mission Museum, San Gabriel ... 52802
San Juan Bautista Mission Museum, San Juan Bautista ... 52812
San Miguel Mission Church, Santa Fe ... 52887
Santuario de Nuestra Senora de Guadalupe, Santa Fe ... 52889
Scotland Heritage Chapel and Museum, Scotland ... 52972
Shaker Heritage Society Museum, Albany ... 46163
Sherwin Miller Museum of Jewish Art, Tulsa ... 53611
Siskin Museum of Religious and Ceremonial Art, Chattanooga ... 47413
Skirball Museum, Cincinnati ... 47563
Spertus Museum, Chicago ... 47514
Temple Museum of Religious Art, Beachwood ... 46658
Texas Baptist Historical Center Museum, Brenham ... 47006
Thomas L. Kane Memorial Chapel, Kane ... 49674
Tome Parish Museum, Tome ... 53527
Trinity Museum of the Parish of Trinity Church, New York ... 51273
Union of American Hebrew Congregation, New York ... 51275
Vasa Lutheran Church Museum, Vasa ... 53711
Washington National Cathedral, Washington ... 53932
Whitefield House Museum, Nazareth ... 50994
William A. Quayle Bible Collection, Baldwin City ... 46541
Wordspring Discovery Center, Orlando ... 51596
World Methodist Museum, Lake Junaluska ... 49886
Yeshiva University Museum, New York ... 51281

Vatican City

Museo Sacro, Città del Vaticano ... 54511

Venezuela

Museo Diocesano, Valle del Espíritu Santo ... 54599
Museo Sacro de Caracas, Caracas ... 54556

Reptiles

Bulgaria

Prirodonaučen Muzej, Černi Osăm ... 05002

Germany

Fossilien-Museum, Fürth ... 18497

Lithuania

Kauno Tado Ivanausko Zoologijos Muziejus, Kaunas ... 30252

United States of America

Imagination Station Science Museum, Wilson ... 54211
The Natural Science Center of Greensboro, Greensboro ... 49067
Rattlesnake Museum, Albuquerque ... 46186
Southern Oregon University Museum of Vertebrate Natural History, Ashland ... 46389
University of North Dakota Zoology Museum, Grand Forks ... 48990

Rescue Equipment → Fire Fighting and Rescue Work

Rocks → Mineralogy

Salt

Austria

Salzwelten Altaussee, Altaussee ... 01709
Salzwelten Hallstatt, Hallstatt ... 02048
Salzwelten Salzburg-Bad Dürrnberg, Bad Dürrnberg ... 01742

Bosnia and Herzegovina

Muzej Solane Tuzla, Tuzla ... 04291

Canada

Malagash Area Heritage Museum, Malagash ... 06222

China

Zigong Salt Industry History Museum, Zigong ... 08518

Colombia

Museo de la Sal, Zipaquirá ... 08791

France

Musée Claude Nicolas Ledoux Saline Royale, Arc-et-Senans ... 11661
Musée Départemental du Sel, Marsal ... 13985
Musée des Marais Salants, Batz-sur-Mer ... 11862
Musée du Sel, Salies-de-Béarn ... 15597

Germany

Alte Saline Bad Reichenhall, Bad Reichenhall ... 16806
Besucherbergwerk Bad Friedrichshall-Kochendorf, Heilbronn ... 19095
Borlach-Museum, Bad Dürrenberg ... 16718
Deutsches Salzmuseum - Industriedenkmal Saline Lüneburg, Lüneburg ... 20096
Forum Fränkischer Hof, Bad Rappenau ... 16805
Museum Obere Saline mit Bismarck-Museum, Bad Kissingen ... 16759
Museum Salz und Moor, Grassau ... 18740
Saline- und Heimatmuseum, Bad Sulza ... 16849
Salinenmuseum Unteres Bohrhaus, Rottweil ... 21428
Salzbergwerk Berchtesgaden, Berchtesgaden ... 17004
Salzmuseum, Bad Sooden-Allendorf ... 16842
Stadthistorisches Sole, Salz und Kali-Bergbaumuseum, Bad Salzdetfurth ... 16820
Technisches Halloren- und Salinenmuseum, Halle, Saale ... 18905

Italy

Museo Storico delle Saline, Margherita di Savoia ... 26604

Japan
Shio-no-michi Hakabutsukan, Itoigawa 28600
Tabako to Shio no Hakubutsukan, Tokyo 29479

Mexico
Museo Comunitario de la Sal, Armería 30724

Netherlands
Zoutmuseum Delden, Delden 32055

Panama
Museo de la Sal y el Azúcar, Aguadulce 34059

Portugal
Museu Municipal de Alcochete-Núcleo do Sal, Alcochete 35398
Núcleo Museológico das Salinas Naturais de Rio Maior, Rio Maior 35891

Russia
Bereznikovskij Istoriko-kraevedčeskij Muzej, Berezniki 36803

Spain
Gatzaren Ekomuseoa, Leintz-Gatzaga 39450
Museo del Mar y de la Sal, Torrevieja 40160
Museo Etnográfico de el Cerrp de Andèvalo, El Cerro de Andévalo 39163

Switzerland
Mines de Sel de Bex, Bex 41226
Musée Suisse du Sel, Aigle 41095
Salzkammer, Pratteln 41734

Taiwan
Tainan Salt Museum, Cigu Township 42166

United Kingdom
Lion Salt Works, Northwich 45272
Salt Museum, Northwich 45273

United States of America
Salt Museum, Liverpool 50147

Sarcophagi → **Tombs**

Scientific Apparatus and Instruments

Argentina
Exploratorio, San Isidro 00580
Museo de Ciencia y Técnica, Buenos Aires 00190
Museo de los Niños Abasto, Buenos Aires 00214
Museo Interactivo de Ciencias-Puerto Ciencia, Paraná 00487
Museo Meteorológico Nacional Dr. Benjamín A. Gould, Córdoba 00334
Museo Participativo de Ciencias, Buenos Aires 00269

Australia
Museum of the History of Science, Sydney 01556
Physics Museum, Saint Lucia 01485
Sydney Children's Museum, Merrylands 01299

Austria
Alchimistenmuseum, Kirchberg am Wagram 02165
Museum Sender Dobl, Dobl 01822
Photohistorische Sammlung, Penk 02484
Sanitärhistorisches Museum, Wien 03067

Brazil
Espaço Museu da Vida, Rio de Janeiro 04691

Canada
The Canadian Clock Museum, Deep River 05687
Saskatchewan Science Centre, Regina 06731

China
Nanjingdi Zhen Science Museum, Nanjing 08047

Croatia
Tehnički Muzej, Zagreb 09012

Czech Republic
Technické Muzeum v Brně, Brno 09440

Denmark
Steno Museet, Århus 09966

Egypt
Science Museum, Alexandria 10478

Finland
Helsingin Yliopistomuseo Arppeanum, Helsinki 10785

France
Musée de l'Horlogerie et du Décolletage, Cluses 12563
Musée du Temps, Besançon 11959

Germany
Comtoise Uhren Museum, Düsseldorf 17986
Deutsches Museum, München 20433
Mathematikum, Gießen 18640
Mathematisch-Physikalischer Salon, Dresden 17959
Optisches Museum der Ernst-Abbe-Stiftung Jena, Jena 19385
Otto-von-Guericke-Museum in der Lukasklause, Magdeburg 20128
Philipp-Matthäus-Hahn-Museum, Albstadt 16458
Rhein-Museum Koblenz, Koblenz 19588
Science Center Spectrum, Berlin 17255
Uhrenmuseum Lorenz-Berlin, Berlin 17275
Wettermuseum Lindenberg, Tauche 22012

Hungary
Eötvös Loránd Emlékkiállítás, Budapest 23397

India
Bardhaman Science Centre, Bardhaman 24076
Dhenkanal Science Centre, Dhenkanal 24165
District Science Centre, Purulia 24364

Israel
National Museum of Science, Planning and Technology, Haifa 24966

Italy
Museo degli Strumenti per il Calcolo, Pisa 27241
Museo Galileo, Firenze 26192

Korea, Republic
National Science Museum of Korea, Daejeon 29802

Latvia
Latvijas Republikas Meliorācijas un Zemkopības Muzejs, Mālpils 30047

Mexico
Museo Universitario de Ciencias, Zacatecas 31583

Netherlands
Historical Collection Electronical Engineering, Delft 32057
Museum Boerhaave, Leiden 32484
NEMO, Amsterdam 31875
Teylers Museum, Haarlem 32302

Philippines
Philippine Science Centrum, Manila 34452

Portugal
Museu da Ciência e Indústria, Porto 35844
Museu de Física, Coimbra 35533
Museu dos Capelinhos, Horta 35645

Romania
Muzeul Ştiinţei şi Tehnicii Ştefan Procopiu, Iaşi 36426

Russia
Angarskij Muzej Časov, Angarsk 36760
Biljarskij Gosudarstvennyj Istoriko-Archeologičeskij i Prirodnyj Muzej-Zapovednik, Biljarsk 36806
Centr Istorii Aviacionnych Dvigatelej im. N.D. Kuznecova, Samara 37611
Centralnyj Muzej Svjazi im. A.S. Popova, Sankt-Peterburg 37630
Dom-muzej K.È. Ciolkovskogo, Kaluga 36994
Muzej Istorii Razvitija Mediciny, Astrachan 36784
Muzej Istorii Vyčislitelnoj Techniki, Moskva 37301
Muzej K. È. Ciolkovskogo, Aviacii i Kosmonavtiki, Kirov 37040
Muzej M.V. Lomonosova, Sankt-Peterburg 37697
Muzej Meteorologii Glavnoj Geofizičeskoj Observatorii im. A.I. Voejkova, Sankt-Peterburg 37698
Muzej Meterologičeskaja Stancija Simbirska, Uljanovsk 37903
Muzej Moskovskogo Gosudarstvennogo Techičeskogo Universiteta im. N.È. Baumana, Moskva 37337
Muzej Nižegorodskaja Radiolaboratorija, Nižnij Novgorod 37462
Muzej Radio im. A.S. Popova, Ekaterinburg 36898

Serbia
Muzej Nikole Tesle, Beograd 38071

Singapore
Singapore Science Centre, Singapore 38180

Spain
Colecciòn dos Gabinetes de Física, Química e Historia Natural do Antigo, Santiago de Compostela 40016
CosmoCaixa Madrid, Alcobendas 38822
Museo-Colección del Observatorio Astronómico Nacional, Madrid 39529
Museo Nacional de Ciencia y Tecnología, Madrid 39570
Museu d'Història de Cambrils, Cambrils 39110
Parque de las Ciencias, Granada 39370

Sweden
Bil Teknikhistoriska Samlingarna, Köping 40645
Bramtidsmuseet, Borlänge 40475
Observatoriemuseet, Stockholm 40923

Taiwan
921 Earthquake Museum, Kengkou 42251
Museum of Meteorology, Tainan 42363

United Arab Emirates
Sharjah Science Museum, Sharjah 43230

United Kingdom
Barometer World Museum, Merton 45132
Cavendish Laboratory, Cambridge 43726
Collins Gallery, Glasgow 44278
Hunterian Museum, Glasgow 44285
Museum of the History of Science, Oxford 45334
Saint Andrews University Museum Collections, Saint Andrews 45557
Satrosphere, Aberdeen 43248
University of Dundee Museum Collections, Dundee 44052
Whipple Museum of the History of Science, Cambridge 43736
Whowhatwherewhenwhy W5, Belfast 43449

United States of America
A.W.A. Electronic-Communication Museum, Bloomfield 46853
American Museum of Science and Energy, Oak Ridge 51456
Arizona Science Center, Phoenix 51900
Baltimore Public Works Museum, Baltimore 46553
Berea College Burroughs Geological Museum, Berea 46746
Black Hills Museum of Natural History, Hill City 49283
Bowers Science Museum, Cortland 47829
Brazosport Museum of Natural Science, Brazosport 46999
California Science Center, Los Angeles 50215
Carnegie Science Center, Pittsburgh 51944
Catawba Science Center, Hickory 49266
The Children's Science Center, Cape Coral 47232
Cincinnati Museum Center, Cincinnati 47557
Coastal Discovery Museum, Hilton Head Island 49294
Columbia River Exhibition of History, Science and Technology, Richland 52307
Cranbrook Institute of Science - Natural History Collections, Bloomfield Hills 46857
Dakota Science Center, Grand Forks 48986
Discovery Center of Idaho, Boise 46891
Discovery Center Science Museum, Fort Collins 48606
Discovery Museum, Sacramento 52506
Discovery Park, Safford 52513
Discovery Science Center of Central Florida, Ocala 51482
Discovery World - The James Lovell Museum of Science, Economics and Technology, Milwaukee 50707
Don Harrington Discovery Center, Amarillo 46252
East Tennessee Discovery Center, Knoxville 49816
Edison and Ford Winter Estates, Fort Myers 48661
Everhart Museum of Natural History, Science and Art, Scranton 52980
Explora Science Center and Children's Museum, Albuquerque 46177
The Exploratorium, San Francisco 52773
Flandrau Science Center, Tucson 53589
Fort Worth Museum of Science and History, Fort Worth 48713
The Franklin Institute, Philadelphia 51836
Gateway to Science, Bismarck 46838
Geology Museum, Huntington 49434
George Washington Carver House, Diamond 48137
Harbor Branch Oceanographic Institution, Fort Pierce 48667
Houston Museum of Natural Science, Houston 49390
Insights - El Paso Science Center, El Paso 48324
Iredell Museum of Arts Heritage, Statesville 53274
Kauai Children's Discovery Museum, Kapaa 49706
Lakeview Museum of Arts and Sciences, Peoria 51773
Louisiana Arts and Science Center, Baton Rouge 46633
McAllen International Museum, McAllen 50339
McWane Science Center, Birmingham 46823
Mary Brogan Museum of Art and Science, Tallahassee 53426
Maryland Science Center, Baltimore 46578
Miami Museum of Science, Miami 50633
Monte L. Bean Life Science Museum, Provo 52187
Museum at Southwestern Michigan College, Dowagiac 48173
Museum of Arts and Sciences, Macon 50367
Museum of Discovery and Science, Fort Lauderdale 48639
Museum of Geology, Rapid City 52253
Museum of Science, Boston 46941
Museum of Science and History of Jacksonville, Jacksonville 49584
Museum of the Center for Meteorite Studies, Tempe 53470
National Inventors Hall of Fame, Akron 46147
National Museum of Natural History, Washington 53904
New Detroit Science Center, Detroit 48128
New York Hall of Science, Flushing 48578
North Carolina Museum of Life and Science, Durham 48223
The North Museum of Natural History and Science, Lancaster 49929
Omnisphere and Science Center, Wichita 54148
The Oyster and Maritime Museum of Chincoteague, Chincoteague 47539
Pacific Science Center, Seattle 53010
Paterson Museum, Paterson 51730
Petaluma Wildlife and Natural Science Museum, Petaluma 51791
Porter Thermometer Museum, Onset 51567
Reading Public Museum and Art Gallery, Reading 52265
Red River Valley Museum, Vernon 53723
Reuben H. Fleet Science Center, San Diego 52743
Rochester Museum, Rochester 52398
Saint Louis Science Center, Saint Louis 52598
Sci-Port Discovery Center, Shreveport 53089
Sci-Tech Center of Northern New York, Watertown 53973
Science Center of Connecticut, West Hartford 54061
Science Center of Iowa, Des Moines 48101
Science Learning Center, Upton 53668
Science Museum of Long Island, Manhasset 50430
Science Museum of Virginia, Richmond 52338
Science Museum of Western Virginia, Roanoke 52382
Science Spectrum, Lubbock 50313
Science Station Museum, Cedar Rapids 47307
Sciencenter, Ithaca 49550
Sony Wonder Technology Lab, New York 51259
South Dakota Discovery Center, Pierre 51916
Southwest Museum of Science and Technology, Dallas 47941
Union Station Kansas City, Kansas City 49705
Virginia Living Museum, Newport News 51337
Weather Discovery Center, North Conway 51390
William McKinley Presidential Library and Museum, Canton 47229

Scientific Expeditions

Australia
Australian Antarctic Division Display, Kingston 01203
Bligh Museum of Pacific Exploration, Bruny Island 00912
Charles Sturt Museum, Grange 01129

S

Austria

Heinrich Harrer-Museum, Hüttenberg 02103

Belgium

Koninklijk Museum voor Midden-Afrika, Tervuren 04058

Canada

Champlain Trail Museum and Pioneer Village, Pembroke 06580
Heritage Centre Canada - HVACR, Mississauga 06311
Maison Léon-Provancher, Cap-Rouge 05550
TELUS World of Science, Edmonton 05755

China

Dufu Caotang Museum, Chengdu 07616

Czech Republic

Památník Dr. Emila Holuba, Holice v Čechách 09522

Denmark

Hauchs Physiske Cabinet, Sorø 10280
Knud Rasmussens Hus, Hundested 10104

Dominican Republic

Museo Faro a Colón, Santo Domingo 10355

France

Musée des Sciences et Techniques, Laval (Mayenne) 13540

Germany

Abenteuermuseum Saarbrücken - Sammlung Heinz Rox-Schulz, Saarbrücken 21453
Alfred Wegener Museum, Rheinsberg 21302

Italy

Museo Aerospaziale Monte di Apollo, Perugia 27146

Kyrgyzstan

Oblastnoj Muzej im. N.M. Prževalskogo, Prževalsk 29964

Latvia

Gaidu un Skautu Muzejs, Ogre 30051

New Zealand

Canterbury Museum, Christchurch 33069

Norway

Ishavsmuseet Aarvak, Brandal 33452
Olav Bjaaland Museum, Morgedal 33681
Polarmuseet, Andenes 33393
Roald Amundsens Hjem Uranienborg, Svartskog 33894
Roald Amundsens Minne, Torp 33912

Peru

Museo Antonio Raimondi, Lima 34224

Portugal

Casa Museu Cristóvão Colombo, Porto Santo 35870

South Africa

Hartenbos Museum, Hartenbos 38639

Spain

Casa Museo de Colón, Valladolid 40228
Monasterio de Santa María de la Rábida, Palos de la Frontera 39804
Museo Casa de Colón, Las Palmas de Gran Canaria 39799

Sweden

Andréemuseet, Gränna 40562
Ostindiefararen Götheborg, Göteborg 40553

United Kingdom

Captain Cook Birthplace Museum, Middlesbrough 45139
Captain Cook Memorial Museum, Whitby 45994
Captain Cook Schoolroom Museum, Great Ayton 44352
David Livingstone Centre, Blantyre 43542
Gilbert White's House and the Oates Museum, Selborne 45632
John McDouall Stuart Museum, Kirkcaldy 44608
Royal Research Ship Discovery, Dundee 44050
Scott Polar Research Institute Museum, Cambridge 43732

United States of America

Captain Nathaniel B. Palmer House, Stonington 53311
John Woolman Memorial, Mount Holly 50879
Johnson Collection of Photographs, Movies and Memorabilia, Chanute 47341
Lawrence L. Lee Scouting Museum, Manchester 50422
Lewis and Clark Boat House and Nature Center, Saint Charles 52534
Martin and Osa Johnson Safari Museum, Chanute 47342
Stephen Phillips Memorial House, Salem 52661

Uruguay

Museo de Descubrimiento, Montevideo 54397

Sculpture

Andorra

Museu Viladomat d'Escultura, Escaldes-Engordany 00094

Argentina

El Fogon de los Arrieros, Resistencia 00521
Museo Casa de Yrurtia, Buenos Aires 00173
Museo de Bellas Artes al Aire Libre Caminito, Buenos Aires 00186
Museo de Bellas Artes de la Boca Quinquela Martín, Buenos Aires 00187
Museo Municipal de Bellas Artes Lucas Braulio Areco, Posadas 00500
Museo Nacional de Bellas Artes, Buenos Aires 00259

Armenia

Ara Sargsyan and Hakob Kojoyan House-Museum, Yerevan 00725
Ervand Kochar Museum, Yerevan 00734
Khoren Ter-Harutyan Museum, Echmiadzin 00697
Wood Carving Museum, Yerevan 00758
Yervand Kochar Museum, Yerevan 00762

Australia

City of Melbourne Collection, Melbourne 01277

Austria

Anton Hanak-Museum, Langenzersdorf 02260
Archäologische Sammlung der Universität Graz, Graz 01963
Atelier Jaindl, Villach 02848
Belvedere - Augarten Contemporary, Wien 02950
Dokumentation des ehemaligen Heilbades, Bad Pirawarth 01761
Essl Museum - Kunst der Gegenwart, Klosterneuburg 02200
Freilichtmuseum Römersteinbruch, Sankt Margarethen 02664
Galerie im Bildhauerhaus und Skulpturenwanderweg, Einöde 01853
Gironcoli Museum, Sankt Johann bei Herberstein 02655
Künstlergarten Wietzen - Alfred Kurz, Lichtenau 02297
Künstlerhaus Büchsenhausen, Innsbruck 02120
Liechtenstein Museum, Wien 03024
Museum des Augustiner Chorherrenstifts, Reichersberg 02572
Schloß Museum, Pram 02524
Schloßmuseum Niederleis, Niederleis 02446
Schwarzenberger Skulpturenpark, Schwarzenberg am Böhmerwald 02721
Siegfried Charoux Museum, Langenzersdorf 02262
Stadtturmgalerie, Innsbruck 02132
Stahlpark Riedersbach, Sankt Pantaleon 02676
Tiroler Künstlerschaft - Kunstpavillon, Innsbruck 02135
Weitwandermuseum, Eibiswald 01850

Azerbaijan

Rustam Mustafayev ad. Azerbaycan Dövlet İncesenet Muzeyi, Bakı 03169

Belgium

BAM - Beaux Arts Mons, Mons 03879
Gallery Sculpturama, Bornem 03384
Krippana - Exposition Internationale de Crèches d'Église, Büllingen 03504
Middelheimmuseum, Antwerpen 03292
Musée Communal du Comte de Jette, Bruxelles 03459
Musée Constantin Meunier, Bruxelles 03460
Musée de l'Art Wallon, Liège 03808
Musée de l'Assistance Publique, Bruxelles 03470
Musée des Beaux-Arts Charleroi, Charleroi 03516
Musée du Vieux-Cimetière, Soignies 04036
Musée en Plein Air du Sart-Tilman, Liège 03813
Musées Royaux d'Art et d'Histoire, Bruxelles 03495
Museum Charles Delporte, Damme 03534
Museum Herman De Cuyper, Blaasveld 03371
Museum Mayer van den Bergh, Antwerpen 03296
Schatkamer van de Sint-Salvatorskathedraal, Brugge 03427
Stedelijk Museum Het Toreke, Tienen 04070
Stedelijk Museum Hof van Busleyden, Mechelen 03862
Steenmuseum, Gent 03645
Stichting George Grard, Gijverinkhove 03656
Taxandriamuseum, Turnhout 04102

Brazil

Museu de Arte Contemporânea da Universidade de São Paulo, São Paulo 04879
Museu de Artes Plásticas, Taubaté 04925
Museu Internacional de Arte Naïf do Brasil, Rio de Janeiro 04754

Bulgaria

Sofijska Gradska Chudožestvena Galerija, Sofia 05203

Cambodia

National Museum of Cambodia, Phnom Penh 05287

Canada

Art Gallery of York University, North York 06485
Canadian Sculpture Centre, Toronto 07099
Galerie Colline, Edmundston 05763
John Weaver Sculpture Museum, Hope 05995
Latcham Gallery, Stouffville 07016
Lieu Historique National du Canada de la Bataille-de-la-Châteauguay, Ormstown 06516
Maison Médard-Bourgault, Saint-Jean-Port-Joli 06827
Mill of Kintail Conservation Area, Almonte 05323
Mind and Matter Gallery, South Surrey 06996
Musée d'Art Inuit Brousseau, Québec 06686
Toronto Sculpture Garden, Toronto 07150

China

Art Museum of Dazu Rock Carvings in Chongqing, Chongqing 07628
Shuanglinsi Painted Sculpture Art Museum, Pingyao 08091

Congo, Democratic Republic

Musée des Beaux-Arts, Kinshasa 08798

Croatia

Atelier Ivan Meštrović, Zagreb 08987
Galerija Antuna Augustinčića, Klanjec 08888
Galerija Skulptura Ivan Sabolić, Peteranec 08928
Hrvatska Akademija Znanosti i Umjetnosti - Gliptoteka, Zagreb 08993
Ivan Meštrović Gallery, Split 08957
Kaštelet Crikvine, Split 08958
Memorijalna Galerija Aleksandar Rukavina, Brtonigla 08850
Muzej za Umjetnost i Obrt, Zagreb 09008
Umjetnička Galerija, Dubrovnik 08874

Cuba

Museo Artes Plásticas, Pinar del Río 09247
Museo Oscar Fernandez Morera, Sancti-Spíritus 09290
Museo Taller de Escultura, Victoria de las Tunas 09350

Cyprus

Limassol Municipal Art Gallery, Lemesos 09388
Pierides Sculpture Garden, Larnaka 09370

Czech Republic

Městské Muzeum a Galerie, Hořice 09524
Městské Muzeum a Galerie Otakara Spaniela a Josefa Wágnera, Jaroměř 09550
Muzeum Antického Sochařství a Architektury, Litomyšl 09633
Muzeum umění a designu, Benešov 09402
Státní Galerie Výtvarného Umění, Cheb 09474

Denmark

Faaborg Museum for Fynsk Malerkunst, Faaborg 10008
Hammerichs Hus, Ærøskøbing 09959
Louisiana Museum of Modern Art, Humlebæk 10103
Ny Carlsberg Glyptotek, København 10145
Randers Kunstmuseum, Randers 10230
Rudolph Tegners Museum, Dronningmølle 09993
Statens Museum for Kunst, København 10152
Thingbaek Kalkminer Bundgaards Museum, Skørping 10273
Thorvaldsens Museum, København 10155
Thorvaldsens Samlingen på Nysø, Vordingborg 10321
Vejen Kunstmuseum, Vejen 10309

Egypt

Mukhtar Museum, Giza 10527

Finland

Emil Wikström Museo, Valkeakoski 11435
Gösta Serlachiuksen Taidemuseo, Mänttä 11093
Huittisten Museo, Huittinen 10842
Lapinlahden Taidemuseo ja Eemil Halosen Museo, Lapinlahti 11044
Mikkelin Taidemuseo, Mikkeli 11115
Wäinö Aaltonen Museo, Turku 11408
Walter Runebergin Veistoskokoelma, Porvoo 11231
Yrjö Liipolan Taidekokoelma, Koski 10985

France

Atelier Manoli-Musée et Jardin de Sculpture, La Richardais 13409
Centre Artistique Jean Baffier, Sancoins 15622
Centre d'Art Contemporain, Istres 13253
Chapelle des Visitandines, Auxerre 11773
Château de Fleurigny, Thorigny-sur-Oreuse 15901
Château-Musée, Saumur 15661
Collections de la Fondation de Coubertin, Saint-Rémy-lès-Chevreuse 15494
Collections de l'Ecole Nationale Supérieure des Beaux-Arts, Paris 14590
Dalí - Espace Montmartre, Paris 14594
Fondation Le Corbusier, Paris 14601
Fondation Musée-Volti, Villefranche-sur-Mer 16192
Fondation Van-Gogh Arles, Arles 11682
Galerie Beaubourg, Vence 16097
Halle Saint-Pierre Museum, Paris 14611
Maison et Sculptures de Jean Cacoud, Fransèches 12998
Musée Adzak, Paris 14628
Musée André Abbal, Carbonne 12279
Musée Bartholdi, Colmar 12582
Musée Boleslaw Biegas, Paris 14634
Musée Bouchard, Paris 14635
Musée Bourdelle, Paris 14636
Musée Calvet, Avignon 11789
Musée Costi, Bergerac 11943
Musée d'Art et d'Archéologie, Aurillac 11754
Musée d'Art et d'Histoire, Meudon 14103
Musée d'Art et d'Histoire de Puisaye, Villiers-Saint-Benoît 16244
Musée David, Angers 11607
Musée de Faykod, Aups 11748
Musée de la Bohème, Yviers 16311
Musée de la Pierre Écrite, Vézelay 16136
Musée de la Société Archéologique et Historique de la Charente, Angoulême 11619
Musée de la Vie Romantique, Paris 14665
Musée de Plein Air, Gueugnon 13163
Musée de Provins et du Provinois, Provins 14951
Musée de Région Auguste Chabaud, Graveson 13124
Musée de Statues d'Enfants, Les Mages 13704
Musée Départemental d'Art Ancien et Contemporain, Épinal 12840
Musée Départemental de l'École Publique, Chevregny 12509
Musée Départemental de l'Oise, Beauvais 11910
Musée des Amis de Thann, Thann 15874
Musée des Années 30, Boulogne-Billancourt 12090
Musée des Augustins, Toulouse 15939
Musée des Beaux-Arts, Angers 11609
Musée des Beaux-Arts, Arras 11698
Musée des Beaux-Arts, Béziers 11976
Musée des Beaux Arts, Lons-le-Saunier 13817
Musée des Beaux-Arts, Marseille 14002
Musée des Beaux-Arts, Pau 14757
Musée des Beaux-Arts, Valenciennes 16043
Musée des Bois Sculptes, Cancale 12263
Musée Despiau Wlérick, Mont-de-Marsan 14158
Musée du Centre de Recherches sur les Monuments Historiques, Paris 14694
Musée du Luxembourg, Paris 14699
Musée du Prieuré, Charolles 12416
Musée du Prieuré de Graville, Le Havre 13596
Musée du Santon, Pierrefeu-du-Var 14817
Musée Étienne Martin, Noyal-sur-Vilaine 14478
Musée François Pompon, Saulieu 15656
Musée Henri Chapu, Le Mée-sur-Seine 13613
Musée Ivanoff, La Borne 13318
Musée-Jardin Paul Landowski, Boulogne-Billancourt 12091

S

Musée Jules Desbois, Parçay-les-Pins 14563
Musée Jules Garinet, Châlons-en-Champagne 12363
Musée Lapidaire, Avignon 11793
Musée Lapidaire de Mozac, Mozac 14321
Musée Lapidaire et Exposition Permanente sur la Restauration, Carcassonne 12286
Musée Le Petit Louvre, La Pacaudière 13399
Musée Louis Cornu, Wassigny 16289
Musée Magnin, Dijon 12742
Musée Moralès, Port-de-Bouc 14914
Musée Morice Lipsi, Rosey 15106
Musée Municipal de la Chartreuse, Douai 12775
Musée Municipal de l'Évêché, Limoges 13770
Musée Municipal de Vendôme, Vendôme 16100
Musée Municipal Nicolas Poussin, Les Andelys 13683
Musée National Auguste Rodin, Meudon 14104
Musée P. Dubois-A. Boucher, Nogent-sur-Seine 14464
Musée René Davoine, Charolles 12417
Musée Rude, Dijon 12743
Musée Santons du Monde, Clermont-l'Hérault 12553
Musée Sculpture sur Bois, Lans-en-Vercors 13516
Musée Zadkine, Paris 14740
Villa Arson, Nice 14442

Germany

1822-Forum Frankfurter Sparkasse, Frankfurt am Main 18352
Akademisches Kunstmuseum - Antikensammlung der Universität, Bonn 17442
Albrechtsburg Meißen, Meißen 20263
Antiken- und Abgußsammlung des Archäologischen Seminars, Marburg 20185
Antikensammlung der Friedrich-Alexander-Universität Erlangen-Nürnberg, Erlangen 18219
Arp Museum Bahnhof Rolandseck, Remagen 21259
Bildhauermuseum Prof. Wandschneider, Plau am See 21051
Bode-Museum - Skulpturensammlung und Museum für Byzantinische Kunst, Berlin 17052
Böke Museum, Leer 19868
Edwin Scharff Museum, Neu-Ulm 20587
Ernst Barlach Atelierhaus und Ausstellungsforum-Graphikkabinett, Güstrow 18827
Ernst Barlach Haus, Hamburg 18926
Ernst Barlach Museum, Ratzeburg 21196
Europäischer Skulpturenpark, Willebadessen 22582
Felix-Müller-Museum im Zehntspeicher, Neunkirchen am Brand 20638
Forum Konkrete Kunst Erfurt, Erfurt 18202
Freies Museum Berlin, Berlin 17085
Fritz-Best-Museum, Kronberg im Taunus 19724
Galerie am Alten Markt, Rostock 21391
Gemälde- und Skulpturensammlung, Nürnberg 20763
Gerhard-Marcks-Haus, Bremen 17551
Gertrudenkapelle, Güstrow 18828
Herbert-Gerisch-Stiftung, Neumünster 20635
Klostergalerie, Zehdenick 22769
Kulturzentrum Sinsteden des Rhein-Kreises Neuss, Rommerskirchen 21367
Kunst- und Skulpturenmuseum, Heilbronn 19097
Kunstforum Ostdeutsche Galerie, Regensburg 21228
Kunsthalle Bremen, Bremen 17556
Kunsthalle im Schloss, Isny im Allgäu 19366
Kunsthalle Mannheim, Mannheim 20166
Kunsthalle Schweinfurt, Schweinfurt 21683
Kunsthalle Würth, Schwäbisch Hall 21657
Kunstmuseum Kloster Unser Lieben Frauen, Magdeburg 20124
Kunstsammlung Lorenzkapelle Rottweil, Rottweil 21425
Kunstsammlungen der Ruhr-Universität Bochum, Bochum 17402
Kunststätte Bossard, Jesteburg 19395
Kunstverein Konstanz e.V., Konstanz 19680
Lapidarium, Berlin 17175
Lechner Museum, Ingolstadt 19343
Mainfränkisches Museum Würzburg, Würzburg 22713
Museum am Ostwall, Dortmund 17924
Museum Bad Arolsen, Bad Arolsen 16668
Museum Bad Arolsen, Bad Arolsen 16669
Museum Fritz von Graevenitz, Stuttgart 21954
Museum für Abgüsse Klassischer Bildwerke, München 20465
Museum für Kunst in Steatit, Frankfurt am Main 18384
Museum Künstlerhaus Gasteiger, Utting am Ammersee 22194
Museum Starnberger See, Starnberg 21864
Museum und Stiftung Anton Geiselhart, Münsingen 20501
Museum Würth, Künzelsau 19736
Nationalgalerie Friedrichswerdersche Kirche - Schinkelmuseum, Berlin 17222
Sammlung Mittelalterliche Kunst in Thüringen, Eisenach 18097
Sammlung Scharf-Gerstenberg, Berlin 17244
Sammlungen des Archäologischen Instituts der Universität, Göttingen 18700
Schloss Seehof, Memmelsdorf 20278
Skulpturenmuseum Glaskasten Marl, Marl, Westfalen 20224
Skulpturenmuseum im Hofberg, Landshut 19805
Skulpturenpark, Bad Nauheim 16792
Skulpturensammlung, Dresden 17970
Skulpturenweg, Emmendingen 18149
Stadtmuseum Simeonstift Trier, Trier 22105
Städtische Galerie im Rathauspark, Gladbeck 18651
Städtische Galerie Liebieghaus, Frankfurt am Main 18394
Stiftsmuseum, Aschaffenburg 16593
Stiftung Wilhelm Lehmbruck Museum, Duisburg 18027
Suermondt-Ludwig-Museum, Aachen 16408
Syker Vorwerk, Syke 21992
Verein für Heimatpflege, Viersen 22233
Von der Heydt-Museum, Wuppertal 22748
Walhalla, Donaustauf 17893
Wilhelm-Hack-Museum Ludwigshafen am Rhein, Ludwigshafen am Rhein 20058

Greece

A. Sohos Museum, Tinos 23225
Christos Capralos Museum, Aegina 22866
Christos Capralos Museum, Agrinion 22874
Historical Museum of Crete, Iráklion 23019
Monemvasia Archaeological Collection, Monemvasia 23100
Museum Alex Mylona, Athinai 22928
Museum of Giannoulis Halepas, Pyrgos 23162
Museum of Sculpture and Figurines of Loukia Georganti, Athinai 22938
Museum of Tinian Artists, Pyrgos 23163
N. Hadjikyriakos-Ghikas Gallery, Athinai 22941
Nikolaos Perantinos Museum of Sculpture, Athinai 22945
Skironio Centrum Kifissa, Néa Kifissia 23123
Yiannoulis Halepas Museum and Museum of Tinian Artists, Panormos Tinou 23133

Guam

Isla Center for the Arts at the University of Guam, Mangilao 23287

Hungary

Amerigo Tot Múzeum, Pécs 23757
Borsos Miklós Múzeum, Győr 23580
Budapest Galéria Kiállítóháza, Budapest 23389
Budapest Galéria Varga Imre Állandó Kiállítóház, Budapest 23391
Finta Múzeum, Túrkeve 23939
Hincz Gyula Állandó Gyűjtemény, Vác 23944
Martyn Ferenc Múzeum, Pécs 23763
Memento Park Budapest, Budapest 23449
Nagy Ferenc Galéria, Tab 23908
Schaár Erzsébet Gyűjtemény, Székesfehérvár 23858
Schaár Erzsébet - Utca, Pécs 23775

Iceland

Ásmundarsafn, Reykjavík 24015
Listasafn Einars Jónssonar, Reykjavík 24021
Listasafn Sigurjóns Ólafssonar, Reykjavík 24023

India

Archaeological Museum, Konarak 24284
Archaeological Museum, Vaisali 24421
Archaeological Museum, Varanasi 24424
Bangiya Sahitya Parisad Museum, Kolkata 24269
Belkhandi Museum, Belkhandi 24081
Bhagalpur Museum, Bhagalpur 24082
Birla Museum, Bhopal 24090
Bundelkhand Chhatrasal Museum, Banda 24070
District Museum, Guntur 24185
District Museum, Shivpuri 24391
Dr. Raj Bali Pandey Puratatva Sangrahalaya, Deoria 24157
Government Museum, Bharatpur 24084
Government Museum, Kodagu 24263
Government Museum, Mathura 24309
Government Museum, Mount Abu 24312
Government Museum, Udaipur 24414
Government Museum and Art Gallery, Chandigarh 24109
Himachal State Museum, Shimla 24389
History Museum, Ahmednagar 24053
I.V.K. Rajwade Sanshodhan Mandal Museum, Dhule 24166
Khajana Buildings Museum, Hyderabad 24213
Khiching Museum, Khiching 24260
Mahatma Gandhi Hindi Sangrahalaya, Kalpi 24250
Mehrangarh Museum, Jodhpur 24246
Mutua Museum, Imphal 24226
Puratatva Sangrahalaya, Gorakhpur 24180
Purvattatva Sangrahalaya Madhya Pradesh Shasan, Damoh 24124
Rani Laxmi Bai Palace Sculpture Collection, Jhansi 24244
Rastrakavi Govind Pai Museum, Udupi 24416
Sahitya Parishad Museum, Midnapore 24311
Sikar Museum, Sikar 24393
Sir Choturam Memorial Museum, Sangaria 24376
State Kala Akademi Museum, Imphal 24229
State Museum, Bhopal 24096
Thanjavur Art Gallery, Thanjavur 24401
TTD Museum, Tirupati 24409
Tulsi Sangrahalaya Ramvan, Satna 24382
Watson Museum, Rajkot 24370
Zilla Samgrahasala, Purulia 24365

Ireland

Bogwood Sculpture Artists, Newtowncashel 24885

Italy

Antiquarium Sestinale, Sestino 27821
Casa Buonarroti, Firenze 26143
Casa Museo Fantoni, Rovetta 27572
Centro d'Arte Verrocchio, Casole d'Elsa 25708
Cripta e Museo di Sant'Anastasio, Asti 25312
Fondazione Magnani Rocca, Mamiano di Traversetolo 26580
Galleria Comunale d'Arte, Cagliari 25578
Galleria d'Arte Moderna, Firenze 26156
Galleria dell'Accademia, Firenze 26159
Galleria Regionale di Palazzo Bellomo, Siracusa 27853
Galleria Rinaldo Carnielo, Firenze 26162
Giardino di Boboli, Firenze 26163
Gipsoteca F. Jerace, Catanzaro 25778
Musei Civici, Pavia 27120
Musei Civici di Padova, Padova 27034
Museo Archeologico e della Collegiata, Casole d'Elsa 25709
Museo Arcivescovile, Ravenna 27370
Museo Barracco, Roma 27463
Museo Berenziano, Cremona 26002
Museo Canonicale, Verona 28231
Museo Civico Antonio Cordici, Erice 26062
Museo Civico d'Arte, Pordenone 27297
Museo d'Arte Antica, Milano 26694
Museo de Santuario di Montevergine, Santuario di Montevergine 27736
Museo degli Affreschi G.B. Cavalcaselle, Verona 28233
Museo del Duomo, Milano 26698
Museo della Figurina, Modena 26750
Museo delle Statue - Stele Lunigianesi, Pontremoli 27293
Museo dell'Opera di Santa Maria del Fiore, Firenze 26176
Museo di Castelvecchio, Verona 28235
Museo di Sant'Agostino, Genova 26309
Museo di Sculture Iperspaziali, Bomarzo 25491
Museo Dupré, Fiesole 26128
Museo e Galleria di Villa Borghese, Roma 27503
Museo e Galleria Mozzi Bardini, Firenze 26189
Museo L. Minguzzi, Milano 26709
Museo Lapidario e del Tesoro del Duomo, Modena 26753
Museo Lapidario Marsicano, Avezzano 25335
Museo Lodovico Pogliaghi, Varese 28163
Museo Michelangelo, Caprese 25664
Museo Nazionale, Ravenna 27374
Museo Nazionale del Bargello, Firenze 26196
Museo Pietro Canonica, Roma 27525
Museo Raffaello Pagliaccetti, Giulianova 26347
Museo Ugo Carà, Muggia 26876
Parco Monumentale di Pinocchio, Collodi 25942
Pinacoteca dell'Accademia di Belle Arti, Carrara 25691
Pinacoteca e Museo delle Arti di Locri Epizephiri, Locri 26506
Pinacoteca Giovanni e Marella Agnelli, Torino 28031
Raccolta G. Manzù, Ardea 25261
Sala Pietro Canonica, Stresa 27913
Villa Carlotta, Tremezzo 28053
Villa San Michele, Anacapri 25224

Japan

Asakura Sculpture Gallery, Tokyo 29370
Chokoku no Mori Bijutsukan, Hakone 28476
Gendai Chokoku Bijutsukan, Tokyo 29393
Hikarinotani Metaru Bijutsukan, Imba 28570
Ibara Kyodo Denchu Bijutsukan, Ibara 28551
Iwate-kenritsu Hakubutsukan, Morioka 28935
Rokuzan Bijutsukan, Hotaka 28549
Sapporo Museum of Sculpture, Sapporo 29223
Takamura's Cottage Memorial Museum, Hanamaki 28494
Utsukushigahara Kogen Bijutsukan, Takeshi 29346
Vangi - The Sculpture Garden Museum, Nagaizumi 28950

Kazakhstan

Častnyj Muzej Figurskoj Kultury, Almaty 29676

Korea, Republic

Mok Am Museum of Art, Koyang 29848
Moran Open-air Museum, Namyangju 29857

Kyrgyzstan

Muzej-masterskaja Gapara Ajtieva, Biškek 29958
Muzej Skulptury pod Otkrytym Nebom, Biškek 29959

Latvia

Tēlnieka Voldemāra Jākobsona Memoriālāis Muzejs, Bebri 29990

Lebanon

Écomusée Rachana, Rachana 30171

Lithuania

Antano Mončio Namai-Muziejus, Palanga 30327
Druskininkų Ž. Lipšico Žydų Muziejus, Druskininkai 30223
Europos Parkas, Vilnius 30403
Juozo Zikaro Memorialinis Muziejus, Kaunas 30246
Juozo Zikaro Memorialinis Muziejus, Paliukai 30330
Kazio Varnelio Namai-Muziejus, Vilnius 30411
Kretingos Muziejus, Kretinga 30288
Rokiškio Krašto Muziejus, Rokiškis 30346
Skulptoriaus Broniaus Pundziaus Memorialinis Muziejus, Pieveniai 30337
Vinco Grybo Memorialinis Muziejus, Jurbarka 30234
Žemaičių Dailės Muziejus, Plungė 30338

Malaysia

Ibrahim Hussein Foundation Museum, Pulau Langkawi 30612

Mexico

Museo de la Escultura Mexica Eusebio Dávalos, Acatitlán 30693
Museo Escultórico de Geles Cabrera, México 31112
Museo Universitario Alejandro Rangel Hidalgo, Nogueras 31216

Morocco

Forbes Museum, Tanger 31695

Nepal

National Wood Working Museum, Bhaktapur 31771

Netherlands

Koninklijk Paleis te Amsterdam, Amsterdam 31855
Museum Beelden aan Zee, Den Haag 32084
Museum Jacobs van den Hof, Amersfoort 31815
Museum Nanning Hendrik Bulthuis, Leeuwarden 32476
Museum Theo Jans, Klarenbeek 32450
Odapark, Venray 32885
Het Oude Slot Heemstede, Heemstede 32328
Pier Pander Museum, Leeuwarden 32477

Nigeria

National Museum Esie, Esie 33321

Northern Cyprus

Taş Eserler Müzesi - Lapidari, Lefkoşa 33373

Norway

Anders Svors Museet, Hornindal 33561
Anne Grimdalens Minne, Dalen i Telemark 33462
Erkebispegården, Trondheim 33923
Nils Aas Kunstverksted, Inderøy 33575
Vigeland-Museet, Oslo 33758

Peru

Museo de Arte Colonial Pedro de Osma, Lima 34231
Pinacoteca Municipal de la Provincia de Corongo, Corongo 34154

Poland

Centrum Rzeźby Polskiej, Orońsko 34989
Dom Artysta Plastyka, Warszawa 35221
Kordegarda, Warszawa 35232
Muzeum Kamienica Łozińskich, Kraków 34801
Zachęta - Narodowa Galeria Sztuki, Warszawa 35290

Portugal

Atelier-Museu António Duarte, Caldas da Rainha 35493
Casa Dorita Castel-Branco, Sintra 35943
Museu de Alberto Sampaio, Guimarães 35639

Museu de Pintura e Escultura Martins Correia, Golegã 35628

Puerto Rico

Museo de Arte de Ponce, Ponce 36049
Popular Arts Museum, San Juan 36078

Romania

Colecţia de Artă Plastică Fr. Storck şi Cecilia Cecilia Cuţescu Stock, Bucureşti 36191
Muzeul Cornel Medrea, Bucureşti 36199
Muzeul de Artă, Drobeta-Turnu Severin 36359
Muzeul de Artă, Târgu Jiu 36664
Muzeul de Artă Ion Irimescu, Fălticeni 36367
Muzeul de Sculptură Ion Jalea, Constanţa 36306

Russia

Chudožestvennyj Muzej im. M.S. Tuganova, Vladikavkaz 37928
Dom-muzej Skulptora A.S. Golubkinoj, Zarajsk 37985
Gosudarstvennyj Muzej-masterskaja Skulptora A.S. Golubkinoj, Moskva 37200
Gosudarstvennyj Muzej Vadima Sidura, Moskva 37203
Memorialnyj Muzej-Masterskaja Skulptora S.T. Konenkova, Moskva 37246
Meždunarodnyj Centr-Muzej im. N.K. Rericha, Moskva 37249
Muzej Skulptury, Sankt-Peterburg 37708
Muzej Skulptury S.T. Konenkova, Smolensk 37771
Muzej-usadba im. N.K. Rericha, Izvara 36958
Park-muzej im. A.K. Tolstogo, Brjansk 36829
Permskaja Gosudarstvennaja Chudožestvennaja Galereja, Perm 37550

Serbia

Galerija Matice Srpske, Novi Sad 38120
Galerija Srpske Akademije Nauka i Umetnosti, Beograd 38045
Savremena Galerija Umetničke Kolonije Ečka, Zrenjanin 38167

Singapore

Dr. Kui-Fang Lai Art Museum, Singapore 38173

Slovakia

Galéria Jána Koniarka v Trnave - Kopplov Kaštiel', Trnava 38379
Galéria Jozefa Kollára, Banská Štiavnica 38190
Galéria Mesta Bratislavy, Bratislava 38211
Múzeum Vojtecha Löfflera, Košice 38272
Starý Zamok Banská Štiavnica, Banská Štiavnica 38194

Slovenia

Gorenjski Muzej Kranj, Kranj 38437
Sadnikarjeva Muzejska Zbirka v Kamniku, Kamnik 38428

Spain

Casón del Buen Retiro, Madrid 39507
Colección de Escultura Contemporánea, Madrid 39512
MUSAC, León 39453
Museo Antón, Candás 39116
Museo Chillida Leku, Hernani 39387
Museo de Escultura al Aire Libre, Madrid 39539
Museo de la Abadía del Sacro-Monte, Granada 39366
Museo de la Catedral, Astorga 38908
Museo Lorenzo Coullaut-Valera, Marchena 39623
Museo Municipal, Ayllón 38926
Museo Municipal Mateo Hernández, Béjar 39014
Museo Nacional de Escultura, Valladolid 40240
Museo Salzillo, Murcia 39702
Museu d'Art, Girona 39344
Museu Frederic Marès, Barcelona 38995
Museu Nacional d'Art de Catalunya (MNAC), Barcelona 39001

Sudan

Merowe Museum, Merowe 40395

Sweden

Carl Eldhs Ateljémuseum, Stockholm 40889
Dalahästmuseum, Dala-Järna 40483
Döderhultarmuseet, Oskarshamn 40802
Gustav III:s Antikmuseum, Stockholm 40897
Millesgården, Lidingö 40673
Skulpturens Hus, Stockholm 40935
Vitlycke Museum, Tanumshede 40969

Switzerland

Antikensammlung, Bern 41202
Ateliermuseum und Skulpturengarten Erwin Rehmann, Laufenburg 41532
August-Suter-Museum, Eptingen 41354
Bruno Weber Skulpturenpark, Dietikon 41329
Kirchner Museum Davos, Davos Platz 41321
Kunsthaus Zürich, Zürich 42083
Kunstmuseum Solothurn, Solothurn 41885
Kunstmuseum Winterthur, Winterthur 42029
Lapidarium der Stiftsbibliotheh St. Gallen, Sankt Gallen 41806
Museo Vincenzo Vela, Ligornetto 41566
Museum der Klosterruine mit Lapidarium, Rüeggisberg 41780
Museum Kleines Klingental, Basel 41172
Paul Gugelmann-Museum, Schönenwerd 41842
Schweizerisches Nationalmuseum, Zürich 42105
Sculpture at Schoenthal, Langenbruck 41524
Skulpturhalle Basel, Basel 41184

Taiwan

Chen Jonghui Memorial Museum, Yuanli 42476
Chien Dai Sculpture Art Museum, Sanyi 42315
Dong Zuo Inkslab Art Museum, Ershuei 42185
Gourd Sculpture Museum, Shanlin 42323
Hualien County Stone Sculpture Museum, Hualien 42215
Juming Museum, Chin-shan 42163
Loshi Stone Art Museum, Ershuei 42186
Ping Yang Wood Sculpture Museum, Hualien 42217
Wang Hsiu Chi Sculpture Park, Taipei 42438
Yo Sin-ci Stone Sculpture Museum, Sioulin 42341
Yuan Shuo Shi Han Lin Cong Huei Stone Sculpture Museum, Sincheng 42332

Turkey

II. Yesemek Heykel Atölyesi ve Açikhava Müzesi, İslahiye 42798
Resim ve Heykel Müzesi, İstanbul 42835

Ukraine

Muzej Sakral'noï Skul'ptury XVIII st. - Tvorčist' Ivana Pinzelja, Lviv 43135

United Kingdom

Andrew Logan Museum of Sculpture, Welshpool 45961
Barbara Hepworth Museum and Sculpture Garden, Saint Ives, Cornwall 45575
Bolton Museum and Art Gallery, Bolton 43553
Burton Art Gallery and Museum, Bideford 43477
Cafe Gallery Projects, London 44811
Cass Sculpture Foundation, Goodwood 44328
Courtauld Institute Gallery, London 44830
Henry Moore Foundation, Much Hadham 45183
Henry Moore Institute, Leeds 44661
The Hepworth Wakefield, Wakefield 45908
New Art Centre, East Winterslow 44096
The New Art Gallery Walsall, Walsall 45920
Penwith Galleries, Saint Ives, Cornwall 45576
Royal British Society of Sculptors, London 44969
William Lamb Memorial Studio, Montrose 45174
Winchester Cathedral Triforium Gallery, Winchester 46036
Wrest Park, Silsoe 45680
Yorkshire Sculpture Park, West Bretton 45966

United States of America

Archie Bray Foundation, Helena 49239
The Artists Museum, New York 51126
Arvada Center for the Arts and Humanities, Arvada 46367
Atrium Gallery, Lewiston 50036
Bass Manor Museum, Malta 50409
Boston Sculptors Gallery, Boston 46924
Brookgreen Gardens, Murrells Inlet 50920
Center Gallery, Demarest 48052
Central Washington Gallery, Ellensburg 48356
City of Brea Gallery, Brea 47000
Coos Art Museum, Coos Bay 47794
DeCordova Museum and Sculpture Park, Lincoln 50090
Dowd Fine Arts Gallery, Cortland 47831
Emerson Gallery, McLean 50362
Emerson Gallery, Clinton 47659
Gallery of Art, Fort Valley 48695
Gallery of Art, Overland Park 51632
Gallery of Art, Saint Louis 52578
Ginny Williams Family Foundation, Denver 48078
Grounds for Sculpture, Hamilton 49133
Herter Art Gallery, Amherst 46267
Hirshhorn Museum and Sculpture Garden, Washington 53877
The Hollycroft Foundation, Ivoryton 49552
International Sculpture Center, Hamilton 49134
International Sculpture Park, Schaumburg 52957
Iris and B. Gerald Cantor Art Gallery, Worcester 54299
Irving Arts Center, Irving 49533
Isamu Noguchi Garden Museum, Long Island City 50182
James and Meryl Hearst Center for the Arts, Cedar Falls 47293
Johnson Atelier, Mercerville 50597
Joseph A. Cain Memorial Art Gallery, Corpus Christi 47821
Krasl Art Center, Saint Joseph 52558
Laumeier Sculpture Park, Saint Louis 52584
Marshall M. Fredericks Sculpture Museum, University Center 53655
Merwin and Wakeley Galleries, Bloomington 46862
Mitchell Museum at Cedarhurst, Mount Vernon 50892
Muse Gallery, Jackson 49572
Museum of Antropology, History and Art, Rio Piedras 52355
Myra Powell Art Gallery and Gallery at the Station, Ogden 51502
Nathan Manilow Sculpture Park, University Park 53656
National Liberty Museum, Philadelphia 51860
Northern Clay Center Galleries, Minneapolis 50741
Opus 40 and the Quarryman's Museum, Saugerties 52927
Plattsburgh Art Museum, Plattsburgh 51984
Putney School Gallery, Putney 52204
Pyramid Hill Sculpture Park and Museum, Hamilton 49139
Robert E. Wilson Art Gallery, Huntington 49424
Rodin Museum, Philadelphia 51876
Rubelle and Norman Schafler Gallery, Brooklyn 47082
Runnels Gallery, Portales 52058
Ruth Chandler Williamson Gallery, Claremont 47576
Santarella Museum, Tyringham 53637
Sculpture Center, Long Island City 50187
Sculpture Galena, Galena 48829
The Smithy-Pioneer Gallery, Cooperstown 47792
Socrates Sculpture Park, Long Island City 50188
Storm King Art Center, Mountainville 50898
Synagogue for the Arts, New York 51266
Umlauf Sculpture Garden and Museum, Austin 46522
University at Buffalo Anderson Art Gallery, Buffalo 47125
University College Gallery, College Park 47687
University Galleries, Normal 51372
Virginia Groot A. Foundation, Chicago 47522
Whitney Museum of American Art at Altria, New York 51280
Woolaroc Museum, Bartlesville 46620

Uruguay

Museo de Artes Plásticas, Madonado 54378

Uzbekistan

Qo'qon Viloyat Ülkashunoslik Muzeyi, Qo'qon 54458

Vatican City

Museo Gregoriano Profano, Città del Vaticano 54505

Venezuela

Museo de Arte Contemporáneo Francisco Narvárez, Porlamar 54593

Vietnam

National Museum of Cham Sculpture, Da Nang 54604

Shells

Australia

Big Shell Museum, Tewantin 01577
Griffiths Sea Shell Museum and Marine Display, Lakes Entrance 01214
Port Stephens Shell Museum, Corlette 01001
Rosewall Memorial Shell Museum, Port Lincoln 01432

Brazil

Museu Oceanográfico Prof. Eliézer de C. Rios, Rio Grande 04775

China

Lushun Snakes Dao Natural Museum, Dalian 07642

Croatia

Malakološki Muzej, Makarska 08908

France

Maison de l'Amiral, Pouldreuzic 14929
Museum du Coquillage, Les Sables-d'Olonne 13716

Germany

Buddelschiff-Museum, Wedel 22352
Internationales Muschelmuseum, Wangerland 22329
Kurioses Muschel-Museum Wremen, Wremen 22701
Muschel- und Schneckenmuseum, Norden 20729

Greece

Corfu Museum of Shells, Kastellanoi Mesis 23040
Museum of Shells of the Greek Seas, Néa Moudiana 23124

Italy

Museo Conchiliologico e della Carta Moneta, Bellaria Igea Marina 25388
Museo Provinciale di Storia Naturale, Foggia 26212

Japan

Kasori Kaizuka Hakubutsukan, Chiba 28395

Mozambique

Museu Ferreira de Almeida, Nampula 31715

Netherlands

Gloria Maris Schelpengalerie, Giethoorn 32253
Miramar Zeemuseum, Vledder 32893
Schelpen Museum, Zaamslag 32974

New Zealand

Picton Community Museum, Picton 33182

Oman

Oman Natural History Museum, Muscat 33991

Taiwan

Shell Exhibit at Chi Chin Seashore Park, Kaohsiung 42243

United Kingdom

Glandford Shell Museum, Glandford 44274

United States of America

Dakota Dinosaur Museum, Dickinson 48138

Shipping → Boats and Shipping

Shoes

Austria

Stiefelmachermuseum, Rechnitz 02568

Belgium

Nationaal Schoeiselmuseum, Izegem 03738

Canada

Bata Shoe Museum, Toronto 07093

Czech Republic

Obuvnické Muzeum, Zlín 09942

France

Musée de la Cordonnerie, Alby-sur-Chéran 11542
Musée de l'Industrie Régionale de la Chaussure, Saint-André-de-la-Marche 15166
Musée du Sabot, Buironfosse 12205
Musée du Sabot, Etang-sur-Arroux 12872
Musée du Sabotier, Soucht 15792
Musée International de la Chaussure, Romans-sur-Isère 15088

Germany

Deutsches Schuhmuseum Hauenstein, Hauenstein 19045
Deutsches Schustermuseum, Burgkunstadt 17659
DLM - Deutsches Ledermuseum/ Schuhmuseum Offenbach, Offenbach am Main 20881
Heimatmuseum, Buchenberg 17616
Historisches Schuhmuseum, Landsberg am Lech 19795
Museum Altes Rathaus, Pirmasens 21041

Museum Weißenfels Schloß Neu-Augustusburg, Weißenfels 22457
Schuhmachermuseum, Ladbergen 19758

Italy

Museo della Bonifica, San Donà di Piave 27631
Museo della Calzatura, Sant'Elpidio a Mare 27729
Museo della Calzatura d'Autore di Rossimoda, Strá 27908
Museo della Calzatura P. Bertolini, Vigevano 28266

Japan

Japan Footwear Museum Japan Folk Toy and Doll Museum, Fukuyama 28447

Mexico

Museo del Calzado El Borceguí, México 31099

Namibia

Leather Museum, Swakopmund 31756

Netherlands

Klompenmuseum De Platijn, Best 31970
Klompenmuseum Gebr. Wietzes, Eelde 32173
Klompenmuseum 't Oale Ambacht, Goor 32262
Klompenmuseum 't Schöpke Enter, Enter 32215
Museum de Klompenmaker, Keijenborg 32444
Nederlands Leder- en Schoenenmuseum, Waalwijk 32918
Scherjon's Klompenmakerij en Museum, Noardburgum 32597

Spain

Museo del Calzado, Elda 39272

Sweden

Skoindustrimuseet, Kumla 40658

Switzerland

Bally Schuhmuseum, Schönenwerd 41841
Kulturpfad Root, Root 41774

United Kingdom

Northampton Museum and Art Gallery, Northampton 45270
Rossendale Footware Heritage Museum, Rawtenstall 45466
Shoe Museum, Street 45790
Souter Johnnie's Cottage, Kirkoswald 44614

Silk

China

China Silk Museum, Hangzhou 07759
China Silk Museum, Shanghai 08160
Suzhou Silk Museum, Suzhou 08260

France

Atelier Municipal de Passementerie, Lyon 13879
Atelier Musée de la Soie, Montboucher-sur-Jabron 14186
Atelier-Musée de la Soie, Taulignan 15866
Chemin de la Soie, Molière-Cavaillac 14143
Musée de la Soie, Monoblet 14149
Musée de la Soierie, Charlieu 12411
Musée des Vallées Cévenoles, Saint-Jean-du-Gard 15337
Musée du Mont-de-Piété et de la Condition des Soies, Avignon 11791
Musée du Tissage et de la Soierie, Bussières 12210
Musée Maison de la Soie, Saint-Hippolyte-du-Fort 15312
Secrets de Soie Musée, Campel 12258
Soieries C.J. Bonnet - Musée de la Soie, Jujurieux 13292

Georgia

Georgian Silk Museum, Tbilisi 16361

Germany

Haus der Seidenkultur Paramentenweberei Hubert Gotzes e.V., Krefeld 19704
Stadtmuseum Pirna, Pirna 21046

Greece

Silk Museum, Soufli 23193

India

Mysore Silk Museum, Bangalore 24074

Italy

Museo della Seta Abegg, Garlate 26274
Museo dell'Arte della Seta, Catanzaro 25780
Museo dell'Artigianato Tessile e della Seta, Reggio Calabria 27381
Museo Didattico della Seta, Como 25952

Japan

Kenpu Hakubutsukan, Komoro 28768
Kenpu Hakubutsukan, Yokohama 29600
Okaya Museum, Okaya 29097
Tango Chirimen Rekishi Hakubutsukan, Nodagawa 29080

Korea, Republic

Silk Museum, Gangnae 29805

Netherlands

Zijdemuseum, Meliskerke 32548

Spain

Museo del Tejido Sedero Valenciano, Valencia 40209

Sweden

Sidenväveri Museum, Stockholm 40932

Taiwan

Taiwan Silk Culture Museum, Gongguan 42193

United Kingdom

Macclesfield Silk Museum, Macclesfield 45065
Paradise Mill Silk Industry Museum, Macclesfield 45066
Silk Mill - Derby's Museum of Industry and History, Derby 43970
Whitchurch Silk Mill, Whitchurch 45999

United States of America

Wistariahurst Museum, Holyoke 49328

Silver → Gold and Silver

Skeletons → Bones

Social Conditions

Angola

Museu da Escravatura, Luanda 00111

Argentina

De los Viejos Colonos, Bariloche 00143
Museo Comunitario de Villa Regina, Villa Regina 00684
Museo Histórico Comunal y de la Colonización Judía Rabino, Moisés Ville 00470
Museo Social Argentino de Universidad, Buenos Aires 00274

Australia

Hahndorf Academy Public Art Gallery and Museum, Hahndorf 01139
Hou Wang Chinese Temple and Museum, Atherton 00817
Immigration Museum, Melbourne 01285
Leonora Gwalia Historical Museum, Leonora 01229
Meroogal House, Nowra 01383
Mulgrave Settlers Museum, Gordonvale 01121
National Museum of Australia, Canberra 00947
National Pioneer Women's Hall of Fame, Alice Springs 00800
Old Parliament House, Canberra 00949
Old Town Hall History House, Portland 01439
Vaucluse House, Vaucluse 01616
Wangaratta Exhibitions Gallery, Wangaratta 01627

Austria

Erstes Österreichisches Museum für Alltagsgeschichte, Neupölla 02442
Franztaler Heimatstube, Mondsee 02404
Frauenmuseum, Hittisau 02084
Hans Gross Kriminalmuseum, Graz 01972
Museum Arbeitswelt Steyr, Steyr 02769
Nadelburgmuseum, Lichtenwörth 02298
Oberösterreichische Landesmuseen, Linz 02322
Österreichisches Gesellschafts- und Wirtschaftsmuseum, Wien 03048

Belgium

Archief en Museum van het Vlaams Leven te Brussel, Bruxelles 03432
Daensmuseum en Archief van de Vlaamse Sociale Strijd, Aalst 03258
Gasthuismuseum Sint-Dimpna, Geel 03613
Musée de la Fourche et de la Vie Rurale, Mortier 03894
Museum van het Vlaams Studentenleven, Leuven 03788
Pelgrom Museum Poorterswoning, Antwerpen 03309

Benin

Musée d'Histoire, Ouidah 04185

Brazil

Memorial Pontes de Miranda da Justiça do Trabalho em Alagoas, Maceió 04541
Museu da Terra e da Vida, Mafra 04550
Museu das Reduções, Ouro Preto 04602
Museu Histórico Visconde de São Leopoldo, São Leopoldo 04842

Bulgaria

Okrăžen Istoričeski Muzej, Plovdiv 05130

Burkina Faso

Musée Provincial du Poni, Gaoua 05273

Canada

Hanna Pioneer Museum, Hanna 05967
Lennoxville Ascot Historical and Museum Society, Lennoxville 06158
Motherwell Homestead, Abernethy 05308
Musée McCord, Montréal 06371
The Salvation Army Museum, Toronto 07136
Westfield Heritage Village, Rockton 06768

China

Museum of Dr. Sun Yat-Sen, Zhongshan 08506

Croatia

Muzej Grada Splita, Split 08960

Czech Republic

Městske Muzeum, Rtyně v Podkrkonoší 09829
Muzeum a Pojizerská Galerie, Semily 09836

Denmark

Arbejdermuseet, København 10123
Kvindemuseet i Danmark, Århus 09964
Odsherreds Kulturhistoriske Museum, Nykøbing Sjælland 10206
Socialmuseet Det Gamle Syge-Plejehjem, Hinnerup 10078

Estonia

Heino Lubja Kaalumuuseum, Mustvee 10596

Finland

Amurin Työläismuseokortteli, Tampere 11348
Kauppilanmäen Museo, Valkeakoski 11436
Pelastusarmeijan Museo, Helsinki 10808
Rakennuskulttuuritalo Toivo ja Korsmanin talo, Pori 11221
Teollisuustyöväen Asuntomuseo, Imatra 10858
Tullimuseo, Helsinki 10831
Työväenmuseo Werstas, Tampere 11367
Vaajakosken Työläiskotimuseo, Jyväskylä 10899

France

CEDIAS Musée Social, Paris 14578
Familistère de Guise, Guise 13169
Maison de l'Outil et de la Pensée Ouvrière, Troyes 15996
Musée de la Cohue, Vannes (Morbihan) 16067
Musée de la Famille, Chasselay 12428
Musée de l'Erotisme, Paris 14669
Musée d'Histoire Jean Garcin 1939-1945, Fontaine-de-Vaucluse 12959
Musée du Travail Charles Peyre, Montfermeil 14206
Musée le Mémoire d'Oyé, Oyé 14549

Germany

Badisches Landesmuseum Karlsruhe, Karlsruhe 19437
Bergbaumuseum, Deuben 17843
DDR-Museum Radebeul, Radebeul 21159
Donauschwäbisches Zentralmuseum, Ulm 22160
Hans-Böckler-Geburtshaus, Trautskirchen 22080
Haus der Geschichte Wittenberg, Lutherstadt Wittenberg 20114
Heimatmuseum, Friedland 18472
Heimatmuseum, Herbrechtingen 19134
Historisches Museum Bielefeld, Bielefeld 17340
Historisches Zentrum Engelshaus, Wuppertal 22742
Löhe-Zeit-Museum, Neuendettelsau 20607
LWL-Industriemuseum, Glashütte Gernheim, Petershagen 21016
LWL-Industriemuseum, Henrichshütte Hattingen, Hattingen 19038
LWL-Industriemuseum, Schiffshebewerk Henrichenburg, Waltrop 22319
LWL-Industriemuseum, Textilmuseum in Bocholt, Bocholt 17395
LWL-Industriemuseum, Westfälisches Landesmuseum für Industriekultur, Dortmund 17921
LWL-Industriemuseum, Zeche Hannover, Bochum 17403
LWL-Industriemuseum, Zeche Nachtigall, Witten 22618
LWL-Industriemuseum, Ziegeleimuseum in Lage, Lage 19764
Museum der Arbeit, Hamburg 18946
Museum Karl-Marx-Haus, Trier 22100
Museum Der Kunststall, Dahlem 17769
Museum Malerwinkelhaus, Marktbreit 20215
Museum Schloss Lichtenberg, Fischbachtal 18322
Rheinisches Landesmuseum für Industrie- und Sozialgeschichte, Ratingen 21193
Sächsisches Industriemuseum Chemnitz, Chemnitz 17703
Schnarch-Museum, Alfeld 16468
Schulze-Delitzsch-Haus, Delitzsch 17813
Speicherstadtmuseum, Hamburg 18959
Stadt- und Industriemuseum Rüsselsheim, Rüsselsheim 21438
Werra-Kalibergbau-Museum, Heringen (Werra) 19143
Willi Dickhut Museum, Gelsenkirchen 18580
Zeche Zollern, Dortmund 17930

Hungary

Nógrádi Történeti Múzeum, Salgótarján 23791

Ireland

Adare Heritage Centre, Adare 24727
Custom House Visitor Centre, Dublin 24782
Irish Labour History Society Museum, Dublin 24800
The Kerry Bog Village Museum, Glenbeigh 24838

Italy

Museo del Lavoro, Castelmagno 25742
Museo delle Scienze della Terra e del Lavoro Contadino, Piandimeleto 27191
Museo dell'Emigrante Casa Giannini, Favale di Malvaro 26091
Museo dell'Emigrazione G.B. Scalabrini, Francavilla Angitola 26244
Museo dell'Uomo e dell'Ambiente, Terra del Sole 27966

Japan

Nagaoka Municipal Science Museum, Nagaoka 28960

Mauritius

Folk Museum of Indian Immigration, Moka 30678

Mexico

Museo Nacional de Culturas Populares, México 31133
Museo Niño Minero, Vetagrande 31539

Moldova

Muzeul Pedagogic Republican, Chişinău 31603

Namibia

Alte Feste, Windhoek 31762
Owela Display Centre, Windhoek 31766

Netherlands

Burger Weeshuis, Zierikzee 33002
De Dubbelde Palmboom, Rotterdam 32704
Erotisch Museum, Amsterdam 31840
Museum Het Schip, Amsterdam 31863
Museum Jannink, Enschede 32211
Museum voor Communicatie, Den Haag 32091
Nationaal Museum Verpleging en Verzorging, Zetten 32997
Nationaal Vakbondsmuseum De Burcht, Amsterdam 31872
Sexmuseum Amsterdam Venustempel, Amsterdam 31887
Westzaanse Bodemvondsten Kok-Voogt, Westzaan 32944

New Zealand

Catlins Historical Museum, Owaka 33167
Engine House, Dunedin 33091
Museum of Wellington, City and Sea, Wellington 33251
Otago Settlers Museum, Dunedin 33096
South Canterbury Museum, Timaru 33212

Norway

Funnefoss Industriarbeidermuseum, Årnes 33399
Labråten, Hvalstad 33573
Norsk Industriarbeidermuseet, Rjukan 33785
Trastad Samlinger, Borkenes 33448
Vesterålsmuseet, Melbu 33664

Peru

Museo de la Inmigración Japonesa, Lima 34243
Museo de la Nación, Lima 34244

Poland

Galicja Muzeum, Kraków 34781
Muzeum Okregowe Mazowsza Zachodniego, Żyradów 35381
Muzeum Tradycji Niepodległościowych, Łódź 34885

Portugal

Museu do Trabalho Michel Giacometti, Setúbal 35937

Puerto Rico

Museo de Nuestra Raíz Africana, San Juan 36069

Romania

Punct Muzeal Gospodăria Tărănească, Enisala 36362

Russia

Centralnyj Muzej Vserossijskogo Obščestva Slepych, Moskva 37152
Muzej Nižegorodskoj Intelligencii, Nižnij Novgorod 37464
Muzej Obščestvennogo Pitanija, Moskva 37341
Muzej Rossijskogo Krasnogo Kresta, Moskva 37352
Ogni Moskvy - Muzej Istorii Gorodskogo Osveščenija, Moskva 37391

Senegal

Musée de la Femme Henriette Bathily, Gorée 38024

Slovenia

Muzej Ljudske Revolucije Trbovlje, Trbovlje 38528

South Africa

Hector Pieterson Museum and Memorial, Johannesburg 38656
Kimberley Mine Museum, Kimberley 38669
Red Location Museum, New Brighton 38694
Slave Lodge, Cape Town 38590

Spain

Museo Comarcal, Zalduondo 40331

Sweden

Arbetets Museum, Norrköping 40769
Centrum för Arbetarhistoria, Landskrona 40667
Emigrantmuseet, Kisa 40644
Gråbo Minnenas Museum, Karlskoga 40630
Siggebohyttans Bergmansgård, Nora 40765
Statarmuseum Mejeriet, Örebro 40790
Utvandrarnas Hus, Växjö 41043

Switzerland

Wohn- und Porzellanmuseum Agentenhaus, Horgen 41489

Turkey

Türkiye Çocuk Esirgeme Kurumu Müzesi, Ankara 42714

United Kingdom

Atholl Country Life Museum, Blair Atholl 43536
Bexhill Museum of Costume and Social History, Bexhill-on-Sea 43472
Blaise Castle House Museum, Bristol 43631
Buckleys Yesterday's World, Battle 43419
Cogges Manor Farm Museum, Witney 46052
Cookworthy Museum of Rural Life in South Devon, Kingsbridge 44591
East Carlton Steel Heritage Centre, East Carlton 44083
Elizabethan House, Great Yarmouth 44357
Epping Forest District Museum, Waltham Abbey 45924
Foundling Museum, London 44851
Highland Museum of Childhood, Strathpeffer 45788
Imperial War Museum North, Manchester 45086
Iona Heritage Centre, Isle-of-Iona 44539
Lancaster City Museum, Lancaster 44633
Landmark Forest Heritage Park, Carrbridge 43765
Langton Matravers Museum, Langton Matravers 44640
Montrose Museum and Art Gallery, Montrose 45173
Museum of Liverpool Life, Liverpool 44742
Newart's Millgate Museum, Newark-on-Trent 45214
Pembroke Lodge Patterson Heritage Museum and Family History Centre, Birchington 43488
People's History Museum, Manchester 45096
People's Story Museum, Edinburgh 44144
Pickford's House Museum, Derby 43967
Powysland Museum and Montgomery Canal Centre, Welshpool 45963
Prescot Museum, Prescot 45442
Ragged School Museum, London 44963
Rugby Art Gallery and Museum, Rugby 45527
Saint Andrews Museum, Saint Andrews 45555
Salvation Army International Heritage Centre, London 44982
Scaplen's Court Museum, Poole 45408
South Wales Miner's Museum, Port Talbot 45418
St John's House, Warwick 45939
Summerlee Heritage Park, Coatbridge 43864
Swaledale Museum, Reeth 45479
Tain and District Museum, Tain 45822
Town House Museum of Lynn Life, King's Lynn 44589
Ulster-American Folk Park, Omagh 45317
Wardown Park Museum, Luton 45048
Waterfront Museum, Poole 45410
West Berkshire Museum, Newbury 45221
Weymouth Museum, Weymouth 45991

United States of America

Birmingham Civil Rights Institute Gallery, Birmingham 46821
Black Heritage Museum, Miami 50622
Chesapeake and Ohio Canal Tavern Museum, Sharpsburg 53054
Clara Barton National Historic Site, Glen Echo 48918
The David Davis Mansion, Bloomington 46859
George Meany Memorial Exhibition, Silver Spring 53106
GLBT Historical Society Museum, San Francisco 52775
Prudence Crandall Museum, Canterbury 47220
The Sandor Teszler Gallery, Spartanburg 53195
Stone's Tavern Museum, Ligonier 50081
Tunica-Biloxi Native American Museum, Marksville 50492
Wapello County Historical Museum, Ottumwa 51631
Wenham Museum, Wenham 54038

Speleology

Austria

Höhlenmuseum, Obertraun 02462
Höhlenmuseum Eisensteinhöhle, Bad Fischau-Brunn 01745
Höhlenmuseum in der Lurgrotte, Peggau 02482
Obir-Tropfsteinhöhlen, Bad Eisenkappel 01743

Belgium

Ardoisalle, Alle-sur-Semois 03263
Grotte Scladina et Centre Archéologique, Sclayn 04006
Natuurhistorisch Museum Boekenbergpark, Deurne 03548

Bulgaria

Muzej na Rodopskija Karst, Čepelare 05001

China

Art Museum of Dazu Rock Carvings in Chongqing, Chongqing 07628

France

Grotte de Blanot, Blanot 12017
Grotte de la Vache, Alliat 11557
Grotte de Pair non Pair, Prignac-et-Marcamps 14947
Grotte de Rouffignac, Rouffignac-Saint-Cernin 15124
Grotte d'Osselle, Saint-Vit 15540
Grottes d'Arcy-sur-Cure, Arcy-sur-Cure 11667
Grottes Préhistoriques de Sare-Lezea, Sare 15630
Historial du Gouffre, Padirac 14551
Musée Archeologique, Azé 11812
Musée de Castel-Merle, Sergeac 15722
Musée de la Fédération Nationale de Spéléologie, Revel 15028
Musée de la Grotte de Cougnac, Payrignac 14767
Musée de la Pétrification, Savonnières 15686
Musée de l'Ours des Cavernes, Entremont-le-Vieux 12826
Musée de Spéléologie, Fontaine-de-Vaucluse 12958
Musée Français de la Spéléologie, Courniou-les-Grottes 12652
Musée Grotte de Limousis, Limousis 13772
Musée Préhistorique, Marsoulas 14016
Parc Préhistorique des Grottes de Fontirou, Castella 12306

Germany

Höhlenkundemuseum Dechenhöhle, Iserlohn 19358
Höhlenmuseum, Frasdorf 18409
Museum für Höhlenkunde, Laichingen 19779
Saalfelder Feengrotten, Saalfeld/Saale 21451

Italy

Antiquarium delle Grotte di Catullo, Sirmione 27856
Frasassi-Le Grotte, Genga 26286
Museo Civico, Garessio 26273
Museo Civico Speleo-Archeologico, Nuoro 26964
Museo di Paleontologia e Speleologia E.A. Martel, Carbonia 25676
Museo di Speleologia, Borgo Grotta Gigante 25503
Museo di Speleologia e Carsismo A. Parolini, Valstagna 28151
Museo di Speleologia Vincenzo Rivera, L'Aquila 26456
Museo Speleopaleontologico, Genga 26288
Museo Speologico, Sgonico 27833

Malta

Ghar Dalam Cave and Museum, Birzebbuga 30627

Mexico

Museo de la Cueva del Señor del Calvario, México 31083

New Zealand

Waitomo Museum of Caves, Waitomo Caves 33230

Portugal

Museu Vulcano-Espeleológico Machado Fagundes, Angra do Heroísmo 35440

Slovakia

Slovenské Múzeum Ochrany Prírody a Jaskyniarstva, Liptovský Mikuláš 38288

Slovenia

Notranjski Muzej Postojna, Postojna 38496

South Africa

Cango Caves Interpretive Centre, Oudtshoorn 38698

Spain

Museu Arxiu Comarcal, Moià 39657

Switzerland

Grottes de Vallorbe, Vallorbe 41975
Höhlenmuseum, Sundlauenen 41922
Musée Suisse de Spéléologie, Chamoson 41289

Thailand

Phathai Cave, Lampang 42576

United Kingdom

Dan-yr-Ogof Showcaves Museum, Abercrave 43231
Llechwedd Slate Caverns, Blaenau Ffestiniog 43533
Wookey Hole Cave Diving and Archaeological Museum, Wookey Hole 46068

Spirits → Wines and Spirits

Sports and Recreation

Argentina

Museo de la Pasión Boquense, Buenos Aires 00205
Museo del Deporte Pierre de Coubertin, La Falda 00402
Museo Municipal del Deporte, San Fernando del Valle de Catamarca 00574

Australia

Australasian Golf Museum, Bothwell 00880
Australian Gallery of Sport and Olympic Museum, East Melbourne 01052
Australian Golf Museum, South Melbourne 01515
Mareeba District Rodeo Museum, Mareeba 01262
Museum of Western Australian Sport, Claremont, Western Australia 00977
National Motor Racing Museum, Bathurst 00840

Austria

Bergbahnmuseum, Kitzbühel 02176
Eisstockmuseum, Sankt Georgen im Attergau 02645
Jenbacher Museum, Jenbach 02141
Oesterreichischer Alpenverein-Museum, Innsbruck 02127
Salzburger Landes-Skimuseum, Werfenweng 02924
Ski- und Heimatmuseum, Sankt Anton am Arlberg 02635
Winter!Sport!Museum!, Mürzzuschlag 02415

Belgium

Sportmuseum Vlaanderen, Hofstade 03708

Brazil

Museu de Esportes Mané Garrincha, Rio de Janeiro 04726

Bulgaria

Nacionalen Muzej na Fizičeskata Kultura i Sport, Sofia 05195

Canada

Alberta Sports Hall of Fame and Museum, Red Deer 06706
Aquatic Hall of Fame and Museum of Canada, Winnipeg 07355
British Columbia Golf House Museum, Vancouver 07202
British Columbia Sports Hall of Fame and Museum, Vancouver 07205
Canada's Sports Hall of Fame, Toronto 07097
Canadian Baseball Hall of Fame Museum, Saint Marys 06868
Canadian Football Hall of Fame and Museum, Hamilton 05952
Canadian Golf Hall of Fame and Museum, Oakville 06494
Canadian Lacrosse Hall of Fame, New Westminster 06435
Canadian Scouting Museum, Ottawa 06537
Canadian Ski Museum, Ottawa 06538
Glengarry Sports Hall of Fame, Maxville 06253
Hockey Hall of Fame, Toronto 07116
Ingersoll Cheese Factory Museum and Sports Hall of Fame, Ingersoll 06012
International Ice Hockey Federation Museum, Kingston 06078
Manitoba Baseball Hall of Fame Museum, Morden 06390
Manitoba Sports Hall of Fame and Museum, Winnipeg 07371
Musée du Ski des Laurentides, Saint-Sauveur 06877
Musée Gilles-Villeneuve, Berthierville 05422
New Brunswick Sports Hall of Fame, Fredericton 05837
Northwestern Ontario Sports Hall of Fame, Thunder Bay 07077
Olympic Hall of Fame and Museum, Calgary 05530
Royal Saint John's Regatta Museum, Saint John's 06856
Saint John Sports Hall of Fame, Saint John 06840
Saskatchewan Baseball Hall of Fame and Museum, Battleford 05401
Saskatchewan Sports Hall of Fame and Museum, Regina 06732
Skate Canada Hall of Fame Museum, Ottawa 06556
Trail Museum and Sports Hall of Memories, Trail 07159

China

China Sports Museum, Beijing 07533

Croatia

Hrvatski Sportski Muzej, Zagreb 09000

Cuba

Museo Casa Natal Juan Manuel Márquez, La Habana 09127
Museo de História del Deporte, La Habana 09134
Museo del Deporte, Cienfuegos 09065
Museo del Deporte, Holguín 09182

S

Cyprus

Cyprus Olympic Committee Museum, Lefkosia 09377

Czech Republic

Oddělení dějin tělesné výchovy a sportu, Praha 09785

Estonia

Eesti Spordimuuseum, Tartu 10648

Finland

Hiihtomuseo, Lahti 11026
Suksimuseo ja Pesäpallomuseo, Vimpeli 11468
Suomen Jääkiekkomuseo, Tampere 11358
Suomen Jalkapallomuseo, Valkeakoski 11438
Suomen Nyrkkeilymuseo, Tampere 11360
Suomen Urheilumuseo, Helsinki 10822
Varalan Liikuntamuseo, Tampere 11369

France

Espace Neige et Montagne, Bonneval-sur-Arc 12056
Musée de la Natation, Chissey-les-Mâcon 12519
Musée des Ballons, Balleroy 11832
Musée des Cultures Taurines, Nîmes 14451
Musée des XVIes Jeux Olympiques d'Hiver, Albertville 11534
Musée du Cerf-Volant, Perrigny-lès-Dijon 14796
Musée du Ski, Besse-en-Chandesse 11962
Musée du Ski et de la Tradition Rousselande, Les Rousses 13712
Musée National du Sport, Paris 14730

Germany

Bergringmuseum, Teterow 22035
Deutsches Boxsport-Museum, Sagard 21468
Deutsches Skimuseum, Planegg 21048
Deutsches Sport- und Olympia Museum, Köln 19597
Eishockey-Museum, Augsburg 16616
Fischinger Heimathaus mit Schimuseum, Fischen im Allgäu 18323
Frankfurter Sportmuseum, Frankfurt am Main 18371
Friedrich-Ludwig-Jahn-Museum, Freyburg (Unstrut) 18458
Golfmuseum, Regensburg 21223
Grünauer Wassersportmuseum, Berlin 17120
Heimat- und Ski-Museum, Braunlage 17516
Das Kleine Museum, Gelsenkirchen 18575
Motor-Sport-Museum Hockenheimring, Hockenheim 19206
Museum im Goetz-Haus, Leipzig 19923
Museum im Klösterle, Peiting 21004
Musik- und Wintersportmuseum, Klingenthal 19572
Schwäbisches Schützenmuseum, Kronburg 19727
Ski- und Heimatmuseum Upfingen, Sankt Johann, Württemberg 21502
Skimuseum, Oberreute 20827
Sportmuseum Berlin, Berlin 17257
Sportmuseum der Stadt Frankfurt/Oder, Frankfurt (Oder) 18407
Sportmuseum Leipzig, Leipzig 19936

Hungary

Czibor Zoltán Emlékszoba, Komárom 23676
Hajózástörténeti Múzeum, Zebegény 23976
Megyeháza Börtönépület, Eger 23541
Sportmúzeum, Kaposvár 23636
Testnevelési és Sportmúzeum, Budapest 23483
Testnevelési és Sportmúzeum Kiállítóterme, Budapest 23484

India

Mountaineering Museum, Darjeeling 24127
National Sportsmuseum, Patiala 24344

Israel

International Jewish Sports Hall of Fame, Netanya 25073
Pierre Gildesgame Maccabi Sports Museum, Ramat Gan 25086

Italy

Museo dei Campionissimi, Novi Ligure 26961
Museo della SAT, Trento 28057
Museo Nazionale della Montagna Duca degli Abruzzi, Torino 28023

Japan

Chichibunomiya Kinen Supohtsu Hakubutsukan, Tokyo 29376
Japan Football Mueum, Tokyo 29402
Japanese Ski Memorial Museum, Joetsu 28619
Sapporo Winter Sports Museum, Sapporo 29227
Shin-Yokohama Ramen Hakubutsukan, Yokohama 29606
Sports Car Museum of Japan, Gotemba 28459
Sumo Hakubutsukan, Tokyo 29477
Yakyu Taiiku Hakubutsukan, Tokyo 29506

Korea, Republic

Gukgiwon Taekwondo Museum, Seoul 29879
Korean Ski Museum, Ganseong 29810
Seoul Olympic Museum, Seoul 29922

Latvia

Latvijas Sporta Muzejs, Rīga 30085
Tenisa muzejs, Rīga 30113

Liechtenstein

Skimuseum, Vaduz 30201

Lithuania

Lietuvos Sporto Istorijos Ekspozicija, Kaunas 30257
Vilniaus Sporto Istorijos Muziejus, Vilnius 30441

Mexico

Museo Casa de la Bola, México 31058
Museo Comité Olímpico, México 31063
Museo del Deporte, Pabellón de Arteaga 31240
Salón de la Fama del Beisbol Profesional de México, Monterrey 31190

Netherlands

Ajax Museum, Amsterdam 31824
Het Eerste Friese Schaatsmuseum, Hindeloopen 32390
Groninger Schaatsmuseum, Sappemeer 32744
Kaatsmuseum, Franeker 32231
Max Euwe Centrum, Amsterdam 31858
Nationaal Schaakmuseum, Den Haag 32095

New Zealand

Bowls New Zealand Museum, New Plymouth 33155
National Scout Museum of New Zealand, Kaiapoi 33132
New Zealand Cricket Museum, Wellington 33254
New Zealands Rugby Museum, Palmerston North 33174
Olympic Museum, Wellington 33257

Nicaragua

Salón de la Fama Nicaragüense, Rivas 33302

Norway

Kongsberg Ski Museum, Kongsberg 33600
Norges Olympiske Museum, Lillehammer 33638
Skimuseet, Oslo 33749
Sondre Nordheimstova, Morgedal 33682

Poland

Muzeum Sportu i Turystyki, Łódź 34883
Muzeum Sportu i Turystyki, Warszawa 35270
Muzeum Sportu i Turystyki Regionu Karkonoszy, Karpacz 34715
Muzeum Sportu i Turystyki Ziemi Gdańskiej, Gdańsk 34646

Portugal

Museu Fundação Laura Santos, Gouveia 35633
Museu Nacional do Desporto, Lisboa 35720
Núcleo Museológico do Lagar de Azeite, Mafra 35752
Sala de Troféus do S.L. Benfica, Lisboa 35734

Puerto Rico

Museo Pancho Coimbre, Ponce 36053
Museo Puertorriqueño del Deportes, Salinas 36055

Romania

Muzeul Sportului din România, Bucureşti 36226

Russia

Istoriko-sportivnyj Muzej, Moskva 37220
Muzej Fizičeskoj Kultury i Sporta, Moskva 37276
Muzej Grebnych Vidov Sporta, Moskva 37283
Muzej Istorii Poljarnych Olimpiad, Murmansk 37435
Muzej Sporta v Lužnikach, Moskva 37362
Muzej Sportivnogo Obščestva Dinamo, Moskva 37363
Muzej Sportivnoj Slavy CSKA, Moskva 37364
Muzej Ugolok Deduški Durova, Moskva 37377

Serbia

Muzej Fizičke Kulture, Beograd 38063
Muzej FK Crvena Zvezda, Beograd 38064

Spain

Museo del Fútbol Aguileño, Águilas 38795
Museu Futbol Club Barcelona President Núñez, Barcelona 38996
Museu i Centre d'Estudis de l'Esport Dr. Melcior Colet, Barcelona 38997

Sweden

Borås Idrottsmuseum, Borås 40467
Degerfors Fotbollsmuseum, Degerfors 40484
Helsingborgs Idrottsmuseum, Helsingborg 40587
Idrottsmuseet, Malmö 40720
Nordiska Travmuseet i Årjäng, Årjäng 40430
Riksidrottsmuseet, Stockholm 40930
Sveriges Tennis Museum, Båstad 40454
Västerbottens Museum, Umeå 41000
Vasalopps Museet Mora, Mora 40747
Vasalopps Museet Sälen, Transtrand 40979

Switzerland

Alpines Ballonsport-Museum, Mürren 41663
Le Musée Olympique Lausanne, Lausanne 41550
Schweizerisches Alpines Museum, Bern 41217
Schweizerisches Schützenmuseum, Bern 41218
Sportmuseum Schweiz, Basel 41185
Wintersport-Museum, Davos Platz 41323

Taiwan

Janfusun Museum, Gukeng 42198

Turkey

Turkish Sports Museum, İstanbul 42846

United Kingdom

Anfield Museum, Liverpool 44730
Arsenal Museum, London 44786
Artemis Archery Collection, Oldland 45315
British Cycling Museum, Camelford 43737
British Golf Museum, Saint Andrews 45552
Deep Sea Adventure and Diving Museum, Weymouth 45987
Donington Grand Prix Collection, Castle Donington 43769
Jim Clark Room, Duns 44065
Leicestershire CCC Museum, Leicester 44677
Manchester United Museum and Tour Centre, Manchester 45091
MCC Museum, London 44915
Millwall FC Museum, London 44918
River and Rowing Museum, Henley-on-Thames 44449
Rugby School Museum, Rugby 45528
Scottish Football Museum, Glasgow 44303
Scottish Rugby Union Museum, Edinburgh 44158
Somerset Cricket Museum, Taunton 45830
Unicorn Trust Equestrian Centre, Stow-on-the-Wold 45772
Wimbledon Lawn Tennis Museum, London 45017
World Rugby Museum, Twickenham 45890

United States of America

Aiken Thoroughbred Racing Hall of Fame and Museum, Aiken 46141
American Quarter Horse Heritage Center and Museum, Amarillo 46251
American Sport Art Museum, Daphne 47965
Babe Didrikson Zaharias Museum, Beaumont 46672
Babe Ruth Birthplace Museum, Baltimore 46548
Black Legends of Professional Basketball Museum, Detroit 48106
B'S Ballpark Museum, Centennial 47311
College Football Hall of Fame, South Bend 53156
Colorado Ski Snowboard Museum and Hall of Fame, Vail 53680
Golf Museum, Far Hills 48506
Golf Museum, Newport News 51333
Green Bay Packers Hall of Fame, Green Bay 49037
Greyhound Hall of Fame, Abilene 46121
Harness Racing Museum and Hall of Fame, Goshen 48978
Huntington Beach International Surfing Museum, Huntington Beach 49439
Indiana Basketball Hall of Fame, New Castle 51041
Indianapolis Motor Speedway Hall of Fame Museum, Indianapolis 49500
International Bowling Museum and Hall of Fame, Saint Louis 52580
International Motor Racing Research Center, Watkins Glen 53981
International Sport Aviation Museum, Lakeland 49900
International Swimming Hall of Fame, Fort Lauderdale 48637
International Tennis Hall of Fame Museum, Newport 51321
Kentucky Derby Museum, Louisville 50287
The Lacrosse Museum and National Hall of Fame, Baltimore 46573
Legends of the Game Baseball Museum, Arlington 46352
Lou Holtz and Upper Ohio Valley Hall of Fame, East Liverpool 48255
Louisville Slugger Museum, Louisville 50291
Michigan Sports Hall of Fame, Detroit 48124
Muhammad Ali Center, Louisville 50294
Naismith Memorial Basketball Hall of Fame, Springfield 53227
National Art Museum of Sport, Indianapolis 49505
National Baseball Hall of Fame and Museum, Cooperstown 47791
National Football Museum, Canton 47228
National Fresh Water Fishing Hall of Fame, Hayward 49231
National Museum of Polo and Hall of Fame, Lake Worth 49897
National Museum of Racing and Hall of Fame, Saratoga Springs 52922
National Museum of Roller Skating, Lincoln 50102
National Museum of the Morgan Horse, Shelburne 53064
The National Soccer Hall of Fame, Oneonta 51564
National Softball Hall of Fame and Museum Complex, Oklahoma City 51520
National Wrestling Hall of Fame, Stillwater 53297
New England Ski Museum, Franconia 48731
Oregon Sports Hall of Fame and Museum, Portland 52084
Paul W. Bryant Museum, Tuscaloosa 53620
Peter J. McGovern Little League Baseball Museum, South Williamsport 53184
Pro Rodeo Hall of Fame and Museum of the American Cowboy, Colorado Springs 47706
San Diego Hall of Champions Sports Museum, San Diego 52748
Sanford Museum, Sanford 52847
Schwartz Collection of Skiing Heritage, Tiburon 53505
South Carolina Tennis Hall of Fame, Belton 46726
South Dakota Amateur Baseball Hall of Fame, Lake Norden 49888
The Sports Museum of New England, Boston 46951
Texas Sports Hall of Fame, Waco 53781
United States Hockey Hall of Fame, Eveleth 48459
United States National Ski Hall of Fame and Museum, Ishpeming 49538
Virginia Sports Hall of Fame and Museum, Portsmouth 52112
Water Ski Hall of Fame, Polk City 52014
Western Ski Sport Museum, Donner Pass 48151
World Figure Skating Museum and Hall of Fame, Colorado Springs 47708
World Golf Hall of Fame, Saint Augustine 52529
Yogi Berra Museum, Little Falls 50135

Uzbekistan

Muzej Olimpijskoj Slavy, Toshkent 54478

Venezuela

Museo del Beisbol, Caracas 54551

Stained Glass → Painted and Stained Glass

Stelae → Tombs

S

Stoves

Belgium

Musée Gaumais, Virton 04116

Germany

Egge-Museum Altenbeken, Altenbeken 16492
Eisenkunstgußmuseum, Witzenhausen 22633
Museum für Eisenkunstguß, Hirzenhain 19200

Netherlands

Haags Openbaar Vervoer Museum, Den Haag 32075
Haardplatenmuseum Klarenbeek, Klarenbeek 32448

Stucco

Germany

Schloss Salem, Salem, Baden 21472
Stuck-Museum, Freiburg im Breisgau 18439

Sugar

Argentina
Museo de la Industria Azucarera Obispo Colombres, San Miguel de Tucumán ... 00599

Australia
Fairymead House Sugar Museum, Bundaberg 00920

Austria
Österreichisches Zuckermuseum, Tulln ... 02832

Barbados
Sir Frank Hutson Sugar Museum, Saint James 03241

Belgium
Maison du Sucre, Frasnes-lez-Anvaing ... 03609

Canada
Redpath Sugar Museum, Toronto ... 07131

Cuba
Museo de la Industria Azucarera, Abel Santamaría ... 09020

France
Musée de l'Art du Sucre, Cordes-sur-Ciel ... 12624
Musée du Sucre d'Orge, Moret-sur-Loing ... 14269

Germany
Zucker-Museum, Berlin ... 17284

Guadeloupe
Ecomusée de Marie Galante, Grand Bourg ... 23274

Indonesia
Museum Gula, Klaten ... 24515

Panama
Museo de la Sal y el Azúcar, Aguadulce ... 34059

Portugal
Museu a Cidade do Açúcar, Funchal ... 35610

Puerto Rico
Museo Castillo Serrallés, Ponce ... 36048

Taiwan
Sugar Industry Museum, Ciaotou ... 42165
Taiwan Sugar Industry Museum, Tainan ... 42370

United States of America
The Judah P. Benjamin Confederate Memorial at Gamble Plantation, Ellenton ... 48362
New England Maple Museum, Pittsford ... 51970

S

T

Tapestries

Belgium
Koninklijke Manufactuur De Wit, Mechelen ... 03859
Maison Jonathas, Enghien ... 03581
Stedelijk Museum, Oudenaarde ... 03942

France
Château-Musée de la Grange, Manom ... 13939
Château-Musée du XVe et XVIIIe Siècle, Commarin ... 12593
Exposition de Tapisseries, Felletin ... 12919
Galerie Nationale de la Tapisserie et de l'Art Textile, Beauvais ... 11909
Musée de la Tapisserie Contemporaine, Angers ... 11608
Musée de la Tapisserie de Bayeux, Bayeux ... 11874
Musée Départemental de la Tapisserie, Aubusson ... 11734
Musée du Tapis et des Arts Textiles, Clermont-Ferrand ... 12551
Musée Jean-Lurçat, Angers ... 11612
Musée Leblanc-Duvernoy, Auxerre ... 11777
Musée Municipal, Hesdin ... 13210
Musée Sobirats, Carpentras ... 12298
Villa et Jardins Ephrussi de Rothschild, Saint-Jean-Cap-Ferrat ... 15317

Germany
Schloss Paretz, Ketzin ... 19523

Italy
Museo Pinacoteca della Santa Casa, Loreto ... 26519
Museo Storico della Tappezzeria, Bologna ... 25477
Pinacoteca Civica Bruno Molajoli, Fabriano ... 26071

Lithuania
Janinos Monkutės-Marks Muziejus-Galerija, Kėdainiai ... 30270

Malaysia
Muzium Tekstil, Kuala Lumpur ... 30583

Morocco
Musée de Dar Jamaï, Meknès ... 31681

Netherlands
Museumboerderij New Greenwich Village, Zuidwolde, Drenthe ... 33015
Vierschaar Museum, Veere ... 32879

Spain
Casa Aymat, Sant Cugat del Vallès ... 39978
Museo Catedralicio, Zamora ... 40332
Museo Catedralicio de La Seo de Zaragoza, Zaragoza ... 40341
Museo de Tapices, Oncala ... 39732
Museo-Exposición Permanente Diocesano, Albarracín ... 38809
Museo Parroquial, Pastrana ... 39815
Real Fábrica de Tapices, Madrid ... 39587

Switzerland
Schloßmuseum Thun, Thun ... 41938

Tunisia
Musée des Musiques Arabes et Méditerranéennes Ennejma Ezzahra, Sidi Bou Saïd ... 42654

United Kingdom
Hampton Court Palace, East Molesey ... 44093
Knutsford Heritage Centre, Knutsford ... 44623
Lanhydrock, Bodmin ... 43549
Lyme Hall, Disley ... 43983
Occupation Tapestry Gallery and Maritime Museum, Saint Helier ... 45573
Quilters' Ruild Resource, Halifax ... 44387
Upton House and Gardens, Banbury ... 43371

Vatican City
Galleria degli Arazzi, Città del Vaticano ... 54499

Tea and Coffee

Bolivia
Museo de la Coca, La Paz ... 04215

Brazil
Museu do Café Cel. Francisco Schmidt, Ribeirão Preto ... 04676

China
China Tea Museum, Hangzhou ... 07760
China Tea Museum, Shanghai ... 08161
Macao Tea Culture House, Macau ... 07996
Mingshan Chashi Museum, Mingshan ... 08018

France
Musée du Thé, Paris ... 14702

Georgia
Nobel Technical Museum, Batumi ... 16336

Germany
Museum für Kaffeetechnik, Emmerich am Rhein ... 18150
Museum Zum Arabischen Coffe Baum, Leipzig ... 19927
Das Ostfriesische Teemuseum und Norder Heimatmuseum, Norden ... 20730

Guadeloupe
Musée-Maison du Café, Vieux-Habitants ... 23283

Iran
Muséyé Chay-e Lahijan, Lahijan ... 24627

Ireland
Bewley's Café Museum, Dublin ... 24779

Japan
Kitami Peppermint Museum, Kitami ... 28717
Onishi Seiwemon Museum, Kyoto ... 28859
Tea Ceremony School Museum, Kyoto ... 28869
Uji Kanbayashi Kinenkan, Uji, Kyoto ... 29560

Korea, Republic
Busan Woman's College Tea-ceremony Museum, Busan ... 29776

Mauritius
Tea Museum, Bois-Cheri ... 30676

Mexico
Museo del Café, San Cristóbal de las Casas ... 31328

Netherlands
Geels Co. Koffie- en Theemuseum, Amsterdam ... 31843
Koffie- en Theemuseum, Amsterdam ... 31853
Koffie- en Winkelmuseum, Pieterburen ... 32669
Museum Joure, Joure ... 32432
De Theefabriek, Houwerzijl ... 32418

Nicaragua
Museo de Inmigrantes de Alemania, Mataglpa 33295
Museo del Café, Mataglpa ... 33297

Portugal
Museu do Café de Novadelta, Campo Maior ... 35502

Switzerland
Johann Jacobs Museum, Zürich ... 42078

Taiwan
Jeoufen Art Gallery, Rueifang ... 42307
Lugu Tea Museum, Lugu ... 42261
Pinglin Tea Museum, Pinglin ... 42287

United Kingdom
Bramah Tea and Coffee Museum, London ... 44798

Technology

Argentina
Museo Faryluk, Aristóbulo del Valle ... 00127
Museo Houssay de Historia de la Ciencia y la Técnología, Buenos Aires ... 00247
Museo Tecnológico Ingeniero Eduardo Latzina, Buenos Aires ... 00275

Australia
Excell Blacksmith and Engineering Workshop Museum, Tumby Bay ... 01605
Museum Victoria, Carlton South ... 00957
Powerhouse Museum, Ultimo ... 01609
Questacon-The National Science and Technology Centre, Canberra ... 00951
Scienceworks Museum, Spotswood ... 01521

Austria
Erdefunkstelle Aflenz, Au bei Aflenz ... 01737
Erstes österreichisches Computermuseum, Bad Pirawarth ... 01762
Erstes Österreichisches Tonbandmuseum, Wien ... 02983
Heizungs- und Schulmuseum Wien, Wien ... 03002
krupp stadt museum, Berndorf ... 01780
Radiomuseum, Grödig ... 01993
Technisches Museum Wien, Wien ... 03079

Bangladesh
Science Museum, Dhaka ... 03228

Belgium
Centre d'Animation Permanente du Rail et de la Pierre, Jemelle ... 03744
Computermuseum, Roeselare ... 03975
Museum voor de Oudere Technieken, Grimbergen ... 03661
Museum voor Industriele Archeologie en Textiel, Gent ... 03638
Museum Zwijgershoek, Sint-Niklaas ... 04022
Technopolis, Mechelen ... 03864

Brazil
Museu da Energia, Itu ... 04502
Museu da Escola de Engenharia, Rio de Janeiro ... 04710
Museu de Ciência e Tecnologia, Salvador ... 04798
Museu do Gás, Rio de Janeiro ... 04736
Museu Municipal, Cachoeira do Sul ... 04378
Museu Naval de Amazônia, Belém ... 04340
Museu Técnico-Cientifício do Instituto Oscar Freire, São Paulo ... 04910

Bulgaria
Nacionalen Politechničeski Muzej, Sofia ... 05197

Canada
Alexander Graham Bell National Historic Site, Baddeck ... 05369
Bell Homestead National Historic Site, Brantford ... 05466
Calgary Science Centre, Calgary ... 05516
Canada Science and Technology Museum, Ottawa ... 06534
Canadian Broadcast Museum Foundation, Toronto ... 07098
Eagle Valley Museum, Sicamous ... 06968
Heart's Content Cable Station, Heart's Content 05982
London Regional Children's Museum, London 06187
Montreal Science Centre, Montréal ... 06350
Musée Historique des Communications Eberdt, Sutton, Québec ... 07042
Newfoundland Science Centre, Saint John's ... 06850
Northern Lights Centre, Watson Lake ... 07291
Ontario Science Centre, Toronto ... 07126
Science North, Sudbury ... 07029
Society for Arts and Technology, Montréal ... 06381
Telecommunications Museum, LaSalle ... 06142

China
Bailian Dongdongkong Science Museum, Liuzhou ... 07965
Hong Kong Science Museum, Hong Kong ... 07804
Science Technology Museum, Beijing ... 07565
Shanghai Science Technology Museum, Shanghai ... 08189
Tianjin Science Museum, Tianjin ... 08288

Colombia
Museo de Ciencia y Juego, Santa Marta ... 08750
Museo de la Ciencia y el Juego, Bogotá ... 08567
Museo de los Niños, Bogotá ... 08568
Museo Mundocaña, Cali ... 08613
Museo Mundocaña, Palmira ... 08702
Museo Siderúrgico de Colombia, Nobsa ... 08697

Croatia
Tehnički Muzej, Zagreb ... 09012

Czech Republic
Hornické Muzeum OKD, Ostrava ... 09697
Hornický Skanzen důl Mayrau, Vinařice u Kladna ... 09908
Národní Technické Muzeum, Praha ... 09780
Technické Muzeum, Petřvald u Karviné ... 09711
Technické Muzeum Tatra, Kopřivnice ... 09596
Textilní Muzeum, Česká Skalice ... 09460

Denmark
Carlsberg Museum Valby, Valby ... 10305
Danmarks Tekniske Museum, Helsingør ... 10063
Elmuseet, Bjerringbro ... 09976
Experimentarium, Hellerup ... 10057

Egypt
Science Museum, Giza ... 10530
Science Museum, Alexandria ... 10478

El Salvador
Museo de los Niños Tin Marin, San Salvador ... 10553
Museo y Biblioteca de Ciencias Stephen Hawking, San Salvador ... 10554

Estonia
Õisu Tehnikumi Muuseum, Halliste ... 10567

Finland
Altia Oyj, Rajamäki ... 11252
Heureka - The Finnish Science Centre, Vantaa ... 11446
Polyteekkarimuseo, Espoo ... 10732

Strömforsin Teollisuusmuseo, Ruotsinpyhtää 11281
Tekniikan Museo, Helsinki 10829
Viestikeskus Lokki, Mikkeli 11118

France

Cité des Sciences et de l'Industrie, Paris 14582
Maison-Musée des Sciences, Lettres et Arts de Cholet et de sa Région, Cholet 12521
Musée de la Monnaie de Paris, Paris 14655
Musée de l'Automate et de la Robotique, Souillac 15795
Musée de l'Holographie, Paris 14672
Musée des Arts et Metiers, Paris 14683
Musée des Automates, Neuilly-sur-Seine 14403
Musée des Egouts de Paris, Paris 14685
Musée des Galvachers, Anost 11633
Musée des Sciences et Techniques, Laval (Mayenne) 13540
Musée Municipal de Millau, Millau 14120
Muséum Mécanique Naturelle, Giverny 13070
Palais de la Découverte, Paris 14742

Georgia

Nobel Technical Museum, Batumi 16336

Germany

Alte Mühle Donsbrüggen, Kleve 19564
Alte Schmiede, Bedesbach 16976
Amüseum - das Saarburger Museum am Wasserfall, Saarburg 21461
Arithmeum, Bonn 17445
Astronomisch-Physikalisches Kabinett, Kassel 19458
Automatik-Museum, Leipzig 19897
Besucherbergwerk Rischbachstollen, Sankt Ingbert 21498
Buderus Zentralheizungsmuseum, Lollar 20029
Carl Bosch Museum, Heidelberg 19070
Deutsches Museum, München 20433
Deutsches Technikmuseum, Berlin 17074
Eisenbahn- und Heimatmuseum Hochdahl, Erkrath 18217
Eisenmuseum Jünkerath, Jünkerath 19406
Erlebniswelt Grundwasser, Langenau 19817
Erlebniswelt U-Boot, Sassnitz 21518
Euro Tecnica Losheim, Hellenthal 19118
Feuerschiff, Wilhelmshaven 22576
Feuerwehr Museum, Lahr 19770
Feuerwehrmuseum, Altenkirchen, Westerwald 16501
Feuerwehrmuseum, Asperg 16600
Feuerwehrmuseum, Hermeskeil 19149
Feuerwehrmuseum, Sachsenheim 21465
Feuerwehrmuseum, Tuttlingen 22133
Feuerwehrmuseum Gahlenz, Gahlenz 18530
Flugausstellung Hermeskeil, Hermeskeil 19150
Friedrich-Gottlob-Keller Museum Krippen, Bad Schandau 16825
Hanf-Museum, Berlin 17124
Haus zur Geschichte der IBM Datenverarbeitung, Sindelfingen 21776
Hebezeug-Museum, Witten 22617
Heinz Nixdorf MuseumsForum, Paderborn 20969
Historisches Museum Bielefeld, Bielefeld 17340
Hütten- und Technikmuseum Ilsenburg, Ilsenburg 19322
Kinder-Akademie Fulda Werkraummuseum, Fulda 18511
Küfer-Museum, Langenau 19819
Kustodie, Dresden 17955
Landesfeuerwehrmuseum Meetzen Mecklenburg-Vorpommern, Meetzen 20254
Luftfahrttechnisches Museum, Rechlin 21207
Märchenofen, Neu-Ulm 20589
Militär-Museum Kossa, Kossa 19693
Motorrad-Museum, Otterbach 20948
Museum 3. Dimension, Dinkelsbühl 17868
Museum der Freiwilligen Feuerwehr, Heiligenhaus 19106
Museum Dingolfing-Herzogsburg, Dingolfing 17865
Museum für Landtechnik und Landarbeit, Emmerthal 18154
Museum für Landwirtschaftliche Maschinen, Erbach, Alb-Donau-Kreis 18188
Museum für Optik, Meßtechnik und mathematische Geräte, Nothweiler 20750
Museum für Waage und Gewicht, Balingen 16901
Museum Kesselhaus Herzberge, Berlin 17208
Museum Ober-Ramstadt, Ober-Ramstadt 20798
Museum Strom und Leben, Recklinghausen 21209
Museumsverein Feuerwehr Obererbach, Obererbach 20805
Nicolaus August Otto Museum, Holzhausen an der Haide 19258
Pfälzisches Bergbaumuseum, Imsbach 19330
phæno gGmbH, Wolfsburg 22666
Rechnermuseum, Furtwangen im Schwarzwald 18521
Rechnermuseum der Gesellschaft für wissenschaftliche Datenverarbeitung, Göttingen 18698
Rheinisches Landesmuseum für Industrie- und Sozialgeschichte, Ratingen 21193
Rundfunkmuseum der Stadt Fürth, Fürth 18499
Sächsisches Feuerwehrmuseum Zeithain, Zeithain 22774
Sägemühle, Thallwitz 22046
Schiffbau- und Schifffahrtsausstellung Roßlau, Roßlau 21383
Schiffmühle Minden, Minden 20338
Silberwaren- und Bijouteriemuseum, Schwäbisch Gmünd 21651
Stadt- und Industriemuseum Rüsselsheim, Rüsselsheim 21438
Technik anno dazumal - Museum Kratzmühle, Kinding 19539
Technik-Modell-Museum Rügen e.V., Samtens 21481
Technik-Museum, Soest 21798
Technik Museum Speyer, Speyer 21838
Technik- und Verkehrsmuseum, Stade 21853
Technikmuseum, Magdeburg 20131
Technikmuseum Hugo Junkers, Dessau 17835
Techniksammlung Backnang, Backnang 16663
Technikschau, Wismar 22614
Technische Sammlungen der Stadt Dresden, Dresden 17974
Technisches Landesmuseum Mecklenburg-Vorpommern, Wismar 22615
Technoseum, Mannheim 20174
Tüftler-Werkstatt-Museum, Frickingen 18465
Universitäts- und Landesbibliothek Darmstadt, Darmstadt 17795
Universitätssammlungen - KunstTechnik der TU Dresden, Dresden 17975
Weltkulturerbe Völklinger Hütte, Völklingen 22249
Werksmuseum der Firma Linke-Hofmann-Busch, Salzgitter 21477
Westfälisches Feuerwehrmuseum, Hattingen 19043

Ghana

Museum of Science and Technology, Accra 22847

Greece

Science Center of Thessaloniki and Technology Museum, Thessaloniki 23220

Greenland

Tele-Museum, Qaqortoq 23266

Hungary

Dunaújvárosi Papírgyár Üzemtörténeti Gyűjteménye, Dunaújváros 23532
Hungarian Museum for Science, Technology and Transport - Industrial Collection, Budapest 23414
Millenaris- Jövő Háza Kiállítás, Csodák Palotája, Budapest 23450

India

Alpha Museum, Bangalore 24071
Bardhaman Science Centre, Bardhaman 24076
Birla Industrial and Technological Museum, Kolkata 24271
Birla Museum, Pilani 24349
District Science Centre, Purulia 24364
Goa Science Centre, Panaji 24340
Kurukshetra Panorama Science Centre, Kurukshetra 24287
Museum of the College of Engineering, Chennai 24118
National Science Centre, Delhi 24151
Nehru Science Centre, Mumbai 24321
Regional Science Centre, Tirupati 24408
Sanjay Sharma Museum, Jaipur, Rajasthan 24235
Visvesvaraya Industrial and Technological Museum, Bangalore 24075

Indonesia

Museum Science and Technology, Jakarta 24498

Iran

Museum of Natural History and Technology, Shiraz 24647
Time Museum, Tehrān 24697

Ireland

College Museum, Maynooth 24875
Ireland's Historic Science Centre, Birr 24745

Israel

Bloomfield Science Museum Jerusalem, Jerusalem 24981
Havayeda Museum, Holon 24976
National Museum of Science, Planning and Technology, Haifa 24966
Nature Park and Galleries, Jerusalem 25007

Italy

Museo degli Strumenti della Riproduzione del Suono, Roma 27471
Museo della Tecnica e del Lavoro MV Agusta, Samarate 27604
Museo dell'Informazione, Senigallia 27802
Museo Leonardiano, Vinci 28285
Museo Nazionale della Scienza e della Tecnica Leonardo da Vinci, Milano 26713
Museo Postale e Telegrafico della Mitteleuropa, Trieste 28095

Japan

Ehime-kenritsu Kagakukan, Niihama 29065
Gakumon Kagaku Hakubutsukan, Tokyo 29389
Hamamatsu-shiritsu Kagakukan, Hamamatsu 28486
Hirano Soundscape Museum, Yamato-koriyama 29585
Hiroshima Children's Museum, Hiroshima 28534
Kagaku Gijutsukan, Tokyo 29404
Kawaguchi-shiritsu Kagakukan, Kawaguchi 28685
Kawasaki Juvenile Science Museum, Kawasaki 28691
Kobe-shi Kagakukan, Kobe 28734
Kokuritsu Kagaku Hakubutsukan , Shinju-ku, Tokyo 29417
Kokuritsu Miraikan, Tokyo 29418
Koriyama-shi Fureai Kagakukan, Koriyama 28771
Kouseiroudoushou Sangyo Anzen Gijutsukan, Tokyo 29422
Kyoto-shi Seishonen Kagaku, Kyoto 28846
Kyoto Yuzen Bunka Kaikan, Kyoto 28849
Muroran-shi Seishonen Kagakukan, Muroran 28947
Nagaoka Gear Museum, Nagaoka 28959
Nagasaki Science Museum, Nagasaki 28982
Nagoya-shi Kagakukan, Nagoya 29008
Niigata Science Museum, Niigata 29056
NTT InterCommunication Center, Tokyo 29453
Osaka-shi Kagaku Hakubutsukan, Osaka 29150
Otaru-shi Seishonen Kagakugijutsukan, Otaru 29162
RiSuPia Panasonic Digital Network Museum, Tokyo 29460
Sendai Music Box Museum, Sendai 29245
Shiseido Art House, Kakegawa 28638
Tepia, Kikai Sangyo Kinenkan, Tokyo 29485
Tokyo Noko Daigaku Kogakubu Fuzoku Sen'i Hakubutsukan, Koganei 28759
Toshiba Kagakukan, Kawasaki 28696
Toyota Commemorative Museum of Industry and Technology, Nagoya 29013
Uchu Kagakukan, Ikoma 28565
The University Museum, Tokyo 29502
Yamaguchi Hakabutsukan, Yamaguchi 29580
Yamanashi-kenritsu Kagaku Senta, Kofu 28755

Korea, Republic

Museum of Chungju National University, Chungju 29791
Seoul National Science Museum, Seoul 29920

Latvia

Minimotor Muzejs, Riga 30096
Rīgas Motormuzejs, Rīga 30108

Liechtenstein

Museum für historische Schreib- und Rechenmaschinen, Schaan 30193

Mexico

Centro de Ciencia y Tecnología, Guadalajara 30887
Descubre Museo Interactivo de Ciencia y Tecnología, Aguascalientes 30697
Imagina-Museo Interactivo Puebla, Puebla 31264
Museo Casa de la Tecnología, Monterrey 31177
Museo de Ciencia, Cuernavaca 30816
Museo de Ciencia y Tecnología del Estado de Veracruz, Xalapa 31558
Museo de Ciencias de Ensenada, Ensenada 30858
Museo El Chapulín, Saltillo 31316
Museo Histórico de la Educación Tecnológica en México, Chilpancingo 30774
Museo Interactivo de Ciencia, Tecnología y Medio Ambiente Sol del Niño, Mexicali 31035
Museo Interactivo del Centro de Ciencias de Sinaloa, Culiacán 30828
Museo Interactivo El Rehilete, Pachuca 31244
Universum, Museo de las Ciencias de la UNAM, México 31161

Morocco

Musée de Maroc Telecom, Rabat 31689

Netherlands

De Dijk Te Kijk, Petten 32668
Explorion, Heerlen 32336
Expo Haringvliet, Stellendam 32806
Modelbouwmuseum, Wassenaar 32927
Molenmuseum De Wachter, Zuidlaren 33013
Museum De Cruquius, Cruquius 32039
Nationaal Baggermuseum, Sliedrecht 32781
NEMO, Amsterdam 31875
Nieuw Land Poldermuseum, Lelystad 32495
Poldermuseum Den Huijgen Dijck, Heerhugowaard 32335
Radiomuseum, Borculo 31985
Techniek Museum Delft, Delft 32066
Telefoonmuseum, Drouwen 32160

New Zealand

East Coast Museum of Technology, Gisborne 33108
Ferrymead Heritage Park, Christchurch 33073
Kaikohe Pioneer Village and Museum, Kaikohe 33133
Museum of Transport and Technology, Auckland 33056
Te Manawa Life Mind, Palmerston North 33176

Norway

Cementmuseet Slemmestad, Slemmestad 33843
Kongsberg Våpenfabrikks Museum, Kongsberg 33601
Norsk Teknisk Museum, Oslo 33744
Norsk Telemuseum - Kristiansand, Kristiansand 33611
Norsk Telemuseum - Lødingen, Lødingen 33642
Norsk Telemuseum - Sørvågen, Sørvågen 33856
Norsk Telemuseum - Stavanger, Stavanger 33872
Norsk Telemuseum - Trondheim, Trondheim 33926

Pakistan

National Museum of Science and Technology, Lahore 34025

Peru

Museo de Ciencias, Huánuco 34179

Philippines

Lapu-Lapu City Science Museum, Lapu-Lapu 34401
Normisist Folk Inventions Museum, Butuan 34355
Philippine Science Centrum, Marikina 34461
Rizal Technological and Polytechnic Institute Museum, Morong 34462
The Science Museum, Cebu 34374

Poland

Muzeum Budownictwa i Techniki Wiejskiej, Bogdaniec 34567
Muzeum Drukarstwa Warszawskiego, Warszawa 35239
Muzeum Gazownictwa, Paczków 35003
Muzeum Gazownictwa, Sobótka 35115
Muzeum Gazownictwa przy Gazowni Warszawskiej, Warszawa 35243
Muzeum Inżynierii Miejskiej, Kraków 34800
Muzeum Kolei Wąskotorowej w Wenecji, Wenecja 35297
Muzeum Politechniki Warszawskiej, Warszawa 35265
Muzeum Pożarnictwa Ziemi Olkuskiej, Olkusz 34974
Muzeum Techniki w Warzawie, Warszawa 35273
Muzeum Wikliniarstwa i Chmielarstwa, Nowy Tomyśl 34964
Muzeum Zagłębia Staropolskiego w Sielpi, Sielpia 35104
Statek-Muzeum Dar Pomorza, Gdynia 34662
Statek-Muzeum Sołdek, Gdańsk 34651
Zabytkowa Kuźnia Wodna, Gdańsk 34653

Portugal

Museu das Crianças, Lisboa 35689
Museu Nacional da Ciência e da Técnica, Coimbra 35540
Núcleo Museológico da Faculdade de Engenharia, Porto 35863
Pavilhão do Conhecimento, Lisboa 35733
Visionarium, Centro de Ciência do Europarque, Santa Maria da Feira 35906

Romania

Expoziția Memorială Traian Vuia, Traian Vuia 36702
Muzeul Memorial Aurel Vlaicu, Geoagiu 36381
Muzeul Național al Petrolului, Ploiești 36536
Muzeul Sării, Slănic Prahova 36626
Muzeul Științei și Tehnicii Ștefan Procopiu, Iași 36426

Russia

Memorialnyj Muzej-laboratorija A.S. Popova, Kronštadt 37085
Memorialnyj Muzej A.S. Popova, Sankt-Peterburg 37664

T

Memorialnyj Muzej A.S. Popova - Izobretatelja Radio, Krasnoturinsk 37081
Muzej Radio im. A.S. Popova, Ekaterinburg 36898
Politechničskij Muzej, Moskva 37401

Serbia

Muzej Nauke i Tehnike, Beograd 38070
Spomen Kuća Laze K. Lazarevića, Beograd 38088

Singapore

Singapore Science Centre, Singapore 38180

Slovakia

Slovenské Technické Múzeum, Košice 38273

Slovenia

Tehnični Muzej Rudarstva, Zagorje ob Savi 38538
Tehniški Muzej Slovenije, Bistra pri Vrhniki 38399
Tehniški Muzej Slovenije, Ljubljana 38467

South Africa

Museum of Science and Technology, Tshwane 38751
South African National Museum of Military History, Johannesburg 38664

Spain

Domus - Casa del Hombre, A Coruña 39215
Miramón Kutxa Espacio de la Ciencia, Donostia-San Sebastián 39250
Museo Aeronáutico Municipalo, Los Alcázares 38821
Museo de la Ciencia y el Cosmos, La Laguna de Tenerife 39440
Museo de la Real Academia Nacional de Farmacia, Madrid 39548
Museo de l'Enginyeria de Catalunya, Barcelona 38967
Museo Elder, Las Palmas de Gran Canaria 39801
Museo Lara, Ronda 39919
Museo Nacional de Ciencia y Tecnología, Madrid 39570
Museu de la Ciència i de la Tècnica de Catalunya, Terrassa 40118
Museu de les Ciències Príncipe Felipe, Valencia 40220

Sweden

Dalénium - Gustaf Daléns Upptäckarcentrum, Stenstorp 40881
Dalénmuseet, Stenstorp 40882
Innovatum Kunskapens Hus, Trollhättan 40982
Kareum - Gamla Bilsalongen, Sparreholm 40877
Radiomuseet i Jönköping, Jönköping 40615
Säffle Marinmotor Museum, Säffle 40821
Sveriges Rundradiomuseum, Motala 40759
Technichus, Härnösand 40573
Teknikens Hus, Luleå 40700
Teknikens och Sjöfartens Hus, Malmö 40727
Tekniska Kvarnen, Örebro 40791
Tekniska Museet, Stockholm 40945
Universeum, Göteborg 40557
Xperiment Huset, Växjö 41045

Switzerland

CERN - Microcosm, Genève 41409
Einstein-Haus, Bern 41205
Musée d'Histoire des Sciences, Genève 41422
Museum Telefonzentrale Magden, Magden 41602
Swiss Science Center Technorama, Winterthur 42039

Taiwan

National Science and Technology Museum, Kaohsiung 42241
Science Hall of Penghu County, Magong 42268

Thailand

Museum of Imaging Technology, Bangkok 42546
Science Centre for Education, Bangkok 42559

United Kingdom

Abbey Pumping Station Museum, Leicester 44672
Alderney Railway - Braye Road Station, Alderney 43268
Amberley Museum and Heritage Centre, Amberley 43297
At-Bristol, Bristol 43630
Bletchley Park, Bletchley 43543
Bressingham Steam Museum, Diss 43985
British Balloon Museum, Newbury 45220
British Engineerium, Hove 44492
Cambridge Museum of Technology, Cambridge 43724
Charney Bassett Mill, Wantage 45928
Discovery Museum, Newcastle-upon-Tyne 45226
Eureka! The Museum for Children, Halifax 44384
Great Orme Tramway, Llandudno 44758
Inspire - Hands-on Science Centre, Norwich 45279
Ironbridge Tollhouse, Ironbridge 44532
Jodrell Bank Visitor Centre, Macclesfield 45064
John King Workshop Museum, Pinxton 45383
Kirkaldy Testing Museum, London 44896
Llangollen Railway, Llangollen 44771
Long Shop Museum, Leiston 44689
Look Out Discovery Centre, Bracknell 43573
Museum of Communication, Burntisland 43682
Museum of Science and Industry in Manchester, Manchester 45092
National Space Centre, Leicester 44680
National Vintage Wireless and Television Museum, Harwich 44412
Oliver Cromwell's House, Ely 44173
Oswestry Transport Museum, Oswestry 45323
Porthcurno Telegraph Museum, Porthcurno 45421
Prickwillow Drainage Engine Museum, Prickwillow 45449
Railway Preservation Society of Ireland, Whitehead 46002
Royal Air Force Air Defence Radar Museum, Norwich 45283
Royal Engineers Museum, Chatham 43798
Science Museum, London 44983
Science Museum Wroughton, Wroughton 46086
Techniquest, Cardiff 43754
Thinktank, Birmingham 43513
Tower Bridge Exhibition, London 44998
Wilton Windmill, Wilton, Marlborough 46018

United States of America

A.W.A. Electronic-Communication Museum, Bloomfield 46853
American Museum of Radio and Eletricity, Bellingham 46711
American Museum of Science and Energy, Oak Ridge 51456
American Precision Museum, Windsor 54228
Ann Arbor Hands-on Museum, Ann Arbor 46303
Arkansas Museum of Natural Resources, Smackover 53132
The B O Railroad Museum, Baltimore 46547
Baker-Cederberg Museum, Rochester 52392
The Bakken, Minneapolis 50729
Baltimore Museum of Industry, Baltimore 46552
Beall Center for Art and Technology, Irvine 49529
Betty Brinn Children's Museum, Milwaukee 50705
California Museum, Sacramento 52500
California Science Center, Los Angeles 50215
Carnegie Science Center, Pittsburgh 51944
Catawba Science Center, Hickory 49266
Charles River Museum of Industry, Waltham 53808
Children's Discovery Museum of Central Illinois, Bloomington 46858
Children's Museum of History, Natural History and Science at Utica, New York, Utica 53675
The Children's Science Center, Cape Coral 47232
Children's Science Center, Carbondale 47240
Cle Elum Telephone Museum, Cle Elum 47604
Coca-Cola Memorabilia Museum of Elizabethtown, Elizabethtown 48339
Columbia River Exhibition of History, Science and Technology, Richland 52307
COSI Columbus, Columbus 47753
Cosi Toledo, Toledo 53521
CTS Turner Museum, Elkhart 48345
Cumberland Science Museum, Nashville 50966
Discovery Center, Fresno 48789
Discovery Center of Springfield, Springfield 53231
Discovery Center Science Museum, Fort Collins 48606
Discovery Museum, Sacramento 52506
The Discovery Museums, Acton 46133
Discovery Park, Safford 52513
Discovery Place, Charlotte 47388
Discovery Science Center of Central Florida, Ocala 51482
Discovery World - The James Lovell Museum of Science, Economics and Technology, Milwaukee 50707
Don Harrington Discovery Center, Amarillo 46252
Eli Whitney Museum, Hamden 49131
Elwood Haynes Museum, Kokomo 49825
Essex Shipbuilding Museum, Essex 48423
Explora Science Center and Children's Museum, Albuquerque 46177
Fort Discovery, Augusta 46477
The Franklin Institute, Philadelphia 51836
Franklin Mint Museum, Franklin Center 48752
Frontiers of Flight Museum, Dallas 47933
Gateway to Science, Bismarck 46838
Great Explorations, Saint Petersburg 52639
Great Lakes Science Center, Cleveland 47631
Gulf Coast Exploreum, Mobile 50767
The Hagley Museum, Wilmington 54193
Hagley Museum and Library, Wilmington 54194
Hands on Children's Museum, Olympia 51544
Healdton Oil Museum, Healdton 49235
Hidalgo Pumphouse Heritage, Hidalgo 49271
Hiller Aviation Institute, San Carlos 52735
Historical Electronics Museum, Linthicum 50123
Hoover Dam Museum, Boulder City 46963
Hoover Historical Center, North Canton 51386
Hudson Valley Children's Museum, Nanuet 50949
Imaginarium Discovery Center, Anchorage 46289
Imaginarium Hands-On Museum, Fort Myers 48663
Imagination Station Science Museum, Wilson 54211
Impression 5 Science Center, Lansing 49941
Intel Museum, Santa Clara 52862
The Invention Factory, Trenton 53559
Jefferson Telephone Museum, Jefferson 49606
John Gorrie Museum, Apalachicola 46327
John Rivers Communications Museum, Charleston 47368
Kalamazoo Valley Museum, Kalamazoo 49667
Lawrence Hall of Science, Berkeley 46757
Long Island Museum of Science and Technology, Garden City 48858
Louisiana Tech Museum, Ruston 52491
Louisville Science Center, Louisville 50290
McWane Science Center, Birmingham 46823
Margaret Hutchinson Compton Gallery, Cambridge 47181
Maryland Science Center, Baltimore 46578
Memphis Pink Palace Museum, Memphis 50579
Mercer Museum of the Bucks County Historical Society, Doylestown 48180
Mid-America Science Museum, Hot Springs 49373
Milton J. Rubenstein Museum of Science and Technology, Syracuse 53400
Minnesota Children's Museum, Saint Paul 52622
MIT Museum, Cambridge 47183
Motorola Museum, Schaumburg 52958
Mount Vernon Museum of Incandescent Lighting, Baltimore 46581
Museum of Anthracite Mining, Ashland 46390
Museum of Art and History, Weston 54102
Museum of Independent Telephony, Abilene 46122
Museum of Natural History and Primitive Technology, Somers 53144
Museum of Radio and Technology, Huntington 49437
Museum of Science, Boston 46941
Museum of Science and Industry, Chicago 47497
Museum of Science and Industry, Tampa 53434
National Canal Museum, Easton 48270
National Great Rivers Museum, Alton 46241
National Inventors Hall of Fame, Akron 46147
New England Wireless and Steam Museum, East Greenwich 48241
New Mexico Museum of Space History, Alamogordo 46151
New York Hall of Science, Flushing 48578
Oregon Museum of Science and Industry, Portland 52083
Orlando Science Center, Orlando 51594
Outagamie Museum, Appleton 46331
Pacific Science Center, Seattle 53010
Patent Model Museum, Fort Smith 48686
The Petroleum Museum, Midland 50674
Real World Computer Museum, Boothwyn 46912
Remington Firearms Museum, Ilion 49467
Reuben H. Fleet Science Center, San Diego 52743
Rothschild Petersen Patent Model Museum, Cazenovia 47288
Saint Louis Science Center, Saint Louis 52598
Schenectady Museum, Schenectady 52962
Sci-Port Discovery Center, Shreveport 53089
Sci-Tech Center of Northern New York, Watertown 53973
Science and Technology Museum of Atlanta, Atlanta 46451
Science Center of Connecticut, West Hartford 54061
Science Center of Iowa, Des Moines 48101
Science Center of West Virginia, Bluefield 46877
Science Discovery Center of Oneonta, Oneonta 51565
The Science Factory Children's Museum, Eugene 48438
Science Museum of Virginia, Richmond 52338
Science Spectrum, Lubbock 50313
Sciencenter, Ithaca 49550
Scitech, Aurora 46494
Sciworks, Winston-Salem 54250
See Science Museum, Manchester 50425
Sloane-Stanley Museum and Kent Furnace, Kent 49737
Sloss Furnaces, Birmingham 46824
Sony Wonder Technology Lab, New York 51259
South Dakota Discovery Center, Pierre 51916
South Florida Science Museum, West Palm Beach 54075
Southwest Museum of Science and Technology, Dallas 47941
Space Center Houston, Houston 49406
Spartanburg Science Center, Spartanburg 53197
Springfield Science Museum, Springfield 53229
Steam Railroading Institute, Owosso 51644
Stevens County Gas and Historical Museum, Hugoton 49416
The Tech Museum of Innovation, San Jose 52811
Telephone Pioneer Museum of New Mexico, Albuquerque 46187
Thomas Edison House, Louisville 50298
Thomas Newcomen Museum, Exton 48467
Titan Missile Museum, Sahuarita 52518
United States Army Aviation Museum, Fort Rucker 48677
United States Army Communications-Electronics Museum, Fort Monmouth 48657
US Patent and Trademark Office - Museum, Arlington 46359
Vestal Museum, Vestal 53730
The Waltham Museum, Waltham 53812
Weeks Air Museum, Miami 50636
Western Museum of Mining and Industry, Colorado Springs 47707
Wheels O' Time Museum, Peoria 51775
The Works, Newark 51295

Venezuela

Museo de Ciencias, Caracas 54545

Telecommunications → Postal Services

Television → Mass Media

Terracotta

Argentina

Museo de Bellas Artes al Aire Libre Caminito, Buenos Aires 00186

Brazil

Museu da Bacia do Paraná, Maringá 04563

Italy

Collezione di Ceramiche Mostra Permanente, Fiorano Modenese 26139
Mostra Permanente della Ceramica, San Lorenzello 27655
Museo Civico delle Ceramiche, Santo Stefano di Camastra 27733
Museo della Tecnica e del Lavoro MV Augusta, Trequanda 28063
Museo Ocarina e degli Strumenti in Terracotta, Budrio 25560

Spain

Terracota Museu de Terrissa i Ceràmica Industrial, La Bisbal d'Empordà 39051

Switzerland

Bibel Orient Museum, Fribourg 41383

United States of America

Tile Heritage Foundation, Healdsburg 49234

Vatican City

Museo Gregoriano Etrusco, Città del Vaticano 54504

Textiles

Australia

Ararat Regional Art Gallery, Ararat 00809
Frances Burke Textile Resource Centre-Collections, Melbourne 01278
Tamworth Regional Gallery, Tamworth 01569

Austria

Handarbeitsmuseum, Traunkirchen 02826
Hermesvilla, Wien 03003
Lebendes Textilmuseum, Groß-Siegharts 01998
Stadtmuseum und Tuchmachermuseum Pinkafeld, Pinkafeld 02499
Stickereimuseum Lustenau, Lustenau 02342
Textildruckmuseum Mittelweiherburg, Hard 02050

Belgium

Centre Touristique de la Laine de la Mode, Verviers 04106
Chantilly Museum, Geraardsbergen 03649
Kant- en Textielcentrum, Diest 03551
Musée de la Rubanerie Cominoise, Comines 03525
Musée du Costume et de la Dentelle, Bruxelles 03483

Museum van Hedendaagse Kantwerken, Sint-Truiden 04028
Stedelijk Textielmuseum, Ronse 03987

Bhutan

Textile Museum, Thimphu 04204

Bulgaria

Muzej na Tekstilnata Industrija, Sliven 05168

Canada

Barrington Woolen Mill Museum, Barrington 05389
Chapel Gallery, Bracebridge 05452
Mississippi Valley Textile Museum, Almonte 05324
Musée Beaulne, Coaticook 05626
Musée Marsil - Musée du Costume et du Textile du Québec, Saint-Lambert 06861
Red Deer and District Museum, Red Deer 06709
Textile Museum of Canada, Toronto 07142
Ukrainian Museum of Canada, Saskatoon 06925
Le Vieux Moulin, Sainte-Jeanne-d'Arc 06895
Wile Carding Mill Museum, Bridgewater 05475

China

Cotton Spinning Weaving Workshop, Shanghai 08164
Nantong Textile Museum, Nantong 08061
Yun Brocade Museum, Nanjing 08051

Czech Republic

Městské Muzeum, Moravská Třebová 09662
Městské Muzeum, Krnov 09609
Muzeum Zábřeh, Zábřeh 09924
Textilní Muzeum, Česká Skalice 09460

Denmark

Textilforum, Herning 10074

Estonia

Heimtali Koduloomuuseum, Heimtali 10569

Finland

Kotiseutumuseo Pellava ja Sahti, Alfred Kordelinin Ajokalumuseo, Lammi 11042
Längelmäen Pellavamuseo, Länkipohja 11024

France

Atelier-Musée des Tisserands et de la Charentaise, Varaignes 16072
Ecomusée du Roannais, Roanne 15058
Maison de Pays, Sainte-Marie-aux-Mines 15565
Maison du Lin, Routot 15131
Musée de la Draperie, Vienne 16151
Musée de la Soierie, Charlieu 12411
Musée de la Viscose, Échirolles 12807
Musée de l'Impression sur Etoffes, Mulhouse 14324
Musée Departemental du Textile, Lavelanet 13551
Musée des Tapisseries, Aix-en-Provence 11515
Musée des Tissus, Lyon 13893
Musée du Costume Comtois, Ornans 14539
Musée du Fer et du Fil, Dompierre (Orne) 12762
Musée du Feutre, Mouzon 14319
Musée du Textile des Vosges, Ventron 16107
Musée du Textilen, Labastide-Rouairoux 13463
Musée Municipal de la Toile de Jouy, Jouy-en-Josas 13286
Musée Souleïado Charles Demery, Tarascon (Bouches-du-Rhône) 15856
Musée Textile de Haute-Alsace, Husseren-Wesserling 13232
Musées de La Poëslerie et de la Dentelle, Villedieu-les-Poêles 16185

Germany

Angermuseum - Margaretha-Reichardt-Haus, Erfurt 18197
Brandenburgisches Textilmuseum Forst, Forst, Lausitz 18341
Deutsches Damast- und Frottiermuseum, Großschönau 18805
Deutsches Textilmuseum Krefeld, Krefeld 19703
Erstes Deutsches Strumpfmuseum, Gelenau 18567
Flachsmuseum Wegberg-Beeck, Wegberg 22359
Heimatmuseum Fronfeste, Limbach-Oberfrohna 19992
Kreismuseum Zons, Dormagen 17904
Leinen- und Spitzenmuseum Haigerseelbach, Haiger 18867
Maschenmuseum, Albstadt 16454
Museum auf dem Burghof, Springe 21846
Museum Haus Martfeld, Schwelm 21690
Museum Voswinckelshof, Dinslaken 17870
Museum Weiler Textilgeschichte, Weil am Rhein 22377
Museumshof, Rahden 21167
Oberfränkisches Textilmuseum, Helmbrechts 19119
Rheinisches Industriemuseum, Euskirchen 18283
Schraube Museum, Halberstadt 18880
Textilmuseum Brennet, Wehr 22362
Textilmuseum Die Scheune, Nettetal-Hombergen 20579
Textilmuseum Sandtnerstiftung, Mindelheim 20332
Textilsammlung Max Berk, Heidelberg 19085
tim - Bayerisches Textil- und Industriemuseum, Augsburg 16638
Tuch und Technik Textilmuseum Neumünster, Neumünster 20636
Tuchmacher Museum Bramsche, Bramsche 17500
Webereimuseum, Sindelfingen 21779
Webermuseum, Biberach an der Riß 17323
Westsächsisches Textilmuseum Crimmitschau, Crimmitschau 17750
Wiesenthäler Textilmuseum, Zell im Wiesental 22781

Greece

Corgialenios Historical and Cultural Museum, Argostolion 22890

Guatemala

Museo Ixchel del Traje Indigena, Guatemala 23300

Hungary

Textil- és Textilruházati Ipartörténeti Múzeum, Budapest 23485

Iceland

Heimilisidnardarsafnid Halldórustofa, Blönduós 23988

India

Calico Museum of Textiles, Ahmedabad 24046
Governmental College of Engineering and Textile Technology Museum, Hooghly 24208
Kachchh Museum, Bhuj 24099
Sardar Vallabhbhai Patel Museum, Surat 24399
State Kala Akademi Museum, Imphal 24229
TTD Museum, Tirupati 24409

Indonesia

Museum Batik, Pekalongan 24542
Museum Takstil, Jakarta 24502

Italy

Museo dei Damaschi, Lorsica 26526
Museo del Tessile, Chieri 25866
Museo del Tessuto, Prato 27335
Museo di Palazzo Mocenigo, Venezia 28195

Japan

Bunka Gakuen Fukushoku Hakubutsukan, Tokyo 29374
Kenji Igarashi Memorial Museum of Laundry, Tokyo 29411
MIYAI Furoshiki and Fukusa Gallery, Kyoto 28851
Nishijin Textile Center, Kyoto 28856
Oriamu Museum, Izumiotsu 28615
Orinasukan, Kyoto 28860

Malaysia

Muzium Tekstil, Kuala Lumpur 30583

Mexico

Museo Textil La Trinidad, Santa Cruz Tlaxcala 31377

Netherlands

Art of Velvet Gallery De Stierenstal, Montfoort 32560
Nederlands Textielmuseum, Tilburg 32829
De Wascht en Strekt, Gilze 32256

Norway

Norsk Trikotasjemuseum, Salhus 33807
Råde Bygdetun, Råde 33774

Poland

Centralne Muzeum Włókiennictwa, Łódź 34860

Portugal

Museu da Indústria Têxtil, Calendário 35498
Museu de Francisco Tavares Proença Júnior, Castelo Branco 35515
Museu de Lanifícios, Covilhã 35553

Romania

Expoziţia Muzeală Istoricul Industriei Textile, Cisnădie 36282

Russia

Dalnevostočnyj Chudožestvennyj Muzej, Chabarovsk 36858
Muzej Chudožestvennych Tkanej, Moskva 37267
Muzej Ivanovskogo Sitca, Ivanovo 36950

Spain

Centre de Documentació i Museu Tèxtil, Terrassa 40116
Fundación Museo de las Ferias, Medina del Campo 39638
Museo Etnográfico Textil Pérez Enciso, Plasencia 39844
Museo Etnológico de Morella y del Meastrazgo, Morella 39685
Museo Nacional de Artes Decorativas, Madrid 39569
Museo Pedagógico Textil, Madrid 39577
Museu de l'Estampació de Premià de Mar, Premià de Mar 39869
Museu Textil i d'Indumentària, Barcelona 39005

Sweden

Dock- och Textilmuseum, Katrineholm 40640
Lövångers Sockenmuseum, Lövånger 40692
Norrköpings Stadsmuseum, Norrköping 40774
Regionmuseet Kristianstad, Kristianstad 40655
Rydals Museum, Rydal 40819
Skattkammaren - Uppsala Domkyrkas Museum, Uppsala 41011
Textilarkivet Västernorrland, Sollefteå 40869
Textilmuseet, Borås 40471

Switzerland

Freiämter Strohmuseum, Wohlen 42044
Musée des Toiles Peintes, Colombier (Neuchâtel) 41310
Museum Bellerive, Zürich 42092
Museumsspinnerei Neuthal, Bäretswil 41150
Ortsmuseum Diessenhofen, Diessenhofen 41328
Textil- und Heimatmuseum, Sennwald 41862
Textilmuseum, Hauptwil 41472
Textilmuseum, Sankt Gallen 41809

Turkey

Aydın Müzesine Bağlı Örenyerleri, Aydın 42722
Halı, Kilim ve Düz Dokuma Yaygılar Müzesi, İstanbul 42820

United Kingdom

Axminster Museum, Axminster 43341
Bankfield Museum, Halifax 44382
Childhood and Costume Museum, Bridgnorth 43604
Colour Museum, Bradford 43578
Falkland Palace and Garden, Falkland 44207
Friary Art Gallery, Lichfield 44703
Guildford House Gallery, Guildford 44370
Helmshore Mills Textile Museum, Helmshore 44441
Heriot-Watt University Museum, Edinburgh 44127
Irish Linen Centre and Lisburn Museum, Lisburn 44725
Lewis Textile Museum, Blackburn, Lancashire 43527
National Wool Museum, Llandysul 44761
Newarke Houses Museum, Leicester 44682
Paulise de Bush Costume Collection, Broadclyst 43643
Ruddington Framework Knitters' Museum, Ruddington 45524
Saddleworth Museum and Art Gallery, Uppermill 45900
Textile Conservation Centre, Winchester 46034
Totnes Costume Museum, Totnes 45871
Trowbridge Museum, Trowbridge 45878
Verdant Works, Dundee 44053
Weavers' Triangle Visitor Centre, Burnley 43680
The Whitworth Art Gallery, Manchester 45099

United States of America

Ace of Clubs House, Texarkana 53488
American Textile History Museum, Lowell 50303
Big Red Barn Gallery, Clearmont 47608
Dar Museum First Ladies of Texas Historic Costumes Collection, Denton 48060
The Design Center at Philadelphia University, Philadelphia 51827
Elizabeth Sage Historic Costume Collection, Bloomington 46863
The Fabric Workshop and Museum, Philadelphia 51833
FIDM Museum and Galleries, Los Angeles 50220
Georgia State Cotton Museum, Vienna 53743
The Goldstein Museum of Design, Saint Paul 52616
Grovewood Gallery, Asheville 46373
Gustafson Gallery, Fort Collins 48608
Helen Louise Allen Textile Collection, Madison 50390
Henry Art Gallery, Seattle 53000
The Lace Museum, Sunnyvale 53372
Liberace Museum, Las Vegas 49974
Mattye Reed African Heritage Center, Greensboro 49066
Maxwell Museum of Anthropology, Albuquerque 46181
Milan Historical Museum, Milan 50680
Moravian Museum of Bethlehem, Bethlehem 46781
Museum of the American Quilter's Society, Paducah 51669
New England Quilt Museum, Lowell 50306
Quilters Hall of Fame, Marion 50484
Reuel B. Pritchett Museum, Bridgewater 47022
Rocky Mountain Quilt Museum, Golden 48961
San Jose Museum of Quilts Textiles, San Jose 52810
Scalamandre, Long Island City 50186
Slater Mill Historic Site, Pawtucket 51738
South Carolina Cotton Museum, Bishopville 46834
The Textile Museum, Washington 53921
Wilmette Historical Museum, Wilmette 54186
Windham Textile and History Museum, Willimantic 54179

Theatre → Performing Arts

Tin Figures → Lead and Tin Figures

Tobacco

Andorra

Museu del Tabac - Antiga Fàbrica Reig, Sant Julià de Lòria 00105

Australia

Texas Historical Museum and Tobacco Museum, Texas 01580

Austria

Heimatmuseum, Sankt Aegyd am Neuwalde 02632
Pfeifenmuseum, Sankt Aegyd am Neuwalde 02633
Pfeifenmuseum, Scheifling 02705
Pfeifenmuseum, Windigstein 03115
Stadtmuseum Wienertor Hainburg, Hainburg an der Donau 02039
Tabakmuseum, Oberzeiring 02468

Belgium

Historisch Pijp en Tabaksmuseum, Sint-Niklaas 04019
Maison de la Ruralité, Thuillies 04063
Musée du Vieux Nimy, Mons 03887
Nationaal Tabaksmuseum, Wervik 04147
Sigaar Musea, Geraardsbergen 03653
Stedelijk Museum voor Pijp en Tabak, Harelbeke 03680
Tabak Musea, Geraardsbergen 03654

Canada

Delhi Ontario Tobacco Museum Heritage Centre, Delhi 05691

Cuba

Museo Casa de Puerto Rico y Sala del Tabaco, La Habana 09121

France

Musée d'Intérêt National du Tabac, Bergerac 11945

Germany

Deutsches Tabak- und Zigarrenmuseum, Bünde 17634
Deutsches Tabakpfeifenmuseum, Oberelsbach 20804
Oberrheinisches Tabakmuseum, Mahlberg 20133
Schnupftabakmuseum, Grafenau, Kreis Freyung-Grafenau 18728
Tabak-Museum der Stadt Hockenheim, Hockenheim 19207
Tabakhistorische Sammlung Reemtsma im Museum der Arbeit, Hamburg 18962
Tabakmuseum, Lorsch 20036

Japan

Tabako to Shio no Hakubutsukan, Tokyo 29479

Netherlands

Amerongs Historisch Museum/ Tabaksmuseum, Amerongen 31807
Kamper Tabaksmuseum, Kampen 32437
Museum De Moriaan, Gouda 32268

Museum Joure, Joure 32432
Niemeyer Tabaksmuseum, Groningen 32286
Pijpenkabinet Smokiana, Amsterdam 31880

Puerto Rico

Museo del Tabaco Herminio Torres Grillo, Caguas 36032

Slovakia

Starý Zamok Banská Štiavnica, Banská Štiavnica 38194

Slovenia

Tobačni Muzej, Ljubljana 38468

Sweden

Tobaks- och Tändstickmuseum, Stockholm 40947

Switzerland

Tabak- und Zigarrenmuseum, Menziken 41626

United Kingdom

Clay Tobacco Pipe Museum, Broseley 43658
Sutton Windmill and Broads Museum, Sutton, Norwich 45806

Tombs

Australia

Sexton's Cottage Museum, Crows Nest, New South Wales 01012

Austria

Bestattungsmuseum der Bestattung Wien GmbH, Wien 02953
Freilichtmuseum spätbronzezeitlicher Hügelgräber, Siegendorf 02740
Kaisergruft, Wien 03010

China

Beijing Dabao Taixi Western Han Tomb Museum, Beijing 07512
Jiuquan Museum, Jiuquan 07904
Lei Cheng Uk Han Tomb Museum, Hong Kong 07808
Wangjian Tomb Museum, Chengdu 07622

Croatia

Muzej Hvarske Baštine, Hvar 08880

Cuba

Museo Chorro de Maita, Yaguajay 09353

Denmark

Ladbyskibsmuseet, Kerteminde 10117

Germany

Bajuwarenmuseum Waging am See, Waging am See 22271
Grabkapelle auf dem Württemberg, Stuttgart 21936
Heimatkundliche Sammlung Burg Rheinfels, Sankt Goar 21496
Mausoleum im Schlossgarten Charlottenburg, Berlin 17185
Museum Burg Linn, Krefeld 19706
Museum für Sepulkralkultur, Kassel 19470
Museum Kloster Asbach, Asbach 16579

Greece

Feres Archaeological Site, Velestino 23233

Hungary

Kegyeleti Szakgyűjtemény, Budapest 23419

Israel

Dominus Flevit, Jerusalem 24985

Italy

Ipogeo dei Volumni, Ponte San Giovanni 27286
Museo Civico, Troia 28102
Museo Civico Aufidenate De Nino, Alfedena 25195
Museo della Badia Benedettina, Cava dei Tirreni 25786
Museo di Archeologia Sacra e Profana, Roma 27492
Museo Nazionale Archeologico, Tarquinia 27943
Tombe Reali di Casa Savoia, Torino 28032

Mexico

Museo de la Muerte, San Juan del Río 31338
Museo de Momias de Celaya, Celaya 30754

Northern Cyprus

Canbulat Türbe ve Müzesi, Gazimağusa 33350

Spain

Museo Provincial, Lugo 39499
Museu i Necròpolis Paleocristians, Tarragona 40104

Sweden

Sundsvalls Museum, Sundsvall 40960

Switzerland

Museo Plebano, Agno 41092
Museum im Turmhof, Steckborn 41908
Sammlung Friedhof Hörnli, Riehen 41765

Turkey

Çifte Medrese Müzesi, Kayseri 42873
Selçuk Müzesi, Konya 42890
Sırçalı Medrese Müzesi, Konya 42891
Türbeler Müzesi, İstanbul 42844

United Kingdom

Iona Abbey and Nunnery, Isle-of-Iona 44538

Vatican City

Museo Pio Cristiano, Città del Vaticano 54509

Tools

Australia

Excell Blacksmith and Engineering Workshop Museum, Tumby Bay 01605
Norseman Historical and Geological Collection, Norseman 01371

Austria

Hobel- und Werkzeugmuseum, Laxenburg 02273
Museumsdorf Trattenbach, Ternberg 02803

Canada

City of York Museum, York 07413

Denmark

Give-Egnens Museum, Give 10037
Mariager Museum, Mariager 10183

Finland

Aitonevan Turvemuseo, Kihniö 10954
Billnäsin Kirvesmuseo, Billnäs 10717
Jokioisten Naulamuseo, Jokioinen 10875
Rosenlew-museo, Pori 11222

France

Ecomusée de Saint-Dégan, Brec'h 12151
Maison de l'Outil et de la Pensée Ouvrière, Troyes 15996
Musée de la Pince, Montécheroux 14196

Germany

Deutsches Werkzeugmuseum, Remscheid 21264
Heimat- und Schmiedemuseum, Widdern 22536
Heimatmuseum Hossingen, Meßstetten 20308
Heimatmuseum in der Scheune, Mössingen 20365
Holzgerätemuseum, Schlüchtern 21577
Museum Würth, Künzelsau 19736
Resenhof, Bernau im Schwarzwald 17292

Guadeloupe

Musée Municipal de Saint-Barthélemy, Saint-Barthélemy 23280

Hungary

Korom Gyula Fodrászati Eszközök Gyűjtemény, Budapest 23427

Ireland

Macroom Museum, Macroom 24872
Tullow Museum, Tullow 24906

Israel

Man and His Work Center in Memory of S. Avitsur, Tel Aviv 25115

Italy

Museo del Chiodo, Forno di Zoldo 26235
Museo della Civiltà Contadina, Serrastretta 27816
Museo dell'Opificio delle Pietre Dure, Firenze 26177
Museo dell'Utensileria, Lanzo Torinese 26453

Japan

Hida Minzoku-mura, Takayama 29338
Kyoto Chopstick Culture Museum, Kyoto 28835
Sakai-shiritsu Hamono Hakubutsukan, Sakai 29200
Takenaka Carpentry Tools Museum, Kobe 28739

Netherlands

Gereedschap Museum Mensert, Delft 32056
Koperslagersmuseum Van der Beele, Horst 32415
Stedelijke Oudheidkamer, Genemuiden 32243

New Zealand

Museum of Woodwork and Ornamental Turning, Ashburton 33038

Norway

Roparshaugsamlinga, Isdalstø 33578

Romania

Muzeul Tehnic Prof. Ing. Dimitrie Leonida, București 36229

Sweden

Kvarnbacken Prylmuseum, Sandviken 40833
Penselmuseet i Bankeryd, Bankeryd 40452
Skogsmuseum, Sunne 40961

United Kingdom

Forge Museum, Much Hadham 45182
Longdale Craft Centre and Museum, Ravenshead 45464
Moira Furnace, Moira 45162
Museum of Bath at Work, Bath 43406
Museum Workshop and Stores, Grangemouth 44343
Pinchbeck Marsh Engine and Land Drainage Museum, Spalding 45722
Ryedale Folk Museum, Hutton-le-Hole 44505
Swinford Museum, Filkins 44228
Usk Rural Life Museum, Usk 45901
Weald and Downland Open Air Museum, Chichester 43832

United States of America

Audie Murphy/American Cotton Museum, Greenville 49080
Moravian Museum of Bethlehem, Bethlehem 46781
Sloane-Stanley Museum and Kent Furnace, Kent 49737
Southern Highland Craft Guild at the Folk Art Center, Asheville 46376
World O' Tools Museum, Waverly 53997

Tourism

Argentina

Exposición Roberto J. Mouras, Carlos Casares 00293

Austria

Montafoner Tourismusmuseum Gaschurn, Gaschurn 01927
Museum Schloß Lackenbach, Lackenbach 02250

Finland

Suomen Matkailumuseo, Imatra 10857

Italy

Museo del Turismo dell'Alta Pusteria, Villabassa 28274

Poland

Gabinet Prof. Kazimierza Sosnowskiego, Kraków 34772

Russia

Muzej Istorii Turizma, Sankt-Peterburg 37687

Sweden

Hamnmuseét, Varberg 41047

Switzerland

Bergführer-Museum, Sankt Niklaus 41814
Hotelmuseum, Flims Waldhaus 41375
Touristik-Museum, Unterseen 41960

United States of America

Eagle Tavern Museum, Watkinsville 53982
Guest House Museum, Fort Bragg 48596
Hotel de Paris Museum, Georgetown 48889

Vietnam

Crazy House Museum, Đà Lat 54602

Toys and Games

Algeria

Musée de l'Enfance, Alger 00040
Musée du Mont Riant, Alger 00042

Australia

Australian Toy Museum, Collingwood 00994
Doll and Bear Museum, Manunda 01260
Jigsaw Gallery and Bridgetown Heritage Museum, Bridgetown 00894

Austria

Badener Puppen- und Spielzeugmuseum, Baden bei Wien 01773
Museum für Billard- und Kaffeehauskultur, Wien 03036
Österreichisches Spiele Museum, Leopoldsdorf im Marchfelde 02295
Salzburg Museum, Salzburg 02625
Spielkartensammlung Piatnik, Wien 03077
Spielzeugmuseum, Wolfurt 03121
Spielzeugmuseum Salzburg, Salzburg 02627
Stadtmuseum, Traiskirchen 02817

Belgium

Archeologisch Streekmuseum Het Labyrint, Kemmel-Heuvelland 03751
Musée du Jouet, Bruxelles 03484
Speelgoedmuseum Mechelen, Mechelen 03860

Canada

Aunt Margaret's Museum of Childhood, Dufresne 05719
Elliott Avedon Museum and Archives of Games, Waterloo 07287
Old Strathcona Model and Toy Museum, Edmonton 05751

Cuba

Museo Gran Casino, Bayamo 09035

Czech Republic

Městské Muzeum a Galerie, Hořice 09524
Muzeum Hraček, Praha 09770
Sex Machines Museum, Praha 09793
Východočeské Muzeum v Pardubicích, Pardubice 09707

Estonia

Tartu Mänguasjamuuseum, Tartu 10658

Finland

Linnanmäen Museo, Helsinki 10800
Nukkemuseo Suruton, Savonlinna 11302
Suomenlinnan Lelumuseo, Helsinki 10826

France

Espace-Jouets, Firminy 12938
Jouëts d'Autrefois Rêves d'Aujourd'hui, Azay-le-Rideau 11810
Musée-Animé du Jouet et des Petits Trains, Colmar 12581
Musée Comtois, Besançon 11955
Musée d'Allard, Montbrison 14188
Musée de la Figurine, Tulette 16005
Musée de la Figurine Historique et du Jouet Ancien, Le Val 13661
Musée de la Poupée et du Jouet Ancien, Wambrechies 16283
Musée du Jeu Ancien, Saint-Hilaire-de-Brethmas 15305

Musée du Jeu de l'Oie, Rambouillet ... 14981
Musée du Jouet, Moirans-en-Montagne ... 14138
Musée du Jouet, Poissy ... 14866
Musée du Jouet Ancien "Atlantrain", Saint-Just-Luzac ... 15352
Musée du Modèle Réduit, Varces ... 16074
Musée la Vie des Jouets, Mauleon ... 14035
Musée Louis Lutaud, Ferrière-la-Grande ... 12923
Musée Municipal Département Jouet, La Ferté-Macé ... 13360
Musée Rambolitrain, Rambouillet ... 14983
Musées de Vieille Forge, de la Traversée du Siècle et des Automates, Fréland ... 13007
Nef des Jouets, Soultz-Haut-Rhin ... 15802

Georgia

T. Tumanishvili Children Toy Museum, Tbilisi 16384

Germany

Boxenstop Auto- und Spielzeugmuseum, Tübingen ... 22112
Deutsch-Französisches Puppen und Spielzeugmuseum, Schweigen-Rechtenbach ... 21681
Deutsches Spiele-Archiv, Nürnberg ... 20757
Deutsches SPIELEmuseum e.V, Chemnitz ... 17693
Deutsches Spielzeugmuseum, Sonneberg ... 21818
Erzgebirgisches Spielzeugmuseum, Seiffen / Erzgebirge Kurort ... 21730
Fränkisches Spielzeugmuseum, Gößweinstein 18687
Heimatmuseum, Rieneck ... 21328
Historisches Puppen-, Spielzeug- und Ikonenmuseum, Bernkastel-Kues ... 17297
Kindergartenmuseum, Bruchsal ... 17597
Kindermuseum im FEZ-Berlin, Berlin ... 17154
Kindheitsmuseum, Marburg ... 20190
Das Kleine Museum, Hagnau am Bodensee ... 18863
Das kleine Spielzeugmuseum, Baden-Baden ... 16882
Kleine Welt, Freiburg im Breisgau ... 18431
Märklin-Museum, Göppingen ... 18676
Margarete-Steiff-Museum, Giengen an der Brenz ... 18632
Mitspielzeugmuseum, Leipzig ... 19920
Modellbahnmuseum-Muggendorf, Wiesenttal 22562
Museum für Puppen, Bären und Spielzeug, Solingen ... 21804
Museum im Frey-Haus, Brandenburg an der Havel ... 17511
Museum im Schloß, Frohburg ... 18487
Museum Obere Saline mit Bismarck-Museum, Bad Kissingen ... 16759
Museum Ravensburger, Ravensburg ... 21204
Museum Schloss Tenneberg, Waltershausen ... 22318
Museum und Galerie in Ugendorf, Unlingen ... 22174
Musikinstrumenten- und Puppenmuseum, Goslar ... 18715
Puppen- und Spielzeugmuseum, Lichtenstein/Sachsen ... 19983
Puppen- und Spielzeugmuseum, Rottweil ... 21427
Puppen- und Spielzeugmuseum Sammlung Katharina Engels, Rothenburg ob der Tauber ... 21409
Puppenmuseum, Bad Schönborn ... 16829
Puppenmuseum, Künzelsau ... 19737
Rügener Puppen- und Spielzeugmuseum, Putbus ... 21141
Schachmuseum Ströbeck, Halberstadt ... 18879
Schlossmuseum Aulendorf-Kunst des Klassizismus-Altes Spielzeug, Aulendorf ... 16643
Spielmuseum Soltau und Stiftung Spiel, Soltau ... 21815
Spielzeug anno dazumal - Museum Beilngries, Beilngries ... 16984
Spielzeugmuseum, Alsfeld ... 16482
Spielzeugmuseum, Bad Lauterberg ... 16776
Spielzeugmuseum, Görlitz ... 18682
Spielzeugmuseum, Hemau ... 19126
Spielzeugmuseum, Hinterzarten ... 19196
Spielzeugmuseum, Schkeuditz ... 21560
Spielzeugmuseum im Alten Rathausturm, München ... 20485
Spielzeugmuseum im Alten Schloß, Sugenheim ... 21975
Spielzeugmuseum im Stadtmuseum, Ingolstadt ... 19345
Spielzeugmuseum - Museum Lydia Bayer, Nürnberg ... 20790
Spielzeugmuseum Solz, Bebra ... 16969
Spielzeugmuseum Trier, Trier ... 22103
Stadtmuseum Sindelfingen, Sindelfingen ... 21778
Städtisches Heimatmuseum Lippstadt, Lippstadt ... 20012
Städtisches Museum Zirndorf, Zirndorf ... 22810
Steinhuder Spielzeugmuseum, Wunstorf ... 22735

Greece

Museum of the History of Children's Games and Books, Myrina ... 23106

Hungary

Szórakaténusz Játékmúzeum, Kecskemét ... 23655

Ireland

The Berkeley Costume and Toy Museum, New Ross ... 24884

Italy

Museo del Cavallo Giocattolo, Grandate ... 26362
Museo del Giocattolo, Palermo ... 27069
Museo del Giocattolo, Sezze ... 27832

Japan

Erzgebirge Toys Museum Museum of Picture Books, Karuizawa ... 28668
Gendai Gangu Hakubutsukan, Mimasaka ... 28909
Little Sweets Museum, Osaka ... 29132
Nihon Gangu Hakubutsukan, Kodera ... 28753
Nihon Kyodogangu-Kan, Kurashiki ... 28788
Osaka Lottery Dream Museum, Osaka ... 29146
Sekai Gangu Hakubutsukan, Karuizawa ... 28670
Sekai Oota Kokaikan, Yokaichi ... 29590
Tateshina World Arts Crafts Museum, Tateshina ... 29352
Tin Toy Museum Motion Display Museum, Yokohama ... 29609
Warabekan, Tottori ... 29517

Luxembourg

Musée du Jouet, Clervaux ... 30461
Musée Folklorique et Viticole à Possen, Bech-Kleinmacher ... 30454

Malta

Pomskizillious Museum of Toys, Xaghra ... 30659
Toy Museum, Valletta ... 30648

Mexico

Museo del Juguete Tradicional Mexicano, San Francisco de los Romo ... 31331

Mongolia

Mo Tu Uv - Museum of Mechanical Toys, Ulaanbaatar ... 31621

Netherlands

Koffie- en Winkelmuseum, Pieterburen ... 32669
Matchboxmuseum Latent, Prinsenbeek ... 32671
Museum Kinderwereld, Roden ... 32696
Museum voor Klederdracht en Oud Speelgoed, Warnsveld ... 32924
Het Nationaal Autominiaturen Museum, Asperen ... 31934
Nationaal Modelspoor Museum, Sneek ... 32790
Nederlands Kansspelmuseum, Den Haag ... 32096
Oude Ambachten en Speelgoedmuseum, Terschuur ... 32819
Poppen- en Speelgoedmuseum, Tilburg ... 32831
Poppenhuis Carmen, Utrecht ... 32855
Radio- en Speelgoed Museum, Onstwedde ... 32622
Speelgoedmuseum De Kijkdoos, Hoorn ... 32412
Speelgoedmuseum Deventer, Deventer ... 32121
Toy-Toy Museum, Rotterdam ... 32730
TV Toys Dieren, Dieren ... 32127

Norway

Norsk Barnemuseum, Stavanger ... 33868

Poland

Muzeum Zabawek ze Zbiorow Henryka Tomaszewskiego, Karpacz ... 34716
Muzeum Zabawkarstwa, Kielce ... 34745

Portugal

Museu do Brinquedo, Évora ... 35588
Museu do Brinquedo de Sintra, Sintra ... 35947
Museu do Jogo, Oeiras ... 35788

Russia

Chudožestvenno-pedagogičeskij Muzej Igruški - Kulturno-delovoj Centr, Sergiev Posad ... 37762

Slovakia

Múzeum Bábkarských Kultúr a Hračiek Hrad Modry Kameň, Modrý Kameň ... 38307

South Africa

Toy and Miniature Museum, Stellenbosch ... 38739

Spain

Museo-Exposición Permanente de Juegos Tradicionales, Campo ... 39111
Museu de la Jugueta de Mallorca, Sa Pobla ... 39847
Museu del Joguet de Catalunya, Figueres ... 39299
Museu Municipal de la Festa d'Elx, Elche ... 39270
Museu Valencia del Joguet, Ibi ... 39405

Sweden

Hobby- och Leksaksmuseum, Stockholm ... 40900
Museet Leksaksland, Malmö ... 40725
Rallarmuseet, Moskosel ... 40749
Smålands Bil, Musik och Leksaksmuseum, Rydaholm ... 40818

Switzerland

Krippen- und Spielzeugmuseum, Bubendorf ... 41262
Musée Suisse du Jeu, La Tour-de-Peilz ... 41945
Puppen- und Spielzeugmuseum, Solothurn ... 41887
Puppenhausmuseum, Basel ... 41179
Sammlung historischer Biscuitdosen, Basel ... 41181
Spielzeug-Museum, Aeugstertal ... 41091
Spielzeugmuseum, Davos Platz ... 41322
Spielzeugmuseum, Dorf- und Rebbaumuseum, Riehen ... 41766
Spielzeugmuseum - Sammlung Depuoz, Zürich ... 42106
Teddybär Museum, Luzern ... 41599
Teddybärmuseum, Sempach ... 41861
Zürcher Spielzeugmuseum, Zürich ... 42120

Taiwan

Dongshan Kite Museum, Dongshan ... 42181
Jioufen Kite Museum, Rueifang ... 42308

Turkey

Oyuncak Müzesi, Ankara ... 42709

United Kingdom

Arundel Toy Museum Dolls House, Arundel ... 43324
Bear Museum, Petersfield ... 45372
Broadway Magic Experience, Broadway ... 43647
The Corgi Heritage Centre, Heywood ... 44464
Cumberland Toy and Model Museum, Cockermouth ... 43867
Eureka! The Museum for Children, Halifax ... 44384
Hollytrees Museum, Colchester ... 43874
House on the Hill Museums Adventure, Stansted Mountfitchet ... 45738
Judges' Lodgings, Lancaster ... 44631
Lilliput Antique Doll and Toy Museum, Brading ... 43588
Museum of Childhood, Sudbury, Derbyshire ... 45796
Museum of Childhood Memories, Beaumaris ... 43427
Penshurst Place and Toy Museum, Penshurst 45358
Pollock's Toy Museum, London ... 44957
Romney Toy and Model Museum, New Romney ... 45207
St John's House, Warwick ... 45939
VA Museum of Childhood, Bethnal Green, London ... 45001
Vina Cooke Museum of Dolls and Bygone Childhood, Newark-on-Trent ... 45217

United States of America

Angels Attic Museum, Santa Monica ... 52897
Creatabilitoys! - Museum of Advertising Icons, Coral Gables ... 47800
Crossroads of America, Bethlehem ... 46776
The Doll and Toy Museum of NYC, Brooklyn ... 47071
Eugene Field House and Saint Louis Toy Museum, Saint Louis ... 52576
Heritage Museum of Fine Arts for Youth, Detroit ... 48120
House of a Thousand Dolls, Loma ... 50163
Lancaster Antique Toy and Train Collection, Portsmouth ... 52108
Matchbox Road Museum, Newfield ... 51313
Milan Historical Museum, Milan ... 50680
Milano Model and Toy Museum, Elmhurst ... 48377
Museum of Carousel Art and History, Sandusky ... 52837
National Farm Toy Museum, Dyersville ... 48229
The National Toy Train Museum, Strasburg ... 53328
Pauline E. Glidden Toy Museum, Ashland ... 46385
Strong National Museum of Play, Rochester ... 52400
Toy and Miniature Museum of Kansas City, Kansas City ... 49703
Trainland U.S.A., Colfax ... 47684
Washington Dolls House and Toy Museum, Washington ... 53931
Wenham Museum, Wenham ... 54038
World Kite Museum and Hall of Fame, Long Beach ... 50180
Yesterday's Children Antique Doll Toy Museum, Vicksburg ... 53736
Yesteryears Doll and Toy Museum, Sandwich 52843

Trades and Guilds

Australia

Brennan and Geraghty's Store Museum, Maryborough, Queensland ... 01267

Austria

Alte Huf- und Wagenschmiede, Dietmanns ... 01818
Alte Schmiede - Kunstverein Wien, Wien ... 02930
Bienen- und Wagnereimuseum, Tollet ... 02814
Erstes Kärntner Handwerksmuseum, Baldramsdorf ... 01779
Erstes Österreichisches Tischler-Museum, Pöchlarn ... 02504
Freilichtmuseum Handwerkerhaus Stegwagner, Windhaag bei Freistadt ... 03109
Heimat- und Ständemuseum, Tragöß ... 02815
Heimatmuseum und Apothekenmuseum, Mauthausen ... 02372
Kaufmannsmuseum, Haslach ... 02055
Museum Steinmetzhaus, Eggenburg ... 01844
Museumsdorf Trattenbach, Ternberg ... 02803
Pulkauer Gewerbemuseum, Pulkau ... 02538
Sägehammer Hofwies, Windhaag bei Freistadt 03113
Sammlungen des Marktamtes, Wien ... 03066
Schmiede im Hammergraben, Helfenberg ... 02068
Schmiedegesellenhaus, Göstling an der Ybbs ... 01956

Belgium

Koninklijke Hoofdgilde Sint-Sebastiaan, Brugge ... 03419
Museum van de Kruisbooggilde Sint-Joris, Brugge ... 03422

Canada

Commanda General Store Museum, Commanda ... 05636
Dickson Store Museum, Dickson ... 05701
Forwarders' Museum, Prescott ... 06657
Hyman and Sons General Store, Gaspé ... 05853
Site-Historique du Banc-de-Paspébiac, Paspébiac ... 06573

China

Akesu Zone Museum, Akesu ... 07477
Ankang Zone Museum, Ankang ... 07479
Ganzhou Zone Museum, Ganzhou ... 07712
Heze Zone Museum, Heze ... 07788
Kashi Zone Silu Museum, Kashi ... 07914
Liaocheng Zone Museum, Liaocheng ... 07934
Linfen Zone Museum, Linfen ... 07942
Nansong Guanstove Museum, Hangzhou ... 07765
Quanzhou Overseas-Relations History Museum, Quanzhou ... 08132
Shangqiu Zone Museum, Shangqiu ... 08202
Shangrao Zone Museum, Shangrao ... 08204
Songzi Fuwu Guan, Songzi ... 08246
Xinzhou Zone Museum, Xinzhou ... 08408
Yili Zone Museum, Yili ... 08446
Yuncheng Zone Hedong Museum, Yuncheng ... 08472

Czech Republic

Expozice Řemesel, Moravské Budějovice ... 09663

Denmark

Roskilde Museums Købmandsgård, Roskilde 10248
Slagelse Museum for Handel, Håndværk og Industri, Slagelse ... 10276
Slagterbutikken anno 1920, Roskilde ... 10249

Finland

Hotelli- ja Ravintolamuseo, Helsinki ... 10788
Jakobstads Museum, Jakobstad ... 10865
Kurikan Museo, Kurikka ... 11014
Naantalin Museo, Naantali ... 11133
Suomen Värimuseo, Kemi ... 10942

Germany

Bäckereimuseum, Rimpar ... 21334
Bauern- und Handwerker-Museum, Malgersdorf ... 20157
Christophs Friseur-Museum, Leipheim ... 19888
Deutsches Fleischermuseum, Böblingen ... 17417
Drechsler- und Metalldrücker-Museum, Wendelstein ... 22475
Fränkisches Museum, Feuchtwangen ... 18313
Gildehaus, Bardowick ... 16927
Handwerk Gewerbe Museum Sagar, Krauschwitz ... 19700
Handwerksmuseum am Mühlenberg, Suhlendorf ... 21978
Haus Peters, Tetenbüll ... 22034

Heimat-und Handwerkermuseum, Leutershausen-Wiedersbach 19970
Heimatmuseum, Calau 17674
Heimatmuseum, Zöblitz 22814
Heimatmuseum der Stadt Herrnhut, Herrnhut 19162
Heimatmuseum der Stadt Ketzin, Ketzin 19522
Historische Gesellenherberge, Blankenburg 17372
Historischer Saal der Fischerzunft, Würzburg 22711
Historisches Museum Bamberg, Bamberg 16910
Kreismuseum Finsterwalde, Finsterwalde 18321
Krell'sche Schmiede, Wernigerode 22496
Lechflößermuseum, Lechbruck am See 19867
Museum der Stadt Mittweida "Alte Pfarrhäuser", Mittweida 20346
Museum in der Burg, Coppenbrügge 17725
Museum in der Osterburg, Weida 22365
Museum Wurzen mit Ringelnatzsammlung, Wurzen 22751
Museumsstadl, Bernried, Kreis Deggendorf 17303
Schmiedemuseum Arfeld, Bad Berleburg 16694
Schwarzwald-Museum, Triberg im Schwarzwald 22092
Stadtmuseum Pforzheim, Pforzheim 21030
Steinhauermuseum, Randersacker 21177
Umstädter Museum Gruberhof, Groß-Umstadt 18778

Ireland
Findlater Museum, Dublin 24789

Italy
Museo del Cappello, Montappone 26800
Museo di Merceologia, Roma 27496

Japan
Kakunodate Museum of Traditional Crafts and Industry, Kakunodate 28639
Matsusaka Oroshi Kinenkan, Matsusaka 28897
Nihon Biyoin Hakubutsukan, Kyoto 28855

Korea, Republic
Museum of Korean Commercial History, Yongin 29944

Malta
Gharb Folklore Museum, Gharb 30628

Netherlands
Stadhuismuseum, Zierikzee 33005
Tattoo Museum, Amsterdam 31893

New Zealand
Briar Herb Factory Museum, Clyde 33080

Pakistan
Mughal Museum, Lahore 34024

Russia
Muzej Kupečeskogo Byta, Ekaterinburg 36896

Saudi Arabia
National Museum for Trading, Riyadh 38018

Slovakia
Múzeum Obchodu, Ilava 38264
Múzeum Obchodu Bratislava, Bratislava 38225
Považské Múzeum, Žilina 38391

Spain
Fundación Museo de las Ferias, Medina del Campo 39638

Switzerland
Wertpapierwelt, Olten 41710

Taiwan
Customs Museum, Taipei 42387
Sweet Shop Museum, Donghe 42179

United Kingdom
Guildhall Museum, Rochester 45507
Her Majesty Customs and Excise National Museum, Liverpool 44735
How we lived then - Museum of Shops, Eastbourne 44098
Museum of Net Manufacture, Bridport 43614
Orkney Museum, Kirkwall 44615
Riverside Museum at Blake's Lock, Reading 45471
Stephens Collection, London 44992
Stockwood Park Museum, Luton 45047
Mr. Straw's House, Worksop 46077
W.H. Smith Museum, Newtown, Powys 45259

United States of America
Blacksmith Shop Museum, Dover-Foxcroft 48171
Forts Folle Avoine Historic Park, Danbury 47954
Grand Island Trader's Cabin, Munising 50907
Hubbell Trading Post, Ganado 48853
Jefferds Tavern, York 54334
John Hancock Warehouse, York 54335
Old Market House, Galena 48828
Phelps Store Museum, Palmyra 51686
Salem Maritime Houses, Salem 52658
South Sutton Old Store Museum, South Sutton 53182
Southern Highland Craft Guild at the Folk Art Center, Asheville 46376
Wells Fargo History Museum - Old Sacramento, Sacramento 52511
Zippo and Case Visitors Center Company Museum, Bradford 46989

Transport

Argentina
Museo de Transportes, Luján 00434
Museo Histórico del Transporte Carlos Hillner Decoud, Quilmes 00510
Museo Rural de la Posta de Sinsacate, Jesús María 00394

Australia
Australian Electric Transport Museum, Saint Kilda, South Australia 01477
Brisbane Tramway Museum, Ferny Grove 01079
Museum of Transportation and Rural Industries, Boyanup 00886
Port of Morgan Historic Museum, Morgan 01321
Victoria's Tramway Museum, Bylands 00930

Austria
Bergbahnmuseum, Kitzbühel 02176
Faszination Zugspitze, Ehrwald 01847
Mödlinger Stadtverkehrsmuseum, Mödling 02394
Weinviertler Oldtimermuseum Poysdorf, Poysdorf 02520
Wiener Tramwaymuseum, Wien 03091

Belgium
Musée des Camions de Brasserie, Romedenne 03983
Musée des Transports en Commun du Pays de Liège, Liège 03810
Musée du Transport Urbain Bruxellois, Bruxelles 03487
Vlaams Tram- en Autobusmuseum, Berchem 03360

Brazil
Museu do Bonde, Rio de Janeiro 04734
Museu do Trem da Cidade de Recife, Recife 04670
Museu dos Transportes Públicos Gaetano Ferolla, São Paulo 04898

Bulgaria
Nacionalen Muzej na Transporta i Săobščenijata, Ruse 05154

Canada
1910 Boomtown, Saskatoon 06909
Canadian Canoe Museum, Peterborough 06591
Canadian Transportation Museum and Historic Village, Kingsville 06092
Canal-de-Carillon, Saint-André-d'Argenteuil 06794
Cumberland House, Regina 06715
History of Transportation, Moose Jaw 06385
Musée Maritime de Charlevoix, Saint-Joseph-de-la-Rive 06860
Pouce Coupe and District Museum, Pouce Coupe 06652
Rideau Canal Museum, Smiths Falls 06987
Sault Sainte Marie Canal, Sault Sainte Marie 06933
Spruce Woods Provincial Heritage Park Museum, Carberry 05557
Yukon Transportation Museum, Whitehorse 07329

China
Changdao Voyage Museum, Changdao 07591
Xian Highway Display Center, Xian 08365

Czech Republic
Muzeum MHD - vozovna Střešovice, Praha 09772
Muzeum Průmyslových Železnic Brno, Zbýšov 09931
Železniční Muzeum Výtopna Jaroměř, Jaroměř 09553

Denmark
Lolland-Falsters Traktor- og Motormuseum, Eskildstrup 10004

Finland
Keitele-Museo, Suolahti 11337
Keski-Suomen Tieliikennemuseo, Kintaus 10961
Mobilia-Auto- ja Tiemuseo, Kangasala 10910
Raitioliikennemuseo, Helsinki 10811
Saimaan Kanavan Museo, Lappeenranta 11049
Varistaipaleen Kanavamuseo, Varistaipale 11451

France
Espace du Cheminot, Nîmes 14447
Mémoire de la Nationale 7, Piolenc 14821
Musée de l'Automobile de Vendée, Talmont-Saint-Hilaire 15846
Musée des Arts et Metiers, Paris 14683
Musée des Télécoms, Pleumeur-Bodou 14842
Musée des Transports, du Tourisme et des Communications, Toulouse 15940
Musée du Canal, Ecuisses 12811
Musée Provençal des Transports Urbains et Régionaux, La Barque 13310

Germany
Achtertalbahn, Ottenhöfen im Schwarzwald 20945
Automuseum Oldie Monschau, Monschau 20372
Bergische Museumsbahnen e.V, Wuppertal 22738
Brandenburgisches Museum für Klein- und Privatbahnen, Gramzow 18735
Dauerausstellung Württembergische Schwarzwaldbahn, Althengstett 16508
DB Museum, Nürnberg 20756
Deutsches Straßenmuseum, Germersheim 18604
Deutsches Technikmuseum, Berlin 17074
Eisenbahnmuseum, Blumberg 17390
Eisenbahnmuseum, Gerstetten 18616
Eisenbahnmuseum der Köln-Bonner-Eisenbahn-Freunde, Wesseling 22512
Eisenbahnmuseum Lehmann, Hofheim in Unterfranken 19227
Eisenbahnverein "Hei Na Ganzlin", Röbel/Müritz 21349
Fahrzeugmuseum, Frankenberg, Sachsen 18345
Flößerei- und Verkehrsmuseum, Gengenbach 18582
Flößereimuseum, Lychen 20118
Gundelsheimer Oldtimermuseum, Gundelsheim, Württemberg 18835
Hannoversches Straßenbahn-Museum, Sehnde 21728
Heimat- und Flößermuseum Calmbach, Bad Wildbad 16859
Historisches Straßenbahn-Depot Sankt-Peter, Nürnberg 20770
Historisches Straßenbahndepot, Halle, Saale 18897
Kutschenkabinett, Landau in der Pfalz 19790
Kutschenmuseum, Neustadt, Dosse 20677
Museum Bahnhof Bexbach, Bexbach 17317
Museum für Landwirtschaftliche Maschinen, Erbach, Alb-Donau-Kreis 18188
Museum für Nummernschilder, Verkehrs- und Zulassungsgeschichte, Großolbersdorf 18799
Museum Industriekultur mit Motorradmuseum, Nürnberg 20784
Museumsdampfzug Rebenbummler, Freiburg im Breisgau 18435
Museumsverein Bahnhofmuseum e.V, Benneckenstein 16994
Museumszug Hespertalbahn, Essen 18265
Oldtimer Museum, Sternberg 21902
Oldtimermuseum Meßkirch, Meßkirch 20306
Profil Auto- und Technikmuseum, Mittelbiberach 20341
Sächsisches Nutzfahrzeugmuseum, Hartmannsdorf 19020
Straßenbahn-Museum Thielenbruch, Köln 19633
Süddeutsches Eisenbahnmuseum, Heilbronn 19102
Technik Museum Speyer, Speyer 21838
Technik- und Verkehrsmuseum, Stade 21853
Verkehrsmuseum, Frankfurt am Main 18400
Verkehrsmuseum Dresden GmbH, Dresden 17976
Verkehrsmuseum Karlsruhe, Karlsruhe 19452
Westerwald Museum Motorrad Technik, Steinebach 21885

Hungary
Közlekedési Múzeum, Budapest 23426
Közlekedési Múzeum - Kiskőrösi Közúti Múzeum, Kiskőrös 23663
Magyar Motorok Múzeuma, Bodrogkeresztúr 23374
Magyar Vasúttörténeti Park, Budapest 23446
Millenniumi Földalatti Vasúti Múzeum, Budapest 23451

Israel
Egged Museum of Passengers Traffic, Holon 24975

Italy
Museo dei Trasporti, Ranco 27362

Japan
Chikatetsu Hakabutsukan, Tokyo 29378
Kotsu Kagaku Hakubutsukan, Osaka 29131
Umekoji Steam Locomotive Museum, Kyoto 28872

Lithuania
Dviračių Muziejus, Šiauliai 30354
Geležinkelio Muziejus, Šiauliai 30357

Luxembourg
Musée des Tramways et Autobus, Luxembourg 30481

Mexico
Museo de Historia del Transporte, Guadalupe 30907

Namibia
TransNamib Museum, Windhoek 31768

Netherlands
Amsterdams Openbaarvervoer Museum, Amsterdam 31827
Museum Stichting van het Rotterdamsche Tramweg Maatschappij, Ouddorp 32649
Nationaal Smalspoormuseum - Stoomtrein Valkenburgse Meer, Valkenburg, Zuid-Holland 32867
Openbaar Vervoer Museum, Doetinchem 32137
Openbaar Vervoer Museum, Rotterdam 32724
Tram Museum Rotterdam, Rotterdam 32731

New Zealand
East Coast Museum of Technology, Gisborne 33108
Ferrymead Heritage Park, Christchurch 33073
Museum of Transport and Technology, Auckland 33056
Yaldhurst Museum of Transport and Science, Christchurch 33079

Norway
Birkenes Bygdemuseum, Birkeland 33437
Haldensvassdragets Kanalmuseum, Ørje 33709
Kristianhus Motormuseum, Vik i Sogn 33978
Norsk Emballasje Museum, Oslo 33739
Norsk Vegmuseum, Fåberg 33486
Vegmuseet - Rogaland, Stavanger 33877

Peru
Associón Museo del Automóvil, Lima 34205

Poland
Skansen Lokomotyw, Zduńska Wola 35368

Portugal
Museu da Carris, Lisboa 35676
Museu do Carro Eléctrico, Porto 35850
Museu dos Transportes e Comunicações, Porto 35854
Museu dos Transportes Urbanos, Coimbra 35537

Romania
Muzeul Căilor Ferate Române, Bucureşti 36197

Russia
Konjušenny Muzej-Muzej Avtomobilnogo Transporta, Sankt-Peterburg 37654
Muzej Avtomobilnogo Transporta, Kursk 37096
Muzej Istorii Černomorskogo Flota, Moskva 37289
Muzej Istorii Moskovskogo Metropolitena, Moskva 37292
Muzej Morskogo Flota, Moskva 37333

Slovakia
Múzeum Dopravy, Bratislava 38221

South Africa
James Hall Museum of Transport, Johannesburg 38657
Outeniqua Railway Museum, George 38625

Sweden
Holmgrens Volkswagenmuseum, Pålsboda 40806
Kanalmuseum Skantzen, Hallstahammar 40576
Norrbottens Järnvägsmuseum, Luleå 40698

Switzerland

Fondazione Dazio Grande, Rodi-Fiesso 41770
Verkehrsdrehscheibe Schweiz und unser Weg zum Meer, Basel 41186
Verkehrshaus der Schweiz, Luzern 41600

Taiwan

Children's Museum of Transportation and Communication, Taipei 42382

Turkey

Rahmi M. Koç Müzesi, İstanbul 42834

United Kingdom

Aston Manor Transport Museum, Birmingham 43496
Bournemouth Transport Museum, Bournemouth 43566
Brattle Farm Museum, Staplehurst 45740
Bury Transport Museum, Bury, Lancashire 43692
Castle Point Transport Museum, Canvey Island 43747
Cobham Bus Museum, Cobham 43866
Cotswold Motoring Museum and Toy Collection, Bourton-on-the-Water 43568
Dover Transport Museum, Whitfield 46003
East Anglia Transport Museum, Carlton Colville 43761
Grampian Transport Museum, Alford, Aberdeenshire 43277
Ipswich Transport Museum, Ipswich 44531
Kennet and Avon Canal Trust Museum, Devizes 43978
Launceston Steam Railway, Launceston 44647
Lincolnshire Road Transport Museum, Lincoln 44713
Llangollen Motor Museum, Llangollen 44770
London's Transport Museum, London 44909
Museum of Transport, Glasgow 44292
Museum of Transport - Greater Manchester, Manchester 45093
Museum Workshop and Stores, Grangemouth 44343
National Motor Museum, Beaulieu 43424
North West Museum of Road Transport, Saint Helens 45567
Nottingham Transport Heritage Centre, Ruddington 45523
Oxford Bus Museum Trust, Long Hanborough 45026
Pendon Museum, Long Wittenham 45028
Snibston Discovery Park, Coalville 43863
South Yorkhire Transport Museum, Aldwarke Rotherham 43274
Stondon Museum, Lower Stondon 45040
Streetlife Museum, Kingston-upon-Hull 44600
Trolleybus Museum at Sandtoft, Doncaster 43995
Ulster Folk and Transport Museum, Holywood 44482
War Room and Motor House Collection, Harrogate 44403
Wirral Transport Museum and Bierkenhead Tramways, Birkenhead 43494

United States of America

American Bicycle and Cycling Museum, Santa Fe 52871
American Truck Historical Society, Kansas City 49684
Atchinson, Topeka and Santa Fe Depot, Alden 46191
Baltimore Streetcar Museum, Baltimore 46554
Carriage House Museum, Colorado Springs 47698
Central Texas Museum of Automotive History, Rosanky 52455
Chesapeake Children's Museum, Annapolis 46317
Cole Land Transportation Museum, Bangor 46596
Connecticut Trolley Museum, East Windsor 48264
Cookeville Depot Museum, Cookeville 47785
Delaware and Hudson Canal Museum, High Falls 49274
Estes-Winn Antique Automobile Museum, Asheville 46372
Forest Park Museum, Perry 51780
Forney Transportation Museum, Denver 48077
Genesee Country Village and Museum, Mumford 50901
Great Overland Station, Topeka 53534
Great Plains Transportation Museum, Wichita 54143
The H. H. Franklin Museum, Tucson 53591
Hollenberg Pony Express Station, Hanover 49160
Indiana Transportation Museum, Noblesville 51357
Lake Waccamaw Depot Museum, Lake Waccamaw 49894
Larz Anderson Auto Museum, Brookline 47064
Long Island Museum of American Art, History and Carriages, Stony Brook 53313
Michigan Transit Museum, Mount Clemens 50873
Mifflinburg Buggy Museum, Mifflinburg 50679
Minnesota Transportation Museum, Saint Paul 52626
Miracle of America Museum, Polson 52015
Museum of Alaska Transportation and Industry, Wasilla 53954
Museum of Bus Transportation, Hershey 49264
Museum of Transportation, Saint Louis 52593
National Road/Zane Grey Museum, Norwich 51441
New York Museum of Transportation, West Henrietta 54062
New York Transit Museum, Brooklyn 47080
Orange Empire Railway Museum, Perris 51778
Owls Head Transportation Museum, Owls Head 51641
Pate Museum of Transportation, Cresson 47876
Patee House Museum, Saint Joseph 52563
Pennsylvania Trolley Museum, Washington 53943
Pioneer Auto Museum, Murdo 50910
Pioneer Museum of Motorcycles, Tacoma 53413
Portholes Into the Past, Medina 50558
R.E. Olds Transportation Museum, Lansing 49945
Railways to Yesterday, Rockhill Furnace 52419
Schoharie Crossing, Fort Hunter 48627
Seashore Trolley Museum, Kennebunkport 49729
Shore Line Trolley Museum, East Haven 48250
Stanley Museum, Kingfield 49780
Station 15, Norton 51429
Steam Railroading Institute, Owosso 51644
Steamtown National Historic Site, Scranton 52984
Studebaker National Museum, South Bend 53160
Swigart Museum, Huntingdon 49423
Travel Town Transportation Museum, Los Angeles 50266
United States Army Transportation Museum, Fort Eustis 48620
Virginia Museum of Transportation, Roanoke 52383
Wade House Stagecoach Inn and Wesley Jung Carriage Museum, Greenbush 49045
Wells Fargo History Museum, Sacramento 52510
West Chicago City Museum, West Chicago 54051
Wheels O' Time Museum, Peoria 51775
Whitewater Canal Historic Site, Metamora 50614
Wisconsin Automotive Museum, Hartford 49201

Uzbekistan

Muzej Železnodorožnoj Techniki, Toshkent 54482

Venezuela

Museo del Transporte Guillermo José Schael, Caracas 54553

Travel → Tourism

Treasuries

Austria

Schatzkammer und Museum des Deutschen Ordens, Wien 03069

Belgium

Basilica Museum, Schatkamer van de Onze-Lieve-Vrouwbasiliek, Tongeren 04074
Krypte en Schatkamer van de O.L.V. Basiliek, Halle 03669
Musée du Chapitre, Soignies 04035
De Schatkamer van de Onze-Lieve-Vrouwekerk, Sint-Truiden 04030
Schatkamer van de Sint-Salvatorskathedraal, Brugge 03427
Schatkammer Sint-Catharina, Maaseik 03838
Stedelijk Museum Schatkamer van Sint-Pieter, Leuven 03790
Trésor de la Cathédrale de Liège, Liège 03818
Trésor de la Cathédrale Notre-Dame, Tournai 04089
Trésor de la Collégiale de Visé, Visé 04119
Trésor de l'Eglise Primaire Saint-Sebastien, Stavelot 04051
Trésor du Frère Hugo d'Oignies, Namur 03907
Trésor et Musée de la Collegiale Sainte-Begge, Andenne 03275

Canada

Basilica Museum, Saint John's 06843
Musée de Sainte-Anne-de-Beaupré, Sainte-Anne-de-Beaupré 06888

Croatia

Riznica Franjevačkog Samostana Split, Split 08963
Riznica Stolne Crkve u Dubrovniku, Dubrovnik 08873

Denmark

Rosenborg Slot, København 10150

France

Musée-Abbaye de Charroux, Charroux 12419
Musée du Suaire, Cadouin 12227
Musée Napoléon Ier et Tresors des Eglises, Brienne-le-Château 12177
Musées de Sens, Sens 15719
Palais du Tau, Reims 15006
Salles du Tresor, Prades 14937
Trésor de la Cathédrale Notre-Dame, Moulins (Allier) 14308
Trésor de Pleaux, Pleaux 14836
Trésor de Sainte Foy, Conques 12615
Trésor d'Église, Narbonne 14381
Tresorama, La Ferté-Macé 13361

Georgia

Kutaisi History Museum, Kutaisi 16343

Germany

Domschatz, Trier 22097
Domschatzmuseum, Regensburg 21219
Kirchenschatz im Sankt Fridolinsmünster, Bad Säckingen 16815
Merseburger Domschatz in der Südklausur und Kapitelhaus, Merseburg 20294
Naumburger Domschatzgewölbe, Naumburg, Saale 20554
Schatzkammer der Residenz München, München 20482

Italy

Museo del Tesoro del Duomo, Berceto 25397
Museo del Tesoro del Duomo, Monza 26864
Museo del Tesoro del Duomo, Rieti 27403
Museo del Tesoro del Duomo, Veroli 28226
Museo del Tesoro della Cattedrale, Anagni 25226
Museo del Tesoro della Cattedrale, Aosta 25248
Museo del Tesoro della Cattedrale, Savona 27777
Museo del Tesoro della Cattedrale di San Lorenzo, Genova 26302
Museo del Tesoro Santuario della Misericordia, Savona 27778
Museo Diacesano, Enna 26057
Museo Lapidario e del Tesoro del Duomo, Modena 26753
Museo Tesoro del Tempio della Beata Vergine, Reggio Emilia 27391
Museo Tesoro della Collegiata di San Lorenzo, Chiavenna 25864
Tesoro del Duomo, Messina 26670
Tesoro del Duomo, Milano 26725
Tesoro del Duomo, Vigevano 28267
Tesoro della Santa Maria della Scala, Chieri 25868
Tesoro di Santa Maria La Stella, Militello in Val di Catania 26732

Japan

Honda Museum, Kanazawa 28655
Meiji Jingu Homotsuden, Tokyo 29434

Montenegro

Riznica Katedrale Svetog Tripuna, Kotor 31648

Netherlands

Schatkamer van de Kathedrale Basiliek Sint-Bavo, Haarlem 32300

Portugal

Tesouro-Museu da Catedral, Viseu 36025

Russia

Almaznyj Fond, Moskva 37140

Spain

Museo de la Catedral, Granada 39369
Museo Diocesano y Cedralicio, Calahorra 39094
Museo Parroquial, Pastrana 39815
Museo-Tesoro Catedralicio, Toledo 40139
Museo-Tesoro de la Cofradía de la Expiración, Málaga 39606
Museo Tesoro Parroquial, Lebrija 39447
Museu-Tresor Parroquial, Olot 39729
Tresor de la Catedral, Girona 39350

Sweden

Skattkammaren, Stockholm 40934

Switzerland

Domschatz Sankt Ursen Kathedrale, Solothurn 41881
Domschatzmuseum, Chur 41302
Kirchenschatz, Bremgarten (Aargau) 41253
Kirchenschatz, Glarus 41436
Kirchenschatz, Sursee 41923
Kirchenschatz des Chorherrenstiftes, Beromünster 41224
Kirchenschatz-Museum, Baden 41146
Kirchenschatz-Museum Sankt Martin, Altdorf 41105
Kirchenschatz Sankt Pelagius Kirche, Bischofszell 41240
Klostermuseum und Caspar Wolf-Kabinett, Muri 41666
Musée et Trésor de l'Hospice, Bourg-Saint-Pierre 41250
Museum Münster, Münster 41662
Trésor de la Cathédrale Notre-Dame-du-Glarier, Sion 41878
Trésor de la Cathédrale Saint-Nicolas, Fribourg 41392

Turkey

Topkapı Sarayı Müzesi, İstanbul 42843

United Kingdom

Christ Church Cathedral Treasury, Oxford 45329
Durham Cathedral Treasures of Saint Cuthbert, Durham 44069
Guildford Cathedral Treasury, Guildford 44369
Saint Asaph Cathedral Treasury, Saint Asaph 45559

Typography → Books, Book Art and Manuscripts

Uniforms, Military

Argentina

Museo de la Casa Rosada, Buenos Aires 00200

Austria

Pioniermuseum, Klosterneuburg 02203

Belgium

Bastogne Historical Center, Bastogne 03345
Canada War Museum, Adegem 03261
Musée de la Bataille des Ardennes, La Roche-en-Ardenne 03769
Museum Slag der Zilveren Helmen, Halen 03666

Brazil

Museu da Força Expedicionária Brasileira, Rio de Janeiro 04715
Museu do Monumento Nacional aos Mortos da Segunda Guerra Mundial, Rio de Janeiro 04739

Canada

48th Highlands Museum, Toronto 07089
Association des Artilleurs de la Garnison, Lévis 06165
Black Watch of Canada Regimental Memorial Museum, Montréal 06329
Canadian Force Base Gagetown Military Museum, Oromocto 06517
Jamieson Museum, Moosomin 06389
Lord Strathcona's Horse Museum, Calgary 05523
Military Communications and Electronics Museum, Kingston 06083
Musée les Voltigeurs de Québec Manège Militaire, Québec 06693
Queen's Own Cameron Highlanders of Canada Regimental Museum, Winnipeg 07374
The Royal Canadian Regiment Museum, London 06190
Royal Military College of Canada Museum, Kingston 06090

Denmark

Forsvarsmuseet på Bornholm, Rønne 10243
Tøjhusmuseet, København 10156

Finland

Mannerheim museo, Helsinki 10804
Ratsuväkimuseo, Lappeenranta 11048

France

Musée de la Cavalerie, Saumur 15662
Musée de la Gendarmerie, Melun 14068
Musée de la Guerre, Fromelles 13015
Musée de la Légion Étrangère, Aubagne 11719
Musée de la Seconde Guerre Mondiale, Tourcoing 15952
Musée de l'Uniforme Légionnaire, Puyloubier 14961
Musée des Sapeurs-Pompiers, Thônes 15892
Musée du Costume Militaire, Sainte-Mesme 15575

Musée International des Polices et des Gendarmeries, Charvieu-Chavagneux 12427

Germany

Mindener Museum für Geschichte, Landes- und Volkskunde, Minden 20336
Museum der Belgischen Streitkräfte in Deutschland, Soest 21796
Museum im Mönchenkloster, Jüterbog 19407

Ghana

Ghana Armed Forces Museum, Kumasi 22856

India

Army Medical Corps Centre Museum, Lucknow 24289
Fort Saint George Museum, Chennai 24114

Indonesia

Museum Abri Satriamandala, Jakarta 24477

Iran

Military Museum, Tehrān 24680

Ireland

Mullingar Military Museum, Mullingar 24881

Italy

Museo Civico della Navigazione Fluviale, Battaglia Terme 25384
Museo della Battaglia, San Martino della Battaglia 27661
Museo delle Armi e delle Uniformi, Urbisaglia 28129
Museo Storico della Fanteria, Roma 27530
Museo Storico dell'Arma del Genio, Roma 27535

Malta

War Museum, Valletta 30649

Morocco

Museo de la Legión, Ceuta 31664

Netherlands

Historisch Museum Grenadiers en Jagers, Arnhem 31924
Marinemuseum, Den Helder 32110
Museum Verbindingsdienst, Ede 32170
Slag van de Somme Museum, Schagen 32749
Traditiekamer Regiment Stoottroepen, Assen 31940

New Zealand

Queen Elizabeth II Army Memorial Museum, Waiouru 33222

Pakistan

Pakistan Army Museum, Rawalpindi 34041

Poland

Muzeum Bitwy Grunwaldzkiej, Stębark 35134

Portugal

Museu de Escola Prática de Artilharia, Vendas Novas 35981

Russia

Centralnyj Voenno-morskoj Muzej, Sankt-Peterburg 37632
Voenno-istoričeskij Muzej Tichookeanskogo Flota, Vladivostok 37952

Serbia

Muzej Stara Livnica, Kragujevac 38107

Slovenia

Vojaški Muzej Tabor, Lokev 38473

South Africa

Castle Military Museum, Cape Town 38581

Spain

Museo de Farmacia Militar, Madrid 39540

Sweden

Fästningsmuseet Karlsborg, Karlsborg 40627
Garnisonsmuseet i Skövde, Skövde 40850

Switzerland

Polizeimuseum, Basel 41178

Thailand

Royal Thai Navy Museum, Samut Prakan 42598

Ukraine

Nacional'nyj Muzej Zbrojnych Syl Ukraïny, Kyïv 43108

United Kingdom

Aldershot Military Museum and Rushmoor Local History Gallery, Aldershot 43271
Argyll and Sutherland Highlanders Regimental Museum, Stirling 45747
Bankfield Museum, Halifax 44382
Border Regiment and King's Own Royal Border Regiment Museum, Carlisle 43757
Cheshire Military Museum, Chester 43816
Durham Light Infantry Museum and Durham Art Gallery, Durham 44071
Fusiliers' Museum Lancashire, Bury, Lancashire 43693
Household Cavalry Museum, Windsor 46042
Kent and Sharpshooters Yeomanry Museum, Edenbridge 44113
King's Own Scottish Borderers Regimental Museum, Berwick-upon-Tweed 43461
Military Museum, Bodmin 43550
Military Museum of Devon and Dorset, Dorchester 44001
National War Museum of Scotland, Edinburgh 44140
The Queen's Own Hussars Museum, Warwick 45937
Regimental Museum of The Royal Welsh (Brecon), Brecon 43599
Regimental Museum The Highlanders, Inverness 44526
Royal Irish Fusiliers Museum, Armagh 43320
The Royal Scots Regimental Museum, Edinburgh 44150
Royal Signals Museum, Blandford Forum 43541
Royal Ulster Rifles Regimental Museum, Belfast 43447
Royal Wiltshire Yeomanry Museum, Swindon 45818
Somerset Military Museum, Taunton 45831
Winchcombe Folk and Police Museum, Winchcombe 46023
Worcestershire Yeomanry Cavalry Museum, Worcester 46074

United States of America

Admiral Nimitz National Museum of the Pacific War, Fredericksburg 48761
American Red Cross Museum, Washington 53844
Fort Lowell Museum, Tucson 53590
Military Historians Museum, Westbrook 54085

Uzbekistan

Centralnyj Muzej Vooružennych Sil Uzbekistana, Toshkent 54469

Vehicles

Andorra

Museu de les dues rodes, Canillo 00087
Museu Nacional de l'Automòbil, Encamp 00090

Argentina

Museo Argentino de Motos Antiguas, Mendoza 00451
Museo del Automóvil, Buenos Aires 00222
Museo del Automóvil, Luján 00435
Museo del Automóvil Club Argentino, Buenos Aires 00223
Museo del Automóvil de Santa Teresita, Santa Teresita 00627
Museo del Automovilismo Juan Manuel Fangio, Balcarce 00138
Museo Ramos Generales, Villa María 00682

Australia

Australian Electric Transport Museum, Saint Kilda, South Australia 01477
Binalong Motor Museum, Binalong 00867
Camperdown Buggy Museum, Camperdown 00942
Canberra Bicycle Museum, Dickson 01038
Charlie's Arthurs Seat Auto Museum, Arthurs Seat 00816
Chiltern Motor Museum, Chiltern 00971
Cobb and Co. Museum, Toowoomba 01592
Gippsland Heritage Park, Moe 01309
Latrobe Bicycle Race Club Museum, Latrobe 01218
National Automobile Museum of Tasmania, Launceston 01222
National Motorcycle Museum, Nabiac 01344
Powerhouse House and Car Museum, Portland 01441
Raverty's Motor Museum, Echuca 01058
Serpentine Vintage Tractors and Machinery Museum, Serpentine 01500

Austria

Achenseer Museumswelt, Maurach am Achensee 02368
American Motorcycle Museum, Unterweitersdorf 02845
Automobil-Museum Siegfried Marcus, Stockerau 02781
Automobil- und Motorradmuseum Austria, Mitterndorf an der Fischa 02389
Automobilmuseum, Aspang 01732
Automobilmuseum, Stainz 02758
Autosammlung, Linz 02309
Dorf- und Kutschenmuseum, Eisgarn 01864
Fahrradmuseum, Ybbs 03122
Kutschenmuseum, Laa an der Thaya 02248
Manro-Classic - Auto Musik Museum, Koppl 02216
Motorrad-Museum Krems-Egelsee, Krems 02230
Motorradmuseum, Neunkirchen 02441
Motorradmuseum, Sulz 02791
Museum Fahrzeug-Technik-Luftfahrt, Bad Ischl 01756
Museum Fahrzeug - Technik - Luftfahrt, Lauffen 02267
Museumstramway Mariazell-Erlaufsee, Mariazell 02358
Nostalgie auf Rädern - Fahrzeugmuseum, Großklein 02005
Oldtimermuseum, Blindenmarkt 01789
Oldtimermuseum, Kleinwetzdorf 02195
Oldtimermuseum Kröpfl, Hartberg 02052
Oldtimermuseum Stift Ardagger, Ardagger 01726
Porsche Automuseum Helmut Pfeifhofer, Gmünd, Kärnten 01942
Radmuseum anno dazumal, Altmünster 01717
Rolls-Royce Museum, Dornbirn 01828
Tauernstraßen-Museum, Eben im Pongau 01832
Traktorveteranensammlung Dorf an der Pram, Wendling 02919
Villacher Fahrzeugmuseum, Villach 02862
Vötters Oldtimermuseum, Kaprun 02154
VW-Käfer Museum, Gaal 01916

Barbados

Mallalieu Motor Museum, Christ Church 03239

Belgium

Automuseum Old Timer, Lo-Reninge 03825
Autoworld, Bruxelles 03435
De Velodroom, Snellegem 04033
First Wing Historical Center (1WHC) The Golden Falcon, Beauvechain 03351
Mahymobiles - Musée Automobile, Leuze-en-Hainaut 03791
Musée du Circuit de Spa-Francorchamps, Stavelot 04048
Musée du Cycle, Huccorgne 03721
Musée du Cyclisme, Lavaux-Sainte-Anne 03781
Nationaal Wielermuseum, Roeselare 03976

Bulgaria

Muzej Ford i Bălgarija, Sofia 05183

Canada

Canadian Automotive Museum, Oshawa 06521
Car Life Museum, Charlottetown 05590
Cosmodôme, Laval 06146
Gervais Wheels Museum, Alida 05316
Heaman's Antique Autorama, Carman 05567
Manitoba Automobile Museum, Elkhorn 05772
Musée Automobile, Saint-Jacques 06825
Musée de Saint-Eustache et de ses Patriotes, Saint-Eustache 06816
Ontario Electric Railway Museum, Milton 06291
Reynolds-Alberta Museum, Wetaskiwin 07312
Smithson International Truck Museum, Rimbey 06751
Spoke Wheel Car Museum, Charlottetown 05598
Trev Deeley Motorcycle Museum, Richmond 06747

Chile

Auto Museum Moncopulli, Osorno 07442

China

Grand Prix Museum, Macau 07993

Cuba

Depósito del Automóvil, La Habana 09101

Czech Republic

Automuseum Praga, Zbuzany 09930
Muzeum Historických Motocyklů, Kašperské Hory 09580
Muzeum Motocyklů, Lesná u Znojma 09622
Old Racing Collection, Dvůr Králové nad Labem 09501
Škoda Auto Museum, Mladá Boleslav 09658
Technické Muzeum Tatra, Kopřivnice 09596

Denmark

Aalholm Automobil Museum, Nysted 10207
Danmarks Cykelmuseum, Ålestrup 09960
Egeskov Veteranmuseum, Kværndrup 10170
Jysk Automobilmuseum, Gjern 10038
Radio- og Motorcykel Museet, Stubbekøbing 10287

Estonia

Palmse Mõis, Viitna 10673

Faroe Islands

Bilmuseum, Vatnsoyrar 10702

Finland

Espoon Automuseo, Espoo 10726
Helsingin Automuseo, Helsinki 10778
Kotiseutumuseo Pellava ja Sahti, Alfred Kordelinin Ajokalumuseo, Lammi 11042
Oulun Automuseo, Oulu 11171
Tiemuseo, Helsinki 10830
TS-Auto- Viestintämuseo, Turku 11401

France

Automusée du Lyonnais, Seyssuel 15754
Collection de l'Aventure Automobile à Poissy, Carrieres-sous-Poissy 12299
Collection Schlumpf, Mulhouse 14323
Espace Automobiles Matra, Romorantin-Lanthenay 15090
Galerie des Transports, Marseille 13992
Manoir Automobile, Lohéac 13802
Musée Auto Moto Vélo, Châtellerault 12462
Musée Automobile, La Clayette 13346
Musée Automobile, Saint-Germain-Laval 15292
Musée Automobile Bellenavois, Bellenaves 11929
Musée Automobile de la Sarthe, Le Mans 13606
Musée Automobile Espera Sbarro, Pontarlier 14897
Musée Barthélemy Thimonnier de la Machine à Coudre et du Cycle, Amplepuis 11593
Musée de la Chartreuse et Fondation Bugatti, Molsheim 14144
Musée de l'Automobile, Orgon 14527
Musée de l'Automobile, Valençay 16036
Musée de l'Automobile, Velaine-en-Haye 16092
Musée de l'Automobile de Vendée, Talmont-Saint-Hilaire 15846
Musée de l'Automobile Française, Reims 15000
Musée de l'Automobile La Belle Époque, Pont-l'Évêque 14891
Musée de l'Automobile Miniature, Fontenay-sur-Mer 12971
Musée de l'Automobile Sportive, Sanary-sur-Mer 15620
Musée de l'Automobiliste, Mougins 14297
Musée de l'Estérel, Les Adrets-de-l'Estérel 13682
Musée de Voitures Anciennes et Sellerie, Menetou-Salon 14075
Musée de Voitures Hippomobiles Anciennes, Tour-en-Sologne 15949
Musée des Arts et Metiers, Paris 14683
Musée du Cycle et de la Moto, Lunéville 13866
Musée du Souvenier, Seillans 15702
Musée du Vélo, Cormatin 12628
Musée du Vélo, Periers 14775
Musée du Vélo et de la Moto, Domazan 12757
Musée du Vélo Le Belle Échappée, La Fresnaye-sur-Chédouet 13371
Musée du Vélocipède, Cadouin 12228
Musée et Salon Permanent de l'Automobile de Sport Prestige et Collection, Chauffailles 12480
Musée Henri Malartre, Rochetaillée-sur-Saône 15075
Musée Hippomobile, Saint-Michel-des-Andaines 15417
Musée National de la Voiture et du Tourisme, Compiègne 12599
Le Petit Musée de la Bicyclette, La Batie-Montgascon 13314

Germany

Audi Museum Mobile, Ingolstadt 19337
August Horch Museum Zwickau, Zwickau 22831
Ausstellung für Kraftwagen und Krafträder, Satteldorf 21522
Auto Museum Dr. Carl Benz, Ladenburg 19761
Auto-Museum Fritz B. Busch, Wolfegg 22648
Auto Technik Museum Sinsheim e.V, Sinsheim 21783
Auto- und Motorradmuseum, Witzenhausen 22632
Automobil-Museum, Asendorf 16598
Automobil Museum, Dortmund 17914
Automuseum Engstingen, Engstingen 18174

Automuseum Störy, Bockenem 17408
AutoMuseum Volkswagen, Wolfsburg 22658
AutoStadt ZeitHaus, Wolfsburg 22659
Auwärter-Museum, Landau an der Isar 19785
BMW Museum, München 20428
Boxenstop Auto- und Spielzeugmuseum, Tübingen 22112
DDR-Motorradmuseum, Borna 17483
Deutsches Automuseum Schloss Langenburg, Langenburg 19823
Deutsches Museum - Verkehrszentrum, München 20434
Deutsches Traktoren und Modellauto Museum e.V., Paderborn 20967
Deutsches Zweirad- und NSU-Museum, Neckarsulm 20565
EFA Museum für Deutsche Automobilgeschichte, Amerang 16537
ErfinderZeiten, Schramberg 21630
Erstes Deutsches Motorroller-Museum, Kaub 19483
Erstes Niederbayerisches Automobil- und Motorrad-Museum, Adlkofen 16432
Fahrradmuseum Zumhaus und Brauereimuseum, Feuchtwangen 18312
Fahrzeugmuseum Marxzell, Marxzell 20231
Fahrzeugmuseum Suhl, Suhl 21976
Gottlieb Daimler-Gedächtnisstätte, Stuttgart 21934
Gottlob Auwärter Museum, Stuttgart 21935
Heinrich-Büssing-Haus, Wolfsburg 22662
Historische Fahrzeugsammlung, Simmelsdorf 21768
Horster Motorrad-Museum, Gelsenkirchen 18574
Kutschenkabinett, Landau in der Pfalz 19790
Landwirtschaftliches Museum Wetzlar, Wetzlar 22528
Luftfahrt- und Technik- Museumspark Merseburg, Merseburg 20293
Mercedes-Benz Museum, Stuttgart 21952
Motor-Sport-Museum Hockenheimring, Hockenheim 19206
Motor Technica Museum, Bad Oeynhausen 16799
Motorrad-Museum, Augustusburg 16640
Motorrad Museum, Ibbenbüren 19303
Motorrad-Träume, Zschopau 22825
Motorrad-Veteranen- und Technikmuseum, Großschönau 18806
Museum, Großpösna 18801
Museum Autovision, Altlußheim 16512
Museum für Kutschen, Chaisen, Karren, Heidenheim an der Brenz 19091
Museum im Porsche Werk Leipzig, Leipzig 19925
Nicolaus August Otto Museum, Holzhausen an der Haide 19258
Niederrheinisches Motorradmuseum, Moers 20363
Niedersächsisches Kutschenmuseum, Lilienthal 19990
Oldenburger Fahrradmuseum, Oldenburg 20899
Porsche Museum, Stuttgart 21957
Rübesams Da Capo Oldtimermuseum, Leipzig 19929
Sächsisches Nutzfahrzeugmuseum, Hartmannsdorf 19020
Schnauferlstall, Ruhpolding 21449
Strassenbaumuseum, Sayda 21527
Technik u. Zweiradmuseum Dargen/Usedom, Dargen 17783
Vehikelsammlung Eppelheim, Eppelheim 18180
Westfalia-Auto-Museum, Rheda-Wiedenbrück 21294
Zündapp-Museum der Brauerei Zoller-Hof, Sigmaringen 21766
Zweirad- und Technikmuseum, Werder 22488
Zweiradmuseum, Berlin 17285
Zweiradmuseum, Pleidelsheim 21057

Hungary

A Tegnap Autói, Tatabánya 23920
Autómatuzsálem Gyűjtemény, Bordány 23377

Indonesia

Museum Mobil, Bogor 24458

Iran

Air Force Museum, Tehrān 24656

Ireland

An Dun Transport and Heritage Museum, Ballinahown 24734
Celbridge Motor Museum, Celbridge 24759
Fethard Park and Folk Museum, Fethard 24833
Museum of Irish Transport, Killarney 24851
National Transport Museum, Dublin 24812

Israel

Israeli Air Force Museum, Hatzerim 24971
Yad Lashiryon Museum, Latrun 25060

Italy

Archivio Storico Fiat, Torino 27996
Museo del Motociclo, Rimini 27406
Museo del Motociclo Moto Guzzi, Mondello del Lario 26777
Museo dell'Automobile, San Martino in Rio 27664
Museo dell'Automobile Carlo Biscaretti di Ruffia, Torino 28008
Museo dell'Automobile e della Tecnica, Farra d'Isonzo 26089
Museo dell'Automobile L. Bonfanti, Romano d'Ezzelino 27551
Museo delle Carrozze, Trani 28048
Museo Storico Alfa Romeo, Arese 25263

Japan

Bicycle Culture Center, Tokyo 29372
Ferrari Museum of Art, Gotemba 28458
Fukuyama Auto and Clock Museum, Fukuyama 28441
Gallery Abarth, Yamanakako 29582
Honda Collection Hall, Motegi 28941
Jitensha Bunka, Tokyo 29403
Jitensha Hakubutsukan Saikuru Senta, Sakai 29196
Mazda Motor Museum, Hiroshima 28540
Mitsubishi Auto Gallery, Okazaki 29104
Nasu Classic Car Museum, Nasu 29048
Nihon Jidosha Hakubutsukan, Futatsunashi 28455
Sports Car Museum of Japan, Gotemba 28459
Toyota Automobile Museum, Nagakute 28952
Toyota Kuragaike Commemorative Hall and Art Salon, Toyota 29533
Yokohama-shi Sutori-Toka-Hakubutsukan, Yokohama 29619

Malta

Malta Classic Car Collection, Qawra 30635

Mexico

Museo de Autos y del Transporte Humberto Lobo Villarreal, Monterrey 31179
Museo del Automóvil, México 31098
Museo del Automóvil Puebla, Puebla 31284
Museo del Transporte, Garza García 30872

Monaco

Collection de Voitures Anciennes de S.A.S. Le Prince de Monaco, Monaco 31605

Netherlands

American Motorcycle Museum, Raalte 32675
Het Amsterdams Automuseum, Zwanenburg 33022
Automusa, Bergeijk 31965
Automuseum Deventer, Deventer 32118
Automuseum Histo-Mobil, Giethoorn 32251
Automuseum Schagen, Schagen 32746
Autotron, Rosmalen 32701
Bromfietsmuseum De Peel, Ospel 32638
DAF Automobiel Museum, Eindhoven 32185
Eerste Nederlandse Opel Automuseum, Tijnje 32826
Den Hartogh Ford Automobielmuseum, Hillegom 32384
Indian Museum, Lemmer 32497
Kinderwagenmuseum, Nieuwolda 32583
Klein Saab Museum d'Oude Bolneus, Woerden 32958
Louwman Collection - Het National Automobilmuseum, Raamsdonksveer 32677
Nationaal Fietsmuseum Velorama, Nijmegen 32592
Noordelijk Busmuseum, Winschoten 32953
Oldtimercentrum Select, Joure 32433
Oldtimermuseum De Rijke, Oostvoorne 32631
Oldtimermuseum De Ronkel, Kloosterburen 32451
Portanje's Vespa Scooter en Nostalgie Collectie, Bunnik 32019
Rijwiel- en Bromfietsmuseum, Zoutkamp 33010
Tweewielermuseum Tankstop, Workum 32966
Wagenmakersmuseum, Leiden 32490

New Zealand

Ashburton Vintage Car Club Museum, Tinwald 33213
Dr. Heins Classic Car Collection, Christchurch 33072
Shantytown, Greymouth 33115
Southward Museum, Paraparaumu 33179

Norway

Flyhistorisk Museum, Sola 33848
Hølonda Skimuseum of Bygdasamling, Gåsbakken 33510
Horten Bilmuseum, Horten 33562
Motormuseet, Strømmen 33888
Norsk Luftfartsmuseum, Bodø 33446
Rogaland Bilmuseum, Vigrestad 33977
Søndhordland Motormuseum, Valevåg 33956

Poland

Muzeum Dworu Polskiego, Plochocin 35012
Muzeum Motoryzacji, Otrębusy 34999
Muzeum Motoryzacji Automobilklubu Wielkopolskiego, Poznań 35040
Zbiory Sprzętu Ratowinctwa Kolejowego, Poznań 35050

Portugal

Museu do Automóvel, Fafe 35593
Museu do Automóvel, Vila Nova de Famalicão 36008
Museu do Automóvel Antigo, Paço d'Arcos 35801
Museu do Caramulo, Caramulo 35506

Russia

Konjušenny Muzej-Muzej Avtomobilnogo Transporta, Sankt-Peterburg 37654
Lomakovskij Muzej Starinnych Avtomobilej i Motociklov, Moskva 37225
Muzej Avtomobilnogo Transporta, Kursk 37096
Muzej Zavoda Moskvič, Moskva 37385

San Marino

Museo Auto d'Ecopa, San Marino 38003

Serbia

Muzej Automobila, Beograd 38061

Slovakia

Múzeum Dopravy, Bratislava 38221

Spain

Museo Eduardo Barreiros, Valdemorillo 40187
Museu de Cera, Barcelona 38974
Museu de l'Automòbil, Barcelona 38981
Museu De L'automoció Roda Roda, Lleida 39480

Sweden

Arvika Fordonmuseum, Arvika 40436
Eds MC- och Motormuseum, Ed 40490
Johanna Museet, Skurup 40855
Laganland Bilmuseum, Lagan 40664
Målilla Hembygdspark, Målilla 40713
Målilla Motormuseum, Målilla 40714
MC-Museum Gyllene Hjulet, Surahammar 40963
Motala Motormuseum, Motala 40755
Motor Nostalgimuséet, Grängesberg 40561
Saab Museum, Trollhättan 40985
Skokloster Motormuseum, Skokloster 40853
Smålands Bil, Musik och Leksaksmuseum, Rydaholm 40818
Spårvägsmuseet, Stockholm 40936
Svedinos Bil- och Flygmuseum, Slöinge 40858
Torsby Fordonsmuseum, Torsby 40977
Vännäs Motormuseum, Vännäs 41027
Vagnshistoriska Museet, Fristad 40527

Switzerland

FBW-Museum, Wetzikon 42009
Historicum, Sargans 41817
Monteverdi Car Collection, Binningen 41234
Motorradmuseum, Gossau (Sankt Gallen) 41444
Musée de l'Auto, Moto et Cycle Jean Tua, Genève 41419
Musée de l'Automobile, Martigny 41611
Musée de l'Automobile, Plan-les-Ouates 41724
Musée International de l'Automobile, Le Grand-Saconnex 41448
Saurer Museum, Arbon 41123

United Kingdom

Alex Brown Cycle History Museum, Thornhill, Dumfriesshire 45854
Atwell-Wilson Motor Museum, Calne 43715
Automobilia Transport Museum, Hebden Bridge 44434
Battlesbridge Motorcycle Museum, Battlesbridge 43420
Benson Veteran Cycle Museum, Benson 43452
Betws-y-coed Motor Museum, Betws-y-Coed 43466
British Commercial Vehicle Museum, Leyland 44701
British Cycling Museum, Camelford 43737
Brooklands Museum, Weybridge 45985
Bugatti Trust, Cheltenham 43808
C.M. Booth Collection of Historic Vehicles, Rolvenden 45510
Cars of the Stars Motor Museum, Keswick 44561
Coventry Transport Museum, Coventry 43900
Donington Grand Prix Collection, Castle Donington 43769
First Penny Farthing Museum, Knutsford 44622
Geeson Brothers Motor Cycle Museum, South Witham 45699
Glenluce Motor Museum, Glenluce 44312
Haynes International Motor Museum, Sparkford 45724
Heritage Motor Centre, Gaydon 44266
Hooton Park Exhibition Centre, Ellesmere Port 44167
Isle of Wight Wax Works, Brading 43587
Jaguar Daimler Heritage Trust, Coventry 43903
Lakeland Motor Museum, Grange-over-Sands 44341
London Motorcycle Museum, Greenford 44363
Military Vehicle Museum, Sunderland 45798
Motor Museum, Filching 44226
Mouldsworth Motor Museum, Mouldsworth 45181
Murray's Motorcycle Museum, Laxey 44651
Myreton Motor Museum, Aberlady 43253
National Cycle Collection, Llandrindod Wells 44756
National Motor Museum, Beaulieu 43424
National Motorcycle Museum, Bickenhill 43475
Newburn Hall Motor Museum, Newburn 45219
Patrick Collection Motor Museum, Birmingham 43509
REME Museum of Technology, Arborfield 43310
Romany Folklore Museum, Selborne 45633
Royal Signals Museum, Blandford Forum 43541
Sammy Miller Motorcycle Museum, New Milton 45205
Sandringham House Museum, Sandringham 45614
Shackerstone Railway Museum, Shackerstone 45643
Shuttleworth Collection, Old Warden 45313
Sir Henry Royce Memorial Foundation, Paulerspury 45347
Sir Joseph Banks Conservatory, Lincoln 44716
Worcestershire County Museum, Kidderminster 44573
World of Country Life, Exmouth 44198
Yorkshire Museum of Carriages and Horse Drawn Vehicles, Wensleydale 45965

United States of America

Americana Museum, Terra Alta 53479
Antique Auto and Race Car Museum, Bedford 46686
Ark-La-Tex Antique and Classic Vehicle Museum, Shreveport 53083
Auburn-Cord-Duesenberg Museum, Auburn 46466
Auto Museum, Escondido 48417
Automobile Driving Museum, Los Angeles 50210
Automotive Hall of Fame, Dearborn 48005
Belair Stable Museum, Bowie 46968
Bicycle Museum of America, New Bremen 51026
Blackhawk Museum, Danville 47958
Boyertown Museum of Historic Vehicles, Boyertown 46979
C. Grier Beam Truck Museum, Cherryville 47427
California Automobile Museum, Sacramento 52499
Canton Classic Car Museum, Canton 47226
Car and Carriage Museum, Pittsburgh 51941
Classic Cars International, Salt Lake City 52690
Crawford Auto-Aviation Museum, Cleveland 47628
Darlington Raceway Stock Car Museum, Darlington 47968
Dixie Gun Works' Old Car Museum, Union City 53653
Ellingson Car Museum, Rogers 52441
Elwood Haynes Museum, Kokomo 49825
Fort Bedford Museum, Bedford 46690
Four Wheel Drive Foundation, Clintonville 47662
Gilmore Car Museum, Hickory Corners 49269
Hall of Fame and Classic Car Museum, Weedsport 54019
Hartung's Auto and License Plate Museum, Glenview 48938
Henry Ford Museum, Dearborn 48009
Henry H. Blommel Historic Automotive Data Collection, Connersville 47780
International Motor Racing Research Center, Watkins Glen 53981
Justice Brothers Racing Museum, Duarte 48184
Mac's Antique Car Museum, Tulsa 53609
Memoryville USA, Rolla 52446
Miracle of America Museum, Polson 52015
Motorcycle Hall of Fame Museum, Pickerington 51913
Museum of Automobile Art and Design, Syracuse 53401
Museum of Automobiles, Morrilton 50848
National Automobile Museum, Reno 52287
National Automotive and Truck Museum of the United States, Auburn 46467
National Corvette Museum, Bowling Green 46974
National Packard Museum, Warren 53825
National Sprint Car Hall of Fame and Museum, Knoxville 49809
Northeast Classic Car Museum, Norwich 51440
Pacific Northwest Truck Museum, Brooks 47088
Petersen Automotive Museum, Los Angeles 50258
Pioneer Museum of Motorcycles, Tacoma 53413
Roaring Twenties Antique Car Museum, Hood 49358
S. Ray Miller Auto Museum, Elkhart 48350
Saint Louis Car Museum, Saint Louis 52597
San Diego Automotive Museum, San Diego 52746
Sarasota Classic Car Museum, Sarasota 52915
Sloan Museum Longway Planetarium, Flint 48560
Volo Antique Auto Museum, Volo 53768
Walter P. Chrysler Boyhood Home and Museum, Ellis 48368
Walter P. Chrysler Museum, Auburn Hills 46474
Wells Auto Museum, Wells 54030

Uruguay
Museo del Automóvil, Montevideo 54398

Veterinary Medicine

Austria
Museum für Beschirrung und Besattelung, Hufbeschlag und Veterinär-Orthopädie, Wien 03035

France
Musée Fragonard de l'Ecole Nationale Vétérinaire d'Alfort, Maisons-Alfort 13924

Germany
Veterinärmedizinhistorisches Museum der Tierärztlichen Hochschule Hannover, Hannover 19007
Zoologisches und Tiermedizinisches Museum der Universität Hohenheim, Stuttgart 21971

India
Museum of Animal Husbandary and Veterinary, Guwahati 24193
Museum of Veterinary, Guwahati 24194

Italy
Museo Anatomico Veterinario, Pisa 27240
Museo di Anatomia Patologica e Teratologia Veterinaria, Ozzano dell'Emilia 27026

Norway
Norsk Veterinærmedisinsk Museum, Oslo 33746

Poland
Muzeum Weterynarii, Ciechanowiec 34609

Portugal
Museu do Lar dos Veteranos Militares, Torres Vedras 35971

South Africa
Gertrud Theiler Tick Museum, Onderstepoort 38695

Spain
Museo Colección de Veterinaria Militar, Madrid 39528

Sweden
Veterinärhistoriska Museet, Skara 40847

Thailand
Veterinary Parasitology Museum, Bangkok 42564

United Kingdom
Army Medical Services Museum, Mytchett 45187
Veterinary Museum, London 45003

Viticulture → Wines and Spirits

Voyages of Discovery → Scientific Expeditions

Wallpaper

France
Maison du Patrimoine de Saint-Chef, Saint-Chef 15201
Musée du Papier Peint, Rixheim 15057

Germany
Deutsches Tapetenmuseum, Kassel 19461

Italy
Museo Bottega della Tarsia Lignea, Sorrento 27880

United Kingdom
The Whitworth Art Gallery, Manchester 45099

Water

Austria
Hochschwabmuseum, Sankt Ilgen 02651
Wasserleitungsmuseum, Reichenau an der Rax 02570
Wasserleitungsmuseum, Wildalpen 03103

Belgium
L'Amusette, Mesvin 03872
Centre Didactique de l'Eau, Seneffe 04007
Musée de la Fontaine et de l'Eau, Genval 03648

Canada
Cascapedia River Museum, Cascapedia-Saint-Jules 05573
Odyssée des Bâtisseurs, Alma 05322
Rideau Canal Museum, Smiths Falls 06987
Site Historique de l'Île-des-Moulins, Terrebonne 07068

China
Huanghe Museum, Zhengzhou 08499

Czech Republic
Ekotechnické Museum, Praha 09743

Finland
Maanmittausmuseo, Helsinki 10802
Saimaan Kanavan Museo, Lappeenranta 11049
Varistaipaleen Kanavamuseo, Varistaipale 11451

France
Centre de la Mer, Paris 14579
Maison de la Rivière, Chinon 12514
Maison de l'Eau, Auxerre 11775
La Maison de l'Eau, Mallièvre 13933
Maison de l'Eau, Neuvy-sur-Barangeon 14413
Maison de l'Eau et de la Pêche, Besse-en-Chandesse 11961
Maison de l'Eau et de la Pêche, Neuvic (Corrèze) 14404
Maison de Loire, Belleville-sur-Loire 11932
Maison de Pays, Hede 13198
Maison des Eaux Minérales, Vic-sur-Cère 16142
Moulin à Eau Maître Marcel, Sainte-Agathe-en-Donzy 15545
Musée de l'Eau, Pont-en-Royans 14886
Musée de l'Hydraviation, Biscarrosse 11999
Musée du Touage et Canal de Saint-Quentin, Riqueval 15049

Germany
Aquarius Wassermuseum, Mülheim an der Ruhr 20409
EAM-live-Museum Wasserkraftwerk Merkenbach, Dillenburg 17859
EAM-live-Museum Wasserkraftwerk Wülmersen, Trendelburg 22088
Inn-Museum, Rosenheim 21376
Museum der Stadtwerke, Pirmasens 21042
Museum im Wasserkraftwerk, Wahlsburg 22274
Museum im Wasserwerk, Berlin 17207
Technisches Museum Alte Wasserkunst, Bautzen 16942
Wasserkunst von 1535, Bad Arolsen 16674

Greece
Open-Air Water-Power Museum, Dimitsana 22984

Hungary
Duna Múzeum - Magyar Környezetvédelmi és Vízügyi Múzeum, Esztergom 23551

India
Regional Science Centre, Lucknow 24299

Iran
Iranian National Museum of Water, Tehrān 24673

Israel
Inland-Water Ecological Laboratory Collection, Jerusalem 24991

Italy
Vintschger Museum, Sluderno 27858

Japan
Ice Museum, Kamikawa 28647
Osaka-furistu Sayamaike Hakubutsukan, Osakasayama 29159
Sapporo Waterworks Memorial Museum, Sapporo 29226

Korea, Republic
Dongjin Irrigation Folk Museum, Gimje 29814
Tong-Jin Irrigation Folk Museum, Kimje 29846

Mexico
Museo Casa del Agua, Tehuacán 31417
Museo Hidráulico del Sistema de Drenaje Profundo, México 31120
Museo la Antigua Casa del Agua, Hunucmá 30953

Netherlands
Milieucentrum De Grote Rivieren, Heerewaarden 32334
Museum De Cruquius, Cruquius 32039
Nederlands Waterleidingmuseum, Utrecht 32853
Poldermuseum, Puttershoek 32674
Poldermuseum de Hooge Boezem Achter Haastrecht, Haastrecht 32306

Norway
Fetsund Lenser Fløtingsmuseum, Fetsund 33491
Hydro Industripark-Museum, Porsgrunn 33769
Norsk Bremuseum, Fjærland 33493
Norsk Vasskraft- og Industristadmuseum, Tyssedal 33944
Vassdragsmuseet Labro, Skollenborg 33840

Portugal
Museu da Agua da Epal, Lisboa 35674
Museus Municipal de Almada, Almada 35415

Russia
Muzej Mir Vody Sankt-Peterburga, Sankt-Peterburg 37699
Muzej Vody, Moskva 37383

Spain
Museu de l#Aigua, Lleida 39479

Sweden
Tom Tits Experiment, Södertälje 40864

Switzerland
BKW-Museum, Mühleberg 41654
Zürcher Abwassermuseum Kläranlage Werdhölzli, Zürich 42119

Taiwan
Museum of Drinking Water, Taipei 42408

United Kingdom
Basingstoke Canal Exhibition, Mytchett 45188
Brindley Bank Pumping Station and Museum, Rugeley 45529
Coultershaw Beam Pump, Petworth 45376
Finch Foundry Working Museum, Okehampton 45311
Kew Bridge Steam Museum, Kew 44569
Lagan Lookout Centre, Belfast 43441
Strathpeffer Spa Pumping Room Exhibition, Strathpeffer 45789
Water Supply Museum, Weymouth 45990
Yorkshire Waterways Museum, Goole 44330

Yemen
The Tanks of Aden, Aden 54644

Wax, Molded

Australia
World in Wax Museum, Echuca 01059

Austria
Figurencabinett Madame Lucrezia, Rosegg 02598

Canada
Louis Tussaud's Waxworks Museum, Niagara Falls 06442
Movieland Wax Museum of the Stars, Niagara Falls 06445
Royal London Wax Museum, Victoria 07259

Czech Republic
Muzeum Voskových Figurín České Historie, Karlštejn 09579
Muzeum Voskových Figurín České Historie, Praha 09774
Wax Museum Český Krumlov, Český Krumlov 09469

Denmark
Louis Tussaud Wax Museum, København 10138

France
Musée Grévin, Paris 14710
Musée Historique, Mont-Saint-Michel 14162

Germany
Lebzelterei- und Wachsziehereimuseum, Pfaffenhofen an der Ilm 21019

Hungary
Huszármúzeum, Balatonszentgyörgy 23356
Népviseletes Babamúzeum - Történelmi Panoptikum, Keszthely 23662

Ireland
Wax Museum, Dublin 24822

Italy
Museo delle Cere, Roma 27484

Lebanon
Musée de Cire Historique, Byblos 30168

Mexico
Museo Ceramoteca, Salamanca 31310
Museo de Cera, Guadalajara 30892
Museo de Cera de la Ciudad de México, México 31072
Museo de Cera de la Villa, México 31073
Museo de Cera de Tijuana, Tijuana 31456
Museo de La Cerería, Tlayacapan 31480

Monaco
Musée de Cires, Monaco 31608

Netherlands
Madame Tussaud's Amsterdam, Amsterdam 31856

Portugal
Museu da Cera, Fátima 35601

Russia
Muzej Voskovych Figur, Sankt-Peterburg 37711
Otraženie - Muzej Voskovych Figur, Moskva 37393

United Kingdom
Isle of Wight Wax Works, Brading 43587
Madame Tussaud's, London 44910

United States of America
Great Blacks in Wax Museum, Baltimore 46566
Hollywood Wax Museum, Branson 46996
Hollywood Wax Museum, Los Angeles 50234
John Brown Wax Museum, Harpers Ferry 49175
Madame Tussaud's Wax Museum, New York 51207
National Presidential Wax Museum, Keystone 49772
Old West Wax Museum, Thermopolis 53493

Weapons → Arms and Armour

Weaving

Armenia
Middle East Museum, Yerevan 00743
State Museum of Wood Carving, Yerevan 00756

Australia
Elizabeth Farm, Rosehill 01468
National Wool Museum, Geelong 01104
Tasmanian Wool Centre, Ross 01470

Austria

Erstes Waldviertler Webereimuseum, Waidhofen an der Thaya 02884
Museum Alte Textilfabrik, Weitra 02907
Webereimuseum, Haslach 02058
Webermuseum, Heidenreichstein 02064

France

Musée du Tissage et de la Soierie, Bussières 12210
Le Musée du Tisserand Dauphinois, La Batie-Montgascon 13313

Germany

Deutsches Teppichmuseum, Oelsnitz 20871
Fischer- und Webermuseum Steinhude, Wunstorf 22733
Handweberei Henni Jaensch-Zeymer, Schwielowsee 21706
Handwebmuseum Rupperath, Bad Münstereifel 16788
Heimatmuseum, Sohland an der Spree 21800
Heimatmuseum Alte Weberstube, Niedercunnersdorf 20703
Heimatstube Wilthen, Wilthen 22592
Historische Schauweberei, Braunsdorf 17532
Stadtmuseum Povelturm, Nordhorn 20743
Textilmuseum Die Scheune, Nettetal-Hombergen 20579
Weberei- und Heimatmuseum Laichingen, Laichingen 19780
Webereimuseum, Breitenberg 17539
Weberhausmuseum, Schauenstein 21535
Weberstube Jonsdorf, Jonsdorf, Kurort 19401

Greece

Ioannina Museum, Ioannina 23014

Italy

Museo dell'Artigianato Tessile e della Seta, Reggio Calabria 27381

Japan

Shiozawa Tsumugi Commemorative Museum, Shiozawa 29278

Netherlands

Museumboerderij, Staphorst 32801
Museumboerderij Vreeburg, Schagen 32747
Weverijmuseum, Geldrop 32240

Norway

Sjølingstad Uldvarefabrik, Mandal 33660

Poland

Muzeum w Bielsku-Białej, Bielsko-Biała 34562

Sweden

Textilmuseet, Borås 40471

Switzerland

Saurer Museum, Arbon 41123
Weberei- und Heimatmuseum Ruedertal, Schmiedrued 41839

Taiwan

Weaving Craft Museum, Fengyuan 42190

United Kingdom

Bancroft Mill Engine Trust, Barnoldswick 43383
Mill Trail Visitor Centre, Alva 43295
Newtown Textile Museum, Newtown, Powys 45256
Paisley Museum and Art Galleries, Paisley 45345
Weaver's Cottage, Kilbarchan 44575

United States of America

Handweaving Museum and Arts Center, Clayton 47601
Watkins Woolen Mill, Lawson 49998

Weights and Measures

Argentina

Museo de la Balanza, Buenos Aires 00198

Germany

Adam-Ries-Museum, Annaberg-Buchholz 16552
Stadt-und Wagenmuseum Oschatz, Oschatz 20918
Zirkelmuseum, Wilhelmsdorf 22571

Netherlands

Weegmuseum de Oude Waag, Someren 32795

Poland

Muzeum Miar Głównego Urzędu Miar, Warszawa 35260

Portugal

Núcleo Museológico de Metrologia, Évora 35590

Spain

Fundación Museo de las Ferias, Medina del Campo 39638

United Kingdom

Avery Historical Museum, Smethwick 45691
Westgate Museum, Winchester 46035

United States of America

Degenhart Paperweight and Glass Museum, Cambridge 47190
Hollywood Guinness World of Records Museum, Los Angeles 50232

Whaling → Fishing and Fisheries

Wines and Spirits

Australia

Grapevine Museum, Chiltern 00972
Pioneer Rum Museum, Beenleigh 00853
Tasmania Distillery Museum, Hobart 01158

Austria

Bezirksmuseum Döbling, Wien 02956
Erstes Oberösterreichisches Schnapsmuseum, Sankt Oswald bei Freistadt 02671
Erstes Südburgenländisches Schnapsbrennerei-museum, Kukmirn 02244
Fassbinder- und Weinbaumuseum, Straß im Straßertale 02785
First Austrian Scotch Whisky Museum, Steyregg 02776
Heimatmuseum Grafenwörth, Grafenwörth 01960
Loisium, Langenlois 02257
MostBirnHaus Stift Ardagger, Ardagger 01725
Mostmuseum, Neumarkt im Mühlkreis 02437
Mostmuseum, Sankt Leonhard am Forst 02661
Pulkautaler Weinbaumuseum, Hadres 02035
Retzer Erlebniskeller, Retz 02580
Schlumberger Wein- und Sektkellerei, Wien 03071
Stadtmuseum mit Weinmuseum, Bad Vöslau 01769
Weinbau- und Fahnenschwingermuseum, Neckenmarkt 02424
Weinbau- und Landwirtschaftsmuseum, Hadersdorf 02032
Weinbaumuseum, Sitzenberg-Reidling 02746
Weinmuseum, Güssing 02016
Weinmuseum Burgenland - Eisenstadt, Eisenstadt 01863
Weinmuseum Kitzeck, Kitzeck im Sausal 02179
Weinmuseum Prellenkirchen, Prellenkirchen 02531
Weinmuseum Schloß Gamlitz, Gamlitz 01922
Weinstadtmuseum Krems, Krems 02232

Belgium

Nationaal Jenevermuseum, Hasselt 03686
Wijnkelder Soniën, Overijse 03946

Canada

British Columbia Wine Museum, Kelowna 06051

China

Macau Wine Museum, Macau 07999

Cuba

Museo del Ron, Santiago de Cuba 09319
Museo del Ron Havana Club, La Habana 09141

Finland

Altia Oyj, Rajamäki 11252

France

Caveau-Musée Les Vignerons de Roueïre, Quarante 14963
Château du Clos de Vougeot, Vougeot 16273
Château-Musée, Boën-sur-Lignon 12037
Château-Musée Verdus, Saint-Seurin-de-Cadourne 15512
Collection d'Outillage Ancien de Tonnellerie, Cussac-Fort-Médoc 12689
Collection Vini-Viticole et de Tonnellerie, Riquewihr 15050
Collection Viti-Vinicole, Banyuls-sur-Mer 11841
Collections Claeyssens, Wambrechies 16282
Distillerie du Périgord, Sarlat-la-Canéda 15635
Ecomusée de la Pomme au Calvados, Le Sap 13649
Ecomusée de la Vigne et du Vin, Gradignan 13092
Ecomusée du Libournais, Montagne 14165
Ecomusée du Pays de la Cerise de Fougerolles, Fougerolles 12989
Ecomusée du Vignoble Drouais et des Artisans, Dreux 12791
Exposition du Château, Cadillac-en-Fronsadais 12226
Exposition Universelle des Vins et Spiritueux, Bandol 11837
Hameau en Beaujolais, Romanèche-Thorins 15084
Maison de la Vigne, Essoyes 12867
Maison des Liqueurs de l'Auzonnet, Saint-Florent-sur-Auzonnet 15268
Maison du Cidre et du Terroir, Le Hezo 13600
Moulin des Grands Vignes, Périssac 14781
Musée Animé du Vin et de la Tonnellerie, Chinon 12515
Musée-Château de Pommard, Pommard 14873
Musée de la Baleine, Port-la-Nouvelle 14917
Musée de la Cave des Champagnes de Castellane, Épernay 12832
Musée de la Distillerie, Pontarlier 14898
Musée de la Distillerie Combier, Saumur 15663
Musée de la Mémoire de la Vigne et du Vin, Saint-Pierre-de-Mézoargues 15461
Musée de la Pomme et du Cidre, Pleudihen-sur-Rance 14839
Musée de la Pomme et du Cidre, Vaudeurs 16083
Musée de la Pomme et du Cidre et des Métiers Traditionnels, Rosay 15101
Musée de la Tonnellerie, Cadillac 12225
Musée de la Tonnellerie, Generac 13039
Musée de la Tonnellerie et du Vin, Aloxe-Corton 11560
Musée de la Tradition Champenoise, Épernay 12833
Musée de la Vie Rurale, Desaignes 12715
Musée de la Vigne, La Roche-Vineuse 13420
Musée de la Vigne, Verzenay 16132
Musée de la Vigne et du Patrimoine, Brioude 12184
Musée de la Vigne et du Vin, Aubière 11728
Musée de la Vigne et du Vin, Boën-sur-Lignon 12038
Musée de la Vigne et du Vin, Cabara 12217
Musée de la Vigne et du Vin, Cérons 12348
Musée de la Vigne et du Vin, Gabarnac 13022
Musée de la Vigne et du Vin, Gigondas 13057
Musée de la Vigne et du Vin, Gruissan 13153
Musée de la Vigne et du Vin, Le Mesnil-sur-Oger 13615
Musée de la Vigne et du Vin, Lézignan-Corbières 13736
Musée de la Vigne et du Vin, Pierreclos 14814
Musée de la Vigne et du Vin, Vaux-en-Beaujolais 16088
Musée de la Vigne et du Vin, Portets 14923
Musée de la Vigne et du Vin d'Anjou, Saint-Lambert-du-Lattay 15354
Musée de la Vigne et du Vin de Franche-Comté, Arbois 11658
Musée de la Vigne et du Vin-Galerie d'Art, Lods 13801
Musée de la Vigne-Moulin du Haut-Benauge, Gornac 13086
Musée de l'Abbaye, Gaillac 13025
Musée de l'Alambic, Saint-Desirat 15237
Musée de l'Armagnac, Condom 12609
Musée départemental du Revermont, Treffort 15979
Musée des Arts du Cognac, Cognac 12566
Musée des Arts et Métiers de la Vigne et du Vin, Moulis-en-Médoc 14309
Musée des Contenants Vinaires, Pouzols-Minérvois 14934
Musée des Distilleries Limougeaudes, Limoges 13767
Musée des Poupées d'Antan et de la Tonnellerie, Maisons-lès-Chaource 13926
Musée des Vieux Métiers et des Eaux-de-Vie, Dabo 12695
Musée des Vins de Touraine, Tours 15965
Musée du Calvados et des Métiers Anciens de Pont-l'Évêque, Pont-l'Évêque 14892
Musée du Château, Margaux 13967
Musée du Château, Pauillac 14762
Musée du Château de Monbazillac, Monbazillac 14146
Musée du Cidre, Domagné 12756
Musée du Cidre, Montaure 14177
Musée du Cidre Breton, Argol 11677
Musée du Cidre du Pays d'Othe, Eaux-Puiseaux 12803
Musée du Donjon de la Toque, Huriel 13231
Musée du Mont Corbier, Saint-Jean-de-Maurienne 15331
Musée du Pays du Der, Sainte-Rémy-en-Bouzemont 15578
Musée du Vieux Pressoir et de la Vigne, Coulanges-la-Vineuse 12642
Musée du Vigneron, Buxy 12214
Musée du Vigneron, Chaumont-le-Bois 12485
Musée du Vigneron, Les Marches 13705
Musée du Vigneron, Verdigny 16111
Musée du Vigneron, Wuenheim 16303
Musée du Vigneron Paul Coulon et Fils, Rasteau 14987
Musée du Vignoble et des Vins d'Alsace, Kientzheim 13302
Musée du Vignoble Nantais, Le Pallet 13628
Musée du Vin, Cheignieu-la-Balme 12496
Musée du Vin, Cissac-Médoc 12531
Musée du Vin, Paris 14703
Musée du Vin, Saint-Hilaire-de-Loulay 15306
Musée du Vin, Châteauneuf-du-Pape 12447
Musée du Vin, Saint-Cirq-Lapopie 15212
Musée du Vin dans l'Art, Pauillac 14763
Musée du Vin de Bourgogne, Beaune 11904
Musée du Vin de Champagne Piper-Heidsieck, Reims 15003
Musée du Vin de la Batellerie et de la Tonnellerie, Bergerac 11946
Musée du Vin et de Saint-Jacques de Compostelle, Pomerol 14872
Musée du Vin-Mimoria di Vino, Luri 13868
Musée Historique du Cidre, Melleray-la-Vallée 14067
Musée Lillet, Podensac 14864
Musée Marc Mureau, Restigné 15021
Musée-Privé de Cire du Vieil, Albi 11539
Musée Regional de la Vigne et du Vin, Montmélian 14227
Musée Régional du Cidre et du Calvados, Valognes 16060
Musée Vigneron, Vendres 16101
Musée Viticole, Saint-Yzans-de-Médoc 15544
Musée Viticole Charles Bernard, Fixin 12942
Planète Bordeaux, Beychac-et-Caillau 11970
Terra-Vinea, Portel-des-Corbières 14922
Vinorama - Musée du Vin, Bordeaux 12076

Germany

Baierweinmuseum, Bach an der Donau 16652
Brennereimuseum Haselünne, Haselünne 19023
Dampfkornbranntweinbrennerei-Museum, Wildeshausen 22568
Deutsches Weinbaumuseum, Oppenheim 20906
Dorfmuseum Hanweiler, Winnenden 22599
Dorfmuseum Pfaffenweiler, Pfaffenweiler 21022
Frankfurter Äpfelwein-Museum, Frankfurt am Main 18368
Grenzwald-Destillation Museum, Crottendorf 17752
Heimat-Museum Maintal, Maintal 20140
Heimatmuseum, Dornstetten 17906
Historisches Museum der Pfalz Speyer, Speyer 21834
Hohenloher Freilandmuseum, Schwäbisch Hall 21655
Kaiserstühler Weinbaumuseum, Vogtsburg im Kaiserstuhl 22251
Kelter- und Weinbaumuseum, Niederstetten 20712
Kölner Weinmuseum, Köln 19611
Kupferberg-Museum, Mainz 20145
Markgräfler Museum Müllheim, Müllheim 20417
Mittelmoselmuseum im Barockhaus Böcking, Traben-Trarbach 22075
Museum Bürgstadt, Bürgstadt 17639
Museum für Weinbau und Stadtgeschichte, Edenkoben 18055
Museum für Weinkultur, Deidesheim 17809
Rheingauer Weinmuseum Brömserburg, Rüdesheim am Rhein 21436
Schnapsmuseum, Kötzting 19669
Schwäbisches Schnapsmuseum Bönnigheim, Bönnigheim 17428
Sieben-Keltern-Museum, Metzingen 20315
Stadtmuseum Herrenmühle, Hammelburg 18973
Tauberländer Dorfmuseum, Weikersheim 22373
Vinarium, Cleebronn 17711
Wein- und Heimatmuseum, Durbach 18030
Weinbau in Mundelsheim, Mundelsheim 20528
Weinbau- und Heimatmuseum Klingenberg, Klingenberg am Main 19570
Weinbaumuseum, Asperg 16601
Weinbaumuseum, Flörsheim-Dalsheim 18331
Weinbaumuseum, Ortenberg, Baden 20913
Weinbaumuseum Alte Kelter, Erlenbach, Kreis Heilbronn 18230
Weinbaumuseum im Herrenhof, Neustadt an der Weinstraße 20674
Weinbaumuseum Meersburg, Meersburg 20252
Weinbaumuseum Stuttgart-Uhlbach, Stuttgart 21967
Weinmuseum, Vaihingen an der Enz 22200

Weinmuseum Schlagkamp-Desoye Senheim, Senheim ... 21745
Winzermuseum Rauenberg, Rauenberg ... 21199

Guadeloupe

Musée du Rhum, Sainte-Rose ... 23282

Hungary

Borgazdasági Múzeum, Tolcsva ... 23935
Nemzeti Bormúzeum, Eger ... 23542
Tokaji Múzeum, Tokaji ... 23934
Zwack Gyümölcspálinka -Fozde és Kiállítás, Kecskemét ... 23657
Zwack Unicum Múzeum, Budapest ... 23495

Ireland

Desmond Castle, Kinsale ... 24856
Locke's Distillery Museum, Kilbeggan ... 24845
The Old Jameson Distillery, Dublin ... 24814

Italy

Enoteca Regionale Permanente la Serenissima, Gradisca d'Isonzo ... 26355
Museo Agricolo e del Vino Ricci Curbastro, Capriolo ... 25668
Museo Bersano delle Contadinerie e Stampe antiche del Vino, Nizza Monferrato ... 26929
Museo del Vino, Caldaro ... 25595
Museo del Vino, Torgiano ... 27993
Museo del Vino e della Moda, Ciliverghe di Mazzano ... 25888
Museo della Liquirizia, Rossano ... 27564
Museo della Vite e del Vino, Carmignano ... 25681
Museo della Vite e del Vino, Rufina ... 27578
Museo della Viticoltura, Prarostino ... 27327
Museo dell'Arte del Vino, Staffolo ... 27897
Museo dell'Enoteca Regionale Piemontese, Grinzane Cavour ... 26369
Museo di Storia dell'Agricoltura, Cesena ... 25834
Museo Enologico Grasso, Milazzo ... 26728
Museo Martini di Storia dell'Enologia, Pessione ... 27177
Museo Ratti dei Vini d'Alba, La Morra ... 26430
Museo Sella e Mosca, Alghero ... 25199
Poli Museo della Grappa, Bassano del Grappa ... 25382

Japan

Dewanoyuki Sake Museum, Tsuruoka ... 29544
Fureai Minatokan, Osaka ... 29127
Gekkeikan Okura Kinenkan, Kyoto ... 28811
Kunizakari Sake Museum, Handa ... 28496
Nambu Toji Denshokan, Ishidoriya ... 28585
Ponshu-kan-Echigo Sake Museum, Yuzawa ... 29636
Sake Museum, Sakata ... 29207
Yoshinogawa Sake Museum Hisago-tei, Nagaoka ... 28966

Luxembourg

Freilichtmuseum Schwebsingen, Bech-Kleinmacher ... 30453
Musée du Vin, Ehnen ... 30471

Martinique

Musée du Rhum, Sainte-Marie ... 30673

Mexico

Museo de la Sidra Copa de Oro, San Andrés Cholula ... 31321
Museo Francisco Javier Sauza, Tequila ... 31450

Netherlands

Doesburgsch Mosterd- en Azijnmuseum, Doesburg ... 32132
Het Gedistilleerd Museum, Schiedam ... 32752
Nationaal Likeur- en Frisdrankenmuseum Isidorus Jonkers, Hilvarenbeek ... 32387
Nederlands Wijnmuseum, Arnhem ... 31931
Het Wijnkopersgildehuys, Amsterdam ... 31902
Wijnmuseum Maastricht, Cadier en Keer ... 32033

Poland

Muzeum Ziemi Lubuskiej, Zielona Góra ... 35372

Portugal

Museu da Fundação da Casa de Mateus, Vila Real ... 36011
Museu da Vinha e do Vinho de Reguengos, Reguengos de Monsaraz ... 35884
Museu do Vinho, Funchal ... 35615
Museu do Vinho, Pico da Pedra ... 35823
Museu do Vinho dos Biscoitos, Biscoitos ... 35468
Museu Nacional do Vinho, Alcobaça ... 35395
Museu Rural e do Vinho do Concelho, Cartaxo ... 35511
Museu Temático do Vinho, Almeirim ... 35420

Puerto Rico

Museo Castillo Serrallés, Ponce ... 36048

Romania

Expoziţia de Viticultură, Teremia Mare ... 36683
Muzeul Podgoriei Drăgăşani, Drăgăşani ... 36355
Muzeul Viticol, Huşi ... 36409
Muzeul Viticol, Odobeşti ... 36496

Slovakia

Malokarpatské Múzeum v Pezinku, Pezinok ... 38324
Vinohradnícka Expozícia Apponyiho Palác, Bratislava ... 38236

South Africa

Groot Constantia Manor House and Wine Museum, Constantia ... 38597
Groot Constantin Homestead and Wine Museum, Constantia ... 38598

Spain

Cortijo Bacardi, Málaga ... 39595
Estación Enológica de Haro y Museo del Vino, Haro ... 39384
Museo de la Sidra de Asturias, Nava ... 39711
Museo del Castillo de Peralada, Peralada ... 39836
Museo del Vino, Villafamés ... 40298
Museo del Vino, Peñafiel ... 39827
Museo del Vino de la D. O. Campo de Barja, Vera de Moncayo ... 40264
Museo del Vino de Montilla, Montilla ... 39678
Museo del Vino Valdepeñas, Valdepeñas ... 40190
Museo Etnográfico e do Viño de Cambados, Cambados ... 39109
Museos de la Atalaya, Jerez de la Frontera ... 39427

Sweden

Vin- och Sprithistoriska Museet, Stockholm ... 40950

Switzerland

Argauisch Kantonales Weinbaumuseum, Tegerfelden ... 41929
Gesellschaft für das Weinbaumuseum am Zürichsee, Au ... 41133
Haus zum Torggel, Berneck ... 41221
Heimat- und Rebbaumuseum, Spiez ... 41892
Mazot Musée, Troistorrents ... 41951
Musée de la Vigne et du Vin, Aigle ... 41094
Musée de la Vigne et du Vin, Boudry ... 41248
Musée Régional d'Histoire et d'Artisanat du Val-de-Travers, Môtiers ... 41649
Musée Valaisan de la Vigne et du Vin, Sierre ... 41867
Orts- und Weinbaumuseum, Neftenbach ... 41673
Ortsmuseum Höngg, Zürich ... 42101
Rebbaumuseum am Bielersee-Musée de la Vigne au Lac de Bienne, Ligerz ... 41565
Schaffhauser Weinbaumuseum, Hallau ... 41469
Walliser Reb- und Weinmuseum, Salgesch ... 41799
Weinbau-Museum, Chur ... 41305

Taiwan

Chia Chi Lan Wine Museum, Yilan ... 42466
Chiayi Distillery Museum, Minsiong ... 42276
Puli Wine Museum, Puli ... 42297
Taichung Distillery Museum, Taichung ... 42354

United Kingdom

Cider Museum and King Offa Distillery, Hereford ... 44453
Dufftown Whisky Museum, Dufftown ... 44032
Glenfarclas Distillery Museum, Ballindalloch ... 43359
Mill House Cider Museum and Dorset Collection of Clocks, Owermoigne ... 45326
Scotch Whisky Heritage Centre, Edinburgh ... 44154
Sheppy's Farm and Cider Museum, Bradford-on-Tone, Taunton ... 43584

United States of America

Arkansas Historic Wine Museum, Paris ... 51699
Browntown Museum, Lake City ... 49880
Historic Hermann Museum, Hermann ... 49261
Oscar Getz Museum of Whiskey History, Bardstown ... 46608
Rodgers Tavern, Perryville ... 51784
Wine Museum of Greyton H. Taylor, Hammondsport ... 49144

Wood → Forest Products

Wood Carving

Armenia

National Folk Art Museum of Armenia, Yerevan ... 00750
Wood Carving Museum, Yerevan ... 00758

Bulgaria

Muzej na Rezbarskoto i Zografsko Izkustvo, Trjavna ... 05230

Canada

Cedar Ridge Creative Centre, Scarborough ... 06936
MacLachlan Woodworking Museum, Kingston ... 06081
Windsor Wood Carving Museum, Windsor, Ontario ... 07346

France

Ferme des Bois Nois, Saint-Eloy La-Glaciere ... 15245
Galerie Europeenne de la Forêt et du Bois, Dompierre (Orne) ... 12761
Musée de la Boissellerie, Bois-d'Amont ... 12046
Musée de l'Outil des Métiers du Bois, Montjoie-Saint-Martin ... 14219
Musée des Tourneurs, Aiguines ... 11507
Musée Étienne Martin, Noyal-sur-Vilaine ... 14478
Musée National Picasso La Guerre et la Paix, Vallauris ... 16051
Musée Rural du Bois, Coupiac ... 12647

Germany

Städtische Galerie, Traunstein ... 22079

Ireland

Clonfert Museum of Ecclesiastical Art, Loughrea ... 24871

Italy

Mostra Permanente di Xilografie di Pietro Parigi, Firenze ... 26165
Museo della Xilografia, Carpi ... 25686
Museo dell'Arte della Tornitura del Legno, Pettenasco ... 27181
Museo Diocesano e Collezione dei Presepi, Bressanone ... 25544
Museo Etnografico del Pinerolese e Museo del Legno, Pinerolo ... 27223

Japan

Nihon Ukiyo-e Hakubutsukan, Matsumoto ... 28896

Netherlands

Museumboerderij New Greenwich Village, Zuidwolde, Drenthe ... 33015

New Zealand

Museum of Woodwork and Ornamental Turning, Ashburton ... 33038
Puke Ariki, New Plymouth ... 33158

Nigeria

National Museum Oron, Oron ... 33338

Russia

Architekturno-ėtnografičeskij Muzej Derevjannogo Zodčestva Chochlovka, Perm ... 37543
Muzej Narodnogo Derevjannogo Zodčestva Vitoslavlicy, Velikij Novgorod ... 37921
Permskaja Gosudarstvennaja Chudožestvennaja Galereja, Perm ... 37550
Tomskij Muzej Derevjannogo Zodčestva, Tomsk ... 37857

Switzerland

Ausstellung der Kantonalen Schnitzerschule, Brienz (Bern) ... 41255

Taiwan

Miaoli Wood Sculpture Museum, Sanyi ... 42317
Paiwan Carving Gallery, Pingtung ... 42292
Wood Carving Museum and Taiwan Sericultural Museum, Zaociao ... 42488

Thailand

Kamthieng House, Bangkok ... 42544

Turkey

Vakıf Inşaat ve Sanat Eserleri Müzesi, İstanbul ... 42847

United Kingdom

Little Moreton Hall, Congleton ... 43883
Thomas Bewick's Birthplace, Mickley ... 45136
William Lamb Memorial Studio, Montrose ... 45174

United States of America

Alaska Indian Arts, Haines ... 49124
Barnegat Bay Decoy and Baymen's Museum, Tuckerton ... 53581
Crafts Museum, Mequon ... 50594
Hoo-Hoo International Forestry Museum, Gurdon ... 49107
Jim Savage Art Gallery and Museum, Sioux Falls ... 53120
Museum of Woodcarving, Shell Lake ... 53069
National Museum of Woodcarving, Custer ... 47916
Shawano County Historical Society Museum, Shawano ... 53056
Trinity Heights - Saint Joseph Center-Museum, Sioux City ... 53115
Wendell Gilley Museum, Southwest Harbor ... 53190

Zoology

Argentina

Museo de Animales Venenosos, Buenos Aires ... 00178
Museo de Ciencias Naturales de Especialidad Zoológica Médica, Buenos Aires ... 00191
Museo de Zoología, Córdoba ... 00328
Museo del Colegio San José, Buenos Aires ... 00226
Museo Histórico Municipal Alfredo E. Múlgura, General Belgrano ... 00377
Museo Municipal de Ciencias Naturales Lorenzo Scaglia, Mar del Plata ... 00447
Museo Natural Dr. Carlos A. Marelli, Colón ... 00309

Australia

Rodney Fox Shark Experience-Museum, Glenelg ... 01115
Tasmanian Museum and Art Gallery, Hobart ... 01159
Western Australian Museum Perth, Perth ... 01412
Wildlife Wonderlands-Giant Earthworm Museum, Bass ... 00835
Zoology Museum, Saint Lucia ... 01486

Austria

Biologiezentrum der Oberösterreichischen Landesmuseen, Linz ... 02310
Naturhistorisches Museum, Admont ... 01694
Naturkundemuseum, Dobersberg ... 01821
Naturwissenschaftliche Sammlungen am Landesmuseum Joanneum, Graz ... 01981
Robert Hytha-Museum, Oberwölbling ... 02465

Belgium

Aquarium-Muséum, Liège ... 03793
Musée d'Histoire Naturelle de Mons, Mons ... 03885
Musée d'Histoire Naturelle et Vivarium, Tournai ... 04088
Muséum de Zoologie et d'Anthropologie de l'ULB, Bruxelles ... 03496
Naturzentrum Ternell, Eupen ... 03591

Brazil

Inpa Coleções Zoológicas, Manaus ... 04552
Museu de Arte e Cultura Canjira, Itaobim ... 04498
Museu de Zoologia, São Paulo ... 04891
Museu Regional da Fauna e Flora do Itatiaia, Itatiaia ... 04500

Canada

Aasland Museum Taxidermy, Cranbrook ... 05657
Collection Entomologique Ouellet-Robert, Outremont ... 06559
Cowan Vertebrate Museum, Vancouver ... 07210
Grizzly Bear Prairie Museum, Wanham ... 07277
International Fox Museum, Summerside ... 07033
Stewart Hay Memorial Museum, Winnipeg ... 07385
UBC Fish Museum, Vancouver ... 07224
University of Calgary Museum of Zoology, Calgary ... 05533

China

Muaseum of Nature and Agriculture, Macau ... 08001
Paleozoological Museum of China, Beijing ... 07560

Colombia

Museo de Ciencias Naturales del Colegio de San José de La Salle, Medellín 08677
Museo de Historia Natural, Pereira 08720
Museo de História Natural, Tunja 08775
Museo de Historia Natural, Bucaramanga 08600
Museo Vitaliano Zuccardi, Sincelejo 08756

Czech Republic

Biskupský Dvůr, Brno 09420
Blatské muzeum - Rožmberský Dům, Soběslav 09842
Galerie Pravěk Očima Zdeňka Buriana, Dvůr Králové nad Labem 09499
Lesnické, Myslivecké a Rybářské Muzeum, Hluboká nad Vltavou 09516
Lovecko Lesnicke Muzeum, Úsov 09891
Zámek Budišov, Budišov u Třebíče 09447

Denmark

Naturama, Svendborg 10291
Zoologisk Museum, Grindsted 10045
Zoologisk Museum, København 10158

Egypt

Giza Zoological Museum, Giza 10526
Scientific Researches Museum, Cairo 10518

Estonia

Tartu Ülikooli Zooloogiamuuseum, Tartu 10662

Ethiopia

Zoological Museum, Addis Ababa 10682

Finland

Eläinmuseo, Helsinki 10774
K.E. Kivirikon Lintu- ja Nisäkäskokoelma, Helsinki 10789
Luontotalo Arkki, Pori 11219
Ostrobothnia Australis, Vaasa 11425
Oulun Yliopiston Eläinmuseo, Oulu 11173
Turun Biologinen Museo, Turku 11402

France

Espace de Découvert du Cheval, Hennebont 13201
Maison de la Baie du Mont Saint-Michel, Vains 16029
Maison de la Faune, Vaujany 16085
Musée Animalier à Jougne, Jougne 13282
Musée-Aquarium d'Arcachon, Arcachon 11662
Musée Césaire Phisalix, Mouthier-Haute-Pierre 14315
Musée d'Histoire Naturelle, Nantes 14366
Musée du Loup, Le Cloître-Saint-Thégonnec 13572
Musée et Atelier d'Art Animalier, La Chapelle-Thouarault 13338
Musée Lorrain du Cheval, Gondrecourt-le-Château 13079
Musée Vivant de la Faisanderie, Milly 14121
Musée Zoologique de l'Université Louis Pasteur et de la Ville de Strasbourg, Strasbourg 15834
Muséum-Aquarium de Nancy, Nancy 14357
Muséum d'Histoire Naturelle, Nice 14440
Muséum National d'Histoire Naturelle, Paris 14741
Parc Zoologique Arche de Noé et Amazonia, Saint-Clément-des-Baleines 15217

Germany

Brehm-Gedenkstätte, Renthendorf 21279
Burgauer Tiermuseum, Burgau 17648
Dinosaurier-Freilichtmuseum Münchehagen, Rehburg-Loccum 21240
Dr.-Carl-Haeberlin-Friesen-Museum, Wyk auf Föhr 22759
Ernst-Haeckel-Haus, Jena 19377
Fehn- und Schiffahrtsmuseum, Rhauderfehn 21291
Forstmuseum Heringen, Heringen (Werra) 19142
Haus der Natur - Cismar, Cismar 17707
Jagd- und Naturkunde Museum, Niederstetten 20711
Münchner Tiermuseum, München 20463
Museum der Natur, Gotha 18719
Museum der Stadt Alfeld (Leine), Alfeld 16467
Museum für Naturkunde, Berlin 17205
Museum für Naturkunde der Stadt Gera, Gera 18595
Natureum am Schloß Ludwigslust, Ludwigslust 20061
Naturhistorisches Museum, Mainz 20149
Naturkunde-Museum Bamberg, Bamberg 16915
Naturkundemuseum, Bayreuth 16963
Naturkundemuseum, Potsdam 21098
Naturkundemuseum Erfurt, Erfurt 18209
Naturkundemuseum im Tierpark Hamm, Hamm, Westfalen 18972
Naturkundemuseum Ostbayern, Regensburg 21230
Phyletisches Museum, Jena 19387
Rhöner Naturmuseum, Tann 22005
Senckenberg Naturhistorische Sammlungen Dresden, Dresden 17969
Staatlicher Schloßbetrieb Schloß Nossen / Kloster Altzella, Nossen 20748
Staatliches Naturhistorisches Museum, Braunschweig 17529
Tiermuseum, Lenggries 19955
Tierpark Sababurg Museum, Hofgeismar 19225
Zoologische Sammlung der Universität Rostock, Rostock 21396
Zoologische Sammlung des Zentralmagazins Naturwissenschaftlicher Sammlungen, Halle, Saale 18907
Zoologische Schausammlung, Tübingen 22128
Zoologisches Forschungsinstitut und Museum Alexander Koenig, Bonn 17469
Zoologisches Museum der Universität Göttingen, Göttingen 18702
Zoologisches Museum der Universität Hamburg, Hamburg 18965
Zoologisches Museum der Universität Heidelberg, Heidelberg 19088
Zoologisches Museum zu Kiel, Kiel 19536
Zoologisches und Tiermedizinisches Museum der Universität Hohenheim, Stuttgart 21971

Ghana

Zoology Museum, Legon 22862

Greece

Animal and Fish Museum, Evros 22996
Zoological Museum, Zografou 23250

Guatemala

Museo Nacional y Zoológico La Aurora, Guatemala 23304

India

Agharkar Museum, Allahabad 24058
Biodiversity Museum, Dehradun 24129
Government J.T.C. Museum, Faizabad 24173
Government Museum, Chennai 24115
Madras Christian College Museum, Chennai 24117
Mahatma Gandhi Museum, Mangalore 24307
Museum of Arthropoda, Pune 24361
Zoological Museum, Faizabad 24174
Zoological Museum, Gorakhpur 24181
Zoological Museum, Hardwar 24205
Zoological Museum, Jaunpur 24240
Zoological Museum, Meerut 24310
Zoology and Botany Museum, Ernakulam 24169
Zoology Museum, Annamalai Nagar 24068
Zoology Museum, Kanpur 24254
Zoology Museum, Muzaffarnagar 24327

Indonesia

Museum Kebun Raya Bogor, Bogor 24457
Museum Komodo, Jakarta 24489
Museum Taman Laut Ancol, Jakarta 24503
Zoological Museum, Pematangsiantar 24546

Iran

Zoological Museum at Agricultural College, Karaj 24615

Ireland

National Museum of Ireland - Natural History, Dublin 24810
University College Zoological Museum, Cork 24770
Zoological Museum, Dublin 24825

Israel

Bet Gordon, Deganya Aleph 24941
Ussishkin House, Kibbutz Dan 25035
Zoological Museum, Tel Aviv 25124

Italy

Gabinetto Geologico e Botanico Piccinini, Pergola 27138
Museo Alpino Duca degli Abruzzi, Courmayeur 25997
Museo Apistico, Bologna 25444
Museo Civico di Storia Naturale, Ferrara 26108
Museo Civico di Storia Naturale Giacomo Doria, Genova 26297
Museo Civico di Zoologia, Roma 27468
Museo dell'Orso, Villavallelonga 28282
Museo di Anatomia Comparata Battista Grassi, Roma 27488
Museo di Scienza della Terra, Bari 25365
Museo di Storia Naturale, Parma 27104
Museo di Storia Naturale, Parma 27105
Museo di Storia Naturale A. Stoppani, Venegono Inferiore 28177
Museo di Storia Naturale dell'Accademia dei Fisiocritici, Siena 27844
Museo di Storia Naturale - Sezione Zoologica La Specola, Firenze 26186
Museo di Zoologia, Bari 25366
Museo di Zoologia, Napoli 26902
Museo di Zoologia, Bologna 25467
Museo di Zoologia, Roma 27502
Museo di Zoologia e Mineralogia, Siddi 27835
Museo di Zoologia Pietro Doderlein, Palermo 27074
Museo di Zoologico e di Anatomia, Torino 28017
Museo Didattico Zoologico, Oria 26992
Museo Zoologico, Catania 25776
Museo Zoologico, Padova 27049
Museo Zoologico degli Vertebrati, Avellino 25332
Museo Zoologico dell'Istituto Nazionale per la Fauna Selvatica, Ozzano dell'Emilia 27027
Museo Zoologico G. Scarpa, Treviso 28074

Jamaica

Zoology Museum, Kingston 28336

Japan

Meguro Kiseichukan, Tokyo 29429

Jordan

Animal Museum, Amman 29643

Kyrgyzstan

Zoologičeskij Muzej, Biškek 29960

Laos

Dinosaur Museum, Savannakhet 29970

Latvia

Latvijas Universitātes Zoologijas Muzejs, Rīga 30092

Lithuania

Kauno Tado Ivanausko Zoologijos Muziejus, Kaunas 30252
Vilniaus Universiteto Zoologijos Muziejus, Vilnius 30446

Mexico

Museo de la Fauna, Aguascalientes 30700
Museo de la Fauna de Veracruz, Xalapa 31559
Museo de la Fauna y Ciencias Naturales, Monterrey 31181
Museo de la Vida Animal Herpetario, Puebla 31280
Museo de Zoología Alfonso L. Herrera, México 31097
Museo Zoologico César Dominguez Flores, Tuxtla Gutiérrez 31518

Netherlands

Kikkermuseum, Den Haag 32077
Museonder, Otterlo 32643
Schoonewelle, Zwartsluis 33023
Zoölogisch Museum Amsterdam, Amsterdam 31904

New Zealand

Kahutara Canoes and Taxidermy Museum, Featherston 33105
Otago Museum, Dunedin 33095

Nigeria

Zoology Museum, Ibadan 33324

Norway

Naturhistorisk museum, Oslo 33737

Pakistan

Islamia College Museum, Peshawar 34034
Zoological Museum, Lahore 34029

Peru

Museo de Zoología Juan Ormea, Trujillo 34322
Museo Forestal de la Policía Ecológica, Jesús María 34197
Museo Mirador de Quistococha, Iquitos 34193

Philippines

Quest - Center for Earth Science And Discovery, Makati 34417
Zamboanga National Museum, Zamboanga 34527

Poland

Muzeum i Instytut Zoologii PAN, Warszawa 35248
Muzeum Lasu i Drewna przy Lesnym Zakładzie Doświadczalnym SGGW, Rogów 35083
Muzeum Przyrodnicze Instytutu Systematyki i Ewolucji Zwierząt PAN, Kraków 34808

Portugal

Museu Bocage da Zoológico e Antropológico, Lisboa 35671
Museu de História Natural, Porto 35847
Museu Zoológico, Coimbra 35544
Museu Zoológico, Pombal 35826
Sala de Zoologia, Porto 35867

Romania

Complex Muzeal de Ştiinţele Naturii, Galaţi 36378
Muzeul Zoologic al Universităţii Babeş-Bolyai, Cluj-Napoca 36292

Russia

Gosudarstvennyj Muzej Životnovodstva im. Akademika E.F. Liskuna, Moskva 37206
Muzei Permskogo Gosudarstvennogo Universiteta, Perm 37547
Muzej Archeologija, Ėtnografija i Ėkologija Sibiri Kemerovskogo Gosudarstvennogo Universiteta, Kemerovo 37033
Muzej Instituta Biologii Morja, Vladivostok 37944
Muzej Koški, Moskva 37310
Muzej Priroda Morja i eë Ochrana, Vladivostok 37947
Novosibirskij Muzej Prirody, Novosibirsk 37487
Zoologičeskij Muzej, Vladivostok 37953
Zoologičeskij Muzej, Kazan 37029
Zoologičeskij Muzej, Samara 37623
Zoologičeskij Muzej, Syktyvkar 37822
Zoologičeskij Muzej im. N.M.Kulagina, Moskva 37430
Zoologičeskij Muzej Kafedry Zoologii Pozvonočnych Sankt-Peterburgskogo Gosudarstvennogo Universiteta, Sankt-Peterburg 37736
Zoologičeskij Muzej Moskovskogo Gosudarstvennogo Universiteta im. M.V. Lomonosova, Moskva 37431
Zoologičeskij Muzej Rossijskogo Gosudarstvennogo Pedagogičeskogo Universiteta im. A.I. Gercena, Sankt-Peterburg 37737
Zoologičeskij Muzej Rossijskoj Akademii Nauk, Sankt-Peterburg 37738
Zoologičeskij Muzej Tomskogo Gosudarstvennogo Universiteta, Tomsk 37860

Slovakia

Múzeum Tatranského Národného Parku, Tatranská Lomnica 38370
Ponitrianske Múzeum, Nitra 38315

Slovenia

Muzej Polovič, Dobova 38410

South Africa

Albany Museum, Grahamstown 38629
Life Sciences Museum, Johannesburg 38659
Port Elizabeth Museum, Port Elizabeth 38712
South African Museum, Cape Town 38592
Transvaal Museum, Tshwane 38759

Spain

Museo de Ciencias Naturales, Pamplona 39807
Museo de Historia Natural, Santiago de Compostela 40021
Museo de Zoología Comparada de Vertebrados, Madrid 39554
Museo Nacional de Ciencias Naturales, Madrid 39571

Sweden

Ånge Zoologiska Museum, Ånge 40424
Evolutionsmuseet, Uppsala 41005
Vänermuseet, Lidköping 40676
Zoologiska Museet, Lund 40709
Zoologiska Utställingar, Uppsala 41018

Switzerland

Hasenmuseum, Bubikon 41263
Musée Cantonal de Zoologie, Lausanne 41543
Musée des Grenouilles, Estavayer-le-Lac 41362
Tierwelt-Panorama, Ebikon 41337
Zoologisches Museum der Universität Zürich, Zürich 42118

Taiwan

Animal Ecology and Dinosaur Museum, Taipei 42373
Research Museum of the Institute of Zoology, Taipei 42422
Sperm Whale Museum, Tainan 42368
Yuanyehao Animal Specimen Museum, Puli 42299
Zoology Museum of National Taiwan University, Taipei 42442

Uganda

Department of Zoology Museum, Kampala 42989
Murchison Falls National Park Museum, Murchison Falls 42996
Queen Elizabeth National Park Museum, Lake Katwe 42994
Wildlife Education Centre, Entebbe 42988

Ukraine

Zoolohičnyj Muzej imeni Benedikta Dybows'koho, Lviv 43143

United Kingdom

Bell Pettigrew Museum, Saint Andrews 45551
Cole Museum of Zoology, Reading 45467
Grant Museum of Zoology and Comparative Anatomy, London 44861
Hunterian Museum, Zoology Section, Glasgow 44286
Loch Ness 2000 Exhibition, Drumnadrochit 44026
Moray Firth Wildlife Centre, Spey Bay 45726
Natural History Museum, Bangor, Gwynedd 43377
Natural History Museum, Belfast 43442
Oxford University Museum of Natural History, Oxford 45336
University Museum of Zoology, Cambridge 43735
Walter Rothschild Zoological Museum, Tring 45877

United States of America

Charles R. Conner Museum, Pullman 52197
Chesapeake Children's Museum, Annapolis 46317
Danville Science Center, Danville 47964
Las Cruces Museum of Natural History, Las Cruces 49958
Mount Rainier National Park Museum, Ashford 46378
Museum of Science at the Toledo Zoo, Toledo 53522
Museum of Vertebrate Zoology, Berkeley 46759
Museum of Zoology, Ann Arbor 46310
National Museum of Wildlife Art, Jackson 49573
North Carolina Museum of Natural Sciences, Raleigh 52234
Olympic National Park Visitor Center, Port Angeles 52027
Oregon Museum of Science and Industry, Portland 52083
Philip L. Wright Zoological Museum, Missoula 50757
Space Farms Zoological Park and Museum, Sussex 53386
University of North Dakota Zoology Museum, Grand Forks 48990
University of Northern Iowa Museum Museums Collections, Cedar Falls 47294
University of Wisconsin Zoological Museum, Madison 50395

Uruguay

Museo Zoológico Dámaso A. Larrañaga, Montevideo 54422

Uzbekistan

Natural History Museum, Nukus 54450

Zimbabwe

Natural History Museum of Zimbabwe, Bulawayo 54666